User's guide

HOW TO USE THIS DIRECTORY

Please read these instructions carefully before using this directory to ensure you get the maximum benefit from it.

There are 4 main ways to use this directory, depending on your requirements:

1. If you know the name of the company/organization you are looking for:

- Use the **QUICK REFERENCE** (*first set of green pages*). This is an alphabetical index of all the companies and organizations included in this edition. Abbreviations (for example, NATO) are also included as far as possible.
- This QUICK REFERENCE will provide you with the page number and reference number of the organization, which refer to the full contact details listed in the CONTACT REFERENCE.

2. If you know the type of organization and country you are looking for:

- The **CONTACT REFERENCE** (*first set of white pages*) provides the main contact information for all organizations indexed in the QUICK REFERENCE. Listings are arranged by type of organization within each country section. The ACTIVITY KEY overleaf identifies each organization category used within the country sections.

3. If you are looking for a selection of companies that provide a particular type of service or product:

- The **PRODUCT/SERVICE INDEX** (*second set of green pages*) is an alphabetical list of products and services included in this directory. Identify the product/service you are interested in. Make a note of the code number.
- Turn to the PRODUCT SUPPLIER LINK (*second set of white pages*), and locate the code number you noted earlier. The companies listed under that code number are suppliers of that product or service. A page number and reference number directs you back to the CONTACT REFERENCE for the main directory entry.

Example entry:

Index reference number — Company name & abbreviation/acronym

80-19050 Oceano Instruments (UK) Ltd
(a subsidiary of Mors SA)
9 New Broom Park, Granton Park Ave Industrial Estate, Edinburgh, Scotland EH5 1RS — Address details
☎ (+44 131) 552 03 03 [E] sales@oceano.demon.co.uk
[W] http://www.oceano.demon.co.uk
⊖ (+44 131) 552 66 19
♣ £ 1,000,000 (1996) 5 (1997) 200 m2 — Electronic contact & financial information
Managing Dir: R.K. Riddell
Sales & Marketing Mgr: Steve Palmer — Named individuals
■ Sales & service of underwater acoustic & oceanographic instrumentation. Long & short baseline acoustic positioning. Acoustic telemetry. Acoustic command (releases, valves, explosives, pyrotechnics). Acoustic control of mine disposal devices. Acoustic beacons & pingers for torpedo/ missile trajectography & relocation. Sound velocimeters. Currentmeters, tidegauges, metocean buoys, etc. → 75 [98] — Description of company activities

Cross reference to other activity types — Year of update

4. If you are looking for an e-mail address or web site for a specific company:

- The **COMPANY ELECTRONIC ADDRESS INDEX** *(the last set of pages)* is an alphabetical list of all the organisations in the directory which have e-mail and web site details. Under each name are listed all the electronic addresses attributed to that organisation, plus page and reference numbers to direct you back to the CONTACT REFERENCE for the main directory entry.

BY LAND AND BY SEA
RECONNAISSANCE SIMBA WARRIOR
POLICE
GKN Defence
GKN Defence
GKN Defence Ltd. Hadley Castle Works,
P.O. Box 106, Telford, Shropshire TF1 4QW.
England. Tel: +44(0) 1952 224500.
Fax: +44(0) 1952 243910.

Contents

Quick Reference

Contact Reference

Product/ Service Index

Product-Supplier Link

Company Electronic Address

Jane's

INTERNATIONAL DEFENCE DIRECTORY 1998

ISBN 0 7106 1550 7

Jane's Information Group
Sentinel House, 163 Brighton Road, Coulsdon, Surrey CR5 2NH, UK
☎ (+44 181) 700 3700 Fax (+44 181) 763 1006, 700 3816
Telex 916907 JANES G
Internet Email: directories@janes.co.uk
Jane's Homepage: http://www.janes.com

Distributed in the Americas by:
Jane's Information Group Inc.,
1340 Braddock Place, Suite 300, Alexandria VA, 22314-1651, USA
☎ (+1 703) 683 3700 Fax (+1 703) 836 1593

Editor - Simon Durham
☎ (+44 181) 700 3809 E simon.durham@janes.co.uk

Financial Editor - Katie Brown
☎ (+44 181) 700 3808 E katie.brown@janes.co.uk

Regional Editor: United Kingdom, Ireland & Scandinavia - James Hales
☎ (+44 181) 700 3717 E james.hales@janes.co.uk

Regional Editor: Northern Europe - Antonio Santos
☎ (+44 181) 700 3810 E antonio.santos@janes.co.uk

Regional Editor: Southern & Central Europe, CIS, Middle East & Africa - Ben Sheppard
☎ (+44 181) 700 3813 E ben.sheppard@janes.co.uk

Regional Editor: Northern USA & Canada - David Shipton
☎ (+44 181) 700 3811 E david.shipton@janes.co.uk

Regional Editor: Southern USA, Central & South Americas - Gwen Mostyn
☎ (+44 181) 700 3812 E gwen.mostyn@janes.co.uk

Regional Editor: Australia & Pacific Rim - Katherine Simmons
☎ (+44 181) 700 3712 E katherine.simmons@janes.co.uk

Administration
Managing Director: Alfred Rolington
Publishing Director: Bob Hutchinson
Publisher: Alan Condron
Managing Editor: Ian Tandy
☎ (+44 181) 700 3807 E ian.tandy@janes.co.uk

The Jane's International Defence Directory is published under licence by
Jane's Information Group Ltd, a member of the Thomson Corporation.
Typeset and bound by Unwin Brothers Ltd, Old Woking, Surrey

Jane's

INTERNATIONAL DEFENCE DIRECTORY 1998

Free Listing Questionnaire

Is your company listed in this directory? If not, please photocopy this page, complete the questionnaire below, and return to the address shown.

Name of your company/organisation:

Name of parent company/holding company:

Full address:

Country:	EMail:	Internet Homepage:
Telephone:	Fax:	Telex:
Free phone:	Teletex:	Cable address:

Ordinary capital/annual sales turnover:

Total working area:

Number of employees:

Key personnel

- Managing Director/President/General Manager:
- Marketing/Sales Director:
- Purchasing Director/Manager:
- Advertising Manager:
- Public Relations Manager:
- Production Director/Head of Production/Manufacturing

Please provide details of any divisions, plants/facilities, field offices on a separate sheet.

Please describe your defence related products/services and send your latest corporate brochure/annual report.

PLEASE FORWARD COMPLETED QUESTIONNAIRE TO:

Simon Durham, Jane's Information Group, Sentinel House,
163 Brighton Road, Coulsdon, Surrey CR5 2NH, UK
Tel: (+44 181) 700 3809 Fax: (+44 181) 700 3816 Tlx: 916907 JANES G
EMail: directories@janes.co.uk

ADVERTISERS

For information on advertising in this product, please tick this box: ☐

BOOK ORDER

For information on ordering further copies of this product, or any other Jane's product, please tick this box: ☐

Introductory Notes/Foreword

Welcome to the 13th edition of the *Jane's International Defence Directory.* This latest edition of the directory sees the introduction of a new section to its many pages - the *Company Electronic Address Index.*

This new section contains an alphabetical listing of all the organisations in the directory for which we currently hold e-mail and web site addresses. With the ever increasing use of the internet for contact and research purposes, we hope that this index will serve as a valuable tool to help you communicate more effectively with the organisations and members of the defence community.

As in previous editions, the information contained in the IDD is largely compiled from details supplied by the companies and organisations themselves, in response to our requests to update and expand the information we hold on them. In instances where we have been unable to elicit a response in time for this edition any information which has not been validated within the last 18 months has been omitted from the directory. This should ensure the accuracy of the information listed and reflects our continued unwillingness to compromise on the quality of the information we provide.

Should you find that your organisation is not listed please forward your details to us for inclusion by either making use of the *Free Listing Questionnaire* included here or through the electronic version available on the Jane's homepage on the internet. Please remember - inclusion in this directory is *still* totally free of charge.

In closing, I should like to thank all those responsible for this latest edition of the IDD. As ever, many thanks go to my editorial team for their persistence and hard work in the pursuance of their research, but also to all the companies and organisations who responded to our information requests. My grateful appreciation goes to them for making the compilation of this latest edition of the IDD possible.

Simon Durham, Editor
September 1997

Index to Advertisers

Countries in this edition

Activity key

Entries in the Contact Reference section are grouped under the following activity key numbers:

10	Head of State
11	Prime Minister
12	Ministries
15	Parliament
20	Defense Administrations/Services/Agencies
25	Defense Forces
26	Police (Civil)
28	Coastguard, Border Protection Forces
29	Customs & Excise
30	Civil Defense
31	Embassies & Attachés
32	Military Liaison Offices
33	Foreign Forces
34	Multinational Intergovernmental Organizations
40	Government Defense Sales/Procurement Organizations
55	Government-owned Industry & Production Facilities
60	Research Organizations, Institutes & Test Establishments
65	Industry, Trade & Armed Forces Associations
70	Financial Holding Groups, Parent Companies
74	Consortia, Joint Venture Companies
75	Manufacturing Companies
80	Service Companies & Organizations
85	Stockists, Distributors & Spare Parts Suppliers
90	Sales Agents, Sales & Liaison Offices

Abbreviations

✈	Fleet (Fixed Wing)	🚁	Fleet (Rotary Wing)
☎	Telephone	♧	Capital
✆	Tollfree phone	♣	Turnover
✉	Cable address	👥	Workforce
TX	Telex	🏠	Roofed area
⊜	Telefax	■	Description of activities
→	See relevant activity section	†	Previous name of company
®	Distributorships/Representatives	E	Electronic mail (e-mail)
97	Update/verification 1997	W	Internet address
98	Update/verification 1998		

A

Company/Organization Name	Country	Page	Key No.
A. Boët SA	France	133	75-06945
A E Atherton & Sons Pty Ltd	Australia	12	75-01570
A. Eickhorn GmbH & Co für Schneidwaren & Waffen (AES)	Germany	168	75-07400
A G Electro-Optics Ltd	UK	520	90-00044
A Goninan & Co. Ltd	Australia	16	75-07760
A Haberkorn GmbH & Co KG	Austria	33	75-01140
A M Corporation	Korea (ROK)	266	80-00010
A. Noble & Son Ltd	Australia	19	75-14690
A P Imports	Australia	29	90-18820
A S Yakovlev Design Bureau	Russia	339	75-78000
A. Smith Gt Bentley Ltd	UK	484	75-54540
A&P Consultants (Pvt) Ltd	Sri Lanka	370	80-00100
A&T = Analysis & Technology Inc	USA	631	80-02600
A&T International Corporation	USA	629	80-00025
A-Mold Corp	USA	553	75-06640
AAAI = Association of Australian Aerospace Industries	Australia	10	65-01000
AAC = American Aerospace Controls Inc	USA	552	75-05450
AACI = Advanced Applications Consultants Inc	USA	630	80-00250
AACO Inc	USA	545	75-00010
AADI = American Aerospace & Defense Industries Inc	USA	651	90-01115
AAE = Aircraft Appliances & Equipment Ltd	Canada	68	75-00430
AAEM = Association of Aero Engine Manufacturers	Russia	336	65-01000
AAF International	USA	545	75-00015
AAF International BV	Netherlands	291	75-00005
Aage Christensen A/S	Denmark	106	90-03340
AAI Corp	USA	545	75-00020
AAI/ACL Technologies Inc	USA	545	75-00360
AAR Advanced Structures	USA	545	75-00120
AAR Allen Airmotive	Netherlands	296	80-00050
AAR Allen Group International	Netherlands	296	80-00050
AAR Corp	USA	629	80-00035
AAR Defense Systems	USA	629	80-00040
AAR International Inc	UK	520	90-00011
AAR Technical Service Center Corp	USA	629	80-00045
Aarding BV	Netherlands	291	75-00020
Aardvark Active Barriers Ltd	UK	426	75-00032
Aardvark Clear Mine Ltd	UK	426	75-00034
Aardvark Holdings Ltd	UK	421	70-00030
Aardvark Tactical Inc	USA	545	75-00122
Aardvark Training Schools Ltd	UK	496	80-00200
AAS = Agusta Aerospace Services SA	Belgium	48	85-00100
AASI = Advanced Aerodynamics & Structures Inc	USA	546	75-00790
AAV = Arab American Vehicles Co	Egypt	109	74-00100
AB Carl Munters	Sweden	377	75-08230
AB Connectors Ltd	UK	426	75-00040
AB Electronic Assemblies Ltd	UK	426	75-00050
AB Electronics (Rogerstone) Ltd	UK	426	75-00080
AB Elektronik GmbH	Germany	164	75-00020
AB Mikroelektronik GmbH	Austria	32	75-00050
AB Volvo Penta	Sweden	379	75-11850
AB Wibe	Sweden	379	75-13000
Abacus Technology Corp	USA	629	80-00050
ABAP = Association Belge pour l'Avion Patrouilleur	Belgium	44	65-00080
ABB = Asea Brown Boveri AB	Sweden	374	70-00200
ABB = Asea Brown Boveri A/S Head Office	Norway	308	75-00110
ABB Asea Brown Boveri Ltd	Switzerland	385	74-00100
ABB Asea Skandia A/S	Norway	308	75-00110
ABB Control A/S	Norway	308	75-00110
ABB Daimler-Benz Transportation (Norway) A/S	Norway	308	75-00120
ABB Dalmer Benz Transportation (TIBB)	Italy	230	75-00010
ABB Financial Services A/S	Norway	308	75-00110
ABB Flexible Automation A/S	Norway	308	75-00110
ABB Industri A/S	Norway	308	75-00110
ABB Installasjon A/S	Norway	308	75-00110
ABB Komponent A/S	Denmark	106	90-01130
ABB Kraft A/S	Norway	308	75-00110
ABB Leitungsbau GmbH	Germany	164	75-00055
ABB Lummus Global Inc	USA	650	90-00040
ABB Miljø Environmental Services A/S	Norway	308	75-00110
ABB Nera Ltd	UK	517	85-01570
ABB Norsk Kabel A/S	Norway	308	75-00110
ABB Offshore Technology A/S	Norway	308	75-00110
ABB Seatec Ltd	UK	520	90-00012
ABB Service A/S	Norway	308	75-00110
ABB Teknologi A/S Corporate Research	Norway	308	75-00110
ABB Trallfa Robot A/S	Norway	308	75-00110
ABB Vetco Gray Inc	USA	650	90-00050
Abbey Electronics	UK	426	75-00140
Abbey Equipment Pty Ltd	Australia	11	75-00100
Abbott Electronics Inc	USA	545	75-00150
Abbott Electronics Ltd	UK	426	75-00180
Abbott Interfast Corp	USA	545	75-00180
Abbott Transistor Laboratories Inc	USA	545	75-00150
Abeking & Rasmussen	Germany	164	75-00060
Abex Aerohydraul UK Office	UK	520	90-00015
Abex GmbH Aerohydraul	Germany	164	75-00105
ABG-SEMA SA	France	138	75-24950
ABIM Anadolv Bilig islem Merkezi ve Genel Danmismanlik Ltd	Turkey	402	75-00010
ABLE Corp	USA	545	75-00260
ABNL = HQ Admiral Benelux Netherlands	Netherlands	288	25-40000
Abrams Instrument Corp	USA	545	75-00280
Abrasive Developments Ltd	UK	427	75-00280
ABS = Aerospace Bearing Support Inc	USA	645	85-00040
Absolute Security	USA	629	80-00080
AC Inc	USA	545	75-00300
AC Tridon - North America	USA	545	75-00460
Academia da Força Aerea	Brazil	54	25-29100
Acal Auriema AB	Sweden	380	90-00050
ACAL Auriema Ltd Top Optic Group	UK	427	75-00290
Acal Electronics Ltd	UK	520	90-00018
ACAL GmbH - Transtech	Germany	180	90-00030
ACAL ITALIA Srl	Italy	242	90-00040
acbLIPS SA	France	129	74-00050
ACC = Ateliers de Constructions du Centre	France	132	75-03700
Accademia Aeronautica	Italy	227	25-29200
Access International Ltd	UK	427	75-00300
Access International Ltd Andover Division	UK	427	75-00300
Accles & Pollock Ltd	UK	427	75-00320
ACCO Foreign Shipping Inc	USA	629	80-00100
Accumulatorenfabrik Sonnenschein GmbH	Germany	164	75-00110
Accuracy International Ltd	UK	427	75-00360
Accurate Energetic Systems, LLC	USA	545	75-00330
ACEC	Belgium	44	75-00060
ACEL = Alan Cobham Engineering Ltd	UK	441	75-14030
Acer Belgium NV	Belgium	49	90-00010
Acer Computer BV	Netherlands	298	90-00005
Acer Computer GmbH	Germany	180	90-00047
Acer Computer Handels GmbH	Austria	34	90-00019
Acer Computer Iberica SA	Spain	368	90-00020
Acer Computer Representative Hungary	Hungary	201	90-00100
Acer Consultants Limited	UK	504	80-12900
Acer France Sarl	France	152	90-00053
Acer Incorporated	Taiwan	393	90-00020
Acer Italy Srl	Italy	242	90-00050
Acer Scandinavia A/S	Denmark	106	90-01200
Acer Scandinavia A/S Norwegian Liaison Office	Norway	311	90-00020
Acer UK Ltd	UK	520	90-00019
ACH	France	130	75-00150
ACI Aviation Consultants Int'l	Australia	23	80-00300
Acieries Aubert & Duval SA	France	132	75-04515
ACL Engine Parts	Australia	11	75-00140
Acofab SA	France	130	75-00200
Acorn Hardas Ltd	UK	427	75-00410
Acousticom Corp	USA	545	75-00380
ACR Systems Inc	USA	650	90-00150
Acro Aerospace	Canada	78	80-00280
Actil	Australia	11	75-00175
Action Information School (AIS)	Denmark	100	25-43000
Active Memory Technology Inc	USA	562	75-16480
Actronics Inc	USA	545	75-00560
Actus Corporation	Japan	250	75-00100
Acutronic Control Sarl	France	131	75-00220
Acutronic Deutschland GmbH	Germany	164	75-00120
Acutronic Ltd	UK	427	75-00520
Acutronic (Switzerland) AG	Switzerland	385	75-00050
Acutronic USA Inc	USA	545	75-00590
AD Aerospace Finishes	Singapore	350	90-00020

Quick Reference

Quick Reference

AOtec Inc.	USA	*554*	75-07940
A.P. Metalcraft Ltd.	UK	*430*	75-03410
AP Precision Hydraulics (Bolton) Ltd	UK	*431*	75-03415
AP Precision Hydraulics Ltd	UK	*431*	75-03415
A.P. Publications Ltd.	UK	*497*	80-01253
APC = Aircraft Parts Corp.	USA	*548*	75-03440
APD	USA	*554*	75-07980
APE = Automated Production Equipment Corp.	USA	*556*	75-10380
Apex Broach & Machine Co.	USA	*554*	75-08000
Apex Storage Systems Ltd	UK	*431*	75-03420
Apex Storage Systems Ltd	UK	*431*	75-03420
API = Image Interpretation Systems Inc.	USA	*584*	75-46110
API-Harowe	USA	*554*	75-08015
APM Hexseal	USA	*554*	75-08030
APNV = Advanced Products NV	Belgium	*45*	75-00075
Apollo Metal SA	France	*151*	85-00300
Apollo Metals (Birmingham) Ltd.	UK	*515*	85-00085
Apollo Metals (International) Ltd	UK	*515*	85-00085
Apollo Metals (London) Ltd	UK	*515*	85-00085
Apollo Metals (Manchester) Ltd.	UK	*515*	85-00085
Apollo Metals (Oxford) Ltd.	UK	*515*	85-00085
Apollo Metals plc.	UK	*515*	85-00085
Apollo Metals (UK) Ltd.	UK	*515*	85-00085
Apparatebau Gauting GmbH, AOA	Germany	*165*	75-00900
Appareillages et Matériels de Servitudes SA (AMS)	France	*132*	75-03050
APPH (Bolton) Ltd	UK	*431*	75-03415
APPH Ltd	UK	*431*	75-03415
Apple Computer IMC	Germany	*177*	80-00200
Appletec Ltd	Israel	*226*	90-00130
Applications des Techniques Nouvelles en Electronique	France	*149*	80-00170
Applied Data Sciences Inc.	USA	*554*	75-08050
Applied Dynamics International (ADI)	USA	*554*	75-08100
Applied Dynamics International Ltd (ADI)	UK	*521*	90-00150
Applied Electro Mechanics Inc. (AEM)	USA	*554*	75-08130
Applied Holographics plc	UK	*431*	75-03520
Applied Measurement Australia Pty Ltd	Australia	*12*	75-01420
Applied Microengineering Ltd	UK	*431*	75-03530
Applied Microsystems Ltd	Canada	*68*	75-00720
Applied Microwave Energy Inc.	USA	*554*	75-08165
Applied Power Australia Ltd	Australia	*12*	75-01500
Applied Science Laboratories (ASL)	USA	*554*	75-08200
Applied Signal Technology	USA	*554*	75-08210
Applied Technology Services (ATS)	Israel	*226*	90-00135
APPSS = The Association of Police and Public Security Suppliers	UK	*420*	65-00100
APR = Aero Products Research Inc.	USA	*546*	75-01113
APREL Laboratories	Canada	*66*	60-00050
APT Radar Systems Ltd	UK	*431*	75-03560
APTEC Surface Engineering Ltd.	UK	*497*	80-01254
APV Baker Ltd	UK	*456*	75-27420
Aqua Bell Ltd	UK	*431*	75-03600
Aquabeam Ltd	UK	*431*	75-03590
Aquarius Fabrique International Pvt Ltd.	India	*204*	75-00160
A.R. Fabb Bros. Ltd	UK	*449*	75-22160
AR Technologies Division	Canada	*77*	80-00080
ARA = ARA Inc.	USA	*554*	75-08370
ARA Inc. (ARA)	USA	*554*	75-08370
Arab American Vehicles Co. (AAV)	Egypt	*109*	74-00100
Arab British Helicopters Co.	Egypt	*110*	74-06000
Arab Commercial Agencies Centre	Egypt	*110*	90-01810
Arab Defence Journal	UK	*497*	80-01255
Arab Engineering Centre.	Egypt	*110*	90-01810
Arab International Optronics Co.	Egypt	*109*	74-00500
Arab Organization for Industrialization (AOI)	Egypt	*109*	55-01700
Arabian Aviation Corp EC	Bahrain	*36*	90-00150
Aracor = Advanced Research & Applications	USA	*546*	75-00940
ARAM	S. Africa	*357*	75-00795
Arbau-Klaus GmbH & Co Fahrzeug und Maschinenbau.	Germany	*165*	75-00930
Arbeitsgruppe für Luft-und Raumfahrt (ALR)	Switzerland	*384*	60-00020
ARC = Allied Research Corp.	USA	*549*	75-04670
ARC = Atlantic Research Corp.	USA	*555*	75-10150
ARC Industries Inc.	USA	*554*	75-08380
Arca Systems	USA	*631*	80-03100
Arca-Regler.	Germany	*165*	75-00950
Arcadia Engineering Co.	UK	*465*	75-36870
Arcane Balistic	France	*138*	75-24010
Arcata Associates Inc.	USA	*554*	75-08400
Archer Mullins Limited	UK	*497*	80-01257
Architektur Städtebau Infrastruktur (ASI)	Germany	*177*	80-00210
Arco Electrical Industries Ltd	Israel	*217*	75-01680
Arco Solar Inc.	USA	*614*	75-80570
Arcom Control Systems Ltd	UK	*431*	75-03700
Arcus Co.	Bulgaria	*61*	55-00060
Arde Inc.	USA	*554*	75-08500
ARDEC = US Army Tank-Automotive and Armaments Command	USA	*535*	25-38980
Ardennlame SA	France	*147*	75-43670
Aremco Products	UK	*431*	75-03705
Arendnoe NPO Zavod Imeni Popova (ZIP ANPO)	Russia	*338*	75-16400
ARES Inc.	USA	*555*	75-08600
Arestel Componentes Equipamentos Electrónicos SA	Portugal	*328*	90-00280
ARF Vola	Bulgaria	*62*	55-02490
Arga Controls Inc.	USA	*555*	75-08620
Argo-Tech Corp.	USA	*555*	75-08630
Argos CLS	France	*132*	75-03320
Arhimidous & Ippokratous (Stratopedoi Siliotopoulou)	Greece	*185*	20-05000
Arianespace	Japan	*255*	80-00050
Aries Ingenieria y Sistemas SA	Spain	*367*	80-00200
Arinc	USA	*631*	80-03130
Arinc Research Corp.	USA	*631*	80-03130
Arjo Ltd.	UK	*431*	75-03715
Arkonia Systems Ltd	UK	*431*	75-03703
Armada Nacional Columbia	Colombia	*91*	25-30000
Armalon Ltd	UK	*431*	75-03729
Armament Documentation Centre	France	*121*	20-04000
Armaz (Bagira) International Consulting Ltd	Israel	*223*	80-00180
Armed Forces Communications & Electronics Association (AFCEA)	USA	*541*	65-00130
Armed Forces Communications and Electronics Association (AFCEA)	Belgium	*44*	65-00050
Armed Forces General Headquarters	Syria	*391*	25-10000
Armed Forces Headquarters	Belgium	*39*	25-10000
Armed Forces Headquarters	Germany	*157*	25-10000
Armed Forces Headquarters	Indonesia	*208*	25-10000
Armed Forces Institute of Pathology	USA	*527*	20-04372
Armed Forces Medical College	Germany	*157*	25-19500
Armed Forces Police Command	Taiwan	*392*	25-31000
Armed Forces Reserve & Coast Guard Command	Taiwan	*392*	28-10000
Armed Forces Staff College	USA	*528*	25-19100
Armed Forces University Hamburg	Germany	*157*	25-19200
Armed Forces University Munich	Germany	*157*	25-19100
Arméns Materielunderliallscentrum	Sweden	*372*	25-31000
Arméns Tekniska Centrum (ATC)	Sweden	*372*	25-31000
Armored Vehicle's Spares Ltd (AVS)	Israel	*217*	75-01700
Armour of America Inc.	USA	*555*	75-08850
Armour Technology International Ltd	UK	*431*	75-03790
Armourshield Limited	UK	*431*	75-03775
ARMS = Atlantic Research Marketing Systems Inc.	USA	*631*	80-03365
Arms Corporation of the Philippines (ARMSCOR)	Philippines	*320*	75-00150
ARMSCOR	S. Africa	*356*	40-01000
ARMSCOR = Arms Corporation of the Philippines	Philippines	*320*	75-00150
Armstrong Fastening Systems	UK	*431*	75-03800
Armstrong Monitoring Corp., The.	Canada	*68*	75-00740
Armstrong Monitoring Corp., The.	USA	*651*	90-01300
Armtec Defense Products Co.	USA	*555*	75-08950
Armtec/Ragen Inc.	USA	*595*	75-59842
Army Air Corps (AVES)	Italy	*227*	25-32000
Army and Air Force Commando Headquarters	Austria	*31*	25-20000
Army Aviation Headquarters	Argentina	*3*	25-32100
Army Aviation School	Belgium	*40*	25-39200
Army Board, Defence Council	UK	*409*	12-10300
Army Headquarters	Argentina	*3*	25-30000
Army Headquarters	Belgium	*40*	25-30000
Army Headquarters	Colombia	*91*	25-30000
Army Headquarters	Greece	*185*	25-30000
Army Headquarters	Luxembourg	*274*	25-10000
Army Headquarters	Myanmar	*285*	25-30000
Army Headquarters	Portugal	*326*	25-30000
Army Headquarters	Singapore	*346*	25-30000
Army Helicopter Forces Command	Spain	*362*	25-31000
Army Institute for Research and Control - National Proof Testing Organization	Chile	*85*	60-00950
Army Materiel Command	Denmark	*100*	25-32100
Army Merchants	Cyprus	*96*	75-00010
Army Operational Command	Denmark	*100*	25-30000
Army Signal School and Technical College for Electrical Engineering	Germany	*157*	25-39200
Army Technology & Engineering Agency	Australia	*7*	25-37100

Quick Reference

B

Quick Reference

Quick Reference

Bofors Ant Armour Systems	Sweden	376	75-02975
Bofors Canada Ltd	Canada	81	90-00200
Bofors Carl Gustaf AB	Sweden	376	75-02950
Bofors Deutschland GmbH	Germany	181	90-00260
Bofors Explosives AB	Sweden	376	75-03000
Bofors Industries AB	Sweden	375	74-00020
Bofors LIAB AB	Sweden	376	75-03025
Bofors Missiles	Sweden	376	75-03030
Bofors Norge A/S	Norway	312	90-00300
Bofors SA Marine AB	Sweden	376	75-03040
Bofors (UK) Representative Office	UK	521	90-00255
Bofors Underwater Systems AB	Sweden	376	75-03050
Bofors Underwater Systems AB	Sweden	376	75-03052
Bofors Weapon Systems	Sweden	376	75-03075
Böhler Bleche GmbH	Austria	32	75-00650
Böhler Schmiedetechnik GmbH	Austria	32	75-00650
Bolivian Army Staff College	Bolivia	52	25-39100
Bollé, Ets	France	133	75-06960
Bollinger Machine Shop & Shipyard Inc.	USA	560	75-14600
Bollinger Shipyard Inc.	USA	560	75-14600
Bolloré Protection	France	133	75-06970
Bolvinkel ApS, Erik	Denmark	106	90-02670
Boma Bisma Indra PT	Indonesia	209	55-00050
Bombardier Eurorail Industrial Vehicles Division	Belgium	45	75-01650
Bombardier Inc. Defence Services	Canada	69	75-01400
Bombardier Inc Defence Systems Division	Canada	69	75-01900
Bombardier-Rotax GmbH	Austria	32	75-00660
BOMISA	Italy	233	75-03120
Bondina NBC Materials	UK	435	75-07600
Boneschi Carrozzeria Sarl	Italy	233	75-05000
Boneschi Centro Sud	Italy	233	75-05000
Bonetti Aircraft Supports SpA	Italy	243	90-00250
Booth Industries Limited	UK	435	75-07670
Booyco Engineering	S. Africa	357	75-02400
Booz, Allen & Hamilton Inc.	USA	632	80-04700
Border Holdings (UK) Ltd	UK	435	75-07740
Bosch Telecom GmbH	Germany	166	75-03410
Bose BV European HQ	Netherlands	291	75-00790
Bose Corporation	USA	560	75-14700
Bosma Machine and Tool Corporation	USA	560	75-14750
Bosman Powersource Support BV	Netherlands	292	75-00800
Bospec Neue Technologien	Germany	166	75-03420
Boss Group Limited	UK	435	75-07750
Boston Whaler Inc. Commercial Products	USA	560	75-14760
Botelho	Brazil	59	90-13850
Botswana High Commission	UK	413	31-00220
Bottoni Minuterie SpA	Italy	233	75-03120
Bougas Panagiotis Ltd	Greece	187	75-00600
Boughton Group	UK	422	70-00247
Boulder Scientific Co.	USA	560	75-14780
Bowers Group	UK	435	75-07770
Bowmar Instrument Corp.	USA	560	75-14860
Bowtech Products Ltd	UK	521	90-00270
Bowthorpe plc	UK	435	75-07780
Bowthorpe Thermometrics	UK	435	75-07785
Bozza SA, José Murillia	Brazil	55	75-02220
BP Chemicals Ltd Advanced Materials Division	UK	435	75-07790
BP Solar	Philippines	321	90-00200
BP Solar Arabia Limited	Saudi Arabia	343	90-00250
BP Solar Australia Pty Ltd	Australia	13	75-02630
BP Solar c/o PT Nusa Cerindo Utama	Indonesia	210	90-00050
BP Solar East Africa Limited	Kenya	263	90-05000
BP Solar España SA	Spain	365	75-01660
BP Solar Inc	USA	651	90-02600
BP Solar Limited	UK	435	75-07800
BP Solar Malaysia	Malaysia	278	90-00255
BP Solar Southern Africa	S. Africa	361	90-01000
BP Solar Vietnam	Vietnam	660	90-00500
BP Thai Solar Corporation Ltd	Thailand	396	90-00250
BPD Difesa e Spazio SpA	Italy	234	75-09112
BPIS = The Agency for Strategic Industries of the Republic of Indonesia	Indonesia	210	65-00100
BPL Healey Instruments Ltd	UK	434	75-06195
Bradford Cylinders Ltd	UK	435	75-07850
Bradley Doublelock Ltd	UK	435	75-07855
Bradley Electronics Ltd	UK	435	75-07860
Brake & Clutch Industries Australia Pty Ltd	Australia	20	75-16245
Brandt Fijnmechanische Industrie BV	Netherlands	292	75-00850
Brassey's (UK) Ltd	UK	499	80-03800
Brassinter SA Indústria e Comércio	Brazil	55	75-02290
Bravo & Filhos, Alberto Maria	Portugal	328	90-00270
Brazan SA	Spain	364	55-00140
Brazilian Embassy	Italy	227	31-00230

Break-Free Inc.	USA	560	75-14990
Breeze Eastern Corp.	USA	560	75-15110
Breitling Ltd	Switzerland	385	75-01400
Breitling Watch Co Ltd	Switzerland	385	75-01400
Bremer Vulkan Marineschiffbau GmbH	Germany	166	75-03600
Bremer Vulkan Werft und Maschinenfabrik GmbH	Germany	166	75-03600
Bren-Tronics Inc.	USA	560	75-15170
Brenco Thermal Coating Pty Ltd	Australia	13	75-02740
Brennan & Co (Pte) Ltd, W.H.	Singapore	350	90-00265
Brennan Inc., Leo J.	USA	632	80-04800
Bressel SA	Spain	366	75-01670
Brett Harris Ltd	UK	435	75-08000
Bretta Spain	Spain	366	75-01700
Bridgestone Corp Head Office	Japan	250	75-00700
Bridon International Ltd	UK	435	75-08350
Bridport Aviation Products	UK	435	75-08400
Bridport-Gundry plc	UK	422	70-00270
Bridport-Quin Pty Ltd	Australia	13	75-02746
Brigade and Battle Group Trainer (North) (BBGT(N))	UK	412	25-33000
Brighton Sheet Metal Ltd	UK	436	75-08550
Brimar Inc.	USA	560	75-15210
Brimech Industries	Australia	20	75-18300
Brimotor Ltd	UK	436	75-08600
Bristol Aerospace Ltd (BAL)	Canada	69	75-01500
Bristol Uniforms Ltd	UK	436	75-08760
Bristoll Group	Canada	69	75-02250
Bristow Helicopters Ltd	UK	499	80-03950
Britannic Aviation	UK	435	75-08400
British Aerospace Airbus Limited	UK	436	75-08882
British Aerospace Australia Ltd Operating Headquarters (BAeA)	Australia	13	75-02750
British Aerospace (Consultancy Services) Ltd	UK	499	80-03990
British Aerospace Dynamics	UK	425	74-06850
British Aerospace Flight Training (UK) Ltd	UK	499	80-03995
British Aerospace Germany GmbH	Germany	181	90-00270
British Aerospace Holdings Inc	USA	542	70-02600
British Aerospace Military Aircraft	UK	436	75-08886
British Aerospace plc (BAe)	UK	422	70-00275
British Aerospace Royal Ordnance	UK	481	75-50620
British Aerospace Simulation Ltd	UK	478	75-48380
British Aerospace (Sweden) AB	Sweden	381	90-00500
British Aerospace (Systems & Equipment) Ltd (BASE)	UK	436	75-08890
British Aerospace Systems and Services	UK	499	80-04000
British Aluminium Plate	UK	436	75-09000
British Aluminium Speciality and Aerospace	UK	429	75-02455
British Aluminium Tubes	UK	436	75-09020
British American Security Information Council	USA	542	65-00200
British American Security Information Council (BASIC)	UK	419	60-00250
British Association of Women Police (BAWP)	UK	420	65-00200
British Electrical Machines	UK	437	75-10450
British Filters Ltd	UK	436	75-09030
British International Helicopters	UK	499	80-04090
British International Industries Ltd (BII)	UK	436	75-09150
British Marine Industries Federation	UK	420	65-00250
British National Space Centre (BNSC)	UK	409	12-30000
British Naval Equipment Association (BNEA)	UK	420	65-00280
British Nuclear Fuels plc (BNFL)	**UK**	**499**	**80-04175**
British Rollmakers Corporation	UK	436	75-09500
British Sarozal Ltd	UK	436	75-09570
British Security Industry Association (BSIA)	UK	421	65-00285
British Steel (Asia) Ltd	Hong Kong SAR	198	90-00180
British Steel Belgium Sprl	Belgium	49	90-00265
British Steel Canada Inc	Canada	81	90-00225
British Steel Denmark A/S	Denmark	106	90-02300
British Steel Deutschland GmbH	Germany	181	90-00275
British Steel Engineering Steels	UK	436	75-09600
British Steel Engineering Steels Sales Inc	USA	651	90-02750
British Steel España SA	Spain	369	90-00200
British Steel Finland Oy	Finland	120	90-00250
British Steel France SA	France	152	90-00260
British Steel Ireland Ltd	Ireland	215	90-00150
British Steel Liaison Services (India) Limited	India	207	90-00200
British Steel Norge AS	Norway	312	90-00310
British Steel (South East Asia) Pte Ltd	Singapore	351	90-00275
British Steel Svenska AB	Sweden	381	90-00360
British Underwater Test and Evaluation Centre (BUTEC)	UK	410	20-00200
British Vita plc	UK	437	75-09750
Britorion Ltd	UK	515	85-00235

Brittain Engineering (Guildford) Ltd	UK	*521*	90-00280
Brittain Engineering Ltd	UK	*521*	90-00280
Britte SA	Belgium	*45*	75-01900
Britton, G.B.	UK	*437*	75-09900
Broadcast Electronics Inc.	USA	*560*	75-15240
Broadcast Microwave Services Inc.	USA	*560*	75-15250
Brochier SA/Ciba Composites	France	*133*	75-07300
Brock's Explosives Ltd	UK	*437*	75-09980
Broco Inc	USA	*560*	75-15260
Brodoprojekt	Croatia	*95*	80-00100
Brodribb Pty Ltd, M	Australia	*13*	75-02760
Bromma Inc.	USA	*560*	75-15270
Brooke Marine Shipbuilders Ltd (see FBM Marine Ltd)	UK	*450*	75-22485
Brookes & Gatehouse Ltd	UK	*437*	75-10100
Brookhouse Paxford Ltd	UK	*437*	75-10103
Brown Boveri, Tecnomasio Italiano SpA	Italy	*230*	75-00010
Brown, David John Welsh Custom Build Pty Ltd	Australia	*13*	75-02805
Brown Ltd, S.G.	UK	*437*	75-10330
Brownell & Co. Inc.	USA	*560*	75-15298
Brownell Climatic Controls Ltd	UK	*437*	75-10385
Brownell Defense & Aviation Products	USA	*560*	75-15298
Brownell Ltd	UK	*437*	75-10385
Brownells Inc.	USA	*560*	75-15300
Browning SA	Belgium	*45*	75-01950
Bruce D. Vallillee Electronics Ltd	Canada	*80*	80-02210
Bruce Industries Inc.	USA	*561*	75-15360
Bruck Textiles plc	Australia	*13*	75-02810
Brüel & Kjaer Belgium NV	Belgium	*49*	90-00275
Brüel & Kjaer Canada Ltd	Canada	*81*	90-00250
Brüel & Kjaer Instruments Inc.	USA	*651*	90-02800
Brügger & Thomet Manufacturing Co	Switzerland	*385*	75-01450
BRUHN NewTech A/S	Denmark	*104*	80-00050
BRUHN NewTech Inc	USA	*632*	80-04940
Brunswick Corp.	USA	*620*	75-86764
Brunton Co.	USA	*561*	75-15460
Brunton's Propellers Ltd	UK	*437*	75-10400
Brush Electrical Machines Ltd (BEM)	UK	*437*	75-10450
BS Hansen Trade & Consult A/S	Norway	*312*	90-00700
BS Hansen Trading AB	Norway	*312*	90-00700
BSD Steel Service Centre	UK	*515*	85-00240
BSD Tubes	UK	*515*	85-00240
BSIA = British Security Industry Association	UK	*421*	65-00285
BT (Marine) Ltd	UK	*499*	80-05280
BT Sales & Services Ltd	UK	*508*	80-20450
BTI BV = Besthoma Trading International BV.	Netherlands	*299*	90-00220
BTR Aerospace Australia	Australia	*13*	75-02815
BTR Engineering (Australia) Ltd	Australia	*26*	85-02900
BTR Environmental Pty Ltd	Australia	*14*	75-02830
BTR plc	UK	*422*	70-00290
BTR Silvertown Ltd.	UK	*437*	75-10830
BTR Technology Services Ltd	UK	*499*	80-04550
Bucher-Guyer Ltd, Vehicles	Switzerland	*385*	75-01580
Buck Chemisch-Technische Werke GmbH & Co	Germany	*162*	70-00050
Buck Werke GmbH & Co	Germany	*162*	70-00050
Budenberg Gauges (Australia) Pty Ltd	Australia	*14*	75-02860
Buergerrecht und Polizei	Germany	*161*	60-00250
Bullet Guard Corp.	USA	*561*	75-15620
Bulova Technologies L.L.C.	USA	*561*	75-15600
Bülow Air Target Systems AB	Sweden	*376*	75-03600
Bumjin Corp	Korea (ROK)	*267*	90-00270
Bunce (Ashbury) Ltd	UK	*438*	75-10870
Bundesakademie für Wehrverwaltung und Wehrtechnik (BAKWVT)	Germany	*156*	20-19100
Bundesamt für Wehrtechnik und Beschaffung (BWB)	Germany	*161*	40-01000
Bundesanstalt für Materialforschung und Prufung (BAM)	Germany	*161*	60-00600
Bundeskanzleramt	Germany	*156*	11-10000
Bundeskriminalamt	Germany	*161*	60-00300
Bundesministerium der Finanzen	Germany	*156*	12-60000
Bundesministerium für Inneres	Austria	*30*	12-40000
Bundesministerium fur Landesverteidigung	Austria	*30*	12-10000
Bundesministerium für Landesverteidigung	Austria	*31*	25-20000
Bundesministerium für Verkehr (BMV)	Germany	*156*	12-95000
Bundespolizeidirektion Wien	Austria	*31*	26-10000
Bundespräsidialamt	Germany	*156*	10-10000
Bundesrat	Austria	*30*	15-10000
Bundesrat	Germany	*156*	15-10000
Bundesrat	Switzerland	*381*	10-10000
Bundesverband der Deutschen Industrie e.V. (BDI)	Germany	*162*	65-00150
Bundesverband der Deutschen Luft-und Raumfahrt-Industrie e.V. (BDLI)	Germany	*162*	65-00100
Bundesverband der Deutschen Luftfahrt, Raumfahrt und Ausrüstungsindustrie e.V.	Germany	*162*	65-00100
Bundeswehrverwaltungsamt (BWVA)	Germany	*156*	20-11000
Bunnell Plastics	USA	*561*	75-15630
Burdeshaw Associates, Ltd (BAL)	USA	*632*	80-04960
Burfield & Co. (Gloves) Ltd	UK	*438*	75-10900
Burkhardt GmbH	Germany	*164*	75-00300
Burkhart Grob Luft und Raumfahrt GmbH & Co KG	Germany	*163*	74-04300
Burlington Microelectronics, SST	USA	*643*	80-29650
Burmarc Ltd	UK	*521*	90-00290
Burr-Brown International Ltd	UK	*438*	75-11032
Burton Electrical Engineering	USA	*561*	75-16005
Busak + Shamban Canada	Canada	*81*	90-00280
Busak + Shamban France SA	France	*152*	90-00270
Busak + Shamban Inc. Shamban Seals Division	USA	*561*	75-16006
Busak + Shamban K.K.	Japan	*257*	90-00240
Busak + Shamban Ltd	UK	*438*	75-11050
Busak + Shamban SA	Belgium	*46*	75-02000
Busak + Shamban West Export Group	USA	*561*	75-16006
Bushmaster Firearms Inc.	USA	*561*	75-16007
BUTEC = British Underwater Test and Evaluation Centre	UK	*410*	20-00200
Butyl Products Ltd	UK	*438*	75-11090
BVR Technologies Ltd	Israel	*218*	75-03350
BWB = Bundesamt für Wehrtechnik und Beschaffung	Germany	*161*	40-01000
BWVA = Bundeswehrverwaltungsamt	Germany	*156*	20-11000

C

C & R Systems Inc.	USA	*561*	75-16016
C. Gavriil SA	Greece	*189*	75-01294
C. Raymond Hunt Associates Inc.	USA	*637*	80-14900
C. Rokas SA	Greece	*191*	75-02310
C W Price & Co. (Pty) Ltd	S. Africa	*362*	90-16900
C&E	Israel	*224*	80-00275
C&S Antennas Inc.	USA	*652*	90-03760
C-CUBED Corp.	USA	*633*	80-05260
C-RAN Corp.	USA	*566*	75-22400
C-Tech Inc.	USA	*651*	90-03000
C-Tech Ltd	Canada	*70*	75-03800
C-Way Industries Inc.	USA	*646*	85-00300
CAA = Civil Aviation Administration	Finland	*116*	20-00300
Cabinet du Premier Ministre	France	*120*	11-10000
Cabiran (1991) Ltd	Israel	*218*	75-03400
Cable and Wireless (Marine) Ltd	UK	*499*	80-05280
CAC Systèmes	France	*133*	75-07680
Cadex Electronics Inc.	Canada	*69*	75-01750
Cadillac Gage Textron	Singapore	*352*	90-02025
Cadillac Plastics Australia	Australia	*26*	85-03000
Cadorath Aerospace Inc.	Canada	*77*	80-00242
Cadorath Plating Co. Ltd.	Canada	*77*	80-00242
CAE	UK	*438*	75-11125
CAE Aviation Ltd	Canada	*78*	80-00245
CAE Electronics Ltd	Canada	*69*	75-01770
CAE Elektronic GmbH	Germany	*166*	75-04500
CAE Inc.	Canada	*67*	70-03100
CAE Invertron Ltd	UK	*438*	75-11145
CAE MRad Pty Ltd	Australia	*24*	80-02800
CAEPE = Centre d'Achèvement et d'Éssais des Propulseurs et Engins	France	*122*	20-40000
CAESA = Compañia Aeronàutica Española SA	Spain	*368*	85-00400
Cagiva Motor SpA	Italy	*229*	65-00450
Cagiva TRADING SpA	Italy	*229*	65-00450
CAIA = Confederation Amicale des Ingenieurs de l'Armement	France	*128*	65-00340
Caina Company (Canada)	Canada	*80*	85-00450
Calbrooke Marketing Inc.	Canada	*80*	85-00500
Caledonian Mining Co. Ltd	UK	*438*	75-11250
Caley Ocean Systems	UK	*438*	75-11260
Calgon Carbon Corp.	USA	*561*	75-16210
CALHAVEN = Callington Haven Pty Ltd.	Australia	*14*	75-03100
Calian Communications Systems Ltd	Canada	*69*	75-01800
Calian Technology Ltd	Canada	*69*	75-01800
Caliber Sales Engineering Inc.	USA	*646*	85-00273
Calif Gasket & Rubber Corp.	USA	*561*	75-16230
California Avi-Tron Corp.	USA	*561*	75-16260
California Eastern Laboratories	USA	*646*	85-00274
California Instruments Corp.	USA	*561*	75-16280

Centre d'Achèvement et d'Éssais des Propulseurs et Engins (CAEPE)	France	*122*	20-40000
Centre d'Electronique de l'Armement (CELAR)	France	*121*	20-16000
Centre d'Enseignement Supérieur Aérien (CESA)	France	*122*	25-29200
Centre d'Essais de Fauga-Mauzac	France	*127*	60-05000
Centre d'Éssais de le Méditerranée (CEM)	France	*122*	20-40000
Centre d'Essais de Modane-Avrieux	France	*127*	60-05000
Centre d'Éssais des Landes (CEL)	France	*122*	20-40000
Centre d'Études de la Défence	Belgium	*39*	20-01000
Centre d'Études de la Plice National	France	*126*	60-00560
Centre d'Études et de Recherches de Toulouse (CERT)	France	*127*	60-05000
Centre d'Études et de Recherches sur la Police	France	*126*	60-00500
Centre de Documentation de l'Armement (CEDOCAR)	France	*121*	20-04000
Centre de Recherche pour la Défense Valcartier (CRDV)	Canada	*67*	60-00450
Centre de Recherches de Politique Criminelle	France	*126*	60-00650
Centre de Recherches Sociologiques sur le Droit et les Institutions Penales	France	*126*	60-00670
Centre de Sciences Criminelles	France	*126*	60-00570
Centre for Criminal Justice Studies	UK	*419*	60-00300
Centre for European Law Policy	Germany	*161*	60-00400
Centre for Interdisciplinary Research	Germany	*161*	60-00430
Centre National d'Instruction de la Protection Contre l'Incendie	France	*123*	30-00100
Centre of Criminal Sciences	France	*126*	60-00570
Centre of Recherche for Criminal Policy	France	*126*	60-00650
Centre of Sociological Research on the Penal Law	France	*126*	60-00670
Centro De Paracaidismo	Spain	*367*	80-00060
Centronix PT	Indonesia	*210*	75-00200
Century Electronics	USA	*562*	75-17385
Century International Arms Inc.	USA	*633*	80-05400
Century International Arms Ltd	Canada	*81*	90-00325
Cenzin Co. Ltd Foreign Trade Enterprise	Poland	*324*	75-00200
CEOA = Central Europe Operating Agency	France	*125*	34-00010
CEP SA	Belgium	*49*	90-00320
Ceramic Devices Inc.	USA	*563*	75-17397
Ceramicast (Precision Investment Castings) Ltd	UK	*439*	75-12410
Ceramtech Ltd	Australia	*13*	75-02740
Cerberus AG	Switzerland	*385*	75-01700
Cerberus Pyrotronics	USA	*651*	90-03280
Cerberus-Guinard SA	France	*133*	75-08800
Cercom Inc.	USA	*563*	75-17400
Ceridian Corporation	USA	*542*	70-02650
Cermat Ltd	Israel	*223*	80-00250
CERME	France	*133*	75-08810
CERN = European Laboratory for Particle Physics	Switzerland	*384*	34-10000
CERT = Centre d'Études et de Recherches de Toulouse	France	*127*	60-05000
Certacom Ltd.	UK	*515*	85-00320
Certified Aircraft Parts Inc	USA	*646*	85-00277
CESA = Centre d'Enseignement Supérieur Aérien	France	*122*	25-29200
CESELSA	Spain	*366*	75-03800
Ceska Zbrojovka	Czech Republic	*98*	75-00300
Cessna Aircraft Company	USA	*563*	75-17750
CET Technologies	Singapore	*348*	75-01080
Cexec Inc.	USA	*633*	80-05425
CFI = Chartered Firearms Industries Pte Ltd	Singapore	*348*	75-01150
CFM GmbH Chemo-technische Produkte	Germany	*166*	75-04800
CFM International Inc. (CFMI)	USA	*544*	74-00400
CFMI = CFM International Inc.	USA	*544*	74-00400
CFSCE = Canadian Forces School of Communications and Electronics	Canada	*65*	25-19200
CGF Gallet SA	France	*133*	75-08840
CGG = Compagnie Générale de Géophysique Head Office	France	*134*	75-11335
CGI Systems BV	Netherlands	*296*	80-03500
CGWIC = China Great Wall Industry Corp.	China PR	*87*	55-00100
Chadwick Downs Pty Ltd	Australia	*26*	85-03500
Chairman of the Council of Ministers	Kyrgyzstan	*269*	11-10000
CHAM (Concentration, Heat & Momentum) Ltd	UK	*499*	80-05900
Chamber of Deputies	Chile	*83*	15-20000
Chamber of Deputies	Luxembourg	*274*	15-20000
Chamber of Representatives	Belgium	*39*	15-05000
Chamberlain Manufacturing Corp.	USA	*563*	75-17950
Champion Friction Co.	USA	*563*	75-17975
Champion Spark Plug Co.	Canada	*81*	90-00330
Champion Spark Plug Co. Aviation Products	USA	*563*	75-17980
Champion Spark Plug Division	Canada	*81*	90-00330
Channel Industries Inc. (CII)	USA	*563*	75-18050
Chantiers de l'Atlantique	France	*133*	75-08850
Chantiers OCEA SA	France	*133*	75-08950
Chapman & Smith Ltd	UK	*439*	75-12500
Chapy Corporation	USA	*563*	75-18065
Charcoal Cloth (International) Ltd	UK	*439*	75-12550
Charles A. Hulcher Co. Inc.	USA	*583*	75-45000
Charlotte Aircraft Corp.	USA	*651*	90-03360
Chartered Ammunition Industries Pte Ltd	Singapore	*348*	75-01090
Chartered Firearms Industries Pte Ltd (CFI)	Singapore	*348*	75-01150
Chartered Industries of Singapore	Singapore	*347*	70-01900
Chase Inc.	USA	*633*	80-05450
Chase Initiatives	USA	*633*	80-05450
Chase Management Systems (Pty) Ltd	S. Africa	*360*	80-03000
CHC Helicopter Corporation	Canada	*78*	80-00280
Checkmate Industries Ltd	Hong Kong SAR	*198*	75-00200
Checkmate Industries of EEC Ltd	Denmark	*102*	75-03300
Checkmate Industries of USA Ltd	USA	*563*	75-18100
Chelsea Instruments Ltd	UK	*439*	75-12600
Cheltenham Technical Publications Ltd	UK	*500*	80-07650
Chelton (Electrostatics) Limited	UK	*439*	75-12700
Chelton Inc.	USA	*563*	75-18120
Chelton/Rayan	France	*152*	90-00326
Chem - Tech	Greece	*188*	75-00750
Chemische Fabrik Freising GmbH	Germany	*166*	75-04800
Chemring Countermeasures	UK	*439*	75-12800
Chemring Group plc	UK	*422*	70-00320
Chemviron Carbon	Belgium	*46*	75-02260
Cherry Aerospace Fasteners	Israel	*226*	90-00319
Cherry Division of Textron Inc	Germany	*181*	90-00330
Cherry Division of Textron Inc.	UK	*522*	90-00335
Cherry Textron	USA	*622*	75-88656
Cheshire Constabulary Headquarters	UK	*412*	26-30000
Chesterfeild Cylinders Ltd	UK	*490*	75-61780
Chicago Army Navy Surplus Co.	USA	*646*	85-00278
Chicago Bullet Proof Equipment Co.	USA	*563*	75-18200
Chicago Bullet Proof Systems	USA	*563*	75-18200
Chicopee Manufacturing Limited	Canada	*70*	75-03200
Chief Belgian Army Staff	Belgium	*40*	25-32200
Chief of Defence Denmark, Headquarters	Denmark	*100*	25-10000
Chief of the General Staff	UK	*411*	25-10100
Chilean Naval Academy	Chile	*85*	60-00500
Chilean Navy Telecommunications Service	Chile	*84*	25-44000
Chilport (U.K) Limited	UK	*499*	80-06000
China Aerospace Corporation	China PR	*86*	12-70000
China Eastern AlliedSignal Aircraft Wheels and Brakes	China PR	*89*	80-00370
China Great Wall Industry Corp. (CGWIC)	China PR	*87*	55-00100
China National Aero-Technology Import & Export Corp. (CATIC)	China PR	*88*	55-00200
China National Precision Machinery Import & Export Corp. (CNPMIEC)	China PR	*88*	55-00265
China North Industries Corp. (NORINCO)	China PR	*88*	55-00280
China Precision Machinery Import & Export Corp.	China PR	*88*	55-00265
China Shipbuilding Trading Co. Ltd (CSTC)	China PR	*88*	55-00360
China Steel Corporation	Taiwan	*392*	75-04700
China Steel Corporation Osaka Office	Japan	*255*	80-00115
China Steel Corporation Singapore Office	Singapore	*349*	80-00430
Chincul SACAIFI	Argentina	*4*	75-03451
Chinese Society of Aeronautics and Astronautics	China PR	*88*	65-00050
Chinn & Co. Ltd, J.S.	UK	*439*	75-13045
Chinn Engineering Co. Ltd, J.S.	UK	*440*	75-13050
Chinn Holdings Ltd, J.S.	UK	*422*	70-00345
Chloride Batteries New Zealand	New Zealand	*302*	75-00750
Chloride Industrial Batteries Ltd	UK	*457*	75-28490
Chomerics	USA	*563*	75-18275
Chomerics Europe Ltd	UK	*473*	75-43880
Chopperline Flight Training	Australia	*24*	80-03400
CHR Industries Inc.	USA	*576*	75-36400
Christensen A/S, Aage	Denmark	*106*	90-03340
Christie Electric Australia	Australia	*26*	85-01000
Chromalloy France	France	*133*	75-09650
Chromalloy Gas Turbine Corp.	USA	*563*	75-18360
Chromalloy Holland	Netherlands	*297*	80-04000
Chromalloy Precision Products	USA	*580*	75-41900
Chronopoulos Ltd	Greece	*188*	75-00800
Chrysanthacopoulos SA	Greece	*188*	75-00910
Chrysanthacopoulos SA, G	Greece	*188*	75-00910
Chrysler Corp.	USA	*542*	70-02700

Quick Reference

Quick Reference

Comepa	France	134	75-10175
Comercial Donald Grant SL	Spain	369	90-00370
Comet GmbH Pyrotechnik-Apparatebau	Germany	166	75-05000
Comex Pro	France	134	75-10200
Comité Richelieu	France	128	65-00330
Comlab Telecommunications Inc.	Canada	70	75-03540
Command & Staff College	Australia	6	25-19300
Command, Control and Communications Systems Depot	Finland	116	25-20000
Command, Control and Communications Systems School	Finland	116	25-20000
Commandancia de la Aviación Naval	Chile	83	25-43000
Commandancia General de la Infanteria de Marina	Spain	362	25-42000
Commandement de L'Armée	Luxembourg	274	25-10000
Commandement de la Force Aérienne de Combat	France	122	25-22300
Commandement des Écoles de l'Armée de l'Air (CEAA)	France	122	25-22500
Commander British Forces Gibraltar	Gibraltar	185	25-40000
Commander Standing Naval Force Atlantic (STANAVFORLANT)	UK	418	34-20050
Commerce International Group Ltd (CIG)	Spain	368	85-00300
Commerce Overseas Corp.	USA	651	90-03640
Commercial Contracts Division	UK	409	12-12000
Commercial Hydraulics GmbH	Germany	166	75-05050
Commercial Hydraulics Keelavite Ltd	UK	441	75-14410
Commercial Hydraulics Kontak Ltd	UK	441	75-14415
Commercial Hydraulics SA	Luxembourg	275	75-00350
Commercial Intertech Corp.	USA	564	75-20050
Commercial Intertech SA Astron Building Systems	Luxembourg	275	75-00350
Commercial Marketing Assistance International (CMA INTL)	USA	633	80-05855
Commercial Shearing Inc.	USA	564	75-20050
Commerciale Avio Marina SAS (CAM)	Italy	243	90-00341
Commissariat for Defence Materiel	Switzerland	382	25-10000
Commissariat Général des Expositions et Salons du GICAT (COGES)	France	128	65-00730
Commission of Science, Technology and Industry (COSTIND)	China PR	86	12-10000
Commodore Aviation Inc.	USA	633	80-05860
Commonwealth Trading Corporation	USA	646	85-00285
Commslogic Ltd	UK	500	80-06400
Communication Control Systems, CCS	UK	522	90-00325
Communications & Machinery Corporation	Pakistan	315	85-01000
Communications & Measurement Technologies Ltd	UK	441	75-14460
Communications & Power Industries	Canada	70	75-03550
Communications & Power Industries	USA	565	75-20393
Communications & Power Industries International Inc	Switzerland	385	75-01900
Communications Instruments Inc. (CII)	USA	565	75-20300
Communications-Applied Tech.	USA	564	75-20290
Communitron (Australia) Pty Ltd	Australia	14	75-03860
COMNAVACTUK = US Naval Activities, United Kingdom	UK	418	33-45000
Compac	USA	565	75-20400
Compact A/S	Norway	309	75-00540
Compagnia Importazioni & Esportazioni Rappresentanze	Italy	243	90-00340
Compagnie d'Études, de Réalisations et d'Installations de Systèmes	France	134	75-12450
Compagnie Française de Manutention	France	152	90-00350
Compagnie Générale de Géophysique Head Office (CGG)	France	134	75-11335
Compagnie Industrielle des Lasers (CILAS)	France	134	75-10240
Compagnie Industrielle des Reflecteurs Radar (CIRRA)	France	134	75-11450
Compagnie Industrielle Radioélectrique SA (CIR)	Switzerland	385	75-01800
Compair (Australasia) Ltd	Australia	14	75-03870
CompAir BroomWade Ltd	UK	442	75-14520
CompAir Holman Iberica SA	Spain	366	75-01900
CompAir Luchard SA	France	134	75-10250
CompAir Reavell Ltd	UK	442	75-14530
Companhia Brasileira de Cartuchos (CBC)	Brazil	55	75-03150
Companhia Comércio e Construções (CCC)	Brazil	56	75-03230
Compañia Aeronàutica Española SA (CAESA)	Spain	368	85-00400
Compañia de Ingenieria de Sistemas y Desarrollos Functionales Ltda (SISDEF)	Chile	85	80-01940
Complas Limited	UK	442	75-14543
Component Repair Technologies Inc. (CRT)	USA	633	80-05896
Composants et Systèmes de Précision	France	150	80-00375
Composite Limited	Israel	226	90-00340
Composites Aquitaine	France	134	75-10265
Compounding Ingredients Ltd	UK	442	75-14545
Compstock International	UK	515	85-00350
CompStream Corporation	China PR	89	90-00400
CompStream Corporation	Indonesia	210	90-00100
Comptek Federal Systems Inc.	USA	565	75-20650
Compton Webb Ltd	UK	442	75-14550
Computer Optics Inc.	USA	565	75-20680
Computer Power Group Ltd	Australia	24	80-03730
Computer Resources International A/S (CRI)	Denmark	104	80-00100
Computer Sciences Corp	UK	500	80-07500
Computer Sciences Corp. Corporate Headquarters (CSC)	USA	565	75-20720
Computerised Training Systems Ltd (CTS)	UK	442	75-14650
Computing Applications Software Technology Inc. (CAST)	USA	651	90-03665
Computing Devices Canada	Canada	70	75-03600
Computing Devices Canada Ltd.	Canada	70	75-03600
Computing Devices Canada Ltd.	UK	442	75-14620
Computing Devices Company Ltd	UK	442	75-14700
Computing Devices Hastings (UK) Ltd UK Office	UK	522	90-00364
Computing Devices International	USA	565	75-20730
Comrod A/S	Norway	309	75-00550
Comstream Corporation	Canada	81	90-00420
ComStream Corporation	UK	522	90-00367
Comstream Corporation	USA	565	75-20780
Comstream Corporation, Andes Region Branch	Colombia	91	90-00400
Comtech Systems Inc.	USA	565	75-20810
Comtronic GmbH	Germany	166	75-05070
Conap Inc.	USA	565	75-20825
Conax Florida Corp.	USA	565	75-20833
Concentration, Heat & Momentum/CHAM.	UK	499	80-05900
CONCO Inc.	USA	566	75-20835
Concoat Ltd Materials and Machinery Division	UK	442	75-14780
Concorde Battery Corp.	USA	566	75-20845
Concurrent Computer Corporation Ltd	UK	442	75-14810
Concurrent Computer GmbH	Germany	181	90-00350
Condor Enterprises Inc.	USA	633	80-06200
Condor Systems Inc.	USA	566	75-20950
Confederation Amicale des Ingenieurs de l'Armement (CAIA)	France	128	65-00340
Confederation of Danish Industries (DI)	Denmark	102	65-00100
Confederation of Norwegian Business and Industry (NHO)	Norway	308	65-00050
CONFORMISC = État-Major du Commandement des Formations Militaires por la Sécurité Civile	France	123	30-00100
Congreso de los Diputados	Spain	362	15-20000
Congress of Deputies	Spain	362	15-20000
Conjay Arms Co. Ltd.	UK	442	75-14855
Connecticut Machine Products Inc. (CMP)	USA	566	75-21050
Connector Microtooling Systems Inc.	USA	579	75-40355
Connei SpA	Italy	233	75-05380
Connettori Elettrici Industriali	Italy	233	75-05380
Connex Italiana	Italy	243	90-00344
Conplan AB	Sweden	380	85-03920
Conplan International AB	Sweden	380	85-03920
Conrac Elektron GmbH	Germany	166	75-05100
Conrac Systems Inc	USA	566	75-21520
Conrac UK Limited	UK	522	90-00372
Conseil des Industries de Défense Françaises (CIDEF)	France	128	65-00400
Conseil National des Transports	France	121	12-70500
Consolidated Cybernetics Controls Ltd	Canada	78	80-00360
Consolidated Manufacturing Industries Limited	Australia	15	75-03925
Consortium Aviaspectrans JSC (AST)	Russia	336	74-01200
Construcciones Aeronáuticas SA (CASA)	Spain	364	55-00120
Constructeurs Associés de Matériels d'Incendie, Voirie, Aviation (CAMIVA)	France	134	75-10280
Construction Association of Korea	Korea (ROK)	264	65-00500
Construction Véhicules Industriels	France	139	75-25200
Constructions Industrielles de la Méditerranée	France	133	75-09990
Constructions Mécaniques des Vosges Marioni (CMW)	France	134	75-12300
Consulplano SA	Portugal	327	80-00320
Consultancy in Advanced Technology	S. Africa	357	75-03100
Consulting Group Headquarters	USA	565	75-20720
Contahal Ltd	Israel	224	80-00300
Container Research Corp. (CRC)	USA	566	75-21700

Quick Reference

Contemporary Systems Design Ltd (CSD)	UK	442	75-14860
Conteneurs d'Alsace, Les	France	138	75-24590
Continental AG	Germany	166	75-05150
Continental AG Werk Uniroyal	Germany	176	75-32090
Continental Electronics Corp	USA	566	75-21800
Continental Microwave and Toll Co Inc	USA	566	75-21830
Continental Microwave Technology Ltd	UK	470	75-42318
ContiTech Holding GmbH	Germany	166	75-05150
Contract Professionals, Inc. (CPI)	USA	633	80-06470
Contraves GmbH	Germany	166	75-05200
Contraves Inc	USA	566	75-21850
Contraves Italiana SpA	Italy	237	75-16800
Control Data Systems Canada Ltd	Canada	68	75-00480
Control Products Corp	USA	566	75-21928
Control Techniques Dynamics Ltd	UK	442	75-14875
Control Yuan	Taiwan	392	15-20000
CONTROP Precision Technologies Ltd	Israel	218	75-04170
Conventional Munitions Systems Inc.	USA	564	75-19170
Cooksey Consulting Pty Ltd	Australia	25	80-06500
Cooper Aviation International Inc.	USA	652	90-03680
Coopers Filters Ltd	UK	442	75-14950
Coors Ceramics Co. Structural Products Group	USA	566	75-22010
Coperimprel SA	Peru	319	90-00350
Cordant, Inc.	USA	644	80-33154
Cordwell Investigations	Australia	24	80-03780
Corina Corporation AG	Switzerland	389	90-00360
CORIS	France	134	75-12450
Cornes & Co Ltd	Japan	257	90-00300
Corrigan Technologies Ltd	Canada	78	80-00380
Corrintec Ltd	UK	442	75-15120
Cortana Corp	USA	540	60-03500
Cortex Vicsion Systems, Inc	USA	566	75-22135
Cosens & Co Ltd	UK	443	75-15140
Cosens Engineering Ltd	UK	443	75-15140
COSM Ltd Equipamento Aeronáutico e Maritimo	Portugal	327	80-00400
COSMOS SpA	Italy	233	75-05700
Cossor Electronics Ltd	UK	478	75-48367
Costain Civil Engineering Ltd	UK	443	75-15213
Costain Group PLC	UK	422	70-00360
Costain Marine International Ltd	UK	443	75-15213
Costain Middle East	UAE	407	75-00300
COSTIND = Commission of Science, Technology and Industry	China PR	86	12-10000
Cosworth Engineering Inc	USA	652	90-03690
Cosworth Racing Inc	USA	652	90-03690
Cotag International	UK	443	75-15235
Cotag International	USA	566	75-22200
Cougar Limited	UK	443	75-15250
Council of Defense & Space Industry Associations (CODSIA)	USA	542	65-00300
Council of Europe Crime Problems Division	France	125	34-00080
Counterrisk Inc	Canada	78	80-00385
Courtalds Aerospace	USA	625	75-93450
Courtaulds Aerospace	UK	443	75-15330
Coutant Electronics Ltd	UK	443	75-15390
Coutant-Lambda Ltd	UK	443	75-15390
Covergie SA	France	134	75-12400
Cow Industrial Polymers Ltd	UK	443	75-15650
Cox & Company Inc	USA	566	75-22300
CP Cases	UK	443	75-15680
C.P. Mega	UK	467	75-38330
CPAD Holding Limited	Canada	78	80-00386
CPAD Technologies Inc	Canada	78	80-00386
CPI = Contract Professionals, Inc.	USA	633	80-06470
CPI Europe Ltd	Sweden	381	90-02200
CPI International Inc	Switzerland	389	90-00365
CPP = Cambridge Parallel Processing	USA	562	75-16480
CQC Plc	UK	443	75-15700
Craig Systems	USA	566	75-22350
Crane Aluminium Extrusion	Australia	15	75-03954
Crane Co. Hydro-Aire Inc	USA	566	75-22452
Crane Fruehauf Ltd	UK	443	75-15750
Crane Inc., John	USA	566	75-22500
Cranfield Aerospace Ltd	UK	500	80-07200
Cranfield Precision	UK	443	75-15772
Cranfield Precision Engineering Ltd	UK	443	75-15772
Cranfield University	UK	419	60-00350
Crawford Collets	UK	443	75-15870
Cray Communications Ltd	Australia	11	75-01293
Cray Marine	UK	443	75-15915
Cray Research Inc	USA	566	75-22570
Cray Systems Ltd	UK	497	80-01240
Crayford Special Equipment Co. Ltd	UK	515	85-00370
CRC = Container Research Corp	USA	566	75-21700
CRDV = Centre de Recherche pour la Défense Valcartier	Canada	67	60-00450
Crellon Microsystems	UK	516	85-00375
Crescent Aviation Services	Pakistan	316	90-00430
Crescent International	USA	652	90-03740
Crestbury Ltd	UK	443	75-16030
Crew Services Ltd	UK	500	80-07370
Crewsaver	UK	443	75-16050
CRI = Computer Resources International A/S	Denmark	104	80-00100
Crime Prevention Council	Germany	161	60-00450
Criminal Intelligence Administration	USA	633	80-06730
Criminal Research Products Inc	USA	566	75-22590
Criminological Research Institute of Lower Saxony	Germany	162	60-01080
Criminology Institut	Spain	364	60-02000
Criminology Institut of Barcelona	Spain	364	60-01900
Critec Pty Ltd	Australia	16	75-07600
CRL = Central Research Laboratories Ltd	UK	499	80-05895
CRL Technologies	Canada	70	75-03760
CRM Fabbr. Motori Marini e Industriali	Italy	233	75-06300
CRM SpA Motori Marini	Italy	233	75-06300
CRM Technologies Ltd	Israel	226	90-00360
Crompton Eternacell Ltd	UK	457	75-28500
Crompton Greaves Ltd Registered Office	India	205	75-00280
Crompton Instruments (Aust) Pty Ltd	Australia	28	90-03750
Crompton Instruments BV	Netherlands	292	75-01220
Crompton Instruments Inc	USA	652	90-03750
Crompton Instruments Ltd	Hong Kong SAR	198	90-00300
Crompton Instruments Ltd	UK	443	75-16070
Crompton Instruments Ltd Canada	Canada	81	90-00450
Crompton Instruments (S.E. Asia) Pty Ltd	Singapore	351	90-00350
Crompton Messinstrumente GmbH	Germany	181	90-00375
Cronite Precision Castings Ltd	UK	443	75-16136
Crosfield Group	UK	444	75-16150
Crouse Hinds Australia Pty Ltd	Australia	15	75-03963
CRP Marine	UK	443	75-16140
CRT = Component Repair Technologies Inc.	USA	633	80-05896
CryoService Ltd	UK	516	85-00380
Crypto AG	Switzerland	385	75-02400
Crystallume	USA	563	75-18425
CR2A-DI	France	134	75-12600
CSC = Computer Sciences Corp. Corporate Headquarters	USA	565	75-20720
CSC Artemis Limited	UK	497	80-01259
CSC Computer Sciences Ltd	UK	500	80-07500
CSC Financial Services Headquarters	USA	565	75-20720
CSC (Specialised Vehicles) Ltd	UK	444	75-16220
CSD = Contemporary Systems Design Ltd	UK	442	75-14860
CSE Aviation Ltd	UK	500	80-07600
CSI = Circuiti Stampati Italia	Italy	233	75-05183
CSIP	UK	444	75-16250
CSIST = Chungshan Institute of Science and Technology	Taiwan	392	60-01000
CSTC = China Shipbuilding Trading Co. Ltd	China PR	88	55-00360
CTA Inc	USA	633	80-06810
CTF Systems Inc	Canada	70	75-03810
CTL = Cincinnati Testing Laboratories Inc.	USA	633	80-05490
CTL-Aerospace Inc	USA	567	75-22700
CTP Logistics Ltd	UK	500	80-07650
CTS = Computerised Training Systems Ltd	UK	442	75-14650
CTS Corporation	USA	567	75-22800
Cuartel General de la Armada	Spain	362	25-40000
Cubic Communications Inc. OAR DF Products Group	USA	567	75-22850
Cubic Corp	Japan	257	90-00380
Cubic Defense Systems Inc	USA	567	75-22900
Culligan & McCallum	USA	567	75-23145
Cummins Diesel NV	Belgium	48	85-00500
Cummins Engine Co. Inc	USA	567	75-22980
Curaçao National Guard	N. Antilles	300	25-40000
Curran Ltd, John	UK	444	75-16430
Currie & Brown (Australia) Pty Ltd	Australia	24	80-03850
Currie & Brown, Cameron & Middleton Pty Ltd	Australia	24	80-03850
Curtis Instruments (UK) Ltd	UK	444	75-16440
Curtiss-Wright Corp	USA	567	75-23050
Curtiss-Wright Flight Systems Inc	USA	567	75-23055
Curtiss-Wright Flow Control Corporation	USA	619	75-86366
Custom Cameras Ltd	UK	444	75-16444
Custom Electronics Inc	USA	567	75-23085
Custom Electronics Inc	USA	652	90-03800
Customs & Excise	Austria	31	29-00500
Customs & Excise, H.M.	UK	412	29-00500

Quick Reference

Quick Reference

E

Quick Reference

Quick Reference

Quick Reference

F

Quick Reference

FLAG = Flemish Aerospace Group	Belgium	44	65-00300
Flag Officer Commanding Fleet	Chile	83	25-42800
Flag Officer Sea Training	UK	412	25-42500
Flagship Training	UK	502	80-10500
Flame Enterprises Inc.	USA	647	85-00600
Fleet Industries Ltd	Canada	71	75-05500
Flemish Aerospace Group (FLAG)	Belgium	44	65-00300
Flexibox Ltd	UK	451	75-23800
Flexibulb Plastics Inc.	Canada	71	75-05550
Flight Components AG	Switzerland	389	90-00640
Flight Data Company Ltd	UK	502	80-10600
Flight International Inc. Group Headquarters	USA	635	80-11600
Flight Management International Corporation	USA	635	80-11630
Flight Operations Ltd.	UK	502	80-10650
Flight Products Inc.	USA	575	75-34700
Flight Refuelling Limited FR-HiTEMP Division.	UK	451	75-23850
Flight Suits Ltd	USA	575	75-34800
Flightcases International A/S	Denmark	102	75-02000
Flightline Aircraft Instruments	USA	575	75-34840
FlightSafety International Inc.	USA	635	80-11650
FlightSafety International Inc. deHavilland Learning Center	Canada	78	80-00660
FlightSafety International Inc. Montreal Learning Center	Canada	78	80-00650
FlightSafety International Inc Paris Learning Center	France	150	80-00600
Flodins Filter AB	Sweden	377	75-05300
Flottweg GmbH	Germany	168	75-09250
Floturn Inc.	USA	576	75-35060
FLS Aerospace (Light Aircraft) Ltd	UK	451	75-23900
Flugmálastjórn	Iceland	201	20-00100
Flugzeugwerke Altenrhein AG, FFA	Switzerland	386	75-04350
Fluid Gravity Engineering Ltd	UK	502	80-10660
Fluor Daniel Inc.	USA	635	80-11720
Fluorglas	USA	576	75-35200
Fluorocarbon Company Limited	UK	451	75-24020
Fluorocarbon Seals Ltd	UK	451	75-24020
Flygtekniska Försöksanstalten	Sweden	374	60-00100
FMC-Nurol Savunma Sanayii (FNSS)	Turkey	402	74-00680
FMI Composites Ltd	UK	451	75-24040
FN Herstal SA	Belgium	46	75-03600
FN Manufacturing Inc (FNMI)	USA	576	75-35320
FN Moteurs	USA	656	90-29250
FNMI = FN Manufacturing Inc.	USA	576	75-35320
FNSS = FMC-Nurol Savunma Sanayii.	Turkey	402	74-00680
FOA = Försvarets Forskningsanstalt	Sweden	374	60-01400
Focal Displays Ltd	UK	451	75-24095
Focal Signs Ltd.	UK	451	75-24095
Fockas Pantelis Ltd	Greece	189	75-01293
Foden Trucks	UK	451	75-24100
Foerster Instruments Inc.	USA	576	75-35330
Foerster Italia Sarl	Italy	234	75-09300
Foerster UK Ltd	UK	451	75-24150
Fokker Aerostructures BV	Netherlands	292	75-02305
Fokker Aviation NV	Netherlands	292	75-02300
Fokker Defence Marketing	Netherlands	292	75-02300
Fokker Elmo BV	Netherlands	292	75-02320
Fokker Services BV	Netherlands	297	80-06600
Fokker Space & Systems	Netherlands	293	75-02325
Fokker Special Products BV	Netherlands	293	75-02330
Fonderies d'Ussel, Société	France	146	75-40400
Fono	West Samoa	341	15-10000
Forbairt	Ireland	213	20-00600
Forcas Armadas de Defesa de Moçambique (FADM)	Mozambique	285	25-10000
Force Aerienne de la Côte d'Ivoire	Cte d'Ivoire	93	25-20000
Force Aérienne du Burkina	Burkina Faso	63	25-20000
Force Computers GmbH European Headquarters	Germany	168	75-09270
Force Computers (UK) Ltd.	UK	451	75-24155
Forces Alliées de l'Atlantique (GQG)	USA	539	34-20000
Forecast Associates	USA	636	80-11760
Forecast International/DMS	USA	636	80-11760
Foreign Affairs Committee	Austria	30	15-10000
Foreign Legion First Foreign Regiment	France	122	25-35000
Foremost Industries Inc.	Canada	71	75-05620
Foreningen for Danske Aerospace Industri (FDAI)	Denmark	102	65-00050
Forged Rolls (UK) Ltd	UK	452	75-24165
Forgemasters Steel & Engineering Ltd	UK	452	75-24168
Forges de Zeebrugge SA	Belgium	46	75-03750
Forjas Taurus SA.	Brazil	56	75-06640
Formsprag-Warren.	USA	576	75-35650
Forrester Stephen Aviation	Australia	28	90-06830
Förster Imaden	Brazil	56	75-06650
Forster-France Sarl	France	136	75-18300
Forsvarets Bygningstjeneste	Denmark	100	20-04300
Försvarets Forskningsanstalt (FOA)	Sweden	374	60-01400
Forsvarets Forskningsintitutt	Norway	307	60-01400
Forsvarets Høyskole	Norway	307	25-19300
Forward Industries Ltd.	UK	452	75-24200
Fotec Inc.	USA	576	75-35680
Fothergill Engineered Fabrics Ltd	UK	452	75-24380
Fotona D.D.	Slovenia	355	75-00400
Fourth Foreign Regiment	France	122	25-35000
Fourth Naval Zone Headquarters	Chile	83	25-42400
Fox & Sons Ltd, C.J.	UK	516	85-00620
Foxtronics Inc.	USA	576	75-35752
FP Diesel LP	USA	576	75-35753
FP Diesel Ltd	UK	516	85-00630
FPE Fischer GmbH	Germany	168	75-09300
FPI Inc.	USA	576	75-35755
FPLM = People's Forces Headquarters General Headquarters of FADM	Mozambique	285	25-10000
FPT Inc.	USA	576	75-35760
FPT Industries Ltd	UK	452	75-24400
FR Aviation Limited	UK	503	80-10850
FR Group	UK	422	70-00350
F.R. Ordnance International Ltd	UK	452	75-24480
Framatome	France	136	75-18405
Framatome Connectors Belux	Belgium	49	90-00690
Framatome Connectors International (FCI)	France	136	75-18405
Framatome Connectors Nederland BV	Netherlands	299	90-00630
Francis & Lewis International Ltd	UK	452	75-24500
Francis B. Willmott Ltd.	UK	494	75-66310
Francis Lord Optics	Australia	18	75-12550
Francisco Eduardo Müller Botelho	Brazil	59	90-13850
Franco-German Brigade	Germany	160	34-41000
Franerex BV	Netherlands	293	75-02335
Franz Inc.	USA	576	75-35850
Franz-V.-Liszt-Haus	Germany	161	60-00490
Frasca International Inc.	USA	576	75-35900
Fraser Pty Ltd., H.I.	Australia	16	75-06835
Fraser-Volpe Corp.	USA	576	75-35950
Frazer-Nash Consultancy Limited	UK	503	80-10950
Frazer-Nash Engineering Ltd	UK	452	75-24660
Frazer-Nash (Midhurst) Ltd	UK	452	75-24660
Frazier Aviation Inc.	USA	576	75-35970
Fred V. Fowler Co Inc.	USA	653	90-06700
Frederick Duffield Pty Ltd	Australia	16	75-06840
Frederick Electronics SFA/DataComm Inc.	USA	576	75-36000
Freeman & Proctor Ltd	UK	452	75-24700
Freeman Marine Equipment Inc.	USA	576	75-36100
French Aeronautics and Space Industry Group	France	128	65-00740
French Air Force Training Command	France	122	25-22500
French-German Research Institute of Saint Louis (ISL)	France	122	20-35000
Frequency Electronics Inc. (FE)	USA	576	75-36190
Fresard Microtechniques SA	France	135	75-13680
Freudenberg Technical Products LP	UK	452	75-24730
Friskem Division, Infinetics Inc.	USA	584	75-46700
FRIWO SILBERKRAFT Gesellschaft für Batterietechnik mbh	Germany	175	75-28000
Frochot Representaciones Industrial, Michel.	Venezuela	659	90-01300
Frontec Corporation	Canada	78	80-00670
Frontier Guard Headquarters	Finland	116	28-10000
Frost & Sullivan Ltd	UK	503	80-10970
Frost Engineering	USA	576	75-36300
Froude Consine	UK	452	75-24800
Froude Engineering Inc.	USA	576	75-36340
Froude France	France	152	90-00665
Frunze M.V., Arsenal Design Bureau	Russia	333	55-01300
FSI = Fastening Systems International Inc.	USA	575	75-33870
FT Technologies Ltd	UK	452	75-24880
FTL Fluid Seals Ltd.	UK	452	75-24870
FTL Seals Technology Limited	UK	452	75-24870
FUCHS Electronics.	S. Africa	358	75-06000
Fuehrungsakademie der Bundeswehr	Germany	161	60-00700
Fuelquip Services Ltd	New Zealand	302	75-00700
Fuerza Aérea Argentina	Argentina	3	25-20000
Fuerza de Naval Aviación	Peru	318	25-34000
Fuerzas Aeromoviles del Ejército de Tierra (FAMET)	Spain	362	25-31000
Fugro NV	Netherlands	291	70-01000
Fugro-UDI Ltd	UK	452	75-24885
Führungsstab der Marine	Germany	157	25-40000
Fuji Electric Co Ltd	Japan	251	75-01050
Fuji Heavy Industries Ltd (FHI)	Japan	251	75-01100
Fuji Industries Co Ltd	UK	522	90-00700

Fuji Industries Co Ltd (FIC)	Japan	*257*	90-00600
Fuji Industries Corp Los Angeles Office	USA	*653*	90-06830
Fuji Jukogyo KK	Japan	*251*	75-01100
Fujisawa Manufacturing Division	Japan	*251*	75-01200
Fujitsu Ltd	Japan	*251*	75-01400
Fullerton, Sherwood Engineering Ltd	Canada	*71*	75-05625
Furneaux Riddall & Co. Ltd	UK	*452*	75-24890
Furno Co. Inc.	USA	*577*	75-36500
Furon	Belgium	*49*	90-00700
Furon	France	*136*	75-18650
Furon	Germany	*168*	75-09600
Furon	Netherlands	*299*	90-00650
Furon CHR	USA	*576*	75-36400
Furon Co. Holmesville Site	USA	*577*	75-36550
Furon Company	USA	*576*	75-36380
Furon Forming Center of Excellence	USA	*547*	75-02450
Furon Limited	UK	*453*	75-24900
Fusion Inc.	USA	*653*	90-06870
Futronix Inc.	USA	*577*	75-36650
Future Systems Group	UK	*510*	80-28600
FWD International Inc.	USA	*577*	75-36680
FXC Corp.	USA	*577*	75-36700

G

G E Capital Spacenet Services, Inc	USA	*636*	80-13310
G Kull AG	Switzerland	*386*	75-07200
G Kull AG, Lunor	Switzerland	*386*	75-07200
G L Rexroth Ltd.	UK	*466*	75-37100
G P Embelton & Co. Pty Ltd	Australia	*16*	75-05740
G. Shiposh, Engineering Works Ltd	Israel	*222*	75-20250
G&H Technology Inc.	USA	*578*	75-39550
GAAM Fire & Rescue	Australia	*16*	75-07000
Gabi Shoef Ltd	Israel	*219*	75-07650
Gabriel Electronics Inc.	USA	*577*	75-36715
GAF Support Command	Germany	*157*	25-22200
GAIC = Guizhou Aviation Industry Corp.	China PR	*88*	55-00800
Gaimusho	Japan	*245*	12-20000
Gajon Institute of Technology SpA	Italy	*229*	60-00080
Galileo Corp.	USA	*577*	75-36755
Galileo Electro-Optics Corp.	USA	*577*	75-36755
Gallay Ltd.	UK	*453*	75-25100
Gallet CGF SA	France	*133*	75-08840
Gallet SA	France	*133*	75-08840
Gamatronic Electronic Industries Ltd	Israel	*219*	75-07680
GAMCO = AlliedSignal Aerospace Marketing Sales & Service	UAE	*408*	90-00100
GAMESA = Grupo Auxiliar Metalurgico SA	Spain	*366*	75-03650
Gamrat SA	Poland	*324*	75-01000
Gandalf Canada Ltd	Canada	*71*	75-05650
Gandalf Technologies Inc.	Canada	*71*	75-05650
GAP Instrument Corp.	USA	*577*	75-36900
Garbarino Pompe SpA	Italy	*234*	75-09530
Garber International Associates Inc. (GIA)	USA	*636*	80-12150
Garde Inc.	USA	*590*	75-55060
Garden Reach Shipbuilders & Engineers Ltd	India	*203*	55-00300
Garmin (Europe) Ltd	UK	*453*	75-25150
Garmin International	USA	*577*	75-36920
Garofoli Costruzioni Meccaniche SpA	Italy	*235*	75-09550
Garphyttan Industrier	UK	*516*	85-00660
Garrard & Co Ltd	UK	*453*	75-25205
Garret GmbH	Germany	*164*	75-00610
Garrett Engine Division, AlliedSignal	USA	*550*	75-04700
Garrett Metal Detectors	USA	*577*	75-36950
Garrett Singapore Pte Ltd	Singapore	*348*	75-00550
Garsite Inc.	USA	*579*	75-39950
Gates Hydraulics Ltd	UK	*453*	75-25210
Gavriil C.	Greece	*189*	75-01294
Gayrobot Undersea Technology	Switzerland	*386*	75-04700
G.B. Britton	UK	*437*	75-09900
GCE = General Consulting & Engineering Srl	Italy	*240*	80-00735
GD Engineering	UK	*475*	75-46050
GD Protective Clothing and Equipment Limited	Ireland	*214*	75-00120
GDI = Graphic Datakits International	USA	*579*	75-40355
GDI Simulation	France	*137*	75-19700
GE = General Electric Co.	USA	*578*	75-38400
GE Aircraft Engines	Germany	*181*	90-00740
G.E. Mortley, Sprague & Co. Ltd	UK	*469*	75-41000
GE Silicones	UK	*516*	85-00690
GEBECOMA = Groupement Belge des Constructeurs de Matériel Aérospatial	Belgium	*44*	65-00060
GEC = The General Electric Company plc	UK	*423*	70-00740
GEC Aerospace Ltd	UK	*453*	75-25300

GEC ALSTHOM Diesels Ltd	UK	*453*	75-25220
GEC ALSTHOM Engineering Systems Ltd	UK	*453*	75-25225
GEC Alsthom French Head Office	France	*130*	74-00650
GEC ALSTHOM Gears Ltd	UK	*453*	75-25230
GEC ALSTHOM International Inc.	USA	*653*	90-06887
GEC ALSTHOM Large Machines Ltd	UK	*453*	75-25240
GEC ALSTHOM Large Steam Turbines	France	*136*	75-19040
GEC ALSTHOM Large Steam Turbines	UK	*453*	75-25242
GEC ALSTHOM Medium Turbo Machines	UK	*453*	75-25243
GEC ALSTHOM Paxman Diesels Ltd	UK	*453*	75-25245
GEC ALSTHOM Regulateurs Europa Ltd	UK	*453*	75-25250
GEC ALSTHOM Ruston Diesels Ltd.	UK	*453*	75-25255
GEC ALSTHOM Sales Network Inc.	USA	*653*	90-06887
GEC ALSTHOM T&D Power Electronic Systems Ltd	UK	*453*	75-25260
GEC ALSTHOM Turbine Generators Ltd	UK	*453*	75-25243
GEC ALSTHOM UK Head Office	UK	*425*	74-06550
GEC Avery Ltd	UK	*432*	75-04560
GEC Electronics Division	Australia	*16*	75-07350
GEC Malaysia Sdn Bhd	Malaysia	*278*	90-00730
GEC Marconi Hazeltine Corp Electronic Systems Division	USA	*578*	75-37415
GEC of Singapore Pte Ltd	Singapore	*351*	90-00710
GEC Plessey Semiconductors	France	*152*	90-00720
GEC Plessey Semiconductors	UK	*454*	75-25363
GEC Plessey Semiconductors	USA	*578*	75-37425
GEC Zimbabwe (Pvt) Ltd.	Zimbabwe	*663*	90-01950
GEC 3 Ltd	Greece	*189*	75-01300
GEC-Marconi	USA	*653*	90-06890
GEC-Marconi Aerospace Inc.	USA	*577*	75-37412
GEC-Marconi Aerospace Ltd	UK	*453*	75-25300
GEC-Marconi Aerospace Systems	UK	*453*	75-25310
GEC-Marconi Avionics Inc.	USA	*577*	75-37413
GEC-Marconi Avionics Ltd.	UK	*454*	75-25320
GEC-Marconi Communications Ltd	UK	*454*	75-25335
GEC-Marconi Dynamics Ltd (GMDL)	UK	*454*	75-25337
GEC-Marconi Electronic Systems Corp.	USA	*577*	75-37414
GEC-Marconi Hazeltine Corp.	USA	*577*	75-37414
GEC-Marconi Infra-Red Limited	UK	*454*	75-25338
GEC-Marconi Ltd.	UK	*422*	70-00735
GEC-Marconi Materials Technology Ltd	UK	*419*	60-00600
GEC-Marconi Materials Technology Ltd	UK	*454*	75-25340
GEC-Marconi Materials Technology Ltd Hirst Division	UK	*419*	60-00550
GEC-Marconi (New Zealand) Ltd	New Zealand	*304*	90-00730
GEC-Marconi Research Centre	UK	*419*	60-00700
GEC-Plessey Telecommunications Ltd	UK	*425*	74-06600
GECI Ltd	UK	*503*	80-11000
GECO Defence A/S	Norway	*309*	75-00875
GED (Southampton) Ltd	UK	*503*	80-11650
GEFA-FLUG = Gesellschaft zur Enticklung und Forderung Aerostatischer Flugsysteme	Germany	*169*	75-10700
Geier & Bluhm Inc.	USA	*578*	75-37427
Geislinger & Co Schwingungstechnik GmbH	Austria	*32*	75-00990
GEL Engineering Ltd	UK	*454*	75-25500
Gelbyson Srl	Italy	*241*	85-00600
Gelvenor Textiles	S. Africa	*358*	75-06100
GEM = Graff Electronic Machines Ltd.	UK	*455*	75-26450
GEM Elettronica Srl	Italy	*235*	75-10100
Gemini International Corp.	USA	*647*	85-00710
Genco Mobile Systems BV	Netherlands	*293*	75-02340
GenCorp	USA	*543*	70-05050
Gendarmerie, The	Belgium	*43*	60-00300
General Atomics	USA	*578*	75-37447
General Aviation Sarl	France	*153*	90-00732
General Cables Orfandis Brs SA	Greece	*189*	75-01297
General Connectors Corp.	USA	*556*	75-11100
General Consulting & Engineering Srl (GCE)	Italy	*240*	80-00735
General Customs Administration (Administracion General de Aduanas)	Costa Rica	*92*	29-10000
General Directorate of Civil Protection	Spain	*363*	30-00500
General Directorate of Customs	Czech Republic	*97*	29-01000
General Directorate of Customs	France	*123*	29-01000
General Directorate of Customs	Iceland	*201*	29-01000
General Directorate of Customs & Excise	Germany	*158*	29-01000
General Dynamics Ordnance Systems	USA	*535*	25-36500
General Electonics	Greece	*189*	75-01300
General Electric Co. (GE)	USA	*578*	75-38400
General Electric Co. of Bangladesh Ltd	Bangladesh	*37*	90-00730
General Electric Co of Hong Kong, The	Hong Kong SAR	*199*	90-00730
General Electric Company of Singapore Pte Ltd	Singapore	*351*	90-00710
General Electric Company plc	UK	*423*	70-00740

Quick Reference

H

HAECO = Hong Kong Aircraft Engineering Co. Ltd	Hong Kong SAR	*198*	80-00200
Hagen	UK	*456*	75-27220
Hagen Morfax Ltd	UK	*456*	75-27220
Hagenuk GmbH	Germany	*170*	75-14800
Hägglunds GmbH	Germany	*181*	90-00805
Hägglunds Liaison Office	Canada	*81*	90-00800
Hägglunds Liaison Office	Korea (ROK)	*267*	90-00811
Hägglunds Moelv AS	Norway	*309*	75-00880
Hägglunds Vehicle AB	USA	*653*	90-08035
Hägglunds Vehicle AB Head Office	Sweden	*377*	75-05800
Hägglunds Vehicle GmbH	Germany	*181*	90-00805
HAGS = Hellenic Army General Staff	Greece	*185*	25-30000
HAI = Hellenic Aerospace Industry Ltd	Greece	*192*	80-00750
HAIC = Hermetic Aircraft International Ltd Corp	USA	*581*	75-43100
Hainsworth & Sons Ltd., A.W.	UK	*456*	75-27300
HAL = Hindustan Aeronautics Ltd	India	*203*	55-01000
HAL Communications Corp.	USA	*580*	75-41450
Hale Group International	USA	*636*	80-13600
Hale Hamilton (Valves) Ltd	UK	*456*	75-27350
Haley & Weller Ltd	UK	*456*	75-27400
Haley Industries Ltd	Canada	*72*	75-06600
Halifax Shipyard Limited	Canada	*76*	75-12300
Halifax Shipyard Limited	Canada	*78*	80-00810
Hall & Watts (Defence Optics) Group	**UK**	***456***	**75-27650**
Hall Limited, J&E	UK	*456*	75-27420
Hallam Engineering (Leicester) Ltd, E C	UK	*456*	75-27430
Halmatic Ltd	UK	*456*	75-27700
Halocarbon Products Corp.	USA	*580*	75-41630
Halter Marine Group	USA	*580*	75-41700
Hamech Ltd	UK	*456*	75-27730
Hämmerli AG Target Arms Factory	Switzerland	*386*	75-05850
Hammonds Technical Services Inc.	USA	*580*	75-41800
Hamworthy Marine	UK	*457*	75-27850
Hamworthy Pumps & Compressors Ltd	UK	*457*	75-27850
Han Sun International Inc.	Korea (ROK)	*265*	75-01480
Hanjin Heavy Industries Co. Ltd (HHIC)	Korea (ROK)	*265*	75-01505
Hanjin Heavy Industries Co Ltd London Office	UK	*522*	90-00820
HANOMAG Aktiengesellschaft	Germany	*171*	75-16185
Hans Harboe-Hansen	Denmark	*105*	80-00800
Hansen Trading AS, BS	Norway	*312*	90-00700
Hanshin Diesel Works Ltd	Japan	*251*	75-01820
Harbin Dongan Engine Company	China PR	*88*	55-00850
Harboro Rubber Co. Ltd	UK	*457*	75-27970
Harco Manufacturing Co. Inc.	USA	*580*	75-41870
Hardcote Technologies	USA	*580*	75-41900
Hardie Engineering, James	New Zealand	*303*	80-00900
Hardigg Cases	USA	*580*	75-41930
Hardware Specialty Co. Inc. (HS)	USA	*647*	85-00800
Harkness Hall Ltd	UK	*457*	75-28050
Harowe Servo Controls	USA	*554*	75-08015
Harris & Associates, Arthur	USA	*653*	90-08055
Harris Brett Ltd	UK	*457*	75-28140
Harris Corporation Corporate Office	USA	*581*	75-42050
Harris GmbH Computer Systems Division	Germany	*169*	75-11900
Harrison & Sons, T S	UK	*457*	75-28150
Hart & Son Inc., Robert W.	USA	*581*	75-42450
Hartron Corporation	Ukraine	*406*	75-03000
Hartron Research & Production Association	Ukraine	*406*	75-03000
Harvey J. McGeorge & Associates Inc.	USA	*641*	80-24955
Haslar Hydrodynamic Test Centre	UK	*411*	20-01015
Hatcham Rubber Ltd.	UK	*457*	75-28300
Hatco A.S.	Turkey	*404*	90-00810
Hatjiantoniou SA	Greece	*189*	75-01650
Hatlapa Uetersener Maschinenfabrik GmbH & Co	Germany	*169*	75-11735
Hausler Scientific Instruments (Pty) Ltd.	S. Africa	*358*	75-06800
Hausmann GmbH Nietwerkzeuge, Edgar	Germany	*181*	90-00820
Haux-Life-Support GmbH	Germany	*169*	75-11970
Havis-Shields Equipment Corp.	USA	*581*	75-42700
Hawke Cable Glands Ltd.	UK	*457*	75-28475
Hawke Connection Systems.	UK	*457*	75-28480
Hawker Batteries Ltd.	UK	*457*	75-28490
Hawker de Havilland Limited	Australia	*17*	75-08200
Hawker Energy Products Ltd	UK	*457*	75-28490
Hawker Eternacell Ltd	UK	*457*	75-28500
Hawker Siddeley Canada Inc.	Canada	*79*	80-01450
Hawker Siddeley Japan KK	Japan	*257*	90-00812
Hawkeye Surveillance Systems Ltd	UK	*503*	80-12240
Hayes Targets	USA	*603*	75-69135
Haynes & Cann Ltd	UK	*457*	75-28750
Haynes International Inc.	USA	*581*	75-42740
Haynes International Ltd.	UK	*457*	75-28760
Haynes International Sarl	France	*153*	90-00810
Haynes International SRL	Italy	*243*	90-00800
Hazeltine Corp.	USA	*578*	75-37415
Hazmat	S. Africa	*358*	75-07000
HB Flugtechnik GmbH	Austria	*33*	75-01225
HB-Aircraft Industries	Austria	*33*	75-01225
H.C. Slingsby plc	UK	*484*	75-54430
H.C. Starck	USA	*617*	75-83957
HCC Industries Inc.	USA	*581*	75-42770
HCC Industries International	UK	*516*	85-00760
HCH Electronics	Egypt	*110*	90-08180
HCH Supply Co., Inc.	USA	*653*	90-08098
HCH Supply Co. Ltd	Egypt	*110*	90-08180
HDA Forgings Ltd	UK	*457*	75-28770
HDL Research Laboratory Inc.	USA	*581*	75-42780
HDOS = Hughes Danbury Optical Systems Inc.	USA	*582*	75-44857
HDW = Howaldtswerke Deutsche Werft AG	Germany	*170*	75-13400
Headquarters Air Cadets	UK	*411*	25-29000
Headquarters Allied Forces North Europe	Norway	*306*	25-12000
Headquarters Australian Defence Force	Australia	*6*	25-10000
Headquarters Chief of Defence Denmark	Denmark	*100*	25-10000
Headquarters Commander-in-Chief Fleet	UK	*412*	25-42100
Headquarters Defence Command Norway Air Force Staff	Norway	*306*	25-13000
Headquarters Defence Command Norway Army Staff (K-13)	Norway	*307*	25-13000
Headquarters Defence Command Norway (DEFCOMNOR)	Norway	*306*	25-13000
Headquarters Defence Command Norway Home Guard Staff	Norway	*306*	25-13000
Headquarters Flag Officer Naval Aviation	UK	*412*	25-42300
Headquarters Flag Officer Surface Flotilla	UK	*412*	25-42320
Headquarters Flagofficer Netherlands Forces Caribbean	Netherlands	*288*	25-40000
Headquarters Land Command	UK	*412*	25-32100
Headquarters Naval Home Command	UK	*412*	25-42400
Headquarters Philippine Army	Philippines	*320*	25-30000
Headquarters, Southern Joint Command	Sweden	*372*	25-32100
Headquarters Strike Command, Royal Air Force	UK	*411*	25-22100
Headquarters Training Command (Army)	Australia	*7*	25-32300
Headquarters United Kingdom Support Command (Germany) (HQ UKSC (G))	Germany	*160*	34-25000
Headquarters US Pacific Fleet	USA	*528*	25-15000
Headquarters 1st (United Kingdom) Armoured Division	Germany	*160*	34-24500
Headquarters 22d Area Support Group	Italy	*227*	25-45000
Headquaters Aviation Support Group	Australia	*7*	25-32100
Heath Tecna Aerospace Co.	USA	*582*	75-43320
Heatly & Gresham (I) Ltd	India	*205*	80-00500
HeavyLift Aircraft Engineering Ltd	UK	*503*	80-12300
HeavyLift Cargo Airlines Ltd	UK	*503*	80-12300
Hebron & Medlock Engineering Ltd	UK	*457*	75-28860
HEC (Aerospace) Ltd	UK	*457*	75-28880
Heckler & Koch GmbH Defence Division (HK)	Germany	*169*	75-11972
Heckler & Koch Inc.	USA	*581*	75-42860
Heckler & Koch SA.	Switzerland	*389*	90-00830
Heckler & Koch UK.	UK	*522*	90-00830
Hedemora Diesel AB.	Sweden	*377*	75-06030
HEEQAC = Hellenic Electronics Equipment Quality Assurance Centre	Greece	*187*	60-00100
Heeqac SA	Greece	*193*	80-00760
Heeresamt	Germany	*157*	25-30100
Heeresfliegerbrigade 3	Germany	*158*	25-60100
Heeresfliegerkommando 3	Germany	*158*	25-60100
Heflin Steel Co.	USA	*581*	75-42880
Heinsoldt Systemtechnick GmbH	Germany	*171*	75-17800
Heiser Josef	Austria	*33*	75-01230
Héli Union	France	*150*	80-00820
Heliavia Nusantara PT	Indonesia	*210*	90-00825
Helicopter Services Group ASA (HSG)	Norway	*308*	70-00800
Helicópteros do Brasil S/A	Brazil	*56*	75-08250
Heliflite Pty Ltd	Australia	*28*	90-08300
Heliflyg AB	Sweden	*379*	80-00800
Helikopter Service AS (HS)	Norway	*311*	80-00800
Helio Mirror Co. Ltd	UK	*457*	75-28900
Heliodinâmica SA	Brazil	*56*	75-08270
Heliponents Inc.	USA	*636*	80-13910
Heliquip Pty Ltd	S. Africa	*361*	85-01300
Helitech Pty Ltd	Australia	*25*	80-08400
Hella Aerospace GmbH	Germany	*169*	75-11975
Hella KG Hueck & Co	Germany	*169*	75-11975
Hella Manufacturing Co. Pty Ltd	Australia	*17*	75-08370
Hellenic Aerospace Industry Ltd (HAI)	Greece	*192*	80-00750

Quick Reference

I

ICDO = International Civil Defence Organisation	Switzerland	*384*	34-10500
Icea Sud SpA.	Italy	*233*	75-05150
Icelandic Civil Aviation Administration	Iceland	*201*	20-00100
Icelandic Coast Guard	Iceland	*201*	28-00500
Icelandic Meteorological Office	Iceland	*201*	20-00200
ICEP SA	Romania	*330*	60-00800
ICI Devoe Coatings	USA	*584*	75-45895
ICL DESC	UK	*501*	80-08400
ICL Enterprises	UK	*504*	80-13100
ICL Technical Plastics Ltd	UK	*460*	75-31060
Icom (UK) Ltd	UK	*460*	75-31070
Icore International GmbH	Germany	*181*	90-00920
Icore International Inc.	USA	*584*	75-45890
Icore International Ltd	UK	*460*	75-31080
IDA = Institute for Defense Analyses	USA	*540*	60-09100
I.D.E.	USA	*651*	90-01090
IDE = Interactive Development Environments Inc.	USA	*637*	80-15950
IDE = Interactive Development Environments Ltd	UK	*504*	80-13330
IDE Technologies Ltd	Israel	*219*	75-09170
Ideal Aerosmith Inc.	USA	*584*	75-45900
Ideal Electric Co.	USA	*584*	75-45920
Ideal Knitwear Manufacturing Ltd	Canada	*72*	75-07720
Idecanarias SA	Spain	*369*	90-01000
IDIC = Instituto de Investigaciones y Control del Ejercito	Chile	*85*	60-00950
IDM Electronics Ltd	UK	*460*	75-31100
IDS = International Data Sciences Inc.	USA	*585*	75-47865
IDS Ingegneria dei Sistemi SpA Head Office	Italy	*240*	80-00920
IEP Airfoils AMTech Division	UK	*446*	75-18420
IEP Airfoils Monk Bridge Division	UK	*446*	75-18450
IEP Airfoils Moorside Division	UK	*446*	75-18455
IEP Airfoils Renston Division	UK	*446*	75-18460
IEP Doncasters Blaenavon Division	UK	*446*	75-18430
IEP Structures Bramah Division	UK	*446*	75-18465
IFI Srl	Italy	*235*	75-11220
IFO International (UK) Ltd	UK	*523*	90-00930
IGM = Instituto Geográfico Militar	Chile	*85*	60-00700
ign Robotersysteme AG	Germany	*174*	75-25200
IHEDN = Institut des Hautes Études de Défense Nationale	France	*127*	60-02800
IHESI = IHESI Department d'Études et Recherches	France	*127*	60-02500
IHESI Department d'Études et Recherches (IHESI)	France	*127*	60-02500
IHI = Ishikawajima-Harima Heavy Industries Co. Ltd	Japan	*251*	75-02800
II-VI Inc.	USA	*624*	75-91355
IIBR = Israel Institute for Biological Research	Israel	*216*	60-00960
IISS = International Institute for Strategic Studies	UK	*419*	60-00900
IKA Works Inc	USA	*584*	75-46020
Ika-Maschinenbau Janke & Kunkel GmbH & Co KG	Germany	*170*	75-13900
IL = Instytut Lotnictwa	Poland	*324*	60-02000
Il Diritto Aero	Italy	*240*	80-00800
ILC Data Device Corp.	UK	*522*	90-00425
ILC Data Device Corp.	USA	*584*	75-46050
ILC Dover Inc.	USA	*584*	75-46050
ILC Technology Inc.	USA	*584*	75-46070
ILC-DDC Elektronik GmbH	Germany	*181*	90-00925
Ilco Unican Inc.	Canada	*72*	75-07725
ILEE AG.	Switzerland	*386*	75-06300
Ilex Systems Inc.	USA	*584*	75-46080
ILG = International Logistics Group	USA	*585*	75-48120
ILI = Intamar Logistics Inc.	USA	*585*	75-47050
Iliakopoulos Ltd	Greece	*190*	75-01850
Ilmavoimien Esikunta	Finland	*116*	25-20000
Ilmavoimien Viestikoulu	Finland	*116*	25-20000
ILS Oy	Finland	*119*	80-00900
Ilyushin Aviation Complex	Russia	*333*	55-09000
Ilyushin Design Bureau	Russia	*333*	55-09000
Image Interpretation Systems Inc. (API)	USA	*584*	75-46110
ImageBase Technology Ltd	UK	*516*	85-00769
Imagenex Technology Corp.	Canada	*72*	75-07728
IMBEL = Indústria de Material Bélico do Brasil	Brazil	*56*	75-09190
Imco Industries Ltd	Israel	*219*	75-09400
IMFL = Institut de Mécanique des Fluides	France	*127*	60-05000
Imgraf SA	Dominican R.	*107*	90-00300
IMI = Israel Military Industries Ltd	Israel	*220*	75-10300
IMI Marston Ltd	UK	*460*	75-31155
IMI Norgren Ltd	UK	*460*	75-31158

IMI plc	UK	*423*	70-00950
IMI Trading USA Inc.	USA	*637*	80-15080
IMI-Tech Corp.	USA	*584*	75-46140
Imo Industries Inc.	USA	*584*	75-46150
IMO Industries Inc. USA	UK	*469*	75-40945
Imo Industries Ltd	UK	*489*	75-60960
Imo Industries Sarl	France	*153*	90-00920
Imo Industries Srl	Italy	*243*	90-00895
Imo Pump AB	Sweden	*377*	75-06500
IMP Aerospace Limited	Canada	*72*	75-07730
IMP Group Ltd	Canada	*67*	70-07000
iMPath Networks Inc.	Canada	*77*	75-46151
Imperial House	Japan	*245*	10-10000
Impex, Sojuzkarta	Russia	*340*	80-02000
Importécnica SA	Brazil	*59*	90-10000
Impreglio SpA	Italy	*240*	80-00790
Impresa Forniture Industriali	Italy	*235*	75-11220
Imrex Aviation Co. Inc.	Singapore	*351*	90-00910
Imrex Company Inc.	USA	*647*	85-00835
I.M.S. Stainless Fasteners Ltd	UK	*517*	85-01690
Imtra Corp.	USA	*653*	90-08990
IN-LHC	France	*137*	75-22320
INC = Ishikawajima Noise Control Co. Ltd	Japan	*252*	75-02820
INC Corp. Pty Ltd	Australia	*17*	75-09250
Inco Alloys Ltd	UK	*460*	75-31260
Inco Engineered Products Ltd	UK	*446*	75-18410
Indal Technologies Inc.	Canada	*72*	75-07740
Indamer Co Pvt Ltd	India	*207*	90-00940
INDEP = Indústrias de Defesa SA	Portugal	*327*	55-00950
Independant Syndicat of the National Police	France	*129*	65-01780
Independent Instruments (Aviation)	Australia	*28*	90-09900
Indian Air Force Western Air Command Headquarters	India	*202*	25-22100
Indian Avitron	India	*206*	85-01000
Indiana Rubber	Netherlands	*293*	75-03540
Indivers NV/Interturbine	Netherlands	*293*	75-03550
INDRA	Spain	*366*	75-03800
Inductoheat Pty Ltd	Australia	*17*	75-09500
Inductosyn International	USA	*575*	75-33800
Industri Pesawat Terbang Nusantara North America Office (IPTN)	USA	*653*	90-09020
Industri Pesawat Terbang Nusantara PT	Indonesia	*209*	55-00300
Industria Aeronautica Neiva S.A.	Brazil	*56*	75-09460
Indústria Brasileira de Infláveis Nautika Ltda	Brazil	*57*	75-14150
Industria Condensatori Applicazioni Elettroelettroniche SpA (ICAR)	Italy	*235*	75-11600
Industria de Fundicao Tupy Ltda	Brazil	*56*	75-06840
Indústria de Material Bélico do Brasil (IMBEL)	Brazil	*56*	75-09190
Industria de Turbo Propulsores SA Head Office and Manufacturing Plant (ITP)	Spain	*366*	75-03870
Industria Metalurgica del Norte Ltda	Chile	*85*	75-00300
Industria SA	France	*137*	75-22300
Industrial & Airborne Systems Handelsgesellschaft mbH, (IAS)	Germany	*178*	80-00800
Industrial Acoustics Co Inc	Germany	*181*	90-00935
Industrial Acoustics Company Ltd (IAC)	UK	*460*	75-31300
Industrial College of the Armed Forces	USA	*528*	25-19100
Industrial e Mercantil de Artefatos de Ferro, Cia	Brazil	*56*	75-03310
Industrial Engineering Designers Ltd	Ireland	*215*	90-00950
Industrial Engineering Services Ltd	Bangladesh	*37*	90-00850
Industrial Laser & Electronic Engineering	Switzerland	*386*	75-06300
Industrial Protector Limited	Hong Kong SAR	*199*	90-00943
Industrial Technology & Machines AG	Switzerland	*385*	75-00950
Indústrias de Defesa SA (INDEP)	Portugal	*327*	55-00950
Indústrias e Participações de Defesa SA	Portugal	*327*	55-00950
Indústrias Nacionais de Defesa EP	Portugal	*327*	55-00950
Industrieanlagen-Betriebsgesellschaft mbH (IABG)	Germany	*178*	80-00945
Industrigruppen JAS AB	Sweden	*375*	74-00100
INESC = National Institute of Public Safety Studies	France	*123*	30-00100
Infantry Export Support Team	UK	*419*	40-10500
Infinetics Inc.	USA	*584*	75-46700
Infocom Systems A/S	Denmark	*103*	75-09200
Infodex Incorporated	USA	*585*	75-46701
Infodoc Sarl Registered Office	Italy	*240*	80-00922
InfoLink Technologies Inc.	USA	*585*	75-46703
Informasjonskontroll A/S	Norway	*307*	60-00300
Information Handling Services Inc.	USA	*637*	80-15530
Information Resources Ltd	Netherlands	*299*	90-01250
Information Resources Management College	USA	*528*	25-19100
Information Security Systems Ltd	UK	*461*	75-31430
Infotel Technologies (Pte) Ltd	Singapore	*351*	90-00940

Quick Reference

International Hospitals Group	UK	*505*	80-13410
International Industrial Co. Ltd	China PR	*90*	90-00900
International Institute for Strategic Studies (IISS)	UK	*419*	60-00900
International Logistics Group (ILG)	USA	*585*	75-48120
International Loss Prevention Systems	Canada	*72*	75-07775
International Machine & Tool Corp.	USA	*585*	75-46860
International Maritime Organization	UK	*418*	34-25000
International Police Association British Section	UK	*421*	65-00600
International Polymer Corp.	USA	*585*	75-48200
International Precision Products BV	UK	*523*	90-00965
International Procurement Services Inc.	USA	*653*	90-09550
International Procurement Services Ltd.	UK	*461*	75-31820
International Professional Security Association (IPSA)	UK	*421*	65-00700
International Promotion Services Sarl	Italy	*243*	90-00950
International Radiator Services Ltd	UK	*461*	75-31840
International Rectifier Co.(GB) Ltd	UK	*461*	75-31860
International Research Institute Inc.	USA	*637*	80-16100
International Security & Defense Systems Ltd (ISDS)	Israel	*224*	80-00960
International Submarine Engineering Ltd (ISE)	Canada	*73*	75-07780
International Technical Acoustic Company	Switzerland	*388*	80-00955
International Technological Applications	Greece	*193*	80-00795
International Technologies Inc.	USA	*585*	75-48330
International Technology SA (ITSA)	Spain	*366*	75-03950
International Trade Contracting & Services Company (ITCS)	Saudi Arabia	*343*	80-01600
International Trade Operations Inc.	USA	*637*	80-16050
International Trading Establishment	Jordan	*260*	90-00800
International Transducer Corp. (ITC)	USA	*585*	75-48650
International Turbine Engine Corporation	Taiwan	*393*	80-00140
Interoperability Systems Interntional Hellas SA (ISI)	Greece	*190*	75-01962
Interpoint A Crane Company	USA	*585*	75-46870
Interpoint Corporation	USA	*585*	75-46870
Interpoint (UK) Ltd	UK	*516*	85-00775
Intersec	UK	*514*	80-55300
Interservice Territorial Command	Belgium	*40*	25-32200
Interspiro	USA	*585*	75-48680
Interspiro AB	Sweden	*377*	75-06950
Interspiro AG	Switzerland	*390*	90-00950
Interspiro Gesmblt	Austria	*33*	75-01420
Interspiro GmbH	Germany	*170*	75-14620
Interspiro Ltd	UK	*461*	75-31945
Interspiro NL	Netherlands	*293*	75-03570
Interstate Electronics Corp.	USA	*586*	75-48700
Interstate Technical Products	UK	*461*	75-31946
Interstate Wire Co. Inc.	USA	*647*	85-00990
INTERTECHNIK Techn. Produktionen GmbH & Co KG	Austria	*33*	75-01430
Intertechnique (Aerospace) Ltd	UK	*461*	75-31948
Intertechnique SA Head Office	France	*137*	75-23000
Interturbine Compressor Technology	Singapore	*349*	80-01000
Interturbine Germany GmbH	Germany	*170*	75-14630
Interturbine Singapore Pte Ltd	Singapore	*349*	80-01000
Intracom SA	Greece	*190*	75-01964
Intramet SA	Greece	*190*	75-01965
Intramet Steel Structures & Electromechanical Equipment S.A.	Greece	*190*	75-01965
Intro Corporation	USA	*647*	85-01000
Intronic GmbH	Germany	*178*	80-00975
Introtec GmbH	Germany	*178*	80-00975
Invar Manufacturing Ltd	Canada	*73*	75-08220
IOC = US Army Industrial Operations Command	USA	*534*	25-33000
ION Track Instruments	UK	*461*	75-32000
Ionix High Performance Wiring Systems	UK	*461*	75-32007
Ipeco Europe	UK	*461*	75-32040
IPECO Inc.	USA	*637*	80-16800
IPL = Instro Precision Ltd	UK	*461*	75-31630
IPSA = International Professional Security Association	UK	*421*	65-00700
IPTN = Industri Pesawat Terbang Nusantara North America Office	USA	*653*	90-09020
IPTN = PT Industri Pesawat Terbang Nusantara	Indonesia	*209*	55-00300
Iran Faravar Co. Ltd	Iran	*212*	90-00950
Iran Helicopter Support & Renewal Inc. (ISHRI)	Iran	*212*	80-09600
Irenco	S. Africa	*358*	75-08300
IRI - Istituto per la Ricostruzione Industriale	Italy	*230*	70-09070
Irish Defence Forces	Ireland	*213*	25-10000
Irish Naval Service	Ireland	*214*	25-41000
Irkutsk Joint Aviation Industrial Association	Russia	*333*	55-09200
IRT Infrarot-Technik Eiselt	Germany	*170*	75-14640
Irvin Aerospace Canada Limited	Canada	*73*	75-08300
Irvin Aerospace Canada Limited	USA	*653*	90-09690
Irvin Aerospace Limited	UK	*462*	75-32050
Irvin Aerospace Ltd	UK	*423*	70-00975
Irvin Aerospace SpA	Italy	*235*	75-12200
Irvin Group Ltd	UK	*423*	70-00975
Irvin Industries Canada Ltd	USA	*653*	90-09690
Irvin Industries Inc.	USA	*586*	75-48960
Irvin Manufatture Industriall SpA	Italy	*235*	75-12200
Irwin Industries Canada Ltd	Canada	*73*	75-08300
Iscan Inc.	USA	*586*	75-49100
ISDEFE = Ingenieria de Sistemas	Spain	*367*	80-00900
ISDS = International Security & Defense Systems Ltd	Israel	*224*	80-00960
ISE = International Submarine Engineering Ltd	Canada	*73*	75-07780
Iseco Safety Shoes	Canada	*81*	90-00940
Ishikawajima Noise Control Co. Ltd (INC)	Japan	*252*	75-02820
Ishikawajima-Harima Heavy Industries Co. Ltd (IHI)	Japan	*251*	75-02800
ISHRI = Iran Helicopter Support & Renewal Inc.	Iran	*212*	80-09600
ISI = Interoperability Systems Interntional Hellas SA	Greece	*190*	75-01962
Iskra Agency Ltd	UK	*462*	75-32060
ISL = French-German Research Institute of Saint Louis	France	*122*	20-35000
Island Commander Faeroes	Denmark	*100*	25-35000
Ismailji & Sons, M	Pakistan	*316*	90-01200
Isoli SpA	Italy	*235*	75-12250
I.Spiewak & Sons Inc.	USA	*617*	75-83300
ISPRA = Israel Product Research Co. Ltd.	Israel	*219*	75-10000
Israel Aerospace and Defence (IADE)	Israel	*224*	80-00970
Israel Aircraft Industries International Inc.	USA	*653*	90-09760
Israel Aircraft Industries Liaison Office	Singapore	*350*	85-01000
Israel Aircraft Industries Ltd (IAI)	Israel	*219*	75-10200
Israel Aircraft Industries-Europe European Purchasing Mission (IAI)	France	*153*	90-00970
Israel Cartridge Co. Ltd	Israel	*225*	85-00500
Israel Desalination Engineering	Israel	*219*	75-09170
Israel Institute for Biological Research (IIBR)	Israel	*216*	60-00960
Israel Military Industries Ltd (IMI)	Israel	*220*	75-10300
Israel Military Industries Ltd Swiss Liaison Office	Switzerland	*390*	90-00990
Israel Military Industries Rocket Systems Division	**Israel**	***220***	**75-10310**
Israel Product Research Co. Ltd (ISPRA)	Israel	*219*	75-10000
Israel Shipyards Ltd	Israel	*220*	75-10600
ISSMI = Istituto Superiore di Statu Maggiore Interforze	Italy	*229*	60-00100
IST = Integrators of Systems Technology Pty (Ltd)	S. Africa	*360*	80-09500
Istec Consultants	UK	*505*	80-13500
Istituto Superiore di Statu Maggiore Interforze (ISSMI)	Italy	*229*	60-00100
ISVR = Institute of Sound & Vibration Research	UK	*419*	60-00870
ISVR Consultancy Services	UK	*505*	80-13520
Italian Institute of External Trade	Hungary	*200*	31-00900
Italtel SIT	Italy	*235*	75-13205
Italtel Spa Headquarters	Italy	*235*	75-13205
I.T.B. Comércio Internacional Lda	Portugal	*327*	85-05000
ITB Comércio Internacional Lda.	Portugal	*327*	85-05000
ITC = International Transducer Corp.	USA	*585*	75-48650
ITCS = International Trade Contracting & Services Company	Saudi Arabia	*343*	80-01600
ITEC = AlliedSignal Engines	Taiwan	*393*	80-00140
ITEC = AlliedSignal Engines International Turbine Engine Company	Czech Republic	*99*	80-00100
ITEC Ltd	UK	*505*	80-13550
Itochu (Europe) plc.	UK	*523*	90-00980
ITP = Industria de Turbo Propulsores SA Head Office and Manufacturing Plant	Spain	*366*	75-03870
I.T.P. Ltd	Israel	*216*	65-00050
ITP Ltd	Israel	*220*	75-10800
ITPN Europe Aircraft & Logistics GmbH	Germany	*182*	90-01000
ITSA = International Technology SA	Spain	*366*	75-03950
ITT Avionics	USA	*586*	75-49455
ITT Cannon	USA	*586*	75-49452
ITT Cannon Benelux Office	Belgium	*49*	90-00970
ITT Cannon Japan	Japan	*252*	75-02850
ITT Cannon Korea Ltd	Korea (ROK)	*267*	90-00970

ITT Cannon RF Products	UK	*462*	75-32170
ITT Cannon/Sealectro	UK	*462*	75-32170
ITT Corp.	USA	*543*	70-06000
ITT Defense & Electronics	Hong Kong SAR	*199*	90-00960
ITT Defense & Electronics	USA	*586*	75-49455
ITT Defense International	Hong Kong SAR	*199*	90-00960
ITT Defense International (Bahrain)	Bahrain	*36*	90-00700
ITT Defense International (Saudi Arabia)	Saudi Arabia	*344*	90-00600
ITT Defense International UK	UK	*505*	80-13570
ITT Defense Ltd	UK	*505*	80-13570
ITT Electro-Optical Products	USA	*586*	75-49455
ITT Federal Services Corp.	USA	*586*	75-49455
ITT Gilfillan	USA	*586*	75-49455
ITT Industries Inc.	USA	*543*	70-06000
ITT Night Vision	USA	*586*	75-49455
IUSD Publishing Company	UK	*497*	80-01253
Ivar Rivenaes Ltd	UK	*479*	75-49820
IVECO Fiat SpA	Italy	*235*	75-13230
Iveco Pegaso SA	Spain	*364*	55-00250
Iver Precision Engineers Ltd	UK	*462*	75-32230
IVEX Corp.	USA	*586*	75-49950
Izhevsk Mechanical Plant	Russia	*333*	55-09500
Izhevskiy Mekhanicheskiy Zavod	Russia	*333*	55-09500
I2E = Études Electroniques Informatiques	France	*150*	80-00540

J

J & E Hofmann Engineering Pty Ltd	Australia	*17*	75-08630
J Blaschke Wehrtechnik GmbH	Austria	*32*	75-00560
J L Weibull AB	Sweden	*379*	75-12100
J T Inglis & Sons Ltd	UK	*461*	75-31500
J&D Wilkie Ltd	UK	*494*	75-66240
J&E Hall Limited	UK	*456*	75-27420
J-Tec Associates Inc.	USA	*587*	75-51250
Jacman Aircraft Inc.	USA	*654*	90-09900
Jacques Cable Systems Ltd	UK	*462*	75-32480
JADI = Japan Association of Defence Industry	Japan	*249*	65-10200
JAE = Japan Aviation Electronics Industry Ltd	Japan	*252*	75-03300
JAI A/S	Denmark	*103*	75-09400
Jaivision Ltd UK	UK	*523*	90-00995
Jakab Industries Pty Ltd	Australia	*17*	75-10170
Jalex BV	Netherlands	*297*	80-10000
James Greaves & Co.	UK	*522*	90-00780
James Hardie Building Services Ltd	Australia	*17*	75-10178
James Hardie Engineering	New Zealand	*303*	80-00900
James North & Sons Ltd	UK	*471*	75-42730
James Walker & Co. Ltd	UK	*492*	75-64120
Jan Staubo A/S	Norway	*312*	90-02000
Jane's Information Group Ltd (JIG)	**UK**	***505***	**80-13700**
Jane's Information Group US Operation (JIG)	**USA**	***637***	**80-17080**
Japan Aircraft Manufacturing Co Ltd	Japan	*257*	90-00945
Japan Aircraft Manufacturing Co. Ltd (NIPPI)	Japan	*252*	75-03200
Japan Association of Defence Industry (JADI)	Japan	*249*	65-10200
Japan Aviation Electronics Industry Ltd (JAE)	Japan	*252*	75-03300
Japan Industrial Technology Association (JITA)	Japan	*249*	65-10400
Japan Machinery Exporters' Association (JMEA)	Japan	*249*	65-10500
Japanese Maritime Safety Agency Headquarters (MSA)	Japan	*246*	28-10000
Jaques	Australia	*17*	75-10200
Jardine Matheson (S) Ltd Aviation Division	Singapore	*351*	90-01010
Jarlsø Fabrikker A/S	Norway	*310*	75-02260
Jarvis Manufacturing Co.	UK	*462*	75-32330
JAS Industry Group	Sweden	*375*	74-00100
Jasmin plc	UK	*462*	75-32370
Jasmin Simtec Ltd	UK	*462*	75-32370
Jason's Cradle	UK	*462*	75-32296
Javac Pty Ltd	Australia	*18*	75-10310
Jay-El Products Inc.	USA	*586*	75-50100
Jaybeam Ltd	UK	*462*	75-32400
JC Bamford Excavators Ltd (JCB)	UK	*462*	75-32297
JCB = JC Bamford Excavators Ltd	UK	*462*	75-32297
JCB Sales Ltd	UK	*523*	90-01050
JDK Controls Inc.	USA	*586*	75-50200
JDR Cable Systems	France	*154*	90-11200
JDR Cable Systems BV	Netherlands	*293*	75-03600
JDR Cable Systems Inc	USA	*654*	90-10100
JDR Cable Systems Ltd	UK	*462*	75-32480
Jean Tanazacq, Ets	France	*147*	75-43670
Jecco Aviation Ltd	UK	*428*	75-01530
Jeckells Sails Ltd	UK	*462*	75-32490
Jehier SA	France	*137*	75-23200
Jennings Technology Corp.	USA	*586*	75-50350
Jenoptik Impulsphysik GmbH	Germany	*170*	75-14800
Jensen Tools Inc.	USA	*647*	85-01010
Jepico Corp	Japan	*257*	90-01030
Jeppesen & Co GmbH	Germany	*170*	75-15050
Jeppesen UK Ltd	UK	*505*	80-13850
Jered Brown Brothers Inc.	USA	*586*	75-50380
Jessop Saville Ltd	UK	*462*	75-32500
Jet Aviation Basel	Switzerland	*388*	80-01030
Jet Aviation Cologne	Germany	*178*	80-00985
Jet Aviation Dusseldorf	Germany	*178*	80-00990
Jet Aviation Geneva	Switzerland	*388*	80-01030
Jet Aviation Hannover	Germany	*178*	80-00995
Jet Aviation Kassel	Germany	*178*	80-01000
Jet Aviation Management AG	Switzerland	*388*	80-01030
Jet Aviation Munich	Germany	*178*	80-01010
Jet Aviation Saarbrucken	Germany	*178*	80-01020
Jet Aviation Saudi Arabia Co Ltd Head Office and Jeddah Station	Saudi Arabia	*343*	80-01700
Jet Aviation Saudi Arabia Co Ltd Riyadh Station	Saudi Arabia	*343*	80-01700
Jet Aviation Zürich AG	Switzerland	*388*	80-01030
Jet Engineering Inc.	USA	*647*	85-01050
Jet Maintenance Pte Ltd.	Singapore	*349*	80-01050
Jet Propulsion Laboratory, NASA (JPL)	USA	*540*	60-10250
Jet Support Ltd	UK	*516*	85-00900
Jetway Associates Ltd.	UK	*462*	75-32520
JFS Inc.	USA	*586*	75-50920
JGW International Ltd	USA	*654*	90-10130
J.I. Case Co.	USA	*562*	75-16950
Jiei Kantai Shireibu	Japan	*246*	25-42100
JIG = Jane's Information Group Ltd	**UK**	***505***	**80-13700**
JIG = Jane's Information Group US Operation	**USA**	***637***	**80-17080**
JITA = Japan Industrial Technology Association	Japan	*249*	65-10400
J.M. Cantor	Honduras	*197*	90-00100
JMEA = Japan Machinery Exporters' Association	Japan	*249*	65-10500
JML Optical Industries Inc.	USA	*587*	75-50940
JMS Electronics Inc.	USA	*647*	85-01080
J.N.S. Electronics Industries	Australia	*18*	75-10350
John Brown Engineering Ltd	UK	*437*	75-10250
John C. Small & Tidmas plc	UK	*487*	75-58080
John C Stennis Space Center	USA	*529*	25-19900
John Chatillon & Sons Inc.	USA	*651*	90-03400
John Crane Inc.	USA	*566*	75-22500
John Crane Marine	USA	*566*	75-22500
John Curran Ltd	UK	*444*	75-16430
John Heine & Son Pty Ltd	Australia	*17*	75-08350
John J. McMullen Associates Inc.	USA	*639*	80-20800
John Jovino Co. Inc.	USA	*647*	85-01150
John Laing International Ltd	UK	*505*	80-13870
John Lawson Display & Technical Models	UK	*464*	75-34875
John Lawson Modelmakers	UK	*464*	75-34875
John M. Davis & Associates	USA	*634*	80-07200
John M. Piris & Sons Ltd	Gibraltar	*185*	75-01000
John Reed (Specialist Market Research & Analysis)	UK	*505*	80-13900
John Trelawny Ltd	UK	*489*	75-61000
John Valves Pty Ltd	Australia	*18*	75-10400
John W. Drake Transportation Consultant	USA	*634*	80-08600
John Welsh Pty Ltd	Australia	*13*	75-02805
Johnson Firearm Specialties Inc.	USA	*586*	75-50920
Johnson John Company, The	USA	*587*	75-51000
Johnson Matthey Plc Noble Metals Business Unit	UK	*462*	75-32815
Joint Medical Materiel Service	Norway	*306*	20-15000
Joint Norwegian Medical Service	Norway	*306*	20-15000
Joint Services Command and Staff College (JSCSC)	UK	*419*	60-01100
Joint Services Survival, Evasion, Resistance and Escape Agency	USA	*529*	25-19920
JONA Studio Consulenza	Italy	*240*	80-04000
Jonas Aircraft & Arms Co. Inc.	USA	*647*	85-01120
Jonathan Arthur Ciener Inc.	USA	*564*	75-18590
Jones & Rickard Pty Ltd	Australia	*18*	75-10750
Jordan Valley Applied Radiation Ltd (JVAR)	Israel	*220*	75-10850
Jørgen Andersen Ingeniørfirma A/S	Denmark	*103*	75-09400
Jørgen Anderson Ingeniorfima	UK	*523*	90-00995

Quick Reference

Kawasaki Motors NV	Netherlands	293	75-03802
Kawasaki Motors Pty Ltd	Australia	18	75-10950
Kawasaki Motors Pty Ltd	Australia	18	75-10955
Kawasaki Motors (UK) Limited	UK	463	75-33250
Kawasaki Motors (UK) Ltd	UK	463	75-33252
Kawasaki Precision Machinery (UK) Ltd.	UK	463	75-33255
Kawasaki Robotics (UK) Ltd	UK	463	75-33260
Kay & Associates Inc.	USA	637	80-17660
Kazan Aircraft Production Association	Russia	333	55-10340
Kazan Aviation Industrial Association	Russia	333	55-10340
Kazan Design Bureau	Russia	333	55-10350
Kazan Helicopters JSC	Russia	337	75-11000
KB "Arsenal"	Russia	333	55-01300
KB Electronics (1989) Ltd	Canada	73	75-08710
KBI	USA	588	75-51670
KBK = Kyokuto Boeki Kaisha Ltd Bangkok Liaison Representative	Thailand	396	90-01150
KBK = Kyokuto Boeki Kaisha Ltd Beijing Office	China PR	90	90-01000
KBK = Kyokuto Boeki Kaisha Ltd Head Office	Japan	258	90-01220
KBK = Kyokuto Boeki Kaisha Ltd London Branch Office	UK	523	90-01180
KBK = Kyokuto Boeki Kaisha Ltd Shanghai Office	China PR	90	90-01005
KBK Inc	USA	654	90-11106
KBOM = Design Bureau of General Machine Building	Russia	336	75-03150
KBP Instrument Design Bureau	Russia	335	60-02100
KCI Carruthers Ltd	UK	463	75-33300
KD Manitou Inc.	USA	654	90-11110
KDI Precision Products Inc.	USA	588	75-51705
KDIA = Korea Defense Industry Association	Korea (ROK)	264	65-11300
ke autoeletric GmbH	Germany	170	75-15750
Kearfott Guidance & Navigation Corp.	Belgium	49	90-00175
Kearfott Guidance & Navigation Corp.	USA	588	75-51760
Kearsley Airways Ltd	UK	463	75-33370
KEC Limited	UK	463	75-33700
Keelavite Commercial Hydraulics Ltd	UK	441	75-14410
Keiaisha NMB Co. Ltd	Japan	253	75-04500
KEL Aerospace Pty Ltd	Australia	18	75-11440
Kelan Circuits Ltd	UK	463	75-33400
Kelsey Instruments Ltd	UK	463	75-33455
Keltec Florida Inc.	USA	588	75-51950
Keltron Corp.	USA	588	75-52000
Keltron Power Systems	USA	588	75-52000
Kelvin Hughes A/S (KH)	Denmark	106	90-11270
Kelvin Hughes Ltd	UK	463	75-33453
Kelvin Hughes (Singapore) Pte Ltd	Singapore	351	90-01130
Kembrey Wiring Systems Ltd	UK	463	75-33525
Kementerian Dalam Negeri	Malaysia	276	12-40000
Kempton Ltd, T.W.	UK	463	75-33600
Kenmax Special Products Pty Ltd (KSP)	Australia	28	90-11800
Kennedy Aerospace Corp.	USA	540	60-11111
Kennedy Marketing Communications Corp.	USA	637	80-17680
Kenrich Petrochemicals Inc.	USA	588	75-52310
Kent Aerospace	UK	428	75-01040
Kent Aerospace plc	UK	431	75-04000
Kentavros-Kouloussiou Bros SA	Greece	190	75-01981
Kentree Ltd	Ireland	215	75-00250
Kentron	S. Africa	358	75-09000
Kenya Air Force Headquarters	Kenya	261	25-20000
KEP Beynelmilel AS	Turkey	404	90-01130
Kerametal Co. Ltd	Slovakia	353	55-01000
Kern Electrical Components Ltd.	UK	463	75-33700
Kern Engineering & Manufacturing Corp.	USA	588	75-52330
Kester Solder	USA	588	75-52350
Kestronics (M) Sdn Bhd	Malaysia	278	90-01030
Kevlin Corporation	USA	588	75-52450
KeyMed (Medical & Industrial Equipment) Ltd	UK	472	75-43020
Keystone Helicopter Corp.	USA	637	80-17780
KFN = Kriminologisches Forschungsinstitut Niedersachsen e.V.	Germany	162	60-01080
KFW Canada	Canada	73	75-08740
KH = Kelvin Hughes A/S	Denmark	106	90-11270
KHD	Germany	170	75-15900
KHD Great Britain Ltd	UK	523	90-01110
KHI = Kawasaki Heavy Industries Ltd Shanghai Office	China PR	90	90-00960
KHI (Dalian) Computer Technology Co Ltd	China PR	89	80-00800
KHI Europe Finance BV	Netherlands	297	80-11000
Khrunichev State Research & Production Space Center	Russia	334	55-10400
Kidde Aerospace & Defence Far East	Singapore	351	90-01180
Kidde Aerospace, Walter	USA	588	75-52500
Kidde Dexaero	France	138	75-23700
Kidde Graviner Australia	Australia	18	75-11495
Kidde-Fenwal	USA	588	75-52510
Kidde-Graviner Ltd	UK	463	75-33880
Kiev Aircraft Repair Plant	Ukraine	406	55-01500
Kimura Corporation	Japan	258	90-01160
Kinemetrics/Truetime	USA	624	75-91110
Kinetic Systems Corp.	USA	588	75-52660
Kinetics Ltd	Israel	220	75-11320
King Fifth Wheel Co.	USA	589	75-52670
Kinhill Engineers Pty Ltd	Australia	25	80-11000
Kioleides SA	Greece	190	75-01982
Kirasa Co. Ltd	Russia	337	75-11350
Kirov Research Production Corp, Perm	Russia	338	75-15975
Kirovo Chepetsky Electro Machine-Building Plant	Russia	337	75-11370
Kjeld Dannemand A/S	Denmark	106	90-04500
KJM Fabritek Inc.	USA	589	75-52877
KJM Fabritek Inc. Aeroballistics Group	USA	589	75-52877
Klaus Transportsysteme GmbH & Co KG	Germany	165	75-00930
Klaus-Maffei Automationstechnik GmbH	Germany	171	75-16295
Klein Associates Inc.	USA	589	75-52900
Klimov Design Bureau	Russia	337	75-11400
Klöckner-Humboldt-Deutz AG	Germany	167	75-06350
KMA = Royal Military Academy for Army and Air Force	Netherlands	288	25-39100
K.M.Fuze Engineering	Israel	220	75-11350
Knabenhans	Switzerland	386	75-07100
Knesset	Israel	215	15-10000
Knight Strip Metals Ltd	UK	517	85-01160
KNMI = Koninklijk Nederlands Meteorologisch Instituut	Netherlands	287	20-12000
Knox Ltd, W&J	UK	463	75-34050
KNT Engineering & Trading Pte Ltd	Singapore	351	90-01238
Knürr-Mechanik für die Elektronik AG	Germany	171	75-16150
Kobelt Manufacturing Co. Ltd	Canada	73	75-08770
Kockums Computer Systems (UK) Ltd	UK	505	80-14180
Kockums Holdings	Sweden	375	70-00450
Kockums Industries AB	Sweden	377	75-07500
Kockums Submarine Systems AB	Sweden	377	75-07505
Kocoverk International AB	Sweden	377	75-07520
Kodak Brasileira Com Ind. Ltda	Brazil	56	75-11700
Kohla Max KG Sportgeraeteerzeugung	Austria	33	75-01570
Kohlbrat & Bunz GmbH	Austria	33	75-01575
Kollmorgen Artus	France	138	75-23820
Kollmorgen Hightech Limited	UK	523	90-01162
Kollsman	USA	589	75-53150
Kollsman Inc.	USA	589	75-53150
Kolon International Corp. Montevideo Branch	Uruguay	657	75-05000
Komatsu Hanomag Aktiengesellschaft	Germany	171	75-16185
Komatsu-Howmet Ltd	Japan	250	74-00600
Kombinat Przemyslu Narzedziowego VIS	Poland	324	75-02520
Kombinat PZL-Hydral SA	Poland	324	75-02500
Kommando Fliegerdivision	Austria	31	25-32200
Kompanija Sloboda D D (SDD)	Yugoslavia	661	75-00475
Kone Carruthers Ltd	UK	463	75-33300
Kongsberg Aerospace	Norway	309	75-01045
Kongsberg Gruppen ASA	Norway	308	70-01000
Kongsberg Gruppen ASA	UK	523	90-01165
Kongsberg Informasjonskontroll A/S Defence Systems Group	Norway	307	60-00300
Kongsberg Norcontrol AS	Norway	309	75-01047
Kongsberg Simrad Inc	USA	589	75-53160
Kongsberg Simrad Ltd	UK	463	75-34100
Kongsberg Simulation & Training Defence Systems Group	Norway	309	75-01060
Kongsberg Spacetec A/S Aerospace Group	Norway	311	75-02640
Koninklijk Nederlands Meteorologisch Instituut (KNMI)	Netherlands	287	20-12000
Koninklijke Nederlandsche Springstoffenfabrieken NV	Netherlands	294	75-04600
Kontor Norge	Norway	312	90-00320
Kontron Electronique	France	153	90-01160
Kooi BV	Netherlands	293	75-04000
Koor Industries Ltd	Israel	220	75-11800
Kopex International Ltd	UK	463	75-34150
Korea Defense Industry Association (KDIA)	Korea (ROK)	264	65-11300
Korea Express USA Inc.	USA	638	80-17900
Korea Shipbuilding & Engineering Corp.	Korea (ROK)	265	75-01505
Korea Technology Bank	Korea (ROK)	266	80-11800
Korps Politie Suriname	Suriname	371	26-00000
Korr. Dätwylet AG	Switzerland	390	90-00900
Korry Electronics Co.	USA	589	75-53250
Korth GmbH	Germany	171	75-16200
Korth GmbH & Co KG	Germany	171	75-16200

KOSI = Kaiser Optical Systems Inc.	USA	*587*	75-51355
Kratos Pressure Products	USA	*589*	75-53300
Krauss-Maffei Aktiengesellschaft	Germany	*163*	70-00350
Krauss-Maffei Asia Pacific Pte Ltd.	Singapore	*351*	90-01245
Krauss-Maffei Chemiomeccanica SrL.	Italy	*243*	90-01170
Krauss-Maffei Corp	USA	*654*	90-11700
Krauss-Maffei Dienstleistung GmbH	Germany	*171*	75-16300
Krauss-Maffei France Sarl	France	*153*	90-01100
Krauss-Maffei Kunstofftechnik GmbH	Germany	*171*	75-16310
Krauss-Maffei (UK) Ltd.	UK	*523*	90-01170
Krauss-Maffei Verfahrenstechnik GmbH	Germany	*171*	75-16320
Krauss-Maffei Verkerhrstechnik GmbH	Germany	*171*	75-16330
Krauss-Maffei Versicherungsdienst GmbH	Germany	*171*	75-16340
Krauss-Maffei Wehrtechnik GmbH	Germany	*171*	75-16345
Krautkramer Japan Co Ltd.	Japan	*250*	75-00860
Krigsflygskolan, F5	Sweden	*372*	25-29400
Kriminalforsogens Personaleskole	Denmark	*102*	60-02000
Kriminalistisches Institut des Bundeskriminalamtes	Germany	*162*	60-01120
Kriminalistisk Institut	Denmark	*102*	60-01000
Kriminalpolitisk Forening	Denmark	*102*	60-01800
Kriminologische Abteilung der Philippps Universität	Germany	*162*	60-01140
Kriminologische Forschungsgruppe der Bayerischen Polizei	Germany	*162*	60-01160
Kriminologische Forschungstelle der Universität zu Köln	Germany	*162*	60-01180
Kriminologische Seminar der Universität Bonn	Germany	*162*	60-01200
Kriminologische Zentralstelle e.V. (KrimZ)	Germany	*162*	60-01300
Kriminologischer Dienst bei der Justizvollzugsanstalt Hannnover	Germany	*162*	60-01110
Kriminologisches Forschungsinstitut Niedersachsen e.V. (KFN)	Germany	*162*	60-01080
KrimZ = Kriminologische Zentralstelle e.V.	Germany	*162*	60-01300
Krishna Rao, M/S E.	India	*207*	90-01240
Kristal SA	France	*151*	85-01200
KRZ Flotsky Arsenal	Bulgaria	*62*	55-02500
KSP = K S Paul Products Ltd	UK	*464*	75-34160
KSP = Kenmax Special Products Pty Ltd	Australia	*28*	90-11800
Kubo-Tech AG	Switzerland	*390*	90-01100
Kuhler Hellas SA	Greece	*190*	75-01989
Kuhn & Siefert	Germany	*179*	85-01500
Kuhn GmbH Bauelemente der Elektrotechnik	Germany	*182*	90-01180
KUKA Wehrtechnik GmbH	Germany	*171*	75-17200
Kulite International Sarl	France	*153*	90-01170
Kulite Sensors Ltd	UK	*464*	75-34180
Kull AG, Lunor	Switzerland	*386*	75-07200
Kulon Scientific Research Institute	Russia	*335*	60-02600
Kumar Udyog, M.	India	*205*	75-02100
Kurgan Machine Building Plant Production Association	Russia	*337*	75-11850
Kurganmashzavod JSC	Russia	*337*	75-11850
Kutaisi Automobile Plant Production Association	Georgia	*156*	55-05000
Kvaerner ASA	Norway	*309*	75-01090
Kvaerner Energy AS	Norway	*309*	75-01093
Kvaerner Energy Limited, Thermal Power Division	UK	*437*	75-10250
Kvaerner Eureka AS	Norway	*309*	75-01095
Kvaerner Mandal A/S	Norway	*309*	75-01100
Kvaerner Masa-Yards Inc	**Finland**	***118***	**75-01370**
Kvaerner plc	UK	*423*	70-00980
Kvant Scientific Research Institute	Ukraine	*406*	60-00880
KVH Europe A/S	Denmark	*103*	75-09500
KVH Industries Inc.	USA	*589*	75-53670
Kylar Inc	USA	*654*	90-11900
Kyokuto Boeki Kaisha (KBK) GmbH	Germany	*182*	90-01190
Kyokuto Boeki Kaisha Ltd Bangkok Liaison Representative (KBK)	Thailand	*396*	90-01150
Kyokuto Boeki Kaisha Ltd Beijing Office (KBK)	China PR	*90*	90-01000
Kyokuto Boeki Kaisha Ltd Head Office (KBK)	Japan	*258*	90-01220
Kyokuto Boeki Kaisha Ltd London Branch Office (KBK)	UK	*523*	90-01180
Kyokuto Boeki Kaisha Ltd Shanghai Office (KBK)	China PR	*90*	90-01005
Kyokuto Boeki Kaisha Ltd Taipei Branch Office	Taiwan	*393*	90-01150
Kyriazis Ltd	Greece	*190*	75-01986
KYTE SA	Greece	*190*	75-01995
KZB Technical Rubber Products (1987) Ltd	Israel	*220*	75-12190

L

L'Air Liquide	France	*131*	75-01200
L'Équipement et la Construction Electrique	Germany	*167*	75-07200
L'Garde Inc.	USA	*590*	75-55060
L'hotellier Bourgogne Industries	France	*138*	75-24890
L'hotellier Montrichard.	France	*138*	75-24900
L&F Industries	USA	*589*	75-53690
L&S Machine Co. Inc.	USA	*589*	75-53700
La Armada de Chile	Chile	*83*	25-30500
La Macarena	Argentina	*5*	90-12000
La-Vey Security International Ltd	UK	*506*	80-14215
Labbe SA, Société	France	*146*	75-40500
Labinal	France	*138*	75-23900
Labinal Aero Defense Systems Inc	USA	*638*	80-18100
Labinal Components & Systems GmbH	Germany	*182*	90-01200
Labinal K.K.	Japan	*258*	90-01230
Laboratoire Arcane	France	*138*	75-24010
Laboratoire de Recherches Balistiques et Aérodynamiques (LRBA)	France	*122*	20-40000
Laboratoire de Salon de Provence	France	*127*	60-05000
Laboratoire Européen pour la Physique des Particules	Switzerland	*384*	34-10000
Labtech Ltd	UK	*464*	75-34280
Lacamiron International Ltd	Colombia	*91*	85-00300
Lacombe Indústria e Comércio de Turbos Ltd	Brazil	*56*	75-12050
Ladish Co.Inc.	USA	*589*	75-53850
LADS Corporation	Australia	*18*	75-12100
Lagardère	France	*138*	75-24100
Laing International Ltd, John	UK	*505*	80-13870
LAIR = Laser & Atomic Research & Development	Croatia	*94*	60-00450
Lambert Aircraft Engineering	Belgium	*46*	75-04750
Lanarealm Ltd	UK	*464*	75-34430
Lancaster Machining Inc.	USA	*589*	75-53935
Land and Marine Products Ltd	UK	*462*	75-32296
Land Headquarters	Australia	*7*	25-32200
Landair International Ltd.	UK	*505*	80-14200
LANDCENT = Allied Land Forces Central Europe Headquarters	Germany	*160*	34-03000
Landes Test Centre	France	*122*	20-40000
Landeskriminalamt.	Germany	*162*	60-01600
Landesverteidigungsausschuss	Austria	*30*	15-10000
Landhelgisgaelan	Iceland	*201*	28-00500
LANDSOUTH = Allied Land Forces Southern Europe Headquarters	Italy	*229*	34-32100
Lane Electronics Ltd, F.C.	UK	*517*	85-01210
Lansdowne Technologies Inc	Canada	*78*	80-01205
Lantor (UK) Ltd.	UK	*464*	75-34665
Lapland Air Command	Finland	*116*	25-20000
LAPTC = Lucas Aerospace Power Transmission	USA	*592*	75-56265
Lapua Oy	Finland	*118*	75-01430
LARIMART SpA	Italy	*236*	75-13500
LAS = Light Aero Spares Ltd	UK	*517*	85-01240
Laser & Atomic Research & Development (LAIR)	Croatia	*94*	60-00450
Laser Electronics (Operations) Pty Ltd	Australia	*26*	85-10000
Laser Fiber Optics Canada Ltd	Canada	*73*	75-08890
Laser-Scan Ltd.	UK	*506*	80-14210
Lasso	S. Africa	*360*	80-25000
Latécoère Société Industrielle d'Aviation	France	*138*	75-24200
Laurel Technologies	USA	*570*	75-27380
Law Enforcement International Ltd	UK	*464*	75-34870
L.B. Holliday & Co. Ltd	UK	*458*	75-29650
L.B. Parkes & Co Ltd.	UK	*473*	75-43950
LBG Agencies (1979) Ltd	Israel	*224*	80-00980
Le Magnésium Industriel	France	*137*	75-21750
Le Protecteur	France	*138*	75-24440
Leach International Europe SA	France	*139*	75-25350
Leach International Germany GmbH	Germany	*182*	90-01230
Leach International UK Ltd	UK	*517*	85-01230
Leafield Engineering Ltd	UK	*464*	75-34910
Leafield Holdings Ltd	UK	*423*	70-00990
Leafield Logistics & Technical Services Ltd	UK	*506*	80-14218
Lear Astronics Corp.	USA	*589*	75-54390
Lear Siegler Power Equipment Corp.	USA	*592*	75-56260
Learjet Inc.	USA	*590*	75-54430
Lee Hydraulische Miniaturkomponenten GmbH	Germany	*182*	90-01233
Lee James Electronics Ltd.	UK	*470*	75-42270
Lee Products Ltd.	UK	*523*	90-01235
LegCo = Legislative Council	Hong Kong SAR	*197*	15-10000

Quick Reference

Logica Informatique SA	Switzerland	390	90-01300
Logica Limited	Hong Kong SAR	198	80-00300
Logica (Malaysia) Sdn Bhd	Malaysia	278	80-00250
Logica Middle East	UAE	408	90-00900
Logica New Zealand Limited	New Zealand	303	80-01200
Logica plc	UK	423	70-01100
Logica Pte Ltd	Singapore	349	80-01100
Logica Pty Ltd	Australia	25	80-12500
Logica SA/NV	Belgium	48	80-01000
Logica s.r.o.	Czech Republic	99	90-01500
Logica Svenska AB	Sweden	379	80-01200
Logica UK Limited	UK	506	80-14350
Logicasiel SpA	Italy	230	74-10000
Logicon Geodynamics Inc.	USA	638	80-18625
Logistic Projects	S. Africa	358	75-09800
Logistic Services International	USA	654	90-13000
Logistics Management Engineering Inc.	USA	638	80-18330
Logistics Pty Ltd	Australia	25	80-12700
Logistics Services International Inc.	USA	638	80-18800
Logtek (Logistic Technologies)	S. Africa	360	80-09950
Lohmann GmbH & Co KG	Germany	171	75-18910
Lohr Défense	France	139	75-25200
Lohr Industrie	France	139	75-25200
Londesborough Security Equipment plc	UK	465	75-36300
Lopez Montes y Associados AP	Mexico	282	90-02000
Loral International GmbH	Germany	182	90-01280
Loral Vought International	Belgium	49	90-01300
Loral Vought Systems	Belgium	49	90-01300
Lord SA	France	150	80-01350
Lordco CWG (NZ) Limited	New Zealand	303	85-00200
LORI Heat Transfer Operations	USA	592	75-56090
Lorraine Electronics Head Office	UK	465	75-36500
Losinjska Plovidba-Brodogradiliste Mali Losinj	Croatia	94	75-06500
L.P. Foreman & Sons Ltd	UK	452	75-24160
LRBA = Laboratoire de Recherches Balistiques et Aérodynamiques	France	122	20-40000
LRC Products Ltd	UK	465	75-36530
LRDE = Electronics and Radar Development Establishment	India	204	60-00550
LRE Relä + Elektronik AB	Sweden	381	90-01230
LRE Relais et Electronique Sarl	France	139	75-25350
LRE Relays & Electronics Ltd	UK	517	85-01230
LSI = Life Support International Inc.	USA	590	75-55163
LSI = Luminescent Systems	Belgium	48	80-01400
LSI Computer Systems Inc.	USA	592	75-56150
Lucas Aerospace	Brazil	57	75-12970
Lucas Aerospace	France	139	75-25390
Lucas Aerospace	UK	465	75-36650
Lucas Aerospace - Australia	Australia	25	80-12850
Lucas Aerospace Cargo Systems	USA	592	75-56250
Lucas Aerospace Customer Support Americas	USA	638	80-18900
Lucas Aerospace Engine Control Systems	Canada	73	75-09140
Lucas Aerospace FCS Liaison Office	Indonesia	210	90-00350
Lucas Aerospace FCS Paris	France	139	75-25390
Lucas Aerospace France Corporate Headquarters	France	153	90-01290
Lucas Aerospace GmbH Power Generation Regional Office	Germany	182	90-01285
Lucas Aerospace Power Equipment Corp.	USA	592	75-56260
Lucas Aerospace Power Transmission (LAPTC)	USA	592	75-56265
Lucas Aerospace Representative Japan	Japan	258	90-01260
Lucas Aerospace Singapore Pte Ltd	Singapore	349	80-01200
Lucas Aftermarket Operations	UK	465	75-36655
Lucas Assembly & Test Systems	UK	465	75-36655
Lucas Control Systems Products	UK	465	75-36675
Lucas Industries plc	UK	424	70-01280
Lucas Management Systems Deutschland GmbH	Germany	164	75-00230
Lucas Management Systems Italia Srl	Italy	240	80-00100
Lucas Management Systems Ltd	UK	497	80-01259
Lucas NSF Ltd	UK	465	75-36675
Lucas Weinschel	USA	627	75-95544
Lucas Western Inc.	USA	592	75-56300
Lucas Yuasa Batteries Ltd	UK	495	75-67560
LucasVarity plc	UK	424	70-01280
Luchaire Défense SA	France	139	75-25400
Lucznik SA	Poland	324	75-02550
Ludlow Corp.	USA	592	75-56350
Ludlum Measurements Inc.	USA	592	75-56360
Luen Hing Taiwan Co. Ltd	Taiwan	393	90-01290
Luftfartsverket	Norway	306	20-04000
Luftschifftechnik GmbH	Germany	177	75-34780
Luftwaffenamt	Germany	157	25-21000
Luftwaffenunterstützungskommando	Germany	157	25-21100
Luftwaffenunterstützungskommando	Germany	157	25-22200
Lukassen Flughafentechnic GmbH (LFT)	Germany	171	75-18950
Lumenition Div. Autocar Electrical Equipment Ltd	UK	432	75-04440
Luminator Aircraft Products	USA	593	75-56450
Luminescent Systems Inc.	USA	593	75-56480
Luminescent Systems (LSI)	Belgium	48	80-01400
LUTAB	Sweden	379	80-01280
Luwa Bahnson Inc.	USA	654	90-13020
Luxtron Inc. Emergency Lighting Div.	USA	593	75-56700
Lyncole XIT Grounding	USA	593	75-56705
Lynwood	UK	465	75-36700
Lynwood Scientific Developments Ltd	UK	465	75-36700
Lynx Corp Pty Ltd	Australia	25	80-12980
Lynx Engineering UK Ltd	UK	465	75-36710
Lyon & Brandfield Ltd	UK	465	75-36720
Lyons Associates Inc., Richard T	USA	641	80-26280
Lyulka-Saturn Inc.	Russia	337	75-12000

M

M & J Vardas Co.	Greece	194	90-02100
M Brodribb Pty Ltd	Australia	13	75-02760
M International Export (MINTEX)	USA	639	80-19780
M Ismailji & Sons Machinery and Equipment	Pakistan	316	90-01200
M. Kumar Udyog	India	205	75-02100
M. Paz Logistics Ltd	Israel	224	80-01600
M. Rosenblatt & Son Inc.	USA	641	80-26450
M W Microwave Corp.	USA	597	75-63495
M&N Aerospace	USA	597	75-62240
M&N Plastics	USA	597	75-62240
M&T Management	Canada	80	85-02000
M-Systems Flash Disk Pioneers	Israel	220	75-12600
M.A. Bigland (Preparations) Ltd	UK	434	75-06700
MA COM	UK	523	90-01290
M/A COM Limited	UK	523	90-01290
Maaseik Police	Austria	32	34-04610
MABAK = Markas Besar Kepolisian R I	Indonesia	208	26-00000
Mabey & Johnson Ltd	UK	465	75-36830
MAC Aerospace Corporation	USA	647	85-01240
MACAN = Mando Aéreo de Canarias	Spain	362	25-22400
MacArtney A/S	Denmark	103	75-12800
MacArtney AS	UK	523	90-01295
Maccaferri Ltd	UK	465	75-36850
MacDonald Dettwiler	Malaysia	278	80-00270
MacDonald Dettwiler and Associates Ltd	Canada	78	80-01310
MacDonald Dettwiler Limited	Australia	25	80-12990
MacDonald Dettwiler Ltd	UK	506	80-15100
Macedonia Machine Works SA	Greece	190	75-01998
MacGregor (Aus) Pty Ltd	Australia	18	75-13000
MacGreor-Navire (Aus) Pty Ltd	Australia	18	75-13000
Machine Technology Ltd	UK	465	75-36870
Machinery Oy	Finland	120	90-01310
Mack Electric Devices Inc.	USA	593	75-56850
Mackenzie Tribbeck Associates (MTA)	UK	506	80-15150
MacNaught Pty Ltd	Australia	18	75-13140
Macneal-Schwendler Corporation	USA	593	75-56930
MacNeal-Schwendler European Headquarters	Germany	178	80-01280
MA/COM Antenna & Cable Div.	USA	593	75-56953
MacTaggart Scott & Co. Ltd	UK	466	75-36900
MacTaggart Scott & Co. Ltd	USA	593	75-57000
Mäder Lacke AG Lach and Farbenfabrik	Switzerland	386	75-08000
MAEST = Mando Aereo del Estrecho	Spain	362	25-22200
MAFO Howden BV	Netherlands	294	75-04350
Mag Instrument Inc.	USA	593	75-57020
Magal Security Systems Ltd	Israel	221	75-12650
Magellan Group Limited	Australia	18	75-13150
Magellan Systems Corp.	USA	593	75-57050
Magellan Systems Corporation	UK	523	90-01300
Magnaghi Milano SpA Head Office and Works	Italy	236	75-13980
Magnaghi Napoli SpA	Italy	236	75-14000
Magnaghi Oleodinamica SpA	Italy	236	75-13980
Magnavox Adv. Products & Systems Co.	USA	593	75-57120
Magnavox Electronic Systems Company West Coast Division	USA	593	75-57120
Magnesium Elektron	UK	466	75-37000
Magnesium Elektron Inc.	USA	593	75-57330
MagneTek Euroatlas GmbH	Germany	168	75-08680
MagneTek Tempe	USA	548	75-03100
Magnetic Seal Corp.	USA	593	75-57380

Quick Reference

Michel Frochot Representaciones Industriales	Venezuela	*659*	90-01300
Michelin Italiana SpA Aircraft Tyre Product Line	Italy	*237*	75-15425
Michelin Italiana SpA Lighe Produit Avion	France	*140*	75-26600
Michelin Sime Aircraft Tyre Co. Sdn Bhd	Malaysia	*278*	75-01800
Michell Bearings	UK	*468*	75-38935
Michigan Wheel Corp.	USA	*596*	75-60930
Michsaf	Israel	*221*	75-14200
Micrel Inc.	USA	*596*	75-60950
Micro Communications Inc.	USA	*596*	75-60980
Micro Data Base Systems Inc.	USA	*639*	80-21080
Micro Design A/S	Norway	*310*	75-01260
Micro Metalsmiths Ltd.	UK	*468*	75-39010
Micro Processor Systems A/S (MPS)	Norway	*309*	75-01060
Micro Robotics Center	USA	*597*	75-63150
Microelectronics and Technology Center	USA	*549*	75-04700
Microelectronics, Guidance and Electrooptics Division	Turkey	*402*	55-00050
Microelectronics Ltd	Israel	*221*	75-14250
Microinfo Ltd	UK	*506*	80-16500
Microkim Ltd	Israel	*221*	75-14350
Microlease plc	UK	*506*	80-16560
Micron Instruments Private Limited	India	*205*	75-01200
Micron Metals	USA	*609*	75-75480
Micronel AG	Switzerland	*387*	75-08490
Micronel AG	Switzerland	*390*	90-00900
Micronic Devices	India	*207*	90-01340
Microphase International	UK	*524*	90-01346
Microsemi Corp.	USA	*596*	75-61090
MicroSpire Paris	France	*153*	90-01340
MicroSpire SA	France	*140*	75-26780
Microtec AG, MB	Switzerland	*386*	75-08485
Microtecnica-Torino SpA	Italy	*237*	75-15500
Microturbo Inc	USA	*596*	75-61170
Microturbo Ltd	UK	*468*	75-39160
Microturbo North America Inc.	USA	*596*	75-61170
Microturbo SA	France	*140*	75-27000
Microwave Engineering Group (MEG)	Israel	*217*	75-01400
Microwave Power Devices Inc.	USA	*597*	75-63130
Midcon Cables Co.	USA	*596*	75-61300
Middle East Marketing Services Ltd.	Jordan	*260*	80-05000
Middlesex Group Limited	UK	*468*	75-39330
Midland Rollmakers Ltd	UK	*436*	75-09500
MIFA Aluminium BV	Netherlands	*294*	75-04525
MIG, Moscow Aircraft Production Organization	Russia	*334*	55-13050
Migada Ltd Medical Equipment Ind.	Israel	*217*	75-00250
Mikes Microwave Electronic Systems Inc.	Turkey	*403*	75-01450
Mikoyan Aviation Scientific-Industrial Complex MIG	Russia	*334*	55-13050
Mikro-Lambda GmbH	Germany	*182*	90-01368
Mikron	Russia	*337*	75-13250
Mikron Ltd	Greece	*191*	75-02050
MIL Davie Inc.	Canada	*70*	75-04040
MIL Moscow Helicopter Plant	Russia	*337*	75-13300
MIL Systems	Canada	*79*	80-01360
MIL Systems Engineering	Canada	*79*	80-01360
Mil Tec = Military Technology Pty Ltd.	Australia	*29*	90-13620
MIL-BASE Industries Inc.	USA	*596*	75-61530
Milan Army Ammunition Plant	USA	*535*	25-36500
Milano Brothers International Corp.	USA	*654*	90-13440
MILCOM International Ltd	Greece	*191*	75-02060
Miles Air Products	Canada	*80*	85-01330
Miles Associates Inc	Canada	*80*	85-01330
Milfa Lda	Portugal	*328*	90-01340
Military Agency for Standardization (MAS)	Belgium	*43*	34-05000
Military Air School	Colombia	*91*	25-29100
Military Aircraft Spares Ltd	UK	*506*	80-16780
Military Development Committee	Madagascar	*275*	20-10000
Military Engine Business	UK	*480*	75-50265
Military Industry Complex and Conversion, Ministry of Engineering	Ukraine	*405*	12-60000
Military Logistics MFC	USA	*596*	75-61550
Military Ordnance Defence Ltd	Spain	*368*	85-00300
Military Survey	UK	*411*	20-01125
Military Systems Group Inc.	USA	*596*	75-61570
Military Technology Pty Ltd (Mil Tec)	Australia	*29*	90-13620
Militech International	USA	*647*	85-01343
Milldale Limited Pulvermatic Sales	UK	*468*	75-39500
Miller Industries Towing Equipment Inc.	USA	*596*	75-61655
Millitech Corp.	USA	*596*	75-61665
Mills-Tui Ltd	New Zealand	*302*	75-01350
Milo Holdings Ltd	UK	*506*	80-16850
Milo S = Södra Militärområdet	Sweden	*372*	25-32100
Milper Ltd	Israel	*226*	90-01310
Mils Elektronik GmbH & Co KG	**Austria**	***33***	**75-01700**
Milsearch Pty Ltd.	Australia	*25*	80-13800
Miltec SA Administration and Marketing Office	Switzerland	*387*	75-08498
Miltope Asia-Pacific	Singapore	*351*	90-01330
Miltope Corp.	USA	*596*	75-61810
MIMA = Malaysian Institute of Maritime Affairs	Malaysia	*277*	60-12000
Minebea Co. Ltd Head Office	Japan	*253*	75-05600
Minebea Co. Ltd Hong Kong Branch	Hong Kong SAR	*198*	75-01000
Minebea Co. Ltd Kuala Lumpur Branch	Malaysia	*277*	75-00200
Minebea Co. Ltd Penang Office	Malaysia	*277*	75-00200
Minebea Co. Ltd Shanghai Office	China PR	*90*	90-05000
Minebea Electronics (Thailand) Ltd	Thailand	*396*	75-01300
Minebea Electronics (UK) Ltd	UK	*468*	75-39700
Minebea Tehnologies (Pte) Ltd Taiwan Branch	Taiwan	*393*	90-01325
Minebea Thai Ltd	Thailand	*396*	75-01400
Mineral Derivatives Ltd	UK	*468*	75-39800
Mining Resource Engineering Ltd (MREL)	Canada	*74*	75-09770
Ministère de l'Intérieur	France	*121*	12-40000
Ministère des Affaires Etrangères	France	*121*	12-20000
Ministerie van Buitenlandse Zaken	Netherlands	*287*	12-20000
Ministerie van Defense	Netherlands	*287*	12-10000
Ministerie Van Economische Zaken	Netherlands	*287*	12-12000
Ministerio de Asuntos Exteriores	Cuba	*95*	12-20000
Ministerio de Defensa Nacional	Colombia	*90*	12-10000
Ministerio de Desarrollo Económico	Colombia	*90*	12-16000
Ministerio de Justicia y del Derecho	Colombia	*90*	12-50000
Ministerio do Exercito	Brazil	*53*	12-13000
Ministerio Relagses Exteriores	Brazil	*53*	12-20000
Ministirio da Aeronautica	Brazil	*53*	12-12500
Ministry Communications/Transportation	Slovakia	*353*	12-70000
Ministry for Foreign Affairs	Tonga	*397*	12-20000
Ministry of Aeronautics	Brazil	*53*	12-12500
Ministry of Civil Defence	New Zealand	*300*	12-40000
Ministry of Commerce	Thailand	*395*	12-30000
Ministry of Commerce, Industry, Trade and Public Enterprises	Fiji	*114*	12-30000
Ministry of Communications	Colombia	*90*	12-15000
Ministry of Communications	Singapore	*346*	12-80000
Ministry of Defence	Austria	*30*	12-10000
Ministry of Defence	Bangladesh	*36*	12-10000
Ministry of Defence	Cambodia	*63*	12-10000
Ministry of Defence	India	*202*	12-10000
Ministry of Defence	Luxembourg	*274*	12-15000
Ministry of Defence	Mongolia	*283*	12-10000
Ministry of Defence	Myanmar	*285*	12-10000
Ministry of Defence	Netherlands	*287*	12-10000
Ministry of Defence	New Zealand	*300*	12-10000
Ministry of Defence	Singapore	*345*	12-10000
Ministry of Defence	Spain	*362*	12-10000
Ministry of Defence	Thailand	*395*	12-10000
Ministry of Defence	Tonga	*397*	12-10000
Ministry of Defence	UK	*410*	20-00700
Ministry of Defence and National Security	Maldives	*279*	12-10000
Ministry of Defence Defence Information Division (D INFO D)	UK	*409*	12-11500
Ministry of Defense	Nepal	*286*	12-10000
Ministry of Economic Affairs	Netherlands	*287*	12-12000
Ministry of Economic Affairs	Taiwan	*391*	12-30000
Ministry of Economy and Transport	Austria	*30*	12-60000
Ministry of Electronics Industry	China PR	*86*	12-80000
Ministry of External Affairs	India	*202*	12-20000
Ministry of External Economic Affairs	Korea (DPR)	*263*	12-50000
Ministry of Finance	Indonesia	*208*	12-60000
Ministry of Finance	Korea (ROK)	*263*	12-50000
Ministry of Finance	Singapore	*346*	12-60000
Ministry of Finance	Taiwan	*391*	12-50000
Ministry of Finance	Thailand	*395*	12-50000
Ministry of Finance Customs Department	Bermuda	*51*	29-01000
Ministry of Foreign Affairs	Bangladesh	*36*	12-20000
Ministry of Foreign Affairs	Bhutan	*52*	12-20000
Ministry of Foreign Affairs	Brazil	*53*	12-20000
Ministry of Foreign Affairs	China PR	*86*	12-20000
Ministry of Foreign Affairs	Croatia	*94*	12-20000
Ministry of Foreign Affairs	Cuba	*95*	12-20000
Ministry of Foreign Affairs	France	*121*	12-20000
Ministry of Foreign Affairs	Japan	*245*	12-20000
Ministry of Foreign Affairs	Korea (DPR)	*263*	12-20000
Ministry of Foreign Affairs	Korea (ROK)	*263*	12-20000
Ministry of Foreign Affairs	Laos	*269*	12-20000
Ministry of Foreign Affairs	Maldives	*279*	12-20000

N

Quick Reference

Polycast Technology Corp.	USA	*604*	75-70858
Polyformes Ltd	UK	*475*	75-46348
Polymathic Engineering Co. Ltd	UK	*475*	75-46350
Polyot Research and Production Co.	Russia	*338*	75-14700
Polyships SA	Spain	*367*	75-05000
Polytec-PI SA	France	*141*	75-30450
Polytronic AG	Switzerland	*387*	75-10250
Polytronic Inc.	USA	*604*	75-70870
Polytronic International Ltd	Switzerland	*387*	75-10250
Polytronic-Jung GmbH	Germany	*173*	75-23850
Polywarm Products Ltd	UK	*475*	75-46356
Polyziv Plastic Products	Israel	*221*	75-17580
Pont-à-Mousson SA	France	*141*	75-30520
Poole and Co (Stainless) Ltd, R.A.	UK	*517*	85-01690
Popov Communications Equipment Plant JSC	Russia	*338*	75-16400
Popov Plant Scientific Production Association	Russia	*338*	75-16400
Port of Singapore Authority	Singapore	*346*	28-10000
Portescap	Switzerland	*387*	75-10320
Portescap Deutschland GmbH	Germany	*182*	90-01665
Portescap France SA	France	*153*	90-01669
Portescap Japan Ltd	Japan	*258*	90-01675
Portescap Scandinavia AB	Sweden	*381*	90-01630
Portescap (UK) Ltd	UK	*475*	75-46420
Portescap US Inc.	USA	*655*	90-16690
Positioning Resources Ltd	UK	*517*	85-01677
Potter & Brumfield	USA	*614*	75-80565
Power Components Ltd	UK	*516*	85-00630
Power Conversion Inc.	USA	*604*	75-71070
Power Spectra Inc.	USA	*605*	75-71085
Power Technology Inc.	USA	*605*	75-71090
Power Ten ICN	USA	*605*	75-71095
Powerent Ltd	UK	*509*	80-23850
Powerex Inc.	USA	*604*	75-71080
Powerpack Representaçoes e Comércio Ltda	Brazil	*59*	90-17000
Powersport International Ltd	UK	*475*	75-46500
PowerTech Inc.	USA	*605*	75-71087
PPA = Pay & Personnel Agency	UK	*411*	20-01145
PPG Industries Glass SA	France	*153*	90-01680
PPI = Pacific Propeller Inc.	USA	*640*	80-23060
PPIHP = Prva Petoletka Industrie Hidraulike i Pneumatike	Yugoslavia	*661*	75-00610
PPO Elektromekhanika	Russia	*337*	75-05500
Pradhan Mantri Karyalaya	India	*202*	11-10000
Pratt & Whitney	USA	*605*	75-71150
Pratt & Whitney Canada (A'Asia) Pty Ltd	Australia	*25*	80-16950
Pratt & Whitney Canada Inc (P&WC)	Canada	*75*	75-11400
Praxair Surface Technologies Inc.	USA	*640*	80-24680
Praxair Surface Technologies K.K.	Japan	*256*	80-00800
Praxair Surface Technologies Ltd	UK	*509*	80-23925
Praxis Critical Systems Ltd.	UK	*509*	80-23930
Praxis plc	UK	*509*	80-23930
Präzisionstechnik Gedern	Germany	*173*	75-24370
PRB Associates Inc.	USA	*640*	80-24700
PRB, Société Nouvelle	Belgium	*47*	75-06050
PRC Inc.	USA	*640*	80-24800
Pre-Mac (Kent) Ltd.	UK	*476*	75-46660
Precimetal Fonderie de Précision SA	Belgium	*46*	75-05500
Precise Flight Inc.	USA	*605*	75-71200
Precision Castparts Corp. (UK) Ltd	UK	*439*	75-12280
Precision Component & Equipment Pte Ltd	Singapore	*352*	90-01670
Precision Filters Inc.	USA	*605*	75-71330
Precision Machining Inc.	USA	*605*	75-71350
Precision Sensors	USA	*605*	75-71360
Precision Standard Inc. (PCSN)	USA	*605*	75-71365
Precision Technologies Pte Ltd	Singapore	*352*	90-01670
Precision Tube Co. Inc. Coaritube Division	USA	*605*	75-71370
Précitube SA	France	*141*	75-30890
Préfecture Maritime Zone Maritine Atlantique	France	*122*	25-41200
Premaberg Manufacturing Ltd	UK	*475*	75-46600
Premier Fuel Systems Limited	UK	*475*	75-46655
Premier Hazard	UK	*476*	75-46690
Premier Hazard Systems Inc.	USA	*605*	75-71395
Premier Hazard Systems Pty Ltd	Australia	*20*	75-16865
Premier Hazard Systems (UK) Ltd	UK	*476*	75-46690
Premier Microwave Corp.	USA	*605*	75-71400
Premier Product Support Services Ltd	UK	*509*	80-23995
Presentey Engineering	Canada	*75*	75-11500
Presidency of Hellenic Republic Military Office	Greece	*185*	10-10000
President's Secretariat	India	*202*	10-10000
Presidential Secretariat	Sri Lanka	*369*	10-10000
PressCut Industries Inc.	USA	*605*	75-71436
Pressure Systems Inc.	USA	*605*	75-71438
Presto Casting Co.	USA	*605*	75-71440
Preussag AG	Germany	*163*	70-00900
Prexer-projekt sp zo o	Poland	*324*	75-02630
PRI South Pacific Inc.	USA	*559*	75-13774
Price & Co. (Pty) Ltd, C W	S. Africa	*362*	90-16900
Price & Orphin Limited	UK	*509*	80-24250
Price Western Leather Co. Ltd	UK	*476*	75-46755
Prikladnoi Mechaniki, NPO	Russia	*338*	75-14600
Prima Armi	Italy	*241*	85-00900
Primary Image Ltd	UK	*476*	75-46790
Prime Minister's Office	New Zealand	*300*	11-10000
Prime Minister's Office	UK	*408*	11-10000
Primex Defence Products Ltd	Canada	*79*	80-01700
Primex Physics International	USA	*605*	75-71460
Primex Technologies, Inc	USA	*544*	70-17000
Primex Technologies, Inc	USA	*606*	75-71472
Prior Data Sciences	UK	*509*	80-24300
Prior Data Sciences Ltd	Canada	*75*	75-11520
Prism Color Group	USA	*640*	80-24875
PRL = Programming Research Ltd	UK	*509*	80-24700
PRM Précimétal	France	*141*	75-30750
Pro Active Sales	USA	*655*	90-16722
Proaction	USA	*641*	80-24900
Procal SA	Greece	*191*	75-02235
Procurement Executive, Ministry of Defence	UK	*409*	12-11000
Product Risk Associates	UK	*509*	80-24400
Prof Sten Luthander Ingenjörsbyrå AB	Sweden	*379*	80-01280
Professional School of Security (PSOS Ltd)	Israel	*216*	60-15000
Profile Security Services Ltd	UK	*509*	80-24500
Programmed Composites Inc.	USA	*606*	75-71500
Programming Research Ltd (PRL)	UK	*509*	80-24700
Progress-Werk Oberkirch AG (PWO)	Germany	*173*	75-24320
Proje Yonetim AS	Turkey	*403*	80-01200
Projects International Development Corp. (PIDC)	USA	*655*	90-16725
Projects Unlimited Inc.	USA	*606*	75-71560
Promavia SA	Belgium	*50*	90-01670
Prometey Central Research Institute of Structural Materials	Russia	*335*	60-03555
Pronal SA	France	*141*	75-31100
Pronest AB	Sweden	*381*	90-01660
Pronit SA	Poland	*324*	75-02640
Proparms Ltd	Canada	*75*	75-11540
Propex Research & Engineering	USA	*584*	75-46020
Propex Research & Engineering GmbH	Germany	*170*	75-13900
Proprietary Software Systems	USA	*641*	80-24930
Proquimed SA	Spain	*366*	75-04930
Prosig Computer Consultants Ltd	UK	*476*	75-46865
Prosig USA Inc	USA	*655*	90-16727
Prosser/Enpo	USA	*574*	75-31670
Prostab International Ltd.	UK	*476*	75-46870
Protechnik Laboratories	S. Africa	*359*	75-17000
Protecteur, Le	France	*138*	75-24440
Protection Development International Corp. (PDIC)	USA	*606*	75-71670
Protector Safety Limited	UK	*476*	75-46890
Protector Technologies Ltd	UK	*476*	75-46890
Prototypa-ZM, s.r.o.	Czech Republic	*98*	60-01000
Protovale Electronic Research Instruments	UK	*476*	75-47000
Protovale (Oxford) Ltd	UK	*476*	75-47000
Protrade Ltd	Israel	*224*	80-01760
Prva Petoletka Industrie Hidraulike i Pneumatike (PPIHP)	Yugoslavia	*661*	75-00610
Prvni Brnenska Strojirna (PBS)	Czech Republic	*98*	55-01500
Przemyslowy Instytut Telekomunikacji	**Poland**	***325***	**75-02650**
PSG = Paint Services Group Limited	UK	*524*	90-01575
PSI = Pathfinder Systems Inc.	USA	*603*	75-68995
PSOS Ltd = Professional School of Security	Israel	*216*	60-15000
PT Boma Bisma Indra	Indonesia	*209*	55-00050
PT Centronix	Indonesia	*210*	75-00200
PT Havilindo Utama	Indonesia	*210*	90-00810
PT Heliavia Nusantara	Indonesia	*210*	90-00825
PT Industri Pesawat Terbang Nusantara (IPTN)	Indonesia	*209*	55-00300
PT Kawasaki Motor Indonesia	Indonesia	*210*	75-00300
PT Kemenagen	Indonesia	*210*	90-00050
PT Metraco Jaya	Indonesia	*210*	90-01330
PT Multi Dharma Guna	Indonesia	*211*	90-01335
PT Noble Denton Utama	Indonesia	*211*	90-01400
PT PAL Indonesia (PERSERO) Head Office (PAL)	Indonesia	*210*	75-00500
PT Panorama Timur Jaya	Indonesia	*211*	90-01500
PT Petrosea	Indonesia	*211*	90-01600
PT Stimec Elcom	Indonesia	*211*	90-01970

Q

R

Raikka Oy	Finland	*118*	75-01800
Railko Ltd	UK	*477*	75-47850
Rajavartiolaitoksen esikunta	Finland	*116*	28-10000
Rajkumar International	India	*205*	75-01780
Raksha Mantralaya	India	*202*	12-10000
RAL = Rutherford Appleton Laboratory	UK	*420*	60-01850
Ralbo Inc.	USA	*607*	75-72690
RAM System GmbH	Germany	*164*	74-18100
Ramapo Helicopters Inc	USA	*641*	80-25450
RAMC = Royal Army Medical Corps	UK	*412*	25-39150
RAMFAN Corporation	USA	*607*	75-72700
RAMO Manufacturing Inc.	USA	*607*	75-72810
Ramsey Winch Co.	USA	*607*	75-72830
RAN Maritime Headquarters	Australia	*7*	25-41000
RAND	USA	*540*	60-17000
Rand Instrument Corp.	USA	*607*	75-72900
Randolph Products Co.	USA	*607*	75-72905
Rank Brimar Inc.	USA	*560*	75-15210
Rantec Microwave & Electronics Inc.	USA	*607*	75-72915
RAO = Royal Army of Oman	Oman	*313*	25-30000
Rapid Acquisition Targeting Systems (RATS)	USA	*607*	75-72920
Rapid Data Ltd	UK	*524*	90-01810
Rapid Deployment Towers Inc.	USA	*607*	75-72925
Rapiscan Security Products Ltd	UK	*477*	75-48305
RAS = Research Associates of Syracuse	USA	*641*	80-25900
Rashtriya Metal Industries Ltd.	India	*205*	75-01810
RATEC = Raufoss Technology AS	Norway	*310*	75-02200
Ratier-Figeac SA	France	*141*	75-31800
Rational Software Corporation	USA	*641*	80-25600
Ratmil RA	Romania	*331*	75-01000
RATS = Rapid Acquisition Targeting Systems	USA	*607*	75-72920
Raufoss ASA (RA)	Norway	*308*	70-02200
Raufoss France Sarl	France	*141*	75-31850
Raufoss Technology AS (RATEC)	Norway	*310*	75-02200
Ray Engineering Company	UK	*478*	75-48350
Raychem AG	Switzerland	*390*	90-01820
Raychem (Canada) Ltd	Canada	*81*	90-01808
Raychem GmbH Bereich Elektronik	Germany	*182*	90-01818
Raychem SA	Belgium	*50*	90-01810
Raydex/C.D.T. Ltd	UK	*477*	75-48340
Raymark Friction Co.	USA	*624*	75-91382
Raytheon Aerospace	USA	*641*	80-25650
Raytheon Aircraft Company	USA	*607*	75-73200
Raytheon Anschütz GmbH	Germany	*173*	75-24650
Raytheon Anschütz GmbH	Germany	*182*	90-01820
Raytheon Canada Ltd	Canada	*75*	75-11700
Raytheon Cossor Electronics Ltd	UK	*478*	75-48367
Raytheon E-Systems	Belgium	*50*	90-01815
Raytheon E-Systems Corporate Headquarters	USA	*608*	75-73302
Raytheon E-Systems DBA	USA	*608*	75-73302
Raytheon International Inc.	Belgium	*50*	90-01815
Raytheon International Inc.	UK	*524*	90-01825
Raytheon Systems Co.	Japan	*258*	90-01820
R.B. Tennent Ltd	UK	*436*	75-09500
RBC Transport Dynamics Corp.	USA	*608*	75-73380
RBDF = Royal Bahamas Defence Force	Bahamas	*35*	25-11000
RBR Armour Limited	Hong Kong SAR	*199*	90-01830
RBR Armour Ltd	UK	*478*	75-48370
RBTS = Royal Brunei Technical Services	Brunei	*60*	40-01000
RCM Computers Ltd	Israel	*226*	90-01800
RCM Ltd	Israel	*226*	90-01800
RDAE = R&D Aeronautical Engineering Co. Inc.	USA	*607*	75-72430
RDI = Reutech Defence Industries	S. Africa	*359*	75-18400
RDM Submarines BV	Netherlands	*294*	75-05770
RDM Technology BV	Netherlands	*294*	75-05780
RDNMC = Naval Materiel Command	Denmark	*100*	25-42100
RDS Servo Components	UK	*525*	90-01826
RDTE = Research, Development, Testing & Evaluation Division	USA	*536*	25-41300
RE = Research Engineers	UK	*478*	75-49250
RE Components Ltd	UK	*478*	75-48400
Reade Advanced Materials	USA	*608*	75-73860
Reade Communications Group	USA	*608*	75-73860
Real Santa Rita Equipamentos de Segurança Ltda	Brazil	*57*	75-18300
Recoil Pty	Australia	*20*	75-18300
Recon/Optical Inc. Corporate Headquarters	USA	*608*	75-73930
Recorder Co.	USA	*579*	75-40360
Recrusul SA	Brazil	*57*	75-18340
RECTA SA	Switzerland	*387*	75-10900
Recticel NV	Belgium	*46*	75-05585
Red Wings Scientific JSC	Russia	*338*	75-21050
Redifon MEL Limited	UK	*478*	75-48530
Redpoint Thermalloy Ltd.	UK	*478*	75-48960
Reflectone Inc	USA	*608*	75-74150
Reflectone UK Ltd	UK	*478*	75-48380
Refractory Metals Association	USA	*542*	65-18250
Regent Mfg Inc.	USA	*609*	75-74200
Regis SpA	Italy	*238*	75-20110
Rehovoth Optics Ltd	Israel	*223*	75-22504
Reil Industrial Enterprises Ltd	Canada	*75*	75-12000
Reime Coatection A/S	Norway	*310*	75-02250
Reime Jarlsø A/S	Norway	*310*	75-02260
Reims-Aviation SA	France	*141*	75-32100
Reinhard Oil A/S	Denmark	*103*	75-18100
Relcom Agencies Ltd	Israel	*226*	90-01840
Reliance Comm/Tec Corp.	USA	*609*	75-74285
Reliant Design Development Co Ltd	UK	*478*	75-49170
Relli Technology Inc.	USA	*609*	75-74280
RELTEC Corporation	USA	*609*	75-74285
R.E.M.F. Radiocoms-Avionics	France	*141*	75-32200
Remmele Engineering Inc.	USA	*609*	75-74400
REMOTEC Inc.	USA	*609*	75-74430
Remsdaq Ltd	UK	*478*	75-49210
Remtech SA	France	*141*	75-32230
Renair Antenna	UK	*478*	75-49225
Renault Véhicules Industriels SA (RVI)	France	*141*	75-32300
RENCOL = The Ray Engineering Company	UK	*478*	75-48350
Rendar Ltd	UK	*478*	75-49226
RENK AG	Germany	*173*	75-24700
Reno Jet Center	USA	*639*	80-20870
Renown Engineering Ltd.	UK	*478*	75-49227
REOMIE BV	Netherlands	*297*	85-05000
REOSC	France	*151*	80-01700
Repair Yard Sibenik	Croatia	*95*	90-05000
Repaircraft plc	UK	*517*	85-01700
Repairtech International Inc.	USA	*641*	80-25880
Repetti & Montiglio SpA	Italy	*238*	75-20120
Reproduction Technologies Inc.	USA	*609*	75-74550
Republic Arms (Pty) Ltd	**S. Africa**	***359***	**75-18210**
Republic Electronics	USA	*597*	75-63125
Republic Fiji Military Forces Naval Division	Fiji	*115*	25-40000
Republic of Fiji Military Forces Headquarters	Fiji	*114*	25-10000
RES Srl	Italy	*244*	90-01835
Research and Development Institute for Electrical Engineering	Romania	*330*	60-00800
Research Associates of Syracuse (RAS)	USA	*641*	80-25900
Research, Development and Production Enterprise Zvezda JSC	Russia	*336*	60-08350
Research, Development, Testing & Evaluation Division (RDTE)	USA	*536*	25-41300
Research Engineers (RE)	UK	*478*	75-49250
Research Institute de Saint Louis, French-German	France	*122*	20-35000
Research Institute for Aeroelastic Systems (RIAS)	Ukraine	*406*	60-00990
Research Institute for the Study of Juridical Information	France	*127*	60-05500
Research Institute of Gas Discharge Devices Plasma	Russia	*338*	75-21150
Research Triangle Institute (RTI)	USA	*540*	60-18000
Reshef Technologies Ltd	Israel	*222*	75-18600
Resil Industria e Comercio Ltda	Brazil	*57*	75-18400
Resil Minas Industrial e Comercio Ltda	Brazil	*57*	75-18450
Reson A/S	Denmark	*103*	75-18300
RESON Inc.	USA	*609*	75-74610
Reson Offshore Ltd	UK	*478*	75-49260
Reson UK	UK	*478*	75-49260
Resource Consultants Inc.	USA	*641*	80-25920
Rest Assured Limited	UK	*478*	75-49280
Retlif Testing Laboratories	USA	*641*	80-25950
Reumech Austral	S. Africa	*359*	75-18230
Reumech-OMC	S. Africa	*359*	75-18265
Reunert Technology Systems (Pty) Ltd (Reutech)	S. Africa	*359*	75-18300
Reutech	S. Africa	*359*	75-18340
Reutech = Reunert Technology Systems (Pty) Ltd	S. Africa	*359*	75-18300
Reutech Defence Industries (RDI)	S. Africa	*359*	75-18400
Reutech Radar Systems (Pty) Ltd Radar Division	S. Africa	*359*	75-18500
Reutech Systems	S. Africa	*359*	75-18510
Reutech Systems	S. Africa	*360*	80-25000
Revere Transducers Inc.	USA	*609*	75-74850
REVIMA	France	*151*	80-01720
Revue THOMMEN AG	Switzerland	*387*	75-11100

Rohde & Schwarz Canada Inc.	Canada	*79*	80-01825
Rohde & Schwarz France Telecom	France	*142*	75-32900
Rohde & Schwarz GmbH & Co KG (R&S)	Germany	*174*	75-25400
Rohde & Schwarz Italia SpA.	Italy	*244*	90-01850
Rohr - San Marcos	USA	*611*	75-76200
Rohr Aero Services Inc.	USA	*641*	80-26400
Rohr Europe	France	*142*	75-33000
Rohr France SA	France	*142*	75-33000
Rohr Inc.	USA	*611*	75-76150
Rohr Industries Inc.	USA	*611*	75-76150
Roje Telecomunicazioni SpA	Italy	*244*	90-01850
Rojone T/A Head Office	Australia	*29*	90-18820
Rokar International Ltd.	Israel	*222*	75-18805
Roke Manor Research Ltd (RMR)	UK	*420*	60-01800
Rolco Products Pty Ltd.	Australia	*20*	75-18750
Rollab Aktiebolag.	Sweden	*380*	80-01850
Rollalong	UK	*479*	75-50180
Rollet H. & Co. Ltd	UK	*479*	75-50208
Rolls-Royce & Associates Ltd (RRA)	UK	*510*	80-29100
Rolls-Royce Aero Engine Services Ltd	UK	*480*	75-50258
Rolls-Royce (Canada) Ltd	Canada	*75*	75-12120
Rolls-Royce Commercial Aero Engines Ltd	UK	*480*	75-50255
Rolls-Royce (Far East) Ltd	Japan	*258*	90-01860
Rolls-Royce (France) Ltd.	France	*154*	90-01850
Rolls-Royce International Ltd	Australia	*29*	90-18850
Rolls-Royce International Ltd	France	*154*	90-01850
Rolls-Royce International Ltd Tokyo Branch Office	Japan	*258*	90-01860
Rolls-Royce MatEval	UK	*480*	75-50260
Rolls-Royce Military Aero Engines Ltd	UK	*480*	75-50265
Rolls-Royce Motor Cars Ltd Specialist Engines	UK	*480*	75-50270
Rolls-Royce North America Inc.	USA	*655*	90-18630
Rolls-Royce plc.	UK	*479*	75-50250
Rolls-Royce Turbomeca Ltd	UK	*426*	74-18000
Romtehnica	Romania	*330*	55-01000
Rood Technology UK Ltd	UK	*510*	80-29200
ROR Rockwell	Germany	*174*	75-25540
ROR Rockwell B V Heavy Vehicle Systems	Netherlands	*294*	75-05850
ROR Rockwell España SA	Spain	*368*	85-00950
ROR Rockwell Intalia Srl Heavy Vehicle Systems	Italy	*238*	75-20820
ROR Rockwell (Ireland) Ltd Heavy Vehicle Systems	Ireland	*215*	75-00520
ROR Rockwell SA	France	*142*	75-33250
Ros Roca SA	Spain	*367*	75-05030
Rose Bearings Ltd	UK	*480*	75-50400
ROSE Elektrotechnik GmbH & Co KG	Germany	*174*	75-25630
Rosemount Aerospace Inc.	USA	*611*	75-76650
Rosenbauer International Aktiengesellschaft AG	Austria	*34*	75-02132
Rosenblatt & Son Inc., M.	USA	*641*	80-26450
Rosker Ltd	UK	*489*	75-60460
ROSM = Royal Ordnance Speciality Metals	UK	*481*	75-50630
Ross & Catherall Ltd	UK	*480*	75-50555
Ross Video Limited	Canada	*75*	75-12150
Rotadata Ltd	UK	*481*	75-50570
Rotadata Telecommunications Inc	USA	*648*	85-01522
Rotax Aviation Corp	India	*205*	80-01900
Rotax GmbH	Austria	*32*	75-00660
Rotech Components Ltd.	UK	*481*	75-50575
Rotek Inc..	USA	*611*	75-76750
Rotherham Precision Engineers Ltd.	UK	*481*	75-50590
Rotortech Ltd	UK	*510*	80-29325
Rotrax Cars International	S. Africa	*359*	75-18700
Rottefella A/S	Norway	*310*	75-02350
Rotterdamsche Droogdok Maatschappij BV	Netherlands	*294*	75-05770
ROTZLER GmbH & Co.	Germany	*174*	75-25800
Roux Systems Analysis and Test	S. Africa	*360*	80-30000
Rowe Industries Inc.	USA	*611*	75-76900
Rowlinson Packaging Ltd	UK	*481*	75-50600
Rowlinson (Wirebound Containers) Ltd.	UK	*481*	75-50600
Royal Advisory Council.	Bhutan	*52*	10-30000
Royal Air Force Benevolent Fund	UK	*421*	65-01100
Royal Army Medical Corps (RAMC)	UK	*412*	25-39150
Royal Army of Oman (RAO)	Oman	*313*	25-30000
Royal Australian Air Force Association	Australia	*10*	65-09000
Royal Bahamas Defence Force (RBDF)	Bahamas	*35*	25-11000
Royal Bahamas Police Force	Bahamas	*35*	26-00000
Royal Brunei Armed Forces, Combat Engineer Squadron, Support Battalion	Brunei	*60*	25-10000
Royal Brunei Navy	Brunei	*60*	25-40000
Royal Brunei Technical Services (RBTS)	Brunei	*60*	40-01000
Royal College of Defence Studies	UK	*420*	60-01810
Royal Danish Court.	Denmark	*99*	10-10000
Royal Danish Navy	Denmark	*100*	25-40000
Royal Fiji Police Force	Fiji	*115*	26-00000
Royal Hong Kong Auxiliary Air Force	Hong Kong SAR	*197*	25-05000
Royal Hong Kong Police Force Headquarters.	Hong Kong SAR	*197*	26-00000
Royal Hong Kong Regiment (The Volunteers)	Hong Kong SAR	*197*	65-10000
Royal Jordanian Air Academy	Jordan	*260*	80-29100
Royal Logistic Corps Headquarters	UK	*412*	25-33950
Royal Military Academy for Army and Air Force (KMA)	Netherlands	*288*	25-39100
Royal Military College	Australia	*7*	25-39100
Royal Military College of Science (RMCS)	UK	*420*	60-01820
Royal Netherlands Air Force Headquarters (RNLAF)	Netherlands	*287*	25-20000
Royal Netherlands Army Headquarters (RNLA)	Netherlands	*288*	25-30000
Royal Netherlands Meterological Institute	Netherlands	*287*	20-12000
Royal Netherlands Navy Centre	Netherlands	*288*	25-49200
Royal Netherlands Navy Headquarters (RNLN)	Netherlands	*288*	25-40000
Royal Netherlands Navy Naval Air Service	Netherlands	*288*	25-40130
Royal New Zealand Air Force Air Staff (RNZAF)	New Zealand	*300*	25-20000
Royal New Zealand Army	New Zealand	*300*	25-30000
Royal New Zealand Navy (RNZN)	New Zealand	*300*	25-40000
Royal Nooteboom Trailers BV	Netherlands	*294*	75-06100
Royal Norwegian Navy	Norway	*307*	25-42000
Royal Ordnance	UK	*481*	75-50620
Royal Ordnance Speciality Metals (ROSM)	UK	*481*	75-50630
Royal Papua New Guinea Constabulary	Papua N.G.	*317*	26-08000
Royal Schelde Koninklijke Schelde Groep BV	Netherlands	*294*	75-06150
Royal School of Military Engineering	UK	*420*	60-01830
Rozner Sales	USA	*655*	90-18800
RRA = Rolls-Royce & Associates Ltd	UK	*510*	80-29100
RSL Electronics Ltd	Israel	*222*	75-18900
RTE Relli Technology Ersatec GmbH	Germany	*179*	80-01800
RTE-Relli Engineering und Vertriebs GmbH.	Germany	*179*	80-01800
RTI = Research Triangle Institute	USA	*540*	60-18000
RTK Marine	UK	*481*	75-50650
Ruart Bearing Company	USA	*645*	85-00040
Rubb Buildings Ltd	UK	*481*	75-50750
Rubber Industrie Soest NV	Netherlands	*294*	75-06155
Rubery Owen Group Services Ltd.	UK	*510*	80-29350
Ruggedtronics	USA	*611*	75-77015
Ruggieri Technologie	France	*142*	75-33500
Ruska Instrument Corp.	USA	*611*	75-77040
Russian Institute of Remote Sensing Methods for Geology	Russia	*334*	55-21000
Ruston Diesels Ltd.	UK	*453*	75-25255
Ruston Gas Turbines Inc.	USA	*653*	90-05840
Rutherford Appleton Laboratory (RAL)	UK	*420*	60-01850
RVI = Renault Véhicules Industriels SA	France	*141*	75-32300
R.X. Products Ltd	UK	*481*	75-51020
Rybinsk Engine-Building	Russia	*337*	75-09090
Rybinsk Engine-Building Design Bureau	Russia	*337*	75-09090
Ryco Hydraulics (see Australasian Steel Products Pty Ltd)	Australia	*12*	75-01720

S

S Kimura & Co. Ltd	Japan	*258*	90-01160
S M Engineering	Korea (ROK)	*267*	90-01955
S&D Group	UK	*510*	80-32000
S&D Security (Equipment) Ltd & STI Ltd.	UK	*510*	80-32000
SA Marine AB	Sweden	*376*	75-03040
Saab AB Gripen	Sweden	*378*	75-09200
Saab Aircraft AB	Sweden	*378*	75-09000
Saab Combitech AB	Sweden	*378*	75-09050
Saab Dynamics AB	Sweden	*378*	75-09100
Saab Ericsson Space AB	Sweden	*378*	75-09150
Saab Group Head Office	Sweden	*375*	70-09000
Saab Training Systems AB	Sweden	*378*	75-09250
SAAF = South African Air Force Headquarters	S. Africa	*355*	25-20000
SAAF Flying School	S. Africa	*356*	25-29100
Saarstahl AG	Germany	*174*	75-25860
SABCA = Société Anonyme Belge de Constructions Aéronautiques	Belgium	*46*	75-05590
Sabharwal's Bulk Drugs Ltd	India	*205*	75-01880
Sabo France	France	*142*	75-33650
Sabre Computers International Ltd	UK	*481*	75-51150
Sabreliner Corp.	USA	*641*	80-26500

Quick Reference

Quick Reference

SAC = Shenyang Aircraft Corp.	China PR	88	55-01950
SAC Hitec Ltd	UK	509	80-27840
SAC Technographic	UK	510	80-30100
SAC Technology Ltd.	UK	509	80-27820
Sächsische Spezialkonfektion GmbH (Spekon)	Germany	174	75-25880
SACLANT = Supreme Allied Command Atlantic.	USA	539	34-20000
SACLANT Undersea Research Centre (SACLANTCEN)	Italy	229	34-42350
SACLANTCEN = SACLANT Undersea Research Centre.	Italy	229	34-42350
Sacm Diesel	France	149	75-47090
Saco Defense Inc.	USA	612	75-77340
Sadler Aircraft Corp.	USA	612	75-77345
SAE = Society of Automotive Engineers	USA	542	65-20000
SAE Alsetex	France	142	75-33810
Saetong Corporation.	Korea (ROK)	267	90-01895
SAF Military Academy	Sweden	372	25-29550
Safare-Crouzet SA	France	142	75-33900
Safariland Ltd Inc.	USA	612	75-77350
Safe Flight Instrument Corp.	USA	612	75-77400
Safe Flight Instrument Ltd	UK	525	90-01885
Safe Training Systems Ltd.	UK	481	75-51300
Safety & Reliability Consultants Ltd	UK	512	80-46200
Safety Air Services Ltd.	UK	482	75-51830
Safety Consulting Engineers Inc. (SCE)	USA	641	80-26680
Saft	France	142	75-33920
Saft America Inc	Mexico	282	75-18000
Saft America Inc..	USA	612	75-77550
Saft Australia Pty Ltd.	Australia	20	75-18950
Saft Ferak A/S	Czech Republic	99	90-03000
Saft Hong-Kong	Hong Kong SAR	199	90-01900
Saft Korea Co Ltd	Korea (ROK)	266	75-04300
Saft Nife AB	Sweden	378	75-09850
Saft Nife Argentina SA.	Argentina	4	75-09000
Saft Nife AS	Norway	310	75-02400
Saft Nife Brasil Sistemas Eléctricos Ltda	Brazil	57	75-19045
Saft Nife BV	Netherlands	294	75-06160
Saft Nife Corp	Canada	75	75-12290
Saft Nife GmbH	Austria	34	75-02142
Saft Nife GmbH	Germany	174	75-25901
Saft Nife Iberica SL	Spain	367	75-05100
Saft Nife Ltd	Hong Kong SAR	199	90-01910
Saft Nife Ltd	UK	481	75-51470
Saft Nife M.E. Ltd	Cyprus	96	75-05000
Saft Nife (México) SA de CV	Mexico	282	75-18100
Saft Nife Oy	Finland	118	75-01960
Saft Nife Power Systems Australia Pty Ltd	Australia	20	75-18950
Saft Nife Power Systems Pte Ltd	Singapore	348	75-03400
Saft Nife Power Systems Sdn Bhd	Malaysia	278	75-00900
Saft Nife SpA.	Italy	238	75-21350
Saft Singapore Pte Ltd.	Singapore	348	75-03400
Safta	Belgium	50	90-01910
SAFTI Military Institute Library	Singapore	347	60-01000
Safway Steel Products Inc.	USA	642	80-26700
Sage Laboratories Inc..	USA	612	75-77610
SAGEM Nederland BV.	Netherlands	295	75-06180
SAGEM SA	**France**	**143**	**75-33950**
SAGEM-B	Belgium	47	75-05595
Sageware Inc.	USA	612	75-77670
SAI = Swedish Aerospace Industries Association	Sweden	374	65-01900
SAI Industries Corp.	USA	612	75-77700
SAIC Ltd	UK	510	80-30220
Saint John Naval Systems.	Canada	75	75-12300
Saint John Shipbuilding Limited.	Canada	75	75-12300
SAIT Systems SA	Belgium	47	75-05600
SAJ-Froude Test Plant Pvt Ltd	India	204	74-01000
Sako Ltd Riihimäki Works	Finland	118	75-01980
Sako-Valmet Ltd	Finland	118	75-01980
Sakr Factory for Developed Industry	Egypt	109	55-01700
SAL = Standard Aero Ltd	Canada	79	80-01970
Sala International Pty Ltd	Australia	21	75-18980
Saladin Holdings Ltd.	UK	510	80-30300
Saladin Security Ltd	UK	510	80-30300
Salchow Elektronik AB.	Sweden	381	90-01913
Salver srl Ciba Composities	Italy	235	75-10900
Salyut DB = Salyut Design Bureau	Russia	338	75-30000
Salyut Design Bureau (Salyut DB).	Russia	338	75-30000
Sama Electronics Corporation.	Korea (ROK)	267	90-01910
Sama Trading Company Head Office	Korea (ROK)	266	85-00900
SAMAERO Company Pte Ltd	Singapore	352	90-01900
Samas Roneo Ltd	UK	481	75-51650
Samas Vickers Ltd	UK	481	75-51650
Samet Consultants.	UK	525	90-01895
Samitac.	UK	510	80-30800
SAMM = Société d'Applications des Machines Motrices	France	143	75-34300
Samson Distraco U.K. Ltd.	UK	517	85-01850
Samsung Aerospace Industries Ltd Head Office.	Korea (ROK)	266	75-04630
Samsung Semiconductor & Telecommunications Co Ltd (SST)	Korea (ROK)	266	75-04650
Samsung Shipbuilding & Heavy Industry Co Ltd	Korea (ROK)	266	75-04660
Samuel Groves & Co. Ltd	UK	495	75-67700
SAN Elektroheat	UK	525	90-01900
Sandeep Metalcraft Pvt Ltd (SM)	India	205	75-01900
Sanders.	USA	612	75-77850
Sandhurst Mfg Co Ltd	UK	481	75-51750
Sandow-Technic SA.	France	143	75-34312
Sandvik AB	Sweden	378	75-10150
Santa Barbara SA	Spain	364	55-01000
Santierul Naval Constanta SA.	Romania	330	75-00750
Sanvas SA Indústria Metal-Mecânica.	Brazil	57	75-19140
Sapac International Co. Ltd	Thailand	396	90-01910
S.A.P.H.I. Group	S. Africa	359	75-19160
S.A.P.H.I. 2000	S. Africa	359	75-19160
SAR-ASSIST	USA	642	80-26850
Sarakakis Pros Aebe	Greece	194	90-01910
Sarantopoulos SA, D.F.	Greece	193	85-00900
Saratov Electric Instrument Manufacturing Plant after Sergo Ordzhonikidze.	Russia	339	75-31075
Saratov Industrial Automation Design Bureau	Russia	337	75-03200
Sargent Fletcher Inc.	USA	612	75-78150
SARMA.	France	143	75-34330
Sarnoff Research Center, David.	USA	540	60-19150
Sartank Industrial & Computer Supply	Germany	183	90-01875
Sarum Systems	UK	525	90-01945
SAS Communications Ltd	UK	482	75-51800
SAS Limited	UK	482	75-51830
Sas R&D Services Inc.	USA	642	80-26870
Sas Research & Development Services Inc.	USA	642	80-26870
Sascal Displays	UK	482	75-51820
Satair A/S	Denmark	105	80-03500
Satair ASIA Pte Ltd	Singapore	350	85-03800
Satair China Representative Office	China PR	89	85-05000
Satair Malaysia Sdn Bhd Representative Office.	Malaysia	278	85-00600
Satakunta Air Command.	Finland	116	25-20000
Sattler Texilwerke KG	Austria	34	75-02155
Saudi Arabian Parsons Limited	Saudi Arabia	344	90-01900
Saudi Naval Support Co. Ltd	Saudi Arabia	343	80-01600
Saudi Vosper Thornycroft Company Limited, The.	Saudi Arabia	343	80-02000
Sauer & Sohn GmbH, J.P.	Germany	174	75-26050
Saunders-Roe Ltd (SRL).	UK	482	75-51850
SAV = Standard Aero V.o.F.	Netherlands	295	75-07000
Saval BV	Netherlands	295	75-06200
Saval NV	Belgium	47	75-05650
Saviat Special Aviation Technology.	Russia	339	75-31100
SBAC = The Society of British Aerospace Companies Ltd	UK	421	65-01950
SBS Engineering Inc.	USA	612	75-78330
SBS Technologies Inc..	USA	612	75-78330
SCA Distributors Pty Ltd.	Australia	16	75-07600
Scalar R F I Pty Ltd.	Australia	20	75-18500
Scaled Composites Inc.	USA	541	60-19160
SCAM SpA.	Italy	238	75-21530
Scan Para ApS	Denmark	103	75-12000
ScanAvia	Norway	311	80-17000
ScanAvia A/S	Denmark	105	80-03550
ScanAvia, Scandinavian Aviation Consultants Sweden AB	Sweden	380	80-01860
Scancopter A/S	Norway	312	90-01920
Scandinavian Aerospace Industry.	Sweden	381	90-01940
Scandinavian Aviation Consultants	Denmark	105	80-03550
Scandinavian Aviation Consultants Group	Norway	311	80-17000
Scandinavian Avionics A/S	Denmark	105	85-01500
Scania AB Head Office and Technical Centre	Sweden	378	75-10200
Scania Argentina SA.	Argentina	4	75-19500
Scania de México SA de CV.	Mexico	282	75-18200
Scania Kapena SA	Poland	325	75-02730
Scania Latin America Ltda.	Brazil	58	75-19180
Scania Production Angers SA.	France	143	75-34380
SCAT = South China Aero Technology Ltd.	Hong Kong SAR	198	80-00850

Quick Reference

Senate	Belgium	*39*	15-10000
Senate	Bermuda	*51*	15-10000
Senate	Canada	*64*	15-10000
Senate	France	*121*	15-10000
Senate	Jordan	*259*	15-10000
Sencea Ltd	UK	*482*	75-52730
Sener Ingenieria y Sistemas SA	Spain	*367*	75-05200
Sensidyne Inc.	USA	*613*	75-79320
Sensitised Coatings Ltd	UK	*482*	75-52800
Senso-Metrics Inc.	USA	*613*	75-79340
Sensor Electronics Limited	UK	*482*	75-52820
Sensor Systems Inc.	USA	*613*	75-79400
Senstar Corporation	Canada	*76*	75-12760
Senstar (UK) Ltd	UK	*525*	90-01937
Senstar-Stellar Corp	Canada	*76*	75-12760
Senstar-Stellar Inc.	USA	*655*	90-19250
SEOS Displays	USA	*613*	75-79450
SEOS Displays Ltd	UK	*482*	75-52830
SEP = Société Europeenne de Propulsion Head Office	France	*144*	75-34850
Sequa Corp.	USA	*613*	75-79490
SERAT = Société d'Études, Réalisations et d'Applications Techniques	France	*144*	75-34885
SERCEL	France	*144*	75-34887
Serck Aviation Div.	UK	*482*	75-52860
Serck Heat Transfer	UK	*483*	75-52880
Serck Heat Transfer	USA	*655*	90-19270
Serck Marston	UK	*461*	75-31840
Serck Services	UK	*461*	75-31840
Serck Services International	UK	*461*	75-31840
Serco Aviation Services	UK	*511*	80-34150
Serco-IAL Limited	UK	*511*	80-34150
Sermatech International Inc	USA	*613*	75-79530
Servaux SA	France	*144*	75-34900
Service Aviation International	France	*151*	85-01900
Service Cantonal de la Protection Civile	Switzerland	*388*	80-03000
Service Hydrographique et Océanographique de la Marine (SHOM)	France	*123*	25-42500
Service Technique des Systèmes de Missiles Tactiques (STSMT)	France	*122*	20-40000
Service Technique des Systèmes Stratégiques et Spatiaux	France	*122*	20-40000
Services Sound & Vision Corp. (SSVC)	UK	*511*	80-34200
Servicio de Telecomunicaciones de la Armada de Chile (BIBLIOTECA)	Chile	*84*	25-44000
Servicios Industriales de la Marina - SIMA PERU	Peru	*319*	55-01950
Servo Corp	Switzerland	*390*	90-01930
Servo Corp. of America	USA	*613*	75-79650
Servogear A/S	Norway	*310*	75-02435
Servotecnica	Italy	*244*	90-01923
Servotecnica SRL	Italy	*241*	85-01000
Servotron SA	Argentina	*5*	90-19300
SETAF = United States Army in Europe Southern European Task Force	Italy	*229*	33-34010
SETEC Pty Ltd	Australia	*21*	75-19415
Seton Healthcare Group plc	UK	*483*	75-53000
Sevastopol JSC Marine Plant	Ukraine	*407*	75-06500
Sevastopol S. Ordzhonikidze Marine Plant	Ukraine	*407*	75-06500
Severe Environment Systems Co.	USA	*623*	75-89630
Seward Medical Limited	UK	*483*	75-53050
SEXTANT Avionique	China PR	*90*	90-18200
Sextant Avionique Asia Pte Ltd	Singapore	*349*	80-01800
SEXTANT Avionique GmbH Büro Deutschland	Germany	*179*	85-02500
SEXTANT Avionique Inc Beacon Centre	USA	*655*	90-19310
SEXTANT Avionique Italia SpA	Italy	*244*	90-01925
SEXTANT Avionique SA Head Office	France	*144*	75-35000
SEXTANT Avionique UK Branch Office	UK	*483*	75-53100
SEXTANT Holdings Inc	USA	*544*	70-18500
SFIM	**France**	***144***	**75-35080**
SFIM Inc	USA	*614*	75-79678
SFIM Industries Deutschland GmbH	Germany	*175*	75-27495
SFIM ODS Head Office	France	*144*	75-35085
SFIM (UK)	UK	*525*	90-01945
SFMI = Société Française de Moteurs à Induction SA	France	*145*	75-35100
SFTE = Society of Flight Test Engineers	UK	*420*	60-02000
SFU = Société des Fonderies d'Ussel	France	*146*	75-40400
S.G. Brown	UK	*437*	75-10330
SGAP (Malaysia) Sdn Bhd	Malaysia	*279*	90-01900
S.G.D. Engineering Ltd	Israel	*224*	80-01930
SGDN = Secrétariat Général de la Défense Nationale	France	*121*	12-10050
SGS Yarsley International Certification Services Ltd	UK	*511*	80-34550
SGS Yarsley Technical Services	UK	*511*	80-34550
SH = Schuberth Helme GmbH	Germany	*174*	75-26800
Shaanxi Aircraft Manufacturing Co.	China PR	*88*	55-01900
Shade (Computer Services) Ltd	UK	*511*	80-34570
Shakespeare Co.	USA	*614*	75-79800
Shakespeare Products Group Electronics & Fibreglass Division	USA	*614*	75-79800
Shalon Chemical Industries Ltd	Israel	*222*	75-20000
Shamban & Captain Inc.	Japan	*257*	90-00240
Shamban Belux SA	Belgium	*46*	75-02000
Shamban Canada Inc.	Canada	*81*	90-00280
Shamban Europa A/S, W.S.	Denmark	*106*	90-19410
Shamban Europa (UK) Ltd	UK	*438*	75-11050
Shamban International	USA	*561*	75-16006
Shanghai Daewoo Center Development Co Ltd	China PR	*89*	75-01910
SHAPE = Supreme Headquarters Allied Powers Europe	Belgium	*43*	34-12100
Sharp Inc., George G.	USA	*642*	80-28000
Shaw Aero Development Inc.	USA	*614*	75-79900
Shaw Aero Devices Inc.	USA	*614*	75-79950
Shaw Packing Ltd	UK	*511*	80-34700
Sheffield Forgemasters Ltd	UK	*483*	75-53200
Sheldahl Inc.	USA	*614*	75-80150
Shenango Steel Buildings Inc.	USA	*614*	75-80205
Shenyang Aircraft Corp. (SAC)	China PR	*88*	55-01950
Sheraton Sofia Hotel Balkan JSC (DAESO)	Bulgaria	*62*	75-03000
Sherwin Inc.	USA	*614*	75-80300
Sherwin Williams do Brazil Indústria e Commércio Ltda	Brazil	*58*	75-19540
Sherwood International Export Corp.	USA	*640*	80-22000
SHI = Sumitomo Heavy Industries Ltd Head Office	Japan	*254*	75-08000
Shield International Ltd	UK	*511*	80-35500
Shimadzu Corporation	Japan	*254*	75-07600
ShinMaywa (America) Ltd	USA	*656*	90-19320
ShinMaywa (America) Ltd Los Angeles Branch	USA	*656*	90-19320
ShinMaywa (Europe) GmbH	Germany	*183*	90-01932
ShinMaywa Ground Services Ltd	Japan	*256*	80-02000
ShinMaywa Industries Ltd	Japan	*254*	75-07800
Ship Analytics Inc.	USA	*614*	75-80455
Shipbuilding Association of Canada	Canada	*67*	65-00300
Shipham & Co. Ltd	UK	*483*	75-53350
Shiposh Engineering Ltd	USA	*656*	90-19330
Shipyard Beograd	Yugoslavia	*661*	75-00690
Shock-Tech Inc.	USA	*614*	75-80457
SHOM = Service Hydrographique et Océanographique de la Marine	France	*123*	25-42500
Shooting Systems Group Inc.	USA	*614*	75-80460
Shorrock Integrated Systems	UK	*483*	75-53450
Shorrock Major Systems Ltd	UK	*483*	75-53450
Shorts Missile Systems Ltd (SMS)	**UK**	***483***	**75-53550**
Shorts Support Services Division (Airwork Ltd)	UK	*511*	80-35600
Shoshin Corp.	Japan	*258*	90-01960
Shoval-Saar Ltd (SSL)	Israel	*225*	80-01938
Shredaway Ltd	UK	*511*	80-35700
Shur-Lok International SA	Belgium	*47*	75-05900
SI = Spectral Intelligence Corp.	USA	*642*	80-29250
SIA = Società Italiana Avionica SpA	Italy	*239*	75-23880
SIAI Marchetti Sarl	Italy	*238*	75-21950
Siam Aviation	Thailand	*396*	90-01695
SIBAT - Foreign Defense Assistance & Defense Export	Israel	*216*	40-01900
SICAMB SpA Head Office and Plant	Italy	*238*	75-22020
Sicaprep (Madeira) Lda	Portugal	*328*	90-01970
Sidem (UK) Ltd	UK	*511*	80-35850
SIDES = Société Industrielle pour le Développment de la Sécurité	France	*146*	75-40470
Siebe Gorman & Co. Ltd	UK	*483*	75-53600
Siebe North Inc.	USA	*614*	75-80555
Siebe plc	UK	*424*	70-01940
Siecor Corp.	USA	*614*	75-80560
Sieger Ltd	UK	*495*	75-67600
Sielman SA	Greece	*191*	75-02360
Siemens AB	Sweden	*378*	75-10300
Siemens Aktiengesellschaft Group Headquarters	Germany	*175*	75-27700
Siemens A/S Defence Electronics Division	Norway	*310*	75-02450
Siemens Atea NV	Belgium	*47*	75-05950
Siemens Egypt Ltd	Egypt	*111*	90-19300

Siemens Electromechanical Components Inc.	USA	*614*	75-80565
Siemens Environmental Systems Ltd Radiation Monitoring	UK	*483*	75-53700
Siemens Ltd Melbourne Head Office	Australia	*21*	75-19450
Siemens Nixdorf Informationssysteme AG	Germany	*179*	80-01955
Siemens plc	UK	*483*	75-53750
Siemens Plessey Controls Ltd	UK	*483*	75-53700
Siemens Plessey Electronic Systems Pty Ltd	Australia	*21*	75-19475
Siemens Plessey Systems	UK	*483*	75-53770
Siemens SA	Brazil	*60*	90-19400
Siemens Showa Solar Pte Ltd	Singapore	*352*	90-01928
Siemens Solar GmbH	Germany	*183*	90-01935
Siemens Solar Industries	USA	*614*	75-80570
Siemens Switzerland Ltd	Switzerland	*387*	75-11980
Siemens-Plessey Assessment Services Ltd	UK	*497*	80-01500
Sierra Research	USA	*614*	75-80640
Sierra Technologies Inc. Sierra Research Division	USA	*614*	75-80640
Sierracin/Magnedyne	USA	*614*	75-80651
Sierracin/Sylmar Corporation	USA	*614*	75-80650
SIFT SAS di Cutore & Pitotti	Italy	*239*	75-22150
SIFT snc di Cutore & Co.	Italy	*239*	75-22150
SIGARMS Inc.	USA	*614*	75-80700
Sigma Industries Pty Limited (Hong Kong)	Hong Kong SAR	*198*	75-00080
Sigma Industries Pty Ltd	Australia	*11*	75-00450
Sigma International Inc	USA	*548*	75-03690
Sigma Security Devices Ltd	UK	*483*	75-53778
Sigma Security Services Ltd	UK	*483*	75-53778
Sigma Signals Corporation	USA	*615*	75-80790
Sigma 3 SA	Greece	*191*	75-02380
Signaal Communications	Netherlands	*295*	75-06350
Signaal Special Products	Netherlands	*295*	75-06400
Signaal USFA	Netherlands	*295*	75-06500
Signal Computing Ltd	UK	*511*	80-36050
Signal Processors Ltd	UK	*483*	75-53780
Signal Technology Corp Olektron Operation	USA	*615*	75-80795
Signal Technology Corp. Systems Operation	USA	*618*	75-84240
Signature Industries Ltd	UK	*484*	75-53785
Signature Management	UK	*484*	75-53790
SIHI Australia Pty Ltd	Australia	*21*	75-19505
Sikorsky Aircraft	USA	*615*	75-80900
Sikorsky United Technologies Int'l Ops	Turkey	*404*	90-01990
Silberkraft Leichtakkumulatoren GmbH	Germany	*175*	75-28000
Silent Partner Body Armor Inc. Personal Armour Systems Division (SPBA)	USA	*615*	75-80930
Silicone Engineering	UK	*511*	80-36070
Sillinger	France	*145*	75-35660
Silvani Antincendi SpA	Italy	*239*	75-22250
Silver Arrow	Israel	*222*	75-20400
Silver Eagle Manufacturing Co.	USA	*615*	75-81020
Silver Star International Co	Hong Kong SAR	*198*	85-00500
Silverstone Electronics Ltd	UK	*484*	75-53830
SIM Security & Electronic System GmbH	Germany	*175*	75-28010
SIMA PERU	Peru	*319*	55-01950
Simad SpA Commercial Office	Italy	*230*	74-23350
Simagine Simulation Systems	Netherlands	*295*	75-06600
SIMC Sarl Administration Office	Italy	*244*	90-01930
Sime Aircraft Tyre Co. Sdn Bhd	Malaysia	*278*	75-01800
SIME s.a.s. di Paolo de Gaetano & C. Commercial Office	Italy	*244*	90-01930
Simeca SA	Greece	*191*	75-02400
Simetra Inc.	Canada	*79*	80-01940
Simo Loukarelli Ltd	Greece	*191*	75-02410
Simon Abbey Pty Ltd	Australia	*11*	75-00100
Simon Access (UK) Ltd	UK	*484*	75-53908
Simon Gloster Saro Ltd	UK	*484*	75-53908
Simonneau Marine International	France	*146*	75-40350
Simpson Lawrence	UK	*517*	85-01910
Simrad (Asia) Pte. Ltd	Singapore	*352*	90-01935
Simrad Inc	USA	*589*	75-53160
Simrad Ltd	UK	*463*	75-34100
Simrad Mesotech Ltd	Canada	*76*	75-12795
Simrad Mesotech Systems Ltd	Canada	*76*	75-12795
Simrad Norge A/S	Norway	*310*	75-02630
Simrad Optronics ASA	Norway	*310*	75-02620
Simrad Optronics Ltd	UK	*525*	90-01957
Simrad Peru SA	Peru	*319*	75-16000
Simrad Shipmate AS	Denmark	*103*	75-19000
SimuFlite Training International	USA	*642*	80-28160
Simula Government Products, Inc.	USA	*615*	75-81250
Simunition	Canada	*67*	74-19100
Singapore Aerospace International	USA	*656*	90-19420
Singapore Aerospace Systems Pte Ltd	Singapore	*350*	80-02005
Singapore Aerospace (UK) Pte Ltd	UK	*525*	90-01959
Singapore Automotive Engineering	Singapore	*348*	75-04280
Singapore British Engineering Pte Ltd.	Singapore	*350*	85-04000
Singapore Electronic & Engineering Ltd	Singapore	*349*	80-01960
Singapore Engineering Software Pte Ltd	Singapore	*349*	80-01965
Singapore Ordnance Engineering Pte Ltd	Singapore	*349*	80-01970
Singapore Technologies Aerospace Ltd	Singapore	*348*	75-03905
Singapore Technologies Automotive Ltd (STA)	Singapore	*348*	75-04280
Singapore Technologies Holdings Pte Ltd	Singapore	*347*	70-01900
Singapore Technologies Pte Ltd	Singapore	*347*	70-01900
Sino Swearingen Aircraft Company	USA	*615*	75-80880
Sintef Group	Norway	*308*	60-01600
S.I.P. Services International Ltd	UK	*511*	80-36500
Sipka Manufacturing (Pvt) Ltd	Pakistan	*315*	75-01970
SIPRI = Stockholm International Peace Research Institute	Sweden	*374*	60-01900
Sir J.H. Biles (Naval Services) Ltd	UK	*498*	80-03350
Sira Ltd Electro-Optics Division	UK	*484*	75-54250
SIRN = Società Impianti Riparazioni Navali, Per Azioni	Italy	*240*	80-01950
SIRN-Società Imp.Riparazioni Navali	Italy	*240*	80-01950
Sirples Sarl	Italy	*239*	75-22580
SIRS Navigation Ltd	UK	*484*	75-54300
SISDEF = Sistemas y Desarrollos Functionales Ltda	Chile	*85*	80-01940
SISTEL-Comunicações, Automação e Sistemas SA	Portugal	*327*	75-18300
Sisu Defence Oy	Finland	*118*	75-01710
SITEP SpA	Italy	*239*	75-22595
SITO = Security Industry Training Organisation Ltd	UK	*511*	80-33500
SIXATAF = Sixth Allied Tactical Air Force	Turkey	*402*	34-22100
Sixth Allied Tactical Air Force (SIXATAF)	Turkey	*402*	34-22100
Sixth Foreign Regiment Engineering	France	*122*	25-35000
SJA International Inc.	USA	*642*	80-28200
SJAC = The Society of Japanese Aerospace Companies Inc.	Japan	*249*	65-19000
SK-IMPEX JSC	Russia	*340*	80-02000
Skanacid A/S	Denmark	*105*	80-03650
Skandinavisk Teleindustri Skanti A/S	Denmark	*106*	90-19450
SKANTI - Skandinavisk Teleindustri A/S	Denmark	*106*	90-19450
SKB GeoKos Special Design Bureau	Russia	*338*	75-14315
SKC Inc.	USA	*615*	75-81480
SKF	Sweden	*378*	75-10340
SKF do Brasil Ltda	Brazil	*58*	75-19680
SKF USA Inc.	USA	*615*	75-81500
Skilled Engineering Ltd.	Australia	*25*	80-19350
Skipasmidastodin hf	Iceland	*202*	75-05000
Skurka Engineering Company	USA	*615*	75-81520
Skyfox Aviation	Australia	*17*	75-08300
Skyfox Aviation Ltd.	Australia	*17*	75-08300
Skyline Aviation BV	Netherlands	*297*	80-19000
Skyline Industries Inc.	USA	*615*	75-81550
Skytech Aviation Services Ltd	UK	*511*	80-37000
Skyways Inc.	USA	*642*	80-28250
Sleeman Engineering	UK	*484*	75-54400
SLI International	USA	*648*	85-01990
Slingsby Aviation Ltd.	UK	*484*	75-54410
Slingsby Engineering Ltd	UK	*484*	75-54420
Sloboda DP	Yugoslavia	*661*	75-00475
Slocomb Company, J.T.	USA	*615*	75-81700
Slovenian Territorial Defence Staff	Slovenia	*354*	25-10000
SM = Sandeep Metalcraft Pvt Ltd.	India	*205*	75-01900
S.M. Alexander (Plastics) Ltd	UK	*429*	75-02580
SMA = Scientific Management Associates Pty Ltd	Australia	*25*	80-19100
SMA Wescare, Protection Safety Div.	UK	*429*	75-02580
SMC = South Midlands Communications Ltd	UK	*485*	75-55213
SMD = Soil Machine Dynamics Ltd	UK	*485*	75-54870
SMG = Société Métallurgique de Gerzat SA Factory and Commercial Services	France	*146*	75-40570
SMIN Consortium	Italy	*230*	74-03700
Smith & Nephew Consumer Products Limited	UK	*484*	75-54620
Smith & Nephew Group Research Centre	UK	*484*	75-54620
Smith & Nephew Healthcare Limited Medical Products	UK	*484*	75-54620
Smith & Nephew Healthcare Surgical Products	UK	*484*	75-54620
Smith & Nephew Homecraft	UK	*484*	75-54620
Smith & Nephew Medical Fabrics	UK	*485*	75-54620
Smith & Nephew Medical Limited	UK	*484*	75-54620
Smith & Nephew plc.	UK	*484*	75-54620

Smith & Prince Ltd	UK	*485*	75-54630
Smith & Wesson	USA	*615*	75-82060
Smith Engineering Pty Ltd, W.E.	Australia	*21*	75-19545
Smith Group, H.R.	UK	*484*	75-54590
Smith Gt Bentley Ltd, A.	UK	*484*	75-54540
Smith Industries Defense Systems International	USA	*648*	85-01990
Smith System Engineering Limited	Belgium	*48*	80-01950
Smith System Engineering Ltd	Japan	*256*	80-00950
Smith System Engineering Ltd	UK	*512*	80-39500
Smith's Electric Vehicles Ltd	UK	*485*	75-54650
Smith-Clayton Forge (Aerospace)	UK	*484*	75-54570
Smiths Industries Aerospace	UK	*485*	75-54700
Smiths Industries Aerospace & Defense Systems Inc	USA	*615*	75-82090
Smolensk Aircraft Plant	Russia	*339*	75-32800
Smolensk Aviation Plant	Russia	*339*	75-32800
SMP = Special Melted Products Ltd	UK	*485*	75-55240
SMR Technologies Inc	USA	*615*	75-82100
SMS = Shorts Missile Systems Ltd	**UK**	***483***	**75-53550**
SMS AG	Germany	*175*	75-28030
Snap-on Tools International Ltd	USA	*656*	90-19580
SNAPC = Syndicat National Autonome des Policiers en Civil	France	*129*	65-01800
SNAT = Syndicat National Autonome de la Tenue	France	*129*	65-01900
SNC = Syndicat National des Commissaires	France	*129*	65-01950
SNC Industrial Technologies Inc	Canada	*76*	75-12960
SNC-Lavalin Group Inc., The	Canada	*79*	80-01840
SNE = Syndicat National des Enquêteurs	France	*129*	65-02000
SNECMA = Société Nationale d'Études et de Constructions de Moteurs d'Aviation	France	*145*	75-36305
SNFA	France	*145*	75-36306
SNIPAT = Syndicat National Indépendant des Personnels Administratifs et Techniques de la Police National	France	*129*	65-02050
SNO = Syndicat National des Officiers	France	*129*	65-02100
Snorkel	USA	*616*	75-82170
SNPE = Société Nationale des Poudres et Explosifs	France	*145*	75-36315
SNUIP = Syndicat National Unitaire des Enquêteurs de Police	France	*129*	65-02300
SNUIP = Syndicat National Unitaire des Inspecteurs de Police	France	*129*	65-02320
SNUPAT = Syndicat National Unifié des Personnels Administratifs et Techniques de la Police National	France	*129*	65-02270
SOAC = Special Operations Acquisition Center	USA	*528*	25-12550
Sobeih Trade & Enterprises	Egypt	*111*	90-19590
SOC = Sparton of Canada Ltd	Canada	*77*	75-13100
SOCATA = Société de Construction d'Avions de Tourisme et d'Affaires	France	*145*	75-36350
SOCHATA Headquarters	France	*145*	75-36400
Sociedade Zickermann SA	Portugal	*328*	90-03850
Società di Condensazione ed Applicazioni Meccaniche	Italy	*238*	75-21530
Società Esercizi Aeroportuali SpA	Italy	*240*	80-01925
Società Esplosivi Industriali SpA Difesa Division (SEI)	Italy	*239*	75-23500
Società Finanziaria Telefonica	Italy	*230*	70-19400
Società Impianti Filtrazione Trasferimento Fluidi	Italy	*239*	75-22150
Società Impianti Riparazioni Navali, Per Azioni (SIRN)	Italy	*240*	80-01950
Società Italiana Avionica SpA (SIA)	Italy	*239*	75-23880
Società Italiana Chimici	Italy	*230*	65-02000
Società Italiana Costruzioni Aeronautiche Martin-Baker	Italy	*238*	75-22020
Società Italiana Materiali & Construzioni	Italy	*244*	90-01930
Società Italiana Telecomunicazioni SpA	Italy	*235*	75-13205
Società per Azioni Sistemi Elettronici Avanzati	Italy	*239*	75-22595
Société Anonyme Belge de Constructions Aéronautiques (SABCA)	Belgium	*46*	75-05590
Société Anonyme Belge de Mécanique et d'Armement SA	Belgium	*46*	75-04930
Société Anonyme d'Études et Réalisations Nucléaires (SODERN)	France	*145*	75-37100
Société Chiarelli d'Études et Réalisations Mécaniques et Electroniques	France	*133*	75-08810
Société d'Applications des Machines Motrices (SAMM)	France	*143*	75-34300
Société d'Applications Générales d'Electricité et de Mécanique	**France**	***143***	**75-33950**
Société d'Armement et d'Études Alsetex	France	*142*	75-33810
Société d'Études et de Fabrications Aéronautiques (SEFA)	France	*143*	75-34810
Société d'Études, Réalisations et d'Applications Techniques (SERAT)	France	*144*	75-34885
Société d'Études, Recherches & Constructions Electroniques	France	*144*	75-34887
Société d'Exploitation des Matériels Martin-Baker (SEMMB)	France	*130*	74-01800
Société d'Exploitation du Conditionnement Dydra	France	*146*	75-39455
Société d'Exploitation et de Constructions Aéronautiques	France	*151*	80-01775
Société de Construction d'Avions de Tourisme et d'Affaires (SOCATA)	France	*145*	75-36350
Société de Construction des Avions Hurel-Dubois	France	*146*	75-40375
Société de Constructions Mécaniques Panhard et Levassor	France	*140*	75-28750
Société de Fabrication d'Instruments de Mesure	**France**	***144***	**75-35080**
Société de Medecine Legale et de Criminologie de France	France	*128*	60-06000
Société de Promotion Aéronautique	Belgium	*50*	90-01670
Société des Céramiques Techniques (SCT)	France	*146*	75-39510
Société des Fonderies d'Ussel (SFU)	France	*146*	75-40400
Société des Tuyaux Bonna	France	*146*	75-41520
Société Dinac	Egypt	*110*	90-04000
Société Europeenne de Propulsion Head Office (SEP)	France	*144*	75-34850
Société Française d'Exportation de Systèmes d'Armement	France	*126*	40-01900
Société Française de Formation et de Conseil Aéronautiques	France	*149*	80-00120
Société Française de Moteurs à Induction SA (SFMI)	France	*145*	75-35100
Société Generale des Prisons et de Legislation Criminelle	France	*127*	60-05900
Société Girondine d'Entretien et de Réparation de Matériel Aéronautique	France	*146*	75-42030
Société Industrielle pour le Développment de la Sécurité (SIDES)	France	*146*	75-40470
Société Labbe SA	France	*146*	75-40500
Société Métallurgique de Gerzat SA Factory and Commercial Services (SMG)	France	*146*	75-40570
Société Nationale d'Études et de Constructions de Moteurs d'Aviation (SNECMA)	France	*145*	75-36305
Société Nationale de Construction Aerospatiale	Belgium	*47*	75-06100
Société Nationale des Poudres et Explosifs (SNPE)	France	*145*	75-36315
Société Navale Française de Formation et de Conseil	France	*150*	80-01420
Société Nouvelle des Ateliers et Chantiers du Havre	France	*130*	75-00150
Société Nouvelle Labbe	France	*146*	75-40500
Société Nouvelle PRB	Belgium	*47*	75-06050
Société pour le Perfectionnement des Matériels et Équipements Aérospatiaux	France	*151*	80-01997
Société Rennaise d'Electronique Professionnelle SA	France	*146*	75-42400
Société SMAC	France	*146*	75-40700
Société Suisse des Constructeurs de Machines	Switzerland	*384*	65-01500
Société Suisse des Explosifs	Switzerland	*387*	75-13500
Société Technique d'Application et de Recherche Electronique	France	*147*	75-42775
Society for Underwater Technology	UK	*421*	65-01900
Society of Automotive Engineers (SAE)	USA	*542*	65-20000
Society of British Aerospace Companies Ltd	UK	*421*	65-01950
Society of Flight Test Engineers (SFTE)	UK	*420*	60-02000
Society of Japanese Aerospace Companies	Japan	*249*	65-19000
Society of Legal Medicine and Criminology of France	France	*128*	60-06000
SOCOM = US Special Operations Command	USA	*528*	25-12500
Sodankyla Geophysical Observatory	Finland	*117*	60-03500
Soderberg Manufacturing Co. Inc	USA	*616*	75-82200
SODERN = Société Anonyme d'Études et Réalisations Nucléaires	France	*145*	75-37100
Södra Militärområdet (Milo S)	Sweden	*372*	25-32100
Soevaernets Kampinformationskursus	Denmark	*100*	25-43000
Sofinfra	France	*122*	20-60000
SOFRESA	France	*126*	40-01900
SOFRESA Riyadh	Saudi Arabia	*343*	40-01970

Name	Country	Page	Code
Software Center of Excellence (SCOE)	USA	*644*	80-33154
Sogerep Aviation Support	Saudi Arabia	*344*	90-01980
Sogerep Logistics Center	Saudi Arabia	*344*	90-01980
SOGERMA Maintenance Group Headquarters	France	*146*	75-42030
Sogitec Electronic Division	France	*146*	75-42050
Soil Machine Dynamics Ltd (SMD)	UK	*485*	75-54870
Sojuzkarta - Impex	Russia	*340*	80-02000
Solar Turbines	USA	*616*	75-82300
Solarex Pty Ltd	Australia	*21*	75-19565
Solid State Testing Inc.	USA	*643*	80-29650
Solomat	USA	*616*	75-82330
Solomat Ltd	UK	*485*	75-55000
Solvay Duphar BV	Netherlands	*295*	75-06850
Somchem	S. Africa	*359*	75-19200
Somchem	S. Africa	*361*	85-01900
Somer Electronics Ltd	UK	*525*	90-01963
Somers Forge Ltd	UK	*485*	75-55150
Sommer Allibert (UK) Ltd	UK	*429*	75-02745
SONACA SA	Belgium	*47*	75-06100
Sonak Systems & Software SA	Greece	*193*	80-01970
Sonar Research & Development Ltd	UK	*485*	75-55170
Sonda Aviation Enterprises Limited	UK	*512*	80-40000
Sonic Communications (Int.) Ltd	UK	*485*	75-55202
Sonnenschein Lithium GmbH Co-ordination Office	France	*154*	90-01983
Sonnenschein Lithium GmbH Head Office	Germany	*175*	75-28050
Sonomed Inc.	USA	*616*	75-82360
Sonovision ITEP Technologies	France	*151*	80-01850
Sonsub International Ltd.	UK	*512*	80-41000
Sonsub North Sea Ltd	UK	*512*	80-41000
Sony Corporation	Japan	*251*	75-00975
Sony Trading International Corporation	Japan	*256*	85-01800
Sooa Corporation	Korea (ROK)	*267*	90-01960
SOPEMEA	France	*151*	80-01997
Sorensen Company	USA	*573*	75-30720
Sorep	France	*146*	75-42400
Soucy International Inc.	Canada	*76*	75-12980
Sound Attenuators Ltd	UK	*485*	75-55208
Souter Shipyard Ltd	UK	*485*	75-55212
South African Air Force Headquarters (SAAF)	S. Africa	*355*	25-20000
South African Army	S. Africa	*356*	25-30000
South Central Ordnance Co.	USA	*616*	75-82400
South China Aero Technology Ltd (SCAT)	Hong Kong SAR	*198*	80-00850
South Midlands Communications Ltd (SMC)	UK	*485*	75-55213
South Military Region Command	Sweden	*372*	25-32100
Southcombe Brothers Ltd	UK	*485*	75-55214
Southern Ammunition Co. Inc.	USA	*616*	75-82440
Southern Joint Command	Sweden	*372*	25-32100
Southern Oceanics (Pty) Ltd.	S. Africa	*360*	80-41000
Southwest Aerospace	USA	*616*	75-82470
Southwest Research Institute (SRI)	USA	*541*	60-19200
Sovaernets Materielkommando	Denmark	*100*	25-42100
Sowerby Research Centre British Aerospace (Operations) Ltd	UK	*420*	60-02030
Soyuz Aviamotors Scientific Technical Complex	Russia	*339*	75-33050
SP aerospace and vehicle systems bv	**Netherlands**	***295***	**75-06860**
SP Radio A/S	Denmark	*103*	75-19600
SPA Ferrite Domain	Russia	*333*	55-07300
Space & Naval Warfare Systems Command Headquarters (SPAWAR)	USA	*535*	25-41300
Space & Naval Warfare Systems Washington Office	USA	*536*	25-41300
Space & Strategic Systems Technical Department	France	*122*	20-40000
Space Electronics Inc.	USA	*642*	80-29050
Space Hellas SA	Greece	*191*	75-02430
Space Machine & Engineering Corp.	USA	*616*	75-82700
Space Systems Finland Ltd (SSF)	Finland	*118*	75-02100
Spacerep Srl	Italy	*244*	90-01965
Spacetec A/S	Norway	*311*	75-02640
Spanish Association of Armament and Defence Material Manufacturers	Spain	*365*	65-00120
Spanish Fleet Commands	Spain	*362*	25-41000
Spanos	Greece	*191*	75-02420
SpanSet Inter AG	Switzerland	*387*	75-13550
SpanSet Ltd	UK	*485*	75-55230
Spantech Pty Ltd.	Australia	*21*	75-19573
Spar Aerospace Ltd Corporate Office	Canada	*76*	75-13000
Spar Aerospace (UK) Ltd	UK	*525*	90-01970
Spar Aerospace U.S. Limited	USA	*656*	90-19700
Spar Aviation Services Godfrey Aerospace	Canada	*76*	75-13020
Sparks & Ellis Uniforms	S. Africa	*359*	75-19390
Spartanics	USA	*616*	75-82850
Sparton Corp.	USA	*616*	75-82900
Sparton Electronics	USA	*616*	75-82900
Sparton Engineered Products Inc. Flora Group	USA	*616*	75-82930
Sparton of Canada Ltd (SOC)	Canada	*77*	75-13100
Sparton Southwest Inc.	USA	*616*	75-82950
Sparton Technology Inc.	USA	*616*	75-82950
SPAT = Syndicat des Personnels Administratifs et Techniques de la Police Nationale	France	*129*	65-02400
Spaulding Composites	USA	*616*	75-82970
SPAWAR = Space & Naval Warfare Systems Command Headquarters	USA	*535*	25-41300
SPBA = Silent Partner Body Armor Inc. Personal Armour Systems Division	USA	*615*	75-80930
SPC International Corp.	USA	*642*	80-29200
Spearhead Exhibitions Ltd.	UK	*512*	80-45100
Spec Tool Co.	USA	*616*	75-83050
SPECAC Ltd	UK	*455*	75-26640
Special Electronic Security Products Ltd	Israel	*222*	75-21050
Special Melted Products Ltd (SMP)	UK	*485*	75-55240
Special Operations Acquisition Center (SOAC)	USA	*528*	25-12550
Special Surface Warfare Vehicle Group	China PR	*89*	75-01450
Specialised Electro-Optical Services Ltd	UK	*482*	75-52830
Specialist Engines/Rolls-Royce Motors Cars	UK	*480*	75-50270
Specialist Training & Technical Services Ltd (STATS)	UK	*512*	80-45700
Specialty Bar Products	USA	*616*	75-83005
Specialty Plastic Products Inc.	USA	*616*	75-83035
Specialty Steel & Forge	USA	*616*	75-83040
Specmat Ltd	UK	*484*	75-54590
Specmat Technologies Inc.	USA	*616*	75-83055
Spectral Dynamics Inc.	USA	*616*	75-83075
Spectral Dynamics (UK) Ltd	UK	*525*	90-01973
Spectral Intelligence Corp. (SI)	USA	*642*	80-29250
SpecTran Corp.	USA	*616*	75-83090
Spectrex Inc.	USA	*617*	75-83095
Spectrol Reliance Ltd	UK	*485*	75-55555
Spectrolab Inc.	USA	*617*	75-83100
Spectron Optik GmbH	Germany	*166*	75-03420
Spectronic Denmark	Denmark	*103*	75-19650
Spectronix Ltd	Israel	*222*	75-21100
Spectrum Associates Inc.	USA	*642*	80-29350
Spectrum Sciences Institute	Canada	*66*	60-00050
Spectrum Technologies Ltd	UK	*486*	75-55620
Spekon = Sächsische Spezialkonfektion GmbH	Germany	*174*	75-25880
Spencer Clark Metal Industries plc	UK	*486*	75-55700
Spencer Moulton	France	*132*	75-04750
SPER	France	*128*	65-00800
Sperry Marine AS	Denmark	*106*	90-19500
Sperry Marine A/S	Norway	*312*	90-01950
Sperry Marine Asia	Singapore	*352*	90-01960
Sperry Marine Asia Inc.	Japan	*258*	90-01985
Sperry Marine BV	Netherlands	*299*	90-01960
Sperry Marine GmbH	Germany	*183*	90-01950
Sperry Marine Inc.	USA	*617*	75-83250
Sperry Marine Limited	UK	*525*	90-01975
Sperry Marine Sarl	France	*154*	90-01985
Sperry Marine SpA	Italy	*239*	75-24400
Spesaero Engineering	S. Africa	*360*	80-42000
SPI = Suspension & Parts Industries Ltd	Israel	*222*	75-21320
Spice Communications	UK	*512*	80-45800
Spider N. Petsios & Sons, Metal Ind.	Greece	*191*	75-02450
Spiral Flex	Greece	*192*	75-02465
Spiral System Instruments Inc.	USA	*617*	75-83330
Spire Corp.	USA	*643*	80-29400
Spitz Inc.	USA	*617*	75-83350
SPL Ordnance Ltd	Israel	*225*	80-01935
SPLAV = State Research and Production Association SPLAV	Russia	*334*	55-18325
SPLAV, State Research and Production Enterprise	Russia	*334*	55-18325
SPM Instrument UK Ltd.	UK	*486*	75-56050
Sportartikelfabrik Wolfgang Grabner	Austria	*33*	75-01100
Spot Image Corp.	USA	*648*	85-02050
Sprecher + Schuh, Industrial Control Group	Switzerland	*387*	75-13590
Sprecher + Schun Niederspannung GmbH	Austria	*33*	75-02128
Sprint	USA	*643*	80-29500
Sprint/Carolina Telephone	USA	*643*	80-29500
Sprung Instant Structures Ltd	Canada	*77*	75-13200
Squadron International Ltd	USA	*643*	80-29550

Quick Reference

Quick Reference

Quick Reference

U

Quick Reference

Quick Reference

X

Y

YDAB	Sweden	*379*	75-15000
Yewlands Engineering Co. Ltd	UK	*495*	75-67520
Young Co., R.M.	USA	*628*	75-98020
Yuasa Automotive Batteries (Europe) Limited	UK	*495*	75-67560
Yugoimport - SDPR	Yugoslavia	*661*	40-00600
Yuzhnoye SDO	Ukraine	*406*	55-02500

Z

Z. Mantos SA	Greece	*190*	75-02007
Z. Rabiner Ltd	Israel	*221*	75-17840
Zaklady Mechaniczne BUMAR-LABEDY SA	Poland	*325*	75-11000
Zaklady Starachowickie Star SA	Poland	*325*	75-11500
Zapp Werkstofftechnik GmbH & Co, Robert	Germany	*184*	90-02600
ZARC International Inc.	USA	*628*	75-98050
Zargal Sarl	France	*154*	90-25000
Závody Tazkého Strojárstva (ZTS)	Slovakia	*353*	75-01000
Zbrojovka Vsetin	Czech Republic	*99*	75-03100
Zeiss-Eltro Optronic GmbH	Germany	*177*	75-34700
Zellweger Analytics Ltd	UK	*495*	75-67600
Zen Technologies Ltd	India	*205*	75-03000
Zenair Ltd.	Canada	*77*	75-17200
Zenon SA	Greece	*192*	75-05100
Zentralstelle für Gesamtverteidigung	Switzerland	*382*	20-10100
Zentralverband Elektrotechnik-und Elektronikindustrie eV (ZVEI)	Germany	*162*	65-02600
Zephyr Flags & Banners	UK	*495*	75-67680
Zeppelin Luftschifftechnik GmbH	Germany	*177*	75-34780
Zeppelin Mobile Systeme GmbH Headquarters	Germany	*177*	75-34800
Zernow Technical Services Inc.	USA	*645*	80-36650
Zero Cases (Europe)	UK	*495*	75-67700
Zero Enclosures	USA	*629*	75-98150
Zeta	USA	*629*	75-98190
Zettler GmbH, Kommunikation und Sicherungftechnik	Germany	*177*	75-34930
Zettler GmbH Kommunikations und Sicherungstechnik	Germany	*177*	75-34930
Zeus Industrial Products Inc.	USA	*629*	75-98210
ZF Friedrichshafen AG	Germany	*177*	75-35000
ZF Gears (GB) Ltd	UK	*495*	75-67720
ZF Great Britain Ltd	UK	*495*	75-67720
ZF Luftfahrttechnik GmbH	Germany	*177*	75-36000
Zickermann SA, Sociedade	Portugal	*328*	90-03850
Zika Ltd, Electrode Works	Israel	*223*	75-23800
Zimbabwe High Commission	Australia	*9*	31-02050
Zinc Alloy Ltd	UK	*434*	75-07500
Zincolcelere Sarl	Italy	*240*	75-32000
Zinn Trading AB	Sweden	*381*	90-04000
ZIP ANPO = Arendnoe NPO Zavod Imeni Popova	Russia	*338*	75-16400
Zippertubing Co.	USA	*629*	75-98250
Zitel Corporation	USA	*629*	75-98260
Znamya Production Group	Ukraine	*407*	75-10000
Zodiac International	France	*149*	75-47310
Zodiac SA/Holding	France	*129*	70-30000
ZPRGB Ovetch	Bulgaria	*62*	55-02480
ZRRT Bozhuriste	Bulgaria	*61*	55-02110
ZRST Razgrad	Bulgaria	*62*	55-02310
ZTS = Závody Tazkého Strojárstva	Slovakia	*353*	75-01000
ZTS TEES Martin s.p.	Slovakia	*353*	75-01500
Zubair Enterprises LLC	Oman	*313*	80-06000
Zuken-Redac Ltd	UK	*495*	75-67800
ZVEI = Zentralverband Elektrotechnik-und Elektronikindustrie eV	Germany	*162*	65-02600
Zvezda JSC, Research, Development and Production Enterprise	Russia	*336*	60-08350
Zvezda-Strela State Research & Production Centre	Russia	*336*	60-08400
ZVS Dubnica Jas	Slovakia	*353*	75-02000
Zwick Energy Research Organization Inc.	USA	*629*	75-98400
Zwicky Engineering Ltd	UK	*495*	75-67900

NUMERIC

21st Century Airships Inc	Canada	*67*	75-00010
3 Sigma SA	Greece	*191*	75-02380
3-D Structures	USA	*613*	75-79450
3C Systems, Inc.	USA	*632*	80-05000

Quick Reference

Notes

Notes

Notes

Albania, Republic of

Head of State:
Rexhep Mejdani

10 Head of State

10-10000 Office of the President
Rr Abdyl Frashëri 1, Tirana
☎ (+355 42) 285 07; 284 20 ☒ 4298
⊖ (+355 42) 337 61; 235 48; 284 20
President of the Republic: Rexhep Mejdani 98

11 Prime Minister

11-10000 Office of the Chairman - Council of Ministers
Martyrs Bulvar, Tirana
☎ (+355 42) 279 47; 282 10 ☒ 4201
⊖ (+355 42) 278 88
Prime Minister, Council of Ministers: Fatos Nano
Deputy Prime Minister: Bashkin Fino 98

12 Ministries

12-10000 Ministry of People's Defense
Martyrs Bulvar, Tirana
Defense Minister: Sabit Brokaj 98

29 Customs & Excise

29-00500 Customs Administration
Rs. Sami Frasheri, Tirana 98

31 Embassies & Attachés

31-00690 Embassy of Greece
Rruga Frederick Shiroka 3, Tirana
☎ (+42) 342 90 ☒ 34443 98

31-01840 Embassy of the Republic of Turkey
Rrug e Kavajes, No 31, Tirana
☎ (+355 42) 333 99 ⊖ (+355 42) 327 19
Ambassador: H E Mr Ahmet K Okcun
Deputy Chief of Mission: Ms Reyhan Akant
Third Secretary: Kazim Batmazoglu 98

31-01890 Embassy of the United Kingdom
Rruga Skenderbeg 12, Tirana
☎ (+355 42) 349 73; 349 74; 349 75
⊖ (+355 42) 349 73; 349 74; 349 75
Ambassador: H E Mr H A P Tesoriere
Deputy Ambassador: G S Briggs
Defence and Military Attaché: Capt G Wiltshire
Management Officer: Dave Biker 98

31-01900 Embassy of the United States of America
Rruga E Elbansanit 103, Tirana
☎ (+355 42) 328 75; 335 20 ⊖ (+355 42) 322 22
Deputy Chief of Mission: Douglas Smith
Commercial Officer: Frank Yacenda
Defence Attaché: Cdr Stephen Fox
Public Affairs Officer: Charles Walsh
☎ (+355 42) 332 46; 297 54 97

Algeria

Head of State:
Gen. Liamine Zeroual

10 Head of State

10-10000 Office of the President
El Mouradia, Algiers
☎ (+213 2) 60 03 60; 60 03 66 ☒ 66044
President: Gen Liamine Zeroual
Secretary General to the Presidency: Said Bouchair 98

11 Prime Minister

11-10000 Office of the Prime Minister
32 Souidani Boudjemaa Avenue, Algiers
☎ (+213 2) 60 94 99; 60 98 07 ☒ 66217 to 66221
Prime Minister: Ahmed Ouyahia 98

12 Ministries

12-10000 Ministry of National Defence
Avenue Ali Khodja, Les Tagarins, PO Box 246, Algiers
☎ (+213 2) 61 15 15; 63 14 76; 63 17 65
☒ 67172; 66420
Minister of National Defence: Liamine Zeroual 98

12-20000 Ministry of Foreign Affairs
6 Rue Ibn Badran, El Mouradia, Algiers
☎ (+213 2) 60 47 44 ☒ 52794 ⊖ (+213 2) 60 11 64
Minister: Ahmed Attaf
Vice-Minister: Mohamed Aberkane 98

12-40000 Ministry of Interior
Government House, Algiers
☎ (+213 2) 63 23 40 ☒ 52073 ⊖ (+213 2) 73 43 68
Minister: Mostefa Benmansour 98

31 Embassies & Attachés

31-00250 Embassy of the Republic of Bulgaria
13 Boulevard Colonel M'hamed Bougara, El Biar, Alger
☎ (+213 2) 23 00 14; 23 01 21 ☒ 66087 bgm dz
⊖ (+213 2) 23 05 33
Deputy Chief of Mission: Nikolov Stanislav 98

31-00660 Embassy of the Federal Republic of Germany
165 Chemin Sfindja, Alger-Gare, PO Box 664, Algiers
☎ (+213 2) 74 19 56; 74 20 47 ☒ 67343; 0408/56/043
⊖ (+213 2) 74 05 21
Ambassador: H E Dr Joachim Broudé-Gröger 98

31-00690 Embassy of Greece
31 rue A les Crétes-Paradou, 16035 Hydra, Alger
☎ (+213 2) 60 08 55; 600855
✉ AMBASSADE GRECE Alger ☒ 66071 98

31-00890 Embassy of Italy
18 rue Mohamed Ouidir Amellal, El Biar, Alger
☎ (+213 2) 92 23 30; 92 25 50 ☒ 61357 itald dz
⊖ (+213 2) 92 59 86
Ambassador: H.E. Dr Antonio Badini
Military and Air Attaché: Col. F. Merlino 98

31-01130 Embassy of the Republic of Mali
Cité DNC, Villa No.15, Chemin du Kaddous, Hydra, Alger
☎ (+213 2) 69 13 51 ☒ 66109 ⊖ (+213 2) 69 20 82
Ambassador: H.E. Mr Cheick S. Diarra
Deputy Chief of Mission: Sekou Kasse
Public Affairs Officer: M. Toure 97

31-01370 Embassy of Pakistan
14 Avenue Souidani Boudjemaa, PO Box 404, Algiers
☎ (+213 2) 69 22 12; 69 37 81; 69 31 94
✉ PAREP Alger ☒ 66277 ⊖ (+213 2) 69 22 12
Ambassador: H E Mr Muhammad Aslam Rizvi
Head of Chancery: Malik Muhammad Nawaz
☎ (+213 2) 69 34 93 98

31-01890 Embassy of the United Kingdom
PO Box 43, Alger-Gare, 16000 Algiers
Residence Cassiopée, Batiment B, 7 Chemin des Glycines, Algiers
☎ (+213 2) 23 00 40; 23 00 68 ✉ PRODROME ALGIERS
☒ 66151 prdrm DZ ⊖ (+213 2) 23 00 67
Ambassador: H E Mr Francois Gordon
Deputy Head of Mission: Jeremy MacAdie
Administrative Attaché: Mrs Laura Page
Pro-Consul: Ms N Abid 98

31-01950 Embassy of Venezuela
3 rue Impasse Ahmed Kara, PO Box 297, 16000 Algiers Gare
☎ (+213 2) 69 38 46; 69 23 84 ✉ EMBAVENEZ Alger
☒ 66642 ⊖ (+213 2) 69 35 55
Ambassador: H E Mr Heliodoro L Claverie R
First Secretary: Simon Becerra 98

31-02010 Embassy of the Federal Republic of Yugoslavia
PO Box 366, Algiers
7 Rue des Frères Benhafid, Hydra
☎ (+213 2) 69 12 18; 69 23 26; 69 27 04 ☒ 66076
⊖ (+213 2) 69 34 72
Ambassador, Chargé d'Affairs: H E Mr Babic Dimitrije 98

Contact Reference

Angola, Republic of

Head of State:
Gen. José Eduardo dos Santos

10 Head of State

10-10000 Office of the President
Palacio do Povo, Luanda
⊠ 3345 gabpres an, 3072 lugol an
President of the People's Republic:
Gen. José Eduardo dos Santos [98]

11 Prime Minister

11-10000 Office of the Prime Minister
Luanda
Prime Minister: Fernando Franca van Dunem [98]

12 Ministries

12-10000 Ministry of Defense
Ministerio Da Defesa
Rua 17 De Setembro, Luanda
☎(+244 2) 33 49 12 ⊠ 3138 def an, 3259 mindef an
⊖ (+244 2) 33 42 76
Minister: Gen Pedro Sebastio
Vice Minister and Commander FAPLA-DA:
Gen Antonio dos Santos França [98]

12-20000 Ministry of Foreign Affairs
Avenida Comandante Jika, Luanda
☎(+244 2) 32 06 70 ⊠ 3127; 3128 mirex an
⊖ (+244 2) 39 51 26
Minister: Venacio Da Moura [98]

31 Embassies & Attachés

31-00660 Embassy of the Federal Republic of Germany
Avenida 4 de Fevreiro 120, PO Box 1295, Luanda
☎(+244 2) 33 47 73; 45 16 ⊠ 2096 aaluan an
⊖ (+244 2) 33 45 16
Ambassador: H.E. Helmut Van Edig [97]

31-00890 Embassy of Italy
Rua Dos Enganos 1, PO Box 6220, Luanda
☎(+244 2) 33 12 45/6 ⊠ 3265 italdip an
⊖ (+244 2) 33 37 43
Ambassador: H.E. Mr Franco Micieli De Biase
First Secretary (Commercial): Dr Stefano Queirolo [98]

31-01270 Royal Netherlands Embassy
Avda 4 de Fevreiro 42-6, Edif Secil, PO Box 3624, Luanda
☎(+244 2) 33 35 40; 33 35 44 ⊖ (+244 2) 33 36 99
Ambassador: H E Drs H Kroon
Second Secretary, Commercial and Public Information:
Drs R F N W Wols [98]

31-01680 Embassy of Spain
PO Box 3061, Luanda
Avda 4 de Fevreiro 95, P1, Salas 11/12, Luanda
☎(+244 2) 39 11 66; 39 11 87 ⊠ 2621 espanha an
⊖ (+244 2) 39 11 88
Ambassador: H E Mr Rafael Fernandez-Pita
Deputy Chief of Mission: Luis Zaballa
Commercial Counsellor: Francisco Balaquer Lara [98]

31-01890 Embassy of the United Kingdom
Rua Diogo Cão, Caixa Postal 1244, Luanda
☎(+244 2) 33 45 82; 39 29 91; 33 45 83; 39 76 81
⊖ (+244 2) 33 33 31
Ambassador: H E Mr Roger Hart
Consul and Deputy Head of Mission: F J Martin
First Secretary Chancery: D G Cox
Second Secretary, Commercial: C Gedge
Second Secretary, Aid: E Rich
Third Secretary: S Digsby
British Liaison Officer: Lieut Col J D A Galt [98]

31-01900 Embassy of the United States of America
Rua Major Kanhangulo, 134, Luanda
☎(+244 2) 39 24 98 ⊖ (+244 2) 39 05 15
Ambassador: Donald Steinberg
Deputy Chief of Mission: John Hartley
Public Affairs Officer: Phillip Ives [98]

Antigua & Barbuda

Head of State:
H.M. Queen Elizabeth II

Governor-General:
Sir James Beethoven Carlisle

10 Head of State

10-10000 Office of the Governor General
Government House, St Johns
☎(+1 268) 462 03 04 ⊖ (+1 268) 462 25 66
Governor General: H E Sir James Beethoven Carlisle [98]

11 Prime Minister

11-10000 Office of the Prime Minister
Factory Road, St Johns
☎(+1 268) 462 49 56 ⊠ 2127 ofprem ak
⊖ (+1 268) 462 32 25
Prime Minister: Hon Lester B Bird [98]

12 Ministries

12-10000 Ministry of Defence
Factory Road, St Johns
☎(+1 268) 462 07 73 ⊠ 2127 ofprem
⊖ (+1 268) 462 32 25
Minister of Defence: Hon Lester B Bird [98]

12-20000 Ministry of External Affairs
New Administration Building, St Johns
☎(+1 268) 462 07 87 ⊠ 2122 extrnal ak
Minister of External Affairs: Hon Lester Bird [98]

12-30000 Ministry of Health and Home Affairs
St John's Street, St Johns
☎(+1 268) 462 08 60 ⊖ (+1 268) 462 50 03
Minister of Health and Home Affairs: Sam Aymer
Minister of Labour, Home Affairs and Citizen Services:
Adolphus Freeland [98]

12-35000 Ministry of Agriculture, Trade, Commerce, Industry and Consular Affairs
Temple and Nevis Streets, St Johns
☎(+1 268) 462 10 07; 462 10 08; 462 10 09
⊖ (+1 809) 462 61 04
Minister of Trade, Industry and Consumer Affairs:
Hilroy Humphreys
Minister of Telecommunications, Civil Aviation, International Transportation and Gaming:
Hon Lester Bird [98]

25 Defense Forces

25-10000 Antigua Barbuda Defence Force
Camp Blizard, Coolidge, St Georges
☎(+1 268) 462 33 63; 426 59 86
⊖ (+1 268) 462 27 16 (force headquarters)
Commander: Lieut Col Trevor A Thomas
Deputy Commander, Operations Director:
Maj Edward H Croft
Procurement and Logistics Director: Lieut A Nicholas [98]

31 Embassies & Attachés

31-01890 United Kingdom High Commission Antigua and Barbuda
11 Old Parham Road,
PO Box 483, Price Waterhouse Centre, St Johns
☎(+1 268) 462 00 08; 462 00 09; 463 00 10
⊖ (+1 268) 462 28 06
High Commissioner: H.E. Mr R. Thomas
(CMG, Resident in Bridgetown)
Deputy High Commissioner: P J Mathers
(Resident in Bridgetown)
Defence Adviser: Capt. I. Hime
(RN, Resident in Bridgetown)
Counsellor: S.M. Boldt [97]

Argentine Republic

Head of State:
Carlos Saúl Menem

10 Head of State

10-10000 Office of the President
Casa de Gobierno, Balcarce 50, 1064 Buenos Aires
☎(+54 1) 374 98 41 ⊠ 21464 copre ar
⊖ (+54 1) 331 13 98
President: H E Carlos Saúl Menem
☎(+54 1) 343 30 51; 343 30 59
Vice President: Carlos Ruckauf
☎(+54 1) 331 03 65; 331 63 04
Cabinet Chief: Jorge Rodríguez
☎(+54 1) 331 51 23; 342 62 86 [98]

10-10500 Office of the Vice President
Hipolito Yrigoyen 1835-49, 1089 Buenos Aires
Vice President: Carlos Ruckauf [98]

12 Ministries

12-10000 Ministry of Defence
Avenida Paseo Colón 255, 1063 Buenos Aires
☎(+54 1) 331 99 11; 343 15 61; 331 15 60; 331 15 69; 342 21 41 ⊠ 222000 ⊖ (+54 1) 343 42 38
Minister of Defence: Jorge Dominguez
Undersecretary Military Affairs: Juan Pihno 98

12-20000 Ministry of Foreign Relations and Worship
Arenales 761, Palacio San Martin, 1061 Buenos Aires
☎(+54 1) 311 00 71
⊠ 21194; 21330; 21339; 22214 relba ar
⊖ (+54 1) 311 57 08
Minister: Guido Di Tella
Director: Enrique de la Torre 98

12-40000 Ministry of Interior
Balcarce 524, 1064 Buenos Aires
Minister: Dr Carlos Corach 98

15 Parliament

15-10000 Senate
Camara de Senadores
Hipólito Yrigoyen 1835, 1089 Buenos Aires
☎(+54 1) 953 30 81
President: Dr Carlos Ruckauf 98

15-20000 Chamber of Deputies
Camara de Diputados
Rivadavia 1864, 1033 Buenos Aires 98

25 Defense Forces

25-19100 National Defence School
Escuela de Defensa Nacional
Maipu 262, 1084 Buenos Aires
E escdef@datamarkets.com.ar
Vice Director: Col Eugenio R Diaz Jausoro (Army)
☎(+54 1) 326 12 73
Research and Publications: Col Hugo G Sarno (Army Rtd)
☎(+54 1) 326 13 18
Inst. Relations and Special Courses Department: Col Carlos Deltorchio (Army)
Senior Course Director: Capt Carlos V Bahnson (Navy, retd) ☎(+54 1) 326 82 67
Formation course: Cdr Oscar V Bulfon (Navy)
Applicative Course: Cdre Homero Vartorelli (Air Force) 98

25-20000 Air Force Headquarters
Fuerza Aérea Argentina
Edificio "Condor", Avenida Comodoro Pedro Zanni 250, 1104 Buenos Aires
☎(+54 1) 312 91 70; 312 34 14 ⊖ (+54 1) 312 82 40
Chief of Staff Air Force: Brig Gen Juan Daniel Paulik
Deputy Chief of Staff: Brig Tomas A Rodríguez
Chief Air Force Operations Command: Brig Hector Ricardo Gilobert
Chief Air Regions Command: Brig Alberto Alegria
Chief Materièl Command: Brig Hector Maria Sambrizzi
Chief of Materièl Area Cordoba: Brig Roberto José Engroba ☎(+54 1) 312 91 70 ext. 2491
Chief of Staff Training: Brig Julio Horacio Caviglia
Chief of Staff Personnel: Brig Roberto José Noe ☎(+54 1) 312 91 70 ext. 2156; 2991
Chief of Staff Materièl: Brig Arturo Herberto Ay
Chief Staff Operations: Cdr Raul Quiroga
Chief Protocol Department: Vice Cmdr Raul E Gonzalez
Chief Air Force Intelligence: Cdr Mario Alfredo Laporta
Chief Air Force Planning Division: Brig Manuel Rafael Villagran
Chief Air Force Logistics Division: Cdr Enrique Josi Pessana
Chief Press Department: Cdr Samuel Efrain Echenique
Chief of Public Relations Department: Cdr Augusto Raul Fieroli
Military Chief of Air Force General Staff: Cdr Jorge R Arguello ☎(+54 1) 312 91 70 ext. 2208
Deputy Chief Press Department: Enrique Rearte
Director General Personnel: Brig Romero Gallo
Director General Accounting and Finance: Cdr Eraldo Avila
Director General Aeronautics Materièl: Cdr Eduardo Josi Giannastasio
Director General of Supply: Cdr Benito Gerardo Wierna
Director General of Infrastructure: Cdr Ernesto Victor Michelli ☎(+54 1) 312 91 70 ext. 2991
Director General Health Services: Cdr Oscar J Trama
Director General Maintenance: Brig Eduardo Martinez
Director Air Force Commissariat: Cdr Juan Josi A Garcia de Diego
Director Air Force Hospital: Cdr Nestor Gerardo Rodriguez Correa
Director Air Transit: Cdr Dante Rene Vesaccia
Economics Director: Cdr Tito German Oviedo
Director of Publications: Cdr José Humberto Aguirre
Communications Director: Cdr Jacobo Edmundo Peil
Deputy Secretary General of the Air Force: Walter Felix Videla ☎(+54 1) 312 91 79 ext. 2509
Commander of Training: Brig Henry R Graton
Information Contact: Cdr Eugenio Miari 98

25-30000 Army Headquarters
Comando General del Ejercito
Azopardo 250, Edificio Libertador, 1328 Buenos Aires
☎(+54 1) 331 94 46; 331 46 01 ⊠ 22200
⊖ (+54 1) 331 94 46
Commander
Deputy Commander, Chief of Staff: Lieut Gen Martín Antonio Balza Gen Anibal U Laino ☎(+54 1) 331 68 27
Operations Director: Gen Ricardo Brinzoni
Procurement and Logistics Director: Gen Pacifico Britos ☎(+54 1) 331 71 06
Army Secretary General: Ernesto Juan Bossi 98

25-32100 Army Aviation Headquarters
Azopardo 250 Piso 13, 1328 Buenos Aires
☎(+54 1) 343 8885; 334 3173; 343 3011/19
⊖ (+54 1) 343 8885; 334 3173
Commander: Brig.Gen. Arturo Emilio Grandinetti
Chief of Staff: Lt Col Hugo Argentino Magnaghi
Logistics Dir: Major Horacio Arturo Sabin Paz 97

25-40000 Navy Headquarters
Comando General de la Armada
Estado Mayor General de la Armada
Comodoro Py y Corbeta Uruguay 2055, Edificio Libertad, 1104 Buenos Aires
☎(+54 1) 312 50 01; 313 09 33; 313 10 45
⊠ 21338 racel ar ⊖ (+54 1) 313 89 39
Chief of Staff Navy: Vice Adm Enrique Emilio Molina Pica
Deputy Commander in Chief: Vice Adm Maximo Rivero Kelly
Chief Operations Department: Rear Adm Adolfo Ardouino
Commander Naval Aviation: Rear Adm Hector Martini 98

31 Embassies & Attachés

31-00070 Armenian Diplomatic Mission
Avenida Pte Rouge Saenz Pena 570, Piso-3, 1035 Buenos Aires
☎(+541) 31 45 18 82 ⊖ (+541) 343 24 67
Chargé d'Affaires: Ara Akopian 98

31-00090 Australian Embassy
Villaneva 14000, 1426 Buenos Aires
☎(+54 1) 777 65 80/85 ⊖ (+54 1) 772 33 49
Ambassador: H.E. Mr Warwick Weemaes
First Secretary: Peter M. Heyward
Trade Commissioner: Cesar Fernández 97

31-00100 Embassy of Austria
Calle French 3671, PO Box 4889, 1000 Buenos Aires
☎(+54 1) 802 71 95; 802 70 96; 802 14 00
⊖ (+54 1) 805 40 16
Ambassador: H E Dr Wolfgang Kriechbaum
Deputy Chief of Mission: Mag Dorothea Auer
Commercial Office
Cerrito 1294, 15th Floor, Buenos Aires
☎(+54 1) 816 15 46; 816 04 79; 816 03 98; 814 36 69
⊖ (+54 1) 814 36 70
Commercial Counsellor: Josef Schwald 98

31-00380 Embassy of Colombia
Carlos Pellegrini 1363, 3er. piso, 1011 Buenos Aires
☎(+54 1) 325 11 06; 325 04 94; 325 02 58
E embacolbuca@datamarkets.com ⊖ (+54 1) 322 93 70
Ambassador: H E Mr Víctor G Ricardo
Deputy Chief of Mission: Jario Montes Moreno
Defense Attaché: Jorge Pineda Carvajal
Secretary to the Ambassador: Ximena Roldàn
Commercial Counsellor's Office
Carlos Pellegrini 1363-3er.piso, Buenos Aires
☎(+54 1) 312 55 38; 312 54 46 ⊖ (+54 1) 313 85 63
Counsellor: Miguel Muñoz Paladines 98

31-00535 Embassy of Estonia
Malaver 1180, 1638 Vincente Lopez
☎(+54 1) 794 53 53 ⊖ (+54 1) 794 53 55
Honorary Consul: Peet Pulisaar 97

31-00660 Embassy of the Federal Republic of Germany
PO Box 2979, 1000 Buenos Aires
Calle Villanueva 1055, 1426 Buenos Aires
☎(+54 1) 778 25 00 ⊠ 21668 aabai ar
⊖ (+54 1) 775 25 50
Ambassador: H E Dr Wiegand Pabsch
Deputy Chief of Mission: Dr Horst Heubaum
Defense Attaché: Franz Berger (Air Force)
Commercial Counsellor: Dr Woerner 98

31-00810 Embassy of the Republic of Hungary
Coronel Diaz 1874, 1425 Buenos Aires
☎(+54 1) 822 07 67; 824 58 45; 826 41 32
⊠ 22843 ungdi ar ⊖ (+54 1) 805 39 18
Ambassador: H E Dr Károly Misley
Commercial Officer: Ferenc Herbert
Public Information Officer: Ferenc Balogh
Commercial Office
Paraguay 577, Buenos Aires
☎(+54 1) 311 25 18; 313 02 10 ⊠ 21 439 ungex ar
⊖ (+54 1) 311 06 39; 35 67 98

31-00870 Embassy of Ireland
Suipicha 1380 Piso 2, 1011 Buenos Aires
☎(+54 1) 325 85 88; 325 08 49 ⊖ (+54 1) 325 75 72
Secretary of Embassy: David Noonan 98

31-00890 Embassy of Italy
Calle Billinghurst 2577, 1425 Buenos Aires
☎(+54 1) 802 00 71/74 ⊠ 21961; 28527 embit ar
⊖ (+54 1) 804 49 14
Ambassador: H.E. Giuseppe Maria Borga ☎(+54 1) 802 99 82
First Secretary: Giuseppe Scognamiglio ☎(+54 1) 802 63 77
Deputy Chief of Mission: Enrico Guicciardi ☎(+54 1) 802 63 77
Commercial Counsellor: Franco Giordano ☎(+54 1) 802 96 96
First Counsellor: Ms Nicola Manduzio ☎(+54 1) 802 84 53 ☎(+54 1) 805 32 54 Fax
Armed Forces Office
3 de Febrero 1365, 1426 Buenos Aires
☎(+54 1) 787 21 45; 787 22 35 ⊖ (+54 1) 787 21 45
Deputy Naval Attaché: Capt. Nicola De Natale ☎(+54 1) 787 21 35; 20 35 ☎(+54 1) 787 21 35 Fax 97

31-01020 Embassy of Lebanon
Av.Libertador Gen.San Martin 2354, 1425 Buenos Aires
☎(+54 1) 802 04 66; 44 92 ⊠ 22 866 liban ar
⊖ (+54 1) 802 29 09
Ambassador: H.E. Mr Riad Kantar 97

31-01270 Royal Netherlands Embassy
Avenida de Mayo 701, piso 19, 1084 Buenos Aires
☎(+54 1) 334 34 74; 334 40 00; 334 33 60
✉ HOLLANDIA BAIRES ⊠ 21824 eholn ar
⊖ (+54 1) 334 27 17
Ambassador: H E Mr Albert J A M Nooij
Deputy Chief of Mission: Dr Johannes W G Jansing

Argentine Republic

First Secretary: Dr Kees T Smit Sibinga
Agricultural Attaché: Ing Philippus J M de Jong
Second Secretary, agricultural affairs: Ms Anne Rauwerda [98]

31-01430 Embassy of the Republic of Poland
Avda Alejandro Maria de Aguado 2870,
1425 Buenos Aires
☎(+54 1) 802 96 81; 802 96 82
[E] ambbuenos@datamarkets.com.ar
⊖ (+54 1) 802 96 83
Ambassador: H E Dr Eugeniusz Noworyta
Councellor (Deputy Chief of Mission) and Public Information Officer: Andrzej Kulesza
First Secretary, Chief of Consular Department: Dr Andrzej W Czub
Second Secretary: Zbigniew Salyga
Second Secretary: Zielinski Zbigniew

Commercial Department
3147, Virrey del Pino Str, Buenos Aires
☎(+54 1) 551 54 97; 551 50 97; 551 53 97
Telex 18255 brh ar ⊖ (+54 1) 551 50 97
Commercial Counsellor: Wlodzimierz Gurnik
Commercial Attaché: Ferdynand Szwed [98]

31-01480 Embassy of Romania
Arroyo 962, 1007 Buenos Aires
☎(+54) 322 26 30; 326 58 88 Telex 24 227; 24301
⊖ (+54) 322 26 30
Ambassador: H E Mr Stelian Oancea
Political Counsellor: Gheorghe Dima
Second Secretary, Consular Section: Viorel Ipate
Economic Counsellor: Ion Plopeanu
First Secretary, Economic Section: Ovidiu Curea
Third Secretary, Culture, Press, Protocol: Ms Cornelia Avram [98]

31-01600 Embassy of Saudi Arabia
Avda Alejandro Maria de Aguado 2881,
1425 Buenos Aires
☎(+54 1) 802 07 60; 802 33 75
Telex 23291 njdia ar, 18933 njdia ar ⊖ (+54 1) 802 75 73
Ambassador: H E Mr F A Nazir [98]

31-01650 Embassy of the Slovak Republic
Embajada de la Republica Eslovaca
Avenida Figueroa Alcorta 3240, 1425 Buenos Aires
☎(+54 1) 801 39 17; 801 38 04
[E] embresvl@unterprov.interprov.com
⊖ (+54 1) 801 46 54
Ambassador: H E Dr Marian Masarik
First Secretary: Dr Michal Biganic
Commercial Counsellor: Ing Jozef Sauer
Administrative Attaché: Vladimir Latta [98]

31-01655 Consulate of the Republic of Slovenia (Honorary)
Camara de Com Esloveno Latinoamer
Moreno 2763, 1752 Lom del Mirador, Buenos Aires
☎(+54 1) 654 64 38 ⊖ (+54 1) 441 17 19
Honorary Consul: Herman Zupan [98]

31-01890 Embassy of the United Kingdom
Dr Luis Agote 2412/52, Capital Federal,
1425 Buenos Aires
☎(+54 1) 803 70 70; 803 70 71
⊖ (+54 1) 807 31 57; 803 17 31
Ambassador: H E Mr William Marsden
Deputy Head of Mission: D N Reddaway
Defence and Military Attaché: Col R J Lawson
Naval and Air Attaché: Gp Capt D K L McDonnel
First Secretary Head of Political Section: D D W Martin
First Secretary, Commercial: H Wiles
Second Secretary, Information Officer: T Kennedy [98]

55 Government-owned Industry & Production Facilities

55-06160 Fabrica Militar "Frey Luis Beltran"
Ruta 11-km 323, 2156 Frey Luis Beltran, Santa Fe
☎(+54 41) 91 26 38 [E] myaflb@redigital.com.ar
⊖ (+54 41) 91 66 08
♣ US$ 5,000,000 (1996) 👥 268 (1997) 🏭 43,000 m2
Managing Director: Lieut Col Luis A Rechiman
Operations Director: Mario A Macagno
Sales and Marketing Director: Ing Oscar Gorosito
Purchasing Manager: Isidoro Faletto
■ Cartridges for military and civil use. Defence ammunition and weapon systems. [98]

75 Manufacturing Companies

Lockheed Martin Aircraft Argentina . . 80-08000

75-00200 Aero Boero SA
Brasil y Alem, 2421 Morteros, Cordoba
☎(+54 562) 226 90; 221 21
Socio Gerente: Hector A Boero
Managing Director: Hector A Zinoni
Technical Director: Ing.Aer. Juan Ferreyra Tarrags
■ Manufacture light aircraft for training, ambulance, agriculture and liaison roles. [98]

75-03451 Chincul SACAIFI
25 de Mayo 489, Piso 6°, 1339 Buenos Aires
☎(+54 1) 312 56 71; 312 56 75; 312 54 58; 311 00 24
[E] macarena@century.com.ar ⊖ (+54 1) 311 57 42
♧ US$ 20,000,000 (1995) ♣ US$ 35,000,000 (1995)
👥 500 (1995) 🏭 20,000 m2
President: José Maria Beraza
Managing Director: Oscar Prieto
Marketing Director: Fernando Barberio
Operations and Technical Director: Ing. Jaime Ferreyra
Purchasing Manager: Enrique Vazquez
■ Produces the following aircraft in military and civil versions: PA 18-150 Super Cub, PA 28-161 Warrior II, PA 28-181 Archer II, PA 28R-201 Arrow, PA 28R-201T Turbo Arrow, PA 28-236 Dakota, PA 31-350 Navajo Chieftain, PA 31T Cheyenne I, PA 31T Cheyenne II-XL, PA 42 Cheyenne IIIA, PA 42-1000 Cheyenne IV.

Plant 1
Calle Mendoza entre 6 y 7, PO Box 80, 5400 San Juan
☎(+54 1) 312 56 71; 312 56 75; 312 54 58
[E] macarena@century.com.ar Telex 59143
⊖ (+54 1) 311 57 42
👥 330 🏭 14,000 m2
Works Director: Ing. Jaime Ferreyra

Plant 2
Diaz Velez 1034, 1702 Ciudadela (Bs As)
☎(+54 1) 312 56 71/5 ⊖ (+54 1) 311 57 42
Works Dir: Eng. Juan C. Aguero
■ Vertically integrated to produce parts, assemblies and subassemblies for aircraft and helicopters, toolroom machines, fixtures, tool holders, upgrade and major modifications for aircraft and helicopters. Manufacture under licence of Piper aircraft and own design 2-seat Chincul trainer.

Timen SA
Maintenance and Repair
Aerodromo Internacional Don Torcuato, CC N° 7,
Don Torcuato Pcia
☎(+54 1) 748 32 60 ⊖ (+54 1) 741 38 00
Works Director: Ing. Pedro Beraza
Maintenance Chief: Ing. Leonel Santorsola
Sales Manager: Ing. Cesar Elizalde
Technical Manager: Ing. Augusta Roberts
■ Maintenance facility for Bell helicopters, Piper, Learjet, Pratt & Whitney, Bendix/King, Sperry, Mooney. [97]

75-09000 Saft Nife Argentina SA
(a subsidiary of Saft SA, France)
Jose Leon Suarez 2244, PO Box 1440, Buenos Aires
☎(+54 1) 684 19 95; 684 19 98 ⊖ (+54 1) 684 19 24
Managing Director: Norberto Traverso [98]

75-19500 Scania Argentina SA
Casilla de Correos Nro.3, Correo Central,
4000 San Miguel de Tucumán
☎(+54 81) 50 90 00 ⊖ (+54 81) 50 90 01 [98]

75-30000 VEC Srl
Special Battery Div.
Cazadores de Coquimbo 3122, 1605 Munro (Bs As)
☎(+54 1) 762 39 99; 756 04 68 Telex 26211 iatear
⊖ (+54 1) 756 04 61
♧ $ 300 000 (1995) ♣ $ 450 000 (1995) 👥 20
🏭 1 000 sq m
Managing Dir: Eric E. Heuser
Sales & Marketing Dir: Mariano M. Heuser
Operations/Technical Dir: Elbio E. Ojoda
Purchasing Mgr: Javier S. Heuser
■ Manufacture, reconditioning, re-build & distribution of Ni/Cd battery packs for military & industrial use. [97]

80 Service Companies & Organizations

80-01000 AlliedSignal Aerospace Marketing Sales and Service
(a subsidiary of AlliedSignal Aerospace, USA)
Buenos Aires
Gerenacia de Maintenimiento, Aeropuerto
Ministro Pistarini, 1802 Suc. Aeropuerto Ezeiza
☎(+54 1) 480 09 15 (recording) ⊖ (+54 1) 480 09 15
Field Service Engineer: Ricardo Amoretti
☎(+54 1) 440 57 43
■ On-site customer service for AlliedSignal Aerospace [98]

80-08000 Lockheed Martin Aircraft Argentina (LMAASA)
(† Fabrica Militar de Aviones (FMA))
Avenida Fuerza Aerea Argentina, Km 5500,
5010 Cordoba (Cba)
☎(+54 51) 65 17 06 ⊖ (+54 51) 65 06 98; 68 17 06
👥 1,100 (1997) 🏭 259,000 sq m
President: Harry Q Radcliffe
☎(+54 51) 65 44 48 (also fax)
Technical Director: Juan Lucini
☎(+54 51) 65 06 20 (also fax)
Operations Director: James Barnes
Sales and Marketing Director: Oscar Bendetti
General Manager of Operations: Alberto Osvaldo Buthet
☎(+54 51) 65 06 20
Marketing Manager: Victor Hugo Almiron
Purchasing Manager: Horatio Gallardo
■ Aircraft design, manufacture, maintenance and repair. IA 63 Pampa basic/advanced jet trainer. Upgrading of IA 58 Pucara close support aircraft. → 75

Ministry of Defence
Paseo Colón 255, 1063 Buenos Aires
☎(+54 1) 342 37 93 ⊖ (+54 1) 334 59 71 [98]

80-20000 Schottel Sudamericana I.C.S.A.
Ave Montes de Oca 2044, 1271 Buenos Aires
☎(+54 1) 303 00 12 15 Telex 22847 schotel ar
⊖ (+54 1) 303 00 10
President: Ricardo F. Place
■ Shipbuilding, installation, maintenance repair & overhaul services. Marine diesel maintenance, overhaul & repair. Ship reconditioning, refit & conversion.
® Germany: Schottel-Werft [97]

90 Sales Agents, Sales & Liaison Offices

Companies represented:

ADB Alnaco, USA	90-05000
Aeroquality Sales, USA	90-05000
AlliedSignal, USA	90-12000
Andrex Radiation Products, Denmark	90-00750
Ateliers de la Haute Garonne, France	90-05000
Balzers AG, Liechtenstein	90-00750
Bell Helicopter Textron Inc, USA	90-12000
Bendix/King, USA	90-12000
Bet-Air, USA	90-20100
Bihler Ltd, USA	90-00750
Comelec SA, Belgium	90-05000
Endevco, USA	90-00750
Endress & Hauser, Germany	90-19300
Exide, USA	90-19300
FMC Jetway Systems, USA	90-05000
Institut Dr Foerster, Germany	90-00750
Krautkraemer GmbH, Germany	90-00750
Lamaco, USA	90-05000
Lockheed Aircraft, USA	90-20100
Measurements Group Inc, USA	90-00750
MTS Systems Corp, USA	90-00750
MTU Friedrichschafen GmbH, Germany	90-13000
N.V. ADB SA, Belgium	90-05000
Oshkosh Truck Corp., USA	90-05000
Schottel-Werft, Germany	80-20000

Smiths Industries, USA 90-05000
Textron Lycoming, Stratford Division, USA 90-12000
Textron Lycoming, Williamsport Division, USA 90-12000
Trace Worldwide Corp., USA 90-05000
Trimos-Sylvac Task Force, Switzerland . 90-00750
Trimos-Sylvac Task Force, Switzerland . 90-20300
Van Hool, Belgium 90-05000

90-00750 ARO SA
Avenida Belgrano 369, 1092 Buenos Aires
☎ (+54 1) 331 39 18; 331 45 03; 331 45 04
⊖ (+54 1) 331 35 72
♧ peso 300,000 ♣ peso 5,000,000 ⅲ 45 ⌂ 280 m2
President: J Antonio Gómez
Vice President: Ing. Ricardo N Brugger
Operations and Technical Director: O Boniti
Purchasing Manager: A Ripari
® Denmark: Andrex Radiation Products
Germany: Institut Dr Foerster, Krautkraemer GmbH
Liechtenstein: Balzers AG
Switzerland: Trimos-Sylvac Task Force
USA: Bihler Ltd, Endevco, MTS Systems Corp, Measurements Group Inc 98

90-01700 Daewoo Corp Sucursal Buenos Aires
Cerrito 1294, Piso 17 - Cap Fed, Buenos Aires
☎ (+54 1) 816 32 01 ▣ 17152 dwba ar
⊖ (+54 1) 815 11 58 98

90-05000 Equipos de Abordo SA
Avenida Paraguay 610, piso 11, 1057 Buenos Aires
☎ (+54 1) 311 39 41 ✉ EQUIBORDO ▣ 24472
⊖ (+54 1) 311 90 14
♧ US$ 100,000 ♣ US$ 3,000,000 ⌂ 120 m2
President and General Manager: Ing Ricardo E Pérez
Operations and Technical Director: Ricardo E Pérez (Jr)
® Belgium: Comelec SA, N.V. ADB SA, Van Hool
France: Ateliers de la Haute Garonne
USA: ADB Alnaco, Aeroquality Sales, FMC Jetway Systems, Lamaco, Oshkosh Truck Corp., Smiths Industries, Trace Worldwide Corp. 98

90-12000 La Macarena
Calle 25 de Mayo 489, Piso 6, 1339 Buenos Aires
☎ (+54 1) 312 56 71; 312 56 75
Ⓔ macarena@century.com.ar ▣ 27238 macub ar
⊖ (+54 1) 311 57 42
♧ US$ 2,000,000 (1996) ♣ US$ 10,000,000 (1996)
ⅲ 60 (1996) ⌂ 1,000 m2
President: José Maria Beraza
Managing Director: Oscar Prieto
Sales and Marketing Director: Fernando Barberio
Operations and Technical Director: Ing. César Elizalde
Purchasing Manager: Enrique Vazquez
■ Sale of aircraft, engines, spares, helicopter equipment and accessories for general and military aviation.
® USA: AlliedSignal, Bell Helicopter Textron Inc, Bendix/King, Textron Lycoming, Stratford Division, Textron Lycoming, Williamsport Division 97

90-13000 MTU Argentina SA
Casilla de Correo n°12, 1618 El Talar. Pcia, Buenos Aires
☎ (+54 1) 736 73 37 ⊖ (+54 1) 11 13 84
Managing Director: Mr Mlekusch
® Germany: MTU Friedrichschafen GmbH 98

90-19300 Servotron SA
Avenida Belgrano 768,5 Piso, 1092 Buenos Aires
☎ (+54 1) 343 63 23/13 10; 331 48 84
Ⓔ servoci@microstar.com.ar ⊖ (+54 1) 334 01 04
♣ US$ 12 000 000 ⅲ 60 ⌂ 1 000 sq m
President: Bruno Capra
® Germany: Endress & Hauser
USA: Exide
Fábrica
Zufriategui 4921, 1063 Villa Martelli, Pcia
☎ (+54 1) 760 97 94; 761 36 30/43
Ⓔ 75074.566@compuserve.com
⊖ (+54 1) 760 97 94; 761 36 30/43 97

90-20100 Alejandro y Rodolfo Taylor Srl
Corrientes 534, Piso 12, 1043 Buenos Aires
☎ (+54 1) 325 17 14; 394 37 51 ✉ TAYLOR
⊖ (+54 1) 394 37 51
♣ US$ 500,000 ⅲ 5 ⌂ 100 m2
Managing Partner: Alejandro H Taylor
■ Sale of aircraft and components to Argentine armed forces.
® USA: Bet-Air, Lockheed Aircraft 98

90-20300 TEC SRL
Avda. Maipu 3902-2P-DTO B, Pcia Buenos Aires
☎ (+54 1) 794 64 12 ⊖ (+54 1) 794 19 74
Managing Dir: Oscar Gazzotti
Marketing Dir: Jorge Pérez
Technical Dir: Ricardo Agnoletti
® Switzerland: Trimos-Sylvac Task Force 97

Armenia, Republic of

President:
Levon Ter-Petrosyan

10 Head of State

10-10000 Office of the President
26 Marshall Baghramyan Street, 375012 Yerevan
⊖ (+374 8852) 15 11 52
President: Levon Ter-Petrosyan ☎ (+7 8852) 52 02 04 98

11 Prime Minister

11-10000 Office of the Prime Minister
1 Government Building, Republic Square, 375010 Yerevan
⊖ (+374 8852) 15 10 36
Prime Minister: Robert Kocharian
☎ (+7 8852) 52 03 60; 52 82 50 98

12 Ministries

12-10000 Ministry of Defence
Vagarsh Vagarshyan 26, Stepanakert 374430, Nagorno Karabakh Republic
☎ (+374 8852) 35 78 81; (+374 89322) 436 92
Ⓦ http://www.nic.nk.am/artsakh/army.html
ⅲ 1,000
Defence Minister: Vazgen Sarkisyan 98

12-20000 Ministry of Foreign Affairs
10 Marshall Baghramyan Street, 375012 Yerevan
☎ (+374 8852) 52 35 31; 58 88 29
⊖ (+374 8852) 15 10 42; 52 70 22
Minister of Foreign Affairs: Alexander Arzoumanian
Chief Consular and Legal Section: Minna Gurgenyan
☎ (+7 8852) 58 88 49 98

12-30000 Ministry of Trade
69 Teryan Street, 375009 Yerevan
☎ (+374 8852) 56 25 91 ⊖ (+374 8852) 58 35 21
Minister: Garnik Nangulian 98

12-40000 Ministry of Internal Affairs
130 Nalbandyan Street, 374430 Stepanakert, Nagorno-Karabakh Republic NK
☎ (+374 8852) 56 09 08 Ⓔ news@mia.nk.am
⊖ (+374 8852) 52 93 88
ⅲ 1,000
Minister: Serzhik Sarkisyan 98

12-50000 Ministry of Economy
1 Government Building, Republic Square, 375010 Yerevan
☎ (+374 8852) 57 73 42 ⊖ (+374 8852) 15 10 69
Minister: Vahram Avanesyan 98

12-60000 Ministry of Industry
2 Government Building, Republic Square, 375010 Yerevan
☎ (+374 8852) 52 18 77; 52 05 79
⊖ (+374 8852) 15 10 84; 52 35 64
Minister of Industry: Ashot Safaryan
Deputy Minister: Meron Shishmanyan
Foreign Relations Dir: Yuri Gasparyan
Chief of Staff: Artour Grigoryan 98

12-70000 Ministry of Communication
28 Nallbardyu Street, 375050 Yerevan
☎ (+374 8852) 52 66 32
⊖ (+374 8852) 15 14 46; 53 86 45
Minister: Grigor Poghpatyan 98

15 Parliament

15-00050 Government of Nagorno-Karabakh Republic
374430 Stepanakert
☎ (+7 89) 32 24 32 90 Ⓔ root@gov.nk.am
⊖ (+7 8852) 28 18 97 97

31 Embassies & Attachés

31-00660 Embassy of the Federal Republic of Germany
Erlwan Hrazdan Hotel, 7th Floor, 72 Pioneerakan Str, 375109 Yerevan
☎ (+374 8852) 53 67 74 ▣ 64243138 aaeri su
⊖ (+374 8852) 15 11 12
Ambassador and Deputy Chief of Mission: Ms Carola Müller-Holthemper
Commercial Officer: Marion Ramackers
Public Affairs Officer: Ingrid Delfs 98

31-00690 Embassy of Greece
Erevan
☎ (+374 2) 53 00 51; 53 71 03 ▣ 536754
⊖ (+374 2) 505 62 17 98

31-01900 Embassy of the United States of America
No 18 General Bagramian, 375019 Yerevan
☎ (+374 2) 15 15 51; 52 46 61 ▣ 243137 amemy
⊖ (+374 2) 15 15 50
Ambassador: H E Mr Peter Tomsen
Deputy Chief of Mission: Louis Licht
Political, Economic Section: Susan Thornton
Commercial Officer: Andrew Hovannisyan
Public Affairs Officer: Morgan Liddick
Defence Attaché's Office: Lieut Col Robert Brown 98

Australia

Head of State:
H.M. Queen Elizabeth II

Governor-General:
Sir William Deane

10 Head of State

10-10000 Governor-General's Office
Government House, Canberra, ACT 2600
☎ (+61 6) 283 35 33 ⊖ (+61 6) 281 37 60
Governor-General: Sir William Deane
Official Secretary: R D Sturkey
Household and Property Manager: David Gordon Mair 98

11 Prime Minister

11-10000 Department of the Prime Minister
Parliament House, Room M94, Canberra, ACT 2600
☎ (+61 6) 277 77 00 ⊠ 61616 primin aa
Prime Minister: John Howard
Deputy Prime Minister: Timothy Fischer
Minister Assisting the Prime Minister for Public Service Matters: Peter McGauran 98

12 Ministries

12-10000 Department of Defence
Russell Offices, Canberra 2600
☎ (+61 6) 265 91 11

Ministers Office
Suite MF 46, Parliament House, Canberra, ACT 2600
☎ (+61 6) 277 78 00; 265 29 99 (public relations)
⊖ (+61 6) 273 41 18; 265 20 99 (public relations)
Minister for Defence: Ian McLachlan
Minister for Defence Industry, Science and Personnel: Ms Bronwyn Bishop
Defence Secretary: A J Ayers ☎ (+61 6) 265 28 50
Chief of Defence Force: Gen John Baker ☎ (+61 6) 265 29 76
Vice Chief of the Defence Force: Rear Adm Chris Barrie ☎ (+61 6) 265 10 20
Chief of Navy: Vice Adm Don Chalmers ☎ (+61 6) 265 10 20
Chief of Army: Lieut Gen John Sanderson ☎ (+61 6) 265 41 98
Chief of Air Force: AM Les Fisher ☎ (+61 6) 265 54 75
Deputy Secretary Strategy and Intelligence: Hugh White ☎ (+61 6) 265 28 48
Chief Defence Scientist: R Brabin Smith ☎ (+61 6) 266 61 13
Head Joint Education and Training: Rear Adm John Lord ☎ (+61 6) 266 50 30
Head Personnel Executive: Maj Gen Peter Dunn ☎ (+61 6) 265 69 02
Head Corporate Support: Peter Sharp ☎ (+61 6) 266 61 30
Commander Support Australia: Maj Gen Des Mueller ☎ (+61 6) 266 67 09
Defence Estate Manager: Rod Corey ☎ (+61 6) 265 43 00
Head Finance: Frank Lewincamp ☎ (+61 6) 265 53 74
Inspector General: Mike McNamara ☎ (+61 6) 266 43 95
Head Corporate Information: Phil Huntley ☎ (+61 6) 266 89 44
Deputy Secretary Aquisition: Gary Jones ☎ (+61 6) 265 39 22
First Assistant Secretary Capital Equipment Program: Ms M McPherson ☎ (+61 6) 265 30 09
Acquisition Corporate Management: C Campbell ☎ (+61 6) 265 47 10
Acquisition Finance and Reporting: G Watters ☎ (+61 6) 265 66 01
Head Industry/Procurement Infrastructure: S Pearce ☎ (+61 6) 266 34 77
Head Systems Acquisition - Electronic Systems: R Bonighton ☎ (+61 6) 266 69 03
Head Systems Acquisition - Maritime and Ground: Rear Adm P Purcell ☎ (+61 6) 266 20 34
Head Systems Acquisition - Aerospace: AVM D Riding ☎ (+61 6) 265 30 51
Director General Public Information: Col Andrew Reynolds ☎ (+61 6) 265 29 29 98

12-10400 Deputy Secretary Acquisition
Department of Defence, Russell Offices - F-2-66, Canberra, ACT 2600
☎ (+61 6) 265 37 42 ⊖ (+61 6) 265 12 14

Capital Equipment Program Division
Russell Offices, F-1-40, Canberra, ACT 2600
☎ (+61 6) 265 30 09 ⊖ (+61 6) 265 11 42 98

12-20000 Department of Foreign Affairs and Trade
Parliament House - Room MG 68, Canberra, ACT 2600
☎ (+61 6) 277 75 00 ⊠ FOREIGN Canberra
W http://www.dfta.gov.au ⊖ (+61 6) 273 41 12
Minister of Foreign Affairs: Alexander Downer
Minister for Trade: Senator the Hon Bob McMullan
Minister for Pacific Island Affairs: The Hon Gordon Bilney
High Comissioner: H E P D Allars
Senior Adviser to Minister of Foreign Affairs: Geoff Bentley
First Secretary: Mr C Gallagher
Second Secretary: Leslie Harrison
E leslie.harrison@dfat.gov. au
Secretary: Michael Costello
☎ (+61 6) 261 91 11; 273 20 81 fax 98

15 Parliament

15-10000 Senate
Parliament House, Canberra, ACT 2600
☎ (+61 6) 277 71 11 E clerk-sen@senate.aph.gov.au
W http://senate.aph.gov.au ⊠ 62326 senate aa
⊖ (+61 6) 277 33 85; 277 33 87
President: Senator the Hon Margaret Reid ☎ (+61 6) 277 33 00
Deputy President: Senator Mal Colston ☎ (+61 6) 277 36 90
Minister for Justice and Attorney General: The Hon Daryl Williams
Attorney General and Police Minister Australian Central Territory: Gary Humphries
Clerk of the Senate: Harry Evans ☎ (+61 6) 277 33 50
Clerk Assistant Management Office: Peter O'Keeffe ☎ (+61 6) 277 33 99
Clerk Assistant Committee Office: Cleaver Elliott ☎ (+61 6) 277 33 71

Foreign Affairs, Defence and Trade Legislation Committee
Parliament House, Canberra, ACT 2600
☎ (+61 6) 277 35 35
Chairman: Senator Judith Troeth ☎ (+61 6) 277 30 02
Secretary: Paul Barsdell ☎ (+61 6) 277 35 35

Foreign Affairs, Defence & Trade References Committee
Parliament House, Canberra, ACT 2600
☎ (+61 6) 277 35 35
Chairman: Senator Michael Forshaw ☎ (+61 6) 277 38 05
Secretary: Paul Barsdell 98

15-20000 House of Representatives
Parliament House, Canberra, ACT 2600
☎ (+61 6) 277 71 11 ⊠ 61884 pro ⊖ (+61 6) 277 20 58
Speaker: Hon R G Halverson (OBE MP) ☎ (+61 6) 277 40 00
Deputy Speaker: G B Nehl (MP)
Clerk of the House: L M Barlin ☎ (+61 6) 277 41 11
Chamber Research Office Director: R Webber ☎ (+61 6) 277 46 70

Joint Committee on Foreign Affairs, Defence and Trade
Parliament House, Canberra, ACT 2600
☎ (+61 6) 277 23 13 ⊖ (+61 6) 277 22 21
Chairman: Rt Hon I M Sinclair (MP) ☎ (+61 6) 277 40 64
Deputy Chairman: Hon L J Brereton (MP)
Secretary: Ms J Towner 98

25 Defense Forces

25-10000 Headquarters Australian Defence Force
Russell Offices B-4-22, Canberra, ACT 2600
☎ (+61 6) 265 29 99 ⊖ (+61 6) 265 10 99
Chief of the Defence Force: VAdm. Alan Beaumont ☎ (+61 6) 265 28 58
ADC to Chief of Defence Force: Flt Lt P. Tolhoek ☎ (+61 6) 265 46 44
Staff Officer Operations to Chief of Defence Force: Lt Col Leslie ☎ (+61 6) 265 28 45
Chief of Staff: Ms Amanda Forman
Asst Chief of Defence Force Development: RAdm. K.A. Doolan ☎ (+61 6) 265 23 32
Asst Chief Defence Force Operations: AVM L.B. Fisher ☎ (+61 6) 265 46 49
Asst Chief of Defence Force Logistics: Maj.Gen. S.N. Gower ☎ (+61 6) 266 34 77
Chief of Staff, Joint Exercise Planning Staff: Brig. Chris Roberts ☎ (+61 6) 265 36 79
Asst Chief of Defence Force Personnel: RAdm. E. Stubington ☎ (+61 6) 265 69 02
Dir General, Joint Operations & Plans: Brig. Fred Pfitzner ☎ (+61 6) 265 29 57
Dir Joint Communications-Electronics: Air Cdre N.P. Middleton ☎ (+61 6) 65 24 14
Dir General, Service Personnel Policy: Cdre. A.L. Thompson ☎ (+61 6) 265 48 52
Dir General, Military Strategy & Concepts: Air Cdre B.D. O'Loghlin ☎ (+61 6) 265 31 53
Surgeon General SGADF: AVM M.D. Miller ☎ (+61 6) 266 39 19

Central Staff Human Resources & Management Div.
Russell Offices, Canberra, ACT 2600
First Asst Secretary: A.G. Thompson ☎ (+61 6) 266 86 22

Public Information Office
Russell Offices 3-4-22, Canberra, ACT 2600
☎ (+61 6) 265 29 29 ⊖ (+61 6) 265 10 99
Dir General Public Information: Brig. A.S. D'Hage

Force Development & Analysis Div.
Russell Offices, Canberra, ACT 2600
☎ (+61 6) 265 31 09; 29 22
First Asst Secretary: M. Brady ☎ (+61 6) 265 30 21 97

25-19300 Command & Staff College
Fort Queenscliff, Queenscliff, Victoria 3225
☎ (+61 3) 52 58 07 11 ⊖ (+61 3) 52 58 07 73
Commandant: Brig P McNamara
Commanding Officer Support Wing: Maj Tim Kemp
Commanding Officer Instructional Wing: Col R Shoebridge
Commanding Officer Development Wing: Col L T Long
Library Manager: Ms Cheryle Marriner ☎ (+61 3) 525 87 31
Library Officer: Ms Patricia Schoning 98

25-19500 Australian Defence Force Academy
Northcott Drive, Campbell, ACT 2600
☎ (+61 6) 268 86 11 ⊖ (+61 6) 268 86 48
Commander: Maj.Gen. F.J. Hickling
Deputy Commander: Gp Capt. J. Cole
Personal Staff Officer: Sqn Ldr G.J. Flint 97

25-22100 RAAF Air Command Headquarters
Air Command Headquarters, Glenbrook, New South Wales 2773
☎ (+61 47) 37 70 00 ⊖ (+61 47) 37 70 14
Air Commander: AVM P G Nicholson (AM)
Chief of Operations: Air Cdre P J Criss (AM AFC)
Chief of Support: Air Cdre R J Gibson
Director of Technical and Logistics Services: Gp Capt T A Roediger
Aide-de-Camp to Air Commander: D M Yarnall
PSO: Wg Cdr D K Harrison 98

25-22200 RAAF Logistics Command Headquarters
RAAF Williams Building 474-A-G, Laverton, Victoria 3027
☎ (+61 3) 697 32 35 ⊠ 31655 airsos aa
Air Officer Commanding: AVM E M Weller (AM) ☎ (+61 3) 92 56 30 00
Chief of Staff: Air Cdre C Hingston ☎ (+61 3) 92 56 30 01

Aide-de-Camp to Air Officer Commanding:
Flt Lieut M Ranunica ☎(+61 3) 92 56 31 82
Director General Logistics Operations:
Air Cdre I H Ashbrook ☎(+61 3) 92 56 30 04
Director General Engineering and Logistics Support, Controller Logistics Systems Agency: Air Cdre N. Smith ☎(+61 3) 92 56 30 03
PSO to Air Officer Commanding: Sqn Ldr R S Barnes ☎(+61 3) 92 56 31 84
Secretary: Cpl D Timms [98]

25-22300 HQ Air Lift Group
RAAF Base, Richmond, New South Wales 2755
☎(+61 45) 70 20 20 ⊖ (+61 45) 70 29 03
Commander: Air Cdre D C Chipman ☎(+61 45) 70 20 00
Staff Officer Plans and Preparedness: Wg Cdr P C Shinnick ☎(+61 45) 70 20 01
Staff Officer Preparedness: Sqn Ldr R Drinkwater ☎(+61 45) 70 20 21
Officer Commanding 86 Wing: Gp Capt C G Spence
Officer Commanding 84 Wing: Wg Cdr C Richards
Officer Commanding 303 Air Base Wing: Gp Capt G Smith
Command Support: W Silverman [98]

25-22400 Maritime Patrol Group
RAAF Base, Edinburgh, South Australia 5111
☎(+61 8) 83 93 40 00 ⊖ (+61 8) 83 93 40 03
Commander Maritime Patrol Group:
Air Cdre J R McCulloch
Officer Commanding 92 Wing: Gp Capt D Long
■ The Maritime Patrol Group exercises control over 92 Wing, which in turn controls 10 Sqn, 11 Sqn, 292 Sqn, 492 Sqn and Detachment A (Malaysia). Other detachments are formed as required. [98]

25-32100 Headquaters Aviation Support Group
(Directorate of Aviation-Army)
Army Aviation Centre, Oakey, Queensland 4401
☎(+61 76) 91 90 11 ⊖ (+61 76) 91 90 10
Commander Aviation Support Group: Brig W J A Mellor (DSC AM) ☎(+61 76) 91 90 06
Director of Aviation - Army: Col R E Walford (AFC) ☎(+61 76) 91 90 04
Staff Officer Grade One Operations and Finance:
Lieut Col N J Crowther ☎(+61 76) 91 90 20
Staff Officer Grade One Logistics: Lieut Col R Aberle ☎(+61 76) 91 90 50
Staff Officer Grade One Training and Personnel:
Lieut Col G F Rutter ☎(+61 76) 91 90 30
Staff Officer Grade One Aviation Psychology:
Lieut Col R Collyer ☎(+61 76) 91 90 41
Staff Officer Grade One Aviation Medicine:
Lieut Col J Turner ☎(+61 76) 91 90 40 [98]

25-32200 Land Headquarters
Victoria Barracks, Paddington, NSW 2021
☎(+61 2) 339 04 55 ⊖ (+61 2) 339 31 45
Land Commander Australia: Maj.Gen. P.M. Arnison (AO)
Chief of Staff: Brig. P.S. O'Sullivan (MBE)
Military Asst: Lt Col Colley
DBRM LC: D. Leary-Smith
Colonel Operations: Col M.H. Hyde
Staff Officer 1 Operations: Lt Col P.L. Schmitt
Staff Officer 1 Plans: Lt Col R.A. Tiller
Staff Officer 1 Communications: Lt Col P. Straughair
Staff Officer 1 Survey: Lt Col R.A. Coote
Staff Officer 1 Organisation: Col K. Ashman
Staff Officer 1 Training: Lt Col P.J. Charlesworth
Colonel Development: Col J. Cox
Staff Officer 1 Plans & Admin: Lt Col G.C. Cunningham
Commander Artillery: Col A. English
Staff Officer Artillery: Lt Col G.D. Mulherin
Staff Officer 1 Intelligence: Lt Col S.R. Dowse
Colonel Admin: Col P.P. Martyn
Commander Medical Services: Col W.P. Ramsey (CSC, AM)
Commander Engineering: Col R.J. McKinnon
Staff Officer 1 Engineering: Lt Col P. Van der Moezel [97]

25-32300 Headquarters Training Command (Army)
Suakin Drive, Georges Heights, NSW 2088
☎(+61 2) 99 60 94 11
[E] so1coord@trg.army.defence.gov.au
⊖ (+61 2) 99 60 94 04
General Officer Commanding: Maj Gen M J Keating (AM)
Chief of Staff: Brig P R Kilpatrick
SO1 Co-ordination: Lieut Col D Smith
Co-ordinator CPL: A Todd [98]

25-37100 Army Technology & Engineering Agency
(† Engineering Development Establishment (Army))
Raleigh Road, Maribyrnong, Private Bag 12, Ascot Vale, Vic 3032
☎(+61 3) 319 55 11 ⊠ ENGDEV Melbourne
⊖ (+61 3) 318 34 21
Commander: Brig. M.P.J. O'Brien
Technical Dir: K.J. Crawford
Engineering Dir: G. Van Ree
Quality Dir: D. Heath
Laboratories Dir: D. Long
Commercial Mgr: D. Collins
Staff Officer Grade 1 Operations: Maj. R.D. McCallum → 60 [97]

25-39100 Royal Military College
Duntroon, Canberra, ACT 2600
☎(+61 6) 275 91 11 ⊖ (+61 6) 275 90 40
Commandant: Brig. Brian Stevens
Director of Military Art: Col D.J. Harris
SO2 Co-ordination: Maj Michael Woods
Military Assistant to Commandant: Capt. Damian White [97]

25-40000 Department of Defence Navy Office
Queen Victoria Terrace, PO Box E33, Canberra, ACT 2600
☎(+61 6) 265 91 11 ⊠ DEFNAV Canberra
⊖ (+61 6) 265 47 90
Chief of Naval Staff: VAdm. R.G. Taylor (AO RAN) ☎(+61 6) 265 51 63
Staff Officer Research to Chief of Naval Staff:
Lt R.C.A. Leahy (RAN)
Deputy Chief of Naval Staff: RAdm C.A. Barrie (AM) ☎(+61 6) 265 51 58
Maritime Commander: Rear Adm Bill Schofield
Dir General Maritime Studies Program: Capt. J. McCaffrie ☎(+61 6) 266 61 14
Naval Scientific Adviser: Dr A. Steele ☎(+61 6) 265 50 10
Director General Naval Policy & Warfare:
Cdre D.J. Shackleton ☎(+61 6) 265 50 65
Submarine Policy & Warfare Dir: Capt. J.A. Diercks ☎(+61 6) 265 51 52
Naval Warfare Dir: Capt. B.L. Adams ☎(+61 6) 265 51 16
Dir General Facilities Branch: Capt. R. Cook ☎(+61 6) 266 43 04
Dir General Naval Engineering Services: H.R. Machin ☎(+61 6) 266 30 20
Asst Chief Naval Staff Personnel: RAdm M.B. Forrest ☎(+61 6) 265 32 51
Information Systems Dir: Cdr M.A. Rennie ☎(+61 6) 265 51 52
Public Information Dir: Cdr S.E. Jones ☎(+61 6) 265 64 09

Naval Materiel Division
Russell Offices, Canberra, ACT 2600
☎(+61 6) 265 35 00
Asst Chief Naval Staff Materiel: RAdm. P.T. Purcell ☎(+61 6) 266 20 34
Dir General Naval Engineering Requirements:
Cdre C.J. Elsmore ☎(+61 6) 266 21 04
Dir General Naval Production: Cdre E.M. Asker ☎(+61 6) 266 31 89
Dir General Naval Project Provisioning: G.K. Tuisk ☎(+61 6) 266 41 01
Dir General Logistics Policy: Cdre D.G. Walkington ☎(+61 6) 266 41 02
Dir General Naval Logistics Policy: Cdre P.F. McGuire

Hydrographer Office
118 Walker Street, North Sydney, NSW 2059
☎(+61 2) 925 42 09 ⊖ (+61 2) 925 48 35
Hydrographer: Cdre J.W. Leech

Naval Health Services
Campbell Park Offices, Campbell, ACT 2601
☎(+61 6) 266 38 55
Dir General Naval Health Services: Cdre M.H. Dowsett [97]

25-41000 RAN Maritime Headquarters
Wylde Street, Potts Point, New South Wales 2011
☎(+61 2) 563 44 44 [E] fpao@mhqe.mc.gov.au
⊖ (+61 2) 563 46 36
Maritime Commander: Rear Adm C Ritchie
Chief of Staff: Cdre G R Walpole
Chief Staff Officer (Engineering): Capt P Hatcher
Chief Staff Officer (Operations): Capt R Moffitt
Chief Staff Officer (C3I): Capt. Shackleton
Chief Staff Officer (Support): Capt R Crane
Staff Officer to Chief of Staff: Lieut L Wray
Fleet Public Affairs Officer: Lieut Carmen Marshall [98]

25-47000 Defence Science & Technology Organisation
(a Department of Defence Establishment)
PO Box 44, Pyrmont, New South Wales 2009
☎(+61 2) 96 92 15 09 ⊖ (+62 2) 96 92 15 60
Chief: R Creaser
Librarian: Ms Elizabeth Wright [E] liz.wright@dsto.defence.gov.au
Assistant Secretary, Science, Industry and External Relations: Mrs Karen Nagle
■ The DSTO Sydney carries out research in support of Australian Defence Force maritime operations. [98]

26 Police (Civil)

26-10000 Australlian Federal Police Head Office
PO Box 401, Canberra, ACT 2601
☎(+61 6) 256 77 77 ⊖ (+61 6) 275 70 58
Commissioner and Chief Police Officer Australian Central Territory Region: Mick Palmer
Deputy Commissioner: Adrien Whiddett
Deputy Commissioner: James Allen
Asssistant Commissioner Australian Central Territory Region: William Stoll
General Manager Eastern Region: Bob McDonald
General Manager Southern Region: Nigel Hadgkiss
General Manager Northern Region: Mike Keelty
General Manager Central Region: Ms Sandi Peisley
General Manager Western Region: Andy Wells
General Manager National Operations: Alan Mills
General Manager Protective Services: John Davies
General Manager Professional Development: Rod Leffers
General Manager Corporate Strategies and Communications: Chris Whyte
General Manager Scientific and Technological Systems and Management: Phil Baer
Director, Media and Public Relations: Steve Jiggins
National Media Liaison Officer: Tracey Dickerson
Media and Public Relations Branch: Ms Ally Mercer
■ National law enforcement agency. [98]

31 Embassies & Attachés

31-00100 Embassy of Austria
PO Box 3375, Manuka, Canberra, ACT 2603
12 Talbot Street, Forrest, Canberra, ACT
☎(+61 6) 295 15 33; 13 76 ⊠ 62726 oecnb aa
⊖ (+61 6) 239 67 51
Ambassador: H.E. Dr Stephan Toth
First Secretary: Dr Anna Jankovicer

Foreign Trade Office
1 York Street, Sydney, NSW 2000
☎(+61 2) 247 85 81 ⊠ AUSTROTRAD Sydney
⊖ (+61 2) 251 10 38
Commercial Consul: Dr Gunter Schimmel
Vice Consul: Mag. Adelheid Hobart [97]

31-00130 High Commission for the People's Republic of Bangladesh
35 Endeavour Street, PO Box 5, Red Hill, Canberra, ACT 2603
☎(+61 6) 295 33 28 ⊠ 61729 bdtaus aa
⊖ (+61 6) 295 33 51; 295 00 32
High Commissioner:
H E Maj Gen Moinul Hussain Choudury
Deputy High Commissioner: S A L M Mustakim
Second Secretary: Sheikh Mohammad Belal
Commercial Counsellor: Muhammad Abul Kalam Azad [98]

31-00230 Embassy of Brazil
PO Box 1540, Canberra, ACT 2601
19 Forster Crescent - Yarralumla, Canberra, Australian Central Territory 2600
☎(+61 6) 273 23 72 [E] aasbrem@interconnet.com.au
⊠ 62327 brasem aa ⊖ (+61 6) 273 23 75
Ambassador: H E Mr Renato Prado Guimaraes

Contact Reference

Counsellor Head of Trade Section and Head of Chancery: Julio Celso Ramos
Science and Technology Affairs: Mrs Maria Helena Nicoll
[98]

31-00310 Canadian High Commission
Commonwealth Avenue, Yarralumla, Canberra, ACT 2600
☎(+61 6) 273 38 44 ⊖ (+61 6) 273 32 85
High Commissioner: H E Mr B Schumacher
Deputy High Commissioner: Gardiner Wilson
Commercial/Economic Counsellor: John Donaghy
Canadian Forces Adviser: Capt Steve Foldesi
[98]

31-00360 Embassy of the People's Republic of China
15 Coronation Drive, Yarralumla, Canberra, ACT 2600
☎(+61 6) 273 47 80 (chancery); 273 47 83 (consular office); 273 47 84 (office of the Defence Attaché); 273 47 85 (commercial office); 273 47 86 (scientific office); 273 44 71 (consular enquiry)
Telex 62489 chiem aa
⊖ (+61 6) 273 48 78; 273 51 89 (consular office)
Ambassador: H E Mr Hua Junduo
Counsellor: Zhao Xianling
Minister Counsellor, Commercial: Zhu Zhenyuan
☎(+61 6) 273 47 85
Counsellor, Consular: Tang Qinlheng
Scientific Counsellor: Prof Fan Mingyi
☎(+61 6) 273 47 86
Defence Attaché: Col Yu Huadong
Deputy Military Attaché: Col Yan Shifa
First Secretary: Gu Sicong
First Secretary: Zhou Jinfu
First Secretary: Zhang Kanshan
First Secretary, Administration: Zhang Wenming
Secretary (Scientific): Luo Dongfeng

Cultural Office
6 Terrigal Crescent, O'Malley, ACT 2606
☎(+61 6) 286 24 65
Counsellor, Cultural: Mao Xinli
First Secretary, Cultural: Ms Ru Xian

Education Office
2 Dalman Crescent, O'Malley, ACT 2606
☎(+61 6) 286 43 51
Counsellor, Education: Liu Dianqiu
[98]

31-00450 Embassy of the Czech Republic
38 Culgoa Circuit, O'Malley, ACT 2606
☎(+61 6) 290 13 86; 290 00 10 ⊖ (+61 6) 290 00 06
Ambassador: H E Dr Jaroslav Suchanek
☎(+61 6) 290 13 86
Counsellor: Ing. Eduard Metela ☎(+61 6) 290 00 10
Secretary: Nadezda Boubalova ☎(+61 6) 290 13 86 [98]

31-00510 Embassy of the Arab Republic of Egypt
1 Darwin Ave, Yarralumla, ACT 2600
☎(+61 6) 273 44 37/8 Telex 62497 boustan aa
⊖ (+61 6) 273 42 79
Ambassador: H.E. Mr Ali El-Sherei
Minister: Abdel Fattah M. Ezzeldin
First Secretary: Mohamed Abdelkader Elkhashab
Admin Attaché: Ragab Saber Ibrahim Gouda
Third Secretary: Maged Nafea Mosleh

Press and Information Bureau
1 Darwin Ave, Yarralumla, ACT 2600
☎(+61 6) 273 46 29 Telex 61443 aa ⊖ (+61 6) 273 14 43
Press & Information Attaché: Magdi Ramzi S. Girgis
Press & Information Attaché: Magdi Abdel Kader [97]

31-00535 Embassy of Estonia
86 Louisa Road, Birchgrove NSW 2041
☎(+61 2) 810 74 68 ⊖ (+61 2) 818 17 79
Honorary Consul General: Malle Tohver [97]

31-00570 Embassy of the Republic of Fiji
Queen Victoria Terrace, PO Box E159, Canberra, ACT 2600
☎(+61 6) 239 68 72/73 ⊖ (+61 6) 295 32 83
Ambassador: H.E. Dr T.M. Biumaiwai
First Secretary: Vincent Thomas Rajalingam

Fiji Consul Sydney
Level 12 St Martin's Tower, Sydney, NSW 2000
☎(+61 2) 267 88 55 ⊖ (+61 2) 267 11 95 [98]

31-00580 Embassy of Finland
10 Darwin Avenue, Yarralumla, Canberra, ACT 2600
☎(+61 6) 273 38 00 ✉ FINLANDIA Canberra
Telex 62713 finla ⊖ (+61 6) 273 36 03
Ambassador: H.E. Mr Charles Murto
Counsellor: Pekka Hyvonen
Attaché: Ms Taina Kettula

Commercial Section
537 New South Head Road, Double Bay, NSW 2028
☎(+61 2) 327 79 04 Telex 123388 aa ⊖ (+61 2) 327 75 28
Commercial Counsellor: Jarmo Turtiainen [97]

31-00690 Embassy of Greece
9 Turrana Street - Yarralumla, Canberra, ACT 2600
☎(+61 6) 273 30 11; 273 31 58
✉ GREEK EMBASSY Canberra Telex 62724 gremb aa
⊖ (+61 6) 273 26 20
Ambassador: H E Mr George C Constantis
First Secretary: Adam-Georges Adamides
Attaché (Administration): Michael Kassimatis
Attaché (Consular): Evangelos Athanassiou

Commercial Office
37 Pitt Street - 3rd Floor, Sydney, New South Wales
☎(+61 6) 247 23 34 ⊖ (+61 6) 251 72 08
Commercial Consul: Christos Lelempsis [98]

31-00810 Embassy of the Republic of Hungary
17 Beale Crescent, Deakin, ACT 2600
☎(+61 6) 282 32 26 ⊖ (+61 6) 285 30 12
Ambassador: H E Dr Gyorgy Varga
Counsellor: Bela Fazekas
Secretary to the Ambassador: Mrs Judy Pummeroy

Commercial Consulate of the Republic of Hungary
Suite 201 - Picadilly Court,
356-366 New South Head Road, Double Bay, NSW 2028
☎(+61 2) 327 50 92/63 35 Telex 123262 aa
⊖ (+61 2) 327 43 11
Consul (Commercial): Janos Ligeti [98]

31-00830 High Commission of India
3-5 Moonah Place - Yarralumla, Canberra, ACT 2600
☎(+61 6) 273 39 99; 273 37 74; 273 38 75
Telex 62362 hcomind aa ⊖ (+61 6) 273 13 08; 273 33 28
High Commissioner: H E Mr G Parthasarathy
☎(+61 6) 273 19 20
Deputy High Commissioner: B B Tyagi
☎(+61 6) 273 11 06
Counsellor: Asok Kumar ☎(+61 6) 273 13 65
Defence Adviser: Capt V S Chaudari (IN)
☎(+61 6) 273 38 49
Assistant, Defence: Alok Singh [98]

31-00850 Embassy of the Islamic Republic of Iran
25 Culgoa Circuit, O'Malley, ACT 2606
☎(+61 6) 290 24 21 Telex 62490 iran aa
⊖ (+61 6) 290 24 31
Ambassador: H.E. Mr Mohammad Roohi-Sefat
Economic Attaché: Ashgar Manoussi [97]

31-00920 Embassy of Japan
112 Empire Circuit - Yarralumla, Canberra, ACT 2600
☎(+61 6) 273 32 44 Telex 62034 taisi aa
⊖ (+61 6) 273 18 48
Ambassador: H E Mr Yukio Satoh
Minister: Akio Suda
Defence Attaché: Capt Takahide Yamamoto
First Secretary (Scientific Affairs): Mr Saito
Counsellor (Economic): Kazutomo Irie
Second Secretary (Press and Information): Masaki Tani
Personal Assistant to Ambassador: Ms Lin Calver [98]

31-00990 Embassy of the Republic of Korea
113 Empire Circuit - Yarralumla, Canberra, ACT 2600
☎(+61 6) 273 30 44 ⊖ (+61 6) 273 48 39
Ambassador: H E Mr Moon Dong Suk
Minister Counsellor (Designate): Hyo Seung Ahn
Defence Attaché: Col Kye Sung Lee ☎(+61 6) 273 38 59
Counsellor: Eon Kee Baek
Counsellor: Bong Ryull Yang
Counsellor Cultural: Kyu Euk Chung
Counsellor Commercial: Tae Yong Lee
First Secretary: Han Taek Im
First Secretary - Consular: Yong Kyu Kwon
First Secretary Agriculture: Ji Bong Song
Third Secretary: Joo Young Kim
Personal Assistant to the Ambassador: Ms J Battersby
[98]

31-01010 Embassy of the Lao People's Democratic Republic
1 Dalman Crescent, O'Malley, Canberra, ACT 2606
☎(+61 6) 286 45 95, 286 69 33 Telex 61627 emlaos aa
⊖ (+61 6) 290 19 10
Ambassador: H E Mr Phantong Phommahaxay
First Secretary: Akhom Saignasith [98]

31-01020 Embassy of Lebanon
27 Endeavour Street, Red Hill, Canberra, ACT 2603
☎(+61 6) 295 73 78 ✉ AMBALIBAN Canberra
Telex 61762 lebano aa ⊖ (+61 6) 239 70 24
Ambassador: H.E. Dr Latif Abul-Husn
First Secretary: Nidal Yehya [97]

31-01140 Malta High Commission
2 Red Hill Court, 97 Monaro Crescent, Red Hill, ACT 2603
☎(+61 6) 295 15 86; 295 16 45 [E] maltahc@cs.net.au
⊖ (+61 6) 239 60 84
First Secretary Information and Trade Divisions: Raymond C Xerri
[98]

31-01270 Royal Netherlands Embassy
120 Empire Circuit, Yarralumla, Canberra, ACT 2600
☎(+61 2) 62 73 31 11 [E] nlgovcan@ozemail.com.au
Telex 62047 nethemb aa ⊖ (+61 2) 62 73 32 06
Ambassador: H E Mr Roelof R Smit
Counsellor: John P Roosegaarde Bisschop
First Secretary: Robert Jumelet
Secretary to the Ambassador: Ms Diana Parks-Bouwer
[98]

31-01300 New Zealand High Commission
Commonwealth Avenue, Canberra, ACT 2600
☎(+61 6) 270 42 11 ⊖ (+61 6) 273 31 94
High Commissioner: H E Mr Peter Hamilton
First Secretary: A Windmill
Defence Adviser: Wg Cdr Peter Randerson (resident in Wellington) [98]

31-01330 High Commission of the Federal Republic of Nigeria
PO Box 241, Canberra, ACT 2608
7 Terrigal Crescent, O'Malley, ACT 2606
☎(+61 6) 286 13 22; 10 44; 19 66
✉ NIGERIAN Canberra Telex 62778 niganb
⊖ (+61 6) 286 53 32
High Commissioner: H.E. Prof. Jonathan Osmam Ndagi
Minister Head of Chancery: John O. Oboh
Defence Adviser: Capt F E Agbiti (Resident in New Delhi)
First Secretary Information: Frank Nmadu [97]

31-01440 Embassy of Portugal
PO Box 92 Deakin, Canberra, ACT 2600
23 Culgoa Circuit, O'Malley, ACT 2606
☎(+61 6) 290 17 33 ✉ LUSITANIA Canberra
Telex 62649 lusa aa ⊖ (+61 6) 290 19 57
Ambassador: H.E. Dr Rui Goulart de Ávila
First Secretary: Dr José Frederico Ludovice [97]

31-01640 High Commission of the Republic of Singapore
17 Forster Crescent - Yarralumla, Canberra, ACT 2600
☎(+61 6) 273 39 44 Telex 62192 singawa aa
⊖ (+61 6) 273 32 60; 273 03 01
Defence Adviser: Loor Han Seng [98]

31-01655 Embassy of the Republic of Slovenia
Level 6 - Advance Bank Center,
PO Box 284 - 60 Marcus Clarke Street,
Canberra, ACT 2601
☎(+61 6) 243 48 30 ⊖ (+61 6) 243 48 27
Chargé d'Affaires: Aljaz Gosnar

Consulate of the Republic of Slovenia
PO Box 188, Coogee, Sydney NSW 2034
☎(+61 2) 314 51 16 ⊖ (+61 2) 399 62 46

Consulate of the Republic of Slovenia (honorary)
PO Box 5, Smithfield NSW 2164

☎(+61 2) 604 51 33 ⊜ (+61 2) 604 00 96
Honorary Consul: Dusan Lajovic 98

31-01670 South African High Commission
Corner State Circle & Rhodes Place, Yarralumla, Canberra, ACT 2600
☎(+61 6) 273 24 24 ✉ SAKOM Canberra
⊜ (+61 6) 273 35 43
Counsellor: P. Gerber 97

31-01680 Embassy of Spain
PO Box 9076, Deakin, ACT 2600
15 Arkana Street,
Yarralumla, Australian Central Territory 2600
☎(+61 6) 273 35 55 ⊜ (+61 6) 273 39 18
Ambassador: H E Mr Emilio Fernández-Castano
Deputy Chief of Mission: Agustín Santos Maraver
Attaché (Administration): Miguel Angel Carrasco
Commercial Office
Edgecliff Centre - Suite 408, 203 New South Head Road, Edgecliff, New South Wales 2027
☎(+61 2) 362 42 12; 362 42 13; 362 42 14
⊠ 22194 ofcome aa ⊜ (+61 2) 362 40 57
Commercial Counsellor: José Antonio García López 98

31-01730 Swedish Embassy
5 Turrana Street - Yarralumla, Canberra, ACT 2600
☎(+61 6) 273 30 33 ✉ SVENSK Canberra
E sweden@netinfo.com.au ⊠ 62303 svensk aa
⊜ (+61 6) 273 32 98
Ambassador: H E Mr Göran Hasselmark
☎(+61 6) 273 23 36
Counsellor Deputy Chief of Mission: Grels Zakrisson
Defence Attaché: Col Lars Olof Nilsson
Second Secretary: Ms Lena Wedén
Swedish Trade Council
44 Market Street, National Mutual Building 25th floor, Sydney, New South Wales 2000
☎(+61 2) 92 62 10 77 ⊠ 178166 svensk aa
⊜ (+61 2) 92 90 39 07
Trade Officer: Christian Olsson 98

31-01740 Embassy of Switzerland
7 Melbourne Avenue, Forrest, Canberra, ACT 2603
☎(+61 6) 273 39 77 ⊠ 62275 amswiss aa
⊜ (+61 6) 273 34 28
Ambassador: H.E. Dr Bernard Marfurt
Deputy Chief of Mission: Dr Jean Jacques de Dardel
Attaché & Head of Chancery: Ms Christine Bertschinger 97

31-01780 Royal Thai Embassy Chancery
111 Empire Circuit, Yarralumla, ACT 2600
☎(+61 6) 273 11 49; 273 29 37 ✉ THAIDUTO Canberra
⊠ 62533 tduto aa ⊜ (+61 6) 273 15 18
Ambassador: H E Mrs Lanxanachantorn Laohaphan
Minister (Deputy Chief of Mission): Mrs Sasitorn Sintusingha
Minister-Counsellor: Sukasem Yothasamutr
Minister-Counsellor (Education): Mrs Pra-on Komolpith
Minister-Counsellor (Agriculture): Sanchai Tontyaporn
Minister-Counsellor (Commercial): Miss Saipin Chantanaruman
Counsellor: Thakul Disatha-Amnarj
Counsellor (Education): Mrs Chantanee Powichit
First Secretary: Biravij Suwanpradhes
First Secretary: Krairawee Sirikul
First Secretary (Agriculture): Chan Chiumkanokchai
Third Secretary: Muhamed Luftfee Useng
Third Secretary: Vittaya Srichainont
Third Secretary: Saowanee Teerapongsakorn
Office of the Army Attaché
5 Stonehaven Crescent, Red Hill, Canberra, ACT 2603
☎(+61 6) 273 32 71
Army Attaché: Col Sirapong Boonpat
Office of the Defence & Air Attaché
6 Titheradge Place, Chapman, Canberra, ACT 2611
☎(+61 6) 288 75 66
Defence and Air Attaché: Gp Capt Weracoop Srisongkram
Office of the Naval Attaché
127 Hawkesbury Crescent, Farrer, Canberra, ACT 2607
☎(+61 6) 286 55 13
Naval Attaché: Capt Wallop Kerdphol (RTN)
Office of the Commercial Affairs
1 Phipps Place (Unit 3A), Deakin, ACT 2600
☎(+61 6) 285 23 28 ✉ THAIDUTO Canberra
⊜ (+61 6) 282 26 13
Commercial Counsellor: Miss Rachakorn Saskri 98

31-01890 United Kingdom High Commission
Commonwealth Avenue, Yarralumla, Canberra, ACT 2600
☎(+61 6) 270 66 55 ✉ UKREP Canberra
⊠ 62222 ukrep aa ⊜ (+61 6) 270 66 16
High Commissioner: H E Sir R Carrick
Deputy High Commissioner: A J Pocock
Defence and Naval Adviser and Head of British Defence Liaison Staff: Cdre P C Wykeham-Martin (RN)
Air Adviser: Col M A Radforth (RAF)
First Secretary Information: R Holland
First Secretary Defence Research: G Stott
First Secretary Defence Export Services: D Richardson
Personal Assistant to Commodore P C Wykeham-Martin: Maureen Clemence 98

31-01900 Embassy of the United States of America
Moonah Place, Canberra, ACT 2600
☎(+61 6) 270 50 00 ⊠ 62104 usaemb aa
⊜ (+61 6) 270 59 70
Ambassador: H.E. Mr Edward J. Perkins
Chargé d'Affaires: Kaarn J. Weaver ☎(+61 6) 270 58 43
Defense Attaché: Col Stephen Barneyback (USAF)
Political Affairs Counsellor: Roger McGuire (Jr)
Admin Affairs Counsellor: Marshall F. Atkins
Public Affairs Officer: Ms Shiela Austrian
Economic Section: Ralph Moore
Melbourne
553 St Kilda Road, PO Box 6722, Melbourne, Vic 3004
☎(+61 3) 526 59 00 ⊜ (+61 3) 510 46 46
Consulate General: Ross Wilson
Political Section: James McNaught
Commercial Section: Kenneth L. Norton
Consular Section: Paul Fitzgerald
Perth
13th Fl. 16 St Georges Terrace, Perth, WA 6000
☎(+61 9) 231 94 00 ⊜ (+61 9) 231 94 44
Consulate General: Nicholas Sherwood
Consular Section: Daniel P. Claffey
Sydney
59th Fl. MLC Centre, 19-29 Martin Place, Sydney, NSW 2000
☎(+61 2) 373 92 00 ⊜ (+61 2) 373 91 25
Consulate General: Jerome Tolson
Political Section: Ms Kristine Pelz
Commercial Section: John Bligh
Consular Section: Teddy Gong
Branch Public Affairs Officer: Christopher Fitzgerald 97

31-01910 Embassy of Uruguay
1st Floor - Suite 7 MCL Tower, Woden, Canberra, ACT 2606
☎(+61 6) 282 48 00 ✉ URUCAN Canberra
⊠ 61486 urucan aa ⊜ (+61 6) 282 43 35
Ambassador and Deputy Chief of Mission: H E Mr Pablo Sader 98

31-01950 Embassy of Venezuela
MLC Tower - 1st Floor, PO Box 37, Woden, Canberra, ACT 2606
☎(+61 6) 282 48 27; 282 48 28
✉ EMBAVENEZ Canberra ⊠ 62110 embave aa
⊜ (+61 6) 281 19 69
Ambassador: H E Mr Antonio Rodríguez Yturbe 98

31-02010 Embassy of the Federal Republic of Yugoslavia
11 Nuyts Street - Red Hill, Canberra, ACT 2603
☎(+61 6) 295 14 58; 295 71 62
E yuembau@ozemail.com.au
W http://www.ozemail.com.au/~yuembau ⊠ 62317 aa
⊜ (+61 6) 239 61 78
Chargé d'Affaires: Dusan Vukasinovic (acting)
Attaché: Trivun Savic 98

31-02050 Zimbabwe High Commission
11 Culgoa Circuit - O'Malley, Canberra, ACT 2606
☎(+61 6) 286 27 00 ⊠ 62211 aa ⊜ (+61 6) 290 16 80
High Commissioner: H E Prof Hasu H Patel
Counsellor: Wenceslaus Mudyanadzo (acting)
First Secretary: Ms Dumisani Nhongo
Second Secretary: Ms Placidia Savanhu
Third Secretary: Ms Dorothy Chikoore 98

55 Government-owned Industry & Production Facilities

55-01000 Australian Defence Industries Ltd (ADI)
Level 22 - Plaza II, Grafton Street,
Bondi Junction, NSW 2022
☎(+61 2) 93 65 93 00 ⊜ (+61 2) 93 69 24 04
Managing Director: Ken A Harris
Group General Manager Business Development: L Targ
Information Services Manager: Ms Susan Tennant
■ Areas of activity include ordnance manufacture, large calibre guns, small arms, ammunition, explosives and propellants, engineering services, machining, large scale finishing, large scale fabrication, repair and maintenance of naval and commercial ships. Design and manufacture of live fire target systems, eye safe laser rangefinder and high speed engineering vehicles.
Operations Group Head Office
Level 12 - 815 Pacific Highway,
Chatswood, New South Wales 2067
☎(+61 2) 99 04 71 00 ⊜ (+61 2) 99 04 71 02
Chief Executive Officer: Steven Milner
Ordnance
Level 1 - Norwich House, 509 St Kilda Road,
Melbourne, Victoria 3004
☎(+61 3) 98 68 61 00 ⊜ (+61 3) 98 68 61 46
■ Design, development and manufacture of a comprehensive range of ammunition (5.56mm to 5/54 Naval), single, double and triple base military and sporting propellants, high explosives, rocket motors, grenades, firing devices and submarine launched flares. Manufacture of military rifles, light machine guns, investment casting and other defence related products.
Marine
Garden Island, Sydney, New South Wales 2000
☎(+61 2) 95 62 33 33 ⊜ (+61 2) 95 62 32 87
■ Repair, refit and modernisation of naval and commercial ships. Resources include dry dock, electronic and missile support facility, computed aided design and engineering systems, optical disk based technical information conversion, storage and management systems.
Engineering
Finn Street, Bendigo, Victoria 3550
☎(+61 3) 54 40 43 11 ⊜ (+61 3) 54 41 67 64
■ Design, development, manufacture, refurbishment and sale of defence related commercial products for light, medium, heavy and precision engineering. Manufacture of marine engines, turbine engine components, submarine equipment, high mobility wheeled military vehicles and weapons systems.
Systems Group Head Office
4 Talavera Road, North Ryde, New South Wales 8113
☎(+61 2) 93 25 15 00 ⊜ (+61 2) 93 25 16 00
Chief Executive Officer: David Robinson
■ Activities centre on the development and support of complex command, control and communications (C3) Systems.
Systems Support
Unit 5 - Silver State, 107 Carnarvon Street,
Silverwater, New South Wales 2128
☎(+61 2) 93 25 19 06 ⊜ (+61 2) 93 25 19 05
Combat Systems
PO Box 60, Garden Island, NSW 2000
☎(+61 2) 562 33 33 ⊜ (+61 2) 56 23 07
Command and Control Information Systems
20-22 Stirling Highway,
Nedlands, Western Australia 6009
☎(+61 89) 333 88 81 ⊜ (+61 89) 333 88 89
Technology Group Head Office
Level 6 - Westfield Towers, 100 William Street,
East Sydney, New South Wales 2011
☎(+61 2) 99 16 49 00 ⊜ (+61 2) 93 68 09 75
Chief Executive Officer: John Barclay
■ Minehunter construction and mine countermeasures, environmental remediation, ship maintenance management programmes, logistic support (defence), calibration and technical laboratory services for defence and commercial sectors.
Refit Planning and Logistic Support Services
Building 89, Level 3, Bay E, Garden Island, NSW 2000
☎(+61 2) 562 28 85 ⊜ (+61 2) 562 26 52
Environment
77 Parramatta Road, Siverwater, New South Wales 2141
☎(+61 2) 93 50 92 00 ⊜ (+61 2) 93 50 92 74
Services (Technical and Logistic)
Ground Floor - 77 Parramattta Road,
Silverwater, New South Wales 2141
☎(+61 2) 93 50 92 00 ⊜ (+61 2) 93 50 92 75

Contact Reference

Minehunter Construction Facility
Fitzroy Street, Carrington,
Newcastle, New South Wales 2294
☎ (+61 49) 24 50 00 ⊖ (+61 49) 24 50 05 [98]

60 Research Organizations, Institutes & Test Establishments

Army Technology & Engineering Agency 25-37100
Colan Products 75-03650

60-00300 Cypher Research Laboratories
58-60 Cinderalla Street,
Machans Beach, Queensland 4878
☎ (+61 70) 55 91 38 [E] cypher@internetnorth.com.au
⊠ 48314 aa ⊖ (+61 70) 55 02 74
♣ A$ 1,200,000 (1992/93) 👥 10
Research Director: John Rogge
■ Manufacture frequency agile HF radio equipment & encrypted data systems. Also repairs military and civil avionics equipment. Fibre optic modems and remote surveillance systems. [98]

60-00400 Defence Science and Technology Organisation (DSTO)
APW1-3-03 Anzac Park West Offices,
Canberra, ACT 2600
⊠ 62063 aa ⊖ (+61 6) 266 63 20
👥 2,400
Chief Defence Scientist: Dr R G Brabin-Smith
☎ (+61 6) 266 61 11
Administrative Officer, Corporate Communications: Christine Grecl
Science Policy Division
APW1-3-12 Anzac Park West Offices,
Canberra, ACT 2600
☎ (+61 6) 266 60 02 ⊠ 62063 aa ⊖ (+61 6) 266 63 20
Director General Science Policy Development: Dr R Spear
☎ (+61 6) 266 66 61
First Assistant Secretary: Dr D Sinnott
☎ (+61 6) 266 60 00
Science Corporate Management Branch
APW1-3-11 Anzac Park West Offices,
Canberra, ACT 2600
☎ (+61 6) 266 60 46 ⊠ 62063 aa ⊖ (+61 6) 266 63 20
Assistant Secretary: M Domney ☎ (+61 6) 266 60 46
Science, Industry and External Relations Branch (part of Science Policy Division)
APW 2-3-16 Anzac Park West Offices,
Canberra, ACT 2600
☎ (+61 6) 266 62 81 ⊖ (+61 6) 266 60 30
Corporate Communications Co-ordinator: G Thompson
☎ (+61 6) 266 60 27; 266 66 54 (fax)
Assistant Secretary: Ms K Nagle ☎ (+61 6) 266 62 81
Aeronautical and Maritime Research Laboratory
506 Lorimer Street, Fishermens Bend, Victoria 3207
☎ (+61 3) 96 26 70 00 ⊠ 82799 aa
⊖ (+61 3) 96 26 70 81
Director: Dr W Schofield ☎ (+61 3) 96 26 74 00; 96 26 70 84 (fax)
Chief of Air Operations Division: Dr C Guy
☎ (+61 3) 96 26 77 00; 96 26 70 84 (fax)
Chief of Airframes and Engines Division: Dr P Preston
☎ (+61 3) 96 26 76 66; 96 26 70 89 (fax)
Chief of Weapons Systems Division: Dr N Nandagopal
☎ (+61 8) 82 59 51 38; 82 59 70 80 (fax)
Chief of Maritime Platforms Division: Dr A McLachlan
☎ (+61 3) 96 26 82 00; 96 26 82 10 (fax)
Chief of Maritime Operations Division: Dr R Creaser
☎ (+61 8) 82 59 51 63; 82 59 61 04 (fax)
General Manager DSTO Business Office: P Quested
☎ (+61 3) 96 26 74 18; 96 26 70 91 (fax)
Manager Corporate Communications: B Humphreys
☎ (+61 3) 96 26 76 22; 96 26 71 33 (fax)
■ Aircraft structures, materials, flight mechanics and propulsion, avionics, aircraft systems and guided weapons. Survivability of ships and submarines, signature management, vulnerability, health and safety, underwater weapons and countermeasures, personnel protection, ballistic weapons and propulsion.
Electronics & Surveillance Research Laboratory
PO Box 1500, Salisbury, SA 5108
☎ (+61 8) 259 55 55 ⊠ 82799 aa ⊖ (+61 8) 259 65 57
Director: J.S. Allison ☎ (+61 8) 259 55 86
Chief of Communications Division: Dr A. Bedford
☎ (+61 8) 259 56 54
Chief of Electronic Warfare Division: Dr M. Lees
☎ (+61 8) 259 52 90
Chief of Information Technology Div.: N. Bryans
☎ (+61 8) 259 59 16
Chief of High Frequency Radar Division: Dr M. Golley
☎ (+61 8) 259 68 11
Chief of Microwave Radar Division: Dr D. Gambling
☎ (+61 8) 259 61 80
Chief of Land, Space & Optoelectronics: Dr R. Lough
☎ (+61 8) 259 51 84
Mgr DSTO Business Office: R. Reichelt
☎ (+61 8) 259 57 42
Corporate Communications Mgr: G. Smith
☎ (+61 8) 259 69 23
■ High level command & control, information technology, communications, signals intelligence & electronic warfare, combat systems & guided weapons. Electromagnetic surveillance & reconnaissance, microwave & high-frequency radar & opto-electronics. [98]

65 Industry, Trade & Armed Forces Associations

65-01000 Association of Australian Aerospace Industries (AAAI)
PO Box 817, 2601, Australian Capital Territory Canberra
214 Northbourne Ave,
Canberra, Australian Capital Territory 2600
☎ (+61 6) 247 22 33 ⊖ (+61 6) 248 61 57
Deputy Chairman: Ian Meibusch
Executive Director: Geoffrey J Mitchell
■ Provides a national aerospace industry grouping for the harmonisation of Australian government/industry and export development. Promotes Australian aerospace related industries and activities through exhibitions, promotions and other marketing activities.
Metal Trades Industry Association (MTIA)
PO Box 817, Canberra, ACT 2601
☎ (+61 6) 247 22 33 ⊖ (+61 6) 248 61 57
Executive Director: L W Purnell
Defence Manufacturers Council (DMC)
PO Box 817, Canberra, ACT 2601
☎ (+61 6) 247 22 33 ⊖ (+61 6) 248 61 57
Chairman: John Jeremy
Executive Director: L W Purnell
■ Represents companies dealing with the Commonwealth on matters of defence procurement. Membership of about 230 companies. [98]

65-01500 Australian Electrical & Electronic Manufacturers Association Ltd (AEEMA)
Suite 404 - Optus House, 10 Moore Street,
Canberra, ACT 2601
☎ (+61 6) 247 46 55 ⊖ (+61 6) 247 98 40
Chairman Defence Electronics Division: Peter Rowland
Chief Executive: Alex Gosman
Deputy Director: Bryan Douglas
■ Represents electrical, electronic and telecommunications industries. Specialist Defence Electronics Division and C3I export. → 75
Members of Defence Electronics Division:
AC & E Sales Pty Ltd
Amtron Australia
Andrew Australia
Aust Technical Publications
ADI Limited
British Aerospace Australia
Cablex Pty Ltd
CAE MRad Pty Ltd
Celsius Tech Australia
Codan Pty Ltd
Compucat Research Pty Ltd
CSC Australia Pty Ltd
CSSIP
Cypher Research
Defence Liaison Services
Design & Industry Pty Ltd
DSTO - Business Development Unit
Ericsson Defence Systems
GEC Marconi Systems Pty Ltd
Graeme H Wallace & Associates
Hewlett-Packard Aust Ltd
JNS Electronics Pty Ltd
Mayer Krieg & Co
Morris Productions Pty Ltd
NEC Australia Pty Ltd
Philips Electronics Australia Ltd
Philips Traffic & Engineering Systems
Raytheon International
R F I Industries Pty Ltd
Siemens Plessey Electronic
Standard Communications
Transfield Defence Systems
Utilux Pty Ltd
Vicom Australia
Vision Abell Pty Ltd
Wormald Technology [98]

65-01700 Australian Industry Defence Network (ASBDN)
PO Box 2538, North Parramatta 2151
☎ (+61 2) 96 30 49 88 ⊖ (+61 2) 96 30 46 41
President: Graham Biesse
Vice President: Kevin Bridger
Executive Director: Kevin Carroll
■ Non-profit organisation assisting small and medium sized businesses to obtain Department of Defence contracts either individually or from within a member network. [98]

65-02000 Defence Industries Council of Western Australia (DIC of WA)
190 Hay Street, PO Box 6209,
East Perth Western Australia 6892
☎ (+61 8) 93 65 75 12 [E] gatt@cciwa.zsn.au
⊠ 93609 chaco aa ⊖ (+61 8) 93 65 75 50
Chairman: Graham Mahony
Executive Officer: James G (Jim) Gatt
■ Council promoting the involvement of Western Australian industry as material and service suppliers to the Australian armed forces. Encourages defence self-sufficiency by developing a viable industry base and two coast defence policy. [98]

65-09000 Royal Australian Air Force Association
222 Pitt Street, Sydney, New South Wales 2000
☎ (+61 2) 92 64 60 07 ⊖ (+61 2) 92 61 88 48
National President: Air Cdre G G Michael
National Secretary: Peter C Alexander
■ Welfare Association. [98]

70 Financial Holding Groups, Parent Companies

70-16200 Orbital Engine Corporation Limited Registered Office
1 Whipple Street, Balcatta, Western Australia 6021
☎ (+61 9) 441 23 11 ⊖ (+61 9) 441 21 33
Chairman: Ross William Kelly
Chief Executive Officer: Kim Christopher Sclunke
Director: John William Beech
Director: Alan James Castleman
Director: Ian Roy Egan
Director: Robert Slade Forbes
Director: Kenneth Noel Johnsen
Director: John Richard Marshall
Director: John Grahame Young
Company Secretary: Ms Deana Cesari [98]

70-16420 Philips Electronics Australia Ltd
(† Philips Industries Holdings Ltd)
15 Blue St, PO Box 1138, North Sydney, NSW 2060
☎ (+61 2) 925 33 33 ✉ ARGENTA Sydney
⊠ 20165; 73670 philind aa ⊖ (+61 2) 929 47 84
Chairman & Managing Dir: J. Veeneklaas
☎ (+61 2) 99 25 32 31
Divisional Mgr Traffic & Engineering Systems: I.B. Buchanan ☎ (+61 2) 600 55 15
Marketing Exec.: O. Storchegger ☎ (+61 2) 600 54 10
Traffic & Engineering Mgr: M. Battaglia
☎ (+61 2) 600 54 15
Divisions:
Philips Traffic & Engineering Systems
Philips Components Pty Ltd [97]

74 Consortia, Joint Venture Companies

74-01500 Bliss-Fox Corporation Limited
Locked Bag N°5, Wetherill Park 2164
106 Woodpark Road, Smithfield, New South Wales 2164
☎ (+61 2) 96 32 58 88 ⊖ (+61 2) 96 81 54 64

Bliss-Fox Manufacturing Pty Ltd
(a manufacturing joint venture)
(a subsidiary of Bliss Corporation Limited)
106 Woodpark Road, Locked Mailbag N°5,
Wetherill Park, NSW 2164
☎ (+61 2) 96 32 58 88 ⊜ (+61 2) 96 81 54 64
115
Group General Manager: William Green
Manager, Ground Systems Equipment: N McPherson
Chief Engineer: R Abela
■ Design and manufacture of mining machinery, materials handling systems and special purpose vehicles. Metal stamping presses, coil processing equipment and GSE equipment. → 75

Cardiff Office
PO Box 335, Cardiff, New South Wales 2285
19 Aruma Close
☎ (+61 49) 56 75 33 ⊜ (+61 49) 56 79 96

Preston Office
8 Fink Street, Preston, Victoria 3072
☎ (+61 3) 94 80 42 22 ⊜ (+61 3) 94 80 45 82

Woodville Office
853-867, Woodville, South Australia 5011
☎ (+62 8) 93 47 41 11 ⊜ (+62 8) 93 47 03 96 [98]

75 Manufacturing Companies

75-00100 Abbey Equipment Pty Ltd
(† Simon Abbey Pty Ltd)
98 Fairbank Rd, PO Box 309 RMDC, Clayton, Vic 3169
☎ (+61 3) 551 88 66 ⊜ (+61 3) 551 57 23
♣ A$ 12 000 000 100
Managing Dir: I. Stewart
Sales & Marketing Mgr: Bernie J. Kiekebosch
■ Manufacture a complete range of vehicle mounted, self-propelled & trailer type access platforms having working heights from 31m. Also manufacture firefighting & rescue aerial equipment up to 31m high. [97]

75-00140 ACL Engine Parts
(a subs. of Automotive Components Ltd)
PO Box 243, Footscray, Vic 3011
☎ (+61 3) 93 17 18 88 ⊜ (+61 3) 93 17 31 98
♧ A$ 1 500 000 (1995) ♣ A$ 34 000 000 (1995) 231
General Mgr: Ewen Kilpatrick
Manufacturing Mgr: Alan Goodwin
Purchasing Mgr: Alex Simonic
Marketing Mgr: David Elliott
■ Manufacture of pistons and piston rings. [97]

75-00175 Actil
(a subs. of Textile Ind. Australia Ltd)
Actil Ave, Woodville, SA 5011
☎ (+61 8) 321 61 25 ⊜ (+61 8) 321 62 91
740
Production Mgr: Geoff Hodgson
Sales & Marketing Mgr: David Robinson
■ Textiles, woven fabrics - sheeting, towels, canvas, towelling, flat cloth dyed and/or printed. [97]

75-00180 Adam Technology
Enterprise Unit 1 - Hayman Road, Technology Park,
Bentley, Western Australia 6102
☎ (+61 8) 94 70 23 22 [E] adam@adamtech.com.au
⊜ (+61 8) 94 70 23 03
♧ A$ 2,000,000 (1997) ♣ A$ 4,000,000 (1997)
24 (1997) 750 m2
Managing Director: Yutaka Nagayasu
Operations and Technical Director: Hans Schoch
Sales and Marketing Manager: Michael E Goble-Garratt
Purchasing Manager: Robert Etienne
■ Design and manufacture of mapping equipment. [98]

75-00200 Advanced Hybrid Aircraft Ltd
10/6 Cameron Road, Mount Barker, SA 5251
☎ (+61 8) 398 36 20 ⊜ (+61 8) 398 36 20
Manager: Bruce N. Blake [98]

75-00310 Aerolyte Industries Pty Ltd
647-649 Waterdale Road, West Heidelberg, Vic 3081
☎ (+61 3) 94 59 27 70 ⊜ (+61 3) 94 59 43 18
♧ A$ 600 000 ♣ A$ 1 500 000 14 4 500 sq m
Managing Dir: D.J. Beresford-Long
☎ (+61 3) 94 59 87 68
Sales Mgr: P. Williams ☎ (+61 3) 94 59 58 43
■ Aluminium shipping & storage containers for technical electronics, audio-visual & camera equipment, computer software & hardware, communications & related equipment, spare parts & tool kits, rifles & guns. [97]

75-00330 Aeraustral Trading
(† Aeronautique Equipment)
Toowoomba Div.
16 Partridge Street, Toowoomba, Qld 4350
☎ (+61 76) 38 19 44 ⊜ (+61 76) 38 19 44
♧ A$ 175 000
Managing Dir: J.A. Bou
■ Safety equipment; pilot, aviation restraint systems, rescue equipment; parachutes; parachute harnesses; canvas products; flight clothing; body armour, helicopter cargo nets & all types of helicopter lifting slings. Individual equipment, webbing suspension, combat belts, field packs large & small, individual shelters, mosquito nets, ballistic vests, etc. [97]

75-00340 AeroSpace Technologies of Australia Ltd (ASTA)
(a division of Boeing Australia Ltd)
226 Lorimer Street, Private Bag No 4,
Port Melbourne, Victoria 3207
☎ (+61 3) 647 31 11
[W] http://www.asta.bra.boeing.com/
⊜ (+61 3) 646 22 53
♧ US$ 200,000,000 (1997) 800 (1997)
Chairman: Ken Peacock
General Manager ASTA Components: Mark Jenkins
Operations Director: David Main
Business Development, Planning Manager and Marketing Director: Ms Jo Staines
Purchasing Manager: Peter Higginbotham

ASTA Components
226 Lorimer Street, Fishermens Bend, Melbourne 3207
☎ (+61 3) 96 47 31 11
⊜ (+61 3) 96 45 34 24; 96 46 22 53
530
General Manager: Mark Jenkins
Human Resources Manager: Chris Sealie
Engineering Manager: Agnan Raghado
Program Manager: Al Tyler
Program Manager: Wal Mykytenko
Controller: John Johnson
■ Manufacture of aerospace components for Aerospatiale, Boeing and McDonnell Douglas including A330/340 floor support structure, main and central landing gear doors for B737-700, B747-400 Krueger flaps, B757 and B777 rudder, B767 Slat Wedges and F/A-18 C/D trailing edge flap.

ASTA Aircraft Systems Division
Private Bag N°4, Lara, Victoria 3212
☎ (+61 3) 52 27 94 44 ⊜ (+61 3) 52 82 33 45
250
■ Design, manufacture and maintenance of military aerospace equipment. [98]

75-00400 Air Affairs Australia Pty Ltd
PO Box 3089, Nowra, New South Wales 2541
☎ (+61 44) 23 67 55 [E] sales@airaffairs.com.au
⊜ (+61 44) 23 12 33
♧ A$ 5,000,000 14 (1997) 13,250 m2
Managing Director: Stephen A Lill
Operations and Engineering Director: Christopher J Sievers
Quality Assurance Director: Jan Ujma
Engineering Design Manager: Andrew Keleher
■ Manufacture of aerial tow targets under licence from Hayes Targets, USA. Operators of UMA, target drones and E/W systems.
® UK: Phoenix Simulation, Racal E/W
USA: Southwest Aerospace Inc. [98]

75-00450 Air International Transit
(† Sigma Industries Pty Ltd)
4 Bachell Ave, PO Box 209, Lidcombe, NSW 2141
☎ (+61 2) 330 71 00 ⊜ (+61 2) 330 71 99
250
Managing Dir: K.R. Allen
Divisional Mgr Defence Products: A.C. Milroy
Divisional Mgr Marine Products: G. Werdier
☎ (+61 3) 94 82 45 11
■ Military standard, railcar, locomotive & mining air conditioning equipment. Design & manufacture of equipment suitable for high temperature, contamination & vibration applications. Standard equipment available for hard track or tyred mobile plant & equipment, military maintenance shelters, communications shelters & field hospital installations.

Subsidiary:
Sigma Air Conditioning Inc. (USA)
Sigma Industries Pty Ltd (Hong Kong)

Queensland Branch
101-105 Randolph St, Rocklea, Qld 4106
☎ (+61 7) 274 19 22 ⊜ (+61 7) 274 01 46
Branch Mgr: T. Mamo

South Australia Branch
2 Rosberg Road, Wingfield, SA 5013
☎ (+61 8) 349 84 88 ⊜ (+61 8) 349 82 82
Branch Mgr: D. Goess

Western Australia Branch
3 Tipping Road, Kewdale, WA 6105
☎ (+61 9) 353 45 44 ⊜ (+61 9) 353 44 67
Branch Mgr: B. Cook

Air International Marine Projects
260A Darebin Road, Fairfield, Vic 3087
☎ (+61 3) 94 82 45 11 ⊜ (+61 3) 94 82 72 07
Branch Mgr: G. Werdier
→ 80

Victoria Branch
26 Industrial Drive, Sunshine, Vic 3020
☎ (+61 3) 93 64 84 38 ⊜ (+61 3) 93 64 88 51
♣ A$ 60 000 000 250
Managing Dir: K.R. Allen ☎ (+61 2) 330 71 00
Divisional Mgr Defence Products: A.C. Milroy
☎ (+61 2) 330 71 00
Divisional Mgr Marine Products: G.F. Werdier
☎ (+61 3) 482 45 11
■ Design & manufacture complete air conditioning systems including refrigeration & HVAC equipment for navy & commercial seagoing vessels, as well as army & other defence related projects. All navy & defence related equipment are shock & vibration proof to the Australian Defence Standard. Nationwide branch facilities provide system & product support including sales & service. [97]

75-01293 Anite Networks Ltd
(† Cray Communications Ltd)
12 Rodborough Road - Frenchs Forest,
Sydney, New South Wales 2086
☎ (+61 2) 99 28 00 00 ∅ (+61 1800) 65 27 29
[W] http://www.anite.com.au ⊜ (+61 2) 99 28 00 11
♧ A$ 30,000,000 (1997) ♣ US$ 40,000,000 (1997)
135 (1997)
Chief Executive Officer: David Arlidge
☎ (+61 2) 99 28 01 01
Marketing Director: Gordon Vick ☎ (+61 2) 99 28 04 51
Finance Director: Bernard Carbone
☎ (+61 2) 99 28 04 01
Purchasing Officer: Danny Lewington
Marketing Communications Manager: Mrs Fiona Martin
☎ (+61 2) 99 28 04 52
■ Network integration company working with third party suppliers manufacturing a comprehensive range of LAN, WAN and voice products.

Victoria Office
Level 3 - 553 St Kilda Road, Melbourne, Victoria 3004
☎ (+61 3) 95 29 76 44 ⊜ (+61 3) 95 29 83 03

Queensland Office
Suite 1 Ground Floor, North Tower,
John Oxley Centre 339 Coronation Dr,
Milton, Queensland 4064
☎ (+61 7) 32 17 55 99 ⊜ (+61 7) 32 17 53 22

Western Australia Office
Level 1 - 47 Burswood Road, Victoria Park,
Western Australia 6100
☎ (+61 9) 472 10 60 ⊜ (+61 9) 472 36 59

Australian Capital Territory
Unit 5, Kingston Manor, 10 Kennedy Street,
Australian Central Territory 2604
☎ (+61 6) 239 54 49 ⊜ (+61 6) 239 56 86

South Australia Office
39 Dequetteville Terrace,
Kent Town, South Australia 5067
☎ (+61 8) 364 17 44 ⊜ (+61 8) 364 35 38

Tasmania Office
19 Brisbane Street, Hobart, Tasmania 7000 [98]

Contact Reference

75-01420 Applied Measurement Australia Pty Ltd
PO Box 159, Oakleigh, Vic 3166
14 Dalgety St, Oakleigh, Vic 3166
☎ (+61 3) 95 68 05 88 ⊖ (+61 3) 95 68 07 87
♣ A$ 2 000 000 ⋔ 12
Managing Dir: John W. Ralton
Operations & Technical Dir: Edward B. Hall
Sales Mgr: Russ Monger
■ Industrial electronics, industrial & scientific measurement, transducers & systems. Test stands for military & civil aircraft components (actuators & controls). → 85 → 90
● Suite 6, 124 Rowe Street, Eastwood, NSW 2112
☎ (+61 2) 858 21 11 ⊖ (+61 2) 858 34 47 [97]

75-01500 Applied Power Australia Ltd
(a subsidiary of Applied Power Inc, USA)
PO Box 261, Regents Park, NSW 2143
☎ (+61 2) 97 43 89 88 ⊖ (+61 2) 97 43 86 48
⋔ 36 (1997)
Managing Director: A E Caldersmith
Sales and Marketing Director: R Coulson
■ Manufacture and supply of Enerpac powered maintenance lifting equipment, torque wrench and bolt tensioning equipment and rescue products. Also supply of computer furniture for LAN and WAN systems, laboratory equipment and electric workstations. [98]

75-01530 ASC Shipbuilding Pty Limited
(† ASC-Newcastle Pty Limited)
(a div. of Australian Submarine Corp.)
Old Punt Rd, Tomago, NSW 2322
☎ (+61 49) 64 80 71 ⊖ (+61 49) 64 83 16
Managing Dir: H. Ohff
Shipyard Operations Mgr: Peter Higgins
■ Builders of amphibious logistic ships, minehunters, patrol boats, landing craft, tugs & support vessels up to 10,000 tonnes. Construction in steel, aluminium, GRP or GRP foam sandwich.
● Old Punt Rd, Tomago, NSW 2322
☎ (+61 49) 64 80 71 ⊖ (+61 49) 64 83 16
■ Builder of the "Bay" class GRP foam sandwich catamaran minehunter, landing craft and rigid inflatable boats. [97]

75-01560 Associated Controls (Australia) Pty Ltd
PO Box 482, Matraville, New South Wales 2036
29 Smith Street, Hillsdale, New South Wales 2036
☎ (+61 2) 93 11 32 55 [E] michael@ascom.com.au
⊖ (+61 2) 93 11 37 79
♣ A$ 2,500,000 ⋔ 24 ⌂ 1,115 m2
Managing Director: Eric Mansfield
General Manager Marketing: R Mansfield
General Manager Manufacturing: G Henderson
Sales Manager: Michael Mansfield
Engineering Manager: B J Petrich
■ Electronic control devices, printed circuit boards, flow soldering and test capabilities, microprocessor design hardware and software. Mil-Spec rotary switches and relays. [98]

75-01570 A E Atherton & Sons Pty Ltd
PO Box 116, Northcote, Victoria 3070
364 Darebin Road, Thornbury 3071
☎ (+61 3) 94 97 25 00 ⊖ (+61 3) 94 97 11 90
♧ A$ 3,000,000 (1996) ♣ A$ 20,000,000 (1996)
⋔ 200 (1996)
Managing Director: S R Atherton
General Manager: M Atherton
■ Pressure vessels, sterilizers, heat exchangers, electric steam generators, commercial catering equipment, water stills, exhaust hoods, mechanical services, hospital equipment to all services on continuous basis. [98]

75-01710 Austral Specialised Vehicles Manufacturing Division
(a division of the Clifford Corporation)
PO Box 422, Virginia, Queensland 4014
356 Bilsen Road, Geebung, Queensland 4034
☎ (+61 7) 32 65 05 55 ⊠ 42256 auspec aa
⊖ (+61 7) 32 65 35 04
♧ A$ 36,600,000 ♣ A$ 15,000,000 ⋔ 377
Queensland State Manager: William M Raper
Contracts and Sales: Eddie Fabian
■ Design and manufacture of fire appliances and equipment. Crash, fire and rescue equipment and specialised vehicles. Firefighting vehicles, buses and coaches. [98]

75-01720 Australasian Steel Products Pty Ltd
19 Whitehall Street, Footscray, Victoria 3011
☎ (+61 3) 96 89 82 22 ⊖ (+61 3) 96 89 77 28
⋔ 200 (1997)
Managing Director: Leigh Morrison
General Manager: Chris James
■ Specialises in the design, development, manufacture and supply of hydraulic hose assemblies, hose couplings, fittings, adaptors, quick release couplings, flanges, specialised filter heads, bodies, pneumatic couplings and adaptors. Accredited to AS3901/ISO9001. Additional capabilities include CNC machining, CAD design and drafting. Facilities for repetition, mass production to close tolerances, high pressure hose testing, impulse testing and hydrostatic external cyclic presssure testing. [98]

75-01780 Australian AMP Pty Ltd
13 Hudson Ave, Castle Hill, NSW 2154
☎ (+61 2) 680 33 77 ⊠ 122195 ⊖ (+61 2) 634 61 88
Managing Dir: Bruno Palmano
Sales Dir: Christopher Bran
Sr Product Mgr: Zoran Markovic
■ Electrical & electronic connectors, cables & cable assemblies.
® USA: AMP [97]

75-01785 Australian Calibrating Services (A'sia) Pty Ltd
126 Oxford St, Collingwood, Vic 3066
☎ (+61 3) 94 17 56 88 ✆ (800) 33 41 83
⊖ (+61 3) 94 17 15 78
♣ A$ 3 000 000 ⋔ 30 ⌂ 8 000 sq m
Director: J.T. Kennon
Director: A.D. Ainsworth
■ Manufacturer of laboratory equipment, including force measuring systems, specialised load cells, strain rate indicator systems, universal testing machines.
Sydney
Commercial Measurement Laboratory
23/31 Wentworth St, Greenacre, NSW 2190
☎ (+61 2) 642 65 88 ⊖ (+61 2) 742 64 54
Adelaide
242 Glen Osmond Rd, Fullarton, SA 5063
☎ (+61 8) 338 13 05 ⊖ (+61 8) 338 15 91
Brisbane
3/36 Achievement Crescent, Acacia Ridge, Qld 4110
☎ (+61 7) 839 33 00 ⊖ (+61 7) 417 15 78 [97]

75-01830 Australian Hard Materials
PO Box 292C, Newcastle, New South Wales 2300
☎ (+61 49) 24 21 00 ⊖ (+61 49) 24 21 75
♧ A$ 15,000,000 ⋔ 50
Director: D MacDonald
Sales and Marketing Director: N Kennewell
■ Supply of tungsten alloy penetrators to the Department of Defence. Supplier of tungsten carbide products to various government departments. [98]

75-01910 Australian Submarine Corporation
GPO Box 2472, Adelaide, South Australia 5001
Mersey Road, Osborne, South Australia 5017
☎ (+61 8) 83 48 70 00 ⊖ (+61 8) 83 48 70 01
♣ A$ 300,000,000 ⋔ 800 ⌂ 20,000 m2
Managing Director: Hans J Ohff
Business Development Manager: R H Milton
■ Manufacture of submarines and surface vessels including ferries, tugs, naval ships, passenger cruise ships and special purpose vessels. Engineering and logistical support, independent engineering projects. [98]

75-01920 Australian Timken Pty Ltd
(a subsidiary of Timken Co, USA)
PO Box 240, Ballarat, Victoria 3350
☎ (+61 3) 53 38 10 55 ⊖ (+61 3) 53 38 11 86
⋔ 265 (1996)
Managing Director: J E Hoover
Sales and Marketing Director: P R Diener
Operations and Technical Director: P J Turner
Purchasing Manager: G C Dodds
Distribution and Customer Support Manager: Noel Trengove
■ Mechanical engineering - bearings, including tapered roller bearings to approx 150 mm diameter. [98]

75-01995 Avtronics (Australia) Pty Limited
(† Filmtronics (Australia) Pty Ltd)
33 Higginbotham Road,
Gladesville, New South Wales 2111
☎ (+61 2) 98 07 14 44 ⊖ (+61 2) 98 09 71 36
♧ A$ 58,000 ♣ A$ 3,000,000 ⋔ 15 ⌂ 10,000 sq ft
Chairman: F. Lord
Director: Jack Cairns
Director: Alan Fry
Dir & Service Mgr: John Weismantel
■ Manufacture of precision optical components. Supply of aircraft Ni-Cad batteries, borescopes, diesel and electric GPUs, frequency changers and rechargeable hi-intensity flashlights. → 85 [98]

75-02020 Barracuda Technologies Pty Ltd
(a subsidiary of Barracuda Tech AB, Sweden)
PO Box 6538, Baulkham Hills, NSW 2153
☎ (+61 2) 98 99 83 66 ⊖ (+61 2) 98 94 61 69
⋔ 10
Managing Director: Robert S Fuller
Director: H E Nilsson
■ Manufacture and supply of Barracuda camouflage systems. [98]

75-02050 Barrett Communications Pty Ltd
PO Box 198, Hamilton Hill, WA 6163
10 Port Kembla Drive, Bibra Lake, Western Australia 6163
☎ (+61 9) 94 34 17 00 ✆ (+61 800) 99 95 80
[W] http://www.iinet.net.au/~barrett
⊖ (+61 9) 94 18 67 57
♣ A$ 7,000,000 ⋔ 29
Managing Director: Tom Barrett
General Manager: Philip Bradshaw
National Sales Executive: Martin Johnson
■ Manufacturers specialising in turnkey radio communications: single-side band transceivers for land mobiles and marine services. HF fax and data systems, HF/telephone interconnect equipment. [98]

75-02425 Bisalloy Steels Pty Ltd
PO Box 1246, Unanderra, New South Wales 2526
☎ (+61 42) 72 04 44 ⊠ 29227 bisloy aa
⊖ (+61 42) 72 04 52
♣ A$ 50,000,000 ⋔ 55
Managing Director: Roy Rogers
Operations and Technical Director: Greg Leary
Controller: Michael Bermejo
Marketing and Sales Manager: Scott Hawkins
■ Manufacturer of abrasion resistant quenched and tempered steel plate, including armour plate, HY-80 and HY-100 steel plate. Induction bending of pipe and structural sections. [98]

75-02490 Blundstone Pty Ltd
88 Gormanston Road, PO Box 316,
Moonah, Tasmania 7009
☎ (+61 2) 82 72 30 00 [E] info@blundstone.com.au
[W] http://www.blundstone.com.au ⊖ (+61 2) 82 72 39 21
⋔ 300
Chief Executive Officer: T Stacey
Marketing Manager: B Smith
■ Manufacture defence uniform boots, safety boots and shoes. [98]

75-02500 Boeing Australia Limited Head Office - Sydney
(† Rockwell Australia Ltd)
(a subsidiary of The Boeing Company, USA)
Level 6 - 3 Thomas Holt Drive, PO Box 165,
North Ryde, New South Wales 2113
☎ (+61 2) 98 05 55 55 ⊖ (+61 2) 98 05 55 99
⋔ 1,800 (1997)
Chairman: Ken Peacock
Managing Director: David M Gray
General Manager Marketing: C W Walford
☎ (+61 6) 257 44 41
General Manager Operations and Maintenance: A Swansson
General Manager Communication and Combat Systems: L M Pears
General Manager Aircraft Systems: C D Giles

Manager, Public Affairs: A E Hill
Marketing Support Co-ordinator: Ms Leanne Vroonland
Melbourne Office
(a subsidiary of The Boeing Company, USA)
Corner Maroondah Highway/Sheep Road, PO Box 200,
Lilydale, Victoria 3140
☎ (+61 3) 97 27 97 11
⊖ (+61 3) 97 26 59 20; 27 13 57
Manufacturing Manager: D P Bowden
■ Design, development, manufacture and installation and support of defence, combat and communication systems.
Canberra Office
First Floor - 99 Northbourne Avenue, Canberra, ACT 2600
☎ (+61 6) 257 44 41 ⊖ (+61 6) 257 47 74
Manager: C W Walford [98]

75-02630 BP Solar Australia Pty Ltd
(a subsidiary of BP Solar Ltd, UK)
1/100 Old Pittwater Road, PO Box 519,
Brookvale, New South Wales 2100
☎ (+61 2) 99 38 51 11 ⊖ (+61 2) 99 39 15 48
♣ US$ 30,000,000 (1996) 👥 100 (1996)
Managing Director: D J Shah
Managing Director: K Brown
Operations Director: J Driver
Sales and Marketing Director: D Evans
Sales Manager: A C Stocken
■ Solar electric generating systems. Purpose designed portable solar modules for manpack radios and other battery charging applications. Custom-built solar plants for remote area power generation. Hazardous area approved (zone 1 and 2) solar panels for offshore platforms and fuel depots. → 90 [98]

75-02740 Brenco Thermal Coating Pty Ltd
(† Ceramtech Ltd)
PO Box 173, West Footscray, Victoria 3012
171-173 Fairbairn Road, Sunshine, Victoria 3020
☎ (+61 3) 93 10 19 77 [E] thermal@ozemail.com.au
⊖ (+61 3) 93 10 19 55
♣ A$ 5,000,000 (1997) 👥 20 (1997) ⌂ 2,500 m2
Managing Director: D Browne
Research and Development Engineer: A Sicinski
■ Production of plasma sprayed coatings and HVOF coatings for protection of parts from wear and corrosion. → 80 [98]

75-02746 Bridport-Quin Pty Ltd
PO Box 384, Port Adelaide, SA 5015
77-89 St. Vincent Street, Port Adelaide, SA 5015
☎ (+61 8) 47 12 77 ✆ (800) 81 13 03 ⌧ 88009 seas aa
⊖ (+61 8) 341 05 57
👥 14
Managing Dir: S.R. Quin
Sales Mgr: A.H. Quin
Purchasing Mgr: B. McBride
Wholesale Sales: R.J. King
■ Camouflage aircraft retaining barriers, cargo restraint nets, camouflage & concealment systems. [97]

75-02750 British Aerospace Australia Ltd Operating Headquarters (BAeA)
(a subsidiary of British Aerospace plc)
PO Box 180, Salisbury, South Australia 5108
14 Park Way - Technology Park,
The Levels, South Australia 5095
☎ (+61 8) 82 90 88 88 [E] enquiries@baea.com.au
⊖ (+61 8) 82 90 88 00; 82 90 88 33
♣ A$ 260,000,000,000 (1996) 👥 1,700 (1996)
Group Chief Executive: Robin Southwell
Managing Director: Barry Murphy
Head of International Marketing: Tony Ennis
☎ (+61 2) 93 58 29 00; 93 58 48 16 (fax)
Head of Corporate Affairs and Communications: Ms Julie Sheam
Head of Sales: Ron Tayles
Head of Marketing: Peter Rowland
Head of Business Improvement: John Devine
General Manager - Systems: Nigel Hennessy
General Manager - Manufacturing and Military Vehicles: Phil McAllan
General Manager - Support Programmes: Mike Brownhill
General Manager - Human Resources and Administration: Ms Delma Bowen
Commercial Director: Mark Hender
Finance Director: Dougie McTaggart
Director - Lead-In Fighter: Bill Gardner
Company Secretary and Legal Counsel: David Mahony
Marketing Manager: John Hamilton
Marketing Manager: Mark Melvin
Marketing Manager: Kelvin Kitto
Marketing Manager: Mike Fullgrabe
Marketing Manager: Keith Puyenbroek
Marketing Manager: Paul Fisher
Corporate Communications Officer: Ms Odette Elix
☎ (+61 8) 82 90 82 27 [E] oelix@baea.com.au
■ Aircraft systems integration, military vehicles, missiles, decoys and targets, electronic warfare, electro-optics, communication, command and control. Shorland series infantry mobility vehicles, Nulka hovering rocket (decoy), Forward Looking Infra Red Multi Role (FLIR), TIS heat-sensitive thermal imaging system, passive radar identification system (PRISM). Compact electronic support measures systems (ESM). Muzzle Velocity System; small lightweight tactical radar system. MODIR infra red jamming system, Parakeet high capacity land-based satellite communications system. ALR-2001 electronic support measures system, ALR-2002 radar warning receiver (replacement receiver designed specifically for RAAF F-111C aircraft).
Services: aircraft systems and avionics design and integration; design and integration of electronic and software systems; military vehicle assembly and manufacture; industrial and mechanical engineering; customer support for military; aerospace and commercial activities; electronics systems manufacture for defence and civil customers; civil and military facilities management.
Capabilities: aircraft systems integration; naval systems integration; military vehicles and systems; decoys, missiles and targets; surveillance; electronic warfare; secure communications; electro-optics; weapons systems engineering; C2 systems; command and control; simulation; radar; digital systems processing.
Research and development activities: Information systems and technologies; computer hardware and software; applied sciences and technologies; aerospace technologies; manufacturing and process technologies; mechanical, industrial, electrical and electronic engineering; electro-optic technologies.
Support Programs:
Aircraft Component Maintenance - Nowra, New South Wales
Australian Centre for Remote Sensing - Canberra, Australian Central Territory
Australian Centre for Remote Sensing - Alice Springs, Northern Territory
Canberra Deep Space Communications Complex - Tidinbilla, Australian Central Territory
Deeper Maintenance - Adelaide, South Australia
ESSO Bass Strait Oilfields - Longford, Victoria
Jervis Bay Range Facility - Jervis Bay, New South Wales
Mobile Laser Ranging Facility - Dongara, Western Australia
Mulloka Sonar System - Adelaide, South Australia
National Transmission Authority - South Australia and Northern Territory
Project Al-Yamamah - Kingdom of Saudi Arabia
Rapier Base Repair Facility - Adelaide, South Australia
Regional Hawk Support - South East Asia
Singapore Armed Forces Support - Rockhampton, Queensland
Sea King Helicopter Simulator - Nowra, New South Wales
TACAN Beacon System - Adelaide, South Australia
Canberra Office
5th Floor - 12 Moore Street, Canberra, ACT 2600
☎ (+61 6) 257 66 66 ⌧ 61793 baecan aa
⊖ (+61 62) 47 27 01
Marketing Dir: A.K. Jaugietis
International Marketing
Suite 2 - Level 3 Westfield Towers, 100 William Street, Sydney, New South Wales 2011
☎ (+61 2) 93 58 29 00 ⊖ (+61 2) 93 58 48 16 [98]

75-02760 M Brodribb Pty Ltd
68 Cape Street, Heidelberg, Victoria 3084
☎ (+61 3) 457 37 01 ⊖ (+61 3) 458 29 73
👥 18
Managing Director: L A Brodribb
Director: R A Brodribb
■ Electrical engineering including: industrial battery chargers, voltage stabilisers, mains isolating transformers, regulating transformers, uninterruptible power supplies, stepdown transformers and reactors. Cathodic protection equipment. [98]

75-02770 The Broken Hill Proprietary Co. Ltd (BHP)
BHP Tower 40th Floor - Bourke Place, 600 Bourke Street, Melbourne, Victoria 94104
☎ (+61 3) 96 09 33 33 [W] http://www.bhp.com.au
⊖ (+61 3) 96 09 30 15
♧ A$ 30,287,000,000 ♣ A$ 15,270,500,000
👥 55,000
Chairman: Brian T Loton
Chief Executive Officer: John B Prescott
Chief Financial Officer: G E Heeley
Group General Manager Assisting the Chief Executive Officer: Ian McMaster
General Manager, Investor Relations and Company Secretary: J R McGregor
Public Affairs Director: Ms Diane M Plummer
Human Resources: C W Mudge
Business units:
BHP Minerals
BHP Petroleum
BHP Steel
BHP Service Companies [98]

75-02800 David Brown Gear Industries Ltd
(a division of David Brown Group plc, UK)
13-19 Franklin Avenue, Bulli, New South Wales 2516
☎ (+61 42) 83 03 00 ⊖ (+61 42) 83 03 45
♣ A$ 40,000,000 👥 300
Managing Director Asia Pacific Division: J Madge
Managing Director: Tom Cross
Operations Director: M Hulme
Sales and Marketing Director Asia Pacific Division: B Robarge
Director Service and Sales: Bill Stevens
General Manager Finance and Administration: D Sirotich
General Manager Application Engineering and Development: T D Hall
Business Manager Couplings: Tony Green
Sales Support Team: Ms Elizabeth King
■ Computer design of gearing, development of complete transmission systems. Design and development of transmission test facilities. Mechanical transmissions systems and components. Service and rebuild facilities. CDG conveyor drives, propulsion and mill gears, wind turbine and agitator drives, pump drives and mobile equipment drives and transmissions. [98]

75-02805 David Brown John Welsh Custom Build Pty Ltd
(† John Welsh Pty Ltd)
12-20 Miller Street, PO Box 77,
East Brunswick, Victoria 3057
☎ (+61 3) 93 87 41 33 ✉ WELSH GEAR Melbourne
⊖ (+61 3) 93 87 06 39
♣ A$ 10,000,000 👥 65 ⌂ 6,000 m2
Managing Director: Thomas Cross
Operations Director: Greg Mautin
Director Service and Sales: Bill A Stevens
☎ (+61 41) 933 42 81 (mobile)
Purchasing Manager: Barry Wood
■ Manufacture of reduction gearboxes; mill type, turbine custom built, bevel; helical, gearing to 8.0 metre diameter flexible couplings. [98]

75-02810 Bruck Textiles plc
PO Box 4254, Sydney, New South Wales 2001
10-14 Waterloo Street,
Surrey Hills, New South Wales 2010
☎ (+61 2) 93 21 61 20 ⌧ 120452 aa
⊖ (+61 2) 93 21 62 90
♣ A$ 80,000,000 (1996) 👥 645 (1996)
Managing Director: John B Harvey
Operations Director: Max White
Sales and Marketing Director: Steven Arch
Manager Defence - Sales, Marketing and Export: A Eastwood ☎ (+61 3) 94 17 71 77
■ Fabric for combat (DPM) & general clothing, high technology applications (ballistic protection, flame retardants) & other applications (flag bunting, sleeping bags, wet & cold weather clothing, camouflage fabrics).
● PO Box 129, Abbotsford, Victoria 3067
159 Victoria Parade, Collingwood, Victoria 3066
☎ (+61 3) 94 17 71 77 ⊖ (+61 3) 94 19 74 20
♣ A$ 80,000,000 👥 645
Manager Defence - Sales, Marketing and Export: A Eastwood [98]

75-02815 BTR Aerospace Australia
(a div. of BTR Marketing Pty Ltd)
836 Mountain Highway, Locked Bag 39,
Bayswater, Victoria 3153
☎ (+61 3) 97 20 61 66 [E] btrsales@btraero.com.au
⊖ (+61 3) 97 20 49 96
♣ A$ 10,000,000 (1997) 👥 85 (1997) ⌂ 6,000 m2
General Manager: Dieter Glenk
Sales and Marketing Manager: Rick Young

Contact Reference

Manufacturing Manager: Graham Paull
Quality and Technical Manager: Richard Doig
Financial Controller: Michael Giurovich
Sales and Production Control Manager: Peter Ablett
■ Precision engineering, overhaul and testing of aircraft components for civil and defence applications. Aircraft, wheel and brake assemblies, brake control equipment, landing gear components, aircraft engine, hydraulic and fuel seals, gaskets, fuel cell mouldings and hose assemblies. Hydraulic, pneumatic and solenoid operated valves. Rubber components: O-rings, seals, gaskets, special mouldings and bonded to metal seals. Missile components and assemblies and launcher equipment.
® UK: Dunlop Aviation Division [98]

75-02830 BTR Environmental Pty Ltd
(† DCE Vokes Pty Ltd)
(a subsidiary of BTR plc, UK)
1/1 The Crescent, PO Box 325,
Kingsgrove, New South Wales 2208
☎ (+61 2) 91 50 03 01 [E] dcedust@localnet.com.au
[E] dceair@localnet.com.au
[W] http://www.aec.com.au/dcevokes/
[W] http://www.giga.aet.au/vokes ⊖ (+61 2) 95 02 34 74
👥 49 (1997) 🏠 4,000 m2
Managing Director: J M Budden
Marketing Officer: Ms Julie A Brown
■ Fabrication of equipment and filters for filtering applications.
Affiliates:
DCE Group Ltd, UK
Delta Neu, France
NAFCO, USA
Neotechnik, Germany
Vokes Air Filter Ltd, UK
Vokes Ltd, UK [98]

75-02860 Budenberg Gauges (Australia) Pty Ltd
(a subsidiary of Budenberg Gauge Co. Ltd, UK)
PO Box 337, Ringwood, Victoria 3134
☎ (+61 3) 98 70 06 66 ⊖ (+61 3) 98 79 50 55
⊕ A$ 700,000 (1997) ♣ A$ 2,500,000 (1997)
👥 18 (1997) 🏠 4,000 m2
Managing Director: R J Britton
■ Pressure and temperature measuring instruments; dead-weight testers. NATA certification of pressure gauges transducers and dead weight testers. Third party accredited ISO 9002. [98]

75-03100 Callington Haven Pty Ltd (CALHAVEN)
2 Euston Street, Rydalmere, New South Wales 2116
☎ (+61 2) 96 84 16 66 ⊖ (+61 2) 96 84 42 15
Managing Director: Nicholas J Schoulal
Sales and Marketing Director: J Hill
Operations and Technical Director: P Fernandes
Purchasing Manager: M Pryor
■ ISO9002 accredited and NATA (Defence) laboratory certified chemical manufacturer to retail, industry, defence and aviation. Enviro and safe solvents for cleaning and surface preparation. Methylene chloride and benzyl alcohol paint strippers. Detergent cleaners and degreasers. Immersion chemicals and weapon cleaning compounds. Deoxidising compounds. Aerosol and residual insecticides. [98]

75-03170 Capral Aluminium Ltd
(† Alcan Australia Ltd)
PO Box 12, Granville, NSW 2142
☎ (+61 2) 682 07 11 ✉ ALUMIN SYDNEY ▣ 120281
⊖ (+61 2) 682 07 77
⊕ A$ 178 000 000 (1995) ♣ A$ 831 000 000 (1995)
👥 2 915
Managing Dir: Ian F. Edwards
Sales & Marketing Dir: Bronte G. Duffy
☎ (+61 2) 682 09 81
Operations & Technical Dir: William J. Wright
☎ (+61 2) 682 00 18
Public Relations Officer: Paul Mills ☎ (+61 2) 682 07 66
Purchasing Mgr: David Tran ☎ (+61 2) 682 06 19
■ Production of aluminium ingot & semi-fabricated products including plate, sheet, foil & extrusions & a range of added value aluminium products for packaging, transport & construction. [97]

75-03230 Caterpillar Elphinstone Pty Ltd
(a joint venture between Elphinstone Pty Ltd and Caterpillar Inc)
2-8 Hopkinson Street, South Burnie, Tasmania 7320
☎ (+61 4) 64 31 23 11 ▣ 59164 aa
⊖ (+61 4) 64 31 23 11
👥 380 (1997) 🏠 15,000 m2
Managing Director: D B Elphinstone
Resources Manager and Director: P Robotham
Sales and Marketing Manager: S Poke
General Manager: A Ransley
Parts and Supply Manager: D Batt ☎ (+61 3) 64 31 93 63
■ Design and manufacture of underground mining equipment. [98]

75-03240 Caterpillar of Australia Ltd
(a subs. of Caterpiller Inc., USA)
Private Mail Bag 4, Tullamarine, Vic 3043
☎ (+61 3) 93 39 93 33 ▣ 30240 aa
⊖ (+61 3) 93 35 33 66
⊕ A$ 4 047 500 (1995) ♣ A$ 920 377 000 (1995)
👥 413
Managing Dir: R.D. Nitto
Sales Development Mgr: N.D. McKay
Manufacturing Resources Mgr: J.D. Donovan
Engines & Product Support Services Mgr: G.S. Winnett
Commercial Mgr Mining: D.J. Mills
Commercial Mgr Construction, Quarry & Forestry: R.L. Fraser
Commercial Mgr Growth Industries: J.M. Dynan
Business Services Mgr: M.C. Tong
Defence Accounts Mgr: David Purdie
■ Diesel powered earthmoving machines. Diesel engines for medium & heavy trucks, mobile plant, marine craft & power generation & lift trucks. [97]

75-03323 CelsiusTech Australia Pty Ltd
(a subsidiary of CelsiusTech AB, Sweden)
Endeavour House, Fourth Avenue, Technology Park, SA 5095
☎ (+61 8) 83 43 38 00 [E] xxxx@celsiustech.com.au
⊖ (+61 8) 83 43 37 77
Managing Director: Nick Hammond [98]

75-03400 Chubb Australia Ltd
Fire Div.
Private Bag 2, Mordialloc, Vic 3195
☎ (+61 3) 92 64 97 00 ⊖ (+61 3) 92 64 97 50
👥 350
General Mgr Chubb Fire: P. Sutherland
Commercial Mgr: G. Twaddell
Government & Defence Mgr: K. Whyte
■ Extinguishers: fire portable/mobile & fixed type. Water, foam, dry chemical powder, carbon dioxide, metal fire, fat fire, fire blankets, fire hose reels. Fire, smoke & thermal detectors. Beam & explosion suppression systems. Foam & foam generating equipment. Fire extinguishing chemicals, portable filling rigs, hydrostatic test rigs. FM200 systems. [97]

75-03450 Circle-C Transformers
(† Transformer Products Pty Ltd)
6 Grieve Close, West Gosford, NSW 2250
☎ (+61 43) 24 61 77 ⊖ (+61 43) 23 27 47
♣ A$ 1,900,000 (1997) 👥 19 (1997)
Director: Alan Cadwallader
Sales and Marketing Director: Harry Patel
■ Manufacturer of power and toroidal core power transformers. Current and voltage transformers to 33 kV rating. HV test equipment to 300 kV. [98]

75-03480 Climax Engineering Pty Ltd
14-18 Quinn Street, Preston, Victoria 3072
☎ (+61 3) 480 50 77 ⊖ (+61 3) 484 20 45
⊕ A$ 2,000,000 (1997) ♣ A$ 1,500,000 (1997)
👥 20 (1997) 🏠 1,700 m2
Managing Director: B F Stevens
Chief Engineer: I Pieroni
■ Mechanical engineering, machining and fabrication, including components for government aircraft factories and tent components for the Department of Defence. [98]

75-03600 CMI - Ballarat
(a division of Consolidated Manufacturing Industries)
903 Latrobe Street, PO Box 198, Ballarat, Victoria 3350
☎ (+61 3) 53 35 81 21 [E] ballarat@cmil.com.au
[W] http://www.cmil.com.au ⊖ (+61 3) 53 35 54 46
♣ A$ 14,000,000 (1997) 👥 150 (1997)
Executive Manager: Dan J Gallagher
Quality Control Manager: Paul Wager
Manufacturing Manager: Bill Brookes
National Sales and Export Manager: John Bramley
Materials Manager: R Verlin
■ Small hardware for defence including utensils (army can opener, spoon combination) clothing, arms componentry, specialist spring steel fastenings and foodstuffs. [98]

75-03620 Codan Pty Ltd
PO Box 96, Campbelltown, SA 5074
81 Graves Street, Newton, South Australia 5074
☎ (+61 8) 83 05 03 11 [E] tadcom@codan.com.au
[E] satcom@codan.com.au [W] http://www.codan.com.au
⊖ (+61 8) 83 05 04 11
👥 340 (1997) 🏠 10,000 m2
Chairman: J A Uhrig
Managing Director: M K Heard
Director: I J Bettison
Director: I B Wall
Radcom Marketing Manager: D J Segredos
Satcom General Manager: Y A Gobolos
■ Land and marine radio transceivers (HF). HF fax and data systems. Mobile systems and antennas. Satellite earth station transceivers and Ku and C Band suitable for simple and multicarrier systems and DAMA applications. Manufacture of printed circuit boards. [98]

75-03650 Colan Products
6 Woods Close, PO Box 809, Huntingwood, NSW 2148
☎ (+61 2) 96 72 78 88 ▣ 72697 ⊖ (+61 2) 96 72 76 66
👥 45
Managing Dir: G.R.J. Coghlan
General Mgr: Peter J. Love
National Sales & Export Mgr: Andrew J. Lucas
■ Engineered textiles of glass fibre, aramids, polyolefins & other synthetic fibres either coated, laminated or resin compatible. Ballistic panels, fireblankets, protective clothing, welding blankets, ballistic fabrics, reinforcement fabrics & tapes for marine, transport & container construction. Thermal insulation fabrics & safety fabrics. → 60 → 80 → 85 → 90 [97]

75-03750 Combat Simulation Systems (Aust) Pty Ltd
Lot 72, St Andrews Road,
Leppington, New South Wales 2171
☎ (+61 2) 96 06 40 11 ⊖ (+61 2) 96 06 40 83
⊕ A$ 150,000 (1997) ♣ A$ 2,000,000 (1997)
👥 14 (1997) 🏠 5,250 m2
Managing Director: Rod Morris
Sales and Marketing Director: Terry Dinneen
Purchasing Manager: Jenni Ellis
■ Pyrotechnics and military explosives for military training and operational applications. [98]

75-03860 Communitron (Australia) Pty Ltd
PO Box 1040, Meadowbank, New South Wales 2114
Unit 1/5 11 Mellor Street,
West Ryde, New South Wales 2114
☎ (+61 2) 98 09 70 77
[E] communitron@speednet.com.au
⊖ (+61 2) 98 07 37 17
⊕ A$ 10,000 (1997) ♣ A$ 7,500,000 (1996)
👥 45 (1997) 🏠 3,000 m2
Chairman: D Toth
General Manager: G K Weisske
Manufacturing Manager: George Ting
Senior Purchasing Officer: Dimitry Broyda
Sales and Customer Support: Peter Groth
■ Manufacture of electronics and telecommunications including small systems, PCB assemblies, cable assemblies, wirewrapping. Test capabilities.
Production of Datagate RS 232 port switch and DPE 2120 data protection equipment.
PCB assembly, mechanical assemblies, cable harnesses, in-circuit testing, functional and environmental testing, turnkey projects, project management and production engineering. [98]

75-03870 Compair (Australasia) Ltd
(a subsidiary of Siebe plc, UK)
34-40 Ricketts Road, PO Box 126,
Mount Waverley, Victoria 3149
☎ (+61 3) 95 44 07 77 ✉ COMPAIRAUS Melbourne
⊖ (+61 3) 95 62 88 94

♧ A$ 36,000,000 ♣ A$ 80,000,000 (1991) 👥 330
Managing Director: R L Bridgewater
Finance Director: I A Ashman
Sales and Marketing Director: D Daniels
Supply Manager: W Sunbloom
Regional Manager Defence: F J Morris
■ Warship compressors, high pressure compressors, gases, breathing air sets, diving compressors, rotary compressors and exhausters, centrifugal compressors and boosters, industrial compressors, reciprocating, screw and vane, portable compressors, pneumatic tools, drills, screwdrivers, impact wrenches, breakers, pneumatic control equipment, dryers. → 80 [98]

75-03925 Consolidated Manufacturing Industries Limited
(† CMI Repetition Engineering)
9 Chalmers Crescent, Mascot, NSW 2020
☎ (+61 2) 669 59 66 ⊖ (+61 2) 669 34 16
♣ A$ 2 200 000 👥 33 🏠 12 000 sq ft
Exec. Chairman: Max Hofmeister
Manager: George Terawsky
Chief Quality Controller: D. Parmley
■ High volume turning of precision components & fasteners up to bar capacity of 60mm dia. [97]

75-03954 Crane Aluminium Extrusion
(† Alumex)
(a division of Crane Bufield Metals plc, Australia)
PO Box 53, Huntingdale, Victoria 3166
Edward Street, Huntingdale, Victoria 3166
☎ (+61 3) 95 50 62 00 ⊖ (+61 3) 95 63 15 24
👥 200
Managing Director: Don Wilson
Operations Director: Peter Bradhurst
Sales and Marketing Director: Peter Darwell
Purchasing Manager: Chris Alway
Secretary: Ms Lauren Wild
■ Aluminium extrusions and rods. [98]

75-03963 Crouse Hinds Australia Pty Ltd
Block X 391 Park Road, PO Box 257,
Regents Park, NSW 2143
☎ (+61 2) 97 43 70 00 [E] sales@crouse-hinds.com.au
⊖ (+61 2) 97 43 70 69
♧ A$ 9,600,000 (1997) 👥 78 (1997) 🏠 4,422 m2
Operations Manager: S Hood
Marketing Manager: B Mohan
Company Secretary: Warren T Jurd
Factory Manager: D Oehlman
■ Manufacture of electrical equipment for hazardous areas, with specialised electrical equipment for marine and adverse environments. [98]

75-04020 Daewoo Engineering & Construction Australia Ltd
(a subsidiary of Daewoo Corporation, South Korea)
Level 5, 52 Phillip Street, Sydney, New South Wales 2000
☎ (+61 2) 251 77 99 ⊖ (+61 2) 251 58 00 [98]

75-04220 Davies Craig Pty Ltd
PO Box 176, Port Melbourne, Victoria 3207
Corner Beacconsfield Parade and Esplanade East
☎ (+61 3) 646 30 51 [E] dcfans@ozemail.com.au
[W] http://www.ozemail.com.ao/~dcfans
⊖ (+61 3) 646 26 32
♧ A$ 3,000,000 (1997) 👥 15 (1997) 🏠 11,000 m2
Managing Director: Richard D Davies
Sales and Marketing Director: Robert Thompson
Operations and Technical Director: Derek Johns
Purchasing Manager: W Hidson
■ Thermatic engine cooling fans, automotive air-conditioner condenser fans. [98]

75-04450 Detroit Diesel-Allison Australia Pty Ltd
Lockeg Bag 98, Moorebank, New South Wales 2170
78-82 Riverside Road, Chipping Norton, New South Wales
☎ (+61 2) 97 94 26 00 [E] sydneyddaa@ddaa.com.au
⊖ (+61 2) 97 94 26 77
♧ A$ 13,500,000 ♣ A$ 152,000,000 (1997)
👥 620 (1997) 🏠 36,930 m2
Managing Director: Leo Tutt
Managing Director: Don Cox
Operations Manager: James Tutt
Sales and Marketing Manager: Bruce Collins
☎ (+61 8) 83 00 81 00
National Product Support Manager: Andrew Culley
☎ (+61 8) 83 00 81 00
Defence Products Manager: Michael Evans
☎ (+61 2) 97 94 26 03
■ Manufacture of power generators and components. → 90
® Germany: MTU Diesel Engines
UK: Perkins Engine Company
USA: Detroit Diesel Corporation, General Motors, Allison Transmission Div, General Motors, Electro Motive Division

Gepps Cross Branch
569 Grand Junction Road,
Gepps Cross, South Australia 5094
☎ (+61 8) 83 00 81 00 ⊖ (+61 8) 83 00 81 10

Willnellie Branch
9 km Stuart Highway, Willnellie, NT 0820
☎ (+61 8) 89 84 31 55 ⊖ (+61 8) 89 84 43 02

Bassendean Branch
174 Railway Parade, Bassendean, WA 6054
☎ (+61 8) 92 73 77 77 ⊖ (+61 8) 92 73 77 20

Altona North Branch
488 Blackshaws Road, Altona North
☎ (+61 3) 97 03 14 44 ⊖ (+61 3) 97 69 33 11

Richlands Branch
147 Archerfield Road, Richlands, Queensland 4077 [98]

75-04580 The Dindima Group Pty Ltd
10 Argent Place, Ringwood, Victoria 3134
☎ (+61 3) 873 44 55 [E] dindima@ozemail.com.au
⊖ (+61 3) 873 47 49
Managing Director: Peter A R Wingett
Marketing Director: Peng Chew
■ Video image processing systems for scientific, industrial and security applications including temporal filter/image store systems with real time noise reduction capability, real time image capture and slow scan transmission systems, motion detecting image capturing equipment, multi-satellite standard scan converters (to accept multiple computer graphics raster and composite TV video signals of disparate standards) and to "window" them into one raster standard video output, interactive image and data analysis systems for scientific and industrial applications including remote sensing. Object recognition "machine vision" systems, video equipment and instrumentation. Variable bit rate real-time TV signal compression equipment/software. [98]

75-04600 D.M. Hull & Co. Pty Ltd
(a subs. of Delta Group Aust. Pty Ltd)
PO Box 587, Cheltenham, Vic 3192
☎ (+61 3) 553 16 99 ⊖ (+61 3) 553 23 64
♧ A$ 250 000 (1995) ♣ A$ 7 155 000 (1995) 👥 76
Managing Dir: F.W. Fewster
Financial Dir: E.C. Wallace
Sales Mgr: P. Rafferty
■ Precision machined components & sub-assemblies. [97]

75-04720 Dorman Equipment Pty Ltd
(a subsidiary of Inductoheat Pty Ltd)
PO Box 170, Seaford, Victoria 3198
☎ (+61 3) 786 66 22 [E] inducto@ozemail.om.au
[W] http://www.ozemail.com.au/~inducto
☒ 32136 inducto aa ⊖ (+61 3) 786 85 22
👥 87 (1996)
Managing Director: J K Robson
Sales Director: P W Burren
Product Manager: G W Fielding
■ Standard & special purpose resistance welding machines & associated electrical controls, jigs & fixtures. Hot dimpling machines. [98]

75-04750 Downee RMD
(† RMD Press Co Pty Ltd)
6-10 Market Road, Sunshine, Vic 3020
☎ (+61 3) 93 64 82 88 ⊖ (+61 3) 93 11 20 30
👥 100 🏠 1 Hectare
Managing Dir: K.J. Doidge
Sales & Marketing Dir: P. Cummins
Purchasing Mgr: M. Cooper
■ Metal pressing & associated assembly, including fuel tanks & stainless steel mufflers for the Department of Defence. Fire extinguisher containers. Pressure vessel ends. [97]

75-04930 Dunlop & Olympic Tyres South Pacific Tyres
(† Pacific Dunlop Ltd)
Private Bag 12, Campbellfield, Victoria 3061
Hume Highway, Somerton, Victoria 3062
☎ (+61 3) 93 05 02 22 ☒ 36019 ⊖ (+61 3) 93 05 34 79
👥 1,726
Managing Director: J M Gurrieri
General Manager: C Fritz
Marketing Manager: Kevin Halpin
Corporate Affairs and Marketing Services Manager: Chris Meehan
■ Retread tyres for civil and military jet aircraft to international standards.

Industrial Division
838 Mountain Highway, PO Box 41,
Bayswater, Victoria 3153
☎ (+61 3) 721 61 11 ☒ 3643 ⊖ (+61 3) 720 42 24
👥 350
General Manager: R Robinson
■ Production of rubber products.

Somerton Office
Hume Highway, Private Bag 12, Somerton, Victoria 3062
☎ (+61 3) 305 02 22 ✉ AUTOTYRES ☒ 31486
⊖ (+61 3) 305 36 02
👥 2 335
Managing Director: J M Gurrieri
Australian Marketing Manager: Mark Stewart
● Cross Street, PO Box 1, West Footscray, Vic 3012
☎ (+61 3) 680 37 77 ✉ OLYMPTYRE ☒ 34588
⊖ (+61 3) 689 56 28
Technical Mgr: Alan Hepworth [98]

75-04980 The Duriron Group Pty Ltd
(† Valtek Australia Pty Ltd)
(a subsidiary of The Duriron Company Inc, USA)
14 Dalmore Drive, Scoresby, Victoria 3179
☎ (+61 3) 764 85 22 ⊖ (+61 3) 764 00 13
👥 60 (1997)
Managing Director: Robert Thompson
Operations Director: Henryk Plucinski
Sales Manager: Zbigniew Tarka
Purchasing Manager: Peter Hooper
Administrator: Ms Olga Fraser
■ Valves, actuators, pumps, filters and related products. [98]

75-05000 Eagle Aircraft
PO Box 586, Fremantle, WA 6160
Clarence Industrial Estate, Cockburn Rd,
Henderson, WA 6106
☎ (+61 9) 410 10 77 ⊖ (+61 9) 410 24 30
👥 75 🏠 10 000 sq m
CEO: Nawi Bin Abdullah
COO: Manshor Ghafar
Sales & Marketing Dir: Ron Scherpenzeel
Purchasing Mgr: Terry Ariss
■ Eagle aircraft; composite industries. [97]

75-05260 Ebs-Ray Pumps Pty Ltd
628 Pittwater Rd, Brookvale, NSW 2100
☎ (+61 2) 99 05 02 34 ⊖ (+61 2) 99 38 38 25
👥 70
Managing Dir: W.A. Ebasry
Sales & Marketing Mgr: Ian Graham
Purchasing Mgr: Peter Goulding
■ Production of a comprehensive range of gear & vane type pumps, regenerative turbine & vacuum pumps; contract machine work. [97]

75-05350 EG&G Sealol Pty Ltd
(a subs. of EG&G Sealol Inc., USA)
18 Gertrude St, Arncliffe, NSW 2205
☎ (+61 2) 597 24 77 ☒ 22624 ⊖ (+61 2) 597 78 01
General Mgr: Mike Lazarevski
■ Manufacture, design, installation and servicing of mechanical seals for submarine, shipping, aircraft and general industry. Specialised valves and bellows devices for ancillary equipment. → 80 [97]

75-05520 Elastic Webbing (Aust) Ltd
390 Settlement Rd, Thomastown, Vic 3074
☎ (+61 3) 94 64 08 33 ☒ 39412 aa
⊖ (+61 3) 94 65 65 31
♣ A$ 7,500,000 👥 5,664
Managing Dir: Kerry Wallace
Sales Mgr: Harry Baker

Contact Reference

■ Textiles, including elastics, blood filters, kevlar bearing tape, industrial, automotive, seat belts & military specification tapes & webbings. [97]

75-05630 Elf Atochem
(a division of Clyde Industries Limited)
PO Box 6509, Silverwater, New South Wales 2128
Gate 5 - Factory Street, Granville, New South Wales 2142
☎ (+61 2) 682 14 00 ✆ (+61 800) 63 93 76
⊖ (+61 2) 682 67 71
Sales and Marketing Director: Michael Renehan
Product Manager: Anthony Symons
☎ (+61 2) 99 12 27 22
Production Manager: Anup Shah
Aviation Manager: Ben Edwards ☎ (+61 2) 99 12 29 13
■ Design and development of chemicals for aircraft and industrial applications. Original equipment manufacturers. [98]

75-05680 Email Air Handling
(† Email Westinghouse Pty Ltd)
30 Auburn Rd, Regents Park, NSW 2143
☎ (+61 2) 644 26 41 ▣ 24758 ⊖ (+61 2) 644 74 45
240 5 000 sq m
General Mgr: R.J. Nella
Mgr Contamination Control Lab.: Chris Christoforidis
■ Precision machining sheet metal fabrication, air conditioning equipment. HEPA air filters. Laminar flow safety cabinets. Clean rooms.
® USA: Flanders Filters Inc. [97]

75-05740 G P Embelton & Co. Pty Ltd
(a division of Embelton Ltd)
PO Box 207, Coburg, Victoria 3058
☎ (+61 3) 93 50 28 11 ⊖ (+61 3) 93 50 17 05
♣ A$ 10,000,000 (1996) 60
Chairman and Managing Director: G R Embelton
Group Commercial Manager: D J Taylor
■ Pads, energy dissipating, aerial delivery, honeycomb "Log Type", manufactured in accordance with Australian Army specification for use in aerial delivery of munitions and supplies. Ammunition, gun plugs for naval gun mounts 105-130 mm. [98]

75-05760 Emco Wheaton Australia Pty Ltd
(a Syltone company)
100 Station Street, Nunawading, Victoria 3131
☎ (+61 3) 98 74 29 00 ✆ (+61 800) 62 62 65
⊖ (+61 3) 98 73 24 22
♣ A$ 5,000,000 (1996) 21
Managing Director: James R Quinn
Operations Manager: A Reid
Sales and Marketing Manager: R Mackenzie
■ Fluid handling equipment. Marine loading arms, fuel couplers and nozzles. Aviation and vehicle refuelling. → 90 [98]

75-05820 Ericsson Defence Systems Pty Ltd
(a subs. of Ericsson Australia Pty Ltd)
202 Bell St, Preston, Vic 3072
☎ (+61 3) 92 30 33 60 ⊖ (+61 3) 92 30 33 56
♧ A$ 1 000 000 20 1 000 m2
Managing Dir: John T. Scanlon
■ Marketing of communications, radars, electro-optics & military electronic systems. → 90
® Sweden: Ericsson Microwave Systems AB, L.M.Ericsson [97]

75-06360 Fesca Qualos Gears
Hi-Tech Drives plc
12 Steane Street, Fairfield, Victoria 3078
☎ (+61 3) 489 78 33 Ⓔ sales@fpg.com.au
⊖ (+61 3) 489 29 04
♧ A$ 4,000,000 ♣ A$ 5,500,000 (1997) 50 (1997)
3,000 m2
General Manager: Bert Van Dijk
Sales and Marketing Manager: Jim Thomson
■ Spur and helical machine cut and precision ground gears, spiral and straight bevel gears; worms and worm wheels, gear componentry. CAA approved and defence approved. Specialised gear boxes (planetary, reduction, splitter), gear box refurbishment, replacement gears, in-house case carburising, standard gears and gear racks. [98]

75-06400 Fibreoptic Lightguides
7/109-111 Hunter Street,
Hornsby, New South Wales 2077
☎ (+61 2) 94 82 32 16 ⊖ (+61 2) 94 82 32 41
4
Managing Director: Ms Elizabeth Cowley
Sales and Marketing Director: John Cowley
■ Manufacture of fibreoptic cables for light and sensor applications including microscopy; endoscopy; industrial sensors and control. [98]

75-06835 H.I. Fraser Pty Ltd
PO Box 885, Mona Vale, NSW 2103
Unit 6/5 Vuko Place, Warriewood, NSW 2102
☎ (+61 2) 99 70 73 22 ⊖ (+61 2) 99 13 72 07
♧ A$ 500 000 30
Managing Dir: Mark Kirkby
■ Manufacture of high pressure pipe couplings & fittings & custom items on CNC machining equipment. Service & repair of Hale Hamilton & other gas & liquid control valves. Clean room to AS1386 class 3500. Accredited to ISO 9000.
® UK: Hale Hamilton (Valves) Ltd, Lindapter Bolt Adaptors, Orseal Valves [97]

75-06840 Frederick Duffield Pty Ltd
15-17 Beaumont Road, PO Box 38,
Mount Ku-ring-gai, NSW 2080
☎ (+61 2) 94 57 96 88 ⊖ (+61 2) 94 57 84 59
210
Managing Director: F R Duffield
General Manager: S F Tonks
Sales Manager: A R McLean
Product Development Manager: Ross Bruwel
■ Manufacture of components for pneumatic and hydraulic applications specialising in high pressure hose and fittings, associated assembling machines and test equipment. Code matched hose and fittings, Stocksaver swaged system, wear resistant two wire braid hose and pressure gauges. [98]

75-07000 GAAM Fire & Rescue
(† Wajax (Australasia) Pty Ltd)
(a subsidiary of Tyco International Pty Ltd, Australia)
29 Temple Drive, Thomastown, Victoria 3074
☎ (+61 3) 94 66 12 44 ✉ PUMPSFIRE Melbourne
⊖ (+61 3) 94 66 47 43
♧ A$ 400,000 (1997) ♣ A$ 8,000,000 (1997)
18 (1997) 1,200 m2
General Manager: Barry J Kincade
Advertising Manager: Brian Campisi
Sales and Marketing Manager: I M Price
Purchasing Manager: A Mitchell
■ Vehicle mounted fire fighting pumps. Diesel and petrol powered fire fighting pump sets. Fire fighting hoses, fire helmets, nozzles. Rescue equipment and air bags. [98]

75-07350 GEC Electronics Division
Unit 1, 38 South Street, Rydalmere, NSW 2116
☎ (+61 2) 96 38 18 88 ⊖ (+61 2) 96 38 17 98
Chief Exec.: G. Hole
Representing:
EEV Ltd, UK [98]

75-07520 General Motors-Holden's Automotive Ltd
(a subs. of General Motors Corp., USA)
GPO Box 1714, Melbourne, Vic 3001
241 Salmon Street,
Fisherman's Bend, Port Melbourne 3207
☎ (+61 3) 96 47 11 11 ▣ 30168 ⊖ (+61 3) 96 47 25 50
♣ A$ 2 750 000 000 5 600
Managing Dir: W.J. Hamel
Sales & Marketing Dir: K.E. Wale
Operations Dir: R.M. Keane
Purchasing Mgr: R.A. Burkhart
■ Automotive components including plastics-injection mouldings 50-9500 gm; sheet metal large stampings & assemblies; hardware automotive door frames, small pressings. Recent contracts include tooling of F18 & frigate propeller blades. [97]

75-07600 Global Lightning Technologies Pty Ltd
(† Critec Pty Ltd)
GPO Box 536, Hobart, Tasmania 7001
Technopark, Dowsings Point, Hobart, Tasmania 7010
☎ (+61 3) 62 37 32 00 Ⓔ info@gtt.com.au
Ⓦ http://www.gtt.com.au ▣ 58260 crtec aa
⊖ (+61 3) 62 73 03 99
♧ A$ 8,000,000 (1997) ♣ A$ 20,000,000 (1996/1997)
135 (1997) 3,800 m2
Chairman: J R Gumley
Managing Director: Dr S J Gumley
Group Sales Director: W D Temple
National Sales Manager: R Chapman
Production Manager: R Eveleigh
Technical Marketing Manager: Dr A J Surtees
Principal Engineer: P Hollingsworth
Principal Engineer: P Jones
Principal Engineer: P Johnson
Marketing Co-ordination Officer: C S Stopp
Marketing Co-ordination Officer: Ms Elizabeth Gibbs
■ Design, consulting, and manufacture of all types of surge and transient protection devices, line conditioners, mini ups; surge reduction filters, filters Movtec Shunt surge diverters, local area network protectors. Nuclear electromagnetic pulse software (NEMP) research and products. Contract manufacturing of electronic PCB based products. Consulting and field survey work on all aspects of lightning, surge and NEMP protection. Lightning protection for structures, systems and equipment. → 80

New South Wales Branch Office
PO Box 6245, Baulkham Hills, NSW 2153
33 Brookhollow Avenue
☎ (+61 2) 98 99 31 11 ⊖ (+61 2) 98 99 77 11

Victoria Branch Office
9 Lillian Street, Bulleen, Victoria 3105
☎ (+61 3) 98 52 16 55 ⊖ (+61 3) 98 52 11 19

Australian Capital Territory Office
Suite 16/71 Constitution Avenue, Campbell, ACT 2612
☎ (+61 6) 257 30 55 ⊖ (+61 6) 257 31 27

South Australia Branch Office
PO Box 171, Stepney, South Australia 5069
239 Magill Road, Maylands, South Australia
☎ (+61 6) 83 66 65 55 ⊖ (+61 6) 83 66 65 56

Western Australia Branch Office
Unit 4 - 129 Welshpool Road,
Welshpool, Western Australia 6106
☎ (+61 9) 358 12 33 ⊖ (+61 9) 358 14 04

SCA Distributors Pty Ltd
178 Wecker Road, PO Box 2305,
Mansfield, Queensland 4122
☎ (+61 7) 38 49 50 77 ⊖ (+61 7) 38 49 70 35

Australian Power Protection
9/99 Reichardt Road, PO Box 2305, Winnellie, NT 0821 [98]

75-07700 GNB Australia Limited
(† Pacific Dunlop Batteries)
(a division of Dunlop & Pacific Tyres, Australia)
PO Box 141, Bankstown, New South Wales 2200
55 Bryant Street, Padstow, New South Wales 2211
☎ (+61 2) 774 05 00 ⊖ (+61 2) 774 29 66
280
General Manager: J E Hodgkinson
Works Manager: G R Green
Industrial Marketing: G F Scott
■ Electrical batteries.

Elizabeth Plant
Peachy Road and Barfield Crescent,
Elizabeth, South Australia 5112
■ Lead acid batteries including main propulsion batteries for diesel electric submarines, charging systems and UPS equipment.

● 395 Toorango Road, Hawthorne East, Victoria 3123
☎ (+61 3) 270 01 00 ⊖ (+61 3) 270 02 99
General Mgr: M.G. McColl [98]

75-07760 A Goninan & Co. Ltd
(a subsidiary of Howard Smith Limited, Australia)
Broadmeadow Road, PO Box 21,
Broadmeadow, New South Wales 2292
☎ (+61 49) 23 50 00 ▣ 28061 ⊖ (+61 49) 23 50 01
♣ A$ 350,000,000 (1996) 2,200 (1996)
Chief Executive: John G Fitzgerald
General Manager Mining and Industrial Services: Philip Blanch
Business Development Manager: Mike Peachey
Purchasing Manager: Bill Stuart
■ Manufacture and general engineering covering design, fabrication, machining, assembling and installation. Products include tracked armoured vehicles, locomotives, communication systems and equipment. → 80 [98]

75-07885 GUD Manufacturing Co. Pty Ltd
(a division of GUD Holdings Limited)
PO Box 62, Sunshine, Victoria 3020
☎ (+61 3) 92 43 33 33 ✆ (+61 1800) 80 45 41
[E] apaj@ozemail.com.au ⊖ (+61 3) 92 43 33 66
♣ A$ 60,200,000 (1997) 👥 650 (1997)
Managing Director: Eric Rainsford
Operations Manager: Trevor Fell
National Marketing and Sales Manager: Colin E Tidball
Purchasing Manager: Roger Aske
■ Manufacture of automatic filters, heavy duty air filters, fuel pumps and fuel pump repair parts. [98]

75-08110 Haden Drysys
(a div.of Haden Engineering Pty Ltd)
PO Box 222, Moorabbin, Vic 3189
11-15 Christensen St, Cheltenham, Vic
☎ (+61 3) 95 55 32 22 ⊖ (+61 3) 95 53 31 31
♧ A$ 3 000 000 (1995) ♣ A$ 125 000 000 (1995)
👥 550
Director: R.J. Simcock
General Mgr: G. Whittaker
Business Development Mgr: Ian McLeod
Sales Mgr: R. Lees
■ Complete plants for metal treatment & paint finishing, fume incinerators & heat recuperators, pollution control systems, paint spray booths, drying ovens, spray painting equipment (automatic & manual), materials handling equipment, heating equipment, industrial washing equipment & parts washers.
Affiliates:
Haden Drysys Int'l Ltd (UK)
Haden Schweitzer Corp. (USA) [97]

75-08200 Hawker de Havilland Limited
(a subsidiary of BTR plc, UK)
361 Milperra Road, PO Box 30,
Bankstown, New South Wales 2200
☎ (+61 2) 97 72 82 84 ☒ 20719 ⊖ (+61 2) 97 92 36 04
👥 1,000 (1997)
Director: C R Burns
Director: J A N Burggraaff
Director: H E Walters
Marketing Manager: F J Makin
New Projects Manager: K C Sayers
☎ (+61 2) 97 72 87 77
■ International airframe and aircraft component design and manufacture, defence aircraft production.
Airframe Design and Manufacture
361 Milperra Road, PO Box 30,
Bankstown, New South Wales 2200
☎ (+61 2) 97 72 83 08 ⊖ (+61 2) 97 73 09 01 [98]

75-08300 Skyfox Aviation Ltd
PO Box 3561, Caloundra, Queensland 4551
☎ (+61 7) 54 91 53 55 [E] skyfoxpl@ozemail.com.au
[W] http://www.ozemail.com.au/%7 eskyfoxpl/
⊖ (+61 7) 54 91 82 37
♧ A$ 1,200,000 (1997) ♣ A$ 4,000,000 (1997)
👥 35 (1997)
Company President: Sir Joh Bjelke-Petersen
Sales and Marketing Director: Fraser Anning
Company Secretary: Doug Docherty
Business Development Manager: David Anning
Purchasing Manager: Rod Brittain
■ Manufacturer of Skyfox CA-25N two-seat cabin light aircraft for flight training. [98]

75-08350 John Heine & Son Pty Ltd
273 Edgar Street, Locked Bag 30,
Bankstown, New South Wales 2200
☎ (+61 2) 97 71 22 00 ⊖ (+61 2) 97 92 13 29
👥 160
Managing Director: R D Vanderfield
■ Machine tool manufactures - presses up to 250 tonnes. Full range of medium machining capacity. High alloy iron and steel casting to 12 tonnes. → 80 [98]

75-08370 Hella Manufacturing Co. Pty Ltd
(a subsidiary of Hella KG Hueck & Co, Germany)
Southern Road, PO Box 89, Mentone, Victoria 3194
☎ (+61 3) 95 81 93 33 ☒ 31877 hella aa
⊖ (+61 3) 95 84 17 41
♧ A$ 13,000,000 (1997) ♣ A$ 90,000,000 (1997)
👥 571
Chairman and Chief Executive Officer: W W J Uhlenbruch
Managing Director: Egon Vetter
Finance Director: P F Doyle
Sales Director: J Reddaway
Technical Director: K D Sigge
Advertising Manager: S Camm
Purchasing Manager: Jack Craven
■ Automotive headlamps and signal lamps, decorated and undecorated plastic components (including trim panels, fascia panels, instrument covers, gearbox consoles, switches and relays. [98]

75-08400 Suspension Components Australia Pty Ltd
83 Alfred Street, PO Box 270,
North Melbourne, Victoria 3051
☎ (+61 3) 93 28 24 12 ⊖ (+61 3) 93 29 04 37
👥 120 (1997)
General Manager: Alan Thom
Export Manager: Alan Denehey
Purchasing Manager: Alan Tustin
Sales Manager: Robert Ross
■ Forgings to 5 kg, stabiliser bars, suspension strut arm supports, wheel wrenches. [98]

75-08630 J & E Hofmann Engineering Pty Ltd
3 Alice St, Bayswater, WA 6053
☎ (+61 9) 279 55 22 ☒ 95061 ⊖ (+61 9) 279 93 86
♣ A$ 21 000 000 (1995) 👥 220 (1996)
🏭 14 500 m2 (1996)
Managing Dir: John Hofmann
Operations/Technical Dir: Leighton White
Technical Sales Mgr: Kevin Hatten
Purchasing Mgr: Terry Craen
■ Manufacture of portable line boring equipment, transmissions & gear boxes. → 80 [97]

75-08720 Howard Porter Pty Ltd
PO Box 76, Hamilton Hill, WA 6163
Murphy Street, O'Connor, Western Australia
☎ (+61 8) 93 37 35 33 ⊖ (+61 8) 9331 29 31
♣ A$ 17,000,000 (1996) 👥 160 (1996) 🏭 12,000 m2
Managing Director: Colin Stewart
Director: Thomas G Porter
Operations Director: Matt Lazzarich
Sales and Marketing Director: Neil Porter
Purchasing Manager: Colin Cassidy
■ Manufacturers of engineered products for road transport and materials handling industries. [98]

75-08740 Howden Fan Company Pty Ltd Godfrey Howden Division
(a Howden Group Company)
Matthews Avenue - Airport West, PO Box 84,
Niddrie, Victoria 3042
☎ (+61 3) 93 38 38 11 ⊖ (+61 3) 93 38 61 36
♣ A$ 6,000,000 (1997) 👥 22 (1997)
🏭 6,387 m2 (1997)
General Manager: P H Egan
Engineering Manager: B Bianchi
Marketing and Sales Manager: John S Finch
■ Supply of airborne components for pressurisation, environmental control, hydraulic and electrical systems. Air and gas handling equipment. Roots blower packages. → 80 → 90
® Netherlands: MAFO-Howden BV, Rotterdamsche Droogdok BV
UK: Airscrew Howden Ltd, Howden Air Control Ltd, Howden Wade Ltd, McTaggart Scott & Co. Ltd, Thermal Control Division
USA: Dresser Industries, Roots Division, Kunkle Valve Co. Inc, Noah Howden Inc, Pneu Devices Inc, Vickers-Fluid Controls & Actuator, Vickers-Systems Monitoring Division, Western Design Corp. [98]

75-08800 Hycast Metals Pty Ltd
404 Woodpark Road, PO Box 2144,
Smithfield, NSW 2164
☎ (+61 2) 725 10 11 ⊖ (+61 2) 725 25 58
👥 68
Managing Dir: J.F. Kell
Sales & Marketing Dir: J. Blain
■ Mechanical engineering & foundry. Production of precision, lost wax investment castings as used in a wide range of industries and equipment. [97]

75-08820 Hy-Q International (Australia) Pty Ltd
Rosebank MDC, Private Bag 100, Clayton, Vic 3169
17 Winterton Rd, Clayton, Vic 3168
☎ (+61 3) 562 82 22 ⊖ (+61 3) 562 90 09
👥 35
Divisional Mgr: P. Cooper
Sales Mgr: Rob Gornowicz
■ Complete design & manufacturing facilities for all aspects of frequency management. Quartz crystal units; quartz crystal filters; quartz crystal blanks; temperature compensated crystal oscillators (TCXOs); quartz crystal oscillators.
Affiliates:
Hy-Q International (UK) Ltd
Hy-Q International (Singapore) Pte Ltd
HY-Q International (USA) Inc [97]

75-09250 INC Corp. Pty Ltd
22 Cleeland Road, South Oakleigh, Vic 3167
☎ (+61 3) 95 43 28 00 ⊖ (+61 3) 95 43 81 08
♧ A$ 4 000 000 👥 42
Managing Dir: M.W. Coates
Operations Dir: A. Mutimer
■ Noise control materials & specialist damping materials to all types of vehicles, aircraft & ships. Also bonding of materials for clothing. [97]

75-09500 Inductoheat Pty Ltd
62 Bardia Avenue, PO Box 170, Seaford, Victoria 3198
☎ (+61 3) 97 86 80 00 [E] inducto@ozemail.com.au
[W] http://www.ozemail.com.au/~inducto
☒ 32136 inducto aa ⊖ (+61 3) 97 86 85 22
👥 87 (1996)
Managing Director: J K Robson
South East Asia Sales Director: P W R Burren
Executive Director: G Jans
Welding Sales Manager: F A Barker
Technical Director: I Hartman
Export Director: J R Powell
Product Manager: G W Fielding
■ Manufacture of induction heating and resistance welding, spare parts. Accredited to AS/NZS and ISO9001. [98]

75-10170 Jakab Industries Pty Ltd
(a subsidiary of Rosemont Investments P/L)
Cook Street, PO Box 614,
Tamworth, New South Wales 2340
☎ (+61 67) 65 41 55 [E] jakab@mpx.com.au
⊖ (+61 67) 62 15 46
♧ A$ 2,000,000 (1997) ♣ A$ 18,000,000 (1997)
👥 160 (1997) 🏭 8,400 m2
Managing Director: Karl Jacoby
Financial Controller: Ms Wendy Willis
Sales and Marketing Director: J McGranagham
Export Sales Manager: Jamie Groth
Engineering, Research and Development Manager: Geoff Botfield
Production Manager: Gordon Botfield
■ Civilian and military ambulances, medical modules and paramedic services. [98]

75-10178 James Hardie Building Services Ltd
(a subs. of James Hardie Industries)
17-19 Orion Road, Lane Cove, NSW 2066
☎ (+61 2) 99 28 70 00 ✆ (800) 81 65 47
⊖ (+61 2) 99 28 71 11
♧ A$ 36 000 000 (1995) ♣ A$ 533 000 000 (1995)
👥 2 000
Managing Dir: John Moller
Marketing Mgr: Paul Magee
Marketing Co-ordinator: Charlie O'Hanlon
■ Design & manufacture of electronic access control & security equipment. Full turnkey systems, ranging from single door applications to sophisticated multi building management systems. Fire protection & detection systems for key land based & marine applications. Explosion suppression systems for APCs, tanks & other vehicles, damage control systems & engine fire suppression systems. [97]

75-10200 Jaques
1 Palmer Street, PO Box 90, Richmond, Victoria 3121
☎ (+61 3) 420 94 44 [E] jaques@sprint.com.au
⊖ (+61 3) 429 84 47
♣ A$ 110,000,000 (1997) 👥 250 (1997) 🏭 7,000 m2
General Manager - Jaques: Graeme Phipps
General Manager - Mining Division: Ian Kitchen
General Manager - Quarrying: Adrian J Town
Manufacturing Manager: T Gardiner
Purchasing Manager: R Vilsten

■ Heavy engineering, construction, mining quarrying equipment. Apron and vibrating feeders, screens, jaw and cone crushers including primary gyratory for mining and quarrying. Jaw crusher laboratory. Heavy engineering on project/jobbing basis. Design and supply on turnkey basis for crushing and screening plants for mining and quarrying.
Major plant equipment and facilities: heavy engineering machining, 50 tonne maximum lift in main assembly. [98]

75-10310 Javac Pty Ltd
54 Rushdale Street, Knoxfield, Victoria 3180
☎ (+61 3) 97 63 76 33 [E] javac@compuserve.com.au
[W] http://ourworld.compuserve.com/ homepages/javac
⊖ (+61 3) 97 63 27 56
♧ A$ 1,800,000 (1997) ♣ A$ 7,000,000 (1997)
👥 42 (1997)
Chairman: R F Davies
Managing Director: A F Davies
■ Electronic and mechanical engineering; high vacuum engineering, including ultra high vacuum systems for specialised vacuum coating and associated techniques. High vacuum pumps and ACCs. Refrigerant recovery and recycling equipment. [98]

75-10350 J.N.S. Electronics Industries
(a subs of JNS Electronic (Aust) Pty Ltd)
37-39 Greenway St, Bulleen, Vic 3105
☎ (+61 3) 98 52 21 00 ⊖ (+61 3) 98 52 09 77
♧ A$ 5 000 000 (1996) ♣ A$ 5 000 000 (1996) 👥 15
🏭 6 000 sq m
Managing Dir: J.N. Stannard
Operations & Technical Dir: R.N. Thompson
■ Manufacturers of special prototype equipment. → 80 [97]

75-10400 John Valves Pty Ltd
(a subs. of Australian Eureka Holdings)
PO Box 279, Ballarat, Vic 3350
☎ (+61 53) 33 07 77 ⊖ (+61 53) 38 17 71
♧ A$ 27 000 000 ♣ A$ 25 000 000 👥 260
Managing Dir: Tony Klein
Operations & Technical Dir: Mike Cuttle
National Sales Mgr: Kevin Eddy
Purchasing Mgr: Rob Wilkie
■ Valves for project or service (including FFG-03 Frigate & FOD Frigates Project). Casting & machining of general engineering components. [97]

75-10750 Jones & Rickard Pty Ltd
34 Bay Street, PO Box 218, Botany, NSW 2019
☎ (+61 2) 666 56 11 ⊖ (+61 2) 666 67 69
♣ A$ 10 000 000 👥 82
Managing Dir: J.R. Rickard
Engineering Services Mgr: N. Tinker
■ Aircraft ground power units, frequency changers & rotary power supplies, computer & simulator power supplies. 400 Hz, 60 Hz & 50 Hz. → 80 [97]

75-10780 Jotun Australia Coatings Pty Ltd
(a subsidiary of Jotun A/S Norway)
PO Box 105, Altona North, Victoria 3025
9 Cawley Road, Brooklyn Via North, Victoria 3025
☎ (+61 3) 93 14 07 22 ⊖ (+61 3) 93 14 04 23
♧ A$ 4,500,000 (1997) ♣ A$ 18,000,000 (1997)
👥 60 (1997)
Managing Director: J S Aubert
Operations Manager: I Campbell
Sales Mgr: K. Gaskell
Technical Manager: E Riding
■ Marine and protective coatings. Zinc and epoxy technologies including Dimet Australia powder coatings for electrostatic applications and Dimetote inorganic zinc silicate coatings. [98]

75-10950 Kawasaki Motors Pty Ltd
63-65 Burrows Road, Alexandria, New South Wales 2015 [98]

75-10955 Kawasaki Motors Pty Ltd
(a subsidiary of Kawasaki Heavy Industries Ltd, Japan)
Unit Q, 10-16 South Street,
Rydalmere, New South Wales 2116
☎ (+61 2) 684 25 85 ⊖ (+61 2) 684 45 80 [98]

75-11440 KEL Aerospace Pty Ltd
231 High Street, Ashburton, Victoria 3147
☎ (+61 3) 98 89 00 22 [E] kelaero@kel.com.au
[W] http://www.kel.com.au ⊖ (+61 3) 98 89 00 06
♧ A$ 10,000,000 (1997) ♣ A$ 2,000,000 (1997)
👥 20 (1997) 🏭 500 m2
Chief Executive Officer: T D Kelly
■ HF amplifiers and mixers; over-the-horizon radar receivers; ionospheric monitoring systems; high voltage electrical transient monitoring system; digital ionosondes; automatic ionogram interpretation and scaling. [98]

75-11495 Kidde Graviner Australia
(a div. of Williams Holdings plc)
68 Dunn Crescent, Dandenong, Vic 3175
☎ (+61) 37 93 15 40 ⊖ (+61) 37 06 80 06
👥 10 🏭 8 000 sq m
General Mgr: P. Nelson
Sales & Marketing Dir: P. Harrington
Operations Dir: J. Trewin
Purchasing Mgr: A. Nelson
■ Support & repairs of divisional products in Australia & New Zealand. [98]

75-12100 LADS Corporation
(a division of Vision Systems Ltd)
Second Avenue - Technology Park,
The Levels, South Australia 5095
☎ (+61 8) 83 00 44 47 ⊖ (+61 8) 83 49 75 28
General Manager: Tom Spurling
■ Advanced airborne laser applications, including aerial hydrography. Products include laser airborne depth sounder in service with the Royal Australian Navy. Design and development of laser and optical systems. Systems integration of airborne equipment. [98]

75-12370 Lennys Commercial Kitchens Pty Ltd
PO Box 123, Mount Hawthorn, WA 6016
☎ (+61 8) 94 46 63 22 [E] lennys@ca.om.au
⊖ (+61 8) 94 46 41 56
♧ A$ 100,000 (1996) ♣ A$ 3,000,000 (1996)
👥 28 (1996) 🏭 2,000 m2
Managing Director: Peter Lenny
Sales and Marketing Director: Bert Stray
Operations and Technical Director: Bob Eagle
Purchasing Manager: Gary Thomas
■ Custom built stainless steel products for hospital, building and catering industries. Custom built stainless steel handrails for ships. Custom built stainless steel kitchen equipment. Distributor of commercial kitchen equipment. → 80 → 85 [98]

75-12400 Lightforce Australia Pty Ltd
16 East Terrace, Cleve, South Australia 5460
☎ (+61 8) 86 28 21 81 ⊖ (+61 8) 86 28 25 38
Managing Director: Dr Raymond L Dennis
Operations Director: David Holmes
Sales Director: Paul Petherick
■ High performance 12 volt portable spotlights and related accessories. [98]

75-12410 The Lincoln Electric Co. (Aust) Pty Ltd
PO Box 17, Revesby, New South Wales 2212
35 Bryant Street, Padstow, New South Wales 2211
☎ (+61 2) 97 72 72 22 ⊖ (+61 2) 97 92 13 87
♣ A$ 80,000,000 👥 300
Managing Director: Robert J Lee
Sales Director: Robert B Hannah
Sales Manager Sydney: John W Selkold
Technical Manager: R Sim
■ Welding machines and supply of consumables, equipment and technology. [98]

75-12550 Francis Lord Optics
33 Higginbotham Rd, Private Mail Bag,
Gladesville, NSW 2111
☎ (+61 2) 807 14 44 ⊠ 25629 meopt aa
⊖ (+61 2) 809 71 36
♧ A$ 100 000 ♣ A$ 300 000 👥 5 🏭 3 000 sq ft
Chairman: Francis Lord
Managing Dir: P.N. Griffiths
Sales & Marketing Dir: Phil Griffiths
Optics Div. Mgr: Alan Fry
■ Manufacture of optical components and/or precision components from glass, quartz & ceramics; repair of optical components, lenses & lens systems. → 80
Affiliates:
Filmtronics (Aust)
Avtronics P/L
Christie Electric (Aust)
Meopta (Aust) [97]

75-13000 MacGregor (Aus) Pty Ltd
(† MacGreor-Navire (Aus) Pty Ltd)
(a subsidiary of MacGregor-Navire)
PO Box 76, Balmain, New South Wales 2041
Unit 13 - 5/13 Parsons Street,
Rozelle, New South Wales 2039
☎ (+61 2) 810 05 57 ⊠ 123691 ⊖ (+61 2) 818 18 17
👥 7 🏭 4,000 sq ft
General Manager: John Smith
Financial Controller: R Poynter
Service Manager: M Stacey
■ Cargo handling and ro-ro equipment, deck machinery, container lashing and securing devices, fittings, service and repair. Hagglunds cranes. [98]

75-13140 MacNaught Pty Ltd
(a subsidiary of MacNaught Holdings)
41-49 Henderson Street, Turella, New South Wales 2205
☎ (+61 2) 95 67 04 01 [E] sales@macnaught.com.au
⊖ (+61 2) 95 97 77 73
👥 120 (1997)
Managing Director: Peter C MacNaught
Operations Director: Keith Daly
Marketing Director: Roger A Yexley
Purchasing Manager: Robert Lee
■ Manufacturer of portable hand/air operated greasing systems, grease guns, hand/power operated drum pumps and flowmeters. Retractable hose reel assemblies. [98]

75-13150 Magellan Group Limited
(a subsidiary of Magellan Corporation (Australia) Pty Ltd)
PO Box 1403, Canning Vale, WA 6970
☎ (+61 8) 94 55 22 31 [E] magtech@iinet.com.au
⊖ (+61 8) 94 55 22 43
♧ A$ 5,000,000 (1996) ♣ A$ 3,000,000 (1997)
👥 25 (1997) 🏭 2,200 m2
General Manager: John A Henderson
Operations Manager: Jonathon S Wilson
Projects and Customer Service Manager: Graham Lewis
Purchasing Officer: Brett Pill
■ Design and manufacture of major electronic sub-systems of weapons and procurement of Mil-Spec components, ISO 9001 approved QA system. "Local content". Second or third tier contractor. [98]

75-13160 Malco Engineering Pty Ltd
(† Malco Industries Ltd)
Foundry Division
Cnr Rich St & Victoria Rd, PO Box 41,
Marrickville, NSW 2204
☎ (+61 2) 560 81 44 ⊖ (+61 2) 560 74 14
👥 60
Managing Dir: C.R. Storey
Director: R.A. Nevison
General Mgr: R.A. Stephens
Sales Mgr: C.H. Hill
■ Production of wheelspiders (trucks), brake drums.
Engineering Div.
PO Box 31, Welland, SA 5007
121 Regency Rd, Croydon Park SA 5008
☎ (+61 8) 340 27 88 ⊖ (+61 8) 340 10 61
👥 330
General Mgr: D.G. Salveson
Sales Mgr: J.R. Tregoning
■ Materials handling equipment; vibrating feeders; belt conveyors; reclaim equipment; screw conveyors; bucket elevators; slat conveyors.
® Switzerland: Georg Fischer Ltd
UK: Gough Holdings
USA: Aero-Motive Mfg Co., Carthage Machine Tool, Dresser Inc., Lubriquip Div of Houdaille
● 14 Helen Street, Newstead, QLD 4006
☎ (+61 7) 32 52 58 71 ⊖ (+61 7) 32 52 57 27
● 30 Pickering Road, Mulgrave, Vic 3170
☎ (+61 3) 95 01 03 44 ⊖ (+61 3) 95 61 77 10 [97]

75-13165 Mason & Cox Pty Ltd.
PO Box 83, Torrensville, SA 5031
☎ (+61 8) 352 48 33 ⊖ (+61 8) 354 09 53
♣ A$ 14 000 000 (1995/1996) 👥 125 🏭 4 500 sq m
Joint Managing Dir: G. French
Joint Managing Dir: P.J. Lander
National Sales Mgr: J. Ferencz
■ Stainless steel for stem tube castings, rudder keel castings, valve castings, watertight door castings, castings for tanks, truck & trailer suspension products. [97]

75-13170 Master Instruments Pty Ltd
Corner Sloane and Saywell Street, PO Box 177, Marrickville, NSW 2204
☎ (+61 2) 95 19 12 00 ⊖ (+61 2) 95 19 46 04
👥 50
Managing Director: John Dunn
Operations Director: Joseph B Dunn
Marketing Director: Ron Dunn
Purchasing Manager: Paul Burrows
■ Electrical indicating instruments, miniature ruggedised D'Arsonval and digital meters. Process control and environmental systems. Rechargeable battery packs for all battery types. General purpose and specialised test sets: multimeters; resistance bridges; capacitance boxes. Specialised telecommunications test equipment. [98]

75-13355 Mercedes-Benz (Australia) Pty Ltd
(a subsidiary of Daimler-Benz AG, Germany)
PO Box 4214, Mulgrave, Victoria 3170
☎ (+61 3) 95 61 70 88 ⊖ (+61 3) 95 61 70 88
♧ A$ 138,000,000 👥 491
Managing Director: B W Schlickum
Manager Special Projects: Jack C Wilson
[E] wilson.jack@mbau.mercedes- benz.com
■ Trucks, coaches, cross-country vehicles, industrial engines, Unimogs and related spare parts. [98]

75-13490 Minitronics Pty Ltd
(a subs. of Saft SA, France)
204 Harbord Rd, Brookvale, Sydney, NSW 2100
☎ (+61 2) 99 39 22 44 ⊖ (+61 2) 99 05 18 39
Managing Dir: Eddy Bidder [97]

75-13495 MM Cables
(a division of Metal Manufactures Ltd)
1 Heathcote Road, Liverpool, New South Wales 2170
☎ (+61 2) 96 00 03 43 ⊖ (+61 2) 96 00 03 14
♣ A$ 5,000,000 (1996) 👥 800 (1997) 🏭 70,000 m2
Group General Manager: Nigel Pringle
Marketing Manager Defence: Alan Saunders
[E] alan_saunders@mmcables. com.au
Purchasing Manager: Peter Hall

Clayton Division
PO Box 220, Clayton, Victoria 3168
☎ (+61 3) 95 44 15 11 ⊖ (+61 3) 95 44 31 36
👥 250
General Manager Communication Products: Bruce Williamson
Technical Marketing Manager: L H Ding
Commercial and Marketing Manager: G Devery
■ Communication optic fibre cables, single and multi mode; steel armoured optic cables; communication wire and cables; solid and cellular plastic insulated local and trunk cable dry core and jelly filled cellular plastic insulated cable. Lightweight assault cable. [98]

75-13500 Modra Electric Power Pty Ltd
(† Winslade & Co. Pty Ltd)
PO Box 421, Cloverdale, WA 6105
527 Abernethy Road, Kewdale, WA
☎ (+61 9) 353 33 11 ⊖ (+61 9) 353 34 74
♣ A$ 5 000 000 👥 5
CEO: M. Winslade
Sales Mgr: Bob Lee
■ Specialised generating sets ground power units & portable engine-driven lighting towers to RAAF & Ministry of Defence, Singapore. [97]

75-13770 Moonraker Australia Pty Ltd
Tasmania Technopark, Dowsing Point, Tasmania 7010
☎ (+61 3) 62 73 15 33 [E] radiocom@moonraker.com.au
⊖ (+61 3) 62 73 17 49
👥 20 🏭 1,000 m2
Managing Director: C W Edwards
Sales and Marketing Director: G L Edwards
Operations and Technical Director: R Logan
Purchasing Manager: P J Collinson
■ Design and distribution of antenna systems and locating beacons for MF, HF and VHF. HF frequency agile antenna systems for land mobile, marine antenna systems, base station antennas and satellite tracked data collection platforms. D/GPS fleet management tracking systems and remote radio monitoring/control systems. [98]

75-14300 Nautronix Ltd
PO Box 1352, Fremantle, Western Australia 6160
☎ (+61 8) 94 30 59 00
[E] defence.enquiries@nautronix.com.au
[W] http://www.nautronix.com.au ⊖ (+61 8) 94 30 59 01
♧ A$ 9,654,000 (1997) ♣ A$ 19,699,000 (1996)
👥 271 (1997)
Chairman and Managing Director: A W Tribe
General Manager: B O'Reilly
Business Development Manager Defence: Dr J H Ranicar
■ Design and manufacture of ASK 4000 series vessel dynamic positioning systems including ASK 4000JS joystick control model. ATS and RS5D/RS 910 series underwater acoustic tracking and positioning systems, including ultra short, short and long baseline systems. HyPro series hydrographic data acquisition and processing systems, including multibeam echosounder compatible systems. Underwater sound ranges and hydroacoustic information links; pingers, call pingers, pinger interrogators and receivers. Diver communication systems and helium speech unscramblers. Maintenance and repair of acoustic and sonar equipment. Calibration of underwater instrumentation. Leasing of acoustic test tank facilities.

Subsidiary Companies:
Nautronix Ltd (UK)
Nautronix Inc, USA
Nautronix (Singapore) Pte Ltd, Singapore
Qubit (Australia) Pty Ltd, Australia
McCorkills Marine Pty Ltd, Australia
Seismic Supply International Pty Ltd, Australia
Seismic Supply Pte Ltd, Singapore
Seismic Supply Hong Kong Ltd, Hong Kong [98]

75-14680 Nilsen Electric (SA) Pty Ltd
Engineering/Switchboard Manufacturing Division
PO Box 111, Kilkenny, South Australia 5009
100 Regency Road, Ferryden Park, South Australia 5010
☎ (+61 8) 82 68 23 00 [E] nilsensa@nilsen.com.au
[W] http://www.nilsen.com.au ⊖ (+61 8) 83 47 03 47
♣ A$ 25,000,000 (1997) 👥 250 (1997) 🏭 7,000 m2
Managing Director: John Nilsen
General Manager: Peter Van Den Heuvel
Sales and Marketing Director: David Hardaker
Engineering Division Manager: B Tetley
Purchasing Manager: John Doyle
Company Secretary: D J Plant
■ Design and manufacture of sheetmetal cabinets, switchboards, motor control centres, control desks, panels etc. for naval and submarine applications, quality accreditation ISO 9001. [98]

75-14690 A. Noble & Son Ltd
PO Box 130, Blair Athol, SA 5084
☎ (+61 8) 260 66 88 ⊖ (+61 8) 349 62 60
♧ A$ 800,000 (1996) ♣ A$ 6,000,000 (1996) 👥 145
🏭 12,000 sq m
Managing Dir: J.N. Rogers
SA State Mgr: T.J. Miller
Company Secretary: I.A. Thorpe
■ Wire rope assemblies, pulley blocks & all other lifting tackle requiring type & proof testing, mooring chains & anchors. [97]

75-14870 Normalair-Garrett Australia
Normalair-Garrett Manufacturing Pty Ltd
(a subs. of Normalair-Garrett Ltd, UK)
45 King Street, PO Box 131, Niddrie, Vic 3042
☎ (+61 3) 93 30 15 11
⊖ (+61 3) 93 35 18 84 Product Support;
38 82 82 Manufacturing
♣ A$ 14 150 000 (1995) 👥 102 🏭 5 700 sq m
Managing Dir: Paul Coleman
Product Support Mgr: Ralph Herron
Aerospace Manufacturing Mgr: David Mayne
Turbochargers Mgr: Tony Keeler ☎ (+61 2) 775 33 11
Support Business Development Mgr: Mike Tobin
Manufacturing Business Development Mgr: Adriaan Maan
■ Manufacture, overhaul, repair & test of military & civil aerospace, marine & ground equipment including pneumatic, hydraulic, fuel, pressurisation, air conditioning, starting, aircrew breathing, auxiliary power & structural parts.
® UK: Normalair-Garrett
USA: AlliedSignal

Affiliates:
Thompson Valves

Turbocharger Division
14 Alfred Road, Chipping Norton, NSW 2170
☎ (+61 2) 755 33 11 ⊖ (+61 2) 724 79 09
General Mgr: Tony Keeler
■ Supply, repair & overhaul of Garrett Turbochargers throughout Australia. [97]

75-14882 Norwellan Ltd
Playford Street, Stawell, Vic 3380
☎ (+61 53) 58 81 00 ⊖ (+61 53) 58 38 73
♧ A$ 3 804 027 ♣ A$ 19 000 000 👥 170
Managing Dir: P.R. Kerr
Marketing Dir: J.T. Bennett
Development Mgr: R. Hamilton
■ Uniform fabrics of all types; blanket fabrics; technical fabrics; transport fabrics; projectile resistant fabrics; military apparel fabrics. [97]

75-14885 Novamarine Instruments Pty Ltd
30 Downie Street, PO Box 96, Wickham, New South Wales 2293
☎ (+61 49) 69 44 77 ⊖ (+61 49) 62 12 10
♣ A$ 3,000,000 (1997) 👥 12 🏭 660 m2
Director: Murray Slana
■ Design & development, electronic repairs, servicing & overhaul of instruments. Software development. [98]

75-14890 NQEA Australia Pty Ltd
(† N Queensland Engineers & Agents Pty Ltd)
60-92 Cook Street, Portsmith, PO Box 1105, Cairns, Queensland 4870
☎ (+61 70) 52 72 22 [E] nqea_cns@tpgi.com.au
⊖ (+61 70) 35 25 20; 35 28 12
♧ A$ 13,500,000 (1996) ♣ A$ 54,300,000 (1996)
👥 300 (1997) 🏭 28,000 m2
Chairman and Owner: D G Fry
General Manager: E W Graham
Technical Support Manager: G E Mears
Marketing: P Harrison
■ Manufacture and design of sophisticated vessels and shipbuilding for commercial and naval requirements. Contractors to Australian Department of Defense. [98]

75-15640 Olex Cables
(a division of Dunlop & Olympic Tyres, Australia)
207 Sunshine Road, Tottenham, Victoria 3012
☎ (+61 3) 92 81 44 44 ⊠ 30498 olex aa
⊖ (+61 3) 93 14 03 83
♧ A$ 35,100,000 👥 1,200
Managing Director: Ian Campbell
Operations and Technical Director: Ken Barber
Group General Manager, Energy and Communications Division: A W Stobart
Group General Manager, Construction Division: T Wingrove
National Sales and Marketing Manager: G Simpson
Purchasing Manager: Tony Kerr
Market Manager: Tony Kiernan
■ Power cables suitable for voltage ranges up to 220 kV with paper, PVC, polyethylene (including cross-linked PE) and EPR insulations, with various outer protections. Communication cables (paper and plastic voice frequency), high frequency cables, low voltage instrumentation cables, rubber insulated and sheathed mining cables, optical fibre cables, fire rated cables. [98]

75-15642 Olin Australia Ltd Winchester Division
(a subsidiary of Olin Corp, USA)
PO Box 776, Geelong, Victoria 3220
☎ (+61 3) 52 48 20 00 ⊖ (+61 3) 52 48 24 09
♧ A$ 8,000,000 (1997) ♣ A$ 30,000,000 (1997)
👥 125 (1997)
General Manager: J J Canning
Project Manager: Wayne M Pilkington
■ Arms & ammunition, including F5 .22 rimfire ammunition to Australian army & special performance centrefire rifle ammunition to special operations groups. [98]

Contact Reference

75-15645 Oliver Footwear
(† Oliver & Stevens)
Corner Prest and Humffray Street, Ballarat, Victoria 3350
☎ (+61 3) 53 31 26 88 ⊖ (+61 3) 53 33 33 10
♤ A$ 1,300,000 👥 110
Managing Director: A O Oliver
Marketing Manager: G B Roberts
■ Occupational footwear including general purpose boots, firemans boots, flying boots and shoes. Manufacturer for defence contracts. [98]

75-15890 Overall Forge Pty Ltd
PO Box 5275, Aubury, New South Wales 2640
☎ (+61 60) 25 67 77 ⊖ (+61 60) 25 79 81
♣ A$ 11,100,000 (1995/96) 👥 65
Managing Director: Lawrie Chadwick
Operations Director: Ric Toseland
Sales Manager: Glen Vitullo
Technical Services Manager: P Howard
■ Steel forgings, rough machined, heat treated and tested to client specifications. Typical products - gear rims, hubs, slabs, shafting, flanges, ship rudder stock, trackwheels, steel tyres, tube sheets, hollow bar. Seamless and profiled rings. [98]

75-16200 Pains-Wessex (Australia) Pty Ltd
(a subsidiary of Chemring Group plc)
PO Box 25, Glen Iris, Victoria 3146
Suite 4 - 1955 Malvern Road, Malvern East 3145
☎ (+61 3) 98 85 04 44 ⊖ (+61 3) 98 85 55 30
♣ A$ 4,000,000 (1995) 👥 14
Managing Director: C H Tegner
Operations Director: P Deegan
Company Secretary: G S Moorhouse
■ Filling and assembly of pyrotechnic stores; marine distress flares, smoke generators, training devices, windproof and waterproof matches. Imported Pains-Wessex Schermuly products. Military and civilian pyrotechnics for signalling illumination, simulation, training, search and rescue, law enforcement and countermeasures.
® Denmark: Quitzau Industry
UK: Chemring plc, Concept Engineering plc, Haley & Weller Ltd, McMurdo Ltd, Octavius Hunt Ltd, Pains-Wessex, Vacuum Reflex Ltd
USA: Alloy Surfaces Inc., Omniglow [98]

75-16245 PBR Automotive Pty Ltd
(† Brake & Clutch Industries Australia Pty Ltd)
(a division of Pacific BBA Ltd, Australia)
264 East Boundary Road, PO Box 176,
East Bentleigh, Victoria 3165
☎ (+61 3) 95 75 22 00 ⊖ (+61 3) 95 75 25 91
👥 1,000
Managing Director: John MacKenzie
Operations Director: Richard Waldron
Sales and Marketing Manager: Chris Nicoll
Sales and Marketing Director: Philip Mason
Marketing Manager: Graeme Heine
Purchasing Manager: Phil Matthias
■ Manufacturer of complete brake and clutch systems for Australian and overseas vehicle assemblers (passenger and commercial applications).
Affiliates:
Automotive Products (UK) Ltd
Bendix Mintex [98]

75-16460 Perry Engineering
(a div. of Boral Johns Perry Industries)
GPO Box 1838, Adelaide, SA 5001
Railway Terrace, Mile End South, SA 5031
☎ (+61 8) 352 17 77 ⊖ (+61 8) 234 18 96
♤ A$ 145 000 000 ♣ A$ 25 000 000 👥 220
🏭 58 000 sq m
General Mgr: A. Nuttall
Marketing Mgr: R. Deane
■ Manufacture pendulum borers; articulated cranes; aircraft tow tugs; airfield rescue & fire fighting vehicles; aerial stores lift truck; ship repair; oily water separators; heavy duty trenching machines. Taclop (truck, aircraft, loading, off-pavement), loaders, aircraft cargo, self-propelled, self-loading. Manufacture armoured infantry mobility vehicles. Provision of medium/heavy engineering & manufacturing services including fabrication, forging, machining & fitting. [97]

75-16490 Philips Components Pty Ltd
(† Philips Electronic Components & Materials)
(a div. of Philips Electronics Australia Limited)
34 Waterloo Road, Ryde, NSW 2113
☎ (+61 2) 805 44 55 ⊖ (+61 2) 805 44 66
👥 200 🏭 2 800 sq m
General Mgr: M.P. McKinlay
Sales & Marketing Mgr: Tom Knox
Purchasing Mgr: K. Murphy
■ Design & manufacture of custom microelectronic components including integrated circuits, thick film hybrid circuits, surface mounted assemblies. Also rigid & flexible PCBs.
Adelaide Microelectronics Facility
PO Box 2001, Alberton, SA 5014
☎ (+61 8) 348 52 00 ⊖ (+61 8) 243 10 48
Factory Mgr: J. Simpson [97]

75-16550 Philips Traffic & Engineering Systems
(a div. of Philips Electronics Ltd)
2 Greenhills Ave, PO Box 269, Liverpool, NSW 2170
☎ (+61 2) 600 54 44 ⊠ 20165 ausmope aa
⊖ (+61 2) 601 76 65
👥 200
Sales Mgr: M. Battaglia ☎ (+61 2) 600 54 14
■ Radar display consoles, military airborne radar; depot level maintenance digital control systems; aerospace equipment; naval combat systems test program sets; calibration laboratory; environmental testing; engineering development; full quality systems accreditation. Automatic test equipment. [97]

75-16750 Plenty Uniquip Pty Ltd
61 Alleyne Street, Chatswood, New South Wales 2067
☎ (+61 2) 417 44 33 ⊠ 121687 syd
⊖ (+61 2) 417 51 46 [98]

75-16865 Premier Hazard Systems Pty Ltd
107-121 Howard Street, North Melbourne, Victoria 3051
☎ (+61 3) 328 34 44 ⊖ (+61 3) 326 53 39
♣ A$ 300 000 👥 3
Contact: Darryl Ballinger
■ Manufacture of a complete line of emergency vehicle warning equipment including strobe & halogen lightbars, beacons, sirens, speakers, switch control & auxiliary lighting products. [98]

75-17000 PWB Anchor Ltd
(† Pitt Waddell Bennet Chains Ltd)
(a division of Email Ltd, Australia)
441 Grimshaw Street, PO Box 77,
Bundoora, Victoria 3083
☎ (+61 3) 94 67 29 33 ⊖ (+61 3) 94 67 72 90
♣ A$ 29,000,000 (1996) 👥 110 (1996)
Managing Director: W L Cooper
Marketing Manager: Ian R Allan
Operations and Technical Manager: G K McCredie
Purchasing Manager: M Tan
■ Chain, chain accessories, chain assemblies, chain slings, hoists, winches, clamps, pulley and chain blocks. [98]

75-17200 Queensland Rubber Co. Pty Ltd
515 Zillmere Road, PO Box 484,
Virginia, Queensland 4014
☎ (+61 7) 32 63 41 44 ⊖ (+61 7) 32 63 49 12
♣ A$ 10,500,000 👥 70
Managing Director: I M Russell
Technical Director: Dr G E Russell
Marketing Manager: Bruce Cage
■ Marine rubber fenders, elastomeric bearings, elastomeric seals and gaskets, mouldings, extrusions and sheet rubber using natural rubber and synthetic elastomers. Acoustic vibration bearings, rubber hydraulic seals, electrical insulation matting, fabric-reinforced rubber dams and dragline buffers. [98]

75-18100 Radstone Technology Pty Ltd
1a First Avenue, PO Box 59,
Eastwood, New South Wales 2122
☎ (+61 2) 98 74 91 43 Ⓔ unitsyd@unitronix.com.au
⊖ (+61 2) 98 74 91 50
Managing Director: Philip C Cohen
■ Manufacture of standard and custom board level, systems and software products for military and industrial applications. Product support open bus architecture (VME/Multibus) including processors, memories, I/O and enclosures. [98]

75-18300 Recoil Pty
(† Brimech Industries)
(a subsidiary of Kaynar Technologies)
PO Box 24, Oakleigh, Victoria 3166
☎ (+61 3) 95 63 15 00 Ⓔ sales@recoil.com.au
⊖ (+61 3) 95 63 19 80
🏭 2,200 m2
Managing Director: Ken Jones
National Sales Manager: Shane Dougherty
Technical Manager: Q Matriske
■ Mechanical engineering: fasteners, stainless steel recoil-helically wound thread inserts to MIL, DoD, SBAC standards.
Branches in UK, USA and Belgium. [98]

75-18500 RFI Industries Pty Ltd
(† Scalar R F I Pty Ltd)
(a subsidiary of Townsend Group Pty Ltd)
54 Holloway Drive, PO Box 254, Bayswater, Victoria 3153
☎ (+61 3) 97 62 67 33 Ⓔ rfind@enternet.com.au
⊖ (+61 3) 97 62 85 01
♣ A$ 5,000,000 (1997) 👥 32 (1997)
Managing Director: J H Pluck
General Manager: D R Johnston
Technical Services Manager: R Davies
Sales Engineer: Eric J Buggee
■ Radio frequency shielded enclosures; R F I Accessories. Manufacture and supply of all sizes of EMI enclosures together with associated accessories. EMI gasket materials including elastomers, mesh, beryllium copper ferrite absorbers. EMI/EMC compliance resins.
Represented by:
Cosmotec Enterprises, Singapore
RFI Electronic Industries, India
IES Electronic Agencies (1986) Ltd, Israel
Components and Instrumentation, New Zealand
Telesplicing Enterprise Co Ltd, Taiwan
Sangho Lead Industry, Korea
Mortdale Office
50-56 Barry Avenue, Mortdale, New South Wales 2223
☎ (+61 2) 95 33 40 66; (+61 41) 931 35 79
⊖ (+61 2) 91 53 85 04 [98]

75-18650 R&I Instrument & Gear Co. Pty Ltd
Unit 1/26 Laser Drive, Rowville, Victoria 3178
☎ (+61 3) 97 64 06 44; 97 59 91 41
⊖ (+61 3) 97 64 01 23
♤ A$ 10,000 ♣ A$ 500,000 👥 8
Managing Director: Jeff P Rowell
General Manager: Andrew Rowell
Operations Manager: John Cubitt [98]

75-18750 Rolco Products Pty Ltd
13 Cahill Street, Dandenong, Victoria 3175
☎ (+61 3) 794 78 33 ⊖ (+61 3) 794 88 52
♤ A$ 800,000 (1997) ♣ A$ 2,400,000 (1997)
👥 26 (1997)
Managing Director: Gary Luxford
Operations Director: David Horne
Purchasing Manager: Lou Masin
■ Bronze and aluminium castings, valves fittings, bearings and brackets, including pump housings for decanting water from fuel bunkers (army, navy, airforce) and bronze flanges (navy). Vacuum and rota moulding tools for plastics. [98]

75-18950 Saft Nife Power Systems Australia Pty Ltd
Unit 5 - Block Q, Princes Road East, PO Box 156,
Regents Park, NSW 2143
☎ (+61 2) 738 42 22 ⊖ (+61 2) 738 42 42
Managing Director: Richard Jensen
Marketing Services Co-ordinator: Toni Cutjar
Saft Australia Pty Ltd
Unit 5 - Block Q, Princes Road East, PO Box 156,
Regents Park, NSW 2143
☎ (+61 2) 738 42 20 ⊖ (+61 2) 743 78 35 [98]

75-18980 Sala International Pty Ltd
(† P A Safety Systems Pty Ltd)
(a subsidiary of BTP plc, UK)
55 Philpott Street, Marrickville, NSW 2204
☎ (+61 2) 95 19 13 33
[E] 100017.1506@compuserve.com
⊖ (+61 2) 95 16 12 68
♣ A$ 6,000,000 (1997) 👥 30 (1997) 🏭 15,000 sq ft
Managing Director: B Johnson
Purchasing Manager: S Finch
Operations and Technical: S Zenos
■ Textile design engineers specialising in parachutes and safety harnesses. Fall arrest products. Approved manufacturer to Australian, USA, French and German standards. [98]

75-19200 Scitec Limited
Corporate Head Office
2 Apollo Place, Lane Cove, NSW 2066
☎ (+61 2) 94 28 95 55 [E] sales@scitec.com.au
[W] http://www.scitec.com.au ⊖ (+61 2) 94 28 99 88
♣ A$ 43,000,000 (1996)
Chairman: Christopher C Golis
Managing Director: Paul Magee
General Manager, Marketing: Colin Kempter
General Manager, Operations: Adrian Siebel
General Manager, Sales and Service: Laurie Stevens
General Manager, Finance and Administration: Dwin Tucker
Company Secretary: John Giddings
■ Research, development and manufacture in the fields of communications and networking technology. Integration of the following technology areas: frame relay; bandwidth management; voice, data, fax and local area networks; T1 and E1 network access equipment; integrated network management and ATM. → 80 [98]

75-19350 Sea Tiger Marine Pty Ltd
PO Box 157, Mordialloc, Victoria 3195
☎ (+61 3) 95 80 15 00 ⊖ (+61 3) 95 80 07 01
♧ A$ 150,000 ♣ A$ 9,000,000 (1996/97) 👥 4
Managing Director: Colin E Pratt
■ Design, manufacture of closed cell buoyancy types of RIBs 4 meters to 9 meters with outboard, sterndrive or jet drives, for rescue assault or shipboard use.
® USA: Stewart and Stevenson Tactical Vehicles [98]

75-19360 SEC Plating Pty Ltd
105 Lakemba St, PO Box 231, Belmore, NSW 2192
☎ (+61 2) 750 80 11 ⊠ 20149 ⊖ (+61 2) 750 65 65
♧ A$ 1 500 000 👥 75 🏭 5 500 sq m
Managing Dir: W.R. Jacobs
General Mgr: R.G. Jacobs
Financial Controller: A.C. Jacobs
Production Mgr: D.E. Jacobs
Technical Mgr: R. Fong Yan
Purchasing Mgr: A.D. Jacobs
■ Automatic rack plating, barrel plating, selective plating, high speed plating, ultra sonic processing, programmed pulse power plating & continuous plating. [97]

75-19380 See-Ezy Products Pty Ltd
6 Chalder Street, Marrickville, NSW 2204
☎ (+61 2) 95 57 20 61 ⊖ (+61 2) 95 16 28 76
♧ A$ 50,000 👥 6
Managing Director: C Harris
Sales and Marketing Director: Paul de Rome
■ Manufacture of lead lights, guard lamps and damage lights, light extensions for Army. Incandescent and fluorescent handlamps. [98]

75-19415 SETEC Pty Ltd
(† Scientific Electronics)
6 Holloway Drive, PO Box 127, Bayswater, Victoria 3153
☎ (+61 3) 97 62 57 77 ⊖ (+61 3) 97 62 91 97
♧ A$ 1,000,000 (1996) ♣ A$ 14,000,000 (1996)
👥 150 (1997)
Managing Director: P D Lloyd
[E] pdlloyd@compuserve.com
Technical Director: K P Smith
Purchasing Manager: Peter Saingles
■ Design and development of switch mode and linear power supplies for OEM, defence and laboratory use. [98]

75-19450 Siemens Ltd Melbourne Head Office
(a subsidiary of Siemens AG, Germany)
544 Church Street, Richmond, Vic 3121
☎ (+61 3) 429 71 11 ⊠ SIEMENS Melbourne ⊠ 30425
⊖ (+61 3) 420 73 09
♣ A$ 250,000,000 👥 1,100
Managing Dir: K.O. Lahr
General Mgr, Telecommunications: Dr O. van Zyl
General Mgr, Electrical Engineering: J.R. Coulton
General Mgr, Commerce: B.K. Roberts
Defence Mgr: J.B. Mason
■ Teleprinters, telegraph transmission & switching systems, telecom transmission equipment, PABXs, special telephone handsets, measuring instruments, electronic components, spectrometry equipment, telemetry & remote control equipment, power supply & distribution equipment.
® Germany: Siemens AG [97]

75-19475 Siemens Plessey Electronic Systems Pty Ltd
(† Plessey Australia Pty Ltd)
(a subsidiary of Siemens AG, Germany)
Faraday Park, Railway Road,
Meadowbank, New South Wales 2114
☎ (+61 2) 98 07 04 00 [E] spspes@siemens.spes.com.au
⊖ (+61 2) 98 07 06 00; 98 08 27 87
♧ A$ 78,509,000 (1995/1996) ♣ A$ 100,000,000
👥 470
Managing Director: E J Matiuk
Business Administration Director: G O Wilcke
Engineering Director: H Tacey
Sales and Marketing Executive: M Jaggard
Corporate Communications Officer: H Davis
Operations: G Baumgartel
■ Capabilities: Research, development, design, test, evaluation, manufacture and total logistic support of complex C3S systems. Manufacture of electronic assemblies. Software development. Systems engineering. Metallurgical and analysis laboratories. Automated test systems. Electronic maintenance. Shelterisation and other remote site installation of electronic systems. Project management.
Products include: RAVEN HF and VHF combat net radios, Mercury military messaging terminal, Jupiter and Saturn spectrum management systems, personal locator beacon, Secure videoconferencing system.
Quality Assessment: ISO 3001.
® UK: Denro Inc, Siemens Plessey Systems.
Bruce Office
Thynne Street, Fern Hill Park, Bruce, ACT 2616
☎ (+61 62) 51 49 88 ⊖ (+61 62) 51 60 92
General Manager: L Farrelly [98]

75-19505 SIHI Australia Pty Ltd
(a subsidiary of Sihi International)
820 Mountain Highway, Bayswater, Victoria 3153
☎ (+61 3) 97 20 15 00 [E] sfsaus@ozemail.om.au
⊖ (+61 3) 97 20 40 76
👥 18
Managing Director: J L Parker
Operations and Technical Director: G J Napper
Sales and Marketing Director: M G Baldock
■ Pump and ancillary equipment for defence projects (Army, Navy and Airforce). Leak proof pumps, plants for oil treatment, fuel pumping, vacuum, water purification and firefighting. Vacuum sewage systems, steam and steam recovery plants. [98]

75-19545 W.E. Smith Engineering Pty Ltd
(a subs. of Heat Exchangers Int. Ltd)
Hamilton Drive, Boambee via, PO Box 274,
Coffs Harbour NSW 2450
☎ (+61 66) 58 34 66 ⊖ (+61 66) 58 34 99
♧ A$ 4 000 000 (1996) ♣ A$ 20 000 000 (1995)
👥 140 (1996) 🏭 7 000 sq m (1996)
Managing Dir: W. Schlechtriem ☎ (+61 66)58 34 66
Operations/Technical Dir: John Morrison
Finance Controller: E. Tschaut
Sales Mgr: L.P. Douglass ☎ (+61 66) 58 34 37
Purchasing Mgr: Peter Sherman
■ Heavy fabrication & machining, pressure vessels & heat exchangers. [97]

75-19565 Solarex Pty Ltd
(a subsidiary of Solarex Corporation)
PO Box 204, Chester Hill, NSW 2162
☎ (+61 2) 97 27 44 55 ⊖ (+61 2) 97 27 74 47
♧ A$ 5,000,000 ♣ A$ 9,239,568 (1996) 👥 65 (1996)
Managing Director: D A Bird
Regional Sales Manager: E L Poole
Executive Secretary: Ms Margaret Griffiths
■ Design and production of military solar power panels. [98]

75-19573 Spantech Pty Ltd
PO Box 622, Broadbeach, Queensland 4218
36 Hutchinson Street,
Burleigh Junction, Queensland 4220
☎ (+61 75) 593 44 49 ⊖ (+61 75) 593 44 48
♧ A$ 25,000,000 ♣ A$ 15,000,000 👥 40
Managing Director: R J Lucas
Export Manager: T J Watts
■ Aircraft hangar construction and design, military explosive storehouses, hardened aircraft shelters, bunker storage, nuclear blast and chemical resistant command centres, barracks and buildings, underground personnel shelters, and defence buidings. [98]

75-19690 Stanilite Electronics Pty Ltd
(a subs. of Stanilite Pacific Ltd)
424 Lane Cove Road, Ryde, NSW 2113
☎ (+61 2) 325 15 00 ⊖ (+61 2) 325 16 00
♧ A$ 193 000 000 (1995) ♣ A$ 125 000 000 (1995)
👥 500 🏭 1 000 sq m
CEO Defence & Aerospace Div.: R.E. Foster
Exec. Dir: J.R. Harriss
Exec. Dir: R.W. Harriss
Exec. Dir: D.W. Fayle
Marketing Mgr Defence & Aerospace Div.: G.J.S. Kendall
■ Capabilities include facilities for electronics equipment design, testing & final product assembly. Specialised facilities exist for computer aided drafting, automated battery testing & certification, microprocessor development systems, calibration of communication equipment & radar to NATA traceable standards. Also applications orientated mechanical design & packaging with established marketing & distribution network. Total integrated logistics support services available. [97]

Contact Reference

75-19780 Static Engineering Pty Ltd
6 Oldham Road, PO Box 95,
Elizabeth, South Australia 5112
☎ (+61 8) 82 55 23 55 ⊠ STATIC Adelaide
⊖ (+61 8) 82 55 44 04; 82 55 77 25
♧ A$ 110,000 (1997) ♣ A$ 5,500,000 (1997)
👥 65 (1997) 🏭 8,000 m2
Managing Director: Dr-Ing. Walter Verschoor
Sales Manager: Jack Middleton
■ Design and manufacture of aircraft ground support equipment and other equipment. Specialising in TALU and TASLU air-transportable cargo loading vehicles to service from C-130 to B-747 aircraft. Also elevating work platforms, bomb-lifters, tail-servicing platforms. [98]

75-20250 Tecalemit (Australasia) Pty Ltd
(a subs. of Siebe plc, UK)
Head Office
PO Box 103, Woodville, SA 5011
6 Sheffield St, Woodville North SA 5012
☎ (+61 8) 243 52 00 ⊖ (+61 8) 45 98 15
♣ A$ 30 000 000 👥 260 🏭 12 000 sq m
Managing Dir: Dennis A. Overton
Sales & Marketing Mgr: Ron E. Carter
■ Lubrication equipment. Air operated pumps; hose reels; hose assemblies; control valves; hand pumps for lubricants, fuels water & air; vehicle hoists; automotive OE components; grease guns & nipples; industrial & vehicle lubrication systems; field service units & equipment.
Brisbane
1 Success Street, Archerfield, Qld 4108
☎ (+61 7) 32 74 37 00 ⊖ (+61 7) 32 74 30 80
Melbourne
Unit 6/25-31 Fenton Street, Huntingdale, Vic 3166
☎ (+61 3) 544 40 77 ⊖ (+61 3) 562 97 94
Perth
Unit 6, 54 Crocker Drive, Malaga, WA 6062
☎ (+61 9) 249 57 00 ⊖ (+61 9) 249 59 11
Sydney
115-121 Ballandella Rd, Pendle Hill, NSW 2145
☎ (+61 2) 636 97 77 ⊖ (+61 2) 636 90 27 [97]

75-20300 T E D Engineering Australia Limited

(a subsidiary of Hawker Richardson Ltd)
292-298 Bay Road, PO Box 612,
Cheltenham, Victoria 3192
☎(+61 3) 95 84 98 76 [E] tedeng@ozemail.com.au
⊖ (+61 3) 95 84 90 57
A$ 20,000,000 ♣ A$ 20,000,000 ⁂ 130
20,000 m2
Executive Chairman and Managing Director: T Fried
Sales Manager: George Derrick
■ Engineering designs, tool and die making, jigs, fixtures, fabrication, machine construction. Aircraft components and tooling. [98]

75-20400 Tesko Engineering & Tooling

28 Wren Road, Moorabbin, Vic 3189
☎(+61 3) 95 55 79 33 ⊖ (+61 3) 95 53 31 52
♣ A$ 1 335 000 ⁂ 13
Managing Dir: F.W. Fewster
Marketing/Sales Mgr: T. Stanyer
■ Aerospace components machining. Tool & die manufacture. Machining components for Government organizations. [97]

75-20480 Thomas Electronics of Australia Pty Ltd

13 Larkin Street, PO Box 291,
Riverwood, New South Wales 2210
☎(+61 2) 91 53 07 21 [E] teo@thomas.com.au
⊖ (+61 2) 95 34 21 71
A$ 2,500,000 ⁂ 3 5,000 m2
Production Manager: Robert Lum
Managing Director: William A Hutchinson
■ Cathode ray display tubes including electrostatic and magnetic deflection. Ruggedised tubes to military and avionic specifications and high resolution tubes for data display. Display monitor sales and service. [98]

75-20540 Thorn DNT Airfield Lighting Pty Ltd

(† DNT Europhane Pty Ltd)
7-9 Newcastle Road, PO Box 548,
Bayswater, Victoria 3153
☎(+61 3) 97 20 32 33 ⊖ (+61 3) 97 20 82 33
♣ A$ 10,000,000 (1996) ⁂ 30
Managing Director: Hank Steuten
Financial Controller: Tony Diepenhorst
Sales and Marketing Manager: Ms Meryl Ellsmore
■ Airfield systems and equipment. Visual and navigational aid systems. → 85 [98]

75-20550 THORN Lighting Pty Ltd

(a subsidiary of THORN Lighting Ltd UK)
13 Cooper Street, Smithfield, New South Wales 2164
☎(+61 2) 604 43 00 ⊖ (+61 2) 604 45 88
Managing Director: John Quinn
Operations Director: Steve Wyatt
Purchasing Manager: Arnold Peters
Marketing Manager: Brett Pieper [98]

75-20680 Timcast Pty Ltd

(a subsidiary of A I Engineering Corp Ltd, Australia)
97 Hector Street, Osborne Park, WA 6017
☎(+61 8) 94 46 28 44 ⊖ (+61 8) 94 45 38 28
A$ 3,000,000 (1996) ♣ A$ 7,000,000 (1996) ⁂ 50
2,400 m2
Chairman: Rob Martin
General Manager: Ivan W Irving
Quality Manager: A Ayrey
Technical Manager: I Burgess
■ Suppliers of finished and semi-finished castings in copper, nickel, chromium and cobalt based alloys, nickel aluminium bronze speciality products. Assessed Bureau Veritas to ISO9002 No per 491279 [98]

75-20805 Transfield Fabrication Head Office

PO Box 470, North Sydney, New South Wales 2059
100 Arthur Street, North Sydney, New South Wales 2060
☎(+61 2) 99 29 86 00 ⊖ (+61 2) 99 54 05 27
■ Fabricated structures, plant and equipment material for maritime and related industries.

Seven Hills Facility
Locked Bag 11, Seven Hills, New South Wales 2147
15 Powers Road
☎(+61 2) 624 44 00 ⊖ (+61 2) 624 72 84

Newcastle Facility
64 Greenleaf Road, Koorangang Island,
Newcastle, New South Wales 2304
☎(+61 49) 28 25 33 ⊖ (+61 49) 20 10 30

Coniston Facility
PO Box 205, Port Kembla, New South Wales 2505
Springhill Road, Coniston, New South Wales 2500
☎(+61 42) 76 27 77 ⊖ (+61 42) 74 39 55

Richlands Facility
479 Freeman Road, Richlands, Queensland 4076
☎(+61 7) 33 75 35 77 ⊖ (+61 7) 33 75 59 82

Traralgon Facility
PO Box 680, Traralgon, Victoria 3844
Eastern Road - Industrial Park
☎(+61 51) 74 76 66 ⊖ (+61 51) 74 50 50

Whyalla Facility
PO Box 293, Whyalla, South Australia 5600
Jacob Street
☎(+61 86) 44 00 99 ⊖ (+61 86) 45 38 78

Kwinina Facility
PO Box 39, Kwinina, Western Australia 6165
Corner Beard and Morley Streets,
Naval Base, Western Australia 6165
☎(+61 9) 410 10 66 ⊖ (+61 9) 410 14 03 [98]

75-20810 Transfield Pty Ltd

30 Alfred Street, Miwsons Point, New South Wales 2071
☎(+61 2) 99 29 86 00; 99 97 882 08
[W] http://www.transfield.com.au ⊠ 21396
⊖ (+61 2) 99 54 49 89
♣ A$ 1,500,000,000 (1997) ⁂ 6,000 (1997)
28,000 m2
Founding Chairman: Franco Belgiorno-Nettis
Deputy Chairman Elect: Gerry Gleeson
Managing Director: Marco Belgiorno-Zegna
Chief Executive Officer Eng/Construction: Peter Gregg
Chief Executive Officer, Operations and Maintenance: George Thomas
Chief Executive Officer Project Development - Australia: Tony Shepherd
Chief Financial Officer: Brian Eslick
Executive Director - Development: Guido Belgiorno-Nettis
Regional Director - Asia: Luca Belgiorno-Nettis
Business Development Manager: David Dick-Smith
[E] dicksmithd@transfield.com. au
Engineering and Construction: Peter Gregg
■ Manufacture of folding-floating bridges and communication bridges, armoured personnel carriers, turrets (M113), reinforced composites and ballistic armour. Infrastructure construction for airports, wharfs, offshore oil platforms, railways and power stations. Infrastructure operations and maintenance. → 80 [98]

75-20811 Transfield Corrosion Protection Head Office

73 Station Road, Seven Hills, New South Wales 2147
☎(+61 2) 838 73 44 ⊖ (+61 2) 624 25 48
General Manager: Enrico Piccioli
■ Corrosion protection for infrastructure, shipbuilding, power, petrochemical, industrial and defence contracts. Hot dip galvanising, industrial blasting and painting, hot wire metallising, design, manufacture and installation of fibre reinforced plastic (FRP) components and structures. [98]

75-20812 Transfield Defence Systems Pty Ltd

(† Transfield Shipbuilding Pty Ltd)
(a subsidiary of Transfield Holdings)
Nelson House, Nelson Place, Williamstown, Victoria 3016
☎(+61 3) 92 44 40 00 ⊖ (+61 3) 92 44 44 00
⁂ 850
Chief Executive: Andrew Johnson
Group Managing Director: Paul Salteri
General Manager Services: Jim Kelly
General Manager Business Development: Garry Seaborne
General Manager Products: Ian McDougall
General Manager Commercial: Harry Kinread
Deputy General Manager Engineering and Technical: Mick Aylward
Systems Engineering Manager: William Parkins
Test and Trials Manager: Gary Reid
Integrated Logistics Support Manager: Graeme Lloyd
Training Systems Manager: Brian Archer
Industry Development Manager: Peter Black
Public Relations Manager: Pat Wilson
Communications: Ms Cath Templeton
■ Owner and operator of the Williamstown naval marine facility. Construction of ANZAC class frigates for the Royal Australian and New Zealand Navies. Defence project management; ship design; system design and integration; integrated logistics support and systems operations. Repair and refit. Technical audit services and industry development training.

Australian Capital Territory
Level 3 - AMA Building, 42 Macquarie Street,
Barton, ACT 2600
☎(+61 6) 273 42 99; 273 42 90 ⊖ (+61 6) 273 42 96
Business Development Manager: Peter Bowler

Western Australia
PO Box 516, Rockingham, WA 6168
Corner Sparks and Russell Road West,
Henderson, Western Australia 6166
☎(+61 9) 437 03 57; (+61 18) 32 50 65 (mobile)
⊖ (+61 9) 410 15 96
Manager: Kerry Britton

Northern Territory
Office 4 - Ground Floor, 68 Cavenagh Street,
Darwin, Northern Territory 0800
☎(+61 89) 41 15 30 ⊖ (+61 89) 41 25 88
Manager: Graham Johnston

Transfield Shipbuilding Victoria
Locked Bag 1, Williamstown, Victoria 3016
Nelson House - Nelson Place, Williamstown, Victoria 3016
☎(+61 3) 92 44 44 00 ⊖ (+61 3) 92 44 40 00
General Manager: Don Tellman

Transfield Shipbuilding WA
PO Box 206, Hamilton Hill, WA 6163
775 Cockburn Road, Henderson, Western Australia 6166
☎(+61 8) 94 37 04 37 [E] tranship@iinet.net.au
⊖ (+61 8) 94 10 20 65
⁂ 400 27 hectares
General Manager: Chris Evans
Technical Manager: Martin Hartman
Purchasing Manager: John Williams
Secretary, Marketing Department: Judith Oetelmans
■ Warships, landing craft, auxiliaries, coastguard vessels, workboats, submarine systems & equipment.

Transfield Shipbuilding FOSA
PO Box 2058, Fortitude Valley, Queensland
Unit 9, 43-45 Butterfield Street, Herston, Queensland
☎(+61 7) 252 33 36 ⊖ (+61 7) 252 33 71
Manager: Harry Warnick [98]

75-20813 Transfield Technologies Pty Ltd

38 South Street, Rydalmere, New South Wales 2116
☎(+61 2) 638 02 10 ⊖ (+61 2) 638 01 66 [98]

75-20840 Trevin Products Pty Ltd

46 Bay Road, Taren Point, NSW 2229
☎(+61 2) 525 25 77 ⊖ (+61 2) 540 14 43
A$ 2 500 000 ⁂ 33
Managing Dir: W.S. Vine
Manager: C. Mustica
■ Industrial type control & protection panels. Load banks for testing, AC & DC switchboards & PLC cabinets. [97]

75-20850 Trimble Navigation Australia

Level 1, 123 Gotha St., Fortitude Valley, Qld 4004
☎(+61 7) 216 00 44 ⊖ (+61 7) 216 00 88
Regional Mgr: Peter O'Connor [97]

75-20910 Tubemakers of Australia Ltd

(a subsidiary of BHP, Australia)
4th Floor, 37-49 Pitt Street, Sydney 2000
☎(+61 2) 92 55 09 00
[E] wogig!milsonspt!bartlem@toa.corp. attmail.com
⊖ (+61 2) 92 55 09 01
Managing Director: W G Johnston
Operations Director: H Terry
Sales and Marketing Director: M G Keegan
Purchasing Manager: C Leckie
Marketing Services Co-ordinator: Ms Melinda Bartlett
■ Manufacture and marketing of spun ductile iron pressure pipe, ductile iron flanged pipe, cast pipe fittings, general and automotive castings, large diameter steel coated and lined water pipe, API line pipe, piling, casing, fabricated fittings, pressure vessel fabrication, special fabricated products and tubeline, RHS and victaulic pipe. Turnkey treatment plants for domestic and industrial waste water.

Chief Administrative Office
1 York Street, PO Box 536,
Sydney, New South Wales 2001
☎(+61 2) 92 39 66 66 ✉ TUBEMAKERS Sydney
⊠ 22849 ⊖ (+61 2) 92 51 30 42
⁂ 5 093

Group General Manager and Chief Operating Officer: G M Parker

Structural Products Division
PO Box 156, Newcastle, New South Wales 2300
Industrial Drive, Mayfield, New South Wales 2304
☎(+61 49) 41 17 11 ⊖ (+61 49) 41 17 05
238
Group General Manager: L B Daley
Marketing Manager: G W Downing
■ Manufacture and marketing of welded carbon steel pipes, RHS and fittings. 98

75-21750 Universal Safety Supply Pty Ltd
1A Redbank Road, Northmead, NSW 2152
☎(+61 2) 840 07 01 ⊖ (+61 2) 683 64 89
A$ 5 400 000 84
Managing Dir: L. Weavers
Marketing & Sales Mgr: F. Shirlow
Purchasing Mgr: R. Breakwell
■ Specific development & production in the field of respiratory protection & complete body protection. Air line systems, static & portable. 97

75-22100 Varian Australia Pty Ltd
(a subsidiary of Varian Associates, USA)
679 Springvale Road, Mulgrave, Victoria 3170
☎(+61 3) 95 60 71 33 TECHNICAL Melbourne
http://www.varian.com 31229
⊖ (+61 3) 95 60 79 50
460
Managing Director: Greg Davis
Managing Director: Philip E Thomas
Operations and Technical Director: G C Davis
Operations Director: Kirsty Sheppard
Sales and Marketing Director: Michael Knowles
Sales and Marketing Director: Philip Binns
Sales and Marketing Director: Alan Marks
Purchasing Manager: Ms Mary Leah
Business Manager: M Knowles
■ Manufacture of analytical instrumentation for materials characterisation including AA, ICP-ES, ICP-MS and UV-Vis/NIR for elemental and molecular determinations. Manufacture of optical components, lenses, mirrors and optic systems. → 90

Affiliates:
Varian Associates, USA

● 6/81 Frenchs Forest Road (East), French Forest, NSW 2086
☎(+61 2) 451 91 33 24503 ⊖ (+61 2) 452 26 00 98

75-22450 Veetol Helicopters Pty Ltd
PO Box 5195C, Newcastle, New South Wales 2302
☎(+61 49) 43 53 48 maloneyb@bigpond.com
⊖ (+61 49) 28 43 01
3 200 m2
Chairman of Directors: Duan A Phillips
■ Formed in 1991 to carry former VTOL Aircraft Pty Ltd Phillicopter programme into production stage.
Manufacture two-seat light helicopters. 98

75-22890 Vision Systems Ltd
Second Avenue, Technology Park, The Levels, South Australia 5095
☎(+61 8) 83 00 44 00 http://www.vsl.com.au
⊖ (+61 8) 83 49 59 89
A$ 60,000,000 (1996) A$ 84,800,000 (1996)
500 (1997)
Managing Director: Dr James Fox
Executive Director: Dr Chris Abell
Financial Manager: M J Peacock
Company Secretary: Euan Pizzey
■ Qualified to ISO 9001, AS3901 and AS3563. Incorporates NATA registered test facility.

Vision Abell
Second Avenue - Technology Park, The Levels, South Australia 5095
☎(+61 8) 300 44 00 ⊖ (+61 8) 349 74 20
Managing Director: Lloyd Groves
■ Research and development contractor to defence and industrial clients in specialised areas: defence equipment development, manufacture, systems engineering, systems integration, project management and logistics support for airborne, naval and land based platforms. Design, manufacture and through life support of electro-optical surveillance systems, detection and image processing systems for military and industrial applications. Advanced research and development in: electro-optical and laser systems, electronic warfare systems, radar threat simulators, thermal signature simulators, electronics hardware and software development, signal and high speed data processing systems, mechanical and process control systems, aeronautical systems. Technology consulting and testing services. Products include RAPTOR military land based surveillance system comprising IR and radar sensors and data fusion technology.

Vision Products
Second Avenue - Technology Park, The Levels, South Australia 5095
☎(+61 8) 300 44 00 ⊖ (+61 8) 300 44 22
Managing Director: Peter Murphy
■ Research, development, manufacture and marketing of advanced closed circuit television systems, high speed image and audio processing systems, advanced electronic security systems, specialised image processing software. Products include video motion detection systems and video transmission systems for telephone, two way radio, cellular telephone, digital and spread spectrum radio for protection of civil and military assets.

Vision Invetech
96 Ricketts Road, Mount Waverley, Victoria 3149
☎(+61 3) 92 65 88 88 ⊖ (+61 3) 92 65 88 81
Managing Director: Peter Pick
■ Business and technolgy consultants. Business and technology strategy, market strategy and research. Consumer products and packaging development. Industrial and scientific product development. Advanced manufacturing systems and automation. Human resources management. 98

75-23090 Walkers Ltd
(a subs. of Evans Deakin Ltd)
23 Bowen St, PO Box 211, Maryborough, Qld 4650
☎(+61 71) 20 81 00 ITOLZAK 49718 itolzak aa
⊖ (+61 71) 22 44 00
A$ 50 000 000 700
Group General Mgr: R. Hardy
Commercial Mgr: G. Nelson
General Mgr, Rolling Stock: P. Bennett
General Mgr, General Engineering: J.S. Pricket
■ Railway equipment including locomotives, rolling stock & components & suburban railcars. Large naval propulsion components. Material handling systems, dragline & shovel buckets, off highway coal haulers & similar vehicles. Sugar cane processing equipment. Heavy gearing. Special castings. 97

75-23300 Weir Engineering Pty Ltd
(a subs. of The Weir Group PLC, UK)
14/25 French's Forest Rd, PO Box 115, Brookvale, NSW 2100
☎(+61 2) 99 50 51 00 BONWEIR Sydney
⊖ (+61 2) 99 50 51 01
US$ 65 000 000 (1996) 220
Managing Dir: K.B. Wilson
Operations & Technical Dir: V. Kinder
Sales & Marketing Mgr: W. McAndrew
■ Manufacture pumps, turbines & heat exchange equipment, repair & overhaul services & pump testing facilities. → 80 → 90
® UK: Weir Pumps 97

75-25000 Vision Products Pty Ltd
(a subsidiary of Vision Systems Ltd)
Second Avenue - Technology Park, The Levels, South Australia 5095
☎(+61 8) 300 44 00 http://vsl.com.au/adpro
87587 adpro aa ⊖ (+61 8) 300 44 22
A$ 50,000 A$ 12,000,000 40 5,000 sq m
Managing Director: David Lloyd
Operations and Technical Director: Rob Galic
Product Marketing Manager: Bob Thomas
Purchasing Manager: Andrew Martin
■ Microprocessor based electronics designed to be incorporated with conventional CCTV surveillance systems. Alerts guard to change in video pictures by application of high speed digital image processing techniques and products. Transmission of video across standard telephone or digital networks.

Sydney Office
Unit 6 - 192 Kingsgrove Road, Kingsgrove, New South Wales 2208
☎(+61 2) 95 54 40 00 ⊖ (+61 2) 95 54 40 42

Melbourne Office
15-17 Normanby Road, Clayton, Victoria 3168
☎(+61 3) 95 44 84 11 ⊖ (+61 3) 95 62 61 97 98

80 Service Companies & Organizations

80-00300 ACI Aviation Consultants Int'l
(a subsidiary of Handled Buy Pty Ltd, Australia)
431 Civic Square, PO Box 431, Canberra, ACT 2608
☎(+61 2) 62 47 19 88 ⊖ (+61 2) 62 57 34 94
A$ 1,000,000 4 (1997)
Managing Director: Hugh A Ritchie
h.ritchie@effect.net.au
Marketing and Sales Director: Ms Linda Parkinson
■ Consultancy services. Feasibility studies to governments, manufacturers, operators and airports throughout South East Asia. Adviser for aviation defence requirements and communications. 98

80-00650 Aircraft Equipment Overhauls & Sales (NSW) Pty Ltd (AEOS)
PO Box 294, Riverwood, New South Wales 2210
29-33 Norman Street, Peakhurst 2210
☎(+61 2) 95 33 14 44 ⊖ (+61 2) 95 34 36 30
A$ 3,000,000 (1997) 25 3,100 m2
Director: Ross J Jarman
Overhaul Director: G Jarman
Quality Director: M Jarman
Technical Services Manager: C Rolfi
■ Maintenance, repair and overhaul. → 80 → 90
® France: Auxilec, Dexaero, Equipements Aeronautiques
Germany: Elektro-Metall Export
USA: Abex Corp, American Fuel Cell Co., Aro Corporation, Essex Cryogenics/Missouri, Jet Electronics & Technical, Kidde Inc, Lear Romec Corp, Litton Instruments, Lucas Aerospace, Marathon Power Technologies, Puritan-Bennett, Scott Aviation 98

80-00720 Air Communications Pty Ltd
Hangar 4 - First Street, Moorabbin Airport, Mentone, Victoria 3194
☎(+61 3) 580 37 34; 580 95 27 ⊖ (+61 3) 580 74 05
13
Managing Director: J A Taylor
Operations and Technical Director: J McDonald
Purchasing Manager: J Taylor
■ Installation and maintenance of radios, HF and VHF (including portable), navigational equipment (airborne); weather radar; intercom systems (airborne). Global positioning systems (satellite) and stormscope systems. 98

Contact Reference

80-00840 Air Target Services Pty Ltd
PO Box 511, Nowra, New South Wales 2541
☎ (+61 44) 23 67 99 [E] admin@airaffairs.com.au
[W] http://www.airaffairs.com.au ⊜ (+61 44) 22 17 72
♣ A$ 1,600,000 (1997) 👥 8
Managing Director: Stephen A Lill ☎ (+61 68) 42 10 33
Operations Director: Jan M Ujma ☎ (+61 44) 47 85 54
Technical Director: Christopher J Sievers
☎ (+61 44) 22 16 96
■ Operation of scoring systems, targets for air-to-air and surface-to-air missiles, guns. Operation and maintenance of towing systems and RPVs for the Australian Defence Forces, under contract. [98]

80-00905 Analysis & Technology Australia Pty Ltd
(a subsidiary of Analysis & Technology Inc, USA)
Unit 4, 55 Townsville Street, Fyshwick, ACT 2609
☎ (+61 6) 280 73 47 [E] ataust@tpgi.com.au
⊜ (+61 6) 280 73 46
♧ A$ 100,000 (1996/97) ♣ A$ 300,000 (1996/97)
🏠 100 m2
Managing Director: Paul Fothergill
Operations and Technical Director: Mike Curry
■ Computer based training and defence consultancy. [98]

80-00910 Andersen Consulting
(Member firm of Andersen Worldwide)
PO Box 4329, Sydney, NSW 2001
☎ (+61 2) 99 27 50 00 ⊜ (+61 2) 99 27 50 12
♣ A$ 1 000 000 000 (1995) 👥 6 000
Managing Partner - Government: David Hunter
Sales & Marketing Dir: Chris Burrows
Operations/Technical Dir: Steve Farmer
Purchasing Mgr: Jonathan Brown
Defence Industries Team: Roger Dooley
■ Systems development, artificial intelligence applications, strategic planning, education & training, productivity & effectiveness reviews, organisational studies & project management assistance. [97]

80-01000 ANI Engineering Services Pty Ltd
(a subsidiary of Australian National Industries Limited)
Level 5 - Merlin Centre, 235 Pyrmont Street,
Pyrmont, New South Wales 2009
☎ (+61 2) 95 77 67 00 ⊜ (+61 2) 95 77 68 88
Chief Exec.: J.C. Jeremy [98]

80-01010 Ansett Australia Avionics
(† Ansett Technologies)
Airport West, Victoria 3042
Corner Matthews Avenue and McIntosh Street
☎ (+61 3) 93 39 56 00 ⊜ (+61 3) 93 38 39 30
👥 160 🏠 2,000 m2
General Manager: Michael Brown
Financial Controller: Sean Kelley
Sales and Marketing Manager: B W Lloyd
Technical Librarian: P Fanning
■ Maintainance of avionic test equipment; repair and overhaul of instruments, electrical radio and radar equipment; training of personnel on aircraft equipment, repair and overhaul. [98]

80-01150 Aspect Computing Pty Ltd
551 Glenferrie Rd, Hawthorn, Vic 3122
☎ (+61 6) 247 76 77 ⊜ (+61 6) 249 16 20
♣ A$ 75 000 000 (1995) 👥 700 (1996)
Managing Dir: Mrs Lyndsey Cattermole
☎ (+61 3) 98 18 06 04
Operations/Technical Dir: David Jones
Sales & Marketing Dir: Ian Farrington
Purchasing Mgr: Brad Reeves
Melbourne Contact: J. Thomas ☎ (+61 3) 98 18 06 04
Sydney Contact: M. Adams ☎ (+61 2) 99 57 31 88
Canberra Contact: G. Rhodes ☎ (+61 6) 247 76 77
Adelaide Contact: C. Anderson ☎ (+61 8) 379 04 66
■ All types of ADP systems development & modification working with defence organisations or their suppliers. [97]

80-01170 ASTA Defence
Beach Road, Lara, Victoria 3212
☎ (+61 3) 52 82 29 88 ▣ 35658 astaav aa
⊜ (+61 3) 52 82 33 45
General Manager: Colin Giles
Human Resources Manager: Steve Kenedy
Commercial Manager: Terry Brown
Controller: Leon Fitzgerald
Project Manager: George Greaves
Project Manager: Brian Robinson
Project Manager: Jeff Walton
Project Manager: Alan Hickey [98]

80-01180 Asteg SA Pty Ltd
416 Churchill Road, Kilburn, SA 5084
☎ (+61 8) 349 44 11 ✆ (800) 88 85 48
⊜ (+61 8) 349 81 21
Managing Director: Darrly Cowan
Marketing Director: Greg Murphy
Product Manager: Warwick Neil
® Switzerland: Trimos-Sylvac Task Force [98]

80-01880 Australian Technical Publications (ATP)
PO Box 146, St Agnes, SA 5097
Building 85, Contractors' Area, Defence
Science & Technology Organisation, Salisbury, SA
☎ (+61 8) 255 50 10 ⊜ (+61 8) 287 13 70
General Mgr: Peter Haley
IT Mgr: Eric Youle
Engineering Mgr: John Bell
■ Provides technical support to defence & industry. Experienced, equipped & accredited to undertake all phases of technical publication production to defence & commercial standards.
The company can provide experienced staff to undertake technical writing & drawing production, development of on-line publishing & information systems & a range of engineering & quality assurance/control support services. [97]

80-02800 CAE MRad Pty Ltd
(† MRad Pty Ltd)
(a division of CAE Electronics (Aust) Pty Ltd)
Innovation House, Technology Park,
The Levels, South Australia 5095
☎ (+61 8) 82 60 89 42 [E] mrad@mrad.com.au
[W] http://www.cae.ca ⊜ (+61 8) 82 60 89 80
👥 50 🏠 900 m2
Operations Manager: Tom Jonhammar
Adelaide Marketing Manager: Peter Tilsley
■ Radar environment generation and sensor stimulation products and systems. Supplied for F-18 radars, naval shipborne radars, air defence radars, electronic support measures and guided missile stimulation. [98]

80-03100 CEA Technologies Pty Ltd
CEA Research & Development Pty Ltd
65 Gladstone Street, PO Box 1095, Fyshwick, ACT 2609
☎ (+61 6) 280 79 32 [E] cea@cea.com.au
[W] http://www.cea.com.au ⊜ (+61 6) 280 48 36
♧ A$ 394,000 (1996) ♣ A$ 12,000,000 (1996)
👥 90 (1996) 🏠 1,600 m2
Managing Director: David J Gaul
Research and Development Director: Ian Croser
Marketing Engineer: Mark Foster
Marketing Assistant: Ms Angela Petty
■ Products: advanced radar processing systems, vessel traffic services (VTS) and coastal radar surveillance systems, communications ESM and DF systems, ACRISS air traffic control system, active communications antennas, advanced multi-sensor data fusion and display systems and software, solid state continuous wave illuminator transmitter, sips integrated communication system, 3D radar, equipment shelters and towers, specialised liquid cooling systems and hydraulic/pneumatic system components.
Capabilities: research and development, system design and integration for multi-discipline and complex systems including electronic, marine engineering and software components. Advanced modelling and simulation tools including finite element analysis. Advanced 3D mechanical CAD systems. ILS capabilities. Manufacture of prototype electronic, electro-mechanical and hydraulic/pneumatic components. Defence accredited to ISO9001.

Melbourne Office
25 Cumberland Drive, Seaford, Victoria 3198
☎ (+61 3) 97 76 96 77 ⊜ (+61 3) 97 76 97 78
Director: Peter N Croser

Adelaide Office
5/77 Fullarton Road, Kent Town Centre, SA 5071
☎ (+61 8) 83 31 80 44 ⊜ (+61 8) 83 31 28 11
Director: Michael G Atchison [98]

80-03400 Chopperline Flight Training
12 Pathfinder Drive, Caloundra West, Queensland 4551
☎ (+61 754) 91 85 88
✆ (+61 800) 07 21 18 (Australia only)
⊜ (+61 754) 91 85 19
♣ A$ 1,900,000 👥 9
Managing Director: B W Costa
Sales and Marketing Director: B Webb
Chief Flying Instructor/Chief Pilot: H M Matheson
■ Military and civil fixed and rotary wing pilot school. [98]

80-03600 CMPS&F Pty Limited (CMPS)
PO Box 201, Chatswood, New South Wales 2057
South Tower, The Interchange 67 Albert Avenue
☎ (+61 2) 94 12 99 99 [W] http://www.cmpsf.com.au/
⊜ (+61 2) 94 12 98 55
♣ A$ 126,000,000 (1996) 👥 1,250 (1997)
Managing Director: J D Cooper
General Manager International: M Ponsonby
General Manager Oil and Gas: S Hollis
General Manager Infrastructure: M Peergrane
General Manager Mining and Manufacture: R E Spence
General Manager Environmental: J R Moody
Public Relations Manager: Ms Kay E Turnbull
■ Professional engineering services: telecommunications, mechanical, electrical, fibre optics instrumentation and control, civil, structural, chemical, computers; project planning, scheduling, estimating, cost and progress control; project management, quality assurance inspection services, configuration management, integrated logistic support services, environmental services, transportation studies, contracts claims analysis. [98]

80-03730 Computer Power Group Ltd
43 Geils Court, West Deakin, ACT 2600
☎ (+61 6) 283 67 77 [E] pille@cbr.cpg.com.au
[W] http://www.cpg.com.au ⊜ (+61 6) 281 57 89
♧ A$ 59,000,000 (1996) ♣ A$ 165,200,000 (1996)
👥 1,400 (1997)
Managing Director: Peter James
Operations Director: Peter Madden
Business Development Manager: Colin Pilliner
Purchasing Manager: Ms Deborah Louden
■ Strategic consulting services, computer systems and programming analysis. Project management, quality management, systems integration, ILS and CALS.

Subsidiaries:
CP International Inc (Hong Kong)
Computer Power Training Institute
MTE (Australia)
Computer Power Group Pty Ltd (Singapore) [98]

80-03780 Cordwell Investigations
94 Glen Dhu Street, Launceston, Tasmania 7249
☎ (+61 3) 63 44 44 87
[E] ac-cordwell@postoffice.stpatricks.tas. edu au
⊜ (+61 3) 63 44 44 87
♣ A$ 62,000 (1995) 👥 2 🏠 100 m2
Managing Director: Tony Cordwell
■ Provides investigation and counter industrial espionage services to industry. [98]

80-03850 Currie & Brown (Australia) Pty Ltd
(† Currie & Brown, Cameron & Middleton Pty Ltd)
135 Wickham Terrace, PO Box 161,
Spring Hill, Queensland 4004
☎ (+61 7) 38 31 32 61 [E] cbbris@curriebrown.aust.com
⊜ (+61 7) 38 32 03 41
👥 11
Director: Robert I McCallum
■ Specialists in contract documentation and cost management for support facilities. Services include: quantity surveying, contract engineering, cost engineering, cost planning and estimating, valuations, feasibility studies.

Melbourne Office
441 St Kilda Road, Melbourne, Victoria 3004
☎ (+61 3) 98 67 30 11 ✉ CEEMEL ▣ 33181
⊜ (+61 3) 98 67 59 50
Director: Keith Irwin
Director: Peter Byrne

Chatswood Office
2 Thomas Street, Chatswood, New South Wales 2067
☎ (+61 2) 94 15 16 00 ⊜ (+61 2) 94 15 14 43
Director: Eric Liddell
Director: Mike O'Shea

Adelaide Office
Level 18 - State Bank Centre, 91 King William Street, Adelaide, South Australia 5000
☎ (+61 8) 231 72 88 ⊖ (+61 8) 211 87 19
Director: Lloyd Goudge

Perth Office
Level 4 - 19 Pier Street, Perth, Western Australia 6000
☎ (+61 9) 325 80 99 ⊖ (+61 9) 321 80 59
Director: Phil Angel [98]

80-04000 Domain Defence Systems (Aust) Pty Ltd
PO Box KL 921, Kings Langley, NSW 2147
☎ (+61 2) 96 74 48 93 ⊖ (+61 2) 96 74 38 47
👥 15 (1997)
Chairman of Directors: J D G Marsh
Operations Director: J P Marsh
Sales and Marketing Director: L Breen
Purchasing Manager: R Hancox
■ Single and three-axis measurement magnetometers; EOO magnetometers, airborne weapons launch systems, cartridge and propellant activated devices. [98]

80-06500 Evro Pacific Partners Pty Ltd
(† Cooksey Consulting Pty Ltd)
Suite 705 - Barrack House, 16-20 Barrack Street, Sydney, New South Wales 2000
☎ (+61 2) 92 99 89 99 ⊖ (+61 2) 92 99 81 05
Executive Chairman: Robert J Cooksey [98]

80-07500 Gutteridge Haskins & Davey Pty Ltd
39 Regent Street, PO Box K839, Haymarket, New South Wales 1238
☎ (+61 2) 96 90 70 70 [E] sydneymail@ghd.com.au
[W] http://www.ghd.com.au ⊖ (+61 2) 96 98 17 80
♣ A$ 98,900,000 (1997) 👥 900 (1997)
Chairman: Henry Adcock ☎ (+61 7) 32 58 36 00
Managing Director: Peter Manger ☎ (+61 7) 92 78 22 00
Sales and Marketing Director: Ms Evelyn Raedler
Marketing Services Co-ordinator: Ms Kellie Rhyll Parkin
■ Engineering design, project management, feasibility studies, project planning and economic assessments, design, documentation, construction management, training, operations and maintenance, asset management, quality management to ISO 9001. [98]

80-08400 Helitech Pty Ltd
(† BH Pacific Pty Ltd)
(a subsidiary of Helitech Industries Pty Ltd, Australia)
PO Box 138, Hamilton Central, Queensland 4007
☎ (+61 7) 32 68 40 77
[E] 1002530433@compuserve.com
⊖ (+61 7) 32 68 78 15
👥 75
Managing Director: Joe V Moharich
General Manager: Robert M Millar
Sales and Marketing Manager: Peter Crook
Purchasing Manager: Ms Pam Harris
■ Helicopter repair, maintenance and overhaul. → 90
® USA: Bell Helicopter Textron [98]

80-08800 Hughes Australia International Pty Ltd
MLC Centre, Level 66, Martin Place, Sydney, NSW 2000
☎ (+61 2) 238 21 23 ⊖ (+61 2) 235 12 48 [98]

80-11000 Kinhill Engineers Pty Ltd
(a subs. of Kinhill Ltd)
Head Office
168 Greenhill Rd, Parkside, SA 5063
☎ (+61 8) 301 12 34 ⊠ 82364 aa ⊖ (+61 8) 301 13 01
♣ A$ 105 000 000 (1995) 👥 1 300
Exec. Chairman: Malcolm Kinnaird
☎ (+61 8) 301 12 34; 12 01 (fax)
Managing Dir: David Klingberg
Federal Government Business Dir: AVM H.F. Roser (RAAF Retd) ☎ (+61 2) 257 52 45; 52 58 (fax)
■ Design & construction management airfields, harbours, factories, defence infrastructure development. Environmental services, feasibility & economic studies, quality assurance service, consultancy services, supply support services, defence acquisition, logistics & project management services. [97]

80-12500 Logica Pty Ltd
(a subsidiary of Logica plc, UK)
30 Collins Street, Melbourne, Vic 3000
☎ (+61 3) 92 03 43 00 ⊖ (+61 3) 92 03 43 33
Managing Director: Ms Carmel Gray
Communications Manager: Ms Carolyn Chaplin
■ Operational analysis and system design, modelling, simulation, command and control systems, knowledge based systems, system evaluation and procurement, hardware and software engineering and project management.

North Sydney Office
76 Berry Street - Level 5, North Sydney, NSW 2060
☎ (+61 2) 92 02 81 00 ⊖ (+61 2) 92 02 81 00 [98]

80-12700 Logistics Pty Ltd
PO Box 848, Glen Waverley, Victoria 3150
☎ (+61 3) 98 88 23 44 [W] http://www.logistic.com.au
⊖ (+61 3) 98 88 23 55
👥 8 (1997)
Managing Director: Adrian Stephan
Business Manager: Ms Jill Batkin
■ Consulting logistics, including preparation of management plans for logistics services; technical writing; life cycle cost and trade-off analysis. Analytical software tools. [98]

80-12850 Lucas Aerospace - Australia
(a division of Lucas Varity Ltd)
PO Box 65, Waterloo, New South Wales 2017
84 Epsom Road, Zetland, Sydney, New South Wales 2017
☎ (+61 2) 93 15 70 00 ⊖ (+61 2) 93 13 76 76
♣ A$ 18,000,000 (1997) 👥 90 (1997) 🏭 3,000 m2
Director and General Manager Asia Pacific: John R Harper
Business Manager Sydney: John W Mason
Sales Manager: Ross Whitmarsh
■ Aerospace repair, overhaul and test of aircraft systems. [98]

80-12980 Lynx Corp Pty Ltd
Locked Bag 71, Fivedock, New South Wales 2046
21 Bibby Street, Chiswick, New South Wales 2046
☎ (+61 2) 97 13 22 22 ⊖ (+61 2) 97 13 22 11
👥 20 (1997) 🏭 2,000 m2
Managing Director: John Bruderlin
Director: P Hayes
■ Reconditioning and overhaul of large industrial, automotive, marine, diesel and petrol engines, compressors, associated plant and equipment. Complete machining facilities and dynomometer test cells. [98]

80-12990 MacDonald Dettwiler Limited
North Sydney, NSW 2059
Level 9, 60 Miller Street, North Sydney, NSW 2060
☎ (+61 2) 956 79 66 ⊖ (+61 2) 956 78 46
👥 42 [97]

80-13800 Milsearch Pty Ltd
PO Box 1339, Tuggeranong, ACT 2900
☎ (+61 6) 293 29 99 ⊖ (+61 6) 293 19 66
♣ A$ 2 500 000 (1995/96) 👥 26
Managing Dir: David Halmarick
Dir Operations: Ron Furze
Operations Mgr LAO PDR: Ronald Hawkins
■ Consultancy in the rehabilitation of land contaminated by military activity. Consultancy in all aspects of explosive ordnance safety and explosive ordnance storage (including regulations & requirements). Detection and clearance of UXO on land and in water, utilising a full kit of EOD tools including digital imaging magnetometers and ground probing radar. [97]

80-15830 Orbital Engine Company (Australia) Pty Ltd
(a subsidiary of Orbital Engine Corp Ltd)
1 Whipple Street, Balcatta, Western Australia 6021
☎ (+61 9) 441 23 11 [E] oemai@orbeng.com.au
[W] http://www.orbeng.com.au ⊠ 94415
⊖ (+61 9) 441 23 45
♧ A$ 61,563,592 (1995) ♣ A$ 20,867,998 (1995)
👥 250 (1997) 🏭 10,818 m2
Managing Director: B A Fitzgerald
Sales and Marketing Director: Ken Johnsen
Purchasing Manager: John Cusack
Public Affairs Co-ordinator: Michael Aspinall
Operations and Technical Engineer: Ken Seeber
Sales and Marketing: Darren Smith
■ Engines for marine and automotive applications. Orbital combustion process light 2-cycle engine. Orbital pneumatic injection system. [98]

80-16000 The OSIRIS Group
(a division of Leading Edge PC Consultancy Pty Ltd, Australia)
Leading Edge PC Consultancy Pty Ltd, Abignano House 7/19-23 Bridge Street, Pymble, New South Wales
☎ (+61 2) 94 88 74 71 [E] leading@magna.com.au
[W] http://www.leadingedge.net.au ⊖ (+61 2) 94 88 74 38
👥 30
General Manager: Timothy Scott
[E] tscott@laurel.ocs.mq.edu. au
Advertising Director: Edward Pollitt
■ Information warfare, information security and intelligence services. [98]

80-16490 Philips Calibration Services
2 Greenhills Avenue, Moorebank, NSW 2170
☎ (+61 2) 600 54 06 ⊖ (+61 2) 600 56 70
♣ A$ 2 000 000 (1995) 👥 21
Managing Dir: I.B. Buchanan
Operations Dir: Ian Kinnard
Sales & Marketing Dir: M. Battaglia
Calibration Business Mgr: Richard Soulie
Purchasing Mgr: H. Proctor
Public Relations Mgr: Ms Catherine Flood
☎ (+61 2) 600 54 11
■ Calibration & maintenance of electrical & metrology equipment. Training of technical staff, procedure development, calibration management of customer sites. Standards laboratory offers traceability to national & international standards in compliance to:- ISO 9001, Australian standards AS2415, AS3901 & AS1822. Environmental testing of large devices in the areas of thermal cycling vibration. [97]

80-16700 PM2 Systems Pty Ltd
Level 9 - 228 Queen Street, Melbourne, Victoria 3000
☎ (+61 3) 96 42 85 00 ⊖ (+61 3) 96 42 85 85 [98]

80-16710 PM2 Systems Pty Ltd
Level 2 Tandem House, 76 Berry Street, North Sydney, NSW
☎ (+61 2) 99 56 82 22 ⊖ (+61 2) 99 56 84 70
Country Manager: Roy Edwards [98]

80-16950 Pratt & Whitney Canada (A'Asia) Pty Ltd
(a subsidiary of Pratt & Whitney Canada Inc)
PO Box 758, Hamilton Central, Queensland 4007
30 Industry Court, Eagle Farm, Queensland 4009
☎ (+61 7) 32 68 51 99 ⊖ (+61 7) 32 68 52 09
Regional Manager: Stephen Bell [98]

80-19100 Scientific Management Associates Pty Ltd (SMA)
200-202 Gladstone St, PO Box 774, Fyshwick, ACT 2609
☎ (+61 62) 39 11 33 ⊠ 61444 ⊖ (+61 62) 80 52 04
👥 210
Group Managing Dir: Keith E. Snell
General Mgr: V.C. Brice
General Mgr: O.F. Culley
General Mgr: W.L. King
■ Engineering & project management support; integrated logistic support; documentation/technical manuals; QA/QC assessments & reporting; contract administration; AIP representation, planning & implementation; training analysis & design; training implementation.
® USA: Integrated Support Systems, Republic Electronics
Subsidiary Companies:
SMNZ Limited (Auckland, NZ)
SMA (UK) Limited (Fareham, UK) [97]

80-19350 Skilled Engineering Ltd
45 Buckhurst Street, South Melbourne, Victoria 3205
☎ (+61 3) 96 99 41 99 [W] http://www.skilled.aust.com
⊖ (+61 3) 96 96 38 23
♧ A$ 21,000,000 ♣ A$ 350,000,000 👥 6,000
Managing Director: F Hargrave
General Manager: C Dyball
National Marketing Manager: Mrs S Weber
Finance Manager: G Buchanan

Contact Reference

■ Project management: defence, facility-total engineering maintenance programs. PC-Maintenance: Computerised maintenance management system communications, facilities management. [98]

80-20000 Technical & Field Surveys Pty Ltd
Friesdale Park, 1562 Limekilns Road, Bathurst, NSW 2795
☎ (+61 63) 37 66 60 ⊜ (+61 63) 37 66 67
👥 7
Managing Dir: M.C. Aubrey
Director: B. Aubrey
■ Geological, terrain & environmental studies; remote sensing applications, regional interpretation of overseas geography & geology resource inventories. Cartographic mapping & land cover assessment. [97]

80-20600 Technology Australasia Pty Ltd
1010 Doncaster Road, Doncaster East, Victoria 3109
☎ (+61 3) 98 41 97 33 [E] taa@taa.ccom.au ☒ 38615
⊜ (+61 3) 98 41 83 74
👥 20 🏠 50 m2
Managing Director: Robert J Halligan
Sales and Marketing Director: Alan Conrad
■ Service company to the defence industry, providing market research, program profiles, government liaison and in-country project management, technical specialists. [98]

80-20700 Transfield Power Services
38 South Street, Rydalmere, New South Wales 2116
■ Plant maintenance for power generators. [98]

80-20900 TSB International Pty Limited
Falcon Business Centre, 47 Falcon Street,
Crows Nest NSW 2065
☎ (+61 2) 99 23 13 35
Contact: Stuart Robins [97]

80-22580 Vipac Engineers & Scientists Ltd
Victorian Technology Centre, 275 Normanby Road,
Port Melbourne, Victoria 3207
☎ (+61 3) 96 47 97 00 ⊜ (+61 3) 96 46 43 70
♣ A$ 12,000,000 (1996) 👥 90 (1997) 🏠 8,000 m2
Managing Director: M J Smith
General Manager: John C Simmons
■ Engineering consulting and applied research services specialising in acoustics, vibration, fluid dynamics and dynamics. Quiet vessel/vehicle, sonar and aerospace structure design.
® Japan: Ono Sokki
USA: Larson Davis, DLI

South Australia Office
21 King William Street, Kent Town, South Australia 5067
☎ (+61 8) 362 54 45 ⊜ (+61 8) 362 07 93
👥 10
Managing Director: M J Smith
South Australia Manager: D Rennison
■ Consulting services in the field of noise and vibration control. [98]

85 Stockists, Distributors, Spare Parts Suppliers

Applied Measurement Aust Pty Ltd	75-01420
Avtronics (Australia) Pty Limited	75-01995
Colan Products	75-03650
Forrester Stephen Aviation	90-06830
Lennys Commercial Kitchens Pty Ltd	75-12370
Thorn DNT Airfield Lighting Pty Ltd	75-20540
Vaisala Pty Ltd	90-21500

85-00500 Aviaquip Pty Ltd
41 Kembla Street, Cheltenham, Victoria 3192
☎ (+61 3) 95 85 12 11 ⊜ (+61 3) 95 85 18 37
General Manager: R D Darbyshire
■ Military and defence suppliers. [98]

85-01000 Avtronics Pty Ltd
(† Christie Electric Australia)
33 Higginbotham Road, Gladesville,
Sydney, New South Wales 2111
☎ (+61 2) 98 07 14 44 (7 lines) ⊜ (+61 2) 98 09 71 36
♧ A$ 100,000 (1997) ♣ A$ 3,000,000 (1997)
👥 15 (1997) 🏠 10,000 sq ft (1997)
Aviation Director: W Jack Cairns
Advertising Manager: Jack Cairns
■ Aviation GSE, hi-intensity searchlights and ground power units. Diesel and electric, flashlights, Ni-Cad batteries, chargers and analysers, borescopes, infra red equipment and precision optical component manufacturing. → 80

Subsidiary companies:
Francis Lord Optics
Filmtronics Australia
Angenieux Australia
Meopta Australia [98]

85-02900 BTR Engineering (Australia) Ltd
(a subsidiary of BTR Nylex Ltd)
Sales & Service Division
21 Browns Rd, PO Box 173, Clayton, Vic 3168
☎ (+61 3) 95 44 48 88 ☒ 31502 aa
⊜ (+61 2) 95 43 45 53
♣ A$ 50 000 000 👥 548 👥 260
General Mgr: N. Wilson
National Commercial Mgr: B. McConnell
Financial Controller: G. Smith
Regional Mgr (Vic/SA/Tas): R. Tidswell
■ Supplier of valves & controls, maintenance products, pumps, fasteners, industrial composites, power transmissions, shaft couplings, bearing systems, backstops & camshaft clutches, industrial bearings, water lubricated bearings, marine bearing material, marinised diesel engines & gearboxes. Diesel engines & generators. Mobile & marine transmissions, linear actuators. Life jackets & survival suits. Safety flooring. [97]

85-03000 Cadillac Plastics Australia
(a subs. of M.A. Hanna Corp.)
PO Box 145, Ermington, NSW 2115
☎ (+61 2) 748 37 22 ⊜ (+61 2) 648 43 73
♧ A$ 7 000 000 ♣ A$ 60 500 000 (1995) 👥 132
Managing Dir: M.J. Binnie
Engineering Plastics Marketing Mgr: David Evans
Aerospace & Transportation Marketing Mgr: Peter Heidecker
Finance & Admin Group Mgr: L. Ward
General Mgr Australia: A.J. Mouw
■ Distribution of aerospace & engineering plastics for military & domestic applications. Product range includes composite processing materials, such as Richmond aircraft products & composite processing material. Texstar aircraft transparencies, Polycast, military specification acrylics, Teklam composite panels, Orcon Corporation insulation films, tapes & systems, Wichitech portable hot bonding units & a comprehensive range of aircraft & industrial grade thermoplastic sheets, plus a comprehensive range of engineering plastics. [97]

85-03500 Chadwick Downs Pty Ltd
(a member of the Air Affairs Australia Pty Ltd Group)
PO Box 205, Coonabarabran, NSW 2357
☎ (+61 68) 42 17 57 [E] cdpl@ozemail.com.au
⊜ (+61 68) 42 22 87
♧ A$ 8,000,000 ♣ A$ 5,000,000 (1997) 👥 16
🏠 28,500 sq ft
Managing Director: Stephen A Lill ☎ (+61 68) 42 10 33
Director: Ms Elaine Lill ☎ (+61 68) 42 10 33
Company Secretary, Finance Director: Stephen Thom ☎ (+61 3) 98 36 19 66
Sales and Marketing Manager: P Guy
Purchasing Manager: I Bassett
■ Suppliers of equipment and technical services to the Australian DoD. Hayes Targets, aircraft modifications, target reeling machines, towlines, radar and acoustical scoring systems, Target and surveillance drones.
® UK: Ampfield Power, Racal Radar Defence Systems, Sabre Computer Scoring Ltd, Target Technologies Ltd
USA: Cartwright Electronics Inc, Hayes Targets-Pemco Aeroplex, Kaiser Marquardt Co., South West Aerospace Inc

Associated companies:
Air Affairs, New Zealand Ltd Air Target Services Pty Ltd, Australia Air Affairs (Australia) Pty Ltd [98]

85-10000 Laser Electronics (Operations) Pty Ltd
PO Box 359, Southport, QLD 4215
9 Price St, Nerang, QLD 4211
☎ (+61 7) 55 96 01 77 ⊜ (+61 7) 55 96 35 30
👥 6
Managing Dir: N. Walden
Sales & Marketing Mgr: C. Holberton
Administration: R. Walden
■ Distributor/representative of laser optics, electro-optics and micro-positioning equipment. [97]

85-14000 Outboard Marine Corporation (Australia) Pty Ltd.
(a subs. of Outboard Marine Corp, USA)
56 Canterbury Rd, Private Bag # 92,
Bankstown, NSW 2200
☎ (+61 2) 794 66 00 ⊜ (+61 2) 794 66 97
♧ A$ 19 800 000 (1995) ♣ A$ 64 000 000 (1995)
👥 140
General Mgr: G. McFarlane
Marketing Dir: G. Henniker
Advertising Mgr: G. Henniker
Secretary/Treasurer: B.L. Hazell
Purchasing Mgr: Harry Dawson
■ Supply of Johnson & Evinrude outboard motors for military & other uses. Also Haines Hunter Fibreglass runabouts & stacer aluminium boats. [97]

85-16000 Racal Australia Pty Ltd
3 Powells Road, Brookvale, New South Wales 21002131
☎ (+61 2) 936 70 00 ⊜ (+61 2) 936 70 36
♣ A$ 10,000,000 👥 20
Director and General Manager: Phil Taylor
Sales Manager: L Solvyns
■ Marketing, sales and support for RF communications, radar avionics, electronic warfare, electronic surveillance, electro-acoustics, data communications, voice and data security, voice and data logging recorders, test and measurement instrumentation, products and systems. → 90
® UK: Racal Electronics Group Plc
USA: Antenna Products Corp., Racal Communications Inc. [98]

90 Sales Agents, Sales & Liaison Offices

Companies represented:

3M, Australia	90-20150
Aaxico/Negretti, UK	90-06830
Abex Corp, USA	80-00650
ACR Electronics Inc, USA	90-20600
Aermacchi SpA, Italy	90-14650
Aero-Motive Mfg Co., USA	75-13160
Air-A-Plane Corp, USA	90-06830
Airscrew Howden Ltd, UK	75-08740
Alfred Karcher GmbH & Co, Germany	90-11000
ALL States, USA	90-20150
Allen ACFT Products Inc, USA	90-20600
Allen-Bradley Pty. Ltd, Australia	90-20150
Allied-Signal Aerospace Co., USA	90-00650
AlliedSignal, USA	75-14870
Alloy Surfaces Inc., USA	75-16200
American Fuel Cell Co., USA	80-00650
AMP, USA	75-01780
Ampfield Power, UK	85-03500
Amplica Inc, USA	90-18820
Anglo SwissOptical Co. plc, UK	90-16320
Antenna Products Corp., USA	85-16000
Apcom Inc, USA	90-18800
Apollo Spray Equipment, USA	90-08300
Appa Technologies, Taiwan	90-20150
Armaturen Gesellschaft mbH, Austria	90-16320
Aro Corporation, USA	80-00650
Assurance Technologies Inc, USA	90-06830
Autoflug GmbH, Germany	90-12600
Auxilec, France	80-00650
Aviation Tech & Resources, USA	90-06830
Barringer Instruments Ltd, Canada	90-14650
Basewest Inc, USA	90-20600
Bell Helicopter Textron, USA	80-08400
Belling Lee Ltd, UK	90-20150
Berkefeld Filter Anlagenbau GmbH, Germany	90-05600
Bernhardt Apparatebau, Germany	90-12600
Beta Aktiengesellschaft, Liechtenstein	90-16320
BFGoodrich, USA	90-20600
Blucher GmbH, Germany	90-12600
Buck, Germany	90-13620
Bulova Technologies Inc, USA	90-05600
Bussmann Manufacturing, USA	90-20150
CAE-Electronics Ltd, Canada	90-14650
Cam Lock (UK) Ltd., UK	90-12600

Camberley Rubber Mouldings, UK 90-12600
Carthage Machine Tool, USA 75-13160
Cartwright Electronics Inc, USA 85-03500
Centronic Industries, Taiwan 90-20150
Chemring plc, UK 75-16200
Conax Florida Corp., USA 90-20600
Concept Engineering plc, UK 75-16200
Continental Microwave & Tooling Co., USA 90-18820
Control Products Corp., USA 90-20600
Cosmos SpA, Italy 90-05600
Crompton Instruments Ltd, UK 90-03750
Crouzet, France 90-06830
Dawe Co., UK 90-20150
Dayton Granger Inc., USA 90-20600
Denon Instruments Co., Japan 90-20150
Denro Inc, UK 75-19475
Detroit Diesel Corporation, USA 75-04450
Dexaero, France 80-00650
Dowty Armourshield, UK 90-06830
Drello, Germany 90-13620
Dresser Inc., USA 75-13160
Dresser Industries, Roots Division, USA 75-08740
Dunlop Aviation Division, UK 75-02815
Dynalec Corporation, USA 90-02000
Dynamit Nobel Graz (DNG), Austria 90-05600
Dynapower-Strato Power Inc., USA 90-20600
Eeco Ltd, USA 90-20150
EG&G Astrophysics, USA 90-14650
EG&G Wakefield, USA 90-20150
Electro-Adapter, USA 90-18820
Electro-Metrics, USA 90-18820
Electrochem Ind., USA 90-20150
Elektro-Metall Export, Germany 80-00650
Equipements Aeronautiques, France 80-00650
Ericsson Microwave Systems AB, Sweden 75-05820
Essex Cryogenics/Missouri, USA 80-00650
Fairchild Scandinavian Bellyloading Co., Sweden 90-02000
Filotex, France 90-18820
Flanders Filters Inc., USA 75-05680
Flight Components, Switzerland 90-06830
Fluke, USA 90-20150
FMA, Canada 90-06830
G&H Technology Inc., USA 90-20600
G L S Gesellschaft für Logistischem Svc, Germany 90-05600
GEC Marconi Aerospace, USA 90-20600
General Motors, Allison Transmission Div, USA 75-04450
General Motors, Electro Motive Division, USA 75-04450
Genyex Corp., USA 90-20600
Georg Fischer Ltd, Switzerland 75-13160
Glückauf Logistik, Germany 90-05600
Gough Holdings, UK 75-13160
Gulton Industries, USA 90-20150
Hale Hamilton (Valves) Ltd, UK 75-06835
Haley & Weller Ltd, UK 75-16200
Hall & Watts Defence Optics Ltd, UK 90-16320
Harris Manufacturing Co. Inc, USA 90-02000
Hayes Targets-Pemco Aeroplex, USA 85-03500
Heinrich C. Kosmeier, Germany 90-20150
Helmets Ltd., UK 90-12600
Herkules Fahrzeuglogistik GmbH, Germany 90-05600
Hino Motors Ltd, Japan 90-05000
Hirtenberger AG, Austria 90-16320
Hokuriku Company, Japan 90-20150
Howden Air Control Ltd, UK 75-08740
Howden Wade Ltd, UK 75-08740
Hughes Treitler Inc., USA 90-20600
Hycor Inc., USA 90-12600
I.B.D., Germany 90-13620
ILC Dover Inc., USA 90-20600
IMC Magnetics Inc., USA 90-20600
Innovative Industrial, Taiwan 90-20150
Integrated Support Systems, USA 80-19100
International Power, USA 90-20150
Jet Electronics & Technical, USA 80-00650
Kaiser Marquardt Co., USA 85-03500
Kay Electrics, USA 90-20150
Kidde Inc, USA 80-00650
Kunkle Valve Co. Inc, USA 75-08740
L.M.Ericsson, Sweden 75-05820
Larson Davis, DLI, USA 80-22580
Laser Products Inc, USA 90-11800
Lear Romec Corp, USA 80-00650
Lindapter Bolt Adaptors, UK 75-06835
Litton Aero Products, USA 90-04050
Litton Guidance and Control Systems, USA 90-04050
Litton Industries Inc, USA 90-12650
Litton Instruments, USA 80-00650
Lockheed Martin, USA 90-14650
Lockheed Martin Overseas Corp., USA 90-19900
Lord Corporation, USA 90-06830
Lubriquip Div of Houdaille, USA 75-13160
Lucas Aerospace, USA 80-00650
Lucas Weinschel Engineering, USA 90-18800
Mace Security Industries Inc, USA 90-11800
MAFO-Howden BV, Netherlands 75-08740
Manroy Engineering, UK 90-16320
Marathon Power Technologies, USA 80-00650
Marathon Power Technologies, USA 90-20600
Marquardt & Schulz GmbH, Germany 90-12600
Matra Australia Pty Ltd, Australia 90-01800
Matra Hachette, France 90-13130
Matra Marconi Space, France 90-13130
Matra Marconi Space, UK 90-01800
Matra Marconi Space France, France 90-01800
Matra Transport International SA, France 90-01800
Matra Transport International SA, France 90-13130
McMurdo Ltd, UK 75-16200
McTaggart Scott & Co. Ltd, UK 75-08740
Menlo Industries, USA 90-18820
Meridian Ocean Systems, USA 90-02000
Moldip SpA, Italy 90-12600
Moog Inc, USA 90-13750
MTU Diesel Engines, Germany 75-04450
Mustang Survival, USA 90-20600
Narda, USA 90-18820
Nico Pyrotechnik, Germany 90-11800
Noah Howden Inc, USA 75-08740
Normalair-Garrett, UK 75-14870
NWM de Kruithoorn BV, Netherlands 90-16320
OAR Corporation, USA 90-18800
Octavius Hunt Ltd, UK 75-16200
OFEMA, France 90-15400
Omni Yig, USA 90-18820
Omniglow, USA 75-16200
On Board Systems, USA 90-20600
Ono Sokki, Japan 80-22580
Orseal Valves, UK 75-06835
Otehall, UK 90-20150
Oxley Developments Co., UK 90-20150
Pacaero, USA 90-02000
Pains-Wessex, UK 75-16200
Pascall, USA 90-18820
Perkins Aerospace Products, USA 90-08300
Perkins Engine Company, UK 75-04450
Phoenix Simulation, UK 75-00400
Pneu Devices Inc, USA 75-08740
Polynor AS, Norway 90-05600
Potter & Brumfield, USA 90-20150
Prestincert, UK 90-20150
Professional Aviation Associates Inc, USA 90-02000
Puritan-Bennett, USA 80-00650
Pye-Electro Devices, UK 90-20150
Quitzau Industry, Denmark 75-16200
Racal Communications Inc., USA 85-16000
Racal E/W, UK 75-00400
Racal Electronics Group Plc, UK 85-16000
Racal Radar Defence Systems, UK 85-03500
Radiall, France 90-18820
Raufoss AS, Norway 90-05600
Remploy, UK 90-12600
Republic Electronics, USA 80-19100
Rheinmetall GmbH, Germany 90-05600
Robinson Helicopter Company, USA 90-08300
Rohde & Schwarz GmbH & Co KG, Germany 90-18800
Rolls-Royce plc, UK 90-18850
Rotterdamsche Droogdok BV, Netherlands 75-08740
Rustrak, USA 90-20150
S E I SpA, Italy 90-05600
Saati SpA, Italy 90-12600
Sabre Computer Scoring Ltd, UK 85-03500
Schiebel Elektronische Geraete GmbH, Austria 90-16320
Schottel-Werft, Germany 90-01000
Scott Aviation, USA 80-00650
Sekur SpA, Italy 90-12600
SFA Datacomm Inc, USA 90-18800
Siemens AG, Germany 75-19450
Siemens Plessey Systems., UK 75-19475
Sigarms Inc, USA 90-11800
Simmel Difesa SpA, Italy 90-05600
Simpson Electric, USA 90-20150
Sistima Compositi SpA, Italy 90-12600
SL Auburn Inc, USA 90-02000
SNC Industrial Technologies Inc., Canada 90-11800
Solent Energy, UK 90-06830
Sonicor, USA 90-20150
South West Aerospace Inc, USA 85-03500
Southwest Aerospace Inc., USA 75-00400
Stainless Steel Products, USA 90-20600
Stewart and Stevenson Tactical Vehicles, USA 75-19350
Steyr-Daimler-Puch AG, Austria 90-16320
STN ATLAS Elektronik GmbH, Germany 90-19960
Submersible Systems, USA 90-12600
Submersible Systems Inc., USA 90-20600
Systems Research Laboratories Inc., USA 90-20600
Target Technologies Ltd, UK 85-03500
Tecom, USA 90-18820
Teledyne Inc., USA 90-20300
Thermal Control Division, UK 75-08740
Thompson Saginaw Inc., USA 90-20600
Thomson-CSF, France 90-20450
Times Microwave, USA 90-18820
Trak Microwave, USA 90-18820
Trimos-Sylvac Task Force, Switzerland 80-01180
TRM Inc, USA 90-18820
Truetime, USA 90-18800
TRW Components Int'l Inc., USA 90-01800
Turbo Dynamics Corporation, USA 90-02000
Tynemouth, UK 90-06830
Unicorn Int'l Pte Ltd, Singapore 90-16320
Vacuum Reflex Ltd, UK 75-16200
Vaisala Oy, Finland 90-21500
Vallon, Germany 90-13620
Van Halteren, Netherlands 90-13620
Vernon Gauging Systems, UK 90-06830
Vickers-Fluid Controls & Actuator, USA 75-08740
Vickers-Systems Monitoring Division, USA 75-08740
Weber Aircraft Int'l Support Group, USA 90-06830
Wegmann, Germany 90-13620
Wegmann, USA 90-13620
Weir Pumps, UK 75-23300
Western Design Corp., USA 75-08740
Yaskawa Electric Co., Japan 90-20150
Zengrange Defence Systems, UK 90-16320

Sales agents/representatives listed elsewhere:

Aircraft Equipment Overhauls & Sales 80-00650
Applied Measurement Aust Pty Ltd 75-01420
BP Solar Australia Pty Ltd 75-02630
Colan Products 75-03650
Detroit Diesel-Allison Australia Pty 75-04450
Emco Wheaton Australia Pty Ltd 75-05760
Ericsson Defence Systems Pty Ltd 75-05820
Helitech Pty Ltd 80-08400
Howden Fan Company Pty Ltd 75-08740
Racal Australia Pty Ltd 85-16000
Varian Australia Pty Ltd 75-22100
Weir Engineering Pty Ltd 75-23300

90-00650 AlliedSignal Aerospace Pty Ltd

(a subsidiary of AlliedSignal Aerospace, USA)
Unit B, 10-20 McEvoy Street, Waterloo, NSW 2017
☎ (+61 2) 699 12 44; (+61 1) 589 61 38 (emergency hotline); 897 92 84 (emergency hotline)
⊜ (+61 2) 319 71 21
Director and Manager: Robert Fenton
® USA: Allied-Signal Aerospace Co.

AlliedSignal Aerospace Marketing Sales and Service
Mascot, NSW 2020
Quantas Airways Ltd, Quantas Jet Base, Administration Building, MB2, Floor 2, North Wing, Quantas Drive, PO Box 213
☎ (+61 2) 691 58 08; 693 79 88 (recording) ⊠ K80935
⊜ (+61 2) 693 52 93
Administrator: Ms Julie Probert
Field Service Engineer: Steve Stolle
■ On-site service for the region of Australia and New Zealand.

AlliedSignal Aerospace Marketing Sales and Service

Contact Reference

Unit 5, 217-219 Mickleham Road, PO Box 125, Tullamarine, Victoria 3034
☎(+61 3) 93 34 55 42 ⊖ (+61 3) 93 34 55 46
Senior Field Service Engineer, AlliedSignal Aircraft Landing Systems: Keith Van Berkel
Senior Field Service Engineer: Mike Gardner
☎(+61 3) 93 34 55 43
Field Service Engineer: Steve Murphy
☎(+61 3) 93 39 51 46; 93 39 27 53; 93 35 27 62 (fax)
[98]

90-01000 Analytical Control Engineering
Unit 27, 17 Lorraine Street, PO Box 94, Peakhurst, NSW 2210
☎(+61 2) 234 15 33 ⊖ (+61 2) 534 32 01
Director: Keith Burnham
® Germany: Schottel-Werft
[97]

90-01800 Austernetics Pty Ltd
PO Box 503, Manly, NSW 2095
☎(+61 2) 99 77 11 94 ⊖ (+61 2) 99 77 45 82
Managing Dir: P.C.F. Morgan (FRAeS)
■ High technology consultants in the field of space & defence. Project management, design services. → 80
® Australia: Matra Australia Pty Ltd
France: Matra Marconi Space France, Matra Transport International SA
UK: Matra Marconi Space
USA: TRW Components Int'l Inc.
[97]

90-01970 Aviall Australia Pty Ltd
(a subsidiary of Aviall Inc, USA)
PO Box 42, Niddrie, Victoria 3042
☎(+61 3) 93 38 11 55 (sales); 93 30 37 11 (administration)
⊖ (+61 3) 93 38 97 73 (sales); 93 30 31 94 (administration)
👥 36
Managing Director: David N Crouch
Operations Manager: Paul Fletcher
Airline and Military Sales Manager: Tony Zarka
■ Manufacture, distribution and overhaul of turbine aero engines. Distributor of aircraft engines and airframe parts.

Sydney Customer Service Office
453 Airport Avenue, Bankstown Airport, NSW 2200
☎(+61 2) 97 91 05 11 ⊖ (+61 2) 97 91 01 02
Manager: Craig Lockerbie

Brisbane Customer Service Centre
9 Grenier Drive, Archerfield Airport, Archerfield, Queensland 4108
☎(+61 7) 32 77 83 22 ⊖ (+61 7) 32 74 11 70
Manager: Bob Armstrong

Melbourne Customer Service Centre
8-10 Halsey Road, Airport West, Victoria 3042
☎(+61 3) 93 38 11 55 ⊖ (+61 3) 93 38 97 73
Manager: Ian Wilson

Cairns Customer Service Centre
Hanger 9 - Cairns Airport, Cairns, Queensland 4870
☎(+61 70) 35 99 50 ⊖ (+61 70) 35 99 49
Manager: David Adams

Moorabbin Customer Service Centre
Northern Avenue, Moorabbin Airport, Victoria 3194
☎(+61 3) 95 80 76 44 ⊖ (+61 3) 95 80 62 67
Manager: Mike Dickson

Perth Customer Service Centre
Hanger 126 - 11 Eagle Drive, Jandakot Airport, WA 6164
☎(+61 9) 332 78 55 ⊖ (+61 9) 332 88 58
Manager: Stephen Bennett
[98]

90-01980 Avio Consultants Pty Ltd
(an Avio Group Company)
Unit 3f - 169 Newcastle Street, Fyshwick, ACT 2609
☎(+61 6) 239 21 55 [E] avioccon@esprit.com.au
⊖ (+61 6) 239 17 44
♧ A$ 100,000 (1997) 👥 3 (1997)
General Manager: Ron Jemeson
■ Consultancy and other assistance for companies developing business relationships with the Department of Defence and other government departments/authorities.
[98]

90-02000 Banner Aero (Australia) Pty Ltd
(† Banner Aeronautical Corp)
PO Box 92, Niddrie, Victoria 3042
☎(+61 3) 93 74 22 66 [E] banner@interspace.com.au
[W] http://www.banner.com ⊖ (+61 3) 93 74 15 11
Managing Director: David Burton

® Sweden: Fairchild Scandinavian Bellyloading Co.
USA: Dynalec Corporation, Harris Manufacturing Co. Inc, Meridian Ocean Systems, Pacaero, Professional Aviation Associates Inc, SL Auburn Inc, Turbo Dynamics Corporation
[98]

90-03700 Clark Equipment Australia Pty Ltd
30 Salisbury Road, PO Box 50, Hornsby, New South Wales 2077
☎(+61 2) 94 77 84 44 ⊠ 70189 ⊖ (+61 2) 94 76 22 41
♧ A$ 16,200,000 👥 161
Managing Director: R Hammond
Technical Director: C Hancox
Marketing Manager: A Foulkes
■ Forklift trucks, reach stackers, terminal tractors and skidsteer loaders.
[98]

90-03750 Crompton Instruments (Aust) Pty Ltd
PO Box 492, Campbelltown, NSW 2560
☎(+61 2) 603 20 66 ⊖ (+61 2) 603 93 35
♣ £ 2 000 000 👥 18
General Mgr: T. Shead
® UK: Crompton Instruments Ltd
[98]

90-03890 Daewoo Australia Pty Ltd
(a subsidiary of Daewoo Corp, South Korea)
Suite 405 Victoria Cross Building, 60 Miller Street, North Sydney, NSW 2060
☎(+61 2) 99 57 55 99 ⊠ 24890 daewoo aa
⊖ (+61 2) 99 25 04 58
[98]

90-04050 Dale Budd & Associates Pty Ltd
Kingston, PO Box E340, Canberra, ACT 2604
☎(+61 6) 257 25 54; 295 38 40 ⊖ (+61 6) 247 89 89
Managing Director: Dale Budd
® USA: Litton Aero Products, Litton Guidance and Control Systems
[98]

90-05000 Diesel Drive Pty Ltd
353 Newbridge Road, Moorebank, New South Wales 2170
☎(+61 2) 96 01 11 33 ⊖ (+61 2) 96 01 82 27
♧ A$ 10,000 (1995) 👥 15 (1995) 🏠 4,500 m2
Director: Brian Ramsey
National Sales Manager: Peter Ramsey
■ Supply of diesel engines for automotive, industrial and marine use.
® Japan: Hino Motors Ltd
[98]

90-05600 Elvish & Associates
29 Magnolia Drive, Lower Templestowe, Victoria 3107
☎(+61 3) 98 48 67 92 ⊖ (+61 3) 98 48 67 92
President: Reginald T Elvish
Associate: N Alexander ☎(+61 6) 299 90 83
Associate: M H MacKenzie-Orr ☎(+61 6) 285 35 29
® Austria: Dynamit Nobel Graz (DNG)
Germany: Berkefeld Filter Anlagenbau GmbH, G L S Gesellschaft für Logistischem Svc, Glückauf Logistik, Herkules Fahrzeuglogistik GmbH, Rheinmetall GmbH
Italy: Cosmos SpA, S E I SpA, Simmel Difesa SpA
Norway: Polynor AS, Raufoss AS
USA: Bulova Technologies Inc
[98]

90-06830 Forrester Stephen Aviation
(a division of Roband Engineering Distributors Pty Ltd, Australia)
Roband House - 179 Park Street, South Melbourne, Victoria 3205
☎(+61 3) 96 96 52 55 [E] roband@netspace.net.au
⊖ (+61 3) 96 96 55 82
♣ A$ 1,000,000 (1997) 👥 12 (1997) 🏠 144 m2
Chief Executive Officer: Robert F Stephen
Director New Zealand: Mal Gunton
Engineering Director: M R Willis
Purchasing Manager: F Fernandes
■ Major supplier of aircraft spares to the Department of Defence. Parts supplied from manufacturers and distributors in USA, Canada, UK, Ireland, France, Switzerland, Germany and Italy. → 85
® Canada: FMA
France: Crouzet
Switzerland: Flight Components
UK: Aaxico/Negretti, Dowty Armourshield, Solent Energy, Tynemouth, Vernon Gauging Systems

USA: Air-A-Plane Corp, Assurance Technologies Inc, Aviation Tech & Resources, Lord Corporation, Weber Aircraft Int'l Support Group
[98]

90-08100 Hadland Photonics Pty Ltd
19A Hampshire Rd, PO Box 323, Glen Waverley, Victoria 3150
☎(+61 3) 560 23 66 ⊖ (+61 3) 560 84 02
Managing Dir: H.E. Biram
Dir Technical: L.V. Jones
■ Reconnaissance, strike, flightpath, high speed & HUD/HDD film cameras. High resolution, HUD/HDD & night sight TV cameras, airborne video tape recorders. High speed video cameras/recorders to 1000PPS (also continuous remote controlled systems for spontaneous events). Film interpretation, assessment & image analysis systems.
[98]

90-08300 Heliflite Pty Ltd
Baulkham Hills Business Centre, PO Box 6080, Sydney, New South Wales 2153
☎(+61 2) 96 80 15 11 [E] heliflite@onaustralia.com.au
⊖ (+61 2) 98 99 13 15
♣ A$ 5,000,000 👥 8
Managing Director: Lyndsay Edmonds
Sales and Marketing Director: Bill Miller
Engineer: Peter McGee
Secretary to the Managing Director: Ms Carolyn Butler
■ Sale of small helicopters. Service and pilot training. → 80
® USA: Apollo Spray Equipment, Perkins Aerospace Products, Robinson Helicopter Company
[98]

90-09900 Independent Instruments (Aviation)
First Avenue - Moorabbin Airport, Mentone, Victoria 3194
☎(+61 3) 95 80 14 44 ⊖ (+61 3) 95 87 38 20
♧ A$ 300,000 (1997) 👥 7
Managing Director: Patrick Miles
■ Test equipment, aircraft instruments and electrical components; auto-pilot systems; industrial instruments.
[98]

90-11000 Karcher Pty Ltd
2/158 Browns Road, Noble Park, Melbourne, Victoria 3174
☎(+61 3) 97 95 44 22 ⊖ (+61 3) 97 01 03 34
Managing Director: Robert Alan Stamley
® Germany: Alfred Karcher GmbH & Co
[98]

90-11200 Kawasaki Heavy Industries Ltd Sydney Office
c/o Kawasaki Motors Pty Ltd, Unit Q, 10-16 South Street, Rydalmere, New South Wales 2116
☎(+61 2) 684 25 85 ⊖ (+61 2) 684 45 80
[98]

90-11800 Kenmax Special Products Pty Ltd (KSP)
PO Box 2009, North Parramatta, NSW 2151
☎(+61 2) 98 90 84 44 ∅ (+61 1800) 24 37 22
⊖ (+61 2) 98 90 34 44
♧ A$ 100,000 (1997) ♣ A$ 1,400,000 (1997)
👥 5 (1997) 🏠 250 m2
Managing Director: Ms Susan J Jenkins
Operations Director: Kenneth J McBroom
Sales and Marketing Director: Alan B James
Company Secretary: Ms Marie Corney
■ Represents foreign defence manufacturers in Australia. Manufactures and sells domestically and exports military pyrotechnics, training and simulation devices.
® Canada: SNC Industrial Technologies Inc.
Germany: Nico Pyrotechnik
USA: Laser Products Inc, Mace Security Industries Inc, Sigarms Inc
[98]

90-12600 Light Aircraft Pty Ltd
PO Box 471, Brookvale, New South Wales 2100
16 Beatrice Street, Clontarf, New South Wales 2093
☎(+61 2) 99 07 92 08 ⊠ EJECTIONS Sydney
⊖ (+61 2) 99 07 93 04
Managing Director: Bruce E Patterson
Director: Michael S Patterson
→ 80
® Germany: Autoflug GmbH, Bernhardt Apparatebau, Blucher GmbH, Marquardt & Schulz GmbH
Italy: Moldip SpA, Saati SpA, Sekur SpA, Sistima Compositi SpA

UK: Cam Lock (UK) Ltd., Camberley Rubber Mouldings, Helmets Ltd., Remploy
USA: Hycor Inc., Submersible Systems [98]

90-12650 Litton Precision Products International Inc
414 Gardeners Road, Rosebery, PO Box 253,
Sydney, NSW 2018
☎(+61 2) 96 69 61 11 ⊖ (+61 2) 96 93 50 48
General Manager: Ron Robertson
® USA: Litton Industries Inc [98]

90-13100 Martin International
Suite 2 - 86 Giles Street, Kingston, ACT 2604
☎(+61 6) 295 38 88
[E] martin.international@compuserve.com
⊖ (+61 6) 295 34 58
👥 4 (1997)
General Manager: Paul L Martin
Consultant: Garry H Jesser
Consultant: Ian Gammage
■ Representing a wide range of companies and products including electronic equipment, ammunition, spare parts, radio equipment, wire, trackway, connectors, filters and oscillators, microwave components, avionic power equipment, security monitoring and testing equipment, military equipment.
Alcatel SEL AG:
Alelco Italy
American Science & Engineering
ASELSAN Turkey
B E Meyers & Co Inc
Delifn Systems Inc
Elisra Electronic Systems (Israel)
Faun (Laird) Ltd UK
FirstFire Ltd UK
General Kinetics Inc
Global Water Technologies Inc
IDM Electronics Ltd UK
Kevlin Corp (Microwave Division) Inc
Litton Systems (Solid State) Inc
Lockheed Martin Fairchild Systems Inc
NAPCO International Inc
Omega Atomizers Inc
Patten Co Inc
Portable Power Systems Inc
Precision Echo Inc
REA Int Ltd
Sparton Electronics Inc
ST Research Inc
SystemWare Inc
Texas Instruments Defense Systems and Electronics Inc
Zarc Inc
Zandex Inc [98]

90-13130 Matra Australia Pty Ltd
PO Box 503, Manly, NSW 2095
☎(+61 2) 99 77 11 94 ⊖ (+61 2) 99 77 45 82
Director: P.C.F. Morgan
® France: Matra Hachette, Matra Marconi Space, Matra Transport International SA [97]

90-13620 Military Technology Pty Ltd (Mil Tec)
PO Box 577, Willoughby, NSW 2067
☎(+61 2) 99 75 72 03 ⊖ (+61 2) 99 75 72 04
👥 4
Managing Dir: K.K. Reimann
Secretary: M.G. Reimann
■ Armoured vehicles & spare parts, M.G. mounts, 76mm smoke launchers, decoy pyrotechniques (Navy), international contacts, transmitting valves.
® Germany: Buck, Drello, I.B.D., Vallon, Wegmann
Netherlands: Van Halteren
USA: Wegmann [97]

90-13750 Moog Australia Pty Ltd
(a subsidiary of Moog Inc, USA)
14 Miles Street, Mulgrave, Victoria 3170
☎(+61 3) 95 61 60 44 ☒ 36375 ⊖ (+61 3) 95 61 55 04
♣ US$ 2,133,000 (1995) 👥 17
Managing Director: R Park
Operations Director: R Melik
Sales and Marketing Director: P Heitmann
Purchasing Manager: R Sridharan
Branch Office
Unit 5/5-9 Hunter Street,
Parramatta, New South Wales 2150
☎(+61 2) 98 91 41 66 ⊖ (+61 2) 98 91 31 85
Manager: Peter Heitmann ☎(+61 2) 756 42 00
® USA: Moog Inc
Representing:
Novotechnik
Elap
Quin [98]

90-13900 MTU Australia Pty Ltd
(a subsidiary of MTU Friedrichshafen GmbH, Germany)
PO Box 6703 Delivery Centre,
Blacktown, New South Wales 2148
☎(+61 2) 96 71 35 55
[E] miu-australia@onaustralia.com.au
⊖ (+61 2) 98 31 19 02
♧ A$ 4,549,000 (1996) ♣ A$ 60,000,000 (1996)
👥 41 (1996) 🏭 3,500 m2
Managing Director: Hans J Walloschek
Financial Controller: Brian Mellander
Operations and Technical Manager: Joerg Semmelroggen
Sales Manager: Richard George
Materials Manager: Alen Fak [98]

90-14650 Nicoll-Cooke Associates Pty Ltd
5/175 Collins Street, Melbourne, Victoria 3000
☎(+61 3) 96 54 17 33 ⊖ (+61 3) 96 54 13 49
Managing Director: D R Lovett
Director: A R Freeman
Office Manager: Ms Jo Pryde
® Canada: Barringer Instruments Ltd, CAE-Electronics Ltd
Italy: Aermacchi SpA
USA: EG&G Astrophysics, Lockheed Martin [98]

90-15400 OFEMA Australia
(a subsidiary of OFEMA, France)
10 Rudd Street - 3rd Floor, PO Box 2712,
Canberra, ACT 2601
☎(+61 6) 248 68 66 ⊖ (+61 6) 247 22 39
General Manager: Brendan Roberts
® France: OFEMA [98]

90-16320 Pentarch Pty Ltd
123 High Street, Prahran, Victoria 3181
☎(+61 3) 95 25 16 33 ⊖ (+61 3) 95 10 96 17
♧ A$ 20,000 (1997) ♣ A$ 5,000,000 (1997)
👥 8 (1997) 🏭 300 m2
Managing Director: Malcolm McComb
Operations Director: Peter Bellew
Sales and Marketing Director: Christopher Deighton
■ Rifles, ammunition and handling of various defence-related agencies.
® Austria: Armaturen Gesellschaft mbH, Hirtenberger AG, Schiebel Elektronische Geraete GmbH, Steyr-Daimler-Puch AG
Liechtenstein: Beta Aktiengesellschaft
Netherlands: NWM de Kruithoorn BV
Singapore: Unicorn Int'l Pte Ltd
UK: Anglo SwissOptical Co. plc, Hall & Watts Defence Optics Ltd, Manroy Engineering, Zengrange Defence Systems [98]

90-18800 Rohde & Schwarz (Australia) Pty Ltd
(a subsidiary of Rohde & Schwarz GmbH & Co KG, Germany)
63 Parramatta Road, Silverwater, NSW 2141
☎(+61 2) 97 48 01 55 ⊖ (+61 2) 97 48 18 36
👥 17
Managing Director: A Bean
Advertising Manager: Simon Haynes
■ Suppliers of SIGINT, COMINT, ELINT, COMSEC, TRANSEC, HF/VHF/UHF communications equipment for airborne, shipborne and ground applications.
® Germany: Rohde & Schwarz GmbH & Co KG
USA: Apcom Inc, Lucas Weinschel Engineering, OAR Corporation, SFA Datacomm Inc, Truetime
Melbourne Office
320 St Kilda Road - 4th Floor, Melbourne, Vic 3004
☎(+61 3) 699 79 22 ⊖ (+61 3) 696 21 00
Manager, Radiomonitoring/Comms: Peter Watschur [98]

90-18820 Rojone T/A Head Office
A P Imports
61 Aero Road, Ingleburn, New South Wales 2565
☎(+61 2) 98 29 15 55 [E] rojone@ozemail.com.au
⊖ (+61 2) 96 00 88 12
♣ A$ 6,000,000 (1997) 👥 33 (1997)
Managing Director: Ms Livia Grabowski
Operations Director: Ms Dianne Price
Technical Director: Ian Janoska
Sales and Marketing Director: Graham Sansom
Purchasing Manager: Darren Murphy
■ Military components: cables, wires, antennas and systems, connectors, cable assemblies, microwave and RF active/passive components. → 75
® France: Filotex, Radiall
USA: Amplica Inc, Continental Microwave & Tooling Co., Electro-Adapter, Electro-Metrics, Menlo Industries, Narda, Omni Yig, Pascall, TRM Inc, Tecom, Times Microwave, Trak Microwave
Brisbane Branch
18 Lorraine Crescent, Lawnton, Queensland 4501
☎(+61 7) 32 85 57 77 ⊖ (+61 7) 32 85 56 80
Melbourne Branch
30 Highland Avenue, Croydon, Victoria 3136
☎(+61 3) 97 25 21 72 ⊖ (+61 3) 97 25 01 65
Perth Branch
Unit 5 - 1914 Beach Road,
Malaga, Western Australia 6062
☎(+61 9) 249 63 00 ⊖ (+61 9) 249 69 49 [98]

90-18850 Rolls-Royce International Ltd
(a subsidiary of Rolls-Royce plc, UK)
39 Waterloo Road, Ryde, New South Wales 2113
☎(+61 2) 98 78 62 11 ⊖ (+61 2) 98 78 60 30
👥 20
Chairman: K J Kirby
Regional Executive: R Pratt
Company Secretary: J Hewitt
■ Marketing and technical service support for Rolls-Royce products in Australasia. → 80
® UK: Rolls-Royce plc [98]

90-19900 Air Cdre Gordon Steege, RAAF (Retd)
120 Pacific Road, Palm Beach, New South Wales 2108
☎(+61 2) 99 74 50 26 ⊖ (+61 2) 99 74 53 18
Consultant: Air Cdre Gordon H Steege
→ 80
® USA: Lockheed Martin Overseas Corp. [98]

90-19960 STN ATLAS (Australia) Pty Ltd
(a subsidiary STN ATLAS Elektronik GmbH)
PO Box 718, Artarmon, New South Wales 2064
39 Herbert Street, St Leonards, New South Wales 2065
☎(+61 2) 437 45 77 [E] atlas@stn-atlas.com.au
☒ 24815 ⊖ (+61 2) 439 75 76
👥 120
Managing Director: J H Hazell
Engineering Manager: G Thomson
Marketing Manager: G Pothoven
Marine Systems and Equipment Manager: B H Groves
Manager, Financial Services: Ms Rosalind Gall Hunt
■ Capability for software development, electronic hardware design and mechanical design. → 80
® Germany: STN ATLAS Elektronik GmbH
Affiliates:
STN ATLAS Elektronik GmbH (Germany) [98]

90-20150 Tecnico Electronics
(a division of Philips Components Pty Ltd, Australia)
34 Waterloo Road, Ryde, New South Wales 2113
☎(+61 2) 805 44 22 ☒ 20165 ⊖ (+61 2) 805 44 33
👥 25
Product Manager, Instrumentation: Raoul Punt
■ Import & distribution of electrical/electronic components & electronic & scientific instruments.
® Australia: 3M, Allen-Bradley Pty. Ltd
Germany: Heinrich C. Kosmeier
Japan: Denon Instruments Co., Hokuriku Company, Yaskawa Electric Co.
Taiwan: Appa Technologies, Centronic Industries, Innovative Industrial
UK: Belling Lee Ltd, Dawe Co., Otehall, Oxley Developments Co., Prestincert, Pye-Electro Devices
USA: ALL States, Bussmann Manufacturing, EG&G Wakefield, Eeco Ltd, Electrochem Ind., Fluke, Gulton Industries, International Power, Kay Electrics, Potter & Brumfield, Rustrak, Simpson Electric, Sonicor [98]

90-20300 Teledyne Industries International Inc.
Ground Floor, 285 Whitehorse Rd, Balwyn, Vic 3103
☎(+61 3) 98 36 71 77 ☒ 38264 telcio
⊖ (+61 3) 98 36 19 64
® USA: Teledyne Inc. [97]

Contact Reference

90-20450 Thomson-CSF Pacific Pty Ltd
16 National Circuit, Barton, ACT 2600
☎(+61 6) 273 32 66 ⊖ (+61 6) 273 16 97
4
Managing Dir: O. Le Marechal
Business Development Mgr: Neil Searle
France: Thomson-CSF [97]

90-20600 Transaero Australia
(a subsidiary of Transaero Inc, USA)
PO Box 1241, Glen Waverley, Victoria 3150
☎(+61 3) 98 87 90 76 ⊖ (+61 3) 98 03 18 80
Vice President: Charles Kane
Regional Manager: Martin Magee
USA: ACR Electronics Inc, Allen ACFT Products Inc, BFGoodrich, Basewest Inc, Conax Florida Corp., Control Products Corp., Dayton Granger Inc., Dynapower-Strato Power Inc., G&H Technology Inc., GEC Marconi Aerospace, Genyex Corp., Hughes Treitler Inc., ILC Dover Inc., IMC Magnetics Inc., Marathon Power Technologies, Mustang Survival, On Board Systems, Stainless Steel Products, Submersible Systems Inc., Systems Research Laboratories Inc., Thompson Saginaw Inc. [98]

90-21000 UBE Industries Ltd
Sydney Representative Office
Level 18 Royal Insurance House, 1 York Street, Sydney, New South Wales 2000
☎(+61 2) 92 51 41 24; 92 51 41 26 74203
⊖ (+61 2) 92 51 42 93 [98]

90-21500 Vaisala Pty Ltd
(a subsidiary of Vaisala Oy, Finland)
3 Guest Street, Hawthorn, Victoria 3122
☎(+61 3) 98 18 42 00 (+61 800) 33 58 40
⊖ (+61 3) 98 18 45 22
♣ A$ 5,700,000 8
Managing Director: Robert Caplikas
Sales Manager: Simon Harrod
■ Atmospheric sounding equipment comprising Omega based radiosondes and MARWIN family of military ground processing equipment. Tactical wind measurement sensors and displays. → 80 → 85
Finland: Vaisala Oy [98]

Austria, Republic of

Federal President:
Thomas Klestil

10 Head of State

10-10000 Office of the Federal President
Hofburg - Bellariator, PO Box 10, A-1014 Vienna
☎(+43 1) 53 42 20 133484 praes a
⊖ (+43 1) 535 65 12
Federal President: Thomas Klestil [98]

11 Prime Minister

11-10000 Office of the Federal Chancellor
Ballhausplatz 2, A-1014 Vienna
☎(+43 1) 661 50; 53 11 50 115585 bkabpd a
Federal Chancellor: Dr Viktor Klima
Vice Chancellor: Wolfgang Schuessel [98]

12 Ministries

12-10000 Ministry of Defence
Bundesministerium fur Landesverteidigung
Dampfschiffstrasse 2, A-1030 Vienna
☎(+43 1) 51 59 50 112145 bmlv a
⊖ (+43 1) 515 95 34 01; 515 95 21 26 (cabinet)
Minister of Defence: Dr Werner Fasslabend
Chief of Cabinet: Raimund Schittenhelm
Director of Press and Public Information Office: MinRat Dr Klaus Sartorius-Thalbo [98]

12-20000 Federal Ministry for Foreign Affairs
Ballhausplatz 2, A-1014 Vienna
☎(+43 1) 53 11 50 ⊠ AUSSENAMT Wien
W http://www.gov.austria-info.at/foreign affairs
13710 aawn a ⊖ (+43 1) 531 85 20; 535 45 30
Federal Minister and Vice Chancellor: Wolfgang Schuessel
State Secretary for Foreign Affairs: Dr Benita Ferrero Waldner
Director of Cabinet of the Minister: Dr Wolfgang Loibl
Spokesman of the Minister: Dr Florian Krenkel
Secretary General for Foreign Affairs: Dr Albert Rohan
Director of Foreign Relations: Dr Peter Hohenfellner
Chief of Press and Information Division: Dr Gerald Kriechbaum
Chief of Security Policy Division: Dr Thomas Mayr Harting
Chief of Southern Countries Division: Dr Walter Hagg
Chief of Eastern Countries Division: Dr Klas Daublebsky
Chief of Africa, Asia and Oceania Division: Dr Marius Calligaris
Chief of Western Countries Division: Dr Christian Prosi [98]

12-40000 Federal Ministry for Interior
Bundesministerium für Inneres
Herrengasse 7, PO Box 100, A-1014 Vienna
☎(+43 1) 53 12 60 115695 bmfijd a
⊖ (+43 1) 531 26 22 11
Federal Minister of Defence: Karl Schloegl
Head of Department I/2: Dr Herbert Stammer [98]

12-60000 Ministry of Economy and Transport
Oberste Zivilluftfahrtbehörde (OZB)
Radetzkystrasse 2, A-1030 Vienna
☎(+43 1) 02 22; 71 16 20 ⊠ CIVILAIR
3221155 wverkm a ⊖ (+43 1) 02 22; 713 03 26
Minister of Economy and Transport: Johann Farnleiter

Aviation Department
Radetzkystrasse 2, A-1030 Vienna
☎(+43 1) 711 62 91 05 ⊖ (+43 1) 713 03 26
Director General of Civil Aviation: Dr Gerhard Stadler
■ Personnel management, annual estimates, administration and co-ordination, press department, aviation. [98]

15 Parliament

15-10000 Austrian Parliament
Parlamentsdirecktion
Dr Karl Renner Ring 3, A-1017 Vienna
☎(+43 1) 40 11 00 ⊖ (+43 1) 401 10; 401 25 37
Director: Dr Heinz Wasserbauer ☎(+43 1) 401 10/22 15
Deputy Director: Dr Sigurd Bauer ☎(+43 1) 401 10/22 27
Deputy Director: Dr Adolf Klausgraber ☎(+43 1) 401 10/22 16
Director of Secretariat: Dr Christian Müller ☎(+43 1) 401 10/22 33

National Council
Nationalrat
Dr Karl Renner Ring 3, A-1017 Vienna
☎(+43 1) 40 11 00 115231 parld a
⊖ (+43 1) 401 10/25 37
President: Dr Heinz Fischer ☎(+43 1) 401 10/22 01
Secretary of the President: Dr Georg Posh ☎(+43 1) 401 10/24 56
Second President: Dr Heinrich Neisser ☎(+43 1) 401 10/22 02
Third President: Prof.-Dr Willi Brauneder ☎(+43 1) 401 10/22 03

Defence Committee
Landesverteidigungsausschuss
Dr Karl Renner Ring 3, A-1017 Vienna
☎(+43 1) 40 11 00 115231 parld a
⊖ (+43 222) 401 10 25 37
Chairman: Herbert Scheibner ☎(+43 1) 401 10/56 81
Deputy Chairman: Anton Gaál ☎(+43 1) 401 10/36 35
Deputy Chairman: Dr Karl Maitz ☎(+43 1) 401 10/49 11
Deputy Chairman: Ing Gerald Tychtl ☎(+43 1) 401 10/35 51
Secretary: Dipl.-Ing Werner Kummerer ☎(+43 1) 401 10/36 76
Secretary: Wolfgang Jung ☎(+43 1) 401 10/46 10
Secretary: Willi Sauer ☎(+43 1) 401 10/48 51

Foreign Affairs Committee
Aussenpolitischer Ausschuss
Dr Karl Renner Ring 3, A-1017 Vienna
☎(+43 1) 40 11 00 115231 parld a
⊖ (+43 1) 401 10/24 36
Chairman: Peter Schieder ☎(+43 1) 401 10/36 30
Deputy Chairman: Michael Spindelegger ☎(+43 1) 401 10/42 48
Deputy Chairman: Dr Josef Cap ☎(+43 1) 401 10/42 48
Deputy Chairman: Dr Jörg Haider ☎(+43 1) 401 10/56 39
Secretary: Rudolf Schwarzböck ☎(+43 1) 401 10/49 24
Secretary: Dr Willi Fuhrmann ☎(+43 1) 401 10/33 90

Federal Council
Bundesrat
Dr Karl Renner Ring 3, A-1017 Vienna
☎(+43 1) 40 11 00 115231 parld a
⊖ (+43 222) 401 10/24 35
President: Prof.-Dr Herbert Schambeck ☎(+43 1) 401 10/22 09
Vice President: Ms Anna Elisabeth Haselbach ☎(+43 1) 401 10/22 08
Vice President: Jürgen Weiss ☎(+43 1) 401 10/43 55

Foreign Affairs Committee
Aussenpolitischer Ausschuss
Dr Karl Renner Ring 3, A-1017 Vienna
☎(+43 1) 40 11 00 115231 parld a
⊖ (+43 1) 401 10/24 35
Chairman: Dr Manfred Mautner Markhof ☎(+43 1) 401 10/44 14
Deputy Chairman: Albrecht Konecny ☎(+43 1) 401 10/37 07
Deputy Chairman: Alfred Gerstl ☎(+43 1) 401 10/44 14
Secretary: Ferdinand Gstöttner ☎(+43 1) 401 10/35 25
Secretary: Gerhard Tusek ☎(+43 1) 401 10/44 14

Legal Affairs Committee
Rechtsausschuss
Dr Karl Renner-Ring 3, A-1017 Vienna
☎(+43 1) 40 11 00 115231 parld a
⊖ (+43 222) 401 10/24 35
Deputy Chairman: Dr Milan Linzer ☎(+43 1) 401 10/46 10
Deputy Chairman: Albrecht Konecny ☎(+43 1) 401 10/37 03
Secretary: Dr Kurt Kaufmann ☎(+43 1) 401 10/44 15
Secretary: Herbert Platzera ☎(+43 1) 401 10/34 71 [98]

20 Defense Administrations/Services/Agencies

20-00100 Federal Armament Office
Vorgartenstrasse 225, AG-Vorgartenstrasse, A-1024 Vienna
☎(+43 222) 217 61 ⊖ (+43 222) 217 61 61 46
Director: Dipl.-Ing F Dechant
Chief Co-ordination and Equipment Development Department: Col R E Hofer
Chief of Armament Ammunition Department: Brig Dipl.-Ing.Dr W Pexa
Chief of Electronics Department: Dipl.-Ing F Macek ☎(+43 222) 727 62 52
Chief Civil Engineering and Pioneer Equipment Department: Brig Dipl.-Ing F Faltus
Chief of Engineering Department: Brig Dipl.-Ing G Hohl ☎(+43 222) 727 62 37
Chief of Chemical and Physical Engineering Department: Brig Dipl.-Ing Hans Klaus Wieser ☎(+43 222) 727 62 33 [98]

25 Defense Forces

25-19100 National Defence Academy
Institute for Strategic Research
Stiftgasse 2a, A-1070 Vienna
☎(+43 222) 523 84 87 ⊖ (+43 222) 523 84 87
Commander of National Defence Academy: Lieut Gen Erich Eder
Deputy Commander: Brig Gen G Ebner
Director of Institute for Strategic Research: Dr Rudolf Hecht
■ The Institute consists of 4 scientfic departments, strategic research, international security, society and forces, conflict and international use of forces. [98]

25-20000 Army and Air Force Commando Headquarters
Bundesministerium für Landesverteidigung
Dampfschiffstrasse 2, A-1030 Vienna
☎(+43 1) 515 95 ⊠ 112145 bmlv a
⊖ (+43 1) 515 95 34 01
Commander: Dr Werner Fasslabend
Chief of Staff: Gen Karl Majcen
Director of Aviation Logistics: Brig Wolfgang Katter
Director of Press and Information: Dr Klaus Sartorius-Thalborn
Director of Aviation Planning Department: Brig Josef Bernecker
Head of Aviation Material Branch: Brig Dipl.-Ing Manfred Munzer
Chief of Logistic and Procurement Department: Gen Dr Peter Corrieri
Air Division Command
Kommando Fliegerdivision, A-3425 Langenlebarn
☎(+43 2272) 23 85
Commander: Othmar Pabisch
Military Press and Information Officer: Col Wolf-Diettich Tesar [97]

25-32200 Air Division Command Headquarters
Kommando Fliegerdivision
Fliegenhorst Brumowski, A-3425 Langenlebarn
☎(+43 2272) 62 30 82 26 [E] fldiv@bmlv.gv.at
[W] http://www.bmlv.gv.at ⊠ 111655 mcomc a
⊖ (+43 2272) 62 30 83 43
Commander: Maj Gen Othmar Pabisch
Tactical Commando: Lieut Gen Peter Kolecko
Personnel and Logistics Commando: Lieut Gen Johann Rathgeb
Technical Commando: Lieut Gen Manfred Mümzer
Press and Information Officer: Col Wolf Dietrich Tesar [98]

26 Police (Civil)

26-10000 Federal Police Headquarters
Bundespolizeidirektion Wien
Schottenring 7-9, A-1010 Vienna
☎(+43 222) 313 10; 72 32; 72 78
⊖ (+43 222) 313 10; 79 65
Head Commissioner: Dr Walter Türk [97]

29 Customs & Excise

29-00500 Customs & Excise
c/o Bundesministerium für Finanzen,
Himmelpfortgasse 4/8, A-1015 Vienna [98]

31 Embassies & Attachés

31-00085 Armenian Diplomatic Mission
Neubaugasse 12-14 1/16, A-1070 Vienna
☎(+43 1) 522 74 79 ⊖ (+43 1) 522 74 81
Ambassador: H E Mr Arman Navasardian [98]

31-00230 Embassy of Brazil
Lugeck 1/5/15, A-1010 Vienna
☎(+43 1) 512 06 31 ⊠ BRASEMB Wien
⊖ (+43 1) 513 83 74
Ambassador: H E Mr Affonso Celso de Ouro Preto
Minister and Counsellor: Jayro Coelho
First Secretary: José Fiuza Neto
Counsellor: Ms Sónia Maria de Castro
Third Secretary: Roberto Teixeira de Avelar
Attaché: Mrs Regina B Flores [97]

31-00350 Embassy of Chile
Lugeck 1/III/10, A-1010 Vienna
☎(+43 1) 512 92 08; 512 33 70; 513 10 58; 512 52 26
⊠ 115952 emchi a ⊖ (+43 1) 512 92 08 33
Ambassador: H E Dr Osvaldo Puccio-Huidobro
Counsellor: Jaime Acuña-Pimentel
Scientific Counsellor: Claudio Silva-Hennings
First Secretary and Commercial Attaché: Luciano Parodi-Gambetti
Third Secretary: C R A Donoso Maluf
Third Secretary: Juan Carlos Salazar Álvarez [97]

31-00430 Embassy of the Republic of Cuba
Himmelhofgasse 40 a/c, A-1130 Vienna
☎(+43 1) 877 81 98 ⊠ 131398 emcuba a
⊖ (+43 1) 877 77 03
Ambassador: H E Mr Alberto San Jose Velazco
Commercial Counsellor: Alfredo Puig Pino
☎(+43 1) 877 21 59 [97]

31-00510 Embassy of the Arab Republic of Egypt
32 Raffelspergergasse, A-1190 Vienna
☎(+43 1) 478 88 00 ⊠ BUSTAN Wien
⊠ 1010/115623 bustan a ⊖ (+43 1) 478 88 00 27
Ambassador: H E Dr Mostafa El-Feki
Minister and Deputy Chief of Mission: Mona Seaudy
Minister, Commercial Affairs: Mohamed El Mantawi
Counsellor: Wagdy Abouzeid
Defence Attaché: Mohamed Ahmed Mahmoud Aly (resident in Prague)
Third Secretary: Hossam Aly [98]

31-00535 Embassy of Estonia
Marokkanergasse 22/6, A-1030 Vienna
☎(+43 1) 718 07 29 [E] saatkond@estemb.or.at
⊖ (+43 1) 718 07 29 15
Ambassador: H E Mr Toivo Tasa
Honorary Consul General
Reichenhallerstraße 10a, A-5020 Salzburg
☎(+43 662) 848 49 61 ⊖ (+43 662) 848 49 64
OSCE Representative
Marokkanergasse 22/6, A-1030 Vienna
☎(+43 1) 718 07 29 ⊖ (+43 1) 718 07 29 15
Ambassador: H E Mr Vaino Reinart [98]

31-00860 Embassy of the Republic of Iraq
Johannesgasse 26, A-1010 Vienna
☎(+43 1) 73 81 95; 73 13 33; 73 82 08; 73 91 96
⊠ 135397 irak a ⊖ (+43 1) 713 82 08
Ambassador: H E Mr Neama Faris [97]

31-00880 Embassy of Israel
Anton Frank Gasse 20, A-1180 Vienna
☎(+43 1) 470 47 41
[E] embassyofisrael-vienna@mailbox.telecom .at
⊖ (+43 1) 470 47 46; 470 57 75
Ambassador: H E Mr Yoel Sher
Counsellor: Doron Mordechai Grossmann
First Secretary: Eitan Levon
Minister-Counsellor, Economic Affairs: Jakob Fass
Minister and Counsellor: Yehoshua Krith-man
Second Secretary and Head of Consular Department: Avraham Koren
Military and Defence Attaché: Col Yitzak Abercohen (resident in Switzerland)
Counsellor: Ammon Noy (resident in Germany) [98]

31-00890 Embassy of Italy
Metternichgasse 13, A-1030 Vienna
☎(+43 1) 712 51 21 ⊠ 132620 itadpa; 134843 itatta
⊖ (+43 1) 713 97 19
Ambassador: H E Dr Joseph Nitti
Deputy Chief of Mission: Dr Liana Marolla
Military and Air Attaché: Col Giovanni Nitti
Commercial Counsellor: Dr Luca Franchetti Pardo
Press and Information Officer: Dr Alberto Bertoni [98]

31-00920 Embassy of Japan
Japanische Botschaft
Argentinierstrasse 21, A-1040 Vienna
☎(+43 1) 50 17 10
Ambassador: H E Mr Atsuhiko Yatabe
Minister and Deputy Chief of Mission: Chihiro Tsukada [97]

31-01020 Embassy of Lebanon
Oppolzergasse 6, A-1010 Vienna
☎(+43 1) 533 88 21; 533 88 22 ⊖ (+43 1) 533 49 84
Ambassador: H E Mr Samir N Hobeica [98]

31-01420 Embassy of the Philippines
Billrathstrasse 2/2/21, A-1190 Vienna
☎(+43 1) 368 55 19 [E] jase@ping.at
⊠ PHILEMBASSY Wien ⊠ 132740 filvn a
⊖ (+43 1) 318 71 38
Ambassador: H E Mr Jose A Zaide (Jr)
Deputy Chief of Mission and Consul General: Mrs Victoria S Bataclan
Third Secretary, Vice-Consul and Administrative Officer: Julius Caesar A Flores
Assistant Security Officer: Felix de Leon (Jr) [97]

31-01440 Embassy of Portugal
Operngasse 20b, A-1040 Vienna
☎(+43 1) 586 75 36 ⊠ 113237 ⊖ (+43 1) 587 58 39
Ambassador: H E Mr Álvaro Mendonça e Moura
☎(+43 1) 513 32 97
Deputy Chief of Mission: José Manuel Santos Braga
☎(+43 1) 505 59 71
First Secretary: Hugo Solano Cabral de Moncada
☎(+43 1) 502 92 22
Counsellor: Mrs Leonor Maria Seixas Ribeiro da Silva
Counsellor: Frederico Alcantara de Melo
☎(+43 1) 505 79 64
Commercial Department
Stubenring 16, A-1010 Vienna
☎(+43 1) 513 26 70 ⊠ 114339 icep a
⊖ (+43 1) 512 88 28
Commercial Counsellor: Dr Manuel Silva Alexandre
☎(+43 1) 712 79 33 [97]

31-01485 Embassy of the Russian Federation
Reisnerstrasse 45-47, A-1030 Vienna
☎(+43 1) 712 12 15; 713 12 29 ⊖ (+43 1) 712 33 88
Ambassador: H E Mr Wladimir Michailowitsch Grinin
Deputy Chief of Mission: Wladimir M Polenow
Counsellor, Science and Technical: Michail G Solovjev
Defence Attaché: Arkadij A Bobrow
Commercial Department
Argentinierstrasse 27, A-1040 Vienna
☎(+43 1) 505 75 89 ⊠ 131848 vnt a
⊖ (+43 1) 505 81 98
Head of Commerce: Gennadij I Filshin
☎(+43 1) 505 44 58 [98]

31-01655 Embassy of the Republic of Slovenia
Nibelungengasse 13, A-1010 Vienna
☎(+43 1) 586 13 09 ⊖ (+43 1) 586 12 65
Ambassador: H.E. Dr Katja Boh
Consulate of the Republic of Slovenia
Bahnhofstrasse 22/2, A-9020 Klagenfurt
☎(+43 463) 541 50 ⊖ (+43 463) 50 40 71
Consul General: Jozef Jeraj
Consulate of the Republic of Slovenia
Fredrichgasse n°6/IX/37, A-8010 Graz
☎(+43 316) 82 30 45
Honorary Consul: Dr Anton Keru [98]

31-01730 Swedish Embassy
Obere Donaustrasse 49-51, PO Box 18, A-1025 Vienna
☎(+43 1) 214 77 01 ⊠ SVENSK Wien ⊠ 114720 svns a
⊖ (+43 1) 214 50 44
Ambassador: H E Mr Björn Skala
Military Attaché: Brig Gen Rune Hall
[E] rune.hall@foreign.ministry .se
Commercial Office
Universitätsstrasse 5, A-1010 Vienna
☎(+43 1) 402 35 15 ⊖ (+43 1) 402 35 15 24 [98]

Contact Reference

Austria, Republic of

31-01890 Embassy of the United Kingdom
Jaurègasse 12, A-1030 Vienna
☎ (+43 1) 716 13 22 17 ☒ 133410 britem a
⊖ (+43 1) 716 13 29 99
Ambassador: H E Sir Anthony Figgis (CMG)
Deputy Head of Mission and Head of Commercial: I C Cliff (OBE)
Defence Attaché: Lieut Col A R Manton
First Secretary, Economic: W R Brandon
Public Information Officer: Miles Fisher [98]

31-01900 Embassy of the United States of America
Boltzmanngasse 16, A-1091 Vienna
☎ (+43 1) 313 39
W http://www.usa.gov/posts/vienna.html
⊖ (+43 1) 310 06 82
Ambassador: H E Mr Swanee G Hunt
Defence Attaché: Col John Fairlamb
Air Attaché: Col Albert Chapman
Commercial Officer: Stephen Craven
Assistant Army Attaché: Maj Michael Hallisey
Public Affairs Officer: Ms Helena Kane Finn
Customs Service: John Kellogg [98]

34 Multinational Intergovernmental Organizations

34-04610 Maaseik Police
Boomgaardstraat 9, A-3680 Maaseik
☎ (+32 89) 56 75 36
Chief: L Haine [98]

34-10000 International Atomic Energy Agency (IAEA)
Wagramerstrasse 5, PO Box 100, A-1400 Vienna
☎ (+43 1) 206 00 W http://www.iaea.or.at/worldatom
☒ 112645 atom a ⊖ (+43 1) 206 07
Director General: Dr Hans Blix
Deputy Director General for Safeguards: Bruno Pelland
Deputy Director General, Administration: David B Waller
Deputy Director General, Research and Isotopes: Sueo Machi
Deputy Director General, Nuclear Energy: Victor Mourogov
Deputy Director General, Technical Safety: Jihuí Qian
Deputy Director General, Nuclear Safety: Zygmund Domaratzk
Press and Information Director: David Kyd
Press and Information Officer: Dr Hans-Friedrich Meyer
■ Intergovernmental agency with 125 member countries throughout the world. [98]

60 Research Organizations, Institutes & Test Establishments

60-05000 Osterreichisches Forschungszentrum Seibersdorf
A-2444 Seibersdorf
☎ (+43 2254) 78 00
W http://www.arcs.ac.at/fzs/jb95.html ☒ 14353 fzs a
⊖ (+43 2254) 740 60
♣ Sch 694,000,000 (1995) ⅏ 537 (1996) ⌂ 50,000 m2
Managing Director: Prof.-Dr Franz Leberl
Head of Information: Dr Nevyjel
■ Faxprint 20 facsimile printers. Hard copy printers (modification). Meteorological data acquisition systems. Airborne radiation detectors. Radiation protection dosimeters. (Thermoluminescence dosimeters). Radiation protection measuring equipment. Systems for EMC testing, early warning systems and networks for radiation monitoring. [98]

70 Financial Holding Groups, Parent Companies

70-01100 Steyr-Daimler-Puch AG
PO Box 62, A-1011 Vienna
☎ (+43 1) 531 44 27 00 ☒ 131706
⊖ (+43 1) 531 44 27 80
♧ Sch 1,000,000,000 (1996)
♣ Sch 11,402,000,000 (1996) ⅏ 6,037 (1996)
President: Prof.-Dr Rudolf Streicher
Sales and Marketing Director: Dr Martin Platzer
Company Secretary: Dr Josef Dellinger
■ Manufacture of all kinds of vehicles. Military products include all-terrain vehicles, tracked vehicles and tanks, firearms.

Subsidiaries with defence business:
Steyr-Daimler-Puch Fahrzeugtechnik GmbH → 75
Steyr Mannlicher AG → 75
Steyr-Daimler-Puch Spezialfahrzeug AG → 75 [98]

75 Manufacturing Companies

75-00050 AB Mikroelektronik GmbH
(a subsidiary of AB Electronic Products Group, UK)
Josef Brandstätter Strasse 2, PO Box 18, A-5013 Salzburg
☎ (+43 662) 44 99 10 ☒ 622213
⊖ (+43 662) 449 91 10
Managing Director: B Plank
■ Design and manufacture of thick film networks and hybrid micro-circuits.
® UK: AB Electronic Products [98]

75-00075 Akustische und Kino-geräte GmbH (AKG)
(a subsidiary of Harman International, USA)
Lemböchgasse 21-25, A-1230 Vienna
☎ (+43 1) 86 65 40 ⊖ (+43 1) 86 65 42 08
♣ Sch 600,000,000 ⅏ 450 ⌂ 17,000 m2
Managing Director: Mag Hendrik Homan
Sales Director: Heinrich Zant
Operations Director: Dr Alexander Fritz
Technical Director: Dr Alexander Persterek
Press and Public Relations Officer: Ms Christina Burkhardt
Purchasing Manager: Mag Christian Linasi
■ Cabled and wireless headphones, microphones, headphone-combinations. Electro-acoustic components. Special applications. [97]

75-00560 J Blaschke Wehrtechnik GmbH
Wienerbergstrasse 42-44, A-1120 Vienna
☎ (+43 1) 810 09 09 ⊖ (+43 1) 810 09 09 33
♣ Sch 17,000,000 (1997) ⌂ 1,500 m2
Managing Director: Ing Volker Haager
■ Camouflage systems: special weapon camouflage; mock-bodies; battlefield camouflage systems; multispectral, active dummies of camouflaged weapons and systems; windscreen camouflage; folding-tanks made of decontaminable material, retention time up to 2,000 hours; NBC-protective clothing made of rubberised material, retention time up to 2,000 hours; NBC-ventilating system for heavy protective suits; light protective suits; tents made of decontaminable material, retention up to 2,000 hours.
Casualty bags: casualty half bags, ventilating systems for casualty bags; transport bags for dead bodies; NBC-sample collecting sets; decontamination collecting trough (size variable from 5 to 500 m2). [98]

75-00650 Böhler Schmiedetechnik GmbH
(a subsidiary of Böhler Uddeholm AG, Austria)
Mariazeller Strasse 25, A-8605 Kapfenberg
☎ (+43 3862) 20 71 24; 20 66 78 ☒ 36612 bok a
⊖ (+43 3862) 20 75 70; 20 20 78
♧ Sch 70,000,000 (1996) ♣ Sch 603,000,000 (1996)
⅏ 330 ⌂ 25,000 m2
Managing Director: F H Romen-Kierner
Managing Director, Finance: H Weigand
Head of Marketing: W Schnedl
Head of Sales: Herbert Puschnik
E herbert.puschnik@buag.co
Purchasing Manager: Erich Kukla
■ Precision forged and machined parts for engines, compressors and aerospace applications.

Böhler Bleche GmbH
Bleckmanngasse 10, A-8680 Murzzuschlag
☎ (+43 3852) 55 53 19 ☒ 36612 boc a
⊖ (+43 3852) 55 53 24
■ Sheets and plates for aerospace applications in high alloyed special steel grades and titanium alloys. [97]

75-00660 Bombardier-Rotax GmbH
(† Rotax GmbH)
Welser Strasse 32, PO Box 5, A-4623 Gunskirchen
☎ (+43 7246) 60 10 ✉ BOMBROTAX ⊖ (+43 7246) 370
♣ Sch 3,140,426,000 (1996) ⅏ 1,436 (1997)
⌂ 604,000 sq ft
Managing Director: Dipl.-Ing Hans Joachim Lucae
Operations Director: Dipl.-Ing Harald Plockinger
Finance Director: Klaus Peter Müller
Product Manager for Aircraft Engines: Ing Josef Fürlinger
☎ (+43 7246) 27 12 32
Purchasing Manager: Altenhofer Hubert
■ Engines for ultra-light experimental aircraft (2 and 4 stroke). Reduction gearboxes for ultralight, certified and experimental aircraft. Four stroke aircraft engines. JAR-22 and FAR-33 approved. [98]

75-00750 W&H Dentalwerk Bürmoos GesmbH
5111 Bürmoos bei Salzburg, Salzburg
☎ (+43 6274) 623 60 E office@wnh.co.at
W http://www.whdental.com ☒ 633346
⊖ (+43 6274) 62 36 55
♧ Sch 3,500,000 ♣ Sch 370,000,000 ⅏ 330
⌂ 3,027 m2
Chief Executive Officer: Dr Bernd Rippel
Operations and Technical Director: Peter Malata
Purchasing Manager: Wolfgang Oberreiter
■ Manufacture of military dental clinics: turbine contra-angles, handpieces for micro motors, standard contra-angles and handpieces for Slip-Joint, laboratory handpieces, electro-compact-handpiece ELCO Plus with micro motor and electronic control box for dental laboratories, ELCOmed-special operation system for surgery and implantology, as well as surgical handpieces. ASSISTINA cleaning and lubricating unit for all contra-angles and handpieces with DIN-connection and W&H turbine contra-angles (also possible for other coupling systems with adaptor). [98]

75-00760 Diamond Aircraft Industries GmbH
(† Hoffman Aircraft Ltd (HOAC))
N.A. Ottostrasse 5, A-2700 Wiener Neustadt
☎ (+43 2622) 267 00 ⊖ (+43 2622) 267 80
President: Christian Dries
Managing Director: Wolfgang Grummeth
Chief Designer: Martin Volck
Director of Sales: Michael Feinig
■ Manufacture primary training aircraft.
Company formed 1981 in Carinthia, southern Austria as Hoffman Flugzeubau-Friesach GmbH; reformed after bankruptcy as Hoffman Aircraft Ltd; relocated to new facility at Wiener Neustadt in 1987. New name HOAC Austria Flugzeugwerk Wiener Neustadt GmbH adopted in 1990 after management buy-out. Renamed Diamond Aircraft Industries GmbH in March 1996. [98]

75-00800 Dynamit Nobel Graz GmbH (DNG)
Annenstrasse 58, PO Box 1124, A-8020 Graz
☎ (+43 316) 80 36 32 70; 83 26 31;
⊖ (+43 316) 803 69 47
⅏ 107
Managing Director: Dr Lohberger
Technical Director: Dipl.-Ing Anetshofer
Sales and Marketing Director: Holger Zülsdorff
Sales and Marketing Manager: Harald Mayr
Purchasing Manager: Mr Strasser
■ Electronic anti-tank mines of the latest generation with self-neutralising units; anti-helicopter and anti-vehicle mines; hollow/cratering charges; plastic blank cartridges; mortar bombs; 155 mm calibre ammunition; pyrotechnical products; boosters; plastic explosives; mechanical ignition systems (NON-EL); 120 mm calibre multiple mortar systems; demilitarization and disposal of ammunition and mines; project management. Complete 81 mm and 120 mm mortars. Small arms ammunition. [98]

75-00920 Empl Fahrzeugwerk GesmbH
PO Box 41, A-6272 Kaltenbach
☎ (+43 5283) 224 80 E empl@telebox.ada.at
⊖ (+43 5283) 22 48 26
♣ Sch 200,000,000 (1996) ⅏ 140 ⌂ 5,000 m2
Managing Director: Herbert Empl
Operations Director: Josef Empl
Sales and Marketing Director: Klaus Reisinger
Purchasing Manager: Franz Kerschhaggl
■ Ambulance shelters, field ambulances. Mobile workshops. Towing and salvaging units. Lubrication units. Lorry mounted platforms for main carriage. [98]

75-00990 Dr-Ing Geislinger & Co Schwingungstechnik GmbH
Hallwanger Landesstrasse 3, A-5300 Hallwang/Salzburg

☎ (+43 662) 66 99 90 🄴 info@geislinger.co.at
☒ 75314424 geis a ⊖ (+43 662) 669 99 40
ꝑ 170 (1997)
Managing Director: C Geislinger
Sales Manager: Uwe Merl
■ Manufacture of couplings and dampers for the marine industry. 98

75-01060 Gold-Eck Austria
Seebacher Strasse 11-13, A-9871 Seeboden
☎ (+43 4762) 510 10 ☒ 48181 ⊖ (+43 4762) 51 01 18
♧ Sch 1,000,000 ♣ Sch 120,000,000 ꝑ 80
General Manager: August Mayer
■ Manufacture of sleeping bags. 97

75-01100 Grabner GmbH
(† Sportartikelfabrik Wolfgang Grabner)
Weistracherstrasse 11, A-3350 Haag
☎ (+43 7434) 42 25 10
⊖ (+43 7434) 422 51 66; 42 25 13
Managing Director: Wolfgang Grabner
Export Manager: Mrs Gertraud Pfaffenbichler
■ Manufacture of inflatable boats, life jackets, rucksacks, isolation bottles, gaiters and packing bags. 98

75-01140 A Haberkorn GmbH & Co KG
Werndl Strasse 3, PO Box 85, A-4240 Freistadt
☎ (+43 7942) 771 77 ☒ 21355 hacof a
⊖ (+43 7942) 771 80
♧ Sch 6,000,000 ♣ Sch 370,000,000 (1995) ꝑ 175
⌂ 15,000 sq m
Managing Director: Mag Karl Haberkorn
■ Trouser belts and body belts, military harnesses, carrying harnesses, gun belts, carrying belts, webbings, braids and ropes in military colours. Fire hoses, climbing harnesses, cargo lashings, lifting slings. Rescue equipment. 97

75-01225 HB Flugtechnik GmbH
(† HB-Aircraft Industries)
Dr Adolf Schärf Strasse 42, PO Box 74,
A-4053 Haid Ansfelden
☎ (+43 7229) 791 04; 791 17
⊖ (+43 7229) 791 04 15; 791 17 15
Managing Director: Ing Heino Brditschka
■ Manufacture two seat light training aircraft. 98

75-01230 Heiser Josef
PO Box 1, A-3291 Kienberg-Gaming
☎ (+43 7485) 972 00 ✉ JOSHEIZER ☒ 19258
⊖ (+43 7485) 972 27
ꝑ 350 ⌂ 20,000 sq m
Managing Director: C L Reitlinger
Sales and Marketing Director: Dr-Ing Timm Friedel
Operations Director: G Konul
Purchasing Manager: K Seydali
■ Gas cylinders for use in industry, hospitals, medical work, beverage industry, fire extinguishers. Cylinders for diving, breathing, resuscitation apparatus and porous mass for acetylene escape-sets, for welding. Blank and semi-finished pressings for projectiles, rocket-drives. Speciality: extremely lightweight cylinders, in particular composites for higher pressures (300 bar). 97

75-01240 Hexcel Composites Austria
(† Danutec Werkstoff GmbH)
Industriestrasse 1, A-4061 Pasching
☎ (+49 7229) 77 20 ⊖ (+49 7229) 77 22 99
♣ Sch 30,000,000 ꝑ 105
Managing Director: Hans Schmid
General Manager: Dr Gerhard Hofinger
■ Polymer composites and polyurethane formulations. 98

75-01250 Hirtenberger AG
Leobersdorferstrasse 33, A-2552 Hirtenberg
☎ (+43 2256) 81 18 40 ☒ 14447 hpag a
⊖ (+43 2256) 818 07
♧ Sch 35,000,000 (1996) ♣ Sch 450,000,000 (1996)
ꝑ 400 (1996) ⌂ 151,000 sq m
Managing Director: Heinz Traeder
Sales Manager: Rudolf Doerr
■ Pistol and revolver ammunition production: 6.35 mm; 7.65 mm; 9 mm Parabellum; 0.45 ACP; 0.32 calibres. S&W Long; 0.38 S&W Special; 0.357 Magnum.
Rifle ammunition: 7.62 mm x 51 mm; 5mm x 45mm; 0.30; versions: ball, steel core, sniper, blank, tracer).
Mortars (60, 81, 82 and 120 mm). Mortar ammunition (60, 81, 82 and 120 mm; versions: HE, smoke, illuminating, practice).
Howitzer shells (105, 130 and 155 mm). Electronic sensors for off-route anti-tank and anti-helicopter systems components. Turnkey projects and technology. 98

75-01280 Hitzinger Gesellschaft GmbH
Helmholtzstrasse 56, PO Box 5000, A-4021 Linz
☎ (+43 732) 38 16 81 ⊖ (+43 732) 381 68 15
♣ Sch 280,000,000 (1996) ꝑ 190
Managing Director: Mr Platzer
Managing Director: Mr Roland
Managing Director: Mr Gatterbauer
Executive Vice President: Peter K Unterweger
Sales and Marketing Director: Mr Pinter
Purchasing Manager: Mr Kolmbauer
Sales Manager, Airport Equipment and Converters: Ing J Mair
■ Centralised 400Hz power system. 97

75-01420 Interspiro Gesmblt
(a subs. of Interspiro AB, Sweden)
Mühlgasse 128, A-8200 Gleisdorf
☎ (+43 31) 12 46 46; 12 44 17 ☒ 131180
⊖ (+43 31) 12 44 17 22
ꝑ 5 98

75-01430 INTERTECHNIK Techn. Produktionen GmbH & Co KG
Industriezeile 56, PO Box 100, A-4040 Linz
☎ (+43 732) 78 92 🄴 info@intertechnik.at ☒ 221522
⊖ (+43 732) 78 92 13
ꝑ 60 (1997) ⌂ 6,000 m2
President: Ing E Hainzl
Managing Director: Ing Reinhard Zierler
Sales and Marketing Director: Helmut Possegger
Operations and Technical Director: Ing Helmut E Kern
Purchasing Manager: Otto Schrittwieser
■ Design and manufacture of integrated systems in weaponry, military hardware and munitions sectors. 98

75-01470 Just Leitern AG
Heiligenstädterstrasse 40, A-1190 Vienna
☎ (+43 1) 368 12 70 ⊖ (+43 1) 368 16 98
Managing Director: Ewald Tuwora
Marketing Director: R Leistner-Mayer
■ Manufacture of ladders (in wood, aluminium, fibre glass). Pit ladders, escape ladder system, mast girder type ladders, industrial stairways, movable building cleaning systems, staircases for aircraft boarding, tank stairways, suspension platforms, loading ramps.

Zeiselmauer Plant
Bundesstrasse B-14, A-3424 Zeiselmauer
☎ (+43 2242) 70 40 80 ⊖ (+43 2242) 725 07

Graz-Strassgang Branch
Mühlfelderweg 40, A-8055 Graz-Strassgang
☎ (+43 316) 29 74 00 ⊖ (+43 316) 29 72 29

Innsbruck-Völs Branch
Innsbrucker Strasse 68, A-6176 Innsbruck-Völs
☎ (+43 512) 30 42 30 ⊖ (+43 512) 30 26 59

Villach Branch
An d. Südwest-Umfahrung, Untere Fellacher Strasse 58, A-9500 Villach
☎ (+43 4242) 580 28 ⊖ (+43 4242) 565 76

Salzburg-Gnigl Branch
Lerchenstrasse 121, A-5023 Salzburg-Gnigl
☎ (+43 662) 66 04 81 ⊖ (+43 662) 66 04 81 98

75-01500 Kapsch AG
Wagenseilgasse 1, A-1121 Vienna
☎ (+43 1) 81 11 10 🅆 http://www.kapsch.net
☒ 131299 kag w a ⊖ (+43 1) 81 11 18 88
♧ US$ 130,000,000 (1995)
♣ US$ 280,000,000 (1995) ꝑ 2,000 (1995)
Managing Director: Georg Kapsch
Member of the Board of Directors: Gerhard Lippburger
Operations Director: Wilhelm Kapsch
Marketing: G Binder
■ Manufacture of telecommunications equipment. 98

75-01570 Kohla Max KG Sportgeraeteerzeugung
Rossaug 28, A-6021 Innsbruck
☎ (+43 512) 335 10 ⊖ (+43 512) 34 42 10
■ Carrying frames. Rucksacks. Alpine equipment and skiing accessories. 98

75-01575 Kohlbrat & Bunz GmbH
4-8 Loretostrasse, A-5550 Radstadt
☎ (+43 6452) 719 30 ☒ 67746 jonsra a
⊖ (+43 6452) 71 93 13
♣ Sch 100,000,000 ꝑ 25
President: Ms Christine Dzionara
Sales and Marketing Director: Ms Rosa Bunz
■ Chainsaw winch; UT 2000 universal transporting and rescue device. 97

75-01615 Lem Norma GmbH
(† Norma Goerz Instruments GmbH)
Palmersstrasse 2, A-2351 Wiener Neudorf
☎ (+43 2236) 69 10; 69 81 50
⊖ (+43 2236) 624 74; 630 80; 69 14 15
♣ Sch 200,000,000 ꝑ 140
Information Contact: Ing I V Gschmeidler
■ Eye safe laser rangefinders for observation or high power laser rangefinder for sniper rifles and infantry observation. Optic aiming devices such as panoramic sights for howitzers and accessories. Reflex sights for anti-aircraft guns. Electronic ballasting machines. Measuring instruments such as precision power meters, voltmeters, ammeters, recorders, spectrum and power analysers, safety testers, insulation testers. 98

75-01625 Liebherr-Werk Nenzing GmbH
PO Box 10, A-6710 Nenzing
☎ (+43 5525) 60 60 🄴 lwn@liebherr.vol.at ☒ 52167
⊖ (+43 5525) 606 20
♧ Sch 220,000,000 (1995)
♣ Sch 2,443,322,000 (1996) ꝑ 832 (1997)
⌂ 61,000 m2
Managing Director: R Copony
Sales and Marketing Director: R Nasahl
Operations and Technical Director: W Lange
Purchasing Manager: T Fritz
■ Deck cranes. 98

Contact Reference

75-01700
Mils Elektronik GmbH & Co KG
Dorfstrasse 20, A-6060 Mils
☎ (+43 5223) 57 71 00 🄴 consult@mils.com (enquiries)
🄴 support@mils.com (customers)
🅆 http://www.mils.com ☒ 533009 mils a
⊖ (+43 5223) 57 71 01
General Manager: Werner Kraetschmer
■ Development and production of cipher equipment for tactical and strategic security. Cipher systems on Personal Computers (text and image) with implemented random One Time Key (OTK) cipher method and/or Mils proprietry stream cipher algorithm. Secure Telephones, Random Key Generators, Network Managers and Customised Security Solutions. Rugged and Tempest Computer Equipment. 98

75-01820 Osterreichiche Automobilfabrik Graf & Stift (OAF)
(a subsidiary of MAN Nutzfahrzeuge AG, Germany)
Brunner Strasser, A-1231 Vienna
☎ (+43 1) 86 63 10 ⊖ (+43 1) 86 63 12 78
♣ Sch 2,214,000,000 (1996) ꝑ 990 (1996)
⌂ 58,000 m2
Managing and Operations Director: Dr Richard Daimer
☎ (+43 1) 86 63 11 00
Export Manager, Truck Division: Heinrich Schiefer
☎ (+43 1) 86 63 11 95
Purchasing Manager: Dr Rudolf Wild
☎ (+43 1) 86 63 12 50
■ Chassis for catering vehicles and for fire fighting trucks. 98

75-02128 Rockwell Automation Ges mbH
(† Sprecher + Schun Niederspannung GmbH)
(a subsidiary of Rockwell Automation, USA)
Bäckermühlweg 1, A-4030 Linz
☎ (+43 732) 38 90 90 ⊖ (+43 732) 389 09 61
ꝑ 71 (1997)
Regional Sales Director: Christoph Becher
🄴 cbecher@eu1.ra.rockwell. com
Human Resources: Ms Faye Macherhammer
® USA: Rockwell Automation 98

75-02132 Rosenbauer International Aktiengesellschaft AG
Paschinger Strasse 90, A-4060 Leonding
☎ (+43 732) 67 94 E rbi@rosenbauer.co.at
Tx 227271 rosba a ⊖ (+43 732) 67 94 88
♣ Sch 1,813,000,000 ⁂ 871
President: Julian Wagner
Managing Director: Helmut Gumbsch
General Manager, Export: Wolfram Mücke
Advertising and Marketing Manager: Manfred Schwedz
■ Crash tenders, dry powder units, foam trucks and trailers, runway foam units, wide range of portable pumps, combined HP pumps. Fire fighting equipment. [98]

75-02142 Saft Nife GmbH
(a subsidiary of Saft SA, France)
Eduard Kittenberger Gasse 56, A-1235 Vienna
☎ (+43 1) 865 93 68 ⊖ (+43 1) 869 12 28
Managing Director: Erwin Matzner [98]

75-02155 Sattler Texilwerke KG
PO Box 6, A-8041 Graz-Thondorf
☎ (+43 316) 40 15 51 ⊖ (+43 316) 40 34 07
♣ Sch 860,000,000 ⁂ 500 ⌂ 60,000 m2
Managing Director: Dkfm.Dr Werner Tessmar-Pfohl
Technical Director: Dipl.-Ing Franz-Wilhelm Mohné
Marketing Director: Ing-Mag.Dr Ernesto Scholbesberger
Purchasing Manager: Erich Guttmann
■ Manufacture of high tech textile and textile systems. Ballistic protection fabrics and composites for personnel and vehicles. Textile coverings and ageing protection coverings for arms and tanks. Military tents and tarpaulins for vehicles, made of synthetic and natural fibres and military bags. [97]

75-02180 Schiebel Elektronische Geraete Ges.mbH
(a subsidiary of Schiebel GmbH)
Margaretenstrasse 112, A-1050 Vienna
☎ (+43 1) 54 62 60 Tx 136741 schwk a
⊖ (+43 1) 545 23 39
⁂ 100
Marketing Director: Dipl.-Ing Hans G Schiebel
■ Manufacture of AN/PSS-12 Handheld Metallic Mine Detector for the US Army. Mine detection equipment and UXO detection equipment. [97]

75-02195 Schrack Aerospace GmbH
(a subsidiary of Saab Ericsson Space and Ericsson, Austria)
Breitenfurterstrasse 106/108, A-1120 Vienna
☎ (+43 1) 80 19 90 ⊖ (+43 1) 80 19 96 95
♣ US$ 8,500,000 (1996) ⁂ 61
Managing Director: Ulf Hansson
Managing Director: Mario Rohracher
Marketing Manager: Bernhard Eichinger
■ Instruments and sub-systems for communications, earth observation and scientific satellite missions, electronic satellite check out equipment. [97]

75-02800 Steyr Antriebstechnik
(a division of Steyr-Daimler-Puch AG)
Schönauerstrasse 5, A-4400 Steyr
☎ (+43 7252) 580 25 20 ⊖ (+43 7252) 433 00
♣ Sch 1,100,000 (1995) ⁂ 800
Director: Prof-Dr Rudolf Streicher
Director: Dipl.-Ing Rudolf Aita
Director: Dr A Koch
Sales and Marketing Director: Dipl.-Ing Joachim Bohnen
Head of Engineering Department: Dr Wilfried Eichelseder
Purchasing Manager: Rudolf Pinter
■ Design and development of automotive components/military utility vehicles. [98]

75-02805 Steyr-Daimler-Puch Fahrzeugtechnik AG & Co KG
(a subsidiary of Steyr-Daimler-Puch AG)
Liebenauer Hauptstrasse 317, PO Box 823, A-8011 Graz
☎ (+43 316) 40 40
Tx 311315 stdpwg a; 311577 stpwg a
⊖ (+43 316) 404 29 86
♣ Sch 4,227,000,000 (1996) ⁂ 3,872 (1996)
⌂ 260,000 m2
General Manager: Prof.-Dr Rudolf Streicher
Member of the Executive Board: Dipl.-Ing Rudolf Aita
Member of the Executive Board: Dr Alfred Koch
Marketing Director: Dr Siegfried Gabler
Marketing Director: Friedrich Rohr
■ Manufacturer of Pinzgauer light wheeled cross-country vehicle and Mercedes-Benz G/Puch G. [98]

75-02810 Steyr-Daimler-Puch Spezialfahrzeug AG & Co KG
(a subsidiary of Steyr-Daimler-Puch AG)
2 Haidequerstrasse 3, PO Box 100, A-1111 Vienna
☎ (+43 1) 76 45 11 ✉ STEWESIMM Wien
Tx 131810 stdpsi a ⊖ (+43 1) 769 81 49
♣ Sch 925,000,000 (1996) ⁂ 292 (1996)
⌂ 65,000 sq m
President and Chief Executive Officer: Dipl.-Ing Hans M Malzacher
Executive Vice President: Dipl.-Ing Wolfgang Riedl
Communications Manager: G Binder
■ Design and manufacture of systems integration for wheeled and tracked armoured vehicles. [98]

75-02820 Steyr Mannlicher AG & Co KG
(a subsidiary of Steyr Daimler Puch AG)
Mannlicher Strasse 1, PO Box 1000, A-4400 Steyr
☎ (+43 7252) 89 60 Tx 613732224 smgv a
⊖ (+43 7252) 786 21
♧ Sch 50,000,000 (1997) ♣ Sch 320,000,000 (1996)
⁂ 198 (1997) ⌂ 14,700 m2
Member of the Executive Board: Ing Heinz Hambrusch
Member of the Executive Board: Gerhard Leitner
Sales Director: Wolfgang Stadler
Operations Director: Karl Kattnigg
Purchasing Manager: Wolfgang Holzmann
■ Manufacture of small arms, automatic and sniper rifles, light machine guns, submachine guns, hunting and sporting rifles, CO2 air pistols/rifles. [98]

75-03040 Swarovski Optik KG
Swarovskistrasse 70, A-6060 Absam
☎ (+43 5223) 51 12 30 ⊖ (+43 5223) 418 60
Managing Director: Gerhard Swarovski
Sales Manager: Albert Wannenmacher (Mag. BA)
Marketing Department: C Carpenter [97]

75-03150 Teufelberger GmbH
Vogelweiderstrasse 50, PO Box 294, A-4600 Wels
☎ (+43 7242) 41 30 Tx 25530 teuf a
⊖ (+43 7242) 41 31 69
Public Relations Director: M Teufelberger
■ Manufacture of ropes of steel, nylon and kevlar. Alpine equipment. [97]

75-03160 THORN Licht GmbH
(a subsidiary of THORN Lighting Ltd)
Erzherzog-Karl-Straße 57, A-1220 Vienna
☎ (+43 1) 221 51 10 Tx 136128 ⊖ (+43 1) 22 99 40
Chief Executive: W Fabich [98]

75-03330 Voere Kufsteiner Gerätebau und Handels Ges.mbH
Untere Sparchen 56, PO Box 416, A-6333 Kufstein
☎ (+43 5372) 625 47 ⊖ (+43 5372) 657 52
♧ Sch 8,900,000 ♣ Sch 66,000,000 ⁂ 60
⌂ 5,700 sq m
Vice President: Dipl.-Ing S Ruhland
Managing Director: Ing Alfons Ruhland
Operations and Technical Director: Ing Werlberger
Purchasing Manager: Dipl.-Ing S Rutland
■ Semi-auto high power hunting rifle model 2185 in most big and small game calibre. Police Swat team semi-auto sniper rifle weapon system. Model VEC 91 high power calibre 5.7 mm UCC bolt action repeating rifle firing caseless ammunition. 5-6 and 7 mm UCC high power caseless ammunition. [98]

90 Sales Agents, Sales & Liaison Offices

Companies represented:

AB Electronic Products, UK 75-00050
Acer Inc, Taiwan 90-00019
Alfred Karcher GmbH & Co, Germany 90-01140
Break-Free Inc, USA 90-00350
Cannon Electric GmbH, Germany 90-00320
Hayes Int'l Targets, USA→ UK 80-07600
Rockwell Automation, USA 75-02128
Schottel-Werft, Germany 90-00400
Sonnenschein Lithium GmbH, Germany 90-00670

90-00019 Acer Computer Handels GmbH
Jochen Rindt Strasse 25, A-1230 Vienna
☎ (+43 1) 615 08 20 ⊖ (+43 1) 615 08 20 50
® Taiwan: Acer Inc [97]

90-00320 Cannon Electric GmbH
PO Box 146, A-1021 Vienna
Afrikanergasse 3/1, Vienna
☎ (+43 1) 216 09 47 ⊖ (+43 1) 216 09 48
Area Manager: Ing G B Junger
® Germany: Cannon Electric GmbH [98]

90-00350 CLP Chemicals Handels GmbH
(a daughter company of Break-free Inc, USA)
Hauptstrasse 17, A-3012 Wolfsgraben
☎ (+43 22) 33 77 43 ⊖ (+43 22) 33 77 45
Managing Director: Henning Hauff
Sales and Marketing Director: K Bialowas
® USA: Break-Free Inc [98]

90-00370 Daewoo Corp Vienna Branch
Am Concorde - Park 1/B3, A-2320 Vienna
☎ (+43 1) 706 38 28 Tx 134359 dwvina
⊖ (+43 1) 706 38 26 [98]

90-00400 Deutz Motoren Vertriebsgesellschaft mbH
Dampfmühlgasse 6-8, A-1112 Vienna
☎ (+43 222) 74 16 71 ⊖ (+43 222) 74 16 71 12
Director: Johann Marterbauer
® Germany: Schottel-Werft [97]

90-00670 Exide Batteriewerke GmbH
Franz Schubert Gasse 7, A-2345 Brunn, Geb
☎ (+43 222) 363 56 10 40 ⊖ (+43 222) 363 56 10 45
Agent: H Weiss
® Germany: Sonnenschein Lithium GmbH [98]

90-01140 Alfred Karcher GmbH
Obachgasse 26, A-1220 Vienna
☎ (+43 1) 25 06 00 Tx 047132969 ⊖ (+43 1) 250 60 30
Managing Director: H H Konhausner
® Germany: Alfred Karcher GmbH & Co [98]

Azerbaijani Republic

President:
Heydar Aliyev

10 Head of State

10-10000 Office of the President
19 Istiglaliyat Street, 370066 Baku
☎ (+994 12) 98 33 65 ⊖ (+994 12) 98 33 28
President: Heydar Aliyev [98]

11 Prime Minister

11-10000 Office of the Prime Minister
Lermontova Kuc, 63, Baku
☎ (+994 12) 92 72 32
Prime Minister: Artur Rasizade
First Deputy Prime Minister: Abbas Abbasov

First Deputy Prime Minister: Vahid Akhmedov
First Deputy Prime Minister: Samed Sadykov
First Deputy Prime Minister: Elchin Efendiyev
Deputy Prime Minister: Izzet Rustamov
Deputy Prime Minister: Tofiq Azizov
Deputy Prime Minister: Abid Sherifov 98

12 Ministries

12-10000 Ministry of Defence
Prospekt Azizbekova 3, 601 Baku
☎(+994 12) 39 31 08
Minister: Lieut Gen Safar Abiyev 98

12-20000 Ministry of Foreign Affairs
Controlny Pereulok 2, 370005 Baku
☎(+994 12) 93 82 31 ⊖ (+994 12) 93 82 43
Minister of Foreign Affairs: Hasan Hasanov
Minister of Foreign Economic Relations (acting): Qudrat Quliyev 98

12-30000 Ministry of Internal Affairs
Husu Hajiyev Kuc, Baku
☎(+994 12) 92 57 54
Minister: Ramil Usubov 98

12-40000 Ministry of National Security
Prospect Azizbekova 2, 601 Baku
☎(+994 12) 93 10 00
Minister: Namiq Abasov 98

12-70000 Ministry of Communications
Prospekt Azerbaijana 33, 370139 Baku
☎(+994 12) 93 00 04 ✉ SAhIL AI 142492
⊖ (+994 12) 98 42 85
Minister: Siruz Abasbeyli
Deputy Minister: Omaz Husseynov
Director of Foreign Relations: Fizuli Melikov
Chief of Staff: Masuma Ibrahimora 98

15 Parliament

15-10000 National Assembly
Ezizbayov Prospekti 1, 370152 Baku
☎(+994 12) 39 97 50; 98 99 58 ▣ 142294 ganun
⊖ (+994 12) 93 49 43; 98 02 42
Chairman: Murtuz Aleskerov ☎(+994 12) 98 09 92
First Deputy: Arif Rahimzade ☎(+994 12) 92 10 60
Deputy: Yashar Aliyev ☎(+994 12) 92 63 48
Chairman of the Foreign Affairs Commission: Yusif Vekilov ☎(+994 12) 93 74 15
Chairman of Legal Policy and State Building Commission: Zakir Zeynalov ☎(+994 12) 98 97 30
Chief of International Relations Department: Namig Akhundov ☎(+994 12) 39 97 50 97

20 Defense Administrations/Services/Agencies

20-01000 Azerbaijan National Aerospace Agency
159 Azadlyg prospect, 370106 Baku
☎(+994 12) 62 93 87 Ⓔ mekhtiev@anasa.baku.az
▣ 142288 space ai ⊖ (+994 12) 62 17 38
General Director: Arif Sh Mekhtiev ☎(+994 12) 62 93 87
Deputy General Director: Rustam Rustamov ☎(+994 12) 62 17 22
■ State policy and co-ordination of national aerospace and space programmes. 98

29 Customs & Excise

29-01000 State Customs Committee
Inshaatchilar Prospecti 2, 370600 Baku
☎(+994 12) 39 73 94; 38 74 97 ⊖ (+994 12) 38 80 63
Chairman: Ibrahim Ali Oglu Valiyev 98

31 Embassies & Attachés

31-00690 Embassy of Greece
Residence "Republica", Baku
☎(+994 12) 98 91 84; 92 89 10 ▣ 24383
⊖ (+994 12) 65 60 08 98

31-01890 Embassy of the United Kingdom
c/o Outward Bagroom, British Embassy, (Baku), King Charles Street, SW1A 2AH London
2 Izmir St, 370065 Baku
☎(+994 12) 98 55 58; 92 48 13; 92 27 39
☎(+873) 144 64 55 (satellite) Ⓔ office@britemb.bakuaz
⊖ (+994 12) 39 13 41; 92 27 39
⊖ (+873) 144 64 56 (satellite)
Ambassador: H E Mr Robert Thomas
Deputy Chief of Mission: M K Oliver
Commercial Officer: N Babayey
Defence Attaché: Col R G Greenham (resident in Moscow) 98

Bahamas

Head of State:
H M Queen Elizabeth II

Governor-General:
Sir Orville Alton Turnquest

10 Head of State

10-10000 Office of the Governor General
Government Hill, PO Box N-8301, Nassau
☎(+1 242) 322 10 21
Governor General: Sir Orville Alton Turnquest 98

11 Prime Minister

11-10000 Office of the Prime Minister
Sir Cecil V Wallace-Whitfield Centre, PO Box N7147, Nassau
☎(+1 242) 327 58 26; 327 58 29
⊖ (+1 242) 327 58 06; 327 58 06
Prime Minister: The Hon Hubert A Ingraham
Deputy Prime Minister: Frank Watson
Minister of Foreign Affairs: Ms Janet Bostwick
Minister for Finance and Planning: William Allen 98

12 Ministries

12-40000 Ministry of National Security
(† Ministry of Public Safety and Immigration)
East Hill Street, PO Box 4891, Nassau
☎(+1 242) 356 68 01; 356 68 02; 356 68 03; 356 68 04; 356 67 93
Minister of National Security: Frank Watson
Ⓔ watsondpm@batelnet.bs
Permanent Secretary: Ms Idris Reid
Deputy Permanent Secretary: Peter-Deveaux Isaacs
■ Related departments: Royal Bahamas Police Force; Royal Bahamas Defence Force; Her Majesty's Prisons; Department of Road Traffic; Port Department. 98

15 Parliament

15-10000 Senate
PO Box N-1365, Nassau
☎(+1 242) 322 20 41 ⊖ (+1 242) 322 11 18 98

15-20000 House of Assembly
Rawson Square, Nassau
☎(+1 242) 322 20 41
⊖ (+1 242) 328 82 94 (c/o Cabinet) 98

25 Defense Forces

25-10000 Defence Force Headquarters
Coral Harbour Base, PO Box No 3733, Nassau
☎(+1 242) 322 19 94; 323 36 91
⊖ (+1 242) 328 89 12; 362 13 74
Commander Defence Force: Cdre Leon L. Smith ☎(+1 809) 322 1994; 323 3741
Principal Staff Officer: Lt Cdr G. Rolle ☎(+1 809) 322 2938
Personnel/Recruiting Officer: Lt F.A. Clarke ☎(+1 809) 322 1994; 323 3691
Staff Intelligence Officer: Lt Cdr E. Rolle ☎(+1 809) 809 322 2726
Public Relations Officer: Lt L. Klein ☎(+1 809) 322 1994; 323 3741 97

25-11000 Royal Bahamas Defence Force (RBDF)
Coral Harbour Base, PO Box No 3733, Nassau
☎(+1 242) 323 36 91 ⊖ (+1 242) 362 13 74
Commanding Officer (Admin) Coral Harbour: Cdr D. Rolle ☎(+1 809) 325 8632
Squadron Commanding Officer (SCO): Cdr Anthony Allens ☎(+1 809) 325 8663
Squadron Commanding Officer (ASCO): Lt Cdr A. Armbrister ☎(+1 809) 322 4436
Air Wing Commanding Officer (AWCO): Lt Cdr M. Williams ☎(+1 809) 362 1473
Military Operations Platoon Officer (MOPO): Snr Lt W.I. Rolle ☎(+1 809) 322 3741
Base Exec. Officer (BXO): Lt P. Clarke ☎(+1 809) 322 7463
Base Engineering Officer (BEO): Lt Cdr P. McNiel ☎(+1 809) 325 4527
Supply Officer (SO): Sr Lt S. Plakaris ☎(+1 809) 362 1381
Training Officer (TO): Lt Cdr D. Duncombe ☎(+1 809) 362 1722
Staff Intelligence Officer (SIO): Lt Cdr E. Rolle ☎(+1 809) 322 2726 97

26 Police (Civil)

26-00000 Royal Bahamas Police Force
East Street, PO Box No 458, Nassau
☎(+1 242) 322 44 44; 322 44 45
⊖ (+1 242) 328 29 91
Commissioner of Police: B.K. Bonamy ☎(+1 809) 325 7602; 322 6250
Sr Deputy Commissioner: Errol Farquharson ☎(+1 809) 322 1200
Asst Comm. CID: N.J. Rolle ☎(+1 809) 322 2311
Asst Comm. Out Island Divisions: A. Miller ☎(+1 809) 322 2487
Asst Comm. Uniformed Divisions: B. Dean
Director Fire Services: Asst Supt H. Douglas ☎(+1 809) 325 8528
Internal Security Div.: Supt Rubin Smith ☎(+1 809) 323 5362
Staff Officer: Det.Supt H. Thurston

Contact Reference

Police College: Asst Comm. Keith V. Mason (Police Reserve) ☎ (+1 809) 325 4346
Drug Enforcement Unit: Asst Supt R. Ferguson ☎ (+1 809) 393 7202
Road Traffic Department: Supt R. Butler ☎ (+1 809) 322 4379 [97]

31 Embassies & Attachés

31-01890 United Kingdom High Commission
PO Box N 7516, Nassau
BITCO Building (3rd floor), East Street
☎ (+1 242) 325 74 71 ✉ UKREP NASSAU
⊖ (+1 242) 323 38 71
High Commissioner: H E Mr P M H Young
Deputy High Commissioner: J P M Culligan
Defence Adviser: Col A L Moorby (resident in Kingston)
Commercial Officer: Mrs Jane Lloyd [98]

Bahrain, State of

Head of State:
Sheikh Isa Bin Salman Al-Khalifa

10 Head of State

10-10000 Office of the Amir
Amiri Court, PO Box 555, Manama
☎ (+973) 66 66 66 ⊠ 8666 qasar bn
Amir of Bahrain: H H Sheikh Isa bin Salman Al-Khalifa [98]

11 Prime Minister

11-10000 Office of the Prime Minister
Government House, PO Box 1000, Manama
☎ (+973) 27 14 96; 26 22 66 ⊠ 9336 prom bn
Prime Minister: H H Sheikh Khalifa bin Salman Al-Khalifa [98]

12 Ministries

12-10000 Ministry of Defense
PO Box 245, Manama
☎ (+973) 66 55 99 ⊠ 8429 deffaa bn
⊖ (+973) 65 03 46
Minister of Defense: Maj Gen Khalifa Bin Ahmad Al-Khalifa [98]

12-20000 Ministry of Foreign Affairs
Government House, PO Box 547, Manama
☎ (+973) 26 22 77 ⊠ 8228 karjia bn
⊖ (+973) 25 61 99, 21 46 16
Minister:
H H Sheikh Muhammad bin Mubarak Al-Khalif Al Khalifa [98]

12-40000 Ministry of Interior
(Public Security Headquarters)
Salmaniya, PO Box 13, Manama
☎ (+973) 27 21 11 ⊠ 9572 psmkt bn
⊖ (+973) 26 21 69; 27 04 63
Minister:
H H Sheikh Muhammad bin Khalifa bin Hamid Khalifa
Deputy Dir General of Public Security:
Maj.Gen. Sheikh Ibrahim Bin Mohamed Al Khalifa [98]

12-60000 Ministry of Transportation and Communications
PO Box 10325, Manama
☎ (+973) 53 45 34 ⊠ 8989 mincom bn
⊖ (+973) 53 75 37
Minister: H E Mr Ali bin Khalifa bin Salman Al Khalifa [98]

12-70000 Ministry of Commerce and Agriculture
PO Box 5479, Manama
☎ (+973) 53 15 31 ⊠ 9171 tejara bn ⊖ (+973) 53 04 55
Minister: Ali Salih Abdullah al Salih [98]

31 Embassies & Attachés

31-00830 Embassy of India
Building No 182, Road 2608 Area 326, PO Box 26106, Gudaibiya
☎ (+973) 71 26 83; 71 27 85 ✉ INDEMBASSY Bahrain
🄴 indemb@batelco.com.bh ⊠ 9047 indemb bn
⊖ (+973) 71 55 27
Ambassador: H E Mr Rajanikanta Verma
☎ (+973) 71 26 49
Deputy Chief of Mission: Madhava Chandra
Commercial Officer: K L Arora
Public Information Officer: Jagdish Parshad [98]

31-00850 Embassy of the Islamic Republic of Iran
PO Box 26365, Manama
☎ (+973) 72 24 00 ⊠ 8238 Iran BN ⊖ (+973) 72 21 01
Chargé d'Affaires: Seyed Mohammed Ahmadi
Third Secretary and Head of Consular Affairs:
Abdul Rasool Dokouhaki [98]

31-00990 Embassy of the Republic of Korea
Bldg 69, Rd 1901, Block 319, PO Box 11700, Manama
☎ (+973) 29 16 29 ✉ GONGKWAN ⊠ 8736 rokbah bn
Ambassador: H.E. Mr Soo Whan Lee
Counsellor & Deputy Chief of Mission: Young Won Kim [97]

31-01830 Embassy of Tunisia
PO Box 26911, Al-Adliya, Manama
☎ (+973) 71 41 49 ⊠ 7136 ambatn bn
⊖ (+973) 71 57 02
Ambassador: H E Mr Mohsen Frini ☎ (+973) 71 41 49
First Secretary: Abdelhamid Cherif ☎ (+973) 71 09 87
Administrative Attaché: Lotfi Trabelsi
Administrative Attaché: Mohamed Ali Oueslati
☎ (+973) 29 55 36 [98]

31-01890 Embassy of the United Kingdom
21 Government Avenue, PO Box 114, 306 Manama
☎ (+973) 53 44 04; 53 48 65/6
✉ PRODROME BAHRAIN ⊠ 8213 prodro bn
⊖ (+973) 53 12 73; 53 33 07
Ambassador: H E Mr D I Lewty
First Secretary Commercial, Consul and Deputy Head of Mission: Barry Seddon
First Secretary Political: Norman Cameron
Defence Attaché: Brig B Strong (Resident in Riyadh)
Defence Co-operation Officer: Cdr B Reeves
Second Secretary Commercial: Ian Lewis
Second Secretary, Works: B M Ward
Vice-Consul: Peter Langham
Third Secretary Political/Information: Ms Caroline Alcock [98]

90 Sales Agents, Sales & Liaison Offices

Companies represented:

Bell Helicopter Textron, USA	90-00150
ITT Defense, USA	90-00700
Sierra Research/LTV, USA	90-00150
The Expanded Metal Co.Ltd, UK	90-00080

90-00080 A.K. Almoayed
PO Box 363, Manama
☎ (+973) 21 11 55 🅆 http://www.almoayed.com
⊖ (+973) 21 19 99
Managing Director: Khalid Almoayed
Sales Manager: Yaseen Alsayed
Secretary, Contact: Horiya Ali
® UK: The Expanded Metal Co.Ltd [98]

90-00150 Arabian Aviation Corp EC
(a subsidiary of Hawker Pacific Pty Ltd)
PO Box 10476, Manama
☎ (+973) 53 23 08 🄴 aviation@balelco.com.bh
⊠ 9289 heli bn ⊖ (+973) 53 00 32
General Manager: Peter A Smales
Sales Manager: Alan L Parsons
Corporate Secretary/Finance Manager:
Ms Linda A H Smythe
Mechanical Manager: Paul M Giles
■ Bell Helicopter Independent representatives for the Arabian Gulf countries
® USA: Bell Helicopter Textron, Sierra Research/LTV [98]

90-00650 Gulf Safety & Engineering Establishment
PO Box 655, Manama
☎ (+973) 25 40 38 ⊖ (+973) 21 06 88 [97]

90-00700 ITT Defense International (Bahrain)
Villa 799, Road 3315, Manama
☎ (+973) 53 17 01 ⊖ (+973) 72 02 56
Middle East Operations Director: Lee W Eyer
® USA: ITT Defense [98]

Bangladesh

Head of State:
Ahmed Shahabuddin

10 Head of State

10-10000 Office of the President
Bangabhaban, 1000 Dhaka
☎ (+880 2) 41 01 41 49 ⊖ (+880 2) 86 63 00
President: Ahmed Shahabuddin [98]

11 Prime Minister

11-10000 Office of the Prime Minister
Gono Bhaban, Sher-e-Banglanagar, Dhaka
☎ (+880 2) 32 82 92; 41 89 89
Prime Minister: Ms Sheikh Hasina Wajed [98]

12 Ministries

12-10000 Ministry of Defence
Old High Court Building, Dhaka 2
Prime Minister and Minister of Defence:
Ms Sheikh Hasina Wajed [98]

12-20000 Ministry of Foreign Affairs
Topkhana Road, Dhaka
☎ (+880 2) 23 32 18; 23 40 90
⊠ 642200; 642222 pama bj
Minister: Abdus Samad Azad [98]

12-40000 **Ministry of Home Affairs**
School Building, 2nd and 3rd Floors, Dhaka
☎(+880 2) 40 41 38; 40 41 42
Minister: Rafiqul Islam 98

31 Embassies & Attachés

31-00090 **Australian High Commission**
184 Gulshan Avenue, Gulshan, Dhaka
☎(+880 2) 60 00 91/95 ✉AUSTCOM ▣ 642317 ahc bj
⊖(+880 2) 88 37 94
High Commissioner: H.E. Mr K.W. Aspinall
First Secretary: B. Farmer
Commercial Officer Austrade Officer: Mr Fasihuzzaman
Public Information Officer: Hifzur Rahman 97

31-00840 **Embassy of the Republic of Indonesia**
75 Gulshan Ave, 1212 Dhaka
☎(+880 2) 60 01 31; 60 01 32
✉PERWAKILAN INDONESIA Dhaka ▣ 65639 rida bj
⊖(+880 2) 88 53 91
Ambassador: H.E. Mr Hadi Ahmad Wayarabi Alhadar
Defense Attaché: Col Soeharnanto Hadisoedarmo
Commercial Counsellor: Muhammad Ischak 97

31-01220 **Embassy of the Kingdom of Morocco**
PO Box GN 6112, Gulshan, Dhaka
House No.44, United Nations Road,
Baridhara Model Town, Dhaka
☎(+880 2) 88 31 76; 61 03 29 ▣ 632302 sifma bj
⊖(+880 2) 87 00 18
Ambassador: H.E. Mr Hassan Mohamed Daoud
Counsellor: Omar Berechid
First Secretary: Mohamed El Ghroussi
Attaché: Mustapha Sadiq
Financial Attaché: Abdelkrim Guedida 97

31-01370 **High Commission for Pakistan**
House NE(C)2, Road 71, Gulshan, Dhaka
☎(+880 2) 88 53 88/9 ✉PAREP Dhaka
▣ 642335 parep bj ⊖(+880 2) 88 36 77
High Commissioner: H.E. Mr Karam Elahi
Defence Attaché: Col Zahid Hussain Khan
Deputy High Commissioner & Head of Chancery: Mrs Rifat Iqbal
Counsellor: Syed Sharfuddin
Counsellor: Hafeezur Rahman
Counsellor: Khurshid Anwa Butt
Counsellor: Abdul Hayee Qureshi
First Secretary: Riaz Ahmad
Commercial Secretary: Sarwar Brohi
Second Secretary: Muhammad Yaqoob 97

31-01430 **Embassy of the Republic of Poland**
House 12A, Road 86, Gulshan, PO Box 6089, 1212 Dhaka
☎(+880 2) 87 14 58 ▣ 632301 pol bj; 642316 dhan bj
⊖(+880 2) 88 73 13
Chargé d'Affaires a.i.: Piotr Opalinski
☎(+880 2) 87 14 58
Commercial Counsellor: Roman Lysakowski
☎(+880 2) 60 03 65
Deputy Commercial Attaché: Krzysztof Splawiec
☎(+880 2) 60 14 81
Chancery
House No. 111, Road No. 4, PO Box 6089,
1212 Banani, Dhaka
☎(+880 2) 60 85 03; 60 60 98 ▣ 632301 pol bj
⊖(+880 2) 87 14 58
Commercial Office
House No. 11, Road No. 77, PO Box 6074,
1212 Gulshan, Dhaka
☎(+880 2) 60 03 71; 60 03 69; 60 03 65
▣ 642316 dhan bj ⊖(+880 2) 88 73 13
Commercial Counsellor: Waldemar Markuszewski
Commercial Attaché: Krzysztof Splawiec 97

31-01730 **Swedish Embassy**
House 1, Road 51, GPO Box 304, Gulshan, Dhaka
☎(+880 2) 88 47 61/4 ✉SVENSK Dhaka
▣ 642 303 sven bj ⊖(+880 2) 88 39 48
Ambassador: H.E. Mr Björn Sternby
Deputy Chief of Mission: Mrs Britt F. Hagström
First Secretary Commercial & Cultural: Lennart Karlson
First Secretary Admin: Lars Wallén 97

31-01890 **United Kingdom High Commission**
United Nations Road, PO Box 6079, Dhaka 12
United Nations Road, Baridhara Dhaka
☎(+880 2) 88 27 05 ⊖(+880 2) 88 34 37; 88 36 66
High Commissioner: H E Mr David C Walker
Deputy High Commissioner: Miss M R McIntosh
Counsellor, Aid: K L Sparkhall
Defence Adviser: Col J M Phillips
First Secretary, Medical Officer: Dr H M Carpenter
First Secretary, Management: D J Fisher
First Secretary, Immigration and Consular: G Fairhurst 98

90 Sales Agents, Sales & Liaison Offices

Companies represented:

Bell Helicopter Textron, USA→ Singapore	90-00220
Capintec Inc, USA	90-01550
Crompton Eternacell, UK	90-00850
Erma Inc., Japan	90-01550
Farnell Int'l, UK	90-01550
Fluka Chemie AG, Switzerland	90-01550
Kodak Clinical Diagnostics Ltd, UK	90-01550
N.E.Technology Ltd, UK	90-01550
Ogawa Seiki Co.Ltd, Japan	90-01550
Sartorius AG, Germany	90-01550
The General Electric Co. plc, UK	90-00730
Thomas Scientific, USA	90-01550
Varian Associates, USA	90-01550

90-00100 **Daewoo Corp Dhaka Branch**
GPO Box 2579, Dhaka
Brac Tower 356 Mohakhali, Commercial Area
(6th and 7th Floor), Dhaka
☎(+880 2) 988 13 16 ▣ 642641 dwd bj
⊖(+880 2) 88 94 84 98

90-00730 **The General Electric Co. of Bangladesh Ltd**
72 Dilkhusha Commercial Area, PO Box 123, Dhaka 1000
☎(+880 2) 956 60 31-37 ✉GECDAC dhaka
▣ 642774 gecb bj ⊖(+880 2) 956 45 17
Managing Dir: M.S. Haque
® UK: The General Electric Co. plc 97

90-00850 **Industrial Engineering Services Ltd**
PO Box 2218, 1000 Dhaka
National Scout Bldg., 7th Floor Inner Circular Rd,
1000 Kakrail, Dhaka
☎(+880 2) 41 70 93 ⊖(+880 2) 86 37 15
Managing Dir: M.A. Choudhury
Operative Dir: S.A.Y. Saber
Sales & Marketing Dir: Mrs Z. Saber
® UK: Crompton Eternacell 97

90-01550 **Moonbeam Corp.**
89 Motijheel Commercial Area, 1st Fl, GPO Box 2101,
Dhaka 2
☎(+880 2) 956 74 18; 83 07; 82 39
✉MOONBEAM DHAKA ▣ 632174 aeol bj
⊖(+880 2) 955 65 70; 65 71; 86 20 70
♣ US$ 500 000 ⛬ 10 ⌂ 800 sq ft
Proprietor: M.A. Razzak
Purchasing Mgr: Noor Ullah
® Germany: Sartorius AG
Japan: Erma Inc., Ogawa Seiki Co.Ltd
Switzerland: Fluka Chemie AG
UK: Farnell Int'l, Kodak Clinical Diagnostics Ltd, N.E.Technology Ltd
USA: Capintec Inc, Thomas Scientific, Varian Associates 97

Barbados

Head of State:
H.M. Queen Elizabeth II

Governor-General:
Sir Clifford Husbands

10 Head of State

10-10000 **Office of the Governor General**
Government House, St Michael
Governor General: Sir Clifford Husbands 98

11 Prime Minister

11-10000 **Office of the Prime Minister**
Government Headquarters, Bay Street, St Michael
☎(+1 809) 436 64 35 ⊖(+1 809) 436 92 80
Prime Minister: Rt Hon Owen S Arthur
Deputy Prime Minister: Hon Billie A Miller
Minister of State: Hon Ronald Toppin 98

12 Ministries

12-20000 **Ministry of Foreign Affairs**
CDC Building, 1 Culloden Road, St Michael
Minister of Foreign Affairs: Hon Billie A Miller
Minister of Foreign Trade and International Business: Hon Phillip C Goddard 98

15 Parliament

15-10000 **Senate**
Senate Chambers, Public Building, Trafalgar Street, Bridgetown
☎(+1 246) 426 53 31; 426 37 12
▣ WB 2222 foreign barbados
⊖(+1 246) 427 74 74; 436 41 43
President of the Senate: His Honour Senator Sir Fred W Gollop
Deputy President of the Senate: His Honour Senator C L Bolden
Leader of the Senate: Honour Reginald Farley
☎(+1 246) 426 44 52
Clerk of Parliament: George Branker
Secretary of Parliament: Ms Diana M Ellis 98

15-20000 **House of Assembly**
Public Building, Trafalgar Street, Bridgetown
☎(+1 246) 426 37 17; 427 20 19
▣ WB 2222 foreign barbados ⊖(+1 246) 436 41 43
Speaker: His Honour Ishmael A Roett
Deputy Speaker: His Honour Rommell D Marshall
Leader of the House of Assembly: Honour Miss Billie A Miller
Deputy Leader of the House of Assembly: Honour D A C Simmons
Clerk of Parliament: George E T Branker
Secretary of Parliament: Ms Diana M Ellis 98

Contact Reference

31 Embassies & Attachés

31-00310 Canadian High Commission
Bishop's Court Hill, St Michael, PO Box 404, Bridgetown
☎ (+1 809) 429 3550 ⊖ (+1 809) 429 3780
High Commissioner: H.E. Mrs Colleen Swords
Counsellor: Kenneth Harley
Counsellor Commercial: Michael Wondergem [97]

31-01890 United Kingdom High Commission
Lower Collymore Rock, PO Box 676, Bridgetown
☎ (+1 809) 436 6694 ⊠ UKREP BRIDGETOWN
⊡ 2219 ukrep bri wb ⊖ (+1 809) 436 5398; 426 7916
High Commissioner: H.E. Mr R. Thomas
Deputy High Commissioner: P.J. Mathers
Defence Adviser: Capt. I.M. Hime (RN)
■ The British High Commission in Barbados is the UK representative to the Regional Security System Member Nations: Barbados, St Lucia, Dominica, St Kitts and Nevis, Grenada, Antigua & Barbuda and St Vincent & The Grenadines. DA also accredited to 5 British Dependent Territories namely Anguilla, Cayman Islands, British Virgin Islands, Montserrat, Turks & Caicos Islands. [97]

31-01900 Embassy of the United States of America
PO Box 302, Bridgetown
☎ (+1 246) 436 49 50 [E] jacbar@usa.gov
⊡ 2259 usemb bg1 wb ⊖ (+1 246) 429 52 46
Ambassador: H E Ms Jeanette W Hyde
Deputy Chief of Mission: Don Holm
Political/Economic Section: Steve Snow
Economic Section: Leo Gallaghar
Consular Section: Philip Jones
Administrative Section: Earl Ferguson
Defense Attaché: Lt Col Donald Robinson (USA)
Military Liaison Officer: Cdr Charles Dickens (USN)
Public Affairs Officer: Ms Jennifer A Clark
☎ (+1 246) 436 63 00; 429 53 16 (fax) [98]

Belarus, Republic of

President:
Aleksandr Lukashenko

10 Head of State

10-10000 Office of the President
Dom Pravitelstva, pl Nezavisimosti, 220010 Minsk
☎ (+375 172) 29 60 12
President: Aleksandr Lukashenko [98]

11 Prime Minister

11-10000 Office of the Prime Minister
Dom Pravitelstva, 220010 Minsk
☎ (+375 172) 29 60 43; 29 60 07
Prime Minister: Sergey Ling
First Deputy Prime Minister: Petr Prakapovich
Deputy Prime Minister: Vladimir Garkun
Deputy Prime Minister: Vasiliy Dolgalev
Deputy Prime Minister: Valeriy Kokarev
Deputy Prime Minister: Vladimir Rusakevich [98]

12 Ministries

12-10000 Ministry of Defence
ulitsa Kommunisticheskaya 1, 220034 Minsk
☎ (+375 17) 276 89 74 ⊖ (+375 17) 276 89 74
Minister of Defence: Col Gen A Tchoumakov
Deputy Minister: Maj Gen Oleg V Yatskevitch
Chief of Staff: Lieut Gen M Kozlav
Director of Foreign Affairs: Col S Zhitikhin (acting)
Deputy Chief of Department of Bilateral Co-operation, Nato: Col Y Epoletov [98]

12-20000 Ministry of Foreign Affairs
19 Lenin Straat, 220030 Minsk
☎ (+375 172) 27 29 22 ⊡ 252285 mena su
⊖ (+375 172) 27 45 21
Minister for Foreign Affairs: Ivan Antanovich
Director International Security and Disarmament Affairs: Mikalai M Lepyashko
Director Public Information: Vladzimir S Paulovich [98]

12-30000 Ministry of Trade
ulitsa Kirova 8 Dom 1, 220050 Minsk
☎ (+375 172) 27 08 97; 27 48 02
⊖ (+375 172) 27 61 21; 27 03 82
Minister of Trade: Petr Kazlov
Minister of Foreign Economic Relations: Mikhail Marinich [98]

12-40000 Ministry of Internal Affairs
ulitsa Uritskogo 5, 220050 Minsk
☎ (+375 172) 29 75 96; 27 21 23; 29 78 08
⊖ (+375 172) 26 12 47
Minister of Internal Affairs: Valentin Agolets [98]

12-50000 Ministry of Industry
2/4 Partyzansky prospect, 220033 Minsk
☎ (+375 172) 27 95 95; 27 69 20
⊖ (+375 172) 24 87 84
Minister of Industry: Anatol Kharlap [98]

12-60000 Ministry of Finance
Dom Pravitelstva, 220010 Minsk
☎ (+375 172) 29 61 37; 20 21 72; 29 69 83
⊖ (+375 172) 29 66 40; 29 62 83
Minister of Finance: Mikalay Pyatrovich Korbuit [98]

20 Defense Administrations/Services/Agencies

20-01000 State Security Committee (KGB)
F Skoryny Avenue 17, 220623 Minsk
☎ (+375 172) 29 94 01 ⊖ (+375 172) 26 00 38
Chairman: Col Gen Vladimir Matskevich [98]

29 Customs & Excise

29-01000 State Customs Committee
11 Kamunistychaya Street, 220029 Minsk [98]

31 Embassies & Attachés

31-00535 Embassy of Estonia
Varvas'eni 17, Minsk
☎ (+375 17) 234 59 65 ⊖ (+375 17) 210 12 60
Third Secretary Consular: Vladimir Redpap [98]

31-00920 Embassy of Japan
Hotel Oktyabrskaya Room N° 303, 13 Engels Street, 220030 Minsk
☎ (+375 172) 27 47 18; 26 03 32; 26 13 90
⊖ (+375 172) 27 43 19 [97]

31-01700 Armenian Diplomatic Mission
Kiev
☎ (+7 0172) 29 30 72
Chargé d'Affaires: Spartak Kostanian (acting) [98]

31-01890 Embassy of the United Kingdom
37 Karl Marx Street, 220030 Minsk
☎ (+375 172) 29 23 03; 29 23 04; 29 23 05
☎ (+375 172) 29 23 10 (visa section)
⊖ (+375 172) 29 23 06
⊖ (+375 172) 29 23 11 (visa section)
Ambassador: H E Miss Jessica Pearce M
Deputy Head of Mission: M J Pethick
Defence Attaché: Capt C A M Parrish (resident in Moscow)
Second Secretary: P Waterhouse (resident in Warsaw)
Third Secretary: L A Beaton [98]

31-01900 Embassy of the United States of America
Starovilenskaya #46, Minsk
☎ (+375 17) 231 50 00 ⊖ (+375 17) 234 78 53
Ambassador: H E Mr Kenneth Spencer Yalowitz
Chargé d'Affaires: John J Boris
Administration Section: William Orr
Consular Section: Ms Patricia Gu
Public Information Officer: Diana Moxhay
Commercial Officer: Anne Carson
Defence Attaché: Maj Robert Hand [98]

75 Manufacturing Companies

75-00200 Beloma
Makayonok Street 23, 220836 Minsk
☎ (+375 17) 263 55 47 ⊘ (+375 17) 263 85 86
[E] market@belomo.minsk.by ⊖ (+375 17) 263 44 47
👥 10,000
General Director: Vyacheslav A Bursky
☎ (+375 17) 264 11 90
Commercial Director: Vladimir Kiskin
☎ (+375 17) 263 65 50
Technical Director: Victor Khadkevich
Manager of Foreign Relations: Igor Kuchura
☎ (+375 17) 263 55 50
Marketing Manager: Ivan Zhukovsky
☎ (+375 17) 263 55 47
■ Space photographic systems, photogrammetric instruments and techniques, stabilized night vision devices and binoculars. [97]

75-00400 Ekran Belarusskoye PO
N-Neman 167, Borisov-4, Minsk Region
☎ (+375 1777) 342 01 ⊘ 206 62
⊖ (+375 1777) 228 82
♣ Rb 130,000,000 (1997) 👥 3,500 (1997)
🏭 5.32 hectares (1997)
Managing Director: Yuri Blinov
Sales and Marketing Director: Ivan Shedov
☎ (+375 1777) 206 61
Operations Director: Yuri Berestenko
Technical Director: Ms Valery Kaspersky
☎ (+375 1777) 218 01
Purchasing Manager: Ms Valery Kiselevick
☎ (+375 1777) 209 64
■ Development and production of aircraft. [98]

Belgium, Kingdom of

Head of State:
H M King Albert II

10 Head of State

10-10000 Office of H M the King
Rue de Brederode, Palais Royal, B-1000 Brussels
☎(+32 2) 513 07 70
King of Belgium: H M King Albert II [98]

11 Prime Minister

11-10000 Office of the Prime Minister
16 rue de la Loi, B-1000 Brussels
☎(+32 2) 501 02 11 ⊠ 62400 primin b
⊖ (+32 2) 513 72 62
Prime Minister: Jean Luc Dehaene
Vice Prime Minister and Minister of Finance and External Commerce: Philippe Maystadt
Vice Prime Minister and Minister of the Interior: Johan Vande Lanotte
Vice Prime Minister and Minister of Economy and Telecommunications: Elio Di Rupo
Vice Prime Minister and Minister of Budget: Herman Van Rompuy
Minister of Political Science: Yvan Ylieff
Minister of Public Health and Pensions: Marcel Colla
Minister of Foreign Affairs: Erik Derycke
Minister of Employment and Labor: Miet Smet
Minister of Social Affairs: Ms Magda De Galan
Minister of Agriculture: Karel Pinxten
Minister of Transport: Michel Daerden
Minister of Justice: Stefaan De Clerk
Minister of National Defence: J P Poncelet
State Secretary for Co-operation and Development: Réginald Moreels [98]

12 Ministries

12-03000 Ministry of Communications
General Information Services
155 rue de la Loi, B-1040 Brussels
☎(+32 2) 287 47 80 ⊖ (+32 2) 230 51 62
Minister of Communications: Elio Di Rupo
Information Contact: Patrick Anselot
Aeronautical Administration
Centre de Communications Nord,
Rue du Progrès 80 - Boîte 5, B-1030 Brussels
☎(+32 2) 206 32 11 ⊖ (+32 2) 203 15 28
Director and Co-ordinator: Paul Verlaenen
Administration of Maritime Affairs and Navigation
Rue d'Arlon 104, B-1040 Brussels
☎(+32 2) 233 13 21 ⊖ (+32 2) 230 30 02
Industrial Engineer: Joseph Lacquet [98]

12-05000 Ministry of Economic Affairs
Square de Meeûs 23, B-1000 Brussels
☎(+32 2) 506 55 00; 506 55 08 ⊖ (+32 2) 506 54 58
Minister of Economic Affairs: Elio Di Rupo
Director of Information: Maurice Charles
Director of Information: Ms Gilberte de Wever [98]

12-05500 Ministry of Finance and Foreign Trade
2 rue des Quatre Bras, B-1000 Brussels
☎(+32 2) 516 83 11 ⊠ 25183 vercom b
⊖ (+32 2) 512 72 21
Minister of Finance and Foreign Trade: Philippe Maystadt [98]

12-06000 Ministry of Foreign Affairs
2 rue des Quatre Bras, B-1000 Brussels
☎(+32 2) 501 35 11 ⊠ 23979 belext b; 21376 belext b
⊖ (+32 2) 514 30 67
Minister of Foreign Affairs: Eric Derycke
Secretary of State for Co-operation and Development: Reginald Moreels [98]

12-07000 Ministry of Interior
Rue Royale 66, B-1000 Brussels
☎(+32 2) 504 85 11; 511 06 60; 500 21 25
⊠ 21762 mintbr b
⊖ (+32 2) 504 85 00; 504 85 80; 500 21 28
Minister of Interior: Johan Vande Lanotte
Chief of Cabinet: Johan Vande de Berghe
Inspector General: Henri Maes
[E] henri.maes@cominformatics.fgov.be
Press Attaché: Ms Marie-Paule Bribosia
Secretariat for Security
Avenue Gallilée 5 - 10th Floor, B-1210 Brussels
☎(+32 2) 210 19 11 ⊖ (+32 2) 217 33 28
State Secretary for Security, Social Integration and Environment: Jan Peeters [98]

12-08000 Ministry of Justice
Boulevard de Waterloo 115, B-1000 Brussels
☎(+32 2) 504 65 11 ⊖ (+32 2) 538 07 67; 512 08 63
Minister of Justice: S de Clerck [97]

12-10000 Ministry of National Defence
Rue Lambermonstraat 8, B-1000 Brussels
☎(+32 2) 507 66 11; 550 28 11
⊖ (+32 2) 507 66 51; 55 28 09; 550 29 19
Minister of National Defence: Jean-Pol Poncelet
Chief of Cabinet: Maj Gen Joohn Schellemans
General Civil Administration
Quartier Reine Elisabeth - Bloc 4B, Rue d'Evere 1, B-1140 Brussels
☎(+32 2) 701 40 91 ⊖ (+32 2) 701 41 48
Director of Information: Mme Françoise Thoams
National Defence Information Service
Rue d'Evere 1, Quartier Reine Elisabeth - Bloc 5, B-1140 Brussels
☎(+32 2) 701 47 71 [E] sidmanager@mil.be
⊖ (+32 2) 701 39 56; 701 44 40 [98]

15 Parliament

15-05000 Chamber of Representatives
Place de la Nation 2, Palais de la Nation, B-1008 Brussels
☎(+32 2) 549 81 11; 549 81 36 (public and international relations)
⊖ (+32 2) 549 83 77 (Presidence); 512 65 33 (legislation)
President: Raymond Langendrics ☎(+32 2) 549 81 02
Director of Cabinet: Patrick Dubois
National Defence Commission
Place de la Nation 2, B-1008 Brussels
☎(+32 2) 549 80 81 ⊖ (+32 2) 512 65 33
President: Jean Paul Henry
Administration Secretary: Guy Bracaval
Foreign Relations Commission
Commission des Relations Extérieures
Place de la Nation 2, B-1008 Brussels
☎(+32 2) 549 81 38 ⊖ (+32 2) 512 65 33
President: François Xavier de Dornvéa
Administration Secretary: Ms Mireille Pöttgens [98]

15-10000 Senate
Place de la Nation, Palais de la Nation, B-1009 Brussels
☎(+32 2) 515 82 11 ⊖ (+32 2) 514 06 85
President: Frank Swaelen [98]

20 Defense Administrations/Services/Agencies

20-01000 Defence Study Centre
Centre d'Études de la Défence
Dreve Sainte-Anne 90, B-1020 Brussels
☎(+32 2) 474 47 90; 474 47 91
⊖ (+32 2) 478 97 78; 478 97 37
Director: Col R Laloux (BEM)
Deputy Director: Maj J L Pire
Chief of Staff: Col W van Hoeck [98]

25 Defense Forces

25-10000 Armed Forces Headquarters
Quartier Reine Elisabeth, Rue d'Evere 1, B-1140 Brussels
☎(+32 2) 701 31 50; 701 31 52 ⊠ 21339 bephg b
⊖ (+32 2) 701 66 25
Director: VAdm Willy Maurits Herteleer
Chief of Personnel Division: Maj Gen Maurice Derycker
Chief of Operations Division: Maj Gen Jean Pierre Roman
Chief of Material Division: Maj Gen François Cornelis
Chief of Budget Division: Maj Gen Henri Tytgat
Chief of Infrastructure Division: Maj Gen Alain Devignon
Chief of Procurement Services: Maj Gen Francies Corbisier
Chief of Intelligence Division: Maj Gen Guillaume Simons
Chief of Army Staff: Lieut Gen Roger Maes
Chief of Air Staff: Lieut Gen Guido Van Hecke
Chief of Naval Staff: RAdm M Verhulst
Chief of Medical Services and Surgeon General: Maj Gen Marc Jean de Coninck
Chief of Naval Staff: Rear Adm M Verhulst
Information and Public Affairs Office
Service de l'Information
Quartier Reine Elisabeth, Rue d'Evere 1, B-1140 Brussels
☎(+32 2) 243 39 55 ⊖ (+32 2) 243 48 62
Director: Col A Vereecke [98]

25-22200 Training and Support Command Headquarters
70 rue de la Fusée, Quartier Roi Albert 1ère, B-1130 Brussels
☎(+32 2) 701 54 00 ⊖ (+32 2) 701 59 20
Commander: Maj Gen K Vervoort
Commander of Training and Logistics: Maj Gen M Mandl
Chief of Staff: Col F Tutelaars ☎(+32 2) 701 54 02 [97]

25-29100 Air Force Royal Technical School
Luikersteenweg 371, Kazerne Kol - Vlieger Renson, B-3800 Sint-Truiden
☎(+32 11) 70 22 50 ⊖ (+32 11) 70 26 00
Commander: Col Rene Claes ☎(+32 11) 70 22 10
Director of Operations: Lieut Col Warnier
Director of Procurement: Maj Cappan
Press and Information Officer: Olt Janssen [98]

Contact Reference

Belgium, Kingdom of

25-30000 Army Headquarters
Rue d'Evere, Quartier Reine Elisabeth, B-1140 Brussels
☎(+32 2) 701 34 45 ⊖ (+32 2) 701 67 92
Chief of Staff: Lieut Gen Maes
Aide de Camp to Chief of Staff: Cdt Savelkoels
Deputy Chief of Staff: Maj Gen van Steenlandt
Deputy Chief of Staff: Maj Gen Fairon
Chief of Personnel (GI): Col Lejolyy
Chief of Logistics (G4): Col Folens
Chief of Plans and Programs (P): Col Delcour
Chief of GST: Col Robyn
Chief Army Surgeon (GDM): Col Dr Lecocq
Chief of Information and Public Relations: Col van Put
GS-IRP
Rue d'Evere, Quartier Reine Elisabeth, B-1140 Brussels
☎(+32 2) 701 34 43 ⊖ (+32 2) 701 67 92
Public Affairs Officer: Lieut Col Teulingkx [98]

25-30700 Test and Evaluation Center (CEM)
Kwartier Oost, B-2130 Brasschaat
Commander: Lieut Col Roothrofd
Head of Technical Department: Maj J M Vanbockstaele [98]

25-32200 Chief Belgian Army Staff
(† Interservice Territorial Command)
Quartier Reine Elisabeth, Rue d'Évere, B-1140 Brussels
☎(+32 2) 701 31 26 ⊖ (+32 2) 701 69 99
Commander: Lieut Gen R Maes
Deputy Commander: Maj Gen van Steenlandt
Aide de Camp: Cmdt L Savelkoels [97]

25-39200 Army Aviation School
Brasschaat Air Base, B-2930 Brasschaat
☎(+32 3) 630 27 30 ⊖ (+32 3) 630 27 39
Commander: Col van Raemdonck ☎(+32 3) 630 27 00
Deputy Commander, Instruction and Operations:
Maj Lenaerts ☎(+32 3) 630 27 01
Logistics Director: Maj Verhelle ☎(+32 3) 630 27 02
Operations Director: Maj Meskens ☎(+32 3) 630 27 50
Deputy Director, Operations: Col Gielis
☎(+32 3) 630 27 52 [98]

25-40000 Naval Force Headquarters
Rue d'Evere, Quartier Reine Elisabeth, B-1140 Brussels
☎(+32 2) 701 33 06 ⊖ (+32 2) 701 48 41
Chief of Naval Staff: Adm M Vermulst
Flag Lieut to the Chief of Naval Staff: Lieut W de Baes
Chief Naval Surgeon: Lieut Col Dr Stubbe
Chief Naval Surgeon: Col Dr T Joens
Director of Information Service: L Delprat
Commander: C Thibaut de Maisieres
Director of Procurement/Logistics: L Vermaere [98]

25-41000 Naval Command Logistics and Supply
Esplanadestraat 1, B-8400 Oostende
☎(+32 5) 956 35 03 ⊖ (+32 5) 956 36 29
Commander: Cdre J Wouters ☎(+32 5) 956 36 00
Chief of Staff: Capt J van den Berghe
☎(+32 5) 956 36 05
Director of Procurement/Logistics: Capt J Devos
☎(+32 5) 956 36 07
Operations Director: Mr Grandchamps [97]

29 Customs & Excise

29-00500 Administration of Customs & Excise
Administration des Douanes et Accises
c/o Ministère des Finances - Cité,
Administrative de l'État - Tours Finances,
Boîte 37- Boulevard du Jardin Botanique 50,
B-1010 Brussels [98]

29-01000 World Customs Organisation (WCO)
Rue de l'Industrie 26/38, B-1040 Brussels
☎(+32 2) 508 42 11 (switchboard)
http://www.bodnet.g-net.be ⊠ 61597
⊖ (+32 2) 508 42 40 (enquiries)
62 (technical experts); 56 (administration)
Secretary General: James Shaver ☎(+32 2) 508 44 00
Deputy Secretary General: Leonid Lozbenko
☎(+32 2) 508 44 01
Director Customs Techniques: Douglas Tweddle
☎(+32 2) 508 43 50
Communications and Public Relations: David Brakemore
☎(+32 2) 508 44 30
■ Studies questions related to co-operation in customs matters and examines technical aspects of customs, with a view to attaining the highest possible degree of uniformity. Preparation of conventions and recommendations, ensuring uniform interpretation and application of customs conventions (on valuation, tariff and statistical nomenclature and customs procedures), conciliatory action in case of dispute, circulation of information and advice regarding customs procedures and regulations. → 34 [98]

31 Embassies & Attachés

31-00085 Armenian Diplomatic Mission
Rue Franz Merjay 157, 1060, Brussels
☎(+32 2) 346 56 67 ⊖ (+32 2) 346 56 67
Deputy Chief of Mission: Grigor Badalian [98]

31-00090 Embassy of Australia
6/8 rue Guimard - Guimard Center, B-1040 Brussels
☎(+32 2) 231 05 00 ✉ AUSTEMBA Bruxelles
⊠ 21834 austem b ⊖ (+32 2) 230 68 02
Ambassador: H E Mr E R Pocock (AM)
Minister and Deputy Chief of Mission:
Ms J Drake-Brockman
Public Affairs Counsellor: P Fuller
Customs Affairs Counsellor:
Mrs Christine Marsden-Smedley
Senior Trade Commissioner: Mrs Sallyanne Atkinson (resident in Paris)
Industry, Science and Technology Counsellor:
Mrs Linda Hall
Administrative Assistant: P Minos [97]

31-00100 Embassy of Austria
Rue de l'Abbaye 47, B-1050 Ixelles, Brussels
☎(+32 2) 649 91 70 ⊖ (+32 2) 640 98 27
Ambassador: H E Prof. Dr Winfried Lang
Deputy Chief of Mission: Mag Walter Grahammer
Counsellor: Dr Eva Maria Ziegler
Defence Attaché: Maj Gen Richard Bondi
Deputy Defence Attaché: Adolf Pöschl
Secretary (NATO): Dr Andreas Riecken
Trade Commission
479 Avenue Louise, B-1050 Brussels
☎(+32 2) 648 21 11 ⊠ 22138 autrad b
⊖ (+32 2) 640 12 69
Trade Commissioner: Dr Rupert Roth [98]

31-00200 Embassy of Bolivia
176 Avenue Louise, PO Box 6, B-1050 Brussels
☎(+32 2) 627 00 10 ⊖ (+32 2) 647 47 82
Ambassador: H E Mr Alvaro Moscoso Blanco
First Secretary: Mrs Rut Corina Ybarnegaray Urquidi [98]

31-00230 Embassy of Brazil
350 Avenue Louise, Boîte 5, B-1050 Brussels
☎(+32 2) 640 20 15 ⊖ (+32 2) 640 81 34
Ambassador: H E Mr Bernardo Pericás Neto
Minister and Counsellor:
Washington Luis Pereira de Sousa [98]

31-00250 Embassy of the Republic of Bulgaria
58 Avenue Hamoir, B-1180 Brussels
☎(+32 2) 374 59 63 ⊖ (+32 2) 375 84 94
Ambassador: H E Mr Boyko Noev
Defence Attaché: Petar Strantchevski
Commercial Officer: Guavraïl Kovatchev
Public Information Officer: Evgueni Stoytchev [98]

31-00260 Embassy of Burkina Faso
16 Place Guy d'Arezzo, B-1180 Brussels
☎(+32 2) 345 99 12 ⊠ 22252 amb bfb
⊖ (+32 2) 345 06 12
Ambassador: H E Mr Youssouf Ouedraogo
Second Counsellor: Raymond Balima
Economic Counsellor: Ambroise Balima
Economic Counsellor: Richard Hien
Finance Attaché: Ouanza Maurice Domboue
Attaché: Mme Catherine Lengani [98]

31-00280 Embassy of the Republic of Burundi
46 Square Marie-Louise, B-1000 Brussels
☎(+32 2) 230 45 35; 230 45 48 ⊠ 23572 arundi b
⊖ (+32 2) 230 78 83
Ambassador: H E Mr Léonidas Ndoricimpa
Deputy Chief of Mission and Commercial Counsellor:
Nduwimana Idolephonse
Commercial Counsellor: Firmin Sindayihebura [98]

31-00340 Embassy of the Republic of Chad
52 Boulevard Lambermont, B-1030 Brussels
☎(+32 2) 215 19 72; 215 19 75; 245 54 99 ⊠ 63380
⊖ (+32 2) 216 35 26
Ambassador: H E Mr Ramadane Barma
Deputy Chief of Mission: Mahamat Warou Zegbadaï
First Secretary: Mahamat Abderamane Haggar
Second Counsellor: Idriss Adjideye [98]

31-00350 Embassy of Chile
Rue Montoyer 40, B-1000 Brussels
☎(+32 2) 280 16 20 ⊖ (+32 2) 280 14 81
Ambassador: H E Mr Sergio Pizarro Mackay
Commercial Officer: Fernando Morales Barria
First Secretary: Juan Pablo Hiriat
Counsellor: Ricardo Ortiz Vidal [98]

31-00430 Embassy of the Republic of Cuba
77 rue Roberts Jones, Uccle, B-1180 Brussels
☎(+32 2) 343 71 46 ⊖ (+32 2) 344 96 91
Ambassador: H E Mr René J Munica
Commercial Office
77 rue Roberts Jones, B-1180 Brussels
☎(+32 2) 346 29 03 ⊠ 65465 oficom b
⊖ (+32 2) 346 27 08
Commercial Attaché: Francisco Alfonso Ledesma [98]

31-00440 Embassy of Cyprus
2 Square Ambiorix, B-1000 Brussels
☎(+32 2) 735 35 10 ⊠ 25172 embacy b
⊖ (+32 2) 735 45 52
Ambassador: H E Mr Michalis M Attalides
Counsellor 'A': Ms Kalliopi Avraam
Counsellor 'B': Mr Georges Zodiates
Secretary 'A': Spyros Attas
Secretary 'B': Andreas Hadjichrysanthou
Financial and Administrative Affairs: Leontios Leontiou
Commercial Counsellor: Ioonnis Shekeris
☎(+32 2) 735 54 09
Attaché, Tourism: Costas Kyprianides
Press Counsellor: Demetris Komodromos
☎(+32 2) 735 90 15
Press Attaché: Ikaros Moushouttas [98]

31-00450 Embassy of the Czech Republic
555 rue Engeland, B-1180 Uccle, Brussels
☎(+32 2) 374 12 03; 375 93 34; 375 59 51
⊖ (+32 2) 375 92 72
Ambassador: H E Mr Jaroslav Sedivy
Minister and Counsellor: Pavel Vacek
Commercial and Economic Counsellor:
Mme Bretislav Kalusek
First Secretary: Mme Hana Hubáckova
First Secretary: Richard Huk
Second Secretary, Culture and Press: Jim Dolezel
Military and Air Attaché: Lieut Col Pavel Baborsky
Deputy Military and Air Attaché: Maj Pavel Zuna
First Secretary, Consul: Ms Magda Kunclova
Attaché, Military Liaison with NATO & UEO:
Col Jaroslav Skopek
Attaché, Military Liaison with NATO & UEO: Maj. Jan Vana
Counsellor: Gen Maj Pavel Jandacek
Attaché, Military Liaison with NATO & UEO:
Maj. Martin Dohnal
Attaché, Military Liaison with NATO & UEO:
Maj. Ivan Dvorak
Military Attaché: Lieut Col Jiri Soldat
Administration Attaché: Petr Jelinek
Administration Attaché: Josef Kordac [98]

31-00510 Embassy of the Arab Republic of Egypt
44 Avenue Léo Errera, B-1180 Brussels
☎ (+32 2) 345 52 53; 345 50 15
▣ 64565 ambinf b; 23716 bostan b ⊖ (+32 2) 343 65 33
Ambassador: H E Mr Muhammad Chabane
Deputy Chief of Mission: Mohamed El Zorkany
Attaché of Administration: Ahmed Metwally El Madbouh
Commercial Section
522 Avenue Louise, B-1050 Brussels
☎ (+32 2) 647 32 27; 647 27 78 ▣ 64809 comrau b
⊖ (+32 2) 646 45 09
First Secretary: Afifi Mahmoud Mohamed Selim
Press Section
42 Boulevard Général Jacques, B-1050 Brussels
☎ (+32 2) 640 35 01 ⊖ (+32 2) 646 45 32
Counsellor: Mrs Laila Soliman [98]

31-00535 Embassy of Estonia
Avenue Isidore Gerard 1, B-1160 Brussels-Auderghem
☎ (+32 2) 779 07 55 [E] saatkond@estemb.be
⊖ (+32 2) 779 28 17
Ambassador: H E Mr Clyde Kull
Honorary Consul
Charlottalei 34, B-2018 Antwerp
☎ (+32 3) 231 16 60 ⊖ (+32 3) 233 76 60
Honorary Consul: Anthony Brouwers [98]

31-00570 Embassy of the Republic of Fiji
66 Avenue de Cortenberg, PO Box 7, B-1000 Brussels
☎ (+32 2) 736 90 50; 736 09 15; 732 66 07
⊖ (+32 2) 736 14 58
Ambassador: H E Mr Kaliopate Tavola
Counsellor (Investment, Political and Institutions): Mrs Ross Ligairi
Counsellor (Agriculture and Trade): Nidhendra Singh
Second Secretary, Tourism: Mrs Salote Radrodro
Second Secretary, Administration: Suliasi Turagabeci
Counsellor, Commercial and Commodities: Sung Kangwai (Resident in London)
Research Assistant: B Puplett [98]

31-00630 Embassy of the Gabonese Republic
112 Avenue Winston Churchill, B-1180 Uccle, Brussels
☎ (+32 2) 340 62 10 ▣ 23383 ambaga b
⊖ (+32 2) 346 46 69
Ambassador: H E Mr Jean Robert Goulongana
Deputy Chief of Mission: François Ebibi-MBA
Commercial Counsellor: William Anguile
Defence Attaché: Lieut Col Jean Baptiste Lamou [98]

31-00640 Embassy of Gambia
126 Avenue Franklin Roosevelt, B-1050 Brussels
☎ (+32 2) 640 10 49 ▣ 24344 gamext b
⊖ (+32 2) 646 32 77
Ambassador: H E Mr Ismaila Briama Ceesay
☎ (+32 2) 351 40 38
Counsellor: Dr Omar Alieu Touray ☎ (+32 2) 372 00 14
First Secretary: Mrs Salimatta Ebironke Toure
☎ (+32 2) 372 00 14
Financial Attaché: Waka Mass Mbai ☎ (+32 2) 375 51 98 [98]

31-00690 Embassy of Greece
2 Avenue Franklin Roosevelt, B-1050 Brussels
☎ (+32 2) 648 17 30; 648 33 02
✉ AMBASSADE GRECE Bruxelles ▣ 25521 grinfb
⊖ (+32 2) 647 45 25
Ambassador: H E Mr Adamantios Vacalopoulos
Deputy Chief of Mission: Denosttiene Stoidis
Defence Attaché: K Karageorgiou
Commercial Counsellor: Mr Ananiadis
☎ (+32 2) 648 15 26 [97]

31-00770 Embassy of Guyana
12 Avenue du Bresil, B-1050 Brussels
☎ (+32 2) 675 62 16 ⊖ (+32 2) 675 63 31
Chargé d'Affaires: Samuel Mann (acting)
First Secretary, Commerce: James Ramsahoye
Finance Attaché: Mrs Dianne Abel
Administration Attaché: Devely B Peters Kissoon [97]

31-00790 Embassy of Honduras
3 Avenue des Gaulois, B-1040 Brussels
☎ (+32 2) 374 00 00 ⊖ (+32 2) 735 26 26
Ambassador: H E Mr Ivan Romero Martinez
Minister-Counsellor and Consul General: Ms Elizabeth Gamundi de Duve
Counsellor: Eduardo Enrique Reina [98]

31-00820 Embassy of Iceland
Rue de Treves 74, B-1040 Brussels
☎ (+32 2) 286 17 00 ▣ 29459 ⊖ (+32 2) 286 17 70
Ambassador: H E Mr Gunnnar Snorri Gunnarsson
Minister-Counsellor and Deputy Chief of Mission: Grétar Már Sigurdsson
Counsellor: Jóhann R Benediktsson
First Secretary: Harald Aspelund
Attaché: Jóna Gróa Valdimarsdóttir
Attaché: Ms Ingibjörg Thorsteinsdóttir
Counsellor, Fisheries: Ms Arndís Ármann Steinthórsdóttir
Counsellor, Science and Education: Rögnvaldur Ólafsson
Counsellor, Science and Education: Hans Gudmundsson
Counsellor, Environment, Industry and Commerce: Thórir Ibsen
Counsellor, Transport and Telecommunications: Jóhann Gudmundsson
Counsellor, Health, Labour and Social Affairs: Mrs Vilborg Hauksdóttir
Counsellor, Judicial Affairs: Högni S Kristjánsson [98]

31-00850 Embassy of the Islamic Republic of Iran
415 Avenue de Tervuren, B-1150 Brussels
☎ (+32 2) 762 37 45 ⊖ (+32 2) 762 39 15
Ambassador: H E Mr Hamid Aboutalébi
Counsellor: Eslamizad Hamid
First Secretary: Khan Belverdi Khosrow
Second Secretary: Talebi Far Abbas
Second Secretary: Gharanfoli Masoud
Second Secretary: Mashayekhi Hamid Reza
Third Secretary: Toufigh Khoshkhou
Attaché: Abroon Ahmad [98]

31-00860 Embassy of the Republic of Iraq
131 Avenue de la Floride, Uccle, B-1180 Brussels
☎ (+32 2) 374 59 92 ▣ 26414 amiraqb
⊖ (+32 2) 374 76 15
Ambassador: H E Mr Said S Ahmad [97]

31-00890 Embassy of Italy
28 rue Emile Claus, B-1050 Brussels
☎ (+32 2) 649 97 00; 649 97 06 ▣ 23950 itadpl b
⊖ (+32 2) 648 54 85
Ambassador: H E Mr Francesco Corrias
First Counsellor: Umberto Colesanti
Counsellor, Commercial Affairs: Ernesto Massimo Bellelli
Counsellor, Social Affairs: Giovanni Adorni Braccesi
Scientific Attaché: Giovanni Briganti
Cultural Attaché: Panagiotis Kizeridis
Finance Attaché: Aldo Pelosio
Military Attaché: Gen Antonio Lombardo
Air Attaché: Col Ercolano Annicchiarico
Naval Attaché: Capt Gianluca Assettati [97]

31-00910 Embassy of Jamaica
Jamaican Mission to the European Communities
2 Avenue Palmerston, B-1000 Brussels
☎ (+32 2) 230 11 70; 230 13 17
[E] emb.jam.brussels@skynet.be ⊖ (+32 2) 230 37 09
Ambassador: H E Mr Douglas C Saunders
Deputy Chief of Mission: Ms Claudia Barnes
First Secretary: Ms Cheryl Spencer
Consular and Finance Attaché: Ms Marva Taylor
Attaché: Ms Adrienne Austin
Attaché: Ms Ann Marie Steele
Trade Department
Avenue Palmerston 2, B-1000 Brussels
☎ (+32 2) 231 01 62 ⊖ (+32 2) 231 02 27
Trade Commissioner: Ms Diane Edwards
Commercial Attaché: Mrs Martine Maitrejean [98]

31-00920 Embassy of Japan
58 Avenue des Arts, PO Box 17/18, B-1040 Brussels
☎ (+32 2) 513 23 40 ✉ TAISHI Bruxelles ▣ 22174 taisi b
⊖ (+32 2) 513 15 56
Ambassador: H E Mr Jumpei Kato
Deputy Chief of Mission: Toshio Watanabe
Defence Attaché: Col Kenichiro Hokazono (JASDF) [97]

31-00930 Embassy of the Hashemite Kingdom of Jordan
104 Avenue Franklin D Roosevelt, B-1050 Brussels
☎ (+32 2) 640 77 55 ▣ 62513 ambjor b
⊖ (+32 2) 640 27 96
Ambassador: H E Dr Umayya Toukan
First Secretary: Omar Al-Nadif [98]

31-01000 Embassy of the State of Kuwait
43 Avenue Franklin Roosevelt, B-1050 Brussels
☎ (+32 2) 647 79 50 ▣ 62904 koutuhb
⊖ (+32 2) 646 12 98
Ambassador: H E Mr Ahmad A Al-Ebrahin
Counsellor and Deputy Chief of Mission: Mohammed Khalaf
Second Secretary: Muteb Al-Mutoteh
Third Secretary: Deyab Al-Rachid
Administration Attaché: Zakaria Al-Khamis [97]

31-01070 Embassy of Luxembourg
211 rue du Noyer, B-1000 Brussels
☎ (+32 2) 733 99 77 ▣ 22974 rplux b
⊖ (+32 2) 732 20 99
Ambassador: H E Mr Paul Schuller
Consul General: Ms Camille Weis
Consul: Victor Clement [98]

31-01090 Embassy of the Republic of Madagascar
276 Avenue de Tervuren, B-1150 Brussels
☎ (+32 2) 770 17 26 ▣ 61197 telmad b
⊖ (+32 2) 772 37 31
Ambassador: H E Mr Jean Omer Beriziky
First Counsellor and Chief of Protocol: Pierre Rabarivola
Customs Attaché: Eugene Rakotomalala
Customs Attaché: Jean Adalphe Ramilison [98]

31-01130 Embassy of the Republic of Mali
487 Avenue Molière, B-1060 Brussels
☎ (+32 2) 345 74 32; 345 75 89 ⊖ (+32 2) 344 57 00
Ambassador: H E Mr N'Tji Laico Traoré [98]

31-01220 Embassy of the Kingdom of Morocco
Boulevard Saint Michel 29, B-1040 Brussels
☎ (+32 2) 736 11 00 ▣ 21233 sifmar b
⊖ (+32 2) 734 64 68
Ambassador: H E Mr M Mohammed Rachad Bouhlala
☎ (+32 2) 736 54 43
Economic First Counsellor: M Farhat Bouazza
☎ (+32 2) 736 97 67
Military, Naval and Air Attaché: Col Tayeb Bainouti
☎ (+32 2) 736 98 51
First Secretary: A Sabri [98]

31-01230 Embassy of Mozambique
97 Boulevard Saint Michel, B-1040 Brussels
☎ (+32 2) 736 25 64; 736 00 96 ▣ 65478 emobru b
⊖ (+32 2) 735 62 07
Ambassador: H E Mr Álvaro Ó da Silva
First Secretary: Ermindo A Ferreira
Second Secretary: C M C Rodrigues da Costa
Atttaché: Ms Maria Julia [98]

31-01300 Embassy of New Zealand
47-48 Boulevard du Régent, B-1000 Brussels
☎ (+32 2) 512 10 40 ⊖ (+32 2) 513 48 56
Ambassador: H E Mr Derek W Leask
Deputy Head of Mission: Nigel Fyfe
Second Secretary (Science and Technology): Ms Penny Macdonald
Secretary of Customs Affairs: Merv Kemp [98]

31-01330 Embassy of the Federal Republic of Nigeria
288 Avenue de Tervuren, B-1150 Brussels
☎ (+32 2) 762 52 00; 762 98 31; 762 98 32
▣ 22435 ngeria b ⊖ (+32 2) 762 37 63; 762 39 10
Minister, Deputy Chief of Mission and Chargé d'Affaires: S A D Olisa
Head of Chancery: Danjuma N Sheni
Commercial Attaché: U H Orjiako
Information Officer: B O Okujagu
Counsellor: Ms Vivian N R Okeke-Anunobi
Counsellor: Joseph I Okoro [98]

31-01340 Royal Norwegian Embassy
Avenue Louise 130a, B-1050 Brussels
☎ (+32 2) 646 07 80 ⊠ 62563 noramb b
⊖ (+32 2) 646 28 82
Ambassador: H E Mr Tor B Naess
Counsellor and Deputy Chief of Mission:
Mrs Lise Kleven Grevstad [98]

31-01370 Embassy of Pakistan
57 Avenue Delleur, B-1170 Brussels
☎ (+32 2) 673 80 02 to 07 ✉ PAREP BRUSSELS
⊠ 61816 parep b
⊖ (+32 2) 675 31 37 (economic section);
675 83 94 (diplomatic section)
Ambassador: H E Mr Riaz Mohammad Khan
Minister of Economic Affairs: Ch Muhammad Ilyas
Counsellor: Syed Hasan Javed
Counsellor: Mohammad Sadiq
Third Secretary: Murad Ashraf Janjua [98]

31-01380 Embassy of the Republic of Panama
Boulevard Brand Whitlock 8, B-1150 Brussels
☎ (+32 2) 733 36 61; 733 90 89 ✉ c
[E] panama@antrasite.be ⊖ (+32 2) 733 77 79
Ambassador and Deputy Chief of Mission:
H E Ms Vilma Ramirez (CE, CECA, CEEA)
Minister and Counsellor: Ms Barletta de Nottebohm
Counsellor, Political Affairs: Luis Torres Herrera
Counsellor, Economic Affairs: Ms Juana Galvan Murillo [98]

31-01410 Embassy of Peru
179 Avenue de Tervueren, B-1150 Brussels
☎ (+32 2) 733 33 19; 733 31 85
[E] embassy.of.peru@unicall.be ⊖ (+32 2) 733 48 19
Ambassador: H E Mr José Antonio Arrospide
Counsellor: Eric Anderson
Commercial Section
5 rue Joseph II, PO Box 2, B-1040 Brussels
☎ (+32 2) 230 82 85 ⊠ 25079 comper b
Commercial Counsellor: Eric Anderson [98]

31-01420 Embassy of the Philippines
83 rue Washington, B-1050 Brussels
☎ (+32 2) 533 18 11 ✉ PHILEMBASSY BRUSSELS
⊖ (+32 2) 538 35 40
Ambassador: H E Mr Alberto A Pedrosa
Minister and Counsellor: Julius D Torres
Commercial Attaché: Romeo G Borrilo
Second Secretary, Consul: Eduardo E Malaya
Foreign Information Officer: Meneleo D Hernandez [97]

31-01430 Embassy of the Republic of Poland
29 Avenue des Gaulois, B-1040 Brussels
☎ (+32 2) 735 72 12; 735 73 81; 735 21 99
⊖ (+32 2) 736 18 81
Ambassador: H E Mr Andrzej Krzeczunowicz
Deputy Chief of Mission: Janusz Niesyto
Defence, Military, Naval and Air Attaché: Col Jan Debski
Assistant Armed Forces Attaché: Capt Artur Krawczyk
Office of Commercial Counsellor
33 rue Jules César, B-1150 Brussels
☎ (+32 2) 771 67 54; 771 68 15 ⊠ 21562 morhan b
Commercial Counsellor: Dariusz Salata
Deputy Commercial Counsellor: Ms Marian Skowerski [97]

31-01440 Embassy of Portugal
Avenue de la Toison d'Or 55, B-1060 Brussels
☎ (+32 2) 539 35 21; 539 38 50; 539 37 13
⊠ 24178 embrux b ⊖ (+32 2) 539 07 73
Ambassador: H E Mr João da Rocha Páris
Deputy Chief of Mission: Ms Fátima Perestrelo
Defence Attaché: Cdt Pedro Miguel P Correia do Amaral
Press Affairs Counsellor: Luis Amaro
Commercial Office
5 rue Joseph II, PO Box 3, B-1040 Brussels
☎ (+32 2) 230 96 25 ⊠ 24570 icep b
⊖ (+32 2) 231 04 47
Commercial Counsellor: António Carlos Silva [98]

31-01460 Embassy of the State of Qatar
71 Avenue Franklin D Roosevelt, B-1050 Brussels
☎ (+32 2) 640 29 00 ⊠ 63754 qatbru
⊖ (+32 2) 648 40 78
Chargé d'Affaires: Mohamed Al-Haiyki [98]

31-01480 Embassy of Romania
105 rue Gabrielle, B-1180 Uccle, Brussels
☎ (+32 2) 345 26 80; 343 69 35; 347 53 38
⊠ 21859 romamb b ⊖ (+32 2) 346 23 45
Chargé d'Affaires: H E Mr Ion Corneliu Iorgulescu
(acting Ambassador)
Counsellor: Petru Pavel Gravilescu
Military, Naval and Air Attaché: Steliar Georgescu
First Secretary, Economic Affairs: Mugur Popovici
First Secretary: Viorel Carata
Minister and Counsellor: Gheorghe Rotaru
First Secretary (SHAPE): Lieut Col Marius Chiotea
First Secretary (SHAPE): Cmdt Dan Predatu
First Secretary (NATO): Iordache Olaru
First Secretary (NATO): Ms Silvia Mariana Badescu
Second Secretary: Victoria Gavrilescu
Second Secretary: Ion Serpe
Press Attaché and Second Secretary:
Carmen Liliana Podgorean
Attaché: Petru Grigoras
Attaché: Gheorghe Dragomir
Economic Section
105 rue Gabrielle, B-1180 Uccle, Brussels
☎ (+32 2) 345 00 40 ⊠ 21859 romamb b
⊖ (+32 2) 346 23 45
Secretary, Economic Affairs: Niculae Dinu Carata
Secretary, Economic Affairs: Mugur Popovici [98]

31-01490 Embassy of the Republic of Rwanda
1 Avenue des Fleurs, B-1150 Brussels
☎ (+32 2) 763 07 02; 763 07 05; 763 07 21; 763 07 38
[E] rwanda.info@a-1.be ⊖ (+32 2) 763 07 53
Ambassador: H E Mr Manzi Bakuramutsa
Public Information Officer: Viviane Dassart [98]

31-01640 Embassy of the Republic of Singapore
198 Avenue Franklin Roosevelt, B-1050 Brussels
☎ (+32 2) 660 29 79 [E] amb.eu@singembbru.be
⊠ 26731 absingem b ⊖ (+32 2) 660 86 85
Ambassador: H E Dr Pang Eng Fong
Counsellor: Miss Joyce Davamoni
First Secretary: Syed Raziff Aljunied
First Secretary (Commercial): Lok Hwee Chong
Second Secretary: Munwar Basha [98]

31-01650 Embassy of the Slovak Republic
195 Avenue Molière, B-1050 Brussels
☎ (+32 2) 346 43 42; 346 42 61 ⊖ (+32 2) 346 63 85
Ambassador: H E Mr Július Hauser
Counsellor, Political and Cultural Affairs: Ms Elena Letková
Counsellor, Economic and Commercial Affairs:
Ms Jana Kotová
Second Secretary, Consular and Visas Affairs:
Ms Marta Domoková
Defence Attaché: Lieut Col Jaroslav Tomas
Counsellor for NATO: Vladimir Jakabcin
Counsellor for NATO: Dusan Zoller
Counsellor for NATO: Jan Prochyra
First Secretary for NATO: Gabriel Kopecky
First Secretary for NATO: Rastislav Kácer
First Secretary for SHAPE: Dusan Ptácek
Second Secretary for NATO: Peter Sedlácek
Attaché for NATO: Branko Toman [98]

31-01655 Embassy of the Republic of Slovenia
179 Avenue Louise, B-1050 Brussels
☎ (+32 2) 646 90 99 ⊖ (+32 2) 646 36 67
Ambassador: H E Mr Jasa L Zlobec
Mission of the Republic of Slovenia to the European Union
30 Avenue Marnix, B-1000 Brussels
☎ (+32 2) 512 44 66 ⊖ (+32 2) 512 09 97 [98]

31-01680 Embassy of Spain
19 rue de la Science, B-1040 Brussels
☎ (+32 2) 230 03 40
⊠ 22092 embesp b; 23250 alebel b
⊖ (+32 2) 230 93 80

Office of the Commercial Counsellor
21/22 Avenue des Arts, PO Box 8, B-1040 Brussels
☎ (+32 2) 230 14 65 ⊠ 21995 ofcome b
⊖ (+32 2) 230 61 04 [97]

31-01730 Embassy of Sweden Defence Section
148 Avenue Louise, B-1050 Brussels
☎ (+32 2) 289 57 80 ⊠ 21148 svensk b
⊖ (+32 2) 289 57 81
Ambassador: H E Mr Göran Berg
Defence Attaché: Brig Gen Anders Sifvertsson
Assistant: A Palin
Swedish Trade Council
33 Boulevard de la Cambre, B-1050 Brussels
☎ (+32 2) 648 46 34 ✉ SWEDTRADE Brussels
⊠ 21112 swetra b ⊖ (+32 2) 648 80 71
Trade Commissioner: Ronald Wintzéus [98]

31-01740 Embassy of Switzerland
26 rue de la Loi, PO Box 9, B-1040 Brussels
☎ (+32 2) 285 43 50 ✉ 100634.361@compuserve.com
⊖ (+32 2) 230 37 81
Ambassador: H E Mr Pierre-Yves Simonin
Head of Chancery: Urs Bühler
Defence Attaché: Col EMG Daniel Combernous
Counsellor and Deputy Chief of Mission: Robert Steiner [98]

31-01790 Embassy of the Republic of Togo
264 Avenue de Tervueren, B-1150 Brussels
☎ (+32 2) 770 17 91 ⊖ (+32 2) 771 50 75
Minister, Counsellor and Chargé d'Affaires:
Kati Ohara Korga
First Secretary, Commercial Affairs: Komlan Afoto
Attaché Commercial and Consular Affairs:
Ms Akuto Kafui Agbossou
Finance Attaché: Alangba Talboussouma
Attaché: Klusseh G Denke
Attaché: Dzigbodi Gogan [97]

31-01870 Embassy of Ukraine
Avenue L Lepoutre 99-101, B-1060 Brussels
Customs Directorate
Avenue L Lepoutre 99-101, B-1060 Brussels
Special Counsellor: O Koval [98]

31-01890 Embassy of the United Kingdom
Rue d'Arlon 85, B-1040 Brussels
☎ (+32 2) 287 62 62; 287 62 11 (switchboard)
⊖ (+32 2) 287 63 55; 287 62 30 (commercial)
Ambassador: H E Mr David Colvin (CMG)
Counsellor and Deputy Head of Mission: E Glover
Commercial Counsellor: Mervyn Jones
First Secretary, Commerce: I McCrory
Defence Attaché: Gp Capt D C Hencken (RAF)
Public Information Officer: Patrick O'Brien [98]

31-01950 Embassy of Venezuela
Franklin Rooseveltlaan 10B, B-1050 Brussels
☎ (+32 2) 639 03 40 ⊠ 61742 emvene b [98]

31-02010 Embassy of the Federal Republic of Yugoslavia
11 Avenue Emile de Mot, B-1000 Brussels
☎ (+32 2) 647 57 81; 647 26 51; 647 29 41
⊠ 26156 yuambb ⊖ (+32 2) 647 29 41
Chargé d'Affaires a.i.: Dragan Momcilovic
Consul Attaché: Milija Ilic [97]

31-02040 Embassy of the Republic of Zimbabwe
21-22 Avenue des Arts, B-1040 Brussels
☎ (+32 2) 762 58 08 ⊠ 24133 zimbru b
⊖ (+32 2) 762 96 05
Ambassador: H E Mr Simbarashe Mumbengegwi
Minister and Deputy Chief of Mission:
Jonathan Wutawunashe
Counsellor of Economic Affairs: Ezekiel Chitsunge [98]

33 Foreign Forces

33-10000 European Logistical Support Office US Embassy Branch Office (ELSO)

Noorderlaan 147, PO Box 12a, B-2030 Antwerpen
☎(+32 3) 542 47 75 ⊠ 34964 elso b
⊖ (+32 3) 542 65 67
Director: Robert Sarofeen
Director: Olga P Murphy
■ Diplomatic Branch Office. 98

34 Multinational Intergovernmental Organizations

World Customs Organisation 29-01000

34-01000 North Atlantic Treaty Organization (NATO)

Organisation du Traité de l'Atlantique Nord
Organisation du Traité de l'Atlantique Nord (OTAN)
B-1110 Brussels
☎(+32 2) 707 41 11 ⊠ 23867 ⊖ (+32 2) 707 45 79
Deputy Secretary General: Ambassador Sergio Balanzino (Italy)
Assistant Secretary General for Political Affairs: Ambassador Gebhardt von Moltke (Germany)
Assistant Secretary General for Defence Planning and Operations: Anthony Cragg (UK)
Assistant Secretary General for Defence Support: Norman W Ray (USA)
Assistant Secretary General for Infrastructure, Logistics, and Civil Emergency Planning: VAdm Herpert van Foreest (Netherlands)
Assistant Secretary General for Scientific and Environmental Affairs: Dr Jean-Marie Cadiou (France)
Executive Secretary: Leo Verbruggen (Netherlands)
Director of the Private Office: Jorge Domecq (Spain)
Director of Press and Information: C F Prebensen (Norway)
Spokesman: Dr Jamie Shea (UK)

Member States:
Belgium, Canada, Denmark, France, Germany, Greece, Iceland, Italy, Luxembourg, The Netherlands, Norway, Portugal, Spain, Turkey, UK and USA.

Council
B-1110 Brussels
Honorary President: Eric Derijcke (Belgium)
Chairman: Javier Solana (Spain) (Secretary General of NATO)
Deputy Secretary General: Sergio Balanzino (Italy)
Permanent Representative: Jean De Ruyt (Belgium)
Permanent Representative: John Anderson (Canada)
Permanent Representative: Gunnar Riberholdt (Denmark)
Permanent Representative: Gérard Errera (France)
Permanent Representative: Dr Hermann Freiherr von Richthofen (Germany)
Permanent Representative: Georges Savvaides (Greece)
Permanent Representative: Thorsteinn Ingólfsson (Iceland)
Permanent Representative: Giovanni Jannuzzi (Italy)
Permanent Representative: Paul Schuller (Luxembourg)
Permanent Representative: Lambert Veenendaal (The Netherlands)
Permanent Representative: Leif Mevik (Norway)
Permanent Representative: António Martins da Cruz (Portugal)
Permanent Representative: Javier Conde de Saro (Spain)
Permanent Representative: Tugay Özçeri (Turkey)
Permanent Representative: Sir John Goulden (United Kingdom)
Permanent Representative: Robert E Hunter (United States)

Military Committee
B-1110 Brussels
Chairman: Gen Klaus Naumann (Germany) (Army)
Deputy Chairman: Lieut Gen N B Kehoe (USA) (Air Force)
Military Representative: Lieut Gen G Bastien (Air Force) (Belgium)
Military Representative: Lieut Gen P G Addy (Army) (Canada)
Military Representative: Lieut Gen O L Kandborg (Army) (Denmark)
Military Representative: Maj Gen P Wiroth (Air Force) (France)
Military Representative: Lieut Gen K Wiesmann (Germany) (Army)
Military Representative: Vice Adm A Vennis (Greece) (Navy)
Military Representative: Lieut Gen G Marrafa (Italy) (Army)
Military Representative: Lieut Col G Lenz (Luxembourg) (Army)
Military Representative: Lieut Gen H W M Satter (Air Force) (The Netherlands)
Military Representative: Lieut Gen D P Danielsen (Army) (Norway)
Military Representative: Lieut Gen M J Alvarenga Santos (Portugal) (Air Force)
Military Representative: Vice Adm J Poblaciones (Navy) (Spain)
Military Representative: Vice Adm A Kiyat (Turkey) (Navy)
Military Representative: Sir John Cheshire (Air Force) (United Kingdom)
Military Representative: Lieut Gen T M Montgomery (United States) (Army)
International Military Staff Director: Lieut Gen G J Folmer (The Netherlands) (Army)

International Military Staff
B-1110 Brussels
Director of the International Military Staff: Lieut Gen G L Folmer (The Netherlands)
Assistant Director, Intelligence Division: Maj Gen F Kikiras (Greece)
Assistant Director, Plans and Policy Division: Maj Gen H Kujat (Germany)
Assistant Director, Operations Division: Rear Adm A Ross (United Kingdom)
Assistant Director, Armament, Logistics and Resources: Maj Gen A Grønheim (Norway)
Assistant Director, Communications and Information Systems: Lieut Gen L O Feliu (Spain)
Assistant Director, Co-operation and Regional Security: Maj Gen J Marchand (France)
International Military Staff: Air Cdre P O Sturley (United Kingdom)
Representative of SACEUR (SACEUREP): Maj Gen C C Ahnfeldt-Mollerup (Denmark)
Representative of SACLANT (SACLANTREPEUR): Vice Adm M P Gretton (United Kingdom)

French Military Representative
B-1110 Brussels
☎(+32 2) 707 74 70
Chief General of Aerial Division: Pierre Wiroth

NATO Situation Center
B-1110 Brussels
☎(+32 2) 707 51 00 ⊖ (+32 2) 707 41 17
Chief: Col L Frank (USAF)

Office of Information and Press (NATIP)
B-1110 Brussels
☎(+32 2) 707 11 11 ⊠ 23867 ⊖ (+32 2) 707 45 79
Director of Information and Press: C F Prebensen
Deputy Director and Spokesman: Dr Jamie Shea
Head of Publications: N W G Sherwen

Allied Communications and Computer Security Agency (ACCSA)
Brussels
☎(+32 2) 707 55 34 ⊖ (+32 2) 707 41 17

Allied Long Lines Agency (ALLA)
Brussels
☎(+32 2) 707 55 17 ⊖ (+32 2) 707 41 17

Allied Radio Frequency Agency (ARFA)
Brussels
☎(+32 2) 707 55 28 ⊖ (+32 2) 707 41 17

Allied Tactical Communications Agency (ATCA)
Brussels
☎(+32 2) 707 54 19 ⊖ (+32 2) 707 41 17

Allied Data Systems Interoperability Agency (ADSIA)
Brussels
☎(+32 2) 707 55 49 ⊖ (+32 2) 707 41 17 98

34-05000 Military Agency for Standardization (MAS)

NATO Headquarters, B-1110 Brussels
⊖ (+32 2) 707 87 77
Chairman: Maj Gen G B Ferrari (Army) 97

34-06000 NATO Air Command and Control System Management Agency

8 rue de Genève, B-1140 Brussels
☎(+32 2) 707 41 11 ⊖ (+32 2) 707 87 77
General Manager: R A Giacomo 97

34-06050 NATO Consultation Command and Control Agency (NC3A)

8 rue de Genève, B-1140 Brussels
☎(+32 2) 707 41 11 ⊖ (+32 2) 707 87 70
General Manager: Loren D Diedrichsen
Deputy General Manager: RAdm E Happack (GE Navy)
Executive: Tim Waugh 98

34-12100 Supreme Headquarters Allied Powers Europe (SHAPE)

B-7010 Shape
☎(+32 65) 44 71 11 Ⓔ shapepio@shape.nato.int
⊖ (+32 65) 44 35 44
SACEUR: Gen George A Joulwan (US Army)
Deputy SACEUR: Gen Sir Jeremy Mackenzie (UK Army)
Assistant to SACEUR for International Affairs (INTAF): Min Michael L Durkee (US)
Executive to SACEUR: Brig Gen James Solignan (US Army)
Chief of Staff: Gen. Peter H Carstens (GE Army)
Chief of Staff (ACE Reaction Force Planning Staff): Lieut Gen P Huysman (NL Army)
Deputy Chief of Staff: VAdm L Lillo (IT Navy)
Assistant Chief of Staff, Operations and Logistics: Maj Gen C Heflebower (USAF)
Assistant Chief of Staff, (Communications and Information Systems): Maj Gen Y Karagoz (TU Army)
Assistant Chief of Staff, Policy and Requirements: RAdm P K Haddacks (UK Navy)
Assistant Chief of Staff, Intelligence: Maj Gen G Bastiaans (NL Army)
Assistant Chief of Staff, ACE Resources: Maj Gen Schade (GE Army)
Chief of Public Information: Col Konrad Freytas (GEAF)
Public Information Executive Officer: Lieut Col David Cannon (USAF)
Administration Assistant: Van der Eeckon 98

34-50000 Western European Union (WEU)

Union de l'Europe Occidentale (UEO)
4 rue de la Régence, B-1000 Brussels
☎(+32 2) 500 44 55 ⊖ (+32 2) 511 35 19
Secretary General: José Cutileiro
Public Information Officer: Myriam Sochacki
■ European intergovernmental defence and security organisation created by the Modified Brussels Treaty in 1954. Member nations: Belgium, France, Germany, Italy, Luxembourg, the Netherlands, Portugal, Spain, UK and Greece. 98

55 Government-owned Industry & Production Facilities

Shur-Lok International SA 75-05900

60 Research Organizations, Institutes & Test Establishments

60-00300 The Gendarmerie Bureau Central des Recherches

Central Research Bureau
47 rue Fritz Toussaint, B-1050 Bruxelles
☎(+32 2) 642 61 11 ⊖ (+32 2) 642 69 55
Commander: Lieut Col Berkmoes 97

60-00550 Institut National de Criminalistique

Chaussée de Vilvoorde 98/100, B-1120 Bruxelles
☎(+32 2) 245 13 69 ⊖ (+32 2) 245 09 25 97

60-00600 Instituut voor Politiele en Gerechtelijke Organisatie

KU Leuven - Hooverplein 10, B-3000 Leuven
Contact: Jef Thys 98

60-01000 Permanent Secretariat for Prevention Policy

Vast Secretariat voor Het Preventiebeleid (VSPP)
(a subordinate of the Ministry of the Interior)
Rue de la Loi 26/28, B-1040 Brussels
☎(+32 2) 500 24 47 Ⓦ http://www.rspp.fgov.be
Ⓔ info@vspp.fgov.be ⊖ (+32 2) 500 24 47
Permanent Secretary: K van Limbergen
☎(+32 2) 500 24 45
Deputy Permanent Secretary: M van Hoecke
☎(+32 2) 500 24 40

Contact Reference

Deputy Permanent Secretary: C Beaupere
☎(+32 2) 500 24 41
■ Deals with crime prevention. Also contributes to the realisation of analysis of criminal phenomena. [97]

60-02000 Scientific Centre for Police Statistics
Rue Royal 47, B-1000 Brussels
Contact: Jan Capelle [98]

60-02800 Vakgroep Strafrecht en Kriminologie
Universiteitsstraat 4, B-2000 Gent
☎(+32 9) 274 69 58 ⊖ (+32 9) 264 69 88
7 (professional); 2 (clerical)
Director (Criminology, Sociology of Law, Crime Prevention): Prof.-Dr Patrick Hebberecht
■ Research, teaching and documentation on victimization studies, police studies (including organisations with special police functions); crime prevention evaluation research and criminal statistics. Established in 1986. [98]

60-03000 Vrije Universiteit Brussel
Pleinlaan 2, B-1050 Brussels
Department of Criminology: Els Enhus [98]

65 Industry, Trade & Armed Forces Associations

65-00050 Armed Forces Communications and Electronics Association (AFCEA)
(a subsidiary of AFCEA International, USA)
European Office
6 rue de Genève, PO Box 2, B-1140 Brussels
☎(+32 2) 705 27 31 [E] afcea@infoboard.be
⊖ (+32 2) 705 28 94
General Manager: Maj Gen E Greindl (retd)
Office Manager: Mrs P Sementilli-Mortier
■ Organises symposia, exhibitions. Provides assistance to AFCEA's chapters in Europe. [98]

65-00060 Belgian Aerospace Industries Association (BAIA)
Groupement Belge des Constructeurs de Matériel Aérospatial (GEBECOMA)
Rue de Strasbourg 13, B-1130 Brussels
☎(+32 2) 729 56 13 ⊖ (+32 2) 729 55 92
3 (1997)
Chairman and President: Marc Goddaer
General Delegate: Marcel Terrasson
Secretary General: J P Vlerick
■ Independent association of Belgian aerospace industry of high standards (AQAP and ISO 9000) dealing with airframes, engines, avionics and accessories for military and civilian applications. Belgian member of AECMA.
Members:
Alcatel Bell SDT
Alcatel Bell
Egic
Formétal
Fabrisys
Metal Forming
Sabena Technics
SABCA
Shur-Lok International
SONACA SA
Structural Design & Analysis
Technical Airborne Components SA
Techspace Aero
Thomson CSF Electronics Belgium [98]

65-00080 Belgian Association for Patrol Aircraft
Association Belge pour l'Avion Patrouilleur (ABAP)
(a joint venture between Techspace Aero, SABCA and SONACA)
1470 Chaussée de Haecht, B-1130 Brussels
☎(+32 2) 729 58 02 ⊖ (+32 2) 216 15 70
President: Remo Pellichero ☎(+32 2) 729 59 19
Administration Manager: Henri Vanhaesendonck
■ Partner in Breguet Atlantic production and ATL 2 program.
Member companies:
Techspace Aero
SABCA
SONACA [97]

65-00090 Belgian Business International
13 rue du Ruchaux, B-1490 Court Saint Etienne
☎(+32 10) 45 50 60 [E] alain.plavard@infoboard.be
59200 bbint b ⊖ (+32 10) 45 50 20
BFr 1,750,000
Managing Director: Alain Planard [97]

65-00110 Belgian Defence and Security Industry Group (BDIG)
c/o Fabrimetal, Rue des Drapiers 21, B-1050 Brussels
☎(+32 2) 510 25 20 21078 fabrim b
⊖ (+32 2) 510 24 56
2 (1997) 70 sq m
President: René Hannon
Vice President: Remo Pellichero
Director: R Godechoul [E] roger.godechoul@fabrimetal .be
■ Professional organisation of the Belgian defence industry. Membership of 25 companies. [98]

65-00140 European Defence Industries Group (EDIG)
94 Gulledelle, B-1200 Brussels
☎(+32 2) 775 81 10 ⊖ (+32 2) 775 81 31
5 560 m2
President and Chairman of the Board: H J Kränzle
Vice President and Vice Chairman of the Board: J Rasquin
Secretary General: G Woodcock
Technical Secretary: Mrs Myriam Croon-Janssens
Chairman of Economic and Legal Committee: D Scillitoe
Chairman of Technical Committee: B Theile
■ Association composed of 12 national (defence) trade associations from: Belgium, Denmark, France, Germany, Greece, Italy, Netherlands, Norway, Portugal, Spain, Turkey and United Kingdom. [98]

65-00170 The European Association of Aerospace Industries (AECMA)
Gulledelle 94, B-1200 Brussels
☎(+32 2) 775 81 10 ⊖ (+32 2) 775 81 11
President: Sir Richard Evans
President Elect: Serge Dassault
Former President: Manfred Bishop
Secretary General: Dr Peter Fichtmüller
Policy Director and Deputy Secretary General: David Ashworth
Operations and Public Relations Director: Fabio Dani
■ Represents the national aerospace industry associations of nine European countries.
Associates:
BDLI (Bundesverband der deutschen Luft und Raumfahrt)
GEBECOMA (Groupement Belge des Constructeurs de Matériel Aérospatial)
FDAI (Foreiningen for Danske Aerospace Industri)
ATECMA (Agrupación Tecnica Española des Constructores de Material Aerospacial)
GIFAS (Groupement des Industries Françaises Aéronautiques et Spatiales)
AIA (Associazione Industrie Aerospaziali)
NAI (Netherlands Aerospace Industries)
SBAC (Society of British Aerospace Companies)
SAI (Swedish Aerospace Industries) [98]

65-00300 Flemish Aerospace Group (FLAG)
Vlaamse Vereniging van Lucht en Ruimtevaart
Brouwersvliet 5, PO Box 4, B-2000 Antwerpen
☎(+32 3) 202 44 07 72893 ⊖ (+32 3) 226 43 34
President: Tony Vangrieken
Managing Director: A Van Hove
■ Non-profit organisation representing companies in the field of defence offsets, civil and military aerospace. Membership of about 90 companies. [98]

65-01000 Trans European Policy Studies Association
11 rue d'Egmont, B-1050 Brussels [98]

70 Financial Holding Groups, Parent Companies

70-00400 Dassault Belgique Aviation SA Head Office
100 Boulevard du Souverain, B-1170 Brussels
☎(+32 2) 672 73 75 ⊖ (+32 2) 673 52 84
President and Chief Executive Officer: Serge Dassault
General Delegate, European Affairs: Guy Servolle
Information Contact: Gabriel Herman
■ Holding company (Dassault Industries Group) and representative office of Dassault Aviation to UE.
Majority shareholdings in:
SABCA, France
Industrial Procurement Services Inc, USA
Falcon International SA, Switzerland [98]

70-01000 National Federatie van Commissarisen & Adj. Commissarisen van Politie
500 Noorderlaan, Antwerpen
Chief: Charles Nuyts [98]

74 Consortia, Joint Venture Companies

74-00500 BAMS Export Office
(† BAMS Association Momentanee)
c/o Sait Systems, Chaussée de Ruisbroek 66, B-1180 Brussels
☎(+32 2) 370 55 08 61511 phemb b
⊖ (+32 2) 376 68 73
President: Raymond Lefevre ☎(+32 3) 829 55 66
Export Manager: Wim Casteleyn
Program Manager: Roger Vanheertum ☎(+32 3) 829 52 37
■ New generation of EW protected VHF combat net radio system providing various manpack, vehicular and helicopter configurations. VHF handheld radioset. → 75 [98]

75 Manufacturing Companies

75-00055 Alcatel Bell
(† Alcatel Bell Telephone)
(a member of Alcatel NV)
Francis Wellesplein 1, B-2018 Antwerpen
☎(+32 3) 829 51 11 [W] http://www.alcatelelecom.be
72128 bella b ⊖ (+32 3) 240 99 96; 829 55 79
BFr 41,100,000,000 (1996) 7,591
President and Chief Executive Director: Julien de Wilde
General Manager, Radio, Space and Defence: Noël Parmentier
Communications Director: Piet Van Speybroeck
Defence and Avionics Director: R Lefevre
Purchasing Director: Luc Van Beek
Information Contact: E Van Tolhuysen
■ Defence communications networks, combat net radio and secure teleprinters. Avionic subsystems and navaids. Air defence and missile electronics. Simulators.
Subsidiaries with defence business:
Alcatel Bell-SDT → 75 [98]

75-00060 Alcatel Bell-SDT Head Office
(† ACEC)
(a subsidiary Alcatel Telecom)
101 rue Chapelle Beaussart, B-6032 Mont sur Marchienne
☎(+32 71) 44 22 09; 44 22 11 51766 belsdt b
⊖ (+32 71) 44 22 56; 44 22 00
BFr 1,250,000,000 BFr 4,500,000,000 (1994)
800 250,000 sq m
General Manager: R Hannon
Publicity Department: B Van Der Maren

■ Electronic assemblies and subassemblies for VHF combat net radio systems, HF modems equipment and terminals, fast battery chargers, tactical message terminals, peripheral electronics for IR sensors, power conditioning, hybrid circuits (signal and power) and ASICs.

Defence Division
PO Box 4008, B-6000 Charleroi
☎ (+32 71) 44 22 89 ▣ 51766 belsdt b
⊜ (+32 71) 44 22 56
Defence Division Manager: M Milecan [98]

75-00070 ADB NV/SA
Etablissements Adrien De Backer SA
(a subsidiary of Siemens AG, Germany)
Leuvensesteenweg 585, B-1930 Zaventem
☎ (+32 2) 722 17 11 ✉ ADB B ▣ 22154
⊜ (+32 2) 722 17 64
♧ BFr 218,000,000 (1996)
♣ BFr 1,408,278,000 (1996) 👥 280
Chief Exec. Officer: M. Bruns
Group Vice President: E. Toussaint
Asst to Group Vice President: A. Mortier
General Manager: Herbert Hess
General Manager Aviation Lighting: Alain Mortier
Marketing Manager: Roland Maes
■ Aviation lighting systems design, manufacture, installation and supervision.
Subsidiaries:
ADB SA France → 90
ADB Inc. USA → 90
ADB UK → 90 [97]

75-00075 Advanced Products NV (APNV)
(a subsidiary of Advanced Products Co, USA)
Rupelweg 9, Industrieterrein Krekelenberg, B-2850 Boom
☎ (+32 3) 880 81 50 ✉ METOSEAL
⊜ (+32 3) 888 48 62
♧ BFr 5,000,000 👥 49 (1997) 🏭 5,000 sq m
Managing and Sales/Marketing Director: M S Nedée
Technical Director: Robert Hermans
Administration and Finance Manager: D Lemmens
Contact: X Lagae
■ Manufacture seals for extreme environmental conditions. Markets: nuclear, aircraft and aerospace, automotive, chemical process and equipment manufacturing industries.
Subsidiary companies:
Advanced Products (Seals & Gaskets) Ltd (UK)
Advanced Products Dichtungen GmbH (Germany)
Advanced Products France Sarl (France) [98]

75-00100 Agfa-Gevaert NV
Septestraat 27, B-2640 Mortsel
☎ (+32 3) 444 41 60 ▣ 73055 agnv b
⊜ (+32 3) 444 76 96
International Marketing Manager, Industrial Imaging Systems: Frans Van Camp
■ Full range of camera films and printing material for aerial photography, photogrammetry and cartography.

Business Unit NDT
Septestraat 27, B-2640 Mortsel
☎ (+32 2) 444 82 41 ⊜ (+32 2) 444 82 43
General Manager: H Deschaumes
■ Non destructive testing systems, industrial x-ray films and processors. Holography. [98]

75-00520 ARSOC Sprl
146 rue du Roi Albert, B-4680 Oupeye
☎ (+32 4) 240 21 01 ▣ 41096 maestr b
⊜ (+32 4) 240 22 02
♣ BFr 50,000,000 (1996) 👥 3 (1997)
Director: F V Maes
■ Laser pointers, ballistic equipment, ammunition and ammunition manufacturing machines. Provision of shooting training. [98]

75-00550 ASCO Industries NV/SA
(† ASCO NV/SA)
Weiveldlaan 2, B-1930 Zaventem
☎ (+32 2) 716 06 11 ⊜ (+32 2) 716 07 70
♧ BFr 300,000,000 (1996)
♣ BFr 1,700,000,000 (1996) 👥 328 (1996)
🏭 18,509 m2
Managing Director: Christian Boas
Organisation and Audit Manager: Emile Boas
Commercial Manager, Aeronautics: Jozef Maes
Marketing Manager, International Prospection: Frans Peeraer
Commercial Manager, Civil and Defence: Jean Claude Vandeputte
Human Resources Manager: Fernand Vandewijngaerden
Finance and Administration Manager: Ivan Van Malderen
Quality Assurance Manager: Bert Vanreyten
Operations Manager: Serge Wintgens
Marketing Manager: Christian van Crombrugge
Security Equipment Trade Manager: Joseph Roonacker
■ Production, maintenance and overhaul of light armoured vehicles and components for missile systems, aircraft and nuclear equipment. Subcontractor manufacturing and refurbishing special metal assemblies, components and spares for civil and military manufacturers and major subcontractors. Involved in the manufacture and assembly of various aerostructures in special alloys for the Airbus programmes (A310, A320, A330 and A340), Boeing programmes and other major aircraft subcontractors. [98]

75-00900 Atlas Copco Airpower NV
Boomsesteenweg 957, B-2610 Antwerpen
☎ (+32 3) 870 21 11 ✉ ATLASAIR ▣ 73007 copap b
⊜ (+32 3) 870 26 55
♧ BFr 2 000 000 000 ♣ BFr 22 703 000 000 (1995)
👥 2 000 🏭 132 580 m2
Managing Dir: G. Mazzalupi
Finance Dir: J. Petersson
Corporate Services: Ms Paula Bracke
■ Portable & stationary air compressors, turbo compressors, dryers, stationary gas compressors. Aircraft ground power units & starters.
Associated companies:
Atlas Copco (Sweden) [97]

75-00925 Aydin Corporation NATO Programs
Bilzersteenweg 126/1, B-3700 Tongeren
☎ (+32 41) 78 05 68 ⊜ (+32 41) 78 05 78
Director of NATO Programs: Charles J Steiner
Office Manager and Marketing Assistant: Ms Marie Dominique Tounissou [98]

75-00950 Automation NV
86 Bevrijdingslaan, B-1640 Saint Genesius Rode
☎ (+32 2) 358 35 75 Ⓔ automation@automation.be
⊜ (+32 2) 358 36 09
♧ US$ 170,000 (1996) ♣ US$ 8,000,000 (1996)
👥 85 (1997) 🏭 4,000 m2
Managing Director: Frans Snyers
Technical Director: Frank Vanlocke
Sales and Marketing Director: Patrick Collet
Purchasing Manager: Bruno Wauters
■ Production of power converters, transformers, rectifiers and battery chargers. [98]

75-01120 Barco NV
President Kennedypark 35, B-8500 Kortrijk
☎ (+32 56) 26 26 11 ▣ 85842 barind b
⊜ (+32 56) 26 22 62
♣ BFr 18,920,000,000 👥 3,210
President and Chief Executive Officer: H R Vandamme
Senior Vice President and Chief Operating Officer: Erik Dejonghe
■ Visualisation: Projection systems: large screen projectors. Display systems: public information displays, hi-res colour displays for pre-press and process control, avionics displays, ruggedised and military standards displays, graphic display generators. Communication systems: broadcast and cable products. Automation: sensors, machine controllers and CIM systems. Graphic systems: graphic design, production and imaging systems. Specialised sub-contracting: thick-film and hybrids, printed circuits, surface mounting, plastic parts, moulds and transformers.
Specialised Sub-contracting:
Barco International Technologies
Barco Special Components [98]

75-01130 Barco Display Systems
(a subsidiary of Barco NV, Belgium)
Th Sevenslaan 106, B-8500 Kortrijk
☎ (+32 56) 23 32 11 Ⓦ http://www.barco.com
▣ 85842 barind b ⊜ (+32 56) 23 34 62; 23 34 80
♣ US$ 550,000,000 (1996) 👥 3,300 (1996)
Vice President and General Manager: Luc Vandenbroucke
Product Manager, Avionics and Rugged Displays: Guy Putman
Head of International Sales: Danny Sergeant
Product Line Manager, Display Systems: Geert Callens
Marketing Communications Manager: Mrs Ann Galland
■ Ruggedised full colour CRT displays (10" up to 29") for aircraft, surface ships, submarines and vehicles, 10" ruggedised LCD display, designed to meet airborne requirements.
Avionics displays and control display units (CDU's).
High performance, real-time display generators and ultra high resolution colour monitors for command and control (C3I), air traffic control, vessel traffic systems, process control, simulation in severe environments. [98]

75-01300 NV Bekaert SA
President Kennedypark 18, B-8500 Kortrijk
☎ (+32 56) 23 05 11 ✉ BEKAERT ⊜ (+32 56) 23 05 43
♣ BFr 61,816,000,000 👥 11,706
President: T Verhaeche de Nayer
Corporate Vice President and General Secretary: W Hallemeesch
Corporate Executive Officer: R Decaluwé
Corporate Communications Officer: W Snaet
■ Metal fibres and conductive surfaces for EMI, ESD and thermal applications. Wire products, steel cord for rubber and plastic re-inforcement, "dramix" for airfield pavements, military protective structures and industrial floors.

Bekaert Fibre Technologies
Bekaertstraat 2, B-8550 Zwevegem
☎ (+32 56) 76 61 11 ⊜ (+32 56) 76 79 66
General Manager: Pol Huysentruyt
Business Development Manager: R de Bruyne
■ Ultra fine metal and metallised fibres for electrical and thermal conductivity. High grade steel fibres for high temperature applications and ESD/EMI applications.

Documentation Centre
B-8550 Zwevegem
☎ (+32 56) 76 61 11 ⊜ (+32 56) 76 70 99
Information Manager: Jo Vandenhende [98]

75-01320 Belgian Advanced Technology Systems SA (BATS)
Parc Ind de Recherches Sart Tilman,
Avenue des Noisetiers, B-4031 Angleur, Liège
☎ (+32 4) 367 08 88 ⊜ (+32 4) 367 13 14
♧ BFr 140,000,000 ♣ BFr 300,000,000 👥 27
🏭 1,250 sq m
General Manager: M Tadmor
Buyer: Eric Philipper
Sales and Marketing: Jean-Claude Griess
■ Design, manufacture, assembly and test of ground-based battlefield surveillance radar, radar for detection of low-flying aircraft and hovering helicopter, land navigation system (inertial), radar training simulator, ground-based data handling and display system. [98]

75-01650 Bombardier Eurorail Industrial Vehicles Division
(† BN Division of Bombardier Eurorail)
Vaartdykstraat 7, B-8200 Brugge
☎ (+32 2) 535 55 11 ⊜ (+32 2) 539 04 28
Managing Director: W Broynooghe
Division Manager: M van Vooren [98]

75-01900 Britte SA
27 rue de Cheratte, B-4683 Vivegnis, Oupeye
☎ (+32 41) 64 44 70 ▣ 41439 ⊜ (+32 41) 64 08 63
♧ BFr 45 000 000 (1996) ♣ BFr 289 750 000 (1996)
👥 85 🏭 4 500 m2
President: Louis Dessart
Managing Dir: Vincent Pissart
Operations & Sales & Marketing Dir: J.M. Agnessen
Purchasing Mgr: Ghislaine Andrien
General Manager: Serge Bohyn
Sales Engineer: Charles Ledocq
Project Engineer: David Eloy
■ Machining of high precision mechanical parts for defence, space & aerospace applications. [97]

75-01950 Browning SA
(a subsidiary of FN Herstal SA, Belgium)
Parc Industriel des Hauts-Sarts, 3ème Avenue,
B-4040 Herstal
☎ (+32 41) 64 65 55 ▣ 41223 fabna b
⊜ (+32 41) 48 14 90
♣ BFr 10,658,000 👥 1,727
■ Guns and ammunition for hunting and leisure shooting, gun accessories, anti-riot ammunition, helmets, shields, bludgeons, holsters, bullet-proof vests, training and shooting range equipment. [98]

Contact Reference

75-02000 **Busak + Shamban SA**
(† Shamban Belux SA)
Chaussee de Louvain 334, B-1300 Wavre
☎(+32 10) 22 57 50 ▣ 59486 ⊖ (+32 10) 22 69 31
General Manager: Gabriela Brondorf
® USA: Shamban & Co. 98

75-02200 **Canons Delcour SA**
(a subsidiary of FN Herstal SA, Belgium)
1 rue J B Cools, B-4040 Herstal
☎(+32 41) 40 61 40 ⊖ (+32 41) 40 22 88
♣ BFr 450,000,000 (1995) 👥 150
Managing Director: G Claisse
Sales Director: José Serrano
■ Manufacture of machine gun barrels (0.5 cal) including stellite lined barrels, ballistic test barrels and sniper rifle barrels. 98

75-02260 **Chemviron Carbon**
(a subs. of Calgon Carbon Corp., USA)
Blvd de la Woluwe 60, Bte 7, B-1220 Brussels
☎(+32 2) 773 02 11 ⊖ (+32 2) 770 93 94
Managing Dir: D.L. Rees
Marketing Dir: E.F. Legros
Business Development Mgr: J.F. Marc
Advertising & Publicity Mgr: Ms Dorite Blanco
■ High quality impregnated activated carbons for the removal of chemical warfare agents by absorption, chemisortion or catalytic destruction. Emergency water treatment service (ETS) for gound water contamination, accidental spills.
® USA: Calgon Carbon Corp. 97

75-02410 **Cockerill Mechanical Industries Defence Division (CMI)**
(a subsidiary of Cockerill-Sambre)
Avenue A Greiner 1, B-4100 Seraing
☎(+32 41) 30 21 11 ▣ 41093 ⊖ (+32 41) 30 25 26
♧ BFr 1,250,000,000 (1996)
♣ BFr 8,500,000,000 (1996) 👥 1,500 (1996)
🏭 55,000 sq m
Managing Director: C Dermiens
Commercial Director: J F Levaux
Contracts Manager: H Art
Engineering Manager: Ph Laurent
■ Manufacture and development of 90 mm gun/turret systems Mk IIIM; VIII, 25 and 30 mm automatic canon turret for LAV and AIFV. Turret and gun modifications. 98

75-02500
Delft Sensor Systems
(† OIP Optronic Instruments & Products)
(a Delft Instruments company)
Westerring 21, B-9700 Oudenaarde
☎(+32 55) 33 38 11 E fdg@oip.be
W http://www.innet.net. oip/ ⊖ (+32 55) 31 68 95
♧ BFr 200,000,000 ♣ BFr 1,000,000,000 👥 120
Managing Director: Freddy Versluys
Finance and Administration Director: M Allegaert
Operations Director: J C Mattelaer
Sales Director: G J Vinke
Research and Development Director: R Slegtenhorst
Purchasing Manager: D Van den Berghe
Marketing Director: A Boumdouma
Quality Assurance Manager: G Brandt
Public Relations: Ms Francoise De Groote
■ Portable display systems; thermal observation equipment (handheld and vehicle mounted); night vision systems; fire control systems for armoured vehicles; space projects; advanced and holographic optics; customised research and development projects; equipment qualification. → 80 97

75-02900 **Emerson and Cuming Microwave Products NV**
(† Grace NV)
Nijverheidsstraat 7a, B-2260 Westerlo
☎(+32 14) 56 25 00 E ecmwprod@innet.be
⊖ (+32 14) 56 25 01
♣ BFr 310,000,000 (1996) 👥 41 (1997) 🏭 6,000 m2
Managing Director: A J Geurts
Sales and Marketing Director: G Timmerman
Personnel and Communications Officer: T Anthonis 98

75-03600 **FN Herstal SA**
33 Voie de Liège, B-4040 Herstal
☎(+32 4) 240 81 11 E info@fnherstal.com
▣ 41223 fabna b ⊖ (+43 4) 240 86 79 (marketing and management)
♧ BFr 3,100,000,000 (1996)
♣ BFr 4,641,000,000 (1996) 👥 1,099 (1996)
Chairman and Managing Director: Jacques Gentgen
Marketing and Programmes Manager: Francis Bossut
Publicity Manager: Chantal Regibeau
■ Research, development and sale of small arms and ammunition (9 mm, 5.7 mm, 5.56 mm, .50 calibre). Mounted weapon systems for helicopter, subsonic aircraft, land vehicles and small ships. Threat simulators from 9 mm up to 120 mm. Weapon maintenance. Military engineering. 98

75-03750 **Forges de Zeebrugge SA**
(a subsidiary of TDA Armements SA, France)
Rue en Bois 63, B-4040 Herstal
☎(+32 4) 248 20 77 ⊖ (+32 4) 248 25 10
♧ BFr 250,000,000 (1996) ♣ BFr 454,000,000 (1996)
👥 36 (1996)
Managing Director: Léon-Maurice Haulet
Finance Director: Ms Étienne van Bossche
Operations and Purchasing Director: Daniel Drion
Marketing and Communications: Ms Mireille Dantine
■ Complete rocket systems: 2.75"/5" ZUNI, 68 mm/100 mm launchers. Motors and warheads. Gun rocket control systems. Submunition warheads. 98

75-04000 **Genk Technical Products NV**
(a sister company of Clark Masts Teksam Ltd and Clark Masts Teksam NV, Belgium)
(a subsidiary of Evergreen Holdings Ltd, UK)
Woudstraat 31, B-3600 Genk
☎(+32 89) 38 08 31 ⊖ (+32 89) 38 61 41
♧ BFr 15,000,000 👥 45
Managing Director: A N Clark
Purchasing Manager: P Herremans
■ Fast erecting transportable mast systems for telecommunications, surveillance and DME equipment. The company works exclusively for the member organisations of Teksam Group. 98

75-04300 **Herbelith NV**
Stationstraat 30, B-8830 GITS Hooglede
☎(+32 51) 22 76 00 ⊖ (+32 51) 22 79 95
Manager: C Vanden Bulchke
■ Epoxy resin castings, composites for radar and electronics. 98

75-04650
INIMEX
Intercontinentalimex
R W van Havrelaan 111-B5, B-2900 Schoten, Antwerp
☎(+32 3) 658 80 52 E defence@inimex.be
▣ 72888 inimex b ⊖ (+32 3) 658 80 53
Managing Director: Pierre Winckel (Jr)
■ Belgium arms exports (authorized by Belgian Authorities). Ammunition from 5.56 mm to 203 mm, armament from 5.56 mm to 155 mm, small arms, rifles (AK-47/74), (G3A3/A4 and M16A2), submachine guns (MP5A3), aircraft practice bombs (BDU-33 D-B), 2.75" rocket assembly, flares, communications and ECM/ESM-optical and thermal hand-held to ship-based imagers. Suppliers to Belgian Air Force and Army, naval and air forces worldwide. Stockists of Ex-DDR naval ships. → 90 98

75-04750 **Lambert Aircraft Engineering**
Lupinestraat 5, B-8500 Kortrijk
☎(+32 56) 21 33 47 ⊖ (+32 56) 21 33 47
Director: Filip Lambert 98

75-04930 **MECAR SA**
Société Anonyme Belge de Mécanique et d'Armement SA
(a subs. of Allied Research Corp., USA)
50 rue Grinfaux, B-7181 Petit Roeulx lez Nivelles
☎(+32 67) 87 64 11 ▣ 57438
⊖ (+32 67) 21 63 07; 21 18 23
♧ BFr 50 000 000 ♣ BFr 1 900 000 000 👥 267
Chairman: J. Sculley
Managing Dir, Financial & Contracts Dir: J.F. Vezina
☎(+32 67) 87 65 50
General Mgr, Industrial & R&D Dir: C. Adant
☎(+32 67) 87 65 00
Sales Dir: J. Bastogne ☎(+32 67) 87 64 30
Commercial & Marketing Products Dir: M. Bely
☎(+32 67) 87 64 40
Marketing & Product Development Dir: L. Palacio
☎(+32 67) 87 64 50
Admin Dir: S. Faccin ☎(+32 67) 87 65 60
Budget Controller: M. Nihoul ☎(+32 67) 87 65 54
Marketing Mgr: A.N. Coates ☎(+32 67) 87 64 32
■ Ammunition ranging from 25, 90, 100, 105, 120, 125, 155mm calibres, particularly APFSDS. 81 & 120mm mortar ammunition. Hand grenades & bullet trap rifle grenades. Anti-Tank rockets & launchers. → 80 97

75-04971 **Meuse & Sambre**
Rue de Namur 16, B-5000 Namur
☎(+32 81) 22 02 33 ⊖ (+32 81) 22 59 20
♧ BFr 42,500,000 (1996) ♣ BFr 225,000,000 (1996)
👥 50 (1996) 🏭 60,000 sq m
Director: T Van Frachen
Export Manager: M Vynckier
■ River bridging craft for the Belgian Army. Patrol boats for the Indonesian Customs. Constructional metalwork and shipbuilding services. 98

75-04998 **MOL Cy NV**
Diksmuidesteenweg 63, B-8830 Hooglede
☎(+32 51) 70 16 81 ▣ 81949 ⊖ (+32 51) 70 30 38
Managing Dir: P. Vanhoorne
Public Relations: Johan Lermytte
■ Manufacture of towing tractors, aircraft refuellers, tank transporters, fuel tanks, anti-riot vehicles & overhead cranes. 97

75-05300 **NOF Europe NV**
(† Awlgrip NV)
(a subs. of US Paint, USA)
Bouwelven 1, IZ Klein-Gent, B-2280 Grobbendonk
☎(+32 14) 23 00 01 ⊖ (+32 14) 23 08 80
Sales Mgr: A. Smit
■ Complete paint systems for aircraft coating, original constructors specified colours, military specifications. 97

75-05500 **Precimetal Fonderie de Précision SA**
Parc Industriel, B-7180 Seneffe
☎(+32 64) 54 05 00 ⊖ (+32 64) 54 05 07
♧ BFr 19,000,000 (1996) ♣ BFr 480,000,000 (1996)
👥 165 (1996) 🏭 6,000 sq m
Managing and Operations Director: Paul Verhaegen
Sales and Marketing Director: Claude Lefèvre
Purchasing Manager: Jean Claude Scorey
■ Investment castings of components weighing from 1 g to 30 kg in all ferrous alloys, including stainless steel alloys, nickel, cobalt and copper alloys. 98

75-05585 **Recticel NV**
Damstraat 2, B-9230 Wetteren
☎(+32 9) 368 92 11 ⊖ (+32 9) 369 64 99
♣ BFr 4,762,920 (1996) 👥 668 (1996)
Managing Director: M Clockaerts
Sales and Marketing Director: H Vliegen
Purchasing Manager: G van de Sijpe
Sales Manager: M Gorré
■ Safety foam to military specifications for suppression of fuel systems explosions. 98

75-05590 **Société Anonyme Belge de Constructions Aéronautiques (SABCA)**
1470 Chaussée de Haecht, B-1130 Brussels
☎(+32 2) 729 55 11 ▣ 21237 sabush
⊖ (+32 2) 705 15 70; 729 58 96
♧ BFr 500,000,000 (1996)
♣ BFr 8,264,000,000 (1995) 👥 1,493 (1996)
🏭 882,000 sq ft
Chairman: Remo Pellichero ☎(+32 2) 729 59 20
Technical Director: Marc Duray ☎(+32 2) 729 55 40
Finance Director: R Dedobbeleer ☎(+32 2) 729 55 20
Commercial Director: Marc Humblet ☎(+32 2) 729 58 00
Marketing Manager, Electronic Defence Products: Pierre Pellegrin ☎(+32 2) 729 56 39
Marketing Manager, Space and Flight Control Systems: Ranbir Bhatti ☎(+32 2) 729 55 54
Purchasing Manager: C Anthonis ☎(+32 2) 729 55 72
■ Design, development and manufacture of a wide range of aerospace equipment.

Haren/Brussels Works
1470 Chaussée de Haecht, B-1130 Brussels
☎(+32 2) 729 55 11 ⊠ 21237 sabca b
⊖ (+32 2) 705 15 70
Works Manager: D Blondeel ☎(+32 2) 729 57 00
■ Design, development and manufacture of aircraft components: airframe structures, hydraulic systems, high precision mechanisms. Design, manufacture and repair of electronic airborne equipment, tank fire control systems, new materials in carbon fibre. Development and environmental testing. Surface and heat treatments. Final assembly, systems integration, ground and flight testing of aircraft, helicopters and accessories, integration of modifications.

Gosselies Works
Aérodrome de Gosselies, B-6200 Gosselies
☎(+32 71) 25 42 11 ⊠ 51251 sabgo b
⊖ (+32 71) 34 42 14
Works Manager: J M Lefevre

Limburg N J
Dellestraat 32, B-3560 Lunmen
☎(+32 13) 53 01 11 ⊖ (+32 13) 52 27 84
👥 100
Works Manager: Philippe Colin
Marketing Manager: D Hoff
■ Development and manufacture of civil, military and industrial high performance composite applications as a spin-off from space technology. Pre-integration of aerospace structures. 98

75-05595 SAGEM-B
(a subsidiary of SAGEM SA, France)
Interleuvenlaan 62, B-3001 Leuven
☎(+32 16) 40 04 28 ⊖ (+32 16) 40 00 44
♣ BFr 26,000,000 (1996) 👥 5 (1996)
Chairman: D Dupuy
Operations Director: Paul Venken
■ Software development for aeronautical applications. 98

75-05600 SAIT Systems SA
(a subsidiary of SAIT-RadioHolland, Belgium)
Chaussée de Ruisbroek 66, B-1180 Brussels
☎(+32 2) 370 53 11 W http://www.saitrh.com
⊖ (+32 2) 376 68 73
♧ BFr 350,000,000 (1992) ♣ BFr 822,442,114 (1995)
👥 134
President: Guy Seutin
Managing Director: R Landrie
Public Relations Officer: Ms Marie-Rose Texidor
■ Ground-based and naval telecommunications systems, navigation aids, industrial TV controls, airport displays for passenger and management information. Real-time and supervision systems. Network management. → 80 98

75-05650 Saval NV
Griffinstraat 7, B-2170 Merksem
☎(+32 3) 645 09 99 ⊖ (+32 3) 644 34 44
👥 8
Managing Director: J F Burns
General Manager: M Kimpe
■ Fire fighting equipment. 98

75-05695 Sector 6
(a division of NV Comauto SA)
Noordkustlaan 12 (Noordzone), B-1702 Groot-Bijgaarden
☎(+32 2) 481 00 20 ⊖ (+32 2) 466 37 45
♣ BFr 400,000,000 👥 95
Managing Dir: Mme J. Verset
Marketing & Sales Dir: A.F. Docking
■ Portable x-ray inspection systems for security use. Purpose built computer driven and battery powered systems. 97

75-05900 Shur-Lok International SA
(a sister co. of Shur-Lok Corp., USA)
Parc Industriel de et à, B-4800 Petit Rechain
☎(+32 87) 32 07 11 ⊖ (+32 87) 32 07 12
♧ BFr 7 900 000 ♣ BFr 220 000 000 (1995) 👥 55
Managing Dir: Peter Grefe
Dir General: Günther Pfeiffer
■ Specialised fastening systems and components for the aircraft industry. Inserts, barrel-nuts, self-locking nuts, position locking systems, expandable diameter fasteners. → 55 → 85
® USA: Shur-Lok Corp., Techeon Fiber-Resin 97

75-05950 Siemens Atea NV
(† ATEA NV)
(a subsidiary of Siemens AG, Germany)
Industriepark Klein Gent, Atealaan 34, B-2200 Herentals
☎(+32 14) 25 20 47 E p80921@vnet.atea.be
⊠ 33695 ateahe b ⊖ (+32 14) 25 32 92
♧ BFr 600,000,000 (1996)
♣ BFr 11,000,000,000 (1996) 👥 1,460 (1996)
🏭 58,300 sq m
President: Jacques Desmedt
Operations Director: F Coosemans
Sales and Marketing Director: T Verbeke
Sales and Marketing Manager, Defence and Space: Vic Kemps
Advertising Manager: Ms Vera Janssens ☎(+32 14) 25 29 82
Purchasing Manager: C Bekaert
■ Voice-data communications, PABXs (ISDN) networks, mobile PABXs, telephone sets, virtual private networks (VPN) systems integrator, AWHQ communications, ASM communications, CCIS and C3I communications, optical fibre LANs for headquarters, peace-keeping systems and shipboard communications. → 90 98

75-06050 Société Nouvelle PRB
(a subsidiary of Giat Industries, France)
Avenue de Tervueren 168, PO Box 7, B-1150 Brussels
☎(+32 2) 774 24 11 ⊖ (+32 2) 763 02 05
♧ FFr 777,000 ♣ BFr 2,500,000 👥 3
Chairman: Jean Claude Allard
■ Manufacture of ammunition, artillery and tank ammunition. Propelling charges. 98

75-06100 SONACA SA
Société Nationale de Construction Aerospatiale
Parc Industriel - Route Nationale 5, B-6041 Gosselies
☎(+32 71) 25 51 11 ⊠ 51241 sonaca b
⊖ (+32 71) 34 40 35
♧ BFr 900,000,000 (1996)
♣ BFr 3,375,000,000 (1996) 👥 1,075 (1996)
🏭 100,000 m2
Chairman: Pierre Sonveaux
Managing Director: Christian Jacqmin
Commercial Director: Marcel Devresse
Finance Manager: Christian de Groote
Technical Manager: Claude Nyssen
Production Manager: J N Guillou
Quality Assurance Manager: Michel Sander
Purchasing Department: Ph Hoste
■ Research, development and manufacture of advanced composite elements, aerospace structures and toolings. Final assembly. Co-producer of the F-16. Belgium prime contractor for Airbus A310, A320, A330 and A340. Participation in space programme (Hermes, ARD, CTV). Maintenance, overhaul and repair of military aircraft. Risk-sharing partner in Embraer EMB-145 programme. → 80 98

75-06200 Techspace Aero SA
(a subsidiary of SNECMA, France)
121 route de Liers, B-4041 Herstal Milmort
☎(+32 41) 78 81 11 ⊠ 42075 techsp b
⊖ (+32 41) 78 52 07
♧ BFr 2,515,000,000 (1997) ♣ BFr 5,800,000 (1996)
👥 1,180 (1997) 🏭 63,000 m2
Chairman: Louis de Spirlet ☎(+32 41) 78 81 40
Chief Executive Officer: Emeric d'Arcimoles ☎(+32 41) 78 84 53
General Manager, Commerce and Programs: Jacques Cipriano ☎(+32 41) 78 87 02
Deputy General Manager, Commmerce and Programs: Jean Christophe Dalla Toffola
General Manager, Space Research and Development: Roger Drosson ☎(+32 41) 78 85 91
General Manager, Finance and Budget: Claude Lemaire
General Manager, Production Control: Jean Marie Gely
General Manager, Human Resources: Willy Kina
General Manager, Economics and Cost Management: Jean Claude Morin
General Manager, Quality: Pierre Parmentier
General Manager, Components Manufacturing: Yves Prete
General Manager, Overhaul Repair and Tests: Jean Luc Royen
General Manager, Purchasing: Pierre Vierset
Manager, Advanced Propulsion Programs: Pierre Lemaire
Communications Mgr: Ferdinand Limpens ☎(+32 41) 78 86 02
■ Design, development and production of components and equipment for aircraft engines: lubrication systems, boosters (low pressure compressors), critical engine components (casings, bearing supports, disks, drums and spool assemblies, turbine blades and vanes); for space propulsion systems: valves (cryogenic, hydraulic, control and gases), regulators and leak control systems. Assembly and testing of complete turbo engines; repair (engineering) and overhaul of aircraft engines; test facilities for turbo engines and equipment. 98

75-06310 Thomson-CSF Electronics Belgium
(† MBLE Defence)
(a subsidiary of Thomson-CSF, France)
28 rue des Frères Taymans, B-1480 Tubize
☎(+32 2) 391 22 11 E teb@thomson-csf.fr
⊠ 21154 thbel ⊖ (+32 2) 391 23 00
♧ BFr 351,000,000 (1996) ♣ BFr 908,000,000 (1996)
👥 160 (1997) 🏭 8,750 m2
Managing Director: Philippe Johansen
Operations Director: Paul Martinet
Sales and Marketing Director: Pierre de Ponthihre
Purchasing Manager: Pierre Delbecque
Sales and Marketing Manager: Janick Legros
International Marketing Manager: Christian Legrand
Head of Technical Department: Michel Francis
Financial Manager: Michel Hautot
■ Radio communications, radars, ECM, electronic equipment for defence, professional multi-layer, flexible and single-sided printed boards. 98

75-06320 Twin Disc International SA
(a subsidiary of Twin Disc Inc, USA)
Chaussée de Namur 54, B-1400 Nivelles
☎(+32 67) 88 72 11 W http://www/twindisc.com
⊠ 57414 ⊖ (+32 67) 88 73 33
♣ BFr 1,628,633,103 (1996)
♣ BFr 289,000,000 (1997) 👥 330 (1996) 🏭 12,652 m2
Managing Director: Ph Pécriaux
Sales and Marketing Director: H C Farby
Purchasing Manager: A Vandenhaute
■ Marine transmissions from 100 to 3,600 kW. Variable speed drives for industrial and marine applications from 200 to 3,000 kW. Arneson surface drives from 200 to 7,000 kW. Electronic controls for up to 6 stations. Powershift transmissions from 200 to 620 kW. 98

75-06345 US Army Equipment Guns and Ammo
14 rue de Bost, PO Box 24, B-3300 Tirlemont
☎(+32 16) 81 33 72 ⊖ (+32 16) 82 54 12
♧ US$ 120,000 ♣ US$ 800,000 👥 2 🏭 300 sq m
Gun Manufacturer: Edgard J M Lemmens
■ Sniper rifles. Semi-auto, commando daggers, inflatable tents, assault rifle scopes. 98

75-06373 UTEXBEL SA/NV
Avenue C Snoecklaan 30, B-9600 Renaix Ronse
☎(+32 55) 23 12 11 ⊠ 85290 utbtis b
⊖ (+32 55) 21 03 46
♧ BFr 232,514,028 (1996)
♣ BFr 4,471,998,009 (1996) 👥 1,118 (1996)
🏭 191,266 sq m
Product Manager: P Merchiers
Contact: H Vandendriessche
■ Military fabrics for all types of uniforms, battledress, workwear, camouflage and protective wear. Fabrics produced to NATO ISO9002 specification. Infra-red reflective fabrics available on request. 98

75-06375 Vandeputte NV
Binnensteenweg 160-172, B-2530 Boechout
☎(+32 3) 460 02 45 ⊖ (+32 3) 460 02 40
♣ BFr 1,300,000,000 (1994) 👥 250
Commercial Director: Herman Dufoer
General Manager: Walter Vandeputte ☎(+32 3) 460 02 55
Export Director: Raoul de Coene
■ Manufacture of protective clothing, safety spectacles, respiratory protection, safety shoes, first-aid supplies, gloves, fire fighting tunics and equipment. 98

75-07000 Velleman Switch NV
(a subsidiary of Sidinvest NV, Belgium)
Legen Heirweg 37, B-9890 Gavere
☎(+32 9) 384 73 73 E velleman.switch@skynet.be
⊖ (+32 9) 384 73 88

Contact Reference

♧ BFr 36,642,000 (1996) ♣ BFr 123,845,658 (1995)
👥 29 🏭 1,600 sq m
Managing Director: Piet Blomme
Production, Research and Development Manager: Luc Demets
Sales Manager Non-Benelux: Danny Marquenie
Sales Manager Benelux: Piet Lietaer
■ Design and manufacture of custom made keypads. Custom-made membrane switches (for military and other purposes), circuitry printing on polyester, silicone keypads, custom-made touch screens. Own design services. [98]

75-08000 VSK Electronics

Venetielaan 39, B-8530 Harelbeke
☎ (+32 56) 21 99 51 ⊖ (+32 56) 20 05 20
♧ BFr 1 000 000 ♣ BFr 575 000 000 👥 119
■ Development, manufacture, installation, service & retailing of security systems. [97]

75-40000 Wolfsberg Aircraft Corporation NV

(† Triloader Aircraft Corporation NV)
Woudstraat 23, B-3600 Genk
☎ (+32 89) 38 08 31 ⊖ (+32 89) 38 61 41
Managing Director: Alex N Clark
■ The company was formed in 1995 to design and develop the Raven 2F small freight aircraft programme. [98]

80 Service Companies & Organizations

80-00100 AlliedSignal Aerospace Marketing Sales and Service

(a subsidiary of AlliedSignal Aerospace, USA)
Delta Air Transport, Building 117,
Brussels Airport, 1820 Melsbroek, B-SK7 1QR Brussels
☎ (+32 2) 754 17 97 ⊖ (+32 2) 754 19 26
Field Service Engineer: Graeme Davidson
■ On-site customer service support for AlliedSignal Aerospace → 90

● Sabena Belgium World Airlines,
Brussels National Airport, B-1930 Brussels
☎ (+32 2) 723 47 31 ⊠ 22-512 SNTEBRB
⊖ (+32 2) 723 42 00
Manager, Material Sales and Consignment Stores: Jan De Vierman
■ AlliedSignal Aerospace warehouse facility. [98]

80-00500 European Aviation Training Center NV (EATC)

Airport Building 119 - Block 3, B-1820 Melsbroek
☎ (+32 2) 752 94 30 ⊖ (+32 2) 751 43 00
👥 23 (1997)
Chairman: André Vanhecke ☎ (+32 2) 752 94 31
Sales and Marketing Director: Kristin Van Beek
Purchasing Manager: Joost de Cocker
Technical Manager: Piet Van de Ven ☎ (+32 2) 752 94 38
Pilot Training Manager: Franco Dorigo ☎ (+32 2) 752 94 34
■ Basic and advanced training for pilots, cabin crew and ground personnel for civil and military aviation. [98]

80-00600 E2S NV Software Engineering Division

Technologiepark 5, B-9052 Zwijnaarde
☎ (+32 9) 221 03 83 Ⓔ e2s@e2s.be
⊖ (+32 9) 220 31 91
♧ BFr 42,000,000 (1996) ♣ BFr 63,000,000 (1995)
👥 12 (1996) 🏭 800 m2
Managing Director: Raoul Majelyne
Sales and Marketing Director: Jacques Dewulf
Research and Development Director: Michel Huybrechts
Administration and Finance Manager: Pascal de Cloedt
■ Development and commercialisation of CASE (Computer-Aided Software Engineering) tools. Consultancy in software engineering, application of software engineering methods, techniques and tools by co-realisation custom software systems. Supported by continuous research and development. → 75

Space and Avionics Division
Technologiepark 5, B-9052 Zwijnaarde
☎ (+32 9) 221 03 83 ⊖ (+32 9) 220 31 91
👥 10
Division Manager: Johan Galle
■ Development of software and systems for space and avionics applications. Realises projects for the European Space Agency to develop and execute test procedures for satellites. Also co-develops the European Space Software Development Environment. Participation in several avionics programs. Production of embedded systems, support systems and test programs. [98]

80-00900 LMS International

Interleuvenlaan 68, B-3001 Leuven
☎ (+32 16) 38 42 00 Ⓦ info@lms.be/lmsintl.com
⊖ (+32 16) 38 43 50
♣ US$ 65,000,000 (1996) 👥 240
President: Dr V Vandeurzen
Operations Director: Dr J Leuridan
Partnering and Business Development Director: T Luniewski
Technical Director: H Wynendaele
Sales Director: T Edwards
Marketing Director: D Roesems
Marketing Communications Manager: P Dercon
■ Development and marketing of systems for acoustics, structural dynamics and durability, combining test and prediction for design optimization. Consulting and engineering services.

Subsidiaries:
LMS France, LMS UK, LMS Japan, LMS Korea, LMS Italiana, LMS Belgium, LMS North America, LMS Numerical Technologies, LMS Skalar, LMS Difa Instruments

LMS Numerical Technologies
Interleuvenlaan 68, B-3001 Leuven
☎ (+32 16) 38 45 00 ⊖ (+32 16) 38 45 50
General Manager: H Wymendaele [98]

80-01000 Logica SA/NV

(a subsidiary of Logica plc, UK)
Naamsestraat/Rue de Namur 72-74, B-1000 Brussels
☎ (+32 2) 512 99 76 ⊖ (+32 3) 512 98 69
■ Systems and software house; ground control systems, remote sensing, communications, data handling, monitoring and control systems.
® UK: Logica plc [98]

80-01400 Luminescent Systems (LSI)

Van Kerckhoven Straat 10, Bus 204, B-2880 Bornem
☎ (+32 3) 890 47 40 Ⓔ lsi.europe@glo.be
Ⓦ http://www.lumsys.com ⊖ (+32 3) 890 47 44
👥 2 🏭 60 m2
Senior Applications Engineer: Paul De Hous
Senior Applications Engineer: R J De Vos Dirk
■ Main activities include: electroluminescent emergency egress lighting systems, night vision goggle compatible lamps, internal cockpit and cabin lighting, formation lights, avionics panels, key switches, ruggedised keyboards.
® USA: Luminescent Systems [98]

80-01500 North Atlantic Consult SA

Kwabeekstraat 4, B-3370 Boutersem
☎ (+32 16) 72 00 71 ⊖ (+32 16) 72 00 78
♧ BFr 2,500,000 ♣ BFr 10,000,000 👥 1 🏭 30 sq m
President: Mrs C Fraiture [97]

80-01940 Security Management Consulting

35 Ijzerstraat, B-1560 Holeilaart
☎ (+32 2) 657 38 44 ⊖ (+32 2) 657 50 02
♧ BFr 1,250,000 (1996) ♣ BFr 6,000,000 (1996)
👥 3 (1996) 🏭 150 m2
General Manager: Maj Gen G Saint Viteux
Operations Director: R Bath
Purchasing Manager: A Nieberding
■ Security consultants. [98]

80-01950 Smith System Engineering Limited

Avenue des Arts, 53, Kuntslaan, B-1000 Brussels
☎ (+32 2) 502 89 40 ⊖ (+32 2) 502 89 35 [98]

80-02000 Transammo NV

Amerikalei 186, B-2000 Antwerpen
☎ (+32 3) 216 00 78 ⊠ 35362 ammo b
⊖ (+32 3) 248 50 33
Managing Director: A Braet
General Manager: P Braet
■ Forwarding and shipping of ammunition and explosives. [97]

80-02100 TSB International NV/SA

(† Systems Reliability (Europe) NV)
Bedrijfspark, Parc Industriel Keiburg,
B-1930 Excelsiorlaan, Zaventem
☎ (+32 2) 720 70 00 ⊠ 4662595 ⊖ (+32 2) 720 56 05
Contact: J P Deschuyter [97]

85 Stockists, Distributors, Spare Parts Suppliers

85-00100 Agusta Aerospace Services SA (AAS)

(a subsidiary of Agusta, Italy)
Rue Derain Patar 1, Aéroport de Liège,
B-4460 Grace Hollogne
☎ (+32 4) 234 23 23; 234 35 91 ⊠ 42820 aas
⊖ (+32 4) 234 19 45; 234 19 46
👥 30 🏭 6,500 sq m
Managing Director: Dr Maria Grazia Boetti
Operations Director: Luigi Ielmini
Purchasing Manager: Giorgio Volpi
■ Maintenance, repair and overhaul of Agusta helicopters, parts and components.
® Italy: Agusta SpA [98]

85-00500 Cummins Diesel NV

(a subsidiry of Cummins Engine Company, UK)
Blarenberglaan 4, Industriepark Noord 2,
B-2800 Mechelen
☎ (+32 15) 28 92 11 ⊖ (+32 15) 28 92 90
Government Business Director: Klaus D Kusatz [97]

85-02000 Trade Mark Electronics BVBA (TME)

Merksemsteenweg 194, B-2100 Deurne, Antwerpen
☎ (+32 3) 325 19 10 Ⓔ info@tme.be
⊖ (+32 3) 326 05 10
♧ US$ 300,000 (1996) ♣ US$ 5,500,000 (1996)
👥 15 (1997) 🏭 1,300 sq m (1997)
Director: P Vermeulen
Administration Director: A Beerens
Advertising Manager: E van Bocklaer
Sales: Lucas Duym
■ Marketing of electronic components, modular microprocessor systems for military and industrial applications.
® USA: Data Key, ILC Data Devices, Microsemi, Paravant Computers, Silicon General, Troyan Systems [98]

90 Sales Agents, Sales & Liaison Offices

Companies represented:

Connectral, France 90-00690
Cortini, Moramdo Coremu, FPT, Tacchi,, Italy 90-01330
Crawford-Swift, UK 90-01330
CSE, UK 90-00050
Daimler-Benz Aerospace AG, Germany . 90-00440
Data Key, USA 85-02000
DK-Codim, France 90-01330
Duracell, Belgium 90-00690
ECE, France 90-00690
EDAK, Switzerland 90-00690
ELNO, France 90-00690
Encore Computer Corp., USA 90-00570
Filotex/Cabeltel, France 90-00690
Furon, USA 90-00700
Graseby Ionics, UK 90-00050
H&S Aviation, UK 90-00050
Hellermann Electric, UK 90-00690
Hughes Electronics Corp, USA 90-00883
Hunting Aviation, UK 90-00050
Ibag, Kelllemberger, Switzerland 90-01330
ILC Data Devices, USA 85-02000
Interturbine, Netherlands 90-00050
Labotek, Boetherm, Denmark 90-01330
Lhotellier Montrichard, France 90-00690
Lockheed Martin Corp., USA 90-00050
Logica plc, UK 80-01000
Loral Corp., USA 90-01300
Luminescent Systems, USA 80-01400
Manitou BF, France 90-01304
Microsemi, USA 85-02000
Northop Grumman International Inc, USA 90-01400
OFEMA, France 90-01450
Paravant Computers, USA 85-02000
Radiall, France 90-00690
Raychem Corp., USA 90-01810
Raytheon Inc, USA 90-01815
Rockford Powertrain Inc., USA 90-01830
Rockwell Collins, USA 90-00050
Royal Ordnance, UK 90-00050
Saft SA, France 90-01910
Shamban & Co., USA 75-02000
Shur-Lok Corp., USA 75-05900
Sierracin-Sylmar, USA 90-00050
Silicon General, USA 85-02000
Société Nouvelle de Connexion, France 90-00690
Sonnenschein, Germany 90-00690
Souriau, France 90-00690
Superflexit, France 90-00690
SW, Reform, Wera, Germany 90-01330
Techeon Fiber-Resin, USA 75-05900
Teledyne Inc., USA 90-02030
Transrack, France 90-00690
Troyan Systems, USA 85-02000
UPM, UK 90-01330
Waverley Electronics, UK→ Netherlands 80-18300
Westinghouse Electric Corp., USA . . . 90-00050
Westland Helicopters, UK 90-00050

Sales agents/representatives listed elsewhere:

AlliedSignal Marketing Sales 80-00100
INIMEX 75-04650
Siemens Atea NV 75-05950

90-00010 Acer Belgium NV

Coremansstraat 34, B-2600 Antwerpen
☎ (+32 3) 230 50 32 ⊖ (+32 3) 281 33 25
® Taiwan: Acer Inc.
[97]

90-00050 Aero Systems SA

10 Chemin des Chasseurs, B-1380 Ohain
☎ (+32 2) 652 22 22 ⊖ (+32 2) 653 15 17
♧ BFr 1 250 000 (1995) 👥 6 ⌂ 200 m2
Managing Dir: Sir Dennis Mitchell
Sales & Marketing Dir: Michael Mitchell
Technical Dir: André Pierre
Purchasing Mgr: Ms Françoise Bindelle
® Netherlands: Interturbine
UK: British Aerospace, CSE, Graseby Ionics, H&S Aviation, Hunting Aviation, Royal Ordnance, Westland Helicopters
USA: Lockheed Martin Corp., Rockwell Collins, Sierracin-Sylmar, Westinghouse Electric Corp.
[97]

90-00170 ASE Europe NV Defence and Security Division

De Keyserlei 58, Bus 1, B-2018 Antwerpen
☎ (+32 3) 234 06 66 ✉ ASEUROP ANTWERP
▣ 31143 acmet b ⊖ (+32 3) 226 22 99
♧ BFr 158,000,000 (1995)
♣ BFr 3,200,000,000 (1995) 👥 80 (1995)
General Manager: G van Ende
Commercial Manager: A Stuer
■ Trade of special vehicles and very high-speed seagoing vessels, mobile medical units and installation. Decontamination equipment for personnel, riot control vehicles and equipment, gensets, flexible tanks, bullet proof and anti-flagration vests, helmets (standard, ballistic, tankcrew, aircrew), parachutes. Bullet proof glass and wall systems, technical shelters, parts.
Offices in:
Europe, Hong Kong, India, Kenya, Nigeria, Saudi Arabia, Thailand, Turkey.
[98]

90-00175 Astronautics Corporation of America

(† Kearfott Guidance & Navigation Corp.)
20 Ave des Tarius, B-1301 Bierges
☎ (+32 10) 41 46 26 ⊖ (+32 10) 41 48 33
Managing Dir: Geoffrey B. Symonds
■ Marketing Office.
® USA: Astronautics Corporation of America
[98]

90-00265 British Steel Belgium Sprl

(a subsidiary of British Steel Engineering Steels, UK)
Avenue Chantecler 33, B-1420 Braine L'Alleud
☎ (+32 2) 384 60 57 ⊖ (+32 2) 384 25 79
[98]

90-00275 Brüel & Kjaer Belgium NV

Ambachtelijke zone De Vunt 17, B-3220 Holsbeek
☎ (+32 16) 44 92 25 ⊖ (+32 16) 44 92 30
Admin Mgr: Arie J.J. Vis
® Denmark: Brüel & Kjaer
[97]

90-00320 CEP SA

Rue de Seraing 33, B-4020 Liege
☎ (+32 41) 43 83 95 ⊖ (+32 41) 41 28 58
[97]

90-00333 Clark Masts Teksam NV

(† Genk Technical Products NV)
(a subsidiary of Evergreen Holdings, UK)
Woudstraat 21, B-3600 Genk
☎ (+32 89) 38 44 72 ▣ 39589 ⊖ (+32 89) 38 61 41
♧ BFr 15,000,000 ♣ BFr 120,000,000 👥 20
Managing Director: A N Clark
Masts Sales Manager: E de Tollis
Teklite Sales Manager: R Masselin
■ Fast erecting portable and transportable masts for military communication systems. Portable and mobile emergency floodlighting sold under the registered trademark TEKLITE.
[98]

90-00440 Daimler-Benz Aerospace AG

(† Dornier GmbH)
(a subsidiary of Daimler-Benz Aerospace AG, Germany)
Avenue de Tervueren 273, B-1150 Brussels
☎ (+32 2) 762 22 75 ⊖ (+32 2) 770 04 07
Director: Dr Alfred Böswald
Contact: Ms Andrea Weber
® Germany: Daimler-Benz Aerospace AG
[97]

90-00570 Encore Computer Belgium

Ave Reine Astrid 1, B-1430 Wauthier-Braine
☎ (+32 2) 366 17 52 ⊖ (+32 2) 366 18 79
® USA: Encore Computer Corp.
[98]

90-00575 Europavia Belgium SA

35 Ave des Courses, B-1000 Brussels
☎ (+32 2) 647 66 35 ⊖ (+32 2) 647 64 04
Managing Dir: Gen. A. Jacobs
[97]

90-00690 Framatome Connectors Belux

(† S.E.B. Souriau)
(a subs. of Framatome Connectors International, France)
8 Quai des Usines, B-1000 Brussels
☎ (+32 2) 204 07 00 ⊖ (+32 2) 203 41 29
♧ BFr 5 000 000 (1995) ♣ BFr 485 104 000 (1995)
👥 37 ⌂ 2 000 sq m
Managing Dir: Michel Schepens
Sales & Marketing Dir: J. Kerkhofs
® Belgium: Duracell
France: CEAC (Tudor, Fulmen), Citel, Connectral, ECE, ELNO, Filotex/Cabeltel, Lhotellier Montrichard, Radiall, Société Nouvelle de Connexion, Souriau, Superflexit, Transrack
Germany: Sonnenschein
Switzerland: EDAK
UK: Hellermann Electric
[97]

90-00700 Furon

(a subs. of Furon Company, USA)
Chaussée de Tirlemont, 100, B-5030 Gembloux
☎ (+32 81) 62 62 11 ⊖ (+32 81) 61 57 89
General Mgr: L. Hansen
■ Pressure-sensitive tapes, coated fabrics, silicone rubber; solid, sponge & foam EMF/RFI shielding materials. Rulon Bearings & Omniseal Spring Energized seals. Complete product line of fluid handling components. Hydraulic & high pressure hose & tubing, wire & cables.
® USA: Furon
[97]

90-00756 Giravia Sarl

183 Avenue Milcamps, Boîte 12, B-1040 Brussels
☎ (+32 2) 653 34 34 ⊖ (+32 2) 652 17 22
General Manager: J D Goffard
[98]

90-00780 The Goodyear Tire & Rubber Company

Spoorwoglaan 3, B-2610 Wilrijk, Antwerp
☎ (+32 3) 820 32 60
[98]

Contact Reference

90-00883 Hughes Europe SA/NV Headquarters for Europe and Africa

5 Avenue Ariane, B-1200 Brussels
☎ (+32 2) 778 49 11 ⊖ (+32 2) 778 49 99
Chairman and Managing Director: Robin L Beard
Vice President, Business Development: W Richard Frederick
Vice President, Telecommunications and Space: Dr Eric J Novotny
Director of Aerospace and Defence: Fred J Fiorenza
Director of Air Defence and Air Traffic Control Systems: Thomas J Harvey
Director of Missile Systems: James E Monroe
Director for Russia and Former Soviet Union: Alex Okun
Vice President, Space and Communications Marketing: William R Schnicke
Director of Tactical Missile Systems: R M Wright
Public Affairs Representative for Europe, Middle East and Africa: Paul Stefens
Representative Delco Systems Operations: John J Frantz
® USA: Hughes Electronics Corp
Hughes Europe Systems Division
1 Berkenlaan, B-1831 Diegem
☎ (+32 2) 725 60 50 ⊖ (+32 2) 725 62 62
General Manager: Robert L Puich
[98]

90-00970 ITT Cannon Benelux Office

Rue Col Bourg Straat 105a - B3, B-1140 Brussels
☎ (+32 2) 726 75 94 ⊖ (+32 2) 726 92 01
Sales Manager: P N Joukes
® Germany: Cannon Electric
[98]

90-01120 Karcher Benelux

(a subsidiary of Alfred Karcher GmbH, Germany)
Industrieweg 12, B-2320 Hoogstraten
☎ (+32 3) 340 07 11 ▣ 4671585 ⊖ (+32 3) 340 08 40
Manager: H Wally Wagtmans
® Germany: Alfred Karcher GmbH & Co
[98]

90-01300 Loral Vought Systems

(† Loral Vought International)
15b Ave Herrmann Debroux, B-1160 Brussels
☎ (+32 2) 660 70 54 ⊖ (+32 2) 660 12 09
Contact: Erich von Marbod
® USA: Loral Corp.
[97]

90-01304 **Mantrec**
(a subsidiary of Manitou BF, France)
Chaussée de Wavre, B-1360 Perwez
☎(+32 81) 65 55 01 ⊜ (+32 81) 65 57 37
® France: Manitou BF [98]

90-01330 **MEC SA**
Mechanical Equipment Company SA
(a subsidiary of Treco, Belgium)
Rue de la Senne 21, B-1000 Brussels
☎(+32 2) 512 91 35 ⊜ (+32 2) 512 54 16
♧ BFr 16,500,000 (19959 ♣ BFr 180,000,000 (1995)
👥 10 (1995) 🏭 2,000 m2
Manager Director: Roland Van Maele
■ Machine tools, robots and plastics machinery.
® Denmark: Labotek, Boetherm
France: Braillon, CMB, Clextral, DK-Codim
Germany: SW, Reform, Wera
Italy: Cortini, Moramdo Coremu, FPT, Tacchi,
Switzerland: Ibag, Kelllemberger
UK: Accrapak, Cincinnati-Milacroy, Conair Churchill, Crawford-Swift, UPM
USA: Cincinnati-Milacron [98]

90-01400 **Northrop Grumman International Inc**
(a subsidiary of Northrop Grumman International Inc, USA)
12 Avenue de Broqueville, PO Box, B-1150 Brussels
☎(+32 2) 772 04 09 ⊜ (+32 2) 771 62 17
👥 6
Vice President, Europe: Darcy Tom
[E] t.darcy@pophost.eunet.be
Office Manager: Dieuwke Rethmeier
® USA: Northop Grumman International Inc [98]

90-01450 **OFEMA**
(a subsidiary of OFEMA, France)
28 Boulevard du Régent, BP 12/13, B-1000 Brussels
☎(+32 2) 511 67 87 ⊜ (+32 2) 514 33 74
Technical Delegate: Col Pattyn
® France: OFEMA [98]

90-01670 **Promavia SA**
Société de Promotion Aéronautique Headquarters
Chaussée de Fleurus 181, Gosselies Airport, B-6041 Gosselies
☎(+32 71) 35 08 29; 34 15 11 ⊜ (+32 71) 35 79 54
♧ BFr 211,000,000 🏭 3,000 m2
Chairman and CEO: André L. Delhamende
Deputy Chairman: Philippe Delhamende
Financial Director: M F Ruppen
Marketing Manager: Joseph Bernas
Contact: Raymond Di Luciano
■ Two seat primary and basic jet trainers. [98]

90-01810 **Raychem SA**
Diestsesteenweg 692, B-3010 Kessel-Lo
☎(+32 16) 27 17 11 ⊜ (+32 16) 27 17 27
Sales Dir: Luc Mergeay
® USA: Raychem Corp. [97]

90-01815 **Raytheon International Inc**
(† Raytheon E-Systems)
(a subsidiary of Raytheon Inc, USA)
Vandammestraat 7, Bus 5a, B-1560 Hoeilaart
☎(+32 2) 657 70 28 ⊜ (+32 2) 657 60 74
Regional Director: Robert A Norman
Administration Manager: Diane J Salt
® USA: Raytheon Inc [98]

90-01830 **Rockford Powertrain Inc.**
81 rue Franz Merjay, B-1060 Brussels
☎(+32 2) 346 37 10 ⊠ 22961 ⊜ (+32 2) 346 33 76
■ Vehicle transmissions, driveshafts, U-joints, fan drives, mechanical clutches.
® USA: Rockford Powertrain Inc. [98]

90-01910 **Safta**
(a subsidiary of Saft SA, France)
8 rue des Deux Gares, B-1070 Brussels
☎(+32 2) 556 44 00
[E] erik.deschuyffeleer@saft.alcatel.be
⊜ (+32 2) 520 16 84
♧ BFr 3,600,000 (1997) ♣ BFr 267,500,000 (1996)
👥 16 (1997) 🏭 1,200 m2
Managing Director: Gérard Mathis
Sales and Marketing Director: Pierre Vanossel
■ Nickel-cadmium batteries, lithium batteries, air-depolarized cells, converters and power supplies, advanced battery systems, rectifier chargers, nickel-cadmium batteries for space applications.
® France: Saft SA [98]

90-02030 **Teledyne Belgium SA**
European Marketing Office
Airport Business Center, Vuurberg 80, B-1831 Diegem
☎(+32 2) 721 52 52 ⊠ 25881 tdybel b
⊜ (+32 2) 721 53 40
Vice President Europe: James H. Baillie
Managing Dir: Mathieu Tijskens
® USA: Teledyne Inc. [97]

90-02150 **Universal Refining NV**
(a subsidiary of Daewoo Corp, Republic of Korea)
Haven 279 - Beliweg 20, B-2030 Antwerp
☎(+32 3) 541 36 00 ⊠ 71494 unidae b
⊜ (+32 3) 541 29 14 [98]

Belize

Head of State:
H.M. Queen Elizabeth II

Governor:
Dr Colville Norbert Young

10 Head of State

10-10000 **Office of the Governor-General**
Belize House, Belmopan
☎(+501 8) 225 21 ⊜ (+501 8) 220 50
Governor General: Sir Colville Norbert Young [98]

11 Prime Minister

11-10000 **Office of the Prime Minister**
East Bloc, Belmopan
☎(+501 2) 757 21
Prime Minister: Manuel Esquivel [98]

11-20000 **Office of the Deputy Prime Minister**
Belmopan
Deputy Prime Minister: Dean Barrow [98]

12 Ministries

12-20000 **Ministry of Foreign Affairs**
New Administration Building, Independence Hill, Belmopan
☎(+501 8) 221 67 ⊠ 102 foreign bz ⊜ (+501 8) 228 54
Minister of Foreign Affairs: Dean Barrow [98]

12-40000 **Ministry of Home Affairs**
Belmopan
Minister of Home Affairs and Labour: Hubert Erlington [98]

12-50000 **Ministry of National Security**
Belmopan
☎(+501 8) 222 25 ⊜ (+501 8) 226 15
Prime Minister and Minister of Defence: Manuel Esquivel
Minister of National Security: Dean Barrow
Minister of National Co-ordination and Mobilisation: Ruben Campos [98]

25 Defense Forces

25-10000 **Belize Defence Force Headquarters (BDF)**
PO Box 141, Belize City
☎(+501 2) 21 71; 21 72; 21 74; 21 76
⊜ (+501 2) 25 21 75
Commandant: Brig Gen Earl E Arthurs
☎(+501 2) 025 21 74
Adjuntant: Maj Lloyd Gillett
Operations Director: Maj Peter Parachue
Deputy Commander/Chief of Staff: Col Robert Garcia
Logistics and Procurement Director: Maj L Castillo
Procurement Director: Maj Jean Lucas
Air Wing
PO Box 141, Belize City
☎(+501 2) 20 98
Commanding Officer: Capt John M A Flowers
Maritime Wing
PO Box 141, Belize City
Commanding Officer: Maj Stephen Heusner
2nd Infantry Batallion: Lt Col Glen S Arthurs [98]

26 Police (Civil)

26-00000 **Belize Police Force**
Office of the Commissioner, PO Box 245, Belmopan, Cayo District
☎(+501 8) 222 24 ⊜ (+501 8) 232 91
Commissioner of Police: Sherman Zuniga
Deputy Commissioner: Ornel Brooks
Asst Commissioner Personnel & Training: Alvan Rowland
Asst Commissioner Operations: Carmen Zetina
Asst Commissioner Administration: Hughington Williams
Superintendent National Crimes Investigation: Raymond Thimbriel
Superintendent Administration: Gerald Bernard
Assistant Superintendent Administration: Gilroy Nicholas
Superintendent Head Special Branch: Melicio Uck
Assistant Inspector Joint Intelligence Co-ordinating Centre: Justo Augustine
Asst Superindent Dragon Unit: Gerald Westby
Senior Superintendent and Commander Eastern Police Division: Lester Garnett Racoon Street Belize City
Asst Superintendent Corozal District: Robert Rosalez
Superintendent Orange Walk District: Bernard Lino
Superintendent Cayo District: Edwardo Wade
Inspector Stann Creek District: Martin Yama
Asst Superintendent Toledo District: Clifford Martinez [97]

31 Embassies & Attachés

31-01760 **Embassy of the Republic of China**
PO Box 1020, Belize City
3rd Fl. James Blake Bldg, Corner Hutson/Eyre Streets, Belize City
☎(+501 2) 318 62 ⊜ (+501 2) 318 90
Ambassador: H.E. Mr Shu-Chi Chang
First Secretary: Benjamin T.H. Ho
Second Secretary: Yin-Lung Kao [97]

31-01890 **United Kingdom High Commission**
PO Box 91, Belmopan
☎(+501 8) 221 46; 221 47 [E] brithicom@bti.net
⊜ (+501 8) 227 61
High Commissioner: H E Mr G M Baker
Deputy High Commissioner: T D Stew
Counsellor for Regional Affairs: J Headley (resides in Mexico City)
Defence Adviser: Col S R Daniell (resides in Mexico City)
Second Secretary: M Dubbey (resides in Miami)

Third Secretary (Aid): G D O'Brien
Management and Consular Officer: Mrs L M O'Brien
Archivist: Miss Lisa Harrison [98]

31-01950 Embassy of Venezuela
18-20 Unity Boulevard, PO Box 49, Belmopan
☎ (+501 8) 223 84 ⊖ (+501 8) 220 22z
Ambassador: Christiaan van der Ree [97]

Benin

Head of State:
Mathieu Kerekou

10 Head of State

10-10000 Office of the President
PO Box 2028, Cotonou
☎ (+229) 30 00 90; 30 10 45; 30 12 56
5222/33 presirep ctnou
President of the People's Republic: Mathieu Kerekou
Minister Parliamentary Communication: Fransisco Marius [98]

11 Prime Minister

11-10000 Office of the Prime Minister
Cotonou
Prime Minister: Adrien Houngbedji [98]

12 Ministries

12-10000 Ministry of Defence and Armed Forces
PO Box 2028, Cotonou
☎ (+229) 30 00 90; 30 04 12; 30 06 23
Minister Delegate to the President Charged with National Defence: Severin Adjovi [98]

12-20000 Ministry of Foreign Affairs and Co-operation
PO Box 318, Cotonou
☎ (+229) 30 04 00; 30 09 29; 30 18 70
5200 minafet ctnou ⊖ (+229) 30 02 45
Minister: Pierre Osho [98]

12-40000 Ministry of Interior and Public Security
Cotonou
☎ (+229) 30 11 06; 30 11 54 5065 ministerieur
Minister: Theophile N'da [98]

25 Defense Forces

25-40000 Naval Forces Command
PO Box 163, Cotonou
☎ (+229) 31 57 24; 31 24 69
Commander: Béré Prosper Kiando
Chief of Staff: Soulemane Marcos
Operations Dir: Anatole Djossou
Procurement/Logistics Dir: Randolphe Houenou [97]

31 Embassies & Attachés

31-01890 Consulate of the United Kingdom
Lot 24, Patte d'Oie, Cotonou
☎ (+229) 30 11 20 972 50 47 (a/b) 5047 stin ctnou)
Honorary Consul: C Barnes [98]

31-01900 Embassy of the United States of America
rue Caporal Anani Bernard, PO Box 2012, Cotonou
☎ (+229) 30 06 50; 30 05 13; 30 17 92
⊖ (+229) 41 15 22
Ambassador: H E Mr John Melvin Yates [98]

Bermuda

Head of State:
H.M. Queen Elizabeth II

Governor:
H E Mr J T Masefield

10 Head of State

10-10000 Office of the Governor
11 Langton Hill, Hamilton HM 13
☎ (+1 441) 292 36 00 E depgov@ibl.bm
⊖ (+1 441) 295 38 23
Governor: H E Mr J T Masefield
Deputy Governor: P Willis
Minister: The Hon Pamela Gordon
Deputy Minister: The Hon Jerome Dill
Aide-de-Camp: Capt William White
Registrar: P J Dryden [98]

11 Prime Minister

11-10000 Office of the Premier
105 Front Street, The Cabinet Building, Hamilton HM 12
☎ (+1 441) 295 51 51 3775 modus ba
⊖ (+1 441) 292 83 97
Premier: The Hon Pamela Gordon
Deputy Premier: Hon Jerome Dill
Minister of Finance: Grant Gibbons
Secretary to the Cabinet: Leopold N Mills

Department of Information Services
Global House, 43 Church Street, Hamilton HM 12
Minister of Technology and Information: John Barritt [98]

15 Parliament

15-10000 Senate
The Cabinet Building, 105 Front Street, Hamilton HM 12
☎ (+1 441) 292 74 08 ⊖ (+1 441) 292 20 06
President of the Senate: Senator the Hon A.S. Jackson (CBE, JP)
Speaker House of Assembly: Hon. E.D. DeCouto (JP, MP) ☎ (+1 441) 292 7408
Minister for Home Affairs, Labour & Public Affairs: Hon. G.L. Edness (CBE, JP, MP)
Clerk to the Senate: Mrs Y.M. Roach
Chairman Defense Board: Col Michael J. Darling (OBE, ED, JP) ☎ (+1 441) 238 1045
Clerk to the Legislature: James E. Smith [97]

26 Police (Civil)

26-00000 Bermuda Police Department
Police Department
10 Headquarters Hill, PO Box HM 530,
Prospect, Devonshire DV 02
☎ (+1 441) 295 00 11 ✉ COMPOL Bermuda
3442 compl ba ⊖ (+1 441) 299 43 64
$ 34,047,510 83 (civilian), 434 (uniformed)
Commissioner of Police: Colin P Coxall
Deputy Commissioner: Jean Jacques Lemay
Procurement Director: Stuart Innes
Finance and Administration Manager: Robert L Jardine [98]

29 Customs & Excise

29-01000 Ministry of Finance Customs Department
Hamilton Hall, 40 Front Street, Hamilton HM 11
☎ (+1 441) 295 48 16 ⊖ (+1 441) 295 53 92
Collector of Customs: Gerald Ardis
P C O: Ken Stone [98]

31 Embassies & Attachés

31-00460 Royal Danish Consulate
PO Box HM 666, Hamilton HM CX
Clarendon House, 2 Church Street, Hamilton HM 11
☎ (+1 441) 295 14 22 ⊖ (+1 441) 292 47 20
Consul: Nicholas B Dill [98]

31-00590 Consulate of France
PO Box HM 282, Hamilton HM AX
Frith Liquors Building, 57 Front Street, Hamilton, HM 11
☎ (+1 441) 295 35 44 ⊖ (+1 441) 295 10 49
Consul: William De V Frith [98]

31-00690 Consulate of Greece
Fair Haven, 20 Astwood Road, Pagst DV04
☎ (+1 441) 236 74 11
Honorary Consul: J Macrae Mitchell [98]

31-00890 Consulate of Italy
PO Box 3313, Hamilton HM PX
☎ (+1 441) 295 19 62 ⊖ (+1 441) 295 19 62
Honorary Consul: Romano Erba
Vice Consul: Ing. Micca Ferrero [98]

31-01270 Consulate of the Royal Netherlands
PO Box HM 1561, Hamilton HM FX
Milner House, 18 Parliament Street, Hamilton HM 12
☎ (+1 441) 295 46 30 ⊖ (+1 441) 292 78 80
Consul: William M. Cox (MP) [97]

31-01340 Royal Norwegian Consulate
PO Box HM 564, Hamilton HM CX
☎ (+1 441) 295 55 19
Consul: C N A Butterfield (Jr) [98]

31-01440 Consulate of Portugal
PO Box HM 1726, Hamilton HM GX
Melbourne House, 11 Parliament Street, Hamilton HM 12
☎ (+1 441) 292 10 39 ⊖ (+1 441) 292 01 37
Vice Consul: Ms Maria B Machado
Secretary: Ms Maria de Fatima Carvalho [98]

31-01730 Swedish Consulate
PO Box HM 1554, Hamilton HM FX
Boyle Building, 31 Queen St, Hamilton, HM 11
☎ (+1 441) 292 8369 ⊖ (+1 441) 292 9740
Consul: W. Neville Conyers
Chancellor: Ms Judith A. Craig [97]

Contact Reference

31-01740 **Consulate of Switzerland**
3 Spice Wood Drive, Forest Hill, Warwick WK 08
☎ (+1 441) 236 29 85 ⊖ (+1 441) 236 29 85
Honorary Consul of Switzerland: Leo Betschart [98]

31-01900 **Consulate of the United States of America**
PO Box HM 325, Hamilton HM BX
Crown Hill, 16 Middle Road, Devonshire DV 03
☎ (+1 441) 295 13 42 ⊖ (+1 441) 295 15 92
Consul General: Robert A Farmer
Deputy Principal Political and Administration Officer: Bruce Berton
Customs Service: Robert Colbert ☎ (+1 441) 293 03 53
Consular Officer: Edmund Leather
NASA Representative: Steven Stompf ☎ (+1 441) 293 11 42
INS: Edward Moore ☎ (+1 441) 293 04 67
USDA: Tom Andre ☎ (+1 441) 293 27 52
Political/Economic Assistant: Ms Astrid C Black ☎ (+1 441) 295 13 42 [E] blackac@hamilwpoa.us-state.gov [98]

33 Foreign Forces

33-20000 **US Coast Guard**
Tudor Hill, Southampton [97]

Bhutan

Head of State:
H.M. King Druk Gyalpo Jigme Singye Wangchuck

10 Head of State

10-10000 **Office of H M the King**
Tashicho Dzong, Thimphu
King of Bhutan: H M King Druk Gyalpo Jigme Singye Wangchuck [98]

10-30000 **Royal Advisory Council**
Tashicho Dzong, Thimphu
Chairman: Kungang Tangbi [98]

12 Ministries

12-20000 **Ministry of Foreign Affairs**
Tashicho Dzong, Thimphu
Minister of Foreign Affairs: Lyonpo Dawa Tsering [98]

12-40000 **Ministry of Home Affairs**
Tashicho Dzong, Thimphu
Minister: Dago Tsering [98]

Bolivia

Head of State:
Gonzalo Sánchez de Lozada Bustamente

10 Head of State

10-10000 **Office of the President**
Palacio de Gobierno, Plaza Murillo, La Paz
☎ (+591 2) 37 40 30; 35 97 76 ▣ 3560031
⊖ (+591 2) 37 13 88
President of the Republic: Gonzalo Sánchez de Lozada Bustamente
Vice President: Victor Hugo Cárdenas Conde [98]

12 Ministries

12-10000 **Ministry of National Defence**
Avenida 20 de Octobre - Izquierdo, Plaza Avaroa, La Paz
☎ (+591 2) 37 13 65; 35 94 61
⊖ (+591 2) 35 31 56; 31 35 56
Minister of National Defence: Alfonso Kreidler [98]

12-20000 **Ministry of Foreign Affairs and Worship**
Cancillería de la Republica de Bolivia
Calle Ingavi esq. Junin No 1067, La Paz
☎ (+591 2) 37 11 50; 36 76 46 ▣ 3325, 2320 minrel bv
⊖ (+591 2) 37 11 55; 37 24 69
Minister of Foreign Affairs and Worship: Antonio Aranibar Quiroga [98]

25 Defense Forces

25-39100 **Bolivian Army Staff College**
"Mariscal Andres de Santa Cruz", PO Box 501, Cochabamba
☎ (+591 42) 557 89; 557 87 ⊖ (+591 42) 557 89
School Commander: Hugo Alvarez Jiménez
Deputy Commander: Justo Gareca G
Director of Operations: Manuel Cordozo P
Director of Procurement and Logistics: José Muñiz Vaca [98]

31 Embassies & Attachés

31-00150 **Embassy of Belgium**
Avenida H Siles, 5290-Obrajes, PO Box 2433, La Paz
☎ (+591 2) 78 58 26; 78 58 27; 78 49 25
✉ AMBABEL La Paz ▣ 3274 ambabel bv
⊖ (+591 2) 78 67 64
Ambassador: H E Mrs Christine Stevens
Head of Chancery: Marcel Hardy
Head of Co-operation Section: Joseph Engelen [98]

31-00890 **Embassy of Italy**
Avenida 6 de Agosto 2575, PO Box 626, La Paz
☎ (+591 2) 36 42 11; 36 11 29 ▣ 2654 italdip bv
⊖ (+591 2) 39 10 75
Ambassador: H.E. Dr Enric'Angiolo Ferroni Carli
First Secretary: Dr Domenico La Spina [97]

31-01680 **Embassy of Spain**
Avenida 6 de Agosto, 2827, PO Box 282, La Paz
☎ (+591 2) 43 01 18 ⊖ (+591 2) 43 27 52
Ambassador: H E Mr Manuel Viturro De La Torre
Deputy Chief of Mission and First Secretary in charge of Political Affairs: Luis Prados Covarrubias
First Secretary, Consular Affairs: Enrique Ojeda Vila
Defence Attaché: Col Manuel Montes de Undabetia
Commercial and Economic Counsellor: Javier Pareja Cerdó
Counsellor: Nicolas Vasquez de Castro
General Co-ordinator: Carlos Garcia-Tornel Florensa
Director of Latinamerican Centre: Juan Ignacio Pita Rodrigañez
Administrative Counsellor: Ms Amparo Lastagaray Guerrero
Commercial Analyst: Ramon Ynaraja [98]

31-01890 **Embassy of the United Kingdom**
Ave Arce 2732-2754, PO Box 694, La Paz
☎ (+591 2) 43 34 24 ⊖ (+591 2) 43 10 73
Ambassador: H E Mr D F C Ridgeway
Deputy Head of Mission and Second Secretary, Commercial: J Gardner
Commercial Officer: E Suarez
Third Secretary: S A Hunt [98]

31-01900 **Embassy of the United States of America**
Avenida Arce 2780, PO Box 425, La Paz
☎ (+591 2) 43 02 51 ▣ 3268 amemb bv
⊖ (+591 2) 43 39 00
Ambassador: H E Mr Curtis W Kamman ☎ (+591 2) 43 09 42
Deputy Chief of Mission: Robert Perry ☎ (+591 2) 43 09 42
Defense Attaché: Col Gregory H Landers (USAF)
Head of Military Group: Col Rodger Slaughter (US Army)
Economic and Commercial Attaché: Paul Larsen
Public Affairs Officer: Phillip T Parkerson ☎ (+591 2) 43 26 05 [98]

31-01950 **Embassy of Venezuela**
Avenida Arce 2678, Edificio Illimani, PO Box 960, La Paz
☎ (+591 2) 43 13 65; 43 20 23 ✉ EMBAVENEZ La Paz
▣ 2383 embvenz bv ⊖ (+591 2) 43 23 48
Ambassador: H E Mr Héctor M Lirer
Deputy Chief of Mission: Luis Navarro Fagúndez
Military, Naval and Air Attaché: Col Cruz Modesto Alfonso Perdon
Deputy Attaché Armed Forces: Lieut Col Jesus Machado B [98]

Bosnia and Herzegovina

Head of State:
Alija Izetbegovic

10 Head of State

10-10000 **Office of the President**
B Putnika 3, 71000 Sarajevo
☎ (+387 71) 61 73 25; 27 16 66
President Chairman: Alija Izetbegovic
Presidency Member: Momcilo Krajisnik
Presidency Member: Kresimir Zubak [98]

11 Prime Minister

11-10000 **Office of the Chairman of the Council of Ministers**
Vojvode Putnika 3, 71000 Sarajevo
☎ (+38 71) 21 37 77 ⊖ (+38 71) 27 28 77
Federation Prime Minister: Edhem Bicakcic
Federation Deputy Prime Minister: Drago Bilandzija
Co-Chairman of the Council of Ministers: Haris Silajdzic
Co-Chairman of the Council of Ministers: Boro Bisoc [98]

12 Ministries

12-10000 Ministry of National Defence
Vojvode Putnika 3A, 71000 Sarajevo
☎ (+38 71) 354 27 ⊖ (+38 71) 65 35 92
Federation Minister: Ante Jelavic
Federation Deputy Minister: Fikret Muslimovic [98]

12-20000 Ministry of Foreign Affairs
Vojvode Putnika 3, 71000 Sarajevo
☎ (+38 71) 66 38 13 ⊖ (+38 71) 65 35 92
Minister: Jadranko Prlic
Deputy Minister: Dragan Bozanic [98]

12-40000 Ministry of Internal Affairs (MUP)
Borise Kovacevica 7, 71000 Sarajevo
☎ (+38 71) 51 28 87 ⊖ (+38 71) 65 35 92
Federation Minister: Mehmed Zilic
Deputy Minister: Leutar Jozo [98]

31 Embassies & Attachés

31-01890 Embassy of the United Kingdom
8, Tina Ujevica, Sarajevo
☎ (+387 71) 66 39 22; 66 40 85; 47 13 24
⊖ (+387 71) 66 61 31
Ambassador: H E Mr Charles Crawford
Deputy Head of Mission: S J Potter
Commercial Counsellor: B A M Marshall
First Secretary: I D Marsh
First Secretary, Aid: D Houston
First Secretary, British Council Director: Ms S Barnes
Second Secretary, Management: M H Bowe
Second Secretary, KHF: Miss F O Conner
Third Secretary: K McGurgan [98]

Botswana

Head of State:
Sir Ketumile Masire

10 Head of State

10-10000 Office of the President
State House, Private Bag 001, Gaborone
☎ (+267) 35 08 00; 35 08 44; 35 08 50 ⊠ 2655 pula bd
⊖ (+267) 35 08 88; 35 08 58; 35 78 00
President of the Republic: Sir Ketumile Masire
Vice President and Minister of Finance and Development Planning: F G Mogae
Minister for Presidential Affairs and Public Administration: Pontashego Kedikilwe
Permanent Secretary to the President: E W M J Legwaila
Permanent Secretary (Political Affairs): M L Selepeng
Chief of Protocol: S C George
Senior Private Secretary to the President: B K Sentle
Administrative Secretary: T D Mogami [98]

12 Ministries

12-20000 Ministry of External Affairs
Private Bag 00368, Gaborone
☎ (+267) 360 21 00 ⊠ 2414 pula bd ⊖ (+267) 31 36 00
Minister External Affairs: Lieut Gen Mompati Merahfe (retd)
Deputy Minister External Trade: K Sebele
Minister Presidential Affairs and Public Administration: P H K Kedikilwe
Foreign Relations Director Acting: C Manyeula
Commander: Lieut Gen S K L Khama
Staff Officer: Capt M Kewakae [98]

25 Defense Forces

25-10000 Defence Force Headquarters
Sir Seretse Khama Barracks, Private Bag X06, Gaborone
☎ (+267) 35 11 21; 35 64 13 ⊠ 2400 bdfhq bd
⊖ (+267) 37 36 00
Commander Defence Forces: Lieut Gen S K I Khama
Deputy Commander: Maj Gen Moeng Pheto [97]

31 Embassies & Attachés

31-01730 Swedish Embassy
The Mall, Development House, Private Bag 0017, Gaborone
☎ (+267) 35 39 12 ✉ SVENSK Gaborone
[E] swembgab@global.co.za ⊠ 2421 swede bd
⊖ (+267) 35 39 42
Ambassador: H E Ms Christina Reulen [98]

31-01890 United Kingdom High Commission
Private Bag 0023, Gaborone
☎ (+267) 35 28 41 ⊖ (+267) 35 61 05
High Commissioner: H E Mr D C B Beaumont
Deputy High Commissioner: A R M Dent
Defence Attaché: Col A Reed Screen (Resident in Harare)
Third Secretary Development and Information: S Cartwright [98]

31-01900 Embassy of the United States of America
PO Box 90, Gaborone
☎ (+267) 35 39 82 ⊖ (+267) 35 69 47
Ambassador: H E Mr Robert Krueger
Deputy Chief of Mission: Ms Gilian Milovanovic [98]

Brazil

Head of State:
Fernando Henrique Cardoso

10 Head of State

10-10000 Office of the President
Palácio do Planalto, Praça dos Três Poderes 3° andar, 70150-900 Brasilia, DF
President of the Republic: H E Mr Fernando Henrique Cardoso
Vice President: Marco Maciel
Secretary General: H E Mr Eduardo Jorge Caldas Pereira
Head Press Department: Ms Ann Tavares

Military Household
Chief, Military Household of the President: Gen Alberto Cardoso

Civilian Household
Chief, Civil Household of the Presidency: Clovis de Barros Carvalho [98]

10-20000 Office of the Vice President
Edificio Banco do Brazil, Sede 3, 18 Andar, 70083 Brasilia DF
☎ (+55 61) 224 46 29 ⊠ 611040 vice br
Vice President: Marco Maciel [98]

12 Ministries

12-12100 Grupo de Transporte Especial (GTE)
(Ministry of Aeronautics)
Base Aérea de Brasilia, 71600-900 Brasilia, DF
☎ (+55 61) 365 10 12 ⊠ 611267 maer br
⊖ (+55 61) 365 15 21
Commander of Group de Transport Especial: Lieut Col Louis Jackson Josuá Costa
■ Official transportation of Government VIPs. [97]

12-12500 Ministry of Aeronautics
Ministirio da Aeronautica
Bloco M, 1° Andar, Esplanada dos Ministerios, 70045-900 Brasilia, DF
☎ (+55 61) 223 10 99; 313 21 20
⊖ (+55 61) 224 22 10
Chief of Air Staff: Lieut Brig João Felippe Sampaio Lacerda ☎ (+55 61) 223 64 09
Deputy Chief of Air Staff: Maj Brig Nelson de Souza Taveira ☎ (+55 61) 223 74 09
Deputy Chief of Logistics Department: Maj Brig Astor Nina de Carvalho Neto ☎ (+55 61) 223 23 59
Deputy Chief of Planning and Doctrine: Maj Brig Frederico de Queiroz Veiga ☎ (+55 61) 223 54 09
Deputy Chief of Legislation: Brig Wellington Godeiro Fernandes ☎ (+55 61) 313 22 40
Deputy Chief Operational Planning: Brig Sérgio Candiota da Silva ☎ (+55 61) 313 24 92
Commander of Air Defense and Commander III FAe (Brasilia): Maj Brig Manoel Carlos Pereira ☎ (+55 61) 365 14 35; 384 65 77
Commander II FAe (Joint Naval Operations, Rio de Janeiro): Brig Alvaro Moreira Pequeno ☎ (+55 21) 462 08 17
Commander V FAe (Air Transport Command, Rio de Janeiro): Brig Nelson Teixeira Pinto ☎ (+55 21) 390 14 07
Chief of Public Relations Division: Leiut Col Jader da Silva Garcia [98]

12-13000 Ministry of the Army
Ministerio do Exercito
70630-901 Brasilia DF
Setor Militar Urbano, Bloc A 4° Andar
Quartel General do Exército
☎ (+55 61) 415 52 00; 415 52 20 ⊠ 611145
⊖ (+55 61) 415 54 89
Minister of State: Gen Zenildo Gonzaga Zoroastro de Lucena ☎ (+55 61) 315 52 00 [98]

12-14000 Ministry of the Navy
Bloco N, 2° Andar, Esplanada dos Ministerios, 70055 Brasilia, DF
☎ (+55 61) 225 14 05 ⊠ 611133/166/392 mmar br
Minister of Navy: Adm Mauro Cesar Rodrigues Pereira [98]

12-20000 Ministry of Foreign Affairs
Ministerio Relagses Exteriores
Esplanada dos Ministerios - Bloco H, Palacio do Itamaraty - 20 Andar, 70170-900 Brasilia, DF
☎ (+55 61) 211 62 40; 211 62 42
⊠ 611311 mnre br; 611319 mnre br
⊖ (+55 61) 223 23 92
Minister of Foreign Affairs: Luiz Felipe Lampreia
Deputy Minister and Secretary-General: Ambassador Sebastão do Rego Barros
Head of Trade Promotion Department: Ambassador Celso Marcos Vieira de Souza
Chief Information Officer: Verra Barouin Machado [98]

12-40000 Ministry of Justice
Esplanada dos Ministerios, Bloco T, 4° andar, 70064-900 Brasilia, DF
☎ (+55 61) 218 30 00 ⊠ 108812313
⊖ (+55 61) 224 09 54
Minister of Justice: Hon Nelson Jobim
Secretary General: Milton Seligman
Director General Federal Police: Vicente Chelotti [98]

Contact Reference

15 Parliament

15-10000 National Constituent Assembly
Praça dos Tres Poderes, 70160 Brasilia
Federal Senate
Praça dos Tres Poderes, Edificio Principal,
70160 Brasilia DF
Federal Chamber of Deputies
Praça dos Tres Poderes, Edificio Principal,
70160 Brasilia DF [98]

25 Defense Forces

25-29100 Air Force Academy
Academia da Força Aerea
13630-000 Pirassununga, SP
☎ (+55 19) 561 31 42 ⊜ (+55 19) 561 93 50
Commandant: Brigadier José Carlos Pereira
☎ (+55 19) 561 34 55
Operations Director: Francisco José Da Silva Lobo
☎ (+55 19) 561 22 33
Chief of Staff: Colonel Antonio J L Velardi
☎ (+55 19) 561 31 42 [98]

25-45000 Marine Corps Headquarters
Praça Barão de Ladário, s/no-Ilha das Cobras,
20091 Rio de Janeiro-RJ
☎ (+55 21) 216 66 27 ⊜ (+55 21) 216 66 28
Commandant of Marine Corps: Adm Valdir Bastos Ponte
Deputy Commander and Chief of Staff:
Capt Luiz Roberto De Moraes Passos
Procurement and Logistics Director:
Capt Celso Antonio Junqueira Resende
Operations Director: Capt Alvaro Augusto Dias Monteiro
Assistant Officer: Antonio H M De Souza [98]

31 Embassies & Attachés

31-00100 Embassy of Austria
Avda das Nações, Lote 40, Quadra 811,
70426-900 Brasilia, DF
☎ (+55 61) 443 31 11; 443 33 73; 443 34 21
⊠ AUSTROAMB Brasilia ▣ 611 202 oebr br
⊜ (+55 61) 443 52 33
Ambassador: H E Dr Manfred Ortner [98]

31-00380 Embassy of Colombia
SES Avenida das Naçoes, Lote 10, PO Box 07 04 43,
70444 Brasilia, DF
☎ (+55 61) 226 89 02; 226 89 97
⊜ (+55 61) 224 47 32
Ambassador: H E Mr Mario Galofre
Counsellor: Juan Bernal Roa
Second Secretary: Ms Maria del Pilar Adames
Commercial Counsellor: Alvaro Rey Rey [98]

31-00450 Embassy of the Czech Republic
Avenida das Nações, Lote 21, 70414 Brasilia DF
☎ (+55 61) 242 77 85; 242 79 05
⊜ (+55 61) 242 78 33
Chargé d'Affaires a.i.: Dr Josef Adamec
First Secretary: Ondrej Kasina [98]

31-00460 Royal Danish Embassy
PO Box 0484, 70359-970 Brasilia, DF
SES - Avenida das Nações, Lote 26,
70416-900 Brasília, DF
☎ (+55 61) 242 81 88 ⊠ AMBADANE Brasilia
▣ 611494 ambd br ⊜ (+55 61) 244 52 45
Ambassador: H E Mr Johannes Dahl-Hansen
First Secretary: Lars Steen Nielsen [98]

31-00500 Embassy of Ecuador
SHIS QI 11, Cj. 9, Casa 24, 71600 Brasilia DF
☎ (+55 61) 248 55 60; 248 56 60 ▣ 611290 eeqd br
⊜ (+55 61) 248 12 90
Ambassador: H.E. Sr Cesar Valdivieso Chiriboga
Counsellor: Rafael Paredes
Military and Air Attaché: Col Alfredo Fiallos
☎ (+55 61) 248 53 60
Naval Attaché: Capt. Gonzalo Toledo
☎ (+55 61) 248 42 19
Secretary: Alvaro Garcia
Secretary: Tomas Larrea [97]

31-00510 Embassy of the Arab Republic of Egypt
Avda das Nações, SEN, Lote 12, 70435 Brasilia, DF
☎ (+55 61) 225 85 17/83 42 ▣ 611 387 eegi br
⊜ (+55 61) 223 58 12
Ambassador: H.E. Dr Mohamed Ahmed Abdel Wahab [97]

31-00580 Embassy of Finland
Avda das Nações, Lote 27, 70417 Brasilia, DF
☎ (+55 61) 242 85 55 [E] suomi@tba.com.br
▣ 611155 efin br ⊜ (+55 61) 244 96 20
Ambassador: H E Mr Pertti Harvola
Deputy Chief of Mission and First Secretary:
Erik af Hällström [98]

31-00590 Embassy of France
Avenida des Naçoes, Lote 4, PO Box 07-665,
79359-970 Brasilia DF
☎ (+55 61) 312 91 00 [E] france@brnet.com.br
▣ 611491; 611078; 611755 efra br
⊜ (+55 61) 312 91 08
Ambassador: H E Mr Phillipe Lecourtier
Deputy Chief of Mission: Patrick Nicoloso
Defense and Air Attaché: Col Yves Rocquancourt
☎ (+55 61) 225 23 22
Scientific Attaché: Roger Frety
Public Information Officer: C Geschwind
Commercial Office
PO Box 07-0499, 70000 Brasilia DF
☎ (+55 61) 225 07 22 ⊠ COMATTA Brasilia
▣ 1720 efra br ⊜ (+55 61) 225 03 28
Commercial Counsellor: Philippe Delleur [98]

31-00630 Embassy of the Gabonese Republic
SHIS, QI 09, Cj 11, Casa 24, 61615-300 Brasilia, DF
☎ (+55 61) 248 35 36; 248 35 33 ▣ 611 710 emga br
⊜ (+55 61) 248 22 41
Ambassador: H E Mr Marcel Odongui-Bonnard [98]

31-00770 Embassy of Guyana
SBN Qd. 02, B1. 'J', 13 Andar, Salas 1310/15,
70438-900 Brasilia, DF
SBN Qd. 02, Bl. 'J', Edificio Eng. Paulo Mauricio
☎ (+55 61) 224 92 29; 224 97 25 ▣ 611268 egui br
⊜ (+55 61) 226 30 22
Ambassador: H E Mr Ivan B Evelyn
First Secretary: Mrs Audrey Jardine-Waddel [98]

31-00810 Embassy of the Republic of Hungary
SES 805, Avda das Nações, Quadra 805, Lote 19,
70413-900 Brasilia, DF
☎ (+55 61) 243 08 22/36/54 ▣ 611285 erph br
⊜ (+55 61) 244 34 26
Ambassador: H.E. Mr Gábor Tóth
Commercial Office
SHIS-QI 23, CONJ. 02, CASA 07, 71660 Brasilia, DF
☎ (+55 61) 366 33 63; 366 28 60 ▣ 611 253 erph br
⊜ (+55 61) 366 39 27; 366 288 80
Commercial Counsellor: Sándor Bago [97]

31-01270 Royal Netherlands Embassy
Via L-3 Sul, Quadra 801, Lote 05, PO Box 0098,
70359 Brasilia DF
☎ (+55 61) 321 47 69 ▣ 611492 nedb br
⊜ (+55 61) 321 15 18
Counsellor: A C Brouwer
Economic Section: H P F A Sleegers [98]

31-01480 Embassy of Romania
Avda das Nações, Lote 6, 70456 Brasilia DF
☎ (+55 61) 226 07 46 ▣ 611283 lrom br
⊜ (+55 61) 226 66 29
Ambassador: H E Mr Ion Bar
Deputy Chief of Mission: Constantin Rusei
☎ (+55 61) 226 24 81
Counsellor, Economic Affairs: Ion Nito
First Secretary, Cultural Affairs: Romulus Roman
Second Secretary, Consular Affairs: Aurel Plaiesu [98]

31-01710 Embassy of the Republic of Suriname
SHIS, QI 7-Cj 1, Casa 6, 70457-900 Brasilia DF
☎ (+55 61) 248 64 06; 248 35 95 ▣ 611414 embs br
⊜ (+55 61) 248 37 91
Ambassador: H E Mr Robert L Christopher
Counsellor: Roy Wong Lun Hing
Military Assistant: SMA Marto [98]

31-01820 Embassy of Trinidad & Tobago
SHIS QI 8, Conj. 4, Casa 5, 71620-245 Brasilia DF
☎ (+55 61) 248 19 22; 248 12 68; 248 12 86
⊜ (+55 61) 248 15 33
Ambassador: H E Mr Robert Torry [98]

31-01890 Embassy of the United Kingdom
Setor de Embaixadas Sul, Quadra 801,
Caixa Postal 07-0586, 70359 Brasilia, DF
☎ (+55 61) 225 27 10 ⊠ PRODROME BRASILIA
▣ 611360 eing br ⊜ (+55 61) 225 17 77
Ambassador: H.E. Mr P.W. Heap (CMG)
Deputy Head of Mission: P.R. Jenkins
First Secretary Commercial: P.B. Yaghmourian
First Secretary Political: I.N. Anthony
Defence, Military & Air Attaché: Col D.M. Black
Naval Attaché: Capt. R.J. Tempest (RN) [97]

31-01900 Embassy of the United States of America
SES; Avenida das Nações, Lote 3, 70403-900 Brasilia, DF
☎ (+55 61) 321 72 72 ⊜ (+55 61) 321 28 33
Ambassador: H E Mr Melvyn Levitsky
Deputy Chief of Mission: Ms Lacy Wright
Head of Commercial Section: Miguel Pardo de Zela
(Resident in São Paulo)
Scientific Attaché: Xenia Wilkinson
Public Affairs Officer: John Dwyer
Defense Attaché: Col Layton Dunbar (USA)
Military Liaison Office
Brasilia
☎ (+55 61) 364 15 71 ⊜ (+55 61) 364 10 71
Military Liaison Officer: Col James Bjork (USAF)
Plans and Programs Co-ordinator: Michael Marshall [98]

55 Government-owned Industry & Production Facilities

55-05000 Embraer
Empresa Brasileira de Aeronáutica SA
Avenida Brig Faria Lima 2170, PO Box 343,
12227-901 São José dos Campos (SP)
☎ (+55 12) 345 10 00
⊜ (+55 12) 341 18 53 (public relations);
345 10 90 (sales)
♧ US$ 1,600,000 ♣ US$ 355,000,000 (1994) 👥 3,976
🏠 256,093,87 m2
Chairman: Gilberto Gabriel Garby
President and Chief Executive Officer:
Maurício Novis Botelho ☎ (+55 12) 345 12 15
President Embraer Aircraft Corp. (USA): Samuel D Hill
☎ (+1 954) 359 37 00; 359 37 01 (fax)
Executive Vice President and Chief Executive Officer:
António L P Manso ☎ (+55 12) 345 12 16; 322 60 70 (fax)
Executive Vice President Commercial: Walter L Leusin
☎ (+55 12) 345 12 15; 341 67 02 (fax)
Executive Vice President Planning, Organisational Development and Quality: Federico F Curado
☎ (+55 12) 345 13 78; 321 53 39
Executive Vice President Industrial: Juarez S B Wanderley
☎ (+55 12) 345 12 00; 321 28 18 (fax)
Senior Vice President Special Programs: Walter Bartels
General Counsel: Marcelo M C Prais
☎ (+55 12) 345 12 14; 321 19 33 (fax)
EMB-145 Programme Director: Satoshi Yokota
☎ (+55 12) 345 10 59
Engineering Director: Luis Carlos Affonso
☎ (+55 12) 345 12 27
Manufacturing Director: Artur Aparecido Valério Coutinho
☎ (+55 12) 345 12 00; 321 28 18 (fax)
Public Relations Director: Cláudio A J Moreira
☎ (+55 12) 345 10 59
Senior Commercial Sales Manager: Ricardo Pesce
☎ (+55 12) 345 17 11

Senior Government Sales Manager:
Eduardo Muñoz de Campos ☎(+55 12) 345 14 00
Marketing Division: Ms Da Lila Andrade
■ EMB-111 Bandeirante twin turboprop maritime patrol aircraft. EMB-312 Tucano twin turboprop trainer (T-27 military desig.). EMB-120 Brasilia pressurized 30-seat twin turboprop. EMB-110P1A Bandeirante. EMB-110P2A Bandeirante twin turboprop general purpose transport. AMX tactical jet fighter/ground attack aircraft jointly developed with Aeritalia/Aermacchi. EMB 145 50-seat turbofan-powered regional transport aircraft. Principal sub-contractor Super Tucano primary aircraft trainer with Northrop Grumman. → 75

EDE Equipment Div.
Rua Itabaiana, 40 Parque Industrial,
12237-540 São José dos Campos, SP
☎(+55 12) 345 10 00 ⊜ (+55 12) 331 13 00
♣ US$ 7 000 000 ♚ 158
General Mgr: Luiz Cláudio Sigaud Ferraz
☎(+55 12) 345 25 10
■ Landing gears, brakes, wheels, valves, actuators, shock absorbers, anti skid systems, pylons, eject release units. 98

74 Consortia, Joint Venture Companies

74-00100 AMX International Ltd
c/o Embraer, PO Box 343,
122277-901 São José dos Campos (SP)
Av. Brigadeiro Faria Lima, 2170,
12225 São José dos Campos, SP
☎(+55 12) 345 13 11 ⊜ (+55 12) 341 18 54
Public Relations Dir: Cláudio A. Moreira
■ Marketing of AMX ground attack aircraft. A joint programme with Embraer (Brazil), Alenia/Aeritalia & Aermacchi (Italy). 97

74-00905 Artefatos Eléctricos e Mecânicos de Aeronaútica Ltda (AEMA)
Rua Paraibuna 838, 12245 São José dos Campos, SP
☎(+55 12) 322 70 77 ⊜ (+55 12) 321 58 47
♧ US$ 100,000 ♣ US$ 16,000,000 ♚ 220 ⌂ 6,500 m2
Managing Director: Durval Gonçalves
Operations and Technical Director:
Dorival Ferreira Gonçalves
Sales and Marketing Manager: G Schlickmann
■ Aerospace products: latches, clamps, hose clamps, seat belt buckles, static dischargers, ground supply DC sockets and plugs, terminal block assemblies, recirculating toilets. Defence products: cluster bombs, parachute retarded bombs, 70mm rockets and launchers, fuzes and practice bombs. 97

75 Manufacturing Companies

Embraer 55-05000
Hispano-Suiza do Brazil Equipamentos 90-08100
Rockwell Do Brasil Ltda 90-18809

75-00585 Aeroelectrónica Industria Componentes Aviónicos SA
(a subsidiary of Aeromot Aeronaves e Motores SA)
Avenida Sertório, 4400, PO Box 8031,
91040-620 Porto Alegre, RS
☎(+55 51) 361 12 22 ⊠ 515026 aemt br
⊜ (+55 51) 361 27 73
♧ US$ 140,000 ♣ US$ 8,000,000 (1995) ♚ 120
⌂ 1,500 m2
President: Cláudio B Viana
Managing Director: Luis Salazar
Commercial Director: Vitor Neves
Operations and Technical Director: Joao C Jotz
Purchasing Manager: Enió Baumgardt
■ Design, manufacture and certification of digital and analogue avionics for civil and military aircraft including reconnaissance systems, communication systems, inertial navigation systems, command and control systems. 97

75-00630 Aeromot Group SA
(† Aeromot Aeronaves e Motores SA)
PO Box 8031, 90201-970 Porto Alegre, RS
Aeroporto Internacional Salgado Filho
☎(+55 51) 371 16 44 ⊠ 515026 aemt br
⊜ (+55 51) 371 16 55
♣ US$ 18 000 000 (1995) ♚ 120
President: Cláudio B. Viana
■ Aircraft sales & maintenance, overhaul & major repairs. 97

75-00635 Aeromot Indústria Mecânico-Metalürgica SA
(a subsidiary of Aermot Aeronaves e Motores SA)
Avenida Das Industrias 1210, PO Box 8031,
90200-290 Porto Alegre, RS
☎(+55 51) 337 13 44; 337 29 77 ✉AEROMOT
⊠ 515026 aemt br ⊜ (+55 51) 343 36 55; 337 12 87
♣ US$ 7,000,000 ♚ 130 ⌂ 3,100 m2
President: Claudio B Viana
Commercial Director: Vitor J P Neves
■ Aircraft and helicopter seats, motorgliders, towable targets and drones. 98

75-00650 Agro Química Maringá SA
(a subs. of Oxmar Oxford Maringá Industrias Quimicas SA)
Rua Alvares Cabral 1210, 09980-160 Diadema, São Paulo
☎(+55 11) 456 16 44 ⊠ 44440 aqms br
⊜ (+55 11) 456 63 55; 749 12 20
♧ US 25,000 (1995) ♣ US$ 31,000 (1995) ♚ 180
⌂ 40,000 sq m
Managing Director: Umberto Silvio Mosseri
Executive Director: Jacques Mosseri
Operations and Technical Director: Eduardo Nakadaira
Sales and Marketing Director: Selim Ezra Setton
Purchasing Manager: Hélio Sergio Carelli
■ Manufacturer of chemical products. 98

75-00910 Teadit Juntas SA
(† ASA Vedações Industriais SA)
Avenida Mercedes Benz 390, PO Box 819,
13055-720 Campinas, SP
☎(+55 19) 225 55 66; 225 56 77
E teadit@dglnet.com.br ⊜ (+55 19) 225 56 14
♣ US$ 8,000,000 ♚ 110 (1997) ⌂ 10,000 m2
Managing Director: José Carlos Carvalho Veiga
Sales Manager: Gilberto Silva
■ Design and production of exhaust manifolds, metallic bellows, expansion joints and gaskets for aircraft and military vehicles. 98

75-00968 Avibras Aeroespacial SA
Avenida Brigadeiro Faria Lima, No 3.305,
Parque Martin Cerere,
12225-000 São José dos Campos, SP
☎(+55 12) 351 66 44
⊠ 1233844 aiae br; 1233845 aiae br
⊜ (+55 12) 351 62 77
♧ US$ 150,000,000 (1995) ♣ US$ 18,000,000 (1996)
♚ 400
President and General Manager: Joao Verdi Leite
Vice President, Official Relations Mgr:
Rubens Domingues Porto
Purchasing Manager: Bartolomé Romero Comas
Media Co-ordinator: Ms Denise Pragana Videira
■ Ammunition. Research, meteorological sounding rockets. Wheeled armoured vehicles, personnel carriers, command post, radio vehicles. Utility, Gendarmerie vehicles.
Rocket research and sounding systems and equipment. Ground-based weapon fire control systems. Data processing systems. Missiles and missile systems. Missile warheads. Rocket systems. Weapon handling systems. Fusing systems, explosives, explosive devices. Propellants, propellant charges.
● Rodovia dos Tamoios, Km 14, PO Box 278,
12300 Jacarei, SP
☎(+55 123) 51 66 44 ⊠ 1233844; 1233845 aiae br
⊜ (+55 123) 51 62 77
■ Rockets for scientific and military applications, unguided rockets and launchers, armament for helicopters, missiles, radars, complete defense systems, engineering expertise, tailor-made systems, inertial navigational systems, microelectronics, fibre optics, satellite communications antennas. 98

75-00975 Aviquei Aerospace
(a division of Aviquei Produtos Hidraulicos E Pneumaticos Ltda, Brazil)
Rua Brigadeiro Galvão, 130, 01151-001 São Paulo, SP
☎(+55 11) 826 85 74 E aviquei@amcham.com.br
W http://www.amcham.com.br/homepages/837 .html
⊜ (+55 11) 66 54 63
♧ US$ 450,000 (1997) ♣ US$ 4,000,000 (1997)
♚ 50 (1997) ⌂ 2,500 sq ft
President, Director: J Quieroz Ferreira
Import and Export Manager: Dr Eduardo Queiro
■ Flexible tubes and joints for aircraft use. 98

75-02120 Bernardini SA Indústria e Comércio
Ordnance Division
Rua Hipólito Soares 79, 04201 São Paulo, SP
☎(+55 11) 273 33 23; 274 80 33
⊜ (+55 11) 274 85 67
♣ US$ 33,000,000 ♚ 800 ⌂ 30,000 m2
President: Richard B Cury
Vice President: Flavio M Bernardini
■ Armoured vehicles and systems. Armoured vehicle bridge-launching unit. Military jeeps. Updating of tanks and military trucks. Parts, casings, nozzles, nutation dampers and magnetic coils for satellite attitude control systems. 98

75-02220 José Murillia Bozza SA Com. e Ind.
R. Tiradentes 931, PO Box 433,
09700 São B. do Campo SP
☎(+55 11) 756 99 66 E bozza@regard.com.br
⊠ 4259 jmbs br ⊜ (+55 11) 756 14 99
♚ 140 (1997) ⌂ 7,150 m2 (1997)
Managing Director: Ms Elizabeth B Bozza
Sales and Marketing Director: Gerson Oritz
Purchasing Manager: Hércules G Negrão
■ Lubricating equipment for military vehicles. 98

75-02290 Brassinter SA Indústria e Comércio
Avenida des Nações Unidos 21344,
04795-911 São Paulo, SP
☎(+55 11) 548 37 11 ⊜ (+55 11) 246 24 90
♧ US$ 20 000 000 (1995) ♣ US$ 50 000 000 (1996)
♚ 500 ⌂ 20 000 sq m
Managing Dir: José Roberto Torres
Technical Dir: Valter Unterberger
Marketing Dir: Ary Frederico Torres
Purchasing Mgr: Wladimir Beneveli
■ General use powder metallurgy products: Heavy metal components, rods, balls, Tungsten carbide - cylinders, balls, or any shape. Discs or pads for heavy duty brakes or clutches, including aircraft brakes. Bronze & iron/steel sintered parts, including self lubricated bearings. 97

75-02700 IBQ Industrias Químicas Ltda
Head Office
Rodovia Régis Bittencourt Km 71,
83420 Quatro Barras, Paraná
☎(+55 41) 772 12 11 E ibq@embratel.com.br
⊠ 415513 ⊜ (+55 41) 772 20 31
♧ US$ 2,000,000 (1995) ♚ 670 ⌂ 22,000 sq m
Technical Director: Gonzalo Orlando Claure Mendez
Commercial Director: Allyrio de Jesus Dipp Filho
Financial Director: Marco Antonio Cassou
Advertising Manager: Marco Antonio Salcedo
■ Explosives manufacturing. 98

75-03030 Carbrasmar Ind. Com Ltda
Rua do Alho 145/700, 21012-350 Rio de Janeiro, RJ
☎(+55 21) 280 44 43
♚ 200 ⌂ 36 000 sq m
■ Patrol boats. 97

75-03150 Companhia Brasileira de Cartuchos (CBC)
Avenida Humberto de Campos, 3220, PO Box 182,
09400-000 Ribeirão Pires, SP
☎(+55 11) 742 75 00
⊜ (+55 11) 459 19 33; 73 85; 32 96
♧ US$ 31 423 433 ♣ US$ 56 055 500 (1995) ♚ 1 037
⌂ 1 850 000 sq m
President: Antonio Marcos Moraes Barros
Dir Operations: Marco Aurelio Salvany
Dir Exports: Gilberto P. Salm

Contact Reference

- Manufacture ammunition for light & medium calibre weapons, both bolt action rifles & pump action shotguns. [97]

75-03230 Companhia Comércio e Construções (CCC)

Rua Eng. Antonio Penido 124, 12700 Cruzeiro SP
☎ (+55 125) 44 17 32; 44 37 92; 44 36 04
⊜ (+55 125) 44 13 29
350 540
Superintendent: Carlos Alberto Bühler
- Special railroad trucks for tank transport, foldable ramps for tanks. [98]

75-03310 Cia Industrial e Mercantil de Artefatos de Ferro (CIMAF)

Avenida Marechal Rondon, 915 Centro,
06093-900 OSASCO, São Paulo SP
☎ (+55 11) 704 88 30 ⊜ (+55 11) 701 70 01
US$ 44,720,000 (1996) ♣ US$ 75,349,000 (1996)
624 (1996) 64,813 m2
President: José Pedote
Administrative and Financial Director: Carlos Camargo
Marketing, Import and Export Manager: Valberto Chrisofani
E christof@amcham.com.br
- Aircraft components. Bolts, Steelwire rope. Fishing ropes. Anchor lines for offshore platforms. [98]

75-05250 Elastic Indústria de Artefatos de Borracha

R. Fabio Eduardo Ramos Esquivel 840,
09900 Diadema, SP
☎ (+55 11) 445 10 33 ⊜ (+55 11) 456 85 85
Cru 8 210 000 000 ♣ US$ 12 000 000 285
President: Percy Putz
Managing Dir: João Alfredo Soliani
Sales & Marketing Dir: Valmir Aparecido Torres
Operations/Technical Dir: Jorge L. Alves
Purchasing Mgr: Alberto Erbert [97]

75-05350 Villares Metals SA

(† Electrometal SA Metals Especials)
(a subsidiary of Agos Villares SA)
Avenida Eletrometal 1100, PO Box 146,
13177-900 Sumari SP
☎ (+55 192) 64 80 00; 64 82 08 191963 elsm br
⊜ (+55 192) 64 10 84; 64 83 90;
(+55 11) 211 19 71 (export department)
US$ 60,000,000 ♣ US$ 97,000,000 (1994) 1,400
720,000 sq ft
President: José Francisco Martins de Viveiros
Managing Director: Paulo Villares Musetti
General Export Manager: Arturo Maceira
Technical Assistant Export Department: Eric Muhne
- Semi-finished barrels, breeches & other gun components, armour plate for combat cars (including bi-metallics). Specialty alloys for use in satellite launchers, missiles & rockets. Ultra high strength steels & specialty alloys for the aerospace industry. [97]

75-06240 FEVA - Maquinas Ferdinand Vaders SA

Rodovia Raposo Tavares 27580, 06700 Cotia, São Paulo
☎ (+55 11) 792 91 33 ⊜ (+55 11) 492 27 72
Cru$ 1 453 520 (1995) ♣ US$ 11 411 107 (1995)
170 10 000 m2
Managing Dir: Monica Mora Vaders
Technical Dir: Richard Vaders
Marketing Dir: Odair Cardoso
Purchasing Mgr: Ricardo Lino Cardoso
- Vehicle winches. [97]

75-06640 Forjas Taurus SA

(† Taurus SA)
Avenue do Forte 511, PO Box 44, 91360 Porto Alegre, RS
☎ (+55 513) 40 22 44 FORJAS 513690 plta br
⊜ (+55 513) 40 49 81
President: Carlos Alberto P Murgel
Export Manager: Nelson Maltchik
- Riot shields, protective helmets, body armour jackets - lightweight bullet-proof.

Subsidiary company:
Taurus Blindagens Ltda → 75
- Rau Kari, 234, Guarulhos, SP
☎ (+55 11) 209 91 66 ⊜ (+55 512) 40 49 81
General Manager: Rogerio Rigon [98]

75-06650 Förster Imaden

Rua dos Feltrins 321, PO Box 532,
São Bernardo do Campo, SP
☎ (+55 11) 419 98 33 1144191
⊜ (+55 11) 419 94 43
65
Managing Director: Carlos Buch Serrano
Sales and Marketing Director: Alejandro Spoerer
Operations Director: Roberto Prado
Purchasing Manager: Carlos Pastoriza
- Production of the Förster-Imaden DM-1000 used in detection of mines, bombs, ammunition and war material containing metallic components.

® Germany: Institut Dr Förster GmbH & Co KG [98]

75-06840 Industria de Fundicao Tupy Ltda

(a subs. of Tupy SA)
Rua Albano Schmidt 3400, PO Box D 301 and 303,
89200-900 Joinville SC
☎ (+55 474) 441 81 81 ⊜ (+55 474) 441 81 41
♣ US$ 314,061,040 (1996) 4,100 (1997)
172,000 m2
Chief Executive Officer: Mário Egerland
Industrial Director: Walmor Krause
Marketing Director: Antonio C Galvão
Finance Director: Alceu D Calciolari
Sales Director: Pedro Duran
Sales Director: Luis C Guedes
Purchasing Director: Almir J de Aguiar
Process and Product Development Director: Wilson L Guesser
Commercial Analyst: Ricardo Ferro Borini
- Cylinder blocks, cylinder heads, bearing caps, crankshafts, brake discs and drums, supports, calipers, camshafts, wheel hubs, steering systems, housings, exhaust manifolds, intake manifolds, connecting rods, knuckles. Rough and machined parts. Pipe fittings, metal abrasives and continuous castings. [98]

75-07350 Gespi Industria e Com. de Eqt Aeronauticos Ltda

Rua Guaratingueta No. 55,
12231-150 S José dos Campos, SP
☎ (+55 12) 341 16 80 ⊜ (+55 12) 341 16 80
US$ 150 000 (1996) 80 1 400 sq m
President: Engo João Batista Mesquita Scarparo
Sales Dir: Carlos A. Picolini
Technical Dir: Sebastião Barbosa [97]

75-07600 Goodyear do Brasil Ltda Aviation Products Division

Sales Office
Rua da Intendencia 91, 030015-010 São Paulo SP
☎ (+55 11) 948 72 74; 608 72 74 GOODYEAR
63335 ⊜ (+55 11) 948 74 44; 608 74 44
6,000
General Manager: Vanderlei Nazareth
Sales and Customer Service Manager: Celso P Camargo
- Retreading of aircraft tyres and inner tube repair. [97]

75-08250 Helicópteros do Brasil S/A

(a subsidiary of Eurocopter, France)
Rua Santos Dumont 200, Dist Ind, PO Box 184,
37500-000 Itajuba, Minas Gerais
☎ (+55 35) 623 20 00 ⊜ (+55 35) 623 20 01
200 (1996) 11,000 m2
President: José Albuquerque
Managing Director: Michel Meylan ☎ (+55 11) 251 17 22
Sales and Marketing Director: Patrick de La Revelière
☎ (+55 11) 251 17 22
Operations and Technical Director: João Bosco
☎ (+55 11) 03 56 23 20 05
Marketing Manager: Luís H Testa ☎ (+55 11) 251 17 22
- Assembly and overhaul of aircraft produced under licence to Eurocopter: Esquilo AS350 BA, AS 350 B2, single engine and civil versions. Esquilo AS355 F2 and AS355 N twin engine and civil versions. Fennec AS555 U2 and AS555 A2 twin engine military version. Dauphin AS365 N2 twin engine, civil version. Panther AS565 (Dauphin military version). BK117 C-1, twin-engine civil aircraft.

Main customers: Brazilian army, navy and air force. State governments, offshore transport and medical organisations. [98]

75-08270 Heliodinâmica SA

Rod. Raposo Tavares km 41, PO Box 111,
06730-970 Vargem Grande Paulista, SP
☎ (+55 11) 79 60 35 11 (800) 14 37 55
⊜ (+55 11) 79 60 37 55
US$ 27,000,000 (1994) ♣ US$ 1,600,000 (1994)
77 (1994) 2,902 m2
President: Bruno Topel
Sales and Marketing Director: Daniel Topel
- Solar cells, silicium cells for the electronic industry, complete photovoltaic systems. [98]

75-08290 Hellermann do Brasil Industria e Comercio Ltda

Vila Anastácio, 05120 São Paulo SP
Avenida Marginal Direita do Tiete 800
☎ (+55 11) 831 81 11 1183096
⊜ (+55 11) 260 80 50
190
Managing Director: M da Costa
- Manufacture and marketing of cable care and heatshrinkable insulation and protection products for the Latin American military and aerospace markets. [98]

75-08600 Hobeco Sociedade Consignatária Ltda

Rua Sacadura Cabral 236, 20221 161 Rio de Janeiro, RJ
☎ (+55 21) 233 91 77 23292 hobc br
⊜ (+55 21) 263 90 67
US$ 360,000 (1997) ♣ US$ 3,500,000 (1997)
34 (1997) 1,100 sq m
President: Manuel Carlos Mendes Pereira
Technical Director: Gilson L Feitosa
Marketing Director: Diego Ribeiro Gomes
Purchasing Manager: Niza Carmo
- Meteorological stations. [98]

75-09190 Indústria de Material Bélico do Brasil (IMBEL)

Avenida 15 de Março, Casa 1, Vila Estrela,
12620-000 Piquete, SP
☎ (+55 12) 556 20 11 ⊜ (+55 12) 556 14 77
Commercial Director: Remy de Andrade
Commercial Administration Department: Paulo Roberto Diaz
- Development and manufacture of rifles, sub-machine guns and pistols for use by Brazilian and other military forces, the Brazilian Police and for competition shooting. [98]

75-09460 Industria Aeronautica Neiva S.A.

Avenida Alcides Cagliari, 2281, Aeroporto,
18608-900 Botucatu, SP
☎ (+55 14) 821 21 22 142736 ⊜ (+55 14) 822 12 85
US$ 17,000,000 (1996) ♣ US$ 13,000,000 (1996)
210 (1996) 20,580 m2
President and Director: Maurício Novis Botelho
☎ (+55 12) 345 12 15
Executive Director: Paulo Urbanavicius
☎ (+55 14) 822 28 55
Production Manager: Carlos Daiti Nakamura
☎ (+55 14) 822 25 99
Commercial Department: Luiz Fabiano Zaccarelli Cunha
☎ (+55 14) 822 50 77
- A range of piston-powered singles and twins for the internal market plus the EMBRAER designed ag-plane EMB-202 Ipanema and structural parts for other civil and military EMBRAER programmes. [98]

75-11410 Kawasaki do Brasil Indústria e Comércio Ltda

(a subsidiary of Kawasaki Heavy Industries Ltd, Japan)
Avenida Paulista 1294/1318-5, CEP 01310, São Paulo
☎ (+55 11) 289 23 88 ⊜ (+55 11) 289 27 88 [98]

75-11700 Kodak Brasileira Com Ind. Ltda

(a subs. of Eastman Kodak Company)
Av Maria Cosiho do Agular, 215,
Bloco "E" 6 Andar - Jarim São Luiz,
05804-900 São Paulo, SP
☎ (+55 11) 848 60 00 ⊜ (+55 11) 848 61 76 [97]

75-12050 Lacombe Indústria e Comércio de Turbos Ltd

Rua Isaias Regis de Miranda 145, 81500 Curitibas PR
☎ (+55 41) 376 15 12 ⊜ (+55 41) 276 90 33
♣ US$ 1,500,000 20 1,000 sq m
President, Owner and Mechanical Engineer: Antonio Cavalconti LaCombe
Controller: Igor Kosmini

■ Turbocharger conversions and repairs. Diesel engine upgrading through turbocharging and aftercooling. [98]

75-12420 Líder Táxi Aéreo SA
Aeroporto Pampulha, Avenida Santa Rosa 123,
31270-750 Belo Horizonte, MG
☎ (+55 31) 448 47 00; 441 49 62 ⊠ 0311275; 391629
⊖ (+55 31) 443 60 22; 443 41 79
♧ Cr 32,465,342,00 ♣ US$ 60,000,000 👥 580
President: Capt. José Afonso Assumpção
Chief Operating Officer: Eduardo Vaz
Financial Director: Alexandre Almeida
Legal Director: Ricardo Alvarenga
Maintenance Director: Irineu Vieira Bueno
Helicopter Director: Geraldo Strambi
Air Taxi Director: Paulo Batista Assuncco Filho
Assistant to Chief Executive Officer: Wilson G Cavalcanti
■ Air taxi, charter freight, maintenance, aircraft and equipment sales, emergency medical services. Maintenance of military helicopters, helicopter tail booms and components.
® USA: Bell Helicopter Textron, Bombardier Amphibious Division, Bombardier Business Aircraft Division, Bombardier Regional Aircraft Division, Flight Safety International, Garrett AlliedSignal, General Electric, Global Systems, Learjet, Raytheon Company (ATC Products)
Subsidiaries:
APEL Aero Participagoes & Empreendimentos Ltda (Holding Co.)
Lider Amazonia Taxi Aireo SA
Lider Representagoes Aeronauticas Ltda
Lideravia Corretora de Seguros Ltda
Lideravia Ltda; 92
Lider Transportes Aereos-Air Brazil SA
Rotorbras-Comercio e Industria de Helicopteros SA [97]

75-12970 Lucas Aerospace
rua Voluntarios da Patria, 45, sala 1005,
22270-000 Botafogo, Rio de Janeiro
☎ (+55 21) 286 3480 ⊖ (+55 21) 286 7625
Comercial Mgr Latin America: John Gattasse [97]

75-13260 Mambrini Equipamentos Rodoviarios Ltda
(a partner of Alcoa Aluminio SA, Brazil)
Rodovia Amaral Peixoto Km 26, São Joaquim,
24800 Itaborai RJ
☎ (+55 21) 635 30 00; 635 25 50; 635 21 24
⊖ (+55 11) 747 62 05; (+55 21) 635 20 50
👥 60 🏭 15,000 m2
Managing and Sales and Marketing Director: Mario V Q de Toledo
Technical and Purchasing Director: Eduardo A A Toledo
Marketing Director: Eustáquio Ferreira
Purchasing Manager: Mario A Silva
■ Military truck bodies and equipment for all trucks. [98]

75-13900 MTU do Brasil Ltda
(† MTU Motores Diesel Ltda)
PO Box 11791, 05090-970 São Paulo, SP
☎ (+55 11) 847 14 99 ⊖ (+55 11) 847 14 58
■ Marine engines 100-600 HP. Marine generating sets 100-5,000 kV. Stationary generating sets 100-5,000 kVA.
Representing:
MTU Friedrichschafen GmbH, Germany [98]

75-13980 MWM Motores Diesel Ltda
Avenida des Nações Unidas, 22.002,
04795 São Paulo SP
☎ (+55 11) 822 32 00 Ⓦ http://www.mwm.com.br
⊠ 1157563/1153838 mwm br ⊖ (+55 11) 822 35 73
♣ US$ 3,000,000 👥 1,335 🏭 45,000 sq ft
President: Thomas A P Wagner
Managing Director: Jurgen Ostermann
Operations Director: Benedicto C D Silva
Technical Director: Manfred Ypma
Sales and Marketing Director: Marcelo Geoffroy
■ Diesel engines for military vehicles and light ships. [98]

75-14000 Nansen SA
Rua José Pedro Araujo 960 Centro, PO Box 213,
32341560 Contagem MG
☎ (+55 31) 359 20 00
Ⓔ nansen@unix.horizontes.com.br
⊖ (+55 31) 359 21 25
♧ US$ 15,000,000 (1996) ♣ US$ 39,700,000 (1996)
👥 666 (1996)
Managing Director: Manoel Ricardo P B Da Costa
Operations Director: Tarcisio Da Costa Lage
Sales and Marketing Director: Lairson Pesce
Purchasing Manager: Claudio Augusto Braga
■ Meteorological balloon-carried radiophones. [98]

75-14100 Nardini Indústria SA
Av. Campos Salles 1735, 13470 Americana, SP
☎ (+55 194) 60 40 26 ⊠ 1551 inna br
⊖ (+55 194) 60 17 59
■ Special lathes for gun tube drilling & machining. [97]

75-14150 Nautika Ltda
Indústria Brasileira de Infláveis Nautika Ltda
R. Santana de Ipanewa, 450, 07220-020 Enarulhos, SP
☎ (+55 11) 912 06 77 ⊖ (+55 11) 912 43 18
♧ US$ 1 000 000 (1995) ♣ US$ 18 000 000 (1995)
👥 250 🏭 9 000 sq m
■ Life vests, inflatable rubber boats. [97]

75-14400 Nife Brasil Sistemas Eletricos Ltda
Av. Pires do Rio, 4615-Itaquera,
08240-000 São Paulo, SP
☎ (+55 11) 996 00 33 ⊠ 62429 ⊖ (+55 11) 205 03 80
♣ US$ 23 900 000 👥 531
Managing Dir: Dilio Sergio Penedo
■ Industrial NiCad and Lead-Acid batteries, AC & DC power electronics, resale of SAFT portable & advanced batteries (for various applications incl. military), emergency lighting. [97]

75-14780 Novatração Artefatos de Borracha SA
Avenida Dep Cantidio Sampaio, 4714, Parada de Taipas,
02860-000 São Paulo, SP
☎ (+55 11) 841 23 22 ⊠ NOVATRACAO São Paulo
⊠ 1181079 ntspbr ⊖ (+55 11) 841 32 41
♧ US$ 1,190,000 (1996) ♣ US$ 2,100,000 👥 91
🏭 9,000 m2
President: Josef Soucek
Purchasing Manager: Ms Wagner Rosa Matiolli
Sales Manager: Joseph Roland Ladislav Soucek
■ Tyres for wheeled armoured vehicles, logistics vehicles and aircraft. Run-flat tyres for airport security vehicles. Tank tracks, road wheels and pneumatic lifts. Elastic containers for fuel, maintenance of flexible Pronal fuel tanks for helicopters. → 80 [98]

75-14900 Nutrimental SA/Comexport
Av. Nacoes Unidas No.989, 04578-000 São Paulo SP
☎ (+55 11) 829 78 88 ⊠ 1121502
⊖ (+55 11) 829 77 59
♣ US$ 70 000 000 👥 2 000
Marketing Dir: Roberto Junqueira
Public Relations Mgr: Duncan Potter
■ Field food for military use. [97]

75-16200 Petroflex Indústria e Comércio SA
Rua Marumbi, 600-Campos Elíseos,
25221-000 Duque de Caxias RJ
☎ (+55 21) 776 42 87; 776 15 37
Ⓔ petroflex@petroflex.com.br
⊖ (+55 21) 776 39 84; 776 40 46
♧ US$ 147,400,000 (1996)
♣ US$ 384,500,000 (1996) 👥 606 (1996)
Chairman: Max Feffer
Vice President: Otto Vicente Perrone
Managing Director: Isaac Plachta
Finance Director: Pedro Paulo Da Poian
Commercial Manager: Ms Maria Locatelli
Purchasing Manager: Ms Hélia Maria Souza Pamplona Cotia
■ Chemical components for missile fuel. [98]

75-18300 Real Santa Rita Equipamentos de Segurança Ltda
Rua Benedito Capistrano, 150 deAlckmin,
37540 Sta Rita de Sapucaí
☎ (+55 35) 471 14 55; 471 12 44
⊖ (+55 35) 471 14 61
♧ US$ 18,000,000 ♣ US$ 20,000,000 👥 400
🏭 12,500 sq m
President: Ronaldo Bauer
Managing Director: Valter Bacarin
Operations Director: C F Silva
Sales and Marketing Director: L A Capelli
Purchasing Manager: L Gallo
■ A full range of protective clothing including boots, special fireproof suits, cold resistant suits for polar use, high visibility jackets, hardhats, face masks, ear defenders, goggles and gloves made from leather, split leather PVC and cotton. [98]

75-18340 Recrusul SA
Avenida Luiz Pasteur 1020,
93212-360 Sapucaia do Sul, RS
☎ (+55 51) 474 12 33 Ⓔ recrusul@recrusul.com.br
⊠ 55 511355 ⊖ (+55 51) 474 18 97
♧ US$ 47,000,000 (1996) ♣ US$ 125,000,000 (1996)
👥 762 (1996) 🏭 34,900 sq m
President: Hélio Wosiack
Operations and Technical Director: Rasso C Lamprecht
Sales and Marketing Director: Alarico Carneiro Cuhna
Purchasing Manager: Nelson Silveira
■ Semi trailers for military trucks, armoured truck bodies. Industrial refrigerators. [98]

75-18400 Resil Industria e Comercio Ltda
Avenida Prestes Maia 775,
09930-900 Diadema, São Paulo
☎ (+55 11) 713 80 39 ✆ (+55 11) 713 80 00
Ⓔ resil@regra.com.br ⊖ (+55 11) 713 80 78
👥 1,300 (1997) including Plant at Minas
President: Stefan Sachs
Managing Director: Cláudio Sachs
Operations Director: Enrico E Galliena
Commercial Director: Alberto Perestrelo
Purchasing Manager: Luis Almeida
■ Fire extinguishers (vehicular and industrial applications) and components. [98]

75-18450 Resil Minas Industrial e Comercio Ltda
(a partner of Resil Industria e Comercio Ltda)
Av. Jorge Sachs 650 (Esq.c/av.Resil),
32900-000 São Joaquim de Bicas, Minas
São Joaquim de Bicas, Minas Gerais
☎ (+55 31) 534 12 33 ⊖ (+55 31) 534 13 65
Managing Director: Sergio León Sachs
■ Original Equipment Manufacturer - autoparts. Seat systems, metallurgical products. [98]

75-18610 Rockwell Do Brasil Ltda Braseixos Division
(a subsidiary of Rockwell International Corp., USA)
Avenida João Batista 824, 06097 900 Osasco SP
☎ (+55 11) 704 65 10 ⊠ 71569 bsxs
⊖ (+55 11) 704 66 00
♣ US$ 150,000,000 👥 3,100
Operations Manager: Luiz A Arruda
■ Axles and carriers for cars, pick-ups, trucks and construction machinery. [98]

75-18780 Amadeo Rossi SA
Rua Amadeo Rossi, 143, PO Box 28,
93030 São Leopoldo, RS
☎ (+55 51) 592 22 12 ⊠ 524011 armm br
⊖ (+55 51) 592 59 89
👥 1,000 🏭 88,000 m2
President: Norberto N Rossi
Managing Director: Claudio A Rossi
Director: Luciano Rossi
Sales and Marketing Director: Vincent Daudt
Purchasing Manager: Nestor Scherer
■ Revolvers, shotguns, rifles and carbines. [98]

75-19045 Saft Nife Brasil Sistemas Eléctricos Ltda
(a subsidiary of Saft SA, France)
Avenida Pires do Rio - 4615 Itaquera,
08240-000 São Paulo
☎ (+55 11) 686 00 33 ⊠ 62429 ⊖ (+55 11) 205 03 80
♣ US$ 20,000,000 (1995) 👥 300
Managing Director: Clas Wanning
Marketing Director: Roberto Mayo [98]

75-19140 Sanvas SA Indústria Metal-Mecânica
Avenida Almirante Barroso 63 GR.1004,
20.031.003 Rio de Janeiro, RJ CEP

☎(+55 21) 553 20 23 E pwrofice@pobox.com
⊜ (+55 21) 240 79 05
119 (1997) 32.734 m2
Director: Raul Sanson
■ Military trailers. 98

75-19180 Scania Latin America Ltda
Caixa Postal 188, 09810-902 São Bernardo do Campo-SP
☎(+55 11) 752 93 33 ⊜ (+55 11) 451 26 59
■ Military trucks, truck & tank engines. 98

75-19240 Schrader do Brasil SA
Avenida Malek Assad, 1600, Parque Mcia Lua, 12300-000 Jacarei, São Paulo
☎(+55 12) 351 65 44 ✉ cfavaro@home.tecsat.com.br
⊜ (+55 12) 351 65 47; 351 17 99
US$ 9,065,822 ♣ US$ 25,000,000 419 (1997)
6,737 m2
President: R H Ferrero
Managing Director: Nelson Scola
Financial Director: Eduardo Favaro
Sales Manager: Paulo Novoa
Industrial Manager: Fernando Leite
■ Tyre valves and accessories, tyre gauges and tools. 98

75-19300 Seeger-Reno Indústria e Comércio Ltda
(a subsidiary of TTC, USA)
Avenida Prestes Maia 230 Vila Marina, 09930-270 Diadema, SP
☎(+55 11) 713 31 33 ✉ seeger@snet.com.br
⊜ (+55 11) 713 44 12
US$ 5,308,432 (1997) ♣ US$ 8,631,000 (1996)
140 (1997) 8,410 m2
Managing Director: João Scivoletto
Industrial Manager: João Carlos Portescheller
Sales and Marketing Manager: Adolfo Pimentel Filho
Purchasing Chief: Roberval Maimone
Costs Budget and Accounts Chief: Edmilson Polo Favaretto
■ Retaining rings and precision spacers. 98

75-19540 Sherwin Williams do Brazil Indústria e Commércio Ltda
(a subsidiary of The Sherwin Williams Co., USA)
Avenida Ibirama, 480 S J Tadeu, 06785-300 Taboão da Serra, SP
☎(+55 11) 79 68 50 00 ⊜ (+55 11) 79 68 50 13
■ Special paints for aircraft and airport use. 98

75-19680 SKF do Brasil Ltda
(a subs. of SKF, AB)
Rod. Pres. Dutra km 223 Cumbica, PO Box 1745, 07034-900 Guarulhos, SP
☎(+55 11) 912 04 88 ☒ 33977 ⊜ (+55 11) 912 43 45
1 094
President: Liberto Sanz
■ Bearings. 97

75-20000 Textil Tabacow SA
Rua José Tabacow 131 Tatuapé, 03409 São Paulo SP
☎(+55 11) 942 33 11 W http://www.tabacow.com.br
⊜ (+55 11) 942 33 44
♣ US$ 90,000,000 (1996) 1,200 (1996)
70,000 m2
Operations Director: Eduardo Kauffmann
General Communications Manager: M Daniel Gallarreta
■ Fireproof carpets for military aircraft. 98

75-20110 Teadit Industria e Comercio Ltd
Avenida Automóvel Clube 8939, 21530-010 Rio de Janeiro RJ
☎(+55 21) 471 25 00 ☒ 23720 asbe br
⊜ (+55 21) 471 20 50
Managing Director: Luigi Biancheri
Sales and Marketing Director: J C van Hemelryck
Export Manager: Domingos Senra
■ Kevlar for bullet proof vests. 98

75-20140 Tecelagem Lady Ltda
R. Leopoldo Couto Magalhães Jr 758, Itaim Bibi, 04542-000 São Paulo, SP
☎(+55 11) 820 04 33 ☒ 1122973
⊜ (+55 11) 820 71 52
Managing Dir: A. Lowy
General Mgr: E. Lowy
■ Velcro hook and loop tape fasteners. 97

75-20145 Tecelinco Tecnologia Electronica Ltda
463 Rua Major Sertoria, 4° andar, 01222 São Paulo
☎(+55 11) 257 36 45 ⊜ (+55 11) 256 64 46
President: A Megnis
Representing:
EEV Ltd, UK 98

75-20205 TECTRAN
Enganharia, Indústria e Comercio SA (a subsidiary of Avibras Aeroespacial SA)
Rodovia Presidente Dutra km 155/156, PO Box 165, 12240-420 São José dos Campos, SP
☎(+55 12) 331 82 00 ⊜ (+55 12) 331 87 75
US$ 10,163,089 (1996) ♣ US$ 20,968,000 (1996)
243 (1997) 13,500 m2
President: J Verdi Leite
Managing Director: Reubens Domingues Porto
Operations Director: Arnaldo N Fleury
Purchasing Manager: José Carlos Machado
Sales and Marketing Manager: Marco Stéphany
■ Special vehicles for military and civil applications. 98

75-20520 Timken do Brazil commercial e Industrial Ltda
(a subsidiary of The Timken Company, USA)
Rua Eng. Mesquita Sampaio 714, Sto Amaro, 04711 São Paulo SP
☎(+55 11) 247 12 33 ⊜ (+55 11) 547 96 33
12,200 m2
President and Director: Robert W Logston
Vice President and Director: Hubert Safranek
Sales and Marketing Manager: Flavio B Libonati
Purchasing Manager: Jorge Steschenko
■ Bearings for military vehicles. 98

75-20780 Treu SA Máquinas e Equipamentos
Avenida Brazil 21000, 21510 Rio de Janeiro
☎(+55 21) 372 66 33 E treurj01@unijet.com.br
☒ 2121089 treu br ⊜ (+55 21) 372 31 96
US$ 2,500,000 ♣ US$ 7,500,000 200
12,000 sq m
President: Curt Treu
Administrative and Commercial Director: Franz Treu
■ Equipment for chemical, oil and related industries and for nuclear power plants. Mixers for rocket fuel production, vacuum chambers for drying rockets cores, mixers for explosives, microniser mills for ammonium perchlorate. 98

75-22230 Veeder-Root do Brasil
Rua Ado Benatti 92, 05037 Sao Paulo
☎(+55 11) 861 21 55 W http://www.veeder-root.com
⊜ (+55 11) 861 19 82
President Veeder-Root do Brazil: Daniel Lassner
® USA: Veeder-Root 98

80 Service Companies & Organizations

Hispano-Suiza do Brazil Equipamentos 90-08100
Novatração Artefatos 75-14780

80-00100 Aeropartes Indústria Aeronáutica
Av Ayrton Senna, 2541 Hangar Jato, 22775-000 Rio de Janeiro RJ
Avenida Ayrton Senna, 2541 Hangar Jato
☎(+55 21) 325 10 22 (Marketing)
E aeropart@embratel.net.br ⊜ (+55 21) 325 67 50
42 2,400 m2
Managing Director: Waldemar Botelho (Jr)
Marketing Director: Edecio Luna Freise
■ Aircraft Maintenance, overhaul of avionics, instruments and accessories. Aircraft parts and sales. Distribution. Representation of aviation companies. → 90 → 85 98

80-00300 AlliedSignal Commercial Avionics Systems
Rua Assis Brazil 143 BL, 1 Apt 1005, Copacabana, 22020-010 Rio de Janeiro
☎(+55 21) 295 74 42 ⊜ (+55 21) 275 49 43
Field Service Engineer: Abiel de Barros
■ Field service engineering office. 98

80-01000 Comaf Indústria Seronautica Ltda
Avenida João Ribeiro N. 472, 20750-091 Pilares, Rio de Janeiro, RJ
☎(+55 21) 591 59 94
E cmfnco@//br.homeshopping.com.bra
⊜ (+55 21) 592 47 77
US$ 100,000 (1996) ♣ US$ 2,600,000 (1996)
43 (1997) 1,500 m2
Managing Director: Nelson C De Oliveira
Marketing Director: Ms Marcia S Melo
Purchasing Manager: Claudia de Sa Sena
■ Overhaul centre and authorised service centre, LAPEC. 98

80-11200 Karcher Indústria e Comércio Ltda
(a subsidiary of Alfred Karcher GmbH & Co, Germany)
Avenida Professor Benedicto, Montenegro 419 - Betel, 13140000 Paulinia, São Paulo
☎(+55 19) 884 75 15 ☒ 038197520 kaicbr
⊜ (+55 19) 884 71 10
Manager: Johannes Karcher
® Germany: Alfred Karcher GmbH & Co 98

80-20000 Motores Rolls-Royce Ltda
(a subsidiary of Rolls-Royce plc, UK)
Rua Dr Cincinato Braga 47, PO Box 404, 09895 São Bernardo do Campo, SP
☎(+55 11) 759 77 22 ☒ 1144115
⊜ (+55 11) 759 80 71
186 12,000 m2
Managing Director: Anthony John Ward
Finance and Administration Director: Paul A Ratcliffe
Industrial Director: Carlos Roberto T Levy
Commercial Director: Flavio de Lacerda Abreu
■ Aeronautical and industrial engine component supplier, overhaul and repair.
® UK: Rolls Royce plc 97

85 Stockists, Distributors, Spare Parts Suppliers

Aeropartes Indústria Aeronáutica . . . 80-00100

85-02000 TRACE Brazil
Rue Prof. Gastao Bahiana 615/801, 22071-030 Lagoa, Rio de Janeiro
☎(+55 21) 521 70 71 ☒ 33779 jsta
⊜ (+55 21) 521 70 71
Director: Wilson Figueiredo
® USA: Trace Worldwide Corp. 98

90 Sales Agents, Sales & Liaison Offices

Companies represented:

Alfred Karcher GmbH & Co, Germany . . 80-11200
Aviall, USA 90-03100
Avtech, USA 90-07000
Bell Helicopter Textron, USA 75-12420
Bombardier Amphibious Division, USA 75-12420
Bombardier Business Aircraft Division, USA 75-12420
Bombardier Regional Aircraft Division, USA 75-12420
Bowers Internal Gauge, UK 90-10000
Chelton, UK 90-07000
Circle Seals Controls International, USA 90-01240
CV Instruments, UK 90-10000
Deutsch, USA 90-07000
Diehl, Germany 90-07000

Sales agents/representatives listed elsewhere:

90-01240 Aeron International

20020.080 Rio de Janeiro, RJ
Avenida Marechal Camara 160, Grupo 1214
☎ (+55 21) 262 1831 [E] aeron@domain.com.br
⊜ (+55 21) 262 1831
President: Ms Elisabeth Machado
® USA: Circle Seals Controls International [98]

90-01300 AlliedSignal Aerospace Marketing Sales and Service

(a subsidiary of AlliedSignal Aerospace, USA)
Rua 18 de Novembro, 800, Hangar 4,
90240-040 Porto-Alegre, RS
☎ (+55 51) 343 46 41; 357 73 78 ⊠ K60695
⊜ (+55 51) 342 56 56
Field Service Engineer: Cesar Basler
■ On-site service for AlliedSignal customers. [98]

90-01700 Allied-Signal Aerospace

Avenida São Gabriel, 333 cj 131, 01435-001 São Paulo
☎ (+55 11) 280 24 11 ⊜ (+55 11) 280 28 53
Regional Manager, South America: Ralph Diaz [98]

90-03100 Ciageral Comercio e Participacoes Ltda

Avenida Beira Mar, 216/1102, 20021-060 Rio de Janeiro
☎ (+55 21) 220 31 67 ✉ CIAGERAL ⊠ 2138260 ciag br
⊜ (+55 21) 220 67 17
♧ Cr 500,000 ♣ US$ 8,000,000 👥 12 🏭 180 m2
Managing Director: Paulo B Macedo
Sales and Marketing Director: Luiz A B Macedo
General Manager: Haroldo Buarque de Macedo
® USA: Aviall, Global Marine, Helicopter Service Inc., Magnetek, Magneter Defense, Microfilite Simulators, Redco Engineering, Schweizer Ag-Cat H-300, Teledyne Thermatics [98]

90-05000 Daewoo do Brasil Importa e Exporta Ltda

Avenida Paulista 777 13, Conj 172 Cerq,
01311-100 Cesar São Paulo
☎ (+55 11) 285 56 99 ⊠ 1135890 dwsp
⊜ (+55 11) 285 15 35 [98]

90-07000 GFC Aerospaço e Equipamentos Ltda

Rua da Gloria - 344 Suite 903,
20241-180 Rio de Janeiro, RJ
☎ (+55 21) 242 98 13; 242 33 68
[E] gfc@easynet.com.br ⊜ (+55 21) 224 58 27
♧ R$ 450,000 (1997) 👥 10 (1997)
Managing Partner: George Grolman
Managing Partner: Luiz Claudio Coutinho Marquez
Managing Partner: Fernando Arcoverde
■ Marketing and sale of products for aerospace manufacturers including aircraft, avionics, instruments, aircraft components, laboratory test equipment, defence equipment for ground vehicles including tanks, aircraft, boats, missiles and rockets. Avionics overhaul shop.
® Germany: Diehl, Junghans, Otto Fuchs, Teldix
UK: Chelton, GEC Sensors Ltd, Graseby Dynamics, Meggitt Avionics (ex-Negretti), Strachan & Henshaw, Truflo
USA: Avtech, Deutsch, Loral Fairchild, MacDonald Dettwiler Assoc., McDonnell Douglas [98]

90-08100 Hispano-Suiza do Brasil Equipamentos Ltda

Avenida Rio Branco n° 311, 9° Andar-Centro,
20040 Rio de Janeiro
☎ (+55 21) 532 11 20 ⊠ 21989 ⊜ (+55 21) 262 70 10
♧ FFr 23,000,000 (1997) ♣ FFr 5,000,000 (1997)
👥 30 (1997) 🏭 3,000 m2
President: Patrice Sabahi
Operations Director: Jean Pierre Sallé
Commercial Manager: Jacques Fremau
Finance Manager: Mrs Mireille Hassoun
■ Maintenance & repair of gas turbines, production of turbocharger parts. → 75 → 80 [98]

90-10000 Importécnica SA

Rua Baroneza De Porto Carreiro 352, São Paulo
☎ (+55 11) 862 43 55 [E] importec@wm.com.br
⊜ (+55 11) 826 32 47
♧ US$ 1,600,000 ♣ US$ 2,400,000 (1997)
👥 35 (1997) 🏭 2,150 (1997)
Marketing Director: Hugo Varela
Technical Director: Nelson Fidelis
® Germany: Helios, Kordt
Switzerland: Proceo SA, Trimos-Sylvac Task Force, Wyler AC
UK: Bowers Internal Gauge, CV Instruments [98]

90-13280 Vectra Defence Ltda (VDL)

(† Vectra Representaçoes Ltda)
Praça Floriano 19-290 andar,
20031-050 Rio de Janeiro - RJ
☎ (+55 21) 532 06 72; 533 86 30
[E] vectradefence@visualnet.com.br
⊜ (+55 21) 533 68 42
♣ US$ 27,000,000 (1996)
👥 2 fixed, 8 consultants (1997) 🏭 75 m2 (1997)
President: Capt Ervé Nogueira (retd)
Vice President: Capt Carlos Victor Barbosa (retd)
Director: Eng Cdr Mauricio Selbach Pereira (retd)
Director, Operations: Paulo de Figueiredo Ferraz (Junior)
■ Sales and marketing of aviation products: including strategic planning anf advice, proposal preparation, financial and contractual advice, legal support.
Representing:
Alfred Karcher GmbH & Co, Ger
Barco Display Systems, Be
Barracuda Technologies, Swe
Bristow Helecopters UK
Cardion/Siemens, USA
Devonport Management Ltd, UK
Ellectronica, It
Fiar, It
GEC-Marconi Defence/Avionics/Materials, UK
Hughes Aircraft Co, USA
Marion Composites, USA
Matra-British Aerospace Dynamics, UK
Matra Marconi Space, UK
Officine, It
Racal Radar Defence Systems, UK
Redifon MEL, UK
Schill Reglerteknik, Swe
Siemens Environmental Systems, UK
Trimble Aerospace Ltd, UK
Vickers Defence Systems, UK
São Paulo Office
São Paulo
☎ (+55 12) 322 32 15 office; 937 63 99 celular
✉ vectrasjk@netvale.com.br
Director: Eng° Mauricio Selbach Pereira
■ Sales & marketing of aviation products [98]

90-13780 Moog do Brasil Controles Ltda

(a subsidiary of MOOG Inc., USA)
Rua Prof. Campos de Oliveira 338, CEP 04675-100,
04675 Jurubatuba, Santo Amaro SP
☎ (+55 11) 523 80 11 [W] http://www.moog.com
⊜ (+55 11) 524 21 86
♧ US$ 1,560,000 ♣ US$ 2,639,000 (1996)
👥 17 (1997) 🏭 1,160 m2
General Manager: Mario Valdo
[E] mvaldo.brazil@moog.com
■ Actuators, valves and flight control equipment.
® USA: Moog Inc [98]

90-13850 F E Müller Botelho

(† Francisco Eduardo Müller Botelho)
Avenida Presidente Vargas 463, 4 andar,
20071-003 Rio de Janeiro, RJ
☎ (+55 21) 509 37 71; 509 76 22 ⊠ 22093 cflcl
⊜ (+55 21) 221 79 54
Managing Director: Francisco Eduardo Müller Botelho [98]

90-15500 OFEMA Brasil Serviços Ltda

(a subsidiary of OFEMA, France)
Rua Mexico 21-GR 1602 - Centro,
20031-144 Rio de Janeiro
☎ (+55 21) 533 32 55; 533 37 38
⊜ (+55 21) 262 47 38
👥 5 (1997) 🏭 110 m2
Managing Director: M Loic Piedevache
Equipments Delegate: M Bernard Joly
® France: OFEMA [98]

90-17000 Powerpack Representaçoes e Comércio Ltda

Rua General Rabelo, 52 - Gávea,
22451-010 Rio de Janeiro, RJ
☎ (+55 21) 239 40 41 ✉ POWERPACK
⊠ 2122841 powe br ⊜ (+55 21) 294 87 95
♧ US$ 20 000 (1996) 👥 15 🏭 1 650 sq ft
Manager: Roberto de Souza Dantas (Jr)
Manager: Fernando de Souza Dantas
® Canada: Pratt & Whitney Canada
USA: Hamilton Standard, Sikorsky Aircraft, United Technologies [97]

90-18300 Radiall do Brasil Componentes Electrónicos Ltda

(a subsidiary of Radiall, France)
Avenida Beira Mar 262 - 6° Andar, 20021 Rio de Janeiro
☎ (+55 21) 282 13 25 ⊠ 2121968 pepr br
⊜ (+55 21) 220 36 68
Managing Director: Hubert Marceul
® France: Radiall [98]

90-18809 Rockwell Do Brasil Ltda Light Vehicles Systems

(a subsidiary of Rockwell International Corp, USA)
Ave Major José Levy Sobrinho, 2700, PO Box 452,
13486-925 Limeira, SP
☎ (+55 194) 40 21 00 [W] http://www.rockwell.com.br
⊠ 192882 fuma br ⊜ (+55 194) 40 22 22
👥 985 (1996) 🏭 18,000 m2
President: Juan L De La Riva
[E] jdlariva@auto.rockwell.com
Operations Director: A Guerreiro
Sales and Marketing Director: J Bahia
Purchasing Manager: B Simão
→ 75
® USA: Rockwell International Corp. [98]

Contact Reference

90-18850 Rockwell Do Brasil SA Collins Commercial Avionics
Rua Sebastião Humel, 395,
12210-200 Sao José dos Campos, SP
☎(+55 123) 21 16 33 ⊖ (+55 123) 21 19 01
Service Center Exec. Mgr: Nelson Aquino
■ Avionics & marine inertial navigation & telecommunications systems
® USA: Rockwell International Corp. [97]

90-19100 Schottel do Brasil Propulsões Maritimas Ltda
(a subs. of Schottel-Werft Germany)
Ave Tancredo Neves 600, Distrito Industrial,
94930-540 Cachoeirinha, RS
☎(+55 51) 470 17 15 ⊖ (+55 51) 470 32 00
Managing Dir: Harold Herzog
® Germany: Schottel-Werft [97]

90-19400 Siemens SA
Sede Central, Avenida Mutinga 3650, Pirituba,
05150-900 São Paulo SP
☎(+55 11) 836 22 11 ▣ 1181006 siem br
⊖ (+55 11) 836 26 31
■ Electronic systems, telex systems. [97]

90-20000 Teledyne do Brasil
(a subsidiary of Teledyne Inc)
Rua da Assembléia 10, Sala 1124,
20011-000 Rio de Janeiro
☎(+55 21) 531 17 22 ⊖ (+55 21) 531 12 77
7 (1997) 240 m2
Managing Director: James D Martins
® USA: Teledyne Inc. [98]

90-20190 Thomson-CSF Electronica Profissional Ltda
(a subsidiary of Thomson-CSF)
Avenida Presidente Wilson 231-Gr2504,
20030 Rio de Janeiro
☎(+55 21) 240 27 07 ▣ 31322 tcsf br
⊖ (+55 21) 220 26 32
♣ US$ 40,000,000 (1995) 5 (1995)
Director General: Daniel Henner
Sales and Marketing Director: Ivan Castro Da Rocha
® France: Thomson-CSF [98]

Brunei

Head of State:
HM Sultan Sir Bolkiah Hassanal

10 Head of State

10-10000 Office of the Sultan
Istana Nurul Iman, Bandar Seri Begawan
☎(+673 2) 22 23 01 ▣ 2234 istana bu
Sultan of Brunei: H M Sultan Sir Bolkiah Hassanal [98]

11 Prime Minister

11-10000 Office of the Prime Minister
Istana Nurul Iman, Bandar Seri Begawan
☎(+673 2) 22 99 88
Prime Minister: HM the Sultan Sir Muda Bolkiah Hassanal [98]

12 Ministries

12-10000 Ministry of Defence
Istana Nurul Iman, Bandar Seri Begawan
☎(+673 2) 33 05 31 ▣ 2683 paduka bu
⊖ (+673 2) 33 03 95
Minister of Defence and Supreme Commander of the Royal Brunei Armed Forces: H M Sultan Sir Bolkiah Hassanal [98]

12-20000 Ministry of Foreign Affairs
Kementerian Hal Ehwal Luar Negeri
Jalan Subok, 1120 Bandar Seri Begawan
☎(+673 2) 24 11 77; 24 02 81
▣ 2292; 2411 dipserv bu
⊖ (+673 2) 22 99 04; 22 47 09
Minister: HRH Prince Muda Mohamed Bolkiah [98]

12-40000 Ministry of Home Affairs
Kementerian Hal Ehwal, Dalam Negeri,
Bandar Seri Begawan
☎(+673 2) 22 32 25 ▣ 2234 istana bu
Minister: Isa bin Ibrahim [98]

25 Defense Forces

25-10000 Royal Brunei Armed Forces, Combat Engineer Squadron, Support Battalion
Penanjorg Garrison, 5049 Tutong, Brunei Darussalam
☎(+673 2) 33 05 31 ext 314 ⊖ (+673 2) 33 01 10
Commander Royal Bunei Armed Forces:
Maj Gen Pehin Dato Husin
Director of Operations: Col Dato Ibrahim
Chief of Staff: Brig Gen Rhin Shain
Director of Procurement: Col Pewin Yusof
Public Relations Officer: Dyg Hajjah Fatimah Bti Hj Hassan
ADC to Commander RBAF: Lieut Akhmal
Commanding Officer: Mag Mohamad Arrifin [98]

25-40000 Royal Brunei Navy
Muara Base, Bandar Seri Begawan
☎(+673 2) 77 07 94 ⊖ (+673 2) 77 01 45
Commanding Officer: Col Abdul Jalil Bin Hj Ahmad
☎(+673 2) 77 01 33
Deputy Commander: Maj. Mahaleh Bin Hj Majid
Staff Officer Operations & Planning:
Maj. Jorsni Bin Hj Lamat [97]

31 Embassies & Attachés

31-00090 Australian High Commission
PO Box 2990, Bandar Seri Begawan
4th Floor, Teck Guan Plaza,
Cnr Jalan Sultan & Jalan McArthur,
Bandar Seri Begawan 1929
☎(+673 2) 22 94 35/6 ⊖ (+673 2) 22 16 52
High Commissioner: H.E. Mr N. P. Davis (MBE)
First Secretary: Mark Barber
Defence Attaché: Capt Garry Kennedy
(resident in Singapore) [97]

31-01350 Embassy of the Sultanate of Oman
PO Box 2875, 1928 Bander Seri Begawan
No 35, Simpang 100, Kampong Pengkaln,
Bandar Seri Begawan
☎(+673 2) 44 69 53; 44 69 54; 44 69 58; 44 69 57
⊖ (+673 2) 44 96 46
Chargé d'Affaires: H E Mr Ahmed Moh'd M Al-Riyami
First Secretary: Saif Said Ali Al-Mawali
Second Secretary: Mabrook Mubarak Al-Hinai [98]

31-01370 High Commission for Pakistan
House No 5 Simpang 396/128, Bandar Seri Begawan
Kampong Sungai Akar, Jalan Kebang Saan,
Bandar Seri Begawan
☎(+673 2) 33 97 97; 33 49 89
✉PAREP Bandar Seri Begawan ⊖ (+673 2) 33 49 90
Head of Chancery and Deputy Chief of Mission:
Amjad Majid Abbasi [98]

31-01420 Embassy of the Philippines
Badi'ah Building Jalan Tutong, Bandar Seri Begawan
☎(+673 2) 24 14 65; 21 14 66 ✉AMBAPHIL Brunei
▣ 2672 bu ⊖ (+673 2) 23 77 07
Ambassador: H E Mr Ramon Tirol
Defense Attaché: Col Jaime M Viernes
☎(+673 2) 23 42 20 [98]

31-01640 High Commission of the Republic of Singapore
PO Box 2159, 1921 Bandar Seri Begawan
5th Floor, RBA Plaza, 2085 Jalan Sultan,
☎(+673 2) 22 75 83 ▣ 2385 ab sinhc bu
⊖ (+673 2) 22 09 57
High Commissioner: H.E. Mr Anthony Chng
Second Secretary: Lee Thian Chye
Defence Adviser: Col Lau Kee Siong [97]

31-01890 United Kingdom High Commission
BFPO 11, Bandar Seri Begawan
Negra Brunei Darussalam 2.01, 2nd Floor, Block D
Komplexs Bangunan Yayasan,
Sultan Haji Hassanaal Bolkiah
☎(+673 2) 22 22 31; 22 31 21 ⊖ (+673 2) 22 60 02
High Commissioner: H E Mr Ivan R Callan
Deputy High Commissioner: C R Boardman
Defence Adviser: Gp Capt J M Collier
(resident in Singapore)
First Secretary Defence Co-operation: J C A Clephane
Second Secretary Commercial: My Moon
Second Secretary Chancery and Information:
Philip Malone
Third Secretary Management Consular: P Gilbert [98]

40 Government Defense Sales/Procurement Organizations

40-01000 Royal Brunei Technical Services (RBTS)
PO Box 308, 1903 Bandar Seri Begawan
4th Floor, RBA Plaza,
3180 Jalan Gadong, Bandar Seri Begawa
☎(+673 2) 24 27 00 (6 Lines) ▣ 2707 rbts g bu
⊖ (+673 2) 24 37 67; 24 37 97
Managing Director:
Pg Hj Tengah bin Pg Dato Paduka Haji Metassan
Deputy General Manager: Awang Othman Maidin
Procurement Services Officer:
Mdm Hjh Khasijah Hj Kassim
Commercial Officer: Mohd Hakashah Hj Abd Samad
Senior Cargo Services: Pg Abd Malek Pg Hj Omar
Special Project Officer: Hj Yakub Hj Bagol
■ RBTS is a wholly owned company of His Majesty's Government. Authorised sole agent for procurement of sensitive and strategic related items and services. Authorised representative for worldwide manufacturers and suppliers. [98]

90 Sales Agents, Sales & Liaison Offices

Companies represented:

Bell Helicopter Textron, USA→ Singapore 90-00220
Telemit GmbH, Germany→ Netherlands . 80-18300

Bulgaria

Head of State:
Petar Stoyanov

10 Head of State

10-10000 Office of the President
2 Dondoukov Boulevard, 1123 Sofia
☎(+359 2) 85 01 ☒ 803418
President: Petar Stoyanov
Vice-President: Todor Kavaldzhiev [98]

11 Prime Minister

11-10000 Office of the Chairman of the Council of Ministers
Ministerski Savet
1 Dondoukov Boulevard, 1194 Sofia
☎(+359 2) 85 01; 87 24 44
☒ 22281 msavet bg, 22378 msavet bg
⊖ (+359 2) 87 08 78
Prime Minister: Ivan Kostov
Deputy Prime Minister: Aleksandur Dimov Bozhkov
Deputy Prime Minister: Evegniy Bakurdzhiev
Deputy Prime Minister: Veselin Metrodiev [98]

12 Ministries

12-10000 Ministry of National Defense
3 Levski St, 1000 Sofia
☎(+359 2) 54 60 01 ☒ 22649 ⊖ (+359 2) 87 57 32
Minister: Georgi Gerbanov Ananiev [98]

12-20000 Ministry of Foreign Affairs
Ministerstvo na Vanshnite Raboti
2 Alexander Zhendov Straat, 1113 Sofia
☎(+359 2) 72 34 46 ☒ 22530; 22529
⊖ (+359 2) 971 20 12; 971 20 33; 971 24 13
Foreign Minister: Stoyan Zhivkov Stalev [98]

12-40000 Ministry of Interior
29 September 6th Street, PO Box 192, 1000 Sofia
☎(+359 2) 82 25 74; 82 26 89 ☒ 23435
⊖ (+359 2) 82 40 47; 87 79 67
Minister: Bogomil Angelov Bonev [98]

12-50000 Ministry of Finance
102 G S Rakovski Street, 1000 Sofia
☎(+359 2) 87 06 22; 86 93 55 ☒ 22727
⊖ (+359 2) 80 11 48
Finance Minister: Muravei Radev [98]

15 Parliament

15-10000 National Assembly
Narodno Sabranie
2 Narodno Sabranie Square, 1169 Sofia
☎(+359 2) 84 01 ⊖ (+359 2) 981 31 31
President: Yordon Sokolov
Secretary General of the National Assembly: Valentin Georgiev
Vice President: Ivan Kurtev
Vice President: Blagovest Sendov
Vice President: Petya Shopova
International Relations and Protocol Department
2 Narodno Sabranie Square, 1169 Sofia
☎(+359 2) 980 89 16; 840 23 26
⊖ (+359 2) 980 89 16
Head of International Relations and Protocol Department: Mitto Nikolov [98]

25 Defense Forces

25-40000 Navy Headquarters
16 Preslav Street, 9009 Varna
☎(+359 52) 23 56 44 ⊖ (+359 52) 60 32 59
Commander-in-Chief: Vice Adm Christo Knotrov
☎(+359 52) 22 85 70
First Deputy Chief of the General Navy Staff: Rear Adm Konstantin Bugdanov
Deputy Commander in Chief, Training: Capt Milan Milanov
Deputy Commander in Chief, Procurement: Capt Neiro Atanasov
Director of Operations: Capt Peter Petrov [98]

29 Customs & Excise

29-01000 Customs General Directorate
Direction Générale des Douanes
Rue Aksakov 1, 1040 Sofia [98]

31 Embassies & Attachés

31-00075 Armenian Diplomatic Mission
Sofia
☎(+810) 40 13 21 56 79
☒ Website http://wotan.wiwi.hu-berlin.de/-levon/botsch.html 18/2/97 ⊖ (+810) 35 92 54 79 70
Honarary Consul: Sevda Sevan [98]

31-00150 Embassy of Belgium
Veltchova Zavera No 1, Sofia
☎(+359 2) 963 36 22; 963 36 62
☒ AMBASSADE BELGIQUE Sofia ☒ 22455
⊖ (+359 2) 963 36 38 [98]

31-00510 Embassy of the Arab Republic of Egypt
5 Shesti Septemvri St, 1000 Sofia
☎(+359 2) 87 02 15; 88 15 09; 87 64 68
☒ 22270 bostan bg ⊖ (+359 2) 88 14 49
Ambassador: H.E. Mrs May Aboul-Dahab
Second Secretary: Gamal Tawfik
Second Secretary: Mrs Amal Salamah
Administrative Attaché: Ahmed Monem Said
Administrative Attaché: Salah Youssef
Administrative Attaché: Mrs Samira Abu Zeid [97]

31-00690 Embassy of Greece
Boulevard Klement Gotvald 68, Sofia
☎(+359 2) 44 37 70 ☒ LEGOGRECE Sofia
☒ 22 458 greekemb sof ⊖ (+359 2) 44 65 07 [98]

31-00840 Embassy of the Republic of Indonesia
53 Simeonovsko Shosse Res. No.4, 1126 Sofia
☎(+359 2) 68 32 20
Ambassador: H.E. Mr Samsubahri Siregar [97]

31-00920 Embassy of Japan
Ul Lyulyakova Gradina 14, 13008 Sofia
☎(+359 2) 72 39 84; 72 39 86; 71 90 22; 71 81 94
☒ TAISHI Sofia ☒ 22397 taisi bg ⊖ (+359 2) 492 10 95
Ambassador: H.E. Mr Takashi Tajima
Deputy Chief of Mission: Akio Wada
First Secretary Commercial: Toshihiro Araki [97]

31-01730 Swedish Embassy
Alfred Nobel 1, PO Box 620, Sofia
☎(+359 2) 72 04 20; 72 38 71 ☒ SVENSK Sofia
☒ 23956 svensk bg ⊖ (+359 2) 70 55 31
Councellor: T Leijonhufvud [98]

31-01890 Embassy of the United Kingdom
Boulevard Vassil Levski 38, 1000 Sofia
☎(+359 2) 980 12 20 ⊖ (+359 2) 980 12 29
Ambassador: H E Mr R G Short
Deputy Head of Mission: M Tatham
Defence Attaché: Col M I V Dore
Cultural Attaché, British Council Director: K Lewis
First Secretary, Management and Consul: P F C Massey
First Secretary, Political: J Kidner
First Secretary, Commercial: M J Carbine
Second Secretary: R N G Sinclair
Second Secretary: M Campbell
Second Secretary, Political and Information: Ms J Govier
Assistant Cultural Attaché, British Council: David Kirwan
Third Secretary and Vice Consul: Ms Sue Morrell
Third Secretary, Immigration/Political: D Pert
Chaplin: Rev John Poole (resides in Bucharest)
Embassy Medical Adviser: Dr P F C Massey (resides in Warsaw) [98]

55 Government-owned Industry & Production Facilities

55-00060 Arcus Co.
PO Box 6, 5140 Liaskovetz
☎(+359 619) 231 34 ☒ 66608 ⊖ (+359 619) 221 23
♣ US$ 90,000,000 (1996) ⁂ 1,600 (1996)
Managing Director: Banko Bankov
Deputy Executive and Finance Director: Dotcho Karaivanov
Marketing Director: Ivailo Velitchov
Operations Director: Trifon Trifonov
Purchasing Manager: Yolko Rashev
■ Development and production of fuzes, shell bodies and rounds for ground, artillery, rocket systems and small arms - 9mm pistol and revolver. [98]

55-01500 Beta Company
5980 Tcherven Briag
☎(+359 659) 22 17; 27 61; 35 61 ☒ 34565
⊖ (+359 659) 69 66; 29 31
♧ US$ 170,000,000 (1996)
♣ US$ 240,000,000 (1996) ⌂ 274,800 m2
Managing Director: Petar Petrov
Deputy General Manager and Sales Director: Ventzislav Ivanov ☎(+359 659) 29 31
Operations Director: Ivan Kabakchiev
■ Production of MT-LB armoured vehicles: MT-LB-201005; MT-LB-V 21006; MT-LB VM-201013; MT-LB MR HR-204011; MT-LB-U and MT-LB-MTP; 122mm self-propelled howitzer 2S1 Gvozdika; hulls of BMP-23/30 and 120mm self-propelled mortar Tundja. [98]

55-01800 Dunarit Ltd
(a subs. of Metalchim Holding Inc.)
PO Box 12, 7000 Rose
☎(+359 82) 44 68 71; 44 68 91 ⊖ (+359 82) 44 68 26
Managing Dir: Stayan Sabotinov
Head of Marketing Dept: K. Kirova
■ Development & production of: combat & practice high explosive aviation bombs; high explosive mine shells & shaped artillery shells; anti-tank & anti-infantry mines. [97]

55-02100 MRF "Georgi Benkovsky"
(a division of VRZ TEREM)
Tzar Simeon Str 1, 4000 Plovdiv
☎(+359 32) 22 89 84 ⊖ (+359 32) 23 21 75
Director: Mr Dragan Draganov
■ Overhaul of MiG-21, MiG-23, Mi-24, Su-25 A/C and belonging engines, MiG-29 engines. Manufacturer of spare parts. [98]

55-02110 MRF "Gen V Zaymov"
(† ZRRT Bozhuriste)
(a member of TEREM)
2227 Bozhuriste
☎(+359 2) 25 37 23 ⊖ (+359 2) 26 11 59
Managing Director: Banko Bankov
■ Overhaul of radars, subsystems, generators and power plants. [98]

Contact Reference

Bulgaria

55-02300 MRF "Ivaylo"
(† EMZ Ivaylo)
(a member of TEREM)
5000 Veliko Tarnovo
☎(+359 62) 268 61 ⊖ (+359 62) 216 26
Managing Director: Stanislav Rashkov
■ Overhaul of artillery, missiles, infantry weapons and telescope systems. [98]

55-02310 MRF "Tzar Kaloyan"
(† ZRST Razgrad)
(a member of TEREM)
7200 Razgrad
☎(+359 84) 321 91 ⊖ (+359 84) 235 13
Managing Director: Roumen Dobrev
■ Overhaul of radio communication equipment. [98]

55-02350 MRF "Lulin"
(† ARSZ Lulin)
(a member of TEREM)
Sofia
☎(+359 2) 23 93 21 ⊖ (+359 2) 22 90 87
Managing Director: Ivan Balov
■ Manufacture of parts for the MT-LB armoured vehicle and its versions, the T-55 MBT and BMP-23/30 IFV. Production of support equipment for land forces. Overhaul of heavy duty vehicles. [98]

55-02400 MRF "Khan Asparukh"
(† ARZ Khan Asparukh)
(a member of TEREM)
9300 Dobritch
☎(+359 58) 274 49 ⊖ (+359 58) 247 65
Managing Director: Ilya Chalackov
■ Overhaul of MiG-21 fighters, L-29 and L-39 trainers. Manufacture of aircraft spares and ground support equipment. [98]

55-02410 MRF "Khan Kroum"
(a member of TEREM)
7700 Targovishte
☎(+359 601) 453 24; 452 85 ⊖ (+359 601) 466 96
Managing Director: Svetlin Todorov
■ Overhaul of main battle tanks, spare parts production. Production of BMP-23/30 infantry fighting vehicle and 120mm Tundga self-propelled mortar. [98]

55-02450 MRF "Tzar Samuil"
(a member of TEREM)
2030 Kostenetz
☎(+359 7142) 23 20; 28 32 ⊖ (+359 7142) 32 01
Managing Director: Tzvetan Iliev
■ Production of training equipment and ammunition. [98]

55-02460 MRF "Letetz"
(† VRZ Letetz)
(a member of TEREM)
Sofia Airport - Northern Side, Sofia
☎(+359 2) 71 01 39 ⊖ (+359 2) 71 01 19
Managing Director: Alexi Andonov
■ Overhaul of Mil Mi-24, Mi-17, Mi-2, Mi-8 helicopters and piston aero engines. [98]

55-02480 MRF "Ovetch"
(† ZPRGB Ovetch)
(a member of TEREM)
9200 Provadia
☎(+359 518) 21 41; 22 69 ⊖ (+359 518) 22 78
Managing Director: Stoyan Stoyanov
■ Overhaul of all types of wheeled vehicles MTLB IFV, diesel engines. [98]

55-02490 MRF "Vola"
(† ARF Vola)
(a member of TEREM)
3000 Vratza
☎(+359 92) 527 53 ⊖ (+359 92) 527 15
Managing Director: Peter Petrov
■ Overhaul of wheeled armoured vehicles and trucks. Spare parts for armour and automotive equipment. [98]

55-02500 KRZ Flotsky Arsenal
(a member of TEREM)
Varna
☎(+359 52) 77 65 01
✆ (+359 52) 77 80 39 (Marketing) ⊠ 77539
⊖ (+359 52) 77 65 61
Managing Director: Fillip Markovski
Sales Director: Veselin Nikolov ☎(+359 52) 77 80 61
Operations Director: Kiril Anguelov
Finance Director: Petyo Georgiev ☎(+359 52) 77 57 75
■ Construction of small craft up to 100 tonnes displacement. Refitting of warships. Production of spare parts and subsystems. Repair of ships up to 8,000 tonnes light weight. [98]

55-03300 Opticoelectron Ltd
(a subsidiary of Electron Co.)
4500 Pangyurishte
☎(+359 357) 33 67; 41 15; 11 54 ⊠ 46575
⊖ (+359 357) 32 47
👥 2,000 (1997) 🏭 160,000 sq m
General Manager: Ing Mr Tcholakov
Marketing Director: Ing Mr Markov
Operations Director: Ing Mr Ziapkov
■ Manufacture of infantry, artillery and air-defence optical sights, fire control systems, launchers for Factoria, Fagot and Konkours anti-tank missiles, night-vision devices. [98]

55-03500 PIMA Ltd
76 Industrial Street, 3400 Montana
☎(+359 96) 235 59 ✆ (+359 79) 95 41 07; 96 57 60
⊠ 38530 ⊖ (+359 96) 200 51
♧ Lv 500,000,000 (1996) ♣ US$ 9,700,000 (1996)
👥 1,140 (1997) 🏭 9,300
Managing Director: Michael Iordanov
Operations Director: Tsvetan Andreev
Marketing Director: Christo Ratchev
■ Manufacture of 120mm and 82mm mortars, components for infantry weapons and shell bodies. [98]

55-03900 MRF TEREM
34 Totleben Blvd, 1606 Sofia
☎(+359 2) 881 12 01; 51 72 23
⊖ (+359 2) 54 58 51; 54 59 51
President: Maj Gen D Ivanov
■ Overhaul of all modifications of MiG-21. [98]

55-04000 Trema JSC
(† Trema Ltd)
(a subsidiary of Metalchim Holding Inc.)
1 Bedek Straat, 5350 Triavna
☎(+359 677) 24 85 ⊠ 67557 ⊖ (+359 677) 34 11
👥 1,500 (1997) 🏭 80,000
Managing Director: Eng Nikola Draganov
Marketing and Finance Director: Kolyo Slavov
☎(+359 677) 34 15
Operations Director: Slavtcho Dimiev
☎(+359 677) 34 34
Purchasing Manager: Eng Ivan Ivanov
■ Manufacture of 122mm, 115mm rounds. Full charge 122mm. Reduced charge 125mm. High explosive shells. [98]

60 Research Organizations, Institutes & Test Establishments

60-00800 Institute of Metal Science Bulgarian Academy of Sciences
(a subsidiary of the Bulgarian Academy of Sciences)
67 Shipchenski Prohod, 1574 Sofia
☎(+359 2) 70 35 32 [E] stvims@bgcict.acad.bg
⊠ 22150 ⊖ (+359 2) 70 32 07
👥 450 (1996) 🏭 14,000 sq m
Director: Prof Y Arsov ☎(+359 2) 70 11 09
Deputy Director: St Vodenicharov ☎(+359 2) 70 35 32
Finance Director: Ts Keratsova ☎(+359 2) 71 00 25
Marketing Director: G Kaytazov ☎(+359 2) 70 31 88
Research Director: Dr St Vodenicharov
☎(+359 2) 70 35 32
Technical Director: Dr D Nenov ☎(+359 2) 70 33 78
Purchasing Manager: Dr S Todorov
☎(+359 2) 714 24 12
Public Relations Officer: Dr N Popov ☎(+359 2) 72 35 55
■ Development and manufacture of combined armour-plating resistant to kinetic cumulative projectiles for armoured vehicles, the PDM-1/2 anti-assault mines and anti-personnel directional fragmentation mines PMN-150/250. Specialised computing unit for main battle tank artillery fire control systems. [98]

75 Manufacturing Companies

75-03000 Sheraton Sofia Hotel Balkan JSC (DAESO)
(a subsidiary of Daewoo Corporation, South Korea)
5 Steva Nedelya Square, 1000 Sofia
☎(+359 2) 980 97 57; 981 65 41
⊖ (+359 2) 981 00 10 [98]

75-05000 Varna Shipyard
9000 Varna
☎(+359 52) 22 24 58 ⊠ 77514 ⊖ (+359 52) 23 72 22
👥 4 207 🏭 428 969 sq m
Managing Dir: Georgi Dichev
Commercial Dir: Kostadin Gandozov
☎(+359 52) 60 20 56
Technical Dir: Aleko Angelov
☎(+359 52) 60 32 27; 22 27 37
■ Ship construction & repair. [97]

90 Sales Agents, Sales & Liaison Offices

90-05000 Daewoo Corp Sofia Branch
Vaptzarov Boulevard, EXPO 2000, Sofia
☎(+359 2) 963 04 29 ⊠ 24362 dw sof
⊖ (+359 2) 963 00 64 [98]

Burkina Faso

Head of State:
Capt. Blaise Compaoré

10 Head of State

10-10000 Office of the President
Présidence du Faso
PO Box 7030, Ouagadougou
☎(+226) 30 66 30; 30 66 31; 30 66 32
⊠ 5221 pregouv bf ⊖ (+226) 31 05 78
President of the National Council of the Revolution:
Capt Blaise Compaoré [98]

11 Prime Minister

11-10000 Office of the Prime Minister
Ouagadougou
☎(+226) 31 47 17
Prime Minister: Kadré Desiré Ouedraogo [97]

12 Ministries

12-10000 Ministry of Popular Defence
PO Box 496, Ouagadougou
☎(+226) 30 72 14 ☒ 5297; 5333 minidef bf
⊖ (+226) 31 00 86
Minister: Albert D Millogo [97]

12-20000 Ministry of External Relations
PO Box 7038, Ouagadougou
☎(+226) 30 73 14 ☒ 5222 minaffet bf
⊖ (+226) 30 87 92
Minister: Ablassé Ouedraogo [98]

12-40000 Ministry of Territorial Administration
PO Box 7034, Ouagadougou
Minister: Yero Boly [98]

25 Defense Forces

25-20000 Air Force Headquarters
Force Aérienne du Burkina
PO Box 96, Ouagadougou
☎(+226) 30 70 04 ⊖ (+226) 31 33 65
Chief of Staff: Lieut Col Ali Traore [98]

31 Embassies & Attachés

31-00660 Embassy of the Federal Republic of Germany
PO Box 600, Ouagadougou
☎(+226) 30 67 31; 30 67 32 ☒ 5217 diplgerma bf
⊖ (+226) 31 39 91
Ambassador: H E Ms Doretta Loscheleren
Deputy Chief of Mission: Andreas Gögze [98]

31-01890 Consulate of the United Kingdom
No 306 Rue 13,04, BP 3769, Ouagadougou
☎(+226) 31 11 37; 30 02 20 ⊖ (+226) 31 05 03
Honorary Consul: M Jones [98]

31-01900 Embassy of the United States of America
BP 35, Ouagadougou 01
☎(+226) 30 67 23; 30 67 25 ☒ 5290 amemb bf
⊖ (+226) 31 23 68
Ambassador: H E Ms Sharon Wilkinson
Deputy Chief of Mission: John Jone [98]

Burundi

Head of State:
Pierre Buyoya

10 Head of State

10-10000 Office of the President
Bujumbura
☎(+257) 260 63 ☒ 5036 presibu bdi
President of the Republic: Maj Pierre Buyoya
Minister of the Interior and Public Security: Lieut Col Epitace Bayaganakandi [98]

11 Prime Minister

11-10000 Office of the Prime Minister
Bujumbura
☎(+257 22) 63 39
Prime Minister: Pascal-Firmin Ndimira [98]

12 Ministries

12-10000 Ministry of National Defense
Ministère de la Défense Nationale
Bujumbura
☎(+257) 260 63 ☒ 5049 mdn bdi
Minister of National Defense: Lieut Col Firmin Sinzoyiheba [98]

12-20000 Ministry of External Relations and Cooperation
PO Box 1840, Bujumbura
☎(+257) 221 50 ☒ 5065 minirex bdi
Minister of Foreign Affairs and Co-operation: Luc Rukingama [98]

12-40000 Ministry of Interior
Bujumbura
☎(+257) 242 42
Minister of Interior and Public Security: Lieut Col Epitace Bayaganakandi [98]

31 Embassies & Attachés

31-00660 Embassy of the Federal Republic of Germany
22 rue du 18 Septembre, PO Box 480, Bujumbura
☎(+257 2) 264 12 ☒ 5068 aabuj bdi
Ambassador: H E Mr Berrnt Moast [98]

31-01485 Embassy of the Russian Federation
9 Avenue de l'Uprona, PO Box 1034, Bujumbura
☎(+257 2) 260 98 ⊖ (+257 2) 229 84
Ambassador: Igor Liakim-Frolol [98]

31-01890 Embassy of the United Kingdom
43 Avenue Bubanza, PO Box 1344, Bujumbura
☎(+257 2) 237 11 ☒ 5126 intac bdi
Ambassador: H E Miss K Oliver (resident in Kampala)
Deputy Head of Mission: G Scott (resident in Kampala)
Defence Adviser: Lieut Col N P C Lewis (resident in Kampala) [98]

31-01900 Embassy of the United States of America
1720 Avenue des Etats-Unis, BP 1720, Bujumbura
☎(+257) 22 34 54 ⊖ (+257) 22 29 26
Ambassador: H E Mr Morris Hughes
Deputy Chief of Mission: James Hallse [98]

Cambodia

Head of State:
H.R.H. Samdech Norodom Sihanouk

10 Head of State

10-10000 Supreme National Council
Khemarin Palace, Phnom Penh
President of the National Assembly: Chea Sim
First Prime Minister: Prince Norodom Ranariddh
Second Prime Minister: Hun Sen
Deputy Prime Minister and Minister of State: Ing Kiet
Deputy Prime Minister: Sar Kheng
Minister of State: Keat Chhon
Minister of State: Ung Phan
Minister of State: Van Molivann
Minister of State: Chem Sngoun
Minister in the Cabinet of Ministers: Veng Sereivut
Minister in the Cabinet of Ministers: Sok An
State Secretary in the Cabinet of Ministers: Sum Manit
State Secretary in the Cabinet of Ministers: Neou Kanon [98]

12 Ministries

12-10000 Ministry of Defence
Pochetong, Phnom Penh
Co-Minister of National Defence: Tie Banh
Co-Minister of National Defence: Gen Tea Chamrath
State Secretary for National Defence: Ek Sereywath
State Secretary for National Defence: Cheang Sayun
State Secretary for National Defence: Peng Pat [98]

31 Embassies & Attachés

31-01890 Embassy of the United Kingdom
29 Street 75, Phnom Penh
☎(+855 23) 42 71 24 ⊖ (+855 23) 42 71 25
Ambassador: H E Mr George Edgar
Deputy Head of Mission: Lawrence Pickup
Commercial and Public Relations Officer: P Freer [98]

31-01900 Embassy of the United States of America
27 EO St 240, Phnom Penh
☎(+855 23) 42 64 36; 42 64 38 ⊖ (+855 23) 42 64 37
Ambassador: H.E. Mr Kenneth M. Quinn
Deputy Chief of Mission: Robert Porter
Commercial Officer: Ms Helen Hudson
Public Affairs Officer: Frank Huffman [97]

Contact Reference

Cameroon

Head of State:
Paul Biya

10 Head of State

10-10000 Office of the President
Yaoundé
☎(+237) 23 40 25, 22 04 55 ▣ 8207 presid kn
President of the Republic: Paul Biya [98]

11 Prime Minister

11-10000 Office of the Prime Minister
Yaoundé
☎(+237) 23 57 63 ▣ 8257 kn
Prime Minister: Peter Mafany Musonge
Vice Prime Minister, Housing and Town Planning: Hamadou Moustapha
Vice Prime Minister, Territorial Administration: Gilbert Andze Tsoungui [98]

31 Embassies & Attachés

31-00340 Embassy of the Republic of Chad
PO Box 506, Yaoundé
☎(+237) 21 06 24; 20 15 46 Residence
✉POB606 YAOUNDE ▣ ambatchad kn88352
Ambassador: H E Mr H Ouangmotching
Counsellor: Mahamaabdelkerim Addei
First Secretary: M A Tidjani
Attaché: Dono Moywegar [98]

31-00510 Embassy of the Arab Republic of Egypt
PO Box 809, Yaoundé
☎(+237) 20 39 22 ▣ 88360 ⊖ (+237) 20 26 47
Ambassador: H E Mr Nofal Ibrahim El Sayed
Counsellor: Sanber Mansour
Attaché: Ms Laila El Sherbiny
Attaché: Atef Abdel Salam [98]

31-00660 Embassy of the Federal Republic of Germany
PO Box 1160, Yaoundé
☎(+237) 20 05 66; 21 00 56 ▣ 88238 kn
⊖ (+237) 20 73 13
Ambassador: H E Mr Klaus Holderbaum
Deputy Chief of Mission: Ralf Mildebrath
Commercial Attaché: Ms Susan Tapella
Consular Affairs: Oliver Fixson [98]

31-00890 Embassy of Italy
Quartier Bastos, PO Box 827, Yaoundé
☎(+237) 21 21 98; 20 33 76 ▣ 8305 itadipl kn
⊖ (+237) 21 52 50
Ambassador: H E Dr Pietro Lonardo
First Secretary: Dr Enrico Vicenti
Commercial Secretary: Dr Roberta Cocconi [98]

31-01220 Embassy of the Kingdom of Morocco
PO Box 1629, Yaoundé
☎(+237) 20 50 92 🄴 sifamayoun@camnet.cm
▣ 88347 KN ⊖ (+237) 20 37 93
Ambassador: H E Mr Mohamed Benomar
First Secretary, Deputy Chief of Mission: Mohamed Maliki
Second Secretary, Commercial Attaché: Abderrahman Madani
Third Secretary: Mustapha El Marouani [98]

31-01890 British High Commission
Ave Winston Churchill, PO Box 547, Yaoundé
☎(+237) 22 05 45; 22 07 96 ⊖ (+237) 22 01 48
High Commissioner: H E Mr N M McCarthy
Deputy Head of Mission and First Secretary Commercial: Peter Broom
Vice Consul: Ms Sarah O'Donnell
Information Officer: A Thygesen [98]

31-01900 Embassy of the United States of America
Rue Nachtigal, PO Box 817, Yaoundé
☎(+237) 23 40 14 ▣ 8223 amembya kn
⊖ (+237) 23 07 53
Ambassador: H E Mr Charles Twining
Deputy Chief of Mission: Mark Boulware
Economic/Commercial Section: Leroy Smith
Public Affairs Officer: Gerald McLoughlin
☎(+237) 23 04 16
Defense Attaché: Lieut Col James L Cobb [98]

Canada

Head of State:
H.M. Queen Elizabeth II

Governor General:
Romeo Le Blanc

10 Head of State

10-10000 Office of the Governor General
Rideau Hall, 1 Sussex Drive, Government House, Ottawa, Ont K1A 0A1
☎(+1 613) 993 8200 ▣ 533280 rideauhall ott
⊖ (+1 613) 990 7636
Governor General: H.E. Rt Hon. Romeo Le Blanc [98]

11 Prime Minister

11-10000 Office of the Prime Minister
80 Wellington Street,
Langevin Block, Parliament Buildings,
Ottawa, Ont K1A 0A2
☎(+1 613) 992 4211 ▣ 533208 pmo pco ott
⊖ (+1 613) 995 0101
Prime Minister: Hon. Jean Chrétien
Deputy Prime Minister: Herb Gray [98]

12 Ministries

12-10000 Department of National Defence (NDHQ)
101 Colonel Drive, Major General G R Peakes Building, Ottawa, Ont K1A 0K2
☎(+1 613) 922 45 81; 922 45 82
▣ 0534218 comm sqn ott ⊖ (+1 613) 992 45 82
Minister: Arthur Eggleton
Deputy Minister: Ms Louise Fréchette
☎(+1 613) 992 4258
Vice Chief of Defence Staff: Vice Adm L E Murray
☎(+1 613) 992 6052
Deputy Chief of the Defence Staff: Lieut Gen J A Roy
☎(+1 613) 992 3355
Assistant Deputy Minister Policy and Communications: Dr K J Calder ☎(+1 613) 992 34 58
Assistant Deputy Minister Personnel: Lieut Gen David N Kinsman ☎(+1 613) 992 75 82
Assistant Deputy Minister Matériel: P L Lagueux
☎(+1 613) 992 66 22
Assistant Deputy Minister Finance and Corporate Services: Robert M Emond ☎(+1 613) 992 03 59
Assistant Deputy Minister Defence Information Services: Maj Gen J G Leech ☎(+1 613) 995 20 17
Assistant Deputy Minister Infrastructure and Environment: J L Adams ☎(+1 613) 945 75 45
Associate Assistant Deputy Minister Finance and Corp. Services: Rear Adm B M Keeler
Associate Assistant Deputy Minister Personnel: D J Lindley ☎(+1 613) 992 74 43
Associate Assistant Deputy Minister Policy and Communications: Rear Adm J A King
☎(+1 613) 992 27 69
Chief Reserves and Cadets: Maj Gen E W Linden
☎(+1 613) 995 98 02
Chief Review Services: Rear Adm K J Summers
☎(+1 613) 995 85 61
Judge Advocate General: Brig Gen J A G P Boutet
☎(+1 613) 992 30 19
Defence Science Advisory Board: W R Bullock
☎(+1 613) 992 40 73 [97]

12-50000 Ministry of Industry Aerospace and Defence Branch
235 Queen St, Ottawa, Ont K1A 0H5
☎(+1 613) 954 3343 ▣ 534123 ⊖ (+1 613) 941 2379
Minister: John Manley
Director General: Ms Catherine A. Kerr [97]

15 Parliament

15-10000 Senate
Parliament Building, Wellington Street, Ottawa, Ont K1A 0A4
☎(+1 613) 995 19 00 ⊖ (+1 613) 995 03 20
Speaker of the Senate: Hon. Fildas Molgay
☎(+1 613) 992 44 16
Leader of the Government: Hon. Joyce Fairbairn
☎(+1 613) 996 43 82
Clerk of the Senate and Parliament: Paul Bélisle
☎(+1 613) 992 24 93

Senate Communications
Parliament Building, Wellington St 300-VB, Ottawa, Ont K1A 0A4
☎(+1 613) 992 11 49 ⊖ (+1 613) 995 49 98
Director Communications: Mark Audcent (Acting) [98]

15-20000 House of Commons
Parliament Building, Wellington St, Ottawa, Ont K1A 0A6
☎(+1 613) 992 4793 ⊖ (+1 613) 995 4253
Speaker: Hon. Gilbert Parent
Government House Leader: Herb Gray
Deputy Speaker: David Kilgour
Clerk of the House: Robert Marleau
Principal Clerk Research & Exchanges: William C. Corbett

Standing Committee on Foreign Affairs & International Trade
Parliament Building, Wellington St, Ottawa, Ont K1A 0A6
☎(+1 613) 992 3150 ⊖ (+1 613) 996 1626
Chairman: Bill Graham

Standing Committee on National Defence & Veterans Affairs
Parliament Building, Wellington St, Ottawa, Ont K1A 0A6
☎(+1 613) 992 3150 ⊖ (+1 613) 996 1626
Chairperson: Ms Mary Clancy [97]

20 Defense Administrations/Services/Agencies

20-05000 Director Cataloguing and Initial Provisioning (DCIP)
National Defence Headquarters,
Major General Pearkes Building, Ottawa, Ont K1A 0K2
☎(+1 819) 994 9401 ⊖ (+1 819) 994 9398
Director: R.H. Grove
- Provides support to DND capital acquisition programme by analyzing & determining the spares requirements associated with major equipment or systems being introduced in the Canadian Forces; provision of NATO cataloguing services to the Department as well as other Federal Government Depts, NATO allied countries & friendly foreign nations.

Mission:
To provide timely & cost-effective support to the material & inventory management needs of client organisations. [97]

25 Defense Forces

25-19100 Canadian Land Force Command & Staff College
Fort Frontenac, Kingston, Ont K7K 5L0
☎ (+1 613) 541 5010 Ext. 5821 ⊖ (+1 613) 541 5944
Commander: Col J.S Labbé
Deputy Commander: Col J.C.M. Gauthier
Chief of Staff: Lt Col R.M. Gervais
Dir Operations: Maj. A.F. Robertson
Dir Procurement/Logistics: Capt. W.F. Thomas [97]

25-19200 Canadian Forces School of Communications and Electronics (CFSCE)
Stn Forces, PO Box 1700, Kingston, Ont K7K 7B4
☎ (+1 613) 541 52 00 ext. 5243 ⊖ (+1 613) 541 44 89
Commandant: Lt Col A F Hamel ☎ (+1 613) 541 52 00
Chief of Staff: Maj. J G C Lareau
☎ (+1 613) 541 50 10 ext. 5203
Officer Commanding A Squadron: Maj. R M Lilenthal
☎ (+1 613) 541 50 10 ext. 5675
Officer Commanding B Squadron: Maj. M C Burke
☎ (+1 613) 541 50 10 ext. 5230
Officer Commanding G Squadron: Maj. D B Gravel
☎ (+1 613) 541 50 10 ext. 5266
Officer Commanding E Squadron: Capt. C E Penney
☎ (+1 613) 541 50 10 ext. 5615
Officer Commanding Training Support Squadron:
Capt. G A Ritchie ☎ (+1 613) 541 50 10 ext. 5253
Adjutant: Capt. P Laudry ☎ (+1 613) 541 52 16
School Chief Warrant Officer: CWO J D S Lebrun
☎ (+1 613) 541 50 10 ext. 5209 [98]

25-21000 Air Command Headquarters (AIRCOM)
Westwin, Manitoba R3J 0T0
☎ (+1 204) 833 25 00 Ext. 5524 ⊖ (+1 204) 833 20 31
Commander: Lt Gen. A.M. DeQuetteville
Deputy Commander: Maj.Gen. B.G. Horseman
Chief of Staff Operations: Brig.Gen. W R C Little
Chief of Staff Support: Brig.Gen. P. Gartenburg
Chief of Staff Personnel: Brig.Gen. E A Findlay
Chief of Staff Reserves: Brig.Gen. J G Blouin
Commander Maritime Air Group: Brig.Gen. B.N. Cameron
Commander Air Transport Group: Brig.Gen. J.R.B. Proulx
Commander Fighter Group: Brig.Gen. D.M. Jurkowski
Commander 10 Tactical Air Group: Brig.Gen. K.R. Pennie
Command Information Officer: Lt Col R. Adam [98]

25-24100 Aerospace Engineering Test Establishment
Stn Forces, PO Box 6550, Cold Lake, AB T9M 2C6
☎ (+1 403) 840 80 00; 86 96
[E] co@aete.coldlake.dnd.ca ⊖ (+1 403) 840 73 80
Commander: Col D. Matthews
Deputy Commander: Lt Col R.P. Cowden
Technical Officer: Lt Col D.I. Tudor
Dir Support: Lt.Col. K.E. Gladstone
→ 60 [98]

25-25000 Air Cadet League of Canada
Constitution Building, 313 Rideau Street,
Ottawa, Ontario K1N 5Y4
☎ (+1 613) 991 43 49 [E] leaguehq@iosphere.net
⊖ (+1 613) 991 43 47
President: Peter Prescott
Executive Director: Richard Logan
Secretary and Treasurer: Mrs Chantal Fauvelle [97]

31 Embassies & Attachés

31-00130 High Commission for the People's Republic of Bangladesh
85 Range Road, Suite 402, Ottawa, Ont K1N 8J6
☎ (+1 613) 236 0138-39 ⊖ (+1 613) 567 3213
High Commissioner: H.E. Mufleh R. Osmary
Deputy High Commissioner: Mr Anwar-ul-Alam
First Secretary: Mr Mahbubuzzanan [97]

31-00260 Embassy of Burkina Faso
48 Range Road, Ottawa, Ont K1N 8J4
☎ (+1 613) 238 4796 ⊠ 534413 ambabf ott
⊖ (+1 613) 238 3812
Ambassador: H.E. Mr Mouhoussine Nacro
First Counsellor: Eric Y. Tiare
Commercial Attaché: Mrs Mâ Ouedraogo
Attaché: Ms Eve Sanon
Financial Attaché: Hamado Ouedraogo [97]

31-00420 Embassy of the Republic of Côte d'Ivoire
9 Marlborough Avenue, Ottawa, Ont K1N 8E6
☎ (+1 613) 236 99 19 ⊠ 533794 ambivoire ott
⊖ (+1 613) 563 82 87
Ambassador: H E Jean Obeo Coulibaly
Deputy Chief of Mission: Kouamé Konan
Counsellor Commercial Affairs: Kouadio Kouakou [98]

31-00430 Embassy of the Republic of Cuba
388 Main Street, Ottawa, Ont K1S 1E3
☎ (+1 613) 563 01 41 [E] cuba@iosphere.net
⊠ 533 135 cubaemba ott ⊖ (+1 613) 563 00 68
Ambassador: H E Mr Bienvenido García
☎ (+1 613) 563 03 26
Deputy Chief of Mission: Jorge Lamadrid
☎ (+1 613) 563 0209
Public Information Officer: Yuri Gala [98]

31-00450 Embassy of the Czech Republic
541 Sussex Drive, 2nd Floor, Ottawa, Ont K1N 6Z6
☎ (+1 613) 562 3875 [E] czemb@ottawa.net
⊖ (+1 613) 562 3878
Ambassador: H.E. Mr Stanislav Chylek
Counsellor: Václav Prosec
Third Secretary Economic & Trade Affairs: Martin Tlapa
Third Secretary Cultural, Press & Consular Affairs:
Mrs Nora Jurkovicová
Defence, Military and Air Attaché: Maj Gen Jiri Giesl
(resident in Washington DC) [97]

31-00535 Honorary Consul of Estonia
958 Broadview Avenue, Toronto, Ontario M4K 2R6
☎ (+1 416) 461 07 64
[E] 75144.2750@compuserve.com
⊖ (+1 416) 461 04 48
Honorary Consul: Ilmar Heinsoo [98]

31-00580 Embassy of Finland
55 Metcalfe Street, Suite 850, Ottawa, Ont K1P 6L5
☎ (+1 613) 236 2389 ⊠ 534462 finamb ott
⊖ (+1 613) 238 1474
Ambassador: H.E. Mr Veijo Sampovaara
First Secretary Deputy Head of Mission: Roy Eriksson
Military, Naval and Air Attaché: Col Pertti Suominen
(resident in Washington DC)
Assistant Military Naval and Air Attaché:
Col Kari Juhani Kokkonen (resident in Washington DC)
Attaché Information, Culture & Admin: Ms Outi Polón
Consulate General of Finland (Toronto)
1200 Bay Street, Suite 604, Toronto, Ont M5R 2A5
☎ (+1 416) 964 0066 ⊠ 0622513 finconsul tor
⊖ (+1 416) 964 1524
Consul: Mrs Helena Lappalainen
Trade Commission (Toronto)
1200 Bay Street, Suite 604, Toronto, Ont M5R 2A5
☎ (+1 416) 964 7400 ⊖ (+1 416) 964 7400
Consul & Sr Trade Commissioner: Ms Mirja Maja
Consulate General of Finland (Vancouver)
1188 West Georgia Street, Suite 1100,
Vancouver, BC V6E 4A2
☎ (+1 604) 688 4483 ⊠ 0455703
⊖ (+1 604) 687 8237 [97]

31-00590 Embassy of France
42 Sussex Drive, Ottawa, Ont K1M 2C9
☎ (+1 613) 789 17 95 [E] presse@amba-ottawa.fr
[W] http://www.amba-ottawa.fr ⊠ 533564 ambafrance ott
⊖ (+1 613) 562 37 04
Ambassador: H E Mr Loic Hennekinne
Deputy Chief of Mission: Jean-François Valette
Defense Attaché (Ground, Naval, Air): Col Jacques Michel
☎ (+1 613) 789 40 07/08
Counsellor (Commercial and Economic): Pierre Letocart
Counsellor (Cultural Affairs): François Trecourt
Scientific Affairs Section
464 Wilbrod Street, Ottawa, Ont K1M 6M8
☎ (+1 613) 238 10 51/52 ⊖ (+1 613) 238 78 84
Counsellor (Scientific Affairs): Bernard Bach [98]

31-00660 Embassy of the Federal Republic of Germany
PO Box 379, Postal Station 'A', Ottawa, Ont K1N 8V4
275 Slater Street, 14th Floor, Ottawa, Ont K1P 5H9
☎ (+1 613) 232 1101/05
[E] 100566.2620@compuserve.com
[E] bn555@freenet.carleton.ca
[W] http://www.docuweb.ca/Germany/
⊠ 534226 aa ott ca ⊖ (+1 613) 594 9330
Ambassador: H.E. Dr Hans-Guenter Sulimma
Deputy Chief of Mission: Dr Juergen Hellner
Defence Attaché: Lt Col Christian Ibrom
Asst to Defence Attaché: Thomas Kaup
Counsellor (Economic and Commercial Affairs):
Mathias von Polenz
First Secretary (Economic & Commercial Affairs):
Joachim Holzenberger
First Secretary (Press): Ulrich Koehn [97]

31-00690 Embassy of Greece
76-80 MacLaren Street, Ottawa, Ont K2P 0K6
☎ (+1 613) 238 62 71/72/73 [E] grnewsca@globalx.net
⊠ 053 3852 greek emb ott ⊖ (+1 613) 233 56 76
Ambassador: H E Mr John Thomoglou
Deputy Chief of Mission: Constantine Glovas
Defence Attaché: Cdr John Kidoniefs
☎ (+1 613) 233 23 83
Commercial Officer: E Thanassas ☎ (+1 416) 515 01 35
Public Information Officer: George Dardavillas
☎ (+1 613) 232 67 96 [98]

31-00810 Embassy of the Republic of Hungary
299 Waverley Street, Ottawa, Ont K2P 0V9
☎ (+1 613) 230 27 17
[W] http://www.docuweb.ca/hungary
⊖ (+1 613) 230 75 60
Ambassador: H E Mr Károly Gedai ☎ (+1 202) 230 27 17
Deputy Chief of Mission: Istvan Torzsa
Counsellor Economic Affairs Press, Cultural Attaché:
Norbert Konkoly ☎ (+1 613) 230 4859
Military Attaché: Col Tamás Tóth
(resident in Washington DC) ☎ (+1 202) 362 6730 [98]

31-00830 High Commission of India
10 Springfield Road, Ottawa, Ont K1M 1C9
☎ (+1 613) 744 37 51 [E] hicomind@ottawa.net
[W] http://www.docuweb.ca/india ⊠ 534172 hicomind ott
⊖ (+1 613) 744 09 13
High Commissioner: H E Mr Prem K Budhwar
Deputy High Commissioner: A K Banerjee
Commercial Secretary: Shri Chand [98]

31-00840 Embassy of the Republic of Indonesia
55 Parkdale Avenue, Ottawa, Ont K1Y 1E5
☎ (+1 613) 724 11 00 ✉ INDONESIA Ottawa
⊠ 533119 indonesia ott ⊖ (+1 613) 724 11 05
Ambassador: H.E. Mr Benjamin Parwoto
Minister Counsellor: Ms Sulastri S Martosoendjojo
Commercial Attaché: Doddy Hidajat
Defence Attaché: Brig.Gen. Sudradjat
Economic Counsellor: Chaidir Siregar
Information Officer: Faisal Bafadal [98]

31-00850 Embassy of the Islamic Republic of Iran
245 Metcalfe Street, Ottawa, Ont K2P 2K2
☎ (+1 613) 235 47 26; 233 47 26 (Consulate)
[E] iranemb@salamiran.org [W] http://www.salamiran.org
⊠ 534229 iran ott
⊖ (+1 613) 232 57 12; 236 47 26 (Consulate)
Ambassador: H E S M H Adeli
Commercial Officer: M S Ghaemi Nadhafi
Public Information Officer: A Sadeghi
Attaché for Administration: I Anssani [98]

Contact / Reference

31-00910 High Commission of Jamaica
275 Slater Street, Suite 800, Ottawa, Ont K1P 5H9
☎(+1 613) 233 93 11/13/14 ⊠ 533287 jamhicom ott
⊖ (+1 613) 233 06 11
High Commissioner: H E Ms Maxine Eleanor Roberts
Deputy High Commissioner: Mrs Ann Scott
First Secretary: Miss Caroline Blake
Attaché: Mrs Allison Oliver [98]

31-00930 Embassy of the Hashemite Kingdom of Jordan
100 Bronson Avenue, Suite 701, Ottawa, Ont K1R 6G8
☎(+1 613) 238 80 90 ⊠ 534538 urduniyah ott
⊖ (+1 613) 232 33 41
Ambassador: H.E. Mr Hani Khalifeh [98]

31-00990 Embassy of the Republic of Korea
151 Slater Street, 5th Floor, Ottawa, Ont K1P 5H3
☎(+1 613) 232 1715 ⊖ (+1 613) 232 0928
Ambassador: H.E. Mr Kee Bock Shin
Minister: Mr Young-Jo Jung
Defence Attaché: Col Oh Hyun Joh
Counsellor Political Affairs: Mr Byung Jin Min
Counsellor Cultural & Press: Mr Young Ho Cho
Counsellor Economic: Mr Jong Kon Yoon
Counsellor Commercial: Mr Hae jung Oh [97]

31-01020 Embassy of Lebanon
640 Lyon Street South, Ottawa, Ont K1S 3Z5
☎(+1 613) 236 5825/55 ⊠ 533571 ambaliban ott
⊖ (+1 613) 232 1609
Ambassador: H.E. Dr Assem Salman Jaber
Counsellor: Michel Haddad [97]

31-01110 Malaysian High Commission
60 Boteler Street, Ottawa, Ont K1N 8Y7
☎(+1 613) 241 51 82 ✉ MALAWAKIL Ottawa
⊠ 533520 ⊖ (+1 613) 241 52 14
High Commissioner:
H.E. Dato' Abdullah Zawawi bin Haji Mohamed
Second Secretary: Jojie Samuel
Second Secretary (Administration): Abu Bakar bin Hussein
Defence Attaché: Col Chia Chan Sing
(resident in Washington DC)
Assistant Defence Attaché: Maj Mohd Nor Dand
(resident in Washington DC) [98]

31-01190 Embassy of Mexico
45 O'Connor Street, Suite 1500, Ottawa, Ont K1P 1A4
☎(+1 613) 233 89 88; 92 72; 99 17
[E] nstn0281@fox.nstn.ca ⊠ 534520
⊖ (+1 613) 235 91 23
Ambassador: H E Mrs Sandra Fuentes
Deputy Chief of Mission: Francisco Olguín
Minister Economic Affairs: Minister Alonso de Gortari

Office of Military & Air Attachés
45 O'Connor Street, Suite 1500, Ottawa, Ont K1P 1A4
☎(+1 613) 232 64 78 ⊖ (+1 613) 232 01 12
Military and Air Attaché: Col Daniel Velasco
Assistant Military and Air Attaché: Lt Col Sergio Martínez [98]

31-01270 Royal Netherlands Embassy
Suite 2020, 350 Albert Street, Ottawa, Ont K1R 1A4
☎(+1 613) 237 5030/36 ⊖ (+1 613) 237 6471
Ambassador: H.E. Mr J.H.W. Fietelaars
Deputy Chief of Mission: A.H. Huitzing
Defence Attaché: Col W. Hageman
Head of Commercial Department: Mr Fred H. Olthof
Press & Cultural Affairs: Mrs M. Bos [97]

31-01300 New Zealand High Commission
99 Bank Street, Metropolitan House, Suite 727,
Ottawa, Ont K1P 6G3
☎(+1 613) 238 59 91 [E] nzhcott@istar.ca
⊖ (+1 613) 238 57 07
High Commissioner: H E Hon Jim Gerard
Deputy High Commissioner: Stephen Jacobi
Defence Adviser: Gp Capt. Peter Stockwell
Public Affairs Officer: W Alexander

New Zealand Consulate General
Suite 1200, 888 Dunsmuir Street,
Vancouver, BC V6C 3K4
☎(+1 604) 684 73 88 ⊖ (+1 604) 684 73 33
Trade Commissioner: Stephen Bryant [98]

31-01370 High Commission for Pakistan
151 Slater Street, Suite 608, Burnside Bldg,
Ottawa, Ont K1P 5H3
☎(+1 613) 238 7881-83 ✉ PAHIC Ottawa
⊠ 534428 parep ott ⊖ (+1 613) 238 7296
High Commissioner: H.E. Dr Farouk Rana
First Secretary: Muhammed Sarfraz A. Khamzada
First Secretary: Abdul Latif
Press Attaché: Khalid Sarwar
Defense & Military Attaché: Brig. Javed Hussain
Air Attaché: Gp Capt. Syed Ghazanfar Hussain

Trade Commision
3421 Peel Street, Montreal H3A 1W7
☎(+1 514) 845 2297-98 ⊠ 055 62154
⊖ (+1 514) 845 1354
Trade Commisioner: Hassan Wasim Afzal [97]

31-01480 Embassy of Romania
655 Rideau Street, Ottawa, Ont K1N 6A3
☎(+1 613) 789 3709; 789 5345 ⊠ 533101 romamb ott
⊖ (+1 613) 789 4365
Counsellor and Chargé d'Affaires: Radu Mateescu
Counsellor Economic Affairs: Viorel Onel
Defence, Military, Air and Naval Attaché: Col Vasile Ilies
(resident in Washington DC)
Second Secretary: Gabriel Coropciuc
Second Secretary: Nicanor Teculescu
Third Secretary: Nicolae Liute
Administration Attaché: Dan Constantinescu [97]

31-01655 Embassy of the Republic of Slovenia
150 Metcalfe Street, Suite 2101,
Ottowa, Ontario K2P 1P1
☎(+1 613) 565 57 81; 565 57 82
⊖ (+1 613) 565 57 83
Ambassador: H E Marijan Majcen
Minister: Mitja Strukelj
Defence Attaché: Col Mitja Miklavec [98]

31-01670 South African Embassy
15 Sussex Drive, Ottawa, Ont K1M 1M8
☎(+1 613) 744 0330 ⊖ (+1 613) 741 1639
Ambassador: H.E. Mr B.I.L. Modise
Counsellor (Information): Ms J.M. Kenny [97]

31-01690 High Commission for the Democratic Socialist Republic of Sri Lanka
333 Laurier Avenue West, Suite 1204,
Ottawa, Ont K1P 1C1
☎(+1 613) 233 8449 ⊖ (+1 613) 238 8448
High Commissioner: H.E. Mr A.C. Goonasekera
Minister Counsellor: S.B. Weragama
First Secretary (Commercial): Mrs Sonali Wijeratne
Third Secretary: A.L. Ratnapala [97]

31-01730 Swedish Embassy
377 Dalhousie Street, Ottawa, Ont K1N 9N8
☎(+1 613) 241 8553 ✉ SVENSK Ottawa
⊠ 533331 svensk ott ⊖ (+1 613) 241 2277
Ambassador: H.E. Mr Ian Stahl
Deputy Chief of Mission: Lennart Linnér
Defence Attaché: Capt. Björn Hamilton (Navy)
Press Attaché: Lisette Lindahl Owens [97]

31-01820 High Commission for Trinidad & Tobago
75 Albert Street, Suite 508, Ottawa, Ont K1P 5E7
☎(+1 613) 232 24 18/19 [E] tthcotta@travel-net.com
⊖ (+1 613) 232 43 49
Deputy High Commissioner: Stephen Kanoal [98]

31-01840 Embassy of the Republic of Turkey
197 Wurtemberg Street, Ottawa, Ont K1N 8L9
☎(+1 613) 789 40 44; 34 40 ⊠ 534716 turkel ott
⊖ (+1 613) 789 34 42
Ambassador: H E Mr Ömer Ersun
Counsellor: Hasan Servet Öktem
Military, Naval and Air Attaché: Col Ergin Ersoy
Commercial Counsellor: Kadie Bal
Deputy Counsellor (Economic): M B Tütüncü [98]

31-01890 United Kingdom High Commission
80 Elgin Street, Ottawa, Ont K1P 5K7
☎(+1 613) 237 15 30 ✉ BRITMILREP OTTAWA
⊠ 533318 ukrep ott ⊖ (+1 613) 232 40 30
High Commissioner: H E Anthony Goodenough
Deputy High Commissioner: Miss L Duffield
Counsellor (Economic and Commercial): B McCleary
Naval and Air Adviser: Gp Capt. T J Willams
Head of BDLS and Defence Attaché: Brig. E P O Springfield
Head of Secretariat BDLS: Lt Cdr R G Blair [98]

31-01990 Embassy of the Republic of Yemen
350 Sparks Street, Suite 1100, Ottawa, Ont K1R 7S8
☎(+1 613) 232 8525 ⊖ (+1 613) 232 8276
Ambassador: H.E. Dr Mohamed Saed Ali
Minister, Deputy Chief of Mission: Ismail Al-Kibsi
Counsellor: Ahmed Ghaleb
First Secretary: Abdullah Fadhel [97]

31-02010 Embassy of the Federal Republic of Yugoslavia
17 Blackburn Avenue, Ottawa, Ont K1N 8A2
☎(+1 613) 233 6289 ⊖ (+1 613) 233 7850
Chargé d'Affaires a.i.: Aleksandar Mitic
Attaché: Dragoslav Stojanovic
Counsellor: Mrs Ljiljana Pavicevic
Military Attaché: Col Stanko Nesic
(resident in Washington DC) [97]

31-02040 High Commission of the Republic of Zimbabwe
332 Somerset Street West, Ottawa, Ont K2P 0J9
☎(+1 613) 237 43 88; 237 44 84
[E] zim.highcomm@sympatico.ca
[W] http://www.docuweb.ca/zimbabwe
⊖ (+1 613) 563 82 69
High Commissioner: H E Ms Lillie Chitauro
Counsellor: S R Marufu
Defence Attaché: Lieut Col L G Masina
(resident in Washington DC) ☎(+1 202) 332 71 00
First Secretary (Commercial): Wilbert C T T Dumba [98]

60 Research Organizations, Institutes & Test Establishments

Aerospace Engineering Test 25-24100
CRL Technologies 75-03760
Racal Filter Technologies Ltd 75-11600

60-00050 APREL Laboratories
51 Spectrum Way, Nepean, Ont K2R 1E6
☎(+1 613) 820 27 30 [E] info@aprel.com
[W] http://www.aprel.com ⊖ (+1 613) 820 41 61
Chairman and Chief Executive Officer: Dr Jacek J Wojcik
President: Ms Kathy MacLean
■ An independent research laboratory providing engineering & test services, specializing in electromagnetics, acoustics & electroacoustics.

NCL Calibration Laboratories
51 Spectrum Way, Nepean, Ont K2R 1E6
☎(+1 613) 820 49 88 ⊖ (+1 613) 820 41 61
Chairman and Chief Executive Officer: Dr Jack Wojcik
Presdient: Ms Kathy Maclean
Operations/Technical Director: Dr Paul Cardinal
■ Provides comprehensive calibration services for electronic and other instrumentation, particularly EMI/EMC and radio instrumentation, shielding field testing.

Spectrum Sciences Institute
51 Spectrum Way, Nepean, Ont K2R 1E6
☎(+1 613) 820 64 71 ⊖ (+1 613) 820 41 61
Chairman and Chief Executive Officer: Dr Jacek Wojcik
Director Business Development: Brian J Hall
■ The Institute brings together professionals from industry, academia and government, who are concerned with non-ionzing radiation phenomena, applications, hazards and protection/mitigation.
Provides research, engineering and testing tools through its associated laboratories. Conference and meeting facilities are available. [98]

60-00300 Canadian Institute of Strategic Studies
2300 Yonge Street, Suite 402, PO Box 2321,
Toronto, Ont M4P 1E4
☎ (+1 416) 322 81 28 E info@ciss.ca
W http://www.ciss.ca ⊖ (+1 416) 322 81 29
Executive Director: Alex Morrison
Associate Executive Director: Jim Hanson
Director of Finance and Administration: Mark Larsen 98

60-00410 Defence Research Establishment Atlantic (DREA)
(a division of the National Defence Headquarters)
PO Box 1012, Dartmouth, NS B2Y 3Z7
☎ (+1 902) 426 31 00 W http://www.crad.dnd.ca
⊖ (+1 902) 426 96 54
200 (1997)
Managing Director: Dr P Bhartia
SSO Plans: B A Trenholm 98

60-00420 Defence Research Establishment Ottawa (DREO)
Department of National Defence, Ottawa, Ont K1A 0Z4
☎ (+1 613) 998 2303 E mbd@dreo.dnd.ca
⊖ (+1 613) 991 2964
Director General: Dr L. John Leggat
Deputy Dir: Dr Raymond Carbonneau
General Chief Scientist: Gordon Marwood
☎ (+1 613) 998 2569; 998 2237
Manager: W. Roy McPherson
Business Development: Ms S. McIntyre
☎ (+1 613) 998 2203; 998 2657 97

60-00440 Defence Research Establishment Suffield (DRES)
PO Box 4000, Medicine Hat, Alta T1A 8K6
☎ (+1 403) 544 46 63 ⊖ (+1 403) 544 33 88
Library Manager: Jerry Fitzgerald E jfitzg@dres.dnd.ca
Head of Information Services Group: Jerome Fitzgerald 98

60-00450 Defence Research Establishment Valcartier (DREV)
Centre de Recherche pour la Défense Valcartier (CRDV)
2459 Boulevard Pie-XI Nord, Val-Belair, QC G3R 1X5
☎ (+1 418) 844 42 55; 844 44 96 E info@drev.dnd.ca
W http://www.drev.dnd.ca ☒ 513098 dre vctr
⊖ (+1 418) 844 47 00
US$ 38,000,000 (1997) 340 25,000 m2
Chief Defence Research Establishment Valcartier: Dr Jacques Gilbert
Director General: Dr Robert S Walker
Operations Director: Jean-François Drolet
Research Director: Dr Dennis Faubert
Plans and Programs Office Director: Lieut Col Maurice Coderre
Chief Scientist: Maurice Laviolette
Public Relations Officer: Mrs Lise Ladouceur
■ R&D in fields of ballistics, ordnance, command & control, surveillance and remote sensing, lasers, rocketry, explosives, pyrotechnics, guidance and control, systems analysis. 98

60-00500 Esquimalt Defence Research Detachment (EDRD)
(† Defence Research Establishment Pacific)
(a div. of Defence Research Establishment Atlantic)
FMO, Victoria, BC V0S 1B0
☎ (+1 604) 363 28 67 ⊖ (+1 604) 363 28 56
60
Director: Dr K.N. Street 97

60-01300 Mustang Survival Corp.
(† META Research Inc)
101-3840 Jacombs Road,
Richmond, British Columbia V6V 1Y6
☎ (+1 604) 270 36 31 ⊖ (+1 604) 273 64 66
25
President: Dwight J Davies
Vice President Government Products: Robert Askew
Project Engineering Manager: S F Portman
■ Services related to military personnel, safety and survival equipment. Basic and applied research, product development, prototyping and testing. Pattern design and grading, preparation of manufacturing, drawings and specifications, preparation of user manuals. Research, development, technical investigations and technical data packaging. 98

60-01500 Operational Research Division (ORA)
(† Operational Research & Analysis)
Maj.Gen. George R. Pearkes Bldg., National Defence HQ, Ottawa, Ont K1A 0K2
☎ (+1 613) 995 0418 ⊖ (+1 613) 992 3342
Dir General Operational Research: P.R. Anderson
☎ (+1 613) 992 6866 97

65 Industry, Trade & Armed Forces Associations

65-00100 Aerospace Industries Association of Canada (AIAC)
Association des Industries Aerospatiales du Canada
60 Queen Street, Suite 1200, Ottawa, Ontario K1P 5Y7
☎ (+1 613) 232 42 97 W http://www.aiac.ca
⊖ (+1 613) 232 11 42
Chairman: Robert Brown
President and Chief Executive Officer: Peter R Smith
Operations Director: Peter S Boag
Public Affairs Director: Ray Windsor
■ National trade organisation of Canadian companies involved in the aerospace sector.
Committees:
Airworthiness
Contracts and Finance
Customs and Traffic
Electronics Systems
Human Resource Development
International Exhibitions
Product Support
Science and Technology
Standards and Quality
Trade and Marketing 98

65-00300 Shipbuilding Association of Canada
Association de la Construction Navale du Canada
222 Queen Street, Suite 1502, Ottawa, Ont K1P 5V9
☎ (+1 613) 232 71 27 ⊖ (+1 613) 238 55 19
President: Peter Cairns
■ Represents the interests of Canadian companies involved in: shipbuilding & ship repair, manufacture of marine equipment and products. Ship design, marine electronic and electrical engineering services. Systems integration & project management services. 98

70 Financial Holding Groups, Parent Companies

70-03100 CAE Inc
Royal Bank Plaza - Suite 3060, PO Box 30,
Toronto, Ontario M5J 2J1
☎ (+1 416) 865 00 70 ✆ (800) 760 06 67
W http://www.cae.ca ☒ 623053 ⊖ (+1 416) 865 03 37
C$ 142,046,000 (1997) ♣ C$ 867,344,000 (1997)
6,400 (1997)
Chairman: David H Race
President and Chief Executive Officer: John E Caldwell
Vice President Finance and Chief Executive Officer: Paul G Renaud
Vice President Human Resources: Ms Ruth Brothers
Vice President Marketing and Corporate Relations: Robert E Waite
Vice President Business Development Affairs: Allan M Bignell
■ Flight simulators and visual systems: supervisory control and data acquisition systems; airborne magnetic anomaly detection equipment and magnetometers; aircraft systems training devices; maintenance; repair; overhaul and calibration of electronic and electro-mechanical instruments and systems. Aircraft maintenance modification, repair and overhaul. Aircraft interiors and avionics.
Subsidiary companies and divisions:
CAE Electronics Ltd
CAE Electronics GmbH, Germany
CAE Aviation Ltd
CAE Electronics Australia (Pty)
CAE Ransohoff Inc
CAE Vanguard
CAE Machinery Ltd
CAE Screenplates
CAE Trislot
CAE Electronics Inc 98

70-07000 IMP Group Ltd
Suite 400, 2651 Dutch Village Road, Halifax, NS B3L 4T1
☎ (+1 902) 873 22 50 E impaero@impgroup.com
W http://www.impgroup.com ☒ 1921668 imp leckie hfx
⊖ (+1 902) 873 22 90
1,000 (1997) 250,000 sq ft
President: Kenneth C Rowe
Director International Marketing: Derek W Kinsman 98

70-09000 Northern Telecom Ltd
Corporate Headquarters
2920 Matheson Boulevard East,
Mississauga, Ont L4W 4M7
☎ (+1 903) 238 70 00 ⊖ (+1 905) 238 73 50
65,000
Chief Executive Officer: Jean C Monty
President and Chief Operating Officer: John A Roth
Senior Vice President Corporate: C Scott
■ Leading supplier of fully digital telecommunications switching equipment. 97

74 Consortia, Joint Venture Companies

74-01000 The Aerospace Consortium Inc.
730 Gana Court, Mississauga, Ont L5S 1P1
☎ (+1 905) 564 66 01 ⊖ (+1 909) 670 16 95
President: William H Reil
Vice President Business Development: F Davenport
Secretary: Paul MacPherson
■ Aerospace & defence components, subsystems & systems: machined components & assemblies, military communications & electronics equipment, whip antennas, composite parts, repair & overhaul, programme management, meteorological equipment. → 75 → 80 98

74-19100 Simunition
(a division of SNC Industrial Technologies Inc.)
5 Montée des Arsenaux, Le Gardeur, Que J5Z 2P4
☎ (+1 514) 581 59 99 ✆ (800) 465 82 55
20
Vice President International Marketing and General Manager: Robert Leboeuf
North America Sales Director: Brian Berger
☎ (+1 860) 589 94 18 ☎ (+1 860) 589 94 21 (fax)
PR & Advertising Mgr/Sr Co-ordinator Int'l Sales: Ms Lise Laberge
Co-ordinator Sales US: Ms Kim Hodnett
Training Co-ordinator: Denis Renaud
■ Small calibre training ammunition for safer and more realistic training. Marking cartridges for man on man/force on force training. Short range training ammunition. Non-penetrating ammunition for use in improvised facilities. Frangible, lead free ammunition for use on steel targets. → 75 98

75 Manufacturing Companies

Aero Machining Ltd 80-00020
Array Systems Computing Inc. 80-00060
Rohde & Schwarz Canada Inc 80-01825
SED Systems Inc 80-01937
Simunition 74-19100
The Aerospace Consortium Inc. 74-01000

75-00010 21st Century Airships Inc
180 Main Street, PO Box 177,
Newmarket, Ontario L3Y 4X1
☎ (+1 905) 898 62 74 ⊖ (+1 905) 898 72 45
President and Chief Executive Officer: Hokan Colting
■ Research, development and manufacture of airships and related products. Spherical airships. 98

75-00300 Aerotech International Inc
100 Eagle Drive, Winnipeg, Manitoba R2R 1V5
☎(+1 204) 633 19 99 ✆ (800) 486 43 28
W http://www.aerotech-herman-nelson.com
⊖ (+1 204) 694 16 12
C$ 12,000,000 (1997) ♣ C$ 6,000,000 (1997)
40 (1997) 45,000 sq ft
President: Paul Sigurdson III
Vice President Finance: Fran Koch
Vice President Sales: Peter Marykuca
Production Manager: Mike Stanley
■ Supplier, manufacturer and military contractor of portable heaters, canvas ducting and aerospace and airline ground support equipment for military applications including inventories and Herman Nelson air filter material.
® USA: Herman-Nelson [98]

75-00430 Aircraft Appliances & Equipment Ltd (AAE)
150 East Drive, Bramalea, Ontario L6T 1C1
☎(+1 905) 791 16 66 ⊖ (+1 905) 791 72 18
♣ C$ 10,000,000 (1995) 80 80,000 sq ft
President and Chief Executive Officer: Bryan T Dawson
Sales and Marketing Director: Andrew Willnecker
Materials Director: Teng Carino
■ Repair and overhaul of military and commercial aircraft equipment, ground support equipment, power supplies, generator test stands and all types of test equipment. Design and fabrication of specialty test equipment power supplies and load banks. Design and development of coalescers for the separation of water and solid contaminants from lubricating and fuel systems for marine gas turbine power plants. Micronic filters. Design and manufacture of airborne tachometer generators, aircraft windshields. Surveillance support equipment.
® USA: Cortec Corporation, GEC Plessey Aerospace, Keco Industries, Pilkington Aerospace Inc, Rapid Power Technologies Inc, Roller Bearing Corporation of America, Smiths Industries [97]

75-00470 AlliedSignal Aerospace Canada
(a division of AlliedSignal Canada Inc)
255 Attwell Drive, Etobicoke,
Etobicoke, Ontario M9W 6L7
☎(+1 416) 675 14 11 ▣ 06989142
⊖ (+1 416) 675 40 21
♣ C$ 235,000,000 (1997) 1,400 (1997)
34,340 m2
President: James Close
Vice President Business Development and Government Affairs: David O'Blenis
Engineering Director: Len Grove
Sales and Marketing Director: John Hawkes
Site General Manager: George Nguyen
Public Relations and Communications: Ms Jo-Anne O'Quinn
■ Manufacture of electronic environmental control systems, missile control actuation systems and customer support. Emergency locator transmitters and ice detection/protection. → 80 → 90

AlliedSignal Aerospatiale Canada Inc
(a divsion of AlliedSignal Aerospace, Canada)
200 Marcel-Laurin Boulevard,
St Laurent, Québec H4M 2L5
☎(+1 514) 744 28 11; 744 74 85 ⊠ AVIATION
▣ 5826688 ⊖ (+1 514) 342 37 95
♣ C$ 90,000,000 550
Site General Manager: Alan Smith
Procurement Manager: John Cousineau
Public Relations Co-ordinator: Ms Griba Furtado
■ Manufacture electronic and mechanical aerospace products, fuel and advanced defence electronic products utilizing electro-optics, thermal imaging, aircraft engine fuel control systems and accessories. Services include overall product life cycle management, including ILS product support, repair and overhaul.
● 255 Attwell Drive, Etobicoke, Ont M9W 6L7
☎(+1 416) 798 67 24 ▣ K3477 ⊖ (+1 416) 798 68 46
Manager, Customer Support: Norman MacPherson
■ On-site service for AlliedSignal Aerospace customers.

Ocean Defense Systems
670 Queen Street, Suite 1200, Ottawa, Ont K1P 5Y7
☎(+1 613) 236 04 41 ⊖ (+1 613) 232 11 42
Regional Director: J David O'Blenis
■ Marketing, sales, services and Government relations.

Atlantic Support Division
Slemon Park, PO Box 194, Prince Edward Island COB 2A0
☎(+1 902) 436 17 60 ⊖ (+1 902) 436 05 70
♣ C$ 7,000,000 50
Site General Manager: Kenneth Forse
■ Repair and overhaul of regional and commuter aircraft engine accessories and equipment, design and manufacture of cable harnesses.

AlliedSignal Aerospatiale Canada Inc
PO Box 2140, Montreal, Quebec H4L 4X8
200 Marcel-Laurin Boulevard
☎(+1 514) 744 28 11 ⊖ (+1 514) 342 37 95
Director, Site Operations: Dr Alan Smith
☎(+1 514) 744 74 10; 748 44 20 (fax)
■ Manufacture and service defence related products.

AlliedSignal Aerospace Marketing Sales and Service
c/o Air British Columbia, 5520 Miller Road,
Richmond, BC V7B 1L9
☎(+1 604) 270 82 64 ⊖ (+1 604) 270 32 70
Field Service Engineer: Carl W Rockel
■ On-site product services and marketing.

AlliedSignal Aerospace Marketing Sales and Service
49 Capistrano Road, Dartmouth, NS B2X 3N8
☎(+1 902) 873 39 38 ⊖ (+1 902) 873 39 19
Field Service Engineer: Jack Wolf
● 500 Cote Vertu Boulevard West, Dorval 005-3,
Dorval, Quebec H4S 1Y9
☎(+1 514) 855 71 93 ⊖ (+1 514) 855 77 37
Field Service Representative: Mark Ostheimer

AlliedSignal Aerospace Marketing Sales and Service
Air Canada, 83 Montreal International Airport,
Dorval, Quebec H4S 1C2
☎(+1 514) 422 73 37 ⊖ (+1 514) 422 69 42
Field Service Engineer: Didier Portela

AlliedSignal Aerospace Marketing Sales and Service
12 Riverwood Grove, Kirkland, Quebec H9J 2X2
☎(+1 514) 694 57 20 ⊖ (+1 514) 694 57 23
Account Manager: Loring McKenzie [98]

75-00475 Alpha Zaicon Tech. Inc.
35A Weston Road, Toronto, Ont M6N 3P1
☎(+1 416) 763 51 58 ⊖ (+1 416) 762 10 54
15
President: Arnold Friedman
Operations Director: Reg Kerr
Sales and Marketing Director: Allan Friedman
Purchasing Manager: Ted Rutledge
■ Manufacture security taut wire fence for perimeter. Microphonic cable security systems for fence, floor concrete buildings & roof areas. Platform Intrusion Detection Systems (PIDS) for subway & light rail transits. [98]

75-00480 Ameridata Canada Ltd
(† Control Data Systems Canada Ltd)
(a subs. of Ameridata Technologies Inc.)
One Antares Drive, Suite 400, Nepean, Ont K2E 8C4
☎(+1 613) 723 1174 ⊖ (+1 613) 225 7971
130
President: Jan Kaminski
Finance & Administration Dir: A. Blais
National Sales Mgr: Blair Nichols
■ Systems integrator providing products & services. Services include network design & implementation, help desk, remote network management & educational services. [97]

75-00500 Anachemia Canada Inc.
PO Box 147, Lachine, Que H8R 1A3
255 Norman, Ville St-Pierre, Que H8R 1M3
☎(+1 514) 489 57 11 W http://www.anachemia.com
⊖ (+1 514) 363 52 81
♣ C$ 40,000,000 (1997) 250 (1997) 15,000 m2
Exec. Vice President: Martin Robinson
VP Operations & Administration: G.R. Quenneville
Manufacturing Dir: Ms J. Moussa
Purchasing Mgr: Andrew Solty
Operations & Administration Mgr: G.R. Quenneville
■ Chemical warfare detection devices: liquid chemical agent detector papers, nerve agent vapour detectors, the C-2, M256, M8, M9, 3-way liquid M28, M29 and M34 chemical agent test kits, various decontamination chemicals and devices. Different configurations available. [98]

75-00600 Andrew Canada Inc
(† Andrew Antenna Co. Ltd)
(a subsidiary of Andrew Corp USA)
606 Beech Street, Whitby, Ontario L1N 5S2
☎(+1 905) 668 33 48 ⊠ ANDCORP whitby
⊖ (+1 905) 430 39 64
♣ C$ 37,000,000 (1995) 98 (1996) 9,000 m2
President: T E Charlton ☎(+1 708) 349 33 00
Vice President Government Antenna Systems: G Tong
Engineering Manager: R Boyko
Accounting Manager: L Ashmore
■ Microwave terrestrial antennas, HELIAX coaxial cable and elliptical waveguide, earth station antennas, radar and navigation aid systems, HF antennas, direction finding tactical antennas, weather radar. Radiax cables, towers, weather sensors and broadcast antennas.
® USA: Andrew Corp [97]

75-00720 Applied Microsystems Ltd
Oceanographic & Geophysical Instr.
2035 Mills Rd, Sidney, BC V8L 3S1
☎(+1 604) 656 0771 ⊖ (+1 604) 655 3655
15 20 000 sq ft
President: Danford Bodrug
Dir Engineering: Tom Dakin
Marketing Mgr: Mike Penny
■ Design & manufacture of state-of-the-art oceanographic, geophysical & water quality monitoring instrumentation. [97]

75-00740 The Armstrong Monitoring Corp. (AMC)
215 Colonnade Road South, Nepean, Ont K2E 7K3
☎(+1 613) 225 95 31 ✆ (800) 465 57 77
E info@armstrongmonitoring.com
W http://www.armstrongmonitoring.com
⊖ (+1 613) 225 69 65
32
President: Lawrence J Armstrong
Chief Executive Officer: Philip Payette
Vice President and General Manager: William L Armstrong
Vice President Technology: Don Segall
Sales and Marketing Manager: Yves Grandmaitre
■ Complete manufacturing of portable & fixed gas detection systems for such gases as propane, methane, hydrogen, chlorine, ammonia, carbon monoxide, carbon dioxide, gasoline & diesel fume emissions, oxygen deficiencies. [98]

75-00750 Atlantis Aerospace Corp. Avalon Information Technologies Inc.
1 Kenview Boulevard, Brampton, Ont L6T 5E6
☎(+1 905) 792 19 81 E info@atlantis.com
W http://www.atlantis.com ▣ 6986766
⊖ (+1 905) 792 72 51
♣ C$ 21,000,000 (1996) 200 (1997) 85,000 sq ft
Managing Dir: Dale Simmons
Operations Director: Richard Turnock
Marketing Programs Manager: Chris Lehman
Purchasing Manager: Larry Dean
■ Training systems/simulators for operational & maintenance training, avionics test equipment such as MIL-STD-1553 bus testers & custom instrumentation/control systems. Digital guidance & control systems for RPVs & target drones. [98]

75-00752 ATS Aerospace Inc.
(† Ballistech Systems)
1250 Marie-Victorin Blvd, St Bruno, Que J3V 6B8
☎(+1 514) 441 9000 ⊖ (+1 514) 441 6789
♣ C$ 10 000 000 45
President: William H. Friend
Marketing Mgr: William LaBerge
Marketing Co-ordinator: C. Tartagia
■ R&D & fabrication of supersonic ballistic radar targets, ATC radar environment, visual tower and communications simulators, ATC signal light gun, special instrumentation including stand-alone digital recorders & blast wave instrumentation. [97]

75-00800 Autonav Marine Systems Inc
50 Electronic Avenue, Port Moody V3H 2R8
☎(+1 604) 461 4061 ⊖ (+1 604) 461 4062
Managing Director: Paul Wagner
Operations Manager: Shafiq Gulamani
■ Manufacture of marine steering systems and autopilots. [98]

75-01080 Barrday Inc.
75 Moorefield Street, PO Box 790,
Cambridge, Ont N1R 5W6
☎(+1 519) 621 36 20 E abeth@barrday.com
⊖ (+1 519) 621 41 23
President: Michael J. Buckstein
Marketing Mgr: Brian Scherer

■ Manufacture of protective armour: soft body armour, military ballistic protective armour, military helmets, load carriage for combat troops (personal webbing), riot shields. 98

75-01090 BatteryPRO Systems Inc.

(† Batterypro Centres Inc.)
203-1 Avenue NW, PO Box 160,
Black Diamond, Alta TOL OHO
☎ (+1 403) 933 53 93 ✆ (800) 661 94 01
E sales@batterypro.com W http://www.batteryprom.com
⊖ (+1 403) 933 32 33
♣ C$ 2,400,000 (1997) 👥 18 (1997)
President: Frank O Holemans
■ Battery analyzers. Battery reconditioners. Battery chargers. 98

75-01130 Bell Helicopter Textron Canada

(a division of Textron Canada Ltd)
12800 rue de l'Avenir, Mirabel, Que J7J 1R4
☎ (+1 514) 437 27 64 ⊖ (+1 514) 437 20 06
👥 1,900 (1997) 🏠 50,000 m2
President: Robert MacDonald
Vice President: Paul Costanzo
Operations Director: Jean-Marc Lariviere
Sales and Marketing Director: Dennis Lacroix
Purchasing Manager: François Chagnon
■ Totally integrated helicopter manufacture facility producing 206B JetRanger III, 206L LongRanger IV, 407 light singles, 430 intermediate twin, 212, 412 and 412-CF medium twins. 98

75-01400 Bombardier Inc. Defence Services

Corporate Office
10000 Cargo A-4 Street, Montreal International Airport, Mirabel, Que H3B 1Y8
☎ (+1 514) 476 40 00 W http://www.bombardier.com
⊖ (+1 514) 476 44 67
♣ C$ 8,000,000,000 (1996) 👥 41,000 (1996)
Chairman and Chief Executive Officer: Laurent Beaudoin
President and Chief Operating Officer: Anthony I Kalhok
Corporate Public Relations Director: Ms Linda Coates
Corporate Secretary: Roger Carle
■ Technical services for various aircraft, special mission aircraft, aviation related training services, design and manufacture of CL-327 unmanned aerial vehicle systems, facilities management for customers, aircraft conversion and modification, support services including full turnkey operations for both civil and military customers, simulation and training equipment and aerial targets. → 80

Aerospace/defence business units:
Canadair Canada
Short Brothers UK
Learjet USA
de Havilland Canada 98

75-01500 Bristol Aerospace Ltd (BAL)

(a Magellan Aerospace Company)
660 Berry Street, PO Box 874, Winnipeg, Man R3C 2S4
☎ (+1 204) 775 83 31 ⊖ (+1 204) 775 74 94
♣ C$ 120,000,000 (1996) 👥 950 🏠 700,000 sq ft
VP Marketing Contracts: W.A. Matthews
Marketing Dir: E. McCullins
Public Relations Mgr: R.C. Walker
■ CRV-7, 70 mm (2.75 inch) air-to-air, air-to-ground rocket weapon system, Black Brant family of sub-orbital research rocket vehicles; Wire Strike Protection Systems (WSPS) for helicopters; large airframe structures, composites and gas turbine engine "hot section" components.
Repair, overhaul, modification of military helicopters, high performance fixed wing fighter aircraft and gas turbine engine tail pipes, afterburners and thrust augmentors. → 80 98

75-01750 Cadex Electronics Inc.

111-7400 MacPherson Avenue, Burnaby, BC V5J 5B6
☎ (+1 604) 451 79 00 ✆ (800) 565 52 28
E analyzers@cadex.com W http://www.cadex.com/cadex
⊖ (+1 604) 451 79 91
👥 32 🏠 14,000 sq ft
President & CEO: Isidor Buchmann
Sales and Marketing Director: James Lindgren
■ Manufacture battery analysers to test & restore rechargeable batteries for portable radios, notebook computers, cell phones & other portable devices.
Overseas offices:
Cadex Electronics (Switzerland) → 75 98

75-01770 CAE Electronics Ltd

(a subsidiary of CAE Inc.)
8585 Cote de Liesse Road, Saint-Laurent, Que H4T 1G6
☎ (+1 514) 341 67 80 W http://www.cae.ca
TX 5824856 cae mtl; TWX 610 422 3063
⊖ (+1 514) 341 76 99
👥 3,400 🏠 68,500 sq m
President: J Cherry
Vice President Sales and Marketing: A Morris
Vice President Simulation Engineering and Manufacturing: H Macramallah
Vice President Business Development: M Rutherford
Director Communications: Luc Ferraudec
Vice President Program Management: R Glass
Director Government Affairs: Q Nadcau
Vice President Systems Engineering: L H White
Vice President Systems Engineering and Development: B Sinott
Sales and Marketing Director: Alberto Costa
Purchasing Manager: Ted Bronk
Trade Show Manager: John Paterson
■ Design & manufacture commercial, military flight simulators, visual systems nuclear & fossil power plant simulators, magnetic anomaly detection systems, control systems for hydro, nuclear, air traffic, marine & space applications. 98

75-01800 Calian Technology Ltd

(† Calian Communications Systems Ltd)
(a subsidiary of Calian Technology Ltd)
300 Legget Drive, Kanata, Ont K2K 1Y5
☎ (+1 613) 599 72 00 W http://www.calian.com
TX 534164 ⊖ (+1 613) 592 77 71
♣ C$ 56,000,000 (1996) 👥 700+ (1997)
President: Larry O'Brien
Vice President: Ed Lambert
Director of Corporate Finance: Shaun McEwan
Director of Public and Government Relations: Greg Strong
■ Provides advanced technology services and products to the satellite communications, space and defence industries.
Designs and manufactures advanced communications systems and equipment for military and commercial applications. The advanced systems division performs system design studies analyses and simulations, mainly in communications and electronic warfare. The monitoring systems division produces computer-controlled instrumentation for communications monitoring, research and simulation (ECM, ECCM, ESM) The telemetry systems division produces DATA downlink systems designed to transmit high speed information in real time from an aircraft to a ship or ground station. 98

75-01900 Bombardier Inc Defence Systems Division

(a member of Bombardier Inc., Services Group)
10,000 Cargo A-4 Street, Montreal International Airport, Mirabel, Que J7N 1H3
400 Chemin de la Côte Vertu, Dorval, Qué H4S 1Y9
☎ (+1 514) 476 40 00 ⊖ (+1 514) 476 44 60
👥 1,000
President: David Huddleston
Vice President Business Development and Administration: Jean MacGibbon
Public Relations: Richard Descarries
■ Provides a wide range of aircraft systems support and services to military customers including system and structural engineering, special mission aircraft, aircraft maintenance, logistics services and aviation training. It has developed a VTOL unmmaned air vehicle.

Defence Systems Division
Canadian Aviation Training Centre,
Barker Avenue, 6th Avenue East,
Southport, Man ROH 1NO
☎ (+1 204) 428 40 00 ⊖ (+1 204) 428 43 76
Director: Ken Carr
■ The Center offers a comprehensive international aviation training programme for military operators. Courses include: primary flying training, multi-engine training, rotary wing training & continuation flying training. 98

75-02250 Canadian Body Armour Ltd

(† Bristoll Group)
8056 Torbram Rd, Brampton, Ont L6T 3T2
☎ (+1 905) 799 2999 ⊖ (+1 905) 799 2890
♧ $ 60 000 ♣ $ 10 000 000 👥 25 (1996) 🏠 500 sq m
President: G. Isberg
Director: C. Bristoll
■ Body armour, safety wear, police, military, clothing. 97

75-02600 Canadian Marconi Co. (CMC)

(a subsidiary of The General Electric Company plc, UK)
Executive Office
600 Dr Frederik Philips Boulevard,
Ville Saint-Laurent, Que H4M 2S9
☎ (+1 514) 748 30 46 W http://www.marconi.ca
⊖ (+1 514) 748 31 84
♧ C$ 274,552,000 (1994) ♣ C$ 250,353,000 (1995)
👥 1,750 🏠 125,000 sq m
Chairman CMC Montreal: William I Mackenzie Turner (Jr)
President and Chief Executive Officer Montreal: Carmen Lloyd
Vice President Business Development and Strategic Planning: Gregg Fawkes
Vice President Human Resources Montreal: Jean-Claude Gauthier
Vice President Commercial Communications Division Kanata: Bryan Locker ☎ (+613) 592 65 00
Vice President Aerospace: Sol Rauch
Vice President and Chief Financial Officer: Greg Yeldon (Acting)
Secretary-Treasurer Montreal: Ms Marcia McKenzie
Public Relations Manager: Ms Janka Dvornik
Government & Public Relations Co-ordinator: P M Rogers
■ AVIONICS:
Navigation & Landing Systems (Doppler, Omega, GPS, MLS); Intelligent Monitoring and Display Systems; Fibre-optics Sensors; Adaptive and Phased-Array Antenna Systems; Airborne Satellite Communications Antenna Systems; Systems Engineering.
COMMERCIAL COMMUNICATIONS:
Sales & service of Marine Electronics, Land Communications Products, Electronic Instruments and Security Systems; Calibration, Repair and Overhaul Services.
COMMUNICATIONS SYSTEMS:
Line-of-sight Tactical Radio Sets with conventional and ECCM capabilities; Multiplexers; Switchboards; Shelter-mounted Transportable Communications Systems; Air Defence Operations Centres; Command, Control and Communications Systems; Packet Switchpads; Tactical Digital Switches; Systems Engineering & Integration.
CUSTOMER SUPPORT:
Technical Publications; Training; Field Engineering; Automated Test Equipment; Logistics Systems Engineering; Spares and Repair & Support Planning services to CMC's worldwide commercial & military aviation customers.
DATACOMM PRODUCTS:
Turnkey Messaging Networks integrating Telex, Facsimile and Data Communications into a single interconnected service; Telex and Facsimile Switching Systems & Conversion Products; Time Division Multiplexers; Packet Assembler Disassemblers; Data Communications Systems Software Development.
COMPONENTS:
Microelectronics; Power Conversion and Magnetics; Specialised Electronic Components; Display Components and Machined Parts; Contract Manufacturing; Circuit Packaging.
GROUND SYSTEMS:
Ground-based Navigational Aids including ILS, VOR and RMMS.
RADAR SYSTEMS:
Navigation Radars; Air & Surface Surveillance Radar Systems; Vessel Traffic Services.

● 415 Leggett Drive, PO Box 13330,
Kanata, Ont K2K 2B2
☎ (+1 613) 592 65 00 ⊖ (+1 613) 592 74 27

Avionics (Montreal)
600 Dr Frederik Philips Boulevard,
Ville Saint-Laurent, Que H4M 2S9
☎ (+1 514) 748 31 48 TX 05827822
⊖ (+1 514) 748 31 00
General Manager: Mark Wasserman
Marketing Services Manager: Ms Tina Verni

Avionics (Kanata)
415 Leggett Drive, Kanata, Ont K2K 2B2
☎ (+1 613) 592 65 00 TX 0534805
⊖ (+1 613) 592 74 27
Vice President Aerospace: Sol Rauch

Components
600 Frederik Philips Boulevard,
Ville Saint-Laurent, Que H4M 2S9
☎ (+1 514) 748 31 48 TX 05827822
⊖ (+1 514) 748 30 17
General Manager: Robert Tanquay

Commercial Communications
415 Leggett Drive, Kanata, Ont K2K 2B2
☎ (+1 613) 592 65 00 TX 0534805
⊖ (+1 613) 592 74 27
Vice President: Bryan Locker

Communications Systems
600 Dr Frederik Philips Boulevard,
Ville Saint-Laurent, Que H4M 2S9
☎(+1 514) 748 31 48 ✉ARCON Montreal ☒ 05827822
⊜ (+1 514) 748 30 35
Marketing Manager International: Jean Akel
Marketing Manager North America: Antoine Cortezi
Radar Systems/Ground Systems
415 Legget Drive, PO Box 13330, Kanata, Ont K2K 2B2
☎(+1 613) 592 65 00 ☒ 0534805
⊜ (+1 613) 592 74 27
Group Manager Radar Systems: Bruce Bailey
Customer Support
600 Dr. Frederik Philips Boulevard,
Ville Saint-Laurent, Que H4M 2S9
☎(+1 514) 748 31 48 ⊜ (+1 514) 748 30 14
General Manager: Alan Barker
98

75-02900 Carl Zeiss Canada Ltd
(a subsidiary of Carl Zeiss)
45 Valleybrook Drive, Don Mills, Ont M3B 2S6
☎(+1 416) 449 46 60 ⊜ (+1 416) 449 35 24
General Manager: Jim Mahar
■ Aerial cameras, photogrammetric instruments, telescopes, military optics, light microscopes, electron microscopes, epi technoscopes. Coordinate measuring instruments, geodetic instruments, image analysis systems, confocal laser scan microscopes. 98

75-02950 Carleton Life Support Technologies Ltd
(a subsidiary of Carleton Technologies Inc.)
1-1200 Aerowood Drive, Mississauga, Ont L4W 2S7
☎(+1 905) 629 32 45 ⊜ (+1 905) 629 83 06
♣ C$ 3,000,000 (1996) 👥 14 🏭 2,400 sq m
Managing Director: Frank Sova Ⓔ sova@carletech.com
Operations Director: Gary Machin
Purchasing Manager: Ms Heather Fieldhouse
■ Manufacture and supply aeronautical life support equipment, including liquid oxygen converters, cryogenic equipment, test stands, oxygen regulators, valves, hoses, chemical defence accessories and equipment, test equipment and reducers. Personal cooling equipment. 98

75-03200 Chicopee Manufacturing Limited
975 Wilson Avenue, Kitchener, Ont N2C 1J1
☎(+1 519) 893 75 75 ⊜ (+1 519) 893 59 52
👥 85 🏭 100,000 sq ft
President & General Mgr: Mrs Betty Sims
Marketing/Sales Dir: Tim Rueffer
Business Development Consultant: Bruce MacRitchie
Purchasing: D. Moore
■ Actuators, aircraft spares, components, computer aided design, manufacturing. Landing gear machined components, machined metals, metal fabrication, metal processing, non-destructive test, precision machining. 98

75-03400 Codalex Ltd
5780 Decelles Avenue, Montreal, Que H3S 2C7
☎(+1 514) 731 32 51 Ⓔ codalex@odyssee.net
⊜ (+1 514) 738 19 70
👥 25 (1996/97) 🏭 10,000 sq ft
President: R.A. Henry
Marketing Dir: P.G. Agnew
Purchasing Mgr: Ms Esther Ohayon
■ Tactical frequency division multiplex (FDM) systems. 98

75-03540 Comlab Telecommunications Inc.
2300 Leon-Harmel, Suite 220, Quebec, Quebec G1N 4L2
☎(+1 418) 682 33 80 Ⓔ sales@comlab.com
Ⓦ http://www.comlab.com ⊜ (+1 418) 682 89 96
♣ C$ 950,000 (1996) 👥 19
President: John F Ahern
■ Radar and EW training systems. → 80 98

75-03550 Communications & Power Industries
(† Varian Canada Microwave Products)
45 River Drive, Georgetown, Ontario L7G 2J4
☎(+1 416) 877 01 61 ⊜ (+1 416) 877 53 27
👥 200
President: Joe Caldarelli
Marketing Manager: Klaus Beecker
■ Klystron tubes for communication, x-ray subsystems and power supplies and amplifiers for terrestrial wireless communication. 98

75-03600 Computing Devices Canada Ltd
(a subsidiary of the Ceridian Corporation, USA)
3785 Richmond Road, PO Box 8508,
Ottawa, Ontario K1G 3M9
☎(+1 613) 596 70 00
Ⓦ http://www.computingdevices.com
⊜ (+1 613) 596 73 96
♣ C$ 375,000,000 (1996) 👥 1,100 🏭 300,000 sq ft
President: D E Scott
Sales and Marketing Director: E W Trischuk
Advertising Manager: Ms Suzanne Belanger
Ⓔ suzanne.belanger@canada.cd ev.com
■ Integrated digital voice and data distribution; acoustic signal processing; high resolution tactical digital displays; C3 systems design and integration multi-sensor scan conversion; all weather, multi spectral; surveillance for air, sea or land; integrating Commercial Off-The-Shelf (COTS) technology; information technologies; software development; communications engineering; sensor integration; research and development. Approved to ISO 9001. → 80
Computing Devices Canada
(a subsidiary of the Ceridian Corporation, USA)
3785 Richmond Road, PO Box 8508,
Nepean, Ont K2H 5B7
☎(+1 613) 596 70 00
Ⓦ http://www.computingdevices.com
☒ 534139; TWX 610 563 1632 ⊜ (+1 613) 820 50 81
♣ C$ 376,533,000 (1996) 👥 1,200
🏭 300,000 sq ft (Ottawa); 🏭 140,000 sq ft (Calgary)
President: David Scott ☎(+1 613) 596 7098
VP Business Development: Ernie Trischuk
Marketing Communications Mgr: Ms Suzanne Belanger
☎(+1 613) 596 7059
■ Direct & indirect fire control systems, vehicle electronics, reconnaissance vehicle surveillance system, computerized laser sight for anti-tank weapons, tactical communications systems, headquarters information distribution system, tactical voice & data distribution system, acoustic signal processing, ASW mission systems, sonobuoy processors, active sonar systems, towed array sonar systems, tactical acoustic trainer, MIL-SPEC electroluminescent displays, large multi-sensor displays, coastal intrusion detection systems, fibre-optic distribution systems, contract manufacturing. 98

75-03760 CRL Technologies
(Division of Forexport Inc., a subsidiary of Le Groupe Forex)
55 Lindsay Street, Dorval, Que H9P 2S6
☎(+1 514) 636 99 21 ✆ (800) 263 99 21
Ⓔ crl.technologies@sympatico.ca ⊜ (+1 514) 636 04 76
♣ C$ 2,000,000 (1997) 👥 40 (1997) 🏭 6,000 sq m
Executive Vice President and General Manager: Pierre Desjardins
■ Night vision compatible lighting systems. Type III, IV, V, VI integrally lit panels, keyboards and switch panels. All types of bezels. Caps. → 60 98

75-03800 C-Tech Ltd
525 Boundary Road, PO Box 1960,
Cornwall, Ont K6H 6N7
☎(+1 613) 933 79 70 ⊜ (+1 613) 933 79 77
👥 100 🏭 110,000 sq ft
President: D Derouin
Marketing Manager: Bob Fraser
■ Design & manufacture underwater acoustics encompassing both military & commercial applications. Primary emphasis in light hull mounted sonar, harbour surveillance & mine countermeasure sonar. Design & manufacture custom electronic equipment, transducers in compliance with military standards & Standard Electronic Modules. (SEM) 98

75-03810 CTF Systems Inc.
15-1750 McLean Avenue, Port Coquitlam, BC V3C 1M9
☎(+1 604) 941 85 61 Ⓔ sales@ctf.com
Ⓦ http://www.ctf.com ⊜ (+1 604) 941 85 65
👥 85 (1997) 🏭 2,800 m2
Managing Director: Dr M B Burbank
Marketing Manager: Gordon Haid Ⓔ gordonh@ctf.com
■ Research, develop and manufacture in the fields of magnetic anomaly detection and magnetoencephalography. SQUID magnetometers. 98

75-04040 Davie Industries Inc.
(† MIL Davie Inc.)
(a subsidiary of Dominion Bridge Corp.)
22 George D. Davie Street, PO Box 130,
Levis, Que G6V 6N7
☎(+1 418) 837 58 41 ⊜ (+1 418) 835 10 17
👥 650 🏭 870,000 sq ft
President: J Arthur Gélinas
Vice President Marketing: Michael G Ayre
Vice President Operations: Jean Yves Rhéaume
Purchasing Manager: Jean Claude Houle
■ Products offered include military & commercial vessels on a "Turnkey" basis, offshore structures & a wide range of industrial equipment as well as equipment for the nuclear industry. Design & other engineering services are also provided. Including maintenance, repair & overhaul services. 98

75-04050 Davis (W.R.) Engineering Ltd
1260 Old Innes Rd, Ottawa, Ont K1B 3V3
☎(+1 613) 748 5500 ☒ 533862 ⊜ (+1 613) 748 3972
👥 30 🏭 40 000 sq ft
President: W.Roland Davis
Engineering Mgr: Glenn Long
Exec. Asst: Ms Carol I. Anderson
■ Consulting & manufacturing engineering services in the mechanical, vehicle, propulsion & electronics fields. Short run manufacturing capability. Infra-red signature suppression systems & technology for land, air & sea applications. Active shaft grounding system for naval applications. Military vehicle design, fabrication, prototyping. Wave generation systems for vessel design & research. 97

75-04100 Designed Precision Castings Inc.
75 Eastern Avenue, Brampton, Ont L6W 1X9
☎(+1 905) 453 04 21 ⊜ (+1 905) 453 36 94
👥 50 🏭 42,000 sq ft
Vice President: Mike Holland
Vice President: Mike Podrebarac
Quality Control Mgr: Terry Pell
■ Ferrous & non-ferrous investment castings in a size range from ounces to 50 lbs, for defence, aerospace & commercial industries. Full in-house non-destructive testing capability. 98

75-04300 Devtek Corporation
(† Devtek Aerospace Company)
100 Allstate Parkway, Suite 500, Markham, Ont L3R 6H3
☎(+1 905) 477 68 61 ⊜ (+1 905) 477 04 81
♧ C$ 77,900,000 (1996) ♣ C$ 328,755,000 (1996)
👥 2,000 (1996) 🏭 400,000 sq ft
Chairman: Helmut Hofmann
President and Chief Executive Officer: James N Renner
Executive Vice President and Chief Financial Officer: Murray G Kingsburgh
Vice President Finance: Peter F Monsberger
Director Corporate Communications: Ms Jan Bornstein
■ Develop & manufacture systems & products for the automotive, aerospace & telecommunications markets. The corporation consists of eleven manufacturing companies with products including aircraft subassemblies, hydraulic & mechanical actuation systems & small arms weapon systems.
Manufacturing subsidiaries:
Diemaco Inc.
West Heights Mfg
Hochelaga Aerospace
Devetek Aerospace
West Heights Mfg, 1665 Highland Road West,
Kitchener, Ont. N26 3W7
☎(+1 519) 576 89 10 ⊜ (+1 519) 576 51 19
👥 144
President: H. Noesgaard
VP Finance: P. Andrews
■ Manufacturer of large landing gear components & sub assemblies & small components for both commercial & military aircraft. Large structural components such as cylinders, pistons, torsion links, drag braces & side braces. Small components such as brake housings, yokes links & spindles.
Hochelaga Aerospace
(a subsidiary of Devtek Aerospace)
3675 Industrial Boulevard, Laval, Que H7L 4S3
☎(+1 514) 629 34 54 ⊜ (+1 514) 629 16 55
👥 91
Vice President and General Manager: B Beckett
Sales & Marketing Dir: Paul Roussea

■ Design, development, manufacturing, testing & overhaul of hydraulic & mechanical actuation & locking systems, flight controls, landing gear attachment & other aircraft actuation systems for civil & military aircraft & helicopters.

Devtek Electronic Packaging Systems (DEPS)
1401 Huntingwood Boulevard,
Scarborough, Ont M1S 3R9
☎(+1 416) 754 21 33 ⊜ (+1 416) 754 26 39
250
President: M Clarke
Sales Manager: G Helary
Marketing Manager: B Bergey
■ Products include chassis, modules, racks & enclosures for the telecommunications, electronic & digital markets.

Precision Systems Division
1480 Birchmount Road, Scarborough, Ontario M1P 2G2
☎(+1 416) 754 21 33 ⊜ (+1 416) 754 26 39
Operational Manager: Wolfgang Mildenberger
Plant Manager: G Haase
Marketing Manager: R C Bergey
■ Manufacture of precision aerospace products including electronic packaging, avionic chassis and heat exchangers, electro-mechanical assemblies, shipboard cabinets and ATC consoles. Machined metal parts.

Deimaco
1036 Wilson Avenue, Kitchener, Ont N2C 1J3
☎(+1 519) 893 68 40 06955164
⊜ (+1 519) 893 68 40
General Mgr: Brian Smith
Dir Sales: B. van Kere Broeck
Contracts Administrator: Ms Sandra Davies
■ Supply & centre of excellence for engineering & development, manufacturing, repair & overhaul of military small arms.

GFI Control Systems
100 Hollinger Crescent, Kitchener, Ont N2K 2Z3
☎(+1 519) 576 42 70 ⊜ (+1 519) 576 40 10
100
President: L. Austin
Marketing & Sales Mgr: T. McIver
■ Manufacturer of natural gas conversion systems technology for use in vehicles. [98]

75-04650 DY-4 Systems Inc.
21 Fitzgerald Road, Nepean, Ont K2H 9J4
☎(+1 613) 596 9911 ⊜ (+1 613) 596 0574
♣ $ 55 000 000 (1995) 198 73 000 sq ft
President: Danny Osadca
VP Marketing: Ed Hennessy
Administrative Assistant: Ms Lisa Hall
☎(+1 613) 596 0574 ☎(+1 613) 596 0574 (fax)
E lhall@dy4.com
■ Design, manufacture & integration of commercial, ruggedised & MIL-SPEC computer systems based on the VMEbus open-system architecture. Conduction cooled cards & chassis for harsh military environments. Air-cooled cards for military, critical & civil applications & prototyping. Provision of board level products, chassis and software modules. Program management, life cycle support, QA and configuration management. [97]

75-04685 Eberhard Hardware Manufacturing Ltd
(a subsidiary of The Eastern Co., USA)
PO Box 367, Tillsonburg, Ont N4G 4H8
☎(+1 519) 842 84 57 ⊜ (+1 519) 688 05 75
42 (1997) 2,300 sq m
Managing Dir: Robert G. Alexander
Operations & Technical Dir: Robert J. McAra
Sales Mgr: Cheryl Buchner
■ Latches, hinges, handles & special products for military purposes. Seats, folding locks, one, two & three point. [98]

75-04688 EDO Canada Ltd
1940 Centre Avenue N.E., Calgary, Alberta T2E 0A7
☎(+1 403) 569 5400 ⊜ (+1 403) 569 5499
115 65 000 sq ft
Acting General Mgr: J. Douglas Moore
■ Manufacture advanced materials components (composites, fuel cylinders, airframe components). [97]

75-04700 EEV Canada Ltd
67 Westmore Drive, Rexdale, Ont M9V 3Y6
☎(+1 905) 678 98 11 ✉ ENELECAN Toronto
⊜ (+1 905) 678 77 26
General Manager: David Soward
Representing:
EEV Ltd, UK [98]

75-05050 Eurocopter Canada Ltd
(† MBB Helicopter Canada Ltd)
(a subs. of Eurocopter SA)
PO Box 250, Fort Erie, Ont L2A 5M9
☎(+1 905) 871 7772 ✆ (800) 267 4999
⊜ (+1 905) 871 3320
♣ $ 30 000 000 (1995) 120 (1995) 7 900 m2
President & CEO: Willy Heidbuechel
Vice President & COO: Michael D. Lavoy
■ Structured as a total design, engineering, manufacturing, flight test, system certification & Eurocopter product support company. [97]

75-05270 Far West Industries Inc.
1701 Kosmina Road, Vernon, BC V1T 8T1
☎(+1 250) 542 02 37 E farwest@junction.net
W http://www.farwestcanada.com ⊜ (+1 250) 545 20 88
♧ C$ 5,000,000 (1997) ♣ C$ 10,500,000 (1997)
100 (1997) 60,000 sq ft
President: Gary Pratt
Controller: D E MacKenzie
Operations Director: John Mussenden
Sales Manager: M Wardlow
Purchasing Manager: Ms Debbie Frigon
■ Functional clothing for all-climate protection. Special purpose garments for military, search & rescue or police use. [98]

75-05400 Firan Corp.
353 Iroquois Shore Rd, Oakville, Ont L6H 1M3
☎(+1 905) 844 2870 ⊜ (+1 905) 844 9247
♧ C$ 21 000 000 (1995) ♣ C$ 202 789 000 (1995)
906
President & COO: Curtis S. Bossi
Exec. Vice President: Thomas G. Ivaskiv
■ Main areas of activity include air traffic control communication systems, ground navigation systems, printed wiring & edge-lit panels.
Subs. with defence business:
Graphico → USA 75
Denro → USA 75
Fernau → UK 75 [97]

75-05500 Fleet Industries Ltd
(a subsidiary of Magellan Aerospace Corp)
1011 Gilmore Road, Fort Erie, Ont L2A 5N3
☎(+1 416) 871 21 00 ⊜ (+1 416) 871 27 22
♣ C$ 40,000,000 (1996) 620 (1997)
170,000 sq m
President & General Mgr: Ernie Dueck
Mgr New Business Dev.: Brian M. Oakley
☎(+1 905) 871 2100 Ext 301
☎Fax: (+1 905) 871 2722
■ Aerospace manufacture of complex structural assemblies, specialising in flight control components with advanced metal to metal & composite bonding techniques. [98]

75-05550 Flexibulb Plastics Inc.
9000 Parent Boulevard, PO Box 635,
Trois Rivieres, Que G9A 5J3
☎(+1 819) 374 9250 ⊜ (+1 819) 374 5143
22 28 000 sq ft
CEO/President: Pierre Tellier
VP Sales/Marketing: Louis Martin Tellier
Head Engineering: Lyne Levesque
Production Mgr: Robert Maynard
■ Vacuum thermoforming of thermoplastics & thermosets. Vacuum metallising. Aircraft interiors & enclosures. Large dimension windshield & glazing. Composite parts & composite tooling. Clear ballistic screen protective structures. [97]

75-05620 Foremost Industries Inc.
1225-64 Ave. NE, Calgary, Alta T2E 8P9
☎(+1 403) 295 5800 3822772
⊜ (+1 403) 295 5810
♣ C$ 40 000 000 150 (Approx.)
Vice President: W.L. Pusch
Marketing Mgr: Ms Barbara Bryde
■ Design and manufacture of heavy-duty off-highway vehicles, both wheeled and tracked, with load capacities ranging from 5 to 70 tonnes. [97]

75-05625 Fullerton, Sherwood Engineering Ltd
6450 Van Deemter Court, Mississauga, Ont L5T 1S1
☎(+1 905) 670 06 56 E info@fseng.com
W http://www.fseng.com ⊜ (+1 905) 670 83 18
President: David J Fullerton
Sales and Marketing Director: Doug Elsey
■ Clearance divers' rebreather apparatus & recompression chamber systems. [98]

75-05650 Gandalf Technologies Inc.
130 Colonnade Road South, Nepean, Ont K2E 7M4
☎(+1 613) 274 6500 ⊜ (+1 613) 274 6505
♣ US$ 120 000 000 (1995/96) 650 (worldwide)
President & CEO: Tom Vasfiliades
VP & Sales Marketing: Ms Joceline Lemux
■ Worldwide manufacturer & supplier of remote access solutions.

Gandalf Canada Ltd
130 Colonnade Rd S., Nepean, Ont K2E 7M4
☎(+1 613) 274 6508 ⊜ (+1 613) 274 6805
President: Tom Vasfiliades
Marketing Dir: Ms Pam Minelli [97]

75-05890 General Kinetics Engineering Corp.
110 East Drive, Bramlea, Ont L6T 1C1
☎(+1 905) 458 01 88 Ext. 227
E 70673.3204@compuserve.com
⊜ (+1 905) 458 75 66
♣ US$ 10,000,000 (1997) 60 (1997) 5,100 sq m
Managing Director: Robert Cathcard
Vice President: Charles Williamson
Operations Director: Nigel Sizmur
Purchasing Manager: Ms Lynda Roselle
■ Manufacture of heavy duty shock absorbers designed explicitly for military applications. [98]

75-05900 General Motors of Canada Ltd
(a div. of General Motors Corp. USA)
Diesel Division
PO Box 5160, London, Ont N6A 4N5
☎(+1 519) 452 5184; 452 5054
⊜ (+1 519) 452 5688; 452 5688
2 000 (1995)
General Dir: W.J. Kienapple
Marketing & Programmes Dir: W.J. Pettipas
Purchasing Mgr (Defence): Alan L. Green
Int'l Sales Mgr (Export): W.L. Claggett
Marketing, Advertising, Exhibits & Defence Operations: Ms Hazell M. Booker
■ Development and production of wheeled armoured vehicles. [97]

75-06050 GIII Ltd
274 Jarvis Avenue, Winnipeg, Man R2W 3A5
☎(+1 204) 586 84 61 E g3@cybersp.mb.ca
⊜ (+1 204) 589 38 83
General Mgr: Ms Mavis Goddard
Controller: Herb Singer
■ Comprehensive line of sleeping bags. Products can also be manufactured to customer specifications. [98]

75-06200 Global Thermoelectric Inc.
#9, 3700-78th Avenue SE, Calgary, Alta T2C 2LB
☎(+1 403) 236 55 56 W http://www.globalte.com
⊜ (+1 403) 236 55 75
♣ C$ 9,600,000 (1997) 74 (1997) 4,500 sq m
President: Jim Perry ☎(+1 403) 720 11 75
Vice President, Generators: Bernie LeSage
☎(+1 403) 720 11 76
Manager Heater Products: Daryl Marling
☎(+1 403) 720 11 86
Marketing Manager, Military Sales: John Howard
☎(+1 403) 720 11 79
Sales and Marketing Generators: Eric Potter
☎(+1 403) 720 11 83
■ Thermoelectric power generation devices. Self-powered vehicle engine heaters. Fuel fired air heaters. [98]

75-06460 Graphico Electronics Group
(a div. of Firan Corporation)
1100 Bellamy Rd North, Scarborough, Ont M1H 1H2
☎(+1 416) 438 6076 ⊜ (+1 416) 438 5961
♣ C$ 15 000 000 (1995) 137 65 000 sq ft
President: Maher Ishak
Sales Mgr: Greg Smith

■ Manufactures military & aerospace printed wiring boards of multilayer, double-sided & teflon configuration. Qualified to MIL-P-55110D, AQAP-4, AQAP-6, MIL-I-45208, WS6536E. Edge-Lit panels & Dead-Front panels MIL P 7788, MIL L 27160, MIL C 25050, MS 33558 (ASG). [97]

75-06500 Greening Donald Co. Ltd
16 Commerce Rd, Orangeville, Ont L9W 2X7
☎(+1 519) 941 1920 ⊜ (+1 519) 942 8357
136 115 000 sq ft
President: Lawrence H. Fort
Sales & Marketing Dir: Kirk D. Miller
Mgr Manufacturing: Jim Ouellette
■ Manufacture woven wire cloth, stainless steel wire, specialty alloy wire & perforated tubes for automotive airbags.

Erin Facility
Highway 24, Erin, Ont N0B 1T0
☎(+1 519) 833 9632 ⊜ (+1 519) 833 2213
13
Specialty Wire Superintendent: Werner Stadler
Marketing Mgr: Kirk D. Miller
■ Supplies stainless steel and other special alloy wires.

Wire Cloth
Commerce Rd, Orangeville, Ont L9W 1X7
☎(+1 519) 941 1920 697851 ⊜ (+1 519) 941 9683
84
Product Mgr: Ms Donna Spike
■ Produces wire cloth & wire screens for applications such as filtering, screening & protective grillework & airbag components. [97]

75-06530 Grow Group Canada Ltd Devoe Coatings, Division
55A MacDonald Avenue,
Dartmouth, Nova Scotia B3B 1T9
☎(+1 902) 468 98 88 ⊜ (+1 902) 468 98 89
Managing Director: Richard Keen
■ Manufacture of high performance paints and coatings. [98]

75-06570 Guildline Instruments Ltd
(a subs. of Guildline Instruments Inc.)
21 Gilroy Street, PO Box 99, Smiths Falls, Ont K7A 4S9
☎(+1 613) 283 3000 ⊜ (+1 613) 283 6082
President & CEO: Anthony Anderson
International Sales Contact: David Moat
■ Electrical and temperature measuring instruments and calibration standards & oceanographic instrumentation. Trading for 40 years. [97]

75-06600 Haley Industries Ltd
Haley Station, Haley, Ont K0J 1Y0
☎(+1 613) 432 88 41 [W] http://www.haley.on.ca
⊜ (+1 613) 432 07 43
♣ C$ 48,700,000 (1996) 480 (1997) 140,000 m2
President and Chief Executive Officer: Dave Gorman
☎(+1 613) 432 88 41
Chairman: Jim Meekison
Vice President and General Manager: Mike McLaughlin
Senior Vice President Marketing: Harvey W Murray
[E] harvey.murray@haley.on.ca
Vice President Finance: Jim Lemenchick
Sales Director: Randy Joe
Purchasing Manager: Pat Gillespie
■ Aluminium and magnesium sand castings for large and small aircraft. Aerospace subsystems. In-house capabilities: pattern shop; CNC finishing department; x-ray. [98]

75-07100 Henotex (1981) Inc
1625 Chabanel West - Suite 483,
Montreal, Québec H4N 2S7
☎(+1 514) 382 61 20 ⊜ (+1 514) 383 65 23
♣ C$ 1,200,000 5 1,300 sq ft
President: André Hénault
■ Hand-held metal detectors LYMPS HS+ and OMNI used by airports, army, embassies, customs and security forces. [98]

75-07300 Hermes Electronics Ltd
(a subsidiary of Ultra Electronics Holding Ltd, UK)
40 Atlantic Street, PO Box 1005,
Dartmouth, Nova Scotia B2Y 4A1
☎(+1 902) 466 74 91 ✆ (800) 491 77 54 01921744
⊜ (+1 902) 463 60 98
♣ C$ 226,005,000 (1994) 180 (1997)
150,000 sq ft
President: Wayne Trowse
Operations Director: Derrick Aucoin
Marketing Director: Rene Cousineau
[E] rcousino@hermes.ns.ca
Business Development Director: Paul Adlakha
Purchasing Manager: Leo Mackenzie
Marketing: Ms Andrea Hollett
■ Design and manufacture of passive sonobuoys, towed arrays, bathythermograph buoys, resonant power supplies, electronic ballasts, underwater acoustic monitoring, HF loop antennae, sea water batteries, hydrophones, monitor and control systems, oceanographic and meteorological drifting buoys. Tactical HF antenna systems. [98]

75-07400 Heroux Inc
755 Thurber Street, Longueuil, Québec J4H 3N2
☎(+1 514) 679 54 50 05560630
⊜ (+1 514) 679 23 76
♧ C$ 44,000,000 ♣ C$ 100,000,000 (1996)
700 (1996) 19,000 m2
President and Chief Executive Officer: Gilles Labbe
Vice President Marketing: Jacques L Bonaventure
Managing Director: L Zakaib
Operations Director: G Demers
Purchasing Manager: A Vaillancourt
■ Development, design, manufacture and overhaul of aircraft landing gear for KC-135, KC-135R, C-130, C-5, A-10, P-3 and F-18. Manufacture of jet engine components for F-100, F-110, F-404.
Flight controls, hydraulic valves, servos, actuators and cylinders. [98]

75-07500 Hewitt Equipment Ltd Industrial Division
4000 Trans Canada Highway, Pointe Claire, Que H9R 1B2
☎(+1 514) 426 30 00 ⊜ (+1 514) 426 52 24
105 7,427 sq m
General Manager: Marc Métivier
Sales Manager: Michel Durand
Sales Representative: Rémy Boulianne
■ Manufacture, supply & installation of MIL-SPEC fixed & mobile diesel engine generator sets ranging from 50 to 5720 kW, in addition to switchgear & electrical apparatus for low & medium voltage up to 34.5kV, 1000 MVA. [98]

75-07520 Hewlett-Packard (Canada) Ltd
(a subs. of Hewlett-Packard Co., USA)
2670 Queensview Drive, Ottawa, Ont K2B 8K1
☎(+1 613) 726 2116 ⊜ (+1 613) 726 2118
♣ C$ 1 250 000 000 (1995) 1 650 (1995)
President: Dan Branda
Business Development Mgr Canadian Defence: Vince Terpstra
■ Design & manufacture electronic products & systems for test, measurement & computation. Acoustic data processing systems, signal development systems for Elint. [97]

75-07630 Howmet-Cercast Group
(a division of Howmet Corp.)
3905 Industrial Boulevard, Montreal North, Que H1H 2Z2
☎(+1 514) 322 23 71 ⊜ (+1 514) 322 13 40
♣ C$ 15,000,000 (1995) 200 (1996)
100,000 sq m
President: Henri M Fine
Vice President Sales and Marketing: Richard Durant
Sales Manager: Dieter Rupp
■ Manufacturer of aluminium & copper-base alloy investment castings. [98]

75-07650 Hughes Elcan Optical Technologies
(† Hughes Leitz Optical Technologies Ltd)
(a subsidiary of Hughes Aircraft Company)
Leitz Road, Midland, Ontario L4R 5B8
☎(+1 705) 526 54 01 ⊜ (+1 705) 526 58 31
♣ C$ 100,000,000 625 200,000 sq ft
President: Dr Joseph De Remigis
Vice President Sales and Marketing: Dr John W Mactaggart
Vice President Marketing: Cameron Perry
■ Opto-electronic and opto-mechanical products. High precision lens systems and optical asssemblies (visible and infrared). Day and night sights, aiming circles and mortar sights.
Photographic and aerial lenses, fire control instruments and underwater optics. [97]

75-07720 Ideal Knitwear Manufacturing Ltd
8955 Blvd Saint-Laurent, Ste 201,
Montreal, Que H2N 1M5
☎(+1 514) 381 4496 ✆ (800) 363 7803
⊜ (+1 514) 381 8792
♣ C$ 4 000 000 75 (Approx.) 45 000 sq ft (1994)
President: Isaac Kupperstein
Managing Dir: Steve Mormina
Sales/Marketing Dir: S. Mulman
■ Knitted headwear & scarves. [97]

75-07725 Ilco Unican Inc.
(a subs. of Unican Security Systems Inc.)
7301 Decarie Blvd, Montreal, Que H4P 2G7
☎(+1 514) 735 5411 ⊜ (+1 514) 735 8707
♣ C$ 118 146 000 1 600
International Sales Mgr: Christopher Beames
■ Manufacture of mechanical, changeable pushbutton access controls, electronic locks (standalone) & hard wired card access systems. [97]

75-07728 Imagenex Technology Corp.
209-1875 Broadway Street, Port Coquitlam, BC V3C 5W2
☎(+1 604) 944 8248 ⊜ (+1 604) 944 8249
7 (1995) 360 sq m
Managing Dir: Willy Wilhelmsen
VP Sales & Marketing: Alan Mulvenna
Purchasing Mgr: Gordon Kristensen
■ Manufacture colour imaging sonar systems & marine radar repeaters & target extractors. [97]

75-07730 IMP Aerospace Limited
Suite 400, 2651 Dutch Village Road, Halifax, NS B3L 4TI
☎(+1 902) 873 22 50 [E] impaero@impgroup.com
[W] http://www.impgroup.com ⊜ (+1 902) 873 22 90
1,000 (1997) 250,000 sq ft
President: Kenneth C Rowe
Executive Vice President: M Phillips
Director International Marketing: Derek W Kinsman
■ Aircraft repair & overhaul. Aircraft component production; extensive wire harness capability; bilingual technical publications. Precision machine parts. → 80 [98]

75-07740 Indal Technologies Inc
(a subsidiary of Indal Limited)
3570 Hawkestone Road, Mississauga, Ontario L5C 2V8
☎(+1 905) 275 53 00 ✆ (800) 263 73 40
INDALTECH MSGA ⊜ (+1 905) 273 70 04
♧ C$ 18,000,000 (1995) ♣ C$ 35,700,000 (1996)
150 215,000 sq ft
President and Chief Executive Officer: Vincent Lacey
Senior Vice President Operations: Denis Gillen
Vice President International Sales: E G Howard
Operations Director: Colin Reid
Contracts and Program Management Director: Ms Joan Patrick
Manufacturing Director: Derek Orr
Sales and Marketing Director: Jay Doyle
Marketing Director (USA): Terry McCloskey
Area Sales Manager, Aero Marine Systems: Gene Joelson
Sales Manager Cable Handling Systems: Tadeusz Slupski
Public Relations Co-ordinator: Ms Janine LeForestier
■ Shipboard helicopter recovery assistance, secure and traverse (RAST) systems, aircraft ship, integrated secure and traverse (ASIST) systems, aluminium telescopic helicopter hangars, curtain and sliding doors, light airborne messenger winches, visual landing aids, unmanned vehicle handling systems, variable depth sonar handling systems, lightweight dipping sonar winch systems and tactical towed line array handling systems. Licenced sub-contractors, manufacturing and system integration, integrated logistics support services.
Frangible ILS antenna supports, cable handling systems, fairings and lightweight dipping sonar systems. [98]

75-07775 International Loss Prevention Systems
1350 East 4th Avenue, Vancouver, BC V6M IL8
☎(+1 604) 255 50 00 [E] ilps@mindlink.bc.ca
[W] http://www.ilps.com ⊜ (+1 604) 254 25 75
♣ C$ 7,800,000 25 15,000 sq ft
President: Ian J. Abrasmson
Managing Dir: Randy Scott

Sales/Marketing Dir: Ken Morey
■ Security products to prevent the theft of office equipment. [98]

75-07780 International Submarine Engineering Ltd (ISE)
1734 Broadway Street, Port Coquitlan, BC VC3 2M8
☎ (+1 604) 942 52 23 [W] http://www.ise.bc.ca
⊖ (+1 604) 942 75 77
♣ C$ 10,700,000 (1997) 👥 100 (1997)
President: James R McFarlane
Operations Director: Mike Macdonald
Sales and Marketing Director: Owen Williams
Purchasing Manager: Don Muth
■ Designers & builders of manned, tethered & autonomous undersea vehicles, imaging sonars, remotely controlled manipulators, remotely controlled land vehicles. [98]

75-08220 Invar Manufacturing Ltd
(a subsidiary of Linamar Corp)
1 Parry Drive, Batawa, Ontario K0K 1E0
☎ (+1 613) 398 61 06 ⊖ (+1 613) 966 79 32
♧ C$ 25,000,000 ♣ C$ 25,000,000 👥 200
🏠 200,000 sq ft
General Manager: E McGregor
Sales Administration Manager: Cam E Nardocchio
Purchasing Manager: Craig Baker
■ Wing components, undercarriage parts, hydraulic assemblies, aircraft structure machine parts. [97]

75-08300 Irvin Aerospace Canada Limited
(† Irwin Industries Canada Ltd)
479 Central Avenue, PO Box 280,
Fort Erie, Ontario L2A 5M9
☎ (+1 905) 871 65 10 ⊖ (+1 905) 871 65 34
♣ C$ 11,500,000 👥 175 🏠 66,000 sq ft
President and Chief Executive Officer: Christopher L Corbin
Marketing and Sales Manager: Ms Elizabeth Matchett
Purchasing: Edward J Johnson
Marketing and Sales: Ms Pam Dobrensky
■ Parachutes for personnel, cargo, weapon and aircraft retardation. Recovery systems for missiles and space vehicles, cargo delivery systems. Safety harnesses and air drop sequence control devices. Inflatable survival products and systems.
Anti-g suits, paratrooper parachutes, missile recovery parachutes, flare chutes, jump tower chutes, safety harnesses. [98]

75-08708 Canadian Kawasaki Motors Inc
(a subsidiary of Kawasaki Heavy Industries Ltd, Japan)
25 Lesmill Road, Don Mills, Toronto, Ontario M3B 2T3
☎ (+1 416) 445 77 75 ⊖ (+1 416) 445 53 89 [98]

75-08710 KB Electronics (1989) Ltd
150 Bluewater Rd, Bedford, Nova Scotia B4B 1G9
☎ (+1 902) 835 7268 ⊖ (+1 902) 835 6026
♣ C$ 5 000 000 👥 50
President & CEO: Paul S. Cugno
Sr Vice President: Sandy Barnhill
■ Design & manufacture fully hardened & ruggedised static power conversion equipment in the 1-10 kW power range for military & high end commercial applications.
Products include uninterruptible power supplies; frequency changers; inverters; converters; battery chargers; helicopter starters & embedded power supplies.
The company also provides complete integrated logistic support as well as repair & overhaul services. The company's quality system meets the requirements of ISO 9001 quality system. [97]

75-08740 KFW Canada
(a member of the TI Group plc, UK)
RR#1 20 Progress Drive, Orillia, Ont L3V 6H1
☎ (+1 705) 326 6611 ⊖ (+1 705) 326 8688
👥 80
General Mgr: Robert Finlayson
Controller: Ian Jones
■ Manufacture precision components for the gas turbine industry. Machining is performed on state-of-the-art equipment. Can provide rapid turn around of quality machined parts. [98]

75-08770 Kobelt Manufacturing Co. Ltd
8238 129 Street, Surrey, BC V3W 0A6
☎ (+1 604) 590 73 13 ⊖ (+1 604) 590 83 13
President: Jack Kobelt ☎ (+1 604) 572 39 35
Managing Director: Thomas Kobelt
Operations Director: John Purnell
Sales and Marketing Director: Valerie Simmons
Purchasing Manager: Kurt Dreger
■ Manufacture & supply industrial disc brakes & marine controls. [98]

75-08890 Laser Fiber Optics Canada Ltd
Head Office
30 Concourse Gate Unit 40, Nepean, Ont K2E 7V7
☎ (+1 613) 723 74 97 ⊖ (+1 613) 723 77 94
♣ C$ 100,000 (1996) 👥 7 (1997) 🏠 3,500 m2
President: Darshan Kant
Marketing Dir: Hari Kant
Edmonton Office
10325A 61 Avenue, Edmonton, Alta T6H 1K9
☎ (+1 613) 723 7497 [T] 534 370 ⊖ (+1 613) 723 7794
Montreal Office
Montreal, Que
☎ (+1 613) 723 7497 [T] 534370 ⊖ (+1 613) 723 7794
■ Line of sight LAN and WAN and wireless communication system FDDI, SONET. [98]

75-09090 Litton Canada Ltd Litton Marine Systems Division
7510 Airport Road, Mississauga, Ont L4T 2H5
☎ (+1 905) 405 11 44 ⊖ (+1 905) 405 11 50
👥 31 (1997) 🏠 2,044 sq m
General Manager: Patrick Warr
Insulators Manager: John Molloy-Vickers
Marine Marketing Manager: Michael Ennis
■ Manufacture and supply marine navigation systems, radar, instrumentation and antenna products, RF insulators and isolation transformers for use on MF, LF and VLF antenna systems. Other marine products also include Gyro Compass, Steering Control Systems, Voyage Management Systems, Integrated Navigation Systems, Doppler Speeds Logs and GMDSS. → 90
® UK: C-Plath, Decca, Sperry Marine [98]

75-09100 Litton Systems Canada Ltd
(a subsidiary of Litton Industries Inc., USA)
25 City View Drive, Etobicoke, Ontario M9W 5A7
☎ (+1 416) 249 12 31 ⊖ (+1 416) 245 03 24
♣ C$ 109,405,000 👥 1,144 🏠 61,915 sq yd
President: Timothy O Westover
Vice President of Engineering: Avi Zisman
Vice President of Marketing: D A Fuller
Press and Public Relations Director: Peter Broadhurst
■ Inertial navigation systems, airborne surveillance radar, flight inspection systems. LED and liquid crystal avionic displays, tactical data management systems, ship command and control systems, large scale system engineering, naval systems integration. [98]

75-09105 LNS Systems Inc
(† Leigh Navigation Systems Inc (LNS))
3003 Etingin Street, Ville St-Laurent, Québec H4S 1Y7
☎ (+1 514) 337 72 21 [T] 05821529 lns canada pclr
⊖ (+1 514) 337 66 84
👥 35 🏠 27,500 sq ft
President: Fred Welton
Executive Vice President and Marketing Director: George Sinoyannis
Operations Director: D Butcher
Marketing Administration Manager: Ms J Garner
■ Manufacturer of mobile, fixed and transportable air traffic control systems, mobile runway lighting systems and radio spectrum monitoring systems. [98]

75-09110 Lockheed Martin Canada Inc.
3001 Solandt Road, Kanata, Ont K2K 2M8
☎ (+1 613) 599 32 70 ⊖ (+1 613) 599 32 82
♣ C$ 100,000,000 👥 357
President: Larry A Ashley
Business Operations Director: Lee C Gartley
Operations and Product Support Director: Ian Malcolm
Engineering Director: Ed Derbyshire
Business Development Director: Darrel Newman
Acoustics and Security Systems Director: Luc Maglierl
Information Systems and Services Director: Andy Bond
EW Programmes Director: Larry Conway
EST Programmes Director: Keith Tucker
■ ECM and ESM systems, ASW systems, AN/SLQ-501 (CANEWS) ESM, AN/SLQ-503 (RAMSES) ECM, SCIMITAR naval ECM. Engineering design, support and systems integration for military electronic systems. Repair and overhaul. [98]

75-09120 Lockheed Martin Electronic Systems Canada
6111 Royalmount Avenue, Montreal, Québec H4P 1K6
☎ (+1 514) 340 83 10 ⊖ (+1 514) 340 83 18
President: Paul Manson
■ Manufacture of electronic surveillance monitors (ESM), electronic countermeasures (ECM) and anti-submarine warfare systems (ASW). AN-SLQ501 (CANEWS) ESM. Engineering design support and systems integration for military electronic systems. Repair and overhaul. [97]

75-09140 Lucas Aerospace Engine Control Systems
5595 Royalmount Ave, Montreal, Que H4P 1J9
☎ (+1 514) 735 1536 ⊖ (+1 514) 342 2047
VP & General Mgr: David Ashton
Engineering Mgr: Alex Howells
Human Resources Mgr: Bruna Crecco
■ Manufacture, repair & overhaul turboprop & turbofan engine accessories, fuel control, pumps, hydraulics & actuation. [97]

75-09215 Manitoba Aerospace Initiative
(† Manitoba Aerospace Technology Program)
155 Carlton Street - 4th Floor,
Winnipeg, Manitoba R3C 3H8
☎ (+1 204) 945 23 97; 945 87 41
⊖ (+1 204) 945 75 92; 945 75 92
Senior Development Officer: Robert Jack
Senior Development Officer: Robert Manson
[E] rmanson@itt.gov.mb.ca
■ Composite structural components, aerial targets, solid propellant rocket systems, jet-engine components, large airframe structures, wire strike protective systems for helicopters, overhaul & repair of turbine engines, structural composite helicopter components, precision machining & sheet metal fabrication, defence electronics. [97]

75-09250 Manta Industries Ltd
1010 Logan Avenue, PO Box 3657/Stn B,
Winnipeg, Man R2W 3R4
☎ (+1 204) 783 70 16 ⊠ MANTENT Winnipeg
⊖ (+1 204) 775 21 22
👥 55 🏠 35,000 sq ft
President and Chief Executive Officer: Barry Guberman ☎ Ext 25
Vice President and Sales Manager: Fraser Kulba ☎ Ext 23
Consultant: J Cohen
■ Manufacture canvas products for military & domestic purposes including military bags, cases & covers; chemical warfare detector kits; electric blankets for search air & rescue flights; heaters for jeep seats; medical stretchers; insect barriers; ruck sacks; tarpaulins; tents; load carrying vests. [98]

75-09507 Matrix Systems Ltd
161 Alden Road, Markham, Ont L3R 3W7
☎ (+1 905) 447 44 42 ⊖ (+1 905) 477 36 06
👥 10 🏠 1,590 sq m
President: Allan Guttman
■ Manufacture space, defence, communications, information & electronic ATC systems. [98]

75-09520 McDonnell Douglas Canada Ltd
PO Box 6013, Toronto AMF, Toronto, Ontario L5P 1B7
☎ (+1 905) 677 43 41 [T] 968825
⊖ (+1 905) 673 43 21; 673 43 22
[T] 6104924341 twx ♣ US$ 200,000 (1994)
👥 1,800 (1996) 🏠 200,000 sq m
President: Kenneth R Reed
■ Manufacture of wings for MD-80 Twinjet, MD-90 and MD-11 Trijet commercial aircraft. Components for CN military transport.
Ottawa Office
50 O'Connor Street, Suite 1424, Ottawa, Ont K1P 6L9
☎ (+1 613) 236 97 01 [T] 534915 ⊖ (+1 613) 563 98 23
Vice President, Ottawa: Ronald G Slaunwhite
■ Manufacture wings for the KC-10 tanker cargo aircraft, components for the CF-18 fighter aircraft for the Canadian Forces and F/A-18 for the US Navy and allied nations.
® USA: McDonnell Douglas [98]

75-09540 Mechron Power Systems Inc
2437 Kaladar Avenue, Ottawa, Ontario K1V 8B9
☎ (+1 613) 733 38 55 ⊖ (+1 613) 733 81 97

⊕ C$ 5,000,000 (1996) ♣ C$ 12,000,000 (1996)
👥 62 (1997) 🏭 10,000 m2
Managing Director: Jean Dumont
Operations Director: Robert Roe
Sales and Marketing Director: George Mayo
Purchasing Manager: Shiela Grubb
■ Production, installation, servicing and maintenance of custom-engineered power generating systems. Standby power and rotary uninterruptible power supply (UPS) systems, DC power systems and related components. Communications rectifiers (AC and DC power distribution), thermoelectric generators, renewable source electric generators, cycle charge power systems. [98]

75-09550 **Med-Eng Systems Inc.**
2470 Don Reid Dr., Ottawa, Ont K1H 8P5
☎ (+1 613) 739 9646 ⊖ (+1 613) 739 4536
👥 40 (1996) 🏭 700 sq m
President: Richard J. L'Abbé
VP Operations: Vince Crupi
Advertising Mgr: Richard J. L'Abbé
Purchasing Mgr: Richard Douglas
■ Explosive disposal helmet/suit with state of the art protection against fragmentation, overpressure, impact & heat. Also producers of EOD shields, telescopic/robotic manipulators & body cooling suits and communication systems. → 80 [97]

75-09600 **Menasco Aerospace Ltd**
(a division of Coltec Aerospace Canada Ltd)
1400 South Service Road West, Oakville, Ontario L6L 5Y7
☎ (+1 905) 827 77 77 ✆ (800) 784 7009
[W] http://www.coltecind.com ⊖ (+1 905) 825 15 83
👥 640 (1997) 🏭 345,000 sq ft
President: P Wright
Vice President Engineering: Gerry Kouverianos
Vice President Program Management: Steve Chalmers
■ Development, design, manufacture, testing and overhaul of fixed-wing and rotary landing gear systems. Electro-hydraulic and hydro-mechanical systems related to primary and secondary flight controls; fly-by-wire electro-hydraulic flight controls; ground steering, including steer-by-wire; aircraft hydraulic systems; variable wing and wing sweep actuation and machining of aircraft and helicopter components. Ramp and door actuation, field service, spares support. Certification testing. → 80 [98]

75-09650 **Messier-Dowty Electronics**
(† Dowty Aerospace Peterborough)
2000 Fisher Drive, PO Box 4525,
Peterborough, Ont K9J 7B1
☎ (+1 705) 743 6903 ⊖ (+1 705) 745 1394
👥 50 🏭 1 900 m2
General Manager: G Novacek
Assistant General Manager: J Hall
Financial Controller: L Donker
■ Custom designed, microprocessor based controllers & monitors for flight controls, landing gear & ground steering systems. [98]

75-09655 **Messier-Dowty Inc.**
(† Dowty Aerospace Landing Gear (Montreal))
(a member of the TI Group)
13000 du Parc, St Janvier (Mirabel), Que J0N 1L0
☎ (+1 514) 434 3400 ⊖ (+1 514) 434 3410
👥 100 🏭 15 000 m2
Vice President & General Mgr: G. Giard
■ Manufacture and testing of commercial and military landing gear equipment.

Messier-Dowty Inc
(† Dowty Aerospace Toronto)
(a member of the TI Group plc)
574 Monarch Avenue, Ajax, Ontario L1S 2G8
☎ (+1 905) 683 31 00 ⊖ (+1 905) 686 29 14
👥 386 🏭 19,000 m2
President: Ken Laver
Vice President Manufacturing Operations: Tony Wood
Vice President Finance: Bryan Teed
Vice President Engineering: Dr Ralph Darlington
Procurement Director: Kevin Alexander
Human Resources Director: Barry Wohl
Contracts Director: Allan Harvey
■ Design, develop & manufacture fully integrated landing gear systems. Capability covers small to ultra large landing gear. Design & manufacture critical electronic controls to support landing gear systems & other applications. [98]

75-09770 **Mining Resource Engineering Ltd (MREL)**
1555 Sydenham Road, RR #8, Kingston, Ont K7L 4V4
☎ (+1 613) 545 04 66 [E] mre@limestone.kosone.com
⊖ (+1 613) 542 80 29
General Mgr: A.W. (Bill) Bauer
Managing Dir: W.A. Crosby
Explosives R&D Mgr: C.John Anderson
■ Develop and manufacture liquid explosive foam (LEXFOAM) which is suitable for EOD operations, rigid and flexible solid explosive foam (R&F-SEF) sheets. [98]

75-09800 **Robert Mitchell Inc.**
350 Decarie Boulevard,
St Laurent, Montreal, Québec H4L 3K5
☎ (+1 514) 747 24 71 [T] 5825854
⊖ (+1 514) 747 77 12
👥 460 (1995) 🏭 25,000 m2
President: G H Holland
Vice President Sales: Paul Emile Dostie
Vice President Secretary and Treasurer: D St Jean
Manager Foundry Division: G Dufresne
■ Aluminium and magnesium sand castings, machine shop, sheet metal fabrication, passenger loading bridges, airport ground equipment. Galley and cabin service trucks, tow bars, lavatory trucks, mobile passenger stairs, baggage conveyors, aircraft de-icers. [97]

75-09855 **MPB Technologies Inc**
151 Hymus Boulevard, Pointe Claire, Québec H9P 1J1
☎ (+1 514) 694 87 51 [E] tds@mpbtech.qc.ca
⊖ (+1 514) 695 74 92
👥 200 🏭 30,000 sq ft
President: Dr M P Bachynski
Lasers and Electro-optics Director: W R L Clements
Purchasing Manager: N Garnett
Public Relations Manager: Ms J Bachynski
■ Specialises in high technology products and systems, research, development and measurement services. Activities include: communications; electromagnetics (radar, defence electronics, microwave systems); electronic systems; fusion technology; lasers and electro-optics; space technology. Inter-satellite links, ASIC design and qualification, MMIC design, fast processing, laser flourescence spectroscopy, high speed communications, multiplexing, atmospheric laser communication system for surveillance. → 80 [97]

75-09860 **Mustang Survival Corp.**
(† Mustang Industries Inc)
3810 Jacombs Road,
Richmond, British Columbia V6V 1Y6
☎ (+1 604) 270 86 31 ⊖ (+1 604) 270 04 89
♣ C$ 25,000,000 (1996) 👥 300 (1997) 🏭 75,000 sq ft
President: Dwight J Davies
Vice President Government Products: Robert J Askew
Operations and Technical Director: Gerald Harder
Manager Government Sales: Jim Hartt
[E] jhartt@mustangsurvival.com
■ Personnel safety and survival equipment for land, sea and air operations. Numerous forms of immersion suits, anti-exposure coveralls, inflatable life preservers, 'g' suits, life vests and buoyant coats form basics of product line. Cold water protective devices and anti-g assemblies.
Subsidiary:
Mustang Survival (USA) Corp [98]

75-09950 **Nordic Systems Inc**
1044 Rangeview Road, Mississauga, Ontario L5E 1H3
☎ (+1 905) 278 33 31 ✆ (800) 268 17 56
[T] 6961146 nordic msga ⊖ (+1 905) 278 51 97
♣ C$ 7,000,000 (1996) 👥 48 (1997) 🏭 3,250 m2
President: George F Hrubecky
Operations and Technical Director: John Nyhuis
Sales Manager: John R Bain
Purchasing Manager: Jane Elcome
■ Manufacture of crash fire rescue vehicles and components. Fire suppression systems, dual agent skid units and refueller reels. [98]

75-10145 **Northern Airborne Technology Ltd (NAT)**
(a subsidiary of Chelton Group Company Ltd)
1925 Kirschner Road - Unit 14,
Kelowna, British Columbia V1Y 4N7
☎ (+1 604) 763 22 32 ⊖ (+1 604) 762 33 74
⊕ US$ 9,000,000 (1996) 👥 75 (1997) 🏭 2,300 m2
President: Geoff Cooper
Vice President Finance: Mark Phillips
Vice President Marketing: Robert Sibson
General Manager: Steve Brunsden
Quality Assurance Manager: Ray Lewis
Sales Manager: John Dendekker [E] johnd@natech.com
■ Helicopter audio controllers, airborne loudhailer and siren systems, navigation interfaces. Tactical FM communication system including voice encryption and muliple transceivers. 2-axis tactical multi-band airborne DF system.
Aircraft cockpit audio systems; intercoms, alerting systems, navigational interfaces and tactical FM communications transceiver and radio controls. AM/FM direction finder. [98]

75-10300 **Novatronics Inc**
(† Novatronics of Canada Ltd)
677 Erie Street, PO Box 610, Stratford, Ontario N5A 6V6
☎ (+1 519) 271 38 80 ✆ (800) 892 38 80
⊖ (+1 519) 271 97 81
♣ C$ 10,000,000 (1997) 👥 100 (1996) 🏭 30,000 sq ft
President: Don McDougall
General Manager: Peter Vandrunen
Technical Director: Mark Bastani
Product Assurance Director: Mike Kolbnasnik
Purchasing Manager: Jim Steward
Sales Manager: Paul Smith
Sales Administrator: Ms Lise St Jean
■ Research, development and manufacture of stepping motors, synchros, servometers, digital indicators (mag wheels), DC brushless motors, surface control position transmitters, RVDTs, LVDTs and cockpit instruments. Precision electro-mechanical devices and related electronics for sensing, actuation and indication of motion/position for aerospace and defence industries. [98]

75-10400 **Ontario Drive & Gear Ltd (ODG)**
220 Bleams Road, PO Box 280,
New Hamburg, Ont N0B 2G0
☎ (+1 519) 662 28 40 [E] sales@odg.com
[W] http://www.argoatv.com ⊖ (+1 519) 662 24 21
👥 90 (1997) 🏭 6,000 sq m
President: Joerg Stieber
Vice President Marketing: Robert C Archer
Purchasing Manager: Randy Knichewsky
■ Argo line of lightweight multiwheel, amphibious off-road vehicles. [98]

75-10420 **Optotek Ltd**
62 Steacie Drive, Kanata, Ontario K2K 2A9
☎ (+1 613) 591 03 36 ✆ (800) 361 29 11
[E] optotek@optotek.com [W] http://www.optotek.com
⊖ (+1 613) 591 05 84
👥 12 🏭 37,000 sq ft
President: Dr David I Kennedy
Operations Director: Gunnar Wareberg
Purchasing Manager: Steve Wilson
■ Development and manufacture of LED materials and devices based on Group III-V semiconductors, custom LED arrays for military and industrial applications, display-related electronic subsystems and test equipment, gallium arsenide field-effect transistors and monolithic microwave integrated circuits, microwave and monolithic integrated circuit analysis and design software. [98]

75-10430 **Oracle Telecomputing Inc.**
25 Industrial Avenue, Carleton Place, Ont K7C 3V7
☎ (+1 613) 257 44 25 [E] oratel@entrenet.com
⊖ (+1 613) 257 77 64
President: James Harvey
Marketing Manager: Don Anderson
Simulation and Training Marketing: Al Hamblin
■ Radar processing and display systems, ATC Simulation. Aviation Information Systems. PSO (Project Single Option) System which monitors flight plans to identify suspicious aircraft for drug interdiction. [98]

75-10770 **Pan Atlantic Aerospace Corporation**
PO Box 599, Station B, Ottawa, Ontario K1P 5P7
☎ (+1 613) 772 14 54 ⊖ (+1 613) 722 16 91
President: Nick Baumberg
■ Specialist in advanced lighter-than-air technology for surveillance, transportation and telecommunications applications. [98]

75-11035 **Pelorus Navigation Systems Inc.**
5428-11 Street North East, Calgary, Alberta T2E 7E9
☎ (+1 403) 730 55 55 W http://www.pelorus.ca
⊖ (+1 403) 730 55 11
♣ C$ 5,095,461 (1996) 👥 30 (1997) 🏭 10,000 m2
Chief Executive Officer: Edward G Fitzhenry
Corporate Secretary and Senior Vice President: Mrs Shirley A Fitzhenry
Vice President Marketing: Peter W Bolton
E pbolton@pelorus.ca
■ Design, development and manufacture of distance measuring equipment (DME), and local area differential global positioning systems (LADGPS). International distribution and support services for engineering, installation and maintenance. Navaids for airports. 98

75-11300 **Pointer Avionics**
(a division of Pointer Technologies (Canada) Ltd)
1235 Shawson Drive, Unit 11b,
Mississauga, Ontario L4W 1C4
☎ (+1 905) 564 16 00 E pointer@idirect.com
⊖ (+1 905) 564 16 01
🏭 21,000 sq ft
President: Leo Bronee
Quality Control: Jan Jensen
■ Aircraft emergency locator transmitters. Pointer ELT service and overhaul. → 80
® USA: Pointer ELT 98

75-11400 **Pratt & Whitney Canada Inc (P&WC)**
(a subsidiary of United Technologies Corp, USA)
1000 Boulevard Marie Victorin,
Longueuil, Québec J4G 1A1
☎ (+1 514) 677 94 11 (global services); 442 80 80
✆ (800) 268 80 00 (international help desk)
✉ PRATTWHIT MTL
📠 5267509 prattw mtl; 8104223872 twx
⊖ (+1 514) 647 36 20
▥ TWX 8104223872 PRATT WHIT CANADA
♣ C$ 1,700,000,000 (1995) 👥 8,200
🏭 2,800,000 sq ft
Chairman and Chief Executive Officer: L David Caplan
President and Chief Operating Officer: Gilles P Ouimet
Vice President Operations and Design Engineering: J Blackie
Vice President Quality Assurance: Michael J David
Vice President Communications: Ms Francine Osborne
Vice President Engineering: Brian H Sanders
Vice President Service Centres: Jean Colpin
Vice President Operations: Louis R Chênevert
Vice President Marketing Corporate and Military Aircraft: F John Wright
Vice President Human Resources: Brian F Duffy
VP Marketing International Business Dev: Joseph N Torchetti
Vice President Finance: R F Steers
Vice President Counsel and Corporate Secretary: C J C Pascoe
Vice President Materials and Procurement: Claude Paquette
Vice President Engineering Product Integrity: Gordon M Hogg
Vice President Industrial and Marine Division: J Brian Haworth
Vice President Marketing Regional Transport and Helicopter Engines: Kevyan Fard
Vice President Customer Support: C Dickman
Controller: Miguel C Doyon
Treasurer: Ms Wendy Conn
Public Relations Officer: Sylvie Savard
■ Aircraft propulsion (turbofan, turboshaft, turboprop); aircraft auxiliary power unit (APU); marine and land-based gas turbine propulsion; gas turbine maintenance services and overhaul. 97

75-11500 **Presentey Engineering**
2784 Fenton Road, PO Box 919 - RR 5,
Gloucester, Ontario K1T 3T7
☎ (+1 613) 822 12 51 E pepl@presentey.com
W http://www.presentey.com ⊖ (+1 613) 822 12 56
♣ C$ 2,000,000 (1997) 👥 15 (1997) 🏭 10,000 m2
President: Ms Nicole Presentey
Sales Director: Kevin Mercer
Operations and Technical Director: Mike Stashin
■ Research, development, consulting and production for electronics, telecommunications and applied science. Brushless DC motors, transmissometers, runway visual range sensor systems and visibility sensors. Wind and altimeter digital display systems, navigation aids. Comsec and Tempest communications equipment, Tempest enclosures and MIL-SPEC patch panels. Electromechanical devices, remote control signalling systems and power amplifier. 98

75-11520 **Prior Data Sciences Ltd**
(a subsidiary of Spar Aerospace Ltd)
240 Michael Cowpland Drive, Kanata, Ont K2M 1P6
☎ (+1 613) 591 72 35 ⊖ (+1 613) 591 03 43
♣ C$ 28,136,000 (1996) 👥 270
President: Larry Humphries
Managing Director: Andre Stavri
Sales and Marketing Director: Robert Moon
Marketing Co-ordinator: Bev Chartrand
■ Oceanic flight data systems, domestic flight data systems, radar data processing, ATM displays and functional workstations, flow management, electronic flight strip automation. 98

75-11540 **Proparms Ltd**
2930 Ch Suite Thérèse, Carignan J3L 2B2
☎ (+1 514) 658 52 07 ⊖ (+1 514) 447 27 27
👥 16 (1997) 🏭 6,600 sq ft
President: Josef Elsener
VP Marketing & Sales: Ms Lesley Rust
Operations/Technical Dir: Pierre Grenier
Purchasing Mgr: Joan Wiener
■ Design and manufacture explosive ordnance equipment: Neutrex Waterjet disrupter, MKII, 29 mm disrupter. Pl exploder generator power source. 12.75 mm disrupter, EOD tools. Products for Police and Military. 98

75-11590 **Racal Canada Inc**
(a division of Racal Electronics plc, UK)
38 Auriga Drive - Unit 12, Nepean, Ontario K2E 8A5
☎ (+1 613) 225 46 40 E sales@racalcanada.com
W http://www.racal.ca ⊖ (+1 613) 225 46 41
♣ C$ 10,000,000 (1997) 👥 50
President and Managing Director: Richard F Howes
Vice President Business Development: Don S Poole
Sales and Marketing Director: B Tuyssuzian
Operations and Technical Director: N R Massey
Purchasing Manager: G Mior
■ Supplier of radio and data communications equipment, defence electronics, avionics systems and accessories for strategic, tactical and civil applications. Products: HF radios; frequency-hopping radios; electro-acoustics; headsets; command and information systems; communication security; electronic warfare systems; unattended ground sensor systems; command and control systems; antennas; defence radar and tactical radio communications.
® UK: Park Air Electronics, Racal Electronics plc Group
USA: Antenna Products Corp, Antenna Research Associates

Branch Office
1806 Woodward Drive, Ottawa, Ontario K2C 0P7
☎ (+1 613) 225 46 40 ⊖ (+1 613) 225 46 41 98

75-11595 **Racal Pelagos Canada, Ltd**
(a subsidiary of The Racal Corporation, USA)
209 Acrotech Drive, Units 2A and 3A,
Enfield, Nova Scotia B2T 1K3
☎ (+1 902) 873 22 82 ⊖ (+1 902) 873 20 67
■ Supplier of high resolution geophysical surveys, seabed mapping and oceanographic programs. 98

75-11600 **Racal Filter Technologies Ltd**
(a subsidiary of The Racal Corporation Canada Inc.)
1360 California Avenue, Brockville, Ont K6V 5V8
☎ (+1 613) 345 01 11 ⊖ (+1 613) 345 26 39
👥 120
President: Phil Doutre
Operations and Technical Director: Pierre Legare
■ Manufacturing of NBC filters, canisters, and gas mask filters and canisters. Research and testing of advanced filter and canister design. → 60 98

75-11700 **Raytheon Canada Ltd**
(a subsidiary of Raytheon Company, USA)
400 Phillip Street, PO Box 1619,
Waterloo, Ontario N2J 4K6
☎ (+1 519) 885 01 10 📠 6955431
⊖ (+1 519) 885 86 41; 885 11 78
▥ TWX 6103653469 ♣ C$ 70,000,000 (1994)
👥 300 (1995) 🏭 129,000 sq ft
President and General Manager: R P Gaffney
Marketing Director: Moe Vyas
Marketing Manager: C W Mayo
■ Turnkey air traffic control systems and air traffic control systems with primary and secondary radar. Ground control approach radar with ASR and PAR, RDPS, FDPS and displays, data processing. Military satellite earth terminals (SATCOM), fast frequency-hopping synthesizers, HF surface wave radars and troposcatter communication systems.
® USA: Raytheon 97

75-12000 **Reil Industrial Enterprises Ltd**
730 Gana Court, Mississauga, Ont L5S 1P1
☎ (+1 905) 670 10 70 ⊖ (+1 905) 670 16 95
♣ C$ 3,000,000 👥 35 🏭 23,000 sq ft
President: William H. Reil
■ Precision CNC machine aerospace components to MIL-SPEC. 98

75-12120 **Rolls-Royce (Canada) Ltd**
(a division of Rolls-Royce Industries Canada)
9500 Cote de Liesse Road, Lachine, Québec H8T 1A2
☎ (+1 514) 631 35 41 ☎ C$ 26,000 ✉ ROYCAR
📠 058221882 ⊖ (+1 514) 636 99 69
👥 890 🏭 350,000 sq ft
President: Jean Guilbault
Vice President Commercial: W J Cooney
Vice President Industrial: Charles McIvor
Vice President Aero: T D Cribbes
Vice President Finance: A S Fleming
Vice President Finance/Administration: Paul Bourgon
Vice President New Build: Alex Best
■ Repair, overhaul and service support for aero and industrial gas turbines. Manufacturers of aircraft parts, industrial gas turbines and off-engine support systems. Service agreements.
® UK: Rolls-Royce 97

75-12150 **Ross Video Limited**
PO Box 220, Iroquois KOE 1KO
■ Video production switchers/mixers. Distribution amplifiers. 98

75-12290 **Saft Nife Corp**
(a subsidiary of Saft SA, France)
125 Nantucket Boulevard, Scarborough, Ont M1P 2N8
☎ (+1 416) 757 51 51 ✆ (800) 263 95 93
⊖ (+1 416) 752 45 14
♣ C$ 25,000,000 👥 140 🏭 6,690 sq m
Managing Director: Paul Terrien
Technical Director: Austin Drohan
Marketing Director: Ms Joanna Kurylowicz
Purchasing Manager: Bruce Rennie 98

75-12300 **Saint John Shipbuilding Limited**
300 Union Street, 10th Floor, PO Box 970,
Saint John, NB E2L 4E5
☎ (+1 506) 632 59 39 ✉ DRYDOCK
⊖ (+1 506) 632 59 12
👥 1,000
President: John F Thomas
Senior Manager, Marketing: J Struthers
Purchasing Manager: Norman Speake
Sales and Marketing New Business Development: D Ian Jones
■ Ship design, construction & repair. Programme management, quality assurance, integrated logistic support, procurement & technology transfer. → 80

Saint John Naval Systems
10 Morris Drive, Suite 6, Building B,
Dartmouth, NS B3B 1K8
☎ (+1 902) 494 57 00 ⊖ (+1 902) 494 57 92
General Manager: Norman Brown
■ Management, design and production of major communication items for commercial and government sector integrated logistics support service.

Dartmouth Marine Slips
1 King Street, PO Box 380,
Dartmouth, Nova Scotia B2Y 3Y5
☎ (+1 902) 494 53 00 ⊖ (+1 902) 494 52 03
General Manager: John Salvage
■ Ship repair services.

East Isle Shipyard Ltd
115 Water Street, PO Box 220,
George Town, P.E.I. COA 1LO
☎(+1 902) 652 22 75 ⊖ (+1 902) 652 27 64
General Manager: William Hind
■ Construction of small vessels.

Halifax Shipyard Limited
3099 Barrington Street North, PO Box 9110,
Halifax, Nova Scotia B3K 5M7
☎(+1 902) 423 58 18 ⊖ (+1 902) 423 92 71
General Manager: Robert Shepard
■ New construction, ship repair, refits and major vessel overhauls. [98]

75-12550 Scintrex Ltd

222 Snidercroft Road, Concord, Ontario L4K 1B5
☎(+1 905) 669 22 80 [E] scintrex@scintrexltd.com
✉ GEOSCINT TORONTO ⊖ (+1 905) 669 51 32
C$ 13,430,000 (1996) ♣ C$ 18,421,000 (1995)
150 (1997) 70,000 sq ft
Chairman: H O Seigel
President and Chief Executive Officer: Abe Rolnick
Vice President Operations: Adrian Beale
Vice President and General Manager Airborne Systems and Surveys Division: Terry McConnell
Vice President Sales and Marketing Analytical and Security Products: Hugh Andrew
Vice President Finance: Mario Scarcelli
Operations and Technical Director, Nuclear and Analytical Products: Larry McNelles
Engineering Director Geophysical Instruments: Ivo Brcic
Sales Manager: Phil Hembruff
Marketing Communications Specialist: Ms Dolores Cece
■ Radiation monitor and automatic alarm systems and gamma survey meters. Magnetic underwater surveillance array, explosive ordnance detector magnetometers, hand held explosive and narcotics detectors.
* Aerial, vehicle & fixed radio sets. [98]

75-12600 Securesearch Inc.

3500 Pharmacy Ave, Unit 4, Scarborough, Ont M1W 2T6
☎(+1 416) 492 5349 ⊠ 06986391 tor
⊖ (+1 416) 492 3656
2 1 200 sq ft
President: Bruce B. Koffler
■ Explosives recognition training aids for security/military/law enforcement. Security publications, videos on X-ray identification of bomb components, drug smuggling, letter bombs, parcel bombs, bomb threat management; grenade fuze removal tool (remote); security awareness posters on explosive devices; aviation security training/test kit. Narcotics smuggling training aids for customs/drug officers. Destructive testing of light vehicle armour, ballistic glazing materials, body armour. Pipe bomb decapper (remote).

Overseas Representatives:
L.E.A. (USA)
Scopex (France, Belgium, Luxembourg, Switzerland, Spain, North Africa, W.Africa)
Swiss-Impex Trading & Consulting Co. Ltd (Hong Kong, China, Macau, Taiwan, Korea)
Ontario Security Contractors Ltd. (United Arab Emirates).
The North Group (Philippines)
Nichimen Aviation Corporation (Japan)
Herasco ApS (Denmark, Norway, Sweden, Finland, Estonia, Latvia, Lithuania)
Unicorn International Pte Ltd (Singapore, Brunei, Indonesia)
Nassar Al-Shuraian's Sons Gen. Trading & Contracting Co. W.L.L. (Kuwait)
Rotax Electronics Private Limited (India)
Fahad International Co. (Saudi Arabia) [97]

75-12750 SEI Industries Ltd

7400 Wilson Avenue, Delta, British Columbia V4G 1E5
☎(+1 604) 946 31 31 ✉ SPECTRAL
[E] seisales@sei-ind.com ⊠ 4355779 sei ind vcr
⊖ (+1 604) 940 95 66
50
President: Mark McCooey
Managing Director: Gary Wilson
Sales and Marketing Director: Greg Emry
Sales Representative: Rob Ferguson
Marketing Co-ordinator: Ms Zabrina Chew
■ Manufacture of lightweight, portable products including fully-collapsible helicopter firefighting bucket (Bambi Bucket), fully collapsible free-standing fuel storage containers for helicopters and external load fuel tanks. Open top water tanks for firefighting. Fabric storage bladders up to 50,000 gallons for storage of chemicals, fuel and water. [97]

75-12755 Sellick Equipment Ltd

358 Erie Street North, PO Box 1000,
Harrow, Ontario NOR 1GO
☎(+1 519) 738 22 55 ⊖ (+1 519) 738 34 77
82 (1997) 80,000 sq ft
Chairman: Walter R Sellick
President: Howard R Sellick
Vice President: David M Sellick
Operations Manager: Ronald Wood
Purchasing Manager: John Kok
■ Rough terrain forklifts, telescopic material handlers and truck mounted forklifts. [98]

75-12760 Senstar-Stellar Corp

(† Senstar Corporation)
Box 13430, Kanata, Ont K2K 1X5
☎(+1 613) 839 55 72 [E] info@senstar.ca
[W] http://www.senstar.ca ⊖ (+1 613) 839 58 30
100
Managing Director: Ron Clifton
Director Business Development and Marketing: Julian Romeskie
Director International Sales: John Sherren
Operations Director: Ms Eleanor Hodgson
Marketing Manager: Michael Rack
Marketing and Sales Administration: Ms Shirley Reddick
■ Manufacture outdoor security products including PerimiTRAX, a covert perimeter sensor; DAVID 300 an outdoor Video Intrusion Detection; REPELS, a portable perimeter sensor providing 100 meters of protection which can be set up in fifteen minutes. Senstar also supplies colour graphic annunciators and communications systems for intrusion detection applications. → 80

Overseas offices:
Senstar Inc. (USA) → 90
Senstar UK Ltd (UK) → 90 [98]

75-12795 Simrad Mesotech Ltd

(† Simrad Mesotech Systems Ltd)
(a subs. of Simrad A/S, Norway)
1598 Kebet Way, Port Coquitlam, BC V3C 5W9
☎(+1 604) 464 8144 ⊖ (+1 604) 941 5423
40 22 000 sq ft
Exec. Vice President: Hans Gray
Sales Mgr: David Adams
■ Design & manufacture underwater acoustic equipment, including minehunting & surveillance sonar systems. [97]

75-12960 SNC Industrial Technologies Inc.

Les Technologies Industrielles SNC Inc.
(a subsidiary of The SNC-Lavalin Group Inc.)
Head Office
5 Montée des Arsenaux, Le Gardeur, Que J5Z 2P4
☎(+1 514) 581 30 80
⊖ (+1 514) 581 02 31; 585 73 02
600
President and Chief Executive Officer: Anthony Rustin
Vice President National Sales: Claude Daigneault
Vice President Marketing: Robert Leboeuf
Vice President Manufacturing: André Breton
Vice President Technologies: Paul Cobetto
Program Director International Marketing: Charles Pommet
Director National Sales: Claude Tassé
Director International Marketing: Marc Delahaut
Senior Co-ordinator International Simunition Sales: Lise Laberge
■ Manufacturer of conventional & training ammunition as well as protective equipment to military, special forces & police in Canada & abroad.

SIMUNITION
5 Montée des Arsenaux, Le Gardeur, Que J5Z 2P4
☎(+1 514) 581 30 80 ⊖ (+1 514) 581 02 31
500
Vice President Manufacturing: André Breton
■ Manufacture of small to large calibre ammunition, protection equipment.

St Augustin Plant
40 d'Anvers Street, Saint-Augustin, Que G3A 1S4
☎(+1 418) 878 44 80 ⊠ 524642 acldr vldg
⊖ (+1 418) 878 43 37
150
Plant Manager: Adrien Pelletier

Ottawa Office
170 Laurier Avenue West, Ottawa, Ont K1P 5V5
☎(+1 613) 567 70 19 ⊖ (+1 613) 567 55 09
Manager: Rémi Vézina [98]

75-12980 Soucy International Inc.

5450 St Roch, Drummondville, Que J2B 6W3
☎(+1 819) 474 66 66 ⊖ (+1 819) 477 94 23
♣ C$ 25,000,000 (1995) 200 (1995) 20,000 sq m
President: Gilles Soucy
Vice President Marketing and Business Development: Jack W Jennings
Operations and Technical Director: Yves Deschenaux
Purchasing Manager: Yvon Granger
■ Design & manufacture rubber tracks, road wheels & drive sprockets for military vehicles. [98]

75-13000 Spar Aerospace Ltd Corporate Office

121 King Street West, Suite 1800,
Toronto, Ontario M5H 4C2
☎(+1 416) 682 76 00 [W] http://www.spar.ca
⊖ (+1 416) 682 76 01
♣ C$ 538,000,000 (1996) 2,600 (1996)
Chairman of the Board: Earl H Orser
Vice Chairman: Philip A Lapp
President and Chief Executive Officer: Colin D Watson
Senior Vice President and Chief Financial Officer: Mark C Steinman
President Spar Space Systems: D F Massotti
Vice President and Legal Counsel, Secretary: Sheldon Polansky
Vice President International Relations: Karsten Westphal
President Comstream: Frederick D Lawrence
Vice President Public Affairs: Ms Deborah M Allan
☎(+1 416) 682 76 19; 682 76 21 (fax)
■ Design, development, manufacture and servicing of systems for aviation, communications, space and defence markets. Systems and services include remote manipulator systems for space and ground applications, satellites, satellite subsystems, land-based earth stations, electro-optic systems, IR surveillance systems, gearboxes, transmission systems and equipment. Aircraft repair and overhaul, helicopter services. → 80

Government Relations Office
222 Queen Street, Suite 402, Ottawa, Ont K1P 5V9
☎(+1 613) 563 02 30 ⊖ (+1 613) 563 42 84
Government Relations Dir: Ken Mackay

Aviation Services Division
7785 Tranmere Drive, Mississauga, Ont L5S 1W5
☎(+1 905) 673 60 00 ⊠ 6524240 sparcal tor
⊖ (+1 905) 671 58 02
President: Tony Caputo
■ Repair & overhaul of electrical, mechanical & navigational components for fixed & rotary wing aircraft.

Satellite and Communications Systems Group
21025 Trans-Canada Highway,
St Anne-de-Bellevue, PQ H9X 3R2
☎(+1 514) 457 21 50 ⊠ 05822792
⊖ (+1 514) 457 27 24
President: David F Masotti
■ Satellite systems, subsystems and payloads. Satellite earth stations, components and subsystems, time division multiple access systems (TDMA).

Spar Space Systems
9445 Airport Road, Brampton, Ont L6S 4J3
☎(+1 905) 790 28 00 ⊖ (+1 905) 790 44 00
Vice President and General Manager: Steve Sorocky
■ Advanced manipulators for space & industrial application, deployable solar arrays, infra-red surveillance systems for land, sea & air applications.

Spar Applied Systems Group
365 March Road, Kanata, Ont K2K 1X3
☎(+1 613) 592 34 30 ⊠ 053-4714 sparby ott
⊖ (+1 613) 592 44 86
President: David W Stapley
■ Electro-optical defence products, electronic assemblies for satellites and defence equipment, product support services.

Spar Applied Systems
115 Emily Street, Carleton Place, Ont K7C 4J5
☎(+1 613) 253 30 20 ⊖ (+1 613) 253 30 33 [98]

75-13020 Spar Aviation Services Godfrey Aerospace

(† Godfrey Aerospace Inc.)
(a subsidiary of Spar Aerospace Ltd, Canada)
70 Hymus Boulevard, Pointe Claire, Que H9R 1C9
☎(+1 514) 426 43 00 ✆ (888) 780 74 84
⊖ (+1 514) 426 34 99
♣ C$ 12,000,000 140 100,000 sq ft
Managing Director: Vincent Ko
Operations Director: Stephane Allard
Sales Manager: Claude Lambert
Purchasing Manager: Michael Casey

■ Design, fabrication of land, naval and airborne support equipment; repair and overhaul of airborne accessories; design and fabrication of galley equipment. → 80 [98]

75-13100 Sparton of Canada Ltd (SOC)
(a subsidiary of Sparton Corp, USA)
99 Ash Street, London, Ontario N5Z 4V3
☎ (+1 519) 455 63 20 Ø (800) 663 00 11
⊖ (+1 519) 452 39 67
👥 100 (1997)
Vice President and General Manager: W David Weind
Acoustics Director: Linas Siurna
Engineering Director: Steve Elder
Director of Oceanography: Colin McCrae
Controller: Michael Woods
Purchasing Manager: George Ryan
Marketing: Ms Josie Graham
■ Active and passive sonobuoys, microbuoys, low frequency sound projectors and controlling sub-systems, bathythermographics (ship and air launched) and controlling sub-systems. Sonar research projects, calibration systems and training. Electronic contract manufacturing, power supplies and oceanographic instrumentation.
® USA: Sparton [98]

75-13200 Sprung Instant Structures Ltd
1001-10th Avenue South West, Calgary, Alberta T2R 0B7
☎ (+1 403) 245 33 71 Ø (800) 661 11 63
W http://www.sprung.com ⊖ (+1 403) 229 19 80
👥 100 🏭 1,000,000 sq ft
Managing Director: P Sprung
Vice President: Philip Sprung (Jr)
Sales Manager: Jim Avery
Purchasing Manager: George Laliberte
Advertising Co-ordinator: Christine Harper
■ Manufacturer and supplier of air-transportable military shelters for operations in severe climates. [98]

75-13510 Tecmotiv Corporation
191 Caldari Road, Unit 9, Concord, Ont L4K 4A1
☎ (+1 905) 669 59 11 E info@tecmotiv.com
W http://www.tecmotiv.com ⊖ (+1 905) 669 94 42
Chairman: Henry Lubaszka
President: Arthur Hayden
Vice President Finance and Administration: Ms Sally Ann Look Yan
Vice President Sales and Marketing: Gordon Forbes
■ Manufacture of diesel engine conversion kits for tracked armoured vehicles & military trucks. Manufacture & supply of maintenance spare parts for tracked armoured & wheeled military vehicles. → 85 [98]

75-13540 Tektronix Canada Inc.
(a subs. of Tektronix Inc.)
555 March Road, Kanata, Ont K2K 2M5
☎ (+1 613) 599 4511 Ø (800) 267 7016
⊖ (+1 613) 599 4513
👥 90
General Mgr: Phil McDonall
Government Accounts Mgr: Rick Weller
Marketing Communications Mgr: Chuck Glassier
■ Test & measurement instrumentation, computer graphics & communications.
● 785 Arrow Road, Weston, Ont M9M 2L4
☎ (+1 416) 747 5000 ⊖ (+1 416) 747 9799 [97]

75-13600 H I Thompson
(a division of Valcom Ltd)
175 Southgate Drive, Guelph, Ontario N1H 6J9
☎ (+1 519) 824 32 20 ⊖ (+1 519) 824 34 11
♧ C$ 10,000,000 ♣ C$ 15,000,000 👥 100 🏭 7,000 m2
General Manager: Bob Jeffery
Technical Director: Alan Henry
■ Design and manufacture of thermal insulating heat shields and precision sheet metal components for aerospace, military and marine markets. [97]

75-13770 Thought Technology Ltd
2180 Belgrave Avenue, Montreal, Que H4A 2L8
☎ (+1 514) 489 82 51 Ø (800) 361 36 51 ✉ THOTTECH
E mail@thoughttechnology.com
W http://www.thoughttechnology.com
⊖ (+1 514) 489 82 55
👥 30
President: Dr Hal Myers
VP Sales: Lawrence Klein
Marketing Dir: Mark Schwartz
Purchasing Mgr: Bran Rosenblatt
■ Audio heart rate monitor that permits monitoring of the heart through 6 layers of protective clothing & displays the heart rate on a digital display. [98]

75-14435 Vancouver Shipyards Co. Ltd
(a subsidiary of Seaspan International Ltd, Canada)
50 Pemberton Avenue, North Vancouver, BC V7P 2R2
☎ (+1 604) 988 63 61
℡ 4352532 seaspan vcr vanshipyd vcr
⊖ (+1 604) 990 32 90
Operations/Technical Dir: Rob Miller
Marketing Mgr: Bill Armstrong
Purchasing Mgr: Bill Fryer
■ New ship construction, ship repair. [98]

75-14540 VHB Industrial Batteries Ltd
(† VARTA Industrial Batteries Ltd)
(a subsidiary of BTR plc, UK)
72 Devon Road - Unit 22, Brampton, Ontario L6T 5B4
☎ (+1 905) 790 07 30 Ø (888) 842 22 88
E varta@vartabattery.com
W http://www.vartabattery.com ⊖ (+1 905) 790 07 44
♣ US$ 4,000,000 (1995) 👥 6 🏭 752 m2
General Manager: Neil Hale
Manager: Ravi Menon
■ Manufacture of NiCad batteries for aircraft. Submarine and vehicle batteries. Lips, telecommunications, solar and motive power. Standby power systems for telecommunications, switchgear, photovoltaic and UPS applications.
® Germany: VARTA Batterie AG
Representing:
VHB Industriebatterien, Germany [97]

75-15000 Victrix Ltd
416 Elizabeth St, Guelph, Ont N1E 2Y1
☎ (+1 519) 836 1480 ⊖ (+1 519) 836 4693
♣ £ 4 000 000 👥 20
President: P.T. Valeriote
Sales Mgr: Keith Dunn
■ Modular practice bombs, signal cartridges, telescopic antenna masts, military hardware & target systems. [97]

75-15200 Walbar Canada
(a subsidiary of Coltec Inc, USA)
1865 Sharlyn Road, Mississauga, Ontario L4X 1R2
☎ (+1 905) 602 40 41 ⊖ (+1 905) 625 83 60
℡ TWX 6104924384 ♣ C$ 35,000,000 (1996) 👥 250
🏭 38,000 m2
General Manager: Michael Savilo
Finance Director: Peter Mammone
Human Resources Director: Nick Johnson
Contracts Administrator: T Whitehouse
■ Aircraft jet engine turbine blades, vanes, discs. Electro-chemical machined compressor blades and vanes. Fuel adaptors and other components. [97]

75-15220 Weatherhaven
(† Weatherhaven Resources Ltd)
5700 Marine Way, Burnaby, BC V5J 5C8
☎ (+1 604) 451 8900 ℡ 0454247 vcr
⊖ (+1 604) 451 8999
👥 100
Managing Dir: Jim Allen
Sales Dir: Glen V. Thorne
Marketing Services Co-ordinator: Ms Louise Labelle
■ Turnkey camps, command centres, mobile field hospitals, workshops, kitchens, wash houses, sleeping quarters - private & semi-permanent, barracks/dorms, warehouses, aircraft hangars, ISO container shelters, modular structures & air transportable buildings & camps. [97]

75-17200 Zenair Ltd
Huronia Airport, PO Box 235, Midland, Ontario L4R 4K8
☎ (+1 705) 526 28 71 E zenair@zenithair.com
W http://www.zenithair.com/ch2000
⊖ (+1 705) 526 80 22
👥 35 🏭 20,000 sq ft
President: Christophe J Heintz
Production Manager: Bruce Barker
Marketing Manager: Mathiem Heintz
■ Design and manufacture [98]

75-46151 iMPath Networks Inc.
(† Fiberlign)
65 Auriga Drive, Suite 100, Nepean, Ont K2E 7W6
☎ (+1 613) 226 40 00 Ø (888) 467 28 49
W http://www.impathnetworks.com
⊖ (+1 613) 226 46 02
👥 35 (1997) 🏭 20,000 sq ft
President: Gregory Boyle
Managing Director: James Paul
Sales Manager: Jean-Paul Clermont
Purchasing Manager: Len Churchward
■ Design & manufacture of fibre optic communications systems & installation products. [98]

80 Service Companies & Organizations

80-00020 Aero Machining Ltd
Les Ateliers d'Usinage Aéro Limitée
(in partnership with CAM-TAC Industries Inc)
5411 Industrial Boulevard,
Montréal North, Québec H1G 3H7
☎ (+1 514) 324 42 60 ⊖ (+1 514) 324 99 97
👥 70 🏭 25,000 sq ft
President: Gilles Demers
General Manager: Marcel Desjarlais
■ Machine shop specialising in components for aircraft structures, missiles, engine parts, hinges and landing gear parts. Medium-size mechanical assemblies. Assembly and testing of mechanics and hydraulic aircraft components. → 75 [98]

80-00060 Array Systems Computing Inc.
1120 Finch Avenue West, 8th Floor,
North York, Ont M3J 3H7
☎ (+1 416) 736 09 00 W http://www.array.ca
⊖ (+1 416) 736 47 15
👥 108 🏭 35,000 sq ft
President: Stuart Berkowitz
Finance Dir: Ronald Jewel
Business Development Mgr Radar: Robert Bruce
Business Development Manager Sonar: Bruce Eidsvik
Marketing and Sales: Ms Leanne K Mielczarek
■ Development & integration of 'commercial off the shelf' (COTS) solutions for signal & image processing systems. Products include sonar, radar, remote sensing & damage control. → 75 [98]

80-00080 AR Technologies Division
(a division of Orenda Aerospace Corporation, Canada)
Unit 220, 13155 Delf Place, Richmond, BC V6V 2A2
☎ (+1 604) 273 17 17 Ø (800) 661 16 33
W http://www.orenda.com ⊖ (+1 604) 273 12 63
♣ C$ 3,200,000 (1996) 👥 30 (1996)
Managing Director: Stuart Davidson
Director Marketing and Sales: Chris Trsek
E ctrsek@orenda.com
■ Repair & overhaul facility for jet engine components, specialising in hot section. [98]

80-00242 Cadorath Aerospace Inc
(† Cadorath Plating Co. Ltd)
(a division of Cadorath Plating Co. Ltd)
2150 Logan Avenue, Winnipeg, Manitoba R2R OJ2
☎ (+1 204) 633 27 07 Ø (800) 470 70 69
⊖ (+1 204) 632 76 63
👥 45 (1995)

Contact Reference

President: Gerry J Cadorath
Vice President Marketing and Operations: Gerry O Cadorath
Quality Control Manager: Thane Allen
■ Turbine engine component overhaul, special processes for aerospace and defence. [97]

80-00245 CAE Aviation Ltd
(† Northwest Industries Ltd)
(a subsidiary of CAE Inc, Canada)
Edmonton International Airport, PO Box 9864, Edmonton, Alberta T5J 2T2
☎ (+1 403) 890 63 00 ⊖ (+1 403) 890 65 50
♣ C$ 75,000,000 ⚉ 650 ⌂ 400,000 sq ft
President: L H Prokop
Vice President Finance and Administration: B Hook
Vice President Engineering: D Kobayashi
Vice President Aircraft Maintenance: H Harper
International Marketing: Terry MacEwan
Executive Assistant Marketing and Business Development: Ms Coleen Schmidt
■ Aircraft maintenance, repair, overhaul and modification. Manufacture of major aircraft components, instrument repair and overhaul. NDT, engineering support and services, avionics. Aircraft painting and furnishings. Technical publications in English, French and Spanish. [98]

80-00250 Cametoid Ltd
1449 Hopkins Street, Whitby, Ontario L1N 2C2
☎ (+1 416) 666 34 00 [E] coatings@cametoid.com
[W] http://www.cametoid.com ⊖ (+1 416) 666 34 13
⚉ 30 ⌂ 35,000 sq ft
President: D G Newman
Vice President: Mrs M A Newman
General Manager: Lawrence H Covens
Engineering and Quality Manager: David Pile
■ Aerospace coatings: anodizing, IVD electroplating, dry film lubricants, vacuum coatings, optical coatings, aluminium coating lubricants. Inorganic/organic paint.
Divisions:
Cametoid Technologies Inc, USA [97]

80-00270 Canadian Commercial Corporation
50 O'Connor Suite 1100, Ottawa, Ont K1A 0S6
☎ (+1 613) 996 00 34 ✆ (800) 748 81 91 ⊠ CCC-OTT
⊖ (+1 613) 995 21 21
♣ C$ 1,000,000,000 (1995/96) ⚉ 85
Executive Vice President and Chief Operating Officer: Douglas Patriquin
Operations Division Director: Mike Chatzipantazis
Director Business Development Division: Stephen Bigsby
Director External Relations Division: Glen Nichols
■ Functions as the contracting agency for foreign governments & international agencies wishing to purchase goods or services from Canada on a government-to-government basis. [98]

80-00280 CHC Helicopter Corporation
(† Okanagan Helicopters)
(a subs. of CHC Helicopter Corp.)
Hangar No.1, St John's Airport, PO Box 5188, St John's, NF A1C 5V5
☎ (+1 709) 570 0700 [E] webmaster@stjohns.chc.ca
[W] http://www.chc.ca ⊖ (+1 709) 570 0506
Chairman: Craig L. Dobbin
President & CEO: Rudy Palladina
CFO: Christopher P. Lloyd
Sr Vice President: Mark D. Dobbin
■ Helicopter operation & technical support services.
International Division
4391 Agar Drive, Richmond, BC V7B 1A5
☎ (+1 604) 276 7500 ⊖ (+1 604) 279 2474
President: Sylvain Allard
Western Division
12021 - 121 Street, Edmonton, Alta T5L 4H7
☎ (+1 403) 429 6900 ⊖ (+1 403) 452 2574
President: Don Wall
Eastern Division
1215 Montée Pilon, CP 188, Succ. Dorion, Les Cèdres, Que J7V 5W1
☎ (+1 514) 455 5687 ⊖ (+1 514) 452 2472
President: Jean Pierre Blais
Acro Aerospace
(a subs. of Canadian Helicopters Limited)
4551 Agar Drive, Richmond, BC V7B 1A4
☎ (+1 604) 276 7600 ⊖ (+1 604) 276 7675
President: Jurgen Funke

■ Provides: Dynamic components, power plants, accessories, structures, parts sales, blades, manufacturing & parts restorations.
Atlantic Turbines Inc.
Hangar #8, Slemon Park, PO Box 150, Summerside, PEI C1N 4P6
☎ (+1 902) 436 1333 ⊖ (+1 902) 436 0777
President: John MacDougal [97]

80-00360 Consolidated Cybernetics Controls Ltd
3488 Côte-des-Neiges - Suite 111, Montréal, Québec H3H 2M6
☎ (+1 514) 934 00 00 ⊠ CYBERNETICS
[E] cyber@cam.org ⊠ 0524452 ⊖ (+1 514) 932 77 76
♧ C$ 4,700,000 ⚉ 12 ⌂ 3,800 m2
Chief Executive Officer: Dr R A Javitch
Vice President and Chief Engineer: Prof. G G Bach
Chief Test Pilot: Andre Giacommetti
Aerospace Sciences Manager: Dr Phillipe de la Savanne (Jr)
Military Systems Co-ordinator: Dr R H Platz
Aeronautical Certification Specialist: T Jagninsky
■ Research and development in general aviation technology. [97]

80-00380 Corrigan Technologies Ltd
59 Sinclair Avenue, Georgetown, Ontario L7G 4X4
☎ (+1 905) 873 06 68 ⊠ 697575 ⊖ (+1 905) 877 41 91
♣ C$ 5,000,000 (1995) ⚉ 40 ⌂ 2,300 m2
Managing Director: Joe Corrigan
Technical Director: Pentti Makela
Sales and Marketing Manager: Nick Law
Purchasing Manager: Art Murphy
■ CAN-X-SCAN x-ray baggage screening systems. [97]

80-00385 Counterrisk Inc.
Mont-Royal Postal Station, PO Box 220, Montreal, Que H3P 3C4
☎ (+1 514) 341 56 84 ⊖ (+1 514) 341 29 00
♣ $ 500,000 (1997) ⚉ 5 ⌂ 610 sq m
Managing Dir: Desmond S. Clarke
Technical Dir: John Drolet
Marketing Dir: Jack Dunn
■ Security consulting services. [98]

80-00386 CPAD Technologies Inc.
(† CPAD Holding Limited)
40 Camelot Drive, Ottawa, Ont K2G 5X8
☎ (+1 613) 224 9939 ⊖ (+1 613) 224 2418
⚉ 50 ⌂ 10 000 sq ft
President: Dr Mariusz Rybak
Exec. Vice President: Andy A. Rybak
Executive Assistant: Ms Debby Hynes
■ Explosives detection, explosives ordnance disposal, detection systems, security systems, surveillance equipment, counter-terrorist systems, bomb detection. [97]

80-00650 FlightSafety International Inc. Montreal Learning Center
9555 Ryan Avenue, Dorval, Que H9P 1A2
☎ (+1 514) 631 20 84 ✆ (800) 573 40 25
⊖ (+1 514) 631 22 63
Contact: Doug Ware [98]

80-00660 FlightSafety International Inc. deHavilland Learning Center
95 Garrett Boulevard, Downsview, Ont M3K 2A5
☎ (+1 416) 638 93 13 ⊖ (+1 416) 638 33 48
Director: Jon Pollack [98]

80-00670 Frontec Corporation
10035 - 105 Street, PO Box 2426, Edmonton, Alberta T5J 2V6
☎ (+1 403) 420 71 12 ⊖ (+1 403) 420 34 26
♣ C$ 80,000,000 ⚉ 500
President: George N Paicu
Vice President Marketing Development: Ian D Hargrave
Vice President Controller: Harish K Mohan
Support Services and Contracts Director: W F Christensen
■ Air and ground services including surface maintenance, navigation aid maintenance, radar operation and maintenance, emergency response. [97]

80-00810 Halifax Shipyard Limited
(a div. of Irving Shipbuilding Group)
3099 Barrington Street North, PO Box 9110, Halifax, NS B3K 5M7
☎ (+1 902) 423 9271 ⊖ (+1 902) 422 5253
President: W.E. Haggett ☎ (+1 506) 633 4444
General Mgr: Robert Shepherd ☎ (+1 902) 492 5802
■ Engineering, design, construction & repair of ships up to Panamax size. [97]

80-00900 HVR Canada Inc
350 Sparks Street #701, Ottawa, Ontario K1R 7S3
☎ (+1 613) 230 50 04 ⊖ (+1 613) 230 05 59
♣ C$ 200,000 ⚉ 2 ⌂ 500 sq m
President: Harry V Rogers
Executive Vice President: Peter D Cook
Vice President Operations: Carl A Dalton
■ Risk assessment and equipment cost analysis for government and defence related industry, operational analysis and modelling studies, software development and analysis. [98]

80-00960 Innotech Aviation Ltd
(a division of IMP Aerospace Group Ltd)
595 Stuart Graham Boulevard, Dorval, Québec H4Y 1E3
☎ (+1 514) 636 84 84 ✆ (800) 463 96 25 ⊠ 05821856
⊖ (+1 514) 636 88 87
♧ C$ 15,500,000 ♣ C$ 60,800,000 (1996) ⚉ 346
⌂ 137,516 sq ft
Chairman: Kenneth C Rowe
Executive Vice President and Chief Operating Officer: Stephen K Plummer
Vice President Engineering: Tom Galley
Vice President Maintenance: Yvon Gagnon
Vice President Maintenance Division: Ken Dandy
Sales and Marketing Manager: Jerry Sosontovich
Advertising: Ms Tracy Biggs
■ Airframe modification, engineering, certification, custom avionics systems integration, prototypes, ECM, EW systems integration, flight inspection systems installation, remote sensing operations, aircraft maintenance, aircraft sales.
® Canada: Cessna Citation
USA: Bendix King, Global Wulfsberg Systems, Honeywell, Rockwell Collins, Sunair Electronics [98]

80-00980 InterCon Consultants
275 Slater St, Suite 1003, Ottawa, Ont K1P 5H9
☎ (+1 613) 236 4451 ⊖ (+1 613) 230 8707
⚉ 8
Partner: R. Campbell
Partner: E.B. Creber
Sr Associate: G.H. Craven
Associate: D.E. Samson
Associate: T.A. Irvine
Associate: P.R. Munro
Associate: P. Gameron
■ Provides service & assistance to foreign or Canadian firms wishing to do business with Canadian Federal or Provincial government or the Canadian private sector with special reference to defence, aerospace, communications, nuclear energy. [97]

80-01205 Lansdowne Technologies Inc
(a subsidiary of Canadian Shipbuilding & Engineering Ltd)
1001-275 Slater Street, Ottawa, Ontario K1P 5H9
☎ (+1 613) 236 33 33 ⊖ (+1 613) 236 44 40
♣ C$ 1,500,000 ⚉ 12
President: James A Faulkner
Sales and Marketing Director: Michael Washer
■ Project management, planning and control, proposal and logistic management for defence annd space industries. Non-tactical software development services. [98]

80-01250 Logica Inc
55 University Avenue - Suite 1710, Toronto, Ont M5J 2H7
☎ (+1 416) 594 66 61 ⊖ (+1 416) 594 16 67 [98]

80-01310 MacDonald Dettwiler and Associates Ltd
(a subsidiary of Orbital Sciences Corporation)
13800 Commerce Parkway, Richmond, British Columbia V6V 2J3
☎ (+1 604) 278 34 11 [E] sales@mda.ca
[W] http://www.mda.ca ⊠ 4355599
⊖ (+1 604) 278 12 85; 278 18 37

♣ C$ 101,400,000 ♠ C$ 101,000,000 ⚑ 775
⌂ 27,900 m2
President and Chief Executive Officer: Daniel E Friedmann
Chairman: John MacDonald
Vice President Geo-Information Systems: Bernie Clark
Vice President Space and Defence Systems: David N Caddey
Aviation Systems Director: Hans Baelli
Sales and Marketing Director: Alan Vlemmiks
Public and Investor Relations Director: Alan Bayless
Purchasing Manager: T W Piche
Advertising: Ms Deborah D'Aguiar
■ Optical, radar and sonar signal and image processing systems for surveillance, reconnaissance and mine countermeasures. Systems integration, command and control, EW subsystem development. Software development to military, NASA or ESA standards.

● Ste 290-1101 Prince of Wales Drive, Ottawa, Ont K2C 3W7
☎ (+1 613) 226 2701 ⊜ (+1 613) 226 2271
Defence Marketing Mgr: Don McClure

● 1000 Windmill Road, Ste 60, Dartmouth, Nova Scotia B3B 1L7
☎ (+1 902) 468 3356 ⊜ (+1 902) 468 2278
Marketing Mgr: Terry Doyle
97

80-01320 Martin Marietta Canada Ltd
(a subsidiary of Lockheed Martin Corp)
50 O'Connor Street - Suite 1450, Ottawa, Ontario K1P 6L2
☎ (+1 613) 232 64 30 ⊜ (+1 613) 232 66 98
♠ US$ 30,000,000 (1994) ⚑ 130 (1995)
President: Gerry Zutler
New Business Initiatives Director: Ian Bayne
■ Systems management of large, complex, leading edge technology control systems in areas of transportation, defence, avionics, space, energy and communications.
® USA: Martin Marietta Corporation
98

80-01360 MIL Systems
(† MIL Systems Engineering)
(a division of the Cedar Group (Canada) Inc.)
1150 Morrison Drive, Suite 200, Ottawa, Ont K2H 8S9
☎ (+1 613) 726 05 00 E quality@milsystems.com
W http://www.milsystems.com ⊜ (+1 613) 726 02 52
♠ C$ 10,000,000 (1997) ⚑ 100 (1997) ⌂ 30,000 sq ft
President: J C Keast
Vice President Finance and Administration: G M Skinner
Marketing Manager: R Perks
■ Full-service engineering consultant: ship design, naval architecture, marine systems engineering, production engineering support. Assistance in developing new technology applications for marine systems. Experienced in program management, life cycle, technical & integrated logistics support. Design & support of HVAC, NBC & environmental control systems. Documentation & publication. Quality assurance to AQAP-1/ISO 9001.

MSEI Pacific
Victoria, BC
☎ (+1 604) 385 62 39 ⊜ (+1 604) 385 80 17
General Mgr: B. Hodgkin
■ Supplementary design office.
98

80-01380 MSEI Services Inc.
(a division of MIL Systems)
1150 Morrison Drive, Ottawa, Ont K2H 8S9
☎ (+1 613) 727 69 48 ⊜ (+1 613) 727 98 38
⚑ 20
President: J.C. Keast
General Mgr: G. Lanigan
■ Technical consulting services: life cycle material management (LCMM), technical investigations, integrated logistic support (ILS), technical publications, data management, test & trials & field support for ships, systems & equipment.
98

80-01400 Oerlikon Aerospace Inc ADATS Assembly and Testing Plant
(a subsidiary of Oerlikon-Bührle Holding AG)
225 Boulevard du Séminaire Sud, St-Jean-sur-Richelieu Québec J3B 8E9
☎ (+1 514) 358 20 00 ⊜ (+1 514) 358 17 44
♠ C$ 33,093,000 (1996) ⚑ 240 (1997) ⌂ 32,000 m2
President and Chief Executive Officer: John F Lemieux
Vice President Advanced Programs Management: Gerhard Hoy
Vice President Human Resources: Normand Bellemare
Vice President Business Development: Henning Jacobsen
Vice President Project Management: Jean-Pierre Autotte
Vice President Business Management: Bruce Fitzsimmons
Corporate Communications Director: Norbert Cyr
■ World mandate for the production and marketing of the ADATS missile system. Design, integration, test and support of advanced defence, communications and electronic systems. Integrated logistics support and turnkey project management. Shelter-based configuration of ADATS without radar for integration with other life control units. Low-level air defence systems (CF LLADS) and related support equipment. Systems and software engineering, integration and testing, customised life-cycle support.
98

80-01450 Orenda Aerospace Corporation
(† Hawker Siddeley Canada Inc.)
3160 Derry Road East, Mississauga, Ont L4T 1A9
☎ (+1 905) 677 32 50 ⊜ (+1 905) 678 15 38
♠ C$ 55,000,000 (1996) ⚑ 400 (1997) ⌂ 430,000 sq ft
Vice President and General Manager: R A Neill
Director Finance: L Winegarden
General Manager Repair and Overhaul: F G Button
General Manager Component Manufacturing: D McIntosh
Manager Repair and Overhaul Marketing: John W Armstrong
■ Maintenance, repair and overhaul of J85 and F404 engines, modules and components, engineering support, sub-contract manufacture of gas turbine engine components, engineering, graphics and laboratory testing services. Manufacture 600 hp aviation reciprocating engine.
98

80-01650 Peacock Inc.
(a subs. of The Weir Group PLC, UK)
Station 'A', CP 1040, Montreal, Que H3C 2X5
☎ (+1 514) 366 5900 ⊜ (+1 514) 366 5904
⚑ 500
President: Randy Hammell
■ Supply & servicing of fluid handling & mechanical equipment for process, mining, metallurgical, marine, pulp, paper & utility industries. Staffs & manages the Naval Engineering Test Establishment (NETE) for the Department of National Defence.
97

80-01700 Primex Defence Products Ltd
275 Slater Street, Suite 1002, Ottawa, Ont K1P 5H9
☎ (+1 613) 238 39 70 ⊜ (+1 613) 238 48 57
⚑ 8
President: T B Morrice
■ International marketing & management services.
98

80-01800 Quantum Inspection & Testing Ltd
(a subsidiary of Westinghouse Canada Inc)
916 Gateway, Burlington, Ontario L7L 5K7
☎ (+1 905) 632 58 69 ⊜ (+1 905) 632 37 41
♠ C$ 4,500,000 ⚑ 40
President: Bert Hildebrand
Marketing Manager: Scott Brown
■ A third party services organisation which provides dedicated quality specialists to clients requiring non-destructive testing Mil Std 410 certified, NATO AQAP 6 calibration laboratory for mechanical test instrumentation, recognised by Department of National Defence. Quantum's non-destructive test facility is the 'only' approved NASA/ Rockwell International fracture critical laboratory in Canada. Government approvals include Standards Council of Canada, Department of National Defence, Transport Canada Airworthiness section & 32 additional accreditations by major aerospace & defence contractors including Pratt & Whitney, General Electric. Recent addition of an Electric/Electronic calibration & certification laboratory for measurement & test equipment. ISO 9003/ISO Guide 25 calibration certification & non-destructive test centre.
97

80-01825 Rohde & Schwarz Canada Inc
(a subsidiary of Rohde & Schwarz GmbH & Co)
555 March Road, Kanata, Ontario K2K 2M5
☎ (+1 613) 592 80 00 W http://www.rsd.dc
⊜ (+1 613) 592 80 09
President: David G Stephenson
Vice President Sales and Marketing: Peter H Foulger
■ Remotely controlled, near real time, networked RF monitoring systems. → 75
Representing:
Rohde & Schwarz GmbH & Co KG, Germany
98

80-01840 The SNC-Lavalin Group Inc.
(† The SNC Group Inc.)
2 Place Félix Martin, Montreal, Que H2Z 1Z3
☎ (+1 514) 393 10 00 ✉ SCINC Montreal
⊜ (+1 514) 866 07 95
⚑ 6,000
Chairman: Guy Saint-Pierre
President and Chief Executive Officer: Jacques Lamarre
Senior Vice President Finance and Planning: Pierre Robitaille
Vice President Public Affairs: Robert Racine
■ Defence products & engineering-construction services.
Subs. with defence business:
SNC Industrial Technologies Inc. → 75
Securiplex → 75
98

80-01850 Stan Reynolds Sales Ltd
4110-57 Street, Wetaskiwin, Alta T9A 2B6
☎ (+1 403) 352 5201 ⊜ (+1 403) 352 4666
President: Stanley G. Reynolds
Aviation Dept Mgr: Byron E.J. Reynolds
■ Purchase & repair of pre-1946 aircraft, engines & parts, especially antique & military aircraft.
97

80-01937 SED Systems Inc
(a subsidiary of Calian Technologies Ltd)
18 Innovation Boulevard, PO Box 1464, Saskatoon, Sask S7K 3P7
☎ (+1 306) 931 34 25; 933 14 46
W http://www.sedsystems.ca ✉ 742495 sedsystems skn
⊜ (+1 306) 933 14 86
♠ C$ 35,000,000 (1996) ⚑ 270 (1997) ⌂ 1,400 m2
President and Chief Operating Officer: Dugald Buchanan
Chief Executive Officer: Larry O'Brien
Finance Director: Ray Basler
Business Development Director: Kent McKerlie
Head of Engineering and Technical: Jim Rennie
Business Director Defence and Government Systems: Robert D Herbert E herbert@sedsystems.ca
Communications Administrator: Ms Karen Davidson
Purchasing Manager: Mike Tourigny
■ Space and communications systems design, engineering and integration specialising in satellite ground terminals, defence communication systems, satellite test and control and mobile satellite communication systems.
Major products: INMARSAT coast-earth stations; satellite in-orbit test systems; satellite monitoring systems; defence communication systems; satellite instruments for resource and environmental monitoring; remote control systems; HF adaptive beam forming receiver systems. → 75
98

80-01940 Simetra Inc.
8-6 Antares Drive, Nepean, Ont K2E 8A9
☎ (+1 613) 225 4573 ⊜ (+1 613) 225 2964
⚑ 8
Managing Dir: Ron Warburton
■ Provision of marketing, technical & support services for the Artemis project management organisations that need enterprise-wide, departmental or divisional support systems for programme & multi-project management in Canada.
97

80-01970 Standard Aero Ltd (SAL)
(a subsidiary of BTR plc, UK)
Winnipeg International Airport, 33 Allen Dyne Road, Winnipeg, Manitoba R3H 1A1
☎ (+1 204) 775 97 11 W http://www.standardaero.ca
✉ 757878 standaero wpg ⊜ (+1 204) 885 22 29
♠ C$ 210,000,000 (1996) ⚑ 1,075 (1997)
⌂ 545,000 sq ft
President: David Shaw
Vice President Finance & Administration: Brad Bertoville
Vice President Engineering and Quality Control: Ron Yanos
Vice President Marketing and Business Development: Tim Jordan ☎ (+1 204) 788 27 54
Vice President Human Resources: Bruce Clarke
Director ALEPU: Lee Beaumont
Director A250: Robert Stone
Director PT6: Edward Richmond
Marketing Assistant: Ms Kimberley Forsen
■ Overhaul, recondition, repair and test aircraft engines, accessories, components. Technical publications and translations.
98

80-02020 Thomson-CSF Systems Canada Inc
(a subsidiary of Thomson-CSF, France)
49 Auriga Drive, Nepean, Ontario K2E 8A1

☎(+1 613) 723 70 00 E thomcsf@thomson-csf.ca
W http://www.thomson-csf.ca ⊖ (+1 613) 723 56 00
♲ C$ 500,000 ♣ C$ 12,000,000 (1996) 👥 80 (1997)
🏭 24,000 sq ft
President & CEO: Norman R.A. Smyth
VP Finance/Administration: Jean A. Rochefort
Business Development Mgr: Ken Bowering
Contracts Mgr: Kevin Dalliday
Engineering Systems Mgr: Barry Brown
Engineering Software Mgr: Ray Carbonneau
Engineering Logistics Mgr: Rick Temelini
Personnel: Stan Janas
Purchasing Mgr: Ted Kelly
■ Systems integration company with expertise in mission systems design, systems engineering, integration and delivery; command, control and communications; systems technology development; countermine technology and systems development; space systems and robotics technology development and integrated logistics support.
Complex turnkey systems, mission systems, trainers, simulators, minefield clearing systems, space and robotics technology, command and control technology. Systems technologies for management, design and development of complex systems including project management, concept definition, systems, analysis, computer engineering, software development, computer simulation and modelling. Integrated logistics support (ILS) services for projects and products. ILS training, tools and services. 98

80-02100 TriLink Technologies Group Inc.
(† M.K. Venables & Associates Inc.)
85 Albert Street, Suite 1620, Ottawa, Ont K1P 6A4
☎(+1 613) 567 95 75 ⊖ (+1 613) 746 30 49
👥 6
President: M K Venables E mikeve@tenebris.com
■ Consulting in: engineering project management, marketing/sales planning & support, engineering studies & analysis, flight operations. As applied to aviation & aerospace, computer applications. 98

80-02120 TSB International
5399 Eglinton Avenue West, Ste 115, Ont M9C 5K6
☎(+1 416) 622 7010 ⊖ (+1 416) 622 3540
Corporate & Marketing Communications Mgr: Michael Maltby
■ The company's product & services manage network planning, performance & cost.
Spectif PBX data management hardware & software collect CDR, alarms & traffic from multi-vendor switches. ORBi-TEL network management applications provide management information for voice networks of all sizes. 97

80-02210 Bruce D. Vallillee Electronics Ltd
36 Trawley Cres., Ajax, Ont L1S 5X8
☎(+1 905) 420 6786 ⊖ (+1 905) 420 8199
👥 3 🏭 3 400 sq ft
President: Bruce Vallillee
Vice President: Shawn D. Vallillee
■ Marketing consultants. 97

80-02230 Vanwell Publishing Ltd
1 Northrup Crescent, PO Box 2131,
St Catharines, Ont L2R 7S2
☎(+1 905) 937 3100 ✆ (800) 661 6136
⊖ (+1 905) 937 1760
👥 18 (1996) 🏭 1 500 sq m
Director: Ben Kooter
Sales Dir: Simon Kooter
Operations Mgr: Ms Diane Kooter
Circulation Mgr: Ms Linda Telega
→ 90
Agents for:
Jane's Publications
Ian Allan Publications
Naval Institute Press Publications 97

85 Stockists, Distributors, Spare Parts Suppliers

Tecmotiv Corporation 75-13510

85-00050 Rockwell Automation Canadian Headquarters and Plant
(† Allen-Bradley Canada Ltd)
(a division of Rockwell International, Canada)
135 Dundas Street, Cambridge, Ont N1R 5X1
☎(+1 519) 623 18 10 ⊖ (+1 519) 623 89 30
President: Charles A Cipolla
E ccipolla@cam1.ra.rockwell. com
® USA: Rockwell International Corp. 98

85-00200 Aviation & Defence Products Support
(a division of American Aviation Parts & Service (CDA) Ltd)
6461 Northam Drive, Mississauga, Ont L4V 1J2
☎(+1 905) 677 53 72 ⊖ (+1 905) 677 10 89
Comptroller: Ms Susan Wilson
■ Full line of military parts. 98

85-00450 Caina Company (Canada)
1547 Merivale Road, PO Box 5521, Ottawa, Ont K2C 3MI
☎(+1 613) 723 19 21 ⊖ (+1 613) 224 69 58
Managing Director: Christopher S Baxter
■ Distributors and sub contractor of small arms/firearms, ammunition and related equipment, both new and surplus/used/obsolete. Refurbishment and upgrade specialities include: 7.62 & 7.92 MG-42/MG-3 series guns; Bren Guns; surplus ammunition; used revolvers; sniper/match rifles; magazines/"clips". Countertrade, accepting used/obsolete equipment in exchange for new materiel. → 80 → 90 98

85-00500 Calbrooke Marketing Inc.
PO Box 792, Arthur, Ont N0G 1A0
☎(+1 519) 848 3810 ✆ (800) 265 7283
⊖ (+1 519) 848 2555
♣ C$ 2 500 000 👥 4
President: John Waters
Dir Sales & Marketing: Jeff Huys
Operations/Technical Dir: Patricia Waters
■ Precision, resistors, test points, Emirfi filters, connectors, switchmode power supplies, inductors, computer systems, capacitors, MIL-SPEC semis, LEDs, PIRs, contract manufacturing, PC Board relays, resistors, sockets, transformers, buzzers, pxe device. Shielding materials. 97

85-00600 EBH Trading Corp.
7040 Torbram Road, Unit 18,
Mississauga, Ontario L4T 3Z4
☎(+1 905) 677 18 61 E ebh@ican.net
⊖ (+1 905) 677 18 67
♣ US$ 3,000,000 👥 10
President: Edward H Badalian
■ Aircraft parts sales. 98

85-01000 MHD Int'l Aviation Parts Inc.
1830 Marie Victorin, St Bruno, Que J3V 4P6
☎(+1 514) 441 51 11 E mhd_intl@sypatico.ca
⊖ (+1 514) 441 49 74
♣ C$ 7,500,000 (1997) 👥 15 (2997) 🏭 15,000 sq ft
Vice President: Art Thorburn
Operations Director: Bryan Dollimore
Quality Assurance/Material Manager: Paul Furion
Sales/Government Manager: John O'Brien
■ Distribution & supply of aircraft parts. Representative to Canadian & US manufacturers. Consulting on military contracts. 98

85-01330 Miles Air Products
(† Miles Associates Inc)
30 Fallingbrook Road, Scarborough, Ontario M1N 2T4
☎(+1 416) 690 44 04 ⊖ (+1 416) 698 49 27
👥 4
President: John A. Miles
■ Distributors of aviation components & accessories.
® UK: Clement Clarke Int'l 98

85-01900 Rockwell Heavy Vehicle Systems
Worldwide Aftermarket, Canadian Distribution Centre
Units 3 and 4, 80 Van Kirk Drive, Brampton, Ont L7A 1B1
☎(+1 905) 840 23 73 E brampton@auto.rockwell.com
⊖ (+1 905) 840 23 78
Facility Manager: Kevin W Little
■ Aftermarket parts and services.
Canadian Regional Sales
Units 3 and 4, 80 Van Kirk Drive, Brampton, Ont L7A 1B1
☎(+1 905) 840 05 08 ⊖ (+1 905) 840 73 54
Manager: Roger C Clark
■ Canadian regional sales.
● PO Box 2500, Tilbury, Ont N0P 2L0
☎(+1 519) 682 17 40 ⊖ (+1 519) 682 35 86
Site Manager: Wayne A Watson
E wawatson@auto.rockwell.com
Rockwell International Suspension Systems Company
105 St George Street, Chatham, Ont N7M 4P3
☎(+1 519) 354 11 00 ⊖ (+1 519) 354 53 25
Plant Manager: Gerald Anderson
E ganders@auto.rockwell.com
● 150 Steeles Avenue, Milton, Ont L9T 2Y5
☎(+1 905) 878 23 95 ⊖ (+1 905) 878 12 90
Plant Manager: John G Gates
E jggates@auto.rockwell.com
Switching Systems Division (Canada)
Suite 620, 200 Yorkland Boulevard,
North York, Ont M2J 5C1
☎(+1 416) 502 01 81 ⊖ (+1 416) 502 03 93
Operations Manager: Philip Ridge 98

85-02000 Vallillee Wide Format Products Limited
(† M&T Management)
Tribute Corporate Center, 1815 Ironstone Manor, Unit 13, Pickering, Ont LIW 3W9
☎(+1 905) 420 6786 ✆ (800) 423 9013
⊖ (+1 905) 420 8199
👥 11 🏭 3 400 sq ft
Vice President: Shawn D. Vallillee
■ Sales & distribution of US & Canadian military product lines including the WIDEfax large format document handling system & Miltrace 36" tactical field fax, copier, scanner, plotter for traces & maps. High speed large format scanners & plotters. 97

90 Sales Agents, Sales & Liaison Offices

Companies represented:

A&M, UK 90-00450
Alfred Karcher GmbH & Co, Germany . . 90-01000
Andrew Corp, USA 75-00600
Antenna Products Corp, USA . . . 75-11590
Antenna Research Associates, USA . . 75-11590
Bendix King, USA 80-00960
Bofors AB, Sweden 90-00200
British Steel Engineering Steels, UK . . 90-00225
C-Plath, UK 75-09090
Cessna Citation, Canada 80-00960
Champion Spark Plug, USA 90-00330
Clement Clarke Int'l, UK 85-01330
Cortec Corporation, USA 75-00430
Crompton Instruments Ltd, UK . . . 90-00450
Decca, UK 75-09090
Detroit Armor Corp., USA 90-01940
Eje Translite, Canada 90-01850
GEC Plessey Aerospace, USA . . . 75-00430
Global Wulfsberg Systems, USA . . . 80-00960
Hägglunds Vehicle AB, Sweden . . . 90-00800
Herman-Nelson, USA 75-00300
Honeywell, USA 80-00960
Howell Exports, USA 90-01850
Ideal Aerosmith, USA 90-01850
Keco Industries, USA 75-00430
Kollmorgen Motion Technologies Group, USA 90-01850
Leach International, USA 90-01850
Leland Electrosystems, USA . . . 90-01850
LRE Relais + Elektronik, Germany . . . 90-01850
Martin Marietta Corporation, USA . . 80-01320
McDonnell Douglas, USA 75-09520
Modutec, UK 90-00450

Newport, UK 90-00450
Park Air Electronics, UK 75-11590
Pilkington Aerospace Inc, USA 75-00430
Pointer ELT, USA 75-11300
Racal Electronics plc Group, UK 75-11590
Rapid Power Technologies Inc, USA . . 75-00430
Raychem Corp., USA 90-01808
Raytheon, USA 75-11700
Rockwell Collins, USA 80-00960
Rockwell International Corp., USA . . . 85-00050
Roller Bearing Corporation of America, USA 75-00430
Rolls-Royce, UK 75-12120
Shamban & Co., USA 90-00280
Smiths Industries, USA 75-00430
Sparton, USA 75-13100
Sperry Marine, UK 75-09090
Sunair Electronics, USA 80-00960
Transaero Inc., USA 90-02100
VARTA Batterie AG, Germany 75-14540
Westland Group, UK 90-02330

Sales agents/representatives listed elsewhere:

AlliedSignal Aerospace Canada 75-00470
Caina Company (Canada) 85-00450
Litton Canada Ltd 75-09090
Vanwell Publishing Ltd 80-02230

90-00200 Bofors Canada Ltd
(a subsidiary of Bofors AB, Sweden)
130 Albert Street, Suite 1915, Ottawa, Ont K1P 5G4
☎(+1 613) 238 87 84 ⊖ (+1 613) 238 19 49
President: G.J. Hollink
■ Suppliers of defence ordnance.
® Sweden: Bofors AB
98

90-00225 British Steel Canada Inc
(a subsidiary of British Steel Engineering Steels, UK)
Suite 405-6505, Trans Canada Highway,
St Laurent, Quebec
☎(+1 514) 855 01 01 ⊖ (+1 514) 855 02 74
® UK: British Steel Engineering Steels
98

90-00250 Brüel & Kjaer Canada Ltd
90 Leacock Rd, Pointe Claire, Que H9R 1H1
☎(+1 514) 695 8225 ⊠ 821691 ⊖ (+1 514) 695 4808
97

90-00280 Busak + Shamban Canada
(† Shamban Canada Inc.)
211 Advance Boulevard, Units 5 & 6,
Brampton, Ont L6T 4S8
☎(+1 416) 213 9444 ⊠ 0697574
⊖ (+1 416) 213 9462
General Mgr: Bruce Hasler
® USA: Shamban & Co.
98

90-00320 CATIC Development Inc
6530 Dunbar Street, Vancouver, BC V6N 1X8
☎(+1 604) 263 1039 ⊖ (+1 604) 261 5086
97

90-00325 Century International Arms Ltd
7035 Trans-Canada Highway, Montreal PQ H4T 1A2
☎(+1 514) 333 9933 ⊖ (+1 514) 333 99 40
President: Brian Sucher
■ Rifles, handguns, ammunition and bayonets.
98

90-00330 Champion Spark Plug Division
(† Champion Spark Plug Co.)
(a division of Cooper Industries Canada Inc.)
336 Courtland, Vaughan, Ontario L4K 4Y1
☎(+1 905) 793 26 85 ✆ (800) 563 85 07 ✉ CHAMPION
⊖ (+1 905) 761 56 00
♣ C$ 100,000,000 👥 200 🏠 250,000 sq ft
President: E Grant
Sales and Marketing Director: S Stone
Marketing Director: R G Armstrong
Customer Services Manager: Ms Maria Cutrara
Purchasing Manager: K McKee
® USA: Champion Spark Plug
98

90-00420 Comstream Corporation
20201 Clark Graham Avenue, Baie d'Urfe, Que H9X 3T5
☎(+1 514) 457 0444 ⊖ (+1 514) 457 0459
97

90-00450 Crompton Instruments Ltd Canada
(a subs. of BTR)
2717 Rena Road, Mississauga, Ontario L4T 3K1
☎(+1 905) 671 2253 ⊖ (+1 905) 671 2617
👥 20
General Manager: L Shean
® UK: A&M, Crompton Instruments Ltd, Modutec, Newport
98

90-00470 Daewoo Canada Ltd Toronto Office
Suite #403 - 1500 Don Mills Road,
North York, Ontario M3B 3KD
☎(+1 416) 510 28 86 ⊖ (+1 416) 510 09 08
98

90-00475 Daewoo Corp Vancouver Branch
3820 Jacombs Road - Suite 103, Richmond V6V 1Y6
☎(+1 604) 270 30 03 ⊖ (+1 604) 270 30 22
98

90-00800 Hägglunds Liaison Office
222 Queen Street - Suite 1502, Ottawa, Ont K1P 5B9
☎(+1 613) 237 7496 ⊖ (+1 613) 238 5519
Liaison Officer: Svante Anderson
■ Hägglunds liaison office. Products represented include the Bv206 All Terrain Carrier, Combat Vehicle 90 (CV90) and recovery units.
® Sweden: Hägglunds Vehicle AB
98

90-00940 Iseco Safety Shoes
2640 Argentia Rd, Mississauga, Ont L5N 6C5
☎(+1 542) 33 00 ⊖ (+1 542) 37 84
97

90-01000 Karcher Canada Inc
(† Karcher Cleaning Systems Inc)
1770 Alstep Drive, Mississauga, Ont L5S 1W1
☎(+1 905) 672 82 33 ⊖ (+1 905) 672 31 55
Manager: Wolfgang Mors
® Germany: Alfred Karcher GmbH & Co
98

90-01340 Mitsubishi Canada Ltd
Ste 2181, Commerce Court West,
PO Box 17 Commerce Court PS, Toronto, Ont
☎(+1 416) 362 67 31 ⊠ 6524288
⊖ (+1 416) 365 13 84
Marketing Mgr: S. Takase
98

90-01380 NMB Precision Inc
370 Britannia Road - East #5,
Mississauga, Ontario L4Z 1X9
☎(+1 905) 890 0570 ⊖ (+1 905) 890 9279
👥 8
President: J Duggan
■ Sale of ball bearings, keyboards, fans, speakers, motors, R&S bearings.
98

90-01480 OFEMA Ottawa Office
(a subsidiary of OFEMA, France)
275 Slater Street - Suite 401, Ottawa, Ont K1P 5H9
☎(+1 613) 238 23 31 ⊖ (+1 613) 238 85 38
Director: Howard Kneen
98

90-01802 Racal-Datacom (Canada) Inc.
2595 Skymark Avenue, Mississauga, Ontario L4W 4L5
☎(+1 905) 602 77 55 ⊖ (+1 905) 629 16 55
Manager: M Oke
98

90-01805 Racal Protection Canada
(a division of Racal Filter Technologies, Canada)
1175 California Avenue, Brockville, Ont K6V 5V8
☎(+1 613) 345 13 49 ✆ (800) 267 44 14 (Canada only)
Ⓔ rpc@recorder.ca ⊖ (+1 613) 345 08 26
👥 13 (1996)
Operations Director: John Halliday
Sales and Marketing Director: Jamie Price
Administration and Marketing Assistant: Tressa Gray
98

90-01808 Raychem (Canada) Ltd
(a subsidiary of Raychem Corp.)
650 - 32nd Avenue, Suite 505, Lachine, Quebec H8T 3K5
☎(+1 514) 634 88 55 ✆ (800) 561 07 95
⊖ (+1 514) 634 88 09
Canadian Sales Manager: Rene Cournoyer
■ Manufacture heat-shrinkable tubing, molded parts and gels, wire and cables, interconnect devices and identification products.
® USA: Raychem Corp.
98

90-01810 Racal Recorders (Canada) Inc
(a subsidiary of Racal Recorders Ltd, UK)
635 Forth Line - Unit 11, Oakville, Ontario L6L 5W4
☎(+1 905) 844 84 47 ⊖ (+1 905) 844 11 66
98

90-01850 Robin Aerospace Products Ltd
2150 Thurston Drive, Ottawa, Ont K1G 5T9
☎(+1 613) 739 9245 ⊖ (+1 613) 739 9641
👥 14 🏠 650 sq m
President: Cliff Squires
Sales Dir: Art Maw
Customer Support Dir: Jake Duret
® Canada: Eje Translite
Germany: LRE Relais + Elektronik
USA: Howell Exports, Ideal Aerosmith, Kollmorgen Motion Technologies Group, Leach International, Leland Electrosystems
97

90-01940 Barry Shane & Associates
PO Box 120, Dewinton, Alta TOL OXO
☎(+1 403) 938 5348 ⊖ (+1 403) 938 5274
Representative: Barry Shane
® USA: Detroit Armor Corp.
97

90-02100 Transaero Canada
216 Migneron St, St Laurent, Que H4T 1Y7
☎(+1 514) 735 6790 ☎(+1 518) 483 4654
⊖ (+1 514) 735 0210 ⊖ (+1 518) 483 4572
Regional Mgr: Bengt Ohman
® USA: Transaero Inc.
97

90-02330 Westland Technologies Canada Ltd
(a subsidiary of GKN plc, UK)
Royal Bank Plaza, PO Box 38, Toronto, Ont M5J 2J7
☎(+1 613) 782 22 41 ⊖ (+1 613) 782 29 01
👥 6 (1996) 🏠 2,000 sq ft
Vice President: Frank Bamford
Vice President and Commercial: Jeffrey R Fausett
Sales and Marketing Director: Clive Soord
® UK: Westland Group
98

Cape Verde

Head of State:
Antonio Monteiro Mascarenhas

10 Head of State

10-10000 Office of the President
Rua Guerra Mendes 14, Praia
☎(+238) 260 ⊠ 51 prep cv
President of the Republic: Antonio Monteiro Mascarenhas [98]

11 Prime Minister

11-10000 Office of the Prime Minister
Praça 12 de Septembro, Praia
☎(+238) 248 ⊠ 52 gover cv
Prime Minister: Carlos Wahnon Veiga
Deputy Prime Minister: Jose Antonio dos Reise [98]

12 Ministries

12-10000 Ministry of Defense and National Security
Rua Andrade Corvo, PO Box 29, Praia-Santiago
☎(+238) 61 16 50 ⊠ 6077 empf cv
Minister: Ulpio Napoleao Fernandes [98]

12-20000 Ministry of Foreign Affairs
Praça 1 de Maio, Praia-Santiago
☎(+238) 61 57 33; 61 67 35; 61 47 74; 61 47 74
✉ NESTRANGEIROS ⊠ 70 mne cv
⊖ (+238) 61 39 52; 61 19 60
Minister: Amilcar Spencer Lopes
Secretary of State for Foreign Affairs and Co-operation: Jose Luis de Jesus [98]

31 Embassies & Attachés

31-01890 Consulate of the United Kingdom
Shell Cabo Verde Sarl, Avenue Amilcar Cabral CP4, Sao Vicente
☎(+238) 31 44 70; 31 46 05; 31 42 32 ⊠ 3082 CV
⊖ (+238) 31 47 55
Ambassador: H E Mr A E Furness (resides in Dakar)
First Secretary and Deputy Chief of Mission: R W Hyde (resides in Dakar)
Education Attaché: R H Budd (resides in Dakar)
Third Secretary (Management) and Vice Consul: I L McKinlay (resides in Dakar)
Third Secretary, Chancery and Information: Miss S G Dobson (resides in Dakar)
Honorary Consul: A Cannuto [98]

31-01900 Embassy of the United States of America
Rua Abilio Macedo 81, CP 201, Praia
☎(+238) 61 56 16/17 ⊠ 6068 amemb cv
⊖ (+238) 61 13 55
Deputy Chief of Mission: Theorphilus J. Rose
Office of the Defense Attaché: Lieut Col Paul S Cariker (resident in Dakar)
Admin Section: Ms Lisa Gamble-Barker [97]

Central African Republic

Head of State:
Ange Félix Patasse

10 Head of State

10-10000 Office of the President
Palais de la Renaissance, Bangui
☎(+236) 61 03 23
⊠ 5201 presicabb rc, 5300 presicaba rc
President of the Republic: Ange Félix Patasse [98]

11 Prime Minister

11-10000 Secretariat-General of the Government
Bangui
Prime Minister: Michel Gbezzera-Bria [98]

12 Ministries

12-10000 Ministry of National Defense & Veterans' Affairs
Bangui
☎(+236) 61 46 11/30 11 ⊠ 5298 mindca rc
Minister of Defence: Dr Pascal Kado [98]

12-20000 Ministry of Foreign Affairs and International Co-operation
PO Box 836, Bangui
☎(+236) 61 15 74; 61 00 85; 61 35 60
⊠ 5213 diploma rc
Minister of Foreign Affairs: Jean Mette-Yapende [98]

31 Embassies & Attachés

31-00660 Embassy of the Federal Republic of Germany
Avenue du Presidt Gamal Abdel Nasser, PO Box 901, Bangui
Avenue du President Gamal Abdel Nasser
☎(+236) 61 47 65 ⊠ 5219 aabgui rc ⊖ (+236) 61 19 89
Ambassador: H E Mr Reinhard Bucholz
Deputy Chief of Mission: Ms Laura Matten [98]

31-01890 Consulate of the United Kingdom
PP Box 977, Bangui
☎(+236) 61 83 38 ⊖ (+236) 61 80 29
Honorary Consul: J Y Lauge [98]

Chad

Head of State:
Lieut Gen Idriss Deby

10 Head of State

10-10000 Office of the President
N'Djamena
☎(+235) 32 15; 32 80 ⊠ 5201 presirep kd
President of the Republic: Lieut Gen Idriss Deby [98]

11 Prime Minister

11-10000 Office of the Prime Minister
N'Djamena
Prime Minister: Nassour Guelengdouksia Ouaidou [98]

12 Ministries

12-10000 Ministry of Armed Forces
PO Box 916, N'Djamena
☎(+235) 32 15; 31 58
Minister of Armed Forces: Mahamat Nimir Hamata [98]

12-20000 Ministry of Foreign Affairs and Co-operation
N'Djamena
☎(+235) 23 13; 51 57 20 ⊠ 5238 minaffet kd
⊖ (+235) 51 56 08
Minister of Foreign Affairs: Mahamat Saleh Annadif [98]

31 Embassies & Attachés

31-00590 Embassy of France
Rue du Lieutenant Franjoux, PO Box 431, N'Djamena
☎(+235) 52 25 75/6 ⊠ 5202 ambafran kd
⊖ (+235) 52 28 55
Ambassador: H.E. Mr André Janier
First Counsellor: Christian Testot
First Secretary: Thomas Wagner
Defence Attaché: Col Guillou [97]

31-01890 Consulate of the United Kingdom
Avenue Charles de Gaulle, BP 877, Ndjamena
☎(+235) 51 60 64 ⊠ 5235 act kd
Ambassador: H E Mr McCarthy (resident in Abuja)
Deputy Chief of Mission: G S Hand (resident in Abuja)
Honorary Consul: Mrs E Abtour [98]

31-01900 Embassy of the United States of America
Avenue Felix Eboue, PO Box 413, N'Djamena
☎(+235) 51 70 09; 51 90 52; 51 92 33
⊠ 5203 amemb kd ⊖ (+235) 51 33 72; 51 56 54
Ambassador: H E Mr Abavavid Hallsce [98]

Chile

Head of State:
Eduardo Frei Ruiz-Tagle

10 Head of State

10-10000 Office of the President
Palacio de la Moneda, Santiago
☎(+56 2) 671 41 03; 690 40 00
⊖ (+56 2) 697 32 62; 698 41 30
President of the Republic: H.E. Eduardo Frei Ruiz-Tagle
98

11 Prime Minister

11-10000 Office of the Minister (Secretary General of Government)
Palacio de la Moneda, Santiago
☎(+56 2) 671 16 80; 690 41 60 ⊠ 24012
⊖ (+56 2) 699 16 57
Minister, Secretary General of Government:
H E Jose Joaquin Brunner
Minister, Secretary General of the Presidency: Juan Villarzu
98

12 Ministries

12-10000 Ministry of National Defence
Plaza Bulnes, Santiago
☎(+56 2) 671 05 51; 696 52 71 ⊠ 40537
⊖ (+56 2) 34 53 39
Minister of Defence: Edmundo Perez Yoma
98

12-20000 Ministry of Foreign Relations
Morande 441/Ex Edificio del Congreso, Santiago
☎(+56 2) 698 25 01; 679 42 00 ⊠ 40595
⊖ (+56 2) 699 42 02
Minister of Foreign Relations: José Miguel Insulza
98

12-40000 Ministry of the Interior
Palacio de la Moneda, Santiago
☎(+56 2) 671 70 54; 41 03 ⊠ 240273
⊖ (+56 2) 696 87 40
Minister: Carlos Figueroa Serrano
Chief of Staff: Gustavo Villalobos
98

15 Parliament

15-10000 Senate
Pedro Montt s/n, Edificio del Congreso, Valparaiso
98

15-20000 Chamber of Deputies
Cámara de Diputados
Pedro Montt s/n, Congreso Nacional, Valparaiso
☎(+56 32) 23 09 95; 697 00 22 E oic@congreso.cl
⊖ (+56 32) 23 05 08
President: Gutenberg Martinez ☎(+56 32) 23 45 36
First Vice President: Armando Arancibia
☎(+56 32) 23 47 62
Second Vice President: Jaime Rocha
☎(+56 32) 23 49 74
Secretary: Carlos Loyola ☎(+56 32) 23 45 59
98

25 Defense Forces

25-22400 Fourth Air Brigade
Punta Arenas
Brigade Commander in Chief:
Brig.Gen. Eitel Yon Muhlenbrock
97

25-30500 Navy Headquarters
La Armada de Chile
Plaza Sotomayor 594, Correo Naval, Valparaiso
☎(+56 32) 25 86 69 ⊖ (+56 32) 25 27 13
Commander in Chief: Adm. Jorge Martínez Busch
☎(+56 32) 25 86 69 ☎(+56 32) 21 37 13 (fax)
General Secretary of the Navy: VAdm. Federico Klein Koch
☎(+56 32) 25 28 18 ☎(+56 32) 21 37 13 (fax)
Chief of Naval General Staff: VAdm. Juan Mackay Barriga
☎(+56 32) 25 51 52 ☎(+56 32) 21 37 13 (fax)
Deputy Chief of Naval General Staff:
Capt. Alex Waghorn Jarpa ☎(+56 32) 21 82 88
☎(+56 32) 21 37 13 (fax)
General Director of Logistics: Adm. Sergio Jarpa Gerhard
☎(+56 32) 25 23 79 ☎(+56 32) 21 99 79 (fax)
General Director of Personnel: Adm Jorge Arancibia Reyes
☎(+56 32) 25 20 94 ext. 6629; 25 54 99 (fax)
Director of Naval Intelligence:
RAdm. Andrés Swett Serrano
☎(+56 32) 28 18 45 ext. 7325
☎(+56 32) 28 19 80 (fax)
Director of Asmar: Adm. Jorge Swett Browne
☎(+56 32) 25 94 11 ☎(+56 32) 21 46 27 (fax)
Flag Officer Commanding Fleet:
Adm. Germán Goddard Dufeu
☎(+56 32) 28 18 45 ext. 6040
☎(+56 32) 21 22 04 (fax)
Deputy Chief of Navy Public Relations:
Lt Cdr Juan Pablo Barros Torres
Chief of Naval Mission in London:
RAdm. Jorge Arancibia Clavel ☎(+44 171) 734 41 76
☎(+44 171) 434 07 93 (fax)
Chief of the Naval Mission in Washington:
RAdm. Onofre Torres Colvin ☎(+1 202) 466 2505
☎(+1 202) 224 9733 (fax)
Commander in Chief of Submarine Force:
RAdm. Arturo Oxley Dueñas ☎(+56 41) 50 50 75
☎(+56 41) 54 21 83 (fax)
Chief of Staff of the Fleet: Capt. Rodolfo Codina Díaz
☎(+56 32) 25 71 63 ☎(+56 32) 21 22 04 (fax)
Commanding General Marine Corps.:
RAdm. Miguel Alvarez Ebner
☎(+56 32) 25 20 94 ext. 6652
☎(+56 32) 25 85 59 (fax)
Director Naval Engineering:
RAdm. Sergio Martínez González ☎(+56 32) 25 56 49
☎(+56 32) 21 47 85 (fax)
Accounting & Supplies Dir: VAdm. Santiago Díaz Torres
☎(+56 32) 25 20 94 ext. 6211
☎(+56 32) 21 99 73 (fax)
General Director of Maritime Territory & Merchant Marine:
VAdm. Ariel Rosas Mascaró ☎(+56 32) 20 80 00
☎(+56 32) 25 25 39 (fax)
Education Dir: RAdm. Enrique Caselli Ramos
☎(+56 32) 68 92 22 ext 8801
☎(+56 32) 97 13 30 (fax)
Naval Welfare Dir: RAdm. Pedro Veas Diabuno
☎(+56 32) 25 20 94 ext. 6704
☎(+56 32) 25 96 16 (fax)
Naval Weapons Dir: Capt. José Marchant Ortega
☎(+56 32) 25 20 94 ext. 6612
☎(+56 32) 21 24 05 (fax)
Naval Health Dir: RAdm. Jaime Guzman Jara
☎(+56 32) 28 18 45 ext. 7663
☎(+56 32) 25 47 87 (fax)
Deputy Chief Public Relation Service:
Lt Cdr Juan Pablo Barros Torres

Public Relations Service
Plaza Sotomayor 592, Valparaiso
☎(+56 32) 25 20 94 ext. 6438 ⊖ (+56 32) 25 21 03
Chief Public Relations: Capt. Cristian Cifuentes Cabello
Deputy Chief of Public Relations:
Lt Cdr Juan Pablo Barros Torres
97

25-30800 Air Force Headquarters
Plaza Bulnes, José Ignacio Zenteno 45 piso 2, Santiago
☎(+56 2) 671 05 51 ⊖ (+56 2) 633 15 15
Commander in Chief Air Force:
Gen Fernando Rojas Vender
Chief of Air Staff: Gen Maximo Venegas Fuentes
Secretary General of the Air Force:
Col José Ignacio Concha Besa
Inspector General: Gen Raul Tapia Esdale
Operations Director: Marcos Meirelles Guizman
Intelligence Director: Gen Bruno Ugatti Geminiani
Chief of the Logistics Command:
Gen Adolfo Von Muhlenbrock Hevia
Head of the Logistics Command Weapons Systems:
Gen Patricio Rios Ponce
Director of Air Defence and Telecommunications:
Gen Humberto Ianiszewski Nogues
Director of Health Department:
Gen Sergio Olmedo Droguett
Chief of Personnel Command:
Gen Enrique Montealegre Jullian
Director General of Aeronautics:
Gen Gonzalo Miranda Aguire
Administration Director: Gen Paulino Villafaena Morales
Chief of Communications Department:
Col Mario Gonzales Ossa
General Auditor: Gen Adolfo Celedon Sandoval
Commander in Chief of Third Air Brigade (PTO Montt):
Gen Ricardo Bermudez Sanhueza
Commander in Chief of First Air Brigade (Iquique):
Gen Alejandro Huerto Forch
Commander in Chief of Second Air Brigade (Santiago):
Gen Patricio Rios Braga
Commander in Chief of Fourth Air Brigade (Punta Arenas):
Jorge Sandoval Bravo
Commander in Chief of Fifth Air Brigade (Antofagasta):
Gen Nelson Sanhueza Lagos
Commander of General Air Supplies:
Gen Luis Berrios Carrasola
Aeronautical Attaché: Gen Ettore Rossi Zerega

Public Relations Officer's Office
Bulnet 147, Santiago
☎(+56 2) 696 16 46 ⊖ (+56 2) 694 34 77
Chief of Communications Department:
Col Mario Gonzalez Ossa
98

25-42100 First Naval Zone Headquarters
Correo Naval, Valparaiso
☎(+56 32) 25 02 07
Commander of First Naval Zone: V D G Goddard
Chief of Staff: Capt Rudolfo Soria-Galvorro
J.Dept A-3 Operations: Cdr Roberto Leniz
Director General of Personnel:
Vice Adm Hernán Couyoum Djian
Director of Ordinance: RAdm José Marchant
98

25-42200 Second Naval Zone Headquarters
Correo Naval, Talcahuano
☎(+56 41) 50 50 07 ⊖ (+56 41) 54 14 60
Commander Second Naval Zone:
RAdm. Jorge Balaresque Walbaum
97

25-42300 Third Naval Zone Headquarters
Correo Naval, Punta Arenas
☎(+56 61) 22 20 21 ⊖ (+56 61) 22 43 94
Commander in Chief Third Naval Zone:
RAdm. Roman Fritis Pérez
97

25-42400 Fourth Naval Zone Headquarters
Av. A. Prat 706, Iquique
☎(+56 57) 42 12 19 ⊖ (+56 57) 42 90 06
Commander: Cdre Pedro Frioli Otonel
Chief of Staff: Capt. Raul Silva Gordon
97

25-42800 Flag Officer Commanding Fleet
Correo Naval, Valparaíso
☎(+56 32) 81 10 71 ext. 8201 ⊖ (+56 32) 81 14 34
Commander: VAdm. Jorge Llorente Domínguez
Chief of Staff: Capt. Miguel Vergara Villalobos
97

25-43000 Naval Aviation Command Headquarters
Commandancia de la Aviación Naval
Correo Naval, Valparaiso
☎(+56 32) 81 15 97 ⊖ (+56 32) 81 14 34
Commander: RAdm. Eduardo Schnaidt Parker
Chief of Staff: Capt. Gudelio Mondaca Oyarzún
Operations Dir: Cdr Roberto Ramírez Von Holle
Director of Logistics: Cdr Juan Gibson Orueta
HA-1 Squadron Commander:
Capt. Eugenio Arellano Palma
VP-1 Squadron Commander:
Capt. Percy Richter Silberstein
VT-1 Squadron Commander: Cdr Eduardo Caprile Febres

VC-1 Squadron Commander: Cdr Federico Jara Cartes
HU-1 Squadron Commander: Cdr Daniel Boisset Encina
[97]

25-44000 Chilean Navy Telecommunications Service
Servicio de Telecomunicaciones de la Armada de Chile (BIBLIOTECA)
Av. Gran Bretaña 980, Playa Ancha, Valparaiso
☎ (+56 32) 28 19 81 ⊖ (+56 32) 28 71 72
Head of Service: Capt. Miguel Vasquez (Navy) [97]

31 Embassies & Attachés

31-00090 Australian Embassy
Gertrudis Echeñique 420, Las Condes,
PO Box 33, Correo 10, Santiago
☎ (+56 2) 228 50 65 ⊠ AUSTEMBA Santiago
⊖ (+56 2) 208 17 07
Ambassador: H E Mrs Susan Tanner
First Secretary and Counsellor: Mrs Raewyn Henelius [97]

31-00100 Embassy of Austria
Barros Errázuriz 1968, Piso 3, 16196 Santiago 9
☎ (+56 2) 223 47 74; 223 42 81
⊠ AUSTROAMB Santiago 9 ▣ 240528 oestg cl
⊖ (+56 2) 204 93 82
Ambassador: H E Dr Horst-Dieter Rennau
☎ (+56 2) 208 02 28
First Secretary: Ms Gertraud Borges da Gama
☎ (+56 2) 228 27 20
Commercial Section
Isidora Goyenechea 2934, Oficina 601, PO Box 165-D, Santiago
☎ (+56 2) 233 05 57 ▣ 240866 austr cl
⊖ (+56 2) 233 69 71
Commercial Counsellor: Josef Hofer
Commercial Attaché: Heinz Thenmeyer
☎ (+56 2) 242 61 86 [98]

31-00500 Embassy of Ecuador
Avenida Providencia 1979, Piso 5, Santiago
☎ (+56 2) 231 50 73 ⊠ MECUADOR SANTIAGO
⊖ (+56 2) 232 58 33
Ambassador: H.E. Rodrigo Valdez [97]

31-00510 Embassy of the Arab Republic of Egypt
Roberto del Rio 1871, PO Box 9904, Santiago
☎ (+56 2) 274 88 81; 274 63 34 ▣ 440156 bostn cz
⊖ (+56 2) 274 63 34
Ambassador: H E Mr Talaat Selmy [98]

31-00660 Embassy of the Federal Republic of Germany
PO Box 9949, Santiago
Calle Agustinas 785, Piso 7, Santiago
☎ (+56 2) 633 50 31 to 35 ▣ 240583 aasgo cl
⊖ (+56 2) 633 61 19
Ambassador: H E Dr Werner Reichenbaum
Deputy Chief of Mission: Dr Alexander Mühlen
Defense Attaché: Lt Col Norbert Hartwig
Economic Counsellor: Eugen Wollfarth
Public Information Officer: Bernd Weidlich [98]

31-00740 Embassy of Guatemala
Monseñor Sótero Sanz 55, Piso 8, Provedincia, Santiago
☎ (+56 2) 335 15 65 ⊖ (+56 2) 335 12 85
Ambassador: H E Mrs Ana Lucrecia Rivera de Ampuero [98]

31-00780 Embassy of the Republic of Haiti
Calle Zurich 255, Officina 21, Las Condas, Santiago
☎ (+56 2) 231 09 67 ⊖ (+56 2) 231 09 67
Ambassador: H E Mr Guy Pierre Andre
Counsellor: Pétuel Raymond [98]

31-00830 Embassy of India
Triana 871, PO Box 10 433, Santiago
☎ (+56 2) 235 20 05; 235 26 33
[E] embindia@chilesat.net ▣ 340046 indemb ck
⊖ (+56 2) 235 96 07
Ambassador: H E Mr Ram Mohan ☎ (+56 2) 235 26 30
Commercial Officer: K K Sachdev [98]

31-00840 Embassy of the Republic of Indonesia
Avenida Nueva Costanera 3318, Vitacura, Santiago
☎ (+56 2) 207 62 66 [E] kbristgo@iusanet.cl
▣ 244068 indon cl ⊖ (+56 2) 207 99 01
First Secretary of Administration: Hindaryudo Soerkarman [98]

31-00890 Embassy of Italy
Clemente Fabres 1050, PO Box 3114, Santiago
☎ (+56 2) 225 90 29; 225 92 12; 225 94 39
▣ 440321 itdpl cz ⊖ (+56 2) 223 24 67
Ambassador: H E Mr E Costa
Counsellor: Francesco Barbado
Commercial Officer: Alessandre Mooliano
Public Information Officer: Luis Perna [98]

31-00930 Embassy of the Hashemite Kingdom of Jordan
Los Militares 4280, Las Condes, PO Box 10431, Santiago
☎ (+56 2) 228 89 89; 208 45 95 ⊖ (+56 2) 228 87 83
Ambassador: H E Mr Atef Halasa
Plenipotentiary Minister: Fouad Atieh
First Secretary and Consul: Muwaffaq Ajlouni [98]

31-01020 Embassy of Lebanon
Alianza 1728, Vitacura, Santiago
☎ (+56 2) 218 28 35 [E] maalouf@iusanet.cl
▣ 440118 amlib cz
Ambassador: H E Mr Massoud Maalouf
☎ maalouf@iusanet.cl FAX [98]

31-01270 Royal Netherlands Embassy
Las Violetas 2368, PO Box 56-D, Santiago
☎ (+56 2) 223 68 25 [E] hollenb@thm.cl
▣ 30520 nedamb ck ⊖ (+56 2) 225 27 37
Ambassador: H E Mr S E Ramondt
First Secretary: K van der Tempel
Second Secretary: J Roodenburg
Administrative and Consular Affairs Attaché: L J G A Felten [98]

31-01670 South African Embassy
Avenue Once de Septembre 2353, 16th Floor, Santiago 9
☎ (+56 2) 231 28 60; 231 28 61; 231 28 62; 231 28 63
⊖ (+56 2) 231 31 85
Ambassador: H E Mr Casper S C Venter
First Secretary: M Jurgens
First Secretary: E O Janse van Rensburg
Armed Forces Attaché: Col J J van Heerden
Counsellor: T Beukes [98]

31-01840 Embassy of the Republic of Turkey
Nuncio Sótero Sanz 136, Providencia, Santiago
☎ (+56 2) 231 89 52
⊠ EMBAJADA DE TURQUIA SANTIAGO
▣ 340278 turkem ck ⊖ (+56 2) 231 77 62
Ambassador: H E Mr Sadi Calislar
First Secretary: Ayhan Enginar [98]

31-01890 Embassy of the United Kingdom
Avenue El Bosque Norte 0125, Casilla 72-D, 16552 Santiago 9
☎ (+56 2) 231 37 37 ▣ 340483 britemb ck
⊖ (+56 2) 234 97 71 (Commercial/Consulate); 234 25 66 (Management/Information); 234 17 50 (Chancery); 234 73 75 (British Council)
Ambassador: H E Miss Glynne Evans
Deputy Head of Mission: D G Roberts
Defence Attaché: Capt P J Ellis (RN)
First Secretary, Commercial: T A Torlot [98]

31-01900 Embassy of the United States of America
Avenida Andrés Bello 2800, Santiago
☎ (+56 2) 232 26 00 [W] http://www.rdc.cl/~usemb
⊖ (+56 2) 330 31 60
Ambassador: H E Mr Gabriel Guerra-Mondragon
☎ (+56 2) 330 33 43
Deputy Chief of Mission: Charles S Shapiro
☎ (+56 2) 330 33 37
Counsellor for Commercial Affairs: Carlos F Poza
☎ (+56 2) 330 33 16
Defense and Naval Attaché: Capt Thomas L Breitinger (USN) ☎ (+56 2) 330 33 31
Public Affairs Officer: Ms Barbara C Moore
☎ (+56 2) 330 33 52
Counsellor for Administrative Affairs: Charles E Nichols [98]

31-02010 Embassy of the Federal Republic of Yugoslavia
Calle Exequias Allende 2370, Providencia, Santiago
☎ (+56 2) 223 05 10
Chargé d'Affaires a.i.: Svetislav Rajevic [98]

55 Government-owned Industry & Production Facilities

55-00150 Astilleros y Maestranzas de la Armada
ASMAR Shipbuilding & Docking Co.
PO Box 150-V, Valparaiso
Prat 856, Edificio Rapa-Nui, Floor 13, Valparaiso
☎ (+56 32) 26 05 00; 26 05 09 ⊠ ASMAR
⊖ (+56 32) 26 01 57 ⊖ (+56 32) 21 46 27; 23 12 97
♣ US$ 100,000,000 (1996) 👥 3,500
Managing Director: RAdm Jorge Swett
Commercial Director: Emilio Cabrera
■ Services include: repair, maintenance and modernisation of naval and merchant ships. Fishing vessel construction and the production of naval equipment. Onshore work for companies and establishments who have particular requirements. Design, manufacture and sale of military equipment. → 75 → 80
Valparaiso Yard
Avda Altamirano 1015, PO Box 150-V, Valparaiso
☎ (+56 32) 27 20 00 ⊖ (+56 32) 27 20 99
Talcahuano Yard
Naval Base Talcahuano, PO Box 9-D, Talcahuano
☎ (+56 41) 50 41 00; 50 42 00
⊖ (+56 41) 50 40 01; 50 40 04
Magallanes Yard
21 de Mayo Number 2421, PO Box 65-D, Correo Naval, Punta Arenas
☎ (+56 61) 26 13 15 ⊖ (+56 61) 26 13 99 [98]

55-00550 Empresa Nacional de Aeronáutica de Chile (ENAER) (ENAER)
Grand Avenida J M Carrera 11087, Paradero 36 1/2, El Bosque
☎ (+56 2) 528 27 35; 528 28 23; 528 25 99
[E] infanaer@/iusanet.cl
⊖ (+56 2) 528 26 99 (commercial); 528 28 15
👥 1,200 (1997) 🏭 47,000 m2
President: Fernando Rojas V
Executive Director: Fernando Silva C
Commercial Manager: Jaime Estay V
Financial Manager: Ms Martha Sarmiento G
Marketing Manager: Alejandro Vargas B
Aircraft Manufacture Manager: Simón Adjemian
■ Manufacture of T-35 Pillan primary trainer, T-35 DT Pillan Turbo primary trainer and Ñamcu light two-seater aircraft. A-36 Halcon (C-101) advanced trainer co-produced in agreement with CASA (Spain). Maintenance, overhaul, repair and upgrading for aircraft, engines and accessories. Risk-sharing partner Embraer EMB-145 50 seat commercial aircraft. Upgrade programme for Mirage and F-5 aircraft for Chilean Air Force. → 75 → 80
Desarrollo de Tecnologias y Sistemas Ltda (DTS)
Rosas 1444, Santiago
☎ (+56 2) 697 09 91 ⊖ (+56 2) 699 33 16
President: Fernando C Silva
■ Development and manufacture of electronic warfare systems. [98]

55-00695 Fábricas y Maestranzas del Ejercito (FAMAE)
Pedro Montt 1606, Santiago
☎ (+56 2) 551 32 05; 556 91 31 ✆ 556 10 11
▣ 242346 famae cl ⊖ (+56 2) 551 18 94

General Director: Luis Iracal Lobo
Sales and Marketing Director: Gustavo Latorre Vásquez
■ Land-based weapons and weapon systems. [98]

60 Research Organizations, Institutes & Test Establishments

60-00500 Chilean Naval Academy
Casilla 23-v, Valparaiso
☎ (+56 32) 28 10 04 [E] esnaval@chilesat.net
⊖ (+56 32) 28 10 04
General Manager: Claudio Arancibia Tavra
☎ (+56 32) 67 13 22 [98]

60-00700 Instituto Geográfico Militar (IGM)
Nueva Santa Isabel, 1640 Santiago
☎ (+56 2) 696 82 21 [E] igm@reuna.al ▣ 441677 igmcz
⊖ (+56 2) 698 82 78
21,081 m2
Director: Brig Gen Enrique Gillmore Callejas
Research Director: Brig Sergio Matus Martinez-Conde
Operations and Technical Director: Col Pablo Gran Lopez
Commercial Subdirector: Lieut Col Sergio Urrejola Dastres
Public Relations Officer: Lieut Col Alfredo Gonzalez Silva
■ Aerial photography, maps, atlas. [98]

60-00950 Instituto de Investigaciones y Control del Ejercito (IDIC)
Army Institute for Research and Control - National Proof Testing Organization
Avda Pedro Montt 2050, Santiago
☎ (+56 2) 556 60 31; 556 80 58 ⊖ (+56 2) 555 09 45
Director: Col Juan A. Carvajal ☎ (+56 2) 556 80 58
Deputy Dir: Col Carlos N. Domínguez
☎ (+56 2) 556 61 51
Assistant Dir: Lt Col Ricardo Riquelme
Dir Research: Maj. Rodrigo Ortíz
Operations & Technical Dir: Maj. Francisco Greve
■ Quality assessment, inspection, testing for homemade or imported products, including ammunition, weapons, powders, explosives, electronics, clothing, foodstuffs etc. As Proof House, controls small arms and its ammunition to guarantee users' security. Affiliated to CIP. [97]

75 Manufacturing Companies

ASMAR . . . 55-00150
Empresa Nacional de Aeronáutica . . . 55-00550

75-00300 Industria Metalurgica del Norte Ltda
Head Office and Defense Division
Exequil Fernández 3397, Macui, Santiago
☎ (+56 2) 221 71 26 ⊖ (+56 2) 221 00 06
President: Dr Eng Carlos Cardoen
General Manager: René Cardemil ☎ (+56 57) 41 18 16
Commercial Counsellor: Cristian Silva
☎ (+57 2) 221 71 62
■ Aircraft bombs (cluster, G.P. and others), small explosive weapons for infantry.
Public Relations Dept
Monseñor Escrivá de Balaguer 6173, Vitacura, Santiago
☎ (+56 2) 219 99 03 ⊖ (+56 2) 218 99 10 [98]

80 Service Companies & Organizations

Aerotech . . . 90-00030
ASMAR . . . 55-00150
Empresa Nacional de Aeronáutica . . . 55-00550

80-01940 Compañia de Ingenieria de Sistemas y Desarrollos Funcionales Ltda (SISDEF)
(a subsidiary of General Electric Company, UK)
Los Fresnos 400, Miraflores, PO Box 150-V, Viña del Mar
☎ (+56 32) 67 37 75; 67 40 92; 67 81 73
[E] sisdef@entelchile.net ⊖ (+56 32) 67 30 71
£ 300 000 45 1000 sq m
President: Claudio Niadal
Sales and Marketing Manager: Eduardo Morales
■ Hardware and Software development and support for defense equipment. Command and control systems, mission control systems, ESM, ELINT and COMINT systems.
® UK: GEC [98]

90 Sales Agents, Sales & Liaison Offices

Companies represented:

ADB/Siemens Airport Lighting, Belgium . . . 90-00030
Aero Union Corporation, USA . . . 90-00030
Aerotec, USA . . . 90-00030
Air Taser Inc, USA . . . 90-00030
Airport Equipment Company (AEC), Sweden . . . 90-00030
AM General Div./LTV, USA . . . 90-00070
Beretta, Italy . . . 90-00120
Beretta, USA . . . 90-00120
CAE Aviation Ltd, Canada . . . 90-00150
Canadian Marconi Co., Canada . . . 90-00150
CEIA, Italy . . . 90-00030
Cossor Electronics, UK . . . 90-00150
Dranetz Technologies Inc., USA . . . 90-00320
E&P Telecommunication, Spain . . . 90-00030
EEV Ltd, UK . . . 90-00700
Euro-Diesel SA, Belgium . . . 90-00030
Field Aviation, Canada . . . 90-00150
GEC, UK . . . 80-01940
Gulfstream Aerospace, USA . . . 90-00150
Hunting Engineering, UK . . . 90-00150
IDC/Kollsman, USA . . . 90-00150
Infoplan, S. Africa . . . 90-00120
Kuiken Hytrans BV, Netherlands . . . 90-00030
Litton, USA . . . 90-00120
Lockheed Martin Tactical Aircraft, USA . . . 90-00150
Loral Electronic Systems, USA . . . 90-00150
Maule Air, USA . . . 90-00150
McDonnell Douglas, USA . . . 90-00120
Norquip, UK . . . 90-00030
Osborne Rescue Boats Ltd, UK . . . 90-00030
Raytheon, USA . . . 90-00120
Schmid Telecommunication, Switzerland . . . 90-00030
Signal, Netherlands . . . 90-00120
Tekno Industries Inc, USA . . . 90-00030
Tracor Airspace, USA . . . 90-00120
Trase/Somati, Belgium . . . 90-00030
TSS Ocean Survey Systems, UK . . . 90-00030
Van Hool, Belgium . . . 90-00030
Vinten Military Systems, UK . . . 90-00150
Zarc International Inc, USA . . . 90-00030

90-00030 Aerotech
Hernando de Aguirre 194, PO Box 206, Correo 35, Santiago
☎ (+56 2) 232 67 57 ⊖ (+56 2) 231 35 30
6 (1997) 70 m2 (1997)
Executive Director: Mathias C Pairoa
Technical Director: Hernan V Goas
Projects Development Director: René N Pairoa
■ Representatives, aeronautical agents and consultants. → 80
® Belgium: ADB/Siemens Airport Lighting, Euro-Diesel SA, Trase/Somati, Van Hool
Italy: CEIA
Netherlands: Kuiken Hytrans BV
Spain: E&P Telecommunication
Sweden: Airport Equipment Company (AEC)
Switzerland: Schmid Telecommunication
UK: Norquip, Osborne Rescue Boats Ltd, TSS Ocean Survey Systems
USA: Aero Union Corporation, Aerotec, Air Taser Inc, Tekno Industries Inc, Zarc International Inc [98]

90-00070 Alberto Grünwald y Cia Ltda
Americo Vespucio No. 1730, Quilicura, Santiago
☎ (+56 2) 603 54 75 ⊖ (+56 2) 603 61 79
US$ 3 000 000 (1996) ♣ US$ 25 000 000 (1996)
100 3 000 sq m
President: J.C. Grünwald
Technical Dir: Camilo Galvez
Marketing Dir: Fernando Moro
® USA: AM General Div./LTV [97]

90-00120 Benjamin Berkovits Graf
Avenida Las Americas 700 Cerrillos, Casilla 14103, Correo 21, Santiago
☎ (+56 2) 557 47 30 [E] nimrod@chilesat.net
⊖ (+56 2) 557 06 56
General Manager: Benjamin Berkovits G
® Italy: Beretta
Netherlands: Signal
S. Africa: Infoplan
USA: Beretta, Litton, McDonnell Douglas, Raytheon, Tracor Airspace [98]

90-00150 Thomas Bertie & Co. Ltda
(† Tattoo Ltda)
Luis Thayer Ojeda 95, Of 204, Casilla 178, Correo 35, Santiago
☎ (+56 2) 232 68 10 [E] tbertie@netup.cl
⊖ (+56 2) 232 76 47
President: Thomas Bertie
® Canada: CAE Aviation Ltd, Canadian Marconi Co., Field Aviation
UK: Cossor Electronics, Hunting Engineering, Vinten Military Systems
USA: Gulfstream Aerospace, IDC/Kollsman, Lockheed Martin Tactical Aircraft, Loral Electronic Systems, Maule Air [98]

90-00320 Coasin Chile S A
Correo 21, 14588 Casilla, Santiago
☎ (+56 2) 225 06 43 ⊖ (+56 2) 274 45 88
♣ US$ 62,000,000 (1996) 350
Managing Director: Raul Ciudad De La Cruze
Marketing Director: Guillermo Aguire
Commercial Director: Alvaro Croquevielle
® USA: Dranetz Technologies Inc. [98]

90-00400 Daewoo Corp Santiago Office
El Bosque Norte 0440 Oficina 407, Edificio Atlantis, Las Condes, Santiago
☎ (+56 2) 246 64 80 ▣ 340497 daewoo ck
⊖ (+56 2) 246 64 83 [98]

90-00700 Morgan and Cia SA
Almirante Pastene 256, PO Box 1-T, Tajamar, Providencia, Santiago
☎ (+56 2) 236 12 78 ⊖ (+56 2) 235 75 52
General Director: A Morgan
® UK: EEV Ltd [98]

90-01050 Rafael Armament Development Authority
San Sebastian 2839, Oficina 504, Las Condes-Santiago
☎ (+56 2) 233 63 88 ⊖ (+56 2) 231 10 62
General Mgr: Ami Manor [97]

China, People's Republic

Head of State:
Jiang Zemin

10 Head of State

10-10000 Office of the President
Great Hall of the People, Beijing
☎ (+86 106) 66 73 80
President: Jiang Zemin
Vice President: Rong Yiren [98]

10-20000 State Council Office of the Premier
State Council Secretariat, Beijing
☎(+86 106) 66 64 53
Premier: Li Peng
Vice Premier: Chunyan Jiang
Vice Premier: Lanqing Li
Vice Premier: Qian Qichen
Vice Premier: Bangguo Wu
Vice Premier: Rongji Zhu
Vice Premier: Jiahua Zou
State Councillor: Jungsheng Chen
State Councillor: Haotian Chi
State Councillor: Ismail Amat
State Councillor: Li Guixian
State Councillor: Li Tieying
State Councillor: Peng Peiyun
State Councillor: Song Jian
State Councillor and Secretary General: Luo Gan [98]

10-30000 Central Military Commission
Chairman: Jiang Zemin [98]

10-50000 State Planning Commission
Yuetan Nan Jie, Beijing
Minister: Chen Jinhua [98]

12 Ministries

12-10000 Ministry of National Defense
Dianmen, Beijing
Minister: Chi Haotian
Commission of Science, Technology and Industry (COSTIND)
PO Box 1092, 100034 Beijing
Minister: Song Jian [98]

12-10050 Ministry of State Security
Dongchangan Street, Beijing
Minister: Jia Chunwang [98]

12-20000 Ministry of Foreign Affairs
225 Chaoyangmennei Dajie, Beijing
Minister: Qian Qichen
Bureau of Cultural Relations
Director: Sun Weixue [98]

12-30000 Ministry of Foreign Economic Relations and Trade Commission
2 Changan East Road, Beijing
Minister: Wu Yi [98]

12-40000 Ministry of Public Security
East Chang'an Street, Beijing
Minister: Tao Siju [98]

12-70000 China Aerospace Corporation
China National Space Administration
No 8 Fucheng Road, Beijing
☎(+86 106) 837 01 56 ⊖ (+86 106) 837 08 49
Minister: Liu Jiyuanng
Vice Minister: Wang Liheng
Foreign Relations Dir: Meng Bo
Chief of Staff: Zhang Baoqian
Science and Technology Committee
No 8 Fucheng Road, Beijing
Chairman: Sun Jiadong [97]

12-80000 Ministry of Electronics Industry
49 Fuxing Lu, Beijing
☎(+86 1) 81 07 31; 81 05 61 ▣ 22383 mei cn
Minister: Hu Qili [98]

31 Embassies & Attachés

31-00020 Embassy of the People's Republic of Albania
28 Guanghua Lu, Jian Guo Men Wai, Beijing
☎(+86 1) 65 32 11 20 ⊖ (+86 1) 65 32 54 51
Ambassador: H E Mr Tahir Elezi ☎(+86 1) 65 32 11 19 [98]

31-00090 Australian Embassy
21 Dongzhimenwai Dajie - Sanlitun, 100600 Beijing
☎(+86 10) 523 23 31; 523 23 37
⊖ (+86 10) 532 46 05
Ambassador: H E Mr R C Smith
Minister (Deputy Chief of Mission): Ms Penny Richards
Defence Attaché: Col Paul McGrane
☎(+86 10) 532 46 05 (fax) ▣ 22263 austm cn
Senior Trade Commissioner: Ms Maurine Chong
Counsellor (Administrative): Martin Fitzpatrick [98]

31-00100 Embassy of Austria
5 Xiushui Nan Jie, Jianguomenwai, Beijing
☎(+86 106) 532 20 61/62/63 ✉ AUSTROAMB Beijing
▣ 22 258 oebpk cn ⊖ (+86 106) 532 15 05
Ambassador: H.E. Dr Dietrich Bukowski
Deputy Chief of Mission: Dr Josef Müllner
☎(+86 10) 532 20 61 [97]

31-00150 Embassy of Belgium
6 San Li Tun Lu, Beijing
☎(+86 106) 532 17 36/37/38 ✉ AMBABEL Beijing
▣ 22260 ambel cn ⊖ (+86 106) 532 50 97
Ambassador: H.E. Mrs Claire Kirschen
First Secretary & Acting Deputy Head of Mission: Peter Lescouhier
Commercial Attaché Flanders Region: P. Verplancken
Defence Attaché: Capt Daniel de Ruytter
☎(+86 106) 973 64 01
Commercial Attaché Walloon Region: Claude Vincent
Consul & Second Secretary: Guido Grauwels
Second Secretary Public Information Officer: Michel Malherbe [97]

31-00250 Embassy of the Republic of Bulgaria
4 Xiu Shui Bei Jie, Jianguomenwai, 100600 Beijing
☎(+86 106) 65 32 11 79; 65 32 19 46; 65 32 19 16
▣ 210640 tispk cn ⊖ (+86 106) 65 32 45 02
Ambassador: H E Mr Stefan Gaberov
☎(+86 10) 65 32 27 71
Counsellor/Public Information Officer: Panteley Spassov
☎(+86 10) 65 32 19 46
Economic and Commercial Counsellor: Vesselin Rachev
☎(+86 10) 65 32 29 69
Second Secretary: Feodor Pavlov
Attaché: Peter Danailov Komitcki [98]

31-00350 Embassy of Chile
1 Sanlitun Dong Sijie, 100600 Beijing
☎(+86 1) 65 32 16 41; 65 32 15 91;
65 32 20 75; 65 32 15 22 ✉ EMBACHILE Beijing
🄴 echile@public3.bta.net.cn ▣ 22252 en chile
⊖ (+86 1) 65 32 71 30
Ambassador: H E Mr Eduardo Arriagada
Deputy Chief of Mission: Franscisco Ossa
☎(+86 19) 65 32 17 59
Defence Attaché: Col Mario Arce ☎(+86 1) 65 32 33 45
Commercial Officer: Felipe Artal ☎(+86 1) 65 32 12 87
Consulate Officer: José Jaliliye
Cultural Officer: Marco Aguayo ☎(+86 10) 65 32 43 76 [98]

31-00430 Embassy of the Republic of Cuba
1 Xiushui Nan Jie, Jian Guo Men Wai, Beijing
☎(+86 106) 532 28 22; 66 56 ▣ 22249 comek cn
⊖ (+86 106) 532 28 70; 67 03
Ambassador: H.E. Mr José A.Guerra Menchero
Deputy Chief of Mission: Ms Gilda Lopez
Military, Naval & Air Attaché: Col Jesus Morejon Morales
☎(+86 10) 532 28 22
Counsellor: Alberto J. Blanco Silva
Economic Counsellor: Ramon Cortada Jacome
First Secretary: Juan M. Rodriguez
Commercial Counsellor: Raul Leonardo Penalver Martinez
☎(+86 10) 532 12 43 [97]

31-00460 Royal Danish Embassy
1 Dong Wu Jie - San Li Tun, Beijing
☎(+86 1) 532 24 31; 532 24 32; 532 24 37; 532 24 38
🄴 amleadan@public.bta.net.cn
🅆 http://www.gcinfo.com/ambadan.cn
▣ 22255 amdan cn ⊖ (+86 1) 532 24 39
Ambassador: H E Mr Christopher Bo Bramsen
Counsellor: Franz-Michael Mellbin
Counsellor Commercial and Economic Affairs: Steven Foster
Agricultural Counsellor: Steen Thorsted
Attaché: Michael Petersen
Secretary: Morten Jespersen
Commercial Secretary: Torben Iversen
Secretary-Archivist: Ms Lise-Lotte L Nielsen
Secretary-Archivist: Ms Jane G Hansen Nielsen [98]

31-00670 Embassy of Ghana
8 Sanlitun Lu, Sanlitun N, Beijing
☎(+86 106) 65 32 13 19; 65 32 15 44; 65 32 20 12
✉ GHANAEMB Peking ▣ 210462 ghp cn
⊖ (+86 106) 65 32 36 02
Ambassador: H E Mr Kojo Amoo-Gottfried
☎(+86 106) 65 32 20 48
Head of Chancery: Ferdinand O Somuah
☎(+86 106) 65 32 20 48
Counsellor: Joseph Annim ☎(+86 106) 65 32 21 28
Second Secretary: Godwin Kosi Seade
☎(+86 106) 65 32 21 03 [98]

31-00690 Embassy of Greece
19 Guanghua Lu, Jianguomenwai, Beijing
☎(+86 1) 532 19 93 ✉ GREEKEMBA Beijing
▣ 22267 gremb cn ⊖ (+86 1) 532 12 77
Ambassador: H.E. Mr Panayotis Rellas
First Counsellor: Leonidas Chryssanthopoulous
First Secretary: George Veis [98]

31-00810 Embassy of the Republic of Hungary
10 Dongzhimenwai Dajie, Sanlitun, Beijing
☎(+86 1) 65 32 14 31 ⊖ (+86 1) 65 32 50 53
Ambassador: H E Mr Ottó Juhász
Deputy Chief of Mission: Duruasi István
Military and Air Attaché: Col Dr Lászlo Török
Commercial Counsellor: Péter Vértes
☎(+86 1) 65 32 38 45
Second Secretary: Ms Andrea Vajda [98]

31-00960 Embassy of Kenya
4 Xi Liu Jie - San Li Tun, 100600 Beijing
☎(+86 10) 65 32 33 81, 65 32 24 73
🄴 koenyla@iuol.cn.net ▣ 22311 kerep cn
⊖ (+86 10) 65 32 17 70
Ambassador: H E Mr James Simani
☎(+86 10) 65 32 33 25
Counsellor: Hezborn A Lidambiza
☎(+86 10) 65 32 38 22 [98]

31-01010 Embassy of the Lao People's Democratic Republic
11 Sanlitun Dongsi Jie, Sanlitun N, Beijing
☎(+86 1) 65 32 12 24; 65 32 54 45
🄴 laoemcn@mailhost.cinet.co.cn ▣ 22144 alaop cn
⊖ (+86 1) 65 32 67 48
Deputy Military and Air Attaché: Bounthy Vongthachanh [98]

31-01210 Embassy of Mongolia
Shiu Peichienjian, Guomenwai 2, Beijing
☎(+86 10) 532 12 03; 532 19 52
⊖ (+86 10) 532 50 45
Ambassador: H.E. Mr Nyamyn Luvsanchultem [98]

31-01260 Royal Nepalese Embassy
1 Sanlitun Xiliujie, Beijing
☎(+86 10) 65 32 17 95; 65 32 27 39
✉ NEPEMBASSY Peking ⊖ (+86 10) 65 32 32 51
Ambassador: H E Prof Yubaraj Singh Pradhan
Minister-Counsellor: Bhagirath Basnet
Military Attaché: Lieut Col Netra Bahadur Thapa
Second Secretary: Ramesh Prasad Khanal
Third Secretary: Pushpa Raj Ghimire
Attaché: Madan Shrestha
Attaché: Dwarika Das Shrestha [98]

31-01270 Royal Netherlands Embassy

Liangmahe Nanlu 4, 100600 Beijing
☎ (+86 10) 532 11 31 ⊠ 210759 nethl cn
⊜ (+86 10) 532 46 89
Ambassador: H E Mr D J van Houten
Deputy Chief of Mission: Jan Pieter Kleiweg de Zwaan
Agricultural Counsellor: Pieter J Noordman
Counsellor Economic and Commercial: Johannes S Kramer
First Secretary Economic Affairs: Simon H de Muinck
First Secretary Development: E H W Van der Akker
First Secretary Political Affairs: Paul J A M Peters
First Secretary Commercial: K W Ruyg (Hong Kong)
Second Secretary Consular and Administrative: Jos A P M Vroom
Third Secretary Economic Affairs: Wicher J Slagter
Third Secretary Commercial: Guido J L Tielman
Attaché for Administrative & Consular Affairs: P.G. Rövekamp
Agricultural Attaché: A G van Arnhem
Assistant Attaché Agricultural: F A G M Claassen
Budgetary Affairs Attaché: Rene Rotteveel
Secretary-Archivist: Ms Christina Kimmel
Secretary-Archivist: Ms Elisabeth Van Meurs
Secretary-Archivist: Ms Christina Westplate [98]

31-01300 New Zealand Embassy Chancery

N°1 Dong Er Jie - Ri Tan Lu, 100600 Beijing
☎ (+86 10) 65 32 27 31; 65 32 27 32; 65 32 27 33
E nzemb@eastnet.co.cn ⊜ (+86 10) 65 32 43 17
Ambassador: H E Mr Christopher Elder
Minister: Warwick Hawker
Commercial Counsellor: Jonathan Watt
Counsellor Immigration: Wayne Levick
First Secretary: Lester Kok Wai Kong
First Secretary: Mrs Pam Dunn
Second Secretary: Matthew Dalzell
Second Secretary: Ms Jillian Dempster
Attaché: Miss Janet Morris
Attaché: David Nicol
Attaché: Mrs Sarah Doyle [98]

31-01420 Embassy of the Philippines

23 Xiu Shui Bei Jie, Jianguomenwai, 100600 Beijing
☎ (+86 10) 65 32 18 72; 65 32 24 51
E ph@mailost.cinet.co.cn ⊜ (+86 10) 65 32 37 61
Ambassador: H E Mr Romualdo A Ong
Deputy Chief of Mission: Diosdado Orocio (Jnr)
First Secretary: Mrs Farita A Aguilucho-Ong
Third Secretary: Alex G Chua
Third Secretary: Roy B Ercraela
Defence and Armed Forces Attaché: Capt Rafael R Vilches (PN) ☎ (+86 10) 65 32 69 86 [98]

31-01480 Embassy of Romania

Ri Tan Lu Dong Er Jie, Jianguomenwai, Beijing
☎ (+86 106) 532 34 42; 532 38 79 ⊠ 22250 rocom cn
⊜ (+86 106) 532 57 28
Ambassador: H E Mr Virgil Constantinescu
Minister Counsellor: Ms Marian Chirila
Counsellor: Mrs Georgeta Molosaga
☎ (+86 10) 532 33 15
Military, Naval and Air Attaché: Lieut Col Dumitro Stanescu ☎ (+86 10) 532 54 17
First Secretary: Constantin Atudosiei
☎ (+86 10) 532 36 16
First Secretary Economic Commercial: Victor Moldoveanu
☎ (+86 10) 532 33 15
First Secretary Economic Commercial: Emil Timofei
☎ (+86 10) 532 33 15
Second Secretary: Calin Stoica [98]

31-01600 Embassy of Saudi Arabia

8-1-51 Tayuan Diplomatic Appartments, Beijing
☎ (+86 1) 65 32 58 60 ⊜ (+86 1) 65 32 58 59
Military Attaché: Col Mohammed I Al-Khodair [98]

31-01630 Embassy of Sierra Leone

7 Dongzhimenwai Dajie, Sanlitun N, 100600 Beijing
☎ (+86 106) 532 12 22; 21 74 ✉ REPLEONE BEIJING
⊜ (+86 106) 532 37 52
Ambassador: H.E. Dr Mohamed Lamin Kamara
Counsellor & Head of Chancery: Dominic Ibrahim Sayo [97]

31-01650 Embassy of the Slovak Republic

Ri Tan Lu - Jian Guo Men Wai, 100600 Beijing
☎ (+86 106) 532 15 31 ⊜ (+86 106) 532 48 14
Ambassador: H E Mr Vladimir Viliam Klimo
Counsellor: Michel Kottman
Defence Attaché: Maj Gen Rudolf Zidek
☎ (+86 10) 532 41 78
Second Secretary Economic and Commercial: Michal Vrábel
Third Secretary: Igor Niepel
Administration Attaché: Ms Vanda Zamarovà
Attaché: Ladislav Patucz
Attaché: Mrs Darina Patuzova
Contact: Mrs Helena Niepelova
Contact: Pavel Zamora [98]

31-01655 Embassy of the Republic of Slovenia

23 Jian Guo Men Wai Da Jie,
3-53 Jian Guo men Wai Dipl. Res, 100600 Beijing
☎ (+86 1) 532 63 56; 532 63 57 ⊜ (+86 1) 532 63 58
Ambassador: H E Mr Ivan Senicar [98]

31-01700 Embassy of the Republic of the Sudan

1 Sankitun Donger Jie, Sanlitun N, 100600 Beijing
☎ (+86 10) 65 32 37 15; 65 32 22 05
⊠ 22116 sudan cn ⊜ (+86 10) 65 32 12 80
Ambassador: H E Dr Ali Yousif Ahmed
Deputy Chief of Mission: Ahmed M Shiekh Mohammed
Military, Naval and Air Attaché: Brig Gen Harahim Ahmed Taseen
☎ (+86 10) 65 32 21 27; 65 32 15 56 [98]

31-01750 Embassy of the Syrian Arab Republic

6 Sanlitun Dongsi Jie - Sanlitun, 100600 Beijing
☎ (+86 106) 65 32 13 72; 65 32 13 47
⊠ 22138 ostlm cn ⊜ (+86 106) 65 32 15 75
Ambassador: H E Dr Loutof Haydar
☎ (+86 10) 65 32 15 63
Deputy Chief of Mission: Suliman Aad
☎ (+86 10) 65 32 14 29
Counsellor and Commercial Officer: Mohamad Bassam Imadi
Attaché: Ismail Souweidan
Attaché: Youssef Mehrez [98]

31-01840 Embassy of the Republic of Turkey

Dong Wu Jie 9, Sanlitun, 100600 Beijing
☎ (+86 106) 65 32 23 47; 26 50 ⊠ 210690 tupek cn
⊜ (+86 106) 65 32 54 80
Ambassador: H.E. Mr Berhan Ekinci
☎ (+86 10) 65 32 26 50
Counsellor: Mehmet Bilir ☎ (+86 10) 65 32 27 12
Military, Naval & Air Attaché: Lt Col Fahir Altan
☎ (+86 10) 65 32 12 58
Economic & Commercial Counsellor: Hüseyin Dönmez
☎ (+86 10) 65 32 38 46 ☎ (+86 10) 65 32 32 68 (fax)
Third Secretary: Ms Deniz Eke ☎ (+86 10) 65 32 23 47
Political Attaché: Ms Ozlem Demir
☎ (+86 10) 65 32 23 47
Admin Attaché: Ayse Kilic [97]

31-01890 Embassy of the United Kingdom

11 Guang Hua Lu, Jian Guo Men Wai, 100 600 Peking
☎ (+86 10) 65 32 19 61; 65 32 19 62;
65 32 19 63; 65 32 19 64; 65 32 68 95;
65 32 68 96; 65 32 68 97; 65 32 68 98
⊜ (+86 10) 65 32 19 37; 65 32 19 38;
65 32 19 38; 65 32 19 39; 65 32 19 30 (consular)
Ambassador: H E Mr Tony Galsworthy
Minister, Consul-General and Deputy Head of Mission: A D Sprake
Counsellor, Political/Economic: S M Featherstone
Counsellor, Commercial: C M J Segar
Counsellor, Cultural (British Council Director): M D Davidson
Counsellor, Public Affairs: K I Malin
Defence, Military and Air Attaché: Brig K O Winfield
Naval Attaché: Capt B N Rayner (RN)

Cultural and Education Section
4th Floor Landmark Building, 8 North Dongsanhuan Road, 100026 Chaoyong District, Peking
☎ (+86 10) 65 01 19 03 ⊜ (+86 10) 65 01 19 77

British Consulate-General
Suite 301 - Shanghai Centre 1376,
200040 Nanjing Xi Lu, Shanghai
☎ (+86 21) 62 79 76 50; 64 71 48 49 (British Council)
⊜ (+86 21) 62 79 76 51 (consulate);
64 33 31 15 (British Council)
Consul-General: W D Townend
Deputy Consul-General and Consul (Commercial): S J Smith
Consul (Cultural and Educational) British Council: Ms B Wickham
Consul (Commercial and Economic): P R Davies
Consul (Commercial): I Richards
Vice-Consul: Miss S Perry

British Consulate-General
Suite 5303-7, Guandong International Hotel,
510098 Guangzhou
☎ (+86 20) 83 31 18 88 ext. 5305; 5306; 5307
⊜ (+86 20) 83 33 64 85
Consul-General: I B Wellfare
Deputy Consul-General: A W J Prime
Vice Consul: G K Hoar [98]

31-01950 Embassy of Venezuela

14 Sanlitun Lu, Sanlitun N, Beijing
☎ (+86 10) 532 12 95; 532 26 94;
65 32 12 95; 65 32 26 94; 65 32 36 54
✉ EMBAVENEZ Beijing E emvenez@mailhost.cinet.co.cn
⊠ 22137 emven cn ⊜ (+86 10) 532 38 17
Ambassador: H E Mr Maglio E Montiel
First Secretary, Commercial Officer: Ms Ana F Cazzadore
Third Secretary: Pedro M Madrid [98]

31-01990 Embassy of the Republic of Yemen

5 Dong San Jie - San Li Tun, Chao Yang District,
100600 Beijing
☎ (+86 10) 65 32 15 58; 65 32 16 88; 65 32 33 46
⊠ 210297 ⊜ (+86 10) 65 32 43 05
Ambassador: H E Mr Mohamed Hadi Awad
Minister: Kasem A Abdullah Radhwan
Counsellor: Nagib Mohamed A Al-Gobani
First Secretary: Mohamed O D Al-Mikhlafi
First Secretary: Gamal Saeed Shamsan Saeed
Second Secretary: Hamood M A Al-Ghazali
Attaché: Kamal O Hetari [98]

31-02010 Embassy of the Federal Republic of Yugoslavia

1 Dong Liu Jie - San Li Tun, Beijing
☎ (+86 10) 65 32 35 16; 65 32 16 93;
65 32 30 16; 65 32 54 13 E yuamba@public.bta.net.cn
⊠ 22403 ambyu cn ⊜ (+86 10) 65 32 12 07
Ambassador: H E Dr Slobodan Unkovic
Minister-Counsellor: Dusan Gajic
Defence Attaché: Capt Dr Slobodan Tejic (Navy)
Counsellor, Economic: Srdjan Simic
Counsellor, Scientific and Technological: Dragoslav Avramovic
Counsellor: Dr Vidan Markovic
Third Secretary, Consular Affairs: Ivan Kandijas
Third Secretary, Economic: Goran Gvozdeovic
Third Secretary, Cultural/Press Affairs: Dejan Hinie
Attaché: Dragan Madlo [98]

31-02020 Embassy of Zaire

6 Dong Wu Jie, San Li Tun, 100600 Beijing
☎ (+86 106) 532 13 60 ✉ AMBAZAIRE BEIJING
⊠ 22273 ambaz cn ⊜ (+86 106) 532 13 60
Ambassador: H E Mr Lombo Lo Mangamanga
Minister Counsellor: Bukasa Mulumba
Second Counsellor: Mabela Lukoki Mayala
Second Counsellor Economic Affairs: Alluma Idolua Mulamba
First Secretary: Nzapalaye Ndagbia [98]

55 Government-owned Industry & Production Facilities

55-00100 China Great Wall Industry Corp. (CGWIC)

No. 22 Fu Cheng Rd, HaiDianQu, 100036 Beijing
☎ (+86 106) 842 91 13; 837 02 85; 06 25
✉ 22484 CPMC CN ⊠ 222777 httx cn
⊜ (+86 106) 842 92 17; 92 17
Chairman: Guo Guozheng
President: Zhang Tong
Vice President: Yu Fu Sheng
Vice President: Xu Yinfu
Vice President: Liu Zhi Xiong
Public Relations Mgr: Zhang Xiaolin

■ Deals mainly with international co-operation on commercial satellites, carrier rockets & accepting orders from foreign customers for satellite launching services. Handles the import & export of space technology & equipment, precision machinery, electronic products, instruments & meters, special vehicles, medical apparatus, broadcast & communication equipment, general machinery & tools, magnetic materials, appliances, measuring tools & chemicals. → 75 [97]

55-00200 China National Aero-Technology Import & Export Corp. (CATIC)

CATIC Plaza, 18 Beichen East Rd, Chaoyang District, 100101 Beijing
☎ (+86 106) 494 22 55 ✉ CAID Beijing
⊠ 22318 aerot cn ⊖ (+86 106) 494 10 88; 06 58
Chairman: Liu Guomin
President: Yang Chunshu
Vice President: Li Zhoushu
VP Export: Wang Dawei
VP International Co-operation: Yan Tiannan
VP Export: Ding Shiqing
Chief Financial Officer: Shan Zumao
Chief Economist: Zhang Siyuan
■ Authorized agent for import/export business of Chinese aviation industry. Products include high performance fighters, attack aircraft, bombers, primary & advanced trainers, transport planes, helicopters & aircraft engines, missiles & aircraft components.

Catic Guangzhou Corporation
39-1 Zhu Si Gang, 2 St, Dongshan, 510080 Guangzhou
☎ (+86 20) 775 49 73 ⊠ 44434 catic cn
⊖ (+86 20) 776 56 19

Catic Shenzhen Company
Catic Zone, Shennan Central, 518041 Shenzhen
☎ (+86 755) 336 36 98; 336 64 27; 336 64 25
⊠ 420200 asitc cn ⊖ (+86 755) 336 67 32; 332 47 36

Catic Shanghai Company
8th Fl, 583 Lingling Rd, 200030 Shanghai
☎ (+86 21) 439 04 36; 439 28 00 ⊠ 33390 aeros cn
⊖ (+86 21) 439 01 66; 439 03 54
Mgr Business Dept: Cai Hong Kui

Catic Fujian Company
Zhong Hang Xincun, Yingji, Gulou, 350003 Fuzhou
☎ (+86 591) 781 21 70 ⊠ 92178 catic cn
⊖ (+86 591) 783 75 11

Catic Xiamen Company
4 Yuehua Road, Huli, 361006 Xiamen
☎ (+86 592) 62 13 10; 62 13 19 ⊠ 93080 axitc cn
⊖ (+86 592) 62 15 74; 23 54 39

Catic Zhuhai Company
Catic Building, Jiuzhou Boul., Ji Da Qu, 519015 Zhuhai
☎ (+86 756) 33 28 11; 33 47 11 ✉ 5300
⊠ 456234 catic cn ⊖ (+86 756) 33 27 85

Catic Harbin Company
4 Hongjiun St, 150001 Nangangqu, Harbin
☎ (+86 451) 362 71 31 ⊠ 87328 hchtc cn
⊖ (+86 451) 362 16 60

Catic Hangzhou Company
93 Tiyuchang Rd, 310003 Hangzhou
☎ (+86 571) 57 17 49; 57 22 12 ⊖ (+86 571) 57 20 53

Catic Beijing Company
10th Floor, Huibin Office Building, No.8 Beichendonglu, 100101 Chaoyangqu, Beijing
☎ (+86 10) 491 11 45; 491 06 12 ⊠ 210705 catic cn
⊖ (+86 10) 499 39 91

Catic Dalian Company
3 Xian St, Xi Gang Qu, 116011 Dailan
☎ (+86 411) 363 81 78 ⊠ 86141 booth cn
⊖ (+86 411) 363 99 06 [97]

55-00265 China National Precision Machinery Import & Export Corp. (CNPMIEC)

(† China Precision Machinery Import & Export Corp.)
22 Fu Cheng Lu, PO Box 129, 100036 Beijing
☎ (+86 106) 837 16 50; 842 91 26 ⊠ 22484 cpmc cn
⊖ (+86 106) 842 91 17
♣ US$ 275 700 000 👥 400
President: Jin Xueknan
General Mgr Marketing Dept: Zhang Zhimin
General Mgr Marketing Dept: Luo Zhinging
General Mgr Commercial Dept: Luo Zhiying
■ Import & export of tactical defensive missiles & related equipment; space equipment; precision machinery; electronic products; optical instruments. CPMIEC also deals with joint ventures, compensation trade, contracts for the design & construction of industrial & civil projects. [97]

55-00280 China North Industries Corp. (NORINCO)

(a member of The Xinshidai Group)
12 A Guang An Men Nan Jie, 100053 Beijing
☎ (+86 106) 347 33 22 ✉ NORINCO Beijing
⊠ 22339 cnic cn ⊖ (+86 106) 340 11 16
👥 1,000,000
Chairman: Zhang Junjiu
Vice Chairman: Tian Ruizhang
President: Wang Liansheng
Vice President: Li De
Vice President: Wang Donquing
Vice President: Qin Jianqing
Director: Li Zhanmin
Director: Lu Zhongyin
Director: Sun Chengtao
Director: Gu Yongchun
Director: Li Changbin
Director: Zhao Shaoliang
Senior Expert: Zhao Jiazheng
Senior Expert: Huang Ruzhen
Senior Business Project Manager: Shi Shangyu
■ Manufacture and development of armoured vehicles, howitzers, mortars, rocket launchers, anti-aircraft weapons, anti-tank missile systems, small arms, ammunition, fire control systems, sighting and aiming devices, high-powered engines and NBC protection systems. → 75 [97]

55-00360 China Shipbuilding Trading Co. Ltd (CSTC)

(a subsidiary of China State Shipbuilding Corp)
10 Yuetan Beixiaojie, Beijing
☎ (+86 10) 68 58 25 60; 62 ✉ CSTC BEIJING
⊠ 22335 cssc cn ⊖ (+86 10) 68 58 33 80; 24 20
☼ Yuan 200 000 000 ♣ US$ 400 000 000 (1995) 👥 160
Managing Dir: Li Jian
Sales & Marketing Dir: Chen Shu Sheng
Operations & Technical Dir: Zhang Xiao Shan
Purchasing Mgr: Zhu Hua Xin
■ Export of ships & marine equipment, import of materials & expertise, & undertaking of construction of shipbuilding facilities & other buildings. Also production of pontoon river bridges. [97]

55-00800 Guizhou Aviation Industry Corp. (GAIC)

PO Box 38, 561000 Anshun, Guizhou
☎ (+86 853) 55 10 27 ⊠ 66018 aimga cn
⊖ (+86 853) 55 12 36
General Mgr: Sun Ruisheng
■ Design & manufacture of military trainer aircraft, turbojet engines, air-to-air missiles & rocket launchers. → 75 [97]

55-00850 Harbin Dongan Engine Company

PO Box 51, 150055 Harbin, Heilongjiang
☎ (+86 451) 860 21 20 ⊠ 87131 hdec cn
⊖ (+86 451) 860 22 66
■ Design & manufacture of amphibious aircraft, general purpose transport aircraft & helicopters. → 75 [97]

55-00900 Nanchang Aircraft Company

PO Box 5001, 330024 Nanchang, Jiangxi
☎ (+86 791) 45 18 33 ⊠ 95068 namc cn
⊖ (+86 791) 45 14 91
President: Li Wan Xin
Deputy President: Ho Fukang
Dir Sales Division: Nang Wang Yanxiang
■ Design & manufacture of military combat aircraft. → 75 [97]

55-01900 Shaanxi Aircraft Manufacturing Co.

PO Box 35, 723215 Chenggu, Shaanxi Province
☎ (+86 916) 21 49 74 ⊠ 70141 sac cn
⊖ (+86 916) 20 20 31
President: Wang Wengfeng
Marketing Dept Mgr: Li Yousheng
■ Manufacture of military/civil transport aircraft. Non-aerospace products include coaches & small trucks. → 75 [97]

55-01950 Shenyang Aircraft Corp. (SAC)

PO Box 328, 110034 Shenyang, Liaoning
☎ (+86 24) 684 26 80 ⊠ 80018 samc cn
⊖ (+86 24) 686 26 89
■ Design & manufacture of military combat aircraft. → 75 [97]

55-02400 Xian Aircraft Company (XAC)

PO Box 140 Extension 84, 710089 Shaanxi, Xian
☎ (+86 29) 684 50 65; 684 60 65
⊖ (+86 29) 620 37 07
👥 20,000
General Manager: Gao Da Cheng
Manager of Marketing Department: Wang Zhi Gang
■ Design and manufacture of military bomber, fighter and transport aircraft (Y7 serial). → 75 [98]

55-02500 Yan Yun Company

4th Fl, 9 Anwaixiaoguandongjie, Chaoyang District, 100029 Beijing
☎ (+86 106) 491 63 19 ⊖ (+86 106) 491 63 19 [97]

65 Industry, Trade & Armed Forces Associations

65-00050 Chinese Society of Aeronautics and Astronautics

N°5 Liangguocchang Road, 100010 Dongchseng District, Beijing
☎ (+86 10) 64 02 14 16 ⊖ (+86 10) 64 02 14 13
👥 40 (1997)
President: Zhu Yu-Li
Director International Affairs Department: Wang Xiaozhou [98]

75 Manufacturing Companies

China Great Wall Industry Corp.	55-00100
Guizhou Aviation Industry Corp.	55-00800
Harbin Dongan Engine Company	55-00850
Nanchang Aircraft Company	55-00900
NORINCO	55-00280
Shaanxi Aircraft Manufacturing Co.	55-01900
Shenyang Aircraft Corp.	55-01950
Xian Aircraft Company	55-02400

75-00200 Beijing Lufthansa Center Co Ltd (BLC)

(a subsidiary of Daewoo Corporation, South Korea)
50 Liangmaqiao Road, 100016 Chaoyang District, Beijing
☎ (+86 10) 64 65 10 15; 64 65 10 18
⊖ (+86 10) 64 67 83 32 [98]

75-00320 Daewoo Corp Beijing Office

C-#712 Beijing Lufthansa Center, 50 Liangmaqiao Road, 100016 Chaoyang District, Beijing
☎ (+86 10) 64 65 19 48; 64 65 19 52
⊖ (+86 10) 64 65 19 83 [98]

75-00325 Daewoo Corp Construction Division Shanghai Office

Room 313/308 Haihua Garden N°8, Dapu Road, 200025 Shanghai
☎ (+86 21) 64 66 50 72 ⊖ (+86 21) 64 66 50 72 [98]

75-00345 Daeyan Center Co Ltd

(a subsidiary of Daewoo Corp, South Korea)
272 Juzi Street, Yanji, Jilin
☎ (+86 433) 265 77 00
⊖ (+86 433) 265 77 05; 265 88 84 [98]

75-00350 Dalian Marine Diesel Works (DMD)

N°1 Hai Fang Street, PO Box 116021, Dalian
☎ (+86 411) 441 72 73 ⊠ 86262 dmdcn
⊖ (+86 411) 441 74 99
☼ Rmb 260,000,000 (1995)
♣ Rmb 250,000,000 (1995) 👥 1,100 (1995)
🏠 82,000 m2
President: Zou Zhi-Ming ☎ (+86 411) 441 94 30

Technical Director: Lu Shi-Chang ☎(+86 411) 441 77 24
Sales and Marketing Director: Lu Chao-dong ☎(+86 411) 441 77 24
Purchasing Manager: Li Ru-ruei ☎(+86 411) 441 15 58
Head of General Office: Wang Gang-gu ☎(+86 411) 441 72 73
■ Manufacture of marine diesel engines. 98

75-00400 Guangzhou Shipyard International Co Ltd (GSI)

2 Fang Cun Road, Guangzhou
☎(+86 20) 81 89 17 12 ⊠ 44418 gzsy cn
⊖ (+86 20) 81 89 63 41; 81 89 15 75
♣ Rmb 2,061,349,000 (1996) Ⅿ 6,072 (1996)
General Manager: Hu Guo Liang
Sales and Marketing Director, Vice General Manager: Li Guanghui
Manager Business Department: Wu De Bao
Purchasing Manager: Peng Sheng
■ Shipbuilding, ship-repairing and ship-breaking services. Container manufacturing, mechanical engineering, specialised vessels and shipping. International standard/non-standard specialised steel dry cargo containers. Steel structures for bridges and high rise construction. Pressure vessels for chemical engineering, fabric reinforced plastic life boats and high speed boats, foils and studs for hydrofoil, terrestrial machinery including port and hydraulic machines, cranes. 98

75-00450 Guilin Daewoo Hotel Co Ltd

(a subsidiary of Daewoo Corp, South Korea)
Bin Jiang Nan Road, Guilin Guangxi
☎(+86 773) 282 55 88 ⊖ (+86 773) 283 55 98 98

75-00890 Hudong Shipyard

Dingningsi Pudong, Shanghai
☎(+86 21) 58 71 32 22 ⊠ 33025 shdsy cn
⊖ (+86 21) 58 71 26 03
♣ Rmb 277 000 000 ⌂ 870 000 sq m
Managing Dir: Zhou Zhenbo
Operations/Technical Dir: Wang Hengyuan
Sales & Marketing Dir: Hu Minghe
Purchasing Mgr: Gu Miaoyan
■ Shipbuilding & manufacture of marine diesel engines (licence from SEMT, France & B&W, Denmark) 97

75-01450 Number 605 Research Institute of AVIC

Special Surface Warfare Vehicle Group
(† Huahang Airship Development Group)
(a subsidiary of Aviation Industries of China)
PO Box 307, 448035 Jingmen, Hubei
☎(+86 724) 233 25 21; 233 25 25
⊖ (+86 724) 233 25 51
Chairman: Wang Fu Ling ☎(+86 724) 234 50 92
Operations Director: Song Min Do
Sales and Marketing Director: Xiu Feng Tian ☎(+86 724) 233 25 21 ext. 2213
Purchasing Manager: Chen Kun
Information Officer: Dong Xue Ping ☎(+86 724) 233 25 21 ext. 2261
■ Design and manufacture of rigid and non-rigid airships, special shaped balloons and tethered aerostats. 98

75-01910 Shanghai Daewoo Center Development Co Ltd

(a subsidiary of Daewoo Corporation, South Korea)
Room 1303-06 Rui-jin B/D 205, Mao-Ming Road (South), Shanghai
☎(+86 21) 64 12 45 61 ⊖ (+86 21) 64 72 04 37 98

75-01950 Tangram Electronic Engineering Co Ltd

Rm 105-107, Hua Bai Building,
No A-1, Bei Tai Ping Zhuang Road,
100088 Haidan District, Beijing
☎(+86 10) 62 06 11 01 ⊖ (+86 10) 62 06 11 00
Sales Manager: R Shen Wei
Representing:
EEV Ltd, UK 98

75-02000 Wuhan Kawasaki Marine Machinery Co Ltd

(joint venture with Kawasaki Heavy Industries Ltd, Japan)
N°9 Wudong Street Qingshan, Wuhan
☎(+86 27) 659 06 26 ⊖ (+86 27) 659 06 27 98

80 Service Companies & Organizations

80-00050 AlliedSignal China Inc. Marketing Sales and Service

(a subsidiary of AlliedSignal Aerospace, USA)
23-03 Landmark Building,
8 North Dongsanhuan Road, Chaoyang, 100004 Beijing
☎(+86 10) 65 01 21 47; 65 01 21 48; 65 06 62 42; 65 06 62 43; 65 06 80 17
⊖ (+86 10) 65 04 53 61; 65 04 53 62
Vice President: Francis Yuen ☎(+86 10) 65 06 80 16
Account Team Leader: Ms Anne Tan
Field Service Manager: Soon Fai Wong
Field Service Engineer: Michael Haray
Field Service Engineer: Ye Wang
Field Service Engineer: Walter Wu
■ Representation for AlliedSignal Aerospace products in China. → 90

● Wanguan Hotel, Shuangliu Airport,
Chengdu, Sichuan Province
☎(+86 28) 554 88 89 ext. 4293 ⊖ (+86 28) 556 53 70
Field Service Engineer: Jason (Guang Chun) Liu

Guangzhou Representative Office
BaiYun International Airport Hotel, Room 1575,
510405 Guangzhou Province
☎(+86 20) 86 65 90 25 ⊠ K82880
⊖ (+86 20) 86 65 90 29
Field Service Engineer: Robert (Jian Cong) Zheng

Haikou International Commercial Centre
Room 701, Da Tong Road, Haikou, Hainan Province
☎(+86 898) 679 69 99 (room 701)
⊖ (+86 898) 677 47 51
Field Service Engineer: Kam Chuen Chan

● No 8, Niu Dun Lu, Zhang Jiang,
Hi-Tech Park, Pudong New Area, 201203 Shanghai
☎(+86 21) 58 55 40 20 ⊖ (+86 21) 58 55 40 30
Account Team Leader: Jeff (Zhenning) Song

● 200335 Shanghai
c/o China Eastern Airlines, Hongquiao
International Airport, Office Building No 4, Room 317
☎(+86 21) 62 68 77 88 ext. 7955
⊖ (+86 21) 62 19 38 47
Field Service Engineer: Hua Hee Tan ☎(+86 21) 62 68 03 82

● c/o Sheraton Xian Hotel,
12 Feng Gao Road, Rooms 1330, 1332,
200335 Xian, Shanxi Province
☎(+86 29) 728 33 50; 426 18 88 (room 1332)
⊖ (+86 29) 728 33 50; 426 21 88 (room 1332)
Field Service Engineer: Michael Haray 98

80-00060 AlliedSignal Commercial Avionics Systems

(a subsidiary of AlliedSignal Aerospace, USA)
Unit 23-03 Landmark Building,
8 North Dongsanhuan Road, Chaoyang, 100004 Beijing
☎(+86 10) 65 06 62 42; 65 01 62 43; 65 01 66 88 ext. 22303 ⊠ K80882
⊖ (+86 10) 65 04 53 61; 65 04 53 62
Sales Representative: William Shao ☎(+86 10) 65 01 21 47
■ Representation of AliiedSignal Aerospace products in China. 98

80-00370 China Eastern AlliedSignal Aircraft Wheels and Brakes

(a subsidiary of AlliedSignal Aerospace, USA)
Long Hua Airport, Building 24, 200232 Shanghai
☎(+86 21) 64 57 67 86 ⊖ (+86 21) 64 57 56 29
General Manager, CEA JV: Edward (Ted) Leahy ☎(+86 21) 58 55 40 20; 58 555 40 30 (fax) 98

80-00400 Dantsin Consulting SA

N°2 Cui Wei Road - 4th Floor, Yinshua Kemao Building,
100036 Beijing
☎(+86 106) 68 21 99 66 ⊖ (+86 106) 68 15 28 77
Ⅿ 6 (1996) ⌂ 150 sq m
Managing Director: Daniel Brandenburger
Marketing Director: Feng Zhinong
Technical Director: Gu Ming
® Switzerland: Trimos-Sylvac Task Force 98

80-00600 Hughes International

Room 502 Int'l Club Office Building,
21 Jianguomenwai Dajie, 5th Floor, 100020 Beijing
Room 502 International Club Office
Building, 21 Jianguomenwai Dajie, 5th Floor
☎(+86 106) 532 16 31 ⊖ (+86 106) 532 18 98 98

80-00800 KHI (Dalian) Computer Technology Co Ltd

(a subsidiary of Kawasaki Heavy Industries Ltd, Japan)
Building Foreign Specialist Dalian,
Maritime University - Ling Shui Qiao, Dalian
☎(+86 411) 467 25 07 ⊖ (+86 411) 467 24 59 98

85 Stockists, Distributors, Spare Parts Suppliers

85-05000 Satair China Representative Office

(a subsidiary of Satair AS, Denmark)
Room No BO712, 7th Floor, Hui Bin,
Office Building, No 8 Beichendong St,
100101 Chaoyang District, Beijing
☎(+86 10) 64 99 35 37 ⊖ (+86 10) 64 99 35 35
■ Offers stocks of factory new parts for military & commercial aircraft. 98

90 Sales Agents, Sales & Liaison Offices

Companies represented:

Astrophysics, USA	90-00530
Berthold, Germany	90-00530
Chandler, USA	90-00530
EG&G Inc., USA	90-00530
Flow, USA	90-00530
Gamma Scientific, USA	90-00530
Lakeshore, USA	90-00530
Marine Instruments, USA	90-00530
Moisture & Humidity, USA	90-00530
Opteolectronics Group, USA	90-00530
Ortec, USA	90-00530
Par, USA	90-00530
Sema Group plc, UK	90-18000
SEXTANT Avionique SA, France	90-18200
Trimos-Sylvac Task Force, Switzerland	80-00400
Vaisala Oy, Finland	90-21500
Zengrange Ltd, UK→ Netherlands	80-18300

Sales agents/representatives listed elsewhere:

AlliedSignal China Inc.	80-00050

90-00400 CompStream Corporation

Consultec Building, Ste 3113, Yi12 Guang Hua Road,
100020 Beijing
☎(+86 106) 505 13 84 ⊖ (+86 106) 505 13 80 97

90-00401 Daewoo Corp Beijing Representative Office

C-701 Beijing Lufthansa Centre, 50 Liangmaqiao Road,
100016 Chaoyang District, Beijing
☎(+86 10) 465 22 26 ⊠ 21052 dwbjo cn
⊖ (+86 10) 465 15 31 98

90-00403 Daewoo Corp Dalian Representative Office

Dalian International Hotel #911-914, N°9 Renmin Road,
Dalian
☎(+86 411) 263 82 38 ⊠ 860557 dwdlc cn
⊖ (+86 411) 280 72 72 98

Contact Reference

China, People's Republic

90-00405 Daewoo Corp Guangzhou Branch
Room 1301 Gitic Plaza Main Building,
339 Huanshi Dong Road, Guangzhou
☎ (+86 20) 83 31 13 70 ⊠ 440850 dwgz cn
⊖ (+86 20) 83 31 27 89 [98]

90-00406 Daewoo Corp Qingdao Office
Ba-Da-Guan Hotel, 15 Shaoguan Road, Qingdao
☎ (+86 532) 386 16 88 ⊠ 321256 dwtao cn
⊖ (+86 532) 387 32 78 [98]

90-00408 Daewoo Corp Wuhan Branch
707 Wuhan International Trade Center,
Commerce Centre, N°297 Xinhua Road, Wuhan
☎ (+86 27) 577 09 93; 577 09 94
⊖ (+86 27) 577 09 92 [98]

90-00412 Daewoo Corp Shenyang Branch
2107 Shenyang Sankei High Tech Tower,
N°262a Shifu Avenue, Shenhe District, Shenyang
☎ (+86 24) 279 00 98 ⊠ 804243 dwshy cn
⊖ (+86 24) 279 00 99 [98]

90-00413 Daewoo Shanghai Office
Room 1303-1307 1318,
Ruijin B/D 205 Mao Ming Road South, Shanghai
☎ (+86 21) 64 72 42 80 ⊖ (+86 21) 64 72 04 37 [98]

90-00414 Daewoo Group Shantou Representative
B1/15th Floor Evergreen Building, East Jinsha Road,
Shantou
☎ (+86 754) 816 34 42 ⊠ 454182 dwsnt cn
⊖ (+86 754) 816 34 45 [98]

90-00420 Daicel Chemical Industries Ltd Beijing Office
910 Beijing Fortune Building 5, Dong San Huan Bei Lu,
Chao Yang District, Beijing
☎ (+86 10) 501 05 26 ⊖ (+86 10) 501 05 28
Shanghai Office
c/o Sheraton Hotel Shanghai, 1200 Cao Xi Bei Lu,
Shanghai
☎ (+86 21) 439 98 12 ⊖ (+86 21) 481 01 34 [97]

90-00530 EG&G
China Representative Office
Ste 408, Scite Tower, No.22 Jianguomen Wai Da Jie,
Beijing
☎ (+86 10) 512 40 79; 512 22 88
⊖ (+86 10) 512 40 84
Manager China Operations: John Lewis
® Germany: Berthold
USA: Astrophysics, Chandler, EG&G Inc., Flow, Gamma Scientific, Lakeshore, Marine Instruments, Moisture & Humidity, Opteolectronics Group, Ortec, Par [98]

90-00900 International Industrial Co. Ltd
(a rep. of Recon/Optical Inc.)
Nanking East Road, 4th Section, 10th Floor No. 25,
Taipei, Taiwan
☎ (+886 2) 718 26 18 ✉ INTINCO ⊠ 11116
⊖ (+886 2) 715 43 32
♲ US$ 2 000 000 ♣ US$ 18 000 000 👥 10 ⌂ 600 sq m
CEO: Rudy Chu [97]

90-00950 Kawasaki Heavy Industries Ltd Beijing Office
Room N°2602 China World Tower, China,
World Trade Center, N°1 Jian Guo Men Wai, Avenue,
100004 Beijing
☎ (+86 10) 65 05 13 50 ⊖ (+86 10) 65 05 13 51 [98]

90-00960 Kawasaki Heavy Industries Ltd Shanghai Office (KHI)
1307 Shanghai Int'l Trade Centre, 2200 Yan An Xi Road,
Shanghai
☎ (+86 21) 62 70 19 68 ⊖ (+86 21) 62 70 19 69 [98]

90-01000 Kyokuto Boeki Kaisha Ltd Beijing Office (KBK)
Room 607 Scite Tower, Jianguo Men Wai Da Jie, Beijing
☎ (+86 10) 512 34 66 ⊖ (+86 10) 512 47 69 [98]

90-01005 Kyokuto Boeki Kaisha Ltd Shanghai Office (KBK)
Room 2011 Ruijin Building, 205 Mao Ming Road South,
Xu Hui District, Shanghai
☎ (+86 21) 64 72 08 72 ⊖ (+86 21) 64 72 56 29 [98]

90-05000 Minebea Co. Ltd Shanghai Office
Room 2311 - Rui Jin Building,
N° 205 Mao Min Road South, 200020 Shanghai
☎ (+86 21) 472 33 09 ⊖ (+86 21) 472 94 63
👥 9
Office Manager: M Sasaki
Electronics and Hi-Tech Components
Xichen, Qingpu, Shanghai
☎ (+86 21) 926 76 80 ⊖ (+86 21) 926 76 97
President: H Kobayashi [98]

90-18000 Sema Group
(a division of Sema Group)
N°107 Yan Dang Road - Room 7F, 200020 Shanghai
☎ (+86 21) 63 72 57 51 ⊖ (+86 21) 63 72 46 46
Managing Director: Bernard Gagou
Permanent Representative: Dominique Chevet
® UK: Sema Group plc [98]

90-18200 SEXTANT Avionique
(a subsidiary of SEXTANT Avionique SA, France)
Tr A Beijing Kelun Building,
Room 1006 - 12a Guanghua Lu, 100020 Beijing
☎ (+86 10) 65 93 29 61; 65 93 29 62
⊖ (+86 10) 65 93 29 63
General Representative: Marc de Montessus
® France: SEXTANT Avionique SA [98]

90-19000 UBE Industries Ltd Beijing Representative Office
Room 25-10, China World Trade Center, 100004 Beijing
☎ (+86 10) 505 13 27; 505 13 28 ⊠ 22515
⊖ (+86 10) 505 13 29 [98]

90-21500 Vaisala Beijing Representative Office
Wangfujing Grand Hotel, Room 518-520,
57 Wangfujing Street, 100006 Beijing
☎ (+86 10) 65 22 40 50 ⊖ (+86 10) 65 22 40 51
Director: Knut Saxen
® Finland: Vaisala Oy [98]

Colombia

Head of State:
Ernesto Samper Pizano

10 Head of State

10-10000 Office of the President
Palacio de Nariño, Carrera 8, 7-26, Bogotá, DE
☎ (+57 1) 284 33 00; 336 44 66 ⊠ 396 44984
⊖ (+57 1) 266 74 34
President of the Republic: H E Ernesto Samper Pizano
Vice President: H E Carlos Lemos Simmonds [98]

12 Ministries

12-10000 Ministry of National Defence
Ministerio de Defensa Nacional
Avenida El Dorado, CAN Carrera 52,
Santa Fé de Bogotá, DE
☎ (+57 1) 222 04 99; 266 93 00 ⊖ (+57 1) 222 18 74
Minister of National Defence: Gilberto Echeverri Mejia
Minister of the Navy: Adm Mauro Pereira
Secretary General: Maj Gen Luis A Rodriguez
Commander of the Armed Forces: Gen Harold Bedoya
Army Commander: Gen Manuel José Bonett Lacarno
Navy Commander: Adm Edgar Romero [98]

12-11000 Ministerio de Comercio Exterior
Bogotá, DE
Calle 28 13A-15 Pisos 5,6,7 and 9
Edificio Centro de Comercio Internacional
⊖ (+57 1) 336 24 93
Minister of Economic Development:
Orlando Cabrales Martinez
Minister of Foreign Trade: Carlos Ronderos Torres [98]

12-15000 Ministry of Communications
Edificio Murillo Toro, Carretera 7 and 8, Calle 12A and 13,
Bogotá, DE
☎ (+57 1) 286 69 11 ⊖ (+57 1) 341 57 03; 286 11 85
Minister of Communications: Saulo Arboleda Gomez [98]

12-16000 Ministerio de Desarrollo Económico
Cra 13 28-01 Ps 5,6,7 & 8, Bogotá, DE
☎ (+57 1) 320 00 77 ⊖ (+57 1) 287 45 83
Minister: Rodrigo Marín Bernal
Contact: Ms Ma. del Pilar Cortaza [97]

12-20000 Ministry of Foreign Affairs
Ministerio de Relaciones Exteriores
Palacio de San Carlos, Calle 10 N° 5-51,
Santa Fé de Bogotá, DE
☎ (+57 1) 282 78 11; 287 68 00 ⊠ 44394
⊖ (+57 1) 284 20 99
Minister of Foreign Affairs: Ms Maria Emma Mejia [98]

12-40000 Ministry of Interior
Ministerio de Gobierno
Calle 13, No 8-38, Santa Fé de Bogotá, DE
☎ (+57 1) 284 02 14 ⊠ 45406 mingo co
⊖ (+57 1) 281 58 84
Minister: Hon Carlos Holmes Trujillo [98]

12-50000 Ministerio de Justicia y del Derecho
Av Jiménez 8-89, Bogotá, DE
☎ (+57 1) 286 02 11 ⊖ (+57 1) 281 55 79; 281 63 84
Contact: Clara Rodríguez [97]

15 Parliament

15-10000 Senate (Senado de la Republica)
Calle 10 Carreras 7 y 8a, Capitolio Nacional,
Santa Fé de Bogotá, DC
☎ (+57 1) 283 84 11 [98]

15-20000 House of Representatives (Camara de Representantes)
Calle 10 Carreras 7 y 8a, Capitolio Nacional, Bogotá, DE
☎ (+57 1) 243 05 06
President: Alvaro Benedetti Vargas [98]

25 Defense Forces

25-29100 Military Air School
Escuela Militar de Aviación
Base Aérea Marco Fidel Suárez, PO Box 1233, Santiago de Cali
☎ (+57 2) 441 88 07; 444 22 51 ⊜ (+57 2) 443 13 00
Commander: Brig Gen Edgar Lesmes A
Chief of Staff, Deputy Commander: Col Jesus Alvarez A
Chief of Staff, Financial Office: Lt Col Rafael Saldarriaga ☎ (+57 2) 443 12 48
Commander, Cadets Group: Lt Col Rommel Zapata C ☎ (+57 2) 441 58 42
Director of Operations: Maj Carlos Montealegre
Director of Procurement and Logistics: Maj Juan Garcia ☎ (+57 2) 443 12 48
Wing Commander: Capt Jaime Garzon ☎ (+57 2) 443 14 35
Press and Information Officer: Capt Jaime Herrera ☎ (+57 2) 443 13 00
Public Relations: Lt Carlos Enrique Herrera Cabra ☎ (+57 2) 443 13 00 [98]

25-30000 Armada Nacional Columbia
Army Headquarters
Centro Administrativo Nacional, Eldorado, Santa Fé de Bogotá, DE
☎ (+57 1) 222 27 09 ☒ 20429 ⊜ (+57 1) 222 26 59
Commander: Vice Adm Edgar Moreno Vasquez
Assistant to the Vice Admiral: Capt Rodolfo Guillen [97]

31 Embassies & Attachés

31-00880 Embassy of Israel
Calle 35 No 7-25, P14, Edif. Caxdac, Bogotá
☎ (+57 1) 287 78 08 E embisrae@latino.net.co
⊜ (+57 1) 287 77 83
Ambassador: H E Mr Abraham Hadad
Deputy Chief of Mission, Consul: Peleg P Lewi
Defense Attaché: Alex Harel [98]

31-00920 Embassy of Japan
Carrera 7 71-21 Torro B Piso 11, PO Box 7407, Bogotá
☎ (+57 1) 317 47 93; 317 50 01 ✉ TAISHI Bogota
☒ 43 327 tais co ⊜ (+57 1) 317 49 49
Ambassador: H E Mr Makoto Asami ☎ (+57 1) 317 47 93 - 82
First Secretary, Economics Section: Eijiro Nagamori
Defence Attaché: Tadashi Abe
Public Information Officer: Atsushi Yamaguchi [98]

31-01890 Embassy of the United Kingdom
Torre Propaganda Sancho, Calle 98 No 9-03 Piso 4, Bogotá
☎ (+57 1) 218 51 11 ☒ 44503 brit co
⊜ (+57 1) 218 24 60
Ambassador: H E Mr A L S Coltman
Deputy Head of Mission: D S Gordon MacLeod
Defence Attaché: Col P C Maynard
First Secretary, Commercial: S J Davey
Second Secretary, Chancery: Dr J A Hobbs
Third Secretary and Vice Consul: K P Faulkner
Third Secretary, Commercial: R D Shackleton [98]

31-01900 Embassy of the United States of America
Calle 22 D bis No. 4751, PO Box A A 3831, Santafé de Bogotá, DE
☎ (+57 1) 315 08 11 ☒ 44843 ameb co
⊜ (+57 1) 315 21 97
Ambassador: H E Mr Myles R R Frechette
Deputy Chief of Mission: Oliver P Garza
Commercial Counsellor: Ms Dorothy Lutter ☎ dorothylutter@usfcsbogota@doc
Public Affairs Officer: Terence Kneebone
Defense Attaché: Col William C Spracher
Head of Military Asst Advisory Group: Col Robert Newmann [98]

80 Service Companies & Organizations

80-00100 Aero-Industrias Leaver & Cia SA
Carrera 3 No 56-19, PO Box 6781, Bogotá, DE
☎ (+57 1) 211 81 00 ⊜ (+57 1) 212 89 52
👥 100
President: James G Leaver
■ Authorised service, repair and overhaul for: Piper Aircraft; McDonnell Douglas Helicopter; AllliedSignal Aerospace; Bendix King; Global Wulfsberg Systems; Tectron Lycoming and Teledyne Continental. [98]

80-00500 AlliedSignal Aerospace Marketing Sales and Service
(a subsidiary of AlliedSignal Aerospace, USA)
Santade de Bogata, DC
Oficina de AlliedSignal, Representative Tecnico, Avanca Hangar No 1, Piso 2, Puente Aereo
☎ (+57 1) 413 99 03 ⊜ (+57 1) 413 99 03
Field Service Engineer: Dave Heilman [98]

85 Stockists, Distributors, Spare Parts Suppliers

85-00300 Lacamiron International Ltd
Avda 19 No. 118-95 Ste 513, PO Box 37819, Bogotá, DE
☎ (+57 1) 620 88 48; 620 92 05 ☒ 42442 camir co
⊜ (+57 1) 620 88 48
♣ US$ 15 000 000 👥 20
President: Eng. Mario Ortegón
Vice President: General Luis E. Morillo A. (retd)
Dir Sales & Marketing: Luz A. Roa
Purchasing Mgr: Jacobo Pulido
■ Special forces equipment, communications, ammunition, armament, aircraft, boats and vehicles. [97]

90 Sales Agents, Sales & Liaison Offices

Companies represented:

Bendix Aerospace, USA	90-00025
Global Navigation, USA	90-00025
King Radio, USA	90-00025
MDD Hughes Helicopters, USA	90-00025
Piper Aircraft, USA	90-00025

90-00025 Aero-Mercantil SA
Guyamaral Airport, PO Box 6781, Bogotá, DE
☎ (+57 1) 211 81 00; 676 08 83 ⊜ (+57 1) 212 89 52
President: James G Leaver
® USA: Bendix Aerospace, Global Navigation, King Radio, MDD Hughes Helicopters, Piper Aircraft [98]

90-00050 Alberto Duque R & Cia Ltda
(in partnership with Avocet Aviation, Canada)
Carrera 7a N°65-74, Oficina 904, Apartado Aéreo 11494, Bogotá, DE
☎ (+57 1) 211 61 67 ⊜ (+57 1) 211 61 67
♧ US$ 30,000 (1996) ♣ US$ 1,000,000 (1996)
👥 5 (1996) 🏠 300 sq m
President: Alberto R Duque
Sales and Marketing Director: Luis E Duque
Operations and Technical Director: Juan F Duque
Purchasing Manager: Ms Monica Posada
■ Parachute manufacture and general aeronautical supplies. [97]

90-00400 Comstream Corporation, Andes Region Branch
(a subsidiary of the Spar Company)
Calle 98 Nos 15-17 Office 207, Santa Fé de Bogotá, DE
☎ (+1 57) 16 21 36 51; 16 21 36 81
W http://www.comstream.com ⊜ (+1 57) 16 21 37 96
Andean Region Director: Ivan Dario Almeyda [98]

90-00430 Daewoo Corp Oficina Santafe de Bogata
Cra 7 N°32-33 Ofi 1603-4, Edificio Seguros Fenix, Santafe de Bogota
☎ (+57 1) 284 02 04 ☒ 42326 dwbog
⊜ (+57 1) 285 35 94 [98]

Comoros

Head of State:
Mohamed Taki Abdoulkarim

10 Head of State

10-10000 Office of the President
PO Box 421, Moroni
☎ (+269) 22 32; 22 17 ☒ 233 presirep ko
President of Federal Islamic Republic: Mohamed Taki Abdoulkarim [98]

11 Prime Minister

11-10000 Office of the Prime Minister
Moroni
Prime Minister: Ahmed Abdouh [98]

12 Ministries

12-20000 Ministry of Foreign Affairs
PO Box 428, Moroni
☎ (+269) 21 60; 23 06 ☒ 219 maerfic ko
Minister of Foreign Affairs: Mouhtare Ahmed Charif [98]

12-40000 Ministry of Interior and Administration Reform
Moroni
☎ (+269) 24 45
Minister: Achirafi Said Hachim [98]

15 Parliament

15-10000 Federal Assembly
PO Box 447, Moroni
☎ (+269) 74 40 13; 74 40 00 ⊜ (+269) 74 42 67
General Secretary: Aboubakar Abdou M'sa
President and Chief of Staff: Salim Djabir Salim
Secretary: Aboubakar Abdou M'Sa
Chairman Foreign Affaires de Defense Committee: Adamo Mohamed ☎ (+269) 73 08 92
Vice Chairman: Attoumane Dhoifir ☎ (+269) 73 08 92
Telltale: Cheikh Mahamoud
Member of the Defence, Foreign Affairs, Coorperation and State Safety Committee: Mohamed Abdouwahab
Member of the Defence, Foreign Affairs, Cooperation and State Safety Committee: Mze Abdou Soule Abdoulbak
Member of the Defence, Foreign Affairs, Cooperation and State Safety Committee: Dhoubidane Ahmed
Member of the Defence, Foreign Affairs, Cooperation and State Safety Committee: Ahmed Oussene
■ The Defence, Foreign Affairs, Cooperation and State Safety Committee activities are: Cooperation, conventions, treaty, international agreements and agreement with the structural adjustment programme; safety and general organisation of the defence; politics of cooperation and military assistance; military planning in the long run; rural police and military pensions. [98]

31 Embassies & Attachés

31-01890 Consulate of the United Kingdom
Henri Fraise Fils and Ocean Indien, PO Box 986, Moroni
☎(+269) 73 31 82 ⊖ (+269) 73 31 82
Honorary Consul: J de Jaegar [98]

31-01900 Embassy of the United States of America
PO Box 1318, Moroni
☎(+269) 73 22 03; 29 22 ⊠ 257 amemb ko
⊖ (+269) 73 22 03
Ambassador: H.E. Mr Kenneth N. Peltier
Deputy Chief of Mission: Karl I. Danga
Defense Attaché: Cdr Richard W. Sim
(Resident in Antananarivo) [97]

Congo

Head of State:
Pascal Lissouba

10 Head of State

10-10000 Office of the President
Palais du Peuple, Quartier Plateau, Brazzaville
☎(+242) 81 07 37; 81 30 37
⊠ 5210 precongo kg, 5325 presirep kg
President of the People's Republic: Pascal Lissouba [98]

11 Prime Minister

11-10000 Office of the Prime Minister
Palais de la Primature, PO Box 2096, Brazzaville
☎(+242) 81 11 24/25 ⊠ 5210 precongo kg
Prime Minister and Chief of Armed Forces:
David Charles Ganao [98]

12 Ministries

12-10000 Ministry of Defense
Quartier General, Brazzaville
☎(+242) 81 42 25 ⊠ 5210 precongo kg
Minister of Defense: Francois Ayayene [98]

12-20000 Ministry of Foreign Affairs and Co-operation
PO Box 2070, Brazzaville
☎(+242) 81 06 48 ⊠ 5210 precongo kg
Minister: Destin-Arsene Tsaty-Boungou [98]

31 Embassies & Attachés

31-00890 Embassy of Italy
2-3 Boulevard Lyautey, PO Box 2484, Brazzaville
☎(+242) 83 25 82; 83 40 47 ⊖ (+242) 83 00 11
Chargé d'Affaires a.i.: Dr Vittorio Fumo [98]

31-01890 British Consulate
Cote de L'Hotel Meridien, Pointe Noire
Cote de L'Hotel Meridien, Rue Lyantey 26
☎(+242) 83 85 27 ⊖ (+242) 83 72 57
Honorary Consul: D Picard [98]

31-01900 Embassy of the United States of America
Avenue Amilcar Cabral, BP 1015, Brazzaville
☎(+242) 83 26 42; 83 20 70
Ⓔ les.mcbee@dos.us-state.gov ⊠ 5367 kg
⊖ (+242) 83 46 90
Deputy Chief of Mission: Vicente Valle
Consular Section: Ava Rogers
Economic Section: William Rowland
Public Affairs Officer: Leslie W McBee ☎(+242) 83 42 64
USIS Librarian: Ms Marie-Léontine T Bilombo [98]

Congo, Dem. Republic of

President and Head of State:
Laurent Desire Kabila

10 Head of State

10-10000 Office of the President
Mont Ngaliema, Kinshasa
☎(+243 12) 313 12, 599 03
⊠ 21155 spc zr, 21200 bs pres zr
President and Head of State: Laurent Desire Kabila [98]

12 Ministries

12-10000 Ministry of National Defence
Batiment Administratif de la Defence,
Mont-Ngaliema, Kinshasa
☎(+243 12) 331 82; 331 83
President and Head of Defence: Laurent Desire Kabila [98]

12-20000 Ministry of Foreign Affairs
Place de l'Independance, BP 7100, Gombe, Kinshasha
Building des Affaires Etrangères 7-8 éme étage-Gombe, Kinshasa
☎(+243 12) 330 76; 342 55; 334 83
Minister of Foreign Affairs: Brizima Karaha
Minister of International Co-operation: Thomas Kanza [98]

31 Embassies & Attachés

31-00690 Embassy of Greece
72 Avenue des 3Z, PB 478, Kinshasa
☎(+243 12) 331 69 [98]

31-00890 Embassy of Italy
8 Avenue de la Mongala, Gombé, PO Box 1000, Kinshasa 1
☎(243 12) 210 18; 211 08 ⊠ 21 650 ambitaz kin zr
Ambassador: H.E. Dr Vittorio Amedeo Farinelli
Commercial Secretary: Dr Giuseppe Mistretta [98]

31-01680 Embassy of Spain
PO Box 8036 Kin 1, Kinshasa
Boulevard 30 Juin, Building de la Communauté Hellénique
☎(+243 12) 218 44, 217 59, 218 81 ⊠ 21401 span zr
⊖ (+243 12) 431 95 [98]

31-01890 Embassy of the United Kingdom
Avenue de Trois Z, BP 8049, Gombe, Kinshasha
☎(+243 12) 347 75; 331 30; 337 72; 334 53
⊖ (+243 12) 87 11 44 54 70
Ambassador: H E Mr Marcus Hope
Consul Deputy Head of Mission: A L G Juleff [98]

Costa Rica

Head of State:
José Maria Figueres Olsen

10 Head of State

10-10000 Office of the President
Casa Presidencial, PO Box 10089, 1000 San José
☎(+506) 234 23 10 ⊠ 2106 preside
President: Ing. José Maria Figueres Olsen
First Vice President: Rodrigo Oreamuno Blanco
Second Vice President: Ms Rebecca Grynspan Mayufis [98]

12 Ministries

12-20000 Ministry of Foreign Affairs and Worship
Apartado 10027, 1000 San José
☎(+506 2) 23 60 06; 221 10 91; 233 11 85
⊠ 2107 rree ⊖ (+506 2) 23 93 28
Minister: Fernando Naranjo Villalobos
Vice Minister: Rodrigo Carreras Jiménez
Chief of Cabinet: Fernando Díaz [98]

12-30000 Ministry of Foreign Commerce
PO Box 96-1005, San José Montes de Oca
La Llacuna 12° Avenida Central, Calle 5
Minister of Foreign Commerce:
Jose Manuel Salazar Xirinachs [98]

12-50000 Ministry of Public Security
Apartado 4768, San José
Centro Commercial del Sur, Frente al Castro Madrid
Minister of Public Security: Ms Laura Chinchilla
Director, Social Security Board: Alvaro Salas Chavez [98]

12-60000 Ministry of Science & Technology
San Pedro de Montes de Oca de la,
Agencia del Blanco Nacional,
100 Este y 350 Sur, San José
Minister of Science and Technology:
Jose Leon Desanti Montero [98]

15 Parliament

15-10000 National Legislative Assembly
Apartado 10162, 1000 San José
President, Legislative Assembly: Alberto Cañas [98]

29 Customs & Excise

29-10000 General Customs Administration (Administracion General de Aduanas)
Calle 14, Avenidas 1-3, San José
☎(+506) 233 95 25 ⊖ (+506) 223 73 34
Director General: Gerardo Bolaños [97]

31 Embassies & Attachés

31-01190 Embassy of Mexico
Avenida 7 A No 1371, Calle 13, PO Box 1017, San José
☎ (+52) 222 06 33 ⊖ (+52) 222 60 80
Ambassador: H.E. Mr Emilio Mújica Montoya
Counsellor: Luis Efren Bauza Garcia-Cano
Head Consular Section: Alfredo Gaytan Hinojosa
Office of the Commercial Counsellor
3a Entrada Los Yoses,
310 Metros Al Sur, PO Box 141-1007, San José
☎ (+52) 234 24 66; 234 96 01 ⊖ (+52) 234 96 13
Commercial Counsellor: Samuel Lara Sánchez
Asst Commercial Counsellor: Ms Gabriela Hernández
Office of the Centro Cultural De Mexico
3a Entrada Los Yoses, PO Box 10107-1000, San José
☎ (+52) 283 23 33 ⊖ (+52) 283 24 06
Cultural Counsellor: Carlos Ortega Guerrero 97

31-01270 Royal Netherlands Embassy
Oficentro Ejecutivo La Sabana, PO Box 10285,
1000 San José
☎ (+506) 296 14 90; 48 96; 15 02; 35 95
✉ HOLANDA San José ▣ 2187 holanda cr
⊖ (+506) 296 29 33
Ambassador: H.E. Mr F.B.A.M. van Haren
☎ (+506) 289 53 72
Minister Counsellor: Lic. J.F.L. Blankenberg
☎ (+506) 228 94 62
First Secretary: Mrs Ch. W. van Kooten
☎ (+506) 228 21 12
First Secretary: Dr K.G. Wit ☎ (+506) 282 71 57
First Secretary: Ing. J.A. Bauer ☎ (+506) 232 52 45
First Secretary: Ing. C.D. Konstapel
First Secretary: Mrs Th. van Osch ☎ (+506) 228 18 44
First Secretary: E. van der Sleen ☎ (+506) 228 25 23
Second Secretary: Mrs Hannah A. Tijmes 97

31-01410 Embassy of Peru
Barrio Pops de Curridabat,
Del Indoor Club 100m Sur y 75m Oeste, San José
☎ (+506) 25 91 45 ⊖ (+506) 53 04 57
Ambassador: H E Mr Alberto Varillas
Minister Counsellor: José Luis Perez
Consul: José I Mariategui 98

31-01650 Embassy of the Republic of Slovakia
PO Box 3910, 1000 San José
☎ (+506) 224 64 67 ⊖ (+506) 224 91 84
Chargé d'Affaires: Vladimir Grácz 97

31-01890 Embassy of the United Kingdom
1007 San José
Edificio Centro Colón, 11th Floor, San José
☎ (+506) 221 55 66; 221 57 16; 221 58 16
▣ 2169 prodrom cr ⊖ (+506) 233 99 38
Ambassador: H E Mr R M Jackson
First Secretary (Commercial) Director of Trade Promotion for Central America: Alan S Green
Third Secretary (Technical Assistance Management Officer) and Vice-Consul: J P Miller 98

Côte d'Ivoire

Head of State:
Henri Konan Bedie

10 Head of State

10-10000 Office of the President
Boulevard Clozel, PO Box 1354, Abidjan
☎ (+225) 32 02 22 ▣ 23754 cotivoir, 3311 cotivoir
President of the Republic: Henri Konan Bedie 98

11 Prime Minister

11-10000 Office of the Prime Minister
Government Buildings, Abidjan 01
Prime Minister: Daniel Kablan Duncan 98

12 Ministries

12-10000 Ministry of Defense and Maritime Affairs
Blvd Nangui Abrogoua, PO Box V-11, Abidjan
☎ (+225) 32 02 88
Minister: Bandama N'Gata 98

12-20000 Ministry of Foreign Affairs
PO Box V-109, Abidjan
Building Angoulvant, Bloc Ministériel
☎ (+225) 32 08 88 ▣ 22118 minafet abidjan
⊖ (+225) 33 23 08
Minister: Amara Essy
Director of Cabinet: Dassys Claudie
☎ (+225) 32 08 88 ext 4272
Chief of Cabinet: Kouame N'Goran
Director International Cooperation: Tanoe Fatoumata
☎ (+225) 32 08 88 ext 4222; 4203
Technical Counsellor: Attoungbre Gervais
☎ (+225) 32 08 88 ext 4267 98

12-40000 Ministry of the Interior
Boulevard Angoulvant, PO Box V 121, Abidjan
☎ (+225) 32 08 88 ▣ 22296 mininter abidjan
⊖ (+225) 32 62 35
Minister: Emile Constant Bombet 98

12-50000 Ministry of Security
Angoulvant Boulevard, BP V241, Abidjan 01
☎ (+225) 22 19 50; 22 19 54 ▣ 23873 minisi ci
Minister of Security: Marcel Dibona Kone 98

25 Defense Forces

25-20000 Force Aerienne de la Côte d'Ivoire
c/o Ministere de la Defense, Abidjan
☎ (+225) 27 90 27 ▣ 43399 ⊖ (+225) 27 88 03
Commander Air Force: Col Eittien Koffi
Deputy Commander: Lieut Col Raymond Zomy Gueu
Director of Operations: Lieut Col Mamadou Comara
Director of Logistics: Lieut Col Brédou Angoua 98

31 Embassies & Attachés

31-00380 Embassy of Colombia
PO Box 3874, Abidjan 01
☎ (+225) 33 12 44 ⊖ (+225) 32 47 31
Ambassador: H E Mr Carlos Adolfo Arenas Campos
Deputy Chief of Mission: Ms Maria Eugenia Olarte
First Secretary: Rubén Dario Agudelo Herrera 98

31-00660 Embassy of the Federal Republic of Germany
Av.Nogues, Bld Botreau Imm.'Le Mans', PO Box 1900, 01 Abidjan
Avenue Nogues Boulevard Botreau Roussel Le Mans'
☎ (+225) 21 47 27 ▣ 23 642 dipger ci
⊖ (+225) 32 47 29
Ambassador: H E Dr Hans Albrecht Schraepler
Commercial Counsellor: M Manngast 98

31-00890 Embassy of Italy
16 Rue de la Canebière, Cocody, PO Box 1905, Abidjan 01
☎ (+225) 44 61 70; 44 63 61 ▣ 26123
⊖ (+225) 44 35 87
Ambassador: H.E. Dr Raffaele Campanella
Deputy Chief of Mission: Dr Giovanni Iannuzzi
Commercial Secretary: Dr Vera Costantino 98

31-01330 Embassy of the Federal Republic of Nigeria
Nigeria House, PO Box 1906, Abidjan 01
☎ (+225) 21 19 82; 22 30 82 ▣ 23532
⊖ (+225) 21 38 17
Ambassador: H.E. Mr J.O. Coker
Minister & Head of Chancery: D. Egwudobi
Defence Attaché: Col M.A. Garba (resident in Cotonou)
Counsellor Political: I.A.D. Nuhu
First Secretary Commercial: O.A. Akerele
First Secretary: D.H. Alhassan
First Secretary: Miss U.E. Ohadiugha
Second Secretary: U. Mbanaso
Financial Attaché: I.S. Nasiru 97

31-01340 Royal Norwegian Embassy
PO Box 607, Abidjan 01
Immeuble N'Zarama, Boulevard Lagunaire
(a droite de l'hotel Sofitel), 5éme étage, Abidjan 01
☎ (+225) 21 21 66; 22 25 34
E noramb@africaonline.co.ci ▣ 23355
⊖ (+225) 21 91 99
Ambassador: H E Mr Jan Naerby
Second Secretary: Stig Travik
Administration Affairs: Tove Laukli 98

31-01890 Embassy of the United Kingdom
PO Box 2581, Abidjan 01
3rd Floor, Immeuble "Les Harmonies", Angle
Boulevard Carde et Ave Dr Jamot, Plateau, Abidjan
☎ (+225) 22 68 50; 22 68 51; 21 82 09 22 68 52
▣ 23706 prdrme ci ⊖ (+225) 22 32 21; 22 04 38
Ambassador: H E Miss M I Rothwell
First Secretary Commercial and Deputy Chief of Mission: F G Geer
Defence Attaché: Lieut Col A R Gales (resident in Accra)
Second Secretary and Information Officer: A T Richmond
Third Secretary and Vice Consul: Mrs J Hemmingway
Third Secretary, Vice Consul: Miss L Renfrew
Third Secretary, Commercial: W P Hemmingway 98

75 Manufacturing Companies

75-05000 Daewoo Corp Abidjan Office
PO Box DAE, Côte d'Ivoire 01
BPV 135, Villa N°20 Sise, Rue des Alizes
Marcopy Residentiel, Abidjan
☎ (+225) 32 37 34 ⊖ (+225) 22 10 88 98

Contact Reference

90 Sales Agents, Sales & Liaison Offices

Companies represented:

Piper Aircraft, USA 90-12000

90-12000 Air Inter Ivoire
07 B P 62, 07 Abidjan
☎(+225) 27 84 65; 27 72 13 ✆ (+225) 07 76 75
✉07 bp62 abidjan 07 ivory coast ⊖ (+225) 27 71 65
♲ CFAFr 70,000,000 ⚉ 30 (1997) ⌂ 6,000 m2
Director General: Jacques L Bergounioux
Director: Patrick Dufaud
Sales and Marketing Director: Ms Helen L Bergounioux
■ Air transport and aerial work. Medical air evacuation. Fixed and rotary wing. Authorised service for manufacturers represented. Leasing.
® USA: Piper Aircraft [98]

Croatia

Head of State:
Franjo Tudjman

10 Head of State

10-10000 Office of the President
Presidential Palace, Pantovcak 241, 41000 Zagreb
☎(+385 1) 46 51 92; 44 40 00
⊖ (+385 1) 44 30 75; 46 51 88
President: Franjo Tudjman [98]

11 Prime Minister

11-10000 Office of the Prime Minister
Trg Sv Marka 2, 41000 Zagreb
☎(+385 1) 44 40 00; 44 46 66
⊖ (+385 1) 43 20 41; 22 70 82
Prime Minister: Zlatko Matesa
Deputy Prime Minister, Economics: Borislav Skegro
Deputy Prime Minister, Refugee Issues: Mate Granic
Deputy Prime Minister, Humanitarian Affairs: Ivica Kostovic
Deputy Prime Minister, Social Affairs: Ljerka Mintas-Hodak
Deputy Prime Minister: Jure Radic [98]

12 Ministries

12-10000 Ministry of Defence
Ministarstvo Obrane
Trg Kralja Petra Kresimira 4, Opaticka 1, 41000 Zagreb
☎(+385 1) 46 71 11 ⊖ (+385 1) 45 11 05; 43 24 15
Minister: Gojko Susak
Department of Aviation
Maksimirskn 63, Zagreb
☎(+385 1) 486 10 22 Ⓔ kopriv@zvonmir.morh.tel.hr
⊖ (+385 1) 486 10 97 Goran Redzepovic [98]

12-20000 Ministry of Foreign Affairs
Trg Nikole Subica Zrinskoga 7-8, 41000 Zagreb
☎(+385 1) 456 99 64 ⊖ (+385 1) 455 17 95
Minister: Dr Mate Granic
Deputy Minister: Dr Ivan Simonovic
Secretary of State Ministry: Zoran Piculjan
Assistant Minister: Hidajet Biscevic
Assistant Minister: Frane Krnic
Assistant Minister: Jaksa Muljacic
Assistant Minister: Spomenka Cek
Assistant Minister: Miroslav Medimorec
Chief of Staff: Dr Vladimir Matek
Communications & Information Systems Department
Trg Nikole Subica Zrinskoga 7-8, Zagreb
☎(+385 1) 456 98 96 ⊖ (+385 1) 456 99 93
Head of Department: Mladen Gavrancic [98]

12-70000 Ministry of Transportation and Communication
Mesnicka 23, 41000 Zagreb
☎(+38 41) 61 20 17 ⊖ (+38 41) 51 81 13
Minister: Zeljko Luzavec [98]

12-80000 Ministry of Science and Technology
Strossmayerov trg. 4, 10000 Zagreb
☎(+385 1) 459 44 31 Ⓔ office@science.hr
⊖ (+385 1) 459 44 39
Minister: Prof Dr Ivica Kostovic
Deputy Minister: Prof Dr Nikola Ruzinski
Director Foreign Relations: Prof Dr Radovan Fuchs
Assistant to Minister: Predrag Pale [98]

25 Defense Forces

25-40500 Naval Medical Institute of the Croatian Navy
Institut Pomorske Medicine HRM
PO Box 196, 21000 Split
☎(+385 21) 35 46 77 ⊖ (+385 21) 35 46 78
Director: Brig.Prof. Darko Ropac (MD, DSc, CroNav MC)
Head of Dept of Undersea & Hyperbaric Medicine: Lt Cdr Nadan M. Petri (MD, DMO)
Head of Dept of Occupational Medicine: Drago Pretkovic (MD)
Head of Dept of Edipemiology & Tropical Diseases: Ms Rosanda Mulic (MD, DSc)
Head of Dept of Sanitary Chemistry & Toxicology: Ms Visnja Katalinic (DSc)
Head of Dept of Medical Selection: Col Jakov Lukac (MD)
Head of Dept of Microbiology and Medical Oceanography: Col Nikola Radosevic (MD)
■ Preventative medicine & medical research. → 60 [97]

29 Customs & Excise

29-01000 Customs Directorate
(a subordinate of the Ministry of Finance)
c/o Ministry of Finance, Bogoviceva 1a, 10000 Zagreb [98]

31 Embassies & Attachés

31-00100 Embassy of Austria
Jabukovac 39, 41000 Zagreb
☎(+385 1) 42 68 59 ⊖ (+385 1) 42 69 01
Ambassador: H.E.Mr Andreas Berlakovich
First Secretary & Deputy to Ambassador: Andreas Stadler
Consul & First Secretary: Hartwig Pomper
Defence Attaché: Brig. Gen. Jürgen Köberl
Assistant Defence Attaché: Warrant Officer Rudolf Langbrugger
Cultural Affairs Attaché: Erich Campestrini
Cultural Affairs Councellor: Dr Walter Maria Stojan
Trade Relations Attaché: Stanislaw Pawlicki
Trade Relations Counsellor: Dr Gunter Mühlberger [97]

31-00660 Embassy of the Federal Republic of Germany
Ulica Grada Vukovaraa 64, 10000 Zagreb
☎(+385 1) 615 81 50 ⊖ (+385 1) 615 81 03 [98]

31-00690 Embassy of Greece
Zagreb
☎(+385 1) 64 12 22; 52 00 11 ⊖ (+385 1) 52 46 59 [98]

31-01480 Embassy of Romania
Srebrnjak 150A, 41000 Zagreb
☎(+385 1) 233 60 91 ⊖ (+385 1) 233 91 70
Ambassador: H.E. Dr Vasile Leca
Deputy Chief of Mission: Nicolae Perhaita
Defence Attaché: Col Tomuta Iulian [97]

31-01655 Embassy of the Republic of Slovenia
Savska cesta 41/IX, 41000 Zagreb
☎(+385 1) 612 15 03; 612 14 66; 51 74 01
⊖ (+385 1) 51 78 37; 62 15 03
Ambassador: H.E. Matija Malesic ☎(+358 1) 53 55 08
Economic Affairs Counsellor: Anton Grabeljsek
First Secretary & Consul: Mrs Alenka Jerak
Third Secretary & Vice Consul: Ms Jasna Sivec [97]

31-01740 Embassy of Switzerland
Bogoviceva 3, 41000 Zagreb
☎(+385 1) 42 15 73 ⊖ (+385 1) 42 59 95
Ambassador: H.E. Mr Petar Troendle [97]

31-01890 Embassy of the United Kingdom
Vlaska 121/III, PO Box 454, 10000 Zagreb
2nd Floor, Astra Tower, Tratinska 5
☎(+385 1) 455 53 10 Ⓔ british-embassy@39.tel.hr
⊖ (+385 1) 455 16 85
Ambassador: H E Mr C A Munroe
Deputy Chief of Mission: M D MacPherson
Defence Attaché: Wg Cdr A J Fovargue
First Secretary Commercial: P J Ivey [98]

31-01900 Embassy of the United States of America
Andrije Hebranga 2, Zagreb
☎(+385 1) 455 55 00 ⊠ 21180 amemb yu
⊖ (+385 1) 455 85 85
Ambassador: H E Mr Peter W Galbraith
Deputy Chief of Mission: Robert P Finn
Defence Attaché: Lieut Col John Sadler
Public Affairs Officer: Douglas Davidson
Commercial Officer: Patrick Hughes [98]

60 Research Organizations, Institutes & Test Establishments

Naval Medical Institute of 25-40500

60-00450 Laser & Atomic Research & Development (LAIR)
(a subsidiary of the Ruder Bosovic Institute)
Bijenicka c 54, PO Box 1016, 10000 Zagreb
☎(+38 51) 27 27 59 Ⓔ persin@olimp.irb.hr
⊖ (+38 51) 42 48 69
♲ US$ 4,200,000 (1997) ♣ US$ 3,000,000 (1997)
⚉ 60 (1997) ⌂ 700 sq m
Managing Director: Dr Sc Antun Persin
■ Research and development in the field of opto-electronics and lasers. Production of systems and equipment based on own Research and Development including training equipment, weapons simulators, warning and detection systems, night sights and laser mine blasters. [98]

75 Manufacturing Companies

75-06500 Losinjska Plovidba-Brodogradiliste Mali Losinj
Losinskih Brodograditelja 66, 51550 Mali Losinj
☎(+385 51) 23 13 22 ⊖ (+385 51) 23 10 74
⚉ 150 (1997)
Managing Director: V Muzic
Sales Director: K Toic
Technical Director: D Jankovic
■ Ship repair services. [98]

75-07800 Trogir Shipyard
Put Brodograditelja 16, 58220 Trogir
☎(+385 58) 88 15 55 ⊖ (+385 58) 88 18 81
⚉ 1,300
Managing Director: Josko Buble

Sales Director: Vedrana Vukmann
Newbuilding Director: Mateo Tramontana
Ship Repair Director: Niksa Tomas
■ Ship construction and repair, two slipways, 20 m wide and 47 m wide, slipways can accommodate ships up to 60,000 tdw ranging from oil tankers, cargo ships, ferry boats, supply vessels, tugs, rescue vessels as well as floating docks of 60,000 tons lifting capacity built in sections and thereafter connected afloat. The shipyard has its own design department and ship repair works which dispose of one floating dry dock of 10,000 tons lifting capacity and 705 m outfitting and repairing wharves and piers are able to perform all kinds of works on shore and dry dock, reconstructions, conversions and jumboizing of all kinds of ships. [98]

75-07900 Uljanik
Flciusova 1, PO Box 208, 52100 Pula
☎ (+385 52) 343 22; 243 22 ⊠ 211655
⊖ (+385 52) 21 16 55
♣ US$ 100,000,000 ⚉ 3,700
President of Business Board: K Radolovic
Sales Director: D Marchi
Shipbuilding Director: R Batelic
President of B B: V Matosevic
President of B B: R Fonovic
Director of Special Equipment: S Bursic
☎ (+385 52) 27 23 51
■ Ship building facility, mechanical and electrical engineering, and welding machines. [98]

75-08000 Viktor Lenac Ship Repair
Martinscica, PO Box 210, 51001 Rijeka
☎ (+385 51) 21 62 55 ⊠ 24194
⊖ (+385 51) 21 70 33/35
♣ US$ 32 000 000 ⚉ 650 ⌂ 20 000 m2
Managing Dir: D. Vrhovnik
Deputy Managing Dir & Technical Dir: O. Antunac
Operations Dir: F. Skoyahdic
Sales & Marketing Dir: R. Cvecic
Purchasing Mgr: F. Kukic
■ Provision of a wide range of ship repair & related services. Conversion, modification, inspection & maintenance. [97]

80 Service Companies & Organizations

80-00100 Brodoprojekt
Ciottina 17a, 51000 Rijeka
☎ (+385 51) 21 10 78 ⊖ (+385 51) 21 13 18
⚉ 80
Managing Director: Vesna Stamenkovic
Manager: David Vukelic
Manager: Mladen Merlak
Manager: Slobodan Lalic
■ Ship design, marine engineering and consulting. [98]

90 Sales Agents, Sales & Liaison Offices

90-05000 Repair Yard Sibenik
(† Velimir Skorpik Navy Yard)
Port Facility, Sibenik
⊠ (+385 22) 33 33 99; 33 38 90
⊖ (+385 22) 33 35 05; 33 36 66
⚉ 760
General Manager: Vladimir Lipovic
Chief Engineer: Ivo Baljkas
■ Capacity to repair 16 vessels of 900-1,500t displacement simultaneously. Operates a 1,500t capacity floating dock. [98]

Cuba

Head of State:
Dr Fidel Castro Ruz

10 Head of State

10-10000 Office of the President
Ciudad de la Habana, Habana
⊠ 511383
President of the Council of State: Dr Fidel Castro Ruz
Minister Secretary of the Council of State:
José M Miyar Barruecos [98]

10-20000 Office of the First Vice-President
Ciudad de la Habana, Habana
⊠ 511151 minfar hab
First Vice-President and Minister of Revolutionary Armed Forces: Gen Raúl Castro Ruz
Vice President of the Council of State:
Juan Almeida Bosque
Vice President of the Council of State:
Gen Abelardo Colome Ibarra
Vice President of the Council of State: Carlos Lage Davila
Vice President of the Council of State:
Esteban Lazo Hernandez
Vice President of the Council of State:
José Ramon Machado Ventura
Minister Secretary of the State Council:
Jose M Miyar Barruecos
Minister for Communications:
Brig Gen Silvano Colas Sanchez [98]

12 Ministries

12-10000 Ministry of the Revolutionary Armed Forces
Plaza de la Revolución, Ciudad Habana
⊠ 511151
Minister of the Revolutionary Armed Forces:
Gen Raúl Castro Ruz [98]

12-20000 Ministry of Foreign Affairs
Ministerio de Asuntos Exteriores
Calzada YG, Quinta y Paseo Vedado, Havana
☎ (+53 7) 30 50 31 ⊠ 511122 minrex hab
Minister of Foreign Relations: Roberto Robaina Gonzalez
Minister of Foreign Investment and Economic Co-operation: Ibrahim Ferradaz [98]

12-30000 Ministry of Foreign Trade
Infanta y Paseo Vedado, Habana
☎ (+53 7) 70 93 41 ⊠ 511174/175/165 mincex hab
Minister of Foreign Trade: Ricardo Cabrisas Ruiz [98]

12-40000 Ministry of the Interior
Ciudad de La Habana, Habana
Minister of Interior: Gen Abelardo Colome Ibarra [98]

15 Parliament

15-10000 National Assembly of People's Power
Asamblea Nacional del Poder Popular
Calle 42 No 2308 e/23 y 25, Playa Ciudad, Habana [98]

31 Embassies & Attachés

31-00100 Embassy of Austria
Calle 4 No 101 e/1ra y 3ra, Miramar, Habanna
☎ (+53 7) 24 28 25; 24 23 94 ⊖ (+53 7) 24 12 35
Ambassador: H E Mr Yuri Standenat
Commercial Counsellor: Walter Koren
(resident in Mexico City)
Attaché, Consular and Administrative Affairs:
Thomas Hollensteiner [98]

31-00590 Embassy of France
Calle 14 No 312, Miramar, PO Box 293, Habana
Ambassador: H.E. Mr Philippe Peltier
Commercial Office
Calle 21, No 140 (entre L y M), Habana 4
☎ (+53 7) 32 99 17 ✉ COMATTA La Havane
⊠ 511 884 comat cu
Commercial Counsellor: Bernard Paulien [98]

31-01890 Embassy of the United Kingdom
Calle 34 No 702/4 entre 7ma Avenida, Miramar, Ciudad de La Havana
☎ (+53 7) 24 17 71; 24 17 72; 24 10 49; 24 92 31 (Chancery/Commercial); 24 12 86; 24 12 99 (Consular/Management)
⊖ (+53 7) 33 81 04 (Consular/Management); 24 92 14 (Commercial/Information)
Ambassador: H E Mr P A McLean
First Secretary and Deputy Head of Mission: N P Kay
Defence Attaché: Col L M C Watson
(resident in Mexico City)
Second Secretary, Commercial: T Brownbill [98]

Contact Reference

Cyprus

Head of State:
Glafcos Clerides

10 Head of State

10-10000 Office of the President
Nicosia
☎ (+357 2) 45 13 33 ⊖ (+357 2) 44 50 16
President of the Republic: H E Glafcos Clerides
Under Secretary to the President: Pantelis Kouros
Government Spokesman: Manolis Christobis
☎ (+357 2) 51 09 30 [98]

12 Ministries

12-10000 Ministry of Defense
4 Emm Roïdi Street, Nicosia
☎ (+357 2) 30 31 89 ⊖ (+357 2) 36 26 05
Minister of Defense: Costas Eliades ☎ (+357 2) 47 30 75
Permanent Secretary: George Hadjianastassio
☎ (+357 2) 51 06 57
Chief of the National Guard: Lieut Gen Nicolaos Vorvolukos [98]

Cyprus

12-20000 Ministry of Foreign Affairs
Government House 18-19, Demosthenis Severis Avenue, Nicosia
☎ (+357 2) 30 21 22 ⊖ (+357 2) 45 18 81; 36 53 13
Minister of Foreign Affairs: Yiannakis "John" Kassoulides
Permanent Secretary: Vanias Markides [98]

12-40000 Ministry of the Interior
Dem Severis Avenue, Ex-Secretarial Compound, Nicosia
☎ (+357 2) 30 26 17; 30 21 05
⊖ (+357 2) 45 34 65; 45 34 65
Minister: Dinos Michaelides ☎ (+357 2) 51 02 22
Permanent Secretary: Th Michael ☎ (+357 2) 51 08 28
Press and Information Office Director: George Hadlisavvas ☎ (+1357 2) 51 09 50
Press and Information Office: John Solomou ☎ (+357 2) 44 84 42 [98]

12-50000 Ministry of Finance
Dem Severis Avenue, Ex-Secretariat Compound, Nicosia
☎ (+357 2) 30 21 64 ⊖ (+357 2) 36 60 80
Minister: Christodoulos Christodoulou
Permanent Secretary: Antonios Malaos [97]

12-80000 Ministry of Justice & Public Order
12 Helioupoleos Street, Nicosia
☎ (+357 2) 30 21 27 ⊖ (+357 2) 46 14 27
Minister: Nicos Koshis
Permanent Secretary: George Anastasiades [97]

26 Police (Civil)

26-00000 Cyprus Police Headquarters
(a subordinate of the Ministry of Justice and Public Order)
PO Box 1478, Nicosia
☎ (+357 2) 30 50 00 ⊖ (+357 2) 31 18 56
👥 3,752 (uniformed); 280 (special constables); 33 (civilian)
Chief Commissioner: Andreas Potamaris
Director of Administration Department: A Lemonaris ☎ (+357 2) 30 50 00 ext. 2339
Assistant Director of Administration Department: G Panayiotou ☎ (+357 2) 30 50 00 ext. 2236
Senior Administration Officer for Permanent Secretary: Th Kazamias [97]

29 Customs & Excise

29-01000 Department of Customs and Excise
(a subordinate of the Ministry of Finance)
c/o Ministry of Finance, Katsoni's Street 29, Nicosia [98]

30 Civil Defense

30-00100 Civil Defence Force Headquarters
(a subordinate of the Ministry of Interior)
39 Kennedy Avenue, PO Box 3830, Nicosia
☎ (+357 2) 42 22 62; 49 92 57 ⊖ (+357 2) 31 56 38
Civil Defence Commissioner: Modestos Pentaliotis [98]

31 Embassies & Attachés

31-00090 Australian High Commission
4 Annis Komninis St cnr Stassinos, PO Box 1331, Nicosia
☎ (+357 2) 47 30 01 ⊠ AUSTCOM Nicosia
⊖ (+357 2) 36 64 86
His Excellency the High Commissioner:
High Commissioner: H E Mr J W Sullivan
Deputy High Commissioner, First Secretary: J G Judge
Attaché and Vice Consul: A A Morris
Commercial Attaché: Mrs Anita Aaritou [98]

31-00250 Embassy of the Republic of Bulgaria
Constantinou Paleologou Street, Nicosia
Constantinou Paleologou Street, No 13, Engomi
☎ (+357 2) 47 24 86; 47 27 40 ⊖ (+357 2) 45 65 98
Ambassador: H E Mr Alexei Ivanov ☎ (+357 2) 44 91 85
Counsellor: Ivan Bakalov
Defence Attaché: Iovko Tsarkov
First Secretary, Press Attaché: Alexei Alexandrov
First Secretary, Commercial Attaché: Assen Babansky
First Secretary, Consular Affairs: Georgui Georguiev [98]

31-00660 Embassy of the Federal Republic of Germany
10 Nikitaras Street, PO Box 1795, CY-1513 Nicosia
10 Nikitaras Street, Ay Omologhitae
☎ (+357 2) 44 43 62; 44 43 63; 44 43 64; 44 43 68
⊠ 2460 aanico cy ⊖ (+357 2) 36 56 94
Ambassador: H E Dr
Head of Administration: Wolfgang Hentschel [98]

31-00690 Embassy of Greece
8/10 Byron Avenue, PO Box 199, Nicosia
☎ (+357 2) 44 18 80 ⊠ 2397 gremb cy
⊖ (+357 2) 47 39 90 [98]

31-00890 Embassy of Italy
25th March Street No 11, POB 7695, 2408 Engomi Nicosia
☎ (+357 2) 35 76 35; 35 76 36; 35 82 58
⊠ 3847 italdipl cy ⊖ (+357 2) 35 76 16
First Secretary: Dr Raffaele De Benedictis [98]

31-01650 Embassy of the Slovak Republic
4 Kalamatas Street, PO Box 1165, Nicosia
4 Kalamatas Street, 2002 Strovolos
☎ (+357 2) 31 16 81; 31 16 83 ⊖ (+357 2) 31 17 15
Chargé d'Affaires: Dusan Rozbora [98]

31-01890 United Kingdom High Commission
BFPO 567, PO Box 1978, Nicosia
☎ (+357 2) 47 31 31; 47 31 37 ⊠ 2208 ukrepnic cy
⊖ (+357 2) 36 71 98
High Commissioner: H E Mr D C A Madden
Counsellor and Deputy High Commissioner: J E Buck
Defence Adviser: Col A C Taylor
First Secretary Commercial: W T Preston
First Secretary Information: G J Ley [98]

31-01900 Embassy of the United States of America
Metochiou & Ploutarchou Str., Nicosia
☎ (+357 2) 47 61 00 ⊠ 4160 amemy cy
⊖ (+357 2) 46 59 44
Ambassador: H.E. Mr Richard A. Boucher
Deputy Chief of Mission: Alejandro Wolff
Economic/Commercial Section: Patricia Nelson-Douvelis
Office of the Defense Attaché: Lt Col David N. Fetter (USA)
Public Affairs Officer: Ms Judith Raine Baroody [97]

75 Manufacturing Companies

75-00010 Army Merchants
16 Zthermopylon Street, 8010 Pafos
☎ (+357 6) 25 40 16 ⊖ (+357 6) 25 40 16
♣ US$ 675,000 👥 9 🏠 350 sq m
Managing Director: Nicos Hadjipetrou
■ Manufacture military uniforms, light self defence and hunting equipment. [98]

75-05000 Saft Nife M.E. Ltd
(a subsidiary of Saft SA, France)
PO Box 4307, Limassol
☎ (+357) 532 24 35 ⊠ 5421 cy ⊖ (+357) 532 96 77
♣ US$ 7,000,000 👥 14 🏠 500 sq m
Managing Director: Stuart Chinn
Sales and Marketing Director: H Schlenk
■ Nickel-Cadmium batteries, lithium batteries, air-depolarized cells, emergency lighting, converters and power supplies, advanced battery systems, rectifier chargers, nickel-cadmium batteries for space applications.
® France: Saft SA [98]

75-05500 Telectronics Communications Ltd
Head Office
PO Box 1573, Limassol
34 King George 11, Telecoms Court, Limassol
☎ (+357 5) 33 22 90 [E] telectronics@cytanet.com.cy
⊖ (+357 5) 33 24 04
⊕ US$ 4,000,000 (1997) ♣ US$ 5,000,000 (1996)
👥 30 (1997)
Director: Dozos Christofi
Marketing Director: Christoe Poulli
Operations Director: Narius Alexandrou
■ Specialists in radio communication systems. [98]

80 Service Companies & Organizations

80-00200 ESL Limited
(† Emerson Systems Ltd)
50 Arch Makarios III Ave, 204 Alpha House, 1065 Nicosia
☎ (+357 2) 45 56 12; 45 57 21
[E] eslcy@spidernet.com.cy ⊖ (+357 2) 47 63 37
President: Wm J Enloe
Vice President and General Manager: Richard A Haber
■ Independent consultant to and distributor for ESCO Electronics Corporation and others. → 85
® USA: Esco Electronics [98]

80-02000 Famalift Shipyard Ltd
PO Box 1903/1558, Limassol
☎ (+357 5) 35 52 93; 39 20 60
⊠ FAMALIFT LIMASSOL Cyprus ⊠ 4223; 3160
⊖ (+357 5) 37 15 54; 39 33 62
⊕ C£ 2,000,000 ♣ C£ 600,000 👥 65 🏠 7,500 m2
Managing Director: Andreas Tzortzis
Operations Director: Tomis Tzortzis
Marketing Manager: Doros Papaicovou
Technical Manager: Renos Phokas
■ Floating dock 1,300 tons lifting capacity, grid-blasting, water blasting, painting, plating, shaft's cladding. [98]

85 Stockists, Distributors, Spare Parts Suppliers

ESL Limited 80-00200

85-02000 Goliber Trading Ltd
PO Box 7524, 3316 Limassol
☎ (+357 5) 74 57 54 ⊠ 4181 goliber cy
⊖ (+357 5) 74 59 02
⊕ US$ 5,500,000 (1996) ♣ US$ 6,500,000 (1996)
👥 5 (1996) 🏠 200 sq m
Managing Director: Leonid Bravinski
Marketing and Sales Director: Matvei Lvine
Operations and Technical Manager: Michael Rodionov
■ Supply of spare parts for all types of aircraft and helicopters made in Russia or the former Soviet Union, including both brand-new and overhauled engines, aircraft tyres, electronic devices, hydraulic pumps and fuel-regulating equipment. → 90 [98]

90 Sales Agents, Sales & Liaison Offices

Companies represented:

Esco Electronics, USA 80-00200
Saft SA, France 75-05000

Sales agents/representatives listed elsewhere:

Goliber Trading Ltd 85-02000

Czech Republic

Head of State:
Václav Havel

10 Head of State

10-10000 Office of the President
119 08 Prague 1
☎ (+420 2) 33 37 11 11 ⊖ (+420 2) 33 37 33 00
President of the Republic: Václav Havel [98]

11 Prime Minister

11-10000 Office of the Prime Minister
Nabr E Benese 4, 125 09 Prague 1
☎ (+420 2) 24 00 21 11
⊖ (+420 2) 24 81 02 31; 24 81 07 40
Prime Minister: Vaclav Klaus
Deputy Prime Minister: Jiri Skaliky
Deputy Prime Minister and Minister of Agriculture: Josef Lux
Deputy Prime Minister: Josef Zieleniec [98]

12 Ministries

12-10000 Federal Ministry of Defence
Federalnm Ministerstvo Obrany (MNO)
Délostrelecká 11, 160 02 Prague 6
☎ (+420 2) 32 23 50; 34 10 12
⊖ (+420 2) 32 23 50; 34 14 33
Minister of Defence: Miroslav Vyborny
Deputy Minister of Defence: Miroslav Kalousek
Deputy Defence Minister: Vladimir Suman
Spokesman: Vaclav Tichy [98]

12-20000 Ministry of Foreign Affairs
Loretanske nam 5, 118 00 Prague 1
☎ (+420 2) 24 18 11 11 ⊖ (+420 2) 24 18 20 44
Minister: Josef Zieleniec
Director of Security Policy Dept: Hynek Fajmon [98]

12-30000 Ministry of Industry and Trade
Na Frantisku 32, 110 15 Prague 1
☎ (+420 2) 24 06 21 45; 24 06 11 11; 24 32 03 30
⊡ 225578 kkm h ⊖ (+420 2) 285 22 27; 26 54 13
Minister: Karel Kuhnl [98]

12-35000 Ministry of Finance
Letenska 15, 118 10 Prague 1
☎ (+420 2) 514 11 11 ⊖ (+420 2) 514 27 88
Minister: Ivan Pilip [98]

12-40000 Ministry of the Interior
Nad Stolou 3, 170 00 Prague 7
☎ (+420 2) 33 51 11 11 ⊖ (+420 2) 37 82 16
Minister: Jan Ruml [98]

12-60000 Ministry of Transport
Nabr L Svobody 12, 110 15 Prague 1
☎ (+420 2) 23 03 11 11 ⊖ (+420 2) 231 40 15
Minister: Martin Riman [98]

12-80000 Ministry of Justice
Vysehradska 16, 128 10 Prague 2
☎ (+420 2) 29 45 45 ⊖ (+420 2) 20 90 64
Minister: Vlasta Parkanova [98]

29 Customs & Excise

29-01000 General Directorate of Customs
Budejovicka 7, 14096 Prague 4 [98]

31 Embassies & Attachés

31-00310 Embassy of Canada
Mickiewiczova 6, 12533 Prague 6
☎ (+420 2) 24 31 11 08 (ext 3805)
E michael.jay@prgue01.x400.gc.cz
⊖ (+420 2) 24 31 28 25
Ambassador: H E Mr Alain Dudict
First Secretary: Yvan Jobin
Canadian Forces Attaché: M E Jay
Assistant to the Defense Attaché: Rich Chagkuwski
Commercial Counsellor: Royer Belenyer [98]

31-00580 Embassy of Finland
Drevná 2, 225 01 Prague 2
☎ (+420 2) 24 91 35 94; 24 91 35 95
⊖ (+420 2) 420 2) 29 03 63
Ambassador: H E Mr Esko Rajakoski
☎ (+420 2) 20 51 36 66
Counsellor/Commercial Affairs: Jorma Meronen
☎ (+420 2) 36 12 12 ⊡ 122 571 fico c
Counsellor: R S Jussila ☎ (+420 602) 34 34 01 (mobile)
Military and Naval Attaché: Lieut Col Jyrki Laakso (resident in Germany) ☎ (+49 228) 38 29 80 [98]

31-00660 Embassy of the Federal Republic of Germany
Vlasská 19, POB 88, 118 01 Prague 1
☎ (+420 2) 57 32 01 90 ⊡ 122814 aapg c
⊖ (+420 2) 57 32 00 43
Ambassador: H E Dr Anton Roßbach
Deputy Chief of Mission: Heinrich Horsten
Military Attaché: Col Franz-Xaver Lauterer
Commercial Counsellor: Dr Maximilian Hurnaus
Chancellor: Hans 1Schonlau
Counsellor Press Affairs: Dr Thomas Bagger [98]

31-00690 Embassy of Greece
Na Orechovce 19, 16000 Prague 6
☎ (+420 2) 35 42 79 ⊡ 121545 ⊖ (+420 2) 3121849 [98]

31-00830 Embassy of India
Valdstejnska 6, Mala Strana, 12528 Prague 1
☎ (+420 2) 57 32 02 66; 57 32 02 60
E indembprague@bohem-net.cz ⊡ 121901 inde c
⊖ (+420 2) 53 94 95
Ambassador: H E Mr Gisish Dhume
☎ (+402 2) 57 32 00 72
Counsellor (Commercial) and Head of Chancery: T Topden
☎ (+402 2) 53 91 00
Military Attaché: Gp Capt J S Gujral ☎ (+402 4) 53 91 04
Head of Chancery: Y C Nazang
Attaché: V N Mehta ☎ (+402) 37 85 84
Attaché: S Raghavan ☎ (+402) 311 19 46
Attaché: Muhemler Singh Kanyal ☎ (+402) 37 22 73
Attaché: V K Bansal ☎ (+402) 37 68 27 [98]

31-00840 Embassy of the Republic of Indonesia
Nad Bud'ankami II/7, Smichov, 150 00 Prague 5
☎ (+420 2) 52 28 20; 52 60 41; 52 60 42; 52 60 43
⊠ INDONESIA Praha ⊡ 121 443 indo c
⊖ (+420 2) 57 21 21 05
Ambassador: H E Mr Leornard Tobine
Commercial Counsellor: Utsnu Edi Prtionyo
Consul: H P Harjanto
Public Information Officer: Krishna Djelani [98]

31-00890 Embassy of Italy
Nerudova 20, 110 00 Prague 1
☎ (+420 2) 57 32 00 11 ⊖ (+420 2) 57 32 02 57
Ambassador: H E Mr Maurizio Moreno
Deputy Chief of Mission: Puccio Francesco
Military and Air Attaché: Gialomo Carulli
Commercial Secretary: Andrea Orizo [98]

31-00920 Embassy of Japan
Maltézské námesti 6, 11000 Mala Straana, Prague 1
☎ (+420 2) 57 32 05 61 ⊖ (+420 2) 53 99 97 [97]

31-01190 Embassy of Mexico
Nad Kazankou 8, 171 00 Prague 7
☎ (+420 2) 855 55 54; 855 00 90; 855 15 39; 854 14 71 ⊖ (+42 2) 855 04 77
Ambassador: H E Mr Gonzalo Enrille ☎ (+42 2) 37 61 59
Second Secretary, Political and International Affairs: Porfirio Martinez Morales
Attaché, Secretary: Mrs Araceli Otero Diaz
Attaché, Press Affairs: Denis Reynaud Pulido [98]

31-01270 Royal Netherlands Embassy
Maltézské nám 1, 22540 Prague 1
☎ (+420 2) 57 32 00 68 ⊠ HOLLANDIA Praha 1
⊡ 122643 neda c ⊖ (+420 2) 57 32 00 94
Ambassador: H E Mr P F C Koch ☎ (+420 2) 37 64 41
Counsellor: Ms Renée V M Jones-Bos
☎ (+420 2) 57 31 16 20
First Secretary, Press and Cultural Attaché: Z R Wijchers
☎ (+420 2) 472 74 48
First Secretary, Environmental Affairs: Thymen 1Kouwenaar ☎ (+420 2) 24 51 09 13
Second Secretary, Commercial Affairs: Wouter W 1Plomp
☎ (+420 2) 35 23 01
Attaché, Administrative and Consular Affairs: Fred Kunst
☎ (+420 2) 581 56 66
Assistant Attaché: Hans D A Derksen
☎ (+420 2) 52 68 63
Archivist: Arno V M Heuschen ☎ (+420 2) 581 07 66
Archivist: Ms Carola B T Hakhof ☎ (+420 2) 301 59 22
Archivist, Secretary: Ms Esther M Bouwmeester
☎ (+420 2)316 34 37
Secretary to the Defence Attaché: Mrs Anita Winterink
☎ (+420 2) 20 92 14 34
Counsellor, Agricultural Section: Henderikus R Toxopeus (resides in Vienna)
Agricultural Section: Mrs Ingrid Hamid-Hardenberg
Counsellor, Transportation Section: Michael J Goppel (resides in Warsaw)
Military, Air and Ground Forces Attaché: Lieut Col E A Imkamp ☎ (+420 2) 402 28 88 [98]

31-01655 Embassy of the Republic of Slovenia
Pod Hradbami 15, 16000 Praha 6
☎ (+420 2) 32 08 72; 343 14 31; 32 58 14
⊖ (+420 2) 32 08 64; 24 31 41 06
Ambassador: H.E. Mr Zvone Dragan [98]

31-01680 Embassy of Spain
Pevnostni 9, 162 00 Prague 6
☎ (+420 2) 24 31 14 41; 24 31 12 22 ⊡ 121974 spain c
⊖ (+420 2) 312 17 70; 24 31 15 76 (military office); 24 31 13 18 (press office)
⊖ (+420 2) 312 08 13 (cultural office)
Ambassador: H.E. Mr Roberto Bermúdez
☎ (+42 2) 24 31 13 63; 32 49 37
Minister Counsellor: Antonio F. Mazarambroz Bernabéu (resides in Bratislava)
Political Affairs Counsellor: Milagros Hernando Echevarría
Defence & Military Attaché: Col Miguel Giménez Martínez
Cultural & Administrative Attaché: Ms Itziar TaboadaAquerreta
Press Affairs Counsellor: Javier Melendo Gaspar
Tourism Affairs Counsellor: José Maria Bethencourt Fontenla

Commercial Section
Stepánská 33, 225 03 Prague 1
☎ (+420 2) 24 21 34 61; 24 23 51 21; 26 46 35
⊡ 122673 ⊖ (+420 2) 24 21 34 61
Commercial and Economic Affairs Counsellor: Inigo Gil-Casares [97]

31-01740 Embassy of Switzerland
Pevnostni 7, 162 00 Prague 6
☎ (+420 2) 24 31 12 28 ⊠ AMBASUISSE Prague
⊖ (+420 2) 24 31 13 12
Ambassador: H E Mr Walter Fetscherin
Third Secretary: Lorella Renggli Odin [98]

31-01890 Embassy of the United Kingdom
Thunovská 14, 118 00 Prague 1
☎ (+420 2) 57 32 03 55 ⊡ 121011 prdm c
⊖ (+420 2) 57 32 10 23
Ambassador: H E Sir David Broucher

Contact Reference

Deputy Head of Mission: J M Cresswell
Defence Attaché: Col W E Nowosielski-Slepowron
First Secretary, Political: A J R Jackson
First Secretary, Commercial: M L Conner
Second Secretary, Commercial: M P Twigg [98]

31-01900 **Embassy of the United States of America**
Trziste 15, 11801 Prague 1
☎ (+420 2) 24 51 08 47 ⊠ 212196 amembc
⊖ (+420 2) 24 51 10 01
Ambassador: H E Ms Jenonne Walker
Deputy Chief of Mission: Eric Edelman
Head of Commercial Section: Ms Kathleen Kriger (acting)
Defence Attaché: Col David Potts (USAF)
Public Affairs Officer: Leonado Williams [97]

55 Government-owned Industry & Production Facilities

55-00200 **Aero Vodochody Ltd**
(an associated company of Aero Holding)
250 70 Odolena Voda
☎ (+420 2) 688 09 71; 688 11 40; 66 41 00 41; 688 00 41 ⊠ 121169 aero c
⊖ (+420 2) 82 31 72
♤ CSk 4,200,000,000 (1994)
♣ CSk 4,291,000,000 (1994) ♚ 2,330
President: Zdenik Chalupnmk
Vice President Sales and Marketing: Zdenik Prokop
Technical Director: Adam Stranák
Marketing Director: Milos Valis
Chief Designer for Military Programs: Dobroslav Rak
Chief Designer for Civil Programs: Jan Mikula
■ Production of the training and light attack aircraft Aero L-39, L-59 and L-139. Development of the L-159 training and light multi-role combat and Aero Ae-270 utility aircraft. → 75 [97]

55-00850 **Letov Prague**
(an associated co. of Aero Holding)
Beranovych 65, 199 02 Prague 9
☎ (+420 2) 66 31 19 33; 66 31 19 02 ⊠ 11430 letp c
⊖ (+420 2) 858 71 75
♣ US$ 6,500,000 (1994) ♚ 1 118
Managing Director: Vaclav Matousek
Commercial Director: Zdenek Badura
Aviation Division Director: Ivan Dubsky
Electronics Division Director: Miroslav Vodicka
■ Production of flight simulators, flight training devices, ejection seat simulators, ULM aircraft. [97]

55-01500 **Prvni Brnenska Strojirna (PBS)**
Vlkovska 279, 595 01 Velka Bites
☎ (+420 619) 96 27 11 ⊖ (+420 619) 96 25 95
Managing Director: Jan Solar ☎ (+420 619) 96 27 50
Finance Director: Vitezslav Musil ☎ (+420 619) 96 26 07; 96 26 07 (fax)
Director of Aviation Division: Zdenek Katolicky ☎ (+420 619) 96 27 52; 96 29 34 (fax)
Head of Sales Department: Zdenek Kralik ☎ (+420 619) 96 29 34; 96 29 34 (fax)
■ Manufacturer of aviation and special accessories equipment for turbine engines. [98]

55-05000 **VOP 025 Novy Jicín, s.p.**
Dukelská 102, 74242 Senov u Noveho Jicina
☎ (+42 656) 78 33 16 ⊘ (+42 656) 78 31 11 ⊠ 52196
⊖ (+42 656) 70 17 34
♤ Kcs 380,958,000 (1995) ♣ Kcs 483,866,000 (1995)
♚ 750 ⌂ 179,200 sq m
Managing Director: Ostar Sima
Technical Director: Adolf Vermirovsky
Sales Director: Lumir Hajny
Production Director: Miroslav Lesenar
Advertising Manager: Miroslav Adamcik
Purchasing Manager: Vrazel Svatopluk
■ Manufacture of armoured recovery vehicles based on the T-55 chassis and modernisation of T-72 tanks.
Products: T-34, T-54, T-55, T-72 tanks and their modifications; modernised tanks T-72 and T-55 AM2; bridge laying tank MT-55A; recovery tank VT-55A; field engineer tanks; dynamic protection of tanks T-55A & T-72 (explosive reactive armour); bulldozer attachments; chemical decontamination vehicles; mobile workshops for repair of tanks, automobiles, infantry combat vehicles' electrical equipment and artillery material; spare parts for tanks; special tank-based vehicles (extinuishing vehicles, snow ploughs, earth moving vehicles).
Services: Know-how of T-55 & T-72 repair; repair and modernisation of tanks; training in military materiel exploitation; scrapping and discarding of military materiel. [98]

55-06000 **Walter Prague**
(f Motorlet Prague)
(a subsidiary of the Aero Holding)
Jinonická 329, 158 01 Prague 5 - Jinonice
☎ (+420 2) 52 96 11 11 [E] marketing@walter.cz
[W] http://www.pressline/com/cz/walter _as
⊖ (+420 2) 52 60 60
♤ US$ 50,000,000 (1996) ♣ US$ 17,600,000 (1996)
♚ 1,100 ⌂ 40,000 m2
Managing Director: Václav Vanek
Commercial Director: Michal Cerny
Research and Development Director: Oldrich Matousek ☎ (+420 2) 57 21 63 14
Marketing Manager: Michal Ptacnie
■ Production and repair of turboprop engines Walter M-601 and M-602; piston engines Walter M-202. → 75 [98]

60 Research Organizations, Institutes & Test Establishments

60-00150 **Aeronautical Research & Test Institute (VZLU)**
Vyzkummy a Zkusebni Letecky Ustav (VZLU)
(an associated company of Aero Holding)
Beranovych 130, 199 05 Prague 9 Letnany
☎ (+420 2) 66 31 13 97 [E] vzlu@vzlu.anet.cz
[W] http://www.ms.anet.cz/~vzlu ⊠ 121169 vzlu c
⊖ (+420 2) 859 06 53; 66 31 05 18
♤ Csk 750,000,000 (1995) ♣ Csk 186,000,000 (1995)
♚ 361 (1997) ⌂ 40,000 m2
Managing Director: Ing Milan Holl
Commercial Director: Stanislav Cekal
Marketing Director: Stamislav Kavale
Purchasing Manager: Marie Breussova
Information Manager: Ing Ladislav Vymetal
■ Research, development and testing in the fields of civil and military aviation. Basic and applied research and testing of aeronautical products. [98]

60-01000 **Prototypa-ZM, s.r.o.**
Hudcova 78, 61200 Brno
☎ (+420 5) 41 51 36 25 ⊖ (+420 5) 41 51 36 81
♤ US$ 52,000 (1997) ♣ US$ 1,300,000 ♚ 38 (1997)
⌂ 2885 m2
Managing and Operations Director: Dipl Ing Jiri Nepovím
Sales and Purchasing Director: Dipl Ing Zdenek Horak
Public Relations Manager: Dipl Ing Jirí Klecka, CSc
■ Research, development, design and testing of weapons and ammunition up to 30 mm, anti-tank weapons and severe climatic testing of arms and ammunition, limited deliveries of boosting rocket motors, parachute recovery systems and launching ramps on trucks for the UAV SOJKA system, special anti-terrorism means, ballistic testing barrels for military hunting rounds in standard calibers. [98]

60-02000 **Synthesia a.s - Vúpch**
532 17 Pardubice
☎ (+420 40) 682 57 00 [E] vupch@pce.czn.cz
⊠ 194235syntc ⊖ (+420 40) 682 29 35
Director: Miroslav Horacek ☎ (420 40) 682 57 00
■ Basic and applied research activities. [98]

70 Financial Holding Groups, Parent Companies

70-00100 **Aero Holding a.s. Headquarters**
(The Group of Czech Aeronautical Companies)
Beranovych 130, 199 04 Prague 9 - Letnany
☎ (+42 2) 66 31 07 27; 88 40 65; 88 27 47
⊖ (+42 2) 88 65 81; 88 13 40; 88 27 47
♤ US$ 385,000,000 (Group 1994)
♣ US$ 370,000,000 (Group 1994) ♚ 7,126 (Group)
Chairman of the Board: Radim Kriz
General Director: Frantisek Petrásek
Technical and Commercial Affairs Director: Jan Barton
Public Relations: Zdenek Burian
■ Parent company of group manufacturing: aircraft, aircraft engines, aircraft instruments, hydraulic equipment, landing gears and aircraft accessories. [98]

75 Manufacturing Companies

Aero Vodochody Ltd 55-00200
LET a.s. 75-00550
Walter Prague 55-06000

75-00080 **Artisys s.r.o.**
Tumova 44, 616 00 Brno
☎ (+420 5) 41 32 12 34; 41 24 39 80 [E] art@traveller.cz
⊖ (+420 5) 44 21 76 06
♤ Cks 102,000 (1995) ♣ Cks 2,300,000 (1996) ♚ 8
⌂ 150 sq m
Managing Director: Dipl Ing Dana Brehelova
Operations and Technical Director: Jiri Sasek
Systems Specialist: Petr Bahula
Representative: Dipl Ing Dana Brhelová
■ Real-time information and control systems, focused on aeronautics. Products include: ATC Simulator RYS; technical/control monitoring workstations for civil and military applications; AMIS - system for radar data visualization and data exchange; 3D radar data information system - secondary and tertiary radar data processing; off-line mapper for creation of machine videomaps with joined data import. → 80
Representing:
Lynx Real-Time Systems Inc, USA
Kinesix Corporation, USA [97]

75-00300 **Ceska Zbrojovka**
Sv Cecha 1283, 688 27 Uhersky Brod
☎ (+420 633) 65 11 11 [E] marketing@czub.cz
[W] http://www.czub.cz ⊠ 60200 ubr cz
⊖ (+420 633) 63 36 65
♤ US$ 22,900 (1996) ♣ US$ 46,200(1996)
♚ 2,700 (1996)
President: Dr Ing Karel Kopp
Managing Director: Ing Jiri Martinec
Operations Director: Ing Roman Vaclevik
Marketing Director: Ing Jiri Koutny
Marketing Manager: Ing Miroslav Vesely
■ Manufacture guns, ammunition, aircraft engines, cutting and special tools. [98]

75-00380 **Eldis s.r.o.**
Prumyslová 387, 530 03 Pardubice
☎ (+420 40) 54 52 82; 54 82 83
⊖ (+420 40) 208 80; 661 12 47
Managing Director: Petr Krejcar
Marketing Director: Frantisek Remta
Signal Processing: Martin Brát
■ Products and services include: approach control centre equipment; air defence centre equipment; airport tower equipment; radar modernisation; training simulators.
Sub-systems comprise: radar display systems; radar data digital logging systems; multi-channel voice communication; digital recorders; radar signal processors; radar data converters; remote monitoring and control systems; unified time distribution systems. [97]

75-00400 **ERA a.s.**
Prumyslová 387, 530 03 Pardubice
☎ (+420 40) 224 19 [E] era.pce@era.anet.cz
⊖ (+420 40) 224 19
♤ Kcs 1,000,000 ♣ Kcs 10,000,000 ♚ 41 ⌂ 400 sq m

Managing Director: M Sotona
Technical Director: V Kubecek
Marketing Director: M Bernard
Marketing: Jan Hrabánek
Project Manager: Pavel Sterba
■ ATC systems; airport surface movement controls; low level non-radar surveillance systems; voice and data recording and logging systems for ATC. 97

75-00550 LET a.s.
(a subsidiary of Aero Holding)
Kunovice 1177, 686 04 Kunovice
☎ (+420 632) 51 11 11 E let@let.cz
W http://www.let.cz ☒ 60387 let c ⊖ (+420 632) 613 52
1,723
Managing Director: Josef Skrásek ☎ (+420 632) 611 23
General Director: Zdeněk Pernica ☎ (+420 632) 611 20
Technical Director: Jan Fridrich ☎ (+420 632) 616 17
Finance Director: Frantisek Laga ☎ (+420 632) 615 16
Commercial Director: Zbynek Sabata ☎ (+420 632) 615 18
Chief Designer: Miroslav Pesák ☎ (+420 632) 614 80
Purchasing Director: Frantisek Ancik ☎ (+420 632) 626 48
Production Director: Miroslav Zamecnik ☎ (+420 632) 611 24
Chief Pilot: Jaromír Novák
Marketing Manager: Milan Morkus
■ Development, production, assembly, maintenance, repair, sales and marketing and product support of L410 and L610 twin turboprop commuter aircraft and L23 and L33 sailplanes. → 75 98

75-00700 Nowatron Elektronik s.r.o.
Herspicka 6, 639 00 Brno
☎ (+420 5) 43 16 93 61; 43 16 93 62
⊖ (+420 5) 43 16 93 54
Group Manager: Dipl Ing Robert Odehnal
Representing:
Barco Monitors
Barco Projectors
Kontron Elektronik
Teknor
Getac
Emerson Computer Power 97

75-00750 Omnipol Ltd
(† Omnipol Foreign Trade Corp)
Nove Mesto, Nekazanka 11, 112 21 Prague 1
☎ (+42 2) 24 01 11 11 ☒ OMNIPOL Praha
E omnipol@gls.cz ☒ 121297 ⊖ (+42 2) 24 01 22 41
♤ Kcs 1,616,000,000 (1996)
♣ Kcs 2,552,000,000 (1996) 263
General Director: Pavel Musela
Chief of Board: Richard Hava
Export Division Director: Petr Vesely
Import Division Director: Michal Hon
Marketing Director: Peter Jacina
Public Relations Manager: Alexander Petrzelka ☎ (+420 2) 24 01 21 49 ☎ (+420 2) 24 01 21 18 (fax)
■ Trading company for defence materials, aircraft and spare parts. Also modernise and lease aerospace products. → 90 98

75-00820 Rockwell LVS Liberec as
Ceske mladeze 452, 460 08 Liberec 8
☎ (+420 48) 522 91 11 ⊖ (+420 48) 48 54 32
Plant Manager: Petr Hora E pehora@auto.rockwell.com
® USA: Rockwell International Corp. 98

75-00900 Sellier & Bellot a.s. Joint Stock Company
Lidika 667, 25813 Vlasim
☎ (+420 303) 59 1111 ⊖ (+420 303) 432 83
♣ US$ 25,000,000 2,200
General Manager: Ing. Karel Kopecek
Public Relations and Marketing Director: Ing Milos Dolezal
Operations Director: Ing Ludek Tichy
Sales Manager: Miroslav Kurilla
■ Ammunition for hand arms, pistols and revolvers for infantry, sport and hunting purposes.
Pyro products: initiators, fuses, detonating fuses, machines, lines and tools for the production of ammunition, explosion technology equipment. 98

75-01000 Synthesia a.s.
(† Vchz Synthesia)
(a division of Synthesia a.s.)
Explosia Division, 532 17 Pardubice
☎ (+420 40) 682 11 11 ⊖ (+420 40) 682 29 40
♤ Kcs 1,036,000 (1996) ♣ Kcs 895,000,000 (1996)
1,240 (1996)
Director of Division: Karel Lorenz
Operations Director: Jaroslav Nimrichter
Sales Director: Jan Rez
Purchasing Manager: Jaroslava Matulova
■ Solid propellants, smokeless powders, pyrotechnical charges and explosives. 98

75-01050 TATRA JSC
Stefánikova 1163, 74221 Koprivnice
☎ (+420 656) 72 11 41 ☒ 527716
⊖ (+420 656) 89 44 81
♣ Kcs 5,200,000,000 (1995) 8,500
Managing Director: Karel Beneda
Sales Director: Zbynek Keisler
Operations Director: Jan Vala
Advertising and Public Relations Manager: Ms Erika Síkorová
Purchasing Manager: Jirí Plíhal
■ Manufacture military transport vehicles. 97

75-01075 Technometra Radotín
(a subs. of Aero Holding)
Vrázská 239, 153 05 Prague 5
☎ (+42 2) 54 92 41 ☒ 122307 tmra c
⊖ (+42 2) 54 95 27
♤ US$ 5,994,258 (1996) ♣ US$ 3,431,900 (1996)
285
Managing Director: Stanislav Vránek ☎ (+420 2) 59 44 24
Commercial Director: Milan Pulkrábek ☎ (+420 2) 59 44 24
Technical Director: Vladimír Andrlik ☎ (+420 2) 59 44 24
Purchasing Manager: Zdenka Seidlova
■ Production of landing gears for L-39, L-59, L-139, Ae-270, L-23 and L-33, aircraft hydraulic systems. 98

75-01100 THORN Lighting CS spol sro
Bila Hora, Chynska 9, 6 Prague
☎ (+420 2) 302 12 40 ⊖ (+420 2) 302 10 44
Chief Executive: Pavel Sedekk 98

75-03100 Zbrojovka Vsetin
Jasenice 1254, 755 37 Vsetin
☎ (+420 6) 57 27 68 ⊖ (+420 6) 57 42 64; 57 47 71
♣ US$ 36 430 000 (1994) 4,000
Chairman: Pavel Novosad
Vice Chairman: Milevoj Slacálek
Managing Dir: Jiri Mikes
Commercial Dir: Karel Dancák
Financial Dir: Radomír Fic
Dir of Controlling: Michal Pilarsky
Personnel Dir: Josef Svácek
Technical Dir: Vlastimil Stradej
Production Dir: Stanislav Skabraha
Board Member: Jirí Vanek
Board Member: Zdenek Raich
Board Member: Radek Vana
■ Manufacture of arms & anti-terrorist devices. 97

80 Service Companies & Organizations

Artisys s.r.o. 75-00080

80-00100 AlliedSignal Engines International Turbine Engine Company (ITEC)
(a subsidiary of AlliedSignal Aerospace, USA)
Aero Vodochody, 67-20/ITEC, 250 70 Odolena Voda
☎ (+420 2) 685 75 79 ⊖ (+420 2) 685 75 79
Manager, L159 Technical Liaison: Peter T Klima
■ Provide technical support for engine aircraft certification. 98

90 Sales Agents, Sales & Liaison Offices

Companies represented:

Alfred Karcher GmbH & Co, Germany . . 90-01100
Rockwell International Corp., USA . . . 75-00820
Saft SA, France 90-03000
Trimos-Sylvac Task Force, Switzerland . 90-00800

Sales agents/representatives listed elsewhere:

Omnipol Ltd 75-00750

90-00200 Beatronic Supply
(a subsidiary of Beatronic Supply Aps)
Sumavska 31, 612 54 Brno
☎ (+420 5) 41 23 55 17 ⊖ (+420 5) 41 23 51 92
Managing Director: Jan Webber 98

90-00500 Daewoo Corp Prague Branch
Snemovni 9, 110 00 Prague 1
☎ (+42 2) 24 51 18 00 ☒ 123563 dwph c
⊖ (+42 2) 24 51 17 90 98

90-00800 V. Fialová s.r.o.
Pod kastany 23, 602 00 Brno
☎ (+42 5) 74 13 32 ⊖ (+42 5) 74 13 32
Managing Director: Vlasta Fialová
Marketing Director: Tomás Fiala
® Switzerland: Trimos-Sylvac Task Force 98

90-01100 Karcher CR Spol s.r.o.
Livanska 1880, 19300 Prague 9
☎ (+42 2) 652 16 65 ⊖ (+42 2) 652 16 65
® Germany: Alfred Karcher GmbH & Co 98

90-01500 Logica s.r.o.
(a subsidiary of Logica plc, UK)
Vaclavske nam 66, Mezibranska 23, 110 00 Prague 1
☎ (+420 2) 24 22 51 39 ⊖ (+420 2) 24 22 54 31 98

90-03000 Saft Ferak A/S
(a subsidiary of Saft SA, France)
Okr Frydek mistek, 73907 Raskovice
☎ (+420 658) 932 21; 932 25 ⊖ (+420 658) 932 26
General Manager: Jean Pierre Cittanova
® France: Saft SA 98

Denmark

Head of State:
H.M. Queen Margrethe II

10 Head of State

10-10000 Royal Danish Court
Det Gule Palae, DK-1256 Copenhagen K
☎ (+45) 33 40 10 10 ☒ 27585 Ambhf dk
⊖ (+45) 33 93 05 60
Queen of Denmark: H.M. Queen Margrethe II
The Lord Chamberlain's Department
Det Gule Palae, Amaliegade 18, DK-1256 Copenhagen K
☎ (+45) 33 40 10 10 ☒ 27585 Ambhf dk
⊖ (+45) 33 93 05 60
Lord Chamberlain: Søren Haslund-Christensen
Master of Ceremonies: Col Christian Eugen-Olsen
Royal Guard
☎ (+45) 33 13 56 03 ⊖ (+45) 33 15 04 06

Royal Guard Commander in Chief: Col Jens Greve
☎(+45) 42 81 36 88 [98]

11 Prime Minister

11-10000 Office of the Prime Minister
Statsministeriet
Christiansborg, Prins Joergens Gaard 11,
DK-1218 Copenhagen K
☎(+45) 33 92 33 00 ⊠ 27027 stat dk
⊖ (+45) 33 11 16 65
Prime Minister: Poul Nyrup Rasmussen
Permanent Secretary of State: Ulrik Federspiel
Asst Secretary of State: Soren Thorup
Asst Secretary of State: Karsten Dybvad
Principal Private Secretary to the Prime Minister:
Peter Lauritzen [98]

12 Ministries

12-09000 Ministry of Business and Industry Agency for Trade and Industry
Tagensvej 137, DK-2200 Copenhagen N
☎(+45) 35 86 86 86 ⊠ 22373 indumi dk
⊖ (+45) 35 86 85 75
Minister of Business and Industry: Jan Trójborg
Head of Department: Bent L Andersen
Counsellor, Industrial Participation: Lars-Bjørn Larsen [98]

12-10000 Ministry of Defence
Holmens Kanal 42, DK-1060 Copenhagen K
☎(+45) 33 92 33 20 ⊠ 27190 fmn dk
⊖ (+45) 33 32 06 55
Minister for Defense: Hans Haekkerup
Head of Department Planning and Maintenance:
Henrik H Ekmann
Secretary to the Minister: N H Hedegaard
Permanent Under-Secretary of State: Anders Troldborg
Director for Materiel Procurement: Jørn Olesen
Head of Procurement and Logistics Division: E T Pedersen
Information & Welfare Services
Ryvangs Alle 1, PO Box 857, DK-2100 Copenhagen 4
☎(+45) 39 27 68 00 ⊖ (+45) 39 27 61 60
Director: Preben Dich
Information Officer: Maj. L.F.S. Nielsen [98]

12-20000 Royal Danish Ministry of Foreign Affairs
Udenrigsministeriet
2 Asiatisk Plads, DK-1448 Copenhagen K
☎(+45) 33 92 00 00 ✉ ETRANGERES Copenhagen K
⊠ 31292 etr dk ⊖ (+45) 31 54 05 33
Minister for Foreign Affairs: Niels Helveg Petersen
Minister for Development Co-operation: Poul Nielson
Permanent Secretary, Director:
Ambassador Mr Henrik Wøhlk
State Secretary, the North Group:
Ambassador Mr J Ørstrøm Møller
State Secretary, the South Group:
Ambassador Mrs Ellen Margrethe Løj
Under-Secretary, Secretariat for Foreign Trade:
Ambassador Mr Ulrik Helweg-Larsen
Under-Secretary, Joint Secretariat:
Ambassador Mr Hans Michael Kofoed-Hansen
Commercial Counsellor: Ditte Jessing [98]

20 Defense Administrations/Services/Agencies

20-04300 Defense Construction Service (DCS)
Forsvarets Bygningstjeneste
Frederiksholms Kanal 30, DK-1220 Copenhagen K
☎(+45) 33 13 47 01 ⊖ (+45) 33 32 10 29
Director: B. Frank [98]

25 Defense Forces

25-10000 Headquarters Chief of Defence Denmark
PO Box 202, DK-2950 Vedbaek
☎(+45) 42 89 22 55 ⊖ (+45) 42 89 07 48
Chief of Staff: Gen. (AF) Christian Hvidt
Chief of Defence Staff:
Maj.Gen. (A) Ove Hoeegh-Guldberg Hoff
Deputy Chief of Staff, Personnel: RAdm. Poul Schriver
Deputy Chief of Staff, Control & Audit:
Maj.Gen. (A) Jens Johansen
Deputy Chief of Staff, Operations, Budgets & Finance:
Maj.Gen. (AF) Klaus L. Axelsen
Dpty Chief of Staff, Materiel Infrastructure & Logistics:
Maj.Gen. (A) Jens C. Frandsen
Asst Chief of Staff, Materiel Division:
Capt. (N) Paul B. Sørensen
Asst Chief of Staff, Infrastructure & Logistics:
Col (AF) Jens Sannom
Asst Chief of Staff, Operations & Management Div.:
Capt. (N) Kurt B. Jensen
Asst Chief of Staff, Budget & Finance Division:
Col. (A) Gunnar Johansen
Chief of Information Branch:
Lt Col (A) Arne Beier-Holgersen [97]

25-20000 Tactical Air Command
Koelvraa, DK-7470 Karup J
☎(+45) 99 62 49 50 ⊖ (+45) 99 62 49 55
Commander: Maj.Gen. Laurits Tophøj [97]

25-22100 Air Materiel Command (AMC)
Flyvematerielkommandoen (FMK)
PO Box 130, DK-3500 Vaerloese
☎(+45) 44 78 22 55 ⊖ (+45) 44 66 25 33
Chief: Maj Gen Viggo D Nielsen
Chief, Public Information: Lieut Col Ole Hertel [98]

25-30000 Army Operational Command
Headquarters
PO Box 59, DK-7470 Karup J.
☎(+45) 99 62 49 70 ⊖ (+45) 99 62 49 65
Commanding General: Maj.Gen. H.G.G. Grüner
Chief of Staff: Col L. Scheel
Asst Chief of Staff Plans Division: Lt Col H.P. Christensen
Asst Chief of Staff Management Division: Lt Col S. Krarup
Asst Chief of Staff Operations & Production Division:
Col E. Kusk
Chief of Administration: Maj. H.E. Andersen
Press & Information Officer: Maj. P.E. Christiansen
Det Kgl Garnisons Bibliotek
Kastellet 42, DK-2100 Copenhagen Ø
☎(+45 1) 11 22 33 [97]

25-32100 Army Materiel Command
Arsenalvej 55, DK-9800 Hjoerring
☎(+45) 99 24 11 11 ⊖ (+45) 99 24 11 19
Chief Army Materiel Command: Maj Gen Henrik H Ekmann
Chief of Staff: Col J Jensen
Director, Army Maintenance and Supply Service:
Col S Krarup
Director, Army Technical Service: T Hansen
Director, Army Depot Area: Col B B Nielsen
Director, Army Main Workshop: E T Østergaard
Director, Ammunition Arsenal: F J Andersen [98]

25-32200 General Medical Services Armed Forces Denmark
PO Box 149, DK-2820 Gentofte
☎(+45) 39 68 16 55 ⊖ (+45) 39 76 04 65
Director: Maj.Gen. Knud Jessen (Medical Corps) [97]

25-35000 Island Commander Faeroes
PO Box 195, DK-110 Thorshavn, Faroes
☎(+298) 114 20 ⊖ (+298) 194 20
Commander: Capt Jesper Dinesen
Deputy Commander/Chief Staff Officer:
Cdr Finn Christensen
Director of Operations: Lieut Cdr Steen Petersen
Director of Procurement/Logistics: Lieut Cdr Peter Gram [98]

25-40000 Royal Danish Navy
Admiral Danish Fleet Headquarters
PO Box 483, DK-8100 Aarhus C
☎(+45) 89 43 30 99 ⊖ (+45) 89 43 31 41
Commander, Admiral Danish Fleet:
RAdm. Kristen Husted Winther ☎(+45) 89 43 30 01
Chief of Staff: Capt N A Binder ☎(+45) 89 43 30 56
Commandant: Cdr L. Hoel ☎(+45) 89 43 31 00
Assistant Chief of Staff, Operations Division:
Cdr s.g. P Cortes ☎(+45) 89 43 30 07
Point of Contact, Operations Division: B. Kristjansson
☎(+45) 89 43 31 15
Asst Chief of Staff, Material & Info Systems Division:
Cdr O. Saugbjerg ☎(+45) 89 43 30 20
Point of Contact, Material & Info Systems Division:
H. Kristensen ☎(+45) 89 43 31 17
Asst Chief of Staff, Planning Division:
Cdr L. Wille-Jørgensen ☎(+45) 89 43 30 30
Point of Contact, Planning Division: H. Drescher
☎(+45) 89 43 31 16
Assistant Chief of Staff, Budget and Finance Division:
Cdr s.g. N J Priisholm Hansen ☎(+45) 89 43 30 50
Point of Contact, Budget & Finance Division: T. Hulgaard
☎(+45) 89 43 31 18
Assitstant Chief of Staff, Personnel & Training Division:
Capt E Schramm ☎(+45) 89 43 30 70
Point of Contact, Personnel and Training Division: I Hansen
☎(+45) 89 43 30 77
Point of Contact, Personnel and Training Division:
T Johansen [98]

25-42100 Naval Materiel Command (RDNMC)
Sovaernets Materielkommando
Danneskiold-Samsoes Allé 1, Holmen,
DK-1434 Copenhagen K
☎(+45) 32 66 32 66 ⊠ 22372 smk dk
⊖ (+45) 32 66 32 99
Commander RDNMC: RAdm. N. Mejdal [98]

25-43000 Action Information School (AIS)
Soevaernets Kampinformationskursus
Naval Base Frederikshavn, DK-9900 Frederikshavn
☎(+45) 99 22 30 25 [E] tvais@inet.uni-g.dk
⊖ (+45) 99 22 30 35
Commander, Chief of Materiel and System Section:
Cdr Danny W Qvist Nepper [98]

31 Embassies & Attachés

31-00310 Embassy of Canada
Kr Bernikowsgade 1, DK-1105 Copenhagen K
☎(+45) 33 12 22 99 ✉ DOMCAN Copenhagen
⊠ 27036 dmcnc dk ⊖ (+45) 33 14 05 85
Ambassador: H E Mr Brian E Baker
Canadian Forces Attaché: Col Leslie A White
(based in Oslo)
Commercial Counsellor: Guy Cadieux
Attaché and Vice Consul: Jocelyne Carmen Fortier [98]

31-00350 Embassy of Chile
Kastelsvej 15/3, DK-2100 Copenhagen K
☎(+45) 31 38 58 34 [E] chiledkáinet.uni-c.dk
⊠ 15099 chile dk ⊖ (+45) 31 38 42 01
Ambassador: H E Raul Schmidt
First Secretary: Enrique Barriga
Commercial Counsellor: Victor Olmedo
(resident in Stockholm)
Labour Attaché: Aldo Verdugo [98]

31-00450 Embassy of the Czech Republic
Ryvangs Allé 14, DK-2100 Copenhagen Ø
☎(+45) 39 29 18 88 ⊖ (+45) 39 29 09 30
Ambassador: H.E. Mr Alois Buchta
Deputy Chief of Mission: Jan Misiarz
Commercial Office
Ryvangs Allé 14, DK-2100 Copenhagen Ø
☎(+45) 39 28 18 88 ⊖ (+45) 39 29 09 30
Commercial Counsellor: Martin Vitek [97]

31-00535 Embassy of Estonia
Aurehojvej 19, DK-2900 Hellerup-Copenhagen
☎(+45) 39 40 26 66 [E] seketar@estemb.dk
⊖ (+45) 39 40 06 88 (consular office)
Ambassador: H E Mr Arvo-Juergen Alas

Honorary Consulate of Estonia
4 Havnegarde, DK-8000 Aarhus C
☎ (+45 86) 13 95 55
Honorary Consul: Soren Claus Kjaer [98]

31-00580 Embassy of Finland
Skt. Annae Plads 24, DK-1250 Copenhagen K
☎ (+45) 33 13 42 14 ▣ 27084 finamb dk
Ambassador: H.E. Mr Ralf Friberg
Deputy Chief of Mission: Bo Lindroos
Commercial Secretary: Antti Saarinen [97]

31-00660 Embassy of the Federal Republic of Germany
Stockholmsgade 57, DK-2100 Copenhagen Ø
☎ (+45) 35 26 16 22 ▣ 27166 aakhgn dk
⊖ (+45) 35 26 71 05
Ambassador: H E Johann Dreher
Deputy Chief of Mission: Mrs Karin E Blumberger
Defense Attaché: Cdr S G Jochen Krambeer
Commercial Attaché: Bernhard A Zobel
Press Officer: Boris Ruge [98]

31-00880 Embassy of Israel
Lundevangsvej 4, DK-2900 Hellerup
☎ (+45) 39 62 62 88 ⊖ (+45) 39 62 19 38
Ambassador: H E Mr Abraham Setton
First Secretary: Aliza Bin-Noun Binnewies
Defence and Armed Forces Attaché: Col Itzchak Ganor
Counsellor (Administrative Affairs): Hanoch Shem-Tov
Counsellor (Scientific Affairs): Baruch Raz [98]

31-00890 Embassy of Italy
Gammel Vartov Vej 7, DK-2900 Hellerup
☎ (+45) 39 62 68 77 ▣ 27078 italdi dk
⊖ (+45) 39 62 25 99
Ambassador: H.E. Dr Giacomo Ivancich Biaggini
Counsellor Deputy Chief of Mission:
Dr Francesco Calogero
First Secretary: Dr Bruno Archi [98]

31-00920 Embassy of Japan
Pilestræde 61, DK-1112 Copenhagen K
☎ (+45) 33 11 33 44 (9.00-17.00);
33 11 30 40 (17.00-9.00) ⊠ TAISHI Copenhagen
▣ 27082 taishi dk ⊖ (+45) 33 11 33 77
Ambassador: H E Mr Takaya Suto
Counsellor (Deputy Chief of Mission): Masamichi Ishikawa
Secretary (Political Affairs): Toshio Ohmura
Press Secretary: Saburo Sato [98]

31-01050 People's Bureau of the Socialist People's Libyan Arab Jamahiriya
Rosenvaengets Hovedveg 4, DK-2100 Copenhagen Ø
☎ (+45) 35 26 36 11 ▣ 22652 alibya dk
⊖ (+45) 35 26 56 06
Chargé d'Affaires a.i.: Fauzi Milad Abu-Argub
Attaché: Abdulwahab Salem Fathallah
Administrative Attaché: Lutfia H Ghanimi
Financial Attaché: Mohamed Ali Al-Shakrbie [98]

31-01220 Embassy of the Kingdom of Morocco
Øregaards Allé 19, Hellerup, DK-2900 Copenhagen
☎ (+45) 39 62 45 11 ▣ 22913 maamb dk
⊖ (+45) 39 62 24 49
Ambassador: H E Mr Jamal Eddine Ghazi
Counsellor: M'hammed Chafik El Khatib
First Secretary: Mahfoud Bahbouhi
First Secretary: Lahssen Remmane
Second Secretary: Bahija Erroumanjou [98]

31-01340 Royal Norwegian Embassy
Amaliegade 39, DK-1256 Copenhagen K
☎ (+45) 33 14 01 24 ⊖ (+45) 33 14 06 24
Ambassador: H E Thorvald Stoltenberg
Deputy Chief of Mission: Bjørn Blokhus
Commercial Counsellor: Jon Strømberg [98]

31-01440 Embassy of Portugal
Hovedvagtsgade 6, DK-1103 Copenhagen K
☎ (+45) 33 13 13 01 ▣ 16586 emport dk
⊖ (+45) 33 14 92 14
Ambassador: H.E. Mr Rui de Meira Ferreira
Deputy Chief of Mission: Francisco Duarte Lopes
Economic Attaché: Ms Maria Fernanda Horta

Commercial Department
Gammel Torv 4, DK-1457 Copenhagen K
☎ (+45) 33 12 76 13/ 32 ▣ 22936 icep dk
⊖ (+45) 33 93 88 85
Commercial Counsellor: J. Estrela Leão [97]

31-01480 Embassy of Romania
Strandagervej 27, DK-2900 Hellerup
☎ (+45) 39 40 71 77; 39 40 70 02 (Commercial section)
▣ 27017 rocom ⊖ (+45) 31 62 78 99
Ambassador: H E Dr Ion Diaconu
Second Secretary Political Affairs: Vasile Cringanu
Second Secretary, Consul: Dorin Bálánicá
Defence, Military, Air and Naval Attaché: Gheorghe Nosca
First Secretary Economic Affairs: Costin Lianu
☎ (+45) 39 40 70 02
First Secretary Economic Affairs: Petru Toderici
Attaché: Laurentiu Vasile [98]

31-01655 Consulate of the Republic of Slovenia
Frederiksgade 50, DK-8000 Aarhus C
☎ (+45) 86 19 54 24 ⊖ (+45) 86 19 79 80
Honorary Consul: Knud Kristensen [98]

31-01680 Embassy of Spain
Upsalagade 26, DK-2100 Copenhagen Ø
☎ (+45) 33 42 47 00 ▣ 27145 esco dk
⊖ (+45) 31 26 30 99
Ambassador: H.E. Mr Antonio Oyarzabal
Deputy Chief of Mission: Miguel Bauza
Defence Attaché: Capt. Juan Luis Garcia Moreno
(resident in The Hague,) (Netherlands)

Commercial Office
Vesterbrogade 10, 3, DK-1620 Copenhagen V
⊠ OFCOMES Copenhagen V ▣ 27322 minco dk
⊖ (+45) 31 21 33 90
Commercial Officer: Leon Herrera [97]

31-01730 Swedish Embassy
Sankt Annae Plads 15A, DK-1250 Copenhagen K
☎ (+45) 33 36 03 70 ▣ 22933 svensk dk
⊖ (+45) 33 36 03 95
Ambassador: H E Mr Håkan Berggren
Deputy Chief of Mission: Bruno S Beijer
Military, Naval and Air Attaché: Brig Kaj Nielsen
☎ (+45) 33 14 22 42
Counsellor Administrative and Consular Affairs:
Mrs Anne Stymne
Information Secretary: Peter Stenbak

Office of the Commercial Attaché
Toldbodgade 18, 5, DK-1253 Copenhagen K
☎ (+45) 33 15 55 22 ⊠ SWEDCOM Copenhagen
⊖ (+45) 33 32 94 99
Commercial Counsellor: Per Arlborg [98]

31-01780 Royal Thai Embassy
Norgesmindevej 18, DK-2900 Hellerup
☎ (+45) 39 62 50 10; 52 57; 55 75 ▣ 16216 thaic dk
⊖ (+45) 39 62 50 59
Ambassador: H.E. Dr Rangsan Phaholyothin
Minister Counsellor: Prasittidesh Vichitsorasatra
First Secretary: Mrs Jirawal Wattanapul
First Secretary: Surapon Petra-vra
Attaché: Wanlop Eiamchim

Office of Commercial Affairs
H.C. Andersens Boulevard 39, DK-1553 Copenhagen V.
☎ (+45) 33 13 53 54 ▣ 27172 thaitr dk
⊖ (+45) 33 91 12 62
Minister Counsellor Commercial: Vorathep Supadulya
First Secretary Commercial: Mrs Daranee Klavongse [97]

31-01840 Embassy of the Republic of Turkey
Rosbaeksvej 15, DK-2100 Copenhagen Ø
☎ (+45 31) 20 27 88 ▣ 27476 turkel dk
⊖ (+45 31) 20 51 66
Ambassador: H E Mr Turan Morali
Minister-Counsellor: Mehmet Taser
Counsellor for Labour and Social Security:
Mehmet Ridvan Selçuk
Counsellor for Information: Fatma Bilgi Sakarya

Commercial & Economic Department
Borgergade 42, DK-1300 Copenhagen K
☎ (+45 33) 12 29 20 ⊖ (+45 33) 14 63 46
Commercial Counsellor: Ms Nejla Gonca [98]

31-01890 Embassy of the United Kingdom
Kastelsvej 40, DK-2100 Copenhagen
☎ (+45) 35 44 52 00; 35 26 63 75
W http://www.brit-emb@post.tele.dk
⊖ (+45) 35 44 52 93
Ambassador: H E Mr Andrew Bache
Deputy Head of Mission: F Gallagher
Counsellor: R L Owen
Defence Attaché: Cdr W T Wiseman (RN)
First Secretary, Commercial: D T Cox [98]

31-01900 Embassy of the United States of America
Dag Hammerskjölds Allé 24, DK-2100 Copenhagen Ø
☎ (+45) 31 42 31 44 E ukh@usis.dk
W http://www.usis.dk ▣ 22216 amemb dk
⊖ (+45) 35 43 02 23
Ambassador: H.E. Mr Edward E. Elson
Counsellor: Jimmy Joseph Kolker
Commercial Attaché: R Christian Reed
Chief, Office of Defence Co-operation:
Col Lawrence S Hagenauer
Air Attaché: Lieut Col Thomas W Shubert
Army Attaché: Lieut Col Steven M Czepiga
Defence Attaché: E Carl Swanson (Jr)
Counsellor Political and Economic Affairs:
Mrs Patricia D Hughes
Counsellor: Michael P Morgan
Counsellor Scientific & Technological Affairs:
Stephanie Smith Kinney
Public Affairs Officer: Stephen Strain [98]

31-01950 Embassy of Venezuela
Holbergsgade 14, 3th, DK-1057 Copenhagen K
☎ (+45) 33 93 63 11; 33 93 64 76
⊠ EMBAVENEZ Copenhagen V ⊖ (+45) 33 15 69 11
Ambassador: H.E. Mr Hugo Alvarez Pifano
Third Secretary: Jose Claudio Nino
Attaché: Ms Matilde Villa Blanco [97]

34 Multinational Intergovernmental Organizations

34-00000 Allied Forces Baltic Approaches Headquarters (BALTAP)
DK-7470 Karup J.
☎ (+45) 86 61 51 11 ⊖ (+45) 87 26 10 21
Commander BALTAP: Vice Adm K Borck (DA N)
☎ (+45) 86 61 51 11 ext. 2300
Chief of Staff COSBALTAP: Rear Adm Dr S Hess (GE N)
☎ (+45) 86 61 51 11 ext. 2305
Assistant Chief of Staff, Plans and Intelligence:
Col B W Carmichael (US F)
☎ (+45) 86 61 51 11 ext. 2501
Assistant Chief of Staff, Operations and Exercise:
Brig Gen H H Scholz (GE F)
☎ (+45) 86 61 51 11 ext. 2401
Assistant Chief of Staff, Logistics, Manpower and Support:
Brig Gen O L Jargensen (DA A)
☎ (+45) 86 61 51 11 ext. 2328
Assistant Chief of Staff, Communications and Information Systems: Gp Capt P C Badcock (UK F)
☎ (+45) 86 61 51 11 ext. 2320
FC: H B Petersen (DA Civ) ☎ (+45) 86 61 51 11 ext. 2747
Chief Public Information Officer: Cdr Per H Jönsson (DA N)
☎ (+45) 86 61 51 11 ext. 2301
Secretary to Chief of Staff SECCOS: Lieut Col H P Sand (GE A) ☎ (+45) 86 61 51 11 ext. 2304

■ Safeguarding of the security and territorial integrity of the NATO member states in the Baltic area. Responsibility to conduct and to contribute to Peace Support Operations outside the NATO area. [98]

60 Research Organizations, Institutes & Test Establishments

60-00600 Det Kriminal Praeventive Råd
Hoerhusvej 7, DK-2300 Koebenhavn [98]

60-00800 Direktorate for Kriminalforsorgen
Klarebodgade 1, DK-2500 Koebenhavn [98]

60-01000 Institut of Criminology and Criminal Law
Kriminalistisk Institut
Skt. Peders Straede 19, DK-1453 Koebenhavn K
☎ (+45) 35 32 26 26 ⊖ (+45) 35 32 33 34
👥 12
Director: Prof. Vagn Greve
Criminology and Sociology of Law: Flemming Balvig
Criminology and Sociology of Youth: Joi Bay
Criminology: Ms Britta Kyvsgaard
Criminology: Ulla Bondensen
Criminology: Ms Annalise Kongstad
Criminal Law: Vagn Greve
Criminology: Ms Margaretha Järvinen
Criminology and Psychology of Law: Berl Kutchinsky
Sociology of Law: Ms Rhubya Mehdi
Criminology and Sociology of Law: Annika Snare
Criminal Law and Law of Psychiatry: Joern Vestergaard
Contact: Ms Rie Iversen
■ Research and teaching activities in the field of Criminology. Established in 1957. [98]

60-01050 Institute of Criminal Law and Criminology
University of Copenhagen, Sct. Peters Straede 19, DK-1453 Copenhagen
Director: Dr Vagn Greve [98]

60-01200 Institute of Forensic Genetics
University of Copenhagen, 11 Frederik den Femtes Vej, DK-2100 Copenhagen O
☎ (+45) 35 32 61 10 ⊖ (+45) 35 32 61 20
👥 8 (academics); 30 (technicians)
Director: MD DSc Niels Morling
Academic: Jack B Cowland
Academic: Jorgen Dissing
Academic: Birthe Eriksen
Academic: Hanna E Hansen
Academic: Lars Nellemann
Academic: Ole Svensmark
Academic: Ms Mariann Thymann
Academic: Soren Norby
■ Research, teaching, genetic investigations in crime, paternity and immigration cases.
Established in 1982. [98]

60-01500 Juridisk Institut Afdeling for Proces-og Kriminalviodenskab
Århus Universitet - Bartholins Alle, Bygn. 340, DK-8200 Århus N
☎ (+45) 89 42 13 63 ⊖ (+45) 86 19 44 63
👥 4
Director: Prof. Gorm Toftegaard Nielsen
Criminal Law and Criminology: Prof. Beth Grothe Nielsen
Criminology: Prof. Jorgen Jepsen
Criminal Law: Prof. Anette Storgaard
■ Research and teaching in criminal science (law). Projects on economic crime; environmental crime, police and social disturbances; domestic violence; juvenile institutions; juvenile delinquency; incest. [98]

60-01800 Kriminalpolitisk Forening
Aalborgvej 39, DK-9560 Hadsund
☎ (+45 98) 57 31 65 ⊖ (+45 98) 57 31 65
Chairman: Carsten Glargaard [98]

60-02000 Kriminalforsogens Personaleskole
Faengselsvej 1, DK-2500 Koebenhavn [98]

65 Industry, Trade & Armed Forces Associations

65-00050 Association of Danish Aerospace Industries
Foreningen for Danske Aerospace Industri (FDAI) (a member of AECMA)
H C Andersens Boulevard 18, DK-1787 Copenhagen, V
☎ (+45) 33 77 33 77 ▣ 112217 di dk
⊖ (+45) 33 77 33 00
President: Aage Madsen
Head of Defence Procurement Services: Ena Bjerregaard
■ Organisation set up to establish co-operation in the aerospace area. Membership of six companies.
Members:
Per Udsen Co. Aircraft Industry A/S
Alcatel Kirk A/S
Computer Resources International A/S
Danish Aerotech A/S
Terma Elektronik A/S
Printca A/S [98]

65-00100 Confederation of Danish Industries (DI)
Dansk Industri
H C Andersens Boulevard 18, DK-1787 Copenhagen V
☎ (+45) 33 77 33 77 ⊖ (+45) 33 77 33 00
Chairman: Svend-Aage Nielsen
Managing Director: Hans Skov Christensen
Head of Defence Procurement Services: Ena Bjerregaard
■ The Confederation is concerned with all aspects of Danish industry including labour agreements. [98]

75 Manufacturing Companies

75-01170 Ambu International A/S
Sdr Ringvej 49, PO Box 215, DK-2600 Glostrup, Copenhagen
☎ (+45) 43 63 01 11 ⊖ (+45) 43 63 26 00
VP Sales & Marketing: Lars Bergelius
Managing Dir: Vivi Kjaer
Head of Admin.: Erik Alslund
■ Field medical supplies & equipment. [97]

75-01180 Ammunitionsarsenalet
Tuenvej 120, Elling, DK-9900 Frederikshavn
☎ (+45) 98 48 11 00 ▣ 67175 amaell dk
⊖ (+45) 98 48 04 25
👥 240
Managing Director: Frans Andersen
Production Mgr: Søren Hansen [98]

75-02000 Pierre Beauvais Flightcases International A/S
Flightcases International A/S
95a Vesterbrogade, DK-1620 Copenhagen V
☎ (+45) 31 21 44 44 ⊖ (+45) 31 23 45 67
👥 10 (1997) 🏭 900 m2 (1997)
■ Manufacture and sale of flight cases of wood, formica/aluminium construction for military and medical applications including: document and PC cases, communication equipment, explosives, high-technology components (electrical, electronic, hydraulic), hospital supplies (instruments, glass ampoules), data equipment, mechanical and electronic spare parts, test and measuring equipment, tools and workshop accessories, weapons and components. Water, air and dust-tight cases (ABS) with de-compression valves. [98]

75-02790 A/S Brüel & Kjaer
Skodsborgveg 307, DK-2850 Naerum
☎ (+45) 42 80 05 00 ▣ 37316 ⊖ (+45) 45 80 14 05
👥 1 300
Managing Dir: Peter Morch Eriksen
■ Precision electronic measuring instruments and transducers. [97]

75-03230 Celwave R.F. A/S
30 Frejasvej, DK-3400 Hillerød
☎ (+45) 48 22 36 36 ⊖ (+45) 48 22 37 37
👥 130
Managing Director: J A Leibovici
Deputy General Manager: P Thiele
Sales Manager: F Laval
■ Production, R&D & sales of filters, antennas & antenna combining systems for land-mobile communications. [98]

75-03300 Checkmate Industries of EEC Ltd
Ordrup Jagtvej 48E, DK-2920 Charlottenlund
☎ (+45) 39 63 99 76 ⊖ (+45) 39 64 55 84
👥 12
President: Vartan Keshishian
Sales & Marketing Dir: Ibrahim Arel
■ Military & police uniforms, footwear, webbing equipment, protective bulletproof vests, protective bulletproof helmets, transport equipment & vehicles, anti-riot equipment, uniforms, military tents & camouflage equipment. [97]

75-04000 DAB Silkeborg A/S
Postboks 309, DK-8600 Silkeborg
☎ (+45 86) 82 33 00 ⊖ (+45 86) 81 56 54 [98]

75-04030 Danfoss System Hydraulik A/S
(a subsidiary of Danfoss A/S)
Aaderupvej 41, DK-4700 Naestved
☎ (+45) 55 78 72 00 ⊖ (+45) 55 78 72 72
♣ DKr 160,000,000 👥 185
Managing Director: O L Sørensen
Sales & Marketing Dir: P. Simonsen
Operations & Technical Dir: I.N. Jensen
Purchasing Mgr: P. Bank
Sales Promotion Planning: Claus Christensen
■ Remote control systems for valve operation. [98]

75-04050 Danish Camp Supply
CLT Dept
Virkelyst 11, DK-9400 Nr Sundby
☎ (+45) 98 19 13 00 ▣ 69925 camp dk
⊖ (+45) 98 19 07 00
♣ DKr 100,000,000 (1996) 👥 24 (1996)
Managing Dir: Michael Fossing
Export Sales Mgr: Flemming Pedersen
■ Container load trailers with mobility in all kinds of terrain & loading/unloading capacity of all types of ISO containers & shelters (trailers, railway wagons, aircraft, etc.). [98]

75-04100
A/S Dantherm

Dantherm®
Environmental Air Management

Jegstrupvej 4, DK-7800 Skive
☎ (+45) 97 52 41 44 ⊖ (+45) 97 52 61 34
👥 300 🏭 29 500 m2
Managing Director: Alex Nielsen
Finance Director: Torben Brandstrup
Sales Manager (Military Division): Eskild Andersen
Marketing Assistant: Ms Carsten R Jensen
■ Manufacture of equipment for heating, de-humidification and heat-recovery ventilation to requirements of NATO armed forces.
Subsidiaries:
A/S Dantherm, Tønsberg, Norway
Dantherm AB, Sweden
Dantherm GmbH, Germany
Dantherm Ltd, Clevedon, UK [98]

75-04300 Danyard A/S
(a subs. of J. Lauritzen Group)
Kradegholmen 4, PO Box 719, DK-9900 Frederikshavn
☎ (+45) 98 42 22 99 ✉ DANYARD ▣ 67117 danyrd dk
⊖ (+45) 98 43 29 30
♣ DKr 1 496 213 👥 2 200
President & CEO: Torben Meonertsen
Exec. VP Sales & Marketing: Gunnar Lage
Exec. VP Production, Design & Purchasing: Jens Viskinge Jensen
Defence Products Sales Mgr: Ove Terp

■ Osprey 55 vessel designs which form the basic platforms to be fitted out for a variety of roles, such as patrol vessels, coastguard vessels, customs & police craft. Vessels can be constructed with helicopter landing pad & with stern ramp for launching & recovery of boarding & rescue boat. Standard Flex 300, Multi-Role Naval Ship, designed for surveillance, surface combat, MCM minelaying. Full load displacement 450 tonnes. Standard Flex 1500 Offshore Patrol Vessel. Full load displacement 1320 tonnes. [97]

75-05700 E. Falck Schmidt A/S

Tolderlundsvej 106, DK-5000 Odense C
☎ (+45) 66 13 11 00 ☒ 59886 falken dk
⊖ (+45) 66 13 11 11
♣ DKr 100,000,000 (1996) ⋔ 140 (1996) ⌂ 10,000 m2
Managing Dir: Jan Falck Schmidt
Marketing Mgr: Poul Nord Nielsen ☎ (+45) 98 92 53 92
■ Contractor and system supplier of army equipment. Production of hydraulic equipment and electrical systems. Machine processing and welding of large steel and aluminium products. Modernisation of existing weapons systems and tracked vehicles. Overhaul and rebuild of weapon systems and subassemblies. Overhaul and repair of engines, transmissions and final drives. [98]

75-07000 GW Sprinkler A/S

(a subs. of Williams Holdings plc)
Kastanievej 15, DK-5620 Glamsbjerg
☎ (+45) 64 72 20 55 ⊖ (+45) 64 72 22 55
⋔ 35
Managing Dir: S.B. Poulsen
Sales & Marketing Dir: M. Strangeways
■ Fire protection equipment. [97]

75-08860 Hundested Motor & Propeller Fabrik A/S

Skansevej 1, DK-3390 Hundested
☎ (+45 42) 33 71 17 [W] http://www.hundestedpropeller
⊖ (+45 42) 33 99 02
⋔ 30 (1996) ⌂ 4,000 m2 (1996)
Managing Director: Bo Joergensen
Operations Director: Poul Erik Jensen [98]

75-09000 A/S Hydrema Danmark

Gl. Kirkevej 16, DK-9530 Støvring
☎ (+45) 98 37 13 33 ⊖ (+45) 98 37 19 96
♣ DKr 200 000 000 ⋔ 280 ⌂ 22 000 sq m
Managing Dir: K.W. Jensen
Sales & Marketing Dir: Bruno Winther
Sales & Marketing Dir: Jan W. Jensen
Public Relations: Allan Molgaard
■ Manufacture & marketing of high technology construction machines including pivot steered loader/excavators & pivot steered dumptrucks. Military customers include the Danish & British army.
Overseas agents:
Hydrema AB, Sweden
A/S Hydrema, Norway
Hydrema Baumaschinen GmbH, Germany
Hydrema (UK) Ltd, UK [97]

75-09200 Infocom Systems A/S

(a subsidiary of Infocom A/S)
Ellegårdvej 25, DK-6400 Sønderborg
☎ (+45) 74 43 34 34 ⊖ (+45) 74 43 34 31
♧ DKr 2,100,000 (1997) ♣ DKr 75,000,000 (1997)
⋔ 80 (1997) ⌂ 2,984 m2
Managing Dir: Bent Kristensen
Export Mgr: Frank S. Hansen
■ Integrated Communication System (ICS) for naval platforms and ground installations. Internal communication systems for naval vessels. Automated message handling systems for naval platforms. Radio TEMPEST compliant Crypto management systems for naval platforms. Digital Switching Systems for audio (wideband link compatible) and data. Training systems for defence forces. → 85 → 90 [98]

75-09400 JAI A/S

(† Jørgen Andersen Ingeniørfirma A/S)
Head Office
Produktionsvej 1-3, DK-2600 Glostrup, Copenhagen
☎ (+45) 42 91 88 88 ⊖ (+45) 42 91 32 52
♣ DKr 350,000,000 ⋔ 110 ⌂ 9,000 m2
Managing Director: Jørgen Kjeld Andersen
Operations Director: Jan Sindesen
Sales and Marketing Director: Villads Sondergaard
Research and Development Manager: Karsten Sjxrslev
Purchasing Manager: Claus Oldrup
Marketing Manager: Lise L Jensen
Public Relations Manager: Claus Starck
■ Systems for high security surveillance applications, day & night, man operated or automatic, for mobile or fixed locations. Electro-optical target location of tracking systems, day light/low light. [97]

75-09500 KVH Europe A/S

Ved Klaedebo 12, DK-2970 Hoersholm
☎ (+45) 42 86 82 89 ⊖ (+45) 42 86 70 77
⋔ 10
Managing Dir: M. Bjerre-Petersen [98]

75-12000 Lite Flite Aps

(† Scan Para ApS)
Kolding Airport, PO Box 55, DK-6580 Vamdrup
☎ (+45) 75 58 37 37 ☒ 31159 isa cph
⊖ (+45) 75 58 37 58
♧ DKr 200,000 (1997) ♣ DKr 4,500,000 ⋔ 5 ⌂ 350 m2
President: John C. Holstein
■ Manufacture military, civilian parachutes, aircraft safety belts, cargo nets and helicopter rescue equipment. Sales, service and repair. [98]

75-12800 MacArtney A/S

Gl Guldagervej 48, DK-6710 Esbjerg V
☎ (+45) 75 11 66 77 ☒ 54272 mac dk
⊖ (+45) 75 11 72 20
⋔ 26
Managing Dir: Martin H. MacArtney
Sales Mgr: Anders B. Andersen
General Mgr: Niels-Erik Hedeager
Technical Mgr: Lars Hansen
■ MCM ROV systems, fibre optic telemetry and control systems, electrical & fibre optic slip rings, dynamic electrical & fibre optic cables & connectors, towed steerable undersea system (FOCUS 400), oceanographic winch systems. [98]

75-13060 Marimatech Aps

Samsovej 30, DK-8382 Hinnerup
☎ (+45) 91 22 55 ⊖ (+45) 91 22 88 [98]

75-13070 Markdan Ocean Aps

Kystvejen 21, DK-8000 Aarhus C
☎ (+45) 86 19 27 19 ⊖ (+45) 86 13 21 01 [98]

75-14090 Navitronic Systems AS

Jegstrupvej 54, DK-8361 Hasselager
☎ (+45 86) 28 82 44 ☒ 68728 navico dk
⊖ (+45 86) 28 82 66
♧ DKr 500 000 (1995) ♣ DKr 15 000 000 (1995) ⋔ 12
Managing Dir: Knud Helkjaer
Marketing Mgr: Chief Ing. T. Bundgård
■ Manufacturers and designers of hydrographic equipment, software & systems. Suppliers of echo sounders, data processing equipment and sound velocity probes. Systems for harbour surveying, dredging and general mapping of seabeds are supplied with volume calculation, profile plotting etc.
Worldwide co-operation with ATLAS ELEKTRONIK GmbH [98]

75-14300 Nea-Lindberg A/S

Industriparken 39-43, PO Box 226, DK-2750 Ballerup
☎ (+45) 44 97 22 00 [E] aj@cybernet.dk
[W] http://www.rdt.dk ⊖ (+45) 44 66 09 10
Managing Director: Kurt S Ottesen
Export Manager: Per Friis
■ Manufacture of aviation equipment, aeroplanes, plants and installations. [98]

75-18100 Reinhard Oil A/S

(a subsidiary of Synlube ApS)
8 Helleruplund Alle, DK-2900 Hellerup
☎ (+45) 39 62 40 33 ⊖ (+45) 39 62 74 31
♧ DKr 500,000 (1997) ♣ DKr 450,000 (1997)
⋔ 3 (1997)
Managing Director: Steen Ovist
■ Manufacture a line of rust preventive & a line of MOLY Products brand name: "MolyDane". Also stock Monsanto traction fluids & heat transfer fluids for distribution.
Distributor/Agent for:
Sun Oil Co., USA & Belgium
Millers Oils, UK
Ilex Compounds, UK
GE Silicones, UK [98]

75-18300 Reson A/S

Fabriksvangen 13, DK-3550 Slangerup
☎ (+45) 47 38 00 66 [E] reson@reson.dk
⊖ (+45) 47 38 00 22
Marketing Co-ordinator: Ms Marianne Raun
Area Sales Manager: Hans F Lauszus
■ Underwater acoustic equipment. Manufacture sonars, transducers and hydrophones for sonar systems, echo sounders, sidescans, velocity profiles, acoustic telemetry links and seabed penetrating systems.
Subsidiaries:
Reson Inc, USA
Reson Offshore Ltd, UK [98]

75-18400 Roblon A/S

Fiskerihavngade 23, DK-9900 Frederikshavn
☎ (+45) 98 42 14 66 ☒ ROBLON ⊖ (+45) 98 42 15 69
♧ DKr 94,890 (1995/6) ♣ DKr 148,860 (1995/96)
⋔ 160 ⌂ 14,000 sq m
Managing Dir: Anders Petersen
Technical Dir: John Olesen
Area Sales Mgr: Kim Larsen
■ Non-metallic ballistic protection modular systems, synthetic ropes. [98]

75-18500 A/S Roulunds Fabriker

Hestehaven 51, DK-5260 Odense S
☎ (+45) 63 11 50 00 ☒ 59873 ⊖ (+45) 66 11 23 80
♧ DKr 30 000 000 (1995) ♣ DKr 827 746 000 (1994)
⋔ 1 040 ⌂ 60 000 sq m
Export Manager Automotive Products: Gert Hansen
Product Manager Ballistic Protection Systems: Ove Henriksen
Marketing Manager: F R Hansen
Sales Manager Airport Equipment: Lars Kock
■ Production of disc brake pads, drum brake linings, V-belts, radiator hoses, transmission belts, fan belts & ballistic protection systems. [98]

75-19000 Simrad Shipmate AS

Østre Allé 6, DK-9530 Støvring
☎ (+45) 98 37 34 99 ☒ 69838 ⊖ (+45) 98 37 38 07
⋔ 100
Managing Dir: John Larsen
Operations & Technical Dir: Odin Sletten
Purchasing Mgr: Jorgen Johansen [97]

75-19600 SP Radio A/S

Porsvej 2, PO Box 7071, DK-9200 Aalborg SV
☎ (+45) 98 18 09 99 [E] sailor@sailor.dk ☒ 69789
⊖ (+45) 98 18 67 17
Sales Director: Sven E Rasmussen
Sales Manager Europe: Kjeld Skovgaard
Sales Manager Overseas (East): Philip Dixon-Carter
Sales Manager Overseas, West: Allan Lydersen
■ Maritime radio communications equipment, e.g. VHF & SSB radiotelephones, radiotelex & scrambler equipment for use with VHF, SSB & domestic telephone network & land-mobile equipment. Satellite communication equipment B,M&C. Portable VHF/UHF program. [98]

75-19650 Spectronic Denmark

Skindbjergvej 44, PO Box 63, DK-8500 Grenaa
☎ (+45) 86 38 72 22
[E] grenaa@spectronic-denmark.com ☒ 60320
⊖ (+45) 86 38 77 04
⋔ 22 ⌂ 995 m2
Sales & Marketing Dir: Claus Aabo
Operations & Technical Dir: Peter Kristiansen
Purchasing Mgr: Stig Bjerring
■ Manufacture of electronic intelligence systems. Wired room monitoring systems including passive lines (AWM system), active telephone lines (RMM/RFM systems), active 220/110V main (ACC system). Wireless monitoring system: INCA remote controlled monitoring system for single or multi-target applications. Cellular Monitor System: mobile telephone and cascade coupled multi channel receiver/ monitor systems (NMT 450/900, Z-SAT Immersat A). PSTN Monitoring Systems. Single and multi line fax and voice monitoring, MPLX-600 with client server PC for communication and storage services. [98]

75-19700 svend a nielsen a-s Engineering Co Ltd

Gillelejevej 30, DK-3230 Graested
☎(+45) 42 29 02 22 Ⓔ sanas@inet.uni-c.dk
⊜ (+45) 42 29 04 34
55 (1996) 3,900 m2
Managing Dir: C.B. Nielsen
Technical Dir: J. Kam
Sales Manager: Hans-Joergen Schmidt
■ Development & production of electrical heating equipment, working in accordance with the instructions of ISO 9001. [98]

75-19800 Svendborg Vaerft A/S

(member of Naval Team Denmark)
DK-5700 Svendborg
☎(+45) 62 21 22 21 ✉ FLYDEDOKKEN Svendborg
⊜ (+45) 62 22 45 46
Managing Director: Holger Pedersen
Production Manager: Jorn Neilsen
Operations and Technical Director: Leif I Madsen
Purchasing Manager: Jørgen Reinholdt
■ All types of ship & machinery repairs & conversions. Shipbuilding of all types up to 115 x 20m, 7000 TDW. Steel construction of all kinds. [98]

75-20000 Tempress AS

Nordlandsvej 64-66, DK-8240 Risskov
☎(+45) 89 32 52 00 68778 tpress dk
⊜ (+45) 89 32 52 13
70 (1997) 3,500 m2 (1997)
Managing Director: Henrik Dybbroe
Sales Manager: Peter Bo Jørgensen
Export Manager: Arne K Skjonnemann
Marketing Manager: Ole Nielsen
■ Pressure, temperature and level measurement & control. [98]

75-20300 TERMA Elektronik AS

Hovmarken 4, DK-8520 Lystrup
☎(+45) 86 22 20 00 ✉ TERMA 68109 terma dk
⊜ (+45) 86 22 27 99
♣ DKr 495,100,000 (1995/96) 500 21,000 sq m
Chairman: Kaj Juul-Pedersen
President: Johannes Jacobsen
Sales & Marketing Dir: Jørgen P. Andersen
Sales & Marketing Asst: Ms Elsebeth Banke
Sales and Marketing Assistant: Ms Charlotte Larsen
■ Radar systems for communication, navigation, surveillance & velocity measurements, satellite equipment, antennas, telecom systems. Countermeasures dispenser systems.
Subsidiaries:
Terma Electronics Inc. (USA) → 90 [98]

75-21100 Timco Electronic

44 Moellegade, Holm, DK-6430 Nordborg
☎(+45) 74 45 31 27 ⊜ (+45) 74 45 16 16
Public Relations Manager: Ms Mona Mai
■ Autopilots for ships, servo-controls and end-stops. [97]

75-21300 Per Udsen Co. Aircraft Industry A/S

Fabrikvej 1, DK-8500 Grenaa
☎(+45) 86 32 19 88 Ⓔ pucpuc@ibm.net
⊜ (+45) 86 32 14 48
♧ DKr 25,000,000 (1997) ♣ DKr 160,000,000 (1997)
240 (1997) 27,294 m2 (1997)
President and General Manager: Aage Madsen
Financial Director: Poul Reinholdt Jensen
Product & Sales Mgr, Co- production & Special Programs: Kurt Padkaer
Product & Sales Mgr, Military Programs: John Rygaard
Product & Sales Mgr, Space & Technology Programs: Jørgen Schroll
Production Manager: Dennis Ipsen
Purchasing: Henning MØller
Purchasing: Brian Mikkelsen
Purchasing: Kent Teglgard
Quality Assurance Mgr: F.Ed.ng Thomas
■ Design, manufacture & assembly of aircraft parts (airframe). Graphite composite parts. → 80 [98]

75-22000 Viking Life-Saving Equipment A/S

(† Nordisk Gummibadsfabrik A/S)
Seadding Ringvej, PO Box 3060, DK-6710 Esbjerg V
☎(+45) 76 11 81 00 ✉ VIKINGBOAT 54114 vikng dk
⊜ (+45) 76 11 81 01
450 26 000 sq m
Managing Dir: Ms Katty Toernaes
International Sales Mgr: Erik Soendergaard
Marketing Asst: Ms Anne Pernille Kragh
■ Viking life-saving equipment manufactured & distributed by a worldwide organisation. The product programme includes inflatable liferafts & slides, rescue boats, marine evacuation survival & fire protection suits, lifejackets & other life-saving equipment. [97]

75-22500 Wallac Denmark AS

Gydevang 30, DK-3450 Allerød
☎(+45) 48 16 90 00 ☎(+45) 48 14 60 00 [98]

75-23300 A/S Wejra

Anker Engelundsvej 3, DK-9200 Aalborg SV
☎(+45) 98 18 12 22 ⊜ (+45) 98 18 89 22
Managing Dir: Uggi Mersholm
Secretary: Bente Nalholm
■ Manufacturer & supplier of weapons & ammunition for all types of TAMPELLA mortar & howitzer systems. [98]

80 Service Companies & Organizations

Metier Scandinavia AS 90-12000
Per Udsen Co. Aircraft Industry A/S . . 75-21300
Rockwell Automation A/S 90-18450

80-00050 BRUHN NewTech A/S

Vasekaer 12, DK-2730 Herlev
☎(+45) 44 88 22 55 Ⓔ nwt@newtech.dk
Ⓦ http://www.newtech.dk ⊜ (+45) 44 53 17 55
♣ DKr 12,900,000 (1994/95) 20
Chairman: Joern Meldgaard
Deputy Chairman: Anders Cold
President CEO: Lars Bruhn
Vice President COO: Per Junker Thiesgaard
Sales & Marketing Dir: Ulrik Gjellerup
Technical Dir: Lasse Christiansen
■ Production of operational software systems for hazard prediction, warning and reporting of the effects of nuclear, biological and chemical incidents; administrative and operational software systems for fire services, environmental protection and emergency preparedness agencies. [98]

80-00100 Computer Resources International A/S (CRI)

Bregnerodvej 144, DK-3460 Birkerod
☎(+45) 45 82 21 00 ⊜ (+45) 45 82 12 75
♧ DKr 12 500 000 ♣ DKr 570 000 000 500
15 000 sq m
President: Steen Hansen
Division Mgr: Soren Ipsen
Marketing Mgr: Jens Minnet
■ Independent Danish Software Development & System Integration House offering a range of products & services within advanced knowledge-based & information processing systems.
System development & integration services within the following market segments.
Aerospace: Mission control systems, ground check-out systems, micro satellites & on-board software as well as high-speed telemetry systems.
Software Systems: Software development environments, document handling systems & workflow management systems.
Defence: Data communication, information management, decision support, military message handling, C3I & air traffic control.
Capabilities range from consultancy services to complete turn-key projects, including the integration of other vendors products & advanced subsystems such as radars & radiocommunications. [97]

80-00250 Danish Aerotech A/S

(† ADE-HML Ground Equipment Div.)
(a member of the FLS Aerospace Group)
Karup Air Base, PO Box 53, DK-7470 Karup
☎(+45) 99 62 62 62 Ⓔ aerotech@pip.dknet.dk
60262 datech ⊜ (+45) 99 62 63 64
♧ DKr 50,000,000 ♣ DKr 70,000,000 (1995) 160
15,000 m2
President and Chief Executive Officer: Kurt Ottesen
Sales and Marketing Director: Jørgen Lund
Division Mgr Workshops: Henrik Therkelsen
Management Asst: Sydney S. Jensen
Division Mgr Calibration of Measuring Equipment: Bendix Christensen
Aviation Services Division Mgr: Per Smedegaard
Quality Assurance Mgr: Bent Koed Doktor
Financial Mgr: Erik Christensen
■ Maintenance, modernisation and repair of aircraft and aircraft equipment ranging from calibration of measuring equipment to sale, production and servicing of ground equipment.
® UK: ADE-HML Ltd [98]

80-00330 Danvik

Lykkeholms Alle 102, DK-8260 Viby J
☎(+45 86) 14 66 01 Ⓔ tayo@post3.tele.dk
⊜ (+45 86) 14 51 27
♣ DKr 26,000,000 (1995)
Managing Director: Eric Andersen
Marketing Director: Tue Lorentzen
Public Relations: Carsten Damgaard
■ Installation, repair and servicing of naval electronic equipment. [98]

80-00400 DDC-I A/S

Gl. Lundtoftvej 1B, DK-2800 Lyngby
☎(+45) 45 87 11 44 Ⓔ sales@addci.dk
Ⓦ hhtp://www.ddci.com ⊜ (+45) 45 87 22 17
♧ US$ 2,145,000 (1996) ♣ US$ 2,825,000 (1996)
40
Managing Director: Morten Sølling
Operations Director: Kenneth Plesner
Purchasing Manager: Lane Egsdal
Marketing Co-ordinator: Morten Scelling
■ Develops and markets high performance Ada cross compiler systems and tools to common microprocessor targets for the development of real-time safety critical embedded systems. DDC-I Ada is used in products from many leading military suppliers, including AlliedSignal, Ferranti, Lockheed and Martin Marietta. [98]

80-00450 DEMEX Consulting Engineers A/S

26, 3rd floor Hejrevej, DK-2400 Copenhagen NV
☎(+45) 38 10 89 70 Ⓔ info@demex.dk
⊜ (+45) 38 33 13 17
♣ US$ 2,000,000 (1996) 21
Managing Director: Erik K Lauritzen
Technical Director: Niels K Madsen
Marketing Manager: Ms Camilla Kirkegaard
■ Independent Danish consulting and engineering company which has developed a strong expertise in the fields of demolition, building waste management recycling and demining, including mine awareness, Explosive Ordnance Disposal (EOD) and clearing after war and disasters in general. [98]

80-00460 Dynatech A/S

Marielundvej 35, DK-2730 Herlev
☎(+45) 44 53 41 00 ⊜ (+45) 44 53 41 41
♧ DKr 1,500,000 (1994) ♣ DKr 12,000,000 20
500 m2
Managing Dir & Marketing: Jørgen M. Lundbeck
R&D Department Mgr: Steen K. Hansen
Quality Assurance Mgr: I.B. Woulf-Jensen
Production Manager: Finn Holm
■ Design and development of power supplies and power systems required in electronic systems placed in a harsh environment, utilising many years of practical hands-on experience combined with modern CAD tools.
Production and maintenance of power supplies and power systems up to a rate of 150 units by month using highly skilled personnel, burn in an automatic test equipment combined with 3 sigma statistical control methods of key parameters. (D1-9000). Quality system certified against AQAP 110. [98]

80-00500 Eiva A/S

Teglbaekvej 8, DK-8361 Hasselager
☎(+45) 86 28 20 11 Ⓔ eiva_as@pip.dknet.dk
⊜ (+45) 86 28 21 11
♣ DKr 30,000,000 (1996) 20 (1996)
Managing Dir: Niels Jorgen Vase
Sales Mgr: Palle Herskind

■ Complete range of software products, including the NaviPac on-line navigation & data acquisition software. NaviBat multibeam echosounder data acquisition software as well as the Naviline Digital Terrain Modelling (DTM) & charting software. Furthermore, Eiva develops customized systems solutions, eg integrated command & control system for mine countermeasure vessels. 98

80-00700 GTA
Enghavevej 9a, DK-7800 Skive
☎ (+45) 97 51 15 77 E gerry@login.dkknet.dk
W http://www.nsads.net/fmkb ⊖ (+45) 97 51 15 77
3 (1997) 300 m2
Managing Director: Ing. Alfred G Davison (CEI) FIEIE
■ Wireless data network communications software and tactical remote wireless control solutions. 98

80-00800 Hans Harboe-Hansen
Hjortevaenget 230, DK-2980 Kokkedal
☎ (+45) 42 24 77 82 ⊖ (+45) 42 24 77 82
1
Naval Consultant: Capt Hans Harboe-Hansen (R.Dan.N. Rtd)
■ Consultancy services. 98

80-00900 Integra Consult A/S
Trorodvej 63B, DK-2950 Vedbaek
☎ (+45) 45 66 00 44 E integra@pip.dknet.dk
⊖ (+45) 45 66 05 10
DKr 1,200,000 (1997) ♣ DKr 11,000,000 (1997)
20 (1997)
Managing Director: Peter Thorsen
Marketing and Sales Director: Michael Niels Thorsen
■ Consulting company providing advice within project management, quality assurance and logistics in connection with the specification, tendering, design, development and implementation of new advanced and complex information systems. The company has a large with CCIS and is used to work in an international multicultural environment. Services include independent verification and validation, development of specifications, audit/review of projects status and support to local project management. 98

80-01350 P. Monsted Chartering A/S
6 Vestre Kaj, DK-4700 Naestved
☎ (+45) 55 72 20 00 46220 ⊖ (+45) 55 77 17 05
♣ US$ 20,000,000 (approx. 1997) 6
President & Managing Dir: Peter Monsted
Chartering: Stig Hansen
Chartering: Preben Riis
Chartering: Torben Regaard
■ Independent shipping agents specialising in transport worldwide by sea & air of defence equipment including IMCO-classified material. Always minimum 20 modern vessels en-route to sensitive areas worldwide on behalf of most major producers & traders, including governmental bodies. 98

80-03500 Satair A/S
Amager Landevej 147A, DK-2770 Kastrup
☎ (+45) 32 47 01 00 31214 sat dk
⊖ (+45) 32 51 34 34
US$ 22,000,000 (1997) ♣ US$ 75,000,000 (1997)
158 (1997)
Managing Director: John Staer
Vice President Sales: Morten Olsen
Vice President Products and Marketing: Jens P Pedersen
Executive Vice President Business Development: Svend Jespersen
Executive Vice President Special Sales: Klavs Larsen
■ Distribution of aircraft equipment. Aircraft standards hardware AN, MS, NAS, MIL, Douglas and Boeing standards. General aviation aircraft engines, propellers, components, instruments and spare parts. Aircraft tools and shop equipment. Special support inventory: Fokker F-27/F-28, Shorts 330/360, BAe 146, Dash 8. DHC-6 Twin Otter, Beechcraft, airframe, engines, propellers, components, instruments and avionics, spare parts, tools and workshop equipment. → 85
® Canada: AlliedSignal, Pratt & Whitney Canada
France: Aerazur, Blanc Aero
Germany: Carl W. Kopperschmidt
UK: Brown Frank & Son, Jane's Information Group, Rivenaes Ivar
USA: AIM, APME, ATI Industries, Aero Instruments, Aeroflash, Air Maze, Air Tuf Products, Airborne, Aircraft Braking Systems, Aircraft Instruments & Dev., Aircraft Parts Corp, Aircraft Porous Media, Aircraft Technical Publishers, Airpath, Alcor, Armstrong Tools, Artex, Avibank, Barfield, Barry Controls, Beechcraft, Borroughs Tools & Equip, Camloc, Craig Tools, Curtiss, Cutler-Hammer, Daniels Mfr Corp DHC, Dayton-Granger, Deutsch Am. Pneumatic Tool, Dzus Fasteners, Electrosystems, Fafnir, Fenwall, E.C.E., Flex-tab, Gardner-Denver, General Electric, Gettelman, Goodyear, Hi-Shear, Honeywell, Howell, Imperial Eastman, KS Avionics, Kell-Strom, Lewis, Marathon Flite-Tronics, Marvel-Schebler, McCreary, Milbar Corp, Miraj, Mitsubishi, Narco Avionics, New York Twist Drill, Norwich, PPG Industries, Porter HK, Precision Airmotive, Prestolite, Puritan-Bennett, Revere, Shamban WS, Sigma-Tek, Slick Aircraft Products, Sunair Electronics, Teledyne Battery Products, Teledyne Continental Motors, Telex Communications, Texas Instruments, Textron Lycoming, Timken, Titeflex, Torrington, Unison Industries, W&J Risbridger, Wemac, Whelan, Wilson, Woodward, Zephyr Manufacturing Co 98

80-03550 ScanAvia A/S
Scandinavian Aviation Consultants
Troredvej 63B, DK-2950 Vedbaek
☎ (+45) 45 66 00 44 E scanavia@pip.dknet.dk
⊖ (+45) 45 66 05 10
DKr 600,000 (1996) ♣ DKr 50,000 (1996) 1 (1997)
25 m2
Managing Director: Bent B Heurlen
■ Consultancy services within civil aviation, including airport planning and development, civil aviation administration and airline organisation and operations. 98

80-03650 Skanacid A/S
32 Bredgade, DK-1260 København K
☎ (+45) 33 13 13 30 E skanacid@post4.tele.dk
⊖ (+45) 33 13 13 09
General Mgr: J. Brøndum 98

80-04000 Systematic Software Engineering A/S
Frichsparken, Søren Frichs Vej 38K, DK-8230 Åbyhøj
☎ (+45) 89 43 20 00 E sse@systematic.dk
W http://www.systematic.dk ⊖ (+45) 89 43 20 20
110 (1996) 2,000 m2
Managing Dir: Michael Holm
■ Defence systems integration - specialising in CCCIS, interoperability, EW Library Generation Systems. Certified to NATO AQAP-13 and ISO9001. 98

85 Stockists, Distributors, Spare Parts Suppliers

Infocom Systems A/S	75-09200
Metier Scandinavia AS	90-12000
Satair A/S	80-03500

85-01500 Scandinavian Avionics A/S
Lufthavnsvej 45, Billund Airport, PO Box 59, DK-7190 Billund
☎ (+45) 75 33 84 22 60107 scanic dk
⊖ (+45) 75 33 24 07
75 (1997) 2,000 sq m
President and Sales Director: H B Truelsen
Technical and Production Manager: P Albertsen
Marketing Director: John Mandrup
Quality Manager: K Poulsen
Instrument Service Manager: E Ubbesen
Military Representative: J. Thalund
■ Sales, installation, repair and overhaul of HF, VHF, UHF, communications and navigation, MDE, transponder, radio/radar altimeters autopilots, flight directors, voice/flight data recorder, inertial navigation, GPS, VLF/Omega, instrument, gyro, ground proximity and aircraft electrical systems for civil and military purposes. 98

85-02000 TRACE Scandinavia
Ibsivej 7, Studstrup, DK-8541 Skoedstrup
☎ (+45) 86 99 20 98 E f-vaaben@traceww.com
⊖ (+45) 86 99 30 67
Managing Director: Finn Vaaben 98

90 Sales Agents, Sales & Liaison Offices

Companies represented:

Acer Inc., Taiwan	90-01200
ADE-HML Ltd, UK	80-00250
Advanced Products NV, Belgium	90-08010
Aerazur, France	80-03500
Aero Instruments, USA	80-03500
Aeroflash, USA	80-03500
Aerospatiale, France	90-03340
AIM, USA	80-03500
Air Maze, USA	80-03500
Air Tuf Products, USA	80-03500
Airborne, USA	80-03500
Aircraft Braking Systems, USA	80-03500
Aircraft Instruments & Dev., USA	80-03500
Aircraft Parts Corp, USA	80-03500
Aircraft Porous Media, USA	80-03500
Aircraft Technical Publishers, USA	80-03500
Airpath, USA	80-03500
Alcor, USA	80-03500
AlliedSignal, Canada	80-03500
APME, USA	80-03500
Armstrong Tools, USA	80-03500
Artemis Management Systems Ltd, UK	90-12000
Artex, USA	80-03500
ATI Industries, USA	80-03500
Avibank, USA	80-03500
Barfield, USA	80-03500
Barry Controls, USA	80-03500
Beechcraft, USA	80-03500
Blanc Aero, France	80-03500
Borroughs Tools & Equip, USA	80-03500
Brown Frank & Son, UK	80-03500
Camloc, USA	80-03500
Cannon Electric GmbH, Germany	90-03050
Carl W. Kopperschmidt, Germany	80-03500
Craig Tools, USA	80-03500
Curtiss, USA	80-03500
Cutler-Hammer, USA	80-03500
Daniels Mfr Corp DHC, USA	80-03500
Dassault, France	90-03340
Dayton-Granger, USA	80-03500
Deutsch Am. Pneumatic Tool, USA	80-03500
Dzus Fasteners, USA	80-03500
EEV Ltd, UK	90-01130
Electrosystems, USA	80-03500
Eurocopter, France	90-03340
Fafnir, USA	80-03500
Fenwall, E.C.E., USA	80-03500
Flex-tab, USA	80-03500
FPT Industries, UK	90-18390
Gardner-Denver, USA	80-03500
General Electric, USA	80-03500
Gettelman, USA	80-03500
GIAT Industries, France	90-03340
Goodyear, USA	80-03500
Hadland, UK	90-02200
Haute-Garonne SA, France	90-18390
Hi-Shear, USA	80-03500
Honeywell, USA	80-03500
Howell, USA	80-03500
ICORE International, UK	90-02200
Ilex Systems Inc., USA	90-20250
Imperial Eastman, USA	80-03500
Indal Technologies Ltd, Canada	90-18390
Jane's Information Group, UK	80-03500
Kell-Strom, USA	80-03500
Kelvin Hughes, UK	90-11270
Kollsman Instruments, USA	90-02670
KS Avionics, USA	80-03500
Lewis, USA	80-03500
Lockheed Sanders, USA	90-02670
Lynwood Scientific Developments Ltd, USA	90-20250
Marathon Flite-Tronics, USA	80-03500
Marvel-Schebler, USA	80-03500
Matra, France	90-03340
McCreary, USA	80-03500
Milbar Corp, USA	80-03500
Miraj, USA	80-03500
Mitsubishi, USA	80-03500
Narco Avionics, USA	80-03500
New York Twist Drill, USA	80-03500
Norwich, USA	80-03500
OFEMA, France	90-03340
Optical Data Systems Inc., USA	90-20250

Contact Reference

Denmark

Porter HK, USA 80-03500
PPG Industries, USA 80-03500
Pratt & Whitney, USA 90-02670
Pratt & Whitney Canada, Canada . . . 80-03500
Precision Airmotive, USA 80-03500
Prestolite, USA 80-03500
Puritan-Bennett, USA 80-03500
Revere, USA 80-03500
Rivenaes Ivar, UK 80-03500
Rockwell International Corp., USA . . . 90-18450
secure Systems Group, USA 90-20250
Shamban WS, USA 80-03500
Sigma-Tek, USA 80-03500
Sikorsky Aircraft, USA 90-02670
Slick Aircraft Products, USA 80-03500
SNPE-Melco, France 90-03340
Specialty Cable Corp, USA 90-18390
Sunair Electronics, USA 80-03500
Teledyne Battery Products, USA 80-03500
Teledyne Continental Motors, USA . . 80-03500
Telex Communications, USA 80-03500
Terma Elektronik, Denmark 90-02200
Texas Instruments, USA 80-03500
Textron Lycoming, USA 80-03500
Thomson CSF Communications, France 90-03340
Timken, USA 80-03500
Titeflex, USA 80-03500
Torrington, USA 80-03500
Trimos-Sylvac Task Force, Switzerland . 90-04500
Turbomeca, France 90-03340
Unison Industries, USA 80-03500
W&J Risbridger, USA 80-03500
W.S. Shamban & Co., USA 90-19410
Wemac, USA 80-03500
Whelan, USA 80-03500
Wilson, USA 80-03500
Woodward, USA 80-03500
Zephyr Manufacturing Co, USA 80-03500

Sales agents/representatives listed elsewhere:

Infocom Systems A/S 75-09200

90-01130 ABB Komponent A/S
(† Classen-Smith Import A/S)
Helgeshoej Alle 8-10, DK-2630 Taastrup
☎(+45) 43 99 91 11 ☒ 33609 csi dk
⊖ (+45) 43 99 92 80
Managing Director: N Solgaard
® UK: EEV Ltd [98]

90-01200 Acer Scandinavia A/S
Kongevejen 62A, DK-3460 Birkercd
☎(+45) 45 82 10 00 ⊖ (+45) 45 82 10 72
® Taiwan: Acer Inc. [97]

90-02200 Beatronic Supply Ltd
Industrivenget 13, DK-3400 Hillerod
☎(+45) 41 23 55 17 [E] beatroni@post3.tele.dk
beatsupp@iqnet.cz ⊖ (+45) 41 23 51 92
♤ DKr 200,000 ⅲ 12 (1997)
President: Bent Gothenborg
Operations Director: Jan Weber
Sales Mgr: Steen Lauritzen
■ Test and measuring instruments, components and systems. Climatic chambers, vacuum chambers, test systems for vibration and shock, gyro and test unit, digital multimeters and calibrations equipment. Digital air data computers, airspeed and altitude transducers, accelerometers, signal conditioner, portable data recorder, pitot static test equipment, engine trim equipment, engine trim and tracking equipment, servo hydraulic components, servo valve, actuators, electronic modules for flight simulators. Rotary and linear potentiometers, tilt sensors, joysticks, tracker balls and keyboards.
® Denmark: Terma Elektronik
UK: Hadland, ICORE International [98]

90-02300 British Steel Denmark A/S
(a subsidiary of British Steel Engineering Steels, UK)
Hans Edvard Teglers Vej 7, DK-2920 Charlottenlund
☎(+45) 39 96 09 00 ⊖ (+45) 39 96 09 49 [98]

90-02670 Erik Bolvinkel Consulting ApS
Poppelhøj 21, DK-2900 Hellerup
☎(+45) 39 62 41 48 ☒ 19801 ebol dk
⊖ (+45) 39 62 46 93
Director: Erik A. Bolvinkel
Exec.Secretary: J. Bolvinkel
® USA: Kollsman Instruments, Lockheed Sanders, Pratt & Whitney, Sikorsky Aircraft [97]

90-03050 Cannon Electric GmbH
Park Allé 278a, DK-2605 Brondby
☎(+45) 43 45 52 88 ⊖ (+45) 43 43 58 58
Area Manager: C Stroeh
® Germany: Cannon Electric GmbH [98]

90-03340 Aage Christensen A/S
Skelmosevej 10, DK-2500 Copenhagen Valby
☎(+45) 36 44 24 44 ☒ 22207 chris dk
⊖ (+45) 36 44 20 24
♤ DKr 600,000 (1996) ⅲ 17
Chairman: H. de Neergaard
Director: A. Krebs
® France: Aerospatiale, Dassault, Eurocopter, GIAT Industries, Matra, OFEMA, SNPE-Melco, Thomson CSF Communications, Turbomeca [98]

90-04500 Kjeld Dannemand A/S
Bjerringbrovej 147, DK-2610 Rødovre
☎(+45) 42 84 55 77 ⊖ (+45) 42 84 43 41
Managing Director: Kjeld Dannemand
Marketing Director: Henning Andersen
Technical Director: Finn Nielsen
® Switzerland: Trimos-Sylvac Task Force [98]

90-08010 Haagensen Gunnar A/S
Hejrevang 6, DK-3450 Allerød
☎(+45) 48 17 65 00 ⊖ (+45) 48 17 17 37
Managing Director: Niels Haagensen
Managing Director: Christian Haagensen
® Belgium: Advanced Products NV [98]

90-11270 Kelvin Hughes A/S (KH)
(a subs. of Kelvin Hughes Ltd, UK)
Marselis Blvd 175, DK-8000 Aarhus C
☎(+45) 86 13 77 88 ☒ 64710
Director: Willy Hanson
® UK: Kelvin Hughes [98]

90-12000 Metier Scandinavia AS
(† Metier Management Systems Denmark A/S)
Strandvejen 203 1 Floor, DK-2900 Hellerup
☎(+45 39) 40 11 33 ⊖ (+45 39) 40 20 08
■ Supplier of Artemis project management software. → 85 → 80
® UK: Artemis Management Systems Ltd [98]

90-18390 V. Reitz & Co.
Vangedevej 106B, DK-2820 Gentofte
☎(+45) 39 69 16 55 ⊖ (+45) 39 69 13 26
Managing Dir/Owner: Mrs Else Reitz
® Canada: Indal Technologies Ltd
France: Haute-Garonne SA
UK: FPT Industries
USA: Specialty Cable Corp [97]

90-18450 Rockwell Automation A/S Industrial Automation Systems
(† Allen-Bradley Denmark A/S)
(a division of Rockwell International, US)
Herstedoestervej 27-29, DK-2620 Albertslund
☎(+45) 43 46 60 00 ⊖ (+45) 43 46 60 01
ⅲ 22
Regional Manager: Jorgen Pedersen
[E] jpedersen@eu1.ra.rockwell.com
■ Sales, support and service. → 80
® USA: Rockwell International Corp. [98]

90-19410 W.S. Shamban Europa A/S
Fabriksvej 17, PO Box 310, DK-3000 Helsingør
☎(+45) 49 27 02 14 ⊖ (+45) 49 27 03 74; 42 22 52 09
ⅲ 250
Managing Director: Claus Baroe
General Manager: Steen Ishoy
Aerospace Co-ordinator: Georg Krebbs
■ Sealing systems for hydraulic systems, shock struts, recoil mechanisms and reservoirs. High temperature gas seals and seals for NBC protection of vehicles. Static seals, piston seals, rod seals and slide strips for aircraft and aerospace hydraulic and pneumatic systems.
® USA: W.S. Shamban & Co. [98]

90-19450 Skandinavisk Teleindustri Skanti A/S
(a subsidiary of EuroCom Industries AS)
34 Kirke Vaerloesevej, DK-3500 Vaerloese
☎(+45) 42 48 25 44 [E] skanti@skanti.dk
[W] http://www.skanti.dk ☒ 37292 skanti dk
⊖ (+45) 42 48 45 09
Sales Dir: Jørgen F. Mikkelsen
Advertising Mgr: Ms Birgit Mulbjerg
■ Sales of high quality HF & VHF marine radios. The Skanti HF/VHF products are widely used on naval ships & on ground due to high reliability, advanced features for remote control & a favourable price/performance ratio. [98]

90-19500 Sperry Marine AS
Hoejnaesvej 46, DK-2610 Roedovre
☎(+45) 36 72 20 77 ☒ 19526 ⊖ (+45) 36 72 21 20
♣ US$ 60,000,000 (1995) ⅲ 25 ⌂ 745 m2
Vice President and General Manager of European Operations: H E Rasmussen [98]

90-20250 Tempest Europe A/S
Marielundvej 46A, DK-2730 Herlev
☎(+45) 44 94 89 02 [E] je@temest.dk
⊖ (+45) 44 94 87 39
Marketing Director: Jan Engelund
■ Distributor of Tempest equipment in Europe for US manufacturers.
® USA: Ilex Systems Inc., Lynwood Scientific Developments Ltd, Optical Data Systems Inc., secure Systems Group [98]

90-21000 THORN Lighting AS
Brogrenen 8, DK-2635 Ishøj
☎(+45) 43 54 06 77 ⊖ (+45) 43 54 45 95
Chief Executive: Arne Tholle [98]

Djibouti

Head of State:
Hassan Gouled Aptidon

10 Head of State

10-10000 Office of the President
PO Box 6, Djibouti
☎(+253) 35 01 77; 35 02 01 ☒ 5871 presiden dj
President of the Republic: Hassan Gouled Aptidon [98]

11 Prime Minister

11-10000 Office of the Prime Minister
PO Box 2086, Djibouti
☎(+253) 35 14 94 ☒ 5871 presiden dj
Prime Minister: Barkat Gourad Hamadou [98]

12 Ministries

12-10000 Ministry of National Defense
Boulevard de La Republique, PO Box 42, Djibouti
☎ (+253) 35 20 34; 35 20 62 ✉ 5871 presiden dj
Minister: Abdallah Chirwa Djibril [98]

12-20000 Ministry of Foreign Affairs and Co-operation
PO Box 1863, Djibouti
☎ (+253) 35 24 71 ✉ 5871 presiden dj
Minister: Mohamed Moussa Chehem [98]

12-40000 Ministry of Interior
PO Box 33, Djibouti
☎ (+253) 35 07 91 ✉ 5871 presiden dj
Minister: Idris Harbi Farah [98]

31 Embassies & Attachés

31-01890 Consulate of the United Kingdom
Inchcape Shipping Services et Cie,
PO Box 81, rue de Geneve, Djibouti
☎ (+253) 35 38 44; 35 38 36 ✉ 5843 dj
⊖ (+253) 35 06 55
Honorary Consul: P Lambrecht [98]

31-01900 Embassy of the United States of America
BP 185, Djibouti
Plateau du Serpent, Boulevard Marechal Joffre
☎ (+253) 35 39 95 ⊖ (+253) 35 39 40
Chargé d'Affaires: Ms Teri Robl [98]

Dominican Republic

Head of State:
Leonel Fernández Reyna

10 Head of State

10-10000 Office of the President
Palacio Nacional, Calle Moisés García, Santo Domingo
☎ (+1 809) 686 47 71 ⊖ (+1 809) 682 23 41
President of the Republic: Leonel Fernández Reyna
Secretary of State for the Presidency: Danilo Medina
Administrative Secretary of the Presidency: Diandino Pena
Technical Secretary to the Presidency: Eduardo Selman [98]

10-20000 Office of the Vice President
Palacio Nacional, Calle Moisés García, Santo Domingo
☎ (+1 809) 686 47 71 ✉ 4299 secpre dr
Vice President: Jaime David Fernandez Mirabal [98]

12 Ministries

12-10000 Secretariat of the Armed Forces
Avenida 27 de Febrero, Plaza de la Independencia, Santo Domingo
☎ (+1 809) 530 51 93 ✉ 0652 fmssefa
⊖ (+1 809) 531 04 61
Secretary of State for the Armed Forces:
Adm Ruben Paulino Alvarez [98]

12-40000 Secretariat of Interior and Police
Edificio Gubernamental-El Huacal,
Piso 3, Avenida Leopaldo Navarro, Santo Domingo
☎ (+1 809) 688 9565; 686 5923
Secretary of State for Interior and Police:
Norge Botello Fernandez
Attorney General: Abel Rodriguez Del Orbe [98]

31 Embassies & Attachés

31-00310 Embassy of Canada
Maximo Gomez 30, PO Box 2054, Santo Domingo 1
☎ (+1 809) 685 11 36 ⊖ (+1 809) 682 26 91
Ambassador: H E Mr Yves Gagnon (resident in Caracas)
Chargé d'Affaires: Pierre Giroux [98]

31-00990 Embassy of the Republic of Korea
Avenida Sarasota 98, PO Box 20221, Santo Domingo
☎ (+1 809) 532 43 14 E e.korea@codetel.net.do
✉ 4368 gksd dr ⊖ (+1 809) 532 38 07
Ambassador: H E Mr Chang Ho Lee
Deputy Chief of Mission: Sang Il Kim
Commercial Counsellor: Young Suh Kim [98]

31-01890 Embassy of the United Kingdom
Edificio Corominas Pepin, Avenue 27 de Fabrero No 233, Santo Domingo
☎ (+1809) 427 71 11; 427 79 05 ✉ 23468 proca ve
⊖ (+1809) 472 75 74
Ambassador: H E Mr Dick Thompson
Consulate of the United Kingdom
Saint George School, Abraham Lincoln 552, Santo Domingo
☎ (+1809) 540 31 32 ✉ 3460781 prodrom
⊖ (+1809) 562 50 15
Honorary Consul: Ms M Tejeda [98]

31-01900 Embassy of the United States of America
Cnr of Calle Cesar Nicolas,
Penson and Calle Leopoldo Navarro, Santo Domingo
☎ (+1 809) 221 21 71 ✉ 3460013 amemb dr
⊖ (+1 809) 686 74 37
Ambassador: H E Mrs Donna Hrinak
Deputy Chief of Mission: Cristóbal Orozco
Head of Commercial Section: Robert J Bucalo
Public Affairs Officer: César D Beltrán
Defense Attaché: Lieut Col Jeffrey Kammerer
Head of Military Assistance Advisory Group:
Lieut Col Michael F Fukey [98]

31-01910 Embassy of Uruguay
Avda Mexico 169, Ensanche Esperilla, Santo Domingo
☎ (+1 809) 565 2669 ✉ URUDOMI Santo Domingo
✉ 0442 urudomi ⊖ (+1 809) 687 2167
Ambassador: H.E. Mr Honorio Barrios Tassano [97]

90 Sales Agents, Sales & Liaison Offices

Companies represented:

Detroit Armor Corp., USA 90-00300

90-00300 Imgraf SA
(a division of Ramon L Baez CMA)
Apartado Postal 1960, Santo Domingo
☎ (+1 809) 567 95 41 E rabaez@codetel.net.do
⊖ (+1 809) 541 26 97
President: Ramón P Báez
® USA: Detroit Armor Corp. [98]

Ecuador

Head of State:
Fabian Alarcon Rivera

10 Head of State

10-10000 Office of the President
Palacio de Gobierno, García Moreno 1043, Quito
☎ (+593 2) 21 63 00 ✉ 23751 prerep ed
⊖ (+593 2) 58 07 35
President of the Republic: Fabian Alarcon Rivera [98]

10-10500 Office of the Vice President
Quito
Arenas y Manuel Larrea Edificio Conade,
Provincial de Pichincha 21° Piso
☎ (+593 2) 56 36 66 ✉ 22058; 22375 prerep ed
⊖ (+593 2) 50 33 79; 50 49 34
Vice President: Ms Rosalia Arteaga Serrano de Cordova [98]

12 Ministries

12-10000 Ministry of National Defence
Ministerio de Defensa Nacional
Exposición 208, Quito
☎ (+593 2) 57 25 80; 51 09 75; 51 28 03
✉ 3986 mindef ed; 22703 mindef ed; 23986 mindef ed
⊖ (+593 2) 58 04 31; 57 07 04
Minister for National Defence: Gen Ricaurte Yanez Ramiro [98]

12-20000 Ministry of Foreign Relations
Ministerio Relaciones Exteriores
Avenida 10 de Agosto y Carrión, Quito
✉ 22142 mmre ed
Minister of Foreign Relations: Jose Ayala Lasso [98]

12-40000 Ministry of Government & Justice
Espejo y Benalcázar, Quito
Minister of Government, Police and Municipality:
Cesar Verduga [98]

15 Parliament

15-10000 National Congress
Congreso Nacional
Palacio Legislativo, Avenida 6 de Diciembre y Piedrahita, Quito
☎ (+593 2) 52 62 51
Chamber of Representatives
Camara Nacional de Representantes
Avenida 6 de Diciembre y Piedrahita, Palacio Legislativo, Quito
☎ (+593 2) 23 02 00
✉ 22 579 preleg ed, 22 788 adpalg ed [98]

31 Embassies & Attachés

31-00080 Embassy of the Argentine Republic
Avenida Amazonas 477, Piso 5, PO Box 2937, Quito
☎ (+593 2) 56 22 92 ✉ 22 136 embarg ed
⊖ (+593 2) 56 81 77
Ambassador: H.E. Mrs Maria Esther Bondanza

Contact Reference

Deputy Chief of Mission: Horacio Alberto Amoroso
Military & Air Attaché: Col Hugo Juan Molinari
Office of the Commercial Counsellor
PO Box 9187, Quito
☎ (+593 2) 52 00 69; 76 24
Commercial Counsellor: Norberto T. Aguerre [97]

31-00150 Embassy of Belgium
Calle Juan Leon Mera 863 y Wilson, PO Box 17 21 532, Quito
☎ (+593 2) 54 53 40; 54 34 48; 56 76 33
⊠ AMBABEL Quito ⊞ 22767 ambel ed
⊖ (+593 2) 50 73 67
Ambassador: H E Mr Joris Couvreur
Head of Chancery: André Vancolen
Commercial Adviser: William Stock [98]

31-00890 Embassy of Italy
Calle La Isla, 111 y Humb. Albornoz, PO Box 17 03 72, Quito
☎ (+593 2) 56 10 77; 56 10 74 ⊞ 22715 itdipl ed
⊖ (+593 2) 50 28 18
Ambassador: H E Dr Ferruccio Marri Caciotti
First Secretary: Dr Carmella Maseccia
Commercial Attaché: Ms Elena Marinelli [98]

31-00920 Embassy of Japan
J L Mera 130 y Avenida Patria, PO Box 17-21-01518, Quito
☎ (+593 2) 56 18 99; 56 31 27;56 30 11; 50 13 62; 56 95 48 ⊞ 22 185 taisi ed
⊖ (+593 2) 50 36 70
Ambassador: H E Mr Hajime Nishimiya
Special Officer for Economic Affairs: Katsuya Naruse [98]

31-01410 Embassy of Peru
Avdas Colón y Amazonas, Edif. Espana, PO Box 338 057, Quito
☎ (+593 2) 55 41 61; 52 01 34; 54 16 25; 56 23 00
⊠ LEPRU Quito ⊖ (+593 2) 56 23 49
Ambassador: H.E. Mr Hugo Palma Valderrama
Deputy Chief of Mission: Raul Patino Alvistur
Military Attaché: Col. Julio Salazar Encinas
☎ (+593 2) 50 98 30
Air Attaché: Col José Luis Malpartida Del Pino
☎ (+593 2) 50 60 06
Naval Attaché: Capt. Rafael Loayza Regalado
☎ (+593 2) 56 96 82 [97]

31-01890 Embassy of the United Kingdom
Calle Gonzalez Suarez 111, PO Box 314, Quito
☎ (+593 2) 56 06 69; 56 07 55; 23 68 67
56 06 70; 56 07 71; 56 03 09; 23 60 73;
23 68 67; 54 15 52; 54 14 98; 22 30 88
⊞ 22138 prodqt ed ⊖ (+593 2) 56 07 29; 56 07 30
Ambassador: H E Mr J W Forbes-Meyer
Deputy Head of Mission: A R Weeks
Defence Attaché: Capt K Ridland (RN)
Second Secretary Commercial and Technical Cooperation: Miss J A Ashdown
Second Secretary: M J Hiller [98]

31-01900 Embassy of the United States of America
PO Box 538, Quito
Avenida 12 de Octubre y Avenida Patria, Quito
☎ (+593 2) 56 28 90 ⊖ (+593 2) 50 20 52
Ambassador: H E Mr Leslie Alexander
Deputy Chief of Mission: James C Struble
Chief of Military Group: Col Michael Merz (USA)
Defense Attaché: Col Ovidio Perez (USA)
Public Affairs Officer: Mark B Krischik
Commercial Section: Ms Janice Corbett [98]

90 Sales Agents, Sales & Liaison Offices

Companies represented:

Transaero Inc, USA 90-02000

90-02000 Transaero SA
Apartado 21-018, Quito
☎ (+593) 244 32 74 ⊖ (+593) 244 13 02
Manager: Charles Crawford
® USA: Transaero Inc [97]

Egypt

Head of State:
Mohammed Hosni Mubarak

10 Head of State

10-10000 Office of the President
Oruba Palace, Oruba Street, Heliopolis, Cairo
☎ (+20 2) 83 07 88 ⊞ 331 psi un; 92144 arabic un
President: Mohammed Hosni Mubarak [98]

11 Prime Minister

11-10000 Office of the Prime Minister
Magles Al Shaab Street, Cairo
☎ (+20 2) 264 57, 273 70 ⊞ 93794 wazra un
Prime Minister: Kamal Ahmed al Ganzouri
Deputy Prime Minister: Yousef Amin Wally [98]

12 Ministries

12-10000 Ministry of Defense and Military Production
23rd July Street, Kobry Al Kobba, Cairo
☎ (+20 2) 83 43 45, 83 83 51
⊞ 23258 ared un; 94295 rdma un; 92129 national
Minister of Defense and Deputy CinC: Gen Mohammed Hussein Tantawi [98]

12-30000 Ministry of Economy and Foreign Trade
96 Ahmed Orabi Street, 12411 Mohandessin, Giza
☎ (+20 2) 91 96 61, 91 92 78 ⊞ 92164 national
Economic and Commmercial Counsellor: Yousef Boutros-Ghali [98]

12-40000 Ministry of Interior
El Sheikh Rihan Street, Cairo
☎ (+20 2) 354 58 97
Minister: Hasan al- Alfi [98]

12-60000 Ministry of Information
Television Building, Corniche El Nile, Maspero, Cairo
☎ (+20 2) 75 95 70
Minister: Mohamed Safwat al Sherif [98]

12-70000 Ministry of Planning
Salah Salem Street, Nasr City, Cairo
☎ (+20 2) 60 36 84
Minister of Planning: Kamal Ahmed al- Ganzouri [98]

12-80000 Ministry of Tourism
Misr Travel Building, Abbasseia, Cairo
☎ (+20 2) 282 84 39
Minister: Mamdouh al-Beltagi [98]

31 Embassies & Attachés

31-00085 Armenian Diplomatic Mission
12 Lufti Hassuna Dokki, Cairo
☎ (+20 2) 360 03 17 ⊖ (+20 2) 341 97 84
Ambassador: H E Mr Edward Nalbandian [98]

31-00090 Australian Embassy
11th Floor, World Trade Centre, Corniche El Nil, Boulac, Cairo
☎ (+20 2) 575 04 44 [E] austremb@sc.soficom.com.eg
⊞ 92257 aust un ⊖ (+20 2) 578 16 38
Ambassador: H E Mr Mike Smith
Counsellor: Ms Sandra Vegting
Consul: Lex Bartlen
Press Officer and Third Attaché: Simon Starr [98]

31-00130 Embassy of the People's Republic of Bangladesh
47 Ahmed Heshmat Street, Zamalek, Cairo
☎ (+20 2) 340 21 01; 341 26 45; 341 26 42
⊞ 22326 doot un ⊖ (+20 2) 341 26 31
Ambassador: H E Mr Ruhul Amin
Counsellor, Deputy Chief of Mission: Golam Mohammad [98]

31-00150 Embassy of Belgium
20 Kamel El Shennawi St, Garden City, PO Box 37, Cairo
☎ (+20 2) 354 74 94; 354 74 96 ⊠ AMBABEL Cairo
⊞ 92264 ambel un ⊖ (+20 2) 354 31 47
Ambassador: H E Mr Gastron van Duyse-Adam
☎ (+20 2) 354 74 94
Head of Chancery: Mr Parys
Counsellor Chargé d'Affaires ai: Mr X Leblanc
Defence Attaché: Col Erik Sanders ☎ (+20 2) 354 04 75 [98]

31-00200 Embassy of Bolivia
5 Dar El Shifa St, Garden City, PO Box 182 Dawaween, Cairo
☎ (+20 2) 354 34 22; 86 43; 355 09 17
⊖ (+20 2) 355 09 17; 354 34 22; 86 43
Ambassador: H.E. Mr Hernando Velasco
☎ (+20 2) 375 27 44
First Secretary: David Carlo Gutierrez
☎ (+20 2) 355 76 11
Public Information Officer: Mrs Nagah A. Kassem [97]

31-00580 Embassy of Finland
3 Abu El Feda Street, 13th Floor, Zamalek, Cairo
☎ (+20 2) 341 37 22; 340 28 01 ⊞ 92328 fiemb un
⊖ (+20 2) 340 51 70
Ambassador: H E Mr Garth Castrén
Deputy Chief of Mission: Ms Sanna Halinen
Economic Co-operation Counsellor: Olli Sotamaa
Commercial Office
Kasr el Nil Street 34, Cairo
☎ (+20 2) 392 68 44; 19 05 ⊞ 92268 finta un
⊖ (+20 2) 392 48 39
Commercial Attaché: Pekka Hyytinen [98]

31-00590 Embassy of France
29 Avenue de Giza, Cairo
☎ (+20 2) 72 82 75; 570 39 16/20 ⊞ 92032 france un
⊖ (+20 2) 571 02 76
Ambassador: H.E. Mr Patrick Leclercq
Deputy Chief of Mission: Patrice Paoli
Armed Forces Attaché: Col Olivier Urvoy de Portzamparc
Asst Military Attaché: Lt Col Jean Pierre Beltr
Public Information Officer: C. Farnaud
Commercial Office
10 rue El Azziz Osman, Zamalek, Cairo
☎ (+20 2) 345 56 94/95; 341 57 01/02
⊞ 92540 comata un
Chief Commercial Officer: David Appia [97]

31-00660 Embassy of the Federal Republic of Germany
8A Hassan Sabry Street, Zamalek, Cairo
☎ (+20 2) 339 96 00 ⊠ 92023 aacai un
⊖ (+20 2) 341 05 30
Ambassador: H E Dr Wolf-Dietrich Schilling
Defense Attaché: Col Friedrick von Richter [98]

31-00690 Embassy of Greece
63 Avenue Alexandre le Grand, Chatby, Alexandrie
☎ (+20 3) 483 84 55; 483 84 54; 483 99 07
⊠ 0091-55989 greec un ⊖ (+20 3) 482 58 96 [98]

31-00880 Embassy of Israel
6 Ibn Malek Street, Giza, Cairo
☎ (+20 2) 361 05 28; 361 03 80; 361 05 45
⊠ 93363 misrl un ⊖ (+20 2) 361 04 14
Ambassador: H E Mr Zvi Hazel
Deputy Chief of Mission: Yaacov Brosh
Commercial Counsellor: Reuven Azar
Press and Information Counsellor: Lior Ben Dor [98]

31-00890 Embassy of Italy
Shara Abdel Rahman Fargmi 15, Garden City, Cairo
☎ (+20 2) 354 31 94/5 ⊠ 94229 itadi un
⊖ (+20 2) 354 06 57
Ambassador: H.E. Mr Alberto Leoncini Bartoli
Minister Deputy Chief of Mission:
Dr Alessandro Fallavollita
Military Naval & Air Attaché: Col. Franco Ganguzza [98]

31-01020 Embassy of Lebanon
5 Ahmed Nessim Street, Guizeh, Cairo
☎ (+20 2) 361 04 74; 06 23 ✉ AMBALIBAN
⊠ 92227 liban ⊖ (+20 2) 361 04 63
Ambassador: H E Mr Hisham Dimashkieh [98]

31-01210 Embassy of Mongolia
3 Midan El Nasr Gametdowal, El Ababia Street, Dokki
☎ (+20 331) 346 06 70 ⊠ 042200656
⊖ (+20 331) 46 05 30 16 [98]

31-01270 Royal Netherlands Embassy
18 Hassan Sabri Street, Zamalek, Cairo
☎ (+20 2) 340 64 34 ⊠ 92028 nedam un
⊖ (+20 2) 341 52 49
Ambassador: H.E. Dr N. van Dam
Defence Attaché: Col C.B.P. Donia
Commercial Officer: Dr R.C. Vos
Public Information Officer: A. Stok [97]

31-01430 Embassy of the Republic of Poland
5 Aziz Osman Street, Zamalek, Cairo
☎ (+20 2) 340 54 16 ⊖ (+20 2) 340 54 27
First Secretary: Ochodek Boguslaw
Counsellor: Jozef Osas
Commercial Counsellor: Daniel Wasilewski
Defence Attaché: Brig Gen Jerzy Slowinski
First Secretary: Kazimierz Wera [98]

31-01655 Centre for International Coordination and Development
Magless El Shaab Office, 5 Midan El Saraya El Koubra,
PO Box 124 Garden City, Cairo
☎ (+20 2) 355 57 99 ⊖ (+20 2) 355 87 81
Ambassador: H E Mr Peter Tos [98]

31-01730 Swedish Embassy
13 Mohamed Mazhar Street, PO Box 131,
11211 Zamalek, Cairo
☎ (+20 2) 341 41 32; 341 14 84; 341 02 59 341 91 69
[E] sveamcai@link.com.eg ⊠ 92256 svens un
⊖ (+20 2) 340 43 57
Ambassador: H E Mr Christer Sylvéu
Deputy Chief of Mission: Jan Thesleff
Third Secretary: Ulla Jacobson [98]

31-01740 Embassy of Switzerland
10 Sharia Abdel Khalek Sarwat, PO Box 633, Cairo
☎ (+20 2) 575 81 33; 77 05 45
[E] swiemcai@soficom.com.eg ⊖ (+20 2) 574 52 36
Ambassador: H E Mrs Blaise Godet
Deputy Chief of Mission: Ulrich Lehner
Defense Attaché: Col René Käser
First Secretary and Consul: Pascal Decosterd
Attaché and Vice-Consul: Hans-Peter Heiniger
Attaché: Thomas Fischer
Attaché: Ms Gabriela Hess
Attaché: Ms Daisy Fontannaaz
Attaché: Hans Ulrich Roth
Attaché: Ms Marie Helene Pitteloud-Turner
Attaché: Rudolf Müller [98]

31-01890 Embassy of the United Kingdom
7 Ahmed Raghab Street, Garden City, Cairo
☎ (+20 2) 354 08 50; 354 08 52/58
⊖ (+20 2) 354 08 59
Ambassador: H E Mr D E S Blatherwick
Deputy Chief of Mission: R E Makepeace
Head of Commercial Section: A D F Henderson
Defence and Military Attaché: Col A W Snook
Naval Attaché: Cdr N P Smith (RN)
Public Information Officer: E W F Webb
Management Officer: M G Claughton [98]

31-01910 Embassy of Uruguay
6 Loutfallah Street, Zamalek, Cairo
☎ (+202) 341 51 37; 340 35 89 ✉ URUEGIP Cairo
[E] uruemb@idsc.gov.eg ⊖ (+202) 341 81 23
Ambassador: H E Mr Julio Cesar Franzini
Consul: Dr Jorge Dotta Ibarra [98]

31-01990 Embassy of the Republic of Yemen
12 Amin El Rafu Street, Dakki, Guizeh, Cairo
☎ (+20 2) 361 42 24; 361 42 25; 361 42 26
[E] yemenemb@pacc.com.eg ⊠ 20542
⊖ (+20 2) 360 48 15
Ambassador: Ahmed Mohamed Lugman
Deputy Chief of Mission: Abel Malik Said Abda
Commercial Counsellor: Abdel Rahman Shogaa
Scientific Attaché: Abdul Malik Baadani
Defence Attaché: Abdal Ramin Abdel Elah
Press and Information Officer: Muhammed Jamila Ali Rajaa [98]

31-02030 Embassy of the Republic of Zambia
25 Abd el Moneium Riad Str, Mohandessine, Cairo
☎ (+20 2) 361 02 81/8 ⊖ (+20 2) 361 08 33
Chargé d'Affaires: Lt Col. Chongo Shula (Rtd)
First Secretary: M.B. Tembo
First Secretary: E.B. Mudiyo
First Secretary: Mrs B.C. Kalulu
Second Secretary: J. Mwale [97]

55 Government-owned Industry & Production Facilities

55-01700 Arab Organization for Industrialization (AOI)
2D Midan al-Abbasia, PO Box 770, Cairo
☎ (+20 2) 482 33 77; 93 28 22; 93 24 83
✉ SONARAB Cairo ⊠ 92090; 92014 aoi un
⊖ (+20 2) 482 60 10
👥 17,000
Chairman: Lieut Gen Salah Halaby
Managing Director: Eng Mohamed Hassan el Khatteeb
Technical Director: Dr Nabil Salama
Operation and Marketing Director: Mahmoud El Refai
→ 75

Sakr Factory for Developed Industry (a subsidiary of AOI)
Suez Road, PO Box 33, Heliopolis, Cairo
☎ (+20 2) 415 02 50; 415 03 80 ⊠ 92175 cerva un
⊖ (+20 2) 290 19 78
Chairman: Dr Farouk Fayed
Marketing Manager: Assem Hatem
■ Manufacture of guided missile systems, unguided rocket systems and launchers.

Aircraft Factory
PO Box Hammat Helwan, Helwan, Cairo
☎ (+20 2) 555 39 46; 555 39 48
⊠ 92191; 23183 nasr un
⊖ (+20 2) 555 95 62; 555 95 63; 78 24 08
Chairman: Eng Ahmed El Sayed
☎ (+20 2) 78 02 40; 78 96 00
■ Manufacture of jet fighter aircraft and advanced jet trainers. Currently producing 'Alpha Jet', advanced jet trainer of Tucano, basic turbo prop trainer including assembly, testing and component manufacture.

Engine Factory
PO Box 12, Helwan, Cairo
☎ (+20 2) 554 60 92; 554 60 93 ⊠ 20862 enfac un
⊖ (+20 2) 78 12 36
Chairman: Eng Yosri Abu Amer ☎ (+20 2) 78 13 18
■ Produces jet aero engines and spare parts. Capability for overhaul and repair operations.

Kader Factory for Developed Industries
2 El Tayaran Street, Nasr City, PO Box 287,
Heliopolis, Cairo
☎ (+20 2) 402 43 19; 24; 402 43 25 ⊠ 92136 kadfa un
⊖ (+20 2) 260 87 18
Chairman: Eng Abd El Hamid Wasfy ☎ (+20 2) 403 69 72
■ Manufacture of armoured vehicles.

Electronics Factory
Almaza Km 4.5 Suez Road, PO Box 84, Heliopolis, Cairo
☎ (+20 2) 418 10 51; 66 03 09
⊠ 23384; 20278 arlec un ⊖ (+20 2) 418 24 80
Chairman: Eng Mohammed Kamal Abdul Fatah
☎ (+20 2) 418 24 81
■ Military electronic equipment, eg: battlefield telephone and telephone switchboards, airborne radio communication equipment. Also produces sub-assemblies for radio altimeter and airborne radars.

Joint-venture production units:
Arab British Dynamics Co. → 74
Arab British Engine Co. → 74
Arab American Vehicles Co. → 74

Helwan Factory for Developed Industry
PO Box 162, Helwan, Cairo
☎ (+20 2) 78 32 86; 78 85 09 ⊠ 22759 hfdi un
⊖ (+20 2) 78 86 03
Chairman: Dr Salah Zaki
■ Assembly, overhaul and repair of the Gazelle helicopter, type SA 342L airframe. [98]

Contact Reference

74 Consortia, Joint Venture Companies

74-00100 Arab American Vehicles Co. (AAV)
4.5 KMS Suez Rd, PO Box 2419, Horrea/Heliopolis, Cairo
☎ (+20 2) 29 22 85; 02 000 ⊠ 92791 aav un
⊖ (+20 2) 290 35 62
Chairman: Hussien Y. Gad
Managing Dir: Jatinder Singh
Sales & Marketing Dir: Martin Watson
■ Joint production with Chrysler of military jeeps for the armed forces. Also has assembly lines for passenger cars. Export a range of YJ variants throughout the gulf markets. → 75 → 85 [97]

74-00300 Arab British Engine Co. (ABECO) (joint venture with Rolls-Royce Ltd, and Arab Organization for Industrialization (AOI))
PO Box 71, Helwan, Cairo
☎ (+20 2) 78 33 29; 89 80 29 ⊠ 93790 abeco un
⊖ (+20 2) 78 96 52
♣ US$ 32,400,000 👥 520 (1997) 🏭 18,000 m2
Chairman: Eng Adb-el Maguid Dawoud
Marketing Director: Sayed Saoudi ☎ (+20 2) 78 30 79
Financial and Commercial Director: Maher Abd El Monem
Operations Director: Eng Hassaan M El Alfy
Purchasing Manager: Ibrahim Said Oda
■ Repair and overhaul of turboshaft and piston engines of helicopters and industrial turboshaft engines. → 75 → 80 [98]

74-00500 Arab International Optronics Co. (a subs. of Avimo Singapore Ltd)
PO Box 8182, Nasr City, Cairo
El Salam Rd, El Salam City, Cairo
☎ (+20 2) 280 60 78; 280 60 79 ⊠ 93509 aci un
⊖ (+20 2) 280 60 77
♣ US$ 15 000 000 👥 350
Chairman: Brig. Gen. Nabil Amer
Managing Dir: Kevin Burrows
Commercial Dir: Yousef Sherif
■ Individual weapon sights, NVS 900 night observation device, night vision goggles, telescopic laser rangefinders for T-series tanks. Hand held laser rangefinders, night driver viewers, night commander viewers, night gunner viewers, tripods, aiming circles, sighting systems, binoculars. [98]

74-06000 Helwan Factory for Developed Industries (HFDI)
(† Arab British Helicopters Co.)
(a part of the Arab Organisation for Industrialization (AOI))
PO Box 162, 11421 Helwan, Cairo
☎ (+20 2) 78 33 16; 30 36 ⊠ 22759 hfdi un
⊖ (+20 2) 78 86 03
♧ US$ 45 000 000 ♟ 700
Chairman: Dr.Eng. Salah Zaki
Finance & Commercial Dir: Adel Abdel Fattah
Production Mgr: Eng. Mohamed Zaki
■ Repair & overhaul of Gazelle Helicopters, design & manufacturing of aircraft ground equipment, manufacturing of composite material parts. → 75 [97]

75 Manufacturing Companies

Arab American Vehicles Co. 74-00100
Arab British Engine Co. 74-00300
Arab Organization for Ind. 55-01700
Helwan Factory for Developed 74-06000

80 Service Companies & Organizations

Arab British Engine Co. 74-00300
Atef Badawi Consultants 90-02110
Sobeih Trade & Enterprises 90-19590

80-06000 Eyes & Ears
Defense Requirement Services & Facilities
PO Box 5113, 11771 Heliopolis West, Cairo
☎ (+20 2) 244 03 46; 271 62 07 ⊖ (+20 2) 271 62 37
♟ 8
Managing Dir: Gen. Albert R. Georgy
Operations & Technical Dir: Gen. Nabil Guirguis
Sales & Marketing Dir: Dr Nuhad Ghobriel
■ Consulting in defence requirements including optics, night vision sights, fire control systems, air defence systems, laser aiming devices, ammunition, radar, communication systems, fibre optics, NBC detection & protection, GPS receivers for navigation & survey, security systems, military computers & monitors, all kinds of military spare parts, sales agency, consulting services & logistics management services.
® Austria: Reichhold Chemie
China PR: Silver Star Inc.
UK: Conjay Arms Comp. Ltd, Efficienta, Intelsec Systems Ltd, Mayan Airport Ltd., Pre-Mac (Kent) Ltd, Scanna Msc Ltd, Sutcliffe Speakman & Co. Ltd, Wylam Hill Security Ltd
USA: Adelphon Inc., Aydine Computer & Monitor, Deep Ocean Engineering, Fletcher Paper, Fraser-Volpe Corp., Insulgard Corp., International Sensor Technology, Landmark Tower Corp., Mills Manufacturing Corp., Ordnance Logistics Inc., SAS Research & Development Services [97]

85 Stockists, Distributors, Spare Parts Suppliers

Arab American Vehicles Co. 74-00100

90 Sales Agents, Sales & Liaison Offices

Companies represented:

Adelphon Inc., USA 80-06000
Aerospatiale, France 90-00650
Alliant Defence Electronics Systems, USA 90-00815
Amherst System, Inc, USA 90-00815
ATR, France 90-00650
Aydine Computer & Monitor, USA 80-06000
Barry Research Communications, USA 90-02110
Bink Bullows, UK 90-01810
BR Communications, USA 90-02110
Brent Europe Ltd, UK 90-01810
Conjay Arms Comp. Ltd, UK 80-06000
Contraves USA, USA 90-15500
Control Techniques, UK 90-08180
Crouzet SA, France 90-08180
Deep Ocean Engineering, USA 80-06000
DIAB_Barracuda, Sweden 90-00815
E-Systems & Delta Electronics Inc., USA 90-04000
Efficienta, UK 80-06000
EG&G Sealol, USA 90-18900
Electro-Metrics, USA 90-02110
ERG Process Plant Ltd, UK 90-01810
Eurocopter, France 90-00650
Euromissile, France 90-00650
Extronic Electronic, Sweden 90-00815
Fletcher Paper, USA 80-06000
Fraser-Volpe Corp., USA 80-06000
Fritz Emde, Germany 90-00815
GEC Marconi Combat Systems, UK . . . 90-08180
Ginge Kerr, Denmark 90-00815
Gulfstream Aerospace Corporation, USA 90-02110
Hercules, USA 90-00815
Hughes Rediffusion Ltd, UK 90-08180
Insulgard Corp., USA 80-06000
Integrated Power Corporation, USA . . . 90-02110
Intelsec Systems Ltd, UK 80-06000
International Sensor Technology, USA . 80-06000
International Telecom. Systems, USA . 90-02110
Inverto Matia, Switzerland 90-00815
Iver, UK 90-01810
Kelvin Hughes, UK 90-08180
Landmark Tower Corp., USA 80-06000
LERC & AET, France 90-04000
Lockheed Corp., USA 90-10000
Loral Infrared & Imaging Systems, USA 90-02110
Mayan Airport Ltd., UK 80-06000
Mills Manufacturing Corp., USA 80-06000
Oerlikon Contraves SpA, Italy 90-15500
Oerlikon-Contraves, Switzerland 90-15500
Omnipol, Czech Republic 90-08180
Ordnance Logistics Inc., USA 80-06000
Pre-Mac (Kent) Ltd, UK 80-06000
R/M Equipment, USA 90-08180
Racal & Solapak, UK 90-04000
Radio Detection, UK 90-00815
RC Federal Co., USA 90-00815
Reichhold Chemie, Austria 80-06000
SAS Research & Development Services, USA 80-06000
SBFI, UK 90-08180
Scanna Msc Ltd, UK 80-06000
Servo, USA 90-00815
Siemens AG, Germany 90-19300
Silver Star Inc., China PR 80-06000
Skoda Export, Czech Republic 90-08180
Solar Electric Engineering, USA 90-00815
Sutcliffe Speakman & Co. Ltd, UK . . . 80-06000
T.T.L., UK 90-00815
The Expanded Metal Co.Ltd, UK 90-19590
Tracor Aerospace Inc., USA 90-00815
Tracor Flight Systems Inc., USA 90-00815
Vaisala, Finland 90-00815
VEGA, USA 90-00815
Velotrans, Austria 90-00815
Wermmer, Germany 90-00815
William Fairey, UK 90-08180
Wylam Hill Security Ltd, UK 80-06000

90-00650 Aerospatiale
(a subsidiary of Aerospatiale, France)
Liaison Office
5 Dar El Shifa Street, Garden City, Cairo
☎ (+20 2) 354 86 71; 354 26 47 ⊠ 92447 airspa un
⊖ (+20 2) 355 73 17
Director: Mohamed Farahat
® France: ATR, Aerospatiale, Eurocopter, Euromissile [98]

90-00815 Amcon Group Egypt Ltd
23 Shehab Street, 34 Syria Street, Mohandessin, Cairo-Giza
☎ (+20 2) 349 82 14 ⊖ (+20 2) 349 82 17
♧ E£ 600,000 (1996) ♣ E£ 10,000,000 (1996) ♟ 20
⌂ 800 m2
Chairman and Managing Dir: Moustafa El Asmar
Financial Dir: Ashref Ismail
Operations Director: Gen. Farouk Adwy
Sales and Marketing Director: Walid Asmar
Purchasing Manager: Gen. Tarek Zwel
■ Aircraft & related equipment. Airport ground equipment, avionics, PRV's. ECM & ESM systems. IFF systems. Camouflage & decoys. Radar systems and security & alarm systems. Electronic systems. Ordnance. Engine parts & accessories. Consulting services.
® Austria: Velotrans
Denmark: Ginge Kerr
Finland: Vaisala
Germany: Fritz Emde, Wermmer
Sweden: DIAB_Barracuda, Extronic Electronic
Switzerland: Inverto Matia
UK: Radio Detection, T.T.L.
USA: Alliant Defence Electronics Systems, Amherst System, Inc, Hercules, RC Federal Co., Servo, Solar Electric Engineering, Tracor Aerospace Inc., Tracor Flight Systems Inc., VEGA [97]

90-01810 Arab Commercial Agencies Centre
(† Arab Engineering Centre)
33 Iraq Street, Mohandseen, Giza
☎ (+20 2) 348 05 56 ⊖ (+20 2) 349 59 99
Managing Director: Ibrahim I Kamal
Director: Ayman Kamal
Director: Marwa Kamal
■ Cleaning chemicals. Crack inspection and spray equipment.
® UK: Bink Bullows, Brent Europe Ltd, ERG Process Plant Ltd, Iver [97]

90-02110 Atef Badawi Consultants
11 Dr. M.A. El Bakly St, Heliopolis West, 11351 Cairo
☎ (+20 2) 242 76 60; 245 03 22 ⊖ (+20 2) 245 33 46
♧ M$ 500 000 (1995) ♣ M$ 5 000 000 (1995) ♟ 7
⌂ 170 m2
Managing Dir: Ahmed A. Badawi
Sales & Marketing Dir: Abdel Hamid Arozead
Operations & Technical Dir: Hamdy Murafy
Purchasing Mgr: Ms Halla Seada
■ Marketing services for military products specially in the areas of reconnaissance & surveillance, command, control, communication & information, electronic warfare, simulators, power generation, solar energy, fibre optics, antenna towers, EAPBX. → 80
® USA: BR Communications, Barry Research Communications, Electro-Metrics, Gulfstream Aerospace Corporation, Integrated Power Corporation, International Telecom. Systems, Loral Infrared & Imaging Systems [97]

90-03000 CATIC
Representative Office
Flat No.3, Flr.No.10, Bldg No.2, El Abour, Salah Salem St, Nasr City, Cairo
☎ (+20 2) 263 80 70 ⊖ (+20 2) 263 80 80 [97]

90-03500 Daewoo Corp Cairo Liaison Office
Flat N°11, 11th Floor, 48 El Saura Street, St Dokki, Cairo
☎ (+20 2) 337 309 86 ⊠ 20592 dwegt un
⊖ (+20 2) 348 63 81 [98]

90-04000 DINAC - Digest International Agencies & Consultants
(† Société Dinac)
PO Box 8088, 11371 Nasr City, Cairo
☎ (+20 2) 261 99 57 ⊖ (+20 2) 261 99 56
♣ US$ 7 000 000 ♟ 9 ⌂ 100 sq m
President: M.A. Basyouny
Technical Dir: Amro El-Abd
® France: LERC & AET
UK: Racal & Solapak
USA: E-Systems & Delta Electronics Inc. [97]

90-08180 HCH Supply Co. Ltd
(a subsidiary of HCH Supply Co USA & of HCH Supply Co UK)
7 El Shahid Abdel Moneim Riad Street, El Mohandessin, Cairo
☎ (+20 2) 360 38 77; 360 59 50; 361 39 67; 361 39 68
✉ PLESSCO ⊠ 92431 hch un ⊖ (+20 2) 360 80 19
♟ 400 ⌂ 2,000 sq m
President: Samir H Hamman
General Manager: Eng Aly El Sheikh
Administration Manager: Nael Kahil

Sales: Kamal Elmansterly
■ Offers advisory services, representation, technical assistance, after sales service and support.
® Czech Republic: Omnipol, Skoda Export
France: Crouzet SA
UK: Control Techniques, GEC Marconi Combat Systems, Hughes Rediffusion Ltd, Kelvin Hughes, SBFI, William Fairey
USA: R/M Equipment [98]

90-10000 Lockheed Martin International SA
(† Lockheed Corp. International SA)
PO Box 2648, Horreya
19 El-Shahid Helmy El-Masry, Almaza, Cairo
☎ (+20 2) 418 54 20 ⊜ (+20 2) 291 74 93
President Mid-East Africa: R. Lawrence
☎ (+20 2) 418 41 76
VP Mid-East/Africa: R. Atcheson ☎ (+20 2) 418 54 20
Office Mgr: M. Zaklama ☎ (+20 2) 291 44 30
® USA: Lockheed Corp. [97]

90-15500 Oerlikon-Contraves Representative Office
150 El Tayaraan Street, Heliopolis, Nasr City - Cairo
8th Floor Apartment 81, El Tewfik City
☎ (+20 2) 403 30 42; 403 30 43 ▣ 20456 conteun
⊜ (+20 2) 403 30 15
Manager: Maurizio Barnaba
® Italy: Oerlikon Contraves SpA
Switzerland: Oerlikon-Contraves
USA: Contraves USA [98]

90-18900 Sealol Egypt EG&G Sealol
EG&G Sealol
139 Mogamah El Masaneh Street, PO Box 151 M.Farid, Cairo
☎ (+20 2) 426 21 03; 426 24 62
✆ (+201 2) 213 44 99 (mobile) ✉ AMONIEOM ▣ 21375
⊜ (+20 2) 426 20 97
♤ £E 2,148,000 ♣ £E 2,400,000 ⛬ 29 ⌂ 1,390 m2
General Manager: Adel A Monieum
® USA: EG&G Sealol [98]

90-19300 Siemens Egypt Ltd
(a subs. of Siemens AG, Germany)
6th Floor - 15 Abdel Hamid Badawi St, Nozha, 11351 Heliopolis, Cairo
☎ (+20 2) 247 41 31 ⊜ (+20 2) 247 41 32
⛬ 4 ⌂ 500 sq m
Resident Mgr: John R. Osborne
® Germany: Siemens AG [97]

90-19590 Sobeih Trade & Enterprises
28 Talaat Harb, Cairo
☎ (+20 2) 575 60 63 ⊜ (+20 2) 574 05 74
♤ R 500,000 (1995) ♣ R 7,250,000 (1996)
⛬ 45 (1995)
Managing Director: Ahmed Sobeih
Operations Director: Eng Abdel Basset
Sales Director: Saad Hanna
Purchasing Manager: Elro Zabolo
■ Bullet proof protection vests, contractors/medical maintenance services. → 80
® UK: The Expanded Metal Co.Ltd [98]

El Salvador

Head of State:
Armando Calderón Sol

10 Head of State

10-10000 Office of the President
San Salvador
Avenida Cuba, Calle Dario González 806
Barrosan Jacinto, Casa Presidencial
☎ (+503) 271 15 55; 271 01 37 ▣ 20552
⊜ (+503) 271 09 50; 298 01 21
President: Armando Calderón Sol
Vice President and Minister of the Presidency: Enrique Borgo Bustamente
Secretary of Presidency Information: Francisco Imendia [98]

12 Ministries

12-10000 Ministry of National Defence
Km 5 Carretera a Santa Tecla,
Alameda Manuel Enrique Araujo, San Salvador
Minister of Defence and Public Security: Gen Jaime Guzman [98]

12-20000 Ministry of Foreign Affairs
Kilómetro 6 1/2, Alameda Enrique Araújo, San Salvador
☎ (+503) 298 17 43; 298 17 42 ▣ 20179 rree
⊜ (+503) 298 20 30; 298 67 63
Minister of Foreign Relations: Ramón González Giner [98]

12-30000 Ministry of Economy
Carretera Panamericana KM 9 2/2, Ilopango, San Salvador
Carretera Panamericana KM 9 2/2, Ilopango
☎ (+503) 295 04 06 ⊜ (+503) 295 04 43
Minister: Luís Enrique Córdova Macías
Vice Minister: Rene Leon
Director General of Transport: Mayor P A José Roberto Corleto [98]

12-40000 Ministry of Interior and Public Security
Palacio Nacional, Centro de Gobierno, San Salvador
Minister of the Interior: Mario Oertel Acosta
Minister of Public Security: Hugo Barerra [98]

31 Embassies & Attachés

31-00890 Embassy of Italy
PO Box 01-99, San Salvador
Calle La Reforna 158, Colonia San Benito
☎ (+503) 223 51 84 ▣ 21019 italdipsal
⊜ (+503) 298 30 50
Ambassador: H.E. Dr Mario Foresti
Chancellor: Dr Marco Venanzi
Commercial Attaché: Dr Lorena Maschietto Baistrocchi [97]

31-01310 Embassy of Nicaragua
71 Av.Norte y 1a Calle Poniente 164, Colonia Escalón, San Salvador, CA
☎ (+503) 224 09 70; 298 65 49;224 12 23; 223 77 29
⊜ (+503) 223 72 01
Ambassador: H E Mr R Ferrey
Commercial Officer: Lincoln Escobar [98]

31-01890 Embassy of the United Kingdom
PO Box 1591, San Salvador
Edifio Inter-Inversiones, Paseo General Escalón 4828
☎ (+503) 98 17 63; 98 17 69; 98 14 55
▣ 20033 prodrome ⊜ (+503) 98 33 28
Ambassador: H.E. Mr I. Gerken
First Secretary Commercial: A.S. Green (OBE, MVO. Resides in San José)
Defence Attaché: Col S.R. Daniell (Resides at Mexico City)
Third Secretary & Vice Consul: D.M. Spires [97]

Equatorial Guinea

Head of State:
Brig.Gen. Teodoro Obiang Nguema Mbasogo (retd)

10 Head of State

10-10000 Office of the President
Malabo, Bioko
President of the Republic: Brig Gen Teodoro Obiang Nguema Mbasogo (retd)
Minister of State at the Presidency in Charge of Relations with Assemblies and Legal Matters: Antonio Fernando Nve Ngu
Minister of State at the Presidency in Charge of Special Duties: Alejandro Evuna Owono Asangono
Secretary of State for National Security: Capt Manuel Nguema Mba [98]

11 Prime Minister

11-10000 Office of the Prime Minister
Malabo, Bioko
Prime Minister: Serafin Seriche Dougan
Deputy Prime Minister: Francisco Javier Ndongo Mbengono
Minister Secretary General of the Government: Francisco Pascual Eyegue Obama Asue
Minister Political Adviser to the Government: Demetrio Elo Ndong Ngefumu [98]

12 Ministries

12-10000 Ministry of National Defense
Ministerio de la Defensa
Malabo, Bioco
Minister Delegate of National Defence: Col Melanio Ebendeng Nsomo [98]

12-20000 Ministry of Foreign Affairs and Co-operation
Ministerio de Asuntos Exteriores
Malabo, Bioko
Minister: Miguel Oyono Ndong
Minister Delegate of Foreign Affairs in charge of Francophone Affairs: Agustin Nze Nfumu [98]

12-40000 Ministry Interior & Nat'l Security
Ministerio del Interior
Malabo, Bioko
Minister of the Interior: Julio Ndong Ela Mangue
Secretary of State for National Security: Capt Manuel Nguema Mba [98]

Estonia

Head of State:
Lennart Meri

10 Head of State

10-10000 Office of the President
Weizenbergi 39, EE0100 Tallinn
☎ (+372 6) 31 62 02 E sekretar@nw.vpk.ee
W http://www.ee/president ⊖ (+372 6) 31 62 50
President: Lennart Meri
Director of the Office: Jaanus Pikani
Deputy Director of the Office: Enn Markvart
Foreign Relations Adviser: Kaja Tael
Legal Adviser: Mall Gramberg
Domestic Policy Aide: Rain Rosimannus
Assistant to the President: Piret Part
Senior Aide-de-Camp: Rein Puusepp
Press Spokesman: Ms Indrek Treufeldt
E presinfo@nw.vpk.ee 98

11 Prime Minister

11-10000 Office of the Prime Minister
Lossi plats 1a, EE0100 Tallinn
☎ (+372 6) 31 67 01 E valitsus@rk.ee
⊖ (+372 6) 31 67 04
Prime Minister: Mart Siimann
State Secretary: Ülo Kaevats 98

12 Ministries

12-10000 Ministry of Defence
Sakala 1, EE0100 Tallinn
☎ (+372 2) 640 60 10 E info@kmin.ee
⊖ (+372 2) 640 60 01
Minister of Defence: Andrus Öövel ☎ (+372 6) 40 60 10
Adjutant: Tarvi Juus
Permanent Under-Secretary: Toomas Kitsing
☎ (+372 2) 640 60 12
Deputy Under-Secretary, Logistics: Eero Vaarman
☎ (+372 2) 640 60 16
Defence Policy: Margus Kolga ☎ (+372 2) 640 60 20
PRO: Victor Polyakoff ☎ (+372 2) 640 60 78
E polyakoff@kmin.ee 98

12-10500 Ministry of Economics Affairs
Harju 11, EE0001 Tallinn
☎ (+372 2) 44 05 77 E kantselei@mineco.ee
⊖ (+372) 631 36 60
Minister: Jaak Leimann
Permanent Under-Secretary: Margus Leivo
Deputy-Under Secretary: Taivo Kivistik
Deputy Under-Secretary: Signe Ratso
Deputy Under-Secretary: Tarmo Ruben
Counsellor: Tiit Maenniksaar
Counsellor: UElo Sarv
Counsellor: Andres Tint
Counsellor: Rae Vunk
Press Spokesman: Daniel Vaarik
☎ (+372 2) 44 38 14; (+372) 631 36 60 (fax) 98

12-20000 Ministry of Foreign Affairs
Rävala pst 9, EE0100 Tallinn
☎ (+372 6) 31 70 00 E vminfo@vm.ee
TX 173269 evm su ⊖ (+372 6) 31 70 99
Minister of Foreign Affairs: Toomas Hendzik Ilves
Permanent Under-Secretary: Indrek Tarand
Deputy Under-Secretary: Priit Kolbre
Deputy Under-Secretary: Raul Laanemäe
Press Spokesman: Ms Ehtel Halliste
☎ (+372) 631 76 54; 631 76 00; 631 70 99 (fax)
E vmedia@vm.ee
Director of Information Division, Press and Information Department: Triing Rajasalu 98

12-20550 Ministry of Finance
Suur-Ameerika 1, EE0100 Tallinn
☎ (+372) 611 34 45 E admin@fin.ee
⊖ (+372) 631 78 10
Minister: Mart Opmann ☎ (372) 611 34 45
Permanent Under-Secretary: Agu Lelled
☎ (+372) 611 33 48
Deputy Under-Secretary: Peep Lass ☎ (+372) 611 36 10
Counsellor to Ministry: Tiit Saat ☎ (+372) 611 32 79
Press Spokesman: Ali Ohlau
☎ (+372) 611 35 58; 631 78 10 (fax) E aili@fin.ee
Specialist of Public Relations Divison: Ruth Roht 98

12-21500 Estonian Trade and Investment Board
(† Foreign Investment Agency 27.11.95)
Raevala 6-602b, EE0001 Tallinn
☎ (+372) 642 01 66 W http://www.vm.eia.ee
E info@eia.ee ⊖ (+372) 641 03 12
Chairman of the Board of Directors: Jüri Sakkeus
Director General of the Estonian Investment Agency: Agu Remmelg
Director of the Estonian Export Agency: Kristjan Kivpalu

Tourism Board
Pikk 71, EE0001 Tallinn
☎ (+372 2) 60 17 00 E tallinn@tourism.uninet.ee
⊖ (+372 2) 60 27 43
Director: Silvi Bljumosits 98

12-30500 Ministry of Transportation and Communications
Viru 9, EE0100 Tallinn
☎ (+372) 639 76 13 E nei@mail.kbfi.ee
⊖ (+372) 639 76 06
Minister: Raivo Vare
Permanent Under-Secretary: Ms Ruth Martin
Deputy Under-Secretary: Tonu Naestema
Deputy Under-Secretary: Vaeino Soonike
Deputy Under-Secretary: Nathan Tonisson 98

12-40000 Ministry of Internal Affairs Department of Public Relations
Pikk 61, EE0100 Tallinn
☎ (+372) 612 50 07 ⊖ (+372) 612 50 10
Minister of Internal Affairs: Robert Lepikson
☎ (+372 6) 612 50 01
Permanent Under-Secretary: Vahur Glaase
☎ (+372) 612 50 02
Deputy Under-Secretary: Tiit Sepp ☎ (+372) 612 50 05
Head of Public Relation Department: Piret Muerk
☎ (+372) 612 50 30 ☎ (+372) 612 50 77 (fax)
E piret@sisemin.gov.ee 98

12-40500 Regional Affairs
Lossi plats 1a, EE0100 Tallinn
☎ (+372) 631 67 40; 631 67 09 (press contact)
E regi@wp.rk.ee ⊖ (+372) 631 67 40; 631 69 14 (press contact)
Minister Without Portfolio: Tiit Kubri
Counsellor: Peeter Albi
Counsellor: Garri Raagma
Counsellor: Kaie Masing 97

15 Parliament

15-10000 Parliament
Riigikogu
Lossi Plats 1a, EE0100 Tallinn
☎ (+372) 631 63 31 ⊖ (+372) 631 63 34
Speaker: Toomas Savi ☎ (+372 631) 163 01
Deputy Speaker: Ants Käärma ☎ (+372 631) 63 11
Deputy Speaker: Tunne Kelam ☎ (+372 631) 63 21
Chairman, Finance Committee: Olev Raju
☎ (+372 631) 64 32
Chairman, Economics Committee: Tiit Made
☎ (+372 631) 64 32
Chairman, Constitutional Committee: Tilt Käbin
☎ (+372 631) 64 40
Chairman, Legal Committee: Daimar Liiv
☎ (+372 631) 64 52
Chairman, Foreign Affairs Committee: Eino Tamm
☎ (+372 631) 64 60
Chairman, Defence Committee: Peeter Lorents
☎ (+372 631) 64 72
Chairman, Cultural Affairs Committee: Talvi Märj
☎ (+372 631) 64 80
Chairman, Social Affairs Committee: Endel Eero
☎ (+372 631) 64 90
Chairman, Rural Committee: Aavo Mölder
☎ (+372 631)65 00
Chairman, Environmental Committee: Rein Järlik
☎ (+372 631) 65 10
Secretary General: Heiki Sibul ☎ (+372 631) 63 30
Press Spokesman: Tonu Kaljumaee ☎ (+372 631) 63 50
(+372 631) 63 54 (fax)
Press Secretary: Margit Liivoja ☎ (+372 631) 63 53
E margit.liivoja@riigikogu. mailnet.fi
Press Secretary: Kristiina Jusson-Lindsalu
☎ (+372 631) 63 51 E kristniina.jusson@riigikog u.mailnet.fi
Information Officer: Madli Leikop ☎ (+372 631) 63 48
Head of Foreign Relations Department: Matti Maasikas
☎ (+371 631) 63 90 98

25 Defense Forces

25-10000 General Staff of the Defence Forces
Narva mount 8, EE0100 Tallinn
☎ (+372 2) 44 44 06 ⊖ (+372 6) 31 40 99
Commander of the Defence Forces: Col Johannes Kert
☎ (+372 2) 44 05 36
Chief of Defence Staff: Col Ants Laaneots
☎ (+372 2) 44 53 67
Deputy Chief of General Staff and Chief of Operations Department: Col Oskar Mark ☎ (+372 2) 44 44 21

Logistics Department
Suur Sojamae 23, EE0014 Tallinn
☎ (+372 2) 21 56 81
Chief of Logistics: Capt Einar Laigna

Armament Department
Suur Sojamae 23, EE0014 Tallinn
☎ (+372 2) 21 32 00
Head of Armament Dept: Mihkel Salla 98

25-10050 Office of the Chief of Air Force
Magasini 31, EE0001 Tallinn
☎ (+372 6) 46 13 09
Chief of Air Force: Col Teo Krüüner 98

25-10060 Office of the Chief of Navy
Tööstuse 54A, EE0004 Tallinn
☎ (+372 2) 47 20 02 ⊖ (+372 2) 44 53 54
Chief of Navy: Capt Roland Leit 98

25-10080 Defence Forces Headquarters
Marva mnt 8, EE0100 Tallinn
☎ (+372 2) 42 57 07 ⊖ (+372) 631 40 99
Chief of General Staff: Col Vello Loemaa (acting)
Head of Press Department: Harri Rent
Press Secretary: Leelo Linnask
☎ (+372 2) 42 13 79; (+372) 631 35 88 (fax) 98

25-10100 Defence League
Toompuiestee 8, EE0001 Tallinn
☎ (+372 6) 40 86 36 (duty officer) ⊖ (+372 6) 40 96 96
Commander: Col Lui Jaanson
■ The volunteer Defence League presently consists of approximately 6,500 volunteers who subordinate to the General Headquarters of the Defence Forces. These individuals participate in military exercises, assist national rescue and police operations and form the core of the reserve forces. 98

26 Police (Civil)

26-10000 Estonian Police Board
Politseiamet
(sub. of Ministry of Internal Affairs)
Pagari 1, EE0100 Tallinn
☎ (+372 6) 39 97 97 ⊖ (+372 6) 39 97 61
Director General: Ain Seppik ☎ (+372) 612 30 01
Head of Foreign Relations Department: Jaana Napa
☎ (+372) 612 47 97

Press and Information Officer: Aet Truu
☎(+372) 612 48 48 98

28 Coastguard, Border Protection Forces

28-10000 Estonian Border-Guard Dept
(sub. of Ministry of Internal Affairs)
Toompea 1, EE0103 Tallinn
☎(+372) 631 60 02 ⊖ (+372) 631 60 12 98

29 Customs & Excise

29-01000 National Customs Board
Rävala pst. 9, EE0100 Tallinn
☎(+372 2) 44 46 62 ⊖ (+372 2) 44 15 82
General Director: Mati Jürgens
Head of Administration Department: Lembit Tammera
Press and Information Officer: Merike Grünthal
Board of Passports
Lai 40, EE0100 Tallinn
☎(+372 2) 66 43 30; 66 43 81
Director General: Linnar Liivamägi 98

30 Civil Defense

30-00100 Honorary Consulate of Austria
Pikk 58, EE0001 Tallinn 98

30-10000 National Rescue Board
(sub. of Ministry of Internal Affairs)
Raua 2, EE 0001 Tallinn
☎(+372) 628 20 00 ⊖ (+372) 628 20 99
Director General: Harry Hein 98

31 Embassies & Attachés

31-00100 Honorary Consulate of Austria
Pikk 58, EE0001 Tallinn
☎(+372 2) 44 24 28 ⊖ (+372 2) 631 32 94 98

31-00310 Canadian Embassy Office
Toomkooli 13, EE0001 Tallinn
☎(+372 2) 44 90 56; 31 35 70 E canembt@zzz.ee
⊖ (+372 6) 31 35 73
Information Contact: Mrs Marina Asari 98

31-00360 Embassy of the People's Republic of China
Narva mnt 98, EE0100 Tallinn
☎(+372 2) 641 90 41 ⊖ (+372 2) 641 90 44 98

31-00460 Royal Embassy of Denmark
Ravala 9, EE0100 Tallinn
☎(+372 2) 45 46 00; 45 46 11; 631 31 20 ☒ 173262
⊖ (+372 6) 31 33 51
Ambassador: H E Mr Svend Roed Nielsen 98

31-00580 Embassy of Finland
Liivalaia 12, EE0001 Tallinn
☎(+372 6) 31 14 44; 31 35 45 (chancellery);
31 14 41 (visa section) ☒ 173186
⊖ (+372 6) 31 14 46 (chancellery); 31 14 45
(visa section)
Ambassador: H E Mr Jaakko Kaurinkoski
Consular Office
Uus 3-76, EE2400 Tartu
☎(+372 7) 43 12 40 ⊖ (+372 7) 43 31 45 98

31-00590 Embassy of France
Toomkuninga 20, EE0100 Tallinn
☎(+372 6) 31 12 06; 31 13 68 (visas)
⊖ (+372 6) 31 13 85
Ambassador: H E Mr Jacques Faure 98

31-00660 Embassy of Germany
Raevala 9, EE0100 Tallinn
☎(+372 6) 31 39 70; 31 39 71; 31 39 76 (visas)
⊖ (+372 2) 45 58 35
Ambassador: H E Mr Bernd Muetzelburg 98

31-00810 Honorary Consul of Hungary
Estonia pst 3/5, EE0001 Tallinn
☎(+372 6) 31 37 91 (Mrs Hellam);
(+372 7) 43 54 40; 43 34 07 (Mr Kuennap)
E mau@oef.org.ee ⊖ (+372 6) 31 37 96 (Mrs Hellam);
(+372 7) 352 36 (Mr Kuennap)
Executive Director Open Estonia Foundation:
Mrs Mall Hellam
Honorary Consul with jurisdiction over Tartu, Jogeva, Viljandi, Valga, Polva and Voru: Ago Kuennal 98

31-00890 Embassy of Italy
Muuerivahe 3, EE0005 Tallinn
☎(+372 6) 25 64 44 ☒ 0357.173091
⊖ (+372 6) 31 13 70
Ambassador: H E Mr Roberto Martini
Deputy Head of Mission: Sergio Nervi
Aeronautic, Military and Naval Attaché:
Col Pilot Francesco Gianesin (resident in Stockholm) 98

31-00920 Embassy of Japan
3rd Floor - Tallin Business Centre, 6 Harju Street,
EE0001 Tallinn
☎(+372 6) 31 05 31; 31 15 32 ⊖ (+372 6) 31 15 33
Chargé d'Affaires: Mitsuya Tokunaga (acting) 97

31-00990 Honorary Consul of Korea
Vaeike-Ameerika 10, EE0007 Tallinn
☎(+372 6) 31 37 81 ⊖ (+372 6) 31 37 80
Honorary Consul: Priit Maran 98

31-01015 Embassy of Latvia
Tonismagi 10, EE0001 Tallinn
☎(+372 2) 68 16 70 ⊖ (+372 6) 31 13 66
Ambassador: H E Mr Andris Piebalgs 98

31-01060 Embassy of Lithuania
Uus tn 15, EE0001 Tallinn
☎(+372) 631 40 30; (+372 2) 44 94 71 (visas)
⊖ (+372 6) 31 40 53 98

31-01340 Embassy of Norway
Paernu mnt 8, EE0109 Tallinn
☎(+372 2) 44 80 14; (+372 6) 31 30 03
⊖ (+372 6) 31 30 01
Ambassador: H E Mr Kai Olaf Lie 98

31-01430 Embassy of Poland
Pärnu mnt 8, EE0001 Tallinn
☎(+372 2) 631 40 88 ☒ 173061 ampol
⊖ (+372 2) 44 52 21
Ambassador: H E Mr Jakub Wolasiewicz
Defence Attaché: Col Boleslaw Bilinski
(Resides in Helsinki) ☎(+358 9) 684 80 90
Second Secretary, Political, Press and Promotion Affairs:
Andrzej Olborski
The Economic and Commercial Counsellor's Office
Lauteri 1-3, Tallinn
☎(+372) 631 42 22 ⊖ (+372) 631 45 50
Commercial Attaché: Marcin Nurowski 98

31-01485 Embassy of Russia
Pikk 19, Tallinn
☎(+372 2) 44 30 14 ⊖ (+372 2) 44 37 73
Ambassador: H E Mr Alexandre Trofimov 98

31-01680 Honorary Consulate of Spain
Estonia Avenue 7, EE0101 Tallinn, 1
☎(+372) 646 61 66
Honorary Consul with Jurisdiction over Estonia:
Nathan Kukk ☎(+372) 646 61 66 98

31-01730 Embassy of Sweden
Pikk 28, EE0100 Tallinn
☎(+372 6) 40 56 00 ☒ 173124 svensk ee
⊖ (+372 6) 40 56 95
Ambassador: H E Mrs Katarina Brodin 98

31-01875 Embassy of Ukraine
Endla 8, Tallinn
☎(+372) 631 15 55; (+372 2) 44 23 55 (consular office)
⊖ (+372) 631 15 56 (consular office)
Ambassador: H E Mr Yuriy Olenenko 98

31-01890 Embassy of the United Kingdom
Kentmanni 20, EE 0100 Tallinn
☎(+372 6) 31 33 53; 31 34 61; 31 34 62 ☒ 173074
⊖ (+372 6) 31 33 54
Ambassador: H E Mr Tim Craddock
Deputy Head of Mission and Vice Consul: D Vickers
Defence Attaché: Lieut Col Simon Walter Lamb Strickland
(resides in Helsinki) 98

31-01900 Embassy of the United States of America
Kentmanni 20, EE0100 Tallinn
☎(+372 6) 31 20 21; 631 22 10 (USIS)
W http://www.usislib.ee/usislib E tallinn@usislib.ee
⊖ (+372 6) 31 20 25; 631 20 26 (USIS)
Ambassador: H E Mr Lawrence P Taylor
☎(+372 6) 31 20 21
Defence Attaché: Peter Hendrikson
Commercial Officer: David Katz
Public Affairs Officer: Ms Victoria S Middleton
☎(+372 6) 31 22 10 98

34 Multinational Intergovernmental Organizations

34-07000 Organisation for Security and Cooperation in Europe (OSCE)
Raevala 9, EE0100 Tallinn
☎(+372 631) 75 73; 31 75 74 ⊖ (+372 631) 75 75
Ambassador: H E Mr Herbert Grubmayr 98

34-10000 United Nations Development Programme (UNDP)
PO Box 446, EE0100 Tallinn
Kohtu 8 - Room 55, EE0090 Tallinn
☎(+372) 631 14 96 ⊖ (+372) 631 13 99
Resident Representative: Jan Wahlberg 98

Contact Reference

Ethiopia

Head of State:
Negasso Gidada

10 Head of State

10-10000 Office of the President
PO Box 5707, Addis Ababa
☎(+251 1) 11 30 00
President: Negasso Gidada [98]

11 Prime Minister

11-10000 Office of the Prime Minister
PO Box 1013, Addis Ababa
☎(+251 1) 12 34 00
Prime Minister: Meles Zenawi [98]

12 Ministries

12-10000 Ministry of National Defense
Churchill Road, PO Box 1373, Addis Ababa
☎(+251 1) 44 55 55 ⊠ 21261 mond addis
Minister: Walewa Teferra [98]

12-20000 Ministry of Foreign Affairs
Menelik II Avenue, PO Box 393, Addis Ababa
☎(+251 1) 44 73 45 to 49 ⊠ 21050 afgers addis
Minister for Foreign Affairs: Seyoum Mesfin [98]

12-30000 Ministry of Trade
PO Box 2559, Addis Ababa
☎(+251 1) 15 96 71 ⊠ 21320 ⊖ (+251 1) 51 42 88
Minister: Kassahun Ayele [98]

31 Embassies & Attachés

31-00150 Embassy of Belgium
Fikre Mariam St. (Kebena) Higher 16, PO Box 1239, Addis Ababa
☎(+251 1) 61 18 13; 16 43 ✉AMBABEL Addis-Ababa
⊠ 21157 ambabel et ⊖ (+251 1) 61 36 46
Ambassador: H.E. Mr Alain Genot
Head of Chancery: Jean-Paul Charlier [97]

31-00420 Embassy of the Republic of Côte d'Ivoire
Jimma Road, PO Box 3668, Addis Ababa
☎(+251 1) 71 12 13, 71 27 54 ⊠ 21061
Ambassador: H E Mr P L Yere
Charge d'Affairs a i: Dalo Gbouagbre
Counsellor: Ly Djerou Robert
First Secretary: Ms Sanogo Zenitea
Counsellor, Public Information Officer: Akadia Adjobia
Commercial Officer: Kouakou Christophe [98]

31-00660 Embassy of the Federal Republic of Germany
Khabana, PO Box 660 and 1267, Addis Ababa
☎(+251 1) 55 04 33 ⊠ 21 015 aaadd et
⊖ (+251 1) 55 13 11
Ambassador: H E Mrs Wiltrud Holik
Deputy Chief of Mission: Dr Berthold Johannes
Public Information Officer: York Schuegraf [98]

31-00690 Embassy of Greece
Africa Avenue, PO Box 11680, Addis Ababa
☎(+251 1) 51 51 61; 51 67 99 ⊖ (+251 1) 51 03 62 [98]

31-00840 Embassy of the Republic of Indonesia
Mekanissa Rd, PO Box 1004, Addis Ababa
☎(+251 1) 71 01 21 ⊠ 21264 indon et
⊖ (+251 1) 71 08 73
Ambassador: H.E. Mr Rochsjad Dahlan [97]

31-00890 Embassy of Italy
Villa Italia, Kebenà, PO Box 1105, Addis Ababa
☎(+251 1) 55 30 42; 55 30 44 ⊠ 21342 itadi et
⊖ (+251 1) 55 02 18
Ambassador: H.E. Dr Sergio Angeletti
Counsellor, Deputy Chief of Mission: Pierluigi Velardi
☎(+251 1) 55 02 18
Military, Naval & Air Attaché: Col Giuseppe Pepe
Commercial Secretary: Dr Fernando Gentilini
☎(+251 1) 55 01 20 [98]

31-00920 Embassy of Japan
Sunshine Building, Bole Road, PO Box 5650, Addis Ababa
☎(+251 1) 51 10 88 [E] japan-embassy@telecom.net.et
⊖ (+251 1) 351 13 50
Ambassador: H E Mr Yasuhrio Hamada
Deputy Chief of Mission: Kenji Miyata
Head of Accounts Section: Masaru Sawaguri
Public Information Officer: Takeshi Murazawa [98]

31-01270 Royal Netherlands Embassy
Old Airport, Embassy Compound, PO Box 1241, Addis Ababa
☎(+251 1) 71 11 00 ✉HOLLANDIA Addis Abeba
⊖ (+251 1) 71 15 77
Second Secretary: Bob Hensen [98]

31-01730 Swedish Embassy
Ras Ababa Aregaye Avenue, PO Box 1029, Addis Ababa
☎(+251 1) 51 12 55
[E] sweden.embassy@telecom.net.et
⊠ 21039 svensk addis ⊖ (+251 1) 51 58 30
Ambassador: H E Mr Carl Olof Cederblad
First Secretary Commercial Counsellor: David Abresparr
Counsellor, Development and Cooperation: Mats Svensson
First Secretary, Development and Cooperation: Arne Carlsgard
First Secretary, Development Cooperation: Lars Leander
Second Secretary, Political and Commercial Affairs: David Abresparr
Second Secretary, Economist: Ms Izabella Eriksson
Second Secretary, Consular Affairs: Ms Liselotte Aberg
☎(+251 1) 51 10 67; 51 31 37
Third Secretary: Ms Anette Millard
Administration: Bengt Isaksson
Visa Office
Ras Tessema Sefer, Higher 3, K53, House Number 167, Addis Abab
☎(+251 1) 51 10 67; 51 31 37 ⊖ (+251 1) 51 54 08 [98]

31-01890 Embassy of the United Kingdom
Fikre Mariam Abatechan Street, PO Box 858, Addis Ababa
☎(+251 1) 61 23 54 ⊠ 21299 proad et
⊖ (+251 1) 61 05 88
Deputy Head of Mission and First Secretary: Frances Guy
Defence Attaché: Col R D Vellacott (resident in Nairobi)
Second Secretary Commercial: T Rooney
Political/Information Officer: Ms Janet Duff [98]

Fiji

Head of State:
Ratu Sir Kamisese Mara

10 Head of State

10-10000 Office of the President
Government House, PO Box 2513, Suva
☎(+679) 31 42 44 ⊠ 2167 fosec fj ⊖ (+679) 30 16 45
President: Ratu Sir Kamisese Mara [98]

11 Prime Minister

11-10000 Office of the Prime Minister
Victoria Parade, PO Box 2513, Suva
☎(+679) 31 52 69; 31 42 44; 31 49 24 ⊠ 2167 fosec fj
Prime Minister: Sitiveni Rabuka
Deputy Prime Minister: Ratu Timoci Vesikula
Assistant Minister to the Prime Minister's Office: Vincent Lobendahn [98]

12 Ministries

12-20000 Ministry of Foreign Affairs and External Trade
Government Building, PO Box 2220, Suva
☎(+679) 21 17 24 ⊠ 2167 fosec fj ⊖ (+679) 30 17 41
Minister: Filipe Bole [98]

12-30000 Ministry of Commerce, Industry, Trade and Public Enterprises
Suva
Minister: Isimeli Bose [98]

12-40000 Ministry of Home Affairs and Immigration
Government Buildings Victoria Parade, PO Box 2349, Suva
☎(+679) 21 14 04 ⊠ 2342 homesec fj
⊖ (+679) 30 03 46
Minister for Home Affairs: Col Paul F Manueli
☎(+679) 21 14 01
Minister for Justice and Attorney General: Ratu Etuate Tavai ☎(+679) 21 15 80
Permanent Secretary: Apisalome Tudreu
☎(+679) 21 12 10
Foreign Relations Director: Gyani Nand (acting)
Head Commissioner Police Force: Col Isikia Savua
☎(+679) 31 29 99
Chief of Staff: Brig George K Konrote ☎(+679) 38 52 22
Principal Defence Assistant: Peni R T Lomaloma
☎(+679) 21 14 04 [98]

25 Defense Forces

25-10000 Republic of Fiji Military Forces Headquarters
Queen Elizabeth Barracks, PO Box 102 - Maddocks Road, Suva
☎(+679) 38 52 22 ⊠ 2157 hqrfmf fj ⊖ (+679) 37 00 33
Commander: Brig Ratu Epeli G Ganilau
☎(+679) 38 52 22
Chief of Staff: Brig George K Konrote ☎(+679) 38 52 22
Commanding Officer Naval Division: Capt Frank Bainimarama ☎(+679) 31 28 03
Colonel General Staff: Col George Kadavulevu
☎(+679) 38 52 22

Colonel Administration and Quartermaster:
Lieut Col Ulaiasi Vatu 98

25-40000 Republic Fiji Military Forces Naval Division
PO Box 12 387, Suva
☎(+679) 31 25 85 ☒ 2632 ⊖ (+679) 31 47 83
Commander Fiji Navy: J V Bainimarama
Deputy Commander and Chief of Staff:
Cdr Timoci Lesikivatukouala
Director of Operations: E Teleni 98

26 Police (Civil)

26-00000 Royal Fiji Police Force
Ratu Sukuna House, Suva
☎(+679) 31 29 99 ☒ 2158 polsouth fj
⊖ (+679) 30 32 18
Head Commissioner: Col Isikia Savua
Assistant Commissioner of Police, Crime: Moses Driver
Assistant Commissioner of Police Operations:
Qalo Bulatiko
Assistant Commissioner of Police Administration:
Paras Ram
Assistant Commissioner of Police Internal Affairs:
Savenaca Tuivaga 98

26-01000 Fiji Intelligence Service
PO Box 2349, Suva
Government Buildings - Berkley Crescent
☎(+679) 31 37 99
Director: Lieut Col Metuisela Mua 98

31 Embassies & Attachés

31-00090 Australian Embassy
Dominion House, Thomson St, 8th Fl., GPO Box 214, Suva
☎(+679) 31 28 44 ✉AUSTEMBA Suva
☒ 2126 austcom fj ⊖ (+679) 30 10 06
Ambassador: H.E. Mr G. Urwin
Deputy Chief of Mission & Counsellor: J. McAnulty
Defence Attaché (South Pacific): Lt Col D. Phillips (MBE)
Trade Commissioner: S. Ryan
First Secretary Public Affairs: L. Smith 98

31-00360 Embassy of the People's Republic of Fiji
147 Queen Elizabeth Drive, Suva
☎(+679) 30 18 33 ⊖ (+679) 30 09 50 98

31-00590 Embassy of France
Thomson Street, Dominion House, Private Mail Bag, Suva
☎(+679) 31 22 33; 29 25 ☒ 2326 amfrance fj
⊖ (+679) 30 18 94
Deputy Chief of Mission: Alain Molle
Regional Scientific & Co-operation Counsellor:
Pascal Perron
Armed Forces Attaché: Cdr L.C. Cédric Villedieu de Torcy
(resident in Noumea) 97

31-00880 Embassy of Israel
5th Floor Parade Building, 69 Joske Street, Suva
☎(+679) 30 34 20 ⊖ (+679) 30 04 15 98

31-00920 Embassy of Japan
GPO Box 13045, Suva
☎(+679) 30 21 22 ✉TAISHI Suva ☒ 2253 taishi fj
⊖ (+679) 30 14 52
Ambassador: H.E. Mr Yasunori Kikuchi
Deputy Chief of Mission: Yoshiaki Kotaki 98

31-01110 Embassy of Malaysia
5th Floor Pacific House, Butt Street, GPO Box 356, Suva
☎(+679) 31 21 66 ☒ 2295 mwklsuva fj
⊖ (+679) 30 33 50 98

31-01300 New Zealand Embassy
Reserve Bank Building, Pratt Street, PO Box 1378, Suva
☎(+679) 31 14 22; 30 60 33 (immigration)
⊖ (+679) 30 08 42; 30 56 75 (Tradenz);
30 37 92 (immigration)
Ambassador: H E Ms Suzanne Blumhardt
Deputy Head of Mission and Counsellor: Vince McBride
First Secretary: Gerry Cook
Second Secretary: Ms Isabel Calvert
Second Secretary: Ngawini Keelan
Attaché: Ms Kathryn A Brown
Trade Commissioner: Paul Linton
Branch Manager, Immigration: Grant Mudgeway
Defence Attaché: Cdr O J Hanley (RNZN) 98

31-01390 Embassy of Papua New Guinea
Credit Corporation House Gordon St, PO Box 2447, Suva
☎(+679) 30 42 44 ⊖ (+679) 30 01 78 98

31-01890 Embassy of the United Kingdom
PO Box 1355, Suva
Victoria House, 47 Gladstone Road, Suva
☎(+679) 31 10 33 Ⓔ ukrep@is.com.fj
⊖ (+679) 30 14 06
Ambassador: H E Mr Michael John Peart
(Pitcairn Alternate) (Representative for the South)
(Pacific Commission,)
(non-resident High Commissioner)
(to Nauru, Kiribati and Tuvalu)
Deputy Head of Mission and Consul: Vernon Scarborough
(non-resident Ambassador to)
(Micronesia, Marshall Islands,)
(Palau, Dep. High Commissioner)
(to Kiribati, Nauru and Tuvalu.) ☎(+64 4) 472 60 49
Defence Attaché: Col Peter Barry (resident in Wellington)
Commercial Officer: Mrs Sunil Kumar
Third Secretary, Management: Miss K J Smith
Press and Public Affairs Officer: Mrs Vineeta Nand 98

Finland

Head of State: Martti Ahtisaari

10 Head of State

10-10000 Office of the President
Mariank 2, FIN00170 Helsinki 17
☎(+358 9) 66 11 33 ⊖ (+358 9) 63 02 92
President of Finland: Martti Ahtisaari
Secretary-General to the President: Jaakko K A Kalela
Secretary of Chancery: Ms Sirkka M Peisa 98

11 Prime Minister

11-10000 Office of the Prime Minister
Aleksanterinkatu 3 D, PO Box 275, FIN00171 Helsinki 17
☎(+358 9) 16 01 ⊖ (+358 9) 160 21 64
Prime Minister: Paavo Lipponen 98

12 Ministries

12-10000 Ministry of Defence
Puolustusministeriö
Eteläinen Makasiinikatu 8 A, PO Box 31,
FIN00131 Helsinki 13
☎(+358 9) 161 61 ☒ 124667 hkipe sf
⊖ (+358 9) 65 32 54
Minister: Ms Anneli Taina
Aide-de-Camp to the Minister: Maj Esa Vanonen
Secretary to the Minister of Defence: Ms Maarit Koivisto
Permanent Under-Secretary of State (PUSS) for Defence:
Lt Gen Pertti Nykänen
Private Secretary to PUSS for Defence: Ms Satu Arola
Chief of the Central Department: Eero Lavonen
Chief of the Organisation Department: Ilkka Puukka
Chief of the Real Estate Department: Matti Niemi

Military Affairs Department
☎(+358 9) 161 39 41 ⊖ (+358 9) 161 39 93
Chief: Maj Gen Jikka Hollo
Chief of FN Office: Col Kari Höglund 98

12-20000 Ministry of Foreign Affairs
Ulkoasiainministeriö
Merikasarmi, PO Box 176, FIN00161 Helsinki 16
☎(+358 9) 13 41 51 Ⓔ webmaster@formin.fi
☒ 124636 umin fi ⊖ (+358 9) 62 98 40
Minister: Tarja Halonen
Minister of Foreign Trade: Ole Norrback
Secretary of State: Jukka Valtasaari
Under-Secretary (Economic Affairs): Leif Fagernäs
Under-Secretary (Political Affairs): Jaakko Blomberg
Press and Public Information Director: Yrjö Länsipuro
Head of Information Services Unit: Frank Hellsten

Administration Department
PO Box 176, FIN00161 Helsinki 16
☎(+358 9) 13 41 51 ⊖ (+358 9) 13 41 54 27
Chief: Charles Murto

Political Department
PO Box 176, FIN00161 Helsinki 16
☎(+358 9) 13 41 51 ⊖ (+358 9) 13 41 50 09
Chief: Pertti Torstila

Commercial Policy Department
PO Box 176, FIN00161 Helsinki 16
☎(+358 9) 13 41 51 ⊖ (+358 9) 13 41 50 70
Chief: Jorma Julin

Protocol Department
PO Box 176, FIN00161 Helsinki 16
☎(+358 9) 13 41 51 ⊖ (+358 9) 13 41 65 22
Chief: Henry Söderholm
Deputy: Irma Ertman 98

12-40000 Ministry of the Interior
Sisäasiainministeriö
Kirkkokatu 12, PO Box 257, FIN00170 Helsinki
☎(+358 9) 16 01 ☒ 123644 susto sf
⊖ (+358 9) 160 29 27
Minister of Interior, Police Affairs: Jan-Erik Enestam
Second Minister of Interior, Administration:
Jouni Backman 98

12-50000 Ministry of Trade & Industry
Handels och Industriministeriet
Aleksanterinkatu 4, PO Box 230, FIN00171 Helsinki
☎(+358 9) 16 01 ☒ 124645 minco sf
⊖ (+358 9) 160 36 66
Minister: Antti Kalliomaki 98

12-60000 Ministry of Transport & Communications
Liikenneministeriö
Eteläesplanadi 16, FIN00130 Helsinki
☎(+358 9) 173 61 ☒ 125472 limin sf
⊖ (+358 9) 173 63 40; 173 62 92
Minister of Transport and Communications: Matti Aura 98

15 Parliament

15-10000 Parliament
Mannerheimint 30, Eduskunta, FIN00102 Helsinki
☎(+358 9) 43 21 Ⓔ merja.kivinen@edustunta.fi
☒ 121464 ekirj sf ⊖ (+358 9) 432 22 74
Speaker: Mrs Riitta Uosukainen
First Deputy Speaker: Mikko Pesälä
Second Deputy Speaker: Mrs Kerttu Törnqvist
Secretary General: Seppo Tiitinen
Deputy Secretary General: Jouni Vainio
Chief Information Officer: Ilpo Pohjola
Secretary of Foreign Affairs Committee: Jukka Huopaniemi
Chairman Foreign Affairs Committee: Markus Aaltonen
Vice Chairman Foreign Affairs Committee: Mrs Eva Biaudet
Secretary to the Speaker: Mrs Helena Riutta
Assistant Secretary of Foreign Affairs Committee:
Ilkka Salmi

Contact Reference

Secretary of Defense Committee: Alpo Rivinoja
Chairman Defence Committee: Kalevi Lamminen
Vice Chairman Defence Committee: Jaakko Laakso [98]

20 Defense Administrations/Services/Agencies

20-00300 Civil Aviation Administration (CAA)
Ilmailutie 9, FIN01530 Vantaa
☎ (+358 9) 827 71; 615 11 ⊖ (+358 9) 82 77 20 99
♣ FMk 840,045,000 (1996)
Director General: M Talvitie
Deputy Chairman and Senior Ministerial Secretary: Pekka Hurtola
Director Flight Safety Authority: K Salonen
Director Administration and Finance Department: E Pitkänen
Director Air Navigation Services: J Hulkko
Deputy Director General Airports Department: L Vänskä
Chief Economic Planning and Financing Department: A Haapanen
Director, International Affairs: L Lövkvist
Secretary International Affairs: Ms P Eskuri
Chief of Press and Information: Ms Irmeli Paavola
Chief, Airport Operations: Kauko Hakunti
Chief, Administration: J Salmela
Chief Flight Operations and Licensing: R Lamberg
Chief, Systems Development: J Alakoski
■ Operates 25 Finnish Airports:
Helsinki-Vantaa and Helsinki-Malmi, Enontekiv, Halli, Ivalo, Joensuu, Jyvdskyld, Kajaani, Kauhava, Keimi-Tornio, Kittild, Kruunupyy, Kuopio, Kuusamo, Lappeenranta, Mariehamn, Oulu, Pori, Rovaniemi, Savonlinna, Tampere-Pirkkala, Turku, Utti, Vaasa, Varkaus. [98]

25 Defense Forces

25-10000 Defence Staff of the Finnish Defence Forces
Fabianinkatu 2, PO Box 919, FIN00101 Helsinki
☎ (+358 9) 181 01 11 [W] http://www.mil.fi/
⊠ 124667 hkipe fi ⊖ (+358 9) 18 12 24 39
Chief of Defence: Gen Gustav Hägglund
Chief of Defence Staff: Lieut Gen Matti Kopra

Army Staff
PO Box 919, FIN00101 Helsinki
Chief of the Army: Maj Gen Juha Kainulainen
Deputy Chief of the Army: Brig Gen Heikki Tilander

Personnel Staff
PO Box 919, FIN00101 Helsinki
Chief of Personnel: Maj Gen Antti Simola
Deputy Chief of Personnel: Col Hannu Luotola

Operations Staff
PO Box 919, FIN00101 Helsinki
Chief of Operations: Lieut Gen Jussi Hautamäki
Deputy Chief of Operations: Brig Gen Olavi Jäppilä

Logistics Staff
PO Box 919, FIN00101 Helsinki
Chief of Logistics: Vice Adm Sakari Visa

Information Division
PO Box 919, FIN00101 Helsinki
⊖ (+358 9) 18 12 24 39
Chief of Information Division: Capt Matti Mäkinen (Navy)
Deputy Chief of Information Division: Cdr Eero Karhuvaara
Informaton Contact: Capt Jukka Haltia
☎ (+358 9) 18 12 24 20 [98]

25-19200 Finnish Defence Forces Defence Training Development Centre (PvKK)
Technical Development Section
PO Box 5, FIN04401 Järvenpää
☎ (+358 9) 18 16 27 24 ⊖ (+358 9) 18 16 27 22
Contact: Osmo Forstén (B.Sc. Eng.) [98]

25-20000 Air Force Headquarters
Ilmavoimien Esikunta
PO Box 30, FIN41161 Tikkakoski
☎ (+358 14) 181 01 11 ⊠ 28501 ilmave sf
⊖ (+358 14) 181 43 50
Commander-in-Chief: Maj Gen Matti Ahola
Technical Inspector: Col Eng Kimmo Oksanen
Chief Technical Department: Col Eng Markku Ihantola
Public Relations Officer: Maj K Reiman
Information Secretary: Ms Leena Tiainen

Lapland Air Command
PO Box 22, FIN96101 Rovaniemi
☎ (+358 16) 181 01 11 ⊖ (+358 16) 181 31 09
Commander: Col S Ryynanen
Chief of Staff: Lieut Col J Liinamaa

Satakunta Air Command
PO Box 761, FIN33101 Tampere
☎ (+358 3) 18 16 11 11 ⊖ (+358 3) 18 16 03 69
Commander: Col J. Pystynen
Chief of Staff: Lieut Col M Määttänen

Karelia Air Command
PO Box 5, FIN70901 Toivala
☎ (+358 17) 181 51 11 ⊖ (+358 17) 181 42 13
Commander: Col H. Lyytinen
Chief of Staff: Lt Col H. Myöhanen

Air Support Squadron
PO Box 6, FIN41161 Tikkakoski
☎ (+358 14) 181 01 11 ⊖ (+358 14) 181 60 05
Commander: Lieut Col J Partio

Command, Control and Communications Systems School
Ilmavoimien Viestikoulu
(† Air Force Signals School)
PO Box 7, FIN41161 Tikkakoski
☎ (+358 14) 181 01 11 ⊠ 27819 ivik sf
⊖ (+358 14) 181 54 80
Commander: Lieut Col M Lehto

Air Academy
Ilmasotakaoulu, PO Box 5, FIN62201 Kauhava
☎ (+358 6) 181 31 11 ⊖ (+358 6) 181 37 00
Commander: Col Hannes Bjurström
Deputy Commander/Chief of Staff: Maj Ensio Paalijärri
Director of Operations: Maj Martti Rajaniemi
Director of Procurement/Logistics: Maj Jarmo Ahonen

Aircraft and Weapons Systems School
(† Air Force Technical School)
PO Box 5, FIN35601 Halli
☎ (+358 3) 18 16 61 11 ⊠ 27858 haitk sf
⊖ (+358 3) 18 16 61 40
Commander: Lt Col P. Hiltunen

Air Force Depot
PO Box 210, FIN33101 Tampere
☎ (+358 3) 18 16 41 11 ⊠ 27830 treiv fi
⊖ (+358 3) 367 18 66
Commander: Col Eng Veijo Mustonen
Chief of Technical Department: Lieut Col Eng Hannu Lassiln
Chief of Quality Assurance: Lieut Col Eng Hakki Karppinen
Chief of Flight Test Center: Lieut Col Juha Grönmark

Command, Control and Communications Systems Depot
(† Air Force Signals Depot)
PO Box 14, FIN41161 Tikkakoski
☎ (+358 14) 181 41 11 ⊠ 27821
⊖ (+358 14) 181 46 25
Commander: Col Eng. V. Kangaskoski [98]

25-40000 Naval Headquarters
PO Box 105, FIN00201 Helsinki
Pohjoiskaari 36, FIN00200 Helsinki
☎ (+358 9) 18 12 42 11; 18 16 71 11
⊠ 124666 hkipe sf
⊖ (+358 9) 18 12 42 12; 18 16 70 16 (information section)
Commander-in-Chief: Rear Adm Esko Illi
Chief of Staff: Capt Heikki Salmela (Navy)
Head Research and Development Sector: Capt Alpo Tuurnala
Chief of Maintenance: Cdr Juha Rannikko
Inspector Coastal Artillery: Col Jorma Vuohelainen
Information Section: 1st Lieut Kari Keskinen
☎ (+358 9) 18 12 42 34; 18 12 42 31 (fax)

Coastal Artillery
Defence HQ, PO Box 919, FIN00101 Helsinki
Fabianinkatu 2, FIN00100 Helsinki
☎ (+358 9) 181 01 11 [98]

26 Police (Civil)

26-11000 National Bureau of Investigation
PO Box 285, FIN01301 Vantaa
☎ (+358 9) 838 86 61 ⊖ (+358 9) 83 88 65 08
Executive Secretary: Mattila Tuula [97]

26-12000 Security Police Headquarters
PO Box 151, FIN00121 Helsinki
☎ (+358 9) 13 44 71 ⊖ (+358 9) 13 44 78 06
Director of Division: Mahi Simola
Deputy Director of Division: Seppo Jouko J Pylkkanen
Press and Information Officer: Juhani J Vaananen
Chief of Technical Division: Hannu Tapio Haaranen [97]

28 Coastguard, Border Protection Forces

28-10000 Frontier Guard Headquarters
Rajavartiolaitoksen esikunta
Korkeavuorent 21, PO Box 3, FIN00131 Helsinki
☎ (+358 2) 04 10 65 11 ⊖ (+358 2) 04 10 67 55
Chief of Frontier Guard: Lt Gen. Matti Autio
Deputy Chief of Frontier Guard: Maj.Gen. Hannu Ahonen
Chief of Personnel Div.: Col Markku Hedemäki
Chief of Frontier & Coast Guard Div.: Col Matti Sandqvist
Chief of Technical Division: Lieut Col Jorma Kosunen
Commander of Air Patrol Squadron: Lt Col Pertti Ruotsalainen
■ The Frontier Guard is militarily organised & is under the control of the Ministry of the Interior. In time of war, frontier guard troops may come under the jurisdiction of the Defense Forces. [98]

29 Customs & Excise

29-00500 National Board of Customs
Tullihallitus, PL 512, FIN00101 Helsinki [98]

31 Embassies & Attachés

31-00535 Embassy of Estonia
Itaeinen Puistotie 10, FIN00140 Helsinki
☎ (+358 9) 622 02 60 [E] sekretar@estemb.fi
⊖ (+358 9) 62 20 26 10
Ambassador: H E Mr Mati Vaarmann
☎ (+358 9) 62 20 26 12
Counsellor: Avo Suurthal ☎ (+358 9) 62 20 26 12
Counsellor (Economic Affairs): Urve Nõu
☎ (+358 9) 62 20 26 12
Second Secretary (Press and Information): Ms Merle Pajula ☎ (+358 9) 62 20 26 12 [98]

31-00690 Embassy of Greece
Maneesikatu 2 A 4, FIN00170 Helsinki
☎ (+358 9) 278 11 00 [E] grembfi@grembfi.pp.fi
[W] http://www.personal.eunet.fi/pp/grembfi
⊠ 125279 gremb sf ⊖ (+358 9) 278 12 00
Ambassador: H E Mr Athanassios A Theodoracopoulos
First Secretary: Georgios Iliopoulos [98]

31-00850 Embassy of the Islamic Republic of Iran
Bertel Jungin tie 4, FIN00570 Helsinki
☎ (+358 9) 684 71 33 ⊠ 125888 iran sf
⊖ (+358 9) 684 94 12
Ambassador: H E Dr Mahmoud Broujerdi
Chargé d'Affaires a.i.: Hassan Hassani
First Secretary: Ali Golabi [98]

31-00890 Embassy of Italy
Itainen Puistotie 4, FIN00140 Helsinki
☎ (+358 9) 17 51 44 ⊠ 121753 itafi sf
⊖ (+358 9) 17 59 76
Ambassador: H.E. Dr Raniero Avogadro
Deputy Chief of Mission: Dr Roberto Natali [97]

31-00920 Embassy of Japan
Eteläranta 8, FIN00130 Helsinki
☎ (+358 9) 64 42 06; 63 30 11 ✉ TAISHI Helsinki
⊠ 121037 taisi sf ⊖ (+358 9) 63 30 12
Ambassador: H.E. Mr Wataru Miyakawa [97]

31-00990 Embassy of the Republic of Korea

Annankatu 32a, FIN00100 Helsinki
☎ (+358 9) 686 62 30 ⊖ (+358 9) 68 66 23 55
Ambassador: H.E. Mrs In-Ho Lee
Counsellor: Ihl-Song Choi
First Secretary: Jae-Choon Lee
Second Secretary: Jong-Nam Park
Third Secretary & Vice Consul: Nam-Sun Park [98]

31-01270 Royal Netherlands Embassy

PO Box 886, FIN00101 Helsinki 10
Raatimiehenkatu 2 A 7, FIN00140 Helsinki 14
☎ (+358 9) 66 17 37 ⊠ 121779 amned sf
⊖ (+358 9) 65 47 34
Ambassador: H.E. Mr Bartholomeus de Bruyn Ouboter
Counsellor: Erik Boer
Defence, Military & Naval Attaché: Col F.W. Hoogeland (resident in Oslo)
Air Attaché, Asst Defence, Naval & Military Attaché: Lt Col G P Callenbach (resident in Oslo)
Administration & Consular Affairs Attaché: Cor J. Schouten [97]

31-01430 Embassy of the Republic of Poland

Armas Lindgrenintie 21, FIN00570 Helsinki
☎ (+358 9) 684 80 77; 684 81 79; 684 80 90
⊖ (+358 9) 684 74 77
Ambassador: H E Mr Józef Wiejacz
☎ (+358 9) 684 83 69
First Secretary, Deputy Chief of Mission: Tomasz Chlon
☎ (+358 9) 684 87 02
Defence, Military, Naval and Air Attaché:
Lieut Col Boleslaw Bilinski ☎ (+358 9) 684 79 27
Second Secretary, Cultural and Scientific Affairs:
Grzegorz Marczak ☎ (+358 9) 684 80 77
First Secretary, Chief of the Consular Section:
Jacek Erenfeicht ☎ (+358 9) 684 93 28
First Secretary: Sylwester Sadlowski
☎ (+358 9) 684 93 50
Second Secretary, Administrative Affairs:
Miroslaw Skarzynski
Attaché: Jacek Diugolecki
Attaché: Wladyslaw Domasiewicz ☎ (+358 9) 684 80 77

Commercial Counsellor's Office
Risto Rytin tie 7, FIN00570 Helsinki
☎ (+358 9) 684 91 88; 684 91 70
⊖ (+358 9) 684 89 07
Commercial Counsellor: Krzysztof Biskupski
☎ (+358 9) 684 93 97
Commercial Attaché: Krzysztof Karwowski
☎ (+358 9) 684 91 88
Commercial Attaché: Jerzy Rokicki ☎ (+358 9) 684 91 88 [98]

31-01650 Embassy of the Slovak Republic

Annankatu 25, FIN00100 Helsinki 10
☎ (+358 9) 70 01 84 60 [E] skemb.hels@sci.fi
⊖ (+358 9) 64 01 05
Counsellor, Chargé d'Affaires a.i.: Vladimir Halgas
Military and Air Attaché: Col Jozef Nosál
Second Secretary (Political Affairs): Sona Budayova
Second Secretary (Commercial Affairs): Juraj Jurásek [98]

31-01670 South African Embassy

Rahapajankatu 1A5, FIN00160 Helsinki 16
☎ (+358 9) 65 82 88 ⊠ LEGSA Helsinki [E] saemb@icon.fi
⊖ (+358 9) 65 83 65
Ambassador: H E Dr C Streeter
Counsellor: A Figgins
Second Secretary (Political): F Hentschel
Third Secretary (Administration): J Lötter [98]

31-01680 Embassy of Spain

Kalliolinnantie 6, FIN00140 Helsinki
☎ (+358 9) 17 05 05 ⊖ (+358 9) 66 01 10
Ambassador: H E Mr Vicente Blanco ☎ (+358 9) 62 68 00
First Secretary: Juan Arístegui
Economic and Commercial Counsellor: León Herrera
Attaché (Commercial Affairs): Urcilina Gallego

Commercial Section
Simonkatu 12 B 23, FIN00100 Helsinki 10
☎ (+358 9) 694 48 02/92 ⊖ (+358 9) 694 40 49
Commercial Attaché: Mrs Urcilina Gallego [98]

31-01730 Swedish Embassy

Norra Esplanaden 7B, PO Box 329, FIN00171 Helsinki
☎ (+358 0) 65 12 55 ⊠ SVENSK Helsinki
⊠ 121082 svnsk sf, 124538 svnsk sf
⊖ (+358 0) 65 52 85
Ambassador: H.E. Mr Knut Thyberg ☎ (+358 0) 60 76 58
Military, Naval & Air Attaché: Col Vulf Hessulf
Asst Defense Attaché: Maj. Folke Benjamin

Office of the Commercial Secretary
Sjötullsqatan HB, FIN00170 Helsinki
☎ (+358 9) 66 93 69 ⊠ 121082 swetr sf
Commercial Secretary: Per Olof Dahllöf [98]

31-01740 Embassy of Switzerland

Uudenmaankatu 16 A, FIN00120 Helsinki 12
☎ (+358 9) 64 94 22 ⊠ AMBASUISSE Helsinki
⊖ (+358 9) 64 90 40
Ambassador: H.E. Mr S. Meili [98]

31-01890 Embassy of the United Kingdom

Itäinen Puistotie 17, FIN00140 Helsinki
☎ (+358 9) 22 86 52 46 ⊖ (+358 9) 22 86 52 62
Ambassador: H E Mr Gavin Hewitt
Counsellor and Deputy Chief of Mission: D Gowan
Defence Attaché: Lieut Col S W L Strickland
First Secretary Commercial: H Formstone [98]

31-01900 Embassy of the United States of America

Itainen Puistotie 14A, FIN00140 Helsinki
☎ (+358 9) 17 19 31 [E] webmaster@usis.fi
[W] http://www.usis.fi ⊠ 121644 usemb sf
⊖ (+358 9) 17 46 81
Ambassador: H.E. Mr Derek N. Shearer
Deputy Chief of Mission: J Michael Cleverley
Commercial Section: Peter Frederick
Public Affairs Officer: Philippe du Chateau
☎ (+358 9) 17 51 38; 65 68 46 (fax) [98]

31-02010 Embassy of the Federal Republic of Yugoslavia

Kulosaarentie 36, FIN00570 Helsinki
☎ (+358 9) 684 74 66 [E] yuamba@personal.eunet.fi
[W] http://personal.eunet.fi/pp/yuamba ⊠ 122099 yamb sf
⊖ (+358 9) 684 87 83
Chargé d'Affaires a.i.: Mrs Marija Djordjevic [98]

55 Government-owned Industry & Production Facilities

55-00100 Patria Industries Oy

(† Finnish Defence Equipment Ltd)
Kaivokatu 10 A, 4th Floor, FIN00100 Helsinki
☎ (+358 9) 725 13 11 ⊖ (+358 9) 72 51 32 22
♣ FMk 1,020,000,000 (1996) 👥 2,280 (1996)
President and Chief Executive Officer: Jorma Eloranta
☎ (+358 9) 72 51 32 04; 72 51 32 00 (fax)
Manager, Corporate Communications: Ms Tiina Hansson
☎ (+358 9) 72 51 32 10; 72 51 32 22 (fax)
■ Group formed from the defence equipment industries previously owned by the Finnish government. Principal products include aircraft maintenance and structures, military and special vehicles, road and airport maintenance equipment, guns and mortars, cartridges, propellants, explosives and other ammunition. → 70

Group companies:
Finavitec Oy
Lapua Oy
Patria Vehicles Oy
Vammas Oy
Vihtavuori Oy [98]

55-02220 Vammas Ltd

(† Vammaskosken Tehdas)
(part of the Patria Industries Oy group)

PO Box 18, FIN38201 Vammala
☎ (+358 3) 519 71 ⊖ (+358 3) 519 74 66
♣ FMk 248,000,000 (1996) 👥 512 (1997)
President: Markku Murto
Sales and Marketing Director: Altti Seinela
Technical Director: Seppo Nikula
Purchasing Manager: Into Jarvinen
■ Howitzers, mortars and ammunition in 105mm, 122mm, 130mm 152mm and 155mm artillery calibres, mortar ammunition in 81mm, 120mm and 60mm calibres. Medium calibre ammunition fuzes for artillery ammunition - point detonating and proximity. [98]

60 Research Organizations, Institutes & Test Establishments

60-00400 The European Institute for Crime Prevention and Control

(affiliated with the United Nations)
PO Box 34, FIN00931 Helsinki
Turunlinnantie 8, FIN00930 Helsinki
☎ (+358 9) 343 20 77 ⊖ (+358 9) 33 48 21
👥 4 (professional); 2 (support)
Director: Dr Matti Joutsen, MSc Seppo Leppä, LL. M Terhi Viljanen, MSc Kristina Kangaspunta
■ Exchange of information, research and technical assistance.
Established in 1981. [98]

60-02800 National Research Institute of Legal Policy

PO Box 35, FIN00931 Helsinki
Turunlinnantie 8, FIN00930 Helsinki
☎ (+358 9) 343 26 07 ⊖ (+358 9) 343 26 35
👥 10 (professional); 4 (support)
Director: Patrik Törnudd
Information Contact: Ms Terttu Belgasem [98]

60-03500 Sodankyla Geophysical Observatory

FIN99600 Sodankyla
☎ (+358 16) 61 98 11 [E] aarne.ranta@sgo.fi
[W] http://www.sgo.fi ⊖ (+358 16) 61 98 75
👥 30 (1997) 🏠 1,700 m2
Director: Tauno Turunen
Technical Dir: Dr Aarne Ranta
■ Measurement & research of ionosphere, seismicity & magnetic fields. Design & production of magnetometers. [98]

65 Industry, Trade & Armed Forces Associations

65-00100 The Finnish Foreign Trade Association (FFTA)

Suomen Ulkomaankauppaliitto
Arkadiankatu 2, PO Box 908, FIN00101 Helsinki 10
☎ (+358 9) 204 69 51 ⊠ FOREIGNTRADE
⊠ 121696 trade sf ⊖ (+358 9) 204 69 55 52
Managing Director: Pertti Huitu
■ Official export promotion agency to develop Finnish foreign trade. Also serves importers & other parties involved in foreign trade. The Finnish Defence Equipment Manufacturers Export Collaboration Group comes under the auspices of the FFTA. [97]

Finland

70 Financial Holding Groups, Parent Companies

Patria Industries Oy 55-00100

75 Manufacturing Companies

Finavitec Oy 80-00580
MJP Corporation Ltd 80-01000

75-00300 Elesco Oy
(a joint venture of Ahlstrom Machinery Oy and Fisbars Oy AB)
PO Box 128, FIN02201 Espoo
☎ (+358 9) 525 90 60 ⊖ (+358 9) 52 59 06 60
♧ FMk 5,000,000 ♣ FMk 38,000,000 👥 40
⌂ 2,200 sq m
Managing Director: Seppo Riitamaa
Marketing Mgr: Seppo Savikurki
Marketing & Sales Mgr Ground Systems: A. Soininen
Purchasing Manager: Simo Rapeli
■ Coverage of Elesco Systems. Minesweeping-magnetic & acoustic. Degaussing-distributed microprocessor based systems. Signature measurement, ranges-magnetic, acoustic and pressure. Hydroacoustic surveillance systems. [98]

75-00500 Environics Oy
PO Box 349, FIN50101 Mikkeli
Työmiehenkatu 2
☎ (+358 15) 17 70 11 ⊖ (+358 15) 17 70 13
♣ FMk 20 000 000 (1996) 👥 19
Managing Dir: Hannu Salmi
Marketing Dir: Ilkka Sintonen
Operations & Technical Dir: Tarmo Karhapaa
Purchasing Mgr: Jorma Lamplaen
■ Chemical warfare agent, detector systems (M86A & M90). [97]

75-00595 Finavicomp Oy
(† Valmet Avicomp Inc)
(a subsidiary of Finavitec Oy)
PO Box 52, FIN35601 Halli
☎ (+358 3) 58 09 41 ⊖ (+358 3) 580 95 43
♣ FMk 115,000,000 (1995) 👥 320 ⌂ 16,000 sq m
President: Veijo Vartiainen
Technical Dir: Tapani Huikuri
Sales & Marketing Dir: Jukka Holkeri
Structures Director: Veli-Matti Erkkilä
Purchasing Mgr: Martti Talikainen
■ Aircraft parts manufacture. Sheet metal forming, riveting assembly, NC-machining, chemical milling, protective treatment & welding, composite part manufacturing. [98]

75-00600 Finnyards Ltd
PO Box 139, FIN26101 Rauma
☎ (+358 38) 836 11 [E] finnyards.rauma@finnyards.com
⊠ 65112 fyard fi ⊖ (+358 38) 836 23 66
♧ FMk 3,870,000 (1997) ♣ FMk 1,300,000,000 (1996)
👥 1,600 (1997)
Managing Director: Kari Airaksinen
Sales and Marketing Director: Jukka Laitera
Operations and Technical Manager: Hannu Kanerva
Purchasing Manager: Jussi Voima
Publicity Manager: Sinikka Railo
■ Manufacture naval and coastguard vessels.

Finnyards Electronics Ltd
Naulakatu 3, FIN33100 Tampere
☎ (+358 3) 245 01 11 ⊠ 22723 ⊖ (+358 3) 13 01 88
👥 80
■ ASW sonars, multibeam echo sounders, sound sources, temperature profilers, space electronics, stealth materials etc.. [98]

75-01350 Hj. Jousi Oy
PO Box 16, FIN20781 Kaarina
☎ (+358 2) 242 11 11 ✉ JOUSIOY
⊖ (+358 2) 242 11 66
♣ FMk 40,000,000 👥 100
Vice Managing Dir: Heikki Jousi
■ Alloy extrusions, aluminium cartridge cases (illumination) & paper tubes. [97]

75-01370 Kvaerner Masa-Yards Inc
(a member of the Kvaerner Group)
PO Box 132, FIN00151 Helsinki
☎ (+358 9) 19 41 ⊖ (+358 9) 65 00 51
♣ FMk 3 064 000 000 000 👥 4 900
President and Chief Executive Officer: Martin Saarikangas
Marketing and Sales Manager: Mikko Niini
Public Relations Manager: Henrik Segercrantz
■ Operation of Turku New Shipyard, Helsinki New Shipyard & a cabin module factory Piikkiö Works. Production of icebreakers, cable ships, LNG-carriers, cruise liners, mine layers, training ships, mine sweepers, fast corvettes, missile vessels, coast guard vessels. Production of Azipod azimuthing electric propulsion drives. [98]

75-01430 Lapua Oy
Patruunatehdas Lapua Oy
(† Cartridge Factory Lapua Ltd)
(part of the Patria Industries Oy group)
PO Box 5, FIN62101 Lapua
☎ (+358 6) 431 01 11 ✉ PATRUUNA Seinäjoki
⊠ 72126 lpt sf ⊖ (+358 6) 438 89 91
♣ FMk 156,000,000 (1996) 👥 356 (1996)
President: Sakari Kumara
Exec VP: Juha Mikkola
Export Mgr: Horst E. Zimmerbauer
Export Mgr: Henri Soravuo
Export Mgr: Jukka Lumme
Purchasing Mgr: Markku Huumonen
Advertising Matters: Ms Eija Varsala
■ Ammunition, cartridges:- Air pellets, .22 rimfire ammunition, pistol and rifle centrefire ammunition for hunting, match and military (ball, sniper, blanks, AP, tracers) bullets, cartridge cases, 20mm shells, percussion, igniting primers, igniters, base charges. → 55 [98]

75-01480 Metorex International Oy
(† Outokumpu Instruments Oy)
Nihtisillankuja 5, PO Box 85, FIN02631 Espoo
☎ (+358 9) 32 94 11 ⊖ (+358 9) 32 94 13 00
President: Tuula Lukander
Exec. Vice President: Heikki Sipilä
Dir US Operations: Mike Thompson
Export Mgr: Petri Ikonen
Marketing Communications Mgr: K. Kirjasniemi
■ Metal detector gates: Metor 120 general purpose gate, Metor 120 WP weatherproof gate, Metor 124 high sensitivity gate for detecting small items, Metor shadow mobile general purpose gate, Metor-X10 space saving tabletop luggage x-ray, Metor X-21 x-ray detector with high density visual display, Metornet PC-based network for remote monitoring and adjusting multiple Metor gates. Multi-zone detectors, hand-held, rugged metal detectors with adjustable sensitivity, fast reset speed detectors for optimum passenger flow. [98]

75-01500 Metra Corp Central Administration
(† Wärtsilä Group/Lohja Corp.)
John Stenbergin Ranta 2, PO Box 230, FIN00101 Helsinki
☎ (+358 9) 709 51 ⊖ (+358 9) 76 22 78
♣ FMk 11,660,000,000 (1996) 👥 11,986 (1996)
President & CEO: Georg Ehrnrooth
Sr VP Corp. Relations & Administration: Pekka Virtanen
■ Manufacture of diesel engines. [98]

75-01650 Noptel Oy
Teknologiantie 2, FIN90570 Oulu
☎ (+358 8) 551 43 51 [W] http://www.otm.fi/noptel/
⊖ (+358 8) 556 41 01
♣ FMk 18,000,000 👥 30
Managing Director: Juhani Heinula
Marketing Manager: Martti Särkelä
■ Specialises in opto-electronic measurement technology. Developed laser based products for shooting training & analysis. [98]

75-01700 Mock Doors Ltd
(a subsidiary of Brandt Group Oy Ltd)
PO Box 18, FIN00731 Helsinki
☎ (+358 9) 61 55 01 ⊖ (+358 9) 36 20 50
♣ FMk 23,000,000 (1996) 👥 35 (1996)
Managing Director: Christian Doepel
Technical Director: Jorma Tuomi
Marketing: Tero Leimu [98]

75-01710 Patria Vehicles Oy
(† Sisu Defence Oy)
(part of the Patria Industries Oy group)
PO Box 189, FIN13101 Hämeenlinna
☎ (+358 3) 64 51 ⊖ (+358 3) 619 71 30
♣ FMk 177,000,000 (1996) 👥 192 (1996)
President: Markku Lakomaa
Export Dir: Johan Rosberg
Marketing Dir: Heikki Hulkkonen
Chief Engineer Military Vehicles: Simo Nurminen
■ Military vehicles, rubber tracked amphibious all terrain vehicles, mineclearing vehicles, off road trucks and military equipment for all climatic and terrain conditions. Armoured personnel carriers. Vehicle systems integration, target systems. → 55 [98]

75-01720 Planar International Ltd
(a subsidiary of Planar Systems Inc, USA)
Olarinluoma 9, PO Box 46, FIN02201 Espoo
☎ (+358 0) 420 01 [E] olli-pesola@planar.com
[W] http://www.planar.com ⊖ (+358 0) 42 21 43
👥 260
Managing Dir: Graham Rothon
Technology Dir: Arto Pakkala
Materials & Manufacturing Dir: Tom Bruhn
Sales Director: Thomas Westerlund
Communications Officer: Olli Pesola
■ Manufacture of electroluminescent flat panel displays, monitors, terminals & accessories. [98]

75-01780 Rados Technology Oy
(† Alnor)
PO Box 506, FIN20101 Turku
☎ (+358 2) 468 46 00 ⊠ 62376 ⊖ (+358 2) 468 46 01
♣ FMk 23,500,000 👥 40
Managing Dir: Simo Malkanäki
Marketing Mgr: Pertti Rajamaki
■ Survey meters & systems for civil defence & other uses. [97]

75-01800 Raikka Oy
PO Box 30, FIN00251 Helsinki 25
☎ (+358 9) 241 33 44 ⊖ (+358 9) 241 74 65
Managing Dir: Seppo Liikkanen [98]

75-01960 Saft Nife Oy
(a subsidiary of Saft SA, France)
Piispantilankuja 4, PO Box 2, FIN02231 Espoo
☎ (+358 9) 881 30 33 ⊖ (+358 9) 881 30 60
Managing Director: Hans Sandström
■ Nickel-cadmium batteries, lithium batteries, air-depolarized cells, emergency lighting, converters and power supplies, advanced battery systems, rectifier charger, nickel-cadmium batteries for space applications. [98]

75-01980 Sako Ltd Riihimäki Works
(† Sako-Valmet Ltd)
(an affiliate of Valmet Corp and Nokia)
PO Box 149, FIN11101 Riihimäki
☎ (+358 19) 74 31 ⊖ (+358 19) 72 40 46
♣ FMk 152,000,000 (1995) 👥 290
President: Henry Paasikivi
Chairman: Matti Sundberg
Marketing Dir: Christer Lindroos
Financial Dir: Risto Tujula
■ Rifles & small arms ammunition. [97]

75-02100 Space Systems Finland Ltd (SSF)
Keilaranta 8, FIN02150 Espoo
☎ (+358 9) 61 32 86 00 [W] http://www.ssf.fi/
⊖ (+358 9) 61 32 86 99
♧ ECU 200,000 (1995) ♣ ECU 1,000,000 (1995) 👥 21
⌂ 500 sq m
Managing Dir: Dr Seppo Korpela
Finance Dir: Petteri Mussalo
Operations & Technical Dir: Kari Kumpulainen

■ On-board flight software, system design, software & integration for embedded systems, small satellite technology, software simulations, EO instrument & optical system analysis & consultancy. → 60 → 80 [97]

75-02650 Tumo Ltd
(a subsidiary of Helitrading Oy)
PO Box 11, FIN20611 Turku
Vaskitie 4, FIN20660 Littoinen
☎(+358 2) 275 27 00 ⊜ (+358 2) 275 27 01
FMk 200,000 (1995) ♣ FMk 11,000,000 (1995)
30 (1995) 1,500 sq m
Managing Dir: Juha Nummela
Operations Director: Olli Nummela
■ Machining, internal & external grinding & honing. Production includes components for laboratory instruments, aircraft parts. [98]

75-02800 Vaisala Oy
PO Box 26, FIN00421 Helsinki
☎(+358 9) 894 91 ✉VAISALA
http://www.vaisala.com 122832 vsala f
⊜ (+358 9) 894 92 27
♣ FMk 493,000,000 (1996) 697 (1996)
25,000 sq m
Chairman: Raimo Voipio
Vice Chairman: Yrjö Neuvo
President & CEO: Pekka Ketonen
Dir Finance: Olli Karikorpi
Dir Upper Air Div.: Jan Hórhammer
Dir Surface Weather Div.: Hannu Tuominen
Dir Sensor Systems Div.: Kenneth Forss
Dir Public Relations: Marit Finne
Purchasing Mgr: Jyrki Luojumäki
■ Equipment for radio-sonde and radio-wind observation. Humidity, pressure and wind meters measuring systems. Airport automatic weather stations. Instruments for environmental measurements. Ceilometers, RVR transmissometers, visibility meters, icecast runway condition monitoring systems. ATIS speech synthesisers.

Subsidiaries:
Vaisala Inc → USA 90
Vaisala Inc Artais Division → USA 90
Vaisala TMI → UK 90
Vaisala (UK) Ltd → UK 75
Vaisala SA → France 90
Vaisala KK → Japan 90
Vaisala Pty Ltd → Australia 90
Vaisala GmbH → Germany 90 [98]

75-02850 Vihtavuori Oy
(a subsidiary of Patria Industries Oy)
FIN41330 Vihtavuori
☎(+358 14) 377 92 11 http://www.vihtavuori.fi
⊜ (+358 14) 377 16 43; 377 14 60
♣ FMk 150,000,000 (1996) 300 (1997)
Managing Director: Antti Jurvelin
Blasting Explosives Unit Director: Niilo Kaasinen
Marketing Manager: Harri Vouritsalo
Production Manager: Pasi Niinikoski
■ Explosives and explosive related products for both civilian and military uses. Products range from mining explosives and detonators to powders and hand gun primers. Various ammunition components for artillery and mortar ammunition. Technological know-how for producing explosives and explosive-related products. → 55 [98]

80 Service Companies & Organizations

Space Systems Finland Ltd 75-02100

80-00580 Finavitec Oy
(† Valmet Aviation Industries Inc)
(part of the Patria Industries Oy group)
FIN35600 Halli
☎(+358 3) 580 91 [E] firstname.lastname@finavitec.fi
⊜ (+358 3) 580 96 67
♣ FMk 166,000,000 (1996) 600 (1996)
President and Chief Executive Officer: Veijo Vartianinen
Vice President Marketing: Jukka Holkeri
Vice President Maintenance: Pertti Korhonen
Vice President Structures: Veli-Matti Erttild
Communications Coordinator: Raili Saarinen
■ Overhaul, modification and repair of airframes and engines, final assembly. Development and support of aircraft systems. Overhaul of fast diesel engines. Structures for aircraft and spacecraft. Military systems integration. → 75 → 55

Maintenance
PO Box 51, FIN35601 Halli
☎(+358 3) 580 91 ⊜ (+358 3) 580 93 14
Director: Pertti Korhonen
■ Maintenance, overhaul, repair & modifications of aircraft & accessories.

Engine Maintenance
PO Box 10, FIN37241 Linnavuori
☎(+358 3) 341 76 66 ⊜ (+358 3) 341 76 60
General Mgr Engine Maintenance: Ilkka Jaakkola
■ Engine maintenance and manufacture of engine parts. [98]

80-00600 Finnish Shipbuilding Surveyors Ltd (FISS)
Vuorimiehenkatu 3, FIN00140 Helsinki
☎(+358 9) 70 01 73 71 [E] birger.mickwitz@fsc.pp.fi
⊜ (+358 9) 70 01 73 73
Managing Director: Birger Mickwitz
Technical Director: Seppo Lucenius
Sales and Marketing Director: Ralt Eklund
■ Professional maritime surveyor services provided on a continuous stand-by basis. Tightness tests of hatch covers, doors, scuttles etc. for ships and maritime constructions. [98]

80-00900 ILS Oy
Varppeenaukio 11, FIN21200 Raisio
☎(+358 2) 438 94 98 [E] ils@ils.pp.fi
⊜ (+358 2) 438 94 99
♣ FMk 4,000,000 (1996) 15 (1996)
Naval Architect: Timo Siirila
■ Naval architect, consultancy & marine engineering services. [98]

80-00950 Instrumentointi Oy
Sarankulmankatu 20, FIN33900 Tampere 90
☎(+358 3) 265 91 11 22303 ⊜ (+358 3) 265 92 00
Managing Director: Raimo Juntunen
☎(+358 3) 265 92 17
Technical Director: Matti Salonen ☎(+358 3) 265 95 00
Sales Director Avionics Division: Matti Hahka
☎(+358 3) 265 92 17
■ Overhaul & repair of aircraft instruments & avionics.
® USA: Collins Radio, E-Systems, Honeywell Commercial Flight Sys Group, Narco Avionics, Servo Syst. of America

Special Systems Division
Sarankulmankatu 20, FIN33900 Tampere
☎(+358 3) 265 91 11 22303 ⊜ (+358 3) 265 95 01
Division Director: Esa Einola ☎(+358 3) 265 97 00
Telecommunications Manager: Heikki Rautonen
☎(+358 3) 265 97 72 [98]

80-01000 MJP Corporation Ltd
Peltolantie, FIN04400 Järvenpää
☎(+44 802) 62 53 79 [E] mjp@corp.pp.fi
http://www.kingston .ac.uk/~bs_s714/mjpp.html
⊜ (+358 9) 279 29 49
£ 10,000 ♣ £ 100,000 (6 months)
Managing Director: Markku Paananen
Sales Director: Jari Kuusisto
■ Cleaning foam for firearms, small arms to heavy artillery. Research and development and military consulting. → 75 [98]

80-01100 Munters Oy
(a subsidiary of AB Carl Munters)
PO Box 5, FIN00751 Helsinki
☎(+358 9) 35 09 41 51 ⊜ (+358 9) 346 32 65
♣ FMk 55,000,000 (1996) 115 (1996)
1,000 m2 (1996)
Managing Director: Hannu Ojanen
Sales Manager, Defence: Markku Stenfors
Purchasing Manager: Pertti Tovanen [98]

85 Stockists, Distributors, Spare Parts Suppliers

VESTA, Antero Wilenius 90-03090

90 Sales Agents, Sales & Liaison Offices

Companies represented:

Airscrew Howden, UK 90-01310
Alfred Karcher GmbH & Co, Germany . . 90-01190
Anachemia Canada, Canada 90-03090
Andrew AG, Switzerland 90-01450
Andrew Corp., USA 90-01450
Andrew Ltd, UK 90-01450
Andrew SciComm. Inc., USA 90-01450
Beaufort Air-Sea Equipment, UK 90-01310
Bergen Nautik CS AS, Norway 90-00200
Bofors AB, Sweden 90-02000
Brent Europe Ltd, UK 90-01310
British Aerospace, UK 90-01310
Canadian Marconi, Canada 90-01450
CGG, France 90-00200
Charcoal Cloth, UK 90-03090
Civacon Europe bv, Netherlands 90-00200
Collins Radio, USA 80-00950
Cossor Electronics, UK 90-01310
Datron World Communications Inc, USA 90-01450
Delf Sensor Systems, Netherlands 90-00200
E-Systems, USA 80-00950
E-Systems Inc., USA 90-01450
EG&G Astrophysics, UK 90-01310
ELNA GmbH, Germany 90-00200
English Electric Valve Co. Ltd, UK 90-00160
ENRAF bv, Netherlands 90-00200
Enterprise Electronics Corp, USA 90-01310
Environmental Tectonics Corp, USA 90-01310
Federal Laboratories, USA 90-03090
Flowautomation, UK 90-01640
Folien Vogel, Austria 90-03090
Francis Searchlights Ltd, UK 90-00200
GEC Marconi Avionics Ltd, Sensors Div., UK 90-00160
GEC Marconi Sensors Ltd, UK 90-01310
GEC-Marconi Avionics Ltd, UK 90-01310
GEC-Marconi Defence Systems, UK 90-01310
HAL Communications Corp., USA 90-01450
Honeywell Commercial Flight Sys Group, USA 80-00950
Houchin Aerospace Ltd, UK 90-01310
Hunting Engineering Ltd, UK 90-01310
IPH/Kamstrup-Metro, Denmark 90-00200
Japan Radio Co, Japan 90-00200
Kachina Communications Corp., USA 90-01450
Lucas Zeta Inc., USA 90-01450
Marconi Radar & Control, UK 90-01310
McMurdo Ltd, UK 90-00200
ML Aviation Ltd, UK 90-01310
MORS, France 90-00200
MSI Mace Security Int., USA 90-03090
Muirhead Vactric Components Ltd, UK . . 90-00160
Musasino Co Ltd, Japan 90-00200
Narco Avionics, USA 80-00950
NIK Public Safety, USA 90-03090
Norcontrol Automation AS, Norway 90-00200
Nordisk Marine Hydraulic Danmark AS, Denmark 90-00200
Radiall, France 90-01780
Raytheon Co, USA 90-00200
RBR, Italy 90-01640
Rolls-Royce (Aero/Marine/Industrial), UK 90-01310
Scientific Communications, USA 90-01450
Scientific Radio Systems, USA 90-01450
Sercel AS, France 90-00200
Servo Syst. of America, USA 80-00950
Ship Automation AS, Norway 90-00200
Sierra Networks Inc Zeta Division, USA . 90-01450
Simrad, Norway 90-00200
Simulation Systems & Support Inc, USA 90-01450
Sunair Electronics Inc, USA 90-01450
TechComm Inc., USA 90-01450
Thorn Security Ltd, UK 90-00200

Contact Reference

Tracor Inc, USA 90-01450
Trans World Comms Inc, USA 90-01450
Trimos-Sylvac Task Force, Switzerland . 90-02010
Vimex AS, Norway 90-00200
Vingtor Electronics AS, Norway 90-00200
Vogel & Höller, Austria 90-03090
Wikrolux, Finland 90-00200
Yokogawa Denshikiki Co Ltd, Japan . . 90-00200
ZPA, Czech Republic 90-01640

90-00160 **Aseko Oy Electronics**
Finankuja 3, FIN02430 Masala
☎ (+358 9) 22 10 11 Ⓔ elektroniikka@aseko.fi
⊖ (+358 9) 22 10 13 50
▥ 1001 161 ♣ FMk 18,000,000 (1996) ⚉ 80
⌂ 2,500 sq m
Managing Director: Martti Petterson
Sales and Marketing Director: Raimo Wächter
Division Manager Electronics: Seppo Kyttala
® UK: English Electric Valve Co. Ltd, GEC Marconi Avionics Ltd, Sensors Div., Muirhead Vactric Components Ltd
98

90-00200 **AT-Marine Oy**
(a partner of Copertura Oy)
Kutojantie 1, Box 60, FIN02630 Espoo
☎ (+358 9) 54 94 26 00 Ⓔ at-marine@co.inet.fi
⊖ (+358 9) 54 94 27 00
♧ FMk 600,000 (1996) ♣ FMk 14,000,000 (1995) ⚉ 12
⌂ 500 sq m
Managing Dir: Seppo Parkkinen
Project Mgr: Tuomo Jokela
Section Mgr: Risto Isopahkala
■ Import and export of technical equipment and systems for both marine and industrial use. Also provides engineering, installation and service for these systems and equipment.
® Denmark: IPH/Kamstrup-Metro, Nordisk Marine Hydraulic Danmark AS
Finland: Wikrolux
France: CGG, MORS, Sercel AS
Germany: ELNA GmbH
Japan: Japan Radio Co, Musasino Co Ltd, Yokogawa Denshikiki Co Ltd
Netherlands: Civacon Europe bv, Delf Sensor Systems, ENRAF bv
Norway: Bergen Nautik CS AS, Norcontrol Automation AS, Ship Automation AS, Simrad, Vimex AS, Vingtor Electronics AS
UK: Francis Searchlights Ltd, McMurdo Ltd, Thorn Security Ltd
USA: Raytheon Co
98

90-00250 **British Steel Finland Oy**
(a subsidiary of British Steel Engineering Steels, UK)
Arkadiankatu 21 A5, FIN00100 Helsinki
☎ (+358 9) 454 24 50 ⊖ (+358 9) 454 24 52
98

90-01190 **Karcher OY**
Yrittajantie 17, FIN01800 Klaukkala
☎ (+358 9) 87 91 91 ⊖ (+358 9) 87 81 93 01
Managing Director: H Tom Lindman
® Germany: Alfred Karcher GmbH & Co
98

90-01310 **Machinery Oy**
Ansatie 5, PO Box 560, FIN01741 Helsinki
☎ (+358 9) 895 51 ✉ MACHINERY
⊖ (+358 9) 89 55 34 74; 89 55 33 81
♧ FMk 9,000,000 (1995) ♣ FMk 3,300,000 (1996)
⚉ 155 (1997) ⌂ 11,800 m2
Managing Dir: Heikki Timonen
Dept Mgr Special Products Group: Jouko Gullstén
® UK: Airscrew Howden, Beaufort Air-Sea Equipment, Brent Europe Ltd, British Aerospace, Cossor Electronics, EG&G Astrophysics, GEC Marconi Sensors Ltd, GEC-Marconi Avionics Ltd, GEC-Marconi Defence Systems, Houchin Aerospace Ltd, Hunting Engineering Ltd, ML Aviation Ltd, Marconi Radar & Control, Rolls-Royce (Aero/Marine/Industrial)
USA: Enterprise Electronics Corp, Environmental Tectonics Corp
98

90-01450 **Novacom Oy**
Kuintinmäenkaari 14A, FIN02210 Espoo 21
☎ (+358 9) 884 54 85 Ⓔ matti.pihlanto@novacom.inet.fi
⊖ (+358 9) 884 51 45
Managing Dir: Matti Pihlanto
® Canada: Canadian Marconi
Switzerland: Andrew AG
UK: Andrew Ltd
USA: Andrew Corp., Andrew SciComm. Inc., Datron World Communications Inc, E-Systems Inc., HAL Communications Corp., Kachina Communications Corp., Lucas Zeta Inc., Scientific Communications, Scientific Radio Systems, Sierra Networks Inc Zeta Division, Simulation Systems & Support Inc, Sunair Electronics Inc, TechComm Inc., Tracor Inc, Trans World Comms Inc
98

90-01640 **Pilotmarin Oy**
Vanha Rastaalantie 35, FIN02620 Espoo 62
☎ (+358 9) 59 05 77 ⊖ (+358 9) 59 47 27
♧ US$ 3,000 (1997) ♣ US$ 600,000 (1996/97)
⚉ 8 (1997) ⌂ 800 m2
Managing Director: Jorma Lehtonen
Sales and Marketing Director: Sami Lehtonen
Public Relations Manager: Riita Tekin
■ Special valves for submarines. Primary flow devices as venturinozzles, classical venturies and orifice plates, also special valves for various measuring purposes.
® Czech Republic: ZPA
Italy: RBR
UK: Flowautomation
98

90-01780 **Radiall SF**
(a subsidiary of Radiall, France)
Mekaanikonkatu 13, FIN00810 Helsinki
☎ (+358 9) 755 77 40 ⊖ (+358 9) 755 77 95
Managing Director: H Savolainen
® France: Radiall
98

90-02000 **Suomen Bofors Oy AB**
(a subsidiary of Bofors AB, Sweden)
PO Box 96, FIN02630 Espoo
☎ (+358 90) 502 23 00 ⊖ (+358 90) 502 23 10
Managing Dir: Paavo Varjola
® Sweden: Bofors AB
98

90-02010 **Tarkkuustuonti Oy**
Nallekuja 1, PO Box 77, FIN01901 Nurmijarvi
☎ (+358 9) 20 63 55 ⊖ (+358 9) 20 63 58
Managing Director: Tapani Laaksonen
■ Sale of measuring instruments.
® Switzerland: Trimos-Sylvac Task Force
98

90-02020 **Tecnex Oy**
Hietalahdenranta 19, FIN00180 Helsinki
☎ (+358 9) 693 21 55 Ⓔ georg.stunkel@tecnex.inet.fi
⊖ (+358 9) 693 21 90
♧ FMk 15,000 (1996) ♣ US$ 1,000,000 (1996)
⚉ 2 (1997) ⌂ 100 sq m
Managing Dir: Georg R. Stünkel
■ Represents manufacturers with emphasis on defence & defence-related products.
98

90-03090 **VESTA, Antero Wilenius**
PO Box 50, FIN00931 Helsinki
☎ (+358 9) 31 41 64; (+358 40) 548 65 16 (GSM)
✉ LUWI Helsinki ▥ 126116 seaof sf
⊖ (+358 9) 343 29 37
Managing Dir: Antero Wilenius
■ Independent sales agency for weapons, CW, mil. chemicals, electronics, law enforcement and security. → 85
® Austria: Folien Vogel, Vogel & Höller
Canada: Anachemia Canada
UK: Charcoal Cloth
USA: Federal Laboratories, MSI Mace Security Int., NIK Public Safety
98

France

Head of State:
Jacques Chirac

10 Head of State

10-10000 **Office of the President**
Palais de l'Elysée, 55-57 rue du Faubourg Saint Honoré, F-75008 Paris
☎ (+33 1) 42 81 51 00; 45 45 55 95; 44 39 59
▥ 650127 prerf; 648384 pref f
⊖ (+33 1) 47 42 24 65; 47 05 40 91
President of the Republic: Jacques Chirac
Chief of Staff: Vice Adm Jean Luc Delaunay
Director of Cabinet: Bertrand Landrieu
Deputy Director of Cabinet: CGAI Yvon Jouan
Deputy Secretary General: Jean-Pierre Denis
Secretary of State: Bernard Dufour
Counsellor to the Presidency: Michel Dupuch
Assistant to the President: Mrs Elisabeth Hubert
Assistant to the President: François Baroin
Assistant to the President: René Lenoir
President of Space Studies Group: Vice Adm Louzeau
Deputy Chief of Staff (Land): Gen Henri Bentegeat
Assistant Chief of Staff (Air): Gen Bertrand Dumont
Assistant Chief of Staff (Sea): Cdre Jacques Gheerbrant
Navy Commissioner-in-Chief: Jean-Marc Lefranc
Diplomatic Counsellor: Jean-David Levitte
Secretary General: Dominique de Villepin
Aide de Camp: Col Peer De Jong
Aide de Camp: Col Xavier Laure
Spokeswoman: Mrs Catherine Colonna
Press Counsellor: Mrs Manuela Isnard
97

11 Prime Minister

11-10000 **Office of the Prime Minister**
Cabinet du Premier Ministre
57 rue de Varenne, Hôtel Matignon, F-75700 Paris
☎ (+33 1) 45 56 80 00; 45 56 83 02 ▥ 200724 premin f
⊖ (+33 1) 45 44 15 72
Prime Minister: Lionel Jospin
Minister of Labour: Ms Martine Aubry
Minister of Justice: Ms Elisabeth Guigou
Minister of Education: Claude Allegre
Minister of the Interior: Jean Pierre Chevènement
Minister of Foreign Affairs: Hubert Vedrine
Minister of Finance, Economics and Industry: Dominique Strauss-Kahn
Minister of Defence: Alain Richard
Minister of Housing and Transport: Jean Claude Gayssot
Minister of Culture and Government Spokeswoman: Ms Catherine Trautmann
Minister of Agriculture and Fishing: Louis Le Pensec
Minister of Environment and Regional Development: Dominique Voyet
Minister for Relations with Parliament: Daniel Vaillant
Minister of Civil Service, State Reform and Decentralisation: Emile Zuccarelli
Minister for Youth and Sports: Marie-George Buffet
Government Adviser, Defence Affairs: Gen. J P Kelche
Press Attaché: Agatte Sanson
98

12 Ministries

12-05000 **Ministry of Defence**
14 rue Saint Dominique, F-75997 Paris Armées
☎ (+33 1) 45 55 95 20; 45 44 39 59; 42 19 30 11
▥ 270003 defnat f; 202681 infodef f
⊖ (+33 1) 47 05 96 56; 47 05 40 91
Minister: Allain Richard

General Studies Group
Délégation aux Affaires Stratégiques
10 rue Saint Dominique, F-00450 Paris Armées
☎ (+33 1) 42 19 38 02 ⊖ (+33 1) 42 19 51 62

Director: M Mallet
Deputy Director: Gen Benkheiri
Deputy Director, Army: Capt Caron [98]

12-10050 General Secretariat of National Defence

Secrétariat Général de la Défense Nationale (SGDN)
(a subordinate of the Office of the Prime Minister)
51 Boulevard de Latour Maubourg, F-75700 Paris
☎ (+33 1) 44 18 80 11 ⊠ 200019 segedfnat paris
⊖ (+33 1) 44 18 83 50
Secretary General: Ms Isabelle Renouard
☎ (+33 1) 44 18 80 01
(+33 1) 44 18 80 02: Rear Adm Jean Moulin
Director of Strategic Affairs: Ms Bérengère Quincy
☎ (+33 1) 44 18 82 01
Director of Scientific Affairs: Michel Ferrier
☎ (+33 1) 44 18 80 51
Director of Civil Defence: Mr Bergougnoux
☎ (+33 1) 44 18 80 71
Director of Economic Affairs: Jean Luc de Boissieu
☎ (+33 1) 44 18 80 81
Director of Judicial Affairs: Emmanuel Glaser
☎ (+33 1) 44 18 82 91
Director of Financial Affairs: Cdr Gal Jean Louis Stum
☎ (+33 1) 44 18 81 01
Information Counsellor: Col Michel Berlaud
☎ (+33 1) 44 18 82 71 [98]

12-20000 Ministry of Foreign Affairs

Ministère des Affaires Etrangères
37 Quai d'Orsay, F-75700 Paris
☎ (+33 1) 43 17 53 53
⊠ 202329 aestg f; 270819 afeip paris
⊖ (+33 1) 45 55 06 67
Minister of Foreign Affairs: Hubert Vedrine
Chief of Cabinet: Jean-Claude Schlumberger [98]

12-40000 Ministry of the Interior

Ministère de l'Intérieur
Place Beauvau, F-75800 Paris
☎ (+33 1) 45 22 90 90; 49 27 49 27
⊠ 290922 minintr paris
Minister of the Interior: Jean Pierre Chevènement [98]

12-70000 Ministry of Transport, Equipment and Tourism

Arche de la Défense, F-92055 Paris La Défense Cedex
☎ (+33 1) 40 80 21 22 ⊠ 610835 f
⊖ (+33 1) 40 81 31 18
Minister of Transport, Equipment and Tourism: Bernard Pons

Commissariat of Air Transport
93 Boulevard du Montparnasse, F-75270 Paris Cedex 06
☎ (+33 1) 49 54 36 36
Commissioner: Pierre Graff

General Directorate of Civil Aviation
48 rue Camille Desmoulins, F-92130 Issy les Moulineaux
☎ (+33 1) 41 09 43 21
Director: Pierre Graff

Office of the Finance Controller
F-92055 Tour Pascal - La Défense
☎ (+33 1) 40 81 21 22 [98]

12-70500 Conseil National des Transports

34 Avenue Marceau, F-75008 Paris
☎ (+33 1) 47 23 01 25 ⊖ (+33 1) 47 20 39 22
President: Guy Salmon-Legagneur
Vice President: Gaston Bessay
Vice President: Christian Brossier
Secretary General: Jacques Robert
Deputy Secretary General: Serge Blater
Deputy Secretary General: Ms Germaine Crochet [98]

12-71000 Direction Général de l'Aviation Civile (DGAC)

48 rue Camille-Desmoulins, F-92452 Issy les Moulineaux
☎ (+33 1) 41 09 49 45 ⊖ (+33 1) 41 09 43 69
Director General: Pierre Graff
Deputy Director: Jean François Grassineau
Chief of Cabinet: Ms Florence Rousse
Communications Officer: Thierry Reviron
Press Officer: Mrs Laurence Brun Potard [98]

15 Parliament

15-05000 National Assembly

Assemblée Nationale
126 rue de l'Université, Palais Bourbon, F-75355 Paris Cedex
☎ (+33 1) 40 63 60 00; 45 55 95 30 ⊠ 260420
⊖ (+33 1) 42 60 99 05
President: Laurent Fabius
Vice President: Jean Glavany
Vice President: Ives Cochet
Vice President: Pierre Mazeaud
Vice President: Gilles de Robien
Vice President: Michel Péricard
Vice President: André Santini [98]

15-10000 Senate

Sénat
15 rue de Vaugirard, Palais du Luxembourg, F-75291 Paris Cedex 06
☎ (+33 1) 42 34 37 14 [E] presidence@senat.fr
⊖ (+33 1) 42 34 33 44
President: René Monory
President of Civil Aviation Group: Sinateur Marcel Fortier
Vice President: Etienne Dailly
Vice President: Pierre-Christian Taittinger
Vice President: Jean Chamant
Vice President: Michel Dreyfus Schmidt
Pris.Comm.Affaires Etrangères/Défense: Sen. Jean Lecanuet
Pris.Commiti National Aviation Gin.: Sinateur E Cartigny
Air Force Expert to the Senate: Col Robert Didier
Administration Manager of Civil Aviation Group: Nicolas-Jean Brehon

Cabinet of the President
15 rue du Vaugirard, F-75291 Paris Cedex 16
☎ (+33 1) 42 34 37 14; 42 34 33 55
⊖ (+33 1) 42 34 33 44; 42 34 31 68
Director of Cabinet: Jean Dominique Giuliani
☎ (+33 1) 42 34 37 75; 42 34 36 89
[E] jd.giuliani@senat.fr
Chief of Cabinet: Ms Cécile Antonini
☎ (+33 1) 42 34 30 13; 42 34 31 68 (fax)
[E] c.antonini@senat.fr
Chief of the Private Secretariat: Ms Yvette Castan
☎ (+33 1) 42 34 36 77; 42 34 31 27 (fax)
Technical Counsellor: Hervé Gastinel
☎ (+33 1) 42 34 36 68; 42 34 36 47 (fax)
Technical Counsellor: Ms Guenhaëlle Le Dréau
☎ (+33 1) 42 34 25 20; 42 34 36 47 (fax)
Technical Counsellor: Ms Jacqueline Meyson-Arnaud
☎ (+33 1) 42 34 36 50; 42 34 36 47 (fax)
Technical Counsellor: Michel Roger
☎ (+33 1) 42 34 37 06; 42 34 36 47 (fax)
Technical Counsellor: François Schechter
☎ (+33 1) 42 34 36 96
Technical Counsellor: Jean Marin Schuh
☎ (+33 1) 42 34 36 51 [E] jm.schuh@senat.fr
Technical Counsellor: Ms Marie-José Tulard
☎ (+33 1) 42 34 36 07 [E] mj.tulard@senat.fr
Technical Counsellor: Ms Christina Giuliani
☎ (+33 1) 42 34 21 25 [E] c.giuliani@senat.fr
Communication Counsellor: Ms Solange Stricker
☎ (+33 1) 42 34 36 61; 42 34 31 27 (fax)
Quartermaster of the Presidency: Jean Jacques Mathiou
☎ (+33 1) 42 34 26 21; 42 34 35 35 (fax)
Military Command: Gen Paul Moulian
☎ (+33 1) 42 34 21 12; 42 34 28 15 (fax) [98]

20 Defense Administrations/Services/Agencies

École Nationale Supérieure de 60-00850

20-04000 Centre de Documentation de l'Armement (CEDOCAR)

Armament Documentation Centre
(an affiliate of DRET)
F-00460 Armées
☎ (+33 1) 45 52 45 01 ⊖ (+33 1) 45 52 45 72
Director: E Valensi [97]

20-11150 DGA Communication

Délégation Générale pour l'Armement
14 rue Saint Dominique, F-00457 Armées
☎ (+33 1) 42 19 45 46 ⊠ 270003 defnat f
⊖ (+33 1) 42 19 47 95
General Delegate: Jean Yves Helmer
Delegate, Armament Programs: IGA Daniel Reydellet
Delegate, International Affairs: IGA Daniel Berthault
Delegate, Industrial and Technological Strategy: IGA Jean Paul Gillyboeuf
Inspector General, Armaments: IGA Jean Benoît Ramé
Head of Cabinet: IGA Jean Fournet
☎ (+33 1) 42 19 45 45
Air Adviser: Col Adrien Besombes ☎ (+33 1) 42 19 45 49
Army Adviser: Gen Michel Rondeau
☎ (+33 1) 42 19 45 39
Navy Adviser: Capt Jean Louis Battet
☎ (+33 1) 42 19 45 37
Head of Press and Information: Ms Françoise Perret
☎ (+33 1) 42 19 45 37 [97]

20-16000 Direction de l'Electronique et de l'Informatique (DEI)

26 Boulevard Victor, F-00460 Armées
☎ (+33 1) 45 52 45 80 ⊖ (+33 1) 45 52 55 81
Director: IGA Michel Javelot
Deputy Director: IGA Michel Colin de Verdiere
Head of External Relations Office: LCA Laurent Collet Billon

Centre d'Electronique de l'Armement (CELAR)
F-35998 Rennes Armées
☎ (+33 2) 99 42 90 11 ⊖ (+33 2) 99 42 91 01
👥 700
Director: Michel Kervella
Deputy Director: Jean Francois Pacault
Head of Research and Development: Guy Mahi
■ Electronic data processing techniques and material testing. [97]

20-33500 Direction des Constructions Navales Headquarters (DCN)

10 rue Sextius Michel, F-75015 Paris
☎ (+33 1) 40 59 50 00 [W] http://www.dcnintl.com
⊖ (+33 1) 40 59 54 08
👥 20,000
President and Chief Executive Officer: R Grelf
Executive Vice President, Business Development: F Guerrier
Executive Vice President, Research and Development: J Le Tallec
Executive Vice President and Director of Projects: Ph Magnien
Executive Vice President, Production: J M Poimboeuf
Public Relations Director: Jean Lassara
■ Design, construction, outfitting, trials and maintenance of the Navy fleet. Munitions, armament and on-board systems.

DCN International is the export arm of the French Ministry of Defence's Direction des Constructions Navales (DCN).

DCN International Bureau Stratégie
19-21 rue du Colonal Pierre Avia, F-75015 Paris
☎ (+33 1) 41 08 71 71 [W] http://www.dcnintl.com
⊖ (+33 1) 41 08 00 27
Chief Executive Officer: Dominique Castellan
Secretary General: Robert Guhennec
Executive Vice President, Sales: Guy Kurkdjian
Executive Vice President, Marketing: Alain Fougeron
Executive Vice President, Finance: Gérard Philippe Menayas
Communiations Manager: C Girerd
Press: Rémi Calandra

DCN Ingénierie
8 Boulevard Victor, F-75015 Paris
☎ (+33 1) 40 59 20 00 ⊖ (+33 1) 40 59 22 09
Managing Director: J Le Tallec

DCN Cherbourg Shipyard
BP 10, F-50115 Cherbourg Naval
☎ (+33) 33 92 20 20
Director: Alain Tournyol du Clos

DCN Brest Shipyard
F-29240 Brest Naval
☎ (+33 2) 98 22 10 80 ⊖ (+33 2) 98 05 18 10
Director: L Petitois

DCN Lorient Shipyard
BP 41, F-56998 Lorient Naval
☎ (+33 2) 97 12 12 12
Director: X Marchal

DCN Toulon Shipyard
BP 77, F-83800 Toulon Naval
☎ (+33 4) 94 02 00 00
Director: Jean Pierre d'Hérouville

DCN Papeete Shipyard
SP 91381, F-00208 Armées

☎(+33 6) 89 45 53 00 ⊜ (+33 6) 89 45 53 05
Director: François Tandéo

DCN Indret Shipyard
BP 30, F-44620 La Montagne
☎(+33 2) 40 84 85 00 ⊜ (+33 2) 40 65 84 79
Director: Jean Paul Panié

DCN Ruelle Shipyard
BP 30, F-16600 Ruelle
☎(+33 5) 45 24 30 00
Director: Jacques Lavarec [98]

20-35000 French-German Research Institute of Saint Louis (ISL)

5 rue de Général Cassagnou, PO Box 34,
F-68301 Saint Louis Cedex
☎(+33 3) 89 69 50 00 ⊜ (+33 3) 89 69 50 02
Director: IGA Jean Yves Chaumeton
☎(+33 3) 89 69 50 10
Director: Dipl.-Ing Hans Schulte ☎(+33 3) 89 69 51 10
Technical Director: ICA Dominique Haberer
☎(+33 3) 89 69 50 20
Administration Director: Dr Joachim Buck
☎(+33 3) 89 69 51 20

■ Institute placed under the joint authority of both the French Ministry of Defence and the Ministry of Defence of the Federal Republic of Germany, carrying out basic research, particularly in the fields of ballistics, detonatics, aerodynamics and the metrology of fast events. [98]

20-40000 Missile & Space Directorate

Direction des Missiles et de l'Espace
26 Boulevard Victor - Tour DGA, F-00460 Armées
☎(+33 1) 45 52 43 21 ⊜ (+33 1) 45 52 59 51
Director: IGA HC Jean Pierre Rabault
Deputy Director: IGA 2 Jacques Bongrand
Public Relations Officer: ICT Ms Edith Bailly

Space & Strategic Systems Technical Department
Service Technique des Systèmes Stratégiques et Spatiaux
26 Boulevard Victor, F-00460 Armées
☎(+33 1) 45 52 43 21 ⊠ 204648
⊜ (+33 1) 45 52 59 51
Director: IGA 1 Daniel Estournet
Deputy Director: IGA 2 Yves Duhil

Tactical Missile Systems Technical Department
Service Technique des Systèmes de Missiles Tactiques (STSMT)
26 Boulevard Victor, F-00460 Armées
☎(+33 1) 45 52 43 21 ⊠ 204648
⊜ (+33 1) 45 52 59 51
Director: IGA 1 Jean Yves LeLoup
Deputy Director: IGA 2 Denis Plane

Landes Test Centre
Centre d'Éssais des Landes (CEL)
F-40115 Biscarosse Air
☎(+33 5) 58 82 51 97 ⊠ 560763 centes bisca
⊜ (+33 5) 58 82 20 00
Director: IGA 2 Pierre Jamin

Mediterranean Test Center
Centre d'Éssais de le Méditerranée (CEM)
PO Box 43, F-83800 Toulon Naval
☎(+33 4) 94 02 42 22 ⊠ 404009 cemedit
⊜ (+33 4) 94 02 48 25
Director: Mr ICA Desbois

Thrusters, Missiles Completion and Test Center
Centre d'Achèvement et d'Éssais des Propulseurs et Engins (CAEPE)
PO Box 2, F-33160 Saint Médard en Jalles
☎(+33 5) 56 70 56 56 ⊠ 560769 f
⊜ (+33 5) 56 70 58 69
Director: IGA2 Jean Moret

Ballistics and Aerodynamics Research Laboratory
Laboratoire de Recherches Balistiques et Aérodynamiques (LRBA)
PO Box 914, F-27207 Vernon Cedex
☎(+33 3) 32 21 42 00 ⊠ 770817 f
⊜ (+33 2) 32 21 44 45
Director: ICA Lionel Gouedard
Chef Service Approvisionnements: C T A A Bourdet
LRBA Director: ICA Jacques Darricau
Technical Manager: ICA Marc Giraud
Direction Attaché: Christian Vanpouille

■ LRBA is subordinate to the Directorate General for Armament; Directorate for Missiles and Space (DME). Design of tactical and strategic missiles. Analysis of industry design studies. Test and evaluation of equipment. Operational analysis of equipment in service. Design and Testing; Aerodynamic, Electronic and Space. Multi-service agency for inertial guidance and navigation for missiles, ships, aircraft and ground vehicles. Delegate for standardisation and the interservice military standard GAM EG 13 tailoring the environmental tests. [97]

20-60000 Sofinfra

102 Bureaux de la Colline, F-92213 Saint Cloud
☎(+33 1) 49 11 49 11 ⊜ (+33 1) 46 02 76 07
General Secretary: Piera Cavanna
Deputy General Manager, Sales: Marc Blaise [98]

25 Defense Forces

25-22300 Commandement de la Force Aérienne de Combat

Combat Air Force Command
BP 115, F-57998 Metz Air
☎(+33 3) 87 63 99 28 ⊜ (+33 3) 87 38 36 90
Commander: Gen Pierre Peron
Deputy Commander: Gen Jean Michel Nicolas [97]

25-22500 French Air Force Training Command

Commandement des Écoles de l'Armée de l'Air (CEAA)
Base Aérienne 705, Route Nationale 10,
F-37076 Tours Cedex 2
☎(+33 2) 47 54 59 58 ⊜ (+33 2) 47 51 66 85
Commander: Gen Jean J Brun
Deputy Commander: Gen Jean Michel Plasse
Chief of Staff: Col P Alaux
Aide de Camp: Laurent Saucié

■ The CEAA includes 16 schools and instruction centres, 5 maintenance units and a recruiting/selection department. [98]

25-22800 Directorate of Military Air Traffic

Base Aérienne 921, F-95155 Taverny Cedex
☎(+33 1) 30 40 38 08 ⊜ (+33 1) 30 40 65 57
Director: Brig Gen Luc Robin ☎(+33 1) 30 40 67 05
Deputy Director: Col Jacky Cognee
☎(+33 1) 30 40 67 05
Chief of Staff: Col Jean Paul Labit ☎(+33 1) 30 40 39 09 [98]

25-23500 Direction Centrale de l'Infrastructure de l'Air (DCIA)

(† Direction de l'Infrastructure Air)
Avenue de la Porte de Sèvres - 5 bis, F-00460 Armées
☎(+33 1) 45 52 35 01 ⊜ (+33 1) 45 52 23 40
Director: GDA Alain Perrollaz
Civil Deputy Director: ICPC Bernard Urcel
☎(+33 1) 45 52 25 02
Military Deputy Director: GBA Alain Choplin
☎(+33 1) 45 52 27 01
Plans and Programmes Director: Col Gilles Laugaudin
☎(+33 1) 45 52 35 04
Technical Director: Mr Jean Pierre Bardy
☎(+33 1) 45 52 35 12 [98]

25-29200 Centre d'Enseignement Supérieur Aérien (CESA)

BP 43, F-00445 Armées
☎(+33 1) 45 51 11 12 ⊜ (+33 1) 45 55 91 75
Director: Gen BA Serge Raynaud
Chief of Staff: Lieut Col Bernard Perrochon
Director: Col Christian Rodriguez
Director of Centre for Aeronautical and Spacial Studies: Maj Michel Brouchier

■ Training of air force commissioned officers. Supervision of the scientific and technical aspects of the higher education of air force officers. Publishers of official reports, booklets and brochures for the air force. [97]

25-35000 Foreign Legion First Foreign Regiment

Quartier Viennot, F-13400 Aubagne

Third Foreign Regiment Infantry
Quartier Forget, F-97310 Kourou

Fourth Foreign Regiment
Quartier Danjou, F-11407 Castelnaudary Cedex

Second Foreign Regiment Paratrooper
Camp Raffali, F-20260 Calvi

First Foreign Regiment Cavalry
Quartier Labouche, F-84100 Orange

Second Foreign Regiment Infantry
Quartier Vallongue, F-30998 Nîmes Armées

Sixth Foreign Regiment Engineering
Quartier Général Rollet, Camp de l'Ardoise,
F-30290 Laundun [98]

25-40000 Navy Headquarters

Marine Nationale
2 rue Royale, F-00300 Armées
☎(+33 1) 42 92 10 00 ⊜ (+33 1) 40 20 04 90
Chief of Naval Staff: Adm Jean Charles Lefevre
Chief of Cabinet to Chief of Staff: Rear Adm Mac-Grath
Chief of Staff: Vice Adm Moysan
Inspector General: Adm Gazzano
Chief of Documentation Section: Lieut Cdr Calle
☎(+33 1) 42 92 12 11

Operations and Logistics Division
2 rue Royale, F-00350 Armées
Deputy Chief of Staff, Operations and Logistics: RAdm de Penfentenyo
Chief of International Affairs Office: Capt de Metz

Planning Division
2 rue Royale, F-00350 Armées
Deputy Chief of Staff, Planning: Rear Adm Combarieu

Program and Operational Requirements Division
2 rue Royale, F-00350 Armées
Deputy Chief of Staff, Programs: Rear Adm Cunny
Chief of Platform and Propulsion Division: Capt Maurand
Chief of Communications Systems Division: Capt Cluzel
Chief of Air Surface Weapons Systems Division: Capt Lefevre
Chief of Underwater Weapons Division: Capt Jeuffrain
Chief of Submarine Division: Capt Bellot

Naval Air Division
2 rue Royale, F-00350 Armées
Deputy Chief of Staff: Vice Adm Wild
☎(+33 1) 42 92 15 75
Assistant Deputy Chief of Staff: Capt de Balmann
☎(+33 1) 42 92 15 73

Navy Military Personnel Division
2 rue Royale, F-00350 Armées
Chief of Staff: Vice Adm Girard [98]

25-41200 Préfecture Maritime Zone Maritine Atlantique

Le Château, BP 46, F-29240 Brest Naval
☎(+33 2) 98 22 10 80
Commander: Adm Le Dantec
Deputy Commander, Territorial: Rear Adm Filippi
Director of Operations: Rear Adm Martin
Director of Logistics: Rear Adm Arino
Executive Assistant: Mr Lacoin [98]

25-42000 Direction Centrale du Commissariat de la Marine (DCCM)

2 rue Royale, F-00352 Armées
☎(+33 1) 42 92 12 21 ⊜ (+33 1) 42 92 11 02
Director: CG1 Yves Bourdais
Deputy Director: CG2 Jean Claude Boulet
Chief of Cabinet: CP Cyriaque Grapin

DCM de la Marine à Brest
F-29240 Brest Naval
☎(+33 2) 98 22 12 57 ⊜ (+33 2) 98 22 06 97
Director: CRG2 Bernard de Cadenet
Public Affairs Officer: CR1 Laurent Favier

DCM de la Marine à Cherbourg
F-50115 Cherbourg Naval
☎(+33 2) 33 92 61 03 ⊜ (+33 2) 33 92 60 71
Director: CRC1 Jean-Louis Delnas
Public Affairs Officer: CR1 Georges Le Stum

DCM de la Marine à Toulon
PO Box 70, F-83800 Toulon Naval
☎(+33 4) 94 02 03 73 ⊜ (+33 4) 94 02 11 44
Director: CRG2 Dominique Vandier
Public Affairs Officer: CR1 François Trenenbert

DCM de la Marine à Lorient
PO Box 32, F-56998 Lorient Naval
☎(+33 2) 97 21 22 49 ⊜ (+33 2) 97 12 11 00

Director: CRC1 Pierre Labit
Public Affairs Officer: CR1 Didier Lhoste 97

25-42500 Naval Hydrographic and Oceanographic Service

Service Hydrographique et Océanographique de la Marine (SHOM)
3 Avenue Octave Gréard, BP 5, F-00307 Armées, Paris 7
☎ (+33 1) 44 38 41 54 ⊠ 285421 pour le SHOM
⊖ (+33 1) 40 65 99 98
Director: IGA Milard ☎ (+33 1) 44 38 40 76
Deputy Director: IGA Souquière ☎ (+33 1) 44 38 42 94
Director of Brest Main Establishment: IGA Cailliau
☎ (+33 1) 98 22 12 81
Head of Bureau for General Affairs: ICA Bessero
Deputy Head of Bureau for General Affairs: ICA Renhas
☎ (+33 1) 44 38 43 95

Brest Main Establishment
13 rue du Chatellier, PO Box 426, F-29275 Brest Cedex
☎ (+33 2) 98 22 10 80 ⊠ 940568 hydro f
⊖ (+33 2) 98 22 12 08
Director: IGA Cailliau ☎ (+33 2) 98 22 12 81
Head of Military Oceanographic Centre: ICA Frachon
☎ (+33 2) 98 22 19 43
Public Relations Officer: CC Exandier

Mission Océanographique de l'Atlantique (MOA)
CM La Villeneuve, F-29240 Brest Naval
☎ (+33 2) 98 22 05 21 ⊖ (+33 2) 98 22 06 58
Director: ICA Fourgassié

Mission Hydrographique de l'Atlantique (MHA)
CM La Villeneuve, F-29240 Brest Naval
☎ (+33 2) 98 22 19 52 ⊖ (+33 2) 98 22 19 43
Director: IPA Tonchia

Mission Océanographique du Pacifique (NOUMEA)
Nouméa Nouvelle Calédonie, BP 38, F-00300 Armées
☎ (+687) 25 53 99 ⊖ (+687) 25 53 45
Director: IPETA Deniel

Mission Océanographique du Pacifique (PAPEETE)
SP 91647, F-00260 Armées
☎ (+689) 46 52 68 ⊖ (+689) 46 53 13
Technical Director: IPETA Le Squere 98

25-50000 Mission Innovation

3 Avenue Octave Gréard, F-75007 Paris
☎ (+33 2) 41 51 05 43 ⊖ (+33 2) 41 51 05 43 60
Commander: Maj Gen Bonavita 98

29 Customs & Excise

29-01000 General Directorate of Customs

Direction Générale des Douanes et Droits Indirects (a subordinate of the Ministry of Budget)
23bis rue de l'Université, F-75700 Paris 07
☎ (+33 1) 40 04 04 04 ⊠ 220200 douan
⊖ (+33 1) 42 60 21 54
Director General: J L Vialla
Chief of Cabinet and General Affairs: Mr Claustres
Chief of Service: M Pinguet 98

30 Civil Defense

30-00100 Directorate of Civil Defence (DSC)

Direction de la Sécurité Civile
(† Directorate of Civil Defence)
18 rue Ernest Cognacq, F-92300 Levallois Perret
☎ (+33 1) 49 27 49 27; 40 07 60 60
⊖ (+33 1) 47 57 82 67
Director of Civil Protection: Daniel Canepa
☎ (+33 1) 40 87 74 24
Director of Cabinet: Jean François Verdier
Deputy Director of Cabinet: Ms Corinne Crevot
☎ (+33 1) 40 87 72 10
Military Assistant: Col Compain ☎ (+33 1) 40 87 73 15

Inspection Division
18 rue Ernest Cognacq, F-92300 Levallois Perret
☎ (+33 1) 40 87 74 13; 40 87 74 36
⊖ (+33 1) 47 58 17 42
Chief Inspector: Col Jacques Trohel

Directorate of Administration and Modernisation
Direction de l'Administration et de la Modernisation
18 rue Ernest Cognacq, F-92300 Levallois Perret
☎ (+33 1) 40 87 72 93 ⊖ (+33 1) 47 58 17 41
Deputy Director: Patrick Levaye

Directorate for Emergency Plans
Direction de la Prévention et des Plans de Secours
18 rue Ernest Cognacq, F-92300 Levallois Perret
☎ (+33 1) 40 87 73 74 ⊖ (+331) 47 5849 27
Deputy Director: Bertrand Munch

Directorate for Emergency and Fire Services
Direction des Services de Secours et des Sapeurs-Pompiers
18 rue Ernest Cognacq, F-92300 Levallois Perret
☎ (+33 1) 40 87 73 98 ⊖ (+33 1) 47 58 17 42
Deputy Director: Michel Lalande

Directorate for Emergency Operations
Direction des Opérations de Secours
18 rue Ernest Cognacq, F-92300 Levallois Perret
☎ (+33 1) 40 87 74 23 ⊖ (+33 1) 47 57 82 67
Deputy Director: François Xavier Ceccalbi

État-Major du Commandement des Formations Militaires por la Sécurité Civile (CONFORMISC)
18 rue Ernest Cognacq, F-92300 Levallois Perret
☎ (+33 1) 40 87 73 15 ⊖ (+33 1) 47 58 63 60
Director: Col Compain ☎ (+33 1) 40 87 75 14;
47 58 63 60 (fax)

National Institute of Public Safety Studies (INESC)
Institute National d'Études pour la Sécurité Civile
Nainville les Roches, F-31750 Champcueil
☎ (+33 1) 64 98 09 09 ⊖ (+33 1) 64 98 04 32
Director: Col Jean-Paul Savelli

Higher Fire Officers National School
École Nationale Supérieure des Officiers de Sapeurs-Pompiers
Nainville les Roches, F-97750 Champcueil
☎ (+33 1) 64 98 09 09 ⊖ (+33 1) 64 98 04 32
Director: Col Christian Pourny

Centre National d'Instruction de la Protection Contre l'Incendie
26 rue Chaptal, F-75009 Paris
☎ (+33 1) 42 80 40 51 ⊖ (+33 1) 42 80 51 53
Director: Lt Col Claude Peltier 97

31 Embassies & Attachés

31-00090 Australian Embassy

4 rue Jean Rey, F-75724 Paris Cedex
☎ (+33 1) 40 59 33 00 ✉ AUSTEMBA Paris
W http://www.austgov.fr ⊖ (+33 1) 40 59 33 10
Ambassador: H E Mr John Spender
Deputy Chief of Mission: P Shannon
Minister, Commercial Affairs: S Atkinson
Counsellor, Administration: L Donse
Counsellor, Industry, Technology and Science:
M Gallagher
First Secretary, Information: Angus Mackenzie
Defence Attaché: Col C Galvin 98

31-00100 Embassy of Austria

6 rue Fabert, F-75007 Paris
☎ (+33 1) 45 55 95 66; 40 63 30 63
⊠ 200708 oebopa f ⊖ (+33 1) 45 55 63 65
Ambassador: H E Mr Franz Ceska
Commercial Counsellor and Trade Delegate:
Dr Harald Kasper
Defence Attaché: Col Wolfgang Wosolsobe
☎ (+33 1) 40 63 30 82
Economic Counsellor: Christoph Ceska
Counsellor, Administration: Rainer Schmutzer
Press Counsellor: Mrs Helene Lamesch 98

31-00120 Embassy of the State of Bahrain

Place des États Unis - 3 bis, F-75116 Paris
☎ (+33 1) 47 23 48 68 (embassy);
47 23 49 15 (consulate) ⊠ 640091 f
⊖ (+33 1) 47 20 55 75
Ambassador: H E Dr Ali Fakhro
First Secretary: Ahmed Harmas 97

31-00150 Embassy of Belgium

9 rue de Tilsitt, F-75840 Paris Cedex 17
☎ (+33 1) 44 09 39 39 ✉ AMBASSADE BELGIQUE Paris
⊠ 650484 amba f ⊖ (+33 1) 47 54 07 64
Ambassador: H E Mr Alain Rens
Minister and Counsellor: Ms Benedicte Frankinet
Minister and Counsellor (Economic Affairs):
Patrick de Beyter
Military, Air and Naval Attaché: Col Raymond Keymeulen
☎ (+33 1) 44 09 70 95; 44 09 77 34 (fax) 98

31-00200 Embassy of Bolivia

12 Avenue du Président Kennedy, F-75016 Paris
☎ (+33 1) 42 24 93 44; 45 27 84 35
Ambassador: H E Mr Carlos Antonio Carrasco
Social Affairs Attaché: Zulema Seydoux
Public Information Officer: Walter Rico Frontaura
Secretary: Ms Annabel Dorr 98

31-00240 Embassy of Brunei Darussalam

Rue Logelbach n°4, F-75017 Paris
☎ (+33 1) 42 67 49 47 ⊠ 651653 ambbru f
⊖ (+33 1) 42 67 53 65
Ambassador: H E Mr Pengiran Haji Idriss
First Secretary: Ms Haji Ahmad
First Secretary, Administrative and Financial Affairs:
Ramli Haji Ahmad
First Secretary: Darmawati Haji Idriss
Third Secretary, Consular Affairs:
Mohamad Noor Haji Badaruddin
Attaché, Communication: Brahim Haji Abd Rahman 98

31-00290 Embassy of the Republic of Cameroon

73 rue d'Auteuil, F-75016 Paris
☎ (+33 1) 47 43 98 33; 47 43 98 39 ⊠ 640087
⊖ (+33 1) 46 51 24 52
Ambassador: H E Mr Pascal Biloa Tang
Deputy Chief of Mission: Messoro Essomba Thomas
Military, Naval and Air Attaché: Col Ndigo Nzie
Deputy Military, Naval and Air Attaché: Capt Otele

Commercial Department
73 rue d'Auteuil, F-75016 Paris
☎ (+33 1) 47 43 98 33 ⊠ 640087 mecp f
⊖ (+33 1) 46 51 24 52
Commercial Counsellor: Loe Victor 98

31-00310 Embassy of Canada

35/37 Avenue Montaigne, F-75008 Paris
☎ (+33 1) 44 43 29 00 ✉ STADACONA PARIS
⊠ 280806 canad a ⊖ (+33 1) 44 43 29 96
Ambassador: H E Mr Jacques Roy
Deputy Chief of Mission: John Noble
☎ (+33 1) 44 43 22 50
Commercial Affairs Minister: Bertin Coté
☎ (+33 1) 44 43 22 50
Canadian Defence Attaché: Brig Gen Claude Thibault
☎ (+33 1) 44 43 25 50
Military Attaché: Lieut Col J H G Brassard
☎ (+33 1) 44 43 25 54
Naval Attaché: Cdr Roger Drolet ☎ (+33 1) 44 43 25 55
Defence Research Counsellor: Maurice Laviolette
☎ (+33 1) 44 43 25 53
Commercial Councellor: Claude Charland
☎ (+33 1) 44 43 23 60
Communications and Public Affairs Counsellor:
Denis Boulet 98

31-00400 Embassy of the Republic of Congo

37bis rue Paul Valéry, F-75116 Paris
☎ (+33 1) 45 00 60 57 ⊠ 611954 arcongo f
⊖ (+33 1) 40 67 14 44
Ambassador: H E Mr Pierre Michel Nguimbi
Counsellor, Political and Diplomatic Affairs: Grégoire Kaya
Counsellor, Economic and Commercial Affairs:
Paul Adam Dibouilou
First Secretary, Protocol: Dieudonné Moukengue
First Secretary, Consular Affairs:
Armel Sébastien Oyo Pubieley
Second Secretary, Communication and Cultural Affairs:
Joseph Bahoua 98

31-00440 Embassy of Cyprus

23 rue Galilée, F-75116 Paris
☎ (+33 1) 47 20 86 28 E embrecyp@worldnet.fr
⊖ (+33 1) 40 70 13 44
Ambassador: H E Mr Andreas Pirishis
First Counsellor: Leonidas Markides
Second Secretary: Andreas Zenonos
Press Counsellor: Michalakis Christodoulides
Commercial Counsellor: Mrs Zoe Stephanou 98

Contact Reference

31-00450 Embassy of the Czech Republic
15 Avenue Charles Floquet, F-75007 Paris
☎(+33 1) 40 65 13 00 ⊖ (+33 1) 47 83 50 78
Ambassador: H E Mr Petr Lom
Minister-Counsellor and Deputy Chief of Mission: Ms Katerina Lukesová
Counsellor, Political and European Affairs: Petr Janyska
Second Secretary, Political Affairs and Press: Jaroslav Jíru
Cultural Affairs Counsellor: Václav Jamek
Chief of the Consular Section: Ms Ljuba Svobodová
Third Secretary, Consular Affairs: Ms Alexandra Krusová
Defence, Military and Air Attaché: Col Petr Kysilko
Deputy Defence, Military and Air Attaché: Lieut Col Miroslav Vondruska
Third Secretary, Protocol: Ms Viktoria Gallée
Delegate to UNESCO: David Masek ☎(+33 1) 45 68 35 35

Economic and Commercial Section
15 Avenue Charles Floquet, F-75343 Paris Cedex 07
☎(+33 1) 40 65 13 10 ⊖ (+33 1) 47 83 50 78
Economic and Commercial Counsellor: Milan Vondrácek
Economic and Commercial Attaché: Ms Jana Pleschingerová [98]

31-00460 Royal Danish Embassy
77 Avenue Marceau, F-75116 Paris
☎(+33 1) 44 31 21 21 ⊠ AMBADANE Paris
⊖ (+33 1) 44 31 21 88
Ambassador: H E Mr Peter Dyvig
Minister and Deputy Chief of Mission: Jørgen Munk Rasmussen
Military, Naval and Air Attaché: Capt Jess Bønkel Jensen (RDN)
Commercial Counsellor: Karsten Hansen [98]

31-00510 Embassy of the Arab Republic of Egypt
56 Avenue d'Iéna, F-75116 Paris
☎(+33 1) 53 67 88 30; 53 67 88 35 ▣ 645297 bustan f
⊖ (+33 1) 47 23 06 43
Ambassador: H E Mr Aly Maher El Sayed
Deputy Chief of Mission: Mokhless Kotb
Minister, Commercial Affairs: Moustafa El Mahdi ☎(+33 1) 47 20 47 89

Office of the Defence Attaché
53 rue Dumont d'Urville, F-75116 Paris
☎(+33 1) 45 00 94 10 ▣ 642490 hararab f
Attaché for Defence, Naval and Air: Gen Hamdy Talaat Hassanein
Deputy Attaché, Defence Naval and Air: Col Abdel Aziz Ibrahim
Deputy Defence Attaché, Naval and Air: Col Yosri Selim
Deputy Defence Attaché, Naval and Air: Col Moustafa El Taher Ali [98]

31-00520 Embassy of El Salvador
12 rue Galilée, F-75116 Paris
☎(+33 1) 47 20 42 02; 47 23 98 03
⊖ (+33 1) 40 70 01 95
Ambassador: H E Mr Ramiro Zepeda-Roldan
Minister and Counsellor: Mauricio Mena
Second Secretary: Ms Maria Rosario Molina de Huntrods
Secretary, Consular Affairs: Ms Lucie Calderon
Attaché: Ms Patricia Salazar- Calderon [98]

31-00535 Embassy of Estonia
14 Boulevard Montmatre, F-75009 Paris
☎(+33 1) 48 01 00 22 Ⓔ saadik@est-emb.fr
⊖ (+33 1) 48 01 02 95
Ambassador: H E Mr Andres Tomasberg

Estonian Embassy to the Council of Europe
12 Boulevard Paul Derouede, Grand,
Pavois, Office 1503 CE Building, F-67000 Strasbourg
☎(+33 3) 88 41 39 41 ⊖ (+33 3) 88 41 27 07
Ambassador: H E Karin Jaani [98]

31-00670 Embassy of Ghana
8 Villa Saïd, F-75116 Paris
☎(+33 1) 45 00 09 50 ⊠ GHANASAMB Paris
▣ 645084 f ghanasa ⊖ (+33 1) 45 00 81 95
Ambassador: H E Dr Keli Nordor
Deputy Chief of Mission: Adolphus K Arthur
Counsellor, Economic Affairs: Ramses J Cleland [98]

31-00780 Embassy of the Republic of Haiti
10 rue Théodule Ribot, F-75017 Paris
☎(+33 1) 47 63 47 78 ⊖ (+33 1) 42 27 02 05
Ambassador: H E Mr Marc A Trouillot [98]

31-00810 Embassy of the Republic of Hungary
5 bis Square de l'Avenue Foch, F-75116 Paris
☎(+33 1) 45 00 41 59; 45 00 00 29
⊖ (+33 1) 45 01 66 00
Ambassador: H E Mr Béla Szombati
Deputy Chief of Mission: György Szathmáry
Military and Air Attaché: Lieut Col Ferenc Vizi
Counsellor, Science and Technology Affairs: Ms Maria Pánczel

Commercial Department
140 Avenue Victor Hugo, F-75116 Paris
☎(+33 1) 53 70 67 00
Counsellor, Commercial: József Goloncsér
Secretary, Commercial Affairs: Sandor Faragó [98]

31-00870 Embassy of Ireland
4 rue Rude, F-75116 Paris
☎(+31 1) 44 17 67 00 ⊖ (+33 1) 44 17 67 60
Ambassador: H E Mr Patrick O'Connor
Deputy Chief of Mission: Brian Nason
Economic Affairs Secretary: Ms Patricia Cullen [97]

31-00920 Embassy of Japan
Ambassade du Japon
7 Avenue Hoche, F-75008 Paris
☎(+33 1) 48 88 62 00 ⊠ TAISHI Paris
▣ 660 493 taishi paris ⊖ (+33 1) 42 27 50 81
Ambassador: H E Mr Moriyuki Motono
Deputy Chief of Mission: Yoshio Karita
Defence Attaché: Col Seiji Nakayama
Secretary Scientific Affairs: J Shibata
Commercial Counsellor: Naomi Anesaki

General Consulate of Japan
Consulat Général du Japon
Tour Europe - 20 Place des Halles, F-67000 Strasbourg
☎(+33 3) 88 52 85 00 ⊖ (+33 3) 88 21 98 56

General Consulate of Japan
Consulat Général du Japon
70 Avenue de Hambourg, F-13008 Marseille
☎(+33 4) 91 73 45 55 ⊖ (+33 4) 91 72 55 46 [97]

31-00990 Embassy of the Republic of Korea
125 rue de Grenelle, F-75007 Paris
☎(+33 1) 47 53 01 01; 47 20 83 86
⊖ (+33 1) 47 53 72 98; 47 23 58 97
Ambassador: H E Mr Lee See-Young
Deputy Chief of Mission: Jae Sun Park ☎(+33 1) 47 53 69 51
Military, Naval and Air Attaché: Col Lie Yeol Lie ☎(+33 1) 47 53 69 53
Commercial Officer: Kim Chang-Yeob
Defence Attaché: Col Sac Park-Jong
Assistant Military, Naval and Air Attaché: Maj Jin Gyu Lee ☎(+33 1) 47 53 69 68
Assistant Military Attaché: Col Jung Kim ☎(+33 1) 47 53 69 73
Commercial Attaché: Won Ju Hong ☎(+33 1) 47 53 69 60
Counsellor: Byong Wook Min ☎(+33 1) 47 53 69 54
Public Information Officer: Kim Chang-Ho [98]

31-01040 Embassy of the Republic of Liberia
8 rue Jacques Bingen, F-75017 Paris
☎(+33 1) 47 63 58 55; 47 63 23 85
▣ 290288 libem paris ⊖ (+33 1) 47 64 35 50
Ambassador: H E Mr Aaron J George II
Deputy Chief of Mission: Jeff Dowana [98]

31-01110 Embassy of Malaysia
2bis rue Bénouville, F-75116 Paris
☎(+33 1) 45 53 11 85 ⊠ MALAWAKIL Paris ▣ 645848
⊖ (+33 1) 47 27 34 60
Ambassador: H E Mr Mohamed Haron
Counsellor: Syed Munshe Afdzaruddin
Second Secretary, Political Affairs: Nur Ashikin Mohd Taid
Second Secretary, Administrative Affairs: Musa R Omar
Counsellor, Commercial Affairs: Ali Ahmad Fadzil
Counsellor, Economic Affairs: Abdul Kadir Afifuddin [98]

31-01130 Embassy of the Republic of Mali
89 rue du Cherche-Midi, F-75006 Paris
☎(+33 1) 45 48 58 43 ▣ 260002 ambmali paris
⊖ (+33 1) 45 48 55 34
Ambassador: H E Ms Madina Ly-Tall
Deputy Chief of Mission and Commercial Counsellor: Sékou Gaoussou Cisse [98]

31-01140 Embassy of Malta
92 Avenue des Champs Elysées, F-75008 Paris
☎(+33 1) 45 62 68 16; 45 62 53 01
⊖ (+33 1) 45 62 00 36
Ambassador: H E Dr V Camilleri
First Secretary (Political, Cultural Affairs and Information): Clive Agius [98]

31-01190 Embassy of Mexico
9 rue de Longchamp, F-75116 Paris
☎(+33 1) 45 53 76 43 ⊖ (+33 1) 47 55 65 29
Ambassador: H E Mr Jorge Carpizo ☎(+33 1) 45 53 99 34
Deputy Chief of Mission: Francisco Olguin
Military and Air Attaché: Gen Garcia Elizalde ☎(+33 1) 47 04 47 09
Military and Air Attaché: Col Rosario Escalante Yocupicio
Assistant Military and Air Attaché: Brig Gen Francisco Arriola Arriola

Office of the Commercial Counsellor
4 rue Notre Dame des Victoires, F-75002 Paris
☎(+33 1) 40 20 07 31 ⊖ (+33 1) 42 61 52 95
Commercial Counsellor: Eligio Serna [97]

31-01210 Embassy of Mongolia
5 Avenue Robert Schuman, F-92100 Boulogne-Billancourt
☎(+33 5) 46 05 28 12; 46 05 23 18
Ⓔ 106513.2672@compuserve.com
⊖ (+33 5) 46 05 31 16
Ambassador: H E Mr K Bekhbat
Second Secretary: A Nunhbat
Attaché: R Altangerel [98]

31-01230 Embassy of Mozambique
82 rue Laugier, F-75017 Paris
☎(+33 1) 47 64 91 32 ▣ 641527 f
⊖ (+33 1) 44 15 90 13
Ambassador: H E Mr José Rui Amaral
Counsellor: Jorge Sousa Mateus [98]

31-01235 Embassy of the Union of Myanmar
60 rue de Courcelles, F-75008 Paris
☎(+33 1) 42 25 56 95 ▣ 642190 birmani f
⊖ (+33 1) 42 56 49 41
Ambassador: H E Mr Nyunt Tin
Deputy Chief of Mission: U Hpo Htaw
Third Secretary: U Daw Kay Thi Soe
Attaché: U Win Kyi
Attaché: U Khin Maung Oo
Third Secretary: U Aye Lwin [98]

31-01300 Embassy of New Zealand
7 rue Léonard de Vinci, F-75116 Paris
☎(+33 1) 45 00 24 11 ⊖ (+33 1) 45 01 26 39
Ambassador: H E Mr Richard Woods
Deputy Head of Mission: Rachel Fry
Counsellor, Economics: Julian Ludbrook
First Secretary: Pamela Wilkinson
Press Attaché: Ms Mary Oliver
Second Secretary: Ms Philippa Collins
Attaché: Ms Mary Molloy
Trade Commissioner: Ms Jane Sweeney
Defence Attaché: Brig I J Duthie [98]

31-01330 Embassy of the Federal Republic of Nigeria
173 Avenue Victor Hugo, F-75116 Paris
☎(+33 1) 47 04 68 65 ▣ 640781
⊖ (+33 1) 47 04 47 54
Ambassador: H E Prof O Gabriel Olusanya
Minister, Deputy Chief of Mission: Ms O O Obafemi
Counsellor, Economic and Commercial Affairs: D O Obasa
Counsellor, Consular and Education: A B Magashi
Counsellor and Chancellor: R A Shiyanbola
Counsellor, Protocol and Administration: B S Ringim
Counsellor, Political Affairs: H O Offor
Attaché, Financial Affairs: L O Orimogunje

Attaché, Administration: R O Bello-Elegushi
Attaché, Information: E F I Aniegbuna
Attaché, Information: Ms S O Adeyemo [98]

31-01340 **Royal Norwegian Embassy**
28 rue Bayard, F-75008 Paris
☎ (+33 1) 53 67 04 00 ⊖ (+33 1) 53 67 04 40
Ambassador: H E Mr Reginald Norby
Deputy Chief of Mission: Mrs Aase S Danielsen
Defence Attaché: Col Gunnar Mjell
Commercial Counsellor: Jan Dahler-Nilsson
Counsellor, Scientific Affairs: Per Bruun Lie [98]

31-01420 **Embassy of the Philippines**
4 Hameau de Boulainvilliers, F-75016 Paris
☎ (+33 1) 44 14 57 00 ⊠ PHILEMBASSY Paris
⊖ (+33 1) 46 47 56 00
Ambassador: H E Mrs Rora Navarro Tolentino
Minister and Deputy Chief of Mission: Romeo L Manalo
Commercial Counsellor: Roberto Mercado
Defence Attaché: Lieut Col Fernando Mesa [98]

31-01460 **Embassy of the State of Qatar**
57 Quai d'Orsay, F-75007 Paris
☎ (+33 1) 45 51 90 71 ⊠ 270074 qatar f
⊖ (+33 1) 45 51 77 07
Ambassador: H E Mr Abdul Rahman Al-Kahulaifi
First Secretary: Ahmed Al-Sanidi
First Secretary, Commercial Affairs: Abdulaziz Al-Alsubaie [98]

31-01640 **Embassy of the Republic of Singapore**
12 Square de l'Avenue Foch, F-75116 Paris
☎ (+33 1) 45 00 33 61 ⊠ 645994 singemp f
⊖ (+33 1) 45 00 61 79
Ambassador: H E Mr Foo Meng Tong
First Secretary: Reuben Wong Yik Pern
First Secretary: Ms Zarina Mohamed
Administration Secretary: Ms Judy Han [98]

31-01655 **Embassy of the Republic of Slovenia**
21 Bouquet de Longchamp, F-75116 Paris
☎ (+33 1) 47 55 65 90 ⊖ (+33 1) 47 55 60 05
Ambassador: H E Dr Andrej Capuder
Counsellor: Ales Biber
First Secretary: Ms Ana Vilfan
Attaché: Darja Golez

Consulate of the Republic of Slovenia
40 Allé de la Robertsau, F-67000 Strasbourg
☎ (+33 3) 88 36 60 25 ⊖ (+33 3) 88 37 14 44
General Consul: Andrej Novak

Mission of the Republic of Slovenia to the Council of Europe
40 Allée de la Robertsau, F-67000 Strasbourg
☎ (+33 3) 88 37 14 44 [98]

31-01690 **Embassy of the Democratic Socialist Republic of Sri Lanka**
15 rue d'Astorg, F-75008 Paris
☎ (+33 1) 42 66 35 01 E mmenu@iway.fr
⊠ 281224 lankamb f ⊖ (+33 1) 40 07 00 11
Ambassador: H E Mr Sumitra Peries
☎ (+33 1) 42 66 38 56
Second Secretary: A H M P Herath
International Affairs: Ahmed A Jawad
Commercial Attaché: Surin de Silva
Commercial and Press Attaché: J M Menu [98]

31-01730 **Embassy of Sweden**
17 rue Barbet de Jouy, F-75007 Paris
☎ (+33 1) 44 18 88 00 ✉ SVENSK Paris
⊠ 204675 svensk f ⊖ (+33 1) 44 18 88 26
Ambassador: H E Mr Örjan Berner
Deputy Chief of Mission: Magnus Robach
Defence Attaché: Capt Sven Rudberg
☎ (+33 1) 44 18 88 24
Counsellor, Scientific Affairs: Bertil Carstam
Counsellor, Economic Affairs: Anders Wollter

Office of the Commercial Attaché
67 Boulevard Haussmann, F-75008 Paris
☎ (+33 1) 42 66 08 88 ✉ SUECOME Paris
⊠ 216526 suecome f ⊖ (+33 1) 42 66 63 04
Commercial Counsellor: Tomas Fellbom [98]

31-01740 **Embassy of Switzerland**
142 rue de Grenelle, F-75007 Paris
☎ (+33 1) 49 55 67 00 ✉ AMBASUISSE Paris
⊖ (+33 1) 45 51 34 77
Ambassador: H E Mr Bénédict de Tscharner
Minister and Deputy Chief of Mission: Georges Martin
Minister, Cultural and Educational Affairs:
Jean Jacques de Dardel
Counsellor, Space Affairs: Roland Hofmann
Embassy Staff Director: Franz Hunkeler
First Secretary, Economic Affairs: Philippe Guex
First Secretary, Press and Information: Daniel Haener
First Secretary, Consular Affairs: Kurt Magli
First Secretary, Social Affairs: Walter Deplazes
First Secretary: Roger Sordet
First Secretary, Commercial Affairs: Pierre Buchs
Third Secretary, Visa Section: Bernard Reymond
Attaché, Cultural Affairs: Ms Anne Marie Aeschlimann
Attaché, Press and Information: Danielle Pesquier
Attaché, Educational Affairs: Ms Marie Antoinette Chollet
Attaché, Administrative Affairs: Pierre Chabloz
Defence Attaché: Maj Gen Jean Jules Couchepin [98]

31-01780 **Royal Thai Embassy**
12 rue Lord Byron, F-750008 Paris
☎ (+33 1) 42 89 89 24 ⊖ (+33 1) 42 89 89 34
Ambassador: H E Mr Tej Bunnag
Minister: Thana Duangratana
Minister and Counsellor: Suthasanai Vacharasinthu
Minister and Counsellor, Consular Affairs:
Ms Madurapochana Ittarong
First Secretary, Information and Personnel:
Ms Wanna Sudjit
First Secretary, Scientific and Cultural Affairs:
Amornsate Singha
First Secretary, Political Affairs: Pratak Sikkhamoppnton
Second Secretary, Economic Affairs:
Ms Makawadee Sumitmor
Second Secretary, Judicial Affairs:
Ms Chavanart Thangsumphant
Naval Defence Attaché: Capt Dumrongsak Haocharoen
☎ (+33 1) 42 89 86 70; 42 89 86 71 (fax)
Military Attaché: Col Pien Sincharoen
☎ (+33 1) 42 89 88 10; 42 89 88 11 (fax)
Air Attaché: Col Sommai Dabpetch (Resident in Bonn)

Office of the Commercial Counsellor
31 Quai de Grenelle, Immeuble Mercure 1, F-75015 Paris
☎ (+33 1) 44 37 01 71; 44 37 01 72
⊖ (+33 1) 44 37 01 73
Minister of Consular Affairs: Id Ayawongs [98]

31-01840 **Embassy of the Republic of Turkey**
16 Avenue Lamballe, F-75016 Paris
☎ (+33 1) 45 24 52 24
E 106302.662@compuserve.com ⊠ 645326 parelc f
⊖ (+33 1) 45 20 41 91
Ambassador: H E Mr Tansug Bleda
Deputy Chief of Mission and First Counsellor: Dogan Akdur
Military Attaché: Col Yilmaz Korucuoglu
☎ (+33 1) 45 24 52 24 ext 243
Naval Attaché: Capt Recai Karagöl ☎ (+33 1) 42 88 48 29
Commercial Counsellor: Ayse Öktem
☎ (+33 1) 45 25 29 63
Press Attaché: Oya Tatlinar ☎ (+33 1) 45 25 05 62
Counsellor, Economic: Holit Tutak [98]

31-01860 **Embassy of the Republic of Uganda**
13 Avenue Raymond Poincaré, F-75116 Paris
☎ (+33 1) 53 70 62 70 ⊖ (+33 1) 53 70 85 15
Ambassador: H E Dr David Kazungu
First Secretary, Economic: L B Lukwiva
First Secretary, UNESCO: Mrs Noea L Katabarwa
First Secretary, Information: Eric Mbabazi-Araali
Third Secretary, Administration: S Kasajja [98]

31-01890 **Embassy of the United Kingdom**
35 rue du Faubourg St Honoré, F-75383 Paris Cedex 08
☎ (+33 1) 44 51 31 00 ⊠ 650264 inform f
⊖ (+33 1) 44 51 98 96
Ambassador: H E Sir Michael Jay
Defence and Air Attaché: Air Cdre M A Pakenham (RAF)
Military Attaché: Brig A C I Gadsby (ARMY)
Naval Attaché: Capt D J Thompson (RN)
Counsellor, Economic: Dr V Caton
Counsellor, Political: S L Cowper-Coles
Counsellor, Technology: Dr M A Darnbrough
First Secretary, Defence Equipment: J C Palmer
Assistant Air Attaché: Wg Cdr A Gunner (RAF) [98]

31-01900 **Embassy of the United States of America**
2 Avenue Gabriel, F-75382 Paris Cedex 08
☎ (+33 1) 43 12 22 22 ⊖ (+33 1) 43 12 49 32
Deputy Chief of Mission: Donald K Bandler
Defence Attaché: Col Peter F Herrly (US Army)
Commercial Counsellor: Emilio F Iodice
Scientific Attaché: Jerome J Bosken
Public Affairs Officer: Christopher Snow

Office of Defence Co-operation
2 rue Saint Florentin, F-75382 Paris Cedex 08
☎ (+33 1) 43 12 22 22 ⊠ 650074 ramc f
Chief of Defence Co-operation: Capt P M E Morris (US Army) [98]

31-01910 **Embassy of Uruguay**
15 rue Le Sueur, F-75116 Paris
☎ (+33 1) 45 00 81 37 ⊠ 645564 f
⊖ (+33 1) 45 01 25 17
Ambassador: H E Mr Héctor Gros Espiell
Naval Attaché: Capt de Vaisseau Walter Franco
First Secretary, Commercial Affairs: Julio Moreira
First Secretary: Ms Adriana Berro de Piriz [97]

31-01960 **Embassy of the Socialist Republic of Vietnam**
62 rue Boileau, F-75016 Paris
☎ (+33 1) 44 14 64 00
⊖ (+33 1) 45 24 39 48; 44 14 64 42
Ambassador: H E Mr Nguyen Chien Thang
Minister and Counsellor: Nguyen Van Doat ☎ (+33 1)
Minister and Counsellor: Le Manh Hung
☎ (+33 1) 44 32 08 72; 44 32 08 79 (fax)
Military Attaché: Col Tran Dam ☎ (+33 1) 40 50 30 28; 42 30 71 34 (fax)
Counsellor, Consular Affairs: Cao Xuan Than
☎ (+33 1) 44 14 64 20; 44 14 64 18 (fax)
Counsellor: Pham Huy Son ☎ (+33 1) 44 14 64 18
Counsellor, Commercial: Nguyen Van Tuyen
☎ (+33 1) 46 24 85 77; 46 24 12 58 (fax)
Counsellor, Affairs Francophones: Tran Van Hai
☎ (+33 1) 44 32 08 76; 44 32 08 79 (fax)
First Secretary: Nguyen Van Viem ☎ (+33 1) 44 14 64 01
First Secretary, Scientific, Cultural and Technical Affairs:
Dinh Huy Hieu ☎ (+33 1) 44 14 64 30
First Secretary, Consular Affairs: Ta Nguyen Ngoc
Attaché, Commercial: Nguyen Van Mon

Commercial Section
44 Avenue de Madrid, F-92200 Neuilly sur Seine
☎ (+33 1) 46 24 85 77; 46 24 80 76 ⊠ 612922 secovi f
Commercial Attaché: Pham Khac Thich
☎ (+33 1) 46 24 80 76 [98]

34 Multinational Intergovernmental Organizations

34-00005 **Advisory Group for Aerospace Research and Development (AGARD)**
7 rue Ancelle, F-92200 Neuilly sur Seine
☎ (+33 1) 55 61 22 00 ⊖ (+33 1) 55 61 22 99
Director: J H Wild
Director: Dr E A van Hoek [97]

34-00010 **Central Europe Operating Agency (CEOA)**
BP 552, F-78005 Versailles Cedex
☎ (+33 1) 39 24 49 00 ⊖ (+33 1) 39 55 65 39
General Manager: Lieut Gen J Charlier
■ Operation of pipeline networks. [97]

34-00080 **Council of Europe Crime Problems Division**
BP 431, F-67006 Strasbourg Cedex
Information Contact: Apleia Tsitsoura [98]

34-00100 **European Space Agency (ESA)**
8-10 rue Mario Nikis, F-75738 Paris Cedex 15
☎ (+33 1) 53 69 71 67; 53 69 76 54
⊖ (+33 1) 53 69 76 90; 53 69 75 60
Director General: Ms Jean Marie Luton
Science Director: Dr Roger Bonnet
Administration Director: A Plattenteich

■ International organisation with the task to provide and to promote for exclusively peaceful purposes, co-operation among European states in space research, technology and their space applications. The organisation is composed of 14 states: Austria, Belgium, Denmark, Finland, France, Germany, Italy, Ireland, Netherlands, Norway, Spain, Sweden, Switzerland and the United Kingdom. Canada takes part in some projects under a co-operation agreement. [98]

34-00300 Western European Union Assembly (WEU)

Assemblée de l'Union de l'Europe Occidental
43 Avenue du Président-Wilson, F-75775 Paris Cedex 16
☎(+33 1) 53 67 22 00
[E] 100315.240@compuserve.com
[W] http://www.nato.int/related/or-weu
⊖ (+33 1) 47 20 45 43
Chairman, (National Armaments Directors): Emile Blanc
President of the Assembly: Lluis Maria de Puig
Information Counsellor: Yves Robins
Clerk of the Assembly: Henri Burgelin
■ Member Nations: Belgium, France, Germany, Italy, Luxembourg, the Netherlands, Portugal, Spain, United Kingdom and Greece. Parliamentary assembly of the WEU.

WEU Institute for Security Studies
Institut d'Études de Sécurité de l'UEO
43 Avenue du Président-Wilson, F-75775 Paris Cedex 16
☎(+33 1) 53 67 22 00 ⊖ (+33 1) 47 20 81 78
Director: Guiolo Lenzi [98]

34-00400 NATO Hawk Management Office (NHMO)

26 rue Galliéni, F-92500 Rueil Malmaison
☎(+33 1) 47 08 75 00 ⊖ (+33 1) 47 52 10 99
General Manager: Col F Bartone [97]

34-00500 NATO Helicopter for the 1990's Design and Development, Production and Logistics Management Agency (NAHEMA)

Le Quatuor - Bâtiment A, 42 route de Galice, F-13082 Aix en Provence Cedex 2
☎(+33 4) 42 95 92 00 ⊖ (+33 4) 42 64 30 50
General Manager: Vice Adm U Battigelli [97]

40 Government Defense Sales/Procurement Organizations

40-01900 SOFRESA

Société Française d'Exportation de Systèmes d'Armement
(a subsidiary of SOFINFRA, France)
307 Bureaux de la Colline, F-92213 Saint Cloud
☎(+33 1) 41 12 22 00 ⊖ (+33 1) 41 12 22 67
♧ FFr 100,000,000 (1997) 👥 85 (1997)
Chairman and Chief Executive Officer: Michel Hazens
Managing Director: J Y Gourcuff
Sales and Marketing Director: H Castel
Adviser to the Chairman: Ms Arlette Gaillon
■ Export organisation of the DGA (Délégation Générale à l'Armement), representing the major defence industry companies in certain countries of the Middle East. [98]

55 Government-owned Industry & Production Facilities

Aerospatiale 75-00600
SNECMA 75-36305

55-00100 Giat Industries

13 Route de la Minière, F-78034 Versailles Cedex
☎(+33 1) 30 97 37 37 ⊖ (+33 1) 30 97 39 00
♧ FFr 2,930,244,600 (1995)
♣ FFr 8,300,000,000 (1996) 👥 15,000
Chairman and Chief Executive Officer: Jacques Loppion
Group Executive Vice President: Philippe Tenneson
Communications Director: Ms Martine Chayriguès
Press Officer: Ms Nelly Tartivel
■ Giat Industries, designer and producer of systems and equipment for land, air, naval, internal security and police forces. International industrial group located in Europe and on the American continent, Giat Industries produces a turover of 8,300 million French francs, half of which is on the export market. Giat Industries is organised around three operational units: Armoured Systems Unit, Weapon and Ammunition Unit and the Herstal Group and Management Development Unit. → 75

Armoured Systems Division
F-78022 Versailles Cedex
☎(+33 1) 39 49 80 00 ⊖ (+33 1) 39 49 81 18
■ This division groups all the activities connected to armoured vehicles: main battle tanks such as the Leclerc, or light armoured vehicles such as the VAB; the fleet (more than 18,000 Leclerc, AMX 30, VAB, AMX 10 and AMX 13 have been produced). To be added to these are turrets of all calibres, that can be installed on all existing land or naval platforms. The division also acts as prime contractor to provide customers with a complete armoured force system. The logistic support system engineering is also an integral part of this division.

Weapon and Ammunition Systems Division
F-78034 Versailles Cedex
☎(+33 1) 39 49 80 00 ⊖ (+33 1) 39 49 81 18
■ This division groups all the activities connected to weapons of all calibres and associated ammunition, to reactive or active protection means. Such varied activities cover 105 mm to 155 mm field artillery systems with associated ammunition, classical, cargo or intelligent, medium or small calibre weapons and ammunition, tank guns and ammunition, such as the latest version of the APFSDS rounds.
The anti-tank mining systems and mine clearing systems for engineers, infantrymen anti-tank systems and armoured vehicle protection systems complete this range of products.

Business Development Directorate
F-78034 Versailles Cedex
☎(+33 1) 39 49 83 70 ⊖ (+33 1) 39 49 81 37
■ Within the global re-struturing context of the defence industries, this new directorate has the objective of creating employment in the group and external employment in association with Sofred, the subsidiary specialised in reconversion in order to compensate for the general decrease in activities connected with armament. Its actions are carried out along two main axes: implementation of industrial projects, such as the Very Large Telescope (for the Southern Observatory) and development of new activities valorising the know-how and skills of the group. [98]

60 Research Organizations, Institutes & Test Establishments

60-00350 Association de Recherches Penales Européennes

Association of European Penal Research
28 rue Saint Guillaume, F-75007 Paris
☎(+33 1) 42 22 64 11 ⊖ (+33 1) 44 39 86 43
Director: Ms Mireille Delmas-Marty
■ Research, organisation of meetings, exchange with the European Association of Jurists created by the Commission of European Communities in the perspective of European harmonization.
Established in 1991. [98]

60-00500 Centre d'Études et de Recherches sur la Police

Study and Research Centre on Police
Institut d'Études Politiques de Toulouse
2 rue des Puits-Creusés, F-31000 Toulouse
Information Contact: Jean Louis Loubet del Bayle [98]

60-00560 Centre d'Études de la Plice National

Studies Centre of the National Police
Plateau de Moulon, BP 41, F-91192 Gif sur Yvette Cedex
Contact: Daniel Bellet [98]

60-00570 Centre de Sciences Criminelles

Centre of Criminal Sciences
Faculté de Droit, Avenue du Doyen Poplawski, F-64000 Pau
☎(+33 5) 59 80 75 57; 59 80 75 58
⊖ (+33 5) 59 80 75 59
👥 4 (professors/teachers); 27 (researchers); 6 (professional); 44 (students)
Director: Reynald Ottenhof
Documentation: Ms Françoise Croisier
■ Research, teaching and documentation.
Established in 1972. [98]

60-00650 Centre de Recherches de Politique Criminelle

Centre of Recherche for Criminal Policy
28 rue Saint Guillaume, F-75007 Paris
👥 11
Founding President: Marc Ancel
President: P Arpaillange
Vice President: G Levasseur
Vice President: Y Monnet
Director: Philippe Lafarge
Secretary General: J Véri
Assistant Secretary General: L Cobert
Assistant Secretary General: D Salas
Director of Scientific Work: Ms M Delmas Marty
Treasurer: J Tourin
Treasurer (Honorary): Y Marx
Chief Editor: C Lazerges
■ Research in criminal policy, social defence, human rights.
Established in 1972. [98]

60-00670 Centre de Recherches Sociologiques sur le Droit et les Institutions Penales

Centre of Sociological Research on the Penal Law
4 rue de Mondovi, Paris
☎(+33 1) 44 77 78 42 ⊖ (+33 1) 44 77 78 77
👥 16
Director, Police, History of Penal Justice: René Levy
Legislative Sociology, Sociology of Law, Comparative Penal Policies: Ph Robert
Prison, Drug Addiction: C Faugeron
Police, Local Safety Evaluation: R Zauberman
Labour Law, Rules: F Soubiran Paillet
Private Security: F Ocqueteau
Criminal Statistics: B Aubusson de Cavarlay
Criminal Statistics: M S Hure
Criminal Statistics: M L Pottier
Prison Demography: P Tournier
Traffic Offences: C Pérez Diaz
Cost of Crime: B Laffargue
Cost of Crime: T Godefroy
Drug Addiction: M D Barre
Traffic Offences, Representations: J M Renouard
Traffic Offences, Legislation: A Kletslen
Information Contact: M Bierry
■ Sociology of law, penitentiary demography, quantitative approach to the penal system, actors of penal system and research.
Established in 1969 as SEPC and 1983 as CESDIP. [98]

60-00850 École Nationale Supérieure de Techniques Avancées (ENSTA)

(a subsidiary of the DGA, France)
32 Boulevard Victor, F-75739 Paris Cedex 15
☎(+33 1) 45 52 49 85 ⊖ (+33 1) 45 52 55 87
Director: IGA Pierre Sintes
Dean of Studies: R Dalbera
Assistant Director of Studies: Hervé Lebret
Press and Information Officer: Lionel Paretas
■ ENSTA is intended to train military and civil engineers, officiers and officials, French and foreign, to participate in the naval, mechanical, chemical, electronic, oceanological and related industries. → 20 [98]

60-01500 Fédération Internationale des Droits de l'Homme

27 rue Jean Dolent, F-75014 Paris
Information Contact: David Roger [98]

60-02200 Groupe Européen de Recherche sur les Normativities (GERN)

European Research Group
4 rue de Mondovi, F-75001 Paris
☎(+33 1) 44 77 78 52 ⊖ (+33 1) 44 77 78 77
Director: Philippe Robert
Information Contact: Roland Bierry
■ Norms and deviances; comparative European studies; research, teaching and seminars.
European interdisciplinary network consisting of thirty centers of eight West European Countries.
Established in 1985. [98]

60-02500 IHESI
Department d'Études et Recherches (IHESI)
19 rue Peclet, F-75015 Paris
Commissioner: Jean Claude Monet
Chief: Mme Fabrice Fournier Mickewicz, Dominic Monjardet [98]

60-02750 Institut Alexandre Lacassagne
Alexandre Lacassage Institute
Domaine Rockefeller, 8 Avenue Rockefeller, F-69373 Lyon Cedex 08
☎ (+33 4) 78 77 70 00 ⊖ (+33 4) 72 12 00 75
👥 6
Director: Prof. J M Elchardus
Psychiatry: Dr B Hoestlandt
Psychiatry: Dr P Barlet
Psychiatry: Dr D Gonin
Psychology: D Weber
Magistrate and Jurist: Mr Seys
Secretary and Documentalist: Ms Maitrejean
■ Teaching and research in psychiatry forensic medicine, clinical criminology.
Established in 1960. [98]

60-02800 Institute of Advanced National Defence Studies
Institut des Hautes Études de Défense Nationale (IHEDN)
13-21 Place Joffre, F-75700 Paris
☎ (+33 1) 44 42 38 22 ⊖ (+33 1) 44 42 31 87
Director of IHEDN, EMS and CHEM: Gen Janvier
☎ (+33 1) 44 42 43 36
Deputy Director and Director of Studies: Préfet Decharriere ☎ (+33 1) 44 42 43 39
Chief of Staff: Col Duval ☎ (+33 1) 44 42 38 22
Director of Studies CHEM: Gen Dumaz
☎ (+33 1) 44 42 48 20
■ French national war college. [98]

60-02850 Institute of Criminology
Institut de Criminologie
Université de Paris
12 place du Panthéon, F-75231 Paris
Information Contact: Prof. Jean Claude Soyer [98]

60-02880 Institut de Criminologie et d'Études Judiciaries
Institute of Criminology and Judicial Studies
Place Anatole France, F-310542 Toulouse Cedex
☎ (+33 5) 61 63 36 19
👥 4 (professional); 1 (support)
Director: Prof. G Roujou de Boubee, H Guy, C Rapp, S Mirabail, G Ginestet, T Garé
■ Research and teaching.
Established in 1924. [98]

60-02900 Institut de Criminologie et de Sciences Criminelles
Institute of Criminology and Criminal Sciences
BP 169, F-59653 Villeneuve d'Ascq Cedex
Pont de Bois, F-59650 Villeneuve d'Ascq Cedex
☎ (+33 3) 20 05 74 39 ⊖ (+33 3) 20 05 74 03
Director: Henri Fenaux
■ Research and teaching.
Established in 1954. [98]

60-02920 Institut de Droit Comparé de l'Université de Paris II
Institute of Comparative Law
Section de Science Criminelle
28 rue Saint Guillaume, F-75007 Paris
☎ (+33 1) 44 39 86 19 ⊖ (+33 1) 44 39 86 43
👥 12
Director: Prof. Mireille Delmas-Marty
Information Contact: Ms Marina Krivitsky
■ Penal comparative law, human rights, criminal policy and international penal law.
Established in 1936. [98]

60-02950 Institut d'Études Judiciaires de Nancy
Institut of Judicial Studies of Nancy
Department of Criminology
13 Place Carnot, F-54035 Nancy Cedex
☎ (+33 3) 83 35 46 15; 83 32 99 38
⊖ (+33 3) 83 32 48 29
👥 15
Director: Prof. Jean François Seuvic
Emeritus Professor of Penal Law: Prof. Vitu
General and Clinical Criminology: Dr Diligent
Psychiatry: Dr Bocquel
Psychiatry: Dr Dubois-Bonnefont
Psyichiatry: Dr Bickel
Psychiatry: Dr Bujon Pinard
Psychiatry: Dr Klein
Psychiatry: Dr Blanchard
Forensic Medicine: Prof. Coudane
Forensic Medicine: Dr Peton
Attorney General: Mr Lucazean
Penal Judge: Mr Picquart
Police Officer: M Cristini
Police Officer: Mr Bordeau
■ Teaching and research in criminology, general penal law and procedures, scientific police, forensic psychiatry, forensic medicine and penitentiary science. Research on local criminality and criminal policy.
Established in 1955. [98]

60-03100 Institut de Medecine Legale de Lyon
Institute of Legal Medicine of Lyon
8 Avenue Rockefeller, F-69008 Lyon
☎ (+33 4) 78 74 12 97; 78 01 17 89
⊖ (+33 4) 72 23 00 75
👥 20
Director: Prof. Daniel Malicier
Information Contact: Dr Pierre Feuglet
■ Research and teaching on autopsies, toxicological tests, anatomopathology and anthropology.
Established in 1928. [98]

60-03150 Institut de Medecine Legale et de Medecine Sociale
Institute of Legal and Social Medicine
Place Théo Varlet, F-59000 Lille
☎ (+33 3) 20 62 12 20 ⊖ (+33 3) 20 62 12 29
👥 4
Director: Prof. Didier Gosset, V Medouin, D Bailly
Information Contact: Ms Carre
■ Research and teaching on drug addiction, child abuse, alcoholism, traffic offenses.
Established in 1933. [98]

60-03300 Institute of Penal Sciences and Criminology
Université d'Aix-Marseille III
3 Avenue Robert Schumann, F-13628 Aix en Provence
Director: J Borricand [98]

60-04000 Institut de Sciences Criminelles
Institut of Criminal Sciences
Université Montesquieu-Bordeaux IV, Avenue Léon Duguit, F-33604 Pessac Cedex
☎ (+33 5) 56 84 85 86
W http://www.montesquieu.u-bordeaux.fr
⊖ (+33 5) 56 84 26 32
👥 10
Professor: Jean Pierre Delmas Saint-Hilaire
Professor: Philippe Conte
Professor: Mr Labayle
Professor: Mr Trigeaud
Master of Conference: Ms Peano
Professor: Lapoyade Deschamps
Master of Conference: Adémar du Cheyron du Pavillon
Master of Conference: Mr Guyon
Master of Conference: Mr Renovard
■ Awards the Criminal Sciences Certificate for students at the 3rd and 4th years in law studies, the Diploma of Special Studies in Criminal Law and Criminal Sciences. Also provides guidance in Criminal Law and Criminal Sciences theses. [97]

60-04100 Institut de Sciences Criminelles
Institute of Criminal Sciences
Université de Poitiers
43 Place Charles de Gaulle, F-86022 Poitiers Cedex
☎ (+33 5) 49 60 20 15 ⊖ (+33 5) 49 41 84 28
👥 15 (professional)
Director: Jean Pradel
Secretary: Ms Françoise Trocas
■ Research and teaching in penal law, penal procedure, comparative law, criminology, sociology, psychiatry and forensic medicine.
Established in 1951. [98]

60-04120 Institut de Sciences Criminelles de Poitiers
Institute of Criminal Sciences of Poitiers
Faculté de Droit de Poitiers
10 rue de l'Université, F-86000 Poitiers
Director: Jean Pradel [98]

60-04150 Institute de Sciences Criminelles de Montpellier
Institute of Criminal Sciences of Montpellier I
Faculté de Droit et des Sciences Economiques
39 rue de l'Université, F-34000 Montpellier
☎ (+33 4) 67 61 54 20 ⊖ (+33 4) 67 60 42 31
👥 5
Director, Penal Law and Penal Procedure: Didier Thomas
Juvenile Delinquency: Ms Christine Lazerges
Criminal Psychiatry: Dr Michel Ribstein
Forensic Medicine: Dr Jean Pierre Santoro
Commissioner Forensic Police: Jacky François
■ Teaching and training.
Established in 1948. [98]

60-05000 Office National d'Études et de Recherches Aérospatiales (ONERA)
Avenue de la Division Leclerc, PO Box 72, F-92322 Châtillon Cedex
☎ (+33 1) 46 73 40 40 ⊖ (+33 1) 46 73 41 41
♣ FFr 1,360,000,000 (1996) 👥 2,300
Chairman and Managing Director: IGA Michel Scheller
Public Relations and Information Manager: Ms Corine Aubert-Simon
■ Fundamental research, applied research and contribution to development in the fields of: aerodynamics, structures, energetic materials, optics, acoustics, electronics, flight mechanics, ground windtunnel and flight testing. Computer science.

Établissement de Palaiseau
Fort de Palaiseau, PO Box 18, F-91120 Palaiseau
☎ (+33 1) 69 93 60 60 ⊖ (+33 1) 69 93 61 61

Établissement de Chalais-Meudon
8 rue des Vertugadins, F-92190 Meudon
☎ (+33 1) 46 23 50 50 ⊖ (+33 1) 46 23 51 51

Centre d'Essais de Modane-Avrieux
PO Box 25, F-73500 Avrieux
☎ (+33 1) 79 20 21 22 ⊠ 980661 f
⊖ (+33 1) 79 20 21 68
Managing Director: J Laverre

Centre d'Essais de Fauga-Mauzac
F-31410 Mauzac
☎ (+33 1) 61 56 63 00 ⊠ 530880 f
⊖ (+33 1) 61 56 63 63
Managing Director: J M Carrara

Centre d'Études et de Recherches de Toulouse (CERT)
2 Avenue Edouard-Belin, PO Box 4025, F-31055 Toulouse Cedex
☎ (+33 1) 61 55 71 11 ⊠ 521596 f
⊖ (+33 1) 61 55 71 72
Managing Director: François Jouaillec

Institut de Mécanique des Fluides (IMFL)
5 Boulevard Paul-Painlevé, F-59045 Lille Cedex
☎ (+33 1) 20 49 69 00 ⊠ 160010 f
⊖ (+33 1) 20 52 95 93
Managing Director: Francis Dupriez

Laboratoire de Salon de Provence
Base Aérienne 701, F-13661 Salon de Provence Air
☎ (+33 1) 90 17 01 10 ⊖ (+33 1) 90 17 01 09 [98]

60-05500 Research Institute for the Study of Juridical Information
Université de Montpellier I, BP 1017, F-34006 Montpellier Cedex
Director: Prof. J L Bilon [98]

60-05900 Société Generale des Prisons et de Legislation Criminelle
General Society of Prisons and Criminal Legislation
5 Petite Place, F-7800 Versaille
Director: Ivan Zakine
Information Contact: M Marc

■ Research activities.
Established in 1876. [98]

60-06000 Société de Medecine Legale et de Criminologie de France
Society of Legal Medicine and Criminology of France
2 place Mazas, F-75012 Paris
☎ (+33 2) 43 34 43 54
Director: P E Fournier
Secretary General: Prof. Claude Piva
■ Established in 1886. [98]

60-07000 Université de Droit Sceaux Criminal Sciences Department
54 Boulevard Desgrandes, F-92331 Sceaux Cedex [98]

65 Industry, Trade & Armed Forces Associations

65-00330 Comité Richelieu
United Defence Technologies of France
98 Avenue du Général Leclerc, F-92100 Boulogne
☎ (+33 1) 46 04 55 03 ⊜ (+33 1) 46 04 55 38
General Director: Emmanuel Leprince
■ Non-profit making association representing over 130 companies in the commercial and defence sectors.
Members:
3IP, ABH Technologies, ACMH, ACSYS, ADERSA, ADR, ADV Technologies, AETA, Alkan, Alten Industrie, APSYS, APTIME, AQL, Ares Armement, ATECA, Atermes, Auxitrol, Axlog Ingenierie, Axon' Cable, B+ Development, Berenger Rouard, Bertin & Cie, Bracer Cerf, Brime, Bronzavia, CAC Systemes, Castor Technologies, Cenergys, Chorus Systemes, Cimulec, Codechamp, Coframi, Comatlas, Crausaz & Fournier, Cril Ingenierie, Degreane, Digilog, EIA, Elan Informatique, Elics, Estelle France, Euro-Cockpit Technologies, Eurogiciel, Fecom, G2I, GAIA Converter, Gallet, GECI, Getelec, Giravions Dorand Industries, GIST, GLCS, Groupe HEF, Heptasys, Houvenaghel Hennequin, Hyperparallel Technologies, I.Tech, I2E, I2M, IBS, Ideal Visuel, ILOG, Imagine Technologies, Info Elec, Ingenia, Isotop Ingenerie, Khiplus, Kremlin, KTT, Le Piston Français, Lluis, LMB, Medysys, Metravib RDS, NBCD Technology, O2 Technology, Objectif Technologie, Oktal, Optis, Oritel, Pacte Novation, Paul Boye, Photonetics, Plasmion, Pouyet RPN, Proengin, Protectic, Protem, PSTI Evaluation, Qualience, RBC, Realix Technologies, RGA Systemes, Samtech France, Sciences Industries Conseils, SECA, Secmat NT, Sector, Semia, Sereea, Seriem, SESO, Sigma +, Silicomp, Simulog, SITIA, Socotec Industrie, Sodena, Sodima, Sodipe, Soditech, Sofrelog, Softronic, STAS, Syminex, Teamlog, Telmat Informatique, Topware Technologies, Trialog Informatique, UNIC Industries, VELEC, Verilog, VSM, XRS. [97]

65-00340 Confederation Amicale des Ingenieurs de l'Armement (CAIA)
32 Boulevard Victor, F-75015 Paris
☎ (+33 1) 45 52 44 27 ⊜ (+33 1) 45 52 44 33
President: François Lureau
Secretary General: Stanislas Chapron
Permanent Delegate: Ms Maryvonne Terracol
■ Association of the French Defence Engineers Corporations. [98]

65-00400 Conseil des Industries de Défense Françaises (CIDEF)
c/o GIFAS - 4 rue Galilée, F-75782 Paris Cedex 16
☎ (+33 1) 44 43 17 00; 44 43 17 85
⊜ (+33 1) 44 43 17 56; 40 70 91 41
President: Serge Dassault (Dassault Aviation)
☎ (+33 2) 47 95 86 38; 47 95 86 39
Premier Vice President of GIFAS: Bernard Dufour (SNECMA) ☎ (+33 1) 40 60 80 01
President SPER: Jacques Buzenet (SNECMA)
☎ (+33 1) 40 60 80 01
President GICAT: Pierre Poquin (SFIM Industries)
☎ (+33 1) 69 19 66 10
Delegate: Gen Jacques Favin-Leveque (GICAT)
☎ (+33 1) 44 14 58 21; 44 14 58 23
Premier Vice President GICAT: Jacques Gentgen (GIAT Industries) ☎ (+33 1) 30 97 36 23
Delegate: Gen Bernard Nicolas (GIFAS)
☎ (+33 1) 44 43 17 02
Member SPER: Gérard Bouy (Dassault Electronique)
☎ (+33 1) 49 11 86 15
Delegate: Yves Artru (SPER) ☎ (+33 1) 45 05 71 90
Secretary: Xavier Guetin (GIFAS) ☎ (+33 1) 44 43 17 85
Public Relations Officer: P Guerin [98]

65-00580 Fédération Générale des Syndicates CGT de la Police Nationale
General Federation of the Syndicates CGT of the National Police
263 rue de Paris, BP 550, F-93514 Montreuil Cedex [98]

65-00600 Fédération des Industries Ferroviaires
12 rue Bixio, F-75007 Paris
☎ (+33 1) 45 56 13 53 ⊜ (+33 1) 47 05 29 17
Managing Director: Jean Pierre Aujoux
Public Relations: Mrs Maguy Peltrei [98]

65-00670 Fédération Professionelle Indépendante de la Police
11 rue des Ursins, F-75004 Paris [98]

65-00700 Fédération Syndicaliste Force Ouvière de la Police Nationale
6 rue Albert Bayet, F-75013 Paris [98]

65-00730 Groupement des Industries Concernées par les Matériels de Defénse Terrestre (GICAT)
64 rue du Ranelagh, F-75016 Paris
☎ (+33 1) 44 14 58 20 ⊜ (+33 1) 42 30 80 89
President: Pierre Poquin
General Manager: Brig Jacques Favin-Leveque (rtd)
■ GICAT the French Land Defence Manufacturers Association represents 80 companies active in the promotion, research, development, manufacture and maintenance of weapons for land forces.
Commissariat Général des Expositions et Salons du GICAT (COGES)
64 rue du Ranelagh, F-75016 Paris
☎ (+33 1) 44 14 58 10 ⊜ (+33 1) 42 30 70 88
General Commissioner: Brig J M Ruault (rtd)
■ GICAT's specialised subsidiary is mainly dedicated to the organisation of Eurosatory land defence equipment exhibition, COGES is also responsible for the French land defence equipment industry through various defence exhibitions all over the world. [97]

65-00740 Groupement des Industries Françaises Aéronautiques et Spatiales (GIFAS)
French Aeronautics and Space Industry Group
4 rue Galilée, F-75782 Paris Cedex 16
☎ (+33 1) 44 43 17 00 ⊠ 645615 f
⊜ (+33 1) 40 70 91 41
President: Serge Dassault
Honorary Chairman: Jacques Maillet
Honorary Chairman: Jacques Mitterand
Honorary Chairman: Jacques Benichou
Honorary Chairman: Jacques-André Larpent
Honorary Chairman: Henri Martre
Commissioner General, Air and Space International Shows: Yves Michot
First Vice President: Jean Paul Béchat
Vice President: Charles Edelstenne
Vice President: Noël Forgeard
Vice President: Jean Robert Martin
Treasurer: Philippe Lhotellier
General Delegate: Bernard Nicolas
Press and Communications Manager: Patrick Guerin
■ Co-ordinates the industrial and economic activities of its 200 members engaged in the development, manufacture, maintenance and sales of aircraft, helicopters, engines, missiles, space vehicles and their associated components, equipment, accessories, products and operating facilities. [98]

65-00750 GILEP
Groupement Interprofessionnel de Logistique et d'Équipements Pétroliers
Fort de Vanves, PO Box 63, F-92243 Malakoff Cedex
☎ (+33 1) 55 58 80 30 ⊠ 250028 direspa f
⊜ (+33 1) 55 58 80 01
👥 2 (1997)
Chairman: Jean Pierre Salaün
Secretary General: Bruno Waymel
■ GILEP operates with the technical support of the French Armed Forces Petroleum Service (S.E.A.). Its member companies offer the whole range of equipment, products and services needed to operate military petroleum logistics (Army, Air Force, Naval Aviation): rear area storage and transport; - air base petroleum facilities; - mobile tactical storage and fuelling systems for helicopter units and ground combat forces; - products for lubrication; - antipollution services (advisory, R&D, engineering, training).
Member Companies:
Cornut
E.I.C. Industries
Faure Herman
Intertechnique (Dpt RVT Zenith Aviation)
Kleber Industrie (C M P Palport)
Lafon
Mouvex
Nyco
Pernin Equipements
Perolo
Pronal
Filtres Rellumix
Renault V.I.
Sofitam Équipements-Satam
Titan Aviation
Trailor
and 5 Correspondent Members [98]

65-00800 Groupement des Industries de Télecommunications et d'Électronique Professionnelle (GITEP)
(† SPER)
17 rue Hamelin, F-75783 Paris Cedex 16
☎ (+33 1) 45 05 70 70 ⊜ (+33 1) 45 53 09 23
President: Marc Houery
Vice President: Jacques Buzenet
Secretary: Jacques Blanche
Treasurer: Gérard Bouy
Treasurer: Alain Carlotti
Treasurer: Jacques Payer
Communications Manager: Ms Nadine Allain
■ Non-profit trade organisation. Represents manufacturers of defence electronics, civil and military communications equipment, aeronautical and naval electronic systems and associated products.
Membership of about 42 companies.
Permanent Members:
Adept-Discofone; Aerospatiale; Alcatel Cable Interface; Alcatel Telecom/Alcatel Business Systems; Alcatel Telecom/Alcatel CIT; Alcatel Telecom/Alcatel Espace; Alcatel Telecom/Alcatel Telspace; Arnould; Ascom HPF SA; Ascom Monetel SA; Bosch Telecom; Bull SA; Cegelec; Celogic SA; Citel-2CP; Clemessy SA; CNIM; CS Defence; CS Telecom; Dassault Automatismes et Telecommunications; Dassault Electronique; Dassault Sercel NP; Depaepe; Elecma; GEGA Electronique; IBM France; Intertechnique; Landis & Gyr Communications; Lucent Technologies BCS France; Malico; Matra-BAe Dynamics France SA; Matra Communication; Matra Marconi Space; Matra Systemes & Information; Mecelec; Mecelec Electronique; Mecelec Industries; Met Communication; MGP Instruments; Mors; Motorola Semi-Conducteurs SA; Nortel Europe SA; Nortel Matra Cellular; Philips Electronique Grand Public; Groupe Poyet; Prescom SA; Rockwell Collins France; RXS Morel Accessoires de Câbles SA; SAGEM SA; Schlumberger Industries; Schlumberger Transactions Electroniques; Schneider Electric SA; SERPE-IESM; Service Electronique Engineering (SEE); Sextant Avionique; SFIM; Sicame SA; Siemens SA;: Simoco France; Sodern; Sodielec; Sogitec Industries; TDA Armament SAS; Tekelec Temex; Thomcast; Thomson-CSF; Thomson-CSF Airsys; Thomson-CSF Applications Radar; Thomson-CSF Communications; Thomson-CSF Optronique; Thomson Marconi Sonar; 3M France; TRT Lucent Technologies; Wavetek SA; Westinghouse SGTE. [98]

65-01600 Syndicat des Commandants et Officiers de la Police Nationale (SCO)
Syndicat of Commandants and Officers of the National Police
2bis Quai de la Mégasserie, F-75001 Paris [98]

65-01700 Syndicat CGC des Corps en Civil de la Police Nationale (SCCPN)
283 Avenue du Maine, BP 35, F-75755 Paris [98]

65-01750 Syndicat des Gradés de la Police Nationale
Syndicat of the Officers of the National Police
2bis Quai de la Mégasserie, F-75001 Paris [98]

65-01780 Syndicat Indépendant de la Police Nationale
Independant Syndicat of the National Police
2bis Quai de la Mégasserie, F-75001 Paris [98]

65-01800 Syndicat National Autonome des Policiers en Civil (SNAPC)
55 rue de Lyon, F-75012 Paris [98]

65-01900 Syndicat National Autonome de la Tenue (SNAT)
55 rue de Lyon, F-75012 Paris [98]

65-01950 Syndicat National des Commissaires (SNC)
National Syndicat of Commissioners
6 rue de Départment, F-75019 Paris [98]

65-02000 Syndicat National des Enquêteurs (SNE)
8 rue Charles Pathe, F-94300 Vincennes [98]

65-02050 Syndicat National Indépendant des Personnels Administratifs et Techniques de la Police National (SNIPAT)
National Independent Syndicat of Administrative and Technical Personnel
BP 1, F-69396 Lyon Cedex 3 [98]

65-02100 Syndicat National des Officiers (SNO)
National Syndicat of Officers
6 rue de Départment, F-75019 Paris [98]

65-02200 Syndicat National des Personnels de Police Scientifique
National Syndicat of Personnel of the Scientific Police
22 rue Antoine Becker - LIPS, F-13224 Marseille Cedex 2 [98]

65-02250 Syndicat National des Policiers en Tenue
Corps Urbains
27 rue de Dunkerque, F-75010 Paris [98]

65-02270 Syndicat National Unifié des Personnels Administratifs et Techniques de la Police National (SNUPAT)
United National Syndicat of Administrative and Technical Personnel
FASP
6 rue de Départment, F-75019 Paris [98]

65-02300 Syndicat National Unitaire des Enquêteurs de Police (SNUIP)
FASP
6 rue de Départment, F-75019 Paris [98]

65-02320 Syndicat National Unitaire des Inspecteurs de Police (SNUIP)
United National Syndicat of Police Inspectors
FASP
6 rue de Départment, F-75019 Paris [98]

65-02350 Syndicat CFTC des Personnels en Tenue de la Police Nationale
11 rue des Ursins, F-75004 Paris [98]

65-02400 Syndicat des Personnels Administratifs et Techniques de la Police Nationale (SPAT)
Syndicat of Administrative and Technical Personnel of the National Police
2bis Quai de la Mégasserie, F-75001 Paris [98]

65-02500 Syndicat CFTC des Policiers en Civil
11 rue des Ursins, F-75004 Paris [98]

65-02800 Syndicat des Policiers en Tenue
19 rue des Boulets, F-75011 Paris [98]

65-03000 Syndicat d'Union Policière
16 rue de la Victoire, F-76083 Le Havre Cedex [98]

65-03100 Syndicat Unitaire de la Police Républicaine (SUPR)
United Syndicat of the Republican Police
FASP
6 rue de Département, F-75019 Paris [98]

65-05000 UCLAT (UCLAT)
26 rue Cambacérès, F-75008 Paris
Liaison Officer: J Franks [98]

65-05100 Union Fédérale CFDT de la Police National
47-49 Avenue Simon Bolivar, F-75950 Paris [98]

65-05200 Union National Autonome des Policiers en Civil
4 Place de Gambetta, F-75020 Paris [98]

65-05500 Union des Syndicats Categoriels de la Police (USCP)
2bis Quai de la Mégasserie, F-75001 Paris [98]

70 Financial Holding Groups, Parent Companies

70-00050 Alcatel Alsthom
(† Alcatel Alsthom Compagnie Générale d'Electricité)
54 rue de la Boétie, F-75008 Paris Cedex
☎(+33 1) 40 76 10 10 ⊠ 651953 f
⊖ (+33 1) 40 76 14 00
♧ FFr 5,738,387,400 ♣ FFr 162,100,000,000
190,000
Chairman and Chief Executive Officer: Serge Tchuruk
Senior Executive Vice President: Jean Pierre Halbron
President & COO: François de Laage de Meux
Chief Operating Officer, Alcatel Telecom: Jozef Cornu
President, Cables and Components Segment: Gérard Hauser
President, Engineering and Systems Segment: Pierre Bonafe
President, GEC Alsthom: Pierre Bilger
Corporate Secretary: Philippe Bissara
General Administration: Jean Paul Barth
Legal: Pascal Durand Barthez
Technical: Laurent Citti
International: Louis Jacques Companyo
Euro Project: Pierre Le Roux
Human Resources and Communication: Ms Caroline Mille
Government Affairs: François Petit
Purchasing: Jean Marc Pornet
Information Contact: Ms Beatrice Mandine
■ Communications, energy and transportation.
Subsidiaries with defence business:
Alcatel
Cegelec
Saft
GEC Alsthom

Alcatel Alsthom Group Members:
Alcatel Alsthom Recherche
Alcatel Telspace
Alcatel Espace
Alcatel ISR
Alcatel Titn Answare
GEC Alsthom
Cegelec
Cegelec CGA
Cirma Enterprise
Soprano
SAFT
Alcatel Bell Telephone, Belgium
Alcatel Bell-SDT, Belgium [98]

70-18000 Messier-Dowty International
(a joint venture between TI Group, UK & SNECMA, France)
Zone Aéronautique Louis Bréguet, BP 133, F-78148 Vélizy Cedex
☎(+33 1) 46 29 18 10 ⊖ (+33 1) 46 29 18 02
Chairman and Chief Executive Officer: L A Edwards
■ Headquarters and operational direction for global landing gear operations based at Gloucester in UK, Vélizy and Bidos in France, Toronto, Montreal and Peterborough in Canada.
Operating Companies:
Messier-Dowty Ltd, UK
Messier-Dowty SAS, France
Messier-Dowty Inc, Canada
Messier-Dowty Electronics, UK
Messier-Dowty Customer Support Ctr, UK
Messier-Dowty Customer Support Ctr, France
Messier-Dowty Customer Support Ctr, Americas
Messier-Dowty Customer Support Ctr, Asia Pacific [98]

70-20400 Thomson SA
173 Boulevard Haussman, F-75415 Paris Cedex 08
☎(+33 1) 53 77 86 16 ⊠ 616780 thom f
⊖ (+33 1) 53 77 86 88
Chairman and Chief Executive Officer: Thierry Breton
Secretary General: Patrice Maynial
Vice President, Finance: Frank Dangeard
Vice President: Pierre Cabanes
Corporate Communications Department: Jean Patrick Defert [98]

70-30000 Zodiac SA/Holding
48 Boulevard Galliéni, F-92130 Issy les Moulineaux
☎(+33 1) 41 23 23 23 ⊠ 270569
⊖ (+33 1) 46 48 83 87
♧ FFr 47,845,490 (1996) ♣ FFr 3,414,000 (1996)
5,600
President: Jean-Louis Gerondeau
Director of Communications: D Puig
Subsidiaries:
Aérazur → 75
Zodiac International → 75 [98]

74 Consortia, Joint Venture Companies

Aero International (Regional) 80-00180

74-00050 acbLIPS SA
(a joint venture between GEC Alsthom ACB, France and LIPS BV, The Netherlands)
2 rue Léon Durocher, F-44200 Nantes
☎(+33 2) 40 41 16 02 Ⓔ acblips@easynet.fr
⊠ 729376 acblips ⊖ (+33 2) 40 41 16 00
President: Robert A Chevillon
Commercial Director: Gilles de Longeaux
Operations and Technical Director: J P Gangler
■ Design, manufacture and sale of all types of propulsors (monobloc propellers, transverse thrusters, azimuthing thrusters, controllable pitch propellers, thrusters, waterjets, shaftlines, accessories) for the Navy, Coastguard, Maritime Police and Customs. → 75 [98]

74-00100 Airbus Industrie
(Owned jointly by Aerospatiale (37.9%), Daimler-Benz Aerospace Airbus (37.9%), CASA (4.2%), British Aerospace Airbus (20%).
1 Rond Point Maurice Bellonte, F-31707 Blagnac Cedex
☎(+33 5) 61 93 31 99 ⊠ 530526 airbus f
⊖ (+33 5) 61 93 37 92

Contact Reference

Chairman: Dr Hans Frederichs
President and Chief Executive Officer: Jean Pierson
Chairman of the Supervisory Board: Edward Reuter
Chief Operating Officer: Volker Von Tein
Senior Vice President, Commercial: John Leahy
Senior Vice President, Industry and Programs: Rolf Feddersen
Senior Vice President, Technology and New Product: Jean Roeder
Senior Vice President, Product Support: Bernard Catteeuw
Senior Vice President, Engineering: Heinz D Brandt
Vice President, Contracts: Denis Viard
Vice President, Marketing: David Jennings
Vice President, Customer Finance: Laurence Barron
Vice President, Flight Division: Pierre Baud
Vice President, Integrity Division: Wolfgang Didszuhn
Vice President, Development Control: Klaus Buchholz
General Manager, Press and Information Services: Ms Barbara Kracht
Vice President, General Engineering: Ange Ortega
Senior Vice President, Administration: Alberto Fernandez
Senior Vice President, Transport: Dietrich Knospe
Vice President, Human Relations: Thierry Schutte
Vice President, Corporate Communications: Robert Alizart
Head of Large Aircraft Division: Jurgen Thomas
Press Manager, Trade and Technical: David Velupillai
Financial Controller: Ian Massey
■ Multinational consortium for management, sales and support of the Airbus Airliner. Range developed and built by European co-operation; the short to medium haul single aisle 150 seat A320 and 186 seat A321. The medium to long range 218 seat A310 and 267 seat A300-600 widebody twin-jets, the medium to long range 335 seat twin-engined A330 and the very long range four-engined 262-295 seat A340.

Partners:
British Aerospace 20%
Aerospatiale 37.9%
Deutsche Airbus 37.9%
Casa 4.2%

Other Participants:
Netherlands: Fokker
Belgium: Belairbus
Italy: Alenia

Affiliates:
Aeroformation, France

Offices:
Airbus Industrie, Sydney, Australia → 90
Airbus Industrie, Beijing, China → 90
Airbus Industrie, Japan → 90
Airbus Industrie, North America → 90
Airbus Industrie, Dubai, UAE → 90 [98]

74-00500 Eurocopter SA
(a joint venture between Aerospatiale, France and Daimler-Benz Aerospace AG, Germany)
Aéroport International, Marseille-Provence,
F-13725 Marignane Cedex
☎ (+33 4) 42 85 95 24 ⊜ (+33 4) 42 85 95 64
♣ FFr 9,500,000,000 👥 9,500
Chairman of the Management Board: Jean François Bigay
Co-Chairman of the Management Board: Dr Siegfried Sobotta
Group Vice President Executive Office: Jean Pierre Dubreuil
Executive Vice President, Controlling, Budget and Planning: Philllipe Burtin
Executive Vice President, Commercial Action: Philippe Harache
Executive Vice President, Finance: Dr H Mühlhoff
Executive Vice President, Industrial Competence Centres: J P Baudry
Executive Vice President, Commercial Helicopters: Yves Richard
Executive Vice President, Governmental Helicopters: Frank Dorn
Senior Vice President, Quality: Jean Pierre Dedieu
Executive Vice President, Customer Support: Laurent Samson
Group Vice President, Technical: A Yves Cassier
Group Vice President, Strategy: Andreas Loewenstein
Senior Vice President, Flight Testing: B Fouques
Director of Communication: Aude Bourrilhon
■ Civil helicopters: Ecureuil, BO 105, EC 135, Bk 117, Dauphin, Super Puma, Colibri.
Military helicopters: Fennec, Panther, Tiger, Cougar, NH 90.

Marignane Plant
Aéroport International, Marseille Provence,
F-13725 Marignane Cedex
☎ (+33 4) 42 85 85 85 ⊜ (+33 4) 42 85 89 28

La Courneuve Plant
20/22 Avenue Marcel Cachin, PO Box 109,
F-93123 La Courneuve Cedex
☎ (+33 1) 49 34 40 00 ⊜ (+33 1) 49 34 44 47 [98]

74-00530 Euromissile Dynamics Group (EMDG)
28 rue de la Redoute, F-92260 Fontenay aux Roses
☎ (+33 1) 41 13 27 80 ⊜ (+33 1) 41 13 27 01
Marketing Director: Mike Burke
■ Industrial consortium for design of 3rd generation anti-tank guided weapons. Member companies: Aerospatiale, France, Matra-BAe Dynamics, UK and Daimler-Benz Aerospace, Germany. [98]

74-00535 Euromissile GIE
12 rue de la Redoute, F-92260 Fontenay aux Roses
☎ (+33 1) 41 87 14 14 ⊠ 632758
⊜ (+33 1) 46 61 64 67
♣ FFr 964,000,000 (1996) 👥 132 (1996)
Chairman: Francis Dorveaux
Operations Director: Rolland Quere
Sales and Marketing Director: Peter Ibbeken
Strategy and External Relations: Jean Thyrard
■ Industrial consortium for production and sale of Milan, HOT and Roland weapon systems. Formed by Aerospatiale, France and Daimler-Benz Aerospace, Germany. → 75 [98]

74-00600 Eurosam GIE
(a joint venture with Aerospatiale, Alenia and Thomson-CSF)
63 Boulevard de la Liberté, PO Box 5,
F-92323 Châtillon Cedex
☎ (+33 1) 40 84 75 70 ⊜ (+33 1) 40 84 75 58
♣ US$ 270,000,000 👥 96 🏭 2,500 sq m
Managing Director: Etienne Lefort
Commercial Director: Pier Luigi Romagneli
Director of Ground Programmes: André Dupuis
Director of Naval Programmes: Luciano Borsari
Director of External Relations: Gian Paolo Mealli
Public Relations Manager: Gérard Carcy
■ A joint industrial organisation (GIE) set up in June 1989 by Alenia of Italy and Aerospatiale and Thomson-CSF of France, to develop, produce and sell a Future Surface-to-Air Family (FSAF) of missile systems to meet the air defence requirements both for naval and ground applications. [97]

74-00650 GEC Alsthom French Head Office
French Head Office
38 Avenue Kléber, F-75795 Paris Cedex 16
☎ (+33 1) 47 55 20 00 ⊠ 645414 f
⊜ (+33 1) 47 55 26 14
President and Chief Executive Officer: Pierre Bilger
Managing Director: Claude Darmon
Managing Director: Jim Cronin
External Communications Director: Jean Georges Micol
■ Power generation, power transmission and distribution, transport, industrial and marine equipment. → 75 [98]

74-00800 Groupement Turboméca SNECMA (GRTS)
2 Boulevard du Général Martial Valin, F-75015 Paris
☎ (+33 1) 40 60 83 30 ⊜ (+33 1) 40 60 83 12
♣ FFr 136,100,000 (1996)
Managing Director: Alain Amiot
■ Industrial teaming of Turbomeca and SNECMA for production of the Larzac turbojet engine. [98]

74-01200 NH Industries Sarl
(a joint venture between Agusta, Italy; Eurocopter, France and Fokker Aviation, The Netherlands)
Le Quatuor - Bâtiment C, 42 Route de Galice,
F-13082 Aix en Provence Cedex 2
☎ (+33 4) 42 95 97 00 ⊜ (+33 4) 42 95 97 49
👥 31
General Manager: Jean Pierre Barthélemy
Deputy General Manager: Diego Tassinari
Sales and Marketing Manager: Alain Gauthier
☎ (+33 4) 42 95 97 02
Deputy Commercial Manager: Luigi Martina
☎ (+33 4) 42 95 97 31
■ Prime contractor for design and development, industrialisation, production, logistic support, marketing and sale of NH90 helicopters. [98]

74-01800 Société d'Exploitation des Matériels Martin-Baker (SEMMB)
(a joint venture between SNECMA, France and Martin Baker Aircraft Ltd, UK)
20 Boulevard Louis Seguin, F-92700 Colombes
☎ (+33 1) 46 52 23 75 ⊜ (+33 1) 46 52 23 85
✧ FFr 1,200,000 ♣ FFr 100,000,000 (1996) 👥 22
🏭 330 sq m
President: G Sangis
Managing Director: André Margot ☎ (+33 1) 46 52 23 60
Sales and Marketing Director: Hervé Lesueur
☎ (+33 1) 46 52 23 68
Operations Director: Michel Perkins
Purchasing Manager: J F Bontant
■ Ejection seats fitted on all French made combat aircraft. Crashworthy seats developed for Franco-German Tiger combat and NATO NH90 helicopters. → 80 → 85
® UK: Martin-Baker Ltd [98]

74-02200 TDA
(† Thomson Brandt Armement - TBA)
(a joint venture between Thomson-CSF and Daimler Benz Aerospace)
18 Avenue du Marechal Juin,
F-92366 Meudon la Forêt Cedex
☎ (+33 1) 39 45 52 00; 39 45 52 40
⊜ (+33 1) 39 45 52 25; 39 45 52 27
♣ FFr 1,000,000 (1996) 👥 1,000 (1996)
Managing Director: Jean Monfort ☎ (+33 1) 39 45 52 61
Sales and Marketing Director: Daniel Briffaut
☎ (+33 1) 39 45 52 14
Marketing Director: Jean Claude Jouvray
☎ (+33 1) 39 45 52 40
Communication Manager: Mr Oettgen
☎ (+33 1) 39 45 52 77
Manager of Product Promotion and Marketing Analysis: Ms Sophie Moreau ☎ (+33 1) 39 45 52 47
■ Armament and munitions: mortars, bombs, anti-tank systems, rockets, smart ammunition, warheads and fuses.

Subsidiaries:
TDW → 75 Germany
Forges de Zeebruge SA → 75 France [98]

75 Manufacturing Companies

75-00150 ACH
Société Nouvelle des Ateliers et Chantiers du Havre
30 rue J J Rousseau, BP 1390, F-76066 Le Havre Cedex
☎ (+33 2) 35 19 80 00; 35 25 68 00 ⊠ 190322 f
⊜ (+33 2) 35 19 80 40; 35 25 67 67
Chairman: Gilbert Fournier
Managing Director: Alain Tessandier
Managing Director: Gilles Fournier
Sales and Design Manager: Antoine Castetz
☎ (+33 2) 35 25 67 18
Shipbuilding Manager: François Faury
■ Shipbuilding, repair and conversion of a wide range of vessels. Facilities include floating and drydock. Design and consultancy services. [98]

75-00200 Acofab SA
Parc d'Activités de la Danne, BP 47,
F-95612 Cergy Pontoise Cedex
Rue Jacques Brel - Parc d'Activités de la Danne,
F-95610 Eragny sur Oise
☎ (+33 1) 30 37 63 30 ⊜ (+33 1) 30 37 46 36
✧ FFr 250,000 (1996) 👥 12 🏭 500
Managing Director: Gilbert Lepeltier
■ EMI shielding solutions for use in military and aerospace industries. Products: conductive elastomer EMI gaskets; corrosion resistant shielding gaskets; conductive sealants and adhesives; conductive coatings (chemical and abrasion resistance) to protect enclosure flanges from corrosion; conductive foil tapes; wire mesh products; shielded ventilation panels and windows; thermal interface materials; microwave absorbers.

® UK: Chomerics Ltd
USA: Emerson & Cuming 97

75-00220 Acutronic Control Sarl
(a sister company of Acutronic AG, Switzerland)
2 Allée des Aulnes, ZA du Buisson de la Couldre,
F-78190 Trappes
☎ (+33 1) 34 82 68 68 Ⓔ acutronic@wanadoo.fr
⊜ (+33 1) 34 82 68 69
👥 3
Managing Director: R Hoffmann
■ Inertial guidance test instruments, motion simulators and servo systems. 98

75-00500 Aérazur Aeronautical Equipment Division
(a company of the Zodiac Group, France)
58 Boulevard Galliéni, F-92137 Issy les Moulineaux Cedex
☎ (+33 1) 41 23 23 23 ⊠ 631891 aerazur f
⊜ (+33 1) 46 48 74 79
♧ FFr 8,250,000 ♣ FFr 601,000,000 (1994)
President: Jean Louis Gerondeau
Director of Parachutes Division: Michael Roussel
Director of Aeronautical Equipment Division: Michel Roussel
Director of Elastomers Division: J Delibbias
International Marketing Manager: Jean Pierre Fetu
■ Aircraft arresting systems, life-saving and survival equipment, parachutes, technical storage and fluid transfer systems. Flexible tanks for aircraft, helicopter and battle tanks, collapsible storage or transport tanks, "Flexit" electrical systems, aircraft pneumatic de-icers. Inflatable tents, pollution boom, aristock, inflatable seals and airbags. 98

75-00600 Aerospatiale Head Office
37 Boulevard de Montmorency, F-75781 Paris Cedex 16
☎ (+33 1) 42 24 24 24 ✉ AEROSPATIALE Paris
⊠ 620025 f aispa x ⊜ (+33 1) 45 24 54 14; 42 24 26 19
♧ FFr 4,607,850 ♣ FFr 48,563,000,000 👥 38,373
Chairman and Chief Executive Officer: Yves Michot
Executive Vice President, Programs and Development, President Helicopter Business: Jean Louis Fache
Corporate Executive Vice President, President Aircraft Business: Claude Terrazzoni
Corporate Executive Vice President, President Space and Defence Business: Michel Delaye
President and Chief Operating Officer, Missiles Division: Pierre Dubois
Corporate Vice President International and Commercial Affairs: Denis Verret
Corporate Vice President, Strategy Business and Development: Jacques Battistella
Corporate Vice President, Finance and Economics: François Auque
Corporate Vice President, Research and Technology: Serge Catoire
Corporate Vice President, Legal Affairs and General Consul: Philippe Simon
Corporate Vice President, Human Resources: René Chabod
Vice President, Corporate Communications: Patrice Kreis
Special Advisor to the Chairman: Vincent Lanata
Scientific Adviser to the Chairman: Jacques Balazard
Press Relations: Ms Sophie Roukline
■ Manufacture of aircraft, helicopters, tactical and strategic missiles and space systems. → 55

Divisions:
Aircraft Division
Missile Division
Space and Defence Division

Subsidiaries:
Aeroassurances, Aerospatiale Batteries, Apsys, Barfield Avsco, Composites-Aquitaine, Celerg, Eurocopter, Hemet Exploration, Maroc-Aviation, Nucletudes, Socata, Sogerma-Socea-Seca, Serat, Sextant Avionique, Unilaser.

Overseas Offices:
Abu Dhabi, Ankara, Bangkok, Bonn, Brussels, Cairo, Hong Kong, Johannesburg, London, Madrid, Moscow, Ottawa, Peking, Rio de Janeiro, Riyadh, Seoul, Singapore, Tokyo, Washington.

Aircraft Division
37 Boulevard de Montmorency, F-75781 Paris Cedex 16
☎ (+33 1) 42 24 24 24 ⊠ 620059 f aispa x
⊜ (+33 1) 45 24 54 14
Corporate Executive Vice President, President of Aircraft Business: Claude Terrazzoni
■ Airbus A300/600, A310, A319, A320, A321, A330, A340, ATR 42, ATR 72, Socata light aircraft for general aviation (TB, TBM700), VLCT (Very Large Capacity Transport).
Socata military aircraft (Epsilon, Omega), FLA (Future Large Aircraft).
Aircraft maintenance. Automatic test systems, development of software, on-board electronic equipment and Electronic Library Systems Simulation.

Toulouse Plant
316 route de Bayonne, F-31060 Toulouse Cedex 3
☎ (+33 5) 61 93 55 55 ⊠ 531546 f aispa x
⊜ (+33 5) 61 49 70 55
Director: Christian Beugnet

Nantes Plant
Rue de l'Aviation, F-44340 Bouguenais
☎ (+33 2) 40 13 70 00 ⊠ 700559 f aispax x
⊜ (+33 2) 40 13 71 00
Director: Jean Noël Guillou

Saint Nazaire Plant
Boulevard des Apprentis, BP 404,
F-44608 Saint Nazaire Cedex
☎ (+33 2) 40 90 80 00 ⊠ 700909 f aispa x
⊜ (+33 2) 40 90 81 31
Director: Jacques Crusson

Méaulte Plant
PO Box 210, F-80810 Méaulte
☎ (+33 3) 22 74 34 77 ⊠ 140740 f aispa x
⊜ (+33 3) 22 75 37 11
Director: Claude Berlan

Space and Defence Division
37 Boulevard de Montmorency, F-75781 Paris Cedex 16
☎ (+33 1) 42 24 24 24 ⊠ 620059 f aispa x
⊜ (+33 1) 42 24 54 14
Corporate Executive Vice President, President Space and Defence Business: Michel Delaye
■ Industrial prime contractor for the French ballistic missile deterrent systems: SSBS and MSBS. Industrial architect for the French Air Force System for Command and Control of Aerial Operations (SCCOA). Industrial architect and stage contractor for the Ariane launch vehicles. Prime contractor for ARD, Atmospheric Reentry Demonstrator. Prime contractor for telecommunications satellites: Eutelsat, Arabsat, Türksat, Nahuel, Thaïcom 3; civil and military satellites: Meteosat and Helios; scientific: ISO, prime contractor for the interplanetary probe Huygens.

Les Mureaux Plant
BP 3002, F-78133 Les Mureaux Cedex
☎ (+33 1) 34 92 12 34 ⊠ 695988 f aispa x
⊜ (+33 1) 34 75 09 15
Director: Jean-Charles Aldeguer

Châtillon Plant
2 rue Béranger, BP 84, F-92322 Châtillon Cedex
☎ (+33 1) 47 46 21 21 ⊠ 250881 f aispa x
⊜ (+33 1) 47 46 27 77
Director: Jean Destruys

Bourges Plant
8 rue Le Brix, BP 35, F-18001 Bourges Cedex
☎ (+33 5) 48 55 50 01 ⊠ 760035 f aispa
⊜ (+33 5) 48 55 54 94
Director: Daniel Dubreuil

Aquitaine Plant
BP 11, F-33165 Saint Médard en Jalles Cedex
☎ (+33 5) 56 57 30 00 ⊠ 560715 f aispa x
⊜ (+33 5) 56 57 31 50
Director: Jean Yves Birotteau

Cannes Plant
100 Boulevard du Midi, F-064000 Cannes la Bocca
☎ (+33 4) 92 92 70 00 ⊠ 470902 f aerocan
⊜ (+33 4) 92 92 71 90
Director: Jean Zieger 98

75-00800 Agamice
4 rue Philidor, F-75020 Paris
☎ (+33 1) 43 72 10 05 ⊜ (+33 1) 43 72 38 91
♧ FFr 100,000 👥 50
Managing Director: J A Ducros
■ Electronic components; antistatic protection; finger strip; heat resistant gaskets.

Representing:
Altoflex GmbH, Germany
Altoflex Elektro, Germany 98

75-01100 AIR-LB Head Office
240 rue Gabriel Péri, PO Box 1, F-92704 Colombes Cedex
☎ (+33 1) 47 86 85 00 ⊜ (+33 1) 47 82 06 64
♧ FFr 28,000,000 (1996) 👥 300 (1996) 🏭 12,000 m2
Chairman and Chief Executive Officer: Lucien Blanchet
Managing Director: Michel Marchet
Sales and Marketing Director: Joël Chotte
Sales and Marketing Director: Fréderic Blanchet
President, Consul of Administration: L Blanchet
■ Electrical connectors and hydraulic equipment. Installation equipment.

Technical and Commercial Services
21 rue de Seine, PO Box 1, F-92704 Colombes Cedex
☎ (+33 1) 46 52 27 00 ⊜ (+33 1) 47 86 39 63
Sales Manager: Daniel Davranche ☎ (+33 1) 46 52 27 27
Research and Development Manager: Patrick Vantouroux 98

75-01200 L'Air Liquide
75 Quai d'Orsay, F-75321 Paris Cedex 07
☎ (+33 1) 40 62 55 55 ✉ AIRLIQUID ⊠ 270716 alss f
⊜ (+33 1) 40 62 54 77
♧ FFr 4,639,661,320 (1995)
♣ FFr 32,188,700,000 (1995) 👥 27,000
President and Chief Executive Officer: Alain Joly
Managing Director: Gérard Levy
Executive Vice President and Director of Astronautical Services: Michel Bourgarel
■ Production of high and low pressure oxygen, nitrogen (liquid/gas), cryogenic transport stocking distributor, oxygen containers for aircraft, miniature cryostats for IR equipment. Cryogenic turbines. 97

75-01350 Air Précision
5 Avenue Denis Papin, PO Box 36,
F-92350 Le Plessis Robinson
☎ (+33 1) 46 01 21 24 ⊠ 634093 aprec f
⊜ (+33 1) 46 31 85 25; 46 32 52 07
👥 145
President: Colin F Jones
Managing Director: Eiji Kawaishi
Finance Director: John Selman
Information Contact: Ms Veronique Ferre
■ Electrical slip rings, designed to specification or off the shelf: applications in tank turrets, surveillance or fire control radars, missile launchers, gun-sights, sonar winches, periscope mast, helicopter blades electrical de-icing or strain gauge measurement. Aircraft instruments such as anti-collision strobe lights, electronic clocks, turn-and-bank indicators, attitude indicators, side slip indicators, stand-by batteries. These products are standard fit on ATR 42-72, Airbus A310, A320, A330, A340, Canadair RJ, ATL 1/2, EMB 145, Super Puma, Dolphin, Squirrel. JAR 145 qualified repair centre for navigation and panel instruments, gyroscopic equipment and electronics, fuel pumps, metering and gauging, engine accesories. Test benches and Scorsby Tables. Quality assurance: ISO 9001, AQAP 120, FAA. → 80 98

75-01540 AD Aerospace Finishes SA
(a subsidiary of Akzo Nobel Coatings BV)
ZI Les Bas Près, BP 151, F-60761 Montataire Cedex
☎ (+33 3) 44 31 37 75 ⊜ (+33 3) 44 31 38 79
Director: H C de Marcheville
■ Primers, topcoats and all specialities for aerospace coating purposes, maskants for chemical milling, epoxies and polyurethane paint systems approved to American and European specifications. 98

75-01570 Alcatel Espace Headquarters
(part of the Alcatel Alsthom Group)
5 rue Noël Pons, F-92734 Nanterre La Défense
☎ (+33 1) 46 52 62 00 ⊜ (+33 1) 46 52 62 50
♧ FFr 180,000,000 (1996)
♣ FFr 2,800,000,000 (1996) 👥 1,600
President and Chief Executive Officer: Jean Claude Husson
Executive Vice President: Bernard Deloffre
Executive Vice President: Lionel Fleury
Senior Vice President, Business Development: Pierre de Bayser
Senior Vice President, Operations: Henri Paul Brochet
Communications and Public Relations Manager: Thierry Deloye
Military Advisor: Christian Fontaine
General Manager, TESAM: Jean Bernard Lagarde
Satellite Services Manager: Alain Roger
Strategic Plan: Pascale Sourisse
■ Design and development, production and integration of complete systems, satellites and ground segments for telecommunications, Earth observation and science for civilian and military applications.
Communication payloads for worldwide satellite manufacturers. Equipment, instruments and space sub-systems.

Contact Reference

Toulouse Industrial Site
26 Avenue J F Champollion, F-31037 Toulouse
☎(+33 5) 61 19 50 50 ⊖ (+33 5) 61 44 49 90 [98]

75-02300 Altoflex
36 rue des Maraîchers, F-75020 Paris
☎(+33 1) 43 72 81 88 ⊖ (+33 1) 43 72 38 91
FFr 600,000 50
President: J A Ducros
■ EMI-RFI shielding: conductive rubber; electronic components.
® Germany: Altoflex GmbH
Switzerland: Altoflex Elektro
Representing:
Altoflex GmbH, Germany
Altoflex Iberica [98]

75-02700 Angénieux
(a subsidiary of Thomson-CSF, France)
F-42570 Saint Héand
☎(+33 4) 77 30 42 10 330695 f
⊖ (+33 4) 77 30 48 75
FFr 250,000 (1996) ♣ FFr 102,000,000 (1995)
190 18,000 m2
President: Jean François Pernotte
Commercial Director: Pierre Duranton
Chief Executive Officer and General Manager: Denis Suvéran
■ Design and manufacture of satellite optical elements such as Spot's Beam Splitters. [97]

75-03050 Appareillages et Matériels de Servitudes SA (AMS)
12 rue de Chevilly, ZI La Cerisaie Nord, F-94260 Fresnes
☎(+33 1) 46 15 54 00 634809 f ams
⊖ (+33 1) 40 96 95 61
FFr 8,400,000 ♣ FFr 85,000,000 100
Chairman and Chief Executive Officer: Louis Bertrand
Managing Director: Claude Gremy
Technical Director: Olivier de Menibus
Export Manager: G Paillard
Purchasing Manager: Pascal Ponroy
☎(+33 1) 46 15 54 46
■ Hydraulic power packs, hydraulic servicing carts, hydraulic components test benches, engine accessories test benches, lube accessory test benches, ground equipment, tools for maintenance and operations of aircraft, special hardware for high and low pressure liquids, automated machinery lines, industrial NC systems, electronic test equipment.
Subsidiary:
AMS Inc, USA
Chatellerault Plant
Avenue Léonce Duteil, ZI du Sanital, F-86100 Chatellerault
☎(+33 2) 49 02 53 40 ⊖ (+33 2) 49 21 42 18
Director: M Valdelievre [97]

75-03320 Argos CLS
18 Avenue Edouard Belin, F-31055 Toulouse Cedex
☎(+33 5) 61 39 47 00 ⊖ (+33 5) 61 75 10 14
95
General Manager: Michel Taillade
■ Turnkey services, subsurface mooring monitoring, moored buoys monitoring. [98]

75-03700 Ateliers de Constructions du Centre (ACC)
Rue Pré la Reine, PO Box 93,
F-63016 Clermont Ferrand Cedex 1
☎(+33 4) 73 98 38 38 990835 f acc
⊖ (+33 4) 73 98 38 20
FFr 8,000,000 ♣ FFr 119,000,000 (1996)
300 (1997) 40,230 sq m
Chairman: Christian Régnier
Technical Manager: Gérard Delorme
General Secretary: Jean Marc Couderc
■ Fixed and mobile stations for tracking, telemetry for military and civil applications. Automatic antenna test facilities, shelters, turrets and elevators. [98]

75-03800 Ateliers de la Haute Garonne Ets Auriol & Cie (AHG)
ZI de Flourens, F-31130 Balma
☎(+33 5) 61 83 60 35 ⊠ AHG Toulouse
⊖ (+33 5) 61 83 95 05
FFr 715,000 (1997) ♣ FFr 35,000,000 (1996)
90 (1997) 5,000 m2
President: Jean-Marc Auriol
Sales and Export Director: Ms Françoise Montsarrat
Technical Director: Philippe Bornes
■ Solid rivets for aircraft industry (civilian and military). FYBRFAST and FYBRDRIVE (patented) rivet for composites. Aerolock - exclusive licence from Lockheed. HATTRICK dyes (exclusive licence).
Representing:
AHG Inc, USA [98]

75-04450 Auxitrol SA
(a subsidiary of Esterline Technologies)
168 Bureaux de la Colline, F-92213 Saint Cloud Cedex
☎(+33 1) 49 11 65 65 ⊠ MAXITROL 620359 masca f
⊖ (+33 1) 49 11 65 66
FFr 25,000,000 320
President: Henri Vergès
Director, Aerospace Equipment Division: Fabrice Lemarquis
■ Design and production of sensors and measurement systems for the aeronautics, space and defence industries. Thermocouples, total temperature probes for aircraft and engines, RTD's, pressure transducers and switches, optical and acoustic pyrometers and related test equipment, speed sensors, flowmeters, level gauges and detectors, clearance sensors, high temperature connectors and electrical heaters.
Subsidiaries:
Auxitrol Iberico SA
Auxitrol Co
Auxitrol Bourges
78/80 Avenue de la Prospective, BP 1051,
F-18023 Bourges Cedex
☎(+33 2) 48 23 85 85 760418 f
⊖ (+33 2) 48 70 03 95
Production: G Neuvialle [98]

75-04515 Acieries Aubert & Duval SA
41 rue de Villiers, PO Box 120,
F-92202 Neuilly sur Seine Cedex
☎(+33 1) 40 88 20 00 620072 f
⊖ (+33 1) 47 57 69 39
Managing Director: Edouard Duval
Information Contact: A Kalnin
■ Manufacture special steels and superalloys in bars, forgings for land equipment in the aerospace and nuclear industries. Melting by electric arc, vacuum induction, vacuum arc and electroflux remelting; full range of steels from low alloy to maraging steels, nickel-based, cobalt-based and titanium alloys. Product forms: billet bar, wire, sheet, closed/open die forgings, forged rings, rolled frings, static and centri-spun castings. Aerospace applications including structural parts (landing gear, flap tracks, engine supports); engine parts (discs, shafts); parts for mechanical assemblies (gears, helicopter transmissions); bars and sheets for a wide variety of uses.
Subsidiary:
Special Metals Corp, USA
Factory
F-63770 Les Ancizes Comps
☎(+33 4) 73 86 04 21 990679; 990670 adanci f [97]

75-04520 Auverland/Sovamag
(† Autoland SA)
(a division of Sovamag)
98 rue du Point du Jour, F-92100 Boulogne
☎(+33 1) 46 21 21 09 ⊖ (+33 1) 46 08 52 22
1,000 (1997) 8,000 m2
President: François Servanin
Sales and Marketing Director: B Touron
Operations Director: C Breuil
Purchasing Manager: B Guette
Export Manager: G Sevanin
■ Different versions of light cross-country vehicles. Troops transport, tank trucks, ambulances, radio liaison vehicles, workshop vans, police vehicles and transport of equipment.
Saint Germain Laval Plant
PO Box 12, F-42260 Saint Germain Laval
☎(+33 4) 77 65 54 44 310663
⊖ (+33 4) 77 65 49 18 [98]

75-04600 AVIAC Technologies
(a subsidiary of SFIM Industries, France)
BP 848, F-78208 Mantes la Jolie Cedex
15/19 rue Louise Michel, F-78200 Mantes la Jolie Cedex
☎(+33 1) 34 78 96 96 ⊖ (+33 1) 34 78 76 45
FFr 50,000,000 ♣ FFr 190,000,000 (1996) 285
Chief Executive Officer: Jacques Cuvillier
Managing Director: A Raffy
Sales and Marketing Director: F Durdux
■ Design and manufacture of mechanical and electromechanical devices and systems for civilian and military applications, mainly for aeronautics, but also for navy and land vehicles. Main products: flight controls, electromechanical actuators, high power actuators, auxiliary electromechanical equipment. [98]

75-04750 Avon Spencer Moulton
(† Spencer Moulton)
(a subsidiary of Avon Rubber plc, UK)
21 rue de la Gare, F-45330 Malesherbes
☎(+33 2) 38 32 72 29 760918
⊖ (+33 2) 38 34 73 42
♣ FFr 10,000,000 (1995/96) 80
Managing Director: Jean Jacques Arrighi
Operations Director: Patrick Bruno
Sales and Marketing Director, Defence Products: Ms Valerie Picard
Purchasing Manager: Jean Lucan
■ Truck roller and wheel rubberising: running, support, tension. Rubberising of track shoes for all types of tanks. Elastomer acoustic signature reduction material. Any extruded, moulded and hand-made elastomer parts. [98]

75-04800 Axon Cable SA
Route de Châlons en Champagne, PO Box 1,
F-51210 Montmirail
☎(+33 3) 26 81 70 00 [E] axon@axon-cable.com
[W] http:// www.axon-cable.com ⊖ (+33 3) 26 81 28 83
FFr 12,707,400 (1996) ♣ FFr 208,000,000 (1996)
400 (1996) 40,000 sq m
Managing Director: Joseph Puzo
Sales Director: Christian Roux
Technical Director: Frédéric Didier
Finance Director: Josette Ferat
Research, Development and Technical Director: Gérard Biscaras
Technical Director: Gilles Rouchaud
Technical and Engineering Director: Thierry Michaut
Purchasing Manager: Ms Sylviane Paul
Military Sales Manager: Christian Roux
Marketing Attaché: Ms Bettina François
Marketing Co-ordinator: Mrs Véronique Faivre
■ Manufacture of electric wires, cables and interconnect systems. Equipment wires with fluorinated or polyimid insulation. Bus harnesses according to MIL-STD-1553B, special bus harnesses. EMI protected harnesses; optimised shielding; shield connection with stainless steel bands. Micro miniature harnesses. Aluminium conductors. [98]

75-04900 Bachmann SA
69 Avenue Danielle Casanova, BP 15,
F-94201 Ivry sur Seine
☎(+33 1) 46 72 42 69 ⊖ (+33 1) 46 58 91 79
FFr 2,024,000 (1997) ♣ FFr 31,000,000 (1996) 50
6,000 m2
President: L M Janny
Managing Director: Alain Mansion
■ Sealed containers with monitored hygrometry, individual compartments, technical shelters, containers, NBC protective suits, inflatable structures, radomes, waterproof tents, inflatable cushions, obturators, mobile shelters for aircraft, tanks and rocket launchers. → 80
Monteceau les Mines Plant
20 rue de Digoin, F-71300 Montceau les Mines
☎(+33 3) 85 57 23 74 ⊖ (+33 3) 85 58 32 23 [98]

75-04950 Axson France
(† Hexcel France)
ZI des Béthunes, Rue de l'Equerre,
F-95310 Saint Ouen L'Aumone
☎(+33 1) 34 40 34 60 [E] axson@axson.fr
[W] http://www.axson.fr ⊖ (+33 1) 34 21 97 87
FFr 10,500,000 (1996) ♣ FFr 150,000,000 (1996)
150 (1996)
Managing Director: Lionel Puget
Sales Director: Patrick Blosse
Marketing Manager: Ms Florence Aillerie [98]

75-06900 Blanc Aéro Industries
PO Box 10, F-75560 Paris Cedex 12
15 rue Lasson, F-75012 Paris
☎(+33 1) 43 07 00 11; 44 67 85 85 220670 etsblan f
⊖ (+33 1) 43 40 88 76
FFr 10,002,500 650 18,800 sq m
Chairman: Etienne Blavet
Technical Director: Jacky Ordines
Marketing Director: Alain Teyssier

Export Sales Manager, Germany: Jean Marc Schmitt
Sales Manager, Rest of the World: J F du Cleuziou
Export Sales Manager, United Kingdom: Philip Delfour
■ Bolts and screws, engine bolts, controlled pre-load fasteners, (Hi-Lok, Hi-Lite, LGPL Lockbolt, GPL), hydraulic and pneumatic tools for fasteners installation, precision fittings and machined parts, locknuts. Aerospace bolts and screws, special nuts, solid rivets and aerospace coatings. [98]

75-06945 A. Boët SA

26 rue Paul Doumer, PO Box 704,
F-59657 Villeneuve d'Ascq Cedex
☎ (+33 3) 20 41 04 36 ⊠ 160815 boetnor f
⊖ (+33 3) 20 41 31 77
♧ FFr 7,678,000 ♣ FFr 115,646,755 ⚇ 140
Managing Director: Olivier Darrieus
International Business Manager: Daniel Dechy
Purchasing Manager: Alain Carlier
■ Sound proofing treatment, run-up silencers (fixed and mobile), hush houses, jet test cells, turnkey projects.
Subsidiaries:
Stopson Española
Stopson Italiana SpA
Boet American Co [97]

75-06960 Ets Bollé

6 rue Tacon, PO Box 139, F-01104 Oyonnax Cedex
☎ (+33 4) 74 77 21 23 ⊖ (+33 4) 74 73 64 65
Manager: Ms Patricia Bollé
Manager: Franck Bollé
Manager, Safety Division: J P Passaquay
■ Manufacture of glasses and goggles for military purposes and safety in general, computer glasses, hearing protectors. Parachutist goggles, mine clearance goggles, ballistic goggles. [98]

75-06970 Bolloré Protection

76 Boulevard des Américains, F-44300 Nantes
☎ (+33 2) 40 77 82 82 ⊠ 711671 bollex f
⊖ (+33 2) 40 72 84 56
♧ FFr 2,520,000 (1997) ♣ FFr 32,800,000 (1996)
⚇ 25 (1997) ⌂ 1,600 m2
Managing Director: Patrick Bolloré
Purchasing Manager: F Arnoux
Treillières Plant
ZI du Ragon, 14 Avenue Louis Pasteur, F-44119 Treillières
☎ (+33 2) 40 77 82 82 ⊠ 711671 bollex f
⊖ (+33 2) 40 72 84 56
■ Perimeter security and access control management. Bullet-proof mantraps and counters; turnkey security installations. [98]

75-07020 Paul Boyé

(a member of NBC Defense Group)
53 Quai de Bosc, PO Box 112, F-34202 Sète
☎ (+33 4) 67 46 87 87 ⊠ 480087
⊖ (+33 4) 67 74 22 75
♣ FFr 600,000,000 ⚇ 800
Managing Director: Jacques Boyé
Operations Director: Philippe Boyé
Sales and Marketing Director: Jean Louis Bruchig
Purchasing Manager: Ms Marinette Devillière
Export Manager: J L Bruchig
Export Assistant: G Laporte
■ Production of: NBC protective garments, tank crew overalls, aircrew overalls, service uniforms for summer and winter, fire retardant protective suits, ballistic protection suits. Combat clothing and battledress. [97]

75-07300 Hexcel Composites

(† Brochier SA/Ciba Composites)
ZI Les Chartinières, BP 27,
F-01121 Dagneux Montluel Cedex
☎ (+33 4) 72 25 26 27
⊖ (+33 4) 78 06 02 92; 72 25 27 28
♧ FFr 75,000,000 ♣ FFr 388,000,000 (1995) ⚇ 330
President: Pierre Jeune
Communications Department: M G Giacometti
■ Pre-impregnation of high-performance fibres (glass, carbon, aramid) for advanced composites.
Representing:
Hexcel Corporation, USA
Composite Materials Unit
Sales and Marketing Department
Zone Industrielle Les Chartinières, BP 27,
F-01121 Dagneux Montluel Cedex
☎ (+33 4) 72 25 26 27 ⊠ 300371 brochie f
⊖ (+33 4) 78 06 02 92

Sales and Marketing Director, South Europe: Thierry Merlot
Sales and Marketing, France: Bruno Bonnel
Sales and Marketing, France: Henri Le Blan [98]

75-07680 CAC Systèmes

Le Breuil Airport, F-41330 La Chapelle, Vendômoise
☎ (+33 2) 54 52 65 65; 54 52 66 66
[E] cacsystems@hol.fr ⊖ (+33 2) 54 52 65 75
♧ FFr 4,206,400 (1997) ♣ US$ 10,000,000 (1996)
⚇ 80 (1997) ⌂ 6,000 sq m
President: Christophe Corizzi
Technical Manager: Jean M Fournier
Export Manager: Philippe Rowell
Purchasing Manager: Joel Guinet
■ Target and UAV systems department: targets for missile and gun training (Fox TSI, FOX TS3 and Eclipse); surveillance and reconnaissance UAV systems (FOX AT1 and FOX AT2). Electronic warfare drone systems (FIX TS). Target and RPV sub-systems department: remote control, telemetry, video transmission equipment, mini-calculators and software. [98]

75-07780 Campenon Bernard SGE

(a subsidiary of SGE, France)
5 Cours Ferdinand de Lesseps,
F-92851 Rueil Malmaison Cedex
☎ (+33 1) 47 16 47 00 ⊠ 631775 f
⊖ (+33 1) 47 16 33 60
♧ FFr 6,837,400 ♣ FFr 505,750,000 ⚇ 11,960
Chairman: Antoine Zacharias
Managing Director: H Stouff
Operations Director: Bernard Loze
Operations Director: Jean Etienne Treffandier
Sales and Marketing Manager: Jean Marie Aoust
Manager of Quality and Military Affairs: Jean François de Champs
■ Turnkey operations for air, navy and army bases. Construction of permanent and semi-permanent buildings and structures. Operations support facilities and equipment. [98]

75-08800 Cerberus-Guinard SA

(a subsidiary of Cerberus AG, Switzerland)
ZI - Rue Fourny 617, BP 20, F-78531 Buc Cedex
☎ (+33 1) 30 84 66 52 ✉ CERBERUS
[E] haure106170.1673@compuserve.com
⊠ 699976 cerb f ⊖ (+33 1) 39 56 13 64
♧ FFr 22,000,000 (1996) ♣ FFr 943,774,569 (1996)
⚇ 860 (1996) ⌂ 15,300 sq m
Chairman: Patrick Jouglard ☎ (+33 1) 30 84 66 11
Aviation and Export Director: Christian Haure
☎ (+33 1) 30 84 66 48
Sales and Communication Assistant: Anne Bonnet
☎ (+33 1) 30 84 66 52
■ Aircraft smoke detection systems as fitted in Boeing AWACS, 747-400 (option) Grumman J-Star, Airbus, A300, A300-600, A310, A319, A320, A321, A330, A340, ATR42/72, Fokker 50, 100, 70, MD-11. [98]

75-08810 CERME

Société Chiarelli d'Études et Réalisations Mécaniques et Electroniques
(a subsidiary of SFIM Industries, France)
22 rue de la Fontaine Henri-IV, PO Box 29,
F-92370 Chaville
☎ (+33 1) 47 50 08 00 ⊖ (+33 1) 47 50 22 91
♣ FFr 48,000,000 (1996) ⚇ 70
Chief Executive Offficer: Pierre Pinot
Managing Director: François Romac
■ Design and manufacture of high performance electronics and micro data processing systems for civil and defence applications.
Ajaccio Office
ZI du Vazzio, F-20090 Ajaccio
☎ (+33 4) 95 22 03 18 ⊖ (+33 4) 95 20 41 09 [98]

75-08840 CGF Gallet SA

(† Gallet SA)
74 Avenue Kléber, F-75784 Paris Cedex
☎ (+33 1) 53 70 03 45 ⊖ (+33 1) 53 70 03 46
♧ FFr 20,042,400 (1996) ♣ FFr 167,000,000 (1996)
⚇ 270 (1996) ⌂ 7,000 m2
Managing Director: A Gallet
Director of Military Department: J P Gueneau
■ Manufacture of civil and military helmets for police forces and firefighting.
Châtillon Office
ZI du Sud, F-01400 Châtillon sur Chalaronne
☎ (+33 4) 74 55 01 55 ⊠ 340251 gallet
⊖ (+33 4) 74 55 24 80 [98]

75-08850 Chantiers de l'Atlantique

(a subsidiary of GEC Alsthom, France)
38 Avenue Kléber, F-75116 Paris
☎ (+33 1) 47 55 27 54 ⊠ 630042 shipyar f
⊖ (+33 1) 47 55 27 77
♧ FFr 355,000,000 (1995)
♣ FFr 4,700,000,000 (1995) ⚇ 4,200
Managing Director: Jean Noel d'Acremont
Commercial Director: J M de L'Epinois
Operations Director: P Payen
Purchasing Manager: P Person
■ Naval ships, frigates, corvettes, OPV and LSD. Support and supply ships.
Saint Nazaire Shipyard
PO Box 400, F-44608 Saint Nazaire
☎ (+33 2) 40 90 91 00 ✉ SHIPYARD Paris
⊠ 700945 shipyar stnaz ⊖ (+33 2) 40 66 23 20
Operations Manager: Guy Sallenave
Naval Ships Division: F Le Miere
Naval Ships Division: B Peltier [97]

75-08950 Chantiers OCEA SA

Quai Cabaude, F-85100 Les Sables d'Olonne
☎ (+33 2) 51 21 05 90 ⊠ 710775 pouvro f
⊖ (+33 2) 51 32 75 78
⚇ 150 ⌂ 10,000 m2
Managing Director: R Joassard
Sales and Marketing Director: Vincent Fepaud
Purchasing Manager: O Gendry [98]

75-09650 Chromalloy France

(a subsidiary of Chromalloy Gas Turbine Corp, USA)
PO Box 7120, F-95054 Cergy Cedex
Avenue des Gros Chevaux - ZI Vert Galant,
F-95310 Saint Ouen l'Aumône
☎ (+33 1) 34 40 36 36 ⊠ 609464
⊖ (+33 1) 34 21 97 37
♣ FFr 80,000,000 ⚇ 150 ⌂ 8,000 sq m
President: Jean Claude Morisson
Vice President, Mechanical Repairs: M Delmas
Vice President, Processes: A Bertoli
Manager of Quality Control: V d'Auzers
■ Erosion, corrosion, oxidation and wear resistant coatings by intermetallic diffusion or spraying: brazing; production and installation of gas turbine honeycomb or felt seals; specialised heat treatment; atmospheric and low pressure plasma coatings. Repair of gas turbine components. [97]

75-09660 Cime Bocuze

(a subsidiary of Giat Industries, France)
13 route de la Minière, F-78008 Versailles Cedex
☎ (+33 1) 39 49 80 51 ⊖ (+33 1) 39 49 80 14
♧ FFr 25,000,000 ♣ FFr 87,000,000 (1995) ⚇ 97
Chairman: Pierre André Moreau
■ Heavy metal components based on Tungsten. [97]

75-09990 CNIM

Constructions Industrielles de la Méditerranée
35 rue de Bassano, F-75008 Paris
☎ (+33 1) 44 31 11 00 ✉ VASCO PARIS
⊠ 642869 vasco f ⊖ (+33 1) 44 31 11 30
President and Director General: V Dmitrieff
Assistant Director General: F Canellas
Defence Sales Manager: Michel Tauzin
■ River crossing equipment, motorised pontoon bridge PFM Mle F1/TCP - assault treadway bridge/SFC - soldier fighting cover, Bac 20 and Bac 60 ferries.
La Seyne sur Mere Plant
ZI de Bregaillon, PO Box 208,
F-83507 La Seyne sur Mer Cedex
☎ (+33 4) 94 30 30 00 ⊠ 401235 f
⊖ (+33 4) 94 30 31 00 [98]

75-10100 Cofran Lubrification

(a subsidiary of Fuchs Petrolub)
ZI de Chef de Baie, F-17026 La Rochelle Cedex
☎ (+33 5) 46 35 23 22 ⊠ 790792
⊖ (+33 5) 46 42 53 80
♧ FFr 3,200,000 ♣ FFr 100,000,000 ⚇ 90
⌂ 6,400 sq m
Chairman and General Manager: Michel Quinzii
Export Manager: Jean Paul Delestre

Contact Reference

■ Private French company manufacturer of high technology lubricants including special military lubricants under licence of the French Ministry of Defence. [97]

75-10175 Comepa

166 rue du Landy, F-93207 Saint Dennis Cedex
☎ (+33 1) 42 43 50 50 [E] comepa@starnet.fr
⊖ (+33 1) 42 43 64 44
♣ FFr 6,000,000 (1996) ♣ FFr 9,379,300 (1996) 👥 32
🏭 1,356 m2
Managing Director: Bertrand Marsaud
Sales and Marketing Director: Alain de Beauvais
Operations Director: Jean Kuentz
Communication Manager: Mr Birebent
■ Manufacture of thermostats. [98]

75-10200 Comex Pro

36 Boulevard des Oceans, BP 143,
F-13275 Marseille Cedex 9
☎ (+33 4) 91 23 50 00 ⊠ 401755
⊖ (+33 4) 91 40 72 75
President: H G Delauze
Senior Vice President: B Debano
Technical Manager, Submarine Division: André Laurent
■ Design and construction of small to medium size submarines (under 500 tons). Design and construction of diving equipment, decompression chambers, diving simulators. Diving services, wreck recovery.
Subsidiaries:
Comex Industries
Comex Pro [98]

75-10240 Compagnie Industrielle des Lasers (CILAS)

(a subsidiary of Aerospatiale and Atomic Energy Commission (CEA), France)
Route de Nozay, PO Box 27, F-91460 Marcoussis
☎ (+33 1) 64 54 48 00 ⊠ 601862 f
⊖ (+33 1) 69 01 37 39
◈ FFr 81,073,600 (1996) ♣ FFr 183,000,000 (1996)
👥 240 (1996)
President and Managing Director: Jacques Battistella
Deputy Managing Director: Jean Pollono
Sales and Marketing Director: Jacques Bourron
■ Military laser applications: design, development and series production of rangefinders and designators-illuminators, ship-based optical helicopter landing aids.
Orleans Factory
ZI La Source, F-45060 Orleans Cedex
☎ (+33 2) 38 64 15 55 ⊠ 760496
⊖ (+33 2) 38 76 02 49 [98]

75-10250 CompAir Luchard SA

(a subsidiary of Siebe plc, UK)
ZI Vaux le Penil - 31 rue de Justice, BP 514,
F-77015 Melun Cedex
☎ (+33 1) 64 37 91 43 ⊠ 690189
⊖ (+33 1) 64 37 00 36
◈ FFr 58,884,100 (1997) ♣ FFr 155,000,000 (1996)
👥 140 (1997) 🏭 7,700 m2
Managing Director: P Robelin
Finance Director: H Job
Sales, High Pressure: Yves Carteaux
■ Manufacture of marine airforce compressors. [98]

75-10265 Composites Aquitaine

(a subsidiary of SOGERMA, France)
17 route de Lacanau, F-33160 Salaunes
☎ (+33 5) 56 68 55 01 ⊖ (+33 5) 56 58 51 93
♣ FFr 90,000,000 (1996) 👥 160 (1996)
Chairman and Chief Executive Officer:
Jean Charles Joseph
■ Research and manufacture of composite parts for the aeronautic, automobile and rail industries. [98]

75-10280 Constructeurs Associés de Matériels d'Incendie, Voirie, Aviation (CAMIVA)

(a subsidiary of Iveco, Italy)
F-73230 Saint Alban Leysse
☎ (+33 4) 79 75 66 66 ⊠ 320934
⊖ (+33 4) 79 70 54 40
◈ FFr 12,263,400 (1996) ♣ FFr 215,000,000 (1996)
👥 224 (1996) 🏭 22,000 m2
President: Saverio Gaboardi
Managing Director: Gerard Cattaud
Sales Director: Jacques Stinder
Management Attaché: R M Perrier
■ Airport fire and rescue vehicles. [98]

75-11335 Compagnie Générale de Géophysique Head Office (CGG)

1 rue Léon Migaux, F-91341 Massy
☎ (+33 1) 64 47 30 00 [E] mkt@cgg.com
⊠ 602442 cggec f ⊖ (+33 1) 64 47 39 70
◈ FFr 38,717,360 (1997) ♣ US$ 603,000,000 (1996)
👥 3,200 (1996)
Chairman and Chief Executive Officer: Yves Lesage
Director and Member of the Board: Mrs Geneviève Gomez
Director and Member of the Board: Olivier Appert
Director and Member of the Board: Robert Castaigne
Director and Member of the Board: Bernard Duval
Director and Member of the Board: Achille Ferrari
Director and Member of the Board: Daniel Fournie
Director and Member of the Board: Christian de Gournay
Director and Member of the Board: Christian Marbach
Director and Member of the Board: Jean Pierre Marchand
Director and Member of the Board: Henri Moulard
Director and Member of the Board: Gilles Rappeneau
Director and Member of the Board: Jean Vallet
Secretary of the Board: Jean Maurice Dalongeville
Executive Vice President and Adviser to the Chairman:
Jean Paul Voglimaci
Senior Executive Vice and Chief Operating Officer:
Robert Brunck
Senior Vice President: Christian Lerat
Vice President, Land Product Line: Gérard Chambovet
Deputy Vice President, Land Product Line: Didier Treussier
Vice President, Marine Product Line: Yves Meignie
Vice President, Data Processing Product Line:
Jean François Arzel
Vice President, Software Product Line: Ms Annick Laroche
Vice President, Equipment Product Line: Thierry Le Roux
Vice President, General Geophysics: Peer Norgaard
Vice President, Marketing: Patrice Canal
☎ (+33 1) 64 47 35 91
Chief Financial Officer: Michel Ponthus
General Secretary and Vice President, Human Resources and Legal Affairs: Jean François Marquaire
Vice President, Economic, Financial Studies and Audit:
Jean Maurice Dalongeville
Vice President for Europe, Africa and Middle East:
Pierre Benichou
Vice President for Latin America: Bruno Deletang
Vice President for Asia: Dominique Robert
Defence Applications Manager: Jean Paul Carle
☎ (+33 1) 64 47 38 23
Marketing Information Co-ordinator: Daniel Potevin
■ Instrumentation for all types of geo-physical measurements; industrial geology; underwater detection equipment (civil and military).
Affiliated companies:
SERCEL [98]

75-11450 Compagnie Industrielle des Reflecteurs Radar (CIRRA)

285 rue Hélène Boucher, F-78531 Buc Cedex
☎ (+33 1) 39 56 50 09 ⊖ (+33 1) 39 56 29 81
◈ FFr 1,100,000 (1997) ♣ FFr 11,000,000 (1996)
👥 25 (1997) 🏭 1,000 sq m
Managing Director: P A Guinard
General Manager: Michel Laroze
Technical Director: Mr Baroth
■ Chaff payloads, air launched and naval. Manufacture of metallic radar reflectors. [98]

75-12300 Constructions Mécaniques des Vosges Marioni (CMW)

43 Avenue du Général Leclerc,
F-88320 Rozières sur Mouzon
☎ (+33 3) 29 07 33 08 ⊖ (+33 3) 29 07 32 52
◈ FFr 279,000 (1996) ♣ FFr 22,000,000 (1996)
👥 60 (1996) 🏭 3,000 m2
Chairman: François Wildenberg
■ Machining of precision parts for aircraft, military equipment and mould, die and other heavy industrial applications. → 80 [98]

75-12400 Covergie SA

(† ETRI SA)
(a subsidiary of Dynateg, France)
9/11 rue du Moulin des Bruyéres, F-92400 Courbevoie
☎ (+33 1) 49 04 92 00 ⊠ 620361 f
⊖ (+33 1) 47 88 09 67
◈ FFr 12,000,000 (1995) ♣ FFr 230,000,000 (1995)
👥 260 🏭 5,000 sq m
Managing Director: Marc Grosset
Commercial Director: Philippe van Gansen
Marketing Director: Serge Grollier
Purchasing Manager: Jean Boivin
Export Sales Manager: Franck Rosengoltz
■ Cooling fans for land, air and seaborne electronic cooling. Fans for heat exchangers, DC/DC converters switching regulators, power supply equipment, electronic loads. Design and development services. [97]

75-12450 CORIS

Compagnie d'Études, de Réalisations et d'Installations de Systèmes
2 Avenue Descartes, F-92350 Le Plessis Robinson
☎ (+33 1) 41 36 15 00 ⊠ 634011 f
⊖ (+33 1) 46 32 15 64
◈ FFr 13,977,800 ♣ FFr 273,000,000 (1995) 👥 277
Chairman: Jean Marie Brun
Managing Director: Claude Vialatte
■ Design, manufacture, installation and maintenance of electromechanical and electronic systems. Mobile and fixed infra-structures, NBC protection equipment, EMP hardening, security transmission networks, command posts, radio communications, fibre optic systems. [97]

75-12600 CR2A-DI

(a subsidiary of Dassault Electronique SA and CGI Company, France)
19 Avenue Dubonnet, F-92411 Courbevoie Cedex
☎ (+33 1) 47 68 37 00 ⊖ (+33 1) 47 68 87 81
♣ FFr 272,000,000 (1996) 👥 612 (1997)
Chairman of the Supervisory Board: Gérard Jousset
Chairman of the Directorate Committee: Xavier Noury
Managing Director: Philippe Chabalier
Member of the Directorate Committee:
Guy Georges Legrand
Operations Director: Felix Mauguen
[E] mauguen@cr2a-di.fr
■ Information systems for military applications.
Regional Office
Le Pilon du Roy - Bâtiment C,
Rue Pierre Berthier - ZI Les Milles,
F-13854 Aix en Provence Cedex 3
☎ (+33 4) 42 24 43 23 ⊖ (+33 4) 42 39 75 83
Regional Office
Avenue Magudas - Bâtiment B3, Parc Héliopolis,
F-33700 Mérignac
☎ (+33 5) 56 34 00 23 ⊖ (+33 5) 56 34 82 92
Brest Regional Office
1 rue Jean Marie Lebris, F-29200 Brest
☎ (+33 2) 98 44 57 93 ⊖ (+33 2) 98 46 06 66
Rennes Regional Office
8 Avenue Henri Fréville, F-35200 Rennes
☎ (+33 2) 99 53 36 36 ⊖ (+33 2) 99 51 15 45
Toulouse Regional Office
32 rue des Cosmonautes, ZI du Palays, F-31400 Toulouse
☎ (+33 5) 61 34 99 77 ⊖ (+33 5) 61 54 39 41 [98]

75-12910 Cybernétix

Technopole de Chateau Gombert, PO Box 94,
F-13382 Marseille Cedex 13
☎ (+33 4) 91 21 77 00 ⊖ (+33 4) 91 21 77 01
◈ FFr 10,000,000 ♣ FFr 160,000,000 (1995) 👥 230
🏭 3,000 sq m
President: Charles Palumbo
Sales Engineer: Gerard Pedersen
Public Relations: Ms Cécile Diaz
■ Manufacture of robotic and automatic systems, remote controlled security vehicles for bomb disposal. [97]

75-12950

Daimler-Benz Aerospace France SAS

(† AEG France SA)
(a subsidiary of Daimler-Benz Aerospace AG, Germany)

Daimler-Benz Aerospace

48 Avenue Victor Hugo, F-75116 Paris
☎ (+33 1) 45 00 24 04 ⊖ (+33 1) 45 00 25 34

♧ FFr 250,000 (1996) 👥 20 (1996)
Managing Director: Gert Runde ☎ (+33 1) 45 00 30 49
Sales Director: P Loewert ☎ (+33 1) 45 00 24 04
■ Activities: reconnaissance and surveillance systems; command, control and communications systems; electronic warfare systems; radars; command, control and guidance systems; aviation and special systems; space and new technologies; optronic equipment; broadcasting and TV transmitters. → 85
® Germany: Daimler-Benz Aerospace AG [98]

75-12970 **Dassault Automatismes et Télécommunications**
(a subsidiary of Dassault Electronique SA)
9 rue Elsa Triolet - ZI les Gâtines, BP 13,
F-78373 Plaisir Cedex
☎ (+33 1) 30 81 20 00 ⊖ (+33 1) 30 55 19 31
♣ FFr 1,415,000,000 (1996) 👥 828 (1996)
Chairman and Chief Executive Officer: Jacques Mijonnet
Vice Chairman and Managing Director: Gilbert Weill
Vice Chairman and Managing Director: Jean Paul Vautrey
Deputy Managing Director: Serge Goldstein Desroches
Deputy Managing Director: Daniel Pelletier
Company Secretary and General Counsel: François Pontet
■ Business equipment for professionals and consumers in: Electronic Transaction Automation, telecommunications and services. [98]

75-12975 **Dassault Aviation Group**
(a partner of Dassault Industries, France)
9 Rond Point des Champs-Elysées, Marcel Dassault,
F-75008 Paris
☎ (+33 1) 47 95 85 85 [W] http://www.dassault-aviation.fr
⊖ (+33 1) 47 41 67 89
♧ FFr 506,294,850 (1996)
♣ FFr 13,000,000,000 (1996) 👥 12,036 (1996)
President and Chief Executive Officer: Serge Dassault
Executive Vice President, Economic and Financial Affairs: Charles Edelstenne
Executive Vice President, Operations: Michel Herchin
Executive Vice President, Engineering, Research, Development and Co-operation: Bruno Revellin-Falcoz
Senior Executive Vice President, International: Pierre Chouzenoux
Director of GAL Industrial and Production Operations: Christian Decaix
Communications Director: Jean Pierre Robillard ☎ (+33 1) 47 95 86 90
Human Resources Director: Pierre Chasseguet
Chief Test Pilot: Yves Kerherve
■ Aircraft, missiles, electronic and electrical equipment, flight control systems, maintenance and repair of aeronautical material.
Subsidiaries:
Dassault International (100%) → 90
Dassault Systemes (46%) → 75
Sogitec (69%) → 75
Coger (74%)
Dassault Falcon Jet Corp (100%)
Saint Cloud Factory
78 Quai Marcel Dassault, F-92214 Saint Cloud
☎ (+33 1) 47 11 40 00 ⊖ (+33 1) 47 11 56 60
Director: Jean Marie Resnard
Technical General Director: Yves Thiriet
International Technical Director: Thierry Prunier
Dassault Equipment Division (DED)
78 Quai Marcel Dassault, F-92214 Saint Cloud
☎ (+33 1) 47 11 35 39 ⊖ (+33 1) 47 11 33 37
Director: Francis Bachelerie
Istres Factory
PO Box 28, F-13801 Istres Cedex
☎ (+33 4) 42 56 77 77 ⊖ (+33 4) 42 56 77 75
Managing Director: Christian Sasso
Vélizy Factory
PO Box 12, F-78141 Vélizy Villacoublay
☎ (+33 1) 40 83 22 12 ⊖ (+33 1) 40 83 99 00
Director: Yves Petit
Purchasing Director: Jacques Lefeuvre
Argenteuil Factory
1 Avenue du Parc, F-95100 Argenteuil Cedex
☎ (+33 1) 34 11 85 85 ⊖ (+33 1) 34 11 83 91
Director: Jean-Marc Grillet
Biarritz Factory
Route d'Hardoy, PO Box 208, F-64205 Biarritz Cedex
☎ (+33 5) 59 31 22 22 ⊖ (+33 5) 59 31 24 10
Director: Guy Piras
Mérignac Factory
Aérodrome de Mérignac, PO Box 24,
F-33701 Mérignac Cedex
☎ (+33 5) 56 55 22 00 ⊖ (+33 5) 56 47 83 14
Managing Director, Bordeaux Group: Philippe Bertin
Martignas Factory
PO Box 38, F-33127 Martignas sur Jalles
☎ (+33 5) 56 68 85 00 ⊖ (+33 5) 56 78 05 50
Director: Jean-Michel Estrade
Seclin Factory
Zone Industrielle, PO Box 289, F-59472 Seclin Cedex
☎ (+33 3) 20 90 93 00 ⊖ (+33 3) 20 90 24 39
Director: Jean Hamandjian
Administration Director: Jean-Louis Chevert
■ Mechanical parts, structural parts and robotics.
Argonay Factory
PO Box 32, F-74370 Pringy
☎ (+33 4) 50 09 10 00 ⊖ (+33 4) 50 09 10 64
Director: Jean-François Vullierme
Cazaux Factory
Base Aérienne 120, BP 24 Cazaux, F-33260 La Teste
☎ (+33 5) 56 22 44 00 ⊖ (+33 5) 56 22 45 22
Managing Director: Roger Lecointre ☎ (+33) 56 22 44 01
Poitiers Factory
24 Avenue Marcel Dassault, ZI de Larnay, F-86580 Biard
☎ (+33 5) 49 58 26 94 ⊖ (+33 5) 49 37 13 83
Managing Director: Jean Lafond ☎ (+33 5) 49 58 02 02 [98]

75-12980
Dassault Electronique SA (Group)
(a subsidiary of Dassault Industries, France)

DASSAULT ELECTRONIQUE

55 Quai Marcel Dassault, BP 301,
F-92214 Saint Cloud Cedex
☎ (+33 1) 49 11 80 00 [TX] 633299 f
⊖ (+33 1) 46 02 57 58
♧ FFr 376,721,520 (1996)
♣ FFr 4,559,000,000 (1996) 👥 4,091 (1996)
Chairman and Chief Executive Officer: Bertrand Daugny
Vice Chairman and Managing Director: Jacques Mijonnet
Vice Chairman and Managing Director: Jean-Etienne Connet
Managing Director: Jean-Louis Michel
Managing Director: Albert Zylberfajn
Company Secretary and General Counsel: François Pontet
■ Production of state-of-the-art electronic systems, software and information systems for military, space and civilian applications. Electromagnetic seekers for equipping missiles: anti-ship, anti-aircraft, anti-tank, anti-radar. Detection: airborne radars (aircraft, helicopters and drones), ground-based radars (ground-to-air search radars, meteorological radars), laser radars, data links. Ground stations for mission preparation and exploitation.
Information Technology and systems: digital data networks (bus), C3I, simulation. On-board computer (aircraft, missiles, battle tanks), data recording systems.
Electronic Warfare: Airborne integrated self protection systems. Surveillance systems for armies. Naval jammers. Tracking systems.
Measurement and test systems for avionics: Satellite Communication Antenna System (SATCOM). Ground Collision Avoidance (GCAS).
Static Memory Data Recorder (SFDR), extended storage Quick Access Recorder (QAR-DAR), noise reduction, airport equipment.
Space: equipment for satellite and data collection - equipment for satellite telecommunications, mass-memories. Microelectronics and microwave applications.
Navigation positioning, GPS equipment, satellite telecommunications, wideband networks, wireless telecommunications. Transportation management for air, sea, rail and vehicles.
Subsidiaries:
Dassault Automatismes et Télécommunications → 75
D.E.3I Informatique → 80
Sorep → 75
CR2A-DI → 75
Dassault Electronique Development → 75
Dassault Sercel Navigation Positionnement → 75 [98]

75-12985 **Dassault Sercel Navigation Positionnement (DSNP)**
(a subsidiary of Dassault Electronique, France)
16 rue de Bel Air, BP 433, F-44474 Carquefou Cedex
☎ (+33 2) 40 30 11 81; 40 30 59 00
[E] jeanpierre.barboux@sercel.fr [TX] 710695 f
⊖ (+33 2) 40 30 58 92
♧ FFr 70,000,000 ♣ FFr 67,000,000 (1996)
👥 62 (1996)
Chairman of the Directorate Committee: Henri Gaillard
Chairman of the Supervisory Board: Jacques Mijonnet
Vice Chairman of the Supervisory Board: Christian Schuller
Sales Manager: Jean Pierre Barboux
Marketing Manager: Francis Lepalud
■ Design and manufacture of navigation positioning equipment. [98]

75-12990 **Dassault Systèmes**
(a subsidiary of Dassault Aviation)
9 Quai Marcel Dassault, F-92156 Suresnes
☎ (+33 1) 40 99 40 99 ⊖ (+33 1) 40 99 43 76
♧ FFr 513,800 (1996) ♣ FFr 1,394,000 (1996)
👥 1,194 (1996)
President: Charles Edelstenne
Marketing and Communication: M Vesco
■ Computer aided three dimensional interactive applications systems (CATIA). [98]

75-13000 **Davey Bickford S.N.C.**
37 rue Saint Maur, BP 574, F-76006 Rouen Cedex
☎ (+33 2) 35 70 94 84 ⊖ (+33 2) 35 88 42 30
♧ FFr 4,000,000 (1996) ♣ FFr 310,000,000 (1996)
👥 750 (1996) ⌂ 750,000 sq m
Managing Director: Jean Jacques Pinel
Sales and Marketing Director: Alain Duchesne
Technical Director: Michel Guibert
Area Marketing Manager: J C Laforge
■ Research, design and development of initiation compounds and ignition powders, delay and short delay firing compounds, primary explosives, initiators with short operation times, pyrotechnic devices without primary explosives, programmable components. Product applications in: mines, quarries, civil engineering, seismic prospecting: detonators (pyrotechnic, electric and non-electric), detonating cords. Fireworks: electrical igniters, firing line. Aeronautical industry: detonators for firefighting systems, impluse cartridges for jettison of aircraft exterior loads, emergency escape systems (aircraft and helicopter). Weapons systems: delayed initiation systems for grenades, anti-tank mine primer, missile propulsion initiation, thermal battery initiation, initiation devices for infra-red (flare) and electromagnetic passive defence munitions, pyrotechnic firing circuits for artillery shells, mortar shells and rockets.
Hery Factory
F-89550 Hery
☎ (+33 3) 86 47 80 70 ⊖ (+33 3) 86 47 84 33
👥 700 ⌂ 20,000 sq m [98]

75-13160 **Defontaine**
(a subsidiary of Krupp Hoesch Group)
3 rue Louis Renault, PO Box 329,
F-44803 Saint Herblain Cedex
☎ (+33 2) 40 67 89 89 [W] http://www.defontaine.com
[TX] 710698 f ⊖ (+33 2) 40 67 89 00
♧ FFr 30,000,000 ♣ FFr 485,000,000 👥 600
⌂ 34,000 sq m
President: Gilles Ferrvillet
Sales and Marketing Director: Luis Moreno
[E] luis.moreno@defontaine.com
Purchasing Director: D Labovere
Sales Manager: Mr Sancereau
■ Flash-butt welded rings and flanges for jet engines produced in alloy steels, stainless steels, corrosion and heat resistant alloys and titanium alloys. [98]

75-13680 **Dixi-Fresard Microtechniques SA**
(† Fresard Microtechniques SA)
(a subsidiary of Dixi Ltd, Switzerland)
4 Chemin de Palente, BP 889, F-25025 Besançon Cedex
☎ (+33 3) 81 88 02 43; 81 53 24 55
⊖ (+33 3) 81 53 37 95
♧ FFr 7,250,000 (1995) ♣ FFr 25,000,000 👥 38
⌂ 2,000 sq m
Managing Director: Mr Tholomier
Commercial Director: Claude Jacquin
Purchasing Manager: Mr Marguet
■ Study, development and manufacture of: electromagnetic display assemblies, failure indicator flags, tachometric motors and specific micromotors, reducers, altimetric counters, miniature time totalisers, fuses, dot-matrix impact printer heads, impact printers used in colour recorders and thermal printers. [97]

75-13800 Dunlop France
Head Office
13 rue Camille Desmoulins,
F-92441 Issy les Moulineaux Cedex
☎(+33 1) 40 93 12 34 ⊜ (+33 1) 40 93 13 31
♣ FFr 4,817,000,000 (1996) 5,698 (1996)
Aviation Director: J Duruis
Press Officer: Ms Anne Marie Moreau
■ Precision rubber seals for aviation and other applications.

Commerce and Sales Department
ZI Amiens Factory, F-80015 Amiens Cedex
☎(+33 3) 22 52 60 00; 22 52 60 24 ⊠ 150374
⊜ (+33 3) 22 52 60 01
Manager: Mr de Ponnat
Sales Manager: C Galempoix [98]

75-14050 ECA SA
(a subsidiary of STN Atlas Elektronik, Germany)
76 Boulevard de la République,
F-92772 Boulogne Billancourt Cedex
☎(+33 1) 46 10 90 60 ⊜ (+33 1) 46 10 90 70
♧ FFr 7,549,200 (1995) ♣ FFr 107,000,000 (1995)
136 (1996) 500 sq m
Managing Director: Jean Pierre Gorgé
Sales and Export Director: Dominique Pageaud
Communications Manager: Ms Wendy Maroger
Purchasing Manager: Alain Maillard
■ Subsea robotics: remotely operated vehicles PAP 104 (mine identification and destruction system), SAR, Epaulard. Industry automation. Subsea remote control systems for offshore production.

Subsidiaries:
CSIP Ltd, UK

Design Office
ZI Toulon Est - Rue des Fères, Lumière - BP 242,
F-83078 Toulon Cedex 09
☎(+33 4) 94 08 90 00 ⊠ 430063
⊜ (+33 4) 94 08 90 70
2,600 sq m
Development Manager: Christian Vigne [98]

75-14100 ECE
(a subsidiary of Intertechnique SA, France)
157 rue Pelleport, PO Box, F-75960 Paris
☎(+33 1) 47 97 55 36 ⊠ 220496
⊜ (+33 1) 47 97 50 78
♣ FFr 390,000,000 (1996) 921 (1996)
Chairman and Chief Executive Officer: Edmond Marchegay
Sales and Marketing Director: Jean Billioque
Executive Vice President: Jean Pierre Brillant
Marketing Manager: Alain Guerrere
Purchasing Manager: Jean Pierre Dutron

Niort Office
Route de Paris, PO Box 147, F-79005 Niort Cedex
☎(+33 5) 49 33 10 40 ⊠ 790602
⊜ (+33 5) 49 33 24 16 [98]

75-14250 Pel-Job Ecomat Industrie SA
Études et Constructions de Matériels Techniques
(† Ecomat Industrie SA)
(a subsdiary of Pel-Job Groupe)
Avenue Charles de Gaulle, PO Box 119,
F-01303 Belley Cedex
☎(+33 4) 79 81 15 09 ⊠ 980366
⊜ (+33 4) 79 81 39 63
♧ FFr 32,000,000 ♣ FFr 450,000,000 300
Managing Director: P Lavalette
Marketing Director: Pascal Frances
Operations Director: Jacques Fournier
Purchasing Manager: Patrick Debroux
■ Hydraulic excavators, compact crawler excavators, mining equipment. [97]

75-14280 Edel Aviation
Herpé SA
Blvd de Beaubourg, Croissy Beaubourg, PO Box 55,
F-77312 Marne la Vallée Cedex 2
☎(+33 1) 60 06 56 66 ⊠ 694881 frp color
⊜ (+33 1) 60 17 21 18
♧ FFr 1,500,400 (1994) ♣ FFr 130,000,000 100
5,000 sq m
Director: G Leroux
Executive Manager: Christian Fontaine
■ Aerosol paints: nitro-synthetic; acrylic polyurethane; supple polyester polyurethane lacquers; primers and base paints; painting systems for all metallic and composite materials. [97]

75-14330 EG&G Sealol Sarl
Head Office and Factory
(a subsidiary of EG&G Sealol Inc,USA)
18 Ter rue des Osiers, Coignières, BP 54,
F-78311 Maurepas Cedex
☎(+33 1) 30 69 30 00 ⊠ 698581
⊜ (+33 1) 34 61 17 56
Manager: Gerald Posso
■ Mechanical seals for tank track rollers, for helicopter turbines and APU for torpedoes and for all rotating equipment such as propeller shafts on ships, pumps, compressors, turbines, jet engines. [98]

75-14400 ELECMA
(The Electronics Division of SNECMA)
Rond point René Ravaud, PO Box 42,
F-77552 Moissy Cramayel Cedex
☎(+33 1) 60 59 71 23 ⊠ 600700 snecma
⊜ (+33 1) 60 59 84 44
Managing Director: Alain Lepoutre
Deputy Director: Bernard Dorier
Commercial Manager: Jean Louis Sauvetre
Communications Manager: Thierry Fagard
Purchasing Manager: Pascal Chadail
External Relations: Ms Christine Howyan
☎(+33 1) 60 59 92 30
■ Aeronautical equipment for engines, aircraft and helicopters, acquisition and data processing equipment, airborne equipment for missiles and launchers. [98]

75-14700 Elektron SA
Site Agroparc, BP 1212, F-84911 Avignon Cedex
☎(+33 4) 90 22 22 22 ⊠ 409717
⊜ (+33 4) 90 22 71 00
♧ FFr 500,000 ♣ FFr 11,000,000 10 500 sq m
President: Jean Jacques Moreau
Operations Director: Yves Caudron
Sales and Marketing Director: Fabrice Latour
Advertising Manager and Executive Secretary:
Ms Margaret L Taig
■ Manufacture of line monitoring systems (fax, telephone), surveillance and communications equipment, theft prevention systems. Distributors of communications security systems, search instruments, EOD equipment, mobile intruder detection systems. Covert video equipment and night vision cameras. [97]

75-14800 ELICS
112 rue Brancion, F-75015 Paris
☎(+33 1) 44 19 65 80 ⊜ (+33 1) 44 19 65 89
♧ FFr 424,800 (1996) ♣ FFr 6,820,343 (1995) 6
150 sq m
Managing Dir: Mr Girault
■ Manufacturer of Delph acquisition and processing systems for sonar data image enhancement and manipulation. [97]

75-16300 Ets Robert Creuzet
Route de Beyssac, F-47200 Marmande
☎(+33 3) 53 20 45 00 ⊜ (+33 5) 53 20 45 20
♧ FFr 32,000,000 (1997) ♣ FFr 148,936,000 (1996)
435 (1997) 24,000 sq m
President: J Barillot-Creuzet
Production Director: Hervé Ortolan
Engineering and Development Director:
Bernard Costeplane
Sales and Marketing Director: Jacques Labat
Purchasing Manager: Didier Creuzet
Public Relations: Ms Graziani
■ Precision engineering, extrusions, thin tubings and hinges, electro-erosion and electro-chemical machinings, precision aluminium forgings, engine compressor blades for the armament and aerospace industries. [98]

75-16500 Euromeca
ZAC de Valentin, BP 3015, F-25045 Besancon Cedex
☎(+33 3) 81 53 32 66 ⊜ (+33 3) 81 88 32 06
♧ FFr 4,000,000 ♣ FFr 38,000,000 (1996) 75 (1996)
5,400 m2
Managing Director: Georges Afonso
Sales Director: Jean Pierre Marguier
Marketing Director: Alain Dehais
■ Design and manufacture of mechanical and hydraulic assemblies, injectors and injection manifolds. Manufacture of landing gears. Standard range of miniature hydraulic valves. Quality standard to: ISO 9001, Veritas, Siar, Raq 2. [98]

75-17500 Faure Herman SA
(a subsidiary of Intertechnique SA, France)
5 Avenue des Andes, BP 126 Les Ulis,
F-91944 Courtaboeuf Cedex
☎(+33 1) 69 82 77 00 ⊠ 603573 f
⊜ (+33 1) 69 86 18 40
Chairman and Chief Executive Officer: Jean-Pierre Salaün (PDG)
Managing Director: Edouard Marienbach
Technical Director: Didier Pabois
Commercial Director: Claude Michel
Aeronautical Sales Manager: Patrick Maillefert
■ Volumetric and mass flowmeter for aerospace and industrial applications. Test benches. Industrial flowmeters for airports and counting systems with printers for services and refuelling tankers. Densitometers and viscometers for liquids and gases. Electromagnetic flowmeters. [98]

75-17550 Ferma
(a subsidiary of Intertechnique SA, France)
60 rue Étienne Dotet, F-92245 Malakoff Cedex
☎(+33 1) 46 12 58 85 ⊜ (+33 1) 47 35 07 88 [98]

75-18300 Forster-France Sarl
Villa Amélie, 17 rue Lafonge, F-94250 Gentilly
☎(+33 1) 45 46 26 10 ⊜ (+33 1) 45 46 22 60
Managing Director: P H Mélou
■ Mine and bomb detectors. Non-destructive test equipment.

Representing::
Institut Dr Förster, Germany [97]

75-18405 Framatome
Tour Framatome, F-92084 Paris La Défense Cedex
☎(+33 1) 47 96 14 14 ⊠ 630635 frama f
⊜ (+33 1) 47 96 30 31
♣ BFr 17,900,000,000 (1995) 19,097 (1995)
President and Director General: Dominique Vignon
Director of Nuclear Production: Christian Cante
Director, Legal: Pierre Charreton
Deputy Director: Pierre Dollon
Director of Nuclear Services: Daniele Hertzog
Deputy Director General, Nuclear Combustion:
Jean Paul Lannegrace
Director of Financial Affairs: Gilbert Lehmann
Deputy Director General of International Affairs:
Jean Daniel Levi
Director of Social Relations and Communication:
François Nogue
Director of Information: Hervé Nora
Secretary General: Philippe Raulin
Director of Industrial Equipment: Jean François Terrien
Technical and Quality Director: Alain Vallee
Press Officer: Mr Chergui
■ Nuclear power systems, connectors, mechanical engineering and computer-based systems. Defence business: navy steam turbines, navy actuators, ground mobile robots, high-tech sub-assemblies for ship and ground vehicles.

Framatome Connectors International (FCI)
Tour Framatome, F-92084 Paris La Défense Cedex
☎(+33 1) 47 96 39 72 ⊜ (+33 1) 47 96 54 44
♧ FFr 2,385,000
Chairman: Dominique Vignon
■ Electrical, electronic and optical connectors and interconnector systems for telecommunications, automotive, electrical, aerospace, harsh and severe environments, industrial control and instrumentation, data and consumer applications. [98]

75-18650 Furon
Avenue du Bac 45, F-94210 Saint Maure La Varenne
☎(+33 1) 43 97 06 03 ⊜ (+33 1) 42 83 84 16 [97]

75-19040 GEC ALSTHOM Large Steam Turbines
141 rue Rateau, F-93126 La Courneuve
☎(+33 1) 49 92 45 67 ⊠ 235900
⊜ (+33 1) 49 92 47 04 [98]

75-19420 Gerand SA
Société Nouvelle
(† Ets Raymond Gérand)
22 rue Rottembourg, F-75012 Paris
☎(+33 1) 44 74 82 00 ⊜ (+33 1) 44 74 82 03

♧ FFr 250,000 🏭 1,500 m2
President: Mr Alloncle [97]

75-19500 EROS GIE
(a joint venture between Intertechnique SA and SFIM Industries, France)
61 rue Pierre Curie, PO Box 73, F-78373 Plaisir Cedex
☎ (+33 1) 30 54 82 00 ⊠ 696242
⊖ (+33 1) 30 55 71 61
President: Jean Pierre Brillant
Sales and Marketing Director: Denis Taoeb
☎ (+33 1) 30 54 83 52
Technical Manager: François Keller
☎ (+33 1) 30 54 84 94
■ Oxygen systems for all types of civil and military aircraft up to supersonic speed, helicopters and general aviation.
Owned by Intertechnique (75%) and SFIM Industries (25%). [98]

75-19700 GDI Simulation
(† Giravions Dorand Industries)
(a subsidiary of Aerospatiale Group, France)
5 rue Jean-Macé, PO Box 30, F-92151 Suresnes
☎ (+33 1) 46 97 26 00 ⊠ 612961 versur f
⊖ (+33 1) 40 99 06 21
♧ FFr 9,000,000 (1997) ♣ FFr 80,000,000 (1996)
👥 70 (1997) 🏭 2,000 m2
Chairman: M Berger
Managing Director: Jean Michel Chiquier
☎ (+33 1) 46 97 26 01
Operations Director: Pierre Sieli ☎ (+33 1) 46 97 27 50
Sales and Marketing Director: Michel Fichoux
☎ (+33 1) 46 97 26 60
■ Procedure Simulators:
Fire control systems for armoured vehicles. Night anti-tank missile firing.
Firing Instruction Technical Simulators:
Anti-tank missiles, heliborne missiles, surface to air artillery.
Tactical Combat Training Simulators:
Helicopters, Anti-tank weapon systems, tanks, rockets, light weapons, zone effect weapons. [98]

75-19800 Goodyear Aviation Products
(† Airtreads)
(a division of Goodyear Tyre & Rubber Company, USA)
101 Avenue de la Châtaigneraie, PO Box 310,
F-92506 Rueil Malmaison Cedex
☎ (+33 1) 47 16 24 01 ⊠ 634205 f
⊖ (+33 1) 47 16 23 08
♧ FFr 379,134,000
President and Chief Executive Officer: P Palusci
Manager, Aviation Products: J M Champagne
■ Manufacture of tyres and retreads. [98]

75-20900 Hellermann France SA
Miniparc du Manet - 2 rue des Hetres, BP 130,
F-78190 Trappes Cedex
☎ (+33 1) 30 13 80 00 ⊠ 697207 f
⊖ (+33 1) 30 13 80 60
👥 80
Managing Director: J P Biget
■ Marketing company with limited manufacturing facilities for the Hellermann range of cable care and protection products manufactured in the UK and Germany. [98]

75-21300 Hispano Suiza
(a subsidiary of SNECMA, France)
18 Boulevard Louis Seguin, F-92707 Colombes Cedex
☎ (+33 1) 41 30 50 10 ⊖ (+33 1) 41 30 50 20
♧ FFr 182,800,000 ♣ FFr 1,552,00,000 (1995)
👥 2,100
Chairman and Chief Executive Officer: Raymond Poggi
Senior Vice President, Industrial Development and Services: Gérard Le Page
Vice President, Finance: Alex Fain
Vice President, Industrial Products Division: Patrice Sabahi
Vice President, Aerostructures Division: Christian Knapp
Vice President, Aeronautical Equipment Division: Emeric d'Arcimoles
Vice President, Human Resources: Bernard Dinet
Strategic Development: Jean Pierre Guyot
Public Relations Manager: Ms Anne Lacourlie
Marketing Department: Olivier Fagard
■ Production of power transmission systems for aircraft engines. Ejection seats (under Martin Baker's licence).

Le Havre Plant
Rue du Pont VIII, PO Box 91 - Gonfreville l'Orcher,
F-76700 Harfleur
☎ (+33 2) 35 55 47 00 ⊠ 190704 f oravia gonfo
⊖ (+33 2) 35 53 35 06
Vice President, Aerostructures and Plant Manager: Christian Knapp
Engineering Department: Pierre Glowacki
■ Nacelles and thrust reversers. Specialised shops for sheetmetal work and composite materials.

Bois Colombes Plant
Rue du Capitaine Guynemer, BP 60,
F-92270 Bois Colombes
☎ (+33 1) 47 60 51 51 ⊠ 620131 f hispa
⊖ (+33 1) 47 81 31 74
Vice President, Aeronautical Equipment Division: Emeric d'Arcimoles
Vice President, Industrial Products Division: Patrice Sabahi
Sales, Programs and Aeronautical Equipment: Olivier Lambert
■ Fuel control units and pumps; turbochargers for high speed diesel engines. [97]

75-21700 Howden Sirocco
(a subsidiary of Howden Group plc, UK)
47 rue Fourier, PO Box 327, F-59020 Lille Cedex
☎ (+33 3) 20 08 13 00 ⊠ 820257
⊖ (+33 3) 20 92 90 76
♧ FFr 7,000,000 (1997) ♣ FFr 100,000,000 (1996)
👥 130 (1997) 🏭 20,000 sq m
President: Jacques Berry
Information Contact: Mrs D Noel
■ Specialists in design and manufacture of fans, blowers, compressors for special applications (military, industry, mines); ammunition handling systems; environmental control systems. [98]

75-21750 Howmet SA Normandy Casting
(† Le Magnésium Industriel)
ZAC des Grands Près, F-14160 Dives sur Mer
☎ (+33 2) 31 28 29 30 ⊖ (+33 2) 31 24 58 66
■ Castings for aerospace, defence and commercial applications; magnesium sand and die castings; investment castings in magnesium, aluminium and copper; air transport equipment.

Tour d'Asnieres
(† Microfusion SA)
4 Avenue Laurent Cely, F-92606 Asnieres
☎ (+33 1) 40 80 25 01 ⊖ (+33 1) 40 80 25 99
Chairman and Chief Executive Officer: Jean Castillon

Gennevilliers Casting
68-78 rue du Moulin de Cage, F-92230 Gennevilliers
☎ (+33 1) 40 85 36 00 ⊖ (+33 1) 47 99 89 73

Le Creusot Casting
26 rue de Pologne, F-71201 Le Creusot
☎ (+33 1) 85 77 66 00 ⊖ (+33 1) 85 56 36 97 [98]

75-21900 Hutchinson s.n.c.
Avenue des Champs Elysées 124,
F-75360 Paris Cedex 08
☎ (+33 1) 40 74 83 00 ⊠ 280002
⊖ (+33 1) 43 59 97 11
♧ FFr 2,500,000 👥 3,024
President: J P Seeuws
Managing Director: P C Clout
■ Runflat devices for wheeled armoured vehicles. Road wheels, pads, bushings for tracked armoured vehicles.

Defence and Security Division
59 rue Marius Aufan, PO Box 75, F-92303 Levallois Cedex
☎ (+33 1) 40 89 54 42 ⊠ 611162
⊖ (+33 1) 47 57 37 36
Head of Division: Alain Olle Laprune [97]

75-22300 Industria SA
(a subsidiary of Micro Mecanique, France)
28 Avenue Clara, F-94420 Le Plessis Trévise
☎ (+33 1) 45 76 53 78 ⊠ 264849 f
⊖ (+33 1) 45 76 47 44
♧ FFr 1,181,375 (1996) ♣ FFr 41,000,000 (1996)
👥 80 (1996) 🏭 3,600 m2
Managing Director: J M Virepinte
Sales and Marketing Director: Alain Dehais
Administration Manager: D Michel
Purchasing Manager: G Graffin
■ Solenoid valves, pressure switches, valves, micromotors, electric linear actuators, self-sealing couplings, thermocouples for aircraft and space. [98]

75-22320 IN-LHC
(a subsidiary of Intertechnique SA, France)
Route de Jallans, F-28200 Châteaudun
☎ (+33 2) 37 97 60 30 ⊠ 760333 hydrcha f
⊖ (+33 2) 37 97 60 35
Sales Manager: Jean François Bonnin
■ Servo valves and servo actuators, electro valves and hydraulic winches, pumps and electro pumps. [98]

75-23000 Intertechnique SA Head Office
61 rue Pierre Curie, PO Box 1, F-78373 Plaisir Cedex
☎ (+33 1) 30 54 82 00 ⊠ 696242
⊖ (+33 1) 30 55 71 61
♧ FFr 110,271,200 (1995)
♣ FFr 1,484,000,000 (1995) 👥 2,510 (1995)
President and Chief Executive Officer: Edmond Marchegay
Deputy Chief Executive Officer and Vice President Aeronautics Branch: Jean-Pierre Brillant
Chairman's Adviser: Jacques Maillet
Chairman's Adviser: Daniel Blondé
Vice President: Eric Le Mer
Vice President Finance and Deputy Chief Executive Officer: Jean Marie Gosse
Vice President, Marketing and Sales: Jean-Pierre Morel
Vice President, Fuel Circulation Department: Jean Michel Mathey
Vice President, Systems Monitoring and Management Department: Dominique Schmauch
Vice President, Oxygen and Life Support Department: François Keller
Vice President, Optronics and Image Processing Department: Gérard Lemoing
Public Affairs Manager: Gérard David
Public Affairs Department: Rémi Stelescot
■ Fuel Circulation Systems:
Electric motor driven pumps (AC/DC), ejector pumps, motor actuated valves, check valves, refuelling couplings and valves, magnetic level indicators, piping couplings, control valves, pressure relief valves, drain valves, pumps and nozzles on fixed or mobile installations for field refuelling of aircraft, armoured vehicles, ships and special vehicles. Components for hydraulics and coolanol systems (motor actuated valves, pressure relief valves, coolanol pumps and bootstraps).
Aerial Refuelling Systems:
Aerial refuelling pods, high-flow transfer pumps, MA2/MA3 couplings, pressure regulators, break joint couplings.
Control and Monitoring Systems:
Monitoring and surveillance controllers, window heating controllers. Air temperature and pressure control systems. Complete oxygen systems for transport aircraft (Crew and passengers/quick donning masks and protective breathing equipment). Military oxygen regulators and anti-g valves (pneumatic and electronic technology), liquid oxygen converters, quick donning masks, positive pressure breathing systems, NBC systems. Oxygen test benches, oxygen nitrogen GSE, emergency oxygen equipment, on-board generating oxygen systems.
Pneumatic Systems:
On-board inert gas generating systems, oxygen enriched cabin air systems, filtration systems (contaminants, CO2).
Optronics:
Seekers, airborne imaging infrared pods, anti-aircraft and missile defence systems.
Subsidiaries:
ECE, France
EROS (GIE Intertechnique - SFIM)
Faure-Herman, France
IN-Aerospace Ltd, UK
Ferma, France
IN-LHC, France
IIN-FLEX, France [98]

75-23200 Jehier SA
Route de Saint Lézin, BP 29, F-49120 Chemillé
☎ (+33 2) 41 64 54 00 ⊖ (+33 2) 41 64 54 01
♧ FFr 2,002,000 (1996) ♣ FFr 60,000,000 (1996)
👥 95 (1996) 🏭 5,000 m2
President: Félix Jehier
Marketing Director: Jacques Maillard
Marketing and Communication Officer: T Renault
■ Thermal and acoustical insulation for high performance applications: aerospace, marine, military vehicles and industry. Manufacturer of materials and equipment (with metallic or non-metallic cover). [98]

75-23500 Karcher France
(a subsidiary of Alfred Karcher GmbH & Co, Germany)
5 Avenue des Coquelicots, ZA des Petits Carreaux, F-94865 Bonneuil sur Marne Cedex
☎ (+33 1) 43 99 67 70 ⊠ 042264285
⊖ (+33 1) 43 39 48 73
♣ FFr 606,000,000 👥 340
Military Department Manager: Claude Olivier
■ Mobile high pressure cleaning and decontamination systems, water and dust aspirators, mechanical sweepers, mobile field kitchens and air heaters. Water treatment systems.
Representing:
Alfred Karcher GmbH & Co [98]

75-23700 Kidde Dexaero
4 rue Henri Poincare, F-92160 Antony
☎ (+33 1) 46 66 08 08 ⊖ (+33 1) 46 66 23 24
👥 70
Managing Director: A Ziegler
■ Supplier of fire detection and suppression equipment for aircraft, armoured vehicles and naval vessels. [98]

75-23820 Kollmorgen Artus
(† ARTUS SA)
(a division of Kollmorgen Corp, USA)
Chemin du Champ des Martyrs, PO Box 9, F-49241 Avrillé Cedex
☎ (+33 2) 41 33 63 40 ⊠ 720970 artus f
⊖ (+33 2) 41 33 63 63
♧ FFr 31,635,000 (1996) ♣ FFr 160,700,000 (1996)
👥 242 (1996) 🏭 29,280 m2
Managing Director: Robert Perrin
Administration Director: Marc de Roquefeuil
Sales and Marketing Director: Franck Dubey
Industrial Operations Director: Jean Paul Briens
Purchasing Manager: Pascale Dupenloup
■ DC motors, torque motors, AC 400 Hz motors, geared motors. Brushless DC motors. Tachogenerators, synchros, resolvers, RVDT. Electromechanical actuators (rotary or linear). Servo control systems to regulate speed and position; alternators, DC generators, starter/generator, AC starters. Transformers, transformer rectifiers. Static rotative convertors, inverters. ILS/VOR calibration systems. [98]

75-23900 Labinal
PO Box 218, F-70851 Saint Quentin Yvelines Cedex
☎ (+33 1) 30 85 30 85 ⊠ 698330 labina f
⊖ (+33 1) 30 43 41 71
♧ FFr 11,673,000,000 (1996)
♣ FFr 410,757,700 (1995) 👥 18,844 (1996)
Chairman and Chief Executive Officer: Amaury Halna du Fretay
Executive Vice President: Serge Becchetti
Executive Vice President: Guy Chardon
Executive Vice President: Jean-Bernard Cocheteux
Executive Vice President: Georges François
Executive Vice President: Jean Lucien Lamy
Executive Vice President, Human Resources: Daniel Rappanello
Corporate Secretary: Henri-Paul Male
Head of Department: Mr Millepied
■ Design, manufacture and support of electrical wiring harnesses and systems; electronic and electromechanical assemblies; tyre pressure monitoring systems; high performance filters and filter systems for use in aviation, space, defence, communications and other related industries. [98]

75-24010 Laboratoire Arcane
(† Arcane Balistic)
PO Box 20, F-92410 Ville d'Avray
☎ (+33 1) 47 09 20 20 ⊖ (+33 1) 47 09 29 98
President: F Bodet
Marketing Director: F Messaoudene
Technical Director: A Belladen
■ Pistols and special purpose ammunition. Riot and crowd control agents, stopson silencers. [97]

75-24050 Étienne Lacroix Tous Artifices SA
PO Box 213, F-31607 Muret Cedex
6 Boulevard de Joffrey, F-31600 Muret
☎ (+33 5) 61 56 65 00 ⊠ 531478 lacart f
⊖ (+33 5) 61 51 42 77
♧ FFr 11,880,000 (1997) ♣ FFr 309,000,000 (1996)
👥 400 (1997)
Chairman: Roger Bares
Managing Director: Étienne Bares
Operations Director: Jean Claude Fontaine
Divisional Director: Pierre Thebault
Sales and Marketing Director: Philippe Rousseau
Purchasing Manager: Jacky Gauvin
Marketing Executive: Mr Malleja
■ Pyrotechnical reseach, development and manufacturing.
Conventional products: illuminating and smoke-emitting devices, pyromechanical assemblies, signalling and rescue devices, aircraft cartridges, self-defence systems for armoured cars, pyrotechnical delays, infra-red tracers. Countermeasures products for protection of aircraft, helicopters, armoured vehicles, surface vessels against electromagnetic radar and infra-red, optronic missile seekers and defence systems.
Paris Office
18 rue Malher, F-75004 Paris
☎ (+33 1) 48 87 53 20 ⊠ 220672 lacart
⊖ (+33 1) 48 87 21 52 [98]

75-24100 Lagardère
(† Matra SA/Matra Hachette)
4 rue de Presbourg, F-75116 Paris
☎ (+33 1) 40 69 16 00 ⊖ (+33 1) 40 69 16 70
♧ FFr 3,880,494,960 ♣ FFr 56,401,000 (1996)
👥 47,000
Chairman and Chief Executive Officer: Jean-Luc Lagardère
Chairman and Chief Executive Officer, Matra Marconi Space: Armand Carlier
Chairman and Chief Executive Officer, Matra BAe Dynamics: Noël Forgeard
Chairman and Chief Executive Officer, Matra Systémes and Information: Bernard Plano
Chief Executive and Chairman of Finance Committee: Philippe Camus
Chief Executive, Automobile, Telecoms, CAD/CAM/CAE Sectors: Frédéric d'Allest
Chief Executive, Communication and Media Sector: Jacques Lehn
Director and Deputy Chairman for Strategic Co-ordination: Jean-Louis Gergorin
Advisor to the Chairman and Senior Vice President, International Affairs: Jean-Paul Gut
Spokesman of the Chairman and Senior Executive Vice President, Communications and Human Resources: Thierry Funck-Brentano
Corporate Vice President, Public Relations: Roland Sanguinetti
Advertising Director: Bernard Pommier
■ Military: development and production of missiles, aeronautical armaments, counter measures, aerial reconnaissance, C3I and mission planning systems. Space: Prime contractor in space programmes (military and civilian observation and telecommunication satellites, scientific satellites, space transportation, orbital infrastructures and manned flights).
Other activities: telecom and CAD/CAM systems. Automobile and transit systems.
Subsidiaries:
Matra Hautes Technologies
Matra BAe Dynamics
Matra Electronique
Matra Marconi Space
Matra Communication
Matra Datavision
Matra Transport International
Matra Automobile
Direction de l'Information
121 Avenue de Malakoff, F-75216 Paris Cedex 16
☎ (+33 1) 40 69 16 74 ⊠ 650772 f
⊖ (+33 1) 47 23 01 70
Information and Press Contact: Frédéric Aragon [98]

75-24200 Latécoère Société Industrielle d'Aviation
Registered Office
135 rue de Périole, F-31079 Toulouse Cedex
☎ (+33 5) 61 58 77 00 ✉ LATECOERE ⊠ 531714
⊖ (+33 5) 61 58 28 37
♧ FFr 11,055,570 ♣ FFr 812,000,000 (1996)
👥 620 (1996) 🏭 58,000 sq m
President: François Junca
Executive Vice President: François Bertrand
Vice President, Corporate Communications: Jean Pierre Robert
Commercial Director: H Schembri
■ Aircraft and aerospace technologies; human centrifuges; mechanical engineering, testing and environmental equipment; integration of airborne weapons; jet engine parts and cowls; thrust reversers; fuselage, wing and nacelles structure elements; ground and on-board equipment; landscape video camera systems, customisation. → 80 [98]

75-24440 Le Protecteur
(a subsidiary of FICAS SA, France)
12 rue Turpin, PO Box 111, F-26103 Romans Cedex
☎ (+33 4) 75 02 58 57 ⊖ (+33 4) 75 02 42 44
♧ FFr 250,000 (1997) 👥 12 (1997) 🏭 200 m2
Managing Director: Jean Christophe Casset
■ Products for self-defence, CS aerosols, electrical defences, small equipment for police forces. [98]

75-24590 Les Conteneurs d'Alsace
(a subsidiary of L'hotellier Montrichard)
15 rue de Quimper, PO Box 2431, F-68200 Mulhouse
☎ (+33 3) 89 50 57 00 ⊖ (+33 3) 89 51 11 66 [98]

75-24890 L'hotellier Bourgogne Industries
(a subsidiary of L'hotellier Montrichard)
13 Avenue du Bois - Genelard, PO Box 17, F-71420 Perrecy les Forges
☎ (+33 3) 85 79 22 65 ⊖ (+33 3) 85 79 23 59 [98]

75-24900 L'hotellier Montrichard
Saint Julien de Chédon, PO Box 17, F-41401 Montrichard Cedex
☎ (+33 2) 54 71 12 12 ⊖ (+33 2) 54 71 12 01
♧ FFr 50,000,000 (1997) ♣ FFr 406,000,000 (1996)
👥 1,056 (1996) 🏭 107,000 m2
Chairman of the Board of Directors: Philippe L'hotellier
Production Director: B Schillewaert
Commercial and Export Director: Johny Ledys
☎ (+33 1) 39 56 29 87
Purchasing Manager: Gabriel Hohn
Sales Assistant: Ms Karine Bermudez
■ Manufacture of watertight re-usable containers specially equipped for shock and vibration requirements, made of plastic and metallic materials to civil and military standards. Aeronautical parts: all types of composite laminated moulded parts. Packaging of aeronautical spare parts. Shelters. → 80
Buc Plant
147 rue Hélène Boucher, F-78530 Buc
☎ (+33 1) 39 56 50 60 ⊖ (+33 1) 39 56 16 45
Marignane Plant
4ème rue n°4 - ZI des Estroublans, F-13127 Vitrolles
☎ (+33 2) 42 89 04 22 ⊖ (+33 2) 42 89 69 86
Toulouse Plant
5 Avenue Clément Ader - ZI Est, F-31770 Colomiers
☎ (+33 5) 61 15 95 95 ⊖ (+33 5) 61 15 95 89
Buc Office
254/256 rue Fourny, F-78530 Buc
☎ (+33 1) 39 56 29 87 ⊖ (+33 1) 39 56 12 01
Montrichard Plant
ZI de Montrichard, PO Box 17, F-41401 Montrichard Cedex
☎ (+33 2) 54 71 12 12 ⊖ (+33 2) 54 71 12 02
Le Haillan Plant
Avenue Toussaint Catros, F-33160 Le Haillan
☎ (+33 5) 56 34 65 85 ⊖ (+33 5) 56 34 87 42 [98]

75-24950 Liebherr-Aerospace Toulouse SA
(† ABG-SEMA SA)
(a subsidiary of Liebherr-Aerospace Lindenberg GmbH, Germany)
408 Avenue des Etats Unis, BP 2010, F-31016 Toulouse Cedex
☎ (+33 5) 61 35 28 28 ⊖ (+33 5) 61 35 28 00
♣ FFr 422,000,000 👥 409 🏭 227,000 sq m
Chief Executive Officer: Alain Guinandeau
Sales and Marketing Director: Christophe Malrin
■ Bleed air systems, air conditioning and environmental control systems (air cycle and vapour cycle), cabin pressure control systems for aerospace and similar applications. [98]

75-25000 Lloyd Instruments SA
(an associated company of Lloyd Instruments Ltd, UK)
3 Avenue des Coudriers, Zone d'Activité de l'Observatoire, F-78180 Montigny le Bretonneux

☎(+33 1) 30 57 47 74 ☒ 699662
⊖ (+33 1) 30 57 50 33 [98]

75-25100 LMB

(a subsidiary of Dynafan Groupe)
36 Avenue Marie et Pierre Curie,
F-19360 Malemort sur Corrèze
☎(+33 5) 55 92 16 33 ⊖ (+33 5) 55 92 19 00
♧ FFr 15,634,000 ♣ FFr 38,000,000 (1995) 👥 95
🏭 4,000 sq m
Managing Director: René Chapalain
Operations Director: Claude Albier
Sales and Marketing Director: Alain Berger
■ Manufacture of cooling systems, fans, heat exchangers, motors and generators.

Plaisir Division
PO Box 57, F-78371 Plaisir Cedex
☎(+33 1) 30 55 21 20 ⊖ (+33 1) 34 81 18 38 [97]

75-25200 Lohr Industrie

Construction Véhicules Industriels
(† Lohr Défense)
(a division of Lohr SA)
29 rue du 14 Juillet, F-67980 Hangenbieten
☎(+33) 88 38 98 00 ☒ 870082 ⊖ (+33) 88 96 06 36
♧ FFr 76,459,936 ♣ FFr 555,000,000 👥 724
🏭 60,000 sq m
President: R Lohr
Sales and Marketing Director: J C LeMaire
■ Ground vehicles for airborne forces, light armoured tracked or wheeled vehicles, shelters, armament carriers, stabilized platforms, tank transporters, vehicle transporters and recovery trucks, road trailers, spanning equipment.

Duppigheim Works
F-67120 Duppigheim
☎(+33 3) 88 38 98 00 ☒ 870082
⊖ (+33 3) 88 96 06 36 [98]

75-25350 Leach International Europe SA

(† LRE Relais et Electronique Sarl)
(a subsidiary of Leach International UK Ltd)
2 rue Goethe, F-57430 Sarralbe
☎(+33 3) 87 97 98 97 ☒ 860959 f
⊖ (+33 3) 87 97 84 04
♧ FFr 2,000,000 👥 265 (1997) 🏭 3,800 m2
Managing Director: Jean François Evellin
Commercial Director: Jean Michel Sigaud
Operations Director: Laurent Steinmetz
■ Electromechanical relays. Time delay relays, timers, electronic, mechanical and electronic systems, switches, panels, central warning systems. → 90
® Germany: LRE Relais + Elektronik
USA: LEACH International North America [98]

75-25390 Lucas Aerospace FCS Paris

(† Lucas Aerospace)
(a division of Lucas Aerospace France SA)
7/9 Avenue de l'Eguillette, BP 7186,
F-92056 Cergy Pontoise Cedex
☎(+33 1) 34 32 63 00 ☒ luca fcs 606330 f
⊖ (+33 1) 34 32 63 10
♣ FFr 400,000,000 (1995) 👥 300 🏭 12,000 sq m
Managing Director: Laurent Mattiussi
Sales and Marketing Director: Yvon Allain
Technical Director: Bruno Thoraval
Operations Director: F Brachfogel
Purchasing Manager: P Moratelli
Product Support Manager: Dominique Jamaux
■ Primary and secondary flying controls; rescue hoists; electro-hydraulic servo systems. [98]

75-25400 Luchaire Défense SA

(a subsidiary of Giat Industries, France)
13 route de la Menière, F-78007 Versailles Cedex
☎(+33 1) 39 49 80 51 ⊖ (+33 1) 39 49 80 14
♧ FFr 50,000,000 ♣ FFr 270,300,000 (1995) 👥 474
Chairman: Jan Paul Philouze
Chief Financial Officer: Mr Steuman
Marketing Manager: Michel Rogier
■ Manufacture of artillery ammunition and warheads. [97]

75-25500 Mannesmann Rexroth SA

(a subsidiary of Mannesmann Rexroth GmbH, Germany)
91 Boulevard Irène Joliot Curie, PO Box 101,
F-69634 Venissieux Cedex
☎(+33 4) 78 78 52 52 ☒ 380852 hydrex f
⊖ (+33 4) 78 78 52 26
♧ FFr 57,000,000 (1996) ♣ FFr 393,579,000 (1996)
👥 434 (1996) 🏭 20,200 sq m
Managing Director: Hubert Quenette
☎(+33 4) 78 78 52 43
Finance and Administration Director: J P Demant
☎(+33 4) 78 78 52 28
Production Director: D Jop ☎(+33 4) 78 78 53 31
Technical Director: W Kauss ☎(+33 4) 78 78 78 52 61
Operations Director: Dieudonne Jop
Director of Sales, Marketing and Export: Gilles Breau
Manager of Marine/Off-Shore Division: M Rivet
☎(+33 4) 78 78 52 55
■ Hydraulic components (gear pumps, multiple sandwich and remote control valves). Hydraulic power packs, hydraulic engineering systems. [98]

75-25550 Manitou BF
Military Equipment Division

Zone Industriel- Route Châteaubriant, PO Box 249,
F-44158 Ancenis Cedex
☎(+33 2) 40 09 10 11 ☒ 710521
⊖ (+33 2) 40 83 36 88
♣ FFr 2,548,000 (1996) 👥 1,700
President: Marcel Braud
Managing Director: Joël Goulet
Operations and Technical Director: Jacques Guerin
Marketing Director: Marcel Claude Braud
Public Relations Manager: Mrs E Gardes
■ Range of rough terrain forklift trucks. [98]

75-25610 Manoir Industries
Division Bar-Lorforge

2 rue de Saint Petersbourg, F-75008 Paris
☎(+33 1) 44 69 88 10 ⊖ (+33 1) 44 70 09 19
♧ FFr 229,076,700 ♣ FFr 1,800,000,000 (1996)
👥 3,000 (1996)
Chairman of the Supervisory Board: Roger Hubert
Managing Director and Member of the Executive Board: Peter Krudup
■ Open-die and closed-die forgings in steel and all forgeable metals. Special alloys and powder metals. Precision die-forgings.

Bar sur Aube Plant
BP 102, F-10200 Bar sur Aube
☎(+33 3) 25 92 31 41 ⊖ (+33 3) 25 27 36 77
👥 190
Director: J C Robin
Commercial Director: Pierre Brans

Custines Plant
F-54670 Custines
☎(+33 3) 83 49 49 49 ⊖ (+33 3) 83 24 27 41
👥 220
Director: F Maraldi [98]

75-25630 Manurhin Défense

(a subsidiary of Giat Industries, France)
13 route de la Minière, F-78009 Versailles Cedex
☎(+33 1) 39 49 80 51 ⊖ (+33 1) 39 49 80 14
♧ FFr 79,500,000 ♣ FFr 405,000,000 👥 743
🏭 66,233 m2
Managing Director: Michel Pinaud
Sales and Marketing Director: J C Allard
Operations Director: M Le Pape
Finance Controller: Mr Valland
Marketing Manager: Olivier Tissier
■ Portable anti-tank systems and medium calibre ammunition. [97]

75-26000 Matra Défense-Espace

(† Matra Défense)
(a subsidiary of Lagardère Groupe, France)
37 Avenue Louis Bréguet, BP 1,
F-78140 Vélizy Villacoublay Cedex
☎(+33 1) 34 88 30 00 ☒ 698077 f
⊖ (+33 1) 34 65 12 15
♧ FFr 154,000,000 (1996)
♣ FFr 6,400,000,000 (1995) 👥 6,000 (France and UK)
Chairman and Chief Executive Officer: Noël Forgeard
Vice President: Jean Paolorsi
Chief Executive Officer, Matra Marconi Space: Armand Carlier
Chairman and Chief Executive Officer of Matra Cap Systèmes: Yves Verer
Senior Vice President, Finance: Michel Thomas
Senior Vice President: François Heisbourg
Senior Vice President, International Division: Jean Paul Gut
Senior Vice President, Marketing: François Desprairies
Vice President, Communications: Marc Vellay
Vice President and Operations Director: Robert Chauprade
Vice President and General Director: Pierre Bourgeon
Vice President and Technical Director: JeanPaul Genest
Vice President, Sales and Export Director: Jean Pierre Talamoni
Space and Defence Advisor: Adm Jean Brusson
Communications Manager: Pierre Iafrate
■ Activities in the space industry. Development and construction of satellites, space equipment, prototypes, missiles, aircraft armament, countermeasures, aerial reconnaissance systems.

Matra Défense
37 Avenue Louis Bréguet,
F-78146 Vélizy Villacoublay Cedex
☎(+33 1) 34 88 30 00 ⊖ (+33 1) 34 65 12 15
Chairman and Chief Executive Officer: N Forgeard
President: P Bourgeon
Vice President: J Paolorsi
Senior Vice President, Strategy: P Turpin
General Office Services: A M Perus
Computing Administration, Management and Finance: M Thomas
Resources Development: B Touet
Human Relations: P Barillot
Quality: L Ribereau Gayon
Communications: M Vellay
Export and Commercial Management: J P Talamoni
Commercial Management, Contracts and Customer Services: B de Vallee
Marketing International Development: F Despraires
Apache Affairs: F Bregier
Technical Research and Development: J P Genest
Operations, Engineering, Production and Purchasing: R Chauprade
Military and Air Advisor: Gen Jacques Deveaud
Military and Land Advisor: Gen André Mathe
Military and Navy Advisor: Adm Alain Denis

Centre de Vélizy la Source
37 Avenue Louis Bréguet, PO Box 1,
F-78146 Vélizy Villacoublay Cédex
☎(+33 1) 34 88 30 00 ☒ 698077 f
⊖ (+33 1) 34 65 12 15
Managing Director: Patrick Devergies

Centre de Vélizy le Bois
20-22 rue Grande Dame Rose, PO Box 150,
F-78141 Vélizy Villacoublay Cedex
☎(+33 1) 34 88 30 00 ☒ 6952238 f
⊖ (+33 1) 34 88 22 88
Managing Dir: Patrick Devergies
Deputy Managing Dir: Michel Dartayre

Centre de Selles Saint Denis
"La Chaudronne", F-41460 Selles Saint Denis
☎(+33 1) 54 95 59 00 ☒ 751356 f
⊖ (+33 1) 54 95 59 59
Managing Director: Jean François Karcher

Centre de Salbris
24 rue de la Convention, PO Box 29, F-41300 Salbris
☎(+33 2) 54 88 36 00 ☒ 750274 f
⊖ (+33 2) 54 88 36 20
Managing Director: Bernard Ancillon [97]

75-26005 Matra Electronique (MEL)

(a subsidiary of Lagardère Group, France)
Rue Ferdinand Meunier, PO Box 2,
F-60610 La Croix Saint Ouen
☎(+33 3) 44 91 45 00 ☒ 140329f
⊖ (+33 3) 44 91 45 45
♧ FFr 10,000,000 ♣ FFr 270,000,000 👥 335
Managing Director: A Yvon Patissier
Sales and Marketing Director: Philippe Ciocci
Operations Director: A Perreard
Purchasing Manager: F Dubois
■ Manufacture of hybrid micro-electronics, electronics and equipment.

Representing:
Matra Défense, France [97]

75-26015 Matrasur SA

(a subsidiary of Vacu-Blast Ltd, UK)
36 Avenue du 1er Mai, ZI Les Glaizes,
F-91124 Palaiseau Cedex
☎(+33 1) 64 47 02 70 ☒ 602753 f
⊖ (+33 1) 64 47 02 75
♧ FFr 2,700,000 ♣ FFr 50,203,000 👥 56 🏭 3,225 sq m
Managing Director: Mike Riley
■ Automated shot peening, automated and manual grit blasting equipment. Equipment for thermoset reinforced plastics (VENUS-GUSMER) such as spray units, resin injection and resin transfer molding units, filament winding units, cast and pour units, meter mix pumps, accessories/support materials. Plastic media blasting equipment. [97]

Contact Reference

75-26040 Mecalectro SA
8 rue Galvani, F-91745 Massy Cedex
☎ (+33 1) 69 32 74 00 ⊖ (+33 1) 69 20 39 34
♧ FFr 10,000,000 (1996) ♣ FFr 63,600,000 (1996)
👥 110 (1996) 🏭 2,800 m2
Managing Director: J C Chabin
Sales and Marketing Manager: P Gillain
Sales Department: P Maire
■ Solenoids and electro-magnetic components. [98]

75-26400 Messier-Bugatti
(a subsidiary of SEP (SNECMA Group), France)
Zone Aéronautique Louis Bréguet, PO Box 40,
F-78141 Vélizy Cedex
☎ (+33 1) 46 29 81 00 ⊠ 634160
⊖ (+33 1) 46 29 87 00
♧ FFr 134,126,560 (1996)
♣ FFr 1,440,000,000 (1996) 👥 1,170 (1996)
Chairman: Yves Leclere
Vice President, Quality: Maurice Perrault
Vice President, Braking Systems: Jean-Christopher Corde
Vice President, Repair and Overhaul: Benoît Gosset
Vice President, Systems and Hydraulic Equipment: Alain Richard
Vice President, Finance: Ms Sylvia Bellon
Vice President, Human Resources: Alain Coté
Communication Director: Ms Monique Ranvier
☎ (+33 1) 46 29 82 20
■ Research, manufacture, development and technical support of: wheels and brakes, braking systems, hydraulic equipment. Repair and overhaul.

Molsheim Plant
19 rue de la Commanderie, F-67120 Molsheim
☎ (+33 3) 88 38 92 00 ⊠ 870788
⊖ (+33 3) 88 38 93 99
Manager: Marcel Schott [98]

75-26500 Messier-Dowty SAS
(a subsidiary of Messier-Dowty International Ltd, UK)
Zone Aéronautique Louis Bréguet, F-78142 Vélizy Cedex
☎ (+33 1) 46 29 18 00 ⊠ 634156
⊖ (+33 1) 46 29 18 01
♧ FFr 272,000,000 ♣ FFr 1,075,000,000 (1995)
👥 1,022
Chairman: Jacques Bouhelier
Products and Operations Director: François Druesne
Financial Director: Georges de Chaisemartin
■ Research, manufacture, development and technical support of complete landing gear systems.

Bidos Plant
F-64400 Bidos
☎ (+33 5) 59 89 60 00 ⊠ 560811 meshi bidos f
⊖ (+33 5) 59 39 62 14
Manager: André Broudeur [98]

75-26600 Michelin Italiana SpA Lighe Produit Avion
(† Manufacture Francaise des Pneumatiques Michelin)
(an affiliate of Compagnie Générale des Etablissements Michelin)
57 Via Livormo, F-10144 Torino
☎ (+39 11) 432 67 11 ⊖ (+39 11) 432 66 98
Division Manager: Pierre Desmarets
Marketing and Sales: Nicolas Baviere
Marketing and Communications: Marco Cervellati
Research and Development: Jean-Philippe Gorez [98]

75-26780 MicroSpire SA
16 Parc d'Activités du Beau Vallon, F-57110 Illange
☎ (+33 3) 82 56 55 64 ⊠ 930395 f
⊖ (+33 3) 82 51 00 49
♧ FFr 1,134,000 (1996) ♣ FFr 82,000,000 (1996)
👥 280 (1996) 🏭 5,000 m2
President: Yves Nold ☎ (+33 3) 82 59 13 26
Sales and Marketing Director: Thierry Berthier
☎ (+33 3) 82 59 17 59
Operations Director: Francis Leyendecker
Purchasing Manager: Jean Luc Marchal
☎ (+33 3) 82 59 13 30
■ Development and manufacture of transformers and inductors. Linear and toroidal products, hybrid technology for professional and military applications. Passive filters. [98]

75-27000 Microturbo SA
(a subsidiary of Turbomeca SA, France)
Chemin du Pont de Rupé, PO Box 2089,
F-31019 Toulouse Cedex
☎ (+33 5) 61 37 55 00 ⊖ (+33 5) 61 70 74 45
👥 427
Chairman and Chief Executive Officer: Jean Bernard Cocheteux
Managing Director: Yves Paul François
Commercial Director: Yves Mazoyer
Operations Director: Claude Delamasantiere
Marketing Manager: Philippe Berges
Purchasing Manager: Remi About
■ Self contained gas turbine starting systems. Air producers and air starters. Auxiliary power units for aircraft and helicopters. Combined APU/starter systems. Turbojet engines. Auxiliary power units for ground applications: trolleys, armoured vehicles, shelters. Integrated power and environmental control systems (IPECS).

Representing:
Microturbo Inc
Microturbo Ltd
Microturbo GmbH [98]

75-27200 NMB Minebea Sarl
18 rue Jules Vercruysse, Zone Industrielle,
F-95100 Argenteuil
☎ (+33 2) 39 80 34 01 ⊠ 42605084
⊖ (+33 2) 39 80 16 00
👥 23
Managing Director: K Koga [98]

75-27750 NEFAB Sarl
(a subsidiary of NEFAB AB, Sweden)
RN 20 Nord, F-41300 Salbris
☎ (+33 2) 54 96 82 55 [E] nefab@wanadoo.fr
⊖ (+33 2) 54 96 35 77
♧ FFr 5,500,000 (1997) ♣ FFr 44,000,000 (1996)
👥 45 (1997) 🏭 5,000 m2
Managing Director: Noël Parrot
Operations Director: Thierry Ulrich
Marketing and Advertising Manager: Ms Anne Leclair
Purchasing Manager: Ms Maurice Laude
Sales Manager: François Simon
Sales Manager: Olivier de Guitant
■ Development and manufacture of packaging for solid protection of delicate material during transport and storage. The concept is based on folding constructions made of sheet board and Aluzink. [98]

75-27770 Neotronics-Solomat SA
(member of Neotronics Technology Group)
16 rue Jacques Tati, F-91006 Evry Cedex
☎ (+33 1) 69 36 05 50 ⊠ 601441 f
⊖ (+33 1) 60 77 93 73 [98]

75-27785 Nereides
4 Avenue des Indes, F-91969 Courtabouef Cedex
☎ (+33 1) 69 07 20 48 [E] nereides@compuserve.com
⊖ (+33 1) 69 07 19 14
♣ FFr 30,000,000 👥 42
Managing Director: Didier Becq
Sales and Marketing Director: Fabria Descours
Communications Executive: Ms V Sandres
■ Specialists in the marine environment. Production of the Wadibuoy System, wave directional system and oil detector Oilspy. NCA (buoyant cable antenna) for submarine VLF/LF/HF communication and navigation. [98]

75-27800 Nicolas Industrie SA
PO Box 3, F-89290 Champs sur Yonne
☎ (+33 3) 86 53 30 09 ⊖ (+33 3) 86 53 79 74
Managing Director: H Suberg
Sales and Marketing Director: O Amichault
Operations Director: Mr Darlot
Purchasing Manager: Mr Baillet
Communication: Mr Suberginaud
■ Vehicles, semi-trailers, heavy transport equipment, tank transporters, engines, Hy-Spec transport system. [97]

75-28350 Oldham France SA
(a part of Hawker Batteries, UK)
Rue Alexander Fleming, ZI Est, F-62000 Arras
☎ (+33 3) 21 60 25 25 ⊠ 120752
⊖ (+33 3) 21 73 16 51
♣ FFr 704,000,000 👥 1,011
Managing Director: C Dreulle
Sales Manager: M Dehé
Sales Manager Army/Defence: G de Magneval
Export Sales Manager: P Specq
■ Motive power batteries, standby power batteries, solar batteries, chargers and gas detectors for surface and mines. [97]

75-28500 ORITEL SA
45 rue du Moulin des Bruyéres, PO Box 303,
F-92402 Courbevoie Cedex
☎ (+33 1) 47 88 97 80 ⊠ 269613
⊖ (+33 1) 47 89 93 27
♧ FFr 6,400,000 (1996) ♣ FFr 18,700,000 (1996) 👥 20
President: P Bodereau
Technical Director: J Y Fabre
Sales and Marketing Director: M Letailleur
■ Development and manufacture of microwave components, test and measurement systems. Military applications in the fields of aircraft, radar and telecommunications systems testing. [98]

75-28750 Panhard & Levassor SA Head Office
Société de Constructions Mécaniques Panhard et Levassor
(a subsidiary of Peugeot SA Holding)
18 Avenue d'Ivry, PO Box 6, F-75621 Paris Cedex 13
☎ (+33 1) 40 77 40 00 ⊠ 270276 f
⊖ (+33 1) 45 82 73 26
♧ FFr 5,000,000 ♣ FFr 429,000 ,000 👥 297
🏭 77,000 sq m
President: Yves Barbe
Director General: Jacques Texier
Technical Director: Mr Moreau
Sales and Marketing Director: P A Valentini
■ Manufacture of light armoured wheeled vehicles. [98]

75-29100 Paulstra SNC Hutchinson
(a subsidiary of Hutchinson, France)
61 rue Marius-Aufan, PO Box 164,
F-92305 Levallois Perret
☎ (+33 1) 40 89 53 31 ⊖ (+33 1) 47 57 44 20
♣ FFr 387,000,000 (1996) 👥 825 (1996)
Sales Director: Laurent Poirier
Manager, Industrial Division: Jacques Pineri
■ Mechanical applications for elastomers. Anti-vibration mounts, laminated assemblies, special sheet products (polyurethane, silicone, special elastomers), heating systems (anti-freeze and de-icing systems for aircraft and helicopters), equipment for missiles and ballistic weapons.

Aviation Plant
2 rue Turgot, BP 26, F-27150 Etrepagny
☎ (+33 2) 32 27 65 00 ⊖ (+33 2) 32 27 26 40

Aviation Plant
26 Boulevard de Péringondas, F-28200 Chateaudun
☎ (+33 2) 37 44 70 70 ⊖ (+33 2) 37 45 77 91

Aviation Plant Vibrachoc
Parc d'Activités de l'Églantier, CE 2804 Lisses,
F-91028 Évry Cedex
☎ (+33 1) 69 91 50 00 ⊖ (+33 1) 60 86 38 59 [98]

75-29590 Petrosystems
(a subsidiary of Compagnie Générale de Géophysique, France)
1 rue Leon Migaux, F-91302 Massy Cedex
☎ (+33 1) 64 47 30 00 ⊠ 602442 cggec f
⊖ (+33 1) 64 47 39 70
♣ US$ 30,000,000 👥 250
President: R Brunck ☎ (+33 1) 64 47 36 35
Chief Executive Officer: A Laroche ☎ (+33 1) 64 47 37 59
Marketing Manager: J P Marbeau ☎ (+33 1) 64 47 34 64
Sales Manager: L Lhommet
■ Design and sale of geological and geophysical software. [98]

75-29600 Automobiles Peugeot
PO Box 01, F-75761 Paris
☎ (+33 1) 40 66 55 11 ⊠ 610700 peugeot paris
⊖ (+33 1) 40 66 54 14
♣ FFr 107,052,000 👥 67,400 (total)
President: Jacques Calvet
■ P4 tactical cross-country command and liaison vehicles in several versions. [98]

75-29800 PICA
(a subsidiary of Compagnie Générale des Eaux, France)
16 rue Trezel, F-92300 Levallois
☎(+33 1) 49 68 09 29 ⊞ 630428
⊖ (+33 1) 49 68 09 30
✥ FFr 7,996,000 (1996) ♣ FFr 150,000,000 (1996)
👥 170
President: P Boulin
Managing Director: G Dagois
Sales Director: P Montagnon
Communications and Marketing Executive:
Pascale Garnier
■ Activated carbon used in gas masks, filtration systems, polyurethane foam for military suits. [98]

75-30402 Poclain Hydraulics SA Military Applications Division
Route Saint Sauveur, PO Box 106,
F-60411 Verberie Cedex
☎(+33 3) 44 40 77 77 ⊞ 140043
⊖ (+33 3) 44 40 77 99
✥ FFr 52,261,860 👥 700 🏭 26,700 sq m
Chairman: Pierre Bataille
Managing Director: Laurent Bataille
Marketing Assistant: Ms Amanda Etheridge
■ Design and production of hydrostatic power transmission (from 20 kW to 1,000 kW) for vehicles. Also specialising in high performance hydrostatic transmission for military applications. [97]

75-30450 Polytec-PI SA
(† RMP)
32 rue Delizy, F-93694 Pantin Cedex
☎(+33 1) 48 10 39 30 ⊖ (+33 1) 48 10 08 03
✥ FFr 1,000,000 (1996) ♣ FFr 24,000,000 (1996)
👥 16 (1996) 🏭 1,000 sq m
President: Dr Francis Levy
Manager, Robotics Department: M Hartmann
■ IR test equipment (radiometers, spectroradiometers, black bodies) and laser training systems for rifles and light anti-tank weapons; fast motion cameras and low light level cameras. [98]

75-30520 Pont-à-Mousson SA
91 Avenue de la Libération, F-54076 Nancy Cedex
☎(+33 3) 83 95 20 00 ⊖ (+33 3) 83 96 31 32
Secretary General: J L Robaux
Contact: Ms Danielle Cuquilleau

Registered Office
91 Avenue de la Libération, F-54000 Nancy
☎(+33 3) 83 96 81 21 ⊞ 850003 pamsa x

Centre de Recherches
PO Box 28, F-54703 Pont à Mousson [98]

75-30750 PRM Précimétal
(† Ets Labonne Précimétal)
(a subsidiary of Acteurs SA, France)
Zone Industrielle, F-18400 Saint Florent sur Cher
☎(+33 2) 48 55 02 31 ⊞ 782140 f
⊖ (+33 2) 48 55 33 82
✥ FFr 400,000 ♣ FFr 20,000,000 (1995) 👥 52
🏭 3,000 sq m
Chairman: M Veillon
Sales and Marketing Director: J F Ollier
■ Machined parts for the aerospace and defence industries. [97]

75-30890 Précitube SA
10 rue de l'Industrie, PO Box 4, F-18290 Charost
☎(+33 2) 48 26 20 12 ⊞ 760434 f
⊖ (+33 2) 48 26 30 42
✥ FFr 4,000,000 (1996) ♣ FFr 31,000,000 (1995) 👥 64
Managing Director: Patrick Champion
Technical Director: Alain Berger
Sales and Marketing Director: Phil Etheridge
Purchasing Manager: M Guilpin
Sales Department: C Moreau
■ Manufacture of stainless steel precision tubes for all aircraft requirements. High pressure circuit tubing, fluid circuits for applications in engines, cells and motors. Shock absorbers for A/V seats. [97]

75-31100 Pronal SA
BP 18, F-59115 Leers
Rue de Trieu du Quesnoy ZI Roubaix Est
☎(+33 3) 20 99 75 00 ⊞ 132750 pronal f
⊖ (+33 3) 20 99 75 20
✥ FFr 5,000,000 (1996) ♣ FFr 65,000,000 (1996) 👥 98
Sales Manager: Guy Doom
■ Flexible tanks for storage and transport of liquids (Hydrocarbons, water) from 20 to 350,000 litres. Lifting bags. Technical inflatable products. [97]

75-31200 Pyromeca
Quartier Lagoubran, BP 2148, F-83063 Toulon Cedex
☎(+33 4) 94 22 86 86 ⊞ 430957 pyrmeca f
⊖ (+33 4) 94 22 86 99
♣ FFr 76,000,000 👥 150
Chairman: Bernard Zeller
President and Director General: Hubert Zante
Commercial Director: David Balducci
Sales Manager, Space: Christian Fritsch
■ System studies and ordnance pyrotechnics for spacecraft, launch pad equipment. [98]

75-31300 Pyrospace
(a joint venture between OEA, USA and Aerospatiale and SNPE, France)
139 Route de Verneuil, BP 52, F-78130 Les Mureaux
☎(+33 1) 34 92 44 44 ⊖ (+33 1) 34 74 30 93
♣ FFr 89,700,000 (1996) 👥 88 (1996) 🏭 4,400 m2
President: Jean Charles Poggi
General Director: Richard Loinard
Defence Projects Engineer: Alexandre Babin
☎(+33 1) 34 92 44 21
Advertising and Sales Manager: Serge Morlan
Marketing and Communication: Laurence Mitachevitch
☎(+33 1) 34 92 44 22
■ Manufacture and sale of various ranges of pyrotechnic devices and components for space, aeronautic, civil and military programs. [98]

75-31405 Quartz et Silice
(a subsidiary of Saint Gobain, France)
BP 102, F-77793 Nemours Cedex
☎(+33 1) 64 45 45 00 ⊞ 692268 f
⊖ (+33 1) 64 28 45 11
✥ FFr 71,000,000 ♣ FFr 150,000,000 👥 300
Managing Director: Pierre Bony
Commercial Director: Ms Héléne Barzoukas
Operations Director: Hervé Delle
Purchasing Manager: Antonio Yerpes
Product Manager: Jean Marie Harry
☎(+33 1) 64 45 46 06
■ Quartz wool, quartz yarn (Quartzel), optical silica, piezo ceramics, fused silica components. [98]

75-31600 Radiall
101 rue Philibert Hoffmann, ZI Ouest,
F-93116 Rosny sous Bois Cedex
☎(+33 1) 49 35 35 35 ⊞ 235220
⊖ (+33 1) 48 54 63 63
✥ FFr 18,709,000 (1996) ♣ FFr 793,400,000 (1996)
👥 1,500 (1996)
President: Yvon Gattaz
Executive Vice President: Lucien Gattaz
Deputy Director General: Pierre-Michel Churg
Executive Vice President: Pierre Gataz
■ Coaxial connectors, rack and panel connectors, optical connectors, passive microwave devices.

Voiron Plant
81 Boulevard Denfert Rochereau, F-38500 Voiron
☎(+33 4) 76 67 50 00 ⊞ 320601
⊖ (+33 4) 76 05 67 38

La Verpillère Plant
Micronde - ZI de Chesnes Tharabie, F-38290 La Verpillère
☎(+33 4) 74 94 26 78 ⊞ 370642
⊖ (+33 4) 74 94 26 05

Voreppe Plant
Rolyz, F-38240 Voreppe
☎(+33 4) 76 50 00 57 ⊖ (+33 4) 76 56 60 97 [98]

75-31800 Ratier-Figeac SA
PO Box 2, F-46101 Figeac Cedex
Route de Cahors, F-46100 Figeac
☎(+33 5) 65 50 50 50 ⊞ 531977
⊖ (+33 5) 65 34 23 63
✥ FFr 51,000,000 (1996) ♣ FFr 560,000,000 (1996)
👥 930 🏭 48,000 m2
Chairman: Daniel Dewavrin
Managing Dir: Robert Vitrat
Director General Adjoint: Robert Gaiani
Marketing Dir: Michel Ferey
Administration, Finance, Public Relations and Personnel Director: Roland Laporte
Production Director: Reni Privat
Service Commercial: Patrick Laumond
☎(+33) 65 50 50 04
■ Mechanical and hydraulic equipment (cable tension regulators, door and flap actuators, vibrations damper). Flight control equipment (side stick control, pedestal, power throttle quadrant). Design and production of aeronautical ballscrews (all applications). Design and production of composite parts (outer guide vane blades, various radomes for missiles, propeller blades for transall aircraft and commuter aircraft). Design and production of aircraft propellers (atlantic, transall composite bladed propellers, Nord 262, 23LF). Production of a new composite bladed propellers for engines within the power range of 1,500 to 5,000 hp. Sub-contracted machining of structural parts, simple annd complex engine parts. Manufacture and repair of all helicopter parts, repair of all aeronautical equipment. [97]

75-31850 Raufoss France Sarl
Rue de la Hardt, Zone Industrielle, F-67120 Molsheim
☎(+33 3) 88 38 55 33 ⊖ (+33 3) 88 38 41 22 [97]

75-32100 Reims-Aviation SA
Aérodrome de Prunay, PO Box 2745,
F-51062 Reims Cedex
☎(+33 3) 26 48 46 46 ✉ REIMSAVIA
⊞ remavia 830754 f ⊖ (+33 3) 26 49 13 60
✥ FFr 97,106,020 (1997) ♣ FFr 154,600,000 (1996)
👥 320 (1997) 🏭 28,488 m2
President and General Director: Jean Paul Chaufour
Operations Director: Louis-Marie Bourcier
Technical Director: Jean Jacques Detant
Commercial and Marketing Director: Gildas Illien
Head of External Relations: Max Boirame
■ Production of F-406 under two versions:
CARAVAN II for the transport of passengers and cargo.
VIGILANT for special missions including maritime surveillance.
Participant in various civil and military programmes along side major European manufacturers, such as Falcon 10/20/200/50 and 900; Mirage 2000; ATR 42 and 72; Airbus A300/310 and A300/340. Manufacture of the 4 seater (F) 172 and (F) 182 single piston engine under CESSNA's licence. Maintenance services. [98]

75-32200 R.E.M.F. Radiocoms-Avionics
Zac Garossos, Rue du Juncassa, F-31700 Beauzelle
☎(+33 5) 61 59 93 37 ⊖ (+33 5) 61 59 33 85
✥ FFr 2,320,000
President: Michel Fourcade
■ Government and military manufacturer of radiocommunications, avionics systems, equipment and ground station equipment. [98]

75-32230 Remtech SA
2/4 Avenue de l'Europe, PO Box 101,
F-78143 Vélizy Villacoublay Cedex
☎(+33 1) 39 46 59 58 ⊖ (+33 1) 39 46 63 10
✥ FFr 1,200,000 ♣ FFr 6,900,000 (1996) 👥 6 (1997)
President and Chief Executive Officer: Jean Michel Fage
Secretary: Christel Tasso
■ Meteorological and environmental remote sensing systems. Doppler sonar acoustic sounding systems for remote detection of wind speed, turbulence and wind shear. Metering equipment for turbulent heat transfer. RASS, remote measurement of temperature profiles up to 1 km. Integrated systems with sophisticated micro-meteorological software for site studies, risk assessment applications. [98]

75-32300 Renault Véhicules Industriels SA (RVI)
129 rue Servient "La Part-Dieu", F-69431 Lyon Cedex 03
☎(+33 4) 78 63 72 50 ⊖ (+33 4) 78 63 71 60
✥ FFr 3,042,620,000 ♣ FFr 30,500,000,000 (1996)
👥 26,050
Chairman and Chief Executive Officer: Shemaya Lévy
Commercial Director: Pierre Colmant
Director of Research, Development and Military Studies:
Stanislas Wersinoski
Public Relations Director: Frédéric Fischer
■ Fire fighting vehicles.

Main production plants:
Annonay, Bourg-en-Bresse, Bouthéon, Blainville, Limoges, Saint-Priest, Vénissieux, Dunstable (UK), Villaverde (Spain).

Subsidiaries:
SINPAR
SEDEMS
SOMAC

Defence Equipment Division
40 rue Pasteur, PO Box 302, F-92156 Suresnes Cedex
☎(+33 1) 40 99 71 11; 40 99 73 15
⊜ (+33 1) 40 99 71 07
■ Tactical trucks, all wheel drive. Wheeled armoured all terrain and amphibious vehicles. Engines ranging from 60 to 1,000 hp. 98

75-32600 Jules Richard Instruments (JRI)
116 Quai de Bezons, PO Box 85, F-95101 Argenteuil
☎(+33 1) 39 47 09 36 ⊠ 608719 f
⊜ (+33 1) 39 47 00 94
♲ FFr 25,000,000 ♣ FFr 50,000,000 ⁂ 85
General Manager: Michel Fauveau
Sales Manager: Luc Lagorce
Marketing Manager: Georges Radisson
■ Temperature, humidity and pressure indicators, transmitters and recorders. 97

75-32750 Rockwell Light Vehicle Systems
(† Rockwell Automotive Body Systems SA)
(a subsidiary of Rockwell International Corp, USA)
Tour Gan, F-92082 Paris La Défense 2
☎(+33 1) 49 06 39 06 ⊠ 620178
⊜ (+33 1) 49 06 38 50
Vice President, Worldwide Materials and Quality:
Alain Deschamps E adeschamp@auto.rockwell. com
■ Worldwide materials, quality and marketing communications.
Representing:
Rockwell International Corp, USA

Roof Systems
30 rue d'Alsace, F-88360 Rupt sur Moselle
☎(+33 3) 29 23 46 46 ⊠ 960461 rckrm f
⊜ (+33 3) 29 24 39 46
Plant Manager, Roof Systems: Bruno Hirtz
E brhirtz@auto.rockwell.com

Access Control Systems
8 rue Sébastien Lehr, PO Box 153,
F-88104 Saint Die Cedex
☎(+33 3) 29 52 20 20 ⊠ 961473 rcksd f
⊜ (+33 3) 29 52 20 28
Plant Manager: Alain Turkietaub
E alturkie@auto.rockwell.com

Door Systems
105 route d'Orléans, PO Box 28, F-45600 Sully sur Loire
☎(+33 2) 38 29 88 00 ⊠ 760610 rcksl f
⊜ (+33 2) 38 29 89 00
Plant Manager: Denis Viale E dviale@auto.rockwell.com

Electric Motors and Electronic Controls
19 rue de Beauvoir, F-14220 Thury Harcourt
☎(+33 2) 31 30 71 10 ⊜ (+33 2) 31 79 42 93
Managing Director, Electric Motors and Electronic Controls: Thomas F Lewis (located in Troy, MI)
E tflewis@auto.rockwell.com
Plant Manager: Guy Lemoine 98

75-32800 Rockwell-Collins France SA
(a subsidiary of Rockwell International, USA)
6 Avenue Didier Daurat, BP 8, F-31701 Blagnac Cedex 1
☎(+33 5) 61 71 77 00 ⊠ 531569 f
⊜ (+33 5) 61 71 51 69
♲ FFr 10,000,000 (1996) ♣ FFr 311,000,000 (1996)
⁂ 289 (1996) ⌂ 6,000 m2
Managing Director: Bernard Loth
Deputy Managing Director: Jean François Bellocq
Communications Manager: Ms Christiane Groussac
Defence Electronics Division: Pierre Rounagnac
■ Products: radiocommunication, radionavigation, command and visualisation systems, automatic test benches, transervers, rotors, GPS, network products. Systems: acoustic analysis, data transmission, network integration, multimedia networks. Services: repair and distribution of Rockwell-Collins equipment, on-site maintenance, training, logistics support, network services.

Centre Commercial Belle Epine
Tour Europa, F-94532 Thiais Cedex 109
☎(+33 1) 45 12 23 50 ⊠ 263359
⊜ (+33 1) 46 86 54 01
Manager of Avionics Sales: Alain Avellan
■ Avionics and marine inertial navigation and telecommunications systems. → 90
Representing:
Rockwell International Corp, USA 98

75-32840 Rockwell SVI SA Heavy Vehicle Systems
24 rue Scheurer Kestner, BP 540,
F-42007 Saint Étienne Cedex 1
☎(+33 4) 77 92 88 00 ⊠ 300751 rckwell f
⊜ (+33 4) 77 92 88 93
Site Manager: Gary Blake (acting)
Representing:
Rockwell International Corp, USA 98

75-32850 Rockwell Telecommunications Sarl
Tour Gran, F-92082 Paris La Défense 2
☎(+33 1) 49 06 39 80 ⊜ (+33 1) 49 06 39 90
Sales Manager, South Europe: Olivier Robert
■ Modems, modem controllers and global positioning systems.
Representing:
Rockwell International Corp, USA

Sophia-Antipolis Office
PO Box 283, F-06905 Sophia-Antipolis Cedex
Les Taissounières - B1, 1681 Route des Dolines,
F-06560 Sophia-Antipolis, Valbonne
☎(+33 4) 93 00 33 00 ⊜ (+33 4) 93 65 32 80
Executive Director, European Operations:
Guillaume d'Eyssautier 98

75-32900 Rohde & Schwarz France Telecom
Parc Tertiaire de Meudon, 25-27 rue Jeannne Braconnier, F-92366 Meudon la Foret Cedex
☎(+33 1) 41 36 11 13 ✉ RUSELEC Paris ⊠ 265759 f rsf
⊜ (+33 1) 41 36 11 11
Managing Director: P Ducasse
Sales Manager: P Le Caer
Manager: C Perrotin
Secretary General: P De Bussac
■ Electronic measuring instruments and test systems. Radio frequency signal generators, spectrum analyzers 0-26,5 GHz, radio (TX/RX) test sets, EMI/EMS systems, EMI receivers, field strength receivers, audio and video measuring equipment, ruggedized computers for process-control.
Representing:
Rohde & Schwarz GmbH & Co KG, Germany

Aix en Provence Office
Tertia II, F-13851 Aix en Provence Cedex 3
☎(+33 4) 94 07 39 94 ⊜ (+33 4) 94 07 55 11

Lyon Office
Miniparc Gerland Batîment 5,
89 Boulevard du Parc de l'Artillerio, F-69007 Lyon
☎(+33 4) 78 29 88 10 ⊜ (+33 4) 78 29 94 71 98

75-33000 Rohr Europe
(† Rohr France SA)
(a subsidiary of Rohr Inc, USA)
22 Chemin de la Crabe, PO Box 16,
F-31931 Toulouse Cedex
☎(+33 5) 61 30 59 59 ⊠ 530513
⊜ (+33 5) 61 49 52 02
♣ FFr 350,000,000 ⁂ 420
President and Director General: Daniel Abehsera
Vice President, Airbus: Bill Frank
Vice President, European Operations: Jean Luminet
Vice President, Hamburg: Klaus Zimmer
■ Assembly of A300, A310, A300/600, A320 A330 and A340 engine nacelles. Assembly of A310 aft pylon fairings. Painting of composite panels. Electrical wiring. 98

75-33250 ROR Rockwell SA
Zone Industrielle du Moulin à Vent,
9 rue des Frères Lumière, F-77290 Mitry Mory
☎(+33 1) 64 27 44 61 ⊜ (+33 1) 64 27 30 45
Managing Director: Ms Genevieve Auvity
Representing:
Rockwell International Corp, USA 98

75-33500 Ruggieri Technologie
Route de Gaudiès, F-09270 Mazères
☎(+33 5) 61 67 91 61 ⊠ 539000 f -ruggieri+
⊜ (+33 5) 61 67 91 69
♲ FFr 10,771,060 ♣ FFr 190,000,000 (1995) ⁂ 135
⌂ 9,500 m2
Managing Director: C Fournet
Operations Director: D Casenave
Commercial Director: P LeCompte
Commercial Export Manager: J L Arrigoni
Purchasing Manager: G Seillier
■ Manufacture of pyrotechnics and related equipment. 98

75-33650 Sabo France
Avenue Augustin Dumont 57, F-92240 Malakoff
☎(+33 1) 42 53 01 52 ⊜ (+33 1) 40 92 84 46
♲ FFr 500,000 ♣ FFr 60,000,000 ⁂ 50 ⌂ 1,500 sq m
President: Jean Vergez
Sales and Marketing Director: M Strenna
Operations Director: G Quillet
Purchasing Manager: G Roure
■ Manufacture of foam and powder extinguishing agents. Fire fighting equipment.

Naintre Plant
ZI La Naurais Bachaud, F-86100 Naintre
☎(+33 5) 49 19 17 17 ⊜ (+33 5) 49 19 17 00
■ Rapid intervention vehicles, powder trucks, fire fighting equipment, dry powders, foam liquids, gas and fire detection. 98

75-33810 SAE Alsetex
Société d'Armement et d'Études Alsetex
Malpaire, BP 5, F-72304 Precigné
☎(+33 2) 43 92 81 00 ⊜ (+33 2) 43 95 41 13
♲ FFr 14,400,000 (1997) ⁂ 90 (1997) ⌂ 2,800 sq m
Managing Director: J L Pomarède
Export Sales Manager: Xavier Fernier
■ Design and manufacture of pyrotechnical devices and active training ammunition, containing explosive, chemical and smoke producing products, riot control products, CS and CN tear gas grenades and dispersers.
Activities also encompass engineering related to ammunition, design and construction of explosive loading machinery.

Malpaire Factory
F-72304 Precigné
☎(+33 2) 43 92 81 00 ⊜ (+33 2) 43 95 41 13 98

75-33900 Safare-Crouzet SA
(a member of the SEXTANT Avionique SA, France)
98 Avenue Saint-Lambert, F-06105 Nice Cedex 2
☎(+33 4) 92 09 76 76 ⊠ 460813 safare f
⊜ (+33 4) 93 51 56 94
♣ FFr 20,000,000 (1996) ⁂ 140 (1996)
President: François Strippoli
Sales Director: Michel Toma
■ Design and manufacture of control tower installations and equipment, underwater acoustics including sonar intercept receivers for submarines and surface ships, passive sonar torpedo detection for surface ships, underwater wireless telephones for surface ships, submarines and divers. 98

75-33920 Saft
(a division of Alcatel Alsthom, Germany)
156 Avenue de Metz, F-93230 Romainville
☎(+33 1) 49 15 36 00 ⊠ 235566
⊜ (+33 1) 49 15 34 00
♲ FFr 500,011,900 (1995)
♣ FFr 4,400,000,000 (1995) ⁂ 6,821 (1995)
Chairman and Chief Executive Officer: D Gounot
Executive Vice President: F Putois
Chief Financial Officer: F Vincent
Communications Manager: Roland Bourgeois
General Counsel: M Valat
Technical and Purchasing Manager: J Quobex
Research Manager: A de Guibert
Human Resources Manager: F M Sissung
General Manager, Portable Batteries Group: F Putois
General Manager, Advanced and Industrial Batteries Group: J C Dutailly
General Manager, Power Electronics Group: J P Lescieux
Area Manager, North America: F Westfall
Area Manager, Europe: K Klinker
Area Manager, Asia-Pacific: J P Crinelli
■ Nickel-Cadmium batteries for aircraft engine starting and APU. Emergency power units. Emergency lighting units on aircraft. Power supply for navigation aids. Maintenance chargers. Special batteries for satellites, missiles, meteorology. Silver-zinc batteries and accumulators.
Subsidiary companies:
Saft Akkumulatoren und Batterien GmbH, Germany
Saft Nife Argentina SA, Argentina
Saft Australia Pty Ltd, Australia
Saft Nife GmbH, Austria
Safta, Belgium
GS-Saft Ltd, Japan
Saft Nife Brazil Ltda, Brazil

Saft Nife Corp, Canada
Saft Korea Co Ltd, Korea
Saft Nife Iberica, Spain
Saft America Inc, USA
Saft Nife Oy, Finland
Saft Hong Kong, Hong Kong
Saft Nife Ltd, Hong Kong
Saft Nife SpA, Italy
Saft America Inc, Mexico
Saft Nife M.E. Ltd, Middle East
Saft Nife AS, Norway
Saft Nife BV, The Netherlands
Saft Nife Ltd, UK
Saft Singapore Pte Ltd, Singapore
Saft Nife Power Systems Pte Ltd, Singapore
Saft Nife AB, Sweden

Corporate Communications Department
156 Avenue de Netz, F-93230 Romainville
☎(+33 1) 49 15 36 00 ⊖ (+33 1) 49 42 13 10
Corporate Communications Director: Roland Bourgeois
Communications Assistant: C Chevalier

Research Department
Laboratoire de Marcoussis, Route de Nozay, F-91460 Marcoussis
☎(+33 1) 69 63 10 00 ⊖ (+33 1) 69 63 16 31
Research Manager: Ms Anne de Guibert

Portable Batteries Department
ZI de Nersac, F-16640 Roullet Saint Estèpe
☎(+33 5) 45 90 50 26 ⊖ (+33 5) 45 90 50 71
General Manager: François Putois

Industrial Batteries
111 Boulevard Alfred-Daney, F-33300 Bordeaux
☎(+33 5) 56 29 24 00 ⊖ (+33 5) 56 50 10 70
General Manager: Jean Claude Dutailly

Advanced Batteries Department
Rue George Lechanché, PO Box 1039, F-86060 Poitiers Cedex 9
☎(+33 5) 49 55 48 48 ⊖ (+33 5) 49 55 48 50
General Manager: Jean-Claude Dutailly

Power Electronics Department
10 rue Jean-Perrin, BP 359, F-37173 Chambray les Tours
☎(+33 2) 47 80 88 60 ⊖ (+33 2) 47 28 07 19
General Manager: Jean-Pierre Lescieux

Lagord Office
1 rue Henri Freyssinet, ZI des Greffières, F-17140 Lagord
☎(+33 5) 46 00 00 29 ⊖ (+33 5) 46 67 89 28 [98]

75-33950
SAGEM SA

Société d'Applications Générales d'Electricité et de Mécanique

SAGEM
GROUPE SAGEM

6 Avenue d'Iéna, F-75783 Paris Cedex 16
☎(+33 1) 40 70 63 63 ⊠ TELESAGEM Paris ⊠ 205255 f
⊖ (+33 1) 40 70 66 00
⇔ FFr 181,280,000 ♣ FFr 9,349,000,000 (1996)
ꟷ 8,458 (1996) ⌂ 200,000 sq m
Chairman and Chief Executive Officer: Pierre Faurre
Executive Vice President and Chief Financial Officer: Mario Colaiacovo
Vice President and Managing Director, Defence and Security Division: Daniel Dupuy

Defence and Security Division
61 rue Salvador Allende, F-92200 Nanterre
☎(+33 1) 40 70 63 63 ⊠ 205255 f
⊖ (+33 1) 40 70 68 40
Executive Vice President and Managing Director: Daniel Dupuy
Managing Director, Navigation and Missile Guidance Department: C Jaeger
Managing Director, IR Detection and Seekers Department: A Delzongle
Managing Director, Security Morpho Systems Department: M Argouse
Managing Director, Avionics Department: J J Floch
Managing Director, Navy Department: Y Labasque
Managing Director, Land Systems Department: J P Gaucher
Director of Development and Diversification: M Argouse
Director of International Affairs: J Paccard
Deputy Director of International Affairs: Patrick Cheniere
■ Inertial Navigation and Guidance Stabilisation:
Inertial navigation systems for submarines, surface ships, land vehicles, aircraft and helicopters, spacecraft. Inertial guidance systems for tactical and strategic missiles, launch vehicles, torpedoes. Homing equipment. Gyro and accelerometer sensor units for stabilisation and flight control systems. Hybrid navigation and localisation systems, inertial-GPS, inertial-terrain contour matching.
Electro-optic Equipment and Systems:
Air reconnaissance line-scanners. Laser military applications. Day/night electro-optic sighting systems. Observation and fire control systems. Periscopes and electro-optic masts.
Integrated Systems:
Avionic systems, aircraft retrofit. Navigation/attack/reconnaissance systems, submarine centralised control.
Data and Image Processing and Transmissions:
Airborne and shipborne computers, mass memories. Digital cartography. Plans and mission planning systems. Image processing and contour matching. Automated fingerprint identification.
Secure Terminals and Systems:
Terminals for military applications. Network, system and terminal security. Encryption.

Argenteuil Plant
72-74 rue de la Tour Billy, F-95101 Argenteuil
☎(+33 1) 34 26 38 00 ⊠ 605108 f
⊖ (+33 1) 34 26 37 20
General Manager: A Dupuy
■ Research and development centre for navigation and defence. Optical sensors and applied physics. Inertial sensors. Sighting devices and electro-optics. Tactical missiles, stabilisation. Navy equipment.

Eragny Plant
PO Box 51, F-95612 Cergy Pontoise Cedex
☎(+33 1) 34 30 50 50 ⊠ 607387 f
⊖ (+33 1) 34 30 50 55
General Manager: A Dupuy
■ Research and development centre for defence and navigation. Automation. Strategic missiles and space. Aeronautic and land navigation. Aeronautic equipment support. UAV's systems. Avionics retrofit for fighter aircraft. Mission planning systems. Automated fingerprint identification.

Montluçon Plant
Usine de la Côte Rouge, PO Box 3247, F-03106 Montluçon Cedex
☎(+33 4) 70 28 75 00 ⊠ 990451 f
⊖ (+33 4) 70 28 28 01
General Manager: J P Jainsky
■ Manufacturing centre for inertial and electro-optic equipment, optic and electronic telecommunications equipment.

Poitiers Plant
Avenue des Hauts de la Chaume, PO Box 20, F-86280 Saint Benoit
☎(+33 5) 49 37 43 43 ⊖ (+33 5) 49 37 43 01
General Manager: G Cubier
■ Manufacturing centre for electro-optic equipment & systems.

After-Sales Service
10 rue des Beaux-Soleils, F-95520 Osny
☎(+33 1) 34 30 50 50 ⊠ 607631 f
⊖ (+33 1) 34 30 50 13
General Manager: C Chambert [98]

75-34300 Société d'Applications des Machines Motrices (SAMM)
Chemin de la Malmaison, F-91570 Bièvres Cedex
☎(+33 1) 69 35 80 00 ⊠ 604085 f
⊖ (+33 1) 69 35 10 20
⇔ FFr 12,000,000 ♣ FFr 350,000,000 ꟷ 495
Chairman and Chief Executive Officer: Jacques Texier
General Manager: Daniel Rivière
Marketing Director: Alain Baudry
Technical Director: Alain Arnaud
General Manager, Quality: Remy Monel
Product Support Manager: Yves Holvoet
Communications Department: Ms Veronique Montel
Publicity Services: Ms N Gransard
■ Hydraulic and electro-hydraulic servocontrols for aircraft and helicopters, servomotors, electro-pump units, electro-valves, pressure switches, pilot handgrips, pushbuttons, helicopter armament, hydraulic circuits (tanks, electropump units), check valves and relief valves, miniature solenoid valves and plug-in pressure switches. Harpoon for heavy helicopters (Super Puma AS 332). Turrets for light armoured vehicles, hydraulic and electro-hydraulic servocontrols, armoured vehicle suspensions.

Vernon Plant
43 rue des Prés, F-27950 Saint Marcel
☎(+33 2) 32 64 70 00 ⊠ 180313
⊖ (+33 2) 32 51 57 47 [98]

75-34312 Sandow-Technic SA
32 rue Chance Milly, F-92110 Clichy
☎(+33 1) 47 37 04 82 ⊖ (+33 1) 47 37 29 45
⇔ FFr 500,000 ♣ FFr 5,000,000 ꟷ 10
President: Mrs Véronique de Chavagnac
Commercial Manager: J Beord
■ Manufacture of elastic shock cords or bungee cords and elastic endless rings from 3 mm to 30 mm for tensioning, pull back, fastening, catapult launchers and shock absorbers for vibration testing. [98]

75-34330 SARMA
PO Box 29, F-26241 Saint Vallier sur Rhône
1 Avenue Marc Seguin, F-26240 Saint Vallier sur Rhône
☎(+33 4) 75 03 40 40 ⊠ 345818 sarma stval f
⊖ (+33 4) 75 03 40 00; 75 03 40 93
⇔ FFr 60,000,000 (1996) ♣ FFr 243,000,000 (1996)
ꟷ 523 (1996) ⌂ 23,800 m2
Managing Director: Pierre Raynal
Deputy Managing Director: Gérard Castaner
Finance and Legal Director: Etiennne de Quillacq
Systems Division Director: Alain Jeunehomme
Commercial Director: Gérard Roberge
Bearings Division Director: Alain Girousse
Rods Division Director: Paul Faure
Quality Director: Jacques Bonzon
Information Systems Director: Roland Poulet
Sales and Marketing Director: Gérard Roberge
Purchasing Manager: Raymond Flevee
■ Manufacture of mechanical flight control components. Flight control rods (metallic and composite), structural rods, torque tube load limiters, cables and cable tension regulators. Airframe ball bearings and spherical plain bearings, airflon bushes, pulleys, electromechanical equipment and actuators. [98]

75-34380 Scania Production Angers SA
PO Box 846, F-49008 Angers Cedex 01
☎(+33 2) 41 41 20 00 ⊖ (+33 2) 41 41 20 48
Vice President: Bengt Klingberg [98]

75-34432 Schlumberger Industries Data Acquisition & Recording
(a subsidiary of Schlumberger Ltd, UK)
1 rue Nieuport, PO Box 54, F-78141 Vélizy Villacoublay Cedex
☎(+33 1) 30 70 30 70 [E] dar@velizy.inst.slb.com
⊠ 205844 ⊖ (+33 1) 30 70 30 80
Managing Director: Jean-Paul Vernhes
Sales and Marketing Director: Bernard Espannet
Business Development Director: Jean François Sulzer
■ Manufacture of digital flight data and maintenance recorders, cartridge memories for airborne and military computers. Flight data analysis systems, telemetry and data acquisition systems, ground and airborne data recorders. [98]

75-34450 Schottel France Sarl
(a subsidiary of Schottel-Werft, Germany)
2 rue Amélie, PO Box 43, F-94192 Villeneuve, Saint Georges
☎(+33 1) 43 82 31 30 ⊠ 266050 schofra f
⊖ (+33 1) 43 82 69 08
⇔ 900,000 (1996) ♣ FFr 20,000,000 (1996) ꟷ 4
⌂ 600 m2
Managing Director: Thierry Dedeyan
Operations Director: Jean Claude Rodriguez
Representing:
Schottel-Werft, Germany [97]

75-34750 SECAPEM
11 rue Henri Beaudelet, F-77330 Ozoir la Ferrière
☎(+33 1) 64 40 00 57 ⊖ (+33 1) 64 40 21 55
President and Managing Director: J Pasqualini
■ Aerial towed targets, including banner targets for air-to-air and ground-to-air weapons training. [98]

75-34810 Société d'Études et de Fabrications Aéronautiques (SEFA)
Rue Marcel Paul 1bis, ZI de la Bonde, F-91742 Massy Cedex
☎(+33 1) 69 53 12 89 ⊖ (+33 1) 69 53 12 88
⇔ FFr 764,000
Managing Director: Manuel Fernandez
■ Aircraft skis for aircraft and helicopters, wheel units to move skid landing helicopters. Hydraulic tool kits for maintenance of aircraft and helicopters. [98]

75-34843 SEMT Pielstick
2 Quai de Seine, F-93202 Saint Denis Cedex
☎(+33 1) 48 09 76 00 ⊠ 233147 f
⊖ (+33 1) 48 09 78 78
♧ FFr 51,380,000 (1995) ♣ FFr 1,343,040,000 (1995)
👥 978 (1995)
Chairman: P Bousseau
Marketing and Licences Director: J L Cavellat
Assistant Marketing Manager: M F Hylander
■ High and medium speed diesel engines for naval applications in the range of 1,000 kW to 25,000 kW.
Other facilities:
Saint Nazaire
Jouet sur l'Aubois [98]

75-34850 Société Europeenne de Propulsion Head Office (SEP)
(a subsidiary of SNECMA, France)
24 rue Salomon de Rothschild, F-92150 Suresnes
☎(+33 1) 47 28 65 00 ⊠ 630906 sepsurn f
⊖ (+33 1) 47 28 05 36; 47 28 05 80
♧ FFr 100,449,960 (1995)
♣ FFr 5,018,000,000 (1996) 👥 3,326 ⌂ 92,000 m2
President and General Director: Roger Vignelles
☎(+33 1) 47 28 08 76 (fax)
Director General: Pierre Betin ☎(+33 1) 47 28 08 13 (fax)
Director General, Operations: Jacques Rossignol
☎(+33 1) 47 28 08 84 (fax)
Vice President, Public Relations: Claude Jarry
Director of Finance and Economic Affairs: Gilbert Font
Quality Director: Bernard Portejoie
Director of Labour Relations: Alain Aubineau
Advertising Manager: Claude Jarry
Director of Solid Propellant Composites Division: G Corteel
Director of Liquid Propellant Division: M Laroche
Press Officer: Ms Jocelyne Terrien
☎(+33 1) 47 28 66 75; 47 28 05 36 (fax)
■ Design and development of solid and liquid propulsion systems for ballistic missiles, tactical missiles and space launch vehicles, propulsion systems and equipment for satellites and spacecraft. Composite materials (ceramic matrix composites and carbon-carbon). Propulsion of ballistic missiles of the French deterrent force; tactical missile propulsion; pyrotechnic squibs; composite armour parts, satellite propulsion, Ariane engines.
DPPC Solid Propellant Composites Divison
Les Cinqs Chemins, Le Haillan,
F-33165 Saint Médard en Jalles
☎(+33 5) 56 55 30 00 ⊠ 560678
⊖ (+33 5) 56 55 30 01
Director: G Corteel
DGPL Liquid Propellant Division
PO Box G6, F-27207 Vernon
☎(+33 2) 32 21 72 00 ⊠ 770019
⊖ (+33 2) 32 21 27 01
Director: Michel Laroche
Saint Medard Plant
Les Cinq Chemins - Le Haillan, BP 37,
F-33165 Saint Medard en Jalles Cedex
☎(+33 5) 56 55 30 00 ⊠ 560678 fsep f
⊖ (+33 5) 56 55 30 01
Vernon Plant
PO Box 802, F-27207 Vernon
☎(+33 2) 32 21 72 00 ⊠ 770019 sep vernon
⊖ (+33 2) 32 21 27 01
Director of Propulsion and Liquids - Space Division: Michel Laroche
GPPE Small Propulsion and Equipment Division
Aérodrome de Melun-Villaroche,
F-77550 Moissy Cramayel
☎(+33 1) 64 71 46 97
⊖ (+33 1) 64 37 68 27; 64 09 33 38
Director of Small Propulsion and Equipment Division: Jean Lin Fournereaux
Centre d'Istres
PO Box 34, F-13800 Istres
☎(+33 4) 42 47 12 00 ⊖ (+33 4) 42 56 69 79 [98]

75-34885 Société d'Études, Réalisations et d'Applications Techniques (SERAT)
(a subsidiary of Aerospatiale, France)
RN 151, F-18570 Le Subdray
☎(+33 2) 48 55 97 60 ⊠ 640227
⊖ (+33 2) 48 55 97 67
♧ FFr 11,000,000
President: H Berger
Managing Director: D Lursat
Financial Manager: X Fisset
■ Design, development, manufacture and production of armament equipment, pyrotechnics and igniters particularly for engines, hollow charges and missile warheads. [98]

75-34887 SERCEL
Société d'Études, Recherches & Constructions Electroniques
(a subsidiary of Compagnie Générale de Géophysique, France)
16 rue de Bel Air, PO Box 439, F-44474 Carquefou Cedex
☎(+33 2) 40 30 11 81 ⊠ 710695 sercel carqf
⊖ (+33 2) 40 30 19 48
♣ FFr 688,000,000 👥 620
Chief Executive Officer: T Le Roux
Senior Vice President: P Rouiller
Sales and Marketing Manager: A Tisserand
Defence Sales: R Jaulin
■ Design and manufacture of seismic acquisition systems. Towed arrays for ASW sonars; underwater hydrophone arrays; winches for wire-type antennas and arrays. Hydraulic equipment. Shelters and caravans. Tracked vehicles.
US Branch:
SERCEL Inc [98]

75-34900 Servaux SA
(in partnership with AMI, France)
710 Chemin du Littoral, F-131016 Marseille
☎(+33 4) 91 46 60 00 ⊠ 420385 f
⊖ (+33 4) 91 46 40 93
♧ FFr 2,647,000 ♣ FFr 80,000,000 👥 100
⌂ 32,330 m2
President: Alain Esteve
Export Manager: C Valery
Technical Manager: Michel Ramon
■ Drop packs with liferafts. Emergency drop packs for desert and jungle survival and unsafe areas. Rescue, survival and search equipment. Fire-fighting equipment. Inflatable liferafts and jackets for aviation and marine applications. [97]

75-35000 SEXTANT Avionique SA Head Office
(a subsidiary of Thomson-CSF and Aerospatiale)
Aérodrome de Vélizy Villacoublay, BP 200,
F-78141 Vélizy Villacoublay Cedex
☎(+33 1) 46 29 70 00
W http://www.sextant-avionique.com ⊠ sxt 631240 f
⊖ (+33 1) 40 94 02 51
♧ FFr 227,276,650 ♣ FFr 4,900,000,000 (1995)
👥 6,200 (1996)
Chairman and Chief Executive Oficer: Jean Robert Martin (PDG)
Senior Vice President: Guy Barruchel
Senior Vice President: Jean Paul Lepeytre
Vice President Strategy and Communications: Gérard Delalande
Vice President, Manufacturing and Engineering Technology: Hervé Girault
Vice President Finance and Control: Lucien Arbel
Deputy Vice President Corporate Communications: Richard Jullien
Vice President Central Procurement: Daniel Delacour
Press Attaché: Ms Florence Fayolle
■ Fields of activity include flight control, navigation, displays and interfaces, on-board instruments, space electronics, automatic flight test equipment.
Jointly owned by Thomson-CSF (66%) and Aerospatiale (34%).
Subsidiaries:
Safare Crouzet SA → France 75
SEXTANT Avionique Inc → USA 90
SEXTANT Avionique Italia SpA → 90
SEXTANT Avionique Asia Pte Ltd, Singapore
Thomson-LCD (with Thomson-CSF)
Crouzet Automatismes, France
CIR, Switzerland
VDO Luft (with BGT)
Customer and Product Support Division
Aérodrome de Villacoublay, PO Box 59,
F-78141 Vélizy Villacoublay Cedex
☎(+33 1) 46 29 70 00 ⊠ sxt 631240 f
⊖ (+33 1) 46 32 85 96
Vice President, Commercial Air Transport Avionics: Jean Louis Moraud
Vice President and General Manager, Customer Support: Daniel Gruaz
■ Twenty-four hour customer and product support for airlines, armed forces and helicopter fleet operators.
Centre de Logistique International
Rue Marcel Dassault, BP 140,
F-86101 Châtellerault Cedex
☎(+33 5) 49 02 10 00 ⊠ sxt 790521 f
⊖ (+33 5) 49 02 10 90
Director: Christophe Humphry
Helicopter Avionics Division
25 rue Jules Védrines, F-26027 Valence Cedex
☎(+33 4) 75 79 85 11 ⊠ 345824 sxt f
⊖ (+33 4) 75 79 85 50
Vice President and General Manager, Helicopter Avionics: Jean Loïc Galle
Vice President, Space: Henry Col
Military Avionics Division
Rue Toussaint Catros, PO Box 91,
F-33166 Saint Médard en Jalles Cedex
☎(+33 5) 56 13 50 00 ⊠ thom 616780 f
⊖ (+33 5) 56 47 73 22
Vice President and General Manager, Combat Training and Aircraft Avionics: Dominique Simonneau
Vice President and General Manager, Military Transport, Surveillance and Aircraft Avionics: Gill Michielin
Meudon la Foret Office
18 Avenue du Marechal Juin,
F-92366 Meudon la Foret Cedex
☎(+33 1) 46 29 70 00 ⊖ (+33 1) 46 29 54 20
Vice President and General Manager, Regional and Business Aircraft Avionics: Jean Pierre Mortreux [98]

75-35080 SFIM
Société de Fabrication d'Instruments de Mesure

sfim

13 Avenue Marcel Ramolfo Garnier,
F-91344 Massy Cedex
☎(+33 1) 69 19 66 00 ✉ SFIM Massy ⊠ 602164 sfim f
⊖ (+33 1) 69 19 69 19
♧ FFr 118,064,640 (1995)
♣ FFr 1,880,000,000 (1996) 👥 2,760
Chairman and Chief Executive Officer: Pierre Poquin
Chief Operating Officer: Michel Faure
Deputy Chief Operating Officer: J Mahieu
Sales and Marketing Director: Pierre Bloch
Production: Alain Riveron
Quality Assurance and Inspection: B Clement
Land and Naval Warfare Systems: F Faget
Aerospace: R Kremer
International Affairs: Claude Robert
Measurement and Trajectography: François Romac
Product Support: Régis Montet
Public Relations Manager: Mrs Ginette Sarrazin
■ Space and Aeronautics:
Day and night gyrostabilized systems for helicopters; on-board video recording systems; night vision goggles; optics for observation satellites and aerial reconaissance; automatic flight control systems; sensors for flight control, navigation and guidance; actuators and flight control equipment; flight test systems; data acquisition, transmission and processing telecommunication; wiring and electrical distribution.
Land Warfare Systems:
Day and night gyrostabilized panoramic sights for main battle tanks and for anti-aircraft systems; fire control systems for lightweight land vehicles; day/night driverscopes for land vehicles; night vision equipment for infantry; submarine periscopes and vessel optical directors; inertial measurement units; vertical/horizontal aiming box and actuators for MBT turrets; sights; optical subsystems; HF antennas for tactical networks; trajectography systems for firing ranges. [98]

75-35085 SFIM ODS Head Office
(a subsidiary of SFIM Industries, France)
16/18 Avenue de l'Europe,
F-78140 Vélizy Villacoublay Cedex
☎(+33 1) 69 19 97 26 ⊖ (+33 1) 69 19 97 52
♧ FFr 106,531,100 ♣ FFr 160,000,000 (1996) 👥 250
Chief Executive Officer: Jean Renée Mahieu
Managing Director: Michel Iscla

■ II portable equipment, II and IR AFV equipment, ground and on-board video equipment, surveillance of large strategic sites, radar signal processing equipment.

Dijon Office
45 Avenue de Stalingrad, F-21059 Dijon Cedex
☎ (+33 3) 80 74 55 00 ⊠ 350981
⊖ (+33 3) 80 72 41 42 [98]

75-35100 Société Française de Moteurs à Induction SA (SFMI)

(a subsidiary of Auxilec, France)
5 rue du Clos d'en Haut, PO Box 115,
F-78702 Conflans Saint Honorine
☎ (+33 1) 39 19 60 60 ⊠ 695820 f
⊖ (+33 1) 39 19 96 31
♧ FFr 5,000,000 ♣ FFr 19,400,000 ⚉ 34 ⌂ 2,200 m2
Chairman: François Quentin
General Manager: Patrick Philibert
Sales Manager: Didier Froger
■ Special motors for aerospace applications, space and defence (AC/DC BLDC and PM-VR). Stepper motors, actuators and servomechanisms, speed reducers, blowers and fans. [97]

75-35660 Sillinger

BP 23, F-75560 Paris Cedex 12
☎ (+33 1) 43 07 21 55 ⊖ (+33 1) 43 40 70 49
♧ FFr 11,503,800 ⚉ 48 ⌂ 2,800 m2
Director: D Sarran
Sales and Marketing Director, Military Division: A Sillinger
Operations Director: Marcel Chopin
Purchasing Manager: Ms Martine Vee
■ Military inflatable boats from 3m to 6,30 m which can be fitted with 60 mm mortar or 12,7 mm heavy machine gun. Floating booms, floating flexible tanks for transport of hydrocarbons, fire-fighting boats. Anti-pollution craft 'Dynamic Skimmer'. Security equipment. [97]

75-36305 Société Nationale d'Études et de Constructions de Moteurs d'Aviation (SNECMA)

2 Boulevard du Général Martial Valin,
F-75724 Paris Cedex 15
☎ (+33 1) 40 60 80 80 ⊠ 600700 snecma f
⊖ (+33 1) 40 60 81 02
♧ FFr 1,800,615,400 (1996)
♣ FFr 9,145,881,763 (1996) ⚉ 11,013 (1996)
Chairman and Chief Executive Officer: Jean Paul Bechat
Executive Vice President and Chief Operating Officer: Yves Bonnet
Senior Vice President, Finance: Jacques Henry Dufour
Senior Vice President, Strategy: Raymond Poggi
Vice President, General Counsel and Secretary: Alain Bosser
Vice President, Human Resources: Claude Loppinet
Vice President, International Development: Philippe Humbert
Vice President, Public Affairs and Communications: Jean Besse
Vice President, Research Development and Quality: Jacques Caruel
Vice President, Commercial Engines: Pierre Alesi
Vice President, Military Engines: Georges Sangis
Vice President, Services: Pierre Cognet
Vice President, Engineering: Jean Paul Herteman
Vice President, Production and Procurement: Marc Ventre
■ Main products: turbojet and/or components of military aircraft, ATAR, M53, LARZAC, TYNE, M88. Civil aircraft: CFM56, CF6, GE90. → 55

Gennevilliers Plant
291 Avenue d'Argenteuil, PO Box 48,
F-92234 Gennevilliers Cedex
☎ (+33 1) 47 60 72 06 ⊠ 600700 snecma
⊖ (+33 1) 47 60 73 00
Managing Director: Pascal Senechal

Villaroche Plant
Rond point René Ravaud Reau, F-77550 Moissy Cramayel
☎ (+33 1) 60 59 71 23 ⊠ 600700 snecma f
⊖ (+33 1) 60 59 71 36
Managing Director: Jean Paul Herteman
Administration Director: Gérard Foucher

Évry Plant
Route Nationale 7, PO Box 81, F-91003 Évry Cedex
☎ (+33 1) 69 87 92 60 ⊠ 600700 snecma
⊖ (+33 1) 69 87 89 28
Managing Director: Claude Foucher
Administration Director: Marc Ventre

Montereau Plant
Aérodrome de Villaroche, PO Box 1936,
F-77019 Melun Cedex
☎ (+33 1) 64 14 88 00 ⊠ 692577 cfmav
⊖ (+33 1) 64 14 87 65
Managing Director: Jacques Renvier

Le Creusot Plant
Avenue de l'Europe, PO Box 97,
F-71203 Le Creusot Cedex
☎ (+33 3) 85 77 77 12 ⊖ (+33 3) 85 77 77 77
Administration Director: Dominique Vaugier [98]

75-36306 SNFA

Zone Industrielle 2, Batterie 900,
F-59309 Valenciennes Cedex
☎ (+33 3) 27 23 52 12 ⊠ 132276 f
⊖ (+33 3) 27 23 52 97
♧ FFr 5,700,000
President: Jean François Lamit
Director General: Olivier Lamit
Finance Director: Mme Geraldine Conrad
Technical Director: André Lamare
Commercial Director: Paul Prodromides
Industrial Director: Jean Pierre Garbuio
■ High precision ball and cylindrical roller bearings for applications in the aerospace industry. [98]

75-36315 Société Nationale des Poudres et Explosifs (SNPE)

12 Quai Henri IV, F-74181 Paris Cedex 04
☎ (+33 1) 48 04 66 66; 48 04 66 93
W http://www.snpe.com ⊠ 220356
⊖ (+33 1) 48 04 68 87
♧ FFr 368,407,800 (1996)
♣ FFr 4,379,000,000 (1996) ⚉ 5,700
Chairman and Chief Executive Officer: Jean Faure
Explosives and Propellants: Alain Brocart
☎ (+33 1) 64 99 10 02
Director of Advanced Technologies and Propulsion Division: Jacques Cardin ☎ (+33 1) 64 99 10 04
Director of Composite Materials Division: Jean Pierre Vettori ☎ (+33 1) 48 04 67 66
Advertising Manager: Ms Catherine Daudin
■ Design, development and manufacture of solid propellants for missiles, rockets, space launchers and gas generators. Liquid propellants and energetic materials for spacecraft: UDMH, MMH, (SNPE CHIMIE), ammonium perchlorate. Pyrotechnics for defence industry and space sectors. Countermeasures equipment for aircraft and helicopters. Risk modelling, operational safety analysis for industrial facilities. Research and development in energetic materials, new molecules and manufacturing processes.

Space and Defence subsidiaries:
Pyromeca

Composite Materials subsidiaries:
Structil, MBS Composites
Engineering subsidiary:
SNPE Ingenierie SA

Explosives and Propellants Division
9 rue Lavoisier, BP 2, F-91710 Vert le Petit
☎ (+33 1) 64 99 10 02; 64 99 12 34
⊖ (+33 1) 64 99 11 70; 64 99 13 55; 64 99 13 25
Director: Alain Brocart
■ Gun propulsion.

Composite Materials Division
12 Quai Henri IV, F-75181 Paris Cedex 4
☎ (+33 1) 48 04 65 63 ⊖ (+33 1) 48 04 68 60
Director: Jean Pierre Vettoti

Le Bouchet Research Centre
F-91710 Vert le Petit
☎ (+33 1) 64 99 10 02 ⊖ (+33 1) 64 99 11 70
Director of Research and Technology: A Davenas
■ Research and advanced studies on composite materials.

Centre du Bouchet
9 rue Lavoisier, BP 2, F-91710 Vert le Petit
☎ (+33 1) 64 99 10 04 ⊖ (+33 1) 64 99 10 08
Director: J Cardin

Sorgues Plant
F-84700 Sorgues
☎ (+33 4) 90 33 62 00 ⊠ 431090
⊖ (+33 4) 90 39 62 64
Manager: Mr Suroca
■ Military and civilian explosives, chemicals.

Toulouse Plant
F-31078 Toulouse
☎ (+33 5) 62 25 73 00 ⊠ 531913
⊖ (+33 5) 61 25 78 83
Manager: M S Morales
■ Chemicals.

Bergerac Plant
F-24108 Bergerac
☎ (+33 5) 53 63 63 63 ⊠ 560795
⊖ (+33 5) 53 63 62 25
Manager: M Meyniel
■ Nitrocellulose; military powders and combustible cases.

Saint Médard Plant
F-33160 Saint Médard
☎ (+33 5) 56 70 50 50 ⊠ 560784
⊖ (+33 5) 56 70 50 51
Manager: Mr Darney
■ Solid propellants; military powders and lightweight materials.

Angoulême Plant
F-16017 Angoulême
☎ (+33 5) 45 65 18 00 ⊠ 790041
⊖ (+33 5) 45 65 18 02
Manager: Mr Stepman
■ Spherical powders, solid propellants and gun cottons.

Pont de Buis Plant
F-29117 Pont de Buis
☎ (+33 2) 98 86 42 42 ⊠ 940144
⊖ (+33 2) 98 73 08 82
■ Sporting powders; military powders; plastic and pyrotechnical devices. [97]

75-36350 Société de Construction d'Avions de Tourisme et d'Affaires (SOCATA)

(a subsidiary of Aerospatiale)
Le Terminal - Bâtiment 413, Zone d'Aviation d'Affaires,
F-93352 Le Bourget Cedex
☎ (+33 1) 49 34 69 69; 49 34 69 70 ⊠ 532835 f
⊖ (+33 1) 49 34 69 71
♧ FFr 64,487,500 ⚉ 870 ⌂ 72,000 m2
President and Managing Director: Jean Marc de Raffin Dourny
Finance Director: Laurent Hermann
Technical Director: Dominique Deschamps
General Secretary: Patrick de Prevaux
Head of Public Relations: Ms Caroline Vallette
☎ (+33 1) 49 34 69 84
■ Manufacture and marketing of single piston engine 4/5 seat aircraft (Tampico, Tobago, Tobago XL, Trinidad, Trinidad TC), Tangara (twin engine piston aircraft), Epsilon and Omega (military trainers) and TBM 700 (single turbine 6/7 seat multi-mission aircraft).

Tarbes Plant
International Airport Tarbes/Ossun, PO Box 930,
F-65001 Tarbes Cedex
☎ (+33 5) 62 41 73 00 ⊖ (+33 5) 62 41 73 55 [97]

75-36400 SOCHATA Headquarters

(a subsidiary of SNECMA, France)
2 Boulevard du Général Martial Valin, F-75015 Paris
☎ (+33 1) 40 60 80 80 ⊖ (+33 1) 40 60 84 80
♣ FFr 947,000,000 (1996) ⚉ 1,202 (1996)
President: Pierre Cognet
■ Specialises in the maintenance of jet engines. Services provided for the French Air Force and Navy, foreign air forces and major airlines.

Saint Quentin en Yvelines Center
1 rue des Frères Farman, F-78114 Magny les Hameaux
☎ (+33 1) 30 96 28 00 ⊠ 695133 biavia
⊖ (+33 1) 30 96 29 84
Director: P D Lamaziere

Chatellerault Center
ZI Nord - Rue Maryse Bastié, PO Box 129,
F-86101 Chatellerault
☎ (+33 5) 49 21 27 51 ⊠ 790118 sochata
⊖ (+33 5) 49 93 00 59
Director: Michel Bischoff [98]

75-37100 Société Anonyme d'Études et Réalisations Nucléaires (SODERN)

20 Avenue Descartes, F-94451 Limeil-Brévannes Cedex
☎ (+33 1) 45 95 70 00 ⊠ 265543 sodern f
⊖ (+33 1) 45 69 14 02
♧ FFr 2,800,000 (1997) ♣ FFr 307,000,000 (1996)
⚉ 315 (1997) ⌂ 10,000 m2
Chairman and Managing Director: Henri Dugré
Director of Space Activities: Ph Lugherini
Director of Nuclear Activities: B Aubert
Personnel and Communication Manager: J L Guillaume
■ Optics:
Optical systems for space instrumentation (IR, UV), flight simulation, nuclear industry.
Nucleonics:
Neutron emission and detection, miss distance detection, NDT by neutron radiography, detection of explosives, chemical materials and narcotics by neutron activation. Bulk material analysis.

Space:
Satellites and spacecraft attitude measurement, star sensing, earth observation cameras. [98]

75-39455 Société d'Exploitation du Conditionnement Dydra
(a subsidiary of Süd Chemie AG, Germany)
12 rue du Port de la Celle, F-77670 Saint Mammes
☎(+33 1) 64 70 71 00 ⊖ (+33 1) 64 70 71 09
FFr 17,819,000 (1996) ♣ FFr 50,000,000 (1996)
45 (1996) 3,600 sq m
Managing Director: J M Devault
Commercial Director: P Cordonnier
■ Desiccants, silica-gel, clay, molecular sieve, breather valves. [98]

75-39510 Société des Céramiques Techniques (SCT)
(a subsidiary of US Filter)
PO Box 1, F-65460 Bazet
☎(+33 5) 62 38 95 95 ☒ 520194
⊖ (+33 5) 62 38 95 50
FFr 67,000,000 ♣ FFr 114,000,000 222
Managing and Sales Director: M Rigaux
Marketing Director: Ms Sokol
■ Advanced ceramics: ceramic armour tiles, metallised ceramics, ceramic power capacitors, feed-through electronic tube components, ceramic/metal sub-assemblies, diodes and thermistor housings, components for optical fibre, biomaterial ceramics (prosthesis). Ceramic membranes for micro, ultra and nanofiltration, ceramic gas filters. [98]

75-40350 Simonneau Marine International
ZI Saint Médard des Prés, PO Box 28,
F-85201 Fontenay le Comte Cedex
☎(+33 2) 51 69 57 40 ☒ 711549
⊖ (+33 2) 51 51 13 65
FFr 2,500,000 (1996) ♣ FFr 82,856,000 (1996)
40 (1996) 5,000 m2
Managing Director: Marc Neny
Sales and Marketing Director: Pierre Renard
Operations Director: Jean Jacques Tard
Purchasing Manager: Jean Dominique Haag
■ Manufacture of fast and ultra fast patrol craft. [98]

75-40375 Société de Construction des Avions Hurel-Dubois
Head Office
13/21 Avenue du Marèchel Juin, PO Box 6,
F-92360 Meudon la Forêt
☎(+33 1) 46 01 11 11 ☒ 250955
⊖ (+33 1) 46 30 78 58
FFr 81,744,500 (1997) ♣ FFr 830,000,000 (1996)
1,300 (1997) 27,550 m2
President and General Manager: Jacques Dubois
Chief Executive Officer: J P Brusson
Technical Director: A Fournier
Director of Human Resources: Jean Bebing
Commercial Manager: P Moots
■ Design and manufacture of prototype and production aircraft: structural parts, metallic composites and equipment. Design and manufacture of escape systems (military aircraft and aircraft during flight tests). Special studies and production mock-ups, simulators, test beds, toolings. Design and manufacture of nacelle for jet engines, jet engines (with and without thrust reversers. Furnishings for civil transport, aircraft and helicopters.
Affiliates:
Société Lorraine de Construction Aéronautique (SLCA)
Société de Précision Mécanique
Société Internationale de Nacelles Toulouse [98]

75-40400 Société des Fonderies d'Ussel (SFU)
(a subsidiary of the Pechiney Group)
ZI La Petite Borde, PO Box 48, F-19202 Ussel
☎(+33 5) 55 96 39 00 ☒ 590034
⊖ (+33 5) 55 96 39 39
300 (1996)
Chairman: J P Labbe
Managing Director: Antoine Darbois
■ Aluminium sand castings and low pressure precision sand castings for the aircraft and aerospace industries; highly sophisticated casting process for products requiring very accurate components. [98]

75-40470 Société Industrielle pour le Développment de la Sécurité (SIDES)
182 rue de Trignac, F-44600 Saint Nazaire
☎(+33 2) 40 17 18 00 ☒ 700094 f
⊖ (+33 2) 40 17 18 03
FFr 13,608,000 ♣ FFr 332,642,834 (1995) 220
32,000 m2
Chairman: Eric Patry
Managing Director: Roland Griaud
Sales and Marketing Director: Lionel Mollard
Export Sales Manager: Georges Guivarch
Publicity Assistant: Ms Christine Corgnet
■ Crash tenders and rapid intervention vehicles for military and civil aviation.
Le Blanc Mesnil Office
2/4 rue Blaise Pascal, F-93152 Le Blanc Mesnil
☎(+33 1) 49 39 40 00 ☒ 213737 f [98]

75-40500 Société Labbe SA
(† Société Nouvelle Labbe)
(a subsidiary of Ógara-Hess & Eisenhardt Armouring Co, USA)
PO Box 414, F-22404 Lamballe
☎(+33 2) 96 50 12 80 ⊖ (+33 2) 96 34 72 65
FFr 15,000,000 (1996) ♣ FFr 120,000,000 (1996)
180 12,000 m2
Managing Director: M Pecheur
Operations Director: Mr Cotin
Sales and Marketing Manager: M Tible
Purchasing Manager: Mr Seguin
■ Protected cars for VIP's; cash-in-transit security vehicles; refrigerated shelters for use by the French Army; custom built special containers for heavy armoured transport for government agencies. Special vehicles for Medical Corps units and the French Navy. [98]

75-40570 Société Métallurgique de Gerzat SA Factory and Commercial Services (SMG)
(a part of the Péchiney Group)
Rue de l'Industrie, PO Box 7, F-63360 Gerzat
☎(+33 4) 73 23 64 00 ✉ METAGERZAT ☒ 392660
⊖ (+33 4) 73 25 49 92
FFr 5,000,000 (1995) ♣ FFr 121,600,000 (1995)
168
Chairman and Chief Executive Officer: J P Labbé
Operations Director: G Guillet
Sales and Marketing Director: B Florenville
Sales and Marketing Director: M Torond
Purchasing Manager: J Lemaire
■ Aluminium alloy high pressure cylinders. Tubular parts with or without bottoms for missiles, rockets, projectile elements. Hydraulic container, jacks. Liners for fibre-wound cylinders. [97]

75-40700 Société SMAC
(a subsidiary of Société Thomson-CSF, France)
ZI Toulon Est, BP 119, F-83079 Toulon Cedex 09
☎(+33 4) 94 75 24 88 ⊖ (+33 4) 94 75 94 99
FFr 400,000 (1996) ♣ FFr 11,000,000 (1996)
28 (1997) 2,000 m2
Managing Director: Alain Sybilensky
☎(+33 4) 92 96 41 01
Operations Director: Thierry Lemoine
Sales and Marketing Director: Alain Pulido
Manager: Philippe Robert
■ Manufacture of rubber and composite engineering parts for transports and industry, marine and armament, underwater connectors, aircraft and space. [98]

75-41520 Société des Tuyaux Bonna
(a subsidiary of Compagnie Générale des Eaux, France)
33 Place Ronde-Espace 21 - Valmy,
F-92981 Paris La Défense Cedex 81
☎(+33 1) 46 53 24 00 ⊖ (+33 1) 46 53 24 11
FFr 88,395,000 (1995) ♣ FFr 1,577,311,000 (1995)
2,400
President: Michel Rivard
Deputy General Manager: André Bernard
Operations Director: François Leblanc
Foreign Department Manager: René Latinus
Purchasing Manager: Xavier Braquilanges
■ Manufacture of NBC modular shelters. [98]

75-42030 SOGERMA Maintenance Group Headquarters
Société Girondine d'Entretien et de Réparation de Matériel Aéronautique
(a part of Aerospatiale Group, France)
Aéroport de Bordeaux-Mérignac, BP 2,
F-33701 Mérignac Cedex
☎(+33 5) 56 55 40 00 ☒ 540748 sogerma f
⊖ (+33 5) 56 34 37 72
FFr 157,685,050 (1996)
♣ FFr 3,307,000,000 (1996) 3,864 (1996)
Chairman and Chief Executive Officer: Henri Paul Puel
Executive Vice President: Bernard Birotteau
Senior Vice President, Strategy and Finance: Régis Trolliet
Senior Vice President, Industry: Jacques Meron
Senior Vice President, Sales and Support: Emmanuel Huot-Marchand
Vice President, Resources: Ms Anne Marie Roux
Vice President, Finance: Gaëtan de Fourtou
Vice President, Press and Communication: Ms Elizabeth Broge
Company Secretary: Jean Luc Monceaux
Chairman and Chief Executive Officer of REVIMA: Philippe Debrun
Executive Vice President, SECA: Yves Boyer
Chairman and Chief Executive Officer, Composites Aquitaine: Jean Charles Joseph
■ Maintenance, overhaul, repair and transformation of all types of aircraft. Repair of hydraulic, electrical and radio navigational equipment. Manufacture of components, spares and ancillary equipment for aircraft, development and building of prototypes, aeronautical engineering. Integration of electronic defence and weapon systems.
Ancien Arsenal
BP 109, F-17303 Rochefort Cedex
☎(+33 5) 46 83 62 00 ☒ 790413 f socearo
⊖ (+33 5) 46 99 96 23
636 (1996)
■ Crew seats RW and FW (flight and deck cabin). Passenger seats, galleys and stowages. All type of test benches. GPU (1,5 to 120 KVa). Frame manufacture (metallic and composite). [98]

75-42050 Sogitec Electronic Division
(a subsidiary of Dassault Aviation, France)
4 rue Marcel Monge, Immeuble Nobel,
F-92158 Suresnes Cedex
☎(+33 1) 41 18 57 00 ☒ 616664 f
⊖ (+33 1) 41 18 59 09
FFr 30,019,200 (1996) ♣ FFr 410,000,000 (1996)
440 (1996) 14,000 sq m
Chairman and Chief Executive Officer: Yves Fouché
Finance and Accounts Director: Jean Pierre Taquet
Sales and Marketing Director: Joël Clairon
Management Attaché: Ms Brigitte Viaud
Operational Advisor: Bruno Le Cornec
■ Training for armed forces personnel; engineering and development simulators; validation; measurement and test equipment; instruction; computerized maintenance management system; technical documentation; electronic publishing; documentation systems engineering.
Electronics Division
4 rue Marcel Monge, Immeuble Nobel,
F-92158 Suresnes Cedex
☎(+33 1) 41 18 57 00 ☒ 616664 f
⊖ (+33 1) 41 18 57 18
Director: Jean Mayou
Export Sales Manager: Jean Claude de Laforcade [98]

75-42400 Sorep
Société Rennaise d'Electronique Professionnelle SA
(a subsidiary of Dassault Electronique SA, France)
ZI de Bellevue, BP 5, F-35220 Chateaubourg
☎(+33 2) 99 00 94 00
⊖ (+33 2) 99 00 94 57; 99 00 94 58
FFr 11,696,200 (1997) ♣ FFr 197,000,000 (1996)
430 (1996) 8,000 m2
Chairman of the Supervisory and Chief Executive Officer: Jean-Etienne Connet
Vice Chairman of the Supervisory Board: François Pontet
President and Chief Operating Officer: Guy Dadou
Managing Director: Jean Claude Le Bleis
■ Design and manufacture of hybrids, MCM's, RF8 microwave modules, boards and ASIC's. Assemblies and test on printed circuit boards. Packaging technologies (LTCC, Fine Pitch, BGA) for military, civil, aerospace, telecommunications and oil prospection applications. → 80
Sales Office
ZI Bellevue, PO Box 5, F-35220 Chateaubourg

☎ (+33 2) 99 00 94 70 ⊖ (+33 2) 99 00 94 58
Export Sales Manager: Bernard Berne
Export Sales Manager: Mokrane Yaker [98]

75-42775 STAREC
Société Technique d'Application et de Recherche Electronique
(a subsidiary of SFIM Industries, France)
7 Avenue de l'Atlantique, F-91952 Les Ulis Cedex
☎ (+33 1) 69 18 34 00 ⊖ (+33 1) 69 18 34 01
♧ FFr 10,000,000 ♣ FFr 520,000,000 (1996) 👥 70
Chief Executive Officer: P Pinot
Managing Director: Albert Levionnois
Marketing Manager: Daniel Lafargue
■ Research and development in electronics and electromagnetism. Airborne antennas for aircraft, missiles and space vehicles. Ground antennas for Satcom, telemetry, tracking stations HF/VHF/UHF/SHF, infrastructure and tactical communications, electronic warfare. Multicouplers, filters, hybrid junctions.
Production subsidiary:
STAREC Chateauroux
USA representative:
Sabre Communications Corp [98]

75-43150 Swarovski Optik
(a subsidiary of Swarovski Optik KG, Austria)
9 rue du Faubourg Poissonière, F-75009 Paris
☎ (+33 1) 48 01 92 80 ⊖ (+33 1) 48 01 00 57
Director General: Pierre Sévery
■ Manufacture of binoculars and telescopes. [98]

75-43670 Ets Jean Tanazacq
(† Ardennlame SA)
Place de la Mairie, F-08260 Maubert Fontaine
☎ (+33 3) 24 35 22 97
♧ FFr 250,000 👥 12
Director General: J Tanazacq
■ Combat and survival knives, daggers in high carbon stainless steel. Camp knives-machettes. [98]

75-44100 Techniques Michel Brochier SA (TMB)
ZI route de Chartinières, Dagneux, F-01120 Montluel
☎ (+33 4) 78 06 32 22 ⊠ 305221
⊖ (+33 4) 72 25 98 25
♣ FFr 26,395,693 👥 35
Chairman and Managing Director: Michel Brochier
Commercial: Patrick Michaud
■ Manufacture of a range of flexible coated fabrics sealed against toxic chemical agents and capable of being high-frequency welded. Finished products range from garments and tents to full-scale aircraft hangars. Tents for hospital with ventilation/filtration systems. Casualty bags for wounded people. [97]

75-44300 Technofan SA
(a division of Labinal Group, France)
10 Place Marcel Dassault, BP 53, F-31702 Blagnac Cedex
☎ (+33 5) 61 30 92 00 ⊠ 530858 f
⊖ (+33 5) 61 30 02 04
♧ FFr 7,531,920 👥 110 🏭 7,000 m2
President and Chief Executive Officer: Jean Claude Lepage
Vice President: Jacques Mignot
Managing Director: Jean Piquet
Sales and Marketing Director: Thierry Misson
Operations and Technical Director: Philippe Carette
Purchasing Manager: Ms Florence Bergoend
■ High performance fans, air valves, heaters and associated temperature control systems, RPM sensing systems, air sound reducers, mainly for application in aerospace industry and air forces. Complete wind tunnel sets, high RPM machines such as turbo generators. [97]

75-44375 Tecphy
Avenue Jean-Jaurès, F-58160 Imphy
☎ (+33 3) 86 21 32 78 ⊖ (+33 3) 86 21 31 01
♧ FFr 193,183,515 ♣ FFr 632,500,000 (1996)
👥 498 (1996)
Managing Director: Alain Pradoura
Sales and Marketing Manager, Aerospace and Defence: Jean Louis Deneuville
■ Heat resistant alloys and super-alloys. High strength steel and semi-products.
Imphy Plant
Avenue Jean Jaurès, BP 2, F-58160 Imphy
☎ (+33 3) 86 21 32 78 ⊖ (+33 3) 86 21 31 01
Firminy Plant
Rue Colonel-Riez, BP 141, F-42704 Firminy Cedex
☎ (+33 4) 77 40 37 00 ⊖ (+33 4) 77 40 36 99 [98]

75-44470 Teleflex-Syneravia SA
(a part of Teleflex Lionel Dupont Group)
15/19 rue d'Antony - Zone Silic, PO Box 449, F-94152 Rungis Cedex
☎ (+33 1) 45 12 84 84 ⊠ 265924 f
⊖ (+33 1) 45 12 84 48
♣ FFr 131,000,000 (1996) 👥 174
President and General Manager: Francis Parain
Industrial Director: Didier Cote
Technical Director: Jean-Pierre Batisse
Finance Director: Pierre Colombet
Sales and Marketing Director: L Rasmont
Purchasing Manager: P Bouzekri
Product Support: P Pierrard
Export: Pierre Boyer
■ Motorised variable pitch aircraft seat adjustment mechanism. Remote control systems (cable or ball drive). Linear and rotary actuators. Aircraft exterior lights. Windshield wiper systems. [98]

75-44930 THERMOCOAX
10 rue de la Passerelle, F-92150 Suresnes
☎ (+33 1) 41 38 80 80 ⊖ (+33 1) 41 38 80 89
Manager: Claude Capron
Sales and Marketing Manager: C Wolff
Market Communications Co-ordinator: Ms Karin Kuhn
☎ (+33 2) 33 62 81 02; 33 62 81 03 (fax)
■ R&D, manufacturing and marketing of products based on mineral cable technology: Temperature sensors: thermoelectric cables and thermocouples, high temperature. Weigh-in-motion, proximity and displacement (capacitive), overheating and fire, neutron. Heating cables and elements: rods, heating devices (de-icing, de-misting). Signal transmission cables and their applications. [97]

75-45100 Thomson-CSF Corporate Headquarters
(a subsidiary of Thomson SA, 58%)
173 Boulevard Haussmann, F-75415 Paris Cedex 08
☎ (+33 1) 53 77 80 00 ⊖ (+33 1) 53 77 83 00
♧ FFr 2,390,862,380 (1996)
♣ FFr 36,300,000,000 (1996) 👥 46,500 (1996)
Chairman and Chief Executive Officer: Marcel Roulet
Senior Executive Vice President: Daniel Rapenne
Senior Executive Vice President: Bernard Rétat
Executive Vice President, International Business: Jean Paul Perrier
Vice President, Finance: Alain Hagelauer
Vice President and General Secretary: Pierre Cabanes
Vice President, Corporate Communications: Ms Sylvie Dumaine
Chief Press Officer: Christophe Robin
Press Officer: Ms Véronique Trivero
Main commercial delegations and offices:
Europe:
Austria Tel: (+43 1) 796 35 94
Belgium Tel: (+32 2) 391 22 11
Germany Tel: (+49 228) 81 99 40
Greece Tel: (+30 1) 77 07 703
Italy Tel: (+39 6) 841 22 08
Netherlands Tel: (+31 70) 360 19 65
Poland Tel: (+48 22) 625 68 28
Russia Tel: (+7 095) 917 44 24
Spain Tel: (+34 1) 563 85 18
Sweden Tel: (+46 8) 660 35 39
Switzerland Tel: (+41 31) 991 62 66
Turkey Tel: (+90 312) 446 51 97
UK Tel: (+44 171) 355 45 43
North America:
Canada Tel: (+1 613) 723 70 00
USA Tel: (+1 703) 838 96 85
Middle East:
Kuwait Tel: (+965) 240 65 36
Oman Tel: (+968) 56 10 32
Qatar Tel: (+974) 32 87 85
Saudi Arabia Tel: (+966 1) 477 50 97
UAE Tel: (+971 2) 76 28 82
Asia Pacific:
Australia Tel: (+61 6) 273 32 66
China Tel: (+86 10) 65 93 29 74
India Tel: (+91 11) 371 52 62
Indonesia Tel: (+62 21) 725 02 32
Japan Tel: (+81 3) 3264 63 41
Malaysia Tel: (+60 3) 262 26 00
Pakistan Tel: (+92 51) 81 32 69
Philippines Tel: (+63 2) 817 57 36
Singapore Tel: (+65) 299 09 20
South Korea Tel: (+82 2) 790 72 33
Taiwan Tel: (+886 2) 705 05 33
Thailand Tel: (+66 2) 254 34 00
Vietnam Tel: (+84 48) 22 62 80
Africa:
Egypt Tel: (+20 2) 348 52 36
Morocco Tel: (+212) 777 08 79
South Africa Tel: (+27 12) 348 46 00
South America:
Argentina Tel: (+54 1) 394 20 74
Brazil Tel: (+55 21) 240 27 07
Chile Tel: (+56 2) 698 30 71
Venezuela Tel: (+58 2) 261 14 81

Aerospace Group
173 Boulevard Haussmann, F-75415 Paris Cedex 08
☎ (+33 1) 53 77 80 00 ⊖ (+33 1) 53 77 81 18
Group Executive Vice President: Noël Clavelloux
■ Radar and countermeasures, optronic equipment, simulators, missile electronics, armament and propulsion, electron tubes, power generation and conversion systems, microwave micro-electronics, hybrid circuits.
Subsidiaries:
Auxilec → 75
TCAR → 75
TDA → 75
Thomson-CSF Radars & Contre-Mesures
Thomson-CSF Optronique
Angénieux
Thomson Tubes Electroniques, France
Thomson Training & Simulation SA, France
Thomson Training & Simulation Inc, USA
Thomson Training & Simulation Ltd, UK
Pilkington Optronics, UK
Thomson-CSF Missile Electronics SA
Thomson-Thorn Missile Electronics Ltd, UK

Thompson-CSF Radars & Contre-Mesures (RCM)
La Clef de Saint Pierre, 1 Boulevard Jean Moulin, F-78852 Elancourt Cedex
☎ (+33 1) 34 59 60 00 ⊖ (+33 1) 34 59 62 36
Chairman and Chief Executive Officer: Bernard Rocquemont
Deputy Manager and Communications: Patrick Henin
■ Airborne electronics equipment: radars for combat and surveillance aircraft, airborne and marine electronic warfare systems, electronic countermeasures, airborne mission systems.

Thomson-CSF Missile Electronics (TME)
23-27 rue Pierre Valette, F-92245 Malakoff Cedex
☎ (+33 1) 49 65 30 30 ⊖ (+33 1) 49 65 30 60
Chairman and Chief Executive Officer: Stanislas Guérin
Commercial Director: Serge Graveleau

Thomson Training & Simulation SA (TT&S)
PO Box 226, F-95523 Cergy Pontoise Cedex
1 rue du Général de Gaulle ZI Les Beaux Soleils
☎ (+33 1) 34 22 82 00 ⊖ (+33 1) 34 22 87 00
Chairman and Chief Executive Officer: Louis Le Portz
Communications Manager: Ms Brigitte Porée
■ Manufacture of simulators for civil and military applications. Training services.

Communications and Command Group
173 Boulevard Haussmann, F-75415 Paris Cedex 08
☎ (+33 1) 53 77 80 00 ⊖ (+33 1) 53 77 89 65
Group Executive Vice President: Roger Chevrel
■ Communication networks and command systems, radiocommunications, electronic warfare and secure communications, navigation, identification.

Thomson-CSF Communications (TCC)
66 rue du Fosse Blanc, PO Box 156, F-92231 Gennevilliers Cedex
☎ (+33 1) 46 13 20 00 ⊠ 620840 f thom
⊖ (+33 1) 46 13 21 63

Thomson-CSF Communications
BP 82, F-92704 Colombes Cedex
Parc d'Activités Kleber 160 Boulevard de Valmy
☎ (+33 1) 41 30 30 00 ⊖ (+33 1) 41 30 33 57
Chairman and Chief Executive Officer: Roger Chevrel
President: Jean Dufour
President: Jose Massol
Commercial Director: Guillaume Dehollain

Detection & Missile Systems Group
173 Boulevard Haussmann, F-75415 Paris Cedex 08
☎ (+33 1) 53 77 80 00 ⊖ (+33 1) 53 77 88 05
Group Executive Vice President: Jean Loup Picard
■ Surface detection systems, naval combat systems, missile systems, anti-submarine warfare.
Subsidiaries:
Thomson-CSF Airsys
Hollandse Signaalapparaten BV, Netherlands
Thomson-CSF NCS France
Thomson Marconi Sonar (France, UK)
Wilcox Electric Inc, USA

Thomson-CSF Airsys (TCA)
7-9 rue Mathurins, F-92221 Bagneux Cedex
☎(+33 1) 40 84 40 00 ⊖ (+33 1) 40 84 33 81
Chairman and Chief Executive Officer: Jean Loup Picard
President: François Lureau
■ Battlefield detection, test ranges, air defence systems, air traffic control systems, navaids, naval combat systems.

Thomson-CSF NCS France (TNF)
7-9 rue des Mauthurins, F-92221 Bagneux Cedex
☎(+33 1) 40 84 20 00 ⊖ (+33 1) 40 84 33 81

Information Technology Group
173 Boulevard Haussmann, F-75415 Paris Cedex 08
☎(+33 1) 53 77 80 00 ⊖ (+33 1) 53 77 89 82
Group Executive Vice President: Jacques Vannier
■ Consulting engineering, software and systems integration, facilities management, software products, computer products, maintenance, logistics and support services.

Subsidiaries:
Aonix → 75
CETIA → 75
Sodeteg → 75
SYSECA → 75
Thomson-CSF Service Industrie
Thomson-CSF Services & Systèmes Sol Spatiaux

Thomson-CSF Service Industrie (TSI)
105 Avenue du Général Eisenhower, BP 1009,
F-81023 Toulouse Cedex
☎(+33 5) 61 19 40 00 ⊠ tcsf 5314423 f
⊖ (+33 5) 61 19 40 01
Chairman and Chief Executive Officer: Bernard Angotti
President: Michel Hanotte
■ Military computers, ruggedised workstations, data handling and display systems, tactical data processing, software.
Operation support and on-site assistance, maintenance and instrumentation training services, international shipping expertise.
Other activities include silicon and GaAs semiconductors, piezo-electric components, passive components.

Thomson-CSF Service Industrie
13 Avenue du Président, Salvadore Allende,
F-94117 Arcueil Cedex
☎(+33 1) 41 48 40 00 ⊠ tcsf 631096 f
⊖ (+33 1) 41 48 40 03
President: Jean Petiot

Laboratoire Central de Recherches (LCR)
Domaine de Corbeville, F-91404 Orsay Cedex
☎(+33 1) 60 33 00 00 ⊖ (+33 1) 60 33 08 78
Managing Director: Jean Pierre D'Haenens
■ Advanced research into new technologies, material science, optics, liquid crystal displays and rapid semiconductor components. [98]

75-45135 Thomson Sintra Activités Sous-Marine
(a subsidiary of Thomson-CSF, France)
525 route des Dolines, PO Box 157,
F-06903 Sophia Antipolis Cedex
☎(+33 4) 92 96 40 18 ⊠ tomsi x 470111f
⊖ (+33 4) 92 96 40 19
♣ FFr 1,500,000,000 ⍭ 1,500
Chairman and Chief Executive Officer: Denis Ranque
Sales and Marketing Director: André Rivière
Export Sales Manager: Daniel Briffaut
Foreign Operations: Jean Georges Malcor
Marketing Director: Philippe Portalier
External Relations Manager: Michel Trastour
■ Design and manufacture of ASW (including mine warfare) equipment and systems. [97]

75-45137 Thomson-CSF Optronique
(a subsidiary of Thomson-CSF, France)
Rue Guynemer, PO Box 55, F-78283 Guyancourt Cedex
☎(+33 1) 30 96 70 00 ⊠ 616780 thom f
⊖ (+33 1) 30 96 75 50
♲ FFr 200,000,000 (1996)
♣ FFr 1,150,000,000 (1996) ⍭ 1,000 (1996)
⌂ 7,000 sq m
Managing Director: Jean Barbet
Operations Director: Michel Keller
Public Relations Manager: Yves Huguenin
■ Manufacture of airborne optronic systems and equipment. [98]

75-45150 THORN Europhane
(a subsidiary of Thorn Lighting Group plc, UK)
156 Boulevard Haussmann, F-75008 Paris
☎(+33 1) 49 53 62 62 ⊠ 380900; 651061 f
⊖ (+33 1) 49 53 62 40; 49 53 62 89
♲ FFr 81,757,000 ♣ FFr 600,000,000
Chief Executive: J P Achale
International Airport Lighting Manager: Bernard Baeyens
■ Airfield lighting equipment: runway, taxiway, approach light fittings (elevated or inset), PAPI system, obstruction and apron lighting. Electrical equipment for series supply (isolating transformers, constant current regulators, control panels). Helicopter platform and general lighting.

Subsidiaries:
Thorn AFL Department, UK
DNT Europhane Pty Ltd, Australia [98]

75-45175 Timet France Sarl
Immeuble Les Cerclades, 2 Mail des Cerclades - BP 183,
F-95023 Cergy Pontoise Cedex
☎(+33 1) 34 41 63 64 ⊖ (+33 1) 34 41 63 60

Associated company:
Timet, UK [98]

75-45200 Titan Aviation
(† Titan SA)
472 rue Paul Claudel, PO Box 407,
F-69651 Villefranche sur Saône
☎(+33 4) 74 68 24 38 ⊠ 305718
⊖ (+33 4) 74 65 52 97
♲ FFr 3,500,000 ♣ FFr 20,000,000 ⍭ 21 ⌂ 3,200 sq m
Chairman: Guy Vernay
Operations Director: René Vaurille
Sales and Marketing Director: Christophe Dufour
Purchasing Manager: Ms Anne Marie Lacombe
Aviation Division Manager: Mr Klein
■ Aircraft fuellers and dispensers. Fuelling systems. [97]

75-45300 Titeflex Europe
Avenue Maurice Chevalier, PO Box 73,
F-77833 Ozoir la Ferrière Cedex
☎(+33 1) 60 18 52 00 ⊖ (+33 1) 64 40 23 37
♲ FFr 3,000,000 (1996) ♣ FFr 56,000,000 (1996)
⍭ 70 (1996) ⌂ 3,530 m2
Managing Director: Ms Fabienne Mahieu
Sales Director: Jean Yves Kerleau
Technical and Quality Director: Pierre Doucet
Purchasing Manager: Mrs L Berthereau
■ Flexible hoses in teflon and stainless steel; fittings. [98]

75-45400 TOPNAV - Positioning Services
(a subsidiary of SOGERMA, France)
1 rue Léon Migaux, F-91341 Massy Cedex
☎(+33 1) 64 47 36 94 ⊖ (+33 1) 64 47 39 34
Manager: D Paul
Sales: S Florin
■ Positioning, geodesy, surveying by all methods. Fleet management systems. [98]

75-45500 Tramec SA
34 rue Gambetta, PO Box 45, F-92174 Vanves Cedex
☎(+33 1) 46 42 12 16 ⊖ (+33 1) 46 42 47 20
♲ FFr 400,000 (1997) ♣ FFr 3,000,000 (1996)
⍭ 10 (1997) ⌂ 300 m2
Chairman: Jean Baptiste Fattaccini
General Manager: Ms Adeline Mela
Manager: Patrice Raimbault
■ SSB/HF tubular antennas, UHF, VHF and EHF antennas, antenna masts, feed through bulkhead union. Anti-static antenna cable, dual swivels, insulators. Tension adjusters, turnbuckles. Terminal boards, connecting blocks, connection boxes, dimmers. All models of shunts. [98]

75-45800 Trimble Navigation France
9 rue de l'Arpajonnais, F-91160 Saulx les Chartreux
☎(+33 1) 64 54 83 90 ⊖ (+33 1) 69 34 49 73
Sales Manager: Yves Loïc Ruello [97]

75-45850 Trimos-Sylvac Metrologie
ZI des Rechardets, 14 rue du Ballon,
F-93165 Noisy le Grand
☎(+33 1) 43 04 40 10 ⊖ (+33 1) 43 05 60 07
Managing Director: Marc Zimmer
® Switzerland: Trimos-Sylvac Task Force [98]

75-46000 Turbomeca
(a subsidiary of Labinal Group, France)
F-64511 Bordes Cedex
☎(+33 5) 59 12 50 00 ⊠ 560928 f
⊖ (+33 5) 59 53 15 12
♲ FFr 189,247,600 ♣ FFr 2,000,000,000 ⍭ 3,500
Chairman and President: Jean Bernard Cocheteux
Executive Vice President: Raymond Heng
Vice President, Commercial and Customer Support: Jean Claude Leyssieux
Vice President, Production: Jean Pierre Subrenat
Vice President, Engineering: Alain Calemard
Director of Sales, Marketing and Public Relations: Jean François Thomas
■ Gas turbines for aeronautical and industrial applications, from aircraft and missiles to high-speed trains and offshore generating sets. [98]

75-46450 USINOR-SACILOR Head Office
(† Groupe USINOR)
Immeuble Pacific, F-92070 Paris La Défense Cedex
☎(+33 1) 41 25 60 10 ⊠ 614730 usinr f
⊖ (+33 1) 41 25 99 78
■ Production of all types of steels. [97]

75-46900 Vibrachoc SA
(a subsidiary of Hutchinson)
Parc d'Activités de l'Eglantier, CE 2804 Lisses,
F-91028 Evry Cedex
☎(+33 1) 69 91 50 00 ⊖ (+33 1) 64 97 50 28
♲ FFr 40,000,000 (1996) ♣ FFr 139,000,000 (1996)
⍭ 180 (1996) ⌂ 11,531 m2
Chairman: P C Clout
Managing Director: Jacques Maigne
Operations Director: Christian Froger
Financial Manager: Laurent Godbillot
■ Elastomeric and metallic anti-vibration and anti-shock mountings; laminated bearings (elastomeric); racking systems; illuminated panels and key-boards; acoustic engineering.

Foreign subsidiaries:
Stop-Choc GmbH → Germany 75
Stop-Choc Ltd → UK 75

Vibrachoc Division
Parc d'Activités de l'Eglantier, CE 2804 Lisses,
F-91028 Evry Cedex
Operations Director: Christian Froger

Claropan Division
CE 2804 Lisses, F-91028 Evry Cedex
Manager: Bernard Bardy [98]

75-46930 Vickers Systems SA AMD Division
Le Parc Club des Septs Deniers,
78 Chemin des Septs Deniers - Bat. 2, F-31200 Toulouse
☎(+33 5) 61 57 33 33 ⊖ (+33 5) 61 57 87 87
Toulouse Marketing Manager: Philippe Coude du Foresto
■ Design and manufacture fluid power and fuel control systems and components including hydraulic, electrohydraulic and fuel pump equipment, pumps, integrated packages, test stands and ground kart support equipment.

Representing:
Vickers Systems GmbH, Germany
Vickers Aerospace, UK
Vickers Systems Ltd, UK
Vickers Electromech, USA
Vickers Inc AMD, USA
Vickers Sterer [98]

75-46950 Vignal Artru Industries Head Office and Works
Route de Romans, PO Box 1, F-26120 Chabeuil
☎(+33 4) 75 59 04 55 ⊖ (+33 4) 75 59 26 60
♲ FFr 2,000,000 (1996) ♣ FFr 39,000,000 (1996)
⍭ 80 (1996) ⌂ 2,600 m2
Managing Director: Jean Artru
Director, Production and Sales: Gérard Patel
Marketing Director: Daniel Kohler
Administration Director: Gérard Rambaud
■ Manufacture of small precision mechanical parts and assemblies (up to 200 mm). Comprehensive services according to drawings and specifications provided by customers. [98]

75-47050 Warner & Tourco Manufacture and Sales Plant
PO Box 313, F-72007 Le Mans Cedex
Route de Spay, F-72700 Allonnes
☎ (+33 2) 43 43 63 63 ⊠ WARNERFRANCE
⊖ (+33 2) 43 43 63 40
Marketing Manager: L du Besset
■ Clutches and brakes: electro-magnetic, hydraulic, pneumatic, overrunning clutches and ball screws. [98]

75-47060 Wartsila Diesel International Sarl
(a subsidiary of Metra Corp, Finland)
12 Quai Saint-Jean, F-67080 Strasbourg Cedex
☎ (+33 3) 88 52 20 60 ⊖ (+33 3) 88 52 20 70
♣ FMk 7,069,000,000 (Group 1996) 👥 6,083 (1996)
President: Pentti-Juhani Hintikka
Technical Director: Daniel Paro
Corporate Communications Manager: Ms Marianne Kopf
■ Manufacture of diesel engines and related systems for power generation, marine propulsion and traction. Maintenance, repair and overhaul services. [98]

75-47070 Wartsila Diesel France SA Navy Business Office
(a part of Wartsila Diesel International Sarl, France)
3 rue Charcot, F-92200 Neuilly sur Seine
☎ (+33 1) 41 43 25 50 [E] wdnavy@aol.com
⊖ (+33 1) 41 42 25 62
♣ US$ 1,600,000,000 (1995) 👥 6,000 (1995)
Sales and Marketing Director, Naval Business: Bertrand de Noblens
■ Worldwide headquarters for naval projects. → 80 [98]

75-47090 Wartsila SACM Diesel
(a division of Wartsila Diesel International Sarl, France)
1 rue de la Fonderie, BP 1210, F-68054 Mulhouse
☎ (+33 3) 89 66 68 68 ⊖ (+33 3) 89 66 68 10

Wartsila France SA
(† Sacm Diesel)
La Combe, BP 115, F-17700 Surgees
☎ (+33 5) 46 30 31 32 ⊠ 790831
⊖ (+33 5) 46 30 31 59
👥 40 (1997)
Chairman: Matti Tulkki
Vice President, Sales and Marketing: Mikael Mäkinen
Sales and Marketing Director: Carl-Hennik Björk
Purchasing Manager: Frédéric Zimmer
■ Family of high speed diesel engines in the 500 - 4,000 hp power range. Supplied for power, traction and special applications. Military applications include battle tanks.

Courbevoie Plant
40 rue du Moulin des Bruyères, F-92400 Courbevoie
☎ (+33 1) 47 17 11 11 ⊖ (+33 1) 43 34 93 21

Gemenos Plant
Les Baux RN 8, F-13420 Gemenos
☎ (+33 4) 42 32 06 06 ⊖ (+33 4) 42 32 06 66

Seclin Plant
ZIA rue de Lorival, BP 411, F-59474 Seclin Cedex
☎ (+33 3) 20 32 61 00 ⊖ (+33 3) 20 32 51 93

Lyon Plant
76 rue Pierre Delore, F-69008 Lyon
☎ (+33 4) 78 58 99 11 ⊖ (+33 4) 78 61 11 88

Boulogne sur Mer Plant
168 Avenue Sarraz Bourmet, F-62200 Boulogne sur Mer
☎ (+33 3) 21 83 59 09 ⊖ (+33 3) 21 33 61 17 [98]

75-47120 Wuilfert SA
(† Mécanique Prècision Wuilfert SA (MPW))
36 rue Parmentier, PO Box 16,
F-94451 Limeil Brèvannes Cedex
☎ (+33 1) 45 69 16 40 ⊖ (+33 1) 45 69 36 99
♧ FFr 500,000 👥 24 (1997) ⌂ 2,000 m2
President and Chief Executive Officer: Philippe Renevey
Technical Director: Etienne Poirier
■ Automatic test systems and test benches for gyrometers, accelerometers and gyroscopes. Sub-contractors for precision mechanical engineering. [98]

75-47310 Zodiac International
(a subsidiary of Zodiac SA Holding, France)
50 Boulevard Galliéni, F-92130 Issy les Moulineaux
☎ (+33 1) 41 23 23 23 ⊠ 270569
⊖ (+33 1) 46 48 75 30
Marine Division Manager: Pierre Barbleu
Export Manager: François Girault
■ Manufacture of inflatable boats, hurricane semi-rigid craft and SOLAS liferafts. Design and manufacture of inflatable products. Superinsulating materials for satellite thermal control systems. [98]

80 Service Companies & Organizations

80-00120 AIRCO
Société Française de Formation et de Conseil Aéronautiques
(a member of Defense Conseil International Group, France)
2 Place Rio de Janeiro, F-75008 Paris
☎ (+33 1) 44 95 07 89 ⊠ 640101 f
⊖ (+33 1) 42 25 94 46
♧ FFr 20,000,000 👥 70
Chairman: Gen Gautier (retd)
Operations Director: Col Patrick Chaignon
■ AIRCO is appointed by the Ministry of Defence to study, organise, co-ordinate, and control the execution of specialised assistance operations connected with the implementation of aeronautical equipment within a foreign air force. [98]

80-00130 AlliedSignal Aerospace Marketing Sales and Service
(a subsidiary of AlliedSignal Aerospace, USA)
Centreda, Avenue Didier Daurat,
F-31700 Blagnac, Toulouse
☎ (+33 5) 61 71 00 76 ⊠ 530421 BENDTLS
⊖ (+33 5) 61 71 05 90
Director, European Commercial OEM Operations: Jim Shaw ☎ (+33 5) 61 71 06 96
Senior Customer Support Director: Pedro Henriques ☎ (+33 5) 61 71 06 99
■ Sales, marketing and support for AlliedSignal Aerospace products. → 90 (ASASCO)

● 126 rue de Stalingrad, F-93700 Drancy
☎ (+33 1) 43 11 51 22 ⊠ ASASCO 230 055F
⊖ (+33 1) 43 11 62 01; 43 11 62 02
Vice President Marketing, Sales and Support - Europe, Middle East and Africa: Thomas M Culligan ☎ (+33 1) 43 11 51 56
Vice President, Airline Sales/Marketing and Service - Europe, Middle East and Africa: Pierre Darpoux ☎ (+33 1) 43 11 51 60
Human Resources: Ms Elisabeth Berger ☎ (+33 1) 43 11 51 62
External Assistant: M de Vigneral
® USA: AlliedSignal Aerospace Service Corp

● 126 rue de Stalingrad, F-93700 Drancy
☎ (+33 1) 43 11 51 60; 43 11 51 57
⊖ (+33 1) 43 11 62 01
Vice President, Airline Market: Pierre Darpoux

● Domaine de la Baumette, F-13450 Grans
☎ (+33 4) 90 45 07 22 ⊖ (+33 4) 90 45 07 65
Field Service Engineer, Airline Team Leader: Alain Vandenaweele
■ Servicing support for AlliedSignal Aerospace products including APU ECS system wheels/brakes for commercial airlines and ATR and TPE 331 engines for regional airlines.

● BP No 1, F-60173 Ivry-le-Temple
☎ (+33 3) 44 08 79 80 ⊖ (+33 3) 44 08 79 68
Director, European Sales: Gerard Phillipe
■ Technical assistance for aviation and avionics equipment. Sales.

● c/o Dassault Aviation, Aeroport de Merignac, BP 24, F-33701 Merignac, (Bordeaux)
☎ (+33 5) 56 13 20 70; 56 13 99 51
⊠ 042 560957 MER F ⊖ (+33 5) 56 13 95 59
Field Service Engineer: Joseph McDonald
Field Service Engineer: Paul Cook

● c/o Air France, ME-GU, Porte 1.024, Zone Entretien, BP 10260, F-95704 Roissy CDG Cedex
☎ (+33 1) 46 75 68 44; 48 62 67 90
⊖ (+33 1) 49 79 04 33
Field Service Engineer: Jean-Pierre Tanvez
Field Service Engineer: Jean-Pierre Matanowski ☎ (+33 1) 49 34 22 87; 48 62 67 89 (fax)

● Murcier Savigny, F-74520 Valeiry
☎ (+33 4) 50 60 41 03 ⊖ (+33 4) 50 60 43 38
Field Service Engineer: Jean-Pierre Pillard [98]

80-00140 AlliedSignal Commercial Avionics Systems
(a subsidiary of AlliedSignal Aerospace, USA)
95 rue des Bouviers, F-26120 Fauconnieres
☎ (+33 4) 75 58 87 95 ⊖ (+33 4) 75 58 87 95
Director, Sales and Service: Yves Leipert
■ Sales and support. → 90 [98]

80-00150 AOG Air Service
(a subsidiary of SCAC Delmas Vieljeux, France)
3 rue du Remblai, PO Box 10355,
F-95706 Roissy Charles de Gaulle
☎ (+33 1) 49 19 15 20 ⊠ 236952 sataog f
⊖ (+33 1) 49 19 15 22
♧ FFr 2,000,000 ♣ FFr 40,000,000 👥 36
Managing Director: Luc Morin
Sales and Marketing Director: Jean Louis Malardier
■ International freight forwarder aeronautical specialist. [98]

80-00170 ATNR
Applications des Techniques Nouvelles en Electronique
(a subsidiary of SFIM Industries, France)
ZA de Courtaboeuf, 1 Avenue de l'Atlantique,
F-91940 Les Ulis
☎ (+33 1) 60 92 65 00 ⊖ (+33 1) 60 92 65 01
♣ FFr 14,000,000 (1996) 👥 23
Chief Executive Officer: M Faure
Managing Director: C Petrigny [98]

80-00180 Aero International (Regional) (AI(R))
(† Avions de Transport Régional)
(a joint venture between Aerospatiale, Alenia and British Aerospace)
1 Allée Pierre Nadot, F-31712 Blagnac Cedex
☎ (+33 5) 62 21 62 21 ⊖ (+33 5) 62 21 63 18
♧ FFr 300,000 (1997) ♣ FFr 7,000,000,000 (1996)
👥 900 (1997)
President and Chief Executive Officer: Patrick Gavin
Senior Vice President, Finance and Operations: Ciro Cerillo
Senior Vice President, Personnel: Antonio di Blasi
Senior Vice President, Commerce: Alain Brodin
Senior Vice President, Sales and Marketing: Jeff Marsh
Senior Vice President, Customer Support: Bill Black
Senior Vice President, Future Projects and Strategic Marketing: Peter Duffy
General Secretary: Paolo Revelli-Beaumont
Press and Information Director: Giancarlo Fre
Legal Counsel: Francesco Giobbe
■ Sales, marketing and customer support operations of the French-Italian consortium ATR and BAe's regional airliner subsidiaries Jetstream and Avro International Aerospace. Products include Jetstream 41, ATR 42-400, ATR 42-500, ATR 72-210A, RJ85 and RJ100. → 74

Ownership:
Aerospatiale (33,3%), Alenia (33,3%), British Aerospace (33,3%). [98]

80-00190 Aviation Defence Service (AVDEF)
Aéroport de Nîmes Arles Camargue, F-30800 Saint Gilles
☎ (+33 6) 66 70 74 10 ⊖ (+33 6) 66 70 12 22
♧ FFr 3,000,000 (1997) 👥 33 (1997)
Managing Director: H Bret
Sales and Marketing Director: B Lamarque
Operations Director: J P Rageau
Purchasing Manager: C Dallet
Head Secretary: Ms Marie Anne Winstanley [98]

80-00230 BFGoodrich Aerospace Evacuation Systems Division
11 rue de la Grande Borne, F-77990 Le Mesnil Amelot

☎ (+33 1) 64 02 67 67 [E] bfgpsc@worldnet.fr
⊜ (+33 1) 64 02 60 61
Managing Director: J Haldemann
■ Service centre.
Representing:
BFGoodrich Aerospace [98]

80-00310 Cegelec
(a subsidiary of Alcatel Alsthom, France)
2 Quai Michelet, F-92309 Levallois Perret Cedex
☎ (+33 1) 41 49 20 00 ⊠ 620454 f
⊜ (+33 1) 41 49 24 85
✣ FFr 566,338,100 (1996)
♣ FFr 16,779,000,000 (1995) 👥 35,000 (1996)
Chairman and Chief Executive Officer: Pierre Bonafé
Senior Executive Vice President: Michel Gérard
Executive Vice President: Yvon Miran
Executive Vice President: Charles Carlier
Director of Management and Budget Control: Guy Mazas
Secretary General: Yves de la Serre
Sales Manager: Daniel Besson
Manager, Energy Systems Group: Michel Augonnet
Manager, Products and Technology Group: Jean Pierre Hauet
Manager of Contracts Group and Services: Philippe Anglaret
Sales Manager, France: Bernard Turbé
International Sales Manager: Alain Berger
Executive Adviser to the Chairman: Patrick Henry
Communications Manager: Michel Masson
Human Resources Manager: Foucauld Lestienne
■ Contracting and systems engineering. → 75 [98]

80-00375 Composants et Systèmes de Précision
ZAC de Mercières, Rue Clément Ader,
F-60200 Compiegne
Managing Director: Christophe Daugny
■ Research and consultancy in high precison mechanics.
Representing:
Cranmos, UK [98]

80-00400 D.E.3I Informatique
(a subsidiary of Dassault Electronique SA and IBM, France)
ZI Les Gâtines, Rue Jean Lepelletier - BP 501,
F-78375 Plaisir Cedex
☎ (+33 1) 30 81 10 00 ⊜ (+33 1) 30 55 01 91
♣ FFr 82,000,000 (1996) 👥 64 (1996)
Chairman and Chief Executive Officer: Jean-Louis Michel
Managing Director: Henri de Darassus
■ Systems integration and software engineering.
Jointly owned by Dassault Electroninque SA (55%) and IBM (45%). [98]

80-00450 Electronique Midi Pyrenées (EMP)
(a subsidiary of SFIM Industries, France)
ZI du Palays, Rue des Cosmonautes, F-31400 Toulouse
☎ (+33 5) 61 54 14 54 ⊜ (+33 5) 61 54 78 25
♣ FFr 250,000,000 (1996) 👥 41
Chief Executive Officer: M Faure
Managing Director: P Walliman [98]

80-00500 Embraer Aviation International
(a subsidiary of Embraer SA, Brazil)
Aeroport du Bourget, Zone d'Aviation d'Affaires, BP 74,
F-93350 Le Bourget
☎ (+33 1) 49 92 73 00 ⊠ 235697 f
⊜ (+33 1) 48 35 90 76
✣ FFr 69,400,000 ♣ FFr 11,500,000 👥 43 🏭 4,000 m2
Financial and Administration Director: Daniel de Sousa
☎ (+33 1) 49 92 73 04
Customer Support Director: Paulo Serra
☎ (+33 1) 49 92 73 01
Commercial Sales Manager, South America: Ricardo Pesce ☎ (+55 12) 345 17 11
Government Sales Manager: Eduardo Munhoz
☎ (+55 12) 345 14 00
■ Commercial aircraft sales and product support activities for EMB-110 Bandeirante, EMB-120 Brasilia, CBA-123, EMB-312 Tucano and other Embraer aircraft. Technical assistance, warranty administration, technical publications, distribution and spare parts. Territory of activities: Europe, Africa and Middle East. → 90
® Brazil: Embraer [97]

80-00540 Études Electroniques Informatiques (I2E)
Avenue Guillibert de la Lauzière,
ZAC de Pichaury II - Les Milles,
F-13794 Aix en Provence Cedex 03
☎ (+33 4) 42 60 00 25 ⊠ 441204 f
⊜ (+33 4) 42 60 00 26
✣ FFr 9,000,000 (1996) ♣ FFr 120,000,000 (1996)
👥 160 (1996)
Managing Director: J C Marcellet
Operations Director: T Tartary
International Business Director: F Reginaud
Sales and Marketing Director: F Sage
Purchasing Manager: N Poulet
■ Development of VHF/UHF interception and direction finding systems (naval, ground and airborne). Manufacture of electronic ground based systems. Simulators and training equipment, ground based radar, telecommunications and electronic test equipment. Airborne electronic systems, airborne and naval communications and equipment, instruments and sensor displays. [98]

80-00550 Eurolang Services
(a division of Sonovision ITEP Technologies)
2 rue Louis Pergaud, PO Box 35,
F-94701 Maisons Alfort Cedex
☎ (+33 1) 45 13 05 03; 45 13 05 41
⊜ (+33 1) 45 13 05 59
♣ FFr 16,000,000 (1994) 👥 25
Sales and Marketing Director: Patrick Bonelli
Manager: Jean Jacques Perot
Purchasing Manager: Ms Catherine Tavant
■ Translation of documentation and localisation of software, all languages, terminology, term databases and sentence memories, graphic and documentation design and development.

Valbonne Office
Les Cardoulines - Batîment 5,
Route Dolines - Sophia Antipolis, F-06560 Valbonne
☎ (+33 4) 93 95 82 67 ⊜ (+33 4) 93 95 82 71
Branch Manager: Agnès Coirier [97]

80-00570 Eurotunnel Operations Headquarters
BP 69, F-62231 Coquelles
Head of Security: B Boubli [98]

80-00595 Export Trading Services Sarl
89 rue de Sevres, F-75006 Paris
☎ (+33 1) 45 48 91 34 ⊜ (+33 1) 45 48 98 58
✣ FFr 500,000 ♣ FFr 5,000,000 (1996) 👥 4
Managing Director: B Cayron
Operations Director: A Cayron
■ Export consultant for aeronautical and defence products. International and domestic. [98]

80-00600 FlightSafety International Inc Paris Learning Center
BP 23, F-93352 Le Bourget Cedex
Zone d'Aviation d'Affaires - Building 404
Aéroport du Bourget
Contact: Bruno Challe [97]

80-00820 Héli Union
(a subsidiary of Compagnie Nationale de Navigation, France)
4 Avenue de la Porte de Sèvres, Héliport de Paris,
F-75015 Paris
☎ (+33 1) 53 78 08 18 ☒ HELIUNION ⊠ 205897 helia
⊜ (+33 1) 53 78 08 19
✣ FFr 77,400,000 (1997) ♣ FFr 394,625,000 (1995)
👥 210 (1997)
Managing Director: G Gerold
Member of the Management Committee: Christophe de Courlon
Member of the Management Committee: Jean Pierre Duchanoy
Commercial Director: Christopher Grainger
Operations Director: Michel Jaguelin
Finance Director: Alain Gaveriaux
Commercial: Sophie Rajonson
■ Helicopter pilot training: basic and advanced IFR; search and rescue instructor. Mono and twin engine. Helicopter assistance for oil and seismic companies. [98]

80-01000 Hemet Exploration
(a subsidiary of SOGERMA Maintenance Group, France)
31 Avenue Clément Ader, F-31770 Colomiers
☎ (+33 5) 61 15 96 00 ⊜ (+33 5) 61 15 96 96
♣ FFr 125,400,000 (1996) 👥 166 (1996)
Chairman and Chief Executive Officer: Régis Trolliet
■ Aircraft maintenance and standardization. Flight test installation. Modification and transformation for all type of civilian aircraft as well as the assembly line of the Beluga Super Airbus Transporter. [98]

80-01350 Lord SA
(a subsidiary of Lord Corp, USA)
Immeuble Strategy Center, 10 rue des Gaudines,
F-78100 Saint Germain en Laye
☎ (+33 1) 30 87 05 78 ⊠ 176211755 lordsa f
⊜ (+33 1) 30 87 08 27
♣ US$ 3,500,000 👥 7
Managing Director: Michel Froidevaux
Customer Support Manager: J C Barthel
■ Bearings for helicopter rotor heads, engine suspension systems, vibration isolators for avionics and flight systems, noise control products, active noise and vibration reduction systems. → 90
® USA: Lord Aerospace [97]

80-01405 Matra Cap Systèmes
(a subsidiary of Lagardére Group, France)
6 rue Dewoitine, BP 14, F-78142 Vélizy Villacoublay Cedex
☎ (+33 1) 34 63 70 00
⊜ (+33 1) 34 63 70 70; 34 63 70 90
✣ FFr 145,098,000 ♣ FFr 850,000,000 (1996)
👥 900 (1996)
Chairman: Bernard Plano
Technical Director: Bruno Masnou
Sales and Marketing Director: Hervé Garnier
Operations Director: Eugéne Dalle
Purchasing Manager: Jacques Pain
Communication Manager: Christian Badé
■ Airborne reconnaissance, air force mission planning, digital mapping, control systems, C3I, space observation systems, image and signal processing, integrated logistic systems. → 75 [98]

80-01420 NAVFCO
Société Navale Française de Formation et de Conseil (a subsidiary of the Defence Conseil International Group)
2 Place Rio de Janeiro, PO Box 49, F-75008 Paris
☎ (+33 1) 44 95 28 00 ⊠ 640942 f
⊜ (+33 1) 44 95 28 99
✣ FFr 25,000,000
Chairman: Adm Joël Bodard (retd)
Adviser to the Chairman: Capt V Ortoli
Operations and Technical Director: J F Tardiveau
Public Relations: Mrs Annie Mazé
■ Consultancy, training and technical assistance in the naval field. Transfer of French naval know-how to foreign forces. [98]

80-01500 Office Français d'Exportation de Matériel Aéronautique (OFEMA)
58 Avenue Marceau, F-75008 Paris
☎ (+33 1) 44 31 17 00 ☒ EXEARO ⊠ 645130 ofema f
⊜ (+33 1) 44 31 18 00
✣ FFr 9,000,000 👥 80
Chairman and Chief Executive Officer: Ivan Barbot
Managing Director: Guy Rupied
Sales and Marketing Director: P Bonnet
Senior Manager, Services: Gen Blancher (retd)
Finance Manager: P Catherine
■ The objective of OFEMA, a service company, is to market the export of all French Aerospace products. [98]

80-01550 Office Général de l'Air (OGA)
33 Avenue des Champs-Elysées, F-75008 Paris
☎ (+33 1) 40 76 20 01 ☒ AIROFFICIA Paris
⊠ 280123 airo f ⊜ (+33 1) 42 56 20 48
✣ FFr 13,350,000 (1995) 👥 140 (1997)
Chairman and Chief Executive Officer: Philippe Esper
Managing Director: Gérard Drogou
Assistant Managing Director: Patrick Belval
Military Advisor and General Manager, Eastern Europe and USA: Gen Paul Clariond
Public Relations and Adviser: Mrs Dominique Jung
Director of Administration and Finance: Martine de Labrouhe

Director of Legal Affairs and Contracts: Philippe-Jean Terrasse
General Manager, Western Europe: Claude Chambon
General Manager, Far East: Antoine Herment
General Manager, Middle East: Michel Glavany
General Manager, Africa: Philippe Buzenet
General Manager, Industry: Patrick Belval
General Manager, Aviation: Jacques Courtin
Director of OGA Service: Bernard Carayon
Assistant to Director: François Barnabé
Public Relations: Mrs Sophie Pottechmer
■ International consulting company providing services to promote the French Aeronautical Industry, Public Works and Engineering.
Also offers decision making analysis, market studies, marketing and contact negotiations, offset agreements, financing and consulting equity participation, maintenance and services, in particular in aerospace, trade, local logistics. 98

80-01690 Radstone Technology SA
6 Avenue des Andes, F-91952 Courtaboeuf Cedex
☎ (+33 1) 64 46 04 03 ⊖ (+33 4) 69 28 03 40
Managing Director: Michel Beaumont
■ Supplier of standard and custom board level, systems and software products for military and industrial applications. Products support open bus architectures (VME) including processors, memories, I/O and enclosures. 98

80-01700 REOSC
(a subsidiary of SFIM Industries, France)
Avenue de la Tour Maury, F-91280 Saint Pierre du Perray
☎ (+33 1) 69 89 72 00 ⊖ (+33 1) 69 89 72 20
♣ FFr 240,000,000 (1996) 👥 3,000
Chief Executive Officer: Michel Faure
Vice President, Strategy and Business Development: Didier Lecomte
Vice President, Administration and Finance: D Turmel
Vice President, High Performance Optics: R Geyl
■ Design, production and integration of spaceborne mirrors. Professional optics (Kinoptik: Genisson) and coating services (EVAP-services: Poujolat MTO). 98

80-01720 REVIMA
(a joint venture between SOGERMA and Air France)
BP 1, F-76490 Caudebec en Caux
☎ (+33 2) 35 95 70 02 ⊖ (+33 2) 35 90 70 48
♣ FFr 498,200,000 (1996) 👥 580 (1996)
Chairman and Chief Executive Officer: Philippe Debrun
Executive Vice President: Gérard Kowal
■ PWC engine repair (PT6, PW100), APU (Garrett, PW), landing gears (Airbus, Boeing, McDonnell Douglas and Lockheed).
Jointly owned by SOGERMA (70%) and Air France (30%). 98

80-01775 SECA
Société d'Exploitation et de Constructions Aéronautiques
(a subsidiary of SOGERMA, France)
BP 132, F-93352 Le Bourget Cedex
☎ (+33 1) 49 34 54 44 ⊖ (+33 1) 49 34 54 35
♣ FFr 483,000,000 (1996) 👥 470 (1996)
Chairman and Chief Executive Officer: Henri-Paul Puel
Executive Vice President: Yves Boyer
■ Maintenance and modification business and commuter aircraft, helicopters, engine repair (PWC, PW100, JT15, PT6, TFE 731, CF700, Allison 250). Manufacture of seals and component repair. 98

80-01850 Sonovision ITEP Technologies
86 rue Regnault, F-75652 Paris Cedex 13
® France: Eurolang Services 97

80-01997 SOPEMEA
Société pour le Perfectionnement des Matériels et Équipements Aérospatiaux
(a subsidiary of Messier-Bugatti, France)
Zone Aéronautique Louis Bréguet, PO Box 48, F-78142 Vélizy Villacoublay Cedex
☎ (+33 1) 46 30 22 74 ⊠ 634137 sopvil f
⊖ (+33 1) 46 30 54 06
♧ FFr 9,274,800 (1997) ♣ FFr 91,000,000 (1996)
👥 122 (1997)
President: Jacques Paul
Commercial Director: Bernard Juignet
Head of Communication: Ms Anne Petit-Boussard
■ Environmental testing for aerospace equipment. Ground vibration testing for aircraft, rockets and complex structures. Maintenance of test facilities. Design and construction of testing and measurement equipment. Technical assistance, training for engineers and technicians. 98

80-02000 Strat*X SA
Tour Pacific, F-92977 Paris La Défense Cedex
☎ (+33 1) 41 02 00 30 Ⓦ http://www.stratx.com
⊖ (+33 1) 47 78 18 44
👥 12
Managing Partner: Laurent Bonnier
Managing Partner: James Thorne
Marketing Manager: Keir Rothnie
■ Specialises in the development and implementation of tailored management development programs in business and marketing strategy. 98

80-03190 Top Air Concepts Sarl
27 rue de la Ferronnerie, F-75001 Paris
☎ (+33 1) 40 26 27 42 ⊖ (+33 1) 40 26 11 18
General Manager: Jacques Chavonnet
■ Consultancy in strategy and management in the aerospace field: market surveys and research; planning and development of products and markets; technical, commercial, manufacturing contracts management; joint ventures and acquisitions. Purchase and sale/lease of civil and military aircraft (fixed and rotary wing), engines, airframes, equipment and spares. Purchase and sale of avionics and instrumentation for civil and military fixed and rotary wing aircraft. 98

85 Stockists, Distributors, Spare Parts Suppliers

Daimler-Benz Aerospace France SAS . 75-12950
Société d'Exploitation 74-01800

85-00050 Aéro-Stock
Aéroport du Bourget - Zone Nord, Bâtiment 378, F-93350 Le Bourget
☎ (+33 1) 48 35 92 59 ⊠ 236463
⊖ (+33 1) 48 35 95 07
♧ FFr 1,000,000 ♣ FFr 57,000,000 (1996) 👥 28 (1997)
🏭 5,500 sq m
Managing Director: Christian Boulanger
Operations Director: Alain Salmon
Sales Manager: Guy Verhegge
Finance Manager: Eric Boulanger
■ Import and export of aeronautical materials.
® USA: Allied-Signal Bendix/King Division, Alumigrip (US Paint), Becker Avionics Inc, Champion Aviation, General Electric (Light Division), Goodyear, Loral (Aircraft Braking Systems), Telex Communications Inc 98

85-00150 Aeroquip SA Division Aéronautique
(a division of Aeroquip Corp, USA)
14 rue du Morvan, Zone Silic 507, F-94623 Rungis Cedex
☎ (+33 1) 46 86 22 30 ⊖ (+33 1) 46 87 00 27
♧ FFr 83,474,600 (1997) 👥 26 (1997) 🏭 1,000 m2
Managing and Operations Director: François Martinez
Sales Director: Guy Eloy
Technical Manager: Philippe Barbier
Purchasing Manager: Gerara Pelletier
Management: Ms Corinne Ginfray
■ Flexible hose assemblies, self-sealing couplings, V-band clamps, instrument clamps, swivel joints, ducting components for bleed air systems.
® USA: Aeroquip Corp 98

85-00300 Apollo Metal SA
(a subsidiary of Apollo Metals plc, UK)
26 rue Fresnel, F-78310 Coignieres
☎ (+33 1) 30 49 42 43 ⊖ (+33 1) 30 49 42 49
♧ FFr 1,000,000 (1996) ♣ FFr 10,000,000 👥 10 (1996)
🏭 1,920 m2
Managing Director: B Vallee
Operations Director: C Vezinat
Purchasing Manager: R Pruski
■ Processing and high-distribution of aluminium for aerospace, defence and high-technology engineering companies. Plates, sheets, bars and sections. 98

85-00500 Euromip
8 rue des Frèeres Caudron, F-78140 Vélizy
☎ (+33 1) 34 65 33 67 ⊖ (+33 1) 35 65 13 92
♧ FFr 1,200,000 (1996) ♣ FFr 13,000,000 (1996)
👥 7 (1996) 🏭 95 m2
President: François de Bernardy
■ Distributor of products and components for avionics, defence and control systems, anti-aircraft systems, simulators, communications, conventional weaponry to the French electronic/aerospace industry.
® Australia: AEMC, RF Devices
Ireland: Electronic Concepts
Taiwan: Taiwan Semiconducters Co
UK: Arcol, MPE, Oxley, Semelab
USA: AT Wall, Alcatel Network Systems, Buckbee Mears, Caddock, DEI, Fenwall, Isolink, Knox Semiconductor, Lumex, VMI 98

85-01200 Kristal SA
61 rue F Mistral, PO Box 114, F-34402 Lunel Cedex
☎ (+33 4) 67 71 52 63
Ⓔ 106477.2152@compuserve.com
⊖ (+33 4) 67 83 08 61
♧ FFr 251,000 (1997) ♣ FFr 9,000,000 (1996)
👥 3 (1997) 🏭 700 m2
President: Gérard Wallet
■ Supply of aircraft parts, hardware instruments, lights, brake parts, ignition harnesses and engine parts for trainers. → 80 98

85-01400 Messier-Dowty Customer Support Centre - Europe
(a division of Messier-Dowty SA)
Zone Aéronautique Louis Bréguet, PB 10, F-78142 Vélizy
☎ (+33 1) 46 29 18 00 ⊠ 634156 f
⊖ (+33 1) 46 29 18 18 (commercial); 46 29 85 31 (technical)
Commercial Manager: Pascal Lajoie
Technical Manager: Charles Cotte
■ Spare parts, AOG service and technical support for Messier-Dowty equipment in Europe, Middle East and Africa. 98

85-01900 Service Aviation International
Impasse Lantelme, F-06700 Saint Laurent du Var
☎ (+33) 93 31 93 10; 93 31 56 84 ⊖ (+33) 93 14 44 42
Managing Director: Paul Lantelme
■ Overhaul and repair of AH-1S Cobra helicopter airframe and components. Spare parts and maintenance training. Upgrade to Super Cobra. Spare parts for helicopters and aircraft. → 80 98

85-03000 TRACE France
12 Sente des Sablons, F-92420 Vaucresson
☎ (+33 1) 47 41 89 89 ⊖ (+33 1) 47 41 92 90
Business Manager: Luc P Neyrinck 98

90 Sales Agents, Sales & Liaison Offices

Companies represented:

Acer Inc, Taiwan	90-00053
AEL, USA	90-00520
AEMC, Australia	85-00500
Aeroquip Corp, USA	85-00150
Alcatel Network Systems, USA	85-00500
Allied International Corp, USA	90-00150
Allied-Signal Bendix/King Division, USA	85-00050
AlliedSignal Aerospace Service Corp, USA	80-00130
Altoflex Elektro, Switzerland	75-02300
Altoflex GmbH, Germany	75-02300
Alumigrip (US Paint), USA	85-00050
Arcol, UK	85-00500
AT Wall, USA	85-00500
Avantek, USA	90-00520
Becker Avionics Inc, USA	85-00050
Beretta SpA, Italy	90-00230
Beta Transformer Technology Corp, USA	90-00430
Bottlang, USA	90-00732
Buckbee Mears, USA	85-00500
Caddock, USA	85-00500
CATIC, China PR	90-00300

Contact Reference

Champion Aviation, USA 85-00050
Chase Electronics, UK 90-00520
Chelton (Electrostatics), UK 90-00326
Chomerics Ltd, UK 75-00200
Daimler-Benz Aerospace AG, Germany . 75-12950
Dassault Aviation Group, France 90-00410
DEI, USA 85-00500
Dranetz Technologies Inc., USA 90-01160
EEV, UK 90-00513
Electronic Concepts, Ireland 85-00500
Embraer, Brazil 80-00500
Emerson & Cuming, USA 75-00200
Emerson Electric, UK 90-01220
Emerson Electric, USA 90-01220
Encore Computer Corp, USA 90-00580
ERI/Whittaker, USA 90-00520
Eurolang Services, France 80-01850
Farnell International, UK 90-00520
Fenwall, USA 85-00500
Fischer Custom Communication, USA . 90-00520
Froude Consine, UK 90-00665
GEC Plessey Semiconductors, UK . . . 90-00720
General Electric, USA 90-00740
General Electric (Light Division), USA . 85-00050
General Microwave, USA 90-00520
Giga-Tronics, USA 90-00520
Goodyear, USA 85-00050
Greene Tweed & Co, USA 90-00775
Grimes Aerospace Company, USA . . . 90-00777
Haynes International, USA 90-00810
Hewlett Packard, USA 90-00520
Honeywell Inc, USA 90-00850
Hughes Aircraft, USA 90-00520
ILC Data Device Corp, USA 90-00430
Imo Industries Inc, USA 90-00920
Isolink, USA 85-00500
Israel Aircraft Industries, Israel 90-00970
JDR Cable Systems BV, Netherlands . . 90-11200
Jeppesen, Germany 90-00732
Knox Semiconductor, USA 85-00500
Kulite Semiconductor Products Inc,
USA 90-01170
LEACH International North America,
USA 75-25350
Litton Industries Inc, USA 90-01236
LMS International, Belgium 90-01250
Loral (Aircraft Braking Systems), USA . 85-00050
Lord Aerospace, USA 80-01350
LRE Relais + Elektronik, Germany . . . 75-25350
Lucas Aerospace, UK 90-01290
Lumex, USA 85-00500
Manitou BF, France 90-00327
Manitou BF, France 90-00350
Manitou BF, France 90-01260
Manitou BF, France 90-02070
Martin-Baker Ltd, UK 74-01800
Microwave Technology, USA 90-00520
Midwest Microwave, USA 90-00520
Moog Inc, USA 90-01400
MPE, UK 85-00500
Northern Airborne Technology Ltd (NAT), Canada 90-00326
Northrop Grumman Inc, USA 90-01500
Orbit, Israel 90-00520
Oxley, UK 85-00500
Portescap, Switzerland 90-01669
PPG Ampaspace, Italy 90-01680
PPG Industries Aircraft Group, USA . . 90-01680
RF Devices, Australia 85-00500
Rolls-Royce plc, UK 90-01850
Semelab, UK 85-00500
Sonnenschein Lithium GmbH, Germany 90-01983
Sumitomo Corp, Japan 90-02000
Sundstrand Aerospace, USA 90-02005
Taiwan Semiconducters Co, Taiwan . . 85-00500
Telex Communications Inc, USA 85-00050
Teledyne Inc, USA 90-02050
Teradyne Inc, USA 90-02065
Times Microwave Systems, USA 90-00520
Trak Systems, USA 90-00520
Trimos-Sylvac Task Force, Switzerland . 75-45850
Vaisala Oy, Finland 90-21000
VMI, USA 85-00500
Zarges Leichtbau GmbH, Germany . . . 90-25000

Sales agents/representatives listed elsewhere:

AlliedSignal Commercial Avionics . . . 80-00140
AlliedSignal Aerospace Marketing . . . 80-00130
Embraer Aviation International 80-00500
Leach International Europe SA 75-25350
Lord SA 80-01350
Rockwell-Collins France SA 75-32800

90-00053 Acer France Sarl
165 Avenue du Boid de la Pie, BP 40005,
F-95911 Roissy Charles de Gaulle
☎ (+33 1) 48 17 40 40 ⊜ (+33 1) 48 17 40 89
® Taiwan: Acer Inc
[97]

90-00085 Aerotech Aviquipo Sarl
(a subsidiary of Aerotech World Trade Corp, USA)
BP 828, F-20192 Ajaccio, Corsica
☎ (+33 4) 95 21 15 89 ⊜ (+33 4) 95 51 16 29
Managing Director: T Felhi
[98]

90-00150 Allied International Corp
(a subsidiary of Allied International Corp, USA)
64 rue Aratole France, F-92300 Levaillois Perret
☎ (+33 1) 47 57 56 00 ⊜ (+33 1) 47 58 50 31
Manager: J Arthur
® USA: Allied International Corp
[98]

90-00230 Berettarmi SA
52 Cours Fauriel, PO Box 83,
F-42010 Saint Etienne Cedex 2
☎ (+33 4) 77 43 82 00 ⊠ 300726 f
⊜ (+33 4) 77 43 82 01
General Manager: Jean François Finot
■ Sale of small arms.
® Italy: Beretta SpA
[98]

90-00260 British Steel France SA
(a subsidiary of British Steel Engineering Steels, UK)
Parc Saint-Exupéry, 9 rue Maryse Bastié, F-69500 Bron
☎ (+33 4) 72 37 07 57 ⊜ (+33 4) 78 41 19 94
[98]

90-00270 Busak + Shamban France SA
38-46 rue Calmette et Guerin, ZI du Prinay, PO Box 74,
F-78501 Sartrouville Cedex
☎ (+33 1) 30 86 56 00 ⊜ (+33 1) 39 13 56 06
General Manager: François Brault
Marketing: A Trent
[98]

90-00300 CATIC Representative Office
76 Avenue de Gambetta - Porte 801,
F-92400 Courbevoie Paris
☎ (+33 1) 43 33 03 09 ⊠ 612188 catic pa f
⊜ (+33 1) 47 89 05 34
® China PR: CATIC
[97]

90-00326 Chelton/Rayan
(a Chelton Group Company)
7 Chemin de Vaubesnard, F-91410 Dourdan
☎ (+33 1) 60 81 95 95 ⊜ (+33 1) 60 81 97 97
👥 26 🏭 2,000 sq m
President: Dr Ronald C Davidson
Sales Manager: Georges Berret
■ Complete protection systems for electrostatic, lightning and EMP hazards for aircraft and military sites. Consultants for protection in above areas. Supplier of full range of antennas, homing systems and dischargers for aircraft.
® Canada: Northern Airborne Technology Ltd (NAT)
UK: Chelton (Electrostatics)
[98]

90-00327 CIMM
(a subsidiary of Manitou BF, France)
Route de Challain la Potherie, F-49440 Cande
☎ (+33 2) 41 92 94 74 ⊜ (+33 2) 41 92 94 76
® France: Manitou BF
[98]

90-00334 C.M.L. International
Avenue de Verdun 54 bis, PO Box 37,
F-78290 Croissy sur Seine
☎ (+33 1) 34 80 19 89 ⊜ (+33 1) 39 76 94 31
President: Claude Leboeuf
■ Export organisation representing major defence industry companies in Africa, the Middle East, Asia and Turkey. Services company specialising in security, mobile hospitals and medical support. General procurement.
→ 80
[97]

90-00350 Compagnie Française de Manutention
(a subsidiary of Manitou BF, France)
Boulevard Pierre et Marie Curie, BP 75,
F-44152 Ancenis Cedex
☎ (+33 2) 40 09 13 13 ⊜ (+33 2) 40 96 29 82
® France: Manitou BF
[98]

90-00410 Dassault International
(a subsidiary of Dassault Aviation Group, France)
27 rue du Prof Pauchet, PO Box 32, F-92420 Vaucresson
☎ (+33 1) 47 95 85 85 ⊠ 203755
⊜ (+33 1) 47 44 67 89
♧ FFr 10,010,000 ♣ FFr 130,000,000 👥 25
Managing Director: Charles Edelstenne
® France: Dassault Aviation Group
[98]

90-00430 DDC Electronique
(a subsidiary of ILC Data Devices Corp, USA)
10 rue Carle Hebert, F-92400 Courbevoie
☎ (+33 1) 41 16 34 24 ⊜ (+33 1) 41 16 34 25
Managing Director: Michel Cazor
Marketing Communications Manager: Lawrence M Margulies
■ A/C and D/C synchronised converters, angle indicators and simulators, MIL-STD-1553 data bus products, custom hybrid micro circuits.
® USA: Beta Transformer Technology Corp, ILC Data Device Corp
[98]

90-00513 EEV France
(a subsidiary of General Electric Co, UK)
2 rue Henri Bergson, F-92665 Asnières Cedex
☎ (+33 1) 40 80 54 00
[E] 106173.3521@compuserve.com ⊠ 610471 inelm
⊜ (+33 1) 40 80 55 29
♣ FFr 51,000,000 (1995/96) 👥 10
General Director: J L Chinzi
Sales and Marketing Director: D Gratusse
Communications Officer: P Silvestre
■ Microwave tubes. Power tubes. EMC protection CCD sensors and cameras. Solar cells. Fire fighting thermal cameras and LCD displays.
® UK: EEV
[98]

90-00520 Elexience
9 rue des Petits Ruisseaux, F-91370 Verrières le Buisson
☎ (+33 1) 60 11 94 71; 69 53 80 00
✉ FTO ELEXIENCE FR ⊠ 691789 f
⊜ (+33 1) 60 11 98 09
♧ FFr 550,000 👥 45 🏭 1,200 m2
President and General Manager: René Berthier
Managing Director: Christian Benatar
Assistant to the Director: Ms Tomarelli
® Israel: Orbit
UK: Chase Electronics, Farnell International
USA: AEL, Avantek, ERI/Whittaker, Fischer Custom Communication, General Microwave, Giga-Tronics, Hewlett Packard, Hughes Aircraft, Microwave Technology, Midwest Microwave, Times Microwave Systems, Trak Systems
[97]

90-00580 Encore Computer SA
Parc d'Activité du Park de Lac,
Secteur Nord, rue Michael Faraday,
F-78180 Montigny Le Bretonneux
☎ (+33 2) 42 10 04 25 ⊜ (+33 1) 34 60 43 95
Country Manager: Stephane Grodner
® USA: Encore Computer Corp
[97]

90-00665 Froude France
49 rue Lamartine, F-78035 Versailles
☎ (+33 1) 39 67 17 16 ⊜ (+31 1) 39 67 17 05
👥 3
Divisional Manager: J M Bruvier
® UK: Froude Consine
[98]

90-00720 GEC Plessey Semiconductors
Avenue des Andes, PO Box 142, F-91944 Les Ulis Cedex
☎ (+33 1) 64 46 23 45 ⊜ (+33 1) 64 46 06 07

Director: G d'Eyssautier
® UK: GEC Plessey Semiconductors 98

90-00732 General Aviation Sarl

97 rue Mademoiselle, F-75015 Paris
☎ (+33 1) 47 83 72 44 ⊖ (+33 1) 45 67 62 09
♧ FFr 50,000
General Manager: Cl Fourès
Sales Service Department: Y Tourniaires
Documentation Services Department: E Garot
■ Equipment suppliers for aeronautical personnel including windshelters, flight suits, radio accessories and flight bags. Naval instruments, helmets and individual flight equipment.
® Germany: Jeppesen
USA: Bottlang 97

90-00740 General Electric International Aircraft Engine Group

(† General Electric France)
18 rue Horace Vernet, F-92136 Issy les Moulineaux Cedex
☎ (+33 1) 47 65 76 10 ✉ INGENETRIC
⊠ 631354 gepta f ⊖ (+33 1) 46 42 94 03
President, General Electric Aircraft Engine Group: Michel Lagorce
Vice President Marine, and Industrial Engine: J J Ferrera
Vice President, Commercial Engine Marketing: John Mor

Toulouse Division
Victoria Center, 20 Chemin de Laporte, F-31300 Toulouse
☎ (+33 5) 62 12 21 00 ⊠ 520083
⊖ (+33 5) 61 31 94 03
Vice President and General Manager: John M Mann
Sales and Marketing Manager: Robert Lamb
Programs Manager: Richard Hellman
CFM56 Programs Manager: David W Oteri
■ Aircraft and helicopter engines, marine gas turbines.
® USA: General Electric 97

90-00750 Emerson and Cuming Microwave Products Sarl

(† Grace NV)
(a subsidiary of Cumming & Emerson Microwave Products NV, Belgium)
16 rue Kléber, F-92130 Issy les Moulineaux
☎ (+33 1) 41 90 61 80 ⊖ (+33 1) 41 90 61 81
👥 3 🏠 80 sq m
Managing Director: A J Geurts
Sales and Marketing Director: J P Heitzmann
■ Low density polymers (foams, resins). Epoxy casting resins, conductive and non-conductive adhesives, conductive castings. 98

90-00775 Greene Tweed & Co France SA

(a subsidiary of Greene, Tweed & Co, USA)
PO Box 409, F-95527 Cergy Pontoise Cedex
9 Chaussée Jules César, F-95520 Osny
☎ (+33 1) 30 73 54 44 W http://www.gtweed.com
⊠ 609099 gtcfra f ⊖ (+33 1) 30 73 45 75
♧ FFr 1,500,000 (1996)
Managing Director: Pierre Flammant
Marketing: Philippe Allienne
■ Fluid sealing systems.
® USA: Greene Tweed & Co 98

90-00777 Grimes Aerospace Company

(a subsidiary of Grimes Aerospace, USA)
Centreda - Batiment 2, Avenue Didier Daurat, F-31700 Blagnac
☎ (+33 5) 61 77 38 34 ⊖ (+33 5) 61 77 38 33
Managing Director: Michael Dubose
Regional Director: Brian Hall
Sales and Marketing Director: Vince Di Giovanni
® USA: Grimes Aerospace Company 98

90-00810 Haynes International Sarl

(a subsidiary of Haynes International Inc, USA)
BP 7110, F-95054 Cergy Pontoise Cedex
☎ (+33 1) 34 48 31 00 ⊖ (+33 1) 30 37 80 22
♧ FFr 20,150,000 👥 13 🏠 1,350 m2
Managing Director: Pierre Limousin
■ Stockists, cutting and sale of titanium, nickel and cobalt-based alloys. Sheet, bar, tube, wire, welding products and fittings.
Registered trademarks: Haynes; Hastelloy; Multimet.
® USA: Haynes International 97

90-00850 Honeywell SA

(a subsidiary of Honeywell Inc, USA)
Parc Technologique de Saint Aubin, Bâtiment Mercury - BP 87, F-91193 Gif sur Yvette Cedex
☎ (+33 1) 60 19 80 00 ✉ HONEYWELL
⊖ (+33 1) 60 19 81 81
♧ FFr 97,000,000 👥 946
Vice President and General Manager: Patrick Jouglard
Director: Jean Ganivet
Marketing Communications: Annick Drouard
® USA: Honeywell Inc

Avionics Division
BP 14, F-31701 Blagnac Cedex
☎ (+33 1) 62 12 15 00 98

90-00920 Imo Industries Sarl

112 rue des Solets, Silic 138, F-94523 Rungis
☎ (+33 1) 46 87 26 38 ⊠ 265735
⊖ (+33 1) 46 86 20 69
General Manager: William Panciroli
® USA: Imo Industries Inc 97

90-00970 Israel Aircraft Industries-Europe European Purchasing Mission (IAI)

(a subsidiary of Israel Aircraft Industries Ltd)
181 Avenue Charles de Gaulle, F-92200 Neuilly sur Seine, Paris
☎ (+33 1) 46 40 47 47 ⊠ 613061 isgal f
⊖ (+33 1) 47 45 73 32
♣ US$ 1,490,000,000 👥 13,500
Managing Director: S Penetz
Sales and Marketing Director: Dr A Sczapiro
Operations Director: J Chriqui
Purchasing Manager: M Chaqui
® Israel: Israel Aircraft Industries 98

90-01100 Krauss-Maffei France Sarl

(a subsidiary of Krauss-Maffei Wehrtechnik GmbH, Germany)
5 Alleé des Barbanniers, F-92632 Gennevilliers Cédex
☎ (+33 1) 46 85 25 25 ⊠ 612091
⊖ (+33 1) 46 85 25 49
Manager: Jacques Ganem 98

90-01160 Kontron Electronique

(a subsidiary of Kontron Elektronik GmbH, Germany)
6 rue des Frères Caudron, PO Box 99, F-78140 Vélizy Villacoublay
☎ (+33 1) 34 58 12 10 W http://www.kontron.com
⊠ 695673 kontron f ⊖ (+33 1) 34 58 12 00
Managing, Sales and Marketing Director: Jean Michel Briere
Purchasing Manager: M H Varlet
® USA: Dranetz Technologies Inc. 98

90-01170 Kulite International Sarl

(a subsidiary of Kulite Semi Conductor Products)
Aéroport de Toussus Le Noble, F-78117 Châteaufort
☎ (+33 1) 39 56 09 00 ⊠ 698800 f
⊖ (+33 1) 39 56 36 48
♧ FFr 293,000 (1995) ♣ FFr 3,000,000 (1995)
Marketing Director: Stéphane Beaufront
® USA: Kulite Semiconductor Products Inc 97

90-01220 Liebert France SA

(† Emerson Liebert)
(a subsidiary of Liebert Corp, UK)
124 Avenue Galliéni, F-93170 Bagnolet
☎ (+33 1) 43 60 01 77 ✆ (0800) 30 02 00
W http://www.liebert.com ⊖ (+33 1) 43 60 80 97
♧ FFr 1,492,500
Managing Director: Pierre Bouchoucha
Commercial Director: Christian Philippe
Technical Director: Alex Klotz
Head of Advertising and Communications: Mrs Caroline Thuysbaert
■ Manufacture of UPS (Uninterruptable Power Systems). Miscellaneous equipment and materials, electrical systems.
® UK: Emerson Electric
USA: Emerson Electric 98

90-01236 Litton Precision Products International Inc

58 rue Pottier, F-78150 Le Chesnay Cedex
☎ (+33 1) 39 23 83 00; 39 55 21 04 E lppifr@aol.com
⊖ (+33 1) 39 55 50 68
👥 11 (1997) 🏠 250 m2
General Area Manager: Jean-Claude Sibi
■ Electrical components, microwave, tubes, flight instruments and connectors.
® USA: Litton Industries Inc 98

90-01250 LMS France

7 bis Burospace, F-91571 Bievres Cedex
☎ (+33 1) 69 41 24 24 ⊖ (+33 1) 69 41 31 27
Sales Manager: G Nuzzo
→ 80
® Belgium: LMS International 98

90-01260 Loc Manutention

(a subsidiary of Manitou BF, France)
Zone Industriel du Vert Galant, F-95310 Saint Ouen L'Aumone
☎ (+31 1) 34 32 36 66 ⊖ (+31 1) 34 64 71 29
® France: Manitou BF 98

90-01290 Lucas Aerospace France Corporate Headquarters

(a subsidiary of Lucas Aerospace, UK)
11 rue Lord-Byron, F-75008 Paris
☎ (+33 1) 45 61 95 25 ⊠ 651080 lucas f
⊖ (+33 1) 45 63 32 46
♣ FFr 600,000,000 (1996) 👥 828 (1996)
Corporate Director: S D Prioult
Operations Director: R Ysambert
Operations Director: L Mattiusi
Secretary: Ms Veronique Tynan
® UK: Lucas Aerospace

Operations Division
270 Avenue des Gresillons, F-92601 Asnieres Cedex
☎ (+33 1) 47 91 61 11 ⊖ (+33 1) 47 91 61 33
General Manager: R Ysambert

Regional/Liaison Office
Victoria Ctr, Bâtiment Didier Daurat, 20 Chemin de Laporte, F-31300 Toulouse
☎ (+33 5) 61 15 04 04 ⊖ (+33 5) 61 15 08 92
Regional Director, Europe: Roger Virnuls
Regional Director: Colin Sams
Secretary: Ms Jacqueline de Chalvron 98

90-01340 MicroSpire Paris

(a subsidiary of MicroSpire SA, France)
7 rue Pierre Métalrie, ZA du Bel Air, F-78120 Rambouillet
☎ (+33 1) 34 85 56 06 ⊖ (+33 1) 30 41 06 20 98

90-01400 Moog Sarl

(a subsidiary of Moog Inc, USA)
38 rue du Morvan, Silic 417, F-94573 Rungis Cedex
☎ (+33 1) 45 60 70 00 ⊖ (+33 1) 45 60 70 01
♣ US$ 8,585,000 👥 33
Managing Director: Bernard Allain ☎ (+33 1) 45 60 70 00
® USA: Moog Inc 97

90-01500 Northrop Grumman International Inc

(a subsidiary of Northrop Grumman Inc, USA)
2 rue Goethe, F-75116 Paris
☎ (+33 1) 53 23 03 40 ⊖ (+33 1) 40 70 01 40
Regional Vice President: Jacques Pous
® USA: Northrop Grumman Inc 98

90-01669 Portescap France SA

(a subsidiary of Portescap, Switzerland)
Le Trident, 3-5 Voie Félix Eboué, F-94024 Créteil Cedex
☎ (+33 1) 48 98 99 66 W http://www.portescap.com
⊖ (+33 1) 48 98 99 55
Managing Director: N Veignat
Sales and Marketing Director: M A Mathez
® Switzerland: Portescap 98

90-01680 PPG Industries Glass SA

(a division of PPG)
8 Avenue Condorcet, F-69100 Villeurbanne
☎ (+33 4) 78 89 74 37 ⊖ (+33 4) 78 94 93 71
Managing Director: Peter J Beucher
Sales Director, Aircraft (Europe): Jean Marc Coursières
Technical Director: Richard H Gross
Sales and Marketing Director: Thomas E Clark

Contact Reference

■ All transparencies for commercial and military aircraft (glass, plastic and laminated).
® Italy: PPG Ampaspace
USA: PPG Industries Aircraft Group [98]

90-01700 Racal Systems Electronique SA
18 Avenue Dutartre, F-78150 Le Chesnay, Parly II
☎ (+33 1) 39 55 88 88 ⊖ (+33 1) 39 55 67 35 [98]

90-01820 LHC Optronic
(a partner of Recon/Optical Inc, USA)
93 rue Claude Chappe, Quartier "Acqueduc de l'Avre",
F-78370 Plasir
☎ (+33 1) 30 55 42 42 ✉ 699515
⊖ (+33 1) 30 55 44 05
♧ FFr 1,000,000 ♣ FFr 20,700,000 (1995) 👥 20 (1997)
President: Gaëtan Calvano
Operations Director: O Racaud
Sales and Marketing Director: A Corne
Purchasing Manager: G Noye [98]

90-01850 Rolls-Royce International Ltd
(† Rolls-Royce (France) Ltd)
(a subsidiary of Rolls Royce plc, UK)
122 Avenue Charles de Gaulle, F-92522 Neuilly sur Seine
☎ (+33 1) 47 22 14 40 ✉ ROYFRANCE PARIS
⊖ (+33 1) 47 45 77 38
👥 2 (management staff) 🏠 100 m2
Regional Executive Director: Anthony R Charlton
® UK: Rolls-Royce plc [98]

90-01983 Sonnenschein Lithium GmbH Co-ordination Office
18 rue de la Tuilerie, F-38170 Seyssinet
☎ (+33 4) 76 96 03 01 ⊖ (+33 4) 76 96 02 26
Co-ordinator: Patrick Simion
® Germany: Sonnenschein Lithium GmbH [98]

90-01985 Sperry Marine Sarl
6 rue des Freres Cubeddu, F-13014 Marseille
☎ (+33 4) 91 63 04 65 ⊖ (+33 4) 91 63 04 63
Regional Manager: R Haasdijk [98]

90-02000 Sumitomo France SA
87 Avenue de la Grande Armée, F-75782 Paris Cedex 16
☎ (+33 1) 40 67 84 00 ⊖ (+33 1) 40 67 98 61
Machinery Section Manager: Jiro Koda
® Japan: Sumitomo Corp [98]

90-02005 Sundstrand Aerospace Europe Les Algorithmes
(a subsidiary of Sundstrand Corporation, USA)
17 Avenue Didier Daurat, Batîment Thales,
F-31700 Blagnac
☎ (+33 5) 61 30 00 18 ⊖ (+33 5) 61 30 00 50
Marketing: Fabrice Vauversin
■ Airborne constant frequency electrical systems based on constant speed (integrated drive generator) or variable speed generators and associated controls. Hybrid 270 volts DC electrical systems. Electrical distribution systems. Secondary power systems including airframe gear boxes, jet fuel starters and auxiliary power units for military and commercial applications. RAM air turbines, high performance cooling fans, main engine fuel pumps and airframe boost pumps.

Longvic Office
ZI de Dijon Sud, PO Box 30, F-21601 Longvic Cedex
☎ (+33 3) 80 66 81 52 ⊠ 351508 avdijon f
⊖ (+33 3) 80 38 33 91
Managing Director: Claude Guiard
■ Overhaul and repair services. → 80
® USA: Sundstrand Aerospace [98]

90-02050 Teledyne SA France
L'Arche du Parc, 138 rue Yves Kerman,
F-92658 Boulogne Billancourt Cedex
☎ (+33 1) 47 61 08 08 ⊠ 614336 tdysa
⊖ (+33 1) 47 61 97 43
Managing Director: Bernard Dolle
® USA: Teleydne Inc [97]

90-02065 Teradyne SA
(a subsidiary of Teradyne Inc, USA)
165 Boulevard de Valmy, F-92706 Colombes
☎ (+33 1) 46 13 15 00
⊖ (+33 1) 46 13 15 01; 46 13 15 02
👥 50
European Marketing Communications: Ms Régine Chretien
® USA: Teradyne Inc [97]

90-02070 TIE
(a subsidiary of Manitou BF, France)
Le Bois de Roche - ZAC de l'Aéropôle, BP 77,
F-44152 Ancenis Cedex
☎ (+33 2) 51 14 51 00
® France: Manitou BF [98]

90-11200 JDR Cable Systems
(† De Regt Câbles Spéciaux SA)
3 route de Choisy, F-60200 Compiègne
☎ (+33 3) 44 40 80 81 ⊖ (+33 3) 44 40 81 18
Managing Director: Han van Veen
Sales and Marketing Director: Peter van Nimwegen
Area Sales Manager: Étienne Lair
® Netherlands: JDR Cable Systems BV [98]

90-21000 Vaisala SA
(a subsidiary of Vaisala Oy, Finland)
3 Park Ariane, Saint Quentin en Yvelines,
F-78284 Guyancourt Cedex
☎ (+33 1) 30 57 27 28 ⊖ (+33 1) 30 96 08 58
Managing Director: Peter Eriksson
■ Automatic weather stations, ice detection and prediction systems, altimeter setting units, digital barometers, ceilometers, visibility and RVR systems, wind measuring systems.
® Finland: Vaisala Oy [98]

90-25000 Zargal Sarl
(a subsidiary of Zarges GmbH, Germany)
30 rue de la Gare, F-67127 Duppigheim
☎ (+33 3) 88 50 82 07 ⊖ (+33 3) 88 49 10 43
♧ FFr 250,000 (1996) ♣ FFr 40,000,000 (1996)
👥 20 (1996) 🏠 1,000 m2
Managing Director: Guenter Lange
Sales and Marketing Director: Frédéric Levieux-Pourrot
■ Provides a wide range of aluminium boxes and cases to handle and safely protect sensitive and valuable material such as electronic equipment, detection materials, optic appliances and medical equipment.
® Germany: Zarges Leichtbau GmbH [98]

Gabon

Head of State:
El Hadj Omar Bongo

10 Head of State

10-10000 Office of the President
PO Box 546, Libreville
☎ (+241) 72 78 36 ⊠ 5211; 5301 presigab go
⊖ (+241) 77 33 46
President of the Republic: El Hadj Omar Bongo
☎ (+241) 76 33 57; 76 33 58 [98]

11 Prime Minister

11-10000 Office of the Prime Minister
PO Box 91, Libreville
☎ (+241) 76 17 89 ⊠ 5409 pm go
Prime Minister and Head of Government: Paulin Obame [98]

12 Ministries

12-10000 Ministry of National Defense, Public Security and Immigration
Ministere de la Defense Nationale, de la Securite et de l'Immigration
PO Box 13493, Libreville
☎ (+241) 72 71 90; 76 25 90; 32 36; 35 80
⊠ 5453 mindefn go ⊖ (+242) 76 60 47
Minister: Gen Idriss Ngari [98]

12-20000 Ministry of Foreign Affairs, Co-operation and Francophone Relations
Ministere des Affaires Etrangeres de la Co-operation et la Francophonie
PO Box 2245, Libreville
☎ (+241) 73 94 69 ⊠ 5255 diplgab go
⊖ (+241) 73 93 38
Minister: Casimir Oye-Mba [98]

25 Defense Forces

25-20000 Air Force Headquarters
Etat Major de l'Armee de l'Air
PO Box 13398, Libreville
☎ (+241) 76 42 47; 74 22 38 ⊠ 565060 fagema
⊖ (+241) 76 42 47; 74 65 89; 73 43 64
Chief of Air Staff: Gen Badinga Cyriaque
☎ (+241) 76 42 47; 74 22 38
Chief G3: Col Etoumbi Pascal
Chief G4: Lieut Col Ekomi Constant
Chief Technical Services: Maj Moudoki Boulingui [98]

31 Embassies & Attachés

31-00290 Embassy of the Republic of Cameroon
PO Box 14001, Libreville
☎ (+241) 73 28 00; 29 10 ⊠ 5396 ambacam go
Chargé d'Affaires a.i.: Isaac Nyemb Nguene
Commercial Officer: Paul-Pierre Ndengue
☎ (+241) 73 40 47
Defence Attaché: Pierre Koundjo
Public Information Officer: Ms Pauline Nyemb [97]

31-00510 Embassy of the Arab Republic of Egypt
PO Box 4240, Libreville
☎ (+241) 73 25 38 ⊠ 5425 boustane go
⊖ (+241) 73 25 19
Ambassador: H E Mrs Laïla Raphael
Deputy Chief of Mission: Mohamed Abdelaziz M Elewa [98]

31-00660 Embassy of the Federal Republic of Germany
Blvd de l'Indépendance Frangipaniers, PO Box 299, Libreville
☎ (+241) 76 01 88; 74 27 90 ⊠ 5248 aalbvl go
⊖ (+241) 72 40 12
Ambassador: H E Mr Horst Rudolf
Deputy Chief of Mission: Wolfgang Müller [98]

31-00890 Embassy of Italy
Immeuble Personnaz & Cardin, PO Box 2251, Libreville
☎ (+241) 74 28 92/3 ⊠ 5287 italdipl go
⊖ (+241) 74 80 35
Ambassador: H.E. Mr Vittorio Fumo
Deputy Chief of Mission: Dott. Sergio Serpelloni [98]

31-00920 Embassy of Japan
Boulevard du Bord de Mer, PO Box 2259, Libreville
☎ (+241) 73 22 97; 02 37 ✉ TAISHI Libreville
⊠ 5428 taisilv go ⊖ (+241) 73 60 60
Ambassador: H E Mr Tomoji Kawai
Counsellor: Yoshitaka Tomita
Attaché: Makoto Kudo

Second Secretary: Shinichi Fujiwara
Third Secretary: Naohisa Yoshida
Assistant Attaché: Kenji Kuratomi
Assistant Attaché: Atsushi Sasaki
Administrative Assistant: Wataru Yuruzume 98

31-01480 Embassy of Romania
PO Box 557, Libreville
☎(+241) 77 42 25 ⊜ (+241) 77 42 25
Charge d'Affairs: Constantin Ene 98

The Gambia

Head of State:
Capt. Yahya Jammeh

10 Head of State

10-10000 Office of the President
State House, Banjul
☎(+220) 282 60; 282 91 ☒ 2204 presof gv
President of the Republic: Capt Yahya Jammeh 98

10-20000 Office of the Vice President
State House, Banjul
☎(+220) 272 43 ⊜ (+220) 270 34
Vice President: Mrs Isatou Njie Jammeh 98

12 Ministries

12-10000 Ministry of National Defense
State House, The Quadrangle, Banjul
☎(+220) 22 40 10; 22 40 46; 22 85 53 ☒ 2204
⊜ (+220) 22 40 01
President and Minister for National Defence:
Col Yayah A J Jammeh (retired)
Permanent Secretary: A O Njie-Baro
Chief of Staff: Lieut Col Lawrence Jarra 98

12-20000 Ministry of External Affairs
State House, The Quadrangle, Banjul
☎(+220) 274 11 ☒ 2254 gamex gv
Secretary of State: Omar Y Njie 98

12-40000 Ministry of Interior
Banjul
Secretarty of State: Capt Momodou Bojang 98

26 Police (Civil)

26-00000 Police Force Headquarters
Buckle Street, Banjul
☎(+220) 22 49 14; 22 72 22; 22 72 47; 22 72 37
⊜ (+220) 22 65 61; 22 75 53
40 (civilian), 1,500 (uniformed)
Commander, Gambia Field Force: F R I Jammeh
Public Realtions: Inspt A Sanyang
Procurement and Finance Director: Chief Supt A Kuyteh
Administration Officer: Abubakarr S Jeng 98

31 Embassies & Attachés

31-01890 United Kingdom High Commission
PO Box 507, Banjul
☎(+220) 49 51 33; 49 51 34 ☒ 2211 ukrep gv
⊜ (+220) 49 61 34
High Commissioner: H E Mr J Wilde
Deputy High Commissioner: H A J Bradley
Third Secretary, Management: N S Chadwick
Third Secretary, Immigration: Mrs A Rowland
Aid Attaché: Mrs A Taylor 98

31-01900 Embassy of the United States of America
Fajara East, Kairaba Avenue, PMB 19, Banjul
☎(+220) 39 28 56; 58; 19 70 ⊜ (+220) 39 24 75
Ambassador: H.E. Mr Gerald W. Scott
Deputy Chief of Mission: Merritt C. Brown
Head of Commercial Section: Ms Kimberely Kelly
Consulate Section: Thomas Cairns (Resident in Dakar) 97

Georgia

Head of State:
Eduard Shevardnadze

10 Head of State

10-10000 Office of the President
State Council Building, Tbilisi
President: Eduard Shevardnadze 98

11 Prime Minister

11-10000 Office of the Prime Minister
Government House, ulitsa Ingorokva, 380034 Tbilisi
☎(+7 8832) 99 70 00 ☒ 212917 ergi su
⊜ (+7 8832) 93 27 27
Prime Minister: Otar Patsatsia
Deputy Prime Minister: Temuraz Basilia
Deputy Prime Minister: Bakur Gulua
Deputy Prime Minister: Zurab Kervalishvili
Deputy Prime Minister: Avtandil Margiani
Deputy Prime Minister: Tamaz Menagharishvili
Deputy Prime Minister: Tamaz Nadareishvili 98

12 Ministries

12-10000 Ministry of Defence
University Straat 2, 380114 Tbilisi
☎(+995 8832) 30 31 63; 30 31 26; 23 55 47
☒ 212903 avaza ⊜ (+995 8832) 98 39 25; 23 55 47
Minister of Defence: Lieut Gen Vardiko Nadibaidze
First Deputy Minister: Maj Gen Guram Nikolaishvily
First Deputy Minister, Chief of Main Staff:
Maj Gen Nodar Tatarashvily
Deputy Minister, Munitions: Lieut Gen Vano Gurgenidze
Deputy Minister, CIS Relations:
Maj Gen Vladimir Chikovany
Deputy Minister, Inspector of the Armed Forces:
Maj Gen Gudjar Kurashvily
Deputy Minister for Rear: Maj Gen Tariel Khitarishvily
Deputy Minister, Emergency Situations:
Maj Gen Tamaz Dumbadze
Chief of Main Dept for Military Readiness:
Maj Gen Shota Bayadze
Chief of Dept for Mobilization and Deputy Chief of Main Staff: Maj Gen Tengiz Razmadze
Chief of Frontier Forces: Maj Gen Valeri Chkheidze
Chief of Combat Service: Maj Gen Boris Ivanov
Commandant of Military Academy:
Maj Gen Tengiz Shubladze
Chief Personnel Department: Maj Gen Temur Kavtaradze
Chief Foreign Affairs Department: Col David Tevzadze
Chief Medical Department: Col Anton Chanturia
Chief of Naval Forces: Capt 1st A Javakhshvily
Air Force Commander: Col Valeri Nakopia
Air Defence Forces Commander: Col George Kvinikadze
Chief Press Service: Lieut Koba Liklikadze
Senior Officer of Foreign Affairs Department:
Lieut George Borchkhadze 98

12-20000 Ministry of Foreign Affairs
4 Nine April Straat, 380018 Tbilisi
☎(+7 8832) 98 93 77; 98 93 54 ☒ 212184 slava su
⊜ (+7 8832) 98 94 56
Minister of Foreign Affairs: Irakli Menagharishvili 98

12-25000 Ministry of Economy
170 Barnov Straat, 380062 Tbilisi
☎(+7 8832) 23 09 25; 23 53 82 ⊜ (+7 8832) 29 00 63
Minister: Vladimer Papava 98

12-30000 Ministry of Trade and Supply
42 Kazbegi Straat, 380077 Tbilisi
☎(+7 8832) 38 96 52; 38 96 67 ☒ 212146 foton su
⊜ (+7 8832) 38 07 23
Minister: Konstantin Zaldastanishvili 98

12-35000 Ministry of Finance
170 Barnov Straat, 38062 Tbilisi
☎(+7 8832) 23 09 25; 22 68 05 ☒ 212348 mofin su
⊜ (+7 8832) 29 23 68
Minister: Mikhail Chkuasleli 98

12-40000 Ministry of Internal Affairs
ulitsa Bolshaya Alleya 10, 380014 Tbilisi
☎(+7 8832) 99 62 33; 80 50
Minister: Kakha Targamadze 98

12-70000 Ministry of Communication
2 Nine April Straat, 380008 Tbilisi
☎(+7 8832) 95 08 15; 99 96 66 ☒ 212911 ptb su
Minister: Pridon Injia 98

12-70500 Ministry of Transportation
prospect A Kazbegi 12, 380060 Tbilisi
☎(+7 8832) 36 45 27
Minister: Merab Adeishvili 98

29 Customs & Excise

29-10000 Customs Committee
24 Akhvlediani Street, 380103 Tbilisi
☎(+7 8832) 96 96 38; 96 96 27 ⊜ (+7 8832) 96 96 21
Chairman: Victor Lordkipanidze 98

31 Embassies & Attachés

31-00500 Embassy of the Republic of Armenia
(† Armenian Diplomatic Mission)
4 Tetelashvili Straat, Tbilisi
☎(+7 8232) 95 17 23 🄴 aspect@iberiapac.ge
⊜ (+7 8232) 99 01 26
Ambassador: Levon Khachatrian
Second Secretary: Ashat Kirakossian 98

Contact Reference

Georgia

55 Government-owned Industry & Production Facilities

55-05000 Kutaisi Automobile Plant Production Association
(a partner of Kamaz and joint venture of Gaz)
ulitsa Avtostroitelia 88, 384009 Kutaisi
☎ (+995 331) 626 95 ✆ (88331) 641 93 ▣ 160112 rul
⊖ (+995 331) 010 81
US$ 19,634,768 (1997) ♣ US$ 3,238,000
2,140 (1997)
General Director: Amiran D Natsvlishvili
Operations Director: Roman Tskhvediani
Purchasing and Marketing Manager: Teimuraz Jerapov
■ Produces trailers for various ex-CIS military vehicles. Also produces diesel engines, trailers and non-armoured vehicle propulsion. [98]

60 Research Organizations, Institutes & Test Establishments

60-01000 Institute of Space Constructions Georgian Technical University
(a partner of International Association Kosmossviaz)
(a subordinate of the Department of Science & Technologies)
Kostava st. 68B, 380075 Tbilisi
☎ (+7 8832) 99 88 13; 36 77 87; 36 77 86
⊖ (+7 8832) 99 87 63
US$ 7 000 000 (1996) ♣ US$ 350 000 (1996) 275
6 200 m2
General Dir & General Designer: Elguja V. Medzmariashvili
Vice General Dir & General Designer: M. Janikashvili
Dir of Design Bureau: G. Bedukadze
Dir of Testing Complex: N. Khatiashvili
Dir of Military Engineering Centre: N. Khachidze
■ Design & testing of solar batteries, reflectors, force frameworks, orbital stations. Building of ground fast-rebasing antennas, unfolding bridges, large spanning lightweight buildings. Research & development in deployable structures. [97]

Germany, Federal Republic

Head of State:
Roman Herzog

10 Head of State

10-10000 Office of the Federal President of the Federal Republic of Germany
Bundespräsidialamt
Kaiser Friedrich Strasse 16, D-53113 Bonn
☎ (+49 228) 20 00 ▣ 886393 adbpn d; 2283624 bpr a
⊖ (+49 228) 20 02 00
President: Roman Herzog
Chief of Staff and Secretary of State: Wilhelm Staudaucher
Spokesman of the President: Roland Lohkamp
☎ (+49 228) 20 03 20; 20 05 55 (fax)
Deputy Spokesman of the President: Thomas Ellerbeck
☎ (+49 228) 20 03 38; 20 05 55 (fax)
Counsellor of Defence Affairs: Capt Hübertüs V Püttkamer (Navy) [98]

11 Prime Minister

11-10000 Office of the Federal Chancellor
Bundeskanzleramt
Adenauerallee 139-141, D-53113 Bonn
☎ (+49 228) 560 ▣ 886750 bkbn d
⊖ (+49 228) 56 23 57
Federal Chancellor: Dr Helmut Kohl
Vice Chancellor: Klaus Kinkel
Chief of the Federal Chancellery: Friedrich Bohl
Head of Foreign Relations and Defence: Mr M D Bitterlich
Head of Internal Security: Dr Roll
Head of Aerospace: Mr M D Wirmer [97]

12 Ministries

12-10000 Federal Ministry of Defence
Bundesministerium der Verteidigung (BMVg)
PO Box 1328, D-53003 Bonn
☎ (+49 228) 121; 12 94 07 ▣ 886575 bw d
⊖ (+49 228) 12 53 57; 12 53 58; 12 94 06
Minister of Defence: Volker Rühe
Parliamentary Secretary of State: Bernd Wilts
Parliamentary Secretary of State: Ms Michaela Geiger
State Secretary, Armaments: Gunnar Simon
Secretary of State, Armaments: Dr Peter Wichert
Director of Press and Information Office and Ministry Spokesman: Col Notker Siebert [98]

12-20000 Federal Ministry of Foreign Affairs
Auswaertiges Amt
Adenauerallee 99/103, D-53113 Bonn
☎ (+49 228) 170 ▣ 886591 aabn d
⊖ (+49 228) 17 28 45
Foreign Minister: Dr Klaus Kinkel
Minister of State: Helmut Schaefer
Minister of State: Dr Werner Hoyer
Secretary of State: Dr Peter Hartmann
Secretary of State: Dr Hans F von Ploetz
Spokesman: Dr Martin Erdmann
Office of the Minister: Dr Gerd Westdickenberg [98]

12-40000 Ministry of the Interior
Bundesministerium des Innern
Graurheindorfer Strasse 198, D-53117 Bonn
☎ (+49 228) 68 11 ▣ 886896; 886662 bmi d
⊖ (+49 228) 681 46 65
Minister of the Interior: Manfred Kanther
Police Director: Mr Scheefeld [98]

12-60000 Federal Ministry of Finance
Bundesministerium der Finanzen
Graurheindorfer Strasse 108, D-53117 Bonn
☎ (+49 228) 68 20
[W] http://www.bundesfinanzministerium.de ▣ 886645
⊖ (+49 228) 682 44 20
Minister: Dr Theo Waigel (MdB)
Head of Minister's Office: Ms Ida M Aschenbrenner
Minister's Spokesperson: Ms Barbara Eckrich
Information and Communication: Ms Margaret Ladener-Malcher
Parliamentary Secretary of of State for Tax Policy: Hansgeorg Hauser
Parliamentary Secretary of State for Budget: Irmgard Karwatzki
Secretary of State: Manfred Overhaus
Secretary of State: Dr Jürgen Stark
Organisation, Personnel and General Administration: Dirk Kühnau
Financial Policy: Dr Horst Schöberle
Budget: Dr Hans Reckers
Customs and Excise Taxes: Siegmar Kunas
Taxes on Income and Property: Heinrich Rendels
Fiscal Co-ordination with the EEC, the States and Municipalities: Dr Jürgen Quantz
Legal, War Claims: Winfried Baldus
Press Office: Waltraud Blöink [98]

12-80000 Federal Ministry of Justice
Bundesministerium der Justiz
Heinemannstrasse 6, D-53175 Bonn
☎ (+49 228) 580 ▣ 228506 ⊖ (+49 228) 58 45 25
Minister: Prof-Dr Edzard Schmidt-Jortzig (MdB)
Press Officer: Dr Böhm
Head of Section IB 1 (Air Regulations): Edwin Frietsch [98]

12-95000 Federal Ministry of Transport
Bundesministerium für Verkehr (BMV)
Robert Schuman Platz 1, PO Box 200100, D-53170 Bonn
☎ (+49 228) 30 00 ▣ 0885700 bmv d
⊖ (+49 228) 300 34 28; 300 34 29
Minister: Matthias Wissmann ☎ (+49 228) 300 20 00
Secretary of State: Hans Jochen Henke
☎ (+49 228) 300 22 00
Parliamentary Secretary of State: Manfred Carstens
☎ (+49 228) 300 21 00
Parliamentary Secretary of State: Johannes Nitsch
☎ (+49 228) 300 21 50
Director of Multilateral Co-operation in Aviation: Rainer Münz
Library: Mr Köppen [98]

15 Parliament

15-10000 Federal Council
Bundesrat
Görresstrasse 15, Bundeshaus (Nordflügel), D-53106 Bonn
☎ (+49 228) 910 00 ▣ 886841 ⊖ (+49 228) 910 04 00
President: Erwin Teufel ☎ (+49 228) 910 01 00
First Vice President: Dr Edmund Stoiber
Second Vice President: Prof-Dr Kurt Biedenkopf
Third Vice President: Heide Simonis
Director: Georg Berndt Oschatz ☎ (+49 228) 910 01 01
Head of Parliamentary Service: MinRat Dr Horst Risse
☎ (+49 228) 910 01 20

Foreign Affairs Committee
Görresstrasse 15, D-53106 Bonn
☎ (+49 228) 910 04 50 ⊖ (+49 228) 910 04 98
Chairman: Prof-Dr Kurt Biedenkopt

Defence Committee
Görresstrasse 15, D-53106 Bonn
☎ (+49 228) 910 04 50 ⊖ (+49 228) 910 04 98
Chairman: Porf-Dr Rolf Eggert

Finance Committee
Görresstrasse 15, D-53106 Bonn
☎ (+49 228) 910 05 10 ⊖ (+49 228) 910 05 28
Chairman: Heinz Schleusser
Deputy Chairman: Prof-Dr Georg Milbradt
Secretary: MinRat Gerard H Semmlaub [98]

20 Defense Administrations/Services/Agencies

20-11000 Federal Office of Defence Administration
Bundeswehrverwaltungsamt (BWVA)
Bonner Talweg 177, D-53129 Bonn
☎ (+49 228) 94 71; 947 20 95
President: Dr Götz Herrmann
Vice President: Dierk Schröder
Press Officer: Mrs Wiltraud Pilz [98]

20-19100 Federal Academy for Military Administration & Technology
Bundesakademie für Wehrverwaltung und Wehrtechnik (BAKWVT)
(an Agency of the Federal Ministry of Defence)
PO Box 250205, D-68085 Mannheim
Seckenheimer Landstrasse 8/10, D-68163 Mannheim
☎ (+49 621) 41 80 91 ▣ 7826564
⊖ (+49 621) 41 84 63
President: Peter J George ☎ (+49 621) 41 80 91 ext. 200
Head of Aeronautics Section: N Schulz
Press Officer: W Reusch ☎ (+49 621) 41 80 91 ext. 202
■ Training of senior officials in administration and procurement of the German Forces including all areas of military aviation and missile technology. [98]

25 Defense Forces

25-10000 Armed Forces Headquarters
Pascalstrasse 10, PO Box 1328, D-53003 Bonn
☎ (+49 228) 12 92 00; 12 92 01
⊖ (+49 228) 12 74 65 (press and information)
Chief of Staff: Gen Hartmut Bagger
Navy and NATO Affairs Spokesman: Cdr Pickohaus [97]

25-17300 Federal Centre for Explosives and Special Technologies
Wehrtechnische Dienststelle für Sprengmittel und Sondertechnik
Oberjettenberg, D-83458 Schneizlreuth
☎ (+49 8651) 79 12 08 ⊖ (+49 8651) 16 00; 79 13 38
Director: Juergen Ackermann ☎ (+49 8651) 12 00
→ 60 [97]

25-17350 Test Establishment 61 (Aircraft)
Wehrtechnische Dienststelle für Luftfahrzeuge (WTD 61)
Flugplatz, D-8072 Manching
☎ (+49 8459) 801 ⊖ (+49 8459) 80 20 22
Director: Dipl.-Ing Werner Schenk ☎ (+49 8459) 20 05
Flight Test Instrumentation: Mr Zerrath
☎ (+49 8459) 32 00
OTA Director of Institute of Aviation - Medicine Department of Ergonomics: Dr Knöfel ☎ (+49 8459) 31 00
Flight Operations Division: Mr Lauer ☎ (+49 8459) 20 89
Aviation Technology: Mr Biel ☎ (+49 8459) 23 07
Aircraft and Engine Technology: Dr Rosenberg
☎ (+49 8459) 20 80
Systems Technology Guided Missiles: Mr Schweiger
☎ (+49 8459) 26 27
Aircraft Equipment Technology: Mr Mang
☎ (+49 8459) 21 26
Avionics and Aircraft Armament: Mr Tröednce
☎ (+49 8459) 22 70
Avionic and Armament Technology and Certification: Mr Frädrich ☎ (+49 8459) 30 15
→ 60 [97]

25-17380 Test Establishment 71 Ships and Naval Armament
Wehrtechnische Dienststelle 81 für Schiffe und Marinewaffen (Engineering Center for Ships and Naval Weapons)
Berliner Strasse 115, D-2330 Eckernförde
☎ (+49 4351) 40 41
E 100625.3066@compuserve.com
⊖ (+49 4351) 454 00
Director: Dipl.-Ing Klaus Borchardt
Deputy Director, Testing and Measurement Services: Mr Cartensen
Director of Planning, Control Accounts, Training and Public Relations: Mr Busche
Administration Director: Mr Kirkskothen
General Services Director: Mr Schultz
Director of Magnetic Effects and Degaussing: Mr Wolff
Director of Underwater Weapons: Dr Vogel
Director of Naval HM&E Technology and Acoustics: Mr Breltensteln
Director of Naval Command and Control: Mr Reelmann
→ 60 [97]

25-19100 Armed Forces University Munich
Universitaet der Bundeswehr München
Werner Heisenberg Weg 39, D-85577 Neubiberg
☎ (+49 89) 60 04 20 00
E h11cpost@rz.unibw-muenchen.de ⊠ 5215800
⊖ (+49 89) 60 04 20 09
President: Dr Hans Georg Lössl
Press Officer: Lieut Col Norbert Hoerpel
☎ (+49 89) 60 04 20 04 [98]

25-19200 University of the Federal Armed Forces Hamburg
Universität der Bundeswehr Hamburg
(† Armed Forces University Hamburg)
Holstenhofweg 85, D-22043 Hamburg
☎ (+49 40) 65 41 27 74 ⊖ (+49 40) 65 41 27 02
President: Prof Dr Gerhard Strunk
☎ (+49 40) 65 41 27 00
Head of Press Office: Mrs Anne Schulz

■ The university offers academic studies in 8 different courses of study (only to officers or cadets of the Federal Armed Forces) in: history, educational science, political science, national economy, business & administration, mechanical engineering, electrical engineering & economic engineering. [97]

25-19500 Armed Forces Medical College
Akademie des Sanitäts- und Gesundheitswesens der Bundeswehr
Neuherbergstrasse 11, D-80937 Munich
☎ (+49 89) 089 31 68 ⊖ (+49 89) 089 33 95
Commander: Gen Med-Dr Peter Fraps
Director of Operations: FltlAp Shäfer
Librarian: Mr Bredl [98]

25-19600 Federal Armed Forces Intelligence School (FAFIS)
Schule für Nachrichtenwesen der Bundeswehr
Alte Kemmenauer Strasse 17/19/21, D-56130 Bad Ems
☎ (+49 2603) 79 71 ⊖ (+49 2603) 79 72 25
Commander: Col Jörg Rappke ☎ (+49 2603) 797 ext 200 [98]

25-21000 German Air Force Office
Luftwaffenamt
Fliegerhorst Köln-Wahn, PO Box 902500/902501, D-51140 Cologne
☎ (+49 2203) 602 33 00; 602 33 02
⊖ (+49 2203) 602 20 71
Head of Air Force Office: Maj Gen Botho Engelien
☎ (+49 2203) 602 23 31
Chief of Staff: Col Wolfgang Born
☎ (+49 2203) 602 45 60
Deputy Director of Air Force Office: Brig Gen Dieter Kellein
Chief of Armaments Division: Brig Gen Jaunig
☎ (+49 2203) 602 21 37
Chief of Air Security Division: Brig Gen Hannes Block
☎ (+49 2203) 602 23 34
Press Office Director: Lieut Col Wulf Bickenbach
☎ (+49 2203) 602 32 96; 602 22 48 [97]

25-21100 Equipment Office of the GAF
Luftwaffenunterstützungskommando
Postfach 902500, D-51140 Cologne
Heidestrasse 245, D-51147 Cologne
☎ (+49 2203) 602 20 46 ⊖ (+49 2203) 693 32
Director: Brig Gen Wolf Gerhard Stäglich [98]

25-22200 GAF Support Command
Luftwaffenunterstützungskommando
Fliegerhorst Köln-Wahn, PO Box 902500/902502, D-51140 Cologne
☎ (+49 2203) 60 21 ⊖ (+49 2203) 693 32
Commander: Maj Gen Karl-Heinz Richter
☎ (+49 2203) 602 51 80; 602 20 45
Deputy Commander and Commander GAF Service Units: Brig Gen Wolf Gerhard Stäglich
☎ (+49 2203) 602 51 80; 602 20 46
Chief of Staff: Col Klaus Knepper
☎ (+49 2203) 602 51 40
Director of Press and Information: Lieut Col Hanns Joachim Biedermann
☎ (+49 2203) 602 20 48; 602 20 71

GAF Materiel Office
Materialamt der Luftwaffe
PO Box 902500/503, D-51140 Cologne
☎ (+49 2203) 60 21 ⊖ (+49 2203) 602 23 04
Director: Brig Gen Lorenz Huber ☎ (+49 2203) 602 20 01
Deputy Director: Col Klaus Peter Treche
☎ (+49 2203) 602 20 02

GAF Armament Division
General Luftwaffenrustung
PO Box 902500/502/09, D-51140 Cologne
☎ (+49 2203) 60 21 ⊖ (+49 2203) 602 21 41
Director: Brig Gen Peter Heinz Lahl
☎ (+49 2203) 602 21 37; 602 31 68
Deputy Director: Col Georg Rodler
☎ (+49 2203) 602 21 38 [98]

25-29100 German Air Force Officer School
Offizierschule der Luftwaffe
PO Box 1264 A/S, D-82242 Fürstenfeldbruck
☎ (+49 8141) 96 21 66 01 ⊖ (+49 8141) 62 00 80
Commandant: Brig Gen Jarosch
Aide-de-camp: Capt Carsten Spiering [98]

25-30100 German Army Office
Heeresamt
Konrad-Adenauer-Kaserne, Brühler Strasse 300, D-50968 Cologne
☎ (+49 221) 93 71 24 25; 937 10
⊖ (+49 221) 93 71 24 24; 93 71 29 56; 93 71 29 90
Director: Gen Maj Jürgen Reichardt
Deputy Director and Chief of Staff: Brig Gen Henning Brümmer
Director of General Affairs Department I: Brig Gen Helmut Harff
Director of Training Department II: Brig Gen Wilhem Romatzeck
Director of Department III: Brig Gen Gerhard Beenders
Director of Department IV: Brig Gen Hans Jürgen Rennack
Director of Department V: Brig Gen Andreas Wittenberg
Director of Department VI: OiG Lühr
Special Staff Department: OiG Helmut Specovius
Commandant Army School of Signals and Electronics: Brig Gen Konrad Bader
Head of Press Office: Lieut Col Volker Rönnike [98]

25-33000 German Army Forces Command
(† Third Army Corps Headquarters)
Am Wöllershof 12, PO Box 6960, D-56064 Koblenz
Commander: Lieut Gen - Dr Klaus Reinhardt
☎ (+49 261) 307 20 00
Deputy Commander: Maj Gen Hubert Gosh
☎ (+49 261) 307 20 10
Chief of Staff: Brig Gen Axel Bürgens
☎ (+49 261) 307 20 30
Press Officer: Lieut Col Hammerstein
☎ (+49 261) 307 20 50 [98]

25-33100 Territorial Command Schleswig Holstein & DBvBev AFNORTH
Niemannsweg 220, PO Box 1151, D-2300 Kiel 1
☎ (+49 431) 380 61 51 ⊖ (+49 431) 380 61 75
Commander: Maj Gen Manfred Dietrich
☎ (+49 431) 380 50 04
Deputy Commander and Chief of Staff: Rear Adm Otto Ciliax
Press Officer: Cdr Wedigo von Wedel
☎ (+49 431) 380 61 30

Military Region Command HQ (VBK 10)
Sophienstrasse 14, D-2000 Hamburg 13
☎ (+49 40) 415 01 ⊖ (+49 40) 415 02 88
Commander VBK 10: Capt Gerhard Krancke [97]

25-39200 Army Signal School and Technical College for Electrical Engineering
Tutzinger Strasse 46, D-82340 Feldafing
☎ (+49 8157) 272 00 ⊖ (+49 8157) 273 06
Commander and Signal Officer-in-Chief: Brig Gen Konrad Bader
UK Exchange Officer: Maj A J Clark
Deputy Commander and Chief of Staff: Col D Mey
Director of Training: Col S Becker
Director of Force Development: Col H Karon [98]

25-40000 Navy Headquarters
Führungsstab der Marine
PO Box 1328, D-53003 Bonn
☎ (+49 228) 12 00 ⊖ (+49 228) 12 51 34
Inspector of the Navy: Vice Adm Hans Rudolf Boehmer
☎ (+49 228) 12 00 ext. 47 00
Deputy Inspector and Chief of Staff: Rear Adm Klaus Jancke ☎ (+49 228) 12 00 ext. 47 02
Assistant to the Chief of Staff: Cdr Jüren Ehle
☎ (+49 228) 12 00 ext. 47 03
Director of Department I, Manpower, Training/Education and Public Relations: Rear Adm Reinhold Siebert
☎ (+49 228) 12 00 ext.49 40
Director of Department II, Logistics Armaments: Rear Adm Bernd Heisc
Director of Department III, Operations, Plans and Policy: Capt Wolfgang Nolting (Navy)
☎ (+49 228) 12 00 ext. 47 10
Public Information/Relations Branch: Capt Gerd Kiehnle (Navy) ☎ (+49 228) 12 00 ext. 47 43 [97]

25-41000 German Naval Office
Marineamt
Marine Anlage Bordum, PO Box 2005, D-26379 Wilhelmshaven
☎ (+49 4421) 470 ⊖ (+49 4421) 47 22 44
Commander: Rear Adm Jürgen Geier

Contact Reference

Deputy Commander and Naval Training Director:
Rear Adm Hans Achim Romer ☎(+49 4421) 49 48 00
Director of Press Department: Wolfgang Jungmann
Surgeon General of the Navy: Adm Dr D Willers
☎(+49 4421) 49 31 40 [97]

25-44000 Naval Support Command Headquarters (MUKdo)
Marineunterstützungskommando
PO Box 2743, D-26379 Wilhelmshaven
Heppenser Groden, D-26384 Wilhelmshaven
☎(+49 4421) 680
⊖ (+49 4421) 68 45 95; 79 25 46; 475 55
Commander: Rear Adm Horst Sommermeyer
☎(+49 4421) 68 40 00
Deputy Commander and Director of Logistic Support and Armament: Rear Adm Karl Wilhelm Rosberg
☎(+49 4421) 68 40 03
Chief of Staff and Director of Logistic Services:
Rear Adm Detlef Kammhelz ☎(+49 4421) 68 44 00
Chief of Planning and Policy Branch: Capt Lothar Schmitt
☎(+49 4421) 68 44 10
Chief of Material Division: Capt Juergen Muehlenmeister
☎(+49 4421) 475 00
Chief of Systems Division: Capt Helmuth Hauck
☎(+49 4421) 79 25 00
Chief of Equipment Division: Capt Dieter Laabs
☎(+49 4421) 68 45 00
Public Information Office
Information-Zentrum-West (Marine)
Schellingstrasse 19, D-26384 Wilhelmshaven
☎(+49 4421) 49 32 61 ⊖ (+49 4421) 30 54 02
Chief of Public Information Office:
Cdre Wolfgang Jungmann [98]

25-47000 Navy Arsenal
Marinearsenal
Postfach 2752, D-26379 Wilhelmshaven
☎(+49 4421) 49 20 01
Director: Dipl.-Ing Horst Eggert
Information/Public Affairs Office
Dezernat 013, Postfach 2752, D-26379 Wilhelmshaven
☎(+49 4421) 49 20 13
Dezernatsleiter: Dipl.-Ing Olaf Petereit [98]

25-49200 Marinewaffenschule
Am Ort 6, D-24340 Eckernförde
☎(+49 4351) 810 41 ⊖ (+49 4351) 828 96
Commander: Capt Fritz Noblé
Press and Information Officer: Lieut Cdr Kail H Schneider [98]

25-50000 Office of the Surgeon General of Federal Armed Forces
Inspekteur des Sanitätsdienstes der Bundeswehr (InspSan)
PO Box 1328, D-53003 Bonn
☎(+49 228) 12 64 27 ⊖ (+49 228) 12 66 89
Surgeon General: Lt Gen.Dr Gunter Desch
Deputy Surgeon General & Chief of Staff:
Gen. Dr Karl-Wilhelm Demmer ☎(+49 228) 12 64 37
Press Officer: Mrs S. Erichsen [97]

25-55000 US Army
(1st Battalion - 50 1st Aviation Regiment)
CMR 477, PO Box 795, D-09165 Hanau
☎(+49 1149) 618 35 14 98
⊖ (+49 1149) 618 35 14 32
Intelligence Officer: Randall I Haws
■ Provides attack helicopter support to US/NATO and UN Forces in Europe. [98]

25-60100 Heeresfliegerbrigade 3
(t Heeresfliegerkommando 3)
Gunther-Plüschow-Kaserne, D-56740 Mendig
☎(+49 2652) 94 13 00 ⊖ (+49 2652) 94 16 39
Commander: Brig Gen Kams
Deputy Commander and Chief of Staff: Col Granz
Operations Director: Lieut Col Dittmar
Procurement Director: Lieut Col Honnens [98]

29 Customs & Excise

29-01000 General Directorate of Customs & Excise
(a subordinate of the Ministry of Finance)
c/o Bundesministerium der Finanzen, Postfach 1308, D-53003 Bonn [98]

31 Embassies & Attachés

31-00085 Armenian Diplomatic Mission
Viktoriastrasse 15, D-53173 Bonn
☎(+49 228) 35 29 03
Ambassador: H E Mr Felix Mamigonian [98]

31-00090 Australian Embassy
Godesberger Allee 105/107, D-53175 Bonn
☎(+49 228) 810 30 ✉ AUSTEMBA Bonn
Telex 885466 ausem d ⊖ (+49 228) 810 31 44
Ambassador: H E Mr Max Hughes
Deputy Chief of Mission: Dr Janet Gardiner
Defence Attaché: Col Gerald McCornack
Assistant Defence Attaché: Lieut Col Keith Alexander
First Secretary: Greg Rimes
Public Affairs Officer: Günter Schlothauer
Berlin Office
Kempinski Plaza - 4th Floor, Uhlandstrasse 181/183, D-10623 Berlin
☎(+49 30) 880 08 80 Telex 113342 auscg dd
⊖ (+49 30) 88 00 88 99 [98]

31-00100 Embassy of Austria
Johanniterstrasse 2, D-53113 Bonn
☎(+49 228) 53 00 60
E oesterreich-botschaft-bonn@compuserve. com
Telex 886780 oebon d ⊖ (+49 228) 54 90 30
Ambassador: Dr Friedrich Hoess
Minister and Deputy Chief of Mission: Dr Alfons Kloss
Defence Attaché: Brig Gen Peter-Erik Czak
☎(+49 228) 23 20 50
Counsellor, Culture and Science: Dr Christa Sauer
Economic Counsellor: Dr Christoph Thun-Hohenstein [98]

31-00150 Embassy of Belgium
Kaiser Friedrich Strasse 7, D-53113 Bonn
☎(+49 228) 21 39 22; 20 14 50
⊖ (+49 228) 21 04 20; 22 08 57
Ambassador: H E Mr Dominique Struye de Swielande
Minister, Deputy Chief of Mission, Economics and European Union: Bruno Nève de Mévergnies
Military, Naval and Air Attaché: Col André Wauters
Commercial Counsellor: Rainer Palm
First Secretary, Press and Political: Guy Sevrin
Economics and Technology Counsellor:
Christian Meerschman
Administrative Counsellor: Louis van den Putte
Agriculture Counsellor: Romain Devylder
First Secretary, Press & Political: Adrien Theatre
First Secretary, Political: Bruno Georges [98]

31-00170 Embassy of the Republic of Benin
PO Box 200254, D-53132 Bonn
Rüdigerstrasse 10, D-53179 Bonn
☎(+49 228) 94 38 70 ⊖ (+49 228) 85 71 92
Ambassador: H E Mr Corneille Mehissou
☎(+49 228) 943 87 21
Counsellor and Deputy Chief of Mission: Mrs Marian Diallo
☎(+49 228) 943 87 94 [98]

31-00200 Embassy of Bolivia
Konstantinstrasse 16, D-53179 Bonn
☎(+49 228) 36 20 38 ⊖ (+49 228) 36 59 52; 93 44 70
Ambassador: H E Mr Orlando Donoso
Press and Cultural Attaché: Mrs Clemencia del Carpio [98]

31-00240 Embassy of Brunei Darussalam
Kaiser Karl Ring 18, D-53111 Bonn
☎(+49 228) 67 20 44; 67 20 47 ⊖ (+49 228) 68 73 29
Ambassador: H E Mr Pengiran Maidin
Head of Chancery: Pengiran Ali Hashim
Defence Adviser: Lieut Col Sari Ahmad [97]

31-00250 Embassy of the Republic of Bulgaria
Auf der Hostert 6, D-53173 Bonn
☎(+49 228) 36 30 61; 36 30 65 Telex 885739 bulg d
⊖ (+49 228) 35 82 15
Ambassador: H E Dr-jur Stoyan Stalev
Minister and Deputy Chief of Mission: Dr Jvan Dimitrov
First Secretary: Dr Valentin Valev
First Secretary, Press: Atanas Krastin
Second Secretary, Protocol: Nikolay Iliev [98]

31-00260 Embassy of Burkina Faso
Wendelstadtallee 18, D-53179 Bonn
☎(+49 228) 95 29 70 Telex 885508 bfaso d
⊖ (+49 228) 952 97 20
Ambassador: H E Mr Jean Baptiste Ilboudo
Second Counsellor: Ms Jacqueline Zaba
Finance Attaché: Moussa Traore
Cultural Counsellor: Mrs Jeanne Marie Ilboudo [98]

31-00380 Embassy of Colombia
Friedrich Wilhelm Strasse 35, D-53113 Bonn
☎(+49 228) 92 37 00 ⊖ (+49 228) 923 70 37
Ambassador: H E Mr Jorge Juan Bendeck Olivella
Deputy Chief of Mission: Alberto Davilla Suarez
Naval Attaché: Capt José Luis Cuenca Ferrada
Cultural Attaché: Cristina Hoyos Koerbel
Commercial Attaché: Ms Adriana Oyola Vivas
Press Attaché: Ms Marcela Gil Helffhrittz
Chief of Consular Department:
Ms Maria Victoria Torres Orozco [98]

31-00400 Embassy of the Republic of Congo
Rheinallee 45, D-53173 Bonn
☎(+49 228) 35 85 55 Telex 886690 kongo d
⊖ (+49 288) 35 83 55
Ambassador: H E Mr Charles Ngouoto Moukolo
Attaché: Ferdinand Ngo-Ngaka
Second Secretary: Grégoire Ntouta [97]

31-00410 Embassy of Costa Rica
Langenbachstrasse 19, D-53113 Bonn
☎(+49 228) 54 00 40 ⊖ (+49 228) 54 90 53
Ambassador: H E Mr Ekhart Peters
Deputy Chief of Mission: Ms Lydia Peralta
Advisor: Klaus Justus
Secretary: Ms Lucy Aruello
Secretary: Alexis Flores [97]

31-00440 Embassy of Cyprus
Kronprinzenstrasse 58, D-53173 Bonn
☎(+49 228) 36 33 36; 36 35 96 ⊖ (+49 228) 35 26 36
Ambassador: H E Mr Theophilou V Theophilou
First Secretary: Dr Michalis Stavrinos
Second Secretary: Philippos Kritiotis
Press Counsellor: Michalakis Koumides
☎(+49 228) 36 50 33
Press Attaché: Demitris Hasikos ☎(+49 228) 36 50 33
Trade Section
Friedrichstrasse 42/44, D-50676 Cologne
☎(+49 221) 23 51 60; 23 51 69 ⊖ (+49 221) 23 17 13
Commercial Counsellor: Charalambos Rousos
Commercial Attaché: Antonios Antoniades [98]

31-00490 Embassy of the Dominican Republic
Burgstrasse 87, D-53177 Bonn
☎(+49 228) 36 49 56 ⊖ (+49 228) 35 25 76
Ambassador: H E Dr Octavio R Cáceres Michel
Second Secretary: Ms Carolina Cáceres Arango
Attaché: Dr Magaly Bello de Kemper [98]

31-00510 Embassy of the Arab Republic of Egypt
Kronprinzenstrasse 2, D-53173 Bonn
☎(+49 228) 95 68 30; 37 88 11 Telex 885719 egt d
⊖ (+49 228) 36 43 04; 37 39 96
Ambassador: H E Mr Mohab Mokbel
☎(+49 228) 956 83 23
Minister and Deputy Chief of Mission: Omar Metwally
☎(+49 228) 956 83 24

Office of the Defence Attaché
Kennedy Allee 43, D-53175 Bonn
☎ (+49 228) 37 95 35; 37 88 11 ⊖ (+49 228) 37 39 96
Defence Attaché: Brig Gen Mohamed Adel el Sheikh
☎ (+49 228) 37 95 35
Deputy Defence Attaché: Maj Ibrahim Mohamed Ibrahim
☎ (+49 228) 37 88 11
Commercial Department
Paracelsusstrasse 72, D-53177 Bonn
☎ (+49 228) 33 09 20 ▣ 8861131 egth d
⊖ (+49 228) 33 14 84
Minister for Commercial Affairs and Envoy:
Hamdy Ahmed Metwally ☎ (+49 228) 33 09 20
Commercial Counsellor: Tarek Abbas
☎ (+49 228) 33 02 32 [97]

31-00520 Embassy of El Salvador
Adenauerallee 238, D-53113 Bonn
☎ (+49 228) 54 99 13; 54 99 14 ⊠ EMBASALVA Bonn 1
⊖ (+49 228) 54 98 14
Ambassador: H E Dr José Saguer Saprissa
Secretary: Ms Gisela Wöllgens [98]

31-00535 Embassy of Estonia
Fritz-Schaeffer-Straße 22, D-53113 Bonn
☎ (+49 228) 91 47 90 [E] bonn@estebonn.gihub.de
⊖ (+49 228) 914 79 11
Ambassador: H E Mr Tiit Matsulevitsch
Honorary Consul
Badestraße 38, D-20143 Hamburg
☎ (+49 40) 450 40 26 ⊖ (+49 40) 45 04 05 15 [98]

31-00580 Embassy of Finland
Friesdorfer Strasse 1, D-53173 Bonn
☎ (+49 228) 38 29 80 ⊖ (+49 228) 382 98 57
Ambassador: H E Mr Arto Mansala
Minister: Mikko Jokela
Deputy Defence Attaché: Lieut Col Jyrki Laakso
Commercial Counsellor: Ms Katariina Huusko-Niemenmaa [98]

31-00690 Embassy of Greece
An der Marienkapelle 10-12, D-53179 Bonn
☎ (+49 228) 35 70 36; 35 70 37; 35 70 38; 35 70 39
[E] 022883010-0001@t-online.de
▣ 885656 hella d; 8869474 grin d
⊖ (+49 228) 35 24 32; 35 32 84
Ambassador: H E Dr Ioannis Bourloyannis-Tsangaridis
Minister-Counsellor and Deputy Chief of Mission:
Dimitrios Contoumas
Defence and Army Attaché: Brig Gen Niouluios Douvas
☎ (+49 228) 35 70 35
Naval Attaché: Capt Athanasious Papachristodoulm
Air Attaché: Lieut Col Stavros Portoulidis
Commercial Office
Kronprinzenstrasee 17, D-53173 Bonn
☎ (+49 228) 35 60 81; 35 60 82 ▣ 885711 gbha d
⊖ (+49 228) 36 19 41
Minister, Economic and Commercial:
Dr Theodoros Karavias
Press and Information
Weberstrasse 83, D-53113 Bonn
☎ (+49 228) 21 56 54 ⊖ (+49 228) 22 17 05
Information Contact: Andreas Papadatos [98]

31-00740 Embassy of Guatemala
Zietenstrasse 16, D-53173 Bonn
☎ (+49 228) 35 15 79; 35 86 09
[E] embaguate_bonn@compuserve.com
⊖ (+49 228) 35 49 40
Ambassador: H E Mr Jorge Skinner-Klée
First Secretary: Ms Sara Solis Castañeda
Second Secretary: Ms Sylvia Orantes Ruiz
Third Secretary: Ms Guisella Garcia Paz [98]

31-00910 Embassy of Jamaica
Am Kreuter 1, D-53177 Bonn
☎ (+49 228) 35 40 45; 36 33 25 ▣ 885493 jamai d
⊖ (+49 228) 36 18 90
Ambassador: H E Mr Peter C Black
Minister, Counsellor and Deputy Chief of Mission:
Mrs Donna Miller
Attaché, Finance and Counsular: Nigel Smith
Attaché, Information: Miss Angela Edwards
Attaché and Senior Secretary: Miss Draniah Trought [98]

31-00920 Embassy of Japan
Japanische Botschaft
Godesberger Allee 102-104, D-53175 Bonn 2
☎ (+49 228) 819 10 ⊠ TAISHI Bonn 1 ▣ 886878 a
⊖ (+49 228) 37 93 99
Ambassador: H E Mr Tatsuo Arima
Minister and Deputy Chief of Mission: Koichi Takahashi
Defence Attaché: Col Masujiro Uchida
Science Attaché: Yutaka Hishiyama
Counsellor, Economic Affairs: Tetsuhiro Hosono
General Consulate of Japan
Japanisches Generalkonsulat
c/o Deutsch-Japanisches Center, Immermanstrasse 45,
D-40210 Dusseldorf
☎ (+49 211) 16 48 20 ⊖ (+49 211) 35 76 50
General Consulate of Japan
Japanisches Generalkonsulat
Rathausmarkt 5, D-20095 Hamburg
☎ (+49 40) 333 01 70 ⊖ (+49 40) 30 39 99 15
General Consulate of Japan
Japanisches Generalkonsulat
Hamburger Allee 2-10, D-60486 Frankfurt am Main
☎ (+49 69) 970 82 30 ⊖ (+49 69) 77 38 73
General Consulate of Japan
Japanisches Generalkonsulat
Wachtelstrasse 8, D-14195 Berlin
☎ (+49 30) 850 89 30 ⊖ (+49 30) 832 69 67
General Consulate of Japan
Japanisches Generalkonsulat
Prinzregentenplatz 10, D-81675 Munich
☎ (+49 89) 47 10 43; 47 10 44; 47 10 45
⊖ (+49 89) 470 57 10 [97]

31-00930 Embassy of the Hashemite Kingdom of Jordan
Beethovenallee 21, D-5300 Bonn 2
☎ (+49 228) 35 70 46; 35 40 51; 35 70 52
▣ 885401 urdun d ⊖ (+49 228) 35 39 51
Ambassador: H E Mr Hussein Hammami
Minister and Deputy Chief of Mission:
Mohammad Ali Qudah
Press Counsellor: Nayef Mutlaq [97]

31-00990 Embassy of the Republic of Korea
Adenauerallee 124, D-53113 Bonn 1
☎ (+49 228) 26 79 60 ▣ 8869508 kor d
⊖ (+49 228) 22 39 43
Ambassador: H E Mr Sun-young Hong
Minister and Deputy Chief of Mission: Young-Oh Song
Defence Attaché: Col Dr Dongmyung Kim
☎ (+49 228) 21 43 63
Assistant Defence Attaché: Cdr Kyong-Jo Sur
Commercial Attaché: Ki-Sub Lee ☎ (+49 228) 21 89 38 [98]

31-01030 Embassy of the Kingdom of Lesotho
Godesberger Allee 50, D-53175 Bonn
☎ (+49 228) 37 68 60; 37 68 68; 37 68 69
[E] lesemger@aol.com ▣ 8869370 lbbn d
⊖ (+49 228) 37 99 47
Ambassador: H E Ms Lebohang Ntsinyi
Counsellor: King Seati Moloi
First Secretary: Joshua Setipa
Third Secretary: Tseliso Benjamin Mafatle
Attaché: Ms Moreheng Lethunya [98]

31-01100 Embassy of the Republic of Malawi
Mainzer Strasse 124, D-5300 Bonn 2
☎ (+49 228) 94 33 50 ▣ 8869689 mabn d
⊖ (+49 228) 943 35 37
Ambassador: H E Mr R N L Nkomba
Diplomatic Minister: W G Mkochi
Counsellor: T A V Chande
First Secretary, Administration: S N Dzama
Second Secretary, Consular: S D P Kapusa
Third Secretary, General Duties: Mrs B A Kabvina [97]

31-01140 Embassy of Malta
Viktoriastrasse 1, D-53173 Bonn
☎ (+49 228) 36 30 17; 36 30 18
[E] maemba@bldtzheim.com ⊖ (+49 228) 36 30 19
Ambassador: H E Carmel J Aquilina
Deputy Chief of Mission: John Busuttil
First Secretary: Dr Elena Grech
Press Secretary: Dr Leo Said [98]

31-01190 Embassy of Mexico
Adenauerallee 100, D-53113 Bonn 1
☎ (+49 228) 91 48 60 [E] mexico@inf.fu-berlin.de
[E] sreale01@att.com ⊖ (+49 228) 21 11 13; 914 86 19
Ambassador: H E Mr Juan José Bremer
Minister and Deputy Chief of Mission: Eduardo Pena
Marine Attaché: Adm Gerónimo Bringas
Counsellor: Mrs Luz Estela Santos
Commercial Section
Adenauerallee 100, D-53113 Bonn 1
☎ (+49 228) 22 30 21 ▣ 885286 ccmf d
⊖ (+49 228) 26 10 04
Counsellor of Economic Affairs: Lajos Szendro Kassa [98]

31-01220 Embassy of the Kingdom of Morocco
Gotenstrasse 7-9, D-53175 Bonn
☎ (+49 228) 35 50 44; 35 50 45; 35 50 46
▣ 885428 sifam d ⊖ (+49 228) 35 78 94
Ambassador: H E Mr Chawki
☎ (+49 228) 35 42 51; 35 32 42
First Counsellor: Abdelhamid B Benchekroun
☎ (+49 228) 35 42 51
Defence Attaché: Col Moulay Hachem El Amrani
☎ (+49 228) 35 66 14
Commercial Attaché: Hassan Bardach
☎ (+49 228) 35 50 44 [98]

31-01270 Royal Netherlands Embassy
Strässechensweg 10, Bonn
☎ (+49 228) 530 50 ⊠ HOLLANDIA BONN 1
▣ 886826 holbo d ⊖ (+49 228) 23 86 21
Defence, Military, Air and Naval Attaché:
Col Wietse F Bÿlsma [97]

31-01300 New Zealand Embassy
Bundeskanzlerplatz 2-10, Bonn-Center HI 902,
D-53113 Bonn
☎ (+49 228) 22 80 70 ⊖ (+49 228) 22 16 87
Ambassador: H E Mr Gerard F Thompson
Deputy Chief of Mission: Dr Michael J McBryde [98]

31-01340 Royal Norwegian Embassy
Mittelstrasse 43, D-53175 Bonn
☎ (+49 228) 81 99 70 [E] info@norwegen.org
⊖ (+49 228) 37 34 98
Ambassador: H E Mr Kjell Eliassen
Minister: Torbjörn Fröysnes
Defence Attaché: Col Jan Skaar ☎ (+49 228) 37 91 94
Political Counsellor: Jon Ramberg
Press and Cultural Counsellor: Jan G Lassen
Economy Counsellor: Helge Skaara [98]

31-01400 Embassy of Paraguay
Uhlandstrasse 32, D-53173 Bonn
☎ (+49 228) 35 67 27 ⊖ (+49 228) 36 66 63
Ambassador: H E Dr Marcos Martinez Mendieta
First Secretary and Deputy Chief of Mission:
Martin Llano-Heyn
First Secretary: Raúl Florentin
Second Secretary: Dieter Escher-Bächle
Secretary: Ms Mónica Acuna Trappe [97]

31-01430 Embassy of the Republic of Poland
Lindenallee 7, D-50968 Cologne
☎ (+49 221) 93 73 02 10; 38 81 72
⊖ (+49 221) 34 30 89
Ambassador: H E Dr Andrzej Byrt
☎ (+49 221) 93 73 02 10; 38 81 72; 34 30 89 (fax)
Minister and Deputy Chief of Mission: Dr Marek Prawda
☎ (+49 221) 93 73 02 12
Defence Attaché: Col Jerzy Gajewski
☎ (+49 221) 93 73 02 21
Head of the Press Section: Jaroslaw Dziedzic
☎ (+49 221) 93 73 02 21
Head of Economic Section: Mariusz Idzikowski
☎ (+49 221) 93 73 02 14
Commercial Section
An der Alteburger Mühle 6, D-50968 Cologne
☎ (+49 221) 34 99 11; 34 99 12; 37 29 14
▣ 8885382 brh d ⊖ (+49 221) 34 99 10
Minister and Head of Commercial Section:
Dr Jozef Olszynski

Contact Reference

Germany, Federal Republic

Berlin Representative Office
Unter den Linden 72, D-10117 Berlin
☎ (+49 30) 220 25 51 ▣ 114007 amb dd
⊖ (+49 30) 229 14 36
Minister and Chief of Representative Office: Dr Jerzy Sulek [98]

31-01480 Embassy of Romania
Legionsweg 14, D-5300 Bonn 1
☎ (+49 228) 55 58 60
▣ 8869791 amro d; 8869792 arse d
⊖ (+49 228) 68 02 47
Ambassador: H E Prof-Dr-Ing Florea Duditza
Commercial Officer: Nicolae Fiser
Air, Naval and Military Attaché: Aurelian Mihalache
Political Counsellor: Vasile Croicu
Political Counsellor: Ion Neamtu
Science and Technology Secretary: Cristian Radu [97]

31-01490 Embassy of the Republic of Rwanda
Beethovenallee 72, D-53173 Bonn 2
☎ (+49 228) 35 52 28 ▣ 885604 amrwa d
⊖ (+49 228) 35 19 22
Ambassador: H E Mr Bernard Makuza
First Secretary: Eugène Richard Gasana
First Counsellor: Charles Ujisenga [97]

31-01600 Embassy of Saudi Arabia
Godesberger Allee 40-42, D-53175 Bonn
☎ (+49 228) 810 90 ✉ NAJDIAH Bonn 2
▣ 885442 bonn d ⊖ (+49 228) 37 55 93
Ambassador: H E Mr Sheikh Abbas Faig Ghazzawi
Minister and Deputy Chief of Mission: Saleh Abdullatif
Cultural Attaché: Dr Raji Al-Itaibi

Office of the Military Attaché
Hohle Gasse 85, D-53177 Bonn
☎ (+49 228) 32 66 88 ▣ 885284 moda d
⊖ (+49 228) 32 38 69
Military Attaché: Col Abdulaziz Al-Balaa [97]

31-01640 Embassy of the Republic of Singapore
Südstrasse 133, D-53175 Bonn
☎ (+49 228) 95 10 30 [E] sing.emb.bonn@t-online.de
⊖ (+49 228) 31 05 27; 951 03 24
Ambassador: H E Mr Chin Siat-Yoon (BG (NS)
First Secretary, Political and Economic: Dominic Goh
Second Secretary: Ho Cheng Heng [98]

31-01655 Embassy of the Republic of Slovenia
Siegfriedstrasse 28, D-5300 Bonn 2
☎ (+49 228) 85 80 33; 85 80 31 ⊖ (+49 228) 85 80 57
Ambassador: H E Dr Boris Frlec
Defence Attaché: Martin Burjan
Commercial Officer: Janez Rezun ☎ (+49 228) 85 81 06
Public Information Officer: Ms Tatjana Kovacic
☎ (+49 228) 85 80 31 [98]

31-01690 Embassy of the Democratic Socialist Republic of Sri Lanka
Noeggerathstrasse 15, D-53111 Bonn
☎ (+49 228) 69 89 46 ▣ 885612 lanka d
⊖ (+49 228) 69 49 88
Ambassador: H E Mr Gunayavedalage Wijayasiri
Minister, Commercial and Economics:
Manikam Sundaralingam
Second Secretary: C A H M Wijeratne [97]

31-01740 Embassy of Switzerland
Gotenstrasse 156, PO Box 260163, D-53175 Bonn 2
☎ (+49 228) 81 00 80; 81 00 32; 81 00 34
✉ AMBASUISSE Bonn 2 ▣ 885646 amsbo d
⊖ (+49 228) 810 08 19
Ambassador: H E Dr Dieter Chenaux-Repond
Minister and Deputy Chief of Mission: Dr Rodolphe Imhoof
Defence Attaché: Col Heinz Loppacher
Commercial Counsellor: Paul Fivat
☎ (+49 228) 810 08 32
Scientific Counsellor: Reto Dürler [97]

31-01890 Embassy of the United Kingdom
Friedrich Ebert Allee 77, D-53113 Bonn
☎ (+49 228) 916 72 51 ⊖ (+49 228) 916 72 02
Ambassador: H E Paul Lever ☎ (+49 228) 916 71 50
Deputy Head of Mission: R F Cooper (MVO)
☎ (+49 228) 916 72 70
Defence Attaché: Brig B R Isbell (MBE)
☎ (+49 228) 916 72 60
Air Attaché: Gp Capt S Wrigley (RAF)
☎ (+49 228) 916 72 40
Naval Attaché: Capt M D Booth (DSC RN)
☎ (+49 228) 916 72 20
Assistant Defence Attaché: Maj N C W Dunkley (MBE)
☎ (+49 228) 916 71 87
Science and Technology Counsellor: R W Barnett
☎ (+49 228) 916 71 77
Defence Supply Counsellor: P Watkins
☎ (+49 228) 916 72 50 [98]

31-02010 Embassy of the Federal Republic of Yugoslavia
Schloßallee 5, D-5300 Bonn 2
☎ (+49 228) 34 40 51 to 55 ▣ 885530 yuamb d
⊖ (+49 228) 34 40 57; 34 40 56
Ambassador: H E Prof-Dr Zoran Jeremic
☎ (+49 228) 34 40 50
Defence Attaché: Col Milan Karagaca
☎ (+49 228) 34 40 58
Commercial Officer: Milidar Zivanovic
Public Information Officer: Zoran Brajovic
Secretary: Zlatko Bucin

Berlin Office
Taubertstrasse 18, D-1000 Berlin 33
☎ (+49 30) 826 20 91; 826 20 92 ▣ 304652 yuamb
⊖ (+49 30) 825 22 06
Deputy Chief of Mission: Petar Tasic [98]

31-02020 Embassy of the Republic of Zaire
Im Meisengarten 133, PO Box 262, D-5300 Bonn 2
☎ (+49 228) 85 81 60 ▣ 885573 zaire d
⊖ (+49 228) 34 03 98
Ambassador: H E Mr Mabolia Inengo Tra Bwato [98]

33 Foreign Forces

33-38400 Turkish Liaison Office to BWB
Türkisches Verbindungsbüro beim Bundesamt für Wehrtechnik und Beschaffung
Mayeber Strasse 87, PO Box 7360, D-56057 Koblenz
☎ (+49 261) 400 79 99 ▣ 862661 bwb d
⊖ (+49 261) 400 42 24
Chief of Liaison Office: Col Emruhan Yalcin [98]

34 Multinational Intergovernmental Organizations

34-01000 Allied Air Forces Central Europe Headquarters (AIRCENT)
D-66877 Ramstein Air Base
☎ (+49 6371) 47 88 48
[E] 100632.64@compuserve.com ⊖ (+49 6371) 443 29
Commander: Gen Michael E Ryan (USAF)
Deputy Commander: Lieut Gen Peter Vogler (Ger AF)
Operations Director: Maj Gen Thierry Fontaine (BEAF)
Director of Support: Maj Gen Franklin W Oliemans (RNLAF)
Public Information Officer: Lieut Col Walter Schneider (GEAF)
■ AIRCENT was established in July 1993 to replace the previous AAFCE, 2ATAF and 4ATAF organisations. [98]

34-03000 Allied Land Forces Central Europe Headquarters (LANDCENT)
PO Box 102620, D-69016 Heidelberg
Campbell Barracks, Romerstrasse 168,
D-6900 Heidelberg
☎ (+49 6221) 39 83 16 ⊖ (+49 6221) 39 83 49
Commander: Gen William B Crouch (USA)
Deputy Commander and Chief of Staff: Lieut Gen de Kleyn (NL)
Deputy Chief of Staff, Operations: Brig Gen Clemmons (USA)
Deputy Chief of Staff, Logistics: Col van Ulden
Deputy Chief of Staff, Intelligence: Col Rohde (GER)
Director of Support: Maj Gen Drewienkiewicz
Operations Director: Maj Gen van Damme
Press and Information Officer: Lieut Col Philipp (Ger) [97]

34-04000 Allied Rapid Reaction Corps Headquarters (ARRC)
Queens Avenue, BFPO 40, D-41179 Moenchengladbach
☎ (+49 2161) 565 57 16 ⊖ (+49 2161) 565 57 18
Commander: Lieut Gen M D Jackson (CB, CBE, UKA)
☎ (+49 2161) 565 54 50
Deputy Commander: Maj Gen C Cabigiosu (IT)
☎ (+49 2161) 565 54 56
Chief of Staff: Maj Gen T J Sulivan (CBE, UKA)
☎ (+49 2161) 565 54 55
Chief of Combat Support: Maj Gen A G Denaro (CBE, UKA)
☎ (+49 2161) 565 54 42
Chief of Public Information: Lieut Col R C L Clifford (MBE, UKA) ☎ (+49 2161) 565 57 13
Public Information Office: Capt H Cate (USA)
☎ (+49 2161) 565 57 15
Chief Clerk: A Steer [98]

34-24500 Headquarters 1st (United Kingdom) Armoured Division
BFPO 15, Herford
☎ (+49 5221) 87 32 27 (public information); 99 54 31
⊖ (+49 5221) 99 53 17
Commander: Maj Gen J P Kiszely (MC)
Deputy Commander: Col J P Riley (DSO)
Director of Procurement and Logistics: Col D R Burns
SO3 G3: M D Norman [98]

34-25000 Headquarters United Kingdom Support Command (Germany) (HQ UKSC (G))
(under the command of Headquarters Land Command in UK)
PO Box 250214, D-41174 Mönchengladbach
☎ (+49 2161) 47 21 69 [E] bfginfo@koan.de
⊖ (+49 2161) 47 40 12; 55 03 44
Commander: Maj Gen C F Drewry (CBE)
Chief of Staff: Brig M G Taylor (MBE)
Media Operations: Lieut Col G A Attard-Manche (RA) (retd) [98]

34-27010 NATO EF2000 and Tornado Development, Production and Logistics Management Agency (NETMA)
(† NATO MRCA Development & Production Management Agency)
Inselkammerstrasse 12-14, D-82008 Unterhaching
☎ (+49 89) 66 68 00 ▣ 529361
⊖ (+49 89) 66 68 05 55; 66 68 05 56
General Manager: J A Gordon
Deputy General Manager: Dr J Kaempf [97]

34-32300 Allied Land Forces Headquarters Schleswig-Holstein & Jutland (HQ LANDJUT)
Eider-Kaserne, PO Box 921, D-24758 Rendsburg
☎ (+49 4331) 52 61 ⊖ (+49 4331) 573 41
Commandant: Lieut Gen M Gerber
Chief of Staff: Brig Gen P O Topp
Assistant Chief of Staff G-1: Lieut Col F Milde
Assistant Chief of Staff G-2: Lieut Col J P R Spliid
Assistant Chief of Staff G-3: Col H E Möller
Assistant Chief of Staff G-4: Col N F Jöncke
Assistant Chief of Staff CiS: Lieut Col L Grandt
Chief of Administration: Capt P K Sneftrup [97]

34-41000 Franco-German Brigade
Deutsch-französische Brigade
D-79371 Müllheim
☎ (+49 7631) 90 21 00 ▣ 224581
⊖ (+49 7631) 90 21 06
Commander: Gen Hans Otto Budde
Deputy Commander: Col Daniel Bruss
Chief of Staff: Col Bernard Oberto
Director of Operations: Werner Weisenburger
Head of Press and Public Relations: von Platen [98]

40 Government Defense Sales/Procurement Organizations

40-01000 Federal Office for Military Technology and Procurement

Bundesamt für Wehrtechnik und Beschaffung (BWB)
Konrad Adenauer Ufer 2-6, PO Box 7360,
D-56057 Koblenz
☎ (+49 261) 40 01 ⊠ 862661 bwb d
⊖ (+49 261) 400 76 30; 400 79 64
President: Peter Koerner ☎ (+49 261) 40 01 Ext 21 00
Deputy President for Technology: Elmar Göbel
☎ (+49 261) 40 01 Ext 75 00
Office of the President/Press Officer: Achim Cajan
☎ (+40 261) 40 01 Ext 28 70
Aircraft Aeronautical Equipment: Helmuth Heumann
☎ (+49 261) 40 01 Ext 77 00
Automotive and Engineer Equipment, Power Supplies:
Hermann Sitterberg ☎ (+49 261) 40 01 Ext 60 00
Weapons and Ammunition Division: Wof-Rüdier Petereit
☎ (+49 261) 40 01 Ext 66 40
Communications and Electronics Division:
Dr Hans-Werner Fieweger
☎ (+49 261) 40 01 Ext 67 00
Ship and Naval Equipment and Underwater Weapons:
Manfred Weiss ☎ (+49 261) 40 01 Ext 43 00
Petrol, Oil, Lubricants and General Equipment Department:
Hans Josef Molitor ☎ (+49 261) 40 01 Ext 33 00
Quality Assurance: Wolfgang Paasch
☎ (+49 261) 40 01 Ext 79 00
Administration Affairs: Detlev Petry
Technical Affairs: Dr Wilhelm Sommer
☎ (+49 261) 40 01 Ext 24 00
Economic Affairs: Dr Dieter Klinker
☎ (+49 261) 40 01 Ext 23 00
Press Attaché: Mr Weidenfeller
☎ (+49 261) 40 01 Ext 28 56
■ The following facilities, among others, report to the BWB: the Army Test Facilities; the Navy Arsenal including subordinate facilities at Wilhelmshaven and Kiel. [97]

60 Research Organizations, Institutes & Test Establishments

Federal Centre for Explosives 25-17300
Test Establishment 71 25-17380
Test Establishment 61 25-17350

60-00150 Aufbau-und Kontakstudium Kriminologie

Universität Hamburg, Jungiusstrasse 6,
D-20355 Hamburg
☎ (+49 40) 41 23 33 29 ⊖ (+49 40) 41 23 23 28
👥 5 (professional); 6 (support)
Director: Dr Fritz Sack
Director: Dr Sebastian Scheerer
Sociology: Ms Susanne Krasmann
Psychology and Criminology: Werner Lehe
Sociology and Criminology: Henning Schmidt-Semisch
■ Research and teaching activities.
Established in 1984. [98]

60-00250 Buergerrecht und Polizei

Free University of Berlin, Malteserstrasse 74/100,
D-1000 Berlin 46
Contact: Heiner Busch [98]

60-00300 Bundeskriminalamt

Postfach 1820, D-6200 Wisbaden
Contact: Herr Ziegler
Oberregierungsrat: Klaus Fischer
Contact: Max-Peter Ratzel
Contact: Peter Vowé [98]

60-00400 Centre for European Law Policy

Universität Bremen, Bibliotekstrasse, D-2800 Bremen
Director: Dr V Gessner [98]

60-00430 Centre for Interdisciplinary Research

Universität Bielefeld, Universitätstrasse 25,
D-4800 Bielefeld
Director: Prof. Dr Peter Weingart [98]

60-00450 Crime Prevention Council

Gartenstrasse 4, D-2217 Rosdorf
Kriminaldirektor: Helmut Hoetzsche [98]

60-00490 Franz-V.-Liszt-Haus

Juristische Fakultät der Universität,
Halle - Universitätsplatz 6, D-06099 Halle/Saale
☎ (+49 345) 83 25 69 ⊠ 318289 uniha d
⊖ (+49 345) 295 15
👥 5
Director: Prof.Dr-iur. Dieter Rössner,
Dr Britta Bannenberg, Michael Rothkirch, Anja Meyer, Ingo Diedrich
■ Researching and teaching.
Established in 1993. [98]

60-00600 Federal Institute for Materials Research & Testing

Bundesanstalt für Materialforschung und Prufung (BAM)
Unter den Eichen 87, D-12205 Berlin
☎ (+49 30) 810 40 🅆 http://www.bam.de
⊖ (+49 30) 811 20 29
👥 1,700 (1997)
President: Prof.Dr-Ing-hc H Czichos
☎ (+49 30) 81 04 10 00
Vice President: Prof.Dr-rer.nat M Hennecke
☎ (+49 30) 81 04 10 20
Head of the President's Office: Dr W Metasch
☎ (+49 30) 81 04 10 01
Member of the Presidential Board:
Dir.Prof.Dr-Ing K H Habig ☎ (+49 30) 81 04 10 30
Head of Publications: Dr-Ing. J Lexow
☎ (+49 30) 81 04 10 03
■ Main objective is to promote the development of the German economy, being responsible for interaction in the fields of Chemistry, Environment and Safety. Engaged in research and development, testing, analysis, approvals, consultation and information.

Fabeckstrasse Branch
Unter den Eichen 44-46, D-12203 Berlin
☎ (+49 30) 810 40 ⊖ (+49 30) 811 20 29

Adlershof Branch
Rudower Chaussee 5, D-12489 Berlin
☎ (+49 30) 639 20 ⊖ (+49 30) 63 92 57 87 [98]

60-00700 Fuehrungsakademie der Bundeswehr

Manteuffelstrasse 20, D-22585 Hamburg
☎ (+49 40) 866 71 ✉ fueakbwhh@t-online.de
⊖ (+49 40) 86 67 23 10 (library) [98]

60-00800 German Aerospace Research Establishment

Deutsche Forschungsanstalt für Luft und Raumfahrt e.V. (DLR)
Linder Höhe, D-51147 Cologne
☎ (+49 2203) 60 10; 601 27 18 🅆 http://www.dlr.de
⊖ (+49 2203) 673 10; 601 46 91
👥 4,400 (1997)
Chairman: Prof Walter Kröll
Vice Chairman: Prof Jürgen Blum
Contact: F Kierspel
■ National research establishment performing scientific and technical research for the development and utilization of future aircraft and spacecraft. Also research on energy technology. [98]

60-00900 Institut of Criminology

Friedrich-Ebert-Anlage 6/70, D-6900 Heidelberg
Director: Prof.Dr Dieter Dölling [98]

60-00930 Institut for Criminal Law, Law of Criminal Procedure, Comparative Law and Medical Law

Martin Luther Universität, Universitätsplatz 10,
D-06099 Halle/S'
☎ (+49 345) 83 25 64 ⊖ (+49 345) 202 95 15
👥 10
Director: Prof.Dr Hans Lilie
Research Assistant: Detlev Bortfeld
Research Assistant: Rudolph Bernd
Contact: Brit Becker
■ Legal commentary on law regarding fiscal offences. Research on relantionship between police and public prosecution. Expertise on liability of medical practitioners in Europe.
Established in 1964. [98]

60-00950 Institut für Jugendstrafrecht und Kriminologie

Universität Giessen, Licher Strasse 64, D-6300 Giessen [98]

60-00970 Institut für Kriminologie

Justus Liebig Universität Licher, Strasse 64,
D-35394 Giessen
👥 6 (professional); 5 (support)
Director: Prof.Dr Arthur Kreuzer
Criminology and Penal Law: Thomas Elpel
Criminology and Psychology: Thomas Görgen
Criminology and Sociology: Lutz Klein
Criminology and Penal Law: Wolfgang König
Criminology and Penal Law: Jürgen Schramke
Criminology and Penal Law: Jürgen Stock
■ Research, teaching, information and continuation training of judges, district attorneys, police officcers, social workers and other professional groups working in the criminal justice system, political counselling, collaboration with scientific institutions and advisory boards.
Established in 1965. [98]

60-00990 Institut für Kriminologie der Universität Heidelberg

Friedrich-Ebert-Anlage 6/10, D-69117 Heidelberg
☎ (+49 6221) 54 74 89 ⊠ 461515 uni hd
⊖ (+49 6221) 54 74 95
👥 5 (professional); 2 (support)
Director: Prof.Dr Dieter Dölling
Social Work: H Beisel
Criminology, Sociology and Jurisprudence: A Hartmann
Criminology and Sociology: D Hermann
Criminology, Sociology and Jurisprudence: C Simsa
■ Research and teaching activities. Projects on victim-offender-reconciliation; drug prevention; community safety; value orientations and criminality; organised crime; expertises on criminal responsibility and judgments of criminal courts; women in prison; criminal law relating to young offenders.
Established in 1962. [98]

60-01000 Institut für Kriminalwissenschafen

Abteilung Kriminologie, Bispinghof 24/25,
D-4400 Münster
Contact: Carda Reiners [98]

60-01020 Institut für Kriminologie der Universität Tübingen

Correnstrasse 34, D-72076 Tübingen
☎ (+49 7071) 29 29 31 ⊖ (+49 7071) 20 20 41
👥 10
Director: Prof.Dr H J Kerner
Criminologie, Psychiatry and Jurisprudence: H Göppinger
Criminology, Jurisprudence: K Boers
Criminology and Psychology: S Fritz Jansen
Sociology and Informatics: P Kurz
Sociology: V Schindler
Psychology: A Schubert
Criminology: E Weitekamp
Criminology and Social Work: H J Witmann
Sociology: W Wittmann
■ Research in empirical criminology, longitudinal comparative investigations into different groups of offenders. Applied criminology for forensic practice (disgnosis, prognosis, therapy and prophylaxis); criminal career research; social transformation and crime in Eastern Europe; community service and restorative justice.
Established in 1962. [98]

60-01040 Institut für Kriminologie und Wirtschaftsstrafrecht der Universität Freiburg im Breisgau

Erbprinzenstrasse 17a, D-79098 Freiburg
☎ (+49 761) 203 22 14 ⊖ (+49 761) 203 22 19
👥 3 (professors); 5 (assistants); 8 (support); 1 librarian; 3 (administration)
Director: Prof. Dr.h.c.mult. Klaus Tiedemann
Contact: Dr Joahim Vogel

■ Research activities in human rights criminal proceedings; comparison of penal law systems in Europe; sanctions on legal persons in German and foreign penal law.
Established in 1930. 98

60-01060 Juristische Fakultaet
Universität Konstanz, Postfach 5560 - D119, D-78434 Konstanz
☎ (+49 7531) 881; 88 21 81; 88 26 74
0733359 univ d
3 (professional)
Director: Prof.Dr-iur. Wolfgang Heinz
Criminology: Ms Cornelia Beck
Criminology: Gerhard Spiess
■ Criminology and empirical legal research.
Established in 1982. 98

60-01080 Kriminologisches Forschungsinstitut Niedersachsen e.V. (KFN)
Criminological Research Institute of Lower Saxony
Lützerodestrasse 9, D-30161 Hannover
☎ (+49 511) 34 83 60 ⊜ (+49 511) 348 36 10
13 (academic); 16 (support)
Director: Prof.Dr Christian Pfeiffer
Sociology of Law and Biographical Research: Dipl.Päd. Andreasa Böttger, Dipl.Soz. Sylvia Christochowitz
■ Research, consulting and teaching.
Established in 1985. 98

60-01110 Kriminologischer Dienst bei der Justizvollzugsanstalt Hannnover
Postfach 5827, D-30058 Hannover
Schulenburger Landstrasse 145, D-30165 Hannover
☎ (+49 511) 679 61 ⊜ (+49 511) 679 62 91
Director: Dr Rolf Herrfahrdt
■ Criminological and penological research.
Established in 1993. 98

60-01120 Kriminalistisches Institut des Bundeskriminalamtes
Postfach 1820, D-65173 Wiesbaden
Thaerstrasse 11, D-65193 Wiesbaden
☎ (+49 611) 551 ⊜ (+49 611) 55 42 82
45 (police investigation/criminology); 48 (technical research)
Director: Rainer Hofmeyer
Contact: Ms Ursula Raber ☎ (+49 611) 55 23 77
■ Interdisciplinary research in the field of criminology and police science (criminal investigation, crime prevention, police organisation, security measures, crime prognosis, operational equipment, pattern recognition, digital signal processing). Research is carried out in co-operation with institutes and universities.
Established in 1973. 98

60-01140 Kriminologische Abteilung der Philippps Universität
Universitätstrasse 6, D-35032 Marburg
☎ (+49 6421) 28 31 20 482357 umrd
⊜ (+49 6421) 28 31 82
Director: Prof.Dr Dieter Meurer
Criminal Law, Criminal Procedure and Criminology: Prof.Dr Erich Hupe
Criminal Law, Foreign and International Law, Criminology: Prof.Dr Georg Freund
Criminal Law and Criminology: Dr Eva Graub
Criminal, Foreign and International Law: Eberhard Knittel
Legal Psychology and Criminology: Ph.D. Siegfried L Sporer
■ Interaction between police and rape victims; release and probation; identification evidence; judicial decision-making by lay-judges and professional judges; criminality of ethnic minorities in the Federal Republic of Germany.
Established in 1971. 98

60-01160 Kriminologische Forschungsgruppe der Bayerischen Polizei
Postfach 190262, D-80602 München
Maillingerstrasse 15, D-80636 München
☎ (+49 809) 125 13 38 ⊜ (+49 809) 125 17 24
5
Director: Dr Wiebke Steffen
Scientific Collaborator: Siegfried Kammhüber
Scientific Collaborator: Dr Johannes Luff
Police Officer: Siegfried Polz
Police Officer: Hans Joachim Bokowsky
Police Officer: Manfred Gerum
■ Research on crime and juvenile crime in Bavaria; multiple offenders and police information systems; crime prevention by police activities; violent crimes, insult, conflict settlement, wife abuse and the response of the police; crimes committed by foreign minorities in Bavaria.
Established in 1979. 98

60-01180 Kriminologische Forschungstelle der Universität zu Köln
Albertus Magnus Platz, D-50923 Köln
☎ (+49 221) 470 42 81 ⊜ (+49 221) 470 51 47
7
Director: Prof.Dr-iur. Michael Walter
Criminology: H Geiter
Criminology: M Kubink
Psychology: A Wagner
Penal Law: W Schuldzinski
Librarian: S Atteln
Secretary: G Blakenburg
■ Research in juvenile delinquency in different European towns. Handling of difficulties and mediation activities by police officers; dealing with violent crimes of young political extremists groups and individuals.
Established in 1968. 98

60-01200 Kriminologische Seminar der Universität Bonn
Adenauer Alle 24/42, D-53113 Bonn
☎ (+49 22) 73 91 33
4 (professional); 1 (support)
Director: Prof.Dr Helmut Marquardt, Dr Lutz Gero Leky, Dr Werner Ruether, Frank Neubacher
■ Research and teaching on environmental offences and social reactions.
Established in 1943. 98

60-01300 Kriminologische Zentralstelle e.V. (KrimZ)
Adolfsallee 32, D-65185 Wiesbaden
☎ (+49 611) 397 31 ⊜ (+49 611) 308 15 68
7 (professional); 5 (support)
Director (Criminology and Criminal Law): Dr-jur. Jörg Martin Jehele
Director (Legal Psychology): Porf.Dr Rudolf Egg
Sociology and Criminal Law: A Dessecker
Sociology: M Kurze
Social Sciences and Documentation: W Sohn
Library: E Herrmann
■ Research and documentation.
Established in 1986. 98

60-01600 Landeskriminalamt
Beim Strohhause 31, Hamburg
Director: W Sielaff 98

60-01700 Lehrstuhl für Kriminologie, Jugendstrafrecht, Strafvollzug, Wissenschaftliche Einrichtung Strafrecht der Freien
Van't Hoff Strasse 8, D-14195 Berlin
Boltzmannstrasse 3, D-14195 Berlin
☎ (+49 30) 838 47 16 ⊜ (+49 30) 838 28 95
8 (professional); 1 (support)
Director: Prof.Dr-iur. Ulrich Eisenberg
■ Research and teaching.
Established in 1976. 98

60-02000 Polizei Führungsakademie
Zum Roten Berge, D-4400 Münster
Department of Criminalistics and Criminology
Zum Roten Berge, D-4400 Münster
Deputy Kriminaldirektor: Joachim Jaeger 98

60-04500 Westfalia University Department of Criminology
Bispinghof 24/25, D-48143 Münster
☎ (+49 251) 83 27 49 ⊜ (+49 251) 83 48 31
7 (professional); 1 (support)
Director: Prof.Dr Dr.h.c. (PL) Hans Joachim Schneider
Academic Assistant, Law: Ms Sabine Nienstedt-Jost-Westendorf
Academic Assistant, Law: Thomas Peppersack
■ Research, teaching and counselling in dangerous recidivists, victimology, juvenile delinquency, comparative criminology and violent crimes.
Established in 1971. 98

60-05000 Wissenschaftszentrum Berlin
Reichpietschufer 50, D-1000 Berlin 30
Contact: Dr Ernst Heinrich Ahlf 98

65 Industry, Trade & Armed Forces Associations

65-00100 Bundesverband der Deutschen Luft-und Raumfahrt-Industrie e.V. (BDLI)
German Aerospace Industries Association
(† Bundesverband der Deutschen Luftfahrt, Raumfahrt und Ausrüstungsindustrie e.V.)
Konstantinstrasse 90, D-53179 Bonn
☎ (+49 228) 84 90 70 E bdli@t-online.de
⊜ (+49 228) 33 07 78
President: Dr Manfred Bischoff
Presiding Manager: Dr Hans Eberhard Birke
Press and Public Relations Manager: Michael K E Hauger
■ Association representing more than 95% of the German aerospace industry involved in the manufacture of aircraft, space systems, engines, aerospace equipment and new materials. Membership of approximately 100 companies. 98

65-00150 Bundesverband der Deutschen Industrie e.V. (BDI)
Gustav Heinemann Ufer 84-88, D-50968 Cologne
☎ (+49 221) 370 86 38 ⊜ (+49 221) 370 86 41 98

65-00400 Deutsche Gesellschaft fur Wehrtechnik (DWT)
Bertha von Suttner Platz 1-7, D-53111 Bonn
☎ (+49 228) 766 82 82; 766 82 83
⊜ (+49 228) 65 63 66
Chairman: Pràs BWB aD Dr Heinz Glaeser
President: Dr Ernst Grosch
Managing Director: Brig Gen aD Hubert Merkel
■ Voluntary non-profit making organisation supporting military research and design in Germany. 98

65-02400 Union Internationale des Syndicats de Police (UISP)
Fortsrasse 3a, D-4010 Hilden
Contact: Wolfgang Dicke 98

65-02600 Zentralverband Elektrotechnik- und Elektronikindustrie eV (ZVEI)
The German Electrical & Electronic Manufacturers' Association
Stresemannallee 19, D-60596 Frankfurt a.Main
☎ (+49 69) 63 020 ✉ zvei@zvei.org
⊜ (+49 69) 630 23 17
Chairman: Dr Franz Josef Wissing
Director: Dipl.-Kfm Norbert Knaup
Director: Dipl.-Ing. Ingo Rüsch
Director: Dr Werner Schmidt
Press and Public Relations: Dipl.-Ing. Gotthard Grass
■ Non-profit trade association for the German electrical and electronic industry. Membership of 1,400 companies. 98

70 Financial Holding Groups, Parent Companies

70-00050 Buck Werke GmbH & Co
(† Buck Chemisch-Technische Werke GmbH & Co)
Mozartstrasse 2, D-83435 Bad Reichenhall
☎ (+49 8651) 70 20 17 8651801
⊜ (+49 8651) 702 70
17 865 1801
Partner: Dipl.-Ing Gert H Buck
Partner: Dipl.-Ing Hartmund H Buck
Managing Director: Dipl.-Ing Hans Wirtz
Group Director, Sales and Marketing: Walter Hanser

Subsidiaries with defence business:
Buck Werk-Neunenburg GmbH
Depyfag Pyrotronic GmbH
Pyrotechnik Uftrungen GmbH [98]

70-00100 Daimler-Benz AG
D-70546 Stuttgart
☎ (+49 711) 179 33 15 Ⓦ http://www.daimler-benz.com
⊖ (+49 711) 179 45 31
♣ DM 103,549,000,000 (1995) 👥 290,031
Chairman: Jürgen Schrempp
Daimler-Benz Aerospace AG: Dr Manfred Bishopp
Daimler-Benz Interservices (Debis): Dr Klaus Mangold
Finance and Controlling: Dr Manfred Gentz
Passenger Vehicles: Juergen Hubber
Research and Technology: Klaus Dieter Vöhringer
Head of Communications: Dr Christoph Walther
Press and Information: Roland Klein
International Media Relations: Eckhard Zanger
■ Group active in various areas of business linked by similar technologies and system structures. Special significance is placed in the areas of traffic and transport technology. The group is divided into Mercedes-Benz (passenger cars, commercial vehicles); direct reporting businesses, rail systems micro-electronics, diesel engines, energy systems technology and automation; Daimler-Benz Aerospace (aircraft, space systems, defence, civil systems and propulsion systems); Daimler-Benz InterServices (Debis) (systemhaus), financial services, insurance brokerage, trading, marketing services, mobile communications services, real estate management.
Member companies:
Daimler-Benz Aerospace AG
Daimler-Benz InterServices AG (debis) [98]

70-00200 Daimler-Benz Aerospace AG
(† Deutsche Aerospace AG)
(a subsidiary of Daimler-Benz AG)
D-81663 Munich
☎ (+49 89) 60 70 ⊠ 5287130 dasam d
⊖ (+49 89) 60 72 64 81
⇔ DM 873,000,000 ♣ DM 15,037,000,000 (1995)
👥 50,784
President and Chief Executive Officer: Dr Manfred Bischoff
Chairman of the Supervisory Board: Jürgen E Schrempp
Finance and Controlling: Dr Wolfgang Piller
Aircraft Group: Dr Dietrich Russell
Space, Defence and Civil Group: Werner Heinzmann
Senior Vice President, Legal: Dr Heinrich J Rodewig
Senior Vice President, Marketing: Dr Michael Ganal
☎ (+49 89) 60 73 45 45
Senior Vice President, Corporate Communications: Christian Poppe ☎ (+49 89) 60 723 42 50
Press and Communications, Space, Defence and Civil Systems: Andreas Breitsprecher
☎ (+49 89) 60 73 42 35
Press, Aircraft Group: Dr Theodor Benien
☎ (+49 89) 60 73 42 59
Communications, Propulsion Systems: Odilo Mühling
☎ (+49 89) 14 89 26 98
Public Relations: Sonja Kemmer
Aircraft Group::
Daimler-Benz Aerospace Airbus GmbH: civil aircraft (Airbus programme, freighter conversion) planned: Future Large Aircraft; Regional Aircraft Division (Eurofighter 2000, Tornado, XX-31, Ranger 2000); Eurocopter (civil and military helicopters, single and twin engine in the light to medium range).
Space Systems Group:
Dornier Satellitensysteme GmbH: satellite systems for science, communication, navigation; remote-sensing satellites; satellite based communications systems.
Space Infrastructure Division: automated and manned orbital systems, space transportation and propulsion systems, components for European rocket Ariane 4/5.
Defence and Civil Systems Group:
Sensor Systems Division - responsible for development, production and integration of communication and radar systems as well as for electronic command and control systems (EloKa). Equipment and systems for strategic radio reconnaissance, tactical radio application, electronic radio command and control systems. The product spectrum in the radar application sector comprises land and ship-based and mobile radars, airborne radar systems for military aircraft application, radar systems for air and sea surveillance tasks, microwave reconnaissance, electronic radar command and control systems. Radar homing heads for guided missiles, drones and shells, fuse sensors for the artillery, data links for remote control and electronics for launchers.
Information and Communication Systems Division: reconnaissance systems (drone CL289; demonstrator Seamos), command/control and information systems, mission planning systems (Diplas/AFA) support systems for weapon deployment (Ares artillery rocket fire control system for Lars and Mars); digital map systems (NH90, Tiger and Eurofighter 2000); mobile ground systems, instruction and training simulators; systems for aerial target simulation; product support; Eurobridge Mobile Brücken GmbH (Dornier folding bridge for military and civil applications); system planning for German Armed Forces; airport and guidance systems security technology (protection of facilities, tangible property); Nortel Dasa Network Systems GmbH Co KG (communications networks including satellite communications). LFK-Lenkflugkörpersysteme GmbH - guided missiles for air defence and anti-aircraft applications (Stinger, Patriot and Roland systems, ship-based Rolling Airframe Missile and ANNG systems. Meads air defence system; future medium-range air-to-air missiles (FMRAAM); anti-tank applications and combat support (Milan, Hot, Pars 3MR/LR, Polyphem); anti-ship missiles (Kormoran 2); stand-off systems for aircraft weaponry (DWS 39 system, ADFS for F-16; Taurus family); sub-systems and components for inertial systems, location finders and logistic systems; missile detection sensors (Missile Launch Detection System).
Profit Center Technical Services and Logistics (Elekluft GmbH): preparation and maintenance of equipment, systems and installations. Profit Center Disposal: munitions disposal and recycling of electric and electronic equipment.
CMS Inc (Conventional Munition Systems): missiles and missile components and systems.
Joint ventures TDA Armaments (armament systems), Bayern Chemie (missile propulsion systems) together with Thomson CSF and Zeiss Eltro Optronik with Carl Zeiss. [97]

70-00350 Krauss-Maffei Aktiengesellschaft
Krauss-Maffei Strasse 2, D-80997 Munich
☎ (+49 89) 889 90 ⊠ 523163 ⊖ (+49 89) 88 99 22 06
Chairman: Dipl.-Ing Burkhard Wollschläger
Executive Board: Dipl.-Ing Gerhard Haas
Executive Board: Dipl.-Ing Klaus D Vogt
Chairman of the Supervisory Board: Dr Joachim Funk
Vice-Chairman of the Supervisory Board: Peter Krahl [98]

70-00800 MAN Aktiengesellschaft
PO Box 401347, D-80713 Munich
☎ (+49 89) 36 09 80 Ⓦ http://www.man.de
⊠ 5215710 manm d ⊖ (+49 89) 36 09 82 50
⇔ DM 771,000,000 (1995/96)
♣ DM 20,269,926,000 (1995/96) 👥 57,454 (1996)
Chairman of the Executive Board: Dr Rudlof Rupprecht
Executive Board Member: Dr Ferdinand Graf von Ballestrem
Executive Board Member: Dr Klaus von Menges
Executive Board Member: Dr Klaus Schubert
Executive Board Member: Gerhard Thulmann
Executive Board Member: Dr Philipp J Zahn
Information Department: Dieter Gratzel
■ Supply of commercial vehicles, machinery and construction plants and trading. High-tech products and services for transport engineering and materials handling, printing and communications engineering for the chemical and petrochemical as well as basic industries. Power generating sector and environmental protection.
Subsidiary companies:
MAN Nutzfahrzeuge AG → 75
Ferrostaal AG → 75
MAN Roland Druckmaschinen AG
MAN Gutehoffnungshütte AG → 75
MAN B&W Diesel AG → 75
MAN Technologie AG → 75
RENK AG → 75
Deggendorfer Werft und Eisenbau GmbH → 75
Major shareholdings in SMS AG & Schwäbische Hüttenwerhe GmbH [98]

70-00850 Metallgesellschaft AG
Bockenheim Landstrasse 73-77,
D-60271 Frankfurt am Main
☎ (+49 69) 71 19 90 ⊠ 41225 ⊖ (+49 69) 71 19 91 00
⇔ DM 669,602,000 ♣ DM 15,825,000,000 (1995/96)
👥 43,292 (1995/96)
President: Dr Karl Josef Neukirchen
■ Non-ferrous metals, marketing of surface treatment processes and products. → 80 [98]

70-00900 Preussag AG
Karl Wiechert Allee 4, D-30625 Hannover
☎ (+49 511) 566 00 ⊠ 922828 ⊖ (+49 511) 566 19 01
⇔ DM 66,226 ♣ DM 25,043,508,000 (1995/96)
Chairman: Dr Michael Frenzel
Group Accounting Department: Jürgen Mick
■ Main business areas are steel and non-ferrous metals production, energy and commodities, trading, building engineering and components.
Subsidiaries with defence business:
Howaldtswerke-Deutsche Werft AG → 75 [98]

70-01000 Rheinmetall AG
Kennedydamm 17, PO Box 104261, D-40033 Dusseldorf
☎ (+49 211) 473 04 ⊖ (+49 211) 473 29 00
♣ DM 3,417,100 (1995) 👥 14,574 (1995)
Chairman of the Board: Dr Hans U Brauner
Board Member: Dr Ernst-Otto Krämer
Board Member: Dieter Niederste-Werbeck
Public Relations Manager: H G Tebrake
Subsidiary companies with defence business:
Rheinmetall Industrie AG
Rheinmetall Waffe & Munition GmbH
Subsidiary companies:
Europ Scan SA, France
Heimann Systems GmbH, Germany
Mak System Gesellschaft mbH, Germany
Mauser-Werke GmbH, Germany
Nico-Pyrotechnik, Hans-Jürgen Diederichs GmbH & Co KG, Germany
NWM De Kruithoorn BV, Netherlands
Oskar Lünig GmbH, Germany
Pyrotechnik Silberhütte GmbH, Germany
Rheinmetall Entsorgungsgesellschaft Weichensdorf mbH, Germany
RMP Rheinmetall Mess- und Prüftechnik GmbH, Germany
RMV Rheinmetall Machine Vision GmbH, Germany
TZN Forschungs- und Entwicklungs- zentrum Unterlüss GmbH, Germany
WNC-Nitrochemie GmbH, Germany [98]

74 Consortia, Joint Venture Companies

74-04300 Burkhart Grob Luft und Raumfahrt GmbH & Co KG
(† Grob-Werke GmbH & Co)
Flugplatz Mattsies, Lettenbachstrasse 9,
D-86874 Tussenhausen
☎ (+49 8268) 99 80 Ⓦ http://www.grob-aerospace.com
⊖ (+49 8268) 99 81 24
♣ DM 25,000,000 👥 60
President: Burkhart Grob
Managing Director: Dr Gero Matel
Vice President and General Manager: Roland Rischer
■ Manufacture of G115 glassfibre light two seat trainer/tourer, high-altitude aircraft. GF200 business and executive aircraft. [98]

74-05800 Eurocopter Deutschland GmbH
(a subsidiary of Eurocopter SA, France)
D-81663 Munich
☎ (+49 89) 60 00 64 88
⊖ (+49 89) 60 00 49 15; 60 00 90 33
⇔ DM 200,000 (1996) ♣ DM 733,000,000 (1996)
👥 3,000 (1996)
Chairman: Dr Siegfried Sobotta ☎ (+49 89) 60 00 45 48
Press and Information Manager: Ms Christina Gotzhein
■ Industrial concern in charge of development, production and product support of current and future products originating in the German part of the Eurocopter group.
Ottobrunn Plant
Postfach 801140, D-8000 Munich 80
☎ (+49 89) 60 72 ⊖ (+49 89) 60 00; 64 81
Donauwörth Plant
Postfach 1353, D-8850 Donauwörth
☎ (+49 906) 71 00 00 ⊖ (+49 906) 715 25 [98]

74-05850 Eurofighter Jagdflugzeug GmbH
(a joint venture between Alenia (Italy), British Aerospace (UK), CASA (Spain) and DASA (Germany))
Am Söldnermoos 17, D-85399 Hallbergmoos
☎ (+49 811) 800 ⊠ 5213908; 5213744 efam d
⊖ (+49 811) 80 15 57
⇔ DM 5,000,000
Managing Director: B Phillipson
Deputy Managing Director: Pablo de Bergia

Commercial Director: Stephen Ward
Product Support Director: Massimo Tarantola
Technical Director: Martin Friemer
Operations Director: Ramon Madrid
Procurement Director: Franco Sella
Vice President, Marketing and Public Relations: Terry G Marsh
■ Eurofighter Jagdflugzeug GmbH has been formed to manage the European Fighter Aircraft Programme. The company is wholly owned by the four partner companies, Alenia in Italy, BAe in Great Britain, CASA in Spain and DASA in Germany. [98]

74-05900 Eurojet Turbo GmbH
Inselkammerstrasse 5, D-82008 Unterhaching, Munich
☎ (+49 89) 66 69 20
⊖ (+49 89) 66 69 21 39; 66 69 21 62
♧ DM 9,000,000 (1996) ♣ DM 500,099,000 (1996)
👥 39 (1996)
Managing Director: Kenneth James Greenall
Technical Director: Dietrich Schulte
Commercial and Marketing Director: Menotti Zinna
Product Assurance Director: Juan Villate
Public Relations: Ms Jessica M Boriani
■ Consortium responsible for the design, development, production and product support of the EJ200 engine for Eurofighter 2000.
Consortium members:
FiatAvio (Italy), ITP (Spain), MTU (Germany) Rolls-Royce (UK). → 75 [98]

74-10000 Panavia Aircraft GmbH
Airport Business Centre, Am Soldnermoos 17, D-85399 Hallbergmoos
☎ (+49 811) 80 12 38; 80 12 39 ▣ 529825 pana d
⊖ (+49 811) 80 13 86
♧ DM 20,000,000 (1997) ♣ DM 1,200,000 (1996)
👥 135 (1997)
Managing Director: Dipl.-Ing Rheinhold Falthauser
Operations Director: Dennis McCornick
Public Relations Manager: W D Gnamm
Protocol Officer: Ms Hilla Dörfler
■ Design, development, production and industrial product support of the Anglo-German-Italian all-weather combat aircraft Tornado. [98]

74-18100 RAM System GmbH
Otto-Hahn Str. 28/30, Eingang 4, D-85521 Riemerling
☎ (+49 89) 60 80 03-43 ⊖ (+49 89) 60 80 03-16
Managing Director: M Leupold
Managing Director: V Pauly
Marketing Manager: Nikos Argyrakis
■ Consortium formed for the production of the RAM Rolling Airframe Missile ship-to-air missile system.
Member companies:
Bodenseewerk Gerätetechnik
Diehl GmbH & Co
Daimler-Benz Aerospace [97]

74-20000 Turbo-Union Ltd
Inselkammerstrasse 5, D-82008 Unterhaching
☎ (+49 89) 66 69 20 ⊖ (+49 89) 66 69 22 00
Chairman: John Ferrie
Managing Director: Karlheinz Koch
■ Consortium formed by FiatAvio (Italy), MTU (Germany) and Rolls-Royce (UK), for the design, development, production and product support of the Turbo-Union RB199 re-heated military turbofan engine. [98]

75 Manufacturing Companies

Elettronica GmbH 75-07565
Eurojet Turbo GmbH 74-05900

75-00020 AB Elektronik GmbH
(a subsidiary of TT Group plc, UK)
Klocknerstrasse 4, D-59368 Werne
☎ (+49 2389) 78 80 ▣ 820916 abek d
⊖ (+49 2389) 78 81 90
Managing Director: Hans Jurgen Zegula
■ Manufacture of automotive sensors, attenuators, potentiometers and switches.
® UK: AB Electronics

Klingerg Office
Saxony Factory, Salzstrasse 1, D-01738 Klingenberg
☎ (+49 35) 20 25 64 45 ⊖ (+49 35) 20 25 64 01 [98]

75-00055 ABB Leitungsbau GmbH
Asea Brown Boveri AG
Ohmweg 11-15, PO Box 240162, D-68197 Mannheim
☎ (+49 621) 810 12 73 ▣ 462411172 abd
⊖ (+49 621) 810 14 85
♣ DM 320,000,000 👥 1,400
Managing Director: Jochen Lehmann
Managing Director: Dr Hans Hugo Kiefer
Head of Marketing: Dipl.-Ing Horst Rúweler
Contact: Ms Gaby Michels
■ Airfield lighting systems, control and monitoring systems for airfield lighting systems, complete range of airfield lights, apron lighting, taxiway guidance signs, docking guidance systems, medium and low voltage power supply systems. [97]

75-00060 Abeking & Rasmussen
PO Box 1160, D-27805 Lemwerder
☎ (+49 421) 673 30 ⊠ ARWERFT
⊖ (+49 421) 673 31 12
👥 500 🏭 25,000 m2
President: Hermann H Schaedla
Sales and Marketing Director: Michael vom Baur
Operations Director: Dr Klaas Spethmann
Purchasing Manager: Wilhelm Ammermann
■ Mine counter-measure vessels (MCMV's). Patrol craft. [98]

75-00105 Abex GmbH Aerohydraul
Lorenzy Schott Strasse 9, D-55252 Mainz Kastel
☎ (+49 6134) 20 40 ⊠ AEROHYDRAUL ▣ 4182071
⊖ (+49 6134) 20 41 43
👥 150 🏭 11,000 sq m
Vice President and General Manager: G Schrank
Vice President, Sales and Marketing: Wolfgang Jung
Technical Director: R Fidler
Purchasing Manager: J Passera
■ Development, production and support of hydraulic pumps, hydraulic motors, electromotor pumps, servopumps, servomotors, transfer units, hydraulic power packages, hydraulic motor driven generators and electro-hydraulic servovalves. Repair and overhaul of hydraulic aircraft components such as cylinders and valves. [97]

75-00110 Accumulatorenfabrik Sonnenschein GmbH
Thiergartenstrasse, PO Box 1180, D-63652 Büdingen
☎ (+49 6042) 810 ⊖ (+49 6042) 815 38
Managing Director: W Bernardi
■ Dryfit batteries. [98]

75-00120 Acutronic Deutschland GmbH
(a sister company of Acutronic AG, Switzerland)
Bretonischer Ring 15, D-85630 Grasbrunn
☎ (+49 89) 46 30 85 [E] acutronic@t-online.de
⊖ (+49 89) 46 46 77
♧ DM 230,000 (1997) 👥 5
Managing Director and Chief Executive Officer: R Hoffmann
■ Inertial guidance test instruments, gyro test tables, metrology equipment, centrifuges, motion simulators. [98]

75-00230 Artemis Management Systems GmbH
(† Lucas Management Systems Deutschland GmbH)
Klenzestrasse 9, D-85737 Ismaning
☎ (+49 89) 960 70 90 ⊖ (+49 89) 96 07 09 11
Managing Director: David Thomson
Sales and Marketing Director: Holger Umcer
■ Supplier of Artemis project management software.
® UK: Artemis Management Systems [98]

75-00260 Aerea Luftfahrtgeräte GmbH
(a subsidiary of Aerea SpA, Italy)
Gewerbepark Bruckmuehl, Gebaude 17, D-D83052 Bruckmuehl
☎ (+49 8062) 781 04; 781 05 ⊖ (+49 8062) 781 06
Managing Director: Peter Wimmer [98]

75-00290 Aeromaritime Systembau GmbH
(a subsidiary of Aeromaritime Systems Group BV, The Netherlands)
PO Box 500152, D-80971 Munich
Hanauer Strasse 105, D-80993 Munich
☎ (+49 89) 14 90 50
[E] asym_aeromaritime@compuserve.com
⊖ (+49 89) 140 11 10; 14 90 52 13 (sales)
♧ DM 6,000,000 ♣ DM 31,000,000 👥 80
🏭 5,000 sq m
Chairman and Managing Director: Horst Winkler
Sales and Marketing Director: Hans-Joachim Ziemann
Advertising and Public Relations Manager: W Juling
■ Shipborne communications systems and equipment; ground based communications systems and equipment; monitoring systems; secondary radar (IFF); IFF test equipment; antennas; electronic test benches and workshops. Military messaging systems. [98]

75-00300 Burkhardt GmbH
(a subsidiary of Krauss-Maffei AG, Germany)
Industriestrasse 8, D-86720 Nördlingen
☎ (+49 9081) 297 40 ⊖ (+49 9081) 29 74 99
Managing Director: Dipl.-Ing Siegfried Koepp [98]

75-00500 AD Aerospace Finishes GmbH
Akzo Nobel Dexter Aerospace Finishes GmbH
(† Akzo Coatings GmbH)
Rathausallee 70, D-22846 Norderstedt
☎ (+49 40) 526 83 80 ⊖ (+49 40) 52 68 38 38
👥 4 (1997) 🏭 120 sq m
Managing Director: K Fleedwood
Sales and Marketing Manager: H G Kuhtz
■ Primers, topcoats and all specialities for aerospace coating purposes, maskants for chemical milling, epoxies and polyurethane paint systems approved to American and European specifications. [98]

75-00515 Akzo Nobel Fazer AG
(† Akzo Nobel)
PO Box 100149, D-42097 Wuppertal
☎ (+49 202) 32 23 50 ⊖ (+49 202) 32 23 51
■ Production of bullet resistant vests. [97]

75-00530 Albrecht Bender GmbH & Co
Nürnberger Strasse 2, D-91781 Weissenburg
☎ (+49 9141) 905 11 ✆ (+49 9141) 90 50 ▣ 624623
⊖ (+49 9141) 905 60
👥 130 (1995)
Managing Director: Gerhard Fürst
Sales and Marketing Director: Ludwig Pfaller
■ Manufacture of uniform accessories for military, police forces and security services. [97]

75-00535 Alcatel Air Navigation Systems GmbH
PO Box 400749, D-70407 Stuttgart
Holderaeckerstrasse 10, D-70499 Stuttgart
☎ (+49 711) 82 13 46 46 ⊖ (+49 711) 82 13 44 05
General Manager: Klaus Schmidt
Marketing and Project Management Director: Christoph Pfuhl
Sales Director: Dietmar Baer
Advertising Manager: Andreas Ziegele
■ Part of the Alcatel Group. Formed in January 1994, Alcatel Air Navigation Systems GmbH replaced the former Alcatel SEL AG and Alcatel Italia SpA companies. All international marketing and sales activities are centralised in Stuttgart, Germany.
Manufacture of Instrument Landing Systems (ILS); VHF Omnidirectional Radio Range Beacon (VOR); Doppler VOR beacon (DVOR); Non-Directional Beacon (NDB); Distance Measuring Equipment (DME); Tactical Air Navigation beacons (TACAN); Microwave Landing System (MLS); Precision DME (DME/P); Remote Maintenance and Monitoring Systems (RMMS); test equipment and spares; DGPS Ground Reference and Monitor Station.
Supplier of complete electronic airport packages, mobile ATC-Towers and mobile VOR/NDB. [98]

75-00610 AlliedSignal Aerospace GmbH
(† Garret GmbH)
(a subsidiary of AlliedSignal Inc, USA)
Frankfurter Strasse 41/65, PO Box 1150, D-65479 Raunheim

☎ (+49 6142) 40 54 27 ▣ 4182134 garet d
⊖ (+49 6142) 40 52 39
♲ DM 500,000 (1995) ♣ US$ 285,000,000 (1996)
👥 600 (1996) 🏠 18,000 sq m
President: Bernd Kessler
Sales and Marketing Director: Greg Albert
Operations Director: Juergen Heller
Marketing Manager: Andrea Paxmann
Purchasing Manager: Rainer Juergens
Customer Support Manager: Wolfgang Gentzsch
[E] wolfgang.gentzsch@alliedsignal.com
■ Repair and overhaul for APU, accessories and TFE of AlliedSignal Aerospace. Manufacture of jet air starters, generator sets and APU (Airbus A319, A320, A330, A340). Customer support for Europe, Middle East and Africa. Worldwide representation for ground support units. → 80 [98]

75-00625 AlliedSignal ELAC Nautic GmbH
(a subsidiary of AlliedSignal International Inc, USA)
Neufeldstrasse, D-24118 Kiel
☎ (+49 431) 88 30 [E] marketing.elac@alliedsignal.com
▣ 292521 ⊖ (+49 431) 88 34 96; 88 34 93; 88 34 89; 88 32 24; 88 34 25; 88 34 12
♣ DM 30,000,000 (1996) 👥 180 (1996) 🏠 111,341 m2
Managing Director: Dr L Hogrefe
Sales and Marketing Director: W Tietz
Operations and Technical Director: H J Maschmann
Purchasing Manager: D Björnsen
Advertising and Sales Support: A Nicopoulos
■ Underwater acoustics for detection, classification, navigation and communication. [98]

75-00650 Alte Haide Baugesellschaft mbH
(a subsidiary of Krauss-Maffei AG, Germany)
Alte Heide 3, D-80805 Munich
☎ (+49 89) 360 49 60 ⊖ (+49 89) 36 60 79
Managing Director: Dipl.-Ing (FH) Siegfried Lössl [98]

75-00750 AMP Deutschland GmbH
Amperestrasse 7-11, D-63225 Langen bei Ffm
☎ (+49 6103) 70 90 ▣ 415019/043
⊖ (+49 6103) 70 92 23
General Manager: K P Götz
Sales and Marketing Director: Mr Bartels
■ Producer of electrical and electronic terminations, interconnection systems and switches.
® USA: AMP Inc [98]

75-00900 AOA Apparatebau Gauting GmbH
Ammerseestrasse 45-49, D-82131 Gauting
☎ (+49 89) 89 31 70 ▣ 521702 gautg d
⊖ (+49 89) 89 31 72 15
♣ DM 33,500,000 (1996) 👥 208 (1996)
Managing Director: Ms Christa Müller
Sales and Marketing Manager: Peter Hacke
■ Development, production and service of: high performance fans, microprocessor systems, sensor systems, fuel pumps, fuel system components. [98]

75-00930 Klaus Transportsysteme GmbH & Co KG
(† Arbau-Klaus GmbH & Co Fahrzeug und Maschinenbau)
PO Box 1141, D-87681 Memmingen
☎ (+49 8331) 921 60 ⊖ (+49 8331) 92 16 50
♣ DM 5,000,000 (1997) 👥 29 (1997)
Managing Director: Franz Prestel
Managing Director: Robert Prestel
Managing Director: Mark Prestel
Operations Director: Werner Rauth
Sales and Advertising: Ms Karin Wellen
☎ (+49 8331) 92 16 30
■ Handling equipment (sideloader) for containers, flat racks to NATO Defence Standards. [98]

75-00950 Arca-Regler
Kempener Strasse 18, PO Box 2120, D-4154 Tönisvorst 2
☎ (+49 2156) 770 90 [W] http://www.arca-valve.com
▣ 853599 arca d ⊖ (+49 2156) 77 09 55
♣ DM 50,000,000 👥 200 (1997) 🏠 10,000 m2
President: Dr-Ing Rüdiger Kaspers (MBA)
Operations Director: Heinz M Nägel
■ Controllers and control valves for shock, vibration and low noise requirements. Steam conditioning valves; linear and rotary valve actuation with intelligent positioning; BIOVENT for hygienic and sterile application. [98]

75-01150 STN ATLAS Elektronik GmbH
(† ATLAS Elektronik GmbH; ex STN-Systemtechnik Nord GmbH)
(a subsidiary of Bremer Vulkan Verbund AG)
Sebaldsbrücker Heerstrasse 235, D-28305 Bremen
☎ (+49 421) 45 70 ⊖ (+49 421) 457 29 00
President: Hans E W Hoffmann
Board Member: N Anding
Board Member: Dr H W Busch
Board Member: K Kunze
Board Member: K Thiedemann
Board Member: H Baier
Marketing and Communication: H Füssinger
Public Relations: R Hahne
■ Sonar systems, torpedo guidance systems, fire control systems, navigation systems, weapon systems, simulators, systems research and development, hydrographic surveying equipment and systems, marine electronics equipment and systems, process data systems. Environmental products and services for military applications. Helicopter avionics. Submarines, mine hunting and clearance equipment and systems, naval targets. Unmanned aerial vehicles. Engine testing facilities, missile test equipment, command and communication ground stations. Mine clearance vehicles, mine clearance equipment, diver delivery vehicles, program management services in boats systems. [98]

75-01400 Autoflug GmbH & Co
Industriestrasse 1, PO Box 1153, D-25462 Rellingen
☎ (+49 4101) 30 72 15 ⊖ (+49 4101) 30 72 07
♲ DM 11,500,000 (1996) ♣ DM 50,000,000 (1996)
👥 237 (1996) 🏠 6,000 (1996)
Managing Director: Andreas Sedlmayr
Managing Director: Heinrich von Paulgerg
■ Development, manufacture, sales and service of escape and safety equipment for aviation and space applications such as Martin-Baker ejection seats, crashworthy helicopter seats, canopy jettison systems, parachute systems, aircraft brake chutes, recovery systems for flying vehicles including space vehicles, shock-absorbing landing bags for RPV's, seat belts and harnesses, liferafts, life vests, survival kits, pilots personal equipment, anti-g suits, immersion suits, helmets, NBC protection equipment, test equipment and logistic support. Development, manufacture, sales and service of rate gyros, vertical sensors, actuators, mast bending moment systems for helicopters. [98]

75-02040 Bavaria Avionik Technologie GmbH (BAT)
Bayernwerkstrasse 39g, PO Box 1150, D-85757 Karlsfeld
☎ (+49 8131) 39 10 ⊖ (+49 8131) 39 11 00
👥 45 🏠 3,000 m2
Managing Director: Roland Harzer
Director: Hans Lampl
Public Relations Officer: Ms Gabriele Egger
■ Development, manufacture and servicing of avionics, electro optical, reconnaissance and test equipment. → 80 → 90
® USA: JcAir, Kollsman Instruments Inc [98]

75-02050 Bavaria Ketronic Gesellschaft für Flugtechnische Geräte mbH (BKT)
Boschstrasse 23, PO Box 500272, D-22761 Hamburg
☎ (+49 40) 89 69 90 ▣ 213930 bkt d
⊖ (+49 40) 890 30 14
♲ DM 2,000,000 ♣ DM 5,000,000 (1996) 👥 30
🏠 3,200 sq ft
Managing Director: M J Rogerson
Vice President, Operations and Civil Airborne Products: Mathias Köhn
Purchasing Manager: K H Ewert ☎ (+49 40) 89 69 91 14
General Manager: Roland Harzer
Civil Airborne Products: Matthias Köhn
☎ (+49 40) 89 69 91 11
Military Airborne and Naval Products: Wolfgang Sell
☎ (+49 40) 89 69 91 10
■ Design, development, production, training, repair and overhaul. Reconnaissance equipment and sensors, airborne data and video recorders, automatic test equipment. Vacuum lavatory systems for aircraft, high speed trains and armoured vehicles.
® Japan: TEAC
USA: Kratos, Rogerson Aircraft Corp [97]

75-02100 Bayern-Chemie Ges. für Flugchemische Antriebe mbH
PO Box 1131, D-84544 Aschau b. Kraiburg/Inn
☎ (+49 8638) 60 10 ✉ BAYERN-CHEMIE
▣ 56440 bcas d ⊖ (+49 8638) 60 13 99
♲ DM 600,000 ♣ DM 53,000,000 (1996) 👥 250
General Manager: Dipl.-Kfm Manfred Haas
Sales and Marketing Director:
Dipl.-Ing Hubert Vockensperger
☎ (+49 8638) 60 15 43
■ Research, development, testing and manufacture in the areas of: propulsion: solid and air breathing propellant rocket motors, mini-rockets, double base and composite propellants; pyrotechnics - valves, actuators, cutting devices, ejection devices, gas generators, infra-red radiators, igniters, ignition mixtures, expandable cushioning elements; consultancy analysis, environmental testing, x-ray, training on machinery and equipment.
Owned by Daimler-Benz Aerospace AG (50%) and Thomson-CSF (50%). [97]

75-02200 Becker Flugfunkwerk GmbH
Becker Avionics GmbH
Baden AirPark, PO Box 34, D-76549 Huegelsheim
☎ (+49 7229) 30 50 ⊖ (+49 7229) 30 52 17
♲ DM 8,000,000 (1995) ♣ DM 20,000,000 (1996)
👥 130 (1997)
Managing Director: Roland Becker
☎ (+49 7229) 30 52 98
Sales and Marketing Director: Peter Kriening
☎ (+49 7229) 30 52 39
Purchasing Manager: Mr Frass
Public Relations: Ms Carole Gassmann
☎ (+49 7229) 30 52 28
■ Airborne communication and navigation systems, entertainment systems, glider and portable transceivers, airborne HF SSB, VHF and UHF ground stations, search and rescue equipment, homing equipment and moving map.
Affiliates:
Becker Avionics Inc USA → 90
Becker Electronics Taiwan → 90
Becker Electronique France → 90 [98]

75-02470 Berleburger Schaumstoffwerk GmbH
PO Box 1180, D-57301 Bad Berleburg
☎ (+49 2751) 80 30 ⊖ (+49 2751) 80 31 59; 80 31 09
♣ DM 80,000,000 (1994) 👥 280 (1995) 🏠 55,000 sq m
Managing Director: Ulf Pöppel
Managing Director: Rainer Pöppel
Sales Manager: Jürgen Jeschke
■ Safety material for shooting ranges, floorings, walls and ceilings, also butts (Regupol shooting blocks).
® USA: Dodge Regupol [98]

75-02490 Hermann Berstorff Maschinenbau GmbH
(a subsidiary of Krauss-Maffei AG, Germany)
Hannoversche Strasse 149, D-30627 Hannover
☎ (+49 511) 570 20 ▣ 921348 ⊖ (+49 511) 56 19 16
Managing Director: Manfred Reichel
Managing Director: Manfred Schick [98]

75-02800 Blohm + Voss GmbH
(a subsidiary of Thyssen Industrie AG)
Defence Technology Division (Navy)
PO Box 100720, D-20005 Hamburg
☎ (+49 40) 311 90 ▣ 211047 bvd
⊖ (+49 40) 31 19 33 96; 31 19 33 24
👥 1,200 (1996)
Managing Director: Dipl.-Ing Herbert von Nitzsch
■ Frigates, corvettes, patrol boats, supply ships, coast-guard ships, training ships and naval auxiliaries SES-technology - new building, overhauls, repairs, modifications, modernisation of any kind of naval ship. Integrated logistic support for naval ships. [98]

75-03000 Bodenseewerk Gerätetechnik GmbH (BGT)
(a member of the Diehl Group, Germany)
PO Box 101155, D-88641 Ueberlingen
☎ (+49 7551) 890 ✉ BODENSEEWERK ▣ 733924 bgt d
⊖ (+49 7551) 89 28 22
♲ DM 51,000,000 ♣ DM 335,000,000 (1995) 👥 1,350
President and Chief Executive Officer: Hans-Peter Reerink
Managing Director: Rainer Ott

Public Relations: E Kiefer
■ Air-to-air missile systems, guidance and control systems for missiles, seeker heads, terminally guided projectiles. Flight control systems for aircraft and helicopters helicopters, electronic engine control systems, heading and attitude reference platforms, display, artificial horizons, meridian gyros, autonomous navigation systems for aircraft, land vehicle navigation systems, automation and medical technology.
Bonn Office
Ullrich von Hassell Strasse 64, D-53123 Bonn
☎ (+49 228) 25 30 49 ⊖ (+49 228) 25 30 40
Koblenz Office
Ferdinand Sauerbruch Strasse 27, D-56073 Koblenz
☎ (+49 261) 94 70 20 ⊖ (+49 261) 40 99 21 [97]

75-03410 Bosch Telecom GmbH
UC/BF, D-13578 Berlin
☎ (+49 3033) 88 14 85
[E] 101575.1726@compuserve.com
⊖ (+49 3033) 88 16 64
♣ DM 330,000,000 (1996) 🏠 1,500 (1997)
Managing Director: J Theverkorn
Operations Director: G Wenzel
Sales and Marketing Director: G Blessing
International Marketing Manager: R Attewell [98]

75-03420 Bospec Neue Technologien
(† Spectron Optik GmbH)
Sportplatzweg 4, PO Box 1149, D-65375 Oestrich, Winkel
☎ (+49 6723) 50 81 [E] bospec@t-on-.de
⊖ (+49 6723) 79 98
♤ DM 300,000 ♣ DM 4,000,000 🏠 20
🏭 1,000 sq m (1996)
President: W R Bost
■ Optical filters and optics for avionics, development, glass processing, coatings, screen printing. Instrument cover glasses, combiner glasses, contrast enhancement filters, EMI filters, anti-frost coating and night vision filters. [97]

75-03600 Bremer Vulkan Marineschiffbau GmbH
(† Bremer Vulkan Werft und Maschinenfabrik GmbH)
Lindenstrasse 110, D-28755 Bremen
☎ (+49 421) 66 80 ⊠ 244858 brvu d; 246790 bvrep d
⊖ (+49 421) 668 30 82
♤ DM 100,000 ♣ DM 6,000,000 🏠 62 🏭 2,238 sq m
Chairman and Managing Director:
Dipl.-Ing Werner Schwarz
Chief Executive Director: Wolfgang Bernecker
Director of Naval Activities: Werner Böschen
Sales Executive: Heiko Evers
■ Design and construction, conversion, after-sales service and logistic support for surface combat and auxiliary ships, such as frigates, corvettes, helicopter carriers, offshore patrol vessels, training vessels, replenishment vessels, logistic support vessels, landing vessels, submarine rescue vessels and survey vessels. → 80 [98]

75-04500 CAE Elektronic GmbH
(a subsidiary of CAE Inc, Canada)
PO Box 1220, D-52201 Stolberg
☎ (+49 2402) 10 60 [E] info@cae-gmbh.de
[W] http://www.cae.ca/gmbh.htm ⊠ 832220 caeg d
⊖ (+49 2402) 10 62 70
♣ DM 139,000,000 (1996) 🏠 550 (1996) 🏭 12,000 m2
Managing Director: Dieter Wolf
Sales and Marketing Director: Wolf Winkler
Operations Director: Karl Kirn
Senior Marketing Manager: Jobst Ott
Purchasing Manager: Bert Frings
Head of Administration and Information Systems:
Michael Stahlmann
■ Design and production of training simulators systems, part task trainers and maintenance trainers. Maintenance, support and modification of sophisticated flight and tactics trainers. Also production of telex, teletex and telefax automation equipment and data distribution networks. → 80
® Canada: CAE Inc [98]

75-04580 Camloc GmbH
(part of Fairchild Fastener Group, Europe)
Industriestrasse 6, D-65779 Kelkheim
☎ (+49 6195) 80 50 ⊠ 4072315 ⊖ (+49 6195) 20 01
Sales Manager: W Filitz
■ 1/4 turn fasteners, custom latches, tension latches, threaded inserts, high-strength fasteners, gas springs for the aerospace industry. [97]

75-04600 Cannon Electric GmbH
(a subsidiary of ITT Industries, USA)
Cannonstrasse 1, PO Box 1120, D-7056 Weinstadt
☎ (+49 7151) 69 90 ⊠ CANNON GmbH ⊠ 7243430
⊖ (+49 7151) 69 92 17
♤ DM 25,000,000 ♣ DM 145,000,000 🏠 900
🏭 20,000 m2
General Manager: Dieter Pesch
Marketing Director: Wolfgang Roland
■ Manufacture and sale of special connectors and harnesses for all defence applications. [98]

75-04610 Carl Baasel Lasertechnik GmbH
(a subsidiary of Krauss-Maffei AG, Germany)
Petersbrunner Strasse 1b, D-82319 Starnberg
☎ (+49 8151) 77 60 ⊖ (+49 8151) 77 61 59
Director: Dipl.-Ing Carl F Baasel
Director: Dipl.-Ing Josef Settele [98]

75-04800 CFM GmbH Chemo-technische Produkte
(† Chemische Fabrik Freising GmbH)
Am Industriepark 3, D-8260 Mühldorf
☎ (+49 8631) 388 40 ⊖ (+49 8631) 38 84 90
♤ DM 50,000 (1996) ♣ DM 4,000,000 (1996)
🏠 30 (1996) 🏭 5,000 m2
Managing Director: Horst Fleischmann
Sales and Marketing Director: Mrs Renate Schönhuber
Operations Director: Ulrich Bleyl
Purchasing Manager: Mrs Nicole Hauptfleisch
■ Chemical and technical products to military standards and DIN specifications like cleaning, polishing, coating paints, silicone adhesives tapes. [98]

75-04900 Clouth Gummiwerke AG
(a subsidiary of ContiTech Group, Hanover)
Defence and Industrial Products Division
PO Box 600229, D-50682 Cologne
Niehler Strasse 102-116
☎ (+49 221) 777 31 ☎ Cologne ⊠ CLOUTH
⊠ 8885376 cl d ⊖ (+49 221) 77 73 698
🏠 750 🏠 630 🏭 141,000 m2
Chairman: Willi Irlenbusch
Member of the Board: Dr Ullrich Masberg
Member of the Board: Martin Norbert
Contact: Dr Christa Osterburg
■ Road wheels, tracks and track pads for armoured vehicles. Add-on ballistic protection for new and existing tracked vehicles. Passive armour and spall liners for military and civil applications (vehicles, aircrafts, ships). Insert plates for bullet-proof vests. [97]

75-05000 Comet GmbH Pyrotechnik-Apparatebau
(a subsidiary of Diehl GmbH, Germany)
PO Box 100267, D-27502 Bremerhaven
Vieländer Weg 147, D-27574 Bremerhaven
☎ (+49 471) 39 30 ⊖ (+49 471) 393 94
♤ DM 12,000,000 ♣ DM 65,000,000 🏠 340
🏭 180,000 m2
Managing Director: Klaus Merdes
Operations Director: Heinz Arnold Swart
Sales and Marketing Manager: Christian Peipers
Purchasing Manager: Götz Hermann Bräutigam
■ Development and production of military and marine pyrotechnics.
Simulators: cannon fire (tank, artillery); hit, tank (HEAT, APDS, a.o.); impact (grenade and artillery).
Illumination: rockets, handheld, 40 mm cartridges, pistol, 26,5 mm. Signalling: stars, parachute-suspended (rocket/pistol); free-falling (rocket/pistol); smokes, full range; signals, distress marine (full range, SOLAS). Explosive breaching equipment, tactical, AP mines; device, line type portable (1 man); device, ladder type portable (2 man); device, ladder type portable (4 man). Pyro components: igniters, pyrotechnic (mechanical/electrical); pyro-mechanical speciality products engineering. [98]

75-05050 Commercial Hydraulics GmbH
(a unit of Commercial Intertech, USA)
Winsener Strasse 69, D-2100 Hamburg 90
⊖ (+49 40) 763 83 99
General Manager: F Bovë
General Manager, Sales: Friedrich W Klein
■ Hangars and metal buildings. [97]

75-05070 Comtronic GmbH
Levelingstrasse 8, D-81673 Munich 80
☎ (+49 89) 43 20 75 ⊖ (+49 89) 431 41 30
♤ DM 240,000 (1995) ♣ DM 8,500,000 (1995)
🏠 40 (1995) 🏭 1,300 sq m (1995)
Managing Director: Manfred Lenmann
Sales and Marketing Director: Ulrich Schmitz-Gärtner
Purchasing Manager: Ute Spreng
■ Front panels, keyboards, Mil-qualified connectors, edge-light-panels, bezel rings, lighting panels.
Factory
In den Kreuzwiesen 20-26, D-69250 Schönau
☎ (+49 6228) 920 50 ⊖ (+49 6228) 89 91 [97]

75-05100 Conrac Elektron GmbH
PO Box 1151, D-97984 Weikersheim
Industriestrasse 18, D-97990 Weikersheim
☎ (+49 7934) 10 10 [E] conrac@compuserve.com
⊖ (+49 7934) 10 11 01
♤ DM 43,000,000 (1996) ♣ DM 3,833,000 (1996)
🏠 160 (1996) 🏭 7,000 sq m
President: Helmut Phillip
Sales Director: Walter Hammel
Director of Marketing and Product Development:
Christian Henn
Sales Manager, Monitors: Horst Nowak
Sales Manager, Information Displays Systems:
Werner Wezel
Operations Manager: Egon Berg
Purchasing Manager: Herbert Hochbeim
Marketing Department: Ms Petra Goetz
■ High resolution colour and monochrome monitors in various screen sizes for applications such as: radar, simulation and surveillance, shipborne, airborne and mobile ground missions, C3I ruggedised, tempest and GL-specified versions.
Overseas Subsidiaries:
Conrac UK Limited [98]

75-05150 ContiTech Holding GmbH
(† Continental AG)
PO Box 169, D-30001 Hannover
☎ (+49 511) 938 02 [W] http://www.contitech.de
⊠ 92170 con d ⊖ (+49 511) 938 27 66
♣ DM 2,900,000 (1996) 🏠 13,800 (1996)
Managing Director: Dr Haverbeck
Marketing Manager: Mr Lerch
Advertising Manager: Mr Abbes
■ Rubber parts, mobile storage tanks for liquids, tarpaulins, refuelling and chemical hoses, anti-vibration and sealing moulded products, liferafts. [98]

75-05200 Contraves GmbH
(a subsidiary of Oerlikon-Contraves AG, Switzerland)
Winterspürerstrasse 17-19, D-78333 Stockach
☎ (+49 7771) 810 ⊖ (+49 7771) 813 96
♣ DM 30,000,000 (1996) 🏠 130 (1996)
Managing Director: H J Etter
Sales and Marketing Director: Dr Seidenstickers
Operations Director: R Röhrig
Purchasing Manager: W Schäuble
■ Development, production and servicing of fire control systems, digital and analogue computer equipment, tracking and measuring equipment. → 80 [98]

75-05300 Dantherm GmbH
(a subsidiary of Dantherm A/S, Denmark)
Günnigfelderstrasse 2, D-45886 Gelsenkirchen
☎ (+49 209) 231 85 ⊖ (+49 209) 20 62 59 [97]

75-05600 Daimler-Benz AG International Government Sales
Dept VLIR Z503, D-70546 Stuttgart
☎ (+49 711) 179 03 08 ⊖ (+49 711) 179 08 76
Managing Director: Dipl.-Ing Dietmar Steffen
Sales Director: Wolfgang Bofinger
Marketing Director: Klaus Berndt
■ All wheel drive vehicles with 2, 3 and 4 axles from 0.5 up to 110 tonnes payload. → 90 [98]

75-05800 **Deggendorfer Werft & Eisenbau GmbH**
(a subsidiary of MAN AG, Germany)
PO Box 1209, D-94452 Deggendorf
☎(+49 991) 38 10 ☒ 69889 dwe d
⊜ (+49 991) 38 11 56
♣ DM 119,000,000 (1995/96) 👥 490
Speaker of the Management Board: Ing Walter Laber
■ Passenger vessels, ferries, freighters, tankers, special vessels, HYDROKLAPP split-barges, floating cranes, floating dredgers, tube reactors, heat exchangers, autoclaves, pressure vessels, pulp digesters, desalination, concentration and evaporation systems, systems for physical processes, plants and erection projects. 98

75-05900 **Deugra Ges. für Brandschutzsysteme**
Halskestrasse 30, PO Box 101260, D-40880 Ratingen
☎(+49 2102) 40 50 ☒ 8585154
⊜ (+49 2102) 40 51 11
👥 100
Managing Director: N Peitzmeier
■ Explosion suppression and fire extinguishing systems and equipment for armoured fighting vehicles, ships, aircraft and high risk areas. 98

75-06310 **Deutsche Sealpak Gesellschaft für Verpackungssysteme mbH**
(a subsidiary of Lohse Textilbau GmbH, Germany)
Kirchweg 141, PO Box 1304, D-2359 Henstedt-Ulzburg
☎(+49 4193) 40 66 ⊜ (+49 4193) 918 29
♧ DM 200,000 (1997) ♣ DM 5,000,000 (1996)
👥 8 (1997) 🏭 1,000 sq m
Managing Director: Jürgen Lohse 98

75-06320 **Deutsche System-Technik GmbH (DST)**
PO Box 450262, D-28296 Bremen
☎(+49 421) 428 70 ⊜ (+49 421) 40 46 60
♧ DM 35,000,000 (1995) ♣ DM 165,000,000 (1995)
👥 800 (1995)
Managing Director: Hans Jürg Zobel
Sales and Marketing Director: Volkmar Jonk
Operations Director: Rolf Gehrke
Communications Director: R Roesler
Purchasing Manager: Ludger Penkhnes
■ Development, manufacture, logistics and service of electronic high-tech products and systems. Optronics and sensor systems, training and simulation systems, communication systems. 97

75-06350 **Deutz Motor Industrie Motoren GmbH**
(† Klöckner-Humboldt-Deutz AG)
Deutz Mülheimer Strasse 147-149, PO Box 800509, D-51057 Cologne
☎(+49 221) 822 24 91 ☒ 88120
⊜ (+49 221) 822 21 16
President: Juergen Schulze
Technical Director: J Vogler
Sales and Marketing Director: R Bridenbach
Head of Research and Development, Engines: Franz X Moser
Head of Prime Mover Systems Division: Dr-Ing Günther Wagner
Purchasing Manager: J Wiest
■ Manufacturer of Deutz range of air-cooled diesel engines. 98

75-06500 **Diehl GmbH & Co**
Strephanstrasse 49, D-90478 Nurenberg
☎(+49 911) 94 70 ☒ 622591 md
⊜ (+49 911) 947 34 29
♣ DM 2,693,000,000 👥 12,590
President and Chief Executive Officer: Dipl.-Ing Thomas Diehl
Executive Officer, Equipment Division: Dr-Ing. Gerd Gassner
Executive Officer, Ammunition Division: Dipl.-Ing Werner Reinl
Public Relations: Dirk-Michael Zahn
■ Main activities include: conventional ammunition (20 mm to 105 mm), dispenser submunitions, electronic and mechanical fuse systems, intelligent munitions, artillery rockets, tracks of all known design categories and suspension systems, maintenance of tracked and wheeled vehicles, rubber parts, rubber and metal components, flare and signal munitions, mine clearing devices, cartridge links, battery systems, air-to-air missile systems, guidance and control systems for missiles, seeker heads, terminally guided projectiles, electronic engine control systems, autonomous aviation navigation systems, information and monitoring systems. Diehl is a partner in the MLRS and RAM programmes.

Affiliated companies:
Jughans Feinwerktechnik (Schramberg)
Bodenseewerk Garaetetechnik GmbH
Comet GmbH, Pyrotechnik-Apparatebau
Diehl & Eagle Picher GmbH
Diehl Remscheid GmbH & Co
DVG Deutsche Verpackungsmittel GmbH
FFG Flensburger Fahrzeugbaugesellschaft mbH
IWS Autotechnik - Fahrzeuge, Fahrzeugteile, Fahrzeugzubehoer GmbH
MFT Metaalwaren-Fabriek Tilburg, (Netherlands)
Neubrandenberger Fahrzengwerke GmbH
VDO-Luftfahrtgeraete-Werk GmbH
MLRS-Europaeische Produktionsgesellschaft mbH
RTG Ammunition GmbH
Zweigwerk Karl Diehl Mariahütte

Ammunition Division
Fischbachstrasse 16, D-90552 Röthenbach/Peg
☎(+49 911) 957 25 16 ⊜ (+49 911) 957 21 60
Contact: Herr Zorn
■ Conventional ammunition from 20 mm to 105 mm in calibre, electronic and mechanical fuse systems, intelligent munitions, submunitions for dispensers, artillery rockets. Combat engineer equipment and mine clearing charges, battery systems, packaging, training and simulation systems. Engineering services. Identification systems. Laser-optical systems. 98

75-06505 **Diehl Remscheid GmbH & Co Equipment Division**
(a subsidiary of Diehl GmbH & Co)
Vieringhausen 118, D-42857 Remscheid
☎(+49 2191) 97 60 ☒ 8513800
⊜ (+49 2191) 97 62 06
♣ DM 110,000,000 (1995) 👥 300 (1995)
🏭 50,000 sq m (1995)
Managing Director: S P Hofmann ☎(+49 2191) 97 62 12
Sales and Marketing Director: M G Haarmann ☎(+49 2191) 97 62 03
Technical Director: Dr A Vieweg
Purchasing Manager: M Heldt ☎(+49 2191) 97 62 15
■ Tracks of all design categories. Sprockets, road and idler wheels and support rollers, moulded rubber parts, rubber metal components, rubber compounds. 97

75-06510 **Walter Dittel GmbH**
Erpftingerstrasse 36, PO Box 1261, D-86899 Landsberg 1
☎(+49 8191) 335 10 ⊜ (+49 8191) 33 51 49
♧ DM 1,800,000 (1996) ♣ DM 12,000,000 (1996)
👥 90 (1996) 🏭 3,600 sq m
Managing Director: Peter Beyeler
Technical Director: Ludwig Krämer
Sales and Marketing Director: E S Pscherer
Sales Manager: Günter Weikamp
Purchasing Manager: G Zierer
Assistant Sales Manager: Harald Weigelt
■ Compact VHF/UHF AM airband transceivers - airborne, portable, mobile, hand-held; static inverters for military and commercial aviation applications. 98

75-06530 **DLK Ventilatoren GmbH**
Ziegeleistrasse 18, D-74214 Berlichingen
☎(+49 7943) 910 20 ⊜ (+49 7943) 91 02 10
♧ DM 800,000 (1996) ♣ DM 13,000,000 (1995)
👥 50 (1996)
President: H J Leithner
General Manager: R Arnold 97

75-06700 **Dornier GmbH**
(a subsidiary of Daimler-Benz Aerospace AG, Germany)
D-88039 Friedrichshafen
☎(+49 7545) 800
W http://www.dasa.com/g/dasa/06.thm ☒ 734209
⊜ (+49 7545) 844 11
♧ DM 150,000,000 (1997)
♣ DM 2,609,600,000 (1996) 👥 3,992 (1997)
Chairman of the Supervisory Board: Dr Manfred Bischoff
Chairman of the Board of Management: Werner Heinzmann
President of Information and Communication Systems: Dr Mathias Hondl
Vice President Corporate Affairs and General Counsel: Dr Ulrich Goebel
President of Satellite Division: Dr Klaus Enszlin
Head of Press and Information: Michael Hartwig
■ Main areas of business include: aircraft, space systems, defence and civil systems, communications systems, medical technology.

Bonn Office
Friedrich-Ebert Allee 26, D-53113 Bonn
☎(+49 228) 917 86 30 ☒ 886507
⊜ (+49 228) 917 86 39
Regional Manager: Dr Udo Pollvogt

Koblenz Office
Poststrasse 8, D-56068 Koblenz
☎(+49 261) 911 68 22 ☒ 261929 ⊜ (+49 261) 144 74
Regional Manager: Karl Bross 98

75-06850 **Drehtainer GmbH**
Bredowstrasse 19, D-22113 Hamburg
☎(+49 40) 732 02 21
E 1066511272@compuserve.com
⊜ (+49 40) 733 85 73
♧ DM 550,000 (1997) ♣ DM 10,000,000 (1996)
👥 30 (1997) 🏭 3,000 m2
Managing Partner: Helmut Meyer
■ Manufacture rough terrain container handlers. 98

75-06900 **Drello Ing P. Drewell GmbH & Co KG**
Max Reger Strasse 35a, PO Box 500126, D-41171 Mönchengladbach
☎(+49 2161) 90 96 ⊜ (+49 2161) 90 97 00
♧ DM 210,000 ♣ DM 10,000,000 👥 37 🏭 3,000 sq m
Managing Director: Hans G Seibt-Drewell
Engineer: Karl Eugen Mackes
Sales Manager: Klaus Kleffner
■ Ballistics measuring equipment. Stroboscopes, CCTV quality control systems; shadowgraph and high-speed photographic equipment. 97

75-06950 **Druck Messtechnik GmbH**
(a subsidiary of Druck Holdings plc, UK)
Lessingstrasse 12, D-61231 Bad Nauheim
☎(+49 6032) 350 28 ⊜ (+49 6032) 711 23
♣ DM 9,000,000 (1997) 👥 15 (1997)
Managing Director: Norbert Habicht 98

75-07100 **Dynamit Nobel GmbH**
(a subsidiary of Metallgesellschaft AG, Germany)
Kaiserstrasse 1, PO Box 1261, D-53839 Troisdorf
☎(+49 2241) 89 13 00 ✉ DYNAMIT NOBEL
☒ 885666 dnd ⊜ (+49 2241) 89 16 69
♧ DM 843,436,000 (1996)
♣ DM 3,707,000,000 (1995/96) 👥 14,924 (1996)
Chairman: Dr Fritz Lehnen
Senior Sales Manager, Defence Division: Matthias Kuhla
Export Sales Manager, Sales Division: Eric Tappert
Area Sales Manager: Franz Wolfgarten
Purchasing Manager: Michael Patt
■ Manufacture and service of sub-calibre training devices (14,5 mm and 25 mm), blank ammunition calibre (5,56 and 25mm), anti-tank weapons, anti-tank mines, off-route mines, area defence weapons, unguided rocket systems. Fire extinguishing soyus. 98

75-07200 **ECE GmbH**
L'Équipement et la Construction Electrique Entwicklung-Fertigung-Vertrieb GmbH
(a subsidiary of ECE, France)
Priessnitz Strasse 4, PO Box 500221, D-80999 Munich
☎(+49 89) 812 60 41 ⊜ (+49 89) 812 60 43
♧ DM 700,000 ♣ DM 7,000,000 (1996) 👥 6 (1996)
Managing Director: Edmond Marchegay
Managing Director: Richard Grützmacher
Company Secretary: Ms Suzanne Oetker-Russwurm
■ Control and protection units for electric generating systems. Differential AC relays. Overvoltage protection units, hermetically sealed relays. Transformer units. Circuit breakers. Contactors. Switches. Illuminating and indicating modules. Indicator lights. Control panels and warning systems measuring instruments. Time interval counters. Testing equipment. 98

Contact Reference

75-07255 EG & G Heimann Optoelectronics GmbH
Weher Koppel 3, Wezelbaksch Straße 31,
D-65199 Wiesbaden
Postfach 3007, D-65020 Wiesbaden
☎ (+49 611) 49 20 [W] http://www.egginc.com/heimann
⊖ (+49 611) 49 26 26 [98]

75-07400 A. Eickhorn GmbH & Co für Schneidwaren & Waffen (AES)
PO Box 120249, D-42677 Solingen
Lohdorfer Strasse 72, D-42699 Solingen
☎ (+49 212) 660 50; 660 05 ∅ (+49 171) 214 67 07
⊖ (+49 212) 674 37
⌂ 1,600 m2
President and Operations Director: Rolf Jürgen Eickhorn
Managing Director: Mrs Annette Eickhorn
Sales and Marketing Director: Jörg Eickhorn
■ Manufacture of military knives and small arms mountings. [98]

75-07550 Elan GmbH
Fritz-Ullmann Strasse 2, D-55252 Mainz Kastel
☎ (+49 6134) 719 50 ⊖ (+49 6134) 17 92
Chairman: Wilfrid H Burke
General Manager: Mrs M C Leblanc
■ Development, manufacture, maintenance and logistic support of military and civil flight navigation instruments, radio communication equipment, navigation ground beacons, ancillary test systems and calibration equipment (TACAN/DME simulators; TACAN/DE ramp tests; instrument/displays testers). [98]

75-07565 Elettronica GmbH
(a subsidiary of Elettronica SpA, Italy)
Industriepark Kottenforst, Am Hambuch 10,
D-53340 Meckenheim
☎ (+49 2225) 880 60; 88 06 24 ⊖ (+49 2225) 88 06 47
✤ DM 1,400,000 (1996) ♣ DM 18,400,000 (1995)
👥 35 (1996) ⌂ 2,350 sq m
Managing Director: Dr Aldo Bianchi
☎ (+49 2225) 88 06 24
Director of Marketing and Business Development: Manfred Kaiser
Technical Director: Dipl.-Ing Peter Trettin
■ Production and modification of mobile receiving and transmitting systems for information gathering, analysis and storage. Production and logistic support for electronic self-protection systems for aircraft. Development and production of end-to-end test equipment for maintenance and performance checks of active and passive electronic systems. Development and production of simulation systems in radar frequency range for training and software support/maintenance. Documentation, maintenance, calibration and logistic services for electronic equipment. Sensors for monitoring and storage of environmental data. → 75 [98]

75-07570 Elkosta VKS Dr Dörnemann GmbH & Co
(a subsidiary of ELKOSTA GmbH, Germany)
PO Box 511240, D-38242 Salzgitter
☎ (+49 5341) 80 20 ⊖ (+49 5341) 80 24 90
✤ DM 6,000,000 (1997) 👥 2,250 (1997)
⌂ 110,000 sq m
Chairman and Managing Director: Dr Hans Malkmus
Managing Director: Dr Holger Schmidt
Domestic Sales and Marketing: Mr Hohenstein
■ Production of gates, barriers, fence construction components, turnstiles for individual passage, low level tripod-type turnstiles, road blocks, tyre killers, blocking bollards. Special products with long-term protection against corrosion, elkosta TRI-PROTECT and zinkalplast, to protect and secure the boundaries of sites in military facilities. [98]

75-07580 ESG Elektroniksystem und Logistik GmbH (ESG)
Einsteinstrasse 174, D-81675 Munich
☎ (+49 89) 921 60 ⊠ 522594 ⊖ (+49 89) 92 16 26 31
✤ DM 10,100,000 (1995) ♣ DM 180,000,000 👥 763
Managing Director: Dr Norbert Servatius
International Marketing Director: Roland Ziegler
Public Relations Manager: Wolfgang Baumbach
■ Development of complex electronic and information systems, command and control systems. Cost-reducing logistics procedures.

Bonn Office
Borsigallee 2, D-53125 Bonn
☎ (+49 228) 25 20 71 ⊖ (+49 228) 25 20 73
Managing Director: Klaus Assion [98]

75-08100 ENPAR Sonderwerstoffe GmbH
Betriebsweg 10, D-51645 Gummersbach
☎ (+49 2261) 79 80 ⊠ 884766 ⊖ (+49 2261) 798 88
♣ DM 25,000,000 👥 45 (1997) ⌂ 2,000 m2
Managing Director: Ulrich Trenkmann
Operations Director: Hans Vlecken
Sales and Marketing Director: Axel Veller
■ Metal distribution for military and aerospace industries. → 85 [98]

75-08680 EuroAtlas GmbH
(† MagneTek Euroatlas GmbH)
Zum Panrepel 2, D-28307 Bremen
☎ (+49 421) 486 930
⊖ (+49 421) 486 93 41; 486 93 77
✤ DM 2,000,000 (1995) ♣ US$ 16,000,000 (1994)
👥 90 (1995) ⌂ 3,000 m2 (1995)
Managing Director: Klaus H Kahrs
Sales and Marketing Manager: Helmut Ronge
Public Relations Manager: K Nawrocki
■ Power electronics: static inverters and converters for military applications and commercial aircraft. [97]

75-08700 Eurobridge Mobile Brücken GmbH
(a subsidiary of Dornier GmbH and Daimler Benz Aerospace AG, Germany)
D-88039 Friedrichshafen
☎ (+49 7545) 836 11 ⊖ (+49 7545) 890 90
Managing Director: Uwe Nortmann
Head of Sales and Marketing: Siegfried Bäumel
■ Development and manufacture of the Dornier foldable bridge DoFB system and other types of mobile bridges. [98]

75-08720 EvoBus GmbH
(† Karl Kässbohrer Fahrzeugwerke GmbH)
(a subsidiary of Daimler-Benz AG, Germany)
Vaihinger Strasse 131, D-70567 Stuttgart
☎ (+49 711) 179 90 04 [E] info@evobus.com
[W] http://www.evobus.com ⊖ (+49 711) 179 91 01
♣ DM 3,600,000,000 (1996) 👥 11,300 (1996)
Managing Director: Wolfang Diez
Public Relations: Irina Compter
■ Manufacture of buses and coaches of the brands Mercedes-Benz and Setra. [98]

75-08750 Excalor Hitzeschutz GmbH
Zum Frankenbrunnen 1, D-66450 Bexbach, Frankenholz
☎ (+49 6826) 80 00 04 ⊖ (+49 6826) 72 79
President: Herbert Kampmann
Managing Clerk: A Wedel
■ Wide range of products encompassing clothing for partial, body and full protection against fire, heat, solid and liquid chemical gases and steam. Complete range of accessories such as: protective helmets according to DIN 4840, rubber boots according to S10 DIN 4843, fire blankets and containers "Flammtrutz" rescue rope acccording to ISSV-Solas 1960, rescue blankets, welding tarpaulins and curtains, heat protective fabrics. [97]

75-08860 Dornier Luftfahrt GmbH
(a division of Fairchild Aerospace)
PO Box 103, D-82230 Wessling
☎ (+49 81) 533 00 [E] 101660.466@compuserve com.
⊠ 526412 doas d ⊖ (+49 81) 53 30 20 07
👥 1,800
Managing Director: James A Robinson
Vice President Sales: Thomas Brandt
Vice President Support: Franz Zellner
■ Sub-contrator to Deutsche Aerospace Airbus GmbH. Technical and logistic servicing of German Navy Breguet Br 1150 Atlantic 1 aircraft. Service centre for business and general aviation aircraft. [98]

75-08900 Faun GmbH
PO Box 100108, D-91205 Lauf an der Pegnitz
☎ (+49 9123) 18 50 ✉ FAUN Laufpegn ⊠ 626093 faun d
⊖ (+49 9123) 753 20
✤ DM 88,000,000 (1996) ♣ DM 230,000,000 (1996)
👥 600 (1997)
President: Koichi Tadano
Sales and Marketing Director: van Acker
Purchasing Manager: Mr Topolniki
■ Heavy duty vehicles, trucks, tractors, recovery vehicles, fire fighting vehicles, mobile cranes, all-terrain cranes, rough-terrain cranes, torpedo launching tubes for naval surface vessels. [98]

75-09230 Ferrostaal AG
(a subsidiary of MAN AG, Germany)
D-45116 Essen
☎ (+49 201) 818 01 ⊠ 857100 fsd
⊖ (+49 201) 818 28 22
♣ DM 4,668,404,000 (1995/96) 👥 1,611
Chairman of Executive Board: Dr Klaus von Menges
■ Worldwide trading of steel products, machinery, plants, systems and components. Future oriented and economic realisation of industrial projects, also as a general contractor. [98]

75-09250 Flottweg GmbH
(a subsidiary of Krauss-Maffei AG, Germany)
Industriestrasse 6-8, D-84137 Vilsbiburg
☎ (+49 8741) 30 10 ⊠ 58680 ⊖ (+49 8741) 30 13 00
Chairman of the Board of Directors: Dipl.-Vw Peter Bruckmayer
Member of the Board of Directors: Dipl.-Kfm Joachim Weiershaus
Member of the Board of Directors: Dr Ernst Jäger [98]

75-09270 Force Computers GmbH European Headquarters
Professor Messerschmitt Strasse 1,
D-85579 Neubiberg, Munich
☎ (+49 89) 60 81 40 ⊖ (+49 89) 609 77 93
👥 250 ⌂ 8,800 sq m
General Manager: Michael Nick Walker
Marketing Director: Chris Williams
Advertising and Marketing Communications Manager: Ms Ulrike Müller
■ ISO 9001 certified manufacturer of embedded computer systems. Such systems are supplied to the defence industry for applications that include: C4I - naval and ground based systems for radar, sonar, weapon control and navigation; simulation - ship, tank, aircraft and helicopter. [97]

75-09300 FPE Fischer GmbH
Postfach 1148, D-88291 Leutkirch
Nadlestrasse 10, D-88299 Leutkirch
☎ (+49 7561) 981 40 [E] fpe@fischer.dbp.de
⊖ (+49 7561) 723 02
✤ DM 50,000 (1996) ♣ DM 7,500,000 (1996)
👥 7 (1996) ⌂ 450 m2
Managing Director: Ulrich Fischer
Operations Director: B Fischer
Sales and Marketing Director: M Dobler
Purchasing Manager: A Butscher
■ Precision potentiometers, printed circuit boards doublesided, multilayers and hybrids, electronic and mechanical assemblies for military and civil aircraft applications. [98]

75-09570 Otto Fuchs Metallwerke
PO Box 1261, D-58528 Meinerzhagen
Derschlager Strasse 26, D-58540 Meinerzhagen
☎ (+49 2354) 731 ⊖ (+49 2354) 732 01
♣ DM 527,000,000 (1995) 👥 2,062 (1996)
Managing Director: Günter Guth
Technical Director: Dr Tepohl
Sales and Marketing Director: Gerd Kollecker
Procurement Officer: Dr H Lowak
■ Die and hand forgings in titanium and nickel materials, extrusions in aluminium, magnesium, copper and their alloys. [98]

75-09600 Furon
Siemensring 36, Gewerbegebiet Munchheide 1,
D-47877 Willich 1
☎ (+49 2154) 42 97 68 ⊖ (+49 2154) 42 99 79 [97]

75-10300 GEREP GmbH
Alter Postweg 10, D-29556 Suderburg 1
☎ (+49 5826) 96 73 12 ⊖ (+49 5826) 83 13
Managing Director: Juergen Remensperger
Managing Director: Dipl.Kfm. Norbert Remensperger
Managing Director: Dipl.-Ing E U Remensperger

■ Development, design, manufacture, marketing and sale of suspension components, telescopic and rotational shock absorbers, buffers, cylinders and track adjusters for armoured vehicles. Qualified to AQAP-4/6 and ISO 9001. [98]

75-10600 Gesellschaft für Intelligente Wirksysteme mbH (GIWS)
(a joint venture between Diehl GmbH & Co and Rheinmetall Industrie GmbH, Germany)
Bahnhofsplatz 6, D-90443 Nürnberg
☎ (+49 911) 24 44 70 ⊖ (+49 911) 244 47 40
Managing, Sales and Marketing Director: Horst Luczynski
Managing and Operations Director: Klaus Loske
■ Manufacture of sensor fused munition for artillery of 155 mm calibre. [98]

75-10700 Gesellschaft zur Enticklung und Forderung Aerostatischer Flugsysteme (GEFA-FLUG)
Weststrasse 24c, D-52074 Aachen
☎ (+49 241) 88 90 40 ⊖ (+49 241) 889 04 20
DM 50,000 (1997) ♣ DM 2,500,000 (1996)
20 (1997) 600 m2
Managing Director: Karl Ludwig Busemeyer
Sales, Marketing and Operations Director: Jürgen Leister
Purchasing Manager: Wolfgang Mainzer
■ Research and development into, and operation of airships, hot air balloons and remotely controlled aerostats. [98]

75-10875 Gisma Steckverbinder GmbH
Oderstrasse 11, D-2350 Neumünster
☎ (+49 4321) 89 51 ⊖ (+49 4321) 89 53
Managing Director: M Maletaluy
Sales and Marketing Director: Tobias Fredx
Sales Department: A Horn
■ Manufacture and design of underwater pluggable and fibre-optic metal shell connectors for use up to 1,000 bar and from signal level to 6Kv and 400 amps. [97]

75-10900 GKN Westland-Sitec GmbH
(a subsidiary of GKN Westland Group)
Schopfgraben 1, D-83714 Miesbach
☎ (+49 8025) 28 10 526113
⊖ (+49 8025) 39 65; 47 36
Managing Director: A Hilgarth
Sales Manager: C M Summers
■ Design, development and production of valves for aircraft fuel, water and hydraulic systems.
® USA: Circle Seal Controls Inc [97]

75-10950 Gloria Werke H Schulte-Frankenfeld GmbH & Co
PO Box 1160, D-59321 Wadersloh/Westf
☎ (+49 2523) 771 88 ⊖ (+49 2523) 772 54
600 (1997)
Export Manager: Eckhard Bauhoff
■ Fire extinguishers and fire safety containers. [98]

75-11000 Goldhofer Fahrzeugwerk GmbH & Co
Donaustrasse 95, PO Box 1364, D-87683 Memmingen
☎ (+49 8331) 150 ⊖ (+49 8331) 152 39
♣ DM 70,000,000 (1996) 430 (1996) 19,500 sq m
Managing Director: Mrs Karoline Goldhofer-Prützel
Executive Vice President, Sales and Marketing: Ralf Schiller
Operations Director: Manfred Müller
Marketing Manager: Erich Traub
Product Manager: H J Habernegg
Export Manager: Horst Haefele
Executive Vice President: Hubert Bantele
■ Low-loaders, semi-low-loaders, special vehicles, heavy duty platform trailers with payloads up to 10,000 tons. Aircraft tow tractors, aircraft recovery transport systems. [98]

75-11300 Gustav Wolf Seil-und Drathwerke GmbH & Co
Carl Bertelsmann Strasse 21/23, PO Box 3353, D-4830 Gütersloh
☎ (+49 5241) 87 60 ✉ DRAHTWOLF 933847 gwolf d
⊖ (+49 5241) 87 61 80
500
Export Manager: Dr E Wolf
Technical Services Manager: Dr S Tappeiner
■ Aircraft cable to Mil-W-83420, special ropes for winches in helicopters, tow-ropes for gliders. Aircraft equipment, 56 aircraft, miscellaneous equipment. [97]

75-11400 Haacon Hebetechnik GmbH
PO Box 1140, D-97893 Freudenberg
Josef Haamann Strasse 6, D-97896 Freudenberg
☎ (+49 9375) 840 ⊖ (+49 9375) 84 66
DM 2,000,000 170 50,000 m2
Managing Director: Jörg Wenzel
Managing Director: Dr W Dreisigacker
Export Director: Jack A Werner
■ Container lifting equipment, special models, lifting tackle and winches. Lifting and loading systems for shelters and containers (air transport). [98]

75-11735 Hatlapa Uetersener Maschinenfabrik GmbH & Co
PO Box 1153, D-25429 Uetersen Hamburg
☎ (+49 4122) 71 10 218510 htlue d
⊖ (+49 4122) 71 11 04
360
Managing Director: Hubertus M Hatlapa
■ Shipborne compressors for naval applications. Steering gears, deck, machinery. [97]

75-11900 Harris GmbH Computer Systems Division
(a division of Harris Corp, USA)
Königswinterer Strasse 554, D-53227 Bonn
☎ (+49 228) 97 02 30 ⊖ (+49 228) 970 23 50
DM 20,000,000 (1995) ♣ DM 12,000,000 (1995)
12 350 m2
Managing Director: Richard D Weinstock
Sales and Marketing Director: Claus Hansen
Sales and Marketing Director: Axel Goeritz
Operations Director: Ulrich Michaelis
Sales Manager: Dr Werner Gillessen
Purchasing Manager: Ms Angelika Umlauft
Representing:
Harris Computer Systems Corp, USA [97]

75-11970 Haux-Life-Support GmbH
Descostrasse 19, D-76307 Karlsbad, Ittersbach
☎ (+49 7248) 916 00 ⊖ (+49 7248) 91 60 66
Managing Director: Gerhard Haux
Managing Director: Torsten Haux
■ Design, development and production of transportable, mobile and stationary decompression chambers, diving simulators, deep diving systems, test facilities, test chambers, locks for material and persons, medical hyperbaric treatment chambers, life-support systems, peripheral systems and equipment, gas supply systems, pressure walls, pilot test chambers, special breathing equipment. Special components for gas storage systems such as pressure reducers and valves. [97]

75-11972 Heckler & Koch GmbH Defence Division (HK)
(a subsidiary of British Aerospace Defence Ltd, UK)
PO Box 1329, D-78722 Oberndorf/Neckar
☎ (+49 7423) 790 ✉ HEKLERKOCH 760313 huko d
⊖ (+49 7423) 79 24 06
600
Business Unit Director: Phil Lee ☎ (+49 7423) 79 24 10
Finance Director: Wilhelm Haaga ☎ (+49 7423) 79 24 64
Sales and Marketing Director: Gert Pannenbecker ☎ (+49 7423) 79 23 11
Technical Director: Ernst Mauch ☎ (+49 7423) 79 22 17
Personnel and Legal Director: Dr Dirk Holzknecht ☎ (+49 7423) 79 23 88
■ Small arms, machine guns, submachine guns, pistols, mounts, special weapons. [98]

75-11975 Hella Aerospace GmbH
(† Hella KG Hueck & Co)
(a subsidiary of Hella KG Hueck and Co)
Rixbecker Strasse 75, D-59552 Lippstadt
☎ (+49 2941) 38 85 12 ✉ HELLA
[E] hella100751.2114.@compuserve.com
⊖ (+49 2941) 38 84 32
150
President and Chief Executive Officer: Dr H G Focken
Executive Vice President: P Metzger
Sales and Marketing Director: F J Beermann
■ Aircraft lighting: headlights, navigation and position and anti-collision lights. NVG compatible lighting. [98]

75-11977 Paul Hellermann GmbH
Siemensstrasse 5, D-25421 Pinneberg
☎ (+49 4101) 70 11 ⊖ (+49 4101) 713 47
Managing Director: C Niemann
■ Manufacture and supply of cable care and heat shrinkable protection products including cable marking and identification systems. Design and production of specialist application tools. [98]

75-12450 Hensoldt AG
Gloelstrasse 3-5, PO Box 1760, D-35573 Wetzlar
☎ (+49 6441) 40 43 80 ⊖ (+49 6441) 40 43 22
DM 1,955,000 (1996) ♣ DM 140,000,000 (1996)
750
Managing Director: Dr Karl Friedrich
Sales and Marketing Director: Dieter Simonis
Operations Director: Paul Naumann
Sales Manager: Martin Hofmann
Purchasing Manager: Gerold Bamberger
■ Special binoculars, telescopes, target periscopes, panoramic periscopes, target monitoring instruments, special tools, instruments for testing hollow bodies, geodetic instruments, testing and adjusting instruments, workshop equipment. [97]

75-12800 Karlheinz Hinze Optoengineering GmbH & Co
PO Box 540906, D-22509 Hamburg
Kieler Strasse 464-470, D-22525 Hamburg
☎ (+49 40) 54 25 90; 54 25 91 ⊖ (+49 40) 54 35 73
Owner: Karlheinz Hinze
Sales and Marketing Manager: Michael Hinze
■ Manufacture of optical inspection instruments (borescopes, miniature TV cameras) and maintenance of aircraft engines, components, arms, gun-barrels. [98]

75-12950 Holset Kompressorenbau Bannewitz GmbH
(a subsidiary of Holset Engineering Co, UK)
Windberastrasse 45, D-01728 Bannewitz
☎ (+49 351) 40 85 50 ⊖ (+49 351) 408 58 40
140 (1996) 70,000 (1996)
Managing Director: Michael Hoschke
Sales and Marketing Director: Paul Ibbotson
Operations Director: Dr Ulrich Kramer
Purchasing Manager: Isolde Sturm [97]

75-13200 Honeywell Regelsysteme GmbH
Geschaftaberich Sondertechnik
(a division of Honeywell Inc., USA)
Honeywellstrasse 2-6, D-63477 Maintal
☎ (+49 6181) 40 11 ⊖ (+49 6181) 40 17 90; 40 12 89
DM 40,000,000 (1997) ♣ DM 90,000,000 (1996)
600 (1997)
Managing Director: Dr Egon Tyssen ☎ (+49 6181) 40 15 00
Communications Director: H K Dietrich ☎ (+49 6181) 40 14 57
Sales and Marketing Director: Eckehard Keip ☎ (+49 6181) 40 16 81
Finance Director: Gerhard Mürschel
Research and Development: R Mollenhauer ☎ (+49 6181) 40 13 18
Production: W Bergmann
Marketing Manager, Avionics: Dr W Hoeltje ☎ (+49 6181) 40 15 99
Purchasing Manager: R Acker
■ Space and aviation control systems.
® USA: Honeywell Inc

Bonn Office
Reuterstrasse 124/136, Bonn Center HI 401, D-5300 Bonn
☎ (+49 228) 21 38 76; 21 38 21 ⊖ (+49 228) 21 22 98
Manager: P Geu

Aerospace and Defence Group, Europe
Honeywellstrasse, D-6457 Maintal 1
☎ (+49 6181) 40 11 4184820 ⊖ (+49 6181) 454 56
General Manager: Dr Egon Tyssen

Optoelectronics
Stuttgarter Strasse 5, D-8000 Munich 40
☎ (+49 89) 35 03 05 10 5216884
⊖ (+49 89) 359 99 71
Business Unit Manager: Walter Hessenberger [98]

75-13270 Hoogovens Aluminium Walzprodukte GmbH
(a subsidiary of Koninklijke Hoogovens NV)
PO Box 100331, D-56033 Koblenz

Contact Reference

☎ (+49 261) 89 14 42; 89 14 66 ✉ HOOGOVENS
⊖ (+49 261) 80 15 74
♣ DM 400,000,000 👥 750 🏭 130,000 sq m
Managing Director: G Buddenbaum
☎ (+49 261) 89 14 43
Technical Director: O Schmitz ☎ (+49 261) 89 13 27
Sales and Marketing Manager, Aircraft and Aerospace Industry: Erland Bade
■ Manufacture of aluminium plates for aircraft, aerospace and for military applications.
® USA: Hoogovens Aluminium Corp [98]

75-13290 HS-Simulatoren Horstmann GmbH Prüfanlagen zur Umweltsimulation

Greizer Strasse 41/49, D-35447 Reiskirchen-Lindenstruth
☎ (+49 6408) 840 ⊖ (+49 6408) 843 58 [98]

75-13400 Howaldtswerke Deutsche Werft AG (HDW)

(a part of the Preussag Group)
PO Box 6309, D-24124 Kiel
☎ (+49 431) 70 00 ✉ HOWALDTDEUTSCH
▣ 2922880 hdw d ⊖ (+49 431) 700 23 14
♧ DM 140,000,000 (1995/96)
♣ DM 1,666,000,000 (1995/96) 👥 3,530 (1995/96)
Chairman: Dr Jürgen Gollenbeck
Head of Public Relations: Waltraud Löcherer
Senior Sales Manager, Naval Ships: Hans Saeger
Senior Sales Manager, Commercial Ships: Gerhard Böttcher
■ Design, construction and building of submarines, frigates, corvettes, supply vessels, auxiliary and landing craft.

Naval Division
PO Box 6309, D-24124 Kiel
☎ (+49 431) 700 ▣ 2922880 hdw d
⊖ (+49 431) 700 23 12
■ Naval shipbuilding of frigates, corvettes, landing ships and submarines with full service in design, construction, repair, documentation, logistic support, training, delivery of know-how and comprehensive sea trials. [98]

75-13520 Hydro-Gerätebau GmbH & Co KG

PO Box 1247, D-77779 Biberach/Baden
☎ (+49 7835) 78 70 ✉ HYDRO ▣ 17783516
⊖ (+49 7835) 18 23
♧ DM 500,000 ♣ DM 18,000,000 👥 82 (1996)
🏭 3,200 sq m
President: Josef Holzer
Vice President: Ms Hilde Holzer
Managing Director: Dipl.-Ing H H Scharsig
Managing Director: Ms B Huttegger
Technical Sales Director: Thomas Hohl
Engineering Manager: Klaus Frutttiger
■ Manufacture of ground support equipment and maintenance tooling for civil and military aircraft. Electronic jacking and levelling systems. Handling systems for tyres and rims. Hydraulic teststands. Production of jet engine handling equipment including engine positioner MLG and NLG handling trolleys. Tools for Boeing, Airbus, McDonnell Douglas and others. Nitrogen servicing carts. NLC tilting dollies, tow bars, tripod jacks, weighing equipment, wheel and brake removal, installation dollies, galley equipment, components for aircraft final assembly lines. [97]

75-13580 Ibak Helmut Hunger GmbH & Co KG

Wehdenweg 1224, PO Box 6260, D-2300 Kiel 1
☎ (+49 431) 727 00 ▣ 292824 ⊖ (+49 431) 72 62 20
Managing Director: Herwig Hunger
Sales Director: Peter Hurlebusch
■ Naval searchlights; special underwater TV systems; acoustic buoys. [98]

75-13900 Ika-Maschinenbau Janke & Kunkel GmbH & Co KG

(† Propex Research & Engineering GmbH)
PO Box 1165, D-79216 Staufen
Janke-u.-Kunkel Strasse 10
☎ (+49 7633) 83 10 ⊖ (+49 7633) 831 98
♧ DM 1,200,000 (1996) ♣ DM 15,700,000 (1996)
👥 65 (1996) 🏭 10,400
Managing Director: Peter Wanninger
Advertising Department: Ruth Lotz
Public Relations Department: Tanja Broscheit

■ Plant design and equipment for batch processing of solid rocket propellants, gun propellants, high explosives and pyrotechnics. [98]

75-14500 Institut Dr Förster GmbH & Co KG

PO Box 1564, D-72705 Reutlingen
☎ (+49 7121) 14 02 15 ▣ 729781 ifrd
⊖ (+49 7121) 14 03 57
President: Martin Förster
■ Mine and bomb detection instruments (registered trademark Metex, Minex and Ferex), magnetic measurement, degaussing systems (registered trade-mark Magnetomat). [98]

75-14620 Interspiro GmbH

(a subsidiary of Interspiro AB, Sweden)
Postfach 1220, D-7529 Forst/Baden
☎ (+49 7251) 170 31 34 ▣ 7822247
⊖ (+49 7251) 25 98
👥 26
Managing Director: Michael Maar [98]

75-14630 Interturbine Germany GmbH

(a division of Interturbine, Germany)
PO Box 1454, D-24562 Kaltenkirchen
Kisdorfer Weg 36-38, D-24568 Kaltenkirchen
☎ (+49 4191) 80 90 ▣ 2180177 itg d
⊖ (+49 4191) 28 26
♧ DM 1,000,000 (1996) ♣ DM 15,000,000 (1996)
👥 85 (1996) 🏭 6,000 sq m
Managing Director: Uwe Blöcker
Operations Director: Hans Dethlefsen
Sales and Marketing Director: Andreas Kalina
Business Development Manager: G Reid
Purchasing Manager: Jürgen Schreck
■ Overhaul and repair of turbine engine casings and hot section components, thermal processing, plasma coating and chemical plating. Manufacture and reconditioning of thrust reversers and engine components of non-rotating parts, cold and hot section. Engine component capability for JT3, JT8, CF6, CFM1, RB-211, JT9, PT6, TPE 331, ST6, Allison 250, vacuum and hydrogen technologies and epolair repair. → 80 [98]

75-14640 IRT Infrarot-Technik Eiselt

PO Box 500323, D-70333 Stuttgart
☎ (+49 7031) 22 22 12 ⊖ (+49 7031) 22 22 28
Managing Director: Magdalena Eiselt
■ Night vision systems; thermal night vision systems; TV night camera systems; high performance light sources. [98]

75-14800 Jenoptik Impulsphysik GmbH

(† Hagenuk GmbH)
(a subsidiary of Jenoptik AG, Germany)
PO Box 1349, D-22860 Schenefeld
Achter de Weiden 10, D-22869 Schenefeld
☎ (+49 40) 83 90 30 [E] info@impulsphysik.com
IMPULSPHYSIK ⊖ (+49 40) 83 90 31 10
♣ DM 13,100,000 (1996) 👥 60 (1997) 🏭 2,900 m2
Managing Director: Volker Sturm
Managing Director: Uwe Bornhöft
Sales Director: Jürgen Zillmer
Purchasing Manager: Kai Kistenmacher
■ Laser cloud height ceilometers, visibility meters, present weather indicators, RVR systems and transmissiometers, fog detectors, complete AWOS Automated Aerodrome Weather Observation Systems, ATIS speech synthesizers. [98]

75-15050 Jeppesen & Co GmbH

(a subsidiary of TM Holdings Inc, USA)
PO Box 700551, D-60555 Frankfurt
☎ (+49 69) 96 12 40 ▣ 412529 jepco d
⊖ (+49 69) 96 12 48 98; 96 12 48 99
👥 250
Managing Director: Hans Lange
Managing Director: Horst Bergmann
Sales and Marketing Director: Thomas Wede
Marketing Representative: Andrea Stumpf
■ Chart services, NavData electronic databases, computer flight planning and weather services, pilot and maintenance training products, pilot supplies and accessories. [98]

75-15120 Junghans Feinwerktechnik

(a subsidiary of Diehl GmbH & Co, Germany)
PO Box 70, D-78701 Schramberg
☎ (+49 7422) 185 29 ⊖ (+49 7422) 184 00
♣ DM 85,000,000 (1996) 👥 550 (1996)
Managing Director: Hans Kaiser ☎ (+49 7422) 183 40
Marketing and Sales: Peter Becker ☎ (+49 7422) 182 83
Public Relations Officer: Mrs Renate Mann
■ Manufacture of fuses, safety and arming systems. Product range includes fuses for automatic cannon ammunition, mortar fuses, fuses for tank gun ammunition, fuses for missiles and rockets, artillery fuses, fuses for AT weapons, S and A devices. [98]

75-15130 Jung Jungenthal GmbH

Bahnstrasse 1, PO Box 20, D-5242 Kirchen
☎ (+49 2741) 68 30 ▣ 875319 ⊖ (+49 2741) 68 32 46
♧ DM 8,000,000 (1997) ♣ DM 80,000,000 👥 600
🏭 40,000 m2
President: Hans H Fries
■ Military engineering, system development and prototype construction of armoured vehicles. Series production of recovery systems and equipment. Conversion of M60 and Leopard 1 into ARV's, bridge layers and fire fighters. [98]

75-15650 Alfred Karcher GmbH & Co

Alfred-Karcher Strasse 28-40, PO Box 160, D-71364 Winnenden, Stuttgart
☎ (+49 7195) 14 22 62; 14 27 97 ▣ 724432
⊖ (+49 7195) 14 27 80; 14 27 20
♣ DM 1,200,000,000 (1995) 👥 4,300
Head of Public Relations: Helmut Stelzmüller
Sales Manager: Klaus Flad
■ Mobile catering systems: modular field kitchens, tactical field kitchens, environmental protection (NBC) defence, containerised environment protection systems, collapsible water tanks, decontamination systems, field heaters, mediclean, cleaning agents, field showers, decontamination tents, fire brigade suits, underwear, NBC suits. Material maintenance systems: high-pressure steam cleaners, hot wax/hot oil preservation kits, auxiliary de-icing kits, tank degassing kits, water purification systems, field laundries, container field kitchens.

Foreign subsidiaries:
Karcher SA, France
Karcher Benelux, Belgium
Karcher Vaporapid AG, Switzerland
Alfred Karcher GmbH, Austria
Karcher SpA, Italy
Karcher AB, Sweden
Karcher A/S, Norway
Karcher Oy, Finland
Karcher UK, Great Britain
Karcher Indústria e Comércio, Brazil
American Kleaner Manufacturing Co, USA
Karcher (Pty) Ltd, Australia
Karcher Cleaning Systems, Canada
Matsuzaka Karcher Cleaning Systems, Japan
Karcher SA, Spain [98]

75-15675 Kawasaki Heavy Industries GmbH

5th Floor Wehrhahn Center, Oststrasse 10, D-40211 Dusseldorf
☎ (+49 211) 35 04 41 ⊖ (+49 211) 16 18 44 [98]

75-15680 Kawasaki Motoren GmbH

Max Planck Strasse 26, D-61381 Friedrichsdorf
(a subsidiary of Kawasaki Heavy Industries Ltd, Japan)
☎ (+49 61) 72 73 40 ⊖ (+49 61) 72 73 41 60 [98]

75-15750 ke autoeletric GmbH

Vohunstransser Strasse 20, D-92685 Floss
☎ (+49 9603) 200 [E] keauto@t-online.de
⊖ (+49 9603) 27 15
Managing Director: Dr Wolfgang Scheideler
Managing Director: Andreas Wolf [98]

75-15900 Deütz Motor GmbH

(† KHD)
Deutz Muelheimer Strasse 147, D-51057 Cologne
⊖ (+49 221) 822 25 29
Chief Executive Officer: Anton Schneider
☎ (+49 221) 822 20 00
Managing Director: Rainer Breidenbach
☎ (+49 221) 822 26 50

Marketing Manager: Peter Pospiech
☎ (+49 221) 822 25 10
■ Production of marine propulsion units and generators. Construction of diesel engines for any kind of application. [97]

75-16150 Knürr-Mechanik für die Elektronik AG

Schatzbogen 29, PO Box 820369, D-81803 Munich
☎ (+49 89) 42 00 40 ⊖ (+49 89) 42 00 41 18
♣ DM 151,000,000 (1996) 👥 520 (1996)
Chairman and Managing Director: Hans Knürr
Sales and Marketing Director: Alfons Brummer
Purchasing Manager: Dieter Hofer
Export Department: Mr Schoenebaum
■ Racks, cabinets, floor standing consoles, cardframes, accessories, aluminium carrying cases and re-usable containers. Industrial furnishing, technical workstations and electronic tools. Military standards. [98]

75-16185 Komatsu Hanomag Aktiengesellschaft

(† HANOMAG Aktiengesellschaft)
Postfach 911325, D-30433 Hannover
☎ (+49 511) 450 90 ▣ 923381 hmg d
⊖ (+49 511) 450 91 85
♧ DM 59,194,600 (1994) ♣ DM 336,800,000 (1994)
👥 1,033 (1994) 🏠 240,000 sq m
Chief Executive of the Board: Jürgen Seidel
Board Member: K Hoshino
Board Member: Johann Werth Pape
Sales Director: Hans Jürgen Scholz
Export Sales Director: Klaus Ulrich Klinger
Marketing Manager: Frank Baustuck
■ Construction machines: wheel loaders, crawler loaders, crawler dozers, compactors. [97]

75-16200 Korth GmbH

(† Korth GmbH & Co KG)
Rober Bosch Strasse 4, D-23909 Ratzeburg
☎ (+49 451) 499 14 97 ⊖ (+49 451) 499 32 30
President: Nikolas Count Bernstorff
Sales Manager: Silke Musik
■ Automatic pistols and revolvers. [97]

75-16295 Klaus-Maffei Automationstechnik GmbH

(a subsidiary of Krauss-Maffei AG, Germany)
Krauss-Maffei Strasse 2, D-80997 Munich
☎ (+49 89) 88 99 22 11 ▣ 523163
⊖ (+49 89) 88 99 29 44
Managing Director: Dipl.-Ing Sigfried Koepp [98]

75-16300 Krauss-Maffei Dienstleistung GmbH

(a subsidiary of Krauss-Maffei AG, Germany)
Krauss-Maffei Strasse 2, D-80997 Munich
☎ (+49 89) 889 90 ▣ 523163 ⊖ (+49 89) 88 99 22 06
Managing Director: Dr Bernhard Nill [98]

75-16310 Krauss-Maffei Kunstofftechnik GmbH

(a subsidiary of Krauss-Maffei AG, Germany)
Krauss-Maffei Strasse 2, D-80997 Munich
☎ (+49 89) 889 90 ☎ (+49 89) 88 99 32 19
▣ 52316341
Chairman of the Board of Directors:
Dipl.-Wirtsch.-Ing Rolf Lidl
Member of the Board of Directors: Dr Arthur Frieges
Member of the Board of Directors: Graziano Parisi
Member of the Board of Directors:
Dipl.-Ing Manfred Petersen
Member of the Board of Directors: Wilhelm Schröder [98]

75-16320 Krauss-Maffei Verfahrenstechnik GmbH

(a subsidiary of Krauss-Maffei AG, Germany)
Krauss-Maffei Strasse 2, D-80997 Munich
☎ (+49 89) 889 90 ▣ 5216504 ⊖ (+49 89) 88 99 32 99
Director: Dipl.-Ing Peter Orlin
Director: Dipl.-Ing Franz Thurner [98]

75-16330 Krauss-Maffei Verkerhrstechnik GmbH

(a subsidiary of Krauss-Maffei AG, Germany)
Krauss-Maffei Strasse 2, D-80997 Munich
☎ (+49 89) 889 90 ▣ 52316391
⊖ (+49 89) 88 99 33 36
Director: Dipl.-Ing Werner Görlitz
Director: Walter Tichy [98]

75-16340 Krauss-Maffei Versicherungsdienst GmbH

(a subsidiary of Krauss-Maffei AG, Germany)
Krauss-Maffei Strasse 2, D-80997 Munich
☎ (+49 89) 889 90 ▣ 523163 ⊖ (+49 89) 88 99 30 87
Director: Klaus Birkholz
Director: Günther Grabow [98]

75-16345 Krauss-Maffei Wehrtechnik GmbH

(a division of Krauss-Maffei AG, Germany)
Krauss-Maffei Strasse 2, D-80997 Munich
☎ (+49 89) 889 90 ▣ 52316331 ⊖ (+49 89) 812 01 90
♧ DM 50,000 (1995) ♣ DM 481,000,000 (1996) 👥 777
👥 821 (1996) 🏠 250,000 sq m
Chairman: Gerhard Haas
Managing Director, Commercial: Rüdiger Rötzel
Managing Director, Technical: Erich Drosen
Sales and Marketing Director: Franz Stangl
Public Relations Manager: Werner F Zickwolff
■ Main contractor for the weapon systems MBT Leopard 1, Leopard 2, AAT Gepard. Add-on-Armour for armoured vehicles. Second generation training systems and simulators.

Subsidiary:
GLS Gesellschaft für Logistischen Service [98]

75-17200 KUKA Wehrtechnik GmbH

(a subsidiary of IWKA Aktiengesellschaft, Germany)
PO Box 431369, D-86073 Augsburg
Zugspitzstrasse 140, D-86165 Augsburg
☎ (+49 821) 79 70 ▣ 5383840 kuk d
⊖ (+49 821) 797 15 52
♣ DM 150,000,000 (1996) 👥 600 🏠 200,000 sq m
Chairman: Hans Eberhard Soeffge
☎ (+49 821) 797 14 25
Executive Vice President: Wolfgang Gott
☎ (+49 821) 797 21 51
Executive Vice President: Claus Detlef Lehmann
☎ (+49 821) 797 19 37
Executive Vice President: Dieter Ost
☎ (+49 821) 797 12 84
Head of Project Management: Dr J Joerg
☎ (+49 821) 797 14 75
■ Development and manufacture of armoured turrets (AFM, AHK, One Man Turret 605; 606A1; 606A2; 607; E23) mounts for weapon calibres ranging from 7,62 mm up to 60 mm, air defence systems, automation technology, logistics, autoloaders for MBT's and howitzers, retrofit and upgrade of wheeled and tracked combat vehicles, environmental protection components, simulators. [98]

75-17350 Wilh. Lambrecht GmbH

Friedlander Weg 65/67, PO Box 2654,
D-37016 Göttingen
☎ (+49 551) 495 80 [E] wilh.lambrecht@t-online.de
▣ 96862 ⊖ (+49 551) 495 83 12
♧ DM 4,000,000 (1996) ♣ DM 13,00,000 (1996)
👥 75 (1996) 🏠 7,278
Managing Director: Klaus D Wyrich
Sales and Marketing Director: Dieter Gailus
■ Meteorological and climatological instruments for weather services, airports and naval applications. [98]

75-17800 Heinsoldt Systemtechnick GmbH

(† Leica Sensortechnik GmbH)
Ernst Leitz Strasse, PO Box 2949, D-35539 Wetzlar
☎ (+49 6441) 29 25 49 ⊖ (+49 6441) 29 24 00
♣ DM 20,000,000 👥 62
Managing Director: Rolf Weyand
Product Manager: Ralf Gissel
■ Tank periscopes day/night/ranging for LAV/IAFV, sighting systems for artillery, panoramic periscopes, weapon measurement and calibration equipment for all calibres and sensors, special optics and technology, training, on-site service. [98]

75-18200 Lewicki Microelectronic GmbH

(a subsidiary of AGIV-Group)
Allee 35, D-89610 Oberdischingen
☎ (+49 7305) 960 20
[E] lewicki.microelectronic@t-online.de
⊖ (+49 7305) 86 01
♧ DM 3,000,000 🏠 2,000 m2
Managing Director: Dr Edmund Rickus
Sales and Marketing Director: Max Sailer
Commercial Manager: Eberhard Werner
■ Thickfilm hybrid IC's custom design and series production of highly miniaturised, high reliability hybrids for space, aerospace, industrial, military and medical applications. [98]

75-18300 Liebherr-Aerospace Lindenberg GmbH (LLI)

(† Liebherr-Aero-Technik GmbH)
(a subsidiary of Liebherr-Holding GmbH, Germany)
Pfaenderstrasse 50/52, PO Box 1363,
D-88153 Lindenberg
☎ (+49 8381) 460 ▣ 541118 lat lg d
⊖ (+49 8381) 463 77
♧ DM 35,000,000 ♣ DM 225,000,000 (1995)
👥 1,000 (1996)
Managing Director: Dipl.-Ing Frieder H Beyer
Managing Director: Dipl.-Volksw Josef Schlick
Sales and Marketing Director: Jochen Muendelein
Exhibitions and Advertising: Ute Braam
■ Development, design, manufacture and support of: operating systems for flight controls; electro-hydraulic primary flight control actuators; redundant fly-by-wire actuators; hydraulic, mechanical and electro-mechanical slat and flap operating systems, associated analogue and digital electronic control systems. Aircraft hydraulic equipment such as pressure and directional control valves, reservoirs and cylinders. Landing gears including brake control systems; landing gear steering systems and associated electronic control systems. Environmental control systems and associated electronic control systems. [98]

Contact Reference

75-18600

LITEF GmbH

(a subsidiary of Litton Industries Inc, USA)
PO Box 774, D-79007 Freiburg im Breisgau
Loerracher Strasse 18, D-79115 Freiburg im Breisgau
☎ (+49 761) 490 10 ▣ 772669 litf d
⊖ (+49 761) 490 14 80
👥 600 🏠 19,060 m2
Managing Director: Richard T Hopman
Deputy Managing Director: Klaus Ehler
Director of Land Navigation Products: Friedrich Schilly
Director of Airborne Navigation Products:
Dr Norbert Sandner
Director of Naval Navigation Products: John M Stewart
■ Development, production and support of: inertial navigation systems, strapdown doppler navigation systems, strapdown heading and attitude reference systems (AHRS), gyros and accelerometers; mobile digital high performance computers, digital communication equipment; development of special electronic units and systems, as well as application of specific integrated circuits (ASIC's) for airborne and space applications, on-board ships and land vehicles.

Bonn Office
Poppelsdorfer Allee 102, D-53115 Bonn
☎ (+49 228) 65 91 22 ⊖ (+49 228) 985 54 22 [98]

75-18910 Lohmann GmbH & Co KG

PO Box 2343, D-56513 Neuwied
☎ (+49 2631) 990 ✉ LOMANEUW ▣ 17263161 loma
⊖ (+49 2631) 99 64 67
♣ DM 380,066,000 (1994) 👥 1,811 (1994)
Managing Director: C Ricker
Technical Director: H W Müller
Export Sales Director: Willi Wortig
■ Metalline aluminized sterile burn dressing kits, individual first aid dressings, compresses and bandages, metalline hoods, gloves, rescue and carrying blankets, NBC warfare wound protection bandages. [97]

75-18950 Lukassen Flughafentechnic GmbH (LFT)

PO Box 1149, D-77724 Oppenau
Poststrasse 22, D-77728 Oppenau
☎ (+49 7804) 17 91 ⊖ (+49 7804) 36 00
Managing Director: P Wiemann
Technical Director: Volker Briegel

Sales and Marketing Director: Florian Stadler
Sales Manager: Rolf Barth
■ Manufacture of mobile hydraulic testers. ground power units, tripod jacks, axle jacks, towbars, catering trucks, engine handling equipment, MLG and NLG handling equipment, maintenance platforms (also movable for helicopters), service vehicles, stairs, tools, assembling and disassembling units for wheelrims and tyres. [97]

75-19500 MaK Systemgesellschaft mbH
(a subsidiary of Rheinmetall Industrie GmbH)
Falckensteiner Strasse 2, D-24159 Kiel
☎ (+49 431) 39 95 02 ⊖ (+49 431) 39 95 29 66
Director: Horst Schön
Advertising Manager: Ms Birgit Baumgart
Sales and Marketing: Ms Susanne Lauda
■ Air start units driven by gas turbines with performance of 190 ppm or 380 ppm at 55 psia, measured at A/C connector. Units are available in skid version, truck or trailer mounted. [98]

75-19610 MAN B&W Diesel AG
(a subsidiary of MAN AG, Germany)
PO Box 100080, D-86135 Augsburg
☎ (+49 821) 32 20 ⊠ 537960 mand
⊖ (+49 821) 322 33 82
♣ DM 1,529,090,000 (1995/96) 👥 5,339
Chairman of the Executive Board: Gerhard Thulmann
■ Large-bore diesel engines for marine and auxiliary propulsion units, stationary and emergency generator sets. Diesel power stations (4-stroke up to 23 400kW, 2-stroke up to 68 520kW), fuel engines, exhaust gas turbochargers, power turbines, complete propulsion systems, monitoring systems, maintenance, repair, modernisation and spare parts supply for diesel engines, foundry products, oil and gas burners. [98]

75-19615 MAN GHH AG
(a subsidiary of MAN Aktiengesellschaft, Germany)
PO Box 1253, D-65453 Ginsheim-Gustavsburg
Ginsheimer Strasse 1, D-65462 Ginsheim-Gustavsburg
☎ (+49 6134) 552 88
E 100656.273@compuserve.com ⊠ 4182058 man d
⊖ (+49 6134) 553 83
♧ DM 200,000,000 (1995/96)
♣ DM 1,792,000,000 (1995/96) 👥 4,807 (1995/96)
Technical Director: Karl Ernst ☎ (+49 6134) 552 92
Senior Sales Manager: Wilfried Mathes
☎ (+49 6134) 552 88
■ Manufacture of the LEGUAN bridge system and Floating Bridge FFB 2000 designed for military use. [98]

75-19620 MAN Gutehoffnungshütte AG
(a subsidiary of MAN AG, Germany)
PO Box 110240, D-46122 Oberhausen
Bahnhofstrasse 66, D-46145 Oberhausen
☎ (+49 208) 692 00 ⊠ 856691 ghh d
⊖ (+49 208) 66 90 21
♣ DM 1,792,003,000 (1995/96) 👥 4,807
Chairman of the Executive Board: Dr Wolfgang Brunn
■ Commissioned manufacture of compressors and turbines. Civil engineering, building construction, solid wall construction and office buildings. Manufacture of machinery and equipment. [98]

75-19640 MAN Nutzfahrzeuge AG
(a subsidiary of MAN AG, Germany)
PO Box 500620, D-80976 Munich
Dachauer Strasse 667, D-80995 Munich
☎ (+49 89) 15 80 22 95 ⊠ 5232110 mad
⊖ (+49 89) 15 80 44 28
♣ DM 6,946,881,000 (1995/96) 👥 26,592
Chairman of the Executive Board: Dr Klaus Schubert
Authorised Officer: Siegfried Skirde
■ Trucks from 6 to 48 tonnes, scheduled bus services and touring coaches, diesel and gas engines for various purposes, automotive and pressed components, systems counselling, transportation consultancy, full service leasing, servicing, spare parts, drive and save training. [98]

75-19645 MAN Roland Druckmaschinen AG Sheet-Fed Printing Press Division
(a subsidiary of MAN AG, Germany)
PO Box 101264, D-63012 Offenbach/Main
☎ (+49 69) 830 50 ⊠ 4152852 mro d
⊖ (+49 69) 83 05 14 40
♣ DM 2,001,985,000 (1994/95) 👥 9,878
Chairman of the Executive Board: Dr Peter Brütt
■ Single & multi-colour sheet-fed offset presses in small, medium & large format available either in the compact five-cylinder system or the flexible unitised design with sheet-turning facilities for paper & board. Sheet-fed rotogravure presses, web-offset presses for newspapers, publications, cartons & book printings in various formats & lengths of print-run. Feeders & deliveries for sheet-fed presses, reel splicers, folder systems, process electronics concept using centralised control console technology & automation concepts for CIM (Computer Integrated Manufacturing) in the graphic arts industry.

Web-Fed Printing Press Division
PO Box 100096, D-86135 Augsburg
☎ (+49 821) 42 40 ⊠ 537960 man d
⊖ (+49 821) 424 33 03 [97]

75-19650 MAN Technologie AG
(a subsidiary of MAN AG, Germany)
PO Box 1347, D-85751 Karlsfeld
☎ (+49 8131) 89 01 ☎ DM 426,296,000 (1995/96)
⊠ 52321121 ma d ⊖ (+49 8131) 89 19 00
👥 1,092
Speaker of the Executive Board: Dipl.-Ing Horst Rauck
■ Development, production, integration and supply of systems for space and lightweight systems. Defence and equipment technology, electronic systems, materials engineering. Plants and sets for the co-generation of power and heat. [98]

75-20005 Mannesmann Rexroth Brueninghaus Hydromatik GmbH
(a subsidiary of Mannesmann Rexroth GmbH, Germany)
Postfach 2260, D-89012 Ulm
☎ (+49 7308) 820 ⊠ 712538 ⊖ (+49 7308) 72 74
♣ DM 789,000,000 (1996) 👥 2,548 (1996)
Managing Director: H Berthold
Sales and Marketing Director: T Kroll
Defence Marketing: E Bretschneider
■ Hydraulic axial piston pumps and motors.

Horb Plant
D-72154 Horb am Neckar
☎ (+49 7451) 920 ⊠ 765321 ⊖ (+49 7451) 82 21 [98]

75-20006 Mannesmann Stahlflaschen GmbH
Karlstrasse 23/25, D-46535 Dinslaken
☎ (+49 2064) 43 32 06 ⊖ (+49 2064) 43 32 95
President: Dr Ernst Lange
President: Klaus D Michels
Advertising Manager: Jens I Reichwald
Purchasing Manager: Bernd Gehla
■ Seamless pressure vessels for storage of compressed air and breathing gas on submarines, frigates and other surface vessels. [98]

75-20300 Mauser-Werke Oberndorf Waffensteme GmbH
(a subsidiary of Rheinmetall Industrie AG, Germany)
Head Office
PO Box 1349, D-78722 Oberndorf/Neckar
Werkstrasse 2, D-78727 Oberndorf/Neckar
☎ (+49 7423) 700 ✉ MAUSERWERKE Oberndorfneckar
⊖ (+49 7423) 706 70
♧ DM 15,000,000 👥 280 (1996)
Chairman: Hugo Wolf
Director: Dietrich C Klöckner
■ Development, design and production of weapon systems for national and civil defence. Services include maintenance, overhaul and repair. Products: medium calibre automatic cannon, missile defence system, air defence system, aircraft armament, tank training system, armoured 120 mm mortar system, ammunition test equipment and sniper rifles, technology studies and development. [98]

75-20370 Medesign GmbH
(† Medesign Dipl.-Ing Lothar Engel)
Partnachplatz 7-9, D-81373 Munich
☎ (+49 89) 769 52 17 ⊖ (+49 89) 769 84 24
Managing Director: Dipl.-Ing Lothar Engel
■ First aid equipment, rescue bags, mobile hospitals. Helicopter and ambulance outfitting, first aid kit Snake/Scorpion with anti-venom, special emergency ECG, burn case for 10 injured persons. Emergency training, inflatable tents, decontamination shower. → 80 [97]

75-20400 Mehler Technische Textilien GmbH
Edelzeller Strasse 44, D-36043 Fulda
☎ (+49 661) 10 37 43 ⊠ 49852 ⊖ (+49 661) 10 36 58
♣ DM 60,000,000 (1995) 👥 380 (1995)
Managing Director: Elmar Schmidt
Sales Director: Elmar Gärtner
Manager, Personnel Protection Business Unit: Siegfried Will
Marketing: Ms Barbara Lill [98]

75-21100 Messo-Chimietechnik GmbH
(a subsidiary of Krauss-Maffei AG, Germany)
Friedrich-Ebert Strasse 134, D-47229 Duisberg
☎ (+49 2065) 410 40 ⊖ (+49 2065) 41 04 99
Managing Director: Dipl.-Ing Wolfgang Wöhlk [98]

75-21650 Moog GmbH
(a subsidiary of Moog Inc, USA)
Hanns Klemm Strasse 28, D-7030 Böblingen
☎ (+49 7031) 62 20 ⊠ 7265777
⊖ (+49 7031) 62 21 00
♣ US$ 41,300,000 (1995) 👥 254 (1995)
Managing Director: Martin Berardi
Sales and Marketing Director: Dieter Boëhm
Manager, Ground Vehicles: Bernd Stehlin
■ Development and manufacture of electrohydraulic, electropneumatic and electromechanic high performance servocontrols for aerospace and defence. [97]

75-21801 MTU Motoren und Turbinen-Union München GmbH
(a subsidiary of Daimler-Benz Aerospace AG, Germany)
PO Box 500640, D-80976 Munich
☎ (+49 89) 148 90 ⊖ (+49 89) 14 89 55 00
♧ DM 156,000,000 (1996)
♣ DM 1,900,000,000 (1996) 👥 4,980 (1996)
Chairman: Dr Manfred Bischoff
President and Board Member: Rainer Hertrich
Board Member, Development and Manufacture: Dr Klaus Steffens
Board Member, Finance and Control: B Liebau
Board Member, Sales and Marketing: Martin Steinberger
Head of Public Relations: Odilo Mühling
■ Development, manufacture and support of jet and turbo-shaft aero engines, including modules and components, aero engines for spacecraft, gas turbines for industrial applications.

Subsidiaries:
MTU Maintenance GmbH
Ludwigsfelde GmbH
Sulzer-MTU Casting Technology GmbH
MTU Turbomeca Rolls-Royce GmbH
Turbo-Union Ltd
EUROJET Turbo GmbH
IAE International Aero Engines AG [98]

75-21802 MTU Motoren-und Turbinen-Union Friedrichshafen GmbH
D-88040 Friedrichshafen
☎ (+49 7541) 900
W http://www.mtu.friedrichshafen.com ⊠ 7342800 mt d
⊖ (+49 7541) 90 50 00
♧ DM 100,000,000 ♣ DM 1,480,000,000 (1996)
👥 4,900 (1996)
President: Dr Rolf A Hanssen
Senior Vice President, Operations and Development: Dr-Ing Michael Wolters
Senior Vice President, Sales and Marketing Electronics: Hermann Amrein
Senior Vice President, Personnel and Commercial: Werner Speckner
General Manager, Military Vehicle Applications: Edo Gerdes
Advertising Manager: Dr Werner Gamser
Market Research: A Ferdynus

■ Development, production, sale and support power systems, diesel engines, gas turbines, power transmissions. Electronic monitoring and control systems for marine, railroad, electric power generators, vehicular and mobile or stationary industrial applications. [98]

75-22135 Munters Euroform GmbH
(a subsidiary of AB Carl Munters, Sweden)
PO Box 1089, D-52011 Aachen
Süsterfeldstrasse 65, D-52072 Aachen
☎ (+49 241) 890 00 ⊜ (+49 241) 890 01 99
♧ DM 2,000,000 (1996) ♣ DM 28,000,000 (1996)
ሐ 120 (1996)
Managing Director: Sigfried Bulang
Operations Director: Udo Bantke
Sales and Marketing Director: Manfred Schillinger
Purchasing Manager: Dieter Heffels
■ Manufacture and distribution of systems for dehumidification, humidification and energy recuperation. [98]

75-22170 Neotronics GmbH
(a member of Neotronics Technology Group)
Bahnhofstrasse 43, D-40764 Langenfeld
☎ (+49 2173) 120 22; 120 24 ⊜ (+49 2173) 120 29 [98]

75-22500 Nico-Pyrotechnik Hanns-Jürgen Diederichs GmbH & Co KG
(a member of the Rheinmetall Group)
Bei der Feuerwerkerei 4, PO Box 1227, D-22946 Trittau
☎ (+49 4154) 80 50 ⊜ (+49 4154) 24 51
ሐ 250
President: Hanns Jürgen Diederichs
Sales and Marketing Director: H G Morotini
Operations Director: Willi Lübbers
Public Relations Manager: Heiko Burmeister
Purchasing Manager: Ulrich Schacht
■ Pyrotechnics and pyrotechnic devices. Sub-calibre training systems for: mortars of all calibres, anti-tank weapons, signal smokes, screening smokes, signal and illuminating cartridges, alarm and simulation devices, 40 mm training and effect ammunition. [98]

75-22700 Nord-Micro Elektronik Feinmechanik AG
PO Box 630225, D-60352 Frankfurt a.Main
Victor-Slotosch Strasse 20, D-60388 Frankfurt a.Main
☎ (+49 6109) 30 30 ✉ NORD-MICRO Frankfurt/M
☒ 4185909 nm d ⊜ (+49 6109) 30 32 33
♧ DM 25,000,000 (1996) ♣ DM 90,000,000 (1996)
ሐ 360 (1996)
Chairman of the Supervisory Board: Raymond P Kurlak
Deputy Chairman of the Supervisory Board: Dr M Thoma
Chairman of the Executive Board: P Carlo Rabino
Deputy Chairman of the Executive Board: Dr Italo Alfano
Sales and Marketing Manager: Rolf Schäfer
■ Design, development, qualification, manufacture, overhaul, repair and product support of air data systems, digital air data computers, air intake control systems, pressure sensors, flight control systems, valves cabin pressure control systems, avionic cooling systems, actuation systems, fuel control systems, secondary flight control systems. [98]

75-22850 Odenwald Werke Rittersbach (OWR)
D-74834 Elztal-Rittersbach
☎ (+49 6293) 731 ⊜ (+49 6293) 732 19
♧ DM 2,000,000 (1996) ♣ DM 25,000,000 (1996)
ሐ 80 (1996) ⌂ 24,000 sq m
Managing Director: F Peter Reis
Operations Director: Mr Geiler
Purchasing Manager: Mr Gaenzler
■ Manufacture of NBC defence systems: detection of chemical warfare agents, personal protection, medical first aid, decontamination, software. [98]

75-22940 O&K Orenstein & Koppel AG Head Office
PO Box 170218, D-44061 Dortmund
☎ (+49 231) 176 00 ☒ 822222
⊜ (+49 231) 17 25 03; 17 90 91
♧ DM 120,000,000 (1996)
♣ DM 1,530,000,000 (1995) ሐ 3,436 (1996)
Managing Director: Dipl.-Ing Manfred Link
Operations Director: Dipl.-Ing Dieter Brammertz
Sales and Marketing Director: Wolfgang P Starke
Director of Advertising, Public Relations and Trade Fairs: G W von Alten
Purchasing Manager: Wilhelm Paulus
■ Rough-terrain forklift trucks, wheel loaders, wheeled and tracked hydraulic excavators, graders, articulated and rigid dump trucks. [98]

75-23330 Parachutes of Germany
Herbst Fallschirm-Service (HFS)
Lempenseite 56, D-69168 Wiesloch
☎ (+49 6222) 518 33 ⊜ (+49 6222) 25 96
President: Siegfried Herbst
■ Parachutes for commando forces. Paragliders. [98]

75-23350 Pechiney Deutschland GmbH
(† Pechiney Aluminium Halnzeug GmbH)
Postfach 104135, D-40032 Dusseldorf
☎ (+49 211) 523 80 ☒ 8582912
⊜ (+49 211) 523 83 00
Managing Director: Jean Louis Goudillère
Sales and Marketing Director: Hans Joachim Steiwheller
Marketing: Mr Helias [98]

75-23600 PK Electronic International
Marschnerstieg 5-7, D-22081 Hamburg
☎ (+49 40) 291 995 ☒ 2174754 pk d
⊜ (+49 40) 299 63 66
ሐ 50
President: P Klüver
Sales Director: Louthean Nelson
■ Design, development and manufacture of professional electronic security systems. [98]

75-23800 Paul Pleiger Maschinenfabrik GmbH & Co KG
PO Box 3263, D-58423 Witten
☎ (+49 2324) 39 80 ☒ 8229964
⊜ (+49 2324) 39 83 24
Operations Director: Mr Hensel
■ Remote operated valves. Hydraulic position indicators, hydraulic drives. Watertight door panels. Electronic controls, control valves. [97]

75-23850 Polytronic-Jung GmbH
(a subsidiary of Polytronic AG, Switzerland)
D-70376 Stuttgart
☎ (+49 711) 54 11 89 ⊜ (+49 711) 56 90 72 [98]

75-24320 Progress-Werk Oberkirch AG (PWO)
Industriestrasse 8, PO Box 1344,
D-77697 Oberkirch Stadelhofen
☎ (+49 7802) 842 75 ✉ PROGRESSWORK Oberkirch
[E] 100517.6550@ compuserve.com ☒ 752631 pwo d
⊜ (+49 7802) 843 43
♧ DM 10,000,000 (1996) ♣ DM 178,000,000 (1996)
ሐ 700 ⌂ 85,000 m2
Board Member: Dipl.Betriebsw Rainer Molenaar
Managing Director: Dipl.-Ing Karl M Schmidhuber
Sales and Marketing Director: B Conrad
Sales and Marketing Director: Karl Schreyer
Operations Director: Karl Leibrecht
Operations and Technical Director: W Moster
Purchasing Manager: Josef Männle
■ Product Division Components/Systems:
Transfer press parts, blank draw and cold form parts, stamp and fire blanking parts, assembly and weld-parts.
Product Division Sheet Structure/Appliances:
Catering Systems, military equipment, aviation, products and sheet structures. [97]

75-24370 PTG GmbH
Präzisionstechnik Gedern
PO Box 1145, D-63684 Gedern
Am Stillhof 6, D-63688 Gedern
☎ (+49 6045) 961 20 ⊜ (+49 6045) 96 12 20
♧ DM 2,000,000 ♣ DM 10,000,000 ሐ 60
President: Horst Kissel
■ Machined structural components for aircraft. Service for planning and programming - NC-Machining. [97]

75-24500 Radiall GmbH
(a subsidiary of Radiall, France)
Carl Zeiss Strasse 10, PO Box 200143,
D-6074 Rodermark 2
☎ (+49 60) 749 10 70 ☒ 4197648
⊜ (+49 60) 74 91 07 70
Managing Director: Claude Fautre [98]

75-24600 Theodor Rapp GmbH Co
Alte Hausacher Strasse 2-4, PO Box 1364,
D-77712 Haslach iK
☎ (+49 7832) 70 30 ⊜ (+49 7832) 703 22
♧ DM 2,500,000 (1997) ሐ 42 (1997) ⌂ 2,800 m2
Managing Director: Heinz Knoblanch
Technical Director: R Hoferer
■ Production and supply of precision parts and assemblies for different weapon systems, optical and mechanical sighting devices, maintenance of mechanical and optical devices and hydraulic systems, NBC equipment, packing and conservation. Licencee of Bofors AB, Sweden. [98]

75-24650 Raytheon Anschütz GmbH
(a subsidiary of Raytheon Marine Co, USA)
PO Box 1166, D-24100 Kiel
☎ (+49 431) 301 90 ⊜ (+49 431) 301 92 91
♧ DM 28,000,000 (1996) ♣ DM 110,000,000 (1996)
ሐ 550 (1996) ⌂ 29,823 m2
Managing Director: Carsten Peters
Finance Director: Gerhard Erb
Sales and Service Director: Wolfgang Petersen
■ Ship bridge equipment: gyro compass equipment; autopilot systems; heading and track controllers; light-spot dead reckoning plotters; data distribution units.
Control and monitoring systems for submarines: steering control consoles; battery monitoring systems; diesel-engine monitoring control systems; navigation and data management centre; sea-water density and temperature measuring systems; depth measuring equipment; rudder position sensors; marine inertial navigation systems (ring laser platform). [97]

75-24700 RENK AG
(a subsidiary of MAN AG, Germany)
PO Box 102307, D-86013 Augsburg
☎ (+49 821) 570 00 ☒ 53781 ⊜ (+49 821) 570 04 60
♣ DM 406,767,000 (1995/96) ሐ 1,708
Speaker of the Executive Board: Prof. Dr Manfred Hirt
Managing Director: Norbert Schulze
■ Automatic vehicle transmissions, test systems, drive elements, industrial and marine gear units, toothed couplings, clutches, disk brakes. [98]

75-24900 Rheinmetall Industrie Aktiengesellschaft
(† Rheinmetall Industrie GmbH)
PO Box 1663, D-40836 Ratingen
Pempelfurtstrasse 1, D-40880 Ratingen
☎ (+49 2102) 900 ⊜ (+49 2102) 90 24 83
♣ DM 1,100,000,000 (1996) ሐ 4,057 (1996)
Chairman: Dr Ernst Otto Krämer
President: Mario Gabrielli
President: Bernhard Hagen
Vice President: Gert Winkler
Vice President: Detlef Moog
■ Development and manufacture of tank and artillery systems, ammunition, support vehicles, armoured infantry fighting vehicles, light armoured tracked and wheeled vehicles, turret systems, weaponry systems, anti-tank mine systems, modular propelling charge systems for artillery, upgrade of tanks and artillery systems, training and simulation systems. [98]

75-24950 Rheinmetall Waffe & Munition GmbH
(a subsidiary of Rheinmetall Aktiengesellschaft)
Pempelfurtstrasse 1, D-40880 Ratingen
☎ (+49 2102) 900 ⊜ (+49 2102) 47 35 53
President: Detlef Moog
■ Development and production of tank and artillery systems, including ammunition, turret systems, anti-tank mine systems, modular propellant systems for artillery, as well as intelligent lethal systems such as sensor-fuse ammunition for artillery. Supplier of ground forces equipment, upgrade of tanks and artillery systems by optimising a wide range of weapon, ammunition and propellant systems.
Systems and Equipment Plant
PO Box 1127, D-29343 Unterlüss
Heinrich Ehrhardt Strasse 2, D-29343 Unterlüss
Vice President: Gert Winkler
Marketing Director: H D Brambusch [98]

Contact Reference

75-25200 ign Robotersysteme AG
(† Messer Griesheim Schweitechnik GmbH)
Steigerwald Strahltechnik, PO Box 1365,
D-82169 Puchheim
☎(+49 89) 809 20 ▣ 521722 telo d
⊖ (+49 89) 809 22 15
Managing Director: Dipl.-Ing Peter Anderl
Operations and Technical Director: Rolf Ertel
Sales and Marketing Director: Ulrich Dobmeier
Purchasing Manager: Paul Jürgen Friedrich
■ Equipment and processes for electron-beam and laser. [97]

75-25265 Rockwell Semiconductor Systems
(a division of Rockwell International Corp, USA)
Paul Gerhardt Allee 50a, D-81245 Munich
☎(+49 89) 829 13 20 ⊖ (+49 89) 834 27 34
Director, Central Europe Area Sales: Georg Wazinski
[E] georg.wazinski@nb.rockwell .com
■ Telecommunications and semiconductor products.
Representing:
Rockwell International Corp. [98]

75-25400 Rohde & Schwarz GmbH & Co KG (R&S)
Mühldorfstrasse 15, D-81671 Munich
☎(+49 89) 412 90 ✉ ROHDESCHWARZ
[W] http://www.rsd.de ▣ 52370320 rus d
⊖ (+49 89) 41 29 21 64; 41 29 35 63
♣ DM 1,000,000,000 (1996) ✱ 4,500
Managing Director, Communications Division:
Ludwig Lohner
Sales and Marketing Director, Communications Division:
Achim Klein
Public Relations Manager: Manfred Grossmann
☎(+49 89) 41 29 22 30
Corporate Communications and Planning: Ms Nicole Julien
■ System integration of HF, VHF and UHF communications. Global communications and air defence systems. ECM resistant transmission. Radio surveillance, analysis equipment and systems: Receivers, direction finders, analysers, antennas VLF to SHF, jammers. Communications test instrumentation. Test and measurement systems. → 80 [98]

75-25540 ROR Rockwell
(a division of Rockwell International Corp, USA)
PO Box 1730, D-58261 Gevelsberg
Hagner Strasse 20, D-58285 Gevelsberg
☎(+49 2332) 40 77 ⊖ (+49 2332) 102 75
Managing Director: Henk van Ettinger
Managing Director: André van Stokkum
■ Trailer axles, mechanical and air suspension units, spring brakes and landing legs for trailers.
Representing:
Rockwell International Corp, USA [98]

75-25630 ROSE Elektrotechnik GmbH & Co KG
PO Box 1362, D-32439 Porta Westfalica
Erbeweg 13-15, D-32457 Porta Westfalica
☎(+49 571) 504 10 ⊖ (+49 571) 504 11 59; 504 12 25
♣ DM 90,000,000 ✱ 260
Export Manager: K Meinhardt
■ Standard enclosures in aluminium, polyester, polycarbonate and ABS to military standards. RFI screened, Ex-e Terminal and electronic enclosures, switchgear cabinets. Control panel and suspension systems. Cable gland and accessories. Machining, irraditing, chromating and special paints to military standards. [98]

75-25800 ROTZLER GmbH & Co
PO Box 1165, D-79577 Steinen
☎(+49 7627) 70 10 [E] marketing@rotzler.com
[W] http://www.rotzler.com ⊖ (+49 7627) 70 11 98
♣ DM 50,000,000 (1996) ✱ 220 (1997) ⌂ 22,000 m2
President: Jürgen Rotzler
President: Ewald Schulz
Marketing Manager: Egbert Heid ☎(+49 7627) 70 11 95
■ Recovery and self recovery winches for off-road trucks. Winches for heavy machinery and tank transporters. Systems for recovery and tow vehicles. Recovery winches for light armoured, wheeled and tracked vehicles. Winches and winch systems for armoured vehicles. [98]

75-25860 Saarstahl AG
(a subsidiary of Saarstahl Holding GmbH)
Bismarckstrasse 57/59, D-66330 Völklingen
☎(+49 6898) 100 ▣ 4424110 svk d
⊖ (+49 6898) 10 40 01
♧ DM 200,000,000 ♣ DM 1,670,000,000 (1995)
✱ 4,630 (1995)
Chairman of Board of Directors: Otfried Forssman
Sales and Marketing Director: Claude Kintz
Personnel and Communication Director: Horst Rival
Production and Engineering Director:
Hans Günter Herfurth
■ Base, quality and special steels. Product forms: bars, wide flats, wire rod, sections. Conditions: as-rolled, forged, peeled, ground, polished, annealed, quenched-and-tempered, age hardened.
Associated Companies:
Vertriebsgesellschaft Saarstahl mbH
Saarstahl-Export GmbH, Germany
Secosar SA, France
Les Aciers Fins de la Sarre, Belgium
Saarsteel Inc, USA
Acciai della Saar SpA, Italy
Saarstahl AG, Switzerland
Saarstahl (UK) Ltd, UK [97]

75-25880 Sächsische Spezialkonfektion GmbH (Spekon)
Nordstrasse 40, D-02782 Seifhennersdorf
☎(+49 3586) 45 60 ⊖ (+49 3586) 45 61 67
♣ US$ 18,000,000 (1996) ✱ 100 (1996) ⌂ 4,200 sq m
Managing Director: Mr Yegin
Marketing Director: Mr Wilhelm
Technical Director: Mr Richter
Purchasing Manager: Mrs Stübner
■ Development and production of parachute products, technical textiles and official clothing. [98]

75-25901 Saft Nife GmbH
(a subsidiary of Saft SA, France)
Colditzstrasse 36-28, D-12099 Berlin
☎(+49 30) 70 02 35 50 ⊖ (+49 30) 70 02 36 76
General Manager: Klaus Kühler
■ Nickel-Cadium, silver-zinc, silver-iron aviation grade accumulators. Lithium, thermal and alkali-manganese batteries for aircraft, industry and military applications. Silver-chloride batteries for rescue lighting. Also produced to customer specifications and QSF-D requirements.
Mainaschaff Office
Behringstrasse 13, D-63814 Mainaschaff
☎(+49 60) 21 70 70 ⊖ (+49 60) 217 07 85; 217 07 90
General Manager: Christophe Gurtner
Nürnberg Office
Nimrodstrasse 9, D-90441 Nürnberg
☎(+49 911) 942 17 40 ⊖ (+49 911) 42 61 44
General Manager: Dieter Petzold
Arnstadt Office
Bierweg 6, D-99130 Arnstadt
☎(+49 3628) 71 15 67 ⊖ (+49 3628) 71 15 60 [98]

75-26050 J.P. Sauer & Sohn GmbH
(a subsidiary of SIG Swiss Industrial Co)
Sauerstrasse 2/6, D-24340 Eckernförde
☎(+49 4351) 47 10 ⊖ (+49 4351) 47 11 60
♧ DM 1,000,000 (1995) ♣ DM 95,000,000 (1995)
✱ 480 (1996) ⌂ 6,500 sq m
Managing Director: Dipl.-Ing Wulf Kraft
☎(+49 4351) 47 11 11
Marketing Director: H Güelzau ☎(+49 4351) 47 11 28
Purchasing Manager: D Nicke ☎(+49 4351) 47 11 17
■ Manufacture of small arms. [98]

75-26200 Schleifring und Apparatebau GmbH
Schleifring Defence Systems
Am Hardtanger 10, D-8080 Fürstenfeldbruck
☎(+49 8141) 40 30 ▣ 527616 ⊖ (+49 8141) 403 45
■ Turret, radar and periscope slip-rings for defence technology (Leopard 1 and 2, Gepard, Marder, LUCHS, Roland, Armoured Vehicle Howitzer, TRMS, Goalkeeper) for peak current circuits up to 2,500 Amp, data buses, video channels and other signal circuits combined with rotary air-channels, angle coders and 18 GHz rotary joints.
Slip-rings manufactured to environmental requirements: shock, vibration, temperature, salt spray, EMP and EMV. [98]

75-26600 Schopf Maschinenbau GmbH
(a subsidiary of Trinity Holdings plc, UK)
Parkstrasse 21, D-73760 Ostfildern
☎(+49 711) 340 00 00 ▣ 7256651
⊖ (+49 711) 340 00 37 (spare parts) 341 10 87
♧ DM 12,000,000 (1995) ♣ DM 38,000,000 (1996)
✱ 70 (1997)
Managing Director: B Geary
Managing Director: Dr H Brüggemann
Finance Director: W Kendzia
Sales and Marketing Director: E Schlichting
Sales Co-ordination: Ms Petra Kornmeier
■ Ground support equipment; conventional and towbarless aircraft towing tractors for for civil and military aircraft; passenger stairs; fresh water and sewage trucks. Cargo loaders and handlers. [98]

75-26700 Schottel-Werft Josef Becker GmbH & Co KG
D-56322 Spay
☎(+49 2628) 610 ✉ SCHOTTEL ⊖ (+49 2628) 613 00
♧ DM 8,600,000 (1996) ♣ DM 81,000,000 (1995)
✱ 350 (1996) ⌂ 6,614 m2
Chairman: Dietrich W Bork ☎(+49 2628) 612 14
Operations Director: Reinhold Reuter
☎(+49 2628) 612 77
Sales Director: Karl Peter Hammer ☎(+49 2628) 612 34
Manager of Defence Projects: Marius Bos
☎(+49 3179) 361 13 91
Public Relations Manager: Kurt Scholz
☎(+49 2628) 612 31
Purchasing Manager: Michael Mludek
☎(+49 2628) 613 99
■ Rudder-propellers from 15 up to 6,000 Kw, transverse thrusters to meet special military design requirements (regarding noise levels, anti-magnetic properties and shockproofing). Certain thruster sizes can be supplied with controllable pitch propellers. Propulsion and steering units for amphibious vehicles, naval vessels, landing craft, military ferries, bridge erection boats. Schottel pump-jets up to 3,500 Kw for main propulsion and for manoeuvring drives, giving thrust in all directions.
Schottel-Hamburg
Pyramidenweg 2, D-25474 Ellerbek
☎(+49 4101) 387 80 ⊖ (+49 4101) 362 14 [97]

75-26750 Schramm Lacke GmbH
(a member of the Grebe Group)
Kettelerstrasse 100, PO Box 101763,
D-63075 Offenbach/Main
☎(+49 69) 860 30 ⊖ (+49 69) 860 32 29
♣ DM 76,000,000 ✱ 200 ⌂ 25,000 m2
Managing Director: Mr Waasen [98]

75-26800 Schuberth Helme GmbH (SH)
Rebenring 31, D-38106 Braunschweig
☎(+49 531) 38 00 50 ▣ 952894 helme d
⊖ (+49 531) 38 00 51 80; 380 05 20
♣ DM 50,000,000 (1996) ✱ 450 ⌂ 26,000 sq m
President: Christian Zahn
Sales and Marketing Director: Franz Josef Görges
Operations Director: Uwe Barth
Export and Governmental Sales Manager:
Michael H Lütsch
■ Lightweight helmets for combat, anti-riot squad, fire brigade, civil defence, industrial safety and motorcyclists. Ballistic protection to class I (German Police Academy). Noise control and communication systems, custom-made portable radios and microphones, respirators, city and country camouflage covers. The Schuberth Combat 826 is equipped with a special shock-absorbing suspension system. [98]

75-27490 Schwäbische Hüttenwerke GmbH
PO Box 3280, D-73414 Aalen-Wasseralfingen
☎(+49 7361) 50 21 ▣ 713832 shwd
⊖ (+49 7361) 50 24 21
♣ DM 346,043,000 (1995/96) ✱ 1,325
Chairman of Management Board: Dipl.-Kfm Klaus Borusiak
■ Tools, process plant equipment, chilled rolls, pumps, castings, forgings, sintered parts, brake discs, bright steel, shafts, linear guide systems, manual implements. [98]

75-27495 SFIM Industries Deutschland GmbH
(a subsidiary of SFIM Industries SA, France)
Gottlieb Daimler Strasse 60, D-71711 Murr
☎ (+49 7144) 811 40 ⊜ (+49 7144) 81 14 22
♧ DM 3,000,000 (1996) ♣ DM 4,900,000 (1996)
👥 18 (1996) 🏠 880 sq m
Managing Director: Dr W Auch
Technical Director: M Oswald
Commercial Director: G Anschuete
■ Development, manufacture, sales and marketing of optic gyroscopes and other inertial measurement units. Product support. [98]

75-27700 Siemens Aktiengesellschaft Group Headquarters
(a division of Siemens AG, Germany)
PO Box 1661, D-85705 Unterschleissheim
Wittelsbacherplatz 2, D-80333 Munich
☎ (+49 89) 23 40 ⊠ 521000 sie d
⊜ (+49 89) 23 43 42 42
⊞ 898 700 siemcw ♧ DM 61,700,000,000 (1994/95)
♣ DM 91,900,000,000 (1994/95) 👥 373,000
Chairman and Chief Executive Officer: Dr Heinrich von Pierer
Corporate Information Department: Dr H Runge
Business and Financial Press Department: Thomas Weber
Strategic Planner: Peter Heinold ☎ (+49 89) 31 79 20 23

Defence Electronics Group
PO Box 1661, D-85705 Unterschleissheim
☎ (+49 89) 31 79 36 47 ⊠ 5288273
⊜ (+49 89) 31 79 38 58
♣ DM 1,300,000,000 (1994/95) 👥 5,600
Managing Director: Helmuth Wiesner
Sales and Marketing Director: Wolfgang Klenner
Marketing Department: Udo H Zinnegger
■ Communications: HF, VHF, UHF radio equipment, radio relay systems, switching systems, network management, encryption equipment, gound stations for satellite communications. Surveillance radars, IFF equipment, air defence control centres, COMINT & ELINT systems. Air traffic management; primary, secondary and airport surveillance radars, navigation aids, voice and data communication systems, arrival and departure co-ordination systems, air traffic control centres and controlling workstations with associated software, planning and project management for small and medium size airports. Integrated systems such as anti-aircraft tanks and tactical air defence systems. air traffic management: radars, displays, communications, microwave landing systems, complete ATC centres and turnkey systems. [97]

75-28000 FRIWO SILBERKRAFT Gesellschaft für Batterietechnik mbh
(† Silberkraft Leichtakkumulatoren GmbH)
(a division of CEAG Industrie-Aktien und Anlagen AG, Germany)
Meidericher Strasse 6-8, PO Box 100552, D-47005 Duisburg
☎ (+49 203) 300 20 ✉ friwosilberkraft@t-online.de
⊠ 855543 ⊜ (+49 203) 30 02 40
♧ DM 10,500,000 (1997) ♣ DM 45,000,000 (1996)
👥 240 (1997) 🏠 10,000 m2
Managing Director: Dr A Leuschner
Operations and Technical Director: J Cornelissen
Sales and Marketing Director: H Ebert
Sales Manager: Mr Schaust
Purchasing Manager: M Corell
■ Manufacture of batteries from traditional nickel-cadium storage, lithium, silver-zinc, special batteries up to batteries systems with a heavy-duty capability. Also production of electrochemical actuators and cap lamps for mining. [98]

75-28010 SIM Security & Electronic System GmbH
Rosslaufstrasse 1, D-67433 Neustadt a.d.W
☎ (+49 6321) 912 70 [E] sim.electronic@t-online.de
⊜ (+49 6321) 170 59
♣ US$ 12,000,000 (1996) 👥 25 (1997)
President: Hans Georg Schilling
Finance Director: J Nikolic
Sales and Marketing Director: J Gregor
■ Mobile Airport Surveillance Systems (MASS), Border Surveillance System, Video Transmission and Scrambling Systems, Night Vision Systems and Gating Systems, Digital Computer Scope (with integrated Fire control System), Digital Spread Spectrum Covert Radio System. [98]

75-28030 SMS AG
PO Box 230229, D-40088 Dusseldorf
☎ (+49 211) 88 10 ⊠ 8589980 smd
⊜ (+49 211) 881 48 97
♧ DM 377,908,000 (1996)
♣ DM 2,152,241,000 (1996) 👥 6,279
Chairman of the Executive Board: Dipl.-Ing Heinrich Weiss
■ Rolling mill and continuous casting technologies, pressing and forging technologies, plastics processing technologies. [98]

75-28050 Sonnenschein Lithium GmbH Head Office
Industriestrasse 22, D-63652 Bahnstation Büdingen
☎ (+49 6042) 810 ⊜ (+49 6042) 815 23
General Manager: Wolfgang Uhl
Product Marketing Manager: Olaf G Lichtlein
■ Lithium, thionyl, chloride batteries, primary and reserve type. [98]

75-29000 Still GmbH Military Division
PO Box 740720, D-22113 Hamburg
Berzeliustrasse 10, D-22097 Hamburg
☎ (+49 40) 733 90 ⊠ 21513160 st d
⊜ (+49 40) 73 39 15 85
♧ DM 120,000,000 (1997)
♣ DM 1,000,000,000 (1997) 👥 5,500 (1997)
🏠 140,000 m2
General Manager: Reinhard Heilemann
Sales and Marketing Director: Mr Frisch
Operations Director: Mr Pfeiffer
Purchasing Manager: Mr Linke
Sales Manager: Karl Heinz Meyer
■ Electric forklift trucks (1,0 to 4,5 tons), Diesel forklift trucks (1,6 to 8,0 tons), electric towing tractors (6 to 20 tons (4,000 N). In use in all NATO and other Armed Forces. [98]

75-29100 Stop-Choc GmbH & Co KG
(a subsidiary of Vibrachoc SA, France)
Benzstrasse 42, Postfach 42, D-7253 Renningen
☎ (+49 7159) 921 90 ⊜ (+49 7159) 92 19 19
♣ DM 10,000,000 👥 50
Managing Director: Lothar Schaefer
Commercial Projects: Jens Trimper
■ Design and manufacture of all metal anti-vibration mountings. [98]

75-29450 Sulzer Hydro Wyss GmbH
(† Sulzer Escher Wyss GmbH)
(a division of Sulzer AG, Switzerland)
PO Box 1380, D-88183 Ravensburg
☎ (+49 751) 83 00 ⊠ 732901 ⊜ (+49 751) 83 32 74
♣ DM 140,000,000 (1996) 👥 500 (1996)
Sales Manager: Mr Petrich
■ Controllable pitch propellers for naval vessels and Sulzer-Kawasaki waterjets for Europe are. [98]

75-29550 TDW GmbH
(a subsidiary of TDA, France)
Postfach 1340, D-86523 Schrobenhausen
☎ (+49 82) 529 90 ⊜ (+49 82) 52 99 65 27
♣ DM 88,000,000 (1996) 👥 293 (1996)
Managing Director: Klaus Bussé
General Manager: Werner Kaltenegger
Financial Manager: Jean Marie Dhainaut
Marketing Manager: Manfred Horst
■ Warhead systems, including safe and arm devices and fusing sensors; submunitions and ejection systems; autonomous anti-tank weapons; EOD-systems. [98]

75-29600 Technics Plasma GmbH Plasma und Ionenstrahl-Systeme
(a subsidiary of Krauss-Maffei AG, Germany)
Dieselstrasse 22a, D-85551 Kirchheim
☎ (+49 89) 90 50 30 ⊠ 529334
⊜ (+49 89) 90 50 31 00
Managing Director: Dr Berthold Kegel
Managing Director: Wolfgang Paul [98]

75-30100 Teldix GmbH
(a division of Litton Industries Inc, USA)
Grenzhöfer Weg 36, PO Box 105608, D-69046 Heidelberg 1
☎ (+49 6221) 51 20 ⊜ (+49 6221) 51 23 05
♧ DM 10,000,000 (1996) ♣ DM 113,000,000 (1995)
👥 499 🏠 20,000 m2
Managing Director: Richard Hopman
Vice President and Marketing Director: Rainer Götting
Research and Development Director: Dr Oliver Stucky
Production Director: Georg Treinies
Public Relations Manager: Armin Göckel
Purchasing Manager: Jörg Fuchs
■ Computer Products:
Avionics Computer and Display Technology: cockpit management systems, mission management systems, missile control systems, defensive aids systems, interface management systems, display technology for helmets, head-up displays, multifunctions displays, GPS receivers. Repair and overhaul of navigation systems for aircraft and land vehicles, gyro systems. Reference projects for F-104 Starfighter; for Tornado (head-up display HUD, fuel flow indicator, missile control unit, launcher decoder unit, weapon interface unit); for EF 2000 (head-up display control panel, computer symbol generator, cockpit interface unit, wing pylon station unit, attack/navigation computer, interface processor unit, defensive aids computer, integrated helmet system, navigation system). Remote terminal unit, doppler velocity sensor for head-up dispay units.
Space Products:
Specialist in high-precision mechanisms for satellites (ball bearing momentum and reaction whells for advanced satellites, solar array drive assemblies (SADA), antenna pointing mechanisms, systems for optical satellite communication (SILEX). [98]

75-30600 Adolf Thies GmbH & Co KG
PO Box 3536, D-37025 Göttingen
Hauptstrasse 76, D-37083 Göttingen
☎ (+49 551) 79 00 10 [E] info@thiesclima.com
[W] http://www.thiesclima.com ⊜ (+49 551) 790 01 65
Managing Director: Wolfgang Behrens
Sales and Marketing Director: Thomas Stadie
Purchasing Manager: Hans Adams
Export Department: Mrs Krambeer
Sales Engineer: Bernd Nasner
■ Meteorological sensors to measure wind direction and velocity, humidity and temperature, precipitation, radiation and atmospheric pressure. Automatic weather stations and systems for acquisition, processing and display. [98]

75-30950 THORN Licht GmbH
Mohnestrasse 55, PO Box 2580, D-5760 Arnsberg 1
☎ (+49 2932) 20 50 ⊜ (+49 2932) 63 31 00
Chief Executive Officer: Klaus Cremer [98]

75-31000 Thyssen Henschel
(a subsidiary of Thyssen Industrie AG, Germany)
Henschelplatz 1, D-34112 Kassel
☎ (+49 561) 801 01 ⊜ (+49 561) 801 69 07
Sales and Marketing Director: Klaus Sonneck
Public Relations: Peter Wiegelmann
■ Development and manufacture of armoured wheeled and tracked vehicles: Marder 1A3 Infantry Combat Vehicle; TH 495 Armoured Infantry Fighting Vehicle (AIFV); TH 495 Infantry Combat Vehicle (ICV); TPz1 Fox Armoured Vehicle NBC; TPz1 Fox Armoured Vehicle and variants; Red Fox; Artillerie M113 for military and police purposes, spare parts and modernisation of existing systems. Logistics.

Defence Department
Henschelplatz 1, D-34112 Kassel
☎ (+49 561) 801 01 ⊠ 99750 ⊜ (+49 561) 801 67 33 [97]

75-31320 Thyssen Rheinstahl Technik GmbH
(a subsidiary of Thyssen Handelsunion AG)
Thyssen Trade Centre, PO Box 105051, D-40041 Düsseldorf
☎ (+49 211) 96 70 ⊜ (+49 211) 967 55 72
♣ DM 1,100,000,000 👥 350
Managing Director: Jürgen Koopmann
Marketing Director: Christoph Hoenings
■ General contractor for defence systems, shipbuilding and marine construction, shipyards and naval dockyards, offset counter-trade arrangements, financial engineering, consultancy and technical assistance and logistics. [97]

Contact Reference

75-31450 Traffipax-Vertrieb GmbH
(a subsidiary of Robot Foto & Electronic GmbH)
PO Box 130726, D-40557 Dusseldorf
Hildener Strasse 57, D-40597 Dusseldorf
☎ (+49 211) 97 14 50 ⊜ (+49 211) 971 45 69
👥 150 (1996)
General Manager: Bernhard Dohmann
General Manager: Heinrich Reith
Technical Director: Erhardt Milatz
Export Manager: Wolfgang Freudenhammer
Purchasing Manager: Ms Sylvia Tanneberger
■ Mobile and stationary radar and photographic equipment for traffic surveillance and speed control. [98]

75-31500 Trimble Navigation Deutschland
Inner Ring 1, D-6454 Bruchkobel
☎ (+49 6181) 710 53 ⊜ (+49 6181) 778 18
Sales and Marketing Manager: T Wunderlich [97]

75-31700 TST Tele Security Timmann Main Plant and Marketing Department
Hauptstrasse 82, D-82327 Tutzing
☎ (+49 8158) 20 51 ⊜ (+49 8158) 20 53
♁ DM 5,000,000 (1997) ♣ DM 60,000,000 (1997)
👥 60 (1997) 🏭 1,000 sq m
President: Dr-hc Klaus P Timmann
Operations Director: M Graf
Sales and Marketing Director: F Klotz
Purchasing Manager: G Engelmann
■ Manufacture of communications security systems. [98]

75-31900 Turbo Lufttechnik GmbH
Gleiwitzstrasse 7, PO Box 1964, D-66469 Zweibrücken
☎ (+49 6332) 80 80 ⊜ (+49 6332) 80 82 67
▥ TWX 6332902 turbo ♁ DM 10,000,000 (1994/95)
♣ DM 200,000,000 (1994/95) 👥 320
President: Dr Johannes Ohlinger
Vice President: Dr Holzdeppe
Marketing Manager: Dipl.-Kfm Kurt Omlor
■ Turnkey climatic wind tunnels, turnkey aerodynamic wind tunnels (low speed, subsonic, transonic, supersonic and hypersonic). Test stands: rocket and motor stands, climatic and other environmental simulation chambers, axial and radial flow fans and sound protection equipment for ventilation systems for tunnels, shelters, underground storage areas, underground aircraft fields, hush houses. [97]

75-32010 Turco-Chemie GmbH
(an ELF/Atochem company)
Wandsbeker Stieg 23, D-22087 Hamburg 76
☎ (+49 40) 251 60 80 ⊠ TURCO HAMBURG
▥ 213820 turco d ⊜ (+49 40) 25 16 08 50
♁ DM 600,000 ♣ DM 8,000,000 👥 23
Managing Director: Karl Ernst
Marketing Manager: Hans Bernhard Moll
■ Chemical surface treatment materials, crack-testing materials according to military specifications for aerospace applications. [97]

75-32050 Ube Industries Ltd
(a subsidiary of UBE Industries Ltd, Japan)
Immermannstrasse 65b, D-40210 Dusseldorf
☎ (+49 211) 178 83 42 ⊜ (+49 211) 361 32 97
Managing Director: Isao Sato
General Manager: Morito Yasumura [98]

75-32090 Continental AG Werk Uniroyal
(† Uniroyal Englebert Reifen GmbH)
Hüttenstrasse 7, PO Box 410, D-52005 Aachen
☎ (+49 241) 519 35 98 ⊜ (+49 241) 519 34 77
Production Manager: E H Potente
Sales Manager: H Gatzweiler
■ Fuel cells, flexible, self-sealing and crashworthy for aircraft and armoured vehicles. Maintenance and servicing. → 80 [97]

75-32135 Vaisala GmbH
(a subsidiary of Vaisala Oy, Finland)
PO Box 540267, D-22502 Hamburg
☎ (+49 40) 85 80 27 ⊜ (+49 40) 850 84 44
Managing Director: Jens Mogensen

Stuttgart Branch
Zweigniederlassung, Pestalozzistrasse 8, D-70563 Stuttgart
☎ (+49 711) 73 40 57 ⊜ (+49 711) 735 63 40 [98]

75-32150
Vallon GmbH
Im Grund 3, D-72800 Eningen
☎ (+49 7121) 985 50
[E] 101726.1642@compuserve.com
[W] http://www.dsk.de/mgm/vallon ▥ 729924
⊜ (+49 7121) 836 43
♁ DM 500,000 (1997) 👥 40 🏭 2,015 m2
President: Dipl.-Ing Gerhard Vallon
■ Metal detectors for security applications, handheld and walkthrough metal detectors, mine detectors and bomb locators. MCI: computer aided magnetic anomaly detection system for the location and identification of buried ferromagnetic objects or magnetic fields in the ground. [98]

75-32600 VDO-Luftfahrtgeräte Werk GmbH
An der Sandelmühle 13, D-60439 Frankfurt am Main
☎ (+49 69) 580 50 ▥ 411519 vdol d
⊜ (+49 69) 580 53 99
♁ DM 10,000,000 ♣ DM 84,500,000 👥 412
🏭 20,000 m2
President: Dieter Spohn
Vice President, Marketing: Bernd Witterstein
Vice President, Sales: Dieter Krug
Director of Research and Development: Manfred Vollmer
Joint Managing Director: Uwe-Dietmar Gross
Vice President, Production: Wolfgang Morgenstern
Public Relations Manager: Hannelore Becker
■ Instruments and systems for avionics and aviation equipment. Test equipment. Instrumentation for army vehicles and navy. Engine and flight control instruments. Head-up and head-down displays, computer symbol generators.
Owned by Bodenseewerk Gerätetechnik GmbH, Germany (50,5%); Sextant Avionique SA, France (49,5%). [98]

75-32800 Manfred Vetter GmbH & Co
PO Box 1355, D-53905 Zülpich
☎ (+49 2252) 30 08 60 ⊜ (+49 2252) 30 08 70
President: Manfred Vetter
Product Manager: Martin Klug
■ Mini lifting bags for rescue, repair, maintenance. Lifting bags for uprighting and lifting purposes. Pipe sealing bags for internal sealing off of pipes. Leak sealing bags and bandages for sealing of leaks on mobile and stationary tanks and pipes. [98]

75-32830 Vickers Systems
(a part of Vickers Aerospace Marine Defence Group)
AM Joseph 16, PO Box 1241, D-61273 Wehrheim
☎ (+49 6081) 19 32 20 ⊜ (+49 6081) 10 32 29
👥 54
Marketing Manager: K Kahl
■ Design and production of fluid power and fuel control systems and components including hydraulic, electrohydraulic and fuel pump equipment; pumps, motors, electric motor pumps, integrated packages, test stands and ground cart support equipment. [98]

75-32870 Voi-Shan/Diessel GmbH
(a division of Fairchild Fasteners, USA)
Steven 3, PO Box 101320, D-3200 Hildesheim, Bavenstadt
☎ (+49 5121) 762 40 ⊜ (+49 5121) 511 500
♁ US$ 1,750,000 (1996) ♣ US$ 30,000,000 (1996)
👥 210 (1996) 🏭 3,000 (1996)
President: Dr Ernest A Marhenke
■ Specialty fasteners, blind bolts, pin/collar systems, threaded inserts, latches, electronic fastening systems, self-retaining bolts, captive screws, alignment pins, fluid couplings, drawer slides, gas springs, struts, airframe and engine bolts and nuts. [97]

75-32950 Volkswagen AG
Berliner Ring 2, D-38436 Wolfsburg
☎ (+49 5361) 90 ⊠ WOLKSWAGEN ▥ 95860 vww d
⊜ (+49 5361) 92 82 82
♁ DM 8,742,000 (1996) ♣ DM 100,123,000 (1996)
👥 24,200
Chairman of the Management Board: Dr-Dipl.Ing Ferdinand Piëch
Finance and Control Director: Bruno Adelt
Sales and Marketing Director: Dr Robert Büchelhofer
Human Resources Director: Dr Peter Hartz
Director, Asia Pacific: Dr Martin Posth
Director of Group Strategy, Treasury, Legal Matters and Organisation: Dr Jens Neumann
Director of Production Optimization and Procurement: Dr-Ing José Ignacio López de Arriortúa
Chairman of the Supervisory Board: Dr-jur Klaus Liesen
Senior Executive: Dr-jur Manfred Pilgrim
Deputy Chairman: Klaus Zwickel
International Press Manager: Matt Gennrich
■ Manufacture and marketing of vehicles and engines. [98]

75-32970 Vötsch Industrietechnik GmbH
(a subsidiary of Weiss Umwelttechnik GmbH, Germany)
Beethovenstrasse 34, D-72336 Balingen-Frommern
☎ (+49 7433) 30 30 ⊜ (+49 7433) 30 31 12

Lindenstruth Office
Greizer Strasse 41/49, D-35447 Reiskirchen-Lindenstruth
☎ (+49 6408) 84 73 ⊜ (+49 6408) 844 44 [98]

75-33030 VSG Verfahrenstechnik für Schiffsbetrieb GmbH
(a subsidiary of ROM-Schiffbau Dockbautechnik, Germany)
Lademanbogen 12, D-22339 Hamburg
☎ (+49 40) 53 90 29 88 ⊜ (+49 40) 53 90 25 67
♣ DM 3,500,000
Managing and Technical Director: Siegfried Kaltwasser (MSc)
■ Project management, manufacture and servicing of all kinds of marine refrigeration and air conditioning plants. → 80
® USA: Carrier-Transicold [98]

75-33400 Wandel & Goltermann Kommunickationstechnik GmbH
Arbachtalstrasse 22, D-72800 Eningen unter Achalm
☎ (+49 7121) 98 60 ⊜ (+49 7121) 98 61 00
♣ DM 20,000,000 (1996) 👥 52 (1997)
Managing Director: Fariborz Khavand
Sales and Marketing Director: Michael Scharnowsky
■ Intercom systems for helicopters, warning systems for police vehicles and motorcycles, passenger information systems for buses and trains. [98]

75-33500 Warner Electric GmbH
(a subsidiary of Dana Corporation, USA)
PO Box 2008, D-72610 Nürtingen
Nuertinger Strasse 70, D-72649 Wolfschlugen
☎ (+49 7022) 50 40 ⊠ WARNERBRAKE
[W] http://www.warnernet.com ⊜ (+49 7022) 541 68
♁ DM 7,100,000 🏭 20,000 m2
Managing Director: L A Brockmeier
Sales and Marketing Director: W Schwab
Technical Director: H R Schneider
Purchasing Manager: D Lederer
Executive Secretary: R Dachs
■ Production and sale of positioning ballscrews and transport ballscrews (tolerance classes P & T 1-10), controls, step motors, linear actuators, tension control systems, high-frequency converters, frequency converters, self-supporting linear motion systems, AC/DC drives. Sales office for DANA motion control products. → 90 [98]

75-33620 Wavetek GmbH
(a division of Wavetek Corp, USA)
Gutenbergstrasse 2-4, D-85737 Ismaning
☎ (+49 89) 99 64 10 [W] http://www.wavetek.com
⊜ (+49 89) 99 64 14 40
♣ US$ 70,000,000 👥 480 🏭 6,500 m2
Managing Director: Dr Klaus Romanek
Sales and Marketing Director: Will Foster
Operations Director: Winfred Lenne
Purchasing Manager: Rene Bahr
Public Relations Manager: Doms Schwarz
■ Test equipment, radio testing stations, measuring and monitoring receivers. [98]

75-33700 **Wegmann & Co GmbH**
August-Bode Strasse 1, D-34127 Kassel
☎ (+49 561) 10 50 ✉ WEGMANN Kassel ⊠ 99859
⊖ (+49 561) 105 22 08
♣ DM 400,000,000 (1996) 👥 1,000 (1996)
Managing Partner: Dipl.-Kfm Dr Manfred Bode
Managing Partner: Wolfgang Bode
Managing Partner: Dipl.-Ing Dr Werner Zimni
Marketing Director: Dr Guenter Hoenig
■ PzH 2000 155 mm Howitzer. Turrets for the Leopard 1 and 2 MBT and the Swedish MBT 122. 4x4 light reconnaissance vehicle. Light air defence system. Turrets for the Gepard SPAAG. MLRS/MARS and 110 SF rocket launcher system. Training systems and simulators. Smoke, HE and decoy grenade launchers. Gun mounts for military and security purposes. Modification and combat effectiveness improvements. Logistics and maintenance. Equipment and components. → 80 [98]

75-33800 **Weiss Umwelttechnik GmbH Simulationsoin Lagen Messtechnik**
Greizer Strasse 41/49, D-35447 Reiskirchen Lindenstruth
☎ (+49 6408) 840 E info@wut.com
⊖ (+49 6408) 843 41
♧ DM 22,000,000 (1996) ♣ DM 240,000,000 (1996)
👥 1,200 (1996)
Managing Director: Karl Heinz Reichent
Managing Director: Karl Heinz Hehl
Marketing Director: Ms E Biedenkapp
■ Manufacturer in the field of environmental simulation and measurement technology.

Berlin Technical Sales Office
Flottenstrasse 58, D-13407 Berlin-Reinickendorf
☎ (+49 30) 40 99 05 10 ⊖ (+49 30) 40 99 05 30

Hamburg Technical Sales Office
Brauhausstieg 21, D-22041 Hamburg-Wandsbek
☎ (+49 40) 68 59 96 ⊖ (+49 40) 68 52 91

Hannover Technical Sales Office
Walsroder Strasse 149, D-30853 Langenhagen
☎ (+49 511) 728 19 10 ⊖ (+49 511) 728 19 30

Oberhausen Technical Sales Office
Mülheimer Strasse 48, D-46045 Oberhausen
☎ (+49 208) 87 00 71 ⊖ (+49 208) 87 87 24

Stuttgart Technical Sales Office
Zettachring 10, D-70567 Stuttgart-Möhringen
☎ (+49 711) 90 02 45 10 ⊖ (+49 711) 90 02 45 30

Munich Technical Sales Office
Wandlhamer Strasse 31, D-82166 Gräfelfing
☎ (+49 89) 89 80 45 10 ⊖ (+49 89) 80 80 45 30

Netzschkau Technical Sales Office
Reinsdorfer Weg 4, D-08491 Netzschkau
☎ (+49 3765) 49 43 76 ⊖ (+49 3765) 49 45 90 [98]

75-34700 **Zeiss-Eltro Optronic GmbH**
(a joint venture between Carl Zeiss and Daimler-Benz Aerospace AG, Germany)
Carl Zeiss Strasse 22, Postfach 1129,
D-73442 Oberkochen
☎ (+49 7364) 20 01 ⊖ (+49 7364) 20 45 88
♣ DM 120,000,000 (1995) 👥 240
President and Chief Executive Officer: Jürgen Mohilo
☎ (+49 7364) 20 48 53
Managing Director and Executive Vice President: Klaus Leinmüller ☎ (+49 7364) 20 33 62
Sales and Marketing Director: Claus Schütz
☎ (+49 7364) 20 33 18
Chief Financial Officer: Andreas Hack
☎ (+49 7364) 20 48 70
Marketing Manager: Edmund Reitter
Guided Missile Systems Director: Bernd Presser
■ Periscopic sights for armoured vehicles, night vision devices, reconnaissance cameras, search and attack periscopes for submarines, fire control systems for battle tanks, airborne navigation FLIR system, stabilization systems. Thermal imaging equipment for main battle tanks, armoured personnel carriers, scout vehicles and upgrading programs. Gyro-stabilized commander's periscopes with integrated thermal imagers. IR missile guidance systems. Optronic sensors for air defence systems, featuring day and TV channels, thermal cameras and laser rangefinders. Thermal imaging systems for border surveillance. Maintenance and repair facilities. [98]

75-34780 **Zeppelin Luftschifftechnik GmbH**
(† Luftschifftechnik GmbH)
Leutholdstrasse 108, D-88045 Friedrichshafen
☎ (+49 7541) 20 25 15 ⊖ (+49 7541) 20 25 16
Managing Director: Max Mugler
Sales and Marketing Director: Hartwig Ochel
Project Manager: Klaus Hagenlocher
■ Design and manufacture of semi-rigid helium airships. [98]

75-34800 **Zeppelin Mobile Systeme GmbH Headquarters**
(a subsidiary of Zeppelin GmbH, Germany)
Albert Einstein Strasse 2, D-77656 Offenburg
☎ (+49 381) 610 50 ⊖ (+49 381) 61 05 50
♣ DM 49,000,000 (1995) 👥 100 ⌂ 8,000 m2
Managing Director: Eberhard Laabs
Sales and Marketing Director: Rolf C Kahle
Advertising Manager: Ms Michaela Kremm
■ Mobile medical systems, shelters with base integration as sub-systems for mobile communications. Transportation vehicles for off-road use.

Production Plant
PO Box 1946, D-88009 Friedrichshafen [97]

75-34930 **Zettler GmbH Kommunikations und Sicherungstechnik**
Holzstrasse 28/30, D-80469 Munich
☎ (+49 89) 23 72 70 ⊠ 17898656 zett d
⊖ (+49 89) 23 72 76 00
♧ DM 20,000,000 ♣ DM 300,000,000 👥 1,500
Managing Director: Thomas Benz
Managing Director: Dieter Klein
Managing Director: Reinhard Sitzmann
■ Intruder and hold-up alarm systems. Automatic emergency telephone call systems. Fire detection equipment. Watchman's control systems, CCTV surveillance equipment. Computer controlled alarm co-ordination.

Profil PR Agentur für Kommunikation König & Ehresmann
Franz Joseph Strasse 48, D-80801 Munich
☎ (+49 89) 272 33 08 ⊖ (+49 89) 272 35 26
Contact: Ms Sandra Sapic [97]

75-35000 **ZF Friedrichshafen AG**
D-88038 Friedrichshafen
☎ (+49 7541) 770; 77 40 41 ⊠ 734207
⊖ (+49 7541) 77 80 00; 77 40 52
♧ DM 245,000,000 (1995)
♣ DM 7,766,000,000 (1996) 👥 32,889 (1996)
Managing Director: W Vogel
Director of Marketing and Head of Public Relations: D Pfister
Operations and Technical Director: E W Behele
Senior Sales and Service Manager: G W Jockers
Purchasing Manager: M Gmeinder
Corporate Manager: Veronika Strobel
■ Manufacture of gearboxes, steering gears for military trucks, armoured cars and tanks; gears and components for aircraft. [98]

75-36000 **ZF Luftfahrttechnik GmbH**
(a subsidiary of ZF Friedrichshafen AG, Germany)
PO Box 101667, D-34016 Kassel
☎ (+49 5674) 70 10 ⊖ (+49 5674) 70 16 06
♣ DM 81,100,000 (1996) 👥 438 (1996)
Chief Executive Officer: Dipl.-Ing Peter Richter
Group Executive Officer: Dr-Ing. Andreas Wolf
Assistant to the Chief Executive Officer: Dipl.-Ing Frank Mergler
■ Development and manufacture of transmissions, gears, gear boxes precision components for helicopter and fixed-wing aircraft. Maintenance, repair and overhaul of helicopter dynamic components. Development, manufacture and maintenance of appropriate test stand facilities. [98]

80 Service Companies & Organizations

80-00050 **AlliedSignal Aerospace Marketing Sales and Service**
(a subsidiary of Allied Signal Aerospace, USA)
c/o LH-WF 02 - Building 337, D-60546 Frankfurt
☎ (+49 69) 690 60 47 ⊖ (+49 69) 690 60 48
Field Service Engineer: Michael Reinheimer
→ 90

Daimler Benz Aerospace Airbus
Halle 14 Sud - Room 881, D-21129 Hamburg
☎ (+49 40) 74 37 70 41 ⊖ (+49 40) 74 37 70 44
Field Service Technician: Ben Rodewald

Cologne Plant
c/o Lufthansa CityLine GmbH,
Flugenhafen Koln/Bonn Hangar II, D-51147 Cologne
☎ (+49 22) 03 59 62 66
⊖ (+49 22) 03 59 61 66; 03 59 61 71
Field Service Engineer: Dave Moniz

Global Repair and Overhaul
Dreieichstrasse 4, D-64546 Moerfelden
☎ (+48 61) 05 20 02 11 ⊖ (+49 61) 05 20 02 10
Site Leader: Norbert Barwinske

Olching Plant
Feldstrasse 22, D-82104 Olching [98]

Contact Reference

80-00060 **AlliedSignal Aerospace Service Corporation**
(a subsidiary of AlliedSignal Aerospace, USA)
Postfach 1150, D-65475 Raunheim
Frankfurter Strasse 41-65
☎ (+49 61) 42 94 51 12 ⊖ (+49 61) 42 94 51 07
Business Manager APUs: Marc Duvall
Manager, Marketing and Support: Dirk Meir
Marketing Manager: Hubert Mayr
Marketing Manager: Fred Reuter
General Aviation Business Manager: Robert Richardson
MSP Sales/Administration: Ms Inge Ellis [98]

80-00075 **AlliedSignal Deutschland GmbH Air Transport Avionics**
(a subsidiary of AlliedSignal Aerospace, USA)
Hahnstrasse 40, D-D-60528 Frankfurt
☎ (+49 69) 669 01 50 ⊠ 412-441
⊖ (+49 69) 666 85 43
Facility Manager: John Laws ☎ (+49 69) 66 90 15 18
Supervisor, Customer Service: Ms Marlene Tritsch
☎ (+49 69) 66 90 15 26
■ Maintenance of avionics equipment. [98]

80-00200 **Amextra GmbH**
Apple Computer IMC
(a subsidiary of Amextra Ltd, Switzerland)
Hafenstrasse 1, D-60327 Frankfurt a.Main
☎ (+49 69) 255 82 32 E largeaccount@amextra.de
⊖ (+49 69) 255 82 72
👥 35 (1997)
Managing Director: Hans Peter Isler
Sales and Marketing Director: Robin Wukits
Operations Director: Ms Barbara Krähenmann
■ Provides solutions involving Apple computers to all NATO forces, supporting agencies and contractors, including hardware, software, installation, training equipment service, consulting and support. [98]

80-00210 **Architektur Städtebau Infrastruktur (ASI)**
Abt. Flughafenplanung
Hubertusstrasse 10, D-41564 Kaarst 2-Büttgen
☎ (+49 2131) 51 44 07 ⊖ (+49 2131) 51 90 85
👥 2
Chief: Dr-Ing. Lothar Prang
■ Research, planning, consultation, airport planning team. Civil and military airport infrastructure. Traffic, structural and operational analysis. Functional systems of terminals. Environmental airport problems, specifically aircraft and airport noise. [98]

80-00313 Cambridge Consultants Ltd
Gustav Stresemann Ring 1, D-65189 Wiesbaden
☎(+49 611) 97 71 10 ⊜ (+49 611) 97 71 17
Representing:
Cambridge Consultants Ltd [98]

80-00350 Carl Cranz Gesellschaft eV
PO Box 1112, D-82234 Wessling
☎(+49 8153) 28 24 13 ⊜ (+49 8153) 28 13 45
Chairman: Dr H H Homuth
Manager: H G Apelt
■ Intensive short courses in: command control, reconnaissance, aerospace technologies, system dynamics, control information technologies, sensorics, remote sensing, weaponry and ballistics. [97]

80-00537 Elekluft GmbH
Elektronik und Luftfahrtgeräte GmbH
Justus Von Liebig Strasse 18, PO Box 1509, D-53121 Bonn 1
☎(+49 228) 668 10 ▣ 8869621
⊜ (+49 228) 668 17 77
✤ DM 10,000,000 ♣ DM 183,600,000 (1996) ⚉ 820
Chairman of the Board: Dr J L Geisel
Managing Director: K Tiemann
Finance Director: M Storz
Public Relations Director: H P Merker
■ System support for complex ground based electronic systems for air defence, ATC and aerospace. Planning and installation of mobile and static communication systems, including EMP system protection. EMC consulting, tests, engineering support and certification. Technical documentation, customer training, material management. Computer based training. Part task simulation. Facility management, site management. [98]

80-00550 Elintech GmbH
Am Schnepfenweg 41, D-8000 Munich 50
☎(+49 89) 150 20 81 ▣ 5215018
⊜ (+49 89) 150 34 88; 150 20 85
✤ DM 50,000 ♣ DM 1,000,000 ⚉ 4 ⌂ 1,000 m2
Managing Director, Operations: Dr Nils E Hansen
Managing Director, Technical: J F Madsen
■ Consulting and technical services. Program management, proposal preparation, system and project engineering. [97]

80-00560 Engineering & Professional Services Inc
Hans Thoma Strasse 81-83, D-6800 Mannheim 1
☎(+49 621) 41 20 87 ⊜ (+49 621) 41 14 66
Director of Field Services, Europe: J Kessel
■ Field services, involving systems engineering and analysis, operations, maintenance and logistic services, including antennas, airborne radar, secure combat net radio, fibre optics, intrusion alarm and telephone systems. Installation and maintenance of computer facilities. Provides third party maintenance and field training. [97]

80-00750 GLS Gesellschaft für Logistischen Service mbH
(a subsidiary of Krauss Maffei AG, Germany)
PO Box 500231, D-80972 Munich
Krauss-Maffei Strasse 2, D-80997 Munich
☎(+49 89) 889 90 ▣ 5215583 gls d
⊜ (+49 89) 812 40 92
♣ DM 89,000,000 (1996) ⚉ 53 (1996)
Managing Director: Günter Grabow
Sales and Marketing Director: Helmut Lawerenz
Operations Director: Gerald Beer
Public Relations Manager: Mrs Ursula Schmidt
■ Initial spare parts supply for MBT Leopard 1 and 2 and AAT Gepard. Retrofit and updating of tracked armoured vehicles with diesel engines, modern armament systems, night sights, fire control systems, stabilization systems, improvement of suspension systems. Sales of armoured and unarmoured vehicles for police command and intervention vehicles. [98]

80-00800 IAS Industrial & Airborne Systems Handelsgesellschaft mbH
PO Box 500747, D-22707 Hamburg
Königstrasse 30, D-22767 Hamburg
☎(+49 40) 389 29 83 ▣ 2161177 ias d
⊜ (+49 40) 389 29 87
⚉ 5 (1997) ⌂ 600 sq m
Managing Director: Uwe K Lehrke
Public Relations Manager: Ms Marianne Latzel
Secretary: M Latzel
■ Logistic support to high technology products and operations, including engineering, development, consultancy and representative services. [98]

80-00945 Industrieanlagen-Betriebsgesellschaft mbH (IABG)
Einsteinstrasse 20, D-85521 Ottobrunn
☎(+49 89) 608 80 [E] iabg@iabg.de
[W] http://www.iabg.de ▣ 524001
⊜ (+49 89) 60 88 24 17
✤ DM 10,000,000 (1995) ♣ DM 300,000,000 (1995)
⚉ 1,400
Senior Vice President: Wolfgang Raasch
Senior Vice President: Karl Ahrensdorf
Senior Vice President: Dr Lothar Becker
Senior Vice President: Dr Dietrich Burkhardt
Senior Vice President: Dr Peter Ebeling
Senior Vice President: Hans Joachim Gudzent
Senior Vice President: Stephen Keller
Senior Vice President: Uwe Nerlich
Senior Vice President: Peter Rinnelt
Senior Vice President: Adolf Neubecker
Managing Director: Dr Rudolf Schwarz
Managing Director: Dr William E Sweeney
Managing Director: Wolf D Schuck
Purchasing Manager: Gerhard Gruber
Public Relations Manager: Wolfgang Mohr
■ Scientific and technical services including systems engineering, operations research studies, logistics; aircraft and spacecraft technology, qualification and acceptance testing, fatigue and static strength tests, environmental tests, realistic field tests, defence weapons systems simulation. [97]

80-00975 Introtec GmbH
(† Intronic GmbH)
Otto Hahn Strasse 7, D-64546 Mörfelden-Walldorf
☎(+49 69) 38 40 82 ⊜ (+49 6105) 714 64
✤ DM 50,000 (1997) ♣ DM 1,000,000 (1996)
⚉ 4 (1997)
Managing Director: Dudley J A Thomas
Purchasing Manager: Ms Monika Dietz [98]

80-00985 Jet Aviation Cologne
(a subsidiary of Jet Aviation Management AG, Switzerland)
Cologne/Bonn Airport, Hangar 3, D-51147 Cologne
☎(+49 2203) 544 61 ⊜ (+49 2203) 535 72
Station Manager: Alwin Terporten
■ Maintenance of jet and turboprop engines; avionics retrofit. [98]

80-00990 Jet Aviation Dusseldorf
(a subsidiary of Jet Aviation Management AG, Switzerland)
Dusseldorf Airport, Hanger 2, D-40474 Dusseldorf
☎(+49 211) 45 49 70 [E] jdus@jetaviation.com
▣ 8584671 ⊜ (+49 211) 454 34 23
Vice President and General Manager: Klaus Dieter Hessenmüller
Manager, Private Aircraft Handling: Theo Rütten
■ Maintenance of jet and turboprop engines; avionic retrofit; engineering and design; aircraft sales. [98]

80-00995 Jet Aviation Hannover
(a subsidiary of Jet Aviation Management AG, Switzerland)
Hannover Airport, Hanger 11, D-30669 Hannover
☎(+49 511) 73 37 60 ⊜ (+49 511) 72 28 68
Station Manager: Karsten Schäfer
■ Maintenance of jet and turboprop engines; avionic retrofit. [98]

80-01000 Jet Aviation Kassel
(a subsidiary of Jet Aviation Management AG, Switzerland)
Kassel-Calden Airport, D-34379 Calden
☎(+49 5674) 999 50 ⊜ (+49 5674) 49 61
Station Manager: Wolfgang Friz
■ Maintenance of jet and turboprop engines; avionics retrofit; propellor maintenance, overhaul and repair facility. [98]

80-01010 Jet Aviation Munich
(a subsidiary of Jet Aviation Management AG, Switzerland)
General Aviation Centre - Hangar 10, Franz Josef Strauss Airport, D-85356 Munich
☎(+49 89) 970 18 01 ⊜ (+49 89) 970 18 02
Station Manager: Hans Jürgen Olzinger
■ Maintenance of jet and turboprop engines. [98]

80-01020 Jet Aviation Saarbrucken
(a subsidiary of Jet Aviation Management AG, Switzerland)
Saarbrucken Airport, D-66131 Ensheim
☎(+49 6893) 948 80 ⊜ (+49 6893) 51 26
Station Manager: Heiko Menke
■ Maintenance of turboprop engines. [98]

80-01255 Logica GmbH
(a subsidiary of Logica plc, UK)
Wilhelm Leuschner Strasse 255, D-64347 Griesheim
☎(+49 6966) 802 70 ⊜ (+49 6966) 679 86
Managing Director: Michael Mutterer
Finance Director: Ms Heidi Kuhn [98]

80-01280 MacNeal-Schwendler European Headquarters
(a subsidiary of MacNeal-Schwendler Corp, USA)
Innsbrucker Ring 15, PO Box 801240, D-81612 Munich
☎(+49 89) 431 98 70 [W] http://www.macsch.com
⊜ (+49 89) 436 17 16
♣ DM 60,000,000 (1996) ⚉ 160 (1996)
Managing Director: Christopher Saint John
Aerospace Marketing Director: Antoine Reymond
Operations Director: Dr Hans Sippel
Purchasing Manager: Franz Wengenmayer
Manager of European Marketing Communications: Richard Bush
Automotive Marketing: Rudi Ott
■ Engineering software and support services. [98]

80-01300 Masin-Aircraft GmbH
Grüngürtelstrasse 12, D-50996 Cologne
☎(+49 221) 39 14 31; 39 12 03 ⊜ (+49 221) 39 58 93
Managing Director: Dieter K Salomon
■ Transport aircraft, worldwide ferry. Sales of military and civil transport aircraft, related parts and engines. Consultants. → 90 [98]

80-01310 Materialinformationszentrum Gesellschaft für Logistik mbH (MIZ)
PO Box 1965, D-26359 Wilhelmshaven
☎(+49 4421) 39 10 ▣ 253405 ⊜ (+49 4421) 39 13 44
♣ DM 19,400,000 ⚉ 150 ⌂ 5,500 sq m
Managing Director: J Leising ☎(+49 4421) 39 12 20
■ Integrated logistic support services. Organisation and system advice (logistics management) maintenance, overhaul and repair, material supply and handling, training, training documentation. Electronic data processing, configuration management. [97]

80-01370 MTG Marinetechnik GmbH
Wandsbeker Königstrasse 62, PO Box 701249, D-22041 Hamburg
☎(+49 40) 65 80 30 ⊜ (+49 40) 65 80 33 92
⚉ 100 (1997)
Marketing Manager: Rainer P Kröger
■ Planning and design office for surface warships and general naval technology. [98]

80-01374 MTU Motoren und Turbinen-Union Ludwigsfelde GmbH
Brandenburgische Strasse, D-14974 Ludwigsfelde
☎(+49 3378) 824 00 ▣ 361880 mtu d
⊜ (+49 3378) 82 43 00
♣ DM 59,658,000 ⚉ 298
General Manager: Rainer Geisler
Sales Manager: Hans Georg Kwauka
Operations MTU-Programs: Manfred Krebs
Finance and Accounting Manager: Heinz Hoffknecht [98]

80-01375 MTU Maintenance GmbH
(a subsidiary of MTU München GmbH, Germany)
Hannover Airport, Münchner Strasse 31,
D-30855 Langenhagen
☎ (+49 511) 780 60; 780 61 44 ✉ 9230309 mtuh d
⊖ (+49 511) 780 61 11; 780 61 00
♧ DM 30,000,000 ♣ DM 351,000,000 (1996)
👥 720 (1996)
Managing and Operations Director: Paul Egon Grall
Sales and Marketing Director: Franz Weinzierl
Finance and Administration Director: Volker Beckmann
Purchasing Manager: Jörg Socko
Public Relations Department: Gudrun M Sobottka
■ Overhaul and repair of commercial turbofan engines and their industrial derivatives. Overhaul programme for General Electric CF6-50, CF6-80, LM 2500 and LM 5000. Overhaul shop certified by JAA, LBA, FAA, DCA (Malaysia), DOA (Thailand), CAA (Egypt), CAA (China), CAA (Pakistan), CAA (India), DGAC (Oman) and AAC (Macau). [98]

80-01380 Mühlhan + Co International
Schlinckstrasse 3, D-21107 Hamburg
☎ (+49 40) 75 27 10 ✉ 217679 myint d
⊖ (+49 40) 75 27 11 38
👥 1,700 (1997)
Managing Director: Dr W D Greverath
Operations Director: W Welke
Operations Director: K O Boltze
Sales and Marketing Director: D Braack
■ Industrial Services:
Corrosion protection (naval and industry); scafolding.
Production:
Special machinery, lashing material, steel parts and ship sections.
Trade:
Protection of corrosion equipment, work platforms. Import and export. [98]

80-01450 Neubrandenburger Fahrzeugwerke GmbH
(† System Instandsetzungs und Verwertungsgesellschaft)
(a subsidiary of DIEHL GmbH & Co, Germany)
Lindenstrasse 39, D-2000 Neubrandenburg
☎ (+49 395) 358 40 ⊖ (+49 395) 368 43 04; 368 43 50
👥 400 (1997) 🏭 40,000 (1997)
Managing Director: Thomas Bockhold
☎ (+49 395) 358 41 00
Operations Director: Klaus Lilienthal
Purchasing Manager: Harald Wachholz
■ Maintenance, repair and overhaul service for armoured vehicles. Reconditioning, refit and conversion services. [98]

80-01500 Nucletron Electronic Vertriebs GmbH
(a subsidiary of Nucletron Electronic AG)
Gaertnerstrasse 60, D-8000 Munich 50
☎ (+49 89) 14 90 02 20
⊖ (+49 89) 14 90 02 11; 14 90 02 50
♧ DM 750,000 (1995) ♣ DM 11,000,000 (1995) 👥 16
General Manager: Peter Peach
Sales Manager: Robert Tittl
■ Technical consultants and distributors specialised in microwave, EMI and electro optical products. → 90
Ⓡ UK: EEV Ltd
USA: ASE, Andrea Radio, Marlow, UVC [98]

80-01700 Philotech GmbH
Seemüllerstrasse 8, D-81549 Munich
☎ (+49 89) 68 30 78 Ⓔ philotech@t-online.de
Ⓔ philotech@compuserve.com ⊖ (+49 89) 68 30 79
👥 70
Managing Director: Hartwig Dirscherl
Sales and Marketing Director: Stewart Allen
Operations Director: I Balszun
■ Software systems and data base development, S/W reliability, prediction, cost estimation and maintainability including quality, safety and hazard analysis. H/W and S/W analysis to ISO, DIN and Mil-Std's test software (C-ATLAS) production. Training, seminars and production of R, M and T, Logistic Support Analysis, ILS, CE and CALS. Provision of consultancy services to aerospace and defence industries including procurement of support personnel. Independent evaluation, studies and analysis. [98]

80-01800 RTE-Relli Engineering und Vertriebs GmbH
(† RTE Relli Technology Ersatec GmbH)
Karl Arnold Strasse 7, D-47877 Willich
☎ (+49 2154) 42 88 21 ⊖ (+49 2154) 42 72 77
♧ DM 300,000 (1996) ♣ DM 1,000,000 (1996)
👥 2 (1996) 🏭 50 sq m
Managing Director: Stefan Viell
■ Logistical service, avionics, airframe, electronics, optics, hardware, nylon fuel tanks, personnel heaters, vehicles spares. [97]

80-01955 Siemens Nixdorf Informationssysteme AG
Otto Hahn Ring 6, D-81739 Munich
☎ (+49 89) 63 64 43 37; 636 01
⊖ (+49 89) 63 64 21 73; 63 65 20 00
♣ DM 13,000,000,000 (1994/95) 👥 37,200
Senior Executive: Gerhard Schulmeyer
Sales and Marketing Manager: S Pathe
■ Supplier of information technology solutions. [97]

80-01975 Strat*X GmbH
Mooshamerstrasse 7, D-82544 Egling
☎ (+49 8176) 70 63 ⊖ (+49 8176) 13 43
Managing Partner: Joseph Leah
Representing:
Strat*X SA, France [98]

85 Stockists, Distributors, Spare Parts Suppliers

ENPAR Sonderwerstoffe GmbH	75-08100
Robert Zapp Werkstofftechnik GmbH	90-02600
Unimex Handels GmbH	90-02150
Willburger System GmbH	90-02370

85-00054 Adelco Elektronik GmbH
Boxholmstrasse 5, D-25451 Quickborn
☎ (+49 4106) 20 24 ✉ ADELC Ⓔ adelco@t-online.de 101704.210@compuserve.com ⊖ (+49 4106) 38 52
♧ DM 400,000 (1996) ♣ DM 5,000,000 (1996) 👥 12
🏭 200 sq m
Managing Director: Joerg Hesse
Managing Director: Fred Hesse
Sales and Marketing Director: Reinhard Ohland
■ Distributor of commercial and military electronic components.
Ⓡ Germany: Telefunken
USA: General Electric Capacitors, Harris, Mallory, Motorola, NACC, Raytheon, Teledyne Relays & Telcom [98]

85-01500 Kuhn & Siefert
PO Box 1733, D-6350 Bingen
☎ (+49 6721) 120 16 ⊖ (+49 6721) 120 18
👥 30 🏭 3,000 sq m
Managing Director: W Siefert
■ Signal guns and ammunition. Nitrocellulose varnish, black varnish melanoid, green/red transparent varnish, case mouth varnish for ammunition. Ultra-violet curing lacquers. [98]

85-02000 Lhotellier Technische Verpackungen GmbH
(a subsidiary of Lhotellier Montrichard SA, France)
Stäffeleswiesen 8, D-89522 Heidenheim
☎ (+49 7321) 95 34 30 ⊖ (+49 7321) 95 34 11
Managing Director: Mr Döll
Director: Erich Cordes
■ Metal and composite containers for missile and ordnance support.
Ⓡ France: Lhotellier Montrichard
Israel: Aeroequipment Ltd
Italy: Connex Italiana
Norway: Nordby
Spain: Tecnologias Reunidas para la Defensa TRD
Sweden: HEMEC
UK: Lhotellier Montrichard (UK) Ltd [98]

85-02500 SEXTANT Avionique GmbH Büro Deutschland
(a subsidiary of SEXTANT Avionique SA, France)
Gochesberger Allee 90, D-53175 Bonn
☎ (+49 228) 819 94 11 ⊖ (+49 228) 819 94 99
Director of Business Development: Hagen Holzhauser
Ⓡ France: SEXTANT Avionique SA [98]

85-02600 TFC - Neiderlassung Bochum
Arnoldschacht 12, D-44894 Bochum
☎ (+49 234) 26 12 07 ⊖ (+49 234) 26 12 09 [98]

85-02700 TRACE Germany
Regensburger Strasse 1, D-84104 Rudelzhausen
☎ (+49 8752) 699 ⊖ (+49 8752) 73 68
Managing Director: Martin Hacker [98]

85-02800 UV-Electronic Consulting GmbH
Am Hang 1a, D-60437 Frankfurt am Main
☎ (+49 69) 95 04 90 77/87 ⊖ (+49 69) 95 04 90 89
♧ DM 50 000 (1994) ♣ DM 3 000 000 (1995) 👥 5
🏭 400 sq m
Managing Director: Elke Vespermann
Operations Director: Klaus Vespermann
Sales and Marketing Director: Ralf Vespermann
■ Distributor of military qualified components such as: opto-electronics, hybrids, military relays and relay sockets, junction modules, resistors, capacitors, feed through filters, feed relays.
Airborne instruments. [97]

90 Sales Agents, Sales & Liaison Offices

Companies represented:

A/S Hydrema, Denmark	90-00890
AB Electronics, UK	75-00020
Acal plc, UK	90-02000
Aeroequipment Ltd, Israel	85-02000
Aerolyte Systems, USA	90-00343
Aerospace Knob, USA	90-01180
Air Technology Inc, USA	90-00064
AlliedSignal Aerospace Co, USA	90-00080
Aluminium Co of America, USA	90-00068
American Pneumatic Tool, USA	90-00820
AMP Inc, USA	75-00750
Ampex Corp, USA	90-00105
Amphenol Corp, USA	90-00110
Ancra Corp, USA	90-00117
Andrea Radio, USA	80-01500
Artemis Management Systems, UK	75-00230
ASE, USA	80-01500
ATI Industries, USA	90-00820
Behlman, USA	90-02000
Bei Motion Control Inc, USA	90-02342
Best Power Inc, USA	90-02342
Beta Transformer Technology Corp, USA	90-00925
BKC, USA	90-00140
Bofors AB, Sweden	90-00260
Bournemouth Industrial Supplies, UK	90-00745
Brent Europe Ltd, UK	90-00064
Bristol Uniforms, UK	90-00064
British Steel Engineering Steels, UK	90-00275
Broadcast Electonics, USA	90-00220
Caddock Electronics Inc, USA	90-02342
CAE Inc, Canada	75-04500
Canadian Marconi, Canada	90-00064
Carrier-Transicold, USA	75-33030
CATIC, China PR	90-00300
Centrax, UK	90-00064
Cherry Aerospace Fasteners, USA	90-00330
Circle Seal Controls Inc, USA	75-10900
Clemco Industries, USA	90-00343
Components Research Corp, USA	90-00140
Computer Optical Products, USA	90-02000
Connex Italiana, Italy	85-02000
Crescent International Corp, USA	90-00220
Crompton Instruments Ltd, UK	90-00375
Custom Electronics Inc, USA	90-02342
Data Display Products, USA	90-01180
DCS Data Control Systems, USA	90-02000
DEL Electronics, USA	90-02000
Digitran, USA	90-01180
Dodge Regupol, USA	75-02470
Dowty, UK	90-00064
Duncan Electronics, France	90-02370
Eaton Pressure Sensors, USA	90-02000
Eaton Valve & Actuators, USA	90-02000

Sales agents/representatives listed elsewhere:

90-00030 ACAL GmbH - Transtech
(† Transtech Hochfrequenz GmbH)
(a subsidiary of ACAL plc, UK)
Fischraecker 2, PO Box 1251, D-74221 Flein/Heilbronn
☎ (+49 7131) 58 10 ⊖ (+49 7131) 58 12 90
♧ DM 1,200,000 (1997) ♣ DM 30,000,000 (1996)
👥 43 (1997) 🏠 800 m2
Managing Director: Dr Dieter Donninger
Sales and Marketing Manager: Ralf Spundflasche
Purchasing Manager: Heinrich Back
Advertising Manager: Ms Helen Cole
■ Microwave and RF components, instruments and subsystems for radar, communication, navigation and ECM applications. Equipment for broadband cable communication.
Representing:
United Kingdom: Faraday Technology, Filtronic Comtek, GEC-Marconi, Thorn Microwave Devices
Canada: GHz Technologies Inc
USA: MBF Microwave Inc, MPD Republic, Res-Net Microwave, Samsung Microwave, Times Microwave Systems, Trompeter Electronics, Vari-L, Weinschel Associates [98]

90-00047 Acer Computer GmbH
Kornkamp 4, D-22923 Ahrensburg, Hamburg
☎ (+49 4102) 48 80 ⊖ (+49 4102) 48 81 01 [97]

90-00064 Aerotechnik Anhalt GmbH & Co KG
Lenaustrasse 1, D-40470 Düsseldorf
☎ (+49 211) 62 69 82 ⊖ (+49 211) 63 75 27
♧ DM 250,000 (1995) 👥 5 (1995) 🏠 110 sq m
Managing Partner: Hans Jürgen Anhalt
Sales and Commercial Manager: Mrs Helga Konrath
® Canada: Canadian Marconi
UK: Brent Europe Ltd, Bristol Uniforms, Centrax, Dowty, Fairey-Microfiltrex Ltd, Flight Refuelling, GEC Avionics
USA: Air Technology Inc, Loral Fairchild Corp, Rexnord Corp [97]

90-00065 Ahlmann Baumaschinen GmbH
(a subsidiary of Manitou BF, France)
Postfach 964, D-24758 Rendsburg
☎ (+49 31) 35 13 02 ⊖ (+49 31) 35 13 84
® France: Manitou BF [98]

90-00068 ALCOA Deutschland GmbH
Lydwerstrasse 11a, D-6000 Frankfurt am Main 71
☎ (+49 69) 664 10 20 ⊖ (+49 69) 666 60 59
Managing Director: H A Steinkusch
® USA: Aluminium Co of America [97]

90-00080 AlliedSignal Aerospace Service Corp (ASASCO)
Frankfurter Strasse 41-65, D-65475 Raunheim
☎ (+49 6142) 94 50; 94 51 38 ⊠ 4064334
⊖ (+49 6142) 94 51 07
Managing Director: Menhorn Gerd
Marketing Manager: Fred Reuter ☎ (+49 6142) 94 51 47; 94 51 06 (fax)
Marketing Department: Ms Margot Schwarz
■ Provision of marketing, sales and service support for AlliedSignal Aerospace US Business Units: Engines; Aerospace Equipment Systems; Commercial Avionics Systems; Government Electronic Systems; Technical Services, Kansas City Division.
® USA: AlliedSignal Aerospace Co [97]

90-00085 Allied International Corp
(a subsidiary of Allied International Corp, USA)
Adalbert Stifter Strasse 8, D-86929 Penzing
☎ (+49 8191) 98 98 33
E alliedgermany@compuserve.com
⊖ (+49 8191) 98 98 34
Manager: Klaus Bruder [98]

90-00105 Ampex Europa GmbH
Otto Volger Strasse 7c, D-65843 Sulzbach (Taunus)
☎ (+49 61) 967 65 00 ⊖ (+49 61) 967 65 29
♧ DM 1,500,000 (1995) ♣ DM 25,000,000 (1995)
👥 20 (1995) 🏠 1,300 m2
Managing Director: Willy Björklund
Finance Manager: Hermann Volk
Personnel Manager: Mrs Gisela Rosin
® USA: Ampex Corp [98]

90-00110 Amphenol Tuchel Electronics GmbH
(a subsidiary of Amphenol Corporation, USA)
August Häusser Strasse 10, D-74080 Heilbronn/Neckar
☎ (+49 7131) 92 90 ⊖ (+49 7131) 92 94 86
♧ DM 20,100,000 ♣ DM 140,000,000 👥 500
Managing Director: Udo Naujoks
Technical Director: Rainer Schröder
Sales and Marketing Director: Dr Ulrich Meyer
Finance Director: Dieter Beck
Human Resources Director: Horst Kleimacheck
Communications Manager: Ralf Stegmann
Purchasing Manager: Auton Sommer
® USA: Amphenol Corp [97]

90-00117 Ancra Jungfalk GmbH
(† Eric Jungfalk & Co GmbH)
(a subsidiary of Ancra International Corporation, USA)
Richard Stocker Strasse 19, PO Box 1309, D-78230 Engen 1
☎ (+49 7733) 500 20 ⊠ 793876 jungf d
⊖ (+49 7733) 50 02 10; 50 02 47
👥 50
Managing Director: Karl Koller
Export Sales Manager: Otto Horstschutz
Export Sales Manager: Christian Kuhn
■ Air cargo systems, ball mats, strap assemblies, seat tracks, seat and galley fittings. Aircraft interior hardware.
® USA: Ancra Corp [97]

90-00140 Asternetics GmbH
Wettersteinstrasse 2, D-8028 Taufkirchen
☎ (+49 89) 612 90 00 ⊖ (+49 89) 612 81 98
♧ DM 200,000 ♣ DM 2,500,000 (1995) 👥 3 🏠 330 m2
President: Günther Olbrich
■ Military qualified electronic components: semiconductors, capacitors, resistors, hybrid oscillators.

® USA: BKC, Components Research Corp, Knox, Micropac Industries, NES, PPC Semiconductors, Semicon, TRW Resistive Prods 97

90-00220 Baxter Elektro-und Motorentechnik
(† Neuzeitliche Elektro-und Motorentechnik GmbH)
PO Box 1547, D-56015 Koblenz
☎ (+49 261) 310 56 ⊖ (+49 261) 310 57
♧ DM 1,700,000 ♣ DM 1,000,000 (1994) ⅲ 3
Managing Director: Caleb G Baxter
® UK: Granger Associates, Technical Material Corp
USA: Broadcast Electonics, Crescent International Corp, Sunair Electronics 97

90-00235 F.A. Bernhardt GmbH
Postfach 1628, D-083636 Bad Tölz
☎ (+49 8041) 768 90 ⊖ (+49 8041) 76 89 32
Chief Executive Officer: S Sokolov
® UK: EEV Ltd 98

90-00260 Bofors Deutschland GmbH
(a subsidiary of Bofors AB, Sweden)
Plittersdorfer Strasse 95, D-53173 Bonn
☎ (+49 228) 36 75 60 ⊖ (+49 228) 367 56 20
♧ DM 100,000 (1996) ♣ DM 2,500,000 (1996)
ⅲ 2 (1996) ☗ 100 m2
Managing Director: Peter Akesson
® Sweden: Bofors AB 98

90-00270 British Aerospace Germany GmbH
Viktoriastrasse, D-53173 Bonn
☎ (+49 228) 93 58 10 ⊖ (+49 228) 93 58 22 97

90-00275 British Steel Deutschland GmbH
(a subsidiary of British Steel Engineering Steels, UK)
Gartenstrasse 2, D-40479 Dusseldorf
☎ (+49 211) 492 60 ⊖ (+49 211) 492 62 82
® UK: British Steel Engineering Steels 98

90-00300 CATIC Trading Development GmbH
Uhlandstrasse 36a, D-22087 Hamburg
☎ (+49 40) 227 22 00 ⊖ (+49 40) 220 47 89
General Manager: C Huang
® China PR: CATIC 97

90-00330 Cherry Division of Textron Inc
Tulpenstrasse 8, D-41466 Neuss
☎ (+49 2131) 46 09 55 ⊖ (+49 2131) 46 09 96
Regional Sales Manager: Hans Werner Schmitz
® USA: Cherry Aerospace Fasteners 98

90-00343 Clemco International GmbH
Arnikastrasse 6, D-8011 Siegertsbrunn
☎ (+49 8102) 30 58 ⊠ 528009 ⊖ (+49 8102) 56 73
Information Contact: W Woitynek
® USA: Aerolyte Systems, Clemco Industries, Zero Products 98

90-00350 Concurrent Computer GmbH
(† Perkin-Elmer Data Systems)
(a subsidiary of Concurrent Computer Corporation, USA)
Lena Christ Strasse 46, Martinsried, D-82152 Planegg
☎ (+49 89) 85 60 30 ⊖ (+49 89) 85 60 31 50
♧ DM 622,000 (1994) ♣ DM 16,000,000 (1994) ⅲ 21
Managing Director: Kevin Dell
Marketing Manager: Albert Soelter
■ International development and manufacture of event-driven, high-performance computer systems for a variety of real-time applications. 98

90-00375 Crompton Messinstrumente GmbH
Harkorstrasse 35, D-40880 Ratingen
☎ (+49 210) 24 10 63 ⊖ (+49 210) 24 60 01
♣ £ 1,000,000 ⅲ 10
General Manager: H Kievit
® UK: Crompton Instruments Ltd 98

90-00390 Daewoo Handels GmbH
Postfach 710254, D-60492 Frankfurt am Main
☎ (+49 69) 66 40 40 ⊠ 416869 daewoo d
⊖ (+49 69) 66 40 41 12 98

90-00550 Elomex Electronic GmbH
Tiefenbachstrasse 14, D-83098 Brannenburg
☎ (+49 8034) 689 ⊖ (+49 8034) 80 15
☗ 7,000 sq m
Managing Director: Ms Cita Lenhart
Senior Vice President: Herbert Mittermayer
■ Stock material for: Tornado, F-104, F-4, RF-4, Alpha Jet, C-160, Brequet Atlantic 1150, HFB-320, Computer, Gyro, Instruments, Radar APQ-99/-120/-172, Testbenches, Optical Sight, Brakes, Actuators and Parts. 98

90-00557 Encore Computer GmbH
(† Gould Electronics GmbH)
Itterpark 8-10, D-40724 Hilden
☎ (+49 2103) 201 26 ⊖ (+49 2103) 240 428
Managing Director: A Wolfgang Müller
Marketing Manager: Jürgen Poppensieker
Controller: Gilbert Pruss
■ High performance computer systems for real-time applications and Unix based information processing.
Munich Office
Tegernseer Landstrasse 161, D-81538 Munich
☎ (+49 89) 69 77 80 ⊖ (+49 89) 697 78 11
Contact: Dr Dieter Lorenz 97

90-00569 Erichsen Prüftechnik Wuppertal GmbH
Simonshöfchen 31, D-42327 Wuppertal
☎ (+49 202) 73 90 80 ⊖ (+49 202) 73 90 81
® UK: Lloyd Instruments Ltd 98

90-00605 Evans & Sutherland GmbH
(a subsidiary of Evans & Sutherland Corporation, USA)
Martin Kollar Strasse 15, D-81929 Munich
☎ (+49 89) 420 99 00 Ⓦ http://www.es.com
⊖ (+49 89) 42 64 88
ⅲ 12 ☗ 7,000 sq m
Managing Director: Joseph Muellner
Office Manager: Ms Christine Anson
■ Simulators and training equipment. 97

90-00740 GE Aircraft Engines
(† General Electric Technical Services Co Inc)
(a subsidiary of General Electric Co, USA)
Büropark Oktavian, Münchner Strasse 14, D-85744 Munich - Unterföhring
☎ (+49 89) 99 27 85 21 ⊖ (+49 89) 957 70 20
Central European Sales and Marketing Development: Herbert Pokorny
Manager Product Support Europe: W J Boehm
Manager Field Service Europe: J L Stone 97

90-00745 Gesellschaft für Hochdrucktechnik mbh (GHT)
PO Box 1251, D-22887 Tangstedt
Heidestrasse 34d, D-22889 Tangstedt
☎ (+49 4109) 67 89 ⊖ (+49 4109) 67 90
Managing Director: H J Lüders
® UK: Bournemouth Industrial Supplies, Stansted Fluid Power 98

90-00795 Greene, Tweed & Co GmbH
(a subsidiary of Greene, Tweed & Co, USA)
Kirchstrasse 9, PO Box 1211, D-6240 Königstein
☎ (+49 6174) 30 73 74 ⊠ 410872 gtc d
Sales Office Manager: Fernando Georgi
Sales Manager: Gerd Dlugosch 97

90-00797 Gusstechnik Peter GmbH
Stresemannstrasse 10a, Postfach 20, D-61178 Karben 6
☎ (+49 603) 99 10 00 ⊖ (+49 603) 991 00 30
Owner: Elmar W Peter
® UK: PI Castings 98

90-00800 Hadland Photonics GmbH
(a subsidiary of Hadland Photonics Ltd, UK)
Luxemburger Strasse 258, PO Box 1260, D-50329 Hürth-Efferen
☎ (+49 2233) 740 21 ⊖ (+49 2233) 744 55
ⅲ 3 (100 worldwide)
General Manager: P S Berkenberg
■ High speed cameras, electronic imaging systems for ballistic research, development and proofing. Other related imaging and instrumentation systems.
® UK: Hadland Photonics Ltd 98

90-00805 Hägglunds Vehicle GmbH
(† Hägglunds GmbH)
(a subsidiary of Hägglunds Vehicle AB, Sweden)
Ernst Grote Strasse 13, D-30916 Isernhagen 1
☎ (+49 511) 613 73 65 ⊖ (+49 511) 613 73 67
♧ DM 250,000 (1996) ♣ DM 6,200,000 (1996)
ⅲ 3 (1996)
Managing Director: Wulf Rinke
Operations Director: Dieter Ripke
■ Products represented include the Bv 206 All Terrain Carrier and Combat Vehicle 90 (CV90).
® Sweden: Hägglunds Vehicle AB 98

90-00820 Edgar Hausmann GmbH Nietwerkzeuge
Förster Busch Strasse 10, D-34346 Hannover
☎ (+49 5545) 781 ⊖ (+49 5545) 14 48
Managing Director: Edgar Hausmann
■ Manufacture of screws, rivets, cutting tools, mobile aircraft tools and stationary riveters.
® USA: ATI Industries, American Pneumatic Tool, Hi Shear, Lok Fast Inc, Zephyr Tools 97

Contact Reference

90-00825 Karl E. Hengstler GmbH Precision
Vorlandstr. 7, D-77756 Hausach
☎ (+49 7831) 60 66 ⊖ (+49 7831) 76 90
Managing Dir: K.E. Hengstler
■ Aircraft jacks & GSE for military & civil aviation industries.
® USA: Regent Jack Manufacturing Co. 97

90-00850 Hughes International Corporation
PO Box 200615, D-53136 Bonn
☎ (+49 228) 95 68 50 ⊖ (+49 228) 956 85 25
General Manager: Karl Horn 98

90-00890 Hydrema Baumaschinen GmbH
Gewerbepark 5, PO Box 5150, D-D16833 Fehrbellin
☎ (+49 033) 932; 724 90 ⊖ (+49 033) 932; 724 91
General Manager: Bruno Winther
® Denmark: A/S Hydrema 97

90-00920 Icore International GmbH
(a subsidiary of Smiths Industries plc, UK)
Hoehestrasse 31-33, D-61348 Bad Homburg 1
☎ (+49 6172) 240 74; 240 76 ⊠ 415880
⊖ (+49 6172) 67 73 10
General Manager: W Rastätter
■ High specification flexible conduit systems, pressure hoses and end terminations for an extensive range of electrical and mechanical applications.
® UK: Icore International Ltd 98

90-00925 ILC-DDC Elektronik GmbH
Ahornstrasse 2, D-82296 Schoenfeising
☎ (+49 8141) 34 90 87 ⊖ (+49 8141) 34 90 89
Managing Director: Simon Oswald
Marketing Communications Manager: Lawrence M Margulies
■ Components, electrical A/D, D/A-Converter, synchro/resolver-converter, synchro/resolver instruments. MIL-STD-1553 data bus components. Power and motor control hybrids. For military applications all hybrids are available in MIL-STD-883 B (Rev C) screening.
® USA: Beta Transformer Technology Corp, ILC Data Device Corp 98

90-00935 Industrial Acoustics Co Inc
(a subsidiary of IAC Ltd, UK)
Sohlweg 17, D-41372 Niederkruchten

☎(+49 2163) 84 31 ⊠ 852261 iacn d
⊖ (+49 2163) 806 18
Managing Director: Heinz Müller
■ Noise control and suppression in aviation, maritime, shelters, gas turbines and audiometric rooms.
® UK: Industrial Acoustics Co [98]

90-01000 ITPN Europe Aircraft & Logistics GmbH
Heuberg 1, D-20354 Hamburg
☎(+49 40) 357 68 40 ⊖ (+49 40) 35 76 84 44 [98]

90-01145 Just Leitern AG
(a subsidiary of Just Leitern AG, Austria)
Wilhelminenhofstrasse 89a, D-12549 Berlin
☎(+49 30) 53 00 92 62 ⊖ (+49 30) 52 00 92 62
® Austria: Lust Leitern AG [98]

90-01180 Kuhn GmbH Bauelemente der Elektrotechnik
Im Lettenhorst 28, D-67105 Schifferstadt
☎(+49 6235) 34 82 ⊖ (+49 6235) 819 55
👥 10 🏭 8,000 sq ft
Managing Director: Kurt W Kuhn
Sales and Marketing Manager: E Leiss
■ Alphanumeric display systems in plasma, fluorescent and LCD technology. Militarised data entry systems (keyboard and digital switches) MIL-SPEC subminiature lamps. Lighted display pushbutton switches, knobs.
® USA: Aerospace Knob, Data Display Products, Digitran, IEE Inc., Stacoswitch, Stake Fastener Co., Staver, Western Indicator, XCEL [97]

90-01190 Kyokuto Boeki Kaisha (KBK) GmbH
(a subsidiary of Kyokuto Boeki Kaisha, Japan)
Immerstrasse 13, D-40210 Düsseldorf
☎(+49 211) 35 00 15 ⊖ (+49 211) 164 95 31 [98]

90-01200 Labinal Components & Systems GmbH
(a subsidiary of Labinal, France)
Aidenbachstrasse 137, D-81479 Munich
☎(+49 89) 784 98 44 ⊖ (+49 89) 780 99 26
👥 8
Managing Director: Jean Lucien Lamy
® France: Labinal [98]

90-01230 Leach International Germany GmbH
Postfach 1561, D-86715 Nördlingen
Hofer Strasse 5, D-86720 Nördlingen
☎(+49 9081) 80 00 ⊖ (+49 9081) 80 01 30 [98]

90-01233 Lee Hydraulische Miniaturkomponenten GmbH
(a subsidiary of The Lee Co, USA)
Bernerstrasse 77, D-60437 Frankfurt am Main
☎(+49 69) 507 61 11 ⊖ (+49 69) 507 28 53
♧ DM 800,000 (1995) ♣ DM 4,000,000 (1995)
👥 4 (1995)
General Manager: E G Rautenberg
■ Miniature hydraulic components, valves, restrictors, plugs.
® USA: Lee Co [98]

90-01240 Litton Precision Products International Inc
(a subsidiary of Litton Industries Inc, USA)
Oberfoehringer Strasse 8, D-81679 Munich
☎(+49 89) 92 20 41 81 ⊠ 524596 liton d
⊖ (+49 89) 982 96 49
President: Mrs Constance L Zagar
Vice President, Business Development: Paolo M Novazzzi
■ International sales and marketing organisation for Litton Industries representing high technology Litton systems, subsystems and components for the global market suitable for navigation, guidance, electronic warfare, command, communication and control applications.
Sales Office
Oberföhringer Strasse 8, D-81679 Munich
☎(+49 89) 92 20 40 ⊖ (+49 89) 98 51 84
General Manager: Werner Sudars [98]

90-01250 LMS GmbH
Mollenbachstrasse 33-35, D-71229 Leonberg
☎(+49 7152) 97 97 90 ⊖ (+49 7152) 97 97 99
Sales Manager: W Hanisch
→ 80
® Belgium: LMS International [98]

90-01280 Lockheed International GmbH
(† Loral International GmbH)
(a subsidiary of Lockheed Martin, USA)
Defence Systems Liaison Office
Duerenstrasse 48, D-53173 Bonn
☎(+49 228) 35 64 14; 35 64 27 ⊖ (+49 228) 35 59 14
Managing Director: Ms Ursula Schöne
■ Synthetic aperture radar reconnaissance and targeting systems, guidance control and missile warning systems, electronic surveillance systems, simulation and training. [97]

90-01283 Loggers Rubbertechniek
Schauenburgerstrasse 15-21, D-20095 Hamburg
☎(+49 40) 32 81 08 85 ⊖ (+49 40) 32 81 08 14 [98]

90-01285 Lucas Aerospace GmbH Power Generation Regional Office
Zaubzerstrasse 11, D-81677 Munich
☎(+49 89) 470 10 33 ⊖ (+49 89) 470 52 72
Managing Director: Hildebrecht Haehn
Regional Manager: David Daniels
Executive Assistant: C Ganserer
® UK: Lucas Aerospace
USA: Lucas Aerospace [98]

90-01315 Marconi Messtechnik GmbH
Landsberger Strasse 65, D-8034 Germering, Munich
☎(+49 89) 84 93 60 ⊠ 5212642 mmmc d
⊖ (+49 89) 841 91 42
Manager: Gerd Von Garrel
® UK: Marconi Instruments [98]

90-01335 Mauer-Werke Oberndorf Waffensysteme GmbH
PO Box, D-78727 Oberndorf
☎(+49 7423) 701 ⊖ (+49 7423) 706 70
President: Hugo Wolf
Sales Director: Dietrich C Klöckner
® Germany: Rheinmetall Industrie Aktiengesellschaft [98]

90-01368 Mikro-Lambda GmbH
Malsenstrasse 66, D-80638 Munich
☎(+49 89) 15 20 50 ⊖ (+49 89) 15 45 43
Manager: Herbert Schekatz
® USA: Microphase Corp [97]

90-01400 Munters GmbH
(a subsidiary of AB Carl Munters, Sweden)
Hans Duncker Strasse 14, PO Box 800780, D-21007 Hamburg
☎(+49 40) 734 16 01 ⊠ 2174333
⊖ (+49 40) 73 41 61 31
♣ DM 20,400 👥 56
General Mgr: Hans-Jurgen Tiebel
Chief of Defence Division: Hans Werz
☎(+49 2241) 33 52 27 [97]

90-01420 NICO Pyrotechnik
PO Box 1227, D-22943 Trittau
☎(+49 4154) 80 50 ⊖ (+49 4154) 24 51
President: Hanns Jürgen Diederichs
Sales Director: Hans Morotini
® Germany: Rheinmetall Industrie Aktiengesellschaft [98]

90-01490 N.V. Tools GmbH
Wittener Strasse 162, D-5600 Wuppertal-Barmen
☎(+49 202) 66 65 11 ⊠ 8592491 towu d
⊖ (+49 2029) 84 17 03
Managing Director: Cyril F Bennett
Secretary: Ms H Keitel
® UK: N.V. Tools Ltd [97]

90-01520 WNC-Nitrochemie GmbH
(a subsidiary of Rheinmetall Industrie Aktiengesellschaft)
Liebigstrasse 17, D-84544 Aschau a Inn
☎(+49 8638) 680 ⊖ (+49 8638) 681 00
President: Gregor Stockmann
President: Werner Köhn
Sales Director: Manfred Kuhl [98]

90-01550 Oasis GmbH Optics and Systems
(† RMI GmbH)
Zum Panrepel 2, D-28307 Bremen
☎(+49 421) 486 93 65 ⊖ (+49 421) 486 93 76
♧ DM 50,000 (1997)
Managing Director: Hans J Meister
■ Lenses, plastic lenses, prisms, beamsplitters, mirrors, windows, objetives, filters, coatings, beam expanders, precision etched metal parts. Image intensifier tubes. [98]

90-01665 Portescap Deutschland GmbH
(a subsidiary of Portescap, Switzerland)
Gülichstrasse 12, D-75179 Pforzheim
☎(+49 7231) 49 15 ⊖ (+49 7231) 49 16 20
♧ DM 600,000 (1995) 👥 18 🏭 400 sq m
® Switzerland: Portescap [97]

90-01812 Rafael North European Office
(a subsidiary of Rafael - Armament Development Authority)
2 Simrock Allee, D-53173 Bonn
☎(+49 228) 35 22 11 ⊠ 885421 isra d
⊖ (+49 228) 35 75 96
Manager: Eli Levy
® Israel: Rafael [97]

90-01815 Racal Elektronik System GmbH
Frankenforster Strasse 21, D-51427 Bergisch Gladbach
☎(+49 2204) 922 20 ⊖ (+49 2204) 214 91 [98]

90-01818 Raychem GmbH Bereich Elektronik
(a subsidiary of Raychem Corp, USA)
Haidgraben 6, D-85521 Ottobrunn
☎(+49 89) 608 90 ⊠ 5212879
⊖ (+49 89) 608 95 70; 609 89 16
Managing Director: Kurt Kantner
Sales Manager, Electronics Division: Erwin Lauschner
® USA: Raychem Corp [97]

90-01820 Raytheon Anschütz GmbH
(† Anschütz and Co GmbH)
(a subsidiary of Raytheon Company, USA)
Zeyestrasse 16-24, D-24106 Kiel
☎(+49 431) 301 90 [E] receptionde@raykiel.com
[W] http://www.raytheon.com ⊠ 292804 ansch d
⊖ (+49 431) 301 92 91
♣ DM 120,000,000 (1996) 👥 500 (1997)
Managing Director: Dr Carsten Peters
Assistant Managing Director: Gerhard Erb
Sales and Marketing Director: Ulrich Schulster
Marketing Manager: Bernhardt Schell
Marketing Services Manager: Bernhardt Schell
■ Nautical gyro-compasses, self-steering and rudder control systems, servo control systems, adaptive course track controllers, naval plotting and dead-reckoning tables, retransmission units, platform systems (3-axis stabilized). Nautical information displays. Mins-ring laser platform, data distribution units, battery monitoring, main engine control (remote control). Radars (TM/ARPA), GMDSS communication radio stations, ECHO sounders, ECDIS electronic chart and information systems, portable fathometers (GPS/DGPS). [98]

90-01855 Rheinmetall W&M GmbH
(a subsidiary of Rheinmetall Industrie Aktiengesellschaft)
PO Box 1127, D-29343 Unterlüß
☎(+49 5827) 800 ⊖ (+49 5827) 10 90
President: Detlef Moog
President: Bernhard Hagen
Vice President: Helmut Ortmann
Sales Director: Dr Hans Peter Hellmeister

Purchasing Manager: Michael Krämer
® Germany: Rheinmetall Industrie Aktiengesellschaft [98]

90-01865 Rockwell-Collins Deutschland

(a subsidiary of Rockwell International Corp, USA)
PO Box 600251, D-63096 Rodgau
Boschstrasse 8-10, D-63110 Rodgau
☎ (+49 6106) 69 70 [W] http://www.rockwell.com
⊜ (+49 6106) 156 73
♣ DM 20,000,000 ⚉ 55 ⌂ 4,200 sq m
Managing Director: Bernd Felski
Controller: N Ameen
Marketing Manager: G Fecher
■ Distribution, service and repair of communication and navigation equipment for airborne and aerospace platforms. Development, production, distribution and service for integrated communication systems for ship, ground or shelterised applications. → 80
® USA: Rockwell International Corp

Collins General Aviation Division
Kelsterbacher Strasse 20, D-65479 Raunheim
☎ (+49 6142) 992 50 ⊜ (+49 6142) 217 62
Regional Sales Manager: Manfred Jahnel [98]

90-01870 Rockwell International GmbH Rockwell Automation

(a division of Rockwell International Corp, USA)
Frankfurt Sales Office
Haus C1-40G - Robert Bosch Strasse 5, D-63303 Dreieich
☎ (+49 6103) 379 70 ⊜ (+49 6103) 37 97 10
Branch Manager: Harmut Pleyer
[E] hpleyer@eu1.ra.rockwell. com
® USA: Rockwell International Corp

Regional Headquarters
Düsselberger Strasse 15, D-42781 Haan-Gruiten
☎ (+49 2104) 69 00 ⊜ (+49 2104) 69 01 21
Sales Director, German Region: Karl Hugo Sönnecken [98]

90-01875 Sartank Industrial & Computer Supply

Grueneburgweg 101, D-6000 Frankfurt 1
☎ (+49 69) 724 14 00 ⊜ (+49 69) 72 25 02
⚉ 17 ⌂ 300 m2
Managing Director: M Sartank
Sales and Marketing Director: A Sartang
Operations Director: S Jawanisky
Purchasing Manager: R Schmid [98]

90-01932 ShinMaywa (Europe) GmbH

Cantadorstrasse 3, D-40211 Dusseldorf 1
☎ (+49 211) 36 07 27 ⊜ (+49 211) 361 38 54
® Japan: ShinMaywa Industries Ltd [97]

90-01935 Siemens Solar GmbH

Frankfurter Ring 152, D-80807 Munich
☎ (+49 89) 35 00 25 91 ⊜ (+49 89) 35 00 25 73
Sales: Jürgen Duch
■ Representing Siemens Solar in Europe, Africa and Middle East.
® USA: Siemens Solar [97]

90-01941 SEG GmbH

Bahnhopstrasse 6, D-79359 Riegel
☎ (+49 7642) 902 40 ⊜ (+49 7642) 90 24 24
■ Land, sea and air navigation systems, INS gyro system development; design and production of antenna stabilization systems. Painting device deformation measuring systems. Gyro production and test equipment. [97]

90-01950 Sperry Marine GmbH

Kleine Bahnstrasse 6, D-22525 Hamburg
☎ (+49 40) 853 13 30 ⊜ (+49 40) 85 31 33 70
⚉ 10
Vice President and General Manager: H E Rasmussen [98]

90-01960 Sumitomo Deutschland GmbH Head Office

(a subsidiary of Sumitomo Corporation, Japan)
Georg Glock Strasse 14, PO Box 104561,
D-40474 Dusseldorf
☎ (+49 211) 457 00 ⊠ 8584119
⊜ (+49 211) 454 15 01
♧ DM 20,000,000 (1996) ♣ DM 402,000,000 (1995)
⚉ 75 (1996)
President: S Tamura
Aerospace and Electronics Manager: M Inaoka
■ Aerospace, defence and telecommunications equipment.
® Japan: Sumitomo Corp [98]

90-01980 Sundstrand Aerospace

Kreetslag 7, D-21129 Hamburg
☎ (+49 40) 74 37 38 49 ⊜ (+49 40) 74 37 38 49
Marketing: Lutz Werner [97]

90-02000 Technitron-Acal/Auriema GmbH

(a subsidiary of ACAL plc, UK)
Gustav Heinemann Ring 135, D-81739 Munich
☎ (+49 89) 630 22 00 ⊜ (+49 89) 637 65 66
♧ DM 1,250,000 ⚉ 8 ⌂ 300 m2
Managing Director: Dr Dieter Donninger
Avionics Division Manager: Pangels Detler
Manager, Instrumentation: Manfred Wagner
Management Assistant: Ms Petra Sabine Desch
® Denmark: Procom Antennas A/S
UK: Acal plc, Navstar Ltd, Sycos
USA: Behlman, Computer Optical Products, DCS Data Control Systems, DEL Electronics, EMP Electromagnetic Processes, Eaton Pressure Sensors, Eaton Valve & Actuators, FJW Optical Systems, G.L. Collins Corp, IEC Interstate Electronics, Jay-EL Products, Monitron Corp, North Atlantic Instruments, Nurad Technologies, Path Products, Quad-Tron Inc, SBS Engineering Inc, Sabtech Industries Inc, Semco Engineering, Trompeter Electronics, Vega Power Systems, Xitron Technologies [97]

90-02020 Tactron Elektronik GmbH

Einsteinstrasse 35, D-82152 Martinsried
☎ (+49 89) 856 10 71 [E] tactron@compuserve.com
⊠ 898891 ⊜ (+49 89) 857 76 05
♣ DM 5,000,000 (1996) ⚉ 10 (1996) ⌂ 200 m2
Managing Director: Gunther Wolf
General Manager: Georg Schmidt
Sales and Marketing Director: Franz Frierst
Purchasing Manager: Leif Clausen
■ Antennas, power supplies for magnetrons, amplifiers, isolators, circulators, single/multi layer microwave ceramic, porcelain capacitors, DC blocks, self bins and bias filter networks, impedance transformers, coax switches up to 40 GHz / WG and coax filters, PIN dioden components, millimeter wave components, VCO's, DTO's, DRO's, phase shifter, I/Q vector modulators, surface mount chip resistors, SMA-N-TNC attenuators, termination equalizers, microwave testfixtures DC to 60 GHz, ultra broadband couplers, hybrids, detectors, power dividers, diodes, YIG-tunned oscillators, synthesizers, microstrip and stripline, multi-layers substrates, gold-metalized broadband RF power FET's. [98]

90-02025 Teledyne Germany

Abraham Lincoln Strasse 38-42, D-65189 Wiesbaden
☎ (+49 611) 763 60 ⊜ (+49 611) 763 61 55 [97]

90-02045 Texas Instruments Deutschland GmbH

(a division of Texas Instruments Inc, USA)
Haggertystrasse 1, D-85356 Freising
☎ (+49 8161) 80 47 33 [E] ddet@msg.ti.com
⊜ (+49 8161) 80 47 35
Director of Business Development: Richard Rhinehart (Jr)
Operations Manager: A Raderm Acker
® USA: Texas Instruments Inc [98]

90-02050 Tisto

Dinnendahlstrasse 31, D-40235 Dusseldorf
☎ (+49 211) 23 08 50 ⊜ (+49 211) 230 88 40
Sales Manager: W Forkheim
Associated company:
Timet, UK [98]

90-02055 Torrington GmbH European Customer Services

Carlstrasse 50, D-52531 Übach-Palenberg
☎ (+49 2451) 97 14 50 ⊜ (+49 2451) 97 14 80
♧ DM 50,000,000 (1996) ♣ DM 110,000,000 (1996)
⚉ 30 (1996)
Managing Director, ESA Logistics: Gert Snijders
☎ (+49 2451) 97 14 68
Manager of European Service Area Customer Service: H K Vierbuchen
■ Aircraft bearings, track rollers (cam followers), aircraft control bearings.
® UK: Torrington [98]

90-02070 Transaero Southern Europe

Austrasse 46, D-53179 Bonn
☎ (+49 228) 85 80 85 ⊜ (+49 228) 85 86 64
Manager: Nadir D Afghanistan
■ Represents leading manufacturers of life support products and systems.
Companies represented::
ACR Electronics
Aerospace Control Products
Aircraft Pordus Media
Allen Aircraft Products
Breeze-Eastern
Control Products Corporation
Co-Operative Industries Inc
Dayton-Granger
Dynapower/Stratopower
GEC-Marconi Aerospace
G & H Technology
Hayes International
Hughes-Treitler
IMC Magnetics
Marathon Power Technologies
Mectron
Onboard Systems
Pacer Systems
Pall Land and Marine Corp
Precision Kinetics
Scott Aviation
Semco Instruments Inc
Simmonds Precision Engine Systems
Stainless Steel Products
Symetrics Inc
Tedeco
Teledyne Battery Products
Thermoid [97]

90-02120 Turbomeca GmbH

(a subsidiary of Turbomeca SA, France)
Geschaftsfliegerzentrum-Gebaude 347, Flughafen, D-22335 Hamburg
☎ (+49 40) 500 21 60 ⊜ (+49 40) 50 02 16 61
⚉ 7 (1996) ⌂ 200 m2
General Manager: Richard Musil
[E] richard.musil@turbomeca.de
■ Distribution of gas turbines for helicopters.
® France: Turbomeca [98]

90-02125 UBE Europe GmbH

Immerman Hof - Immermannstrasse 65b, D-40210 Dusseldorf 1
☎ (+49 211) 17 88 30 ⊜ (+49 211) 361 32 97 [98]

90-02150 Unimex Handels GmbH

Rabenkopfstrasse 27, D-81545 Munich
☎ (+49 89) 64 81 99 ✆ (0171) 752 79 81
[E] unimex@compuserve.com ⊠ 5215466 unex d
⊜ (+49 89) 642 22 47
♧ DM 1,000,000 ♣ DM 18,000,000 (1995) ⚉ 16
⌂ 4,000 sq m
Managing Director: Alexander Dajani
Sales and Marketing Director: Peter Hoffmann
Operations Director: Ostold Gunther
Purchasing Manager: Mr Till
■ Remote controlled vehicles and equipment for EOD, stethoscopes, bomb detectors, mine detectors, metal detectors, hand held metal detectors DMD, vapour detector UNEX 2500, anti magnetic tools. → 80 → 85 [98]

90-02335 Weber Aircraft Inc

(a subsidiary of Group Zodiac, USA)
Gerowег 10, D-33014 Bad Driburg
☎ (+49) 525 39 88 10 ⊜ (+49) 42 53 98 81 81
Sales Director, Europe: Hans J Zurbrüggen
■ Ejection seats.
® USA: Weber Aircraft [97]

90-02337 Weinberger Deutschland GmbH

(a subsidiary of Weinberger AG, Switzerland)
Stephanienstrasse 92, D-76133 Karlsruhe
☎ (+49 721) 91 32 80 [E] weinberger@ilk.de
[W] http://www.ilk.de/weinberger ⊜ (+49 721) 913 28 30

Managing Director: Volker Schumann
® Switzerland: Weinberger AG [98]

90-02340 WEL Consult GmbH
Graf v Moltke Weg 25, D-61267 Neu Anspach
☎(+49 6081) 94 40 20; 94 40 25
[E] 100542.2707@compuserve.com [W] http://ourworld.compuserve.com/homepages/wel- consult
⊖ (+49 6081) 432 92
♻ DM 50,000 (1995) ♣ DM 6,000,000 (1995) 👥 2
Managing Director: Wulf E Luther
■ Guidance and control systems, gyros, north-finding gyros, INS, antenna stabilization.
® USA: Humphrey Inc, SEG GmbH [97]

90-02342 Weltronic GmbH
Trausnitzstrasse 8, D-81671 Munich
☎(+49 89) 49 20 66; 49 20 68 ⊖ (+49 89) 49 62 34
♻ DM 200,000 (1997) ♣ DM 7,000,000 (1997)
👥 15 (1997) 🏭 400 sq m
Managing Director: W Schwarz
Technical Director: H Braun
Sales and Marketing Director: A Riedel
Purchasing Manager: M Maier
® USA: Bei Motion Control Inc, Best Power Inc, Caddock Electronics Inc, Custom Electronics Inc, Electro Films Inc, Maida Inc, Transtek Inc [98]

90-02370 Willburger System GmbH
Auf der Schuchen 11, D-82418 Seehausen
☎(+49 88) 41 30 28 [E] info@willburger.de
[W] http://www. willburger.de ⊖ (+49 88) 41 51 58
♻ DM 100,000 ♣ DM 14,000,000 👥 6 🏭 150 sq m
Managing Director: Ulrich Willburger
■ All MIL-Capacitors, RFI/EMI-Filters, potentiometers, resistors and thermistors, resolvers, motors, base material for the PCB-industry. → 85
® France: Duncan Electronics, Eurofarad SA, Precilec
USA: Instec, Ketema/Rodan Division, Polara Engineering Inc [97]

90-02500 VLSI Technology GmbH
(a subsidiary of VLSI Technology Inc, USA)
Bayerwaldstrasse 11, D-81737 Munich
☎(+49 89) 92 79 50 ⊠ 5214279
⊖ (+49 89) 62 70 60; 62 70 61 01
♣ US$ 700 000 000 (1995) 👥 2,500
Managing Director: Pascal Langlois
Sales and Marketing Director: Max Lögel
Operations Director: Jean Luc Bouvress
Marketing Engineer: Richard Foster [97]

90-02600 Robert Zapp Werkstofftechnik GmbH & Co KG
PO Box 105545, D-40046 Dusseldorf
Goltsteinstrasse 28, D-40211 Dusseldorf
☎(+49 211) 355 00 ⊠ 8581911
⊖ (+49 211) 355 02 90
Managing Director: Dipl.-Ing T H Meyer
Sales and Marketing Director: I Gumnior
Sales and Marketing Director: R Eisenkölbl
■ Semi-finished products and components made of special stainless steels, nickel and cobalt base alloys, titanium and zirconium and their alloys and tantalum. → 80 → 85
® USA: Elgiloy, Haynes International, High Performance Tube, Industrial Tectonics, OREMET, RMI Co, Valley Metals, Xaloy [98]

Ghana

Head of State:
Flt Lt Jerry John Rawlings (retd)

10 Head of State

10-10000 Office of the President
The Castle, Osu, Accra
☎(+233 21) 66 54 15
President: H E Flt Lieut Jerry John Rawlings (retd) [98]

12 Ministries

12-10000 Ministry of Defence
Burma Camp, Accra
☎(+233 21) 77 61 11
Minister for Defence: The Hon Mahama Iddrisu [98]

12-20000 Ministry of Foreign Affairs
PO Box M.53, Accra
☎(+233 21) 66 49 51; 66 49 53
⊠ 2001 star gh accra; 2665 star gh accra
Minister of Foreign Affairs: Ahwoi Kwamena [98]

12-40000 Ministry of Internal Affairs
PO Box M 42, Accra
☎(+233 21) 66 54 21
Secretary for Internal Affairs: Nii Okaidjah Adamafio [98]

31 Embassies & Attachés

31-00250 Embassy of the Republic of Bulgaria
3 Kakramadu Rd, East Cantonments, PO Box 3193, Accra
☎(+233 21) 77 24 04 ✉ BULEMBASSY, ACCRA
⊠ 2709 bultg gh ⊖ (+233 21) 77 42 31
Chargé d'Affaires a.i.: Alexander Peytchev
Public Information Officer: Mrs Diana Georgieva [97]

31-00310 Canadian High Commission
42 Independence Avenue, PO Box 1639, Accra
☎(+233 21) 22 85 55 ✉ DOMCANADA Accra
[E] accra@accraoi.x400.gc.ca ⊖ (+223 21) 77 37 91
High Commissioner: H E Mr John Schram
Counsellor, Development: Richard Beattie
First Secretary, Immigration: Ms Patti Page
First Secretary, Development:
Chandran Thiruchittampalam
First Secretary, Administration: Stanley Baran
First Secretary, Development: Ms Anne Lavender
First Secretary, Immigration: Ms Lise Pouliot
Third Secretary, Political and Trade: Mario Gariepy
Third Secretary: Serge Deshaies
Attaché: Diane Hosker
Attaché: Barry Rice [98]

31-00660 Embassy of the Federal Republic of Germany
7th Avenue Extension 4, North Ridge, PO Box 1757, Accra
☎(+233 21) 22 13 26 ⊠ 2025 aaaccr gh
Ambassador: H E Mr Christiaw Nakonz
Deputy Chief of Mission First Secretary:
Maij Einig-Heidenhal
Commercial Counsellor: Michael Jürgens
Military Attaché: Lieut Col Dietmar Nickel
Public Information Officer: Dr Holger Ruthe [98]

31-00890 Embassy of Italy
Jawahar Nehru Road, PO Box 140, Accra
☎(+233 21) 77 56 21 ⊠ 2039 italdipl accra
⊖ (+233 21) 77 70 56 [98]

31-01270 Royal Netherlands Embassy
PO Box 3248, Accra
☎(+233 21) 77 36 44 [E] nlgovacc@ncs.co.gn
⊠ 2128 holland gh ⊖ (+233 21) 77 36 55
Ambassador: Drs Hein C R M Princen
First Secretary: Jan W de Waal [98]

31-01680 Embassy of Spain
PO Box 1218, Accra
☎(+233 21) 77 40 04; 77 40 05 ⊠ 2680 embspa gh
⊖ (+233 21) 77 62 17
Ambassador: H E Mr Diego MaSanchez Bustamante
Deputy Chief of Mission: Lourdes Melendez Garcia
Commercial Counsellor: Ramon Muro (Resident in Lagos)
Chancellor: Segundo Garcia Fernandez [98]

31-01740 Embassy of Switzerland
9 Water Rd, North Ridge Area, PO Box 359, Accra
☎(+233 21) 22 81 25; 85 ✉ AMBASUISSE Accra
⊠ 2197 amswis gh ⊖ (+233 21) 22 35 83 [97]

31-01890 United Kingdom High Commission
PO Box 296, Accra
Osu Link, Off Gamel Abdul Nasser Avenue, Accra
☎(+233 21) 22 16 65; 66 95 85 (24 hrs)
⊠ 2323 ukrep gh ⊖ (+233 21) 66 46 52
High Commissioner: H E Mr I Mackley
Deputy High Commissioner: D Wyatt
First Secretary Commercial: W F Somerset
First Secretary Management: B Cope
Defence Adviser: Lieut Col A R Gale [98]

Gibraltar

Head of State:
H.M. Queen Elizabeth II

Governor:
Rt Hon Richard Luce

10 Head of State

10-10000 Office of the Governor
The Convent, Gibraltar
☎(+350) 542 01 ⊖ (+350) 759 08
Governor: Rt Hon Richard Luce
Deputy Governor: M Robinson
Commander British Forces: Cdre S Taylor (RN)
Assistant to the Deputy Governor: Dr S Ellwood [98]

11 Prime Minister

11-10000 Office of the Chief Minister
6 Convent Place, Gibraltar
☎(+350) 700 71 ⊠ 2223 admsec gk ⊖ (+350) 763 96
Chief Minister: Peter Caruana
Chief Secretary: Mrs M Sanchez
Personal Private Secretary: Dennis Figueras [98]

15 Parliament

15-10000 House of Assembly
Gibraltar
☎(+350) 784 20 ⊠ 2223 admsec gk ⊖ (+350) 428 49
Speaker: Hon Judge J E Alcantara
Chief Minster: Hon P R Caruana

Leader of the Opposition: Hon J J Bossano
Clerk of the House: D J Reyes
98

25 Defense Forces

25-20000 RAF Gibraltar
BFPO 52, Gibraltar
☎ (+350) 535 22 ⊖ (+350) 536 00
Station Commander: Wg Cdr R G Fraser
OC Operations Wing: Flt Lieut A Rowley

Gibralter Services Education Centre († Education & Recreation Flight)
BFPO 52, Gibraltar
☎ (+350) 542 60
503 J3 Public Information Coordinator: Capt M Wills Aarel
98

25-30000 The Gibraltar Regiment
Devil's Tower Camp, BFPO 52, Gibraltar
☎ (+350 5) 42 84; 42 48 ⊠ 2355 tpsgib gk
⊖ (+350 5) 43 48
Commanding Officer: Lieut Col Dennin Duarte
Commander HQ Company and Thomson's Battery: Capt Michael Borg
Commander "G" Company: Maj Mark Randall
Commander "I" Company: Maj Patrick Canessa
Commander "B" Company: Maj Philip Canessa
Press and Information Officer: Lieut Colin Risso
98

25-40000 Commander British Forces Gibraltar
HOBF, The Tower, BFPO 52, Gibraltar
☎ (+350) 550 01, 550 02 ⊠ 2331 navgib gk
⊖ (+350) 552 17
Commander British Forces Gibraltar: Cdre A J S Taylor (RN)
Chief Staff Officer (Operations and Plans): Col J D Sankey
Chief Staff Officer (Support): Capt R J Lord (RN)
Military Assistant to Commander British Forces: Sqn Ldr A S Barmby (RAF)
98

29 Customs & Excise

29-01000 Customs Office
Custom House - Waterport Wharf, Gibraltar
☎ (+350) 772 51; 788 79 ⊖ (+350) 783 62
Collector of Customs: Anthony D Lima
☎ (+350) 788 79 ext 3702
98

75 Manufacturing Companies

75-01000 John M. Piris & Sons Ltd
Unit 61, Harbours Deck, The New Harbours, Gibraltar
☎ (+350) 731 71 ⊖ (+350) 443 70
♤ £ 100 ♣ £ 280 000 (1994/95) 9 200 m2
Managing Dir: John M. Piris (Sr)
Electrical Services Dir: John M. Piris (Jr)
Mechanical Services Dir: Joseph Michael Piris
■ Ship repair. Welding, pipe work, boiler repair, mechanical services including engine overhaul.
Member of Electrical Apparatus Service Association (EASA).
97

Greece

Head of State:
Costis Stefanopoulos

10 Head of State

10-10000 Presidency of Hellenic Republic Military Office
Stisichorou Street 17, 10028 Athens
☎ (+30 1) 724 87 20 (military office)
⊖ (+30 1) 724 89 88 (military office)
President of the Hellenic Republic: Costis Stefanopoulos
☎ (+30 1) 724 64 90
Secretary General: E Gikas ☎ (+30 1) 722 69 37
Director Military Affairs Office: Adm P Lambiris
☎ (+30 1) 724 87 40
Aide de Camp to the President: Lieut Col J Baltzois
☎ (+30 1) 724 87 90
98

11 Prime Minister

11-10000 Office of the Prime Minister
House of Parliment, Syntagma Square, 101 80 Athens
☎ (+30 1) 724 90 40 ⊖ (+30 1) 722 77 35
Prime Minister: Konstandinos Simitis
Undersecretary to Prime Minister: Antonios Livanis
Deputy Minister to the Prime Minister's Office: Yeoryios Paskhalidhis
98

12 Ministries

12-10000 Ministry of National Defence
Holargos, 100 20 Athens
☎ (+30 1) 646 54 01 ⊠ 216780 rkga gr
⊖ (+30 1) 644 37 88
Minister of National Defense: Apostolos-Athansios Tsochatzopoulos
Deputy Minister: Dhimitrios Apostolakis
Deputy Minister: Ioannis Stathopoulos
Alternate Minister: Alexandros Papadongonas
98

12-20000 Ministry of Foreign Affairs
Odos Kolokosta 2, Athens
☎ (+30 1) 361 05 81 ⊠ 218213 ypex gr
⊖ (+30 1) 360 97 16
Minister: Theodhoros Pangalos
Deputy Foreign Minister: Ioannis Kranidhiotis
98

12-25000 Ministry of Finance
10 Karageorgis Servias, 10184 Athens
Minister of Finance: Yioannos Papandoniou
Deputy Minister of Finance: Nikolaos Khristodhoulakis
Deputy Minister of Finance: Yeoryios Dhris
98

12-40000 Ministry of Interior
27 Stadiou Street, 101 83 Athens
☎ (+30 1) 322 35 21
⊠ 2161109 veso gr; 216110 yeso gr; 215776; ⊠ 215868 ⊖ (+30 1) 323 32 18
Minister of Interior: Alexandros Papadhopoulos
Deputy Minister: Anastasios Mandelis
Deputy Minister: Lambros Papadhimas
98

12-50000 Ministry of Public Order
1 Katehaki, 10177 Athens
☎ (+30 1) 692 85 10; 691 10 54; 691 63 83
⊠ 218905 ydta gr ⊖ (+30 1) 692 16 75
Minister: Yeoryios Romaios
Dpty Minister of Public Order: Constantinos Geitonas
98

12-60000 Ministry of Transport and Communications
13 Xenophontos Street, 101 91 Athens
☎ (+30 1) 923 03 69 ⊠ 216369 ⊖ (+30 1) 923 33 30
Minister: Kharalambos Kastanidhis
☎ (+30 1) 324 46 62; 324 74 00 (fax)
Deputy Minister: Emmanouil Loukakis
☎ (+30 1) 324 56 11; 323 90 39 (fax)
Secretary General of Transport: Ioannis Constantinidis
☎ (+30 1) 324 49 37; 324 05 06 (fax)
Secretary General of Communications: Panayotis Vournas
☎ (+30 1) 324 05 01
Director General of Transport: Iakovos Avazos
☎ (+30 1) 323 13 70
Director General for Administration: Christos Polydoropoulos
☎ (+30 1) 323 71 40; 324 40 50 (fax)
Head of International Affairs Division: Georgios Patsiavos
98

20 Defense Administrations/Services/Agencies

20-05000 Hellenic Codification Centre (HNDGS)
Arhimidous & Ippokratous (Stratopedoi Siliotopoulou)
11146 Galatsi, Athens
98

25 Defense Forces

25-20000 Air Force Headquarters
136 Messogion, Holargos, 155 61 Athens
☎ (+30 1) 642 84 11; 642 81 01; 642 81 04
⊖ (+30 1) 646 97 33; 642 84 79
Chief of Air Force General Staff: Lieut Gen G Antonetsis
☎ (+30 1) 643 59 40; 672 36 40 (fax)
Deputy Chief of Staff: Maj Gen Gerorge Koletsos
☎ (+30 1) 642 93 37
General Inspector: Brig Gen K Gounaris
☎ (+30 1) 652 76 80
Director Supply Directorate: Brig gen A Sypropoulos
☎ (+30 1) 652 32 17
Director Technical Corps: Maj Gen D Kateloyzos
☎ (+30 1) 642 79 59
Director "A" Branch: Brig Gen E Georgousis
☎ (+30 1) 642 95 74
Director "B" Branch: Brig Gen C H Kalyviotis
☎ (+30 1) 642 94 96
Director "C" Branch: Maj Gen D Katelouzos
☎ (+30 1) 642 79 59
Director "D" Branch: Brig Gen T H Moschos
☎ (+30 1) 651 00 50
Director Flight and Ground Safety Directorate: Brig Gen K Gounaris ☎ (+30 1) 646 29 71
Director Financial Directorate: Brig Gen K Iliopoulos
Chief Secretary: Mazy Sazika
98

25-30000 Army Headquarters
Hellenic Army General Staff (HAGS)
10 Thalou - Pittakou Street, 105 58 Athens
☎ (+30 1) 646 52 01 ⊖ (+30 1) 655 27 01
Director Military Police Division: Maj Gen Ioannis Grammatikos
Director "B" General Staff Directorate: Maj Gen Dimitrios Komninos
Director Supply-Transportation Directorate: Maj Gen Konstandinos Varelas
Director "C" General Staff Directorate: Maj Gen Ioannis Fassaris
Director Ordnance Corps Directorate: Maj Gen George Nikolinakos
Director Medical Corps Directorate: Maj Gen Spyridon Doumas
Director Procurement Directorate: Maj Gen Aristophanis Koromilias
Director Finance Directorate: Maj Gen Stylianos Manidakis
Director "A" General Staff Directorate: Maj Gen Dimitrios Dimou
Director "D" General Staff Directorate: Maj Gen George Liakouris
Director Technical Corps Directorate: Maj Gen Elias Saperas
Director Infantry Directorate: Maj Gen George Lahanas
Director Armour Directorate: Maj Gen George Angelopoulos

Contact Reference

Director Artillery Directorate:
Maj Gen George Mavroyianneas
Director Engineering Directorate:
Maj Gen Theodoros Dimoudis
Director Signals Directorate: Maj Gen Dimitrios Arvanitis
Director Special Forces Directorate:
Brig Gen Theodoros Nikolaidis
Director Army Aviation Directorate:
Brig Gen Michail Papamarkou
Director 2nd Staff Office: Brig Gen Grigorios Tendolouris
Director 3rd Staff Office: Brig Gen Minas Kondrafouris
Director 7th Staff Office: Brig Gen George Vlassis
Director Training Division: Brig Gen Nikolaos Giouvanis
Director Organization Division:
Brig Gen Panayiotis Papadogonas
Director Defense Planning and Programming Division:
Brig Gen Leonidas Economopoulos
Director Computor-Studies Division:
Brig Gen Stylianos Psaroudakis
Director CFE-CSBM Divison: Brig Gen Christos Tsekouras
Director 1st Staff Office:
Brig Gen George Andonakopoulos
Director Military Personnel Division:
Brig Gen Nikolaos Tsagatakis
Director Mobilization Division: Brig Gen Athanassios Grivas
Director Personnel Selection Division:
Brig Gen Nikolaos Pliakos
Director Conscription Division: Brig Gen George Goulas
Director 4th Staff Office: Brig Gen Charalambos Paliouras
Director Infrastructure Division:
Brig Gen Leonidas Demestihas
Director Support Division: Brig Gen Dimitrios Kouyianis
Staff Officer/3rd SO: Capt C Kalogeropoulos
Chief/HAGS: Lieut Gen Konstandinos Voulgaris
Army Inspector General: Lieut Gen Angelos Kedikoglou
"A" Deputy Chief/HAGS:
Lieut Gen Dimitrios Alevromagiros
"B" Deputy Chief/HAGS: Lieut Gen George Thiakos
Director Military History Directorate:
Lieut Gen Ioannis Zozonakis 98

29 Customs & Excise

29-01000 Directorate of Customs & Excise
(a subordinate of the Ministry of Finance)
c/o Ministry of Finance, Karageorgis Servias,
10184 Athens 98

31 Embassies & Attachés

31-00085 Armenian Diplomatic Mission
Athens
☎(+301) 934 57 27; 934 57 27 (consulate)
⊖ (+301) 934 57 27
Ambassador: Arman Kirakossian 98

31-00100 Embassy of Austria
26 Leoforos Alexandras, 10683 Athens
☎(+30 1) 821 68 00; 821 10 36; 882 75 20; 822 30 67
✉AUSTROAMB ⊖ (+30 1) 821 98 23
Ambassador: H E Dr Hans Sabaditsch
☎(+30 1) 821 68 00
Deputy Chief of Mission: Dr Michael Desser
☎(+30 1) 821 68 00
Military Attaché: Brig Josef Klamminger
Commercial Counsellor: Mag Bruno Freytag
☎(+30 1) 884 37 11, 882 79 13 (Fax)
⊠ 216123 oehd gr 98

31-00250 Embassy of the Republic of Bulgaria
33A Stratigou Kallari Street, Paleo Psychico,
15452 Athens
☎(+30 1) 624 81 05; 624 81 08 ⊠ 215873
⊖ (+30 1) 624 81 30
Ambassador: H E Mr Branimir Petrov
Counsellor: Dimitar Arnaoudov
Deputy Chief of Mission: Dimitar Arnaoudov
Military and Naval Attaché: Col Neyko Braynov
Counsellor: Plamen Atanassov
Commercial Counsellor: Latckezar Shefanov
Counsellor, Commercial Affairs Thessaloniki:
Georgy Georgiev
First Secretary, Cultural Affairs: Harizan Harizanov
First Secretary: Mrs Roumiana Mitreva
First Secretary, Consular Affairs: Tochko Tochev
Second Secretary, Press: Lyudmil Anguelov
Second Secretary: Plamen Dascalov
Second Secreatry: Vihren Nikolov
Attaché, Public Information Officer: Emilia Kraleua
Attaché, Administrative Affairs: Valeri Milev
Attaché, Administrative Affairs: Tchavdar Belinski
Secretary of the Defence Attaché: Mgr Roumen Stoyanov
Third Secretary, Commercial and Economic Affiars:
Ms Valery Vladimirov
Air Attaché: Lieut Col Vesselin Ivanov 98

31-00310 Embassy of Canada
4 Ioannou Gennadiou Street, 11521 Athens
☎(+30 1) 725 40 11 ⊖ (+30 1) 725 23 46
Ambassador: H.E. Mr D.R.T. Fraser
Canadian Forces Attaché: Col Lea Vachon
(Resident in Rome) ☎(+39 6) 44 59 81
☎(+39 6) 44 59 87 81 (fax)
Commercial Counsellor: David Cohen 97

31-00450 Embassy of the Czech Republic
6 G. Seferi St, Palaio Psychico, 154 52 Athens
☎(+30 1) 671 37 55; 671 97 01 ⊖ (+30 1) 671 06 75
Ambassador: H.E. Mr Jirí Marvan
Political Affairs Counsellor: Josef Matejka
Commercial & Economic Affairs Counsellor:
Miroslav Cancík
Third Secretary, Consular Affairs: Milan Fiser
Attaché: Mrs Jirina Kvapilová 97

31-00580 Embassy of Finland
1 Eratosthenous Street, 11635 Athens
☎(+30 1) 701 04 44; 701 17 75
⊖ (+30 1) 751 50 64; 701 14 44
Ambassador: H E Mr Antti Lassila
Second Secretary: Renne Klinge
Consul: Ms Marja Korhonen 98

31-00660 Embassy of the Federal Republic of Germany
Karaoli Dimitriou 3, 10675 Athens
☎(+30 1) 728 51 11 ⊖ (+30 1) 725 12 05
Ambassador: H E Dr Friedrich Reiche
Deputy Chief of Mission: Julius Bobinger
Economic and Commercial Counsellor:
Matthias von Kummer
Defense Attaché: Col Gerorge-Wilhelm von Graevenitz 98

31-00690 Embassy of Greece
PO Box 94375, 11693 Riyadh
☎(+966 10) 465 50 26 ⊠ 406322 98

31-00810 Embassy of the Republic of Hungary
16 Kalvou St, Psychico, 15452 Athens
☎(+30 1) 672 37 53; 672 53 37; 672 59 95
⊠ 215006 hexp gr (Commercial Section)
⊖ (+30 1) 674 08 90
Ambassador: H E Mr Janos Hermans
Military and Air Attaché: Lieut Col Andras Nemeth
Press Attaché: Gyorgy Selmeci
Commercil Officer: Laszlo Schiller
First Counsellor: Ferenc Kontra
Counsellor (Commercial Affairs) Head of the Commercial Office: Laszlo Schiller
First Secretary: Katalin Benedick
Commercial Section
11 Vas. Sofias, 3rd Floor, 10671 Athens
☎(+30 1) 364 80 03 ⊠ 215006 hexp gr
⊖ (+30 1) 363 85 83
Commercial Counsellor: Laszlo Schiller 98

31-00890 Embassy of Italy
2 Osos Sekeri, 10674 Athens
☎(+30 1) 361 72 60/3; 72 73/5 ⊖ (+30 1) 361 73 30 98

31-00920 Embassy of Japan
2-4 Messoghion Avenue,
Athens Tower Building - 21st Floor, Athens
☎(+30 1) 775 81 01; 775 81 02; 723 37 32
⊖ (+30 1) 770 59 64
Ambassador: H E Mr Yasumitsu Kiuchi
Minister: Ikuya Yoshimura 98

31-01270 Royal Netherlands Embassy
Leoforos Vassileos Konstantinou 5-7, 106 74 Athens
☎(+30 1) 723 97 01/4 ✉HOLLANDIA Athens
⊠ 215971 neth gr ⊖ (+30 1) 724 89 00
Ambassador: H.E. Mr H. Wagenmakers
First Secretary: G.H.C. van der Lingen 97

31-01300 New Zealand Embassy
15-17 Odos An Tsoha, 115 21 Ambelokipi, Athens
☎(+30 1) 777 06 86; 771 01 12 ✉RAUPO Athens
⊠ 216 630 raup gr ⊖ (+30 1) 777 73 90
Hon Consul General: H E Mr C Cotsilinis 97

31-01370 Embassy of Pakistan
6 Loukianou Sofias, Kolonaki, 10675 Athens
☎(+30 1) 729 01 22; 729 02 14 ✉PAREP Athens
⊠ 221016 parp gr ⊖ (+30 1) 724 72 31
Ambassador: H.E. Mr Rasheed Ahmad
☎(+30 1) 729 01 22
First Secretary: Mohammad Aqil Nadeem 97

31-01480 Embassy of Romania
Odos Emmanuel Benaki 7, Pal.Psychiko, Athens
☎(+30 1) 671 80 20 ⊠ 215301 romc gr
⊖ (+30 1) 671 48 60
Ambassador: H.E. Mr Christian Ionescu
☎(+30 1) 671 80 08
Deputy Chief of Mission: Ion Patrascu
☎(+30 1) 672 51 77
Defence, Military, Naval & Air Attaché:
Colonel Alexandru Videa ☎(+30 1) 647 25 40
Deputy Defence, Military, Naval & Air Attaché:
Laurentiu Anghelescu ☎(+30 1) 647 25 40
First Secretary: Horia Dima
First Secretary: Mihai Rasica
First Secretary: Petre Ghioalda ☎(+30 1) 671 27 19
Commercial Counsellor: Nicholache Blacioti
☎(+30 1) 671 60 06
Commercial Secretary: Radu Hergheligiu
☎(+30 1) 671 60 06
Commercial Secretary: Ion Serban ☎(+30 1) 671 60 06 97

31-01655 Embassy of the Republic of Slovenia
43 Vas Pavlou, 154 52 Athens, Palco Psychico
☎(+30 1) 685 66 55 ⊖ (+30 1) 685 58 19
Ambassador: H E Mr Miha Vrhunec
Consulate of the Republic of Slovenia
Marad Hellas Ltd, 213b Alexsandras Avenue,
115 23 Athens
☎(+30 1) 643 28 56; 641 04 90 ⊖ (+30 1) 642 47 58
Honorary Consul: Evangelos Tziavos 98

31-01780 Royal Thai Embassy
23 Taigetou Street, Paleo Psychico, 154 52 Athens
☎(+30 1) 671 79 69; 671 01 55 [E] thaiath@hol.gn
⊠ 225856 thai gr ⊖ (+30 1) 647 95 08
Ambassador: H E Lieut Ravee Hongsaprabhas
Minister Counsellor: Phairoj Vikiniyatanee
Commercial Counsellor: Sutirak Issadisai
Counsellor, Labour: Songsak Tantayotin
First Secretary: Miss Boonkerd Visudhipailbul
First Secretary: Pairach Chiaraanunt
Attaché: Mrs Sunee Muangnapha
Attaché: Mrs Chanphen Kongdej
Office of Commercial Affairs
119, Panormou Street, 5 FL, 11524 Athens
☎(+30 1) 649 66 71; 649 66 73 ⊖ (+30 1) 649 66 70
Counsellor: Miss Rommanee Chun-Obhas 98

31-01840 Embassy of the Republic of Turkey
Odos Vassileos Gheorgiou B.8, 10674 Athens
☎(+30 1) 721 36 59; 724 59 15; 724 59 17
⊠ 214498 turk gr ⊖ (+30 1) 722 95 97
Ambassador: H.E. Mr Umit Pamir
Deputy Chief of Mission/ Minister Counsellor: Tacan Ildem
Naval Attaché: Capt. Bulent Bostanoglu
Military Attaché: Col Hayati Sevgi
Air Attaché: Col Recai Sevinc ☎(+30 1) 724 05 12
☎(+30 1) 723 13 79 (fax)
Deputy Economic Counsellor: Refik Ozgur
☎(+30 1) 722 21 96 97

31-01890 Embassy of the United Kingdom
1 Ploutarchou Street, 10675 Athens
☎ (+30 1) 723 62 11 ▣ 216440 lion gr
⊖ (+30 1) 725 88 20
Ambassador: H E Mr Michael Llewellyn Smith (CMG)
Deputy Head of Mission: C J A Denne (CMG)
Defence and Military Attaché: Brig W A McMahon
Naval and Air Attaché: Capt W K Benbow (RN)
Commercial Counsellor: G Thomas [98]

55 Government-owned Industry & Production Facilities

55-00530 EBO Group of Companies
Kifissias Ave 160, 115 25 Athens
☎ (+30 1) 647 26 11 ▣ 218562 ebo gr
⊖ (+30 1) 647 27 15
♣ Dr 13 977 742 592 ⅱ 1 626
President: Nikos Papadakis
■ State-owned corporation controlling manufacturing subsidiaries. Product range covers small to large calibre weapons, explosives, propellants, ammunition & integrated weapon systems. [97]

55-00800 Hellenic Arms Industry SA
(part of the EBO Group of Companies)
Head Office
160 Kifissias Avenue, 115 25 Athens
☎ (+30 1) 647 26 11 ▣ 218562 ebo gr
⊖ (+30 1) 647 27 15
♣ Dr 12,000,000,000 ⅱ 1,900
Chairman of the Board: S Vellidis
Managing Director: A Barakos
Operations and Technical Director: G Pintelas
Marketing Director: M Tsontos
Group Commercial Manager: V D Cambas
■ State owned corporation controlling manufacturing subsidiaries. The company operates five plants producing the following:
Small and medium calibre guns, Mortars 81mm 120mm, small and medium calibre gun barrels, 30mm ammunition, Artillery rounds 105/155/155mm Extended range, 203mm (8"). Propellant powders, Propelling charges, Nitrocellulose, TNT, Anti-tank weapons, Anti-aircraft weapons, Aircraft fuel drop tanks, Bomb loaders, Pylons, TER-MER, Ejector units. Casting and processing of non-ferrous metal. Military uniforms.
Subsidiaries:
Hellenic Arms Industry SA
Electromechanica Kymi
Ipirou Metalworks Industry SA
Lavrion Plant
PO Box 29, Lavrion
☎ (+30 292) 230 91/96 ▣ 210909
■ 30mm ammunition, complete artillery rounds, 105mm/155mm, fuzes, propellant charges, nitrocellulose, TNT, single-base powders [97]

55-00850 Hellenic Shipyards Co.
Athens Office
5 Koumbari Street, PO Box 3480, 102 33 Athens
☎ (+30 1) 557 83 15 ✉ SHIPYARD Athens
▣ 215396 hsys gr ⊖ (+30 1) 557 07 00; 557 07 19
ⅱ 2,163 (1997)
Chairman: G Kasmas
Vice Chairman: G Kontakis
Managing Director: D Groves
Financial Director: A Vasishta
Commercial Director: D Blincow (Acting)
Personnel and Administration Director: A Trigonis
Production Director: F Wilcox
Yard Director: P Mademtzoglou
Defence Business Development Director: E Bacopoulos
Technical Director: P Mitchell
Estimating Manager: A Evans
Project Services Manager: R Stone
New Building Project Manager: V Katsikaros
Weapons and Electronics Manager: A Papathanasiou
Repair Manager: E Markousis
Marketing Department: M Constantinides
Public Relations Department: H Karoubi
■ Shipbuilding and repair (naval, merchant) and industrial steel constructions: railcars, electric railcars and heavy metal constructions and offshore structures. Frigates and patrol vessels and replinishment tankers for Hellenic Navy. → 75 → 80 [98]

60 Research Organizations, Institutes & Test Establishments

60-00100 Hellenic Electronics Equipment Quality Assurance Centre (HEEQAC)
6 Ierosolimon Straat, Amerikis Square, 112 52 Athens
☎ (+30 1) 865 56 11; 861 39 28; 861 35 75
▣ 218326 gehr gr ⊖ (+30 1) 867 71 75
♧ US$ 250,000 (1996) ♣ US$ 2,000,000 (1996) ⅱ 50
⌂ 5,000
Managing Director: George Chrysanthacopoulos
Marketing Director: C Cariniotakis
■ Design and manufacture of electronic systems. Quality control and assurance testing of electronic devices. [98]

60-00300 Hellenic Foundation for European & Foreign Policy
4 Xenophontos Street, 10557 Athens
☎ (+30 1) 33 15 22 Ⓔ eliamep@compulink.ge
⊖ (+30 1) 364 21 39
ⅱ 25
Head Documentation Department: Ms Vassi Leontari
■ Research, conferences, publications, teaching (postgraduate), electronic, data processing and documentation. [98]

65 Industry, Trade & Armed Forces Associations

65-00500 Association of Hellenic Defence Manufacturers (SEKPY)
10 Messogion Avenue, 115 27 Athens
☎ (+30 1) 770 24 39 ⊖ (+30 1) 770 24 39
President: George Stavrou
Managing Director: George Tsolkas ☎ (+30 1) 778 89 66
Secretary General: Michael Anglos ☎ (+30 1) 778 89 66
Sales and Marketing Director: Ms Sofia Tsolkas
☎ (+30 1) 770 24 39
■ Non-profit trade organisation. Represents over 90 companies in the Greek defence industry. [98]

65-01000 Federation of Greek Industries
5 Xenofontos Street, 105 57 Athens
⊖ (+30 1) 322 29 29 [98]

75 Manufacturing Companies

Hellenic Shipyards Co. 55-00850
SSA SA 90-01980

75-00030 Advanced Autoelectric Hellas Ltd
(a member of Hellenic Manufacturers of Defence Materials Association)
3rd km Argos-Tripolis, 212 00 Argos
☎ (+30 751) 209 89; 284 62 ⊖ (+30 751) 213 27
■ Manufacturing assembly of vehicle starters and electrical equipment. [98]

75-00050 Advanced Technology Applications
(a member of Hellenic Manufacturers of Defence materials Association)
Volos Industrial Area Road, 385 00 Volos
☎ (+30 421) 955 67; 955 67; 966 68 ▣ 282375 tsia gr
⊖ (+30 421) 955 69
♣ US$ 3,000,000 (1993) ⅱ 20 ⌂ 10,000 sq ft
Marketing/Sales: N Tsianakas
Advertising Manager: G Christogianafis
■ Production of aircraft/engine spare parts and systems development including low-level air defence systems, remotely-piloted vehicles, rocket launchers, chaff and flare dispensers, dummy bombs and maintenance facilities. [98]

75-00200 Anco SA
Head Office
44 Syngrou Avenue, 117 42 Athens
☎ (+30 1) 923 68 61 Ⓔ i4e3lis@athen.compulink.g
▣ 215303 ⊖ (+30 1) 922 88 02
♧ Dr 400,000,000 (1996) ♣ Dr 2,464,000,000 (1996)
ⅱ 130 (1996) ⌂ 4,850 m2
Managing Director: Char Delakouras
Finance Director: K Angelis
Sales and Marketing Director: Michael Dimitriadis
Public Relations: Chris P Hamis
■ Digital telecommunications networks, radio communications networks, telematics products and services, electronic test and data communication networks, measuring instruments, power supplies for telecommunications equipment, iron and steel representations of raw materials and finished products. Closed circuit television (CCTV) power supplies. [98]

75-00280 Astron SA
PO Box 30, 193 00 Aspropyrgos
☎ (+30 1) 557 73 20 ▣ 215827 dmca
⊖ (+30 1) 557 42 88
ⅱ 42
Chairman: Dimitri Kanakis
■ Galvanised metal constructions; underground shelters for aircraft; ammunition depots. [98]

75-00330 Axon Ltd
(a member of Hellenic Manufacturers of Defence Materials Association)
7 Delfon Street, 151 25 Marousi, Athens
☎ (+30 1) 689 70 65; 680 43 68 ▣ 214811 xenagr
⊖ (+30 1) 689 71 65
♧ Dr 3,000,000 ♣ Dr 120,000,000 ⅱ 20 ⌂ 1,000 sq m
Contact: K Papandreou
■ Metal constructions; machining works; spare parts for military and civil use. [98]

Contact Reference

75-00400 Bail Ltd
(a memeber of Hellenic Manufacturers of Defence Materials Association)
38 Plapouta Street, 114 73 Athens
☎ (+30 1) 821 70 77 ⊖ (+30 1) 825 02 37
Contact: Costas Papandreou
■ Systems in informatics and electronic security. [98]

75-00450 Barracuda Hatzistamatiou
(a member of Hellenic Manufacturers of Defence Materials Association)
22 Erinis Avenue, 177 78 Tavros
☎ (+30 1) 341 09 86 ⊖ (+30 1) 345 61 09
Contact: A Hatzistamatiou
■ Construction of inflated boats. [98]

75-00570 BMTE SA
(a member of Hellenic Manufacturers of Defence Materials Association)
2 Anat Thrakis Street, 161 22 Kaisariani
☎ (+30 1) 724 61 82 ⊖ (+30 1) 722 38 31
ⅱ 55
■ Fully automatic technology, quality control laboratory contains transmission spectrograph, metallographic microscope, tensile & bending machine, impact test machine, hardness measuring machines, ultrasonic apparatus, chemical laboratory, exporting bearing blocks & heat resistant materials spare parts. [98]

75-00600 Bougas Panagiotis Ltd
(a member of Hellenic Manufacturers of Defence Materials Association)
20 Dimitros Street, 177 70 Tavros, Attikis
☎ (+30 1) 342 15 40; 345 80 59 ⊖ (+30 1) 342 15 40
ⅱ 5
Managing Director: Pan Bougas
■ Manufacturer of stainless steel and aluminium constructions, fuel tanks, muffers for vehicles and tanks. [98]

75-00700 Catsaros Automation SA
Paparrigopoulou 13, 105 61 Athens
☎ (+30 1) 322 61 09; 323 82 80 ▣ 210357 auto gr
⊖ (+30 1) 322 38 66
♧ US$ 800,000 (1994) ♣ US$ 3 000 000
♣ US$ 3,500,000 (1994) ⅱ 25 ⅱ 24 ⌂ 800 m2
Managing Dir: Dipl.Ing. Byron Catsaros [97]

75-00750 Chem - Tech
(a member of Hellenic Manufacturers of Defence Materials Association)
8 Appianou and Argonauton Street,
121 31 Peristeri, Athens
☎ (+30 1) 574 25 06 ⊖ (+30 1) 576 27 59
Contact: Ch Papadopoulos
■ Manufacture printed circuit boards. [98]

75-00800 Chronopoulos Ltd
(a member of Hellenic Manufacturers of Defence Materials Association)
29 Lechourity Street, 121 32 Peristeri, Athens
☎ (+30 1) 571 77 49 ⊖ (+30 1) 573 52 90
👥 23
Contact: A Katsanos
■ Metal accessories production, spare parts. [98]

75-00910 Chrysanthacopoulos SA
Electronic & Communications Enterprises (ECE)
6 Ierosolymon Street, 112 52 Athens
☎ (+30 1) 865 56 11; 861 39 28; 861 35 75
⊠ 218326 gehr gr ⊖ (+30 1) 867 71 75
♧ US$ 1,500,000 (1996) ♣ US$ 3,000,000 (1996)
👥 500 🏭 3,000
Managing Director: George Chrysanthacopoulos
Assistant General Manager: Constanine Cariniotakis
■ EW Systems; Jammers for radiotele- communications; Early warning receivers (ESM RECEIVERS); Deception jammer systems RGPO/XEYE; Special types of antennas at the frequency band of VHF/UHF/SHF; Mobile, portable, base station transceivers at the frequency band of HF/VHF/UHF; Repeaters, radio links and transporters at the frequency band of VHF/UHF/SHF; Transducer systems for telecommunication applications; Radiobeacons for military and marine use; Marine watch receivers; Automatic recorders with data loggers for telecommunication systems; Educational simulator systems; Auto-visual educational systems; Simultaneous translating systems for conference; Electroacoustic public address systems for use in ships and buildings; Signalling systems for railways; Electronic power supplies, stabilizers and inverters. [98]

75-00915 Colettis J. & T. Metal Works Ltd Main Office & Factory
(a member of Hellenic Manufacturers of Defence Materials Association)
103, 25th Martiou Street, 121 32 Peristeri
☎ (+30 1) 572 36 18 ⊖ (+30 1) 575 93 62
👥 6
Contact: J Colettis
■ Construction of metallic parts. [98]

75-00940 Danezis-Bellos (Tekesyp) SA
(a member of Hellenic Manufacturers of Defence Materials Association)
4-6 Fliountos Street, 104 77 Votanikos
☎ (+30 1) 347 90 35 ⊠ 219492 ⊖ (+30 1) 347 13 51
👥 8
Managing Director: E Danezis
Managing Director: J Bellos
■ Connectors for middle, high and very high pressure, control cables (push-pull), tubes, metal and non-metal, for various applications, assembly of pressure tubes according to specialisation, INOX pressure tubes. [98]

75-00980 Dourou Bros & Co. SA Head Office
(a member of Hellenic Manufacturers of Defence Materials Association)
60 Filonos Street, 18535 Peiraias
☎ (+30 1) 412 46 84 ⊠ 241626 ⊖ (+30 1) 412 98 03
👥 30
Contact: O Douros
■ Manufacture of boilers, metal constructions, machining works, ship general repairs, ship dismantling, diesel fuel containers.
Factory
101 L. Athinon, 18541 Peiraias
☎ (+30 1) 411 56 18/19; 481 25 98
⊖ (+30 1) 412 98 03 [98]

75-00990 Dramaliotis SA
(a member of Hellenic Manufacturers of Defence Materials Association)
4th Km - Ethnikis Road, 40300 Farsalon Mikrothivon
☎ (+30 491) 221 41; 22 22 79 ⊖ (+30 491) 221 42
👥 90
Contact: T Dramaliotis
■ Machining and electrical works. [98]

75-00995 Dromor Yachts SA
(a member of Hellenic Manufacturers of Defence Materials Association)
Kalivia Thorikou GR, 19010 Attiki
☎ (+30 299) 299 00 ⊖ (+30 299) 230 67
👥 75
Contact: N Drossopoulos
■ Fast cruising boats, pilot and patrol boats, inflated and sailing boats for special purpose. [98]

75-01050 EBEH A. Kypriotis SA
(a member of Hellenic Manufacturers of Defence Materials Association)
1st km, Koropiou-Varis Road, 19400 Koropi
☎ (+30 1) 662 46 81 ⊖ (+30 1) 662 45 68
👥 20
General Manager: A Kypriotis
■ Manufacture of induction heating equipment, heat treating, uninterruptable power supplies (UPS). [98]

75-01060 Econ Industries SA
Head Office
32 Kifissias St, 15125 Maroussi
☎ (+30 1) 680 06 20 ⊖ (+30 1) 680 06 60
Managing Dir: D. Economides
Deputy Managing Dir: Alexandros Economides
Sales & Marketing Dir: P. Papantonopoulos
Purchasing Mgr: M. Apartoglou
Plant
PO Box 60, 19003 Markopoulo
☎ (+30 299) 238 02; 31/3 ⊖ (+30 299) 238 30
Operations & Technical Dir: P. Dimakos
Subsidiaries:
Econ Electronics
Econ Mechanics
Econ Optics [97]

75-01070 Econ Optics Mechanics SA
(a member of Hellenic Manufacturers of Defence Materials Association)
PO Box 60, 190 03 Markopoulo
☎ (+30 299) 238 31 ⊖ (+30 299) 238 30
👥 80
Managing Director: D Economidis
■ Manufacture of rangefinders, night vision equipment, spare parts for vehicles and tanks, upgrade kits for M48 tanks. [98]

75-01100 EIS Hellas Ltd
(a member of Hellenic Manufacturers of Defence Materials Association)
64 Louisis Riancor, 115 23 Athens
☎ (+30 1) 691 52 76 ⊖ (+30 1) 691 07 59
👥 12
Contact: Klis Papatsos [98]

75-01127 Electrometal SA
(a member of Hellenic Manufacturers of Defence Materials Association)
Nato Avenue - Spithari, 19300 Aspropirgos
☎ (+30 1) 557 18 22 ⊖ (+30 1) 535 05 49
👥 13
Contact: F Nikolaidis
■ Metal platings in general, anodising of aluminium, plating in cooper, zink and cadium. [98]

75-01130 Elefsis Shipyards SA
(a member of Hellenic Manufacturers of Defence Materials Association)
Elefsina GR, 19200 Elefsis
☎ (+30 1) 553 51 11 ⊖ (+30 1) 554 60 16
👥 1,950
Contact: J Stergiopoulos
■ Shipbuilding, shipconversions, shiprepairing and Heave industrial constructions. [98]

75-01150 Elfon Ltd
(a member of Hellenic Manufacturers of Defence Materials Association)
30 Ikarias Street, 16675 Glifada
☎ (+30 1) 962 82 12; 962 82 34; 962 82 556
⊠ 218243 vaco gr ⊖ (+30 1) 962 85 39
👥 50
Contact: George Chrisos
Contact: M Dikis
Contact: D Kemos
■ Manufacture of electrical and electronic subassemblies. Design and development of high technology material. [98]

75-01200 Elviemek SA
(a member of Hellenic Manufacturers of Defence Materials Association)
32 Kifisias Avenue, Pirgos Atrina, 151 25 Maroussi
☎ (+30) 682 86 01 ⊠ 214258 elv gr
⊖ (+30 1) 684 15 24
👥 145
Contact: S Parlas
Contact: Mr Prokopiou
■ Manufacture of explosives and ammunition. [98]

75-01220 Elvitex SA
(a member of Hellenic Manufacturers of Defence Materials Association)
30-32 El Venizelou Street, 14231 North Ionia, Athens
☎ (+30 1) 277 04 21; 279 97 07 ⊠ 0219207 gepegr
⊖ (+30 1) 277 04 03
♧ Dr 514,335,000 (1996) ♣ Dr 1,600,000,000 (1996)
👥 130 🏭 9,500 m2
Contact: K Partsemlidis
■ Manufacturers of military and working protective garments. [98]

75-01250 Emka SA
(a member of Hellenic Manufacturers of Defence Materials Association)
62 Kifisias Avenue, 151 25 Maurousi, Athens
☎ (+30 1) 689 90 76; 689 90 78; 689 90 79
⊖ (+30 1) 689 90 75
👥 15
Contact: G Stavrou
Contact: St Komnopoulos
■ Spare parts for military vehicles and armoured tanks. [98]

75-01280 Epirus Foundries SA
(a member of Helenic Manufacturers of Defence Materials Association)
PO Box 1025, 45110 Loanninon
☎ (+30 651) 615 03; 626 66 ⊠ 322173 epfo gr
⊖ (+30 651) 291 71; 349 38
👥 64
Contact: Sp Kitsatis
■ High precision investment castings of various metals for vehicles, rifles, machinery and components. [98]

75-01285 Ergon Battaries Ltd
(a member of Hellenic Manufacturers of Defence Materials Association)
11 Praxitelous Street GR, 10562 Athens
☎ (+30 1) 246 30 31 ⊖ (+30 1) 246 94 89
👥 15
Contact: K Strofilas
■ Lead-acid industrial batteries, traction batteries both P2S and P2S-H types, stationary batteries with OP2S cells, battery charger service. [98]

75-01288 Eurodef Aerospace & Defence Industries SA
22 Leof Dimokratias, 154 51 N Psychiko
☎ (+30) 677 82 40; 674 75 34 ⊖ (+30) 16 72 27 94
♧ US$ 1,000,000 ♣ US$ 8,500,000 (1994)
Director: Dr Y Kavaratzis
Marketing Manager: Peter Letros
Marketing and Sales Executive: Ms Theodora Maniati
■ Products: weapon systems design and development, airborne weapons and ordnance, aircraft pylons, ejector racks, LGB laser guided bombs, helicopter and aircraft modifications, gunship conversions. Rocket launchers 2.75 inch, flare launchers SUU-25, helicopter emergency floats, aeroengine hot section brakes for F-4, A-7, Mirage 2000, Mirage F1.
Bosa Weapons Facility
Volos
Alpha Aerospace Facility
385 00 A'Vipe [97]

75-01293 Fockas Pantelis Ltd
(a member of Hellenic Manufacturers of Defence Materials Association)
16 Palaiologou, 16673 Voula, Athens
☎(+30 1) 895 95 24 ⊜ (+30 1) 895 18 42
👥 10
Contact: P Fokas
■ Electronic applications, technical systems for airport lighting, shooting range systems, automatic control with microprocessors and specialised floodlights. [98]

75-01294 C. Gavriil SA
17th Km National Road, Lamia N.K. Kifissia, Athens
☎(+30 1) 807 12 90 ⊜ (+30 1) 620 33 12
US$ 4 000 000 (1995) ♣ US$ 2 000 000 (1995) 👥 45
4 100 sq m
Managing Dir: Christos Gruriil
Operations & Technical Dir: Sokratis Gruriil
■ Production of textile products for special end-uses mainly for the military sector. [97]

75-01295 Eurodef Aerospace & Defence Industries S.A
22 Leof Dimokratis, N Psychiko, 15451 Athens
☎(+30 1) 677 82 40, 674 75 34 ⊜ (+30 1) 672 27 94
US$ 500,000 (1997) ♣ US$ 8,000,000 (1996) 👥 52
10 000 sq m
Managing and Operations Director: Y Kavaratzis
Sales and Marketing Manager: Peter Letros
Purchasing Manager: S Politis
■ Manufacture of: rocket launcher LAU-68, LAU-131; aircraft flare launcher SUU-25F/A; cluster bomb dispenser CBU-71, CBU-82; aircraft incendiary bomb BLU-1/B, M116A2; Sidewinder training missile AIM-9P, AIM-9L; R550 Magic; practice bomb BDU-33, MK106, P11.
Manufacture of aircraft parts: carbon brake (Mirage 2000, F1) HS-M200; multiple and triple ejectors MER/TER; wing and centerline pylons; external fuel tanks; emergency flotation kit (helicopter) EFS.
Service: Aeroengine hot section repair. [98]

75-01297 General Cables Orfandis Brs SA
(a member of Hellenic Manufaturers of Defence Materials Association)
197 Antigonis Street GR, 10443 Athens
☎(+30 1) 513 56 11; 513 56 16
☎(+30 741) 227 74 (factory) ⊜ (+30 1) 512 48 00
US$ 250,000,000 (1996)
♣ US$ 4,600,000,000 (1996) 👥 140 15,000 m2
Managing and Operations Director: C Dounias
Marketing Director: K Malieras
Purchasing Manager: G Keramidiotis
Trading and Financial Advisor: A Dounia
■ Production of cables for electrical and communications use, and optical fibres during end of 1997 and beginning of 1998. [98]

75-01300 General Electronics Communications Consultants Contractors Ltd
GEC 3 Ltd
(† General Electonics)
(a subs. of Milcom International Ltd)
PO Box 51084, 145 10 Kifissia, Athens
☎(+30 1) 801 25 48; 623 33 40 ⊠ 218347 ge gr
⊜ (+30 93) 42 11 42
Dr 30 000 000 (1994) ♣ Dr 150 000 000 (1994)
👥 50 5 500 sq m
General Mgr: Din Psiloyannis
Technical Mgr: F. Georgiades
Sales & Marketing Mgr: E. Theodossiades
Purchasing Mgr: A. Kotsanis
■ Design & manufacture of antenna systems (towers, radomes, reflectors) from low frequency, medium HF, VHF to microwave & satellite, for military customers worldwide. Also manufacture of electronic systems telecommunications equipment & related support systems. Systems & services: frequency management systems for armed forces & government. Surveillance systems, construction & installation (complete engineering). Antenna site acquisition (management, analysis). Broadcast industry consulting & contracting, mobile tracking systems - voice, data. Planning & development. → 80 → 90 [97]

75-01325 Geniki Michanourgiki SA
(a member of Hellenic Manufacturers of Defence Materials Association)
32 Peiraios, 18346 Moschato
☎(+30 1) 481 47 61 ⊜ (+30 1) 482 45 27
👥 8
Contact: J Christofis
■ Die works for tanks, spare parts, jigs, precision machining works. [98]

75-01335 Germanos Batteries SA
(a member of Hellenic Manufacturers of Defence Materials Association)
14452 Athens
12th km National Road Athens - Lamia
☎(+30 1) 284 25 00 ☎(+30 541) 952 40 (factory)
⊜ (+30 1) 284 26 03 ⊜ (+30 541) 954 46 (factory)
👥 150
Contact: M Zachariadis
■ Batteries (zinc-carbon, nickel, cadmioum lead acid) for different technologies, chargers, backup units, power supply units, rectifiers, inverters, special battery packs. [98]

75-01336 GMC SA
(a member of Hellenic Manufacturers Defence Materials Association)
Anatolikis Thrakis 2, Kaisariani
☎(+30 1) 723 93 59
☎(+30 541) 918 02; 918 24; 918 46 (factory)
E misios@xan.forthnet.gr ⊜ (+30 1) 722 38 31
⊜ (+30 541) 918 80 (factory)
4132GMC US$ 2,500,000 (1997)
♣ US$ 6,000,000 (1996) 👥 120 (1997) 12,500 m2
Managing Director: Kabakas Elis
Production Director: Missios Dimitpis
Operations Director: Rofaelos Kostas
Marketing Director: D Mislos
Purchasing Manager: A Patsatzis
■ Steel cast iron castings, machining. [98]

75-01350 Georgakis Anastasios
(a member of Hellenic Manufacturers of Defence Materials Association)
15 Othonos Street, 548 33 Thessalonica
☎(+30 31) 23 81 93 ⊜ (+30 31) 27 00 43
Contact: A Georgakis
■ Manufacturer of electromechanical, electrological and electronic applications. [98]

75-01360 Grammenos Athan & Co. Ltd
(† A.Grammenos)
(a member of Hellenic Manufacturers of Defence Materials Association)
16 Kallidromou N Sepolia, 12133 Peristeri
☎(+30 1) 574 72 45 ⊜ (+30 1) 575 69 46
👥 8
Contact: Ath Grammenos
■ Manufacture of gears of various sizes and specifications, engine parts conforming to military specifications. [98]

75-01400 Grantex Hellenic Brakes Industry SA
(† Grantex SA)
(a member of Hellenic Manufacturers of Defence Materials Association)
Terma Orpheos GR, 122 41 Aegaleo, Athens
☎(+30 1) 346 59 42 ⊜ (+30 1) 347 21 85
👥 60
Contact: N Paskalidis
■ Brake linings, disc brake pands, clutchs, friction materials, brake parts. [98]

75-01500 Greek Powder & Cartridge Co.SA (PYRKAL)
(a member of Hellenic Manufacturers of Defence Materials Association)
1 Ilioupoleos Avenue, 172 36 Hymettus, Athens
☎(+30 1) 971 39 62; ⊠ 221986 eepk gr
⊜ (+30 1) 970 50 09
Dr 65,000,000,000 ♣ Dr 15,000,000,000 👥 1,650
1,065 000 sq m
Contact: M Nikoladis
■ Manufacture of a wide range of ammunition covering the requirements of the Greek Forces and various foreign markets. [98]

75-01650 Hatjiantoniou SA
(a member of Hellenic Manufacturers of Defence Materials Association)
3 Tripoleos Street GR, 18866 N Ikonio, Piraeus
☎(+30 1) 431 17 51 ⊜ (+30 1) 431 26 51
👥 25
Contact: G Hatjiantoniou
■ Internal combution (diesel) engine parts such as cylinder heads, liners, inlet casting, piston rings and manipolds. [98]

75-01700 Hellenic Aerospace Industry SA
(a member of Hellenic Manufacturers of Defence Materials Association)
2-4 Messogiou Avenue GR, 71525 Athens
☎(+30 1) 779 96 78 ⊜ (+30 1) 779 79 41
👥 3,015
Contact: M Grammatopoulos
■ Support services to aircraft and engines. Manufacture of aircraft and engines, electronic products and systems. Training services in all aerospace industry disciplines.
Tanagra Factory
PO Box 23, 32009 Shimatari
☎(+30 1) 883 67 11 ⊜ (+30 1) 883 87 14 [98]

75-01780 Hellenic Vehicle Industry SA (ELBO)
7 Anastaseos Straat, 15561 Holargos, Athens
1 Perikleous St, 15561 Holargos, Athens
☎(+30 1) 652 45 31 ⊠ 222082 elbo gr
⊜ (+30 1) 653 33 78
Dr 2,757,382,000 ♣ Dr 26,000,000,000 👥 1,000
270,000 sq m
President and Managing Director: Likourgos Sakellaris
Managing Director: Elias Efthimiadis
Commercial Director: Ektor Valetopoulos
Finance Director: Dimitris Missios
Technical Director: Vassilis Kazantzoglou
After Sales Director: Dimitrios Keskinis
Marketing Manager: Alkiviadis Kompselidis
Marketing Office
230 Messogion Ave & 1 Perikleous St,
155 61 Holargos, Athens
230 Messogion Avenue and 1 Perikleous Street
☎(+30 1) 652 45 31; 652 21 30 ⊠ 222082 sth gr
⊜ (+30 1) 653 33 78
Managing Director: Elias Eftimiadis
Public Relations: Ath Kotsakis
Head Office Factory
PO Box 10239, 541 10 Thessaloniki
Industrial Area of Thessaloniki, 570 22 Sindos
☎(+30 31) 79 85 02; 79 86 02 ⊠ 412586 elbo gr
⊜ (+30 31) 79 84 26
General Manager: Elias Eftimiadis
■ Production of military trucks, armoured personnel carriers, fire engines, buses and diesel engines. → 90 [98]

75-01805 Hitec - Modern Technology and Computer Applications SA
(a member of Hellenic Manufacturers of Defence Materials Association)
204, Messogion Avenue, 155 61 Holargos, Athens
☎(+30 301) 654 88 73 ⊜ (+30 301) 654 77 48
👥 45
Contact: Mr Kabouroglu
■ Defence material automation systems, electronic information management systems, computer and data networks, system simulators, command and control systems, electronic document archiving, defence material and spare parts sales. [98]

75-01810 Hourdakis SA
(a member of Hellenic Manufacturers of Defence Materials Association)
204 Messogion Avenue, 155 61 Holargos, Athens
☎(+30 1) 662 40 44 ⊜ (+30 1) 622 27 86
👥 132
Contact: M Hourdakis
■ Manufacture printed circuits and electronic telecommunication equipment. [98]

75-01835 Iatriki Techniki
(a member of Hellenic Manufacturers of Defence Materials Association)
5-7 Distomou Street, 151 25 Maroussi, Athens
☎(+30 1) 689 82 02 ⊜ (+30 1) 689 85 55
👥 25

Contact: S Kavvadias
■ Production of surgical instruments machinery works. |98|

75-01850 Iliakopoulos Ltd
(† Athansios Iliakopoulos Workshops)
(a member of Hellenic Manufacturers of Defence Materials Association)
11 Ag Nikolaou, 17455 Alimos
☎(+30 1) 983 19 52 ⊖ (+30 1) 982 40 27
👥 6
Contact: Ath Iliakopoulos
■ Metal machining works (bushings, cylinders, magnetic and non magnetic liners, bolts, couplings, crankshafts, pistons). |98|

75-01960 Interkat SA
(a member of Hellenic Manufacturers of Defence Materials Association)
8th km - Varis-Koropiou Road, 19400 Portsi area
☎(+30 1) 662 36 61 ⊖ (+30 1) 662 42 58
👥 42
Contact: S Gourzoulidis
■ Car seats and mirrors; air and oil filters; tactor fences and side plates; press tooling construction. |98|

75-01962 Interoperability Systems Interntional Hellas SA (ISI)
(a member of Hellenic Manufacturers of Defence Materials Association)
104 Gennimtaa Street, 165 61 Ano Glyfada
☎(+30 301) 964 77 56; 964 77 57
👥 28
Contact: Hirt Alan
■ Design, development and programming of software intensive tactical C3 systems. Provides C3 and interoperability systems for the domestic and international market. Design and deliver multi layer printed circuit cards for implementation of special interface or processing requirements. |98|

75-01964 Intracom SA
(a member of Hellenic Manufacturers of Defence Materials Association)
19002 Paiania Attikis
☎(+30 1) 664 49 61; 884 37 15; 686 00 00
⊖ (+30 1) 664 43 79; 664 37 18
👥 1,400
Contact: G Troulinos
■ Digital switching centres, power supply systems, telecommunications equipment (voice-data-fax), PCM systems, radio links, real time telecommunication equipment software. |98|

75-01965 Intramet SA
(† Intramet Steel Structures & Electromechanical Equipment S.A.)
(a member of Hellenic Manufacturers of Defence Materials Association)
19.5 km - Leof Markopoulou, PO Box 68, 190 02 Paiania Attikis
☎(+30 1) 664 49 61; 884 37 15; 686 00 00
⊠ 6646211; 295421 kats gr ⊖ (+30 1) 664 37 18
👥 112
Contact: A Achouzaridis
Contact: K Stoupis
■ Manufacture of steel towers for overhead power transmission lines, steel structures for power stations and substations, opencast mining constructions (belt conveyors, bucket-wheel, excavators), industrial steel buildings. |98|

75-01970 Kanakas Ltd
(a member of Hellenic Manufacturers of Defence Materials Association)
Chalastra Thessalonikis
☎(+30 31) 79 38 84 ⊖ (+30 31) 79 38 35
👥 8
Contact: Th Kanakas
■ Manufacture of parts for armoured vehicles and trucks. |98|

75-01972 Kapa SA Texoga
(a member of Hellenic Manufacturers of Defence Materials Association)
127 Thessaloniki Street, Moschato, Athens
☎(+30 1) 481 82 24; 481 89 12 ⊖ (+30 1) 481 80 54
👥 11
Contact: G Mitsotakis
■ Manufacture anti-detonating material. |98|

75-01975 Karamanou Bros Ltd
(a member of Hellenic Manufacturers of Defence Materials Association)
A Industrial Area, 38 500 Volos
☎(+30 421) 950 60 ⊖ (+30 421) 399 97
👥 4
Contact: Mr Karamanos
■ Manufacture of sheet metal parts. |98|

75-01978 Karcher Cleaning Systems AE
99 Aristotelous and Ioanninon, 13671 Aharnes
☎(+30 1) 244 13 12 ⊖ (+30 1) 246 31 32
Managing Director: Dimitrios Shistohilis
Representing:
Alfred Karcher GmbH & Co |98|

75-01980 Karipidis Bros. SA
20 Dimaraki St, 118 55 Votanikos, Attikis
☎(+30 1) 345 61 53; 345 70 37 ⊠ 222210 kar gr
⊖ (+30 1) 347 80 14
♲ Dr 50 000 000 (1995) ♣ Dr 687 553 000 (1995)
👥 30 👥 70 🏭 5 000 sq m
Managing Cons.: Kyriakos G. Karipidis
Military Equipment Branch: Yiannis P. Koutoselis
■ Precision rubber & metal rubber parts; car & truck brake repair kits; parts according to customer needs, drawings & specifications. |97|

75-01981 Kentavros-Kouloussiou Bros SA
(a member of Hellenic Manufacturers of Defence Materials Association)
PO Box 40065, 56000 Stravroupoli-Thes/niki
☎(+30 31) 68 35 01 ⊖ (+30 31) 68 19 49
👥 40
Contact: Ath Koulousios
■ Manufacture of rubber products for military and civil use. |98|

75-01982 Kioleides SA
(a member of Hellenic Manufacturers of Defence Materials Association)
22nd km - Athinon Lamias Krioneri, PO Box 51156, 14510 Kifissia
☎(+30 1) 816 19 80; 807 16 00 ⊖ (+30 1) 816 17 39
👥 150
Contact: N. Kioleides
■ Manufacture of heavy trucks & tank carriers. |98|

75-01986 Kyriazis Ltd
(a member of Hellenic Manufacturers of Defence Materials Association)
3 Rodopis Street, 11855 Votanikos
☎(+30 1) 345 48 76 ⊖ (+30 1) 346 56 39
👥 4
Contact: K Kyriazis
■ Production of machinery for bolts and nuts. |98|

75-01989 Kuhler Hellas SA
(a member of Hellenic Manufacturers of Defence Materials Association)
A Industrial Area, 385 00 Volos
☎(+30 421) 951 13; 951 57 ⊖ (+30 421) 951 35
👥 27
Contact: B Theocharakis
Contact: St Papadopoulos
■ Manufacture cooling units for vehicles. |98|

75-01995 KYTE SA
(a member of Hellenic Manufacturers of Defence Materials Association)
Industriel Area Kilkis, PO Box 14, 611 00 Kilkis, Athens
☎(+30 341) 718 80 ⊖ (+30 341) 715 10
👥 12
Contact: A Tsakirdis
■ Manufacture and assembly of electronics, converters, telecommunication systems and security systems. |98|

75-01998 Macedonia Machine Works SA
(a member of Hellenic Manufacturers of Defence Materials Association)
1 Smyrnis Street, 50100 Kozani
☎(+30 461) 242 15 ⊖ (+30 461) 242 16
👥 16
Contact: C Anthropoulos
■ Manufacture of spare parts for tanks. |98|

75-01999 MacOptic SA Head Office
6-8 Fragon St, 54626 Thessaloniki
☎(+31 30) 54 23 22 |E| macoptic@spark.net.gr
⊖ (+31 30) 53 90 09
♲ Dr 550,000,000 (1996) 👥 20 (1996) 🏭 1,800 sq m
President: Sakis Hadjivalasis
Marketing Director: Emmanuel Kyriakidis
Purchasing Manager: John Telioudis
■ Production of vision blocks with infrared filters for Western and Eastern APC's and MBT's. Overhaul of vision blocks. |97|

75-02000 Mandra Plant
(† General Engineering Ltd)
(part of The EBO Group of Companies and a subsidiary of Hellenic Arms Industry)
160 Kifissias Avenue, 115 25 Athens
☎(+30 1) 647 26 11 |E| ebosem@ath.forthnet.gr
⊠ 218562 ebo gr ⊖ (+30 1) 647 27 15
♲ Dr 2,000,000,000 ♣ Dr 20,000,000,000 👥 300
🏭 15,00 m2
Managing Director: Kakinakis Theodore
Operations Director: Karauassilis Dimitri
■ Drop fuel tanks, pylons, external stores, mechanical bomb loaders, A/A systems and hybrid gun/missile systems. |98|

75-02002 Manolessos MFG Co
(a member of Hellenic Manufacturers of Defence Materials Association)
3 Verikokias Street, 185 34 Piraeus
☎(+30 1) 413 60 86 ⊠ 241951 ⊖ (+30 1) 413 18 45
👥 6
Contact: Mr Vretos
Contact: Mr Paviou
■ Metal working ateliers for military uniforms. |98|

75-02007 Z. Mantos SA
(a member of Hellenic Manufacturers of Defence Materials Association)
3 Verkokias Street, 185 34 Piraeus
☎(+30 1) 240 27 91 ⊖ (+30 1) 240 27 90
👥 12
Contact: Z Mantos
■ Manufacture ambulances, spare parts, closets vans, bullet proof vehicles, and trailers. |98|

75-02010 Matren Ltd
(a member of Hellenic Manufacturers of Defence Materials Association)
5 Delfon, 106 80 Athens
☎(+30 1) 362 31 53; 361 81 20 ⊠ 219338 anho gr
⊖ (+30 1) 360 81 28
👥 3
Contact: Michael Agglos
■ Retrofit kits for M48 tanks, spares for M47 and M48 and M113 tanks, multicontact electrical connectors, ammunition boxes, containers for large calibre rounds, propelling charges, ammunition magazines for G3 rifles. Precision stamping parts for aircraft and aluminium fuel tanks. |98|

75-02025 Metal Constructions of Greece SA (METKA)
(a member of Hellenic Manufacturers of Defence Materials Association)
11 Marinou Antypam, N Iraklio, Athens
☎(+30 1) 421 604 61 ⊖ (+30 1) 665 82
👥 410
Contact: Mr Anastassopoulos
■ Manufacture heavy and compound electromechanical equipment for mines, ports, industries, electric power stations. The erection and installation works of heavy metal-mechanical constructions as well as large electromechanical and energy projects. |98|

75-02027 Metelko SA
(a member of Hellenic Manufacturers of Defence Materials Association)
8 Kritis Street, 144 52 Athens
☎(+30 1) 283 05 13 ⊖ (+30 1) 281 09 98

Contact: P Toskos
■ Construction of printed circuit boards. [98]

75-02030 **Mevaco Metalourgiki SA**
(a member of Hellenic Manufacturers of Defence Materials Association)
Prari Moustaki Aspropyrgos, 19300 Attikis
☎ (+30 1) 248 09 34 ⊖ (+30 1) 248 01 02
👥 40
Contact: D Delendas
■ Manufacture of metallic parts for air conditioning and heating systems, solar systems, electronic systems, gas containers, military material. Air pressure vessels, steel boilers, special metallic constructions. [98]

75-02050 **Mikron Ltd**
(a member of Hellenic Manufacturers of Defence Materials Association)
17 Mainemenis, 117 78 Tavros
☎ (+30 1) 347 98 40 ⊖ (+30 1) 347 98 45
👥 15
Contact: A Kantzis
■ Manufacture of special die constructions, spare parts for armoured vehicles, machining works. [98]

75-02060 **MILCOM International Ltd**
PO Box 51086, 145 10 Athens
☎ (+30 1) 623 33 40; (+30 93) 27 67 11 ⊠ 218347 gr
⊖ (+30 1) 807 86 97; 623 33 41
⊕ £ 20,000,000 ♣ £ 80,000,000 👥 25 🏭 2,000 sq m
Managing Director: E Theodossiades
Sales and Marketing Director: N Papageorgiou
Operations and Technical Director: K Spyros
■ Telecommunication products. [97]

75-02080 **Mostro Ltd**
(a member of Hellenic Manufacturers of Defence Materials Association)
105 Vouliagmenis Avenue, 16674 Athens
☎ (+30 1) 963 29 88 ⊖ (+30 1) 963 19 22
👥 40
Contact: K Farmakis
■ Manufacture of military inflatable boats, inflatable floating bridges for tanks, inflatable platforms for helicopter sea landings. [98]

75-02150 **Netcom Ltd**
(a member of Hellenic Manufacturers of Defence Materials Association)
4 Metamorfoseos Street, 15344 Glyka Nera
☎ (+30 1) 665 88 02 ⊖ (+30 1) 665 88 80
👥 76
Contact: A Kousathanas
Contact: K Manthos
■ Electronic material, frequency crystals, crystal filters, filters custom made design. [98]

75-02160 **Nikpa Springs SA**
(a member of Hellenic Manufacuturers of Defence Materials Association)
12 Karaiskaki, 144 51 Metamorfosi
☎ (+30 1) 279 93 18; 279 62 44 ⊖ (+30 1) 282 81 88
👥 25
Contact: P Papadopoulos
■ Manufacture of springs for weapons and weapons systems, for vehicles and electromechanical applications. [98]

75-02165 **Olympic Filter SA**
(a member of Hellenic Manufacturers of Defence Materials Association)
57-59 Palamidou Street, 10441 Athens
☎ (+30 1) 514 52 81; 514 52 83 ⊠ 214708 sapo gr
⊖ (+30 1) 513 26 79
👥 60
Contact: St Samourdanis
■ Manufacture of air and oil engines filters, diesel hydraulic filters according to military specifications. [98]

75-02170 **Olympic Hellas Papapanagiotoy-Vrettos SA**
(a member of Hellenic Manufacturers of Defence Materials Association)
7 L Vouliagmenis Avenue, 17236 Ag Dimitrios
☎ (+30 1) 973 50 60; 975 60 06 ⊖ (+30 1) 971 71 34
👥 45
Contact: M Vrettos
■ Manufacture of military inflatable boats and life rafts. [98]

75-02180 **Olympic Tool Hellas SA**
Industrial Area, 320 11 Inofyta Viotias
☎ (+30 262) 317 81; 323 81 ⊠ 214117 sar gr
⊖ (+30 262) 317 82
⊕ Dr 514,300,000 (1996)
♣ Dr 310,782,000,000 (1996) 👥 37 🏭 7,000 m2
Chairman: C Sarantopoulos
Managing Director: L Anotine
Managing Director: F Devant
Operations Director: A Argyrion
■ Fabrications, reverse engineering and assembly parts including electromechanical assembly, high precision machnining, and harness and cabling manufacturing for aircraft. → 85 [98]

75-02190 **Orfanides SA**
(a member of Hellenic Manufacturers of Defence Materials Association)
Veroias 86 Lonia, 57005 Thes/niki
☎ (+30 31) 51 53 13; 78 15 13 ⊖ (+30 31) 78 15 13
👥 70
Contact: C Orfanides
■ Manufacture of parts for tanks. [98]

75-02192 **Ostraco Armor Products Ltd**
PO Box 10039, 541 10 Thessaloniki
☎ (+30 31) 79 81 02 ⊠ 412420 zura
⊖ (+30 31) 79 96 94
⊕ US$ 20 000 ♣ US$ 260 000 (1995) 👥 8 🏭 200 sq m
Managing Dir: N. Sournopoulos
Sales & Marketing Dir: P. Zouras
Purchasing Mgr: K. Zouray
■ Development, manufacture & marketing of body armour equipment. [97]

75-02195 **Papadopoulos & Co Ltd**
(a member of Hellenic Manufacturers of Defence Materials Association)
48 Peiraios Street, 18346 Mochato
☎ (+30 1) 482 00 15; 481 50 75; 482 48 58
⊖ (+30 1) 482 76 69
👥 25
Contact: N Papadopoulos
■ Armoured vehicle tracking material, weapon parts, heat treatment works. [98]

75-02200 **Petropoulos Pertros SA**
(a member of Hellenic Manufacturers of Defence Materials Association)
39 Lera Odos GR, 10447 Athens
☎ (+30 1) 347 33 00 ⊖ (+30 1) 347 34 04
👥 86
Contact: P Petropoulos
■ Power generating sets, diesel engines, military vehicles, trucks and buses. [98]

75-02230 **Plan Aero Ltd**
(a member of Hellenic Manufacturers of Defence Materials Association)
129 L Kifisias Avenue, 11524 Athens
☎ (+30 1) 648 25 18 ⊠ 225462 plan gr
⊖ (+30 1) 648 25 17
👥 20
Contact: A Savvinidis
Contact: D Rekleitis
Contact: P Noussis
■ Manufacture of brakes and parts for aircraft. [98]

75-02235 **Procal SA**
(a member of Hellenic Manufacturers of Defence Materials Association)
Ag Sotira Acharnes, 13671 Athens
☎ (+30 1) 246 20 56; 246 64 66 ⊖ (+30 1) 246 20 55
👥 55
Contact: A Psichas
■ Manufacture of cooling units. [98]

75-02310 **C. Rokas SA**
46 Patission Street, 106 82 Athens
☎ (+30 1) 884 27 11; 881 72 57; 881 71 61
⊠ 216356 roks gr ⊖ (+30 1) 821 06 85
⊕ US$ 2,300,000 (1997) ♣ US$ 17,800,000 (1997)
👥 170 (1997) 🏭 20,000 m2
Managing Director: C Rokas
Commercial Director: Th Tsantilas
Operations Manager: G Sryrou
■ Lifting equipment and steel structures for aircraft hangars, overhead travelling cranes and monorail systems for the aerospace industry and shipyard cranes. [98]

75-02360 **Sielman SA**
(a member of Hellenic Manufacturers of Defence Material Association)
A Industrial Area Volou, 35800 Volos
☎ (+30 421) 951 18 ⊖ (+30 421) 950 83
⊕ US$ 2,000,000 (1995) ♣ US$ 4,000,000 (1995)
👥 60 🏭 4,000 m2
President: Theo Efremides
Vice President Sales and Marketing: Ms M Efremides
General Manager: Theofilos Efremides
■ Spares and electrical accessories for military vehicles/tanks, automatic control applications and wire harnesses. Brake discs and pads.

Athens Office
Pindou 46, Koliatsou, Athens
☎ (+30 1) 201 40 67 ⊖ (+30 1) 201 33 32
Manager: K. Morakis [98]

75-02380 **3 Sigma SA**
(a member of Hellenic Manufacturers of Defence Materials Association)
Group of Companies
115 27 Athens
2-4 Messogion Avenue, Athens Tower A' Building
☎ (+30 1) 771 661 78 ⊠ 214443 ⊖ (+30 1) 779 94 18
⊕ Dr 32,000,000 (1994) ♣ Dr 1,130,000,000 (1994)
👥 25 🏭 600 sq m (1995)
Managing Director: Elina Sarri
■ Manufacture of RPV and target systems. → 90 [98]

75-02400 **Simeca SA**
(a member of Hellenic Manufacturers of Defence Materials Association)
Portsi, PO Box 101, 19400 Koropi
☎ (+30 1) 662 35 03 ⊖ (+30 1) 662 36 03
👥 15
Contact: Mr Chaftas
■ Manufacture of metallic constructions. [98]

75-02410 **Simo Loukarelli Ltd**
(a member of Hellenic Manufacturers of Defence Materials Association)
12th km - Ethnikis Road, 14455 Athinon Lamias
☎ (+30 1) 281 33 01 ⊖ (+30 1) 281 19 11
👥 20
Contact: Loukarelli Simo
■ Manufacture of electrical equipment, retrofit kits for M48 tanks, electric power generators. [98]

75-02420 **E. Spanos & Sons**
103 D Moutsopoulou Street, 18541 Piraeus
☎ (+30 1) 420 94 96; 420 94 53 ⊖ (+30 1) 420 94 72
🏭 1,919 m2
Managing Director: Constantinos Spanos
■ Manufacturer and supplier of shoes, boots and general leather items for the Greek armed forces, police forces and state services. [98]

75-02430 **Space Hellas SA**
(a member of Hellenic Manufacturers of Defence Materials Association)
302 Mesogion Avenue, 155 62 Holargos
☎ (+30 1) 672 70 08; 653 34 51 ⊖ (+30 1) 651 67 12
👥 75
Contact: D Manolopoulos
■ Military and civilian projects, data communications and offset benefits. [98]

75-02450 **Spider N. Petsios & Sons, Metal Ind.**
(a member of Hellenic Manufacturers of Defence Materials Association)
79 Pireos Street, 18346 Travos
☎ (+30 1) 486 11 91 ⊖ (+30 1) 426 06 04
👥 250
Contact: K Petsios

Contact Reference

■ Manufacture of military vehicles superstructures, cartridge cases for 20mm and 30mm calibre rounds.

Factory
Loannina - near airport
☎ (+30 651) 271 91; 255 14 ⊜ (+30 651) 345 35 [98]

75-02465 Spiral Flex
(a member of Hellenic Manufacturers of Defence Materials Association)
7th km, Road Larissa - Athens, PO Box 1352,
411 10 Larissa
☎ (+30 41) 92 15 84 ⊜ (+30 41) 92 15 83
♤ US$ 400,000 ♣ US$ 1,000,000 👥 10 🏠 500 sq m
Sales and Marketing Director: Vaios Belagias
Operations and Technical Director: Andreas Zisi
■ Production of barbed tape concertina, security fencing. [98]

75-02490 Stylidou Ste Ltd
(a member of Hellenic Manufacturers of Defence Materials Association)
110 Giannitson, 546 27 Thes/niki
☎ (+30 31) 52 74 33; 52 75 20 ⊜ (+30 31 51 43 04
👥 6
Contact: Th Stylidis
■ Gaskets industry. [98]

75-02525 Synergy Systems SA
68 Amerikis, 26441 Patra
☎ (+30 61) 65 30 92 ⊜ (+30 61) 65 38 93
Contact: A Tslodimos
■ Services include microelectronics (ASICS - FPAG's), telematics applications, hardware description language, and digital signal processing. [98]

75-02560 Tekson SA
(a member of Hellenic Manufacturers of Defence Materials Association)
2 Filias, 144 51 Metamorfosi, Attikis
☎ (+30 1) 282 55 68; 282 89 12 ⊜ (+30 1) 282 83 93
👥 40
Contact: P Karamanolis
■ Development of fuses for grenades and mortar bombs. [98]

75-02570 Tele Watt Color SA
(a member of Hellenic Manufacturers of Defence Materials Association)
12th km North Road Larissa, 415 00 Thes/niki
12th km North Road Larissa - Thes/nika
☎ (+30 41) 54 14 42 ⊜ (+30 41) 54 12 07
Contact: J Vogiatzakis
■ Manufacture protective clothing, safety equipment, protective armour and rescue safety equipment. [98]

75-02580 Terry Hellas GHR. Gelatis Co
(a member of Hellenic Manufacturers of Defence Materials Association)
30 Nikis Street GR, 10557 Athens
☎ (+30 1) 321 13 92 ⊜ (+30 1) 324 27 74
👥 25
Contact: Char Charalabidis
■ Manufacture of knithing equipment (socks, sweaters, jackets), protective clothing and safety equipment, protective armour equipment, rescue safety equipment.

Factory
Inofita Viotias GR, 32011 Viotia
☎ (+30 262) 314 48 ⊜ (+30 262) 322 93 [98]

75-02590 Theodoratos Ltd
(a member of Hellenic Manufacturers of Defence Materials Association)
8, D Theodora, 14452 Metamorfosi, Attikis
☎ (+30 1) 281 80 70; 284 50 47 ⊜ (+30 1) 281 80 68
👥 12
Contact: Th Theodoratos
■ Aluminium sheet works, special painting, NATO labels, signs, industrial plates, special parts. [98]

75-02600 Thorax SA
(a member of Hellenic Manufacturers of Defence Materials Association)
Nea Ethniki Road 18, 135 61 Ag Anargyroi
☎ (+30 1) 832 73 10; 832 85 20 ⊜ (+30 1) 832 14 34
👥 14
Contact: E Kokkinelis
■ Manufacture of bullet proof vests, jackets, bullet proofing of cars. [98]

75-02700 Topuzeli Machine Works
(a member of Hellenic Manufacturers of Defence Materials Association)
66 Monastiriou Street GR, 56334 Thessaloniki
☎ (+30 31) 76 30 60 ⊜ (+30 31) 76 37 11
👥 30
Contact: K Topouzelis
■ Precision metal parts for military equipment (vehicle weapon systems). Crane bridges, self propelled conveyor belts (on caterpillars), steel structures. [98]

75-03200 Valpak SA
PP Germanou 7, 10561 Athens
☎ (+30 1) 323 25 38; 325 21 26 ⊠ 226631
⊜ (+30 1) 325 21 53
♣ Dr 500,000,000 (1996) 👥 72 (1996) 🏠 2,000 m2
General Manager: Athanasios Tsoukas
Marketing and Sales Director: Dimitris Tsoukas
Public Relations Manager: Takis Petrou
■ Production and distribution of camouflage netting. [98]

75-03500 Veka Papermills SA
(a member of Hellenic Manufacturers of Defence Materials Association)
145 65 Athens
21st Ethnikis Road athinon - Lamias
☎ (+30 1) 814 26 12; 814 26 34 ⊠ 210281 veka gr
⊜ (+30 1) 814 29 26
♤ US$ 6,000,000 ♣ US$ 4,386,000 (1994)
👥 95 (1995) 🏠 6,600 m2 (1995)
Managing Director: St Arvanitides
Sales Director: E A Vamvakidis
■ Manufacture fibre containers for ammunition, cores and textile, plastic foil, paper drums for chemicals aand composite cans (papert body, metal TSP) [98]

75-04000 Viana SA
(a member of Hellenic Manufacturers of Defence Materials Association)
3rd km Varis - Koropiou, 19400 Athens
☎ (+30 1) 662 39 10; 662 47 00 ⊜ (+30 1) 662 47 24
👥 20
Contact: Mr Anargyrou
■ Manufacture protective articles for the face, body and breathing. [98]

75-04300 Viokat SA
(a member of Hellenic Manufacturers of Defence Materials Association)
Aristeidou 6, 105 59 Athens
☎ (+30 1) 686 67 00 ⊠ 219577 biok gr
⊜ (+30 1) 683 04 54
👥 290
Contact: G Kostantinidis
■ Turnkey projects, heavy structural steel works, maintenance and repair of industrial, mechanical and electrical equipment.

Factory
69th km P.E.O., Athinon-Thivon
☎ (+30 262) 249 41; 251 81 ⊜ (+30 262) 251 82 [98]

75-04450 Vippon-Hellenic Rubber Industry SA
(a member of Hellenic Manufacturers of Defence Materials Association)
9/28th October Street, 182 33 Piraeus
☎ (+30 1) 491 48 46 ⊜ (+30 1) 491 29 76
♣ Dr 450,000,000 (1995) 👥 30 🏠 3,000 m2
Managing Director: Alexander Kalafatis
Sales and Marketing Director: Thanos Kalafatis
■ Rubberisation of track components for tracked armoured vehicles and assembly of completed track systems (trackshoes, pads, pins, bushings, wheels). Refurbishment of track systems for tracked armoured vehicles. Hard rubber battery cases and cell covers (NATO type). gas masks. Certified according to ISO 9002 & AQAP 120. [98]

75-04500 Vissariou K.
(a member of Hellenic Manufacturers of Defence Materials Association)
163 L Athinon, Athens
☎ (+30 1) 345 63 33 ; 342 15 07 ⊜ (+30 1) 347 71 53
Chairman: Konst Vissariou
■ Muffler construction, armoured vehicle spares. [98]

75-05100 Zenon SA
(a member of Hellenic Manufacturers of Defence Materials Association)
Kanari 5, 153 44 Glyka Nera
☎ (+30 301) 604 15 82 ⊜ (+30 301) 604 10 51
Contact: A Economopoulos
■ Specialises in industrial robotics, computer vision and CIM. [98]

80 Service Companies & Organizations

80-00180 Akmon Ltd
(† OLS Ltd)
848 E. Antistaseos Street, 152 31 Halandri, Athens
☎ (+30 1) 677 34 80; 677 34 81 Ⓔ akmon@netor.gr.
⊜ (+30 1) 677 34 82
♤ US$ 50,000 (1997) ♣ US$ 8,000,000 (1997) 👥 10
🏠 200 m2
Co-Chairman: John Rozolis
Co-Chairman: D Kaliakhanis
Sales and Marketing Director: Tassos Rozolis
Operations and Technical Director: G Papadimoy
Purchasing Manager: B Tsakiris
■ Consultants, representative for foreign defence companies. [98]

80-00275 C3 - C2T SA
(a member of Hellenic Mnufacturers of Defence Materials Association)
Kilkis Industrial Area, PO Box 50, 611 00 Kilkis
☎ (+30 341) 719 47; 227 82
⊜ (+30 341) 719 87; 255 60
👥 16
Contact: G Passalidis
■ Services offered are area of measurements, callibration testing, certification and quality control. [98]

80-00650 European Finance Associates Ltd (EFA)
(part of Albion Group International Inc.)
7 Stratigi Street, Psychico, 15451 Athens
☎ (+30 1) 672 86 10 ⊜ (+30 1) 672 86 24
👥 50
General Manager: Christian Hadjiminas
■ Design, negotiation & fulfillment of offset programs in Europe, Middle East & Asia. Emphasis on investment in high technology projects. Fields of expertise in international investments in offset projects related to software, robotics, telecommunications & aerospace. [98]

80-00700 GENME SA Marine & Industrial
6 Naxou 150 Iera Odos Street, 12242 Aigaleo, Athens
☎ (+30 1) 345 31 41; 346 29 67 ✉ GEJONSUR
⊠ 210815 geme gr ⊜ (+30 1) 347 60 41
♤ US$ 1,003,703 (1996) ♣ US$ 1,844,397 (1996)
👥 30 (1997) 🏠 1,500 sq m
Managing Director: John Soursos
Sales and Marketing Director: John Fokas
Public Relations and Advertising Manager: John Fokas
■ Engineering services. Marine main and auxiliary machinery including deck machinery (diesel, steam, gas and other). Surveying, testing, reconditioning, repairing, related component/spares, constructing/fabricating. ISO 9002 certified. [98]

80-00750 Hellenic Aerospace Industry Ltd (HAI)
Tanagra, PO Box 23, 320 09 Schimatari
☎ (+30 1) 883 67 11; 26 25 20 00
Ⓔ hai@hai.e-mail.com Ⓦ http://www.haicorp.com
⊠ 299306 hai gr ⊜ (+30 1) 883 87 14; 26 25 21 70

♧ US$ 863,645,000 (1995)
♣ US$ 114,505,000 (1995) ⚇ 2,869 (1995)
⌂ 140,777 m2
Chairman and Managing Director: Eleftherios Moschopedis
Managing Director: George Goulis
Operations and Technical Director: Evangelos Ladis
Engineering and Research and Development Director: Th Spathopoulos
Sales and Marketing Director: Nick Vassilopoulos
Administration: Michael Grammatopoulos
Purchasing Manager: Spyros Nnasoulas
Head of Public Relations: Ms Katia Apostolopoulou
Public Relations Office: A Bourtzoglou
■ Maintenance, overhaul, modification and testing of aircraft, engines, helicopters and electronics (airborne and ground equipment and systems). Manufacture of aircraft and helicopter structural parts, assemblies and modification kits. Design and manufacture of electronics and telecommunication products. Technical training. [98]

80-00760 Heeqac SA
(a member of Hellenic Manufacturers Defence Materials Association)
6 Ierossolimon Street, 11252 Athens
☎ (+30 1) 861 39 28; 861 35 75; 865 56 11
⊖ (+30 1) 867 71 75
⚇ 20
Contact: G Chrysanthakopoulos
■ Equipment quality assurance centre for a wide range of electric components, devices and systems. Its main function is the certification of technical specifications as well as the calibration of instrumentation and measuring devices. [98]

80-00795 International Technological Applications
67 Daskaroli Street, 16675 Glyfada
☎ (+30 1) 965 20 31 Ⓔ ita@otenet.gr ⊠ 226913
⊖ (+30 1) 965 20 81
♧ Dr 150,000 000 (1997) ♣ Dr 700,000,000 (1997)
⚇ 28 ⌂ 1,000 sq m
President/Managing Director: Anthony Gerassimou
Operations and Technical Director: Costas Stroutzas
Administration Director: Kostas Lambrou
Financial Director: Stelios Andreau
Sales and Marketing Director: Ms Mandalena Gerassimou
Purchasing Manager: Nassos Vitoropoulos [98]

80-01180 Marac Electronics SA
(a member of Hellenic Manufacturers of Defence Materials Association)
455 Dimokratis Street, 188 63 Piraeus
☎ (+30 1)431 43 61 ⊖ (+30 1) 431 42 34
⚇ 89
Contact: N Pitsoulis
■ Represent and manufacture telecom, automation and instrumentation products. [98]

80-01200 Marine Technical Bureau Inc.
(Incorporating E. Vlassopoulos Marine Technical Bureau)
117 Notara Street, 185 35 Piraeus
☎ (+30 1) 428 33 48; 428 33 50 ✉ VLASLE
⊠ 213630 jhev gr ⊖ (+30 1) 428 33 51
Managing Director: Lefteris E Vlassopoulos
Marine Surveyor: Capt Th Lefkantinos
Naval Architect: Evangelos Vas Vassiliou
Marine Engineer: G F Karavias
Marine Engineer/Electrical Engineer: Alex Demokratis
Naval Architect: John Papaioannou
Assistant Naval Architect: Antony Iskos [98]

80-01400 Overseas Maritime Consultants EPE
57 Notara Street, 185 35 Piraeus
☎ (+30 1) 422 27 30; 422 27 38
Ⓦ http://www.hol.gr/business/ overseas
Ⓔ overboard@hol.gr ⊠ 211586; 241306 omc gr
⊖ (+30 1) 422 27 37
⚇ 14 ⌂ 300 m2
Spokesperson: James Walsh
■ Consultancy services. [98]

80-01970 Sonak Systems & Software SA
(a member of the Hellenic Manufacturer of Defence Materials Association)
178 Kifissias Avenue, 152 31 Halandri, Athens
☎ (+30 1) 647 86 55 ⊠ 221992 ⊖ (+30 1) 675 64 00
♣ Dra 162,000,000 (1990) ⚇ 27
Managing Director: Elias P Arkoumaneas
Marketing: Panayiotis Tsotras
■ Electronic systems and software company. Electronic systems integration. Software products are all developed using fourth generation tools and relational database design. [98]

80-02010 Talos Engineering SA
27 Vas. Sophias Avenue, 106 74 Athens
☎ (+30 1) 722 96 76; 722 96 75
Ⓔ 314mk141@otenet.gr ⊖ (+30 1) 722 98 19
♧ Dr 25,000,000 ♣ Dr 200,000,000 ⚇ 20 ⌂ 500 m2
Managing Director: Nicholas Kouleimanis
■ Design and construction supervision of ground-based installations. Consultancy services for warship-related projects. Project management. Security engineering services. [98]

80-03000 Unicon Ltd
14 Leof Demokratias Avenue, 15121 Pefki
☎ (+30 1) 612 06 79; 932 37 22; 801 58 87
Ⓔ unicon@acropolis.gr ⊖ (+30 1) 934 15 62; 805 28 74
Managing Director: Dr Harry Angelopulos
Sales and Marketing: A Tzogios
Purchasing Manager: D Papasotiriou
Public Relations Manager: E Chaldeou
■ Consultancy firm specialising in assisting manufacturers and service companies implement ISO 9000, ISO 14000, ASQP, CALS and CE marketing. Services include implementation planning, preparation of the quality assurance and procedures manual, supplier development training, failure mode analysis (FMEA), implementation of SPC and DOE techniques R&M allocation, prediction and analysis, product assurance , documentation and codification, logistics support analysis and quality engineering services. [98]

85 Stockists, Distributors, Spare Parts Suppliers

Mercury SA 90-01300
Olympic Tool Hellas SA 75-02180

85-00500 Eltron SA
20 Herodotou Street, 106 75 Athens
☎ (+30 1) 721 06 69; 724 95 11; 724 95 15
✉ DALMAR ⊠ 216435 dalm-gr
⊖ (+30 1) 723 95 56; 721 18 60
⚇ 25 (1997) ⌂ 450 m2
Managing Director: Marios Dallegio
Executive Consultant: D A Anetis
® Aruba: National Forge
Australia: MILS Elektronik
Canada: Oerlik Aerospace
France: SFIM Industries
Italy: Agusta SpA, Augusta, Marconi SpA
Russia: Almaz, Rubin
Slovakia: Technopol JSC
Sweden: Barracudaa Technologies AS, Ericsson Radio Systems
Switzerland: Mettler Toledo, Oerlikon-Contraves, SIG Swiss Industrial Co.
UK: Cegelec Project Ltd, Graseby
USA: AlliedSignal Inc., Chadwick Helmuth Co, FLIR Systems, Marsh Aviation, Telephonics Corp. [98]

85-00900 D.F. Sarantopoulos SA
92 Leof. Kifissou, 10443 Athens
☎ (+30 1) 514 64 11; 514 64 12; 514 64 13
Ⓔ dfsara@hol.gr ⊠ 214400 tesb gr ⊖ (+30 1) 514 64 34
♧ US$ 400,000 ♣ US$ 8,000,000 ⚇ 55 ⌂ 1,200 sq m
President: D F Sarantopoulos
→ 80 → 90
® Germany: Eder, Krupp, Sennebogen, Zettelmeyer
Japan: Mitsubishi Heavy Industries
USA: Nitco [98]

85-01000 Survival
10 Nikotsara Str, 14562 Kifisia
☎ (+30 1) 801 17 63; 801 71 97; 801 10 15
Ⓦ http://www.survival.gr ⊠ 224680 yfdk gr
⊖ (+30 1) 801 76 34; 325 08 75
♧ US$ 500,000 (1996) ♣ US$ 1,500,000 (1996)
⚇ 10 (1996) ⌂ 800 m2
Managing Director: Yannis Fanurakis
Marketing Director: Dimitris Katsikas
■ Stockists and distributors of large range of military and civilian survival equipment. [98]

85-02000 TRACE Middle East Regional Liaison Office
(a division of TRACE World Wide Corp)
Pythagora 23, 15562 Holargos, Athens
☎ (+30 1) 651 17 44 Ⓔ jpp@inet.gr
⊖ (+30 1) 654 19 86
Managing Director: John P Pavlanto
® USA: TRACE Worldwide Corp [98]

90 Sales Agents, Sales & Liaison Offices

Companies represented:

Agusta SpA, Italy 85-00500
AlliedSignal Inc., USA 85-00500
Almaz, Russia 85-00500
Augusta, Italy 85-00500
Auxilec, France 90-00400
Barracudaa Technologies AS, Sweden . 85-00500
Beretta SpA, Italy 90-00270
Cegelec Project Ltd, UK 85-00500
Chadwick Helmuth Co, USA 85-00500
Eder, Germany 85-00900
EEV Ltd, UK 90-02000
Ericsson Radio Systems, Sweden . . . 85-00500
Eurocopter, France 90-00400
FLIR Systems, USA 85-00500
Graseby, UK 85-00500
Krupp, Germany 85-00900
Lockheed-Martin, USA 90-01300
Marconi SpA, Italy 85-00500
Marsh Aviation, USA 85-00500
McDonnell Douglas Helicopter Systems, USA 90-01300
Messier-Bugatti, France 90-00400
Mettler Toledo, Switzerland 85-00500
Michelin, France 90-00400
MILS Elektronik, Australia 85-00500
Mitsubishi Heavy Industries, Japan . . 85-00900
National Forge, Aruba 85-00500
Nitco, USA 85-00900
Norsk Forsvarsteknologi A/S, Norway . 90-01400
Oerlik Aerospace, Canada 85-00500
Oerlikon-Contraves, Switzerland 85-00500
Ofema, France 90-00400
Rubin, Russia 85-00500
Schweizer Helicopters, USA 90-01300
Sennebogen, Germany 85-00900
SFIM Industries, France 85-00500
SIG Swiss Industrial Co., Switzerland . 85-00500
Technopol JSC, Slovakia 85-00500
Telephonics Corp., USA 85-00500
TRACE Worldwide Corp, USA 85-02000
Turbomecaa, France 90-00400
Zettelmeyer, Germany 85-00900

Sales agents/representatives listed elsewhere:

3 Sigma SA 75-02380
D.F. Sarantopoulos SA 85-00900
General Electronics Communications . 75-01300
Hellenic Vehicle Industry SA 75-01780

90-00020 ALMI SA
39 Daskaroli Street, 166 75 Glyfada, Athens
☎ (+30 1) 963 32 74; 963 37 72 Ⓔ almi@hol.gr
⊠ 224430 mat gr ⊖ (+30 1) 963 23 85
♧ Dr 20,000,000 ⚇ 7 ⌂ 200 m2
Managing Director: Nick Gavalas [98]

90-00270 Beretta Hellas
(† Beretta Hellas Sarl)
Stassinopoulos Building, 30 Baknana Street, 11744 Athens
☎ (+30 1) 321 81 88 ▣ 214052 stasgr
⊖ (+30 1) 321 05 47
Managing Director: Elias M Stassinopoulos
® Italy: Beretta SpA [98]

90-00400 EKTE-A Ltd
(† OFEMA)
Avenue Kifissias 16, 151 25 Maroussi, Athens
☎ (+30 1) 685 08 20 ▣ 224 232 ofma gr
⊖ (+30 1) 685 20 22
♧ Dr 12,000,000 (1997) 👥 6 (1997) 🏠 121 sq m
General Manager: Constantin Tsalas
® France: Auxilec, Eurocopter, Messier-Bugatti, Michelin, Ofema, Turbomecaa [98]

90-00700 German Genuine Co
86 Filonos St, 18536 Piraeus
☎ (+30 1) 429 46 47; 429 46 48 ▣ 211300 blom gr
⊖ (+30 1) 429 47 05
Managing Director: George Giannakouros [98]

90-01000 Kartalis Near East Agency
21 V Heracleou Street, 546 24 Thessaloniki
☎ (+30 31) 23 95 33 ▣ 419260 kart
⊖ (+30 31) 24 07 10
♣ US$ 1,000 000 (1996) 👥 3 (1996) 🏠 100 sq m
President: Diam Kartalis
Managing Director: Peter Kartalis
Sales and Marketing Director: Ms Marianna Dundon
■ Agency representing overseas manufacturers interests and facilitating their sales to to Greek armed forces and other government agencies.
Trading: Ordnance logistics supplies. Small calibre ammunition and small calibre weapons. [98]

90-01250 MEGADIESEL SA
5/7 Alkamenous Street, 10 439 Athens
☎ (+30 1) 881 02 86; 882 21 21; 882 34 61; 882 66 45
▣ 219394 mega gr
⊖ (+30 1) 821 35 48; 884 13 16; 825 40 57
♧ Dr 60,000,000 (1996) ♣ Dr 500,000,000 (1995)
👥 10 🏠 500 sq m
Managing Director: L Pangalos [97]

90-01300 Mercury SA
(a member of Hellenic Manufacturers of Defence Materials Association)
2 Loulianou, 106 75 Athens
☎ (+30 1) 721 02 41; 724 27 46 ⊖ (+30 1) 721 51 98
♣ US$ 220,000,000 (1992) 👥 15 🏠 300 m2
Managing Director: Peter Spanakos
Operations Director: Bill Houliacos
■ Retrofit kits for M48. Wire harness. → 85
® USA: Lockheed-Martin, McDonnell Douglas Helicopter Systems, Schweizer Helicopters [98]

90-01400 Norsk Forsvarsteknologi A/S
65 Akzi Miaouli, 18536 Piraeus
☎ (+30 1) 418 23 21-4 ⊖ (+30 1) 451 92 10
Director: Nick G. Petmezas
® Norway: Norsk Forsvarsteknologi A/S [97]

90-01910 Sarakakis Pros Aebe
(a member of the Saracakis Group)
41 Athinio Avenu, 101 73 Athens
☎ (+30 1) 346 53 21; 346 70 11 ▣ 2154 osbsa gr
⊖ (+30 1) 346 73 29 (Group 3)
♧ Dr 7,967,000,000 (1994)
♣ Dr 19,592,000,000 (1994) 👥 679 🏠 90,000 sq m
President/Managing Director: D J Saracakis
Vice President/Managing Director: F J Saracakis
Marketing Manager: Th Venardos
■ Local distributors for military vehicles manufactured by Volvo. Local conversion of chassis-cabs and/or vans into military fire extinguishing vehicles and ambulance for the following manufacturers: Volvo, Mitsubishi Motors. [98]

90-01980 SSA SA
Ethnikis Antistaseos 84, 15231 Halandri
☎ (+30 1) 672 51 06; 672 24 19 [E] DM@HOL.GR
▣ 225644 ssa gr ⊖ (+30 1) 672 66 82; 675 66 26
♧ Dr 50,000,000 (1996) ♣ Dr 500,000,000 (1996)
👥 12
Managing Director: Nicholas Broudos
Marketing Director: Seraphim Seferiades
Technical Director: Seraphim Seferiades
■ Electronic components, systems, technical support. → 75 → 80 [98]

90-02000 Telmaco SA
69 George Vlahou Street, 115 25 Athens
☎ (+30 1) 677 74 00 ✉ TELTRAMACO
▣ 219185 telm gr ⊖ (+30 1) 677 74 06
Managing Director: D D Kyreazis
® UK: EEV Ltd [98]

90-02100 M & J Vardas Co.
121 Patission Avenue, 112 51 Athens (813)
☎ (+30 1) 821 62 38; 821 64 43; 322 40 07; 325 53 78
▣ 219475 ⊖ (+30 1) 867 89 89; 323 43 73
👥 9 (1997) 🏠 244
Managing Director: Michael S Vardas
■ Represents the interests of overseas manufacturers and stockists in Greece and assists them in promoting their sales to the Greek Armed Forces and other state owned authorities. [98]

90-02200 Vasilios Alateras
Stadiou Street 5, 10562 Athens
☎ (+30 1) 324 27 37; 644 15 29; 322 16 28; 643 67 54
⊖ (+30 1) 322 90 78; 643 67 54
👥 1
Marketing and Sales Director: Vasilios Alateras
■ Armed forces and law enforcement supplier. → 80 [98]

Grenada

Head of State:
H.M. Queen Elizabeth II

Governor General:
Daniel Williams

10 Head of State

10-10000 Office of the Governor-General
Upper Lucas Street, St George's
☎ (+1 809) 440 24 01 ⊖ (+1 809) 440 66 88
Governor General: Daniel Williams [98]

11 Prime Minister

11-10000 Office of the Prime Minister
Botanical Gardens, Tanteen, St George's
☎ (+1 809) 440 22 55; 440 22 65 ▣ 3458 trade ga
⊖ (+1 809) 440 41 16
Prime Minister: Keith Mitchell
● Botanical Gardens, Tanteen, Saint Georges
☎ (+1 809) 440 22 55 ▣ 3423 grenex ga
⊖ (+1 809) 440 41 16 [98]

12 Ministries

12-10000 Ministry of National Security
St George's
Minister of National Security: Keith Mitchell [98]

12-20000 Ministry of External Affairs
Botanical Gardens, St George's
☎ (+1 809) 440 22 55 ▣ 3423 grenexga
⊖ (+1 809) 440 41 84
Minister of External Affairs: Raphael Fletcher [98]

12-25000 Ministry of Finance
Lagoon Road, St George's
☎ (+1 809) 440 27 31; 440 27 32; 440 27 33; 440 27 34 ▣ 3418 minfinge ⊖ (+1 809) 440 41 15
Minister of Finance Trade and Industry: Patrick Bubb [98]

15 Parliament

15-10000 Senate
York House, Saint Georges [98]

15-20000 House of Representatives
York House, Saint Georges [98]

26 Police (Civil)

26-00000 Grenada Police Force
Police Headquarters, Saint Georges
☎ (+1 809) 440 3999; 0082; 3499
⊖ (+1 809) 440 4139 [97]

31 Embassies & Attachés

31-01890 British High Commission
14 Church Street, St. George's
☎ (+1 809) 440 32 22; 440 35 36
⊖ (+1 809) 440 49 39
High Commissioner: H E Mr R Thomas (Resides at Bridgetown)
Resident Acting High Commissioner: M R Eastburn
Defence Adviser: Capt I M Hime (Resides at Bridgetown) [98]

31-01950 Embassy of Venezuela
598/1 Archibald Avenue, PO Box 201, Saint Georges
☎ (+1 809) 440 17 21; 440 17 22
✉ EMBAVENEZ Saint Georges ⊖ (+1 809) 440 66 57
Ambassador: H E Mr Teofilo R Labrador [98]

Guatemala

Head of State:
Alvaro Arzu Irigoyen

10 Head of State

10-10000 Office of the President
Palacio Nacional, Zona 1, Guatemala City
☎(+502 2) 39 33 33 ☒ 5325 relpub gu
⊖ (+502 2) 51 97 02
President: Alvaro Arzu Irigoyen
Minister of Government: Rudolfo Adrian Mendoza Rosales
98

10-15000 Office of the Vice President
Presidential House, 6a Avenida 4-19, Zone 1, Guatemala City
☎(+502 2) 53 92 03 ☒ 5331 capres gu
Vice President: Luis Flores Asturias 98

12 Ministries

12-10000 Ministry of Defence
Palacio Nacional, Guatemala City
☎(+502 2) 212 12 ☒ 248368
Minister of Defence: Hector Mario Barrios Celada 98

12-20000 Ministry of Foreign Relations
Palacio Nacional, Guatemala City
☎(+502 2) 212 12
Minister of Foreign Relations: Eduardo Stein 98

15 Parliament

15-10000 Congress of the Republic
9a Avenue 9-48, Zona 1, Guatemala City
☎(+502 2) 53 33 33; 53 33 34; 53 33 35; 53 33 56; 53 33 37 98

31 Embassies & Attachés

31-00590 Embassy of France
Imm.Marbella 16 Calle 4-53, Zona 10, A.P. 1252, 01014 Guatemala City
☎(+502 2) 37 36 39/40 80 ☒ 5963 amfran gu
Ambassador: H.E. Mr Paul Poudade
Deputy Chief of Mission: Gilles Bienvenu 98

31-01410 Embassy of Peru
2a Avenida 9-67, Zona 9, Guatemala City
☎(+502 2) 31 85 58 Ⓔ lepruguate@gua.gbm.net
⊖ (+502 2) 34 37 44
Ambassador: H E Mr Julio Florián
Deputy Chief of Mission: Hon Alfredo Tejeda 98

31-01680 Embassy of Spain
6a Calle 6-48, Zona 9, PO Box 669, Guatemala City
☎(+502 2) 334 37 57 (PBX) ⊖ (+502 2) 332 24 56
Ambassador: H E Mr Victor Fagilde González

Commercial Office
Guatemala City
12 Calle 1-25, Z 10-Edificio Géminis 10, officina 1701, Torre Sr Nivel 17
☎(+502 2) 335 30 11; 335 30 14
⊖ (+502 2) 335 30 16
Commercial Counsellor: Jesús Murujosa

Cooperative Technical Office
6a Calle 6-48, Zona 9, PO Box 675, Guatemala City
☎(+502 2) 331 03 12 ⊖ (+502 2) 331 53 03 98

31-01890 Embassy of the United Kingdom
Guatemala City
Edificio Centro Financiero, 7th Floor
☎(+502 2) 32 16 01; 32 16 04; 332 16 02; 332 16 06
⊖ (+502 2) 34 19 04
Ambassador: H E Mr P M Newton
First Secretary, Consul and Deputy Head of Mission: Miss C E McEwen
First Secretary, Commercial: A S Green (Resides at San José)
Third Secretary (Management) and Vice Consul: Miss J M Johnson 98

Guinea

Head of State:
Gen. Lansana Conté

10 Head of State

10-10000 Office of the President
Conakry
☎(+224) 44 11 47; 44 11 22 ☒ 623 prg cky, 2165 cky
President of the Republic: Gen Lansana Conté 98

12 Ministries

12-10000 Ministry of National Defense
Conakry
☎(+224) 44 16 36
Minister: Gen Lansana Conte 98

12-20000 Ministry of Foreign Affairs
Conakry
☎(+224) 40 50 55 ☒ 634 minaf cky
Minister: Lamine Camara 98

12-21000 Ministry of Planning and International Co-operation
Conakry
Minister: Ousmane Kaba 98

31 Embassies & Attachés

31-00660 Embassy of the Federal Republic of Germany
PO Box 540, Conakry
☎(+224) 41 15 06; 44 15 08 ☒ 24779 aacy gui
⊖ (+224) 41 22 18
Ambassador: H E Mr Pius A Fischer
Second Secretary: Mr Thomas Petereit 98

31-00890 Embassy of Italy
Quartier Camayenne, PO Box 84, Conakry
☎(+224) 44 35 88 ⊖ (+224) 44 35 88
Chargé d'Affaires: Dr Roberto Rosellini 98

31-01485 Embassy of the Russian Federation
PO Box 329, Conakry
☎(+224) 41 27 77; 46 37 25 ⊖ (+224) 41 27 77
Ambassador: H E Mr Igor Stoudennikov
Deputy Chief of Mission: Vladimir Timoshenko
Military Attaché: Capt. Guennady Tourkov
Commercial Representative/Secretary: Boris Fomitchev
☎(+224) 46 26 76 98

31-01740 Consulate of Switzerland
Agence Consulaire de Suisse
Corniche Sud, Madina, PO Box 720, Conakry
☎(+224) 46 32 76 Ⓔ adra@adra.org.gn
⊖ (+224) 46 32 76; 46 34 80
Honorary Consul: Colin Richardson ☎(+224) 46 32 76
98

31-01890 Consulate of the United Kingdom
PO Box 834, Conakry
☎(+224) 44 29 59; 41 39 08 ☒ 22294
⊖ (+224) 44 42 15; 41 39 08
Consul: R W Hyde (Resident in Dakar)
Vice Consul: I L McKinlay (Resident in Dakar)
Honorary Consul: Mrs V A Treitlein 98

Guinea Bissau

Head of State:
Gen João Bernardo Vieira

10 Head of State

10-10000 Office of the President
Bissau
☎(+245) 20 11 62 ☒ 251 pce bi ⊖ (+245) 20 14 04
President: Gen João Bernardo Vieira 98

11 Prime Minister

11-10000 Office of the Prime Minister
Bissau
Prime Minister: Manuel Saturnino Da Costa 98

12 Ministries

12-10000 Ministry of Defense
Bissau
☎(+245) 21 28 71 ☒ 249
Minister of Defence: Samba Lamine Mane 98

12-20000 Ministry of Foreign Affairs
Bissau
☎(+245) 21 23 14 ☒ 249 mnestra bi
Minister: Dr Fernando Delfin Da Silva 98

12-40000 Ministry of the Interior
Bissau
Minister: Ms Francesca Pereira 98

31 Embassies & Attachés

31-01890 Consulate of the United Kingdom
Mavegro Int, CP 100, Bissau
☎(+245) 20 12 24; 20 12 16 ☒ 259 mavegro b1
⊖ (+245) 20 12 65
Hyde: R W Hyde (Resident in Dakar)

Contact Reference

Vice Consul: I L McKinlay (Resident in Dakar)
Honorary Consul: J Van Maanen [98]

31-01900 Embassy of the United States of America
Bairro de Penha, CP 297, 1067 Bissau Codex
☎(+245) 25 22 73; 25 22 76 ⊖ (+245) 25 22 82
Ambassador: H E Mr Peggy Blackford [98]

Guyana

Head of State:
Samuel Hinds

10 Head of State

10-10000 Office of the President
New Garden Street, Georgetown
☎(+592 2) 513 30; 715 74; 714 76; 677 25
2205 presoff gy ⊖ (+592 2) 633 95
President of the Republic: Samuel Hinds
First Vice President: Ms Janet Jagen
Second Vice President: Reepu Daman Persaud
Minister of the Office of the President (Public Service): George Fung-On
Head of Presidential Secretariat: Roger Luncheon
Government Information Service Director: Dr Rouin Deodat [98]

11 Prime Minister

11-10000 Office of the Prime Minister
Wight's Lane, Parliament Building, Georgetown
Prime Minister: Ms Janet Jagen [98]

12 Ministries

12-20000 Ministry of Foreign Affairs
Takuba Lodge, Georgetown
254 South Road and New Garden Streets, Georgetown
☎(+592 2) 616 06; 616 09; 564 67 2220 foreign gy
⊖ (+592 2) 25 91 92
Senior Minister of Foreign Affairs: Clement Rohee [98]

12-40000 Ministry of Home Affairs
6 Brickdam, Georgetown
☎(+592 2) 572 70 ⊖ (+592 2) 531 12; 627 40
Minister of Home Affairs: Feroze Mohamed [98]

15 Parliament

15-10000 National Assembly
Parliament Office, Brickdam, Georgetown
☎(+592) 664 67 ⊖ (+592) 513 57
Speaker: Derek C Jagan [98]

25 Defense Forces

25-10000 Defence Force Headquarters
Thomas Lands, Camp Ayanganna, Georgetown
☎(+592 2) 588 66 ⊖ (+592 2) 660 49
Chief of Staff Guyana Defence Force: Brig Gen Joseph G Singh ☎(+592 2) 643 09
Commander Guyana People's Militia: Col Harol B Hinds ☎(+592 2) 660 14
Colonel General Staff: Col Michael Atherly
Colonel Administration and Quatering: Lieut Col Lennox Wilson ☎(+592 2) 725 12
Director General Guyana National Service: Lieut Col Clifford Klass
Commander Officer Coast Guard: Lieut Cdr Bary Best ☎(+592 2) 590 90
Commander Officer Air Corp: Maj Stanislaus Canzius
Staff Officer Grade 1 Operations and Training: Maj J B Lovell [97]

31 Embassies & Attachés

31-00310 Canadian High Commission
High & Young Streets, PO Box 10880, Georgetown
☎(+592 2) 720 81/85 ✉ DOMCAN Georgetown
⊖ (+592 2) 583 80
High Commissioner: H.E. Mr Simon Wade
First Secretary (Development): S. Greaves
Commercial Officer: L. Primo [97]

31-01890 United Kingdom High Commission
44 Main Street, PO Box 10849, Georgetown
☎(+592 2) 658 81; 658 84 ⊖ (+592 2) 535 55; 688 18
High Commissioner: H E Mr David John Johnson
Deputy High Commissioner: C Brazier
Defence Adviser: Col A Moore (Resident in Caracas)
Second Secretary Chancery: D Stewart (Resident in Caracas)
Third Secretary of Management and Consular Affairs: N Floyd
Third Secretary for Immigration: Miss L Donven
Commercial Officer: C McGarrell [98]

Haiti

Head of State:
Rene Preval

10 Head of State

10-10000 Office of the President
Palais National, Port-au-Prince
President of the Republic: Rene Preval [98]

12 Ministries

12-10000 Ministry of National Defense
Palais des Ministères, Port-au-Prince
Secretary of State for Interior and National Defence: Paul Yves Joseph [98]

12-20000 Ministry of Foreign Affairs
Cité de l'Exposition, Port-au-Prince
Minister of Foreign Affairs: Emmanuel Fritz Longchamp [98]

31 Embassies & Attachés

31-00660 Embassy of the Federal Republic of Germany
2 rue Claudinette, PO Box 1147, Port au Prince
☎(+509 57) 72 80 20072 aaprinc hn
⊖ (+509 57) 41 31
Ambassador: H.E. Mr Bernd K. Vierkoetter
Commercial Attaché: Ms Hedwig Matt [97]

31-01190 Embassy of Mexico
Delmas 60 No.2 - Musseau, PO Box 327, Pétion Ville
☎(+509 49) 25 97; 21 00 ⊖ (+509 57) 67 83
Ambassador: H.E. Mr Sergio Romero Cuevas
First Secretary: M.Miguel Garcia Zamudio
Civil Attaché: Gustavo Ernesto Calderon Miranda
Chargé d'Affairs Administration: M.Estban Soberanes Pelcastre [97]

31-01890 Consulate of the United Kingdom
Hotel Montana, PO Box 1302, Port-au-Prince
☎(+509) 57 39 69 ⊖ (+509) 57 40 48
Consul: Miss J MacLeod (resides in Kingston)
Vice-Consul: D T J Bull (resides in Kingston)
Vice-Consul: Mrs M Guercy [98]

Honduras

Head of State:
Carlos Roberto Reina Idiaquez

10 Head of State

10-10000 Office of the President
Casa Presidencial, Avenida 6, Calle 1, Tegucigalpa, DC
☎(+504) 34 49 22; 38 27 10
President of the Republic: Carlos Roberto Reina Idiaquez
First Vice President: Gen Walter López (retd)
Second Vice President: Juan de la Cruz Avelar
Third Vice President: Guadalupe Jerezano [98]

12 Ministries

12-10000 Ministry of Defence and Public Security
Palacio de los Ministerios, Calle 4, Avenida 5, Tegucigalpa
☎(+504) 37 95 51
Minister of National Defence and Public Security: Jose Luis Nunez Bennet [98]

12-20000 Ministry of Foreign Affairs
Ministerio Relaciones Exteriores
Palacio José Cecilio del Valle, Boulevard Juan Pablo II, Tegucigalpa, DC
Minister of Foreign Affairs: Delmar Panting Urbizo [98]

12-40000 Ministry of Interior
Palacio Nacional, Tegucigalpa
☎(+501) 22 81 04
Minister of Government and Justice: Efrain Mondaca Silva
Minister of Industry, Commerce and Tourism: Fernando Garcia [98]

31 Embassies & Attachés

31-00490 Embassy of the Dominican Republic
PO Box 633, Tegucigalpa DC
☎ (+504) 36 56 65 ✉ DICOSA HO ▣ 1258 dicosa ho
⊖ (+504) 36 91 13
Ambassador: H.E. Mr Simon S. Kafié [97]

31-00500 Embassy of Ecuador
PO Box 358, Tegucigalpa, DC
Colina Palmira, Avenida Juan Lindo 122
☎ (+504) 36 59 80 Ⓔ mecuahon@hondutel.hn
⊖ (+504) 36 69 29
Ambassador: H E Dr José Ignacio Jijón Freile
Commercial Counsellor: Ing. V Orlando Paredes [98]

31-01890 Embassy of the United Kingdom
PO Box 290, Tegucigalpa
Edificio Palmira,3er Piso, Colonia Palmira
☎ (+504) 32 54 29; 32 06 12; 32 06 18
▣ 1234 prodrome ho ⊖ (+504) 32 54 80
Ambassador: H E Mr P R Holmes
First Secretary, Commercial: A S Green
Defence Attaché: Col D B Simpson
(resident in Guatemala City)
Third Secretary, Vice-Consul and Deputy Head of Mission: R A Legg [98]

31-01900 Embassy of the United States of America
Avenida La Paz, Apartado Postal No 3453, Tegucigalpa, DC
☎ (+504) 36 93 20 ⊖ (+504) 36 90 37
Ambassador: H.E. Mr James Francis Creagan
Deputy Chief of Mission: Ms Deborah Schwartz
Commercial Section: Donald James McNally (Acting)
Public Affairs Officer: Mark Krischik [97]

90 Sales Agents, Sales & Liaison Offices

Companies represented:

Detroit Armor Corp., USA 90-00100

90-00100 J.M. Cantor
PO BOX 3253, Tegucigalpa, M D C
☎ (+504) 37 40 94; 37 40 95 ⊖ (+504) 32 61 67
Representative: José M Cantor
® USA: Detroit Armor Corp. [98]

Hong Kong SAR

Head of State:
Jiang Zemin

Chief Executive Hong Kong Administrative Region:
Tung Chee Hwa

10 Head of State

10-30000 Executive Council
Government Secretariat, Lower Albert Road, Hong Kong SAR
Governer of the Executive Council: Sir Sze-yuen Chung
Member of Executive Council: Henry Tang
Member of Executive Council: Chun-ying Leung
Member of Executive Council: Anthony Leung [98]

12 Ministries

12-40000 Government Secretariat Security Branch
Lower Albert Road, Main Wing - 6th Floor, Hong Kong SAR
☎ (+852) 28 10 26 42 ⊖ (+852) 28 68 50 74
Secretary for Security: Peter Lai
EO (G&A): Miss Louisa Lewry
Government Security Officer: John Ng
■ Defence related departments: Auxiliary Medical Service, Civil Aid Service, Correctional Services Department, Customs and Excise Department, British Forces, Fire Service Department, Immigration Department, Government Flying Service, Royal Hong Kong Police Force, Royal Hong Kong Auxiliary Police Force. [98]

15 Parliament

15-10000 Legislative Council (LegCo)
8 Jackson Road, Hong Kong SAR
☎ (+852) 28 69 46 16 ⊖ (+852) 28 77 96 00
Chairman: Andrew W F Wong Ⓔ wkwong@hkstar.com [98]

25 Defense Forces

25-05000 Government Flying Service
(† Royal Hong Kong Auxiliary Air Force)
1 Cheung Yip Street, Kowloon Bay
☎ (+852) 23 05 82 50 Ⓔ gfsbrian@hkstar.com
⊖ (+852) 27 53 84 08
Controller: Capt Brian Butt
Chief Aircraft Engineer: Capt Graeme McIntosh
Chief Pilot, Helicopter: Capt Barry Collier
Chief Pilot, Fixed Wing: Capt Ceri Harris
Senior Aircraft Engineer, Quality Assurance: Capt Benny Lee C C
Senior Supplies Officer: Ms Teresa Kong [98]

26 Police (Civil)

26-00000 Hong Kong Police Force Headquarters
(† Royal Hong Kong Police Force Headquarters)
Arsenal Street, Wanchai, Hong Kong SAR
☎ (+852) 28 60 20 00 ▣ 65367 hx
⊖ (+852) 28 65 65 46
Commissioner of Police: K-O Hui ☎ (+852) 860 20 02
Force Librarian: E Cheung
☎ (+852) 86 66 210; 529 84 78 (fax) [98]

31 Embassies & Attachés

31-00100 Consulate General of Austria
34-37 Connaught Road, Central,
Room 2201, Chinachem Tower, Hong Kong SAR
☎ (+852 5) 222 80 86 ✉ AUSTROKO ▣ 86006 austroko
⊖ (+852 5) 221 87 73
Consul General: Dr Gunther Gallowitsch
Vice Consul: Karl Simml
Assistant to Consul General: Ms Mabel Ku
Trade Commission
14th Floor Diamond Exchange Building,
8-10 Duddell Street, Hong Kong SAR
☎ (+852 5) 223 75 55; 222 23 88
⊖ (+852 5) 223 28 10; 223 64 93
Trade Commissioner: Dr Alfred Mayer [98]

31-00880 Consulate General of Israel
Room 701 Admiralty Centre Tower 2, 18 Harcourt Road, Hong Kong SAR
☎ (+852) 25 29 60 91 Ⓔ isrtrhk@asiaonline.net
⊖ (+852) 28 04 27 50
Consul General: Nachman Perel
Consul (Commercial Attaché): Ephraim Aviram [98]

31-00920 Consulate General of Japan
46/47th Floor One Exchange Square,
8 Connaught Place, Central, Hong Kong SAR
☎ (+852) 25 22 11 84; 25 22 11 88 ✉ RYOJI Hong Kong
▣ 73301 ryoji hx ⊖ (+852) 28 68 01 56
Consul General: Hideaki Ueda
Deputy Consul General: Takashi Saito
Head General Affairs Department: Ishikane Kimihiro [97]

31-01110 Commission of Malaysia
47-50 Gloucester Road, Wanchai,
24th Floor Malaysia Building, Hong Kong SAR
☎ (+852) 25 27 09 21; 25 27 09 23 ▣ 74207 mwaki hx
⊖ (+852) 28 65 16 28
Commissioner: Mohamed Khalis [97]

31-01300 New Zealand Consulate General
3416 Jardine House - Central, GPO Box 2790, Hong Kong SAR
☎ (+852) 25 25 50 44 ⊖ (+852) 28 45 29 15
Consul General: Carl Worker
Consul: Kevin Kay
Vice Consul: Ms Stephanie Lee
Attaché: Ms Christine Holah
Trade Commissioner: Donald MacLean
Branch Manager: Kevin Cameron
Consular Assistant: Ms Judy Cameron [98]

31-01890 British Trade Commission
PO Box 528, Hong Kong SAR
9th Floor, Bank of America Tower,
12 Harcourt Road, Central, Hong Kong
☎ (+852) 29 01 30 00 Ⓔ btchk@hkabconet
Ⓦ http://www.webhk.com/btchk ▣ 73031 uktra hx
⊖ (+852) 29 01 30 66
Senior Trade Commissioner: R F Cornish
Trade Counsellor: Andrew Seaton
Trade Counsellor: N N Inkster
Press and Information Officer: Gary Walder [98]

31-01900 Consulate of the United States of America
26 Garden Road, PO Box 66, Hong Kong SAR
☎ (+852) 25 23 90 11 ▣ 63141 usdoc hx
⊖ (+852) 28 45 15 98
Consul General: Richard A Boucher
☎ (+852) 28 41 23 32
Deputy Principal Officer: Stephen A Schlaikjer
☎ (+852) 28 41 22 98
Head of Commercial Section: David Katz
☎ (+852) 25 21 57 53
Public Affairs Officer: Robert B Laing
☎ (+852) 28 41 22 55
Defense Attaché: Capt James R Glover (US Navy)
☎ (+852) 28 41 22 55
US Defense Liaison Office Hong Kong
PSC 464, Box 10, 96522-0002 FPO AP
☎ (+852) 28 41 22 26 Ⓔ tndklap@asiaonline.net
⊖ (+852) 28 41 00 97
14
Operations Officer: Bradley Bennett [98]

65 Industry, Trade & Armed Forces Associations

65-10000 RHKR The Volunteers Association
(† Royal Hong Kong Regiment (The Volunteers))
PO Box 47296, Morrison Hill Post Office, Hong Kong SAR
☎ (+852) 29 66 78 81; 29 66 78 95
⊖ (+852) 29 66 78 92
Chairman: Maj Michael Leung (retd)
First Vice Chairman: Maj Michael Ip (retd)
Second Vice Chairman: Maj S Lord (retd)
Contact: WO2 T Chan (retd)
■ Disbanded 1995. [98]

Contact Reference

Hong Kong SAR

74 Consortia, Joint Venture Companies

South China Aero Technology Ltd . . . 80-00850

75 Manufacturing Companies

75-00080 Air International Pty Limited (Hong Kong)
(† Sigma Industries Pty Limited (Hong Kong))
35 Tin Ha Road, Hung Shui Kiu, Yuen Long, N.T.
☎(+852) 26 17 79 71 ⊖ (+852) 26 17 28 67
Manager: S. Li [97]

75-00200 Checkmate Industries Ltd
1213-1214 New World Office Bldg,
W.Wing, 20 Salisbury Rd, Kowloon
☎(+852) 27 39 67 83 ⊠ 54468 vache hx
⊖ (+852) 27 39 72 49
♣ US$ 10 000 000 ⅏ 40
President: Vartan Keshishian
Sales and Marketing Director: Zarmair Keshishian
■ Military and police uniforms, footwear, webbing equipment, protective bulletproof vests, protective bulletproof helmets, transport equipment and vehicles. Anti-riot equipment, uniforms, military tents and camouflage equipment. → 85 [97]

75-00600 Kawasaki Heavy Industries (HK) Ltd
Room 1619 Jardine House, Connaught Road, Central, Hong Kong SAR
☎(+852) 25 22 35 60 ⊖ (+852) 28 45 29 05 [98]

75-01000 Minebea Co. Ltd Hong Kong Branch
Room 1416 - Starhouse 3, Salisbury Road,
Tsim Sha Tsui, Kowloon
☎(+852) 730 99 13 ⊖ (+852 3) 735 45 35
⅏ 13
Branch Manager: O Misawa [98]

75-02000 THORN Lighting Hong Kong
PO Box 51 / GPO, Hong Kong SAR
19th Floor Jardine Eng House, 260 Kings Road
☎(+852) 578 43 03 ⊖ (+852) 887 02 47
Chief Executive: Steve Grao [98]

75-08000 Wyldar Machine Tool Ltd
Fabrico Ind Building 1st Floor, 78-84 Kwai Cheong Road,
Kwai Chung, NT
☎(+852) 24 89 98 81 ⊖ (+852) 24 80 47 21
Managing Director: Rockey Y K Lo
Technical Director: Owen Y K Lam
Sales Manager: David Lee Pui Tin
® Switzerland: Trimos-Sylvac Task Force [98]

80 Service Companies & Organizations

Swiss-Impex Co Ltd 90-01990

80-00010 AlliedSignal Aerospace Marketing Sales and Service
Hong Kong International Airport, Kowloon City
c/o HAECO Hangar Four, North Annex Office
60, Concorde Road, Hong Kong International Airport,
Kowloon City
☎(+852) 23 82 80 33 ⊖ (+852) 27 16 49 30
Account Team Leader: Leonard Fok
☎(+852) 94 56 81 30 (mobile)
Field Service Engineer: Anthony Chia
Field Service Engineer: John Krings
■ Manages commercial programs, provides technical support for AlliedSignal Aerospace products in Southern China. [98]

80-00150 Cogent Managed Services
34th Floor Central Plaza, 18 Harbour Road, Wanchai
☎(+852) 25 85 28 66 ⊖ (+852) 25 85 22 19
Managing Director: Mike Kelly
Director Global One: Brian Spearpoint
Director India: Simon MacCormac
Director Asia: Malcolm Nicholls
■ Provides a complete service offering to national and international organisations providing alternate networks to established suppliers. This has been provided in the UK, Netherlands, Hong Kong, Japan and internationally. [97]

80-00200 Hong Kong Aircraft Engineering Co. Ltd (HAECO)
(a subsidiary of John Swire & Sons Ltd)
60 Concorde Road, Hong Kong International Airport,
Kai Tak, Kowloon
☎(+852) 27 67 61 42 ✉ AIRENG Hong Kong
⊠ 44664 haeco hx ⊖ (+852) 23 33 45 14
♧ HK$ 185,194,000 ♣ HK$ 2,390,776,000 (1995)
⅏ 5,100 ⌂ 750,000 sq ft
Managing Director: John Slosar
Engineering Director: Keith Law
General Manager Marketing and Business Aquisition: A Hurran
Operations Director, China: P K Chan
Commercial Manager: D Briant
Commercial Manager: Hans Chau
Purchasing Manager: A Sathianathan
■ Maintenance, overhaul, repair, refurbishment and modification of aircraft, engines, components and related equipment.
Subsidiaries and associated companies:
South China Aero Technology (SCAT)
Bridgestone Aircraft Tire Company (Asia) Ltd
Kai Tak Refuellers Company Ltd (KTR)
Lucas TAECO Aerospace (Xiamen) Co Ltd
BF Goodrich Aerospace Asia Pacific Ltd
Taikoo (Xiamen) Aircraft Engineering Co Ltd
AlliedSignal TAECO Aerospace (Xiamen) Co Ltd
Hong Kong Aero Engine Services Ltd [97]

80-00250 Hughes Asia Pacific (Hong Kong) Ltd
(a division of Hughes Electronics Corporation, USA)
Suite 3407 Citibank Tower, Citibank Plaza, 3 Garden Road,
Central, Hong Kong SAR
☎(+852) 28 48 32 88 ⊖ (+852) 25 37 68 83
Managing Director: Chris Lanzit
E crlanzit@ccgate.hac.com [98]

80-00300 Logica Limited
(a subsidiary of Logica plc, UK)
4th Floor, Chung Nam Building, 1 Lockhart Road,
Wanchai, Hong Kong
☎(+852) 28 66 26 63 ⊖ (+852) 25 20 18 06 [98]

80-00410 PMsoft International Limited
Room 710 7th Floor Tai Yau Building, 181 Johnston Road,
Wanchai
☎(+852) 28 36 32 18 ⊖ (+852) 25 91 04 71 [98]

80-00500 Scott Wilson (Hong Kong) Ltd
(† Scott Wilson Kirkpatrick)
38th Floor - Metroplaza Tower 1, 223 Hing Fong Road,
Kwai Fong
☎(+852) 24 28 88 66 E swk@swkhk.com.hk
⊖ (+852) 24 28 99 22
⅏ 2,500 (worldwide)
Chairman: Ron H Rakusen
Managing Director: Martin H C Kwong
Director: Robin Osbourn
Director: Michael L Chalmers
Director: Alan D Morris
Director: Tony S Chan
Director: Albert C H Cheung
Director: Alexander S F Wong
Director: Peter C W Chan
Director: George E Tedbury
Director: Robert J Hunt
■ Consultancy for engineering and related fields, planning, environment, railways, leisure and tourism. [98]

80-00700 Secure Consulting International
FB3, First Floor Garden Mansion, 150 Tai Hang Road,
Tai Han [98]

80-00850 South China Aero Technology Ltd (SCAT)
(a joint venture between HAECO & CATIC)
Hong Kong International Airport, 60 Concorde Road,
Kai Tak, Hong Kong
☎(+852) 27 67 64 29 ⊠ 51644 ⊖ (+852) 23 26 48 70
Director: Joseph Wong
■ Import/export, marketing services and technical support for aero products between PRC and rest of the world. → 74 [97]

85 Stockists, Distributors, Spare Parts Suppliers

Checkmate Industries Ltd 75-00200

85-00500 Silver Star International Co
PO Box 91153 TST, Hong Kong SAR
☎(+852) 26 04 97 14 ⊖ (+852) 26 92 44 19
Chief Executive Officer: Stephen Lau
■ Distributor of Chinese arms and ordnance to clients worldwide. [98]

90 Sales Agents, Sales & Liaison Offices

Companies represented:

Alfred Karcher GmbH & Co, Germany . . 90-01100
Bell Helicopter Textron, USA→ Singapore 90-00220
Crompton Instruments Ltd, UK 90-00300
Farnell International, UK 90-00550
ITT Defense, USA 90-00960
Racal Electronics Group, UK 90-01810
Radiall, France 90-01812
Rockwell-Collins, USA 90-01850
Sait Radio-Holland Group, Netherlands . 90-01813
Sema Group plc, UK 90-01920
Teledyne Inc., USA 90-02015
The General Electric Co., UK 90-00730
Trimos-Sylvac Task Force, Switzerland . 75-08000

90-00150 Aviall Airstocks
604 New Bright Building,
11 Sheung Yuet Road - Kowloon Bay, Kowloon
☎(+852) 23 18 02 33 ⊖ (+852) 23 31 92 22
Manager: Yenho Tree [98]

90-00180 British Steel (Asia) Ltd
(a subsidiary of British Steel Engineering Steels, UK)
6th Floor Jardine Engineering House,
260 Kings Road, North Point, Hong Kong SAR
☎(+852) 28 07 01 96 ⊖ (+852) 28 07 18 05 [98]

90-00300 Crompton Instruments Ltd
Aberdeen, Hong Kong SAR
11th Floor Kingsun Comouter Industry
Building, 40 Shek Pai Wan Road, Tin Wan,
Aberdeen, Hong Kong SAR
☎(+852) 25 52 02 67 ⊖ (+852) 28 73 14 76
♣ £ 1,000,000 ⅏ 10
General Manager: E Choi
® UK: Crompton Instruments Ltd [98]

90-00395 Daewoo Hong Kong Ltd
(a subsidiary of Daewoo Corp, South Korea)
2301-2309 23rd Floor Natwest Tower,
Times Square, 1 Matheson Street, Causeway Bay
☎(+852) 25 99 57 00 ⊠ 75274 dwhkg hx
⊖ (+852) 25 06 06 66 [98]

90-00400 Daicel (Hong Kong) Ltd
Suite 915, World Shipping Centre, Harbour City, Kowloon
☎(+852) 27 30 21 29 ⊖ (+852) 23 75 30 90 [97]

90-00550 Euro Tech (Far East) Ltd
18th Floor Gee Chang Hong Centre,
65 Wong Chuk Hang Road, Hong Kong SAR
☎(+852 5) 28 14 03 11 ▣ 72449 efeld hx
⊖ (+852 5) 28 70 04 79
⚉ 60 ⌂ 2,000 sq m
Managing Director: T C Leung
Sales and Marketing Director: C P Kwan
Operations and Technical Director: Alex Sham
Office Manager: Ms Nancy Wong
■ Supplier of process control, scientific, telecommunication, electronic test and measurement instruments, equipment and systems.
® UK: Farnell International [97]

90-00730 The General Electric Co of Hong Kong Ltd
PO Box 15 GPO, Hong Kong
☎(+852) 29 19 82 82 ▣ 73098 ghong hx
⊖ (+852) 28 34 57 73
Managing Dir: J. Chiu
® UK: The General Electric Co. [97]

90-00943 Industrial Protector Limited
Rm 1007/8 Penisula Square, 18 Sung On Street,
Hunghom, Kowloon
☎(+852) 27 64 82 72 (5 lines) ▣ 38954 isafe hx
⊖ (+852) 27 64 40 63; 23 03 06 89 [97]

90-00960 ITT Defense International
(† ITT Defense & Electronics)
705 Bank of America Tower, 12 Harcourt Road Central,
Hong Kong SAR
☎(+852) 25 84 61 20 ⊖ (+852) 25 37 56 55
Asia Pacific Operations Director: Tom Roberts
® USA: ITT Defense [98]

90-01100 Karcher Ltd
Unit 14 5th Floor Hewlett Centre, 52-54 Hoi Yuen Road,
Kwun Tong, Kowloon
☎(+852) 23 57 58 63 ⊖ (+852) 23 57 56 32
Managing Director: Dr Bernhard Graf
® Germany: Alfred Karcher GmbH & Co [98]

90-01400 Moog Controls Hong Kong Ltd
Unit 7-6 Floor, Block B, New Trade Plaza, 6 On Ping Street,
Shatin, N.T.
☎(+852) 26 35 32 00 ⊖ (+852) 26 35 45 05
♣ US$ 4,100,000 ⚉ 11
General Manager: M Ohno
Deputy General Manager: Samuel Lo [97]

90-01810 Racal Electronics Group China Liaison Division
(a division of Racal Electronics plc, UK)
Sun House, 13th Floor, 90 Connaught Road,
Central Hong Kong SAR
☎(+852) 25 43 06 63 ▣ 83556 rdehk hx
⊖ (+852) 25 44 94 06
Consultant: Brian Stewart
® UK: Racal Electronics Group
Racal-Datacom (China Region) Ltd
Sun House, 13th Floor, 181 Des Voeux Road,
Central, Hong Kong SAR
☎(+852) 28 15 86 33 ⊖ (+852) 28 15 81 41
General Manager: Steve Bossom
General Sales Manager: Eddie Ho [98]

90-01812 Radiall Electronics (Asia) Ltd
Unit 1 6th Floor Charm Centre, 700 Castle Peak Road,
Kowloon
☎(+852) 29 59 38 33 ▣ 57072 bisha hx
⊖ (+852) 29 59 26 36
♣ HK$ 13,000,000 (1995) ⚉ 10 (19979 ⌂ 150 m2
Managing Director: Charles Wu
Sales and Marketing Director: E Hung
® France: Radiall [98]

90-01813 Radio Holland BV
806 Join-in Hang Sing Centre, 71-75 Container Port Road,
Kwai Chung, N.T.
☎(+852) 24 23 90 07 ▣ 50881 ⊖ (+852) 24 80 58 98
♣ US$ 15,000,000 (1995) ⚉ 38 ⌂ 1,800 sq m
General Manager: V G Hoyle
Sales and Marketing Director: Paul Smulders
Operations and Technical Director: Andy Richardson
■ Marine communications, land communications, satellite communications, radar, navigational aids, electro-optic surveillance systems and night vision devices.
® Netherlands: Sait Radio-Holland Group
Argos Vision Systems, Flir Systems Inc,:
ITT, Motorola, S.P. Radio, Furuno & Jotron. [97]

90-01830 RBR Armour Limited
(a subsidiary of RBR Armour, UK)
1st Floor N°1 Sixth Lane, Sha Kok Mei, Saikung
☎(+852) 27 92 47 54
🄴 100452.3251@compuserve.com
⊖ (+852) 27 91 74 18
⚉ 1
Director Asia Pacific: John Bowman
■ Supply of personal protection products. [98]

90-01850 Rockwell International (Asia Pacific) Limited
(a division of Rockwell International Corp., USA)
13th Floor, Suites 8-10,
Harbour Centre, 25 Harbour Road, Wanchai
☎(+852) 28 27 01 81 ⊖ (+852) 28 27 64 88
Office Manager: Richard Yu
® USA: Rockwell-Collins [98]

90-01900 Saft Hong-Kong
(a subsidiary of Saft SA, France)
Unit 7-10, 23/f Apec Plaza,
49 Hoi Yuen Road, Kwun Tong, Kowloon
☎(+852) 27 96 99 32 ⊖ (+852) 27 98 06 19
Managing Director: Mrs Grace Chiam [98]

90-01910 Saft Nife Ltd
Unit 5-6, 3/f Corporation Square, 8 Lam Lok Street,
Kowloon Bay
☎(+852) 27 95 27 19 ⊖ (+852) 27 98 05 77
Managing Director: Ng King [98]

90-01920 Sema Group (HK) Ltd
Rm 2201 Windsor Hse, 311 Gloucester Road,
Causeway Bay
☎(+852) 28 81 68 60 ⊖ (+852) 25 04 45 73
® UK: Sema Group plc [98]

90-01970 Stratech (Hong Kong) Ltd
35th Floor, Central Plaza, 18 Harbour Road,
Hong Kong SAR
☎(+852) 25 93 11 12 ⊖ (+852) 25 93 12 22
Managing Director and Chief Executive Officer:
Dr David K M Chew
Executive Director, Technology: Dr Kennedy K M Chew
Executive Director, Corporate and Legal Affairs:
Chew-Leong Sook Ching
Director: Ms Lucy Ng [97]

90-01990 Swiss-Impex Co Ltd Head Office
(† Swiss-Impex Trading & Consulting Co Ltd)
GPO Box 943, Central, Hong Kong SAR
6th Floor, Block E, Shing Lee Commercial
Building, 6-12 Wing Kut Street, Hong Kong SAR
☎(+852) 25 45 51 13 ▣ 72722 sitc hx
⊖ (+852) 25 42 32 69
⚉ 8 (1995) ⌂ 1,000 sq ft
Managing Director: Urs Baeumle
■ Specialising in anti-guerrilla warfare and anti-terrorism equipment including combat simulators, detection systems, military transceivers, active/passive night vision equipment, infantry equipment and NBC suits. Supplier of customised intelligence gathering devices. → 80 [97]

90-02015 Teledyne Industries International Inc.
RM 918 New World Centre (East Wing),
24 Salisbury Rd, Tsimshatsui, Kowloon
☎(+852) 23 68 52 31 ▣ 30240 tdyhx
⊖ (+852) 23 69 89 78
® USA: Teledyne Inc. [97]

90-02100 UBE (Hong Kong) Ltd
Rooms 2622-4 Sum Hung Kai Centre, 30 Harbour Road,
Hong Kong SAR
☎(+852) 28 77 16 28 ⊖ (+852) 28 77 12 62 [98]

Hungary

Head of State:
Arpad Göncz

10 Head of State

10-10000 Office of the President
Kossuth L tér 3-5, H-1055 Budapest
☎(+36 1) 268 40 00 ⊖ (+36 1) 268 48 75
President: Arpad Göncz
Presidential Adviser Military Affairs: Col Robert Pick [98]

11 Prime Minister

11-10000 Office of the Prime Minister
Kossuth L tér 1-3, H-1055 Budapest
☎(+36 1) 268 40 00 ▣ 225547 mtbp h
⊖ (+36 1) 153 00 30; 268 47 37
Prime Minister: Gyula Horn
Deputy Prime Minister: Gabor Kuncze
Spokeswoman Council of Ministers: Ms Judit Fuhász [98]

12 Ministries

12-10000 Ministry of Defence
Balaton ulitsa 7-11, H-1055 Budapest
☎(+36 1) 311 96 35; 132 25 00
⊖ (+36 1) 312 82 46; 132 25 00
Minister of Defence: György Keleti
Political State Secretary: Dr István Fodor
Administrative State Secretary: Dr József Fehér
Deputy State Secretary: Tibor Tóth
Deputy State Secretary: László Borsits
Deputy State Secretary: Dr Károly Janza
Deputy State Secretary & Chief of Private Office:
Csaba Liszkay
Speaker of the Ministry: Col Lajos Erdélyi
Public Relations Dir: Anna Koós
Head of Press Office: Col László Tikos [98]

12-15000 Ministry of Environment and Regional Policy
Föutca 44-50, H-1011 Budapest
☎(+36 1) 201 41 33
Minister: Ferenc Baja [98]

12-18000 Ministry of Finance
József nádor ter 2-4, H-1051 Budapest
☎(+36 1) 118 20 66; 138 26 33 ⊖ (+36 1) 118 25 70
Minister of Finance: Peter Medgyessy [98]

12-20000 Ministry of Foreign Affairs
Külügyminisztérium
PO Box 423, H-1394 Budapest 62
Bem rkp. 47, Budapest II
☎(+36 1) 156 80 00 ▣ 225571 digudh
⊖ (+36 1) 212 59 18; 155 96 93
Minister: László Kovács ☎(+36 1) 156 56 84
Political State Secretary: István Scent-Iványi
☎(+36 1) 201 95 39
State Secretary Administration: Ferenc Somogyi
☎(+36 1) 201 95 38

Deputy State Secretary International Organisations, Press: André Erdős ☎(+36 1) 201 42 57
Deputy State Secretary Advanced Countries: István Pataki ☎(+36 1) 201 86 15
Deputy State Secretary Third World: Zoltàn Pereszlényi ☎(+36 1) 201 80 93
Deputy State Secretary East-Central Europe: János Tóth ☎(+36 1) 201 78 57
Spokesman's Office, Spokesman: Gábor Szentivànyi ☎(+36 1) 202 49 79; 202 50 98 (fax)
Government Spokesperson: Ms Evelyn Forró ☎(+36 1) 269 48 41; 268 47 37 (fax) 98

12-25000 Ministry of Industry and Trade
Honvéd utca 13-15, H-H-1880 Budapest
☎(+36 1) 302 23 55 W http://www.ikm.hu
⊖ (+36 1) 302 23 94
225578
Minister of Industry, Trade and Tourism: Dr Szabolcs Fazakas
Press and Public Relations Director of the Ministry: Dr Judit F Gyarfas
Minister's Office
Honvéd utca 13-15, H-1880 Budapest
☎(+36 1) 302 23 55 ⊖ (+36 1) 302 23 94 98

12-40000 Ministry of Interior
József Attila utca 2-4, H-1051 Budapest
☎(+36 1) 112 17 10; 131 37 00; 132 57 90
⊖ (+36 1) 118 97 19
Minister of the Interior: Gabor Kuncze 98

12-45000 Ministry of Justice
Szalay utca 16, H-1055 Budapest
☎(+36 1) 132 53 30; 132 61 70
Minister of Justice: Pal Vastagh 98

12-50000 Ministry of Labour
Roosevelt tér 7-8, H-1051 Budapest
☎(+36 1) 132 21 00; 111 02 00 ⊖ (+36 1) 131 63 99
Minister of Labour: Peter Kiss 98

12-60000 Ministry of Transport, Communication and Water Affairs
Dob utca 75-81, H-1077 Budapest
☎(+36 1) 122 02 20; 142 05 20 ⊖ (+36 1) 122 86 95
Minister: Lotz Karoly 98

15 Parliament

15-10000 National Assembly
Kossuth L tér 1-3, H-1055 Budapest V
☎(+36 1) 268 40 00 (Parliament Building)
☎(+36 1) 268 50 00 (MP's Building)
W http://www.mkogy.hu 225547 ⊖ (+36 1) 268 48 00
Speaker: Zoltán Gál ☎(+36 1) 268 44 02
Deputy Speaker: Ms Agnes G Nagy Macz ó ☎(+36 1) 268 48 09
Deputy Speaker: Ms Mária Kóródi ☎(+36 1) 268 42 90
Deputy Speaker: Tobor Füzessy ☎(+36 1) 268 48 08
Secretary General: István Soltész ☎(+36 1) 268 44 50
President of the Foreign Committee: Istvan Szent-Ivaanyi ☎(+36 1) 268 46 26
Vice President of the Foreign Committee: András Bársony ☎(+36 1) 268 56 07
Vice President of the Foreign Committee: György Csóti ☎(+36 1) 268 51 37
President of the Defence Committee: Imre Mécs ☎(+36 1) 268 48 14
Vice President of the Defence Committee: Ferenc 1Juhasz ☎(+36 1) 268 55 10
Vice President of the Defence Committee: Béla Gyuricza ☎(+36 1) 268 58 54
Head of the Press Department: Mrs Emőke Kulcsár ☎(+36 1) 268 42 75 98

25 Defense Forces

25-10000 Hungarian Home Defence Forces
Budapest
☎(+36 1) 111 96 35 ⊖ (+36 1) 112 82 46
Chief of General Staff: Lt Gen. Sándor Németh
First Deputy Chief of General Staff: Maj.Gen. Ferenc Végh
Commander of 4 Mechanized Corps: Maj.Gen. Ambrus Preininger
Commander of 2 Air & Air Defence Corps: Maj.Gen. Attila Kositzky
Commander of 1 Mechanized Division: Brig.Gen. László Tánczos
Commander of 2 Mechanized Division: Brig.Gen. Mihály Cserhalmi
Commander of 3 Mechanized Division: Brig.Gen. Ferenc Bátor
Head of Intelligence Branch: Maj.Gen.Dr Géza Stefán
Head of Material-Technical Branch: Eng.Maj.Gen. Nándor Hollósi
Head of Personnel Branch: Maj.Gen.Dr József Holló 97

25-20000 Air Forces & Air Defence Command Headquarters
Schweidel Ulitsa 2-4, Bp. Pf 25, H-1885 Budapest
☎(+36 1) 166 74 38 ⊖ (+36 1) 166 98 20
Chief of Aviation, Hungarian Home Defence Forces: Maj Gen György Feminger
Deputy Commander and Chief of Staff: Col János Sági
Operations Director: Col József Babos
Procurement Director: Col István Jobbik 97

29 Customs & Excise

29-01000 Customs & Finance Guard
PO Box 109, H-1450 Budapest 98

31 Embassies & Attachés

31-00310 Embassy of Canada
Budakeszi Utca 32, H-1121 Budapest
☎(+36 1) 275 12 00 ✉ CANADA Budapest
224588 canada h ⊖ (+36 1) 275 12 10
Ambassador: H E Mr Susan Cartwright
First Secretary: Guy Archambaunt
Commercial Counsellor: W G McEwen
Counsuler Secretary: Nemeth Zita 98

31-00460 Royal Danish Embassy
Határör út 37, H-1122 Budapest XII
☎(+36 1) 155 73 20 E danishembassy@pronet
224137 amda h ⊖ (+36 1) 175 38 03
Ambassador: H E Mr Erik Skov
Deputy Chief of Mission and First Secretry: Michael Bennet Jensen ☎(+36 1) 326 04 34
Secretary of the Embassy Commercial: Ms Anne Marie Tyndeskov Voetmann ☎(+36 1) 202 14 44
Attaché: Ms Susanne Baden Jorgensen ☎(+36 1) 325 76 34
Agricultural Counsellor: Carsten Philipsen 98

31-00590 Embassy of France
Lendvay utca 27, H-1062 Budapest VI
☎(+36 1) 332 49 80 225165 ⊖ (+36 1) 111 82 91
Ambassador: H E Mr Paul Poudade
First Counsellor Deputy Chief of Mission: Alain Holleville
First Secretary: Ms Elisabeth Barsacq
Military Section
Lendvay utca 27, H-1062 Budapest VI
☎(+36 1) 269 18 15 ⊖ (+36 1) 269 18 16
Armed Forces Attaché: Col Philippe Toussaint
Air Attaché: Lieut Col Guy Planet
Commercial Section
Petőfi Sandor utca 11, H-H-1052 Budapest (V)
☎(+36 1) 266 12 40 225143 ⊖ (+36 1) 266 12 48
Commercial Counsellor: Paul Hunsinger
Commercial Attaché: Patrick de Cazenove
Cultural Section
Fő utca 17, H-1011 Budapest (1)
☎(+36 1) 202 11 33 223558 ⊖ (+36 1) 202 13 23
Cultural Counsellor: Henri Lebreton 98

31-00690 Embassy of Greece
Szegfu utca 3, H-1063 Budapest
☎(+36 1) 322 80 04; 322 80 65 224113 98

31-00830 Embassy of India
Buzavirag Utca 14, H-1025 Budapest II
☎(+36 1) 325 77 42; 325 77 43
E chancery@indemb.datanet.hu 226374 indem h
⊖ (+36 1) 325 77 45
Ambassador: H E Mr Satnam Jit Singh
Defence Attaché: Col B V Nair (Residing in Prague,) (concurrently accredited to) (Hungary)
Counsellor (Commercial Information and Political): Ms Anita Nayer 98

31-00840 Embassy of the Republic of Indonesia
Varosligeti fasor 26, H-1068 Budapest VI
☎(+36 1) 342 85 08; 342 85 49; 342 85 85
✉ INDONESIA Budapest 227808 indon h
⊖ (+36 1) 322 86 69
Ambassador: H E Mr Hasan Abduldjalil
First Secretary and Public Information Officer: Djoko Harjanto
Commercial Officer: Ade Petranto 97

31-00890 Embassy of Italy
Stefania Ulitsa 95, H-1143 Budapest
☎(+36 1) 343 60 65; 343 60 57 225294 itabp h
⊖ (+36 1) 343 60 58
Ambassador: H E Dr Pietro Ercole Ago
Deputy Chief of Mission, First Counsellor: Dr Gianni Ghisi
Military and Defense Attaché: Col Piero Lucchetti
Scientific Attaché: Dr Gianfranco Cicognani
First Secretary for Commercial Affairs: Dr Emanuela d'Alessandro
Cultural Affairs
Brody Sandor utca 8, Budapest
☎(+36 1) 266 75 55 ⊖ (+36 1) 266 01 71
Cultural Attaché: Dr Giuseppe Manica 98

31-00900 Italian Institute of External Trade
Rakoczi ulitsa 1/3, Budapest
☎(+36 1) 266 75 55 ⊖ (+36 1) 266 01 71
Director of the Italian Institute of External Trade: Dr Luiga Liberati 98

31-00920 Embassy of Japan
1024 II Rómer Flóris Utca 56-58, PO Box 78, H-1525 Budapest
☎(+36 1) 156 45 33; 156 46 48; 156 44 81
⊖ (+36 1) 212 26 83 97

31-01190 Embassy of Mexico
Vérhalom ulitsa 12-16, Ep 3, 31/33, H-1023 Budapest
☎(+36 1) 326 04 47 ⊖ (+36 1) 326 04 85
Ambassador: H E Mr Margaarita Gonzhlez Gamio ☎(+36 1) 326 04 87
Deputy Chief of Mission: Eduardo Moguel
Commercial Counsellor: Roberto Fagoaga
Second Secretary Consular Affairs: Raul Villanueva
Third Secretary: Edgar Galindo Cota
Cultural Attaché: Silvia Ramirey R 98

31-01665 Embassy of the Republic of Slovenia
Lendvay Utca 23, H-1026 Budapest
☎(+36 1) 153 23 74; 132 71 78 ⊖ (+36 1) 153 47 30
Ambassador: H.E. Mr Ferenc Hajós
Counsellor: Milivoj Butinar
Military & Air Attaché: Col Bogdan Koprivnikar
First Secretary, Commercial Affairs: Iztok Grmek
Attaché: Ms Ksenija Skrilec 98

31-01740 **Embassy of Switzerland**
Stefania ut. 107, H-1143 Budapest
☎(+36 1) 343 18 38; 343 94 91
E 100634.3571@compuserve.com
⊖ (+36 1) 343 94 92; 344 05 04
Ambassador: H E Mr Claudio Caratsch
Deputy Chief of Mission: Johannes Kunz
Defense Attaché: Col Thomas Mathys 98

31-01890 **Embassy of the United Kingdom**
Harmincad Utca 6, H-1051 Budapest V
☎(+36 1) 266 28 88 ⊖ (+36 1) 266 09 07
Ambassador: H E Mr C W Long
Deputy Head of Mission: C N R Prentice
Defence Military and Air Attaché: Col H V C Stephenson
First Secretary (Commercial): S C Martin 98

75 Manufacturing Companies

75-00100 **Army-Coop**
XIV ker Hermina út 17, H-1146 Budapest
☎(+36 1) 343 12 36 ⊖ (+36 1) 343 18 44
US$ 5,000,000 68 37,562 m2
Managing Director: Sandor Veress
Technical Director: Dr Tibor Elek
Sales and Marketing Director: Jozsef Bernoth
Commercial Director: Ferenc Lovas
■ Manufacture of: 81mm automatic mortars; 82(81)mm self-propelled automatic mortars; type DE 82 towed variation 82mm automatic mortars; 120mm mortars; 2A31 122mm howitzers. 97

75-00150 **Daewoo Engineering and Construction (Hungary) Co Ltd**
(a subsidiary of Daewoo Corporation, South Korea)
48 BAJCSY Zsiliszksy Street, 2nd Floor, H-1054 Budapest
☎(+36 1) 302 25 92 ⊖ (+36 1) 302 27 58 98

90 Sales Agents, Sales & Liaison Offices

Companies represented:

Acer Inc., Taiwan 90-00100
Alfred Karcher GmbH & Co, Germany. . 90-00800

90-00100 **Acer Computer Representative Hungary**
Dayka G.u.3. Bl-1, H-1118 Budapest
☎(+36 1) 209 20 56 ⊖ (+36 1) 248 13 68
® Taiwan: Acer Inc. 97

90-00250 **Daewoo Corp Budapest Represetative Office**
Bajcsy-Zsilinszky Ut 12,
International Trade Center Room 304, H-1051 Budapest
☎(+36 1) 266 37 37 223719 dwbud h
⊖ (+36 1) 266 38 38 98

90-00800 **Karcher Hungaria KFT**
Gyepsor u 1, H-1751 Budapest
☎(+36 1) 276 03 52 ⊖ (+36 1) 276 05 51
® Germany: Alfred Karcher GmbH & Co 98

Iceland

Head of State:
Olafur Grimsson

10 Head of State

10-10000 **Office of the President**
v/Laekjargötu, IS-150 Reykjavik
☎(+354) 562 26 00 ⊖ (+354) 562 40 14
President: Olafur Grimsson 98

11 Prime Minister

11-10000 **Office of the Prime Minister**
v/Laekjargötu, IS-101 Reykjavik
☎(+354) 552 50 00
Prime Minister: David Oddsson 98

12 Ministries

12-10000 **Ministry of Communications**
Hafnarhusinu v/Tryggvagotu, IS-150 Reykjavik
☎(+354) 560 96 30 ⊖ (+354) 562 17 02
Minister of Communications: Halldór Blöndal
Secretary General: Jón Birgir Jónsson
Deputy Secretary General: Halldór S Kristjánsson 98

12-20000 **Ministry for Foreign Affairs**
Raudarárstígur 25, IS-150 Reykjavik
☎(+354) 560 99 00 E postur@utn.stjr.is
2050; 2225 extern is ⊖ (+354) 562 23 86; 562 23 73
Minister of Foreign Affairs: Halldór Asgrimsson
Chairman of Defence Division: Thordur A E Oskarsson
Director of Culture, Press and Information: Bjarni Sigtryggsson ☎(+354) 560 99 18
■ Liaison between NATO forces and government.
Information Office
Raudarárstígur 25, IS-150 Reykjavik
☎(+354) 560 99 65 ⊖ (+354) 562 31 52
Press Officer: Bjarni Sigtryggsson
Defence Department
Raudarárstígur 25, IS-150 Reykjavík
☎(+354) 560 99 50 ⊖ (+354) 551 56 80 98

12-50000 **Ministry of Justice and Ecclesiastical Affairs**
Dóms-Og Kirkjumálaráduneytid
Arnarhvoll, IS-150 Reykjavik
☎(+354) 560 90 10 ⊖ (+354) 552 73 40
Minister of Justice: Thorsteinn Pálsson 98

15 Parliament

15-10000 **Parliament**
Althing
IS-150 Reykjavik
☎(+354) 563 05 00 ⊖ (+354) 563 05 20
Speaker of the Althing: Ólafur G Einarsson
Secretary-General: Fridrik Ólafsson
Foreign Affairs Committee
IS-150 Reykjavik
☎(+354) 563 05 00 ⊖ (+354) 563 06 10
Chairman: Geir H Haarde 98

20 Defense Administrations/Services/Agencies

20-00100 **Icelandic Civil Aviation Administration**
Flugmálastjórn
Reykjavik Airport, PO Box 350, IS-121 Reykjavik
☎(+354) 569 41 00 ✉ CIVILAIR 2250 falcon is
⊖ (+354) 562 36 19
Director General: Thorgeir Pálsson
Deputy Director of Civil Aviation and Ground Facilities: Haukur Hauksson
Airport Division Director: Johann H Jonsson
International Civil Aviation Director: Thordor O Sigurdsson
Administration Department Director: Sigrun Traustadottir
Flight Safety Department Director: Gudmundur Matthiasson
Air Traffic Services Division Director: Asgeir Palsson
Manager Keflavik Airport: Pitur Gudmundsson 97

20-00200 **Icelandic Meteorological Office**
Vedurstofa Islands
Bustadavegi 9, IS-150 Reykjavik
☎(+354) 560 06 00 ✉ METEO E office@vedur.is
⊖ (+354) 552 81 21
Director: Magnús Jónsson
■ 24 hour weather forecasts and observation at Reykjavik and Keflavik airports. 97

28 Coastguard, Border Protection Forces

28-00500 **Icelandic Coast Guard**
Landhelgisgaelan
Seljavegi 32, PO Box 7120, IS-127 Reykjavik
☎(+354) 511 22 22 (main); 511 33 33 (operations/SAR)
E sar@skima.is 2048 varsk is ⊖ (+354) 511 22 44
US$ 11,000,000 128
General Director: Hafsteinn Hafsteinsson
Finance Manager: Ms Maria Solbergsdottir 97

29 Customs & Excise

29-01000 **General Directorate of Customs**
Tollhúsinu - Tryggvagötu 19, IS-150 Reykjavik 98

31 Embassies & Attachés

31-01340 **Royal Norwegian Embassy**
Fjólugötu 17, Reykjavik
☎(+354) 551 30 65 ⊖ (+354) 552 95 53
Ambassador: H.E. Mr Nils O. Dietz
Deputy Chief of Mission: Olav Myklebust 97

31-01890 **Embassy of the United Kingdom**
PO Box 460, IS-121 Reykjavik
Laufasvegur 49
☎(+354) 550 51 00 ⊖ (+354) 550 51 05
Ambassador and Consul General: H E Mr J McCulloch
Deputy Head of Mission and Consul: Miss J S Wills
Commercial Officer: O Valdimarsson 98

31-01900 **Embassy of the United States of America**
Laufasvegur 21, IS-101 Reykjavik
☎(+354) 562 91 00 E amemb@itn.is
W http://www.nyherji.is/america ⊖ (+354) 562 91 39
Ambassador: H E Mr Day Olin Mount
Deputy Chief of Mission: Thomas Scaletta
Chief Economic/Commercial Secretary: Olaf North Otto (III)
Public Affairs Office
US Information Service, Laugavegur 26, IS-101 Reykjavik
☎(+354) 562 91 00 ⊖ (+354) 562 91 39
Public Affairs Officer: Walter Douglas ☎(+354) 562 10 20 98

75 Manufacturing Companies

75-00500 Hugrún h/f
Sidumuli 13, IS-108 Reykjavik
☎(+354) 568 10 91 [E] hugran@centrum.is
[W] http://www.centrum.is/hugrun ▣ 2307 isinfo is
⊜ (+354) 568 90 61
Managing Director: Hjalti Hardarson
Operations and Technical Director: Kjartan Hardarson
Marketing Director: Steingrimur Gunnarsson
Purchasing Manager: Magnus H Steinarsson
■ Manufacture of underwater temperature, depth and salinity measuring loggers. Sea-level, tide and wave measuring systems. Tension recorders. Air temperature recorders. [97]

75-05000 Skipasmidastodin hf
6 Sudurtangi, IS-400 Isafjordur
☎(+354) 456 38 99 [E] skipas@sne.pa.is
⊜ (+354) 456 44 71
♧ Ikr 10,000,000 ♣ Ikr 150,000,000 (1996)
👥 25 (1997) ⌂ 2,000 m2
Managing Director: Sigurdus Jonsson
■ Provision of ship repair services & facilities. [98]

India

Head of State:
Shankar Dayal Sharma

10 Head of State

10-10000 President's Secretariat
President's Estate 4, New Delhi 110004
☎(+91 11) 301 53 21
President: Shankar Dayal Sharma [98]

11 Prime Minister

11-10000 Office of the Prime Minister
Pradhan Mantri Karyalaya
South Block, Gate N° 6, New Delhi 110 011
☎(+91 11) 301 23 12
Prime Minister: H D Deve Gowda
Principle Secretary: T R Satish Chandran [98]

12 Ministries

12-10000 Ministry of Defence
Raksha Mantralaya
South Block 11, New Delhi 110 011
☎(+91 11) 301 23 80; 301 30 33
Minister: Mulayam Singh Yadav
Minister of State for Defence: N V N Somu [98]

12-20000 Ministry of External Affairs
Videsh Mantralaya
South Block 11, New Delhi 110 011
☎(+91 11) 37 23 18; 37 45 31 ▣ 313192 frgn in
Minister for External Affairs: Inder Kumar Gujral
Foreign Secretary: Shri Salman Haider [98]

12-40000 Ministry of Home Affairs
Grih Mantralaya
North Block 1, New Delhi 110 011
☎(+91 11) 37 19 89; 37 32 35
Minister: Indrajit Gupta
Minister of State for Home Affairs:
Mohammed Maqboop Dar [98]

25 Defense Forces

25-22100 Indian Air Force Western Air Command Headquarters
Subroto Park, New Delhi 110 010
☎(+91 11) 329 32 33; 548 39 00
⊜ (+91 11) 329 32 22
Air Officer Commander-in-Chief: AM Vinod Patney
Senior Air Staff Officer: AM V K Bhatia
Staff Officer to Air Officer Commander-in-Chief:
Wg Cdr A Bhakoo [98]

25-42500 The Commanding Officer Senior MBT Officer
Ins Dega, NAD (Post), Visakhapatnam 530009
☎(+91 891) 55 84 59
Commanding Officer: Capt M V K Rau
Executive Officer: Lieut Cdr Balerau
Senior MBT Officer: Cdr C B Rao
Flight Commander: Lieut Cdr M S Mahesh [98]

31 Embassies & Attachés

31-00090 Australian High Commission
1/50-G Shantipath, Chanakyapuri, PO Box 5210,
New Delhi 110 021
☎(+91 11) 688 82 23; 688 56 37 ☒ AUSTCOM
▣ 3182001
⊜ (+91 11) 687 41 26; 687 22 28 (Defence)
High Commissioner: H.E. Mr D.P. Gribble
Deputy High Commissioner: Rakesh Ahuja
Defence Adviser: Capt. Cris F. George (AM RAN)
Defence Admin Officer: Woff T. Polley
Minister Counsellor Commercial: R. Bayliss
Public Affairs Counsellor: Ms Gai Brodtmann [97]

31-00100 Embassy of Austria
EP-13 Chandragupta Marg,
Chanakyapuri, New Delhi 110021
☎(+91 11) 688 90 50 ⊜ (+91 11) 688 69 29
Ambassador: H E Mr Karl Peterlik
Minister Counsellor and Deputy Chief of Mission:
Peter Launsky-Tieffenthal
Second Secretary Administration and Vice Consul:
Robert Sinnhuber
Second Secretary: Ms Elisabeth Koegler
Attaché Administration and Consular Affairs: Edwin Ferner
Secretary to the Ambassador:
Ms Anneliese Baumschlager
Office of the Commercial Counsellor
12A Amrita Shergill Marg, New Delhi 110 003
☎(+91 11) 461 83 95; 461 83 97; 469 08 67
▣ 3162259 ahnd in ⊜ (+91 11) 461 87 42
Commercial Counsellor and Trade Commissioner:
Dr Heinz Rampitsch
Commercial Attaché and Deputy Trade Commissioner:
Ms Martina Lempach [98]

31-00150 Embassy of Belgium
50-N Shanti Path, Chanakyapuri, New Delhi 110 021
☎(+91 11) 60 82 95; 688 98 51; 688 92 04; 687 57 28
▣ 82008 abel in ⊜ (+91 11) 688 58 21
Ambassador: H E Mr Guillaume Metten
Counsellor Deputy Chief of Mission: Willy De Buck
First Secretary: Miss Anne Marchal
Attaché and Consul: Peter Ponjaert
Attaché, Technical: Paul Baeten
Attaché: Miss Isabelle Forrest
Office of the Economic and Trade Commissioner of the Flemish Region
50n Shantipath, Chanakypuri, New Delhi 110021
☎(+91 11) 611 72 39 ☎ 3172008 abel in
⊜ (+91 11) 611 02 58
Office of the Economic and Trade Commissioner of the Walloon Region
50n Shantipath, Chanakyapuri, New Delhi 110021
☎(+91 11) 611 02 59 [E] waldel@del2.vsnl.net.in
▣ 3172008 abel in ⊜ (+91 11) 611 78 72
Trade and Economic Commissioner of the Walloon Region:
Pierre Henusse [98]

31-00535 Embassy of Estonia
M-2 Hemkunt Tower, 98 Nehru Place, New Delhi 110 019
☎(+91 11) 641 45 75; 641 10 07; 643 10 67
⊜ (+91 11) 644 46 42
Honorary Consul General: Kasturi Lal Vig [98]

31-00580 Embassy of Finland
Nyaya Marg, Chanakyapuri, New Delhi 110 021
☎(+91 11) 611 52 58; 611 80 96
⊜ (+91 11) 688 67 13; 688 53 80 (commercial)
Ambassador: H E Mr Benjamin Bassin
Counsellor: Heimo Laakkonen
Second Secretary: Ms Terttu-Liisa Aho
Commercial Secretary: Erkki Salminen
Archivist: Aila Oikarinen [98]

31-00660 Embassy of the Federal Republic of Germany
PO Box 613, New Delhi 110 001
N°6 Shantipath, Chanakyapuri, New Delhi 110 021
☎(+91 11) 687 18 31; 687 18 32; 687 18 33;
687 18 34; 687 18 35; 687 18 36; 687 18 37
▣ 3182077 ⊜ (+91 11) 687 31 17
Ambassador: H E Mr Frank Elbe
Deputy Chief of Mission: Ute Banerjea-Komers
Minister: Jürgen Elias
Defence Attaché: Capt Klaus-Dieter Pusch (GE Navy)
Minister Counsellor: Karl-Andreas Freiherr Von Stenglin
Minister Counsellor: Dr Peer C Stanchina
Counsellor: Gerd Krösing
Counsellor: Dr Günther Theisen
Counsellor: Peter Von Wesendonk
Counsellor: Hermann Kern
Counsellor: Dr Stephan Freiherr Von Welck
Counsellor: Wolfgang Freiherr Von Edrfa
First Secretary: Hubert Schmalz
First Secretary: Alexander Stedfeld
First Secretary: Konrad Arz Von Straussenburg
First Secretary: Karl-Heinz Biging [98]

31-00670 Ghana High Commission
50-N Satya Marg, Chanakyapuri, New Delhi 110021
☎(+91 11) 688 32 98; 33 38 ☒ GHANACOM DELHI
▣ 3172484 ghcd in ⊜ (+91 11) 688 32 02
High Commissioner: H.E. Mr Anthony Forson
Minister Counsellor: Kwabena Baah-Duodu
Head of Chancery Counsellor: C. Mintah-Agyemang
Second Secretary: Ms Beatrice Okwei
Third Secretary: Ms Irene Nartey [97]

31-01020 Embassy of Lebanon
10 Sardar Patel Marg, Chanakyapuri, New Delhi 110 021
☎(+91 11) 301 31 74; 36 37 ▣ 3161161 lban in
⊜ (+91 11) 301 55 55
Ambassador: H.E. Mr Michel Bitar [97]

31-01210 Embassy of Mongolia
34 Archbishop Makarios, Marg, New Delhi 11003
☎(+91 11) 463 17 28; 461 79 89
⊜ (+91 11) 463 32 40 [98]

31-01300 New Zealand High Commission
50-N Nyaya Marg Chanakypuri, New Delhi 110 021
☎(+91 11) 688 31 70 ⊜ (+91 11) 687 23 17
High Commissioner: H E Mr Adrian Simlock
Deputy High Commissioner: Ms Catherine McGregor
Second Secretary: Ms Priscilla Clark
First Secretary Immigration: Mrs Justine Muriwai
Trade Commissioner: Ian Stockwell [98]

31-01420 Embassy of the Philippines
50-N Nyaya Marg, Chanakyapuri, New Delhi 110 021
☎(+91 11) 60 11 20; 60 35 11; 60 69 86; 60 89 24
☒ AMBAPHIL New Delhi ⊜ (+91 11) 687 64 01
Ambassador: H.E. Mr Juanito P. Jarasa
Commercial Officer & Third Secretary: Wenceslao Gayola
Defence & Armed Forces Attaché: Capt. Anselmo Legazpi [97]

31-01690 High Commission for the Democratic Socialist Republic of Sri Lanka

27 Kautilya Marg, Chanakya Puri, New Delhi 110 021
☎ (+91 11) 301 02 01; 301 01 03 ▣ 3161162 slhc in
⊖ (+91 11) 301 52 95
High Commissioner: H E Mr Mangala Moonsinghe
☎ (+91 11) 301 12 26
Deputy High Commissioner: C R Jayasinghe
☎ (+91 11) 301 43 31
Minister: M Sumanadasa

Trade Office
Sri Lanka House, 34 Romi Mody Street, Bombay 400 023
☎ (+91 22) 204 58 61; 204 83 03 ▣ 113 510 slco in
Trade Commissioner: S Udegedara
☎ (+91 22) 204 23 39 [98]

31-01890 United Kingdom High Commission

Chanakyapuri, New Delhi 1100021
☎ (+91 11) 687 21 61 ▣ 03165125 bhc in
⊖ (+91 11) 687 28 82
High Commissioner: H E Sir David Gore-Booth
Minister and Deputy High Commissioner: Dr D Carter
Defence and Military Adviser: Brig R A Draper
☎ (+91 11) 687 21 61 ext. 2405
Naval Adviser: Capt J McLees (RN)
☎ (+91 11) 687 21 61 ext. 2516
Air Adviser: Gp Capt R E Wedge (RAF)
☎ (+91 11) 687 21 61 ext. 2576
Counsellor, Economic and Commercial: Warwick Morris
☎ (+91 11) 687 21 61 ext. 2508
Counsellor: K M Williams
Counsellor and Head of Chancery: A M Thomson
Counsellor, Management: A Holmes

Office of the Deputy British High Commissioner
PO Box 11714, Bombay 400 021
Maker Chambers IV, 222 Jamnalal Bajaj Road, Nariman Point, Bombay
☎ (+91 22) 28 30 51; 283 23 30; 283 36 02
⊖ (+91 22) 202 79 40
Deputy High Commissioner: M C Bates
First Secretary, Commercial: P O D McCoy
First Secretary (Cultural Affairs) British Council Director: J W Edmundson

Office of the British Deputy High Commissioner
1 Ho Chi Minh Sarani, Calcutta 700 071
☎ (+91 33) 282 51 71; 282 51 72; 282 51 73; 282 51 74; 282 51 75; 282 65 36 ▣ 215983 ukcl in
⊖ (+91 33) 242 34 35
Deputy High Commissioner: S M Scaddan
First Secretary (Cultural Affairs) British Council Director: A P Thomas
Second Secretary (Commercial): P J Whitehead
Third Secretary (Consular): D R Belgrove

Office of the British Deputy High Commissioner in Southern India
Commercial Section
24 Anderson Road, Chennai 600 006
☎ (+91 44) 827 31 36; 827 31 37; 825 74 22; 825 74 33 ⊖ (+91 44) 826 90 04
Deputy High Commissioner: S H Palmer
Director, South India (British Council Division): D J Utley
Second Secretary (Commercial): Mrs C M McNeill
Attaché (Immigration): M G Morris
Attaché (Consular and Immigration): Miss C S Lewis

Consular and Immigration Section
111th Floor Kakani Towers, 15 Khader Nawaz, Khan Road, Chennai 600 006
☎ (+91 44) 827 06 58; 825 97 12
⊖ (+91 44) 827 51 30 [98]

55 Government-owned Industry & Production Facilities

55-00200 Bharat Electronics Ltd (BEL)

Registered Office
N° 116/2 Race Course Road, Bangalore 560001
☎ (+91 80) 226 73 22 (10 lines) ⊠ BEHO
▣ 8452477; 8458650 be in ⊖ (+91 80) 225 84 10
♧ Rs 800,000,000 ♣ Rs 10,640,000,000 (1995/96)
👥 17,044
Chairman and Managing Director: Rear Adm J J Baxi
Finance Director: Goutam Sengupta
Sales and Marketing Director: Dr V K Kosky
General Manager, International Marketing: Col K M Jayaprakasu
General Manager, Public Relations: J S Isaac
Senior Public Relations Officer: B A Lazar

■ Manufacture of communication equipment of diverse types including transportable and static communication systems, troposcatter, ship-based & mobile/land based communication systems, encryption/secrecy devices, satcom equipment, telecommunication equipment both switching and transmission, equipment for sound and vision broadcast including studio equipment. Radars for primary and secondary surveillance, low level detection fire control and meteorology purposes. Tank fire control systems and night vision equipment. Electronic components, electronic warfare systems, simulators, airport modernisation equipment such as displays, radars and ATC. → 75

Bangalore Complex
Jalahalli PO, Bangalore 560 013
☎ (+91 80) 334 57 41 (12 Lines) ⊠ BETRON
▣ 8452244; 8458379 be bg in
⊖ (+91 80) 334 31 19; 29 95; 26 52; 59 27; 81 35; 62 80
Executive Director: Y Gopala Rao
General Manager, Components: Dr G Manohar

Ghaziabad Unit
PO Box Bharat Nagar, Ghaziabad 201 008
☎ (+91 86) 76 82 22 ⊠ BELGAD ▣ 592207 belg in
⊖ (+91 575) 73 35 83
Executive Director: V Ammineedu
General Manager: O P Monga

Panchkula Unit
Industrial Area, Phase III, Plot 405, Panchkula, Haryana 134 109
☎ (+91 172) 56 15 20 (10 lines) ⊠ BEPANCH
▣ 395247 be in; 3950367 pco in ⊖ (+91 172) 56 05 48
General Manager: S C Khanna

Taloja Unit
MIDC Plot, L-1, Ind.Area Raigad, Dist. Maharashtra, Taloja 410 208
☎ (+91 215) 742 24 86; 742 01 46 ⊠ BETOJA
▣ 133211 be in ⊖ (+91 215) 42 28 88
General Manager: Wg Cdr R Soundararajan

Machilipatnam Unit
Ravindranath Tagore Rd, PO Box 26, Machilipatnam 521 001
☎ (+91 86) 35 81; 35 83; 32 54; 35 26 ⊠ BEOPTEL
▣ 4704201 be in ⊖ (+91 86) 72 26 40
General Manager: K S Ramanujam

Kotdwara Unit
Balbhadrapur, Kotdwara, Pauri Dist., Garhwall UP 246 149
☎ (+91) 224 66; 224 68 ⊠ BEKOT
⊖ (+91) 013 82; 224 64
General Manager: Lieut Col B S Gosain (retd)

Tank Electronics Support Centre
Madras Unit
Nandambakkam, PO Box 981, Chennai 600 089
☎ (+91 44) 234 69 06 (6 lines) ⊠ BEMAELECT
▣ 26184 be in ⊖ (+91 44) 234 69 05
General Manager: K S Ramanujam

Hyderabad Unit
Electro-Magnetics Division, Nacharam Industrial Estate, Hyderabad 501 507
☎ (+91 40) 85 01 14; 85 01 15 ⊠ BEHYD
▣ 4257025 BE IN ⊖ (+91 40) 85 40 06; 85 14 06
General Manager: V V C Rao

Pune Unit
NDA Road, Pashan, Pune 411 008
☎ (+91 212) 33 59 58; 33 80 22; 33 12 00; 33 22 84
⊠ BEBIT ▣ 145305 be in ⊖ (+91 212) 33 95 77
General Manager: Ch Anandam

New Delhi
Chandralok, 2nd Floor, 36 Janpath, New Delhi 110 001
☎ (+91 11) 332 21 36; 332 21 30; 332 13 73
⊠ LOBELTRONS ▣ 3166106 lobl in
⊖ (+91 11) 332 20 58; 322 20 58
General Manager: Col S Dowan (retd)

Bombay
World Trade Centre, 9th Fl, Unit 1, Cuffe Parade, Colaba, Bombay 400 005
☎ (+91 22) 218 28 72; 30 43; 53 82 ⊠ ROBETRON
▣ 1184802 robe in ⊖ (+91 22) 218 86 70; 30 43
General Manager: Cmde Alan Quadros (VSM, retd)

Calcutta
3/F Block A2-A8, Chatterji Int'l, Centre - 33a Jawaharlal Nehru Road, Calcutta 700 071
☎ (+91 33) 29 92 60; 29 40 53 ⊠ ROBE ▣ 215758 be in
⊖ (+91 33) 29 23 80
General Manager: P K Guha Ray

Madras
SIET Building, 312 Mount Road, Chennai 600 018
☎ (+91 44) 45 16 60; 45 08 80 ⊠ ROBEMAS
▣ 416267 robe in

Hyderabad
5-9-241, Abid Road, Hyderabad 500001
☎ (+91 40) 20 27 84; 20 16 96 ⊖ (+91 40) 20 16 98

Central Research Laboratory
25 2nd Fl, Shankarnarayah Building, M.G. Road, Bangalore 560 001
☎ (+91 80) 558 43 40; 584 49 11; 558 49 11
⊠ BELSEARCH ▣ 0845 8951 crlb in
⊖ (+91 80) 558 42 23
Chief Scientist: H Ramakrishma

Software Export Dept
Subharam Complex, 144 M.G. Road, Bangalore 560 001
☎ (+91 80) 558 23 67; 38 51; 38 53 ▣ 08458486
⊖ (+91 80) 558 36 75 [97]

55-00300 Garden Reach Shipbuilders & Engineers Ltd

43-46 Garden Reach Road, Calcutta 700 024
☎ (+91 33) 45 78 30 ⊠ COMBINE
▣ 0218514; 0218556
⊖ (+91 33) 469 81 50; 469 81 44; 479 24 16
♣ Rs 1,861,500,000 (1995/96) 👥 8,882
Chairman and Managing Director: VAdm R B Vohra (IN retd)
Director, Shipbuilding: Cmdr J S Brar (IN retd)
General Manager, Engineering: R K Roy
Assistant Public Relations Officer: K Nandy

■ Shipbuilding, repair and manufacture of diesel engines and generating sets, deck machinery items, general engineering products.
Glass fibre reinforced patrol boats and marine sewage treatment plants.
Naval ships: frigate, corvette, fleet tanker, landing ship tank, survey vessels, marine acoustic research ships.
Coast guard ships: fast patrol vessels, inshore patrol vessels.
Harbour craft: tugs, barges, motor launch, offshore platform support/standby vessels, dredgers, fire floats, fishing trawlers, fibre glass reinforced patrol boats.
Diesel engines: GRSE Bergen collaboration propulsion 1200 to 5400 HP, diesel generator 500 to 3800 kw. GRSE-mtu collaboration, propulsion 665 to 5600 HP diesel generator 350 to 1600 kw.
Commercial vessels up to 28,000 dwt. Repairs and refits for all naval, coast guard and merchant ships.
Shipboard engineering equipment: steering gears, winches, capstans, boat devits, aluminium telescopic helicopter hangar, towed decoy handling equipment, sonar handling equipment, electric anchor windlass, hatch cover.
Portable steel bridges: fixed spun steel bridges up to 200 ft. Bailey type modular steel bridges including suspension type brdges up to 400 ft span.
Vehicle recovery equipment up to 15 tonnes capacity. Agricultural pumps. [98]

55-00400 Goa Shipyard Ltd

Vasco Da Gama, Goa 403 802
☎ (+91 834) 51 21 52; 51 21 56; 51 39 54 ⊠ GOAYARD
▣ 0191218; 0191283
⊖ (+91 834) 51 38 70; 51 39 43; 51 32 60
♧ Rs 129,338,000 (1995/96)
♣ Rs 2,791,800,000 (1996/97) 👥 2,171 (1997)
🏭 10,000 m2
Chairman and Managing Director: RAdm B R Menon
☎ (+91 834) 51 39 43
Sales and Marketing Director: Capt C A Manuel
Company Secretary: Shri R C Asukar
☎ (+91 834) 51 39 43
Purchasing Manager: Cdr S K Mutreja
☎ (+91 834) 51 24 94
Executive Assistant to Chairman and Managing Director: John K James

■ Manufacture of missile craft, advanced offshore patrol vessels, landing craft, seaward defence boats, survey craft, tugs, harbour craft and sail training ships. Survey vessels, surface effect ships, hovercraft, pollution control vessels, deep sea trawlers, fish factory vessels, catamarans, pleasure submarines, offshore supply/standby vessels, passenger ferries, multipurpose tankers. Vessels up to 105m in length and 3,000 tonnes launch weight. General engineering repairs to all types of vessels. [98]

55-01000 Hindustan Aeronautics Ltd (HAL)

15/1 Cubbon Road, PO Box 5150, Bangalore 560001
☎ (+91 80) 286 67 01 ⊠ AERO HIND BANGALORE-HAL
▣ 8452266 ⊖ (+91 80) 286 87 58; 71 40
♧ Rs 406,000,000 (1995/96)
♣ Rs 1,589,000,000 (1995/96) 👥 35,245
Chairman: R N Sharma ☎ (+91 80) 286 31 05
Director Corporate Planning: S N Sachindran
☎ (+91 80) 286 74 65
Director Personnel: N R Mohanty ☎ (+91 80) 286 84 09
Director Finance: A K Zutshi ☎ (+91 80) 286 40 50

General Manager Marketing: K Umamaheswar
☎(+91 80) 286 51 97; 286 71 40 (fax)
Deputy General Manager Public Relations: Cdr M Nirmal
☎(+91 80) 286 86 29
■ Design, development and manufacture of fixed and rotary wing aircraft, avionics and accessories. Manufacture, maintenance, repair and overhaul of fighter, transport and trainer aircraft, helicopters, aero-engines, avionics, accessories, and ground support equipment. Structural components for satellites and launch vehicles, software development for aerospace applications, ground support equipment and design consultancy. → 75

Bangalore Complex
PO Vimanapura, Bangalore 560017
☎(+91 80) 526 52 01 ▣ 8452234
⊖ (+91 80) 526 51 88
Managing Director: Dr C G Krishnadas Nair

Aircraft Division
PO Box 1788, Bangalore 560017
☎(+91 80) 526 52 01 ▣ 8452234
⊖ (+91 80) 526 51 88
General Manager: L M Bharadwaj
■ Manufacture of defence and civil aircraft airframes.

Engine Division
PO Box 7579, Bangalore 560075
☎(+91 80) 528 23 61 ⊖ (+91 80) 528 13 97
General Manager: Mr Beharilal
■ Adour, Orpheus, Dart, Artouste IIIB manufacture and overhaul, aero-engine components.

Foundry & Forge Division
PO Box 1791, Bangalore 560017
☎(+91 80) 526 52 01 ⊖ (+91 80) 526 55 21
General Manager: B C Chaterji
■ High precision castings, semi-finished products.

Helicopter Division
PO Box 1790, Bangalore 560017
☎(+91 80) 526 52 01 ▣ 8452764
⊖ (+91 80) 526 47 17
General Manager: Ashok Baweja
■ Manufacture and overhaul of Chetak, rotables, spare parts.

Overhaul Division
PO Box 1786, Bangalore 560017
☎(+91 812) 57 01 81 ▣ 8452234
⊖ (+91 80) 527 24 77
General Manager: A P Aneja
■ Overhaul of aircraft including Caribou, Otter, Canberra, C119-G, C-47. Piston engine overhaul.

Design & Development Complex
PO Box 1789, Bangalore 560017
☎(+91 80) 225 80 03; 225 30 05
⊠AERO DESIGN BANGALORE HAL ▣ 8458083
⊖ (+91 80) 526 30 96

Accessories Complex
PO HAL, Lucknow 226016
☎(+91 522) 38 39 67 ▣ 5352428
⊖ (+91 522) 38 39 73
Managing Director: H C Kholay

Hyderabad Division
PO HAL, Hyderabad 500042
☎(+91 80) 272 82 81 ▣ 4258016 ⊖ (+91 40) 27 83 89
General Manager: Wing Cdr Roy Choudhary
■ Avionic equipment, airport radars, ceramic pads, connectors, silver zinc batteries. Battlefield surveillance radar. Combined VHF/UHF transreceivers. IFF transponders.

Kanpur Division
PO Box 225, Kanpur 208008
☎(+91 512) 35 17 49 ▣ 325243 ⊖ (+91 512) 35 05 05
General Manager: T K Dhar
■ Manufacture of Dornier aircraft and manufacture/overhaul of HS-748, HPT-32 ab-initio piston engine trainer.

Korwa Division
PO HAL Korwa, Amethi, Sultanpur 227412
☎(+91 536) 50 27 75 (SUL); 82 22 05 (AMT)
⊠SUPERSONIC AMETHI ▣ 0845 094811
⊖ (+91 536) 50 25 18 (SUL); 82 21 86 (AMT)
Additional General Manager: Gp Capt S P Sharma
■ Advanced navigation systems.

Lucknow Division
PO HAL, Lucknow 226013
☎(+91 9522) 38 39 67 ▣ 5352428
⊖ (+91 522) 38 39 73
■ Manufacture and overhaul of accessories for aircraft and engines including landing gears, wheels, brake units, ejection seats, instruments, fuel pumps, trolleys, custom made test rigs, gyro instruments and ground support equipment for civil and military applications.

MIG Complex
Ojhar Township, Nasik, Maharashtra
☎(+91 2533) 758 21; 781 96 ▣ 752241
⊖ (+91 2533) 758 25
Managing Director: N R Mohanty

Nasik Division
Ojhar Township, Nasik, Maharashtra
☎(+91 253) 781 17 ▣ 752241 ⊖ (+91 253) 37 58 25
General Mgr: A.K. Saxena
■ MiG aircraft manufacture & overhaul.

Koraput Division
Sunabeda 9, Koraput, Orissa 763002
☎(+91 6853) 202 00 ▣ 6601206
⊖ (+91 6853) 200 04
General Manager: K Sham Sunder
■ Manufacture and overhaul of MiG aero-engines, forgings and castings.

Liaison Office
101 Aurobindo Place, Hauz Khas, New Delhi 110016
☎(+91 11) 66 46 04 ▣ 3173374 ⊖ (+91 11) 686 46 83
Resident Manager: Gp Capt Shukla

Bombay
New Kamani Chambers, 32, Ramjibhai,
Kamani Marg - Ballard Estate, Bombay 40001
☎(+91 22) 261 03 29; 261 62 76 ▣ 118443
⊖ (+91 22) 266 48 90
Resident Manager: A C Takkar

Madras
N°15a CASA Major Road Egmore, Chennai 600008
☎(+91 44) 828 41 80 ▣ 417699
Liaison Officer: S Ramachandran

Visakhapatnam
Opposite C B Mission Compound,
Waltair Uplands Srirampuram, Visakhapatnam 580003
☎(+91) 55 54 64; 55 43 76 ▣ 495248
Officer-in-charge: R Patnaik [98]

55-01200 Mazagon Dock Ltd

Dockyard Road, Mazagon, Bombay 400 010
☎(+91 22) 376 05 61 (20 lines) ⊠MAZDOCK BOMBAY
▣ 1175568; 1175814
⊖ (+91 22) 373 83 40; 373 81 47; 373 81 59
♧ RS 1,992,000 ♣ RS 8,342,100,000 👥 10,500
Chairman and Managing Director: Rear Adm D V Taneja
Shipbuilding and Offshore Director: Capt S V Nair
Corporate Planning Director: D J Correa
Public Relations Officer: A M Dhuru
■ Construction of warships, missile boats, destroyers, submarines, corvettes, offshore patrol vessels, offshore platforms, coast guard vessels. [97]

55-01300 Mishra Dhatu Nigam Ltd Superalloys Plant

(a Government of India Enterprise)
Kanchanbagh, Hyderabad 500258
☎(+91 40) 23 92 93; 23 97 64
[E] spralloy@hdi.ysnl.net.in [E] mdnexport@hdl.vsnl.net.in
▣ 04256552; 04256494 ⊖ (+91 40) 23 92 14
♧ Rs 1,340,000,000 ♣ Rs 920,000,000 (1996/97)
👥 1,450 (1997)
Chairman and Managing Director: R K Mahapatra
☎(+91 40) 23 92 01
Public Relations Manager: M V Surender
☎(+91 40) 23 90 01
Deputy General Manager Exports: V K Khare
Deputy Manager Exports: C Sabita
■ Manufacture of super alloys, titanium and titantium alloys, Molybdenum special purpose steels. Controlled expansion alloys, soft magnetic alloys, heat resistant alloys, electrical resistance alloys. Products supplied in mill forms as hot/cold rolled billets, bars, sheets, strip wires, welded tubes, cores, laminations, forged products, investment castings, titanium coatings and forged components. Defence related products including forgings for gun barrels and other ordnance spares, trail tubes for field guns. Aero-engine materials for military aircraft, titanium and titanium alloys, special steels for naval, aerospace, chemical and power applications. [98]

60 Research Organizations, Institutes & Test Establishments

60-00550 Electronics and Radar Development Establishment (LRDE)

DRDO Complex, CV Raman Nagar, Bangalore 560093
⊠DEVELECTRONICS ▣ 0845288 ⊖ (+91 80) 524 29 16
Director: George M Cleetus
Scientist: Y V L N Rao [98]

74 Consortia, Joint Venture Companies

74-01000 SAJ-Froude Test Plant Pvt Ltd

72-76 Mundhawa, Industrial Estate, Pune 411 036
☎(+91 212) 67 02 82; 67 08 82 ⊠DYNASTEEL
▣ 01457495 saj in ⊖ (+91 212) 67 10 28
♧ Rs 22,500,000 ♣ Rs 72,418,000,000 👥 147 (1996)
🏭 16,000 sq m
Managing Director: Prakash Jagtap
Senior General Manager, Marketing: Kishore D Shinde
Operations and Technical Director: A R Gupta
Purchasing Manager: C S Desai
■ Engine and vehicle test equipment.
® UK: AJB Associates Electronics Ltd, Biceri Ltd, Froude Consine Ltd, Gendeal Ltd, Mecelec Developments Ltd, Perkins Technology Ltd, Servo Test, Sound Atenuators (Industries) Ltd, Twiflex Ltd
USA: Intelligent Controls, Stone Bennett Corporation [97]

75 Manufacturing Companies

75-00080 Ailga Rubber Works

84/8 Fazalganj, Kanpur 208012
☎(+91 512) 21 70 68; 21 70 44; 21 62 86
⊠AILRUBBER ⊖ (+91 512) 21 62 13
♧ Rs 50,000,000 (1997) ♣ Rs 29,950,000 (1996/97)
👥 125 🏭 3,100 sq yds
Managing Director: Raj Kumar Tandon
Partner Export Sales: Anand Tandon
Advertising Manager: Sharad Tandon
Public Relations Manager: R N Bhargawa
■ Manufacture and export of rubber moulded, rubber-to-metal bonded and extruded rubber profiles as per customer drawings/specifications. [98]

75-00130 AMA Pvte Ltd

Canada Building Dr D Naoroji Road, PO Box 556,
Bombay 400001
☎(+91 22) 207 52 71; 207 52 72; 207 52 73
⊠AMALIMITED [E] amaafls@giasbm01.vsnl.net.in
⊖ (+91 22) 207 89 10
♧ US$ 75,000 ♣ US$ 1,700,000 👥 54 🏭 15 ,76 sq ft
Managing Director: Ms Minal G Patel
Director: Rahul Patel
Director: Sameer Patel
Airport Lighting Division Manager: G R Nandedkar
■ Elevated runway/taxiway lights, VASI units, PAPI units and obstruction marker lights, constant current regulators, circuit, indicators, isolating transformers, transformer housings, remote control desks, power distribution panels, approach masts, approach lights, illuminated wind cones and H indicators.

Plant
C/1-266/267 GIDC Industrial Estate, Naroda,
Ahmedbad 382330
☎(+91 79) 281 10 49; 281 35 68 ⊠AMALIMITED [98]

75-00150 AMP Tools (India) PVZ Ltd

(a subsidiary of AMP Incorporated)
Plot N°44, Cochin Export Processing Zone,
Kakkanad, Cochin 682030
☎(+91 484) 42 25 97; 42 25 98
[E] amptools@md2.vsml.in ▣ 8856974
⊖ (+91 484) 42 25 93
👥 125 (1997)
Divisional Manager Finance and Operation:
I Sethumadhavan
■ Production of tools, moulds and dies for the manufacture of connectors.
® USA: AMP [98]

75-00160 Aquarius Fabrique International Pvt Ltd

101 Gera Chambers, Off Boat Club Road, Pune 411026
☎(+91 212) 79 01 53; 79 29 26
[E] jopi@giaspn01.vsnl.net.in ⊖ (+91 212) 79 28 10
♣ Rs 10,000,000 (1996) 👥 26 🏭 1,600 m2
Managing Director: Jagdeep Singh
Managing Director: Sandeep Singh

■ Manufacture and export of armament components and ammunition additives for reduction of gun barrel wear. Charge bags for 155 mm propelling charges. Liner assembly M6 for 106 mm RCL ammunition. Copper liners/cones for HEAT ammunition. Bore cleaning fluid (BCF) for cleaning and preservation of large and small calibre weapons. [98]

75-00270 Bharat Heavy Plate & Vessels Limited
(a subsidiary of Bharat Yantra Nigam Limited)
BHPV Post, Visakhapatnam, Andra Pradesh 530 012
☎ (+91 891) 51 73 81; 51 76 21
E bhpvmm/vizag@dartmail.dartnot.com
▣ 0495213294313 bhpvin ⊖ (+91 891) 51 76 26
👥 3,959 (1997) 🏭 97,360 m2
Managing Director: B Venkateswarlu
Chief General Manager, Production: A S R P V S Raju
Chief General Manager, Commercial: B Radhamohan Rao
Purchasing Manager: D R G K Mohan Rao
Manager Publicity and Public Relations: D Varada Reddy
Publicity and Public Relations Officer: P Aravindakshan
■ Manufacture of process plant equipment for core sector industries including oil, gas, petroleum, petrochemical, chemical, steel, aluminium, fertilizer and power. High pressure heat exchangers, multilayer vessels, pressure vessels, storage spheres, air separation plants, columns and systems.
Exporters of pressure vessels, columns, heat exchangers, drums, spherical tanks, skid mounted packages, air fin coolers and evaporators. [98]

75-00280 Crompton Greaves Ltd Registered Office
(a subsidiary of the Thapar Group of Industries)
1 Dr V B Gandhi Marg, PO Box 91,
Mumbai, Bombay 400 023
☎ (+91 22) 27 66 10 ⊖ (+91 22) 267 78 50; 265 28 09 (international operations); 27 22 11
♣ Rs 7,934,000,000
Chairman: N M Wagle
Secretary: W Henrigus
→ 80 → 90
® Austria: Bombardier-Rotax GmbH
Belgium: Skyclimber
France: Aerazur-EFA, Glaser, Schlumberger
Italy: Alenia
New Zealand: CWF Hamilton & Co.Ltd
Sweden: Magnetic AB
UK: Alcatel Multipoint Comm Ltd, Griffon Hovercraft, Martin Baker A/C Co., Mctaggart Scott, Normalair Garrett Ltd, Penny & Giles Pot. Ltd, Smiths Ind.Aerospace, Vickers Systems Ltd
USA: Vickers Inc.(SMD), Webber Aircraft Inc. [98]

75-00300 Daewoo Power India Ltd
(a subsidiary of Daewoo Corporation, South Korea)
Apartment A-4 Quatab Hotel OFF, Sri Aurobindo MARG, New Delhi 110016
☎ (+91 11) 651 43 39; 651 43 59
⊖ (+91 11) 686 86 89 [98]

75-00540 Dynamic (India) Pvt Ltd
E-40/7 Okhla Industrial Area, Phase 2, New Delhi 110020
☎ (+91 11) 643 36 30 ▣ 3163002
Vice President: Wg Cdr P.T. Singh (Rtd)
■ Manufacture of high quality military uniforms for the international market.
® Canada: Istec [97]

75-01200 Micron Instruments Private Limited
4/90 Connaught Circus, New Delhi 110001
☎ (+91 11) 336 33 83; 336 47 22; 373 29 21
⊠ MICROFUZE ▣ 61199 mipl in ⊖ (+91 11) 374 64 49
♧ Rs 13,912,000 (1997) ♣ Rs 50,000,000 (1997)
👥 450 (1997) 🏭 5,000 m2
Managing Director: M Vikram Sahgal
Executive Director: Maj Gen P R Dhawan (retd)
■ Registered with MoD, Government of India for manufacture and supply of armament stores including fuses for all types of projectiles for army, navy and air force. Safe and arm devices. Precision components, assemblies and sub-assemblies for small arms. [98]

75-01780 Rajkumar International
97 Kazi Sayed Street, Amar Chambers, 2nd Floor, Bombay 400 003
☎ (+91 22) 342 08 71 ⊠ KACHABO BOMBAY
▣ 76505 tortin ⊖ (+91 22) 344 38 27
♣ Rs 2,500,000 👥 15
General Manager and Sales and Marketing Director: R S Shah
Advertising Manager: K S Shah
Public Relations Manager: T S Shah
■ Manufacture of gum arabic, gum babul, gum powder, gum guar, gum rosin. Preservative and lubricant grades. [97]

75-01810 Rashtriya Metal Industries Ltd
Sir Mathuradas Vasanji Road JB Nagar, PO Box 7406, Bombay 400059
☎ (+91 22) 832 33 11; 836 63 64 ▣ 1179088 rmil in
⊖ (+91 22) 836 63 66
Managing Director: Balkrishna Binani
General Manager: G D Daga
■ Manufacturers of brass strips, discs and cups for use in cartridge cases. [98]

75-01850 Ring Sights (India) Pvt Ltd
(a subsidiary of Ring Sights Group Ltd, UK)
C-291 Defence Colony, New Delhi 110024
☎ (+91 11) 62 25 32 ▣ 3161922 ⊖ (+91 11) 644 49 69
♧ Rs 2,000,000 ♣ Rs 6,000,000
Managing Director: V Paul
■ Design, manufacture and sale of patented unit power collimator, sights and weapon interfaces for pistols, rifles, SMG and machine guns, under licence from Ring Sights International.
® UK: Electro Primastic Collimators Ltd, Ring Sights Holdings Ltd, Ring Sights Ltd
USA: Electro Prismatic Collimators, Ring Sights Inc [98]

75-01880 Dr Sabharwal's Bulk Drugs Ltd
117/H-2/160, Pandu Nagar,
Plot 24, Bhawanipur, Mandhana, Kanpur Dehat 208005
☎ (+91 512) 21 72 77; 21 73 77
⊖ (+91 512) 21 65 52; 21 27 64
♧ Rs 2,000,000 (1996) ♣ Rs 5,000,000 (1994/95)
👥 95 🏭 30,000 sq ft (1996)
Managing Director: Dr A R Sabharwal
Technical Director: Manish Sabharwal
Marketing Director: Mrs Ajyjana Sabharwal
Public Relations Manager: K Nisha
■ Manufacture of a complete range of dressings and medical supplies.

Manufacturing Laboratories Ltd
Plot n° 9 Bhawanipur, Mandhana, Kanpur Dehat 208005
☎ (+91 512) 21 72 77 ⊖ (+91 512) 21 65 52; 21 27 64
👥 49
Managing Director: Dr A R Sabharwal
Sales Director: Mrs A Sabharwal
Advertising Manager: V K Sabharwal
Public Relations Manager: Y P Bhateja
■ Manufacture of zinc oxide adhesive plaster, B.P. adhesive tape USP. Plaster of Paris bandage B.P. Cotton crepe bandage B.P. [97]

75-01900 Sandeep Metalcraft Pvt Ltd (SM)
D-16, M.I.D.C. Industrial Area, Nagpur 440 028
☎ (+91 7104) 378 78; 367 67 ⊠ METALCRAFT
⊖ (+91 7104) 368 60
♧ Rs 2,600,000 ♣ Rs 500,000,000 👥 114
🏭 89,952 sq m
Managing Director: Shyam Agarwal
Technical Director: M K Nair
■ Manufacture and supply of fuses, primers, tail units, small calibre shells, mortar training device, assemblies for small arms and other precision engineering components and assemblies. [97]

75-01950 Stone India Ltd
PO Box 16731, Calcutta 700 088
☎ (+91 33) 478 46 61; 478 46 68
⊠ STONECO CALCUTTA ▣ 0218148 ston in
⊖ (+91 33) 478 48 86
♧ US$ 2,182,318 (1996) ♣ US$ 7,356,118 (1996)
👥 654 🏭 170,000 sq m
Managing Director: Rajinder Malhan
Vice President, Technical: S Kumar
Vice President, Manufacturing: D Das
Vice President, Marketing: S L Ahuja
General Manager, Materials: S Udaybhanu
■ DC motors, air conditioning, filters, cleaners air starters, generators especially for combat vehicles. Navigational aids mainly buoys and beacons. Rotary convertors and DC contactors. [97]

75-02050 Triventi Plenty Engineering Ltd
(a joint venture with Plenty Ltd, UK)
D196 Okhla Industrial Area, Phase 1, New Delhi 110 020
☎ (+91 11) 681 18 78 ▣ 3175154 pae in
⊖ (+91 11) 681 22 80 [98]

75-02100 M. Kumar Udyog
13 Gandhi Gram G.T. Road, Kanpur 208007
☎ (+91 512) 422 40; 426 13 ⊖ (+91 512) 35 17 99
♧ US$ 100,000 (1994) ♣ US$ 1,250,000 (1994)
👥 200 🏭 2,000 sq m
Managing Director: Manoj Cupta
Sales and Marketing Director: Neeraj Gupta
■ Manufacture clothing, headgear, footwear, insignia, individual equipment, tents and bedding and other hardware. Also acts as a merchandise import/export trading house. [97]

75-03000 Zen Technologies Ltd
40 Radha Swamy Colony, Sikh Road,
Secunderabad, A.P. 500 009
☎ (+91 40) 84 32 79 E zentech@hd1.vsnl.net.in
⊖ (+91 40) 84 32 79
Managing Director: Ashok Atluri
Marketing Director: A V Rao
Vice President: A D Kishore
■ Manufacture small arms training simulator and hand grenade simulator. [97]

80 Service Companies & Organizations

80-00100 AlliedSignal Aerospace Marketing and Sales
Air India Engineering Complex, Old Airport,
Kalina, Bombay 400 098
II-747 Hangar, 3rd Floor, Air India
Engineering Complex, Old Airport,
Santa Cruz (E), 3/25 Golden Valley, Kalina, Bombay
☎ (+91 22) 611 94 50, 611 16 50 ext. 5277
⊖ (+92 22) 611 67 07; 611 94 50
Field Service Engineer: Madan Mohan
☎ (+91 22) 618 02 70; (+91 98210) 155 09 (mobile)
■ On-site customer support for regional area. [98]

80-00105 AlliedSignal India Inc
11th Floor, Dr Gophal Das Bhawan Towers,
New Delhi 110 001
28 Barakhamba Road
☎ (+91 11) 371 80 35; 371 80 36; 373 11 75; 373 11 76
President: Sunder Mulchandani
☎ (+91 11) 373 19 54; 371 80 38 (fax)
Field Service Engineer: Gurbaksh Lal
☎ (+91 11) 565 35 06; 565 35 07 (fax)
Deputy Director, Finance: Lalit Lamba
Field Service Engineer: Madan Mohan
☎ (+98 100) 193 46 (mobile)
Director,Marketing, Sales and Service: Shyam Uttamsingh
☎ (
■ Sales and service of AlliedSignal products. → 90 [98]

80-00500 Heatly & Gresham (I) Ltd
E-47/4 Okhla Industrial Area, Phase-II, New Delhi 1100 20
☎ (+91 11) 683 62 93; 683 19 65; 683 74 04; 691 41 31 ▣ 86041 hag in ⊖ (+91 11) 684 71 71
Managing Director: A K Chowdhry
Chief Executive Officer: N S L Narasimhan
Advisor: V M Tandon
® Switzerland: Trimos-Sylvac Task Force [98]

80-01900 Rotax Aviation Corp
12 Shaheed Bhagat Singh Marg, New Delhi 110001
☎ (+91 11) 336 17 79; 336 47 15 ⊠ ROTAXCORP
⊖ (+91 11) 334 41 74
♣ US$ 50,000,000 👥 28

India

President: Surindar K Gupta
Vice President: Wg Cdr V V Bhupati Rao (retd)
Company Secretary: P S Thakur
■ Marketing consultants for overseas manufacturers of high technology equipment of varied types for aerospace, research and development, avionics, aircraft, electronics and telecommunications applications. [98]

85 Stockists, Distributors, Spare Parts Suppliers

85-01000 Indian Avitron
(a division of BMG Enterprises Ltd)
DCM Building, 1st Floor, 16 Barakhamba Road,
New Delhi 110 001
☎ (+91 11) 332 92 71; 332 95 11
E bmgent@giasd101.vsnl.net.in ▣ 3166307 avia in
⊖ (+91 11) 332 75 12
⚑ 50 (1997)
Chairman: Brij Mohan Gupta ☎ (+91 11) 373 01 55
Managing Director: Harsh Mohan Gupta
☎ (+91 11) 371 32 62
■ Aviation, defence, space and meteorological equipment.
Bangalore Office
N°23 1 Main, Defence Colony, Indira Colony,
Bangalore 560 038
Lucknow Office
8-B, Avadh Bazar, Hotel Clarks Avadh, Lucknow 226 001
☎ (+91 522) 22 17 44 ⊖ (+91 522) 24 17 44
Hyderabad Office
33 Jalvayu Vihar, Kukkatpally, Hyderabad 500 072
☎ (+91 842) 86 26 88 ✉ AVITRONICS [98]

90 Sales Agents, Sales & Liaison Offices

Companies represented:

Accurate Sound Corp, USA . . . 90-00500
Advance Group, USA . . . 90-00500
Aerazur-EFA, France . . . 75-00280
Aero Contracts, UK . . . 90-00005
Air Liquide, France . . . 90-00500
Aircraft Furnishing Ltd, UK . . . 90-00005
AJB Associates Electronics Ltd, UK . . . 74-01000
Ako Regelungstechnik GmbH, Germany 90-01240
Alcatel Multipoint Comm Ltd, UK . . . 75-00280
Alenia, Italy . . . 75-00280
American Switch Corp, USA . . . 90-00500
AMP, USA . . . 75-00150
Amphenol Italia SRL, Italy . . . 90-00040
Amphenol Ltd, UK . . . 90-00040
Amphenol Products, USA . . . 90-00040
Amphenol Socapex, France . . . 90-00040
Amphenol-Tuchel Electronics, Germany 90-00040
AMTS Int'l N.V., Belgium . . . 90-01240
Anorad Corp, USA . . . 90-00500
APV South East Asia Pte Ltd, Singapore 90-01240
Argumens GmbH, Germany . . . 90-00425
Axel Electronics, USA . . . 90-00500
Babcock Displays, USA . . . 90-00500
Banner Aircraft International, USA . . . 90-00835
Beech Aircraft, USA . . . 90-00940
Belar Electronics, USA . . . 90-02500
Belgraver, Netherlands . . . 90-00005
Bendix/King Avionics, USA . . . 90-00940
Biceri Ltd, UK . . . 74-01000
Bombardier-Rotax GmbH, Austria . . . 75-00280
BP Solar Limited, UK . . . 90-01980
Carl Dose GmbH, Germany . . . 90-01240
Cassens & Plath GmbH, Germany . . . 90-01240
Champion Techno, USA . . . 90-00500
Chemtronics Inc, USA . . . 90-00005
Christie Electric Co., USA . . . 90-02500
Collins Avionics, USA . . . 90-00940
Collmer Semicondustor, USA . . . 90-00500
Cryolor SMC TBT, France . . . 90-00500
CWF Hamilton & Co.Ltd, New Zealand . 75-00280
Dai-Ichi Denshi Kogyo K.K., Japan . . . 90-00040
Desarrollo Technicas Ind. de Galicia SA, Spain . . . 90-01240
Dowty Aerospace, Singapore . . . 90-00005
EECO Inc, USA . . . 90-00500
EECO Ltd, UK . . . 90-00500
EEV Ltd, UK . . . 90-00030
EIP Metals Ltd. (Evered Tube Div.), UK. 90-00425
Electro Air, USA . . . 90-00005
Electro Primastic Collimators Ltd, UK . 75-01850
Electro Prismatic Collimators, USA . . . 75-01850
Envirovac Inc, USA . . . 90-01240
Fairchild Loral Data Systems, USA . . . 90-00005
Farnell Int'l, UK . . . 90-02500
Foam Tex, UK . . . 90-00005
Focus Veilig BV, Netherlands . . . 90-01240
Froude Consine Ltd, UK . . . 74-01000
G. Hartner Co GmbH, Germany . . . 90-00835
GAMMA-F Corporation, USA . . . 90-00425
Gendeal Ltd, UK . . . 74-01000
Genisco Technology, USA . . . 90-00500
Glaser, France . . . 75-00280
Griffin Woodhouse Ltd, UK . . . 90-01240
Griffon Hovercraft, UK . . . 75-00280
Gulfstream Aerospace, USA . . . 90-00940
Hackman Metos Ltd, Finland . . . 90-01240
Helkama-Kaapeli, Finland . . . 90-01240
Hexcel Composiites, UK . . . 90-00005
Hi Tech Import & Export, USA . . . 90-00500
Hose-McCann Telephone Co Inc, USA . 90-01240
Hundested Motor & Propeller Fabrik A/S, Denmark . . . 90-01240
IMA, Luxembourg . . . 90-00835
Intelligent Controls, USA . . . 74-01000
Isovolta, Austria . . . 90-00005
Israel Microwave Components Ltd, Israel . . . 90-00425
Istec, Canada . . . 75-00540
Itek Systems (BEI), USA . . . 90-00500
Kimball Physics, USA . . . 90-00500
La Auxiliar Navel SA, Spain . . . 90-01240
Limco Airepair, USA . . . 90-00005
MACOR Marine Systems Int'l GmbH, Germany . . . 90-01240
Magnapower (Pvt) Ltd, S. Africa . . . 90-01240
Magnetic AB, Sweden . . . 75-00280
Mallory Capacitor, USA . . . 90-00840
Mankiewicz, Germany . . . 90-00005
Martin Baker A/C Co., UK . . . 75-00280
Maskinfabrik Acta A/S, Denmark . . . 90-01240
Maskinfabrik EMCE, Netherlands . . . 90-01240
Mctaggart Scott, UK . . . 75-00280
Mecelec Developments Ltd, UK . . . 74-01000
Microphase Corp., USA . . . 90-01340
MTS Mikrotechnik GmbH, Germany . . . 90-00425
Naval Electronics AB, Sweden . . . 90-01240
Norac A/S, Norway . . . 90-01240
Norlau A/S, Denmark . . . 90-01240
Normalair Garrett Ltd, UK . . . 75-00280
Odin Control Systems, Denmark . . . 90-01240
OFEMA, France . . . 90-01540
Optomic Microwave Ltd., Israel . . . 90-00425
Penny & Giles Pot. Ltd, UK . . . 75-00280
Perfection Mica, USA . . . 90-00500
Perkins Technology Ltd, UK . . . 74-01000
Phototonics Tech Systems, USA . . . 90-00500
Podszuck GmbH, Germany . . . 90-01240
Pratt & Whitney, Canada . . . 90-00940
Precision Tube Company, USA . . . 90-00425
Radiall, France . . . 90-01700
Ring Sights Holdings Ltd, UK . . . 75-01850
Ring Sights Inc, USA . . . 75-01850
Ring Sights Ltd, UK . . . 75-01850
Rod-L Electronics Inc, USA . . . 90-00500
Satair, Denmark . . . 90-00005
Schaffner Electronik, Switzerland . . . 90-00500
Schlumberger, France . . . 75-00280
Schomandl Vertriebs, Germany . . . 90-02500
Sembawang Aviation, Singapore . . . 90-00005
Senisys Sensors, USA . . . 90-00500
Servo Test, UK . . . 74-01000
Shield Technology, USA . . . 90-00500
Sifam, UK . . . 90-02500
Simac CSI BV, Netherlands . . . 90-01240
Singer Products, USA . . . 90-00500
Skyclimber, Belgium . . . 75-00280
Smiths Ind.Aerospace, UK . . . 75-00280
Sound Atenuators (Industries) Ltd, UK . 74-01000
Spectrum Elektrotechnik GmbH, Germany . . . 90-00425
Speich Srl, Italy . . . 90-01240
Standard Wire & Cable Co, USA . . . 90-00500
Stone Bennett Corporation, USA . . . 74-01000
Sulcer, S.L., Spain . . . 90-01240
Sunair Electronics, USA . . . 90-00940
Tamagawa Seiki Co, Japan . . . 90-00500
Teledyne Controls, USA . . . 90-00005
Textron Lycoming, USA . . . 90-00940
Thomas Electronics Inc, USA . . . 90-00500
Thorn Security Ltd, UK . . . 90-01240
Tierney Metals UK, Germany . . . 90-00835
Trimos-Sylvac Task Force, Switzerland . 80-00500
Twiflex Ltd, UK . . . 74-01000
Unimatic Engineering, UK . . . 90-00500
Universal Shielding, USA . . . 90-00500
Vickers Inc.(SMD), USA . . . 75-00280
Vickers Systems Ltd, UK . . . 75-00280
Victor Pyrate Ltd, UK . . . 90-01240
Wayne Kerr Electronics Ltd, UK . . . 90-02500
Webber Aircraft Inc., USA . . . 75-00280
WS Wilson, USA . . . 90-00005
Zollner GmbH & Co KG, Germany . . . 90-01240

Sales agents/representatives listed elsewhere:

AlliedSignal India Inc . . . 80-00105
Crompton Greaves Ltd . . . 75-00280

90-00005 Aero-Equipment & Agencies Corp
Old Airport, Santacruz, Bombay 400029
☎ (+91 22) 610 30 41 ▣ 1184263 mack in
⊖ (+91 22) 266 14 40; 611 93 28
♣ US$ 2,000,000 ⚑ 20 ⌂ 1,700 sq ft
Managing Director: Mohit Gunja
Sales Manager: V R Ramchandran
Sales Manager: P K Kutty
® Austria: Isovolta
Denmark: Satair
Germany: Mankiewicz
Netherlands: Belgraver
Singapore: Dowty Aerospace, Sembawang Aviation
UK: Aero Contracts, Aircraft Furnishing Ltd, Foam Tex, Hexcel Composiites
USA: Chemtronics Inc, Electro Air, Fairchild Loral Data Systems, Limco Airepair, Teledyne Controls, WS Wilson [98]

90-00030 Aimil Ltd Electronics Division
Naimex House, A-8 Mohan Co-operative,
Industrial Estate, Mathura Road, New Delhi 110 044
☎ (+91 11) 694 62 81 ⊖ (+91 11) 694 54 56
Director: A Verma
® UK: EEV Ltd [98]

90-00040 Amphetronex Ltd
105 Bhosari Industrial Area, PO Box 1, Pune 411 026
☎ (+91 212) 79 03 63; 79 04 63; 79 13 63
✉ AMFETRONIX ▣ 0146237 ⊖ (+91 212) 79 05 81
♧ Rs 7,040,000 (1994) ♣ Rs 225,570,564 (1994)
⚑ 270 ⌂ 4,500 sq m
Chief Executive: Col Paul Varghese (retd)
☎ (+91 212) 79 04 62
Managing Director: M M Murugappan
Deputy General Manager, Marketing: T K Venogopal
☎ (+91 212) 79 04 81
Deputy General Manager, Development: B B Maindergikar
☎ (+91 212) 79 20 13
Purchasing Manager: S K Murthy
Public Relations Manager: S V Achtani
☎ (+91 212) 79 13 63
■ Manufacture and sale of connectors and interconnection products; cable harness and IC sockets. Approved to defence and aerospace standards. Exporting to different European and Southeast Asian countries.
® France: Amphenol Socapex
Germany: Amphenol-Tuchel Electronics
Italy: Amphenol Italia SRL
Japan: Dai-Ichi Denshi Kogyo K.K.
UK: Amphenol Ltd
USA: Amphenol Products
New Delhi Office
Jeevan Deep Building, 1st Floor, 10 Sansad Marg,
New Delhi 110001
☎ (+91 11) 336 53 64 ⊖ (+91 11) 374 66 78
Bangalore Office
Flat n°4931, XI Floor,
High Point Bldg n°IV, 45 Palace Road, Bangalore 560001
☎ (+91 80) 228 10 02 ⊖ (+91 80) 225 38 95
Secunderabad Office
66 Rashtrapathi Road, PO Box 1553,
Secunderabad 500003
☎ (+91 40) 80 25 07; 80 29 06 [97]

90-00200 British Steel Liaison Services (India) Limited
(a subsidiary of British Steel Engineering Steels, UK)
412 Raheja Chambers, 213 Backbay Reclamation, Bombay 400-021
☎ (+91 22) 282 31 26 ⊖ (+91 22) 287 51 48 [98]

90-00380 Daewoo Corp New Delhi Branch
4-6 Sirifort Institutional Area, First Floor, New Delhi 110049
☎ (+91 11) 622 82 01 ▣ 71202 dwoo in
⊖ (+91 11) 622 33 87 [98]

90-00385 Daewoo Corp Bombay Branch
107 Maker Chamber Vi, Nariman Point, Bombay 400 021
☎ (+91 22) 283 59 47 ▣ 1186688 dwbb in
⊖ (+91 22) 204 44 70 [98]

90-00425 Deltronic Marketing Co.
146-148 Chenoy Trade Centre, Park Lane, Secunderabad 500 003
☎ (+91 40) 81 95 14 ▣ 04252126 ⊖ (+91 40) 84 69 89
⊕ US$ 1 000 000 ♣ US$ 3 600 000 👥 16 ⌂ 1 000 sq m
President: D.H. Kushlani
® Germany: Argumens GmbH, MTS Mikrotechnik GmbH, Spectrum Elektrotechnik GmbH
Israel: Israel Microwave Components Ltd, Optomic Microwave Ltd.
UK: EIP Metals Ltd. (Evered Tube Div.)
USA: GAMMA-F Corporation, Precision Tube Company [97]

90-00450 Dynamic Sales & Service
E-40/7 Okhla Industrial Area, Phase 2, New Delhi 110020
☎ (+91 11) 683 82 87; 683 11 74; 683 46 13
[E] dynaipl@gems.vsnl.net.in ▣ 85014
⊖ (+91 11) 683 15 53
President: Maj S P Khosla
Senior Executive Vice President: Wg Cdr K C Soni
Vice President, Technical: Wg Cdr P T Singha
General Manager: Col D S Alagh
■ Export house. [97]

90-00500 Echbee Corp
38 Cawasji Patel Street, 5th Floor Kamer Building, Bombay 400001
☎ (+91 22) 204 34 89; 204 83 41 ✉ SPACEAGE
▣ 115479 echb in ⊖ (+91 22) 204 34 89
⊕ US$ 1,000,000 (1996) ♣ US$ 4,000,000 (1996)
👥 16 (1996)
Proprietor and Managing Director: Hasmukh B Shah
■ Sales and service engineers, service representative for technological companies.
® France: Air Liquide, Cryolor SMC TBT
Japan: Tamagawa Seiki Co
Switzerland: Schaffner Electronik
UK: EECO Ltd, Unimatic Engineering
USA: Accurate Sound Corp, Advance Group, American Switch Corp, Anorad Corp, Axel Electronics, Babcock Displays, Champion Techno, Collmer Semicondustor, EECO Inc, Genisco Technology, Hi Tech Import & Export, Itek Systems (BEI), Kimball Physics, Perfection Mica, Phototonics Tech Systems, Rod-L Electronics Inc, Senisys Sensors, Shield Technology, Singer Products, Standard Wire & Cable Co, Thomas Electronics Inc, Universal Shielding

Echbee Techno-Legal Aid Centre
38 Cawasji Patel Street, Bombay 2
☎ (+91 22) 204 34 89; 204 83 41
⊖ (+91 22) 262 42 02; 202 42 02
■ Technical and legal assistance regarding Indian customs, imports and foreign exchange regulations. [98]

90-00835 Himalaya Aviation
3272 11th Main, HAL 2nd Stage, Bangalore 560038
☎ (+91 80) 527 51 96; 526 84 98 ▣ 8452943 gana in
⊖ (+91 80) 526 84 98
⊕ £ 10,000 ♣ £ 1,000,000 👥 12 ⌂ 140 sq m
Executive Director: B Seshadri
Sales Manager: R Padmanabhan
■ Raw materials, semi-finished components, tooling items, consumables, avionics equipment, electronic/electrical, hardware.
® Germany: G. Hartner Co GmbH, Tierney Metals UK
Luxembourg: IMA
USA: Banner Aircraft International [97]

90-00840 Hindco Corp
205A Dr Annie Besant Road, Worli Naka, Bombay 400018
☎ (+91 22) 493 15 44 ✉ HINDCOMEL
⊖ (+91 22) 493 69 48
⊕ US$ 40,000 ♣ US$ 7,000,000 (1995) 👥 15
President: Deepak Valia
Public Relations Officer: Jayant Gandhi
■ Publishers and advertising. → 80
® USA: Mallory Capacitor [98]

90-00940 Indamer Co Pvt Ltd
Hangar N°1 Juhu Airport, Bombay 400054
☎ (+91 22) 611 56 61; 611 99 77; 611 78 24; 611 13 52 ✉ ALBALDINI ▣ 1171031 icpl in
⊖ (+91 22) 611 36 26
👥 110 ⌂ 20,000 sq ft
Managing Director: Richard G Koszarek
Director, Sales and Service: A K Murthy
Director, Sales and Service: K R Nandkumar
Director, Finance and Company Matters: S K Iyengar
Director, Operations and Administration: K P Raman
■ Sales and after sales service including overhaul of aircraft, aero-engines and avionics.
® Canada: Pratt & Whitney
USA: Beech Aircraft, Bendix/King Avionics, Collins Avionics, Gulfstream Aerospace, Sunair Electronics, Textron Lycoming [98]

90-01240 M/S E. Krishnarao
PO Box 2055, Suryabaugh, Visakhapatnam 530020
☎ (+91 891) 56 67 39 ✉ GOODLUCK ▣ 0495255 ekrs in
⊖ (+91 891) 56 37 77
⊕ Rs 1,200,000 (1996) ♣ Rs 12,000,000 (1996) 👥 21
⌂ 1,200 sq m
Managing Partner: Emandi Krishna Rao
Sales and Marketing Director: Emandi Srinivas
® Belgium: AMTS Int'l N.V.
Denmark: Hundested Motor & Propeller Fabrik A/S, Maskinfabrik Acta A/S, Norlau A/S, Odin Control Systems
Finland: Hackman Metos Ltd, Helkama-Kaapeli
Germany: Ako Regelungstechnik GmbH, Carl Dose GmbH, Cassens & Plath GmbH, MACOR Marine Systems Int'l GmbH, Podszuck GmbH, Zollner GmbH & Co KG
Italy: Speich Srl
Netherlands: Focus Veilig BV, Maskinfabrik EMCE, Simac CSI BV
Norway: Norac A/S
Singapore: APV South East Asia Pte Ltd
S. Africa: Magnapower (Pvt) Ltd
Spain: Desarrollo Technicas Ind. de Galicia SA, La Auxiliar Navel SA, Sulcer, S.L.
Sweden: Naval Electronics AB
UK: Griffin Woodhouse Ltd, Thorn Security Ltd, Victor Pyrate Ltd
USA: Envirovac Inc, Hose-McCann Telephone Co Inc [97]

90-01340 Micronic Devices
(a division of Advanced Micronic Devices Ltd, India)
Arun Complex, 65 DVG Road, Basavenagudi, Bangalore 560 004
☎ (+91 812) 60 06 31; 60 78 00; 60 45 02
[E] mic.amdbt@gems.vsnl.net-in
⊖ (+91 812) 660 87 85; 558 34 33
⊕ Rs 50,000,000 (1997) ♣ Rs 300,000,000 (1997)
👥 240 (1997) ⌂ 10,000 m2
Director: B S Seshadri
Vice President: J Padmanabhan
® USA: Microphase Corp. [98]

90-01540 OFEMA
(a subsidiary of OFEMA, France)
42 Golf Links Road, New Delhi 110003
☎ (+91 11) 461 84 24; 461 54 09 ✉ EXPORAERO
▣ 3165056 deli in ⊖ (+91 11) 46 26 075
Technical Delegate: M Lannevere
® France: OFEMA

Indiranagar Office
101 Defence Colony - HAL Stage 2, 2nd Main, 3rd Cross, Indiranagar, Bangalore 560038
☎ (+91 81) 58 22 60 ✉ OFEMA ▣ 8458661 ofbg in
⊖ (+91 80) 58 15 43
Technical Delegate: Wing Cdr H R Seetharam (retd)
■ Export liaison office for the French aerospace industry.
® France: OFEMA [98]

90-01650 PARCO
16b Connaught Place, New Delhi 110001
☎ (+91 11) 332 08 77 [E] kapil.parco@axcess.net.in
⊖ (+91 11) 332 06 19
👥 6 (1997) ⌂ 85 m2
Managing Director: Kapil Bhagat
Sales and Marketing Director: K K Gupta [98]

90-01700 Radiall Protection Ltd
(a subsidiary of Radiall, France)
17 Andree Road, Shanthinagar, Bangalore 560027
☎ (+91 80) 223 50 87 ▣ 8452839
⊖ (+91 80) 227 94 59
General Manager: M U Bhandari
® France: Radiall [98]

90-01980 Tata BP Solar India Limited
Plot No. 78 Electonic City, Hosur Road, Bangalore 561229
Kssidc Multistorey Blocks, Hosur Road, Electronic City, Bangalore 562158
☎ (+91 80) 852 00 83 ⊖ (+91 80) 852 09 72
® UK: BP Solar Limited [98]

90-02200 Victoria Enterprises Private Ltd
302 World Trade Centre, Babar Road, New Delhi 110001
☎ (+91 11) 331 73 61; 331 73 62 ✉ ROGENTER
▣ 03162560 com in ⊖ (+91 11) 331 87 26
Group President: Mahendra K. Jajodia
Director: K K Arora ☎ (+91 11) 332 39 13
Director: Amrit Pal ☎ (+91 11) 332 29 96
Aerospace Executive: Wg Cdr N M Sharma (rtd)
☎ (+91 11) 331 73 61
■ Sales representative for foreign manufacturers: optical tracking systems, hovercraft, aerospace metals, film readers, precision rate tables, BITES, astrodomes, ultrasonic flaw detectors. [98]

90-02500 VXL Engineers Limited
F-45 NDSE Part 1, New Delhi 110049
☎ (+91 11) 461 95 30; 469 27 88; 462 87 68
✉ LECFLO ⊖ (+91 11) 462 09 24
⊕ Rs 200 000 000 ♣ Rs 175 000 000 👥 500
⌂ 9 290 sq m
President: Y.N. Bhargava
■ Manufacture RF test and measuring instruments: synthesised signal generators up to 1.1 GHz, RF milli voltmeters, directional wattmeters, modulation meters. PCM test equipment: PCM 30 analysers, PCM terminal test set, VF MUX tester, digital transmission analysers up to 140 MBit/sec. Microwave components & connectors: RF connector, type: N, SMA, DIN, RF attenuators, RF filters, RF loads: dry, oil cooled. Communication systems: 2/15 MARR, 8 MB/sec. OLTE. Personal rescue beacon.
® Germany: Schomandl Vertriebs
UK: Farnell Int'l, Sifam, Wayne Kerr Electronics Ltd
USA: Belar Electronics, Christie Electric Co. [97]

Indonesia

Head of State:
Gen Soeharto (retd)

10 Head of State

10-10000 Office of the President
Bina Graha, Jalan Veteran N°17, Jakarta
☎ (+62 21) 34 10 79; 88 46 56 ▣ 44240 seknek ia
President: Gen Soeharto (retd)
Minister of State and State Secretary: Maj Gen Merdiono
Minister of State and Cabinet Secretary: Saadilah Mursjid [98]

10-20000 Office of the Vice-President
Jalan Merdeka Selatan 6, Jakarta
☎ (+62 21) 58 23 26; 58 27 00
Vice-President: Gen Try Sutrisno (retd) [98]

Indonesia

12 Ministries

12-10000 Department of Defence and Security
Jalan Medan Merdeka Barat 13, 10110 Jakarta Pusat
☎ (+62 21) 345 61 84 ⊖ (+62 21) 384 51 78
Minister: Gen Edi Sudradjat (retd) [98]

12-20000 Department of Foreign Affairs
Jalan Taman Prejambon 6, Jakarta
☎ (+62 21) 36 60 14 ⊠ 44205; 44469 deplu jkt
Minister: Ali Abdullah Alatas [98]

12-40000 Department of Home Affairs
Jalan Merdeka Utara 7, Jakarta
☎ (+62 21) 34 22 22
Minister: Lieut Gen Suardi Memed Yogie (retd) [98]

12-50000 Ministry of State for Research and Technology
Gedung BPP Teknologi Lt.111, Jalan M H Thamrin N°8, Jakarta
☎ (+62 21) 32 47 67; 32 23 02
Minister: Prof Bacharuddin J Habibie [98]

12-60000 Ministry of Finance
Jalan Lapangan Banteng Timur 2-4, Jakarta
☎ (+62 21) 34 89 38; 36 50 79; 37 33 09
⊠ 44319 dpkeu ia; 46415 ditjmon jkt
Minister: Mar'ie Muhammad [98]

25 Defense Forces

25-10000 Armed Forces Headquarters
Cilangkap, Jakarta Timur
☎ (+62 21) 840 12 40
Commander in Chief Armed Forces: Gen Feisal Tanjung [98]

26 Police (Civil)

26-00000 National Police Headquarters
Markas Besar Kepolisian R I (MABAK)
Jalan Trunojoyo 3, Kebayoran Baru, Jakarta
☎ (+62 21) 384 85 37
Chief of Police: Lieut Gen Dibyo Widodo
☎ (+62 21) 721 81 15; 720 72 77
Deputy Chief of Police: M B Hutalung
Deputy Chief of Administration: Maj Gen S H Roesmandhi
Police Inspector General: Maj Gen Ronny Lihawa
Planning and Budgeting: Maj Gen Lufti Dahlan
Co-ordinator of Expert Staff: Maj Gen Karyoso
Director of Intelligence and Security: Brig Gen Guntur
Director of Detective: Brig Gen Roesdihardjo
Director of Alert Unit: Brig Gen S A Suhardi
Director of Community Counselling: Brig Gen Ign Koes
Director of Traffic: Brig Gen R H Soewito
Director Personnel, Planning and Control:
Brig Gen Kadaryanto
Director of Personnel Maintenance: Brig Gen S A Sambas
Director of Education: Brig Gen S A Soepardi
Director of Logistics: Brig Gen Soeseno
Head of Police Information Service: Drs Nurfaizi
[E] nurfaizi@indosat.net.id
National Secretary: Brig Gen Ahwil Lutan
Head of Electronic Communication: Brig Gen H Syahril
Head of Medical and Health Unit:
Brig Gen Pamudji Santoso
Head of Research and Development Unit:
Brig Gen Ishak Kodijat
Head of Legal Unit: Brig Gen S H Anwari
Head of Psychological Unit: Brig Gen Moch Nasichun [98]

31 Embassies & Attachés

31-00080 Embassy of the Argentine Republic
Menara Mulia Suite 1901 - 19th Floor,
Jalan Jend Gatot Subroto Kav 9-11, 12930 Jakarta
☎ (+62 21) 526 56 61; 526 56 62 ⊠ 69151 emarin ia
⊖ (+62 21) 526 56 64
Ambassador: H E Mr Juan Carlos Morelli
Counsellor Political Affairs: Miguel Jorge Nougues
Cousellor Consular and Cultural Affairs:
Mrs Maria Soledad Durini de Nougues
First Secretary Economic and Commercial Affairs:
Luis Maria Villafañe [98]

31-00090 Australian Embassy
Jalan H R Rasuna Said Kav C15-16, 12940 Jakarta Selatan
☎ (+62 21) 522 71 11 ⊖ (+62 21) 522 71 01
Ambassador: H E Mr John McCarthy
Deputy Chief of Mission: Murray Cobban
Defence and Army Attaché: Brig E P Chamberlain
Air Attaché: Gp Capt Gary Dunbar
Naval Attaché: Capt J Jones (RAN)
Army Attaché: Col B Millen
Assistant Defence Attaché: Maj D Rutherford
Third Secretary: J Powell
Third Secretary: Ms Trish McKinnon
Public Information Officer: J Milne
Commercial Office
Jalan M H Rasuna Said Kav C15-16,
12940 Jakarta Selatan
☎ (+62 21) 522 71 11 ⊖ (+62 21) 522 71 03
Minister: B Wallace [98]

31-00100 Embassy of Austria
Jalan Diponegoro 44, PO Box 2746, 10027 Jakarta
☎ (+62 21) 33 81 01; 33 80 90 ⊠ 46387 oebjkt ia
⊖ (+62 21) 390 49 27
Ambassador: H E Dr Hans Demel
First Secretary: Ms Eleonora Windisch
Commercial Office
Jalan Jambu 17, Menteng, Jakarta
☎ (+62 21) 314 52 42; 310 08 93 ⊖ (+62 21) 32 73 42
Commercial Counsellor: Dr Oskar Andesner
Commercial Attaché: Mag. Anton Hölzl [98]

31-00310 Embassy of Canada
PO Box 1052/JKT, Jakarta
☎ (+62 21) 525 07 09 ⊠ 62131 dmcan jkt
⊖ (+62 21) 571 22 51
Ambassador: H E Mr Gary J Smith
Deputy Chief of Mission: R Cremonse
Military Attaché: Col M Murphy
Commercial Counsellor: M Moszczenska [98]

31-00350 Embassy of the Republic of Chile
PO Box 4168, 12041 Jakarta
Bina Mulia Blds, 7th floor Jl. H.R. Rasuna Said Kav. 10,
12950 Kuningan, Jakarta
☎ (+62 21) 520 11 31 ⊠ 62587 chilie ia
⊖ (+62 21) 520 19 55
Ambassador: H.E. Mr Fernando Cousino Besa
Minister Counsellor: Julio H. Riethmüller
Commercial Counsellor: Tresjee Mariman
Third Secretary: Fernando Berguno [97]

31-00450 Embassy of the Czech Republic
Jalan Gereja Theresia No. 20, PO Box 1319, Jakarta Pusat
☎ (+62 21) 390 40 75/7 ⊠ 69241 ⊖ (+62 21) 33 62 82
Ambassador: H.E. Mr Jaroslav Olsa
First Secretary Political/ Deputy Head of Mission:
Petr Hrdlicka
Third Secretary Press, Cultural, Education & Youth:
Martin Curda
Commercial Counsellor: Michal Svoboda
Attaché Administration Section: Tomas Slimacek [97]

31-00460 Royal Danish Embassy
PO Box 4459, 12044 Jakarta
Bina Mulia Building, 4th Floor, Jalan H R
Rasuna Said, Kav 10, 12950 Jakarta
☎ (+62 21) 520 43 50 ✉ AMBADANE Jakarta
[E] dkembjak@dnet.net.id ⊠ 62123 ambdk ia
⊖ (+62 21) 520 19 62
Ambassador: H E Mr Michael Sternberg
Deputy Chief of Mission and Commercial Counsellor:
Thorkild Borre
Attaché and Consul: Folmer Jensen [98]

31-00660 Embassy of the Federal Republic of Germany
Jalan M H Thamrin No 1, Jakarta
☎ (+62 21) 390 17 50 ext. 101/117
✉ DIPLOGERMA Jakarta ⊠ 44333 aajakaia
⊖ (+62 21) 390 17 57
Ambassador: H E Dr Heinrich Seemann
Deputy Chief of Mission: W Grolig
Defence Attaché: B Hasenpusch
Commercial Counsellor: W Busch
Public Information Officer: D Lamlé [98]

31-00810 Embassy of the Republic of Hungary
Jl Rasuna Said, Blok X/3, Kuningan, 12950 Jakarta
☎ (+62 21) 520 34 59; 58 75 25 ⊠ 62835 hungex ia
⊖ (+62 21) 520 34 61; 521 26 83
Chargé d'Affaires a.i.: Arpad Kovacs
Political Counsellor: Sandor Torok
Commercial Counsellor: Balazs Vass [97]

31-00890 Embassy of Italy
Jalan Diponegoro 45, 10310 Jakarta
☎ (+62 21) 33 74 45 (hunting) ⊠ 61546 itadpl ia
⊖ (+62 21) 33 74 22
Ambassador: H.E. Dr Mario Brando Pensa
Counsellor: Dr Gianni Martini
First Secretary Commercial: Dr Stefano Pontesilli
Air, Military & Naval Attaché: Col Lanfranco Laccisaglia
☎ (+62 21) 310 42 18 [97]

31-00990 Embassy of the Republic of Korea
Jl.Gatot Subroto 57, Selatan, Jakarta
☎ (+62 21) 520 19 15 ⊖ (+62 21) 525 41 59
Ambassador: H.E. Mr Hyung Ki Min
Deputy Chief of Mission: Hanjohng Yo
Armed Forces Attaché: Brig.Gen. Sae Ho Suh
Commercial Attaché: Chang Soon Kong
Public Information Officer: Bong Hyup Chung [97]

31-01270 Royal Netherlands Embassy
Jalan H R Rasuna Said S-3, Kuningan, Jakarta
☎ (+62 21) 51 15 15 [E] nljakagr@rad.net.id
⊠ 62414 nedam ia ⊖ (+62 21) 570 07 34
Ambassador: H E Mr P R Brouwer
Deputy Chief of Mission: D C B den Haas
Defence Attaché: Col B T de Mars
First Secretary Press and Cultural Affairs: Dr E J Schmidt
Head of Commercial Section: Dr J Th. M Vervloed
Commercial Secretary: Dr W Th. Mohr [98]

31-01340 Royal Norwegian Embassy
Bina Mulia Building 1, 4th Floor,
Jalan HR Rasuna Said Kav 10 Kuningan, 12950 Jakarta
☎ (+62 21) 525 19 90 ⊠ 62127 ambnor ia
⊖ (+62 21) 520 73 65
Ambassador: H E Mr Jan Wessel Hegg
First Secretary: Ms Tanja Kristine Hegge
Administration Attaché: Ms Wenche S Haga [98]

31-01370 Embassy of Pakistan
Jalan Teuku Umar 50, Jakarta
☎ (+62 21) 310 39 42; 65 47 ✉ PAREP Jakarta
⊖ (+62 21) 310 39 45; 40 75
Ambassador: H.E. Mr Mustafa Kamal Kazi
Deputy Chief of Mission Counsellor:
Mrs Naghmana A. Hashmi
Defence Attaché: Gp Capt. Tariq Mahmood Ashraf
☎ (+62 21) 310 65 47
Commercial Attaché: Gufron Memon
☎ (+62 21) 310 39 47
Press Attaché: Ghulam Qadir Khan ☎ (+62 21) 310 39 46
Third Secretary: Afzal Mahmood Mirza
☎ (+62 21) 314 40 74 [97]

31-01390 Embassy of Papua New Guinea
Panin Bank Centre, 6th floor, 1 Jl. Jendera Sudirman,
10270 Jakarta
☎ (+62 21) 725 12 18 ⊖ (+62 21) 725 10 12
Ambassador: H.E. Mr Paul J. Dala

Head of Chancery: Age Kila
Defense Attaché: Col K. Frank [97]

31-01420 Embassy of the Philippines

Jalan Imam Bonjol 6-8, Jakarta Pusat
☎ (+62 21) 310 03 34; 03 45; 03 02
⊖ (+62 21) 35 11 67
Ambassador: H.E. Mr Eusebio A. Abaquin
Minister Counsellor: Belén Fule Anota
Third Secretary & Vice Consul: Frank R. Cimafranca
Defense & Armed Forces Attaché: Capt. Edgardo M. Israel
Commercial Attaché: Bruno Y. Cuenca (Jr)
Press & Information Attaché: Narciso M. Reyes (Jr) [97]

31-01480 Embassy of Romania

Jalan Teuku Cik Ditiro N°42-A, Menteng, Jakarta Pusat
☎ (+62 21) 310 62 40; 310 62 41 ⊠ 61208 romcom ia
⊖ (+62 21) 390 77 59
Ambassador: H E Mr Dumitru Tancu
First Secretary Political: Ovidiu Dianu
Defence Attaché: Col Ioan Niccului Fratita (residing in Seoul, ROK)
Counsellor: Dragos-Mihai Serbanescu
Second Secretary, Cultural, Press and Consular: Valentin Emil Moraru
Administration Attaché: Valeriu Sotu [98]

31-01485 Embassy of the Russian Federation

Jalan Rasuna Said X-7, 1-2, Jakarta
☎ (+62 21) 522 29 72 ⊖ (+62 21) 522 29 16
Ambassador: H E Mr Nikolai N Soloviev
Deputy Chief of Mission: Vladimir Y Plontnikov
Head of Consular Division: Fedor N Poselianov
☎ (+62 21) 522 51 95
First Secretary Administration: Boris Budanov

Trade Representation
60 Jl Imam Bonjol, Jakarta Pusat
☎ (+62 21) 384 64 23; 390 40 51
⊖ (+62 21) 390 40 50
Trade Representative: Alexei B Shmelev
☎ (+62 21) 722 21 02
Deputy Trade Representative: Sergei A Patrikeev
☎ (+62 21) 384 64 23

Military Attachés Office
Jl Rasuna Said X-7, 1-2, Jakarta
☎ (+62 21) 522 29 12
Military and Naval Attaché: Capt 1st Grade Vladimir P Izgorodin
☎ (+62 21) 720 81 79
Senior Assistant Military and Naval Attaché: Capt 2nd Grade Alexandre Ladygin
☎ (+62 21) 77 04 23

Information Section
25 Jl Indramayu, Menteng, Jakarta Pusat
☎ (+62 21) 33 55 15
First Secretary Information: Youry I Noskov
☎ (+62 21) 386 08 61 [97]

31-01690 Embassy of the Democratic Socialist Republic of Sri Lanka

Jalan Diponegoro 70, Jakarta
☎ (+62 21) 314 10 18; 314 10 21; 316 18 86; 315 16 86 Ⓔ lankaemba@vision.net.id ⊠ 69152 slemb ia
⊖ (+62 21) 310 79 62
Ambassador: H E Mr Ratyne Deshapriya Senanayake
First Secretary: I Ansar
Attaché: P T Kumarage
Clerk: C Elegoda [98]

31-01740 Embassy of Switzerland

Jalan H R Rasuna Saïd, Block X 3/2, Kuningan, 12950 Jakarta Selatan
☎ (+62 21) 525 60 61 ✉ AMBASUISSE Jakarta
⊖ (+62 21) 520 22 89
Ambassador: H E Dr Peter Hollenweger
First Secretary and Deputy Head of Mission: Denis Knobel
Trade Counsellor: Werner Wiedmer
Attaché and Head of Chancery: Christian Winter
Attaché: Urs Müller
Attaché: Mrs Anita Brechbühler Foster
Attaché: Mrs Rita Taskiran Bracher [98]

31-01780 Royal Thai Embassy

Jalan Imam Bonjol 74, 10310 Jakarta
☎ (+62 21) 390 40 52; 390 40 53; 390 40 54; 390 40 55 Ⓔ thaijkt@indo.net.id ⊠ 44035 thai jkt
⊖ (+62 21) 310 74 69
Ambassador: H E Mr Somphand Kokilanon
Minister: Wiwat Kuthonthien
Minister Counsellor: Chalit Manityakul
Counsellor: Klaham Chanvanichpom
First Secretary: Chalee Sakolvari
First Secretary: Pinyo Sakornvanasak
Second Secretary: Suwapong Sirisorn
Second Secretary: Miss Pranee Rerkthanakij
Third Secretary: Sukhum Somprasonk
Third Secretary: Somkiat Pugsuwan
Attaché: Napaphon Roonapai
Attaché: Wuttichai Intrasard

Office of the Defence and Army Attaché
Jalan Taman Radio Dalam, Gandara Utara, Keybayoran Baru, Jakarta Selatan
☎ (+62 21) 723 38 23; 723 38 29
Defence and Army Attaché: Col Taratom Sriyabhandhe

Office of the Air Attaché
Jalan Sinabung II Bawah, Simpruk Kav W2 - Kebayoran Baru, Jakarta Selatan
☎ (+62 21) 739 25 20
Air Attaché: Gp Capt Supol Vidhayavej

Office of the Naval Attaché
Jalan Wuluh Block III N°12, Taman Tomang Raya Indah, Jakarta Selatan
☎ (+62 21) 831 21 31 ⊖ (+62 21) 835 66 23
Naval Attaché: Capt Kanat Thongpool

Office of Commercial Affairs
20th Floor Chase Plaza Tower, Jala Jend Sudiman Kav 21, 12910 Jakarta
☎ (+62 21) 579 11 77; 570 11 75
⊖ (+62 21) 570 13 90
Minister: Dumrong Indharameesup
Minister Counsellor: Sit Sriwitthaya
Counsellor: Suphak Saguandeekul [98]

31-01840 Embassy of the Republic of Turkey

Jalan HR Rasuna Said Kav I, Kunigan, 12950 Jakarta
☎ (+62 21) 525 62 50; 526 41 43
Ⓔ tccakabe@rad.net.id ⊠ 62506 turkel ia
⊖ (+62 21) 522 60 56
Ambassador: H E Mr Sevinc Dalyanoglu
Charge d'Affaires: Raif Karaca
Military Attaché: Capt Seycan Sevincler
Attaché: Hasan Ozdemir
Attaché: Cetin Taner [98]

31-01890 Embassy of the United Kingdom

Jalan M H Thamrin 75, 10310 Jakarta
☎ (+62 21) 315 62 64; 314 42 29 ⊠ 61166 prodro ia
⊖ (+62 21) 314 18 24; 390 27 26; 390 74 93
Ambassador: H E Mr Robin Christopher
Deputy Head of Mission: Quinton Quayle
Political Counsellor: Peter Davies
Commercial Counsellor: Peter Johnstone
Defence and Military Attaché: Col I J Hellberg
First Secretary Defence Supply: John Bewley
PADEFATT: Ms Zoë Gough [98]

31-01950 Embassy of Venezuela

Central Plaza Building - 17th Floor, Jl Jenderal Sudirman Kav 47-48, 12930 Jakarta
☎ (+62 21) 522 95 85; 525 01 95; 525 07 95
Ⓔ evenjakt@indo.net.id ⊠ 62701 embven ia
⊖ (+62 21) 525 01 95
Ambassador: H E Mr Vasco Altuve Febres
☎ (+62 21) 522 95 84; 522 30 19
Counsellor: Pablo Gaudnik [98]

34 Multinational Intergovernmental Organizations

34-01000 Association of South East Asian Nations (ASEAN)

70-A Jalan Sisingamangaraja, PO Box 2072, 12110 Jakarta
☎ (+62 21) 726 29 91; 724 33 72
⊖ (+62 21) 724 35 04; 739 82 34
Secretary-General: Dato Ajit Singh
Deputy Secretary-General Brunei: Mahadi Haji Wasli
Dir Economic Co-operation Malaysia: Dr Chee Peng Lim (Singapore)
AFTA Unit Dir Thailand: Dr Suthad Setboonsarng
Functional Co-operation Dir Philippines: Dr Filehon Uriarte
General Affairs Dir Indonesia: Dr Filino Harahap
Sr Officer Information: Ms Geni Achnas
Public Information Officer: Ms Shamiala Kandiah [97]

55 Government-owned Industry & Production Facilities

55-00050 PT Boma Bisma Indra

Jalan Ngagel 155-157, 60246 Surabaya
☎ (+62 31) 568 52 34; 568 52 36; 567 02 95; 567 02 96 Ⓔ bbimark@radned.id
Ⓔ ptbbi@peterpetraac.id ⊖ (+62 31) 567 10 22
👥 2,260
President: Soetomo Soetrisno
Marketing Director: Mr Widodo
Production Director: Tjokro Suprijadi
General Affairs Director: Moch Toha Dahlan
Director of Technology: Christian Titahgluw
■ Manufacture and sale of diesel engines, generating sets and spare parts. Manufacture of plant equipment (condensers, heat exchangers, pressure vessels, low/high pressure heaters). [98]

55-00300 PT Industri Pesawat Terbang Nusantara (IPTN)

Nusantara Aircraft Industries Ltd
Jalan Pajajaran 154, PO Box 1562 BD, 40174 Bandung
☎ (+62 22) 63 20 43; 600 27 02; 600 34 42
⊠ 28295 iptnbd la ⊖ (+62 22) 63 16 96
👥 16,000 🏭 600,000 m2
President: Prof Dr Ing. B J Habibie
Chief Operating Officer: Ir. Hari Laksono
Senior Vice President Airplane Business Unit: Dr Ing. Ilham A Habibie
Senior Vice President, Weapon Systems, Space System: Ir. S D Yusman
Senior Vice President Technology: Dr Ir. Said D Jenie
Vice President Program N-250: Ir. Djoko Sartono
Vice President Program CN-235: Ir. Hari Purnomo
Vice President, Helicopter: Ir. Bambang Soeryawan
Vice President Weapon System: Ir. Sindu U Tampbolon
Vice President Space System: Ir. F X Sudarmono
Vice President Commercial I Aircraft: Ir. Heru Santoso
Vice President Commercial II, Aircraft: Ir. Aditya Kusuma
Vice President Commercial for Helicopter, Weapon System and Space System: Pudji Sulaksono
Public Relations Manager: M T Satoto
Contact: Putu Darmawan
■ Fixed-wing: design and manufacture of N-250 high speed turboprop commuter aircraft, 64-68 seats, fly-by-wire. CASA/ITPN CN235 jointly designed and manufactured with CASA, multi-role aircraft in military, maritime patrol and civil/passenger transport versions. Design and manufacture of N2130 turbofan powered regional aircraft, 114-132 seater, fly-by-wire. NC-212 low to medium speed, multi-role aircraft.
Rotary-wing: Eurocopter BO-105 light 5 seater utility and observation helicopter (licenced by Eurocopter). Bell 412 15 seat utility helicopter (under licence from Bell Helicopter Textron). AS332L Super Puma utility helicopter, 20-24 seats, under licence from Eurocopter.
Weapon systems: FFAR 2.75" rocket, SUT Torpedo, surface under water target. NDL-40 FFAR Launcher.
→ 75

Head Office
Kanter Pusat
BPP Teknologi Lt 24, PO Box 3752, Jakarta
Jalan MH Thamrin 8, 10310 Jakarta
☎ (+62 21) 32 25 02 ⊠ 46141 iptnjk ia
⊖ (+62 21) 32 81 69

Jakarta Office
Kamor
BBD Plaza 14th Floor, PO Box 3752, 10310 Jakarta
Jalan Imran Bonjol 61
☎ (+62 21) 32 22 47 ⊠ 61248 iptnjk ia; 81393 iptnjk ia
⊖ (+62 21) 310 00 81; 32 63 18

Plant
Jalan Palaran 154, PO Box 1582 BD, 40174 Bandung
☎ (+62 22) 83 39 00; 83 39 11
⊠ 28295 iptnbd ia; 28108 iptnbd ia ⊖ (+62 22) 83 39 12

Divial UMC PT IPTN
PO Box 1582, 40174 Bandung
KP IV Jalan Pajajaran 15-1
☎ (+62 22) 83 10 86; 87 84 95
⊖ (+62 22) 83 18 51; 83 77 47 [98]

Contact Reference

Indonesia

65 Industry, Trade & Armed Forces Associations

65-00100 The Agency for Strategic Industries of the Republic of Indonesia (BPIS)

Arthaloka 3rd Floor Jalan Jend, Sudirman 2,
10220 Jakarta
☎(+62 21) 570 53 34; 570 53 35; 570 53 37
[W] http://www.inn.bppt.go.id/government/ bpis/bpis.html
⊜ (+62 21) 58 44 10
■ Represents strategic Indonesian companies providing products and services in transportation, information, telecommunication and electronics.
Provider of technical support, organisation of policy implementation marketing and management supervision.
Member companies:
PT Barata Indonesia
PT Boma Bisma Indra
PT Dahana
PT INKA (Industri Kerata Api)
PT INTI
PT IPTN (Industri Pesawat Terbang Nusantara)
PT Krakatau Steel
PT Len Industri
PT Pal Indonesia
PT Pindad [98]

75 Manufacturing Companies

PT Industri Pesawat Terbang 55-00300

75-00050 Daewoo Corp Jakarta Office

Lippo Plaza 15th Floor #1501,
Jalan Jend Sudirman KAV 25, Jakarta Selatan
☎(+62 21) 520 34 97 ⊜ (+62 21) 520 34 99 [98]

75-00200 PT Centronix

Defence Group
Jl Matraman Raya 36, 13150 Jakarta
☎(+62 21) 850 98 70/71; 858 19 99 ⊠ 48216 centro ia
⊜ (+62 21) 858 31 86
■ General engineering & electronics. Manufacture of military telecommunications & electronics equipment. Applied research & development. [97]

75-00300 PT Kawasaki Motor Indonesia

(joint venture with Kawasaki Heavy Industries Ltd, Japan)
Jalan Perintis Kemerdekaan, Kelapa Gading,
14250 Jakarta Utara
☎(+62 21) 452 33 22 ⊜ (+62 21) 452 35 66 [98]

75-00500 PT PAL Indonesia (PERSERO) Head Office (PAL)

BPP Teknologi Building - 17th Floor,
Jalan M H Thamrin N°8, 10340 Jakarta
☎(+62 21) 315 68 60 [E] paljkt@ibm.net ⊠ 61725 pailjkt
⊜ (+62 21) 315 68 66; 316 87 29
✤ Rs 1,671,000,000 (1996)
♣ Rs 247,470,000,000 (1996) 👥 4,800 (1997)
🏭 700,000 m2 (1997)
Chairman, President and Chief Executive Officer: Prof Dr-Ing. B J Habibie
Executive Vice President General Affairs: Ir Auzai
Executive Vice President Production: Dr-Ir Adwin H Suryohadiprodjo
Executive Vice President Technology: Ir Arief Aini
Executive Vice President Maintenance: Ir Wulang Widada
Executive Vice President Finance: Ir Agus Supriyadi
Executive Vice President Commercial: Dipl.-Ing. Suleman Wiriadidjaja ☎(+62 21) 316 93 35
Executive Vice President Program: Ir Suhadi
■ Shipbuilding activities include construction of fast patrol boats, frigates, mine hunters, submarines, jetfoils and diesel engines. Comprehensive maintenance and overhaul services also available.
Plant
Ujung, Surabaya, Jawa Timur
☎(+62 31) 329 22 75 ⊠ 31223 pal sb
⊜ (+62 31) 329 24 93 [98]

75-02000 P T Trans-Bakrie

Setiabudi Atrium - 4th Floor, Jalan H R Rasuna Said,
12920 Jakarta [98]

75-05000 Transfield Pty Ltd

Landmark Centre Tower A - 16th Floor,
Suite 1606 - Jl Jend Sudirman Kav 1, 12910 Jakarta
☎(+62 21) 522 34 40 ⊜ (+62 21) 522 34 39 [98]

80 Service Companies & Organizations

80-00500 AlliedSignal Aerospace Marketing Sales and Service

PO Box 303 Bush 19101, Cengkareng
c/o Garuda Maintenance Facilities,
Hangar 11, Room 86a, Soekamo-Hatta Airport
☎(+62 21) 550 81 54 (recording) ⊠ 69053 k
⊜ (+62 21) 550 12 35
Field Service Engineer: Pheng Cheng Sim
☎(+62 816) 85 64 39
■ On-site customer support for AlliedSignal Aerospace. → 90 [98]

80-01000 Hughes International Corporation

Mid Plaza II Building, 22nd Floor,
Jalan Jend, Sudirman, Kav 10-11, Jakarta Pusat
☎(+62 21) 573 41 61 ⊜ (+62 21) 573 41 62
Managing Director: George T Lewis [98]

80-05000 PT Wijaya Purna Bisakti

Jalan Naripan n°85, 40112 Bangung
☎(+62 22) 420 47 77; 43 98 00 ⊜ (+62 22) 43 76 00
Chairman: Ir Soedjana Wire Koesoemah
Managing Director: Benjamin Wijaya
Marketing Director: Anton Bambang Hadiojono
Technical Director: Dip.-Ing. David Harryanto
® Switzerland: Trimos-Sylvac Task Force [98]

90 Sales Agents, Sales & Liaison Offices

Companies represented:

Aero International, USA	90-01330
Aeronautical Communications Equip, USA	90-01500
Aerotech, USA	90-01330
Allied International Corp, USA	90-00040
BR Communications, USA	90-01500
Carlson Communications Inc, USA	90-01500
Chesivale Electronics, UK	90-01500
Clough Engineering Group, Australia	90-01600
Dataradio Inc, Canada	90-01500
De Havilland, Canada	90-00810
Dynatech Tactical Communications Inc, USA	90-01500
EEV Ltd, UK	90-01335
FFA AG, Switzerland	90-00825
Futac International Ltd, Hong Kong SAR	90-01500
GEC-Ferranti, UK	90-02000
General Mgr Electro Motive Div., Canada	90-02000
Hamilton Standard, USA	90-02000
Harris DTS, USA	90-01500
Huber + Suhner AG, Switzerland	90-01500
Marconi Instruments, UK	90-01500
MCL Inc, USA	90-01500
Noble Denton, UK	90-01400
Northrop International Aircraft Inc., USA	90-00825
Pratt & Whitney, Canada	90-02000
Premier Crown, USA	90-01330
Racal Avionics Ltd, UK	90-02000
Racal Comsec Ltd, UK	90-01500
Racal Recorder Ltd, UK	90-01500
Racal Tacticom Ltd (Racal Radio), UK	90-01500
Rockwell Int'l, USA	90-02000
SAT, France	90-01500
Sikorsky Aircraft, USA	90-02000
Sirchie Laboratories, USA	90-01330
SR Telecom Inc (SRT), Canada	90-01500
SSPA Microwave Corp, Canada	90-01500
Sundstrand Power Systems, USA	90-02000
Sundstrand Turbomach, USA	90-02000
Technology For Communications Int'l, USA	90-01500
Thomson-CSF, France	90-02030
Thomson-Tubes Electroniques, France	90-01500
Trimos-Sylvac Task Force, Switzerland	80-05000
Uhn-Rohn Inc, USA	90-01500
Viking International, USA	90-01500
Zengrange Ltd, UK→ Netherlands	80-18300

Sales agents/representatives listed elsewhere:

AlliedSignal Marketing Sales 80-00500

90-00040 Alpha Epsico

Jalan Durentiga Selatan #2, 12760 Jakarta
☎(+62 21) 794 03 16 [E] alpha@cbn.net.id
⊜ (+62 21) 780 48 03
👥 8 🏭 115 m2
Managing Director: Bungki Wicaksono
® USA: Allied International Corp [98]

90-00050 BP Solar c/o PT Nusa Cerindo Utama

(† PT Kemenagen)
Niaga Tower, 17th Floor, Jalan Jendral Sudirman 58,
12190 Jakarta
☎(+62 21) 250 56 37 ⊜ (+62 21) 522 59 28 [98]

90-00100 CompStream Corporation

South Asia-Pacific
Wisma Budi, JL H.R.Rasuna Said,
Kav.C-6 Lantai 7, No.702, 12940 Jakarta
☎(+62 21) 521 32 95 ⊜ (+62 21) 521 33 43 [97]

90-00110 Daewoo Corp Jakarta Branch

Lippo Plaza 7th Floor, Jalan Jend Sudirman Kav 25,
12920 Jakarta
☎(+62 21) 520 45 20 ⊠ 60013 daewoo ia
⊜ (+62 21) 520 34 78 [98]

90-00325 Kawasaki Heavy Industries Ltd Jakarta Office

9th Floor Skyline Building, Jalan MH Thamrin 9,
10340 Jakarta
☎(+62 21) 314 07 37 ⊜ (+62 21) 314 10 49 [98]

90-00350 Lucas Aerospace FCS Liaison Office

c/o 4th Fl, GPT, PT IPTN, Jalan Pajajaran No. 154,
Bandung
☎(+62 22) 63 82 29; 67 86 29 ⊜ (+62 22) 63 82 29
Resident Mgr: Mohammad Idrus [97]

90-00810 PT Havilindo Utama

Jalan Iskandarsyah Raya 7,
Wisma Daria 3rd Floor Kebayoran Baru, 12160 Jakarta
☎(+62 21) 720 80 28 ⊠ 47376 daria ia
⊜ (+62 21) 724 73 54
President Director: S M Herlambang
® Canada: De Havilland [98]

90-00825 PT Heliavia Nusantara

PO Box 3160 JKT, 10002 Jakarta
5th Floor, Wisma Pede, Jalan Haryono MT Kav 17,
12810 Jakarta
☎(+62 21) 829 73 04; 803 33 71; 835 17 78
✉ HELIAVIA ⊜ (+62 21) 830 33 71
Director: A Andoko
■ Distribution of spares for military and civil fixed and rotary wing aircraft.
® Switzerland: FFA AG
USA: Northrop International Aircraft Inc. [98]

90-01330 PT Metraco Jaya

PO Box 4412, 12044 Jakarta
☎(+62 21) 799 40 48 ⊜ (+62 21) 719 73 61
✤ Rs 100,000,000 (1997) ♣ US$ 1,000,000 (1997)
👥 10 (1997) 🏭 200 m2

President: S Adiyana
Director: D Champel
Director: I Ibrahim
Office Manager: Ms E Wisnu Y
® USA: Aero International, Aerotech, Premier Crown, Sirchie Laboratories [98]

90-01335 PT Multi Dharma Guna
PO Box 1499/JKS, 12014 Jakarta
☎ (+62 21) 765 51 76 ⊖ (+62 21) 765 51 33
👥 25
President: Sumito Ibnu Hardjono
Marketing Director: Mrs Denise Soedarso
Marketing Director: Ir Silvester L Loho
■ Microwave links, radio communications for ships, submarines and divers, underwater communication and detection, underwater acoustic and complementary items. Closed circuit video equipment, fire alarms systems, infra-red imaging units. Torpedo systems inclusive exercise items, oil filters and separators. Aircraft spares and units, maritime communication spares, electronic valves, spare parts for Tribal class ships.
® UK: EEV Ltd [98]

90-01400 PT Noble Denton Utama
Jalan Cikini Raya No. 111, 10330 Jakarta
☎ (+62 21) 314 94 04; 314 94 05; 310 36 80
☒ 61127 nodent ia ⊖ (+62 21) 32 40 15
® UK: Noble Denton [98]

90-01500 PT Panorama Timur Jaya
JL Paus 86 D/E, Rawamangun, Jakarta
☎ (+62 21) 475 67 88 ☒ 48363 ⊖ (+62 21) 475 16 88
♧ US$ 1,000,000 ♣ US$ 15,000,000 (1995) 👥 43
🏠 600 m2
Director: Harijono Suwarno
Division Manager: Alvin Linardi
Division Manager: Ramo Sumolang
Purchasing Manager: Ms Diana Matondang
® Canada: Dataradio Inc, SR Telecom Inc (SRT), SSPA Microwave Corp
France: SAT, Thomson-Tubes Electroniques
Hong Kong SAR: Futac International Ltd
Switzerland: Huber + Suhner AG
UK: Chesivale Electronics, Marconi Instruments, Racal Comsec Ltd, Racal Recorder Ltd, Racal Tacticom Ltd (Racal Radio)
USA: Aeronautical Communications Equip, BR Communications, Carlson Communications Inc, Dynatech Tactical Communications Inc, Harris DTS, MCL Inc, Technology For Communications Int'l, Uhn-Rohn Inc, Viking International [97]

90-01600 PT Petrosea
(a subsidiary of Clough Eng Ltd, Australia)
Wisma Anugraha, Jl Taman Kemang n°32b, 12730 Jakarta
☎ (+62 21) 797 50 65 ⊖ (+62 21) 797 50 53
♧ US$ 65,000,000 (1996) ♣ US$ 70,000,000 (1995)
👥 1,800
President: T R Humphry
Sales and Marketing Director: Lindsay Courtis
Operations and Technical Director: Neil Cain
Purchasing Manager: Norm Cannard
■ Engineering and construction company providing contracting services to the following industry sectors: Oil and Gas (onshore and offshore) engineering and construction, mining, construction and contract mining, infrastructure, engineering and construction, services, equipment and personnel hire.
® Australia: Clough Engineering Group [97]

90-01970 PT Stimec Elcom
Jalan Musi N°32, PO Box 4467 Jkt, 10150 Jakarta
☎ (+62 21) 385 83 58 ✉ ELCOSTI Jakarta
🄴 stimec@indosat.net.id
☒ 44989 alxgee ia; 47474 made ia
⊖ (+62 21) 384 50 43
♧ US$ 4,500,000 (1997) ♣ US$ 10,000,000 (1997)
👥 20 (1997) 🏠 300 sq m
President, Director: Made Rahardja
Operations Director: Cmdre Mulyadi (retd)
Sales and Marketing Director: Dipl.-Ing Ong Jan Hoei
Purchasing Manager: Herni K Rahardjo
■ Telecommunications equipment.
Representing:
Aydin Corp High Power Amplifier
ETS Inc (Engineering and Technical Services) Ground Station Equipment Manufacturer
IWatsu Measuring Equipment
Rohill, Radio Trunking System [98]

90-02000 PT Transavia Utama
Jl. Letjen S. Parman No. G 1-2, 11480 Jakarta
☎ (+62 21) 534 20 08; 530 16 75
✉ TRANSAVIA Jakarta ☒ 65420 transa ia
⊖ (+62 21) 548 01 02; 530 27 11
♧ Rs 200 000 000 👥 125
President: Ir. Rizwan Halim
Director: Ir. Sukendro Wardojo
General Mgr: Ir. Johanes Hoetomo
Sales & Marketing Mgr: Ir. Haryadi Adhitjan
Purchasing Mgr: Ir. Mike Widjava
■ The company is engaged in the sales & services of various defence related products to the Armed Forces of Indonesia. Service activities also include after sales support & retrofit.
® Canada: General Mgr Electro Motive Div., Pratt & Whitney
UK: GEC-Ferranti, Racal Avionics Ltd
USA: Hamilton Standard, Rockwell Int'l, Sikorsky Aircraft, Sundstrand Power Systems, Sundstrand Turbomach
Subsidiaries:
Supply Express Pte Ltd, Singapore
Transavia Pacific Corp., USA [97]

90-02015 Strategic Technologies - Indonesia
Wisma GKBI 25th Floor, Suite 2515,
Jalan Jend. Sudirman No. 28, 10210 Jakarta
☎ (+62 21) 574 15 06-8 ⊖ (+62 21) 574 15 09
Managing Dir & CEO: Dr David K.M. Chew
Exec. Dir Technology: Dr Kennedy K.M. Chew
Exec. Dir Corporate/Legal Affairs: Chew-Leong Sook Ching
Director: Lucy Ng [97]

90-02030 Thomson-CSF
Panin Center, 4th Floor, Jalan Jendral Sudirman N°1, 12190 Jakarta
☎ (+62 21) 725 02 32 ⊖ (+62 21) 720 03 41
Representative: Philippe de Pimodan
® France: Thomson-CSF [98]

Iran

Head of State:
Ali Akbar Hashemi-Rafsanjani

10 Head of State

10-10000 Office of the President
Teheran
☎ (+98 21) 61 61
President of the Islamic Republic: Ali Akbar Hashemi-Rafsanjsni
Deputy President for Legal and Parliamentary Affairs: Atallah Mohajerani [98]

12 Ministries

12-10000 Ministry of Defence
Kaboli Street, Resalat Freeway, Seyed Khandan, Teheran
Minister of Defense and Armed Forces Logistics: Mohammad Forouzandeh [98]

12-20000 Ministry of Foreign Affairs
Imam Khomaini Avenue, Teheran
☎ (+98 21) 30 39 16; 32 15 14
☒ 212450; 212600 ext ir
Minister: Dr Ali Akbar Velayati [98]

12-40000 Ministry of the Interior
Dr Fatemi Avenue, Teheran
Minister: Ali Mohammad Besharati-Jahromi [98]

31 Embassies & Attachés

31-00080 Embassy of the Argentine Republic
Plaza Argentina, Edif Alitalia 7, PO Box 15875-4335, Teheran 15134
☎ (+98 21) 871 82 94; 871 62 44; 871 91 89
☒ 215994 emba ir ⊖ (+98 21) 871 25 83
Chargé d'Affaires: Eduardo Lionel DeMayo
Counsellor Economic & Commercial: Martin Rivolta [98]

31-00085 Armenian Diplomatic Mission
Ostad Shariar Street, Corner of,
Razi/Tomhouri/Eslami Aye, Tehran
☎ (+98 21) 67 48 33 ⊖ (+98 21) 67 06 57
Chargé d'Affaires: Vahan Baibourdian [98]

31-00130 Embassy of the People's Republic of Bangladesh
Avenue Gandi, 5th Street, Bldg 14, PO Box 11365-3711, Teheran
☎ (+98 21) 877 29 79; 877 29 80
⊖ (+98 21) 877 82 95
Ambassador: H E Mr Syed Muazzemazi
Deputy Chief of Mission: Mustafa Kamal
Commercial Counsellor: M Gias Uddin
First Secretary Labour: Mohaammed Nasin
First Secretary: Rahman Mujibur Bhuiyan
Public Information Officer: S M Narim [98]

31-00580 Embassy of Finland
Valie Asr Avenue Below Vanak Square,
PO Box 15115-619, corner of Nilon St, Teheran
☎ (+98 21) 877 49 85; 888 91 51 ☒ 212930 emfi ir
⊖ (+98 21) 888 91 07
Ambassador: H E Mr Antti Koistinen
Second Secretary: Ari Maki
Administrative Attaché: Mrs Eila Irawan
Administrative Attaché: Mrs Senja Ryhanen
Commercial Attaché: Juhani Ylikoski
Commercial Attaché: Esa Kinnunen [98]

31-00690 Embassy of Greece
PO Box 11365-8151, Teheran
Africa Expressway, Niloufar Street No 20
☎ (+98 21) 222 05 33; 222 37 84 ☒ 212067
⊖ (+98 21) 222 37 84 [98]

31-00870 Embassy of Ireland
Ave Mirdamad, Khiaban Razan Shomali 10, Teheran 19116
☎ (+98 21) 222 76 72 ☒ 213865 irel ir
⊖ (+98 21) 222 27 31
Ambassador: H E Mr Anthony Edward Mannix
Deputy Chief of Mission: Declan Smyth [98]

31-00890 Embassy of Italy
81 Avenue Neauphle-Le-Chateau, PO Box 11365-7863, Teheran
☎ (+98 21) 649 59 55/57 ☒ 214171 itier ir
⊖ (+98 21) 649 69 61
Ambassador: H.E. Mr Giovanni Castellaneta
Deputy Chief of Mission: Paolo Legnaioli
Counsellor: Mario Panaro
First Secretary & Head of the Commercial Section: Luciano Pezzotti
Consulate of Italy
Avenue Neauphle Le Chateau,
Kutdhe Shahid Amir Said Vaezi n.2, Teheran
☎ (+98 21) 649 69 55/7 ☒ 214150 itco ir
⊖ (+98 21) 649 69 61
Consul: Massimo Rustico [98]

31-01270 Royal Netherlands Embassy
PO Box 138, Teheran 11365
Ave Val y Asr, ave Mottahari,
Sarbedaran Street, Jahansouz Alley No.36

☎(+98 21) 89 60 11; 89 60 12; 89 62 07; 89 23 18
⊠ 212788 holl ir ⊖ (+98 21) 889 20 87
Ambassador: H E Mr M Damme
Deputy Chief of Mission: A van der Schaaf
Commercial Secretary: Mrs Y E D M Stassen
Defence Attaché: Col Jean Cemmen
Secretary: Mrs N H Jurriëns [98]

31-01300 New Zealand Embassy
PO Box 11365-436, Teheran 11
29 Koucheh Mirza Hassani, Avenue Mirza-ye-Shirazi
☎(+98 21) 871 50 61 ⊖ (+98 21) 871 17 15
Ambassador: H E Mr Daniel Richards
Commercial Counsellor: Tim Sheppard (Reside's in Dubai)
☎(+971 4) 31 75 00 (fax)
First Secretary: Godfrey Payne [98]

31-01440 Embassy of Portugal
Vali Asr Avenue, Tavanir Avenue,
Nezami Ganjavi Street 30, Teheran
☎(+98 21) 877 13 80; 21 32 ⊠ 212588 LUSA IR
⊖ (+98 21) 877 78 34; 21 34
Ambassador: H E Dr Manuel Marcelo Monteiro Curto
☎(+98 21) 877 21 32
Chancellor Head of the Consular Section:
Mrs Maria Jose Morgado Fortes Conde
☎(+98 21) 877 78 34; 877 13 80
Commercial Counsellor: Joao G V Ribeiro Rodrigues
☎(+98 21) 888 09 57; 888 09 58 [98]

31-01655 Embassy to the Republic of Slovenia
7th Street N° 10, Khaled Eslamboli Avenue, Tehran
☎(+98 21) 885 79 72
Ambassador: H E Dr Joze Kunic [98]

31-01730 Swedish Embassy
No 2 Nastaran Street, PO Box 19575-458, Teheran
Pasdaran Avenue (North of Dr Lavasahi), Teheran
☎(+98 21) 229 68 02 ✉ SVENSK Teheran
⊠ 212822 svsk ir ⊖ (+98 21) 228 60 21
Ambassador: H E Mr Hans Andersson
Deputy Chief of Mission: Ms Asa Arvidson
Public Information Officer: Dag Juhlin-Dannfelt
Third Secretary: Marianne Bylund [98]

31-01780 Royal Thai Embassy
PO Box 11495-111, Teheran 11547
Baharestan Ave, Park Amin-ed-Dowleh 4, Tehran 11547
☎(+98 21) 753 14 33; 753 77 08
✉ THAI DUTO Teheran ⊠ 214140 thai ir
⊖ (+98 21) 753 20 22
Ambassador: H.E. Mr Prida Apirat
Minister Counsellor: Charnvit Buathongchan
First Secretary: Prasidhi Sochiratna
Interpreter (Local-Iranian): Mrs Nasrin Fard-Tamaddon
Commercial Office
2 Mehrdad Alley, Sayeh St, PO Box 19395-4653,
Africa Ave. Teheran
☎(+98 21) 205 73 78/9 ✉ THAI TRADE Teheran
⊠ 214140 thai ir ⊖ (+98 21) 205 43 34
Minister Counsellor: M.R. Chirakom Kitiyakara [97]

31-01890 Embassy of the United Kingdom
PO Box 11365-4474, Teheran 11344
143 Ferdowsi Avenue, Teheran 11344
☎(+98 21) 67 50 11 ⊠ 212493 prod ir
⊖ (+98 21) 67 80 21
Chargé d'Affaires: N W Brown
First Secretary and Deputy Head of Mission: T A Jones
First Secretary Commercial: A Bedford
First Secretary Management and Consul: E Mearns
Second Secretary Commercial: G Langley [98]

75 Manufacturing Companies

75-00500 Daewoo Construction Tehran Office
(a subsidiary of Daewoo Corporation, South Korea)
PO Box 14155-3531, Tehran
Reza Nayebi Street, Vessal Shirazi Avenue, Tehran
☎(+98 21) 640 34 43; 646 35 71
⊖ (+98 21) 646 54 05 [98]

80 Service Companies & Organizations

80-09600 Iran Helicopter Support & Renewal Inc. (ISHRI)
Mehrabad Airport Rd, Mearadj Ave, Teheran 13-873
☎(+98 21) 600 34 14; 600 39 97
⊖ (+98 21) 601 58 62; 95 30 30
Marketing Supervisor: Mr Mojdehi [98]

80-16100 Pars Link Engineering Co.
No 42, 68th Street, Shiraz Janoub, Kordesten Highway,
Teheran 14369
☎(+98 21) 222 66 42; 804 38 49 ⊠ 222203 link ir
⊖ (+98 21) 803 36 58
👥 20
Managing Director: L Yeghnazar
General Manager: R Minassian
■ Design & execution of airports (civil & military). Electrical, electronic & computer systems. [98]

90 Sales Agents, Sales & Liaison Offices

Companies represented:

Claude Lyons Ltd, UK	90-00950
Crompton Vidor, UK	90-00950
Tungstone Batteries Ltd, UK	90-00950

90-00300 CATIC
Representative Office
67 Boustan, 10 St Pasdaran, Teheran
☎(+98 21) 258 33 98 ⊖ (+98 21) 258 33 98 [97]

90-00350 Daewoo Corp Tehran Branch
PO Box 14155-3531, Tehran
☎(+98 21) 646 35 71 ⊠ 2157055 dntx ir
⊖ (+98 21) 646 98 05 [98]

90-00950 Iran Faravar Co. Ltd
PO Box 15815/1715, Teheran
☎(+98 21) 83 24 87; 882 83 06 ⊠ 222647 prgt ir
⊖ (+98 21) 884 45 43
Managing Director: K Azizi
® UK: Claude Lyons Ltd, Crompton Vidor, Tungstone Batteries Ltd [98]

Iraq

Head of State:
Saddam Hussein

10 Head of State

10-10000 Office of the President
Presidential Palace, Kara'adat Marim, Baghdad
President of the Republic: Saddam Hussein
Minister of State for Military Affairs:
Staff Gen Abdul Al-Jabbar Kahil Shanshal [98]

10-20000 Office of the Vice President
Presidential Palace, Kara'adat Marim, Baghdad
Vice President: Taha Muhyi al-Din Ma'ruf
Vice President: Taha Yasin Ramadan [98]

11 Prime Minister

11-10000 Office of the Prime Minister
Presidential Palace, Kara'adat Marim, Baghdad
⊠ 212299 alqasr ik
Prime Minister: Hussein Saddam
Deputy Prime Minister: Tariq Mikha'il Aziz
Deputy Prime Minister: Taha Yasin Ramadan
Deputy Prime Minister: Muhammad Mamza al- Zubaydi [98]

12 Ministries

12-10000 Ministry of Defense
Bab Al-Muadam, Baghdad
☎(+964 1) 888 90 71 ⊠ 212202 mi def ik
Minister: Gen Adb al-Jabbar Tawfiq Shanshal [98]

12-20000 Ministry of Foreign Affairs
Kara'adat Marim, Baghdad
☎(+964 1) 300 91
⊠ 212201 a mi fa ik, 212213 a kharjia ik
Foreign Minister: Muhammad Sa'id Kazim Sahhaf [98]

12-40000 Ministry of the Interior
Baghdad
Minister: Muhammad Zimam Abd al-Razzaq [98]

12-70000 Ministry of Oil
Hay Al-Mustansiriyah, PO Box 19244, Zayoonah, Baghdad
✉ NAFTIYAH - Baghdad
⊠ 212216, 214223, 212229 mi oil ik
Minister of Oil: Amir Muhammad Rashid al-Ubaydi [98]

31 Embassies & Attachés

31-00580 Embassy of Finland
Hai Babel Hse 64, Zuqaq 17 Mah 929,
PO Box 2041 Alwiyah, Baghdad
☎(+964 1) 776 04 88; 776 62 71
✉ FINLANDIA Baghdad ⊠ 212454 finnemb ik [98]

31-00690 Embassy of Greece
PO Box 10003, Baghdad
Jadriyah University Square 2H, 2H2, 2H3
☎(+964 1) 776 65 72 ⊠ 212 479 [98]

31-01110 Embassy of Malaysia
61B/2/35 Al Karada Al Sharqiya, PO Box 1275, Baghdad
☎(+964 1) 776 56 76; 776 59 09; 776 63 88
⊠ 212 452 malwakil ik ⊖ (+964 1) 217 36 19
Ambassador: H E Mr Abdul Latif Bin Awang [98]

31-01370 Embassy of Pakistan
Al-Mansour House No 14 Street No 7, Mohallah 609, Baghdad
☎(+964 1) 541 03 10; 541 51 20 ✉ PAREP Baghdad
⊠ 213 310 parep ik ⊖ (+694 1) 242 87 07
Ambassador: H E Mr Karanatullah K Ghori
Deputy Chief of Mission: Shankat Mogaddam
Defense Attaché: Col Zafar Ali [98]

Ireland

Head of State:
Mrs Mary Robinson

10 Head of State

10-10000 Office of the President
Phoenix Park, Dublin 8
☎ (+353 1) 677 28 15 ⊜ (+353 1) 671 05 29
President of Ireland: Mary Robinson
Secretary to the President: Peter Ryan
Adviser to the President: B Rosney [98]

11 Prime Minister

11-10000 Department of the Taoiseach
Upper Merrion Street, Dublin 2
☎ (+353 1) 668 93 33; 69 93 33
⊠ Name change from internet, TC comfirmed by calling directly. ⊜ (+353 1) 662 10 07; 660 32 81
Prime Minister (Taoiseach): Bertie Ahern
Deputy Prime Minister and Minister for Enterprise and Employment: Mary Harney
Minister for Foreign Affairs: Ray Burke
Finance Minister: Charlie McCreevy [98]

12 Ministries

12-10000 Department of Defence
Coláiste Caoimhin, Mobhi Road, Glasnevin, Dublin 9
☎ (+353 1) 837 99 11; 804 20 00 [E] defence@iol.ie
⊜ (+353 1) 837 79 93
Minister for Defence: David Andrews
Minister of State (Defence): Seamus Brennan
Secretary, Department of Defence: David O'Callaghan
Chief of Staff: Lieut Gen Gerard McMahon
Assistant Chief of Staff: Brig Gen P F Nowlan
Private Secretary: Barbara Burke
Press Liaison Officer: John Hanney
Contracts Branch
Park House, North Circular Road, Dublin 7
☎ (+353 1) 804 20 00 ⊜ (+353 1) 804 28 00
Contracts Manager: Liam Goggin [98]

12-20000 Department of Foreign Affairs
Roinn Gnóthai Eachtracha
Iveagh House, 80 St Stephen's Green, Dublin 2
☎ (+353 1) 478 08 22 ⊠ 93720 estr ei
⊜ (+353 1) 478 14 84
Minister for Foreign Affairs: Ray Burke
Minister of State: Liz O'Donnell [98]

12-50000 Ministry of Transport, Energy and Communication
44 Kildare Street, Dublin 2
☎ (+353 1) 670 74 44 ⊜ (+353 1) 677 31 69
Minister for Transport, Energy and Communication: Mary O'Rourke [98]

12-60000 Department of Tourism & Trade
23 Kildare Street, Dublin 2
☎ (+353 1) 662 14 44
⊜ (+353 1) 676 61 54 (trade enquiries)
Minister for Tourism and Trade: Chris Flood
Secretary, Department of Tourism and Trade: Margaret Hayes [98]

20 Defense Administrations/Services/Agencies

20-00600 Forbairt
(an Agency of the Department of Enterprise and Employment)
Glasnevin, Dublin 9
☎ (+353 1) 808 20 00 ⊠ 832501 olas ei
⊜ (+353 1) 837 01 78
International Programmes Manager: Dr Brian O'Donnell
■ Established by the Government to facilitate the development of Irish industry, to encourage technology/innovation and provide technology services to all firms. [98]

20-01300 Meteorological Service
(Department of Transport, Energy & Communications)
Glasnevin Hill, Dublin 9
☎ (+353 1) 806 42 00 ⊜ (+353 1) 806 42 47
Director: Declan J Murphy ☎ (+353 1) 842 44 11
Head Administration: Eamon Murphy
■ Meteorological & climatological services. Provides meteorological services at Cork, Dublin & Shannon airports. [98]

25 Defense Forces

25-10000 Irish Defence Forces
Defence Forces Headquarters
Parkgate Street, Dublin 8
☎ (+353 1) 837 99 11; 804 20 00 ⊠ 93689 arm ei
⊜ (+353 1) 677 17 26
Chief of Staff: Lieut Gen G J McMahon
☎ (+353 1) 804 26 00
Personal Staff Officer: Comdt S McCann
☎ (+353 1) 804 26 01
Adjutant General: Maj Gen W O'Dwyer
☎ (+353 1) 804 22 60
Personal Staff Officer: Capt P Murphy
☎ (+353 1) 804 22 61
Quartermaster General: Maj Gen D F Stapleton
☎ (+353 1) 804 22 90
Personal Staff Officer: Capt B MacSweeney
☎ (+353 1) 804 22 91
Asst Chief of Staff: Brig Gen P F Nowlan
☎ (+353 1) 804 26 05
Personal Staff Officer: Capt E Smyth
☎ (+353 1) 804 26 06
Director Operations: Col C Dodd ☎ (+353 1) 804 26 10
Director Planning and Research: Col D Taylor
☎ (+353 1) 804 27 40
Director Training: Col W Fitzgerald ☎ (+353 1) 804 26 65
Director Artillery: Col J Prendergast
☎ (+353 1) 804 20 50
Director Signals: Col S F Brennan ☎ (+353 1) 804 60 40
Director Ordnance Corps: Col T M Clancy
☎ (+353 1) 804 60 50
Director Supply and Transport: Col P T Cobbe
☎ (+353 1) 804 60 60
Director Reserve Forces: Col H Johnson
☎ (+353 1) 804 60 80
Director Medical Corps: Col M M Collins
☎ (+353 1) 804 27 90
Director of Military Police Corps: Col R J Campion
Director Public Relations: Lieut Col M Kealy
☎ (+353 1) 804 26 85
Information Officer, Public Relations Section: Capt R McGrath ☎ (+353 1) 804 20 00 ext. 2688; 677 90 18
OIC Military Archives: Comdt P Young
Defence Forces Press Officer: Comdt D Carbery
☎ (+353 1) 804 26 86
Adjutant General's Branch
Parkgate Street, Dublin 8
☎ (+353 1) 804 22 60 ⊜ (+353 1) 836 78 22
Adjutant General: Maj Gen W O'Dwyer
Personal Staff Officer: Capt P Murphy
☎ (+353 1) 804 22 61
OIC Officers Records Section: Comdt W O'Hara
☎ (+353 1) 804 27 60
OIC Enlisted Personnel Section: Comdt D J Reidy
☎ (+353 1) 804 27 70
Deputy Judge Advocate General: Col P J Ghent
☎ (+353 1) 804 22 65
Psychologist: Lieut Col C J Goggin ☎ (+353 1) 804 22 65
Director Army School of Music: Col C O'Brien
☎ (+353 1) 804 64 20
Head Chaplain to Defence Force: Monsr R Field
☎ (+353 1) 804 22 70
Quartermaster General's Branch
Parkgate Street, Dublin 8
☎ (+353 1) 804 22 90
Quartermaster General: Maj Gen D F Stapleton
☎ (+353 1) 804 22 90
Personal Staff Officer: Capt B MacSweeney
☎ (+353 1) 804 22 91
Deputy Quartermaster General: Col J Sreenan
☎ (+353 1) 804 22 92
OIC Q Administration Section: Lieut Col D Lee
☎ (+353 1) 804 22 95
OIC Supply Section: Lieut Col B Goulding
☎ (+353 1) 804 24 77
OIC Accommodation and Barrack Services Section: Lieut Col E Field ☎ (+353 1) 804 24 95
OIC Canteen Section: Lieut Col A Martin
☎ (+353 1) 804 61 60
Eastern Command
Cathal Brugha Barracks, Rathmines, Dublin 8
☎ (+353 1) 804 20 00 ⊜ (+353 1) 497 40 27
General Officer Commanding: Brig Gen C Mangan
☎ (+353 1) 804 62 00
Second in Command and Executive Officer: Col A Murphy
☎ (+353 1) 804 62 01
Adjutant: Lieut Col N McCann ☎ (+353 1) 804 62 02
Southern Command
Collins Barracks, Cork
☎ (+353 21) 51 40 00 ⊜ (+353 21) 50 26 66
General Officer Commanding: Brig Gen D Taylor
☎ (+353 21) 51 41 00
Second in Command and Executive Officer: Col M Walsh
☎ (+353 21) 51 41 01
Adjutant: Lieut Col S Scanlon ☎ (+353 21) 51 41 03
Western Command
Custume Barracks, Athlone, Co Westmeath
☎ (+353 902) 210 00 ⊜ (+353 902) 942 96
General Officer Commanding: Brig Gen J Martin
☎ (+353 902) 21 10 01
Second in Command and Executive Officer: Col J Quilty
☎ (+353 902) 21 10 03
Adjutant: Lieut Col J Moloney ☎ (+353 902) 21 10 04
Quartermaster: Lieut Col P Freeman
☎ (+353 902) 21 10 18
Curragh Command
Curragh Camp, Co Kildare
☎ (+353 45) 44 50 00 ⊜ (+353 45) 44 16 26
General Officer Commanding: Brig Gen F Colclough
☎ (+353 45) 44 50 01
Second in Command and Executive Officer: Col J Vize
☎ (+353 45) 44 50 02
Adjutant: Lieut Col D O'Keefe ☎ (+353 45) 44 50 03
Quartermaster: Lieut Col W J Hayes
☎ (+353 45) 44 50 10
Air Corps
Parkgate Street, Dublin 8
☎ (+353 1) 804 50 00 ⊜ (+353 1) 838 62 10
General Officer Commanding and Director of Aviation: Brig Gen P A Cranfield ☎ (+353 1) 804 25 05
Executive Officer: Lieut Col M O'Malley
☎ (+353 1) 804 25 06
Staff Officer and Administration: Comdt E Scanlon
☎ (+353 1) 804 25 11
Chief Aeronautical Engineer: Lieut Col Ultan G P Lyons
☎ (+353 1) 804 25 07
Headquarters Air Corps Group
Casement Aerodrome, Baldonnel, Co. Dublin
☎ (+353 1) 59 24 93 ⊜ (+353 1) 459 10 36
Group Commander: Col J O'Brien ☎ (+353 1) 804 65 00
Staff Officer and Administration: Comdt C O'Dwyer
☎ (+353 1) 804 65 01
Naval Service
Parkgate Street, Dublin 8
☎ (+353 1) 804 50 00
Flag Officer Commanding: Cdre J J Kavanagh
☎ (+353 1) 804 27 00
Officer Commanding Naval Base and Dockyard: Capt P J McElhinney ☎ (+353 21) 86 47 01
Executive Officer: Cdr P C O'Donnell
☎ (+353 21) 86 47 02
OIC Naval Supervisory Centre: Lieut Cdr D O'Callaghan
☎ (+353 21) 86 48 30
Marine Engineering Superintendent: Capt J F Troy
☎ (+353 21) 86 47 00
Engineer Manager Dockyard: Lieut Cdr K Daly
☎ (+353 21) 86 48 85
Officer Base Support Department: Lieut Cdr P McNulty
☎ (+353 21) 86 47 14
The Military College
Pearse Barracks, Curragh, Co. Kildare
☎ (+353 45) 50 34 ⊜ (+353 45) 44 10 23
Commandant: Brig Gen P Redmond
☎ (+353 45) 44 50 31

OIC Instructional Administration: Capt J O'Loughlin
☎(+353 14) 544 50 32
Commandant Command and Staff School: Col M Mullooly
☎(+353 14) 544 50 60
Commandant Infantry School: Col D Travers
☎(+353 45) 44 50 50
Commandant Cadet School: Comdt S Downes
☎(+353 1) 804 50 40
A/OC UN Training School (Ireland): Lieut Col O MacDonald
☎(+353 1) 804 50 68

Equitation School
McKee Barracks, Dublin 7
☎(+353 1) 804 50 00
Officer Commanding: Col E Campion
☎(+353 1) 804 61 40

Apprentice School
Devoy Barracks, Naas, Co. Kildare
☎(+353 45) 975 56 ⊜ (+353 45) 87 68 71
Officer Commanding: Lieut Col T Whelan [98]

25-41000 Irish Naval Service
Department of Defence, Colaiste Caoimhin, Glasnevin, Dublin 9
☎(+353 1) 804 27 03 ⊜ (+353 1) 677 24 02
Flag Officer Commanding Naval Service: Cdre J J Kavanagh
Staff Officer Operations: Cdr E P Ryan [98]

29 Customs & Excise

29-01000 Office of the Revenue Commissioners
Dublin Castle, Dublin 2 [98]

31 Embassies & Attachés

31-00090 Australian Embassy
Fitzwilton House, Wilton Terrace, Dublin 2
☎(+353 1) 676 15 17 ✉AUSTEMBA Dublin
⊜ (+353 1) 668 52 66
Ambassador: H.E. Mr Edward J. Stevens
First Secretary: David Morris
First Secretary Consular and Administration: Peter Hurditch
Personal Assistant to the Ambassador: Ms Brenda Waters [98]

31-00100 Embassy of Austria
15 Ailesbury Court, 93 Ailesbury Road, Dublin 4
☎(+353 1) 269 45 77; 269 14 51
⊜ (+353 1) 283 08 60
Ambassador: Dr Michael Breisky
Minister-Counsellor: Dr Michael Postl
Counsellor and Consul: Stefan Bagyura
Military/Defence Attaché: Brig Hans Rüdiger Sulzgruber (Resident in London) ☎(+44 171) 235 37 31
Cultural Counsellor: Dr Emil Brix (Resident in London) ☎(+44 171) 584 86 53

Office of the Commercial Counsellor
Merrion Hall, Strand Road, PO Box 2506, Sandymount, Dublin 4
☎(+353 1) 283 04 88 ⊠ 93334 ahst ei
⊜ (+353 1) 283 05 31
Commercial Counsellor: Dr Gerd Dückelmann-Dublany
Commercial Attaché: Mag. Herwig Schenk [98]

31-00150 Embassy of Belgium
2 Shrewsbury Road, Ballsbridge, Dublin 4
☎(+353 1) 269 20 82 ✉AMBABEL Dublin
⊠ 933 22 belg ei ⊜ (+353 1) 283 84 88
Ambassador: H.E. Louis Fobe
Counsellor: Désiré Sergijsels
Attaché Consular & Administration: Jean-Michel Colas
Attaché Economic & Commercial: Pierre-Jean Van Linthoudt ☎(+353 1) 283 94 02 [97]

31-00310 Embassy of Canada
65 St. Stephen's Green, Dublin 2
☎(+353 1) 478 19 88 [E] cdnembsy@iol.ie
⊜ (+353 1) 478 12 85
Ambassador: H E Michael Phillip
Commercial Counsellor: Mrs Monica Heron
First Secretary: Philip Pinnington [98]

31-00590 Embassy of France
36 Ailesbury Road, Ballsbridge, Dublin 4
☎(+353 1) 260 16 66 ⊜ (+353 1) 283 01 78
Ambassador: H.E. Mr Francois Mouton
Deputy Chief of Mission: Mrs Genevieve Galarneau-Mack
Defence Attaché: Lt Col Claude Bonifaix
First Secretary: Gerard Le Corre

Office of the Commercial Counsellor
Clanwilliam Place, Marine House, Dublin 2
☎(+353 1) 668 07 77 ⊜ (+353 1) 661 72 91
Commercial Counsellor: Francis Chambon
Commercial Attaché: Jean Vaury [97]

31-00660 Embassy of the Federal Republic of Germany
31 Trimleston Ave, Booterstown, Blackrock, Co. Dublin 4
☎(+353 1) 269 30 11 ⊠ 93809 aadu ei
⊜ (+353 1) 269 39 46
Ambassador: H.E. Mr Horst Pakowski
Minister Counsellor/Deputy Chief of Mission: Wolfgang Seiwert
First Secretary Political & Press: Ms Sibylle Katharina Sorg
First Secretary Cultural & Consular: Carsten Meyer-Wiefhausen
Defence Attaché: Gen. Eckart Fischer (Resident in London) [97]

31-00830 Embassy of India
6 Leeson Park, Dublin 6
☎(+353 1) 497 08 06; 08 43; 09 59; 09 84/7
⊠ 30670 dtlx ei ⊜ (+353 1) 497 80 74
Ambassador: H.E. Mr Her Charan Singh Dhody
Counsellor, P&E: Lal Dingliana
Second Secretary, Head of Chancery: M.C. Pandey
Second Secretary, CON: Ram Parkash Sehgal [97]

31-00890 Embassy of Italy
63-65 Northumberland Road, Ballsbridge, Dublin 4
☎(+353 1) 660 17 44 ⊠ 93950 ital ei
⊜ (+353 1) 668 27 59
Ambassador: H E Dr Ferdinando Zezza
First Counsellor: Dr Fabio Cristiani

Italian Trade Office
16 St. Stephen's Green, Dublin 2
☎(+353 1) 676 78 29 ⊜ (+351 1) 676 77 87
Acting Director: Vincent O'Callaghan [98]

31-00920 Embassy of Japan
Nutley Building, Merrion Centre, Nutley Lane, Dublin 4
☎(+353 1) 269 42 44; 269 40 33; 269 29 80
✉TAISHI Dublin 4 ⊠ 93790 tasi ei
⊜ (+353 1) 283 87 26
Ambassador: H.E. Mr Yuzo Hatano
Minister, Counsellor: Minoru Yamashita
First Secretary: Yoshiro Morino
First Secretary: Toyomasa Nishizawa [98]

31-01440 Embassy of Portugal
Knocksinna House, Knocksinna, Foxrock, Dublin 18
☎(+353 1) 289 44 16; 289 33 75 ⊠ 30 777 port ei
⊜ (+353 1) 289 28 49
Ambassador: H E Manuel Lopes da Costa
First Secretary, Deputy Head of Mission: Dr Miguel Alves Rita
Commercial Counsellor: Dr Carlos Costa
Defence Attaché: Cdr António Bettencourt [98]

31-01730 Swedish Embassy
13-17 Dawson Street, Dublin 2
☎(+353 1) 671 58 22 ✉SVENSK Dublin 2
⊠ 93341 swed ei ⊜ (+353 1) 679 67 18
Ambassador: H.E. Mr Per Jodahl
Counsellor and Deputy Chief of Mission: Per Lindstrom
Defence Attaché: Col Gustaf Diurlin (Resident in London)
Third Secretary Admin: Kajsa Al Jaouni [98]

31-01740 Embassy of Switzerland
6 Ailesbury Road, Ballsbridge, Dublin 4
☎(+353 1) 269 25 15; 15 66 ⊜ (+353 1) 283 03 44 [97]

31-01890 Embassy of the United Kingdom
29 Merrion Road, Dublin 4
☎(+353 1) 205 37 00 ⊜ (+353 1) 205 38 85
Ambassador: H.E. Mrs Veronica Sutherland (CMG)
Deputy Head of Mission: R I Clarke
Defence Attaché: Col S D Lambe (CBE)
First Secretary Commercial: R N B Baker
Public Information Officer: Ms C. Tuhey [98]

75 Manufacturing Companies

75-00020 Adams-Russell BV
(an Adams-Russell Co.)
Youngline Centre, Pouladuff Rd, Cork
☎(+353 21) 31 12 66
Managing Dir: Brian McCoy
■ European microelectronics sub-contracting source, producing a number of quality RF/microwave products for several Adams-Russell divisions, as well as offering a cost competitive sub-contracting manufacturing capability. [97]

75-00040 AlliedSignal Ireland Limited Precision Forging and Machining
(a subsidiary of AlliedSignal Aerospace, USA)
Western Extension Industrial Park, Unit 328, Cork Road, Waterford
☎(+35 3) 517 64 11 ⊜ (+35 3) 517 88 18
Managing Director: Jim Collins
■ Manufacture and sale of blades and vanes. → 90 [98]

75-00070 BEFAB Safeland Ltd
(subsidiary of Datron Inc, USA)
Bay 15 Industrial Estate, Shannon, Co. Clare
☎(+353 61) 47 18 44 ✉SAFELAND Shannon ⊠ 72142
⊜ (+353 61) 47 14 63
♣ I£ 3,000,000 (1996) 👥 25 (1997) 🏭 11,000 sq m
Managing Dir: Alec O'Sullivan
Commercial Mgr: John Ryan
Technical Mgr: Richard Chawke
■ Provide high quality engineered products for the aerospace, commercial aviation & industrial markets. Repair & overhaul of aircraft safety equipment.
® USA: Datron Inc. [98]

75-00085 DDC Ireland Ltd
(a subsidiary of ILC Data Device Corporation)
Cork Business & Technology Park, Model Farm Road, Cork
☎(+353 21) 34 10 65 ⊜ (+353 21) 34 15 68
👥 55 🏭 25,000 sq ft
Director: Ms Carmel Mulcahy
Sales and Marketing Director: John Gasking
Marketing Communications Manager: Lawrence M Margulies
■ Manufacture hybrid microelectronics, data converters, MIL-STD-1533 products, power hybrids & solid state power controllers, military sub-systems. [98]

75-00100 ESSCO Collins Ltd
(a subsidiary of Electronic Space Systems Corporation, USA)
Kilkishen, Co. Clare
☎(+353 61) 36 72 44 [E] essco@tiac.net
[W] http://www.esscocollins.com ⊜ (+353 61) 31 10 44
♧ I£ 400,000 (1997) ♣ I£ 3,600,000 (1996)
👥 40 (1997) 🏭 30,000 sq ft
Chairman: Albert Cohen
Managing Director: Louis G Orsatti
Operations Director: Liam Brown
International Sales Mgr: Michael O'Connell
■ Design & manufacture radomes (metal/dielectric space frame, sandwich, solid laminate, air supported, shipboard) & antenna systems (aircraft, communications, radar) for satellite communications, radio astronomy, precision pointing & tracking applications. [98]

75-00120 GD Protective Clothing and Equipment Limited
Tallaght Business Park, Dublin 24
☎(+353 1) 451 14 33 ⊜ (+353 1) 452 42 94
👥 55
Managing Dir: John Byrne

Sales & Marketing Dir: Declan Bradbury
Operations & Technical Dir: Alec Feldman
■ Fire fighting clothing & equipment. Flame resistant station wear clothing. [97]

75-00250 **Kentree Ltd**
Kilbrittain, Co. Cork
☎ (+353 23) 497 08 [E] kentree@indigo.ie
⊖ (+353 23) 498 01
I£ 118,750 (1997) I£ 780,000 (1995) 15 (1997)
1,200 m2
Managing Director: John H McGuire
Operations Director: Elgan Loane
Marketing Manager: John McGuire
Purchasing Manager: Ms Lucy O'Donovan Lynch
■ Manufactures a range of remote control equipment designed originally for explosive ordnance disposal. Techniques employed allow for special features to suit client requirements and development for changing applications in military police situations and hostile environments. [98]

75-00520 **ROR Rockwell (Ireland) Ltd Heavy Vehicle Systems**
(a division of Rockwell International Corp, USA)
Newhall, Naas, Co. Kildare
☎ (+353 45) 43 34 35 ⊖ (+353 45) 43 23 56
Sales and Technical Manager: Ray Lowney
Office/Clerical Specialist: Ms Kate Nolan
® USA: Rockwell International Corp. [98]

85 Stockists, Distributors, Spare Parts Suppliers

85-02000 **THORN Lighting Ireland Ltd**
320 Harolds Cross Road, Dublin 6
☎ (+353 1) 492 28 77 ⊖ (+353 1) 492 27 24
Chief Executive: Sean Hayes [98]

90 Sales Agents, Sales & Liaison Offices

Companies represented:

Datron Inc., USA 75-00070
EEV Ltd, UK. 90-00950
Rockwell International Corp., USA . . . 75-00520

Sales agents/representatives listed elsewhere:
AlliedSignal Ireland Limited 75-00040

90-00150 **British Steel Ireland Ltd**
(a subsidiary of British Steel Engineering Steels, UK)
Leeson Court, 86-88 Lower Leeson Street, Dublin 2
☎ (+353 1) 661 67 73 ⊖ (+353 1) 676 54 13 [98]

90-00950 **Industrial Engineering Designers Ltd**
Hayestown, Navan, Co Meath
☎ (+353 46) 243 09 ⊖ (+353 46) 249 59
Marketing Dir: W. Evans
® UK: EEV Ltd [98]

Israel

Head of State:
Ezer Weizman

10 Head of State

10-10000 **Office of the President**
Beit Hanasi, 3 Hanassi Street, 92188 Jerusalem
☎ (+972 2) 70 72 11 ⊖ (+972 2) 66 04 45
President of the State of Israel: Ezer Weizman
Director General President's Office: Arieh Shomer
Military Assistant: Brig Gen Shimon Heifetz [97]

11 Prime Minister

11-10000 **Office of the Prime Minister**
3 Kaplan Street, 91919 Hakirya, Jerusalem
☎ (+972 2) 70 55 55 ⊖ (+972 2) 66 48 38
Prime Minister: Binyamin Netanyahu
Deputy Prime Minister: Moshe Katsav
Deputy Prime Minister: Rafael Eitan
Deput Prime Minister: David Levy
Deputy Prime Minister: Hammer Zevulun
Finance Minister: Ya'akov Ne'eman
Minister of Transport: Yitzak Levi
Minister of the Interior: Eli Suissa
Trade and Industry Minister: Natan Sharansky
National Infrastructure Minister: Ariel Sharon [98]

12 Ministries

12-10000 **Ministry of Defence**
7 "A" Street, 67659 Hakirya, Tel Aviv
☎ (+972 3) 20 51 44; 20 52 19; 20 52 20
▣ 471434 mod il
Defence Minister: Yitzhak Mordechai [98]

12-20000 **Ministry of Foreign Affairs**
Hakirya, Romema, 91950 Jerusalem
☎ (+972 2) 530 31 11 ▣ 25223; 25250; 25333 hutz il
⊖ (+972 2) 530 33 67
Minister of Foreign Affairs: David Levy
Deputy Chief: Mrs Amira Arnon
Director General: Eytan Bentsur
Spokesman: Daniel Shek [98]

12-40000 **Ministry of the Interior**
2 Kaplan Street, Hakirya, 91061 Jerusalem
☎ (+972 2) 66 01 51 ▣ 25291 ispolil
Minister: Eli Suissa [98]

12-50000 **Ministry of Public Security**
(† Israel Police Ministry)
PO Box 18182, 91181 Jerusalem
☎ (+972 2) 530 80 87 ▣ 25291 ispol il
⊖ (+972 2) 582 67 70
Minister for Police: Avigdor Kahalan
Director General: Emi Palant
Inspector General: Assaf Hefetz
Head Information Service and International Relations Division: Ms Naomi Shapira [98]

12-60000 **Ministry of Science and Technology**
PO Box 7182, 61070 Tel-Aviv
☎ (+972 3) 691 67 86 ⊖ (+972 3) 696 28 38
Minister: Binyamin Netanyahu [98]

15 Parliament

15-10000 **The Knesset**
Hakirya, 91950 Jerusalem
☎ (+972 2) 675 33 69 [E] sifra2@netvision.net.il
⊖ (+972 2) 566 27 33
Speaker: Dan Tichon
Director Speaker's Office: Ms Yardena Meler-Horowitch
☎ (+972 2) 675 33 78
Knesset Library Director: Ms Naomi Kimhi
☎ (+972 2) 675 32 45
Public Relations and Press Officer: Giora Pordes
☎ (+972 2) 675 33 69
Foreign Affairs Defense Committee
Hakirya, 91000 Jerusalem
☎ (+972 2) 675 31 23 ⊖ (+972 2) 675 31 00
Chairman: Vzi Landav [98]

31 Embassies & Attachés

31-00090 **Australian Embassy**
Beit Europa, 37 Shaul Hamelech Bvd, 64928 Tel Aviv
☎ (+972 3) 695 04 51 [E] aussie@inter.net.il
⊖ (+972 3) 696 84 04
Ambassador: H E Mr Ian Wilcock
First Secretary: K Muller
First Secretary and Consul: W Helker
Second Secretary: Ms V Plowman
Attaché: J Hawson [98]

31-00100 **Embassy of Austria**
11 Rehov Herman Cohen, PO Box 11095, 61110 Tel Aviv
☎ (+972 3) 524 61 86 [E] avtambta@netvision.net.il
⊖ (+972 3) 524 40 39
Ambassador: H E Dkfm Dr Herbert Kröll
Office of the Commercial Counsellor
198 Hayarkon Street, Tel Aviv
☎ (+972 3) 522 66 41 ⊖ (+972 3) 522 40 71
Commercial Counsellor: Dkfm Hans Kaufmann [98]

31-00150 **Embassy of Belgium**
266 Rehov Hayarkon, 63504 Tel Aviv
☎ (+972 3) 605 41 64/7 ✉ AMBABELGE Tel Aviv
▣ 34211 amblg il ⊖ (+972 3) 546 53 45
Ambassador: H.E. Mr Marc Otte
First Secretary: Carine Petit
Attaché Consular & Admin Affairs: Leo Cortens
Commercial Canvasser Walloon Region: Frederic Delbart
☎ (+972 3) 604 33 40 ☎ (+972 3) 544 59 33 (fax)
Commercial Canvasser Flemish Region: Edouard Renson
☎ (+972 3) 604 13 78 ☎ (+972 3) 604 20 19 (fax) [97]

31-00310 **Embassy of Canada**
PO Box 6410, 61063 Tel Aviv
220 Rehov Hayarkon, 63405 Tel Aviv
☎ (+972 3) 527 29 29 ⊖ (+972 3) 527 23 33
Ambassador: H.E. Mr David Berger
Deputy Chief of Mission: Dave Viveash
Canadian Forces Attaché: Col T. Humphries
Counsellor Commercial: Ms Ruth Zeisler [97]

31-00350 **Embassy of Chile**
7 Havakook Street, 63505 Tel Aviv
☎ (+972 3) 602 01 30 [E] echileil@inter.net.il
▣ 342189 emchi il ⊖ (+972 3) 602 01 33 [98]

31-00460 **Royal Danish Embassy**
23 Bnei Moshe Street, PO Box 21080, 61210 Tel Aviv
☎ (+972 3) 544 21 44/5 ✉ AMBADANE Tel Aviv
▣ 33514 ambdk il ⊖ (+972 3) 546 55 02
Ambassador: H.E. Mr Jakob Rytter
Minister Counsellor: Finn Abrahamsen
Commercial Officer: Ulla H. Larsen [97]

31-00535 **Embassy of Estonia**
6 Hadassah Street, Tel-Aviv
☎ (+972 3) 523 16 58 ⊖ (+972 3) 527 93 80
Honorary Consul: Joseph Shekel [98]

Contact Reference

31-00580 Embassy of Finland
IBN Gvirol 2, 8th Floor, PO Box 20013, 64077 Tel Aviv
☎ (+972 3) 695 05 27 ☒ 33552 finla il
⊖ (+972 3) 696 63 11
Ambassador: H.E. Mr Arto Tanner
First Secretary: Mrs Heidi Schroderus-Fox
Commercial Attaché: Tamir Rotor
Attaché Consular & Cultural Affairs:
Ms Katrina Siebenberg 97

31-00660 Embassy of the Federal Republic of Germany
PO Box 16038, 61160 Tel Aviv
3 Daniel Frisch Street, 64731 Tel Aviv
☎ (+972 3) 693 13 13 ☒ 33621 germa il
⊖ (+972 3) 696 92 17
Ambassador: H E Dr Theodor Wallau
☎ (+972 3) 693 13 13
Defense Attaché: Col Jürgen Eigenbrod
Assistant Defence Attaché: MSgt Franz Haering
☎ (+972 3) 693 13 47 98

31-00690 Embassy of Greece
35 Sderot Shaul Hemelech, PO Box 33631, 64927 Tel Aviv
☎ (+972 3) 695 97 04; 696 61 27 ☒ 341227
⊖ (+972 3) 695 54 23 98

31-00890 Embassy of Italy
Asia House, 4 Weizman Street, 64239 Tel Aviv
☎ (+972 3) 696 42 23; 696 42 25 ☒ 342664 itadi il
⊖ (+972 3) 691 84 28
Ambassador: H.E. Dr Giuseppe Panocchia
Deputy Chief of Mission: Dr Emanuele Punzo
Military, Naval & Air Attaché: Col Giangabriele Carta
Asst Air Attaché: Lt Col Pierluigi Zanetti
First Secretary: Dr Emanuele Farruggia
Commercial Counsellor: Dr Cristiano Maggipinto 97

31-01190 Embassy of Mexico
3 Bograshov St., 63808 Tel Aviv
☎ (+972 3) 523 03 67/9 ⊖ (+972 3) 523 73 99
Counsellor: Eduardo Giles 97

31-01270 Royal Netherlands Embassy
PO Box 33 379, 61333 Tel Aviv
Beit Asia 4, Rehov Weizman, 64239 Tel Aviv
☎ (+972 3) 695 73 77/9 ☒ 342180 neth il
⊖ (+972 3) 695 73 70
Ambassador: H E Mr C M J Kröner
Defence Attaché: Col H J Vandeweijer 98

31-01340 Royal Norwegian Embassy
PO Box 6800, IL-61067 Tel Aviv
☎ (+972 3) 544 20 30; 544 20 33 ☒ 33417 noramb il
⊖ (+972 3) 544 20 34
Ambassador: H E Mr Svein O Saether
Counsellor: Sten Arne Rosnes
First Secretary: Ms Hilde Haraldstad 98

31-01380 Embassy of the Republic of Panama
10 Rehov Hei Beiyar, Kibar Hamedina, PO Box 21 260, 61210 Tel-Aviv
☎ (+972 3) 696 08 49 ☒ 33511 pancl il
⊖ (+972 3) 691 00 45
Ambassador: H E Prof Paulino Romero C 98

31-01480 Embassy of Romania
24 Rehov Adam Hacohen, Tel Aviv
☎ (+972 3) 523 00 66 ☒ 33867 romcom il
⊖ (+972 3) 524 73 79
Ambassador: H E Dr Cheorghe Popescu
Defence Attaché: Capt Cdr Aurel-ion Stanciu (Airforce)
☎ (+972 3) 523 07 98
Commercial Officer: Dorin Refea 98

31-01485 Russian Embassy
Hayarkon 120, 63573 Tel Aviv
☎ (+972 3) 522 67 33 E embrus@goldnet.net/il
⊖ (+972 3) 522 67 13 98

31-01680 Embassy of Spain
3 Rehov Daniel Frish, The Tower, 16th Floor, 64731 Tel Aviv
☎ (+972 3) 696 52 18; 52 19; 52 20
⊖ (+972 3) 696 52 17
Ambassador: H.E. Mr José Luis Crespo De Vega
Counsellor: Fernando Valderrama
Commercial Officer: José Garcia Nuno 97

31-01730 Swedish Embassy
4 Weizman Street, 64239 Tel Aviv
☎ (+972 3) 695 81 11 E sweenbtlv@trendline.co.il
☒ 33 650 svnsk il ⊖ (+972 3) 695 81 16
Ambassador: H E Mr John Hagard
Second Secretary: Henrik Cederin ☎ (+972 3) 522 96 63 98

31-01890 Embassy of the United Kingdom
192 Hayarkon Street, 63405 Tel Aviv
☎ (+972 3) 524 91 71; 524 91 78
⊖ (+972 3) 524 33 13 (Commercial); 527 85 74
Ambassador: H E Mr David Manning
Deputy Head of Mission and Consul General: C J B White
Defence and Military Attaché: Col C S Wakelin
Naval and Air Attaché: Wg Cdr M Whitehouse (RAF)
First Secretary Commercial: W Magor

Consulate of the United Kingdom
6th Floor Migdalor Building, 1 Ben Yehuda Street, Tel Aviv
☎ (+972 3) 510 01 66 ☒ 371411 brita il
⊖ (+972 3) 510 11 67
British Consul: J Fielder 98

31-01891 British Consulate General
Nashashibi Street 19, PO Box 19690, Jerusalem
Nashashibi Street 19, Jerusalem
☎ (+972 2) 582 82 81; 582 84 82 ☒ 26260 ukjsm il
⊖ (+972 2) 532 23 68
Consul General: R J Dalton
Deputy Head of Mission: C P Winnington-Ingram
HM Consul General: Ms Kim Bowers
Commercial Information Officer: Yacoub Dahdal 98

31-01900 Embassy of the United States of America
71 Hayarkon Street, PSC 98 Box 100, 63405 Tel Aviv
☎ (+972 3) 517 43 38 ⊖ (+972 3) 517 32 27
Ambassador: H E Mr Martin Indyk
Deputy Chief of Mission: James A Larocco
Defense Attaché: Col John V Siebert (USAF)
Commercial Counsellor: Barry Friedman
Scientific Attaché: David W Mulenex 98

31-01950 Embassy of Venezuela
Kaufman 2, Textile Center 16th Floor, 68012 Tel Aviv
☎ (+972 3) 517 62 87; 517 62 88; 517 62 89
E emven@netmedia.net.il ☒ 342 172 venez il
Ambassador: H E Mr Freddy Alvarez Yanes
Counsellor: Jorge Mendez Montilla
Second Secretary: Yaneth Arocha Rivas
Assistant: Ms Ruth Macudu 98

40 Government Defense Sales/Procurement Organizations

40-01900 SIBAT - Foreign Defense Assistance & Defense Export
Ministry of Defense, Hakirya, 61909 Tel Aviv
☎ (+972 3) 697 55 85; 57 62; 55 15
⊖ (+972 3) 697 67 24
👥 120
Director General: David Shoval ☎ (+972 3) 697 55 15
Deputy Dir Marketing: Israel Meiri ☎ (+972 3) 79 01
Deputy Dir Surplus Sales: Alex Goldengorn
☎ (+972 3) 697 72 07
Head of Export Services Center: Rafi Bonovitsky
☎ (+972 3) 697 63 76
Head Latin America Division: Jacob Rubinovitch
☎ (+972 3) 697 79 25
Head North America Division: Haim Barzilai
☎ (+972 3) 697 79 05
Head Far East Division: Amir Gal ☎ (+972 3) 697 79 19
Head Central & Eastern Europe Division: Rudi Schmidt
☎ (+972 3) 697 79 29
Head Africa Division: Arie Nordan ☎ (+972 3) 697 79 11
Head Mediterrenean Countries Division: Izhak Soroka
☎ (+972 3) 697 79 39
Head South East Asia Division: Jacob Winternitz
☎ (+972 3) 697 79 20
Head European Division: Haim Aviv ☎ (+972 3) 697 79 28
Marketing & Advertising Services Dir:
Mrs Rachel Naidek-Ashkenazi ☎ (+972 3) 697 55 15
■ Export licencing, of defence export-systems, consulting, know-how & training services, promotion & marketing activities. Sale of IDF phased-out equipment; quality control of military equipment via the IDF Testing Unit. Quality assurance. 97

55 Government-owned Industry & Production Facilities

Israel Aircraft Industries Ltd 75-10200

55-01800 Rafael Armament Development Authority
PO Box 2250, 31021 Haifa
☎ (+972 4) 877 69 65; 879 54 19 ☒ 471508 vered il
⊖ (+972 4) 879 46 57
♣ US$ 490 000 000 👥 4 600
President: Brig.Gen. Itzachk Gat (Retd)
Asst D-G for Research & Dev.: Dr Naftali Amit
Asst D-G for Marketing & Bus.Dev.: Ran Galli
Asst D-G for Finance & Economics: Dan Schweppe
Spokesman: Noah Shachar
■ Guided & unguided systems (missiles & rockets), special-purpose computers, EW systems, electro-optic & thermal imaging systems & ballistic protection training simulators. → 75 97

60 Research Organizations, Institutes & Test Establishments

60-00960 Israel Institute for Biological Research (IIBR)
PO Box 19, 74100 Ness-Ziona
☎ (+972 8) 938 16 56 E mambi@iibr.gov.il
W http://www.iibr.gov.il ⊖ (+972 8) 940 14 04
👥 320
Director: Dr Avigdor Shafferman ☎ (+972 8) 938 15 95
Operatons and Technical Director: Dr Amram Golombek
■ Manufacture of ETKplus explosives testing kit enabling simple, rapid and reliable detection of persons engaged in preparing, handling or carrying explosive charges and allows for fast screening of suspects in the field. The kit aids in detection, through exposure and proof of contact contamination of various surfaces by explosives. 98

60-15000 Professional School of Security (PSOS Ltd)
(a subsidiary of PubLiss International Trade Co Ltd)
Beit Merkazim, PO Box 12273, 46733 Herzlia Pituach
☎ (+972 9) 55 29 02 ⊖ (+972 9 55 29 93
👥 12
Managing Director: Asher Cohen
Dir Sales & Marketing: José Benauram
Chief Instructor: Nadav Sharow
■ Specialist in training security personnel. → 80 97

65 Industry, Trade & Armed Forces Associations

65-00050 I.T.P. Ltd
25 Lech Street, 51200 Bnei Brak
☎ (+972 3) 578 09 01 E ronyg@itpgroup.com
⊖ (+972 3) 578 09 11
👥 10
General Manager: Ron Grynholc
■ Represents Israeli companies abroad and foreign companies within Israel concerning defence and security equipment. Consultancy and training services. 98

70 Financial Holding Groups, Parent Companies

70-00700 Elron Electronic Industries Ltd
Advanced Technology Center, PO Box 1573, 31015 Haifa
☎ (+972 4) 854 50 00 [E] elron@alron.net
[W] http://www.elron.com ⊜ (+972 4) 855 02 48
♧ US$ 200,000,000 (1996)
♣ US$ 1,200,000,000 (1996)
Chairman and Chief Executive Officer: Uzia Galil
☎ (+972 4) 854 50 00
Senior Vice President Business Development: Dr Jacov Ben-Zvi
Vice President Technologies: Abraham Peri
Vice President Information Technology: Arie Amit
Vice President Finance and Corporate Secretary: Doron Birger
■ Elron's global businesses are conducted through a group of affiliated companies, in the fields of medical diagnostic imaging, advanced defence electronics, communication, information technologies, semiconductors, networking services and internet software products.
Subsidiaries with defence business:
Elbit Ltd → 75 [98]

75 Manufacturing Companies

Rafael 55-01800
Syscon Engineering 80-01940

75-00200 Achidatex Nazareth Elite (1977) Ltd
PO Box 2156, 52121 Ramat Gan
14 Hamefalsim Street, 52121 Petach Tiqva
☎ (+972 3) 921 34 31 ⊜ (+972 3) 921 34 36
♣ US$ 16,000,000 (1996) 👥 140 (1996)
🏭 8,000 (1996)
President and Purchasing Director: Fredy Davidovitz
Managing Director: Abraham Hatzor
Marketing Director: Arik Benari
Marketing Director: Amos Zunrez
Operations Director: Yosef Goldenberg
■ Production of dry storage covers, protective vests, military equipment, tarpaulins and tents. [98]

75-00250 Adam Medical Products Ltd
(† Migada Ltd Medical Equipment Ind.)
(a member of the TEVA Group)
Light Industrial Zone, PO Box 2, 77100 Ashdod
☎ (+972 8) 851 02 22 ⊜ (+972 8) 851 02 80
♣ US$ 60 000 000 👥 400
President: Dr Oded Awerbuch
Sales & Marketing Dir: Zvi Levinhar
Operations & Technical Dir: Mr Shreiber
Business Development Mgr: David Ben Ami
■ Disposable medical products. Disposable solution & blood infusion sets, critical care disposal devices, I.V. cannula, dialysis blood lines. I.V. solutions, pour bottles, surgical sutures, disinfectants. [97]

75-00400 Advanced Technology Ltd
PO Box 58180, 61581 Tel Aviv
☎ (+972 3) 548 35 55 [E] atlaran@inter.net.il
[W] http://www.tekem.co.il ⊜ (+972 3) 649 99 90
♣ US$ 43,500,000 (1995) 👥 500 🏭 6,500 sq m
Managing Director: Yehezkel Zeira
Marketing Director: Zvi Aran
■ Software developer and systems house. Currently involved in the development and deployment of projects in Israel, Europe, USA and the Far East. Main business areas include Management Information Systems; Real Time and Communications Systems; Defence Systems Applications; Representation and Local Support for Leading International Software Vendors. Capabilities of the Real Time and Defence Systems Division includes air defence and air traffic control systems; C3I systems; communications and networking; Geographical Information Systems (GIS), computer graphics and mapping applications; simulators and trainers; operations research and algorithms for ballistics, tracking and simulations; Automated Guided Vehicles (AGV) and port automation systems; embedded computer systems programming, e.g., fire control systems; conversions of legacy software running on older operating systems to modern languages and operating systems; large logistics systems, data storage and retrieval systems.
Certified to ISO 9001 for designing, developing, integration, installation, maintenance and marketing of software and computer embedded systems. → 80 [97]

75-00630 Aitech Software Engineering Ltd
(a subsidiary of Aitech Defense Systems Inc.)
PO Box 2088, 46120 Herzliya B
☎ (+972 52) 55 00 24 ⊜ (+972 52) 54 43 15
President: Tsvi Lavi
■ Real-time embedded board support packaging for VxWORKS, PSOS and ONT. [98]

75-01400 Microwave Engineering Group (MEG)
(a division of RCM Ltd)
31 Habarzel Street, Ramat-Hahayal, PO Box 39844, 61397 Tel-Aviv
☎ (+972 3) 648 78 85 ⊜ (+972 3) 648 72 90
👥 6 (1997) 🏭 250
Managing Director: Eddie Wiegenfeld
Director of Engineering: George Lazar
Marketing Director: Mobhe Szekatch
Purchasing and Operations Director: M Bonjack
■ Consulting, custom products design, project management, manufacturing, preparation of manuals, setting up training and product service support. The areas of activity in the microwave field include: Radar, EW, communications, navigation and fast growing commercial applications. [98]

75-01680 Arco Electrical Industries Ltd
(a subsidiary of the ARIEL Group)
7 Shapiro St, New Industrial Park, PO Box 4565, 75144 Rishon Le-Zion
☎ (+972 3) 963 08 08 ⊜ (+972 3) 961 46 75
🏭 10,000
General Manager: Zeki Balli
■ Plastics injection moulding and polyester pressing, light and heavy metal processing, CNC sheet metal forming and punching, electrical wire assembly, electrostatic painting, coating, electromechanical final assembly, double insulated polycarbonate boxes, Reinforced polyester cabinets, polycarbonate traffic signal lights, 3-pole HRC fuse switches, prism clamps and BusBar terminals, and N-type DO fuse bases. [98]

75-01700 Armored Vehicle's Spares Ltd (AVS)
Industrial Area, PO Box 630, 42815 Pardesia
☎ (+972 9) 64 64 72 ⊜ (+972 9) 64 64 73
♣ US$ 3 500 000 (1995) 👥 50 🏭 1 800 sq m
President: Eli Cohen
Operations & Technical Dir: Benny Posti
Managing Dir: Ariel Hazut
Purchasing Mgr: Avichai Wienstein
Public Relations Mgr: Ms Laura Shuck
■ Track Shoes including T-130 (M-113), T-132 (M-107), T-136 (M-109), T-91 (M-41/59), T-97 (M-48/60), T-142 (M-60), AMX13 and more. [97]

75-01870 Arotec Ltd
33 Bilu St, PO Box 275, 70700 Gedera
☎ (+972 8) 859 18 17 ⊜ (+972 8) 859 69 42
♣ US$ 600 000 (1995) 👥 12 🏭 300 sq m
General Mgr: Oded Arkin
■ Develops and produces modules and subsystems in Hybrid & SMT Technologies in accordance with the customer electronic design in analogue, digital, power and RF ranges. → 80 [97]

75-02100 Ashot Ashkelon Industries Ltd
(a subsidiary of Israel Military Industries Ltd)
PO Box 21, 78100 Ashkelon
☎ (+972 7) 72 15 11 ⊜ (+972 7) 72 81 67
♣ US$ 45,000,000 (1995) 👥 400 (1996)
🏭 300,000 sq ft
President: Dr Robert Gero
Managing Dir: Michael Erlich ☎ (+972 7) 72 91 86
Operations & Technical Dir: Dan Katz
Purchasing Mgr: Danny Rihan ☎ (+972 7) 72 15 60
■ Design, develop & manufacture gear based products for the aerospace & automotive industries for commercial & military applications. Three manufacturing plants - aerospace systems, land systems & sintered (heavy metal) products.
Overseas Offices:
Ashot USA Inc. [97]

75-02400 Astronautics C.A. Ltd
(a subsidiary of Astronautics Corp of America)
23 Hayarkon Street, PO Box 882, 51261 Bnei-Brak
☎ (+972 3) 579 15 55 ⊜ (+972 3) 70 44 04
♣ US$ 2,000,000 👥 390
General Manager: David Ronen
■ Design & manufacture of military electronics systems for airborne, naval & land applications. Display systems (processors & multifunction displays), cockpit indicators, air-data computers, tank fire-control systems, operating control consoles, specialised test equipment.
® USA: Astronautics Corporation of America [98]

75-02500 Automotive Industries Ltd
(a subsidiary of Automotive Equipment Ltd)
PO Box 535, 17105 Nazareth Illit
☎ (+972 6) 655 81 11 ⊜ (+972 6) 655 81 03
👥 350
General Manager: Amos Eshet ☎ (+972 6) 655 58 11
Export and Marketing Manager: Dan Gafni
☎ (+972 6) 655 81 22
Operations Director: Arie Beizer
Purchasing Manager: Yitzhak Sprinzak
■ Manufacture of 4x4 military vehicles M462 "ABIR", M240 multi mission vehicle, Armoured vehicles against all calibres, light weapons and LSVC light surveillance vehicle. [98]

75-02610 Azimuth Ltd
15 Hataasiah Street, PO Box 2497, 43654 Raanana
☎ (+972 9) 760 41 34 [E] mkt@azimuth.co.il
⊜ (+972 9) 742 95 48
President: Shlomo Reisman
Vice President International Marketing and Sales: David Alster
■ Target acquisition systems, GPS navigation systems. [98]

75-02800 Bedek Aviation Group
(a division of Israel Aircraft Industries Ltd)
70100 Ben Gurion Int'l Airport
☎ (+972 3) 971 12 40; 935 37 43; 935 38 43
☒ 381014; 381033; 381002 israv il
⊜ (+972 3) 935 70 66; 935 45 77 (Marketing)
Coperation Vice President and General Manager: D Arzi
■ Repair, overhaul, retrofit, modification, conversion, remanufacture and customising of civil aircraft.
Made up of four autonomous business divisions.
Engines Division
70100 Ben Gurion Int'l Airport
☎ (+972 3) 935 70 64 ⊜ (+972 3) 935 87 40
General Manager: S Shalev
■ Overhaul, repair, retrofit, outfitting and testing of 30 types of civil and military engines including turbo fan and turbo jet: PWA: PT6, JT4A, JT3D, F52, FT8D, JT9D, F100, PW1120. GE: T64, J79, F110. Allison: T56, 250C, 501. SNEMA: ATAR 9B, 9C. Garret: TFE-731. Lycoming: T53.
Aircraft Division
70100 Ben Gurion Int'l Airport
☎ (+972 3) 935 49 30 ⊜ (+972 3) 935 42 79
General Manager: S Shkedi
■ Service and conversion of B747's, including Section 41. Maintenance of heavy transport aircraft conversion of airframes to customer defined transport, tanker, EW and recondition mission configurations. Provide complete 24 hour on-site airline support throughout customer network.
Components Division
70100 Ben Gurion Int'l Airport
☎ (+972 3) 935 35 54 ⊜ (+972 3) 935 33 11

General Manager: E Hattem
■ Maintenance, overhaul, repair, modification and test of 6,000 types of aircraft accessories and instruments. Approved by major civil and military regulatory agencies and OEMS.

Flight Operations
70100 Ben Gurion Int'l Airport
☎(+972 3) 935 30 36 ⊖ (+972 3) 935 89 93
General Manager: A Raziel
■ Test flights of all IAI customer owned aircraft (fixed and rotary wing light and heavy) following maintenance, overhaul, upgrades and conversion programs. Test flights of IAI produced aircraft and international ferrying flights.

Ramta Division
Industrial Zone, PO Box 323, 84102 Beer-Sheeva
☎(+972 7) 627 09 51; 627 09 52; 27 48 51; 27 48 52
☒ 5298 iaibs il ⊖ (+972 7) 627 67 70
General Manager: N Galili
■ Combat engineering equipment; fighting and support vehicles; GSE and aircraft structures including galleys composite materials design and manufacture. 98

75-02850 Bental Industries Ltd
12436 Kibbutz Merom Golan
☎(+972 6) 96 01 99 ⊖ (+972 6) 96 01 87
50 15 000 sq ft
President: Moti Nitzan
Operations & Technical Dir: B. Pharchat
Sales & Marketing Dir: A. Avizemer
Engineering Dir: Arkadi Shore
Purchasing Mgr: M. Yeshuran
■ Specialise in the design & manufacture of high quality & performance electrical motors. Rare earth permanent magnet D.C. & A.C. motors. Approved to ISO 9001. 97

75-03200 Bet Shemesh Engines Ltd
(a subsidiary of Ormat Industries Ltd)
Post Office Haela, 99000 Bet Shemesh
☎(+972 2) 990 94 87 ⊖ (+972 2) 991 19 70
♣ US$ 25,000,000 (1996) 350 32 000 sq m
Managing Director: Avner Shacham
Marketing and Sales Manager: Avigdor Hirsh
Regional Marketing Manager: Zvi Barak
■ Development, manufacture, overhaul and repair of turbojet engines, production of investment castings and manufacture of jet engine parts.

Divisions:
Manufacturing Division
Casting Division
Assembly and Overhaul Division 98

75-03350 BVR Technologies Ltd
1 Korazin Street, 53583 Givataim
☎(+972 3) 571 56 71 ✆ (050) 26 88 46
E zamir@bvr.co.il W http://www.bvr.co.il
⊖ (+972 3) 571 56 68
♤ US$ 10,000,000 ♣ US$ 20,500,000 (1996)
100 (1996) 1,500 m2
Chairman: Yaron Sheinman
Managing Director: Aviv Tzidon
Vice President Marketing: Zamir Avraham
Operations Director: Amiram Eliasaf
Purchasing Manager: Roni Erlich
■ Hotshot family of squadron-deployable combat tactics trainers. Pilot training systems. Large scale "Wargames" simulation for decision support. Advanced ACMI systems, airborne and ground stations. EO/IR simulators for weapons operators training. Naval and army trainee systems. 98

75-03400 Cabiran (1991) Ltd
26120 Kibbutz Cabri
☎(+972 4) 995 26 11 E cabiran@cabri.org.il
W http://www.cabri.org.il/cabiran ⊖ (+972 4) 995 26 12
♣ US$ 4,500,000 (1996) 60 4,000 m2
Managing Director: Ido Rosolio
Operations Director: David Leb
Marketing Manager: Micha Ron
■ Non-ferrous investment casting. 98

75-03450 Carcom
(a subsidiary of Cyclone Aviation Products Ltd)
PO Box 114, 20100 Carmiel
☎(+972 4) 996 07 11 ⊖ (+972 4) 996 22 20
♣ US$ 7,000,000
General Manager: Gideon Goren
Marketing Director: Jacob Ron

■ Production of composite parts and panels and adhesively bonded assemblies. Composite products for aeronautical structures. Armour plates and seats for pilot protection. 98

75-03750 CI Systems Ltd
(† CI Ltd)
(a subsidiary of CI Technologies Inc, USA)
Industrial Park, PO Box 147, 10551 Migdal HaE'mek
☎(+972 6) 644 88 88 E cis@actcom.co.il
⊖ (+972 6) 654 35 70
♤ US$ 16,000,000 (1997) ♣ US$ 6,500,000 (1996)
70 2,000 m2
Chief Executive: Nauum Shy ☎(+972 6) 644 88 11
Research Director: Yaron Ish-Shalom
☎(+972 6) 644 88 37
Marketing Director: Zvi Kopolovitch
☎(+972 6) 644 88 33
Operations Director: Dror Motobilov
Purchasing Manager: Harel Melamed
■ Designs and manufactures electro-optical BITE, simulators and FLIR test equipment and boresight sets. Visual and IR/FLIR system test sets in development, production, field operation. Produces IR spectroradiometers and IR sources. 98

75-04170 CONTROP Precision Technologies Ltd
PO Box 611, 45105 Hod Hasharon
☎(+972 9) 740 70 62 E cntrpnir@netvision.net.il
⊖ (+972 9) 740 53 76
♣ US$ 3,000,000 (1996) 23 (1996)
General Manager: Shlomo Nir
Business Development Director: Eli Ben-Aharon
International Marketing Director: Asher Aber
■ Specialises in the development and production of electro-optical and precision motion control systems. High performance stabilised observation payloads used for day and night surveillance on board of: UAV's, helicopters and light aircraft. Special directional stabilized antenna pedestals for satellite communication systems mounted on UAV's. Special film cameras for aerial reconnaissance. 98

75-04250 Cyclone Aviation Products Ltd
(a subsidiary of Etz Lavud Ltd)
PO Box 114, 20100 Carmiel
☎(+972 4) 996 07 16; 996 87 46
E cyclver@netvision.net.il ⊖ (+972 4) 996 85 67
590 (1997) 20,000 m2
President: Gideon Goren ☎(+972 4) 996 07 01
Vice President and Operations Director: Johathan Glinert
☎(+972 4) 996 07 20
Marketing Director: Jacob Ron ☎(+972 4) 996 07 16
Purchasing Manager: Gaby Fisch
■ Helicopter and light aircraft maintenance, overhaul, repair, rebuild and modification. Aircraft flight surfaces production, repair and rebuild. Structural components production and assembly. Precise and heavy machinery. Composite material component production. → 80

Subs. with defense business:
Carcom → 75 98

75-04310 Dafna Industries
Kibbutz Dafna, 12235 Upper Galilee
☎(+972 6) 681 81 07; 681 81 20 E inmann@shani.co.il
⊖ (+972 6) 690 21 16
140 (1996)
Managing Director: Dov Cabiri
Director Marketing and Sales: Elik Gilad
Operations Director: Ilan Harlam
Purchasing Manager: O N Factor
■ Produces NBC overboots from PVC. 98

75-04450 Datic Cable Works Ltd
(Telecommunication Cables Division of) Cvalim - Electric Wire & Cable Company of Israel Ltd
Beit She'an, PO Box 189, 10900 Beit She'an
☎(+972 6) 648 92 22 ☒ 45103 ⊖ (+972 6) 648 92 50
♣ US$ 40,000,000 160 (1996) 16,000 (1996)
Managing Director: Michael M Goldschmidt
Sales Director: Braumen Meir
Purchasing Manager: Lax Ariea
General Manager: Berman Yoram 98

75-04575 Decision Systems Israel Inc (DSI)
11 Ben Gurion Street, 54017 Givat Shmuel
☎(+972 3) 531 33 33 W http://www.dsi.co.il
⊖ (+972 3) 531 33 22
♤ US$ 12,000,000 (1996) ♣ US$ 12,000,000 (1996)
230 (1997) 2,000 m2
Managing Director: Samuel Fogel
Operations Director: Benny Sela
Marketing Director: Moshe Shan
Purchasing Manager: Ran Avgar
■ Specialising in the development and manufacture of naval, airborne and ground state-of-the-art systems. Sonar systems including torpedo sensors, underwater surveillance systems, command and control systems for various C2 missions, tactical air picture for military and traffic control application, ECM systems and upgrades, simulators (radar, ESM, ECM), integrated mobile navigation and location systems including moving map displays and multi purpose reconfigurable operator consoles. 98

75-05050 Echtman Engineering Corp. Ltd
(† Echtman Hamalchim Engineering Corp.)
PO Box 18015, 61180 Tel Aviv
☎(+972 3) 921 92 94 ⊖ (+972 3) 921 99 55
♣ US$ 12,000,000 120 7,000 sq m
Managing Director: Alex Echtman
Sales Manager: Yigal Kurtz
Publicity Manager: Mrs B Echtman
■ Military aircraft and helicopter positioner-handler tractors, fork-lift trucks, electric tow-tractors.

Plant
New Industrial Zone, 49512 Petach Tikva
New Industrial Zone, 2 Hayetzira Straat
☎(+972 3) 921 92 94 ⊖ (+972 3) 921 99 55 98

75-05300 Elbit Systems Ltd
(† Elbit Ltd)
(an affiliate of Elron Electronic Ind. Ltd)
Advanced Technology Center, PO Box 539, 31053 Haifa
☎(+972 4) 831 53 15 ✉ ELBIT Haifa
W http://www.elbit.co.il
⊖ (+972 4) 855 00 02; 55 16 23
♣ US$ 307,508,000 1,735
Chairman: Ermmanuel Gill
President: Joseph Ackerman
Vice President and Chief Financial Officer: I Forchi
Marketing Promotion Director: Ms Ilana Gelfer
■ Ground Systems: Combat vehicle upgrades, tank fire control systems, electric turret and control systems, artillery command and control systems, day-night observation systems, digital image processing and communications systems, C3 systems, training systems, paramilitary systems.
Airborne Systems: Fixed wing and helicopter upgrades, weapons delivery and navigation systems, airborne computer systems, cockpit display systems, helmet mounted displays, stores management systems, precision guidance systems, helicopter avionic systems, command, control and communication (C3I) systems. Unmanned Airborne Vehicles.
Naval Combat Systems Intergration:
Tactical control systems, C3 systems, trainer and computerised simulation systems, electronic warfare systems, communication systems.

Information Services Center
PO Box 539, 31053 Haifa
☎(+972 4) 831 53 15 ✉ ELBIT Haifa
⊖ (+972 4) 852 00 02
Manager: D Platner 98

75-05755 Cvalim - Electric Wire & Cable Co. of Israel Ltd (EWC)
Haifa Bay, PO Box 400, 27103 Kiriat Blalik
☎(+972 4) 84 66 222 ☒ 45103 ⊖ (+972 4) 846 62 86
♣ NIS 135,000 (1995) 480 (1997)
50,750 m2 (1997)
Managing Director: M Goldschmidt
Vice President Marketing: Braumen Meir
Operations Director: J Waits
Purchasing Manager: Lax Arie
■ Electrical telecommunication and electronic wire and cable products to Israeli, European, American and Australian military specification standards.

Divisions:
Datic Telecommunication Cables Div. → 75
Thermofil-Cvalim Ltd, Electronic Cables Div → 75
Power Cables Division

Power Cables Division
Haifa Bay, PO Box 443, 31003 Haifa
☎(+972 4) 846 62 22 ☒ 45103 ewcco il
⊖ (+972 4) 841 39 54 98

75-05800 Electronics & Aviation Systems Ltd (EAS)
PO Box 265, 81101 Yavne
☎ (+972 8) 943 17 86 ✆ (+972 50) 40 29 58
⊖ (+972 8) 943 17 40
♧ US$ 1,200,000 (1996) ♣ US$ 2,000,000 (1996)
👥 12 ⌂ 500 m2
Managing Director: Joseph Gross
■ Manufacture UAV's and flight control systems. [98]

75-06150 Elisra Electronic Systems Ltd
(a subsidiary of Tadiran Ltd)
Corporation Headquarters
48 Mivtza Kadesh Street, 51203 Bene Baraq
☎ (+972 3) 617 51 11; 754 51 11; 617 52 27
⊠ 033553 ⊖ (+972 3) 617 58 50; 754 54 68
♣ US$ 107,858,000 👥 825 (1997) ⌂ 22,000 m2
President: Avner Raz
Vice President Marketing: Meir Barton
Operations/Technical Director: Manahem Oren
Purchasing Manager: Arie Dolev
Acting Supervisor Advertising Agency: Gil Goldfine
■ EW systems house. Develops and produces the following: A series of combat proven aircraft self-protection systems, for first-line combat aircraft and helicopters, incorporating advanced RWR and laser warning systems, also airborne system simulation, airborne computers, airborne EW and radar simulation system and field loaders. A series of naval ESM/ECM and ELINT systems for a variety of vessels from small patrol boats to frigates and other combat ships. Ground-based, airborne and naval, ELINT systems. Still and video image processing and transmission systems for reconnaissance and intelligence purposes. EW and radar simulators for systems evaluation, maintenance and training. Active and passive microwave components, supercomponents and power amplifiers. [98]

75-06300 EL-OP Electro Optics Industries Ltd
(a subsidiary of Federmann Enterprises)
Advanced Techn.Park, Kiryat Weizmann, PO Box 1165, 76110 Rehovot
☎ (+972 8) 938 62 11 ⊠ 381328 ⊖ (+972 8) 938 62 37
♣ US$ 270,000,000 (1996) 👥 1,700 ⌂ 32,000 sq m
President: Jacob Toren
Marketing Director: Josef Gaspar
Operations and Technical Director: Avi Azrad
Purchasing Manager: Avi Shoham
Marketing Communication Manager: Aviva Haleva
■ Development and manufacture of thermal imaging systems, laser systems, fire control systems, displays for avionic of ground platforms, optronic stabilised payloads. Electro-optical counter measures systems, remote sensing systems. [98]

75-06800 Erez Forensic Technology Ltd
PO Box 35008, 91350 Jerusalem
☎ (+972 2) 653 54 53 ⊖ (+972 2) 651 31 18
♣ US$ 1,000,000 (1995) 👥 5 ⌂ 250 sq m
Director: Menahem Ansbacher
Operations and Technical Director: Shai Amisar
■ Explosive detection kits covering the whole range of explosives including TNT, Semtex, dynamite and those based on inorganic nitrate compounds. Drug detection field kits for the identification of drugs such as cannabinoids (marijuana and hashish mainly), heroin, cocaine and other forensic speciality products. [98]

75-07200 Eyal Engineering & Industrial Co.
Industrial Zone, PO Box 374, 46103 Herzliya
☎ (+972 9) 954 11 11 ⊖ (+972 9) 954 12 17
♣ US$ 10,000,000 (1996) 👥 70 (1996) ⌂ 4,500 sq m
Managing and Operations Director: Israel Ayalon
Sales and Marketing Director: Uria Ayalon
Purchasing Manager: Ms Rachel Ehi [98]

75-07650 Gabi Shoef Ltd
5 Eliau Shamir St, 50297 Mishmar Hashivaa
☎ (+972 3) 960 55 59 ⊖ (+972 3) 960 41 60
♣ US$ 1 200 000 👥 20 ⌂ 20 300 sq ft
Managing Dir: Gabi Shoef
Sales & Marketing Dir: Yossi Shoef
Operations & Technical Dir: Ben Zvi Shraga
Head of Southern Branch: Yossi Bukanet
Head of Northern Branch: Avi Dahan
■ Non-destructive testing, radiography x-ray & isotopes, ultrasonic inspection, die penetrants, eddy current & magnetic particles. Working & certified by international specifications & standards. [97]

75-07680 Gamatronic Electronic Industries Ltd
Science-Based Industries Park, PO Box 45029, 97774 Jerusalem
☎ (+972 2) 586 01 11 [E] gammaa@netvision.net.il
⊖ (+972 2) 582 88 75
♣ US$ 10,000,000 👥 140
President: Yossi Goren
Purchasing and Marketing Director: Garen Sharon
■ PWM zero time UPS. UPS systems from 5KVA to 20KVA, three phase, line voltage stabilizers, power supplies, frequency converters. [98]

75-07950 Goussinsky B. Engineering & Manufacturing Co.Ltd
26110 Haifa Bay, PO Box 10033, Haifa
☎ (+972 4) 841 73 51 ⊖ (+972 4) 872 30 64
👥 35 (1996)
Managing Dir: Doron Goussinsky
Marketing Mgr: M. Goussinsky
General Mgr: Uri Vekstein
■ Gates & barriers for vehicles & pedestrians. Cranes, hoists, materials handling. Gates electric, barriers electric, turnstiles, manual & electric; anti-terrorist. "Defender" line of gates & barriers, bridge cranes; gantry cranes, "No-Rail" patented bridge cranes. [97]

75-08205 Hadaikan Automotive Assemblies
PO Box 13088, 61130 Tel Aviv
☎ (+972 9) 65 52 17/8 ⊖ (+972 9) 65 31 73
👥 40 (1996) ⌂ 5 000 sq m
Managing Dir: Dorom Lachman
■ Manufacturer of automotive parts for passenger, commercial, military & agricultural vehicles. [97]

75-09170 IDE Technologies Ltd
(† Israel Desalination Engineering)
13 Zarchin ST, PO Box 591, 43104 Ra'Anana
☎ (+972 9) 90 97 77 ⊠ 33590 ⊖ (+972 9) 90 97 15
♣ US$ 40 000 000 (1995) 👥 170 (1995)
President: D. Waxman
VP Technology: A. Ophir
VP Finance: M. Goldstein
Marketing Mgr: E. Oliver
Product Mgr Desal: Gustavo Kronenberg
Product Mgr Evaporators & Heat Pumps: Abraham Koren
Product Mgr Water Treatment: Freddi Lokiec
■ Development of saline water desalination equipment & other related technologies, such as purification of water streams. Aquaport-sea water vapour compression distillation units in capacities from 25 to 3 000 cubic m/day per module in static installation & from 25 to 100 cubic m/day in mobile units. [97]

75-09400 Imco Industries Ltd
Industrial Park, PO Box 189, 20302 Tel Hanan
☎ (+972 4) 32 63 22 ⊠ 46400 ⊖ (+972 4) 23 75 08
♧ US$ 6 000 000 (1995) ♣ US$ 10 000 000 (1995)
👥 120 (1996) ⌂ 4 500 sq m
President: J. Yakir
Managing Dir: E. Greenberg
Operations & Technical Dir: H. Geva
■ Military & civilian mechanical, electromechanical & electronics. Electronic packaging, harnesses, cable & cords, printed circuit assemblies, pitot tubes, thermocouples, pilot & commander sticks, delay relays, antennas, rocket launchers, oxygen valves, control panels, transponders for IFF transponders, complete tanks, armoured vehicles, electrical, electronic & vectronics systems. [97]

75-10000 Israel Product Research Co. Ltd (ISPRA)
PO Box 2086, 46120 Herzliya
16 Galgal Haplada St, Ind. Zone, 46722 Herzliya
☎ (+972 9) 955 54 64; 954 04 84; 954 02 24
[E] ispra@netvision.net.il ⊠ 342590 ispra il
⊖ (+972 9) 55 91 46
👥 38 ⌂ 800 m2
Managing Director: Dr A Jacobson
Sales and Marketing Director: Y Stein
Operations and Technical Director: J Bialy
Purchasing Manager: H Michaeli
■ Non-lethal chemical weapons with the use of tear-gas; specifically tear-gas hand grenades, launchers, spray devices and tear-gas protectojet - anti-riot fog ejector. "Fibrespider" fibre optics fences and flexible fibre optic net perimeter protection. [98]

75-10200 Israel Aircraft Industries Ltd (IAI)
70100 Ben Gurion Int'l Airport
☎ (+972 3) 935 31 11; 935 81 11 ✉ ISRAELAVIA
⊠ 381014; 381033; 381002 isravia
⊖ (+972 3) 935 31 31; 971 22 90
♣ US$ 1,500,000,000 (1996) 👥 13,150
⌂ 680,000 sq m
Chairman: Maj Gen Avigdor Yanouch Ben-Gal (retd)
President and Chief Executive Officer: Moshe Keret
Vice President Finance: N Gilad
Vice President Marketing: M Broder
Vice President General Manager Bedek Aviation Group: D Arzi
Vice President General Manager Military Aircraft Group: O Harari
Vice President General Manager Commercial Aircraft Group: M Boness
Vice President General Manager Electronics Group: S Alkon
Vice President Corporate Communications: Doron Suslik
Corporate Advertising and Foreign Press: Ms Ilana Sternfeld ☎ (+972 3) 935 85 14 (+972 3) 935 85 12 (fax)
■ Research, design, development, integration, testing, manufacture, upgrading, support and service of land, sea, air systems and equipment, and space technologies. Military and civil aircraft; UAV's; missiles; hydraulic, electro-mechanical, electronic and optronic systems; ordnance and navigation equipment; patrol boats; mine clearance devices; C3I systems; airborne and ground-based radars; software. → 55

Subsidiaries and associated companies:
Israel Aircraft Industries Int'l Inc, (USA)

Military Aircraft Group
70100 Ben Gurion Int'l Airport
☎ (+972 3) 935 41 36; 971 14 71 ✉ ISRAELAVIA
⊠ 381002; 381014; 381033 israv il
⊖ (+972 3) 935 78 43
Corporation Vice President and General Manager: O Harari
■ Design, manufacture, upgrading of manned and unmanned fixed and rotary wing aircraft.
Made up of three autonomous business divisions.

Malat Division
70100 Ben Gurion Int'l Airport
☎ (+972 3) 935 35 08 ⊠ 381014; 381033 israv il
⊖ (+972 3) 935 41 75
General Manager: E Zuckerman
■ Concept, design integration and assembly of UVA's, airframes, on-board components, operational control systems and launch and retrieval equipment.

Mata Division (Helicopters)
Industrial Zone, PO Box 27/160, 91271 Jerusalem
☎ (+972 2) 584 12 11; 583 43 13 ⊠ 26213 jerai il
⊖ (+972 2) 584 12 10; 584 12 99
General Manager: O Shifris
■ Helicopter structures, systems and components maintenance, crash repair, overhaul, modification and upgrading. CAD/CAM assisted cable and harness design and fabrication. Design and manufacture of aircraft crew and passenger seats, aircraft wheels and cockpit controls.

Lahav Division
70100 Ben Gurion Int'l Airport
☎ (+972 3) 935 80 92 ⊖ (+972 3) 935 70 10
General Manager: D Durgan
■ Retrofit design, integration and assembly of advanced aircraft. Development and manufacture of customer specified advanced combat aircraft. Upgrades of MiG series aircraft. F-4, F-5.

Commercial Aircraft Group
70100 Ben Gurion Int'l Airport
☎ (+972 3) 935 33 37; 935 71 49
⊠ 381033; 381014; 381002 israv il
⊖ (+972 3) 935 42 26
Corporation Vice President and General Manager: M Bones
■ Design, development and manufacture of civil aircraft.
Made up of three autonomous business divisions.

Engineering Division
70100 Ben Gurion Int'l Airport
☎ (+972 3) 935 81 50 ⊖ (+972 3) 935 50 50
General Manager: Z Arazi
■ Analysis, design, development, integration and testing of aircraft systems and material requirements for aerospace clients.

Production Division
70100 Ben Gurion Int'l Airport
☎ (+972 3) 935 31 32 ⊜ (+972 3) 935 31 30
General Manager: G Cohen
■ Substructure and components manufacture for domestic and foreign customers. Special capabilities include the fabrication of standard and advanced material structures, nacelles and other engine parts including hot-end components. Development of future civil aircraft, the production of IAI's new Astra-Galaxy business jet, provision of upgrades for Westwind Aircraft and product support for Westwind and Arava STOL. Aircraft.

SHL Division
Industrial Zone, PO Box 190, 71101 Lod
☎ (+972 8) 23 91 11 ▣ 381520 shld il
⊜ (+972 8) 22 27 92
General Manager: H Porter
■ Design, development and manufacture of military and civil hydraulic systems and components; aircraft landing gear and shock absorbers; servo-hydraulic actuators for flight control systems.

Electronics Group
Yahud Industrial Zone, PO Box 105, 56000 Yahud
☎ (+972 3) 531 55 55 ▣ 341450 mbt il
⊜ (+972 3) 536 52 05; 536 39 75
Vice President and General Manager: Shmuel Alkon
■ Analysis, concept, design, development, integration, manufacture and testing of defence and civil electronic systems.
Made up of four autonomous divisions.

ELTA Electronics Industries Ltd
(a subsidiary of Israel Aircraft Industries Ltd, Electronics Division)
PO Box 330, 77102 Ashdod
☎ (+972 8) 857 23 33 ✉ ELTASHDOD ▣ 381807 elta il
⊜ (+972 8) 856 59 14
⌂ 420,000 sq ft
Managing Director: Mordechai Shmueli
☎ (+972 3) 57 23 27 (+972 3) 56 45 68 (fax)
■ Military electronics systems and equipment including airborne, surface, naval and tactical radars, ELINT/ESM, SIGINT, COMINT/COMJAM, computers, ATE, secure communications equipment, programmable signal processors, artificial intelligence, microwave antennas, high voltage power supplies and components.

MBT Division
Yahud Industrial Zone, PO Box 105, 56000 Yahud
☎ (+972 3) 531 55 55; 531 40 01 ▣ 341450 mbt il
⊜ (+972 3) 536 52 05; 536 33 76
General Manager: I Nissan ☎ (+972 3) 531 40 01 (+972 3) 531 41 10
■ Design and manufacture of smart bombs, fire control systems, offensive and defensive precision tactical battlefield missiles. Training simulator systems, civil satellite integration and group service.

Taman Division
Yahud Industrial Zone, PO Box 75, 56100 Yahud
☎ (+972 3) 531 55 55 ▣ 341948 tmm il
⊜ (+972 3) 531 51 40
General Manager: D Rotem ☎ (+972 3) 536 46 66 (+972 3) 531 51 29
■ Design and manufacture of inertial measurement, stabilisation, navigation and optronic systems & components for land, sea and air, manned & unmanned platforms, specialised civil space components.

MLM Division
PO Box 45, 70350 Beer Yaakov
☎ (+972 8) 927 23 33 ▣ 381014; 381033 israv il
⊜ (+972 8) 927 25 74; 927 27 42
General Manager: I Livnat ☎ (+972 8) 927 24 36 (+972 8) 927 20 20 (fax)
■ Real time video compression systems, image processing and enhancement, tactical switching command and control systems. Development of arrow ATBM missile systems. 98

75-10300 Israel Military Industries Ltd (IMI)
(† TAAS Israel Industries Ltd)
64 Bialik Boulavard, PO Box 1044, 47100 Ramat Hasharon
☎ (+972 3) 548 51 18; 52 07 ⊜ (+972 3) 548 51 29
♣ US$ 509,000,000 (1996) 👥 4,300
Chairman of the Board: Yaacov Lifshitz
President: Shlomo Milo
Chief Engineer: Brig Gen Elazar Barak
Internal Auditor: Mr Jehushua
Legal Advisor: Sanrai Shmuel
Research and Development Strategy: Shmuel Keren
Finance and Economics: Hasid Nehemia
Ashot Ashkelon Ltd: Dr Robert Gerro
Marketing: Nissim Mor
Technologies Complex: Emanuel Harel
Systems Group: Brig Gen Shmuel 1Keren (Rtd)
Ammunition Group: Ehud Frank
Weapons and AFV Systems Group: Dr Avi Felder
Spokesman Manager: Menachem Dotan
Advertising Manager: Avinoam Zafir
Administration and Personnel: Moshe Peer
■ Manufacture of arms, ammunition, explosives and propellants, military support systems and equipment and aircraft accessories. Remote-control mobile robots.
Services division:
Hancal-Systems Engineering 98

75-10310 Israel Military Industries Rocket Systems Division
64 Bialik Boulevard, PO Box 1044, 47100 Ramat Hasharon
☎ (+972 8) 924 26 84; 927 74 47/8/9
⊜ (+972 8) 925 28 96
♣ US$ 50,000,000 (1996) 👥 530
Division Director: Nathan Wechesler
Marketing Director: Israel Weinreb
■ Design, development and manufacture of: all sizes of rocket motors from small diameter to satellite launchers; artillery rocket systems; anti-tank weapon systems; pyrotechnic products and countermeasures (illuminating flares and IR decoy flares for aircraft protection); ballistic armor for personal and vehical protection. Extensive composite materials and metal machining capability. 98

75-10600 Israel Shipyards Ltd
PO Box 10630, 26118 Haifa-Bay
☎ (+972 4) 846 02 45 ⊜ (+972 4) 841 05 72
♣ US$ 40,000,000 (1996) 👥 250 ⌂ 30,000 sq m
President: Ishay Haramaty ☎ (+972 4) 846 02 46
Vice President Marketing: Y Meivar ☎ (+972 4) 846 02 45
Purchasing Manager: Moshe Cohen
☎ (+972 4) 846 03 37
■ Design and construction of a wide variety of naval craft, including the Reshef missile and gun boat, fast patrol boats, tugs and landing craft. Upgrading of naval vessels.
→ 80 98

75-10800 ITP Ltd
25 Lechi Street, 51200 Bnei-Brak
☎ (+972 3) 578 09 01 ▣ 381665 ⊜ (+972 3) 578 09 11
👥 50
Managing Director: Ron Grynholc
Sales and Marketing Director: David Soloducho
■ Bullet proof jackets, personal gear, security electronics, anti terror special equipment, army and police consulting. 98

75-10850 Jordan Valley Applied Radiation Ltd (JVAR)
(a subsidiary of the Clal Group)
PO Box 103, 23100 Midgal Ha'Emek
☎ (+972 6) 654 36 66 Ⓔ jvar@actom.co.il
⊜ (+972 6) 654 74 72
♣ US$ 2,000,000 👥 26
Managing Director: Isaac Mazor
Marketing Director: Michael Kulin
Purchasing Manager: Elisha Grünwald
■ Development and manufacture of Elemental Analysis Equipment for product engineering and quality assurance in armament and explosives production. 98

75-11320 Kinetics Ltd
11 Hamlacha Street, PO Box 10, 60251 Or-Yehuda
☎ (+972 3) 533 50 30 Ⓔ market@kinetics.co.il
Ⓦ http://www.kinectics.co.il ⊜ (+972 3) 533 50 28
👥 110 (1997)
President: Kainan Rafaeli
Sales and Marketing Director: Avi Hadayo
Vice President Research and Development: Dov Kleczewski
Purchasing Manager: Moti Ekstein
■ Development and production of fluid power systems and components for defence applications, including systems and components for armoured fighting vehicles suspension, turret control, fuel, braking, NBC and air-conditioning. 98

75-11350 K.M.Fuze Engineering
(a subsidiary of Koor Electronics Ltd)
PO Box 50, 71100 Lod
☎ (+972 8) 927 31 70 ⊜ (+972 8) 927 31 73
♧ US$ 6,000,000 ♣ US$ 3,000,000 👥 60 ⌂ 1,000 m2
Managing and Marketing Director: Michael Naftali
Operations Director: Sharir Bdaz
■ Development and production of a large variety of electronic fuses and electronic A.T mines. 98

75-11800 Koor Industries Ltd
4 Kaufman Street, 68012 Tel-Aviv
☎ (+972 3) 510 56 81; 519 52 01
⊜ (+862 3) 519 53 53
♣ US$ 3,530,000,000 👥 21,500
President and Chief Executive Officer: Benjamin D Gaon
Deputy President and Chief Financial Officer: Yehuda Milo
Assistant to the President: Lyon Roth
Manager Office of the President: Ms Liraz Biran
■ Telecommunictions and electronics, agrochemicals, building and infrastructure materials.
Telecommunications and Electronics:
Tadiran Ltd
Tadiran Telecomminications Ltd
Tadiran Com Ltd
Hisra Electronic Systems Ltd
Tadiran Appliances Ltd
Tadiran Information Systems Ltd
Tadiran Holdings Ltd
Telrad Telecommunications and Electronics Ltd
Agrochemicals and Other Chemicals:
Makhteshim Chemical Works Ltd
Agan Chemical Manufacturers Ltd
Building and Infrastructure Materials:
Mashav Initiation and Developments Ltd
Nesher Israel Cement Enterprises Ltd
Taavura Cement Containers Ltd
United Steel Mills Ltd
Middle East Tube Co Ltd
Merhav Ceramic and Building Materials Ltd
Granite Hacarmel Investment Ltd
Sonol Israel Ltd
Supergas Israel Gas Distribution
Vulcan Batteries Ltd
Other Businesses:
Shemen Industries Ltd
Henkel Soad Ltd
Galilee Fruits Ltd
Hod Lavan Products Ltd
Soltan Ltd
Koor Metals Ltd Oils Div.
Merkavim Metals Works Ltd
Phoenicia Glass Works Ltd
Koortrade Ltd
Koor Tourism Enterprises Ltd
Koor Capital Markets Ltd
Koor Properties Ltd 98

75-12190 KZB Technical Rubber Products (1987) Ltd
(a subsidiary of Hish Processing and Conveying Technology Ltd)
PO Box 10157, 26111 Haifa Bay
☎ (+972 4) 842 44 99 ▣ 45152 hish il
⊜ (+972 4) 872 21 33
👥 30 (1997)
Managing Director: Eitan Ben Natan
Marketing Director: Yaram Levy
■ Technical moulded rubber products; tract shoe pads, road wheel and track support rollers for battle tanks. 98

75-12200 Lehavot Industries
12125 Lehavot Habashan
☎ (+972 6) 95 21 01 ⊜ (+972 6) 95 21 19
♣ US$ 7,000,000 (1994) 👥 60 ⌂ 6,000 sq m
Export Marketing Director: Benzion Feinstein
General Manager: Giora Salz
■ Fire extinguishers, fire fighting and extinguishing equipment: portable, wheeled, manual and automatic actuated, for use on army bases, naval bases, airfields (flightline types) fire extinguishing systems for armoured vehicles and naval craft. 98

75-12600 M-Systems Flash Disk Pioneers
Atidim Industrial Park, Building 7, PO Box 58036, 61580 Tel-Aviv
☎ (+972 3) 647 77 76 Ⓔ info@m-sys.com
Ⓦ http://www.m-sys.com ⊜ (+972 3) 647 66 68
♧ US$ 18,899,000 (1996) ♣ US$ 11,750,000 (1996)
👥 75 ⌂ 1,210 sq m
Managing Director: Dov Moran
Operations and Technical Director: Reuven Sela
Sales and Marketing Director: Aryeh Mergi
Marketing Manager: Dafna Maor
Purchasing Manager: Atalya Levi

■ Develop, manufacture and marketing of flash data storage solutions.
® USA: M-Systems Inc [98]

75-12650 Magal Security Systems Ltd
Industrial Zone, PO Box 70, 56000 Yahud
☎ (+972 3) 539 14 44 ⊜ (+972 3) 536 62 45
♧ US$ 10 000 000 ⱳ 110 ⌂ 20 000 sq ft
President & Managing Dir: I. Dekel
Financial Mgr: M. Brandt
Marketing Mgr: J. Golan
Airport Systems Mgr: O. Katz
■ Advanced computerised control centres, computer controlled taut wire perimeter intrusion detection systems, computerised electronic anti-intrusion grid. Automatic bomb detection systems. [97]

75-12750 Magnus Engineering & Maintenance Ltd
35 Hayarkon Street, 51204 Bnei-Brak
☎ (+972 3) 578 04 80 ⊜ (+972 3) 578 05 25
ⱳ 35 (1997) ⌂ 1,200
Managing Director: Jacob Ashkenazy
Sales and Marketing Director: Shmaryahu Peer
■ Computerised test cells for: turbojet, turboprop and turboshaft engines. RPV engines test cells, diesel engines test cells; engine monitoring system analysers; engine fleet management software; engine performance analysis, software and various engine portable testers and simulators. → 80 [98]

75-13900 Merkavim Metal Works Ltd
(a subs. of Koor Industries)
94 Jabotinsky St, PO Box 472, 49105 Petah, Tiqva
☎ (+972 3) 922 30 44 ⊜ (+972 3) 921 82 94
♣ US$ 30 000 000 ⱳ 300 ⌂ 24 000 sq m
Managing Dir: Alex Magen ☎ (+972 3) 924 33 27
Sales & Marketing Dir: Moshe Greenfeld
☎ (+972 3) 924 51 97
Operations & Technical Dir: S. Sidi ☎ (+972 3) 924 80 50
Operations & Technical Dir: S. Raz
Purchasing Mgr: Avi Malovani
■ Bus body building. [97]

75-14200 Michsaf
Kibbutz Nir Am, 79155 D N Hof Ashkelon
☎ (+972 7) 680 99 00 ⊜ (+972 7) 680 99 71
General Manager: Tamir Simchi
Operations Director: Patrick Cohen
Marketing Director: Jlan Gilad
Purchasing Manager: Jzchak Hadjes
■ Stainless steel cutlery and camping cutlery sets. [98]

75-14250 Microelectronics Ltd
(a div. of Ariel Group)
1 Sharet St, Rishon Le Zion
☎ (+972 3) 961 23 23 ⊜ (+972 3) 961 23 07
♣ US$ 5 000 000 ⱳ 52 ⌂ 2 000
Managing Dir: Jacob Golany
Operations & Technical Dir: Moshe Binderman
Purchasing Mgr: David Markov
■ Manufacturer of RF & microwave fixed & variable capacitors. Manufacturer of high precision mechanical parts. [97]

75-14350 Microkim Ltd
Scientific Industries Center, PO Box 1797, 31016 Haifa
☎ (+972 4) 855 12 11 ⊜ (+972 4) 855 12 18
♣ US$ 3 000 000 (1995) ⱳ 40 ⌂ 1 000 sq m
President: Joseph Moscovitz
Engineering Dir: Alex Hodisan
■ Microwave components & subsystems including MIC super-components & hybrid sub-assemblies, sources, synthesizers, control devices, low-noise amplifiers, power amplifiers, transmitters, receivers. [97]

75-15200 Moked Engineering (1969) Ltd
PO Box 3454, 91034 Jerusalem
5 Shattner Center, Givat Shaul, 95461 Jerusalem
☎ (+972 2) 652 71 41 ⊜ (+972 2) 651 11 33
ⱳ 30
General Manager: Yigal Arad
Assistant General Manager: Mrs Yael Arad
■ The "Third Eye" and the "Airborne Third Eye" systems, developed and manufactured by Moked Engineering Ltd., are unique warning systems designed for the instantaneous detection of laser rangefinders, designators and infra-red searchlight. The systems were respectively designed for armoured vehicles and aircraft, especially helicopters. [98]

75-15450 MTC Industries & Research Carmiel Ltd
PO Box 232, 20100 Carmiel
☎ (+972 4) 98 77 72/3 ⊠ 46378 mtcca
⊜ (+972 4) 98 11 89
VP Marketing: Yair Avni
Managing Dir: Yechiel Cohen
Operations/Technical Dir: Ephraim Luzon
Purchasing Mgr: Dr Victor Shribman
■ Design, development & manufacture of gyroscopic instruments, torque motors & pickoffs, resolvers, accelerometers & synchros. Full range of customer specified design & development services. Thermal coating service qualified to ISO 9001. [97]

75-15500 MTLM Mechanics Karmiel Ltd
(a subsidiary of Cyclone Aviation Ltd)
PO Box 114, 20103 Carmiel
☎ (+972 4) 996 07 16 [E] cyclver@netvision.net.il
⊜ (+972 4) 996 85 67
ⱳ 500 ⌂ 20,000 m2
General Manager: Gideon Goren
Chief Engineer: J Gleinhart
Marketing Director: Ron Jacob
Purchasing Manager: Gabby Fisch
■ Manufacture, maintenance and repair of aircraft structural assemblies, machined and sheet metal primary parts, aircraft wheel and brake systems, composite structural parts, bonding of metal and composite structural parts. [98]

75-15800 Nice Systems Ltd
Atidim Industrial Park, Building 7, PO Box 58070, Neveh Sharett, Tel-Aviv
☎ (+972 3) 645 38 66; 37 00 ⊜ (+972 3) 647 82 73
♣ US$ 21 000 000 (1995) ⱳ 180 ⌂ 5 600 sq m
President/Managing Dir: Benny Levin
☎ (+972 3) 645 37 33
VP Sales & Marketing: Rami Lazar ☎ (+972 3) 645 37 55
Marketing Co-ordinator: Ms Anat Hochberg
☎ (+972 3) 645 38 66
■ Production of NiceLog Voice Logging System. Digital & analogue recording & fast retrieval system. Provides continuous & unattended recording of telephone & radio communication. Up to 120 channels per single 19" 4U unit. On line storage on disc (up to 1500 hours), archiving to DDS2 cassettes. Units stackable to support very large configurations. Production of the CDF-1500 communications intelligence (COMINT) & direction finding (DF) system. Coverage of 20-1200 MHZ (HF optional), with extensive spectrum surveillance emitter location & monitoring capabilities. Advanced graphic user interface. [97]

75-16200 Opgal Optronic Industry Ltd
(a subsidiary of EL-OP/Rafael-Galram)
Industrial Zone No 5, PO Box 462, 20101 Karmiel
☎ (+972 4) 995 39 03; 995 39 31
⊜ (+972 4) 995 39 00; 995 39 46
♧ US$ 8,675,000 (1996) ♣ US$ 20,000,000 (1996)
ⱳ 120 (1996) ⌂ 2,500 m2
Managing Director: Shalke Orbach ☎ (+972 4) 995 39 03
Sales Director: Silvie Dascalou
Operations and Technical Director: Eitean Zeiler
Purchasing Manager: Ilan Golberg
■ Israeli system house for surveillance and observation systems based on state-of-the-art thermal imaging cameras. [98]

75-16400 Oram Electric Industries Ltd
Industrial Zone A'Ramla, PO Box 126, 71101 Lod
☎ (+972 8) 22 45 91 ⊠ 371131 ⊜ (+972 8) 24 03 83
♣ US$ 6 000 000 ⱳ 130 (1995) ⌂ 10 000 sq m
Managing Dir: Y. Bar-Eli
■ Manufacture of transformers & power supply. [97]

75-16800 ORTEK Ltd
(a subsidiary of El-Op Electro-Optics Ind.Ltd)
PO Box 388, 80100 Sderot
☎ (+972 7) 689 16 91 ⊜ (+972 7) 689 16 95
♣ US$ 3,000,000 ⱳ 20
General Manager: Issak Sheps
■ Night sight for infantry weapons and optical sight for small arms. [98]

75-17000 Palbam Metal Works
18960 En-Harod (Ihud)
☎ (+972 6) 653 17 03 ⊜ (+972 6) 653 19 04
♣ US$ 18,000,000 ⱳ 160 (1997) ⌂ 12,000 m2
General Manager: N Hen
Chief Engineer: C Geyari
Technical Office Department Head: Eli Zamlan
Defence Products Departmant and Head of Special Projects: Yonatan Shani
Chief Financial Manager: Simchon Horesh
Research and Development: Lochay Kimchi
Marketing Manager: Moshe Ben-Ishai
Technical Manager: Ido Nur
Operation and Production Manager: Nissim Coranel
Purchasing Manager: Giora Siebner
Quality Control: Moshe Katz
Production and Personnel: Gil Sohar
■ Stainless steel, aluminium, titanium fabrication combining laser-cutting, forming, welding and machining. Typical products; fuel tanks for tracked and wheeled armoured vehicles, food containers, field kitchens, add-on armour.
Subsidiary:
Ricor Ltd → 75 [98]

75-17500 Pladot
Ein Harod Meuchad
☎ (+972 6) 648 53 50 ⊠ 46347 rod il
⊜ (+972 6) 53 15 41
Managing Dir: Eyal Levite [97]

75-17510 Plasan Sasa
Kibbutz Sasa, 13870 M P Merom Hagalil
☎ (+972 6) 98 85 55; 98 78 80
[E] plasan@net.vision.net.il
⊜ (+972 6) 98 02 12; 98 02 12
ⱳ 50 (1996) ⌂ 2,000 m2
Managing Director: Dan Ziv
Sales and Marketing Director: Oded Sheshinski
Operations and Technical Director: D Benisti
Purchasing Manager: Y Rotem
■ Manufacture of lightweight composite armour for personal protection, vehicles, helicopters and patrol ships. [98]

75-17580 Polyziv Plastic Products
(owned by Kibbutz Gesher Hagiv)
Kibbutz Gesher Haziv, 22815 Western Galilee
☎ (+972 4) 982 57 21 [E] polyziv@netvision.net.il
⊜ (+972 4) 982 57 24
♣ US$ 5,500,000 ⱳ 45 ⌂ 4,000 m2
Managing Director: Peter Braun
Marketing Director: Allan Feldman
Operations Director: Ya'akov Shlesberg
Purchasing Manager: Zvi Elkon
■ A wide variety of plastic products, including cargo pallets and practice bombs. [98]

75-17840 Z. Rabiner Ltd
15 Sederot Chen, Industrial Area, PO Box 1154, 27510 Kiryat Bialik
☎ (+972 4) 876 47 77 ⊠ 46400 bxha il
⊜ (+972 4) 876 26 34
♣ US$ 1 500 000 (1995) ⱳ 35 ⌂ 1 500 m2
President: Zeev Rabiner
General Mgr: Ran Rabiner
■ Precision engineering. Products include: telescope masts (for military & civil communications); moulds for precision castings & aircraft-parts; special injectors (for naval diesel & trucks); aircraft parts' prototypes; precision milling (specialised in manufacture of aircraft parts); & deep precision drilling (from 3mm to 50mm to 1.3m in depth. The company also seeks sub-contracts, joint ventures or distribution rights with overseas companies working in the same field. [97]

75-17850 Rabintex Industries Ltd
28 Lechi St, 51201 Bnei Brak
☎ (+972 3) 570 01 96/7 ⊜ (+972 3) 579 26 66
♣ US$ 30 000 000 ⱳ 450
Managing Dir: Zeev Rabinovitz
Marketing Dir: Amir Haramaty

Contact Reference

■ Bullet proof vests, fire retardent overalls, helmets, tents, fighting order harnesses, passive dry storage systems. Medical safety equipment. [97]

75-17900 Rada Electronic Industries Ltd
12 Medinat Hayehudim, PO Box 2059, 46120 Herzliya-B
☎ (+972 9) 954 21 82 [W] http://www.rada.com
⊜ (+972 9) 955 51 76
♧ US$ 34,244,000 (equity) (1996)
♣ US$ 28,065,000 (1996) 👥 200 (1996) 🏠 3,200 m2
President and Chief Executive Officer: Haim Nissenson
Chairman: Yair Grinberg
Director and Vice President, Head of Marketing the CATS: Brig Gen Meir Shariv (Retired)
President Rada Inc, Head of International Activities: Maj Gen Herzle Bodinger (Retired)
Vice President Engineering: Shlomo Rosen
Operations and Technical Director: Michael Weiss
Marketing Co-ordinator: Mrs Nira Porat Bernstien
Chief Financial Officer: Hermel Meir
Operations and Plan Manager: Michael Weiss
■ Automatic test equipment (ATE) for aircraft and electronic systems. Computerised organisation level testers for several systems. Armament, guided weapon, missiles, cables, fire control, universal 'I' level tester. Avionics and airborne computers including data management and transfer systems. Computers and peripherals. Autonomous combat-maneuvers evaluation systems for flight debriefing.
US Office:
Rada Electronic Industries Inc
Plant
New Industrial Zone, PO Box 54, 10900 Beit-Shean
☎ (+972 6) 658 58 11 ⊜ (+972 6) 658 43 01 [98]

75-18600 Reshef Technologies Ltd
(a subsidiary of Aryt Industries Ltd)
7 Haplada Street, PO Box 696, 60256 Or-Yehuda
☎ (+972 3) 538 86 29 ⊜ (+972 3) 533 92 23
♣ US$ 15,300,000 (1996) 👥 140 (1996)
President: Arieh Geller
Vice President Marketing: Shmuel Shoham
■ Main Products:
Alpha M787 Mortar Proximity Fuse Lambda M760 Mortar Electronic Time Fuse EPD M797 Mortar Electronic Point Detonating Fuse. Delta M137 Artillery Electronic Time Fuse. Epsilon M139 Artillary Proximity Fuse. Omega M127 Tank Electronic Time/Distance Fuse. EPD-M176 electroic point detonating fuse for artillery rockets, KAPPA-M175 proximity fuse for artillery rockets, SIGMA-M174 electronic time fuse for artillery rockets. [98]

75-18800 Ricor Cryogenic & Vacuum Systems
(a subsidiary of Palbam Metal Works)
18960 En-Harod (Ihud)
☎ (+972 6) 653 17 03 ✉ RICOR Enharodihud
[E] pundak@inter.net.il ⊜ (+972 6) 653 24 24
♣ US$ 9,200,000 (1996) 👥 65 🏠 1,250 m2
General Manager: N Pundak
Sales and Marketing: Y Zur
Purchasing Manager: D Robinson
Personnel: I Tsphadieth
Finance: S Choresh
Quality Assurance: G Pasternak
Stockroom: R Pundak
Machine Shop: B Prosper
Assembly Department: A Dagan
G-M Products: Z Ron
■ Cryogenic and vacuum systems. Cryogenic coolers for IR thermal imaging systems. Cryogenic storage and transport tanks.
Exclusive distributors:
USA and Canada: C.E. Corp
UK and France: Coercive Ltd [98]

75-18805 Rokar International Ltd
(a subsidiary of Tracor Inc, USA)
PO Box 3294, 91032 Jerusalem
☎ (+972 2) 532 98 88 ⊜ (+972 2) 582 25 22
♣ US$ 8,000,000 👥 90
President: Moshe Peled
Managing Director: Arie Shmueli
■ Develops, manufacturers and supplies customer support services in the following areas: Electronic airborne threat adaptive countermeasures, radio navigation-GPS and Omega, aerial camera control and mission planning. [98]

75-18900 RSL Electronics Ltd
PO Box 21, 10550 Migdal Haemek
☎ (+972 6) 654 75 10; 654 75 82
✆ (+972 6) 654 31 66 [E] zeevd@zsl.co.il
⊜ (+972 6) 654 75 20
♧ US$ 12,500,000 (1995) ♣ US$ 12,000,000 (1995)
👥 110 (1994) 🏠 4,000 sq m
Managing Director: Abraham Ofek
Technical Director: Amos Oren
Business Development Director: Zeev Degani
Marketing Director: Jacob Bridger
Purchasing Manager: Ari Tabor
■ Fixed wing and helicopter jet engine monitoring and control system, event history recorders for jet engines, muzzle velocity measuring radars. Advanced brake systems controllers for aircraft. [98]

75-19500 Semi-Conductor Devices (SCD)
PO Box 2250, 31021 Haifa
☎ (+972 4) 990 66 36 [E] mar-scd@actcom.co.il
⊜ (+972 4) 990 66 26
♣ US$ 20,000,000 👥 180
Managing Director: Dr Yair Alpern
Sales and Marketing Director: Dr Ehud Navon
Purchasing Manager: S Reiss
■ Manufacture of infra-red detectors and semiconductor lasers. [98]

75-20000 Shalon Chemical Industries Ltd
25 Nachmani Street, 65794 Tel Aviv
☎ (+972 3) 629 12 26; 629 12 27
[W] http://www.doryanet.co.il/shlon17.htm
[E] shalon@netvision.net.il ⊜ (+972 3) 629 16 15
♣ US$ 10,000,000 (1996) 👥 250 (1996)
🏠 10,000 sq m
Managing Director: Ram Baharad ☎ (+972 7) 81 12 71
Sales and Marketing Director: Dr Itai Barel ☎ (+972 7) 81 12 71
Operations and Technical Director: Ken Samet
■ NBC protective equipment for personal and collective applications including ballistic, sun, wind, dust and laser protective goggles. [98]

75-20250 G. Shiposh, Engineering Works Ltd
11 Qiryat Bialik, PO Box 11, 27000 Haifa Bay
☎ (+972 4) 877 28 48; 876 36 19 ✉ 45157 ship il
⊜ (+972 4) 877 31 86
👥 80 🏠 2,000 sq m
President: George Shiposh ☎ (+972 4) 876 36 19
Sales and Marketing Director: Micha Shiposh ☎ (+972 4) 651 44 19
■ Military equipment, rockets and rocket launchers, pre-fab pipe lines military application, pre-fab shelters, pre-fab refuelling station systems.
Overseas Offices:
Shiposh Engineering Ltd → USA 90 [98]

75-20400 Silver Arrow
10 Roujansky St, 75706 Rishon-Lezion
☎ (+972 3) 963 21 11 ⊜ (+972 3) 963 21 12
♣ US$ 14 000 000 (1995) 👥 160
President: Shay Shapira ☎ (+972 3) 963 21 14
VP Engineering: Giora Hertzberg ☎ (+972 3) 963 21 60
Marketing Dir: Mooly Dinnar ☎ (+972 3) 963 21 13
Purchasing Mgr: Israel Lunchner ☎ (+972 3) 963 21 17
Human Resources Mgr: Meirav Rosi-Paz ☎ (+972 3) 963 21 30
Financial Mgr: Opher Bar-ner ☎ (+972 3) 963 21 24
■ Development & production of unmanned aerial vehicles (UAV's) & associated support systems. Products: Micro-V system, Hermes 450 & Hermes 1000. [97]

75-21050 Special Electronic Security Products Ltd
12/B Rahavat-Ilan Street, 54056 Givat-Shmuel
☎ (+972 3) 532 51 32; 532 16 95 [E] tunik@sesp.co.il
[W] http://www.sesp.co.il
⊜ (+972 3) 532 43 11; 532 35 53
♧ US$ 900,000 👥 9
Managing Director: Eng Y Tunik
Sales and Marketing Director: J Malka
Operations and Technical Director: Ms Julia Estrin
Purchasing Manager: T Koren
■ Design and manufacture of surveillance and counter-surveillance equipment. Miniature bugs, wire taps and transmitters for telephones. Transmitter locator to 2 GHz scramblers for telephone lines, fax and rooms. Counter-surveillance equipment for de-activation and detection of bugs/taps. Discreet recording devices, miniature televsion covert video transmitters and boosters to 7.5 GHE. [98]

75-21100 Spectronix Ltd
PO Box 29, 80150 Sperot
☎ (+972 7) 689 91 73 [E] spec@netvision.net.il
⊜ (+972 7) 689 05 80
President: Y Spector
Vice President Marketing: A Gonen
■ Automatic fire extinguishing and explosion suppression systems for military vehicles, naval vessels and aircraft shelters. Damage control systems. [98]

75-21320 Suspension & Parts Industries Ltd (SPI)
(a subsidiary of Urdan Industries Ltd, Israel)
Ha-Yotsrim St. Industrial Zone, PO Box 305, 20101 Carmiel
☎ (+972 4) 998 83 01 ✉ 46326 spi il
⊜ (+972 4) 998 11 13
♣ US$ 8,900,000 (1995) 👥 100 🏠 8,500 m2
President: Shlomo Pazzy
■ Designs, develops, manufactures & markets suspension components for armoured vehicles. Full range of roadwheels, sprockets, arms, rotational shock absorbers, track adjusters, track components & assemblies. APCs, MBTs, IFVs, etc. [97]

75-21500 Tadiran Com. Ltd.
(† Tadiran Communications & Systems)
26 Hashoftim St, PO Box 267, 58102 Holon
☎ (+972 3) 557 42 50 ⊜ (+972 3) 557 53 00
♣ US$ 105,000,000 (1997) 👥 790
President and Chief Executive Officer: Hezi Hermoni
Vice President - Marketing: Lior Almagor
Vice President - Operations: David Cohen
Vice President - Human Resources: Mosh Goldstein
Vice President Finance and Control, Chief Finance Officer: Hehaia Shiff
Vice President - Purchasing: Zeev Kowalski
Vice President - Tactical Radio: Amos Weizman
Vice President - Advanced Data Communications: Shmual Katz
Director of Marketing Communications: Ms Margalit Degani
General Council: Layla Chertow
■ Development, production and marketing of a full range of military and commercial products and systems, as follows: Tactical communications, military telecommunications systems, C3I (Communications, Command, Control and Intelligence), military computers, EW (Electronic Warfare), reconnaissance and surveillance, platform upgrades, simulators, security systems, wireless data communications, spectrum monitoring, vehicle location and fleet management, automatic port systems, digital video compression, electronic toll systems. [98]

75-21550 Tadiran Spectralink
2 Hamerrava Street, PO Box 150, 58101 Holon
☎ (+972 3) 557 31 00
[W] http://www.tadiran.spectralink.com
⊜ (+972 3) 557 31 31
♣ US$ 35,000,000 👥 140 (1997) 🏠 4,000 sq m
Managing Director: Yitzhar Beni
Marketing Director: David Hornstein
Marketing Director: Avi Peleg
■ Supplier of data link systems for a variety of airborne platforms and ground installations. Datalink design, software development, system integration, testing, manufacturing of hardware, technnical support and global marketing. [98]

75-21600 Tadiran Ltd
29 Hamerkava Street, PO Box 150, 58101 Holon
☎ (+972 3) 557 32 00 [W] http://www.tadiran.com
⊜ (+972 3) 557 32 80
♧ US$ 331,000,000 (equity 1996)
♣ US$ 1,117,000,000 (sales revenue 1996)
US$ 370,000,000 (defence sales revenue 1996)
👥 8,200 (1996)
Chairman: Benjammin Gaon
President: Israel Zamir
Senior Vice President: Benny Maiden

■ Tactical communication equipment systems, data links, tactical terminals and electronic warfare systems. C3I and reconnaissance systems, simulators and trainers batteries.

Subsidiaries with defence business:
Tadiran Com Ltd
Elisra Electronic Systems Ltd
Tadiran Electronic Systems Ltd
Tadiran Spectralink Ltd
Advanced Technology Ltd
Simtech Advanced Training and Simulation Systems Ltd
[98]

75-21900 Tamor SMR Ltd
PO Box 19, 10551 Migdal Ha'amek
☎ (+972 6) 657 67 35 [E] tamor@doryanet.co.il
[W] http://www.doryanet.co.il/tamor/i.htm
⊖ (+972 6) 657 67 38
♣ US$ 4,500,000 (1996) 👥 35 (1996) 🏭 2,500 m2
Managing Director: Moshe Mazor
Operations Director: Ronen Mazor
■ Supplier of rubber products for military tracked vehicles to the Israel Defense Forces and other Armies. Rubber and rubber-to-metal bonded parts for tracked armoured vehicles, complete tank tracks, road and tension wheels, endless rubber bands for half tracks. [98]

75-22000 TAT Industries Ltd
(† TAT Aero Equipment Industries Ltd)
PO Box 80, 70750 Gedera
☎ (+972 8) 859 54 11 ⊖ (+972 8) 859 28 31
♧ US$ 23,400,000 ♣ US$ 18,000,000 👥 120
🏭 80,800 sq ft
Chairman: S Ostersetzer
Vice President Marketing: Ami Zelcer
Managing Director: D Zeelim
Marketing Director: H Gazit
■ Airborne systems and equipment: fuel, oxygen, instrumentation and electronic assemblies, environmental control systems, heat exchangers, secondary power systems.

US Office:
Limco Airepair Inc. → 90

Plant
PO Box 80, 70750 Gedera
☎ (+972 8) 859 54 11; 859 54 40
⊖ (+972 8) 859 28 31
■ Airborne systems & equipment: fuel, power, oxygen, environmental control systems, heat exchangers & secondary power systems. [98]

75-22400 Telkoor Ltd
5 Hatzoref St, PO Box 1871, 58102 Holon
☎ (+972 3) 557 04 44 ⊖ (+972 3) 557 04 33
♣ US$ 5 200 000 👥 50 🏭 1 500 sq m
Managing Dir: David Amitai ☎ (+972 3) 557 04 30
Sales & Marketing Dir: Josef Ben-Ami
☎ (+972 3) 557 04 35
Operations & Tecnical Dir: Itzhak Harlap
Purchasing Mgr: Sam Halperin
■ Custom made military power supplies. Low voltage power supplies. DC/DC converters. AC/DC switch inverters. [97]

75-22504 Temmek Optics Ltd
(† Rehovoth Optics Ltd)
PO Box 159, 81101 Yavne
☎ (+972 8) 943 84 08 [E] maresco@crendline.co.il
⊖ (+972 8) 943 83 90
♧ US$ 10,000,000 (1996) ♣ US$ 4,000,000 (1996)
👥 85 (1997) 🏭 5,000 m2 (1997)
Managing Director: Z Maresse
Operations Director: E Maresse
Marketing and Sales Director: O Benyakov
Marketing and Sales Director: Yoran Burger
Purchasing Manager: D Maresse
■ Custom precision optical elements and coatings for the visible and IR spectrums. Precision CNC machining of glass and metals. Optical subassemblies. [98]

75-23000 Thermofil-Cvalim Ltd
(Electronic Cables Div. of The Electric Wire & Cable Co. of Israel Ltd)
Industrial Area Ma'alot, PO Box 38, 24952 Ma'alot
☎ (+972 4) 97 69 62; 97 69 64 ⊠ 45103 ewco il
⊖ (+972 4) 97 69 61
Marketing Manager: Meir Brauman
■ Electronic wires and cables for aviation (aerospace) and computer industry. Cables to Military Specification. [98]

75-23055 Trans Security Systems & Technology Ltd
6 Galgalei Haplada Street, PO Box 2014,
46722 IZ Herzliya Pituach
☎ (+972 52) 58 50 51; 58 50 47
[E] transsec@netvision.net.il ⊖ (+972 52) 58 50 60
Managing Director: Uri Alizi
Marketing Director: Yoram Leshem
Technical Manager: Amnon Shvartz
■ Manufacture of F-5000 fibernet Fibre-optic fence detection system; TEE-4000 monitoring and control system for high security risk installations; F-8000 Marinet-maxine security system; F-7000 Trans-Cable multi purpose perimeter intrusion detection system. [98]

75-23210 Urdan Industries Ltd
(a subsidiary of Clal Industries)
Industrial Center, 42378 Natanya
☎ (+972 9) 33 80 74 ⊖ (+972 9) 61 02 46
♣ US$ 54,000,000 (1995) 👥 615
General Mgr: Israel Sever
■ Suspension parts for MBT, APC; mine clearing system, ammunition trailer for SPG; heavy trailers; upgrading kits for MBT & APC; add on armour. → 80 [97]

75-23238 Vargus Ltd
1 Hayotsrim Street, PO Box 266, 22311 Nahariya
☎ (+972 4) 985 51 11 [E] vargusfr@netvision.net.il
⊖ (+972 4) 982 58 10
♣ US$ 8,000,000 👥 160
Managing Director: Harry Ehrenberg
Operations Director: M Bloch
Marketing Manager: S Palmer
■ Hard metal cutting tools. ISO 9002 certified. [98]

75-23240 Vidco Industries Ltd
2 Ha'ofan St, PO Box 2502, 58125 Holon
☎ (+972 3) 556 95 95 ⊖ (+972 3) 556 81 16
♣ US$ 85 000 000 (1996) 👥 900 (1996) 🏭 12 000 sq m
President: Abraham Inbar
■ Thermal shrouds for gun barrels. [97]

75-23260 Vidisco Ltd
R&D Video & Computer Systems
1 Korazin Street, PO Box 1138, 53111 Givataim
☎ (+972 3) 573 05 92; 573 05 93
[E] vidisco@mail.netvision.net.il ⊖ (+972 3) 571 33 37
Chief Executive Officer: Shlomo Shapira
Managing Director: Michael Hilsenrad
Marketing Manager: Ari Diamond
Managing Engineer of Research and Development: Lior Pick
Production Manager: Eli Dayan
■ Portable real-time video imaging X-ray inspection systems. Bomb disposal, drug search, forensics, customs/smuggling, VIP security and non-destructive testing. [98]

75-23290 Vulcan Batteries Ltd
Tefen Industrial Area, PO Box 17, 24959 Tefen
☎ (+972 4) 987 22 05 ⊖ (+972 4) 987 26 72
♣ US$ 15 000 000 (1995) 👥 145 🏭 12 500 sq m
Managing Dir: David Shainin
Sales & Marketing Dir: Gilad Manor
Operations & Technical Dir: David Ifargan
Purchasing Mgr: Abrasha Malkevich
■ Starter batteries for tanks, armoured vehicles, trucks & other military vehicles. [97]

75-23600 Yariv
46 Ahad Haam Street, 43211 Raanana
☎ (+972 9) 748 14 62 ✆ (+972 52) 45 60 67
[E] garivetn@netvision.net.il ⊖ (+972 9) 748 14 62
Air Defence Consultant: Brig Gen Yariv Eitan [98]

75-23800 Electrode Works Zika Ltd
PO Box 2317, 24122 Acre
☎ (+972 4) 985 18 51 [E] zika1@netvision.net.il
⊖ (+972 4) 985 18 78
♣ US$ 35,000,000 (1997) 👥 180
Managing Director: Israel Shotland
Operations/Technical Dir: U Goldstein
Sales Director: Yaron Danieli
Marketing Executive: M Rotter
Export/Import Manager: U Marom
■ Manufacture of a wide range of arc welding electrodes and welding wire. [98]

80 Service Companies & Organizations

Advanced Technology Ltd	75-00400
AlliedSignal Aerospace Marketing	90-00075
Arotec Ltd	75-01870
Cyclone Aviation Products Ltd	75-04250
Israel Shipyards Ltd	75-10600
Kalia Ltd	85-00500
Magnus Engineering & Maintenance Ltd	75-12750
Mipha International Ltd	90-01350
MTI Engineering Ltd	90-01360
Professional School of Security	60-15000
Urdan Industries Ltd	75-23210

80-00040 Aerospace Design Hugo Marom Ltd (ADHM)
(a subsidiary of Hugo Marom Aviation Consultants)
50 Levi Eshkol Street, 69361 Tel-Aviv
☎ (+972 3) 699 12 24; 699 21 41
⊖ (+972 3) 699 19 57
President: Hugo Marom
■ Engineering of all aircraft and airport ground equipment, aircraft test equipment, engine test cells, sound control equipment, airborne equipment and systems, aircraft installations and modifications. Updating and improving aircraft performance; design of aircraft reconnaissance and installations, collimator facilities; equipment standardisation studies, specification preparation, cost and program management. [98]

80-00100 Airport Engineering Hugo Marom Ltd (AEHM)
(subs. of Hugo Marom Aviation Consultants)
50 Levi Eshkol Street, 69361 Tel-Aviv
☎ (+972 3) 699 12 24; 699 21 41
⊖ (+972 3) 699 19 57
♣ US$ 750,000 (1996) 👥 6 🏭 500 sq m
President: Hugo Marom
■ Preparation of airfield concepts; total master planning of military and civil airports; architectural detailed design of complete airport buildings, hangars and underground aircraft shelters; hardened fighter maintenance facilities, fighter aircraft run-up stands and engine test cells with silencers; specialists in offshore airports and airport automated information and recording systems. [98]

80-00150 Aman-Management Consultants Ltd
(a subsidiary of Amanet Systems Ltd)
34 Yitzhak Sateh St, 67212 Tel Aviv
☎ (+972 3) 638 35 55 ⊖ (+972 3) 537 11 70
♧ US$ 13,000,000 (1995) ♣ US$ 27,000,000 (1995)
👥 300 🏭 1,500 sq m
Chairman: Abraham Assef
Managing Dir: S. Ber Or
Operations & Technical Dir: Amnon Bartal
Sales & Marketing Dir: I. Shaham
■ Strategic planning, project management engineering, civil engineering, economics, organisational consulting, systems analysis & design, programming & computer operation. [97]

80-00180 Armaz (Bagira) International Consulting Ltd
7 Haharuv Street, PO Box 4716, 40800 Rosh Haayin
☎ (+972 3) 938 41 25 [E] armaz@aol.co.il
⊖ (+972 3) 938 41 19
♣ US$ 1,500,000 👥 25
President: Arie Mizrachi
Operations Director: Yaron Mizrach
Marketing Director: Tomer Sharon
■ Provides international consulting services to military-oriented industries, armed forces and ministries worldwide. [98]

80-00250 Cermat Ltd
4A Tsolelet Dakar St, 34862 Haifa
☎ (+972 4) 834 33 37 ✆ (972) 52 53 18 67
⊖ (+972 4) 834 33 37
♣ US$ 400 000 👥 2
Managing Dir: Itamar Israeli
■ Involved in projects in armour protection & suppling special materials for composites: Ceramics, glues, etc. → 90 [97]

Contact Reference

80-00275 C&E
PO Box 2367, 55122 Kiryat-Ono
☎ (+972 3) 532 36 03 ⊜ (+972 3) 532 51 48
General Manager: Baruch Sharon
■ Aeronautical engineering consultants specialising in airbourne structural design, aircraft mechanical systems, systems development, and aircraft and helicopters crashworthy seats development. 98

80-00300 Contahal Ltd
(a member of the Tadiran Group)
1 Corazin Street, 53583 Givataim
☎ (+972 3) 572 52 52 E z_amitai@contahal.co.il
⊜ (+972 3) 571 51 27
♧ US$ 12,500,000 (1996) ♣ US$ 21,000,000 (1996)
👥 300
Managing Director: Dan Kunik
Manager Real-Time and Military Systems: Zvi Amitai
■ Software development for defence and real-time systems, database management, industrial process and manufacturing control, commercial and financial data processing. Marketing of of own and other known "off the shelf" software products. 98

80-00400 Data Systems Designers (DSD)
54 Weizman Street, 44247 Kfar-Saba
☎ (+972 9) 748 02 10; 791 32 97; 749 39 44
E dsdmeir@netvision.net.il ⊜ (+972 9) 760 28 29
👥 9
Managing Director: Meir Davidovitz
Sales and Marketing Director: Israel Gatt
■ Consulting and configuration management. 98

80-00430 DEMCO Consultants Ltd
(† DEMCO Defence & Military Consultants Ltd)
29 Ha'Yetzira Street, PO Box 3107, 52130 Ramat Gan
☎ (+972 3) 752 53 33 ⊠ 341167 coin il, Att: DEM
⊜ (+972 3) 752 76 04
♣ US$ 3,600,000 (1996) 👥 35 (1996) 🏭 550 sq m
President: Gen Y Stern (Retired)
Managing Director: Col N Rafaely (Retired)
Deputy Managing Director: Col A Kayam
Chief Finance Officer: Shai Levy
■ Consultancy services in national defence, military security and civil defence fields. Specialises management of large scale emergencies. Intelligence and criminal intelligence, information to the public, security and protection. Stress prevention and management. Essential services and economy in emergencies. Environmental hazards. Border and coastal defence/protection. Inter service/departmental joint/combined operations. Anti-terrorism warfare. Rescue and emergency medicine. Command and control, communications, dispatch systems for emergency services. Training and exercises. Organisation and human resources development. Integrated logistic support -ILS. 98

80-00480 Eastronics Ltd
11 Rozanis Street, PO Box 39300, 61392 Tel-Aviv
☎ (+972 3) 645 87 77 ⊜ (+972 3) 645 86 66
👥 130
Chairman: Theodore Kenny
Managing Director: Michael Kenny
Vice President Marketing and Sales: Arie Eitan
Marketing Manager: Zvi Goldberg
■ Provides representation, maintenance and consultant services.
® Canada: Litton-Canada, MDA
Germany: Litef
Switzerland: Balzers
USA: Bauirgu, Honeywell, Intel, Intergraph, Magnavox Electronic Systems Co, Sanders (a Lockheed Martin Company), Tektronix 98

80-00500 Eilat Consultants Accidents Prevention Media Ltd
26 Hagdud Haivri Street, 66852 Tel Aviv
☎ (+972 3) 687 96 72 E zb@netvision.net.il
⊜ (+972 3) 687 64 07
Managing Director: Itzchak Lichtenfeld
Program Manager: Zvi Beniamini
■ Offers services in the following areas: Safety analysis and research, safety programs and risk management, professional safety training and education to pilots, field commanders, safety officers and high level command. Accidents investigations. 98

80-00750 Gordon Consulting Engineers Ltd
"Elran Centre", 96 Auhza Street, 43450 Raanana
☎ (+972 9) 743 01 66 E ig@netvision.net.il
⊜ (+972 9) 743 01 62
♣ US$ 2,000,000 (1996) 👥 15 (1996) 🏭 250 m2
President: J Gordon
Marketing and Sales Director: M Golan
■ Consulting, planning, supervising for security and communication systems. 98

80-00880 Hugo Marom Aviation Consultants Ltd (HMAC)
50 Levi Eshkol Street, 69361 Tel-Aviv
☎ (+972 3) 699 12 24; 699 21 41
⊜ (+972 3) 699 19 57
♧ US$ 200,000 ♣ US$ 200,000 (1997) 👥 12
🏭 500 sq m
President: Hugo Marom
Vice President: M Marom
■ Aircraft type and performance evaluation; aircraft and engine overhaul maintenance and manufacturing facilities; master planning complete with economic analysis; cost benefit studies of aerospace components and manufacturing facilities; flight test and certification; aircraft operating manuals; pilot evaluation and grading; airfield security; forecasting and accident investigation.
Associates:
Muller & Schnabel, Structural Engineers
Ophir Ronen, Energy Control Engineers
Kaplan & Navot, Electrical Engineers
Yoshpe Segalis, Sanitary Engineers 98

80-00960 International Security & Defense Systems Ltd (ISDS)
PO Box 2283, 76122 Rehovot
☎ (+972 8) 922 76 99 ⊜ (+972 8) 923 51 04
♣ US$ 4 000 000 (1994) 👥 15
President: Leo Gleser
Sales & Marketing Dir: Arieh Avnat
Operations Dir: Alex Ein-Gil
Advertising Mgr: Shaul Bril
Public Relations Mgr: Julian Rubin
■ Licensed by the Israeli Ministry of Defense as a supplier to I.D.F. and also authorized in handling international negotiations and sales for: Arms and Defense equipment, National Defense and Security consulting, and training services. 97

80-00970 Israel Aerospace and Defence (IADE)
Haneviim 26 Street, 64071 Tel Aviv
☎ (+972 3) 528 21 92 E eshel@shani.net
W http://www.iaw.co.il ⊜ (+972 3) 525 18 39
👥 10
Managing Director: Ron Peer ☎ (+972 52) 56 86 69
Marketing Manager: Erika Zuk
Advertising Director: Tmir Eshel
■ Organisers of aerospace and defence conferences, exhibitions and airshows in Israel. 98

80-00980 LBG Agencies (1979) Ltd
3 Shaar Palmer Str., PO Box 33622, 31339 Haifa
☎ (+972 4) 867 17 19; 864 02 02
E santo@netvision.net.il ⊠ 46549 lbg il
⊜ (+972 4) 864 08 46
👥 27
Managing Dir: Naval Capt. (Res.) Sappir Ben-Zouk
Marketing Mgr: Mrs Lia Gabler
■ Representatives of defence industries in Israel, export of military equipment of the Israeli MOD (SIBAT), light & heavy ammunition, small arms, uniforms, vehicles type jeep & security vehicles, night vision, mines, grenades, shells, mortar bombs, explosives, intelligence systems, scabbard & bayonets, boots & shoes, NBC protective equipment. 97

80-01350 MITM Consulting Engineers
9 Baskind Street, 75706 Rishon Lezion
☎ (+972 3) 951 42 61 E mitm_c@netvision.net.il
⊜ (+972 3) 951 50 02
Managing Director: Ahmed Sobelr Soniol
Operations Director: Eng Abdel Basset
■ Specialising in the design of fixed systems.
® France: Groupe Bacou Comoditex
Hungary: Hydro Export
Italy: Grace International, Hospitalia Int
Switzerland: Dupont de Nemours International
UK: Grace International
USA: Grace International, The Kerite Co, Zarc International 98

80-01400 Nimda Co. Ltd.
PO Box 821, 53108 Givataim
☎ (+972 3) 571 21 61 ⊠ 341457 nim il
⊜ (+972 3) 571 50 22
👥 100
President: Zvi Nir
Vice President Engineering and Projects: Arie Semel
Vice President Finance: Shaike Perl
Vice President Marketing: Rami Ziv
Director of Marketing: Zvi Gera
■ Turnkey projects for production, integration and modernisation of military vehicles. Upgrading of MBTs, APCs, amphibious armoured vehicles and trucks. 98

80-01470 Nogay Ltd
6 Hassolelim Street, PO Box 9274, 61092 Tel Aviv
☎ (+972 3) 562 55 44 E mogay@netvision.net.il
⊜ (+972 3) 562 55 48
♣ US$ 500,000 👥 10 🏭 200 m2
President: Pesach Shachar
Operations Director: Isaac Zakh
■ Communications planning and consultancy. 98

80-01600 M. Paz Logistics Ltd
10 Yad Harutzim Street, PO Box 2242, Kfar Saba
☎ (+972 9) 766 15 55 E pazlog@netvision
⊜ (+972 9) 766 15 50
👥 19 (1997) 🏭 450 sq m
President: Ofer Paz
Executive Vice President and General Manager: Oded Paz
Technical Director: Ran Ofer
Sales and Marketing Director: Yariv Sagi
Operations and Purchasing Manager: Anat Paz
■ High level logistic and technical consulting. Specialise in automative and retrofit projects. Represent foreign firms in Israel. → 85 → 90 98

80-01650 Planning & Research Ltd (P&R)
PO Box 39889, 61398 Tel-Aviv
☎ (+972 3) 649 95 15 E fsjarch@trendline.co.il
⊜ (+972 3) 649 96 15
♣ US$ 3,000,000 👥 25 (1997) 🏭 300 sq m
Executive Director: Shmuel Shilo
■ Comprehensive planning and design of military installations and complexes, including: military regional planning and locational criteria analysis; planning and design programmes; master plans and profile; layout planning; detailed architectural design of training lands and installations; special structures and buildings design; specific installations design. Characterisation and design of fire fighting and rescue training facilities. 98

80-01760 Protrade Ltd
(a subsidiary of Procon Ltd)
11 Hibat Zion Street, 52391 Ramat Gan Israel
☎ (+972 3) 570 11 51 ⊜ (+972 3) 570 23 17
Managing Dir: A. Paldi
■ Surplus military hardware, equipment & systems, ex-stock supplies; firearm reconditioning; special requirements; consultancy - anti-terrorist & security devices. → 85 97

80-01900 Scitronix Ltd
15 Rehov Sherf, Pisgat Ze'ev Mizrach, 97842 Jerusalem
☎ (+972 2) 583 31 51 E scitronx@netvision.net.il
⊜ (+972 2) 583 31 52
♣ US$ 2,000,000 👥 3
Managing Director: David Morris
Sales Director: Batya Engelberg
■ Development and marketing consultants, specialising in aerospace and defence. Services the marketing requirements of defence contractors, and for marketing activities on behalf of overseas companies. Marketing services include representation, market surveys, technology surveys, web page design and promotion. 98

80-01930 S.G.D. Engineering Ltd
41 Nachmani Street, PO Box 36637, 67136 Tel Aviv
☎ (+972 3) 560 41 35 ⊠ 371987 sgd il
⊜ (+972 3) 560 21 79
♧ US$ 500,000 (1996) ♣ US$ 2,500,000 (1996) 👥 40
🏭 250 m2

President: Dov Cohen
Chief Engineer: Dan Valdman
Purchasing Manager: Yarona Marnin
■ Aircraft engineering: modifications and installations, mechanical design of airborne systems and aircraft-related equipment. [98]

80-01935 SPL Ordnance Ltd
10 Yad Harutzim Street, PO Box 2242, Kfar Saba
☎ (+972 9) 766 15 55 ⊖ (+972 9) 766 15 50
👥 10 (1997) 🏭 450 sq m
President: Brig Gen Micha Paz (Retired)
Managing Director: Lieut Col Meir Savariego (Retired)
Marketing Manager: Oded Paz
■ Dedicated to the specialist area of EOD (Explosive Ordnance Disposal), IEDD (Improvised Explosive Devices Disposal) and rescue operations. Expertise in consultancy, training and providing specialised equipment in the above fields. → 85 → 90 [98]

80-01938 Shoval-Saar Ltd (SSL)
7 A. Zarchin St. Beit Shaul, PO Box 2220, 43662 Raanana
☎ (+972 9) 626 31 55 ⊖ (+972 9) 46 11 60
♧ NIS 300 000 (1996) 🏭 300 sq m
Managing Dir: Z. Yanai
Operations & Technical Dir: S. Menikevitch
Security Programmes Dir: S. Hotter
Maritime Security Programmes Dir: N. Stick
Special Naval Warfare Dir: I. Arad
■ Maritime & naval consultencies. Training & doctrine evolution & evaluation. Onboard waste processing by Prolysis Plasma fed systems. Surveillance & security for coastal areas land borders & sites: harbours, terminals, resorts, shipping & offshore installations. [97]

80-01940 Syscon Engineering
PO Box 1241, 52112 Ramat-Gan
46/9 Congress Avenue
☎ (+972 3) 677 99 10 ⊖ (+972 3) 677 05 45
♣ US$ 1,500,000 👥 3
Director: H Lieberman
■ Consulting in high-power transmitters design and applications. Consulting and design of telecommunication supervisory systems. AC/DC and battery systems. → 75
® UK: Farnell Int'l [98]

80-02018 Taran Technologies Ltd
(a subsidiary of Petrus Aviation Ltd)
Hayarkonb Street, No 35 Blei-Brack, 521204 Ramat Gan
☎ (+972 3) 578 04 54; 578 04 55 [E] petruss@inter.net.il
⊖ (+972 3) 618 69 70
👥 5 (1997) 🏭 125 (1997)
Managing Director: Avi Mor
Vice President Sales and Marketing: Nadir Segall
■ Representation of defence related manufacturers, particularly in aviation products and services. Sales of Air Force and Army surplus equipment, aircraft, engines, spare parts, armament and avionics. → 90 [98]

80-02025 Technum Integrated Technologies Ltd
PO Box 3744, 31073 Haifa
☎ (+972 4) 824 63 60 ⊖ (+972 4) 824 40 44
👥 5 🏭 250 sq m
Managing Dir: G. Raz
Sales & Marketing Dir: M/s Rebecca Raz
Operations & Technical Dir: Tuvia Sher
■ Promotion of high-tech projects, including projects with military applications. [97]

85 Stockists, Distributors, Spare Parts Suppliers

M. Paz Logistics Ltd 80-01600
Protrade Ltd 80-01760
SPL Ordnance Ltd 80-01935

85-00500 Kalia Ltd
(† Kalia Israel Cartridge Co. Ltd)
PO Box 18434, 91183 Jerusalem
☎ (+972 2) 535 45 65 [E] kalia@netvision.net.il
⊖ (+972 2) 535 45 66
Managing Director: Zev Kaptowsky
Operations Director: Yosef Lockhart
→ 80 [98]

85-01700 Odmat Rotem
8 Hateena Street, Caesarea
☎ (+972 6) 636 21 24 [E] ad_rotem@netvision.net.il
⊖ (+972 6) 636 30 52
♣ US$ 1,000,000 👥 2
President: Amnon Rotem
Vice President: D Rotem
■ Export and marketing services. Activities include exporting Israeli products, international trade, consulting services, market surveys and joint ventures. → 90 [98]

85-06000 T.I.P.
30805 Kibbutz Maayan ZVI
☎ (+972 6) 39 50 60 ⊖ (+972 6) 39 68 61
General Mgr: Avraham Sagie
■ Marketing of Meprolight products. [97]

85-08000 TRACE Israel
PO Box 1128, 47110 Ramat Hasaron
Managing Director: Joshua Ron
® USA: TRACE Worldwide Corp. [98]

90 Sales Agents, Sales & Liaison Offices

Companies represented:

Aeroflex International, USA 90-01800
Aeroflex Laboratories Inc, USA 90-01800
Aeroflex-Lintek, USA 90-01800
Airborne Product Supply Ltd, USA . . . 90-00200
Allied International Corp., USA . . . 90-01960
Amherst Systems Inc, USA . . . 90-01800
Anghel Laboratories Inc, USA . . . 90-01800
Associated Industries Inc., USA . . . 90-00540
Associated Spring Spec Ltd., UK . . . 90-00430
Astralloy-Wear Technology, USA . . . 90-00120
Astro Gear & Instrument Co., USA . . . 90-00120
Astronautics Corporation of America, USA . . . 75-02400
ATX, USA . . . 90-00135
Ausimag Centrifugal Pumps, Australia . 90-00120
Avio Diepen B.V., Netherlands . . . 90-00120
Aydin Microwave, USA . . . 90-00135
Aydin Monitor & Computer, USA . . . 90-00135
Aydin Vector, USA . . . 90-00135
Balzers, Switzerland . . . 80-00480
Barden Corp., USA . . . 90-00430
Bauirgu, USA . . . 80-00480
Berg, Winfred M., USA . . . 90-00430
Bliss & Laughlin Steel Co, USA . . . 90-00120
BMI, France . . . 90-00135
Boysen GmbH, Germany . . . 90-00120
CAE Link Corp., USA . . . 90-00430
Cannon Electric GmbH, Germany . . . 90-00850
Chemimetal SA Bruxelles, Belgium . . . 90-00200
Cherry Div. Textron Inc., USA . . . 90-00319
CHL, USA . . . 90-01800
Circle Seal Controls, USA . . . 90-00120
Comstron Corp, USA . . . 90-01800
Condor Systems, USA . . . 90-01360
Cubic Communications, USA . . . 90-01800
Custom Electronics Inc, USA . . . 90-00130
D.I.O.P., USA . . . 90-00430
D.J.Mouldings, UK . . . 90-01840
Daimler-Benz Aerospace, Germany . . . 90-00200
Datatape, USA . . . 90-01360
Delfin Systems, USA . . . 90-01800
Diamond Antenna & Microwave, USA . 90-01800
Digitech Industries, USA . . . 90-01800
Dixon, USA . . . 90-00120
Dupont de Nemours International, Switzerland . . . 80-01350
Duracell International, USA . . . 90-00430
E.C.E., France . . . 90-01840
Electro-Dynamics, USA . . . 90-00120
Electrodynamics Inc., USA . . . 90-00120
Electronic Space Systems, USA . . . 90-01360
EMC Rantec Anechoic/Shielding, USA . 90-01800
Ena Ltd, Turkey . . . 90-00200
Epsilon Lambda, USA . . . 90-01800
Eurologic Systems Ltd, Ireland . . . 90-01800
Fag Kugelfischer Georg Schafer AG, Germany . . . 90-00430
Fairchild Data Corp, USA . . . 90-01800
Farnell Int'l, UK . . . 80-01940
FEI Communication Inc, USA . . . 90-01360
Folsom Research, USA . . . 90-01800
Furon-Bunnell Plastics, USA . . . 90-00120
G.T.C. RF Products Group, USA . . . 90-00540
General Electric Co., USA . . . 90-01360
Grace International, UK . . . 80-01350
Grace International, Italy . . . 80-01350
Grace International, USA . . . 80-01350
Groupe Bacou Comoditex, France . . . 80-01350
Hansen, USA . . . 90-00120
Hartwell, USA . . . 90-00120
Herotek Inc, USA . . . 90-01800
Honeywell, USA . . . 80-00480
Hospitalia Int, Italy . . . 80-01350
HR Smith Group, UK . . . 90-01800
Hydro Export, Hungary . . . 80-01350
Iiyama Electric Co, Japan . . . 90-01800
IN-SNEC, France . . . 90-00135
Informate, USA . . . 90-01360
Inmet Corp, USA . . . 90-01800
Intel, USA . . . 80-00480
Intercontinental Microwave, USA . . . 90-01800
Intergraph, USA . . . 80-00480
JFW Industries Inc., USA . . . 90-01360
K&M Electronics Ltd, USA . . . 90-01360
Kearfott Guidance & Navig., USA . . . 90-00430
Leutron Vision AG, Switzerland . . . 90-01800
Librascope Corp., USA . . . 90-00430
Litef, Germany . . . 80-00480
Litton/Airtron, USA . . . 90-01360
Litton Amecom, USA . . . 90-01360
Litton-Canada, MDA, Canada . . . 80-00480
Logimetrics Inc, USA . . . 90-01800
M-Systems Inc, USA . . . 75-12600
Magnavox Electronic Systems Co, USA 80-00480
Math Associates, USA . . . 90-01800
Maxtech Inc, USA . . . 90-01800
MBM Electroforms & Components, UK 90-00135
Merrimac, USA . . . 90-01800
Metheus Corp, USA . . . 90-01800
Microphase Corp, USA . . . 90-01800
Microwave Dynamics, USA . . . 90-01800
Microwave Technology Inc., USA . . . 90-01360
Miltope Corp, USA . . . 90-01310
Mistral, USA . . . 90-00120
Nei Syncrolift, USA . . . 90-01360
Nicolet, USA . . . 90-00135
Ocean Technology, USA . . . 90-00540
Optische Werke G. Rodenstock, Germany . . . 90-00430
Orwin Associates Inc, USA . . . 90-01800
Photonetics, France . . . 90-00135
Power Systems Technology, USA . . . 90-01800
Precilec SA, France . . . 90-01840
Precision Antennas, UK . . . 90-01800
Q-BIT Corp, USA . . . 90-01800
Quintek Ltd, UK . . . 90-01800
Reliance Gear, UK . . . 90-01300
Rugged Computer Systems, USA . . . 90-01360
S.T. Research, USA . . . 90-01800
Sanders (a Lockheed Martin Company), USA . . . 80-00480
Sector Microwave Ind, USA . . . 90-01800
Seetrax CEA, UK . . . 90-01800
Seiko Instruments, Japan . . . 90-01800
SEKAI Electronics of America Inc., USA 90-01360
Sensorex, France . . . 90-01840
Signal Design Inc, USA . . . 90-01800
Skydata, USA . . . 90-01800
Sonix (Sonotek), USA . . . 90-01800
Spectrum GmbH, Germany . . . 90-01800
ST Keltec Corporation, USA . . . 90-01360
ST Microwave Corporation, USA . . . 90-01360
ST Oletron, USA . . . 90-01360
Synrad, USA . . . 90-00135
Taconic Plastics, USA . . . 90-00135
Talley Defense Systems, USA . . . 90-00120
TEAC Corp. of America, USA . . . 90-01360
Tektronix, USA . . . 80-00480
Teledyne, USA . . . 90-02030
Teledyne Cast Product, USA . . . 90-00120
Teledyne Picco, USA . . . 90-00120
Teleflex-Syneravia, France . . . 90-01840
The Kerite Co, USA . . . 80-01350
Tiernay Aerospace, USA . . . 90-00120
Tiernay Metals, USA . . . 90-00120

TRACE Worldwide Corp., USA 85-08000
TRL Technology Ltd, UK 90-01800
TTC, USA 90-00135
United Fastener, USA 90-00120
Veritech Microwave, USA 90-01800
Vierling, Germany 90-00135
Zarc International, USA 80-01350

Sales agents/representatives listed elsewhere:

Cermat Ltd 80-00250
M. Paz Logistics Ltd 80-01600
Odmat Rotem 85-01700
SPL Ordnance Ltd 80-01935
Taran Technologies Ltd 80-02018

90-00075 AlliedSignal Aerospace Marketing Sales & Service
(a subsidiary of AlliedSignal Aerospace, USA)
Lod, Israel
c/o Israel Aircraft Industries, Ben Gurion Airport,
Westwind Project Department 4321
☎(+972 8) 928 66 51 ⊠ 0606 381014 usrav il
⊖ (+972 8) 928 60 58
Field Service Engineer: Mike Calderon
→ 80 [98]

90-00120 Amzad Supplies Ltd
(a member of The Amcor Group)
PO Box 2850, 61027 Tel-Aviv
98 Yigal Allon Street, 67891 Tel-Aviv
☎(+972 3) 562 56 91; 563 95 07; 563 95 10
✉PHILAMCOR ⊠ 33444; 33801 ⊖ (+972 3) 562 33 92
♣ US$ 4,000,000 (1996) 👥 6 (1996) 🏠 100 sq m
Managing Director: Haim Shapira
Sales and Marketing Director: Michael Ringel
■ Exotic metals, hardware, ball bearings, pneumatic and hydraulic components, electronic packagings. Castings, advanced metals.
® Australia: Ausimag Centrifugal Pumps
Germany: Boysen GmbH
Netherlands: Avio Diepen B.V.
USA: Astralloy-Wear Technology, Astro Gear & Instrument Co., Bliss & Laughlin Steel Co, Circle Seal Controls, Dixon, Electro-Dynamics, Electrodynamics Inc., Furon-Bunnell Plastics, Hansen, Hartwell, Mistral, Talley Defense Systems, Teledyne Cast Product, Teledyne Picco, Tiernay Aerospace, Tiernay Metals, United Fastener [98]

90-00130 Appletec Ltd
2 Hamaager Street, PO Box 2288,
43653 Ra'anana Ind. Area
☎(+972 9) 98 73 50 ⊖ (+972 9) 44 35 23
® USA: Custom Electronics Inc [97]

90-00135 Applied Technology Services (ATS)
33 Ben Gurion Street, PO Box 5085, 46101 Herzliya
☎(+972 9) 957 41 11 [E] yossi_v@netvision.net.il
⊠ 341787 ⊖ (+972 9) 957 44 47
♣ US$ 6,000,000 👥 10 (1997) 🏠 240 sq m (1997)
Managing Director: Yossi Reiss
Sales and Marketing Director: Mikail Farfel
Operations and Technical Director: Dori Reiss
Purchasing Manager: Zily Efrat
■ Sales representative of microwave components and telemetry, PC boards and test equipment.
® France: BMI, IN-SNEC, Photonetics
Germany: Vierling
UK: MBM Electroforms & Components
USA: ATX, Aydin Microwave, Aydin Monitor & Computer, Aydin Vector, Nicolet, Synrad, TTC, Taconic Plastics [98]

90-00200 Benram Inc.
2/A Downes Hill, PO Box 7197, 31071 Ahuza, Haifa
☎(+972 4) 834 11 28 ✉BENRAM LTD ⊠ 46400 bxha il
⊖ (+972 4) 834 89 05
👥 5
Managing Director: Ben Hefter
■ Promotion and representative for aircraft surplus and spare parts, aviation, avionics, electronics: components, systems (import/export). Communication equipment exports.
® Belgium: Chemimetal SA Bruxelles
Germany: Daimler-Benz Aerospace
Turkey: Ena Ltd
USA: Airborne Product Supply Ltd [98]

90-00319 Cherry Aerospace Fasteners
Nili 8, 43207 Raanana
☎(+972 9) 90 22 65 ⊖ (+972 9) 91 26 97
Regional Sales Manager for Middle East: Ron Gescheidt
® USA: Cherry Div. Textron Inc. [98]

90-00340 Composite Limited
12 Ha-Gat Street, #27, 75498 Rishon le Zion
☎(+972 3) 951 54 84 [E] composit@inter.net.il
⊖ (+972 3) 951 54 85
♣ US$ 6,500,000 (1995) 👥 31 🏠 300 sq m
Managing Director: Amiram Saban
Marketing Director: Lee Saban
■ Special metal retailer. [98]

90-00360 CRM Technologies Ltd
28 Nachlat Itzhak Street, 67449 Tel Aviv
☎(+972 3) 695 61 33 ⊖ (+972 3) 695 83 77
Director: Yoram Malchi
Director: Meir Carmel
■ Source for supply of personal & unit military gear, military medical supplies, recovery & logistic support, military surplus, etc.. [97]

90-00430 Del-Ta Engineering Equipment Ltd
16 Hagalim Blvd, PO Box 389, 46103 Herzliya B
☎(+972 9) 52 18 88; 52 18 70 ⊠ 361714 delta il
⊖ (+972 9) 52 18 82
Managing Dir: T. Oren
General Mgr: Eliezer S. Kish
General Mgr: Uri Ram
® Germany: Fag Kugelfischer Georg Schafer AG, Optische Werke G. Rodenstock
UK: Associated Spring Spec Ltd.
USA: Barden Corp., Berg, Winfred M., CAE Link Corp., D.I.O.P., Duracell International, Kearfott Guidance & Navig., Librascope Corp. [97]

90-00540 Electronic Microsystems Ltd (EML)
PO Box 3135, 46103 Herzliya B
☎(+972 52) 55 58 95 ⊖ (+972 52) 57 49 32
♣ US$ 4 500 000 👥 6 🏠 4 000 sq ft
General Mgr: A.E. Goldstein
■ Microelectronics, hybrids, microwave amplifiers, communication radio.
® USA: Associated Industries Inc., G.T.C. RF Products Group, Ocean Technology [97]

90-00850 Aro-Tech
14 Hamashof Street, PO Box 4447, 75143 Rishon Lezion
☎(+972 3) 951 30 99 ⊖ (+972 3) 951 31 01
® Germany: Cannon Electric GmbH [98]

90-01300 Medital Ltd
74 Rishonim St, PO Box 1038, 47100 Ramat Hasharon
☎(+972 3) 540 02 51 ⊖ (+972 3) 540 02 52
♣ US$ 5 000 000 👥 10 🏠 200 sq m
Managing Dir: Joseph Issler
Sales & Marketing Dir: Avi Yacobi
Operations & Technical Dir: Arie Klien
® UK: Reliance Gear
Representing:
THK, Keyence, KHK, Sakae, AEG Servo Systems, Alpha, Galil Motion Control, Sunmulon, Elmo Motion Control, Bühler, EMC, BNK, APA, Le Ine & Lindr, Rollix, Motion, Fenner, Elliott, HSI, Shindengen. [97]

90-01310 Milper Ltd
PO Box 147, 53101 Givataim
☎(+972 3) 571 31 88 [E] ugig@milpe.co.il
⊖ (+972 3) 571 29 72
Managing Director: Uzi Goldschmidt
Operations Director: D Alcalay
® USA: Miltope Corp [98]

90-01350 Mipha International Ltd
8 Har Nevo Street, 47225 Ramat Hasharon
☎(+972 3) 540 24 03; 540 23 92
⊖ (+972 3) 549 09 31
Managing Director: Michael Itzhar
■ Electronic intelligence, electronic monitoring equipment, security consultancy and electronic fences. → 80 [98]

90-01360 MTI Engineering Ltd
PO Box 43066, 61430 Tel-Aviv
☎(+972 3) 648 79 01 ⊖ (+972 3) 648 79 40
⊕ US$ 5 000 000 (1995) ♣ US$ 3 500 000 (1995) 👥 46
🏠 1 600 sq m
Chairman: Zvi Borovits
Director: David Marshall
Advertising Mgr: Dan Sperling
■ Consulting & representative organisation specialising in sophisticated systems & components for microwave, electronics & aerospace products. → 80
® USA: Condor Systems, Datatape, Electronic Space Systems, FEI Communication Inc, General Electric Co., Informate, JFW Industries Inc., K&M Electronics Ltd, Litton Amecom, Litton/Airtron, Microwave Technology Inc., Nei Syncrolift, Rugged Computer Systems, SEKAI Electronics of America Inc., ST Keltec Corporation, ST Microwave Corporation, ST Oletron, TEAC Corp. of America
Subsidiary:
MTI Technology & Engineering Ltd [97]

90-01600 Oreet-Marketing Communications
15 Kineret Street, 51201 Bene Barak
☎(+972 3) 570 65 27 [E] areetimic@netvision.net.il
⊖ (+972 3) 570 65 26
👥 7 (1997) 🏠 100
Managing Director: Ms Oreet Ben-Yaacov [98]

90-01800 RCM Ltd
(† RCM Computers Ltd)
31 Habarzel Street, Ramat-Hahayal, PO Box 39844, 61397 Tel-Aviv
☎(+972 3) 648 78 85 ⊖ (+972 3) 648 72 90
♣ US$ 4,800,000 (1996) 👥 17 🏠 350 m2
President: Eddie Wiegenfeld
Vice President, Engineering: Moshe Bonjack
Vice President, Marketing: Moshe Szekatch
® Germany: Spectrum GmbH
Ireland: Eurologic Systems Ltd
Japan: Iiyama Electric Co, Seiko Instruments
Switzerland: Leutron Vision AG
UK: HR Smith Group, Precision Antennas, Quintek Ltd, Seetrax CEA, TRL Technology Ltd
USA: Aeroflex International, Aeroflex Laboratories Inc, Aeroflex-Lintek, Amherst Systems Inc, Anghel Laboratories Inc, CHL, Comstron Corp, Cubic Communications, Delfin Systems, Diamond Antenna & Microwave, Digitech Industries, EMC Rantec Anechoic/Shielding, Epsilon Lambda, Fairchild Data Corp, Folsom Research, Herotek Inc, Inmet Corp, Intercontinental Microwave, Logimetrics Inc, Math Associates, Maxtech Inc, Merrimac, Metheus Corp, Microphase Corp, Microwave Dynamics, Orwin Associates Inc, Power Systems Technology, Q-BIT Corp, S.T. Research, Sector Microwave Ind, Signal Design Inc, Skydata, Sonix (Sonotek), Veritech Microwave [98]

90-01840 Relcom Agencies Ltd
PO Box 3079, 46103 Herzlia
c/o Teperson, 10 Hashaked Street,
46910 Kfar Shmaryahu
☎(+972 9) 955 13 13 ⊖ (+972 9) 951 48 90
♣ US$ 5,800,000 (1996) 👥 5 🏠 100 m2
Managing Director: Ilan Caner
Operations Director: David Arad
® France: E.C.E., Precilec SA, Sensorex, Teleflex-Syneravia
UK: D.J.Mouldings [98]

90-01860 Racom Electronics Co Ltd
PO Box 21120, 61210 Tel-Aviv
☎(+972 3) 577 68 80 ⊖ (+972 3) 579 55 22 [97]

90-01960 SRS Engineering Agencies
1 Haatzmaut Street, PO Box 774, 40500 Eben-Yehuda
☎(+97 2) 98 99 82 17 ⊖ (+97 2) 98 99 82 18
Contact: Eli Shifman
® USA: Allied International Corp. [98]

90-02030 Teledyne Intercontinental Ltd
America Hse, 35 Shaul Hamelech Blvd, PO Box 33044, 64927 Tel-Aviv
☎ (+972 3) 696 28 60 ⊖ (+972 3) 695 25 44
General Mgr: Ori Goodovich
Office Mgr: E. Roberts
® USA: Teledyne [97]

90-02500 Transaero Inc.
13 Shefa-Tal Street, 67013 Tel Aviv
☎ (+972 3) 652 67 77 ⊖ (+972 3) 562 74 60
Regional Mgr: Jacob Dekel [97]

Italy

Head of State:
Oscar Luigi Scalfaro

10 Head of State

10-10000 Office of the President
Palazzo del Quirinale, I-00187 Rome
☎ (+39 6) 469 91 ⊠ 620022 sgpr i ⊖ (+39 6) 678 39 98
President of the Republic: Oscar Luigi Scalfaro [98]

11 Prime Minister

11-10000 Office of the Prime Minister
Piazza Colonna 370, Palazzo Chigi, I-00187 Rome
☎ (+39 6) 67 79 ⊠ 613199 chigi; 614010 precom
⊖ (+39 6) 678 39 98; 679 68 94
Prime Minister: Romano Prodi
Under Secretary to the Prime Minister: Lamberto Cardia [98]

12 Ministries

12-10000 Ministry of Defence
Gabinetto del Ministro
Via XX Settembre 8-11, Palazzo Baracchini, I-00100 Rome
☎ (+39 6) 488 21 26 ⊠ 611438 smd i
⊖ (+39 6) 488 47 59; 488 67 21
Minister of Defense: Beniamino Andreatta [98]

12-20000 Ministry of Foreign Affairs
Ministerio Degli Affari Esteri
Piazzale della Farnesina 1, Foro Italico, I-00194 Rome
☎ (+39 6) 369 11; 387 11 ⊠ 610611
⊖ (+39 6) 322 28 50
Minister of Foreign Affairs: Lamberto Dini
■ Communications and telecommunications, tourism, exhibitions in Italy, maritime transport, general affairs, aero transport. [98]

12-30000 Ministry of Foreign Trade
Viale America 341, I-00144 Rome
☎ (+39 6) 59 93 ⊠ 610083
⊖ (+39 6) 592 15 73; 591 37 51
Minister of Foreign Trade: Augusto Fantozzi [98]

12-40000 Ministry of the Interior
Piazza del Viminale, I-00184 Rome
☎ (+39 6) 466 71 ⊠ 610635 miroma i
⊖ (+39 6) 48 67 59
Minister of the Interior: Giorgio Napolitana
Director General of Civil Aviation: Ms Maria Di Fazio
Information Contact: Ms Nicola Mancino
■ Inspectional service for fire-fighting services on harbours/aiports [98]

12-50000 Ministry of Finance
Viale Europa, I-00144 Rome
☎ (+39 6) 599 71 ⊠ 614460 minseg i
⊖ (+39 6) 59 97 20 67; 59 97 34 92
Minister of Finance: Vincenzo Visco [98]

15 Parliament

15-10000 Senate
Piazza Madama, I-00186 Rome
☎ (+39 6) 670 61
Speaker of the Senate: Nicola Mancino [98]

15-20000 Chamber of Deputies
Piazza Montecitorio, I-00100 Rome
☎ (+39 6) 67 60 ⊠ 612523 camdep i
Speaker: Lucioano Violante [98]

25 Defense Forces

25-22010 Air Force 1st Air Region Headquarters
Piazza Ermete Novelli 1, I-20129 Milan
☎ (+39 2) 738 24 41 ⊖ (+39 2) 738 27 60
Commander 1st Air Region: Gen Gianbortolo Parisi
Chief of Staff: Gen B A Roberts Iacomin
Chief Logistics Division: Sante Palombi
Director of Operations: Col Gesualdo Papaale
PIO: Maj Giovanni Fuochi [98]

25-29200 Accademia Aeronautica
Via Domitiana, I-80078 Pozzuoli
☎ (+39 81) 735 51 11 ✉ AEROACCADEMIA
⊖ (+39 81) 735 52 19
Commandante: Gen Da Leonardo Tricario
Deputy Commander: Col Enrico Camerotto [97]

25-32000 Army Air Corps (AVES)
Aviazione Esercito
Via Nomentana 274, I-00161 Rome
☎ (+39 6) 838 00 18
Inspector General of Army Aviation:
Maj.Gen. Pasqualino Verdecchia [97]

25-45000 Headquarters 22d Area Support Group
Caserma Ederle, PO Box 10, I-36100 Vicenza
☎ (+39 444) 51 70 00; 51 79 88 ⊖ (+39 444) 51 75 43
Commander: Col Scott Conrad
Exec. Officer: Lt Col Karen Day
Public Affairs Specialist: Dorothy Spagnuolo [97]

29 Customs & Excise

29-00500 Directorate of Customs
Via M. Carucci 71, I-00143 Rome [98]

31 Embassies & Attachés

31-00050 Embassy of the Republic of Angola
Via Filippo Bernardini 21, I-00165 Rome
☎ (+39 6) 39 36 69 41; 39 36 69 02; 39 36 65 70
⊠ 614505 errar i ⊖ (+39 6) 63 49 60
Ambassador: H E Mr Antero Alberto Ervedosa Abreu
Counsellor Deputy Chief of Mission:
Rui Orlando Ferreira de Ceita Silva Xavier [98]

31-00080 Embassy of the Argentine Republic
Piazza dell'Esquilino 2, I-00185 Rome
☎ (+39 6) 474 25 51 ⊠ 610 386 embarg i
⊖ (+39 6) 481 47 87
Ambassador: H E Mr Antonio Erman Gonzalez
Deputy Chief of Mission: Alfredo Alcorta
Military Attaché: Col Jorge Luis Zamudio
☎ (+39 6) 487 14 56
Naval Attaché: Capt Alois Esteban Payarola
☎ (+39 6) 487 22 55
Air Attaché: Cdr Victor Hugo Borchert
☎ (+39 6) 488 02 96
Secretary: Ms Lilian Skillan
Commercial Office
Plazza della Rotonda 2, P 3, I-00187 Roma
☎ (+39 6) 65 70 22 ⊠ 616 340 cearg i
Commercial Counsellor: Luis Beltran Martinez Thomas [98]

31-00090 Australian Embassy
Via Alessandria 215, I-00198 Rome
☎ (+39 6) 83 27 21; 85 27 23 57
✉ AUSTEMBA Rome 21600 ⊠ 610165 amsem i
⊖ (+39 6) 85 27 23 60
Ambassador: H E Mr Rory Steele
Deputy Chief of Mission: Gordon Miller
Commercial Counsellor: Gerard Lanzarone
Defence Attaché, Southern Europe: Maj Gen Geoff Carter
Defence Administrative Assistant: T W Sweeney [98]

31-00100 Embassy of Austria
Via G.B. Pergolesi 3, I-00198 Rome
☎ (+39 6) 855 82 41/42/43/44 ✉ AUSTROAMB
⊠ 610139 oebrom i ⊖ (+39 6) 854 32 86
Ambassador: H.E. Dr Emil Staffelmayr
Minister: Arno Riedel
Second Secretary: Mrs Melitta Schubert
Second Secretary: Robert Zischg
Second Secretary, Political Affairs: Mrs Renate Kobler
Second Secretary, Administrative Affairs: Gerhard Gerlza
Administration Attaché: Ms Elisabeth Kriechbaum
Consul Affairs Counsellor: Mrs Franz Wechner
Consular Affairs Attaché: Mrs Helga Kubideck
Military, Air & Naval Attaché: Gen. Uwe Maydell
Administrative Affairs Attaché: Ignaz Hnizdo
Attaché: Alexander Benedict
Cultural Affairs Attaché: Klaus Wölfer
Cultural Affairs Attaché: Werner Almhofer
Cultural Affairs Attaché: Mrs Sieglinde Schmidt
Commercial Affairs Counsellor: Herbert Geissler
Commercial Affairs Attaché: Martin Pilser
Deputy Attaché: Ms Edeltraude Rosemarie Hock [97]

31-00150 Embassy of Belgium
49 Via dei Monti Parioli, I-00197 Rome
☎ (+39 6) 360 95 11 ✉ AMBASSADE BELGIUM Rome
[E] ambelrom@pub.xplore.it ⊠ 62 25 14
⊖ (+39 6) 322 69 35
Ambassador: H E Mr André Onkelinx
Deputy Chief of Mission: Ghislain D'Hoop
Counsellor Military/Scientific Affairs: Michel Peetermans
Economic Counsellor: Paul Sadi Brancart
Commercial Canvasser: G Walaevens
Commercial Canvasser: F Murari [98]

31-00200 Embassy of Bolivia
Via Toscana 30, I-00187 Rome
☎ (+39 6) 481 74 38 ⊠ 620221 ⊖ (+39 6) 482 19 75
Ambassador: H.E. Ms Moira Paz Estenssoro
Second Secretary: Herman Gonzales Daza
Second Secretary: Alvaro Gonzales Quint [97]

31-00230 Brazilian Embassy
Piazza Navona 14, I-00186 Rome
☎ (+39 6) 68 39 81 ⊖ (+39 6) 686 78 58
Administrative Assistant: Ms Monica Costa [98]

31-00350 Embassy of Chile
Via Po, 23, I-00184 Rome
☎ (+39 6) 884 14 33; 884 14 35; 884 14 47
⊠ 611 420 embach i ⊖ (+39 6) 884 14 52
Ambassador: H E Mr Jorge Jimenez de la Jare
Counsellor: Luio Alberto Sepulveud
First Secretary: Mauricio Ugalde Bilbao
First Secretary: Oscar Alcaman Riffo
Third Secretary: Christian Blasco Torres

Contact Reference

Cultural Attaché: Juan Barattini
Attaché: Hector Basualto [97]

31-00450 Embassy of the Czech Republic
Via dei Gracchi, 322, I-00192 Rome
☎(+39 6) 324 44 59 ⊖ (+39 6) 324 44 66
Ambassador: H E Mr Martin Stropnicky
Counsellor: Mrs Ivana Holoubkova
Counsellor: Jiri Novak
Defence Attaché: Col Gustav Helm
Cultural Attaché: Mrs Marie Zabranova
Attaché, Political: Pavel Smetacek
First Secretary, Political: Mrs Hana Cigankova
First Secretary: Tomas Kuchta
First Secretary: Mrs Svatava Pargacova
Third Secretary, Commercial: Mrs Dana Pavlouskova
Administrative Attaché: Bouhmila Vodova [98]

31-00510 Embassy of the Arab Republic of Egypt
Via Salaria 265-267, I-00199 Rome
☎(+39 6) 841 74 20; 855 53 61 ⊠ 616339 safrom i
⊖ (+39 6) 855 44 24
Ambassador: H E Mr Nehad Ibrahim Abdel Latif
Deputy Chief of Mission: Ahmed Rezk Mohamed Rezk
Military, Naval and Air Attaché:
Brig Gen Maged George Elias Ghattas
☎(+39 6) 884 47 44
Military, Naval and Air Deputy Attaché:
Maj Pil Hani Ramzi El-Dib
Minister, Commercial Affairs: Moustafa Karam
☎(+39 6) 841 63 87 [98]

31-00580 Embassy of Finland
Via Lisbona 3, I-00198 Rome
☎(+39 6) 854 83 29 ⊠ 625600 finno i
⊖ (+39 6) 854 03 62
Ambassador: H.E. Mr Matti Häkkänen
Deputy Chief of Mission: Ms Kaija Ilander
Military & Defence Attaché: Lt Col Harri Niipola (resident in Bern)
Commercial Assistant: Ms Susanna Remes

Commercial Section
Via Cerva 31, I-20122 Milan
☎(+39 2) 76 02 02 33 ⊠ 320366 fintra i
⊖ (+39 2) 76 02 02 40
Commercial Counsellor: Pentti Rantalainen [97]

31-00670 Embassy of Ghana
Via Ostriana 4, I-00199 Rome
☎(+39 6) 86 21 56 91; 86 21 93 07; 86 21 93
[E] j.ackan@flashnet.it 610270 ghanaemb i
⊖ (+39 6) 86 32 57 62
Ambassador: H E Mr Therese S Scott
☎(+39 6) 86 21 66 29
Deputy Chief of Mission, Head of Chancery: Kodzo K Alabo
☎(+39 6) 860 22 67
Commercial Secretary, Attaché: Joseph Ackon
☎(+39 6) 86 21 66 63 [98]

31-00690 Embassy of Greece
Via S Mercadante 36, I-00198 Rome
☎(+39 6) 854 96 30; 855 85 89; 841 68 73
✉ AMBASSADE GRECE Rome
⊠ 610416 amgrom i; 610673 grtrom i
⊖ (+39 6) 841 59 27
Ambassador: H E Mr Evangelos Frangoulis
Minister-Counsellor and Deputy Chief of Mission:
Dimitri Tsikouris
First Secretary: D Touloupas
Defence Attaché: Col Christos Kaliviotis
☎(+39 6) 855 31 00; 841 59 27 (fax)
Commercial and Economic Affairs Counsellor:
Vassilis Psonis ☎(+39 6) 855 29 30; 854 00 09 (fax)
Press and Information Officer: Vassilis Notas [98]

31-00830 Embassy of India
Via XX Settembre 5, I-00187 Rome
☎(+39 6) 488 46 42; 488 46 43; 488 46 44; 488 46 45
⊠ 611274 indeb i ⊖ (+39 6) 481 95 39
Ambassador: H E Shri K P Fabian
Deputy Chief of Mission: Shri Butshikan Singh
Military, Naval and Air Attaché: Col S G Chatterji
Commercial Attaché: Shri T C Barupal
Commercial Secretary: Shri A K Jain [98]

31-00870 Embassy of Ireland
Pizza di Campitelli, 3, I-00186 Rome
☎(+39 6) 697 91 21 ⊠ 626 030 hibrom i
⊖ (+39 6) 679 23 54; 69 79 12 31
Ambassador: H E Mr Joseph Small
First Secretary: James Gawley [98]

31-00920 Embassy of Japan
Ambasciata del Giappone
Via Quintino Sella 60, I-00187 Rome
☎(+39 6) 48 79 91 ✉ TAISHI Roma ⊠ 610063 taishi i
⊖ (+39 6) 487 33 16
Ambassador: H.E. Mr Shinichiro Asao
Deputy Chief of Mission: Noritake Kai
Second Secretary (Defense): Teruyoshi Maeda
Commercial Counsellor: Yosiyasu Nao
First Secretary (Press Affairs): Yuri Kodeda

General Consulate of Japan
Consolato Generale del Giappone
Via F Turati 16-18, I-20121 Milan
☎(+39 2) 29 00 23 16; 29 00 23 17 29 00 33 18
⊖ (+39 2) 659 72 01 [98]

31-00990 Embassy of the Republic of Korea
Via Barnaba Oriani 30, I-00197 Rome
☎(+39 6) 808 53 51 ⊠ 610182 korome i
⊖ (+39 6) 808 53 54
Military, Naval & Air Attaché: Col Han Sang-Ki [97]

31-01140 Embassy of Malta
Lungotevere Marzio 12, I-00186 Rome
☎(+39 6) 687 99 90; 687 99 47; 687 76 29
⊠ 611205 malrom i ⊖ (+39 6) 689 26 87
Ambassador: H E Mr Henry De Gabriele
First Counsellor: Alfred Falzon
Second Secretary: Rachel Sapiano
Third Secretary and Consul: Robert Borg
Attaché, Commercial: Andrew Spiteri [98]

31-01270 Royal Netherlands Embassy
Via Michele Mercati 8, I-00197 Rome
☎(+39 6) 322 11 41; 321 50 03; 321 58 27
⊠ 610138 olanda i ⊖ (+39 6) 320 10 96
Ambassador: H E Mr Charles Ronald van Beuge
Minister and Deputy Chief of Mission:
Dr Baron Marnix van Aerssen Beijeren van Voshol
Economic Affairs Counsellor: Dr Johannes M Corijn
Press and Cultural Affairs Counsellor: Dr Jan E Boeles
Agricultural Affairs Counsellor: Harm Boer
Scientific and Technological Affairs Counsellor:
Dr Henk E Hoefdraad
First Secretary, Commerce:
Robert H M M A van Hanswijck de Jonge
Military Attaché: Col Henri J J A M Grootveld
First Secretary, Consular Affairs: Dr Victoria A C Pelkman
First Secretary, Political Affairs: Dr Marie Florence van Es
Second Secretary, Administration Affairs:
Johannes A C Vellekoop [98]

31-01310 Embassy of Nicaragua
Via Bescia 16, I-00198 Rome
☎(+39 6) 884 16 95; 841 46 93 ⊠ 616 295 ambnic i
⊖ (+39 6) 884 16 95
Ambassador: H E Mr Maria Eugenia
Counsellor: Ena Yolanda Jerez [98]

31-01370 Embassy of Pakistan
Via della Camilluccia 682, I-00135 Rome
☎(+39 6) 36 30 17 75; 329 48 36 ✉ PAREP Rome
⊠ 620890 - 1 ⊖ (+39 6) 36 30 19 36
Minister: Athar Mahmood
First Secretary and Public Information Officer: Ghalib Iqbal
Commercial Counsellor: Uaman Hasaan
Agricultural Counsellor: Shahid Rashid
Defence Attaché: Capt Najmul Hasnain
☎(+33 1) 45 62 23 32; 45 62 24 06 [98]

31-01430 Embassy of the Republic of Poland
Via P.P. Rubens 20, I-00197 Rome
☎(+39 6) 322 44 55; 322 45 97; 322 46 95
⊖ (+39 6) 321 78 95
Ambassador: H.E. Mr Jan Ludwik Wdowik
Defence Attaché: Andrzej Kostrzewski
Public Information Officer: Lech Kubiak

Commercial Office
Via Olona 2/4, I-00198 Rome
☎(+39 6) 854 11 28 ⊠ 610325 brhrm i
⊖ (+39 6) 855 33 91
Commercial Counsellor: Tomasz Bartoszewicz [97]

31-01480 Embassy of Romania
Via Nicolo Tartaglia 36, I-00197 Rome
☎(+39 6) 808 45 29 ⊠ 610460 roamb i
⊖ (+39 6) 808 49 95
Ambassador: Constantin Mihail Grigorie
Military Naval and Air Attaché: Col Gheorghe Nicolaescu
☎(+39 6) 807 88 98 [98]

31-01485 Embassy of the Russian Federation
Via Gaeta 5, I-00185 Rome
☎(+39 6) 494 16 80; 494 16 81 ⊖ (+39 6) 49 10 31
Ambassador: H.E. Mr Valery Fiodorovich Keniaikine
Minister & Deputy Chief of Mission: Vladimir Efinov
Military & Air Attaché: Col Valery Baranovsky
Naval Attaché: Capt. Valery Briukhovetsky
First Counsellor: Andrei Maslov
Secretary: A. Grigoriev

Commercial Department
Via Clitunno 46, I-00198 Rome
☎(+39 6) 855 54 56; 884 21 87; 844 30 51
Commercial Counsellor: Dmitry Inkine [97]

31-01655 Embassy of the Republic of Slovenia
Via Leonardo Pisano 10, I-00197 Rome
☎(+39 6) 808 12 72; 808 10 75 ⊖ (+39 6) 808 14 71
Ambassador: H E Mr Peter Andrej Bekes

Consulate General of the Republic of Slovenia
Via S Giorgio 1, I-34123 Trieste
☎(+39 40) 30 78 55 ⊖ (+39 40) 30 82 66
Chargé d'Affaires: Vlasta Valencic Pelikan [98]

31-01670 South African Embassy
Via Tanaro 14, I-00198 Rome
☎(+39 6) 85 25 41 ⊖ (+39 6) 85 25 43 00
Ambassador: H E Dr Khorshed Ginwala
Political Counsellor: Jurgens Young
Commercial Counsellor: Eric A Odendaal
First Secretary, Press and Information: J Ferreira
Defence Attaché: Col Ben Arnoldi
Third Secretary, Politics and Information: Izak W Barnard
Third Secretary, Commerce: Schalk W Visser
Third Secretary, Press and Information:
Ms Linda V Shongwe [98]

31-01690 Embassy of the Democratic Socialist Republic of Sri Lanka
Via Giuseppe Cuboni 6-8, I-00197 Rome
☎(+39 6) 322 42 02; 322 41 62 ⊠ 612602 lanka i
⊖ (+39 6) 322 41 93
Ambassador: H.E. Dr U. Pethiyagoda
☎(+39 6) 321 70 70
Second Secretary & Head of Chancery: A.W. Pathiraja
☎(+39 6) 321 71 47
First Secretary, Commerce: R.D.S. Kumararatne
☎(+39 6) 322 42 81
Administration Attaché: G.C. Amerasinghe [97]

31-01700 Embassy of the Republic of the Sudan
Via Lazzaro Spallanzani 24, I-00161 Rome
☎(+39 6) 440 36 09; 440 30 71 ⊠ 610302 sudani i
⊖ (+39 6) 440 43 77; 440 23 58
Ambassador: H.E. Mr Mahdi Mustafa El Hadi
Deputy Chief of Mission & Minister:
Abuel Gasim Abdel Wahid S. Idris
Minister: Abdelazim Abdalla El Tom
Agricultural Affairs Counsellor:
Mohamed Said M. Ali Harbi
Second Secretary, Cultural & Consular:
Hussein Mohamed Hussein
Administrative Attaché: Mirghani Higazi Mahgoud [97]

31-01740 Embassy of Switzerland
Via Barnaba Oriani 61, I-00197 Rome
☎(+39 6) 808 36 41 ✉ AMBASUISSE Roma
⊖ (+39 6) 808 85 10
Ambassador: H E Mr Dante Martinelli
Deputy Chief of Mission: Mrs Ingrid Apelbaum

Defense Attaché: Col Siro Muschietti
Counsellor Economic Affairs: Thomas Kupfer
Counsellor Cultural and Public Affairs: Jean-Hubert Lebet 98

31-01830 Embassy of Tunisia
Via Asmara 7, I-00199 Rome
☎ (+39 6) 860 30 60; 860 30 68 ✉ 610190 ambtun i
⊖ (+39 6) 86 21 82 04
Ambassador: H.E. Mr Ahmed Friaa
Deputy Chief of Mission: Abderrahmen Ben Mansour
Military, Naval & Air Attaché: Ezzeddine Karoui
Deputy Military, Naval & Air Attaché: Tarek Debal
Commercial Counsellor: Mehrez Ben Rhouma 97

31-01890 Embassy of the United Kingdom
Via XX Settembre 80a, I-00187 Rome
☎ (+39 6) 482 54 41 ⊖ (+39 6) 48 90 42 86
Ambassador: H E Mr T Richardson (KCMG)
Minister: Keith Bloomfield (CMG)
Defence and Military Attaché: Brig J A Anderson
Naval Attaché: Capt G Wiltshire (RN)
Air Attaché: Gp Capt J M Kennell (RAF)
Economic and Commercial Counsellor: T G Paxman 98

31-01910 Embassy of Uruguay
Via Veneto 183, I-00187 Rome
☎ (+39 6) 482 17 76 ✉ URUIT Roma
E emb.uruguay@agora.stim.it ⊖ (+39 6) 482 36 95
Ambassador: H E Mr Carlos Brugnini
☎ (+39 6) 482 10 01
Deputy Chief of Mission, Minister: Joaquin Piriz
☎ (+39 6) 482 17 77

Commercial Department
Piazza Armando Diaz 7, Floor 6, I-20123 Milan
☎ (+39 6) 86 11 00 ✉ 334 363 decmi i
⊖ (+39 6) 869 23 26
Commercial Counsellor: Roberto Merola 98

33 Foreign Forces

33-34010 United States Army in Europe Southern European Task Force (SETAF)
Camp Ederle, PO Box 10, I-36100 Vicenza
☎ (+39 444) 51 78 66; 51 76 18
E robinsoj@email.vicenza.army.mil
⊖ (+39 444) 51 75 43
Commander SETAF: Maj Gen Edwin Smith
Commander SETAF Infantry Brigade: Col Glen L Scott
Commander 1st Battalion 508th Infantry: Lieut Col Arnold Bray
Commander 502nd Aviation: Maj John Woodenberry
Commander 22nd Area Support Group: Col Scott W Conrad
Commander 14th Transportation Battalion: Lieut Col Duane N Sweeney
Commander 509th Signal Battalion: Lieut Col Joseph M Rose 98

34 Multinational Intergovernmental Organizations

34-22110 Fifth Allied Tactical Air Force Headquarters (FIVEATAF)
(Allied Air Forces Southern Europe)
Airport dal Molin, Viale Ferrarin 138, I-36100 Vicenza
☎ (+39 444) 93 11 11 ✉ FIVEATAF
W http://www.facgpidiv.tin.it ⊖ (+39 444) 93 13 70
Commander: Lieut Gen A Vannucchi (ITAF)
☎ (+39 444) 93 12 01
Deputy Commander: Maj Gen T Kinnan (USAF)
☎ (+39 444) 93 12 06
Chief of Staff: Brig Gen G Mariani (ITAF)
☎ (+39 444) 93 12 07
Operations Director: Col T C Jones (USAF)
☎ (+39 444) 93 10 03
Procurement and Logistics Director: Brig Gen S Livatino (ITAF) ☎ (+39 444) 93 10 60
Press and Information Officer: Col Giuseppe Di Chio (ITAF)
☎ (+39 444) 93 15 44
■ Ensure Air Defence in designated area of responsibility. 98

34-32100 Allied Land Forces Southern Europe Headquarters (LANDSOUTH)
Via Roma 31, I-37100 Verona
☎ (+39 45) 809 51 70 ⊖ (+39 45) 209 51 00
Commander: Gen. Cesare Pucci (ITA)
Chief of Staff: Maj.Gen. Salvatore Sabatino (ITA)
Deputy Chief of Staff, Operations: Maj.Gen. Alberto Jannelli (ITA)
Deputy Chief of Staff, Support: Brig.Gen. Charles M. Burke (US Army)
Secretary to the Chief of Staff: Col Antonio Amato (ITA)
Assistant Chief of Staff, Personnel & Administration: Col Mauro Massera (ITA)
Assistant Chief of Staff, Intelligence: Col Alessandro Pompegnani (ITA)
Assistant Chief of Staff, Training & Operations: Col Francesco Apicella (ITA)
Assistant Chief of Staff, Logistics: Col Franco Scaramagli (ITA)
Assistant Chief of Staff, Communication & Info. Systems: Col Flavio Zordan (ITA)
Assistant Chief of Staff, Budget & Finance: Col Vito Pepe (ITA)
Commander of Support Group: Col Vincenzo Lorusso (ITA)
Commander of Joint Signal Group: Lt Col Corrado Nastasi (ITA)
Chief of Public Information: Lt Col Giovanni Bernardi (ITA) 97

34-42100 Allied Naval Forces Southern Europe Headquarters (NAVSOUTH)
Nisida, Naples
☎ (+39 81) 760 11 88 ⊖ (+39 81) 570 21 72
COMNAVSOUTH: Adm G Spinozzi (IT Navy)
COMAIRMED: Rear Adm J Ryan (USN)
COMSUBMED: Rear Adm J Crossenbacher (USN)
Chief of Staff (NAVSOUTH): Vice Adm Michael Moore
Assistant Chief of Staff (Plans and Policy): Rear Adm Sungun (Turkish Navy)
Assistant Chief of Staff (Operations): Cdre G Hatzis (Greek Navy)
Assistant Chief of Staff (Communications): Capt W Griffin (RN)
Assitant Chief of Staff (Logistics and Administration): Rear Adm E Bolongaro (It Navy)
Assistant Chief of Staff (Intelligence): Capt S Richason (USN)
French Liaison Officer: Cdr E Gouraud (French Navy)
Public Information Officer: Lieut E Trundle (RN) 98

34-42350 SACLANT Undersea Research Centre (SACLANTCEN)
Viale San Bartolomeo 400,
I-19138 San Bartolomeo (La Spezia)
⊖ (+39 187) 52 46 00
Director: Rear Adm J Spoelstra 97

34-59100 NATO Defence College
Viale della Civiltà del Lavoro 38, I-00144 Rome
☎ (+39 6) 592 37 41 ✉ 621221 nadefcol
⊖ (+39 6) 591 34 08
Commandant: Lieut Gen R Evraire
Deputy Commandant: Amb. D.C. Arnould
Chief of Cabinet: Col. R.C.P. Hoeben (BE AF)
Head Research Library: C. Grant 97

60 Research Organizations, Institutes & Test Establishments

60-00080 Gajon Institute of Technology SpA
Via Piaggio 9, I-16136 Genoa
☎ (+39 10) 272 46 41 ✉ 281229 ⊖ (+39 10) 21 26 71
♣ US$ 10 000 000 ⛹ 53 ⌂ 5 000 sq m
President: Prof. Dr G.P. Noja
■ Design & development of training plans, training systems. Direct training of military personnel. Design & development of turnkey training centres. → 75 97

60-00100 Istituto Superiore di Statu Maggiore Interforze (ISSMI)
(a subs. of The Ministry of Defense)
Piazza della Rovere 83, I-00165 Roma
☎ (+39 6) 474 59 93 ✆ (+30 6) 46 91 31 01
⊖ (+39 6) 474 63 03
⛹ 165 ⌂ 2 800 sq m
Managing Dir: Gen. B.A. Naldi
Joint Services Staff College Dir: R. Romano
Operations Dir: Cdr Bracco
Sales & Marketing Dir: Capt. Grieco
Purchasing Mgr: Capt. Naridellui
■ Joint services staff courses for 3 services: Lieutenant Commanders, Majors, Commanders & Lieutenant Colonels. Open to foreign officers. Participants: approx. 100 (80 nationals & 20 foreign). Opened in July 1994; first course 1995. Subjects: Military Doctrine, Management, Economics & Law. Subjects: Military Doctrine, Management, Economics & Law. 97

65 Industry, Trade & Armed Forces Associations

65-00150 Aerospace Industries Association (AIA)
Associazione Industrie Aerospaziali (a member of AECMA)
Via Nazionale 200, I-00184 Rome
☎ (+39 6) 488 02 47; 488 02 48; 488 02 49
✉ ASSOAVIO ⊖ (+39 6) 482 74 76
♣ Lit 7,500,000,000 ⛹ 35,000
President: Ing. Paolo Torricelli
Vice President: Ing. Amedeo Caporaletti
Secretary General: Carlo Festucci
■ National trade organisation for constructors of aircraft, engines, missiles & associated equipment.
Defence: Skyshield 35-AHEAD gun & missile; Skyguard twin gun & missile air defence systems; AHEAD-system; Gun-King multy-divergence laser sight system; Shorar/Pagoda serach & acquisition radar; Gun-Star compact fire control system; Seaguard CIWS system; ADATS missile systems; Ranger UAV-system; training systems & simulators. Automatic cannon for air defence, vehicle & naval armament.
Pyrotec: ammunition in the calibre range 12.7/35mm. Production machinery & ballistic measuring systems. Facilities & equipment for disposal of ammunition.
Space: spacecraft structures, fairings for launchers, inflatable space rigidized structures, deployment mechanisms, scientific instruments. 97

65-00450 Cagiva Motor SpA
(† Cagiva TRADING SpA)
(a subsidiary of Cagiva Group)
Via G Macchi 144, I-2110 Varese
☎ (+39 51) 31 12 40 ⊖ (+39 51) 31 34 74
♣ Lit 420,000,000,000 ⛹ 100
President: Claudio Castiglioni
Sales Director: L D Giacometti 98

65-01000 National Association of Electronic Industries Defence, Systems, Space (ANIE)
Associazione Nazionale Industrie Electtroniche Defence, Systems, Space
Via Giulio Caccini 1, I-00198 Rome
☎ (+39 6) 884 57 02 ⊖ (+39 6) 841 70 20
President: Enzo Benigni
Secretary: Roberto Del Toro
■ Trade association representing Italian manufacturers of electronic systems and equipment for military & civil applications.
Members:
Alcatel Italia Spa
Alcatel Air Navigation
Alenia Un'Azienda Finmeccanica
Alenia Difesa
Alenia Aerospazio Div. Spavio
Bull HN ISI Spa
CAP HN ISI Spa
Datamat Ingeneria Dei Sistemi Spa
Elettromeccanica Parizzi Spa
Elettronic Spa
Elmer Spa
Elsag Bailey
Ericsson Telecommunicazioni Spa
Fiar Div. Sistemi Avionci
Finsiel Spa
Italtel Spa
Laben Spa
Larimart Spa
Litton Italia
MAC - Alenia Marconi
Marconi Spa

Nuova Telespazio Spa
Alenia Difesa - Div. Sistemi Evionici
Officine Panerai Spa
Alenia Difesa un'Azienda
Page Europa Spa
Pirelli Cavi Spa
SEPA
SIA Spa
SIRTI Spa
Sperry Marine Italia Srl
Vitrociset Spa
WASS
98

65-01800 Raggrupamento delle Industrie a Tecnologia Avanzata per la Difesa (RITAD)

Grouping of Advanced Technology Industries for Defence
Viale Liegi 44, I-00198 Rome
☎(+39 6) 855 02 08; 855 25 41; 855 80 89
⊖ (+39 6) 855 97 41
Director: Adm Giuseppe Rondonotti
■ Non-profit making association representing Italian companies involved in high technology activities in the defence sector.
Members:
Agusta SpA, A I A, A N I E, A N P A M, Assonave, Confindustria, Fiat SpA, Fincantieri C N I SpA, Finmeccancia, OTO Melara/Breda Meccanica Bresciana SpA, Airmacchi SpA, Alenia Elsag Sistemi Navali Srl, Alfa Romeo Avio Società Aero per Azioni, Elettronica SpA, Elmer SpA, Europa Metalli LMI SpA, Fiar SpA, Fiat Aviazione SpA, Italtel SpA, Iveco Fiat SpA, Litton Italia SpA, Oerlikon/Contraves Italiana SpA, Ind Aereon e Meccanica R Piaggio SpA, Riva Calzoni SpA
98

65-02000 Società Italiana Chimici

Via Luigi Pulci 27, I-00162 Rome
☎(+39 6) 44 24 40 78 ✉ITALCHIMICI
⊖ (+39 6) 44 29 07 75
♧ Lit 99,000,000 (1997) ♣ Lit 1,500,000,000 (1996)
👥 7 (1997) 🏭 200 m2
Managing Director: Guido Promontorio
■ Individual solar which converts seawater into potable water.
98

65-02140 UCIMU-Sistemi per Produrre

Unione Costruttori It. Macchine Utensili, Robot e Automazione
PO Box 11031, I-20110 Milan
Viale Fulvio Testi 128, I-20092 Cinisello Balsamo (MI)
☎(+39 2) 26 25 51 ⊠ 312594
⊖ (+39 2) 26 25 52 14/.349
Chairman: Ezio Colombo
General Mgr: Alfredo Mariotti
Press & Promotion: Mrs Rossella Bernuzzi
■ Italian machine tools, robot and automation manufacturers' association.
97

70 Financial Holding Groups, Parent Companies

70-00010 Aeronautica Macchi SpA

Head Office
Vio Foresio 1, I-21040 Venegono Superiore (VA)
☎(+39 331) 86 59 10 ⊖ (+39 331) 86 59 12
♧ Lit 2,262,605,000
President: Dr Fabrizio Foresio
Vice President: Dr Giorgio Zappa
Associated companies:
Aermacchi SpA → 100
Logic SpA → 100
Sicamb SpA → 70
98

70-00270 Finmeccanica

Piazza Montegrappa 4, I-00195 Rome
☎(+39 6) 32 47 31 E news@finmeccanica.com
⊠ 610371 finmec ⊖ (+39 6) 887 05 87
♧ Lit 1,688,000,000 ♣ Lit 12,844,000,000 👥 57,000
Communications: Dr Pippa ☎(+39 6) 32 47 33 95; 320 86 21 (fax)
■ Financial holding company of IRI group. Operates in the areas of aeronautics, space, defence, energy and transportation.
97

70-09070 IRI - Istituto per la Ricostruzione Industriale

Via Vittorio Veneto 89, I-00187 Rome
☎(+39 6) 472 71 ⊠ 610468 istiri i
⊖ (+39 6) 472 723 08
♣ Lit 75,681,000,000 👥 132,489
Chairman: Gian Maria Gros-Pietro
Director General: Pietro Ciucci
Press Officer Manager: Sergio Brund
98

70-10000 Montedison SpA

Piazzetta M Bossi 3, I-20120 Milan
☎(+39 2) 627 01 ⊖ (+39 2) 62 70 52 68
♣ Lit 23,953,000,000 👥 23,564
Public Relations Manager: Franco Cerabolimi
98

70-19400 Telecom Italia Spa Headquarters

Società Finanziaria Telefonica († STET)
Corso d'Italia 41, I-00198 Rome
☎(+39 6) 858 91 ⊠ 610006 ⊖ (+39 6) 858 94 34
♧ Lit 3,680,000,000,000
♣ Lit 37,373,000,000,000 (1995) 👥 132,548
President: Guido Rossi
Managing Director: Tomaso Tommasi di Vignano
General Director: Umberto De Julio
98

74 Consortia, Joint Venture Companies

74-00100 Alenia Difesa, Divislone Sistemi Navali

(† Alenia Elsag Sistemi Navali SpA)
(a Finmeccanica Company)
Head Office
Via di S Alessandro 28/30, I-60131 Rome
☎(+39 6) 41 88 31 ⊠ 621276 AD/DSN-S
⊖ (+39 6) 41 88 32 97
♣ Lit 460,000,000,000 (1997) 👥 1,005
Managing Director and General Manager: Armando Sodaro
Marketing Director: Antonio Bontempi
■ Radar systems (naval and coastal, surveillance, tracking and multifunctional). Fire control systems (radar and electro-optical). Command and control systems (naval including C3I). Missile systems (surface-to-air missile systems). Electro-Acoustic systems (underwater and hull mounted sonars). Mine identification and disposal systems (ROV). Combat systems definition and integration.
98

74-03700 Consorzio SMIN

(Consortium formed by Elsag SpA and Riva Calzoni SpA)
Via Panama 6/B, I-16154 Genova
☎(+39 10) 654 81 ⊠ 621276 ⊖ (+39 10) 654 85 90
Manager Commercial: Giuseppe Bedeschi
97

74-10000 Logicasiel SpA

(a joint venture with Finsiel)
Via Leone XIII 14, I-20145 Milan
☎(+39 2) 48 59 41 ⊖ (+39 2) 48 00 83 12
Managing Director: Roberto Cassaleggio
Public Relations Manager: Adonella Melis
Bologna Office
Via San Vitale 40/3, I-40125 Bologna
☎(+39 51) 29 44 11 ⊖ (+39 51) 29 44 49
Turin Office
Corso Svizzera 185, I-10149 Turin
☎(+39 11) 742 87 11 ⊖ (+39 11) 771 64 46
Ivrea Office
Via Corte d'Assise 8, I-10015 Ivrea (TO)
☎(+39 12) 54 69 88 ⊖ (+39 12) 54 43 57
98

74-23350 Simad SpA Commercial Office

PO Box 1, I-67061 Carsoli
☎(+39 863) 99 61 39; 99 62 44
W http://www.amicus.it/simad ⊖ (+39 863) 99 63 90
♧ Lit 1,000,000,000 (1997)
♣ Lit 3,500,000,000 (1996) 👥 24 (1997) 🏭 6,000 m2
Chairman: Ugo Giovannoni
Managing Director and General Manager: Andrea Giovannoni
Marketing and Export Director: Giuseppe Appierto
Technical Director: Daniele Bugatti
■ Smoke and illuminating signals, tear gas devices, pneumatic launcher, sport and shotgun cartridges, special cartridges 12 bore, 38mm and 40mm for riot control. Anti-sabotage cartridges. → 75
Oricola (AQ) Plant
PO Box 1, I-67061 Carsoli, AQ
Via Tiburtina Valeria, Km 64, I-67063 Oricola, AQ
☎(+39 863) 99 61 39; 99 62 44 ⊠ 601080 start i
⊖ (+39 863) 99 63 90
Plant Manager: Giuseppe Appierto
■ Signal devices, smoke signals, illuminating signals, tear gas grenades and pyrotechnics.
98

75 Manufacturing Companies

75-00010 ABB Dalmer Benz Transportation (TIBB)

(† Tecnomasio Italiano Brown Boveri SpA)
(a subsidiary of ABB Daimler Benz Transportation)
Head Office
Pal Cedri, VLE Europa 24, I-20090 Segrate
☎(+39 2) 579 71 ⊠ 310153
Works
Via Privata Trento 2, I-17047 Vado Ligure SV
☎(+39 19) 88 29 21 ✉TECNOMASIO ⊠ 213016
● Via dell'Industria 28, I-20010 Vittuone MI
☎(+39 2) 901 50 21 ✉TECNOMASIO ⊠ 330127
■ Naval computers, electrics and electronics.
98

75-00020 AD Aerospace Finishes Srl

(a subsidiary of Akzo Nobel Coatings BV)
Via delle Industrie 22, San Zenone Degli Ezzelini, I-31020 Treviso
☎(+39 42) 396 81 22 ⊖ (+39 42) 396 90 46
General Manager: A M L Derksen
■ Washprimers, primers and topcoats for military, commercial and general aviation. Process coatings, maskants for chemical milling, paint-removers and cleaning solvents. Defence coatings and sealants.
98

75-00280 Aerea SpA

(a partner of ACMA GmbH)
Via Cefalonia 18, I-20156 Milan
☎(+39 2) 38 00 02 18 ⊠ 333494 aerea i
⊖ (+39 2) 33 40 26 76
♧ Lit 1,500,000,000 ♣ Lit 18,000,000,000 👥 84
🏭 6,000 sq m
Chairman: Dr Ing. Marco Bognetti
Managing Director: Dr-Ing. Silvano Mantovani
Operations Director: Danild Rainoldi
Marketing Director: Dr-Ing. Vincenzo Riviello
Public Relations Manager: Francesco Nicotra
Purchasing Manager: Domenico Cantele
■ Design, development and manufacture of ejector release units for high performance aircraft, bomb rack units to NATO standard, 14 inch and 14/30 inch missile rail launchers for AIM-9L missile, chaff, flare and (EFA). Program for design and development of ejector release unit equipment type ALDERU/AHDERU and rail beam for ITSPL. bomb and rocket dispenser, door gun posts, defence conversion systems for helicopter and light aircraft, rocket launchers for rockets from 50mm up to 81mm calibre. Contractor to European Fighter Aircraft. Fuel tanks and containers. Anti-submarine and training bombs. Contractor for airframe support equipment AV-8B plus.
Subsidiaries:
Aerea Luftfahrtgerte GmbH
Aerea Cospal SpA
Acma GmbH
98

75-00500 Aermacchi SpA Commercial Headquarters

(a division of Aeronautica Macchi SpA)
Via Ing. Paolo Foresio 1, I-21040 Venegono Superiore, VA
☎(+39 331) 81 31 11; 81 31 30 (public relations)
✉AERMACCHI Varese ⊠ 380070 aermac i
⊖ (+39 331) 81 34 50; 82 75 95

♲ Lit 85,000,000,000 ♣ Lit 241,800,000,000 (1994)
♟ 1,392 ♜ 112,000 m2
President: Dr Fabrizio Foresio
Vice President: Ing. Pasquale d'Angelo
Managing Director: Giorgio Brazzelli
General Manager: Ing. Bruno Cussigh
Finance and Administration Director: Dr Andrea Tarlazzi
Manufacturing Director: Dr-Ing. Emilio Cancer
Technical Manager: Ing. Massimo Lucchesini
Commercial Manager: Dr Cesare Cozzi
Purchasing Manager: Ing. Osvaldo Pedrocchi
Quality Assurance and Test Manager: Dr-Ing. Gugliemo Rossoni

■ Project, production and overhaul of MB-339A and MB-339C advanced jet trainers and MB-339FD Lead-In Fighter aircraft. Production and support of Redigo L-290 Turbopropeller. Production of AMX fighter aircraft, single and two-seater versions. Design an production of the front fuselage section and assembly of the fuselage of DO 328 commuter aircraft.

Design and production of selected items of Tornado and Eurofighter 2000 aircraft.

Manufacture all aircraft systems and the integration of digital avionics equipment for the YAK/AEM-13 advanced trainer. Specialised in wind tunnel tests: low speed-high angle of attack. Specialised in structural tests: fatigue, vibration and bird impact. Design and production of ground support equipment.

Works Plant
I-21040 Venegono Superiore
☎ (+39 331) 81 31 11

Flight Center Plant
I-21040 Venegono Inferiore
☎ (+39 331) 81 31 11

Research Laboratories Wind Tunnel
I-21040 Venegono Superiore
☎ (+39 331) 81 31 11

Structural Test Center
I-21040 Venegono Superiore
☎ (+39 331) 81 31 11 97

75-01000 Agusta Group IRI Finmeccanica

(a division of Finmeccanica company)
Via G Agusta 520, I-21017 Cascina Costa di Samarate
☎ (+39 331) 22 91 72 W http://www.agustacorp.com
⊖ (+39 331) 22 99 84
♣ U$S 955,000,000 (1996) ♟ 5,094 (1996)
Chairman: Gen Amedro Caporalettio
Operations Director: Mario Pellerei
Quality Director: Ing F Reina
Procurement Director: Ing M Sala
Commercial Director: Alessandro Gustapane
Public Relations and Corporate Affairs Director: Gian Luigi Ghezzi
Government Sales: Ing B Spangnolini
Civil Sales: Ing L Monti
Personnel and Organisation: Mario Po
Planning and Controller: Ing A Gianni
International and Programs: Ing G Orsi

■ Research, development, planning, production, maintenance and marketing of helicopters and associated equipment, components and structural parts (mechanical, electrical, optical and electronic). Software: customer support, technical assistance, spares, upgrading, training, overhaul. Integrated Logistic Support (ILS) system design; technical, financial and administrative management and control of subsidiary companies.

Production:
helicopters (own design):
A109K2 civil, military; A109
Power civil; A119 Koala civil, military
single engine; A129 Mangusta military; A129 International, military.
Helicopters under licence: McDonnell Douglas AMD 500 and AMD 520N; Bell AB206, AB212 ASW/ASUW, AB412.
Other activities: avionic and onboard equipment, onboard computers, systems engineering simulators, integrated logistic support, composite materials, castings.
International Programs
EH101
NH90

Subsidiaries:
Agusta Aerospace Corporation, USA
Agusta Aerospace Services AS/SA, Belgium

Associated Companies:
EH Industries Ltd, UK
EH Industries Inc, USA
NH Industries Sarl, France

Registered Office
Via G Agusta 520, I-21017 Cascina Costa di Samarate VA
☎ (+39 331) 22 91 11 ⊖ (+39 331) 22 25 95

Brindisi Plant
Contrada Santa Teresa Pinti, I-72100 Brindisi
☎ (+39 831) 53 11 ⊖ (+39 831) 45 26 59

Vergiate Plant
Via Romasi, I-21019 Vergiate, VA
☎ (+39 331) 94 01 11 ⊖ (+39 331) 94 79 09

Benevento Plant
Contrada Ponte Valentino San, S S 90bis, I-82100 Benevento
☎ (+39 824) 534 41 ⊖ (+39 824) 534 18

Frosinone Plant
Via G Agusta 1, I-03110 Frosinone
☎ (+39 775) 20 51 ⊖ (+39 775) 20 53 97

Anagni Plant
Localita Paduni, I-03012 Anagni, FR
☎ (+39 775) 76 41 ⊖ (+39 775) 76 85 43 98

75-01160 Alenia Aerospazio

(a Finmeccanica company)

Alenia AEROSPAZIO

92 Viale Maresciallo Pilsudski, I-00197 Rome
☎ (+39 6) 80 77 81; 80 77 87 49
⊠ 611395; 611101 allenia i
⊖ (+39 6) 807 22 15; 807 51 84; 80 77 83 00
♣ US$ 1,848,000,000 ♟ 12,000
President: Giorgio Zappa
Head of Aeronautics Division: Filippo Bagnato
Head of Space Division: Giuseppe Viriglio

■ Aeronautics Division Activities:
International Programs
Eurofighter: air superiority aircraft in co-operation with BAe, Daimler-Benz Aerospace and CASA.
Tornado: multirole combat aircraft in co-operation with Daimler-Benz Aerospace and BAe (Panavia Consortium).
AMX: close interdiction and reconnaissance aircraft, in co-operation with Aermacchi and Embraer (Brazil).
ATR 42-500: last version of the ATR family in the 48 seat configuration.
ATR 72-210: stretched version of the last version of the ATR family in the 66/68 seat configuration.
Boeing B767: design and manufacture of aerostructures and composite mobile surfaces.
McDonnell Douglas MD80/90: design and manufacture of composite control surface and fuselage panels.
McDonnell Douglas MD11: manufacture of fuselage panels, winglets and control surfaces.
Falcon 2000: development and construction in co-operation with Dassault.
Falcon 900E: design and production in co-operation with Dassault.
Airbus A330: development of engine nacelles.
McDonnell Douglas MD95: design and manufacture of fuselage elements.
Boeing B777: design and manufacture of wing sections.
Airbus A321: design and manufacture of fuselage sections.
Harrier AV 8BII: final assembly and final test preparation (Italian Navy).
National Programs
F104 ASA/M: updating of avionics and weapon system of F104S ASA fighter aircraft.
G222: multi-purpose turboprop aircraft for military and civil missions (transport, paratroopers, firefighting, spraying, aeroambulance, environmental surveillance, ECM, navigation aid calibration).
Development
F.L.A. (Future Large Aircraft) new European military transport aircraft; C27-J next generation medium tactical airlifter, developed by Alenia Aerospazio and Lockheed Martin; AIR 70 new twin engine aircraft developed by AI(R) Consortium in the 60/90 seat range.
Overhaul and Maintenance
Overhaul and maintenance of military aircraft (Tornado, G222, F104S ASA, AMX, Atlantic, Awacs, B707); overhaul and maintenance of commercial aircraft (DC8, MD80, DC10, MD11, B727, B707, ATR42 and 72, Airbus); modification of DC8/DC10 aircraft to cargo version; re-engining of B727/100 aircraft; modification of B707 aircraft to tanker/transport version; training; product support; cockpit updating for DC8, DC10, B727, Atlantic and G222 aircraft; design, production and maintenance for thrust reversers "target type".
Space Division Activities
Telecommunication satellites (Artemis, DRS, Sicral, Globalstar, Italsat F2); scientific satellites (TSS 1, Lageos 2, SAX, XMM, Integral, SOHO-UVCS); remote sensing satellite (X-Sar, ERS 2, Envisat, Cassini-Huygens, Helios, Skymed-Cosmo); space infrastructures (Colombus Orbiting Facility, Spacehab, MPLM - Multi Purpose Logistic Module); launch, transport and re-entry systems (Iris, ATV, CRV, Ariane IV); software for space, aeronautics and missile applications; parallel supercomputers; data handling systems for satellites and space platforms; onboard scientific instrumentation for remote sensing and interplanetary probes; electronic control systems for launching and space transport systems.

Finmeccanica companies coordinated by Alenia:
Aerospazio
Laben SpA
Officine Aeronavali Venezia SpA
Quadrics Supercomputers World Ltd
Space Software Italia SpA
The Dee Howard Co. 98

75-01165 Alenia Difesa

(a Finmeccanica company)

Alenia DIFESA

92 Viale Maresciallo Pilsudski, I-I-00197 Rome
☎ (+39 6) 80 77 81; 80 77 87 18
⊠ 611395; 611101 alenia i
⊖ (+39 6) 807 51 84; 807 22 15; 80 77 84 80
♣ US$ 9,070,000,000 ♟ 60,012
President: Ing P F Guarguaglini
Head of Radar Systems Division: Ing C A Penazzi
Head of Naval Systems Division: Ing A Sodaro
Head of Missile Systems Divsion: Ing M Di Donato
Otobreda Division: Ing P F Guarguaglini
Avionic Systems and Equipment Division: Ing G Grasso

■ Radars Systems Division Activities:
Mission critical systems; medium/long range tri-dimensional radars; bi-dimensional surveillance radars for coastal and low-level operations; mobile and transportable command and control systems; radars for air surveillance of the battlefield; integrated command and controls systems; battlefield surveillance systems on airborne platforms (helicopters, aircraft, RPV's); fire automation systems; logistic identification and maintenance systems; meteorological systems; SATCAS systems based on four levels (systems for ATC automated service); primary radars for en-route control; primary radars for terminal area (Phased Arary, 3D, Meteo); secondary radars (surveillance, communications, including Ground Data Link Processor); ATC training simulators; navigation aids (en-route, landing, satellite navigation); FDP systems (Flight Data Processing); message switching systems for AFTN:SAFE (Systems for Aeronautical Fixed message Exchange); workstations for ATC controllers; monitoring and control systems; radar data processors; recording and playback systems; meteorological systems for airports; complete systems for guidance and control of airport ground surface movements (SMGCS); arrival/departure information database for airports (ADID); ADS and ATC integrated systems (ADSATC); router for Aeronautical Telecommunication Network; logistics and maintenance; airport turnkey projects and management (runways, runway lighting, buildings).
METEOR - Contruzioni Aeronautiche ed Elettroniche SpA

Contact Reference

Tactical unmanned aerial vehicles; target drones; automatic systems for civil defence; remotely piloted vehicles for territory surveillance; remotely piloted vehicles operations; training courses for operation of produced systems; computer aided training systems; flight and mission simulators; computer image generation; integrated logistic support for simulators and UAV's.
Naval Systems Division Activities:
Surface systems
Integrated combat management systems or single subsystems; surveillance radars for naval and coastal applications; radar and electro-optical surveillance and fire control systems; Command and Control systems (C3I), based on military and ruggedised computers and consoles for military and paramilitary, naval and landbased applications; naval surface-to-air missile systems and CIWS; vessel traffic control systems.
Underwater Systems
WASS - Whitehead Alenia Sistemi Subacquei SpA
Active and passive sonar systems; heavy and light torpedoes systems; integrated combat systems for submarines; anti submarine low cost weapons.
Missile Systems Division
Multirole, surface-to-air and air-to-air missiles; anti-aircraft, surface-to-air missile systems for strategic target defence; anti-aircraft missile systems for the battlefield; air-to-air new generation missile systems; anti-aircraft missile systems family both ground based and for naval applications; production programmes under licence; anti-ship air-to-surface and anti-tank missile systems.
Otobreda Division
Naval guns; weapon systems for army/aeronautical uses; AF cannons and turrets; ammunition and bombs.
Avionics Systems and Equipment Division
Airborne radars for fire control, early warning maritime patrol, ASW/ASV and rescue; laser rangefinders, TV cameras, thermal cameras, TVRR trackers, infra-red search and track systems, displays; Precision Approach Radar (PAR), minehunting sonar and dipping sonar for use on EH 101 class helicopters; power supply systems, photovoltaic cells GaAs, high voltage power supplies, TWTA's, SSPA's frequency synthesisers, telecommunication stations, telecommunication transponders, mobile terminals, expert systems, mini-satellites, space robotics and automation; apparatus installation and repair technical assistance, integrated logistic support, test equipment and trouble shooting expert systems design and production.

Other Finmeccanica Companies co-ordinated by:
Alenia Difesa
Alelco SpA
Galileo SMA
Gematronik GmbH
GF Sistemi Avionici [98]

75-01170 Alenia SpA Headquarters

(a Finmeccanica Company)
92 Viale Maresciallo Pilsudski, I-00197 Rome
☎ (+39 6) 80 77 81; 80 77 87 74;
80 77 87 18 (press department)
Telex 611395; 611101 alenia i
Fax (+39 6) 807 22 15; 807 51 84; 80 77 83 00;
807 83 89 (press department)
Capital Lit 4,800,000,000 ♣ Lit 5,000,000,000 Employees 20,000
President and Chief Operating Officer: Giorgio Zappa
General Manager Finance: Paolo Micheletta
Vice President Marketing and International Affairs: V Proieti
Marketing Communication Director: D Scurti
Missile Systems Division: Ing. M Di Donato
Radar Systems Division: Ing. C A Penazzi
President Otobreda Division: Ing. P F Guarguaglini
Avionic Systems and Equipment Division: Ing. R Esposito
Naval Systems Division: Ing. A Sodaro
■ Design and manufacture of military and commercial aircraft. Activities cover airframe components, aircraft modification, overhaul and maintenance, aero-engines, command and control systems, environmental protection systems, ATC systems, radar, missiles, RPV's and unmanned vehicles, satellites, ground stations, space stations and orbiting infra-structures. Design and manufacture of civil transport aircraft ATR 42 and 72, military aircraft G222, AMX, Tornado, Eurofighter 2000. Under international contracts for design and manufacture of parts for B767, B777, MD-80, MD-90, MD-11. In the field of aircraft modification, contracts are carried for cargo conversion and re-engining of DC-8, DC-10, MD-11, B707 and B727 aircraft.
The company is active in the field of surveillance, command, control and defence systems where it designs and manufactures integrated systems for air defence such as the RAT31S, RAT31SL, RAT31DL weapon systems, Aspide, Spada and Skyguard missile systems, CATRIN ground command control and communication systems, unmanned air vehicles (Mirach series); support and logistic systems for military applications; telecommunication and navigation systems.
Design and manufacture ATC systems for airports, for air and maritime traffic control, the ALE 9 and ALE 3x5 radar antennas and DDS 80 and CDS 2000 consoles.
Design and manufacture of complete space systems including Italsat, Olympus, Sicral, Globalstar, Sarit, Artemis and DRS telecommunication satellites, Meteosat, ERS-1 and Helios remote-sensing satellites, Tethered, Lageos II and Sax scientific satellites, transport systems and manned space modules including Columbus, Ariane IV, Carina ground stations, orbiting infrastructure and space software.
Design, manufacture and overhaul of turboprop, turbojet, turboshaft, and turbofan engines including T-700 series for the Anglo-Italian EH 101 helicopter programme. Design and manufacture of naval systems including surveillance, command and control systems. Underwater systems for surface and underwater vessels.
Radar systems division: mission critical division, medium/long range tri-dimensional radars, bi-dimensional surveillance radars for coastal and low level operations. Surveillance, target aquisition and command/control systems. Battlefield surveillance systems on airborne platforms (helicopters, aircraft, RPV's), fire automation systems, logistic identification and maintenance systems. Meteorological systems, SATCAS systems based on four levels (systems for ATC automated service). Primary radars for en-route control, primary radars for terminal area (phased array, 3D, meteo). Secondary radars (surveillance, communications, including ground data-link processor). ATC training simulators, navigation aids (en-route, landing, satellite navigation). FDP systems, message shifting systems for AFTN: SAFE (System for Aeronautical Fixed Messages Exchange). Workstations for ATC controllers, monitoring and control systems, radar data processors, recording and playback systems. Meteorological systems for airports. Complete systems for guidance and control of airport ground surface movements (SMGCS). Arrival and departure information database for airports (ADID), ADS and ATC integrated systems (ADSATC). Router for aeronautical telecommunication network, logistics and maintenance. Airport turnkey projects and management (runways, runway lighting, buildings).

Finmeccanica Companies Co-ordinated by:
Alenia Difesa
Alelco SpA
Fiar SpA
Galileo Spa
Gematronik GmbH
GF Sistemi Avionici
Meteor CAE SpA
Wass SpA

Subsidiaries with defence business:
Alenia Spazio SpA → 75
Alfa Romeo Avio SpA → 75
Alenia-OTO Sistemi Missilistici
Elmer SpA → 75
Laben SpA → 75
Meteor SpA → 75
Officine Aeronavali Venezia SpA → 75
Alenia Elsag Sistemi Navali → 74
WASS - Whitehead Alenia Sistemi Subaquei [97]

75-01770 Arteco Srl

Via Poggio Alle Viti 12, I-55054 Massarosa (Lu)
☎ (+39 584) 96 11 66 Fax (+39 584) 96 20 21
Capital Lit 600,000,000 (1997)
♣ Lit 200,000,000,000 (1997) Employees 26 (1997)
Area 1,500 sq m
President: Pier Giorgio Del Carlo
■ Equipment for handling and disposal of waste water, bilge waste. Waste dilacerators and pumps. On-board chlorine production by electrochemical cells. Special amagnetic features. Full maintenance service. [98]

75-01800 Astra Veicoli Industriali SpA

(a division of Iveco SpA)
Defense Division
Via Caorsana 79, I-29100 Piacenza
☎ (+39 523) 54 31 Telex 530148 Fax (+39 523) 59 32 93
Capital Lit 20,000,000,000 (1997)
♣ Lit 242,289,000,000 (1996) Employees 500 (1997)
Area 150,000 sq m
General Manager of Defence Division: S Braco
Operations Director: Ing M Manfioletti
Purchasing Manager: A Risoli
Off Road Vehicles Marketing Manager: A Scognamillc
■ Production of special 6x6 and 4x4 vehicles for Engineer Corps. Manufacturer of industrial off-road vehicles (dump trucks) for the Italian market and for the Middle East, Africa, and Central and South America. [98]

75-02300 Azienda Torinese Elettronica Telecomunicazioni (ATET)

42 Via G. da Verrazzano, I-10129 Turin
☎ (+39 11) 568 32 00 Fax (+39 11) 59 04 93
Managing Dir: Corrado Avagnina
Sales & Marketing Dir: Ms Andrea Avagnina
Operations Dir: Lorenzo Brunasso
Purchasing Mgr: Francesco Mancuso
■ Miniaturised radio transmitters & beacons. Covert surveillance equipment.

Rome Office
132 Via Gregorio VII, I-00165 Rome
☎ (+39 6) 39 37 56 43; 638 02 21
Fax (+39 6) 39 36 63 15 [97]

75-02360 Azimut SpA

(a subsidiary of Benetti Shipyard)
Viale dei Mareschi, 14, I-10051 Avigliana Torino
☎ (+39 11) 936 72 71 Fax (+39 11) 936 72 70
Capital 6,450,000,000 (1996) ♣ Lit 149,000,000 (1996)
Employees 296 (1996) Area 15,500 m2
Managing Director: Paolo Vitelli
Technical Director: Ing Ermammo Porro
Marketing Director: Massimo Perotti
Marketing Coordintor, Public Relations Manager: Dr Mary Brayda-Bruno
■ Production of steel and aluminium vessels from 35 to 70 m. Production of fibreglass vessels from 10 to 30 m, motor cruisers and motor yachts. [98]

75-02640 Belotti SpA Head Office

Via Gramsci 2, I-16010 Manesseno (GE)
☎ (+39 10) 730 81 Fax (+39 10) 71 29 74
Capital Lit 6,050,000,000 (1997) Employees 200 Area 10,000 m2
President: Gian Luigi Carrara
Commercial Director: Pier Luigi Cafiero
Operations Director: G Pastorino
Purchasing Manager: P Carara
■ Mobile cranes for lifting and transporting damaged aircraft, containers and goods weighing up to 100 tonnes.

Manesseno San Olcese Plant
Via Gramsci 2, I-16010 Manesseno San Olcese
☎ (+39 10) 730 81 Fax (+30 10) 71 29 74 [98]

75-02700 Benelli Armi SpA

Via della Stazione 50, I-61029 Urbino, Pesaro
☎ (+39 722) 30 71 E marketing@benelli.it
Fax (+39 722) 32 74 27; 30 72 07
Capital Lit 4,200,000,000 (1996)
♣ Lit 52,600,000,000 (1994) Employees 132 Area 8,275 m2
Managing Director: Ing Luigi Moretti
Production Director: Ing Lucio Burigana
Sales and Marketing Director: Ing Gino Priora
■ Semi-automatic weapons, specialised machine tools, precision fabrication. [98]

75-02800 Pietro Beretta SpA

Fabbrica d'Armi P Beretta SpA
Via Pietro Beretta 18, I-25063 Gardone, VT
☎ (+39 30) 834 11 W http://www.beretta.it
Telex 301523 pb bsi Fax (+39 30) 83 25 41
Capital Lit 10,000,000 (1996)
♣ Lit 215,000,000,000 (1996) Employees 1,050
Area 75,000 sq m (covered area)
President and Managing Director: Ugo Gussalli Beretta
Sales Manager (Defence Division): Michele Micheletti
Sales Manager: Giovanni Locanto
Advertising Office: Alessandro Baratta
■ Manufacture of small arms for military and police forces.

Rome Office
Via Puccini 9, I-00198 Rome
☎ (+39 6) 855 48 70; 855 15 48 Telex 610355
Fax (+39 6) 884 00 78 [98]

75-03120 BOMISA
Bottoni Minuterie SpA
13 via Idiomi, I-20090 Assago MI
☎ (+39 2) 48 85 81 ⊠ 310141 ⊖ (+39 2) 488 22 85
♣ US$ 13,00,000 👥 125
Managing Director: Mauro Corinaldi
Sales Director: Pierluigi Di Pisa
Purchasing Manager: Antonio Volontieri
■ Metal buttons, buckles, insignia, stars, medals, handcuffs, parade swords, belts and souvenir plaques. [98]

75-05000 Carrozzeria Boneschi Sarl
(a subsidiary of Boneschi Centro SUD)
Via Giovanni Boneschi 1, I-20040 Cambiago, MI
☎ (+39 2) 95 30 80 45 [E] boncartrade@pointest.com
[E] boncardesign@pointest.com ⊖ (+39 2) 95 30 82 04
♧ Lit 2,000,000,000 (1996)
♣ Lit 10,000,000,000 (1996) 👥 100 (1996)
🏠 10,000 m2
President: Dr Bruno Pezzaglia
Managing Director: Dr Paolo Pezzaglia
■ NATO standard shelters and cases. Armoured vehicles, mobile HQ vans, ambulances and mobile hospitals. Press work.

Boneschi Centro Sud
Via Negroni 71, Ariccia, Rome
☎ (+39 6) 934 34 26 ⊖ (+39 6) 934 31 22 [98]

75-05150 C.I.A.M.-E.A. Sarl
(† Icea Sud SpA)
Via Astura 34, Loc. Le Isole, I-00040 Lanuvio, Roma
☎ (+39 6) 93 74 85 28 ⊖ (+39 6) 93 74 85 17
♧ Llt 60 000 000 (1996) ♣ Llt 3 000 000 000 (1996)
👥 6 🏠 40 000 sq m
President: Eng. Francesco Rietti
Managing & Operations Dir: Eng. Giuseppe Rocchi
Marketing Dir: Dr Elisabetta Tulli
Purchasing Mgr: Dino Pagnotti
■ Fibre containers for fuses, shells, mortar bombs, rockets & propelling charges of any calibre. Primary cartridge cases for mortar bombs, bags, caps & liners for propelling charges; grommets; parachutes for light flares. Accoutrements & military clothing/bedding items. [97]

75-05183 Circuiti Stampati Italia (CSI)
(a div. of Zinocelere Sarl, Italy)
Via Bertone 10bis, I-13042 Cavaglia, VC
☎ (+39 161) 99 63 56; 99 63 57 ⊖ (+39 161) 99 63 69
♧ Llt 2 000 000 000 (1995)
♣ Llt 11 000 000 000 (1994) 👥 60 🏠 3 000 m2
Managing Dir: Dr A. Bellazzini
General Mgr: Piero Pacetti
Operations Dir: C.A. Caglieri
Marketing & Sales Mgr: Giancarlo Silva
Sales Mgr: A. Garavelli
Purchasing Mgr: G. Turcato
■ Professional printed circuit boards, multilayers (rigid and flexible for defence, space and avionics applications.

Representing:
Zincocelere Sarl, Italy [97]

75-05200 Rockwell International Co. Collins General Aviation Division
(† Rockwell CVC SpA)
(a subsidiary of Rockwell International Corp, USA)
Collins General Aviation Division, Europe, Middle East and Africa Operations
Via Monte Giberto 157, I-00138 Rome
☎ (+39 6) 881 85 55 ⊖ (+39 6) 881 91 29
👥 2
Director: Sergio Masserini [98]

75-05380 Connei SpA
Connettori Elettrici Industriali
(a subs. of Smiths Industries plc, UK)
Via Pillea 14/18, I-16153 Genova Sestri
☎ (+39 10) 60 08 21 ⊖ (+39 10) 650 85 73
♧ Llt 7 650 000 000 (1995)
♣ Llt 18 777 000 000 (1995) 👥 131
President: Dr Martin Hartley Jones
Managing Dir: Ing. Giuseppe Lancella
Sales & Marketing Dir: Mauro Gramaccioni
Technical Dir: Ing. Piero Bozzo
Purchasing Mgr: Alberto Massuco
■ Electrical connectors for electronics & telecommunications.

Representing:
Hypertac Ltd, UK
FRB Connectron, France
Interconnectron GmbH, Germany
SI-TAC Connectors Inc., USA
Hypertronics Corp., USA

Rome Office
Via Marco Valerio Corvo 116, I-00174 Rome
☎ (+39 6) 76 96 58 20 ⊖ (+39 6) 767 42 22

Milan Office
Via Ripamonti 195, I-20141 Milan
☎ (+39 2) 569 40 31 ⊖ (+39 2) 57 40 36 82 [97]

75-05700 COSMOS SpA
Via della Padula 83, I-57124 Livorno
☎ (+39 586) 86 80 11 [E] cosmos@master.etrurianet.it
⊠ 500170 motsot i ⊖ (+39 586) 85 93 73
♧ Lit 500,000,000 (1997) ♣ Lit 10,000,000 (1997)
👥 60 (1997) 🏠 3,000 m2
Managing Director and General Manager: Jucio Moricomi
Commercial Executive Director: Capt Manuel De Domto (retired)
Operations Director: Sergio Carlini
Purchasing Manager: Paolo Vaccai
■ Underwater attack craft such as manned mini-submarines. [98]

75-06300 CRM SpA Motori Marini
(† CRM Fabbr. Motori Marini e Industriali)
Via Marnate 41, I-21053 Castellanza (Va)
☎ (+39 331) 57 26 00 ⊠ 334382 ⊖ (+39 331) 50 55 51
♧ Lit 3,000,000,000 (1997)
♣ Lit 14,304,000,000 (1997) 👥 70 (1997) 🏠 12,600
Managing Director: Giorgio Mariani
Sales and Marketing Director: L Mariani
Technical Director: Bruno Piccoletti
Purchasing Manager: E Marchi
■ Manufacture of diesel engines for naval and special applications, from 900hp to 2100 hp. Gears and drives. Non-magnetic versions also available. Engines for airships. [98]

75-06500 Datamat Ingegneria dei Sistemi SpA
Via Laurentina 760, I-00144 Rome
☎ (+39 6) 502 71 [W] http://www.datauiat.it
⊠ 613436 dtmt i ⊖ (+39 6) 505 14 07
♧ Lit 30,000,000,000 (1996)
♣ Lit 205,000,000,000 (1996) 👥 1,200 (1996)
🏠 20,000 sq m
President: Prof Claudio Bianchi
Managing Director and General Manager: Ing Giancarlo Giglio
Marketing Director: Ing Franco Olivieri
Finance Director: Francesco Mannoni
Defence Business Development Director: Ing Enrico Cuturi
Sales Director: Ing E Cambise
Marketing Director, Defence Systems Division: Ing Gabriele Pitzurra
Technical Director: Ing E Cambise
Quality Assurance Director: A Aquili
Budget Control Director: C Carrara
Director of Personnel and Organization: G De Leonardis
■ C3 systems, training simulators, satellite ground stations, software technologies, system integration, logistic support. [98]

75-06525 Defense Telecommunications Division
(a division of MID/Marconi Italtel Difesa)
Via Castello della Magliana 75, I-00148 Rome
☎ (+39 6) 65 96 01
⊖ (+39 6) 65 96 04 65; 65 96 04 66
Director: Gianfranco Soverini
Commercial Director: Giuseppe Capodici
International Marketing Executive: Francesco Esposito
Marketing Executive: Icilio Di Luzio
■ Strategic and tactical command, control, communications and intelligence systems, joint tactical information distribution systems (JTIDS) and identification friend or foe (IFF) systems for aircraft, helicopters, tanks and ships, including the NATO Identification System; encryption systems for radio, telegraph and data communications; tactical radios and avionic equipment. [98]

75-06550 De Filippi Andrea Sas
Via Garibaldi 124, I-12061 Carru, CN
☎ (+39 173) 751 01 ✉ DE FILIPPI ⊠ 210175 caruca i
⊖ (+39 173) 751 12
♧ Lit 50,000,000 ♣ Lit 1,510,000,000 (1995) 👥 12
🏠 3,100 sq m
Managing Director: Franco de Filippi
■ Trailers, semi-trailers and mechanical constructions in general. [98]

75-06695 Druck Italia Sarl
(a subsidiary of Druck Ltd, UK)
Via Capecelatro 11, I-20148 Milan
☎ (+39 2) 48 70 71 66 ⊠ 351183 druck i
⊖ (+39 2) 48 70 55 68
♧ Llt 20,000,000 (1997) ♣ Llt 9,000,000 (1997) 👥 16
🏠 400 m2
Managing Director: Tiziano Casati
Operations Director: Arena Emore
■ Manufacture of pressure sensors, indicators and calibration equipment. [98]

75-06860 Electronic Marketing Co. SpA Consulting Office (EMC)
(a partner of Dremco Telecom Sarl, Italy)
Piazzale A. Ricco 4, I-41100 Modena
☎ (+39 59) 82 61 60 ⊖ (+39 59) 33 33 42
♧ Llt 450 000 000 ♣ Llt 7 000 000 000 (1995) 👥 30
🏠 1 000 sq m
Managing Dir: Vittorio Alfieri
Operations Dir: Raoul Nicolini
Sales & Marketing Dir: Giulio Alfieri
Consultant: Luciano Zerbini
Purchasing Mgr: Ezio Malavasi
■ Specialized in countermeasures antennas, VHF-UHF transceivers. [97]

75-07000 Elettronica Aster SpA (EA)
Via E. Longoni 102b, I-20030 Barlassina, Milan
☎ (+39 362) 56 06 36; 55 70 46 ✉ EASTER
⊠ 340531 easter i ⊖ (+39 362) 56 11 09
♣ Llt 26 098 000 000 👥 158
Engineering Mgr: Gianantonio Moretto
Product & Support Mgr: Carlo Broglia
■ Airborne radar antennas & receivers. Gyroscopes & accelerometers, microwave components, mission data loaders & recorders. Overhaul of flight & navigation instruments.

Head Office
Corso Vittorio Emanuele II n° 15, I-20122 Milano
☎ (+39 2) 76 00 31 94 ⊖ (+39 2) 76 00 07 97
♧ Llt 1 000 000 000 ♣ Llt 27 500 000 000 👥 175
🏠 6 000 m2
President: Enrico Ripamonti
Managing Director: Francesco Spina
Commercial Director: Carlo Broglia
Accounting Manager: Mrs Ornella Peggiani
Personnel Mgr: Renato Ruvolo
Engineering Manager: Giovanni Moretto [97]

75-07050 Elettronica SpA
Via Tiburtina Valeria - Km 13,700, Loc. Settecamini, I-00131 Rome
☎ (+39 6) 415 41 ⊠ 626527 elti ⊖ (+39 6) 419 28 69
♧ Lit 60,000,000,000 (1995) 👥 850 🏠 35 300
President and Managing Director: Filippo Fratalocchi
Vice President and General Manager: Enzo Benigni
Sales and Marketing Director: Giuseppe Maresca
Public Relations Manager: Gianni Carlini
■ Passive EW equipment for the detection, location, analysis & identification of electromagnetic emissions (ESM/ELINT/COMINT). Active electronic countermeasures equipment (ECM). Integrated ESM/ECM systems for naval, airborne & ground applications. Special microwave components.

Subsidiaries with defence business:
Elettronica (UK) Ltd → UK 75
Elettronica GmbH → Germany 75 [97]

75-08200 EUROCONTROL SpA
Via Varenna 52a, I-16155 Genova
☎ (+39 10) 698 98 77 [E] eurocval@tin.it
⊖ (+39 10) 699 00 01
♧ Lit 2,402,439,000 (1997)
♣ Lit 9,000,000,000 (1996) 👥 45 (1996) 🏠 4,000 m2
Chairman: Ing Sergio Bonfoco
Managing Director and General Manager: Prof Gian Paolo Sacco
Sales and Marketing Director: Ing Carlo Valdettaro

Contact Reference

Quality Manager: Ing Emanuele Franzò

■ Main product lines: power electronics (frequency converter, inverter, servo amplifier and power supply); control electronics (data processing, ballistic computer, training firing simulator, actuator fully digital control unit and multifunction remote console); optronics (electro-optical line of sight telecamera based and stabilised gunner periscope); electromechanics (AC and DC motor and generator, DC brushless motor, servomotor, rotating frequency converter, slip-ring assembly, control panel). [98]

75-08400 Eurovinil Industries SpA

(a subsidiary of Wardle Storey plc, UK)

Via Genova 3, I-58100 Grosseto
☎(+39 564) 48 73 00; 48 73 04 ⊠ 500212 evinil i
⊖(+39 564) 48 72 22
♣ Lit 20,000,000,000 (1994)
♣ Lit 20,000,000,000 (1996) 👥 100 🏭 28,000 m2
President: Ing. Alessandro Giachi ☎(+39 564) 48 72 00
Marketing Dir: Giordano Turati ☎(+39 564) 48 74 00
Sales Mgr: Gianluca Tucci
Purchasing Manager: Giuseppe Bisti

■ Liferafts, inflatables, emergency landing pads, rescue & lifting bags, pneumatic field tents, helicopter floats, lifejackets; pneumatic structures coated with thermoplastic resins & sealed by electronic welding. [97]

75-08550 Fabbrica Italiana Apparecchiature Radioelecttriche SpA (FIAR)

Via Montefeltro 8, I-20156 Milan
☎(+39 2) 357 90; 357 91 ⊠ 331140 fiarmo i
⊖(+39 2) 33 40 09 81
♦ Lit 9,516,650,000 ♣ Lit 126,000,000,000 (1994)
👥 799 🏭 48,719 m2
President: Giorgio Oldoini
Managing Dir: Ing. Raffaele Esposito
Public Relations Mgr: Ing. Giuseppe Russo

■ Design, development, manufacture & installation of high technology systems & apparatus for defence & space. Logistic support.

Rome Office
Sicilia 43, I-00185 Rome
☎(+39 6) 482 64 53 ⊠ 623674 ⊖(+39 6) 487 47 12
Director: Alessandro Torrisi

Defence Division
Via G.B. Grassi 93, I-20157 Milan
☎(+39 2) 35 79 05 37 ⊠ 331265 fiargr i
⊖(+39 2) 355 73 25
Director: O. Bosco
International Marketing Dir: A. Padoa

■ Fire control & multimode, weather & precision approach radars; ground controlled approach systems; mine detecting & classifying sonars; dipping anti-submarine sonar air defence systems; daylight & low light TV systems for ground & naval applications; laser rangefinders & TV trackers; thermal imagers; infra-red searching & tracking systems (IRST/FLIR); fire control scout systems for helicopters; fire training simulators.

Logistic Support Division
Via Milano 255, I-20021 Baranzate, Milan
☎(+39 2) 35 79 01 ⊠ 335674
Director: D. Bollini

■ Customer after-sales service. Installation & repair service; warranty management; renewal parts & replacements; field services; field report analysis; improvement of apparatus & systems. Design & development of test equipment; logistic support. → 80

Space Division
Via Montefeltro 8, I-20156 Milan
☎(+39 2) 35 79 01 ⊠ 331140
Director: L Giardino

■ Space systems & equipment: power systems, high voltage converters (EPC), TWTA frequency synthesizers, millimeter wave apparatus, solid state amplifiers, Telecommunication transponders. telecommunication stations, mobile terminals, ground support equipment. [97]

75-08630 Farid Industrie SA

(† Farid SpA)

Head Office & Main Plant
Corso Savona 39bis, I-10024 Moncalieri (TO)
☎(+39 11) 64 54 54 ⊖(+39 11) 64 19 34
♦ Lit 2,000,000,000 ♣ Lit 29,619,000,000 (1993)
👥 91 🏭 8,000 m2
Managing and Marketing Director: Dr Paolo Martinelli
Purchasing Manager: Pier Giovanni Cerrina

■ Road recovery cranes, special cranes. Recovery wreckers. [98]

75-09110 Fiat Avio SpA Head Office

(a subsidiary of Fiat pA)

C so Ferrucci 112, I-10138 Turin
☎(+39 11) 685 81 11 ⊠ FIATAVIO Torino
⊖(+39 11) 685 84 86
♦ Lit 285,000,000,000 (1996)
♣ Lit 1,648,000,000 (1996) 👥 5,141 (1996)
🏭 13,800,000 m2
President: Ing Paolo Torricelli
Chief Executive Officer: Ing Pier Giorgio Rohiti
Vice President Engineering: Ing Leonards Caroni
Product Quality Director: Ing Amato Fadini
Aviation Business Unit and Industrial Director: Ing Saverio Strati
Marketing and Sales Director: Ing Luciano Maccaferri
Vice President Purchasing: Dr Umberto Catani
Vice President Manufacturing: Franco Davico
Vice President Product Support: Ing Franco Cannici
Vice President Personnel and Organisation: Dr Roberto Ciervo
Vice President Engineering and Power Plants Manufacturing: Ing Giovanni Alloni
Public Relations Manager: Dr Roberto Bonetti
Space Business Unit: Ing Davide Maccagnani
Energy Business Unit: Ing Franco Rodi
Marketing: Ing Sergio Crosta

■ Design, develop and produce engine mecahnical components for modern military helicopters and aircraft, design and produce helicopter transmissions and marine propulsion turbines. Produce auxiliary power units, turbofans. Also involved in space propulsion, seperation motors, orbital propulsion systems, light launchers missile and rocket engines. Production of gas turbines in the 20 MW to 240 MW power range for the production of electric power.

Gas Turbine Div.
Corso Ferruce 112, I-10138 Turin
☎(+39 11) 685 81 11 ⊖(+39 11) 685 84 86
Dir: Ing. Franco Rodi

Brindisi Plant
Zona Punto Franco, I-72100 Brindisi
☎(+39 831) 57 91 ⊠ 860048 fiabri
⊖(+39 831) 57 92 62
■ Aerogine plant.

Gas Turbine Plant
Corso Romania 661, I-10152 Turin
☎(+39 11) 685 81 11 ⊠ 221320 fiatav i
⊖(+39 11) 68 59 77
■ Energy plant.

Test Cells
Strada del Drosso 145, I-10100 Turin
☎(+39 11) 684 61 11 ⊖(+39 11) 347 08 00
■ Energy plant.

Torino
Via Nizza, I-312010127 Torino
☎(+39 11) 685 81 11 ⊠ 221320 ⊖(+39 11) 685 84 86
■ Aeroengine plant.

Pomigliano-Napoli (Alfa Romeo Avio)
Viale Impero, I-80038 Napoli
Pomigliano d'Arco - Napoli
☎(+39 81) 316 11 11 ⊖(+39 81) 884 65 22
■ Aeroengine plant.

Acerra - Napoli (Alfa Romeo Avio)
Localita Calabricito -, I-80011 Acerra - Napoli
Localita Calabricito - Zona Industriale
☎(+39 81) 316 11 11 ⊖(+39 81) 316 40 94
■ Aeroengine plant.

Saluggia - Vercelli
Strada per Crescentino, I-13040 Sluggia -Vercelli
☎(+39 161) 48 04 15 ⊖(+39 161) 48 33 49
■ Energy Plant.

Colleferro - Roma
Corso Garibaldi, I-22-00034 Colleferro - Roma
☎(+39 6) 972 91 ⊖(+39 6) 97 29 22 01
■ Space Plant [98]

75-09112 FiatAvio SpA Business Unit Space (BDP)

(† BPD Difesa e Spazio SpA)

Corso Garibaldi 22, I-00034 Colleferro, Rome

■ Space Products: solid propellant apogee motors. Ariane 4 and 5 boosters and separation motors. Liquid propellant and cold gas propulsion systems. Magneto plasma dynamic electric propulsion. Small launcher (VEGA).

Defence Products:
powders and propellants for missile motors. [98]

75-09250 Fincantieri

Cantieri Navali Italiani SpA

Registered Office
Via Genova 1, I-34100 Trieste
☎(+39 40) 319 31 11 ⊖(+39 40) 319 23 05
♦ Lit 10,155,000,000 ♣ Lit 3,478,500,000,000
👥 10,155
Chairman and Chief Executive Officer: Dr Corrado Antonini
Managing Director: Ing Saverio Di Macco
General Manager: Dr Bernardo Carratu
Vice General Manager: Dr Alessandro Franconi
Joint General Manager: Ing Giuseppe Beruardi

■ Fincantieri is the leading IRI-Group company for the production of naval vessels, merchant ships and offshore units of any type and size for ship conversions.

Rome Office
Via Sardegna 40, I-00189 Rome
☎(+39 6) 482 57 41 ⊠ 610180 fincrm i
⊖(+39 6) 675 12 37
Director: Dr Nazareno Silvestrini

Naval Shipbuilding Division
Via Cipro 11, I-16129 Genoa
☎(+39 10) 599 51 ⊠ 270168 fincge i
⊖(+39 10) 599 53 79; 599 52 72
Division General Director: Ing Mario De Negri
■ Construction of naval vessels of any type, submarines, and fast ferries.

Muggiano
Viale S Bartolomeo 440, I-19024 La Spezia
☎(+39 187) 54 31 11 ⊠ 270397 fincmu i
⊖(+39 187) 54 35 50
■ Building facilities: Slips, 1; maximum length 190 m.

Riva Trigoso
Via Erasmo Piaggio, I-16037 Riva Trigoso (GE)
☎(+39 185) 48 31 ⊠ 270638 fincrt i
⊖(+39 185) 48 25 11
■ Building facilities: Slips, 2; maximum length 174 m.

Merchant Shipbuilding Div. Offices
Passeggio Saint Andrea 6, I-34132 Trieste
☎(+39 40) 319 31 11 ⊠ 460041 fincme i
⊖(+39 40) 37 69 69
Division General Manager: Ing Giaucarlo Testo
■ Construction of merchant ships of any type up to 300,000 tdw and offshore units.

Monfalcone
Piazzale A Cosulich 1, I-34074 Monfalcone (GO)
☎(+39 481) 49 21 ⊠ 460108 fincmo i
⊖(+39 481) 71 14 49
■ Building facilities: Drydock, 1; maximum length 350 m; breadth 56 m. Largest ship: 300,000 tdw.

Venezia-Marghera
Via delle Industrie 18, I-30175 Venezia-Marghera
☎(+39 41) 66 61 11 ⊠ 410106 fincma i
⊖(+39 41) 66 64 15
■ Building facilities: Drydock, 1; maximum length 334 m. Largest ship: 250,000 tdw.

Ancona
Lungomare Vanvitelli 80, I-60100 Ancona
☎(+39 71) 59 51 ⊠ 560003 fincan i
⊖(+39 71) 59 52 00
■ Building facilities: Drydock, 1; maximum length 240 m. Largest ship: 100,000 tdw.

Castellammare di Stabia
Piazza Amendola 5, I-80053 Castellammare di Stabia
☎(+39 81) 872 05 22 ⊠ 710074 fincst i
⊖(+39 81) 871 44 03
■ Building facilities: Slips, 2; max. length 239 m. Largest ship: 80,000 tdw.

Palermo
Via dei Cantieri 75, I-90142 Palermo
☎(+39 91) 620 61 11 ⊠ 911073 fincpa i
⊖(+39 91) 54 72 28
■ Building facilities: Slip 1; max. length 310 m.
Repair facilities: Drydocks, 2; max. length 370 m; breadth 68 m; draft 10.45 m. Largest ship: 400,000 tdw. [98]

75-09300 Foerster Italia Sarl

Via dell'Artigianato 55, I-20083 Vigano di Gaggiano
☎(+39 2) 90 84 16 00 ⊖(+39 2) 908 21 97
Managing Director: Roberto Longo [98]

75-09530 Garbarino Pompe SpA

Via S Lazzaro 3, PO Box 108, I-15011 Acqui Terme
☎(+39 144) 586 71 Ⓔ boy108@acqui.alpicom.it
garbarino@acqui.alpicom.it ⊠ 212158 garpum i
⊖(+39 144) 552 60
♦ Lit 2,000,000,000 (1997)
♣ Lit 16,000,000,000 (1996) 👥 65 (1997)
🏭 10,000 m2

President: Franco Mario Garbarino
Managing Director: Carlo Garbarino
Operations/Technical Director: Franco Parodi
Sales and Marketing Director: Alberto Dallari
Purchasing Manager: Lorenzo Rapetti
■ Manufacture of centrifugal pumps. 98

75-09550 Garofoli Costruzioni Meccaniche SpA

Voc Pantano - Z I Sabbione, I-05100 Terni
☎ (+39 744) 80 00 60 ⊜ (+39 744) 81 18 08
♧ Lit 2,000,000,000 (1996)
♣ Lit 15,000,000,000 (1996) 👥 130 🏭 15,000 sq m
Managing Director: A Garofoli
Sales and Marketing Director: F Garofoli
Operations Director: R Marcelloni
Purchasing Manager: P Garofoli
■ Machining and mechanical construction. Shelters, antenna systems, towers for antennas, mechanical and electromechanical components for aeronautical and missile systems. 98

75-10100 GEM Elettronica Srl

Via Vespucci 9, PO Box 280,
I-63039 San Benedetto del Tronto AP
☎ (+39 735) 55 90 51 ✆ (+39 735) 59 05 26
E gem@jth.it W http://www.jth.if/gem
⊜ (+39 735) 59 05 40
♧ US$ 1,000,000 (1997) ♣ US$ 15,000,000 👥 70
🏭 5,000 sq m
Managing Director, Administration and Marketing Manager: Giuseppe Merlini
Operations Director: Tiziana Ferrante
Marketing Director: Giovanni De Renzis
Production Manager: Piergiorgio Difilippo
Business Development Manager: Robert Erington
Purchasing Manager: Antonio Merlettini
■ Radar, navigation systems. Design and manufacture navigation radars, fire control radars, electronic charting systems for vessel traffic management and coastal surveillance, GPS receivers and automated tracking systems. 98

75-10300 Ghisalba SpA

Via Tevere 15, I-10090 Cascine Vica Rivoli, TO
☎ (+39 11) 959 31 31 ⊜ (+39 11) 959 70 64
♧ Lit 3,000,000,000 (1996)
♣ Lit 23,280,000,000(1996) 👥 137 (1996)
🏭 6,756 sq m
President: Dr Ing Gian Luigi Quario
Operations and Purchasing Director: F Magliana
Sales and Marketing Director: Ing E Falcone
■ Manufacture of electrical control equipment. 98

75-10510 Gilardoni Mandello Lario Plant

Via E Fermi 2, I-22054 Mandello Lario
☎ (+39 341) 73 15 83; 70 51 11 E gx@gilardoni.inet.it
W http://www.smartcom.gilardoni.it ⊠ 380496 raygil i
⊜ (+39 341) 73 50 46
President: Mrs Cristina Gilardoni
Managing Director: Richard Bisson
Operations Director: Andrew Orsini
Sales and Marketing Director: Marco Gilardoni
Purchasing Manager: Ivano Bruseghini
■ X-ray tridimensional micrrodose equipment for hand baggage and air cargo/baggage security inspection. X-ray and ultrasonic equipment for NDT (quality tests and maintenance service).

Rome Office
Via dei Foscari 7, I-00162 Rome
☎ (+39 6) 44 29 12 38; 44 29 07 17
⊜ (+39 6) 44 29 12 94

Motta S Anastasia Plant
Strada Provinciale 13, I-95040 Motta S Anastasia
☎ (+39 95) 713 02 26 ⊜ (+39 95) 713 02 45 97

75-10730 Gruppo Nuova MA.NA.RO.

Zona Industriale Via Fonte Damo, PO Box 36,
I-60035 Jesi (AN)
☎ (+39 731) 619 61 ✉ VEDERE FORGLIO ALLEGATO
E mailbox@nuovamanaro.it
W http://www.nuovamanaro.it ⊠ 561127
⊜ (+39 731) 60 51 67 (engineering);
(+39 731) 60 51 86 (commercial)
♧ Lit 12,000,000,000 ♣ Lit 25,000,000,000 👥 120
🏭 15,000 sq m
Managing Director: Marco Marconi
Technical Director: Igino Pelonzi
Sales and Marketing Director: Claudio Vignoli
☎ (+39 731) 60 51 65 E claudiovignoli@nuovamanro.it
Purchasing Manager: Domenico Zamponi
Advertising Manager: Ms Roberta Marconi
■ Aircraft refueller vehicles and equipment. Storage tanks, microfiltering and pumping units. Firefighting vehicles and equipment. Road tankers, trailers and semi-trailers for liquid goods, transport and distribution. Anti-riot vehicles. 98

75-10900 Hexcel Structures

(† Salver srl Ciba Composities)
Via D Macaluso 2/4, PO Box 232, I-72100 Brindisi
☎ (+39 831) 57 31 45 E salver@weil.future.it
⊜ (+39 831) 57 38 78
♣ US$ 8,000,000 (1997) 👥 165 (1997)
🏭 10,000 (1997)
Managing Director: R Di Francesco
Technical and Marketing Manager: F Innocente
■ Composite fabrications. 98

75-11000 Honeywell SpA

Via Vittor Pisani 13, I-20124 Milan
☎ (+39 2) 677 31 ✉ HONEYWELL milano ⊠ 311092
⊜ (+39 2) 677 32 08
♧ Lit 1,500,000,000
Marketing Director: Ing. G Righi 98

75-11220 IFI Srl

Impresa Forniture Industriali
(a partner of Industria Lubrificanti Speciali SRL)
Via Tiburtina 443, I-00159 Rome
☎ (+39 6) 435 60 21 ⊠ 622415 fornif i
⊜ (+39 6) 439 11 38
♧ Lit 90,000,000 (1996) ♣ Lit 10,000,000,000 (1996)
👥 30 (1996) 🏭 6,000 ,2
President: Renato Mantovani
Managing Director: Pio Mantovani
Marketing Manager: Francesco Mantovani
Technical Department: Sergio Durante
Foreign Department: Ms Maria Mantovani
■ Lubricants, oils, synthetic oil, sealant, electronic SPS wrench bolt tightening equipment for defence and all strategic goods for Air Force, Navy and Army. → 90
® France: Nyco S.A., Nyco SA, Weber Rawflex
UK: Brent International plc, Century Oils Group plc
USA: Chemviron Carbon, Citicorp, EM Corporation, Hodges International Inc., John G. Shelley, Lockhart Chemical, Messner, SPS Technologies, Scott Manufacturing, Technos Int'l Intel. Eqpt
● Via G Asproni 26/28, I-09100 Cagliari
☎ (+39 70) 66 47 26 ⊠ 791051 fornif i
⊜ (+39 70) 66 26 43
Managing Director: Bruno Aresu

Plant
Zona Industriale Elmas SS 554, I-09034 Cagliari
☎ (+39 70) 24 02 43
Managing Director: Bruno Aresu 98

75-11600 Industria Condensatori Applicazioni Elettroelettroniche SpA (ICAR)

Via Isonzo 10, I-20052 Monza, Milan
☎ (+39 39) 83 92 51 ⊠ 333339 icar i
⊜ (+39 39) 83 32 27
♧ Lit 4,000,000,000 (1996)
♣ Lit 62,473,000,000 (1996) 👥 200 🏭 7,770 m3
President: Ms Luisa Portaluppi
Vice President: Ing. Marco Bognetti
Managing Director: Ing. Francesco Folli
Operations Director: G Picci
Secretary: R Bellabio
Contact: R Ballabio
■ Complete range of capacitors. HV generators (DC), delay and pulse networks. Radio interference supression filters. Automatically switched banks with harmonic filtering systems. 98

75-11900 Intermarine SpA

(a subsidiary of Montedison SpA, Italy)
Localita Cà del Sale, I-19038 Sarzana, La Spezia
☎ (+39 187) 61 72 93 E marketing@intermarine.it
⊠ 271062 ⊜ (+39 187) 67 42 49
♧ Lit 39,000,000,000 (1996)
♣ Lit 118,000,000,000 (1996) 👥 400 🏭 70,000 sq m
Managing Director: Eng Giovanni Facchinetti
Commercial Director: Capt Giovanni Carlo Bonetti
Contact: Mara Bortolussi
■ Research, design and production of high performance and specialised military craft (MCMVs, FPBs and OPVs) ecological and hydrographic vessels, yachts, lifeboats and tenders using advanced GRP technology.

Plant
Via Alta, PO Box 91, I-19038 Sarzana, La Spezia
☎ (+39 187) 67 18 00; 61 71 11 ⊠ 271062
⊜ (+39 187) 67 42 49
Technical Director: Eng Giuseppe Marchetti 98

75-12200 Irvin Aerospace SpA

(† Irvin Manufatture Industriall SpA)
(Hunting Engineering Group)
Via delle Valli, I-04011 Aprilia LT
☎ (+39 6) 928 28 46 ✉ IRVIN aprilia ⊠ 680019 irvman i
⊜ (+39 6) 92 72 71 65
♧ Lit 2,800,000,000 ♣ Lit 35,000,000000 (1997)
👥 190 (1997) 🏭 35,000 sq ft (including Sehkur SpA)
President: Roger J Whysall
Managing Director: F Massimo Bombacci
General Manager: Attilio Ciuffi
Research and Development Director: Mario Laurenti
Marketing Manager: Ing Mauro de Leonardis
■ Parachutes for personnel, cargo, weapon and aircraft retardation. Recovery systems for missiles/space vehicles and cargo delivery systems. Safety harnesses and air drop sequence control devices. Inflatable survival products and flexible fuel tanks. Nuclear, biological and chemical equipment.
Camouflage, lifevests, liferafts, aerial delivery platforms and equipment field hospital (air transportable). Aircraft components, platforms, pallets and containers. Aeronautical carpentry (galleys and trolleys).
® USA: Irvin Industries 98

75-12250 Isoli SpA

Head Office & Plant
Via Boschi 44, I-35014 Fontaniva (PD)
☎ (+39 49) 595 76 00 ⊠ 430496 isoli i
⊜ (+39 49) 595 86 45
♧ Lit 1,600,000,000 ♣ LIt 32 000 000 000 (1990)
👥 150 🏭 70,000 m2
Chief Executive Officer: Dott D Nigretti
Sales Director: M Mazzolla
Operations Director: G M Isoli
Purchasing Manager: A Isoli
■ Military and industrial cranes (capacity from 6 to 28 tons), elevating working platforms (working height 12.7m to 44.5m), light and heavy rescue equipment, multipurpose equipment. All to be mounted on commercial chassis. 98

75-13205 Italtel Spa Headquarters

Società Italiana Telecomunicazioni SpA
Via A di Tocqueville 13, I-20154 Milan
☎ (+39 2) 438 81; 43 88 51 13 ✉ ITALTEL
⊠ 314840 sitele i ⊜ (+39 2) 43 88 52 20
♧ Lit 396,000,000,000 ♣ US$ 3,900,000 👥 16,000
President and Chief Executive Officer: Salvatore Randi
Honorary Chairman: Francesco Carassa
Executive Vice President: Giovanni Barberi
Executive Vice President: Fausto Plebani
Director: Aldo de Sario
Director: Alessandro Ovi
Director: Antonio Gardone
Director: Peter Pribilla
Director: Giorgio Scanavaaca
Director: Jost Hammerschmidt
Marketing Communications: Piera Maria Galbiati
Public Relations: Ms Maria Patrizia Grieco
■ Design, manufacture, marketing and installation of equipment for public and private telecommunications systems on a turnkey basis.

L'Aquila
Località Boschetto, I-67100 L'Aquila
☎ (+39 862) 33 51 ⊠ 600032 ⊜ (+39 862) 33 52 53
■ Design, manufacture and marketing of defence telecommunications systems and equipment, including IFF identification systems, tactical radios, avionic equipment, cryptographic equipment, MIDS (Multifuntion Information Distribution System), NIS (NATO Identification system). 98

75-13230 IVECO Fiat SpA

Defence Vehicles Division
Via Volta 6, I-39100 Bolzano (BZ)
☎ (+39 471) 90 51 11 ⊠ 400541; 401504 ivedvd i
⊜ (+39 471) 20 08 18; 90 54 44
👥 38,200

Contact Reference

Marketing Manager: Paolo Sinha
☎(+039 471) 90 54 44 (fax)
Commercial Operations, Export: Olga Ceruetto
☎(+39 471) 90 52 28; 90 54 44 (fax)
■ Military trucks with payloads ranging from 1 to 20 tons. Wheeled armoured vehicles, 4x4, 6x6, 8x8. Special vehicles, e.g. amphibious trucks for transport and rescue operations and armoured buses for police operations. Components for military tracked vehicles: engines, gearboxes, powerpacks, tracks. Kits for refitting and repowering.
Subsidiary company:
IVECO-Magirus AG → Germany 75 [98]

75-13500 LARIMART SpA
Via di Torrevecchia 8/12, I-00168 Rome
☎(+39 6) 30 34 31 [E] larimart@mclink.it
℡ 621298 lrtrom i ⊜ (+39 6) 30 34 33 86
♣ US$ 13,100,000 (1996)
Managing Director: Vittorio Biscarini
Commercial Director: Gianmario Fracassa
General Manager: Nicola Mazzotta
■ VHF/FM military hand-held radios; radio intercom harness systems and related ancillaries for vehicular ground and naval applications. Communications and input-output terminals for military specifcations. Antenna masts for vehicles and ground applications. Tactical and mobile reprogrammable batteries chargers. Management systems. Relevant complete technical support is also provided. [98]

75-13700 Litton Italia SpA (LITAL)
(a division of Litton Industries Inc, USA)
PO Box 10064 Eur, I-00100 Rome
Via Pontina 800 - km 27, I-00040 Pomezia, Rome
☎(+39 6) 91 19 21 ℡ 610391; 616435 lital i
⊜ (+39 6) 912 25 17
♧ Lit 3,000,000,000 (1996)
♣ Lit 50,000,000,000 (1996) ⁂ 270 (1996)
⌂ 16,000 m2
Managing Director and General Manager: Dr Piero Biagetti
Vice President, Business Development: Dr Gilberto Pastori
Vice President, International Sales: Ing Alan Edwin Kaile
Vice President, Technical: Ing Italo Calvelli
Vice President Finance: Franco Collepardo
Marketing Director: Dr Franco Rotanti
Public Relations Director: Adm Enrico Rossi
Marketing Support Manager: Federico di Marzo
Purchasing Manager: Ms Rosina Marani
Quality: Gualtiero Bernardini
■ Inertial navigation systems, attitude and heading reference systems for airborne, shipborne and ground applications using both gimbal and strapdown technology, utilising DTG, laser gyroscopes and fibre optic gyros. Research and development of inertial systems. Tactical command and control, intelligent handheld terminals, artillery automation. Inertial guidance and control systems, integrated inertial/GPS systems, C31 systems, artillery fire control systems, automated meteorological radiosounding systems, automated topographic data processing systems, intelligent portable terminals.
® USA: Litton Industries [98]

75-13800 Logic SpA
(a member of the Aeronautica Macchi Group)
Via Brescia 29, I-20063 Cernusco S/N, Milan
☎(+39 2) 92 10 25 51 ✉ LOGIC Milano ℡ 332846 logic i
⊜ (+39 2) 92 10 25 28
♧ Lit 6,325,000,000 ♣ Lit 17,900,000,000 (1995)
⁂ 98 ⌂ 2,750 m2
President: Bruno Cussigh
Managing Director: Ing. V Cattani
Operations Director: F Vecchio
Sales and Marketing Director: G F Campana
Purchasing Manager: P Pagano
■ Deliverable loads management, stores management systems, cockpit control units. Alarm systems, caution, warning and advisory panels, bus-controlled remote display units, aural warning units, synthetic voice alerting units. Closed loop servo-controls, turbine control systems, steering control systems. Ice protection, anti-icing controllers. Cockpit instrumentation, integrated engine instruments panels, fuel quantity indicators. Data acquisition systems, engine monitoring and recording systems, fatigue data acquisition systems. Fuel gauging, fuel quantity gauging systems, level indicators. Ground support equipment, automatic test equipment, special-to-type test equipment. Anti-collision light systems, anti-collision beacons, power supply units. Miscellanea, generation protection units, cockpit control panels, fire extinguisher panels, light modules, electro-mechanical actuators. Centralised monitoring and measurement systems. Military standard 1553 bus terminals. Central warning systems, fuel measurement, rotor and turbine RPM warning systems for helicopters. Closed loop remote controls. Store management systems. LCD engine instrument systems. [97]

75-13980 Magnaghi Milano SpA Head Office and Works
(† Magnaghi Oleodinamica SpA)
Via Stamira D'Ancona 27, I-20127 Milan
☎(+39 2) 28 45 91 ✉ OLEOMAGNAGHI Milano
℡ 320648 magnmi i ⊜ (+39 2) 282 88 19
♧ Lit 13,000,000,000 ♣ Lit 33,000,000,000 (1996)
⁂ 295 (1996) ⌂ 36,000 m2
Chairman: Rag Aldo Vajani Magnaghi
Vice Chairman: Dr Ing Carmelo Consentino
Managing Director: Dr Ing Giorgio Fassina
Member of the Board: Dr Ing Franco Catella
Member of the Board: Dr Ing Francesco Sirgiovanni
Member of the Board: Dr Arduino Barbacci
Financial Director: Rag Luigi Botta
Commercial Director: Ing Guido Cassani
Production Director: Ing Francesco Rinetti
Technical Director: Dr Ing Francesco Violante De Dionigi
Technical Director: Ing Francesco Violante
Quality Assurance Director: Ing Mauro Benetti
■ Complete hydraulic systems and flight servocontrols for aircraft, helicopters and hydrofoils. Wheels and brakes for aircraft and helicopters. Hydraulic test stands. Anti-ice electronic box.
Landing gears, wheels, brakes, shock absorbers, flight power controls, servo-valves, hydraulic packages, nose wheel steerings, main rotor blade dampers, ground test benches.
Co-operative partnerships with: Hydraulic Research Textron Inc, USA; Lockheed Martin, USA; Bendix Electrodynamics, USA; Dunlop Aviation Division, USA; Dowty Aerospace, UK; Fairey Hydraulics, UK; Messier-Bugatti, France; Liebherr Aerospace-Lindenberg GmbH, Germany.
Brugherio Works
Via S Maurizio al Lambro 20, I-20047 Brugherio MI
☎(+39 39) 287 17 01 ⊜ (+39 39) 88 40 10
Works Manager: F Rinetti [98]

75-14000 Magnaghi Napoli SpA
Via Galileo Ferraris 76, I-80142 Naples
☎(+39 81) 597 71 11 ✉ OLEOMAGNAGHI Napoli
℡ 710403 magnna i ⊜ (+39 81) 734 06 48
♧ Lit 10,000,000,000 (1997)
♣ Lit 31,000,000,000 (1996) ⁂ 232 (1997)
⌂ 28,000 m2
Chairman: Rag Aldo Magnaghi
Vice Chairman: Ing M Pellerei
Managing Director: Dr Ing Franco Catella
Finance Director: Rag Luigi Botta
Production Manager: Ms Carmine Savino
Marketing Director and Commercial Manager: Ing Luigi Marinucci
Quality Assurance Manager: Ing Adamo Esposito
Technical Manager: Ing Bruno Montersarchio
Administration: Rag Pasquale Esposito
■ Manufacture of mechanical and hydraulic equipment for aircraft, landing gears, shock absorbers and hydraulic jacks, steering, servocontrols, oil and fuel tanks, jet tail pipes, blankets, special aeronautical containers.
Associated companies:
Magnaghi Milano [98]

75-14290 Manifattura A. Testori di G. SpA
Via Dante 7, I-20026 Novate Milanese
☎(+39 2) 35 47 21 ℡ 331889 ⊜ (+39 2) 39 10 05 10
♧ Lit 31,000,000,000 ♣ Lit 25,000,000,000 ⁂ 130
⌂ 20,000 sq m
Managing Director: Paolo Testori
Operations Director: Marco Astrua
Sales and Marketing Director: Fausto Balzarini
Purchasing Manager: Mrs Luisa Basilico
■ Fabrics and carpets (flameproofed) for aircraft interiors & components. [98]

75-14400 Marchetti Autogru SpA
Via Caorsana 49, I-29100 Piacenza
☎(+39 523) 59 31 50 ⊜ (+39 523) 59 32 53
♧ LIt 6 500 000 000 ♣ LIt 20 000 000 000 (1995) ⁂ 75
⌂ 16 000 m2
President: Luigi Marchetti
Managing Dir: Roberto Girasoli
Commercial & Sales Mgr: Massimo Marchetti
Export Mgr: Maurizio Minerva
■ Hydraulic telescopic truck-mounted cranes multi-road type, all terrain cranes, rough terrain cranes with lifting capacities ranging from 10 to 200 metric tonnes. Special vehicles. Supplier to the Italian Army. [97]

75-14510 Marconi Industrial Services SpA
Via Leopoldo Pilla, 8, I-46010 Curtatone (MN)
☎(+39 376) 38 14 36 [E] marconi@keplero.mynet.it
℡ 300518 marconi i ⊜ (+39 376) 38 05 62
♧ Lit 2,270,000,000 (1997)
♣ Lit 5,300,000,000 (1996) ⁂ 82 (1997)
⌂ 93,000 sq m (1997)
Managing Director: Stefano Viviani [98]

75-14550 Marconi SpA DTM
(a subsidiary of GEC-Marconi, UK)
Via A. Negrone 1a, I-16153 Genova Cornigliano
☎(+39 10) 600 21 ⊜ (+39 10) 650 48 60
♧ Lit 68,200,000,000 (1995)
♣ Lit 610,000,000,000 (1994) ⁂ 2,800 ⌂ 56,000 sq m
Managing Dir: Ing. Sandro Gualano
Technical Dir: Ing. Carlo Bisio
Sales & Marketing Dir: Ing. Luciano Maciotta
Export Sales Mgr: Poggi Silvano
Purchasing Mgr: Ing. Enrico Castanini
■ Manufacture of multichannel telecommunication equipment for civil & military applications, realisation of turnkey military integrated communication systems, avionic equipment & weapon systems & realisation of turnkey projects. Manufacture of SDH network applications.
Milan Office
Via Palmanova 185, I-20132 Milan
☎(+39 2) 256 31 41/3 ℡ 311467 ⊜ (+39 2) 256 77 45
Cisterna di Latina Plant
Borgo Bainsizza, Km 62/SS, 148 Pontina,
I-04010 Cisterna di Latina
☎(+39 6) 968 52 51/2/3 ℡ 621648 marconi
⊜ (+39 6) 968 55 86
Rome Office
Via Campo Nell'Elba 3, I-00199 Rome
☎(+39 6) 88 69 31 ✉ EXPANSE ℡ 610272
⊜ (+39 6) 886 39 68 [97]

75-14600 Marelli Avio
(a division of Simmel Difesa SpA)
Via Verdi 33, San Giorgio Su Legnano, I-20010 Milan
☎(+39 331) 52 84 46; 52 84 65 ⊜ (+39 331) 40 45 50
♣ US$ 15,000,000 (1997) ⁂ 100 (1997) ⌂ 5,000 m2
Managing Director: G Gignolo (Simmel Difesa SpA)
Operations Director: Paolo Fantini ☎(+39 331) 40 22 16
Purchasing Manager: A Roneo
■ Air and oil cooled generators, starters and starter generators, control units for electric power generating systems, electric motors and electric fans, HE ignition units for jet engines, rotating inverters and transformer rectifier units. [98]

75-15300 Metallurgica Bresciana SpA
Via Brognolo 3, I-25050 Passirano (BS)
☎(+39 30) 685 02 64 ℡ 300041 ghitti i
⊜ (+39 30) 65 36 67
♧ Lit 15,500,000,000 (1995)
♣ Lit 31,000,000,000 (1995) ⁂ 119 ⌂ 13,500 m2
Sales, Marketing and Sole Director: Pierjacomo Ghitti
General Director: Aldo Santus
Sales and Marketing Director: Renzo Del Nevo
Purchasing and Administrative Director: Mrs Lorenza Ghitti
■ Cold, fire and solvent resistant wires and cables for naval and aviation applications. [97]

75-15400 Meteor Costruzioni Aeronautiche e Elettroniche SpA

Head Office
Via Mario Stoppani 21, Zona Industriale di Soleschiano, I-34077 Ronchi dei Legionari
☎ (+39 481) 47 91; 47 81 11 ☒ 460288 metmon i
⊖ (+39 481) 47 83 13; 47 93 00 (sales department)
♧ Lit 19,000,000,000 ♣ Lit 35,000,000,000 ⁂ 347
⌂ 20,500 m2
President: Giuseppe Tonutti
Vice President: Giovanni Sarzotti
General Manager, Marketing and Sales Manager: Ing. Filippo Lanzarotto
Sales Manager Italy: Roberto Lestini
External Sales Manager: Cdr Alessandro di Blase
New Product Development and Marketing: Antonio Cenciarelli
■ Tactical unmanned air and naval target vehicles and drones for training, battlefield surveillance, target surveillance, target acquisition, designation and strike missions. Take-off boosters, towed target radio control systems, servo actuators. Autopilots, miss distance indicators, security self-destruction systems, transponders. Supporting ground control and test equipment. Integrated control, tracking, plotting and telemetry systems. RPV field operation support. Computer-aided training systems, flight and mission simulators, computer image generation and visual systems. Integrated logistic support for simulators and UAV's. → 80

Registered Office
Via Fabio Filzi 4, I-34132 Trieste
☎ (+39 40) 694 79
■ Aerial targets and UMA (Unmanned Aircraft) for air surveillance, targetting, decoy and strike missions. Automatic navigators and autopilots. Integrated control, tracking, plotting and telemetry systems. Artillery fire adjustment field. Operation and service of towed targets and transponders.

Co-production and Co-operation Agreements:
Aeritalia SpA
Agusta SpA
Elettronica SpA
Aerospatiale
Swedair Ltd
Pacific Aerosystem Inc
Dornier GmbH

Plant
Stazione F S Ronchi Nord, I-34074 Monfalcone
☎ (+39 481) 77 11 ☒ 460288 metmon i
⊖ (+39 481) 47 83 13
Plant Manager: Dr Carlo Spano

Flight Test Centre
SS 125 Orientale Sarda Km 78, I-09040 Villaputzu
☎ (+39 70) 999 37; 999 46 ☒ 790076 aviels i
Director: Bruno Pipan

Training Centre
I-34074 Monfalcone
☎ (+39 481) 77 74 41 ☒ 460288 [97]

75-15425 Michelin Italiana SpA Aircraft Tyre Product Line

57 Via Livorno, I-10144 Torino
☎ (+39 11) 432 67 11 ⊖ (+39 11) 432 66 98
Managing Director: Pierre Desmarets
Sales Director, Europe, Middle East and Africa: Nicolas Baviere
Marketing Manager: Marco Cervellati
■ Tyres for military and civil aircraft. [98]

75-15500 Microtecnica-Torino SpA

(a subsidiary of Hamilton Standard, USA)
Via Madama Cristina 147, PO Box 1210, I-10126 Turin
☎ (+39 11) 693 21 ☒ 221190 micro i
⊖ (+39 11) 693 22 42
♧ Lit 50,000,000,000 (1996)
♣ Lit 122,600,000,000 (1996) ⁂ 800 ⌂ 38,000 m2
President and Chief Execcutive Officer: Giorgio Lapidari
Operations Director: Giuseppe Dario
Sales and Marketing Director: Umberto Penco
Deputy General Manager, Engineering: Alberto Suppi
Purchasing Manager: Gianfranco Murador
Advertisement: Maurizio Bello
Vice President: Dr Italo Alfano
■ Design, development and servicing of navigation and guidance systems for ships, aircraft, helicopters and missiles. Primary and secondary flight controls; drive, control and actuation systems; air conditioning, pressurization and heat transfer systems. Power generation and control. On-board power units for missiles; test stands.

Experimental Centre for Aerospace Equipment
I-10062 Luserna S Giovanni, Turin
☎ (+39 121) 90 02 11; 90 02 12
✉ MICROTECNICA luserna ☒ 221190 micro i
⊖ (+39 212) 908 28 [98]

75-15900 Secondo Mona SpA

Via C del Prete 1, I-21019 Somma Lombardo VA
☎ (+39 331) 75 61 11 ✉ MONA sommalombardo
⊖ (+39 331) 25 23 34
♧ Lit 2,385,000,000 (1996)
♣ Lit 27,500,000,000 (1996) ⁂ 189 ⌂ 19,657 m2
Chairman and Managing Director: Dr Roberto Mona
Deputy Chairman and Managing Director: Dr Renato Mona
General Director: Dr Ing Lucio D'Andrea
Quality Manager: Dr Ing Carlo Stella
Marketing and Technical Product Support Manager: Sergio Bogni
Marketing and Sales Executive: Dr Ing Giuliano Modesti
Engineering Executive: Dr Ing Renato Bulgheroni
Production and Operations Executive: Francesco Casale
Deputy Sales Executive: Pierluigi Bertolotti
■ Fuel, hydraulic, lubrication and pneumatic equipment for aircraft, helicopters and engines. Fuel and lubrication equipment for armoured vehicles. General purpose and special-to-type test rigs. Equipment repair and overhaul worldwide. [98]

75-16230 Moog Italiana Srl

(a subsidiary of Moog GmbH, Germany)
Via dei Tre Corsi, Zona Industriale Sud D1, I-21046 Malnate, VA
☎ (+39 332) 42 11 11 ☒ 380582 ⊖ (+39 332) 42 92 33
♧ Lit 1,000,000,000 ♣ Lit 14,000,000,000 (1994)
⁂ 37
Chairman: K G Smith
Managing Director: Ing. Emanuele V Minneci
■ Flow servovalves, brushless servomotors and electronics. Propellant valves, redundant valves and servoactuators.
® USA: Moog Inc [97]

75-16630 Novamarine 2 SpA

Head Office
Via dei Lidi, Zona Industriale, PO Box 96, I-07026 Olbia SS
☎ (+39 789) 585 38 ⊖ (+39 789) 588 56
♧ Lit 3,130,000,000 (1996)
♣ Lit 12,500,000,000 (1996) ⁂ 80 (1997)
⌂ 11,000 sq m
Managing Director: Giuseppe Carlini
Operations Director: Sergio Gaggero
Marketing Director: Pier Paolo Pintus
Purchasing Manager: Mario Pittui
Public Relations Manager: Mrs Anna Carlini
Secretary, Export Department: Barbara Duethorn
■ Supremely strong large superboats which have been conceived, designed and built for arduous professional use. The RH 1000 is especially suitable for missions such as service tender, lifeboat, military assignments, fast communications. [98]

75-16650 Novurania SpA Head Office

Via Circonvallazione 3, I-38079 Tione di Trento
☎ (+39 465) 32 11 62 ✉ NOVURANIA tione
⊖ (+39 465) 32 15 51
♧ Lit 2,070,000,000 ♣ Lit 14,000,000,000 (1997)
⁂ 96 ⌂ 10,000 m2
President: Mrs Gabriella Laupp Pellegrini
Managing Director: Dr Mirco Pellegrini
Marketing Director: Ms Flavia Pellegrini
■ Inflatable boats, rafts, tanks, jackets, and pontoons. Rubber coated fabrics, protective garments, camouflage nets, inflatable tents.

Campo Lomaso Works
I-38070 Campo Lomaso, TN
☎ (+39 465) 70 13 86 ⊖ (+39 465) 70 13 86
General Manager: Giuseppe Stefenelli [98]

75-16700 Nuova Connavi Srl

Località Lagoscuro di Ceparana, I-19020 Vezzano Ligure (SP)
☎ (+39 187) 93 97 77 ⊖ (+39 187) 93 45 66
⁂ 11
Managing Director: Renzo Pagliari
■ Machine and equipment for ammunition production. [98]

75-16750 OCEM SpA Head Office and Plant

Via 2 Agosto 1980 n°11, I-40016 San Giorgio di Piano, BO
☎ (+39 51) 665 01 69 ⊖ (+39 51) 665 00 99
♧ Lit 4,500,000,000 ♣ Lit 20,000,000,000 (1996)
⁂ 80 ⌂ 6,000 m2
President: Mrs Lucilla Anderson
Managing Director: Ing Gian Luigi Basile
Marketing and Engineering Manager: Ing Renato Luigi Berzolla
■ Airfield lighting equipment: full range of lights, isolating transformers, constant current regulators and accessories, control and monitoring. Equipment conform to NATO standards 3316 and to ICAO Annex 14. Search lights: production for military and civil protection use. Visual range from 1.5km to 6km; manual or remote control.
Uninterruptable power supplies. Static frequency changers, computerised remote control system, flashing lights. Identification beacon driver. Runway and taxiway lighting. [98]

75-16800 Oerlikon-Contraves SpA Headquarters

(† Contraves Italiana SpA)
(a subsidiary of Oerlikon-Contraves AG, Switzerland)
Via Affile 102, I-00131 Rome
☎ (+39 6) 436 11 ☒ 610166 conita i
⊖ (+39 6) 413 08 30
♧ Lit 40,000,000,000 (1995)
♣ Lit 96,000,000,000 (1995) ⁂ 527
Managing Director: Dr Ing Leopoldo Balestiero
Operations Director: Ing Luciano Zuccarini
Finance Director: Bruno Bernasconi
Sales Director: Ing Claudio Koporossy
Development Manager: Ing Umberto Fazio
Sales and Public Relations Manager: Ing Michele Antonucci
■ Fire control systems for AA guns and missiles. Naval anti-missile systems. Search and tracking radars for ground and naval use (short range).

Lainate Plant
Viale Italia 60, I-20020 Lainate, Milan
☎ (+39 2) 93 59 71 ⊖ (+39 2) 937 43 27 [98]

75-17500 Officine Panerai SpA

Via Ponte a Iozzi 8/3, Cascine del Riccio, I-50020 Monteoriolo, Impruneta FI
☎ (+39 55) 20 93 51 Ⓔ webcom@panerai.it
⊖ (+39 55) 20 93 50
♧ Lit 1,200,000,000 (1997)
♣ Lit 5,900,000,000 (1996) ⁂ 40 ⌂ 2,000 m2
President: Dr-Ing. Dino Zei
Managing Director: Dr.Ing. Bruno Latini
Sales and Marketing Director: Cdr Franco Zavattaro
Purchasing Manager: Rudi Brusconi
■ Optical landing aids systems (on-shore, platform, ship). Meteorological data monitoring systems. Electro-acoustic equipment and systems. Shipborne alarm systems, submarine trim tank level indicator, gauges and special devices for swimmers. Firing training systems, electro-acoustic equipment and sensors. training systems, electro-acoustic equipment. [97]

75-18100 OTO Breda

(† OTO Melara)
(a subsidiary of Finmeccanica, Italy)
Via Valdilocchi 15, Casella Postalle 337, I-19136 La Spezia
☎ (+39 187) 58 11 11 ✉ OTO MELARA La Spezia
☒ 281101; 270368 oto i ⊖ (+39 187) 58 26 69
♣ Lit 500,000,000,000 (1996) ⁂ 2,000 ⌂ 187,000 m2
Operations Director: Ing Gianpietro Pieracci
Purchasing Manager: Ing Franco Antonini
Public Relations Manager: Dr Attilio di Giovanni
☎ (+39 187) 58 24 13
General Manager: Ing Rinaldo Gonnelli
☎ (+39 187) 58 22 10
Public Relations Assistant: Marina Magnani
■ Naval, air force and army guns, artillery, air-to-surface, surface-to-surface missile systems, self-propelled howitzers and tanks. Logistic support, ammunition (research and production), systems engineering. AF cannons and turrets, ammunition, bombs.

Brescia Plant
Via Lunga 2, I-25100 Brescia
☎(+39 30) 379 11 ⊜ (+39 30) 32 21 15 [98]

75-18700 Partenavia Costruzioni Aernautiche SpA
(a subsidiary of Aerocosmos)
Corso Vittorio Emmanuele 24, I-20122 Milan
☎(+39 2) 76 00 18 47 ✉PARTENAVIA naples
⊜ (+39 2) 78 31 47
70
President and Chief Executive Officer: Dr Luciano Zanotti
General Manager: Dr-Ing. Ciro d'Amato
Marketing and Communications Manager: Dr Barbara Contini
■ P68 Observer (aerial work, patrol etc). P68C/P68C-TC. AP68 TP-600 Viator (aerial work, commuter). [98]

75-18790 Rinaldo Piaggio SpA
Via Cibrario 4, I-16154 Genoa Sestri
☎(+39 10) 648 11 ☒ 270695 aerpia i
⊜ (+39 10) 652 01 60; 648 12 28; 650 33 78
Lit 200,533,520 (1996)
Lit 140,000,000,000 (1996) 1,200 110,000 m2
Government Commissioner: Mr Dott Gian Luigi Francardo
Government Commissioner: Gen Luciano Meloni
Government Commissioner: Dr Vittorio Agostino
Administration Director: Dr Stelio Ferrero
Aircraft Marketing and Commercial Director: Ing Adolfo Drago
Civil Marketing Manager: Paolo Barberis
■ Design, manufacture and overhaul of P.166-DL3 utility aircraft & P.180 Avanti business aircraft. Collaboration for Tornado MRCA and G.222 military aircraft.

Plant
Aeroporto C Columbo, I-16154 Genoa Sestri
☎(+39 10) 648 11

Plant
I-17024 Finale Ligure SV
☎(+39 19) 697 01

Branch Office
Via A Gramsci 34, I-00197 Rome
☎(+39 6) 320 08 50; 320 08 91 [98]

75-19310 Poli Officine Meccaniche di alta Precisione SpA
Head Office & Plant
Via Oberdan 5, I-13019 Varallo Sesia VC
☎(+39 163) 527 42; 527 43; 527 44 [E] poliit@tin.it
⊜ (+39 163) 538 81
Lit 1,350,000,000 (1994) Lit 21,000,000,000
134 21,000 sq m
President and Managing Director: Egidio Poli
General Manager: Rinaldo Gioria
Marketing Manager: Riccardo Poli
■ Tridimensional measuring machines, granite surface plates, universal marking off machines, slip gauges. [98]

75-19650 Radiall Elettronica Srl
(a subsidiary of Radiall, France)
Via Concordia 5, I-20090 Assago, Milan
☎(+39 2) 48 88 25 94 ✉RADIALL ☒ 43350037 eurel i
⊜ (+39 2) 48 84 30 18
Lit 495,000,000 Lit 5,500,000,000 10
Managing Director: Adriano Raso

Rome Office
Via Luigi Capuana 10, I-00197 Rome
☎(+39 6) 827 25 45 ☒ 610358 ⊜ (+39 6) 827 24 39
■ Coaxial connectors, hyper frequency materials, multiple connectors.

Representing:
Radiall, France [98]

75-20110 Regis SpA
Salita Riva 3, PO Box 244, I-13051 Biella, VC
☎(+39 15) 322 39 ☒ 214033 regeli ⊜ (+39 15) 271 46
Llt 1 200 000 000 Llt 5 500 000 000 26
2 380 m2
President: Claudio Regis
Managing Dir: Luciano Zanella
Sales & Marketing Dir: Mauro Bertoncini
Operations Dir: Alessio Cartotto
Purchasing Mgr: Ms Francesca Salza
■ High resolution TV cameras & monitors, frame stores, scan converters, thermal image cameras, solid state TV cameras, TV tracking systems & components digital HUD cameras. [97]

75-20120 Repetti & Montiglio SpA
Via A.Grandi 16, I-15033 Casale Monferrato
☎(+39 142) 45 59 00 ✆ (+39 336) 44 28 84
☒ 215024 repmon i ⊜ (+39 142) 45 44 64
Lit 20,000,000 (1996) Lit 22,000,000,000 (1996)
85 (1996) 10,000 m2
Managing Director: Mairo Terenzio
Marketing Manager: Repetti Stefano
Purchasing Manager: Francia Mauro
Solicitor: Procurtore Enrico
■ Conversion and armouring of all types of vehicles. [98]

75-20500 Riva Calzoni SpA
Via E. Ponente 72, I-40133 Bologna
☎(+39 51) 52 75 11 ☒ 510156 calzon i
⊜ (+39 51) 657 46 55
Llt 54 000 000 000 Llt 99 000 000 000 (1995)
481
General Mgr: Guido Ucelli
Aerospace Dir: Montalto Alessandro
Sales & Marketing Mgr: Claudio Podestà
Purchasing Mgr: S. Luciano
■ Main products: helicopter handling systems, naval autopilots, probes for automated booster inspection, hydraulic actuators & valves, special hydraulic motors. Submarine equipment: snorkels, non-penetrating optronic masts, antenna masts, periscope & fairing hoisting systems, gas exhaust valves, hull valves.
Surface vessel systems: missile reloading systems, torpedo handling systems, helicopter handling systems, mine-hunters auxiliary propulsion systems, ROV for mine identification & neutralisation, recovery systems for ROV's.
Aerospace systems: robotised TV sensor probe for booster field inspection, propellant mixing pots, wheels & brakes handling systems for airliners, hydraulic manifold & valves for aircraft. [97]

75-20800 Rockwell Light Vehicle Systems - Avellino Srl
(a subsidiary of Rockwell International Corp, USA)

Access Control Systems
Via Pianodardine, I-83100 Avellino
☎(+39 825) 61 53 11 ⊜ (+39 825) 61 53 29
Plant Manager: Carlo A Benaglia
[E] cabenagl@auto.rockwell.com

Door Systems
Via Statale Briantea 342, I-22032 Albese Con Cassano (Como)
☎(+39 31) 42 93 11 ☒ 380271 rogoit i
⊜ (+39 31) 42 93 33
Controller: Felice Scanziani
[E] fscanzia@auto.rockwell.com [98]

75-20820 Rockwell CVC SpA Heavy Vehicle Systems
(a subsidiary of Rockwell International Corp, USA)
Strada Provinciale, Cameri-Bellinzago km 5, I-28060 Cameri (Novara)
☎(+39 321) 42 31 ☒ 200255 rocvc i
⊜ (+39 321) 42 32 76
Engineering and Site Manager: Carlo Mondin
® USA: Rockwell International Corp.

ROR Rockwell Intalia Srl Heavy Vehicle Systems
Via Monte Fiorino 23, I-37057 San Giovanni Lupatoto Verona
☎(+39 45) 875 03 99 ⊜ (+39 45) 875 05 13
Managing Director: Tiziano Ciocchetta
® USA: Rockwell International Corp. [98]

75-21350 Saft Nife SpA
(a subsidiary of Saft SA, France)
Viale Cembrano 11, Casella Postale 1283, I-16148 Genova
☎(+39 10) 374 79 11 ☒ 270355 ⊜ (+39 10) 38 62 73
Managing Director: Luciano Trestini [98]

75-21530 SCAM SpA
Società di Condensazione ed Applicazioni Meccaniche
Via Arcivescovado 7, I-10121 Turin
☎(+39 11) 562 24 33 ☒ 221016 scam i
⊜ (+39 11) 562 28 33
Llt 4 000 000 000 (1994) Llt 8 000 000 000 50
3 500 sq m
Managing Dir: Giorgio Barolo
Sales & Marketing Dir: Maurizio Gaidano
Operations & Technical Dir: Riccardo Bogino
■ Salt water purification systems; multi-staged vacuum groups; air & water cooling systems; steam condensers & steam heating systems. Drinking water production plants for ships. [97]

75-21650 Sèkur SpA
(a subsidiary of Irvin Aerospace SpA)
Via Delle Valli, I-04011 Aprilia LT
☎(+39 6) 928 28 46 [E] md3274@mclink.it ☒ 680019
⊜ (+39 6) 92 72 71 65
Lit 300,000,000 (1997)
Lit 14,000,000,000 (1996) 80 (1997)
12,000 m2 (including Irvin)
Managing Director: Massimo Bombacci
Operations Manager: Attilio Ciuffi
Marketing and Sales Manager: Mauro De Leonardis
Research and Development Manager: Marco Adami
■ Safety equipment: gas masks, gas and particulate filters for protection against NBC agents, organic fumes, and carbon monoxide. Emergency respirators, NBC overgarments, open circuit self-contained breathing apparatus, complete filtering units for shelters, casemates and armoured armoured vehicles. Safety tanks: flexible, collapsible, self sealing, crash-resistant fuel tanks for aircraft, helicopters and armoured vehicles. [98]

75-21750 SELIN Sistemi Sarl
(a division of FIR Meva Group)
Via Chiaravagna 28, I-16153 Genoa Sestri
☎(+39 10) 651 40 93; 650 84 59 ✉SELINGE
☒ 270302 selin i ⊜ (+39 10) 650 84 06
Lit 90,000,000 (1996) Lit 8,000,000,000 (1996)
30 4,000 m2
Managing Director: Ing. Marco Marsilii
Operations Director: Ing. C Mengaroni
Technical Director: Ing. A Pallottini
■ Converters and no-break systems for safe supply of electronic microwave equipment: radars, fire control units, computers. Electric motors and generators to military specifications. Servo-systems and electronic data processors for weapon controls. Automatic test stands for aircraft electric generators (IDG). 400cps motor-generators ground power supply of military and civil aircraft equipment, servomotors for traverse/elevation of turrets, ground naval weapons systems, frequency convertors, electric motors for torpedoes, high speed generators, electronic equipment generators for military vehicles, drive systems for electrical vehicles.
® Germany: Euroatlas Lic [97]

75-21950 SIAI Marchetti Sarl
Via Indipendenza 2, I-21018 Sesto Calende, VA
☎(+39 331) 92 91 11 ☒ 331848 siaico i
⊜ (+39 331) 92 25 25
Llt 72 450 000 000 Llt 80 000 000 000 (1994)
450 64 380 sq m
Managing Dir: Bruno Cellemme
Operations Director: Ing. Franco Brembati
Sales and Marketing Director: Filippo Meani
Purchasing Mgr: Giorgio Andreis
Quality Manager: Mario Sessa
■ Co-production of aeronautical components for other manufacturers. Produces the SF260 trainer powered by a single piston engine, S211 & S211A turbofan trainers & the SF600A Canguro twin turboprop for light transport. The SF260 & SF600A aircraft can be produced in different versions for empolyment in civil aviation as well as in the military sector.
Components produced for other companies include the wing root trailing edge, flaps & wing tips for the Panavia Tornado, the tail section of the Alenia G222, flaps & central fuselage section for the Alenia F104S, as well as the Airbus A310 tail cone & inspection door. [97]

75-22020 SICAMB SpA Head Office and Plant
Società Italiana Costruzioni Aeronautiche Martin-Baker
Via Eschido 1, I-04110 Borgo Carso Latina
☎(+39 773) 45 12 62 ✉SICAERO latina
☒ 680557 sicaer i ⊜ (+39 773) 45 12 66; 65 42 00
Lit 6,000,000,000 (1996)
Lit 45,000,000,000 (1996) 343 17,000 m2
15,000 m2
President: Ing. Umberto Klinger
Director: Ing. Maurizio Cerruti
Commercial and Technical Manager: Ing. Alfredo Costa
Administration Manager: Rag. Maurizio Gherardi

■ Manufacture of aircraft structural components and sub-assemblies for civil aircraft including A310, DC-8, MD-11, ATR 42, MD-80, MD-90, DO328, Falcon 2000 and military aircraft MB339, AMX, Breguet Atlantique. Manufacture of ejection seats under licence to Martin Baker. Production of thrust reverser for Falcon 900EX and Falcon 2000. [97]

75-22150 SIFT SAS di Cutore & Pitotti
Società Impianti Filtrazione Trasferimento Fluidi
(† SIFT snc di Cutore & Co.)
PO Box 72, I-00040 Montecompatri, Rome
☎ (+39 6) 948 53 32 [E] sift@mbox.micanet.it
⊖ (+39 6) 948 72 51
♧ Lit 50,000,000 (1997)
Managing Director: Dr Ing Gaetano E Cutore
Sales and Marketing Director: Dr Ing Emanuele Cutore
Operations Director: Dr Ing Marzia Cutore
Purchasing Manager: Mrs Giuliana Pitotti
■ Ground Retractable Integrated Fuelling Unit System (GRIFUS) for military air bases and civil airports. [98]

75-22250 Silvani Antincendi SpA
(a sister company of Kidde Aerospace)
Via Redipuglia 8, I-20010 Bareggio, Milan
☎ (+39 2) 90 36 01 17 ⊠ SILVANI bareggio
⊞ 330125 silant i ⊖ (+39 2) 902 85 31; 90 36 20 32
♧ Lit 1,400,000,000 ♣ L1t 60,000,000,000 (1995)
👥 285 👥 200 🏭 35,000 m2
Managing Director: Dr Ing Vincenzo Nicolosi
Operations Director: C Claudi
Sales Director: G Zonno
Marketing Director: Ing Nico Zorzetto
☎ (+39 2) 90 36 16 63 (fax)
Purchasing Manager: Angelo Torcoli
■ Extinguishing agents for aeronautical use. Foam-producing fire fighting liquids: protein, synthetic, fluorprotein, fluorsynthetic, fluorprotein film, universal. Firefighting equipment for naval use. Modular airborne firefighting systems SAA for G222, C-130 Hercules, SAA-E CH47 Chinook.
® UK: ICI Imp Chemical Ind

Rome Office
Viale del Tintoretto 88, I-00196 Rome
☎ (+39 6) 594 07 95; 541 27 23 ⊞ 614221 silvrm i
⊖ (+39 6) 594 19 04
General Manager: F Stovali [98]

75-22580 Sirples Sarl
Via della Pace 41a, I-20098 S Guiliano Milanese
☎ (+39 2) 98 24 22 62 ⊠ SIRPLES
⊖ (+39 2) 98 24 22 72
♧ Lit 150,000,000 (1996) ♣ Lit 3,000,000,000 (1996)
👥 13 (1996) 🏭 1,000 m2
Chairman: Dr Alberto Maffei
Managing Director: Dr Ing Ivo Caporali
■ Linear and rotary actuators, servo actuators, harnesses, generators, microswitches. Electromechanical actuators, filters.
® France: Labinal
UK: Burgess Micro Switch [98]

75-22595 SITEP SpA
Società per Azioni Sistemi Elettronici Avanzati
Via Vincinella, Loc Ponzano, I-19035 S Stefano Magra SP
☎ (+39 187) 623 51 [E] sitep@mbox.vol.it
⊖ (+39 187) 63 05 03
♧ Lit 1,320,000,000 (1996)
♣ Lit 6,300,000,000 (1996) 👥 38 (1996) 🏭 3,824 m2
President: Luigi Pagni
Managing Director: Cristiana Pagni
Marketing Manager: Valdo Lombardi
Purchasing Manager: Franco Canalini
■ Manufacture of solid-state electronic devices for military use; solid-state anemometers; synchro signals amplifiers; operating data display systems; naval systems and shelterized moving systems.

Commercial Office
Via Flaminia 61, I-00196 Roma
☎ (+39 6) 361 18 21 [98]

75-23500 Società Esplosivi Industriali SpA Difesa Division (SEI)
(a partner of Groupe EPC)
Via Gavardo 3, I-25016 Ghedi (Brescia)
☎ (+39 30) 904 11 ⊠ SEI ⊞ 300489 espsei i
⊖ (+39 30) 903 14 61
♧ Lit 1,000,241,000 👥 124
Chairman: Paul Chatel de Brancion
Managing Director: Ing Marc Sassier
Marketing and Commercial Director: Dr Aurelio Maria Buonanno
Technical Sales: Ferrante Musesti
Technical Sales: Ing Fabio Sgarzi
■ Reserch and develop
GPLD aircraft bombs from 250 to 2000 lbs, sea mines, sea mines countermining charges, land mines clearence systems, explosive filling, demolition and shaped charges and accessories, demilitarisation of ammunitions, and pyrotechnics. [98]

75-23880 Società Italiana Avionica SpA (SIA)
Strada Antica di Collegno 253, I-10146 Turin
☎ (+39 11) 772 01 11 ⊞ 221499 syasys i
⊖ (+39 11) 72 56 79
♧ Lit 600,000,000 👥 120
President: Ing. Antonio Filisetti
Vice President: Ing. Umberto Penco
General Manager: Ing. Enrico Giordani
■ Research, development and specialist support for avionics and related technologies. [97]

75-24000 Faure Herman Socrate SpA
(a subsidiary of Faure Herman SA, France)
Via Allende 3-20090, Segrate (MI)
☎ (+39 2) 26 92 23 15 ⊖ (+39 2) 26 92 23 55 [98]

75-24400 Sperry Marine SpA
Piazzale Kennedy 1, I-16129 Genoa
☎ (+39 10) 56 63 51; 56 63 53 ⊞ 270521
⊖ (+39 10) 53 70 20
👥 25
Contact: Roberto Beneventi
■ Electronic and logistic equipment. [98]

75-24990 Tacchella Macchine SpA
Regione Sant'Anna, I-15016 Cassine AL
☎ (+39 144) 711 21; 711 24 ⊞ 210227 tacmac i
⊖ (+39 144) 71 43 80
♧ Lit 2,269,340,000 (1996)
♣ Lit 43,000,000,000 (1996) 👥 200 🏭 25,000 m2
President: Alberto Tacchella
Administrator: Maurizio Tacchella
■ Tool machines, grinding machines (hydraulic and hand), cutter grinding machines, CNC grinders, grinding centres and custom built machines. [97]

75-25000 Tacconi SpA
(a subsidiary of F Lli Sarchi)
Head Office and Plant
Viale Lodi 59, I-27100 Pavia
☎ (+39 382) 57 58 03 ⊞ 315288 ⊖ (+39 382) 46 03 23
♧ Lit 1,590,000,000 (1996)
♣ Lit 25,000,000,000 (1996) 👥 60
President: P Sarchi
Export Division: Ms Elisabeta Sarctu
■ Safety products. [98]

75-25020 Fabbrica d'Armi Fratelli Tanfoglio SRL
Via Valtrompia 39/41, I-25063 Gardone Val Trompia (BS)
☎ (+39 30) 891 03 61 ⊖ (+39 30) 891 01 83
♧ Lit 97,000,000 (1996) ♣ Lit 13,000,000,000 (1996)
👥 72 (1996)
President and General Manager: Bortolo Tanfoglio
Managing Director: Ms Maria Celsa Tanfoglio
Sales Manager: Massimo Tanfoglio
Sales Manager: Lino Tanfoglio
■ Automatic pistols and components in military and civilian calibres. [98]

75-26200 THORN Europhane SpA
16 Via Andrea Costa, I-40013 Castel Maggiore Bologna
☎ (+39 51) 70 00 53 ⊖ (+39 51) 70 16 52
Chief Executive: Pier Carli [98]

75-26300 Tiesse Robot SpA
(joint venture with Kawasaki Heavy Industries Ltd, Japan)
Via Isorella 24, I-25010 Visano, Brescia
☎ (+39 30) 995 86 21 ⊖ (+39 30) 995 86 77 [98]

75-26450 Trimble Navigation Italy
Via Moncone 23, I-00192 Rome
☎ (+39 6) 86 21 60 70 ⊖ (+39 6) 86 21 79 70
Sales Mgr: Enzo Lanza [97]

75-26600 Turco Italiana SpA
(an ELF/ATOCHEM company)
Via Artigianale 29, I-25010 Montirone, Brescia
☎ (+39 30) 26 74 43; 26 74 45 ⊞ 305352 turco i
⊖ (+39 30) 267 71 37
♧ Lit 1,000,000,000 (1994)
♣ Lit 7,500,000,000 (1996) 👥 15 (1996) 🏭 9,500 m2
Chairman: Dr G Comba
Managing Director: Ing Paolo Fisogni
Sales and Marketing Director: Ing C C Venturini
■ Chemical products for aircraft (military and civil) maintenance and production. Structural sealants, water-based maskant, strippers, cleaners and NDT. [98]

75-26630 Umbra Cuscinetti SpA
Via Piave 12, I-06034 Foligno PG
☎ (+39 742) 613 41 ☎ (+39 742) 34 82 70
⊠ UMBRAROLL [E] umbraera@clilne.it ⊞ 660060 umbrol i
⊖ (+39 742) 548 58 ⊖ (+39 742) 34 82 69
♧ Lit 3,000,000,000 ♣ Lit 47,000,000,000 👥 230
🏭 13,000 m2
Managing Director: Ing. Valter Baldaccini
Operations Director: Ing. Reno Ortolani
Sales and Marketing Manager: Ing. Nicola Bacelli
Purchasing Manager: Gianni Bonucci
■ Ball-screw actuators and servo-controls.
® Germany: FAG Bearings [97]

75-26800 Valsella Meccanotecnica SpA Head Office and Plant
Fascia d'Oro, I-25014 Castenedolo, BS
☎ (+39 30) 213 71 ⊞ 300495 emmeti i
⊖ (+39 30) 273 16 87
♧ Lit 2,000,000,000 ♣ Lit 18,000,000,000 👥 80
🏭 11,500 m2
Managing Director: Dr Giuseppe Pessina
☎ (+39 30) 213 72 21
Operations Manager: Giovanni Borletti
Marketing Manager: Claudio Taiola ☎ (+39 30) 213 72 52
■ Land mines, limpet mines, demolition equipment and charges. Land mine scattering systems for military vehicles and helicopters. Remotely controlled exploders. Interdiction and alarm devices. Logistics handling and transportation, plastic containers for ammunition. Plastic components for civil and military applications. [97]

75-27100 VEAM Srl
(a subsidiary of Litton Industries Inc, USA)
Via Statuto 2, I-20020 Arese, Milan
☎ (+39 2) 93 53 21 ⊞ 332502 veam i
⊖ (+39 2) 938 50 00
President: Sandro Petz
■ Multipin cylindrical connectors for military, public transport, robotics, electronics, lighting, industrial machines and nuclear applications.

Other offices and subsidiaries:
VEAM Elektro-Anschlusstechnik GmbH, Germany
VEAM, USA [98]

75-27150 Vickers Systems Trinova SpA Aerospace Marine Defence
(† Vickers Division Sperry SpA)
(a Trinova SpA company)
Via Monzese 34, I-20060 Vignate, MI
☎ (+39 2) 95 05 82 22 ⊖ (+39 2) 956 67 30
Area Manager: Dett Ing Massimo Zava
■ Hydraulic, electrohydraulic and fuel pump equipment. Pumps, motors, electric motor pumps, integrated power packages, test stands, ground support equipment; fluid and pneumatic controls and actuators. Electromechanical motor and rotary and linear actuator products, valves, AC/DC motors and generators. Fluid systems monitoring and diagnostics. Associated electronic controls. [98]

75-27400 VM MOTORI SpA
Via Ferrarese n°29, I-44042 Cento, FE
☎ (+39 51) 683 75 11 ⊠ VIERRE CENTO
⊞ 510462 viemme ⊖ (+39 51) 683 75 17
♧ Llt 40 400 000 (1995) ♣ Llt 413 111 000 (1995)
👥 932 🏭 45 000 sq m
President: R. Penske
Managing Dir: Vilmo Ferioli

Contact Reference

Commercial Dir: Mario Balbi
■ Manufacture of engines for agricultural, industrial, marine & energy applications. [97]

75-27600 Whitehead SpA
Via di Levante 48-50, I-57128 Salviano
☎ (+39 586) 84 01 11 ⊠ 500192 wmfsal i
⊖ (+39 586) 85 40 60
♣ Lit 67,200,000,000 (1993) ⌂ 136,000 m2
President: Luigi Angelo Bellusci
Managing Director: Enrico Dell'Artino
Commercial Manager: Luciano Bellero
Marketing Manager: Manlio Curia
■ Heavyweight wireguided homing torpedoes and lightweight homing torpedoes (supplied with launchers, control consoles and control networks) for surface ships, submarines, fixed and rotary wing aircraft, shore-based torpedo stations; submarine-borne systems for encapsulated missile control and launching; towed/non-towed devices for underwater countermeasures; 3D underwater tracking systems and test ranges for torpedo evaluation; self-propelled and stationary underwater targets. Naval mines, countermining charges and cutters, sonobuoys, minesweeping systems. ASW missiles, computerised systems for torpedo control and delivery and mine warfare systems. → 80 [98]

75-32000 Zincolcelere Sarl
(† Teknecomp SpA)
Via Abate Bertone 10 bis/12, I-18042 Cavaglià, Bi
☎ (+39 161) 96 91; 99 62 11 ⊠ 200039 zinco i
⊖ (+39 161) 96 60 32
♧ Lit 7 500 000 000 (1995)
♣ Lit 428,358,000,000 (1995) 👥 441 ⌂ 25,000 m2
President and Managing Director: Dr Aldino Bellazzini
General Manager: Ing. Pietro Pacetti
Sales and Marketing Manager: Dr Giancarlo Silva
■ Printed circuit boards, multi-layers and double sided PTH, rigid/flexible (UL, IPC, military, NATO, CECC, CNET and BT approved) for defence, aerospace, telecommunications and information technology. ISO 9002. [97]

80 Service Companies & Organizations

Fabbrica Italiana Apparecchiature 75-08550
LMS Italiana 90-01245
Meteor CAE SpA 75-15400
Whitehead SpA 75-27600

80-00100 Artemis International Spa
(† Lucas Management Systems Italia Srl)
Viale Monza, 259, I-20128 Milan
☎ (+39 2) 27 00 47 79 ⊖ (+39 2) 27 00 48 52
President: Carlo Boldi
■ Supplier of Artemis project management software.
® UK: Artemis Management Systems
● Via Cristoforo Colombo 440, I-00145 Rome
☎ (+39 6) 541 62 16 ⊖ (+39 6) 541 62 43
Country Manager: Luigi Capobianchi [98]

80-00430 ED.A.I. Sarl
Publishing Group
Via Guido Guinicelli 4, I-50133 Firenze
☎ (+39 55) 57 47 74; 57 33 02; 57 01 44
[W] http://www.edai.it/edai [E] edai@edai.it
⊖ (+39 55) 57 01 03; 58 09 46
Managing Director: Mrs Randa Eid
Editor-in-Chief: Paolo Gianvanni ☎ (+39 55) 57 49 20
Advertising Manager: Mrs Vania Pistollozi
☎ (+39 55) 57 46 01
Panorama Difesa
(† JP4 Mensile di Aeronautica)
Via Guido Guinicelli 4, I-50133 Florence
☎ (+39 55) 57 47 74; 57 33 02; 57 01 44
[E] edai@edai.it [W] http://www.edai.it/edai
⊖ (+39 55) 57 01 03; 58 09 46
Editor-in-Chief: Vgo Passalacque ☎ (+39 55) 57 49 20
Managing Director: Randa Eid
Editorial Director: Ruggero Stanglini ☎ (+39 55) 57 46 85
Advertising Manager: Mrs Vania Pistolozzi
☎ (+39 55) 57 46 01
■ Aeronautics, space and aviation monthly magazine (in Italian with English digest). Published by EDAI. [98]

80-00450 Ediconsult Internazionale Srl
Piazza Fontane Marose 3, I-16123 Genoa
☎ (+39 10) 58 36 84 [E] ediconsuer@iol.it
⊖ (+39 10) 56 65 78
Managing Director: Dr Vittorio Negrone
Sales Manager: Mrs Ida de Mari
■ Advertising and communication services.
® UK: Jane's Information Group [98]

80-00735 General Consulting & Engineering Srl (GCE)
Via A Ristori 5, I-00197 Rome
☎ (+39 6) 807 86 60; 807 99 09 ⊖ (+39 6) 808 39 68
♧ Lit 90,000,000 ♣ Lit 1,000,000,000 (1994) 👥 5
⌂ 200 m2
President: Claudio Stolfi
Managing Director: Giovanni Cammarano
Technical Director: Paolo Laurienzo
■ Systems research, technical evaluation, product development, consulting and engineering. Design and technological solutions and support for weapon systems, tanks, helicopters, launch, dispensing and propulsive systems, warheads and kinetic energy ammunition. Liquid and solid motors for space activities, alternative energy sources. → 90
® France: Arianespace
Israel: IAI [97]

80-00790 Impreglio SpA
(† Cogefar-Impresit SpA)
Viale Italia 1, I-20099 Sesto San Giovanni (MI)
☎ (+39 2) 24 42 21 11 [E] impregilo@interbusiness.it
⊠ 353352 iglmi i ⊖ (+39 2) 24 42 22 93
♧ Lit 593,322,252,000 ♣ Lit 2,934,000,000 (1996)
👥 31,450
Chairman: Dr Franco Carraro
Managing Director: Dr Giuseppe Gatto
General Manager: Ing Franco Vischi
General Manager: Dr Vittorio Ferrari
Marketing Director: Ing Carlo Maria Gallizzi
Purchasing Manager: Claudio Saibene
Public Realations Manager: Piero Musilli
■ Civil engineering company. [98]

80-00800 Il Diritto Aero
Via Prisciano No 8, P 4, I-00136 Rome
☎ (+39 6) 35 45 09 55 ⊖ (+39 6) 35 34 34 70
Editor: Adalberto Tempesta
■ Law magazine covering national and international air, maritime, intermodal transports and space law. [98]

80-00920 IDS Ingegneria dei Sistemi SpA
Head Office
(a subsidiary of Finsis SpA)
Via Roma 50, I-56126 Pisa
☎ (+39 50) 50 27 69 [E] idspisa.master@tin.it
⊖ (+39 50) 50 24 70
♧ Lit 1,000,000,000 (1996)
♣ Lit 7,000,000,000 (1996) 👥 50 ⌂ 750 m2
Managing Director: Ing Franco Bardelli
Operations and Technical Director: Ing Giuliano Bettini
Marketing Director: Ing Guido Giacometti
Electromagnetic Engineering Manager: Fulvio Bessi
Purchasing Manager: Ing Fabriao Bernardini
Contracts Officer: Ms Elena Frassi
■ Engineering ATC and radio navigation aids, electromagnetic compatibility, weapons systems performance evaluation, radar scattering analysis and signature management, satellite antennae LEMP and ESD, radar engineering, concept and development of complex environments. Research, development and engineering support in EMC, EMP, EMP/lightning and antenna analysis. Radar engineering and RCS analysis. Sensor and system simulation, tracking, navigation systems, design and evaluation. Single wedge tools and design framework delivery. [98]

80-00922 Infodoc Sarl
Registered Office
Via Agnini 76, I-41037 Mirandola, MO
☎ (+39 535) 261 08 [E] infodo@arcanet.it
[W] http://www.infodoc.it ⊖ (+39 535) 260 21
♧ Lit 99,000,000
General Manager: Ettore Giovanetti
Operations Director: Riccardo Giovanetti
■ Technical information services delivered on microform or electronically.
Supplier of CD-ROM databases including military documents, vendor catalogues, electronic components databases (integrated circuits, discrete semiconductors, passive devices), parts research and logistics, regulatory documents (environmental, safety and health), banking communications and logistics, customized integrated information management and retrieval systems, industry standards, military/federal specifications, parts/logistics data, aviation and aerospace regulations, directives, circulars, certificates and standardisation documents: FAA; AIA; (NAS); SBAC and CAA; AECMA; EUROCAE; MSFC and KSFC (NASA); AS and AMS (SAE); ARINC; RTCA; ICAO; DOD Military Specifications and standards (USA); MOD AVP's (UK); JAR (JAA).
® UK: Information Handling Services, Technical Indexes Ltd
Operations Office
Via Agnini 76, I-41037 Mirandola, MO
☎ (+39 535) 261 08 ⊖ (+39 535) 260 21 [98]

80-01500 Officine Aeronavali Venezia SpA
(a Finmeccanica Company)
Headquarters
Via Triestina 214, I-30030 Venice Tessera
☎ (+39 41) 66 68 11 ✉ AERONAV ⊠ 410446 aeronav i
⊖ (+39 41) 541 55 45
♣ Lit 150,000,000,000 👥 761
President: Ing. Beppe Sacchi
Vice President: Ing. Roberto Mannu
Managing Director: Ing. Sorice
Commercial Director: Ing. Massimo Isidori
General Manager: Ing. Gianfranco Vallerini
■ FAA authorised repair station for maintenance and overhaul of major transport and passenger aircraft, passenger to freighter modification of large aircraft, customised avionics design and installation, paint stripping and repainting, test and repair of hydraulic and electrical components, sheet metal fabrication and new parts manufacture, customer support services.
Capodichino Plant
Via Nuovo Tempio 20, I-80144 Capodichino
☎ (+39 81) 234 48 63 ⊖ (+39 81) 736 55 16
GAT Group Manager: Ing. Nino D'Angelo
Group Deputy Manager: Ing. Michele Priolo
Commercial Director: Ing. Renzo Lunardi [97]

80-01925 SEA SpA
Head Office
Società Esercizi Aeroportuali SpA
Aeroporto, I-20090 Milano Linate
☎ (+39 2) 748 51 ⊠ 321167 sealin i
⊖ (+39 2) 74 85 20 10
♧ Llt 50 000 000 000 ♣ Llt 517 733 000 000 (1993)
👥 4 307
Managing Dir: Massimo Amore
Operations Dir: Esposito Vincenzo
Purchasing Mgr: Giorgio Tattoni
Sales & Marketing Mgr: Schmid Aldo
Press Officer: Dr Claudio del Bianco
■ Ground handling services. [97]

80-01950 Società Impianti Riparazioni Navali, Per Azioni (SIRN)
(† SIRN-Società Imp.Riparazioni Navali)
Via Casaregis 22/1, I-16129 Genova
☎ (+39 10) 58 67 71; 59 30 95 [E] sizn.ge@pn.itnet.it
⊖ (+39 10) 54 26 78
♧ Lit 49,000,000 👥 22 ⌂ 150 m2
President: D Bandoni
Managing Director: C Gravina
→ 90
® UK: NEI Clarke Chapman [98]

80-04000 Studio JONA di Consulenza Aeronautica
Via Camerata Picena 379, I-00138 Rome
☎ (+39 6) 881 16 10 ⊖ (+39 6) 881 93 30
👥 1 (1997) ⌂ 100 m2
Principal: Dr Francesco Jona Lasinio
■ Aeronautical consultants, sales agents, representation of aeronautical equipment. Instrument and accessory manufacturers.
® UK: Aerospace Composite Technologies Ltd, Kearsley Airways Ltd, Weston Aerospace (a division of
USA: Hass Instrument Corporation [98]

80-40000 Vitrociset SpA
Head Office & Plant
Via Salaria 1027, I-00138 Rome

☎ (+39 6) 88 17 01 ⊠ 611422 ciset i
⊖ (+39 6) 88 17 02 00
♧ Llt 2 700 000 000 ♣ Llt 250 000 000 000 (1995)
President & Managing Dir: Ing. Girolamo Cartia
General Mgr: Ing. Alberico Pellacchia
Sales & Marketing Mgr: Ing. Vittorio Vergara
Advertising & Press Department: Renato Favretto
■ Maintenance & technical assistance, in Italy & other countries, of automated control centres, ATC radar systems, navaids (ILS, VOR, DME), communication systems. Logistic systems management from technical assistance to full integrated logistic support; design, improvement, maintenance & management of experimental & training firing ranges; logistic information systems for aircraft & weapons systems; trajectography systems & telemetry ground stations for missiles & aircraft. Technical assistance & management of ESA's principal technology & operating centres (ESTEC, ESOC, ESRIN, KOUROU). Integrated logistics for space, data processing & engineering services for the implementation of the earth segment of ESA's main programmes (Columbus, Hermes, Olympus, ERS-1). EDP ranging from project definition to production, maintenance & application of software. Image satellite remote sensing processing & thematic map production. Logistic support backed up by a complete range of training programmes. Systems engineering & installation, design, planning, implementation & testing of complex integrated systems. Engineering, supply & installation, management of implementation programmes assuming responsibility for turn-key jobs, design of infrastructures. Maintenance, repair, overhaul of electronic equipment. Special studies to expand, integrate & upgrade existing systems. Organisation of training courses, spare parts definition, test instruments calibration, technical documentation, integrated logistics support for defence electronic systems.

Rome Plant
Stabilimento di Via Casale di, Settebagni 5/9, I-00138 Rome
☎ (+39 6) 88 17 01

Cagliari Plant
Sede Secondaria di Macchiareddu, PO Box 495, I-09100 Cagliari
☎ (+39 70) 246 81

Cagliari Plant
Stabilimento di Capo San Lorenzo, Villaputzu, I-09040 Cagliari
☎ (+39 70) 997 71 14 [97]

85 Stockists, Distributors, Spare Parts Suppliers

Bonetti Aircraft Supports SpA 90-00250

85-00600 Gelbyson Srl

Registered Office
Via A Albricci 15, I-00194 Rome
☎ (+39 6) 36 30 47 61; 36 30 49 41
⊖ (+39 6) 329 73 37
♧ Lit 99,000,000 (1997) ♣ Lit 3,556,627,000 (1996)
👥 7 (1997) 🏭 1,500 m2
Managing Director: Mrs Amita Endlich
Marketing and Purchasing Manager: Raffaello Tribioli
Representing:
Marathon Battery, USA
Sundstrand System Turbomach Solar, USA
TAT Group, USA [98]

85-00800 Nicotra Meccanica SpA

Via Bergamina 23, I-20016 Pero, Milan
☎ (+39 2) 33 91 11 52 [E] nicotrareccanica@ga.lactica.it
⊖ (+39 2) 33 91 11 56
♧ Lit 200,000,000 (1994) ♧ Lit 800,000,000 (1997)
♣ Lit 7,000,000,000 (1997) 👥 10 (1997)
Managing Director: Giuseppe Nicotra
Finance Director: R Pagani
Sales Director: S Ravasio
Purchasing Manager: R Ostini
Technical Department: F Locarno

■ Rapid clamping fittings (manual, pneumatic and hydraulic), assembly clamps. Closures and quick-release latches for inspection hatches. Clamping devices for milling, grinding and honing. Block jig systems for working centres; quick die change and fixing systems; pneumatic power clamps; vacuum fixing systems; low profile fixings; levelling and vibration damping feet; modular systems for part supporting on 3D co-ordinate measuring machines; mechanical and hydraulical rollblocs and consoles for die moving; clamp levers; advanced holding systems for machining centres. [98]

85-00900 Prima Armi

Viale Kennedy 8, I-10064 Pinerolo, TO
☎ (+39 121) 32 14 22 ✉ PRIMARMI Pinerolo
⊠ 215166 prima ⊖ (+39 121) 39 87 39
👥 12 🏭 1,200
Managing Director: Giuseppe Garbarino
■ Stockist and broker of obsolete ordnance material: small arms, ammunition and spare parts. Distributor of small arms, ammunition, personal and survival equipment, security equipment and body protection.
® Austria: Karl Kahles
Belgium: Cooppal Cartridges
Brazil: Amadeo Rossi SA
France: Matra Manurhin, Soc. Française de Munitions, Unique MAPF
Italy: Knight Fighter Products
Spain: Astra Unceta y CIA, LLama Gabilondo, Star B. Echeverria
Switzerland: Asai Oerlikon, Sphinx Engineering
USA: Michael of Oregon
Yugoslavia: Zastava Arms [98]

85-01000 Servotecnica SRL

Viale Lombardia 20, I-20095 Cusano Milanino MI
☎ (+39 2) 66 42 01 ⊖ (+39 2) 66 40 10 20
♧ Lit 900,000,000 (1997) ♣ US$ 28,000,000 (1997)
👥 42 (1997) 🏭 1,500 m2
President: Adriano Salgarello
Marketing Director: Gnioni Maurizo
■ Tranducers, electric motors, slip rings and gyros. → 90 [98]

85-99999 TRACE Italy

Via Cufra 17, I-20159 Milano
☎ (+39 2) 69 00 87 67 ⊖ (+39 2) 69 00 87 67
Managing Director: Gianfranco Campana [98]

90 Sales Agents, Sales & Liaison Offices

Companies represented:

Aaxico, USA 90-00250
ABB Servomotors Sarl, Italy 90-01930
Abrasive Technology Inc., USA 90-00114
Acer Inc., Taiwan 90-00050
Admiral Training, UK 90-01320
Advanced Products NV, Belgium 90-00167
Advantest Corp, USA 90-01850
Aerazur, France 90-00105
Aero Hardware, USA 90-00114
Aerospace Composite Technologies Ltd, UK 80-04000
Aerospace Fasteners, USA 90-02020
Aggiungerea, France 90-00105
Air Industries Corp., USA 90-00250
Air Precision SA, France 90-00950
Aircraft Supplies, USA 90-00250
AIT Ltd, UK 90-00950
Alcore, USA 90-00114
Alfred Karcher GmbH & Co, Germany 90-01000
Alimex, Germany 90-00114
Allfast Fastening Syst., USA 90-00250
Allied Bendix Aerospace, USA 90-00340
Allied International Corporation, USA 90-00341
AlliedSignal Aerospace, USA 90-00340
Allison Engine Co., USA 90-01320
Alpha Enterprises Inc, USA 90-00950
Aluminium Menziken, Switzerland 90-00114
Amadeo Rossi SA, Brazil 85-00900
American Aviation International, USA 90-01450
AMS Metal Stockholders, UK 90-00114
Arianespace, France 80-00735
Artemis Management Systems, UK 80-00100
Asai Oerlikon, Switzerland 85-00900
Astra Unceta y CIA, Spain 85-00900
Astro Instrument Corp, USA 90-00950
Ateliers de la Haute Garonne, France 90-02020
ATI Industries Tools, USA 90-02020
ATI-Snapon Tools, USA 90-02020
Aurora Forgings, USA 90-00250
Auxitrol, France 90-00105
Aviojet Corp., USA 90-00195
Bard Manufacturing, USA 90-00950
Barry Controls UK, UK 90-00341
Barry Controls USA, USA 90-00341
Baxter Woodhouse Taylor, UK 90-01450
Beaufort Air Sea Equipment, UK 90-00341
BHS Trenntechnik, Germany 90-01170
BMW Rolls-Royce, Germany 90-01320
Bralco Metals, USA 90-00250
Brent International plc, UK 75-11220
British Aerospace plc, UK 90-01320
British Steel Engineering Steels, UK 90-00114
Buckeye, USA 90-02020
Burbank Aircraft Electro Air, USA 90-00250
Burgess Micro Switch, UK 75-22580
Caloric, Germany 90-01170
Canadair Ltd, Canada 90-00340
Century Oils Group plc, UK 75-11220
Chemviron Carbon, USA 75-11220
Cilas SA, France 90-00950
Ciral Casting, UK 90-00250
Citicorp, USA 75-11220
Click Bond, USA 90-02020
Codechamp, France 90-00950
Commscope Inc, USA 90-00250
COMSAT RSI, USA 90-01930
Cooper Power Tools, USA 90-02020
Cooppal Cartridges, Belgium 85-00900
Courtaulds Group, USA 90-00250
Custom Electronics Inc, USA 90-01965
Daco, UK 90-00950
Delft Research Institute, Netherlands 90-01930
Derco Aerospace Inc, USA 90-00250
Detampel, France 90-00114
Detrampel, France 90-00105
Deutsch/American Pneum. Tools, USA 90-02020
Doler Tools, USA 90-02020
Dunlop Aviation Division, UK 90-00341
Dunlop Precision Rubber, UK 90-00341
Dunlop Tyres, UK 90-00341
Ecet-Labem, France 90-00105
EDO Corp, USA 90-00950
EEV Ltd, UK 90-01300
Electrospace Systems, USA 90-01930
Elektro-Metall, Germany 90-00114
EM Corporation, USA 75-11220
EMDS, USA 90-00950
Encore Computer Corp., USA 90-00500
Engineering Systems International, France 90-01930
Eros, France 90-00105
ESSCO, USA 90-00950
ET EL, Switzerland 90-00950
Euroatlas Lic, Germany 75-21750
Eurocopter Group, Germany 90-00120
Eurocopter Group, France 90-00120
Exotic Metals Forming, USA 90-00114
FAG Bearings, Germany 75-26630
Fansteel/California Drop Forge, USA 90-00250
Farrand Controls, USA 90-01923
Fatigue Technology Inc., USA 90-02020
Fine Tubes, UK 90-00250
Fleetworld, UK 90-00114
Flottiveg, Germany 90-01170
Flugzzeug Union Sud, Germany 90-00250
FMC, USA 90-00340
Forged Metals, USA 90-00114
Gardner Denver, USA 90-02020
GBP Tools, USA 90-02020
GEC-Marconi Avionics, UK 90-01320
General Technology Systems, UK 90-01320
GeoAcoustics, UK 90-01320
Grace NV, Belgium 90-00480
Gueneau Industries, France 90-00105
Hamilton Avenet, USA 90-00250
Hardware Specialty Co. Inc., USA 90-00750
Hartwell, USA 90-02020
Harvey Titanium, USA 90-00250
Hass Instrument Corporation, USA 80-04000
Hitchiner Corporation, USA 90-00750
Hodges International Inc., USA 75-11220
Hughes, Microwave Measurement Systems, USA 90-01850
Hughes Aircraft, USA 90-00950

Contact Reference

Hughes Aircraft Co, USA	90-01850
Hughes Defense Comms, USA	90-00950
Hughes Mahwah Electro-Optics Center, USA	90-00950
Humphrey Inc, USA	90-00950
IAI, Israel	80-00735
ICI Imp Chemical Ind, UK	75-22250
Imco Bergman, USA	90-00250
Imo Industries Inc., USA	90-00895
Indal Technologies, Canada	90-01930
Indal Technologies, USA	90-01930
Industria, France	90-00105
Information Handling Services, UK	80-00922
Innovative Tooling Systems, USA	90-02020
Instrument Technology Inc., USA	90-01450
Int'l Light Metals, USA	90-00250
Intertechnique, France	90-00105
Irvin Industries, USA	75-12200
ITT Neo Dyn, USA	90-01500
Jaivision A/S, Denmark	90-00950
Jane's Information Group, UK	80-00450
Jet Products, USA	90-00114
John G. Shelley, USA	75-11220
Kalker, France	90-00105
Karl Kahles, Austria	85-00900
Kaynar, USA	90-02020
Kearsley Airways Ltd, UK	80-04000
Kirkhill Aircraft, USA	90-00250
Knight Fighter Products, Italy	85-00900
Kollmorgen Artus, France	90-00950
Kollmorgen Inland Motor, USA	90-00950
Krauss-Maffei Wehrtechnik GmbH, Germany	90-02250
Labinal, France	75-22580
Lameco, USA	90-00250
Lawrence Aviation, USA	90-00250
LCD Sources, USA	90-00341
Leach International LRE/GEPE, Germany	90-00340
Lefiell Mfg, USA	90-00250
Lehmann Hein, Germany	90-01930
Lewis Engineering, USA	90-01450
Linde, Germany	90-01170
Litton Industries, USA	75-13700
Litton Industries Inc, USA	90-01240
LLama Gabilondo, Spain	85-00900
LMS International, Belgium	90-01245
Lockhart Chemical, USA	75-11220
Lok-Fast, USA	90-02020
Lord SA, USA	90-00250
LRE Relais + Elektronik, Germany	90-00340
Lucas Aerospace Power Transmission, UK	90-00340
M & M Aerospace, USA	90-00114
M.C. Gill, USA	90-00114
Man Technologie AG, Germany	90-00950
Manitou BF, France	90-01290
Marotta Scientific Controls, USA	90-00341
Marshall Aerospace, UK	90-01320
Matra Manurhin, France	85-00900
Matzen & Timm, Germany	90-00114
Mc Starlite, USA	90-00250
Messner, USA	75-11220
Messo-Chemietechnik, Germany	90-01170
Michael of Oregon, USA	85-00900
Microdot Aerospace, USA	90-02020
Millitech Corp, USA	90-01850
MMI Microwave Measurement Systems, USA	90-01850
Monadock Fasteners, USA	90-02020
Monogram, USA	90-02020
Moog Inc, USA	75-16230
MPB Corporation, USA	90-00750
MTU Friendrichschafen GmbH, Germany	90-01350
Murdock, USA	90-00250
NEI Clarke Chapman, UK	80-01950
Nelson Aero Space, USA	90-00250
Nyco S.A., France	75-11220
Nyco SA, France	75-11220
OCTEC Ltd, UK	90-00950
Olaer Industries, France	90-01500
Omega Technologies, USA	90-02020
Ore International Inc., USA	90-01320
Pacific Air Industries, USA	90-00250
PB Fasteners, USA	90-00250
Peerless Aerospace, USA	90-00250
PI Castings, UK	90-00110
PM Putzmeister, Germany	90-01170
Polycast Technology, USA	90-02020
Precilec SA, France	90-01923
Precision Sensors Inc., USA	90-01450
Precitube, France	90-00114
Quackenbush Dresser, USA	90-00250
Quadrant Simulators Inc., UK	90-01930
Rediffusion Simulaton, UK	90-01930
Reliance Gear, USA	90-00750
Remote Ocean Systems, USA	90-00950
Renk AG, Germany	90-01330
Reynolds Metals, USA	90-00250
Ritter Shaus & Blecher, Germany	90-01170
Rockwell International Corp., USA	75-20820
Rockwell International Corp., USA	90-01845
Rohde & Schwarz GmbH & Co KG, Germany	90-01850
Roll Forming Corp, USA	90-00250
Roll Formings Corp, UK	90-00250
Rolls-Royce plc, UK	90-01320
Schmidding, Germany	90-01170
Schottel-Werft, Germany	90-01900
Scott Manufacturing, USA	75-11220
Seepex, Germany	90-01170
SEXTANT Avionique SA, France	90-01925
Sie Metallurpique, France	90-00250
Signal USFA, Netherlands	90-00950
Smith Industries, UK	90-00341
Soc. Française de Munitions, France	85-00900
Spacematic, USA	90-02020
Sphinx Engineering, Switzerland	85-00900
SPS Technologies, USA	75-11220
SRL Systems Research Laboratories, USA	90-00340
Star B. Echeverria, Spain	85-00900
Starlite Industries, USA	90-02020
Starsys Research Corp, USA	90-00950
Summit Aviation, USA	90-00114
Teaero Corp., USA	90-00114
Technetics Corporation, USA	90-00341
Technical Indexes Ltd, UK	80-00922
Technos Int'l Intel. Eqpt, USA	75-11220
Teledyne, USA	90-00340
Teledyne Advanced Materials, USA	90-00250
Textron Filtration Systems, USA	90-01500
The Austin Co., USA	90-00185
The Young Engineers, USA	90-02020
Thomson Brandt Armements, France	90-00340
Thomson-CSF Simulator Division, France	90-00340
Ticalemit Flexibles, UK	90-00105
Tiernay Group, USA	90-00250
Time Electronics, USA	90-00250
Tracor Aerospace, USA	90-00950
Tracor Flight Systems, USA	90-00950
Trilectron Ind., USA	90-01450
Trimos-Sylvac Task Force, Switzerland	90-21000
Tuthill, USA	90-00750
Ultra Electronics Weapon Systems, UK	90-01320
Unique MAPF, France	85-00900
Universal Alloy Products, USA	90-00114
US Industrial Tools, USA	90-02020
US Petsche Co Inc, USA	90-00250
Valtec Aircraft Supply, USA	90-00250
Vernitron Corp., USA	90-01923
Vibra-Metrics Inc., USA	90-01450
Videospection Inc, USA	90-00950
W.Vinten, UK	90-01320
Wadkin, UK	90-00250
Watkins-Johnson Co, USA	90-02300
Weber Rawflex, France	75-11220
Weldon Tools, USA	90-02020
Wesco Manufacturing Inc, USA	90-00250
West Coast Aerospace, USA	90-00114
Weston Aerospace (a division of, UK	80-04000
Weyers Bros Ltd, UK	90-00950
Wilson Corp, USA	90-00250
Windlink A/S, Norway	90-01930
Zastava Arms, Yugoslavia	85-00900
Zephyr Mfr, USA	90-00250

Sales agents/representatives listed elsewhere:

General Consulting & Engineering Srl	80-00735
IFI Srl	75-11220
Servotecnica SRL	85-01000
SIRN sze	80-01950

90-00040 ACAL ITALIA Srl
(† Technitron Systems Srl)
Viale Milanofiori Palazzo E/1, I-20090 Assago, Milan
☎(+39 2) 824 21 12/15 ⊖ (+39 2) 57 51 17 61
Llt 473 000 000 Llt 7 000 000 000 15
550 sq m
Managing Dir: Luciano Pacinotti
Commercial Dir: Pierangelo Vernizzi
Marketing Dir: Maurizio Gatti [97]

90-00050 Acer Italy Srl
Via Cassanese 210, I-20092 Segrate, Milan
☎(+39 2) 26 92 25 65 ⊖ (+39 2) 26 92 10 21
® Taiwan: Acer Inc. [97]

90-00105 Aermar di Marco Rebasti & c Snc
Via Paolo Mascagni 27, I-50124 Florence
☎(+39 55) 22 24 45; (+39 336) 67 10 79
☎(+39 336) 68 42 62 [E] rebasti@dada.it
⊖ (+39 55) 22 02 93
Managing Director: Marco Rebasti
® France: Aerazur, Aggiungerea, Auxitrol, Detrampel, Ecet-Labem, Eros, Gueneau Industries, Industria, Intertechnique, Kalker
UK: Ticalemit Flexibles [97]

90-00110 Aermec 83
Via Medici 42, I-10143 Turin
☎(+39 11) 776 19 55 ⊖ (+39 11) 749 35 69
Managing Dir: G.C. Carando
® UK: PI Castings [98]

90-00114 Aerospace Engineering Genova Srl
(† Aerospace Engineering Sarl)
Via Rimassa 41/6, I-16129 Genova
☎(+39 10) 553 52 53 (4 lines) [E] aereng@tn.village.it
272589 aereng i ⊖ (+39 10) 53 71 39 (2 lines)
Lit 60,000,000 (1997)
Managing Director: Ing Paolo Bandelloni
Sales Manager: Silvio Benvenuto
■ Sales representative and distributor in aircraft and defence fields for metals, raw materials, airframe and engine spare parts, mechanical and electrical hardware.
® France: Detampel, Precitube
Germany: Alimex, Elektro-Metall, Matzen & Timm
Switzerland: Aluminium Menziken
UK: AMS Metal Stockholders, British Steel Engineering Steels, Fleetworld
USA: Abrasive Technology Inc., Aero Hardware, Alcore, Exotic Metals Forming, Forged Metals, Jet Products, M & M Aerospace, M.C. Gill, Summit Aviation, Teaero Corp., Universal Alloy Products, West Coast Aerospace [98]

90-00117 Aerotech Italiana SRL
(a subsidiary of Aerotech World Trade Corp, USA)
Via Monte Delle Gioie, I-00199 Rome
☎(+39 6) 86 20 54 00 ⊖ (+39 6) 86 20 33 45
Managing Director: P Paolicchi [98]

90-00120 Aersud Elicotteri Sarl
(a subsidiary of Helicopters Italia)
Via Bembo 70, Aeroporto Catullo, I-37062 Dossobuono
☎(+39 45) 860 06 46 ⊖ (+39 45) 861 80 21
President: Dr Ing Riccardo Aichner
Vice President: Ing Vittorio Morrassi
■ Eurocopter distributor in Italy. Sales and services.
® France: Eurocopter Group
Germany: Eurocopter Group [98]

90-00155 AlliedSignal Aerospace Marketing Sales & Service
Casella Postale 58, Aeroporto de Fiumicino, I-00050 Rome
☎(+39 6) 65 01 12 75 ⊖ (+39 6) 52 90 53
Account Manager: Claudio Carratu [98]

90-00167 Angst & Pfister SpA
Viale Teodorico 25, I-20149 Milan
☎(+39 2) 310 61 ⊖ (+39 2) 310 63 99
® Belgium: Advanced Products NV [98]

90-00185 Austin Italia SpA
(an associate with The Austin Co., USA)
Via Romagnoli 6, I-20146 Milan
☎(+39 2) 47 79 31 E austin@enter.it
⊖ (+39 2) 48 95 37 58
♧ Lit 500,000,000 (1996)
♣ Lit 43,000,000,000 (1996) 👥 70 🏠 1,200 sq m
Chairman: Paolo Brischi
Vice President: Sergio Treves
Operations Director: Michele Mogotto
Purchasing Manager: Giorgio Ferrario
® USA: The Austin Co. [98]

90-00195 Aviojet Corp
Via Goffredo Casalis 34, I-10143 Turin
☎(+39 11) 437 39 39 ⊖ (+39 11) 437 52 66
Managing Director: Dr Giovanni Boffetta
® USA: Aviojet Corp. [97]

90-00250 Bonetti Aircraft Supports SpA
Via Sottoripa 1A, I-16124 Genova
☎(+39 10) 235 01 ⊠UT BONETTI
⊖ (+39 10) 235 02 00; 235 02 22
♧ Lit 300,000,000 (1996) ♣ US$ 40,000,000 (1996)
👥 45
President: Carlo Bonetti
Vice President: Marco Canali
Vice President Financial: Ms Nora Canali Bonetti
Honorary President: Pietro Bonetti
Credit and Financial Director: Franca Durante
Management Information Systems Division: Laura Rossotti
Raw Materials, Composite and Chemicals Director: Ing Alberto Innamorati
Director Components, Spare: Lorenzo Carboni
Director Components, Spare Parts and Fasteners North: Paolo Blondet
→ 85
® France: Sie Metallurpique
Germany: Flugzeug Union Sud
UK: Ciral Casting, Fine Tubes, Roll Formings Corp, Wadkin
USA: Aaxico, Air Industries Corp., Aircraft Supplies, Allfast Fastening Syst., Aurora Forgings, Bralco Metals, Burbank Aircraft Electro Air, Commscope Inc, Courtaulds Group, Derco Aerospace Inc, Fansteel/California Drop Forge, Hamilton Avenet, Harvey Titanium, Imco Bergman, Int'l Light Metals, Kirkhill Aircraft, Lameco, Lawrence Aviation, Lefiell Mfg, Lord SA, Mc Starlite, Murdock, Nelson Aero Space, PB Fasteners, Pacific Air Industries, Peerless Aerospace, Quackenbush Dresser, Reynolds Metals, Roll Forming Corp, Teledyne Advanced Materials, Tiernay Group, Time Electronics, US Petsche Co Inc, Valtec Aircraft Supply, Wesco Manufacturing Inc, Wilson Corp, Zephyr Mfr

Rome Office
Via Castro Pretoria 115, I-00100 Rome
☎(+39 6) 446 31 45; 31 65 ⊖ (+39 6) 445 33 64
Military Programs Director: Col G Carlo Quieto

Rome Office
Via Mauro Leone, Pamigliano D'Arco, Naples
☎(+39 81) 803 71 03; 803 71 04; 803 71 61
⊖ (+39 81) 803 71 06
Manager: Ms Rita Sorrentino
■ Raw materials, composites, chemical materials and tooling, mechanical, electrical, rubber components and fasteners. [98]

90-00340 CIER Sarl
Compagnia Importazioni & Esportazioni Rappresentanze
Via Gaetano Donizetti 1, I-00198 Rome
☎(+39 6) 841 75 10; 841 68 14 E ciersrl@mbox.vol.it
⊠ 610639 ⊖ (+39 6) 841 11 74
♧ Lit 200,000,000
President and Managing Director: Dr Roberto Bolla
Sales and Marketing Director: Dr Francesco Bolla
Director: Dr Stefano Bolla
Operations Director: Pasqualino Casieri
® Canada: Canadair Ltd
France: Thomson Brandt Armements, Thomson-CSF Simulator Division
Germany: LRE Relais + Elektronik, Leach International LRE/GEPE
UK: Lucas Aerospace Power Transmission
USA: Allied Bendix Aerospace, AlliedSignal Aerospace, FMC, SRL Systems Research Laboratories, Teledyne

Samarate Office
Via Roma 85 (Varese), I-21017 Samarate
☎(+39 331) 22 83 09 ⊖ (+39 331) 22 83 09 [97]

90-00341 Commerciale Avio Marina SAS (CAM)
San Marco 930, I-30124 Venice
San Marco 930, Nuovo Indirzzo Daz, J Luglio
☎(+39 41) 521 25 92 ⊖ (+39 41) 521 25 96
♧ Lit 20,000,000
President: Dr-Ing Maurizio Klinger
■ Aircraft instruments, flight systems, components, engine accessories, parachutes and equipment for army, navy, rescue and survival.
® UK: Barry Controls UK, Beaufort Air Sea Equipment, Dunlop Aviation Division, Dunlop Precision Rubber, Dunlop Tyres, Smith Industries
USA: Allied International Corporation, Barry Controls USA, LCD Sources, Marotta Scientific Controls, Technetics Corporation [98]

90-00344 Connex Italiana
Piazzale Lugano 9, I-20158 Milan
☎(+39 2) 39 32 22 40; 39 32 11 47 ⊠ 325087 connex i
⊖ (+39 2) 646 41 95
Managing Director: Piercarlo Infrano
Information Contact: Marco Bottani [97]

90-00375 Daewoo Italia Srl
Centro Direzionale Colleoni,
Palazzo Liocomo (A2) Via Paracelso 6,
I-20041 Agrate Brianza
☎(+39 39) 609 18 81 ⊠ 333584 daewmi 1
⊖ (+39 39) 609 18 87 [98]

90-00480 Grace Italiana SpA
(† Emerson & Cuming Italiana SpA)
Via Trento 7, PO Box 108, I-20017 Passirana di Rho, Milan
☎(+39 2) 933 23 37 ⊠ 334221 ⊖ (+39 2) 939 09 07
Manager Polymers Division: E Allievi
■ Low density polymers (foams and resins). Epoxy casting resins, silicone moulding materials, conductive and non-conductive adhesives, conductive castings.
® Belgium: Grace NV [97]

90-00500 Encore Computer Italia SpA
Via A. Volta 16, I-20093 Cologno Monzese (MI)
☎(+39 2) 253 912 21
Sales: E. Conlon
® USA: Encore Computer Corp. [97]

90-00750 Gimar Giordano K L
(† Gimar di DF E M Giardono Sas)
Via Busoni 7a, I-20137 Milan
☎(+39 2) 551 20 04; 551 69 18 ⊖ (+39 2) 546 61 63
Director: Fausto Pasqualucci
Owner: D F Giordano
® USA: Hardware Specialty Co. Inc., Hitchiner Corporation, MPB Corporation, Reliance Gear, Tuthill [98]

90-00760 Encore Computer Italia SpA
(a subsidiary of Encore Computer Corporation, USA)
Via A Volta 16, I-20093 Cologno Monzese, Milano
☎(+39 2) 25 39 12 21 W http://www.encore.com
⊠ 335402 selit i ⊖ (+39 2) 254 98 19
Managing Director: Ellen Conlon
Operations Director: Paolo Proserpio
Purchasing Manager: Cristina Rosso
■ Hardware, software and integration systems. [98]

90-00800 Haynes International SRL
Viale Brianza, 8, I-20127 Milan
☎(+39 2) 26 14 13 31 ⊖ (+39 2) 282 82 73 [97]

90-00892 Hughes International Corporation
44 Lungotevere Mellini, I-00193 Rome
☎(+39 6) 32 14 05 ⊖ (+39 6) 321 39 37
Director: Dr Fausto Pasqualucci [98]

90-00895 Imo Industries Srl
Via Corregio 19, I-20149 Milan
☎(+39 2) 48 00 86 15 ⊖ (+39 2) 48 01 39 28
General Mgr: Riccardo Carella
® USA: Imo Industries Inc. [97]

90-00950 International Promotion Services Sarl
Via Gide 33, I-00143 Rome
☎(+39 6) 500 46 45; 502 03 21 E md1182@mclink.it
⊖ (+39 6) 500 50 69
♧ Lit 99,000,000 👥 7 (1997) 🏠 250 m2
President and Managing Director: Carmelo Caruso
Finance and Administration Director: Francesco Caruso
General Manager: Giorgio Caruso
■ Consulting in technology transfers, joint ventures, administrative and logistic support for aerospace, defence, Italian MoD and industry.
® Denmark: Jaivision A/S
France: Air Precision SA, Cilas SA, Codechamp, Kollmorgen Artus
Germany: Man Technologie AG
Netherlands: Signal USFA
Switzerland: ET EL
UK: AIT Ltd, Daco, OCTEC Ltd, Weyers Bros Ltd
USA: Alpha Enterprises Inc, Astro Instrument Corp, Bard Manufacturing, EDO Corp, EMDS, ESSCO, Hughes Aircraft, Hughes Defense Comms, Hughes Mahwah Electro-Optics Center, Humphrey Inc, Kollmorgen Inland Motor, Remote Ocean Systems, Starsys Research Corp, Tracor Aerospace, Tracor Flight Systems, Videospection Inc
Electro-Mechanic Components: [98]

90-01000 Karcher SpA
(a subsidiary of Alfred Kärcher GmbH & Co)
Via Elvezia 4, I-21050 Cantello, VA
☎(+39 332) 41 74 00; 41 74 19 ⊖ (+39 332) 41 83 50
♧ Lit 950,000,000 (1996) ♣ Lit 17,800,000 (1996)
👥 24 🏠 1,500 m2
Manager: Eugenio Cagna
® Germany: Alfred Karcher GmbH & Co [98]

90-01170 Krauss-Maffei Chemiomeccanica SrL
(a subsidiary of Krauss-Maffei Wehertechnik GmbH)
Piazza Castello 2, I-20121 Milano
☎(+39 2) 869 05 21 23 ⊠ 334089
⊖ (+39 2) 890 02 77
Manager: Dr-Ing Guiseppe Morabito
® Germany: BHS Trenntechnik, Caloric, Flottiveg, Linde, Messo-Chemietechnik, PM Putzmeister, Ritter Shaus & Blecher, Schmidding, Seepex [98]

90-01240 Litton Precision Products SpA
Viale Fulvio Testi 126, I-20092 Cinisello Balsamo, Milan
☎(+39 2) 244 04 21 ⊖ (+39 2) 244 06 69
Managing Director: Paolo Novazzi
General Manager: Mario Spinelli
® USA: Litton Industries Inc

Sales Office
Via Montauro 27, I-00040 Morena, Rome
☎(+39 6) 724 77 13; 724 77 14 ⊖ (+39 6) 724 77 13
General Manager: Giuseppe di Rocco [98]

90-01245 LMS Italiana
Via Alcarotti 1, I-28100 Novara
☎(+39 321) 61 23 01 ⊖ (+39 321) 39 95 57
Sales Manager: M Parodi
→ 80
® Belgium: LMS International [98]

90-01290 Manitou Costruzioni Ind Sarl
(a subsidiary of Manitou BF, France)
Via E Fermi 5, I-41013 Castelfranco
☎(+39 59) 95 05 11 ⊖ (+39 59) 92 26 64
® France: Manitou BF [98]

90-01300 Marconi SpA
Via Palmanova 185, I-20132 Milan
☎(+39 2) 256 31 41 ⊖ (+39 2) 256 77 45
® UK: EEV Ltd [98]

90-01320 Mercantile Italo Britannica Sarl (MIB)
Via Barnaba Oriani 34, I-00197 Rome
☎(+39 6) 808 22 44 ⊠MIBSPA Roma
⊖ (+39 6) 439 63 09
♧ Lit 20,000,000 👥 7
Honorary Chairman: Gen C A Fabio Moizo
Chairman: Dr-Ing. Elio de Sabata (OBE, FRAeS)
Managing Director: A W Mead (OBE, FRAeS)

Contact Reference

Director: Dr-Ing. Giorgio Levantesi
Director and Partner: Richard Martin-Hurst
■ Agents and consultants to foreign companies in Italy.
® Germany: BMW Rolls-Royce
UK: Admiral Training, British Aerospace plc, GEC-Marconi Avionics, General Technology Systems, GeoAcoustics, Marshall Aerospace, Rolls-Royce plc, Ultra Electronics Weapon Systems, W.Vinten
USA: Allison Engine Co., Ore International Inc. [97]

90-01330 Metallco International Ltd Liaison Office Italy
(a subsidiary of Metallco International Ltd)
Via Giulia di Barolo 9, I-10124 Turin
☎ (+39 11) 839 53 78 ⊜ (+39 11) 839 53 79
Managing Director: Gian Carlo Bussei
■ Sales agents and commercial advisors representing petrochemical and power suppliers, machinery and components to contractors in Europe, Middle East and South East Asia.
® Germany: Renk AG [98]

90-01350 MTU Italia Sarl
(a subsidiary of MTU Motoren und Turbinen-Union Friedrichshafen GmbH, Germany)
Via Aurelia Nord, I-19021 Arcola
☎ (+39 187) 95 26 11 ⊜ (+39 187) 95 47 13
♧ Lit 4,750,000,000 (1996) ⛬ 32 ⌂ 1,150 sq m
President: Mr Valentini
Operations Director: Mr Felme
Sales and Marketing Director: Mr Dube
Purchasing Manager: Mr Hundsdoerfer
® Germany: MTU Friendrichschafen GmbH [98]

90-01450 Nordavia
163 Str. Cascinoto, I-10158 Torino
☎ (+39 11) 223 98 39; 223 98 67
⊜ (+39 11) 223 98 27
♧ Llt 120 000 000 (1995) ♣ Llt 500 000 (1994) ⛬ 3
⌂ 250
Managing Dir: Mrs C. Gallo
Marketing Mgr: Sergio G. Assereto
® UK: Baxter Woodhouse Taylor
USA: American Aviation International, Instrument Technology Inc., Lewis Engineering, Precision Sensors Inc., Trilectron Ind., Vibra-Metrics Inc. [97]

90-01500 Olaer Italiana SpA
(a subsidiary of Oiltech Group)
83 Strada Fantasia, I-10040 Leini TO
☎ (+39 11) 991 85 11 ⊜ (+39 11) 998 93 24
♧ Lit 840,000,000 (1997) ♣ Lit 18,000,000 (1996)
⛬ 26 (1997) ⌂ 2,000
Managing Director: G Tosco
Sales Director: C Bongiovanni
Financial Director: A Arietti
Purchasing Director: A Leone
® France: Olaer Industries
USA: ITT Neo Dyn, Textron Filtration Systems [98]

90-01835 RES Srl
(a subsidiary of Racal Recorders Ltd, UK)
Via Crescenzago 105, I-20132 Milan
☎ (+39 2) 26 11 19 42; 26 11 19 52
⊜ (+39 2) 26 11 19 37 [98]

90-01845 Rockwell Automation Srl Regional Headquarters
Viale de Gasperi 126, I-20017 Mazzo di Rho, Milan
☎ (+39 2) 93 97 21 ⊜ (+39 2) 93 97 22 01
Director: Alberto Sicoli
® USA: Rockwell International Corp. [98]

90-01850 Rohde & Schwarz Italia SpA
(† Roje Telecomunicazioni SpA)
(a subsidiary of Rohde & Schwarz GmbH & Co KG, Germany)
Via Tiburtina 1182, I-00156 Rome
☎ (+39 6) 41 59 81 ⊜ (+39 6) 41 59 82 70
Managing Director: Maurizio Leon ☎ (+39 6) 41 59 82 01
Sales Manager: Giovanni Bertolini ☎ (+39 6) 41 59 82 30
Purchasing Manager: Marcello Caciotti ☎ (+39 6) 41 59 82 40
■ Sales and service activities in the following fields: radio monitoring, radio localisation, communication, electronic warfare and measuring equipment.
® Germany: Rohde & Schwarz GmbH & Co KG
USA: Advantest Corp, Hughes Aircraft Co, Hughes, Microwave Measurement Systems, MMI Microwave Measurement Systems, Millitech Corp [98]

90-01900 Schottel Italiana Sarl
(a subs. of Schottel-Werft, Germany)
Corso A. Podestà 1/13, I-16121 Genoa
☎ (+39 10) 54 24 53 ⊜ (+39 10) 53 25 59
Managing Dir: Enrico Brizzolara
® Germany: Schottel-Werft [97]

90-01923 Servotecnica
Viale Lombardia 20, I-20095 Cusano Milanino, Milan
☎ (+39 2) 66 42 01 ⊜ (+39 2) 66 40 10 20
♧ Lit 900,000,000 (1997) ♣ US$ 28,000,000 (1997)
⛬ 42 (1997) ⌂ 1,500 m2
President: Adriano Salgarello
Marketing Director: Gnioni Maurizio
® France: Precilec SA
USA: Farrand Controls, Vernitron Corp. [98]

90-01925 SEXTANT Avionique Italia SpA
(a subsidiary of SEXTANT Avionique SA, France)
38 Via F Cavallotti, I-21029 Vergiate, Varese
☎ (+39 331) 94 61 23 ▣ 315518 ⊜ (+39 331) 94 85 27
Managing Director: Pietro Roveda
® France: SEXTANT Avionique SA [98]

90-01930 SIMC Sarl Administration Office
Societá Italiana Materiali & Construzioni
Via dei Monti Parioli 38, I-00197 Rome
☎ (+39 6) 321 08 10; 321 35 06; 321 34 97 ✉ SIMC
▣ 3210908 ⊜ (+39 6) 321 09 08
Chairman: Adm V Pescatori
Vice President: Dr-Ing. Paolo De Gaetano
Managing Director: Dr Claudio Chiodelli
® Germany: Lehmann Hein
UK: Rediffusion Simulaton
USA: Electrospace Systems, Indal Technologies
SIME s.a.s. di Paolo de Gaetano & C. Commercial Office
Via F.S. Benucci 9, I-00149 Rome
☎ (+39 6) 55 27 08 29; 55 27 30 49
Ⓔ paolo@star.flashnet.it ⊜ (+39 6) 551 56 62
General Dir: Paolo de Gaetano
■ Software FEM Numerical Simulation (crash, shock, CFD simulation). Software for networks & for structural calculation. Simulators & trainers, flight, industrial & defence oriented. Satellite earth stations & component parts. Servomotors & controls. Helicopter landing devices & fire doors for ships.
® Canada: Indal Technologies
France: Engineering Systems International
Italy: ABB Servomotors Sarl
Netherlands: Delft Research Institute
Norway: Windlink A/S
UK: Quadrant Simulators Inc.
USA: COMSAT RSI [97]

90-01965 Spacerep Srl
Via Gonzaga 4, I-20123 Milan
☎ (+39 2) 87 88 94; 87 86 62 ⊜ (+39 2) 86 46 02 19
® USA: Custom Electronics Inc [97]

90-02020 Teknoavio Sarl
Via Cesare Battisti 10, I-28040 Dormeletto, NO
☎ (+39 322) 24 15 61 ⊜ (+39 322) 483 19
♧ Lit 20,000,000 ⛬ 6 (1997)
President and Chief Executive Officer: Sergio Barbieri
■ Miscellaneous equipment and components. Sale, repair and maintenance of tools, hardware and materials.
® France: Ateliers de la Haute Garonne
USA: ATI Industries Tools, ATI-Snapon Tools, Aerospace Fasteners, Buckeye, Click Bond, Cooper Power Tools, Deutsch/American Pneum. Tools, Doler Tools, Fatigue Technology Inc., GBP Tools, Gardner Denver, Hartwell, Innovative Tooling Systems, Kaynar, Lok-Fast, Microdot Aerospace, Monadock Fasteners, Monogram, Omega Technologies, Polycast Technology, Spacematic, Starlite Industries, The Young Engineers, US Industrial Tools, Weldon Tools [98]

90-02030 Teledyne Italy
Corso di Porta Romana 2, I-20122 Milan
☎ (+39 2) 72 00 03 85 ⊜ (+39 2) 72 02 11 01 [97]

90-02250 Veronesi Separatori SpA
Via Don Minzoni 1, I-40050 Villanova di Castenaso (BO)
☎ (+39 51) 605 30 07 ▣ 511029 ⊜ (+39 51) 605 31 83
Manager: Dr-Ing. Graziano Sartori
® Germany: Krauss-Maffei Wehrtechnik GmbH [98]

90-02300 Watkins-Johnson Italiana SpA
(a division of Watkins-Johnson, USA)
Piza G. Marconi 25, I-00144 Rome
☎ (+39 6) 592 45 54; 591 25 15 ⊜ (+39 6) 591 73 42
Director: Sergio Testagrossa
Applications Mgr: Giuseppe Caneglias
® USA: Watkins-Johnson Co [97]

90-21000 Trimos-Sylvac Sarl
Via Cassanese 220, I-20090 Segrate, Milan
☎ (+39 2) 213 77 51 ⊜ (+39 2) 213 77 54
Managing Director: P Balestrini
® Switzerland: Trimos-Sylvac Task Force [98]

90-61350 Stelit
Via Zenodoto di Efeso, I-00124 Rome
☎ (+39 6) 65 09 09 57 ⊜ (+39 6) 65 09 09 56 [97]

Jamaica

Head of State:
H.M. Queen Elizabeth II

Governor General:
Sir Howard Cooke

10 Head of State

10-10000 Office of the Governor General
King's House, Hope Road, Kingston 10
Governor General: Sir Howard Cooke [98]

11 Prime Minister

11-10000 Office of the Prime Minister
1 Devon Road, Jamaica House, PO Box 272, Kingston 6
☎ (+1 809) 927 99 41 ▣ 2398 ⊜ (+1 809) 929 00 05
Prime Minister: Percival James Patterson
Deputy Prime Minister: Seymour Mullings [98]

12 Ministries

12-10000 Ministry of National Security
Office Centre, 12 Ocean Boulevard,
Kingston Mall, Kingston 10
☎ (+1 809) 922 00 80 ⊜ (+1 809) 922 69 50
Minister of National Security and Justice: K D Knight [98]

12-20000 Ministry of Foreign Affairs and Foreign Trade
85 Knutsford Boulevard, Kingston 5
☎ (+1 809) 926 42 20; 926 42 21
▣ 2114 external ja; 2374 foreign ja
⊜ (+1 809) 929 67 33
Minister of Foreign Affairs and Trade: Seymour Mullings
Minister of Industry, Investments and Commerce: Dr Paul Robertson [98]

15 Parliament

15-10000 Senate
Gordon House, Duke Street, Kingston
☎(+809) 922 02 00 [98]

31 Embassies & Attachés

31-00660 Embassy of the Federal Republic of Germany
10 Waterloo Road, PO Box 444, Kingston 10
☎(+1 809) 926 5665; 6728; 6729 ✉ 2146 aakgn ja
⊖ (+1 809) 929 8282
Ambassador: H.E. Dr W. Bolewski
Deputy Chief of Mission: Dr Ulrich Seidinburger
Commercial Counsellor: Robert Müller [97]

31-00830 High Commission of India
4 Retreat Avenue, PO Box 446, Kingston 6
☎(+1 876) 927 31 14; 927 42 70
E highcomindkin@toj.com ✉ (291) 3456 hci kgn
⊖ (+1 876) 978 28 01
High Commissioner: H E Mr Vidya Bhushan Soni
Attaché: Subhash Chander Khurana
Attaché: A V Satyanarayana
Attaché: V K Mehta [98]

31-01820 High Commission for Trinidad & Tobago
Pan Jamaica Building, 3rd Floor, 60 Knutsford Boulevard, 5 Kingston
☎(+1 809) 926 57 30; 926 57 39; 968 05 88
⊖ (+1 809) 926 58 01
High Commissioner: H E Mrs Pearl Wilson
Second Secretary: Ms Sharon Bradshaw
Attaché, Administration: Rohit Beharry
Attaché, Administration: Mrs Cynthia Daniel-Sobion [98]

31-01890 United Kingdom High Commission
PO Box 575, Kingston 10
Trafalgar Rd, Kingston 10
☎(+1 809) 926 90 50 ✉ UKREPKIN JA
✉ 2110 ukrepkin ja ⊖ (+1 809) 929 78 69
High Commissioner: H E Mr A R Thomas
Deputy High Commissioner: Allan White
Defence Adviser: Col Al Moorby
Naval Adviser: Capt I M Himes (Resides at Bridgetown)
Commercial Officer: Mrs Cheryl Nurse
Management Assistant: Ms Valerie F Smith [98]

31-01900 Embassy of the United States of America
2 Oxford Road, 1st Floor, Jamaica Mutual Life Center, Kingston
☎(+1 809) 929 4850/9 ⊖ (+1 809) 926 6743
Abassador: H.E. Mr Gary J. Cooper
Deputy Chief of Mission: John Vessey
Head of Commercial Section: Robert Bucalo (Resident in Santo Domingo) ☎(+1 809) 221 2171
Public Affairs Officer: J.Michael Houlahan
Defence Attaché: Lt Col Norman Wiggins (USMC)
Military Liaison Officer: Lt Col Terry K. DeRouchey (USA) [97]

Japan

Head of State:
His Imperial Majesty Akihito

10 Head of State

10-10000 Imperial House
1-1 Chiyoda-ku, Tokyo 100
☎(+81 3) 32 13 11 11; 32 13 21 11
Emperor of Japan: His Imperial Majesty Akihito [98]

11 Prime Minister

11-10000 Office of the Prime Minister
1-6-1 Nagata-Cho, Chiyoda-ku, Tokyo 100
☎(+81 3) 35 81 23 61
Prime Minister: Ryutaro Hasimoto
Chief Cabinet Secretary: Seiroku Kajiyama
Finance Minister: Hiroshi Mitsuzuka
Minister of International Trade and Industry: Shinji Sato [98]

12 Ministries

12-20000 Ministry of Foreign Affairs
Gaimusho
2-2-1 Kasumigaseki, Chiyoda-ku, Tokyo 100
☎(+81 3) 35 80 33 11 ✉ 22350 gaimu j
Minister of Foreign Affairs: Yukihiko Ikeda
Deputy Foreign Minister: Kazuo Ogura
State Secretary for Foreign Affairs: Masahiko Koumura
Spokesman: Nobuaki Tanaka
Special Assistant to the Foreign Minister: Mitsuro Donowaki
Cultural Affairs Department
Director General: Yasukuni Enoki [98]

12-30000 Ministry of International Trade and Industry (MITI)
Tsusho Sangyo Sho
3-1 Kasumigaseki 1-chome, Chiyoda-ku, Tokyo 100
☎(+81 3) 501 15 11 ⊖ (+81 3) 501 77 75
Minister: Shinji Sato [98]

12-40000 Ministry of Home Affairs
2-1-2 Kasumigaseki, Chiyoda-ku, Tokyo 100
☎(+81 3) 35 81 53 11; 35 81 01 41
Minister of Home Affairs: Katsuhiko Shirakawa [98]

25 Defense Forces

25-22100 Air Defence Command Headquarters (ADC)
1-1850 Sengen-Cho, Fuchu-shi, Tokyo 183
☎(+81 92) 581 34 16 ⊖ (+81 92) 581 34 16
Commanding General: Lt Gen. Shigeru Sugiyama
Commander Western Air Defence Force: Lt Gen. Kazuyoshi Suzuki
Chief of Staff: Yoshimitsu Tsumagair
Chief Defence Division: Maj.Gen. Yasuo Ohkushi
Chief Materiel Division: Col Shinroku Sasaki
Northern Air Defence Force HQ
Misawa Air Base, 125-7 Ushirokubo, Misawa-shi, Aomori-ken 033
☎(+81 176) 53 41 21
Commanding General: Lt Gen. Norio Asakura
Chief of Staff: Maj.Gen. Ikuo Ontake
Western Air Defence Force HQ
Kasuga Air Base, 3-1-1 Haramachi, Kasuga-shi, Fukuoka-ken 816
☎(+81 92) 581 40 31
Commanding General: Lt Gen. Hyoubu Etou
Chief of Staff: Maj.Gen. Teruhiko Shigenaga
Central Air Defence Force HQ
Iruma Air Base, 2-3 Inariyama, Sayama Shi, Saitama-ken 350-13
☎(+81 429) 53 61 31
Commanding General: Lt Gen. Kouichi Sugiura
Chief of Staff: Maj.Gen. Shinichi Ichimura
South Western Composite Air Div.
Naha Air Base, 679 Kagamimizu, Naha-shi, Okinawa-ken 901-01
☎(+81 988) 57 11 91
Commanding General: Lt Gen. Atushi Itou
Chief of Staff: Maj.Gen. Kazuhiko Mori [97]

25-30000 Ground Self-Defense Force (GSDF)
Rikujo Jieitai
7-45 Akasaka 9-chome, Minato-ku, Tokyo 107
☎(+81 3) 34 08 52 11 ext. 2425
E jgsdf@ppp.mindnet.or.jp
⊖ (+81 3) 34 08 52 11 ext. 2427
Chief of Staff: Gen Nobutoshi Watanabe
☎(+81 3) 34 08 52 11 ext. 2400
Vice Chief of Staff: Lieut Gen Tsuneo Isojima
Inspector General: Maj Gen Yoshimitsu Miyatake
Controller: Maj Gen Noboru Yamaguchi
Chief of Defence Planning Division: Col Humio Eto
Chief of Operations Division: Col Makoto Hirose
Chief of Aviation Section: Col Hirohisa Hujimoto
Chief Foreign Liaison: Col Tsuguyu Endo
Chief of Research Division: Col Kenji Tokuda
Commanding General Ordnance Depot: Lieut Gen Akihiko Kurushima
Personnel Director: Maj Gen Hajime Matsusaki
Intelligence Director: Maj Gen Mitsuo Nonaka
Logistics Department
☎(+81 3) 34 08 52 11
Director: Maj Gen Masaaki Matsukawa
Deputy Director: Maj Gen Minoru Maeda
Head Logistics Management Division: Maj Gen Isao Imamura
Head of Ordnance and Chemical Division: Col Mitsunobu Kamimoto
Head Communications and Electronics Division: Col Tadayuki Hayashi
Head Aircraft Division: Col Hiroyuki Sumi
Head Quartermaster Division: Col Mitsuyoshi Ozu
Head Engineering Division: Col Toshihiko Sugaya
Head Materièl Research and Development Division: Col Toshihiro Ishii
Head Plans and Administration Section: Col Takeshi Sakai
Head Aviation Safety Section: Col Tadayuki Iwaki
Head Management and Transportation Division: Col Mamoru Hachisuga
Plans and Operations Department
Director: Maj Gen Takashi Arai
Northern Army
Northern Army Headquarters, Nishi-10 Minami-26 Chuo-Ku, Sapporo-shi Hokkaido 604
☎(+81 11) 511 71 16
Commanding General: Lieut Gen Kaneyuki Okoshi
Chief of Staff: Maj Gen Susuma Iwasaru
North Eastern Army
North Eastern Army, 1 Minamime-Yakata Myagino-ku, Sendai-shi, Miyagi-ken 983
☎(+81 22) 231 11 11
Commanding General: Lieut Gen Yoshihiro Takeda
Chief of Staff: Maj Gen Kenji Tsuchihashi
Eastern Army
Eastern Army Headquarters, Ohizumigakuen-cho - Nerima, Tokyo 178
☎(+81 48) 460 17 11
Commanding General: Lieut Gen Yuji Hujinawa
Chief of Staff: Maj Gen Mitsuo Sakudo
Middle Army
Middle Army Headquarters, 7-1-1 Midorigaoka, Itami-shi, Hyogo-Ken 664
☎(+81 727) 82 00 01
Commanding General: Lieut Gen Toshiaka Miyamoto
Chief of Staff: Maj Gen Takenori Yasumura
Western Army
Western Army Headquarters, 1 Higashi-cho, Kumamoto-Shi, Kumamoto-ken 862
☎(+81 963) 68 51 11
Commanding General: Lieut Gen Tomotoshi Tamura [98]

Contact Reference

25-40000 Maritime Self-Defence Force (MSDF)

Kaijo Jiei Tai
Maritime Staff Office (Kaijo Bakuryo Kanbu)
7-45 Akasaka 9-chome, Minato-ku, Tokyo 107
☎(+81 3) 34 08 52 11; ext. 2794/2925
(public information) E koho-mso@iic.or.jp
⊖ (+81 3) 34 08 52 11 ext. 2925; 34 08 24 40
Chief of Staff: Adm Kazuya Natsukawa
Vice Chief of Staff: Vice Adm Mutsuyoshi Gomi
Inspector General: Rear Adm Kataru Hasegawa
Director Administration Department:
Rear Adm Kanji Kushita
Director Personnel and Education Department:
Rear Adm Kaoru Fukutani
Director Operations and Planning Department:
Rear Adm Tooru Ishikawa
Director Intelligence Department:
Rear Adm Satoshi Takemura
Director Logistics Department: Rear Adm Norifumi Takeda
Director Engineering Department:
Rear Adm Tadashi Murakami
Surgeon General: Rear Adm Tooru Nakajima

Operations and Plans Department
Director: Rear Adm Tooru Ishikawa
Head Systems Programs Division: Capt Tooru Takahashi
Head Operations Division: Capt Yasuo Seki

Logistics Department
Director: Rear Adm Norifumi Takeda
Deputy Director: Rear Adm Naoki Takeda
Head Logistics Planning Division: Capt Tsutomo Takahashi
Head Ships Division: Capt Michiyuki Muneyoshi
Head Aircraft Section: Capt Yoshitaka Mitsunaka
Head Weapons Division: Capt Masatoki Matsumoto

Intelligence Department
Director: Rear Adm Satoshi Takemura

Personnel and Education Department
Director: Rear Adm Kaoru Fukatani

Engineering Department
Director: Rear Adm Tadashi Murakamki 98

25-42100 Self-Defense Fleet Headquarters

Jiei Kantai Shireibu
7-73 Funakoshi-cho, Yokosuka-shi, Kanagawa 237
☎(+81 468) 61 82 81
Commander in Chief Fleet: Vice Adm Seiji Saeki
Chief of Staff: Rear Adm Yuuji Minari 98

28 Coastguard, Border Protection Forces

28-10000 Japanese Maritime Safety Agency Headquarters (MSA)

Kairjo Hoan Cho
2-1-3 Kasumigaseki, Chiyoda-ku, Tokyo
☎(+81 3) 35 91 63 61; 35 91 97 80 (direct)
Commandant: Ysutoshi Tsuchisaka
Vice Commandant: Toshiaki Ohmori
Vice Commandant for Operations: Ryuichi Takei
Director General Administration Department:
Michiaki Tsuji
Director-General Equipment and Technology:
Nobutaka Nanbu
Director-General Guard and Rescue Department:
Kunio Tashima
Director-General Hydrographic Department:
Shoichi Ohshima
Director-General Navigation Aids Department:
Gokichi Ogawa
Superintendant Maritime Safety Academy: Hiroyuki Fujii
Superintendant Maritime Safety School:
Hiroyuki Yoshizumi
Foreign Liaison Officer: Masahiro Izumi

■ Organisation for the promotion of safety measures. Maritime traffic information systems, maritime traffic safety guidance. Information gathering systems for search and rescue. Systems for marine leisure-related rescue operations. Protection of marine environment including response to natural disasters. Law enforcement in territorial seas. Ocenographic studies and surveys. Information service for navigational safety, management and provision of oceanographic data and information.

Navigational aids systems, local weather information, maintenance and improvement of navigational aids. Radio navigation aids, differential GPS. Marine safety academy and safety school.

Public Relations Office
2-1-3 Kasumigaseki, Chiyoda-ku, Tokyo 100
☎(+81 3) 35 91 63 61

Maritime Safety Academy
5-1 Wakabacho, Kure-shi, Hiroshima 737
☎(+81 8) 23 21 49 61
Superintendant: Hiroyuki Fujii
Deputy Superintendant: Hiroshi Motegi

Maritime Safety School
Nagahama, Maizuru-shi, Kukuoka 801
☎(+81 7) 73 62 35 20
Superintendant: Hiroyuki Yoshizumi
Deputy Superintendant: Masahiko Sakamoto

Moji Branch - Maritime Safety School
3-3-1 Shiranoe, Moji-ku, Kitakyushu-shi, Fukoaka 989-24
☎(+81 9) 33 41 81 31
Chief: Tadao Matsumoto

Miyagi Branch - Maritime Safety School
4 Kitanagauma - Shimonogo, Iwanuma-shi, Miyagi 989-24
☎(+81 2) 23 24 23 38
Chief: Kimito Sunohara

1st Regional Maritime Safety Headquarters
5-3 Minatomachi, Otaru-shi, Hokkaido 047
☎(+81 1) 34 32 61 61
Commander: Yasuki Kuwaburu
Deputy Commander: Yukinoba Tomonaga

2nd Regional Maritime Safety Headquarters
3-4-1 Sadayamadori, Shiogama-shi, Miyagi 985
☎(+81 2) 23 63 01 11
Commander: Mitsuharu Kodama
Deputy Commander: Katsuzo Mori

3rd Regional Maritime Safety Headquarters
5-57 Kitanakadori - Naka-ku, Yokohama-shi, Kanagawa 231
☎(+81 4) 52 11 07 71
Commander: Nobuo Chihara
Deputy Commander: Shigeru Kikuchi

4th Regional Maritime Safety Headquarters
2-3-12 Irifune - Minato-ku, Nagoya-shi, Aichi 455
☎(+81 5) 26 61 16 11
Commander: Yuko Adachi
Deputy Commander: Takashi Jinmyo

5th Regional Maritime Safety Headquarters
1-1 Hatobacho - Chuo-ku, Kobe-shi, Hyogo 650
☎(+81 7) 83 91 65 51
Commander: Mitsuo Suzuki
Deputy Commander: Kozo Tanaka

6th Regional Maritime Safety Headquarters
3-10-17 Ujina Kaigan, Minami-ku, Hiroshima-shi, Hiroshima 734
☎(+81 8) 22 51 51 11
Commander: Takashi Suyama
Deputy Commander: Susumu Takashi

7th Regional Maritime Safety Headquarters
1-3-10 Nishikaigan - Moji-ku, Kitakyushu-shi, Fukuoka 801
☎(+81 9) 33 21 29 31
Commander: Tsutomo Tokuno
Deputy Commander: Tadashi Gotoh

8th Regional Maritime Safety Headquarters
901 Shimofukui, Maizuru-shi, Kyoto 624
☎(+81 7) 73 76 41 00
Commander: Eisuke Kudo
Deputy Commander: Tetsuo Imazato

9th Regional Maritime Safety Headquarters
2-2-1 Bandai, Niigata-shi, Nigata 950
☎(+81 2) 52 44 41 51
Commander: Kazuo Tanigawa
Deputy Commander: Yoshio Kubo

10th Regional Maritime Safety Headquarters
4-1 Higasikoorimotomachi, Kagoshima-shi, Kagoshima 890
☎(+81 9) 92 50 98 00
Commander: Yoshishiko Terakawa
Deputy Commander: Nobuaki Okajima

11th Regional Maritime Safety Headquarters
2-11-1 Minatomachi, Naha-shi, Okinawa 900
☎(+61 9) 88 66 00 83
Commander: Masuru Kubota
Deputy Commander: Tsuneo Ogawa
Deputy Commander: Akinori Shima
Deputy Commander: Tadao Tatsuno 98

31 Embassies & Attachés

31-00100 Embassy of Austria

1-1-20 Moto-Azabu, Minato-ku, Tokyo 106
☎(+81 3) 34 51 82 81/2 ✉AUSTROAMB Tokyo
⊖ (+81 3) 34 51 82 83
Ambassador: H.E. Dr Martin Vukovich
Minister Deputy Chief of Mission: Andreas List
First Counsellor Commercial: Wolfgang Küng
First Secretary: Harald Günter
Commercial Attaché: Peter Fuchs 97

31-00250 Embassy of the Republic of Bulgaria

36-3 Yoyogi 5-chome, Shibuya-Ku, Tokyo 151
☎(+81 3) 34 65 10 21 ⊖ (+81 3) 34 65 10 31
Ambassador: H.E. Mr Valentin Gatsinski
DCM Minister Counsellor: Zlatin Trapkov
Counsellor Financial, Science & Technology Consular Affaires: Ivan Dimitrov
Counsellor Head of Commercial Section:
Toma Tchouparov
First Secretaray Press & Culture: Mrs Dobrinka Dancheva
Second Secretary Commercial Affairs: Dimitar Dimitrov
Second Secretary Cultural Exchange:
Ms Youliana Lazarova
Attaché: Boiko Touletchki 97

31-00310 Embassy of Canada

3-38 Akasaka 7-chome, Minato-ku, Tokyo 107
☎(+81 3) 34 08 21 01/08 ⊖ (+81 3) 34 79 53 20
Ambassador: H.E. Mr Donald W. Campbell
☎(+81 3) 34 08 21 01
Minister Economic & Commercial: John Tennant
Canadian Forces Attaché: Capt. Roger Chiasson
Minister Head of Chancery: Joseph Caron
Minister Counsellor Commercial: Mark Romoff
Counsellor Science & Technology: Ms Peggy Tsang
Third Secretary Commercial: Wayne Robson 97

31-00350 Embassy of Chile

1-14 Shiba 3-chome, Nihon Seimei Akabanebashi Bldg 8F, Minato-Ku, Tokyo 105
☎(+81 3) 34 52 75 61; 34 52 75 62; 34 52 75 85
⊠ 24585 embchile j ⊖ (+81 3) 34 52 44 57
Ambassador: H E Mr Jaime Lagos Erazo
Minister Counsellor: Patricio Torres
Third Secretary: Oscar Fuentes Lira
Third Secretary: Mrs Constanza Figueroa Sepulveda
Third Secretary and Consul: Fernando Velasco Parada
Cultural and Press Attaché: Jeff Peet

Office of Commercial Affairs
1-14 Shiba 3-chome, Minato-Ku,
Nihon Seimei Akabanebashi Bldg 8F, Tokyo 105
☎(+81 3) 37 69 05 51; 37 69 05 52
⊖ (+81 3) 37 69 41 56
Counsellor Head of Commercial Office: Dario Guzman Leal
Agriculture Attaché: Pablo Alvarez Ravera 98

31-00380 Embassy of Colombia

10-53 Kami-Osaki 3-chome, Shinagawa-ku, Tokyo 141
☎(+81 3) 34 40 64 51 E emcoltky@j-link.or.jp
⊖ (+81 3) 34 40 67 24
Ambassador: H E Mr Pedro Felipe Valencia
Counsellor: Guillermo Mejia
Second Secretary in charge of Consular Affairs:
Luis Amadeo Hernandez
Second Secretary: Juan Pablo Turbay
Commercial Counsellor: Mrs Nora Olave-Shibusawa
Commercial Counsellor: Ricardo Gutierrez
Commercial Counsellor: Armando Castaño 98

31-00430 Embassy of the Republic of Cuba

4-11-12 Shimomeguro, Meguro-ku, Tokyo 153
☎(+81 3) 37 16 31 12 ⊖ (+81 3) 37 16 43 34
Ambassador: H E Mr Ernesto Meléndez Bachs
Deputy Chief of Mission: Pedro Monzón Barata
Commercial Counsellor: Mrs Niurka Atienza Ambou
Third Secretary: Herminio López 98

31-00460 Royal Danish Embassy

29-6 Sarugaku-cho, Shibuya-Ku, Tokyo 150
☎(+81 3) 34 96 30 01 ✉AMBADANE Tokyo
⊠ 24417 ambadane j ⊖ (+81 3) 34 96 34 40
Ambassador: H.E. Mr Flemming Hedegaard
Deputy Chief of Mission: Hans Schou

Minister Counsellor: Erik Klindt Andersen
Commercial Counsellor: Jens Peder Jensen
First Secretary: Anders Ornemark
Counsellor Information & Consul: Bent Lindblad
Consul: Ms Birgit Vesterholt
Attaché Science & Technology: Ms Eva Moe Jacobsen [97]

31-00500 Embassy of Ecuador
12-24 Nishi-Azabu 4-chome, Minato-Ku,
38 Kowa Bldg, Room 806, Tokyo 106
☎ (+81 3) 34 99 28 00 ⊖ (+81 3) 34 99 44 00
Ambassador: H.E. Mr Marcelo Avila
Counsellor Deputy Chief of Mission: Mario Guerrero
Commercial Counsellor: Rodrigo Paez [97]

31-00520 Embassy of El Salvador
Rowa #38 Building RH803, 12-24 Nishi-Azabu 4-chome, Minato-ku, Tokyo 106
☎ (+81 3) 34 99 44 61 [E] embesal@iac.co.jp
⊖ (+81 3) 34 86 70 22
Ambassador: H E Dr Mynor Gil
First Secretary and Consul General: Noemi Reinosa
Assistant: Ms Maritza Vlate [98]

31-00570 Embassy of the Republic of Fiji
Noa Bldg 14th Fl. 3-5, 2 Chome Azabudai, Minato-ku, Tokyo
☎ (+81 3) 35 87 20 38 ⊖ (+81 3) 35 87 25 63
Ambassador: H.E. Mr Robin Hardy Yarrow
First Secretary: E. Blakelock
Commercial Secretary: M. Sakashita [98]

31-00590 Embassy of France
4-11-44 Minami-Azabu, Minato-Ku, Tokyo 106
☎ (+81 3) 54 20 88 00 [E] ambfrajp@popmail.gol.com
⊖ (+81 3) 54 20 88 47
Ambassador: H E Mr Jean-Bernard Ouvrieu
Minister Deputy Chief of Mission: Jacques Lapouge
Defence Attaché: Capt Antoine de Trogof (Navy)
Counsellor Economic and Commercial Affairs: Hubert Testard
Counsellor Press: Daniel Blaize
Counsellor Industrial Affairs: Jean Barthélemy
Attaché Defence Equipment: Maj Jacques Maleval
Attaché Nuclear: Jean-Jacques Lavigne
Attaché Science and Technology: Henri Angelino
Archivist: Miss Akiyo Sawada

Office of the Commercial Counsellor
Masonic 39 Mori Building 12th Floor,
2-4-5 Azabudai, Minato-ku, Tokyo 106
☎ (+81 3) 54 03 23 33
⊖ (+81 3) 54 03 23 19; 54 03 23 43
Commercial Counsellor: Mrs Florence Jeanblanc Rislei
Commercial Counsellor: Laurent Fiscus [98]

31-00630 Embassy of the Gabonese Republic
1-12-11 Kami-Osaki, Shinagawa-ku, Tokyo 141
☎ (+81 3) 34 48 95 40 ⊠ 24812 ambgabon j
⊖ (+81 3) 34 48 15 96
Ambassador: H E Mr Vincent Boulé
First Counsellor: Ms Aimé Bibang Bi Mba
Secretary: Kayoko Abe [98]

31-00690 Embassy of Greece
16-30 Nishi Azabu, 3 chome, Minato-Ku, Tokyo 106
☎ (+81 3) 34 03 08 71/72 ⊠ 27696
⊖ (+81 3) 402 46 42
Ambassador: H.E. Mr Constantinos Vassis
Counsellor (Deputy Chief of Mission): Constantinos Giovas
Second Secretary: Vassilios Tolios [98]

31-00740 Embassy of Guatemala
12-24 Nishi-Azabu 4-chome, Minato-ku,
Kowa Bldg No 38, Room 905, Tokyo 106
☎ (+81 3) 34 00 18 30 ⊖ (+81 3) 34 00 18 20
Ambassador: H.E. Mr F. González-Davison [97]

31-00790 Embassy of Honduras
12-24 Nishi-Azabu 4-chome, Kowa Building 38, 802, Minato-ku, Tokyo 106
☎ (+81 3) 34 09 11 50 [E] honduras@interlink.or.jp
⊖ (+81 3) 34 09 03 05
Ambassador: H E Mr Edgardo Sevilla Idiáguez [98]

31-00810 Embassy of the Republic of Hungary
2-17-14 Mita, Minato-ku, Tokyo 108
☎ (+81 3) 37 98 88 01/4 ⊖ (+81 3) 37 98 88 12
Ambassador: H.E. Dr Zoltan Südy
Deputy Chief of Mission: Jenö Molnar
First Secretary Press: Dr Gabor Diczhazi
First Secretary Political & Cultural: Gyula Dabronaki

Commercial Counsellor's Office
Mori Bldg 28, 4-16-13 Nishi-Azabu, Minato-ku, Tokyo 106
☎ (+81 3) 34 99 49 51/4 ⊖ (+81 3) 34 99 49 18
Commercial Counsellor: Sándor Kiss [97]

31-00830 Embassy of India
2-11 Kudan-Minami 2-chome, Chiyoda-ku, Tokyo 102
☎ (+81 3) 32 62 23 91/7; 32 39 24 49
⊠ 24850 indembsy j ⊖ (+81 3) 32 34 48 66
Ambassador: H.E. Shri Kuldip Sahdev
☎ (+81 3) 32 65 50 36
Deputy Chief of Mission: Dr S. Jaishankar
☎ (+81 3) 32 62 10 36
Minister Counsellor: Sh Bhushan Jain
Counsellor: Dr V.T. Chitnis ☎ (+81 3) 32 62 26 89
Defence Attaché: Col P.C. Katoch ☎ (+81 3) 32 62 92 54
First Secretary: Sh D.M. Mulay ☎ (+81 3) 32 34 13 09
First Secretary: Sh Krishan Kumar
First Secretary Trade: Dr S.K. Agarwal
☎ (+81 3) 32 61 43 05
Head of Chancery: Sh R.K. Kalra ☎ (+81 3) 32 62 26 88
Admin Attaché: R.G. Nair
Second Secretary: Sh S. Bhagwat ☎ (+81 3) 32 62 23 91 [97]

31-00870 Embassy of Ireland
Ireland House, 2-10-7 Kojimachi, Chiyoda-ku, Tokyo 102
☎ (+81 3) 32 63 06 95 ⊖ (+81 3) 32 65 22 75 [98]

31-00890 Embassy of Italy
5-4 Mita 2-chome, Minato-ku, Tokyo 108
☎ (+81 3) 34 53 52 91 to 96 ⊠ 22433 italdipl j
⊖ (+81 3) 34 56 23 19
Ambassador: H.E. Dr Giovanni Dominedo
Deputy Chief of Mission: Dr Gherardo La Francesca
Military, Naval & Air Attaché: Capt. Enrico Bozzano (Navy)
Commercial Counsellor: Dr Pio Mariani [97]

31-00960 Embassy of Kenya
3-24-3 Yakumo, Meguro-ku, Tokyo 152
☎ (+81 3) 37 23 40 06; 37 23 40 07
⊠ 2422378 kenrep j ⊖ (+81 3) 37 23 44 88
Ambassador: H E Mr Samson K Chemai
Counsellor: Philip Richard Okanda Owade
First Secretary Political: Henry O Ragen
First Secretary Immigration: Elijah J Gittao
Financial Attaché: Abel O Sagini [98]

31-01000 Embassy of the State of Kuwait
13-12 Mita 4-chome, Minato-Ku, Tokyo 108
☎ (+81 3) 34 55 03 61; 34 55 03 63 ⊠ 25501 kuwaitia j
⊖ (+81 3) 34 56 62 90
Ambassador: H E Dr Suhail Khalil Shuhaiber
Second Secretary: Abdullah Al-Askar
Third Secretary: Jasem Al-Bedaiwi [98]

31-01010 Embassy of the Lao People's Democratic Republic
3-3-21 Nishi Azabu, Minato-ku, Tokyo 106
☎ (+81 3) 54 11 22 91/2 ⊖ (+81 3) 54 11 22 93
Ambassador: H.E. Mr Thongsay Bodhisane
First Secretary: Souchay Philathivong
Third Secretary: Khonepheng Thammavong
Third Secretary: Phaylam Oudomicham [97]

31-01020 Embassy of Lebanon
Chiyoda House 5F, 2-17-8 Nagata-cho, Chiyoda-ku, Tokyo 100
☎ (+81 3) 35 80 12 06; 35 80 12 27
[E] ambaliba@japan.co.jp ⊠ 25356 ambaliba j
⊖ (+81 3) 35 80 22 81
Ambassador: H E Mr Samir Chamma [98]

31-01190 Embassy of Mexico
2-15-1 Nagata Cho, Chiyoda Ku, Tokyo 100
☎ (+81 3) 35 81 11 31; 35 81 11 35
[E] embamex@twics.com ⊖ (+81 3) 35 81 40 58
Ambassador: H E Mr Manuel Uribe Castaneda
Deputy Chief of Mission: Cesar Perez Saavedra
Minister Political Affairs: Alejandro Madrigal
Minister Cultural Affairs: Guillermo Eguiarte
Counsellor Press: Enrique Morales
Naval Attaché: Vice Adm Daniel Hernandz Gallegos
Commercial Counsellor: Sergio Inclan
☎ (+81 3) 35 80 08 11; 35 80 08 11
Counsellor Financial Affairs: Tito Ayala
☎ (+81 3) 32 84 03 31
First Secretary Science and Technology: Armando Arriaga [98]

31-01210 Embassy of Mongolia
21-4 Kamiyama-cho, Shibuya-ku,
Shoto Pinecrest, Mansion 21-4, Tokyo 150
☎ (+81 3) 34 69 20 88; 34 69 20 92
⊠ 23757 embamngl j ⊖ (+81 3) 34 69 22 16
Ambassador: H.E. Mr Daramyn Yondon
☎ (+81 3) 34 69 20 91
Commercial & Economic Counsellor: Dashdorjiin Hadhuu
☎ (+81 3) 34 69 20 92 [98]

31-01270 Royal Netherlands Embassy
6-3 Shiba-ken 3-chome, Minato-Ku, Tokyo 105
☎ (+81 3) 54 01 04 11 ⊖ (+81 3) 54 01 04 20
Ambassador: H.E. Mr F.P.R. Van Nouhuys
Minister Deputy Chief of Mission: A. Heldring
Political Counsellor: J. Huber
Commercial Counsellor: J. de Vries
Second Secretary Trade: Ms G. Mulder
Second Secretary Consular Affairs & Administration: T. de Boer
Counsellor Industrial Affairs: R.V. Schipper
☎ (+81 3) 54 01 04 30/3
Scientific Counsellor: E. van Koay
☎ (+81 3) 54 01 04 25/6
Scientific Attaché: Dr A. van Blokland
☎ (+81 3) 54 01 04 25 [97]

31-01340 Royal Norwegian Embassy
12-2 Minami Azabu 5-chome, Minato-ku, Tokyo 106
☎ (+81 3) 34 40 26 11 ⊠ 26440 ambanor j
⊖ (+81 3) 34 40 26 20
Ambassador: H.E. Mr John Bjornebye
Minister Counsellor: Harald Neple
Counsellor: Birgit Schjerven
First Secretary: Gerd Pettersen
Second Secretary: Berit Enge
Attaché: Birgit Brantenberg
Counsellor: Emil Ekker
Counsellor Industry & Technology: Wilhelm S. Boe [97]

31-01370 Embassy of Pakistan
2-14-9 Moto-Azabu, Minato-ku, Tokyo 106
☎ (+81 3) 34 54 48 61/64 ✉ PAREP Tokyo ⊠ 28883 j
⊖ (+81 3) 34 55 14 81
Ambassador: H.E. Mr Mujahid Husain
Counsellor/Head of Chancery: Mohammed Akhtar Tufail
Counsellor: Abdul Malik Abdullah ☎ (+81 3) 34 52 37 91
Minister Technical Affairs: Mian Suhai Aslam
☎ (+81 3) 37 98 09 03
Commercial Counsellor: Shaukat Nawaz Tahir
☎ (+81 3) 34 54 50 88
Press Attaché: Shahzad Sharjeel ☎ (+81 3) 34 55 14 81 [97]

31-01380 Embassy of the Republic of Panama
12-24 Nishi-Azabu 4-chome, Minato-ku,
Room #902 Kowa Building 38, Tokyo 106
☎ (+81 3) 499 37 41; 499 37 41 42
[E] panaemb@japan.co.jp
[W] http://www.teleserve.co.jp/embassy/embassy/embassy/pan/index.htm ⊖ (+81 3) 54 85 35 48
Ambassador: H E Mr José A Sosa
Counsellor, Political: Javier Orillac
Counsellor, Commercial: Mrs Maria Gabriela Méndez de Félix
Tourist Attaché: Mrs Diana Azcárraga [98]

31-01400 Embassy of Paraguay
Kowa 38 Building - Room 701,
4-12-24 Nishi-Azabu, Minatoka, Tokyo 106
☎ (+81 3) 54 85 31 01; 54 85 31 02
[E] embapar@gol.com
[W] http://www.teleserve.co.jp/embassy/embassy/par/index.htm ⊖ (+81 3) 54 85 31 03
Ambassador: H E Mr Miguel A Solano Lopez C

Counsellor: Brigido Lezcan Britos
Second Secretary: Julio C Peralta R [98]

31-01420 **Embassy of the Philippines**
11-24 Nanpeidai-machi, Shibuya-ku, Tokyo 150
☎(+81 3) 496 27 31/6 ✉ PHILEMBASSY TOKYO
⊠ 22694 ambaphil j ⊖ (+81 3) 496 27 34
Ambassador: H.E. Mr Alfonso Yuchengco
Minister Counsellor: Ms Erlinda Q. Gavino
Armed Forces Attaché: Col Marcel A. Panares
First Secretary & Consul: Pedro B. Firmalino
Commercial Attaché: Elmer Nocheseda
First Secretary & Consul: Jerril G. Santos
Second Secretary & Consul:
Miss Maria Cleofe R. Natividad
Second Secretary & Consul: Miss Maria Teresa Taguiang
Public Information Officer: J.J.P. Gimenez [97]

31-01430 **Embassy of the Republic of Poland**
4-5-14 Takanawa, Minato-ku, Tokyo 108
☎(+81 3) 32 80 28 81/2 ✉ POLAMBA Tokyo
⊠ 22314 polamba j ⊖ (+81 3) 32 80 23 62
Ambassador: H.E. Mr Henryk Lipszyc
Minister Counsellor: Jan Janiszewski
Commercial Counsellor: Andrzej Kaczmarek
Military, Naval & Air Attaché: Col Jan Lesinski
Counsellor Science & Technology: Wojciech Bogusz
First Secretary Political Affairs: Dr Jadwiga Rodowicz
First Secretary Public Information: Jacek Gaca [97]

31-01440 **Embassy of Portugal**
3-10-3 Kojimachi, Chiyoda-ku, Tokyo 102
☎(+81 3) 52 12 73 22 (chancery);
52 26 06 14 (consular section);
52 26 06 11 (cultural section)
⊖ (+81 3) 52 26 06 16 (chancery);
52 26 06 16 (consular section);
52 26 06 13 (cultural section)
Ambassador: H E Mr Rui Goulart de Ávila
Deputy Chief of Mission and Counsellor: Luis Lorvão
First Secretary Economic and Consular: Eduardo Ramos
Commercial Counsellor's Office
Regency Shinsaka 101, Akassaka 8-5-8, Minato-ku, Tokyo 107
☎(+81 3) 54 74 44 04 ⊖ (+81 3) 54 74 44 03
Commercial Counsellor: Camilo Martins de Oliveira [98]

31-01460 **Embassy of the State of Qatar**
6-8-7 Akasaka, Minato-ku, Tokyo 108
☎(+81 3) 32 24 39 11/2 ⊠ 24877 kharijia j
⊖ (+81 3) 32 24 39 17
Ambassador: H.E. Mr Mohamed Hassan Al-Jaber
First Secretary: Abdulla A. Al-Fakhroo [97]

31-01480 **Embassy of Romania**
16-19 Nishi Azabu 3-chome, Minato-ku, Tokyo 106
☎(+81 3) 34 79 03 11; 34 79 03 13
[E] romembjp@iac.co.jp ⊠ 22664 romamb j
⊖ (+81 3) 34 79 03 12
Ambassador: H E Dr Eugen Dijmarescu
Minister Counsellor: Ion Scumpieru
☎(+81 3) 34 79 10 98
Counsellor: Ms Marin Valcea
First Secretary: Constantin-Dan Sulugiucc
☎(+81 3) 34 79 10 98
Second Secretary Scientific Affairs: Ion Visoiu
☎(+81 3) 34 79 18 04
Second Secretary Cultural Affairs: Aurelian Neagu
☎(+81 3) 34 79 04 13
Defence, Military, Air and Naval Attaché:
Lieut Col Virgil Ioan Dan ☎(+81 3) 34 79 44 60
Press and Consular Affairs: Stefan Modrea
☎(+81 3) 34 79 18 04
Office of the Economic Counsellor
16-19 Nishi Azabu 3-chome, Minato-ku, Tokyo 106
☎(+81 3) 34 79 03 11 ⊠ 22664 j
⊖ (+81 3) 34 79 03 12
Minister Counsellor Economic: Valentin Pavel
☎(+81 3) 34 79 04 11
Economic Counsellor: Danut Bucur
☎(+81 3) 34 79 04 14
Economic Officer: Paul Mihai Dobreanu
☎(+81 3) 34 79 03 13 [98]

31-01485 **Embassy of the Russian Federation**
2-1-1 Azabudai, Minato-ku, Tokyo 106
☎(+81 3) 35 83 42 24; 59 82 ⊠ 24231 rosposol j
⊖ (+81 3) 35 05 05 93
Ambassador: H.E. Mr Lyudvig Chizhov
Minister Deputy Chief of Mission: Vasilii I. Saplin
Minister-Counsellor Economic: Oleg V. Ivanov
Military & Air Attaché: Gen.Maj. Alexander Evstrakhov
Military & Air Attaché: Capt. Vladislav Prokopenko
Naval Attaché: Lt Col Yuri Kapoustine
Asst Military & Air Attaché: Maj. Oleg Nudnov
Asst Military & Air Attaché: Maj. Sergei Kovalev
Asst Naval Attaché: Capt 3rd Rank Maxim Tkhugo
Office of the Trade Representative
6-4-9 Takanawa, Minato-ku, Tokyo 106
☎(+81 3) 34 47 32 91 ⊖ (+81 3) 34 47 32 21
Trade Representative: Vladimir P. Pavlov
Vice Trade Representative: Mikhail A. Popik [97]

31-01600 **Royal Embassy of Saudi Arabia**
1-53 Azabu Nagasaka-cho, Minato-ku, Tokyo 106
☎(+81 3) 35 89 52 41; 35 89 52 44 ⊠ 25731 najdiah j
⊖ (+81 3) 35 89 52 00
Counsellor, Chargé d'Affaires a.i.:
Abdulrhman Ali Magharabi
Second Secretary: Adnan H H Gedair [98]

31-01655 **Embassy of the Republic of Slovenia**
5-15 Akasaka 7-Chome, Minato-ku, Tokyo 107
☎(+81 3) 55 70 62 75 [E] embslo@gol.com.jp
⊖ (+81 3) 55 70 60 75
Ambassador: H E Mr Janez Premoze
First Secretary: Dr Stanislav Rascan [98]

31-01670 **South African Embassy**
7-9 Hirakawacho 2-Chome - Chiyoda-Ku,
414 Zenkyoren Building, Tokyo 102
☎(+81 3) 32 65 33 66; 32 65 33 69
[W] http://sunsite.sut.ac.jp/embassy/ jasanet
⊖ (+81 3) 32 65 11 08
Counsellor: B Freeman
Counsellor, Economic: B N Roberts [98]

31-01690 **Embassy of the Democratic Socialist Republic of Sri Lanka**
14-1 Akasaka 1-chome, Minato-ku, Tokyo 107
☎(+81 3) 585 74 31; 585 74 32
[E] ldpo3432@niftyserve.of.jp ⊠ 24524 lankaemb j
⊖ (+81 3) 586 93 07
Ambassador: H E Mr N Navaratnarajah
☎(+81 3) 585 74 33
First Secretary: Mrs H M M Dissanayake
First Secretary Commercial: Nimal Karunatilake
☎(+81 3) 585 74 35
Second Secretary: Anura Rajakaruna
Attaché: Jinadasa Panagoda [98]

31-01730 **Swedish Embassy**
1-10-3-100 Roppongi, Minato-ku, Tokyo 106
☎(+81 3) 55 62 50 50; 55 62 50 60 (information)
✉ SVENSK Tokyo [E] swedemb@twics.com
⊠ 24586 svensk j ⊖ (+81 3) 55 62 90 95
Foreign Service Officer: Ms Anna Wollberg
Commercial Office
1-10-3-200 Roppongi, Minato-ku, Tokyo 106
☎(+81 3) 55 62 50 00 [E] stcjapan@gol.com
⊖ (+81 3) 55 62 90 80
Commercial Counsellor: Göran Edman
☎(+81 3) 403 92 41/45 FAX: (+81 3) 408 20 68
Counsellor (Science & Technology): Hans Nilsson
☎(+81 3) 470 41 81/85 ✉ SWTECH Tokyo
FAX: (+81 3) 470 41 85 ⊠ 2424196 swtech j
Science and Technology Office
1-10-3-300 Roppongi, Minato-ku, Tokyo 106
☎(+81 3) 55 62 50 30; 55 62 50 37 (reception)
[E] info@swetech.or.jp
Invest in Sweden Agency
1-10-3-300 Roppongi, Minato-ku, Tokyo 106
☎(+81 3) 55 62 50 14 [E] isatyo@beehive.twics.com
⊖ (+81 3) 55 62 51 30 [98]

31-01830 **Embassy of Tunisia**
1-18-8 Wakaba, Shinjuku-ku, Tokyo 160
☎(+81 3) 33 35 41 11; 33 35 41 13
⊠ 27146 ambtunis j ⊖ (+81 3) 32 25 43 87
Ambassador: H E Mr Noureddine Mejdoub
Counsellor (Deputy Chief of Mission): Ridha Azaiez
Counsellor: Nejib Denguezli
Counsellor: Mohamed Antar
Attaché: Mounir Youssef
Attaché: Bechi Fathalli [98]

31-01880 **Embassy of the United Arab Emirates**
9-10 Nanpeidai-cho, Shibuya-ku, Tokyo 150
☎(+81 3) 54 89 08 04 ⊖ (+81 3) 54 89 08 13
Ambassador: H E Mr Hamad Salem Al-Makami
Deputy Chief of Mission:
Abdulla Moh'd Abdul-Wahed Al-Ma'ainah [98]

31-01890 **Embassy of the United Kingdom**
No 1 Ichiban-cho, Chiyoda-ku, Tokyo 102
☎(+81 3) 52 11 11 00 ⊠ 22755 prodrome j
⊖ (+81 3) 52 75 31 64
Ambassador: H E Sir David J Wright
Minister: C T W Humfrey
Minister, Financial: J E W Kirby
Defence and Naval Attaché: Capt N D V Robertson (RN)
Military and Air Attaché: Gp Capt P Edwards (RAF)
Counsellor, Political: D J Fitton
Counsellor, Political: Dr P B Preece
Counsellor, Commercial: P Bateman
Counsellor, Economic: T C Morris [98]

31-01950 **Embassy of Venezuela**
12-24 Nishi-Azabu 4-chome, Minato-ku,
No. 38 Kowa Bldg 703, Tokyo 106
☎(+81 3) 34 09 15 01/04 ✉ EMBAVENEZ Tokyo
⊖ (+81 3) 34 09 15 05
Ambassador: Carlos R. Omàna
Minister Counsellor: Ms Velis M. Villegas
Commercial Counsellor: Carlos Ponce [97]

31-02030 **Embassy of the Republic of Zambia**
10-2 Ebara 1-chome, Shinagawa-ku, Tokyo 142
☎(+81 3) 34 91 01 21; 34 91 01 22
✉ ZAMBIANS TOKYO ⊠ 25210 zamemtyo j
⊖ (+81 3) 34 91 01 23
Ambassador: H E Prof L P Temtio
Chargé d'Affaires a.i.: Ms Tabiso L Nyemba
Second Secretary Accounts: Imasiku Mwanangumbi
Third Secretary/Personal Secretary: Ms Marjorie Yalobi [98]

31-02040 **Embassy of the Republic of Zimbabwe**
5-9-10 Shiroganedai, Minato-ku, Tokyo 108
☎(+81 3) 32 80 03 31; 32 80 03 32 ⊠ 32975 j
⊖ (+81 3) 32 80 04 66
Ambassador: H E Mr Buzwani Donald Mothobi
Counsellor (Deputy Chief of Mission):
Raphael R S Ndawana
First Secretary Trade and Investment: Phillip Hove
First Secretary Tourism, Information and Culture:
Graham Maenzanise
Second Secretary: Ms Daisy Mhlanga
Third Secretary: Mrs Eve Tsikirayi [98]

33 Foreign Forces

33-20100 **US Army Japan**
Camp Zama, [97]

60 Research Organizations, Institutes & Test Establishments

60-01000 **National Aerospace Laboratory (NAL)**
7-44-1 Jindaijihigashi-machi, Chofu, Tokyo 182
☎(+81 422) 47 59 11 ⊖ (+81 422) 47 32 90
♣ Yen 12 445 000 000 (1995) 431 72 992 m2
Dir General: Kazuaki Takashima
Chief Librarian: Miwa Sugita
■ Government establishment founded in 1955 for research & development in the field of aeronautical & space technology. [97]

65 Industry, Trade & Armed Forces Associations

65-10200 Japan Association of Defence Industry (JADI)

2-21-3 Akasaka, Minato-ku, Tokyo 107
☎ (+81 3) 35 84 67 55 ⊖ (+81 3) 35 84 73 63
Secretary General: Hiromichi Eguchi
☎ (+81 3) 35 84 67 55
Chief Executive Director: Kazutaka Tomofuji
Chief of Secretariat: Maj Gen Yoshimitsu Takeda (retd)
☎ (+81 3) 35 84 67 55
Research Director: Col Daisaku Asakawa (retd)

■ Membership consists of 134 companies active across entire range of defence products and services including firearms, guided missiles, ammunition, explosives, military vehicles, electronics, underwater ordnance, shipboard outfits, naval ordnance, aircraft ASW, aircraft armament and electronics and weapon materials.

Members:
Aerospace Logistics & Electronic Co Ltd
Anritsu Corp
Asahi Chemical Industry Co Ltd
Asahi-Seiki Manufacturing Co Ltd
Chugoku Kayaku Co Ltd
Daido Steel Co Ltd
Daikin Industries Ltd
Daisel Chemical Industries Ltd
Denki Kogyo Co Ltd
C3I Systems Corp
Eagle Industry Co Ltd
Fuji Electric Co Ltd
Fuji Heavy Industries Ltd (FHI)
Fuji Photo Optical Co Ltd
Fujikura Ltd
Fujitsu Ltd
Fujitsu General Ltd
Fujitsu Tokki Systems
The Furukawa Electric Co Ltd
Furuno Electric Co Ltd
The Hakodate Dock Co Ltd
Hino Motors Ltd
Hinode Kogyosho Co Ltd
Hitachi Ltd
Hitachi Zosen Corp
Hosoya Fireworks Co Ltd
Howa Machinery Ltd
Ikegami Tsusinki Co Ltd
Ishikawa Seisakusho Ltd
Ishikawajima-Harima Heavy Industries Co Ltd
Isuzu Motors Ltd
Japan Aircraft Manufacturing Co Ltd (NIPPI)
Japan Aviation Electronics Ind Ltd (JAE)
Japan Radio Co Ltd
The Japan Steel Works Ltd
Joint Systems Service Co Ltd
JRC Tokki Co Ltd
Junkosha Co Ltd
Kaiyo Denshi Kogyo Co Ltd
Kanto Aircraft Instrument Co Ltd
Kawasaki Heavy Industries Ltd (KHI)
Kawasaki Steel Corp
KEIO Diesel Industry Co Ltd
Kitazawa Electric Works Co Ltd
Kobe Steel Ltd
Koden Electronics Co Ltd
Kokusai Electric Co Ltd
Kokusai Electric Engineering Ltd
Komatsu Ltd
Komatsu Tokki Ltd
Kowa Co Ltd
Koyo Seiko Co Ltd
Kyosan Electric Manufacturing Co Ltd
Marine United Inc
Matsushita Electric Industrial Co Ltd
Meisei Electric Co Ltd
MHI Logitec Co Ltd
MHI Oceanics Co Ltd
MHI Special Vehicles Parts Supply & Technical Service Co Ltd
Minebea Co Ltd
Miroku Firearms Manufacturing Co
Mitsubishi Electric Corp (MELCO)
Mitsubishi Electric Logistics Support Co
Mitsubishi Heavy Industries Ltd (MHI)
Mitsubishi Motors Corp
Mitsubishi Nagasaki Machinery Mfg Co Ltd
Mitsubishi Precision Co Ltd
Mitsubishi Space Software Co Ltd
Mitsui & Co Ltd
Mitsui Engineering & Shipbuilding Co Ltd
Morioka Tokki Co
Nagano Japan Radio Co Ltd
Nagoya Electric Works Co Ltd
NEC Corporation
NEC Radio Equipment Engineering Ltd
NEC Radio & Electronics Ltd
New Japan Radio Co Ltd
Nichiyu Giken Kogyo Co Ltd
Nikko Tokki Co Ltd
Niko Electronics Co Ltd
Nikon Corp
Nippon Avionics Co Ltd
Nippon Electric Industry Co Ltd
Nippon Koki Co Ltd
Nippon Steel Corp
Nippon Yakin Kogyo Co Ltd
Nissan Motor Co Ltd
Nisshin Electronics Co Ltd
NKK Corp
NOF Corporation
Ohara Ironworks Co Ltd
Ohkawa Engineering Co Ltd
Oki Electric Industry Co Ltd
Oki Seatec Co Ltd
Richo Elemex Corp
Sampa Kogyo KK
Sasebo Heavy Industries Co Ltd
Shimadzu Corp
Shinko-Seisakusho Co Ltd
ShinMaywa Industries Ltd
Shizuoka Oki Electric Industry Co Ltd
Shounan Seiki Co Ltd
Shouwa Aircraft Industry Co Ltd
Showa Kinzoku Kogyo Co Ltd
Silver Built Co
SPC Electronics Corp
Sumiju Tokki Service Co Ltd
Sumitomo Electric Industries Ltd
Sumitomo Heavy Industries Ltd
Sumitomo Metal Industries Ltd
Suzuki Motor Corp
Taiko Electric Works Ltd
Taiyo Musen Co Ltd
Tamagawa Seiki Co Ltd
Tamura Corp
Tanabe Pneumatic Machinery Co Ltd
Tochigi Fuju Sangyo K K
Tokimec Inc
Tokin Corp
Tokyo Aircraft Instrument Co Ltd
Topcon Corp
Toshiba Corp
Toshiba Electronic Systems Co Ltd
Toshiba Logistic Support Corp
Totsu Denshi Co Ltd
Toyo Communications Equipment Co Ltd
Toyo Seiki Co Ltd
Toyota Motor Corp
The Tsurumi-Seiki Co Ltd
Watanabe Engineering Co
Yoh Engineering K K
Yokogawa Denshikiki Co Ltd
Yokogawa Electric Corp
The Yokohama Rubber Co Ltd [98]

65-10400 Japan Industrial Technology Association (JITA)

Toranomon Ichi-chome,
Mori Building 5F - 1-19-5 Toranomon,
Minato-ku, Tokyo 105
☎ (+81 3) 35 91 62 71 ⊖ (+81 3) 35 92 13 68
Technical Research and Information Department Manager: A Torakani [98]

65-10500 Japan Machinery Exporters' Association (JMEA)

Kikai Shinko Building 5-8, Shibakoen, 3-chome, Minato-ku, Tokyo 105
☎ (+81 3) 34 31 98 08 ⊠ 24744 machinex
⊖ (+81 3) 34 36 64 55
👥 80
Executive Managing Director: Mitsuo Watanabe
Managing Director: Toru Hara
Managing Director: Masahiro Imoto
Chairman Aircraft Committee: Nobuhiko Tatebe
Leader Industry Co-ordination Group: Masakazu Io [98]

65-19000 The Society of Japanese Aerospace Companies Inc (SJAC)

Tokyo-Tameike Building 2F,
1-14 Akasaka 1-chome Minato-leu, Tokyo 107
☎ (+81 3) 35 85 05 11 ⊖ (+81 3) 35 85 05 41
Chairman and Chief Executive Director: Dr Hiroshi Ohba (Kawasaki Heavy Industries Inc)
Vice Chairman: Toshifumi Takei (Ishikawajima-Harima Heavy) (Industries Co Ltd)
Vice Chairman: Kikuo Fujiwara (Shimadzu Coporation)
Vice Chairman: Takashi Kitaoka (Mitsibushi Electric) (Corporation)
President: Toshihiko Nakamura
Senior Vice President: Sanshiro Hosaka
Senior Vice President: Hidejiro Yamada
Vice President: Yoshio Suzuki
Vice President: Kazuyuki Hirose
Vice President: Sadao Suzuki
Vice President International Affairs: Miki Suqiura

■ Representative body of the Japanese Aerospace industry.

Members:
APC Aerospeciality Inc, Aero Asahi Corporation, All Nippon Airways Co. Ltd, ANA Aircraft Maintenance Co. Ltd, Asahi Glass Co. Ltd, Bridgestone Corporation, Chiyoda Corporation, CIBA-GEIGY (Japan) Ltd, Commercial Airplane Company, Daicel Chemical Industries Ltd, Daido Steel Co. Ltd, Daikin Industries Ltd, Eagle Industry Co. Ltd, Fuji Bellows Co. Ltd, Fuji Electric Co. Ltd, Fuji Heavy Industries Ltd, Fujikin International Inc, Fujiwara Co. Ltd,
The Furukawa Battery Co. Ltd, The Furukawa Electric Co. Ltd, Furuno Electric Co. Ltd, Hakko Electric Machine Works Co. Ltd, Hirata Technical Co. Ltd, Hitachi Ltd, Hitachi Cable Ltd, Hitachi Metals Ltd, Hoden Seimitsu Kako Kenkyusho Co. Ltd, Honda Motor Co. Ltd, Ishikawajima-Harima Heavy Industries Co. Ltd, Ishikawa Seisakusho Ltd, Iwatani International Corp, Jamco Corp, Japan Aircraft Development Corp, Japan Aircraft Manufacturing Co. Ltd, Japan Air Lines Co. Ltd, Japan Air System Co. Ltd, Japan Aviation Electronics Industry Ltd, Japanese Aero Engines Corp., Japan Radio Co. Ltd, The Japan Steel Works Co. Ltd, Kaiyo Denshi Kogyo K K (KDK), Kanto Aircraft Instrument Co. Ltd, Kawada Industries Inc, Kawanishi Aero-Parts Products Co. Ltd, Kawasaki Heavy Industries Ltd, Kayaba Industries Co. Ltd, Kitamura Machinery Co. Ltd, Kobe Steel Ltd, Koito Industries Ltd, Koito Manufacturing Co. Ltd, Kokusai Electric Co. Ltd, Komatsu Zenoah Co., Koyo Seiko Co. Ltd, Machida Endoscope Co. Ltd, Matsushita Electric Industrial Co. Ltd, Matsuura Machinery Corp, Minebea Co. Ltd, Mitsubishi Cable Industries Ltd, Mitsubishi Chemical Corp, Mitsubishi Electric Corp, Mitsubishi Heavy Industries Ltd, Mitsubishi Materials Corp., Mitsubishi Precision Co. Ltd, Mitsubishi Rayon Co. Ltd, Mitsubishi Space Software Co. Ltd, Mitsubishi Steel Manufacturing Co. Ltd, Mitsui Toatsu Chemicals Inc, NAC Incorporated, Nachi-Fujikoshi Corp, Nagoya Screw Manufacturing Co. Ltd, NEC Aerospace Systems Ltd, NEC Corporation, NEC Radio Equipment Ltd, NGK Spark Plug Co. Ltd, Nihon Tokushu Toryo Co. Ltd, Nihon Pall Ltd, Nikkiso Co. Ltd, Nippon Avionics Co. Ltd, Nippondenso Co. Ltd, Nippon Paint Co. Ltd, Nippon Steel Corporation, NOF Corporation, Nissan Motor Co. Ltd, NKK Corporation, NSK Ltd, NTN Corporation, Obayashi Corp., OKI Electric Industry Co. Ltd, Omori Seikoki Co. Ltd, OTA Keiki Seisakusho Co. Ltd, Owari Precise Products Co. Ltd, Pacific Hydrogen Co. Ltd, Sakura Rubber Co. Ltd, Seiritsu Industries Co. Ltd, Shimadzu Corporation, Shimizu Corporation, Shinko Electric Co. Ltd, Shinmaywa Industry Co. Ltd, Shounan Seiki Co. Ltd, Showa Aircraft Industry Co. Ltd, Sogo Spring Mfg Co. Ltd, Sony Corporation, STS Corporation, Sumitomo Bakelite Co. Ltd, Sumitomo Electric Industries Ltd, Sumitomo Heavy Industries Ltd, Sumitomo Light Metal Industries Ltd, Sumitomo 3M Ltd, Sumitomo Metal Industries Ltd, Sumitomo Precision Products Co. Ltd, Taisei Corporation, Takenaka Corp, Tamagawa Seiki Co. Ltd, Tatsumi Manufacturing Co. Ltd, Teijin Seiki Co. Ltd, Terauchi Mfg Co. Ltd, Toho Cytec Co. Ltd, Toho Rayon Co. Ltd, Tokimec Inc, Tokyo Aircraft Instrument Co. Ltd, Toshiba Corporation, Toshiba Electronic Systems Co. Ltd, Toyo Communication Equipment Co. Ltd, Yokogawa Denshikiki Co. Ltd, Yokogawa Electric Corporation, The Yokohama Rubber Co. Ltd, Yoshimitsu Industries Inc.

Associate Members:
Asashi Air Supply Inc, Fuji Industries Co. Ltd, High-Reliability Components Corporation, Hoogovens Aluminium Japan Ltd, International Aircraft Development Fund, International Task Force Ltd, The Ishida Corporation, Japan Fleet Service Co. Ltd, Japan Manned Space Systems Corporation, JGC Corporation, Jupitor Corporation, Kanematsu Corporation, Kyokuto Boeki Kaisha Ltd, LTCB Systems Co. Ltd, Mainami Boeki Co. Ltd, Marubeni Corporation, Marubun Corporation, Meitec Corporation, Mikuni Shoko Corporation, Mitsubishi Corporation, Mitsubishi Research Institute Inc, Mitsui & Co. Ltd, Morimura Bros. Inc, National Tsusho Co. Ltd, Nichimen Corporation, Nippon Aircraft Supply Co. Ltd, Nissho Iwai Aerospace Corporation, Nissho Iwai Corporation, Nozaki & Co. Ltd, Okura & Co. Ltd, Rocket System Corporation, Shintoa Corporation, Space Engineering Development Co. Ltd, Sumitomo Corporation, Teledyne Japan K K, The Tokyo Marine and Fire Insurance Co. Ltd, Tomen Corporation, Toyo Engineering Corporation, Yamada Corporation. 98

70 Financial Holding Groups, Parent Companies

70-02000 Mitsubishi Corporation
6-3 Marunouchi 2-Chome, Chiyoda-ku, Tokyo 100-86
☎(+81 3) 32 10 46 04
☎(+81 3) 32 10 46 04; 32 10 45 77 ✉MITSUBISHI JT
▣ 333337 nc tak ⊖ (+81 3) 32 10 47 83
♤ Yen 126,476,216,390
♣ Yen 17,794,929,000,000 (1995/96) 👥 13,100
Chairman: Shinroku Morohashi
President: Minoru Makihara
Managing Director: Hironori Aihara
General Manager Information Systems Services Division C: T Ariga
General Manager Japan F-2 Program Unit: Y Kaito
General Manager Aircraft and Engines Unit: T Matszuzawa
General Manager Electronic Systems Unit: M Nagase
General Manager New Business Initiative Unit: T Sato (acting)
General Manager Commercial Aviation Product Unit: T Sasaki (acting)
Assistant General Manager Space Systems Unit: K Ochi
Assistant General Manager Spacecraft/Information Systems Unit: A Yamaura
Assistant General Manager Helicopter Unit: T Fukishima
Assistant General Manager Commercial Aircraft Leasing Unit: K Arai
Assistant General Manager Airport and Project Unit: R Takeuchi
Assistant General Manager Information Systems and Services Division C: K Kaida
■ Commercial and military helicopters: F-2, military aircraft, aircraft engines, guided missiles, defence electronics, satellites and space systems. Airport related projects, aircraft interior equipment including galley, lavatories and seats. Aircraft finance and leasing.
® Canada: Indal Technologies, Pratt & Whitney Canada, Spar Aerospace
France: Thomson
Germany: Daimler-Benz Aerospace
Sweden: Fairchild Scandinavian Bellyloading
USA: Aerojet, AlliedSignal, BE Aerospace, Honeywell, Hughes, Lockheed Martin, McDonnell Douglas, Menasco, Northrop Grumman, Raytheon, Rockwell International, Science Applications International, Space Systems/Loral, Spacehab, TRW Components International, United Technologies

Commercial Aviation Dept
6-3 Marunouchi 2 chome, Chiyoda-ku, Tokyo 10086
☎(+81 3) 32 10 48 11 ✉MITSUBISHI JK
⊖ (+81 3) 32 10 47 92; 32 10 47 93
General Mgr Commercial Aviation Dept: Kenichiro Yoshiyama
Deputy General Mgr: Tadashi Sasaki
Mgr Project Development Team: Ryoichi Takeuchi
Mgr Aircraft Leasing Team: Kyoichi Arai
■ Aircraft interior equipment, galley, lavatory, seats. Aircraft leasing & finance.
® USA: AlliedSignal, BE Aerospace 98

70-05000 Toshiba Corp
1-1 Shibaura 1-chome, Minato-ku, Tokyo 105-01
☎(+81 3) 34 57 32 45; 34 57 21 05 (corporate communications) ✉ TOSHIBA
W http://www.toshiba.co.jp ▣ 22587 toshiba j
⊖ (+81 3) 34 56 16 31; 34 56 16 32; 34 56 16 33; 34 56 47 76 (corporate communications)
♤ Yen 90,000,000,000
♣ Yen 3,713,000,000,000 (defence 1995)
👥 71,170 (defence business 1996)
Chairman of the Board: Fumio Sato
President and Chief Executive Officer: Taizo Nishimuro
Senior Executive Vice President: Atsumi Uchiyama
Senior Executive Vice President: Hideharu Egawa
Senior Executive Vice President: Masaichi Koga
Senior Executive Vice President: Tetsuya Yamamoto
Senior Executive Vice President: Masanobu Ohyama
Senior Executive Vice President: Takeshi Okatomi
Executive Vice President: Kanichi Ito
Executive Vice President: Susumu Abe
Executive Vice President: Kensuke Fujimatsu
Executive Vice President: Isamu Nitta
Senior Vice President: Terunori Aiga
Senior Vice President: Naohisa Shimomura
Senior Vice President: Tetsuo Machii
Senior Vice President: Tomohiko Sasaki
Senior Vice President: Akinobu Kasami
Senior Vice President: Tadashi Okamura
Senior Vice President: Kozo Wada
Senior Vice President: Mamoru Kitamura
Senior Vice President: Kiyoaki Shimagami
Senior Vice President: Kenichi Mori
Senior Manager Marketing - New Business Electronic Systems Division: Jun Ichinose
☎(+81 3) 34 57 30 61; 34 56 03 59 (fax)
Corporate Communications Office: Ms Naomi Furuyu
E naomi.furuyu@toshiba.co.jp
■ Information and communication systems, electronic devices, medical equipment, power plant systems, transportation equipment, consumer products. Close range SAM (SAM-2, Keiko SAM, SAM-3) for air-to-air and shipborne applications. Short range SAM (SAM-1 TAN SAM, SAM 1-C improved TAN SAM), base air defence. Medium range SAM, hawk and main firing radar. Long range SAM, PATRIOT, module, power plant, radar set and transmitter. Avionics for P-3C, HRN 107 ILS receiver, HSN-3 INS, AN/APS-115C search radar, AN/ASQ-212(J) data processing system. Avionics for SH-60J, HRQ-109 VOR receiver, HYQ-1 HCDS (Helicopter Combat Direction System). Avionics for fighter aircraft F-15 and F-2, AN/ASN-109 INS, lead computing gyro unit, map generator. Avionics for C-1 cargo aircraft, J/AYN-1 map navigation display system, J/APN-93 navigation weather radar. Avionics for AH-1S attack helicopter, FCC Fire Control Computer, JARN-A10 VOR/ILS receiver. Avionics for U-125 rescue aircraft, J/APQ-2 search radar. Avionics for E-2C and E-767 AWACS. Field artillery systems including FADS Field Artillery Data-Processing System, JMPQ-P13 mortar locating radar, JTPS-P16 artillery locating radar. Navigation support systems, GCA Ground Controlled Approach, RIPS Radar Information Processing Systems, ILS Instrument Landing System, relocatable control tower. Weather radars: J/FPH-6(); J/FPH-8() and LPN-10, NLPN-8(). Advanced bridge systems.

Subsidiaries with defence business:
Toshiba Electronic Systems Co. Ltd → 75
Toshiba Logistics Support Corporation

Electronics Systems Division
Head Office
72 Horikawa-cho, Saiwai-ku, Kawasaki-shi, Kawasaki 210
☎(+81 44) 549 30 00
General Manager Electronics Systems Division: Takao Inoue
International Procurement Group: Takashi Yamachika
Assistant Group Executive Electronics and Telecommunications.: Yoshiki Okamoto

Komukai Works
1 Komukai Toshiba-cho, Saiwai-ku, Kawasakie Pref, Kanagawa 210
☎(+81 44) 548 50 03 ⊖ (+81 44) 541 11 25
General Manager: Takao Inoue 98

74 Consortia, Joint Venture Companies

74-00600 Komatsu-Howmet Ltd
(joint venture Komatsu Ltd & Howmet Corp)
1-1 3-chome Ueno, Hirakata-shi, Osaka-fu 573
☎(+81 6) 720 49 31 23-5 ⊖ (+81 6) 720 49 41 28
Technology Director: Roger Hambleton 98

75 Manufacturing Companies

75-00100 Actus Corporation
Bicks-Shinjyuku Building, 2-19-1 Shinjyuku, Shinjyuku-ku 160
☎(+81 3) 52 69 32 01 ▣ 34570 actusgal j
⊖ (+81 3) 52 69 32 00
👥 221 98

75-00385 Daewoo Corp Tokyo Branch
404 Toranomon Mitsui Building 8-1, 3-chome Kasumigaseki, Chiyoda-ku, Tokyo 100
☎(+81 3) 35 97 78 79; 35 97 78 81; 35 97 78 82
⊖ (+81 3) 35 97 90 23 98

75-00390 Daewoo Fukoaka Office
(a subsidiary of Daewoo Corporation, South Korea)
AIT Building 6th Floor, Momochihama 2-2-22, Sawaraku, Fukoaka 814
☎(+81 92) 852 08 00 ⊖ (+81 92) 852 08 01 98

75-00700 Bridgestone Corp Head Office
10-1 Kyobashi 1-Chome, Chuo-ku, Tokyo 104
☎(+81 3) 35 63 65 77; 35 63 65 41 ✉BSTIRE Tokyo
▣ 22217 ⊖ (+81 3) 35 63 65 77; 36 63 68 44
♤ Yen 88,032,000,000 (1995)
♣ Yen 1,686,636,000,000 (1995) 👥 89,418
President: Yoichiro Kaizaki
Executive Vice President: Tadakazu Harada
Executive Vice President: Masatoshi Ono
Executive Vice President: Katsuyoshi Shibata
Executive Vice President International Operations: T Uchiyama
Sales and Marketing Director: K Nomura
General Manager Aircraft Tyre Sales Department: K Iwasa
■ Aircraft tyres and retreading. Portable scales for measurement of total weight and weight distribution of aircraft. 98

75-00860 Krautkramer Japan Co Ltd
(joint venture with Krautkramer Branson, USA)
5-5 Sakuragawa 1-chome, Itabashi-ku, Tokyo 174
☎(+81 3) 39 37 96 54 ▣ 2423507
⊖ (+81 3) 39 37 94 54
♤ Yen 44,550,000 (1997)
♣ Yen 3,300,000,000 (1996) 👥 75 (1997) 🏭 4,357 m2
Chief, Eastern Sales Branch: Osamu Fukada 98

75-00950 Daicel Chemical Industries Ltd
Tokyo Head Office
8-1, Kasumigaseki 3-chome, Chiyoda-ku, Tokyo 100
☎(+81 3) 35 07 32 32 ▣ 2224632 daicel j
⊖ (+81 3) 35 07 31 15
♤ Yen 36 267 000 000 (1995)
♣ Yen 160 237 000 000 (1995) 👥 3 224
President & CEO: Akiro Kojima
Sr Managing Dir: Kikuo Aoki
Sr Managing Dir: Manabu Kai
Managing Dir Int'l Div.: Juro Suzuki
Managing Dir Business Group: Shunichi Kimura
Sr General Mgr Purchasing Group: Masaaki Ichino
Sr General Mgr Planning & Development Group: Keiichi Kayano
Sr General Mgr Technical & Engineering Group: Tamotsu Nakagawa
Sr General Mgr Cellulose Div.: Kunio Kanai
Sr General Mgr Plastics & Resins Div.: Toshikazu Morita
Public Relations: Takashi Matsumoto
■ Manufacture & sale of cellulosic derivatives (acetate tow, cellulose acetate), organic chemicals (acetic acid & its derivates, sorbic acid), plastics & films (ABS resin & its alloy, PS resin, PVDC coated OPP film), functional products (inflator for airbag, chiral column) & propulsion systems (pilot emergency-escape systems).

Overseas subsidiaries:
Daicel (USA) Inc → USA 90
Daicel (Europa) → Germany 90
Daicel Chemical (Asia) → Singapore 90
Daicel Kisco Co → Taiwan 90
Dynepco → USA 90

Sakai Office
1 Teppo-cho, Sakai, Osaka 590
☎(+81 722) 27 31 11 ▣ 5374213 daicel j
⊖ (+81 722) 27 30 00

Osaka Office
3-13 Azuchi-machi 2 chome, Chuo-ku, Osaka 541
☎ (+81 6) 263 48 11 ⊠ 5222987 daicel j
⊖ (+81 6) 263 47 11 [97]

75-00975 Sony Corporation
6-7-35 Kitashinagawa, Shinagawa-ku, Tokyo 141
☎ (+81 3) 54 48 22 00 ⊖ (+81 3) 54 48 30 61
♣ US$ 45,670,000,000,000 (1997) 👥 22,000 (1996)
Chairman and Chief Executive Officer: Norio Ohga
President and Chief Operating Officer: Nobuyuki Idei
Managing Director: Koji Tamiyo
International Public Relations: Gene Huh
E g-huh@ccom.sony.co.jp
■ Manufacture of electronics products including video and audio equipment, televisions, displays, semiconductors, electronic components, computers and computer peripherals, telecommunications equipment.
● 6-7-35, Kitashinagawa, Shinagawa-ku, Tokyo 141
☎ (+81 3) 54 48 21 11 ⊖ (+81 3) 54 48 22 44 [98]

75-01050 Fuji Electric Co Ltd
12-1 Yurakucho, 1-chome, Chiyoda-ku, Tokyo 100
☎ (+81 3) 32 11 71 11 W http://www.fujieletric.co.jp
⊠ 22331 fujielea j ⊖ (+81 3) 53 88 78 04
♧ Yen 47,600,000,000
♣ Yen 560,000,000,000 (1996) 👥 13,200 (1996)
President: Yoshihiko Nakazato
Director Aerospace: Shigeyuki Kobayashi
General Manager: Teruhisa Shiozaki
Manager, Defence Systems Department: Fujita Masanari
■ Electric machinery, appliances, measuring instruments, semi-conductors, power supplies inverters, cryogenic coolers for infra-red sensors. [98]

75-01100 Fuji Heavy Industries Ltd (FHI)
Fuji Jukogyo KK
7-2 Nishishinjuku 1-chome, Shinjuku-ku, Tokyo 160
☎ (+81 3) 33 47 25 25 ⊠ 2322268 fuji j
⊖ (+81 3) 33 47 25 88
♧ Yen 83,617,000,000
♣ Yen 824,734,000,000 (1996) 👥 14,763
🏠 1,394,860 m2
President: Takeshi Tanaka
Senior Managing Director and General Manager Aerospace Division: Yasuyuki Kogure
Director, Deputy General Manager Aerospace Division: Satoshi Idei
General Manager Commercial Programs: Kisaburo Wani
General Manager Marketing and Sales, Defence Programs: Shunji Notake
General Manager Administration Department: Takehiko Sarukawa

Aerospace Division
7-2 Nishishinjuku 1-chome, Shinjuku-ku, Tokyo 160
☎ (+81 3) 33 47 25 25 ✉ FUJIHEAVY Tokyo
⊠ 2322268 fuji j ⊖ (+81 3) 33 47 25 88
Senior Managing Director and General Manager Aerospace Division: Yasuyuki Kogure
Director, Deputy General Manager Aerospace Division: Satoshi Idei
General Manager Marketing and Sales Defence Programs: Shunji Notake
General Manager Commercial Business: Kisaburo Wani
General Manager Administration Department: Takehiko Sarukawa
Manager, Administration Department: Kazuhiro Oka
E okak@sb.hq.subaru-fhi.co. jp

Aircraft Plant
1-11, Yonan 1-chome, Utsunomiya-shi, Tochigi-Ken 320
☎ (+81 28) 658 11 14 ⊖ (+81 28) 659 48 50
General Manager Aircraft Plant: Satoshi Idei
General Manager Research and Laboratory Department: H Kikukawa
General Manager Space Technology: K Tanaka
General Manager Rotary Wing Aircraft Engineering Department: T Miyazaki
General Manager Fixed Wing Aircraft Engineering Department: T Hoshi
General Manager Unmanned Air Vehicles Engineering Department: T Sato
General Manager Electronics Engineering Department: H Kikukawa
■ Production of T-5 trainer, Bell AH-1S helicopter, Bell and utility UH-1J 205B helicopter, Ryan BQM-34AJ, J/AQM-1 target drone MAST GROUP (Patriot), Boeing 737 elevators, Boeing 747 ailerons and spoilers, Boeing 757 outboard flaps, Boeing 767 wing/body fairings and MLG doors, Boeing 777 centre wing and MLG doors, MD-11 outboard ailerons, F-15 airframe components, P-3C wing assemblies, SSAM launcher, SSM canister launcher. [98]

75-01200 Fujisawa Manufacturing Division
1-1, 1 Chome, Katase, Fujisawa City, Kanagawa-Pref 251
☎ (+81 466) 23 21 31 ⊖ (+81 466) 27 26 24
👥 476
Factory Manager: Tetsuo Takahashi [98]

75-01400 Fujitsu Ltd
1015 Kamikodanaka, Nakahara-ku, Kawasaki 211
☎ (+81 3) 32 16 32 11 ✉ FUJITSULIMITED TOKYO
⊖ (+81 3) 32 16 93 52 ⊖ (+81 3) 754 25 90
♧ Yen 223 600 000 000 ♣ Yen 3 257 700 000 000
👥 164 364
Chairman: Takuma Yamamoto
President: Tadashi Sekizawa
Public Relations: Isao Hirano ☎ (+81 3) 32 15 52 36
☎ (+81 3) 32 16 93 65 (fax)
■ Radiocommunications equipment, computers, radar equipment, infra-red night vision equipment & communication systems & equipment.
● 1015 Kamikodanaka, Nakaharaku Kawasakishi, Kanagawa 211
☎ (+81 44) 777 11 11
General Mgr: Hisashi Yajima
● 460 Koyama, Suzaka-Shi, Nagano 382
☎ (+81 262) 45 33 33
General Mgr: Yoshiaki Taniuchi
● 28-1 Jyoto 3, Oyama-Shi, Tochigi 323
☎ (+81 285) 24 22 22
General Mgr: Toshihiko Kawai
● 36 Kitaowaribe Nagano-Shi, Nagano 380
☎ (+81 262) 43 11 21
General Mgr: Kohichi Saito
● 64 Tsukamoto Nishiwaki, Okubo-Cho Akashi-Shi, Hyogo 674
☎ (+81 78) 936 11 21
General Mgr: Masaki Ogana
● 1405 Ohmaru Inagi-Shi, Tokyo 206
☎ (+81 423) 77 41 11
General Mgr: Kenji Akasaka
● 1224 Nakanara Kumagaya-Shi, Saitama 360
☎ (+81 485) 23 11 21
General Mgr: Ehichi Yoshida
● 6-1 Satsuki-Machi, Kanuma-Shi, Tochigi 322
☎ (+81 289) 76 21 21
General Mgr: Kazur Norimoto
● 110 Hachiman Ikki-Machi, Aizu Wakamastsu-Shi, Fukushima 965
☎ (+81 242) 24 21 21
General Mgr: Shanji Kumakula
● 4-2 Moriyama, Nishine, Kanegasaki, Sawa-gun, Iwate 029-45
☎ (+81 197) 44 53 11
General Mgr: Izumi Tanaka
● 3 Kogyodanchi, Monden Machi, Aizu Wakamatsu-Shi, Fukushima 965
☎ (+81 242) 28 61 11
General Mgr: Shunji Kumakusa
● 1388 Shimoishigami, Othawara-Shi, Tochigi 329-26
☎ (+81 287) 29 21 11
General Mgr: Katsura Yamada
● 904 Nobecho, Tatebayashi-Shi, Aunma 370-07
☎ (+81 276) 73 61 11
General Mgr: Ehichi Yoshida
● 140 Miyamoto, Numazu-Shi, Shizuoka 410-03
☎ (+81 559) 23 22 22
General Mgr: Shizuo Abe
● 1500 Mizono, Tado-Cho, Kuwana-gun, Mie 511-01
☎ (+81 594) 48 55 11
General Mgr: Kunihiko Wada [97]

75-01820 Hanshin Diesel Works Ltd
Room N°601, Maronouchi-Building,
Marunouchi, Chiyoda-ku, Tokyo
☎ (+81 3) 32 16 36 01 ⊖ (+81 3) 32 01 76 75
Manager, Overseas Business Section: S Iseda [98]

75-02000 Hitachi Ltd Headquarters
4-6 Kanda Surugadai, Chiyoda-ku, Tokyo 101-10
☎ (+81 3) 32 58 11 11 W http://www.hitachi.co.jp
⊖ (+81 3) 32 58 52 45
👥 330,152 (1995)
President: Tsutomu Kanai
■ Communications equipment, avionics, training and simulation equipment. Space systems and equipment. [98]

75-02390 Hitachi Zosen Co. Ltd Head Office
3-28 Nishikujo 5-chome, Konohana-ku, Osaka 554
☎ (+81 6) 466 75 00 ⊖ (+81 6) 466 75 72
♧ Yen 50,289,000,000 (1996)
♣ Yen 597,856,000,000 (1996) 👥 4,393 (1996)
Chairman and Chief Executive Officer: Yoshihiro Fujii
President: Isoh Minami
Executive Vice President: Kiyoshi Tsuboi
Executive Vice President: Toshihiko Tanikoshi
Executive Vice President: Shogo Furuta
Senior Managing Director: Isao Toda
Senior Managing Director: Tadamaro Fujiwara
Managing Director: Susumu Hirota
Managing Director: Kenzo Hamabata
Managing Director: Takashi Satani
Managing Director: Tsutomu Yokoi
Managing Director: Masakatsu Yoshimura
Managing Director: Akira Ikenaga
Managing Director: Hiroyuki Yamada
■ Activities in environmental technologies, plants, shipbuilding and defence, steel structures, construction machinery, electronics and control systems and$ biotechnical engineering. Main product lines, ships and offshore structures: shipbuilding; ship repairs and conversions; passenger ships; merchant ships; high-speed boats; marine machinery; pleasure boats; FRP boats.
Offshore structures: FPSO's; oil drilling rigs; oil production platforms; plant barges; harbour facilities; undersea equipment; offshore work equipment; oil storage bases.
Naval ships and defence equipment: construction and repair of naval ships; patrol boats; defence equipment.

Overseas Offices and Subsidiaries:
Europe
Hitachi Zosen Corporation, Düsseldorf Office
Hitachi Zosen Europe Ltd (London and Gloucester)
USA
Hitachi Zosen USA Ltd (New York and Chicago)
Asia
Hitachi Zosen Corporation, Beijing Office
Hitachi Zosen Corporation, Shanghai Office
Hitachi Zosen Corporation, Jakarta Office
Hitachi Zosen Corporation, Bangkok Office
Hitachi Zosen Corporation, Taipei Office
Hictahi Zosen Corporation, Seoul Office
Hitachi Zosen Corporation, Ho Chi Minh City Office
Hitachi Zosen Company (HK) Limited, Hong Kong
Hitachi Zosen Engineering Singapore (Pte) Limited
Hitachi Zosen Singapore Limited
Dalian Dali Steel Works Co Ltd

Tokyo Office Ship and Naval Headquarters
7th Floor Palaceside Building, 1-1 Hitotsubashi 1-chome, Chiyoda-ku, Tokyo 100
☎ (+81 3) 32 17 84 18 ✉ SHIPYARD
⊖ (+81 3) 32 17 85 34
♧ Yen 503,000,000,000 (1996) 👥 12,443 (1996)
Assistant Chief: Tetsuro Okumura [98]

75-02550 Howa Machinery Ltd
Shinkawa-cho, Nishikasugai-Gun, Aichi 452
☎ (+81 52) 408 12 31 ⊠ 04439931
⊖ (+81 52) 409 48 55
♧ Yen 9 019 722 882 (1995) ♣ Yen 35 412 349 000
👥 2 228 (1995)
Dir & General Mgr: M. Sagoh
■ Manufacture firearms, construction machinery, cast iron & steel, machine tools & textile machinery. [97]

75-02800 Ishikawajima-Harima Heavy Industries Co. Ltd (IHI)
Shin-Ohtemachi Bldg 2-1,
Ohtemachi 2-chome, Chiyoda-ku, Tokyo 100
☎ (+81 3) 32 44 53 33 ✉ IHICO Tokyo ⊠ 22232 ihihqt j
⊖ (+81 3) 32 44 53 98
♧ Yen 64 924 750 000 (1995)
♣ Yen 1 023 200 000 000 (1995) 👥 15 900
Chairman & CEO: Kosaku Inaba
President & COO: Toshifumi Takei
Exec. Vice President: Seiichi Inoue

Contact Reference

Exec. Vice President: Takao Mikoshiba
Exec. Vice President: Fujio Ito
Exec. Vice President: Shozo Ojimi
Sr Managing Dir: Ukichi Kishino
Sr Managing Dir: Takeshi Yana
Public Relations Dept: F. Usami

Aero Engine & Space Ops
Shin-Ohtemachi Bldg 2-1, 2-Chome, Chiyoda-ku, Tokyo 100
☎ (+81 3) 32 44 53 33 ⊜ (+81 3) 32 44 53 98
♣ Yen 133 200 000 000 (1995) 👥 3 550
Operations President: Tohru Ishikawa
Deputy General Mgr: Motosugu Ito
Deputy General Mgr: Takuji Miyo
General Mgr Planning & Control Dept: K. Ichida

Shipbuilding & Offshore
Tokyo Chuo Bldg, 6-2 Marunouchi, 1-chome, Chiyoda-ku, Tokyo 100
☎ (+81 3) 3286 22 68; 3286 23 22 ⊠ 24104 inisen j
⊜ (+81 3) 3286 24 37
General Mgr: M. Miyazaki
Mgr Naval Div.: S. Morioka
■ Tokyo Shipyard: missile destroyers for the Japanese Maritime Defense Force. Yokohama Shipyard: a weather ship for the Japanese Government & navaid tenders for the Japanese Maritime Safety Agency.

Technical Development
2-16 Toyosu 3-chome, Koto-ku, Tokyo 135
☎ (+81 3) 35 34 22 51 ⊠ 23507 ⊜ (+81 3) 35 34 22 07
General Mgr: Kazumichi Motozuna
General Mgr Advisory Staff Group: K. Abiko
General Mgr, Ordnance Engineering Dept: T. Nakamura

Space Development Division
Tokyo Chuo Bldg, 6-2 Marunouchi, 1-Chome, Chiyoda-ku, Tokyo 100
☎ (+81 3) 32 44 57 70 ⊜ (+81 3) 32 44 53 98
Division Dir: Y. Kitamura
General Mgr Sales & Marketing: T. Shioda
General Mgr Space Experiment Project: Y. Ano
General Mgr Space Station Project: K. Inoue
■ Manufacture liquid propellant engines, attitude control systems, ground test facilities & components.

Land & Marine Gas Turbine Division
Tokyo Chuo Bldg, 6-2 Marunouchi, 1-Chome, Chiyoda-Ku, Tokyo 100
☎ (+81 3) 32 86 22 25 ⊜ (+81 3) 32 86 23 86
Division Dir: T. Ikeda
General Mgr Sales & Marketing: I. Hayami
General Mgr Engineering: H. Sugimoto
Product Support Dept: Y. Satoh
■ Manufacture aero-derivative type gas turbines & small gas turbines. [97]

75-02820 Ishikawajima Noise Control Co. Ltd (INC)

(a subsidiary of Ishikawajima-Harima Heavy Industries Co Ltd)
15-18 Hyakunin-cho 1-chome, Shinjuku-ku, Tokyo 169
☎ (+81 3) 33 60 32 22 ⊜ (+81 3) 33 60 12 90
♧ US$ 2,500,000 (1997) ♣ US$ 50,000,000 (1997)
👥 100 (1997)
Managing Director: Hiroshi Kobayashi
General Manager: Yoshiaki Koubuchi
Sales and Marketing Director: K Matsumoto
Operations and Technical Director: H Tanisue
Purchasing Manager: F Kikuchi
■ Products: engine test cell, engine test cell for turbo shaft engine, jet engine with afterburner, large fan engine, full turnkey bases, plans to completion. Aircraft ground run-up noise suppressor, fence and duct type noise suppressor for B747, DC-10, MD-11. Hush house noise suppressor for KC-135 and fighter aircraft. Consulting services for airport noise control. [98]

75-02850 ITT Cannon Japan

5362-1 5-Chome, Hibarigaoka, Zama-City, Kanagawa 228
☎ (+81 462) 57 20 10 ⊜ (+81 462) 57 16 80
♧ US$ 1,800,000 ♣ US$ 42,000,000 👥 170
Sales Manager: Naotake Kuribayashi [98]

75-03200 Japan Aircraft Manufacturing Co. Ltd (NIPPI)

3175 Showa-machi - Kanazawa-ku, Yokohama 236
☎ (+81 45) 773 51 18 ✉ NIPPI Yokohama
⊜ (+81 45) 771 12 53
♧ Yen 6,023,023,315 (1996)
♣ Yen 32,111,000,000 (1996) 👥 1,611 (1996)
🏭 117,627 m2
President: Kanji Sonoda
Managing Director and General Manager Sales Division: Motonobu Akiyama
General Manager, International Projects Department: Masahiro Kaneshige
■ Airframe components, transmission parts, target systems, space equipment, electronic devices. → 80

Atsugi Plant
28 Soyagi 2-chome, Yamato, Kanagawa 242
☎ (+81 462) 65 20 12 ⊜ (+81 462) 65 20 48
Director and Plant Manager: Yoshimi Hanazono

Yokohama Plant
3175 Showa-machi - Kanazawa-ku, Yokohama 236
☎ (+81 45) 773 51 20 ✉ NIPPI Yokohama
⊜ (+81 45) 771 18 07
Director and Plant Manager: Mitsugi Kiyota
■ Manufacture and repair of aircraft components and parts. Manufacture of FRP fan systems, industrial machines, test equipment and light alloy products. [98]

75-03300 Japan Aviation Electronics Industry Ltd (JAE)

21-2 Dogenzaka 1-chome, Shibuya-Ku, Tokyo 150
☎ (+81 3) 37 80 29 32 ⊜ (+81 3) 37 80 29 45
♧ Yen 10,639,000,000 (1996) 👥 2,400 (1996)
President: H Nagatoshi
Director International Marketing Department: Takeshi Morishita
Marketing Manager: Okada Masafumi
■ Inertial navigation & guidance systems, automatic flight control systems, avionics. Connectors, switches & relays, fibre optic products & electronic equipment for aerospace & related applications. [98]

75-04350 Kawasaki Heavy Industries Ltd

Nissei-Kawasaki Building, 1-18 Nakamachidor1 2-chome, Kobe 650-91
☎ (+81 78) 341 77 31

Tokyo Head Office (KHI)
Kawasaki Jukogyo
2-4-1 Hamamatsu-Cho, World Trade Center Building, Tokyo 105
☎ (+81 3) 34 35 21 11 Ⓦ http://www.khi.co.jp/
⊠ 2424371 kawaju j ⊜ (+81 3) 34 36 30 37
♧ Yen 72,466,000,000 (1996)
♣ Yen 948,972,000,000 (1995) 👥 16,343
Chairman: Hiroshi Ohba
President: Toshio Kame
Assistant Manager - Sales Administration Department, Aerospace Sales Division: Shingo Nakamura
☎ (+81 3) 34 35 24 85; 34 32 39 77 (fax)
■ Main products: aircraft and jet engines, spacecraft and equipment, ships, marine engineering, rolling stock, plant and factory automation systems, construction machinery, small engines, motorcycle and Jet Ski personal watercraft. Aerospace products: aircraft; fuselage of Boeing B777 and B767, fuselage panel of A321, wing rib of B737, outboard flap for B747, flap hinge fairing of MD80. T-4 medium jet trainer and T-4 aerobatics trainer (Blue Impulse), wing, tail and after fuselage of F-15, fuselage of XF-2/F-2, P-3C patrol plane, EP-3. Rotary Wing: BK117 (joint development with MBB (ECD), Germany), H500, KV-107, MD900/901 RSS, XOH-1 New Observation Helicopter, CH-47J, OH-6.
Space related activities: payload fairing for H-11 rocket, geological experiment satellite, engineering test satellite VII, H-11 orbiting plane (in design) and automatic landing flight experiment/hypersonic flight experiment, Japanese experiment module (part of space station Alpha), astronaut training facility.
Engines: T53 and T55-K-712 turboshaft engines, V2500, RB211/Trent, PW400 and auxiliary power units (international joint programmes), transmission for XOH-1, BK117 and MD900, Air-Turbo-Ramjet engine (under development).
Other products: construction of airport facilities for domestic and international airports, maintenance hangars, docks, passenger bridges, aircraft cleaning systems, engine run-up noise supressors, air cargo terminals, cargo storage and distribution systems. Hang-glider, motorcycle and missile simulators.
Overseas offices in Jakarta, Bangkok, Beijing, Manila, Sydney, Taipei, Shanghai, Seoul and Kuala Lumpur.

Subsidiaries:
KHI Europe Finance BV
Kawasaki Heavy Industries (HK) Ltd
Kawasaki Heavy Industries (Europe) BV
Kawasaki Heavy Industries (USA) Inc
Kawasaki do Brasil Indústria e Comércio Ltda
Kawasaki Heavy Industries (UK) Ltd
Kawasaki Heavy Industries GmbH
Kawasaki Industries (Singapore) Pte Ltd
KHI (Dalian) Computer Technology Co Ltd
Kawasaki Rail Car Inc
Kawasaki Precision Machinery (UK) Ltd
Kawasaki Motors Corp USA
Canadian Kawasaki Motors Inc
Kawasaki Motors (UK) Ltd
Kawasaki Motoren GmbH
Kawasaki Motors Pty Ltd
Kawasaki Motors Netherlands NV
Kawasaki Motors Manufacturing Corp USA
Kawasaki Loaders Manufacturing Corp USA
Kawasaki Robotics (USA) Inc
Kawasaki Loaders Inc

Kobe Head Office
Kobe Crystal Tower, 1-3 Higashikawasaki-cho 1-chome, Chuo-ku, Kobe 650-91
☎ (+81 78) 371 95 30 ⊜ (+81 78) 371 95 68

Gifu Works
1 Kawasaki-cho, Kagamihara, Gifu 504
☎ (+81 583) 82 57 12 ✉ KAWASAKI HEAVY Gifu
⊠ 4722368 ⊜ (+81 583) 82 54 04; 82 29 81
👥 3,705 🏭 700,000 m2
Associate Director and General Manager Gifu A/C Division: Mitsuaki Kawafi
■ Aircraft and space systems. Manufacture, repair and overhaul of equipment for commercial aircraft, anti-submarine aircraft, fighters, trainers, helicopters, missiles and space systems.
P-3C ASW patrol aircraft, F-15J fighter, Boeing B767 and B777 passenger aircraft, T-4 medium trainer, T-4 aerobatic aircraft, low-noise STOL experimental aircraft, C-1 medium transport aircraft, T-33A trainer, repair of E-2C early warning aircraft and C-130H transport. Manufacture of Kawasaki BK117 helicopter, CH-47J helicopter, Kawasaki Vertol 107-IIA helicopter, Kawasaki Hughes 369D helicopter. Type 64 anti-tank missile (ATM), type 79 anti-landing craft and anti-tank missile (H-ATM), type 87 anti-tank missile (M-ATM), H-II rocket launch, H-II rocket fairing, geodetic satellite (GS), machinery control and surveillance system for destroyer.

Akashi Works (Jet Engine Division)
1-1 Kawasaki-cho, Akashi-shi, Hyogo 673
☎ (+81 78) 921 13 01; 921 15 01; 921 16 11
✉ KAWASAKI HEAVY Akashi ⊠ 5628951
Director and General Manager - Jet Engine Division: Misao Yurigusa
■ Manufacture of motorcycles, industrial robots, general-purpose engines, jet engines for marine and aircraft applications, gas turbine engines. All-terrain vehicles (ATV's), Jet Ski personal watercraft, engine powered electric generators, transmissions. Transmission for helicopters.

Kobe Shipyards
1-1 Higashikawasaki-cho 3 chome, Chuo-ku, Kobe 650-91
☎ (+81 78) 671 50 01 ⊠ 5622115
⊜ (+81 78) 671 55 97
■ Equipment for ship and maritime applications: ultra high-speed passenger boats (Jetfoil), high speed car ferries (Jet Piercer), submarines, research ships, ship repairs and modifications.
Engines: steam turbines, gas turbines and diesel engines for ground and maritime applications. Large accelerators and decelerators, propellors, water-jet propulsors, air blowers, air compressors, natural gas force-feeding modules, wind tunnels and tunnel ventilation systems.
Factory automation devices: water-jet cutting systems (KAWACS), CO2 laser system, shaped steel fabrication (COM), vibrating machines, physical distribution systems.
Construction of Yuushio and improved Yuushio class submarines.

Nagoya Works 1
11 Kusunoki 3-chome, Yatomi-cho Amagun, Aichi-Pref 498
☎ (+81 567) 68 51 17
■ Manufacture and assembly of structural aircraft components.
Manufacture and assembly of aircraft and equipment for space exploration. Fuselage for Boeing B777.

Nagoya Works 2
Kanaoka N°7 Tobishima Village Amagun, Aichi-Pref 490-14
☎ (+81 5675) 508 00
■ Manufacture and assembly of structural aircraft components.
Assembly and construction of forward and mid fuselage for Boeing B767.

Noda Works
118 Futatsuzuka, Noda, Chiba 278
☎ (+81 471) 24 11 21 ⊜ (+81 471) 24 76 45
■ Steel structures: steel frames, airport facilities and equipment, steel pipe structures, steel towers, bridges, steel hydraulic pipes, steel water pipes, floodgates, LNG tanks, LPG tanks, high pressure gas containers, steel stacks, blast furnaces, alloy hydrogen storage equipment.

Sodegaura Works
50-1 Minaisode, Sodegaura, Chiba 299-02

☎(+81 438) 62 71 27 ⊜ (+81 438) 62 71 29
■ Steel structures: bridges, tank, conventional steel products.

Yachiyo Works
1780 Kamikoya, Yachiyo, Chiba 276
☎(+81 474) 83 11 11 ⊜ (+81 474) 82 32 27
■ Plants, crushers, breakers, sieves and driers. Special cast steel and iron, ceramics and other materials. Wear-resistant special cast-iron products, heat and wear resistant cast-iron products, precision machined cast-iron products.

Chiba Works
1 Niihama-cho, Chuo-ku, Chiba 260
☎(+81 43) 263 22 11 ⊜ (+81 43) 263 55 25

Hyogo Works
1-18 Wadayama-dori 2-chome, Hyogo-ku, Kobe 652
☎(+81 78) 682 31 11 ⊜ (+81 78) 671 57 84
■ Trains: electric locomotives, diesel locomotives, electric cars, passenger coaches, diesel cars, mono-rail cars, Automated Guideway Transit System, freight cars, tank cars.
High-pressure vehicles, straddle cranes (K-lift). aerial ropeways, cable cars, platform screen doors.

Nishi-Kobe Works
234 Matsumoto, Hazetani-cho, Nishi-ku, Kobe 651-22
☎(+81 78) 991 11 33 ⊜ (+81 78) 991 31 86
■ Hydraulic equipment: pumps, motors, valves, rotary actuators, hydraulic units and hydraulics systems for industrial applications.
Maritime equipment: electric hydraulic steering devices, automatic steering machines, deck machinery, deck cranes, fishing machinery.
Precision gear pumps for chemical fibre spinning, test and simulation equipment, manipulators, brushless servomotors.

Seishin Works
8-1 Takatsukadai 2-chome, Nishi-ku, Kobe 651-22
☎(+81 78) 992 19 11 ⊜ (+81 78) 992 19 10
■ Jet engine and gas turbine components.

Banshu Works
2680 Oka, Inami-cho, Kako-gun, Hyogo 675-11
☎(+81 794) 95 12 11 ⊜ (+81 794) 95 12 26
■ Construction machinery and cargo-handling equipment. Wheel loader, tyre rollers, macadam rollers, vibratory rollers, rotary drills, snowploughs, concrete paving machines, concrete mixers, log loaders, fork loaders, towing tractors, load-haul dumps.

Harima Works
8 Niijima, Harima-cho, Kako-gun, Hyogo 675-01
☎(+81 794) 35 21 31 ⊜ (+81 794) 35 21 32
■ Plants, construction machinery and steel structures. Steel frames, airport facilities and equipment, steel pipe structures, bridges, LNG tanks, LPG tanks, high pressure gas vessels, subterranean tunnels, martime structures, steel hydraulic pipes, water gates, steel stacks, launchers, container-handling cranes, rampways.

Sakaide Works
1 Kawasaki-cho, Sakaide, Kagawa 762
☎(+81 877) 46 14 73 ⊜ (+81 877) 45 05 56
■ Equipment for ship and maritime applications: tankers, bulk carriers, LNG carriers, LPG carriers, container ships, car carriers, oil drilling rigs, self-elevated maritime operation platforms, floating platforms, various plant barges, repair and modification of ships.

Akashi Technical Institute
1-1 Kawasaki-cho, Akashi, Hyogo 673
☎(+81 78) 921 16 11 ⊜ (+81 78) 921 16 02
■ Fundamental and applied research in areas of physics, strength, corrosion, materials, machinery, thermal energy and welding.

Gifu Technical Institute
1 Kasaski-cho, Kakamigahara, Hyogo 673
☎(+81 583) 82 52 74 ⊜ (+81 583) 82 51 30
■ Research in areas related to space and aerospace products including aerodynamics, structures, materials, information, electronics, controls and spacecraft.

Kanto Technical Institute
118 Futatsuza, Noda, Chiba 278
☎(+81 471) 24 02 82 ⊜ (+81 471) 24 51 52
■ Research and investigation in areas of advanced and high techologies. 98

75-04500 Keiaisha NMB Co. Ltd
18th Floor Arco Tower, 1-8-1 Shimo-Meguro, Meguro-ku, Tokyo 153
☎(+81 3) 54 34 87 11 ⊜ (+81 3) 54 34 87 00
ﾒ 415
President: Tsuguo Yamamoto 98

75-05600 Minebea Co. Ltd Head Office
4106-73 Miyota-machi, Kitasaku-gun, Nagano 389-02
☎(+81 267) 32 22 00 ⊜ (+81 267) 31 13 30
♧ Yen 63,681,000,000
♣ Yen 239,133,000,000 (1995) ﾒ 3,004 (1995)
Chairman and Chief Executive Officer: Iwao Ishizuka
President: Goro Ogino
Public Relations Manager: Ryusuke Misukami
Corporate Communications Department:
Yasuyuki Sugawara ☎(+81 3) 54 34 86 37; 54 34 86 07 (fax)
■ Multinational, high volume manufacturing company. Main business areas are precision bearings, machinery components, electronic devices and other products including defence equipment. Weapon carrier and release equipment, aircraft barrelled weapons, actuators, special components, electronic, integrated circuits, bearings, fastening systems, fasteners, electronic motors.

Headquarters Factory/Karuizawa Plant
4106-73 Oaza Miyota, Miyota-Machi - Kitasaku-gun, Nagano 389-02
☎(+81 267) 32 22 00 ⊜ (+81 267) 31 13 30

Kyoto Manufacturing Division
55 Tozumitoshiro, Yawata-City, Kyoto-Pref 614
☎(+81 75) 971 19 01 ⊜ (+81 75) 981 27 52

Research and Development Laboratory
1743-1 Asana - Asaba-cho, Iwata-Gun, Shizuoka Pref 437-11
☎(+81 538) 23 70 01 ⊜ (+81 538) 23 70 40

Civil Engineering Instrument Center
1-1-1 Katase - Fujisawa-City, Kanagawa-Pref 251
☎(+81 466) 22 71 61 ⊜ (+81 466) 22 17 01
President: M Nagamine
Managing Director: K Abe 98

75-05700 Mitsubishi Electric Corporation Head Office (MELCO)
Mitsubishi Denki Building, Marunouchi, Tokyo 100
☎(+81 3) 32 18 21 11 ✉ MELCO Tokyo
W http://www.mitsubishielectric.com ⊠ 24532 j
⊜ (+81 3) 32 18 29 24
President: Takashi Kitaoka
■ Missile systems, communication equipment, combat direction systems, EW equipment, radars, fire control systems, signal processing equipment, power equipment, computers and satellites. 98

75-05800 Mitsubishi Heavy Industries Ltd (MHI)
5-1 Marunouchi 2-chome, Chiyoda-ku, PO Box 645, Tokyo 100
☎(+81 3) 32 12 31 11 ✉ HISHIJU Tokyo
⊠ 22282; 22381; 22443; 22874 hishiju j
⊜ (+81 3) 32 12 98 65
♧ Yen 265,390,000,000 ♣ Yen 2,733,756,000,000
ﾒ 41,115
Chairman: Kentaro Aikawa
President: Nobuyuki Masuda
Executive Vice President: Yoshihisa Tsuda
Executive Vice President: Yasuo Nagai
Managing Director and General Manager, Aircraft and Special Vehicle Headquarters: Takashi Nishioka
General Manager Aircraft Department: Hideo Egawa
General Manager Helicopter Department: Yusyo Kinoshita
General Manager Aircraft Equipment Department: Toshinori Nishi
General Manager Guided Weapon Systems Department: Yoshifumi Nozawa
General Manager Space Systems Department: Keiji Yasuki
General Manager Special Vehicles Department: Masahiro Kataoka
General Manager Public Relations: Takamichi Wachi
Public Relations: Yasushi Kawano
Aircraft and Special Vehicle Administration Department: Kiyoshi Kondo

Nagoya Aerospace Systems Works
10 Oye-cho, Minato-ku, Nagoya 455
☎(+81 52) 611 21 21 ✉ HISHIAIR ⊠ 4464561
⊜ (+81 52) 611 93 60
Director and General Manager: Tadayuki Tanioka
■ F-15J jet fighter, SH-60J helicopter, B767 aft fuselage and H-II rocket.

Nagoya Guidance and Propulsion Systems Works
1200 Higashi-Tanaka, Komaki-shi, Aichi 485
☎(+81 568) 79 21 11 ✉ HISHIJU ⊠ 4485850
⊜ (+81 568) 78 25 52
Director and General Manager: Minoru Takakuru
■ Type 80 air-to-ship guided missiles; AIM-9L air-to-air guided missiles; Patriot surface-to-air missile systems. Rocket engines LE-5 and LE-7.

Nagasaki Shipyard and Engine Works
1-1 Akunoura-machi, Nagasaki 850-91
☎(+81 958) 28 41 21 ✉ DOCK Nagasaki
⊜ (+81 958) 28 41 05
Director and General Manager: Syoji Fukushima
■ Builders of 1 Asagiri-class destroyer.

Kobe Shipyard and Engine Works
1-1 Wadasaki-cho 1-chome, Hyogo-Ku, Kobe
☎(+81 78) 672 22 20 ✉ DOCK Kobe
⊜ (+81 78) 672 23 00
Director and General Manager: Takaharu Toyada
■ Builders of Yuushio and improved Yuushio class submarines.

Shimonoseki Shipyard and Engine Works
16-1 Enoura, 6 chome, Hikoshima, Shimonoseki
☎(+81 832) 66 59 78 ✉ DOCK Shimonoseki
⊠ 682284 hishiju j ⊜ (+81 832) 67 70 10
Director and General Manager: Akira Inoue
■ Builders of fast attack craft. 98

75-06000 Mitsubishi Precision Co. Ltd Head Office
Mita 43 Mori Building, 3-13-16 Mita, Minato-ku, Tokyo 108
☎(+81 3) 34 53 64 21 W http://www.mindnet.or.jp/mpc
⊜ (+81 3) 34 53 64 34
♧ Yen 3,167,252,000
President: Enkou Itoh
■ Design, manufacture, repair, maintenance and sale of space, aviation and vehicular equipment and systems. Simulator, trainer and visual systems, information processing systems, traffic control systems.

Osaka Sales Office
Athens Shin-Osaka Building, 6-1-15 Nishi-Nakajima, Yokogawa-ku, Osaka 532
☎(+81 6) 303 22 05 ⊜ (+81 6) 303 03 39

Nagoya Sales Office
Koma Building, 3-8-7 Nishiki, Naka-ku, Nagoya 460
☎(+81 52) 961 35 57 ⊜ (+81 52) 961 34 93

Kyushu Sales Office
Asahi Semei Building, 8-23 Tsunaba-cho, Hakata-ku, Fukuoka 812
☎(+81 92) 282 98 63 ⊜ (+81 92) 282 98 65 98

75-06100 Moog Japan Ltd
(a subs. of Moog Inc., USA)
1532 Shindo, Hiratsuka, Kanagawa 254
☎(+81 463) 55 36 15 ⊜ (+81 463) 54 47 09
♧ Yen 226 000 000 (1996) ♣ US$ 23 500 000 (1995)
ﾒ 130
President: Mitsumasa Ohno ☎(+81 46) 355 36 15
Sales Group Mgr: Mark Sakamoto ☎(+81 46) 355 74 10
■ Manufacturer of precision control components & systems, used in a wide range of high performance aircraft, strategic & tactical missiles, space vehicles & military ground vehicles.
® USA: Moog Inc. 97

75-06350 NEC Corporation
7-1, Shiba 5-chome, Minato-ku, Tokyo 108-01
☎(+81 3) 34 54 11 11; 37 98 66 81 ✉ NEC TOKYO
⊠ 22686 nectok j ⊜ (+81 3) 37 98 15 10; 37 98 15 19; 37 98 66 84
♧ Yen 191,200,000,000 (1996) ﾒ 152,719 (1996)
Chairman: Tadahiro Sekimoto
President: Hisashi Kaneko
Guidance and Electro-Optics General Manager: K Nakada
Radio Applications General Manager: K Sugiuchi
Space Systems Development General Manager: K Murata
Senior Manager Government and Public Sector Sales: Masahiko Hama
Assistant Manager Government and Public Sector Planning: Tadashi Shikimori
■ Radio application equipment, ATC radars, defence electronic systems, Navaids, airborne communication equipment, rocket and missile guidance systems, satellites, satellite transponders, satellite earth stations, autopilot systems, flight simulators and telemetry.

Subsidiaries:
Japan Aviation Electronics Ind. Ltd
Nippon Avionics Co. Ltd
NEC Aerospace Systems Ltd

Europe, Middle East and Africa Division
7-1 Shiba 5-chome, Minato-Ku, Tokyo 108-01
☎(+81 3) 34 54 11 11 ✉ NEC TOKYO ⊠ 22686 nectok j
General Manager, Europe: Takao Nishizawa
General Manager, Middle East and Africa: Noboru Sato

Contact Reference

Fuchu Plant
10 Nisshin-cho, 1-chome, Fuchu, Tokyo 183
☎(+81 423) 64 11 11
General Manager: Kunio Koike

Yokohama Plant
4035 Ikebe-cho, Tsuzuki-ku, Yokohama, Kanagawa 224
☎(+81 45) 932 11 11 ⊠ 3822664 yoknec j
General Manager: Konji Saito [98]

75-06790 Nissan Motor Co. Ltd
17-1 Ginza 6-chome, Chuo-ku, Tokyo 104-23
☎(+81 3) 35 43 55 23 ⊠ 22503 nismo j
⊖(+81 3) 35 46 26 69
⚘ Yen 203,647,000,000 (1996)
♣ Yen 3,690,400,000,000 (1996) 👥 44,782 (1996)
Chairman: Yoshifumi Tsuji
President: Yoshikazu Hanawa
Executive Vice President: Kosei Minami
Executive Vice President: Yoshikazu Kawana
Managing Director: Akio Shimohara
Operations Director: Makoto Takemoto
Sales and Marketing Director: Satoshi Suzuki
Purchasing Manager: Toshio Takamiya
Manager Administration Department, Aerospace Division: Yoshifumi Abe
■ Passenger cars, buses and trucks. Activities in the field of aerospace and marine equipment and industrial machinery.

Aerospace Division
5-1, Momoi 3-chome, Suginami-ku, Tokyo 167
☎(+81 3) 33 01 67 20 ⊖(+81 3) 33 01 67 17
♣ Yen 41,100,000,000 👥 876
General Manager: Akio Shinohara ☎(+81 3) 33 01 66 10
General Manager Sales and Marketing: Satoshi Suzuki
Manager Administration Department: Yoshifumi Abe
Defence Systems Department: Tetsuo Nakamura ☎(+81 3) 33 01 66 80
■ Launching vehicles, sounding rockets, defence rocket systems, rocket launchers, guidance and control systems, test equipment. [98]

75-06800 NMB Electro Precision Inc.
8-4 Chome, Ogi-machi, Miyagino-ku,
Sendai City, Miyagi-Pref 983
☎(+81 22) 232 69 51 ⊖(+81 22) 232 78 04
👥 65
President: M Miyamoto
■ Manufacture of fan blowers. [98]

75-07200 Oki Electric Industry Co. Ltd
4-10-3 Sibaura, Minato-ku, Tokyo 108
☎(+81 3) 54 45 60 80 🅆 http://www.oki.co.jp
⊖(+81 3) 54 45 62 02
⚘ Yen 67,849,000,000 (1996)
♣ Yen 733,200,000,000 (1996)
👥 21,700 (consolidated)
President: Shiko Sawamura ☎(+81 3) 35 01 31 11
General Manager Defense Systems Marketing and Sales: Yasuhiro Ono ☎(+81 3) 54 45 60 80
General Manager Marketing and Sales Department: Takeo Suzuki ☎(+81 3) 54 45 60 81
Senior Manager Defence Systems Marketing and Sales: Takashi Mori ☎(+81 3) 54 45 60 91
■ Sonobuoys, sonars, computer systems, communications equipment, electronic devices. [98]

75-07305 Omori Manufacturing Division
18-18 4 Chome, Omori Nishi, Ota-Ku, Tokyo 143
☎(+81 3) 37 65 11 21 ⊖(+81 3) 37 66 76 37
👥 239
Executive Director and Factory Manager: Saburo Koyama [98]

75-07310 Omori Seikoki Co. Ltd (OSCO)
17-12 Omori Naka 2-Chome, Ohta-Ku, Tokyo 143
☎(+81 3) 37 61 22 97 ⊖(+81 3) 37 62 45 27
⚘ Yen 100,000,000 (1997)
♣ Yen 1,500,000,000 (1996) 👥 150 (1997)
🏭 2,343 m2
President and Chief Executive Officer: Koichi Tokimura
Executive Vice President Plant Operations: Akihiro Tokimura
Executive Director and Sales Manager: Shuichiro Shinohara
Director Engineering Department: Keiji Kooda
Advisor: Hanji Mori
■ Repair and overhaul of aircraft and helicopter accessories, wheels, brakes, hydraulic/electric/electronic control system components, radar, hauldown systems, missile and rocket launchers. Manufacture of aircraft hydraulic units, GSE and test equipment. [98]

75-07600 Shimadzu Corporation
1 Nishinokyo-Kuwabaracho, Nakagyo-ku, Kyoto 604
☎(+81 75) 823 11 01 ✉ SHIMADZU Kyoto
⊠ 05422166 shmds j ⊖(+81 75) 811 31 55
⚘ Yen 16,800,000,000 ♣ Yen 162,921,000,000
👥 4,300
President: Kikuo Fujiwara
General Manager: Shuichi Sato
Marketing Manager: Kazuo Wakasa
Sales and Marketing Department - Aircraft Equipment Division: Yoichiro Ikeda
■ Equipment manufacture for aeromechanics and avionics, GSE, flight control, hydraulics, fuel and oil systems, head-up and radar displays, environmental control systems. Analytical and measurement instruments, testing and inspection machines, process control equipment, medical systems and equipment, biotechnology instruments, industrial equipment, opto-electronic equipment and optical devices.

Tokyo Office
3 Kanda-Nishikicho 1-chome, Chiyoda-ku, Tokyo 101 [98]

75-07800 ShinMaywa Industries Ltd
6-2 Otemachi 2 Chome, Chiyoda-ku, Tokyo 100
☎(+81 3) 32 45 66 11 ⊠ 5644493 smic hoj
⊖(+81 3) 32 45 66 16
⚘ Yen 14 260 173 479 ♣ Yen 152 358 000 000 (1994)
👥 5 161
President: Shiko Saikawa
Sr Exec. Managing Dir: Tamotsu Mineno
Sr Exec. Managing Dir: Iwao Hiraoka
Exec. Managing Dir: Junpei Matsuo
Operations/Technical Dir: Yasuo Takeuchi
Sales & Marketing Dir: Yushi Tanaka
General Affairs Mgr: Yoshimasa Nagasaku
Sales Dept Mgr: Motoaki Mishimoto
■ Design & manufacture flying boats; airport equipment; motors; vacuum equipment; pumps & water disposal related products; special purpose truck bodies & equipment.

Aircraft Division
1-1, Ohgi 1-chome, Higashinada-ku, Kobe 658
☎(+81 78) 412 91 51 ✉ SMICAIR ⊠ 5622157 smic k j
⊖(+81 78) 431 36 95
Exec.Managing Dir & General Mgr: Junpei Matsuo
■ Manufacture of US-1 STOL amphibian aircraft, subassemblies & parts for P-3C, F-15, T-4, SH-60J, CH-47, MD-11, MD-80, MD-90, Gulfstream IV, B747, B757, B767 & B777. Modification of Learjet 36A to U-36A.

Sales Head Office & Sales Dept
Nippon Bldg, 6-2 Otemachi 2-chome, Chiyoda-ku, Tokyo 100
☎(+81 3) 32 45 66 11 ⊖(+81 3) 32 45 66 16
Manager: Yushi Tanaka [97]

75-07950 Sumitomo Electric Industries Ltd
5-33 Kitahama 5 chome, Chuo-ku, Osaka 541
☎(+81 6) 220 41 41 ⊠ 5222187 seihdq j
⊖(+81 6) 222 33 80; 466 57 26
⚘ Yen 79,713,000,000 (1996)
♣ Yen 770,489,000,000 (1996) 👥 14,700
President: Noritaka Kurauchi
■ Electric wires and cables, optical fibres, compound semiconductors, rubber fuel cells, de-icers, radomes, life rafts and jackets.

Hybrid Products Division
1-3 Shimaya 1-chome, Konohana-ku, Osaka 554
☎(+81 6) 466 55 65 ⊠ 05245660 ⊖(+81 6) 461 32 45
General Manager: Kazuoki Hanzawa

Defence Marketing Department
3-12 Moto-Akasaka 1-chome, Minato-ku, Tokyo 107
☎(+81 3) 34 23 50 43 ⊖(+81 3) 34 23 50 43
Marketing Manager: Kenji Kawase [98]

75-08000 Sumitomo Heavy Industries Ltd Head Office (SHI)
Sumitomo Heavy Industries Building,
5-9-11 Kita Shinagawa, Shinagawa-ku, Tokyo 141
☎(+81 3) 54 88 80 00 🅆 http://www.shi.co.jp/
♣ Yen 275,317,000,000 (1995) 👥 5,711 (1996)
■ Rocket launch structures, cryogenic technology. Machinery: steel-making, non-ferrous metal, forging, compression, transportation, cargo handling, physical distribution, accelerator, electron irradiation equipment, superconductivity equipment, medical appliances, car parking facilities, moving sidewalks.
Ships and steel structures: ships, warships, marine structures, marine developing equipment, bridges, steel frames, water gates, shielded evacuators, pressure vessels, agitation tanks, low temperature equipment, defence equipment, aircraft equipment, casted and forged products. [98]

75-08070 Sumitomo Precision Products Co Ltd Head Office
1-10 Fuso-cho, Amagasaki, Hyogo Prefecture 660
☎(+81 6) 482 88 11 ⊖(+81 6) 489 58 01
♣ Yen 33,374,000 (1996) 👥 1,086 (1996)
President: Iwao Takai
Executive Vice President: Toshio Egawa
Executive Vice President: Tetsuguro Ishikawa
Senior Managing Director: Masumi Watanabe
Managing Director: Tadayoshi Gotoh
Managing Director: Haruhisa Nakata
Director: Hiroaki Nakamura
Director: Koujiro Mitsui
Managing Director and General Manager, Microelectronics Technology Division: Mutsuo Mukuda
■ Propellers, landing systems, flight control systems, fuel systems, hydraulic equipment, heat exchangers, environmental equipment.

Tokyo Head Office
Kazan Building 2-4 3-chome, Kasumigaseki, Chiyoda-ku, Tokyo 100
☎(+81 3) 35 81 34 41 ⊖(+81 3) 35 80 22 07
Manager: Hayao Ohsaka
Marketing Dir: Tetsugoro Maruyama [98]

75-08500 TOKIMEC Inc.
Head Office
16-46, Minami-Kamata, 2-Chome, Ohtza-ku, Tokyo 144
☎(+81 3) 37 32 21 11 ✉ TOKIMEC
⊖(+81 3) 37 36 02 61
⚘ Yen 7 217 000 000 ♣ Yen 49 116 000 000 👥 2 289
Chairman: Nobue Hirono
President: Keijiro Morita
Managing Dir: Kiyoshi Fujii
■ Aeronautical instruments, aero hydraulic equipment, air surveillance/control indicators for radar sites, gunsights, radar warning systems, multifunction displays, missile displays, Luneburg lenses, submarine inertial navigation systems, stabilised gyro compasses, ring laser gyros, fibre-optics, microwave devices, radar displays, antenna directing systems, dead reckoning equipment.
® USA: AlliedSignal Inc., Honeywell Inc., Howden Airdynamics Inc., Vickers Inc.

Electronics Systems Division
16-46 Minami-Kamata, 2-Chome, Ohta-Ku, Tokyo 144
☎(+81 3) 37 37 86 41 ⊠ 2466193 tokeit j
⊖(+81 3) 37 37 86 68
Dir & General Mgr: Susumu Ishii
Dept Mgr Marketing: Ken-ichi Waki
Dept Mgr Business Planning: Motoaki Yanatori
Dept Mgr Engineering: Kuniaki Shimizu
Dept Mgr Business Development: Kunio Homma

Tochigi Factory
3-1 Takaku-ko, Nasu-machi, Tochigi 325
☎(+81 287) 63 37 11 ⊖(+81 287) 64 21 40
Dir & Deputy General Mgr: Toru Nakaguchi [97]

75-08600 Tokyo Aircraft Instrument Co Ltd
Tokyo Kokukeiki KK (TKK)
35-1 Izumi-Honcho 1-chome, Komae-Shi, Tokyo 201
☎(+81 3) 34 89 11 25 ✉ TKKWADA Tokyochohu
⊠ 2422246 tkkwad j ⊖(+81 3) 34 88 55 21
⚘ Yen 500,000,000 ♣ Yen 14,189,000,000 (1995)
👥 705 (1995)
President: Atsutake Wada
Managing Director: Goro Kondo
Director and Manager Aerospace Sales: Masayuki Numazawa
Operations and Technical Director: Hiroshi Touhei
Manager Business Development: Tatsuo Miki
Purchasing Manager: Isao Uekusa
Assistant Manager: S Mamiya
■ Automatic flight control systems, aircraft instruments, oxygen generating systems, air data computers and flight data recorders.

Haneda Maintenance Office
1-8-2 Tokyo International Airport, Ohdta-ku, Tokyo 144
☎ (+81 3) 37 47 76 71 ⊖ (+81 3) 37 47 03 74
Manager: K Onishi ☎ (+81 52) 203 18 01

Nagoya Sales Office
Shirakawa Academy Building, 2-11-3 Sakae, Naka-ku, Nagoya 460
☎ (+81 52) 203 18 01
Sales Director: K Eguchi
98

75-10000 Toyo Communication Equipment Co Ltd (TOYOCOM)

12-32 Konan 2-chome, Minato-ku, Tokyo 108
☎ (+81 3) 54 62 96 00; 54 62 96 10
⊠ EXPORTOYOCOM ⊖ (+81 3) 54 62 96 25
♣ Yen 84,200,000,000 👥 4,000
President: Shinichi Itoh
Director and General Manager, Wireless Communication Systems Division: Takuo Shiode
Director and General Manager Purchasing Department: Makoto Tachibana
International Operations Division Manager: Junichi Kagayama
■ Telecommunication equipment, PCM transmission systems, subscriber carrier terminals, modems, paging receivers, cellular telephones, ARGOS platform transmitter terminals, IFF equipment synthetic quartz crystals ATC transponders, fixed/mobile radios, multiplex data transmission radios, satellite communications SATCOM antenna sub-system. Cordless interphone system.
98

75-10100 UBE Industries Ltd UBE Head Office

12-32 Nishihonmachi 1-chome, Ube, Yamaguchi 755
☎ (+81 836) 31 11 22 Ⓦ http://www.ube-ind.co.jp
⊠ 684503 ⊖ (+81 836) 21 22 52
Chairman: Motoo Nakahigashi
President: Maomi Nagahiro
Senior Managing Director: Akira Tanaka
Senior Managing Director: Eisuke Shintani
Senior Managing Director: Goro Kato
Senior Managing Director: Tadahiko Yoshimori
Managing Director: Ken Sakai
Managing Director: Kozo Fujii
Managing Director: Koichiro Miyamoto
Managing Director: Tadae Nakanishi
Managing Director: Minoru Tanifuji
Managing Director: Tetsuo Kadoyama
Managing Director: Toshiro Kawaguchi
Managing Director: Yukuo Suzuki
Managing Director: Tadaaki Hirano
Managing Director: Akio Matsumura
■ Chemicals and plastics, construction materials, machinery, engineering. Research and development.

Tokyo Head Office
UBE Building, 3-11 Higashi-Shinagawa 2-chome, Shinagawa-ku, Tokyo 140

Sapporo Branch
Sapporo Kogin Building, 2 Kitaichijo Nishi 5-chome, Chuo-ku, Sapporo 060
☎ (+81 11) 271 04 56 ⊖ (+81 11) 241 83 52

Tohoku Branch
Sumitomo Seimei Sendai Chuo Building, 6-1 Chuo 4-chome - Aoba-ku, Sendai 980
☎ (+81 22) 262 62 35 ⊖ (+81 22) 262 62 40

Nagoya Branch
Nagoya Kogin Building, 11-18 Nishiki 1-chome, Naka-ku, Nagoya 460
☎ (+81 52) 201 70 71 ⊖ (+81 52) 201 70 75

Osaka Branch
Sakurabashi Toyo Building, 2-16 Sonezakishinchi 2-chome, Kita-ku, Osaka 530
☎ (+81 6) 346 13 61 ⊖ (+81 6) 244 71 76

Hiroshima Branch
Hiroshima Kogin Building, 1-22 Kamiya-cho 2-chome, Naka-ku, Hiroshima 730
☎ (+81 82) 244 71 00 ⊖ (+81 82) 244 71 76

Shikoku Branch
Hyaku-ju-yon Building, 5-1 Kamei-cho, Takamatsu 760
☎ (+81 878) 61 51 00 ⊖ (+81 878) 61 51 09

Kyushu Branch
Fukuoka Sanwa Building, 10-24 Tenjin 1-chome, Chuo-ku, Fukuoka 810
☎ (+81 92) 781 01 11 ⊖ (+81 92) 713 54 69

UBE Research Laboratory
1978-5 Kogushi, Ube, Yamaguchi 755
☎ (+81 836) 31 09 74 ⊖ (+81 836) 31 65 71

Polymer Research Centre
8-1 Goi-Minamikaigan, Ichihara, Chiba 290
☎ (+81 436) 23 51 11 ⊖ (+81 436) 22 51 32

Chiba Petrochemical Factory
8-1 Goi-Minamikaigan, Ichihara, Chiba 290
☎ (+81 436) 23 51 11 ⊖ (+81 436) 22 54 91

Sakai Factory
1 Chikko-Shinmachi 3-chome, Sakai, Osaka 592
☎ (+81 722) 43 51 00 ⊖ (+81 722) 43 51 50

UBE Corporate Service Division
1978-10 Kogushi, Ube, Yamaguchi 755
☎ (+81 836) 31 21 11 ⊖ (+81 836) 21 22 52

UBE Chemical Factory
1978-10 Kogushi, Ube, Yamaguchi 755
☎ (+81 836) 21 22 86

UBE Cement Factory
1978-2 Kogushi, Ube, Yamagushi 755
☎ (+81 836) 31 21 11 ⊖ (+81 836) 21 22 52

Isa Cement Factory
4768 Isa, Isa-cho, Mine, Yamaguchi 759-22
☎ (+81 8375) 212 12 ⊖ (+81 8375) 217 50

Kanda Cement Factory
7 Nagahama-machi, Kanda-cho, Miyako-Fukuoka 800-03
☎ (+81 93) 434 21 11 ⊖ (+81 93) 434 01 79

UBE Machinery & Engineering Works
1980 Okinoyama, Kogushi, Ube, Yamagushi 755
☎ (+81 836) 31 21 11 ⊖ (+81 836) 21 22 52

Coal Center
1980-34 Okinoyama, Kogushi, Ube, Yamagushi 755
☎ (+81 836) 31 59 71 ⊖ (+81 836) 31 59 85
98

75-25000 Yamaha Motor Co Ltd

2500 Shingai - Iwata-shi, Shizuoka-ken 438
☎ (+81 538) 32 11 70 ⊠ 59645 yamaha j
⊖ (+81 538) 37 42 59
♤ Yen 22,822,000,000 ♣ Yen 415,903,000,000
👥 7,354
President: Takehiko Hasegawa
General Manager Aerospace: Masatoshi Ohmi
Assistant Manager Patent Department: Michihisa Kagami
Manager: Takafumi Itagaki Ⓔ itagaki-takafumi@ccgw.yamaha-motor.co.jp
■ Piston engines, engines for target drones.
98

80 Service Companies & Organizations

80-00030 AlliedSignal Aerospace Marketing Sales and Service

Shirikawa Dairoku Building 5F, Room 503, 2-18-5 Nishiki, Naka-ku, Nagoya-shi, Aichi-ken 460
☎ (+81 52) 204 04 39 ⊖ (+81 52) 204 04 67
Field Service Engineer: Stanley Longacre
■ On-site customer support for AlliedSignal Aerospace. → 90
98

80-00032 AlliedSignal Aerospace Marketing Sales and Service

Narita, Chiba 282
New Tokyo International Airport, c/o
Japan Airlines Engine Maintenance Center, 4th Floor, Narita, Chiba 282
☎ (+81 476) 32 42 08; 32 00 96 ⊖ (+81 476) 32 63 47
Field Service Engineer: Marco O Gonzales
■ On-site customer support for AlliedSignal Aerospace. → 90
98

80-00035 AlliedSignal Aerospace Marketing Sales and Service

Japan Airlines, Kiso Building 7F, 1-6-3 Haneda Airport, Ota-ku, Tokyo 144
☎ (+81 3) 37 47 00 96; 37 47 39 84
⊖ (+81 3) 37 47 39 76
Manager, Airlines: Taisuke Okumura
☎ (+81 3) 37 69 85 78; 37 69 50 28 (fax)
Field Service Engineer: Brian Muskus
Field Service Engineer: Shingo Yonezawa
■ On-site customer support for AlliedSignal Aerospace.
98

80-00050 Arianespace

(a subsidiary of Arianespace, France)
Kasumigaseki Building 31st Floor, 3-2-5 Kasumigaseki, Chiyoda-ku, Tokyo 100
☎ (+81 3) 35 92 27 66 ⊖ (+81 3) 35 92 27 68
Head, Tokyo Office: Jean-Louis Claudon
■ Satellite launch services.
® France: Arianespace
98

80-00055 Artemis International Limited

(a subsidiary of Artemis Management Systems Ltd, UK)
20-5 Ichibancho, Chiyoda-ku, Tokyo 102
6th Floor American International Building
☎ (+81 3) 32 62 81 61 ⊖ (+81 3) 32 62 85 50
Regional Director: Katsujiro Fujinami
98

80-00056 Artemis International KK

(a subsidiary of Artemis Management Systems Ltd, UK)
Advance Esaka Building 501, 8-10 Toyotsu-Cho, Suita-City, Osaka Prefecture 564
☎ (+81 6) 821 00 71 ⊖ (+81 6) 821 00 74
98

80-00057 Artemis International Limited

9th Floor N°14 Ocean Building, 3-23-3 Aoi, Higashi-ku, Nagoya, Aichi 461
98

80-00100 Center for Information on Security Trade Control (CISTEC)

Bansuiken Building 6-6 1-chome, Toranomon, Minato-ku, Tokyo 105
☎ (+81 3) 35 93 11 48 ⊖ (+81 3) 35 93 11 38
👥 36
Director Research Department: M Yoneta
Senior Rearcher: S Inaba
■ Study and research. Collection, research and analysis of basic data and information on goods and technologies.
98

80-00115 China Steel Corporation Osaka Office

8th Floor - Nantai Building, 2-20 Minamihonmachi 4-Chome, Chuoku, Osaka 542
☎ (+81 6) 281 10 21 ⊖ (+81 6) 281 10 23
98

80-00125 Defense Technology Foundation

4th Floor AHM's One Building, 1-18-4 Ebisu Nishi, Shibuya-ku, Tokyo 150
☎ (+81 3) 34 63 23 08 ⊖ (+81 3) 34 63 21 49
👥 50 (1997)
Director General: Hajime Seki
Managing Director: Yutaka Abe
■ Publication of monthly "Defense Technology Journal". Contracted study and technical assistance for defence technology matters.
98

80-00180 Hornmark KK

5-8-3 Shiroganedai, Minato-ku, Tokyo 108
☎ (+81 3) 34 47 66 14 ⊖ (+81 3) 34 40 62 98
President: Nils Hornmark
■ Consultant for sales of foreign defence products to the Japanese Defense Agency.
98

80-00250 Hughes International Corporation

Kamiyacho Mori Building, 2 Floor, 3-20 Toranomon 4-chome, Minato-ku, Tokyo 105
☎ (+81 3) 54 70 17 70 ⊖ (+81 3) 35 81 56 68
98

80-00450 Nippon Systems Industries Corp

(a subsidiary of Kyokuto Boeki Kaisha Ltd)
3rd Floor Loyal-Kikaku Building, 1-33 Sakaecho 3-chome - Asaka, Saitama 351
☎ (+81 48) 467 24 91 ⊖ (+81 48) 467 20 96
■ Sale and maintenance of electronic equipment. Development of VHF transceivers, general aviation aircraft receivers and telemetry based automatic scales. Technical support: inspection, maintenance and repair of KBK imported equipment. Authorised Rockwell-Collins service shop. → 90
98

80-00625 Nissan Aerospace Engineering Co. Ltd

(a subsidiary of Nissan Motor Co. Ltd.)
5-1 Momoi 3-chome, Suginami-ku, Tokyo 167
☎ (+81 3) 33 01 67 40 ⊖ (+81 3) 33 99 55 44
♣ Yen 1,149,000,000 👥 120
Managing Director: Shinichi Yamazaki
Marketing and Sales Director: Takuji Murakami
Public Relations Manager: Masaaki Samaru
■ Design support services for rockets & other aerospace components. [98]

80-00800 Praxair Surface Technologies K.K.

7th Fl, Shin-Yokohama 1C Blvd,
18-9 Shin-Yokohama 3-Chome,
Kohoku-Ku, Yokohama 222
☎ (+81 45) 474 30 32
⊖ (+81 45) 474 30 27; 474 30 28 [97]

80-00950 Smith System Engineering Ltd

Akaska Wing Building 4F, 6-6-15 Akasaka, Minato-ku, Tokyo 107
☎ (+81 3) 35 86 04 14 ⊖ (+81 3) 55 63 28 23 [98]

80-02000 ShinMaywa Ground Services Ltd

3-20 2 chome, Nakamachi, Izunjsano-shi, Osaka 598
☎ (+81 6) 64 13 71 ⊖ (+81 6) 64 13 81
♧ Yen 50 000 000 (1996) ♣ Yen 434 000 000 (1996)
👥 170
President: Motohiro Matsushita
Director: Kazumi Ishimoto
Director: Tatsuo Matsuoka
Director: Makoto Shimomura
Director: Katsuo Ohno
■ Aircraft ground handling & maintenance including marshalling, aircraft moving, checking, fuel & oil supply, cabin cleaning, water & toilet services. Baggage, cargo & mail handling. Loading & unloading. [97]

85 Stockists, Distributors, Spare Parts Suppliers

85-01800 Sony Trading International Corporation

(a subsidiary of Sony Corporation)

Headquarters
TS Building 2-13-40, Konan, Minato-ku, Tokyo 108
☎ (+81 3) 34 74 61 95
⊖ (+81 3) 34 74 69 93; 34 74 69 94
♧ Yen 550,000,000 ♣ Yen 229,500,000,000 (1995)
👥 350 (1995)
President: Haruo Akita
Managing Director Aircraft Group: Hitoshi Sunabori
General Manager: Taka-yuki Miyajima

Osaka Office
Sony 2nd Building, 34-23 Shinmachi 1-chome, Nishi-ku, Osaka 550
☎ (+81 6) 541 79 91 ⊖ (+81 6) 531 49 25
■ Import and maintenance of aircraft, avionics and other defence equipment. → 80
® France: ABG-SEMCA, Aerazur, Sully, Eurocopter, SAMM, Sextant
Germany: Eurocopter
Norway: Helicopter Service
UK: Helitune Ltd
USA: Aerial Films Inc, Dassault Falcon Jet

Tokyo Heliport Office
Shinkiba 4-chome, Koto-ku, Tokyo 136
☎ (+81 3) 35 21 74 11 ⊖ (+81 3) 35 21 74 15 [98]

85-01900 Sumitomo Corp

2-2 Hitotsubashi 1-chome, Chiyoda-ku, Tokyo 100-91
☎ (+81 3) 32 17 50 00 ✉ SUMITOMO Tokyo
[W] http://www.sumitocorp.co.jp/index-e. html
⊠ 22202; 22203 ⊖ (+81 3) 36 36 72 20
♧ Yen 109,500,000,000 (1996)
♣ Yen 12,710,652,000,000 (1996)
♣ Yen 15 032 716 000 000 👥 6,249 (1997) 👥 6 876
President: Kenji Miyahata
Executive Vice President Machinery and Electric Group: Tsuneo Iwasaki
General Manager: Masanori Nojima
Deputy General Manager, Planning and Development Department: Naoki Karoda
Assistant Manager Aerospace Department N°2: Shingo Nakashima
■ Trading and distribution of commodities, industrial and consumer goods.

Electronics and Aerospace Division
2-2 Hitotsubashi 1 Chome, Chiyoda-ku, Tokyo 100
☎ (+81 3) 32 17 63 73 ⊠ 22202/3
General Manager: Koichi Narikawa
General Manager Aerospace Department N°2: Makoto Sato
General Manager Aerospace Department N°2: Masanori Nojima
■ Worldwide trading of telecommunication systems, broadcasting systems, data processing equipment, consumer electronic and aerospace products including airport equipment.

Osaka Office
PO Box 600, Higashi, Osaka 540-91
☎ (+81 6) 220 60 00; 220 67 78 ⊠ 63227; 63321
⊖ (+81 6) 220 77 11
Assistant General Manager Aerospace Department N°2: Masanori Nojima

Nagoya Office
1-6, Higashisakura, 1-chome, Higashi-ku, Nagoya 461
☎ (+81 52) 963 22 26 ⊖ (+81 52) 963 22 01
General Manager Aerospace and Electronics Department: Koji Nisigaki [98]

85-02000 TRACE Japan

5-18 Kami-Ohsaki 4-Chome, Shinagawa-ku, Tokyo 141
☎ (+81 3) 34 92 68 21 ⊖ (+81 3) 34 92 63 88
Managing Director: Toshifumi Ego [98]

90 Sales Agents, Sales & Liaison Offices

Companies represented:

ABG-SEMCA, France	85-01800
Aerazur, Sully, France	85-01800
Aerial Films Inc, USA	85-01800
Aerojet, USA	70-02000
Aerospatiale, France	90-00500
Air Target AB, Sweden	90-00560
Airport Technology, Sweden	90-00560
Akzo Nobel Coatings BV, Netherlands	90-00050
ALCOA, USA	90-00120
Alfred Karcher GmbH & Co, Germany	90-01136
Allied International Corp., USA	90-00130
AlliedSignal, USA	70-02000
AlliedSignal Aerospace, USA	90-00135
AlliedSignal Inc., USA	75-08500
Arianespace, France	80-00050
AS&E, USA	90-00300
BE Aerospace, USA	70-02000
Beta Transformer Technology Corp., USA	90-00430
BF Goodrich Aerospace, USA	90-01100
Bofors AB, Sweden	90-00560
British Aerospace Systems/Equipment Ltd, UK	90-00300
CATIC, China PR	90-00245
Celsius Group, Sweden	90-00560
Coppus Engineering Corp, USA	90-01160
Crompton Instruments Ltd, UK	90-00812
Cubic Corp, USA	90-00380
Custom Electronics Inc., USA	90-01030
Daimler-Benz Aerospace, Germany	70-02000
Dassault Falcon Jet, USA	85-01800
EEV Ltd, UK	90-00300
Eurocopter, Germany	85-01800
Eurocopter, France	85-01800
Fairchild Scandinavian Bellyloading, Sweden	70-02000
Forges de Zeebrugge SA, Belgium	90-00300
Garrett, USA	90-00300
GEC-Marconi Avionics Ltd, UK	90-00300
GEC-Marconi Defence Systems Ltd, UK	90-00300
Helicopter Service, Norway	85-01800
Helio, UK	90-00300
Helitune Ltd, UK	85-01800
Honeywell, USA	70-02000
Honeywell Inc., USA	75-08500
Howden Airdynamics Inc., USA	75-08500
Hughes, USA	70-02000
ILC Data Device Corp., USA	90-00430
Indal Technologies, Canada	70-02000
Indal Technologies, Canada→ Canada	90-01340
Jetstream Aircraft Ltd, UK	90-00300
Kaydon Corp, USA	90-01160
Kingsbury Inc, USA	90-01160
Labinal, France	90-01230
Lemo SA, Switzerland	90-01235
Lemo UK, UK	90-01235
Litton Industries Inc, USA	90-01240
LMS International, Belgium	90-01245
Lockheed Martin, USA	70-02000
Lucas Aerospace Ltd, UK	90-01260
Martin Baker Aircraft Co. Ltd, UK	90-00300
Martin Marietta Corporation, USA	90-01280
McDonnell Douglas, USA	70-02000
Menasco, USA	70-02000
ML Aviation Ltd, UK	90-00300
ML Douglas Equipment Ltd, UK	90-00300
Moog Inc., USA	75-06100
Moss Management Consultants, UK	90-00300
Northrop Grumman, USA	70-02000
PG Products Ltd, UK	90-00300
Piepenbrock Pyrotechnik GmbH, Germany	90-00560
Portescap, Switzerland	90-01675
Pratt & Whitney Canada, Canada	70-02000
Pyser-SGI Limited, UK	90-00300
Racal Acoustics Ltd, UK	90-00300
Racal Avionics Ltd, UK	90-00300
Raytheon, USA	70-02000
Raytheon, USA	90-01820
Raytheon Corporate Jets Inc., USA	90-00300
Rexnord Corp, USA	90-01160
Rockwell International, USA	70-02000
Rockwell International Corp., USA	90-01830
Rolls-Royce, UK	90-01860
SAMM, France	85-01800
Science Applications International, USA	70-02000
Sextant, France	85-01800
Shamban & Co., USA	90-00240
Simmonds Precision Engine Systems, USA	90-01100
Space Systems/Loral, USA	70-02000
Spacehab, USA	70-02000
Spar Aerospace, Canada	70-02000
Stewart-Warner Corp, USA	90-01160
Synchro-Start Products Inc, USA	90-01160
Teledyne Inc., USA	90-02025
The Hymatic Engineering Co. Ltd, UK	90-00300
The Timken Co, USA	90-01160
Thomson, France	70-02000
Thomson Brandt Armenents, France	90-00300
Thomson Sintra ASM Sophia, France	90-00300
Thomson Training & Simulation Ltd, UK	90-01100
TRW Components International, USA	70-02000
United Technologies, USA	70-02000
Vaisala Oy, Finland	90-02200
Vickers Inc., USA	75-08500
Vidisco, Israel	90-00300
Westinghouse Electric Corp., USA	90-02400
Whittaker Controls, USA	90-01100
Zengrange Ltd, UK	90-00300

Sales agents/representatives listed elsewhere:

AlliedSignal Aerospace Marketing	80-00030
AlliedSignal Aerospace Marketing	80-00032
Nippon Systems Industries Corp	80-00450

90-00050 AD Aerospace Finishes Japan

c/o Akzo Nobel KK, Bancho Kaikan 12-1 Goban-cho, Chiyoda-ku, Tokyo 102
☎ (+81 3) 52 75 62 81 ⊖ (+81 3) 32 61 88 74
Contact: A Watanabe
® Netherlands: Akzo Nobel Coatings BV [98]

90-00120 ALCOA Japan Ltd

427 Fuji Building 2-3, Marunouchi 3-chome, Chiyoda-ku, Tokyo 100
☎ (+81 3) 32 87 17 51 ✉ ALUMIAMERICA Tokyo
[E] tokcco1.matsu002@ssw.alcoa.com
⊖ (+81 3) 32 87 17 54
♧ US$ 1,000,000 (1996) ♣ US$ 100,000,000 (1996)
👥 10 (1997) 🏭 200 m2
Director, General Manager: Takashi Matsuoka
® USA: ALCOA [98]

90-00130 Allied International Corp.
Aerospace Dept
CPO Box 1524, Tokyo 100-91
☎ (+81 3) 32 17 50 00; 217 63 73 ⊠ 22202/3
Mgr Aerospace Dept: S. Suzuki
® USA: Allied International Corp. [98]

90-00135 AlliedSignal Aerospace Marketing Sales and Service
(a subsidiary of AlliedSignal Aerospace, USA)
AlliedSignal Inc. Japan
Mita 43 Mori Building, 4th Floor,
13-16, Mita 3-chome, Minato-ku, Tokyo 108
☎ (+81 3) 37 69 85 51 ⊖ (+81 3) 37 69 50 28
Vice President Asia Pacific: Alain Monié
☎ (+81 3) 37 69 85 74
Vice President, Asia Pacific, AES: Jim Griffin
☎ (+81 3) 37 69 85 60
® USA: AlliedSignal Aerospace [98]

90-00142 Ancra Japan
(a subsidiary of Ancra International LLC, USA)
Kan-Nai Kowa Building 9th Floor, 2-9-4 Minato-Cho,
Nakaku, Yokohama
☎ (+81 45) 681 81 71 ⊖ (+81 45) 681 81 70
Managing Director: Taizo Nakamura [98]

90-00145 Anritsu Corp.
5-10-27 Minami-Azabu, Minato-Ku, Tokyo 106
☎ (+81 3) 54 20 06 52 ⊖ (+81 3) 54 20 06 58 [97]

90-00160 Asia Pacific Enterprise Inc
8-13 1-chome Kyobashi, PO Box 1068, Chou-ku, Tokyo
☎ (+81 3) 35 61 52 91; 35 61 52 92
⊖ (+81 3) 35 61 52 93
President: Xavier Koike [98]

90-00215 The BFGoodrich Company of Japan Ltd
Akasaka Kurokawa Bldg, 2-5-8 Akasaka, Minato-ku,
Tokyo 107
☎ (+81 35) 563 27 31 ⊖ (+81 35) 563 27 33 [97]

90-00240 Busak + Shamban K.K.
(† Shamban & Captain Inc.)
Otsuka Building, 3-18-5 Chitose, Sumida-Ku, Tokyo 130
☎ (+81 3) 38 46 47 31 ⊠ 2222705
⊖ (+81 3) 36 35 53 48
General Manager: Hiroshi Okazaki
® USA: Shamban & Co. [98]

90-00245 CATIC
Representative Office
6F Hoochobori Office, Bldg 9-7,
3-Chome Haachobori Chuo-Ku, Tokyo 104
☎ (+81 3) 35 51 04 59 ⊖ (+81 3) 35 51 09 46
® China PR: CATIC [97]

90-00280 Columbia Import & Export Co Ltd
2-1 Sarugaku-cho 1-chome, Chiyoda-ku, Tokyo 101
☎ (+81 3) 32 91 06 41 Ⓔ jpncolu@super.win.or.jp
⊖ (+81 3) 32 93 67 47
♧ Yen 5,250,000,000 (1997)
♣ Yen 11,000,000,000 (1997) 👥 130 (1997) 🏠 290 m2
President: Tsutomu Tobita
Managing Director: M Oinuma
Operations Director: K Hayashi
Sales and Marketing Director: F Nakahata
Import Manager: T Kasai [98]

90-00300 Cornes & Co Ltd
2-5-12 Higashi Kanda, Chiyoda-ku, Tokyo 101
☎ (+81 3) 58 21 16 21 ✉ CORNES Tokyo
⊖ (+81 3) 58 21 16 32
👥 340 (1996)
Sales Manager: M Ishino
® Belgium: Forges de Zeebrugge SA
France: Thomson Brandt Armenents, Thomson Sintra ASM Sophia
Israel: Vidisco
UK: British Aerospace Systems/Equipment Ltd, EEV Ltd, GEC-Marconi Avionics Ltd, GEC-Marconi Defence Systems Ltd, Helio, Jetstream Aircraft Ltd, ML Aviation Ltd, ML Douglas Equipment Ltd, Martin Baker Aircraft Co. Ltd, Moss Management Consultants, PG Products Ltd, Pyser-SGI Limited, Racal Acoustics Ltd, Racal Avionics Ltd, The Hymatic Engineering Co. Ltd, Zengrange Ltd
USA: AS&E, Garrett, Raytheon Corporate Jets Inc. [98]

90-00380 Cubic Corp
(a subsidiary of Cubic Corp, USA)
Madre Matsudaa Building, 4-13 Kioi-cho,
Chiyoda-ku, Tokyo 102
☎ (+81 3) 32 22 10 98 Ⓦ http://www.cubic.com
⊖ (+81 3) 32 39 33 00
♣ US$ 400,000,000 (1996)
Vice President International: Michael W David
■ Air and ground combat training systems. Data links and ESM antenna systems.
® USA: Cubic Corp [98]

90-00410 Daewoo International (Japan) Corp
Room 404 Toranomon Mitsui Building,
3-8-1 Kasumigaseki, Chiyoda-ku, Tokyo 100
(a subsidiary of Daewoo Corp, South Korea)
☎ (+81 3) 35 97 78 01 ⊠ 2226754 dwjpn j
⊖ (+81 3) 35 02 38 24
Osaka Branch
Midosuji Mitsui Building 10th Floor,
1-3 4-chome Bingomachi, Chuoh-ku, Osaka 541
☎ (+81 6) 222 08 48 ⊠ 63731 mituchun j
⊖ (+81 6) 222 08 65
Nagoya Branch
10th Floor Nagoya Hirokoji Building, 3-1 Sakae 2-chome,
Naka-ku, Nagoya 460
☎ (+81 52) 222 37 17 ⊠ 4427578 dwngy j
⊖ (+81 52) 222 37 19
Fukuoka Branch
6th Floor Ait Building, 2-2-22 Momochi-Hama,
Sawara-ku, Fukuoka-City 814
☎ (+81 92) 852 58 00 ⊖ (+81 92) 852 58 01 [98]

90-00430 DDC Electronics KK
Koraku 1-Chome Building 7F, 2-8 Koraku 1-Chome,
Bunkyo-Ku, Tokyo 112
☎ (+81 3) 38 14 76 88 ⊖ (+81 3) 38 14 76 89
Managing Director Asia/Pacific: Koji Moriyama
Sales Manager: Kosako Tomita
® USA: Beta Transformer Technology Corp., ILC Data Device Corp. [98]

90-00500 Aerospatiale
(a subsidiary of Aerospatiale, France)
Akimoto Minami Aoyama Building-5FI,
1-9 Minami Aoyama 4-Chome, Minato-Ku, Tokyo 107
☎ (+81 3) 401 80 32 ⊠ 23661;073000
⊖ (+81 3) 408 50 65
® France: Aerospatiale [98]

90-00560 FFV Japan KK
(a subsidiary of Bofors AB, Sweden)
Shinei-Hongo Building, 27-16 Hongo 2-chome,
Bunkyo-ku, Tokyo 113
☎ (+81 3) 58 02 18 01; 58 02 18 04
⊖ (+81 3) 58 02 18 00
♧ Yen 48,000,000 ♣ Yen 1,000,000,000 👥 6 (1997)
🏠 132 m2
President: Ikunosuke Shimura
Assistant to Management: Takaichi Terada
® Germany: Piepenbrock Pyrotechnik GmbH
Sweden: Air Target AB, Airport Technology, Bofors AB, Celsius Group [98]

90-00600 Fuji Industries Co Ltd (FIC)
1 Shimbashi Ekimae Building, 2-20-15 Shimbashi,
Minato-ku, Tokyo 105
☎ (+81 3) 55 68 54 11 ⊖ (+81 3) 35 71 69 66
♧ Yen 99,000,000 (1997)
♣ Yen 10,734,275,000 (1996) 👥 120 (1997)
Executive Director: Hiromu Hanaki
Managing Director: Ryohei Takata
Manager: Hidenori Akimoto
Nagoya Branch
Horiuchi Building, 25-9 Meieki 3-chome,
Nakamura-ku, Nagoya
☎ (+81 52) 581 81 21 ⊠ 4422241
⊖ (+81 52) 561 55 78
Manager: Katsuo Yoshimura
Gifu Branch
Dai-Gifu Building, 25 Kandamachi 9-chome, Gifu
☎ (+81 58) 262 31 91 ⊖ (+81 58) 266 60 18
Manager: Takeyuki Kaneko
■ Import of parts, materials and equipment for aircraft. [98]

90-00812 Hawker Siddeley Japan KK
11-15 Yato-Machi 2-chome, Tanashi, Tokyo
☎ (+81 814) 24 22 62 01 ⊖ (+81 814) 24 22 63 80
♣ £ 500 000 👥 100
Managing Dir: G. Yamagushi
® UK: Crompton Instruments Ltd [98]

90-00900 Intercontinental Marketing Corp.
IPO Box 5056, Tokyo 100-31
1-19-8 Kakigaracho Nihonbashi, Tokyo 103
☎ (+81 3) 36 61 74 58 ⊖ (+81 3) 36 67 96 46
Manager: T. Fujimoto
Representative: Warren E. Ball
Book Dept Mgr: K. Ito
■ Agent for distribution of AIAA journals & US & Asian aerospace & defence publications. [97]

90-00945 Japan Aircraft Manufacturing Co Ltd
Tokyo Kaijo New Building, 2-1 Marunouchi 1-chome,
Chiyoda-ku, Tokyo 100
☎ (+81 3) 32 13 89 81 ⊖ (+81 3) 32 13 89 85
Gifu Office
Kawaju Fudosan Building 2, Kawasaki-cho,
Kagamihara, Gifu 504
☎ (+81 583) 83 59 80 ⊖ (+81 583) 83 46 36 [98]

90-01030 Jepico Corp
Shinjuku Dai-Ichi Seimei Bldg,
Nishi-Shinjuku 2-7-1, Shinjuku-ku, Tokyo 163
☎ (+81 33) 348 06 11 ⊠ 2323167 jepico j
⊖ (+81 33) 348 06 23
Engineering Dept, Sales Div.: Takashi Kishino
® USA: Custom Electronics Inc. [97]

90-01100 Jupitor Corp.
3-17-4, Minami Aoyama 3 Chome, Jupitor Bldg,
Minato-ku, Tokyo 107
☎ (+81 3) 34 03 13 11 ⊖ (+81 3) 34 03 13 19
♧ Yen 110 000 000 ♣ Yen 10 000 000 000
👥 115 (1996)
President: Kaoru Fujimura ☎ (+81 3) 34 03 13 12
VP & Sales Dept Dir: Tetsuya Oda ☎ (+81 3) 34 03 13 13
■ Import of flight simulators & aircraft related parts.
Nagoya Branch
5-1-32 Sakae, Naka-ku, Nagoya City 460
☎ (+81 52) 241 03 89 ⊖ (+81 52) 241 02 76
VP & Dir: Nobuyuki Ohya
Futtsu Plant
1174-2 Nishi Oowada, Futtsu, Chiba 299-14
☎ (+81 439) 65 11 46 ⊖ (+81 439) 65 30 99
General Mgr: Tomoo Ito
■ Manufacture of blade indexing tool.
® UK: Thomson Training & Simulation Ltd
USA: BF Goodrich Aerospace, Simmonds Precision Engine Systems, Whittaker Controls [97]

90-01134 Kanematsu Aerospace Corp (KAC)
(a subsidiary of Kanematsu Corporation, Japan)
Toranomon Takagi Building 6th Floor,
1-7-2 Nishi Shinbashi, Tokyo 105
☎ (+81 3) 35 80 34 81 ⊖ (+81 3) 35 80 34 67
♧ Yen 90,000,000 (1996)
♣ Yen 2,700,000,000 (1996) 👥 40
President: Goro Sato
Sales Division Manager: Masaaki Ohyama
Subsidiary company:
Kanematsu-Gosho (France) SA [98]

90-01136 Karcher (Japan) Co Ltd
(† Karcher Co Ltd)
Miyagi 981-34
Irene Karcher Building - n°2 Matsusaka
Daira 3-chome-Taiwa-cho, Kurokawa-gun
☎ (+81 22) 344 31 40 ⊖ (+81 22) 344 31 41
Managing Director: Hachiro Sato
® Germany: Alfred Karcher GmbH & Co [98]

90-01160 Kimura Corporation
(† S Kimura & Co. Ltd)
CPO Box 461, Tokyo
☎ (+81 3) 32 13 02 51 ⊖ (+81 3) 32 13 04 70
♤ US$ 18,000,000 (1997) ♣ US$ 29,000,000 (1997)
👥 47 (1997)
President: Y Frank Kimura
Executive Vice President: Taisei Ishizu
■ Bearings, airborne lights, heat exchangers, combustion cabin heaters, radar altimeters, HF communication, military and aircraft gauges.
® USA: Coppus Engineering Corp, Kaydon Corp, Kingsbury Inc, Rexnord Corp, Stewart-Warner Corp, Synchro-Start Products Inc, The Timken Co [98]

90-01220 Kyokuto Boeki Kaisha Ltd Head Office (KBK)
CPO Box 330, Tokyo
7th Floor - New Otemachi Building,
2-1 Otemachi 2-chome, Chiyoda-ku, Tokyo 100
☎ (+81 3) 32 44 35 00; 32 44 35 11 ▣ 22440 j
⊖ (+81 3) 32 46 18 52; 32 46 21 48
♤ Yen 450,000,000 ♣ Yen 170,631,000,000 👥 622
Chairman and Chief Executive Officer: Hisago Miyoshi
President and Chief Operating Officer: Motoo Imakita
Executive Vice President: Masaharu Tatsutu
Executive Vice President: Takashi Hiraishi
Managing Director: Noburu Kawai
Managing Director: Kunihiko Irie
Managing Director: Kazuo Muraoka
Director: Kosaku Inaba
Director: Takeshi Matsui
Director: Masayoshi Fujimoto
Director: Takehiko Takatori
Director: Akira Terashima
General Manager Aerospace and Defence Systems Department: Hiroshi Inoue
■ Aircraft and aerospace related products including engines, communications, navigation, ground support equipment, satellite equipment, launch related equipment. Global positioning systems, electronics, measuring and control technologies.
Representing:
Rockwell-Collins
Technical Centre
1st Floor, NF Park Building, 9-15 Futaba 2-chome,
Shinagawa-ku, Tokyo 142
☎ (+81 3) 57 02 31 11 ⊖ (+81 3) 57 02 31 57
Sapporo Branch Office
6th Floor Daimuru Fuji Central Bldg,
3-2 Minami Ichijo Nishi, Chuo-ku, Sapporo 060
☎ (+81 11) 221 36 28 ⊖ (+81 11) 271 42 40
Sendai Branch Office
3rd Floor - Kasai Building, 9-1 Chuo 2-chome, Aoba-ku,
Sendai 980
☎ (+81 22) 222 82 02 ⊖ (+81 22) 227 08 33
Nagoya Branch Office
6th Floor - Tokai Building, 16-30 Meieki Minami 1-chome,
Nakamura-ku, Nagoya 450
☎ (+81 52) 571 25 71 ⊖ (+81 52) 586 46 16
Osaka Branch Office
7th Floor - Mainichi Osaka Kaikan,
Kitakan, 6-16 Doijima 1-chome, Kita-ku, Osaka 530
☎ (+81 6) 344 11 21 ⊖ (+81 6) 347 06 26
Hiroshima Branch Office
5th Floor Hiroshima Chugin Building,
15-6 Hacchobori - Naka-ku, Hiroshima 730
Fukuoka Branch Office
5th Floor Denki Building Annex 1-82,
Watanabe-dori 2-chome, Chuo-ku, Fukuoka 810
☎ (+81 92) 751 03 03 ⊖ (+81 92) 714 68 73 [98]

90-01230 Labinal K.K.
Tohsei Building 4f, 3-8-2 Nishi Shimbashi, Mirato-ku 3306
☎ (+81 3) 34 33 33 06 ⊖ (+81 3) 34 33 63 42
👥 5
® France: Labinal [98]

90-01235 Lemo Japan Ltd
Kada-Yushima Building, Bunkyo-ku, Tokyo 113
☎ (+81 3) 56 88 89 31 ⊖ (+81 3) 56 88 89 30
® Switzerland: Lemo SA
UK: Lemo UK [98]

90-01240 Litton-Westrex Co Japan
Chiyoda Building, 2-1-2 Marunouchi, Central PO Box 760,
Chiyoda-ku, Tokyo
☎ (+81 3) 32 11 67 91 ⊖ (+81 3) 32 11 67 97
General Manager: Yoji Kawashima
® USA: Litton Industries Inc [98]

90-01245 LMS Japan
Benex 5-3 Building-9th Floor - 20-8,
Shin-Yokohama, 3-Chome, Kohoku-Ku, Yokohama 222
☎ (+81 45) 476 00 77 ⊖ (+81 45) 476 08 70
Sales Manager: M Saito
® Belgium: LMS International [98]

90-01260 Lucas Aerospace Representative Japan
(liaison office of Lucas Varity plc)
Lucas Kk, Lucas House,
7-5 Iwamotocho 1-chome, Chiyoda-ku, Tokyo 101
☎ (+81 3) 38 63 51 81 ⊖ (+81 3) 38 66 77 45
♤ Yen 360,000,000 (1997) 👥 35
Manager: Akinobu Sakashita
® UK: Lucas Aerospace Ltd [98]

90-01280 Martin Marietta International, Inc.
Kowa 35 Building, 5F, 1-14-14, Alaska,
Minato-ku, Tokyo 107
☎ (+81 3) 35 88 52 32 ⊖ (+81 3) 35 88 52 62
President of Far East Region: Robert F Young
® USA: Martin Marietta Corporation [98]

90-01453 Nihon Radiall KK
(a subsidiary of Radiall, France)
8 Kajiyama Building - 6F, 2-11-3 Higashi Nihonbashi,
Chuo-Ku, Tokyo 103
☎ (+81 338) 66 23 90 ⊖ (+81 338) 66 23 91 [98]

90-01470 Northrop Grumman International Inc.
No. 45 Kowa Bldg, 9th Floor,
15-9 Kinami-Aoyana, 1-Chome, Minato-ku, Tokyo 107
☎ (+81 3) 34 75 63 30 ⊖ (+81 3) 34 75 63 33
Dir: Kidosan Kito [97]

90-01480 Nozaki & Co. Ltd
2-6 Nihonbashi Bakurocho, 2-chome, Chuo-Ku,
Tokyo 103-81
☎ (+81 3) 56 41 43 30 ⊠ NOZAKI
⊖ (+81 3) 56 41 43 92; 56 41 43 93
♤ Yen 3,201,037,500 ♣ Yen 135,233,000,000 (1996)
👥 360 (1996)
President: Hiroshi Kohyu
Senior Executive Director: Masaomi Inoue
General Manager Aerospace and Machinery Group: Tatsuto Ishioka
Aerospace and Marketing Group General Manager: Norihiro Noto
General Manager Aerospace and Machinery Department: Kazuyuki Tamura
Deputy General Manager Aerospace and Machinery Department: Akira Nishizawa
Manager 1st Secretary Aerospace and Machinery Department: Hiroshi Miyazaki
Manager 2nd Secretary Aerospace and Machinery Department: Shinji Naka
Nagoya Branch
6-5 Nishiki 1-Chome - Naka-ku, Nagoya 460
☎ (+81 52) 222 70 70 ⊖ (+81 52) 222 70 78
Manager: Takashi Asano [98]

90-01600 Parsons Polytech Inc.
(† Parsons Corp.)
Kyobashi Maruki Bldg 3F, 17-9 Kyobashi 2-chome,
Chuo-Ku, Tokyo 104
☎ (+81 3) 35 61 01 20 ⊖ (+81 3) 35 61 01 27
President: T. Orita [97]

90-01675 Portescap Japan Ltd
(a subs. of Portescap, Switzerland)
7/10 Nihombashi-honcho, 4 Chome, Tokyo 103 Chuo-ku
☎ (+81 3) 32 41 02 01 ▣ 23967 ⊖ (+81 3) 32 41 02 21
👥 8
Managing Dir: Akio Kubo
® Switzerland: Portescap [97]

90-01820 Raytheon Systems Co.
Mitsubishi Electric Building,
Room 413, 2-2-3-Marunouchi, Chiyoda-ku, Tokyo 100
☎ (+81 3) 32 12 15 72 ⊖ (+81 3) 32 12 16 46
Area Manager: John Hedley
® USA: Raytheon [98]

90-01830 Rockwell International Japan Co. Ltd
(a division of Rockwell International Corp., USA)
Shinkawa Sanko Building, 9F, 1-3-17 Chinkawa, Chuo-ku,
Tokyo 104
☎ (+81 3) 55 43 35 83 ⊖ (+81 3) 55 43 35 85
President and Chief Representative - Japan: Akira Suzuki
☎ (+81 3) 55 43 35 40
Avionics and Communications Director: Isao Anzai
☎ (+81 3) 55 43 35 28
Marketing Manager - Japan, LUS: Ken Asai
Director, Heavy Vehicle Systems - Far East: Haruyuki Wada
☎ (+81 3) 55 43 35 60 [98]

90-01860 Rolls-Royce International Ltd Tokyo Branch Office
(† Rolls-Royce (Far East) Ltd)
Kasumigaseki Building, Room 3104,
3-2-5 Kasumigaseki, Chiyoda-ku, Tokyo 100
☎ (+81 3) 35 92 09 66 ⊖ (+81 3) 35 92 09 69
Regional Executive: Bob Moore
Product Support
Haneda Airport, PO Box 51 Ohta-ku, Tokyo 149
☎ (+81 3) 37 47 59 98 ⊖ (+81 3) 37 47 00 58
Chief Service Representative: B Gregg
® UK: Rolls-Royce [98]

90-01960 Shoshin Corp
2-4-2 Muromachi - Nihonbashi, PO Box 173,
Chuo-ku, Tokyo 103-91
☎ (+81 3) 32 70 59 21 ⊠ TOSHOSHIN
E shori@shoshin.co.jp ▣ 2223423 shoshin j
⊖ (+81 3) 32 45 03 69
♤ Yen 48,000,000 (1998)
♣ Yen 6,600,000,000 (1998) 👥 100 (1998) 🏭 750 m2
President: Yasusuke Ishizuka
Senior Managing Director: Norio Takahashi
Operations and Technical Director: T Eko
Director: M Johraku
Purchasing Manager: T Ide
■ Representative and distributor for US and European defence products manufacturers.
Overseas offices/agents:
Shoshin USA, Inc
Shoshin (Europe) Ltd
TPS Aviation, Inc
Osaka Branch
2-3-8 Uchihonmachi, Chuo-ku, Osaka 540
☎ (+81 6) 946 77 51 ▣ 5228780 ⊖ (+81 6) 946 77 60
Branch Manager and Director: K Matsunaga
Nagoya Branch
3-2-4 Meieki - Nakamura-ku, Nagoya 450
☎ (+81 52) 581 72 91 ▣ 4444462
⊖ (+81 52) 581 72 90
Branch Manager and Director: Yusei Tazawa
■ Parts and materials for aircraft NB rocket, H-1 rocket, FCS, ECM and radar. [98]

90-01985 Sperry Marine Asia Inc.
Kuga Building, 11-6 Kuramae, 4-Chome, Taito-ku,
Tokyo 111
☎ (+81 3) 38 63 74 01 ⊖ (+81 3) 38 63 74 55
Manager: S Usuda [98]

90-02025 Teledyne Japan KK
1-3-2 Li Dabashi, Chiyoda-ku, Tokyo 102
☎ (+81 3) 239 90 80 ▣ 5106007490
⊖ (+81 3) 239 90 21; 90 85
Director: Michael A. Robbins
Director: Dan T. Yamamura
® USA: Teledyne Inc. [97]

90-02100 United Technologies International
Uchisaiwaicho Dai Building, 7th Fl., 1-3-3 Uchisaiwaicho, Chiyoda-Ku, Tokyo 100
☎(+81 3) 35 81 32 21 ⊖ (+81 3) 35 81 32 26
President: Yoshikuni Nagashima 97

90-02200 Vaisala KK
(a subs. of Vaisala Oy, Finland)
42 Kagurazaka 6-chome, Shinjuku-ku, Tokyo 162
☎(+81 3) 32 66 96 11 ⊖ (+81 3) 32 66 96 10
President: Rauno Sirola
® Finland: Vaisala Oy
Osaka Branch
1-12r-15, Higashimikuni, Yodogawa-ku, Osaka 532
☎(+81 6) 391 24 41 ⊖ (+81 6) 391 24 42 98

90-02400 Westinghouse Electric Asia SA Japan
Rm.1510 N.Wing, Yurakucho Denki Bldg, 1-7-1 Yurakucho, Chiyoda-ku, Tokyo 100
☎(+81 3) 32 87 13 06 ▣ 2224989 welstk j
⊖ (+81 3) 32 13 66 74
Regional VP Marketing: Donald E. Vance
® USA: Westinghouse Electric Corp. 97

Jordan

Head of State:
H.M. King Hussein I

10 Head of State

10-10000 Office of H M the King
The Royal Palace, Amman
☎(+962 6) 63 73 41 ▣ 21332; 21333 ryl plc jo
⊖ (+962 6) 62 74 22
King of Jordan: H M King Hussein I
Chief of Royal Court: H H Awn al- Khasawnah 98

11 Prime Minister

11-10000 Office of the Prime Minister
PO Box 80, Amman
☎(+962 6) 64 12 11 ▣ 21444 pm jo
⊖ (+962 6) 64 25 20
Prime Minister: Abd-al-Salam al Majali 98

12 Ministries

12-10000 Ministry of Defense
PO Box 79, Amman
☎(+962 6) 221 31; 621 91-95; 621 01
▣ 21200 hqjaa jo ⊖ (+962 6) 68 74 20
Minister of Defense: Abd-al-Salam al- Majali 98

12-20000 Ministry of Foreign Affairs
Jabal Amman 3rd Circle, PO Box 1577, Amman
☎(+962 6) 443 61 ▣ 21255 kharjk jo
Minister of Foreign Affairs: Fayez Tarawneh 98

12-30000 Ministry of Industry and Trade
PO Box 2019, Amman
☎(+962 6) 66 31 91
Minister: Hani al Mulki 98

12-40000 Ministry of the Interior
PO Box 100, Amman
☎(+962 6) 388 49; 631 11 ▣ 23162 mister jo
Minister: Nadir Rashid 98

12-50000 Ministry of Communications
PO Box 71, Amman
☎(+962 6) 62 43 01
Minister: Bassam al Sakit 98

12-60000 Ministry of Finance
King Hussein Street, PO Box 85, 11118 Amman
☎(+962 6) 63 63 21 🄴 mdf@amra.nic.gov.jo
▣ 236 financ jo ⊖ (+962 6) 00 96 26; 64 31 21
Minister of Finance: Suleiman Hafez
Secretary-General: Raef Al-Alami
Director Customs: Nathmi Al-Adbullah
Director Foreign Relations: Dr Odeh Al-Suweis
Chief of Staff: Isa Al-Halaiqah
Administrative Advisor to the Minister: Arwa Al-Muzaffar 98

12-70000 Ministry of Transport
PO Box 1929, Amman
☎(+962 6) 64 14 61
Minister for Transport: Bassam al Sakit 98

15 Parliament

15-10000 Senate
PO Box 72, Amman
☎(+962 6) 66 41 21 ▣ 24149 Maglis Jo
⊖ (+962 6) 70 17 82
President: Ahmad Al-Louzi
Secretary General: Zaid Zuraikat 98

31 Embassies & Attachés

31-00090 Australian Embassy
4th Circle, Jabal Amman, PO Box 35201, Amman
☎(+962 6) 67 32 46; 67 32 47 ▣ 21743 austem jo
⊖ (+962 6) 67 32 60
Ambassador: H.E. Ms M. Wickes ☎(+962 6) 67 32 46
Consul: G. Elton
Second Secretary: W. Richardson 97

31-00150 Embassy of Belgium
Djebel Amman near 4th Circle, PO Box 942, 11118 Amman
Djebel Amman near 4th Circle, Boumedian Street
☎(+962 6) 69 28 59 ✉AMBABEL Amman
▣ 22340 ambel jo ⊖ (+962 6) 69 74 87
Ambassador: H E Mr G Courtois
Head of Chancery: Theo Dirix
Commercial Officer: Mrs Linda Nulens 98

31-00350 Embassy of Chile
Suez Street, PO Box 830663 Zahran, 11183 Amman
☎(+962 6) 81 42 63; 82 33 60 🄴 echilejo@go.com.jo
▣ 21 696 chile jo ⊖ (+962 6) 81 42 63
Ambassador: H E Mr Jorge Iglesias ☎(+962 6) 81 41 35
Third Secretary and Consul: Juan Verga Urcelay
☎(+962 6) 82 33 60 98

31-00450 Embassy of the Czech Republic
PO Box 2213, 11181 Amman
☎(+962 6) 85 70 51; 85 70 52 🄴 czech@nets.com.jo
⊖ (+962 6) 85 70 53
Ambassador: H E Mr Tomas Smetanka
Deputy Chief of Mission and Commercial Officer: Frantisek Barak 98

31-00590 Embassy of France
Mutanabi Street, Jabal Amman, PO Box 5348, Amman
☎(+962 6) 64 12 73/74 ▣ 21219 ambfra jo
⊖ (+962 6) 65 96 06
Ambassador: H.E. Mr Bernard Bajolet
First Counsellor: Stéphane Romatet
Defence Attaché: Col Alain Butori
Cultural & Scientific Attaché: Mme Agnes Romatet
Press Attaché: Pierre Alain Cofinice
Vice-Consul Head of Administration: Franck Lavel
Commercial Office
18 Abu Firas Al Hamadani Street, PO Box 5101, Amman
☎(+962 6) 64 45 70/23 79 ✉COMATA Amman
▣ 21192 comata jo ⊖ (+962 6) 64 15 43
Commercial Counsellor: Christian Ponsot
Commercial Attaché: Pierre Boedoz 97

31-00660 Embassy of the Federal Republic of Germany
31 Benghazi Street, PO Box 183, Amman
☎(+962 6) 68 93 51 ▣ 0493 21235 aaamn jo
⊖ (+962 6) 68 58 87
Ambassador: H E Mr Peter Mende
Deputy Chief of Mission - Defence Attaché: Col Ulrich Bothe
Commercial Counsellor: Mr Ohnemueller 98

31-00690 Embassy of Greece
PO Box 35069, Amman
Jabal Amman 4th Circle
☎(+962 6) 67 23 31 ▣ 21566 98

31-00810 Embassy of the Republic of Hungary
PO Box 3441, 11181 Amman
☎(+962 6) 81 56 14 ▣ 21815 ⊖ (+962 6) 81 58 36
Ambassador: H E Mr Zoltan Pereszle'nyi (resident in Damascus) ☎(+963 11) 611 07 87
Chargé d'Affairs and Commercial Counsellor: Dr Attila Szanto ☎(+962 6) 81 56 14 98

31-00890 Embassy of Italy
Hafiz Ibrahim Street, 5/7 Jebal Al Webdeh, Amman
☎(+962 6) 63 81 85; 63 64 13; 62 43 42
▣ 21143 italdi jo ⊖ (+962 6) 65 97 30
Ambassador: H.E. Mr Romualdo Bettini
Deputy Chief of Mission: Dr Diego Brasioli
Military Naval & Air Attaché: Col Danilo Neri
☎(+962 6) 63 04 64
Commercial Secretary: Dr Raffaele De Benedictis
☎(+962 6) 63 02 26 98

31-01270 Royal Netherlands Embassy
22 Embassies Street, 4th Circle, PO Box 941361, 11194 Amman
☎(+962 6) 69 15 25; 69 40 89; 69 35 81
⊖ (+962 6) 69 41 61
Ambassador: B F Tangelder
Deputy Head of Mission: J A Kappert
Head of Administration: Mrs C E Hackert 98

31-01370 Embassy of Pakistan
Opp.Bisharat Church, Jabal Luweibdeh, PO Box 1232, Amman
☎(+962 6) 62 27 87; 46 80; 63 83 52 ✉PAREP Amman
🄴 pakemb@nets.com.jo. ▣ 21014 ⊖ (+962 6) 61 16 33
Ambassador: H E Mr Ali Sarwar Naqvi
Head of Chancery: Muhammad Ziad Khan
Defense Attaché: Col Zafar Ali 98

31-01480 Embassy of Romania
Shmeisani, PO Box 2869, Amman
21 Abduliaii Bin Masoud Street, Amman
☎(+962 6) 66 77 38 ▣ 21860 romcom jo
⊖ (+962 6) 68 40 18
Ambassador: H E Mr Ioan Agaficioaia
Economic Counsellor: Onoriu Nan
Second Secretary - Cultural and Press Affairs: George Bogdan
Third Secretary - Consular Affairs: Sorin Vlasceanu
Military, Air and Naval Attaché: Col Dumitru Dumitrescu (resident in Cairo) 98

31-01730 Swedish Embassy
PO Box 830536, 11183 Amman
12 Embassy St, Jabal Amman, 4th Circle, Opposite Salah el Din Mosque, Amman
☎(+962 6) 66 91 77; 66 91 78; 68 86 21 ▣ 22039 jo
⊖ (+962 6) 66 91 79
Ambassador: H E Mrs Agneta Bohman
Deputy Chief of Mission: Mrs B M Hedin 98

Contact Reference

Jordan

31-01740 Embassy of Switzerland
Jabal-Amman, PO Box 5341, 11183 Amman
☎ (+962 6) 68 64 16; 68 64 17 ⊠ 21237 amswis jo
⊜ (+962 6) 69 86 85
Head of Chancery: Martin von Arx
Public Affairs: P Albanesi 98

31-01890 Embassy of the United Kingdom
PO Box 87, Abdoun, Amman
☎ (+962 6) 82 31 00 ⊠ 22209 prodrm jo
⊜ (+962 6) 81 37 59
Ambassador: H E Mr Christopher Battiscombe
Deputy Head of Mission: S P Collis
Defence, Naval and Military Attaché: Col T R Dumas
Air Attaché: Wg Cdr D M Sedman (RAF)
Second Secretary Commercial: B Winson 98

80 Service Companies & Organizations

80-05000 Middle East Marketing Services Ltd
PO Box 2994, 11181 Amman
☎ (+962 6) 82 62 27 E mms@go.com.jo
⊜ (+962 6) 82 62 28
General Manager: F Abdeen
■ Defence consultants and representatives. 98

80-29100 Royal Jordanian Air Academy
Amman Marka Airport, PO Box 340513, Amman
☎ (+962 6) 89 46 77 ⊠ 21386 aauniv jo
⊜ (+962 6) 89 46 70
Director General: Mohammad F Khawaldeh
Press and Information Officer: Ayed Omary
■ Civil aviation training school. 98

90 Sales Agents, Sales & Liaison Offices

Companies represented:

CATIC, China PR	90-00335
Chori, Japan	90-01600
Farnell Int'l, UK	90-01800
Goodwill, Taiwan	90-01800
Hewlett Packard, USA	90-01800
Loral Terracom (Lockeed Martin), USA	90-01600
Park Air, UK	90-01600
Wavetek, UK	90-01800

90-00335 CATIC
Amman Branch
PO Box 927215, Amman
☎ (+962 6) 61 14 91 ⊠ 22019 catic jo
⊜ (+962 6) 61 14 90
® China PR: CATIC 97

90-00400 Daewoo Corp Amman Branch
3rd Floor Strand Building N°2, Shmeisani PO Box 940667, 11194 Amman
☎ (+962 6) 68 30 58 ⊠ 21370 daewoo jo
⊜ (+962 6) 68 30 59 98

90-00800 International Trading Establishment
PO Box 5182, 11183 Amman
☎ (+962 6) 82 17 00 ⊜ (+962 6) 81 91 23
President: Ramzi Fathi Abu-Taleb
■ Defence & related products & equipment.
Middle East consulting. 98

90-01600 Petra International Corp.
PO Box 665, Amman
☎ (+962 6) 62 78 01 ⊠ HASSANCO JO
E zntram@go.com.jo ⊠ 21964 itco jo
⊜ (+962 6) 62 78 03
♧ US$ 50,000 (1990) ♣ US$ 2,000,000 (1996)
👥 10 (1997) 🏭 240 sq m
Chairman: Mamduh Abu-Hassan ☎ (+962 6) 62 55 14
Sales and Marketing Director: Abdul Rahman Abu Hassan
Technical Director: Eng M Abu Hassan
General Manager: Ismail Haron
Purchasing Manager: Abdul Hadi
■ Defence equipment.
® Japan: Chori
UK: Park Air
USA: Loral Terracom (Lockeed Martin) 98

90-01800 Scientific & Medical Supplies Co.
PO Box 1387, Amman
☎ (+962 6) 62 49 07; 63 99 07 ⊠ 21456
⊜ (+962 6) 62 82 58
♧ US$ 4,000,000 (1997) ♣ US$ 14,000,000 (1997)
👥 121 (1997) 🏭 1,000 sq m
Managing Director: M Shammout
Operations and Technical Manager: Eng Saleh Al Madhoun
Purchasing Manager: Miss May Aweimrin
® Taiwan: Goodwill
UK: Farnell Int'l, Wavetek
USA: Hewlett Packard 98

Kazakstan

President:
Nursultan Nazarbayev

10 Head of State

10-10000 Office of the President
Rezidentsiya Presidenta Respubliki Kazakstan
"Ploshad" Respubliki Square 4, 480091 Almaty
☎ (+7 3272) 62 33 08 ⊠ 251320 splav
⊜ (+7 3272) 63 98 71
President: Nursultan Nazarbayev ☎ (+7 3272) 62 30 16
Vice President: Yerik Asanbayev ☎ (+7 3272) 62 77 58
Presidential Chief of Staff: Sagynbek Tursunov
State Secretary: Abish Kekilbaev ☎ (+7 3272) 62 77 58
Head of President Administration: Oralbai Abdykarimov ☎ (+7 3272) 62 06 30
Director of President's Office: Nurtai Abikayev ☎ (+7 3272) 62 34 96
Chief of the Presidential Republican Guard: Sat Tokpakbayev
Chief of Defence Department: Nurlan Tlemisov ☎ (+7 3272) 65 01 87
Chief of International Department: Konstantin Zhigalov ☎ (+7 3272) 62 33 08
Aide to the President First Deputy Chairman of the Supreme Economic Council: Zeinolla Kakimhanov ☎ (+7 3272) 62 38 57
Aide to the President: Imangali Tassmagambetov ☎ (+7 3272) 62 45 08
Press Secretary: Krymbek Kusherbaev ☎ (+7 3272) 62 06 30 98

11 Prime Minister

11-10000 Office of the Prime Minister
Ploshad Respublika Sqaure 4, 48 0091 Almaty
☎ (+7 3272) 62 79 66
Prime Minister: Akezhan Kazhegeldin ☎ (+7 3272) 62 30 97
First Deputy Prime Minister: Akhmetzhan Esimov ☎ (+7 3272) 63 89 67
Deputy Prime Minister: Nigmatzhan Isingarin ☎ (+7 3272) 63 89 67
Deputy Prime Minister: Zhanybek Karibzhanov ☎ (+7 3272) 53 23 08
Deputy Prime Minister: Dusembai Duisenov ☎ (+7 3272) 53 57 54
Deputy Prime Minister: Imangali Tasmagambetov ☎ (+7 3272) 63 47 88
Deputy Prime Minister: Aleksandr Pavlov ☎ (+7 3272) 63 27 70
Head of Prime Minister's Office: Trek Begahmetov ☎ (+7 3272) 54 35 57 98

12 Ministries

12-10000 Ministry of Defence
53 Dzhandosova Straat, 480091 Almaty
☎ (+7 3272) 21 47 35 ⊜ (+7 3272) 21 49 36
Minister of Defence: Lieut Gen Muhtar Altynbaev ☎ (+7 3272) 21 47 35
First Deputy Minister of Defence: Alihan Dzarbulov ☎ (+7 3272) 21 62 98
First Deputy of Defence: Fedor Sherbakov ☎ (+7 3272) 28 15 50
Chairman of National Security Committee: Alnur Musaev ☎ (+7 3272) 69 37 65 98

12-20000 Ministry of Foreign Affairs
65 Aiteke bi Straat, 480091 Almaty
☎ (+7 3272) 63 25 38 ⊜ (+7 3272) 63 13 87
Minister of Foreign Affairs: Kasymzhumart Tokaev ☎ (+7 3272) 63 25 38
First Deputy Minister of Foreign Relations: Erlan Idrisov ☎ (+7 3272) 63 78 45 98

12-30000 Ministry of Foreign Economic Relations
prospect Abylai Khan 77, Almaty
☎ (+7 3272) 62 08 09 ⊠ 251364 radon
⊜ (+7 3272) 62 08 06
Minister and Vice Prime Minister: Umirzak Shukeyev
■ Management, regulation and co-ordination of all Foreign Economic activities. 98

12-40000 Ministry of Internal Affairs
ulitsa Kalinina 95, 480091 Almaty
☎ (+7 3272) 62 84 57 ⊜ (+7 3272) 69 55 30
Minister of Internal Affairs: Lieut Gen Kairbek Suleymenov 98

29 Customs & Excise

29-00500 Customs Committee
127 Zhlbck Zholy, 480012 Almaty 98

31 Embassies & Attachés

31-00500 Armenian Diplomatic Mission
Alma Ata
☎ (+7 327) 62 86 18 ⊜ (+7 327) 61 93 95 98

31-00920 Embassy of Japan
Samal-1 - Building 36, 4th/5th Floor, Almaty
☎ (+7 327) 581 15 66 ⊜ (+7 327) 581 15 62 97

31-01890 Embassy of the United Kingdom
173 Ulitsa Furmanova, Almaty
☎ (+7 3272) 50 61 91; 50 61 92 ⊜ (+7 3272) 50 62 60
Ambassador: H E Mr Douglas McAdam
Deputy Chief of Mission: Charles Hile 98

Kenya

Head of State:
Daniel T. arap Moi

10 Head of State

10-10000 Office of the President
Harambee House, Harambee Avenue, PO Box 30 510, Nairobi
☎(+254 2) 274 11 ⊠ 22372 defence
President of the Republic and Commander in Chief: Daniel T arap Moi [98]

10-20000 Office of the Vice-President
Jogoo House, Harambee Avenue, PO Box 30520, Nairobi
☎(+254 2) 284 11
Vice-President: George Saitoti [98]

12 Ministries

12-20000 Ministry of Foreign Affairs and International Co-operation
Harambee House, Harambee Avenue, PO Box 30551, Nairobi
☎(+254 2) 33 44 33 ⊠FOREIGN Nairobi
⊠ 22003 foreign nairobi ⊖ (+254 2) 33 54 94
Minister: Stephen Kalonzo Musyoka [98]

12-30000 Ministry of Local Government
Jogoo House A, PO Box 30 004, Nairobi
Minister for Local Government and Physical Planning: Francis Lotodo [98]

12-40000 Ministry of Home Affairs
Jogoo House, Harambee Avenue, PO Box 30520, Nairobi
☎(+254 2) 22 84 11
Minister: William Ole Ntimama [98]

12-50000 Ministry of Finance
Harambee Avenue, Treasury Building, PO Box 30007, Nairobi
☎(+254 2) 338 11 ⊠FINANCE
Minister: Musalia Mudavadi [98]

25 Defense Forces

25-10000 Defence Headquarters
Ulinzi House, PO Box 40668, Nairobi
☎(+254 2) 72 11 00 ⊠DEFENCE Nairobi
⊠ 22372 defence ⊖ (+254 2) 72 58 54
Chief of General Staff: Lieut Gen D R C Tonje
Chief of Staff: Maj Gen J Koech
Chief of Logistics Division: Brig Mithika [98]

25-20000 Kenya Air Force Headquarters
Eastleigh, PO Box 48888, Nairobi
☎(+254 2) 76 44 01 ⊠KAF NAIROBI ⊠ 22473
⊖ (+254 2) 76 43 93
Commander: Maj Gen N L Leshan
STSSO: S G Mohammed [97]

31 Embassies & Attachés

31-00030 Embassy of Algeria
Comcraft House, Haile Selassie Ave, PO Box 53902, Nairobi
☎(+254 2) 21 38 64/6 ⊖ (+254 2) 21 74 77; 33 72 86
Ambassador: H.E. Mr Arezki Cherfa [98]

31-00080 Embassy of the Argentine Republic
6th Floor, Posta Sacco Plaza, PO Box 30283, Nairobi
☎(+254 2) 33 52 42; 33 99 49 ⊠ 22544 argent
⊖ (+254 2) 21 76 93
Ambassador: H E Mr José M Cantilo
Counsellor: Carlos A Pezzoni
Ambassador's Secretary: Ms L Lopez [98]

31-00090 Australian High Commission
ICIPE House, Riverside Drive, off Chiromo Road, PO Box 39341, Nairobi
☎(+254 2) 44 50 34; 44 50 39; 44 48 11
⊠AUSTCOM Nairobi ⊠ 22203 austcom
⊖ (+254 2) 44 46 17; 44 47 18
High Commissioner: H E Mr John Trotter
Deputy High Commissioner: Jim Lindsay
Public Information Officer: Tony Goldner

Commercial Office
ICIPE House, Riverside Drive, PO Box 39585, Nairobi
☎(+254 2) 44 50 34; 44 50 39; 44 48 11
⊠AUSTRADE Nairobi ⊠ 22203 austcom
⊖ (+254 2) 44 46 16 [98]

31-00100 Embassy of Austria
PO Box 30560, Nairobi
☎(+254 2) 22 82 81; 22 82 82 ⊖ (+254 2) 33 27 92
Ambassador: H E Mr Franz Hörlberger
Counsellor and Deputy Chief of Mission: Ronald Sturh [98]

31-00130 High Commission for the People's Republic of Bangladesh
PO Box 41645, Nairobi
Ole Odume Road off Argwings Kodhek Road, Nairobi
☎(+254 2) 56 28 15; 56 28 16 ⊠ 25077
⊖ (+254 2) 56 28 17
Ambassador: Dr Afsarul M Qader
Deputy Chief of Mission: Abdur Rahim [98]

31-00150 Embassy of Belgium
Muthaiga, Limuru Road, PO Box 30461, Nairobi
☎(+254 2) 74 15 64; 74 15 67 ⊠AMBABEL Nairobi
⊠ 22269 ambabel ke ⊖ (+254 2) 74 15 68
Ambassador: H E Mr R Nijskens
Head of Chancery: G Uytterhaegen
Defence Attaché: Lieut Col J L Herÿgers [98]

31-00230 Brazilian Embassy
4th Floor, Jeevan Bharati Building, Harambee Avenue, PO Box 30754, Nairobi
Harambee Avenue, Nairobi
☎(+254 2) 33 26 49; 21 57 55 ⊠ 22498
⊖ (+254 2) 33 62 45 [98]

31-00280 Embassy of the Republic of Burundi
4th Floor Development House, Moi Ave, PO Box 44439, Nairobi
☎(+254 2) 21 84 59 ⊠ 22425 ambabu
⊖ (+254 2) 21 90 05 [98]

31-00310 Canadian High Commission
Haut Commissariat du Canada
6th Floor Comcraft House, PO Box 30481, Nairobi
Haile Selassie Avenue, Nairobi
☎(+254 2) 21 48 04 ⊠DOMCAN Nairobi
⊠ 22198 domcan ⊖ (+254 2) 21 64 85; 22 69 87
High Commissioner: H E Mr Bernard Dussaul
Deputy Chief of Mission: Paul S Haddow
Counsellor (Development): Jacques Bidard
First Secretary (Administration): David Smart [98]

31-00350 Embassy of Chile
5th Floor International House, PO Box 45554, Nairobi
Mama Ngina Street, Nairobi
☎(+254 2) 33 79 34; 33 13 20 ⊖ (+254 2) 21 56 48
Chargé d'Affaires a.i.: Jose Luis Morales [97]

31-00360 Embassy of the People's Republic of China
Woodlands Road, Hurlingham, PO Box 30508, Nairobi
☎(+254 2) 72 25 59 ⊠ 22235 ⊖ (+254 2) 74 64 02 [98]

31-00380 Embassy of Colombia
International House, 8th Floor, PO Box 48494, Nairobi
International House, 8th Floor, Mama Ngina Street
☎(+254 2) 24 67 70; 24 67 71
E embcol@form-net.com ⊖ (+254 2) 24 67 72
Ambassador: H E Mr German Garcia-Duran
First Secretary, Deputy Chief of Mission: Carlos H Valero [98]

31-00410 Embassy of Costa Rica
PO Box 30750, Nairobi
Geosurvey Building, Wilson Airport, Lanagata Road, Nairobi
☎(+254 2) 50 02 26/7; 72 18 45 [98]

31-00440 High Commission of the Republic of Cyprus
5th Flr, Eagle House, Kimathi Street, PO Box 30739, Nairobi
☎(+254 2) 22 08 81; 44 19 54 ⊠ 22436 cyprohc
⊖ (+254 2) 33 12 32 [98]

31-00450 Embassy of the Czech Republic
Embassy House, Harambee Avenue, PO Box 48785, Nairobi
☎(+254 2) 22 34 48 ⊠ 25115 ⊖ (+254 2) 22 34 48
Chargé d'Affaires a i: H E Mr Zenekd Dobiás
Deputy Chief of Mission, Commercial: Ivan Korcak [98]

31-00460 Royal Danish Embassy
PO Box 40412, Nairobi
11th Floor, HFCK Building, Kenyatta Avenue/Koinange Street, Nairobi
☎(+254 2) 33 10 88/9 ⊠AMBADANE Nairobi
⊠ 22216 ambadane ⊖ (+254 2) 33 14 92
Ambassador: H.E. Mr Erik Fiil
Minister (Deputy Chief of Mission): N.M. Munk
First Secretary: Lars Kjellberg [98]

31-00510 Embassy of the Arab Republic of Egypt
Harambee Plaza, 7th Floor, Haile Selassie Ave. PO Box 30285, Nairobi
☎(+254 2) 57 03 60 ⊖ (+254 2) 21 15 60 [98]

31-00532 Embassy of the State of Eritrea
2nd Floor, New Rehema House, Raphta Rd, PO Box 38651, Nairobi
☎(+254 2) 44 31 63 ⊖ (+254 2) 44 31 65 [98]

31-00540 Embassy of Ethiopia
State House Avenue, PO Box 45198, Nairobi
☎(+254 2) 72 30 27 ⊖ (+254 2) 72 34 01
Ambassador: Teshome Toga Chanaka
Deputy Chief of Mission: Tefera Gizaw
Defence Attaché: Zedwu Taye
First Secretary: Fiseha Medhin
First Secretary: Yemane Abadi
Second Secretary, Commercial: Wondimu Asamnew
Second Secretary: Abede Kelemu
Third Secretary: Tilahun Eshetu
Attaché: Etalemahu Belew
Public Affairs Officer: James Asibwa [98]

31-00580 Embassy of Finland
2nd Floor International House, Mama Ngina St. PO Box 30379, Nairobi
☎(+254 2) 33 47 77; 33 44 08 ⊠ 22010 finlandia
⊖ (+254 2) 33 59 86 [98]

31-00590 Embassy of France
9th Flr Barclays Plaza, Loita Street, PO Box 41784, Nairobi
☎ (+254 2) 33 97 83; 33 99 78 ▣ 22279 ambfrance
⊖ (+254 2) 33 94 21
Ambassador: H.E. Mr Michel Rougagnou
Deputy Chief of Mission: Mrs Caroline Dumas-Desselas
Armed Forces Attaché: Cdt Jean-François Robert [98]

31-00630 Embassy of the Republic of Gabon
Hotel Intercontinental-City Hall Way, PO Box 30353, Nairobi
☎ (+254 2) 33 55 50 ext 8900 ⊖ (+254 2) 33 78 54 [98]

31-00660 Embassy of the Federal Republic of Germany
PO Box 30180, Nairobi
Williamson House, 1st, 8th & 9th Floor 4th Ngong Avenue, Nairobi
☎ (+254 2) 71 25 27 ▣ 22221 aa narb
⊖ (+254 2) 71 48 86
Ambassador: H.E. Mr Michael Gerdts
Deputy Chief of Mission: Dr Erich Kristof
Commercial Counsellor: Rainer Münzel [97]

31-00690 Embassy of Greece
IPS Building, Kimathi Street, PO Box 30543, Nairobi
☎ (+254 2) 34 07 22 ▣ 22008 ⊖ (+254 2) 21 60 44 [98]

31-00810 Embassy of the Republic of Hungary
Ole Odume Road, PO Box 61146, Nairobi
☎ (+254 2) 56 00 60; 56 04 53 ▣ 22364
⊖ (+254 2) 56 01 14 [98]

31-00820 Embassy of Iceland
Bendera Lane, Spring Valley Road, PO Box 45000, Nairobi
☎ (+254 2) 52 14 87 ⊖ (+254 2) 52 14 87 [98]

31-00830 High Commission of India
Jeevan Bharati Bldg, Harambee Avenue,
2nd Floor, PO Box 30074, Nairobi
☎ (+254 2) 22 51 04; 22 51 80
[E] hcindia@form-net.com ⊖ (+254 2) 33 41 67 [98]

31-00850 Embassy of the Islamic Republic of Iran
Dennis Pritt Road, PO Box 49170, Nairobi
☎ (+254 2) 72 03 43; 72 07 96 ⊖ 22563
⊖ (+254 2) 33 99 36 [98]

31-00860 Embassy of the Republic of Iraq
Loresho Road, PO Box 49213, Nairobi
☎ (+254 2) 58 02 62; 58 10 73 ⊖ (+254 2) 58 28 80 [98]

31-00880 Embassy of Israel
PO Box 30354, Nairobi
☎ (+254 2) 72 21 82; 72 21 83 ⊖ (+254 2) 71 59 66
Ambassador: H E Mr Menashe Zipori
Deputy Chief of Mission: Jonathan Meidan
Public Affairs Officer: Ms Judith Munyasa [98]

31-00890 Embassy of Italy
9th Flr Int'l House, Mama Ngina St, PO Box 30107, Nairobi
☎ (+254 2) 33 73 56 ▣ 22251 italdipl
⊖ (+254 2) 33 70 56
Ambassador: H.E. Dr Renato Volpini
Deputy Chief of Mission: Dr Michelangelo Pipan
Attaché (International Organisations): Mrs Giovanna D'Ulisse [98]

31-00920 Embassy of Japan
Kenyatta Avenue, PO Box 60202, Nairobi
☎ (+254 2) 33 29 55 ⊠ TAISHI Nairobi ▣ 22286 taisi Ke
⊖ (+254 2) 21 65 30
Ambassador: H.E. Mrs Ginko Sato
Deputy Chief of Mission: M. Horie [98]

31-00990 Embassy of the Republic of Korea
Anniversary Towers, University Way, PO Box 30455, Nairobi
☎ (+254 2) 33 35 81 [E] koremb@form-net.com
▣ 22300 gongkwan ⊖ (+254 2) 33 28 39
Ambassador: H.E. Mr Lee Dong-ik [98]

31-01000 Embassy of the State of Kuwait
Muthaiga Road, House No.38, PO Box 42353, Nairobi
☎ (+254 2) 76 16 14 [E] kuwaitembassy@form-net.com
▣ 22467 ⊖ (+254 2) 76 28 37 [98]

31-01030 High Commission for the Kingdom of Lesotho
4th Floor Int'l House, Mama Ngina St, PO Box 44096, Nairobi
☎ (+254 2) 22 48 76; 21 77 85 ▣ 22489 boemeli
⊖ (+254 2) 33 74 93 [98]

31-01100 High Commission for the Republic of Malawi
Mvuli Church Road, Westlands, PO Box 30453, Nairobi
☎ (+254 2) 44 05 69 (8 lines) ▣ 22749 kwacha ke
⊖ (+254 2) 44 05 68
High Commissioner: H.E. Mr M.V.L. Phiri (Acting)
Defense Attaché: Lt Col A.S.H. Kwalingana [98]

31-01190 Embassy of Mexico
Kibagare Way, Off Loresho Ridge, PO Box 41139, Nairobi
☎ (+254 2) 58 28 50; 58 25 79
[E] embmexke@form-net.com ▣ 23065 embmexnbi
⊖ (+254 2) 58 15 00 [98]

31-01220 Embassy of the Kingdom of Morocco
3rd Flr.Diamond Trust House, Moi Ave, PO Box 61093, Nairobi
☎ (+254 2) 22 23 61: 22 23 64
[E] embassymorocco@form-net.com ▣ 22531
⊖ (+254 2) 22 23 64 [98]

31-01330 High Commission for the Federal Republic of Nigeria
Lenana Road, Hurlingham, PO Box 30516, Nairobi
☎ (+254 2) 56 41 16 ▣ 22194 nigerian
⊖ (+254 2) 56 41 17; 56 27 76 [98]

31-01370 Embassy of Pakistan
St Michael's Road, Westlands, PO Box 30045, Nairobi
☎ (+254 2) 44 39 11; 44 39 12 ⊠ PAREP Nairobi
▣ 25097 parep nbi ⊖ (+254 2) 44 65 07
High Commissioner: H E Mr S Shafqat Kakakhel
First Secretary, Head of Chancery: Moazzam Ahmed Khan
First Secretary: Leena S Moazzam
Commercial Counsellor: Naeem B Shikh
Press Attaché: Jauher Husain [98]

31-01420 Embassy of the Philippines
State House Road, PO Box 47941, Nairobi
☎ (+254 2) 72 17 91 ⊠ PHILEEMBASSY Nairobi
▣ 22925 ⊖ (+254 2) 72 58 97 [97]

31-01430 Embassy of the Republic of Poland
Kabarnet Road, Woodley, PO Box 30086, Nairobi
☎ (+254 2) 56 62 88/9 ▣ 22266 ⊖ (+254 2) 72 77 01 [98]

31-01440 Embassy of Portugal
10th Floor Reinsurance Plaza,
Aga Khan Walk, PO Box 34020, Nairobi
☎ (+254 2) 33 89 90; 33 98 53 ▣ 22634
⊖ (+254 2) 21 47 11 [98]

31-01485 Embassy of the Russian Federation
Lenana Road, PO Box 30049, Nairobi
☎ (+254 2) 72 24 62; 72 87 00 ▣ 25261
⊖ (+254 2) 72 18 88 [98]

31-01600 Embassy of Saudi Arabia
Murthaiga Road, PO Box 58297, Nairobi
☎ (+254 2) 76 27 81/4 ▣ 22990 [98]

31-01650 Embassy of the Slovak Republic
Milimani Road, PO Box 30204, Nairobi
☎ (+254 2) 72 18 96 ⊖ (+254 2) 72 18 98
Ambassador: H E Mr Ivan Zachar
First Secretary Deputy Chief of Mission: Miroslav Hacek [98]

31-01680 Embassy of Spain
5th Floor, Bruce House, Standard St, PO Box 45503, Nairobi
☎ (+254 2) 22 65 68, 33 57 11/12, 33 63 30
⊠ EMBASPAIN Nairobi ▣ 22157 embaesken
⊖ (+254 2) 33 28 58
Ambassador: H.E. Mr Luis Calvo Merino
Deputy Chief of Mission: Javier Herrera García-Canturri [98]

31-01690 High Commission for the Democratic Socialist Republic of Sri Lanka
Rose Avenue, Off Lenana Road, PO Box 48145, Nairobi
☎ (+254 2) 22 75 77; 22 75 78 ▣ 25081 lankacom
⊖ (+254 2) 22 53 91
High Commissioner: H E Dr G G M Wijewardena
☎ (+254 2) 33 07 38
Second Secretary: Y K A Rohanajith
Administration Attaché: D M P Dissanayake [98]

31-01700 Embassy of the Republic of the Sudan
Minet ICDC House, Mamlaka Road, PO Box 48784, Nairobi
☎ (+254 2) 72 08 53; 72 08 89; 72 08 83
⊖ (+254 2) 71 26 10
Ambassador: Mohamed Ahmed
Deputy Chief of Mission, Counsellor: Ali Sadig
Defence Attaché: Col Ali Adam Nagi [98]

31-01720 High Commission for the Kingdom of Swaziland
3rd Flr Transnational Plaza, PO Box 41887, Nairobi
☎ (+254 2) 33 92 31; 22 2817 ▣ 22085 [98]

31-01730 Swedish Embassy
10th Flr Int'l House, Mama Ngina St, PO Box 30600, Nairobi
☎ (+254 2) 22 90 42/5 ⊠ SVENSK Nairobi
▣ 22264 svensk nrb ⊖ (+254 2) 21 89 08; 22 08 63 [98]

31-01740 Embassy of Switzerland
7th Flr Int'l House, Mama Ngina St, PO Box 30752, Nairobi
☎ (+254 2) 22 87 35 ⊠ AMBASUISSE Nairobi
▣ 22181 amswis nrb ⊖ (+254 2) 72 86 23
Ambassador: H.E. Dr Richard Gaechter [98]

31-01770 High Commission for the United Republic of Tanzania
Continental House, Uhuru Highway, PO Box 47790, Nairobi
☎ (+254 2) 33 10 56/7 ⊖ (+254 2) 21 82 69 [98]

31-01780 Royal Thai Embassy
Rose Avenue, Hurlingham, PO Box 58349, Nairobi
☎ (+254 2) 71 42 76 ⊠ THAIDUTO NAIROBI
⊖ (+254 2) 71 58 01
Ambassador: H E Mr Pongsak Disyatat
Minister Counsellor: Kamchat Klansunthorn
First Secretary: Nat Pinyowattanacheep
Third Secretary: Phalsal Suwisisth
Third Secretary: Miss Potjana Satapana
Office of Commercial Counsellor
PO Box 30637, Nairobi
☎ (+254 2) 58 29 67 ⊖ (+254 2) 73 25 02
Commercial Counsellor: Manoo Tianniam [98]

31-01840 Embassy of the Republic of Turkey
Gigiri Road, PO Box 30785, Nairobi

☎ (+254 2) 52 04 04 ✉ 22346 turkel
⊖ (+254 2) 52 12 37 98

31-01860 Uganda High Commission
5th Fl Uganda House, Kenyatta Ave, PO Box 60853, Nairobi
☎ (+254 2) 33 08 01; 33 08 14 ✉ 22732
⊖ (+254 2) 33 09 70
High Commissioner: H.E. Mr Joseph Tomusange
Deputy Chief of Mission: Mrs Cissy Taliwaku
First Secretary: John A. Chede
Third Secretary: Mrs Rossette N. Katungye
Consular Attaché: Basil B. Kabanda
Financial Attaché: Richard Wasswa Kabaale
Admin Attaché: Mrs Fatuma Seruyange 98

31-01890 United Kingdom High Commission
Upper Hill Road, PO Box 30465, Nairobi
☎ (+254 2) 71 46 99 ⊖ (+254 2) 71 91 07
Deputy High Commissioner: A Tucker (Acting)
Defence Adviser: Col R D Vellacott
First Secretary Commercial: S Martin
Public Information: Ms G Wilson 98

31-01950 Embassy of Venezuela
International House, Mama Ngina St, PO Box 34477, Nairobi
☎ (+254 2) 34 01 34/38 ✉ 22671 ⊖ (+254 2) 33 74 87
Ambassador: Alberto Lizarralde 98

31-01990 Embassy of the Republic of Yemen
Ngong/Kabarnet Road, PO Box 44642, Nairobi
☎ (+254 2) 56 43 79; 56 45 17; 56 45 27
⊖ (+254 2) 56 10 71
Ambassador: Ahmed A Maisari 98

31-02020 Embassy of the Republic of Zaire
Electricity House, Harambee Ave, PO Box 48106, Nairobi
☎ (+254 2) 297 71/2 ✉ 22057 98

31-02030 Embassy of the Republic of Zambia
Nyerere Road, PO Box 48741, Nairobi
☎ (+254 2) 72 47 96; 72 48 50 ✉ 22193
⊖ (+254 2) 71 84 94 98

31-02040 Embassy of the Republic of Zimbabwe
6th Fl Minet ICDC House, Mamlaka Rd, PO Box 30806, Nairobi
☎ (+254 2) 72 10 45; 120 49; 10 71 ✉ 25033
⊖ (+254 2) 72 65 03 98

90 Sales Agents, Sales & Liaison Offices

Companies represented:

BP Solar Limited, UK 90-05000
English Electric Valve, UK 90-20310

90-05000 BP Solar East Africa Limited
Argwings Kodhek Road, PO Box 47796, Nairobi
☎ (+254 2) 56 97 80 ⊖ (+254 2) 56 94 30
® UK: BP Solar Limited 98

90-07000 Daewoo Corp Nairobi Branch
PO Box 43739, Nairobi
Posta Sacco Building 11th Floor, University Way, Nairobi
☎ (+254 2) 22 86 86 ✉ 22759 daewoo kn
⊖ (+254 2) 33 51 74 98

90-20310 Technical & Industrial Representations Ltd
PO Box 45375, Nairobi
☎ (+254 2) 22 69 36 ⊖ (+254 2) 34 06 62
7 (1995) 200 sq m (1995)
Managing Dir: S.N. Amin
Sales & Marketing Mgr: M.S. Amin
® UK: English Electric Valve 97

Korea, Dem. People's Rep.

Head of State:
Kim Chong-il

10 Head of State

10-10000 Office of the President
Pyongyang
Supreme Commander: Kim Chong-il
Vice President: Pak Song-Chol
Vice President: Kim Pyong-sik
Vice President: Kim Yong-chu
Vice President: Yi Chong-ok 98

12 Ministries

12-20000 Ministry of Foreign Affairs
Pyongyang
Minister: Kim Yong-nam 98

12-40000 Ministry of Public Security
Pyongyang
Minister: AVM Paek Hak-nim 98

12-50000 Ministry of External Economic Affairs
Pyongyang
Chairman, External Affairs Committee: Song-tae Yi 98

Korea, Republic

Head of State:
Kim Yong-sam

10 Head of State

10-10000 Office of the President
Chong Wa Dae, 1 Sejong-no, Chongno-gu, Seoul
☎ (+82 2) 770 00 55 ⊖ (+82 2) 770 03 44
President of the Republic: Kim Yong-sam
Agency for National Security Planning
70 Sejong-no - Chongno-gu, Seoul
☎ (+82 2) 757 02 35 ⊖ (+82 2) 273 71 43
Director: Kwon Young-hae 98

11 Prime Minister

11-10000 Office of the Prime Minister
77 Sejong-no, Chongno-gu, Seoul
☎ (+82 2) 720 01 06 ⊖ (+82 2) 720 43 20
Prime Minister: Ko Kon 98

11-10100 Office of the Deputy Prime Ministers
1 Chungang-dong, Gwach'on, Kyonggi, Seoul
☎ (+82 2) 503 90 01; 503 71 71 ✉ 23202
⊖ (+82 2) 730 70 15
Deputy Prime Minister, Finance and Economy Minister: Kyong-sik Kang
Deputy Prime Minister, National Unification Minister: Kwon Ok-i 98

12 Ministries

12-10000 Ministry of National Defence
1 Yonsan-dong 3ga, Yongsan-ku, Seoul
☎ (+82 2) 795 00 71 ⊖ (+82 2) 748 70 56
Minister for National Defence: Kim Dong-jin
☎ (+82 2) 748 60 00 98

12-20000 Ministry of Foreign Affairs
77 Sejong-no, Chongno-gu, Seoul
☎ (+82 2) 720 23 01
✉ 24651; 24652; 24653 woimubu k
⊖ (+82 2) 724 82 91
Minister: Chon-ha Yu ☎ (+82 2) 720 23 01; 720 23 02
First Minister of State for Political Affairs: Kyong-sik Sin
Second Minister of State for Political Affairs: Ms Yun-tok Kim
Deputy Minister of Foreign Affairs: Song Young Shik 98

12-41000 Ministry of Information
77 Sejong-no, Chorgno-ku, Seoul
☎ (+82 2) 720 14 51 ⊖ (+82 2) 732 82 21
Minister: Pong-kyun Kang 98

12-50000 Ministry of Finance
1 Chungang-dong, Gwach'on-si, Kyonggi-do
☎ (+82 2) 503 71 71 ⊖ (+82 2) 503 90 33
Minister of Finance and Economy: Kang Kyong-sik 98

12-60000 Ministry of Trade and Industry
Jungang-dong, Kwach'on-si, Kyunggi-do
☎ (+82 2) 503 94 00; 503 94 01
⊖ (+82 2) 503 94 96; 503 31 42
Minister: Chang-yol Yim 98

12-62000 Ministry of Science and Technology
1 Chungang-dong, Kwach'on, Kyonggi-do
☎ (+82 2) 503 76 19 ⊖ (+82 2) 503 76 73
Minister: Suk-il Kwon 98

31 Embassies & Attachés

31-00100 Embassy of Austria
CPO Box 10099, Seoul 100-699
1913, 19th Floor Kyobo Building - Chogra 1-ka, Chongno-gu, Seoul
☎ (+82 2) 732 90 71; 732 90 72 ⊖ (+82 2) 732 94 86
Ambassador: H E Dr Erwald Jaeger
Deputy Chief of Mission: Ernst Wiedermann 98

31-00130 Embassy People's Republic of Bangladesh
7-3 Hannam-1 dong, Yongsan-ku, Seoul
☎ (+82 2) 796 40 56; 796 40 57
E banglaem@ktnet.co.kr ✉ 22804 bdoot k
⊖ (+82 2) 790 53 13
Ambassador: H E Mr Md Fazlur Rahman

Contact Reference

Counsellor and Deputy Head of Mission:
Md Shamsul Haque [98]

31-00250 Embassy of the Republic of Bulgaria
4th floor Byuck San 125, Building #12-5 - Tongja-dong, Yongsan-ku, Seoul 140-709
☎ (+82 2) 727 58 10; 727 58 11
[E] edby@bora.dacom.co.kr ▣ 28911 bultarg k
⊖ (+82 2) 727 58 12
Ambassador: H E Mr Dimiter Ikonomov
Counsellor Deputy Chief of Mission: Atanas Mladenov
Assistant to the Ambassador: Kyueng-eun Lee [98]

31-00310 Embassy of Canada
PO Box 6299, Seoul 100-662
Kolon Building 10th Floor, 45 Mugyo-Dong
☎ (+82 2) 34 55 60 28 ▣ 27425 canada k
⊖ (+82 2) 774 69 89
Ambassador: H E Mr Michel Perrault
Commercial Counsellor: David Collins
Defence Attaché: Col Keith Eddy
Canadian Forces Attaché Assistant, Seoul: D Rees
Public Information Officer: Patrice Cousineau [98]

31-00535 Embassy of Estonia
Cheong Ahm Building 85-3, Seosomun-dong, Chung-gu, Seoul
☎ (+82 2) 771 43 00 ▣ k 24281; k 23506
⊖ (+82 2) 756 82 45
Honorary Consul: Jae Ik Park [98]

31-00580 Embassy of Finland
KPO Box 1518, Seoul 110-615
1602 Kyobo Building, 1-1 Chongro, Chongro-ku, Seoul 110-714
☎ (+82 2) 732 67 37 ▣ 24343 finnemb k
⊖ (+82 2) 723 49 69
Ambassador: H.E. Mr Jorma Julin
Counsellor: Ms Anneli Vuorinen
Consular & Admin Attaché: Mrs Maija Miettinen-Timio
Consular & Admin Attaché: Ms Teija Hyvönen
Commercial Counsellor: Jarmo Karesto
☎ (+82 2) 725 20 76 (+82 2) 737 31 07 (fax) [97]

31-00590 Embassy of France
CPO Box 1808, Seoul
☎ (+82 2) 312 32 72; 312 32 79 ⊖ (+82 2) 393 61 08
Ambassador: H E Mr Jean-Paul Reau
Defence Attaché: Col François Torres
Assistant Defence Attaché: Mr Lacroix
Public Information Officer: Dominique Delpuech

Office of the Commercial Counsellor
Young-Dong, PO Box 2165, Seoul 135-621
8th Floor, Samheung Building, 705-9 Yeoksam-Dong, Kangnan-Ku, Seoul 135-080
☎ (+82 2) 564 90 32 ▣ 27528 k ⊖ (+82 2) 564 98 32
Economic and Commercial Counsellor: Jean Louis Latour [98]

31-00630 Embassy of the Gabonese Republic
4th Floor - Yoosund Building, 738-20 Hannam-dong, Yongsan-ku, Seoul
☎ (+82 2) 793 95 75; 793 95 76 ⊖ (+82 2) 793 95 74
Ambassador: H E Mr Joseph Mamboungou
Counsellor: Apangha Okegue
First Counsellor: Paul-Patrick Biffot
Counsellor: Richard Lekoumba [98]

31-01210 Embassy of Mongolia
Yong San-Ku Italwon, A Building, Dong 1-139 Namsan Village, Apt 302, Seoul 9111
☎ (+82 2) 793 56 11 ⊖ (+82 2) 793 56 11 [98]

31-01300 New Zealand Embassy
CPO Box 1059, Seoul
Rooms 1802-1805 Kyobo Building, 1 Chongno 1-ga, Chongno-gu, Seoul
☎ (+82 2) 730 77 94 ⊖ (+82 2) 737 48 61
Ambassador: H E Mr Gerald McGhie
Counsellor and Deputy Head of Mission:
Ms Heather Riddell
Counsellor Commercial: Jeffrey Shepherd
First Secretary Administration: Ms Sonia Thomasen
Second Secretary: Peter Lund
Second Secretary: Ms Tracey Black
Attaché: Ms Patricia Watson [98]

31-01340 Royal Norwegian Embassy
CPO Box 355, Seoul 100-603
☎ (+82 2) 795 68 50; 795 68 51
[E] noram@bora.dacom.co.kr ▣ 25155 ambanor
⊖ (+82 2) 798 60 72
Deputy Chief of Mission: Tom Tjomsland
Attaché: Ms Kristen B Hills [98]

31-01430 Embassy of the Republic of Poland
1-72 Donbingo-dong, Yongsan-gu, Seoul
☎ (+82 2) 749 96 81; 749 96 83 ▣ 24590 k
⊖ (+82 2) 749 96 80
Ambassador and Deputy Chief of Mission:
H E Mr Janusz Switkowski
Defence Attaché: Stanislav Rozycki
Commercial Officer: Henryk Socha
Public Information Officer: Mrs Urszula Raznowiecka [98]

31-01480 Embassy of Romania
1-42 U.N. Village, Hannam-Dong, Yongsan-Ku, Seoul
☎ (+82 2) 797 49 24 ⊖ (+82 2) 794 31 14
Ambassador: H.E. Mr Nicolae Ropotean
Commercial Counsellor: Andrei Maghiari
Defence Attaché: Col Ioan Fratita [97]

31-01485 Embassy of the Russian Federation
1001-13, 14, 15 Daechi-dong, Kangnam-ku, Seoul
☎ (+82 2) 552 70 94; 552 70 96 ▣ 29500 conssu k
⊖ (+82 2) 563 35 89; 552 70 98
Ambassador: H.E. Mr George F. Kunadze
Minister Deputy Chief of Mission: Georgi D. Toloraya
Military, Air & Naval Attaché: Col Victor D. Belvi
Asst Military, Air & Naval Attaché: Col Valery I. Sizov
Asst Military, Air & Naval Attaché:
Maj. Konstantin A. Charnov
Trade Representative: Serguei A. Petrov [97]

31-01730 Swedish Embassy
KPO Box 1154, Seoul 110-611
12/F Hanhyo Building, 136 Soelin-dong, Chongro-ku, Seoul 110-110
☎ (+82 2) 738 08 46; 738 11 49; 720 47 67; 737 86 72; 737 86 73 (commerical) ▣ 27231 svensk
⊖ (+82 2) 733 13 17
Ambassador: H E Mr Sture Stierlöf
Trade Counsellor: Åke Karlsson
Counsellor: Ms Annika Svahnström
Contact: Park Haegyong [98]

31-01740 Embassy of Switzerland
CPO Box 2900, Seoul 100-629
32-10 Songwol-dong - Chongro-gu, Seoul
☎ (+82 2) 739 95 11; 739 75 14 ✉ AMBASUISSE Seoul
[E] swissemb@sobak.kornet.nm.kr ⊖ (+82 2) 737 93 92
Ambassador: H E Mr Eric N Pfister
Head of Chancery: Bernhard Schurch
First Secretary: Christian Schoenenberger
Commercial Counsellor: Ms Irene Flueckiger [98]

31-01840 Embassy of the Republic of Turkey
726-116 Hannam-dong - Yongsan-gu, Seoul 140-212
☎ (+82 2) 794 02 55; 794 37 78 ⊖ (+82 2) 797 85 46
Ambassador: H E Mr Halil Dag
Counsellor for Commercial Affairs: Ahmet Yakici
Economic Counsellor: Nuri Eran Tortop
Military Attaché: Col Sevki Aksu ☎ (+82 2) 794 92 48
Third Secretary: Miss Meltem Buyukkaras
Third Secretary: Cinar Ergin
Attaché: Adnan Bekcekaral
Attaché: Ozcan Sezar
Attaché: Hakan Gumus [98]

31-01890 Embassy of the United Kingdom
4 Chung-dong, Chung-ku, Seoul 100
☎ (+82 2) 735 73 41; 735 73 42; 735 73 43
▣ 27320 prodrom k
⊖ (+82 2) 725 17 38; 736 62 41 (commercial); 733 83 68 (defence supplies); 738 27 97 (economic section)
Ambassador: H E Mr Stephen D R Brown
Deputy Head of Mission: D R Marsh
Counsellor, Political: S J Buckle
Counsellor, Regional Affairs: P J Aron
Defence and Military Attaché: Brig C D Parr
Naval and Air Attaché: Gp Capt S J Coy
Cultural Attaché (British Council Director): T Toney [98]

65 Industry, Trade & Armed Forces Associations

65-00500 Construction Association of Korea
8th Floor Construction Building, 71-2 Nonhyon-don, Kangnam-gu, Seoul 135-701
☎ (+82 2) 547 61 01; 547 61 07; 548 71 07
✉ CONASKOR Seoul ▣ 28972 cakorea k
⊖ (+82 2) 517 69 79; 542 62 64
President: Choi Won-suk
Director Technical Survey and Research: Soong-kon Kim [98]

65-00700 Electronic Industries Association of Korea (EIAK)
#648 Yeoksam-dong, Kangnam-gu, Seoul
☎ (+82 2) 553 09 41; 553 09 47
[E] eiak@soback.kornet.nm.kr
⊖ (+82 2) 555 61 95; 563 73 71
Chairman (LG Semicon Co Ltd): Cha-Hak Koo
Vice Chairman (Samsung Electronics Co Ltd):
Kwang Ho Kim
Vice Chairman (LG Electronics Co Ltd): John Koo
Vice Chairman (Daewoo Electronics Co Ltd): Jae Yeol Yang
Vice Chairman (Hyundai Electronic Industries Co Ltd):
Joo Young Kim
Vice Chairman (Namsung Corp): Bong Soo Yoon
Vice Chairman (Daeduck Electronics Co Ltd):
Jung Shick Kim
Vice Chairman (Dong Hyeong Magnet Wire Co Ltd):
Ke Dong Park
Vice Chairman (Hansoi Electronics Inc): Don Wan Jo
Vice Chairman (EIAK): Sang Won Rhee
Director (Anam Industrial Co Ltd): In Kil Hwang
Director (Korea Computer Inc): Sung Chai Hong
Director (Autonics Corporation): Hwan Ki Park
Director (Korea Electronics Co): Jung So Kwak
Director (Dae A Leadwire Co Ltd): Sung Bak Hwang
Director (LG Information and Communications Ltd):
Jang Ho Chung
Director (Dae Ryung Ind Inc): Soung Woo Kwon
Director (Orion Electric Co Ltd): Gil Yong Um
Director (Daewoo Telecom Ltd): Ki Bum Yoo
Director (Samhwa Capacitor Co Ltd): Myung Heui Han
Director (Doosan Elecctro-Materials Co Ltd):
Chung Hoon Lee
■ Non-profit making association representing Korean manufacturers and exporters of electrical and electronic products. Membership of 480 companies. [98]

65-11300 Korea Defense Industry Association (KDIA)
13th Floor, Sungwoo Building 51-1, Dohwa 2-dong, Mapo-Ku, Seoul 121-042
☎ (+82 2) 32 70 60 50 ⊖ (+82 2) 32 70 60 60
Chairman: Ryu Chan-U
Executive Vice Chairman: Maj Gen Ahn Byung-Kil (retd)
☎ (+82 2) 716 76 00
Managing Director: Rear Adm Choe Jae Yol (retd)
☎ (+82 2) 32 70 60 04
International Corporation Director: Jou Young-Il
☎ (+82 2) 32 70 60 05
Manager: Sohn Yoon ☎ (+82 2) 32 70 60 35
Assistant Manager: Kim Jung Hwan
■ Non-profit making organisation representing Korean manufacturers of defence equipment. [98]

70 Financial Holding Groups, Parent Companies

70-00100 Daewoo Corporation Trading Division
541 5-Ga Namdaemunno, Chung-gu, CPO Box 2810, Seoul
☎ (+82 2) 759 21 14; 759 27 41; 759 26 71
✉ DAEWOO Seoul [W] http://www.dwc.co.kr
▣ 23341-2 daewoo k ⊖ (+82 2) 759 37 81; 753 94 89
♧ US$ 2,200,945,000 (1996)
♣ US$ 22,521,109,000 (1996) 👥 2,560

Founder and Chairman: Kim Woo-Choong
Managing Director: Kim Sung-Ki ☎ (+82 2) 759 20 43
General Trading Division President: Kang Byong-Ho
Marketing Director: Cho Jae-Woo
Aerospace and Defence Business Team: B W Woo
☎ (+82 2) 759 27 59
■ Parent concern, trading and construction arm for the Daewoo Group of companies. 98

75 Manufacturing Companies

75-00100 Asia Motors Co Ltd
(a subsidiary of KIA Motors Co Ltd, Korea)
15-1 Yoido-Dong, Youngdeungpo-Ku, Seoul
☎ (+82 2) 788 86 63 W http://www.asia.co.kr
24627 asiamco k ⊖ (+82 2) 785 14 85
US$ 261,748,669 (1996)
♣ US$ 1,884,913,224 (1996) 9,196 (1996)
Managing Director: Cho Kyu Ho
Operations Director: Park Ki Young
Sales and Marketing Director: Woo Young Dae
Strategic Planning and Co-ordination: Kim Wook
■ Logistics, auxiliary and liaison vehicles, from 1/4t 4x4 utility vehicle to 5t 6x6 cargo truck. KM90 series armoured personnel carrier. 98

75-00540 Daewoo Corporation Construction Division
84-11 5-ga Namdaemunno, CPO Box 8269, Chung-gu, Seoul
☎ (+82 2) 259 31 14 23341-2 daewoo
⊖ (+82 2) 259 31 13 98

75-00550 Daewoo Electronics Co Ltd
(a member of the Daewoo Group)
612-1 Ahyeon-Dong, CPO Box 8003, Mapo-Gu, Seoul
☎ (+82 2) 360 80 01; 360 80 02; 360 80 03; 360 80 06; 360 80 07; 360 80 08 ✉ DAEWOO ELEC
E a8dooe@web.dwe.co.kr
W http://www.dwe.daewoo.co.kr
28177; 28178; 29549
⊖ (+82 2) 364 55 88; 364 50 95; 364 78 32; 360 78 57 (public relations)
♣ US$ 4,000,000,000 (1995) 14,500
Daewoo Group Chairman: Woo-Joong Kim
Chairman and Chief Executive Officer: Soon-Hoon Bae
President and Chief Operating Officer: Jae-Yeol Yang
Specialist, TV Marketing Team: S M Jin
☎ (+82 2) 360 75 33 E h4gooe@web.dwe.co.kr
■ Manufacturer & supplier of electronics products. 98

75-00600 Daewoo Heavy Industries Ltd
(a member of the Daewoo Group)
Head Office
6 Manseok-dong, dong-gu, PO Box 7955, Inchon 160
☎ (+82 32) 760 11 14 ⊖ (+82 32) 762 15 46
♣ US$ 2 052 069 000 (1994) 19 677
President Representative Dir: Yi Bong-Hui
☎ (+82 2) 726 15 04
Exec. VP Representative Dir: Kim Son-Ik
☎ (+82 2) 779 07 60
Aerospace Division
Marketing & Sales
Daewoo Center Bldg,
541 5-ga, Namdaemunno, Chung-gu, Seoul
☎ (+82 2) 779 16 69 ⊖ (+82 2) 756 26 79
Exec. Managing Dir: Kwak Sang-Ha ☎ (+82 2) 755 02 90
Director: Park Kun-Woo ☎ (+82 2) 771 70 18
Plant
3 Block, Sungju Industrial Complex, Changwon, Kyungnam
☎ (+82 551) 80 65 21 ⊖ (+82 551) 85 23 83
Managing Dir: Lee Kwang-Jin ☎ (+82 551) 80 65 11
Research Dir: Ra Duk-Joo ☎ (+82 551) 80 65 30
■ Manufacture of fuselage, wing & composite components for commercial/military fixed & rotary wing aircraft.
Defense Products Division
Marketing & Sales
Daewoo Center Bldg,
541 5-ga, Namdaemunno, Chung-gu, Seoul
☎ (+82 2) 726 32 63 ⊖ (+82 2) 726 32 69
Managing Dir: Oh Yung-Il ☎ (+82 2) 726 30 17
Marketing & Sales Exec. Dir: Ahn Hyo-Kee
☎ (+82 2) 726 30 15
General Mgr Marketing & Sales: Lee Kyu-Chun
☎ (+82 2) 726 32 60
Plant
3 Block, Sungju Industrial Complex, Changwon, Kyungnam
☎ (+82 551) 81 22 62 ⊖ (+82 551) 84 46 23
Managing Dir R&D Centre: Ahn In ☎ (+82 551) 80 63 07
■ Manufacture special purpose, armoured & tracked combat vehicles. Air defence systems, gun & missile, naval & ground. Upgrading & modification of armoured vehicles & naval guns. Multiple launcher rocket systems. Defence electronics.
Representatives:
Daewoo Int'l (America) Corp.
Daewoo Canada Ltd
Daewoo Handels GmbH
Daewoo UK Ltd
Daewoo France SarL
Daewoo Int'l (Japan) Corp.
Shipbuilding Div.
PO Box 6208, Seoul
☎ (+82 2) 726 19 11 24698 dwokpo k
⊖ (+82 2) 778 54 23
11 000 3 530 000 sq m
Special Ship Marketing Dept Dir: Kim Duk-Joong
☎ (+82 2) 726 19 11
Special Ships Div.
1 Ajoo-dong, Koje si, Kyungnam
☎ (+82 558) 680 21 14 29131/3 dwokpo k
⊖ (+82 558) 681 74 47
■ Design & production of a full range of naval ships, combatant ships, auxiliaries, patrol boats, landing ships & submarines.
Seoul Office
541 Namdaemunno 5-Ga, Chung-Gu, PO Box 5208, Seoul 100
☎ (+82 2) 726 19 11 24698 dwokpo k
⊖ (+82 2) 778 54 23 97

75-01200 Eusung Industrial Co. Ltd
395-166 Seogyo-Dong, Mapo-Gu, Seoul
☎ (+82 2) 324 97 11 25720 esibt k
⊖ (+82 2) 333 38 24
US$ 8 254 512 ♣ US$ 39 000 000 250
Managing Dir: Hwa Young Jung
Sales & Marketing Dir: H.S. Ha
Operations & Technical Dir: Ah Young Ho
Purchasing Mgr: Buyng Ki Tung
■ Ground barrier barbed tape concertina. 97

75-01400 GoldStar Electric Co. Ltd
(a subs. of The Lucky-Goldstar Int'l Corp)
Yoido PO Box 880, Seoul
20 Yoido-Dong, Yeoungdeungpo-Gu, Seoul 100, Seoul 150
☎ (+82 2) 787 11 14 ✉ GOLDWAVE Seoul
28295 gswave k ⊖ (+82 2) 784 35 49
Dir.Exports: Jung Jong-Hyun
■ Communications equipment, minehunting & detection equipment. 97

75-01480 Han Sun International Inc.
PO Box 2953, Seoul
☎ (+82 2) 388 02 57 ⊖ (+82 2) 389 98 25
US$ 370 000 (1996) ♣ US$ 3 500 000 (1996) 8
200 sq m
Managing Dir: S.J. Park
Operations Dir: S.W. Park
Sales & Marketing Dir: S.C. Kang
Purchasing Mgr: B.J. Lee
■ Manufacture fabric for outwear. 97

75-01505 Hanjin Heavy Industries Co. Ltd (HHIC)
(† Korea Shipbuilding & Engineering Corp)
29 5-ka Pongnae Dong, Yongdo-Gu, Pusan 606-065
☎ (+82 51) 410 31 14 53330 hhicp k
⊖ (+82 51) 410 36 67
US$ 200,000,000 (1995)
♣ US$ 390,000,000 (1995) 3,300 180,000 sq m
Managing Director: S S Kim
Assistant General Manager: C S Chung
■ Shipbuilding - military: fast attack craft, corvette, frigates, OPV, supply vessel, oil skimmers, salvage tugs, harbour tugs and other special craft. General ship repairs, engine, electronic and electric repairs. Damage repairs, conversions and tank recoating. → 80
Seoul Office
Marine Center 118 2-Ka, Namdaemun-Ro, Chung-Ku, Seoul
☎ (+82 2) 728 54 41 23669 hhics k
⊖ (+82 2) 756 54 55
Ulsan Shipyard
492 Yongjam-dong, Nam-Ku, Ulsan
☎ (+82 522) 259 31 14 ⊖ (+82 522) 259 32 23
250,000 m2
Youngdo Shipyard
29 5-ka Pongnae-Dong, Youngdo-Ku, Pusan
☎ (+82 51) 410 33 11; 410 33 15 53330 hhicp k
⊖ (+82 51) 410 33 37; 410 33 38
300,000 m2
Dadaepo Factory
370-6 Dadae-dong, Saha-ku, Pusan
☎ (+82 51) 260 31 14 ⊖ (+82 51) 260 31 14 98

75-01540 Hyundai Corp.
140-2 Gye-Dong, Chongro-Ku, Hyundai Bldg, 3rd Floor, Seoul 110-793
☎ (+82 2) 741 23 01 ✉ HDSANSA Seoul
24119; 32101; 23175-7 hdcorp k
⊖ (+82 2) 746 10 93
Chairman: Mongi Koo Chung
President: Se Yong Park
■ Marketing & procurement organisation for all companies of Hyundai Group. Main group activities include naval construction, general contracting & civil engineering, buildings, railway locomotives & rolling stock, ship propulsion machinery, ship repair, materials handling equipment, vehicles, chemicals, military uniforms & equipment, webbing, clothing & footwear. Main battle tanks, various kinds of military defence items. → 90 97

75-01550 Hyundai Electrical Engineering Co. Ltd
337 Ilsan-dong, Ulsan, S.Kyongsang Prov. 690
☎ (+82 522) 542 41 51 52172; 53752
■ Transformers, generators, motors & circuit breakers for marine & industrial use. Export enquiries to Hyundai Corp.
Seoul Office
Seoul
☎ (+82 2) 741 26 11 97

75-01560 Hyundai Electronics Industries Co Head Office
San 136-1, Ami-ri, Bubal-eub, Ichon-kun, Kyoungki-do 467-860
☎ (+82 2) 741 06 61 ⊖ (+82 2) 741 07 37
■ Semi-conductors, computers and telecommunications products. Construction of satellite and auto navigation systems.
Information & Communication Service
66 Chuckseon-Dong, Chongro-Gu, Seoul 110-052
☎ (+82 2) 741 13 11/25 Ext.214 (Direct)
29793; 29794; 28361 hdtard k ⊖ (+82 2) 741 03 17
President: Chung Mong Hun
Marketing Dir: Eoung-Sik Yoo
Senior Mgr, Int'l Sales Dept (Telecomm.): Sung-Young Ann
■ Semiconductors, electronic & information systems. 98

75-01570 Hyundai Engine & Machinery Co. Ltd
1-5 Chonha-dong, Ulsan City 690
☎ (+82 522) 542 41 51-60 52191; 53815
■ Ship propulsion machinery, diesel engines, pumps. 97

75-01600 Hyundai Heavy Industries Co. Ltd
(a subs. of Hyundai Corp.)
Special & Naval Shipbuilding Div.
1 Cheonha-dong, Dong-gu, Ulsan, Kyung-nam 682-992
☎ (+82 522) 30 28 40; 30 28 41
✉ HHIYARD Dong Ulsan 52452 hhiyard k
⊖ (+82 522) 30 34 91
♣ US$ 4 900 000 000 (1995) 27 000 (company total)
1 400 000 m2 (works)
President: Jung Kook Kim
Managing Dir: C.H. Cho
Sales & Marketing Dir: RAdm K.S. Kim (Rtd)
Operations/Technical Dir: D.Y. Kwak
Sales & Marketing Mgr: Ki Young Sung
■ Warship construction-frigates, corvettes, fast patrol boats, landing ships & craft, launches & small high-speed craft. Specialised in design & construction of all kinds of naval ships & special purpose vessels. Export enquiries to Special & Naval Shipbuilding Div. or Hyundai Corp.
Seoul Office
140-2 Kye-dong, Chongro-ku, Seoul 110-270
☎ (+82 2) 746 46 71 ✉ HDYARD 28361 hhiyard k
⊖ (+82 2) 741 11 52 97

75-01602 Hyundai Motor Co. Ltd
140-2, Gye-dong, Chongro-Ku, Hyundai Bldg, 10th Floor, Seoul 110-793
☎ (+82 2) 744 02 11 ✉ HYUNDAIMOTOR Seoul
⊠ 23521 hdmoco k
■ Cargo trucks, dump trucks, special vehicles [97]

75-02010 Kangnam Corporation
(a subsidiary of Kunsul Chemical Industrial Co Ltd)
399 Gupyung-Dong, Saha-Ku, Pusan
☎ (+82 51) 262 97 70
⊖ (+82 51) 262 74 00; 262 97 71
♧ US$ 300,000 (1996) ♣ US$ 82,100,000 (1996)
👥 700 (1997)
President: Capt Jin Ok Lee (rtd)
Senior Manager Overseas Business Development Department: Jae Kwan Park
Technical Director: Dong Jo Lim
Managing Director: Young Tae Ahn
Operations Director: Hyo Su Kim
Sales and Marketing Director: Jae Kwan Park
Purchasing Manager: Jae Sup Shin
■ Mine countermeasure vessels, high speed patrol boats, LCVP (Landing Craft, vehicle and personnel), rigid inflatable boats, enclosed life boats, steel shipbuilding and repairs, steel structures. [98]

75-03600 NMB Korea Co. Ltd
9F Woo Duck Building 832-2, Yeok Sam-Dong, Kang Nam-ku 135-080
☎ (+82 2) 557 44 67 ⊖ (+82 2) 557 44 78
👥 25
President: Y H Lee [98]

75-03750 Optechs Korea Inc
291-1 Gajwa-dong, Seo-ku, Inchon
☎ (+82 32) 578 14 41; 745 14 39
[W] http://ibk.co.kr/company/optechs.htm
⊖ (+82 32) 578 14 42
♧ US$ 862,500 (1997) ♣ US$ 1,625,000,000 (1996)
👥 183 (1997) 🏠 2,995 m2
Managing Director: Nam Jung Kim
Operations Director: Tae Woong Kim
Sales and Marketing Director: Tae Young Kim
Purchasing Manager: Hong Duk Kim
Deputy Manager: Seong-Cheon Park
■ Development and manufacture of precision optics. Laser optics, night vision devices, sights for self-propelled howitzers, fire control equipment, mortar sights, and sights for recoilless rifles. [98]

75-04300 Saft Korea Co Ltd
195-42 Anyang 7-Dong, Manan-Ku, Anyang City, Kyunggi-Do
☎ (+82 343) 41 11 34 ⊖ (+82 343) 41 11 39
Managing Director: Sung Kyu Kim
Industrial Battery Division
501 Ansan Bldg, 773-4 Yeoksam-dong, Kangnam-ku, Seoul 135080
☎ (+82 2) 501 00 31/3 ⊖ (+82 2) 501 00 34 [98]

75-04630 Samsung Aerospace Industries Ltd
Head Office
25th Floor Samsung Life Building, 150 2-ka Taepyung-Ro Chung-ku, Seoul 100-716
☎ (+82 2) 751 88 53; 751 88 54
[W] http://www.ssa.samsung.co.kr/ ⊖ (+82 2) 751 85 90
♣ US$ 850 000 000 (1993) 👥 6 100
Vice Chairman and Chief Executive Officer: Dae Won Lee
Executive Vice President and Chief Executive Officer: Moo Sung Yu
Executive Vice President and Chief Executive Officer: Bok Hyun Ahn
■ Fighter aircraft & trainer aircraft, aircraft engine, aircraft engine overhaul, airframe parts, aircraft engine parts, propulsion systems for rockets & missiles. [98]

75-04650 Samsung Semiconductor & Telecommunications Co Ltd (SST)
Head Office
8/12 Floor Samsung Main Building, 250 2-ka, Taepyung-ro, Chung-ku, Seoul
☎ (+82 2) 727 01 14 [W] http://www.secc.samsung.com
⊖ (+82 2) 753 09 67 [98]

75-04660 Samsung Shipbuilding & Heavy Industry Co Ltd
Dongnam Tower 89-25, Daechi-dong, Kangnam-gu, Seoul 135-280
☎ (+82 2) 34 58 60 30; 34 58 60 43; 34 58 65 40; 34 58 75 70
[W] http://sun.ship.samsung.co.kr/
⊖ (+82 2) 34 58 65 01; 34 58 65 02; 34 58 65 03; 34 58 65 04; 34 58 65 05; 34 58 65 06
👥 10,807 (1996)
President and Chief Executive Officer: Hai Kyoo Lee [98]

80 Service Companies & Organizations

Hanjin Heavy Industries Co. Ltd . . . 75-01505

80-00010 A M Corporation
Juangang Mutual Saving and Finance, Building 8th Floor, #61-5 Nonhyun Dong, Kangnam Ku, Seoul, Kangnam Ku, Seoul
☎ (+82 2) 540 33 63 [E] djl1005@aol.com
⊖ (+82 2) 540 33 65
👥 15
Managing Director: Young Woo Lee
Marketing Director: David J Lee
■ Consulting for defence manufacturers, army, navy, airforce and commercial business, specialising in the South Korean market. [98]

80-00080 AlliedSignal Aerospace Marketing Sales and Service
Kimpo International Airport 7F, Asiana Airlines Hangar, Seoul 157-250
Kimpo International Airport 7th Floor, Asiana Airlines Hangar, 1 Konghang-dong, Kangseo-ku
Field Service Engineer: Byung-Gi (BG) Kang
☎ (+82 2) 661 02 15; 661 02 14 (fax)
Technical Representative: Young T Kim
☎ (+82 2) 664 02 18; 665 42 15 (fax) ⊠ 69491 k
■ On-site customer support for AlliedSignal Aerospace. → 90 [98]

80-00090 AlliedSignal Korea Ltd
Seoul 137-040
6th Floor Sambo Mutual Savings and Finance Building, 44-1 Banpo-dong, Seocho-ku, Seoul 137-040
Account Team Leader: K M (Kevin) Yoo
■ On-site customer support for AlliedSignal Aerospace. [98]

80-00300 Cogent
(a Nortel Professional Services Business)
No. 47, 7th Floor, Hyekang Building, 157-36, Samsung-Don, Kangnam-ku, Seoul 135-090
☎ (+82 02) 569 22 77 ⊖ (+82 02) 569 22 24
Director: E L Henderson
Representing:
Cogent → UK 80 [98]

80-00700 Hughes International Corporation
Kyobo Building, Room 1606, 1-1 Jongro 1-ka, Jongro-ku, Seoul 110
☎ (+82 2) 734 47 82 ⊖ (+82 2) 734 47 85 [98]

80-00725 Genex International Corporation
460-2 Pyungchang-dong, Chongro-gu, Seoul 110-012
☎ (+82 2) 379 3100 ⊖ (+82 2) 394 1786
👥 10
Managing Director: Chris P Bach
Director: Chang E Ju [E] ceju@hitel.kol.co.kr
■ Manufacturers representative and consultants in defence. Commission agent for foreign customers who want to do business in Korea. [97]

80-01700 PMsoft Korea Ltd
(a subsidiary of Artemis Management Systems Ltd, UK)
5th Floor Yongma Building, 668-2 Yoksam-dong, Kangnam-ku, Seoul 135-080
☎ (+82 2) 569 07 30 ⊖ (+82 2) 569 07 33
Country Manager: Young Min Park [98]

80-05000 TSM Corporation
(in partnership with TSM Corporation)
Taegun B/D #201, 1009-1 Bangbae-Dong, Seocho-Ku, Seoul
☎ (+82 2) 586 66 45 [E] tsmc@unitel.co.kr
⊖ (+82 2) 521 09 76
♧ £ 125,000 (1996) ♣ £ 1,500,000 (1996)
👥 21 (1996) 🏠 505 m2
Managing Director: Benjamin Park
Marketing Director: J Park ☎ (+82 2) 583 22 94
Technical Director: J Kim ☎ (+82 2) 597 45 75
Purchasing Manager: K Kim
® Switzerland: Trimos-Sylvac Task Force [98]

80-11800 Korea Technology Bank
FKI Building 1st Floor 28-1, Yoidodong, Youngdeungpogu 150-756
☎ (+82 2) 782 76 00 ext. 470 [W] http://www.ktb.co.kr
⊖ (+82 2) 783 42 10
♣ US$ 542,000,000 (1995)
Senior Consultant: Jongeon Lee [E] joanlee@pis.kpc.or.kr
■ Financial services for research and development. [98]

85 Stockists, Distributors, Spare Parts Suppliers

85-00900 Sama Trading Company Head Office
KPO Box 1006, Seoul
11th Floor Suwoon Holkwan Building, 88 Kyungun-Dong, Jongno-Ku, Seoul
☎ (+82 2) 734 06 31 ⊖ (+82 2) 734 06 47
👥 18
Chairman of the Board and President: Joong S Inn
Vice President: Tae S Hong
Director, Sales and Marketing: Jung T Chung
Manager, Sales: Gil Su Choe
Manager, Sales: Dok Joo Lee
Manager, Sales: C S Shin
■ Importer, distributor, wholesaler and retailer of Valenite cutting tools, audio-visual equipment and consumer electronics. Engineering and service. [98]

90 Sales Agents, Sales & Liaison Offices

Companies represented:

Andrew Corporation, USA 90-01955
Arnold & Richter Cine-Technik, Germany 90-01910
Bofors Weapon Systems, Sweden . . . 90-00270
Cadillac Gage Inc, USA 90-00270
Carl Schenck AG, Germany 90-00270
Cockerill Mechanical Industries, Belgium 90-00270
EEV Ltd, UK 90-01895
Farmont GmbH, Germany 90-00270
GEC Alsthom, UK 90-00270
GEC Alsthom ACB, UK 90-00270
GEC Marconi, UK 90-00270
Grumman Corp., USA 90-01500
Hägglunds Vehicle AB, Sweden . . . 90-00811
Huge Network Systems, USA 90-00270
Klockner Ina, France 90-00270
Kudelski SA, Switzerland 90-01910
LMS International, Belgium 90-01235
Lockheed Martin Telemetry, USA . . . 90-01955
Marconi Command & Control Systems, UK 90-00270
Park Air Electronics, UK 90-01955
Philips Broadcast Television Systems, USA 90-01910
Racal Recorders Ltd, UK 90-01955
Raytheon Company, USA 90-01955
Royal Ordnance, UK 90-00270
Shure Brothers Inc, USA 90-01910
Telecommunications Techniques Corp, USA 90-01910
Teledyne Inc., USA 90-02030
Thorn Security Systems, USA 90-00270
Trimos-Sylvac Task Force, Switzerland . . 80-05000
Watkins-Johnson Company, USA . . . 90-01955
Western Design, USA 90-00270

Sales agents/representatives listed elsewhere:

AlliedSignal Aerospace Marketing . . 80-00080
Hyundai 75-01540

90-00080 Aerotech World Trade Corp
Room 809 Seocho Renaissance Building,
1598-3 Seocho-dong, Seocho-ku, Seoul
☎ (+82 2) 598 86 97 ⊖ (+82 2) 598 86 98
Managing Director: Won Bae Park [98]

90-00270 Bumjin Corp
Bumjin Building Seocho-Dong, Seocho-ku, Seoul
☎ (+82 2) 587 49 00 [E] bbn@bjcnet.com
[W] http://www.bjcnet.com ⊖ (+82 2) 587 49 06
♧ US$ 500,000 (1996) ♣ US$ 120,000,000 (1996)
👥 30 🏭 300 m2
President and Chief Executive Officer: Bong Joon Choi
President and Chief Operating Officer: J H Cho
Managing Director: Frank Choi
Operations Director: K W Han
Sales and Marketing Director: K J Choi
Operations and Technical Director: S H Pack
Purchasing Manager: M W Lee
Purchasing Manager: D K Chung
■ Export, import and consulting in defence and technology.
® Belgium: Cockerill Mechanical Industries
France: Klockner Ina
Germany: Carl Schenck AG, Farmont GmbH
Sweden: Bofors Weapon Systems
UK: GEC Alsthom, GEC Alsthom ACB, GEC Marconi, Marconi Command & Control Systems, Royal Ordnance
USA: Cadillac Gage Inc, Huge Network Systems, Thorn Security Systems, Western Design [98]

90-00600 Federal Trade Inc.
2nd Fl, Baik-Sung Bldg, Simsa-Dong, Kangnam-Ku, Seoul
☎ (+82 2) 544 47 01 ⊖ (+82 2) 544 47 03
♧ US$ 1 000 000 (1996) ♣ US$ 5 000 000 (1995) 👥 9
President: Andrew J. Kim
Managing Dir: Park Duck-Am
Sales Mgr: Stephen Oh [97]

90-00700 GENEX International Corp.
406-2 Pyungchang Dong, Chongro-gu, Seoul 110-012
☎ (+82 2) 379 31 00 ⊖ (+82 2) 374 17 86
👥 5
Managing Dir: Chris P. Bach
Marketing Mgr: Chang E. Ju
■ Represent manufacturers in defence related businesses. Represent clients as an agent, consultant or liaison office specialising in radar, missiles, fire control systems, communications & electronic warfare systems. [97]

90-00811 Hägglunds Liaison Office
3rd Floor, Sung-Bo Bldg 90-2, Nonhyun-DONG, Kangnam-Ku, Seoul
☎ (+82) 25 11 86 35 ⊖ (+82) 25 11 86 34
Liaison Officer: Robert Lee
■ Products represented include the Bv 206 All Terrain Carrier, Combat Vehicle 90 (CV90) and recovery units.
® Sweden: Hägglunds Vehicle AB [98]

90-00970 ITT Cannon Korea Ltd
(a joint venture between ITT Cannon and Korea Electric Devices)
620 Changkang Building - n°22, Dohwa-dong, Mapo-Ku
☎ (+82 2) 702 71 11 ⊖ (+82 2) 717 73 30
♣ US$ 5,000,000 👥 15
Information Contact: D C Cho
■ Represents all ITT Cannon products as well as ITT Cannon Switch, Relay and Data Communications. [98]

90-01010 Kawasaki Heavy Industries Ltd Seoul Office
2302 Changgyo Building, 1 Changgyo-Dong, Chung-Gu, Seoul
☎ (+82 2) 778 66 37 ⊖ (+82 2) 778 66 38 [98]

90-01235 LMS Korea
210-2, 7th Floor-Hapdong Building, Yang Jae-Dong Seo Cho-Ku, Seoul
☎ (+82 2) 571 72 46 ⊖ (+82 2) 574 73 21
Sales Manager: S Y Lee
® Belgium: LMS International [98]

90-01355 Moog Korea
(a subs. of Moog Inc., USA)
505-4, Yoelmi-Ri, Shilchone-myon, Kwangju-Kun, Kyunggi-do
☎ (+82 347) 64 67 11 ⊖ (+82 347) 64 09 29
♣ US$ 2 634 000 (1995) 👥 12
General Mgr: Jung Un Cho ☎ (+82) 347 64 67 11
■ Electro-hydraulic/pneumatic/mechanical servo control systems & components, servo valves, actuators, electric motors, gassing systems. [97]

90-01400 Namsung Traders Co. Ltd
CPO Box 395, Seoul 100603
☎ (+82 2) 322 19 86 ⊖ (+82 2) 322 19 18
♧ Wons 100,000,000 (1995) ♣ US$ 5,455,000
👥 8 (1996) 🏭 167 m2
President: Inheng Rhee
Director: See-Hark Lee
Executive Sales Director: Se-Hun Yang
Product Support Manager: Byung-Soon Kong
Assistant Sales Manager: Gil-Hyun Kim [98]

90-01500 Northrop Grumman Korea Ltd
Room 913, Royal Building, 5 Dangoodong, Chongroku, Seoul 110-071
☎ (+82 2) 723 26 46 ⊖ (+82 2) 722 37 20
Director: Curt Orchard
® USA: Grumman Corp. [97]

90-01895 Saetong Corporation
CPO Box 3590, Seoul
Taeyang Building 207/208 Number 1-521, Yeoido-Dong, Seoul 150
☎ (+82 2) 783 65 51 to 65 55 [TX] 23596
⊖ (+82 2) 784 27 88
Chief Executive Officer: B W Lee
® UK: EEV Ltd [98]

90-01910 Sama Electronics Corporation
KWTC PO Box 110, Seoul
2nd Floor Gu Gu Building, 145-18 Samsung-Dong, Kangnam-Ku
☎ (+82 2) 565 86 41; 565 86 45 ⊖ (+82 2) 565 86 40
👥 28
Chairman of the Board and President: Joong S Inn
President: Ki Whan Inn
Vice President: S Y Kang
Manager, General Affairs: Lee Baek Ho
Manager, Sales: Jae Y Choi
Manager, Sales: Jae S Jung
Manager, Sales: Jin S Lee
Manager, Sales: Yon Bok Chung
Manager, Service and Engineering: In C Jung
■ Agents and representatives for film, broadcasting, audio-visual, test and measuring equipment. Engineering services. Telecommunications test equipment, cameras, telecine, switchers, lighting equipment, tape recorders, microphones, cartridge tele-conferencing system.
® Germany: Arnold & Richter Cine-Technik
Switzerland: Kudelski SA
USA: Philips Broadcast Television Systems, Shure Brothers Inc, Telecommunications Techniques Corp [98]

90-01955 S M Engineering
(a subsidiary of Sama Electronics Corp and Sama Trading Co Ltd, Korea)
10th Floor Suwoon Hoikwan, 88 Kyungun-Dong, Jongno-Ku, Seoul
☎ (+82 2) 738 21 84; 738 21 85; 733 93 36; 733 93 38
[E] smeng@nur.net ⊖ (+82 2) 739 96 98; 733 84 61
♧ US$ 1,279,000 (1997) ♣ US$ 13,039,000 (1997)
👥 23 (1997) 🏭 365 m2
Chairman and Chief Executive Officer: Joong S Inn
Vice President and Managing Director: Ki S Park
Executive Director Sales - Navigation Equipment: Sang Bo Cha
Director, Sales - Communication VTS: Moore I Kim
Director, Sales - Electronic Components: Young M Ha
Marketing Director: D H Sohn
Manager, Contracts: Jae Y Woo
Manager, Sales - Communication: Sae T Kim
Manager, Sales - Navigation Equipment: Hyeuk H Kwon
■ Commission agents and representatives. Engineering, installation and maintenance services. Marine radars, navigational systems and products, electronic warfare systems, RF/microwave/digital communication equipment. RF and M/W devices. Long range surface radar, sonars, surveillance receiving systems, ESM systems, test equipment, microwave transceivers. Transmission cables, air-to-ground communication equipment, multi-channel voice communication recording and reproduction system.
® UK: Park Air Electronics, Racal Recorders Ltd
USA: Andrew Corporation, Lockheed Martin Telemetry, Raytheon Company, Watkins-Johnson Company
Branch Office Engineering Service Station
KPO Box 293, Seoul
Room 305 Dong-Il Building, 84-10 Chungang-Dong, Chung-Ku, Pusan
☎ (+82 51) 469 86 82; 469 86 37
⊖ (+82 51) 465 74 93
Service Manager: Young M Kim [98]

90-01960 Sooa Corporation
502, Yusung Building, 702-22, Yoksam-Dong, Kangman-Ku, Seoul
☎ (+82 2) 561 50 20 [E] sukchul@nownuri.net
⊖ (+82 2) 561 63 94
♧ US$ 250,000 (1996) ♣ US$ 30,000,000 (1996)
👥 25
President: Suk-Chul Youn
Marketing Director: Seog-Bong Youn
Advertising Director: Ho-in Chung
Public Relations Manager: San-Gon Yu
■ Marine diesel engines for high and medium speed patrol boats. Demilitarisation technology and systems. Military propellant and explosives. Ammunition and components. Technology and equipment for the production of military propellant, explosives and devices. [98]

90-02030 Teledyne Industries International Inc.
4th Floor, 144-23 Yundang Building, Kangnam-ku, Seoul 135-909
☎ (+82 2) 538 37 61-63 ⊖ (+82 2) 796 24 68
® USA: Teledyne Inc. [97]

90-02330 Woo Kyung Electro-Optics
#640, Sungkok-Dong, Ansan-City, Kyongki-Do 425-110
☎ (+82 345) 494 68 55 ⊖ (+82 345) 494 68 59
♣ US$ 10,000,000
Contact: Y Hahn [98]

Contact Reference

Kuwait

Head of State:
Sheikh Jabir Al-Ahmad al-Jabir Al Sabah

10 Head of State

10-10000 Office of the Amir
PO Box 799, 13008 Safat
☎ (+965) 243 90 21
Amir of Kuwait: Sheikh Jabir al-Ahmad al-Jabir Al Sabah [98]

11 Prime Minister

11-10000 Office of Heir Apparent and Prime Minister
Sief Palace, PO Box 1397, Safat
☎ (+965) 243 19 01
[TX] 22796 offprim kt, 44343 pm off kt

Kuwait

Heir Apparent and Prime Minister:
Sa'ad al-'Abdullah al-Salim Al Sabah
First Deputy Prime Minister:
Sabah Sa'ad al-Ahmed al-Jabir Sabah
Deputy Prime Minister: Nasir Abdallah al Rawden
Deputy Prime Minister: Salim al-Sabah al-Salim Al Sabah
Minister of Commerce and Industry:
Jasim Abdullah al Mudaf
Minister of Communications, Electricity and Water:
Jasim Muhammad al Awn
Minister of Information: Saud al-Nasir Al Sabah
Minister of Oil: Isa Muhammad al Mazdi
Minister of Planning: Dr Ali Fagd al Zumai [98]

31 Embassies & Attachés

31-00250 Embassy of the Republic of Bulgaria
PO Box 12090, 71651 Shamia
☎ (+965) 531 44 59 ⊠ 31550 bulem kt
⊖ (+965) 532 14 53
Ambassador: H.E. Mr Dimitar Dimitrov
Commercial Officer: Alexandrer Alexandrov
Public Information Officer: Mervat Nasri Abedeh [97]

31-00310 Embassy of Canada
PO Box 25281, 13113 Safat
24 Al-Mutawakel Street, Block 4, Da'iah, Kuwait
☎ (+965) 256 30 25 ⊖ (+965) 256 41 67
Ambassador: H E Mr T W Colfer
Counsellor (Commercial) and Program Manager:
J N Guerin
Commercial Secretary: A E Ferrer
Canadian Forces Attaché: Col Larry Bowen
(resident in Saudi Arabia) [98]

31-00450 Embassy of the Czech Republic
PO Box 1151, 13012 Safat
☎ (+965) 252 90 18 (English); 252 90 15 (Arabic)
[E] czechemb@ncc.moc.kw ⊖ (+965) 252 90 21
Ambassador: Josef Rychtar ☎ (+965) 257 50 18
First Secretary Commercial: Joroslav Kalfirt
☎ (+965) 257 50 19
Attaché: Jaroslav Hriba
Administrative Official: Milos Jaro [98]

31-00460 Royal Danish Consulate General
PO Box 5451, 13055 Safat
Al Mubader Center, Qurtoba, Bloc 2, Plot 496, Street 3, Safat
☎ (+965) 534 10 05; 534 10 06 ⊖ (+965) 534 10 07
Honorary Consul: Abdulla Y Alghanim
Deputy Consul: Diraar Y Alghanim
Deputy Consul General: Oussama Said Salame [98]

31-00660 Embassy of the Federal Republic of Germany
Abdullah Al-Salem District Block 1, PO Box 805, 13009 Safat
Nusf Al-Yousef Street, Ave. 14, Villa 13, 13009 Kuwait City
☎ (+965) 252 08 32; 252 08 27; 252 07 56
⊠ 22097 aa kt
⊖ (+965) 252 07 63; 257 43 91 Commercial
Ambassador: H E Dr Gunter Held
First Secretary and Deputy Head of Mission:
Karl-Heinz Wittek
Second Secretary Commercial Affairs: Carsten Ruepke
First Secretary and Head of Press and Information Department: Franz Horn
First Secretary Head of Administration: Klaus Schregle
Second Secretary Consular Affairs: Ralf Freitag [98]

31-00690 Embassy of Greece
Khaldiya, Block 4, Street 44, HS4, Kuwait City
☎ (+965) 481 71 01; 481 71 02 ⊖ (+965) 481 71 03
Ambassador: H E Mr Stellos Mallikourtis
First Secretary and Deputy Head of Mission:
Gerassimos Davaris
Attaché and Vice Consul: Dimitri Hadolios
Military Liaison Officer: Col Dimitrios Koutoulogenis
(resident in Riyadh)
Attaché: Nikolas Bardis
Attaché: George Doulias [98]

31-00810 Embassy of the Republic of Hungary
Surra Block 3, St 5, House No 14, PO Box 23955, 13100 Safat
☎ (+965) 532 39 01/3 ⊖ (+965) 532 39 04
Ambassador: H.E. Dr Pal Fabian
Deputy Chief of Mission & First Secretary: Sandor Keseru
Press Attaché: Antal Dragos
Office of the Commercial Section
PO Box 5671, 13057 Safat
☎ (+965) 531 26 00 (3 lines) ⊖ (+965) 531 37 82
Commercial Counsellor: Peter Tarnoki [97]

31-01320 Embassy of the Republic of Niger
PO Box 44451, 32059 Hawalli
☎ (+965) 565 29 43 ⊠ 23365 ambnir kt
⊖ (+965) 564 04 78
Ambassador: Seyni Maiga
Deputy Chief of Mission: Mahamaned Yacoubu
Finance Attaché: Sarlislsou Mamane Samaila [98]

31-01330 Embassy of the Federal Republic of Nigeria
PO Box 6432, 32039 Hawalli
☎ (+965) 532 07 94; 532 07 95 ⊠ 22864 nigeria kt
⊖ (+965) 532 08 34
Chargé d'Affairs: Sumaila Adamu
Commercial Counsellor: Olawale Idris Maiyegun [98]

31-01430 Embassy of the Republic of Poland
PO Box 12443, 71653 Shamiya
☎ (+965) 484 52 75; 481 36 41 ⊖ (+965) 484 52 18
Ambassador: H E Mr Jan Natkanski ☎ (+965) 531 15 71
Commercial Counsellor: Zygmunt Janiec
☎ (+965) 484 52 75
First Secretary: Piotr Ciecwierz ☎ (+965) 531 15 71 [98]

31-01680 Embassy of Spain
Surra, Block 3, Street 14, Villa 19, PO Box 22207, 13083 Safat
☎ (+965) 532 58 27; 532 58 29 ⊠ 22341 ESPAN
⊖ (+965) 532 58 26
Ambassador: H E Mr José Luis Rosello
☎ (+965) 533 63 92
First Secretary: Ignacio Escobar
Defence Attaché: Col Vicente Martinez Rodriguez
Economic and Commercial Counsellor:
Federico Povedano Alonso
Attaché: Julio Alameda [98]

31-01740 Embassy of Switzerland
Qortuba, Area 2, Street 1, Hse 122, PO Box 23954, 13100 Safat
☎ (+965) 534 01 75 ⊖ (+965) 534 01 76
Ambassador: H E Mr Rudolf Staub
Counsellor: H P Studer
Commercial Officer: Ziad Tarakji [98]

31-01840 Embassy of the Republic of Turkey
Istiqlal Street Block 16, PO Box 20627, 13067 Safat
☎ (+965) 253 17 85; 243 14 66; 242 42 48
[E] trkemb@ncc.moc.kw ⊠ 44806, 22290 turkof kt
⊖ (+965) 256 06 53
Ambassador: H E Mr Ahmet Ertay ☎ (+965) 256 04 68
Deputy Chief of Mission: Kemal Demirciler
Counsellor for Labour and Social Affairs: A Erol Daglier
☎ (+965) 257 09 06
Counsellor for Information: Mehmet Emin Temel
Commercial Counsellor: Salim Rustu Ocku
☎ (+965) 256 06 58
Deputy Counsel for Information: Mustafa Atakar
Economic Counsellor: Ibrahim Uslu
Second Secretary, Head of Consulate Section:
Miss Kezban Nilvana Darama
Deputy Attaché: Izzet Kamil Giritlioglu
Deputy Attaché of Labour: Erkan Unlu
Attaché: Suleyman Ozkan
Attaché: Erkan Unlu
Attaché: Kenan Oygur
Attaché: Levent Aras
Information Office
Al Duwaliyah Center - Mezzanine, PO Box 15518, 35456 Aldaiyah, Kuwait City
Al Duwaliyah Center - Mezzanine Floor
☎ (+965) 242 42 48; 242 58 04; 242 42 98
(fax with telephone) [98]

31-01890 Embassy of the United Kingdom
Arabian Gulf Street, PO Box 2, 13001 Safat
☎ (+965) 240 33 34; 240 33 36 ⊖ (+965) 242 67 99
Ambassador: H E Mr G H Boyce
Deputy Head of Mission: J Jenkins
First Secretary Commercial: M Hurley
First Secretary Defence Supply: H Snedden
Defence Attaché: Col G Sayle [98]

31-01950 Embassy of Venezuela
PO Box 24440, 13105 Safat
☎ (+965) 532 43 67/9; 533 30 62 ⊠ 22782 embaven kt
⊖ (+965) 532 43 68
Ambassador: H.E. Mr Mario Tepedino Raven [97]

65 Industry, Trade & Armed Forces Associations

Ahmed Al Saleh & Sons Co. 85-00100

80 Service Companies & Organizations

80-00600 Hughes International Corporation
(a subsidiary of Hughes Aircraft Company)
PO Box 2118, Safat
☎ (+965) 243 41 93 [E] gitkwt@kuwait.net
⊖ (+965) 243 42 06
⌂ 150 m2 (1997)
Director: Paul James Pezas
■ Marketing of defence and communication products. [98]

80-01000 National Consulting Bureau
PO Box 5092, 13051 Safat
☎ (+965) 562 07 00 ⊖ (+965) 561 15 21
♧ $ 150 000 ♣ $ 2 000 000 👥 15
Dir Region Int'l/5: Fouad Al-Khadra
® UK: Jane's Information Group [97]

85 Stockists, Distributors, Spare Parts Suppliers

85-00100 Ahmed Al Saleh & Sons Co.
PO Box 340, 13004 Safat
☎ (+965) 245 98 03; 245 98 04 ⊠ 23305 youshaa
⊖ (+965) 245 98 05
👥 300
President: Ahmed Al Saleh
Vice President: Abdulmohsen Al Saleh
Executive Manager: Yacoub Al Saleh
→ 65 → 90 [98]

90 Sales Agents, Sales & Liaison Offices

Companies represented:

CATIC, China PR 90-00200
Jane's Information Group, UK 80-01000

Sales agents/representatives listed elsewhere:

Ahmed Al Saleh & Sons Co. 85-00100

90-00200 CATIC
Representative Office
Al Khaldiya, PO Box 17359, 72453 Al Khaldiya
☎ (+965) 488 93 60 ⊠ 22688 chinaem kt
⊖ (+965) 488 96 30
® China PR: CATIC [97]

Kyrgyzstan

President:
Askar Akayev

10 Head of State

10-10000 Office of the President
Government House, 72003 Bishkek
☎ (+7 3312) 21 24 66
President: Askar Akayev [98]

11 Prime Minister

11-10000 Office of the Chairman of the Council of Ministers
720003 Bishkek
Government House ulitsa Pervomayskaya 57
☎ (+7 3312) 22 56 56 ⊖ (+7 3312) 22 59 50
Prime Minister: Apas Jumagulov
First Deputy Prime Minister: Kemilbek Nanayev
Deputy Prime Minister for Agriculture: Karimshev Abdimomunov
Deputy Prime Minister for Social Issues: Mira Jangaracheva [98]

12 Ministries

12-10000 State Committee for Defence Affairs
Logvinenko Street 26, 720001 Bishkek
☎ (+7 3312) 22 78 79
Chairman: Lieut Gen Myrzakan Subanov [98]

12-20000 Ministry of Foreign Affairs
Abdumomunov Street, 205, 720003 Bishkek
☎ (+7 3312) 22 59 14 ⊖ (+7 3312) 22 57 35
Minister: Muratbek Imanaliyev [98]

12-40000 Ministry of Internal Affairs
ulitsa Frunze 469, 720011 Bishkek
☎ (+7 3312) 22 54 90 ⊖ (+7 3312) 22 32 78
Minister: Maj Gen Omurbek Kutuyev [98]

15 Parliament

15-10000 Supreme Soviet
Jokorku Kenesh
Bishkek
☎ (+7 3312) 27 09 96 ⊖ (+7 3312) 22 39 91
Chairman: Medetkan Sherimkulov
Aide to Chairman: Zhusup Alfeitov [98]

20 Defense Administrations/Services/Agencies

20-01000 State Cartography Agency
Kievskaya Street, 107, 720300 Bishkek
☎ (+7 3312) 21 22 96 [E] agency@geodes.bishkek.su
⊖ (+7 33 12) 21 07 72
Director: Valery Tsurkov [98]

25 Defense Forces

25-50000 National Guard
Kievskaya Street 114, 720001 Bishkek
☎ (+7 3312) 26 27 28 ⊖ (+7 3312) 22 05 94
Commander: Maj Gen Abdygul Chotbaev
Chief of Financial Service: Emil Taranchiev ☎ (+7 3312) 22 47 14
Chief of Staff: Col Vladimir Slavskiy ☎ (+7 3312) 22 08 33
Director of Operations: Lieut Col Kyrochkin
Director of Procurement: Lieut Col Vinnik [98]

Laos

Head of State:
Nouhak Phoumsavan

10 Head of State

10-10000 Office of the President
Vientiane
President: Nouhak Phoumsavan
Vice President: Gen Sisavat Keobounphan [98]

11 Prime Minister

11-10000 Office of the Prime Minister
Vientiane
Prime Minister: Gen Khamtai Siphandon
Deputy Prime Minister: Khamphoui Keoboualapha
Deputy Prime Minister: Boungnang Volachit
Minister Attached to the Office of the Prime Minister: Cheuang Sombounkhan [98]

12 Ministries

12-10000 Ministry of National Defence
Vientiane
☎ (+856) 41 28 05; 41 28 10
Minister of Defence: Lieut Gen Choummali Saignason [98]

12-20000 Ministry of Foreign Affairs
Nongbon Road, Vientiane
☎ (+856) 41 40 08 ✉ 4317, 4320 minae ls
⊖ (+856) 41 23 95
Minister of Foreign Affairs: Somsavat Lengsavad ☎ (+856) 41 40 01
Deputy Foreign Minister: Souban Srithirath ☎ (+856) 41 40 02
Deputy Minister of Commerce: Lian Thikeo ☎ (+856) 41 20 01
Director General of the International Relations Department: Col Sisophone Bang Ornsengdeth [98]

12-40000 Ministry of the Interior
Vientiane
Minister: Maj Gen Asang Laoli [98]

31 Embassies & Attachés

31-01730 Swedish Embassy
Route de Sokpaluang, PO Box 800, Vientiane
☎ (+856 21) 31 37 72; 31 50 18 ✉ SVENSK Vientiane
⊖ (+856 21) 31 50 01
Ambassador: H E Ms Eva Heckscher (resident in Thailand)
Chargé d'Affaires a.i.: Olof Milton
First Secretary Economic: Göran Bergman [97]

31-01890 British Trade Office
PO Box 6626, Vientiane
☎ (+856 21) 41 36 06 ⊖ (+856 21) 41 36 07
Ambassador: H E Sir James Hodge (resident in Bangkok)
Head of British Trade Office: Robin Eyre
Counsellor: P R Sizeland (resident in Bangkok)
Commercial Counsellor: M J Greenstreet (resident in Bangkok)
First Secretary Commercial: P J W Hardman (resident in Bangkok) [98]

75 Manufacturing Companies

75-05000 Transfield Pty Ltd
Rue Wat Nak Noi - Ban Thapalanxay, PO Box 5482, Vientiane
☎ (+856 21) 31 33 30 ⊖ (+856 21) 31 20 78 [98]

90 Sales Agents, Sales & Liaison Offices

90-00300 CATIC
Representative Office
PO Box 5091, Vientiane
☎ (+856) 21 98 79 ⊖ (+856) 21 98 79 [97]

Latvia

Head of State:
Guntis Ulmanis

10 Head of State

10-10000 Office of the President
Latvijas Republikas Augstaka Padome
Pils Lukums 3, LV-1900 Riga 50
☎ (+371) 737 95 19 [E] pnede@mail.parks.lv
⊖ (+371) 783 05 38
President: Guntis Ulmanis
Chairman Supreme Council: Alfreds Cepanis ☎ (+371) 708 71 27
Deputy Chairman: Andris Ameriks ☎ (+371) 708 71 63
Deputy Chairman: Aigars Jirgens ☎ (+371) 708 71 61
Defense and Internal Affairs: Karlis Julijs Druva ☎ (+371) 708 72 45
Foreign Affairs: Indulis Bērzius ☎ (+371) 708 73 01
Minister of Foreign Affairs: Valdis Birkavs ☎ (+371) 701 61 01
Minister of Defence: Talavs Jundzis ☎ (+371) 721 01 24
Foreign Media Relations Officer: Viktors Daugmalis ☎ (+371) 708 72 18
Secretary: Imants Daudiss ☎ (+371) 7078 71 04 [98]

11 Prime Minister

11-10000 Office of the Prime Minister
Government Offices, 36 Brivibas Boulevard, 226170 Riga
☎ (+371) 733 22 32 ⊠ 161104 vald su
⊖ (+371) 728 65 98; 728 04 69
Prime Minister: Andris Skele
Deputy Prime Minister: Anatolijs Gorbunovs
Deputy Prime Minister: Albert Kauls
Deputy Prime Minister: Maris Grinblats
Deputy Prime Minister: Juris Kaksitis
Deputy Prime Minister: Andrejs Krastins
98

12 Ministries

12-10000 Ministry of Defence
Kr Valdemara 10/12, 1473 Riga
☎ (+371) 721 04 19 ⊖ (+371) 724 31 15
Minister of Defence: Andrejs Krastins
☎ (+371) 721 01 24
Advisor to the Minister: Ms Saulcerite Ievina Krumina
☎ (+371) 721 04 54
Minister's Office Manager: Mrs Irisa Matisa
☎ (+371) 724 31 14
State Secretary: Einars Vaivods ☎ (+371) 732 33 80
Parliamentary Secretary: Andris Ligotnis
☎ (+371) 732 30 68
Press Secretary: Janis Podins ☎ (+371) 721 15 64
Administrative Department Director: Mrs Regina Elksne
☎ (+371) 722 37 23
Planning Department Director: Edgars Rinkevics (acting)
☎ (+371) 722 92 41
Finance and Contracts Department Director:
Jevgenijs Bogdans (acting) ☎ (+371) 722 86 69
Finance and Contracts Section Deputy Director:
Mrs Maija Skudra ☎ (+371) 722 12 48
Building and Maintenance Department Director:
Alfons Cupans ☎ (+371) 722 27 32

International Relations Department
Kr. Valdemara 10/12, 1473 Riga
⊖ (+371) 724 31 15
Acting International Relations Dir & Head of Protocol Section: Mrs Elvira Smukste ☎ (+371) 733 61 39
Deputy Head of Protocol Section: Modris Stasuls
☎ (+371) 721 04 19
Senior Desk Officer: Ms Ieva Birzkalne
☎ (+371) 721 04 19
Head of Agreements Section: Ms Inara Leimane
☎ (+371) 733 52 29
Desk Officer: Ms Kristiana Tamane ☎ (+371) 721 04 19
Head of International Military Relations & Education Section: Janis Sarts ☎ (+371) 733 16 33
Senior Desk Officer: Martins Nilsons
Senior Desk Officer: Mrs Ida Mezeniece
Desk Officer: Janis Karlsberg
98

12-15000 Ministry of Finance
Smilsu iela 1, 1919 Riga
☎ (+371) 722 66 72 ⊠ 161232 ⊖ (+371) 722 72 20
Minister: Roberts Zile
98

12-20000 Ministry of Foreign Affairs
36 Brivibas Boulevard, 1395 Riga
☎ (+371) 722 00 79; 701 62 50 [E] mfa.cha@mfa.gov.lv
[W] http://www.mfa.gov.lv
⊖ (+371) 782 81 21; 728 28 82
Minister for Foreign Affairs: Valdis Birkavs
☎ (+371) 722 33 07
State Secretary: Maris Riekstins ☎ (+371) 701 62 03
Head of Security Policy Divison: Peteris Vinkelis
☎ (+371) 701 62 16
Information Department Director: Roberts Stafeckis
☎ (+371) 701 61 15
Head of Information Division: Vilnis Melins
☎ (+301) 701 62 12
Press Secretary: Andreos Pildegovics
☎ (+301) 701 61 17
98

12-40000 Minister of Internal Affairs
6 Raina Boulevard, 226181 Riga
☎ (+371 2) 28 72 60 ⊖ (+371 2) 21 22 55
Minister of Internal Affairs: Dainis Turlais
98

12-50000 Ministry of Transport
3, Gogola Iela, 1743 Riga
☎ (+371) 722 69 22 ⊖ (+371 7) 82 81 20; 21 71 80
Minister: Vilis Kristopans ☎ (+371 7) 22 69 22
Deputy Minister: Uldis Petersons
Chief of Staff: Ms Vera Zidova
Director Foreign Relations: Austris Caunitis
Aviation Director: Arnis Muiznieks ☎ (+371 7) 02 82 09
98

15 Parliament

15-10000 Parliament
Saeima
Jekaba ula 11, LV-1811 Riga
☎ (+371 2) 708 72 65 ⊖ (+371 2) 708 72 44
Speaker: Alfreds Espanis ☎ (+371 2) 708 71 11
Chief Press Service: Viktores Daugmalis
☎ (+371 2) 708 72 18
Foreign Relations Department, Chief of Staff: Silvij Kalnena
☎ (+371 2) 708 73 13
Chairman of the Foreign Relations Committee:
Indulio Bizjino ☎ (+371 2) 708 73 01
Chairman of the Defence and Home Affairs Committee:
Kazlis Diruva ☎ (+371 2) 708 72 09
Deputy Chairman of the Defence and Home Affairs Committee: Ms Janis Kazaks ☎ (+371 2) 708 72 45
Deputy Chairman: Andzis Ameaks ☎ (+371 2) 708 71 17
Deputy Chairman: Aigars Jizgens ☎ (+371 2) 708 71 61
Deputy Chirman of the Foreign Relations Committee:
Alexander Kirsteins ☎ (+371 2) 708 72 05
Secretary: Maris Rudjitis ☎ (+371 2) 708 71 14
Secretary: Imants Daudiss ☎ (+371 2) 708 71 17
98

25 Defense Forces

25-10000 Defence Force Headquarters
Latvijas Republikas Aizsardzibas Spéki
4 Raina Boulevard, 1473 Riga
☎ (+371) 722 12 24 ⊖ (+371) 732 51 56
Commander: Col Juris Dalbins ☎ (+371) 721 37 18
Chief of Staff: Col Juris Arness ☎ (+371) 722 02 76
Deputy Chief of Staff: Col Janis Kasocins
☎ (+371) 733 63 55
Commander Naval Forces: Capt. Gaidis Zeibots
☎ (+371) 702 73 19
Commander Air Forces: Col Karlis Kins
☎ (+371) 702 74 81
Commander Border Guards: Lt Col Leonids Lasmanis (Navy) ☎ (+371) 227 40 55
Commander National Guard: Col Juris Eihmanis
☎ (+371) 246 41 45
97

29 Customs & Excise

29-00500 State Revenue Service
1 Smilsu Street, 1919 Riga
Director General: Aija Poca
98

31 Embassies & Attachés

31-00460 Royal Danish Embassy
L Pils iela 11, 1863 Riga
☎ (+371 7) 21 04 33; 22 61 10; 782 02 35
[E] amdkriga@amdkriga.org.lv
⊖ (+371 7) 22 92 18; 782 02 34
Contact: Rita Jørgensen
98

31-00535 Embassy of Estonia
Skolas 13, 1010 Riga
☎ (+371 7) 22 68 45 (consular department)
☎ (+371 7) 82 04 60 [E] sekretar@riia.vm.ee;
sekretar@com.latnet.lv ⊖ (+371 7) 82 04 61
Ambassador: H E Mr Toomas Tiivel
98

31-00580 Embassy of Finland
Kaloakaa bulvaris 1, 1605 Riga
☎ (+371) 733 20 05;
721 60 43; 721 64 29 (commercial section)
⊖ (+371) 733 35 97
Ambassador: H E Mr Hannu Hämäläinen
Defence Attaché: Lieut Col Harri Pulsa (resides in Tallinn)
Administrative Attaché: Ms Marja-Terttu Eskola
Second Secretary: Mrs Päivi Laine
98

31-00660 Embassy of the Federal Republic of Germany
Basteja Boulevard 14, 1050 Riga
☎ (+371) 722 90 96; 782 02 13 ⊠ 161210 arig lv
⊖ (+371) 782 02 23
Ambassador: H E Dr Horst Wiesel
Deputy Chief of Mission and Commercial Counsellor:
Dr Wolfgang Manig
First Secretary, Scientific: Michael Morgenstein
Defence Attaché: Col Wilhelm Tekath
Attaché: S Stohl
98

31-01430 Embassy of the Republic of Poland
Elizabetes Str. 2, Riga
☎ (+371) 732 16 17; 732 15 93; 732 22 33 ⊠ 161212
⊖ (+371) 783 01 74; 733 81 96
97

31-01730 Swedish Embasssy
A. Pumpura Str. 8, Riga
☎ (+371) 782 80 32 ⊠ 16114 svensk
⊖ (+371) 782 80 31
Ambassador: H.E. Prof. Andreas Ådahl
Counsellor: Roland Nilsson
Defence Attaché: Lt Col Erling Johansson
Cultural Attaché: Mats Sylwan
Customs Attaché: Kalev Valdur Lennart Laupa (resides in Tallinn)
First Secretary: Harry Landau
Second Secretary: Peter Vallaste
Second Secretary: Irene Skahlberg
Third Secretary: Roland Glasfors
Administrative Attaché: Ms Camilla Jaurell
Administrative Attaché: Ms Carin Pullam
Agricultural Counsellor: Jan H. Amberg
(resides in Moscow)
97

31-01760 Taipei Mission in the Rep. of Latvia
2 Elizabetes Street, Room 602, 1340 Riga
☎ (+371 7) 32 12 04; 32 11 66; 32 06 10
[E] tmil@mail.bkc.lv ⊖ (+371 7) 32 32 68
Representative: Daniel Hu
☎ (+371 1) 32 11 66; 32 06 10; 32 12 04
Secretary: Yung-shu Chang
Secretary: Mei-shun Lo
98

31-01890 Embassy of the United Kingdom
5 Alunana Straat, 1010 Riga
☎ (+371) 733 81 26/7/8/9/30/31 ⊖ (+371) 733 81 32
Ambassador: H E Mr Nicholas Jarrold
Deputy Head of Mission and Consular: Alan Anstald
Defence Attaché: Lt Col N G W Lang
Second Secretary Political: Fiona MacCallum
Commercial Counsellor: Arnando Pupots
Public Information Officer: Ms Lelde Pfafrode
98

Lebanon

Head of State:
Ilyas Harawi

10 Head of State

10-10000 Office of the President
Baabda Palace, Baabda
☎(+961 1) 38 72 00
President of the Republic: Ilyas Harawi 98

11 Prime Minister

11-10000 Office of the Prime Minister
Rue Riad Solh, Grand Sérail, Beirut
☎(+961 1) 22 10 00
Prime Minister: Rafik al Hariri
Deputy Prime Minister: Michel al-Murr 98

12 Ministries

12-10000 Ministry of National Defense
Yarzé, Beirut
☎(+961 1) 45 24 00; 42 04 00 ⊠ 20901
Minister: Muhsin Dallul 98

12-20000 Ministry of Foreign Affairs
Achrafieh, Beirut
☎(+961 1) 33 31 00; 32 69 05
Minister: Faris Buwayz 98

12-40000 Ministry of Interior
Grand Sérail, Rue Riad Solh, Beirut
Minister: Michel al Murr 98

25 Defense Forces

25-20000 Air Force Headquarters
Yarzé, Beirut
☎(+961 1) 42 04 00; 45 24 00
Chief of Air Staff: Brig.Gen. Mahmoud Mattar 97

25-40000 Navy Headquarters, Naval Base of Beirut
Beirut
☎(+961 1) 42 45 91, 44 33 58, 42 28 95
44 41 08; 44 60 03 ⊖ (+961 1) 44 47 62; 20 22 61
Chief of Naval Staff: Rear Adm Samir Khadem
Deputy for Logistics: Col Salah Yunis
Director of Operations: Col Ali Mouallem
Director of Procurement: Col Salah Younes
Press and Information Officer: Lieut Col Vazih Baroudi 98

31 Embassies & Attachés

31-00660 Embassy of the Federal Republic of Germany
Rabieh-Mtaileb nahe der Jesus, PO Box 2820, Beirut
Rabieh-Mtaileb nahe der Jesus and Mary High School
☎(+961 1) 40 53 10; 40 69 50 ⊠ 48040 le AABEIR
⊖ (+961 1) 40 53 11
Ambassador: H E Dr Peter Wittig 98

31-00690 Embassy of Greece
Imm Sarraf, 11éme Etage, Achrfieh, Beirut
☎(+961 1) 21 92 17; 20 13 22 ⊠ 40131
⊖ (+961 1) 20 13 24 98

31-00890 Embassy of Italy
Hamra rue de Rome, Beirut, Ovest
☎(+961 1) 34 02 25; 34 26 84; 86 83 01
34 02 27; 34 26 86 [E] ambcalia@cyberianet.lb
⊠ 0049448397 ambita le ⊖ (+961 1) 34 26 86
Ambassador: H E Mr Carlo Calia
First Counsellor: Massimo Roscigno
First Counsellor: Fulvio Rustico

Office of the Armed Forces Attaché
Al Mir Street, Nawar Building, Zouk Mikael
☎(+961 9) 21 55 40; 21 55 45 ⊖ (+961 9) 21 55 53
Military, Naval and Air Attaché: Gen Giuseppe Pavone 98

31-01340 Royal Norwegian Embassy
Taher and Fakry Buldg, Bliss Street,
PO Box 113/7001 Hamra, Ras Beirut
Taher and Fakry Building, Bliss Street
☎(+961 1) 35 37 31; 35 37 32 ⊠ 22 690 noramb le
⊖ (+961 1) 35 37 30
Ambassador: Odd Wibe
Administration Attacé: Gro Snuggerud 98

31-01890 Embassy of the United Kingdom
8th Street, Rabieh
☎(+961 1) 41 70 07; 40 50 70; 40 36 40
⊖ (+961 1) 40 20 32
Ambassador: H E Mr David MacLennan
Deputy Head of Mission: L Banks
Defence Attaché: Lieut Col C J A Wilton
Second Secretary, Commercial: B West
Vice Consul: Miss C Kitsell

Consular, Commercial, Management Sections
Autostrade Jal El Dib, Coolrite Building, Beirut
☎(+961 1) 40 63 30; 40 50 33; 40 20 35 ⊠ 44104
⊖ (+961 1) 40 20 33 98

31-02010 Embassy of the Federal Republic of Yugoslavia
Imm. Daouk, Senayeh, PO Box 742, Beirut
☎(+961 1) 73 96 33 ⊖ (+961 1) 73 96 38
Ambassador: H E Mr Dragoslav Pejic
Chargé d'Affaires: Veibor Dulovic 98

Lesotho

Head of State:
King Letsie III

Prime Minister:
Ntsu Mokhehle

11 Prime Minister

11-10000 Office of the Prime Minister
PO Box 527, 100 Maseru
Prime Minister, Minister of Defense and Internal Security: Ntsu Mokhehle
Deputy Prime Minister: Bethuel Pakalitha Mosisili 98

12 Ministries

12-10000 Ministry of Defence and Internal Security
Private Bag A166, 100 Maseru
☎(+266) 31 65 70 ⊖ (+266) 31 05 18
Deputy Prime Minister: Bethuel Pakalitha Mosisili
Minister: Dr Ntsu Mokhehle
Permanent Secretary: A N Ncholu
Public Relations and Foreign Relations: Ms L Makakole
Procurement: T R Tokho 98

12-20000 Ministry of Foreign Affairs
PO Box 1387, 100 Maseru
☎(+266) 32 38 96 ⊠ 4330 bb
Minister: Kelebone Albert Maope 98

31 Embassies & Attachés

31-01890 United Kingdom High Commission
PO Box Ms 521, 100 Maseru
☎(+266) 31 39 61 [E] hcmaseru@lesoff.co.za ⊠ 4343 lo
⊖ (+266) 31 01 20
High Commissioner: H E Mr P J Smith
Deputy High Commissioner, Defence Attaché and Public Information Officer: G W Wain
Third Secretary: Mrs E A Kehoe 98

31-01900 Embassy of the United States of America
PO Box 333, 100 Maseru
☎(+266) 31 26 66 ⊠ 4506 lo ⊖ (+266) 31 01 16
Ambassador: Mark Bismark Magrick
Deputy Chief of Mission: Raymond Brown 98

Libya

Head of State:
Col Muammar Abu Minyar al-Qadhafi

10 Head of State

10-10000 Office of the Leader of the Revolution
Tripoli
☎(+218 21) 482 10; 482 14 ⊠ 20032
Leader of the Revolution and Secretary of Defense: Col Muammar Abu Minyar al-Qadhafi 98

10-10500 Secretariat of the General People's Congress (GPC)
Tripoli
☎(+218 21) 307 77 ⊠ 20032
Secretary-General of the GPC: Muhammad al-Zanati 98

12 Ministries

12-20000 Secretariat of the People's External Liaison & International Cooperation
(Secretariat of the People's External) Liaison and International Cooperation
Al Fateh Road, Tripoli
☎(+218 21) 416 34 ⊠ 20032
Secretary: Umar Mustafa al Muntasir 98

15 Parliament

15-10000 General People's Congress
Tripoli
Secretary-General: Muhammad al- Zanati
Assistant: Ali Mursi al- Shairi [98]

31 Embassies & Attachés

31-00450 Embassy of the Czech Republic
Sharia Ahmed Lotfi Sayed, Ben Ashour, PO Box 1097, Tripoli
☎ (+218 21) 60 34 44 ⊠ (+218 21) 60 08 39
Ambassador: H.E. Mr Alexandr Karych
Armed Forces Attaché: Lieut Col Milan Bachan (resident in Cairo)
Third Secretary: Petr Hladik
Administration Attaché: Josef Archman [97]

31-00660 Embassy of the Federal Republic of Germany
Sharia Hassan el Mashai, PO Box 302, Tripoli
☎ (+218 21) 333 05 54; 38 27; 444 85 52; 01 53
⊠ 20298 aatrp ly ⊖ (+218 21) 444 89 68
Ambassador: H.E. Mr Peter Kiewitt [97]

31-00670 Embassy of Ghana
G Ben-Mafarah Street, PO Box 4169, Andalus, Tripoli
☎ (+218 21) 477 25 34 ⊠ 20879 ghanemb ly
⊖ (+218 21) 477 35 57
Ambassador: H E Mr Laary Bimi ☎ (+218 21) 90 57 78
Head of Chancery Commercial Attaché: Ernest Y Ayittey ☎ (+218 21) 83 38 92
Consular Affairs: Mohamed Hussein ☎ (+218 21) 483 07 08
Administration: W N Kwashie
Communications: Ms Christiana Koffie ☎ (+218 21) 707 62 [98]

31-00690 Embassy of Greece
Sharia Jalal Bayar 18, PO Box 5147, Tripoli
☎ (+218 21) 396 78 ⊠ 20409 ⊖ (+218 21) 419 07 [98]

31-00810 Embassy of the Republic of Hungary
Sharia Talha Bin Abdullah, PO Box 4010, Tripoli
☎ (+218 21) 60 57 99; 60 49 81 ⊠ 20 055
⊖ (+218 21) 60 82 29
Ambassador: H E Mr dr György Busztin
Deputy Chief of Mission: Ferenc Csillag
Counsellor: Sandor Fasiom ☎ (+218 21) 483 28 34 [98]

31-00890 Embassy of Italy
Sharia Uahran 1, PO Box 912, Tripoli
☎ (+218 21) 333 41 31/3 ⊠ 20602
⊖ (+218 21) 333 16 73
Ambassador: H.E. Dr Giorgio Testori
Deputy Chief of Mission: Dr Giuseppe Cinti
Naval, Military & Defense Attaché: Capt. Michele Carrieri ☎ (+218 21) 341 34
Commercial Counsellor: Dr Giuseppe Cinti ☎ (+218 21) 353 62 [98]

31-01270 Royal Netherlands Embassy
Sharia Gelal Bayar 20, PO Box 3801, Tripoli
☎ (+218 21) 444 15 49/50 ✉ HOLLANDIA Tripoli
⊠ 20279 hollar ly ⊖ (+218 21) 444 03 86
Chargé d'Affaires: B.F. Tangelder
Attaché: A.H. Kloos [97]

31-01420 Embassy of the Philippines
Kilometer 7, Gargarish Road, PO Box 12508, Tripoli
☎ (+218 21) 483 39 66 ⊠ 20304 ambphil ly
⊖ (+218 21) 483 39 66
Ambassador: H E Mr Huhtar M Muallam
Career Minister and Consul General: Hon Ricardo Endaya
Finance Attaché: Marden A Garduque
Administrative Attaché: Benito R Bernabe
Cultural Attaché: Ms Victoria U Salian
Commumication Attaché: Reynaldo N Aquino
Attaché: Matias G Ellema
Attaché: Geranfel T Abrillo
Attaché: Adbullah Cali
Attaché: Ms Victoria Macasalong
Labour, Attaché: Bulyok S Nilong [98]

31-01480 Embassy of Romania
Sharia Ahmed Lutfi El-Sayed, PO Box 5085, Tripoli
☎ (+218 21) 60 79 05; 60 79 04
⊖ (+218 21) 360 75 97
Ambassador: H E Mr Krieg Polejack
Deputy Chief of Mission: Alexander Wittwer
Third Secretary, Head of Chancery: Mrs Sabine Jenkins
Consul: Alexandru-Valentin Popesco [98]

31-01840 Embassy of the Republic of Turkey
Sharia Zawia Dahmani Street, PO Box 947, Tripoli
☎ (+218 21) 333 77 17; 333 77 32
E aerbas@mfa.gov.tr/ ⊠ 20765 turkeyly
⊖ (+218 21) 333 76 86
Ambassador: H E Mr Ates Balkan
First Secretary: Orhn Isik
Chancery: Shari Zaviya Dahmani
Consular Section: Shara Jeraba
Consular Section: Ben Shur
Counsellor for Labour and Social Affairs: Halidun Ercan
Expert Counsellor: Ömer Sayer
Counsellor for Economy and Commerce: Besir Iyidiker
Third Secretary: Murat Mustafa Onart
Third Secretary: Faruk Kaymakci
Deputy Attaché for Labour and Social Affairs: Salih Vecdi Seckin
Attaché: Yaser Kücükoglu
Attaché: Lokman Erdogdu
Attaché: Sinan Yesildag
Attaché: Ali Erbas
Attaché: Remzi Inceler
Attaché: Mehmet Cayir
Attaché: Ali Yilmaz [98]

90 Sales Agents, Sales & Liaison Offices

90-05000 Daewoo Corp Tripoli Office
PO Box 15076, Tripoli
Manara Road Benghasir, Tripoli
☎ (+218 022) 309 86 ⊠ 20726 dwtip ly
⊖ (+218 022) 308 77 [98]

Liechtenstein

Head of State:
H.S.H. Prince Hans Adam II

10 Head of State

10-10000 Office of H S H the Reigning Prince
Schloss Vaduz, FL-9490 Vaduz
☎ (+41 75) 232 12 12 ⊠ 889288 ⊖ (+41 75) 232 68 62
Reigning Prince: H S H Prince Hans-Adam II von und zu Liechtenstein [98]

11 Prime Minister

11-10000 Office of the Head of Government
Government Building, FL-9490 Vaduz
☎ (+41 75) 236 61 11 ⊖ (+41 75) 236 64 60
Head of Government: Dr Mario Frick
Deputy Head of Government and Councillor for Interior: Michael Ritter
Foreign Relations Director: lic rer pol Roland Marxer
Chief of Press and Information Office: Ms Daniela Clavadetscher
Chief of Staff: Peter Mella [98]

12 Ministries

12-10000 Office of the Government Councillor for Finances
Government House, FL-9490 Vaduz
☎ (+41 75) 236 61 11 ⊖ (+41 75) 236 60 22
Government Councillor: Mario Frick [98]

12-20000 Office of the Government Councillor for Foreign Affairs
Heiligkreuz 14, FL-9490 Vaduz
☎ (+41 75) 236 60 58 ⊖ (+41 75) 236 60 22
Government Councillor: Dr Andrea Willi [98]

12-30000 Office of the Government Councillor For Transportation
FL-9490 Vaduz
☎ (+41 75) 236 61 11 ⊠ 889290 repi fl
⊖ (+41 75) 236 60 28
Councillor: Dr Norbert Marxer
Deputy: Reinhert Schaffhauser [98]

15 Parliament

15-10000 Parliament
Liechtensteinischer Landtag
Landtagssekretariat, Aeulestraat 80, FL-9490 Vaduz
☎ (+41 75) 236 61 11 ⊖ (+41 75) 236 65 80
President: Dr Peter Wolff
Vice President: Otmar Hasler [98]

26 Police (Civil)

26-00000 Police Department
Gewerbeweg 4, FL-9490 Vaduz
☎ (+41 75) 236 71 11 ⊠ 889292 ipfl fl
⊖ (+41 75) 236 77 22
Chief of Police: Reto Brunhart [98]

30 Civil Defense

30-10000 Office for Civil Defence and National Supply
Amt für Zivilschutz und Landesversorgung
FL-9495 Vaduz
☎ (+41 75) 236 69 15 ⊖ (+41 75) 236 69 24
Director: Alfred Vogt
Deputy Director: Alois Hoop [98]

31 Embassies & Attachés

31-01890 Consulate of the United Kingdom
Postfach 70, Stadtle 7, FL-9490 Vaduz
☎ (+41 75) 236 14 14 ⊖ (+41 75) 232 87 28

Ambassador: Christopher Hulse (resides in Berne)
Consul-General: D M Bell (resides at Zurich)
Honorary Consul: B Jeeves 98

90 Sales Agents, Sales & Liaison Offices

Companies represented:

Beta Company, USA 90-00230

90-00230 Beta AG
Landstrasse 114, PO Box 30, FL-9495 Triesen
☎ (+41 75) 339 30 87 ⊖ (+41 75) 399 30 89
Managing Director: Ms Silvia Kaiser-Block
Director: A J Block (Jr)
® USA: Beta Company 98

Lithuania

Head of State:
Algirdas Brazauskas

10 Head of State

10-10000 Supreme Council of the Republic of Lithuania
Seimas of the Republic of Lithuania
Gedimino 53, 2002 Vilnius
☎ (+370 2) 62 85 51; 62 84 23 ▣ 261138 ltinf su
⊖ (+370 2) 22 45 36
President: Algirdas Brazauskas
☎ (+370 0) 62 89 86; 22 62 10 (fax)
Chairman: Ceslovas Jursenas
☎ (+370 2) 62 16 32; 62 90 40 (fax)
Deputy Chairman: Aloyzas Sakalas ☎ (+370 2) 62 16 54
Deputy Chairman: Egidijus Bickauskas
☎ (+370 2) 62 75 33
Deputy Chairman: Juozas Bertatonis
☎ (+370 2) 62 67 78
Chancellor: Neris Germanas ☎ (+370 2) 22 78 87
Advisor Secretariat of the Seimas Board:
Arunas Godunavicius ☎ (+370 2) 62 14 74
Advisor Secretariat of the Seimas Board: Virgilijus Liauska
☎ (+370 2) 22 48 11 98

11 Prime Minister

11-10000 Government of the Republic of Lithuania
Gedimino pr 11, 2039 Vilnius
☎ (+370 2) 62 21 01 ▣ 261105 ltmt su
⊖ (+370 2) 22 10 88
Prime Minister: Gediminas Vagnorius
Government Secretary: Algirdas Mikulis
☎ (+370 2) 62 65 95
Deputy Government Secretary: Henrikas Grigelevicius
☎ (+370 2) 62 16 97
Head of Executive Office: Rimantas Mencinskas
☎ (+370 2) 62 90 38
Head of Government Information Centre:
Vilius Kavaliauskas ☎ (+370 2) 62 56 51;
61 34 64 (fax)
International Affairs Advisor: Asta Skaisgiryte
☎ (+370 2) 61 53 34 98

12 Ministries

12-10000 Ministry of National Defence
Totoriu 25-3, 2001 Vilnius
☎ (+370 2) 62 48 21 ▣ 261244 ⊖ (+370 2) 22 60 82
Minister: Ceslovas Stankevicius 98

12-20000 Ministry of Foreign Affairs
Tumo-Vaizganto 2, 2026 Vilnius
[E] chancery@uors.urm.it ▣ 261218 urm lt
⊖ (+370 2) 62 07 52; 61 86 89
Minister of Foreign Affairs: Algirdas Saudargas
☎ (+370 2) 61 85 37
Deputy Minister: Januska Albinas ☎ (+370 2) 62 58 51
Deputy Minister: Jonas Rudalevicias ☎ (+370 2) 22 14 71
Deputy Minister: Arnoldas Milukas
Deputy Minister: Algimantas Rimkunas
☎ (+370 2) 22 23 48
Director of Foreign Relations: Vygaudas Usackas
Press Information Officer: Gintarias Jatkonis
☎ (+370 2) 62 16 57
Informations Officer: Zana Tarasevic 98

12-31000 Ministry of Industry and Trade
Tumo-Vaizganto 8a-2, 2739 Vilnius
☎ (+370 2) 62 88 30 ▣ 261262 egle su
⊖ (+370 2) 22 59 67
Minister: Laima Andrikiene
Minister: Vilius Ziedonis 98

12-40000 Ministry of the Interior
Sventaragio 2, 2026 Vilnius
☎ (+370 2) 62 67 52 ⊖ (+370 2) 69 83 80
Minister: Vidmantas Ziemelis 98

12-41000 Ministry of Justice
Gedimino Avenue 30/1, 2600 Vilnius
☎ (+370 2) 62 46 70 ⊖ (+370 2) 62 59 40
Minister: Vytautas Pakalniskis
Deputy Minister of Justice and Director Foreign Relations:
Gintaras Svedas
Chief of Staff: Ms Danute Gasiuniene 98

12-50000 Ministry of Communications and Information
Vilniaus 33, 2026 Vilnius
☎ (+370 2) 62 04 43 ▣ 261166
⊖ (+370 2) 22 50 70; 62 44 02
Minister: Rimantas Pleikys 98

12-60000 Ministry of Construction and Urban Development
Jaksto 4/9, 2026 Vilnius
☎ (+370 2) 61 05 58 ⊖ (+370 2) 22 08 47
Minister: Algis Caplikas 98

12-61000 Ministry of Finance
Sermuksniu 6, 2696 Vilnius
☎ (+370 2) 62 51 72 [E] finmin@finmin.lt
⊖ (+370 2) 22 63 87
Minister: Algirdas Gediminas Semet
Director Foreign Relations: Ruta Skyriene
Deputy Minister: Rimantas Survila
Chief of Staff: Grazina Urboniene 98

12-62000 Ministry of Economics
Gedimino pr 38/2, 2600 Vilnius
☎ (+370 2) 62 57 30 ⊖ (+370 2) 62 56 04; 62 39 74
Minister of Economy: Vincas Babilius
Head of Public Relations: M Barysas (acting) 98

12-63000 Ministry of Transport
Gedimino pr 17, 2026 Vilnius
☎ (+370 2) 62 14 45 ⊖ (+370 2) 22 43 35
Minister: Algis Zvaliauskis 98

25 Defense Forces

25-40000 Navy Headquarters
Naujoji uosto 24, 5800 Klaipeda
☎ (+370 6) 21 35 06; 21 35 76 ⊖ (+370 6) 21 22 96
Commander in Chief: Cdr Raimundas Saulius Baltuska
Deputy Commander COS: Lieut Cdr Vytautas Urbas
Deputy Commander in Chief: Lieut Cdr Algis Rukas
Deputy Commander in Chief: Antanas Mikutis
Director of Operations: Cdr Arturas Andrusartis
Head ASW and MCM LINAV: Lieut Cdr Eduardas Karlonas 98

29 Customs & Excise

29-10000 The Customs Department
(† (Department of Customs))
Jaksto 1/25, 2600 Vilnius
☎ (+370 2) 82 66 31 ⊖ (+370 2) 22 49 48
Director General: Alvydas Budrys ☎ (+390 2) 22 64 15
Deputy Director: Mrs Laima Leipuyiene (acting)
☎ (+370 2) 61 00 59
Deputy Director: Neringa Motiejunaite (acting)
☎ (+370 2) 61 08 16
Deputy Director: Rimantas Satrauskas (acting)
☎ (+370 2) 62 98 85 98

31 Embassies & Attachés

31-00460 Royal Danish Embassy
Kosciuskos 36, Vilnius
☎ (+370) 903 01 11 ⊖ (+370) 903 01 10
Ambassador: H E Mr Per Carlsen ☎ (+370) 22 45 45
Deputy Chief of Mission: Claus Wintop
Commercial Counsellor: Onute Babraviciene
Commercial Counsellor: Sigits Baltuska
Attaché: Esther A Larsen
Defence Attaché: Steen Bornholdt Andersen 98

31-00535 Embassy of Estonia
Tilto 29, 2001 Vilnius, Lietuva
☎ (+370 2) 22 04 86 ⊖ (+370 2) 22 04 61
Ambassador: H E Mr Alar Olljum
Consul: Andreas Traat ☎ (+370 2) 62 20 30
Commercial Counsellor: Toomas Kahur 98

31-00590 Embassy of France
Didzio 1, 2600 Vilnius
☎ (+370 2) 22 29 79; 22 28 58; 22 24 84
⊖ (+370 2) 22 35 30 (visas); 22 42 11 (chancery)
Ambassador: H E Mr Michael Tourine
Deputy Chief of Mission: Jean-Charles Berthonnet
Counsellor: M Didier Le Bret
Commercial Counsellor: Xavier Delamarre
☎ (+370 2) 22 37 86; 22 29 88 (fax)
Vice Counsul: Ms Marie-Therese Poudade
Defence Attaché: Capt Alain Nicolas-Nicolaz
Cultural Counsellor: Patrick Donabedian
☎ (+370 2) 22 04 60; 22 58 70 (fax)
Cultural Attaché: Corinne Micaelli
Attache, Police: Gérard Lacour (resides in Riga)
Finance Counsellor: Partrick Berger (resides in Vienna)
Attaché: Patrick Deunet (resides in Vienna)
Attaché: Ms Florence Jourdain
Attaché: Thierry Bourguet
Attaché: Ms Flavie Cordaro
Attaché: Norbert Coustaud
Attaché: Patrice Lefol 98

31-01730 Swedish Embassy
Jogailos 8, 2600 Vilnius
☎ (+370 2) 22 64 67 ⊖ (+370 2) 22 64 44
Ambassador: H E Mr Stellan Ottosson
Defence Attaché: Lieut Col Erling Sohansson 98

31-01840 Embassy of the Republic of Turkey
Didzioji Gatve 37, 2001 Vilnius
☎ (+370 2) 22 33 80; 22 28 87; 22 34 40
▣ 261006 tcbvl ⊖ (+370 2) 22 32 77

Contact Reference

Ambassador: H E Mr Erkan Gezer
First Counsellor: Yasar Pinar
Second Secretary: Korkut Güngen
Attaché: Sibel Sahin 98

31-01890 Embassy of the United Kingdom
Antakalnio 2, 2055 Vilnius
☎ (+370 2) 22 20 70; 22 20 70
E be-vilniou@post.omnitel.net ⊖ (+370 2) 72 75 79
Ambassador: H E Mr Thomas Macan
Second Secretary and Deputy Head of Mission and Consul: S N Tarry
First Secretary Political: W J Roebuck
Attaché: Ms Amanda Coopere
Attaché, Political: Mathew Prouten
First Secretary, Political: W J Roebuck
First Secretary, British Council Director: A N Sanderson (resides in Riga)
Defence Attaché: Lieut N G W Lang (resides in Riga)
Commercial Officer: Ms Ieva Binkauskaite
Public Information Officer: Ms Mida Babiliene
Commercial Assistant: Ina Rosenite 98

Luxembourg

Head of State:
H R H Grand Duke Jean

10 Head of State

10-10000 Grand Ducal Court
Palais Grand-Ducal, Luxembourg-Ville
☎ (+352) 22 26 01 ⊠ 1817 ⊖ (+352) 47 18 30
Grand Duke of Luxembourg: H R H Grand Duke Jean
Marshal of the Court: Henri Ahlborn 98

11 Prime Minister

11-10000 Office of the Prime Minister
4 rue de la Congrégation, Hôtel de Bourgogne, L-1352 Luxembourg Ville
☎ (+352) 478 21 06; 478 47 81 ⊠ 2790 etafin lu
⊖ (+352) 46 17 20
Prime Minister: Jean-Claude Juncker
Secretary General: Albert Hansen ☎ (+352) 478 21 03

Office of the Deputy Prime Minister
4 rue de la Congregation, L-2910 Luxembourg-Ville
☎ (+352) 478 21 06 ⊖ (+352) 46 17 20
Deputy Prime Minister and Minister of Foreign Affairs, External Trade and Co-operation: Jacques F Poos

Government of Luxembourg Information and Press Service
43 Boulevard Roosevelt, L-2450 Luxembourg Ville
☎ (+352) 478 21 06; 478 21 81 ⊠ 1727 siplu
⊖ (+352) 47 02 85; 22 00 90
Head of Information Service: Mil Jung
☎ (+352) 478 21 83
Press and Information Service: T H A Pescatore
E theodore.pecatore@sip .etat.lu 98

12 Ministries

12-10000 Ministry of Public Force
Plateau du Saint Esprit, L-2915 Luxembourg
☎ (+352) 474 58 11 ⊖ (+352) 46 26 82
Minister of Public Force: Alex Bodry 98

12-15000 Ministry of Defence
Plateau du Saint-Esprit, L-2915 Luxembourg
☎ (+352) 478 28 06 ⊖ (+352) 46 26 82
Minister: Alex Bodry 98

12-20000 Ministry of Foreign Affairs, External Trade and Co-operation
5 rue Notre-Dame, L-2911 Luxembourg-Ville
☎ (+352) 478 11; 478 23 03
⊠ 3405; 1864; 3412 afetr lu ⊖ (+352) 22 31 44
Deputy Prime Minister and Minister of Foreign Affairs, Trade and Co-operation: Jacques F Poos
State Secretary: Georges Wohlfart
Secretary General: Jean Mischo
Public Relations Director: Victor Weitzel

International Economic Relations
6 rue de la Congregation, L-2910 Luxembourg-Ville
☎ (+352) 478 23 50 ⊖ (+352) 22 20 48
Director: Nicolas Schmit ☎ (+352) 478 23 50

Direction Protocol & Administration Ministry of Foreign Affairs
5 rue Notre-Dame, L-2910 Luxembourg
☎ (+352) 478 23 80 ⊖ (+352) 22 31 44
Director: Jean Welter

Direction Political and Cultural Affairs
5 rue Notre-Dame, L-2910 Luxembourg
☎ (+352) 478 23 10 ⊖ (+352) 22 31 44
Director: Jean Feyder 98

12-40000 Ministry of the Interior
19 rue Beaumont, L-2933 Luxembourg
☎ (+352) 478 46 06; 478 46 95 ⊖ (+352) 24 18 46
Minister of the Interior: Michel Wolter
Secretary General: Claude Frieseisen
Director of Civil Defence: Leon Anen 98

12-50000 Ministry of Finance
3 rue de la Congrégation, L-2931 Luxembourg
☎ (+352) 478 26 06 ⊠ 2790 etafin lu
⊖ (+352) 22 19 80
Minister: Marc Fischbach
Treasury Director: Yves Mersch
General Administrator: Gaston Reinesch 98

12-60000 Ministry of Transport
19-21 Boulevard Royal, L-2938 Luxembourg-City
☎ (+352) 478 44 03 ⊠ CIVILAIR ⊠ 1465 civair lu
⊖ (+352) 46 43 15; 46 77 90
Minister: Mrs Mady Delvaux-Stehres ☎ (+352) 478 44 01
Secretary General: Fernand Kesseler ☎ (+352) 478 44 03
Director of Aviation: Henri Klein ☎ (+352) 478 44 12 98

15 Parliament

15-20000 Chamber of Deputies
19 rue du Marché aux Herbes, L-1728 Luxembourg-Ville
☎ (+352) 22 02 22 ⊖ (+352) 22 02 30
President: Jean Spautz
Secretary General: Guillaume Wagener 98

25 Defense Forces

25-10000 Army Headquarters
Commandement de L'Armée
PO Box 1873, L-1018 Luxembourg
☎ (+352) 48 88 36 ⊠ 2912 comdta lu
⊖ (+352) 478 54 20
Commander: Col M. Gretsch
Deputy Commander and Secretary General: Lieut Col René Alzin
G1 (Personnel) and Public Information Officer: Lieut Col F Guth
Chief Intelligence and Operations G2/G3: Lieut Col P Freichel
Chief Logistics G4: Lieut Col N Ries
Budget Officer: Lieut Col S Luca
Host Nations Support and Assistant Logistics Officer: Capt A Arend
CFE Officer: Maj G Hoffmann
Assistant G2/G3: Gérard Schott
Medical Adviser: Lieut Col Georges Boden
General-Chaplain: Capt Nico Wenner 98

29 Customs & Excise

29-10000 Customs Headquarters
Direction des Douanes et Accises (Subordinate to Ministry of Finance)
PO Box 26, L-2010 Luxembourg
☎ (+352) 290 19 11 ⊠ 3113 douan lu
⊖ (+352) 49 87 90
Director: Marc Schloesser
Chief Personnel and General Affairs: Marc Petit
☎ (+352) 290 19 12 04 98

31 Embassies & Attachés

31-00100 Embassy of Austria
3 rue des Bains, L-1212 Luxembourg
☎ (+352) 22 69 57; 47 11 88 ⊠ AUSTROAMB
⊠ 2530 oelux lu ⊖ (+352) 46 39 74
Ambassador: H E Dr Josef Magerl
Counsellor and Deputy Chief of Mission: Walter Grahammer
First Secretary: Ingrid Koehn 97

31-00150 Embassy of Belgium
4 rue des Girondins, Résidence Champagne, Luxembourg
☎ (+352) 442 74 61 ⊠ 2550 ambbel lu
⊖ (+352) 45 42 82
Ambassador: H E Mr B Kethulle de Ryhove
Deputy Chief of Mission: Ms Marie-Antoinette Pirlot
Director of the Chancellery: Henri Colpaert
Military Attaché: Col Jacques Gilson
Commercial Attaché: Pierre Leleux
Attaché: Jan Verhegge 98

31-00660 Embassy of the Federal Republic of Germany
PO Box 95, L-2010 Luxembourg
20-22 Avenue Emile Reuter, L-2420 Luxembourg
☎ (+352) 453 44 51 ⊠ 3413 aaluxb lu
⊖ (+352) 45 56 04
Ambassador: H E Mr Rolf-Eberhard Jung
Counsellor and Deputy Chief of Mission: Erwin Starnitzky
Counsellor Press and Communications: Hubertus von Römer
Defence Attaché: Lieut Col R F Abele
Second Secretary Administrative Affairs: Kurt Kenning
Commercial Attaché: Ms Daniela Weber 97

31-00890 Embassy of Italy
5-7 rue Marie Adelaïde, L-2128 Luxembourg
☎ (+352) 44 36 44; 44 36 45 ⊠ 2216
⊖ (+352) 45 55 23
Ambassador: H E Mr Giovanni Castellanni Pastoris
Deputy Chief of Mission and Counsellor: Dr Adriano Tedeschi 98

31-00920 Embassy of Japan
17 rue Beaumont, L-1219 Luxembourg
☎ (+352) 46 41 51 ⊖ (+352) 46 41 76 97

31-01655 Consulate of the Republic of Slovenia
27 rue des Glacis, L-1628 Luxembourg
☎ (+352) 46 64 79 ⊖ (+352) 46 64 70
Honorary Consul: Franc Dreu 98

31-01680 Embassy of Spain
4 Boulevard Emmanuel Servais, PO Box 290, L-2535 Luxembourg
☎ (+352) 46 02 55 ⊖ (+352) 47 48 50
Ambassador: H E Mr Ramiro Perez-Maura (Duke of Maura)
Counsellor: Fernando Valderrama
First Secretary: Ms Alicia Rico 98

31-01890 **Embassy of the United Kingdom**
14 Boulevard Roosevelt, L-L-2450 Luxembourg
☎(+352) 22 98 64; 22 98 66; 22 98 65
⊠ 3443 prodrolu ⊖ (+352) 22 98 67 Group 2
Ambassador and Consul General: H E Mr J N Elam
First Secretary and Deputy Chief of Mission: Ms M C Harper
Commercial Counsellor: N R Jarrold (Resident in Brussels)
Defence Attaché: Gp Capt D C Hencken (RAF, Resident in Brussels) [98]

34 Multinational Intergovernmental Organizations

34-07010 **NATO Maintenance and Supply Agency (NAMSA)**
11 rue de la Gare, PO Box 13, L-8302 Capellen
☎(+352) 30 63 65 57; 30 63 65 58; 30 63 65 59
⊠ 2359 ⊖ (+352) 30 87 21; 30 63 66 52
General Manager: R W A Zweerts ☎(+352) 30 63 65 02
Procurement Director: Gary Sager ☎(+352) 30 63 65 65
Director Organisation and Administration: Brian Todd ☎(+352) 30 63 65 76
Finance Director: Elmar Osterloh ☎(+352) 30 63 65 71
Director of Logistics: Joseph S Laposota
Head, Corporate Planning and Marketing: Klaus Broecker ☎(+352) 30 63 65 59 E kbroecker@namsa.nato.int
Public Relations Manager: Mrs Marie-Claire Goller ☎(+352) 30 63 65 57
■ NAMSA is NATO's principal logistics agency and currently supports 30 weapon and equipment systems for 15 NATO nations' armed forces.
Repair, maintenance and supply of spare parts, procurement and demilitarization of ammunition and special engineering support services are the main activities in support of aircraft, missiles, artillery and tracked vehicles, electronics and associated equipment.
NAMSA provides NATO with co-operative logistics services including codification/identification services, NATO Master Cross Reference List (N-MCRL), Stock Holding and Asset Requirements Exchange (SHARE) and NATO Ammunition Database (NADB).
NAMSA's activities are extended to cover service requirements for Partnership for Peace nations and contractual support for NATO's operational logistics.
Logistics Programmes and Operations Directorate
L-8302 Capellen, GD [98]

75 Manufacturing Companies

75-00350 **Commercial Intertech SA Astron Building Systems**
(† Commercial Hydraulics SA)
(a subsidiary of Commercial Intertech Corp, USA)
PO Box 152, L-9201 Diekirch
Route d'Ettelbruck
☎(+352) 80 29 11 ⊠ HYDCOM E astromg@pt.lu
⊖ (+352) 80 34 66
Lfr 207,560,000 (1997) ♣ Lfr 2,838,365,197 (1996)
480 (1997) 28,000 m2
Managing Director: Edward K Barnard
Operations Director: Norbert Holtzen
Sales and Marketing Director: Yvan Reuter
Purchasing Manager: Pete Klein
General Administration: Georges Mersch
■ Aircraft hangars, storage buildings, office buildings, controlled humidity warehouses and steel tunnel supports. [98]

Macedonia

Head of State:
Kiro Gligorov

10 Head of State

10-10000 **Office of the President**
Skopje
President: Kiro Gligorov [98]

11 Prime Minister

11-10000 **Office of the Prime Minister**
Dame Grueva 6, 91000 Skopje
☎(+38 91) 20 12 11 ⊖ (+38 91) 21 13 93
Prime Minister: Branko Crvenkovski
Deputy Prime Minister: Jane Miljovski
Deputy Prime Minister: Ljube Trpevski
Deputy Prime Minister: Bekir Zuta [98]

12 Ministries

12-10000 **Ministry of Defence**
Dimce Mirceva bb, 91000 Skopje
☎(+389 91) 23 73 73
Minister: Lazar Kitanovski
Chief of General Staff: Col Gen Trajche Krstevski [98]

12-20000 **Ministry of Foreign Affairs**
No 6 Dame Grueva, 91000 Skopje
☎(+389 91) 11 03 33; 11 52 66 ⊖ (+389 91) 11 57 90
Minister of Foreign Affairs: Blagoje Handziski [98]

12-70000 **Information**
(† Ministry of Information)
Skopje
☎(+389 91) 11 41 62; 11 55 10
⊖ (+389 91) 11 41 62; 11 74 01
Secretary: Ms Ana Panovska [98]

29 Customs & Excise

29-00500 **Customs Administration**
Lazar Licenovski 13, 91000 Skopje [98]

31 Embassies & Attachés

31-01665 **Embassy of the Republic of Slovenia**
Ulitsa Mile Pop Jordanov 12, 91000 Skopje
☎(+38 91) 38 12 21 ⊖ (+38 91) 38 12 27
Ambassador: H E Mr Jozica Puhar [98]

Madagascar

Head of State:
Adm Didier Ratsiraka (retd)

10 Head of State

10-10000 **Office of the President**
Ambohitsorohitra, Antananarivo
☎(+261 2) 274 74
President: Adm Didier Ratsirakda (retd) [98]

11 Prime Minister

11-10000 **Office of the Prime Minister**
Mahazoarivo, Antananarivo 101
☎(+261 2) 252 58
Prime Minister: Pascal Rakotomavo
Vice President in Charge of Decentralization and Budget: Pierrot Jocelyn Rajaonarivelo
Vice President in Charge of Foreign Affairs: Herzio Razafimahaleo [98]

12 Ministries

12-40000 **Ministry of Interior**
Anosy, Antananarivo 101
☎(+261 2) 214 65 ⊠ 22339 gouvernement tana
Minister: Col Jean-Jaques Rasolondraibe [98]

15 Parliament

15-10000 **National Assembly**
(† People's National Assembly)
Tsimbazaza, Antananarivo
☎(+261 2) 216 37 ⊖ (+261 2) 632 35 [98]

20 Defense Administrations/Services/Agencies

20-10000 **Military Development Committee**
Antanarivo
☎(+261 2) 220 91 ⊖ (+261 2) 219 06
40
Chief of President Military Staff: Brig.Gen. Marcel Ranjeva [97]

31 Embassies & Attachés

31-00830 **Embassy of India**
4 Lalana Rajaonson Emile Tsralalana, PO Box 1787, Antananarivo
☎(+261 2) 233 34; 271 56 ⊠ 22484 indemb mg
⊖ (+261 2) 23 37 90
Ambassador: H.E. Dr Prabhakara Jha (Designate)
First Secretary Deputy Chief of Mission: G.K. Mehta
Military & Air Attaché: Col Owen Pereira
Naval Attaché: Capt. S. Handa
Attaché: Shri Guru Sharah Singh [97]

31-00890 **Embassy of Italy**
22 Rue Pasteur Rabary, PO Box 16, Antananarive
☎(+261 2) 212 17; 215 74; 284 43 ⊠ 22293 itadp mg
⊖ (+261 2) 238 14

Contact Reference

Ambassador: H.E. Dr Francesco Sciortino
Commercial Secretary: Dr Fabrizio Marcelli [98]

31-01890 Embassy of the United Kingdom
BP Box 167, Antananarivo
1st Floor, Immeuble "Ny Havana" Cité de 67 Ha
☎(+261 2) 277 49; 273 70 ⊠ 22459 prodro mg
⊖ (+261 2) 266 90
Ambassador: H E Mr R S Dewar
Deputy Head of Mission: Ms L M Clegg
Cultural Attaché: M J Bottle (resides in Port Luis)
Third Secretary: J. Green
Commercial Officer: M Rakotolehibe [98]

31-01900 Embassy of the United States of America
14/16 Rue Rainitovo, BP 620, Antsahavola, Antananarivo
☎(+261 2) 212 57; 200 89; 207 18; 209 56
⊠ 22202 usaemb mg ⊖ (+261 2) 23 45 39
Ambassador: H E Ms Vicki Huddleston
Deputy Chief of Mission: Howard T Perlow
Commercial Section: Peter Barlerin
Public Affairs Officer: Paul J Saxton
Defense Attaché: Lieut Cdr John Logan (USN) [98]

Malawi

Head of State:
H E Dr Bakili Muluzi

10 Head of State

10-10000 Office of the President
Capital City, Private Bag 301, Lilongwe 3
☎(+265) 73 11 99 ⊠ 4389 pres mi
President: H E Dr Bakili Muluzi
First Vice President: Justin C Malewezi [98]

12 Ministries

12-10000 Ministry of Defence
Capital City, Private Bag 339, Lilongwe 3
Minister of Defence: Joseph Kubwalo [98]

12-20000 Ministry of Foreign Affairs
Capital City, PO Box 30315, Lilongwe 3
☎(+265) 78 22 11 ⊖ (+265) 78 24 34
Minister: Mapopa Chipeta
Secretary of External Affairs: B M Ndisale [98]

12-40000 Ministry of Local Government, Forestry and Natural Resources
Capital City, PO Box 30312, Lilongwe 3
Minister: Chakakala Chaziya [98]

15 Parliament

15-10000 National Assembly
Parliament Building, PO Box 80, Zomba
☎(+265) 52 24 88 🄴 parliament@unima.wn.apc.org
⊠ 446553 ⊖ (+265) 52 32 90
Speaker: Rodwell T C Munyenyembe ☎(+265) 52 29 96
Secretary of Parliament: Roosevelt L Gondwe
Deputy Clerk of Parliament: F A U Assani
Chairman Defense and Security Committee:
Mayinga Mkandawire
Member of the Defence and Security Committee:
M M Kalele
Member of the Defence and Security Committee:
A K M Pillane
Member of the Defence and Security Committee:
J L Makumula
Member of the Defence and Security Committee:
C B Thete
Member of the Defence and Security Committee:
Gwanda Chakuamlsa
Member of the Defence and Security Committee:
V G Chinere
Member of the Defence and Security Committee:
B A M Kalinda Phiri
Member of the Defence and Security Committee:
N K S Mwafulirwa
Member of the Defence and Security Committee:
C W Salima
Member of the Defence and Security Committee:
M A K Mgolomba [98]

31 Embassies & Attachés

31-00690 Embassy of Greece
PO Box 2, Blantyre
☎(+265) 63 10 34 [98]

31-01670 South African High Commission
Mpico Building, Capital City, PO Box 30043, Lilongwe 3
☎(+265 783) 722 ⊠ 44255 saleg mi ⊖ (+265 782) 571
Counsellor: G W E Seate (acting)
Defence Attaché: Col J I de Wet
Public Information Officer, Commercial Attaché and Third Secretary: L D Viljoen [98]

31-01890 United Kingdom High Commission
PO Box 30042, Lilongwe 3
☎(+265) 78 24 00 ⊠ 44727 ukrepli mi
⊖ (+265) 78 26 57; 78 21 53
High Commissioner: H E Mr J F R Martin
Deputy High Commissioner: R W Sindon
Defence Adviser: Col A J Reed-Screen
(Resident in Harare)
First Secretary, Development: R J Leverington
First Secretary, Regional Medical Advisor: Dr J O Llwellyn
Second Secretary, Information and Political: S I Wall
Second Secretary Political Information: Miss M Bennett
Third Secretary, Development: M P Reilly [98]

31-01900 Embassy of the United States of America
PO Box 30016, Lilongwe 3
☎(+265) 78 31 66 ⊠ 44627 amembll mi
⊖ (+265) 78 04 71
First Secretary: Bernia Link
Defense Attaché: Lieut Col Delehine Donneloi (US Army) [98]

Malaysia

Supreme Head of State:
H.M. Sultan Tunku Ja'afar ibni Abdul Rahman

11 Prime Minister

11-10000 Office of the Prime Minister
Jalan Dato'Onn, 50502 Kuala Lumpur
🅆 http://www.mdc.com.my
Prime Minister: YAB Dato' Seri Dr Mahathir bin Mohamad
Deputy Prime Minister: Datuk Seri Anwar Ibrahim [98]

12 Ministries

12-40000 Ministry of Home Affairs
Kementerian Dalam Negeri
Jalan Dato'Onn, 50546 Kuala Lumpur
☎(+60 3) 230 93 44; 293 61 22 ✉KDN Kuala Lumpur
⊖ (+60 3) 230 10 51; 230 12 17
Minister of Home Affairs:
Dato'Seri Dr Mahathir bin Mohamad
Deputy Minister: Datuk Seri Megat Junid bin Megat Ayob
Deputy Minister: Dato' Ong Ka Ting
Secretary General: Datuk Samsudin Bin Osman
Deputy Secretary General I:
Dato'Seri Nik Abidin Bin Nik Omar
Deputy Secretary General II: Dato'Ismail Bin Hamzah
Under Secretary Internal Security and Public Order:
En Hamzah Bin Mohd Rus
Public Relations Officer: Jamilah Mohd Shahid
■ Subordinate defence related departments:
People's Vigilante Corps
Civil Defence Department [98]

12-60000 Ministry of Science, Technology and the Enviroment
Tingkat 14 - Wisma Sime Darby, Jalan Raja Laut, 50662 Kuala Lumpur
☎(+60 3) 293 89 55 🅆 http://masti.gov.my
⊠ 28154 MA ⊖ (+60 3) 293 60 06
Minister: Dato'Lav Hieng Ding
Deputy Minister: Abu Bakar Daud ☎(+60 3) 293 88 66
Malaysian Science and Technology Information Centre
19th Floor - Wisma Sime Darby, Jalan Raja Laut, 50662 Kuala Lumpur
Director: Samsurdin bin Ahamad [98]

15 Parliament

15-10000 Parliament (Senate)
Dewan Negara
Parliament Building, 50680 Kuala Lumpur
☎(+60 3) 232 19 55 ⊖ (+60 3) 230 09 98
President: Tan Sri Dato' Mohamed Yaacob
Secretary to the Senate: Abdullah Abdul Wahab [98]

15-20000 Parliament House of Representatives
Dewan Rakyat
Parliament Building, 50680 Kuala Lumpur
Speaker: Tan Sri Dato'Mohamed Zahir bin Haji Ismail
Secretary to the Speaker: Nordin bin Baba [98]

31 Embassies & Attachés

31-00080 Embassy of the Argentine Republic
PO Box 11200, 50738 Kuala Lumpur
3 Jalan Semantan Dua - Damansura Heights, 50490 Kuala Lumpur
☎(+60 3) 255 01 76; 255 25 64 ⊠ 31854 embarc ma
⊖ (+60 3) 255 27 06
Ambassador: H E Mr Guillermo Hector Renom
First Secretary: Carlos R Chiappe Solozabal
First Secretary: Gustavo Horacio Luis Fazzari
Second Secretary: Augusto Cesar Granada
Administrative Attaché: Ms Maria Ana Garcia Cuerva [98]

31-00090 Australian High Commission
PO Box 10921, 50730 Kuala Lumpur
N°6 Jalan Yap Kwan Seng, 50450 Kuala Lumpur
☎(+60 3) 242 31 22 ✉AUSTCOM Kuala Lumpur
⊠ 30260 austkl ma ⊖ (+60 3) 241 57 73
High Commissioner: H E Mr R L Cotton
Deputy High Commissioner: David Mason
Defence Adviser: Gp Capt Rick Jones
☎(+60 3) 240 67 89
Assistant Defence Adviser: Lieut Col John Ramsay
☎(+60 3) 240 67 92
Minister Commercial and Senior Trade Commissioner:
Kym Hewett ☎(+60 3) 240 67 56

Counsellor Administration and Consul General:
John Anthony Osborne
Counsellor Commercial and Trade Commmissioner:
Brendan Dyson ☎ (+60 3) 240 67 58
Counsellor Industry, Science and Technology:
Ms Robyn Flemming ☎ (+60 3) 240 65 05 [98]

31-00100 Embassy of Austria
PO Box 10154, 50704 Kuala Lumpur
7th Floor, MUI Plaza, Jalan P Ramlee,
50250 Kuala Lumpur
☎ (+60 3) 248 42 77; 248 43 59 ▣ 31263 oeboku ma
⊖ (+60 3) 248 98 13
Ambassador: H.E. Mr Udo Ehrlich-Adam
Deputy Chief of Mission: Ms Leonore Knuttner

Office of Commercial Counsellor
Kompleks Kewangan, 10th Floor, Jalan Raja Chulan,
50200 Kuala Lumpur
☎ (+60 3) 261 47 24; 261 48 36 ▣ 31203 ahstkl ma
⊖ (+60 3) 261 31 30
Commercial Councellor: Hartwig Seuchter [97]

31-00130 High Commission for the People's Republic of Bangladesh
204-1 Jalan Ampang, 50450 Kuala Lumpur
☎ (+60 3) 242 32 71; 248 79 40 ▣ 31593 bdut ma
⊖ (+60 3) 241 33 81

Honorary Consulate General of the People's Republic of Bangladesh
15 Bishop Street, 10200 Penang [98]

31-00150 Embassy of Belgium
GPO Box 10558, 50718 Kuala Lumpur
12 Lorong Yap Kwan Seng, 50450 Kuala Lumpur
☎ (+60 3) 262 57 33; 263 58 64; 262 59 91
✉ BELGE Kuala Lumpur ▣ 30044 ambel ma
⊖ (+60 3) 262 59 22
Ambassador: H E Mr Eric Duchene
Second Secretary and Deputy Chief of Mission:
Marc Mullie
Commercial Attaché: Jean-Jacques Manigart [98]

31-00460 Royal Danish Embassy
86 Jalan Ampang, PO Box 10908, 50728 Kuala Lumpur
22nd Floor, Wisma Denmark, 86 Jalan Ampang,
50728 Kuala Lumpur
☎ (+60 3) 202 20 01 [E] rdembsy@po.jaring.my
⊖ (+60 3) 202 20 15
Ambassador: H E Mr Leif Reimann
Vice Consul: Ms Ulla Johansen
Commercial Officer: Ulrich Serensen
Asst Attaché Commercial: Jens Agerskov
Trade Information Officer: Jaime Yong [98]

31-00570 Embassy of the Republic of Fiji
Suite 2.03, 2nd Floor, Wisma Equity,
No 150 Jalan Ampang, 50450 Kuala Lumpur
☎ (+60 3) 242 84 22 ⊖ (+60 3) 241 56 36 [98]

31-00860 Embassy of the Republic of Iraq
2 Jalan Langgak Golf, Off Jalan Tun Razak,
55000 Kuala Lumpur
☎ (+60 3) 248 05 55 ▣ 30328 iraqkl ma
⊖ (+60 3) 241 43 31
Ambassador: H E Mr Mohammed Rifat Ali Al-Ani
☎ (+60 3) 248 06 50
Consul: Saad Y Dawood
Administration Officer: Raad A M Al-Mudares [98]

31-01300 New Zealand High Commission
PO Box 12003, 50764 Kuala Lumpur
21st Floor Menara IMC - 8 Jalan Sultan Ismail
☎ (+60 3) 238 25 33 ☎ Kuala Lumpur
✉ ARAWA Kuala Lumpur ▣ 30284 arawa ma
⊖ (+60 3) 238 03 87
Defence Adviser: Col B W R I Sherwood [98]

31-01480 Embassy of Romania
114 Jalan Damai, off Jalan Ampang, 55 000 Kuala Lumpur
☎ (+60 3) 242 31 72 ▣ 30362 romcom ma
⊖ (+60 3) 244 87 13
Ambassador: H E Mr Eugen Popa ☎ (+60 3) 241 74 54

Trade Representation
Kuala Lumpur
☎ (+60 3) 248 20 65 ▣ 30362 romcom ma
⊖ (+60 3) 244 87 13
Commercial Counsellor: Ion Sihleanu [98]

31-01640 Embassy of the Republic of Singapore
209 Jalan Tun Razak, 50400 Kuala Lumpur
☎ (+60 3) 261 62 77 [E] tltong@shcom.po.my
⊖ (+60 3) 261 63 43
High Commissioner: H E Mr K Kesavapany
Deputy High Commissioner and Minister Counsellor:
Vanu Gopala Menon
Defence Adviser: Michael Goh
First Secretary: Jaya Ratnam
First Secretary: Yip Wei Kiat
First Secretary: Ms Joanna Foo [98]

31-01740 Embassy of Switzerland
PO Box 12008, 50764 Kuala Lumpur
16 Pesiaran Madge, 55000 Kuala Lumpur
☎ (+60 3) 248 06 22; 07 51; 06 39
✉ AMBASUISSE Kuala Lumpur ▣ 31155 emswis ma
⊖ (+60 3) 248 09 35
Ambassador: H.E. Mr Max B. Dahinden
Counsellor: Markus Hirsiger [97]

31-01840 Embassy of the Republic of Turkey
118 Jalan U Thant, 55000 Kuala Lumpur
☎ (+60 3) 457 22 25 ▣ 31245 turkbe ma
⊖ (+60 3) 457 22 27
Ambassador: H E Mr Ferhat Ataman
Defence Attaché: Capt Oguz Orhun
Commercial Officer: Attila Gokhan Kizilarslan [98]

31-01890 United Kingdom High Commission
PO Box 11030, 50732 Kuala Lumpur
☎ (+60 3) 248 21 22; 248 23 54 (commercial);
248 71 22 ⊖ (+60 3) 244 77 66 (High Commission);
248 08 80 (commercial); 244 96 92 (consular)
High Commissioner: H E Mr David J Moss
Deputy High Commissioner and Commercial Counsellor:
T N Byrne
Counsellor, Political: Dr P G Wood
Counsellor, Commercial and Economic: H Parkinson
Counsellor, Defence Supply: B Neale
Defence Adviser: Col M B Cooper
First Secretary: S J Bridges [98]

31-01900 Embassy of the United States of America
PO Box 10035, 50700 Kuala Lumpur
376 Jalan Tun Razak, 50400 Kuala Lumpur
☎ (+60 3) 248 90 11 ⊖ (+60 3) 242 22 07
Ambassador: H.E. Mr John R. Malott
☎ (+60 3) 248 90 11 ext. 200
Deputy Chief of Mission: Ms Wendy Chamberlin
☎ (+60 3) 248 90 11 ext. 201
Political Section: G.Nicholas Mauger (III)
☎ (+60 3) 248 90 11 ext. 210
Economic Section: Ms Deborah L. Linde
☎ (+60 3) 248 90 11 ext. 251
Commercial Section: B.Paul Scogna
☎ (+60 3) 248 90 11 ext. 258
Defense Attaché: Lt Col Russell Thurber (II)
☎ (+60 3) 248 90 11 ext. 234
Public Affairs Officer: Ms Julie Gianelloni/Connor
☎ (+60 3) 248 90 11 ext. 323 [97]

60 Research Organizations, Institutes & Test Establishments

60-12000 Malaysian Institute of Maritime Affairs (MIMA)
16th Floor - Wisma Sime Darby, Jalan Raja Laut,
50350 Kuala Lumpur
☎ (+60 3) 291 29 60 [E] jnmak@mima.gov.my
⊖ (+60 3) 293 91 17; 293 30 44
Director: Dr B A Hamzah
Director of Research: Joon Num Mak
■ Malaysian government funded research institute assigned to national, regional and global maritime issues. [98]

70 Financial Holding Groups, Parent Companies

GEC Malaysia Sdn Bhd 90-00730

70-00500 Kawasaki Motors Holding (Malaysia) Sdn Bhd
(joint venture with Kawasaki Heavy Industries Ltd, Japan)
Polt Surat N°39, Pelabuhan Kelang,
42000 Selangor Darul Ehsan
Lot 24 Jalan Lebuh Sultan Mohamed,
Kawasan Persushahaan, Bander Sultan
Sulaiman, Polt Surat N°39,
42000 Pelabuhan Kelang, Selangor Darul
☎ (+60 3) 376 21 97 ⊖ (+60 3) 376 13 80 [98]

75 Manufacturing Companies

75-00043 Daewoo Corp Malaysia Branch
Suite 2002 - 20th Floor, Pernas International Building,
50250 Kuala Lumpur
Jalan Sultan Ismail
☎ (+60 3) 261 91 62 ⊖ (+60 3) 261 91 63 [98]

75-00055 Hong Leong-Lürssen Shipyard (1992) Sdn Bhd
4567 Jalan Chain Ferry, PO Box 43, 12700 Butterworth
☎ (+60 4) 332 99 55 ⊖ (+60 4) 332 99 66
♧ M$ 5,000,000 (1997) ♣ M$ 50,000,000 (1997)
👥 250 (1997) 🏭 10,000 m2
General Manager: Chuan Yong Chuah
Operations Manager: Tamjis bin Tahir
Purchasing Manager: Hong Yew Mak
■ Shipbuilding, ship-repairing, overhaul of main engines & accessories. → 80

Kuala Lumpur Office
Level 9, Wisma Hong Leong, 18 Jalan Perak,
50450 Kuala Lumpur
☎ (+60 3) 264 26 31 ⊖ (+60 3) 264 26 67 [98]

75-00060 Malaysian Oxygen Berhad
(Association of Alboc (Jersey) Ltd)
PO Box 10633, 50720 Kuala Lumpur
N°13 Jalan 222, 46100 Petaling Jaya
☎ (+60 3) 755 42 33 ▣ 37663 ma maox p
⊖ (+60 3) 756 63 89
♧ US$ 150,000,000 ♣ US$ 283,000,000
👥 670 (1997)
Managing Director: Rodd Duke
General Manager Commercial: Jit Teng Lim
■ Manufacture and distribution of industrial, special and medical gases, welding electrodes and medical equipment. → 85
® France: Zodiac SA [98]

75-00200 Minebea Co. Ltd Kuala Lumpur Branch
4th Floor West Tower Wisma Consplant,
N°2 Jalan SS 16/4, 47500 Subang Jaya, Salengoor
☎ (+60 3) 731 78 49 ⊖ (+60 3) 201 15 34
👥 10
Branch Manager: M Takata

Penang Office
Suite 10-2 10f Manate Penang Garden,
42-A Jalan Sultan Ahmad Shah, 10050 Penang
☎ (+60 4) 227 56 81 ⊖ (+60 4) 227 58 20 [98]

75-00800 Penang Shipbuilding Corporation
Ming Building 11th Floor, Jalan Bukit Nanas,
50250 Kuala Lumpur
☎ (+60 3) 238 77 70 ⊖ (+60 3) 201 77 69
👥 300
Vice President: A Manap
■ Construction, maintenance, repair and overhaul of patrol and landing ships. Repair and overhaul of marine diesel engines. [98]

Contact Reference

Malaysia

75-00900 Saft Nife Power Systems Sdn Bhd
(a subsidiary of Saft SA, France)
37 Jalan 10/91 Taman Shamelin,
Perkasa Batu 3-1/2 Jalan Cheras, 56100 Kuala Lumpur
☎ (+60 3) 985 29 96 ⊖ (+60 3) 984 49 95
General Manager: Ng King
[98]

75-01800 Sime Aircraft Tyre Co. Sdn Bhd
(† Michelin Sime Aircraft Tyre Co. Sdn Bhd)
(a subsidiary of DMIB Berhad, Malaysia)
4 Jalan Tandang, PO Box 66,
46700 Petaling Jaya, Selangor
☎ (+60 3) 791 44 06; 791 61 17
⊖ (+60 3) 791 44 70; 791 77 58
Managing Director: Mohamed Ishak
Sales and Marketing Director: Yap Kwong Tuck
☎ (+60 3) 791 44 07
Technical Director: Yang Lim Guey
■ Manufacture and retread of commercial, commuter and military aircraft tyres.
[98]

75-05000 Transfield Pty Ltd Regional Office
15th Floor - Menara Multi-Purpose,
Capital Sq - 8 Jalan Munshi Abdullah, 50100 Kuala Lumpur
☎ (+60 3) 294 77 88 ⊖ (+60 3) 294 59 66
Regional Director: Luca Belgiorno-Nettie
[98]

80 Service Companies & Organizations

GEC Malaysia Sdn Bhd 90-00730
Hong Leong-Lürssen Shipyard 75-00055

80-00050 AIROD Sdn Bhd
(a subsidiary of Aerospace Industries Malaysia Sdn Bhd)
Locked Bag 4004, Pejabat Pos Kampung Tunku,
47309 Petaling Jaya, Selangor
☎ (+60 3) 746 51 12 [E] kamil@airod.po.my
[W] http://www.jaring.my/biz/airod ⊠ 37910 ma
⊖ (+60 3) 746 47 09
♤ US$ 6,300,000 (1996) ♣ US$ 52,000,000 (1996)
👥 1,100 (1997) 🏭 77.4 acres
President: Dato'Ahmad Johan
☎ (+60 3) 746 51 12 ext. 111
Vice President: Moghni Rahmat
☎ (+60 3) 746 51 12 ext. 100
Senior General Manager Operations: Ramli Mat Api
☎ (+60 3) 746 51 12 ext. 318
General Manager Marketing: Kamil Aziz
☎ (+60 3) 746 51 12 ext. 25 [E] kamil@airod.po.my
Senior Manager Logistics: Mohd Zaini Yusof
☎ (+60 3) 746 51 12 ext. 132
Purchasing Manager: Amir Hussin Awang
Corporate Marketing: Niozam Norwani
■ Overhaul, repair and modification of civil and military aircraft, engines, components, avionics, electronics. Lockheed authorised repair station for Hercules C-130/L-100. Collins certified service centre. Manufacture of portable airfield lighting systems and BAe Hawk Wing Pylon wirelooms.
[98]

80-00075 AlliedSignal Aerospace Marketing Sales and Service
47200 Subang, Selangor
c/o Malaysia Airways, Complex A-AA 1203,
Hangar 3, 3rd Floor, Subang-Kuala Lumpur
International Airport
☎ (+60 3) 740 26 21; (+65) 861 33 62 (Singapore)
⊖ (+60 3) 746 54 63; (+65) 861 50 73 (Singapore)
Field Service Engineer: Abtar Singh
■ On-site customer support for AlliedSignal Aerospace. → 90
[98]

80-00200 Federal Telecommunications Sdn Bhd
(a subsidiary of FCW Holding Berhad)
Lot 2, Jalan 222, Section 51A, Petaling Jaya,
46100 Selangor Darul Ehsan
☎ (+60 3) 759 67 33 [E] cstan@fedtel.po.my ⊠ 37155
⊖ (+60 3) 759 67 78
♣ US$ 80,000,000 (1995) 👥 250 🏭 10,000 sq ft
Managing Director: Ahmad Hishamna
Sales Director: Choon Sian Tan
■ A leading supplier in the field of telecommunications industries. Branch offices at Penang, Sabah, Johor, Sarawak.
[98]

80-00250 Logica (Malaysia) Sdn Bhd
(a subsidiary of Logica plc, UK)
5th Floor Bangunan Getah Asli, PO Box 30,
50450 Kuala Lumpur
5th Floor Bangunan Getah Asli (Menara)
148 Jalan Ampang
☎ (+603) 262 48 78 ⊖ (+603) 262 45 57
[98]

80-00270 MacDonald Dettwiler
(a division of Orbitaz Sciences Corporation, USA)
Suite 14d - 14th Floor Empire Tower, City Square Centre,
50400 Kuala Lumpur
☎ (+60 3) 264 12 73 [E] http://www.mda.ca
⊖ (+60 3) 264 12 72
♣ M$ 100,000,000 (1996) 👥 750 (1997)
🏭 180,000 m2
Managing Director: Daniel Friedmann
Area Director - Asia: N Snaye
Operations Director: Anil Wiresaka
Sales and Marketing Director: Alan Vlemmiks
Purchasing Manager: Terry Piche
[98]

80-00800 Malaysia Mining Corp Bhd (MMC)
32 Fl Menara PNB 201A Jln Tun Razak, PO Box 10936,
50730 Kuala Lumpur
☎ (+60 3) 261 60 00 ⊠ 31316 ⊖ (+60 3) 261 29 51
Managing Director: Horahim Menudin
Operations Director: Haji Ab Sukor Shahar
Group Company Secretary: Darmwatti Dayari
[98]

80-01500 Rima Enterprise
14 Jalan Sri Hartamas 8, 50480 Kuala Lumpur
☎ (+60 3) 651 41 75 [E] hardback@tm.net.my
⊖ (+60 3) 651 41 60
President: Roziah Hashim
Accounts Officer: Bakhtiar Talhah ☎ (+60 3) 651 56 84
■ Book distributor, magazine subscriptions, publishing.
[98]

80-02000 United Technology
No 3A 1st Floor, Jalan SS21/34, Damansara Utama,
47400 Petaling Jaya
☎ (+60 3) 717 84 46 ⊖ (+60 3) 717 84 39
Managing Dir: Dr Colin Edmonds
Personal Asst: Ms Geraldine de Silva
® UK: Jane's Information Group
[97]

85 Stockists, Distributors, Spare Parts Suppliers

Malaysian Oxygen Berhad 75-00060

85-00600 Satair Malaysia Sdn Bhd Representative Office
(a subsidiary of Satair AS, Denmark)
311 Block E, Kelana Parkview, No 1,
Jalan 556/2, Kelana Jaya, 47301 Selangor Darul Ehsan
☎ (+60 3) 704 66 57; 704 66 70 ⊖ (+60 3) 704 66 91
👥 6
[98]

90 Sales Agents, Sales & Liaison Offices

Companies represented:

Akzo Nobel Coatings BV, Netherlands . . 90-00007
AS&E, USA 90-01030
Bell Helicopter Textron, USA→ Singapore 90-00220
BP Solar Limited, UK 90-00255
GEC-Marconi Communications, UK . . 90-01310
ITT Canon Connectors, USA 90-01030
Jane's Information Group, UK 80-02000
Kestronics, Singapore 90-01030
Marconi Radar Systems Ltd, UK . . . 90-00730
Marconi Secure Radio Sys., UK 90-00730
Meteor CAE SpA, Italy→ Netherlands . . 80-18300
Noble Denton, UK 90-01530
OFEMA, France 90-01580
Sema Group plc, UK 90-01900
Ultramare Australia, Australia 90-03020
United Defense L.P., USA 90-04000
Zodiac SA, France 75-00060

Sales agents/representatives listed elsewhere:

AlliedSignal Marketing Sales 80-00075

90-00007 AD Aerospace Finishes Representative Office Australasia
(a subsidiary of Akzo Nobel Coatings BV, Netherlands)
26 Jalan SS 3/6, Sungai Waya, 47300 Petaling Jaya
☎ (+60 3) 777 88 41 ⊖ (+60 3) 777 87 90
Contact: G A Windsor
® Netherlands: Akzo Nobel Coatings BV
[98]

90-00255 BP Solar Malaysia
Tingkat 37 Menara Maybank, 100 Jalan Tun Perak,
50050 KL Malaysia
100 Jalan Tun Perak, 11081 Peti Surat, Kuala Lumpur
☎ (+60 3) 232 63 22 ⊖ (+60 3) 232 76 42
® UK: BP Solar Limited
[98]

90-00350 Daewoo (M) Sdn Bhd
(a subsidiary of Daewoo Corp, South Korea)
308 3rd Floor Pernas International, Jalan Sultan Ismail,
50250 Kuala Lumpur
☎ (+60 3) 261 80 33 ⊠ 30067 daewoo ma
⊖ (+60 3) 261 57 90
[98]

90-00730 GEC Malaysia Sdn Bhd
Jalan 215 Templer, PO Box 225, Jalan Sultan,
Petaling Jaya, Selangor
☎ (+60 3) 791 13 88 ⊖ (+60 3) 792 13 50
♤ US$ 3 300 000 ♣ US$ 150 000 000 (1995) 👥 450
🏭 6 500 sq m
Managing Dir: J.B. Bulmer
■ Represents GEC-Marconi UK defence related products & services. Electronic systems & equipment for armies, navies & air forces. → 70 → 80
® UK: Marconi Radar Systems Ltd, Marconi Secure Radio Sys.
[97]

90-01020 Kawasaki Heavy Industries Ltd Kuala Lumpur Office
6th Floor UBN Tower, 10 Jalan P Ramlee,
50250 Kuala Lumpur
☎ (+60 3) 230 51 41 ⊖ (+60 3) 230 51 48
[98]

90-01030 Kestronics (M) Sdn Bhd
No 6, Block B, Lot 756, Prime Subang Light Ind. Park,
47500 Selangor
Jalan Subang 6, Off Persiaran Subang,
Subang Jaya, Petaling Jaya, 47500 Selangor Darul Ehsan
☎ (+60 3) 733 48 01 ⊖ (+60 3) 733 48 00
Human Resource Mgr: Anthony Yeo
Sales & Marketing Mgr: Ken Liew
® Singapore: Kestronics
USA: AS&E, ITT Canon Connectors
[97]

90-01310 Marconi (Malaysia) Sdn Bhd
Bangunan Tabung Haji, Jalan Tun Razak,
50400 Kuala Lumpur
☎ (+60 3) 261 61 66 ⊖ (+60 3) 261 03 60
Managing Dir: Dato' Kamaruddin 'Ariff
■ Digital transmission equipment, defence communications & computer communication equipment.
® UK: GEC-Marconi Communications
[97]

90-01530 Noble Denton Malaysia Sdn Bhd
3A, Jalan Wawasan 4/1, Bandar Baru Ampang,
68000 Selangar D.E.
☎ (+60 3) 470 16 98 ⊖ (+60 3) 470 17 06
® UK: Noble Denton
[98]

90-01580 **OFEMA**
(a subsidiary of OFEMA, France)
Pernas International Building,
Suite 1301 - Jalan Sultan Ismail, 50250 Kuala Lumpur
☎ (+60 3) 262 16 00 ⊖ (+60 3) 262 38 75
Delegate, South East Asia: M Olivier Rives Georges
® France: OFEMA 98

90-01850 **Rockwell-Collins International Inc.**
(a subsidiary of Rockwell International Corp, USA)
Suite 14.01, 14th Floor,
Wisma Stephens 88, Jalan Raja Chulan,
50200 Kuala Lumpur
☎ (+60 3) 248 28 12; 248 22 51 ⊖ (+60 3) 242 54 63
Regional Marketing Director South East Asia: Y T Ko
E ytk@cacd.rockwell.com
■ Provide technical services for aviation avionics and ground communications products. 98

90-01900 **SGAP (Malaysia) Sdn Bhd**
Suite 2.11, 2nd Fl Wisma Padicorp,
No. 5-29 Jalan SS 17/7A, 47500 Subang Jaya
☎ (+60 3) 732 08 99 ⊖ (+60 3) 732 11 81
® UK: Sema Group plc 98

90-02010 **Teledyne Malaysia**
(a subs. of Teledyne Ind. Int'l Inc., USA)
4th Floor, Wisma Consplant,
No # 2 Jalan SS16/4 Subang Jaya,
47500 Petaling Jaya, Selangor
☎ (+65 3) 732 06 26; 732 06 36 ⊖ (+65 3) 732 06 46
4 (1995) 1 500 sq ft
Managing Dir: Ooi Ah Ba 97

90-02050 **Transaero Malaysia**
1-3B, Jalan PJS 10/4, Taman Subang Indah,
46000 Petaling Jaya, Selangor
☎ (+60 3) 733 78 26; 733 78 28 ⊖ (+60 3) 734 56 00
Regional Mgr: Major Tahir 97

90-03020 **Ultramare Malaysia Sdn Bhd**
2105 Wisma Stephen, Jalan Raja Chulan,
05-12 Kuala Lumpur
▣ 30594 ⊖ (+60 3) 241 29 13
Director: J Camesi
® Australia: Ultramare Australia 98

90-04000 **United Defense L.P. International**
Suite 19-8-3A Level 8, UOA Centre, 19 Jalan Pinang,
50450 Kuala Lumpur
☎ (+60 3) 262 17 14 ⊖ (+60 3) 262 61 75
Manager: Gerald E. Bernard
® USA: United Defense L.P. 97

Maldives

Head of State:
Maumoon Abdul Gayoom

10 Head of State

10-10000 **Office of the President**
13 Marine Drive, Malé
☎ (+960) 32 03 ▣ 66013 radhun mf; 77001 theem mf
President of the Republic: Maumoon Abdul Gayoom
Minister, President's Office: Mohamed Zahir Hussain 98

12 Ministries

12-10000 **Ministry of Defence and National Security**
Badeyrige, Ameer Ahmed Magu, 20-05 Malé
☎ (+960) 32 26 01 ▣ 66056 aman mf
⊖ (+960) 32 55 25
President and Minister of Defence:
Maumoon Abdul Gayoom
Minister of State for Defence and National Security:
Anbaree Abdul Sattar 98

12-20000 **Ministry of Foreign Affairs**
Marine Drive, 20-05 Malé
☎ (+960) 34 05 ▣ 66008 minex mf
Minister: Fathulla Jameel 98

12-40000 **Ministry of Home Affairs**
Fashana Building, Malé
Minister: Abdulla Jameel 98

12-70000 **Ministry of Planning, Human Resources and Environment**
20-05 Malé
Minister: Ismail Shafeeu 98

15 Parliament

15-10000 **Parliament**
Majilis
Naadi, Ameer Ahmed Magu, Malé
Speaker of Citizens Majilis: Abdulla Hameed 98

31 Embassies & Attachés

31-01690 **High Commission for the Democratic Socialist Republic of Sri Lanka**
'Muraka' 5/44 Orchid Magu, Malé 20-02
☎ (+960) 32 28 45 ▣ 66061 lankcom mf
⊖ (+960) 32 16 52
High Commissioner: H.E. Mr C.S. Poolokasingham
☎ (+960) 32 33 38
Commercial Officer & Second Secretary: Mrs D. Rajapakse
First Secretary: L.H.A.U. Senerath 97

Mali

Head of State:
Alpha Oumar Konare

10 Head of State

10-10000 **Office of the President**
PO Box 1463, Bamako
☎ (+223) 22 24 61; 22 37 72
President: Alpha Oumar Konare 98

11 Prime Minister

11-10000 **Office of the Prime Minister**
PO Box 1463, Bamako
Prime Minister: Ibrahima Boubacar Keita 98

12 Ministries

12-10000 **Ministry of Defence and Internal Security**
PO Box 215, Bamako
☎ (+223) 22 26 17; 22 23 03
Minister of Defense: Mamodou Ba 98

12-20000 **Ministry of Foreign Affairs and International Co-operation**
Koulouba, Bamako
☎ (+223) 22 54 89; 22 56 33 ▣ 560 affet bamako
Minister of Foreign Affairs: Dioncounda Traore 98

31 Embassies & Attachés

31-00590 **Embassy of France**
Square Patrice Lumumba, PO Box 17, Bamako
☎ (+223) 22 29 51; 31 36 ▣ 2569 ⊖ (+223) 22 03 29
Ambassador: H E Mr Gabriel de Bellescize
Deputy Chief of Mission: Francoise Maylie
Defence Attaché: Col Michael Chaillet
Commercial Office
PO Box 101, Bamako
☎ (+223) 22 65 73 ▣ 516 comatta bko
Commercial Attaché: M Raimond (Resident in Dakar) 98

31-00690 **Embassy of Greece**
CP 714, Maputo
☎ (+258 1) 49 04 81 ▣ 6299 ⊖ (+258 1) 49 13 97 98

31-01330 **Embassy of the Federal Republic of Nigeria**
Badalabougou East, PO Box 57, Bamako
☎ (+223) 22 57 71; 46 96 ✉ NIGERIAN Bamako ▣ 505
Chargé d'Affaires: M O Kuforija ☎ (+223) 22 42 58
Head of Chancery: M Ajanwachukwu ☎ (+223) 22 57 71
Counsellor: M T O Osang ☎ (+263) 22 53 03
First Secretary: A T Poko ☎ (+223) 23 13 56
Attaché, Finance: J I Ogbeinde
Attaché, Administrative: Ms Victoria U Okogwu
☎ (+223) 22 46 96
Commercial and Public Information Officer: B A A Adejinle 98

31-01890 **Consulate of the United Kingdom**
BP 2069, Bamako

☎(+223) 22 47 38; 25 30 [E] aisb@maliner.ml
⊖ (+223) 22 08 53; 25 30
Ambassador: H E Mr David Snoxell (non-resident)
Consul: Richard Hyde (Resident in Dakar)
Vice Consul: Sean Burns (Resident in Dakar)
British Honorary Consul: Ms Violet Diallo
Commercial Officer: Ms N Thiam (Resident in Dakar) [98]

Malta

Head of State: Ugo Mifsud Bonnici

10 Head of State

10-10000 Office of the President of Malta
The Palace, Valletta
☎(+356) 22 12 21 ⊖ (+356) 24 12 41
President: Ugo Mifsud Bonnici [98]

11 Prime Minister

11-10000 Office of the Prime Minister
Auberge de Castille, Valletta
☎(+356) 24 25 60 ⊠ 1531 opmsec mw
⊖ (+356) 24 33 03
Prime Minister: Alfred Sant
Deputy Prime Minister: George Vella [98]

12 Ministries

12-20000 Ministry of Foreign Affairs
Palazzo Parisio, Merchants Street, Valletta
☎(+356) 24 21 91 ⊠ 1497 minfa.mw
⊖ (+356) 23 78 22
Minister of Foreign Affairs and Deputy Prime Minister: George Vella [98]

12-30000 Ministry for Home Affairs
House of the Four Winds, Valletta
Minister: Tonio Borg [97]

12-40000 Ministry for Health, Care for the Elderly and Family Affairs
(† Ministry for Social Development)
476 St Joseph High Road, HMR 18 St Venera
☎(+356) 48 51 00 ⊖ (+356) 49 37 44
Minister: Dr Michael Farrugia
Personal Assistant: Ms Amy Genovese [98]

12-60000 Ministry of Transport, Communications and Technology
Lascaris, Valletta
☎(+356) 22 62 63; 22 62 66 ⊖ (+356) 24 37 58
Minister: Joe Debono Grech
Press Secretary: Mario Schiavone [98]

15 Parliament

15-10000 House of Representatives
The Palace, Valletta
☎(+356) 22 22 94 ⊖ (+356) 24 25 52
Speaker: Notary Myriam Spiteri Debono
Deputy Speaker: Dr Michael Bonnici
Clerk of the House: R Cauchi
Chairman: Hon Sandio Schembri Adami
Hembels: Hon George Vella
Hembels: Hon Chris Cardona
Hembels: Hon Jose Herrira
Hembels: Hon Gindo de Marco
Hembels: Hon Joseph Borg
Hembels: Hon John Vella
Private Secretary: Charles Stafrace [98]

25 Defense Forces

25-37000 Composite Regiment
(Armed Forces of Malta)
Luqa Barracks, Luqa
☎(+356) 82 42 12 ⊠ 1489 armfor ⊖ (+356) 80 98 60
Commanding Officer 2nd Regiment AFM: Col A Camilleri (MI ExpE) ☎(+356) 23 67 59
Officer Commanding Maritime Squadron: Capt M Gatt ☎(+356) 22 52 89
Officer Commanding Air Defence Battery: Capt A Muscat ☎(+356) 24 88 83
Officer Commanding HQ Company: Capt A Attard ☎(+256) 82 42 12
Officer Commanding Air Squadron: Maj J Smith ☎(+356) 24 07 87
■ The Composite Regiment consists of the Air Squadron, Maritime Squadron and Air Defence Battery. [97]

26 Police (Civil)

26-00000 Police General Headquarters
(a subordinate to the Ministry for Home Affairs)
Floriana
☎(+356) 22 40 01; 22 40 51 ⊠ 1899
⊖ (+356) 23 54 67
Commissioner of Police: George A. Grech ☎(+356) 22 40 02
Personal Asst to the Commissioner: Inspector A.P. Logan [97]

29 Customs & Excise

29-00500 Customs Department
(a subordinate to the Ministry of Finance)
Customs House, Lascaris Wharf, Valletta
☎(+356) 23 57 21; 23 57 29 ⊠ 1600 cusmal mw
⊖ (+356) 24 61 50
Head of Department: Anthony Pavia [98]

31 Embassies & Attachés

31-00590 Embassy of France
12 Sir Temi Zammit Street, Ta'xbiex, PO Box 408, Valletta
☎(+356) 33 11 07; 58 56 ⊖ (+356) 33 46 40
Ambassador: H.E. Mr Patrick Amiot
Deputy Chief of Mission: Francois Morel

Office of the Commercial Counsellor
Flat B, Dolphin Court, Embassy Way,
PO Box 359 GPO Valletta, Ta'xbiex
☎(+356) 33 48 26; 44 ✉ COMATA Malta
⊠ 1327 comata mw ⊖ (+356) 33 48 44
Commercial Counsellor: Pierre Herly [97]

31-00890 Embassy of Italy
4-5 Vilhena Street, Floriana, PO Box 514, Valletta
☎(+356) 23 02 65; 23 02 66; 23 31 57; 23 31 59
⊠ 1388 italia mw ⊖ (+356) 23 53 39
Ambassador: H E Dr Marco Colombo
Counsellor Deputy Chief of Mission: Dr Pier Attinio Forlano
Personal Assistant: Ms Anna Maria Rosso [98]

31-01050 People's Bureau of the Great Socialist People's Libyan Arab Jamahiriya
Dar il-Jamaharia, Notabile Road, BZN 20 Balzan
☎(+356) 48 63 48 ⊠ 1258 ⊖ (+356) 48 39 39
Secretary of the Libyan People's Bureau: H.E. Dr Hend A. Siala ☎(+356) 48 63 47
Counsellor Economic & Co-operation: Mohamed M. Sweisi ☎(+356) 48 63 47 [97]

31-01655 Consulate of the Republic of Slovenia
38 Magazine Street, Floriana
☎(+356) 23 14 50 ⊖ (+356) 24 43 51
Honorary Consul: Nicholas A Baldacchino [98]

31-01680 Embassy of Spain
145 Tower Road, Flat 10, PO Box 17, Sliema
☎(+356) 31 41 64; 31 41 65 ⊖ (+356) 33 37 32
Ambassador: H E Mr José Manuel Cervera de Góngora ☎(+356) 44 24 88
Deputy Chief of Mission: Juan Armando Andrada-Vanderwilde ☎(+356) 33 58 81
Chancellor: Alfred Cauchi ☎(+356) 37 06 09
Commercial Counsellor: Alejandro Magro Mas (Resident in Tripoli) [98]

31-01830 Embassy of Tunisia
144 Tower Rd, SLM 08 Sliema
☎(+356) 34 58 66 ⊠ 1835 ambtun mw
⊖ (+356) 33 13 45
Ambassador: H.E. Mr Farouk Ladjimi
Counsellor: Mannoubi Fassatoui
Attaché: Mohammed Mouldi Chairi [97]

31-01890 United Kingdom High Commission
PO Box 506, Valletta
7 St Anne Street, Floriana
☎(+356) 23 31 34; 23 31 37 ⊖ (+356) 24 20 01
High Commissioner: H E Mr Graham Archer
Deputy High Commissioner and First Secretary Commercial: R G Lewington
Management Officer: S Richards [98]

31-01900 Embassy of the United States of America
PO Box 535, Valletta
2nd Floor Development House, St. Anne Street, Floriana
☎(+356) 23 59 60 ⊖ (+356) 24 32 29
Ambassador: H.E. Mr Joseph R. Paolino (Jr)
Deputy Chief of Mission: Charles N. Patterson (Jr)
Commercial Section: James Martin Perez
Public Affairs Officer: Ms Linda Cheatham [97]

75 Manufacturing Companies

75-50000 Robots International Limited
(joint venture with Kawasaki Heavy Industries Ltd, Japan)
21 Archbishop Street, Valetta
☎(+356) 32 08 97 ⊖ (+356) 34 60 41 [98]

Mauritania

Head of State:
Col Maaouya Ould Sid'Ahmed Taya

10 Head of State

10-10000 Office of the President
PO Box 184, Nouakchott
☎ (+222) 523 17 ▣ 580
President of the Islamic Republic, Prime Minister and Minister of Defense: Col Maaouya Ould Sid'Ahmed Taya [98]

11 Prime Minister

11-10000 Office of the Prime Minister
Nouakchott
Prime Minister: Cheikh Afia Ould Mohamed Khouana [98]

12 Ministries

12-10000 Ministry of Defence
PO Box 617, Nouakchott
☎ (+222) 533 51 ▣ 566 mitra
Minister of National Defence: Mohamed Yelsen Ould El Vil [98]

12-20000 Ministry of Foreign Affairs and Co-operation
PO Box 230, Nouakchott
☎ (+222) 526 82 ▣ 585 minaf
Minister: Sidi Ould Khalifa [98]

12-40000 Ministry of Interior
PO Box 527, Nouakchott
☎ (+222) 536 01, 520 20
Minister: Kaba Ould Elewa
Minister of Information: Ahmed Ould Khalifa Ould Jiddou [98]

31 Embassies & Attachés

31-00660 Embassy of the Federal Republic of Germany
PO Box 372, Nouakchott
☎ (+222) 517 29, 510 32 ▣ 5555 aanuak mtn
⊖ (+222) 517 22
Ambassador: H E Mr Johannes E Westerhoff
Deputy Chief of Mission: Willi-Hans Wallmichrath
Commercial Secretary/Attaché: Frank Stossel [98]

31-01680 Embassy of Spain
PO Box 232, Nouakchott
☎ (+222) 520 80; 525 79 ▣ 563 ambesp mtn
⊖ (+222) 540 88
Ambassador: H E Mr Juan Maria Lopez-Aguilar
Deputy Chief of Mission: Rafael Molina Llado
Military Attaché: Fernando Pinto
Commercial Secretary: Ms Gloria Minguez [98]

31-01890 Consulate of the United Kingdom
B9, 2069 Nouakchott
Honorary Consul: Mrs N Aberderrahmane [98]

Mauritius

Head of State:
H.M. Queen Elizabeth II

10 Head of State

10-10000 Office of the President
State House, Réduit
☎ (+230) 454 30 21 ▣ 4249 ⊖ (+230) 464 53 70
President: Sir Cassam Uteem
Vice President: Angidi Chettiar [98]

11 Prime Minister

11-10000 Office of the Prime Minister
Government House, Port Louis
Prime Minister, Minister of Defense and Internal Security: Navinchandra Ramgoolam
Deputy Prime Minister: Rajkeshwur Kailash Purryag [98]

12 Ministries

12-10000 Ministry of Defence and Internal Security
Government House, Port Louis
☎ (+230) 201 10 01 ▣ 4249 extern iw
⊖ (+230) 212 93 93
Prime Minister and Minister of Defence: Navinchandra Ramgoolam [98]

12-40000 Ministry of Local Government
3rd Fl. Emmanuel Anquetil Bldg, Port Louis
☎ (+230) 201 12 16 ⊖ (+230) 208 14 50
Minister: Hon James Burty David [97]

26 Police (Civil)

26-00000 Mauritius Police Force
Police HQ, Line Barracks, Port Louis
☎ (+230) 208 12 12 ⊖ (+230) 208 70 22
Commissioner of Police: J.R. Dayal (CSK, QPM, PDSM, FBIM, MAIMS)
Police Quartermaster: Asst Super. S.Dorasamy Naiken ☎ (+230) 208 12 12 [97]

31 Embassies & Attachés

31-00090 Australian High Commission
Rogers House, PO Box 541, Port Louis
☎ (+230) 208 17 00 ▣ 4414 ⊖ (+230) 208 88 78
High Commissioner: H E Mr Chris Marchant
First Secretary: Diana Pearce [98]

31-01890 United Kingdom High Commission
PO Box 1063, Port Louis
7th Floor Les Cascades Building Edith Cavell Street
☎ (+230) 211 13 61; 208 98 50; 208 98 51
⊖ (+230) 211 13 69; 212 84 70
High Commissioner: H E Mr James Daly
Deputy High Commissioner and First Secretary Commercial: T J Andrews ☎ (+230) 212 84 70
Defence Adviser: Lieut Col C Mantell (Resident in London)
Third Secretary Aid and Information: Philip Maw [98]

31-01900 Embassy of the United States of America
Rogers House 4th Floor, John Kennedy Street, Port Louis
☎ (+230) 208 97 63 ⊖ (+230) 208 95 34
Ambassador: H E Mr Walter Geisal
Deputy Chief of Mission: Chat Blakeman [98]

Mexico

Head of State:
Ernesto Zedillo Ponce de Leon

10 Head of State

10-10000 Office of the President
Palacio Nacional, Patio de Honor, 06067 Mexico DF
☎ (+52 5) 522 74 89; 542 88 36
▣ 1774400/4500 sppnme
President: Ernesto Zedillo Ponce de Leon [98]

12 Ministries

12-10000 Secretariat of National Defence
11640 Mexico 10 DF
Boulevard Manuel Avila Camacho Esq,
Avenida Industria Militar, 11640 Lomas de Sotelo
☎ (+52 5) 557 45 00; 557 19 32; 395 59 15
▣ 1776312 sdnme
Secretary of National Defence: Gen Enrique Cervantes Aguirre
■ The Secretary of National Defense controls both the Army & Air Force. However, naval operations come under the jurisdiction of the Secretary of Marine. [98]

12-10400 Secretariat of Navy
Revillagigedo 11, Piso 3, 06050 Mexico DF
☎ (+52 5) 521 62 90; 512 24 57
Secretary of Navy: Adm Jose Ramon Lorenzo Franco
■ The Secretary of Marine controls all aspects of Mexico's maritime operations, including the Navy, Coastal Territorial Commands & the Coastguard. [98]

12-20000 Secretariat of External Relations
Secretaria Relaciones Exteriores
Avenida Ricardo Flores Magón 1, 06995 Mexico DF
☎ (+52 5) 782 39 37; 782 41 09; 782 39 35
Secretary of Foreign Relations: Jose Angel Guirra Trevino [98]

12-40000 Secretariat of the Government
Bucareli 99, Colonia Juárez, 06699 Mexico DF
☎ (+52 5) 535 15 13; 546 34 48
▣ 1774375/376 sgobme
Secretary of Government: Emilio Chuayffet Chemor [98]

12-60000 Secretariat of Commerce and Industrial Development
Avenida Cuauhtemoc 80, 06728 Mexico DF
☎ (+52 5) 588 33 01; 588 58 22
▣ 1775 718 scdrme, 172 086 scdfme
Secretary of Commerce and Industrial Development: Herminio Blanco Mendoza [98]

Contact Reference

15 Parliament

15-20000 Federal Chamber of Deputies (Camara de Deputados)
La Fragua No 3 Piso 3, 06030 Mexico DF [98]

31 Embassies & Attachés

31-00030 Embassy of the Popular and Democratic Republic of Algeria
Sierra Madre 540, Lomas de Chapultepec,
11000 DF Mexico
☎ (+52 5) 520 69 50; 520 86 56 ▣ 1774310 radpme
⊖ (+52 5) 540 75 79
Ambassador: H E Mr Abdelatif Debabeche
Deputy Chief of Mission and Commercial Counsellor: Abdelmadjid Khecha [98]

31-00150 Embassy of Belgium
Avenida Alfredo Musset 41 (Polanco), 11550 Mexico, DF
☎ (+52 5) 280 07 58; 280 10 08 ✉ AMBABEL Mexico
▣ 1771030 ambbme ⊖ (+52 5) 280 02 08
Ambassador: H.E. Mr Nestor Cockx
Head of Chancery: Josse Martin
Commercial Counsellor: G. Cole [97]

31-00450 Embassy of the Czech Republic
Cuvier 22, Colonia Nueva Anzures, 11590 Mexico 5, DF
☎ (+52 5) 531 25 44; 27 77 ⊖ (+52 2) 531 18 37
Ambassador: H.E. Mr Václav Cekan
Commercial Officer: Miroslav Konír [97]

31-00460 Royal Danish Embassy
Calle Tres Picos 43, PO Box 105-105,
11580 Col.Chapultepec Morales
☎ (+52 5) 255 34 05 ✉ AMBADANE Mexico
▣ 1773049 amdkme ⊖ (+52 5) 545 57 97
Ambassador: H E Mr Peter Troels Simon Branner
First Secretary: Lars Rimmer
Attaché/Consul: Ms Brigitte Randmets [98]

31-00590 Embassy of France
Campos Eliseos 339, 11560 Mexico, DF
☎ (+52 5) 282 97 00 ▣ 1771302 aframe
⊖ (+52 5) 282 97 03
Ambassador: H E Mr Bruno Delaye
Armed Forces Attaché: Col Jean-Paul Husté
Press Attaché: Olivier Bove ☎ (+52 5) 282 97 31

Commercial Office
Campos Eliseos 339, Col. Chapultepec Morales,
11560 Mexico, DF
☎ (+52 5) 282 98 20 ✉ COMATTA Mexico
▣ 1775899 comame ⊖ (+52 5) 282 98 12
Commercial Counsellor: Gilles Baudouin [98]

31-00810 Embassy of the Republic of Hungary
Paseo de Las Palmas 2005, Lomas de Chapultepec,
11000 Mexico, DF
☎ (+52 5) 596 05 23 ▣ 1774 503 dimeme
⊖ (+52 5) 596 23 78
Ambassador: H E Dr József Kosárka
Administrative Official: A Busa [98]

31-01370 Embassy of Pakistan
Hegel 512, Colonia Polanco, 11570 Mexico, DF
☎ (+52 5) 203 36 36; 203 12 42 ✉ PAREP Mexico DF
▣ 1763355 eirpme ⊖ (+52 5) 203 99 07
Ambassador: H.E. Mr Manzar Shafiq
Deputy Chief of Mission: Haroon R. Rana [97]

31-01890 Embassy of the United Kingdom
Lerma 71, Col Cuauhtémoc, 06500 Mexico, DF
☎ (+52 5) 207 20 89; 207 21 49 Ⓔ embuk@intmex.com
⊖ (+52 5) 207 76 72
Ambassador: H E Mr Adrian Beamish
Deputy Chief of Mission, Minister-Counsellor: Dr P H Tibbet
Defence Attaché: Col S R Daniell
Cultural Counsellor: A Curry
First Secretary, Commercial: M H McIntosh
Management Officer: M Leach [98]

31-01950 Embassy of Venezuela
Schiller 326, Colonia Chapultepec - Morales,
11570 CP 11570, Ciudad de México
☎ (+52) 203 45 87; 203 42 33; 203 44 35
✉ EMBAVENEZ Mexico 5 ▣ 1763219 emveme
⊖ (+52) 203 86 14
Ambassador: H.E. Mr Fernando Ochoa Antica
Defence Attaché: Col Armando Bruno Suarez
Commercial Officer: Beatriz Barreto De Fritz
Public Information Officer: Magaly Saavedra
Ministro Consejero: Alcides Villalba [97]

75 Manufacturing Companies

75-00100 AlliedSignal de Mexico, SA SME
(a subsidiary of AlliedSignal Aerospace, USA)
PO Box 9001, 92231 Calexico, CA
MSC 772-199, Mexicali, BC
☎ (+52 6) 561 05 75 ⊖ (+52 6) 561 12 02
General Manager: Jerry Newman
Plant Manager: David Ridge [98]

75-00200 Daewoo Ingerieria Y Construcciones SA
(a subsidiary of Daewoo Corporation, South Korea)
De CV (Mexico Orio CRT Site), PO Box 3219,
92232-3219 Calexio CA
☎ (+52 65) 61 77 98 ⊖ (+52 65) 61 82 02 [98]

75-18000 Saft America Inc
Componentes Tecnicos de Baja California
Calle Uno Poniente n°201, Ciudad Industrial, Tijuana, BC
☎ (+52 66) 23 30 11 ⊖ (+52 66) 23 57 40
General Manager: Jean Coibion [98]

75-18100 Saft Nife (México) SA de CV
(a subsidiary of Saft SA, France)
San Luis Tlatilco 25, Apartado Postal 34,
53470 Naucalpan
☎ (+52 5) 301 25 13 ▣ 1777352 ⊖ (+52 5) 301 36 17
♣ US$ 4,500,000 ⚉ 25 ⌂ 1,600 sq m
Managing Director: Roberto Troop
Sales and Marketing Director: Juan de la Cruz
Purchasing Manager: António Hernandes [98]

75-18200 Scania de México SA de CV
Prol Av Industrías No 4640, Esq Eje 134,
78395 San Luis de Potosí
☎ (+52 48) 24 05 05 ⊖ (+52 48) 24 05 04 [98]

80 Service Companies & Organizations

80-00100 AlliedSignal Aerospace Marketing Sales and Service
(a subsidiary of AlliedSignal Aerospace, USA)
Avenida Fuerza Aerea Mexicana 416, Col Federal,
15700 Mexico City, DF
☎ (+52 5) 723 83 91; 723 83 00 ext. 2021
⊖ (+52 5) 503 21 91
Field Service Engineer: Mario Servin
→ 90
● Calle Cerezo 665 OTE, Col Valle de Campestre,
66250 San Pedro Garza Garcia, NL
☎ (+52 8) 378 57 28
⊖ (+52 8) 378 57 28 (prompt, then *51)
Field Service Engineer: John Bancroft [98]

90 Sales Agents, Sales & Liaison Offices

Companies represented:

Alfred Karcher GmbH & Co, Germany . . 90-07400
Indal Technologies Inc., Canada 90-02000
Teledyne Inc., USA 90-20000

Sales agents/representatives listed elsewhere:

AlliedSignal Aerospace Marketing . . 80-00100

90-02000 Ciesa
Lopez Montes y Associados AP
Comania Industrial Especializada, S A
Paseo de las Palmas 731 201, Col Lomas de Chapultepec,
11000 México, DF
☎ (+52 5) 259 11 81; 259 10 64 ⊖ (+52 5) 520 91 79
Managing Director: Otoniel Miranda
Sales and Marketing Director: Rudolfo de Bernardi
Operations Director: José Joaquín Lopez-Mendes
Purchasing Manager: Eduardo Blancafort
® Canada: Indal Technologies Inc. [98]

90-02500 Daewoo Mexico SA de CV
Bosques De Ciruelos N°130-5 Piso,
Col Bosques de las Lomas, CP 11700 Mexico DF
☎ (+52 5) 596 57 57 ▣ 1772303 diclme
⊖ (+52 5) 596 15 85 [98]

90-07400 Kärcher Mexico SA de CV
Calle 4 - n°29-B, Fracc Alce Blanco, 53020 Naucalpan
☎ (+52 5) 357 04 28; 576 51 56
Ⓔ 103503.2251@compuserve.com
Ⓦ http://www.karcher.com.mx ⊖ (+52 5) 576 50 65
♣ US$ 5,000,000 (1997) ⚉ 14 (1997) ⌂ 1,000 m2
Managing Director: Hector M Namihira
■ NBC defence and environmental protection suits and systems.
® Germany: Alfred Karcher GmbH & Co [98]

90-20000 Teledyne Industries International
(a subsidiary of Alleghény Teledyne Inc.)
Andres Bello No.10, Piso 5, 11560 Mexico, DF
☎ (+52 905) 282 90 16 ⊖ (+52 905) 282 90 17
® USA: Teledyne Inc. [98]

Moldova

President:
Petru Lucinschi

10 Head of State

10-10000 Office of the President
Chisinau
☎ (+373 2) 23 79 86 ⊖ (+373 2) 24 26 96; 23 77 42
President: Petru Lucinschi [98]

11 Prime Minister

11-10000 Office of the Prime Minister
Chisinau
☎ (+373 2) 23 34 23
Prime Minister: Ion Ciubuc
Deputy Prime Minister: Ion Gutu
Deputy Prime Minister: Valeriu Bulgari [98]

12 Ministries

12-10000 Ministry of Defence
Soseaua Hîncesti 84, Chisinau
☎(+373 2) 23 45 35 ⊜ (+373 2) 23 35 33
Minister: Maj Gen Valeriu Pasat [98]

12-20000 Ministry of Foreign Affairs
pl Maria Adonari Natsionale, Government House, Chisinau
☎(+373 2) 23 37 28
Minister of Foreign Affairs: Nicolae Tabacaru [98]

12-30000 Ministry of the Interior
Bd Stefan cel Mare 67, Chisinau
☎(+373 2) 23 35 69
Minister of Internal Affairs: Maj Gen Mihai Plamadeala [98]

12-45000 Ministry of National Security
Bd Stefan cel Mare 166, Chisinau
☎(+373 2) 23 93 09
Minister: Todor Botnaru [98]

12-80000 Ministry of Education
1 Piata Marii Adunari Nactionale, 2033 Chisinau
☎(+373 2) 23 35 19; 23 35 15; 23 27 27
⊜ (+373 2) 23 35 15; 23 34 74
Minister: Iacob Popovici
First Deputy Minister: Nicolae Andronache
Foreign Relations Director: Ludmila Rotaras
Chief of Staff: Gheorghe Egorov
Advisor to the Minister: Stefan Tiron [98]

29 Customs & Excise

29-00500 Directorate of Customs
Dom Pravitelstva, House of Government,
277001 Kyshynyov [98]

Monaco

Head of State:
H.S.H. Prince Rainier III

10 Head of State

10-10000 Office of H S H the Prince
Palais de Monaco, PO Box 518, Monte Carlo
Ruler of Monaco: H S H Prince Rainier III [98]

11 Prime Minister

11-10000 Office of the Minister of State
Ministère d'Etat, PO Box 522, Monte Carlo
☎(+377 93) 15 80 00 ▣ 469942 govermo carlo
⊜ (+377 93) 15 82 17
Minister of State: Jacques Dupont [98]

15 Parliament

15-10000 National Council
12 rue Colonel Bellando de Castro, Monte Carlo
☎(+377 93) 30 41 15 ⊜ (+377 93) 25 31 90
President of the National Council: Jean-Charles Rey [98]

31 Embassies & Attachés

31-01890 Consulate of the United Kingdom
33 Boulevard Princesse Charlotte, BP 265,
MC 98005 Monaco
☎(+377) 93 50 99 66 ⊜ (+377) 93 50 14 47
Consul-General: I Davies (resides at Marseille)
Honorary British Consul: E J F Blair [98]

80 Service Companies & Organizations

80-01000 Victoria Maritime Services
20 Ave de Fontvieille, 98000 Monaco
☎(+377 92) 05 58 14 [E] vms@monte-cedo.mc
⊜ (+377 92) 05 58 28
Director: Luca Spinelli
Director: Maurizio Taviani
■ Registro Italiano Navale (RINA) agents and non-exclusive surveyors; nautical inspectors on behalf of Barbados, Commonwealth of the Bahamas, Malta and Nigeria; hull, machinery and cargo surveying of all types of vessel for owners, charterers, underwriters, P&I clubs; valuations; member of the Chamber of Maritime Arbitration of Monaco. Quality system certified in accordance to EN ISO 9002 standards. [98]

Mongolia

Head of State:
Punsalmaagiyn Ochirbat

10 Head of State

10-10000 Office of the President
Presidium of People's Great Hural, Ulan Bator
President: Punslmaagiyn Ochirbat [98]

11 Prime Minister

11-10000 Office of the Prime Minister
Ulan Bator
Prime Minister: Mendsaihan Enhsaihan
Deputy Prime Minister and Co-ordinator for Emergency Matters: Choijilsurengiyn Purevdorj
Deputy Prime Minister and Minister for Administration: Lhamsuren Enebish [98]

12 Ministries

12-10000 Ministry of Defence
Ulan Bator
☎(+976 1) 35 84 77; 508 87; 538 54
⊜ (+976 1) 35 84 77; 35 84 57
Minister of Defence: Dambiin Dorligjav [98]

12-20000 Ministry of Foreign Relations
Ulan Bator
☎(+976 1) 31 10 67; 32 12 74 ⊜ (+976 1) 32 21 27
Minister: Shukher Altangerel [98]

31 Embassies & Attachés

31-01890 Embassy of the United Kingdom
PO Box 703, Ulaanbaatar 13
30 Enkh Taivny Gudamzh
☎(+976 1) 35 81 33; 35 82 38
[E] britemb@magicnet.mn ⊜ (+976 1) 35 80 36
Ambassador: H E Mr John Durham
Deputy Head of Mission: K J Brind [98]

31-01900 Embassy of the United States of America
Big Ring Road, Microdistrict 11, PO Box 1021, Ulaanbaatar
☎(+976 1) 32 90 95; 32 96 06 ⊜ (+976 1) 32 07 76
Ambassador: H.E. Mr Donald C. Johnson
Deputy Chief of Mission: David Williams
Office of the Defence Attaché: Col Rocky Roland (USAF, resident in Beijing)
Economic/Consular Section: C. Michael Konner
Public Affairs Officer: Ms Ann E. Welden
Political Officer: Cheryl A. Martin [97]

Morocco

Head of State:
H.M. King Hassan II

10 Head of State

10-10000 Office of H M the King
Palais Royal, Rabat
☎(+212 7) 601 22; 601 22 24 ▣ 31008 goumaroc
King of Morocco: H M King Hassan II [98]

11 Prime Minister

11-10000 Office of the Prime Minister
Palais Royal, Rabat
☎(+212 7) 645 54, 612 47, 622 50 ▣ 31008
Prime Minister: Abdellatif Filali
Secretary-General of the Government: Rabie Abdessadek [98]

Contact Reference

12 Ministries

12-20000 Ministry of Foreign Affairs and Co-operation
Avenue Franklin Roosevelt, Rabat
☎(+212 7) 628 41; 628 42; 628 43; 628 44
E mext@onpt.net.ma W http://www.minicom.gov.ma
⊠ 31007 aferext m; 31014 aferext m; 31049 aferext m
Minister of State and Foreign Affairs: Abdellatif Filali
Secretary of State Foreign Affairs: Taieb Fassi-Fihri 98

12-60000 Ministry of Finance
Rabat
Minister: Mohamed Kabbaj 98

31 Embassies & Attachés

31-00100 Embassy of Austria
2 Rue Tiddas, PO Box 135, Rabat
☎(+212 7) 76 40 03; 76 16 98 ✉AUSTROAMB Rabat
⊖ (+212 7) 76 54 25
Ambassador: H E Dr Tassilo F Ogrinz
First Secretary and Consul: Ingeborg Lahr 98

31-00660 Embassy of the Federal Republic of Germany
7 Zankat Madnine, PO Box 235, Rabat
☎(+212 7) 70 96 62; 96 85; 96 97 ⊠ 31 026 aarabat m
⊖ (+212 7) 70 68 51
Ambassador: H E Mr Herwig Bartels
Deputy Chief of Mission: Rolf Schumacher
Defense Attaché: Col H Kellerhoff
Press Attaché: Philipp Ackermann 98

31-00690 Embassy of Greece
23 Rue d'Oujda, Rabat
☎(+212 7) 72 38 39; 73 34 46
✉AMBASSADE GRECE Rabat ⊠ 31953 greekemb m
⊖ (+212 7) 70 22 70
Ambassador: H E Mr Dionyssios Koddellas 98

31-00890 Embassy of Italy
2 Rue Idriss al Azhar, PO Box 111, Rabat
☎(+212 7) 70 65 97/8 ⊠ 32731 rabital m
⊖ (+212 7) 70 68 82
Ambassador: H.E. Mr Emilio Franco Destefanis
Counsellor: Gianpaolo Scarante
First Secretary Economic & Commercial: Arturo Luzzi
Attaché Cultural: Paolo Ghera
Attaché: Angelo Greco
Attaché: Massimo Ignesti
Office of the Military Attaché
40 Avenue Des Nations Unies, Agdal, Rabat
☎(+212 7) 77 22 11 ⊠ 32023 militalm
Military, Naval & Air Attaché: Col Raimondo Fanara 97

31-01370 Embassy of Pakistan
11 Rue Azrou, Rabat
☎(+212 7) 76 64 53; 76 24 02 ✉PAREP Rabat
⊠ 31 918 parep m ⊖ (+212 7) 76 67 42
Ambassador: H E Mr Syed Azmat Hassan
First Secretary, Head of Chancery: Inayat Ullah Kakar 98

31-01430 Embassy of the Republic of Poland
23 Zankat Oqbah, PO Box 425, Rabat-Agdal
☎(+212 7) 77 17 91; 77 11 73 E apolgne@onpt.net.ma
⊖ (+212 7) 77 53 20
Ambassador: H E Mr Piotr Szymanowski
☎(+212 7) 77 11 73
Deputy Chief of Mission: Janusz Zebrowski
First Secretary: Ms Lidia Wieczorkiewicz
Military, Naval and Air Attaché: Col Wojciech Myszak (Resident in Paris) ☎(+33 1) 45 51 60 80
Commercial Counsellor: Teofil Stanislawski
22 rue de Khouribga ☎(+212 7) 76 83 68; 76 86 11
⊠ 31995 98

31-01610 Embassy of the Republic of Senegal
17 Av. Cadi Hamadi Sanhaji, Souissi, PO Box 365, Rabat
17 Avenue Cadi Hamadi Sanhaji, Souissi
☎(+212 7) 75 41 71 ⊠ 31048 ambasene m
⊖ (+212 7) 75 41 49
Ambassador: H E Mr Doudou Diop
First Consellor: Ousmane Ba
First Counsellor, Commercial: Mouhamadou Diene
Defence Attaché: Ibra Mime
First Secretary: Mohamed Diakhaté Diaw
First Secretary: Babacar Gueye 98

31-01740 Embassy of Switzerland
Square de Berkane, PO Box 169, Rabat
☎(+212 7) 70 69 74; 75 12 ✉AMBASUISSE Rabat
⊖ (+212 7) 70 57 49
Ambassador: H.E. Mr Henri Cuennet
Deputy Chief of Mission: Jorg Feigenwinter
Defence Attaché: Col Walter Berchtold (Resident in Madrid) 97

31-01890 Embassy of the United Kingdom
17 Boulevard de La Tour Hassan, PO Box 45, Rabat
☎(+212 7) 72 09 06; 73 14 03; 73 14 04; 70 45 32
⊖ (+212 7) 70 45 31
Ambassador: H E Mr William Fullerton
Deputy Head of Mission: L Buchanan
Defence Attaché: Lieut Col Michael Argue
Information Officer: Patrick J Davies 98

31-01900 Embassy of the United States of America
2 Avenue de Marrakech, Rabat
☎(+212 7) 76 22 65 E paorabat@usia.gov
⊖ (+212 7) 76 56 61
Ambassador: H E Mr Marc C Ginsberg
Chargé d'Affaires: Gary S Usrey
Economic Section: Alexandra Sundquist
Consulate Section: Ms Shelly Midura (Resides in Casablanca) ☎(+212 2) 26 45 50
Administration Section: Richard E Kramer
Agency for International Development: Michael Farbman
United States Defense Attaché: Col Alan J Tinder (USAF)
Office of Defence Cooperation (ODC): Col Stephen T Fenyon
Public Affairs Officer: James L Bullock
☎(+212 7) 75 81 81
Commercial Officer: Albert Nahae (Resides in Casablanca) 98

75 Manufacturing Companies

75-05000 Daewoo Rabat Sarl (DAERA)
(a subsidiary of Daewoo Corporation, South Korea)
Rabat Hilton Hotel, Aviation Souissi, Rabat
☎(+212 7) 67 45 69; 67 45 99 ⊖ (+212 7) 67 43 66 98

80 Service Companies & Organizations

80-01000 Maroc Aviation
(a subsidiary of SOGERMA, France)
12 rue de Lécrivain, Casablanca
☎(+212 2) 61 85 85 ⊖ (+212 2) 62 18 44
♣ FFr 34,600,000 (1996) 👥 115
Chief Executive Officer: Jean André Morand
■ Component overhaul and avionics repair (C130, helicopters), manufacture of electric hairnesses. 98

90 Sales Agents, Sales & Liaison Offices

90-05000 Daewoo Corp
Lot 20, Lotissement de la Wilya Souissi, Rabat
☎(+212 7) 65 90 95 ⊖ (+212 7) 65 90 53 98

Mozambique

Head of State:
Joaquím Alberto Chissano

10 Head of State

10-10000 Office of the President
Avenida Julius Nyerere, Maputo
☎(+258) 74 11 21 ⊠ 6243
President of the People's Republic: Joaquím Alberto Chissano
Minister for State Administration: Alfredo Gamito 98

11 Prime Minister

11-10000 Office of the Prime Minister
Maputo
☎(+258 1) 42 68 61; 42 68 63 ⊠ 6154
⊖ (+258 1) 42 68 81
Prime Minister: Pascoal Manuel Mocumbi 98

12 Ministries

12-10000 Ministry of National Defence
Avenida dos Martires de Mueda, Maputo
☎(+258) 74 20 81 ⊠ 6331
Minister: Aguiar Real Mazula
Vice Minister and Chief of General Staff: Lieut Gen Antonio Hama Thay
Vice Minister: Maj Gen Eduardo da Silva Nihia 98

12-20000 Ministry of Foreign Affairs and Co-operation
Julius Nyerere Avenue, PO Box 4, Maputo
☎(+258 1) 49 02 18; 49 02 22; 49 02 23
⊠ 8 418 mines mo ⊖ (+258 1) 49 14 60
Minister of Foreign Affairs and Co-operation: Dr Leonardo Santos Simào
Deputy Minister of Foreign Affairs and Co-operation: Mrs Frances Rodrigues ☎(+258 1) 49 30 31
Deputy Minister of Foreign Affairs and Co-operation: Hipolito Patricio ☎(+258 1) 49 30 85
Secretary of State for Foreign Affairs and Co-operation: Antonio Materrula ☎(+258 1) 49 13 76
Chief of Protocol: Francisco E P Cigaro 98

12-40000 Ministry of Interior
46-48 Avenida da Anchieta, Maputo
☎(+258) 201 30 ⊠ 6419
Minister: Almerinho Menhenje 98

15 Parliament

15-10000 Assembly of the Republic
Assembleia da República
PO Box 1516, Maputo
☎(+258 1) 40 08 26; 40 08 33 ⊠ 6473 asrep mo
⊖ (+258 1) 40 07 11
Secretary General: D Sibambo 97

25 Defense Forces

25-10000 People's Forces Headquarters General Headquarters of FADM (FPLM)
Forcas Armadas de Defesa de Moçambique (FADM)
Defence Forces of Mozambique
Avenida 24 de Julho, Maputo
☎(+258) 40 03 84 ▣ 400350 ⊖ (+258) 40 03 50
Deputy Minister of Defence: Lt Gen. Antonio Hama Thai
Deputy Commander/Chief of Staff: Brig. J.S. Jane
Commander Air Force: Maj.Gen. José B. Chagua
Commander Army: Maj.Gen. Olimpio Cardoso
Commander Navy: VAdm. José Pascoal
Operations Dir: Brig. Aregy
Logistics Dir: Maj.Gen. Teofilo
Chief of General Staff: Lagos Lidimu [97]

31 Embassies & Attachés

31-00230 Brazilian Embassy
Avenue Kenneth Kaunda 296, PO Box 1167, Maputo
☎(+258 1) 49 23 87; 49 23 88; 49 28 63
▣ 6454 bremb mo ⊖ (+258 1) 49 09 86; 49 27 51
Ambassador: H E Mr Luciano Rosa ☎(+258 1) 49 20 32
Deputy Chief of Mission: Paulo Américo Veiga Wolowski
Cultural Affairs: Ms Irene Cezne ☎(+258 1) 42 88 00
Secretary Technology and Science: Pery Machado [98]

31-00460 Royal Danish Embassy
Avenida 24 de Julho 1500, PO Box 4588, Maputo
☎(+258) 42 01 72; 42 01 73 ✉ AMBADANE Maputo
E dkembmap@zebra.vem.m2 ▣ 6164 ambdk mo
⊖ (+258) 42 05 57
Ambassador and Deputy of Mission:
H E Mr Ole Blicher-Olsen ☎(+258) 42 90 54
Charge d'Affaires: Preben Gondolf ☎(+258) 42 90 51
Commercial Officer: Inge Marie Olsen
Secretary: Lourdes Teodora [98]

31-00660 Embassy of the Federal Republic of Germany
Rua Damião de Gois 506, CP 1595, Maputo
☎(+258 1) 49 27 14 ▣ 6489 aalum mo
⊖ (+258 1) 49 28 88 [98]

31-00690 Embassy of Greece
CP 714, Maputo
☎(+258 1) 49 04 81 ▣ 6299 ⊖ (+258 1) 49 13 97 [98]

31-00890 Embassy of Italy
387 Kenneth Kaunda Avneue, PO Box 976, Maputo
☎(+258 1) 49 16 05; 49 22 27/9; 49 02 81
▣ 6442 itadi mo ⊖ (+258 1) 49 20 46
Ambassador: H.E. Dr Manfredo Incisa Di Camerana
Deputy Chief of Mission: Cristiano Gallo
Military Attaché: Col Napoli Nicolo
Commercial Secretary: Roberto Vellano [98]

31-01480 Embassy of Romania
Gorongosa, 59, PO Box 4648, Maputo
☎(+258 1) 49 29 99 ▣ 6397 ⊖ (+258 1) 49 29 99
Chargé d'affaires: H E Mr Valenu Licolae [98]

31-01680 Embassy of Spain
Rua Damiao De Gois 347, PO Box 1331, Maputo
☎(+258) 49 20 25; 27; 30; 48 ▣ 6579 espan mo
⊖ (+258) 49 20 55
Ambassador: H.E. Mr Francisco Viqueira Niel
Deputy Chief of Mission: Juan José Buitrago de Benito [97]

31-01730 Swedish Embassy
Av. Julius Nyerere 1128, PO Box 338, Maputo
☎(+258) 49 00 91 ✉ SVENSK Maputo ▣ 6272 svens mo
⊖ (+258) 49 00 56
Ambassador: H.E. Mr H. Odmark
Deputy Chief of Mission: Mrs Ann Stodberg [97]

31-01890 Embassy of the United Kingdom
PO Box 55, Maputo
Av Vladimir I Lenine 310, Maputo
☎(+258) 42 01 11; 42 01 12; 42 01 05;
42 01 16; 42 01 17 ✉ PRODROME MAPUTO
▣ 6265 proma mo ⊖ (+258) 42 16 66
Ambassador: H E Mr B J Everett
Deputy Head of Mission: G C Liversey
Second Secretary, Commercial and Aid: S Mattinson
First Secretary, British Council Director: R P H Woods [98]

31-02030 Embassy of the Republic of Zambia
Av. Kenneth Kaunda 1286, PO Box 4655, Maputo
☎(+258 1) 49 24 52; 49 13 07 ▣ 6415
⊖ (+258 1) 49 18 93
Ambassador: H.E. Maj.Gen. B.B. Chisuta
Defence Attaché: Col U. Uankinza [97]

Myanmar

Head of State:
Gen. Than Shwe

11 Prime Minister

11-10000 Office of the Prime Minister
Minister's Office, Yangon
Prime Minister and Chairman, State Law and Order Restoration: Gen Than Shwe
First Secretary, State Law and Order Restoration Council: Lieut Gen Khin Nyunt
Second Secretary, State Law and Order Restoration Council: Lieut Gen Tin Oo
Minister in the Office of the First Seretary: Lieut Gen Min Thein
Minister in the Office of the First Secretary: Brig Gen Myo Thant
Deputy Prime Minister: VAdm Maung Maung Khin
Deputy Prime Minister: Lieut Gen Tin Tun
Minister in the Office of the Chairman, State Law and Order Restoration Council: Brig Gen Saw Lwin
Minister in the Office of the Prime Minister: Brig Gen Lun Maung
Minister of the Office of the Prime Minister: Col Pe Thein
Minister in the Office of the Prime Minister: Col Than Shwe
Minister in the Office of the Deputy Prime Minister: Mr Khin Maung Yin
Minister in the Office of the Deputy Prime Minister: Brig Gen Maung Maung [98]

12 Ministries

12-10000 Ministry of Defence
Signal Pagoda Road, Yangon
Minister: Gen Than Shwe [98]

12-20000 Ministry of Foreign Affairs
Pyay Road, Yangon
Minister of Foreign Affairs: Ohn Gyaw [98]

12-40000 Ministry of Home Affairs
Yangon
Minister: Lieut Gen Mya Thin [98]

25 Defense Forces

25-30000 Army Headquarters
Signal Pagoda Road, Yangon
Commander in Chief: Gen Maung Aye [98]

31 Embassies & Attachés

31-00590 Embassy of France
102 Pyidaungsu Yeiktha Road, PO Box 858, Yangon
☎(+95 1) 28 21 22; 28 24 18 ⊖ (+95 1) 28 77 59
Ambassador: H E Mr Bernard Pottier
Deputy Chief of Mission: Olivier Vaysset
Commercial Office
35 Po Sein Road, Yangon
☎(+95 1) 54 07 89; 54 07 91 ▣ 21404 comatt bm
⊖ (+95 1) 54 07 90
Commercial Attaché: Robert Igier [98]

31-00660 Embassy of the Federal Republic of Germany
32 Nat Mauk Street, PO Box 12 GPO, Yangon
☎(+95 1) 54 89 51; 54 89 52; 54 89 53
▣ 21401 aarang bm ⊖ (+95 1) 54 88 99
Ambassador: H E Dr Wolfgang Wiesner
Commercial Counsellor: Dr Michael Koch
Second Secretary: Hans Barner [98]

31-00840 Embassy of the Republic of Indonesia
100 Pyidaungsu Yeiktta Road, PO Box 1401, Yangon
☎(+95 1) 28 17 14, 28 13 58
✉ PERWAKILAN INDONESIA Rangoon
▣ 21355 tindon bm ⊖ (+95 1) 28 26 75
Ambassador: H E Mr A Poerwanto Lenggono
Third Secretary, Information: Mr Budimansyah [98]

31-00890 Embassy of Italy
3 Inya Myaing Road, PO Box 866, Yangon
☎(+95 1) 52 71 00; 52 71 01 ▣ 21317 italtx bm
⊖ (+95 1) 53 36 70
Chargé d'Affaires a.i.: Tony Mori [98]

31-00920 Embassy of Japan
100 Natmauk Road, Yangon
☎(+95 1) 54 96 44; 54 96 45; 54 96 46;
54 96 47; 54 96 48 ⊖ (+95 1) 54 96 43
Ambassador: H E Mr Yoichi Yamagu
Deputy Chief of Mission: Koichi Kawakami
First Secretary and Defense Attaché: Col Akira Sato
Commercial Officer: Ryotaro Suzuki
Public Information Officer: Kazu Mitoro [98]

31-00990 Embassy of the Republic of Korea
97 University Avenue, GPO Box 1408, Yangon
☎(+95 1) 52 71 42; 52 71 43; 52 71 44
⊖ (+95 1) 51 322 86
Ambassador: H E Mr Kyong-bo Choi
Counsellor: S C Chang
Defence Attaché: Col K S Oh
First Secretary: D N Park
Administration Attaché: I Y Lee [98]

31-01110 Embassy of Malaysia
82 Pyidaungsu, Yeiktha Rd, Yangon
☎(+95 1) 310 31; 316 77 ▣ 21321 mwakil bm
Ambassador: H.E. Dato Abdul Wahab Haran
Deputy Chief of Mission: Kennedy Jawan
First Secretary Information & Cultural: Mohd. Radi Zun
Defence Advisor: Lt Col S. Hashim [97]

31-01370 Embassy of Pakistan
A-4 Diplomatic Quarters - Pyi Road, PO Box 581, Yangon
☎(+95 1) 22 28 81; 22 28 82 ✉ PAREP Rangoon
▣ 21309 parep bm ⊖ (+95 1) 22 11 47
Ambassador: H E Mr Javed Hafiz
Chargé d'Affaires a.i.: M F Rahman Akbar [98]

Contact Reference

31-01690 **Embassy of the Democratic Socialist Republic of Sri Lanka**
34 Taw Win Road, PO Box 1150, Yangon
☎(+95 1) 22 28 12 ☒ 21352 lanka bm
⊖ (+95 1) 22 15 09
Ambassador: H.E. Mr K.B. Fernando ☎(+95 1) 22 29 72
Third Secretary: Mrs W.A.T. Gunatillake
☎(+95 1) 52 72 36
Attaché: W.S.M. Somadasa ☎(+95) 28 11 95 [97]

31-01780 **Royal Thai Embassy**
45 Pyay Road, 6 1/2 Mile, Yangon
☎(+95 1) 52 56 70; 53 30 82 ⊖ (+95 1) 22 27 84
Ambassador: H.E. Mr Poksak Nilubol
Minister: Uthel Silthorn
Office of the Naval Attaché
11-B Than Lwin Road, Yangon
☎(+95 1) 53 32 01
Naval Attaché: Capt. Suchart Hongprasong
Office of the Defence, Army & Air Attaché
40 University Ave, Yangon
☎(+95 1) 53 05 74 ⊖ (+95 1) 53 07 61
Defence, Army & Air Attaché: Col Ong-Ard Ratanawichal
Office of Commercial Affairs
73 Than Lwin Road, Yangon
☎(+95 1) 52 52 46; 52 55 18 ⊖ (+95 1) 53 27 68
Commercial Counsellor: Sriwat Suwan [97]

31-01890 **Embassy of the United Kingdom**
80 Strand Road, PO Box 638, Rangoon
☎(+95 1) 29 53 00; 28 17 00
⊖ (+95 1) 29 53 06; 28 95 66
Ambassador: H E Mr Robert A E Gordon
Deputy Head of Mission: Jeremy E Larner
Cultural Attaché: Christopher Harrison
Third Secretary Management and Vice Consul: Miss Doris Davis
Third Secretary Political: Miss Tiffany A White
Attaché: Miss Anne Gill [98]

31-01900 **Embassy of the United States of America**
581 Merchant Street, GPO Box 521, Yangon
☎(+95 1) 820 55; 821 82 ☒ 21230 ambygn bm
⊖ (+95 1) 804 09
Charge d'Affaires: Marilyn A. Meyers
Commercial Section: Ms Laura J. Kirkconnell
Public Affairs Officer: Douglas M. Barnes
Defense Attaché: Col Donald R. Moran (US Army) [97]

90 Sales Agents, Sales & Liaison Offices

Companies represented:

Bell Helicopter Textron, USA→
Singapore 90-00220

90-00300 **CATIC**
Representative Office
No.135(D) Mawyawaddi Rd, 81/2 mile, Pyay Rd, Yangon, Myanmar
☎(+95 1) 620 49 [97]

90-10000 **Myanmar Daewoo Ltd**
(a subsidiary of Daewoo Corp, South Korea)
N°22 Yaw Min Gyi Road, Dagon Township, Yangon
☎(+95 1) 899 48 ⊖ (+95 1) 842 46 [98]

Namibia

Head of State:
Sam Nujoma

10 Head of State

10-10000 **Office of the President**
Windhoek
President: Sam Nujoma [98]

11 Prime Minister

11-10000 **Office of the Prime Minister**
Private Bag 13338, Windhoek
☎(+264 61) 287 20 01 ⊖ (+264 61) 23 19 14
Prime Minister: Hage G Geingob
Deputy Prime Minister: Hendrik Witbooi [97]

12 Ministries

12-10000 **Ministry of Defense**
PO Box 2870, 9000 Windhoek
☎(+264 61) 204 91 11 ⊖ (+264 61) 204 20 22; 20 92
Minister: Philemon Malima [98]

12-20000 **Ministry of Foreign Affairs**
Windhoek
Minister: Theo-Ben Gurirab [98]

12-40000 **Ministry of Home Affairs**
Windhoek
Minister of Home Affairs: Jerry Ekandjo
Minister for Local Government: Nicky Iyambo
Minister of Justice: Ngarikutuke Tjiriange
Minister of Information and Broadcasting: Ben Amathila [98]

31 Embassies & Attachés

31-00890 **Embassy of Italy**
Anna Str. Co. Gevers Str., Ludwigsdorf, Windhoek
☎(+264 61) 22 86 02 ⊖ (+264 61) 22 98 60 [98]

31-01890 **United Kingdom High Commission**
116 Robert Mugabe Avenue, 9000 Windhoek
☎(+264 61) 22 30 22 ⊖ (+264 61) 22 88 95
High Commissioner: H E Mr R H G Davies
Deputy High Commisioner and Second Secretary, Development: N P Dickerson
Defence Advisor: Cdr A D Auld (resides at Pretoria)
Third Secretary, Development: Miss H T Milne
Attaché: L A Salter [98]

31-01900 **Embassy of the United States of America**
Ausplan Building, 14 Lossen St., Private Bag 12029, Ausspannplatz, Windhoek
☎(+264 61) 22 16 01 ⊖ (+264 61) 22 97 92
Ambassador: H.E. Mr Marshall F. McCallie
Deputy Chief of Mission: Ms Katherine H. Peterson
Economic/Commercial Section: Philip R. Drouin
Admin Asst.: Ms Claudia Claasen
Public Information Officer: Dr Herman W. Henning
Defence Attaché: Lou Mazel [97]

Nepal

Head of State:
H.M. King Bir Bikram Shah Dev Birendra

10 Head of State

10-10000 **Office of H M the King**
Durbar Marg, Royal Palace, Kathmandu
☎(+977 1) 22 81 87; 22 87 96 ☒ 2236 rexpal np
⊖ (+977 1) 22 73 95
King: H M Bir Bikram Shah Dev Birendra [98]

11 Prime Minister

11-10000 **Office of the Prime Minister**
Singha Durbar, Kathmandu
☎(+977 1) 21 15 55; 21 59 55
Prime Minister: Lokendra Bahadur Chand
Deputy Prime Minister: Bam Dev Gautam [98]

12 Ministries

12-10000 **Ministry of Defense**
Singha Durbar, Kathmandu
☎(+977 1) 21 60 89
Prime Minister and Minister of Defence: Lokendra Bahadur Chand [98]

12-20000 **Ministry of Foreign Affairs**
Singha Durbar, Kathmandu
☎(+977 1) 21 59 06 ☒ 2224, 2442 nofa np
⊖ (+977 1) 21 59 05
Minister: Kamal Thapa [98]

31 Embassies & Attachés

31-00590 **Embassy of France**
Lazimpath, PO Box 452, Kathmandu
☎(+977 1) 41 23 32 ☒ 2209 ambfra np
⊖ (+977 1) 41 99 68
Ambassador: H.E. Mr Daniel Dupont
Deputy Chief of Mission: Olivier Nicolas
Defense Attaché: Col Georges Laine (Resident in New Delhi)
Consul & Head of Chancery: Gèrard Fromageot
Cultural, Scientific & Technical Attaché: Christian Champon-Chirac [97]

31-01370 **Embassy of Pakistan**
Panipokhari, GPO Box 202, Kathmandu
☎(+977 1) 41 14 21/31 ✉PAREP Kathmandu
⊖ (+977 1) 41 91 13
Ambassador: H.E. Mr Kamran Niaz ☎(+977 1) 41 05 65
Defence Attaché: Col Jamil Halder ☎(+977 1)41 95 06 [97]

31-01780 **Royal Thai Embassy**
Ward N°3, PO Box 3333, Bansbari, Kathmandu
☎(+977 1) 37 14 10; 37 14 11 ☒ 2373 thaidu np
⊖ (+977 1) 37 14 08; 37 14 09
Ambassador: H E Mr Precha Pitisant
☎(+977 1) 41 43 72; 41 44 70 (fax)
Minister Counsellor: Thongbi Sornkaew
☎(+977 1) 37 17 34; 37 117 93
Second Secretary: Somphong Kangthong
☎(+977 1) 41 69 80
Second Secretary: Teeraboon Boonwit
☎(+977 1) 37 00 22

Attaché: Mrs Jirapa Liuchulaschan ☎ (+977 1) 41 66 26
Attaché: Ms Sirijan Thothong ☎ (+977 1) 37 12 19 [98]

31-01890 Embassy of the United Kingdom
PO Box 106, Lainchaur Kathmandu
☎ (+977 1) 41 05 83; 41 12 81; 41 45 88
✉ PRODROME KATHMANDU ⊠ 2343 britem np
⊖ (+977 1) 41 17 89
Ambassador: H E Mr David Fall
Deputy Head of Mission: P R Heigl
Defence Attaché: Col A W Blackett
First Secretary, Development and Commercial: M L Alcock
First Secretary, Cultural (British Council Director): Ms S Ewans
First Secretary: M Gough-Cooper (residing in New Delhi)
Second Secretary, Consular and Management: J S Gibson
Second Secretary, Development: B Mackay
Third Secretary, Development: Miss S M O'Connor
Vice-Consul: B A Davidson [98]

31-01900 Embassy of the United States of America
Pani Pokhari, Kathmandu
☎ (+977 1) 41 11 79 ⊠ 2381 aektm np
⊖ (+977 1) 41 99 63
Ambassador: H.E. Ms Sandra L. Vogelgesang
Deputy Chief of Mission: Peter W. Bodde
Head of Commercial Section: Siddhi B. Amatya
Office of the Defense Attaché: Lt Col James Dunn (USA)
Administration Section: William Campbell
Public Affairs Officer: David A. Queen [97]

90 Sales Agents, Sales & Liaison Offices

90-00300 CATIC
PO Box 11, Maharajgunj, Kathmandu
☎ (+977 1) 41 31 87 ⊖ (+977 1) 41 65 74 [97]

Netherlands

Head of State:
H.M. Queen Beatrix

10 Head of State

10-10000 Office of H M the Queen
PO Box 30412, NL-2500 GK The Hague
Paleis Lange Voorhout, NL-2514 EH The Hague
Queen of the Netherlands: H M Queen Beatrix [98]

11 Prime Minister

11-10000 Office of the Prime Minister
Binnenhof 20, PO Box 20001, NL-2500 EA The Hague
☎ (+31 70) 356 41 00 ⊖ (+31 70) 356 46 83
Prime Minister: Wim Kok
Vice Prime Minister: Hans Dijkstal
Vice Prime Minister: Hans van Mierlo
Advisor on Foreign and Defence Affairs: J P H M Merckelbach
Head of Library: I Wubben
The Netherlands Information Service
Binnenhof 19, PO Box 20006, NL-2500 EA The Hague
☎ (+31 70) 356 40 00 ⊖ (+31 70) 356 46 83
Spokesman and Press Officer to the PM: M J D van der Voet [98]

12 Ministries

12-10000 Ministry of Defence
Ministerie van Defensie
Plein 4, PO Box 20701, NL-2500 ES The Hague
☎ (+31 70) 318 81 88 ⊠ 31337 ⊖ (+31 70) 318 72 64
Minister of Defence: Dr J J C Voorhoeve
Secretary of State of Defence: Jan Gmelich Meijling
Press and Information Directorate
Kalvermarkt 38, PO Box 20701, NL-2500 ES The Hague
☎ (+31 70) 318 81 88 ⊖ (+31 70) 318 65 23
Director of Press and Information: H van den Heuvel ☎ (+31 70) 318 83 05 [98]

12-12000 Ministry of Economic Affairs
Ministerie Van Economische Zaken
Postbus 20101, NL-2500 EC The Hague
Bezuidenhouttseweg 30
☎ (+31 70) 379 89 11 [E] voorlichting@minez.nl
[W] http://www.minez.nl ⊖ (+31 70) 347 40 81
Minister of Economic Affairs: Hans Wijers
Secretary of State: Anneke van Dok [98]

12-20000 Ministry of Foreign Affairs
Ministerie van Buitenlandse Zaken
Bezuidenhoutseweg 67, PO Box 20061, NL-2500 EB The Hague
☎ (+31 70) 348 64 86 ✉ CELER ⊠ 31326 buza nl
⊖ (+31 70) 348 48 48
Minister of Foreign Affairs: Hans A F M O van Mierlo ☎ (+31 70) 348 50 36
Secretary of State: M Patiin ☎ (+31 70) 348 50 25
Minister for Development Co-operation: J P Pronk ☎ (+31 70) 348 50 33
Secretary General: D J van den Berg ☎ (+31 70) 348 50 40
Deputy Secretary General: J G S T M van Hellenberg Hubar ☎ (+31 70) 348 50 37
Director General for Bilateral Relations: T P Hofstee ☎ (+31 70) 348 65 41
Deputy Director General for Bilateral Relations: M E P Dierikx
Director General for Political Affairs: J M Vos Mazairac ☎ (+31 70) 348 51 24
Director General for European Co-operation: S I H Goosess ☎ (+31 70) 348 48 43
Director General for International Co-operation: J N M Richelle ☎ (+31 70) 348 44 16
Inspector General for Development Co-operation: Ms H G C Ajoeb ☎ (+31 70) 348 44 21 [98]

12-40000 Ministry of the Interior
Schedeldoekshaven 200, PO Box 20011, NL-2500 EA The Hague
☎ (+31 70) 302 61 03 ⊠ 32109 ⊖ (+31 70) 302 91 53
Minister: Ed van Thÿn
State Secretary: Jacob Kohnstamm
Director General for Public Order & Security: Harry Borghouts [97]

12-70000 Ministry of Transport, Public Works and Water Management
Plesmanweg 1-6, PO Box 90771, NL-2509 LT The Hague
☎ (+31 70) 351 72 47 ⊖ (+31 70) 351 63 48
Minister of Transport: Mrs Annemarie Jorritsma-Gebbink [98]

12-90000 Ministry of Justice
WODC Bibliotheek, PO Box 20301, NL-2500 EH The Hague
☎ (+31 70) 370 76 17 ⊖ (+31 70) 370 79 02
Minister of Justice: Winnie Sorgdrager [98]

15 Parliament

15-10000 First Chamber of the State General
Eerste Kamer der Staten-Generaal
Binnenhof 22, PO Box 20017, NL-2500 EA The Hague
☎ (+31 70) 362 45 71 ⊖ (+31 70) 365 38 68
President: H D Tjeenk Willink
Vice President: Dr A Posfma
Vice President: H Heyne Makkacel
Secretary General: Dr C L Baljé
Public Relations Officer: M Paulusma
Defence Commission
Binnenhof 22, PO Box 20017, NL-2500 EA The Hague
Chairman: J A van Graafeiland
Foreign Affairs Commission
Binnenhof 22, PO Box 20017, NL-2500 EA The Hague
Chairman: J Verbeck
Vice Chairman: Dr J Gennip
Internal Affairs Commission
Binnenhof 22, PO Box 20017, NL-2500 EA The Hague
Chairman: Dr J T J van den Berg
Vice Chairman: H Wiegel [97]

15-10010 Second Chamber of the State General
Tweede Kamer Staten Generaal
Binnenhof 1a, PO Box 20018, NL-2500 EA The Hague
☎ (+31 70) 318 22 11 ⊠ 32653 ⊖ (+31 70) 365 41 22
Chairman: Dr W J Deetman
Defence Commission
Binnenhof 1a, PO Box 20018, NL-2500 EA The Hague
Chairman: Dr Frits Bolkstein
First Vice Chairman: A B M Frinking
Foreign Affairs Commission
Binnenhof 1a, PO Box 20018, NL-2500 EA The Hague
Chairman: Dr Harry J B Aarts
First Vice Chairman: Maarten van Traa
Foreign Trade Commission
Binnenhof 1a, PO Box 20018, NL-2500 EA The Hague
Chairman: W A Mateman
First Vice Chairman: C N van Dis
Internal Affairs Commission
Binnenhof 1a, PO Box 20018, NL-2500 AE The Hague
Deputy Chairman: Jan Kees Wiebenga [97]

Contact Reference

20 Defense Administrations/Services/Agencies

20-10000 General Planning Construction and Maintenance Bureau for RNLAF Airfields
Kalvermarkt 38, PO Box 20701, NL-2500 ES The Hague
☎ (+31 70) 318 81 88
Director General: Mr Flederus [97]

20-12000 Royal Netherlands Meterological Institute
Koninklijk Nederlands Meteorologisch Instituut (KNMI)
(a subordinate to Ministry of Transport, Public Works and Water Management)
Wilhelminalaan 10, PO Box 201, NL-3730 AE De Bilt
☎ (+31 30) 220 69 11 ⊠ 47096 nl
⊖ (+31 30) 221 04 07
540
General Manager: Dr H M Fijnaut
Head of International Relations: H Daan
Co-ordinator Earth Observation: Dr A P M Baede
Head of Communications Division: W F M Verhoeven
Aeronautical Meteorological Division
PO Box 7625, NL-1118 ZJ Schiphol Airport
☎ (+31 20) 602 31 93 (administration); 602 31 46 (operations) ⊖ (+31 20) 648 47 79
Head of Division: E R Sluiter
■ Meteorological forecasts for aviation. [97]

25 Defense Forces

25-19000 Netherlands Defence College
PO Box 20701, NL-2500 ES The Hague
☎ (+31 15) 215 22 00 [E] alg@idl.dico.disp.mindef.nl
⊖ (+31 15) 215 27 28
Director: Gen Maj C. Homan
Librarian: W R J Rovers [98]

25-20000 Royal Netherlands Air Force Headquarters (RNLAF)
Binckhorstlaan 135, PO Box 20703, NL-2500 ES The Hague
☎ (+31 70) 349 35 91 ⊖ (+31 70) 349 25 00

Commander-in-Chief: Lieut Gen B A C Droste (RNLAF)
Vice Commander-in-Chief: Maj Gen A P A de Jong (RNLAF)
Personnel Director: Maj Gen C J M de Veer
Operations Director: Maj Gen G F A Macco

Editorial Team Herkenning

Defence Periodical
PO Box 20703, NL-2500 ES The Hague
☎(+31 70) 349 35 91 ⊖ (+31 70) 349 25 00
Editor-in-Chief: 1st Lieut R W M van Vuuren
Editor, Aircraftrecognition: R Choufoer
Editor, Aircraftrecognition: SM P H de Bock
Editor, Navalrecognition: SM B Booij
Editor, Navalrecognition: Sgt R Geurts
Editor, Tankidentification: AOO A Jonker
Editor, Tankidentification: 1st Sgt F Cruts

Materiel Directorate
Binckhorstlaan 135, PO Box 20703, NL-2500 ES The Hague
☎(+31 70) 349 35 91 ⊠ 43393 luhvs
⊖ (+31 70) 349 25 00
Director: Maj Gen Drs D Altena

RNLAF Directorate of Personnel
Binckhorstlaan 135, PO Box 20703, NL-25000 ES The Hague
☎(+31 70) 349 35 91
Director: Maj Gen C J M de Veer

RNLAF Directorate of Operations
Binckhorstlaan 135, PO Box 20703, NL-2500 ES The Hague
☎(+31 70) 349 35 35
Director: Maj Gen G F A Macco

Information Service
Kalvermarkt 38, PO Box 20701, NL-2500 ES The Hague
☎(+31 70) 318 83 72; 318 83 76
⊖ (+31 70) 318 83 09
Chief of Public Information Service: Lieut Col J D Piëst
Assistant Chief of Public Information Service: 1st Lieut F J P Logister

Eindhoven AFB
PO Box 49, NL-5120 AA Rijen
☎(+31 40) 50 69 11
Commander: Col J H de Jong

Gilze-Rijen AFB
PO Box 49, NL-5120 AA Rijen
☎(+31 1612) 969 11
Commander: Col J.H. de Jong

Volkel AFB
PO Box 10150, NL-5408 ZW Volkel (N.B.)
☎(+31 4132) 720 51 ⊖ (+31 4132) 740 77
Commander: Col W. Glaser

Twenthe AFB
PO Box 5013, NL-7500 GA Enschede
☎(+31 53) 35 07 05 ⊖ (+31 53) 30 99 54
Commander: Col P J M Godderij

Leeuwarden AFB
PO Box 21050, NL-8900 JB Leeuwarden
☎(+31 58) 34 29 11 ⊖ (+31 58) 13 71 06
Commander: Col G Booij

Deelen AFB
PO Box 9207, NL-6800 HJ Arnhem
☎(+31 85) 71 89 11 ⊖ (+31 85) 43 14 20
Commander: Col J R Zijlstra

Soesterberg AFB
PO Box 160, NL-3769 ZK Soesterberg
☎(+31 3463) 329 11 ⊖ (+31 3463) 546 21
Commander: Col J B Ansink [97]

25-30000 Royal Netherlands Army Headquarters (RNLA)

Thérèse Schwartzestraat 15, PO Box 90711, NL-2509 LS The Hague
☎(+31 70) 316 75 01 ⊖ (+31 70) 316 76 66
Commander-in-Chief Royal Netherlands Army: Lieut Gen H A Couzy
Vice Commander-in-Chief: Maj Gen A P P M van Baal
☎(+31 70) 316 75 05; 316 70 62 (fax)
Deputy Chief of Army Staff, Plans: Brig Gen P J E J Striek
☎(+31 70) 316 75 06
Deputy Chief of Army Staff, Operations: Brig Gen C Nicolai
☎(+31 70) 316 75 07
Chief of Staff, National Territorial Command: Brig Gen A Dijustra ☎(+31 1820) 921 03; 923 01 (fax)
Governor of Military Academy: Maj Gen Vogel Poel
☎(+31 76) 27 30 48; 27 35 34 (fax)

Materiel Directorate
Van der Burchlaan 31, PO Box 90701, NL-2509 LS The Hague
☎(+31 70) 316 41 40 ⊖ (+31 70) 316 46 00
Director: Maj Gen Pacanda ☎(+31 70) 316 41 42
Head of Procurement Division: Mr de Jong
☎(+31 70) 316 41 10
Head of Suppliers Section: Brig Gen Woordhuizen
☎(+31 70) 316 41 06
Head of Technical Division: Brig Gen de Klerk
☎(+31 70) 316 41 38 [97]

25-39100 Royal Military Academy for Army and Air Force (KMA)

PO Box 90154, NL-4800 RG Breda
☎(+31 76) 527 49 11 ⊖ (+31 76) 527 35 21
Governor: Maj Gen F J M Vogelpoel
Public Relations Officer: Capt C G T M Jansen
Assistant Librarian: J de Graaf
Secretary to the Governor: M Frens [98]

25-40000 Royal Netherlands Navy Headquarters (RNLN)

Admiraliteit, PO Box 20702, NL-2500 ES The Hague
☎(+31 70) 316 91 11 ⊠ 31335
⊖ (+31 70) 316 20 12; 318 83 71
Commander-in-Chief: VAdm L Kroon
Deputy Commander-in-Chief: RAdm L L Buffart
Deputy Chief of Naval Staff for Plans and Policy: Cdre R A A Klaver
Deputy Chief for Operations and Flag Officer Naval Airservice: Cdre C D M J Leebeck
Personnel Director: RAdm W J E van Rijn
Material Director: RAdm-Ir J G Nanninga
Economic Management Director: R A F van de Kamp

HQ Royal Netherlands Marine Corps (RNLMC)
250 Noordsingel, NL-3032 BN Rotterdam
☎(+31 10) 443 22 00 ⊖ (+31 10) 443 22 40
Commandant of Royal Netherlands Marine Corps: Maj Gen E C Klop

HQ Admiral Benelux Netherlands (ABNL)
PO Box 10000, NL-1780 CA Den Helder
☎(+31 223) 65 30 00 ⊖ (+31 223) 65 69 49
Admiral, Netherlands Fleet: VAdm Dr G G Hooft

Headquarters Flagofficer Netherlands Forces Caribbean
NAPO 401, NL-3509 VP Utrecht
☎(+59 99) 63 74 01 ⊖ (+59 99) 637 44 49
Flagofficer Netherlands Forces Caribbean: Cdre J W de Jager

Information Service Office
PO Box 20701, NL-2500 ES The Hague
Kalvermarkt 38, NL-2511 CB The Hague
☎(+31 70) 318 83 80 ⊖ (+31 70) 318 83 71
Director of Public Information: Cdr Silvain J H Benistant
☎(+31 70) 318 83 81
Deputy Director of Public Information: Lieut Cdr-Ir A van der Leaaij ☎(+31 70) 318 83 78 [98]

25-40130 Royal Netherlands Navy Naval Air Service

Marineluchtvaardtdienst (MLD)
Van der Burchlaan 31, NL-2597 PC The Hague
☎(+31 70) 316 91 11 ⊖ (+31 70) 316 29 61
Naval Air Flag Officer: Cdr C D M J Leebech
Commanding Officer (Fixed Wing): Capt A van Digh
Commanding Officer (Rotary Wing): Col W F Visee [98]

25-49200 Royal Netherlands Navy Centre

Marinevliegkamp Valkenburg, PO Box 590, NL-2220 AN Katwijk aan Zee
☎(+31 71) 405 25 47 ⊖ (+31 71) 405 26 00
Director: Capt M Bod
Head of Operations: Lieut Cdr F A Hermans
Deputy Director and Head of Research: Ir A C Houtman
Acoustics Specialist, Research Department: Ir J de Graaf [98]

31 Embassies & Attachés

31-00100 Embassy of Austria

PO Box 96855, NL-2509 JG The Hague
Van Alkemadelaan 342, NL-2597 AS The Hague
☎(+31 70) 324 54 70; 324 54 71; 324 54 72
⊖ (+31 70) 328 20 66
Ambassador: H E Dr Alexander Christiani
Deputy Chief of Mission: Dr Gerhard Zettl
Military and Air Attaché: Divisionär Richard Bondi
☎(+32 2) 646 02 78
Deputy Defence Attaché: Lieut Richard Ebster [98]

31-00150 Embassy of Belgium

Lange Vijverberg 12, NL-2513 AC The Hague
☎(+31 70) 312 34 56 [E] ambel-dh@wirehub.nl
⊠ 31035 ambel nl ⊖ (+31 70) 364 55 79
Ambassador: H E Mr Jan F Willems
☎(+31 70) 312 34 25
Minister and Deputy Chief of Mission: P Vercauteren Drubbel ☎(+31 70) 312 34 26
Minister, Political Affairs: A Creusen
☎(+31 70) 312 34 28
Commercial Attaché: C Andoulsi ☎(+31 70) 354 62 86
Commercial Counsellor: W Fimmers
☎(+31 70) 355 52 55
Second Secretary: René Hannaert
Trade Secretary: H van Elsacker ☎(+31 70) 312 34 29

Commercial Department
Lange Voorhout 51, NL-2514 EC The Hague
☎(+31 70) 346 94 07 ⊖ (+31 70) 362 10 60 [98]

31-00290 Embassy of the Republic of Cameroon

Amaliastraat 14, NL-2514 JC The Hague
☎(+31 70) 346 97 15 ⊠ 31257 aclh nl
⊖ (+31 70) 365 29 79
Ambassador: H E Mrs J I M Bassong
Deputy Chief of Mission: Timothée Tabapssi Famndie [98]

31-00310 Embassy of Canada

Sophialaan 7, PO Box 30820, NL-2500 GV The Hague
☎(+31 70) 311 16 00 ⊖ (+31 70) 311 16 82
Ambassador: H E Mrs M Bernard-Meunier
Minister and Commercial Counsellor: Claude N Fontaine
Defence Attaché: Col F Laforge
Third Secretary: J Reeves [98]

31-00320 Embassy of the Republic of Cape Verde

Koninginnegracht 44, NL-2514 AD The Hague
☎(+31 70) 346 96 23 ⊖ (+31 70) 346 77 02
Ambassador: H E Mr Julio Vasco de Sousa Lobo
Counsellor: Daniel A Pereira [98]

31-00350 Embassy of Chile

Mauritskade 51, NL-2514 HG The Hague
☎(+31 70) 364 27 48; 363 98 84 ⊠ 34199 echgv nl
⊖ (+31 70) 345 21 09
Ambassador: H E Mr Luis Pascal
Deputy Chief of Mission: Guillermo Echenique
Secretary: E Mozató

Commercial Office
Stadhouderskade 2, NL-1054 ES Amsterdam
☎(+31 20) 612 47 01 ⊖ (+31 20) 685 24 86
Commercial Attaché: Mrs Jimena Castillo
☎(+31 70) 364 52 52 [98]

31-00360 Embassy of the People's Republic of China

Adrian Goekooplaan 7, NL-2517 JX The Hague
☎(+31 70) 355 15 15 ⊖ (+31 70) 355 16 51
Ambassador: H E Ms Zhu Manli ☎(+31 70) 351 51 10
Counsellor: Zhou Congwu ☎(+31 70) 351 48 78
Commercial Counsellor: Li Kanghua ☎(+31 70) 51 55 59
Military Attaché: Li Jingping ☎(+31 70) 350 69 33 [98]

31-00430 Embassy of the Republic of Cuba

Mauritskade 49, NL-2514 HG The Hague
☎(+31 70) 360 60 61 ⊠ 31318 embcu nl
⊖ (+31 70) 364 75 86
Ambassador: H E Mrs Sonia Diaz LLera [98]

31-00450 Embassy of the Czech Republic

Paleisstraat 4, NL-2514 JA The Hague
☎(+31 70) 346 97 12 ⊖ (+31 70) 356 33 49
Ambassador: H E Jaroslav Mihule
Counsellor, Minister and Deputy Chief of Mission: Jaroslav Horák
First Secretary: Igor Sedo
First Secretary: Pavel Rezác
Second Secretary: Petra Sramotová
Second Secretary: Jan Michal
Attaché: Zdenek Valis
Defence Attaché: Lieut Col Pavel Baborsky
Deputy Defence Attaché: Maj Pavel Zuna [98]

31-00460 Royal Danish Embassy
Koninginnegracht 30, PO Box 85654,
NL-2508 CJ The Hague
☎ (+31 70) 302 59 59 ✉ AMBADANE The Hague
▣ 32075 amdan nl ⊖ (+31 70) 302 59 50
Ambassador: H E Mr Jakob Rytter
Counsellor and Deputy Chief of Mission:
Torben E Simonsen
Commercial Counsellor: Poul-Erik Larsen [98]

31-00535 Honorary Consul General of Estonia
Snipweg 101 - 1118 DP Schipol South, PO Box 7634,
NL-1117 ZJ Schipol Centre
☎ (+31 20) 316 54 40 ⊖ (+31 20) 316 54 54
Honorary Consul General: Wibo Aris
Commercial Officer: Mrs Külli Prosa (MA)

Den Haag Office
Emmastraat 1e, NL-2595 EG Den Haag
☎ (+31 703) 82 12 17 ⊖ (+31 703) 47 81 90 [98]

31-00660 Embassy of the Federal Republic of Germany
Groot Hertoginnelaan 18-20, NL-2517 EG The Hague
☎ (+31 70) 342 06 00 ▣ 31012 aagv nl
⊖ (+31 70) 365 19 57
Ambassador: H E Mr Eberhard von Puttkamer
Deputy Chief of Mission: Jürgen Dröge
Defence, Naval, Air and Military Attaché:
Lieut Col Axel Haase
Economic Affairs Counsellor: Winfried Schlote
Head of Administration: Joachim Dost
Attaché: Ingrid Malcomess
Press and Information Affairs Secretary: Peter Dettmar [98]

31-00830 Embassy of India
Buitenrustweg 2, NL-2517 KD The Hague
☎ (+31 70) 346 97 71
[E] ambsr@indemb.nl; couns@indemb.nl
▣ 33543 indem nl ⊖ (+31 70) 361 70 72; 346 25 94
Ambassador: H E Mr Prabhakar Menon
☎ (+31 70) 365 91 25
Counsellor: Mrs Deepa G Wadhwa
Counsellor: Rajinder Bhagat
First Secretary: Ravinder Khanna
First Secretary and Head of Chancery: Ms M Subashini
First Secretary: Mahesh Dutt Kalra
Attaché: Ms Krishna Bahl
Air Attaché: Cdre Ajit
Military Attaché: Brig Gen Ashok Hukku
Naval Attaché: VAdm Thayi Hari [98]

31-00840 Embassy of the Republic of Indonesia
8 Tobias Asserlaan, NL-2517 KC The Hague
☎ (+31 70) 310 81 00 [E] homhaag@mailserv.caiw.nl
▣ 32367 penri nl; 32356 indon nl ⊖ (+31 70) 364 33 31
Ambassador: H E Mr J B Soedarmanto Kadarisman
☎ (+31 70) 310 81 33
Deputy Chief of Mission: Abdurrachman Mattalitti
☎ (+31 70) 310 81 34
Defence Attaché: Col M Luthfie ☎ (+31 70) 310 81 36
Economic Counsellor: Mr Ladjuris ☎ (+31 70) 310 81 36
Commercial Attaché: Yeru Salimianto
☎ (+31 70) 310 81 45 [98]

31-00880 Embassy of Israel
Buitenhof 47, NL-2513 AH The Hague
☎ (+31 70) 376 05 00 [E] ambisra@xs4all.nl
[W] http://www.xs4all.nl/~ambisra ⊖ (+31 70) 376 05 55
Ambassador: H E Mr Yosef Gal
Deputy Chief of Mission: Yaffa Ben-Ari
Military, Naval and Air Attaché: Lieut Col Avi Ronen
Consular and Administration Officer: Nuriel Yitzchaki
Economic Affairs Attaché: Moshe Shoham
Press, Information and Cultural Attaché: Iddo Moed [98]

31-00920 Embassy of Japan
Tobias Asserlaan 2, NL-2517 KC The Hague
☎ (+31 70) 346 95 44 ✉ TAISHI Hague ▣ 32105 taisi nl
⊖ (+31 70) 310 63 41
Ambassador: H E Mr K Fujita
Deputy Chief of Mission: T Matsumoto
Commercial Secretary: N Yamada [97]

31-01000 Embassy of the State of Kuwait
Carnegielaan 9, NL-2517 KH The Hague
☎ (+31 70) 360 38 13/5 ▣ 32119 kwtlh nl
⊖ (+31 70) 392 45 88
Ambassador: H E Mr Khalaf Abass Khalaf
First Secretary: Abdul Rahman H A Al-Otaibi
Third Secretary: Mohammed H Al-Failakawi
Attaché: Mobarak M F S Al-Hajeri [97]

31-01070 Embassy of Luxembourg
Nassaulaan 8, NL-2514 JS The Hague
☎ (+31 70) 360 75 16; 364 75 89 ▣ 33174 amlux nl
⊖ (+31 70) 356 33 03
Ambassador: H E Mr Hubert Wurth
Minister Counsellor: Jean-Paul Munchen [97]

31-01110 Embassy of Malaysia
Rustenburgweg 2, NL-2517 KE The Hague
☎ (+31 70) 350 65 06; 350 58 55
✉ MALAWAKIL Den Haag ▣ 33024 malay nl
⊖ (+31 70) 350 65 36
Ambassador: H E Mr A Ganapathy
First Secretary: Ahmad Izlan Idris
Commercial Counsellor: Dzulkfle Bin Mahmud
Second Secretary: Abdul Latiff Kader
Information Assistant: Ms Zahariah Kahar [98]

31-01220 Embassy of the Kingdom of Morocco
Oranjestraat 9, NL-2514 JB The Hague
☎ (+31 70) 346 96 17 ▣ 34163 abmar nl
⊖ (+31 70) 361 45 03
Ambassador: H E Mr Abderrahman Benomar
Deputy Chief of Mission: Abdelhamid Mouharrar
Military, Naval and Air Attaché: Col Tayeb Bainouti
(resident in Brussels) [97]

31-01300 Embassy of New Zealand
Carnegielaan 10, NL-2517 KH The Hague
☎ (+31 70) 346 93 24 [E] nzemb@bart.nl
⊖ (+31 70) 363 29 83
Ambassador: H E Ms Hilary Ann Willberg
First Secretary: Guy Sergei Lewis
Second Secretary: Murray John Denyer
Executive Assistant: M Bovenlander [98]

31-01340 Royal Norwegian Embassy
Prinsessegracht 6a, NL-2514 AN The Hague
☎ (+31 70) 345 19 00 ⊖ (+31 70) 365 96 30
Ambassador: H E Mr Arne L Langeland
Counsellor: Knut H Christensen [97]

31-01370 Embassy of Pakistan
8 Amaliastraat, NL-2514 JC The Hague
☎ (+31 70) 64 89 48; 64 89 49; 346 10 72
✉ PAREP The Hague ▣ 33696 parep nl
⊖ (+31 70) 310 60 47
Ambassador: H E Mr Saaed M Khan
Deputy Chief of Mission: Mohsin Razi
Defence Attaché: Col Ayyaz Saleem Rana
☎ (+49 228) 35 20 04 09
Commercial Secretary: Mrs Anjum Asad Amin [97]

31-01420 Embassy of the Philippines
Laan Copes Van Cattenburch 125, NL-2585 EZ The Hague
☎ (+31 70) 360 48 20; 360 48 21
✉ PHILEMBASSY's Gravenhage ▣ 33103 amphl nl
⊖ (+31 70) 356 00 30
Ambassador: H E Mr R S Sanchez

Office of the Commercial Attaché
3rd Floor Philippine Embassy,
Laan Copes van Cattenburch 125, NL-2585 EZ The Hague
☎ (+31 10) 364 12 95 ⊖ (+31 10) 362 25 08
Commercial Counsellor: Mr A G Mamaril [98]

31-01430 Embassy of the Republic of Poland
Alexanderstraat 25, NL-2514 JM The Hague
☎ (+31 70) 360 28 06; 360 28 07
[E] ambhaga@polamb.nl ⊖ (+31 70) 360 28 10
Ambassador: H E Mr Stanistaw Komorowski
Military, Naval and Air Attaché: Col Eugeniusz Krakowiak
First Secretary, Scientific and Cultural Affairs:
Alina Prystrom
Commercial Counsellor: Józef Hoppen

Office of Commercial Affairs
Van Lennepweg 51, NL-2597 LG The Hague
☎ (+31 70) 350 27 81 ▣ 31359 morha nl
⊖ (+31 70) 354 39 66
Commercial Attaché: Tomasz Przygoda
Commercial Attaché: Apolinary Wazbinski
Commercial Attaché: Elzbieta Kocanda
Deputy Commercial Attaché: Mirostaw Kurek [98]

31-01480 Embassy of Romania
Catsheuvel 55, NL-2517 KA The Hague
☎ (+31 70) 354 37 96 ⊖ (+31 70) 354 15 87
Ambassador: H E Mr Marin Buhoara
First Secretary, Consular Affairs: Stefan Stoica
Military, Naval and Air Attaché: Capt Antoniu Gavril Pop
(resident in Brussels)
Second Secretary and Head of Political Section: Ioan Tudor
Second Secretary and Cultural Attaché: Dan Sitaru
Second Secretary, Press and Information:
Mrs Ireny Comaroschi

Office of Economic Affairs
Castheuvel 55, NL-2517 KA The Hague
☎ (+31 70) 354 15 80 ▣ 32189 rocom nl
⊖ (+31 70) 354 15 87
Economic Counsellor: Dan Cioanca
First Secretary, Economic Affairs: Liviu Adrian Nita [97]

31-01485 Embassy of the Russian Federation
Andries Bickerweg 2, NL-2517 JP The Hague
☎ (+31 70) 345 13 00; 345 13 01 ▣ 32176 ussr nl
⊖ (+31 70) 361 79 60
Ambassador: H E Mr Leonid A Skotnikov
Commercial Officer: Alexei N Tsaregorodtsev
Public Information Officer: Vladimir V Naidenov
Attaché: Musslim U Gedigushev
Defence Attaché: Col Anatoly J Pivovarov [98]

31-01600 Embassy of Saudi Arabia
Alexanderstraat 19, NL-2514 JM The Hague
☎ (+31 70) 361 43 91 ▣ 33731 najdiah nl
⊖ (+31 70) 363 03 48
Ambassador: H E Mr Mohamed Nouri Ibrahim [98]

31-01670 Embassy of South Africa
Wassenaarseweg 40, NL-2596 CJ The Hague
☎ (+31 70) 392 45 01 ⊖ (+31 70) 346 06 69
Ambassador: H E Mr C G Niehaus
Counsellor: C J Bezuidenhout
Administration Officer: J C France
Commercial Counsellor: J H C Engelbrecht [98]

31-01740 Embassy of Switzerland
PO Box 30913, NL-2500 GX The Hague
Lange Voorhout 42, NL-2514 EE The Hague
☎ (+31 70) 364 28 31; 364 28 32
✉ AMBASUISSE The Hague ⊖ (+31 70) 356 12 38
Ambassador: H E Mr Alfred Rüegg
Head of Chancery: Urs Müller
Defence Attaché: Col Walter Ritzmann [97]

31-01780 Royal Thai Embassy
Buitenrustweg 1, NL-2517 KD The Hague
☎ (+31 70) 345 97 03; 345 20 88; 345 06 32
▣ 34207 tduto nl ⊖ (+31 70) 345 19 29
Ambassador: H E Mr Prachya Davi-Tavedikul
Deputy Chief of Mission: Nimit Prathomvarl

Commercial Office
123 Laan Copes van Cattenburch, NL-2585 EZ The Hague
☎ (+31 70) 346 17 97; 346 21 75
⊖ (+31 70) 345 93 76
Commercial Counsellor: Miss Sasithorn Saksri
Commercial Affairs Secretary: Wichchakorn Chandrasiri [97]

31-01830 Embassy of Tunisia
Gentsestraat 98, NL-2587 HX The Hague
☎ (+31 70) 351 22 51 ▣ 31271 ambtu nl
⊖ (+31 70) 351 43 23
Ambassador: H E Mr Anouar Berraies
Deputy Chief of Mission: Mohamed Ben Mahmoud
Commercial Counsellor: Tarak Ben Salem [98]

31-01840 Embassy of the Republic of Turkey
Jan Evertstraat 15, NL-2514 BS The Hague

Contact Reference

☎(+31 70) 60 49 12; 60 49 14 ☒ 32623 turko nl
⊖ (+31 70) 361 79 69
Ambassador: H E Mr Baki Ilkin
Deputy Chief of Mission: Derya Kanbay

Commercial Affairs Office
Jan van Nassaustraat 25, NL-2596 BM The Hague
☎(+31 70) 328 13 76 ☒ 34521 ticmu nl
⊖ (+31 70) 324 82 11

Office of the Defence Attaché
Jan Everstraat 15, NL-2514 BS The Hague
☎(+31 70) 360 49 12; 360 49 14
Defence Attaché: Col K Nazif Özdagdeviren [97]

31-01890 Embassy of the United Kingdom
Lange Voorhout 10, NL-2514 ED The Hague
☎(+31 70) 427 04 27 (10 lines) ⊖ (+31 70) 427 03 45
Ambassador: H E Miss Rosemary Spencer
Deputy Head of Mission: P Dimond
Commercial Counsellor: C W Robbins
Defence and Naval Attaché: Capt P J Organ (RN)
Military and Air Attaché: Lieut Col C A Le Hardy [98]

31-01900 Embassy of the United States of America
102 Lange Voorhout, NL-2514 EJ The Hague
☎(+31 70) 310 92 09 ⊖ (+31 70) 361 46 88
Ambassador: H E Mr K Terry Dornbush
Deputy Chief of Mission: William P Pope
Commercial Counsellor: Linda Archer
Public Affairs Counsellor: Yolanda Robinson
Defence Attaché: Capt Donald Roulstone (US Navy)

Office of Defence Co-operation
102 Lange Voorhout, NL-2514 EJ The Hague
☎(+31 70) 310 92 82 ⊖ (+31 70) 364 83 30
Chief of Office of Defence Co-operation: Col Ron Levis [98]

31-01910 Embassy of Uruguay
Mauritskade 33, NL-2514 HD The Hague
☎(+31 70) 360 98 15; 360 98 16
⊖ (+31 70) 356 28 26
Ambassador: H E Mr Juan Andrès Pacheco
Second Secretary: Dr Rossana Rubiños [97]

31-01950 Embassy of Venezuela
Nassaulaan 2, NL-2514 JS The Hague
☎(+31 70) 352 38 51; 352 43 51
✉EMBAVENEZ The Hague ☒ 34058 emven nl
⊖ (+31 70) 365 69 54
Ambassador: H E Dr Carlos E Nones Sucre
Deputy Chief of Mission: Ms Alicia Chacín
Military Attaché: Col Oswaldo Enrique Sosa (resident in Belgium) ☎(+32 2) 640 47 36; 647 88 20 (fax)
Commercial Officer: Roberto Dubuc [97]

31-01990 Embassy of the Republic of Yemen
Surinamestraat 9, NL-2585 GG The Hague
☎(+31 70) 365 39 36 ☒ 33290 ⊖ (+31 70) 356 33 12
Ambassador: H E Dr A A Homarah
Counsellor, Economic Affairs: A A Hamid Saeed Al-Mekhlafi
First Secretary: Mahmood Ibrahim Suliman
Third Secretary: Abdulkader Ali Dahhan Altheeb
Attaché: Abdo Mahmood Humees [98]

34 Multinational Intergovernmental Organizations

34-12100 Allied Forces Central Europe Headquarters (AFCENT)
PO Box 270, NL-6440 Brunssum
☎(+31 45) 526 22 88 ⊖ (+31 45) 526 30 95
Commander-in-Chief (CINCENT): Gen Dieter Stöckmann (GEA) ☎(+31 45) 526 22 88 ext. 2352
Deputy Commander-in-Chief (DCINCENT): ACM Sir Michael Stear (UKAF) ☎(+31 45) 526 22 88 ext. 2610
Chief of Staff (CoS): Lieut Gen R de Clerck (BEA) ☎(+31 45) 526 22 88 ext. 2644
Director of Operations (DoO): Maj Gen R W Roper (Jr) (US Army) ☎(+31 45) 526 22 88 ext. 2620
Director of Support (DoS): Maj Gen M W Gentjes (NLAF) ☎(+31 45) 526 22 88 ext. 2159
Assistant Chief of Staff Resources: Brig G L Kerr (UKA) ☎(+31 45) 526 22 88 ext. 2949
Assistant Chief of Staff Policy Division: Brig Gen E Skriver (DAF) ☎(+31 45) 526 22 88 ext. 2931
Assistant Chief of Staff Intelligence Div. (ACoS INT): Brig Gen M Verschoore (BEAF) ☎(+31 45) 526 22 88 ext. 2221
Assistant Chief of Staff Communications and Information Systems Division: Brig Gen P Bos (NLA) ☎(+31 45) 526 22 88 ext. 3153
Assistant Chief of Staff Operations and Logistics Division: Brig Gen D Farwick (GEA) ☎(+31 45) 526 22 88 ext. 3150
Chief of the French Military Mission (FMM): Maj Gen G Folio (FRA) ☎(+31 45) 526 22 88 ext. 2174
Chief of Public Information: Col F Salis (GEA) ☎(+31 45) 526 22 88 ext. 2288 [97]

34-12400 NATO Airborne Early Warning and Control Programme Management Agency (NAPMA)
Akerstraat 7, NL-6445 CL Brunssum
☎(+31 45) 526 22 22 ⊖ (+31 45) 525 43 73
General Manager: Brig Gen F Lübbe [97]

34-12600 NATO Consultation Command and Control Agency (NC3A)
Oude Waalsdorperweg 61, PO Box 174, NL-2501 CD THe Hague
☎(+31 70) 314 21 00 ⊖ (+31 70) 314 21 11
General Manager: L D Diedrichsen
Head of Information Services: C Walker [98]

60 Research Organizations, Institutes & Test Establishments

60-01000 Maritime Research Institute Netherlands (MARIN)
Haagsteeg 2, PO Box 28, NL-6700 AA Wageningen
☎(+31 317) 49 39 11 ⊖ (+31 317) 49 32 45
♣ DFl 35,000,000 👥 235
Managing Director: Ir Th M Oostinjen
Operations Director: Ir A H Hubregtse
Manager of Strategic Marketing: Ir J N Dekkker [98]

60-01400 National Aerospace Laboratory
Nationaal Lucht en Ruimtevaartlaboratorium (NLR)
Anthony Fokkerweg 2, PO Box 90502, NL-1006 BM Amsterdam
☎(+31 20) 511 31 13 Ⓔ info@nlr.nl Ⓦ http://www.nlr.nl
⊖ (+31 20) 511 32 10
♣ G 140,000,000 (1996) 👥 900 (1996)
Chairman: J van Houwelingen
General Director: Dr-Ir B M Spee
Technical Director: Prof-Ir F J Abbink
Finance Director: Dr L W Esselman
Associate Director: Dr A de Graaff
Public Relations Officer: Mrs J F van Esch
Publications Officer: B J Meijer
Purchasing Manager: J F Post
Fluid Dynamics Division: Prof-Ir J W Sloof
Flight Dynamics Division: Ir J T M Van Doorn
Space Flight Division: Ir B J P Van der Peet
Secretary: E Folkers
Engineering and Technical Sciences: Ir J van Twist
Informatics: Ir W Loeve
General Services: Ir W F Wessels
■ Research and development for military aircraft, helicopters and missile projects (aerodynamics, aero-acoustics, structures and materials, flight dynamics, flight simulation, handling qualities) spacecraft development (fluid dynamics, stability and control, environmental simulation, materials and structures, robotics), remote sensing and verification technology, military aircraft and helicopter operations (evaluation, military operations research, computer-aided mission planning and logistics, incident/accident investigation, certification of weapon systems, helicopter-ship qualification). Supercomputer facility available.
Aerospace research under contract to national and foreign industries. Operators, government and international agencies. Flight, space informatics. Research and test facilities: wind tunnels, flight and ATC simulators. Environmental facilities.

Amsterdam Main Laboratory
PO Box 90502, NL-1006 BM Amsterdam
☎(+31 20) 511 31 13 ✉WINDTUNNEL
⊖ (+31 20) 511 32 10

Noordoostpolder Laboratory
Voorsterweg 31, PO Box 153, NL-8300 AD Emmeloord
☎(+31 527) 24 84 44 ⊖ (+31 527) 24 82 10

Library
Voorsterweg 31, PO Box 153, NL-8300 AD Emmeloord
☎(+31 527) 24 84 44 ⊖ (+31 527) 24 82 10 [98]

60-02000 RAND Europe
Landbergstraat 6, NL-2628 CE Delft
☎(+31 15) 278 54 11 ⊖ (+31 15) 278 17 88 [98]

60-20100 TNO Defence Research
TNO Defensieonderzoek
Schoemakerstraat 97, PO Box 6006, NL-2600 JA Delft
☎(+31 15) 269 69 00; 269 47 26
⊖ (+31 15) 262 73 19
♣ DFl 150,000,000 👥 885
Managing Director: E I L O G Margherita (MSc)
Director of Marketing and Programmes: Dr H J Pasman (MSc)
Public Relations: F G J van Aken (MSc)
■ Applied scientific research. Performs studies on all defence aspects and takes part in technological developments.

TNO Physics & Electronics Laboratory (TNO-FEL)
PO Box 96864, NL-2509 JG The Hague
☎(+31 70) 374 00 00 ⊖ (+31 70) 328 09 61
Managing Director: Dr J W Maas
Research Director: J A Vogel (MSc)
Information Officer: W L Smith
■ Research institute. Current research and development activities are focused on sensor systems and information and communication systems. Training and simulation systems and policy studies.

TNO Prins Maurits Laboratory (TNO-PML)
PO Box 45, NL-2280 AA Rijswijk
☎(+31 15) 284 28 42 ⊖ (+31 15) 284 39 91
Director: R M Lutje Schilpholt (MSc)
Director of Research: M van Zelm (MSc)
Information Officer: T M Groothuizen
■ Chemical research (analytical chemistry, chemical toxicity, protection and risk analysis) and technological research (ballistics and propulsion technology, munition functioning and explosives, weapon effects, pulse physics and explosion prevention).

TNO Human Factors Research Institute
PO Box 23, NL-3769 ZG Soesterberg
☎(+31 346) 35 62 11 ⊖ (+31 346) 35 39 77
Director: Dr A van Meeteren (MSc)
Information Officer: Dr A R A van der Horst (MSc)
■ Rearch and development activities: perception (visual speech, hearing), human information processing, skilled behaviour, and work environment. [97]

65 Industry, Trade & Armed Forces Associations

65-00600 Dutch Gas Turbine Association
Vereniging Gasturbine VGT
PO Box 21000, NL-3802 RA Amersfoort
Stadsring 187, NL-3817 BA Amersfoort
☎(+31 33) 479 87 07 Ⓔ vgt@euronet.nl
⊖ (+31 33) 479 87 09
Managing Director: Ir A J A Mom
Managing Dir: A.J.A. Mom
Executive Assistant: Ms Yvonne van Ginkel
■ Promotion of the technical and economic interests of the Dutch gas turbine industry and its related subcontractors.

Member Companies:
Bosman Powersource Support
Chromalloy Holland
Delaval Stork
Demkolec
ECN - Netherlands Energy Research Foundation
EPON _ Maintenance and Power Generation
EPZ
EZH
Interturbine
Elbar
Eldim
KLM - Engineering and Maintenance
NLR - National Aerospace Laboratory
OPRA Optimal Radial Turbine
Philips Machinefabrieken Nederland Machinefabriek ACHT
PNEM Transport and Opwekking
Pratt & Whitney Overhaul and Repair Centre (Europe)
SSB Energy and Utility Engineering

SP Aerospace and Vehicle Systems
STORK Technical Services
STORK Energy Boilers
STORK RMO
Techno Fysica
Croon Elektrotechniek
Thomassen International
Standard Aero
Urenco Aerospace 98

65-01000 NEDIP BV
Netherlands Defence Industry Promotion
(† Netherlands Naval Industries Group)
Hamonterweg 5, NL-6023 AJ Budel-Schoot
☎ (+31 495) 49 63 60 ⊖ (+31 495) 49 63 60
2 (1997)
General Manager: Ing. E M Verberk
Consultant: Cdre E Alting Siberg (retd)
■ NEDIP is an organisation arising from the Netherlands Naval Industries Group (NNIG) which was formed in 1983. Their primary aim is to assist and support the marketing activities of the Netherlands Defence Industries with particular emphasis on overseas opportunities. Internationally recognized as an experienced and innovative name in organizing Netherlands National Group participation in international exhibitions worldwide. 98

65-01450 Netherlands Aerospace Group (NAG)
Boerhaavelaan 40, PO Box 190, NL-2700 AD Zoetermeer
☎ (+31 79) 353 13 56 E nag@fme.nl
⊖ (+31 79) 353 13 65
Chairman: Th A Bierman
General Manager: Jan J C Bartels
Project Manager: Ms Erika Roodzant
■ Group composed by 40 companies that manufacture sub-assemblies for airframes and engines, aircraft equipment, airport and ground equipment, electrical and electronic equipment, maintenance and overhaul parts, research and development. 98

65-01470 Netherlands Defence Manufacturers Association (NIID)
Prinsessegracht 19, NL-2514 AP The Hague
☎ (+31 70) 364 48 07 E niid@pi.net
⊖ (+31 70) 365 69 33
8 (1997)
Managing Director: J Hans Dibbetz
Business Development Manager: Bill A M Klück
Business Development Manager:
Raymond H van der Meer
Manager of Marketing Support: Eelke M Siersema
■ The NIID aims at optimal participation of the Dutch industry in the field of defence equipment development and production for both national and foreign markets. Special attention given to small and medium-sized companies in the Netherlands to help them obtain subcontracts in the defence market. The primary interest for NIID members is the national defence equipment market. The NIID divides the defence market into three segments: the national defence market, the offset market and subcontracting. NATO equipment and maintenance services are also taken in account.
Marketing support: The NIID provides relevant marketing tools to its members. Data on MOD's possible future purchases and its members' capabilities are collected and matched in an extensive database. Foreign IEPG contracts bulletin are scanned for opportunities and members are informed at an early stage about business opportunities.
Intermediary of offset issues: In close co-operation with the Ministries of Economic Affairs and Defence, NIID assists foreign companies in identifying likely Dutch partners and/or subcontractors they need in fulfilling their offset obligations. 98

65-01480 The Netherlands Industrial Simulator Platform (NISP)
Prinsessegracht 19, NL-2514 AP The Hague
☎ (+31 70) 364 48 07 ⊖ (+31 70) 365 69 33
Chairman: Raymond van der Meer
■ Representing a group of established research institutes and companies active in the simulator market. NISP provides an effective pool of know-how and experience in research and development of systems integration and the production of subsystems. Innovative research into training analysis, ergonomics interfacing, visual presentation, modelling and dynamics carried out at the national research institutes and universities is available to the high-tech companies to assist them in the development and production of affordable, user-friendly trainers and simulators for both civil and military applications. Research capabilities in the institutes and universities pooled with the expertise and know-how of the industrial partner in the NISP is also available for simulator projects abroad. The NISP also directs its attention to international markets and partnerships with foreign industries.

Participants in the NISP:
Cap Gemini (Netherlands)
Data Sciences (Netherlands)
Fokker Aviation (Netherlands)
Fokker Space (Netherlands)
Friendship Simulation Company (Netherlands)
Van Halteren Metaal (Netherlands)
High Tech Automation BV (Netherlands)
Hydraudyne Systems and Engineering BV (Netherlands)
ICT Automation BV (Netherlands)
Logica BV (Netherlands)
MSCN BV (Netherlands)
National Aerospace Laboratory NLR (Netherlands)
Nedinsco BV (Netherlands)
Origin/Nieuwegein BV (Netherlands)
Parker Hannifin BV (Netherlands)
Philips Components (Netherlands)
R & H Systems (Netherlands)
University of Groningen (Netherlands)
Siemens Nederland NV (Netherlands)
Silicon Graphics Computer Systems BV (Netherlands)
Special Products Aerospace and Vehicle Systems BV (Netherlands)
Teelen BV (Netherlands)
Technical University Delft (Netherlands)
TNO-Physics and Electronics Laboratory (Netherlands)
TNO Human Factors Research Institute (Netherlands)
Free Universtity Amsterdam (Netherlands) 98

70 Financial Holding Groups, Parent Companies

70-00300 AEG Eurolec Holding BV
PO Box 9433, NL-1006 AK Amsterdam
Aletta Jacobslaan 9, NL-1066 BP Amsterdam
☎ (+31 20) 487 87 87 98

70-01000 Fugro NV
Veurse Achterweg 10, PO Box 41,
NL-2260 AA Leidschendam
☎ (+31 70) 311 14 22 ⊖ (+31 70) 320 27 03
♧ Dfl 374,900,000 (1995) ♣ Dfl 653,700,000 (1995)
President and Chief Executive Officer: G J Kramer
■ Multi-national group of consulting engineers. Provision of advice, technical applications and surveying operations. Divisions include onshore geotechnical services, offshore geotechnical services, environmental services, onshore surveying services, offshore surveying services and positioning services. → 80 98

70-05000 Stork NV Head Office
PO Box 5004, NL-1410 AA Naarden
7 Amersfoortsestraatweg, NL-1412 KA Naarden
☎ (+31 35) 695 74 11 ⊖ (+31 35) 695 12 58
♣ G 4,916,000,000 (1996) 20,504 (1996)
Chairman: Dr J C M Hovers
Deputy Chairman: H A D van den Boogaard
Deputy Chairman: Prof A W Veenman
Chairman of the Supervisory Board: R Hazelhoff
Deputy Chairman of the Supervisory Board: P Schwencke
Member of the Supervisory Board: Prof V Halberstadt
Member of the Supervisory Board: Prof C J Rijnvos
Member of the Supervisory Board: Prof B P T Veitman
Member of the Supervisory Board: J B T Manschot
Member of the Supervisory Board: Jhr F O J Sickinghe
Member of the Supervisory Board: S D de Bree
■ Supply high-level technological products and services to the industrial community. 98

75 Manufacturing Companies

AAR Allen Group International	80-00050
Radio Holland Electronics BV	80-18300

75-00005 AAF International BV
(a subsidiary of O Y L, Malaysia)
Egelenburg 2, PO Box 7928, NL-1008 AC Amsterdam
☎ (+31 20) 549 44 11 ⊠ 12372 amfinl
⊖ (+31 20) 644 43 98
♣ DFl 80,000,000 260 12,000 sq m
Managing Director: M Ryan
Sales and Marketing Director: P Room
Operations Director: D Porter
Purchasing Manager: R Weima
Marketing Manager: J H Wenzek
Marketing Services Manager: T Quail
Marketing Communications Co-ordinator:
Stephan Pasman
■ Manufacture of NBC air filters, air filter equipment and air pollution control systems. 98

75-00020 Aarding BV
(a subsidiary of Aarding Behect BV)
PO Box 29, NL-8070 AA Nunspeet
Industrieweg 61, NL-8071 CS Nunspeet
☎ (+31 3412) 537 44 E aarding@pi.net
⊠ 47412 aawat nl ⊖ (+31 3412) 513 22
♣ DFl 35,000,000 (1996) 100 (1997) 3,500 m2
Managing Director: G Versteeg
Operations Director: T Napel
Manager, Defence Division: N Kleiberg
Purchasing Manager: W Wetering
■ Ground support equipment for military airbases, test facilities for engine and aircraft, noise suppression equipment for engine test facilities ('hush houses'), hydraulic platforms, steel doors for shelters, winches, watertank trailers and trailers for general purposes. 98

75-00225 Akzo Nobel Coatings BV Business Unit
AD Aerospace Finishes vof
PO Box 3, NL-2170 BA Sassenheim
Rijksstraatweg 31, NL-2171 AJ Sassenheim
☎ (+31 71) 308 29 05 ⊠ 39136 aacsh nl
⊖ (+31 71) 308 20 56
Managing Director: K J Fleetwood
Commercial Manager: W Dÿkhuizen
Technical and Marketing Manager: J Gerrits
Account Co-ordinator: M van Pieterson
Controller: D Betti
■ Infra-red and solar heat reflecting polyurethane, acrylic, epox and conventional camouflage coating systems. Paint removers and cleaning solvents. Maskants for chemical milling. Process and fabrication coatings. 98

75-00510 Ancra CHS BV
(† Ancra International BV)
PO Box 33, NL-5280 AA Boxtel
26 Schouwenooy, NL-5281 RE Boxtel
☎ (+31 411) 67 86 75 ⊖ (+31 411) 68 39 95
♧ US$ 10,000,000 ♣ US$ 70,000,000 300
Managing Director: W J Kuipers
■ Loading bridges, lashing equipment for trucks and aircraft. Cargo tiedowns, nets, seat fittings, galley and panel fittings. Cargo systems and spare parts.
® USA: Ancra Corp 97

75-00790 Bose BV European HQ
(a subsidiary of Bose Corp, USA)
Nijverheidstraat 8, NL-1135 GE Edam
☎ (+31 299) 39 01 11 ⊖ (+31 299) 39 01 14
Managing Director, Europe: Anton Schalkamp
Military Business Development Manager:
Frank Van Heyghen E frank_van_heyphen@bose.com
Manager, Noise Reduction Technology Group:
Alex Rawling
■ Manufacture of active noise reduction headsets for flight and armoured vehicle applications. Supplied to US Air Force, US Army, US Navy, Kuwait, Saudi Arabia, Australian and European forces. → 90 98

75-00800 Bosman Powersource Support BV
Brielselaan 45, NL-3008 AG Rotterdam
☎ (+31 10) 423 98 24 ⊖ (+31 10) 484 15 64
👥 70 🏭 100,000 sq ft
Managing Director: H G A Overwater
Area Sales Manager: Johan T van Ammelrooy
Quality Manager: G J van Mourik
■ FAA/JAA approved repair station for aircraft and gas turbine engine components. Repair and manufacture of engine hot-section components used in aero engines and stationary gas turbines. Repair and manufacture of aircraft ducting, exhaust systems and general sheetmetal components. Processes: forming, machining, fusion and resistance welding, torch and vacuum brazing, heat treatments, balancing, shot peening and penetrant inspections. [98]

75-00850 Brandt Fijnmechanische Industrie BV
H J E Wenckebachweg 117, NL-1096 AM Amsterdam
☎ (+31 20) 668 12 81 ⊖ (+31 20) 692 83 54
👥 45
Managing Director: T H Dijk
Contact: Dr-Ing. M H Vlastra
■ Manufacture of precision components, tools and assemblies. Repair and overhaul of mechanical components.
® Germany: Scheer & Co GmbH [98]

75-01130 Cimcool Industrial Products BV
(a subsidiary of Cincinnati Milacron, USA)
PO Box 463, NL-3130 AL Vlaardingen
Schiedamsedijk 20, NL-3134 KK Vlaardingen
☎ (+31 10) 460 06 60 ⊠ 21581 ⊖ (+31 10) 460 32 40
Marketing Manager: H Oostveen
■ Metalworking fluids, rust preventives, tapping compounds.
® UK: Cincinnati Milacron
USA: Cincinnati Milacron [97]

75-01150 Claymount Assemblies BV
(† Clayberg International BV)
(a division of Clayrworld BV)
Marconistraat 25, NL-6942 PX Didam
☎ (+31 316) 52 31 24 [W] http://www.clayrmont.com
⊖ (+31 316) 33 18 59
♧ DFl 3,000,000 (1996) ♣ DFl 8,000,000 (1996)
👥 19 (1997) 🏭 500 m2
President: Robert Tierie
Vice President, Sales and Marketing: C Wen
■ HV interconnection systems and sophisticated HV connectors and cable assemblies for continuous performance and reliability in extreme industrial and military environments. [98]

75-01200 Combimac BV
PO Box 2038, NL-7801 CA Emmen
Kapt Grantstraat 5, NL-7821 AP Emmen
☎ (+31 591) 61 15 31 ⊖ (+31 591) 61 97 30
♣ DFl 9,000,000 (1996) 👥 50 (1996)
Sales Director: H J Robben
Engineering Director: M J T de Zwart
Purchasing Manager: J Smit
■ Design, development and production of shock resistant, low-magnetic, submersible and special electric motors, submersible and immersible pumps, axial and centrifugal fans in accordance with industrial and naval standards. [98]

75-01220 Crompton Instruments BV
PO Box 121, NL-2980 AC Ridderkerk
Noordenweg 79, NL-2984 AG Ridderkerk
☎ (+31 180) 43 20 33 ⊖ (+31 180) 42 56 40
👥 12
General Mgr: H Brandt [98]

75-01400 DAF Trucks NV
Hugo van der Goeslaan 1, PO Box 90065, NL-5600 PT Eindhoven
☎ (+31 40) 214 91 11 ⊠ 51085 daf nl
⊖ (+31 40) 214 43 25
♣ DFl 1,213,000,000 (1994) 👥 3,445
Chairman of the Management Board: Cornelis G Baan
Director: H Wagter
Director: E H M van den Assem
Communications: H Brian
■ Manufacture of medium and heavy trucks for commercial and military use, diesel engines, cabs and axles.

Communications Department
Hugo van der Goeslaan 1, NL-5643 TW Eindhoven
☎ (+31 40) 214 21 04 ⊖ (+31 40) 214 43 17
Communications Director: H Momber [97]

75-01500 Damen Shipyards
PO Box 1, NL-4200 AA Gorinchem
Avelingen West 20, NL-4202 MS Gorinchem
☎ (+31 183) 63 99 11 ⊠ 25335 dame nl
⊖ (+31 183) 63 21 89
♣ DFl 250,000,000 👥 300
Managing Director: Ing. P D H Kalis
☎ (+31 183) 63 94 64
Director of Special Craft: Ing. P H Noordenbos
☎ (+31 183) 63 94 76
Operations Director: J B de Heer
Purchasing Manager: J A de Kuipen
Advertising Manager: Ms D den Tek
☎ (+31 183) 63 94 68
■ Manufacture of patrol, combat and logistic support vessels in aluminium and steel. Other utility vessels: tugs, buoy tenders, landing craft, firefighting craft. [98]

75-01525 Datawell BV
Zomerluststraat 4, NL-2012 LM Haarlem
☎ (+31 23) 531 60 53 ⊖ (+31 23) 531 19 86
■ Wave height and direction buoys. Complete assemblies for the measurement, transmission, logging and analysis of wave data. [97]

75-01575
Delft Electronische Producten (DEP)
Delft Instruments
(a division of Delft Sensor Systems)
Dwazziewegen 2, PO Box 60, NL-9300 AB Roden
☎ (+31 50) 501 88 08 ⊠ 53678 dep nl
⊖ (+31 50) 501 35 10
♧ DFl 1,500,000 ♣ DFl 30,000,000 👥 100
🏭 4,500 sq m
Managing Director: Tjaart W Wiersum
Director Sales and Marketing: Anton D Boswijk
Defence Sales Manager: Henk Arends
Defence Sales Manager: Cor Boot
Industrial Sales Manager: Joost Adema
Industrial Sales Manager: Léon A Bosch
Industrial Sales Manager: Rob J Schomaker
Industrial Sales Executive: Eric Leemhuis
Director Operations: Jan van Spijker
Marketing Communications Manager: Mrs Meta Bartmann van-Dam
■ Image intensifier tubes; (position sensitive) photon counters; auroral images; intensified CCD's. [98]

75-01590 Delft Sensor Systems
(† Delft Instruments Electro-Optics BV)
Röntgenweg 1, PO Box 5083, NL-2600 GB Delft
☎ (+31 15) 269 80 50 ⊖ (+31 15) 269 80 98
♣ BFr 1,000,000,000
Managing Director: Freddy Versluys
■ Portable display systems; thermal observation equipment (handheld and vehicle mounted); night vision systems; fire control systems for armoured vehicles; space projects; advanced and holographic optics; customised research and development projects; equipment qualification. → 80 [97]

75-01650 Draka Kabel BV
(a subsidiary of Draka Holding)
PO Box 1013, NL-1000 BA Amsterdam
☎ (+31 20) 637 99 11 ✉ DRAKA ⊠ 11569 draka nl
⊖ (+31 20) 637 93 63
👥 575
■ Certified manufacturers of electric cables for power and instrumentation. Halogen free, low smoke, flame retardant and oil resistant marine and naval products. Wire and cords, power supply, data transmission, telecommunications, computer systems, instrumentation. Halogen free cables, fire retardant cables and cables for special applications. [98]

75-01730 DSM Head Office
Het Overloon 1, PO Box 6500, NL-6401 JH Heerlen
☎ (+31 45) 578 81 11 ⊖ (+31 45) 578 71 14
♣ DFl 9,822,000 (1995) 👥 16,990 (worldwide)
Chairman: S de Bree
Assistant Director of Defence Affairs: A Nijenhuis
■ Reinforced thermoset/thermoplastic composites, Dyneema fibre (helmets, vests, light weight armour, ropes and cables. Nyrim composites (logistic containers, run flat tyres, armour piercing nozzle protectors) high-grade engineering plastics, plastic packagings, safety shoes, catering.

High Performance Fibers European Office
Eisterweg 3, NL-6422 PN Heerlen
☎ (+31 45) 543 67 67 ⊖ (+31 45) 542 65 38
Marketing Manager: Christiane Thielens [97]

75-02050 Enkev BV
PO Box 3, NL-1130 AA Volendam
☎ (+31 299) 36 43 55 ⊖ (+31 299) 36 84 09
♧ DFl 250,000 (1997) ♣ DFl 8,000,000 (1996)
👥 30 (1997) 🏭 5,000 sq m
Managing Director: W F V Doesburg
Operations Director: J van der Hout
■ Packaging material made of rubberised hair (Hairlok), rubberised coir (Cocolok). Specially suitable for: ammunition and sensitive electronic equipment. [98]

75-02200 Eurometaal NV
Hemkade, PO Box 419, NL-1500 EK Zaandam
☎ (+31 75) 650 42 67 ⊖ (+31 75) 650 42 61
👥 500
President: J J Vonk
Sales and Marketing Director: Paul Kramer
Purchasing Manager: L Broek
■ Ammunition and pyrotechnics, ammunition components, propellant charges. Missile components, high precision aluminium and steel forgings and machined parts. Aircraft ammunition. [98]

75-02300 Fokker Aviation NV
(a member of the Stork Group, Netherlands)
PO Box 75047, NL-1117 ZN Schiphol
☎ (+31 20) 605 48 65 ⊖ (+31 20) 605 48 75
♣ US$ 500,000,000 👥 2,400
Director: C J Berends van Loenen
Secretary: Ms E C Face

Fokker Defence Marketing
PO Box 75047, NL-1117 ZN Schipol
☎ (+31 20) 605 48 50 ⊖ (+31 20) 605 48 75
Managing Director: Kier Vis
Manager, Government Programmes: J Kist
Manager, Industrial Products: J Oosterman
Manager, Customer Engineering: P van der Spek
Manager, Contract Negotiations: N Stageman [98]

75-02305 Fokker Aerostructures BV
(a member of Stork NV, Netherlands)
PO Box 1, NL-3350 AA Papendrecht
Industrieweg 4, NL-Papendrech 3351 LB
☎ (+31 78) 641 93 83 ⊖ (+31 78) 641 97 99
👥 700 (1997)
Managing Director: C A M de Koning
Sales and Marketing Director: D Koen
Purchasing Manager: R P Honig
Marketing Manager: P van Aken
■ Design, development manufacturing and support of airframes structures and structural components for civil and military aircraft and helicopters. Provides high-end specialised know-how and integrated solutions on state-of-the-art airframe technology to aircraft integrators and their major contractors. [98]

75-02320 Fokker Elmo BV
(a member of the Stork Group, Netherlands)
PO Box 75, NL-4630 AB Hoogerheide
Aviolandalaan 33, NL-4631 RP Hoogerheide
☎ (+31 164) 61 70 00 [E] fo-elmo@euronet.nl
⊠ 78472 fbe nl ⊖ (+31 164) 61 77 00
👥 485
President: Gerard A van Beck
Sales and Marketing Director: S Hands
Purchasing Manager: Peter Diks
■ Design, manufacture, installation, maintenance and repair of electrical and electronic systems. Main activities: power distribution equipment, wiring harnesses, electronic control and display systems for aircraft cockpits and electronics for defence applications. [98]

75-02325 Fokker Space & Systems
Newtonweg 1, PO Box 32070, NL-2303 DB Leiden
☎ (+31 71) 524 50 00 ⊜ (+31 71) 524 59 99
♣ DFl 220,000,000 (1996) 460 (1997) 1,500 sq m
President: P G Winters
Vice President, Marketing: B P van Ijsselstein
Finance Manager: P van der Waal
Public Relations: E van Drumpt ☎ (+31 71) 24 51 35
■ Engineering services at system and subsystem level. Production lines with a strong engineering content for space products and structures, missile control subsystems. Design of simulators for development and training. → 80 98

75-02330 Fokker Special Products BV
(a member of the Stork Group, Netherlands)
PO Box 59, NL-7900 AB Hoogeveen
Edisonstraat 1, NL-7903 AN Hoogeveen
☎ (+31 528) 28 50 00 42057 foh nl
⊜ (+31 528) 28 50 10
Managing Director: J A Blikman
Operations Director: S Resink
Sales and Marketing Director: E Noordhof
■ Design and manufacture of a wide range of industrial and defence products and systems. Manufacture of the self supporting mobile system (Air Mobile Hospital System). 98

75-02335 Franerex BV
(a daughter company of Eurometaal NV)
Heimolen 24, NL-4625 DB Bergen op Zoom
☎ (+31 164) 23 39 30 ⊜ (+31 164) 25 42 75
Director: H W Beur 98

75-02340 Genco Mobile Systems BV
(a subsidiary of Gemco Industries BV)
Science Park Eindhoven 5053, NL-5692 EB Son
☎ (+31 499) 46 25 25 W http://www.gms@gemco.nl
⊜ (+31 499) 46 25 00
♧ DFl 13,000,000 (1995) ♣ DFl 50,000,000 (1996) 140
Managing Director: J G V Germert
Manager: P J F C Veltman
Account Manager, Defence Projects: T F Wammes
Account Manager: Mr van Keep
■ Development, design and production of mobile units for on-site service such as vehicle bodies, shelter installations and trailers for maintenance and repair, medical care, rescue and firefighting, chemical ecological rescue, decontamination and cleaning, measuring and control, recovery and salvage, specific application. 98

75-03110 Holec Machines & Apparaten BV
Ringdijk 390, PO Box 4050, NL-2980 GB Ridderkerk
☎ (+31 180) 44 50 00 20377 hslk nl
⊜ (+31 180) 44 58 66
♧ DFl 161,000,000 (1995) ♣ DFl 235,000,000 (1995) 650 (1997) 75,000 sq ft
Managing Director: E C Yntema
Sales and Marketing Director: J G J M Thyssen
Purchasing Manager: J M Schram
Defence Business Manager: J Brink ☎ (+31 1804) 454 05
Advertising Manager: T van der Ploeg
■ Design and manufacture of DC/AC motors and generators; electrical drive systems, static and rotary converters and power supply systems; (diagnostic) maintenance, repair and overhaul of electrical machinery of all makes. 98

75-03112 Holec Power Protection bv
(† Holec Systemen & Components BV)
PO Box 4, NL-7550 GA Hengelo (OV)
☎ (+31 74) 246 28 50 E hppbv@pi.net
W http://www.holecups.com ⊜ (+31 74) 246 28 00
Director: H Koenderink
Marketing Communications: Ms Marion Weinberg
■ Manufacturers and suppliers of dynamic UPS systems. 98

75-03300 Hollandse Signaalapparaten BV
(a subsidiary of Thomson-CSF, France)
Zuidelijke Havenweg 40, PO Box 42,
NL-7550 GD Hengelo Ov
☎ (+31 74) 248 81 11 SIGNAAL Hengelo
44310 sign nl ⊜ (+31 74) 242 59 36
2,600
President: E A van Amerongen
Finance Director: F J Engel
Marketing Director: G J Jonge Poerink
Operations and Technical Director: D Arnold
Purchasing Manager: A P Blom
Commercial Manager: H R Boswijk
Communications Officer: I Edwards
Communications Manager: Ms Wilma Kampman
■ Systems based on in-house design and production of subsystems and components in the field of radar and weapon-control, optronics, sonar, displays, C3I, M3I and communications equipment for combat areas, naval ships and shore stations. Design and production of systems for: naval defence, military telecommunications, ground-based air defence (including ILS and programme management). 98

75-03350 Holmatro Industrial & Rescue Equipment
PO Box 33, NL-4940 AA Raamsdonksveer
Zalmweg 30, NL-4941 VX Raamsdonksveer
☎ (+31 162) 58 92 00 ⊜ (+31 162) 52 24 82
♧ DFl 175,000 (1994) ♣ DFl 30,000,000 (1994) 145
Managing Director: Ing J J Meyer
Advertising Manager: Ad Rombouts
Marketing Department: D Metz
■ Manufacture, sales and service worldwide of industrial and rescue equipment. 98

75-03470 Hydraudyne Systems & Engineering BV
(a subsidiary of Mannesmann Rexroth AG)
PO Box 32, NL-5280 AA Boxtel
Kruisbroeksestraat 1, NL-5281 RV Boxtel
☎ (+31 411) 65 14 51
50308 hydyn nl; 50825 hydro nl
⊜ (+31 411) 65 14 53
♣ DFl 220,000,000 (1996) 640 (1996)
Managing Director: B W Drenth
Sales and Marketing Director: J A C Wels
Public Relations Manager: Bart Dohmen
■ Hydraulic system design, development and production for naval vessels, submarines, shiplifts, armoured vehicles. Anti-magnetic cranes for minesweepers. Test trolleys and arresting gears for aircraft carriers. Hydraulic and electric motion systems. 98

75-03530 Hytech BV
Hyperbaric Technology BV
Brasem 43, NL-4941 SE Raamsdonksveer
☎ (+31 162) 52 22 02 ⊜ (+31 162) 51 90 69
♧ US$ 200,000 ♣ US$ 4,000,000 15
550 m2 (1995)
Managing Director: J Lagrouw
Operations Director: T Jager
■ Manufacture of military diving equipment. Portable, lightweight, self-contained non-magnetic 4-diver decompression chamber and support unit capable of operation at any inlet pressure from 5 to 300 bar. Airportable by helicopter. 97

75-03540 Indiana Rubber
PO Box 48, NL-4700 AA Roosendaal
☎ (+31 165) 59 51 00 ⊘ (+31 165) 59 51 30
⊜ (+31 165) 53 78 05
♣ DFl 14,000,000 70 12,000 m2
Managing Director: R C M van Ginhoven
Operations Director: J Schoemaker
Sales and Marketing Director: B Saly
Purchasing Manager: J Cremers
■ Production and sale of NBC gloves. 97

75-03550 Indivers NV/Interturbine
Arena Office Park, Marathon 9H, NL-1213 PE Hilversum
☎ (+31 35) 642 07 09 58677 ⊜ (+31 35) 642 06 09
■ Manufacture, overhaul and repair of gas turbines and compressors. Engine testing.
Subsidiary:
Elbar BV
Eldim BV
Interturbine Holland 98

75-03570 Interspiro NL
(a subsidiary of Interspiro AB, Sweden)
Asterweg 1, PO Box 37080, NL-1030 AB Amsterdam
☎ (+31 20) 632 29 11 ⊜ (+31 20) 634 01 21
8
Managing Dir: Paul Eissens
■ Breathing apparatus, breathing protection equipment designed for use by military forces, fire services, police & customs, merchant navies & industry worldwide. 98

75-03600 JDR Cable Systems BV
(† De Regt Special Cable BV)
(a subsidiary of BI Group plc, UK)
PO Box 49, NL-2900 AA Capelle aan den Ijssel
☎ (+31 10) 258 68 00 E nl@jdrcables.com
⊜ (+31 10) 442 35 41
♣ £ 7,000,000 60 (1997)
Managing Director: H van Veen
Technical Director: N O'Hear
Marketing Director: P J van Nimwegen
Purchasing Manager: Ms M van Helvoirt
Area Manager, Benelux: Ron Voerman
■ Design and manufacture of customer designed electro-mechanical cables for towed array systems, sonar systems and mine sweeping systems (influential, mechanical) and a wide range of R.O.V. tether cables, diving umbilicals, u/w instrumentation cables. High emphasis on dynamic and/or u/w applications using optical fibres and high tech fibre technologies. 98

75-03780 Kawasaki Heavy Industries (Europe) BV
7th Floor River Staete, Amsteldijk 166,
NL-1079 Amsterdam
☎ (+31 20) 644 68 69 ⊜ (+31 20) 642 57 25 98

75-03800 Kawasaki Motors NV
(a subsidiary of Kawasaki Heavy Industries Ltd, Japan)
Marconiweg 5 Industrie Park, De Biezen, Hagestein 98

75-03802 Kawasaki Motors NV
(a joint venture of Kawasaki Heavy Industries NV, Japan)
Marconiweg 5, NL-4131 Vianen ZH
☎ (+81 3473) 734 44 ⊜ (+81 3473) 764 58 98

75-03805 Kawasaki Motors Netherlands NV
(a subsidiary of Kawasaki Heavy Industries Ltd, Japan)
Diamantlaan 14, NL-2132 Hoofddorp
☎ (+31 23) 567 05 00 ⊜ (+31 23) 563 98 84 98

75-04000 Kooi BV
(an Allegheny Teledyne Co)
Molen 25 - 9077 SW, PO Box 25,
NL-9077 ZV Vrouwenparochie
☎ (+31 518) 40 17 41 ⊜ (+31 518) 40 10 24
♣ US$ 25,000,000 130 7,000 m2
Managing Director: M G de Leeuw
Sales and Marketing Director: G de Haan
Operations Director: W J Westerbeek
Purchasing Manager: P Bakker
■ Manufacture of Kooi-AAP track mounted fork-lift trucks, trucks for (un)loading trucks and handling of palletised goods. Multi-directional fork-lift trucks for distribution and handling missiles in the field. Specifications: lifting capacities from 1 to 4 tonnes, lift-heights up to 3,9 m and engines up to 70hp. 98

75-04200 Lips BV
PO Box 6, NL-5150 BB Drunen
☎ (+31 416) 38 81 15 E lips.drunen@tip.nl
W http://wwwl.tip.nl/~1305591 35185 lips nl
⊜ (+31 416) 37 31 62
♣ £ 200,000,000 (1996) 575 (1996)
Managing Director: Ir I I C Holthaus
Public Relations Manager: Dr A J Konning
Sales Promotions Manager: Annemiek Koning
■ Monobloc propellers, controllable pitch propellers, waterjets, transverse tunnel thrusters, azimuthing thrusters, integrated ship control systems, service and after sales. 98

75-04300 Loggers Rubbertechniek BV
(a part of RVH group of companies)
PO Box 194, NL-3300 AD Dordrecht
Mijlweg 83, NL-3316 BE Dordrecht
☎ (+31 78) 618 28 44 ⊜ (+31 78) 618 06 50
35 2,000 m2

Contact Reference

Managing Director: Stefan Vogelaar
☎ (+31 78) 618 28 44
Operations Director: J P Gitz
Sales and Marketing Director: Dirk F van der Mark
Office Manager: Yvette Schilder Spel
Purchasing Manager: R Tielman
■ Shock mounts, vibration and anti-vibration mounts, cable mounts, air suspension systems, flexible couplings, flexible pipe support systems. [98]

75-04350 MAFO Howden BV
(a subsidiary of Howden Group plc, UK)
Eekboerstraat 65, PO Box 316, NL-7575 AW Oldenzaal
☎ (+31 541) 57 10 00 ⊖ (+31 541) 52 01 45
55 (1997)
Managing Director: H J Lichtenbeld
Defence Business Manager: W M Denkers
Purchasing Manager: R F van der Bles
■ Water/gas tight naval doors and hatches, naval helicopter hangars and hangar doors. Fans for special applications, air-conditioning units, hatches and scuttles, GRP doors and hatches. [98]

75-04525 MIFA Aluminium BV
Deltakade 4-6, PO Box 3111, NL-5902 RC Venlo
☎ (+31 77) 389 88 88 ⊖ (+31 77) 389 89 89
Managing Director: J J Zitman
Sales and Marketing Director: H Pauwels
Operations and Technical Director: A Costermans
Purchasing Manager: J Hendrikx
■ Aluminium extruding, machining, finishing, casting. Precision extruding, CNC machining of extrusion profiles, castings and solids, anodizing, chemical treatments, coatings and mechanical pre-treatment, synergistic and industrial coatings such as Tufram and HCR. Precision casting to the lost-wax method assemblies. [97]

75-04600 Muiden Chemie International BV
(† Koninklijke Nederlandsche Springstoffenfabrieken NV)
Kruitpad 1398 CP, Postbus 1, NL-1398 ZG Muiden
☎ (+31 2942) 647 17 ⊖ (+31 2942) 646 87
80
Managing Dir: Ing. H. Schipper
Commercial Mgr: Mrs M. Paetz
■ All kinds of military propellant, hunting powder, basebleed grains & charges. [98]

75-04700 Nautikaris BV
Hoofdsrtraat 170, NL-2071 Santpoort-Noord
☎ (+31 23) 38 95 02 ⊖ (+31 23) 37 24 15 [98]

75-04800 Nevesbu
PO Box 16350, NL-2500 BJ The Hague [98]

75-05000 Niehuis & van den Berg BV
(a subsidiary of Damen Shipyards Group)
Ship Repair Yard, PO Box 5801, NL-3008 AV Rotterdam
Pastoriestraat 1, NL-3195 HT Pernis TR
☎ (+31 10) 438 11 00 ⊠ 28626 nvdb nl
⊖ (+31 10) 438 19 97
♣ DFl 40,000,000 (1996) 140 (1997)
Managing Director: F Roos
Commercial Director: J den Dunnen
Operations Director: A den Arend [98]

75-05100 NKI
Nederlandse Kunststof Industrie BV
(a subsidiary of Van der Giessen de Noord NV, Netherlands)
PO Box 222, NL-5100 AE Dongen
☎ (+31 1623) 172 00 ⊠ 54578 nki nl
⊖ (+31 1623) 156 57
♣ US$ 15,000,000 80
Managing Director: G E E Aerts
Sales and Marketing Director for Europe, Middle East and Africa: Jan C D M Teulings
Sales Manager, Special Projects: Hans E Scholtens
Sales Support Manager: Loran A H Janssens
■ Glass fibre re-inforced polyester naval craft. Advanced composite lightweight structures for applications in the aerospace and aviation industries. [98]

75-05200 NWM de Kruithoorn BV
(a subsidiary of Rheinmetall Industries GmbH, Germany)
Poeldonkweg, PO Box 1050,
NL-5200 BC 's-Hertogenbosch
Poeldonkweg 5, NL-5216 JX 's-Hertogenbosch
☎ (+31 73) 681 02 22 ⊖ (+31 73) 681 03 10
♧ DFl 15,000,000 ♣ DFl 50,000,000 200
Managing Dir: F. Birkenkamp
Sales & Marketing Mgr: Steef van den Berg
Quality Mgr: C. Taal
■ Manufacturer of break-up ammunition, developed in calibres 20mm x 139mm; 25mm x 137mm; 25mm x 173mm; 35mm x 228mm; an 40mm L-60/L-70. Production of heavy metal penetrators in calibre range 20mm to 120mm related ammunition in calibre range 20mm and to 30mm, such as 25mm APDS-T and 30mm MPDS. Frangible APDS in 20-35mm medium calibre. Ammunition 20mm x 139mm for automatic cannon Rheinmetall RH202, HS820, F2 and Mauser Mk20. Ammunition 25mm x 137mm for automatic cannon Oerlikon KBA, McDonnell Douglas M242 Bushmaster anmd Mauser Mk25. Aircraft ammunition 20mm x 102mm for M61 and M39 cannon. Cartridge cases and synthetic ammunition components.
Member of Rheinmetall Defence Technology Group. [97]

75-05255 Oceanographic Co Netherlands
PO Box 7429, NL-2701 AK Zoetermeer
☎ (+31 79) 342 83 16 ⊖ (+31 79) 341 50 84
Managing Director: F A van Dongen
E f.vandongen@ocn.nl
■ Acoustic current and sediment profilers. High frequency radar current measuring systems in real time. [98]

75-05500 Philips Machinefabrieken Nederland BV
(a subsidiary of Philips Electronics NV)
(Machinefabriek Acht), PO Box 218,
NL-5600 MD Eindhoven
☎ (+31 40) 76 46 90 ⊖ (+31 40) 76 69 54
♧ US$ 30,000,000 ♣ US$ 70,000,000 (1995) 500
Managing Director: Albert Daniëls
Aviation Programs Manager: Ton Staassen
■ Manufacturer of high precision and safety critical parts and assemblies for space and aeronautical equipment. Specialise in sheet metal components for aircraft engines and high speed machining of aluminium aircraft components.
Quality assurance to ISO 9002, Mil-Q-9858A and Mil-I-45208A. [97]

75-05750 Radiall Nederland BV
(a subsidiary of Radiall, France)
Hogebrinkerweg 12, PO Box 64, NL-3870 CB Hoevelaken
☎ (+31 33) 253 40 09 E info@radiall.nl
W http://www.radiall.nl ⊖ (+31 33) 253 45 12
26 (1997) 800 m2
Managing Director: Peter Wigleven
■ Connectors and subsystems for electronic devices, microwave passive products, fibre optic connection systems.
® France: Radiall [98]

75-05770 RDM Submarines BV
(† Rotterdamsche Droogdok Maatschappij BV)
PO Box 913, NL-3000 AX Rotterdam
☎ (+31 10) 487 91 11 ⊠ 20753 rdm nl
⊖ (+31 19) 487 25 00
Managing Director: Ir T Hiltrop
General Manager, Marketing: Dr Ir C A Prins
Sales Manager: G B Spoorenberg
■ Design, construction, refit and maintenance of submarines. Atmosphere Independent Propulsion (Closed Cycle Diesel). [97]

75-05780 RDM Technology BV
(a subsidiary of RDM Technology Holding)
PO Box 1039, NL-3000 BA Rotterdam
21 Heijplaatstraat, NL-3089 JB Rotterdam
☎ (+31 10) 487 27 53 ⊖ (+31 10) 487 22 99
♣ US$ 50,000,000 (1996) 300 (1996) 40,000 m2
Managing Director: W J T H Luijten
General Manager, Sales and Marketing: T Wijker
Sales Manager, Defence and Engineering: J M C Olierook
Sales Manager, Grids and Bore Cleaning Tool: D M Dröge
■ Modification programs for M114/39 155 mm, M101/33 105 mm and M109L47 155 mm howitzers. Modification program of Leopard 1 and 2 Main Battle Tank. Modification, overhaul and maintenance of M109/90 155 mm howitzer. AMX-13 family modernization program.
Flightdeck landing grid systems for shipborne helicopter operations. Secure helicopter restraints. Bore cleaning tool for maintenance of rifled and smooth barrels of both artillery and battle tanks. [98]

75-05790 Rockwell Automation BV
(a subsidiary of Rockwell International, USA)
Amsterdamseweg 15, NL-1422 AC Uithoorn
☎ (+31 2975) 54 35 00 ⊖ (+31 2975) 56 07 01
Country Manager, Sales: Henk Wildeman
® USA: Rockwell International Corp. [98]

75-05800 Rockwell-Collins (UK) Limited Netherlands Branch
(a subsidiary of Rockwell Collins Inc, USA)
PO Box 3074, NL-2130 KB Hoofddorp
Antareslaan 43, NL-2132 JE Hoofddorp
☎ (+31 23) 569 80 10 ⊖ (+31 23) 565 27 84
12 (1997)
Director of EUMEA Operations and Business Development: J H Timmers
Director of Business Development: Otto Blecker
Director for Europe and Africa: Thierry Tosi
Director of Marketing for Europe, Middle East and Africa: H A Ekkelenkamp
Director of Marketing for Europe: T Schiebergen
Director of Marketing for Europe: Ralph S Luijt
Customer Service Engineer: Simon G Stancliffe
■ Avionics and telecommunications systems.
® USA: Rockwell International Inc [98]

75-05850 ROR Rockwell B V Heavy Vehicle Systems
(a subsidiary of Rockwell International Corp, USA)
PO Box 255, NL-5700 AG Helmond
Middendijk 40, NL-5705 CC Helmond
☎ (+31 492) 53 58 05 ⊖ (+31 492) 54 71 75
Managing Director: Andre van Stokkum
Director: Henk van Ettinger [98]

75-06100 Royal Nooteboom Trailers BV
Nieuweweg 190, PO Box 155, NL-6600 AD Wijchen
☎ (+31 24) 648 88 64 ⊖ (+31 24) 648 88 11
300 (1997) 22,000 m2
Managing Director: D Nooteboom
Sales and Marketing Director: C Reuver
Advertising and Public Relations Manager: J van de Water
■ Manufacture of platform trailers, lowbed trailers, semi-lowbed trailers, telescopic flatbed trailers and semi-lowbed drawbar trailers. [98]

75-06150 Royal Schelde Koninklijke Schelde Groep BV
Glacisstraat 165, PO Box 16, NL-4380 AA Vlissingen
☎ (+31 118) 48 28 81 ⊠ 37815 kms nl
⊖ (+31 118) 48 50 10
♣ DFl 750,000,000 2,650
Chairman: R C van den Heuvel
Shipbuilding Director: P E Denis
Manager, Sales and Marketing for Naval Vessels: R Post
■ Design and construction of naval surface vessels such as frigates, corvettes, patrol vessels, support ships, surface effect ships as well as main propulsion gearing. Integrated logistic support. [98]

75-06155 Rubber Industrie Soest NV
PO Box 360, NL-8200 AJ Lelystad
☎ (+31 320) 22 61 15 ⊖ (+31 320) 24 91 70
40 (1997) 11,200 m2
Managing Director: H Rietdijk
Commercial Manager: O C Verbeek
■ Design and supply all types of rubber fenders, moulded articles for ships, locks parts without or bonded to metals and offshore installations. [98]

75-06160 Saft Nife BV
Zijlweg 156, PO Box 3176, NL-2001 DD Haarlem
☎ (+31 235) 15 08 00 ⊖ (+31 235) 32 99 97
♣ DFl 30,000,000 75
Managing Director: Gérard Mathis [98]

75-06180 SAGEM Nederland BV
(a subsidiary of SAGEM SA, France)
Postbus 195, NL-56000 AD Eindhoven
☎ (+31 40) 265 36 24; 265 36 25
⊖ (+31 40) 265 35 44
♣ DFI 1,000,000 (1996) 👥 3 (1996)
Chairman: D Dupuy
Operations Director: Mr Douville
■ Industrial assembly of drone systems. [98]

75-06200 Saval BV
(a subsidiary of Internatio-Müller NV)
Westrik 17, PO Box 100, NL-4800 AC Prinsenbeek
☎ (+31 76) 548 70 00 [E] systems@saval.nl ⊠ 74118
⊖ (+31 76) 541 79 22
👥 200 (1997)
Managing Director: J F Burns
Sales Manager, Portables: V Kimpe
Sales Manager, Systems: J van Straaten
Marketing Services Manager: E Schmitz
■ Portable fire fighting equipment, fire fighting and detection systems for airports, refineries, industrial plants; offshore and marine protection. [98]

75-06280 Schat-Harding BV Head Office
(† Schat-Davit Company BV)
(a division of Umoe Schat-Harding, Norway)
PO Box 9032, NL-3506 GA Utrecht
☎ (+31 30) 264 42 00 ⊠ 47261 davit nl
⊖ (+31 30) 264 42 99
Director: Klaus Flores
Sales and Marketing Manager: J V Drunen
■ Survival and deck equipment, davits and cranes. [98]

75-06300 Scheepswerf & Machinefabriek Akerboom BV
PO Box 260, NL-2300 AG Leiden
☎ (+31 71) 576 16 00 ⊖ (+31 71) 531 29 04
👥 40 (1997)
Managing Director: H A Akerboom
Technical Director: J C Pieterse
■ Custom building and repair of marine vessels. [98]

75-06350 Signaal Communications
(a subsidiary of Thomson-CSF, France)
PO Box 88, NL-1270 AB Huizen
☎ (+31 35) 524 82 48 ⊖ (+31 35) 524 82 42
Managing Director: Adri Blokhuis
Sales and Marketing Director: Rob Lakenman
■ Mobile radio communications, tactical networks and naval communications. [98]

75-06400 Signaal Special Products
(a subsidiary of Hollandse Signaalapparaten Bv)
PO Box 241, NL-2700 AE Zoetermeer
☎ (+31 79) 344 59 99 ⊖ (+31 79) 344 59 38
👥 125 🏭 2,994 m2
General Manager: W E Reesink ☎ (+31 79) 344 59 90
Commercial Manager: G Seppenwoolde
☎ (+31 79) 344 59 92
Commercial Manager: C L de Rek ☎ (+31 79) 344 59 94
■ Design and manufacture of transponder and data link systems; power supplies and inductive components; protective equipment; optical data storage equipment. [98]

75-06500 Signaal USFA
(a subsidiary of Hollandse Signaalapparaten BV)
PO Box 6034, NL-5600 HA Eindhoven
Meerenakkerweg 1, NL-5652 AR Eindhoven
☎ (+31 40) 250 36 03 [E] zewa@usfa.nl
[W] http://www.usfa.nl ⊠ 51732 usfae nl
⊖ (+31 40) 250 37 77
♣ DFI 40,000,000 (1996) 👥 110 (1997) 🏭 10,000 m2
General Manager, Business Development: B Doorenbos
☎ (+31 40) 250 36 60
Finance Manager: H Bosman ☎ (+31 40) 250 36 79
Business Manager, Group Electro-Optics: M Wijnheymer
☎ (+31 40) 250 37 76
Business Manager, Group Cryogenics: D Verbeek
☎ (+31 40) 250 36 64
Business Manager, Group Fuzes: M van Oostrom
☎ (+31 40) 250 36 74
Business Manager, Group Batteries: J Theuns
☎ (+31 40) 250 37 04
Purchasing Manager: Mr Kortlever ☎ (+31 40) 250 36 80
Office Manager: Ms S Bosman
■ Electro-optical and electronic defence systems. Infra-red cameras for surveillance, aiming tracking and observation, CCD-TV cameras for naval /land fire control systems, mission recording cockpit cameras. Proximity fuzes for various applications, stirling-cycle cryogenic coolers, special purpose lithium batteries. [98]

75-06600 Simagine Simulation Systems
PO Box 16068, NL-2500 BB The Hague
Prinses Beatrixlaan 26, NL-2595 AL The Hague
☎ (+31 70) 333 24 54
[E] simulation.systems@gw.siemens.nl
⊖ (+31 70) 333 25 94
Managing Director: P J L van Lieshout
Marketing Contact: Y Kaufman
Marketing: P Bos
■ Design, development and production of training and simulation systems. [98]

75-06850 Solvay Duphar BV
(a part of the Sovay Group, Belgium)
Injectables Division
Veerweg 12, NL-8121 AA Olst
☎ (+31 570) 56 82 04 ⊖ (+31 570) 56 84 71
♣ DFI 900,000,000 👥 3,000
Technical and Marketing Manager: Adriaan D Drewes
■ Internationally operating pharma/chemical company, developing, manufacturing, marketing products for human and animal health. Medical devices specialised on production and marketing of prefilled disposable syringes and automatic injectors particularly suitable for military use. [98]

75-06860
SP aerospace and vehicle systems bv
(† DAF Special Products BV)
PO Box 436, NL-5660 AK Geldrop
Eindhovensweg 120, NL-5660 AK Geldrop
☎ (+31 40) 280 91 91 ⊖ (+31 40) 280 91 23
♣ DFI 100,000,000 👥 250
Managing Director: H Schaeken
Managing Director: J P van Leeuwen
Sales Manager: H Brouwer
Purchasing Manager: H Snyders
■ Defence systems: armoured personnel carriers, drive systems, superstructures for trucks. Aerospace systems: landing gear, helicopter transmissions, structural components. Design and manufacture of aircraft landing gears. Helicopter and aircraft components. [98]

75-07000 Standard Aero V.o.F. (SAV)
(† TOPPS (Turbine Overhaul & Power Plant Support) V.o.F.)
(a division of Standard Aero, Canada)
PO Box 4197, NL-5004 JD Tilburg
☎ (+31 13) 572 85 72 [E] pw100@standardaero.com
[W] http://www.standardaero.com ⊖ (+31 13) 572 85 00
♣ US$ 40,000,000 (1996) 👥 110 🏭 8,000 sq m
Managing Director: T Botha
Operations Director: A Tuck
Sales and Marketing Manager: A C M van der Made
Purchasing Manager: J Kraak
■ Manufacture, repair, overhaul and testing of gas turbines. [98]

75-07200 Stork Bronswerk BV
(a member of the Stork Group)
PO Box 494, NL-3800 AL Amersfoort
☎ (+31 33) 467 83 00 ⊠ BRONSWERK ⊠ 79039 broa nl
⊖ (+31 33) 463 71 61
Managing Director: G A M de Hart
Operations Director: A M Oude Engberink
Communications Manager: E J Peet
Purchasing Manager: M H van der Hoorn
■ Design, engineering, construction and installation of air conditioning, ventilation and refrigeration systems. Products include ground cooling units for aircraft electronics, explosion proof heating and purge units for aircraft. [98]

75-07220 Stork RMO BV
(a subsidiary of Stork NV, Netherlands)
PO Box 1250, NL-1000 BG Amsterdam
☎ (+31 20) 523 37 00 ⊠ 11102 ⊖ (+31 20) 622 06 17
♧ DFI 2,400,000 (1996) ♣ DFI 4,900,000,000 (1996)
👥 300 (1996) 🏭 12,000 m2
Managing Director: J A Pijnappels
■ Research, design, component manufacture and testing of gas turbines and compressors. Design and manufacture of mechanical, special test and simulation equipment for heavy guns. [98]

75-07300 Stork-Wärtsilä Diesel BV Special Products Department (SWD)
(a member of the Wärtsilä Diesel Group)
Assendorperlure 5, PO Box 10608, NL-8000 GB Zwolle
☎ (+31 38) 425 32 53 ⊠ 42116 swdz nl
⊖ (+31 38) 425 33 25; 422 35 64
👥 975
Managing Director: M J Wieringa
General Manager, Special Products: S Miedema
Manager, Marine Sales: T Huisman
Sales Promotion: Ad Smulder
■ Heavy duty, medium speed, four stroke diesel engines ranging from 515 up to 16,290 kW for propulsion, auxiliary and land-based power.

Amsterdam Plant
Czaar Peterstraat 213, PO Box 4196, NL-1009 AD Amsterdam
☎ (+31 20) 520 39 11 ⊠ 14395 swdi nl
⊖ (+31 20) 626 92 14
Projects Manager: N Nörgard
■ Diesel engines. Applications include main propulsion systems for ocean-going vessels, coastal vessels, inland waterway craft, auxiliary power supply aboard craft and land based power generation. Also specialises in system engineering, designing and supplying complete propulsion systems for wide range of vessels including resiliently mounted and shock resistant, non-magnetic applications. Re-engining, supply of complete diesel electric power stations on a turnkey basis and installation of power units for special applications. [97]

75-07350 Sun Electric Systems BV
(a subsidiary of Snap-On Tools Corp, USA)
Nijverheidslaan 17, NL-1382 LG Weesp
☎ (+31 294) 46 55 00 ⊖ (+31 294) 41 50 33
👥 75 (1997)
Managing Director: Ir M Molenaar
Sales and Marketing Director: Dr Jan H Landsaat
■ Aircraft hydraulics testers, hydraulics condition testers, ground power units, special tailor-made hydraulic projects. [98]

75-07400 Techno Fysica Croon Kombinatie
Centre for Measurement Techniques
PO Box 351, NL-2990 AJ Barendrecht
Zuideinde 80, NL-2991 Barendrecht
☎ (+31 180) 62 02 11 ⊜ (+31 180) 62 07 05
20
Managing Director: R W P Leitermarkt
Operations and Technical Director: P Kloppenburg
Sales and Marketing Director: H van Tienhouen
Purchasing Manager: R R Jansen
■ Consultancy, instrumentation, testing and manufacture relating to gas turbine technology. [97]

75-07450 Telpro plus BV
(a daughter company of AEG Eurolec Holding BV)
PO Box 9433, NL-1006 AK Amsterdam
Aletta Jacoslaan 9, NL-1066 BP Amsterdam
☎ (+31 20) 487 87 87 ⊜ (+31 20) 487 87 88
Managing Director: R P Bruggeman
■ Design and manufacture of reconnaissance and surveillance products, command, control and communications, electronic warfare, firepower and fire control, undersea warfare, airborne naval and ground equipment, radar, ground-to-air communication, VLF, HF, UHF, SHF communication, shelters with electronic equipment as well as consoles, panels, power supplies and printed circuit boards.
Representing:
Daimler-Benz Aerospace
ESW Marine and Special Technology
Hagenuk Marine, Radar and Radio Technology
Zeiss Eltro Optronic
Park Air
Denro Air Traffic [98]

75-07500 Ten Cate Protect BV
(a subsidiary of Royal Ten Cate)
Campbellweg 30, PO Box 186, NL-7440 AD Nyverdal
☎ (+31 548) 63 36 23 ⊜ (+31 548) 63 32 56
Managing Director: Ir I de Vrics
Manager of Marketing Support: Mrs Anneke Bergsma
[E] a.bergsma@tip.nl
■ Manufacture of textiles for work and career wear, safety and protection, healthcare and hygiene. Textiles in general. [98]

75-07600 Terberg Benschop BV
PO Box 2, NL-3405 ZG Benschop
☎ (+31 34) 245 92 11 ☒ 40090 ⊜ (+31 34) 245 19 34
♣ DFl 65,000,000 120 20 000 m2
President: G Terberg
Director: G Juberg (Jr)
■ Manufacture of special vehicles including cross-country trucks, vehicles for special applications (4x4, 6x6, 8x8, 10x8), wheeltrains and trailers for special equipment. Revision and overhaul of components. [97]

75-07670 Thomassen International BV
PO Box 70, NL-6990 AB Rheden
☎ (+31 26) 497 51 51 [E] thomassen@thomassen-int.nl
⊜ (+31 26) 495 41 25
♧ DFl 83,800,000 (1996) ♣ DFl 567,000,000 (1996)
800 (1996)
Public Relations Manager: J C Nijland
■ Manufacture, overhaul and repair of gas turbines and compressors. Engine testing. GE overhaul shop, testing of GE LM2500 and LM1600 engines. Gas and steam turbines. Turnkey delivery of powerplants in the 25-250 MW range. [98]

75-07818 Van Beest BV
PO Box 57, NL-3360 AB Sliedrecht
Industrieweg 6, NL-3361 HJ Sliedrecht
☎ (+31 184) 41 33 00 ⊜ (+31 184) 41 49 59
Managing Director: Cees Boer
Commercial Director: Jaap Kleibergen
■ Manufacture and supply of wire rope and chain fittings. → 85 [97]

75-07830 van Buuren-van Swaay BV Marine airconditioning
(a division of Internatio-Müller NV)
Bredewater 24, PO Box 220, NL-2700 AE Zoetermeer
☎ (+31 79) 371 07 10 ☒ 31736 vbvs nl
⊜ (+31 79) 351 04 34
12 1,600 sq m
Managing Director: Ir P L Broekhuijsen
☎ (+31 79) 371 07 50
Project Manager, Navy: G C Annokkee
☎ (+31 79) 371 06 12
Marine Air Conditioning Manager: M A M van Holsteyn
☎ (+31 79) 371 06 84
■ Design, manufacture and installation of CO2-air purification, NBC air filtration, air conditioning, ventilation and refrigerating equipment. [97]

75-07850 Van der Giessen-de Noord Marinebouw BV (GNM)
(a subs. of Van der Giessem de Noord NV)
Schaardijk 23, NL-2921 LG Krimpen Aandan Ijssel
☎ (+31 180) 59 12 00 ⊜ (+31 180) 51 34 64
Managing Dir: J.C.C. Smit
Operations Dir: D. Schonenberg
■ Design, construction, conversion & maintenance of small naval vessels. [97]

75-07870 Van Halteren Metaal BV
PO Box 48, NL-3750 GA Bunschoten
Marconiweg 2, NL-3752 LS Bunschoten
☎ (+31 33) 298 31 24 ☒ 79524 hamet nl
⊜ (+31 33) 298 36 74
Managing Director: I Veerman
Operations Manager: Ed Moorlander
■ Road wheels and suspension parts for M113 and other armoured vehicles. Loading, aiming and firing (ALF) simulator for M109 crew training. Ammunition registration systems and closed loop artillery simulation systems. Numerical control machining. [98]

75-07890 Van Riemsdijk Rotterdam
(a subsidiary of Driessen Aircraft Interiors)
Stolwijkstraat 57, NL-3079 Rotterdam
☎ (+31 10) 479 58 11 ☒ 26401 ⊜ (+31 10) 479 54 78
50 5,000 m2
Managing Dir: Ron E. van Riemsdijk
Sales and Marketing Director: Jaap Marks
Procurement Manager: Rund van Gent
■ Design and manufacture of air cargo restraint equipment. Containers to special sizes. Roller beds in aircraft. [97]

75-07900 Van Rietschoten & Houwens
R&H Systems BV
(a subsidiary of Internatio-Müller NV)
PO Box 5054, NL-3008 AB Rotterdam
Sluisjesdijk 155, NL-3087 AG Rotterdam
☎ (+31 10) 487 19 11 ⊜ (+31 10) 487 17 02
♣ DFl 250,000,000 400 30,000 m2
Managing Director: J T Tromp
Financial Manager: H J J Kooijmans
Public Relations Manager: M J C Griekspoor
■ Main activities include data acquisition, data processing, control and monitoring, automation, communication, training and simulation for onshore and ship-borne installations for military, industrial, utility, offshore, maritime, hydrographic and oceanographic applications. Adaptive roll stabilising autopilots (ARSA), degaussing systems, automated ship-control centres, electronic chart display and information systems (ECDIS), hydrographic data collecting and processing systems, integrated monitoring and control systems (IMCS), steering control systems. SCC (Ship Control Centre) simulator, platform trainers for submarines, damage control trainers, hardware trainers. [97]

75-08100 Verolme Elektra BV
(† Electron Fabrikage BV)
Elektraweg 5, PO Box 33, NL-3140 AA Maassluis
☎ (+31 10) 591 30 33
[E] verolme_elektra@compuserve.com
⊜ (+31 10) 591 91 57
♧ DFl 100,000 (1997) ♣ DFl 7,500,000 (1997)
30 (1997) 10,000 sq ft
Managing Director: A R Koster ☎ (+31 10) 591 30 33
Accounts Manager: J M de Bruin ☎ (+31 10) 593 07 29
Technical and Purchasing Manager: J Verhey
☎ (+31 10) 593 07 68
Marketing and Business Development Manager:
A M C Schram de Jong ☎ (+31 10) 593 07 68
■ Design, engineering and manufacture of cables harnesses, distribution boxes, electrical systems for vehicles, lighting systems, multi-purpose test stands for jet engine and accessories such as electronic pneumatic power generation, fuel, lube oil and hydraulic systems. Ground support equipment. [98]

75-09100 White Products BV
Industrieweg 17, PO Box 31, NL-4330 AA Oudenbosch
☎ (+31 165) 33 66 00 ⊜ (+31 165) 33 66 02
♣ G 60,000,000 200 25,000 m2
Managing Director: B Smit
Operations and Technical Director: E de Wit
Purchasing Manager: K Roovers
Manager of Sales Support and Marketing: O Key
Advertising and Marketing: O Keij
■ Harnesses, battery and starter cables, braided straps and cable systems.
Injection moulding, coax connectors, harmonised and specialised cables. Rubber mouldings, electronic assemblies. Production and repair of military harnesses and cable assemblies.
Associated companies:
USB (Germany)
Cableries de Valenciennes (France)
White Products Ltd (UK)
Cablesies de Valenciennes (France) [98]

80 Service Companies & Organizations

Delft Sensor Systems . . . 75-01590
Fokker Space & Systems . . . 75-02325
Fugro NV . . . 70-01000
Information Resources Ltd . . . 90-01250

80-00050 AAR Allen Group International
AAR Allen Airmotive
(a subs. of AAR Corp., USA)
PO Box 52 ND, NL-2132 Hoofddorp
☎ (+31 20) 655 22 22 ☒ 844 74539
⊜ (+31 20) 653 47 77
180 70 000 sq ft
General Mgr: Jim Clark
VP, Military Programs & Business Development:
Joost Geluk
■ Repair, test & overhaul of aircraft components: avionics, instruments, gyros, electrical systems, pneumatics, hydraulics, jet fuel starters, augment. fuel pumps, convergent engine nozzle controls, accessory drive gearboxes. Design & manufacturer of universal test stands for A/C components:: ADG, CSD, generators, pneumatics, linear actuators, hydraulics, fuel flow transmitters. → 75
Subsidiary companies:
AAR Aviation Services Ltd, UK
AAR Allen Airmotive International Italy SpA
AAR International Inc. Asia-Pacific [97]

80-01100 AlliedSignal Aerospace Marketing Sales and Service
(a subsidiary of AlliedSignal Aerospace, USA)
Dianalaan 7, NL-1183 CX Amstelveen
☎ (+31 20) 640 44 98 ⊜ (+31 20) 640 41 60
Field Service Engineer: Geoffrey Coombe
→ 90
● c/o Fokker Aircraft, Room SO 30-11, BV Schihddyk 231, AB Schiphol
☎ (+31 20) 605 29 79 ⊜ (+31 20) 648 42 87
Field Service Engineer: Rene Veerman
● c/o KLM (Royal Dutch Airlines), PO Box 7700 SPL/TI, NL-1117 ZL Schiphol Airport East
Building 425, Room 046-047, Schiphol
☎ (+31 20) 649 47 13 ☒ K77966
⊜ (+31 20) 648 46 55; 648 42 87 [98]

80-01250 AMC Consultancy
PO Box 97824, NL-2509 GE Den Haag
☎ (+31 70) 652 71 79 40 [E] amccom@worldaccess.nl
⊜ (+31 70) 514 05 48
35 m2
Managing Director: A M C Schram de Jong
■ Assists foreign companies on the Dutch market in particular, with projects for the Ministry of Defence and similar (technical) projects that have been announced in public publications. [98]

80-03500 IBM Nederland NV Education & Training
(† CGI Systems BV)
Winthontlaan 1, PO Box 8303, NL-3503 RH Utrecht
☎ (+31 20) 513 29 40 ⊜ (+31 20) 513 25 85

☗ 70
Managing Dir: Harry Peeters
Sales & Marketing Dir: Harry Julsing
Operations & Technical Dir: Rein Buyze
■ Provides training & education on the Yourdon structured method, OMT-2, Booch, unified-object oriented soft/hardware systems engineering methods applied in embedded software engineering in aviation, telecommunications, simulation, defence & industry. Open & on-site customized training & consulting: structured, object oriented analysis & design for real-time systems, project management, covering Europe, Middle East & Africa.
Training courses offered: Yourdon structured analysis & design for real-time systems; object-oriented analysis design using OMT-2 & Booch; object-oriented analysis & design; object-oriented project management; distributed systems design; object-oriented database concept & implementation. 97

80-04000 Chromalloy Holland
(† Turbine Support Europa BV)
(a subsidiary of Chromalloy Gas Turbine Corp, USA)
Siriusstraat 55, NL-5015 BT Tilburg
☎ (+31 13) 532 84 00 ⊠ 52559 tse nl
⊖ (+31 13) 543 28 33
☗ 400 (1997) ⌂ 15,000 sq m
Managing Director: Ad Appeldoorn
Director of Quality and Customer Support: Adri van Ierland
Operations Manager: Ad Kusters
Operations Director: Benny Appeldoorn
Purchasing Manager: Loek Kuppens
■ Repair, overhaul and modifications of F100, J85 and J79. Turbine engine components such as combustors, turbine blades, nozzles, discs, casings, honeycomb seals and fuel nozzles. 98

80-05500 Eldim BV
(a subsidiary of the Interturbine Group)
Spiktweg 24, PO Box 4341, NL-5944 ZG Arcen
☎ (+31 77) 473 19 19 ⊠ 58184 ⊖ (+31 77) 473 24 85
☗ 250 (1997) ⌂ 10,000 sq ft
Managing Director: Dr G de Vlieger
Business Development Director: John C Ellis
Operations Manager: Jos Bakermans
Sales and Marketing Manager: Jan Rietdijk
Financial Manager: Han Bloemendal
Purchasing Manager: Cor Coopmans
■ Gas turbine blade and vane cooling specialists. EDM-ECM-ECD-Laser processes. Seal manufacturer.
® Netherlands: Indivers NV 98

80-06600 Fokker Services BV
(a member of the Stork Group, Netherlands)
PO Box 3, NL-4630 AA Hoogerheide
☎ (+31 164) 61 80 00; 61 82 07 ⊠ 78063 fow nl
⊖ (+31 164) 61 40 73; 61 86 66
☗ 1,050 (1997)
President: G L M Hamers
Sales and Marketing Director: Erik Goedhart
Operations Director: J A M de Bruyn
Purchasing Manager: W Wesselius
Public Relations Department: Kristine Touw
■ Maintenance, modification, conversion and repair services. Handling of Fokker 27, F28, F50, F60, F70, F100 and Boeing B737. Modification and upgrading of F-16 and Northrop F-5. Also helicopter modification shop. 98

80-07700 Goliath & Co BV
(a subsidiary of Goliath Nederland BV)
Lange Voorhout 72, NL-2514 EH The Hague
☎ (+31 70) 365 49 38 ⊖ (+31 70) 365 70 52
♧ DFI 100,000 (1997) ☗ 7 (1997) ⌂ 200 m2
Managing Director: Ihr J F Hora Siccama
Operations Manager: J B J Walthaus
■ Marketing consultants for military products. 98

80-07900 Herkenning
PO Box 20703, NL-2500 ES The Hague
☎ (+31 70) 349 28 00 ⊖ (+31 70) 349 27 29
Editor-in-Chief: 1st Lieut R W M van Vuwren
Editor "Aircraftrecognition": R Choufoer
Editor "Aircraftrecognition": SM P H de Bock
Editor "Navalrecognition": SM J A F Stassen
Editor "Tankidentification": A Jonker
Editor "Tankidentification": Lieut C A M Meulenbroeks
■ Bi-monthly official magazine of armed forces for aircraft, vessel and tank recognition. 98

80-10000 Jalex BV
(† Technical Consultancy Jalex BV)
20 Luchthavenweg, NL-1786 PP Den Helder
☎ (+31 223) 63 47 99 E skyline.aviation@tip.nl
⊖ (+31 223) 66 02 74
♧ DFI 4,000,000 (1996) ♣ DFI 2,500,000 (1996)
☗ 15 (1997) ⌂ 3,000 m2
Managing Director: Alex van Hoevell
Purchasing Manager: Ruud van Leerdam
Consultant: N Tak
■ Consulting engineers for aviation services, operation and maintenance. Airfield lighting, avionics, EW equipment, thermal imaging, flight safety.
® Belgium: Abelag Aviation SA
UK: Airworthy Ltd, GEC Avionics Ltd, GEC Sensors Ltd, Helmets Ltd 98

80-11000 KHI Europe Finance BV
(a subsidiary of Kawasaki Heavy Industries Ltd, Japan)
Hoekenrode 6, NL-1102 Amsterdam
☎ (+31 20) 644 68 69 ⊖ (+31 20) 629 46 61 98

80-12500 Logica BV
(a subsidiary of Logica plc, UK)
Het Kasteel, PO Box 197, NL-3440 AD Woerden
☎ (+31 348) 57 25 55 ⌀ (0348) 57 26 36
⊖ (+31 348) 52 49 70
♣ G 120,000,000 (1997) ☗ 560 (1997) ⌂ 14,000 m2
Managing Director: Wilfried de Meyere
Operations and Technical Director: Hadi Jassim
Account Manager Defence: W A Wynveldt
Rotterdam Office
Wijnhaven 69, NL-3011 WJ Rotterdam
☎ (+31 10) 433 08 44 ⊖ (+31 10) 433 14 47
Groningen Office
Brugstraat 32b, NL-9711 HZ Groningen
☎ (+31 50) 31 45 66 ⊖ (+31 50) 314 13 09 98

80-18300 Radio Holland Electronics BV Defence Department
(† Radio Holland Sait)
(a div. of SAIT-Radio Holland Group)
Jan Rebelstr. 20, NL-1069 CC Amsterdam
☎ (+31 20) 667 81 00; 667 81 84 ⊠ 13166 rholl nl
⊖ (+31 20) 66 78 318
☗ 100
Managing Dir: W.W.M. Overbeck
Operations Dir: W.G. van Veenen Daal
Sales & Marketing Dir: C. van der Zouwen
Purchasing Mgr: J. Griffioen
■ Activities in the fields of radiocommunications & positioning systems for land mobile purposes. Handling turnkey projects including service & aftersales. → 75 → 90
® Australia: Codan
Denmark: Terma
Finland: Vaisala
Germany: Motorola
Israel: Motorola
Italy: Meteor CAE SpA
Norway: Nera
UK: RAPCO, Redifon, Smith Industries, Waverley Electronics, Zengrange Ltd
USA: Magnavox (Hughes Aircraft), Trimble 97

80-19000 Skyline Aviation BV
(a subsidiary of Jalex BV)
20 Luchthavenweg, NL-1786 PP Den Helder
☎ (+31 223) 66 03 04 E skyline.aviation@tip.nl
⊖ (+31 223) 66 02 74
♣ G 3,500,000 (1996) ☗ 15 (1997) ⌂ 3,200 m2
Managing Director: R A van Hoevell
Operations Director: E J van Wijk
Chief Pilot: J de Visser
■ Special military mission flights, target towing flights, military training flights, air taxi and ambulance flights, aircraft maintenance, sales and trade. Parking facilities for aircraft, sightseeing flights. Sale of Avgas 100LL fuel. 98

80-19500 Stemat Marine Services
PO Box 54511, NL-3008 KA Rotterdam
☎ (+31 10) 299 65 55 E info@stemat.nl
W http://www.stemat.nl ⊖ (+31 10) 299 65 43
Managing Director: H Brigggeman
■ Hire and service of transport pontoons up to 5,500 tons, workboats and tugs, floating/sea going cranebarges, winches, transhipment, storage and water transports inland and offshore. 98

80-20000 W Tielemans, MD
(† B Voorsluÿs, MD)
PO Box 20703, NL-2500 ES's-Gravenhage
☎ (+31 70) 349 24 03 ⊖ (+31 70) 349 34 39
☗ 14
Managing Director: J van den Hoorn (MD)
Operations Director: P de Jager
■ Safety, health and welfare services. 98

80-23300 De Weger Architects and Consulting Engineers
(† FC de Weger International BV)
PO Box 705, NL-3000 AS Rotterdam
☎ (+31 10) 467 80 88 E weger@pi.net
⊠ 24261 weger nl ⊖ (+31 10) 467 45 90
♧ US$ 11,000.000 (1995) ♣ US$ 14,500,000 (1995)
☗ 150 (1997) ⌂ 3,300 sq m
Managing Director: P G de Wit
Operations and Technical Director: P van Putten
Purchasing Manager: H Ligteringen
Public Relations: P van den Harr
■ Civil engineering, architectural design, structural engineering, hydraulic, hydrological and geohydrological engineering. Naval bases, covered berths, drylocks shiplifts. 98

85 Stockists, Distributors, Spare Parts Suppliers

Van Beest BV 75-07818

85-00010 Air-Parts BV
PO Box 255, NL-2400 AG Alphen aan den Rijn
Kalkovenweg 12, NL-2401 LK Alphen aan den Rijn
☎ (+31 172) 42 24 55 ⊠ AIRPARTS
⊖ (+31 172) 42 10 22
☗ 33 (1997)
Managing Director: P de Vries E p.de.pires@central.air-parts.nl
Manager of Aviation/Ordnance Division: G W G M Berendse
Financial Administration: E de Leest
Management and Marketing Assistant: Ms Monique Neve
■ Sale of aircraft parts and accessories, ordnance equipment and parts for military vehicles, tanks and weapon systems. Microwave and RF components and systems for Satam, EW, ECM and communication interception.
® USA: Applied Systems Technologies, Bren-Tronics Inc, Carrier Components Corp, Comstron (AEROFLEX), Condor-Systems, Excalibur Systems Ltd, Giga-Tronics, Godrey Aerospace, Life Support International, Mason Electric, Miteq Hauppage, Natvar Corp, Purdy, Puritan-Bennett Aerosystems, Up-Right 98

85-05000 REOMIE BV
Erlecomsedam 34, NL-6576 JW Ooy
☎ (+31 24) 663 81 00 E reomie@euronet.nl
⊖ (+31 24) 663 26 32
☗ 25 (1997) ⌂ 6,000 m2
Managing Director: Andre Springveld
Sales Director: Andre Kulik
Advertising Manager: Martin van Heteren
■ Import and export of military trucks, parts, equipment and surplus material. US-made tactical vehicles and automotive parts such as M151 series jeeps, M series 2.5 ton trucks, M series 5 ton trucks, M107/109 series howitzers, M113 series armoured carriers, M41/M48/M60 series tanks. Overhaul and retrofit of engines, axles, transmissions and transfers. 98

85-06300 Technisch Bureau H A Muller
PO Box 219, NL-6800 AE Arnhem
☎ (+31 26) 443 37 52 ⊖ (+31 26) 45 84 35
☗ 5
Director: H A Muller
Director: Evelien Kienhorst
■ Specialists in personal protection equipment, special weapons and ammunition, pyrotechnics, commando equipment, military, police and anti-terrorist equipment. 98

Contact Reference

Netherlands

90 Sales Agents, Sales & Liaison Offices

Companies represented:

3Net, UK	90-01150
Abelag Aviation SA, Belgium	80-10000
Aerotel Ltd, Israel	90-01150
AIM, UK	90-00870
Airtour International, UK	90-00185
Airworthy Ltd, UK	80-10000
Allied Signal, USA	90-00870
Amplicilence, Italy	90-01150
Analogic, USA	90-01150
Anaren Microwave, USA	90-00850
Ancra Corp, USA	75-00510
Andrewe Antennas, UK	90-01150
Applied Systems Technologies, USA	85-00010
Artemis Management Systems, UK	90-01300
Austin Co., USA	90-00180
Bell Helicopter, USA	90-00200
Bird, USA	90-00850
Breeze Eastern, USA	90-00870
Bren-Tronics Inc, USA	85-00010
British Aerospace, UK	90-00870
British Steel Engineering Steels, UK	90-00125
C & Gavazzi Soar, Japan	90-01150
Cadex, Canada	90-00850
Cadre, Switzerland	90-01150
Carrier Components Corp, USA	85-00010
CAS, USA	90-00870
Cel Lucas, UK	90-01150
Chase London, UK	90-01150
Cincinnati Milacron, UK	75-01130
Cincinnati Milacron, USA	75-01130
Clement Clarke International, UK	90-00185
Codan, Australia	80-18300
Comstron (AEROFLEX), USA	85-00010
Condor-Systems, USA	85-00010
Conrac, Germany	90-00850
Cooper Aviation International, USA	90-00185
CSA Antennas, UK	90-01150
Dancall, Denmark	90-01150
Danphone, Denmark	90-01150
Datac Aachen, Germany	90-01150
Dataradio, Canada	90-01150
Datasel, Germany	90-01150
Deutsche Aerospace, Germany	90-00100
Dorotech, France	90-01150
E-Systems, USA	90-00870
EEV Ltd, UK	90-00120
EG&G, USA	90-01150
Eicon, Canada	90-01150
Endevco, USA	90-01150
Ensto Connector, Finland	90-00220
Envirovac, USA	90-00870
Excalibur Systems Ltd, USA	85-00010
Farnell, UK	90-01150
Flight Refuelling, UK	90-00870
Framatome, France	90-00630
Fuji, Japan	90-01150
GEC Avionics Ltd, UK	80-10000
GEC Sensors Ltd, UK	80-10000
General Microwave, USA	90-00850
Giga-Tronics, USA	85-00010
Gigatape, Germany	90-01150
Godrey Aerospace, USA	85-00010
Graphtec, Japan	90-01150
Grimes Aerospace Company, USA	90-00800
Guildline Instruments, Canada	90-01150
Harris PRD-Electronics, USA	90-00850
Helmets Ltd, UK	80-10000
HEME, UK	90-01150
Hughes Aircraft, USA	90-01150
Hughes Microelectronics, UK	90-01150
IEM Inc., USA	90-01970
Indivers NV, Netherlands	80-05500
Intamar Logistics Inc, USA	90-00185
Intel, UK	90-01150
Intel, USA	90-01150
IPC Information Systems, UK	90-01150
Jane's Information Group, UK	90-01250
Jovy Atlas, Germany	90-00850
Kenwood, Japan	90-01150
Krenz, Germany	90-01150
Labinal, Germany	90-01150
Lambda, Denmark	90-01150
Lannet, Israel	90-01150
Lear Siegler Inc., USA	90-00192
Lechmotoren, Germany	90-00850
Lernout & Hauspie, Romania	90-01150
Life Support International, USA	85-00010
Ling Dynamic Systems, UK	90-01150
Litton Industries Inc, USA	90-01240
Lucas Aerospace Power Trans., USA	90-00870
Magnavox (Hughes Aircraft), USA	80-18300
Marketing & Technical Serv., UK	90-01250
Mascot, Italy	90-01150
Mason Electric, USA	85-00010
McDonnell Douglas Helicopter, USA	90-01470
Memotec, Canada	90-01150
Meteor CAE SpA, Italy	80-18300
Microphase Corp, USA	90-00850
Millard Towers, Canada	90-00220
Miteq Hauppage, USA	85-00010
Molectron, USA	90-01150
Montclair Enterprises Inc., USA	90-00580
Motorola, Germany	80-18300
Motorola, Israel	80-18300
Motorola GmbH, Germany	90-01150
MWB-EMU Systems, Germany	90-01150
Natvar Corp, USA	85-00010
Nera, Norway	80-18300
Noise, Japan	90-01150
OR-X, Israel	90-01150
Orbit, Israel	90-00850
ORGA, Germany	90-00850
PI Castings, UK	90-02600
Plisch, Germany	90-00850
Purdy, USA	85-00010
Puritan-Bennett Aerosystems, USA	85-00010
PZL, Poland	90-00185
Racal Communications, UK	90-01150
Racal Comsec, UK	90-01150
Racal Milgo, UK	90-01150
Racal Tacticom, UK	90-01150
RAD, Israel	90-01150
Radcom, Israel	90-00850
Radiall, France	75-05750
Radiocom, Switzerland	90-01150
Radius, UK	90-01150
RAPCO, UK	80-18300
Raven, UK	90-01150
RCA Electro Optics, Canada	90-01150
Redifon, UK	80-18300
Research Engineers, UK	90-00220
Rheinmetall Industrie Aktiengesellschaft, Germany	90-01500
RND, Israel	90-01150
Robert E. Hoose Inc., USA	90-00580
Rockwell International Corp., USA	75-05790
Rockwell International Inc, USA	75-05800
Satel OY, France	90-01150
Schaevitz Engineering, USA	90-01150
Scheer & Co GmbH, Germany	75-00850
Schottel-Werft, Germany	90-01900
Scope Communications, USA	90-00850
Shaw, USA	90-00870
Shibasoku, UK	90-01150
Siemens CTE, Germany	90-00850
Sierracin/Sylmar Corp, USA	90-01470
Sikorsky Aircraft, USA	90-00870
Singer Kearfott, USA	90-01150
SIT, France	90-01150
Smith Industries, UK	80-18300
Smiths Industries, USA	90-00870
Spider Systems, UK	90-01150
Spinner, Germany	90-00850
Storno, Denmark	90-01150
Studer, Switzerland	90-00850
Swissphone, Switzerland	90-01150
Taco Tafel, Germany	90-01150
Taiyo Musen, Japan	90-01150
TEAM, France	90-00870
Teledyne Microwave, USA	90-01150
Telefunken Syst Technik GmbH, Germany	90-00100
Terma, Denmark	80-18300
Thandar, UK	90-01150
Thomson CNI, France	90-00870
Thomson CSF, Romania	90-01150
Thorn, France	90-00220
Trend Communications, UK	90-00850
Tricom Borer, UK	90-01150
Trimble, USA	80-18300
TRW, USA	90-01150
Ultra Lastertech, Canada	90-01150
Unison, USA	90-00870
Up-Right, USA	85-00010
Vaisala, Finland	80-18300
Watkins Johnson, UK	90-01150
Watkins Johnson, USA	90-01150
Waveline, USA	90-00850
Waverley Electronics, UK	80-18300
Weinschel (Lucas), USA	90-00850
West, UK	90-01150
World Marine Publishing Ltd, UK	90-01250
XL Microwave, USA	90-00850
Zengrange Ltd, UK	80-18300

Sales agents/representatives listed elsewhere:

AlliedSignal Aerospace Marketing	80-01100
Bose BV	75-00790
Radio Holland Electronics BV	80-18300

90-00005 Acer Computer BV
(a subs. of Acer Incorporated, Taiwan)
European Headquarters
PO Box 70576, NL-5201 CZ's- Hertogenbosch
Europalaan 89, NL-5232 BC's- Hertogenbosch
☎ (+31 73) 645 95 95 ⊜ (+31 73) 645 95 99
♣ US$ 1 898 000 000 (Group)
President Europe: Teddy Lu
[97]

90-00100 AEG Nederland NV
(a subsidiary of AEG Eurolec Holding BV)
PO Box 9433, NL-1006 AK Amsterdam
Aletta Jacobslaaan 9, NL-1066 BP Amsterdam
☎ (+31 20) 487 87 87 ⊘ (+31 20) 487 87 88
[E] sales@aeg-telpro.nl [E] management@aeg-telpro.nl
⊜ (+31 20) 617 00 25
♧ G 27,000,000 (1997) ♣ G 65,000,000 (1996)
92 (1997) 12,000 m2
Managing Director: R P Bruggeman
Operations Director: R Koffeman
Sales and Marketing Director: E E Geelhuysen
Purchasing Manager: A J Wetsteyn
® Germany: Deutsche Aerospace, Telefunken Syst Technik GmbH
[98]

90-00120 Alphatron Professional Components
K.P. van der Mandelelaan 40,
NL-3062 MB Brainpark, Rotterdam
☎ (+31 10) 452 06 00 ⊠ 24009 alpha nl
⊜ (+31 10) 452 12 70
Managing Director: D Slingerland
® UK: EEV Ltd
[98]

90-00125 Anglo Staal Holland BV
(a subsidiary of British Steel Engineering Steels, UK)
Postbus 19088, NL-BB 3011 Rotterdam
Leeuwenstraat 13, NL-3001 Al Rotterdam
☎ (+31 10) 412 88 40 ⊜ (+31 10) 411 68 54
® UK: British Steel Engineering Steels
[98]

90-00180 Austin Nederland BV
(a subs. of The Austin Co., USA)
Airport Office Center, Schipholweg 297,
NL-1171 PK Badhoevedorp
☎ (+31 20) 659 60 66 ⊜ (+31 20) 659 70 81
Managing Dir: Jan Wamsteker
® USA: Austin Co.
[97]

90-00185 Aviation Services International (Benelux) BV (ASI)
PO Box 418, NL-1700 AK Heerhugowaard
Deimoslaan 17a, NL-1702 CK Heerhugowaard
☎ (+31 72) 571 24 67
⊜ (+31 72) 571 52 17; 574 17 17
Managing Director: Mrs M van Gemert
Vice President, Sales and Special Projects: R Kraaipoel
■ Sales of aircraft parts and accessories, commercial and military markets.
® Poland: PZL
UK: Airtour International, Clement Clarke International
USA: Cooper Aviation International, Intamar Logistics Inc
[98]

90-00192 Aviquipo Holland BV
Building 105, PO Box 75052, NL-1117 ZP Schipol Airport
☎ (+31 20) 45 37 84 ⊠ 15578 ⊜ (+31 10) 45 59 59

Manager: Karl Vetter
® USA: Lear Siegler Inc. [98]

90-00200 Bell Helicopter Supply Center
PO Box 7534, NL-1117 ZG Schiphol Airport East
Schipholweg 303, NL-1171 PL Badhoevedorp
☎ (+31 2968) 951 56 ▣ 16325 bell nl
⊖ (+31 2968) 967 77
® USA: Bell Helicopter [97]

90-00220 Besthoma Trading International BV (BTI BV)
(a subsidiary of Besthoma Aviation Holding BV)
PO Box 421, NL-5680 AK Best
De Maas 17a, NL-5684 PL Best
☎ (+31 499) 39 66 48; 39 36 64 [E] bti@btibv.nl
[W] http://www.btibv.nl ⊖ (+31 499) 39 74 39
Managing Director: Huub Hommen
Finance Director: J J M Hommen
Office Manager: C Driedonks
■ Design and supply of airfield lighting systems including elevated and inset lights. Navigational beacons for airports and marine applications. Transformers and connectors, constant current regulators, control systems, taxiway signs, obstruction lighting, retroreflective lighting systems, taxiway markers and windcones.
Infrared cameras with thermal imaging capacity for ground movement control in severe weather conditions, visual support to radar ground movement control.
® Canada: Millard Towers
Finland: Ensto Connector
France: Thorn
UK: Research Engineers [98]

90-00275 Cleco Europe
(a subsidiary of Cleco Limited, UK)
Hanzestraat 4, NL-7006 RH Doetinchem
☎ (+31 8340) 266 99 ⊖ (+31 8340) 241 99 [98]

90-00300 Daewoo Corp Amsterdam Branch
L'Arche Bleue B/D 1st Floor, Overschiestraat 180, NL-1062 XK Amsterdam
☎ (+31 20) 617 81 05 ▣ 18714 daewoo nl
⊖ (+31 20) 617 81 25 [98]

90-00560 Encore Computer Nederland BV
(† Gould SEL Computer Systems BV)
(a subs. of Encore Computer Corporation)
PO Box 1666, NL-3600 BR Maarssenbroek
Computerweg 4, NL-3606 AT Maarssenbroek
☎ (+31 3465) 662 14 ▣ 70667 encore ul
⊖ (+31 3465) 674 23
Marketing: Wytze Van Der Raay
■ Main activities: Computer systems for simulation, telemetry, data acquisition & other real-time applications. Computer systems for software development & OLTP. [97]

90-00580 Eriks BV
Voormeer 33, PO Box 280, NL-1800 BK Alkmaar
☎ (+31 72) 514 15 80 ⊖ (+31 72) 515 56 45
♤ DFI 50 000 000 (1995) ♣ DFI 172 000 000 (1995)
👥 425
Managing Dir: M. Muilwjk
Sales & Marketing Mgr: M. Ebbing
Maritime & Defence Mgr: J.J. Jonker
Public Relations Mgr: R. Roem
Contact: R.V. Weert
® USA: Montclair Enterprises Inc., Robert E. Hoose Inc. [97]

90-00630 Framatome Connectors Nederland BV
(a division of Framatome, France)
Essebaan, PO Box 174, NL-2900 AD Capelle a/d IJssel
☎ (+31 10) 459 63 99 [W] http://www.fciconnect.com
⊖ (+31 10) 458 54 67
Management Assistant: M L van der Wal
® France: Framatome [98]

90-00650 Furon
(a subs. of Furon Company, USA)
Molenweg 7, NL-7447 CW Hellendoorn
☎ (+31 5486) 542 66 ⊖ (+31 5486) 545 73 [97]

90-00800 Grimes Aerospace Company
Saturnusstraat 21, NL-2132 HB Hoofddorp
☎ (+31 2503) 520 60 ⊖ (+31 2503) 519 53
® USA: Grimes Aerospace Company [98]

90-00850 Heynen BV
De Groote Heeze 11, PO Box 10, NL-6590 AA Gennep
☎ (+31 485) 55 09 09 [E] heynen@heynen.com
[W] http://www.heynen.com ⊖ (+31 485) 55 09 00
Managing Director: R Heijnen
Public Relations Manager: J Kieboom
® Canada: Cadex
Germany: Conrac, Jovy Atlas, Lechmotoren, ORGA, Plisch, Siemens CTE, Spinner
Israel: Orbit, Radcom
Switzerland: Studer
UK: Trend Communications
USA: Anaren Microwave, Bird, General Microwave, Harris PRD-Electronics, Microphase Corp, Scope Communications, Waveline, Weinschel (Lucas), XL Microwave [98]

90-00870 Hollinda BV
van der Oudermeulenlaan 1, NL-2243 CR Wassenaar
☎ (+31 70) 517 71 22 ⊖ (+31 70) 511 02 12
♣ US$ 23 000 000 (1995) 👥 5 🏠 300 m2
Managing Dir: A.J.O. van Gennep
Commercial Dir: J.C. Coenraads
Defence Consultant: R. Vinckx
Secretary: H. van Adrichem
® France: TEAM, Thomson CNI
UK: AIM, British Aerospace, Flight Refuelling
USA: Allied Signal, Breeze Eastern, CAS, E-Systems, Envirovac, Lucas Aerospace Power Trans., Shaw, Sikorsky Aircraft, Smiths Industries, Unison [97]

90-01150 Koning en Hartman BV
(a subsidiary of Getronics NV, Amsterdam)
PO Box 125, NL-2600 AC Delft
Energieweg 1, NL-2627 AP Delft
☎ (+31 15) 60 99 06 ✉ EKOHA ▣ 38250
⊖ (+31 15) 61 91 94
👥 350
Managing Director: A H Kersbergen
® Canada: Dataradio, Eicon, Guildline Instruments, Memotec, RCA Electro Optics, Ultra Lastertech
Denmark: Dancall, Danphone, Lambda, Storno
France: Dorotech, SIT, Satel OY
Germany: Datac Aachen, Datasel, Gigatape, Krenz, Labinal, MWB-EMU Systems, Motorola GmbH, Taco Tafel
Israel: Aerotel Ltd, Lannet, OR-X, RAD, RND
Italy: Amplicilence, Mascot
Japan: C & Gavazzi Soar, Fuji, Graphtec, Kenwood, Noise, Taiyo Musen
Romania: Lernout & Hauspie, Thomson CSF
Switzerland: Cadre, Radiocom, Swissphone
UK: 3Net, Andrewe Antennas, CSA Antennas, Cel Lucas, Chase London, Farnell, HEME, Hughes Microelectronics, IPC Information Systems, Intel, Ling Dynamic Systems, Racal Communications, Racal Comsec, Racal Milgo, Racal Tacticom, Radius, Raven, Shibasoku, Spider Systems, Thandar, Tricom Borer, Watkins Johnson, West
USA: Analogic, EG&G, Endevco, Hughes Aircraft, Intel, Molectron, Schaevitz Engineering, Singer Kearfott, TRW, Teledyne Microwave, Watkins Johnson [98]

90-01240 Litton Precision Products International
Benelux Sales Office
Griendstraat 10, NL-2921 LA Krimpen a/d Yssel
☎ (+31 180) 51 85 55; 59 68 88
[E] littonnl@worldaccess.nl
⊖ (+31 180) 51 66 24; 59 68 99
👥 11
Managing Director: Jack Ekelmans
Sales and Marketing Director: W van Drunen
® USA: Litton Industries Inc [98]

90-01250 Information Resources Ltd
(† Marketing & Technical Services Ltd)
Logger 111, NL-1186 RP Amstelveen
☎ (+31 20) 647 44 84 ⊖ (+31 20) 640 23 43
Manager: Dr J.A.K. Henderson
■ Consultancy, market research & marketing information systems. Marine procurement services & publishing. → 80
® UK: Jane's Information Group, Marketing & Technical Serv., World Marine Publishing Ltd [97]

90-01300 Metier Plancon
(a subsidiary of Artemis Management Systems, UK)
Waterlandlaan 42/44, NL-1441 MP Purmerend
☎ (+31 299) 41 81 81 ⊖ (+31 299) 41 81 60
Manager: Paul Phillips
® UK: Artemis Management Systems [98]

90-01470 Noratech BV
North Atlantic Technologies International
Kapelweg 3, NL-2587 BH The Hague
☎ (+31 70) 350 30 38 [E] ahylkema@bart.nl
⊖ (+31 70) 354 29 25
President: Alexander S W Hylkema
® USA: McDonnell Douglas Helicopter, Sierracin/Sylmar Corp [98]

90-01500 NWM de Kruithoorn BV
(a subsidiary of Rheimnetall Industrie Aktiengesellschaft)
PO Box 1050, NL-5200 BC's Hertogenbosch
☎ (+31 73) 681 02 22 ⊖ (+31 73) 681 03 10
President: Friedrich Birkenkamp
Sales Director: Steef van den Berg
® Germany: Rheinmetall Industrie Aktiengesellschaft [98]

90-01900 Schottel-Werft Nederland BV
(a subs. of Schottel-Werft, Germany)
PO Box 477, NL-2700 AL Zoetermeer
Chroomstraat 143, NL-2718 RJ Zoetermeer
☎ (+31 79) 361 13 91 ⊖ (+31 79) 361 14 17
General Mgr, Sales & Administration: Wim Hazelaar
Defence Project Mgr: M. Bos
® Germany: Schottel-Werft [97]

90-01960 Sperry Marine BV
PO Box 274, NL-3130 AG Vlaardingen
James Wattweg 22, NL-3133 KK Vlaardingen
☎ (+31 10) 445 16 00 ⊖ (+31 10) 434 50 15
Regional Manager: R Haasdijk [98]

90-01970 Stellingwerff Beintema BV
Zijlweg 144, NL-2015 BH Haarlem
☎ (+31 23) 531 41 92 ⊖ (+31 23) 532 59 93
General Mgr: S. Stellingwerff Beintema
® USA: IEM Inc. [97]

90-02000 Surcom International BV
Defence & Security Electronics
PO Box 82, NL-3958 ZV Amerongen
☎ (+31 343) 46 00 00 [E] surcom@compuserve.com
⊖ (+31 343) 45 74 44
♤ DFI 40,000 (1996) ♣ DFI 2,500,000 (1996)
👥 3 (1996)
President: Martin P de Zwarte
Manager, Communications Systems: Willem Haverkamp
■ HF, VHF and UHF portable, mobile, fixed site military transceivers; UHF (DAMA) SATCOM Terminals; (Combat) SAR Transceivers; civil and military ATC communications; secure telephone and facsimile; tempest and rugged PC, fax; manpack DF/SIGINT; communication ESM systems; radar transponders and active radar calibration targets; electronic bomb, artillery and mortar fuzes; night vision; thermal imaging; laser ranging, targeting and warning.
Companies Represented:
Motorola Inc, Harris RF Communications Inc, DATRON World Communications Inc, ZETA Inc, Delfin Systems Inc, XL Computing Inc, RICOH Government Systems, Fredwal Inc, Fuchs Electronics Ltd, International Technologies (Lasers) Ltd, Inframetrics Inc, Marconi SpA [98]

90-02217 Varel BV
Positronweg 2, NL-3542 AZ Utrecht
☎ (+31 20) 651 61 71 ⊖ (+31 20) 697 80 84
Managing Director: Tom Rolf [98]

90-02600 Robert Zapp Tooling BV
(† Robert Zapp Holland BV)
Televisieweg 79, NL-1322 AK Almere
☎ (+31 36) 536 05 57 ⊖ (+31 36) 536 02 68
Sales Manager: W Beek
® UK: PI Castings [98]

Netherlands Antilles

Governor:
Dr Jaime Saleh

10 Head of State

10-10000 Office of the Governor
Fort Amsterdam 2, Willemstad, Curaçao
Governor: Dr Jaime Saleh 98

11 Prime Minister

11-10000 Office of the Prime Minister
Fort Amsterdam 17, Willemstad, Curaçao
☎ (+599 9) 461 39 88 ☒ 1079 alsec na
⊖ (+599 9) 461 50 77
Prime Minister: Miguel Pourier 98

12 Ministries

12-20000 Department of Foreign and General Affairs
Curaçao
Minister of Labour and Social Affairs: Jeffrey Corion 98

12-60000 Department of Transport and Communications
Fort Amsterdam 17, Willemstad, Curaçao
☎ (+599 9) 461 39 88 ☒ 1079 alsec na
Minister: Leo Chance 98

25 Defense Forces

25-25000 Netherlands Forces in the Caribbean
Naval Base
Nightingale Way, Parera, Curaçao
☎ (+599 9) 463 74 02
⊖ (+599 9) 463 74 44; 463 74 92
Commander of Caribbean Area: Cdre J W de Jager
Chief of Staff: Col W A J Prins
Operations Director: Cdr J F Geilen-Kirchen
Procurement and Logistics Director: Cdr P J Wich
Public Relations Officer: Flag Lieut J W Klopper 98

25-30000 Marine Corps
Marine-Basis, Parera, Curaçao
☎ (+599 9) 63 73 05; 63 71 00 ⊖ (+599 9) 63 74 44
Commander: Lieut Col A W Kerssemeijer 98

25-40000 Curaçao National Guard
Vrijwilligerskorps Curaçao (VKC)
Curaçao
☎ (+599 9) 37 07 50 ⊖ (+599 9) 63 75 90
Commander: R E L Ursula
Deputy Commander/Chief of Staff: Maj R de Jesus
Operations Director: Capt R Sling
Director of Procurement and Logistics: Lieut B Suemanta
Administration: Capt G Sillé 98

25-45000 Headquarters Flag Officer Netherland Forces in the Caribean
Naval Base Parera, Curaçao
☎ (+599 9) 63 74 30 ⊖ (+599 9) 63 74 44
Commander: Lieut Col A W Kerssemeijer
Commander of Detachment Suffisant: Maj B Van Rijswijk 98

90 Sales Agents, Sales & Liaison Offices

Companies represented:

Radio-Holland Group, Netherlands . . . 90-02000
Waverley Electronics, UK→ Netherlands 80-18300

90-02000 Radio-Holland Caribbean NV
(a division of Radio Holland B.V., NL)
Pletterijweg z/n, PO Box 3894, Willemstad, Curaçao
☎ (+599 9) 461 25 77 ☒ 3901195 rhoca na
⊖ (+599 9) 461 27 23
NAFI 1,000,000 (1997) ♣ NAFI 15,000,000 (1997)
55 (1997) 800 m2
Managing Director: THomas L Merkus
Operations Director: Dennis Heymans
Sales and Marketing Director: Dido Kalma
Sales and Marketing Manager: B Hasham
Purchasing Manager: Richard Doran
■ Any form of electronics equipment for navy, army and air force.
® Netherlands: Radio-Holland Group 98

New Zealand

Head of State:
H.M. Queen Elizabeth II

Governor-General:
H E Sir Micheal Hardie Boys

10 Head of State

10-10000 Office of the Governor-General
Government House, Private Bag, Wellington
☎ (+64 4) 389 80 55 ⊖ (+64 4) 389 55 36
Governor-General:
H E The Right Hon Sir Michael Hardie Boys 98

11 Prime Minister

11-10000 Prime Minister's Office
Executive Wing, Parliament House, Wellington
☎ (+64 4) 471 99 98 ☒ 31702 ⊖ (+64 4) 473 70 45
Prime Minister: James B Bolger
Deputy Prime Minister: Peter Winston 98

12 Ministries

12-10000 Ministry of Defence
Parliament Buildings, Private Bag, Wellington
☎ (+64 4) 496 09 99 ☒ 3513 defcom nz
⊖ (+64 4) 496 08 59
Minister of Defence: Paul East
Chief of Defence: Lieut Gen Anthony Birks 98

12-20000 Ministry of Foreign Affairs and Trade
Private Bag 18-901, Wellington
Stafford House, 40 The Terrace, Wellington
☎ (+64 4) 494 85 00 ☒ 3441 external nz
⊖ (+64 4) 472 95 96
Minister of Foreign Affairs and Trade: Donald McKinnon 98

12-40000 Ministry of Civil Defence
Parliament Buildings, Wellington
☎ (+64 4) 471 99 43 ⊖ (+64 4) 471 29 07
Minister for Civil Defence: Jack Elder 98

25 Defense Forces

25-20000 Royal New Zealand Air Force Air Staff (RNZAF)
HQ New Zealand Defence Force, Private Bag, Wellington
☎ (+64 4) 496 09 99 ☒ 3513 defcom nz
⊖ (+64 4) 496 05 65
Chief of Air Staff: AVM C W Adamson
☎ (+64 4) 496 05 00
Deputy Chief of Air Staff: Air Cdre D M Hamilton
☎ (+64 4) 496 05 03
Assistant Chief of Air Staff Operations: Gp Capt I G Brunton
☎ (+64 4) 496 05 10
Assistant Chief of Staff Personnel: Gp Capt B R Fergusson
☎ (+64 4) 496 05 50
Assistant Chief of Staff Logistics: Gp Capt A M Gill
☎ (+64 4) 496 08 00
Assistant Chief of Staff Programmes and Projects:
Gp Capt G B Lintott ☎ (+64 4) 496 05 46
Personal Staff Officer to CAS: Sqn Ldr S Woodhouse
RNZAF Air Command
RNZAF Base Auckland, Private Bag, Whenuapai, Auckland
☎ (+64 9) 306 77 89
Air Officer Commanding: Air Cdre G J W Goldsmith 98

25-30000 New Zealand Army
(† Royal New Zealand Army)
Army General Staff
HQ NZ Defence Forces, Private Bag, Wellington
☎ (+64 4) 496 09 99 ☒ 3513 defcom nz
⊖ (+64 4) 496 04 93
Chief of General Staff: Maj.Gen. P.M. Reid (CBE)
Deputy Chief of General Staff: Brig. R.R. Ottaway (MBE)
Territorial Advisor: Brig. J.B. O'Brien (OBE ED)
Asst Chief of General Staff Development: Col B.L. Fraher
Asst Chief of General Staff Operations: Col T. O'Reilly
Asst Chief of General Staff Personnel:
Col R. Bywater-Lutman
Asst Chief of General Staff Resources:
Lt Col J.B. Vryenhoek
Director of Logistics Army: Lt Col R.M.S. Johnston
Public Information Officer: Capt. Jody Stewart
☎ (+64 4) 496 02 90 (+64 4) 496 02 95 (fax) 97

25-32200 The Army Combat Centre Army Training Group
Private Bag, PO Box 1702, Waiouru
☎ (+64 6) 387 61 11 Ext 7400 ⊖ (+64 6) 387 60 61
Commander ATG: Col R. Seymour
Commanding Officer: Lt Col T.R. Kuper
☎ (+64 6) 387 61 11 ext. 7400 97

25-40000 Royal New Zealand Navy (RNZN)
Naval Staff
HQ New Zealand Defence Force, Private Bag, Wellington
☎ (+64 4) 496 09 99 ☒ 3513 defcom nz
⊖ (+64 4) 496 03 11
Chief of Naval Staff: RAdm. J.E.N. Welch
Deputy Chief of Naval Staff: Cdre K.R. Moen
Asst Chief of Naval Staff Materiel: Capt. R.H. Keeling
Asst Chief of Staff Operations: Capt. W. Rathbun
Asst Chief of Staff Development: Capt. A.J. Peck
Naval Finance Dir: J.F. Lowen
Asst Chief of Staff Personnel: Cdr J.O. Ladd
Maritime Headquarters
HMNZ Naval Base, Auckland
☎ (+64 9) 444 59 99 ⊖ (+64 9) 445 55 43
Maritime Commander: Cdre J.G. Leonard
Fleet Operations Officer: Cdr G.R. Buchan
Captain Naval Training: Capt. A.D. Clayton-Greene
Captain Fleet Support: Capt. M.N. Franklin
RNZN Hydrographer
HMNZ Naval Base, Auckland
☎ (+64 9) 489 50 61 ⊖ (+64 9) 489 50 63
Hydrographer: Cdr P.F.K. Usher

Defence Scientific Establishment
HMNZ Naval Base, Auckland
☎ (+64 9) 445 59 02 ⊖ (+64 9) 445 58 20
Director: Dr J.J. Buckingham 97

31 Embassies & Attachés

31-00150 Embassy of Belgium
1-3 Willeston Street, PO Box 3379, Wellington
☎ (+64 4) 472 95 58; 472 95 59 ▣ 31452 ambabel nz
⊖ (+64 4) 471 27 64
Ambassador: H E Mr Pierre Vandervoorde
Attaché: Mrs Greta J Fernie-de Brabander 98

31-00310 Canadian High Commission
61 Molesworth Street, PO Box 12049,
Thorndon, Wellington
☎ (+64 4) 473 95 77 ⊖ (+64 4) 471 20 82
Counsellor: Brian Watson
Commercial Counsellor: John Hill ☎ (+64 9) 309 36 90
Canadian Forces Adviser: Capt I S Foldesi
(Resident in Canberra) ☎ (+61 62) 73 38 44
Systems Administrator: Michele Rowe 98

31-00350 Embassy of Chile
1-3 Willeston St, Willis Corroon Hse, PO Box 3861,
Wellington
☎ (+64 4) 472 51 80 ▣ 31034 emchile nz
⊖ (+64 4) 472 53 24
Ambassador: H.E. Mr Demetrio Infante Figueroa
First Secretary: José M. Gonzalez
Second Secretary & Consul: Pablo Arriaràn 97

31-00570 Embassy of the Republic of Fiji
31 Pipitea Street, Thorndon, PO Box 3940, Wellington
☎ (+64 4) 473 54 01/2 ⊖ (+64 4) 499 10 11
Ambassador: H.E. Mr Joji N. Guivalu
Chargé d'Affaires: Abdul Rafik
Second Secretary: Ms Mesema Yabakivitu 97

31-00590 Embassy of France Chancery
Rural Bank Building - Manners Street, PO Box 11-343,
Wellington
☎ (+64 4) 384 25 55; 802 77 78 (defence);
802 77 88 (press)
⊖ (+64 4) 384 25 77 (chancery/defence);
384 25 79 (consular/press)
Ambassador: H E Mr Jacques le Blanc
Counsellor: Xavier Dupont
Defence Attaché: Col Claude Brodin
First Secretary (Consular): Laurent Lagrange
Third Secretary: Christophe Jacquemont

Office of the Commercial Counsellor
34-42 Manners Street, PO Box 11-148, Wellington
☎ (+64 4) 384 50 42; 384 50 43
✉ COMATTA Wellington ⊖ (+64 4) 384 52 98
Counsellor (Economic and Commercial): Franc Secula
Counsellor (Financial): Alain Guillouet (Residing in Tokyo)

Cultural Office
34-42 Manners Street, PO Box 11-281, Wellington
☎ (+64 4) 802 77 78 ⊖ (+64 4) 384 25 78
Counsellor (Cultural and Scientific): Serge Verniau 98

31-00690 Consulate General of the Hellenic Republic
5-7 Willeston Street - 10th Floor, PO Box 24066,
Wellington
☎ (+64 4) 473 77 75; 473 77 76 ⊖ (+64 4) 385 46 08
General Consul: Theodoros W Daskarolis 98

31-00830 High Commission of India
PO Box 4045, Wellington
10th Floor - FAI House, 180 Molesworth Street
☎ (+64 4) 473 63 90; 473 63 91
E hicomind@globe.co.nz ▣ 31676 hicoind nz
⊖ (+64 4) 499 06 65
High Commissioner: H E Mr K M Meena
First Secretary and Head of Chancery: K S Bharti
Attaché: V A John Britto
Defence Adviser: V S Chaudary 98

31-00850 Embassy of the Islamic Republic of Iran
151 Te Anau Road, PO Box 10249 The Terrace,
Hataitai, Wellington
☎ (+64 4) 386 29 83; 386 29 76 ▣ 31226 nz
⊖ (+64 4) 386 30 65
Ambassador: H.E. Mr Abolfazl Khazaee Torshizi
Commercial Counsellor: Reza Allaei Tazaehgheshlagh
Consular Attaché: Seyed Mohammad Sadegh Tabatabaei
Third Secretary: Seyed Majid Tafreshi Khamneh
Attaché: Alireza Safahieh 97

31-00880 Embassy of Israel
13th Floor - DB Tower, PO Box 2171, Wellington
111 The Terrace, Wellington
☎ (+64 4) 472 23 62; 472 23 68
E israel-ask@israel.org.nz
W http://www.iconz.co.nz/~israel ⊖ (+64 4) 499 06 32
Ambassador: H E Mr Nissan Koren Krupsky
Counsellor (Deputy Chief of Mission): Amir Gissin
Second Secretary Political: Yehuda Ya'akov
Second Secretary Administration: Amnon Israel
Public Affairs Assistant: Ms Yvonne Trask
Trade Officer and Paific Islands Co-ordinator:
Dimitri Geidelberg E dimitri.geidelberg@israel. org.nz
Information Officer: Ms Leila Martley 98

31-00890 Embassy of Italy
34-38 Grant Road, PO Box 463, Thorndon, Wellington
☎ (+64 4) 473 53 39 ▣ 31571 italdipl nz
⊖ (+64 4) 472 72 55
Ambassador: H E Dr Antonino Provenzano
Attaché, Consular and Administration: Luciano Capellino 98

31-00920 Embassy of Japan
PO Box 6340, Wellington
Norwich Insurance House, 3-11 Hunter St., Wellington
☎ (+64 4) 473 15 40 ▣ 3544 nz ⊖ (+64 4) 471 29 51
Ambassador: H.E. Mr Endo Tetsuya
Minister Counsellor: Yuzo Ota
First Secretary: Yoshikazu Takeuchi
First Secretary Agricultural & Commercial:
Takashi Hiramatsu
Second Secretary: Kgoru Shimada 97

31-00990 Embassy of the Republic of Korea
ASB Bank Tower-Level 11, PO Box 11-143, Wellington
2 Hunter Street, Wellington
☎ (+64 4) 473 90 73; 473 90 74 ⊖ (+64 4) 472 38 65
Ambassador: H E Mr Yoon-Kyung Oh
Counsellor: Song-Jun Ohm
Second Secretary and Consul: Dong Il Oh
Defence Attaché: Col Kim Sam Gan (Residing in Canberra)
Attaché: Nam Soo Park

Consulate of the Republic of Korea
c/o Price Waterhouse, PO Box 13-250, Christchurch
119 Armagh Street
☎ (+64 3) 379 00 40 ⊖ (+64 3) 379 56 77
Honorary Consul: Clyde L Sugden

Consular Agency of the Republic of Korea
Auckland
Consular Agent: Sang-hoon Kim
Deputy Consular Agent: Min-ho Choo 98

31-01110 Malaysian High Commission
10 Washington Avenue, PO Box 9422,
Brooklyn, Wellington
☎ (+64 4) 385 24 39; 385 20 19 ▣ 30485 nz
⊖ (+64 4) 385 69 73
High Commissioner: H E Datuk Daniel Tajem
Counsellor: Ibrahim Mohamad
Defence Adviser:
Lieut Col Rajah Abdullah bin Rajah Baharuddin
Second Secretary: Teh Eng Beng
Contact: Rohanah Mahmod
Information Assistant: Ms Christina Eguna 98

31-01430 Embassy of the Republic of Poland Office of the Commercial Counsellor
17 Upland Road, Kelburn, Wellington
☎ (+64 4) 471 24 56; 475 94 53
E polishembassy@xtra.co.nz
⊖ (+64 4) 471 24 55; 475 94 58
Ambassador: Agnieszka Morawinska
(Resident in Canberra) ☎ (+61 6) 273 12 08
Counsellor (Commercial), Chargé d'Affaires:
Andrzej Soltysinski ☎ (+64 4) 475 94 53 98

31-01485 Embassy of the Russian Federation
57 Messines Road, Karori, Wellington
☎ (+64 4) 476 61 13 E eor@netlink.co.nz ▣ 31727 nz
⊖ (+64 4) 476 38 43
Ambassador: H E Mr Sergei Belyaev
Minister Counsellor: Anatoly Botov
First Secretary: Andrew Roudenko

Commercial Office
16 Lancaster Street, Karori, Wellington
☎ (+64 4) 476 78 21 ⊖ (+64 4) 476 62 05
Commercial Counsellor: Vadim Lukyanovich 98

31-01570 High Commission of Western Samoa
1a Wesley Road, PO Box 1430, Kelburn, Wellington
☎ (+64 4) 472 09 53; 472 09 54 ▣ 31043 palolo nz
⊖ (+64 4) 471 24 79
High Commissioner: H E Mr Feesago Siaosi Fepuleài
First Secretary: Leilua Tavas Leota 98

31-01655 Consulate of the Republic of Slovenia (honorary)
Easteren Hutt Road, Pomare, Lower Hutt Wellington
☎ (+64 4) 567 00 27 ⊖ (+64 4) 567 00 24 98

31-01740 Embassy of Switzerland
22 Panama Street, Wellington 1
☎ (+64 4) 472 15 93/4 ⊖ (+64 4) 499 63 02
Ambassador: H.E. Ernst Thurnheer
Head of Chancery: Franz Schneider
First Secretary: Walter Simmen 97

31-01890 United Kingdom High Commission
44 Hill Street, PO Box 1812, Wellington 1
☎ (+64 4) 472 60 49 ✉ UKREP WELLINGTON
▣ 3325 ukrep nz ⊖ (+64 4) 473 49 82
High Commissioner: H E Mr R J Alston (CMG)
Deputy High Commissioner: H Salvesen
Defence Adviser & Sr British Defence Liaison Staff:
Col P R Barry (CBE)
First Secretary Commercial: M Capes
First Secretary: R Hargreaves 98

60 Research Organizations, Institutes & Test Establishments

60-00410 Defence Scientific Establishment (DSE)
New Zealand Defence Force
Auckland Naval Base, Private Bag 32901, Auckland
☎ (+64 9) 445 59 02 ⊖ (+64 9) 445 58 90
Director: Dr. J.H. Buckingham 97

75 Manufacturing Companies

75-00110 Albion Clothing Ltd
PO Box 22-646, Christchurch
☎ (+64 3) 379 76 77 ⊖ (+64 3) 365 75 70
♧ NZ$ 192,000 (1997) ♣ NZ$ 2,800,000 (1997)
85 (1997) 900 m2
Managing Director: G M Steele
Operations Director: Mrs R A Steele
Accountant: Peter Stoneman
■ Military uniforms, firefighting suits, bulletproof vests, cold and wet weather suits, jackets, trousers and shirts. 98

75-00150 Anite Networks
Anite House Level 2, 48 Greys Avenue, Auckland
☎ (+64 9) 309 27 29 ✆ (+64 800) 80 27 29
⊖ (+64 9) 309 32 42

Contact Reference

Hamilton Office
NZI House Level 4, Crnr Alexandra St and Garden Place, Hamilton
☎(+64 7) 838 06 25 ⊖ (+64 7) 838 37 56
Wellington Office
Majestic Centre Level 12, 100 Willis Street, Wellington
☎(+64 4) 499 27 29 ⊖ (+64 4) 499 27 39
Christchurch Office
Level 1 - Unit 1, 15 Washington Way, Christchurch
☎(+64 3) 366 55 11 ⊖ (+64 3) 366 25 20
Dunedin Office
80 Filleut Street, Dunedin
☎(+64 3) 479 00 65 ⊖ (+64 3) 477 90 24 [98]

75-00180 Australian Calibrating Services (A'sia) Pty Ltd
State Highway 1, Otaki
☎(+64 6) 364 55 95 ⊖ (+64 6) 364 54 03 [97]

75-00420 Donaghys Industries Ltd
PO Box 15007, Aranui, Christchurch
☎(+64 3) 388 61 86 ⊖ (+64 3) 388 47 48
♣ US$ 187,100,000 (1995) 👥 350
Chief Executive Officer: S Mander
Managing Director: Steve Wilson
Operations and Technical Director: Bill McCook
Purchasing Manager: Les Hill
■ Synthetic tapes, webbing, braids, ropes, shoelaces and cords. [98]

75-00500 Electropar Ltd
PO Box 58623, Auckland
35 Lady Ruby Drive, East Tamaki, Auckland
☎(+64 9) 274 20 00 ⊖ (+64 9) 274 20 01
Managing Director: Grant Wallace
Marketing Manager: Ms Wendy Wallace
■ Manufacturer of electrical accessories for naval vessels including plugs, sockets, junction boxes, lighting accessories and switches. [98]

75-00650 Fitzroy Engineering Group Limited
Private Bag 2053, 4620 New Plymouth
☎(+64 67) 58 61 65 [E] greg.michetski@xttra.co.nz
⊖ (+64 67) 58 17 03
♧ NZ$ 3,000,000 ♣ NZ$ 60,000,000 (1996/97)
👥 290 (1996/97) 🏭 7,000 m2
Managing Director: Peter White-Robinson
Operations and Technical Director: Chas Wilcox
Operations Director: Steve Lutto
Purchasing Manager: Cliff Watson
Business Manager: M Elmes
Marketing Manager: Greg Mischifski
■ Pressure vessels of various sizes to all codes. Heavy plate fabrication, sheet metal fabrication. Consulting in the area of specialist fabrication. → 80 [98]

75-00700 Fuelquip Services Ltd
127 Wexford Road - Miramar, PO Box 15-104, Wellington
☎(+64 4) 802 84 00 ⊖ (+64 4) 388 24 49
👥 300 (1997)
Managing Director: Kevin O'Sullivan
Business Development Manager: Peter Butcher
Marketing Services Manager: Miss A Crawford
■ Aircraft refuellers, mobile lube rigs, fuel pumping and filter systems. Specialised transport engineering and portable bulk refuelling units. → 80 → 85 → 90 [98]

75-00750 GNB Technologies
(† Chloride Batteries New Zealand)
(a division of Pacific Dunlop Holdings (NZ) Limited)
Lower Hutt, PO Box 36-026, Lower Hutt
☎(+64 44) 568 42 46; 568 42 69 [E] gnbn@xtra.co.nz
⊖ (+64 44) 568 66 87
👥 203 (1997)
Managing Director: Mrs Debra J Bright
National Sales Manager: Andrew Duncan
Marketing Manager: Jeff MacGroucutt
Recycling Manager: Grant Bolitho
Factory Manager: Jim Kosega
Engineering and Technical Services Manager: Neville Manson
Finance and Administration Manager: Michael Kavanagh
Manager for People and the Enviroment: Ms Janeen Wood
Logistics Manager: Murray Smith
National Manager - Telecommunications and Power Control: Andy Butchart
National Manager - Electric Vehicles: Rob Bolton
Purchasing Co-ordinator: Neville Ware
Executive Assistant: Ms Lynda McCarroll
Christchurch Branch
PO Box 11-300, Christchurch
☎(+64 3) 348 09 83 ⊖ (+64 3) 348 07 04
Branch Manager: D Beecroft
Auckland Branch
PO Box 76-102, Wiri, Auckland
☎(+64 9) 278 81 79 ⊖ (+64 9) 278 75 03
Branch Manager: B Wilkinson
Hamilton Branch
PO Box 5156, Hamilton
☎(+64 7) 847 51 84 ⊖ (+64 7) 847 21 57
Branch Manager: L Ferguson
Napier Branch
PO Box 626, Napier
☎(+64 6) 843 60 55 ⊖ (+64 6) 843 51 17
Branch Manager: C Macken
New Plymouth Branch
PO Box 525, New Plymouth
☎(+64 6) 757 56 92 ⊖ (+64 6) 757 56 93
Branch Manager: C Rangi
Dunedin Branch
PO Box 720, Dunedin
☎(+64 3) 455 51 79 ⊖ (+64 3) 455 26 10
Branch Manager: W de Vries
Palmerston North Branch
PO Box 1083, Palmerston North
☎(+64 6) 358 20 79 ⊖ (+64 6) 357 08 31
Branch Manager: P Cairney
Nelson Branch
PO Box 489, Nelson
☎(+64 3) 548 18 13 ⊖ (+64 3) 548 08 86
Branch Manager: L Hannah
Invercargill Branch
82 Clyde Street, Invercargill
☎(+64 3) 218 72 49 ⊖ (+64 3) 214 03 36
Branch Manager: R Shaw [98]

75-00780 Guardall Alarms New Zealand Ltd
(† Guardall Reichell)
Symonds Street, PO Box 8462, Auckland 1
5 Carlton Gore Road, Grafton, Auckland
☎(+64 9) 337 34 95 ✆ (800) 50 10 04
⊖ (+64 9) 337 27 15
♧ NZ$ 225 000 ♣ NZ$ 3 500 000 👥 30
Managing Dir: Dennis H. Waldrom
National Sales Mgr: Kerry Penman
Operations: John Wooding
■ Computerised building management systems comprising: access control, CCTV/TLR, intruder & fire detection systems & uninterruptible power supplies. [97]

75-00860 Hutchwilco Limited
(a subsidiary of Chemring Group plc, UK)
10-14 Farmhouse Lane, PO Box 18-064, Glen Innes, Auckland
☎(+64 9) 528 07 88 ✆ (+64 800) 48 82 49
[E] enquiries@hutchwilco.co.nz ⊖ (+64 9) 528 07 22
♧ NZ$ 1,000,000 (1997) ♣ NZ$ 6,500,000 (1996)
👥 45 (1997) 🏭 3,375 m2
Managing Director: Tom Davies
General Manager: Greg Hunt
Operations, Technical and Purchasing Manager: Bernard Orme
Sales and Marketing Manager: Tim Ritchie
■ Lifejackets, lifebuoys and buoyancy aids. Military and marine pyrotechnics. Marine window wiping systems. → 85
® UK: International Safety Group, McMurdo Marine, Mullion, Octavius Hunt, Pains-Wessex, Splashdown, Vacuum Reflex, Wynn Marine Ltd [98]

75-00950 Interlock Group Ltd
PO Box 15-103, Wellington
☎(+64 4) 388 83 55 ⊖ (+64 4) 388 84 81
👥 500 (1997)
Executive Director: D W S Moloney
■ Camouflage net support system. [98]

75-01250 Loadlift Equipment Ltd
(a subsidiary of Zuellig (NZ) Ltd)
PO Box 12-593, Penrose, Auckland
☎(+64 9) 634 37 14 ⊖ (+64 9) 634 00 39
♣ NZ$ 20,000,000 (1996) 👥 33 (1996)
General Manager: Charles Reid
Operations Director: Allan Amers
Sales and Marketing Director: David Stewart
■ Supply, repair and overhaul of forklift trucks, excavators, rollers aand marine cranes. → 80 [98]

75-01300 Marine-Air Systems Ltd
24 Bridge St, PO Box 30-248, Lower Hutt
☎(+64 4) 569 21 70 ⊖ (+64 4) 566 12 47
Managing Dir: N. Jordan
Sales & Marketing Dir Defence: Dr R. Gibbard
Technical Dir: P. Williams
■ Specialises in research & development of electronic/software products for air force, army, navy & government agencies. QA accreditation to AQAP 13 & ISO 9001. Small scale, high quality manufacturing capability.
® Canada: Litton Canada, Raytheon Canada
UK: Cossor Electronics, GEC Ferranti Defence Systems, GEC Plessey Radar, Marconi Underwater Systems
USA: E-Systems Montek, Litton, Raytheon Equipment Division, Texas Instruments [97]

75-01350 Mills-Tui Ltd
(a subsidiary of Fuelquip Services Ltd)
16-38 Pururu Street, PO Box 469, Rotorua
☎(+64 7) 348 80 39 ⊖ (+64 7) 348 68 67
♧ NZ$ 100,000 (1995) 👥 26 🏭 2,000 m2
Managing Director: Kevin O'Sullivan
General Manager: Lloyd A Pickering
Engineer: M H Comer
■ Urban and rural fire appliances. Airport crash rescue vehicles, aluminium boats and specialised transport engineering. → 85 → 90 [98]

75-01550 Oscmar International Ltd
Wellesley St, PO Box 6008, Auckland
25-29 Shaddock Street, Mt Eden, Auckland
☎(+64 9) 379 03 60 ⊖ (+64 9) 307 02 11
Sales & Marketing Dir: T.A. (Tom) Scott
General Mgr: M.W.S. (Mike) Stanbridge
Asst Accountant: Ms J. Higham
■ Military laser based infantry simulation equipment. Industrial engine and generator control systems. [97]

75-01600 Pacific Aerospace Corp. Ltd (PAC)
(joint venture of Aeromotive Ltd and Lockheed Martin Aeronautics International)
Hamilton Airport, Private Bag 3027, Hamilton
☎(+64 7) 843 61 44 [E] pacific@aerospace.co.nz
⊖ (+64 7) 843 61 34
♣ NZ$ 16,000,000 👥 110 (1997) 🏭 104,140 sq ft
Managing Director: Brian Hare
General Manager: Graeme Polley
Quality Assurance Manager: Tony Lea
Company Secretary: Ms Stacey Gray
Manager Policy and Marketing: John McWilliam
Inventory Manager: Murray Dreyer
■ FU24 and Cresco ag/utility aircraft, CT4 Airtrainer civil/military aircraft, overhaul and repair of military aircraft, mechanical components. Manufacture of aircraft components and general engineering to ISO 9002-mil-1-45208 and NZ CAA.
Boeing and MDC QA approved requirements for heat treatment, NDT and surface finishing. → 80
Shareholders:
Aeromotive Ltd 75.1%
Lockheed Martin Aeronautics International 24.9% [98]

75-01620 Pacific Helmets (NZ) Ltd
315 Heads Road, PO Box 866, Wanganui
☎(+64 6) 344 50 19 [E] pacifichelmets@clear.nz
⊖ (+64 6) 344 53 76
👥 40
Managing Director: D W A Bennett
Sales and Marketing Director: T Boswell
■ Safety and protective helmets including fire, rescue and military. Helmets for special applications and customised design. [98]

75-01630 Peco Ltd
95 Moorhouse Avenue, Christchurch 1
☎(+64 3) 366 56 32 ⊖ (+64 3) 366 56 31
👥 20 (19979 🏭 1,000 m2 (1997)
Managing Director: Sydney A Goldsmith
Sales and Marketing Manager: Patrick Fogarty
■ Aircraft ground service equipment, aircraft assembly jig fixtures and tools. Packaging and collating machinery, materials handling equipment. [98]

75-01650 Phillips & Smith Ltd
PO Box 69028, Glendene, Auckland 8
☎(+64 9) 818 80 48 ⊖ (+64 9) 818 44 84
👥 46 (1997)
Managing Director: S R Hampton
Export Operations Manager: Hugh Cronwright
National Sales and Marketing Manager: Warwick Manning
■ Fire fighting equipment. [98]

75-01950 Synapco Industries Ltd
PO Box 36031, Moera, Lower Hutt
☎(+64 4) 568 51 43 ⊖ (+64 4) 568 51 17
♧ NZ$ 310,000 (1997) ♣ NZ$ 800,000 (1997)
👥 8 (1997) 🏭 1,500 m2
Managing Director: David A Ussher
■ Manufacture and refurbishment of armoured vehicle track pads. [98]

75-02050 THORN Lighting (NZ) Ltd
399 Rosebank Road - Avondale, PO Box 71134, Auckland 7
☎(+64 9) 828 71 55 ✆ (800) 80 08 34
⊖ (+64 9) 828 75 91
Chief Executive Officer: Mike Gibbons
Commercial Director: L Woodward
Operations and Technical Director: S Edwards
Secretary: C Hawken [98]

75-02055 Transfield Shipbuilding New Zealand
PO Box 6176, Wellesley Street, Auckland
Level 9 - Lumley House, 7 City Road, Auckland
☎(+64 9) 302 29 44; (+64 25) 98 56 81 (mobile)
⊖ (+64 9) 302 22 38
Manager: Kevan Rogers [98]

75-02060 Transfield Whangerei Ltd
Port Road, PO Box 1864, Whangerei
☎(+64 9) 430 09 02 ⊖ (+64 9) 430 02 97 [98]

75-02080 Trimble Navigation New Zealand
76 Chester Street East, PO Box 13-547 Armagh, Christchurch
☎(+64 3) 371 34 00 ⊖ (+64 3) 371 34 17
👥 76
Managing Dir: Alan Townsend
Sales & Marketing Dir: Terry McDonald
■ GPS software and hardware manufacturing company. [97]

75-02250 Vega Industries Ltd
Heriot Drive, Porirua, Wellington
☎(+64 4) 237 43 93 ⊖ (+64 4) 237 43 92
👥 9
Managing Dir: J.P. Rochfort
General Mgr: C.M. Cook
■ Design & manufacture of lighted navigational aids, searchlights, signalling searchlights, precision approach path indicators & sector lights. [97]

80 Service Companies & Organizations

Fitzroy Engineering Group Limited	75-00650
Fuelquip Services Ltd	75-00700
Loadlift Equipment Ltd	75-01250
Omega DSE Ltd	90-01500
Pacific Aerospace Corp. Ltd	75-01600
Scollay Holdings Ltd	90-01910

80-00050 Air Affairs (NZ) Limited
PO Box 99198, Newmarket, Auckland
☎(+64 9) 524 41 29; 529 09 59 ⊖ (+64 9) 524 46 86
♧ NZ$ 1,000,000 ♣ NZ$ 2,000,000 (1994) 👥 5
Managing Director: Douglas M Pauling
E dpauling@acorn.co.nz
Sales and Marketing Director: Chris Young
Operations and Technical Director: David Stokes
Engineering Manager: Ken Connolly
Office Manager: R V Pauling
■ Contractors to RNZN for target towing. Development and supply of CBT systems for defence. Distributors of Acorn and Tandon computer systems, Westland Systems Assessment UK and Terabyte Interactive Ltd. → 85 [98]

80-00100 Air New Zealand Ltd Engineering Services
Christchurch Airport, PO Box 14005, Christchurch 8015
☎(+64 3) 374 70 02 E brazip@airnz.co.nz
⊖ (+64 3) 374 70 01
👥 2,000 (1997) 🏭 60,000 m2
General Manager Engineering: Ian Diamond
Manager Engineering Operations: Graeme Gilmore
Marketing Manager: Ian V Matthews
Engineering Marketing Representative: Philip A Brazier
Materials Manager: Steve Watts
■ Full overhaul of B767, B737, B727, C-130, and P-3, engines and components. [98]

80-00900 James Hardie Engineering
481 Tremaine Ave, PO Box 517, Palmerston North
☎(+64 6) 356 64 88 ⊖ (+64 6) 356 49 93
♧ NZ$ 12 000 000 (1995) ♣ NZ$ 12 000 000 (1995)
👥 80
General Mgr: M.J. Lewis
■ Sheetmetal products, electrical switchboards control centres, air conditioning plant, light engineering, fire fighting systems, portable fire equipment, fixed special hazard, damage control systems, respiratory equipment.
Design & engineering, project management manufacturing commissioning, testing service & maintenance.

Head Office
50 O'Rorke Road, Penrose, Auckland
☎(+64 9) 579 30 99 ⊖ (+64 9) 579 24 93
Chief Exec.: B. Boniface [97]

80-01150 K S I Agencies Ltd
PO Box 163, Whangarei
☎(+64 9) 438 99 39 ⊖ (+64 9) 438 72 74
👥 114 (1997)
Managing Director: W J Orr
Sales Director: S H Culham
■ Engineering consultancy. Specialised armouring (static and vehicular), body armour. Compass B555 Light Armour Ballistic Plate. [98]

80-01200 Logica New Zealand Limited
(a subsidiary of Logica plc, UK)
Microsoft House - Level 14, PO Box 1436, Wellington, New Zealand
Microsoft House - Level 14, 49 Boulcott Street
☎(+64 4) 499 94 90 ⊖ (+64 4) 499 94 91 [98]

80-02000 Open Technology Limited
Corner Newton and Abbey Streets, Newton, Auckland
☎(+64 9) 378 02 80 ⊖ (+64 9) 378 02 93 [98]

80-02010 Open Tehnology Limited
PO Box 437, Wellington
Corner Wakefield and Taranaki Streets
☎(+64 4) 385 29 69 ⊖ (+64 4) 385 77 64 [98]

80-05000 Wormald
(† Wormald Fire Systems Ltd)
Private Bag, New Lynn, Auckland 7
☎(+64 9) 826 17 00 ⊖ (+64 9) 827 61 33
General Mgr: G. Wallace
Marketing Co-ordinator: J. Edmondson
■ Design, installation & servicing of fire protection & security systems. [97]

85 Stockists, Distributors, Spare Parts Suppliers

Air Affairs (NZ) Limited	80-00050
Fuelquip Services Ltd	75-00700
Hutchwilco Limited	75-00860
Mills-Tui Ltd	75-01350

85-00150 Benfield Marketing Ltd
18 Quetta Street, Ngaio, 6004 Wellington
☎(+64 4) 479 70 95 ⊖ (+64 4) 479 61 71
♧ NZ$ 500,000 (1997) ♣ NZ$ 500,000 (1997)
👥 3 (1997) 🏭 425 m2
Managing Director: Leon Benbaruk
■ Radio VHF transmitters, UHF thermal imaging cameras, IR surveillance and electro-optical surveillance equipment for armed forces. Avionics for Orion P-3 and C-130 EW simulators. 25mm guns for aircraft and navy patrol boats, patrol boats, tankers, assault rifles, sniper rifles. M113 APC armoured personnel carrier upgrade weapon capabilities, 40mm grenade launchers M203/M203PI ammunitions. Chaff & dispensers lorries, ARV, CRV's for army, trailers, inmarsat equipment.
Small arms ammunition: 5.56mm/9mm/7.62mm/.50 cal. Heavy arms ammunition: 105mm/81mm mortar/84 mm Carl Gustav. Overhead weapon systems (OWS) 25mm/.50 cal TI and II: Thermal imaging and Infra-red equipment. V-SAT: small aperture terminal clothing, cloth, uniform, helmets, frag-vests and bulletproof vests. Tents, tarpaulins, water storage tanks, fuel storage tanks, mobile and stationary gas turbine power stations. Medical and hospital equipment. Secondhand army material and equipment (artillery and ammunition). Shelters and electronic repair shelters. EOD/ordnance equipment, signal projectors for navy LRF Laser Range Finders. Global positioning systems for army and air force (GPS). Medium recovery and four-wheel drive vehicles. [98]

85-00200 Lordco CWG (NZ) Limited
PO Box 38-499, Wellington
Unit D-1, 59-61 Marsden Street, Lower Hutt
☎(+64 4) 569 51 08 ⊖ (+64 4) 569 27 07
♧ NZ$ 37,000 (1997) ♣ NZ$ 2,000,000 (1997) 👥 10
🏭 150 m2
Managing Director: D McLean
Puchasing Manager: K A Lee
■ Agents and stockists of
aviation lighting, screened rooms, power supplies, electrical/electronic equipment and components. → 90
® UK: Auto Diesels Edghill, BLP Components, Belling Lee (Intech), LPA Industries, Lindgren Rayproof Ltd, Metalite Aviation Lighting [98]

85-02000 Radiola Corporation Ltd
Wi-neera Drive, Private Bag 50911, Porirua, Wellington
☎(+64 4) 237 01 59 ⊖ (+64 4) 237 12 67
♧ NZ$ 3,000,000 ♣ NZ$ 30,000,000 👥 45
🏭 100,000 sq ft
Managing Director: Brent L Albiston
Sales and Marketing Manager: Richard Thompson
E richardt@radiola.co.nz
■ Specialises in electronic systems and equipment for armed forces, aviation, law enforcement and government agencies. Software capability certified ISO9001. Project management and Through Life Support.
® Australia: AWA Communications, AWA Defence Industries
USA: Airport Systems Int'l Inc, Cardion Inc, Southern Avionics Company [98]

90 Sales Agents, Sales & Liaison Offices

Companies represented:

Airport Systems Int'l Inc, USA	85-02000
Auto Diesels Edghill, UK	85-00200
Aviall Australia Pty Ltd, Australia	90-00150
Aviall Inc., USA	90-00150
AWA Communications, Australia	85-02000
AWA Defence Industries, Australia	85-02000
Belling Lee (Intech), UK	85-00200
BLP Components, UK	85-00200
Cardion Inc, USA	85-02000
Cossor Electronics, UK	75-01300
E-Systems Montek, USA	75-01300
GEC Ferranti Defence Systems, UK	75-01300
GEC Plessey Radar, UK	75-01300
International Safety Group, UK	75-00860
Lindgren Rayproof Ltd, UK	85-00200
Litton, USA	75-01300
Litton Canada, Canada	75-01300
LPA Industries, UK	85-00200
Marconi Underwater Systems, UK	75-01300
McMurdo Marine, UK	75-00860
Metalite Aviation Lighting, UK	85-00200
Mullion, UK	75-00860
Octavius Hunt, UK	75-00860
Pains-Wessex, UK	75-00860

Contact Reference

Raytheon Canada, Canada 75-01300
Raytheon Equipment Division, USA . . 75-01300
Southern Avionics Company, USA . . . 85-02000
Splashdown, UK 75-00860
Texas Instruments, USA 75-01300
The General Electric Co. plc, UK . . . 90-00730
Vacuum Reflex, UK 75-00860
Wynn Marine Ltd, UK 75-00860

Sales agents/representatives listed elsewhere:

Fuelquip Services Ltd 75-00700
Lordco CWG (NZ) Limited 85-00200
Mills-Tui Ltd 75-01350

90-00100 Antares Aerospace
(a division of Kapiti Holdings Ltd)
PO Box 2746, Wellington
☎ (+64 4) 297 09 16 ⊜ (+64 4) 297 19 10
NZ$ 2,500,000 ♣ NZ$ 9,000,000 16
Chief Executive Officer: Murray C Cole
Sales and Marketing Director: L Lissette
Purchasing Manager: N Pashby
■ Importers, exporters, dealers and operators of new and used aircraft and spacecraft. Aviation lighting equipment. Aerial and photogrammetric equipment. Survival and rescue equipment. Airport traffic control systems PAL-PAV monitoring systems. Operators and owners of Paraparaumu Airport Ltd. [98]

90-00150 Aviall New Zealand Pty Ltd
Auckland International Airport, Cyril Kay Road, Auckland
☎ (+64 9) 275 05 71 ⊜ (+64 9) 275 65 69
Manager: Brent Oldfield
® Australia: Aviall Australia Pty Ltd
USA: Aviall Inc. [98]

90-00730 GEC-Marconi (New Zealand) Ltd
(a subs. of the General Electric Co., UK)
Prosser St. Elsdon, PO Box 50-244, Porirua
☎ (+64 4) 237 54 09 ⊜ (+64 4) 237 26 77
♣ NZ$ 70 000 000 300
Managing Dir: Nigel Power
Marketing Mgr GEC-Marconi Div.: A.S. Andrews
® UK: The General Electric Co. plc [97]

90-01470 New Zealand Safety Ltd
401 Great South Rd, Penrose, Auckland
☎ (+64 9) 579 28 80 ⊜ (+64 9) 579 83 35
360
Managing Dir: Andrew Paul
Sales & Marketing Dir: Michael Arnott
Commercial Mgr: Perry Gallagher
Operations & Technical Mgr: Clive Marple
Purchasing Mgr: Glen McWilliams
■ Safety. First Aid. Fire protection. Hazchem. Environmental. Safety specialists.
Kemira (OY):
Bilsom (AA)
Sked Systems
Mada Medical
Stockhausen
Racal
Occumed
Nederman
Moldex Metric
Crews
Bolle
Pulsafe
USSafety
Hornell Elektrooptik AB
Survivair
Drager
Pratt Safety
Kappler
Bel Art USA [97]

90-01500 Omega DSE Ltd
PO Box 5099, Wellington 1
Landcorp House - 101 Lambton Quay, Wellington
☎ (+64 4) 471 06 31 E omega@sequel.co.nz
⊜ (+64 4) 471 02 30
Managing Director: Bob Cooper
Sales and Marketing Director, Naval: Iain Ross
Sales and Marketing Director, Air: John Lawton
Sales and Maketing Director, Land: Todd Cooper
■ Specialist importers and manufacturer's representatives. → 80 [98]

90-01910 Scollay Holdings Ltd
PO Box 2377, Wellington
☎ (+64 4) 472 29 61 E 76267.436@compuserve.com
3461 wiscol ⊜ (+64 4) 472 29 66
AZ$ 750,000 (1997) 25 1,050 m2
Managing Director: C J Scollay
■ Representatives, distributors and consultants. Areas of activity include aerospace, communications, computing and munitions. → 80 [98]

Nicaragua

Head of State:
Arnoldo Aleman

10 Head of State

10-05000 Office of the President
PO Box 2398, Managua
Casa de Gobierno, Avenida Aug C Sandino
☎ (+505 2) 78 62 02; 28 40 15 ⊜ (+505 2) 62 79 11
President: Arnoldo Aleman [98]

10-10000 Office of the Vice President
Casa de Gobierno, PO Box 2398, Managua
☎ (+505 2) 28 26 84; 28 36 74
Vice President: Enrique Bolanos [98]

12 Ministries

12-10000 Ministry of Defence
Casa de la Presidencia, Managua
Minister of Defence: Jaime Cuadra [98]

12-20000 Ministry of Foreign Affairs
Detrás de los Ranchos, Managua
Minister of Foreign Affairs: Emilio Alvarez [98]

12-30000 Ministry of Foreign Co-operation
PO Box 4595, Managua
Minister of Foreign Co-operation: David Robelto [98]

12-40000 Ministry of the Interior
PO Box 68, Managua
Minister of Government: Lorenzo Guerrera [98]

25 Defense Forces

25-40000 Naval Forces
Plaza Espana, 1 Km Al Sur, Managua
☎ (+505 2) 78 45 63 ⊜ (+505 2) 78 45 60
Chief: Lieut Col Manuel Rivas Guatemala
Chief of Staff: Lieut Col Cesar Largaespada
☎ (+505 2) 78 45 61
Director of Operations: Lieut Cdr Manuel Mora Ortíz
Director of Procurement and Logistics:
Lieut Cdr Sergio Lazo Fernández [98]

31 Embassies & Attachés

31-00230 Brazilian Embassy
Km 7 3/4 de la Carretera Sur, PO Box 264, Managua
☎ (+505 2) 65 00 35 ⊜ (+505 2) 65 22 06
Ambassador: H.E. Mr J.G. Marques-Porto
Third Secretary: Patricia Maria O.L. Serapiao [97]

31-00590 Embassy of France
Iglesia El Carmen 1 1/2 C Abajo, A P 1227 Managua
☎ (+505 2) 22 62 10; 22 66 15 1085 amfrance
⊜ (+505 2) 28 31 66
Ambassador: H E Mr Joseph Francis Rapin
Defense Forece Attaché: Lt Col. A Cordier
Commercial Office
1 cuadra Abajo, 10 Varas al Lago, PO Box 2259, Managua
☎ (+505 2) 24 63; 50 95 ✉ COMATTA Managua
1049 comatta ou 1047 [98]

31-00890 Embassy of Italy
De La Rotunda El Geuguense 1, Cuarda Al La Lago, Managua
☎ (+505 2) 66 29 18; 66 29 61; 66 64 86; 66 43 19
E embitalia@tmx.com.ni 1437 itdipl nk
⊜ (+505 2) 66 39 87
Ambassador: H E Mr Nicolo' Goretti De'Flamini [98]

31-01890 Embassy of the United Kingdom
Reparto "Los Robles", Primera Etapa
Entrada Principal de la Carretera a
Masaya, Cuarta Casa a Mano Derecha, Managua
☎ (+505 2) 78 00 14; 78 08 87 E britemb@ibw.com.ni
⊜ (+505 2) 78 40 85
Ambassador: H E Mr Roy Osborne
Deputy Head of Mission: Miss M-L Archer
First Secretary, Commercial: A S Green
(Resident in San José)
Defence Attaché: Col D Simpson (Resident in Guatemala) [98]

31-01900 Embassy of the United States of America
Km 4-1/2 Carretera Sur, Managua
☎ (+505 2) 66 60 10; 66 60 13; 66 60 15; 66 60 18; 66 60 26; 66 60 27; 66 60 32; 66 60 34
W http://www.usia.gov/posts/managua.html
⊜ (+505 2) 66 60 46
Ambassador: H E Mr Lino Gutierrez
Deputy Chief of Mission: Frederick Becker
Political Affairs Officer: Kevin Whitaker
Economic and Commercial Officer: Ms Sandra Dembski
Defense Attaché: Col Lynn Lanzoni (US Army)
Public Affairs Officer: Elizabeth Whitaker [98]

Niger

Head of State:
Col. Ibrahim Bare Mainassaraa

10 Head of State

10-10000 Office of the President
Niamey
☎(+227) 72 23 81 ▣ 5214
President, Niger National Council:
Col Ibrahim Bare Mainassaraa
Prime Minister: Amadou Boubacar Cisse 98

12 Ministries

12-10000 Ministry of National Defence
Niamey
☎(+227) 72 28 40; 72 26 50 ▣ 5291
Minister: Ousmane Issoufou Oubandawaki
Inspector General, Ministry of Defence: Col Nouhou Bako 98

12-20000 Ministry of External Affairs
(† Ministry of Foreign Affairs and Co-operation)
PO Box 396, Niamey
☎(+227) 72 29 07 ▣ 5200 minafet ni
Minister: Ibrahim Hassane Mayaki 98

12-40000 Ministry of Interior
Niamey
☎(+227) 72 21 76; 72 32 10 ▣ 5214 presirep ni
Minister of Interior: Omar Idi Ango 98

31 Embassies & Attachés

31-00310 Embassy of Canada
PO Box 362, Naimey
☎(+227) 75 36 86/7 ▣ 5264 ⊖ (+227) 75 31 07
Consul & Co-operation Counsellor: J. Lepine
Commercial & Public Information Officer: M. Barhouni 97

31-00660 Embassy of the Federal Republic of Germany
71 Avenue du Général de Gaulle, PO Box 629, Niamey
☎(+227) 72 25 34; 72 35 10 ▣ 5223 aa niam ni
⊖ (+227) 72 39 85
Ambassador: H E Mr Angelika Völkel
Counsellor: Jens Kraus-Messé
Commercial Attache: Christel Kolvembad 98

31-01890 Vice-Consulate of the United Kingdom
BP 11168, Niamey
☎(+227) 73 20 51; 73 25 39 ▣ 5244 ni
⊖ (+227) 73 36 92
Honorary Consul: B Niandou 98

31-01900 Embassy of the United States of America
Rue Des Ambassades, BP 11201, Niamey
☎(+227) 72 26 61 ▣ 5444 embnia ni
⊖ (+227) 73 31 67
Ambassador: H E Mr Charles Cecil
Deputy Chief of Mission: Jay O Dehmalow 98

Nigeria

Head of State:
Gen. Sani Abacha

10 Head of State

10-10000 Office of the President
Ribadu Road, Ikoyi Island, State House, Lagos
☎(+234 1) 68 44 00; 68 44 01; 68 19 87 ▣ 21290
President: Gen. Sani Abacha 98

12 Ministries

12-10000 Ministry of Defence
Cabinet Buildings, Tafawa Balewa Square - PO Box 12590, Lagos
☎(+234 1) 63 10 56 ▣ 22610 deptry ng
Minister: Gen Sani Abacha 98

12-20000 Ministry of Foreign Affairs
Republic House, Marina, PO Box 12 600, Lagos
☎(+234 1) 63 52 36; 63 52 58; 63 51 78; 63 40 40
▣ 21236 extnal ng; 21444 extnal ng
Minister: Chief Tom Ikimi 98

12-40000 Ministry of Internal Affairs
Maloney Street, Lagos
☎(+234 1) 68 36 94
Minister: Baba Gana Kingibe 98

31 Embassies & Attachés

31-00660 Embassy of the Federal Republic of Germany
15 Eleke Crescent, Victoria Island, PO Box 728, Lagos
☎(+234 1) 261 10 11;261 10 82
Ⓔ 106655.2050@compuserve.com ▣ 21229 aalags ng
⊖ (+234 1) 261 77 95
Ambassador: H E Mr Johammes Lohse
Military/Defense Attaché: Lieut Col Volkhard Marx
Commercial Counsellor: Ernst Doring
Public Information Officer: Michael Peuss 98

31-00690 Embassy of Greece
Plot 1644 Oko-Awo Close, PO Box 1199,
Victoria Island, Lagos
☎(+234 1) 61 14 12 ▣ 21747 ⊖ (+234 1) 61 48 52 98

31-00810 Embassy of the Republic of Hungary
9 Louis Salomon Close, PO Box 3168,
Lagos, Victoria Island
☎(+234 1) 261 66 02; 261 35 51
Ⓔ huemlgs@infoweb.abs.net ⊖ (+234 1) 261 37 17
Ambassador: H E Dr Janos Balassa
Second Secreataty: Béla Bozsik

Honorary Consul
4, Abahabi Road, Kaduna
☎(+234 062) 23 35 50
⊖ (+234 062) 23 38 24; 21 52 24
Honorary Consul: Endre Déri 98

31-00890 Embassy of Italy
12 Eleke Crescent, Victoria Island, PO Box 2161, Lagos
☎(+234 1) 262 10 46/8 ▣ 21202
⊖ (+234 1) 262 10 50
Ambassador: H.E. Dr Stefano Rastrelli
First Secretary: Dr Antonio Bellavia ☎(+234 1) 61 29 12
Commercial Secretary: Dr Tommaso Procopio
☎(+234 1) 61 76 76/90 68/16 04
☎(+234 1) 61 30 58 (Fax)
Military Attaché: Col Mario Destefanis
☎(+234 1) 68 57 24/07 44 98

31-01340 Royal Norwegian Embassy
3 Anifowoshe St, Victoria Island, Lagos
☎(+234 1) 261 84 67 ▣ 21429 noramb ng
⊖ (+234 1) 261 84 69
Ambassador: H.E. Mr F.H. Nomme
Secretary: Terje T. Nervik 97

31-01420 Embassy of the Philippines
1 Danube Street, Off Babaginda Way,
Maitama District, Abuja
☎(+234 09) 523 76 49 ⊖ (+234 09) 523 76 50
Ambassador: H E Mr M Mas R Umpa
Administrative Officer: Ladrillo D Ordonez
Finance Officer: Alimodin D Eson
Protocol Officer: Abudullah D Rasuman
Consular Officer: Manuel A Duron
Property and Records Officer: Jessie C Esguerra
Secretary: Ms Farida R Tampi
Cultural Information Assistant: Romanoff M Raki-in
Trade Representative: Bukhare Madaya 98

31-01480 Embassy of Romania
Plot 1192D, Olugbosi Close, PO Box 72928,
Victoria Island, Lagos
☎(+234 1) 61 78 06; 61 32 97; 61 82 49
▣ 28828 romcom ng ⊖ (+234 1) 61 82 49
Ambassador: H E Dr Gheorghe Dumitru
Consul and Press Attaché: Silviu Predoiu
Economic Counsellor: Catalin Costache
Political Counsellor: Iacob Prada
First Economic Secretary: Bogdan Dumitrescu
Public Information Officer: Silviu Predoiu 98

Contact Reference

31-01700 Embassy of the Republic of the Sudan
2B Kofo Abayomi St, Victoria Island, PO Box 2428, Lagos
☎(+234 1) 261 58 89; 61 73 11; 61 12 28
⊖ (+234 1) 261 59 46
Ambassador:
H E Brig Gen Al-Awad Mohammed Al-Hassan (Retd)
Minister Plenipotentiary: Isaac C Kenti
Consular: Adil Ibrahim Mustapher
Second Secretary: Mohammed Murtada Mubarak
Administrative Attaché: Mohammed Dahab
Communication Officer: Khalid Hamza
Public Information Officer: Mohammed A A Jagoud 98

31-01840 Embassy of the Republic of Turkey
3 Okunola Martins Close, Ikoyi, PO Box 56252, Lagos
☎(+234 1) 269 11 40 ⊖ (+234 1) 269 30 40
Ambassador: H.E. Mr Numan Hazar 97

31-01890 United Kingdom High Commission
11 Eleke Crescent, Victoria Island,
Private Mail Bag 12136, Victoria Island, Lagos
☎(+234 1) 261 95 31; 261 95 37; 261 95 41
▣ 21247 ukrep ng ⊖ (+234 1) 261 40 21
High Commissioner: H E Mr G S Burton
(also non-resident Ambassador) (to Benin)
Deputy High Commissioner: R A Pullen
Counsellor, Economic and Commercial: D D Pearey
First Secretary, Political: T Clayden
First Secrtetary, Management: C W Corkin
First Secretary, Commercial: P A Stephenson
First Secretary, Immigration: A W McKenzie
First Secretary, Political: G D Beale
First Secretary, Political: A M Sommerlad
First Secretary, Medical Officer: Dr S B Davis
First Secretary, Political: M R Frost
Second Secretary, Political: J Ansell
Second Secretary, Technical Management: K Jacobs
Second Secretary, Management: J C Gardiner
Second Secretary, Management: A Mills
Second Secretary, Works: D G Wright
Second Secretary, Aid: R Watt
Second Secretary, Immigration: W T S Fenton
Second Secretary, Immigration: A Reuter
Second Secretary, Immigration: A P Brown
Second Secretary, Immigration: A P Bell
Second Secretary, Nurse: Ms Y Gibney

Third Secretary, Information: J Wattam
Third Secretary, Technical: J M Dunn
Attaché, Management: Miss A Le Verrier
Attaché, Management: Miss C L Barrow
Attaché, Management: D M A Chambers
Attaché, Management: D L Thomas
Attaché, Management: T Wiltshire [98]

75 Manufacturing Companies

75-05000 Daewoo Nigeria Ltd
(a subsidiary of Daewoo Corporation, South Korea)
GPO Box 8686, Lagos
Plot 1608 Adeola Hopewell Street, Victoria Island, Lagos
☎ (+234 1) 262 21 12; 261 69 57
⊜ (+234 1) 262 21 10 [98]

80 Service Companies & Organizations

80-35000 YASAVA Ltd
27 Boyle Street, Onikan, Lagos
☎ (+234 1) 260 08 45; 61 60 54 ; 262 04 60
⊜ (+234 1) 263 75 47; 269 13 50
10 (1996)
Chief Executive Officer: Christopher N Mbanefo
Executive Director: Capt R E Hayes
■ Specialised flight services for petroleum industry. Contract charter, EMS evacuation flights, aircraft management. [98]

90 Sales Agents, Sales & Liaison Offices

Companies represented:

Indal Technologies Inc., Canada 90-02000

90-02000 Balmoh (Nigeria) Ltd
PO Box 54656, Falomo, Lagos
☎ (+234 1) 84 14 34
Marketing Mgr: Mohammed Alhaji
® Canada: Indal Technologies Inc. [98]

90-03000 Daewoo Nigeria Ltd
Plot 1608 Adeola Hopewell Street, Victoria Island, Lagos
☎ (+234 1) 261 71 08 ⊠ 28647 daewoo ng
⊜ (+234 1) 262 21 10 [98]

Norway

Head of State:
H.M. King Harald V

10 Head of State

10-10000 Office of H.M. the King
Det. Kgl. Slott, N-0010 Oslo 1
☎ (+47) 22 44 19 20 ⊜ (+47) 22 55 08 80
King of Norway: H.M. King Harald V
Private Secretary to HM the King: Magne K. Hagen
Chief of the Court: RAdm. Sivert Farstad
Marshal of the Court: Col Lars Petter Forberg
Aides de Camp Department
Det. Kgl. Slot, N-0010 Oslo 1
Senior Aide de Camp: Col Tore H. Bolstad
Army Representative: Col Bjørn Ruud
Air Force Representative: Col Tore H. Bolstad
Naval Representative: Cdre Petter Andreas Ask (RNN) [97]

11 Prime Minister

11-10000 Office of the Prime Minister
Akersgaten 42, PO Box 8001, Oslo 1
☎ (+47) 22 34 90 90 ⊠ 21565 stm n
⊜ (+47) 22 34 95 00
Prime Minister: Thorbjorn Jagland ☎ (+47 2) 34 40 00; 34 95 00 (fax) ⊠ 21565 stm n [98]

12 Ministries

12-10000 Royal Ministry of Defence
Myntgaten 1, PO Box 8126 Dep, N-0032 Oslo 1
☎ (+47) 23 09 80 00
Minister of Defence: Jorgen Kosmo [98]

12-20000 Royal Ministry of Foreign Affairs
Kgl Norske Utenriksdepartement
7 Juni Plassen 1, PO Box 8114 Dep, N-0032 Oslo 1
☎ (+47) 22 24 36 00
Minister of Foreign Affairs: Bjorn Tore Godal
Minister of Development Aid Co-operation: Ms Kari Nordheim-Larsen
Secretary General: Bjarne Lindstrom
Press Spokesman: Ingvar Havnen [98]

12-40000 Ministry of Justice and Police
Akersgaten 42, PO Box 8005 Dep, N-0030 Oslo 1
☎ (+47) 22 24 90 90 ⊜ (+47) 22 24 95 30
Minister of Justice and Police: Gerd-Liv Valla [98]

12-60000 Ministry of Transport and Communications
Det Kongelige Samferdselsdepartement
PO Box 8010 Dep, N-0030 Oslo 1
☎ (+47) 22 24 90 90 ⊠ 21439 sdep n
⊜ (+47) 22 24 95 71
Minister: Sissel Roenbeck
Secretary General: Ms Karin M Bruzelius [98]

20 Defense Administrations/Services/Agencies

20-04000 Civil Aviation Adminstration
Luftfartsverket
Wergelandsveien 1, PO Box 8124 Dep, N-0032 Oslo 1
☎ (+47) 22 94 20 00 ⊠ 71032 enfb n
⊜ (+47) 22 94 23 90
Director General: Ove Liavaag
Deputy Director General: Ole M Rambech
Director Administration Division: Ragnar Praesttun
Director Technical Division: Bjorn Vinge
Director Aeronautical Inspections Division: Viggo Lofsgaard
Director Airports/Air Navigation Services Division: Geir Ingebrethsen
Head Marketing Department: Lars Mork Gundersen
Public Affairs and Communications Director: Finn Langeland [98]

20-05000 Defence Combined Material Agency (DCMA)
Codification Branch
Oslo Mil/Loeren, N-0018 Oslo
☎ (+47) 23 09 65 48 ⊜ (+47) 23 09 64 57
National Director of Codification: Joar Johanssen
■ National Codification Bureau.
National authority for the implementation of the NATO Codification System. [98]

20-15000 Joint Norwegian Medical Service
(† Joint Medical Materiel Service)
The Library, Oslo Mil/Huseby, N-0016 Oslo
☎ (+47) 23 09 70 52 ⊜ (+47) 23 09 73 35
Commander: Maj Gen Nils H Fagerhaug
Librarian: Steinar Fjelldal [98]

25 Defense Forces

25-12000 Headquarters Allied Forces North Europe
PO Box 8080, N-4003 Stavanger
☎ (+47) 47 51 24 05 ⊜ (+47) 51 57 61 05
Commander Allied Forces North Europe (COMNORTH): Lt Gen. Sigurd Frisvold ☎ (+47) 51 57 24 04
Chief of Staff: Maj.Gen. John F. Miller ☎ (+47) 51 57 24 04
Commader Land Forces North Europe: Maj.Gen. Svein Erik Lysgaard ☎ (+47) 51 57 24 11
Commander Naval Forces North Europe: RAdm. Rino Paus ☎ (+47) 51 57 24 31
Commander Air Forces North Europe: Maj.Gen. Per-Oscar Jacobsen ☎ (+47) 51 57 24 61
Public Information Officer: Lt Col Brynjar Nymo ☎ (+47) 51 57 24 05
■ HQ North is an integrated tri-service NATO & national command with a single staff to perform both NATO & national tasks. It is the only peacetime NATO headquarters in Norway, and the only fully integrated headquarters in NATO.
COMNORTH (Commander Allied Forces North Europe) is dual-hatted. In his NATO capacity he exercises full operational control in all NATO matters, exercise, crisis & war. All NATO force in Norway report through COMNORTH up to the NATO chain of command. During peacetime in his national capacity as COMSONOR (Commander Armed Forces South Norway) he is equivalent to Commander Armed Forces North Norway, both reporting through their national chain of command.
→ 34 [97]

25-13000 Headquarters Defence Command Norway (DEFCOMNOR)
Oslo Mil/Huseby, N-0016 Oslo 1
☎ (+47) 23 09 82 26; 23 09 82 36
⊜ (+47) 23 09 84 83; 23 09 71 66
Chief of Defence: Gen. Arne Solli
Chief of Staff & HQ Defence Command: VAdm. Leonard Revang
Deputy Chief of Staff, Plans & Policy: Maj.Gen. Trond Moltzau
Deputy Chief of Staff, Operations: Cdre Ole Gerhard Røn
Deputy Chief of Staff, Personnel: Brig. Bjørn Ruud
Deputy Chief of Staff, Intelligence: Maj.Gen. Olav Bjerke
Deputy Chief of Staff, Security: Brig. Jan E. Larsen
Deputy Chief of Staff, Logistics: Brig Svein Ivar Hansen
Deputy Chief of Staff, Budget & Finance: Cdt. Petter A. Ask
Deputy Chief of Staff, Communication & Information: Brig. Erik N. Engestøl
Chief of Public Information Division, Spokesman: Brig. Erik lanke
Chief of Staff, Norwegian Army: Maj.Gen. Sverre Økland
Chief of Staff, Royal Norwegian Navy: RAdm. Hans Kr. Svensholdt
Chief of Staff, Royal Norwegian Air Force: Maj Gen Per Oscar Jacobsen
Chief of Staff, Norwegian Home Guard: Maj.Gen. Per Mathiesen
Press and Information Division
Søkedalsveien 148, Oslo Mil/Huseby, N-0016 Oslo 1
☎ (+47) 23 09 82 36 ⊜ (+47) 23 09 84 83; 23 09 71 66
Chief of Press and Information Division, Spokesman: Brig Erik lanke ☎ (+47) 23 09 82 26; 23 09 84 83 (fax)
Chief Spokesman Office: Ltc Knut Roar Bakken ☎ (+47) 23 09 82 34; 23 09 84 83 (fax)
Secretary to the Chief of Staff/HQ DEFCOMNOR: Col Bjarne Syrdal ☎ (+47) 23 09 82 13; 23 09 83 79 (fax)
Home Guard Staff
Oslo Mil/Huseby, N-0016 Oslo 1
☎ (+47) 23 09 86 72 ⊠ 71240 ⊜ (+46) 23 09 71 14
Chief of Staff, Norwegian Home Guard: Maj Gen Per Mathisen
Chief of Staff/Home Guard Staff: Cdre Thor Johansen
Air Force Staff
Oslo Mil/Huseby, N-0016 Oslo 1
☎ (+47) 23 09 88 01 ⊜ (+47) 23 09 75 03
Chief of Staff RNoAF: Maj Gen Per-Oscar Jacobsen
Chief of Staff/Air Force Staff: Brig Åge Hamang
Chief of Planning Division: Col Arne Skogstad
Inspector of Flying: Brig Geir Harildstad
Inspector of Control and Reporting: Brig Lars Myraune

Inspector of Air Defence Artillery: Brig Tor A Sanolí
Inspector of Flight Safety: Lt Col Kjell Tungesvik
Long-Term Planning Section SPLS:
Maj Hans Erik Swendgaard
Aide de Camp to the CoSRNoAF: Capt Atle Bastiansen

Army Staff (K-13) (DEFCOMNOR/Army)
Oslo mil/Huseby, N-0016 Oslo 1
☎ (+47) 22 49 85 03
Chief of Staff Norwegian Army: Maj.Gen. Sverre Økland
Aide-de-camp to the Inspector General:
Capt. Odin Johannessen
Deputy Chief of Staff: Brig. Roar Haugen
Inspector Infantry: Brig. Nils Fosland
Inspector Cavalry: Brig. Johan Brun
Inspector Artillery: Brig. Thorstein Skiaker
Inspector Engineers: Brig. Arne Broberg
Inspector Signals: Brig. Nils Hveberg
Inspector Transport/ Quartermaster Corps & Ordnance:
Brig. Paul Berntsen
Inspector Medical Services: Brig. Svein Ødegarden
Inspector Veterinary Services: Col Odd Knavelsrud
Chief Organisation Div.: Col Torgeir Hagen
Chief Personnel Div.: Col Ole Sannes
Chief Education Div.: Lt Col Asbjørn Schjølberg [98]

25-19300 National Defense College
Forsvarets Høyskole
Oslo Mil/Akershus, N-0015 Oslo 1
☎ (+47) 22 40 37 04 ⊜ (+47) 22 40 33 19
Commandant: Maj.Gen. Gullow Gjeseth
Chief of Staff: E.C. Kjelling [97]

25-23200 Bardufoss Air Station
PO Box 5, N-9201 Bardufoss
☎ (+47) 77 89 50 00 ⊜ (+47) 77 89 61 09
Commander: Col T.M. Arestoel ☎ (+47) 77 89 60 02
Chief of Staff & Information Officer: Maj. B.P. Eilertsen
☎ (+47) 77 89 60 02
Operations Dir: Lt Col B.N. Staff ☎ (+47) 77 89 62 00
Procurement Dir: Lt Col K. Ingebrigtsen [97]

25-42000 Royal Norwegian Navy
Oslo Mil/Huseby, N-0016 Oslo
☎ (+47) 23 09 87 01 ⊜ (+47) 23 09 81 57
Chief of Staff: Rear Adm Hans K Svensholt
ADC to Chief of Staff: John E Johnsen [98]

29 Customs & Excise

29-10000 The Norwegian Directorate of Customs & Excise
Schweigaardsgt. 15, N-0032 Oslo 1
☎ (+47 2) 86 03 00 ⊜ (+47 2) 17 65 24
General Director: Frida Nokken
Director: Finn Fredriksen [97]

31 Embassies & Attachés

31-00100 Embassy of Austria
Thomas Heftyesgt 19-21, N-0244 Oslo
☎ (+47) 22 55 23 48/49 ⊜ (+47) 22 55 43 61
Ambassador: H.E. Mr Harald Wiesner
Deputy Chief of Mission: Dr Thomas Nader

Commercial Section
Oscarsgt 81, N-0256 Oslo
☎ (+47) 22 55 57 30 ⊜ (+47) 22 55 66 22
Counsellor of Commercial Affaires: Robert Luck [97]

31-00250 Embassy of the Republic of Bulgaria
Tidemands Gate 11, N-0244 Oslo 2
☎ (+47) 22 55 40 40 ▣ 79708 bultr n
⊜ (+47) 22 55 40 24
Ambassador: H.E. Mrs Antoinette Primatarov
(Resident in Stockholm) ☎ (+46 8) 723 09 38
Chargé d'Affaires: Mrs Svetla Popova
☎ (+47) 22 44 31 80
Deputy Chief of Mission Attaché: Christo Polendakov
Commercial Counsellor: Christo Karakolev
☎ (+47) 22 43 66 22 [97]

31-00310 Embassy of Canada
Oscarsgate 20, N-0244 Oslo
☎ (+47) 22 46 69 55/58 ⊠ DOMCAN OSLO
▣ 71880 domca n ⊜ (+47) 22 69 34 67
Ambassador: H.E. Mr François Mathys
Counsellor & Head of Chancery: Miles Godfrey
Commercial Counsellor: Ms Catherine Dickson
Canadian Forces Attaché: Col Leslie White
Public Information Officer: Bjørn Petter Hernes [97]

31-00460 Royal Danish Embassy
7 Olav Kyrres Gate, N-0244 Oslo
☎ (+47) 22 54 08 00 ⊠ AMBADANE OSLO
▣ 71275 amdan n ⊜ (+47) 22 55 46 34
Ambassador: H.E. Mr Hans Henrik Bruun
Deputy Chief of Mission: Ole F. Joergensen
Commercial Counsellor: Joachim Rehder [97]

31-00510 Embassy of the Arab Republic of Egypt
90a Drammensveien, N-0244 Oslo 2
☎ (+47) 22 20 00 10; 22 44 77 67 ▣ 78405 faris n
⊜ (+47) 22 56 22 68
Ambassador: H.E. Mr Mahmoud Mubarak
Second Secretary: Amr Moawad [97]

31-00535 Embassy of Estonia
Oscars Gate 216R, N-0244 Oslo
☎ (+47) 22 59 98 02 [E] sekretar@estemb.no
⊜ (+47) 22 59 98 04
Chargé d'Affaires: Enno Tummen
Secretary: Ms Katri-Heléna Mãreka [98]

31-00580 Embassy of Finland
Thomas Heftyesgate 1, N-0244 Oslo
☎ (+47) 22 43 04 00 ⊜ (+47) 22 43 06 29
Ambassador: H.E. Mr Jorma Inki
Minister Counsellor & Deputy Chief of Mission:
Frank Edman
Military, Naval and Air Attaché: Lieut Col Vilijo Hyverinen
Commercial Attaché: Marja Kahra [98]

31-00660 Embassy of the Federal Republic of Germany
Oscarsgate 45, N-0258 Oslo 2
☎ (+47) 22 55 20 10 ▣ 71173 aaosl n
⊜ (+47) 22 44 76 72
Ambassador: H.E. Mr Helmut Wegner
Minister Counsellor: Rolf Dieter Schnelle
First Secretary Commercial Affairs: Joachim Heidorn
Defense Attaché: Lt Col Hans Joachim Hövelmann
Head of Administration: Kanzler Rudolf Berkner [97]

31-00920 Embassy of Japan
33b Parkveien, N-0244 Oslo 2
☎ (+47) 22 55 10 11; 55 10 12 ⊠ TAISHI Oslo
▣ 76085 taisi n ⊜ (+47) 22 44 25 05
Ambassador: H.E. Mr Masaki Seo
Deputy Chief of Mission: Kenro Iino
Defense Attaché: Capt. Hideo Sekiguchi (N) [97]

31-01190 Embassy of Mexico
108b Drammensveien, N-0244 Oslo
☎ (+47) 22 43 11 65; 22 43 14 77 ⊜ (+47) 22 44 43 52
Ambassador: H.E. Mr Antonio Villegas
Deputy Chief of Mission: Bernardo Córdova [97]

31-01480 Embassy of Romania
Oscars Gate 51, N-0258 Oslo 2
☎ (+47 22) 44 15 12; 56 12 37
[W] http://home.sol.no/marascu/romania.htm
▣ 76844 ambro n ⊜ (+47 22) 43 16 74
Ambassador: H.E. Mr Mihai Croitoru
First Secretary: Iacob Olimpiu Radu Seni
Defence Attaché: Col Rudolf Gheorghe Nosca
Commercial Counsellor: Aurel Viorel Marascu
Third Secretary: Florian Gabriel Dragos Medaru
Attaché: Gheorghe Veselu [98]

31-01485 Embassy of the Russian Federation
Drammensveien 74, N-0271 Oslo 2
☎ (+47) 22 55 32 78/9 ⊜ (+47) 22 55 00 70
Ambassador: H E Mr Yuri E Fokine
Deputy Chief of Mission: Vadim V. Rozanov
Defence, Military, Naval and Air Attaché:
Col Gennadi M Savostjuk
Assistant Defence Attaché: Lieut Col Anatoli A Tasolkine
Assistant Defence Attaché: Col Sergej V Volostnov
Counsellor Consular Affairs: Vladimir G Lipaev
Counsellor: Aleksandr F. Monakhov
Counsellor: Serguei V Tchachnikov
First Secretary Admin Affiars: Vladimir N. Eltsjaninov
First Secretary Science & Technology: Igor F. Rumjantsev
Second Secretary Press Affairs: Serguei S Mironov

Commercial Department
Drammensveien 106 B, N-0273 Oslo
☎ (+47) 22 44 40 25 ▣ 74318 ⊜ (+47) 22 55 35 78
Commercial Representative: Vitalij V. Sakharov [98]

31-01740 Embassy of Switzerland
78 Bygdøy Allé, N-0244 Oslo
☎ (+47) 22 43 05 90 ⊜ (+47) 22 44 63 50
Ambassador: H E Mr Gaudenz von Salis
Counsellor: Michel Coquoz
Defence, Military, Naval and Air Attaché:
Col Georg von Erlach (based in Stockholm)
Assistant Defence Attaché: Lieut Col Manfred Lüdi
(based in Stockholm)
First Secretary (Administrative and Consular Affairs):
Beat Moser [98]

31-01890 Embassy of the United Kingdom
8 Thomas Heftyesgate, N-0244 Oslo
☎ (+47) 22 55 24 00 [E] britemb@online.no
⊜ (+47) 22 55 10 41
Ambassador: H E Mr M Elliott (CMG)
Deputy Head of Mission: D Blunt
First Secretary, Economic and Commercial: Dr Sweeny
Defence and Naval Attaché: Lieut Col P Irvine
Military and Air Attaché: Lieut Col H Gullan
Press and Information Officer: M C Peters [98]

31-01900 Embassy of the United States of America
Drammensveien 18, N-0244 Oslo 2
☎ (+47) 22 44 85 50 [E] oslo@usis.no
⊜ (+47) 22 44 04 36
Ambassador: H.E. Mr Thomas A. Loftus
Deputy Chief of Mission: Sharon K. Mercurio
Commercial Attaché: Hilbert Engelhardt
Public Affairs Officer: Michael T. Scanlin
Defense Attaché: Capt. Robert H. Paleck (US Navy)

Office of Defense Co-operation
Drammensveien 18, N-0244 Oslo 2
☎ (+47) 22 44 85 50 ⊜ (+47) 22 44 04 36
Chief of ODC: Col Daniel G. Penny (Jr) USAF [98]

34 Multinational Intergovernmental Organizations

Headquarters Allied Forces 25-12000

60 Research Organizations, Institutes & Test Establishments

60-00300 Kongsberg Informasjonskontroll A/S Defence Systems Group
(† Informasjonskontroll A/S)
(a subs. of Norsk Forsvarsteknologi A/S)
PO Box 265, N-1370 Asker
☎ (+47) 66 79 06 60 ⊜ (+47) 66 90 03 45
♚ 79
Manager: Per Skard
■ Main services and products: software development, system design, signal processing, electronics, communication. [97]

60-01400 Norwegian Defense Research Establishment
Forsvarets Forskningsintitutt
PO Box 25, N-2007 Kjeller
☎ (+47 63) 80 70 00 ⊜ (+47 63) 80 71 15
♣ NKr 330 000 (1995) ♚ 521
Dir General: Nils Holme ☎ (+47 63) 89 71 00

Contact Reference

Chief of Systems Analysis Division: Ragnvald H. Solstrand
☎ (+47 63) 80 77 01
Chief of Underwater Defence Division: Jarl Johnsen
☎ (+47 33) 03 34 23
Chief of Electronics Division: Paul Narum
☎ (+47 63) 80 72 00
Chief of Weapons & Materiel Division: Bjarne Haugstad
☎ (+47 63) 80 75 00
Chief of Environmental Toxicology Division: Frode Fonnum
☎ (+47 63) 80 78 00
Librarian: Mrs Anne Kurland ☎ (+47 63) 80 71 28 [97]

60-01600 **The Sintef Group**

4 Strindveien, N-7034 Trondheim
☎ (+47) 73 59 30 00 [W] http://www.sintef.no/
▣ 55620 sintf n ⊖ (+47) 73 59 24 80
♣ NKr 1,500,000,000 ♟♟ 2,000
Managing Director: Roar Arntzen
Press Officer: Svein Tønseth
■ Contract research & development for industry & technology. [98]

65 Industry, Trade & Armed Forces Associations

65-00050 **Confederation of Norwegian Business and Industry (NHO)**

Middeltuns Gate 27, Oslo
☎ (+47) 22 96 50 00 ⊖ (+47) 22 96 53 71 [98]

65-02000 **Norwegian Trade Council**

(† The Export Council of Norway)
Drammensveien 40, N-0243 Oslo 2
☎ (+47 22) 96 63 00 ▣ 78532 exnor n
⊖ (+47 22) 92 64 00
♟♟ 132 (Head Office)
Managing Dir: Per Andreas Vogt
Sales & Marketing Dir: Bjornar Tinnesand
Operations & Technical Dir: Dag Holler
Information & Public Relations Dir: Helge Lund
■ Organisation handling promotion of exports & industrial co-operation with foreign partners. Acts as a service organ for the Ministry of Defence & Norwegian manufacturers.

Norwegian Defence Industries Export Association (NORDINEX)
Drammesveien 40, N-0255 Oslo 2
☎ (+47 2) 43 77 00 ▣ 78532 n ⊖ (+47 2) 55 26 28
Defence Materiel Co-operation Dir: Theis Uriksen
■ Association for the promotion of Norwegian defence industries on export markets.

Members include:
Dyno Industrier
Helikopter Service
Karl H. Hoie
Kvaerner Eureka
Norsk Forsvarsteknologi
Raufoss
EB Netcom
EB Lehmkuhl
Norcontrol Simulation
Norsk Marconi
Simrad Optronics
STK
Vinghogs Mekaniska Verksted [97]

70 Financial Holding Groups, Parent Companies

Kvaerner ASA 75-01090
Raufoss Technology AS 75-02200

70-00800 **Helicopter Services Group ASA (HSG)**

PO Box 531 Sentrum, N-0105 Oslo
☎ (+47) 22 42 32 70 ⊖ (+47) 22 42 40 01
♣ NKr 2,165,000,000 (1996) ♟♟ 1,806 (1996)
President and Chief Executive Officer: Lars Harlem
Senior Vice President Business Development: Jan Bengtson
Senior Vice President Finance: David A Harrison
Group Secretary: Ingunn Vartdal
■ Global helicopter operator. Helicopter support to the oil and gas industry, emergency services, defence support, contract maintenance and training services.

Operating companies:
Helikopter Service AS
Bond Helicopters Ltd, UK
Lloyd Helicopters Pty Ltd, Australia [98]

70-01000 **Kongsberg Gruppen ASA**

Kirkegårdsvn 45, PO Box 1003, N-3601 Kongsberg
☎ (+47) 32 73 82 00 [W] http://www.kongsberg.com
⊖ (+47) 32 73 85 86
♧ NKr 675,000,000 (1996)
♣ NKr 3,023,000,000 (1996) ♟♟ 3212 (1996)
⌂ 120,000 m2
■ High-technology industrial design and manufacture focused on the defence, maritime, civil aviation and space sectors. Development, production and delivery of advanced systems that integrate electronics, mechanical engineering and information technology. The Group's three business areas are Kongsberg Aerospace, Kongsberg Defence Systems and Kongsberg Norcontrol (Maritime Systems). → 75 → 80

Subsidiaries:
Kongsberg Spacetec AS
Kongsberg Simulation & Training AS
Kongsberg Informasjonskontroll AS
NFT Ericsson Communications ANS
Kongsberg Norcontrol AS
Kongsberg Naeringspark AS [98]

70-02200 **Raufoss ASA (RA)**

Head Office
PO Box 2, N-2831 Raufoss
☎ (+47) 61 15 20 00 ▣ 71444 ra n
⊖ (+47) 61 15 25 99; 61 15 20 94
♧ NKr 2,489,000,000 (book capital) (1996)
♣ NKr 2,664,000,000 (1996) ♟♟ 3,158 (1996)
Chairman: Jan Balstad
Group President: Nils Erik Skarsgård
Chairman of the Board: Hans O Bjøntegaard
■ Research, development and production in the space and automotive sectors. Ammunition and aerospace defence products.

Associated companies/subsidiaries:
Raufoss Technology AS
Raufoss Service AS
Raufoss Metall GmbH, Germany
Raufoss France SARL, France
Raufoss do Brasil Ltda, Brasil
Kongsberg-Raufoss AS [98]

74 Consortia, Joint Venture Companies

74-01000 **NFT Ericsson Communications ANS**

(† EB Ericsson Defence ANS)
(a joint venture between Ericsson AS and Kongsberg Gruppen AS)
PO Box 87, N-1361 Billingstad
☎ (+47) 66 84 24 00 ⊖ (+47) 66 84 82 30
♣ NKr 299,000,000 (1997) ♟♟ 130 (1997)
President: Gunnar Fredrikson
Marketing Dir: Jarl N. Johnsen
Operations Manager: Roar T Karlsen
Purchasing Manager: Jan Antonsen
■ Tactical communications systems, C3 systems. Major products include EriTac tactical multichannel area communication systems for army and air defence applications. On going development of next generation MultiRole Combat Netradio. → 75 [98]

75 Manufacturing Companies

Kongsberg Gruppen ASA 70-01000
Marine Elastomers 85-01000
NFT Ericsson Communications ANS . 74-01000

75-00050 **Alcatel Kitron A/S**

Kilsund, PO Box 13/14, N-4920 Staubø
☎ (+47) 37 05 21 00 ⊖ (+47) 37 05 22 21
♣ NKr 275 000 000 ♟♟ 300
■ Manufacture of crypto electronic systems, radio equipment, infrastructure & tactical communications systems. [97]

75-00090 **Alcatel Telecom Norway A/S**

(a subs. of Alcatel STK A/S)
PO Box 310 Økern, N-0511 Oslo 5
Rislokkveien 2, Okern
☎ (+47) 22 63 88 10 ▣ 74552 atnde n
⊖ (+47) 22 63 79 44
♣ NKr 1 400 000 000 (1995) ♟♟ 1 300
President: Nils Erik Skarsgaard
Marketing Dir: Bernt Mathisen
Marketing and Sales Director: Svein E Dippner
Mgr Defence: Jens Gjerloew
Export Mgr Defence: Bernt Mathisen
Communication & Public Relations Mgr: Helge Qvigstad
■ Manufacture communications systems and equipment, including nodal switching units, encryption hardware, secure telephone and tactical area communication systems. [97]

75-00110 **Asea Brown Boveri A/S Head Office (ABB)**

(a subsidiary of ABB Asea Brown Boveri Ltd, Switzerland)
Bergerveien 12, PO Box 94, N-1361 Billingstad
☎ (+47) 66 84 30 60 [E] info@noabb.abb.no
[W] http://www.abb.no ▣ 71723 teleb n
⊖ (+47) 66 84 35 49
♣ US$ 3,228,000,000 (1996) ♟♟ 12,398 (1996)
President & CEO: Kjell E. Almskog
Senior Vice President Communications: S I Aarshether
■ Supplier of products to the electricity & construction industries. Supplier to the oil & gas industry, nationally & internationally. Cleaning plants for the aluminium industry.

ABB Asea Skandia A/S
Alfasetveien 11, PO Box 205 Alnabru, N-0614 Oslo
☎ (+47) 87 50 00 ⊖ (+47) 87 59 00

ABB Teknologi A/S Corporate Research
Bergerveien 12, PO Box 90, N-1361 Billingstad
☎ (+47) 66 84 30 60 ▣ 71723 teleb n
⊖ (+47) 66 84 35 40

ABB Control A/S
Amtm. Aallsgt. 97, PO Box 100 Sentrum, N-3701 Skien
☎ (+47) 35 58 25 00 ▣ 21524 ⊖ (+47) 35 52 00 10

ABB Financial Services A/S
Bergerveien 12, PO Box 82, N-1361 Billingstad
☎ (+47) 66 84 31 00 ⊖ (+47) 66 98 04 97

ABB Flexible Automation A/S
(† ABB Trallfa Robot A/S)
PO Box 265, N-4341 Bryne
☎ (+47) 51 48 90 00 ⊖ (+47) 51 48 37 25

ABB Industri A/S
Hasleveien 50, PO Box 6540 Rodeløkka, N-0501 Oslo
☎ (+47) 22 87 20 00 ▣ 71105 kurem n
⊖ (+47) 22 87 29 00

ABB Installasjon A/S
Bergerveien 12, PO Box 92, N-1361 Billingstad
☎ (+47) 66 84 30 60 ⊖ (+47) 66 98 19 16

ABB Miljø Environmental Services A/S
Ole Deviksvei 10, PO Box 6260 Etterstad, N-0603 Oslo
☎ (+47) 22 87 40 00 ▣ 71745 vifte n
⊖ (+47) 22 87 49 00

ABB Kraft A/S
Jacob Borchs gate 6, PO Box 470 Brakerøa, N-3002 Drammen
☎ (+47) 32 24 80 00 ▣ 72569 delta n
⊖ (+47) 32 24 80 10

ABB Norsk Kabel A/S
Kjerraten 16, PO Box 369, N-3001 Krammen
☎ (+47) 32 24 90 00 ▣ 78149 ⊖ (+47) 32 83 98 20

ABB Offshore Technology A/S
Luramyrveien 65, PO Box 1200, N-4301 Sandnes
☎ (+47) 51 63 44 00 ⊖ (+47) 51 63 47 10

ABB Service A/S
Jacob Borchsgt 6, PO Box 470 Brakerøa, N-3002 Drammen
☎ (+47) 32 24 84 00 ⊖ (+47) 32 24 84 52 [98]

75-00120 **ABB Daimler-Benz Transportation (Norway) A/S**

(a joint venture between ABB Group and Daimler Benz)
Stasjonsveien 1, PO Box 83, N-2011 Strømmen
☎ (+47) 63 80 96 00 ⊖ (+47) 63 80 96 01
♣ US$ 66,000,000 ♟♟ 288 [98]

75-00520 Centa Transmissioner A/S
(a subsidiary of Centa Antriebe Kirschey GmbH, Germany)
Breiliveien 47B, PO Box 157, N-3201 Sandefjord
☎ (+47) 33 48 31 00 [E] centa@centa.no
⊖ (+47) 33 48 31 10
♧ NKr 500,000 (1997) ♣ NKr 16,000,000 (1996)
ꜩ 5 (1997) ⌂ 535 m2
Managing Director: Börre Karlsrud
■ Manufacture of mechanical power transmission systems. [98]

75-00540 Compact A/S
PO Box 38, N-5212 Softeland Bergen
☎ (+47) 56 30 66 66 ⊠ 40658 comp n
⊖ (+47) 56 30 74 40
♣ NKr 50 000 000 ꜩ 40
Managing Dir: Arne N. Andreassen
Sales & Marketing Dir: Tom Roll
Operations Dir: Jan Are Heldal
■ Military emergency food rations/survival rations. [97]

75-00550 Comrod A/S
PO Box 40, N-4120 Tau
☎ (+47) 51 74 62 00 ⊠ 33206 fiiber n
⊖ (+47) 51 74 51 27
♧ NKr 26,000,000 (1995) ♣ NKr 35,000,000 (1995)
ꜩ 43 ⌂ 3,000 m2
Managing Director: Per Hatlem
Operations Director: Steinar Alme
Purchasing Manager: Otto Lund
Marketing Manager: Tor Reidar Brekke
■ Manufacture of antennas for navaids monitoring and MF/HF/VHF/UHF air ground communication. [97]

75-00600 Dyno Industrier ASA Defence Products
N-3475 Saetre
☎ (+47) 32 79 15 77 ⊠ DYNAMIT ⊖ (+47) 32 22 81 10
♣ NKr 96,000,000 (1996) ꜩ 90 (1996) ⌂ 120,000 sq ft
President and Corporate Chief Executive Officer: Dag Mejdell
Managing Director: Baard Braein
Director Sales, Research and Development: Håvard Theigen
Operations Director: Martin Mellbye
Purchasing Manager: Harald Bergsvåg
■ Manufacture of military high explosives RDX, HMX & compositions derived thereof for pressing & casting in ammunition as well as plastic bonded explosives. Double-base propellants in the form of flakes, strips, tubes & circular sheets for propelling charges for guns & mortar bombs.
Subsidiaries:
Dyno Nobel, Sweden
Dyno Nobel Inc, USA [98]

75-00850 Eidsvoll Electronics AS (EIDEL)
PO Box 38, N-2081 Eidsvoll
☎ (+47 63) 96 42 30 [E] eidel@telepost.no
⊖ (+47 63) 96 20 48
♧ NKr 1,000,000 (1996) ♣ NKr 15,000,000 (1996)
ꜩ 15 (1996) ⌂ 900 m2
Managing Director: Tore Havstein
Product Manager: Ivar Nordby
■ Design & manufacture Pulse Code Modulation (PCM) telemetry equipment incorporating remote control system & time distribution. Encoders/decoders, data acquisition systems, military and professional systems development, solid state recorders for aerospace applications. [98]

75-00860 Ericsson Radar A/S
49 Isebakkeveien, N-1788 Halden
☎ (+47) 69 21 41 00 ⊖ (+47) 69 21 41 88
ꜩ 70
Managing Dir: Johnny Bardal
■ Development & production of military radar systems, including Giraffe 50 AT and ARTHUR (Artillery Hunting Radar). Radar simulators. [98]

75-00875 GECO Defence A/S
(a subsidiary of GECO AS)
Schlumberger House, Solbråveien 23, N-1370 Asker
☎ (+47) 66 78 80 00 ⊖ (+47) 66 78 85 01
Managing Dir: Per Gjestrum
Operations/Technical Dir: Halvor Lunde
■ Acoustic Mine Sweep Systems & Towed Arrays. [98]

75-00880 Hägglunds Moelv AS
(a subsidiary of Hägglunds Vehicle AB, Sweden)
PO Box 244, N-2391 Moelv
☎ (+47) 62 34 60 00 [E] mail@haggmo.no
⊖ (+47) 62 34 60 01
♧ NKr 8,000,000 (1996) ♣ NKr 107,000,000 (1996)
ꜩ 115 (1996) ⌂ 8,000 sq m
Managing Dir: Thormod Bjerke
Sales & Marketing Dir: Sverre Falch
■ Auxiliary equipment for defence applications. Recovery systems for wheeled and tracked vehicles, armoured engineer and armoured mine clearing vehicles, gun mounts, mortar platforms and off-road trailers. Recovery vehicles for recovery of battle tanks/heavy vehicles under difficult field conditions. Armoured engineer vehicles equipped with dozerblade, excavator and winches. Armoured mine clearing vehicles with a flail-based mine clearing unit. Military trailers in various sizes. Ring mounts for 0.50 calibre machine gun or 40mm grenade launcher. Mortar platforms. [98]

75-00890 Hitec A/S
8 Lagerveien, PO Box 178, N-4033 Forus
☎ (+47) 51 81 81 81 ⊖ (+47) 51 80 05 47
♣ NKr 474 527 000 (1995) ꜩ 229
President & CEO: Jon Gjedebo
Vice President: Ola Saetre
Vice President CTO: Arne Smedal
Vice President: Rune Kuernberg
Marketing Mgr: Toni Fadnes
■ Provides systems for automation & remote control in the following business segments: drilling, production units, subsea & environment. [97]

75-00970 Karl H. Høie & Co. A/S
Storgt. 37, N-0182 Oslo
☎ (+47) 22 20 47 79 ⊠ NBCN NORWAY
⊖ (+47) 22 20 28 91
♧ NKr 400 000
Director: Karl H. Høie
■ NBC decontamination, water treatment; cleaning equipment; non-destructive low pressure cleaning field shower equipment. Represented in the USA by Engineered Air Systems Inc. [97]

75-01030 Jotron Electronics A/S
PO Box 85, N-3280 Tjodalyng
☎ (+47) 33 12 45 77 ⊠ 21715 tron n
⊖ (+47) 33 12 67 80
♣ US$ 10,000,000 (1995) ꜩ 50
Chairman: Olaf Stavik
Managing Director: Steiner Saeter
Aeronautical Product Sales Manager: Ole-Steiner Nordhaug [E] ole.nordhaug@pop.marlink
Marine Product Sales Manager: Sten Frizel
Purchasing Manager: Merete Berdal
■ UHF and VHF ground-to-air radio equipment, emergency beacons, transceivers and handhelds. [97]

75-01040 Karmøy Winch A/S
Brygnes, N-4251 Kopervik
☎ (+47) 52 85 68 00 ⊖ (+47) 52 85 68 02
♣ £ 12,000,000 (1996) ꜩ 100 (1997) ⌂ 4,000 m2
Managing Director: Geir Hystad
Sales Manager: Helge Hystad
Purchasing Manager: Imgvar Melkevik
■ Emergency towing systems for tankers. Towing systems for tugs. Provision of 1-400ton winches worldwide. Other deck equipment. [98]

75-01045 Kongsberg Aerospace
(a division of Kongsberg Gruppen ASA)
PO Box 1003, N-3601 Kongsberg
☎ (+47) 32 73 81 10 [W] http://www.kongsberg.com
⊖ (+47) 32 73 93 30
♧ NKr 675,000,000 (Group) (1996)
♣ NKr 3,023,000,000 (Group) (1996)
ꜩ 610; 3,212 (Group) (1996) ⌂ 120,000 m2 (Group)
Managing Director: Tom Gerhardsen
Director Business Development: Odd Kyllingmark
Purchasing Manager: Just Swensen
■ Activities are focused on anti-ship missiles that can be launched from ships, fighter aircraft and helicopters. Develop and manufactures military products based on related propulsion navigation and construction technologies. → 80 [98]

75-01047 Kongsberg Norcontrol AS
(† Norcontrol Seacraft AS)
PO Box 1009, N-3194 Horten
☎ (+47) 33 03 20 00 ⊖ (+47) 33 04 22 50
ꜩ 200 (1997)
Managing Director: Trond Utne
Sales and Marketing Director: Leif Pedersen
Marketing Co-ordinator: Ms Marianne Danielsen [98]

75-01060 Kongsberg Simulation & Training Defence Systems Group
(† Micro Processor Systems A/S (MPS))
(a subs. of Norsk Forsvarsteknologi A/S)
Baneveien 32, N-3600 Kongsberg
☎ (+47) 32 73 57 66 [E] olav.sandnes@ks-t.no
[W] http://www/kongsberg.net/kst/english/
⊖ (+47) 32 73 69 65
ꜩ 47 (1997)
Manager: Torfiur Kiklal
Marketing Director: Olav Sandnes
■ Advanced air defence trainers and simulators, advanced sonar simulator systems, complete tactical trainers and ASW simulators. Combined combat simulation. [98]

75-01090 Kvaerner ASA
Prof. Kohts vei 5, PO Box 169, N-1324 Lysaker
☎ (+47) 67 51 30 00 ⊖ (+47) 67 51 30 10
♣ NKr 58,558,000,000 (1996) ꜩ 60,000 (worldwide)
President and Chief Executive Officer: Erik Tønseth
Executive Vice President Shipbuilding: Diderik Schnitler
Executive Vice President Process: David Moorhouse
Executive Vice President Construction: Keith Clarke
Senior Vice President Finance: Jan Magne Heggelund
Senior Vice President Corporate Development: John Fletcher
Senior Vice President Corporate Communications: Trond Andersen
■ Major engineering and construction projects including shipbuilding, bridge building, mining and tunnelling. Engineering, management and contracting services.
Head Office: Kvaerner plc, UK. → 80 → 70 [98]

75-01093 Kvaerner Energy AS
Enebakkveien 71 c, PO Box 9277, N-0134 Grønland, Oslo
☎ (+47) 22 66 66 66 ⊖ (+47) 22 19 37 65
Manager Sales and Marketing Marine Propulsion: Lieut Cdr Katrine Kierülf (RNON) ☎ (+47) 22 66 65 98
■ Marine propulsion systems. [98]

75-01095 Kvaerner Eureka AS
(a subsidiary of Kvaerner ASA)
PO Box 38, N-3401 Lier
☎ (+47) 32 85 90 00 ⊖ (+47) 32 85 22 55
♣ Nkr 450,000,000 ꜩ 500 ⌂ 31,231 m2
President and Chief Executive Officer: B O Wille
Managing Director, Vice President Defence Division: T Bergheim
Operations and Technical Director, Vice President Manufacturing: Terje Bråthen
Marketing Manager, Defence Division: E R Røssum
Purchasing Manager, Defence Division: T Pedersen
■ Defence Programs:
Building gun turrets for system integration, project management. Building, as sole supplier, MLRS main launcher components for Lockheed Martin Vought Systems. Vehicle refurbishment and modification programs for Norwegian Armed Forces including weapon systems integration. Radar elevation systems to the Norwegian Air Force.
Defence Products:
TOW missile launching turret, radar elevation systems, aircraft shelter doors, general weapon systems integration and programme management. [98]

75-01100 Kvaerner Mandal A/S
(a subsidiary of Kvaerner A/S)
PO Box 283, N-4501 Mandal
☎ (+47) 38 27 92 00 ⊖ (+47) 38 26 03 88
♣ NKr 350,000,000 ꜩ 300
Managing Dir: Fredrik H. Behrens
Operations & Technical Dir: Nere G. Skomedal
Sales & Marketing Director: Petter Andreassen [E] petterandreassen@kman. kvaerner.telemax.no
Purchasing Mgr: Mogens L. Pedersen
■ Manufacture minehunters, minesweeper, fast patrol craft & ASW corvette. [97]

75-01260 Micro Design A/S
Th. Owesens gt. 35c, PO Box 3974 Leangen, N-7002 Trondheim
☎ (+47) 73 82 65 00 ⊜ (+47) 73 82 65 01
👥 70 (1997)
Managing Director: Kai Bogen
Purchasing Manager: Lars Andersen
Vice President, Sales and Marketing: Terje Dahl
■ Security and access control systems. Vehicular access control systems and traffic management systems. → 80 [98]

75-01630 Norcontrol A/S Maritime Systems Group
Bekkajordet 8a, PO Box 1024, N-3194 Horten
☎ (+47) 33 03 20 00 ⊜ (+47) 33 04 34 10
President: Jan Erik Korssdøen
Managing Dir: Steinar Espestøyl
Marketing Dir: Ian C. Vendrell
Marketing Co-ordinator: Ms Rita K. Nyborg
■ Training systems for naval use including tactical trainers, radar navigation & ship handling trainers, radar plotting trainers, engine room training simulators, tactical display units & tactical data links. Coastal & harbour surveillance systems, air traffic control systems. Vessel propulsion plant simulators. [97]

75-01640 Nordic VLSI A/S
Vestre Rosten 81, N-7075 Tiller
☎ (+47) 72 89 89 00 ⊜ (+47) 72 89 89 89
♣ NKr 38 000 000 (1995) 👥 42 (1996)
Managing Dir: Oddbjörn Trööyen
Technical Dir: Torstein Heggebö
Sales & Marketing Dir: Olav L. Lindquist
■ Design of integrated circuits & electronic systems based on ASICs (application specific integrated circuits), ranging from custom to gate arrays. [97]

75-01870 Nortronicom A/S
Components Department
PO Box 33, N-0612 Manglerud, Oslo 6
☎ (+47 2) 268 09 10 ⊜ (+47 2) 268 18 87
Representing:
EEV Ltd, UK [98]

75-01950 Norsk Jetmotor A/S
Kirkegardsveien 45, PO Box 1004, N-3601 Kongsberg
☎ (+47) 32 73 95 00 [E] office@jetmotor.no
[Tx] 79637 jet n ⊜ (+47) 32 73 62 75
♧ US$ 36,000,000 (1997) ♣ US$ 80,000,000 (1997)
👥 400 (1997) 🏠 150,000 sq ft
President: Erik Hernes
Vice President Finance: Svein Midtskog
Vice President Marketing and Programs: Per Hroar Olsen
Purchasing Manager: Trevor G Gosling
■ Manufacture of components and modules in fan, low pressure turbine and exhaust area of military and commercial jet engines, cases, exhaust cases, seals, hubs, rings and vanes. [98]

75-02050 Norwegian Defence Technology A/S
Kongsberg Næringspark, PO Box 1003, N-3600 Kongsberg
☎ (+47) 32 73 82 00 [Tx] 71491 vaapn n
⊜ (+47) 32 73 85 86
♣ NKr 2 200 000 000 (1994) 👥 2 288 (1994)
Chairman: Christian Brinch
President & CEO: Jan T. Jørgensen
Exec. VP Corporate Finance: Arne Solberg
Exec. VP Corporate Staff: Andreas Dahl
Exec. VP & General Mgr Aerospace Group: Tom Gerhardsen
Exec. VP & General Mgr Defence Systems Group: Bjørn Viggo Bjune
Exec. VP & General Mgr Maritime Systems Group: Jan Erik Korssdøen
Information Mgr: Jan Egil Andersen
■ Penguin anti-ship missile, anti-ship missiles for fighter aircraft, helicopters & naval vessels, missile launchers & missile control systems, aerospace structures & mechanical maintenance, electronic ground-based support systems for space operations, environmentally durable electronics for space operations & civil aviation.
Subs.:
Kongsberg Norcontrol A/S 100%, Kongsberg Norcontrol Automation A/S 100%, Kongsberg Norcontrol Systems A/S 100%, Kongsberg Norcontrol Training A/S 100%, Kongsberg Norcontrol Seacraft A/S 100%, NFT Baartek A/S 55.4%, Kongsberg Spacetec A/S 100%, NFT Ericsson Communications ANS 50%, Kongsberg Informasjonskontroll A/S 100%, Kongsberg Simulation & Training A/S 100%, Kongsberg Neringspark A/S 100%, Kongsberg Spedisjon A/S 100%, Basetec A/S 100%. [97]

75-02125 Oceanor
Oceanographic Company of Norway ASA
Pirsenteret, N-7005 Trondheim
☎ (+47) 73 54 52 00 [E] oceanor@oceanor.no
[W] http://www.oceanor.no ⊜ (+47) 73 54 52 01
Managing Director: S A Andersen
Sales and Marketing Director: S Tryggestad
■ Research & development of environmental & performance monitoring systems, used in environmental control, offshore instrumentation & aquaculture operations. [98]

75-02135 OmniTech A/S
Nedre Åstveit 12, N-5083 Ovre Ervik
☎ (+47) 55 19 58 50 ⊜ (+47) 55 19 32 05
■ Production of 3D acoustic cameras. [98]

75-02200 Raufoss Technology AS (RATEC)
PO Box 77, N-2831 Raufoss
☎ (+47) 61 15 20 00 ⊜ (+47) 61 15 27 54
♣ NKr 1,308,000,000 (1996)
President: Nils Erik Skarsgård
Marketing Director: Per Lillehaug
■ Products include: Small arms ammunition; medium calibre 12.7mm to 40mm; anti-ship greandes; anti-armour weapons; artillery ammunition; mortar bombs; pyrotechnical products; 155mm shell bodies; 40mm brass cases; specialised aluminium components for the defence industry; forged steel and aluminium components; rockets; rocket motors; warheads; space products; composite materials and pressure vessels; air brake system couplings for heavy vehicles; ISIFLO couplings. → 70 [98]

75-02250 Reime Coatection A/S
PO Box 55, N-3671 Notodden
☎ (+47) 35 02 55 20; 35 02 55 22 ⊜ (+47) 35 01 55 30
♣ NKr 24 000 000 (1996) 👥 35 (1996) 🏠 3 000 sq m
Managing Dir: Jan Erik Lindvåg ☎ (+47) 35 02 55 24
Sales & Marketing Dir: Jan Petter Lindstad
Technical Dir: Henrik Hognestad
■ Surface treatments ranging from PTFE-coatings on subsea components to chromating of light metal alloy for aircraft. [97]

75-02260 Reime Jarlsø A/S
(† Jarlsø Fabrikker A/S)
PO Box 70 - Jersøy, N-3108 Tønsberg
☎ (+47) 33 32 54 35 ⊜ (+47) 33 32 83 08
♧ NKr 62 700 000 (1995) ♣ NKr 114 400 000 (1995)
👥 150 (1996) 🏠 20 000 sq m (1996)
Managing Dir: Gullik Jensen
Operations Dir: Øivind Skåtun
Sales & Marketing Dir: Lars Strømme
Purchasing Mgr: Roar Bergersen
■ Manufacture of towers, marts & steel structures for high-power transmission & telecommunications. [97]

75-02300 ROFi Industrier AS
(a subsidiary of ROFi Gruppen AS)
Aaroesetervegen 5, N-6400 Molde
☎ (+47) 71 21 11 99 [E] rofi@sn.no ⊜ (+47) 71 21 18 27
♣ NKr 50,000,000 (1996) 👥 54
Managing Director: Berner M Olsen
Purchasing Manager: Oscar Hanson
Marketing Manager: Stein Hagen
■ Manufacture of tents, car covers, bullet and splinter jackets and helmets. Beds, mattresses, stretchers, snow shoes and ski poles. [98]

75-02350 Rottefella A/S
(a subsidiary of Møller Industrier A/S)
Østensjøvn 79, PO Box 23, N-0611 Oslo
☎ (+47) 22 76 00 76 ⊜ (+47) 22 76 00 70
👥 71 (1996)
Managing Dir: Leif Berven
■ Ski bindings for military use. [98]

75-02400 Saft Nife AS
Roa Naeringspark, Aslakveien 14, N-0753 Oslo
☎ (+47) 22 51 15 50 ⊜ (+47) 22 51 15 40
Managing Director: Trond Beyer [98]

75-02410 Seatex A/S
Pirsenteret, N-7005 Trondheim
☎ (+47) 73 54 55 00 [E] firmapost@seatex.no
[W] http://www.seatex.no [Tx] 55102 seatex n
⊜ (+47) 73 51 50 20
👥 70 (1996) 🏠 2,000 m2
Managing Dir: Hans Olav Torsen
Marketing Mgr: Geir H. Mathisen
■ Design & manufacture of advanced navigation & positioning products, subsea inspection, oceanographic buoys & multi-axis motion sensor systems. [98]

75-02435 Servogear A/S
N-5420 Rubbestadneset
☎ (+47) 53 42 73 80 [E] main.office@servogear.no
⊜ (+47) 53 42 77 83
♧ NKr 2,800,000 (1997) ♣ NKr 42,000,000 (1996)
👥 32 (1997) 🏠 1,700 m2
Managing Director: Magne Møklebust
Operations Director: Arvid Kvernøy
■ Manufacture propeller systems for fast craft. [98]

75-02450 Siemens A/S Defence Electronics Division
Østre Aker Vei 90, PO Box 10, N-0518 Veitvet, Oslo
☎ (+47) 22 63 30 00; 22 63 31 97 ✉ SIEMENS OSLO
[E] mja@fs.siemens.no [Tx] 27769 siema e
⊜ (+47) 22 63 41 07
Vice President: F Almaas
Marketing Executive: Moten Jarodd
■ Data distribution and processing, military computers, monitoring and control systems. [98]

75-02620 Simrad Optronics ASA
Ensjøveien 23b, PO Box 6114, N-0602 Etterstad, Oslo
☎ (+47) 22 67 04 90 ⊜ (+47) 22 19 29 91
♧ NKr 70,525,000 (1996) ♣ NKr 133,000,000 (1996)
👥 91 (1996)
Managing Director: Roy Jensen
Marketing Director: Kristian Vikdal
Sales Manager: B Kristensen
Sales Manager: S Hagen
Sales Manager: Ø Sandby
■ Development and manufacture of electro-optical equipment for military range finders, light amplifiers, thermal cameras, laser warning receivers and ceilometers. [98]

75-02630 Simrad Norge A/S
Strandpromenaden 50, PO Box 111, N-3191 Horten
☎ (+47 33) 03 40 00 [Tx] 70391 simhn
⊜ (+47 33) 04 44 24
👥 1 100
Managing Dir: Ernst Eik
Contact: Trond Helland
Marketing Mgr, Naval & Ocean Sc. Dept: Jan Haug Kristensen
■ Manufacturer of hydroacoustic products for naval systems & other applications. For naval purposes: sonars for ASW & MCM, passive listening devices & systems, positioning & tracking systems, hydrographic sounders, dynamic position systems, vessel management systems, ECDIS, combat information systems.
Subs.:
Simrad Srl (Italy)
Simrad (Asia) Pte Ltd (Singapore)
Simrad Far East Ltd (Taiwan)
Simrad Inc. (USA) → 90
Simrad Ltd (UK) → 75
Simrad Mesotech Systems Ltd (Canada) → 75
(† Bentech Subsea A/S)
● Wergelandsveien 23 A, PO Box 55, N-7501 Stjørdal
☎ (+47) 74 83 34 00 ⊜ (+47) 74 83 34 25
♣ NKr 12 700 000 👥 20
Underwater Design Group Mgr: Oddbjorn Malmo
Sub-bottom Profiling Mgr: Johnny Dybedal
■ Parametric sonars, Mine disposal systems, electronic scanning sonars. [97]

75-02640 Kongsberg Spacetec A/S Aerospace Group
(† Spacetec A/S)
(a subs. of Norsk Forsvarsteknologi A/S)
Prestvannveien 38, N-9005 Tromsø
☎ (+47) 77 68 45 00 ⊠ 64025 space n
⊜ (+47) 77 65 58 59
Manager: Bjørn Arne Kanck [97]

75-02700 Bertel O. Steen A/S
PO Box 52, N-1471 Skaarer
☎ (+47 67) 92 60 0 ⊠ 74699 ⊜ (+47 67) 92 62 66
♣ NKr 4 400 000 000 (1994) 👥 1 050
Managing Dir: Erling Ihle
■ Manufacture of emergency power supply generators to customer's specifications, light trailers & assembly of special mounts & equipment in vehicles & vessels. Importer of military & civilian vehicles. → 85 [97]

75-02890 THORN Lighting AS
(† THORN Belysning AS)
Industriveien 11, N-1481 Hagan
☎ (+47) 67 06 04 15 ⊠ 16929 ⊜ (+47) 67 06 03 51
Chief Executive: O Danielsen [98]

75-02950 Trelleborg Viking A/S
(† Viking Mjøndalen A/S)
(a subsidiary of Trelleborg AB, Sweden)
PO Box 55, N-3051 Mjøndalen
☎ (+47) 32 23 21 00 ⊠ 71515 ⊜ (+47) 32 23 22 00
♣ NKr 150,000,000 👥 230
Managing Dir: Erik Leknes
Marketing Mgr: Tore Nikkerud
■ Vehicle tracks, fire protection, fire protected ammunition boxes, ballistic panels. [98]

80 Service Companies & Organizations

Kongsberg Aerospace 75-01045
Kongsberg Gruppen ASA 70-01000
Kvaerner ASA 75-01090
Micro Design A/S 75-01260

80-00800 Helikopter Service AS (HS)
(a subsidiary of Helicopter Services Group ASA)
PO Box 522, N-4055 Stavanger Lufthavn
☎ (+47) 51 65 37 00 ✉ NORCOPTER ⊠ 30366 fwn
⊜ (+47) 51 65 23 62
♣ NKr 1,446,000,000 (1996) 👥 720 (1996)
Chairman: Lars Harlem
Managing Director: Tor Schaathun
Vice President Commercial: Aage Nilsen
Vice President Maintenance: Jan Martinussen
Vice President Flight Operations: Jakob Bae
Information Officer: J P Petersen
Purchasing Manager: T Hoff
■ Offshore helicopter services. Flight training; search and rescue; scheduled services; helicopter and component overhaul.

Bergen Airport Office
Flesland Airport, N-5069 Bergen
☎ (+47) 55 98 31 00 ⊜ (+47) 55 22 91 81

Airport Office
Kristiansund Airport, Kvernberget, N-6500 Kristiansund
☎ (+47) 71 56 51 00 ⊜ (+47) 71 56 51 60

Forwarding Department
Oslo Fornebu Airport, N-1330 Oslo Airport
☎ (+47) 67 12 93 46 ⊜ (+47) 67 12 10 92

Floro Airport Office
PO Box 471, N-6901 Floro
☎ (+47) 57 74 08 20 ⊜ (+47) 67 12 10 92

Bronnoysund Airport Office
PO Box 150, N-8900 Bronnysund
☎ (+47) 75 01 80 80 ⊜ (+47) 75 01 80 90 [98]

80-01400 Det Norske Veritas (DNV)
Veritasveien 1, N-1322 Høvik
☎ (+47) 67 57 99 00 ⊠ 76192
⊜ (+47) 67 57 87 87 (corporate communications)
♣ NKr 3,321,000,000 (1996) 👥 4,000 (1996)
Chief Executive Officer and Chairman of the Executive Board: Sven Ullring
Deputy Chief Executive Officer: Stein Thor Verle
Corporate Management Staff: Miklos Konkoly-Thege
Business Area Maritime Industry: Tor-Christian Mathiesen
Business Area Process Industry: Bjørn Weibye
Division Technology and Products: Terje Staalstrøm
Head of Corporate Communication: Tore Høifødt
☎ (+47) 67 53 13 62
Head of Press Section: Ole Kristian Lunde
☎ (+47) 67 56 20 98
■ Independent provider of safety, quality and environmental management services for industry and shipping. Research and development services. Quality certification services.

Bergen
Johan Berentsensvei 109-111, PO Box 6005, N-5020 Bergen
☎ (+47) 55 94 95 00 ⊜ (+47) 55 94 96 00
Regional Manager: Bjørn Pedersen ☎ (+47) 55 28 55 51
Marketing Manager: Jan V Koren ☎ (+47) 32 84 60 28

Florø (DNVD)
PO Box 240, N-6901 Florø
Strandgaten 43
☎ (+47) 57 74 20 67 ⊜ (+47) 57 74 28 10
Station Manager: Vemund Melvaer ☎ (+47) 57 74 04 54

Fredrikstad
PO Box 304, N-1601 Fredrikstad
Nygårdsgaten 37
☎ (+47) 69 31 99 70 ⊜ (+47) 69 31 86 90
Station Manager: Carl-Morten Gjeldnes
☎ (+47) 69 31 10 30

Førde
PO Box 368, N-6800 Førde
Øyrane 10
☎ (+47) 57 82 31 22 ⊜ (+47) 57 82 01 55

Harstad
PO Box 1047, N-9401 Harstad
Storgt 49
☎ (+47) 77 01 88 30 ⊜ (+47) 77 01 88 40 [98]

80-17000 ScanAvia
Scandinavian Aviation Consultants Group
c/o Norconsult, Fjordvn. 4, N-1301 Scandvika
☎ (+47) 67 57 10 26 ⊜ (+47) 67 54 95 90
Managing Dir: Eilif Ness [97]

80-19000 Selmer ASA
25 St Olavsgt, PO Box 1175 Sentrum, N-0107 Oslo
☎ (+47) 22 03 06 00 [E] firmapost@selmer.telemax.no
⊜ (+47) 22 20 88 30
♧ NKr 2,170,200,000 (1996)
♣ NKr 4,166,100,000 (Group) (1996) 👥 3,391 (1996)
Managing Director: Sigmund Bjorgum
Finance Manager: Espen Gundersen
Purchasing Manager: Leif Dammerud
■ General contractor in heavy construction and building sector. Civil engineering contracts include roads, bridges, harbours, quays, larger industrial development projects, defence projects, hydropower projects, tunnels, rock caverns and dams. [98]

80-30000 Teleplan A/S
PO Box 69, N-1324 Lysaker
☎ (+47) 67 12 70 00 ⊜ (+47) 67 12 72 70
♧ NKr 73,629,510 (1996) ♣ NKr 350,000,000 (1995)
👥 400
President: Asgeir Myhre
Head of Defence Division: Sven-Erik Eriksen
Director System Division: John Vestengen
Director Telematics Division: Bjørn Taale Sandberg
■ Telematics consultancy and systems engineering. Systems design of special applications for EW, communications vulnerability analysis and C3. [98]

80-50000 Vestfold Bil & Karosseri A/S (VBK)
(a subsidiary of Naerings-Invest A/S)
PO Box 98, N-3191 Horten
☎ (+47 33) 07 37 00 [E] vbk-as@vbk-horten.vf.no
⊜ (+47 33) 07 37 05
♧ NKr 22,000,000 (1995) ♣ NKr 90,000,000 (1996)
👥 150 (1995) 🏭 12,500 sq m
Managing Dir: Olav Fossgard
Operations Dir: Arne Ambjørnsen
Purchasing Mgr: Reidar Stenlund
■ Development projects & prototype work. Overhaul, maintenance & repair of aircraft arresting gear, aircraft fuel tanks, firing control units & guns. Production of made-to-measure tarpaulins for military equipment, re-upholstering of aircraft seats & renovation of aircraft interiors. Supply of customised bodywork & interiors for military vehicles. [98]

85 Stockists, Distributors, Spare Parts Suppliers

Bertel O. Steen A/S 75-02700

85-01000 Marine Elastomers
Damsgardsgate 163, PO Box 72, N-5031 Lavsevag/Bergen
☎ (+47) 534 57 70 ⊜ (+47) 534 57 89
Contact: Henrik Bang Andreasen
■ Supply the military with components and systems for underwater applications. Also specialise in applications engineering of connectors, cable and harness systems supported by inhouse moulding facilities. → 75 [98]

90 Sales Agents, Sales & Liaison Offices

Companies represented:

A/S Dantherm, Denmark 90-00400
A/S Hydrema, Denmark 90-00900
A Steinmeyer GmbH & Co KG, Germany 90-00880
Alfred Karcher GmbH & Co, Germany . . 90-01160
Australian Defence Industries Ltd, Australia 90-01370
BDL Systems Ltd, UK 90-01370
Beaufort Air-Sea Equipment Ltd, UK . . 90-01370
Bell Helicopter Textron, USA 90-01920
Bocchi, Italy 90-00880
Bofors AB, Sweden 90-00300
Bowers, UK 90-00880
British Aerospace Defence Ltd, UK . . . 90-01370
British Steel Engineering Steels, UK . . 90-00310
C.E. Johansson, Sweden 90-00880
Cannon Electric GmbH, Germany 90-00320
Chadwick-Helmuth, USA 90-01920
Chemring Ltd, UK 90-01370
Cilas, France 90-01920
Compac SA, Switzerland 90-00880
Conax Florida Corp., USA 90-01370
Diemaco Inc., Canada 90-01370
Dynamit Nobel Graz, Austria 90-01370
Dynasafe AB, Sweden 90-01370
Euroavionics GmbH, Germany 90-01920
Eusung Industrial Co. Ltd, Korea (ROK) . 90-01370
GEC, UK 90-01920
GKN Defence, UK 90-01370
Helmet Integrated Systems Ltd, UK . . 90-01370
Hermes Electronics, Canada 90-02000
Hommelwerke GmbH, Germany 90-00880
Honeywell, USA 90-01920
Hunting Engineering, UK 90-01370
Leitech, Denmark 90-00880
Loral Electro Optical Systems, USA . . 90-01370
Loral Solartron, UK 90-01370
Manroy Engineering Ltd, UK 90-01370
Mowag Motorwagenfabrik AG, Switzerland 90-01370
Oskar Schwenk GmbH & Co, Germany . 90-00880
Pains-Wessex/Schermuly, UK 90-01370
Racal Avionics, UK 90-01920
Richmond Electronics/Engineering Int Ltd, UK 90-01370
Roblon, Denmark 90-00880
Royal Ordnance, UK 90-01370
Scepter Manufacturing Co. Ltd, Canada 90-01370
SEI Industries, Canada 90-01920
Shamban & Co., USA 90-00150
Siebe Gorman, UK 90-01370
Sierra Research, USA 90-01920
Signaal GmbH, Germany 90-01920
Spar Aerospace, Canada 90-01920
Sylvac SA, Switzerland 90-00880
Trimos-Sylvak Task Force, Switzerland . 90-00880
Vickers Defence Systems, ... 90-01370
Westland Helicopters, UK 90-02000
Wylam Hill Ltd, UK 90-00880
Wyler AG, Switzerland 90-00880

90-00020 Acer Scandinavia A/S Norwegian Liaison Office
Bleikerveien 17, N-1370 Asker
☎ (+47) 66 90 10 30 ⊜ (+47) 66 90 02 40 [97]

90-00150 Per-Kr. Asim A/S
Ole Deviks Vei 18, PO Box 33-Bryn, N-0611 Oslo 6
☎(+47) 22 64 60 80 ⊠ 19774 ⊖ (+47) 22 64 87 96
General Mgr: Arild Rognerud
® USA: Shamban & Co. [98]

90-00300 Bofors Norge A/S
Boks 33, Refsted, N-0513 Oslo 5
☎(+47 2) 22 87 90 ⊖ (+47 2) 71 22 60
Managing Dir: Harald Austad
® Sweden: Bofors AB [98]

90-00310 British Steel Norge AS
(a subsidiary of British Steel Engineering Steels, UK)
PO Box 13, N-0212 Skoyen, Oslo
☎(+47 22) 52 69 70 ⊖ (+47 22) 52 34 11
® UK: British Steel Engineering Steels [98]

90-00320 Cannon Electric GmbH
Kontor Norge
Hovfaret 13, N-0275 Oslo 2
☎(+47 2) 50 70 20 ⊖ (+47 2) 50 32 05
Area Manager: F Toemta
® Germany: Cannon Electric GmbH [98]

90-00360 CLP Chemicals A/S
(a subsidiary of Break-Free Inc., USA)
PO Box 44, N-1355 Baerum Postterminal
☎(+47) 67 53 30 80 ⊖ (+47) 67 12 36 95 [98]

90-00400 Dantherm A/S
Postboks 4, N-3101 Tønsberg
☎(+47) 33 38 58 00 ⊖ (+47) 33 38 51 91
Managing Director: Anders Horn
® Denmark: A/S Dantherm [98]

90-00700 BS Hansen Trading AB
(† BS Hansen Trade & Consult A/S)
Kronprinsensgt 3, N-0251 Oslo
☎(+47) 22 83 68 24 [E] hxx004@aft.sn.no
⊖ (+47) 22 83 68 25
👥 1
Managing Director: Bjørn S Hansen
■ Agent for foreign manufacturers of defence equipment and aeronautical equipment. Consultancy services. [98]

90-00880 HTS Vertoy A/S
Ing. Rybergsgt. 97, PO Box 1521, N-3007 Drammen
☎(+47) 32 23 46 50 ⊖ (+47) 32 23 46 65
♧ NKr 1,000,000 (1996) ♣ NKr 7,100,000 (1996)
👥 3 (1997) 🏠 1,500 m2
Managing Director, Purchasing Manager: Edvin Guddal
Sales and Marketing Director: Esben Eliassen
Operations Director: Harald Throndsen
■ Sale of measuring equipment.
® Denmark: Leitech, Roblon
Germany: A Steinmeyer GmbH & Co KG, Hommelwerke GmbH, Oskar Schwenk GmbH & Co
Italy: Bocchi
Sweden: C.E. Johansson
Switzerland: Compac SA, Sylvac SA, Trimos-Sylvak Task Force, Wyler AG
UK: Bowers, Wylam Hill Ltd [98]

90-00885 Hughes International Corporation
Stortings Gata 8 - 8th Floor, N-0161 Oslo
☎(+47) 22 82 84 50 ⊖ (+47) 22 82 84 59
Director: Alden B Billitti [98]

90-00900 Hydrema A/S
Bergermoen, N-3520 Jevnaker
General Mgr: A. Bjørnerud
® Denmark: A/S Hydrema [97]

90-01160 Karcher A/S
Gjerdrumsvei 4, N-0409 Oslo
☎(+47) 22 23 20 80 ⊖ (+47) 22 23 33 60
Managing Director: Tor Sveen
® Germany: Alfred Karcher GmbH & Co [98]

90-01295 Metric A/S
Nordåsveien 5, PO Box 164, N-1203 Holmlia, Oslo
☎(+47) 22 76 40 00 [E] metric@metric.no
[W] http://www.metric.no ⊖ (+47) 22 76 40 50
♣ NKr 60,000,000 (1996) 👥 28 (1997)
Managing Director: Frank De Kinder
Marketing Co-ordinator: Ms Cecilie Holgersen [98]

90-01370 Nauteknik Defence & Security A/S
(a subs. of Detra A/S)
Oscargate 42, PO Box 2521 Solli, N-0203 Oslo
☎(+47) 22 43 82 80 ⊠ 78629 sos n
⊖ (+47) 22 43 42 17
Managing Dir: Flemming Devor
■ Supplier of advanced special equipment for air, sea & land defence forces, police, & professional users of security equipment.
® : Vickers Defence Systems
Australia: Australian Defence Industries Ltd
Austria: Dynamit Nobel Graz
Canada: Diemaco Inc., Scepter Manufacturing Co. Ltd
Korea (ROK): Eusung Industrial Co. Ltd
Sweden: Dynasafe AB
Switzerland: Mowag Motorwagenfabrik AG
UK: BDL Systems Ltd, Beaufort Air-Sea Equipment Ltd, British Aerospace Defence Ltd, Chemring Ltd, GKN Defence, Helmet Integrated Systems Ltd, Hunting Engineering, Loral Solartron, Manroy Engineering Ltd, Pains-Wessex/Schermuly, Richmond Electronics/Engineering Int Ltd, Royal Ordnance, Siebe Gorman
USA: Conax Florida Corp., Loral Electro Optical Systems [97]

90-01920 Scancopter A/S
(a subsidiary of Helikopter Service, Norway)
Oksenøyvn 12, PO Box 58, N-1330 Oslo Lufthavn
☎(+47) 67 12 56 40 [E] copter@online.no
⊖ (+47) 67 12 59 69
♧ NKr 1,000,000 (1997) ♣ NKr 4,000,000 (1997)
👥 4 (1997) 🏠 130 m2
Managing Director: Reidar Andersen
Sales and Marketing Manager: Geir Hamre
Technical Sales Manager: Kjell Håkonsen
■ Distributor/agents for helicopters, helicopter accessories and other defence products.
® Canada: SEI Industries, Spar Aerospace
France: Cilas
Germany: Euroavionics GmbH, Signaal GmbH
UK: GEC, Racal Avionics
USA: Bell Helicopter Textron, Chadwick-Helmuth, Honeywell, Sierra Research [98]

90-01950 Sperry Marine A/S
Eiksveien 115, N-1345 Oesteraas
☎(+47) 22 24 20 40 ⊖ (+47) 22 24 65 05
Vice President: H Rasmussen [98]

90-02000 Jan Staubo A/S
PO Box 1512 Vika, N-0117 Oslo 1
☎(+47) 37 03 41 76; 22 69 59 25
⊖ (+47) 37 03 42 05; 22 46 50 42
Managing Director: Jan Staubo
Sales Director: Tor Nikolausen
Technical Director: Stig Mörck
® Canada: Hermes Electronics
UK: Westland Helicopters [97]

Oman

Head of State:
H M Sultan Qaboos bin Said Al Said

10 Head of State

10-10000 Office of H M the Sultan
The Royal Palace, PO Box 632, Muscat
☎(+968) 73 66 21 ⊠ 5267; 5385 paloff on
⊖ (+968) 73 94 27
Sultan of Oman, Prime Minister, Minister of Defence Minister of Foreign Affairs:
H M Sultan Qaboos bin Said Al Said
Personal Representative of the Sultan:
Thuwaiyni bin Shihab Al Said

Diwan of Royal Court
PO Box 632, 113 Muscat
☎(+968) 73 66 21; 73 66 23 ⊖ (+968) 73 94 27
Minister: H E Sayyid Saif bin Hamed bin Saoud [97]

11 Prime Minister

11-10000 Office of the Prime Minister
The Royal Palace, PO Box 949, Muscat
☎(+968) 74 55 50
Sultan and Prime Minister:
H M Sultan Qaboos bin Said Al Said

Secretariat-General for the Cabinet
PO Box 975, 113 Muscat
☎(+968) 73 85 16 ⊠ 5575 dpmcm on
⊖ (+968) 73 85 17
Deputy Prime Minister for Cabinet Affairs:
H.H. Fahd bin Mahmud Al Said [98]

12 Ministries

12-10000 Ministry of Defence
PO Box 113, 113 Muscat
☎(+968) 70 91 99 ⊠ 5228 defence on
⊖ (+968) 70 25 21
Ruler and Minister: H M Sultan Qaboos bin Said Al Said
Deputy Prime Minister for Security & Defence:
H.H. Sayyid Faher bin Taimur Al Said
Secretary General: Saif bin Hamed Al Batashi
SO1 Market Research Office of COSSAF:
Ahmed bin Said Al-Rawahi
Director Military Intelligence: E W Payne [97]

12-15000 Ministry of Palace Office Affairs
PO Box 2227, 112 Ruwi
☎(+968) 60 08 41 ⊠ 5267 paloff on ⊖ (+968) 60 84 67
Minister and Head of the Office of the Supreme Commander of the Armed Forces:
H E General Ali bin Majid Al Ma'amari [97]

12-20000 Ministry of Foreign Affairs
PO Box 252, 113 Muscat
☎(+968) 69 95 00 ⊠ 5195; 5337 mfa on
⊖ (+968) 69 61 44
Minister: H M Sultan Qaboos bin Said Al Said
Minister of State for Foreign Affairs:
Yousef bin Alawi bin Abdullah
Secretary General: H H Sayyid Haitham bin Tarek Al Said [97]

12-50000 Ministry of Information
PO Box 600, 113 Muscat
☎(+968) 60 32 22 ⊠ 5265 inform on
⊖ (+968) 69 65 92
Minister of Information: Abdul Aziz bin Mohamed Al Rowas

Under Secretary for Admin & Finance:
Ali bin Ahmed Al Ansari
Under Secretary for Information Affairs:
H E Hamed bin Mohamed bin Mohsin Al Rashdi
Director General: Abdullah bin Nasser Al Rhabi [97]

12-60000 Ministry of Communications
PO Box 684, Muscat
☎(+968) 70 22 33 ⊠ 3390 mwasalat on
Minister: Salim bin Abdallah al Ghazali [98]

12-62000 Ministry of Posts Telegraphs and Telephones
PO Box 338, 112 Muscat
☎(+968) 69 70 00 ⊠ 5556 postel on ⊖ (+968) 69 66 70
Minister: Ahmed bin Suwaydan al Baluchi [98]

25 Defense Forces

25-30000 Royal Army of Oman (RAO)
PO Box 721, 111 CPO Seeb
☎(+968) 61 46 00 ⊖ (+968) 61 87 79
Commander Royal Army of Oman:
Maj.Gen. Ali bin Rashid bin Mohammed Al-Kilbani
Principal Staff Officer:
Brig. Abdullah Bin Amor Bin Ali Al Marhuby
Director Logistics: Brig. Saif bin Suleiman Al-Maani
Director Ordnance Services: Col Habib Al-Baluchi
Sultan's Special Force
PO Box 1331, CPO Seeb
☎(+968) 59 01 58/9; 59 09 98; 59 12 33
⊠ 5435 caseeba ⊖ (+968) 59 01 56
Staff Officer Admin & Finance: Lt Col S.B. Nair
Sultan's Special Force
PO Box 18866, Salalah [97]

31 Embassies & Attachés

31-00030 Embassy of Democratic & Popular Republic of Algeria
216 Madinat Al Sultan Qaboos, 115 Muscat
☎(+968) 60 16 98 ⊠ 5054 ambalg on
⊖ (+968) 69 44 19
Ambassador: Brahim Aissa
Deputy Chief of Mission: Abdelkrim Mahieddine [97]

31-00130 Embassy People's Rep. of Bangladesh
PO Box 3959, 112 Ruwi
☎(+968) 56 75 01 ⊖ (+968) 56 75 02
Ambassador: H E Mr Amin Ahmed Chowdhury
Counsellor, Labour: Muhammad Abdul Matin
First Secretary: Khandakar Abdus Sattar [98]

31-00240 Embassy of Brunei Darussalam
Qurum Beach, Way No. 3050 Villa 4062, PO Box 91, 112 Ruwi, Muscat
☎(+968) 60 35 33; 60 42 53 ⊠ 5135 kbomct on
⊖ (+968) 60 59 10
Ambassador: H.E. Mr Pehin Data Haji Mahdini
Deputy Chief of Mission & Second Secretary:
Haji Abdullah bin Haji Saman
Third Secretary Admin Affairs: Mohamad bin Pengas
Third Secretary Consular Affairs: Morni bin Haji Tuah
Attaché: Abd Karim bin Haji Bini [97]

31-00460 Royal Danish Consulate
PO Box 1040, 112 Ruwi-Muscat
☎(+968) 70 32 89; 70 83 04 ⊠ 3214 towel on
⊖ (+968) 79 38 92
Consul: Anwar Ali Sultan [98]

31-00660 Embassy of the Federal Republic of Germany
Next Al-Nahda Hospital, PC 112 Ruwi, Muscat
☎(+968) 70 24 82; 70 21 64 ⊠ 3440 aamuscat on
⊖ (+968) 70 56 90
Ambassador: H.E. Dr Kurt Messer
Head of Chancery: Manfred Reinhardt
First Secretary: Hermann Nicolai ☎(+968) 60 16 84
Military & Defense Attaché: Col Hartmut Schulz
(Resident in Riyadh)
Third Secretary Commercial: Miss Heike Jope [97]

31-00830 Embassy of India
PO Box 1727 Ruwi, 112 Muscat
☎(+968) 70 68 94 ⊠ 3429 indembs on
⊖ (+968) 79 75 47
Ambassador: H E Shri I S Rathore
Defence Attaché: Vimal Narayan (IN)
Commercial and Press Officer: Reddy [98]

31-00890 Embassy of Italy
Way 2411, House No. 842 Qurum, PO Box 3727, Mascate
☎(+968) 56 09 68; 56 48 38 ⊠ 5450 italdipl on
⊖ (+968) 56 48 46
Ambassador: H.E. Dr Sergio Emina
First Secretary: Dr Fabrizio Nicoletti
Naval, Military & Air Attaché: Capt. Maurizio Vacca
(Resident in Abu Dhabi)
Commercial Attaché: Dr Antonio Angelaccio [98]

31-01110 Embassy of Malaysia
PO Box 3939, 112 Ruwi
☎(+968) 69 83 29 ⊠ MALAWAKIL MUSCAT
⊠ 5565 malwakil on ⊖ (+968) 60 50 31
Chargé d'Affaires a.i.: Jamal bin Hassan
☎(+968) 60 71 30
Third Secretary: Walter Fernandis ☎(+968) 69 86 43 [98]

31-01340 Royal Norwegian Consulate General
PO Box 89, 112 Ruwi-Muscat
☎(+968) 70 32 89; 70 83 04 ⊠ 3214 towel on
⊖ (+968) 79 38 92
Consul General: H E Mr Ali Sultan Anwar [98]

31-01690 Embassy of the Democratic Socialist Republic of Sri Lanka
PO Box 95, 115 Muscat
Villa No 701, Way No 2114, Madinat Qaboos West
☎(+968) 69 78 41; 69 78 42 [E] sloman@gto.net.om
⊠ 5158 emblanka on ⊖ (+968) 69 73 36
Ambassador: H E Mr T B Maduwegedera
Second Secretary: N Kamalenthirn [98]

31-01890 Embassy of the United Kingdom
PO Box 300, 113 Muscat
☎(+968) 69 30 77
⊖ (+968) 60 30 87; 31 79; 30 88/9; 30 91
Ambassador: H.E. Mr Richard J. Muir (CMG)
Deputy Head of Mission: Dr Noel J. Guckian
Defence & Military Attaché: Brig. H. Duncan (MBE)
☎(+968) 60 30 89 (fax)
Naval & Air Attaché: Cdr H.S. Clark (DSC RN)
Third Secretary Chancery & Information:
B.E.N. Lyster-Binns ☎(+968) 69 31 79 (fax)
Office of the Commercial Secretary
PO Box 300, 113 Muscat
☎(+968) 69 30 77 ⊖ (+968) 69 30 88
First Secretary, Commercial: P. Williams [97]

31-01990 Embassy of the Republic of Yemen
PO Box 105, 115 Sultan Qabus City, Muscat
☎(+968) 60 08 15; 60 41 72 ⊠ 5109 ghamdan on
⊖ (+968) 60 50 08
Ambassador: H.E. Mr Ahmed Daifallah Al Azeib
Deputy Chief of Mission: Dr Amin Al Yousufi
Administration Financial Counsellor: Hessein Al Rahomi [98]

80 Service Companies & Organizations

OHI Telecommunications Co.LLC . . . 90-01400

80-06000 Zubair Enterprises LLC
PO Box 1127, 112 Muscat
☎(+968) 70 71 47 ⊠ 3258 zubair on ⊖ (+968) 79 69 21
President: Rashad Al Zubair
Executive Vice President Commercial: Nick Marriott
Vice President International: John Fitzgerald
Manager Defence Division: Radha Nair
■ Automotive, avionics, computer systems, logistics and support services, military clothing, munitions, security and surveillance equipment, telecommunications and training simulation. [97]

85 Stockists, Distributors, Spare Parts Suppliers

OHI Telecommunications Co.LLC . . . 90-01400

90 Sales Agents, Sales & Liaison Offices

90-01400 OHI Telecommunications Co.LLC
(a subs. of Oman Holdings International)
PO Box 889, 113 Muscat
☎(+968) 60 36 06 ⊠ 5052 telecom on
⊖ (+968) 60 36 08
General Mgr: Craig Werner
Defence Systems Mgr: David Clements
Marketing & Sales Mgr Communications: Graham Davies
Marketing & Sales Mgr Computer Systems: V. Sharma
■ Activities include: communication equipment, electronic equipment, avionics, weapon systems, optical equipment, ammunition, vehicles, GPS, computer systems, security systems. Workshop facilities. Skilled manpower services. → 80 → 85 [97]

90-02000 Tawoos LLC
PO Box 1676, 114 Muttrah
☎(+968) 79 66 36 ⊠ 3741 tawoos on
⊖ (+968) 70 65 87
Managing Director: Samir Fancy
General Manager, Defence Support: Nigel Harris [98]

Pakistan

Head of State:
Farooq Leghari

10 Head of State

10-10000 Office of the President
Murree Brewery Road, Rawalpindi
☎ (+92 51) 82 38 90; 82 63 25 ▣ 5791 cab pk
President: Farooq Leghari [98]

11 Prime Minister

11-10000 Office of the Prime Minister
Islamabad
Prime Minister: Nawaz Sharif [98]

12 Ministries

12-20000 Ministry of Foreign Affairs
Constitution Avenue, Islamabad
☎ (+92 51) 21 19 42 ▣ 5800 faiba pk
⊖ (+92 51) 21 24 34
Minister: Gohar Ayub Khan [98]

12-40000 Ministry of the Interior
Pak Secretariat, Block R, Islamabad
☎ (+92 51) 82 57 11; 82 06 86; 82 20 26; 82 03 52
▣ 5790 fcr pk
Minister: Chaudhary Shujat Hussain [98]

25 Defense Forces

25-20000 Air Force Headquarters
Chaklala
☎ (+92 51) 58 48 13 ⊖ (+92 51) 56 64 00
Chief of Air Staff: ACM M. Abbas Khattak
Vice Chief of Air Staff: AM Arshad Choudry
Deputy Chief of Air Staff Administration: AVM Javed Hayat Malik
Deputy Chief of Air Staff Training: AVM Amer Sheriff
Deputy Chief of Air Staff Engineering: AVM Afzaal
Deputy Chief of Air Staff Operations: AM Shahid Zulfiqar
Air Secretary: AM Aliuddin
Operations Dir: Gp Capt Ateeb
Procurement Logistics Dir: Gp Capt. Tufail
Inspector General: AVM Imtiaz Haider
☎ (+92 51) 515 84 96 20
Personal Staff Officer: Gp Capt. Raashid Kalim
☎ (92 51) 515 84 96 20 [97]

31 Embassies & Attachés

31-00090 Australian High Commission
Diplomatic Enclave, PO Box 1046, Islamabad
☎ (+92 51) 27 92 23 [E] ahcisb@isb.compol.com
⊖ (+92 51) 27 92 22
High Commissioner: H E Mr Geoffrey C Allen
Deputy Chief of Mission: Geoffrey Brooks
Defence Adviser: Gp Capt Grahame Carroll
Commercial Counsellor: Bill Mahon
First Secretary: B V Brown [98]

31-00150 Embassy of Belgium
N° 2, Street 10, F 6/3, PO Box 1016, Islamabad
☎ (+92 51) 21 00 31; 21 00 32 ⊠ AMBABEL Islamabad
▣ 5865 ambel pk ⊖ (+92 51) 82 23 58
Ambassador: H.E. Mr Wilfried Geens
Deputy Chief of Mission: Peter Claes
Commercial Attaché: Abid Husain [97]

31-00450 Embassy of the Czech Republic
House No. 49, Street No 27, F-6/2, Islamabad
☎ (+92 51) 21 01 95; 22 12 58 ▣ 5705 zamin pk
⊖ (+92 51) 82 53 27
Ambassador: H.E. Mr Jakub Karfík
Counsellor: Jaroslav Domas
Admin Attaché: Karel Vanas
Consul General: Milan Krenek [97]

31-00460 Royal Danish Embassy
Ho 9 90th Street - Ramna 6/3, PO Box 1118, Islamabad
☎ (+92 51) 82 47 22; 82 47 24
⊠ AMBADANE Islamabad
[E] eber@ambadane.sdnpk.undp.org ▣ 5825 ambdk pk
⊖ (+92 51) 82 34 83
Chargé d'Affaires: Kurt Ebert [98]

31-00535 Honorary Consulate of Estonia
22 Nazimuddin Road, F7/1, Islambad
☎ (+92 51) 81 66 03; 922 45 88 ⊖ (+92 51) 27 17 35
Honorary Consul: Malik Ghulam Arbi Khar [98]

31-00590 Embassy of France
Diplomatic Enclave G/5, PO Box 1068, Islamabad
☎ (+92 51) 21 39 81/3 ▣ 5819 frais pk
⊖ (+92 51) 82 53 89; 82 25 83
Ambassador: H.E. Mr Pierre Lafrance
First Secretary: Michel Djokovic
Defence Attaché: Col Philippe Guillaume
Air Attaché: Lt Col Bertrand Zundellier
Commercial Section
PO Box 1068, Islamabad
☎ (+92 51) 21 46 18; 21 90 93 ⊠ COMATTA Islamabad
▣ 5715 comat pk ⊖ (+92 51) 82 22 40
Commercial Counsellor: Ms Dominique Mauppin [97]

31-00690 Embassy of Greece
House No 6, Margalla Road, F-7/3, Islamabad
☎ (+92 51) 82 51 86; 82 25 58; 82 47 74
▣ 54232 gremb pk ⊖ (+92 51) 82 51 61 [98]

31-00810 Embassy of the Republic of Hungary
12 Margalla Road - F 6/3, PO Box 1103, Islamabad
☎ (+92 51) 82 33 52; 82 33 53 ⊖ (+92 51) 82 52 56
Ambassador: H E Mr Janos Barabas [98]

31-00830 High Commission of India
G-5 Diplomatic Enclave, Islamabad
☎ (+92 51) 81 43 71/5 ▣ 825849 indem pk
⊖ (+92 51) 22 42 86
High Commissioner: H.E. Mr Satish Chandra
Deputy High Commissioner: Sharat Sabharwal
Counsellor Information: N. Parthasarathi
Head of Chancery: M.C. Naithani
First Secretary Commerce: M.K. Karnik
First Secretary Economic: M.K. Sinha
Second Secretary Information: N.K. Chopra
Office of the Armed Forces Advisers
G-5 Diplomatic Enclave, Islamabad
☎ (+92 51) 81 43 71 ▣ 5849 indem pk
Defence Adviser: Brig. Ajay Sood
Air Adviser: Gp Capt. Parvez Khokhar
☎ (+92 51) 81 43 71
Naval Adviser: Capt. Rakesh Pandit ☎ (+92 51) 81 43 71 [97]

31-00890 Embassy of Italy
54 Margalla Road, F-6/3, PO Box 1008, Islamabad
☎ (+92 51) 22 29 83; 21 07 91/2 ▣ 5861 itald pk
⊖ (+92 51) 22 29 86
Ambassador: H.E. Mr Pietro Rinaldi
Deputy Chief of Mission: Mrs Maria Accili Sabbatini
Military, Air & Naval Attaché: Gen. Remo Napoletano [98]

31-01430 Embassy of the Republic of Poland
24 Street, G-4 Diplomatic Enclave, PO Box 1032, Islamabad
☎ (+92 51) 27 94 91; 27 94 92
[E] plembpk@biruni.eru.com.pk ▣ 54679 empol pk
⊖ (+92 51) 82 54 42
Ambassador: H E Mr Lucjan Mieczkowski
First Secretary: Jacek Najder
Third Secretary: Janusz Molka
Office of the Commercial Counsellor
8b Embassy Road F-6/4, PO Box 1162, Islamabad
☎ (+92 51) 27 94 95; 82 62 45 ⊖ (+92 51) 27 94 96
Counsellor: Janusz Bilinski [98]

31-01740 Embassy of Switzerland
Street 6 - Dipl Enclave, G-5/4, PO Box 1073, Islamabad
☎ (+92 51) 27 92 90; 27 92 91; 27 92 92
⊠ AMBASUISSE Islamabad ⊖ (+92 51) 27 92 86
Ambassador: M A Salamin
Deputy Chief of Mission: Max Grob
Defence Attaché: Col René Käser [98]

31-01890 United Kingdom High Commission
Diplomatic Enclave - Ramna 5, PO Box 1122, Islamabad
☎ (+92 51) 82 21 31; 82 21 35
⊠ PRODROME ISLAMABAD ▣ 54122 ukemb pk
⊖ (+92 51) 82 34 39
High Commissioner: H E Mr David Dain
☎ (+92 51) 82 21 31; 82 21 35 ext. 202
Deputy High Commissioner: J W Watt
☎ (+92 51) 82 21 31; 82 21 35 ext. 203
Counsellor (Economic/Commercial/Aid): G J Dorey
☎ (+92 51) 82 21 31; 82 21 35 ext. 205
Defence and Military Adviser: Brig B Wheelwright
☎ (+92 51) 82 21 31; 82 21 35 ext. 218
Naval and Air Adviser: Capt P Jackson (RN)
☎ (+92 51) 82 21 31; 82 21 35 ext. 216
First Secretary Commercial/Economic: P Jenkins
☎ (+92 51) 82 21 31; 82 21 35 ext. 227 [98]

31-02010 Embassy of the Federal Republic of Yugoslavia
14 Street 87 - Ramna 6/3, PO Box 1050, Islamabad
☎ (+92 51) 82 95 56; 82 95 57; 82 27 74
[E] slavia@paknet1.ptc.pk ⊖ (+92 51) 82 09 65; 82 23 21
Chargé d'Affaires: Zarko Milosevic (acting)
Administration Attaché: Vladimir Djuric [98]

55 Government-owned Industry & Production Facilities

55-00025 Air Weapons Complex
E-S, Officers Colony, Wah Cantt
☎ (+92 57) 72 28 42 ⊖ (+92 51) 52 42 33
👥 1,000
Chief Executive Officer: Shahid Hamid
General Manager (Manufacturing): Nasir Ali Shah
General Manager (Logistics): Col. Altaf Ur Rehman
Manager Sales Promotion Division: Asad Mir
Manager Commercial Division: Asif Ahmed
Marketing Consultant: A Moeed Khan
Marketing Officer: Ms Naseema Kaleem
■ Air delivered weapons. [97]

55-00100 Pakistan Ordnance Factories (POF)
Wah Cantt
☎ (+92 51) 56 60 30; 56 60 31; 56 60 32; 56 60 33; 56 60 34; 56 60 35; 56 60 36; 56 60 37; 56 60 38; 56 60 39; 56 38 71; 56 38 72; (+92 596) 29 69 97 (29 lines)
▣ 5840; 54178 pofac pk
⊖ (+92 51) 58 41 75; (+92 596) 29 95
♣ US$ 100,000,000 👥 35,000
Chairman: Maj Gen Mahmud Ali Durrani
Exports and Advertising Director: Azhar Nawaz Khan
☎ (+92 51) 56 60 30 ext. 2019
Public Relations Manager: Zafar Mahmood Malik
Exports Manager: Z A Siddiqui
■ State owned ordnance production organisation

Main products: mortar bombs, infantry weapons, small arms ammunition, aircraft and anti-aircraft ammunition, tank and anti-tank ammunition, anti-tank rockets, artillery ammunition, mines, grenades, pyrotechnics, fuzes, detonators, primers, military explosives, propellants and commercial explosives. [98]

55-01600 Pakistan Aeronautical Complex (PAC)
Kamra, District Attock
☎ (+92 51) 580 26 05; 584 16 34; 583 95 12
Telex 5601 pac kamra pk
Fax (+92 51) 58 41 62; 58 38 37; 56 42 07
US$ 150,000,000 (1995/96)
♣ US$ 2,159,000,000 (AMF only 1996/97)
6,350 (1996)
Director General: AVM A Rahim Yousefzai
Managing Director (AMF): Air Cdre Muhammad Younas
☎ (+92 51) 58 42 12
Managing Director (F6RF): Air Cdre Muhammad Salim
☎ (+92 51) 58 41 61
Managing Director (KARF): Air Cdre Qamaruddin
☎ (+92 51) 56 25 90
Managing Director (MRF): Air Cdre M Aslam Ansari
☎ (+92 51) 58 41 60
■ Manufactures 'Mushshak' and 'Shahbaz' for training, forward air control and observation. Progressive manufacture of K-8 Jet Trainer Aircraft. MRF overhauls Mirage III and V aircraft and components. SNECMA ATAR 09C and P&W F-100 engines and accessories. F6RF overhauls, F-6, A-5III, F-7P and FT-5 aircraft, components and instruments. Manufacture of spares and subsonic/super-sonic drop tanks for F-6 and F-7 aircraft. KARF rebuilds. MPDR-45 ground based air defence radars and associated power generators. [98]

75 Manufacturing Companies

M Ismailji & Sons 90-01200

75-00100 Alsons Industries (PVT) Ltd
PO Box 4853, 74000 Karachi
S-18, SITE, Karachi
☎ (+92 21) 256 20 60; 256 38 11
Fax (+92 21) 256 38 58
300
Managing Director: Abdul Rehman Allana
■ Manufacture of fuzes for mortar bombs and other precision components. [98]

75-00500 Daudsons Armoury
Small Industries Estate, Kohat Road, Peshawar
☎ (+92 521) 23 09 04; 23 09 05; 23 09 06; 23 11 35; 23 11 36; 23 11 37 ✉ ARMOURY
E daud@daudson.psw.erum.com.pk. Telex 52369 pcope pk
Fax (+92 521) 27 60 59
Rs 17,397,705 (1994/95)
♣ Rs 24,376,937 (1994/95) 300 30,000 sq m
Managing Director: Nooruddin F. Daud
Sales & Marketing Dir: Hakimuddin F. Daud
Executive: Hasan Gialani
■ MK 82 Low Drag Bomb Bodies 500lbs. Complete. Facility available for producing Forged Empty Bomb Bodies MK 83 & MK 84, complete with parts. Fin Assembly Conical MAU 93-B. Armament accessories like: Arming Wires, Extractor Wires, Ferrules, etc. Practice Bombs Aircraft. Sub Calibre Devices RPG-7, SPG-9, RR, etc. Mortars 120mm, 81mm, 60mm Training Systems. 7-shot 12 bore Pump Action shotguns, called "SHOOTER". [98]

75-00600 Daudsons Industries (Pvt) Ltd
Industrial Estate - Kohat Road, Peshawar
☎ (+92 521) 23 11 35; 23 11 36; 23 11 37; 23 09 04; 23 09 05 E daud@daudson.psw.erum.com.pk
Telex 52369 pcpoe Fax (+92 521) 27 60 59
Contact: Nooruddin F Daud [98]

75-01970 Sipka Manufacturing (Pvt) Ltd
8-D Block-6, PECH Society, PO Box 8290, 75400 Karachi
☎ (+92 21) 44 40 82; 44 13 71 ✉ PLANWELL
Fax (+92 21) 454 55 32; 454 97 39
Rs 6 000 000 (1996) ♣ Rs 50 000 000 (1996) 110
1 250 sq m
Managing Dir: Shahmir Khan
Dir Telecommunications: Sajid Ahmed Khan
Technical Services Dir: John E. George
■ Manufacture of wireless radio equipment & accessories. Sales & service of telecommunication products. → 90
® Germany: Sprecher & Schun
Japan: Nanayo Shoji
Singapore: Motorola Singapore
USA: E.F.Johnson, Telenet Communications
Lahore Office
166-A, Upper Mall, Lahore
☎ (+92 42) 87 81 75 Fax (+92 42) 376 13 58
Islamabad Office
House No.450, Street No 90, Sector G-9/4, Islamabad
☎ (+92 51) 26 35 54; 85 16 90 Fax (+92 51) 26 35 56
■ Import, sales & servicing of telecommunication, electro-medical, defence & electrical products.
Factory
Liaquat Ashraf Colony No.1, Mehmoodabad, Karachi 44
☎ (+92 21) 54 29 55
■ Telecommunication equipment & systems, telephone exchanges, CCTV, antennas, public address systems, security systems, laboratory equipment, criminal research equipment, generating sets, electrical distribution & transmission equipment, electro-medical equipment, hospital equipment & furniture, firearms, protective clothing, cable splicing & joining material. [97]

75-01980 Time & Tune Engineering Division
55/55-1 Shafi Plaza, Canning Road, 46000 Rawalpindi
☎ (+92 51) 56 75 93; 56 37 98 E ikram@infolink.net.pk
Telex 5576 ttism pk Fax (+92 51) 56 84 45
♣ Rs 65,000,000 (1995) 12
General Manager: Munir Hussain
Executive Partner: Mr Moinuddin
Sales Manager: Mustaali Ismailji
Computer Division
37-A Sethi Plaza, Blue Area, Islamabad
☎ (+92 51) 27 32 08; 27 32 09; 27 13 58; 81 01 20
Fax (+92 51) 21 13 56
♣ Rs 25,000,000 16
Executive Partner: Munir Hussain
General Manager: Ejaz Ahmed [98]

75-02300 Wah Nobel (Pvt) Ltd
(a subsidiary of Wah Industries Ltd)
GT RD, Wah Cantt
☎ (+92 51) 56 87 60; 56 60 30 39 ext. 2236 (+92 596) 33 58; 31 11; 34 58 ✉ DYNAMITE Wah Cantt
Telex 5672 comex pk Fax (+92 51) 56 41 80
Rs 50,000,000 (1997) ♣ Rs 250,000,000 (1997)
220 (1997) 46,000 sq ft
Managing Director: S K Mohyuddin
Manager Marketing: Muhammad Nawaz Khan
Purchasing Manager: M Afsar Alam
■ Commercial accessories and blasting services.
® Sweden: Nobel Industries [98]

80 Service Companies & Organizations

NAD International 90-01410

80-00075 AlliedSignal Aerospace Marketing Sales and Service
Dr Daupota Road, Cantt Station, 75520 Karachi
☎ (+92 21) 454 24 29 Fax (+92 21) 454 24 29
Field Service Engineer: William David
■ On-site customer support for AlliedSignal Aerospace. → 90 [98]

80-01000 Eliya Travels & Tours
S-27 Sea Breeze Plaza, Shahrah-e-Faisal, 75530 Karachi
☎ (+92 21) 778 55 36; 778 55 37; 778 55 38
E imranali@paknet3.ptc.pk
Fax (+92 21) 778 55 38; 778 23 62
20
Director General: Najam Ahmed
Advertising Director: Imran Ali Bhatti
E imranali@paknet3.ptc.pk
Marketing Manager: Masood Ahmed Wasi
International Relations Officer: Syed Kashif Qamar
■ Travel, tourism and aviation agency. Overflying permission and technical landings over Pakistan. Air ticket sales for all destinations. [98]

80-01500 Management Equipment Vehicles Co Head Office (MEVCO)
House N°14 Street 2, F-6/3, Islamabad
☎ (+92 51) 82 58 92 Fax (+92 51) 27 03 71
8 2,000 sq ft
President: Lieut Gen Ejaz Azim (retd)
Senior Executive Vice President: Brig Imtiaz Azim (retd)
Executive Vice President: Brig Humayun Butt
Managing Director: Shahid Azim
■ Consultancy, marketing, sales and technical transfers for all defence related products.
® Sweden: Bulow AB
UK: GEC-Marconi Avionics, Thompson Valves
USA: Transaero Inc
Representing:
Oxley Avionics Ltd [98]

85 Stockists, Distributors, Spare Parts Suppliers

M Ismailji & Sons 90-01200
NAD International 90-01410

85-01000 Communications & Machinery Corporation
PO Box 7968, 74400 Saddar, Karachi
22 Farid Chambers, Abdullah Haroon Road
☎ (+92 21) 567 82 52 E zeeshans@paknet1.ptc.pk
Telex 21791 cmcwt pk Fax (+92 21) 568 32 83
♣ US$ 175,000 (1995) 12 500 sq ft
Managing Director: Ahmed Naseem Shamsi
Marketing Director: Zeeshan Ahmed Shamsi
☎ (+91 21) 567 75 07
Advertising Director: Muhammad Akbar
Public Relations: Nauman Ahmed Shamsi
■ Engineering consultancy with Pakistan MoD and turnkey projects. [98]

85-03000 E-Pool Systems
204 Shahid Plaza, Jinnah Avenue, Blue Area, 44000 Islamabad
☎ (+92 51) 82 42 50 E epool@isb.compol.com
W http://www.compol.com/epool Fax (+92 51) 82 42 09
♣ US$ 20,000,000 (1995) 6
Managing Director: Shehzad Ahmed Mir
■ Represents and markets the products of major international manufacturers and suppliers of defence related equipment in Pakistan. [97]

85-07000 Pakistan Business International
603 Kashif Center, Sharea Faisal, 75530 Karachi
☎ (+92 21) 568 72 57; 51 54 11 ✉ OPTIMISTS
Telex 8229839 pbipk Fax (+92 21) 568 43 80
Rs 50 000 000 ♣ Rs 100 000 000 200 sq m
Chief Exec.: M.A. Rasheed
Attorney: M.A. Jabbar
■ Imports, exports, indenting agencies, supplies, liaison services representing variou manufacturers in Asia, Europe, USA & Australia. [97]

90 Sales Agents, Sales & Liaison Offices

Companies represented:

ADI, Taiwan 90-02120
Amerace Corp, USA 90-00950
American Electronic Lab, USA 90-02120
Ampex International, USA 90-01380
Applied Test Systems, USA 90-01410
Bayfern Ltd, UK 90-00950
BF-Goodrich, USA 90-02120
Bulow AB, Sweden 80-01500
Chemring Limited, UK 90-02150
Connevans Ltd, UK 90-00950
Courtaulds Military Association, UK . . 90-00770
Croylek Ltd, UK 90-00950
DX Radio System UK, UK 90-00950
E.F.Johnson, USA 75-01970
E-Systems, USA 90-02120
EEV Ltd, UK 90-01700
Eumoco AG., Germany 90-01200
FMC Airline Equipment Division, USA . 90-01410

Fritz Werner Industrie, Germany 90-01200
GEC-Marconi Avionics, UK 80-01500
Getac, Taiwan 90-02120
GTE Government Systems Corp, USA . 90-00770
H K Porter & Co. Inc, USA 90-00950
Hewlett-Packard, USA 90-01380
Hymatic, UK 90-02120
ITT, USA 90-02120
J Jennings, USA 90-00950
Jenoptik Impulsphysik, Germany 90-00950
Jetstream Aircraft Limited, UK. 90-02150
Lachaussee SA., Belgium 90-01200
Laird (Anglesey) Ltd, UK 90-02150
Litton Industries, USA 90-02120
LNR, USA. 90-02120
Lockheed Martin, USA 90-02120
Macaw Electron, USA 90-00950
Marine Technology, UK 90-00950
Matra MS2i, France 90-02150
MBM Trade (UK) Ltd, UK 90-00950
Motorola Singapore, Singapore 75-01970
Nanayo Shoji, Japan 75-01970
Nobel Industries, Sweden 75-02300
Pilkington Optronics, UK 90-02120
Plessey, UK. 90-01380
PPI Adhesive Products Ltd, Ireland . . . 90-00950
Racal Acoustics, UK 90-02120
Racal Decca Navigator, UK. 90-01380
Ranger Communications, Taiwan 90-02120
Rank Taylor Hobson, UK 90-01410
Redifon SPT Limited, UK 90-02150
Rinaldo Piaggio, Italy 90-01410
Saco Defence, USA. 90-02120
Satel, France 90-00950
Sea Scan, Singapore 90-00950
Seabreeze, Singapore 90-00950
Shakespeare, USA 90-02120
Ship Analytics, USA 90-02120
Siemens Plessey, Australia 90-02120
Sierracin Sylmar, USA 90-02120
Sprecher & Schun, Germany 75-01970
Steiner Offices, Germany 90-02120
Strachan & Henshaw Limited, UK 90-02150
Surcom Associates Inc, USA. 90-00950
Telenet Communications, USA. 75-01970
Thompson Valves, UK 80-01500
Transaero Inc, USA. 80-01500
UCS Engineering, Germany 90-00950
Vectron Lab Inc, USA. 90-00950
Walter Schmaddebreck, Austria 90-00950
Watkins Johnson, USA. 90-02120
Williams Fairey Engineering Ltd, UK . . 90-02150

Sales agents/representatives listed elsewhere:

AlliedSignal Aerospace Marketing . . 80-00075
Sipka Manufacturing (Pvt) Ltd 75-01970

90-00400 CATIC

Representative Office
House No. 151, St. No. 15, E-7 Secter, Islamabad
☎ (+92 51) 82 12 06 ▤ 54206 catic pk islamabad
⊖ (+92 51) 82 62 30
[97]

90-00430 Crescent Aviation Services

(a subsidiary of Crescent Consultants)
Flat N°6 A One Plaza, Markaz F-10, Islamabad
☎ (+92 51) 29 96 05 [E] crescaiva@isb.compol.com
⊖ (+92 51) 29 96 24
♣ US$ 4,000,000 (1996) 👥 10
Chief Executive and Managing Director: Shahid M Rashdi
Marketing Director: Tariq Hamid
Advertising Director: Asma Sheikh
Public Relations: Aalis Rashdi
■ Representation, spare parts supplies, engine overhaul, assessment and financing. Aviation and defence related equipment. Finance for major projects.
Representing:
Piper Aircraft
Fairchild
[98]

90-00440 Daewoo Corp Karachi Liaison Office

11th Floor Shaheen Complex B/D, M R Kayani Road, Karachi
☎ (+92 21) 263 38 63 ▤ 21245 dwkhi pk
⊖ (+92 21) 263 26 74
[98]

90-00770 Greaves Cotton Pakistan (Pvt) Ltd

(a subsidiary of Faruque (Private) Limited, Pakistan)
Modern Motors House, Beaumont Road, PO Box 4908, 75530 Karachi
☎ (+92 21) 568 25 65 ▤ 20815 ⊖ (+92 21) 568 28 39
♧ Rs 20,000,000 (1997) ♣ Rs 125,000,000 (1997)
👥 120 (1997) ⌂ 11,000 sq ft
Managing Director: Mohammed Faruque
Operations Director: Omar Faruque
Sales and Marketing Director: M Q Azam
Purchasing Manager: Iftikhar Ahmad
Rawalpindi
550-C Martin Road, Westridge-1, Rawalpindi
☎ (+92 51) 86 22 80; 47 43 04 ▤ 5564 grvs pk
⊖ (+92 51) 47 43 04
Senior Executive: Saeed Khan ☎ (+91 52) 27 18 86
® UK: Courtaulds Military Association
USA: GTE Government Systems Corp
[98]

90-00950 Interhom (Pvt) Limited

A/4 C C A Shaheen Chambers, PO Box 8008, 75400 Karachi
Block 7 and 8 K C H S, Karachi
☎ (+92 21) 452 15 86 [E] inter@hom.1hr.erum.com.pk
▤ 25226 inter pk; 27563 ⊖ (+92 21) 452 74 33
♣ Rs 25,000,000 👥 150
Chairman: M Amin ☎ (+92 21) 630 50 61; 636 26 25
Managing Director: T M Sheik
☎ (+92 21) 636 88 10; 636 88 63
Director: A Rehman ☎ (+92 21) 636 19 18
Director: Sarmad Amin ☎ (+92 21) 630 50 90
General Manager (Lahore): Tariq Jilani
☎ (+92 45) 41 46 50
Regional Manager (Islamabad): Shakeel Khan
☎ (+92 51) 41 46 50
Branch Manager (Karachi): S M Ishtiaq
® Austria: Walter Schmaddebreck
France: Satel
Germany: Jenoptik Impulsphysik, UCS Engineering
Ireland: PPI Adhesive Products Ltd
Singapore: Sea Scan, Seabreeze
UK: Bayfern Ltd, Connevans Ltd, Croylek Ltd, DX Radio System UK, MBM Trade (UK) Ltd, Marine Technology
USA: Amerace Corp, H K Porter & Co. Inc, J Jennings, Macaw Electron, Surcom Associates Inc, Vectron Lab Inc
Lahore Office
4 Lawrence Road, Lahore
☎ (+92 42) 636 00 09; 630 47 49 ▤ 44580 inter pk
⊖ (+92 42) 636 88 09
Managing Director: Tariq Mahmood Sheikh
☎ (+92 42) 636 89 10; 636 88 63
■ Power generation, telecommunication, oil field equipment, agricultural implements, generators, locomotive parts, defence equipment and steel towers etc.
[98]

90-01200 M Ismailji & Sons Machinery and Equipment

55/55-1 Shafi Plaza, First Floor, Canning Road, Bank Road Square, 46000 Rawalpindi
☎ (+92 51) 56 75 93; 56 37 98 [E] ikram@infolink.net.pk
▤ 5576 ttism pk ⊖ (+92 51) 56 84 45
♣ Rs 75,000,000 (1995) 👥 7 ⌂ 225 m2
Executive Partner: Aman Ali
Sales Manager: Munir Hussain
→ 75 → 85
® Belgium: Lachaussee SA.
Germany: Eumoco AG., Fritz Werner Industrie
[98]

90-01380 Mushko Electronics (Pvt) Ltd

Oosman Chambers, Abdullah Haroon Road, Karachi
☎ (+92 21) 566 04 90; 566 07 73 ⊠ COOPERATOR
▤ 2894 musko pk ⊖ (+92 21) 566 08 01
Managing Director: Faisal A Ali
General Manager (South): Haris Ahmed
General Manager (North): Mr Shafiquddin
Purchasing Manager: Mr Mumtaz
Sales and Marketing: Zahid Uqaili
® UK: Plessey, Racal Decca Navigator
USA: Ampex International, Hewlett-Packard
Representing:
Polar
Rolls-Royce (Industrial and Marine Gas Turbine)
Leica
Wavetek
Bruel & Kjaer
[98]

90-01410 NAD International

Kashif Center - Suite 9/10, 9th Floor Shahrea-Faisal, 75530 Karachi
☎ (+92 21) 56 61 51; 51 87 80 ⊠ LONGVIEW ▤ 20747
⊖ (+92 21) 568 98 73
♣ US$ 5,000,000 (1997) 👥 11 (1997) ⌂ 1,600 sq ft
Chief Executive Officer: Danny Dhalla
General Manager: A Majeed
Operations Director: M Aijaz
Sales and Marketing Director: Steven James
Overseas Sales and Marketing Director: M Shirazi
Sales Engineer: Ghulam Mujtaba
■ Simulators, training equipment, quality control equipment, GSE, electronics, test rigs, wind tunnels, armaments and spares. Aviation, aerospace and defence sales. → 80 → 85
® Italy: Rinaldo Piaggio
UK: Rank Taylor Hobson
USA: Applied Test Systems, FMC Airline Equipment Division
[98]

90-01700 Pearl Engineering (Private) Limited

Head Office
PO Box 8546, Karachi 04
☎ (+92 21) 53 95 05; 53 09 95; 57 64 92
▤ 24028 pearl pk ⊖ (+92 21) 587 04 65
👥 10
President: F.U. Ahmad
Sales & Marketing Dir: Tariq Ashfaq
® UK: EEV Ltd
[98]

90-02120 Uit (Pvt) Ltd

(t United International Technologies)
Omar Plaza, 76-W, Jinnah Avenue, PO Box 1149, Islamabad
☎ (+92 51) 27 36 48; 27 70 38 ✆ (+92 51) 81 81 71
⊠ UIT (Pvt) [E] uitfak@isb.comsats.net.pk
⊖ (+92 51) 27 37 54
♧ US$ 1,000,000 (1996) ♣ US$ 7,000,000 (1997)
👥 9 (1997) ⌂ 4,000 sq ft
Managing Director: Purvis A Hussain
☎ (+92 51) 27 04 37
Sales and Marketing Director: Tahir Javed
Operations and Technical Director: Col Ghani Ur-Rehman (rtd)
Purchasing Manager: Arshad Mehmood
■ Defence marketing consultants to major defence and aerospace companies from USA to Europe.
® Australia: Siemens Plessey
Germany: Steiner Offices
Taiwan: ADI, Getac, Ranger Communications
UK: Hymatic, Pilkington Optronics, Racal Acoustics
USA: American Electronic Lab, BF-Goodrich, E-Systems, ITT, LNR, Litton Industries, Lockheed Martin, Saco Defence, Shakespeare, Ship Analytics, Sierracin Sylmar, Watkins Johnson
[98]

90-02150 United Traders

Haroon Chambers - Altaf Hussain Road, 74000 Karachi
☎ (+92 21) 241 49 16 ✆ (+92 21) 241 57 01
⊠ FOUNDATION ▤ 29356 utx pk ⊖ (+92 21) 241 68 62
👥 13
Managing Director: F H Ansari
Manager: Mohammad Ahmad
Branch Office
Room No 60/A-7 - Khan Chambers, Canning Road, Rawalpindi Cantonment
☎ (+92 51) 56 39 07 ⊠ KEYSTONE Rawalpindi
▤ 29356 utx pk
General Manager: F H Ansari
® France: Matra MS2i
UK: Chemring Limited, Jetstream Aircraft Limited, Laird (Anglesey) Ltd, Redifon SPT Limited, Strachan & Henshaw Limited, Williams Fairey Engineering Ltd
[98]

Panama

Head of State:
Ernesto Pérez Balladares

10 Head of State

10-10000 Office of the President
Palacio Presidencial, Vallija 50, Panama 1
President: Ernesto Pérez Balladares
Minister of the Presidency: Olmedo Miranda [98]

10-15000 Office of the Vice President
Panama
First Vice President: Tomas Altamirano Duque
Second Vice President: Felipe Virzi [98]

12 Ministries

12-20000 Ministry of Foreign Relations
Avenida Perú and Calle 34, La Exposición, Panama 4
Minister of Foreign Relations: Ricardo Alberto Arias [98]

12-40000 Ministry of Government & Justice
Apartado 1628, Panama 1
Minister of Government and Justice: Raúl Montenegro [98]

31 Embassies & Attachés

31-01680 Embassy of Spain
PO Box 1857 Zona 1, Panamá 1
Plaza Belis Porras entre Avenida Peru y Cl 34
☎(+507) 227 51 22; 227 54 72; 227 57 48; 225 05 49
⊖ (+507) 227 62 84
Ambassador: H E Mr Manuel Lorenzo
Deputy Chief of Mission: Francisco Javier Heurgueta
First Secretary: Mrs Ana María Sálomon
Commercial Counsellor: Joaquín Márquez [98]

31-01890 Embassy of the United Kingdom
Torre Swiss Bank, Calle 53, PO Box 889 Zona 1, Panama City
☎(+507) 269 08 66 ⊖ (+507) 223 07 30
Ambassador: H E Mr W B Sinton
Deputy Head of Mission, Third Secretary and Vice Consul: K Burrows
Defence and Military Attaché: Capt K Ridland (Resides in Caracas, Venezuela) [98]

31-01910 Embassy of Uruguay
Cle 32, esq.Avda Justo Arosemena, PO Box 8898 Zona 5, Panamá City
☎(+507 2) 25 00 49; 90 87 ✉URUPANAMA
⊖ (+507 2) 25 90 87
Ambassador: H.E. Mr A. Rodrigueznin
Attaché: Lic. Juan M. Fábrega [97]

75 Manufacturing Companies

75-05000 Daewoo Corp Panama Branch
PO Box 6-2240, El Dorado
Calle Aquilino de la Guardiay, Calle 54
Marbella Torre Banco, General Piso 22, Panama City
☎(+507) 269 45 33 ext. 108; 223 28 57
⊖ (+507) 269 48 11 [98]

80 Service Companies & Organizations

80-00300 Manfredi Aeronautical Industries
PO Box 7528, Panama 5
☎(+507) 223 76 76; 223 19 09 ✉AEROMAN
⊠ 2661 cabpub pg ⊖ (+507) 263 67 43
President: Franz R Manfredi
■ Civil and military aeronautical consultants. Aircraft and related items sales/ distribution. Aircraft inspection/valuation and appraisal. Aviation law. Flight training. International aircraft ferrying. [98]

90 Sales Agents, Sales & Liaison Offices

Companies represented:

Carl Walther GmbH, Germany	90-06000
Dynamit Nobel, Germany	90-06000
Fiocchi of America, USA	90-06000
Interarms, USA	90-06000
J.G. Anschutz GmbH, Germany	90-06000
Raytheon, USA	90-06000

90-03000 Daewoo International (Panama) SA
PO Box 6-2240, El Dorado
Calle Aquilino de la Guardia y Calle 54
Marbella Torre Banco, General Piso 22, Panama City
☎(+507) 269 45 33 ⊠ 8640 daewoo pg
⊖ (+507) 269 48 11 [98]

90-06000 Jorlin SA
Apartado 6-3293, El Dorado, Panama
☎(+507) 223 23 51 ⊖ (+507) 269 38 59
President: Ing. Jorge A Cohen
® Germany: Carl Walther GmbH, Dynamit Nobel, J.G. Anschutz GmbH
USA: Fiocchi of America, Interarms, Raytheon [98]

Papua New Guinea

Head of State:
H.M. Queen Elizabeth II

Governor-General:
Wiwa Korowi

10 Head of State

10-10000 Office of the Governor-General
Government House, Konedobu, PO Box 79, Port Moresby
Governor General: Wiwa Korowi [98]

11 Prime Minister

11-10000 Office of the Prime Minister
Marea Haus, Waigani, PO Box 6605, Boroko
Prime Minister: Bill Skate
Minister for State Assisting the Prime Minister: Sir Pita Lus [98]

26 Police (Civil)

26-08000 Royal Papua New Guinea Constabulary
Police Headquarters, PO Box 85, Konedobu, NCD
☎(+675) 322 61 11 ✉VIGILANCE KONEDOBU
⊠ 22113 police no ⊖ (+675) 322 61 89
♧ kina 100,000,000 ⚉ 8,400 (uniformed); 400 (civilian)
Commissioner: Robert Nenta ☎(+675) 322 61 10
Deputy Commissioner: R Korus
Assistant Commissioner: M Minok
Executive Officer: Andrew Sterns
Assistant Executive Officer: Chief Sgt D Camin
Public Relations and Liaison Officer: Mrs A Gillis [98]

30 Civil Defense

30-00100 Papua New Guinea National Disaster, Surveillance and Emergency Services
PO Box 4970, Boroko, NCT
☎(+675) 301 10 54 ⊠ 70323290 pride ne
⊖ (+675) 325 41 86
Director General: L Anderson ☎(+675) 301 10 53 [98]

31 Embassies & Attachés

31-00590 Embassy of France
PO Box 1155, Port Moresby
Pacific View Apartments 2, 9th Floor 1/84 Pruth Street, Korobosea
☎(+675) 25 13 23; 25 37 40 ⊖ (+675) 25 08 61
Ambassador: H E Mr Pierre Le Gars
Defence Attaché: Capt Robert Alamercery (Resident in Canberra)
Trade Counsellor: Bernard Ould Yahoui (Resident in Sydney)
Financial Counsellor: Alain Guillouet (Resident in Tokyo)
Consul and First Seretary: Mrs Laurence Brun-Walrand [98]

31-00660 Embassy of the Federal Republic of Germany
2nd Floor Pacific View Apartments,
Pruth Street - 3 Mile Hill, Port Moresby
☎(+675) 325 29 71; 325 29 88 ⊠ 23037 aapmby ne
⊖ (+675) 325 10 29
Ambassador: H E Dr Klaus Zeller (Resident in Canberra)
Chargé d'Affaires: Klaus Soering [98]

31-00840 Embassy of the Republic of Indonesia
PO Box 7165, Boroko, NCD
1 2/410 Sir John Guise Drive,
Waigani, National Capital District
☎(+675) 25 31 16; 25 35 44
✉INDONESIA PORT MORESBY
E indoemb@datec.com.au
⊠ 23311 indopom; 22164 indopng ne
⊖ (+675) 25 05 35
Ambassador: H E Maj Gen Abinowo (retd)
First Secretary, Political: Dempo Awang Yuddie
First Secretary, Administration: Sunarto Mangundirdjo
First Secretary, Information: Subari Ardibrata
Second Secretary, Protocol and Consular: Entang Hidayat
Third Secretary, Administration: Deny Tri Basuki
Attaché, Administration: Eustachius Eko Ristiyanto
Attaché, Economic: Widya Parsoaran Gultom
Attaché, Information: Nicolas Hendrik T Manoppo
Defence Attaché: Col Yohanne Albert Ruru
Education and Cultural Attaché: Dr Tandi Roma Andi-Lolo [98]

31-01300 New Zealand High Commission
PO Box 1051, Waigani, NCD
☎(+675) 25 94 44 ⊖ (+675) 25 05 65
High Commissioner: H E Mr John Clarke
Deputy High Commissioner: Neil Robertson
Defence Adviser: Wg Cdr Athol Forrest [98]

31-01420 **Embassy of the Philippines**
PO Box 5916, Boroko, Port Moresby
☎(+675) 325 65 77 ⊜ (+675) 323 18 03
Chargé d'Affaires a.i.: H E Mr Aladin G. Villacorte
☎(+675) 325 06 69 [98]

31-01890 **United Kingdom High Commission**
PO Box 212, 131 Waigani, NCD
Kiroki Street, Waigani, Port Moresby
☎(+675) 325 16 77 ⊜ (+675) 325 35 47
High Commissioner: H E Mr Bob Low
Deputy High Commissioner: M Plumb
Deputy Head of Mission: B P Baldwin
Defence Attaché: Col M Radford (Resident at Canberra)
☎(+61 62) 70 66 66 ext 216
(+61 62) 62 73 32 36 (fax) ⊠ 7162222 [98]

Paraguay

Head of State:
Juan Carlos Wasmosy

10 Head of State

10-10000 **Office of the President**
Palacio de López, Mburbicha Roga, Asunción
☎(+595 21) 44 18 89 ⊜ (+595 21) 49 36 62
President: Juan Carlos Wasmosy
Vice President: Roberto Angel Seifart [98]

12 Ministries

12-10000 **Ministry of National Defence**
Avenida Mariscal Lopez y, Asunción
☎(+595 21) 21 44 77 ⊠ 762 py ccc ffaa
⊜ (+595 21) 21 15 83
Minister: Dr Hugo Estigarribia Elizeche [98]

12-20000 **Ministry of Foreign Relations**
Juan O'Leary 222 y Pte, Franco 707, Piso 4,
Asunción, Paraguay
☎(+595 21) 49 45 93; 49 38 72 ⊠ 111 py relaciones
⊜ (+595 21) 49 39 10
Minister of Foreign Relations: Ruben Melgarejo Lanzoni [98]

12-25000 **Ministry of Interior**
Chile y Manduvira, Asunción
☎(+595 21) 49 36 61; 44 66 64 ⊠ 153 mininterior py
⊜ (+595 21) 44 84 46
Minister of Interior: Miguel Angel Ramirez
Minister of Justice and Labor: Sebastion Gonzalez Insfran [98]

12-30000 **Ministry of Industry and Commerce**
Avenida España 323, Asunción
☎(+595 21) 20 48 33 ⊠ 259 py mic
⊜ (+595 21) 21 05 70
Minister: Atilo Fernandez [98]

15 Parliament

15-10000 **Senate**
Palacio Legislativo, Asunción [98]

15-20000 **Chamber of Deputies**
Palacio Legislativo, Asunción [98]

31 Embassies & Attachés

31-00890 **Embassy of Italy**
Calle Luis Morales 680, PO Box 521, Asunción
☎(+595 21) 20 74 29; 22 59 18
[E] ambasu@quanta.com.py ⊠ 22104 italas py
⊜ (+595 21) 21 26 30
Ambassador: H E Dr Antonio Cavaterra
Deputy Chief of Mission: Ms Lidia Polmonari [98]

31-01410 **Embassy of Peru**
Avenida Mariscal López 648, PO Box 433, Asunción
☎(+595 21) 20 44 56 [E] embperu@pla.net.py
✉LEPRU Asuncion ⊜ (+595 21) 21 29 80
Ambassador: H E Mr Alejandro León Pazos
☎(+595 21) 66 03 07
Deputy Chief of Mission: Ms Amalia Mariategui Succar
☎(+595 21) 60 61 57
Commercial and Public Information Officer:
José David Hurtado Fudínaga [98]

31-01680 **Embassy of Spain**
Calle Yegros 437, p5°y6°, PO Box 6, Asunción
☎(+595 21) 49 06 86; 49 06 87 ⊠ 44160 embespa py
⊜ (+595 21) 44 53 94
Ambassador: H E Mr I García-Valdecasas Fernández
Commercial Office
Calle Presidente Franco, esq.,
Ayolas Edificio Ayfra, ala dcha, p8°, Asunción
☎(+595 21) 49 84 41; 49 84 42 ⊠ 44306 ofcomes py
⊜ (+595 21) 49 84 43
Commercial Consellor/Secretary: Ms Sara Alicaide Guindo [98]

Peru

Head of State:
Alberto Kenyo Fujimori Fujimori

10 Head of State

10-10000 **Office of the President**
Palacio de Gobierno, Plaza de Armas s/n, Lima 1
☎(+51 14) 27 13 66 ⊜ (+51 14) 31 04 27
President of the Republic: Alberto Kenyo Fujimori Fujimori
Minister of the Presidency: Daniel Hokama Tokashiki [98]

11 Prime Minister

11-10000 **Office of the Council of Ministers**
Avenida Inca Garcilaso de la, Vega 1351, Piso 4, Lima
☎(+51 14) 33 41 88 ⊠ 25166 pe precomin
President of the Council of Ministers:
Alberto Pandolfi Arbulu [98]

12 Ministries

12-10000 **Ministry of Defence**
Avenida Arequipa 291, Lima 1
☎(+51 14) 35 51 50; 36 29 12 ⊜ (+51 14) 33 44 70
Minister of Defence: Gen Cesar Saucedo Sanchez [98]

12-20000 **Ministry of Foreign Affairs**
Ministerio Relaciones Exteriores
Jr Ucayali 363, Lima 1
Minister of Foreign Affairs: Eduardo Ferrero Costa [98]

12-40000 **Ministry of Interior**
Ministerio de Interior
Plaza 30 de Agosto 150, San Isidro, Lima 27
☎(+51 14) 41 29 89; 41 19 13 ⊠ 21697
Minister of Interior: Gen José Villanueva Ruesta
Minister of Justice: Alfredo Quispe Correa [98]

15 Parliament

15-10000 **Senate**
Senado de la Republica
Plaza Bolivar, Lima 1
☎(+51 14) 28 31 19 ⊠ 20002 pe senrep [98]

15-20000 **Chamber of Deputies**
Camara de Diputados
Plaza Bolivar, Lima 1
☎(+51 14) 28 79 80 ⊠ 25 311 pe camaradi [98]

25 Defense Forces

25-34000 **Fuerza de Naval Aviación**
Aeroporto Int. Jorge Chávez,
Rampa Sur, Avenida Elmer Faucett S/N, 1 Callao
☎(+51 1) 575 17 21 ⊜ (+51 1) 575 17 21
Commanding Officer: RAdm Frank Boyle Alvarado
Deputy Commander, Chief of Staff:
Capt José Qeuvedo De Lama
Operations Director: Cmdr Enriqué Cebreros D
Director of Procurement and Logistics:
Cmdr Jorge Portocarrero Peñafie [98]

25-41000 **Directorate of Naval Armaments Peruvian Navy**
Avda de La Marina, Cuadra 36, La Perla Callao
☎(+51 511) 420 61 60 ⊜ (+51 511) 420 61 60
Director: Ricardo Fernández [97]

28 Coastguard, Border Protection Forces

28-10000 **Direccion General Capitanias y Guardacostas**
(Subordinate to Ministry of Defence)
102 Daniel Nieto, Plaza Grau, Callao
☎(+51 14) 29 68 98; 29 31 96
[E] postmast@dgcdicap.mil.pe ⊠ 26071 pe dicapi
⊜ (+51 14) 65 10 68; 65 87 70
General Director of Captaincies and Coast Guard:
Rear Adm Gonzallo Rodriguez Maisterrena
☎(+51 14) 65 98 66
Deputy Commander: Rear Adm César Chavez
Operations Director: Capt Alberto Revoredo
Procurement/Logistics Director: Capt Juan Illescas
■ National Maritime Authority. Jurisdiction: Peruvian waters through captaincies in the main maritime ports as well as inland on lakes and rivers. [98]

31 Embassies & Attachés

31-00100 Embassy of Austria
Avenida Central 643 P5, San Isidro, PO Box 853, Lima 100
☎ (+51 1) 442 05 03 ☒ 21128 oelim pe
⊖ (+51 1) 442 88 51
Ambassador: H E Mr Artur Schuschnigg
Second Secretary and Vice-Consul: Walter Rehberger [98]

31-00350 Embassy of Chile
Av. Javier Prado Oeste 790, San Isidro, Lima
☎ (+51 1) 221 20 80; 221 20 81 ☒ 20181 pe emchile
⊖ (+51 1) 221 20 85
Ambassador: H.E. Mr Carlos Martínez Sotomayor
Deputy Chief of Mission: Jorge Montero Figueroa
Air Attaché: Col Rodolfo Acuña Guimpert
Naval Attaché: Capt. Rodrigo Díaz Buzeta
Commercial Attaché: Diego Varas Benavente
☎ (+51 1) 422 27 36 [97]

31-00450 Embassy of the Czech Republic
Baltazar La Torre 398, Casilla Postal 848, San Isidro, Lima
☎ (+51 1) 264 33 81; 264 33 74 ⊖ (+51 1) 264 17 08
Ambassador: H E Dr Lubomír Hladík
Second Secretary, Commercial and Consul: Ing Josef Hlobil
Attaché: Ing Zdenek Kalas [98]

31-00810 Consulate of the Republic of Hungary
Avda Javier Prado Este 1520, San Isidro, Lima
☎ (+51 1) 224 17 80 ☒ 25413 diplima pe
⊖ (+51 1) 36 98 53
Ambassador: H.E. Mr Stefan Hajossy [97]

31-00830 Embassy of India
Av Salaverry 3006, Magdalena del Mar, Lima 17
Avenida Salaverry 3006, Magdalena del Mar
☎ (+51 14) 62 18 40; 62 18 90; 62 90 40
🄴 postmaster@indoperu.org.pe ☒ 25515 pe indebras
⊖ (+51 14) 61 03 74
Ambassador: H E Mrs Nirupama Rao
Third Secretary and Head of Chancery: Sanjiv Ranjan [98]

31-01300 Honorary Consulate of New Zealand
PO Box 3553, Lima 100
Torre Central, Camino Real 390, Piso 17, Lima 27 (San Isidro)
☎ (+51 1) 221 28 33 ⊖ (+51 1) 421 31 94
Ambassador: H E Mr Lawrence Markes (Resident in Mexico)
Honorary Consul: Roberto P Eidson [98]

31-01440 Embassy of Portugal
Avenida Central 643-4TO Piso, San Isidro, Lima 27
☎ (+51 1) 440 99 05 ☒ 25 734 lusitan pe
⊖ (+51 1) 442 96 55
Ambassador: H E Dr José Sarmento ☎ (+51 1) 422 05 19 [98]

31-01890 Embassy of the United Kingdom
PO Box 854, Lima 100
El Pacifico Washington (Piso 12)
Plaza Washington, Avenida Arequipa, Lima
☎ (+51 1) 433 47 38; 433 48 39; 433 49 32
🄴 brit-emb@amouta.rcp.net.pe ⊖ (+51 1) 433 47 35
Ambassador: H E Mr J Illman
Deputy Head of Mission: R G Church
First Secretary, Commercial: S J Collier [98]

31-02010 Embassy of the Federal Republic of Yugoslavia
360 Cle C.Porras Osores, San Isidro, 18-0392 Miraflores, Lima
☎ (+51 14) 421 24 23; 421 24 27
⊖ (+51 14) 421 24 27
Charge D'Affairs a.i.: Zoran Raicevic [98]

55 Government-owned Industry & Production Facilities

55-01950 Servicios Industriales de la Marina - SIMA PERU
Executive Headquarters
Avenida Contralmirante Moira, Base Naval del Peru, Callao
☎ (+51 65) 55 41; 34 20 ✉ SIMAPERU
☒ 26128 pe sima
Chairman: VAdm Augusto Fernandez Hidalgo
Executive Dir.: RAdm Fernando Jimenez Román
Manager SIMA CALLAO: Capt Andrés Chiarella Caballero
Manager SIMA CHIMBOTE: Capt Renán Zuñiga Mossone
Manager SIMA CEFAR: Capt José Silva Ferrer
Manager SIMA IQUITOS: Capt Eugenio Beltran Chávez
Finance Mgr: Javier Luna Samamé
Marketing Mgr: Raúl Zuazo Tolmos
Logistics Mgr: Hector Marrache Palacios
■ Naval repairs, including complete overhauls & modernization of naval surface vessels & submarines. Naval metal & mechanical work, including heavy boiler work, heavy naval structure work & welded heavy profiles work. Naval shipbuilding, with slipways of 65,000 & 27,000 GRT for the construction of frigates & fleet auxiliaries (tankers & freighters) & smaller craft such as fast patrol boats & landing craft. Small bore weaponry. Railway locomotives & rolling stock. → 80 [98]

75 Manufacturing Companies

75-16000 Simrad Peru SA
(a subs. of Simrad Norge)
Larco 730 Of.704 Miraflores, Lima 18
☎ (+51 14) 45 57 60 ⊖ (+51 14) 241 10 67
♧ US$ 50 000 (1996) ♣ US$ 200 000 (1996)
👥 10 (1996) 🏭 200 sq m
Managing Dir: Gunnar Glaatnes
General Mgr: Henri Yurjevic
Marketing Mgr: William Robinson [97]

80 Service Companies & Organizations

85 Stockists, Distributors, Spare Parts Suppliers

90 Sales Agents, Sales & Liaison Offices

Companies represented:

90-00230 Benavides Representaciones Latinoamericanas y Peruanas Srl (BERELAPE)
Malecon de la Marina 150, PO Box 0645, Lima 18
☎ (+51 14) 41 23 23; 41 45 05; 41 92 38
⊖ (+51 14) 41 46 56
♧ US$ 300,000 (1996) ♣ US$ 1,600,000 (1996)
👥 15 (1997) 🏭 1,290 m2
Chairman: Oscar Benavides Gubbins
Commercial Director: Gerardo Chanduui
■ Sale of helicopters, aircraft, components and weapons. Maintenance, repair and overhaul of parts, avionics, radars and electric systems. Sale of components to armed forces. → 80
Representing:
American Equipment Co. USA
Allied Bendix Aerospace USA
Central Engineering Co.
Sikorsky Aircraft Division
Bruggemann and Brand, Ger
Concast Standard AG., Ger
Militair Aviation Limited, UK
Lovaux Ltd, UK
Cobrasma SA, Brazil
Canadian Marconi Co., CAN
Luchaire, FRA [98]

90-00300 Berelape Srl
(† Aviojet Corp)
Malecon de la Marina 150, Miraflorez, Lima 18
☎ (+51 14) 41 45 05 ⊖ (+51 14) 41 46 56
♧ US$ 300,000 (1995) ♣ US$ 1,500,000 (1995)
👥 15 (1995) 🏭 1,290 m2
Managing Director: Oscar Benavides G
Purchasing Manager: Gerardo Chanduvi T
® UK: Militair Aviation Ltd
USA: Aviojet Corp., United Technologies, Aircraft Sales Div. [97]

90-00350 Coperimprel SA
(a subs. of Pietro Schinaia Z.)
Ave Guzman Blanco 269, Lima
☎ (+51 14) 433 85 72; 433 60 85
✉ COPERIMPREL Lima No.174424
⊖ (+51 14) 433 85 72
♧ US$ 100 500 (1995) ♣ US$ 500 557 (1995) 👥 11
🏭 540 sq m
Managing Dir: Pietro Schinaia Z.
Marketing Dir: Blanca Landa
Public Relations Mgr: Pietro Schinaia L.
→ 75 → 80 → 85 [97]

90-00400 Daewoo Corporation Oficina Representiva
Avenida Canaval Y Moreyra N°452, Piso 16, San Isidro, Lima
☎ (+51 1) 221 42 44 ⊖ (+51 1) 421 13 33 [98]

Philippines

Head of State:
Gen. Fidel V. Ramos

10 Head of State

10-10000 Office of the President
Malacanang Park, J P Laurel Street, Manila
☎ (+63 2) 521 23 01; 521 23 10
President of the Republic: Gen Fidel V Ramos
Vice-President: Joseph Estrada
Executive House, Burgos Street Manila
☎ (+63 2) 40 55 33
Executive Secretary: Ruben Torres
Press Secretary: Hector Villanueva
Group Commander, Presidential Security Group: Col Voltaire Gazmin ☎ (+63 2) 50 21 23 01 [98]

12 Ministries

12-10000 Department of National Defence
DND Building - 3rd Floor, Camp E Aguinaldo,
Quezon City, Metro Manila
☎(+63 2) 911 45 29; 911 33 39 ▣ 22471 dnd ph
⊖ (+63 2) 911 71 13
Secretary of National Defence: Renato S de Villa
☎(+63 2) 911 61 83; 911 61 93
Undersecretary of National Defence:
Hon Feliciano M Gacis
☎(+63 2) 911 33 00; 911 61 92
Undersecretary for Civilian Relations: Fernando C Campos
☎(+63 2) 911 62 46
Undersecretary for Veterans and Reserve Affairs, Assistant Secretary for Comptrollership: Vicente R Buenaventura
☎(+63 2) 911 65 61; 911 97 30
Assistant Secretary for Installations and Logistics:
Cesar H Bello ☎(+63 2) 911 62 44
Assistant Secretary for Personnel: Ms Leticia A Gloria
☎(+63 2) 911 43 96
Assistant Secretary for Plans and Programs:
Romeo R Zulueta ☎(+63 2) 911 64 60
Officer for Public Affairs: Ms Lourdes I Ilustre
☎(+63 2) 911 45 29; 911 33 39
Chief Management of Defence Information Systems:
Ramon El Martinez ☎(+63 2) 911 62 62
Chief Office for Strategic Assessments:
Ms Edna L Guevara ☎(+63 2) 911 71 71
Chief Office for Legal Affairs: Ms Iris V Enciso
☎(+63 2) 911 64 13
Chief Office for Legislative Affairs: Manuel A Miguel
☎(+63 2) 911 64 12
Chief of Philippines Armed Forces: Gen Arnulfo G Acedera
☎(+63 2) 911 64 36
Administrator Philippine Veterans Affairs Office:
Umberto A Rodriguez ☎(+63 2) 911 78 78
Director Arsenal: Niceto F Festin ☎(+63 2) 911 65 34
President National Defence College of the Philippines:
Eduardo R Cabanlig ☎(+63 2) 87 74 22
Administrator Office of Civil Defence: Fortunato M Dejoras
☎(+63 2) 911 66 75 [98]

12-20000 Department of Foreign Affairs
MFA Building, Padre Faura, Manila
Secretary of Foreign Affairs: Domingo Siazon [98]

12-30100 Ministry of Trade and Industry
Trade and Industry Building, 361 Buenida Avenue, Makati, Metro Manila
Secretary of Trade and Industry: Ceasar Bautista [98]

12-40000 Department of the Interior and Local Government
PNCC Building, 2nd Floor, EDSA corner, Reliance Street, Mandaluyong, Metro Manila
Secretary of the Interior and Local Government:
Robert Barbers [98]

25 Defense Forces

25-30000 Headquarters Philippine Army
Fort Andres Bonifacio Makati, Makati, Metro Manila 1201
☎(+63 2) 812 74 16; 812 74 31; 812 74 39
Commanding General: Lieut Gen Clemente P Mariano
☎(+63 2) 816 11 62; 813 18 11 (fax)
Vice Commander: Maj Gen Emilio B De Leon
☎(+63 2) 892 06 31
Chief of Staff: Brig Gen Santos B Gabison
☎(+63 2) 892 06 31
Assistant Chief of Staff for Personnel G-1:
Col Delfin M Lorenzana ☎(+63 2) 844 27 77
Assistant Chief of Staff for Security G-2:
Col Gabriel A Habacon ☎(+63 2) 893 28 26
Assistant Chief of Staff for Operations G-3:
Col Generoso S Senga ☎(+63 2) 892 28 84
Assistant Chief of Staff for Logistics G-4:
Col Felicisimo V Lagascca ☎(+63 2) 817 28 59
Assistant Chief of Staff for Plans and Programmes G-5:
Col Reynaldo D Rivera ☎(+63 2) 812 29 35
Assistant Chief of Staff for Comptrollership G-6:
Col Samuel B Bagasin ☎(+63 2) 85 15 65
Assistant Chief of Staff for Civil and Military Operations G-7: Col Rodolfo B Alvarado ☎(+63 2) 844 65 42
Assistant Chief of Staff for Training G-8:
Col Josefino B Manayao ☎(+63 2) 844 79 19
Assistant Chief of Staff for Reservist and Retirees Affairs G-10: Col Froilan M Maglaya ☎(+63 2) 893 47 86
Assistant Chief of Staff for Communication, Electronics and Information System G-11: Col Eliseo E Rio
Commanding Officer, Training and Doctrine Command (TRADOC): Col Roberto P Santiago
Public Information Officer: Lieut Col Antonio V Inciong (Jr)
☎(+63 2) 892 16 93 [98]

31 Embassies & Attachés

31-00590 Embassy of France Chancery
PO Box 1676, Makati, Metro Manila
Pacific Star Building, Corner Gil Puyat and Makati Avenue, Makati, Metro Manila
☎(+63 2) 810 19 81; 810 19 82; 810 19 83;
810 19 84; 810 19 85; 810 19 86;
810 19 87; 810 19 88 ⊖ (+63 2) 817 50 47
Ambassador: H E Mr Samuel Le Caruyer de Beauvais
Deputy Chief of Mission and Counsellor: Yves Lorvellec
Press Counsellor: Jean-Pierre Swyngedauw
Counsellor, Cultural, Scientific and Co-operation:
Marcel Jouve
First Secretary: Ms Claude Abily
First Secretary and Consul: Patrick Vaillant
Armed Forces Attaché: Col François Rozier de Linage

Commercial Section
Unit 34a/b, 34th Floor,
Rufino Pacific Tower 6784 Ayala Ave,
Makati, Metro Manila
☎(+63 2) 811 10 01; 811 10 02; 811 10 03; 811 10 04
⊖ (+63 2) 811 10 33
Counsellor, Commercial: Henri Fremon [98]

31-00830 Embassy of India
PO Box 2123 MCPO, Makati, Metro Manila 1222
2190 Paraiso Street, Dasmarinas Village,
Makati, Metro Manila
☎(+63 2) 643 01 01/2 ⊠ INDEMBASSY MANILA
▣ 63595 ⊖ (+63 2) 815 81 51
Ambassador: H.E. Mr C.P. Ravindranathan
☎(+63 2) 812 41 25
Counsellor: Madhava Chandra ☎(+63 2) 844 14 29
Air & Naval Attaché: Capt. D.K. Joshi
(Resident in Singapore)
Second Secretary: Sanjay Verma [97]

31-01270 Royal Netherlands Embassy
Makati Commercial Center, PO Box 2448,
Makati, Metro Manila 1264
9th Floor, King's Court Building, 2129 Pasong Tamo,
Makati, Metro Manila
☎(+63 2) 812 59 81; 812 59 83 Ⓔ nlgovman@qinet.net
⊖ (+63 2) 815 45 79
Ambassador: H E Mr Eric T J T Kwint
Counsellor and Deputy Chief of Mission: J J J Teunissen
First Secretary Economic and Commercial Affairs:
Ms M A van Drunen Littel
Second Secretary: W W G Oostelbos [98]

31-01300 New Zealand Embassy
PO Box 3228 MCPO, Makati, Metro Manila 1272
23rd Fldoor Far East Bank Center, Gil Puyat Avenue,
Makati, Metro Manila
☎(+63 2) 891 53 58 ⊖ (+63 2) 891 53 57
Ambassador: H E Mr Colin V Bell
First Secretary, Policy: Tim McIvor
First Secretary, Administration: Col Brian Chippindale (rtd)
Defence Attaché: Col Robert Bywater-Lutman [98]

31-01480 Embassy of Romania
1216 Acacia Rd, Dasmarinas Village, Makati, Metro Manila
☎(+632) 843 90 14 ▣ 45460 romcom
⊖ (+632) 843 90 63
Ambassador: H.E. Mr Ioan Emil Vasiltu
Deputy Chief of Mission: Emil Ghitulescu

Commercial Office
1233 Acacia Road, Makati, Metro Manila
▣ 45460 romcom pm
Commercial Counsellor: Gheorghe Balasa [97]

31-01740 Embassy of Switzerland
MCPO Box 2068, Makati Metro, Manila 1260
18th Floor - Solidbank Building, 777 Paseo de Roxas
☎(+63 2) 819 02 02 Ⓔ swissman@webquest.ccom
▣ 22339 esw ph ⊖ (+63 2) 815 03 81
Ambassador: H E Mr Kurt M Höchner
Counsellor and Deputy Head of Mission: Josef W Eisele
Second Secretary and Head of Chancery: Norbert Arnold
Vice Consul: Rudolf Hösli
Commercial Officer: Ms May Olalia
Administration Officer: Ms Marie-José Borghini
Secretary to the Ambassador: Mrs Claudia Richli [98]

31-01890 Embassy of the United Kingdom
PO Box 2927 MCPO, Metro Manila
Fl 15-17 LV Locsin Bldg, 6752 Ayala Ave, Cor Makati Ave,
Makati, Metro Manila 1226
☎(+63 2) 816 71 16 (consular/visa) ▣ 63282 prodme pn
⊖ (+63 2) 819 72 06
(management/information); 810 27 45
(consular/visa); 815 62 33 (commercial/aid)
Ambassador: H E Mr Adrian C Thorpe
Deputy Chief of Mission: M D Reilly
Defence Attaché: Gp Capt P G Wildman (RAF)
First Secretary, Commercial: E J McEvoy
First Secretary, Commercial (British Council Director):
K T Bailey
Second Secretary, Management: A P Newton
Second Secretary: A T Dimbleby
H M Consul: L B Evans [98]

31-01900 Embassy of the United States of America
APO AP 96440, Ermita Manila
1201 Roxas Boulevard
☎(+63 2) 521 71 16 ⊖ (+63 2) 522 43 61
Ambassador: H E Mr Thomas C Hubbard [98]

75 Manufacturing Companies

75-00150 Arms Corporation of the Philippines (ARMSCOR)
Executive Offices
9th Flr, Strata 100 Bldg, Emerald Ave, Ortigas Center,
Pasig City 1600
☎(+63 2) 631 81 56/60 ⊖ (+63 2) 631 28 76
👥 689
Chairman & President: Demetrio R. Tuason
Sr VP & General Mgr: Danilo R. Dela Paz
VP Marketing: Michael P. Melchor
■ Arms & ammunition. Cal .22 Rimfire rifles bolt action/semi-automatic, 12 Ga. pump action shotgun, cal .38 revolvers, cal .22 Rimfire ammunition, Centerfire pistol ammunition cal .38 Auto, .38 Super, 9mm, 32 Auto, .45, 357, 38 Spl, Centerfire rifle ammunition cal 5.56mm, .222, .223, 30 MI Carbine, 12 Ga. shotgun/shotshells. Standard & high capacity 1911 pistols in .45 ACP, .38 super & 9 mm.

Factory:
Fortune Ave. Parang, Marikina, Metro Manila 1800
☎(+63 2) 942 36 53; 941 62 44; 941 42 99
⊖ (+63 2) 942 06 82 [97]

75-00370 Daewoo Corp Manila Branch
15th Floor Unit-C, World Center Building, Makati City
330 Sen Gil J Puyat Avenue
☎(+63 2) 867 82 91; 867 82 92; 867 82 93
⊖ (+63 2) 867 80 44 [98]

75-00800 Far East Industrial Supply Co Inc. (FEISCO)
401-409 Del Monte Ave, Quezon City
☎(+63 2) 35 40 16-18; 361 84 42-43; 361 84 39
⊠ FEISCO MANILA ▣ 63199 etpimo pn
⊖ (+63 2) 361 84 39; 30 46 16
♧ PH.PES 3 000 000 000 (1996)
♣ PH.PES 100 000 000 (1996) 👥 600 (1996)
🏭 200 000 sq m (1996)
President: Danilo Valdez
Operations & Technical Dir: R. Deiparine
■ Manufacturers of steel helmets, ammunition metal boxes, aluminium and steel magazines. [97]

75-01325 **Transfield Pty Ltd**
c/o APTDC - 8 St Peter Street, Cubao, Quezon City 1109
☎ (+63 2) 413 28 92 ⊜ (+63 2) 725 32 09 [98]

75-01350 **Trojan Manufacturing & Marketing Inc (TMMI)**
2109 URC Building, Espana St. Sampaloc, Manila 2806
☎ (+63 2) 741 70 51; 61 70; 70 81 ✉ TROMA MANILA
▣ 40773 pvlmnl pm ⊜ (+63 2) 741 70 81
♧ Ph.Pes 5 029 705 ♣ Ph.Pes 30 200 000 (1995)
180 ; 2 500 sq m
Managing Dir: Miss Fluerdeliz A. Constantine
Sales & Marketing Dir: Leo L. Toledano
Operations & Technical Dir: Jim C. Baquiran
Purchasing Mgr: Alfred S. Constantino
Advertising Mgr: Ms Angelina S. Toledano
■ Manufacturer of clothing, webbing & personal equipment for the Philippine Armed Forces. [97]

80 Service Companies & Organizations

80-00050 **Aerospace & Defense Sales Inc (ADSI)**
Suite 1001, Goldhill Tower Condominium,
Metro Manila 1500
#5 Annapolis Street - Greenhills,
San Juan, Metro Manila 1500
☎ (+63 2) 723 93 89; 724 26 79; 78 71 77
E geo_mark@compass.com.ph ⊜ (+63 2) 723 93 89
♧ peso 2,000,000 (1996) ♣ peso 4,100,000 (1996)
167 28 (1997) 160 m2
Chairman: Bienvenido M Araw II
Vice Chairman: Ms Isabelita R Sy
President: Danilo R Sy
Director: Ms Maureen C Abelardo
Director: Ms Victoria C Delos Reyes
Director: Renato R Sy
Operations Manager: Ms Marie P Atchada
■ Sales, representation and distribution of aviation, maritime and defence equipment, parts, supplies and service. Consultancy and management services through business planning, project implementation, arrangement of joint ventures and countertrade. GIS and aerial photogrammetry services. [98]

80-00100 **AlliedSignal Aerospace Marketing Sales and Service**
Villamor Air Base, Pasay City, Metro Manila
2nd Floor Philippines Airlines Technical Center
☎ (+63 2) 831 02 55 ⊜ (+63 2) 831 75 47
Field Service Engineer: Leo Sebastian
■ On-site customer support for AlliedSignal Aerospace. → 90 [98]

80-01500 **Pacific Mountain Information Ltd**
2/F Wing A - Concorde Condominium, Benavidez Street, Makati, Metro Manila 1200
☎ (+63) 812 81 92; 813 15 24 ⊜ (+63) 813 15 24
President: Prof Grace C Lara
Sales and Marketing Director: Arnold Atienza
Operations Manager: Ms Ofelia P Paulino
Purhasing Manager: Adoracion Oabel
® UK: Jane's Information Group [98]

80-02000 **Philippine Exhibition Services Organization Inc**
Unit 1205-B, Centerpoint Office Condominium,
Metro Manila Pasig 1600
☎ (+63 2) 633 65 61; 633 65 62; 633 65 63
⊜ (+63 2) 633 65 89
♧ peso 1,127,500 (1995) ♣ peso 11,020,252 (1995)
15 175 sq m (1997)
President and General Manager: Danilo R Sy
☎ (+63 2) 633 65 57 59
Assistant General Manager: Ms Marie P Atchada
■ Air, land and sea exhibitions. [98]

90 Sales Agents, Sales & Liaison Offices

Companies represented:

Bell Helicopter Textron, USA→ Singapore 90-00220
Jane's Information Group, UK 80-01500
The Parsons Corp., USA 90-01500
Zengrange Ltd, UK→ Netherlands 80-18300

Sales agents/representatives listed elsewhere:

AlliedSignal Aerospace Marketing . . 80-00100

90-00200 **BP Solar**
(a subsidiary of BP Solar Australia Pty Ltd)
Suite 76, Zeta Building,
191 Salcedo Street - Legaspi Village,
Makati City, Manila 1220
☎ (+63 2) 815 90 36; 894 34 90; 894 85 08
⊜ (+63 2) 817 15 23
Country Manager: Katrina V Ignacio [98]

90-00210 **Daewoo Corp Manila Representative Office**
4th Floor Pacific Star Building, Makati City
Sen Gil Puyat Avenue, Corner Makati Avenue
☎ (+63 2) 818 52 20 ⊜ (+63 2) 817 52 67 [98]

90-00900 **Kawasaki Heavy Industries Ltd Manila Office**
20th Floor Metrobank Plaza Building, Gil J Puyat Avenue, Makati, Metro Manila
☎ (+63 2) 818 27 86 ⊜ (+63 2) 818 27 87 [98]

90-01500 **Parsons International Limited**
(† Parsons Corp.)
10th Fl, Rufino Pacific Tower, 6784 Ayala Avenue, Makati, Metro Manila 1226
☎ (+63 2) 811 01 15 ⊜ (+63 2) 811 00 60
Sr Vice President: Louis G. Ogden
® USA: The Parsons Corp. [97]

Poland

Head of State:
Aleksander Kwasniewski

10 Head of State

10-10000 **Office of the President**
ulitsa Wiejska 10, PL-00-902 Warsaw
☎ (+48 22) 695 38 19 ⊜ (+48 22) 695 38 19
President: Aleksander Kwasniewski [98]

11 Prime Minister

11-10000 **Office of the Prime Minister**
Al Ujazdowskie 1-3, PL-00-583 Warsaw
☎ (+48 22) 694 60 00 ▣ 813446; 815376
⊜ (+48 22) 625 26 37; 628 28 05; 628 42 22
Prime Minister: Wlodzimierz Cimoszewicz
Deputy Prime Minister and Finance Minister: Marek Belka
Deputy Prime Minister: Roman Jagielinski
Deputy Prime Minister: Miroslaw Piertrewicz
Government Spokesperson: Mrs Aleksandra Janubowska
☎ (+48 22) 41 07 22; 694 61 51
☎ (+48 22) 694 68 26 (fax) [98]

12 Ministries

12-10000 **Ministry of National Defence**
Ulitsa Klonowa 1, PL-00-909 Warsaw 60
☎ (+48 22) 621 02 61 ⊜ (+48 22) 45 53 78
Minister of Defence: Stanislaw Dobrzanski [98]

12-10500 **Ministry of Economy**
Warsaw
Minister: Wieslaw Kaczmarek [98]

12-20000 **Ministry of Foreign Affairs**
Al Szucha 23, PL-00-580 Warsaw
☎ (+48 22) 44 16 87 ▣ 814301 ⊜ (+48 22) 623 90 00
Minister of Foreign Affairs: Dariusz Rosati
Deputy Minister: Eugeniusz Wyzner [98]

12-20200 **Ministry of Transport and Maritime Economy**
ulitsa T Chalubifskiego 4/6, PL-00 928 Warsaw
☎ (+48 22) 24 40 00 ⊜ (+48 22) 30 02 61
Minister: Boguslaw Liberadzki [98]

12-20500 **Ministry of Economy**
ulitsa Klonowa 1, PL-00 909 Warsaw
☎ (+48 22) 628 04 61 ⊜ (+48 22) 628 68 08
Minister: Wieslaw Kaczmarek [98]

12-30300 **Ministry of Treasury**
ulitsa Krucza 36, PL-00 522 Warsaw
☎ (+48 22) 695 80 00
⊜ (+48 22) 628 08 72; 621 33 61
Minister: Miroslaw Pietrewicz [98]

12-30500 **Ministry of Communications**
Plac S Malachowskiego 2, PL-00 940 Warsaw
☎ (+48 22) 656 50 00 ⊜ (+48 22) 826 48 40
Minister: Andrzei Zielinski [98]

12-40000 **Ministry of Internal Affairs**
Ministerstwo Spraw Wewnetrznych
Ulitsa Batorego 7/9, PL-02-514 Warsaw
☎ (+48 22) 621 02 51 ⊜ (+48 22) 49 74 94
Minister: Leszek Miller [98]

15 Parliament

15-10000 **Parliament**
Sejm
Ulitsa Wiejska 4-8, PL-00-902 Warsaw
☎ (+48 22) 694 25 00; 28 32 97 ▣ 812544
Speaker: Jozef Zych
Deputy Minister: Wojciech Popkowski
Vice Marshal: Aleksander Malachowski
Vice Marshal: Olga Krzyzanowski
Vice Marshal: Marek Borowski
Chief of Staff: Daria Andrew-Fraczek
Secretary General: Maciej Graniecki
Chairman of the National Defence Committee: Jerzy Szmajdzinski ☎ (+48 2) 694 11 47
Chairman of the Foreign Affairs Committee: Bronistaw Geremek ☎ (+48 2) 694 14 07
Director of Protocol: Dariusz M Chmiel [98]

15-10700 **State Committee of Science and Technology**
ulitsa Badafi Naukowych, PL-00 529 Warsaw
☎ (+48 22) 628 40 71 ⊜ (+48 22) 628 09 22
Head of Committee: Aleksander Luczak [98]

15-20600 **European Integration Committe**
Aleje Ujazdowskie 9, PL-00 582 Warsaw
☎ (+48 22) 694 73 54 ⊜ (+48 22) 629 48 88
Prime Minister and Head of the European Integration Committee: Wlodzimierz Cimoszewicz
Secretary of State: Mrs Danuta Hubner [98]

Contact Reference

20 Defense Administrations/Services/Agencies

20-40000 Government Centre for Strategic Studies
PL-00 926 Warsaw
Rdzadowe Centrum Studiow Strategicznych
ulitsa Wspolna 2/4
☎(+48 22) 661 81 11
Contact: Miroslaw Kuzmiuk [98]

25 Defense Forces

25-40000 Navy Headquarters
Skwer Kosciuszki 1, PL-81-912 Gdynia 12
☎(+48 58) 20 78 70 ⊖ (+48 58) 26 34 87
Commander in Chief: VAdm. Ryszard Lukasik
Chief of Naval Staff: RAdm. Jedrezej Czajkowski ☎(+48 58) 20 20 53
Chief Logistics & Maintenance Div.: RAdm. Jerzy Buczma ☎(+48 58) 20 73 83
Spokesman for the Polish Navy Commander: Lt Janusz Walczak ☎(+48 58) 26 39 19
Commander Naval Aviation PN: Capt. Mieczyslaw Jastrzebski ☎(+48 58) 26 34 71
Head of International Contacts Dept of the Polish Navy: Capt. Jerzy Zubowicz ☎(+48 58) 26 32 80
Head of Analysis Division: Capt. Z. Gromadzki [97]

26 Police (Civil)

26-10000 Police Headquarters
ul. Pulawska 148/150, PL-02-624 Warsaw
☎(+48 22) 43 99 61 ⊖ (+48 22) 43 11 68
Head of Police Force: Nadinsp. Jerzy Stanczyk
Chief of Staff: Insp. Robert Mastalinski ☎(+48 22) 601 23 94
Chief of Investigation Dept: Insp. Wladyslaw Naja
Chief of CID: Insp. Krzysztof Jablonski
Chief of Economic Crime Dept: Insp. Zbigniew Dziemidowicz
Chief of Traffic Police Dept: Insp. Wieslaw Karas
Chief of Prevention Dept: Podinsp. Ireneusz Wachowski
Chief of Forensic Laboratory: Insp. Zbigniew Ruszkowski
Chief of Interpol NCB: Insp. Andrzej Koweszko
Spokesman & Public Relations Officer: Podkom.. Jerzy Kirzynski ☎(+48 22) 601 30 07; 43 11 68 (fax) [97]

29 Customs & Excise

29-00500 Customs Headquarters
Glowny Urzad Cel
(Supervised by the Ministry of Finance)
Ul. Swietokrzyska 12, Skr. Poczt. 10, PL-00-916 Warsaw 51
☎(+48 22) 826 71 55; 826 55 63; 827 30 29
⊖ (+48 22) 827 34 27
814846
President of Central Board: Mr Nogaj
Spokesman: Elizbieta Drabarek ☎(+48 22) 826 89 45
Director of the Bureau of International Co-operation: Jan Zralek [98]

31 Embassies & Attachés

31-00010 Embassy of Afganistan
Ulica Mikolaja Reja 4, PL-02-053 Warsaw
☎(+48 22) 25 22 73 ⊖ (+48 22) 25 16 48 [98]

31-00020 Embassy of Albania
Ulica Sloneczna 15, PL-00-789 Warsaw
☎(+48 22) 49 84 27 ⊖ (+48 22) 48 40 04 [98]

31-00030 Embassy of the Popular and Democratic Republic of Algeria
Ulica Dabrowiecka 21, PL-03-932 Warsaw
☎(+48 22) 617 58 55; 617 59 31 817019
Ambassador: H.E. Mr Brahim Haïssa [98]

31-00080 Embassy of Argentina
Ulica Jana Stykl 17-19, PL-03-928 Warsaw
☎(+48 22) 617 60 28; 617 59 31 812412
⊖ (+48 22) 617 71 62 [98]

31-00090 Embassy of Australia
Ulica Estonka 3-5, PL-02-903 Warsaw
☎(+48 22) 617 60 28; 617 60 25 813032
⊖ (+48 22) 617 67 56 [98]

31-00100 Embassy of Austria
Ulitsa Jurija Gagarina 34, PL-00-748 Warsaw
☎(+48 22) 41 00 81; 41 00 84 ✉ AUSTROAMB
813629 ⊖ (+48 22) 41 00 85
Ambassador: H.E. Dr Gerhard Wagner
Deputy Chief of Mission: Ms Isabel Rauscher
Commercial Counsellor: Dr Rudolf Orisich
Consulate
Ulica Krupnicza 42, PL-31-123 Cracow
☎(+48 12) 21 97 66 326597 ⊖ (+48 12) 21 67 64 [98]

31-00130 Embassy of the People's Republic of Bangladesh
Ulica Rejtana 15, PL-02-516 Warsaw
☎(+48 22) 49 32 00; 48 06 37
E b-doot@wawl-comnet.pl ⊖ (+48 22) 48 49 74
Ambassador: H E Mr Nazem A Choudhury
Second Secretary, Commercial Affairs: M Firoz Hayat [98]

31-00145 Embassy of Belarus
Ulica Atenska 67, PL-03-931 Warsaw
☎(+48 22) 617 39 54 815330 ⊖ (+48 22) 617 84 41 [98]

31-00150 Embassy of Belgium
Ulica Senatorska 34, PL-00-095 Warsaw
☎(+48 22) 27 02 33; 27 02 35 ✉ AMBABEL Warszawa
813340 ⊖ (+48 22) 635 57 11 [98]

31-00230 Embassy of Brazil
Ulica Poselska 11, PL-03-931 Warsaw
☎(+48 22) 617 48 00 813748 ⊖ (+48 22) 617 86 89 [98]

31-00250 Embassy of Bulgaria
Aleje Ujazdowskie 33-35, PL-00-540 Warsaw
☎(+48 22) 29 40 71; 29 40 75 ☎813322
⊖ (+48 2) 625 71 71 [98]

31-00310 Embassy of Canada
Ulica Matejki 1-5, PL-00-481 Warsaw
☎(+48 22) 629 80 51 ✉ DOMCAN Warsaw 813424
⊖ (+48 22) 629 64 57
Ambassador: H E Mr Serge April
Canadian Forces Attaché: Col Alan Stryuadka
First Secretary, Commercial: Ms Linda McDonald
Public Information Officer: Angela Bogdan
Assistant Defence Attaché: D Pierre [98]

31-00350 Embassy of Chile
Ulica Morszynska 71b, PL-02-932 Warsaw
☎(+48 22) 40 90 41; (+48 22) 642 81 55 814542
⊖ (+48 22) 642 80 54 [98]

31-00360 Embassy of the People's Republic of China
Ulica Bonifraterska 1, PL-00-203 Warsaw
☎(+48 22) 31 38 61 813589 ⊖ (+48 22) 635 42 11 [98]

31-00380 Embassy of Columbia
Ulica Zwyciezców 29, PL-03-936 Warsaw
☎(+48 2) 617 09 73 816496 ⊖ (+48 2) 617 66 84 [98]

31-00410 Embassy of Costa Rica
Ulica Kubickiego 17 m. 26, PL-02-951 Warsaw
☎(+48 22) 642 78 32 E embajada.cr@supermedia.pl
⊖ (+48 22) 642 32
Chargé d'Affairs a I: H E Mr Janina Lcon Arayn [98]

31-00425 Embassy of Croatia
Ulica Ignacego Krasickiego 10, PL-02-628 Warsaw
☎(+48 22) 44 12 25; 44 23 93; 44 39 94
E EHMADORA@WAW.PDI.NET ⊖ (+48 22) 44 05 67
Ambassador: H E Dr Ivan Brnelic
Counsellor: Maja Plancic ☎(+48 22) 44 22 70
Counsellor: Damir Tomka
Third Secretary: Josip Spoljaric [98]

31-00430 Embassy of Cuba
Ulica Rejtana 15, PL-02-516 Warsaw
☎(+48 22) 48 17 15; 48 25 73 ⊖ (+48 22) 48 22 31 [98]

31-00450 Embassy of the Czech Republic
Ulica Koszykowa 18, PL-00-555 Warsaw
☎(+48 22) 628 72 21; 628 72 25 ⊖ (+48 22) 29 80 45
Ambassador: H E Mr Karel Stindl
Chargé d'Affaires: Vladimír Sál
Counsellor: Jan Misiarz
Military and Air Attaché: Lieut Col Vladislav Raszka
Consulate
Ulica Pawla Stalmacha 21, PL-40-058 Katowice
☎(+48 32) 51 85 76 ⊖ (+48 32) 51 36 28 [98]

31-00460 Royal Danish Embassy
Ulica Rakowiecka 19, PL-02-517 Warsaw
☎(+48 22) 49 00 56; 48 26 00 ✉ AMBADANE
813387 ⊖ (+48 22) 48 75 80
Ambassador: H E Mr Claus Juul Nielson
Counsellor (Deputy Chief of Mission): Ole Christofferson
Defence Attaché: Col Søren Lyder Jacobson
Commercial Counsellor: Frans J Thorwald Bolvenkel ☎(+48 22) 48 25 40
Assistant Commercial Attaché: Mrs Rikke Mouridsen ☎(+48 22) 48 25 40 [98]

31-00510 Embassy of the Arab Republic of Egypt
Ulica Alzacka 18, PL-03-972 Warsaw
☎(+48 22) 617 90 58; 617 69 73 813605
⊖ (+48 22) 617 90 58 [98]

31-00535 Embassy of Estonia
Karwinska 1, PL-02-630 Warsaw
☎(+48 22) 44 44 80; 44 44 83 (consular)
⊖ (+48 22) 44 44 81
Chargé d'Affaires: Riho Laanemaee [98]

31-00580 Embassy of Finland
Ulica Chopina 4-8, PL-00-559 Warsaw
☎(+48 22) 629 40 91 E Finland@it.com.pl 814286
⊖ (+48 22) 621 60 10
Ambassador: H E Mr Seppo Kuppila
Counsellor: Valtteri Hirvonen
Military, Naval and Air Attaché: Cdr Jari Kahara
Commercial Secretary: Juha Metsäntähti ☎(+48 22) 625 68 65; 21 63 94 (fax) 813731 fint pl [98]

31-00590 Embassy of France
1 Ulica Piekna, PL-00-477 Warsaw
☎(+48 22) 628 84 02; 628 84 09 825580
⊖ (+48 22) 625 20 31
Ambassador: H E Mr Daniel Contenay
1st Counsellor (Deputy Chief of Mission): Michel Gardas
Commercial Counsellor: Michel Bardet ✉ COMATTA Warszawa 813590 comat pl
Armed Forces Attaché: Col Jacques Delaigue
Navy Attaché: CC Gerard Jomini
Scientific Attaché: Dominique Le Masne [98]

31-00660 Embassy of Germany
Ulica Dabrowiecka 30, PL-03-932 Warsaw
☎ (+48 22) 617 30 11; 617 30 12;
617 30 17 (night calls) ⊠ 813455 aaw pl
⊖ (+48 22) 617 35 82; 29 48 03
Ambassador: H E Mr Johannes Bauch
Deputy Chief of Mission: Wolfdietrich Vogel
Armed Forces Attaché: Col Peter Köhler
Commercial Counsellor: Dr Joachim Götz
Consulate
Ulica Stolarska 7, PL-31-043 Cracow
☎ (+48 12) 21 84 73 ⊠ 325690 ⊖ (+48 12) 21 76 28
Consulate
Aleja Zwyciestwa 23, PL-80-219 Gdansk-Wrzeszcz
☎ (+48 58) 41 43 66 ⊠ 512374 ⊖ (+48 58) 41 22 45
[98]

31-00690 Embassy of Greece
Ulica Jana Paska 21, PL-01-640 Warsaw
☎ (+48 22) 33 34 88; 33 34 89; 33 05 32 ⊠ 813692
⊖ (+48 22) 33 17 35
[98]

31-00810 Embassy of the Republic of Hungary
Ulica Chopina 2, PL-00-559 Warsaw
☎ (+48 22) 628 44 51; 628 44 55
⊖ (+48 22) 621 85 61
Cultural Attaché: Ivan Liszaki
[98]

31-00830 Embassy of India
Ulica Rejtana 15, PL-02-516 Warsaw
☎ (+48 22) 49 58 00; 49 62 57; 49 68 50 ⊠ 814891
⊖ (+48 22) 49 67 05
[98]

31-00840 Embassy of the Republic of Indonesia
Ulica Wachocka 9, PL-03-934 Warsaw
☎ (+48 2) 617 29 35; 617 39 17; 617 51 08;
617 51 79; 617 84 51 ⊠ 813680
Ambassador: H E Mr Abdul Salam Gani
Economic Affairs: Temu Alam
Political Affairs: Kardi Ady Zulkarnaen Saputra
Press Affairs: Teguh Wardoyo
[98]

31-00850 Embassy of Iran
Ulica Królowej Aldony 22, PL-03-928 Warsaw
☎ (+48 22) 617 15 85; 617 42 93 ⊠ 813823
⊖ (+48 22) 617 84 52
[98]

31-00860 Embassy of Iraq
Ulica Dabrowiecka 9a, PL-03-932 Warsaw
☎ (+48 22) 617 72 63; 617 57 73; 617 70 65
⊠ 813918 ⊖ (+48 22) 617 70 65
[98]

31-00870 Embassy of Ireland
Ulica Humanska 10, PL-00-789 Warsaw
☎ (+48 22) 49 66 33; 49 66 55; 49 66 80 ⊠ 812305
⊖ (+48 22) 49 84 31
[98]

31-00880 Embassy of Israel
Ulica Krzywickiego 24, PL-02-078 Warsaw
☎ (+48 22) 25 00 28; 25 09 23; 25 63 84 ⊠ 813742
⊖ (+48 22) 25 16 70
[98]

31-00890 Embassy of Italy
Plac Dabrowskiego 6, PL-00-055 Warsaw
☎ (+48 22) 26 34 71; 26 34 73 ⊠ 813742
⊖ (+48 22) 27 85 07
[98]

31-00920 Embassy of Japan
Ulica Grazyny 11, PL-02-548 Warsaw
☎ (+48 22) 49 87 81; 49 87 83 ⊖ (+48 22) 48 14 12
[97]

31-00980 Embassy of the Democratic People's Republic of Korea
Ulica Bobrowiecka 1a, PL-00-728 Warsaw
☎ (+48 22) 48 58 13; 40 57 10 ⊠ 812707
⊖ (+48 22) 40 57 08
[98]

31-00990 Embassy of the Republic of Korea
Ulica Ignacego Krasickiego 25, PL-02-611 Warsaw
☎ (+48 22) 48 33 87; 48 40 75; 48 34 09 ⊠ 817069
⊖ (+48 22) 44 69 50
[98]

31-01010 Embassy of Laos
Ulica Rejtana 15, PL-02-516 Warsaw
☎ (+48 22) 48 47 86; 48 89 49 ⊠ 8135932 ambalao
[98]

31-01015 Embassy of Latvia
Ulica Rejtana 15, PL-02-516 Warsaw
☎ (+48 22) 48 19 47; 48 98 05 ⊖ (+48 22) 48 02 01
[98]

31-01050 People's Bureau of the Socialist People's Libyan Arab Jamahiriya
Ulica Kryniczna 2, PL-03-934 Warsaw
☎ (+48 22) 617 48 83; 617 48 22 ⊠ 825508
⊖ (+48 22) 617 50 91
[98]

31-01060 Embassy of Lithuania
Aleja Jana Chrystiana Schucha 5, PL-00-580 Warsaw
☎ (+48 22) 625 33 68 ⊖ (+48 22) 625 34 40
[98]

31-01110 Embassy of Malaysia
Ulica Gruzinska 3, PL-03-902 Warsaw
☎ (+48 22) 617 31 44; 617 44 13 ⊠ 825568
Ambassador: H E Mr Abdul Rahman Rahim
Deputy Chief of Mission: Nasarudin Salleh
Commercial Counsellor: Samad Kassim
[98]

31-01190 Embassy of Mexico
Ulica Staroscinska 1B m 4/5, PL-02-516 Warsaw
☎ (+48 22) 49 52 50; 49 52 58 ⊖ (+48 22) 48 76 17
Ambassador: H E Mr Jose Gabriel Newman Valenzuela
Counsellor, Press and Information:
Victor Balvanera Miranda
Second Secretary, Commercial Section: Jaime Isla Lope
Third Secretary, Cultural Section: Dolores Repetto
Third Secretary, Visa Section: Maria esther Villanuev
Attaché: Manuel Urraca Gonsen
[98]

31-01210 Embassy of Mongolia
Ulica Rejtana 15 flat 16, PL-02-516 Warsaw
☎ (+48 22) 48 42 64; 48 79 20 ⊠ 814399
⊖ (+48 22) 48 42 64
Ambassador: H E Mr Tumen Budsuren
Attaché: Dorj Barhas
[98]

31-01220 Embassy of the Kingdom of Morocco
Ulica Starosinska 1, PL-02-516 Warsaw
☎ (+48 22) 49 63 41 ⊠ 813740 ⊖ (+48 22) 48 18 40
Ambassador: H.E. Mr Abdeladim Tber
Deputy Chief of Mission: Abdellatif Saâdi
[98]

31-01270 Royal Netherlands Embassy
Ulica Chocimska 6, PL-00-791 Warsaw
☎ (+48 22) 49 23 51; 49 23 52 ⊠ 813666
⊖ (+48 22) 48 83 45
Ambassador: H E Mr S I H Gosses
Deputy Chief of Mission: Mrs A de Bijll Nachenius
Defence, Military, Naval and Air Attaché:
Col C H J Veldkamp
Commercial Secretary: A P Uijterlinde
Press Secretary: Mrs P van der Ijssel
[98]

31-01330 Embassy of the Federal Republic of Nigeria
Ulica Chocimska 18, PL-00-791 Warsaw
☎ (+48 22) 48 69 44; 48 53 08 ⊠ 814675
⊖ (+48 22) 48 53 79
[98]

31-01340 Royal Norwegian Embassy
Ulica Chopina 2a, PL-00-559 Warsaw
☎ (+48 22) 621 42 31; 621 42 33 ⊠ 813738
⊖ (+48 22) 628 09 38
[98]

31-01370 Embassy of Pakistan
Ulica Staroscinska 1, PL-02-516 Warsaw
☎ (+48 22) 49 49 38, 49 48 08 ✉ PAREP Warsaw
⊠ 816063 parep pl
Ambassador: H E Mr S M Inaamullah
[98]

31-01410 Embassy of Peru
Ulica Felinskiego 25, PL-01-555 Warsaw
☎ (+48 22) 39 97 57 ✉ LEPRU warsaw ⊠ 814320
⊖ (+48 22) 39 97 66
Chargé d'Affaires: C Dammert
Deputy Chief of Mission: Aurora Luque de Hjorth
First Secretary Commercial and Consul:
Augusto Cornejo Vásquez
[98]

31-01440 Embassy of Portugal
Ulica Dabrowiecka 10, PL-03-910 Warsaw
☎ (+48 22) 617 60 21; 617 60 22 ⊠ 825509
⊖ (+48 22) 617 44 98
[98]

31-01480 Embassy of Romania
Ulica Chopina 10, PL-00-559 Warsaw
☎ (+48 22) 628 31 56 ⊠ 813420 ⊖ (+48 22) 628 52 64
[98]

31-01485 Embassy of the Russian Federation
Ulica Belwederska 49, PL-00-761 Warsaw
☎ (+48 22) 621 34 53; 621 55 75; 621 59 54
⊠ 813530
[98]

31-01650 Embassy of the Slovak Republic
Ulica Litewska 6, PL-00-581 Warsaw
☎ (+48 22) 628 40 51; 628 40 54 ⊠ 813585
⊖ (+48 22) 628 40 55
[98]

31-01670 South African Embassy
Ulica Koszykowa 54, PL-00-675 Warsaw
☎ (+48 22) 625 62 28; 625 62 33; 625 72 39
⊖ (+48 22) 625 62 70
[98]

31-01680 Embassy of Spain
Ulica Staroscinska 1b, PL-00-516 Warsaw
☎ (+48 22) 49 99 26; 49 99 31; 49 99 67 ⊠ 814515
⊖ (+48 22) 49 12 97
Ambassador: H E Mr Fernando Riquelme
Counsellor Deputy Chief of Mission: José Maria Valdemoro
Armed Forces Attaché: Col Octavio Moneva
[98]

31-01730 Swedish Embassy
Ulica Bagatela 3, PL-00-585 Warsaw
☎ (+48 22) 628 97 00; 646 33 51 ✉ SVENSK warsaw
⊠ 813457 ⊖ (+48 22) 646 52 43
Ambassador: H E Mr Stefan Noreén
Counsellor and Deputy Chief of Mission: Jan Nyberg
Military Attaché: Col Hans Bjernby
Commercial Counsellor: Eva Nilsson
Swedish Trade Council
Ul Pytlasinskiego 13A, PL-00-777 Warsaw
☎ (+48 22) 40 31 03; 40 32 77; 40 56 97 ⊠ 817960
⊖ (+48 22) 41 79 92
[98]

31-01740 Embassy of Switzerland
Aleje Ujazdowskie 27, PL-00-540 Warsaw
☎ (+48 22) 628 04 81; 628 04 82; 625 35 87;
628 21 60 ✉ AMBASUISSE warsaw ⊠ 813528
⊖ (+48 22) 21 05 48
Ambassador: H E Dr Richard Gaechter
Deputy Chief of Mission: Josef Aregger
Defense Attaché: Col Jean Rossier
[98]

31-01750 Embassy of Syria
Ulica Narbutta 19a, PL-02-536 Warsaw
☎ (+48 22) 49 14 54; 49 14 56; 48 48 09 ⊠ 825465
⊖ (+48 22) 49 18 47
[98]

31-01780 Royal Thai Embassy
Ulica Staroscinska 1b, PL-02-516 Warsaw
☎ (+48 22) 49 26 55; 49 14 06; 49 47 30; 49 64 14
[E] thaiwars@it.com.pl ⊠ 825392 ⊖ (+48 22) 49 26 30
Ambassador: H E Mr Nobhorn J Karnkit

Poland

Minister-Counsellor, Deputy Chief of Mission: Asha Dvitiyananda
Minister-Counsellor, Commercial Counsellor: Adisai Dhummakupt
Director of Thai Trade Centre: Worapong Chaiwong
First Secretary, Public Information Officer: Thatree Chauvachata [98]

31-01830 Embassy of Tunisia
Ulica Mysliwiecka 14, PL-00-459 Warsaw
☎(+48 22) 628 25 86; 628 63 30 ⊠ 812827
⊖ (+48 22) 621 62 95 [98]

31-01840 Embassy of the Republic of Turkey
Ulica Malczewskiego 32, PL-02-622 Warsaw
☎(+48 22) 44 32 01; 44 33 67 ⊖ (+48 22) 44 37 37
Ambassador: H E Mr Korkmaz Haktanir
☎(+48 22) 44 77 31
First Secretary, Deputy Chief of Mission: H Avni Aksoy
Armed Forces Attaché: Lieut Col Sabri Dogar
☎(+48 22) 44 10 40
Commercial Counsellor: Yalcin Gübe
☎(+48 22) 44 34 37 [97]

31-01875 Embassy of Ukraine
Aleja Jana Chrystiana Schucha 7, PL-00-580 Warsaw
☎(+48 22) 625 01 27 ⊖ (+48 22) 625 32 30
Second Secretary: Andrii Deshcmytsia [98]

31-01890 Embassy of the United Kingdom
Aleja Róz 1, PL-00-556 Warsaw
☎(+48 22) 628 10 01; 628 10 05 ⊠ 813694
⊖ (+48 22) 21 71 61
Ambassador: H E Mr Christopher Hum
Deputy Chief of Mission: R A E Gordon (OBE)
Defence and Air Attaché: Gp Capt H H Moses (MBE, RAF)
Naval and Military Attaché: Lieut Col R Z A Ciaglinski
First Secretary Commercial: A J Gooch [98]

31-01900 Embassy of the United States of America
Aleje Ujazdowskie 29-31, PL-00-540 Warsaw
☎(+48 22) 628 30 41; 628 30 49 ⊠ 813304
⊖ (+48 22) 628 82 98; 628 93 26
Chargé d'Affaires: Michael M Hornblow
Head of Commercial Section: Ms Joan Edwards
Public Affairs Officer: Ms Anne M Sigmund
Scientific Attaché: James W Chamberlin
Defence Attaché: Col Branko B Marinovich (US Army) [98]

31-01910 Embassy of Uruguay
Ulica Rejtana 15, PL-02-516 Warsaw
☎(+48 22) 49 50 40; 49 56 59; 49 15 90
✉ URUPOL Warsaw ⊠ 814647 urupol pl
⊖ (+48 22) 49 56 59 [98]

31-01950 Embassy of Venezuela
Aleje Jerozolimskie 101, PL-02-011 Warsaw
☎(+48 22) 621 51 45 ✉ EMBAVENEZ Warsaw
⊠ 812788 emve pl ⊖ (+48 22) 628 67 40 [98]

31-01960 Embassy of Vietnam
Ulica Kazimierzowska 14, PL-02-589 Warsaw
☎(+48 22) 44 60 21 ⊖ (+48 22) 44 67 23
Ambassador: H E Mr Vu Duony Huan
Second Secretary - Public Information Officer: Pham Minh Tuan
Commercial Officer: Nguyen Hau Don [98]

31-01990 Embassy of Yemen
Aleja Niepodleglosci 54, PL-02-626 Warsaw
☎(+48 22) 44 02 34; 44 02 36; 628 51 61 ⊠ 814756
⊖ (+48 22) 44 61 29 [98]

31-02010 Embassy of the Federal Republic of Yugoslavia
Aleje Ujazdowskie 23-25, PL-00-540 Warsaw
☎(+48 22) 628 51 61, 25 60 51; 25 60 52; 25 60 53
⊖ (+48 22) 29 71 73 [98]

55 Government-owned Industry & Production Facilities

55-10000 Tlocznia Metali "Pressta"
Bolechowo k/Poznania, Ulica Obornicka 1, PL-62-005 Owinska
☎(+48 61) 12 30 11; 12 32 11 ⊠ 0412501 press pl
⊖ (+48 61) 12 31 92
♣ £ 16 000 000 ⚉ 1 100
Managing Dir: Jerzy Lenartowski
Financial Dir: Ms Wanda Kurylo
Sales & Marketing Dir: Jerzy Stefaniak
■ Artillery & tank ammunition. [98]

55-20000 Warsaw Mechanical Works "Wuzetem"
89/93 Czerniakowska Street, PL-00718 Warsaw
☎(+48 22) 41 06 73; 41 38 61; 41 11 46
⊠ 813711 wzm pl ⊖ (+48 39) 12 03 71; 41 85 68
⚉ 786
Managing Dir: Jaroslaw Lazurko
Marketing/Sales Dir: Janusz Waszak
Advertising/PR Mgr: Tadeusz Janiszewski
■ Fuel injection equipment for diesel engines in military applications. [98]

60 Research Organizations, Institutes & Test Establishments

60-02000 Aviation Institute
Instytut Lotnictwa (IL)
(a subordinate of the Ministry of Trade and Industry)
Aleja Krakowska 110/114, PL-02-256 Warsaw, Okecie
☎(+48 22) 46 00 11 ⊠ 813537 ⊖ (+48 22) 46 44 32
⚉ 530 (1997)
General Manager: Dr Ing Witold Wisniowski
Chief Consultant, Scientific and Technical Co-operation: Dipl-Ing Jerzy Grzegorzewski
Director of Research: Dr-Ing W Potkanski
■ Research and design in high and low speed aerodynamics, static and fatigue testing of aircraft and aero engines, flight instruments and other equipment. Flight testing. [98]

65 Industry, Trade & Armed Forces Associations

65-03000 Navimor
Heweliuska 11, PL-80-90 Gdansk
☎(+48 58) 31 98 52 ⊠ 512453 ⊖ (+48 58) 31 18 31
Deputy General Dir: Zbigniew Towianski
■ Organisation representing member shipyards in international ship building, repair & related service concerns. [98]

75 Manufacturing Companies

75-00200 Cenzin Co. Ltd Foreign Trade Enterprise
2 Frascati Str., PO Box 22, PL-00-489 Warsaw
☎(+48 22) 629 63 96; (+48 22) 621 09 17
⊠ 814505 pl ⊖ (+48 22) 628 63 56
Commercial Dir: I. Grzenkowicz
■ Naval ships & other marine products, tanks, armoured & other vehicles, aircraft, helicopters, anti-aircraft, anti-tank guided missiles, artillery, infantry armament, ammunition, mines, grenades, explosives, air & coastal defence systems/C3I, radar, communication equipment, optoelectronic equipment, military technologies, technical assistance & training for military specialists, overhaul services of radar, tanks, aircraft, vessels. Turnkey projects of air & coastal defence systems, overhaul factories. [97]

75-00250 Dezamet SA
Ulica Szypowskiego 1, PL-39-460 Nowa Deba
☎(+48 15) 46 26 01 ⊖ (+48 15) 46 26 19
■ Manufacture of ammunition, hand grenades, missile and shell fuses and cluster bombs. [98]

75-00300 Elma SA
Ulica Lochowska 69, PL-85-395 Bydgoszcz
☎(+48 52) 79 22 40 ⊖ (+48 52) 73 31 30
■ Manufacture of mines and fuzes. [98]

75-00350 Erg SA
Ulica Chemikow 133, PL-43-150 Bierun
☎(+48 3) 116 09 00 ⊖ (+48 3) 116 03 57
■ Manufacture of explosives and fuzes. [98]

75-01000 Gamrat SA
Ulica Mickiewicza 108, PL-38-200 Jaslo
☎(+48 3) 13 66 20 21 ⊖ (+48 3) 13 66 73 80
■ Propellants for rockets and shells. [98]

75-01700 Huta Stalowa Wola SA
Ulica Kolejowa 57, PL-00-962 Warsaw
☎(+48 22) 32 34 64 ⊖ (+48 22) 32 70 51
■ Manufacture of racked and wheeled combat vehicles, SP howitzers, armoured ammunition carriers and recovery vehicles, tank guns, mortars, SP How 2S1, BWP-40 and AA system Sopel. [98]

75-01800 PCO SA Industrial Optics Centre
Ulica Ostrobramska 75, PL-04-175 Warsaw
☎(+48 22) 13 74 98 ⊖ (+48 22) 13 73 24
■ Manufacture of fire control systems. [98]

75-02500 Kombinat PZL-Hydral SA
ul. Bierutowska 57/59, PL-51-317 Wroclaw
☎(+48 71) 25 18 34; 25 24 57 ⊠ 0712216 wruska pl
⊖ (+48 71) 25 25 65; 24 52 14
Managing Dir: Aleksander Hager (ED)
Business Economy Dir: Augustyn Piwkowski (MaE)
Marketing & Sales Dir: Zbigniew Necki (MaE)
Production Dir: Wladyslaw Markowicz (Eng.)
■ Design & production of turbine engine fuel supply & governing systems & piston engine feed & control systems. Also hydraulic & lubrication systems for fixed & rotary wing aircraft. [97]

75-02520 Kombinat Przemyslu Narzedziowego VIS
Ulica Kasprzaka 29/31, PL-01-234 Warsaw
☎(+48 22) 36 88 21 ⊖ (+48 22) 36 15 10
■ Manufacture of pistols. [98]

75-02550 Lucznik SA
Ulica 1905 Roku 1-9, PL-26-600 Radom
☎(+48) 291 41 ⊖ (+48) 278 23
■ Manufacture of pistols, SMG's, assault and automatic rifles. [98]

75-02600 Morska Stocznia Remontowa SA
ul. Ludzi Morza 16, PL-72602 Swinoujscie
☎(+48 97) 50 06 ⊠ 422341 ⊖ (+48 97) 39 12 45 69
♣ US$ 10 000 000 (1995) ⚉ 550
Managing Dir: Romuald Glowacki
Sales Dir: Zbigniew Nowak
Technical Dir: Zbigniew Miodowski
Contracting Dept Mgr: Jan Podlecki
■ Ship repair services & facilities. [97]

75-02630 Prexer-projekt sp zo o
Ulica Kilinskiego 16, PL-90-209 Lodz
☎(+48 42) 32 70 53 ⊖ (+48 42) 32 17 02
■ Manufacture of bombs and training bombs, telescopic fuzes, cluster bombs and air targets. [98]

75-02640 Pronit SA
Ulica Zakladowa 7, PL-26-940 Pionki
☎(+48) 12 47 23 ⊖ (+48) 12 55 34
■ Manufacture of tank ammunition, line charges, gunpowder, propellants, signal and special ammunition. [98]

75-02650
Przemyslowy Instytut Telekomunikacji
Poligonowa 30 Straat, PL-04-051 Warsaw
☎(+48 22) 10 00 51 Ⓔ office@pit.edu.pl
Ⓦ http://www.pit.edu.pl ⊠ 813231PL
⊖ (+48 22) 10 23 80; 10 23 81
♧ US$ 12,700,000 (1996) ♣ US$ 20,000,000 (1996)
ꝟ 861 (1996) ⌂ 17,679 m2
Managing Director: Roman Dufrene
Marketing Director: Tadevsz Lecki-Orkow
■ Manufacture of fire control systems for armoured vehicles (laser-NV-IR), NV devices, laser warning systems and laser designators. 98

75-02670 **PZL Mielec**
Wojska Polskiego 3, PL-39-000 Mielec
☎(+48 17) 788 72 44; 788 72 43
Ⓔ wskpzlsa@ptcmet.mielec.pl ⊖ (+48 17) 788 65 87
♧ Zl 144,862 ♣ Zl 28,810 ꝟ 8,500 ⌂ 20 ha
Managing Director: Kiestew Pastula
☎(+48 196) 88 70 10; (+48 14) 21 47 85 (fax)
Operations Director: Boguslaw Biernacki
☎(+48 196) 88 79 75; (+48 14) 21 47 85
Marketing Director: Bogden Ostrowski
☎(+48 196) 88 72 44; 88 65 87 (fax)
Financial Director: Tadeusz Szkotnicki
☎(+48 196) 88 70 31; 88 73 41 (fax)
Promotion and Advertising Manager: Wit Blaszczak
■ Manufacture of M20 Mewa, AN28 Skytruck and M96 Iryde Jet. 98

75-02680 **PZL Warszawa-Okecie SA**
Aleja Krakowska 110/1145, PL-0-971 Warsaw
☎(+48 22) 46 09 03 ⊖ (+48 22) 46 27 01
■ Manufacture of PZL 104 Wilga and PZL 130 Orlik. 98

75-02690 **PZL Wola SA**
Ulica Fort Wola 22, PL-00-961 Warsaw
☎(+48 22) 36 36 21 ⊖ (+48 22) 37 45 13 98

75-02705 **Radmor SA**
Ulica Hutnicza 3, PL-81-212 Gdyniia
☎(+48 58) 23 23 71 ⊖ (+48 58) 23 33 00
■ Manufacture of telecommunications equipment. 98

75-02710 **Radwar SA**
Ulica Poligonowa 30, PL-04-051 Warsaw
☎(+48 22) 10 40 61 ⊖ (+48 22) 13 48 84
■ Manufacture of radar systems, command and control systems, IFF systems, fire control systems and air, sea and traffic control systems. 98

75-02730 **Scania Kapena SA**
Grunwaldzka 12, PL-76-200 Slupsk
☎(+48 59) 43 88 71 ⊖ (+48 59) 43 66 01 98

75-02750 **Stocznia Gdanska SA**
Ulica Doki 1, PL-80-958 Gdansk
☎(+48 58) 39 23 17; 39 20 72; 39 22 34; 39 22 94
⊠ 512028 sgpl ⊖ (+48 58) 31 99 25
Managing Director: Jamusz Rymaniak
☎(+48 58) 31 64 20; 31 95 18 (fax) ⊠ 512604 sgpl
Deputy Commercial Director: Zygmunp Czapluk
Technical and Production Director: Jozes Vurcharp
■ Ship building: container vessels, reefers, bulk carriers, ro-ros; also produce marine auxiliary oil fired boilers, spare parts for MAN B&W engines, pressure vessels and other steel constructions. 98

75-02800 **Stocznia Polnocna SA Northern Shipyard**
ul. Marynarki Polskiej 177, PL-80958 Gdansk
☎(+48 58) 39 60 00 ⊠ 512264 sp pl
⊖ (+48 58) 31 69 65
♧ US$ 7 000 000 (1995) ♣ US$ 34 000 000 (1996)
ꝟ 1 700
Chairman of the Board/Managing Dir: Mariusz Leszczynski
☎(+48 58) 31 23 02
Marketing Dir: Czeslaw Jaruga
☎(+48 58) 31 94 94; 39 61 02
Operations/Technical Dir: Andrzej Szwarc
Purchasing Mgr: Marian Krasnopolski
■ Construction of navy craft, (corvettes, landing craft, patrol boats), hydrographic & research vessels, fire fighting & rescue boats & tugs. 97

75-02850 **Stocznia Szczecinska SA**
Hutnicza 1, PL-71-642 Szczecin
☎(+48 91) 22 36 05 ⊠ 0422106 ⊖ (+48 91) 22 24 12
♣ US$ 450 000 000 ꝟ 7 000
Managing Dir: K. Piotrowski
Sales Dir: A. Zarnoch
Technical Dir: R. Kwidzinski
■ Ship building, repair & maintenance of ore & coal carriers, general cargo ships, hydrographs, research vessels, training/cargo ships, bulk carriers, ferries, conventional passenger ships, container ships, tankers, oil drilling platform serving supply/tug boats. Facilities include: slipway cranage & hull prefabrication shops enabling construction of blocks up to 300 Mg weight; 1.5 km outfitting quay; building berths - hull construction centre 'Vulkan' with 2 slipways of 250 x 25.6 m & 248 x 25.6 m, hull construction centre 'Odra' with 3 slipways of 132.8 x 18.8 m, 146 x 19.8 m & 190 x 44 m allowing building of vessels up to 50 000 dwt. 98

75-02900 **Unimor SA**
Ulica Rzeznicka 54/56, PL-80-822 Gdansk
☎(+48 58) 37 57 73 ⊖ (+48 58) 31 60 24
■ Manufacture of electronic and telecommunications equipment. 98

75-02975 **Warel SA**
Ulica Modlinska 6, PL-03-216 Warsaw
☎(+48 22) 11 20 11 ⊖ (+48 22) 11 03 50 98

75-05000 **WSK PZL-Rzeszow**
Ulica Hetmanska 120, PL-35 078 Rzeszow
☎(+48 17) 854 61 00; 854 62 00 ⊠ 0633353
⊖ (+48 17) 62 07 50
♣ Zi 180,000,000 (1996) ꝟ 5,600 (1996)
⌂ 118,000 m2
Managing Director: Tadeusz Cebulak
Sales Director: Wladyslaw Jasiczek
Purchasing Manager: Adam Sienko
Business Development Manager: Andrzej Klimiec
■ Manufacture aircraft and helicopter engines and transmission. 98

75-10000 **WSK-PZL Swidnik SA**
Aleja Lotników Polskich 1, PL-21-045 Swidnik K/Lublina
☎(+48 81) 51 29 61; 51 20 71; 68 09 01
Ⓔ pzlswidn@warman.com.pl ⊠ 0642301 wsk pl
⊖ (+48 81) 51 21 73; 69 09 19; 68 09 18
♧ US$ 150,343,000 (1995)
♣ US$ 147,928,000 (1995) ꝟ 4,387
General Director: Mieczyslaw Majewski
Finance Director: Waldemar Kijanko
Director of Research and Development and Chief Designer: R Hermann
Commercial Director: R Cukierman
Development Director: Boguslaw Maryniak
Production Director: Tadeusz Kochanowski
Marketing Manager: Andrzej Stachyra
Sales Manager: Tomasz Bylica
Public Relations: Ms Elzbieta Wozniak
■ Manufacture of Mi-2, PZL-Kania (Kitty Hawk), PZL-Sokol (PZL Falcon) helicopters, PW-5 glider, assemblies and components for Il-86 and Il-96-300 aircraft in agricultural cargo, crane, medevac, passenger, VIP and armed versions. Components manufacturer for Airbus Industries and Aerospatiale. Operates a training centre for mechanics and pilots. 98

75-11000 **Zaklady Mechaniczne BUMAR-LABEDY SA**
ul.Mechanikow 9, PL-44-109 Gliwice
☎(+48 32) 734 51 11; 734 69 62 ⊠ 316257 zaml pl
⊖ (+48 32) 734 69 66; 734 69 75
♣ US$ 91,500,000 (1996) ꝟ 4,300 ⌂ 346,936 m2
Managing Director: Henryk Pfeifer
Commercial Director: Jozef Jedruch
Marketing: Piotr Oleksy
■ Main battle tanks T-72M1, PT-91-HARD/upgraded T-72M1, armoured repair and recovery vehicle WZT-3 and AVLB PMC-90. 98

75-11500 **Zaklady Starachowickie Star SA**
(† Fabryka Samochodów)
ul. 1-go Maja 12, PL-27 200 Starachowice
☎(+48 47) 73 41 10 ⊠ 0612571 ⊖ (+48 47) 74 70 38
Contact: Anna Jaworska
■ Manufacture of trucks. 97

80 Service Companies & Organizations

80-00200 **BMT-IBW Polska Sp.z.o.o.**
(a subs. of British Maritime Technology Ltd)
Koscierska 7, PL-80-953 Gdansk
☎(+48 58) 52 20 19 ⊖ (+48 58) 52 20 19
Chairman: Andrew Docherty
Managing Dir: Jerzy Graff
■ Port and coastal management, hydroinformatics and application of Geographical Information Systems to the marine environment. 98

80-01000 **Pezetel Foreign Trade Enterprise Ltd**
Pezetel Przedsiebiorstwo Handlu Zagranicznego
Aleja Stanów Zjednoczonych 61, PL-04-028 Warsaw
☎(+48 22) 10 80 01; 13 52 85
⊠ 813314; 814651 pzl pl ⊖ (+48 22) 13 23 56
ꝟ 100
General Mgr: Andrzej Jesionek
Deputy Mgr: Wojciech Staszek
Marketing & Advertising Dir: Witold Kozlowski
■ Majority state-owned trade organisation. Products include light aircraft, trainers, agricultural, transport & sport. Helicopters & gliders. Aircraft engines. Board instruments. Diesel engines.
Ownership: state (51%) aviation & diesel engines industry (49%). 97

90 Sales Agents, Sales & Liaison Offices

Companies represented:
Alfred Karcher GmbH & Co, Germany. . 90-01000
Trimos-Sylvac Task Force, Switzerland . 90-10000

90-00500 **Daewoo Handels GmbH Warsaw Office (Universal B/D)**
Aleja Jerozolimskie 44, PL-00-950 Warsaw
☎(+48 22) 693 71 13 ⊠ 816530 dwwaw pl
⊖ (+48 22) 27 63 59 98

90-01000 **Karcher Poland Ltd Sp.zo.o.**
Ul Godlewskiego 22, PL-30198 Kraków
☎(+48 12) 36 45 65 ⊖ (+48 12) 23 75 65
Managing Director: Jacek Olko
® Germany: Alfred Karcher GmbH & Co 98

90-10000 **Trimos Sylvac SC**
Ulitsa Naleczowska 62/48, PL-02-922 Warsaw
☎(+48 22) 42 15 42; 642 13 30 ⊖ (+48 22) 42 80 05
Director: Andrzej Harenza
Commercial Director: Piotr Ptasinski
® Switzerland: Trimos-Sylvac Task Force 98

Portugal

Head of State:
Jorge Sampaio

10 Head of State

10-10000 Office of the President
Praça Afonso Albuquerque, Palacio Nacional de Belem, P-1300 Lisbon
☎(+351 1) 364 72 57 ⊠ 16733 prerep p
⊖ (+351 1) 363 72 43
President of the Republic: Jorge Sampaio [98]

11 Prime Minister

11-10000 Office of the Prime Minister
Rua Professor Gomes Teixeira, P-1300 Lisbon 3
☎(+351 1) 397 70 01 ⊖ (+351 1) 60 26 00
Prime Minister: António Guterres
Adjunct Minister: Jorge Coelho [98]

12 Ministries

12-05000 Ministry of Finance
Avenida Infante D Henrique 5, P-1100 Lisbon
☎(+351 1) 888 46 75; 888 45 55 ⊠ 12143 minfin p
⊖ (+351 1) 886 00 32; 886 98 38; 886 36 58
Minister: António Sousa Franco [98]

12-20000 Ministry of Foreign Affairs
Ministério dos Negócios Estrangeiros
Largo do Rilvas, Palácio das Necessidades, P-1354 Lisbon Codex
☎(+351 1) 396 50 41; 397 01 00 ⊠ 12276 etran p
⊖ (+351 1) 396 39 84; 60 97 08
Minister: Jaime Gama [98]

12-40000 Ministry of Internal Administration
Praça do Comércio, P-1100 Lisbon
☎(+351 1) 346 45 21 ⊠ 16765 gbmai p
⊖ (+351 1) 342 73 72; 347 96 00
Minister: Alberto Costa [98]

12-50000 Ministry of Public Works, Transport and Communications
Praca do Comercio, P-1100 Lisboa
☎(+351 1) 87 95 41 ⊠ 12251 micon
Minister: Joao Cravinho [98]

20 Defense Administrations/Services/Agencies

20-10000 Directorate of Civil Aviation
Direcção Geral da Aviação Civil
Rua B - Edificio 5/6, Aeroporto de Lisboa, P-1700 Lisbon
☎(+351 1) 842 35 00; 842 35 99 ✉AEROCIVIL lisboa
⊠ 12120 aerciv p ⊖ (+351 1) 847 35 85
General Director: Ing Pedro Reis
Deputy General Director: GostMario Neves
Chief Technical Assistant: Ms Luis Lofs
Deputy Director General: Melo Gureia [98]

25 Defense Forces

25-30000 Army Headquarters
Rua Museu de Artilharia, P-1196 Lisbon Codex
☎(+351 1) 86 71 31/32; 87 36 70 ⊠ 222223
Chief of Staff (CEME): Gen. Octávio G.C. Cerqueira Rocha
☎(+351 1) 886 98 55
Deputy Chief of Staff (VCEME):
Gen. Gabriel Augusto Espírito Santo
☎(+351 1) 887 27 60; 888 17 69 (fax)
General Inspector (IGE):
Gen. José Faria Ribeiro de Oliveira
☎(+351 1) 886 06 90
Military Governor of Lisbon (GML):
Gen. Ramiro José Marcelino Mourato
☎(+351 1) 354 21 84
Dir of the Military High Studies Institute:
Gen. Joaquim Chito Rodrigues ☎(+351 1) 301 17 81
☎(+351 1) 301 24 59 (fax); 301 48 92 (fax)
Logistics Commander (QMG):
Gen. Rui Xavier Lobato de Faria Ravara
☎(+351 1) 887 37 56; 60 70 22
Instruction Commander (CIE): Gen. António Cipriano Pinto
☎(+351 1) 886 95 31; 492 33 30/5
Personnel Commander (AGE): Gen. João Goulão de Melo
☎(+351 1) 886 83 76; 346 87 48; 342 43 82 (fax)
Commander, South Military Region:
Gen. José Eduardo Carvalho de Paiva Morão
(CMDT RMS) ☎(+351 66) 240 69
Land Forces Operational Commander (COFT):
Gen. António Albuquerque
☎(+351 1) 886 43 41; 301 44 97
Commander, North Military Region:
Gen. Manuel de Azevedo Moreira Maia (CMDT RMN)
☎(+351 2) 200 86 45/6/7; 200 89 01
☎(+351 2) 31 55 17 (fax)
Commander, Military Academy (CMDT AM):
Gen. Frutuoso Pires Mateus
☎(+351 1) 53 94 53; 315 03 25 (fax)
Chief of Army Staff Cabinet:
Gen. José Manuel Silva Viegas (CH GAB CEME)
☎(+351 1) 887 27 60; 887 59 73 (fax) [97]

25-50000 National Republican Guard Headquarters
Guarda Nacional Republicana
Largo do Carmo, P-1200 Lisbon
☎(+351 1) 347 56 38 ⊠ 12260 ⊖ (+351 1) 347 48 19
General Commander of Nation Republican Guard:
Gen Henrique Godinho
Deputy General Commander:
Brig Gen Abdel Luis Lemos Caldas
Chief Logistics Division: Col Armenio da Silva Vitoria [98]

29 Customs & Excise

29-00500 Direcção Geral das Alfandegas
Fiscal Guard Headquarters
Rua da Alfandega 5, P-1194 Lisboa Codex [98]

31 Embassies & Attachés

31-00100 Embassy of Austria
Avenida Infante Santo 43-4°, P-1350 Lisbon
☎(+351 1) 395 82 20/2 ⊠ 16768 oebolip
⊖ (+351 1) 395 82 24
Ambassador: H.E. Dr A. Missong
Deputy Chief of Mission: Dr Ulrike Hartmann

Commercial Office
Rua Rodriguez Sampaio 18-5°, P-1100 Lisbon
☎(+351 1) 354 76 09; 70 14; 70 53
⊖ (+351 1) 352 24 89
Commercial Attaché: Dr Dietmar Fellner [97]

31-00310 Embassy of Canada
Ave da Liberdade 144/56, 4th Fl., P-1200 Lisbon
☎(+351 1) 347 48 92 ✉DOMCAN Lisbon
⊠ 12377 domcan p ⊖ (+351 1) 347 64 66
Ambassador: H.E. Ms Patricia Marsden-Dole
Defence Attaché: Lt Col Gilles Brassard (Paris)
Commercial Officer: Louis Gaetan
Public Information Officer: Denys Tessier
Counsellor: Ernest Loignon
Affairs Attaché: John Maffett [97]

31-00460 Royal Danish Embassy
Rua Castilho 14-3, P-1296 Lisbon Codex
☎(+351 1) 54 50 99; 51 24; 52 49; 56 07
✉AMBADANE LISBOA ⊠ 18302 amdan p
⊖ (+351 1) 355 46 15
Ambassador: H E Mr Mogens Isaksen
Deputy Chief of Mission: Allan Christensen
Commercial Counsellor: Jean Christensen [98]

31-00690 Embassy of Greece
Rua do Alto do Duque 13, P-1400 Lisbon
☎(+351 1) 301 69 91 ⊠ 18832 ⊖ (+351 1) 301 12 05 [98]

31-00760 Embassy of the Republic of Guinea Bissau
Rua de Alcolena 17 A, P-1400 Lisbon
☎(+351 1) 303 04 40 ⊠ 14 326 guibis p
⊖ (+351 1) 303 04 50
Ambassador: H E Mr Jose Pereira Batista
Deputy Chief of Mission:
Maria De Lourdes Pimentel Nogueira
Counsellor: Dr Ivo Djalo
First Secretary: João Romano Ferreira
Defense Attaché: Braima Sanha
Commercial Secretary: Ibraima Sano
Attaché: Maria Arminda J C Cabral
Attaché: Cadijato Djalo Gomes Embalo [98]

31-00810 Embassy of the Republic of Hungary
Calçada de Santo Amaro 85, P-1300 Lisbon
☎(+351 1) 364 59 28; 364 59 29 ⊠ 13385 diplom p
⊖ (+351 1) 363 23 14
Ambassador: H E Dr Andras Gulyás
Deputy Chief of Mission: Dr Géza Farkas
Military/Defense Attaché: Col Laszlo Hajdu
Commercial Counsellor: Dr Sandor Szabo [98]

31-00890 Embassy of Italy
Largo Conde de Pombeiro 6, P-1198 Lisbon Codex
☎(+351 1) 354 61 44 [E] embital@individual.EUnet.pt
[W] http://www.embital.pt/embital ⊠ 12336 italia p
⊖ (+351 1) 315 49 26; 354 94 65
Ambassador: H E Dr Antonio C di Melilli
Counsellor: Fiametta Milesi-Ferretti
Military, Naval and Air Attaché: Col Luigi Rizzo
Commercial Secretary: Luca Maestripieri

Office of the Commercial Attaché
Calçada Conde Pombeiro 4, P-1198 Lisbon
☎(+351 1) 354 61 44 ⊖ (+351 1) 352 15 16
Commercial Secretary: Dr Antonio Bellavia [98]

31-00920 Embassy of Japan
Embaixada do Japão
Rua Mouzinho da Silveira N° 11, P-1200 Lisbon
☎(+351 1) 352 34 85 ✉TAISHI Lisboa ⊠ 12378 taisi p
⊖ (+351 1) 353 48 02
Ambassador: H.E. Mr Fumiya Okada [97]

31-01485 Embassy of the Russian Federation
59 Rua Visconde de Santarem, P-1000 Lisbon
☎(+351 1) 846 24 24; 846 44 76 (Consular section); 846 25 24; 846 24 23 ⊠ 60233
⊖ (+351 1) 846 30 08; 847 93 27 (Consular section)
Ambassador: H E Mr Alexandre Smirnov
Deputy Chief of Mission: Serguei Andreev
Military, Naval and Air Attaché: Col Yuri Skibenko
Assistant to Ambassador: Dmitri Potapov

Commercial Office
Avenida das Descobertas 4, P-1400 Lisbon
☎(+351 1) 301 03 81 ⊖ (+351 1) 301 53 10
Chief Commercial Office: Vladimir Riabtchouk [98]

31-01890 Embassy of the United Kingdom
Rua São Bernardo 33, P-1200 Lisbon
☎(+351 1) 392 40 25; 392 41 26; 392 40 27
⊖ (+351 1) 392 41 84
Ambassador: H E Mr R Westbrook
Counsellor and Deputy Chief of Mission: A F Smith
First Secretary, Commercial: P Sinkinson

Public Information Officer: Miss M Astle
Defence Attaché: Cdr R Guddard (RN)
Assistant to Defence Attaché: M Drunnee [98]

31-02010 **Embassy of the Federal Republic of Yugoslavia**
Avenida das Descobertas 12, Restelo, P-1400 Lisbon
☎ (+351 1) 301 53 11; 301 53 12
[E] yuemba@mail.telepac.pt ⊜ (+351 1) 301 53 12
Charge d'Affairs: Vladimir Vilotijevic [98]

55 Government-owned Industry & Production Facilities

55-00950 **Indústrias de Defesa SA (INDEP)**
(† Indústrias e Participações de Defesa SA)
(† Indústrias Nacionais de Defesa EP)
4V de Moscavide, PO Box 8106, P-1885 Moscavide
☎ (+351 1) 944 65 46 ⊜ (+351 1) 944 29 52
♲ Esc 15,450,000,000 (1994) 567 115,500 sq m
President: Dr Alberto Jorge Leitao
Vice President: Dr João Sinôsa De Almeida
Commercial Director: Ing António A L Cristovam
Administrator: Dr Manuel Rosário Sança
Administrator: António D S Oliveira Marques
Administrator: Ing Luis Avelino Roque Esteves
Area Sales Manager: J Páscoa
Area Sales Manager: J Nicolau
Area Sales Manager: H Brandão
■ Automatic rifles, machine-guns and mortars, artillery, mortars and small arms, ammunition. Loading of artillery and mortar ammunition. Laser and optronic systems. Air pressure rifles. Field kitchens. → 75

Commercial Dept
Avenida de Moscavide, P-1802 Lisbon Codex
☎ (+351 1) 944 65 46 ⊠ 16326 indfnm p
⊜ (+351 1) 944 29 52; 944 48 53

Works
Rua Material de Guerra, P-1802 Lisbon Codex
☎ (+351 1) 868 43 70 ⊠ 12514 indfbp p
⊜ (+351 1) 868 23 30. [98]

65 Industry, Trade & Armed Forces Associations

65-01000 **NID/AIP - Defence Industries Association**
(a member of AIP - Portuguese Industrial Association)
Praça das Indústrias, P-1399 Lisbon Codex
☎ (+351 1) 362 01 00 ⊠ 15650 aipfil p
⊜ (+351 1) 363 56 08
General Secretary: Col J L Mingot de Almeida (retd)
■ A branch of the Portuguese Industrial Association (AIP) representing defence related industries. [98]

75 Manufacturing Companies

Indústrias de Defesa 55-00950

75-14000 **Noricum Portugal Division**
Avenida D. Luis I, N°47-1°dto, Alfragide, P-2700 Amadora
☎ (+351 1) 471 13 02 ⊜ (+351 1) 471 14 52
■ Manufacture of heavy artillery systems. [97]

75-15000 **Omnitécnica Sarl**
Estrada de Alfragide, P-2720 Amadora
☎ (+351 1) 471 55 17 ⊠ OMITECNICA
[E] omnitec.ca@mail.telepac.pt ⊠ 13433 omnica p
⊜ (+351 1) 471 36 10
♲ Esc 150,000,000 (1997) ♣ US$ 16,704 062 (1996)
60 (1997) 2,848 m2
President: Dr Luis Filipe Gaspar
Operations and Technical Director: Manuel Maia
General Manager: Dharmsi Murargi
Services for Communications: Luis Alegria
■ Telecommunications equipment. [98]

75-18300 **SISTEL-Comunicações, Automação e Sistemas SA**
Quinta dos Medronheiros, Lazarim, PO Box 9, P-2825 Monte da Caparica
☎ (+351 1) 294 84 00 ⊠ SISTEL Almada
⊜ (+351 1) 295 06 16
♲ US$ 13,800,000 (1997) ♣ US$ 28 000 000 (1997)
208 34,600 m2
Managing Director: Hiroshi Watanabe
Finance Director: Takashi Yahamoto
Manufacturing Director: Eros Dos Reis
Sales and Marketing Director: António Morao
☎ (+351 1) 294 84 26
Export Sales Manager: Fernando Duarte Catulo
Publicity Manager: Ms Ana Paula Oliveira
Purchasing Manager: Carlos Filipe
■ Manufacturer of P/Ble-101 field telephone and H-1184GR handset. [98]

80 Service Companies & Organizations

80-00320 **Consulplano SA**
Avenida Frei Miguel Contreiras 54, 3rd Floor, P-1700 Lisbon
☎ (+351 1) 849 11 87 [E] jbooventoro@consulplano.pt
[W] http://www.consulplano.pt ⊜ (+351 1) 849 67 38
♲ Esc 200,000,000 (1996) ♣ Esc 532,646,000 (1996)
20 250 m2
Managing Director: Ilidio Serôdio
Technical Director: João Boaventura
® UK: Jane's Information Group [98]

80-00400 **COSM Ltd Equipamento Aeronáutico e Maritimo**
Calçada da Picheleira n°166 R/C, P-1900 Lisbon
☎ (+351 1) 80 05 47 ⊜ (+351 1) 847 19 12
Managing Dir: Licinio Lopes Soares
Operations & Sales & Marketing Dir: Óscar Soares
Purchasing Mgr: Dulce Gonçalves
■ Consultancy & sales of aeronautic, naval, civil & military equipment.
® Germany: Walter Dittel GmbH [97]

80-01500 **OGMA - Indústria Aeronautica de Portugal SA**
P-2615 Alverca
☎ (+351 1) 958 10 00 ⊠ 66814 ogma p
⊜ (+351 1) 958 04 01; 958 12 88
♣ Esc 104,700,000 1,679 130,113 m2
President: Gen Castelo Branco
Vice President: Col Nogueira Pinto
Vice President: Dr Barbosa da Silva
Commercial Director: Col Marques Pereira
Production Director: Lieut Col Prata Mendes
Personnel Department: Dr Pedro Medeiros
Marketing Support Manager: Pedro Narciso
■ Servicing, heavy maintenance and modifications of aircraft and helicopter airframe and structures. Engine maintenance and overhaul facilities with test cells for turboprop and turbojet. Electronic repair station and general maintenance. Manufacture of aircraft components and structural assemblies. [98]

85 Stockists, Distributors, Spare Parts Suppliers

Aviquipo de Portugal Ltda 90-00150

85-02000 **Casa Serras E. Dias Serras Ltd**
Rua Augusta 228, P-1100 Lisbon
☎ (+351 1) 342 01 33 ⊜ (+351 1) 347 77 57
Managing Director: Luis Serras
Manager: Jorge Serras
■ Radio communications, avionics, direction range finders, multi-channel log records.
® Belgium: Barco NV
France: SFIM
Switzerland: Kudelski SA
USA: Amphenol Corporation, Belden Electronics, Broadcast Electronics, Rockwell-Collins, Stancil Corporation [98]

85-05000 **I.T.B. Comércio Internacional Lda**
Edficio Alto Amoreiras Pisco 5-A, Rue Joshua Benoliel, P-1250 Lisboa
☎ (+351 1) 383 32 03; 383 32 62; 383 33 91
⊜ (+351 1) 388 02 32
Marketing and Sales Director: Laurent Douillard
Distribution Agreement:
Sig-Sauer
Napco
Helmet Intergrated Systems
Courtaulds Aerospace
CGF Gallet, SA
Simunition
Avon Technicl Products
Famae
Sweedish Body Armour
SPS-Sully Produits Spéciaux
Voss Aerospace
Intertechnique Group:
Eros
Faure Herman
HR-Hidraulique De Regulation
Intertechnique
LHC
Zenith Aviation [98]

85-09000 **Turbomar Comércio e Técnica de Máquinas Lda**
Rua Alexandre Herculano 10, PO Box 9, P-2796 Linda-à-Velha Codex
☎ (+351 1) 419 50 65 ⊠ TURBOMAR
⊜ (+351 1) 419 88 78
♲ Esc 130,000,000 (1994)
♣ Esc 1,000,000,000 (1996) 80 1,800 m2
Chairman: José Conte Silva Morais
Sales and Marketing Director: Luis Morais
Production Manager: Valdemar Fernandes
Purchasing Manager: Bo Dahlin
Sales: A Sardo
■ Aircraft engines, APU vibration isolators, laminated elastomers for helicopter rotor heads, avionic trays, skid mount systems for military shelters, vibration and shock and noise control systems. Sales of ductile iron pipes for water and generating sets; sale and installation of radio telephones systems.
® UK: Liebert Swindon Ltd
USA: Barry Controls
Representing:
VM Diesel Engines - VM Motori Spa, Italy
Newage International, England
Chloride Industrial Batteries, England
Auto Diesels Edghill HML, Uk [97]

90 Sales Agents, Sales & Liaison Offices

Companies represented:

Abex GmbH, Germany 90-00270
Acôme, France 90-02030
Aerazur, France 90-01620
Aerotech/Aviquipo, USA 90-00150
Alden Int'l, USA 90-00270
AlliedSignal Aerospace, USA 90-00270
Amphenol Corporation, USA 85-02000
Arbed, Luxembourg 90-00270
Asea Hafo, Sweden 90-03850
Atis Assmann, Germany 90-03850
Autoflug Gmbh, Germany 90-01620
Barco NV, Belgium 85-02000
Barry Controls, USA 85-09000
Belden Electronics, USA 85-02000
Bendix/King, USA 90-00270
BMW/Rolls-Royce GmbH, Germany . . . 90-00400
Bofors Weapon Systems, Sweden . . . 90-00400
British Aerospace Defence Ltd, UK. . . 90-00400
Broadcast Electronics, USA 85-02000
Bulova Technologies Inc., USA. . . . 90-01620
Chenowth Military Products Inc., USA. 90-01620
Chesilvale Electronics, UK 90-02030
Chesterfield Cylinders Ltd, UK 90-00400
Citel 2CP, France 90-02030
Continental Armor Ltd, Greece 90-01620
Desford Tubes Ltd, UK 90-00400
Deshaies, France 90-02030
Disc Communications, UK 90-00270
E.H. Industries Ltd, UK 90-00400
EEV Ltd, UK. 90-00280

Contact Reference

90-00150 Aviquipo de Portugal Ltda
Avenida Antonio Augusto de Aguiar, N°11/1° Esq, P-1000 Lisbon
☎(+351 1) 352 41 63 ▣ 14500 avikip p
⊜ (+351 1) 352 98 87
Managing Director: J Silva
■ Distribution of airborne equipment and components. Mobile GSE, fixed airport systems, aircraft spares, integrated logistics support and project/program management. → 85
® USA: Aerotech/Aviquipo, Lear Siegler Inc [98]

90-00270 Alberto Maria Bravo & Filhos
Apartado 1373, P-1011 Lisboa Codex
Praça de Londres 3-3°D, P-1000 Lisboa
☎(+351 1) 840 74 48 ⊜ (+351 1) 848 37 14
👥 15
Managing Director: J Simões
Aviation Department Manager: A Alves
Aviation Dept
Praça de Londres 3 - 3°D, PO Box 1373, P-1000 Lisbon
☎(+351 1) 840 74 48 ⊜ (+351 1) 848 37 14
Aviation Department Manager: Antonio A Alves
® France: Thomson-Trt Defense (CNI)
Germany: Abex GmbH
USA: AlliedSignal Aerospace, Bendix/King, Sundstrand Aerospace
Electronics Dept
Praça de Londres 3 - 3°D, PO Box 1373, P-1000 Lisbon
☎(+351 1) 840 74 48 ⊜ (+351 1) 848 37 14
® Germany: Elac Nautik (AlledSignal), Felten & Guilleaume
UK: Disc Communications, Wescom/Rockwell
USA: Alden Int'l, Granger Assoc., RF Communications, Wescom Inc./Rockwell
Steels and Ammunition
Praça de Londres 3 - 3°D, PO Box 1373, P-1000 Lisbon
☎(+351 1) 840 74 48 ⊜ (+351 1) 848 37 14
® Luxembourg: Arbed [98]

90-00280 Arestel Componentes Equipamentos Electrónicos SA
(† Amperel Electrónica Ind Ltda)
Avenida Dr Mario Moutinho, Lt. 1528 c/v, P-1400 Lisbon
☎(+351 1) 301 62 22 ⊜ (+351 1) 301 62 21
Director: A Romao
® UK: EEV Ltd [98]

90-00390 Eastécnica Lda
c/o Luis Santos, P-2702 Carnaxide
☎(+351 1) 416 09 07 ⊜ (+351 1) 416 09 47
⚘ £ 215 000 (1995) ♣ £ 6 000 000 (1995) 👥 28
Managing Dir: Luis Santos
Operations & Technical Dir: João Pessanha
® UK: Farnell Int'l [97]

90-00400 The Engineering Co. of Portugal Lda
Rua São Sebastião da Pedreira 122, PO Box 1053, P-1052 Lisboa Codex
☎(+351 1) 314 70 70 ✉ STILWELL
⊜ (+351 1) 353 10 14
Managing Director: David W Stilwell
Commercial Director: Simon P Stilwell
® Germany: BMW/Rolls-Royce GmbH
Sweden: Bofors Weapon Systems
UK: British Aerospace Defence Ltd, Chesterfield Cylinders Ltd, Desford Tubes Ltd, E.H. Industries Ltd, GEC-Marconi Materials Technology Ltd, GKN-Westland Helicopters Ltd, Phoenix Simulation Ltd, Reflectone UK Ltd, Rolls-Royce plc, Vosper Thornycroft (UK) Ltd, Yorkshire Alloys Ltd [98]

90-00495 Empilhadores de Portugal
(a subsidiary of Manitou BF, France)
Quinta do Cabo - Lote 10, Zona Industrial, P-2600 Vila Franca de Xira
☎(+351 63) 27 16 22 ⊜ (+351 63) 27 16 40
® France: Manitou BF [98]

90-00870 Honos Ltd
(a partner of Monitor Air Supply, USA)
Praceta António Feliciano Catilho, PO Box 14, P-2795 Carnaxide
Praceta António Feliciano Castilho - n°6 2nd Floor
☎(+351 1) 416 69 80 [E] honos@mail.telepac.pt
⊜ (+351 1) 416 69 89
♣ US$ 2,000,000 (1994) 👥 4
General Manager: José Macedo Basto
Operations and Technical Director: José Silva Carvalho
Sales and Marketing Director: Mrs Mida Raichande
Purchasing Manager: Ms Helena Fernandes
■ International supply network for C-130, F-5 and Boeing B707 aircraft parts. [98]

90-01340 Milfa Lda
Avenue Republica 692, P-4450 Matosinhos
☎(+351 2) 938 24 50
⊜ (+351 2) 937 13 05; 938 53 94
⚘ Esc 50,000,000,000 (1997)
♣ Esc 1,400,000,000,000 (1996) 👥 30 (1997)
🏭 1,600 m2
Managing Director: Joaquim Cidade
Communictions and Publicity: Maria Dulce
® USA: Warner Electric Brake & Clutch [98]

90-01620 Paragrup-Importação & Exportação Lda
Rua Gregório Lopes, Lote 1596-a, 1° Esq., P-1400 Lisbon
☎(+351 1) 301 49 62 ⊜ (+351 1) 301 62 19
♣ US$ 1 000 000 (1993) 👥 5
General Mgr: João Manuel de Lemos Costa
Defence Division & Marketing Dir: António dos Santos Frias
Airborne Department: Albano Marques de Carvalho
® Austria: Steyr Mannlicher
France: Aerazur, Parachutes De France, Secapem
Germany: Autoflug Gmbh
Greece: Continental Armor Ltd
Italy: Fiocchi Munizioni SpA
Norway: Raufoss A/S, Simrad Optronics A/S, Vinghogs A/S
Sweden: SAAB Instruments
Turkey: Makina ve Kimya Endustrisi Kurumu
USA: Bulova Technologies Inc., Chenowth Military Products Inc., Mills Manufacturing, Para-Flite Incorporated, XL Vision [97]

90-01970 Sicaprep (Madeira) Lda
Rua Dr Pita 18, Edificio Magnolia, Bloco 2, r/c-E, P-9000 Funchal, Madeira Island
☎(+351 91) 74 26 44 ⊜ (+351 91) 76 19 69
Managing Dir: José Coelho
Marketing Mgr: Adelino Carvalho
Representing:
Fujitsu [97]

90-02030 Teleprinta - Comércio e Tecnologia Electrónica Lda
(an assoc. co. Sicaprep (Madeira) Lda)
Rua Ferreira Lapa 35 r/c, P-1150 Lisbon
☎(+351 1) 352 49 32; 354 89 37; 354 84 23
⊜ (+351 1) 354 84 23
⚘ Esc 2 100 000 ♣ Esc 150 000 000 👥 6 🏭 340 m2
Managing Dir: José Coelho
Operations & Technical Dir: João Sousa Costa
® France: Acôme, Citel 2CP, Deshaies, Pouyet International, Slat
Germany: Krieg
UK: Chesilvale Electronics, Hy-Q, Willsher & Quick [97]

90-03850 Sociedade Zickermann SA
Secção Electrotécnica
Rossio 3 - 3rd Floor, PO Box 2155, P-1103 Lisbon
☎(+351 1) 346 90 65 [E] zickermann.lis@mail.telepal.pt
⊜ (+351 1) 346 91 29
Director: Peter Zickermann
® Austria: EGO
France: Trigofrance
Germany: Atis Assmann, Knürr AG, Schaltbau, Semikron, Wabco Westinghouse
Italy: Signal-Lux
Sweden: Asea Hafo
UK: Elmwood Sensors Ltd, Hamlin Electronics [98]

Qatar

Head of State:
Sheikh Khalifa bin Hamad Al-Thani

10 Head of State

10-10000 Office of the Amir
PO Box 923, Doha
☎(+974) 41 58 88 ▣ 4297 amiri dh
Amir of Qatar: Sheikh Hamad Bin Khalifa Al-Thani [98]

11 Prime Minister

11-10000 Office of the Prime Minister
PO Box 923, Doha
☎(+974) 41 58 88
Prime Minister: Sheikh Abdallah bin Khalifa Al Thani [98]

12 Ministries

12-10000 Ministry of Defense
PO Box 37, Doha
☎(+974) 40 41 11
Minister of Defense and Commander in Chief of Armed Forces: Maj Gen Hamad bin Khalifa Al-Thani 98

12-20000 Ministry of Foreign Affairs
PO Box 250, Doha
☎(+974) 41 50 00 ⊠ 4252 khargi dh
Minister: Hamad bin Jasim bin Jabir Al Thani 98

12-40000 Ministry of the Interior
PO Box 8895, Doha
☎(+974) 33 00 00 ⊠ 4383 ⊖ (+974) 44 37 50
Minister: Abdallah bin Khalifa Al Thani 98

12-50000 Ministry of Communications and Transport
PO Box 3416, Doha
☎(+974) 46 40 00 ⊠ 4800
Minister: Ahmed bin Nasir Al Thani 98

31 Embassies & Attachés

31-00660 Embassy of the Federal Republic of Germany
PO Box 3064, Doha
☎(+974) 87 69 59 ⊠ 4528 aadoha dh
⊖ (+974) 87 69 49
Ambassador: H E Mr Klaus Schröder
Deputy Chief of Mission: Jürgen Handloegsen
Commercial Attaché: Aebert Couslen 98

31-00830 Embassy of India
PO Box 2788, Doha
☎(+974) 67 20 21 ⊠ INDEMBASSY DOHA
indembdh@qatar.qa ⊠ 4646 indemb dh
⊖ (+974) 67 04 48
Ambassador: H E Shri R L Naraayan ☎(+974) 67 20 25
Counsellor: Shri P K Bhutiani ☎(+974) 67 20 23
First Secretary: Shri Amrit Lugun ☎(+974) 67 28 38
Second Secretary, Commercial: Shri Bhuoinder Singh ☎(+974) 67 20 22 98

31-00890 Embassy of Italy
Ali Bin Abi Talib Street 41, PO Box 4188, Doha
☎(+974) 43 68 42/4 ⊖ (+974) 44 64 66 98

31-00920 Embassy of Japan
Diplomatic Area, West Bay, PO Box 2208, Doha
☎(+974) 83 12 24/5 ⊠ TAISHI Doha ⊠ 4339 taisi dh
⊖ (+974) 83 21 78
Ambassador: H E Mr Shinya Nagai
Counsellor: Nobu Ebina
Second Secretary: Ryoji Iwama
Second Secretary: Takeshi Ishiguro
Attaché: Katsuhiro Akutsu
Attaché: Hirotaka Tanaka 98

31-01830 Embassy of Tunisia
PO Box 2707, Doha
☎(+974) 83 26 45; 83 26 46 ⊖ (+974) 83 26 49
Ambassador: H E Mr Janusz Switkowski
Military, Naval and Air Attaché: Col Stanislaw Rozycki
Deputy Chief of Mission: Tarek Letaief
First Secretary: Mrs Urszula Raznowiecka
Attaché: Andrzej Wolfram
Commercial Counsellor: Henryk Socha
Commercial Attaché: Janusz Allina 98

31-01890 Embassy of the United Kingdom
PO Box 3, Doha
☎(+974) 42 19 91 ⊠ PRODROME DOHA
⊖ (+974) 43 86 92
Ambassador: H E Mr David A Wright
Deputy Head of Mission: J Maguire
Second Secretary Commercial: N Abbott
Defence Attaché: Wg Cdr W Southcombe (Resident in Dotta)
British Vice Consul: Ms G Jenkinson
Third Secretary - Political/ Information: D Lelliott 98

31-01900 Embassy of the United States of America
PO Box 2399, Doha
149 Ali bin Ahmed St, Farig bin Omran, Doha
☎(+974) 86 47 01/03 ⊠ 4847 amemb dh
⊖ (+974) 86 16 69
Ambassador: H E Mr Patrick W Theros
Public Affairs Officer: Jeff Hill
Head of US Liaison Office: Lt Col Bruce Deane 98

80 Service Companies & Organizations

80-00700 Gulf Helicopters Co.
(a subsidiary of Gulf Air)
Head Office
Al Areesh Street, Ras Abu Aboud Road, PO Box 811, Doha
☎(+974) 33 38 88 ⊖ (+974) 41 10 04
Chairman: Abdullah Bin Hamad Al Attiya
■ Aviation activities including operations in support of civil engineering projects, oil and gas exploration and production, and executive transport. → 90 98

85 Stockists, Distributors, Spare Parts Suppliers

85-00500 Golden Company
PO Box 12654, Doha
☎(+974) 24 35 00 E golden@qatar.net.qa
♣ US$ 75,100 17
President: Hassan Al-Homoud ☎(+974) 64 42 26
Marketing Director: Saad Al-Rahmmani
Advertising and Public Relations Director: Jone Bit
■ Distributers of computers and computer supplements to the defence industry. 98

90 Sales Agents, Sales & Liaison Offices

Sales agents/representatives listed elsewhere:

Gulf Helicopters Co. 80-00700

Romania

Head of State:
H E Emil Constantinescu

10 Head of State

10-10000 Office of the President
Palatul Cotroceni, Bdul Genuilui nr 1, sect 6, Bucharest
☎(+40 1) 312 11 59; 221 12 59 ⊖ (+40 1) 411 31 31
President and Commander in Chief of Armed Forces: H E Emil Constantinescu
Supreme Council of Defence, Secretary: Div Gen Ion Magdalena
Protection Guard Service Director: Nicu Anghel
Presidential Adviser, Political Analysis Department: Ms Zoe Petre
Presidential Adviser, Defence National Security Department: Dorin Marian
Supreme Council of Defence
Bucarest
Chairman: H E Emil Constantinescu
Secretary: Div Gen Ion Magdalena 98

11 Prime Minister

11-10000 Office of the Prime Minister
Piata Victoriei 1, Sect 1, Bucharest
☎(+40 1) 61 34 00; 230 36 60; 311 05 04
⊖ (+40 1) 222 60 88 (Public Information Department); 311 37 08
Prime Minister: Victor Ciorbea
Secretary General: Remus Opris 98

12 Ministries

12-10000 Ministry of National Defense
Straat Izvor 13-15, Sect 5, Bucharest
☎(+40 1) 410 68 76 ⊖ (+40 1) 410 68 76
Minister of National Defence: Victor Babiuc
Secretary of State, Chief of Defence Policy Department: Constantin Dudu Ionescu
Secretary of State, Chief of General Staff: Div Gen Constantin Degeratu
Secretary of State, Chief of Procurement Department: Div Gen Dan Zaharia
Inspector General: Div Gen Mihai Corneliu Lungu
Chief of Cabinet: Div Gen Gheorghe Andrita
Chief of Human Resources Management Directorate: Brig Gen Neculai Balan
Chief of Military Protection and Safety Directorate: Brig Gen Ion Dohotaru
Defence Policy Department
Straat Izvor 13-15, Sect 5, Bucharest
Chief of Social Services, Military Traditions: Brig Gen Marin Mirea
Chief of Political Military Analysis Directorate: Col Vasile Croitoru
Chief of Interantional Relations Directorate: Col Octavian Dumitrescu
Chief of European and Euro-Atlantic Integration Directorate: Col Ionescu Mihail
Procurement Department
Straat Drumul Taberii 9-11, Sect 6, Bucharest
⊖ (+40 1) 312 74 60 98

12-20000 Ministry of Foreign Affairs
Aleea Modrogan 14, Bucharest
☎(+40 1) 230 20 71 ⊖ (+40 1) 230 74 89
Minister: Severin Adrian
Secretary of State: Dimitru Ceausu
Secretary of State: Lazar Comanescu
Secretary of State: Gabriel Gafita
Secretary of State: Marin Mitris
General Director: Petru Corolos
General Director: Georges Popescu
General Director: Ioan Maxim
General Director: Mihai Costa 98

12-30000 Ministry of Industry and Commerce
Calea Victoriei 152, Bucharest
☎(+40 1) 650 26 57 ⊖ (+40 1) 312 05 13
Minister of Industry and Commerce: Calin Popescu Tariceanu (State Minister)
Secretary of State: Florin Buruiana ☎(+40 1) 650 74 41
Secretary of State: Liviu Marcu ☎(+40 1) 312 70 29 (fax); 312 03 21 (fax)
Secretary of State: Sebastian Vladescu ☎(+40 1) 679 74 24 (fax); 311 07 27 (fax)
Secretary of State: Nicolae Staiculescu ☎(+40 1) 650 21 83 (fax); 650 68 84 (fax)
Secretary of State: Liviu Hagea ☎(+40 1) 615 60 41 (fax); 410 20 69 (fax) 98

12-40000 Ministry of Interior
Straat Mihai Voda nr 8, Bucharest
☎(+40 1) 613 06 25 ⊖ (+40 1) 613 06 25
Minister: Gavril Dejeu
Secretary of State: Brig Gen Teodor Zaharia
Secretary of State: Col Nicolae Langa
Secretary of State: Emanuel Albu 98

Contact Reference

Romania

12-50000 Ministry of Justice
Bdul Regina Elisabeta nr 33, Sect 5, Bucharest
☎(+40 1) 311 22 66 ⊜ (+40 1) 323 61 79
Minister: Valeriu Stoica [98]

12-60000 Ministry of Finance
Bucharest
Minister of Finance: Mircea Ciumara [98]

12-80000 Ministry of Research Technology
Straat Mendeleev nr 21-25, Sect 1, Bucharest
☎(+40 1) 650 06 90 ⊜ (+40 1) 312 14 10
Minister: Bujor Bogdan Teodoriu
Secretary of State: Mircea Valer Pusca [98]

20 Defense Administrations/Services/Agencies

20-01000 Romanian Intelligence Service (SRI)
Bdul Libertatii 14, Bucharest
Sos. Odaii 20, Bucharest
☎(+40 1) 631 42 55 (Press); 633 46 67 (Protocol)
Director: Dr Virgil Magureanu
First Deputy Director: Maj.Gen. Victor Marcu
Deputy Director: Maj.Gen. Ion Popescu
Chief of Counterintelligence Div.(B): Col Nicolae Lupu
Constitution Protection (A): Col Ovidiu Soare
Chief Protection Div. (E): Col Gheorghe Atudoroaie
Chief Economic Counterintelligence: Col Dumitru Badescu
Chief of Anti-terrorist Brigade: Col Nicolae Stefanut
Spokesperson: Nicolae Ulieru [98]

25 Defense Forces

25-10000 Armed Forces Headquarters Ministry of National Defence
(Ministry of National Defense)
Straat Izvor 13-15, Sect 5, Bucharest
☎(+40 1) 312 30 30 ⊠ 11447, 11281, 11167
⊜ (+40 1) 410 68 76
Commander-in-Chief of Armed Forces: H E Emil Constantinescu
Minister of Defence: Victor Babuic
Secretary of State, Chief of General Staff: Dr Div Gen Constantin Degeratu
Secretary of State, Chief of Defence Policy Department: Constantin Dudu Ionescu
Secretary of State, Chief of Procurement Department: Div Gen Dan Zaharia
Inspector General of the General Inspectorate of the MoD: Div Gen Corneliu Mihai Lungu
Commander Military Aviation Command: Maj Gen Stan Ion
Commander Territorial Air Defence: Brig Gen Marineanu Tiberiu
Commander Territorial Air Defence Command: Brig Gen Tiberiu Marineanu
Press and Information Officer: Capt Cdre Amariutii Vasile
Chief of Cabinet: Lt Col Constantin Tomescu [98]

25-19000 The Academy of Advanced Military Studies
Straat Panduri nr 68-72, Sect 5, Bucharest
☎(+40 1) 410 31 00 ⊜ (+40 1) 410 20 72; 410 07 73
Commandant: Army Corps: Gen Dumitri Cioflina
Chief of Staff: Brig Gen Liviu Haoian
Director of Procurement: Col Emil Tanislav [98]

25-19100 The College of National Defence
Straat Panduri nr 68-72, Sect 5, Bucharest
Director: Col Constantin Balaban [98]

25-20000 Air Force Air Defence Headquarters
(Ministry of National Defense)
Sos Bucuresti-Ploiesti km 9, Sect 5, Bucharest
☎(+40 1) 633 74 95 ⊜ (+40 1) 633 24 31
Chief of Staff: Air Flotilla Gen Gheorghe Bucse
Commander Territorial Air Defence Command: Brig Gen Tiberiu Marineanu
Commander Military Aviation Command: Cdr Catrina Gheorghe [98]

25-30000 Army Headquarters
Straat Drumul Taberii nr 9-11, Sect 6, Bucharest
☎(+40 1) 746 01 11 ⊜ (+40 1) 614 33 01
Chief of Staff: Div Gen Dr Nicolae Pastinica [98]

25-40000 Navy Headquarters
(Ministry of National Defense)
Straat Remus Oprescu nr 13, Constantza
☎(+40 1) 61 57 00 ⊜ (+40 1) 61 57 10
Chief of Staff: Rear Adm Traian Atanasiu [98]

29 Customs & Excise

29-00500 Customs Committee
Rue Matei Millo 13 - Sector 1, Officiui Postal 1 - PB 886, R-70704 Bucharest [98]

31 Embassies & Attachés

31-00085 Armenian Diplomatic Mission
Bucharest
☎(+810) 42 13 21 56 79 ⊜ (+810) 40 13 21 59 30 [98]

31-00450 Embassy of the Czech Republic
Str. Ion Ghica 11 sector 3, R-70 418 Bucharest
☎(+40 1) 615 91 42/3; 312 02 68
[E] ambchbuc@tag.vsat.ro ⊜ (+40 1) 312 25 39
Ambassador: H.E. Mr Jaromir Kvapil
Defence Attaché: Col Oldrich Turcinek
Commercial Officer: Karel Lhotsky [97]

31-00460 Royal Danish Embassy
Str. Dr Burghelea 3, R-73102 Bucharest
☎(+40 0) 12 03 52/56 ⊠ 11325 amdan r
⊜ (+40 0) 12 03 58 [97]

31-00590 Embassy of France
13-15 Strada Biserica Amzei, PO Box 143, Bucharest
☎(+40 1) 312 02 17; 312 02 18; 312 02 19
[E] ambafrance.presse@com.pc.net.ro ⊠ 11320 FRABUC
⊜ (+40 1) 312 02 00
Ambassador: H E Mr Bernard Boyer
Deputy Chief of Mission: Liouel Majesté-Larrouy
Commercial Counsellor: Michael Vergnaud
✉ COMATTA Bucuresti ⊠ 10 666 comata r
Armed Forces Attaché: Lieut Col M Laffaille [98]

31-00920 Embassy of Japan
Strada Polona N° 4, Sector 1, Bucharest
☎(+40 1) 21 00 79 02 ⊜ (+40 1) 210 02 72 [97]

31-01050 People's Bureau of the Socialist People's Libyan Arab Jamahiriya
Bulevard Ana Ipatescu 15, Bucharest
Head of the People's Bureau: A S Zintani
Commercial Offier: Ali Abdulaziz Elisaue
First Counsellor Charge d'Affairs A I: Dao Musbah Elegaili
Public Information Officer: Ismael M Salem [98]

31-01190 Embassy of Mexico
Strada Armemeasoa 35, Sect 2, Bucharest
☎(+401) 210 47 28; 210 45 77 [E] eumrm@k.ro
⊜ (+401) 210 47 13
Ambassador: H E Mr Enrique Fernandez Zapata
☎(+401) 210 47 13
Deputy Chief of Mission: Francisca E Mendez De Lozaiyo
☎(+401) 210 47 28 [98]

31-01730 Swedish Embassy
Strada Sofia 11, R-71276 Bucharest
☎(+40 1) 230 21 84 [E] embassy@sweden.sfos.ro
⊠ 11 313 svensk r ⊜ (+40 1) 230 62 56
Ambassador: H E Mr Ragnar Ängeby [98]

31-01890 Embassy of the United Kingdom
24 Strada Jules Michelet, R-70154 Bucharest
☎(+40 1) 312 03 03; 312 03 06; 312 03 04 312 03 05
⊠ 11295 prodrom r ⊜ (+40 1) 312 02 29
Ambassador: H E Mr Christopher Crabbie
Deputy Head of Mission: R M Publicover
Defence, Naval, Military and Air Attaché: Lieut Col R J F Owen
First Secretary, Commercial: N Sheppard
Assistant Defence Attaché: WO R C Guest [98]

55 Government-owned Industry & Production Facilities

55-01000 Romtehnica
Straat Drumul Tabereii 9-11, Sect 6, Bucharest
☎(+40 1) 401 34 05; 745 01 55
⊠ 11 608 TXKCO; 11 640 TXNVS
⊜ (+40 1) 746 66 71; 410 14 67; 746 03 17; 312 74 60
Director: Col Eng Ion Nicolae ☎(+401) 410 34 03
Operations Director: Eugen Lascu
Marketing Manager: George Bajenaru
Commercial Manager: Aurel Nemes
■ Manufacture of military equipment. [98]

60 Research Organizations, Institutes & Test Establishments

60-00800 Research and Development Institute for Electrical Engineering
ICEP SA
Straat Splaiul Unirii nr 313, Sect 3, R-74204 Bucharest
☎(+40 1) 321 63 46; 322 28 13 ⊠ 010486 icpe
⊜ (+40 1) 321 37 69; 321 63 46; 322 37 69
♧ US$ 19,000,000 (1996) ♣ US$ 18,000,000 (1996)
👥 1,400 🏭 6,000 sq m
Managing Director: Dr Vasile Nicolae
☎(+40 1) 322 28 13
Director of Ecology Research: Dipl-Ing Paul Pencioiu
☎(+40 1) 322 01 10
Marketing Director: Traian Balanescu
☎(+40 1) 321 72 48
Director of Research for Special Electric Machines and Avionics: Dr Dipl-Ing Virgil Racicovschi
☎(+40 1) 321 63 46
Finance Director: Ms Maria Burtan Fleicher
☎(+40 1) 322 01 10
Director of Research for Electrical Apparatus: Dipl-Ing Ion Marinescu ☎(+40 1) 322 28 13
Director of Research for Electrical Materials: Dipl-Ing Wilhelm Kappel ☎(+40 1) 322 27 48
Purchasing Manager: Ms Maria Burtan
☎(+40 1) 322 01 10
■ Electrical research/development, engineering and production. Activities structured into four departments:
Special electric machines (micromotors, microturbogenerators, servomotors, synchros, resolvers, tachos) and avionics.
Electrical materials (permanent magnets, insulating materials, electrical contacts and bushes, cryogenics).
Electrical Apparatus and cables.
Ecology (solar and wind equipment for energy conversion, sensors and transducers, ecological electrotechnologies, informatics). [98]

75 Manufacturing Companies

Aerostar SA 80-00100

75-00750 Santierul Naval Constanta SA
Constanta Port, Constanta
☎(+40 41) 61 69 70; 61 48 65 ✆ (+40 41) 61 90 18
⊜ (+40 41) 61 16 51
♧ US$ 58,000,000 (1996) ♣ US$ 46,500,000 (1996)
👥 3,990 (1996) 🏭 198,000 sq m

Managing Director: Aurel Daraban
Sales Director and Purchasing Manager: Nicolae Matei
☎(+40 41) 61 33 98
Technical Director: Paul Hartescu ☎(+40 41) 61 55 86
■ Ship building, repair and overhaul services and facilities. [98]

75-01000 Ratmil RA
Pasajul Victoriei, 48-50 Sector 1, R-70102 Bucarest
☎(+40 1) 312 65 78; 312 26 71 ⊠ 10561
⊖ (+40 1) 312 29 88; 312 26 33
♣ US$ 285,000,000 ☗ 32,000
■ Manufacture pistols, machine guns, sub-machine guns, rifles, light machine guns, anti-aircraft machine guns, portable anti rocket launcher, grenade portable launcher, various rounds, aviation bombs, hand grenades, undirigible sockets, cartridge cases, jet propelled bombs, mines and grenades, mortars, IFV's and APC's [98]

75-01200 Tulcea Shipyard
8 Portului Street, R-8800 Tulcea
☎(+40 40) 55 05 26 ⊠ 52257 sntl r
⊖ (+40 40) 55 03 05
☗ 70 ha ☖ 152 000 sq m
Managing Director: Dumitru Ivanov
Technical Director: Costel Lungu ☎(+40 40) 55 09 63
Sales Director: Silviu Pisica
Purchasing Manager: Mihahi Babdianu
■ Construction, conversion and repair of a wide range of shipping. [98]

80 Service Companies & Organizations

80-00100 Aerostar SA
9 Condorilor Street, R-5500 Bacau
☎(+40 34) 17 50 70; 17 20 06
[E] aerostar@sysnei.02.1cc1.ro ⊠ 21339
⊖ (+40 34) 17 20 23; 17 22 59
♧ US$ 25,000,000 (1996) ♣ US$ 15,000,000 (1996)
☗ 4,538 ☖ 340,000 sq m
President and General Director: Mihai Ciobanu
Finance Director: Doru Damaschin
Commercial Director: Constantin Roventa
☎(+40 34) 17 56 52
Operations and Technical Director: Grigore Filip
Marketing Director: Gheorghe Carnu
■ Manufacture of Yak-52 trainer, avionics, landing gears, hydraulic equipment, engines and reduction gears for aviation, sub-assemblies and spare parts. Military aircraft repairs. → 75 [98]

90 Sales Agents, Sales & Liaison Offices

90-05000 Daewoo Corp Bucharest Branch
(Modern B/D) 14th Floor 44-46, Carol 1, Sector 1, Bucharest
International Trade Center
☎(+40 1) 250 00 55 ⊠ 12405 dwbucr
⊖ (+40 1) 250 03 96 [98]

Russian Federation

President:
Boris Nikolayevich Yeltsin

10 Head of State

10-10000 Office of the President
The Kremlin, Moscow
☎(+7 095) 205 95 10; 205 51 23
President: Boris Nikolayevich Yeltsin
Chief of Staff: Vladimir Babichev
Head of the Presidential Administration: Valenti Yumashev
Chief of Government Press Service: Valentin M. Sergeyev [98]

11 Prime Minister

11-10000 Office of the Prime Minister
Staraya Place 4, Moscow
☎(+7 095) 205 95 10; 206 25 11
Prime Minister: Viktor Stepanovich Chernomyrdin
First Deputy Prime Minister: Anatoly Chubais
First Deputy Prime Minister: Boris Nemstov
Deputy Prime Minister: Vladimir Bulgak
Deputy Prime Minister: Alfred Kokh
Deputy Prime Minister and Economics Minister: Yakov Urinson
Deputy Prime Minister and Minister of Internal Affairs: Gen Anatoly Kulikov
Deputy Prime Minister: Oleg Sysuyev
Deputy Prime Minister, CIS Policy: Ms Valery Serov [98]

12 Ministries

12-10000 Ministry of Defence
37 Miasnitskayaa Straat, 103175 Moscow
☎(+7 095) 293 56 83 ⊖ (+7 095) 296 89 00
Minister of Defence: Col Gen Igor Dimitriyevich Sergeyev [98]

12-20000 Ministry of Foreign Affairs
32/34 Smolenskaye, Sennaya Square, 121200 Moscow
☎(+7 095) 244 16 06 ⊖ (+7 095) 230 21 30
Minister of Foreign Affairs: Yevgeniy Maksimovich Primakov [98]

12-30000 Ministry of Foreign Economic Relations and Trade (MFER)
32/34 Smolenskaya, Sennaya Square, 121200 Moscow
☎(+7 095) 244 14 50 ⊠ 412171 gusk su
⊖ (+7 095) 244 39 81
Minister: Mikhail Yefimovich Fradkov [98]

12-40000 Ministry of Interior
ulitsa Zitnaya, 16, 117049 Moscow
☎(+7 095) 239 01 36 (Public Relations Secretariat); 238 71 11; 239 00 56 (International Relations Secretariate)
Minister: Gen Anatoly Kulikov [98]

12-45000 Ministry of Security (RFMS)
Ministerstvo Bezopasnosti Rossiyskoi Federatsii (MBRF)
ulitsa B Lubyanka 2, Moscow
☎(+7 095) 221 07 62
Minister: Lieut Gen Nikolai Kovalev [98]

12-75000 Ministry of Transport
Sadovo-Samotechnaya ulitsa 10, Moscow
☎(+7 095) 200 08 03; 200 14 85
⊖ (+7 095) 200 14 79
Minister: Nikolay Petrovich Tsakh [98]

12-85000 Ministry of Atomic Energy
ulitsa B Ordinka 26, 101000 Moscow
Minister: Viktor Nikitovich Mikhaylov
Chairman, Federal Inspectorate for Nuclear and Radiation Safety: Yuriy Georgiyevich Vishnevskiy [98]

25 Defense Forces

25-29000 Air Force Museum
141170 Monino, Moscow
☎(+7 095) 584 21 80; 526 33 27
⊖ (+7 095) 526 33 51
Chief: V. Tolkov
Deputy Chief: A. Yablonskih
Chief of Scientific & Display Dept: A. Melovatsky
Chief of Fund Dept: Y. Mamitch
Chief of Museum's Dept: V. Kazashvilly
Chief of Resoration Shop: V. Voronko
■ The museum was founded in 1958 as a structural unit of the Air Force Academy. Nine museum halls are situated in the building of the former aircraft repair plant. Display areas also include a hangar & tarmac.
Exhibits total 39 457 (6 915 on display), including: 164 aircraft (110 aeroplanes, 21 helicopters, 9 gliders & 21 other aircraft); 121 aircraft engines (73 piston & 48 jet); 2 543 pieces of armament, equipment & ammunition; 1 009 pieces of munitions; 184 flags, banners & colours; 6 517 orders, medals, badges, insignia; 2 872 documents; 11 094 negatives; 722 art exhibits. The museum's library possesses more than 10 000 books & 125 films. [97]

28 Coastguard, Border Protection Forces

28-41000 Federal Frontier (Border) Service
ulitsa Lubianka, 2, 101000 Moscow
☎(+7 095) 224 43 03 [98]

29 Customs & Excise

29-00500 State Customs Committee
Pl 1, Komsomolskaya Square, 107842 Moscow
☎(+7 095) 231 76 48; 207 30 72; 244 41 84 [98]

29-40500 Federal Revenue Police Service
ulitsa Marseika, 12, 101968 Moscow
☎(+7 095) 921 21 48 [98]

31 Embassies & Attachés

31-00050 Embassy of the Republic of Angola
ul. Olof Palme 6, Moscow
☎(+7 095) 143 63 24; 143 65 21; 143 65 31
⊠ 413 402 ⊖ (+7 095) 956 18 80
Ambassador: H.E. Mr Luis Doukui Paulo de Castro
☎(+7 095) 956 18 81
First Secretary: Orlando Jorge Encarnação da Cruz Lima
☎(+7 095) 143 66 13
Military Attaché: Col Sebastião Adriano Vizado
☎(+7 095) 143 65 01; 956 10 66 [97]

31-00090 Australian Embassy
Kropotkinsky Pereulok 13, Moscow
☎(+7 095) 956 60 70 ✉ AUSTEMBA Moscow
⊠ 413474 ⊖ (+7 095) 956 61 70
Ambassador: H.E. Mr Geoffery R. Bentley
Political Counsellor & Deputy Head of Mission: Glen Waller
Minister, Commercial Counsellor: Peter Frank

First Secretary, Foreign Policy: Dr Bobo Lo
☎ (+7 095) 232 32 92
First Secretary, Polmil & Defence: Robert Owen-Jones
☎ (+7 095) 232 32 80
First Secretary, Administration: Geoffrey Becher
First Secretary, Immigration: Ms Janette Haughton [97]

31-00130 Embassy of the People's Republic of Bangladesh
Zemledelcheski Per 6-8, Moscow
☎ (+7 095) 246 79 00; 246 73 31 (office);
233 84 11 (residence) ▣ 413196 ⊖ (+7 095) 248 31 85
Ambassador: H.E. Mr Mostafa Faruque Mohammed
☎ (+7 095) 246 78 04
Defence Attaché: Air Cdre Rashid Ahmed
☎ (+7 095) 246 73 31 [97]

31-00150 Embassy of Belgium
ulitsa Malaya Molchanovka 7, 121069 Moscow
☎ (+7 095) 291 60 18; 291 60 27; 05 31; 16 04
⊠ AMBABEL Moscou ▣ 413471 ambel su
⊖ (+7 095) 291 60 05
Ambassador: H E Mr Pierre Championes
Deputy Chief of Mission: Thomas Bachdoult
Economic Counsellor: Ken Adam
Defence Attaché: Brig Gen Marcus
Second Secretary: Marechal Evert [98]

31-00330 Embassy of Central African Republic
Ul. 26-Bakinskikh Komissarov 9, Kv. 124-125, Moscow
☎ (+7 095) 434 45 20 ▣ 413737 amrca
⊖ (+7 095) 434 46 14
Ambassador: H E Mr Claude Bernard Beloum
☎ (+7 095) 434 45 20
First Counsellor: André Doumkodji ☎ (+7 095) 434 43 36
Second Secretary, Finance: Salomon Nganaré
☎ (+7 095) 434 06 48 [98]

31-00535 Embassy of Estonia
Malo Kislovski 5, 103 009 Moskva
☎ (+7 095) 290 50 13; 290 31 78 (consular department)
[E] mskvpost@moskva.vm.ee ⊖ (+07 95) 202 38 30
Ambassador: H E Mr Mart Helme
Consular Department Manager: Jaanus Kirikmae
Chancellor, Press: Peter Miller

General Consulate of Estonia
Bolsaja Monetnaja 14, St Petersburg
☎ (+7 812) 233 55 48 ⊖ (+7 812) 233 53 09
Consul General: Märt Piisrop

General Consulate of Estonia
Narodnaja 25, Pihkva
☎ (+7 112) 44 59 39 ⊖ (+7 112) 15 68 64
Consul: Andres Kolk [98]

31-00690 Embassy of Greece
ulitsa Stanislavskovo 4, Moscow
☎ (+7 095) 290 14 46; 290 10 34 ▣ 413472
⊖ (+7 095) 200 12 52 [98]

31-00830 Embassy of India
ul. Vorontsovo Pole, Building 6-8, 107120 Moscow
☎ (+7 095) 917 16 34 ⊖ (+7 095) 975 23 37
▥ 41352 :NAMOW RU
Ambassador: H E Mr Ronen Sen ☎ (+7 095) 916 05 52
Naval Attaché: Cdr P S Sahi ☎ (+7 095) 917 16 34 [98]

31-00920 Embassy of Japan
Kalashny Pereulok 12, Moscow
☎ (+7 095) 291 85 00; 291 85 01
⊖ (+7 095) 200 12 40

Consulate-General of Japan
Nab Reki Moiki 29, 191065 St Petersburg

Consulate-General of Japan
Ulitsa Pushkina 38a,
680000 Khabarovsk, Khabarovsky Krai
☎ (+7 50985) 210 02; (+7 4212) 33 18 07; 33 19 18
⊖ (+7 50985) 210 03; (+7 4212) 33 19 12 [97]

31-01070 Embassy of Luxembourg
Kroutchevsky Pereulok 3, Moscow
☎ (+7 095) 202 21 71 ;202 53 81 ▣ 413131
⊖ (+7 095) 200 52 43
Ambassador: H.E. Mr Adrien Meisch
Commercial & Public Relations Officer: Guy Diederich [97]

31-01110 Embassy of Malaysia
Mosfilmovskaya Ulitsa 50, Moscow
☎ (+7 095) 147 15 14/15 ⊠ MALAWAKIL
▣ 413 478 malay su ⊖ (+7 095) 147 15 26
Ambassador: H.E. Mr Datuk Yahya Baba
Deputy Chief of Mission: Rosli Ismail
Commercial Counsellor: N. Sivarajah [97]

31-01235 Embassy of the Union of Myanmar
41 Gertsena St., Moscow
☎ (+7 095) 291 01 70 ▣ 413403 umyanru
⊖ (+7 095) 956 31 86
Ambassador: H.E. U Kain Nyunt ☎ (+7 095) 291 05 45
Military, Naval & Air Attaché: Col Soe Nyunt
☎ (+7 095) 291 01 70 [97]

31-01270 Royal Netherlands Embassy
Kalashnyi Pereulok 6, Moscow
☎ (+7 095) 291 29 99; 291 29 48;
291 29 54; 291 29 76 ▣ 413442 hol su
⊖ (+7 095) 200 52 64
Ambassador: H E Baron de Vos van Steemwÿk
Deputy Chief of Mission: J C F von Mühlen
Commercial Counsellor: R E M van Lanschot
First Secretary (Commercial): G P M H Steeghs
Defence Attaché: Brig Gen J Rietvelol
Deputy Defence Attaché: Lieut Col H M L Grothauzen [98]

31-01370 Embassy of Pakistan
Sadova-Kudrinskaya Ulitsa 17, Moscow
☎ (+7 095) 250 39 91; 251 49 24 ⊠ PAREP Moscow
▣ 413 338 parep su, 413 194 pkemb su
⊖ (+7 095) 956 90 97; 232 94 36
Ambassador: H.E. Mr Tanvir Ahmad Khan
Deputy Chief of Mission: Muhammad Younis Khan
☎ (+7 095) 254 39 66
Defence Attaché: Brig. Mohammad Saleem
☎ (+7 095) 254 27 22
First Secretary: Manzoor ul Haq ☎ (+7 095) 254 22 61
Second Secretary: Dr Sohail Khan ☎ (+7 095) 254 16 10
Commercial Counsellor: Muhammad Iqbal Arain
☎ (+7 095) 243 48 02 [97]

31-01480 Embassy of Romania
Mosfilmovskaia Ulitsa 64, Moscow
☎ (+7 095) 143 04 24 ▣ 414355; 414360
⊖ (+7 095) 143 04 49
Ambassador: H E Mr Constantin Girbea
Commercial Counsellor: Liviu Istratoiu
☎ (+7 095) 143 28 09
Economic Counsellor: Liviu Istratoiu
☎ (+7 095) 143 28 09
First Secretary: Vasile Pele
First Secretary: Cristian Botez
First Secretary: Cornel Gheorghias
First Secretary, Economic Counsellors Office:
Mrs Liliana Poida ☎ (+7 095) 143 28 09
First Secretary, Economic Counsellors Office:
Marius Tabacu ☎ (+7 095) 143 28 09
Second Secretary: Liviu Visan
Second Secretary: Remus Ghergulescu
Second Secretary: Cornel Ionescu
Attaché: Geheorghe Nastase
Attaché: Victor Vasilescu
Military, Naval and Air Attaché: Col Radu Ignatescu
☎ (+7 095) 143 04 34
Deputy Military, Naval and Air Attaché:
Lieut Col Florentiu Barbu
Assistant Military, Naval and Air Attaché:
Maj Ing Cristian Ermia [98]

31-01640 Embassy of the Republic of Singapore
Per Kammenaye Sloboda 5, Moscow
☎ (+7 095) 241 37 02 [E] sinemb@online.ru
▣ 413128 sem su ⊖ (+7 095) 241 78 95
Counsellor: Christopher Cheang
First Secretary: Ms Juniper Jim [98]

31-01655 Embassy of the Republic of Slovenia
Gruzinski Pereulok, Dom 3 Kv 41, 123056 Moscow
☎ (+7 095) 254 35 31 ⊖ (+7 095) 254 72 23
Ambassador: H E Mr Sasa Ivan Gerzina
First Consular Secretary: Franc Lukner [98]

31-01730 Swedish Embassy
Ulitsa Mosfilmovskaya 60, Moscow
☎ (+7 095) 956 12 00 ⊠ SVENSK Moscow
▣ 413410 svnsk su ⊖ (+7 095) 956 12 02
Ambassador: H E Mr Sven Hirdman
Deputy Chief of Mission: Lars Fredén
Defence, Military and Naval Attaché: Brig Gen Håkan Linde
Air Attaché: Brig Gen Bo-Göte Eriksson
Commercial Counsellor: John Olof Dahlstein
Administrative Counsellor: Ms May Andersson [98]

31-01840 Embassy of the Republic of Turkey
7 Rostovsky Per 12, Moscow
☎ (+7 095) 246 17 52 [E] turemb@dol.ru
▣ (+7 095) 245 63 48; 245 65 02
▥ 64 413731
Ambassador: H E Mr Bilgin Unan ☎ (+7 095) 247 19 70
Deputy Chief of Mission: Can Altan ☎ (+7 095) 246 28 00
Defence and Naval Attaché: Rear Adm Erdal Bucak
☎ (+7 095) 246 04 89
Air Attaché: Col Sadik Altin ☎ (+7 095) 246 92 52
Army Attaché: Lieut Col Mehmet Adanali
☎ (+7 095) 245 65 01
Chief Counsellor for Commercial Affairs: Tanzer Guven
☎ (+7 095) 246 29 89
Chief Counsellor for Economic Affairs:
Abdurrahman Sensoy ☎ (+7 095) 246 17 52
Press Attaché: Emel Özgür ☎ (+7 095) 246 97 52 [98]

31-01890 Embassy of the United Kingdom
Sofiiskaya Naberezhnaya, Moscow 72
☎ (+7 095) 956 72 31 ⊖ (+7 095) 956 74 35
Ambassador: H E Sir Andrew Wood
Defence and Air Attaché: Air Cdre M L Feenan
Naval Attaché: Capt C A M Parrish
Military Attaché: Col R Greenham
Head of Secretariat, Defence: S Young
Assistant Naval Attaché: Lieut Cdr D G Fields
Assistant Military Attaché: Maj C M Hainge
Assistant Air Attaché: Sqn Ldr A I Farmer (RAF)
First Secretary Commercial: S Trusswell
First Secretary Public Affairs: I Hay-Campbell

Commercial & Scientific Departments
Sofiiskaya Naberezhnaya, Moscow 72
☎ (+7 095) 956 72 00 ⊖ (+7 095) 956 74 20
Counsellor Commercial: A Brenton [98]

31-01910 Embassy of Uruguay
Lomonosovsky Prospekt 38, 1st Fl., 117330 Moscow
☎ (+7 095) 143 04 01; 143 04 04 ⊠ URURS MOSCOW
⊖ (+7 095) 938 20 45
Ambassador: H.E. Mr Pedro Dondo
Consul: Anibal Cabral [97]

40 Government Defense Sales/Procurement Organizations

40-01000 Aviaexport PLC
22 Ulansky per, 101000 Moscow
☎ (+7 095) 207 55 00 ⊠ AVIAEXPORT
▣ 411257; 411555 avex ru ⊖ (+7 095) 207 07 80
General Director: F N Myasnikov
Deputy General Director: A V Pavlov
Deputy General Director: B A Khripunov
Deputy General Director: A V Komissarov
Chief of Public Relations Department: A O Didenko
☎ (+7 095) 207 29 77

■ Export/import of aircraft, helicopters, airborne and ground equipment. Services for aircraft, helicopters and GSE. → 80

● 22 Ulansky Per, 10100 Moscow
☎ (+7 095) 207 55 00 ▣ 411257 avex su
⊖ (+7 095) 207 07 80

■ Authorised by the Committee for Defence Industries of the Russian Federation to represent the aviation industry of the Russian Federation on the world market as well as aircraft manufacturers of the Commonwealth of Independent States on the basis of special agreements. Engaged in export/import/leasing of passenger and cargo aircraft and rendering customer support. Export/import of airfield equipment all types of parachutes, parachute equipment and dropping platforms, simulators, general aviation, deltaplanes, monoparaplanes with fast acting parachute recovery systems, machine tools, metallurgical products. Export/import radio and electrical equipment instruments and spares. Trading in aerospace related licences, know-how and engineering services, marketing in Russia and abroad the results of scientific, technological and intellectual activity, assistance in legal protection in intellectual property including patenting of inventions, registration of trade marks and software in Russia and abroad. Overhaul and repair of aircraft, engines, aviation equipment and aggregates. Operation and post-warranty technical support of aircraft, training of foreign flight and ground personnel, engineering services. Export/import of propulsion and auxiliary engines. Participation of licence production and related services. [98]

55 Government-owned Industry & Production Facilities

55-01000 R E Alekseyev Central Hydrofoil Design Bureau JSC

Alekseyev Design Bureau
51 Svoboda Street, 603003 Nizhniy Novgorod
☎ (+7 8312) 25 10 37 ⊖ (+7 8312) 25 02 48
General Director: Igor M Vasilevsky
Chief of Bureau and Technical Director: Mikhail Andreyevich Suslov ☎ (+7 8312) 25 14 15
Chief Designers: Vldimir Nikolayevich Kirillovykh ☎ (+7 8312) 25 06 22
Chief Designers: Anatoly Nikolaevich Pavlenko ☎ (+7 8312) 25 03 07
Chief Designers: Mikhail Yuryevich Garanov ☎ (+7 8312) 25 12 85
■ Design of hydrofoil ships, air-cavity ships and ekranoplans. [98]

55-01050 All-Russian Institute of Aviation Materials

Vserossiski Institut Aviatsionnykh Materialov (VIAM)
ul. Radio 17, 107005 Moscow
☎ (+7 095) 261 86 77 ⊖ (+7 095) 267 86 09
Director: Radi Yevgenyevich Shalin
First Deputy Dir: Evgeni Borisovich Kachanov
Deputy Dir: Boris Vitalyevich Perov
■ Research & development of aerospace materials including high temperature alloys, metallo-ceramic coatings, metals, plastics & polymers. Specific research includes materials for engine & re-entry systems design, protective coatings for turbine blades & heat-resistant coatings for leading edges of supersonic wings. → 60 [97]

55-01300 Arsenal Design Bureau named after M.V. Frunze

KB "Arsenal"
1/3 Ulitsa Komsomola, 195009 St Petersburg
☎ (+7 812) 542 29 73; 248 98 42 ⊠ 122774 "GROT"
⊠ 121608 stal ru ⊠ i22774 grot ru
⊖ (+7 812) 542 20 60
👥 1,000 (1997) 🏭 20,000 sq m
Director and Chief Designer: Boris T Poletaev
First Deputy Director and Chief Designer: Leonid D Fedotov ☎ (+7 812) 542 18 96
Deputy Chief Designer, Marketing and Communications: Vladimir I Sapojnikov ☎ (+7 812) 248 98 42
■ Satellites. [98]

55-06000 Elektrostal Machine-Building Plant Production Association

PO "Mashinostroitelnyy zavod", POMZ
Elektrostal, 144000 Moscow Oblast
☎ (+7 095) 176 24 55 ⊠ 911669 energ su
⊖ (+7 095) 702 92 21
♧ US$ 640,000,000 (1996)
♣ US$ 106,000,000 (1996) 👥 10,594 (1996)
🏭 559,241 m2
Technical Director: A Panushkin
Marketing Manager: A Suchkov
Purchasing Manager: G Gavrilov
Operations Director: C Potoskaev
■ Uranium for nuclear weapons. [98]

55-06100 Energomash Scientific Production Association

Khimky 1, 141400 Moscow Oblast
☎ (+7 095) 572 22 59; 572 22 00
⊖ (+7 095) 251 75 04
👥 7 850 🏭 282 000 sq m
General Dir & General Designer: Boris I. Katorgin
First Deputy General Dir & General Designer: Vladimir K. Chvanov
First Deputy General Dir & General Designer: Gennady G. Derkach
Deputy General Dir & Foreign Economic Relations: Victor A. Sigaev ☎ (+7 095) 572 22 00
■ Design, test & manufacture of liquid rocket engines for space launch vehicles. [97]

55-07300 Ferrite Domain Scientific Production Association

SPA Ferrite Domain
Ferrite International
Chernigovskaya Ulitsa 8, 196084 St Petersburg
☎ (+7 812) 294 84 11 ⊠ 121014 ferco ru
⊖ (+7 812) 298 37 91
♧ Rb 150,000,000,000 (1996)
♣ Rb 14,000,000,000 (1995) 👥 1,040 🏭 33,000 m2
Dir General: Valery V. Philippov
Managing Director: Vladimir D Voronkov ☎ (+7 812) 294 80 33 (telephone and fax)
Technical Director: Anatoly Gourchenko
Financial Dir: Sergey P. Babenko ☎ (+7 812) 298 22 32
Marketing Director: Vladimir Fjedorov
Marketing Mgr: Andrey V. Nikiforov ☎ (+7 812) 294 71 87
Public Relations Mgr: Dr Rostislav A. Khlebnikov ☎ (+7 812) 298 75 77
■ Microwave ferrite materials & components, microwave & millimeter wave equipment for radar & space communications; soft (poly- & monocrystalline) & hard ferrites; molybdenum permalloy powder cores (MPP-cores). [97]

55-07600 Gidropridor

24 Bolshoi Sumpsonievsky prospect, 194044 St Petersburg
☎ (+7 812) 542 01 47 [E] gidroprb@sovam.com
⊠ 321472 mope ⊖ (+7 812) 542 96 59; 542 95 28
👥 2,000 (1997) 🏭 26,000 m2
Director: Stanislv G Proshkin
Deputy Director: Leonid I Voronin
Commercial Director: Peter D Sukhoparov
Technical Director: Oleg Ivanov [98]

55-09000 Ilyushin Design Bureau

Ilyushin Aviation Complex
Ilyushin Aviation Complex, Leningradsky Prospect 45 "G", 125190 Moscow
☎ (+7 095) 943 83 25; 213 55 70
[E] ilyushin@sovam.com ⊠ 411956 sokol
⊖ (+7 095) 212 21 32; 250 13 34
General Designer & CEO: Genrikh V. Novozhilov
First Deputy CEO & Exec. Dir: Victor Livanov
Chief Designer/Head of Foreign Technical Relations: Igor Katyrev
Chief Designer/Head of Design Bureau: Nikolai Talikov
Asst. General Designer Foreign Economic: Vladimir Beliakov
1st Deputy General Designer and IL-96 Programme Manager: Vyacheslav I Terentyev
Chief Designer and Il-96 Programme Manager: Radi P Papkovsky
■ Design and manufacture of commercial, transport and general aviation light aircraft. → 75

Ilyushin Manufacturing Facility
329029 Voronej
General Director: Albert Georgevic Mikhailov [97]

55-09200 Irkutsk Joint Aviation Industrial Association

3 Novatorov Straat, 664020 Irkutsk
☎ (+7 3952) 39 29 11; 32 29 13; 32 29 42 ✆ 32 29 11
⊠ 231121 ⊖ (+7 3952) 39 29 11; 32 29 13
👥 11,500 (1996)
Director General: Alexei I Federov
Director Marketing, Sales and Product Support: Anatoly A Belov
Finance Director: Ms Valery I Markeev ☎ (+7 3952) 32 29 14
Deputy Director General for Production: Sergi Ivanovich Vassilenko ☎ (+7 3952) 32 29 18
Operations Director: Kovalkov Vladimir
Head of Marketing: Vitaly Zelenkov
Purchasing Manager: Stefanovsky Vladimir
■ Manufactures civil and military aircraft, including the Su-27 UBk, Su-30 family and the Be-200. [98]

55-09500 Izhevsk Mechanical Plant

Izhevskiy Mekhanicheskiy Zavod
Ulitsa Promyshlennaya, 8, 426063 Izhevsk
☎ (+7 3412) 76 58 30 [E] ims@shield.udmurtia.su
⊠ 755111BAIKL RU ⊖ (+7 3412) 76 45 90
General Director: Vasily S Chuguyevsky
Marketing Director: Vladimir Roudakov
Marketing Manager: Victor Gorodilov
■ Manufacture small arms, sporting weapons, anti-tank and surface-to-air missiles. [98]

55-10000 JSC Ulan-Ude Aviation Plant Association

Ulan-Udenskoye APO
(† Ulan-Ude Aviation Production Association)
1, Khorinskaya Street, 670009 Ulan-Ude
☎ (+7 30122) 530 30; 574 75; 526 38; 616 29; 516 29; 535 55 [E] uso4@uuaz.buriatia.ru ⊠ 288110 avia
⊖ (+7 30122) 521 47; 516 29
👥 5,000
Director General: Vladimir F Konstantinov ☎ (+7 30122) 533 86
Technical Director: Serguei Solomin ☎ (+7 30122) 533 86
Director General Deputy on Sales and Marketing: Leonid Belych
Financial Director: Iouri N Gavrilenko ☎ (+7 30122) 523 09
Chief Pilot: Iouri Katchourin ☎ (+7 30122) 552 86
Deputy Chief of Commercial Department: Andrew L Polyutin
■ Manufacturer of the Mi8, Mi-171, KA-62, helicopters and the SU-25UKB trainer/combat aircraft and aviation multi mission strike complex Su-39. [98]

55-10300 Kalujsky Dvigatel

Kaluzhskiy Motorostroitelnyy Zavod
(a joint stock company)
Ulitsa Moskovskaya, 247, 248633 Kaluga
☎ (+7 8422) 240 97 ⊠ 183212 sokol
⊖ (+7 8422) 400 95
👥 9,000
General Director: Y A Leykovskiy
Chief Engineer: V P Slominskiy
Head of Marketing Department: V Lyadov
Purchasing Manager: V P Zlobin
■ Produces Klimov gas turbines for tanks (T-80), and engines for aircraft (Il-114). Gas turbine engines for anti-aircraft artillery, power assemblies for specific machinery.

Design Bureau
Ulitsa Moskovskaya, 247, 248633 Kaluga
☎ (+7 8422) 271 34 ⊠ 183213 ⊖ (+7 8422) 271 74
Director: N V Shleynikov

Experimental Office
Ulitsa Moskovskaya, 247, 248633 Kaluga [98]

55-10340 Kazan Aviation Industrial Association

(† Kazan Aircraft Production Association)
420036 Kazan, Tatarstan
☎ (+7 8432) 54 24 32 ⊖ (+7 8432) 54 02 04
Managing Director: Khairullin Gumerovich
Marketing Director: Shpilkin Ruslan Alexandrovich
■ Manufacture the TU-224, TU-214, TU-324 and TU-330 aircraft, and drying vacuum machines for crome leather production. [98]

55-10350 Kazan Design Bureau

Dementjev Straat, 420036 Kazan, Tatarstan
☎ (+7 8432) 54 15 02 ⊖ (+7 8432) 54 06 43
♣ Rb 500,000,000 👥 700

■ Aircraft instrumentation. Temperature, revolutional speed, torque sensors & relevant instrumentation. Instrument characteristic investigation & testing at gas aerodynamical rig, in cryogenic laboratory & at vibration & gas stands with characteristics near to real conditions as well small-sized thin-film nickel temperature sensors together with a controller designed for use in household appliances. [98]

55-10400 Khrunichev State Research & Production Space Center

Novozavodskaya St 18, 121087 Moscow
☎ (+7 095) 145 89 54; 145 98 27; 145 83 43
(+7 095) 145 98 02 [E] proton@sovam.com
412427 mir su ⊖ (+7 095) 142 59 00
18,000
Director: Anatoliy Ivanovich Kiselev
Prime Deputy Director General: Alexandr A Medvedev ☎ (+7 095) 145 83 43
Deputy Dir: Aleksandr V Lebedev ☎ (+7 095) 145 83 43
Deputy Director General, General Designer: Anatoly K Nedaivoda ☎ (+7 095) 148 08 43
Chief International Cooperation Bureau: Dennis Pivnyuk ☎ (+7 095) 145 98 30
Head of Foreign Relations: Alexksandr S Kondratyev ☎ (+7 095) 145 98 27
Head of Public Relations: Sergei A Jiltsov ☎ (+7 095) 145 98 02
■ Rocket boosters and space vehicles. Mir modules and parts for M-4 Bison space-support aircraft. Rockot lightweight launch vehicle, Proton and Proton-M launch vehicles. [98]

55-10450 Energia Korolev Rocket Space Corporation Korolev Rocket & Space Corporation Energia (RSC Energia)

Energiya Scientific Production Association
4 a Lenina Straat, 141070 Korolev, Moscow region
☎ (+7 095) 513 72 48 [E] rsce.postmaster@rsce.nasa.ru
911518 buran su ⊖ (+7 095) 187 98 77; 513 86 20
20,000 (1997)
Chairman, President and General Designer: Yuriy Pavlovich Semenov ☎ (+7 095) 513 72 48
Prime Vice President and Prime Deputy General Manager: Nickolay I Zelenschchikov ☎ (+7 095) 513 77 64
Vice President and Prime Deputy General Designer: Victor P Legostayev ☎ (+7 095) 513 82 42
Head of International Relations: Aleksander G Derechin ☎ (+7 095) 513 86 52
■ Development and operation of orbital stations, spacecrafts, communication satellites and launchers. Main programs are Mir Space Station, Soyuz-TM and Progress-M Spacecraft, International Space Station Alpha, Sea LAunch and Yamal Satellites for Space Communication Systems. aerospace.

Overseas Offices:
Energia USA
Energia Deutschland [98]

55-13050 Moscow Aircraft Production Organization MIG (MAPO MIG)

(† Mikoyan Aviation Scientific-Industrial Complex MIG)
7 1st Botkinsky Drive, 125040 Moscow
☎ (+7 095) 252 81 41; 252 88 39
⊖ (+7 095) 250 88 19; 956 15 86
25,000
General Dir: Vladimir V. Kuzmin ☎ (+7 095) 252 80 01
General Designer: Rostislav A. Belyakov ☎ (+7 095) 155 23 15
Commercial Dir: Alexander N. Bezrukov ☎ (+7 095) 252 81 41
First Deputy General Dir, Chief Designer: Anatoliy A. Belosvet
General Engineer: Victor M. Puzanov ☎ (+7 095) 252 82 39
Advisor to General Dir on Strategic Analysis: Alexander I. Ageev ☎ (+7 095) 202 35 23
Technical Dir: Ivan I. Butko ☎ (+7 095) 252 81 32
Chief Foreign Economic Relations Department: Vitali Nosov
Chief Advertising & PR Dept: Svyatoslav Yu. Ribas ☎ (+7 095) 252 80 78
Advertising Manager: Ksenija Tolstokhlebova
■ Founded in 1893, MAPO originates from factory No. 1 & the Experimental Design Bureau headed by A.I. Mikoyan. This disassociated from the factory in 1939, & was later officially named the Mikoyan Aviation Scientific-Industrial Complex MIG. In 1995, the Moscow Aircraft Production Organization (MAPO) & the Mikoyan Aviation Scientific-Industrial Complex MIG were merged into the single Moscow Aircraft Production Organization MIG.
MAPO MIG is composed of the main production complex in Moscow, a series-production plant & scientific-industrial complex, & two subsidiaries located in the Moscow region.
Main products:
Military Aircraft: The MIG-29 fighter & its modified versions; upgrade packages supplied to the user. The MIG-AT advanced trainer. Civil Aircraft: the T-101 Grach multi-role aircraft, the AVIATIKA-890 ultralight aircraft, the IL-103 five-seat aircraft, the I-1L two-seat aircraft, etc.
Services: guarantee & post-guarantee services & maintenance. Establishment of maintenance & support centres (joint ventures) on the customer's territory. [97]

55-17900 JSC Samara Scientific-Technical Complex "NK Engines"

(† SSSPE TRUD)
ul S Lazo 1, 443026 Samara
☎ (+7 8462) 50 02 28 [E] sntk@sntk.samara.su
⊖ (+7 8462) 50 12 11
Director General, General Designer: Eugeny A Gritzenko
First Deputy General Designer: S M Ignachkov ☎ (+7 8462) 50 48 27
Production Director: A V Zuev ☎ (+7 8462) 50 54 08
Design Bureau Chief: L M Shirkin ☎ (+7 8462) 50 71 02
■ Development of original an modified aircraft and rocket engines including: NK-12, NK-8-4, NK-8-2U, NK-144, NK-86, NK-86A, NK-88, NK-89, NK-321, NK-31, NK-39, NK-43. The NK-family aircraft engines have been converted into drive engines for modular gas-transfer plants in cross-country gas pipelines, electric generators and other applications. Facilities include a test base for experimental development of engines and assemblies. Equipment includes gas dynamic, strength and vibration test stands and engine test sets. Technology processes developed include production of cast cooled turbine blades and vanes, application of multilayer thermoprotective coatings to turbine blades, compressor blades and discs of titanium alloys, use of electrophysical and electrochemical treatment processes for parts (including parts made of heat-resistant alloys and steels). [98]

55-18320 State Research and Production Enterprise Bazalt

105058 Moscow
☎ (+7 095) 369 01 22 ⊖ (+7 095) 369 24 18
Managing Director: A S Obukhov
Finance Director: M V Artyomov
Marketing Director: V N Sychev
Operations Director: R Sh Rakhmtulin
■ Design of unguided aircraft bombs, anti-tank and anti-saboteur marine grenade launcher systems, hand grenades, mortar artillery shells (for SP guns). Production, delivery, transfer of licences, custom development of new weapon versions. [98]

55-18325 State Research and Production Association SPLAV (SPLAV)

(† State Research and Production Enterprise SPLAV)
4 Shcheglovskaya Zaseka, 300004 Tula
☎ (+7 0872) 44 45 86 ✉ 253183 LES 253110 rtb su
⊖ (+7 0872) 44 14 74; 36 02 87
Director General: Prof Nicolay A Makarovets
Deputy Director General: Rudolf Kobylin
Deputy Director of Economics: Y A Shubkin
Chief of Department of Foreign Relations: A B Kuznetzov
■ Development of multiple-launch artillery rocket systems, including URAGAN and SMERCH. Design and production of anti-submarine weapons and shipborne MLRS, anti-torpedo and other naval defence systems. [98]

55-18500 Sukhoi Design Bureau

23a Policarpova Straat, 125284 Moscow
☎ (+7 095) 945 65 25 414716 suhoi su
⊖ (+7 095) 200 42 43
General Designer: Mikhail P Seemonov
Sales and Marketing Director: Alexei V Komarov
First Deputy General Designer: Mikhail A Pogosian
Deputy General Designer: Alexi I Knychev
Public Relations Manager: Vladimir G Yakovlev
■ Development and manufacture of military aircraft (Su-22, Su-24, Su-25, Su-27, Su-27UB, Su-30, Su-32, Su-33, Su-35); aerobatic sport aircraft (Su-26, Su-29, Su-31) and civil aircraft including supersonic business jets, ground effect vehicles, general purpose aircraft and trainers. [98]

55-19000 Teknomash Scientific Production Association

NPO Tekhnomash
3rd Proyezd, Marinoy Roschi 40, 127018 Moscow
☎ (+7 095) 289 92 08; 289 70 04; 289 11 75; 289 95 31 411351 tent su ⊖ (+7 095) 289 73 45
General Director: V V Bulakin
Director: Vadim A Isachenko
Director of Business Centre: V I Grigoriev
Technical Director: K A Motouznyi
Deputy Director: V Kasayes
Deputy Director: Valentin Kazakov
Engineering Mgr: Vadim B. Glebov
■ Designed equipment used in the production of the Mir space stations, the Energiya space launch vehicle and the Buran shuttle. [97]

55-19450 JSC Tulamashzavod

(† Tula Machine-Building Plant Production Association)
2 Mosin Str., 300002 Tula
☎ (+7 0872) 36 24 65 753114 alfa ru
⊖ (+7 0872) 27 26 20; 36 24 65
⊖ (+7 095) 975 01 40; 975 32 73; 975 20 51
General Dir: Vadim S. Usov
Operations Dir: Evgeni A. Dronov
Sales Dir: Victor I. Aleshin
Foreign Economical Relations Dept Chief: V.N. Izmailov
Purchasing Mgr: Alexander A. Bograd
■ Kashtan air defense system (including both missile & anti-aircraft artillery components); 30mm six-barrel cannon for aircraft & armoured vehicles. [97]

55-19500 Tupolev JSC

Tupolev Aeronautic R&D Complex,
17 Naberejnaia Akademika Tupoleva, 111250 Moscow
☎ (+7 095) 267 24 44 (095) 267 26 44
⊖ (+7 095) 261 08 68; 267 27 33; 261 71 41
General Designer: Dr Alexei A Tupolev
General Director: Valentin T Klimov
Deputy General Director: Oleg Y Alacheyev ☎ (+7 095) 267 24 27
Deputy General Director, Manufacture: Anatoly V Sakharov ☎ (+7 095) 267 27 77
Finance Director: Alexander V Potapov ☎ (+7 095) 267 77 84
Chief Designer: Igor S Kalygine ☎ (+7 095) 263 70 04
Chief Designer: Youry V. Vorobiev ☎ (+7 095) 263 75 56
Chief Designer: Lev A Lanovsky ☎ (+7 095) 263 70 04
Chief Designer: Alexander S Chenguart ☎ (+7 095) 267 28 88
Foreign and Economic Affairs Manager: Serguey M Rajevsky ☎ (+7 095) 267 25 33
■ Design and manufacture of civil and military aircraft. Development of cryogenic fuel systems. [97]

55-21000 VNIIKAM

Russian Institute of Remote Sensing Methods for Geology
Birjevoy proezd 6, 199034 St Petersburg
☎ (+7 812) 218 28 01 ✉ AEROMETODY
121302 limex su ⊖ (+7 812) 218 39 16
♣ Rb 1,500,000,000 (1995) 100 (1996) 3,000 m2
Director: Alexei V Pertsov
Deputy Director: Vladimir L Scherbakov
Deputy Director Science: Alexander A Kirsanov
Deputy Director Science: Vadim S Antipov
Deputy Director Economy: Ms Tatyana S Rodionova
■ Development and installation of remote sensing technology for geological and geoecological tasks. [97]

60 Research Organizations, Institutes & Test Establishments

All-Russian Institute of Aviation . . . 55-01050
NPO Molniya plc 75-13380

60-00090 Ausrire Scientific & Research Institute of Radio Equipment
Shkipersky Protok 19, 199106 St Petersburg
☎ (+7 812) 356 18 34; 351 17 07
⊖ (+7 812) 356 01 38; 850 16 65; 352 37 51
♲ RBI 50 858 000 000 (1995)
♣ RBI 20 950 000 000 (1995) ⚉ 2 278 (1996)
⌂ 90 000 sq m
Managing Dir: Gromov Gennadiy Nikolaev
Finance Dir: Kalchenko Konstantin Evgenjevich
Marketing Dir: Efimkin Evgeniy Petrovich
■ Development of ATC & ATM systems, navigational, meteorological & landing equipment & systems. [97]

60-00300 Central Aerohydrodynamics Institute (TsAGI)
Named after Prof N E Zhukovsky, Russian State Research Center
1 Zhukovsky Street, Zhukovsky, Moscow Region, 140160 Moscow
☎ (+7 095) 911 09 69; 556 41 79; 556 40 21; 271 00 19 ▣ 412573 zakat su
⊖ (+7 095) 911 00 19; 556 43 37; 271 09 69
Director: Prof Vladimir Ya Neyland
Director of TSAGI: Prof German I Zagainov
Deputy Director, Structural Strength and Certification: Dr Yuri A Stoutchalkin ☎ (+7 095) 556 40 35
Deputy Director Aerothermodynamics and Gas Dynamics: Prof. Victor N Gusev ☎ (+7 095) 556 40 16
Deputy Diretor, Flight Dynamics and Control Systems: Dr Valery L Sukhanov ☎ (+7 095) 556 41 55
Deputy Director, Advanced Aerospace Vehicles: Prof. Leonid M Shkadov ☎ (+7 095) 556 44 18
Deputy Director Advanced Test Facilities: Dr Pavel V Miodushevsky ☎ (+7 095) 556 40 27
Deputy Director Aerodynamics: Prof. F A Pavlovets ☎ (+7 095) 556 43 74
Deputy Director Engineering Services: A M Masyuk ☎ (+7 095) 556 42 24
Deputy and Commercial Director: V M Tyurin ☎ (+7 095) 556 40 21
Financial Director: Mrs Olga S Chicherova ☎ (+7 095) 556 42 89
Deputy Director Social Programmes: Mrs Olga A Bessolova ☎ (+7 095) 556 42 62
Head of International Business Department: Dr A A Ionov ☎ (+7 095) 556 40 54
Corresponding Member: Prof. Vladimir Ya Neyland
Industrial Aerodynamics, Hydrodynamics: Prof. Anatoly G Munin
■ Fundamental research:
Aerodynamics of all types of aircraft over full speed range from low subsonic to orbital speeds. Aircraft strength, aeroelasticity, service life, structural integrity, composite materials. Flight dynamics and flight vehicle control systems. Hydrodynamics and solutions to aircraft motion problems in fluid and gas mediums. Aeroacoustics, industrial aerodynamics including wind energy units and fans. Test facilites and data acquisition systems.
Exploratory research
Prediction of developments in aerospace technology, advancement of aircraft concepts.
Applied research:
Design of specific aerospace technology prototypes. Scientific support for projects executed in design bureaus from preliminary design stage to aircraft manufacture and flight testing.
State enterprise:
Resolutions for all aircraft designs, preparation for flight tests and service life limits. Participates in setting aircraft development standards. Testing, certification and development. [98]

60-00400 Central Institute of Aviation Motors (CIAM)
2 Aviamotornaya Straat, 111250 Moscow
☎ (+7 095) 200 22 15; 361 64 81 ▣ 411686 ciam su
⊖ (+7 095) 267 13 54
Director: Prof D Sc Donat A Ogorodnikov ☎ (+1 095) 200 22 15
First Deputy Director: Dr Valentine I Solonin ☎ (+7 095) 261 71 50
Deputy Director: D Sc Sergei A Sirotin ☎ (+7 095) 362 49 28
Deputy Director: Prof D Sc Fiodor N Olifirov ☎ (+7 095) 361 29 42
Commercial Department: Anatoly A Cherkassov ☎ (+7 095) 361 64 81
■ Research, development, test & evaluation of all aeroengine types. [98]

60-01150 Electron Central Scientific & Research Institute
Tsentralny Nauchno-Issledovatelsky Institut Electron
68 Prospekt M. Toreza, 194223 St Petersburg
☎ (+7 812) 552 36 00; 552 71 70
☎ (+7 812) 550 92 08; 550 92 50
⊖ (+7 812) 552 31 97
⚉ 1 030 (1996) ⌂ 35 000 m2
Managing Dir: Ivan Sergeevich Vasilyev ☎ (+7 812) 552 36 00
R&D Dir: Rudolf Mikhailovich Stepanov ☎ (+7 812) 552 28 09; 550 92 91
Dir Marketing: Grigory Isaacovich Vishnevsky ☎ (+7 812) 552 71 70
Dir Business Communications: Igor Alexandrovich Rumyantsev ☎ (+7 812) 550 92 19
Purchasing Mgr: Sergy Alexandrovich Dryablov ☎ (+7 812) 550 60 37
■ Development of solid-state & electronic-beam photo-electronic detectors, for various TV systems & applications. [97]

60-01200 Enics Research Center
2a Academician Pavlova Straat, PO Box 8, 420127 Kazan
☎ (+7 8432) 54 70 43 ✉ KOCMOC
[E] root1@enics.kazan.su ▣ 412062 octet su Box 51012
⊖ (+7 8432) 54 70 43
President: Valery Pobezhimov
Technical Director: Ildar Yakupov
■ Design, production and testing of unmanned aircraft. → 75 [98]

60-02100 Instrument Design Bureau - KBP Tula
ul. Shcheglovskaya Zaseka, 300001 Tula
☎ (+7 872) 41 07 50 ✉ Tula STRELA
⊖ (+7 872) 42 61 39
Dir & General Designer: A.G. Shipunov
Exec. Dir & Chief Engineer: V.P. Tikhonov
Sales & Marketing Dir: S.G. Kuznetsov
■ Armament development in the following areas: Anti-Tank Guided Weapons systems (ATGW); tank-gun launched guided missile systems; guided artillery munitions; short-range anti-aircraft guided weapon systems; IFV weapon systems (including BMP-3); automatic gun systems (naval, air & tank); small arms (rifles, pistols, sub-machine guns); flamethrowers. [97]

60-02600 Kulon Scientific Research Institute
Murmansky Prospekt 14, 129075 Moscow
☎ (+7 095) 287 14 71 ✆ (+7 095) 286 40 27
⊖ (+7 095) 287 76 11
Director: A S Novoselov
Director Research: I V Efanov
Operations and Technical Director: V A Fiks
Public Relations Officer: O A Vladimirov
■ Development and manufacture of equipment for remotely piloted vehicles with airborne and ground-based command and control systems. [98]

60-02730 Morphyspribor
46 Chkalovsky Pr., 197376 St Petersburg
☎ (+7 812) 232 21 05 ✆ (+7 812) 230 39 35
⊖ (+7 812) 235 39 91
Director: Dr Yuri Koriakin
Technical Dir: Dr Konstantin Polkanov
Marketing Dir: Oleg Beliavsky
Head of Foreign Relations Dept & Marketing Mgr: Dr Alla Kalyaeva
■ Research, development & production of surface ship ASW sonar systems. [97]

60-02750 Moscow Institute for Electromechanics & Automatics
(a part of the Aviapribor Corp)
5 Aviatsioniy Pereulok, 125319 Moscow
☎ (+7 095) 152 48 74 ⊖ (+7 095) 152 26 31
Director General: Sergey Krioukov
■ Development, production, integration and aftersales support of flight control, fly by wire and flight management systems, as inertial navigation and reference systems for civil aircraft. [98]

60-02780 Moscow Research Institute Agat
Malaya Grouzinskaya Str. 54, 123557 Moscow
☎ (+7 095) 253 09 96 ▣ 346719 agat su
⊖ (+7 095) 253 65 44
Managing & Technical Dir: Prof. Jozeph G. Akopian ☎ (+7 095) 253 06 66
Chief Engineer: Boris S. Arkhipov ☎ (+7 095) 253 65 44
Deputy Dir Science: Vladimir D. Chernov ☎ (+7 095) 556 81 10
Deputy Dir Finance: Eugeny I. Skorokhodov ☎ (+7 095) 556 72 32
Principal Engineer & Translator: Victor Ph. Molchanov
■ Design & development of radar seekers (semi-active, active, combined).

Regional Office
Zhukovsky 2, 140160 Moscow Region
☎ (+7 095) 556 50 87 ⊖ (+7 095) 556 81 10 [97]

60-03300 OKB Mechanical Engineering
Burnakovski Proezd 15, 603074 Nizhny Novgorod
☎ (+7 8312) 41 89 02 ✆ (+7 8312) 46 21 32; 41 89 90
▣ 151108 blok ru ⊖ (+7 8312) 41 87 72; 24 49 97
⚉ 3,200 (1997) ⌂ 97,780 m2
Managing Director: F M Mitenkov
Sales and Marketing Director: V G Krivenchuk
Advertising Manager: V I Kalentiev
Public Relations Manager: E V Kusmartsev ☎ (+7 8312) 41 72 84 [98]

60-03555 Prometey Central Research Institute of Structural Materials
(a subordinate of the Shipbuilding Department of the State Committee)
49 Spalernaja Straat, 193015 St Petersburg
☎ (+7 812) 274 18 10; 274 37 96
[E] vvv@promety2.spb.su ▣ 322147 alfa ru
⊖ (+7 812) 274 17 07
♲ US$ 45,000,000 (1996) ⚉ 2,000 (1996)
Managing Director: Akad Prof Igor Vasilievich Gorynin
First Deputy, Operations and Technical Director: Prof Oleg Georgievich Sokolov
Sales and Marketing Director: Dr Andrej Vasilievich Dmitriev
Purchasing Manager: Valentin F Sirotkin
■ Development of advanced materials, production processes and manufacturing technologies of metallurgical semifinished products from steel and alloys based on titanium, aluminium and copper. Development of material treatment, including welding and coatings. Research into material behaviour in various operating conditions including materials under radiation influence. Determination of life-time of materials used in structures. [98]

60-03900 Research Institute for Special Equipment of the Ministry of the Interior of the Russian Federation
ulitsa Lubianka, 16/14, 101000 Moscow
☎ (+7 095) 923 60 35 ⊖ (+7 095) 934 27 14 [98]

60-04000 R&D Institute of Mechanical Engineering (R&DIME)
Nizhnyaya Salda, 624610 Sverdlovsk Region
☎ (+7 34345) 217 03; 442 55
✆ (+7 34345) 217 03 41 58 ▣ 348172 omul
⊖ (+7 34345) 222 58; 317 03
General Manager: Boris Alekseevich Nekrasov
Sales and Marketing Director: Yuri Pavlovitch Bondar
Technical Director: Adolf Ivanovich Razhigajev
Low Engine Design Director: Evgeny G Larin
Advertising Manager: Ms Elena Valentinovna Kutueva
■ Development and manufacture of low thrust rocket engines (LTRE) for orientation, stabilization, attitude and motion control systems of the space stations Salyut and Mir, spacecraft Soyuz and Progress, equipment modules Kvant and Kristall, space vehicle Buran and cosmonaut motion control systems. Liquid rocket engines up to 3,000 kn thrust on the cryogenic propellant. [97]

60-04030 National Institute for Aviation Technologies
Nauchno-Issledovatelski Institut Aviatsionnoi Tekhnologii (NIIAT)
(† Scientific Research Institute of Aviation Technology)
ulitsa Petrovka 24, 103051 Moscow

☎(+7 095) 200 76 01; 200 14 89; 200 39 88; 200 34 17 ⊠ 411700 ptb su
⊖ (+7 095) 200 76 01; 200 30 81
♧ US$ 31,000,000 (1997) ♣ US$ 3,000,000 (1997)
👥 1,100 (1997) 🏭 44,000 m2
General Director: Oleg Sergeyevich Sirotkin
☎(+7 095) 200 34 12 (fax)
Managing Director: Vitaly V Plikhunov
☎(+7 095) 200 34 12 (also fax)
Operations Director: Victor Rushkov
☎(+7 095) 200 75 33
Marketing Director: Vladimir V Evstigneev
■ Services offered include welding, casting, composite material, super plasticity shaping, thermoprocesses, and salvage of aviation military and technical facilities. [98]

60-04050 Scientific Research Institute of Radio Equipment

Shkipersky Pzotok 19, RF, 199106 St Petersburg
☎(+7 812) 351 17 07; 356 18 34
⊖ (+7 812) 356 01 38; 352 37 51
♧ R 273,700,000,000 ♣ R 61,000,000,000 👥 2,350
🏭 90,000 m2
Chief Designer: Gennady Gromov
International Marketing Director: Valeriy Jhikarev
■ Research, design and manufacture radio and communications equipment for avionics and ATS systems. [98]

60-04150 State Research Institute of Aviation Systems (GosNIIAS)

(† GosNIIAS)
7 Victorenko Straat, 125319 Moscow
☎(+7 095) 157 70 47; 157 75 22
✆ (+7 095) 157 10 31 Ⓔ czoot@tm.niias.msk.su
⊖ (+7 095) 157 31 27
👥 3,000
Chief Executive Officer: Evgeniy A Fedosov
☎(+7 095) 157 70 17
Chief of Public Relations: V Volodin
Executive Research Director: Pavel V Poznyakov
☎(+7 095) 157 65 82
Executive Commercial Director: Boris S Alyoshin
☎(+7 095) 157 75 21
■ Development and integration of advanced avionics and weapon systems. Certification testing of new airborne equipment. Development of computer-aided design/engineering systems, computer systems and tehnologies for use in air traffic control. Provision of flying testbeds and hardware-in-the-loop simulation and ground testing facilities. → 75 [98]

60-04500 JSC Taganrog Aviation Scientific-Technical Complex named after G M Beriev

Aviatorov Sq. 1, 347923 Taganrog
☎(+7 86344) 498 39; 498 53 ⊠ 298107 beta su
⊖ (+7 86344) 414 54; 448 83
Managing Dir & Designer General: Gennady S. Panatov
☎(+7 86344) 499 01
First Deputy General Designer: Valentin V. Boev
☎(+7 86344) 499 20
First Deputy Designer & Technical Dir: Alexander V. Yavkin
☎(+7 86344) 498 53
Sales & Marketing Dir: Vladimir N. Konoplev
☎(+7 86344) 466 48
Senior Mgr, Marketing Dept: Oleg V. Souspa
☎(+7 86344) 448 83
■ Experimental design bureau, experimental production, flight-test complex, economic support service, financial support service & logistics support service. Also test bases & proving grounds at the Black Sea & the Sea of Azov. Experimental prototypes of amphibious aircraft, wing-in-ground effect vehicles. Design & development of non-conventional aircraft, the test of aircraft & its assemblies in maritime conditions & the training of flying crews & maintenance personnel for seaplane operation. [97]

60-08350 Research, Development and Production Enterprise Zvezda JSC

39 Gogol Straat, 140070 Tomilino, Moscow region
☎(+7 501) 882 22 79 Ⓔ npstar@npstar.msk.su
⊠ 206123 star us ⊖ (+7 95) 557 33 88
👥 2,500
General Director and General Designer: Guy I Severin
☎(+7 95) 557 30 65
Deputy General Director and Deputy General Designer: Vitaly I Svertshek ☎(+7 95) 557 33 77
Deputy General Designer: Vladislav I Kharchenko
☎(+7 95) 557 33 78
Executive Director: Nikoly I Afanasenko
☎(+7 95) 557 16 77
■ Design, development and pilot production of ejection and shock absorbing seats, IVA and EVA space suits, protective gear and oxygen equipment for pilots, inflight refuelling systems, aircraft fire detection and suppression systems, slides, rafts and survival kits for civil aircraft. → 75 [98]

60-08400 Zvezda-Strela State Research & Production Centre

Ilyicha St 7, 141070 Kaliningrad
☎(+7 095) 519 55 01 ⊖ (+7 095) 519 47 22
Chief Designer: Yuri D. Novikov
■ Research, development & production of airborne missiles X-25M, X-31, X-35 & their derivatives. → 75 [97]

65 Industry, Trade & Armed Forces Associations

65-01000 Association of Aero Engine Manufacturers (AAEM)

Glavpochtamt, PO Box 391, 101000 Moscow
☎(+7 095) 207 01 71; 207 09 97
⊖ (+7 095) 207 54 91; 207 57 67
♧ Rb 41,500,000 (1996) ♣ Rb 4,000,000,000 (1996)
👥 28 🏭 295 m2
President and General Director: Victor M Chujko
Vice President and Deputy General Director: Anatoly D Dekhjarenko
Deputy General Director: Valentin M Tolokonnikov
Deputy General Director: Anatoly G Ganzha
Deputy General Director: Vladilen N Razumovsky
Deputy General Director: Nikolay P Kezin
Senior Manager, Foreign Relations: Evgueny A Allyamov
■ Members include aircraft engine designers, manufacturers and users. [97]

74 Consortia, Joint Venture Companies

74-01200 Consortium Aviaspectrans JSC (AST)

Zhukovsky 5, PO Box 230, 140160 Moscow Region
☎(+7 095) 556 56 93; 556 59 93
⊖ (+7 095) 292 65 11 (box 1105) AST
♧ US$ 25,445,000 🏭 832 m2
Board Chairman and General Director: Valentin A Korchagin ☎(+7 095) 556 59 93
Sales and Marketing Director: Svetlana S Baranova
☎(+7 095) 556 56 93
Operations Director: Anatoliy M Kotelnikov
☎(+7 095) 556 56 34
■ Light 18-seat amphibious aircraft "Yamal" intended for operation in unfavourable conditions/undeveloped regions. Six-seat amphibious aircraft, equipment for monitoring, mobile navigation aids. 8-seat deluxe cargo, ambulance and ice reconnaissance. [97]

74-08000 Tula Small Arms Plant Production Association

PO "Tulskiy oruzheynyy zavod"
300002 Tula
☎(+7 0872) 27 05 81; 27 05 92; 27 05 92 27 82 59; 36 41 62 ⊠ 253112 upa
⊖ (+7 0872) 27 34 39
Director: Nikolay D Maslennikov ☎(+7 0872) 27 82 59
Commercial Director: Aleksandr T Vinogradov
☎(+7 0872) 27 05 92
Operations Director: V N Babakhin
Purchasing Manager: Y P Merkulov
■ Manufacture AKC-74Y submachine guns, underbarrel grenade launchers, sniper rifles BCC and special machine carbine AC self loading pistols. [98]

75 Manufacturing Companies

Enics Research Center 60-01200
Ilyushin Design Bureau 55-09000
State Research Institute 60-04150
Zvezda JSC 60-08350
Zvezda-Strela State Research & . . . 60-08400

75-00230 AlliedSignal Commercial Avionics Systems American Russian Integrated Avionics/ARIA

18 Tupoleva Street, Zhukovsky, Moscow
Director General: Eduard P Alexeev
Deputy Director General: Raymond E Hillman
■ Administrative and manufacturing facility for manufacturing avionics for CIS-built aircraft. [98]

75-01040 AOOT OKBM

Experimental Design Bureau for Motor-Building (OKBM)
Voroshilov St 22, 394086 Voronezh
☎(+7 0732) 34 82 96; 77 11 94; 34 83 60
⊠ 412535 motor ⊖ (+7 0732) 33 41 37
👥 960 (1996) 🏭 32 206 sq m (1996)
General Dir: Anatoly Georgievich Bakanov
Deputy General Dir & Sales Mgr: Yuri Borisovich Yurovsky
Operations Dir: Anatoly I. Antonenkov
Sales & Marketing Dir: Valery A. Syssoyev
Purchasing Mgr: Viktor D. Sherstyukov
Chief Information Dept: Stanislav N. Brostovsky
■ Development of medium & low powered engines for aircraft, helicopters & other flight vehicles. [97]

75-01150 Avia Ltd Scientific Industrial Union

Nauchno Proizvodstvennoye Obédinenie Avia Ltd
Kominterna Straat, House 13/4, 129327 Moscow
☎(+7 095) 184 43 77 ⊖ (+7 095) 184 43 77
👥 28 🏭 3,300 sq m
Director General: A Loshkarev
Deputy General Director, Test Flights and Certificate: Mikhail M Pachin
Purchasing Manager: Alexey Stepanov
■ Manufacture Accord multi-purpose aircraft. [98]

75-01200 Aviapribor Corp

5 Aviatzionny Pereulok, 125319 Moscow
☎(+7 095) 152 48 74; 155 34 83
✆ (+7 095) 152 50 21 Ⓔ aomiea@dol.ru
⊖ (+7 095) 152 26 31
♧ Rb 4,727,000,000 ♣ Rb 1,120,000,000,000
👥 16,800 🏭 311,400 m2
President and General Director: Sergey P Krioukov
Sales and Marketing Director: Evgeny L Golosov
Chief Designer: Z Aboutidze
Public Relations Manager: Alexander V Davydov
■ Integrated computer flight-navigation systems for long-range aircraft and helicopters, such as FMS; analogues of FMCS, FCCS, FACS, TCCS and FBW; inertial platforms; strapdown inertial systems; stellar-inertial systems; gyro horizons; vertical gyroscopes; primary data sensors gyrocopes, angular-rate sensors, linear-acceleration transducers, software packages for various applications, air data systems; angle of attack, slip, altitude, velocity Mach indicators; electronic display systems (EFIS, EICAS); warning system annunciators.
Aviapribor Corp Companies:
Moscow Institute of Electromechanics and Automation
Arzamas Design Bureau "Temp"
Ulilanovsk Microinstrument Plant "Signal"
Saratov Electroinstrument
Saratov Industrial Automation Design Bureau
First Moscow Instrument
Micromachina
Ural Instrument Plant
Ulianovsk Instrument Design Bureau
Utios [98]

75-01545 Arzamas Design Bureau "Temp-Avia"

(a part of the Aviapribor Corp)
26 Kirov Straat, Arzamas
☎(+7 83147) 404 35
General Director: Victor Leschiev
■ Manufacture gyrostabilisers, gyro horizons, auto pilots, inertial systems, accerlerometers, free gyros, and angular rate sensors. [98]

75-03150 Design Bureau of General Machine Building (KBOM)

(a subs. of the Russian Space Agency)
Bereshkovskaya naberezhnaya 22, 121059 Moscow

☎ (+7 095) 246 60 44 ⊠ 111515 ⊖ (+7 095) 240 41 96
Chief & General Designer: Prof.Dr Igor V. Barmin
First Deputy Chief & General Designer: Dr Vladimir G. Eliseev
Marketing Dir: Vladimir G. Tarnavski
■ Development, modification & operation of space launch complexes, fuelling stations & material space processing equipment. Main areas of operation: development of technological schemes for space complexes covering the full cycle of launch vehicle operations, maintenance, fuelling & launch; design documentation for launch complex systems, equipment & facilities; author's supervision of manufacture, assembly & test of all site systems & equipment; technical guidance concerning launch complex exploitation; preparation & operation of experiments & test result analysis.
Addition development work includes: development of liquid natural gas transportation & production complexes; development of regional agricultural raw material, product processing & storage complexes.
Operations include participation in the following project & programs:
Russian programs: Vostok, Soyuz, Salyut, Mir, Progress, Cosmos, Proton, Luna, Lunochod, Energiya-Buran, Mars, Venera, Molniya, Ekran, Horizont.
International programs: Soyuz-Apollo, Vega, Phobos, Inmarsat-3, Aster-1E, Tempo, Iridium.
Advanced programs: ISS & Mars-96. → 80 [97]

75-03200 Design Bureau of Industrial Automatics
(† Saratov Industrial Automation Design Bureau) (part of the Aviapribor Corp.)
239 B Sadovaya Straat, 410005 Saratov
☎ (+7 8452) 91 82 61 ⊖ (+7 8452) 24 60 55
General Director: Vladimir Chernishkin
■ Automatic flight control systems for helicopters and general aviation aircraft; automatic control systems for hydrofoil aircraft. [98]

75-05200 Electronintorg Ltd
Vladimir A Sokolov Ul, Usievicha 24/2, 125315 Moscow
☎ (+7 095) 155 40 26; 155 49 12; 155 40 38; 155 40 33 ⊠ 411326 ezp su
⊖ (+7 095) 151 54 41
♣ US$ 15,000,000 (1996) 👥 90 (1996)
General Director: Vladimir A Sokolov
Marketing Director: Alexander Kourliandski
☎ (+7 095) 151 16 39
■ Electronic components, materials and equipment for the electronic industry. [98]

75-05500 PPO Elektromekhanika
Gogol Str. 51/53, 440052 Penza
☎ (+7 8412) 33 23 21 Ø (+7 8142) 33 95 31
⊠ 155211 micron ⊖ (+7 8412) 33 21 29
♧ US$ 5 500 000 (1996) ♣ US$ 6 000 000 (1996)
👥 1 784 (1996) ⌂ 64 237 sq m (1996)
General Dir: Nikolai Ovchinnikov
■ Electromechanical & data processing equipment for space programmes. [97]

75-09090 Rybinsk Engine-Building Design Bureau
(† Rybinsk Engine-Building)
163 Lenin Avenue, 152903 Rybinsk, Yaroslavskaya
☎ (+7 0855) 24 38 27 Ø (0855) 24 31 00
⊠ 412559 topaz ⊖ (+7 0855) 26 23 98
♣ Rb 1,000,000,000,000 👥 23,000
General Director: Ms Valery I Shelgunov
Technical Director, General Designer: Aleander S Novikov
Director of Sourcing, Economics and Finance: Yuri V Lastochkin
Purchasing Manager: Victor A Polyakov
■ Development, manufacture and repair work for various aircraft gas-turbine engines. Development and production of power units based on aircraft, ships, and diesel engines. Production of spare parts for aircraft engines, diesel engines and snowmobiles. [98]

75-10200 Kamov JSC
(† Kamov Helicopter Scientific & Technology Company)
8-go Marta Street, 140007 Lubertsy, Moscow Region
☎ (+7 095) 171 37 43; 700 32 04 ⊠ 206112 kamov
⊖ (+7 095) 700 30 71
👥 3,700
General Designer: Prof Dr Sergei V Mikheyev
First Vice President and Deputy General Designer: Veniamin Kassjanikov
Chief Designer: Juri Sokovikov
Chief Designer: Krygin Vyacheslav
Chief Designer: Evgeny Pak
Spokesman: Yevgeny Sudarev
■ Design and manufacture of KA-50, KA-32, KA-226 and KA-62 helicopters for use in passenger and cargo transport, agriculture and high accuracy flying crane. Development and manufacture of military helicopters for air forces and navies, special purpose and remotely piloted models. [98]

75-11000 Kazan Helicopters JSC
420085 Kazan
☎ (+7 8432) 54 45 52 ⊠ 224848 agat su
⊖ (+7 8432) 54 52 52
General Director: Aleksander P Lavrentiev
☎ (+7 8432) 54 54 73
Marketing Director: M A Tikhonov ☎ (+7 8432) 54 45 52
■ Design and manufacture of Mi-8/Mi-17 family of multi-purpose helicopters.

Moscow Office
Moscow
☎ (+7 095) 207 51 46 ⊠ 812274 wex su
⊖ (+7 095) 207 03 24
■ Design and manufacture of Mi-17 and Mi-8 multi purpose helicopters. [98]

75-11350 Kirasa Co. Ltd
4 Chelyuskintsev Street, 614094 Perm
☎ (+7 3422) 90 39 85; 33 20 37; 90 42 87
⊖ (+7 3422) 34 63 03
♣ US$ 20,000,000 👥 550 ⌂ 10,000 m2
President: Serguei Kashin ☎ (+7 3422) 36 25 26
General Director: Vladimir Koromoushin
☎ (+7 3422) 54 54 73; 49 69 55
Commercial Director: Youri Choubaev
☎ (17 3422) 90 39 86
Technical Director: Albert Kalgin
Overseas Trade Dept: Serguei Nikoulin
☎ (+7 3422) 90 42 87
Advertising Manager: Youri Gribanov
☎ (+7 3422) 90 39 89
■ Design and production of security products. Harpoon vehicle immobilisation unit for vehicle check points, storage areas, drug enforcement operations, anti-terrorist operations, aircraft aprons/airfield security, secure vehicle parking and commercial premises.
Waterclear filtering device for refining water from contaminated sources.
Anti-stab armoured vests for police and prison correction departments, custom made anti-ammunition "Execuvest", Kirasa-2V, Kirasa 3M/h undercover armoured vests, Universal SN 5-2-1 concealable armoured vest from kevlar, terlon, twaron, dyneema, titanium, steel and composite material construction. [98]

75-11370 Kirovo Chepetsky Electro Machine-Building Plant
1b Lenina Street, 613020 Kirovo-Chepetsk
☎ (+7 83361) 381 89; 363 15 Ø (+7 83361) 108 85
⊖ (+7 83361) 135 72
♧ Rb 408,610,000,000 ♣ Rb 17,000,000,000
👥 1,200 (1997) ⌂ 295,400 sq m
Managing Director: Vladimir Shestakov
Sales Director: Viktor Podlevskih ☎ (+7 833 61) 357 06
Technical Director: Vladimir Karpikov
Finance Director: Nikolai Markov ☎ (+7 833 61) 133 24
Purchasing Manager: Sergew Patapov
☎ (+7 83361) 364 75
Marketing Manager: Gennady Moroyovsky
■ Manufacture aircraft equipment including relays, contacts and microcircuits (IC's). [98]

75-11400 Klimov Design Bureau
1 Kantemirovskaya Straat, 184100 St Petersburg
☎ (+7 812) 247 78 61; 245 15 86; 245 80 46; 542 89 75; 542 76 30 ⊠ 121282 jet su
⊖ (+7 812) 245 43 29; 245 33 55; 245 00 65
General Designer: Alexander Sarkisov
Chief Designer: Piotr D. Gavra
Dpty General Mgr: Andrei P. Listratov
■ Manufacture of turbine and turboshaft engines for fixed and rotary wing aircraft including RD-33 for MiG-29 and engines for IL-114 and TV-117. Development of helicopter main gear boxes. [98]

75-11850 Kurgan Machine Building Plant Production Association
Kurganmashzavod JSC
17 Mashinostroitelei Avenue, 640631 Kurgan
☎ (+7 35222) 322 44; 748 33 ⊠ 420263 grad
⊖ (+7 35222) 339 96; 314 97
General Director: Valeriy D Dorodny
Chief Engineer: Dmitry V Knyazev
Operations Director: Anatoly D Zhilyakov
Marketing Director: Sergi V Bocharov
Purchasing Manager: Galina G Ivanova
■ Manufacture BMP-2, BMP-3 IFV's, armoured recovery vehicle BREM-L, repair and modernization of BMP-1, BMP-2. BHP-3, and manufacture of SPTA. [98]

75-12000 Lyulka-Saturn Inc
13 Kasatkin Straat, 129301 Moscow
☎ (+7 095) 283 09 13; 283 94 93; 283 13 74
Ø (+7 095) 283 95 55
⊖ (+7 095) 283 28 63; 286 75 66
👥 2,500
President and Chief Executive Officer: Victor M Chepkin
Marketing and Corporate Communications Director: Victor Nesterov
Managing Director: Valery Lebedev
Sales and Marketing Director: V Boriskov
Purchasing Manager: Ivan Trunov
Advertising Manager: V Plotnikov
■ Military turbojet/turbofan engines, power generation and gas pumping. [98]

75-13250 Mikron
103460 Moscow k-460, Zelenograd
☎ (+7 095) 535 15 09 ⊖ (+7 095) 535 62 64
General Director: Gennady Krasnikov
■ Design, produce intergrated circuits for space apparatus, supercomputers and multi-coodinative control systems. [98]

75-13300 MIL Moscow Helicopter Plant
2 Sokolnichesky Val, 107113 Moscow
☎ (+7 095) 264 90 83; 269 92 19; 264 47 62
⊠ 412144 mil su ⊖ (+7 095) 264 55 71; 264 47 62
👥 3,500
General Designer and General Manager: George A Sinelschikov ☎ (+7 095) 264 90 83
Head of Division: Oleg P Bakhov
■ Design and manufacture of military and civil helicopters. [98]

75-13380 NPO Molniya plc
(† Molniya Scientific & Industrial Interprise)
6 Novoposelkovaya Straat, 123459 Moscow
☎ (+7 095) 493 50 53; 497 47 60
E alex@molniya.msk.ru
⊖ (+7 095) 493 50 53; 492 93 71; 497 47 23
👥 2,500 (1997) ⌂ 100,000 sq m
General Director: Alexander S Bashilov
☎ (+7 095) 493 50 53
Deputy General Director: Alexander B Dolgov
☎ (+7 095) 492 93 61
General Designer: Gleb E Lozino-Lozinskiy
☎ (+7 095) 492 95 55
Marketing Dir: Michael J Gofin ☎ (+7 095) 497 47 60; 497 47 23 (fax)
■ Design and manufacture of: reusable aerospace systems; civil aircraft (6-seater to super heavy); tower type multistorey autoparking. → 60 [98]

75-13400 Moscow Institute for Electromechanics & Automatics
(a partner of Aviapribor Corp.)
5 Aviatsioniy Pereulok, 125319 Moscow
☎ (+7 095) 152 48 74 E comiea@adol.ru
⊠ 113826 anap ⊖ (+7 095) 152 26 31
Director General: Sergey P Krjukov
General Deputy Director: Vyacheslav V Kazakov
■ Development, production, integration and aftersales support of flight control, fly-by-wire and flight management systems, as well as inertial navigation and reference systems for civil aircraft and helicopters. [98]

75-13750 Nakhodka Ship Repair Yard
59 Nakhadka Avenue, Nakhodka 13, 692900 Nakhodka
☎ (+7 42366) 220 37
⊖ (+7 42366) 472 58; (+7 50491) 523 18
👥 1,700
Managing Director: Evgeny Chizh

Contact Reference

Operations Director: Evgeny Klimov
Manager of Marketing Department: Sergey Sazonov
■ Ship repair. Facilities include floating dock. [98]

75-14100 NII Stali
81a Dubninskaya Straat, 127411 Moscow
☎(+7 095) 485 35 10; 485 70 70
⊖ (+7 095) 485 43 95
👥 818 (1997) 🏭 46,600 sq m
Managing Director: Dmitry Rototaev
☎(+7 095) 484 63 61
Deputy Director: Valery Grigorjan ☎(+7 095) 484 62 11
Sales Manager: Nikolai Zhuravlev
■ Design and manufacture of body armour, protective clothing and helmets. [98]

75-14200 NOVO Corporation
6 Dubki Straat, PO Box 34, 127434 Moscow
☎(+7 095) 210 27 60; 210 27 70; 210 27 80; 210 27 90 ⊖ (+7 095) 210 27 71
♧ US$ 800,000 (1995) ♣ US$ 3,500,000 (1995)
👥 50 (1995) 🏭 500 m2
President: Vladimir Philippovsky
Vice President: Serguei M Sorkine
Operations Director: Nicholai B Touloupov
Sales Director: Andrei S Piatakov
■ Night vision systems, surveillance and security equipment. Bug detectors, scramblers and audio surveillance. Bomb, metal and x-ray detection. Bullet-proof clothing, cars and surveillance vehicles. other Police-Gendarmerie equipment. Radio communication, personal protection and Police-Gendarmerie equipment. [97]

75-14300 NPO Energomash
Burdenko Straat 1, 141400 Khimky, Moscow Region
☎(+7 095) 572 22 59; 572 22 00 ☒ 911649 almaz su
⊖ (+7 095) 251 75 04; 573 12 01
👥 7,850 🏭 282,000 m2
General Dir & General Designer: Boris I. Katorgin
First Deputy General Dir & General Designer: Vladimir K. Chvanov
First Deputy General Dir & General Designer: Gennady G. Derkach
Deputy General Dir Foreign Econ.: Victor A. Sigaev
Chief of Information Dept: Vladimir S. Sudakov
☎(+7 095) 572 76 49
■ Design, develop & manufacture powerful liquid propellant rocket engines including RD-170 liquid rocket engines for Zenit & Energia launch vehicles, RD-107 & 108 liquid rocket engines for SOYUZ launch vehicle & RD-253 liquid rocket engine for Proton launch vehicle. [97]

75-14315 NPO Geofizika
SKB GeoKos Special Design Bureau (Scientific and Production Association)
11/17 Irkutskaya Straat, 107497 Moscow
☎(+7 095) 462 03 43
⊖ (+7 095) 462 15 06; 462 06 84
Director General: Aleksey A Fomin
Chief Designer and Deputy Director: Vladimir S Kouzmin
Deputy Chief Designer, International: Alexandre P Dmitriev
☎(+7 095) 462 21 68
■ Design and manufacture of diversified opto-electronics including: automatic opto-electronic instruments for attitude measurement using the Sun, stars, planets; optical visualisation instruments for use by astronauts during manual manoeuvring in daylight and night conditions; instruments for studying the Earth's natural resources and atmosphere surrounding space vehicles (radiometers, photometers, spectroradiometers, spectrophotometers, actiometers, lightmeters and laser measuring devices); ground-based instruments for orientating heliostats towards the Sun; Star, Sun and Earth simulators, radiometers for star simulator calibration, simulators for visualising of external effects during training of astronauts; high precision optical angle-code encoders.
Test facilities include: precise opto-mechanical stands equipped with Sun, star and planet simulators; test centre for vibration, shock, vacuum, acoustic and climatic environment; graduation complex for calibration of spectrometric instrumentation. Photometers, spectrometers, spectrophotometers, actiometers, and lightmeters. [97]

75-14600 NPO Prikladnoi Mechaniki (NPO PM)
Scientific Production Association of Applied Mechanics
660026 Krasnoyarsk 26
☎(+7 39197) 320 32 ⊖ (+7 39197) 226 35
General Designer: Albert G. Kozlov
First Deputy General Designer & General Dir: Valery P. Ganzenko
Head Patents & Information: Roman P. Turkenich
☎(+7 39197) 346 17
■ Design of communication systems, TV-broadcasting, navigation & geodetic satellites. Manufacture & operation of satellites. [97]

75-14700 Polyot Research and Production Co
1 Komosomolskaya Square, GPS-461, 603600 Nizhny Novgorod
☎(+7 8312) 44 24 05; 42 21 04; 42 15 91; 44 21 86; 42 21 86 ☒ 151120 rtb su; 151176 prima
⊖ (+7 8312) 35 64 80; 44 24 05
♣ Rb 6,900,000,000 👥 3,600 🏭 72,500 m2
General Director: Evgenily L Belousov
Deputy General Director: Stanislav T Burtsev
Sales and Marketing Director: Sergey Ya Shishov
Advertising Manager: Pavel G Skachkov
■ Manufacture airborne radiocommunication equipment and avionics systems for civil and military aircraft. Air-to-air and air-to-ground automated communication systems and networks, airborne and ground-based communication complexes for data exchange in a wide frequency range, airborne radios, transmitters and receivers, airborne communication equipment: modems, codecs, computers, control panels, antenna and feeder devices and fibre optic components. [98]

75-15950 Perm Instrument-Building JSC
106, 25th of October Street, GSP-590, 614600 Perm
☎(+7 3422) 45 23 36; 45 35 44 Ⓔ onr@ppk.perm.su
☒ 134840 rotor su ⊖ (+7 3422) 45 12 19; 45 42 64
General Director: A G Andreev ☎(+7 3422) 45 23 36
Operations Director: V Ermkov
Marketing Director: Y N Naugolnykh
☎(+7 3422) 45 45 59
■ Manufacturer of precise inertial navigation systems for aviation and gyrocompasses for sea and river vessels. Manufacturer of DTG, accelerometers, torquers, slip rings, low-voltage micromotors, electronic units. [98]

75-15975 Perm Kirov Research Production Corp
11 Galperina Street, 614113 Perm
☎(+7 3422) 54 84 97 ⌀ (+7 3422) 54 83 21
☒ 134123 kama ⊖ (+7 3422) 55 77 43
♣ US$ 21,600,000 👥 8,000 (1997)
Managing Director: Kusmitskiy Gennadiy
Commercial Director: Romanovich Alexey
Head of Foreign Economical Department of Perm Kirov Plant: Devjatericov Michael
■ Production of industrial explosives, gunpowder for recreational uses, precision-dosage instruments for loose and binding explosives, polyurethanes, varnishes, paints, acrylamide, microcrystalline cellulose, nitrocellulose, colloxylin based granulated compound material. [98]

75-16000 Perm Motors JSC
Permskie Motory JSC
93 Komsomolskiy Prospekt, GSP-621, 614600 Perm
☎(+7 3422) 64 08 61 ☒ 134812 don su
⊖ (+7 3422) 45 13 20; 45 00 21
Dir General: Vladimir Victorovich Kinderkneht
☎(+7 3422) 45 05 24
Marketing, Sales & Commerce Dir: Vladimir Ivanovich Goryainov ☎(+7 3422) 45 11 87
■ Production of the high bypass ratio, turbofan PS-90A engine for passenger & transport aircraft. [97]

75-16200 Pirometer
Bolshaya Monetnaya 16, 197061 St Petersburg
☎(+7 812) 238 72 45 ⌀ (+7 812) 233 74 59 ✉ ARTEK pirometr@sovam.com ⊖ (+7 812) 233 83 06
♧ US$ 33,000,000 (1996) ♣ US$ 16,000,000 (1996)
👥 1,500 🏭 80,000 sq m
President: Sergey D Bodrounov ☎(+7 812) 238 16 40
General Director: Alexander Zakharov
☎(+7 812) 233 74 59
Deputy General Director for Marketing: Anatoly A Samorodov
Sales Director: Alexander Girnov
Operations Director: Sergey Lebedev
☎(+7 812) 233 74 59
Advertising Director: Remezova Irina
☎(+7 812) 238 72 20
■ Manufacture Automatic Control Systems for civil TU-154, Il-62 and military aircraft. Autopilots for pilotless aircraft. Also manufacture test equipment for ground control of the above systems. [98]

75-16400 Popov Communications Equipment Plant JSC
Popov Plant Scientific Production Association Arendnoe NPO Zavod Imeni Popova (ZIP ANPO)
(† Gzas Imeni Popova JSC)
100 Ulitsa Internatinalnaya, 603600 Nizhniy Novgorod 2
☎(+7 8312) 44 41 06; 42 03 59
☒ 151113 ars su; 151145 nebo ⊖ (+7 8312) 42 61 74
♧ Rb 43,002,668,000 (1994)
♣ Rb 11,383,261,000 (1994) 👥 5,000 🏭 100,000 m2
General Manager: Vladimir S Fedorov
Deputy Director: Alexandr Gulyayev
Deputy General Manager Foreign Economic Relations: Gennadiy F Koshelev
■ Ground and airborne aviation radio communication equipment and complexes. Radio centres, MF and HF radio stations for various applications. [98]

75-21050 Red Wings Scientific JSC
Svoboda Straat 100, 347930 Taganrog
☎(+7 86344) 337 03; 444 40 ⌀ (+7 86344) 337 03
⊖ (+7 86344) 440 40
👥 180 (1997) 🏭 900 m2 (1997)
Director General: Oleg N Komarov
Technical Director: Anatoly B Vodolazky
Chief of Production: Alexei A Bondorev
Marketing Director: Vladimir Larin
Operations Director: Anatoly B Vodolzsky
■ Production of ultra-light aircraft. [98]

75-21150 Research Institute of Gas Discharge Devices Plasma
(† Plasma Scientific & Production Corp.)
24 Tsyolkovsky Street, 390023 Ryazan
☎(+7 0912) 44 90 02; 44 90 62 ⊖ (+7 095) 973 00 90
👥 1,000 (1996)
Managing Director: Vladislav 1Samorodov
☎(+7 0912) 44 90 02
Sales and Marketing Director: Vladimir I Aga
☎(+7 0912) 44 90 62
■ Development and manufacture of the following types of products: gas lasers, including atomic lasers (He-Ne), ion lasers (Ar, Kr, Ar + Kr, Ne, He Cd), molecular lasers (CO2,N) and devices using lasers. Plasma display panels and collective usage screens based on them; gas discharge switching devices, including thyratrons, tacitrons and spark gaps. [98]

75-25000 Rockwell International Overseas Corporation
Representative Office
Bolshoi Strochenovsky Pereulok, No 22/25, 4th Floor, Suite 403, 113054 Moscow
☎(+7 095) 230 60 10 ⊖ (+7 095) 230 60 15
Corporate Representative: Gennady Alekseyevich Semchenkov
■ Avionics and marine inertial navigation and telecommunications equipment and systems; industrial automation products, systems and services; semiconductor products; automotive components. → 80
® USA: Rockwell International Corp. [98]

75-30000 Salyut Design Bureau (Salyut DB)
(Affiliated Branch of Khrunichev's State)
18 Novozavodskaya Straat, 121087 Moscow
☎(+7 095) 145 98 84 ⊖ (+7 095) 142 59 21
General Designer: Anatoli K Nedaivoda
Deputy General Designer: Viacheslav N Ivanov
Deputy Director: Nikolai E Kulashkin

■ Design and development of heavy and light launch vehicles. Design and development of 20-t class spacecraft; space stations Salyut and Mir; space vehicles Kosmos-929, 1267, 1443, 1686 series. Kvant, Kvant-2, Kristall, Spectr, Priroda modules. Design and development of upper stages. Design and development of re-entry capsules of various weights. State Programs: Luna, Mars, Venera, Kosmos. International Programs: Vega, Fobes. Orbital space stations: Salyut, Mir. Space vehicles: Astron, Granat, Zond, Gorizont, Raduga, Ekran, Glonass. Modules for space stations: Kvant, Kvant-2, Krystall, Spectr, Priroda.

Rocket Space Plant
18 Novozavodskaya, 121309 Moscow
☎ (+7 095) 145 70 38 ⊠ 412427 su
⊖ (+7 095) 142 59 00
Director General: Anatoli Kalinin ☎ (+7 095) 145 92 98
Director, Engineering: Yuri P Gorodnichev
☎ (+7 095) 145 80 76
Director, Marketing: Eduard V Nikoldev
☎ (+7 095) 145 88 28 [98]

75-31075 Saratov Electric Instrument Manufacturing Plant after Sergo Ordzhonikidze

(a part of the Aviapribor Corp)
239 Bolshaya Sadovaya Straat, 410005 Saratov
☎ (+7 8452) 24 21 32
General Director: Adolf Kazakov
■ Manufature flight and navigation integrated systems for transport aircraft and auto pilots. [98]

75-31100 Saviat Special Aviation Technology

Volokolamskoye Chausse 73, Suite 327, 123424 Moscow
☎ (+7 095) 490 07 72 ⊖ (+7 095) 490 06 15
👥 20
Director: V B Pugachov
President: Vladimiz S Egez
Chief of Marketing Department:
Vladimir Sergeevich Savinich
■ Aircraft design and production including E-1 light multi-purpose aircraft. Design and production of aviation engines from new materials. [98]

75-32800 Smolensk Aircraft Plant

(† Smolensk Aviation Plant)
Frunze St 74, 214006 Smolensk
☎ (+7 08122) 293 13 ⊖ (+7 08122) 293 86; 205 29
Director General: Michail Abramichev
Technical Director: Alexnder Miroshkin
Finance Director: Vitaliy Lomovskih
Purchasing Manager: Vladimir Bavchenkov
■ Production of YAK-18T; SM-92 Finist, M-55 Geophisica; YAK-40; YAK-42 and YAK-112. [98]

75-33050 Soyuz Aviamotors Scientific Technical Complex

Luzhnetskaya Naberezhnaya 2/4, 119270 Moscow
☎ (+7 095) 242 28 62; 242 94 68
⊖ (+7 095) 242 57 02; 242 29 34
🏠 132,000 m2
Director General and General Designer:
Vasiliy K Kobchenko
First General Designer: Evgeniy A Fomin
Marketing Director: Vladislav Belousov
Deputy General Designer: Alexander F Zhirnov
☎ (+7 095) 242 94 44; 242 57 02 (fax)
Operations Director: Mixail N Bychkov
■ Design and production of gas-turbine engines and associated technologies. Thermodynamic and gas-dynamic research. Development of software for SARP, piloting and automatic control programs. [98]

75-33500 Technoavia

Kronshtadsky Bulvar 7a, 125212 Moscow
☎ (+7 095) 452 58 22 ⊘ (+7 095) 452 56 03
⊖ (+7 095) 452 56 94
👥 30 (1997) 🏠 500 m2
President: Viatcheslav Kondratiev
Managing Director: G Makoagonov
Marketing Director: A Konstantiniai
Operations Director: G Egorova
■ Light aircraft manufacture. [98]

75-50000 Ulianovsk Instrument Manufacturing Design Bureau

(part of the Aviapribor Corp.)
10-a Krymov Street, 432021 Ulyanovsk
☎ (+7 842) 231 42 69; 231 42 65 ⊠ 263808
⊖ (+7 842) 231 89 15
♧ Rb 1,720,703,000 ♣ Rb 2,123,439,000 👥 1,178
🏠 9,650 m2
General Director: Nikolay Makarov
■ Electronic Indication System EFIS, EICAS Integrated Electronic Indication and Warning System EIS, Analog/Digital Data Conversion System SPADI, Air Data Computers, Airspeed/Altitude Indicators, Pilot Probes. [98]

75-51000 Ulyanovsk Mechanical Plant

Ulyanovsk Mechanichesky Zavod
94 Moskovskoe Road, 432008 Ulyanovsk
☎ (+7 8422) 31 75 58 ⊘ (+7 8422) 32 61 63
⊖ (+7 8422) 32 61 63
Managing Director: Vyacheslav Abanin
Deputy Director Interational: Igor A Besnosenko
■ Manufacture radar systems and the GANG anti-aircraft missile systems, designed for engagement at high speed and extremely low altitudes. [98]

75-51200 Ulianovsk Plant "Signal"

(a part of the Aviapribor Corp)
36 Prospect Sozidateley, 432064 Ulianovsk
☎ (+7 8422) 20 17 91
General Director: Yuri Babchenko
■ Manufacture parts and assemblies for inertial systems. [98]

75-52000 Urals Instrument Plant

(part of the Aviapribor Corp.)
17 Gorkogo Straat, 620151 Yekaterinburg
☎ (+7 343) 51 17 67; 51 90 94 ⊖ (+7 343) 51 40 11
Director: Vladimir Godlevsky
■ Produces hydroscopic tools such as altitude indicators and hydraulic components. [98]

75-53000 Urals Optical & Mechanical Plant

33b Vostochnaya Street, 620100 Yekaterinburg
☎ (+7 3432) 24 18 03 [E] trank@sin.global-one.ru
⊠ 721764 hvoia su ⊖ (+7 3432) 24 18 44
General Director: Eduard S Yalamov
☎ (+7 3432) 24 17 01
Deputy General Director: Veniamin S Elinson
☎ (+7 3432) 24 18 63
■ Production of optomechanical and telescopic sight systems; geodetic instruments; medical equipment. [98]

75-78000 A S Yakovlev Design Bureau

68 Leningradsky Prospekt, 125315 Moscow
☎ (+7 095) 157 17 34 ⊠ 417575 plaz
⊖ (+7 095) 157 47 26
Chairman and General Designer: Alexander N Dondukov
☎ (+7 095) 157 57 37
President and General Director: Oleg F Demchenko
First Vice President and General Designer:
Vladimir G Dmitriev
Vice President and Deputy General Director External Economic Relations: Arkady I Gurtovoy
Deputy General Director Flignt Test: Andrei A Sinitsin
Vice President and Chief Designer Yak-242, Yak-46:
Vladimir G Dmitriev
Vice President and Chief Designer Yak-130:
Nicolas N Dolzhenkov
Vice President and Chief Designer Yak-148/IAI Galaxy:
Vladimir A Mitkin
Vice President and Chief Designer Yak-42, Yak-142:
Alexei G Rakhimbaev
Chief Designer Yak-3, Yak-40TL, Yak-77:
Sergei A Yakovlev
Chief Designer Yak-54, Yak-55M, Yak-112:
Dmitry K Drach
Chief Designer Yak-18T, Yak-52, Yak-58, UAV:
Yuri I Yankevich
Chief Marketing and Sales Department: Anatoly S Ivanov
Chief Exterior Department: Evgeny M Tarasov
Chief Information and Press Service Department:
Yuri V Zasypkin
Head Information & Press Dept: Yuri V. Zasypkin

■ Design/manufacture of commercial aircraft (Yak-58, Yak-112, Yak-42/42D with Yak-142, Yak-242, Yak-46 under development), piston training aircraft (Yak-52, Yak-54 with Yak- 56 under development) jet training aircraft (UTS/Yak 130), aerobatic aircraft (Yak 55M) and SHMEL (RPV). Bizjets under development: Yak-48, Yak-77. [98]

75-78900 Yantar Shipyard

10 Transportnyi Tupick, 236002 Kaliningrad
☎ (+7 0112) 44 10 83 ⊠ 262158 ⊖ (+7 0112) 44 84 33
👥 5 000 (1996) 🏠 600 000 sq m
Managing Dir: Valery Tsyploukhin
Technical Dir: Boris Krout
Marketing Mgr: Andrey Lysenko
Purchasing Mgr: Nikolay Lysenko
■ Ship construction & repair. [97]

80 Service Companies & Organizations

Aviaexport PLC	40-01000
Design Bureau of	75-03150
Rockwell International Overseas . . .	75-25000

80-00180 JSC Aviadvigatel

(† Motorostroitelnoye Konstruktorskoye Buro)
93 Komsomolsky Prospect, GSP-624, 614600 Perm
☎ (+7 83422) 45 20 19 ✉ PERM 10 LAVA
⊠ 134802 lava su ⊖ (+7 83422) 45 97 77
👥 3 500
General Designer & General Dir: Mikhail L. Kuzmenko
☎ (+7 83422) 45 20 19
Marketing Dir: A.P. Malutin
Head of Foreign Relations Dept: Edward G. Tsyrlin
☎ (+7 83422) 45 81 41
■ Development of a series of engines in the 12-20 ton thrust range, based on the PS-90A engine. Turbojet engines for aviation, industrial & marine applications. Advanced propfans.
Engine production is undertaken by JSC Permskie Motory. [97]

80-00300 Diagnostika JSC

(† Diagnostica Ltd)
Aviamotornaya Straat 2, 111250 Moscow
☎ (+7 095) 362 22 74 [E] nascru@dol.ru
⊖ (+7 095) 362 22 74
♣ US$ 2,000,000 (1996) 👥 18 (1996) 🏠 150 m2
Managing Director: Victor N Kravlsov
■ Computer consultancy services. Software and algorithms for ground and onboard system diagnostics, and prognoses for aviation engines, energetics and turbines. [98]

80-00500 Glavcosmos

9 Ulitsa Krasnoproletarskaya, 103030 Moscow
☎ (+7 095) 972 43 69 ⊠ 411879 ⊖ (+7 095) 288 95 83
Chairman: Alexander Ivanovich Dunaev
☎ (+7 095) 972 43 69
Senior Deputy Chairman: Oleg Firsjuk
☎ (+7 095) 972 47 14
Research and Development Director: Dmitry Poletayev
Head of Marketing Department: Dmitry Poletaev
☎ (+7 095) 972 33 70
Marketing: Yevgeny Sherstobitov ☎ (+7 095) 972 41 30
Head International Dept: Nikolai Semenov
☎ (+7 095) 972 37 55
Contact: Vladimir Martynov
■ Marketing, sale and lease of satellite services, specialising in communications. [97]

80-01000 The Northern Shipping Company Emergency Division

21 Moseyev Ostrov, 163020 Archangel
☎ (+7 8182) 23 18 74; 23 18 57; 23 18 40 23 18 50
✉ morflot archangel ⊠ 24211 mrf ru
⊖ (+7 8182) 43 83 10
General Manager: Machel Suslin ☎ (+7 8182) 43 39 75
Chief of Salvage Dept: Alexander A Nikolayev
☎ (+7 8182) 43 18 57
Chief of Engineering and Technical Department:
Albert V Nekrasov
Manager of Emergency Division: Alexey G Goureyev

Contact Reference

■ Salvage operations, sea and ocean towage of vessels and floating craft, assistance to various ships in sea and river navigation through ice, underwater and diving operations. [98]

80-02000 Sojuzkarta - Impex
SK-IMPEX JSC
(a subsidiary of Sojuzkarta)
45 Volgogradskiy Prospekt, 109316 Moscow
☎ (+7 095) 177 80 13 ☒ 411942 ren su
⊖ (+7 095) 177 51 11; 177 52 01
♣ US$ 1,500,000 (1995) 👥 21 🏠 550 m2
President: Dr Viatcheslav Piskulin
Vice President: Leonid G Suslonov
Managing Director: Rudolf Kazakov
Technical Director: Matveev Victor
Deputy Commercial Dir: Gorokhov Evgueny
Deputy Director: Ms Marina Rezepova
Purchasing Manager: Denisov Iuri
■ Aerial and space photographic, cartographic and geodetic products.
General mapping and topographic surveys; export of map-making equipment and Russian military topographic maps at scales 1/100 000, 1/200 000, 1/500 000; satellite remote sensing by high resolution camera; digital maps and charts at all scales; export of navy nautical charts. [98]

85 Stockists, Distributors, Spare Parts Suppliers

85-02000 TRACE Moscow
(a subsidiary of TRACE Worldwide, USA)
19/18 Bolshaya Filovskaya Ulitsa,
First Korpus, Entrance 7, Apt 190, 121309 Moscow
☎ (+7 095) 956 47 56; 956 47 57
[E] jbarrie@bigfoot.com ⊖ (+7 095) 142 19 14
👥 8 (1997)
Director: Jeff Barrie
Marketing Director: Boris Bobrovich
Operations Director: Ludmilla Mashanoua
Purchasing Manager: Ms Marina Barmouvshaya [98]

90 Sales Agents, Sales & Liaison Offices

Companies represented:

Akzo Nobel Coatings BV, Netherlands	90-00100
Rockwell International Corp., USA	75-25000
Thomson-CSF, France	90-02040

90-00100 Akzo Nobel NV Representative Office
Ulitsa Smolnaya 24d, Meridian Commercial Tower,
125445 Moscow
☎ (+7 095) 960 28 90 ⊖ (+7 095) 960 28 84
Contact: V G Talanovski
® Netherlands: Akzo Nobel Coatings BV [98]

90-00210 AlliedSignal Aerospace Marketing Sales & Service
19-27, Groholsky Pereulok, 129010 Moscow
☎ (+7 503) 956 12 50 ⊖ (+7 503) 956 12 51
Account Team Leader, Site Manager: Sergei Pritulenko
Field Service Engineer: Oleg Alexandrov
Field Service Engineer: Sergei Toropov
■ Provides on-site customer support for AlliedSignal Aerospace [98]

90-00350 Daewoo Corp Moscow Office
Room 1903-04 (19Fl) 1705 (17Fl),
12 Krasnopresnenskaya Nad, 123610 Moscow
Office B/D Sorvincenter
☎ (+7 095) 258 19 81 ☒ 413764 dwmos su
⊖ (+7 095) 258 19 94 [98]

90-00360 Daewoo Corp Vladivostok Branch
5th Floor Sukhanova Street 3a, 690020 Vladivostok
☎ (+7 3712) 39 15 07 ☒ 116580 dwtas su
⊖ (+7 3712) 45 36 56 [98]

90-02020 Teledyne Russia
Laningradsky Prospekt 55, 125468 Moscow
☎ (+7 095) 943 94 02; 943 24 03
⊖ (+7 095) 943 24 03; 943 94 03 [97]

90-02040 Thomson-CSF Delegation
4/17 Pokrovsky Prospekt, Building 3 KV3,
101000 Moscow
☎ (+7 095) 297 44 24; 208 97 16; 206 97 17
☒ 413287 tcsf su ⊖ (+7 095) 230 23 55
® France: Thomson-CSF [97]

90-02600 United Technologies International, Russia
(a division of United Technologies Corporation, USA)
Petrovsko-Razumovsky, Proezd 19/1, 124125 Moscow
☎ (+7 095) 253 4543; 755 83 88 ⊖ (+7 095) 253 4493
President: Geoffrey Kleb
Marketing Director: Mark Pitts
Purcahsing Manager: Douglas McArthur [98]

Rwanda

Head of State:
Pasteur Bizimungu

10 Head of State

10-10000 Office of the President
PO Box 15, Kigali
☒ 517 presirep rw
President of the Republic: Pasteur Bizimungu
Vice President: Maj Gen Paul Kagame [98]

11 Prime Minister

11-10000 Office of the Prime Minister
PO Box 15, Kigali
Prime Minister: Pierre Celestin Rwigema [98]

12 Ministries

12-10000 Ministry of National Defense
PO Box 23, Kigali
☎ (+250) 753 46 ☒ 517 presirep rw
Minister: Mag Gen Paul Kageame [98]

12-20000 Ministry of Foreign Affairs and Co-operation
PO Box 175, Kigali
☎ (+250) 753 39 ☒ 502 minaffet rw ⊖ (+250) 729 02
Minister: Anastase Gasana [98]

12-40000 Ministry of Interior
PO Box 446, Kigali
☎ (+250) 854 77
Minister: Sheik Adbelkarim Harelimana [98]

31 Embassies & Attachés

31-00660 Embassy of the Federal Republic of Germany
8 Rue de Bugarama, PO Box 355, Kigali
☎ (+250) 752 22, 751 41 ☒ 520 aakgl rw
⊖ (+250) 772 67
Ambassador: H E Mr August Hummel
Deputy Chief of Mission: Dr Thomas Zahneisen
Commercial Counsellor: Thomas Lenk
Cultural Attaché: A Pfarrkircher [98]

31-01890 Embassy of the United Kingdom
Parcelle No 1071, Kimihurura, PO Box BP 576, Kigali
☎ (+250) 840 98; 857 71; 857 73 ⊖ (+250) 820 44
Ambassador: H E Miss Kay Oliver
Deputy Head of Mission: G Scott (Resident in Kampala)
Defence Attaché: Lieut Col N P C Lewis [98]

Saint Lucia

Head of State:
H.M. Queen Elizabeth II

Governor General:
George Mallet

10 Head of State

10-10000 Office of the Governor General
Government Building, The Morne, Castries
☎ (+1 809) 452 24 81
Governor General: George Mallet [98]

11 Prime Minister

11-10000 Office of the Prime Minister
New Government Building, Block C, 5th Floor,
Waterfront, Castries
☎ (+1 809) 452 39 80 ☒ 6243 pmslu lc
⊖ (+1 809) 453 73 52
Prime Minister: Kenny Anthony
Senior Minister: Mario Michel [98]

12 Ministries

12-40000 Ministry of Foreign Affairs and Home Affairs
Government Buildings, Block B, Waterfront, Castries
☎ (+1 809) 452 11 78 ⊖ (+1 809) 452 74 27
Minister for Foreign Affairs and International Trade:
George Odlum [98]

15 Parliament

15-10000 Senate
Castries
☎(+1 809) 452 38 56 [98]

15-20000 House of Assembly
Castries
☎(+1 809) 452 38 56; 453 66 50
⊖ (+1 809) 452 54 51 [98]

31 Embassies & Attachés

31-01890 United Kingdom High Commission
2nd Floor, NIS Building, Watefront, PO Box 227, Castries
☎(+1 758) 452 24 84 ⊖ (+1 758) 453 15 43
High Commissioner: H E Mr R Thomas
Acting High Commissioner: H E Mr M J Grolicott (resident in St Lucia)
Defence Adviser: Capt I Hime (RN, resident in Bridgetown)
Information Officer: D Worrell [98]

Western Samoa

Head of State:
H.H. Malietoa Tanumafili II

10 Head of State

10-10000 Office of the Head of State
Vailima, Government House, Apia
☎(+685) 208 40 ☒ 221
Head of State: H H Malietoa Tanumafili II [98]

11 Prime Minister

11-10000 Office of the Prime Minister
PO Box L1861, Apia
☎(+685) 215 00; 236 36; 247 99 ☒ 221 MZLO SX
⊖ (+685) 218 22
Prime Minister: Hon. Tofilau Eti Alesana
Minister of Finance: Hon Tuilaepa Sailele Malielegaoi
Minister of Agriculture, Forestry, Fisheries and Meteorological Services: Teofilo Molio
Minister of Internal Affairs: Leota Lu II
Minister of Post Office and Telecommunications: Tofilau Eti Alesana
Minister of Education: Ms Fiame Naomi Mata'afa
Minister of Public Works: Le'afa Vitale
Minister of Health: Misa Telefoni Retzlaff
Minister of Justice, Land and Titles Court: Mafasolia Papu Vailupe
Minister of Lands, Environment and Government Housing: Tuala Sale Tagaloa Kerslake [98]

15 Parliament

15-10000 Legislative Assembly
Fono
PO Box 1866, Apia
Maoto Fono - Mulinu'u, Apia
☎(+685) 218 11 ⊖ (+685) 218 17
Speaker: Hon Toleafoa Faasifi
Deputy Speaker: Mulitalo Siafausa
Chairman Foreign Affairs Commission: Hon. Tofilau Eti Alesana
Clerk of the Legislative Assembly: Dr Fetuao Toia Alama [98]

31 Embassies & Attachés

31-00090 Australian High Commission
Beach Rd, Fea Gai Ma Leata Bldg, PO Box 704, Apia
☎(+685) 234 11/12 ⊖ (+685) 231 59
High Commissioner: H.E. Mr D.W. Hegarty
Deputy High Comissioner: Ms Melanie Davies
Military/Defence Attaché: Lt Col D. Phillips
Third Secretary: Adam Lees [97]

31-01300 New Zealand High Commission
Beach Road, PO Box 1876, Apia
☎(+685) 217 11 ⊖ (+685) 200 86
High Commissioner: H E Mr Peter Hamilton
Deputy High Commissioner: Ms Penelope Ridings
Defence Adviser: Wg Cdr Peter Randerson (Resident in Wellington)
First Secretary: A Windmill
Second Secretary: Mike Walsh
Attaché: Rodney Lelieveld [97]

San Marino

Captain Regent (Joint Chiefs of State):
Giancarlo Venturini

Captain Regent (Joint Chiefs of State):
Maurizio Rattini

12 Ministries

12-20000 Foreign and Political Affairs Department
Contrada Omerelli, I-47031 San Marino
⊖ (+39 549 378) 99 20 18
Secretary of State for Foreign and Political Affairs: H E Mr Gabriele Gatti
Secretary of State for Finance and Budget: Clara Boscaglia
Secretary of State for Internal Affairs: Antonio Lazzaro Volpinari [98]

15 Parliament

15-10000 Grand and General Council
Consiglio Grande e Generale
San Marino
☎(+378) 88 22 73; 88 23 17 ⊖ (+378) 88 23 89
Secretary of State for Interior Affairs: Antonio Volpinari
☎(+378) 88 24 34
Minister of Foreign Affairs: Dr Gabriel Gatti
☎(+378) 88 22 18
Director: Giovanna Crescentini [98]

31 Embassies & Attachés

31-01890 Consulate of the United Kingdom
Lungarno Corsini 2, I-I-50123 Florence
☎(+39 55) 28 41 33; 28 95 56 (commercial office)
⊖ (+39 55) 21 91 12
resides in Florence: R J Griffiths [98]

31-01900 Consulate of the United States of America
Lungarno Vesvucci, No 38, I-50123 Florence
☎(+39 55) 239 82 76
Consulate: Luis McCall (resides in Florence) [98]

Sao Tomé & Principe

Head of State:
Miguel Trovoada

10 Head of State

10-10000 Office of the President
São Tomé
President: Miguel Trovoada [98]

11 Prime Minister

11-10000 Office of the Prime Minister
São Tomé
Prime Minister: Raul Wagner Conceicao Bragance Neto [98]

12 Ministries

12-10000 Ministry of Defence and Internal Order
São Tomé
Minister: Capt Joao Quaresma Viegas Bexigas [98]

12-20000 Ministry of Foreign Affairs and São Tomean Communities Overseas
São Tomé
Minister: Homero Jeronimo Salvaterra [98]

31 Embassies & Attachés

31-01890 Consulate of the United Kingdom
Residencial Avenida, Avienda Da Indeendencia, CP 257 Sao Tomé
☎(+239 12) 210 26; 210 27 ⊖ (+239 12) 213 72
Honorary Consul: James Gomes [98]

Contact Reference

Saudi Arabia

Head of State:
H M King Fahd bin Abd al-Aziz Al Saud

10 Head of State

10-10000 Office of H M the King
Riyadh
☎(+966 1) 488 22 22
King of Saudi Arabia and Prime Minister: H M Fahd bin Abd al-Aziz Al Saud
First Deputy Prime Minister: H R H Prince Abdullah bin Abdul al-Aziz al-Saud [98]

11 Prime Minister

11-10010 Office of the Second Deputy Prime Minister
(Head of the National Gaurds)
Riyadh
☎(+966 1) 441 60 55 ⊠ 201 064 natgrd sj
Second Deputy Prime Minister: H R H Prince Sultan bin Abd al-Aziz Al Saud [98]

12 Ministries

12-10000 Ministry of Defense and Aviation (MODA)
Airport Road, PO Box 1844, 11165 Riyadh
☎(+966 1) 478 59 00
Minister: H R H Prince Sultan bin Abd al-Aziz Al Saud [98]

12-20000 Ministry of Foreign Affairs
Al Nassiriah Road, Riyadh
☎(+966 1) 405 50 00 ⊠ 401104, 402099/100
⊖ (+966 1) 402 37 07
Minister: H R H Prince Saud al-Faysal bin Abd al-Aziz Saud [98]

12-40000 Ministry of Interior
Airport Street, 11134 Riyadh
☎(+966 1) 401 19 44 ⊠ 203 743/744/745 minint sj
Minister: Nayif bin Abd al-Aziz Al Saud [98]

12-50000 Ministry of Communications
Old Airport Road, 11178 Riyadh
☎(+966 1) 404 30 00 ⊖ (+966 1) 403 57 43
Minister of Communications: Dr Nasir bin Muhammad al- Salum [98]

31 Embassies & Attachés

31-00090 Australian Embassy
Abdullah Bin Hozafa Al-Sahm Avenue, PO Box 94400, 11693 Riyadh
☎(+966 1) 488 77 88 ⊖ (+966 1) 488 79 73
Ambassador: H E Mr Philip Knight
Deputy Chief of Mission: Adrian Morrison
Consul: Hugh D Wilson
First Secretary: Guy Renudiger
Defence Attaché: Gp Capt Grahame Carroll
Australian Trade Commissioner
Riyadh
☎(+966 1) 488 77 88 ⊖ (+966 1) 488 74 58
Counsellor, Commercial: Geoff Rae [98]

31-00100 Embassy of Austria
PO Box 94373, 11693 Riyadh
☎(+966 1) 477 74 45; 476 12 99
⊖ (+966 1) 476 67 91
Ambassador: H.E. Mr Otto Ditz
Deputy Chief of Mission & First Secretary: Friedrich Stift
Commercial Office
PO Box 94 372, 11693 Riyadh
☎(+966 1) 404 10 10; 03 34 ⊖ (+966 1) 404 29 75
Commercial Counsellor: Dkfm. Oskar Smrzka [97]

31-00230 Brazilian Embassy
Diplomatic Quarter, PO Box 94348, 11693 Riyadh
☎(+966 1) 488 00 18; 488 00 25 ⊠ 406711 braemb sj
⊖ (+966 1) 488 10 73
Ambassador: H E Mr Serqio M Thompson-Flores
Head of Chancery, Consular and Commercial Affairs: Affonso Jose Santos [98]

31-00290 Embassy of the Republic of Cameroon
PO Box 94336, 11693 Riyadh
☎(+966 1) 488 00 22; 488 02 03 ⊠ 46688 amcasa sj
⊖ (+966 1) 488 14 63
Ambassador: H E Mohamadou Labarang
☎(+966 1) 488 01 22
Second Counsellor: Jean Jaques Kemadjou
☎(+966 1) 488 16 68
First Secretary: Sduaibu Kingue ☎(+966 1) 488 29 19
Financial Attaché: Alim Mahamoudou
☎(+966 1) 488 27 75 [98]

31-00310 Embassy of Canada
Diplomatic Quarter, PO Box 94321, 11693 Riyadh
☎(+966 1) 488 22 88 ⊠ 404893 domcan sj
⊖ (+966 1) 488 01 37
Ambassador: H.E. Mr Peter Sutherland
Counsellor & Consul: Hartmuth G. Kroll
Defence Attaché: Col L.C. Bowen
Minister Counsellor Commercial: Murray Esselmont
Commercial Counsellor: Edward Shelley
Commercial Counsellor: J. Clapp
Public Information Officer: J. McLaren [97]

31-00540 Embassy of Ethiopia
PO Box 94 341, 11693 Riyadh
☎(+966 1) 477 52 85; 479 16 90 ⊠ 406633 ethemb
⊖ (+966 1) 476 80 20
Ambassador: H E Mr Mohammed Ali ibrahim
Councellor: Guesh Glmariam
First Secretary: Masfin Nigatu
Third Secretary: Asefa Habte Michael
Attaché: Mrs Yetna Yet Ketema [98]

31-00690 Embassy of Greece
PO Box 94375, 11693 Riyadh
☎(+966 1) 465 50 26 ⊠ 406322
Ambassador: H.E. Mr Paul Apostolides [98]

31-00890 Embassy of Italy
Diplomatic Quarter, PO Box 94389, 11693 Riyadh
☎(+966 1) 488 12 12 ⊠ 406188 itdisa sj
⊖ (+966 1) 488 19 51; 488 05 90
Ambassador: H E Dr Marco Sorace Marescaa
☎(+966 1) 488 05 83
Counsellor: Dr Maurizio Pavesi ☎(+966 1) 454 63 13
Finance Manager: G Basile
Defence Attaché: Col Marco Scariatti
Second Secretary, Commercial: Dr Luigi De Chiara
☎(+966 1) 482 09 02
Assistant Military Attaché: Lieut Col Valantino Buttazzo
☎(+966 1) 488 46 93 [98]

31-00960 Embassy of Kenya
PO Box 94358, 11693 Riyadh
☎(+966 1) 488 12 38; 488 24 84
⊠ 406455 KENEMIB ST. ⊖ (+966 1) 488 26 29
Ambassador: H.E. Mr Said Hemed Bat-Haf
Charge d'Affairs: Patrick Rotich ☎(+966 1) 488 12 38
Second Secretary: Salim M Salim ☎(+966 1) 488 12 38
First Secretary Commercial: Michael W Furuha
☎(+966 1) 488 12 38
Second Secretary: Kamatta Muthaa
Immigration: Samuel K Mboje
Financial: Steven M Muturi
Administrative Attaché: William Achieng [98]

31-00990 Embassy of the Republic of Korea
Al-Musa Adna, Mohd Hamza Fatayerji, PO Box 94399, 11693 Riyadh
☎(+966 1) 488 22 11 ⊠ 405858 gksaud sj
⊖ (+966 1) 488 13 17
Ambassador: H E Mr Hyohun Shin
Deputy Chief of Mission: Woong Nam Kim
Commercial Counsellor: Jong In Lee [98]

31-01000 Embassy of the State of Kuwait
Diplomatic Quarter, PO Box 94304, 11693 Riyadh
☎(+966 1) 488 35 00; 488 32 01; 488 34 01
⊠ 401301 ⊖ (+966 1) 488 36 82; 482 12 84
Ambassador: H E Sheikh Sabah Khaled Al-Hama Al-Sabah
First Secretary: Hafith Mohamad Al-Ajmi
Second Secretary: Nasser Abdul Rahman Al-Motairi
Third Secretary: Fahad Meshari Al-Zofairi
Attaché: Meshal Abdul-Aziz Al-Sabah
Attaché: Khaled Al-Armali
Attaché: Hamad Al-Mansour
Chief of Military Liason Office: Col Abdul-Wahab Al-Obaid
Assistant of Chief Military Liason Office: Lieut Col Abdul-Latif Bukreiss
Cultural Attaché: Tamin Abdel-Mohsen Al-Moqateh [98]

31-01140 Embassy of Malta
PO Box 94361, 11693 Riyadh
☎(+966 1) 463 23 45 ⊖ (+966 1) 463 39 93
Ambassador: H E Mr Emannuel Farrugia
Counsellor: Alfred M Falzon
Maltese Consulate
PO Box 7083, 21462 Jeddah
☎(+966 2) 665 79 73 ⊖ (+966 2) 665 79 80
Consul: Dr Mohammed A Mansouri [98]

31-01300 New Zealand Embassy
Diplomatic Quarter, PO Box 94397, 11693 Riyadh
☎(+966 1) 488 79 88 ⊖ (+966 1) 488 79 12
Ambassador: H E Mr David Payton
☎(+966 1) 488 79 88 ext 105
Deputy Chief of Mission: Peter Brown
☎(+966 1) 488 79 88 ext 115
First Secretary: Ms Belinda Brown
☎(+966 1) 488 79 88 ext 111
First Secretary, Administration and Consul: Ian Goodger [98]

31-01330 Embassy of the Federal Republic of Nigeria
PO Box 94386, 11693 Riyadh
☎(+966 1) 482 30 24; 482 39 82 ⊠ 406177 embang sj
⊖ (+966 1) 482 41 34
Ambassador: H E Mr Alhaji Ahmed Al-Gazali
Head of Chancery: H A Opeloyeru
Counsellor: U A Baraya
Counsellor: Sani Bako
Counsellor: M A Musawa [98]

31-01370 Embassy of Pakistan
PO Box 94007, 11693 Riyadh
☎(+966 1) 488 42 22; 488 46 66; 488 41 11; 488 72 72 ✉ PAREP Riyadh 11693 ⊠ 406500 parep sj
⊖ (+966 1) 488 79 53
Deputy Head of Mission: Muhammad Haroon Shaukat
☎(+966 1) 482 39 77
Defense Attaché: Brig Muhammad Tariq
☎(+966 1) 488 79 32
Air and Naval Attaché: Gp Capt Shaheen Hamid Butt
Economic Counsellor: Riaz Bukhari ☎(+966 1) 482 48 70
First Secretary (C): Ayaz Hussain
First Secretary (P): Riaz H Bukhari
Counsellor (Welfare-I): Waheed A Shaikh
Counsellor (Welfare-II): Najeeb Khilji
Military Mission
PO Box 94007, 11693 Riyadh
☎(+966 1) 488 79 32 ⊠ 406500 parep sj
⊖ (+966 1) 488 79 53
⊞ 406500 SJ
Head of Mission: Brig Muhammad Tariq
☎(+966 1) 488 79 32
Air and Naval Attaché: Gp Capt Shaeen Harrnid Burr
☎(+966 1) 488 77 92 [98]

31-01420 Embassy of the Philippines
PO Box 94366, 11693 Riyadh
Olaya Street in front of Al-Owais Market,
King Fahad District, Riyadh
☎ (+966 1) 454 07 77 ✉ AMBAPHIL Riyadh
📠 406377 philem sj ⊖ (+966 1) 454 25 50
Ambassador: H.E. Mr Romulo Espaldon
Minister Counsellor & Consul General: Jesus N. Maravilla
Minister Counsellor & Head of Chancery: Dauday Tago
First Secretary, Consul & Special Asst to Ambassador: Oscar G. Orcine
Third Secretary & Vice Consul Head of Admin Section: Alimatar Garangan
Economic Officer & Attaché: Yusuf J.T. Rasul
Third Secretary & Vice Consul Head of Political Section: Miquel R. Bautista 97

31-01440 Embassy of Portugal
PO Box 94328, 11693 Riyadh
☎ (+966 1) 464 48 53 📠 404477
Ambassador: H.E. Dr Pedro Vasconcelos e Castro
Deputy Chief of Mission: Eduardo Pereira de Sousa 97

31-01680 Embassy of Spain
PO Box 94347, 11693 Riyadh
☎ (+966 1) 488 06 06 ⊖ (+966 1) 488 04 20
Ambassador: H E Mr Francisco Viqueira
First Secretary, Charge d'Affairs a i: Ricardo Sanchez-Blanco
Second Secretary: Fernando Nogales
Defence Attaché: Col Vicente Martinez
Economic and Commercial Counsellor: Federico Povedano
Attaché: Julio Alamed
Head Administrative Officer: Jose Manuel Gil

Commercial Office
Thalaateen Street, PO Box 94372, 116393 Riyadh
☎ (+966 1) 464 51 25; 464 47 94
⊖ (+966 1) 465 04 09 98

31-01730 Embassy of Sweden
Collector Road B, Diplomatic Quarter, PO Box 94382, 11693 Riyadh
☎ (+966 1) 488 31 00 ✉ SVENSK Riyadh
📠 406266 svensk sj ⊖ (+966 1) 488 06 04
Ambassador: H E Mr Lave Johnsson
Counsellor: Leif Ringquist
Commercial Attaché: Adallah Saeed 98

31-01740 Embassy of Switzerland
Diplomatic Quarter, PO Box 94311, 11693 Riyadh
☎ (+966 1) 488 12 91 ⊖ (+966 1) 488 06 32
Ambassador: H E Mr Christian Blickenstorfer
Deputy Chief of Mission and Commercial Counsellor: Etienne Thévoz
Commercial Counsellor: Etienne Thévoz 98

31-01750 Embassy of the Syrian Arab Republic
Diplomatic Quarter, PO Box 94323, 11693 Riyadh
☎ (+966 1) 488 74 81 📠 406677 ⊖ (+966 1) 482 61 96
Ambassador: H E Mr Omar Al Sayyed 98

31-01760 Embassy of the Republic of China
Diplomatic Quarter, PO Box 94393, 11693 Riyadh
☎ (+966 1) 488 19 00 ext 216 📠 406244
⊖ (+966 1) 488 17 16
Ambassador: H E Mr Yeh Chia-Wu
Military Liason Officer: Col Liu-Chi Fung 98

31-01840 Embassy of the Republic of Turkey
Diplomatic Quarter, PO Box 94390, 11613 Riyadh
☎ (+966 1) 482 01 01 (6 lines)
Ⓔ 105504.1521@compuserve.com 📠 406206 tremb sj
⊖ (+966 1) 488 78 23
Ambassador: H E Mr Türkekul Kurttekin
☎ (+966 1) 482 01 01 ext 327
Counsellor: Sadi Altinok ☎ (+966 1) 482 01 01 ext 324
Military Attaché: Col Ömer Esenyel
☎ (+966 1) 482 01 01 ext 381
Press and Information Officer: Vecdi Gedik
☎ (+966 1) 482 01 01 ext 371 98

31-01860 Embassy of the Republic of Uganda
Diplomatic Quarter, PO Box 94344, 11693 Riyadh
☎ (+966 1) 454 49 10 📠 40658 :gemrd sj
⊖ (+966 1) 454 92 60
Ambassador: H E Dr Mohammad Kisuule
☎ (+966 1) 454 11 48
First Secretary/Head of Chancery: John W K Magoba
☎ (+966 1) 450 91 01
First Secretary: A N Kabunga ☎ (+966 1) 454 58 75
Administrative Attaché: Byabajungu Akiiki
☎ (+966 1) 454 92 64 98

31-01890 Embassy of the United Kingdom
PO Box 94351, 11693 Riyadh
☎ (+966 1) 488 00 77 📠 406488 britem sj
⊖ (+966 1) 488 23 73; 488 06 23; 488 12 09
Ambassador: H E Mr A F Green
Deputy Chief of Mission: W C Patey
Counsellor, Commercial: R J Northern
Defence and Military Attaché: Brig W E Strong
Naval and Air Attaché: Cdr I A Vosper (RN)
First Secretary: J A Raine 98

40 Government Defense Sales/Procurement Organizations

40-01970 SOFRESA Riyadh
PO Box 4720, Riyadh
☎ (+966 1) 478 75 28 ⊖ (+966 1) 476 90 34
General Manager: Christian Duplessis
Representing:
SOFRESA, France 98

75 Manufacturing Companies

75-00250 Daewoo Corp Arabia Branch
PO Box 691, 31932 Dharan
10th Floor, Riyadh Tower Building, King Abdul Aziz Street, Al-Khobar
☎ (+966 3) 898 32 75; 646 35 71
⊖ (+966 3) 895 24 29 98

80 Service Companies & Organizations

80-00050 Al Falak Equipment & Supplies Company
(† Metier Management Systems)
Dhahran Airport, PO Box 1663, 31952 Alkhobar
☎ (+966 3) 894 65 60 ⊖ (+966 3) 894 60 32
Middle East Sales Manager: David Blackaby
® UK: Artemis Management Systems 98

80-00800 Hughes Saudi Arabia Ltd
Old Airport Road, Bugshan, PO Box 69357, 11547 Riyadh
Bank Al Saudi Fransi Building, Riyadh
☎ (+966 1) 478 70 11 ⊖ (+966 1) 479 14 85 98

80-01600 International Trade Contracting & Services Company (ITCS)
(† Saudi Naval Support Co. Ltd)
PO Box 3122, 21471 Jeddah
☎ (+966 2) 669 26 68; 669 26 72
⊖ (+966 2) 660 61 64
Managing Director: Mahmoud Bachir
Operation Manager: Edgar A Nemer 98

80-01700 Jet Aviation Saudi Arabia Co Ltd Head Office and Jeddah Station
(a subsidiary of Jet Aviation Management AG, Switzerland)
PO Box 40055, 21499 Jeddah
Private Aviation Terminal
King Abdulaziz International Airport
☎ (+966 2) 685 04 00 Ⓔ jjed@jetaviation.com
📠 607261 ⊖ (+966 2) 685 04 05; 685 04 06 (administration)
Vice President and General Manager: Jürg Reuthinger
Operations Manager: Masood Ali Khan
Chief Inspector and Maintenance Manager: Dirk G de Boer
■ Offers full FBO services, airline services and line maintenance.

Jet Aviation Saudi Arabia Co Ltd Riyadh Station
PO Box 12516, 11483 Riyadh
General Aviation Terminal King Khaled International Airport
☎ (+966 1) 221 42 00; 221 00 25; 221 00 26
Ⓔ jruh@jetaviation.com 📠 405551
⊖ (+966 1) 220 19 27
Operations and Station Manager: Noor Ur Razzak 98

80-02000 The Saudi Vosper Thornycroft Company Limited
PO Box 4900, 11412 Riyadh
☎ (+966 1) 464 97 70 📠 400813 falest sj
⊖ (+966 1) 465 11 39
Contact: B.E. Woodham
■ Maintenance, repair & operation of both military & civil ships, including technical training. 98

90 Sales Agents, Sales & Liaison Offices

Companies represented:

AlliedSignal, USA	90-01980
AlliedSignal/Bendix King, USA	90-01980
Allison, USA	90-01980
Artemis Management Systems, UK	80-00050
Bell Helicopter Textron, USA	90-01980
Bendix King, USA	90-01980
BP Solar Limited, UK	90-00250
Chadwich Helmuth, USA	90-01980
CMI Mechanical Industries, Belgium	90-00400
ITT Defense, USA	90-00600
MFD International Ltd, UK	90-01875
Rockwell-Collins, USA	90-01850
SAIL (Engineering) Consultants, Saudi Arabia	90-01875

90-00095 AlliedSignal Aerospace Marketing Sales & Service
c/o Saudi Arabiaan Airlines, PO Box 167, CC 830, 21231 Jeddah
☎ (+966 2) 684 28 88 ⊖ (+966 2) 685 77 13
Field Service Engineer: John Bartlett 98

90-00100 Advanced Electronics Co. Ltd
PO Box 90916, 11623 Riyadh
☎ (+966 1) 220 13 50 📠 407724
⊖ (+966 1) 220 13 55; 23 28
♲ US$ 29,500,000 (1997) ♣ US$ 125,500,000 (1996)
👥 364 (1997) 🏠 13,800 sq m
President and Chief Executive Officer: Abdulaziz A Sugair
Executive Vice President: Dr Ghassan Shibi
Vice President TQM-HR: Dr Tarik Sulimani
Vice President Research and Development: Dr Khalid Biyari
Vice President Projects: Dr Faisal Sugair
Vice President Business Development: Mike Kelly
Marketing Manager: Khalid Alturki
■ Design, development, manufacture, repair, modification and upgrades of electronic products and systems for military, civil and industrial customers. 98

90-00190 Aujan Industrial Supplies
Al Khobar, PO Box 2297, Kingdom of Saudi Arabia
☎ (+966 3) 857 99 55 📠 801300 aujan sj
⊖ (+966 3) 857 32 18 97

90-00250 BP Solar Arabia Limited
PO Box 85652, 11612 Riyadh
☎ (+966 1) 498 42 79 ⊖ (+966 1) 498 47 96
® UK: BP Solar Limited 98

90-00400 CMI Cockerill Mechanical Industries Middle East Office
PO Box 1555, 21441 Jeddah
☎ (+966 2) 651 20 69 ⊖ (+966 2) 660 78 26
Regional Director: N J Martin
® Belgium: CMI Mechanical Industries 98

Contact Reference

90-00410 Al-Bab Care of Daewoo
(a subsidiary of Daewoo Corp, South Korea)
King Faisal Foundation Building,
South Tower 8th Floor PO Box 89644, 11692 Riyadh
☎ (+966 1) 464 68 14 ▣ 40704 albab sj
⊖ (+966 1) 464 68 14 [98]

90-00600 ITT Defense International (Saudi Arabia)
Al-Wadi Estates, PO Box 61837, 11571 Riyadh
☎ (+966 1) 476 81 75 ⊖ (+966 1) 476 81 75
Vice President, Saudi Arabia Operations:
Raymond E Haddock
® USA: ITT Defense [98]

90-01500 Northrop Grumman International Inc.
Al Yamamah 1, PO Box 61279 Villa 47, 11665 Riyadh
☎ (+966 1) 478 30 22 ⊖ (+966 1) 478 30 22
Director: David G. Jahosky [97]

90-01850 Rockwell Systems International, Inc.
(† Collins Systems International Inc.)
(a division of Rockwell International Corp., USA)
PO Box 2375, 11451 Riyadh
☎ (+966 1) 465 55 49 ⊖ (+966 1) 465 65 63
Middle East Regional Director: Tariq Aziz
■ Telecommunications equipment and systems.
® USA: Rockwell-Collins [98]

90-01875 SAK Engineering Consultants
(† MFD International Ltd)
PO Box 1655, 21441 Jeddah
☎ (+966 2) 671 35 46; 671 14 27 ▣ 405074
⊖ (+966 2) 672 29 60
👥 16
General Manager: David N Castree
■ Consultants for design and management civil, building and industrial projects.
® Saudi Arabia: SAIL (Engineering) Consultants
UK: MFD International Ltd [98]

90-01900 Saudi Arabian Parsons Limited
(† Parsons Group)
PO Box 8037, 21482 Jeddah
☎ (+966 2) 643 70 59 ⊖ (+966 2) 643 75 21
👥 264
Manager: N. Ahmed [97]

90-01980 Sogerep Aviation Support
PO Box 41699, 11531 Riyadh
☎ (+966 1) 464 20 68; 465 53 78
✆ (+966 1) 465 41 55 ▣ 403254 sogrep sj
⊖ (+966 1) 464 41 90
♤ SR 10,000,000 (1997) ♣ SR 50,000,000 (1997)
👥 45 (1997) 🏠 2,000 m2
Vice President: Thomas Thompson
Marketing Manager: James L Beahon
® USA: AlliedSignal, Bell Helicopter Textron, Bendix King

Sogerep Logistics Center
Dhahran Airport, PO Box 490, 31932 Dhahran
☎ (+966 3) 857 65 68 ▣ 801962 sogrep sj
⊖ (+966 3) 857 68 12
Logistics Manager: Frank Andrews
® USA: AlliedSignal/Bendix King, Allison, Bell Helicopter Textron, Chadwich Helmuth [98]

Senegal

Head of State:
Abdou Diouf

10 Head of State

10-10000 Office of the President
Avenue Roume, PO Box 168, Dakar
☎ (+221) 23 10 88 ▣ 258; 537 presid sg
President of the Republic: Abdou Diouf [98]

11 Prime Minister

11-10000 Office of the Prime Minister
Building Administratif, PO Box 140, Dakar
☎ (+221) 21 50 75 ▣ 258; 537
Prime Minister: Habib Thiam [98]

12 Ministries

12-10000 Ministry of Armed Forces
Building Administratif, PO Box 176, Dakar
☎ (+221) 22 56 13 ▣ 483; 485
Minister: Cheikh Hamidou Kane [98]

12-20000 Ministry of Foreign Affairs
Place de l'Indépendance, Dakar
☎ (+221) 21 62 80; 21 10 81 ▣ 482, 483, 485
Minister: Moustapha Niasse [98]

12-40000 Ministry of the Interior
Rond Point de la République, Dakar
☎ (+221) 21 41 51
Minister: Abdourahmane Sow [98]

31 Embassies & Attachés

31-00660 Embassy of the Federal Republic of Germany
PO Box 2100, Dakar
20 Avenue Pasteur, Dakar
☎ (+221) 23 25 19; 48 84 ▣ 21686 aadaka sg
⊖ (+221) 22 52 99
Ambassador: H.E. Dr Eleonore Linsmayer
Deputy Chief of Mission & Commercial Counsellor:
Heinz Wilhelm
Public Information Officer: Lars Gerrit Leymann [97]

31-00890 Embassy of Italy
Rue Alpha Achamiyou Tall, Dakar
☎ (+221) 22 05 78; 22 00 76; 21 65 72
▣ 00 906 51641 ⊖ (+221) 21 75 80
Ambassador: H E Dr Paolo Guido Spinelli
First Secretary: Carla Campanile
Commercial Officer: Patrizia Falcinelli [98]

31-01220 Embassy of the Kingdom of Morocco
Avenue Aheikh Anta Diop, PO Box 490, Dakar
Ambassador: H E Mr Ennahdi Elidrissi ☎ (+221) 24 38 36
Commercial Counsellor: Jaouhar Abdelkader
☎ (+221) 24 69 27 [98]

31-01370 Embassy of Pakistan
Villa No 7602, Sicaap Mermoz, PO Box 2635, Dakar
☎ (+221) 24 61 35; 24 61 38 ✉ PAREP Dakar
▣ 51615 parep dkr sg ⊖ (+221) 24 61 36
Ambassador: H E Mr Saeed Khalid
Deputy Chief of Mission, Head of Chancery:
Manzoor Alam Qureshi [98]

31-01480 Embassy of Romania
Bd de L'Est X Rue 4, BP 3171 et 32 12, Dakar
☎ (+221) 25 19 13; 25 19 13 ▣ 61115 ambrom
⊖ (+221) 25 19 13
Chargé d'Affaires: Ienel Ilie
Commercial Counsellor: Stefan Ochian ▣ 3153 [98]

31-01680 Embassy of Spain
SAIH, 45 Bd de la République, PO Box 2091, Dakar
☎ (+221) 21 30 81; 21 11 78 ▣ 451 ambespa sg
⊖ (+221) 21 68 45
Deputy Chief of Mission:
Santiago Gomez-Acebo Rodriguez-Spiteri
Commercial Secretary: Javier Puebla Rabanal [98]

31-01740 Embassy of Switzerland
Rue R.N'Diaye + R.Seydou Nourou Tall, PO Box 1772, Dakar
☎ (+221) 23 58 48 ⊖ (+221) 22 36 57
Ambassador: H.E. Mr Jean-Claude Richard
Head of Chancery: Jean-Didier Javet
Commercial & Public Affairs: Bernard Pillonel [97]

31-01890 Embassy of the United Kingdom
20 Rue du Docteur Guillet, PO Box 6025, Dakar
☎ (+221) 23 73 92; 23 99 71 ⊖ (+221) 23 27 66
Ambassador: H E Mr David Snoxell (non-resident)
Deputy Head of Mission: R W Hyde
Commercial Officer: Mrs N Thiam
Information Officer / Defence Matters: G T G Collier
■ The Embassy in Dakar is also responsible for British interests in Mali, Guinea, Guinea-Bissau & Cape Verde. [98]

Seychelles

Head of State:
France Albert René

10 Head of State

10-10000 Office of the President
State House, PO Box 655, Victoria, Mahe
☎ (+248) 243 91 ▣ 2217
President of the Republic: France Albert René [98]

12 Ministries

12-20000 Ministry of Foreign Affairs Planning and Environment
(† Ministry of Planning & External Relations)
Maison Queau De Quinssy, PO Box 656, Victoria, Mahe
☎ (+248) 22 46 88 Ⓔ mfapesey@Seychelles
▣ 2260 minae sz ⊖ (+248) 22 48 45
Minister for Ministry of Foreign Affairs Planning and Environment: Joseph Belmont
Principle Secretary: Emmanuel Faure [98]

26 Police (Civil)

26-00000 Seychelles Police Force Headquarters
PO Box 46, Victoria, Mahé
☎(+248) 32 25 91 ⊠ 2298 compol sz
⊖ (+248) 32 30 05
Principle Secretary Department of Internal Affairs: D Quilindo
Commisioner of Police: Andre Quilindo
Deputy Commissioner: Tite Maurice Morin ☎(+248) 32 20 11 ext. 107/108
Assistant Commisioner of Police for Operations: Gaetah Didin ☎(+248) 32 20 11 ext 110
Assistant Commisioner of Police (ADH): Kishnon Labontr ☎(+248) 32 20 11 ext 115
Assistant Commisioner of Police (RHRMoD): Tite Morin ☎(+248) 32 25 91 ext 107
Public Relations Officer: Robert Georgrs 98

31 Embassies & Attachés

31-01890 United Kingdom High Commission
PO Box 161, Victoria, Mahe
3rd Floor, Victoria House, Mahe
☎(+248) 22 52 25; 53 56 E bhcsey@seychelles.net
⊖ (+248) 22 51 27
High Commissioner: H E Mr Peter Thomson
Deputy High Commissioner: T F Robins
Defence Adviser: Lieut Col R C Mantell (Resident in London)
Commercial and Information Assistant: Mrs J C Chetty 98

Sierra Leone

10 Head of State

10-10000 Office of the President
State House, Independence Avenue, Freetown
⊠ 3230 presec sl
President of the Republic: Ahmad Tejan Kabbah
Vice President: Albert Demby
Minister of Presidential Affairs and Public Service: Momodu Koroma 98

12 Ministries

12-10000 Ministry of Defence
Tower Hill, Freetown
Deputy Minister of Defence: Capt Sam Hinga Norman (retd) 98

12-20000 Department of Foreign Affairs
14 Gloucester Street, Freetown
☎(+232 22) 27 77; 22 43 27 ⊠ 3218 external sl
⊖ (+232 22) 56 15
Minister: Ms Shirley Gbjumama 98

12-40000 Ministry of Internal Affairs
State Avenue, Freetown 98

31 Embassies & Attachés

31-01890 United Kingdom High Commission
Spur Road, Wilberforce, Freetown
☎(+232 22) 23 29 61; 23 29 62 ⊠ 3235 ukrep sl
⊖ (+232 22) 22 81 69
High Commissioner: H E Mr P A Penfold
Deputy High Commissioner: C J K Glass
Defence Adviser: Lieut Col Andrew Gale (Resident in Accra)
Third Secretary Consular/Management: D Harris 98

31-01900 Embassy of the United States of America
Corner Walpole & Siaka Stevens St, Freetown
☎(+232 22) 22 64 81 ⊠ 3509 usemb sl
⊖ (+232 22) 22 54 71
Ambassador: H.E. Mr John L. Hirsch
Counsellor & Deputy Chief of Mission: Charles A. Ray
Defence Attaché: Lt Col Billy W. Feltner (Resident in Monrovia)
First Secretary & Public Affairs Officer: Dr Patricia L. Sharpe
First Secretary & Admin Officer: Mrs Elizabeth McGaffey
Second Secretary & Consular Officer: Ms Marlee Anderson
Attaché & Vice Consul: Charles Slater 97

Singapore

Head of State:
Ong Teng Cheong

10 Head of State

10-10000 Office of the President
Istana, Orchard Road, Singapore 0922
☎(+65) 737 55 22 ⊖ (+65) 738 46 73
President of the Republic: Ong Teng Cheong 98

11 Prime Minister

11-10000 Office of the Prime Minister
Istana Annexe, Orchard Road, Istana 0923
✉PRIMIN Singapore
Prime Minister: Goh Chok Tong ☎(+65) 235 85 77 (+65) 732 46 27 (fax)
Deputy Prime Minister: Tony Tan Keng Yam
Deputy Prime Minister: Lee Hsien Loong
Senior Minister: Lee Kuan Yew ☎(+65) 737 51 33 (+65) 734 52 44 98

12 Ministries

12-10000 Ministry of Defence
MINDEF Building - Gombak Drive, Off Upper Bukit Timah Road, Singapore 2366
☎(+65) 768 30 45 ⊠ 21373 singdef rs
⊖ (+65) 764 61 19
Deputy Prime Minister and Minister for Defence: Dr Tony Tan Keng Yam ☎(+65) 768 21 52
Second Minister for Defence: Rear Adm Teo Chee Hean (NS) ☎(+65) 768 21 03
Minister of State for Defence: Matthias Yao Chih ☎(+65) 768 21 03
Permanent Secretary (Defence): Eddie Teo ☎(+65) 768 21 04
Permanent Secretary (Defence Development): Peter Ho ☎(+65) 768 27 97
Chief of Defence Force: Lieut Gen Bey Soo Khiang ☎(+65) 768 21 50
Deputy Secretary (Technology): Brig Gen Wesley D'Aranjo ☎(+65) 768 27 84
Deputy Secretary (Policy): Mrs Chua Siew San ☎(+65) 768 29 20
Deputy Secretary (Administration): Tan Kee Yong ☎(+65) 768 21 84
Chief Defence Scientist: Prof Lui Pao Chuem ☎(+65) 768 27 87
Director Policy: Col Yap Ong Heng ☎(+65) 768 28 14
Director Legal Services: Low Wee Ping ☎(+65) 768 27 99
Director Resource Planning: Lee Chee Teng ☎(+65) 373 45 00
Director Joint Operations and Planning Directorate: Brig Gen Lam Joon Khoi ☎(+65) 768 25 41
Director Joint Intelligence Directorate: Brig Gen Lee Fook Sun ☎(+65) 768 44 20
Director Security and Intelligence Division: Col Choi Sing Kwok ☎(+65) 761 02 01
Head Foreign Military Liaison Branch: Lieut Col Charles Tan ☎(+65) 768 42 95
Director Defence Materièl Organisation (DMO): Quek Tong Boon ☎(+65) 373 46 78
Director Procurement Division: C Manohara ☎(+65) 373 46 78

DSO National Laboratories
20 Science Park Drive, Singapore 118230
Director: Dr Su Guaning ☎(+65) 772 72 35
Divisional Manager (Plans): Tan Soo Kee
Manager, Information Resource Centre: Ms Colleen Lim Tai Siew E ltaisiew@dso.org.sg

AFPN 0007
1 Depot Road #12-05, Singapore 109679
Deputy Director Procurement Division: Col Fred Kang Beng Ho ☎(+65) 373 41 05
Head Public Communications Department: Ms Sandra Buenaventura ☎(+65) 276 84 43

AFPN 0011
303 Gombak Drive #1-26, Singapore 669645
Head Public Relations Branch: Chan Chee Kong ☎(+65) 768 30 56
Head Media Relations Branch: Ms Lydia Lim ☎(+65) 768 30 56
Director Military Security: Lieut Col Jet Kian Yee ☎(+65) 768 31 28
Chief Executive Officer Defence Science Organisation: Dr Su Guaning ☎(+65) 772 72 35

AFPN 0048
20 Science Park Drive, Singapore 118230
Chief of Air Force: Maj Gen Goh Yong Siang ☎(+65) 768 35 01
Chief of Staff (Air Staff): Col Yam Ah Mee ☎(+65) 768 35 02
Head of Air Manpower Department: Lieut Col Chang Long Wee ☎(+65) 768 30 00

AFPN 8055
303 Gombak Drive #2-27, Singapore 669645
Head of Air Intelligence Department: Lieut Col Jimmy Tan Cheng Yaw ☎(+65) 768 31 60
Head of Air Operations: Col Raymond Ng ☎(+65) 768 35 06
Head of Air Plans Department: Lieut Col Chee Wee Kiong ☎(+65) 768 35 64
Commander of Air Defence Artillery: Col Yam Ah Mee ☎(+65) 381 72 11
Head Air Training Department: Col Peter Wong ☎(+65) 768 35 08
Head Air Logistics Department: Col Soh Kong Pheng ☎(+65) 768 31 71
Head Aeromedical Centre: Lieut Col Dr Ng Boon Lui ☎(+65) 381 43 01
Commander Tactical Air Support Command: Col Toh Su Po ☎(+65) 75 94 01
Commander Paya Lebar Airport: Col Loh Kok Hua ☎(+65) 381 31 00
Commander Changi Air Base: Lieut Col Foo Kim Peng ☎(+65) 540 52 13
Commander Tengah Air Base: Col Lim Kim Choon ☎(+65) 761 20 00
Chief of Army: Maj Gen Han Eng Juan ☎(+65) 768 24 14
Assistant Chief of Staff, Personnel: Lieut Col Chin Phei Chen ☎(+65) 768 22 18

AFPN 0002
303 Gombak Drive #02-14, Singapore 669645
Assistant Chief of General Staff, Intelligence: Col Tan Huck Gim ☎(+65) 768 33 99
Assistant Chief of Staff, Logistics: Col Lee Wai Mun ☎(+65) 768 33 99
Assistant Chief of General Staff, Operations: Col Desmond Quek ☎(+65) 768 31 16

Contact Reference

Assistant Chief of General Staff, Plans: Col Phang Hee Hon
☎ (+65) 768 21 63
Assistant Chief of General Staff, Training:
Col Leong Yue Kheong ☎ (+65) 768 21 73
Chief Armour Officer: Col Tan Hang Kee
☎ (+65) 668 62 01
Chief Artillery Officer: Col Philip Su Poon Ghee
☎ (+65) 750 33 01
Chief Engineer Officer: Lieut Col Lim Kok Pheng
☎ (+65) 661 21 00
Chief Infantry Officer: Col Tan Nga Kok ☎ (+65) 860 12 13
Chief Signal Officer: Lieut Col Yun Ta Chun
☎ (+65) 761 10 00
Chief of Navy: Rear Adm Richard Lim ☎ (+65) 768 20 01
Head Naval Personnel Department:
Lieut Col Goh Leong Huat ☎ (+65) 768 20 01

AFPN 6050
303 Gombak Drive #03-05, Singapore 669645
Head Naval Intelligence Department:
Lieut Col Kevin Santa Maria John ☎ (+65) 768 20 80
Head Naval Operations Department: Col Lim Khoon Bock
☎ (+65) 768 31 18
Head Naval Plans Department: Col Simon Ong
☎ (+65) 768 20 08
Fleet Commander: Rear Adm Larry Loon
☎ (+65) 864 20 00
Commander Naval Logistics: Lieut Col Thomas Vergis
☎ (+65) 864 20 01
Commander COSCOM: Col Jimmy Goh
☎ (+65) 371 32 02

AFPN 6504
13 Brani Terminal Avenue #01-03, Singapore 098682
Head of Naval Training Department: Maj Tan Kai Hoe
☎ (+65) 768 20 03

AFPN 6055
303 Gombak Drive #03-03, Singapore 669645 [98]

12-20000 Ministry of Foreign Affairs
250 North Bridge Road, N°7-00, Raffles City Tower, Singapore 179101
☎ (+65) 336 11 77 ◙ 21136; 21242; 20358 foreign rs
⊖ (+65) 339 43 30
Minister for Foreign Affairs and Law:
Shunmugan Jayakumar ☎ (+65) 330 56 01 [98]

12-40000 Ministry of Home Affairs
Phoenix Park, Tanglin Road, Singapore 1024
☎ (+65) 235 91 11 ☒ HOMAFFAIRS ◙ 34363 minha rs
⊖ (+65) 734 44 20
Minister: Wong Kan Seng ☎ (+65) 730 90 00 [98]

12-50000 Ministry of Information and The Arts
460 Alexandra Road, PSA Building, #36-00, Singapore 0511
☎ (+65) 279 97 95 ⊖ (+65) 278 15 26
Minister: George Yeo Yong Boon [98]

12-60000 Ministry of Finance
43rd, 45th, 46th and 50th Storey,
8 Shenton Way, Treasury Building, Singapore 0106
Minister of Finance: Dr Richard Hu Tsu Tau
☎ (+65) 225 99 11 [98]

12-70000 Ministry of Trade and Industry
8 Shenton Way #48-01, Treasury Building, Singapore 0106
☎ (+65) 225 99 11 ⊖ (+65) 320 92 60
Minister for Trade and Industry: Lee Yock Suan [98]

12-80000 Ministry of Communications
460 Alexandra Road, 39th Storey, PSA Building, Singapore 0511
◙ 25500 comms rs ⊖ (+65) 375 77 34
Minister: Mah Bow Tan [98]

12-90000 Ministry of National Development
5 Maxwell Road, #21-00 and #22-00,
Tower Block, MND Complex, Singapore 0106
☎ (+65) 222 12 11 ⊖ (+65) 325 72 54
Minister: Lim Hng Kiang [98]

25 Defense Forces

25-20000 Air Force Headquarters
Paya Lebar Airport, Singapore 1953
☎ (+65) 760 81 88
Chief of Air Staff: Col Soon Eng Boon ☎ (+65) 768 35 02
Head of Air Manpower Dept: Lt Col Kang Beng Ho
☎ (+65) 768 30 00
Head of Air Intelligence Dept: Lt Col Yap Ong Heng
Head of Air Operations Dept: Col Goh yong Siang
☎ (+65) 768 35 06
Head of Air Plans Dept: Lt Col Peter Wong Nin Jee
☎ (+65) 768 35 64
Commander of Air Defence Artillery:
Lt Col Chong Lim Chye ☎ (+65) 282 24 22
Head Air Training Dept: Lt Col Looi Han Seng
☎ (+65) 768 35 08
Head Aeromedical Centre: Lt Col Dr Peng Chung Mien
☎ (+65) 287 61 11 ext 201
Commander Systems Command (AFSC):
Lt Col Mark Wong Lai Choong
☎ (+65) 760 43 11 ext 211
Commander Tactical Air Support Command:
Col K. Kuharajahsingam
Commander Changi Air Base:
Lt Col Timothy James de Souza ☎ (+65) 542 31 13
Commander Sembawang Air Base: Col D.G. Rajoo
☎ (+65) 754 33 11 ext 8000
Commander Tengah Air Base: Lt Col Joe Kua Sian Ten
☎ (+65) 760 33 11 ext 2000 [97]

25-30000 Army Headquarters
Ministry of Defence, MINDEF Building, Gombak Drive, Singapore 2366
☎ (+65) 760 81 88 ◙ 21373 singdef rs
⊖ (+65) 472 33 51
Chief of Army: Brig.Gen. Han Eng Juan
☎ (+65) 768 24 14
Asst Chief of Staff Personnel: Lt Col Chua Chwee Koh
Asst Chief of General Staff Intelligence: Col Tan Nga Kok
Asst Chief of General Staff Logistics: Col Lee Wai Mun
Asst Chief of General Staff Operations:
Lt Col Desmond Kuek ☎ (+65) 768 31 16
Asst Chief of General Staff Plans: Col Pang Hee Hon
Asst Chief of General Staff Training: Col Leong Yue Kheong
Chief Armour Officer: Col Lim Kim Lye
Chief Artillery Officer: Col Philip Su Poon Ghee
Chief Engineer Officer: Lt Col Lee Tong Tai
Chief Infantry Officer: Col Yeo Cheng Ann
Chief Signal Officer: Lt Col Yun Ta Chun
Chief Army Medical Officer: Lt Col Dr Low Wye Mun [97]

28 Coastguard, Border Protection Forces

28-10000 Port of Singapore Authority
PSA Bldg, 460 Alexandra Road, Singapore 119963
☎ (+65) 274 71 11 ☒ TANJONG ◙ 21507 rs
⊖ (+65) 274 42 61; 46 77
Chairman: Dr Yeo Ning Hong ☎ (+65) 279 57 09
CEO: Khoo Teng Chye
Deputy CEO International: Goon Kok Loon
Deputy CEO Operations: Lee Chee Yeng
Finance Dir: Kevin Yap Foo
Engineering Dir: Philip Ng Fook Wah
Operations Dir: Ng Chee Keong
Information Systems Dir: Eric Lui Chew
■ Providing & maintaining adequate & efficient port services & facilities. [97]

31 Embassies & Attachés

31-00090 Australian High Commission
25 Napier Road, Singapore 208507
☎ (+65) 737 93 11 ◙ 21238 rs
⊖ (+65) 733 71 34; 734 42 65 (Commercial Section)
⊖ (+65) 737 54 00 (Public Affairs Section); 737 54 81 (Admin Section)
High Commissioner: H.E. Mr E.F. Delofski
Defence Adviser: Capt. Garry Kennedy
☎ (+65) 733 71 34 (Fax)
Asst Defence Adviser: Sqn Ldr Glen Coy
Counsellor Commercial & Sr Trade Commissioner:
Ms Julie Bayliss
RAN Liaision Officer: Cdr John Brown [97]

31-00130 People's Republic of Bangladesh High Commission
101 Thomson Road #06-07, United Square, Singapore 1130
☎ (+65) 255 00 75; 250 63 23
[E] bdoot@singnet.com.sg ◙ 23312 bangla rs
⊖ (+65) 255 18 24
High Commissioner: H E Mr Ashfaqur Rahman
☎ (+65) 250 55 39
Deputy High Commissioner: Dr M A Samad
First Secretary: Mahbubur Rahman
Second Secretary: Monir ul Islam [98]

31-00150 Embassy of Belgium
10 Anson Road, Int'l Plaza 09-24, Singapore 079903
☎ (+65 220) 76 77/79 ☒ AMBABELGE Singapore
◙ 23301 ambel rs ⊖ (+65 222) 69 76
Ambassador: H.E. Mr Olivier Gilles de Pelichy
Third Secretary: Francois Bontemps
Trade Commissioner: Oscar Cousy [97]

31-00350 Embassy of Chile
105 Cecil Street 14-01/02, The Octagon, Singapore 0106
☎ (+65) 223 85 77/8 ◙ 34187 emchil rs
⊖ (+65) 225 06 77
Ambassador: H.E. Mr Carlos Tudela Aroca
First Secretary: Claudio Rojas
Second Secretary: Enrique Subercaseaux
Second Secretary: Patricio Powell
Third Secretary: Javier Matta

Commercial Department
133 Cecil Street 11-01, Keck Seng Tower, Singapore 06935
☎ (+65) 225 52 44 ◙ 42233 prochile rs
⊖ (+65) 225 52 45
Commercial Attaché: Mrs Rose Marie Bedecarratz

Information Bureau
179 River Valley Road #02-02, Singapore 179033
☎ (+65) 334 43 53 ⊖ (+65) 334 43 52 [97]

31-00460 Royal Danish Embassy
101 Thomson Road, #13-01 United Square, Singapore 307591
☎ (+65) 250 33 83 [E] embassy@denmark.com.sg
◙ 24576 ambadan rs ⊖ (+65) 253 37 64
Ambassador: H E Mr Niels Kaas Dyrlund
Senior Trade Officer: Anthony Leong
Attaché: Per I Johansen [98]

31-00580 Embassy of Finland
101 Thomson Road, #21-03 United Square, Singapore 307591
☎ (+65) 254 40 42 ☒ FINLANDIA Singapore
◙ 21489 finns rs ⊖ (+65) 253 41 01
Ambassador: H.E. Mr Veli J. Ollikainen
(Resident in Jakarta)
Charge d'Affaires: Jukka Leino
Trade Commissioner: Olli Laasanen
Attaché Consular & Admin: Ms Maritaka Maunola [97]

31-00590 Embassy of France
5 Gallop Road, Singapore 258960
☎ (+65) 466 48 61/66; 49 07; 42 58; 42 12; 468 58 80
◙ 21351 amfrans rs ⊖ (+65) 469 09 07; 468 41 90
Ambassador: H.E. Mr Francois Barry Delongchamps
Deputy Chief of Mission: Pierre Levy
Defence Attaché: Capt. Patrick Stervinou
Technical Defence Attaché: Ms Michele Laussucq-Kubler

Commercial Counsellor's Office
331 North Bridge Road, 18-03/06, Singapore 188720
☎ (+65) 339 97 10 ⊖ (+65) 339 31 37
Commercial Counsellor: Guy Lombard [97]

31-00880 Embassy of Israel
58 Dalvey Road, Singapore 259463
☎ (+65) 235 09 66 [E] israel@singnet.com.sg
⊖ (+65) 733 70 08; 737 25 02 (trade department)
Ambassador: H E Mr Daniel Megiddo
Deputy Chief of Mission: Oded Joseph
First Secretary Economic Affairs: Oren Jammari
Counsellor: Ms Louise Shimron
Military Attaché: Adm David Ben-Bashat [98]

31-00890 Embassy of the Republic of Italy
101 Thomson Road 27-02/3, United Square, Singapore 307591

☎(+65) 250 60 22 ✉ITALDIPL Singapore
Telex 21177 italdip rs ⊖ (+65) 253 33 01
Ambassador: H.E. Dr Enrico Augelli
First Secretary: Dr Domenico La Spina
Navy, Military & Aeronautic Attaché:
Col Lanfranco Laccisaglia (Resident in Jakarta)
Cultural Attaché: Dr Gabriele Tombini
Italian Trade Commission
1 Maritime Square #12-05, World Trade Centre, Singapore 0409
☎(+65) 273 14 44 Telex 21865 ittrade rs
⊖ (+65) 278 19 54 [97]

31-00920 Embassy of Japan
80 Anson Road, IBM Towers, #34-00 and #35-00, Singapore 079907
☎(+65) 235 88 55; 235 88 59 Telex 21353
⊖ (+65) 324 26 16
Ambassador: H E Mr Katsuhisa Uchida
Minister: Juichi Takahara
First Secretary and Defense Attaché:
Capt Fumio Suetsugu
First Secretary Commercial: Hideshi Ueda [98]

31-01110 Malaysian High Commission
301 Jervois Road, Singapore 249077
☎(+65) 235 01 11; 02 23; 05 34; 09 90
Telex 21406 surjaya rs ⊖ (+65) 733 61 35
High Commissioner: H E Mr Datuk Emam Mohd Haniff
Deputy High Commissioner: Mohd Fadzil Ajob
Counsellor and Head of Chancery: Hasnudin Hamzah
Economic Counsellor: Ismail Lajen ☎(+65) 235 16 05
Defence Adviser: Col Yusoff Izzudin [98]

31-01270 Royal Netherlands Embassy
Orchard Road, #13-01 Liat Towers, Tanglin PO Box 447, Singapore 23881
☎(+65) 737 11 55 Telex 33815 nethemb rs
⊖ (+65) 737 19 40
Ambassador: H.E. Mr J.G. Wilbrenninck
Defence Attaché: Col B.T. de Mars (Resident in Jakarta)
First Secretary: L.C.J.D. Schillings
Attaché: M.C. Wijnands [97]

31-01300 New Zealand High Commission
391a Orchard Road, #15-06/10 Ngee Ann City, Singapore 238873
☎(+65) 235 99 66 Telex 21244 tainui rs ⊖ (+65) 733 99 24
High Commissioner: H E Mr Geoff Ward
Deputy High Commissioner: Ms Carol Douglas
Defence Adviser: Capt D J Davidson
Regional Trade Commissioner: Chris Girton
Trade Commissioner: Grant Fuller
Travel Commissioner: Fergus Brown [98]

31-01370 High Commission of the Islamic Republic of Pakistan
20a Nassim Road, Singapore 258396
☎(+65) 737 66 21; 737 69 88; ✉PAREP Singapore
⊖ (+65) 737 40 96; 737 12 96 (commercial)
High Commissioner: H E Mr Toheed Ahad
Deputy High Commissioner: Sohail Ittehad Hussain
Commercial Counsellor: Ahmad Farooq
Accountant: M Akram [98]

31-01420 Embassy of the Philippines
20 Nassim Road, Singapore 258395
☎(+65) 737 39 77; 737 33 32; 737 37 54 Telex 34445 rs
⊖ (+65) 733 95 44
Ambassador: H E Mr Alberto A Encomienda
Minister Counsellor: Victoriano A Lecaros
Armed Forces Attaché: Capt Rufino Lopez (Jr)
Commercial Attaché: Eugene Reyes
Attaché and Administrative Officer: Benjamin S Fransisco
Office of the Marketing Representative
20 Nassim Road, Singapore 258395
☎(+65) 235 21 84 ⊖ (+65) 235 95 39
Marketing Representative: Phineas Alburo [98]

31-01430 Embassy of the Republic of Poland
100 Beach Road, Shaw Towers #33-11, Singapore 189702
☎(+65) 294 25 13/4 Telex 26355 morhan
⊖ (+65) 295 00 16
Charge d'Affaires a.i.: Krzysztof Kijak
Commercial Attaché: Marcin Baranski [97]

31-01485 Embassy of the Russian Federation
51 Nassim Road, Singapore 258439
☎(+65) 235 18 34 Telex 23404 suposol rs
⊖ (+65) 733 47 80
Ambassador: H.E. Mr Mikhail M. Bely
Third Secretary Chief of Protocol: Dmitri Lookiqntsev
Military, Air & Naval Attaché: Col Oleg G. Kotkov
Trade Representation
390 Orchard Rd, #13-04 Palais Renaissance, Singapore 238871
☎(+65) 737 62 21 Telex 21346 vnstorg rs
⊖ (+65) 235 07 49
Trade Representative: Anatoli N. Nazarov
Deputy Trade Representative: Mi-khail G. Vaniokov [97]

31-01730 Swedish Embassy
Orchard, PO Box 292, Singapore 912310
111 Somerset Road #05-01, Singapore 238164
☎(+65) 734 27 71 ✉SVENSK Singapore
Telex 23450 svensk rs ⊖ (+65) 732 29 58
Ambassador: H E Mr Krister Isaksson
Counsellor: Mrs Birgitta Malmvall
Defence Attaché: Capt Magnus Waldenström
Public Information Officer: Rune Sjöberg [98]

31-01740 Embassy of Switzerland
Swiss Club Link 1, Singapore 1128
☎(+65) 468 57 88 Telex 21501 amswiss
⊖ (+65) 466 82 45 [98]

31-01840 Embassy of the Republic of Turkey
20B Nassim Road, Singapore 258397
☎(+65) 732 92 11/2; 76 46 Telex 34668 sntrel rs
⊖ (+65) 738 17 86
Ambassador: H.E. Mr Oktay Ozuye
First Secretary: Ms Deniz Gakar
Commercial Officer: Alev Kaymak
Public Information Officer: Nejat Moral
Admin Attaché: Levent Kadioglu [97]

31-01890 United Kingdom High Commission
Tanglin Road, Singapore 247919
☎(+65) 473 93 33; 474 84 66 (visa/consular)
⊖ (+65) 474 19 58 (chancery); 475 97 06 (management); 474 04 68 (consular)
High Commissioner: H E Mr Alan C Hunt
Deputy High Commissioner and Counsellor Economic/Commercial: A J Gooch
Defence Adviser: Gp Capt J M Colllier (RAF)
Cultural and Economic Adviser (British Council Director): Dr J Grote
Head of Chancery: M J K Rickerd
Deputy Cultural and Educational Adviser, Director of English Language Centre: J W Knagg
First Secretary, Management and Consular: P A Ramsay
First Secretary, Commercial: A R Albright
Commercial Department
Tanglin PO Box 19, Singapore 247919
☎(+65) 44 04 61 ✉UKREP SINGAPORE
⊖ (+65) 475 23 20
Commercial and Economic Counsellor: A J Gooch [98]

31-01900 Embassy of the United States of America
27 Napier Road, Singapore 258508
☎(+65) 476 91 00 E mhasing@usia.gov
⊖ (+65) 476 90 35
Ambassador: H E Mr Timothy Chorba
Deputy Chief of Mission: Emil Skodon
Counsellor Public Affairs: Michael H Anderson
Defence Attaché: Capt Terry S Douglas
Army Attaché: Col Chris Hill
Air Attaché: Col Stephen M McClain (USAF)
Commercial Attaché: David Fulton
Economic and Political Officer: Bill Monroe [98]

60 Research Organizations, Institutes & Test Establishments

60-01000 SAFTI Military Institute Library
(Lib) AFPN 1456, 500 Upper Jurong Road, 01-06, Singapore 638364
☎(+65) 799 74 27; 74 29 ⊖ (+65) 792 27 80
Librarian: Ms Ko Yun Chin [98]

70 Financial Holding Groups, Parent Companies

70-01900 Singapore Technologies Pte Ltd
(† Singapore Technologies Holdings Pte Ltd)
Singapore Science Park, Singapore 118258
83 Science Park Drive, #01-01/02 The Curie
☎(+65) 775 32 66 ⊖ (+65) 775 32 33
♣ S$ 4,542,458 (1996) 👥 22,650 (1996)
Chairman: Teo Ming Kian
President and Chief Executive Officer: Ms Ho Ching
Manager, Corporate Communications: Ms Gladys Han
■ Technology-based multinational conglomerate contributing to the development of Singapore through enterprise. Capabilities include research and development, design, engineering, precision manufacturing and infrastructure development. Management of the following five core business groups:-
Engineering
Aerospace, automotive, ordnance, precision engineering, shipbuilding.
Technology
Data storage, electronics, information technology, semiconductor and telemedia.
Infrastructure and lifestyle
Construction, industrial parks, energy, food, leisure and logistics.
Property
Development, investment and management.
Financial Services
Customised financing, venture capital, regional operations.
Chartered Industries of Singapore (a member of Singapore Technologies)
249 Jalan Boon Lay, Singapore 619523
☎(+65) 265 10 66; 261 24 44 ⊖ (+65) 261 69 32
👥 1,900
President: Loh Chuk Yam
■ Integrated range of ordnance products and services including design, development, manufacture, testing and marketing. [98]

74 Consortia, Joint Venture Companies

Siemens Showa Solar Pte Ltd 90-01928

75 Manufacturing Companies

75-00445 AlliedSignal Commercial Avionics Inc
512 Chai Chee Lane #05-01/06, Singapore 469028
☎(+65) 448 61 22 ⊖ (+65) 448 02 73
Managing Director: Michael Penafort ☎(+65) 249 91 81
Director Quality Assurance: P H Chang
☎(+65) 249 91 25
Manager, Materials: Lai Yew Chin ☎(+65) 249 91 05
Manager, Test Engineering: Tan Yong Chuang
☎(+65) 249 91 05
Program Manager: Chua Yak Chye ☎(+65) 249 91 33 [98]

75-00500 Allied Ordnance of Singapore Pte Ltd
(a subsidiary of Chartered Industries of Singapore)
2b/2c Ayer Rajah Crescent, Singapore
☎(+65) 779 79 55 ⊖ (+65) 779 02 84
Vice President and General Manager: Shae Toh Hock
Operations Director: Tjua Jang Long
Senior Marketing Manager: Teddy Oon
Purchasing Manager: Oliver Chong
■ Manufacture of 40mm L70 Air Defence Weapon Systems comprising guns and ammunition. [98]

75-00550 AlliedSignal (Ltd) Singapore
(† Garrett Singapore Pte Ltd)
(a subsidiary of Allied-Signal Aerospace, USA)
7 Joo Koon Circle, Jurong, Singapore 2262
☎ (+65) 861 69 55 ⊖ (+65) 861 42 97
Managing Director: N G Koon Lye ☎ (+65) 861 66 01; 862 23 08 (fax)
■ Manufacture and assembly of precision aerospace component parts and systems. [98]

75-00800 Avimo Electro-Optics Pte Ltd (AVS)
(† Avimo Singapore Ltd)
(an associate of Alvis plc)
14 Fifth Lok Yang Rd, Jurong Town, Singapore 2262
☎ (+65) 265 51 22 ⊖ (+65) 265 14 79
S$ 10 000 000 ♣ S$ 40 000 000 450
1 500 sq m
Managing Dir: Avi Ayalon
VP Operations & Technical: John Yane
Sr Mgr Marketing Business Development: Tan Joon Hong
Purchasing Mgr: Martin Lee
■ Design, development & manufacture optical sub-assemblies & electro-optical instruments (including lens design and coating development). Primary products: night vision systems, sub-assemblies and components. Day/dusk rifle sights. Precision Investment Casting for Aircraft.
Subsidiaries:
Arab International Optronics Co., Egypt
Avimo Ltd
Helio Mirror Co. Ltd [98]

75-01080 CET Technologies
(† CEI Technologies Pte Ltd)
(a member of Singapore Technologies Group of Companies)
Singapore Technologies Building,
100 Jurong East Street 21, Singapore 609602
☎ (+65) 56 76 69 ✉ DOBERMAN Singapore
55992 ceipl ⊖ (+65) 567 63 00
♣ S$ 100,000,000 (1996) 550 (1997)
Vice President and General Manager: Seah Moon Ming
Marketing Director: Khoo Swee Peng
Purchasing Manager: Jeffrey Wong
■ Electronic fuzes (proximity and time) for improved effectiveness of artillery and mortar ammunition. Manufacture of electronic fuzes for mortar and artillery rounds. [98]

75-01090 Chartered Ammunition Industries Pte Ltd
(a subsidiary of Chartered Industries of Singapore)
249 Jalan Boon Lay, Singapore
☎ (+65) 265 10 66 ⊖ (+65) 268 75 79
Assistant Director: Lee Poh Choon
General Manager: Wong Peng Lam
■ Design, develop, manufacture, assembly, refurbishment and modification of ordnance products including small arms, aircraft ammunition, mortar bombs, artillery shells, explosives and pyrotechnics. [98]

75-01150 Chartered Firearms Industries Pte Ltd (CFI)
(a subs. of Chartered Ind. of Singapore)
249 Jalan Boon Lay, Jurong Town, Singapore 619523
☎ (+65) 265 10 66; 261 24 44 ✉ DOBERMAN
38951 cis rs ⊖ (+65) 264 46 06
165 7 799 sq m
Chairman: Lye Fei
Sr Vice President & General Mgr: Wong Peng Lam
Marketing Dir: Tham Mow Siang ☎ (+65) 265 10 66
Sr Engineer Marketing: Miah Wee Lim
■ Manufacture of small arms. [97]

75-02200 Intec Precision Equipment Pte Ltd
N°7 Gul Street 4 - Jurong Town, Singapore 2262
☎ (+65) 862 10 90 ⊖ (+65) 861 16 90
Managing Director: Siew Shie Peng
Marketing Director: Ms Cecelia Ng
® Switzerland: Trimos-Sylvac Task Force [98]

75-02300 Kawasaki Heavy Industries (Singapore) Pte Ltd
6 Battery Road #18-04, Singapore 049909
☎ (+62) 225 51 33 ⊖ (+65) 224 90 29 [98]

75-02720 Metheus Systems (S) Pte Ltd (Asia)
24 Ang Mo Kio St 65, 4th Floor, East Wing, Singapore 569061
☎ (+65) 486 0703 E singapore@metheus.com
⊖ (+65) 482 0715 [97]

75-02740 NMB Singapore Ltd
1 Chai Chee Avenue, Singapore 1646
☎ (+65) 241 10 33 8734331 desin
⊖ (+65) 444 65 27
957
Senior Director: S Fujisaua
■ Manufacture of M&I Bearings. Strain guage lord cell.
Minebea Technologies Pte Ltd
Purchasing Systems and Finance Divisions
1 Chai Chee Avenue, Singapore 469059
☎ (+65) 241 10 33 56630 ⊖ (+65) 444 65 27
Sales Division
PSA Building, 460 Alexandra Road #34-02, Singapore 0511
☎ (+65) 278 85 22 56530 minsin rs
⊖ (+65) 278 84 77 [98]

75-02800 Ordnance Development and Engineering Company of Singapore (1996) Pte Ltd
(a subsidiary of Chartered Industries of Singapore)
15 Chin Bee Drive, Singapore 619863
☎ (+65) 265 53 22 ⊖ (+65) 265 22 85
Senior Exeutive Marketing: Ms Wendy Cheng
■ Manufacture of weapons, 7.62mm machine guns to 60mm, 81mm, 120mm mortars, 155mm Field Howitzers - 39 and 52 calibre, FH88. Light machine guns, multi-purpose grenade launchers, assault rifles. [98]

75-02850 OTO (S.E.A.) Pte Ltd
802 Thong Teck Building, 15 Scotts Rd, Singapore
☎ (+65) 37 79 22 21449
Managing Dir: Ing. Rinaldo Gonnelli
® Italy: OTO Melara [97]

75-03200 Pelmec Industries (Pte) Ltd
28 Gul Way, Jurong Industrial Estate, Singapore 2262
☎ (+65) 861 68 55 8726278 ⊖ (+65) 861 72 12
491
Plant Manager: K Omi
■ Manufacture of ball bearings. [98]

75-03370 Radcoflex Singapore Pte Ltd
(a subs. of Radcoflex Australia Pty Ltd)
Block 8, Pandan Loop, Ste 01-122, Singapore 128326
☎ (+65) 777 26 20 ⊖ (+65) 779 75 31
50 1 000 sq m
Managing Dir: Ian Robinson
Operations & Technical Dir: Richard Ng
Purchasing Mgr: A. Hassim
■ Flexible metal hoses & expansion joints. [97]

75-03400 Saft Singapore Pte Ltd
(a subsidiary of Saft SA, France)
996 Bendemeer Road, 05-01/04,
Kallang Basin Industrial Estate, Singapore 339944
☎ (+65) 296 82 11 36253 saft rs ⊖ (+65) 298 41 01
50
Managing Director: Mrs Grace Chiam
Marketing Co-ordinator: Ms Rosalind Ang
■ Nickel-cadmium sealed, vented and lithium batteries for aircraft.
® France: Saft SA
Saft Nife Power Systems Pte Ltd
54 Genting Lane, # 03-01/03 Hiang Kie Complex 2, Singapore 349562
☎ (+65) 747 53 83 33578 ⊖ (+65) 741 93 65 [98]

75-03905 Singapore Technologies Aerospace Ltd
(a divison of Singapore Technologies Pte Ltd)
540 Airport Road, Paya Lebar 539938
☎ (+65) 287 11 11 43255 samkg
⊖ (+65) 280 97 13; 280 82 13
♣ S$ 523,389,000 (1996) 3,840
Chairman: Ms Ho Ching
Group President and Chief Executive Officer: Boon Swan Foo
Deputy President and President of Military Marketing: Brig Gen Gary Yeo
Executive Vice President and President of Business Development: Tay Kok Khiang
Senior Vice President, Commercial Business Group: Wee Siew Kim
Senior Vice President, Military Business Group: Chang Cheow Teck
Senior Vice President International Marketing: Ooi Ling Heong
Senior Vice President International Business: Lim Lu Hock
Director Business Development: Sew Chee Jhuen
Group Financial Controller: Patrick Lee
Director, Human Resources: Kuan Kwee Jee
Head of Legal Department and Company Seretary: Ms Michelle Neo
Head of Corporate Communications: Ms Shirley Tan
■ Products and services including high echelon aircraft maintenance, structural modification and refurbishment, engine and aircraft component overhaul and repair, supplies, material support and engineering services.
STAe Group:
Military Business Group
ST Aerospace Engineering Pte Ltd
ST Aerospace Engines Pte Ltd
ST Aerospace Systems Pte Ltd
ST Aerospace Supplies Pte Ltd
Commercial Business Group
ST Aviation Services Co Pte Ltd (SASCO)
ST Mobile Aerospace Engineering Inc (MAE)
Airline Rotables Ltd (ARL) UK [98]

75-04280 Singapore Technologies Automotive Ltd (STA)
(† Singapore Automotive Engineering)
5 Portsdown Rd, Singapore 139296
☎ (+65) 473 63 11 ✉ SINAUTENG 25755 sineng rs
⊖ (+65) 471 06 62
S$ 25 000 000 (1995) ♣ S$ 175 89 000 000 (1995)
815 20 000 sq m
Corporate Services Exec.: Sandy Goh
Design & Development Mgr: Fong Saik Hay
Regional Marketing Mgr: Alex Teo
■ Complete facilities for rebuilding, modifying, upgrading & repairing all types of military vehicles from trucks to armoured vehicles. Also customises vehicles for special military use. Installation, systems integrator, automated storage & retrieval system. [97]

75-04400
Strategic Technologies Pte Ltd (STRATECH)
1 International Business Park #05-14, The Synergy, Singapore 609917
☎ (+65) 323 21 88 E stratech@singnet.com.sg
W stratech@pacific.net.sg
⊖ (+65) 323 21 77, 449 39 36
US$ 3,000,000 (1996) ♣ US$ 18,000,000 (1996)
62 5,000 sq ft
Managing Director and Chief Executive Officer: Dr David K M Chew
Executive Director Technology: Dr Kennedy K M Chew
Executive Director Corporate/Legal Affairs: Chew-Leong Sook Ching
Director: Lucy Ng
■ The design, development, manufacture, supply, installation, commissioning & maintenance of weapons scoring systems (both portable & fixed base), weapons ranges & ballistic instrumention for air, sea & land forces, weapons & ammunition manufacture, test & development agencies. Training & simulation systems. Security, law enforcement & covert surveillance & countermeasure systems. Information & advanced technology systems. Total support of air force, army, navy, airports, commercial airlines & aircraft maintenance facilities. ISO 9001 & ISO 9002 certified. [98]

75-04450 Trimble Navigation Singapore Pte Ltd
15 Scotts Rd, 03-01 Thong Teck Bldg, Singapore 0922
☎ (+65) 738 65 49 ⊖ (+65) 738 63 46 [97]

75-04570 Vias Aviation & Engineering Pte Ltd
Block 5095, Ang Mo Kio Ind Park 2, 01-1042/1044, Singapore 569598

☎ (+65) 481 11 36; 11 25 ⊖ (+65) 481 94 94
♧ S$ 1,000,000 (1997) ⛹ 20 (1997) 🏭 5,000 sq ft
Managing Director: Chua Horng
Operations and Technical Director: Lee Cheng Guan
■ Manufacture aircraft spares & GSE, aircraft sectional repair. Research & development of single procurement contact for all approved aerospace material parts for military/civil aircraft. [98]

80 Service Companies & Organizations

Aero Systems Pte Ltd	90-00060
SAMAERO Company Pte Ltd	90-01900
W.H. Brennan & Co. (Pte) Ltd	90-00265

80-00070 AlliedSignal Aerospace Marketing Sales and Service
Core N, 6th Story, SIAEC Hangar, 31 Airline Road, Singapore 1781
☎ (+65) 541 60 73 ⊖ (+65) 542 13 76
Field Service Engineer: Geoffrey Long
Field Service Engineer: Tony Garcia
☎ (+65) 541 60 71; 541 13 76 (fax)
Field Service Engineer: John Garrard
☎ (+65) 541 60 71; 541 39 99 (fax)
■ On-site customer support for AlliedSignal Aerospace. → 90 [98]

80-00072 AlliedSignal Aerospace Marketing Sales and Service
PO Box 166 Juron Town Post Office, Singapore 9161
☎ (+65) 861 45 59; 861 45 33 ⊖ (+65) 861 45 91
Marketing Analyst: Ms Janet Soh
■ On-site customer support for AlliedSignal Aerospace. → 90 [98]

80-00074 AlliedSignal Aerospace Marketing Sales and Service
333 River Valley Road, 07-02 Yong An Park, Singapore 238365
☎ (+65) 734 32 10; 730 46 45
⊖ (+65) 730 40 07; 732 56 95
Field Service, Product Engineer: Gene McCann
■ Marketing, sales, service and field service facility for AlliedSignal Aerospace military and commercial products. [98]

80-00076 AlliedSignal Aerospace Marketing Sales and Service
AlliedSignal Singapore (Pte) Limited
161 Gul Circle, Singapore 629619
☎ (+65) 861 79 12; 869 51 01 ⊠ 25399 garetsn
⊖ (+65) 862 24 73; 861 23 59
Vice President, Repair and Overhaul, Asia and Pacific: Jacques Esculier
Director, Marketing: Francis Lee
☎ (+65) 869 51 07; 862 38 78 (fax)
Customer Support Manager: George Lim
☎ (+65) 869 51 28; 862 38 78 (fax)
Customer Support Manager: Francis Ng
☎ (+65) 869 52 28; 862 38 78 (fax)
Manager, Marketing and Customer Support: Chia Chong Yeng ☎ (+65) 869 51 13; 862 38 78 (fax)
Field Service Engineer: Bill Begue
☎ (+65) 861 18 84; 861 50 73 (fax)
Field Service Engineer: Thye Phing Toh
☎ (+65) 869 18 84; 861 50 73 (fax)
■ Administration, manufacture, sale and servicing of AlliedSignal Aerospace products. [98]

80-00430 China Steel Corporation Singapore Office
#14-01 MAS Building, 10 Shenton Way, Singapore 0207
☎ (+65) 223 87 77 ⊖ (+65) 225 60 54 [98]

80-00450 Dowty Aerospace Aviation Services (1995) Pte Ltd
(a division of TI Group plc, UK)
21 Loyang Crescent, Loyang Industrial Estate, Singapore 508985
☎ (+65) 545 94 55 ✉ DOWTYS ⊖ (+65) 542 39 36
⛹ 125 (1997) 🏭 90,000 sq ft
Managing Director: H H Goh
Finance Director: Ms Mary Lim
Operations Director: C K Yap
Marketing and Sales Director: Albert Tay
■ Repair and overhaul of narrow and wide body airliner landing gears, up to and including B747-400. Overhaul of electrical, hydro-mechanical, pneumatic ducts, propellors, pressure vessels, primary/secondary flight controls, flap carriages/transmissions and accessories. Full landing gear exchange service. [98]

80-00700 Hughes International Corporation
United Square Building #27-01, 101 Thomson Road, Singapore 1130
☎ (+65) 255 50 78 ⊖ (+65) 251 97 17 [98]

80-01000 Interturbine Singapore Pte Ltd
Interturbine Compressor Technology
18 Loyang Crescent, Singapore 1750
☎ (+65) 545 48 66 ⊠ 24565 vacy ⊖ (+65) 542 75 83
♧ S$ 5 000 000 ♣ US$ 11 000 000 ⛹ 150
🏭 4 000 sq m
Managing Dir: Tan Tat Thye
■ Overhaul of turbine engine hot & cold section components. [97]

80-01050 Jet Maintenance Pte Ltd
(a subsidiary of Jet Aviation Management AG, Switzerland)
1075 West Camp Road, Seletar Airport, Singapore 797800
☎ (+65) 481 53 11 🄴 jsin@jetaviation.com
⊖ (+65) 481 83 36
Vice President and General Manager: Ruedi Kraft
Manager, Private Aircraft Handling: Angie Koh
■ Maintenance of jet engines; interior refurbishment and outfitting; avionic retrofit; aircraft management and aircraft charter. [98]

80-01100 Logica Pte Ltd
(a subsidiary of Logica plc, UK)
PIL Building, 140 Cecil Street 16-04, Singapore 0106
☎ (+65) 223 53 43 ⊖ (+65) 223 44 75 [98]

80-01200 Lucas Aerospace Singapore Pte Ltd
(a division of Lucas Aerospace Ltd)
35-37 Loyang Way, Singapore 508733
☎ (+65) 545 99 75 ⊖ (+65) 545 99 65
⛹ 30
Business Manager: Y P Lim
Regional Sales and Marketing Manager: Chan Thud Chee
■ Servocontrols and hydraulic components, helicopter rescue hoists, electrical generation equipment, fuel systems, actuation systems for fixed and rotary wing aircraft. [98]

80-01640 PKM Consultants (S) Pte Ltd
(† PKM Ltd)
101 Cecil Street #10-08/09, Tong Eng Building, Singapore 069533
☎ (+65) 225 36 11; 225 36 12 ⊖ (+65) 225 58 72
⛹ 5 🏭 118 m2
Managing Director: Peter Maier
■ Regional consultants in South East Asia for defence related business.
Consultants to:
Aerea SpA
Autoflug GmbH
Cosmos SpA
Eurotorp SA
Intermarine SpA
Lohr SA
Mak System Gesellschaft mbH
Marconi SpA
Mowag Motorwagenfabrik AG
Noricum
Nico Pyrotechnik
Oto Breda Division of Alenia Difesa
RKS SA
Simmel Difesa
SIG Swiss Industrial Co. [98]

80-01660 PMsoft Pte Ltd
(a subsidiary of Artemis Management Systems Ltd, UK)
89 Science Park Drive, 04-06/07 The Rutherford, Singapore 0511
☎ (+65) 778 16 78 ⊖ (+65) 778 68 84
Regional Director: Johnny Tan [98]

80-01800 Sextant Avionique Asia Pte Ltd
(a subsidiary of SEXTANT Avionique SA, France)
4 Loyang Lane, Singapore 1750-0601
☎ (+65) 542 25 33 ⊠ 35616 ⊖ (+65) 542 96 50
General Manager: Franck Herbert
■ Customer and product support.
® France: Sextant Avionique SA [98]

80-01960 Singapore Electronic & Engineering Ltd
ST Electronic & Engineering Limited
(a subs. of Singapore Tech. Ind. Corp)
24 Ang Mo Kio Street 65, Singapore 569061
☎ (+65) 481 88 88 ⊠ 21901 seelsin
⊖ (+65) 482 10 79; 481 42 72 (Mktg)
♧ S$ 10 000 000 ⛹ 700 🏭 100 000 sq ft
Sr VP & General Mgr: Yap Eng Lip
Marketing Mgr: Fang Yeow Huey
■ Life cycle support for military & industrial electronic instruments, equipment & systems covering system design, installation, maintenance, overhauling & upgrading. → 90
® UK: Farnell Int'l
USA: Craig, Magnavox, Miltope, Watkins Johnson [97]

80-01965 Singapore Engineering Software Pte Ltd
(a joint venture of Singapore Technologies Industrial Corp. & CelsiusTech Systems AB)
73 Science Park Drive, #01-22 Cintech Bldg, Singapore 118254
☎ (+65) 777 88 22 ⊖ (+65) 778 71 71
♧ S$ 1 000 000 ♣ S$ 7 000 000 ⛹ 60
General Mgr: Goh Kee Fong
■ Design, development, production, installation & maintenance of software products & provision of software engineering services. Certified to ISO 9001. Customised Realtime Decision Support Systems. System Integration Services. Software development. Comand & Control Systems. Air & Sea Port Management Systems (traffic control, facility scheduling, resource management). Distributed Logistics Systems. Simulation. Prototyping, Expert Systems. [97]

80-01970 Singapore Ordnance Engineering Pte Ltd
(a subsidiary of Singapore Technologies Automotive Ltd)
Jalan Kayu Post Office, PO Box 91, Singapore 918004
☎ (+65) 481 84 63 ⊖ (+65) 481 91 59
⛹ 25 (1997)
Chairman: Wu Tzu Chien
General Manager: Alex Teo
■ Design, integration and construction of prototypes and speciality mechanical, electrical and communication equipment. Depot maintenance, repair and retrofitting of armoured vehicle systems and equipment. [98]

80-01990 ST Aerospace Engineering Pte Ltd
540 Airport Road, Singapore 539938
☎ (+65) 287 11 11 ✉ SAIRMAINT ⊠ 25507 samair
⊖ (+65) 282 32 36
⛹ 1,011 🏭 61,400 m2
President: Chang Cheow Teck ☎ (+65) 380 65 15
Vice President and General Manager: Ted Chia Chee Kiong
Vice President, Operations: Ho Yuen Sang
Senior Manager: R Balakrishnan
Manager, Customer Relations and Sales: Tang Weng Chuen
■ Maintenance, modification, repair and servicing of civil and military aircraft and helicopters; authorised service centre for Beechcraft, Bell Helicopter Textron, Eurocopter France (Puma and Super Puma), modification centre for Raisbeck Engineering STC kit. [98]

80-02000 ST Aerospace Engines Pte Ltd
(a member of Singapore Technologies)
501 Airport Rd, Paya Lebar, Singapore 539931

☎(+65) 285 11 11 ⊠ 33268 saeng ⊖ (+65) 282 30 10
⚇ 380 (1995) ⌂ 13 984 sq m (1995)
Sr Vice President & General Mgr: Chong Kok Pan
Commercial Mgr: Anthony Tong
Technical Mgr: Chun Mun Cheng
Works Mgr: Choo Han Khoon
■ Overhaul & repair of Pratt & Whitney JT8D, & F100, Allison T56/501, General Electric F404 & J85, AlliedSignal T53, Turbomeca Makila & Arriel engines. [97]

80-02002 ST Aerospace Supplies Pte Ltd
540 Airport Road, Paya Lebar, Singapore 539934
☎(+65) 287 11 11; 380 64 07; 98 29 34 55
⊠ 51158 samkg rs ⊖ (+65) 287 09 10; 284 31 46 [98]

80-02005 ST Aerospace Systems Pte Ltd
(† Singapore Aerospace Systems Pte Ltd)
(a subsidiary of ST Aerospace Ltd)
505a Airport Road, Paya Lebar, Singapore 539934
☎(+65) 287 22 22 ⊠ 55851 saero
⊖ (+65) 284 44 14; 284 02 36
⚇ 320 ⌂ 12,000 m2
Vice President and General Manager: Ang Chye Kiat
☎(+65) 389 46 12; 389 46 13; 284 02 36 (fax)
Manager Product Support and Development: Walter Kan
☎(+65) 380 46 76
■ Maintenance, overhaul and repair of civil and military aircraft components and equipment. Authorised service centre for Aerospatiale, Allied Signal (Bendix/King), Astronautics, ECE, GEC Ferranti, Hydraulic Research Textron, JET Kollsman, Lucas Aerospace, Revue Thommen, Rockwell Collins, Sextant Avionique, Superflexit, Teledyne, Thomson-CSF and Thomson-TRT Defense. [98]

80-02007 ST Aviation Services Company Pte Ltd
(a subsidiary of Singapore Technologies Aerospace Ltd)
11 Changi North Crescent, Singapore 499611
☎(+65) 545 09 88; 540 56 00 ⊖ (+65) 545 67 57 [98]

80-03000 Vipac Engineers & Scientists Pte Ltd Singapore
35 Jalan Peminpin, #03-01 Wee Poh Building, Singapore 2057
☎(+65) 353 87 11 ⊖ (+65) 353 42 11
♣ US$ 1,500,000 ⚇ 10
Managing Director (Asia): I G Jones
■ Consulting engineering services in acoustics, vibration and dynamics. [98]

85 Stockists, Distributors, Spare Parts Suppliers

Ultramare Singapore Pty Ltd 90-02090

85-01000 Israel Aircraft Industries Liaison Office
7 Temasek Boulevard, #09-03a Suntec - Tower One, Singapore 038987
☎(+65) 333 34 56 [E] siairo@pacific.net.sg
⊖ (+65) 338 05 81
Director, Liaison Office: Arie Herzog
Sales Director: Matthew Nursella [98]

85-02000 Messier-Dowty Customer Support Centre - Asia/Pacific
21 Loyang Crescent, Loyang Industrial Estate, Singapore 1750
☎(+65) 543 06 26 ⊠ 21447 rs ⊖ (+65) 542 27 18
Customer Support Dir: Martin Akerman
Commercial Mgr: Johnny Eng
Technical Manager: Steve Barnes
■ Provides spare parts, AOG service & technical support for Messier-Dowty & Dowty Aerospace equipment in service in Asia Pacific regions & Australasia. [98]

85-03800 Satair ASIA Pte Ltd
(a subsidiary of Satair AS, Denmark)
8 Loyang Link, Singapore 1750
☎(+65) 543 09 77 ⊠ 51235 rs ⊖ (+65) 543 07 37
♣ US$ 20,000,000 (1997) ⚇ 23 (1997) ⌂ 1,300 m2
Managing Director: René Frandsen
Operations Manager: Chin-inn Wong
Sales Manager: Michael Leow
Quality Control Manager: Amin Masagoes [98]

85-04000 Singapore British Engineering Pte Ltd
(a joint venture between Singapore Technologies and BAe plc)
435 Orchard Road, #19-02 Wisma Atria, Singapore 238877
☎(+65) 735 83 28 [E] sbesing@mbox2.singnet.com.sg
⊖ (+65) 735 82 33
General Manager: Stanley Kwok
Senior Marketing Manager: K C Tan
■ Marketing of specific BAe defence products and systems. [98]

90 Sales Agents, Sales & Liaison Offices

Companies represented:

Aardvark, UK 90-00265
Aerosonic Corp., USA 90-00060
Akzo Nobel Coatings BV, Netherlands . 90-00020
Alfred Karcher GmbH & Co, Germany . 90-01120
Alliant Techsystems (Kilgore Operation), USA 90-00265
Allied-Signal Aerospace Co., USA . . 90-00085
Altair, USA 90-00060
Astronautics, USA 90-00060
Avon Industrial Polymers, UK 90-00265
Bell Helicopter Textron, USA 90-00220
Bendix/King, USA 90-00060
CMI Mechanical Industries, Belgium . . 90-00335
Collins Avionics, USA 90-00060
Colts Manufacturing Co., USA 90-00265
Companhia Brasileira De Cartuchos (CBC), Brazil 90-00265
Conax Corp., USA 90-00265
Consilium Marine AB, Sweden 90-01130
Cossor Electronics Ltd, UK 90-00940
Craig, USA 80-01960
Crompton Instruments Ltd, UK 90-00350
Cylink Corp., USA 90-00370
Daicel Chemical Industries Ltd, Japan . 90-00400
Eurocopter, Germany 90-01900
Eurocopter, France 90-01900
Farnell Int'l, UK 80-01960
Grumman Corp., USA 90-01480
Hägglunds Vehicle AB, Sweden 90-02210
Hiatts & Co., UK 90-00265
Hughes Training Inc Link Division, USA 90-01010
Kelvin Hughes Ltd, UK 90-01130
KHD, Germany 90-00450
Krauss-Maffei AG, Germany 90-01245
Lektro Inc, USA 90-00060
Lloyd Instruments Ltd, UK 90-01250
Loral Fairchild, USA 90-00060
Magnavox, USA 80-01960
Manitou BF, France 90-01300
Marine Data, UK 90-01130
Miltope, USA 80-01960
MTU Friedrichschafen GmbH, Germany 90-01370
NKI BV, Netherlands 90-01430
Noble Denton, UK 90-01450
OTO Melara, Italy 75-02850
Pains-Wessex Schermully, UK 90-00265
Racal Electronics, UK 90-01700
RFD, UK 90-00265
Ring Sights Holdings Ltd, UK 90-01840
Ring Sights Inc., USA 90-01840
Ring Sights Ltd, UK 90-01840
Rockwell International Corp., USA . . . 90-01850
Rogerson Kratos, USA 90-00060
Rohde & Schwarz GmbH & Kg, Germany 90-00940
S-Tec Corporation, USA 90-00060
Saft SA, France 75-03400
Schottel-Werft, Germany 90-01910
Sema Group plc, UK 90-01915
Sextant Avionique SA, France 80-01800
Showa Solar Energy K.K., Japan . . . 90-01928
Siebe Gorman, UK 90-00265
Siemens Solar Industries, USA 90-01928
SOCATA, France 90-01900
Sperry Marine Inc., USA 90-01960
Sunair Electronics, USA 90-00060
Teledyne Inc., USA 90-02018
Telemit GmbH, Germany→ Netherlands . 80-18300
Textron Marine and Land Systems, USA 90-02025
The General Electric Co., UK 90-00710
Trimos-Sylvac Task Force, Switzerland . 75-02200
Turbomeca, France 90-01900
Watkins Johnson, USA 80-01960
Westland Helicopters, UK 90-01010
William Fairey Engineering, UK 90-00265

Sales agents/representatives listed elsewhere:

AlliedSignal Aerospace Marketing . . 80-00072
AlliedSignal Aerospace Marketing . . 80-00070
Singapore Electronic & Engineering . . 80-01960

90-00020 AD Aerospace Finishes
Representative Office Australasia-Singapore
39 Oriole Crescent, Singapore 288631
☎(+65) 466 04 61 ⊖ (+65) 466 13 26
Contact: R P Wielinga
® Netherlands: Akzo Nobel Coatings BV [98]

90-00060 Aero Systems Pte Ltd
37 Jalan Pemimpin, St.03-01, Singapore 2057
☎(+65) 258 44 77 ⊠ AEROSYSTEM
⊠ 25205 Sales; 37837 Service ⊖ (+65) 258 22 33
Managing Dir: Rama Chandara
Regional Mgr: Frederick Lee
■ Aviation equipment sales & service organisation serving Asia, SE Asia & Australia. FAA certified service facilities at Airport (FAA repair station 651-14F and other approvals). → 80
® USA: Aerosonic Corp., Altair, Astronautics, Bendix/King, Collins Avionics, Lektro Inc, Loral Fairchild, Rogerson Kratos, S-Tec Corporation, Sunair Electronics [97]

90-00075 Aerotech Singapore Pte Ltd
(a subsidiary of Aerotech World Trade Corp, USA)
51 Newton Road, Goldhill Plaza #16-01/03, Singapore 308900
☎(+65) 738 35 11 [E] aerotech@singnet.com.sg
⊖ (+65) 738 25 83
⚇ 6 ⌂ 1,500 sq ft
Managing Director: Baljit Singh
Sales and Marketing Manager: Wee Hock Kee
Administrator: Ms Chandrika Nair [98]

90-00085 Allied-Signal Aerospace
(a subsidiary of AlliedSignal Aerospace, USA)
15-09 Shaw Center, 1 Scotts Road, Singapore 220208
☎(+65) 235 12 43 ⊖ (+65) 732 56 95
Regional Director, South Asia: Henriqué Marques
☎(+65) 730 46 11
Field Service and Product Engineer: Gene McCann
☎(+65) 730 46 45; 734 32 10; 737 56 95 (fax); 734 40 07 (fax)
® USA: Allied-Signal Aerospace Co. [98]

90-00220 Bell Helicopter Asia (Pte) Ltd
(a subsidiary of Bell Helicopter, Textron, USA)
Tampines South Post Office, PO Box 0314, Singapore 9152
50 Loyang Way, Singapore 1750
☎(+65) 542 24 22 ⊠ 50555 bhasia rs
⊖ (+65) 542 73 87; 542 70 38
Marketing Director: Leon Y Lui
Customer Support Director: David P Blake
■ Bell Helicopter sales, service and spare parts for Singapore, Myanmar, Brunei, Bangladesh, Hong Kong, Malaysia, Philippines, Thailand, Sri Lanka, Indonesia, People's Republic of China, Nepal, India, Vietnam and Cambodia.
Wholly owned subsidiary of
® USA: Bell Helicopter Textron [98]

90-00255 Bliss-Fox Manufacturing Pty Ltd
132-a Jalan Dermawan, Hillview Court, Singapore 2366
☎(+65) 763 64 40 ⊖ (+65) 760 01 17 [98]

90-00265 W.H. Brennan & Co. (Pte) Ltd
(a subs. of Boustead Group of Companies)
47 Loyang Way, Singapore 1750

☎(+65) 542 97 22 ⊠ 55539 brenco rs
⊖ (+65) 542 52 46
♁ S$ 3 000 000 (1996) ♣ S$ 16 000 000 (1995) ⚉ 100
⌂ 4 346 sq m
Managing Dir: Choy Cheok Wing
General Mgr Marine Safety & Marketing: Joseph Choo
Marketing Div. Mgr: Tow Peng Soon
Mgr Flight Safety: Mak Kwok Kee
Fire Safety Mgr: Raymond Yap
■ Manufacturers representative. Trading in defence, safety & security equipment. Service & repair facility for evacuation slides, liferaft, lifevests, aircraft/ship fire extinguishing equipment & oxygen systems. Design, supply, install, commission & service fire protection systems. Halon recovery & recycling. → 80
® Brazil: Companhia Brasileira De Cartuchos (CBC)
UK: Aardvark, Avon Industrial Polymers, Hiatts & Co., Pains-Wessex Schermully, RFD, Siebe Gorman, William Fairey Engineering
USA: Alliant Techsystems (Kilgore Operation), Colts Manufacturing Co., Conax Corp. [97]

90-00275 British Steel (South East Asia) Pte Ltd
(a subsidiary of British Steel Engineering Steels, UK)
1123 Serangoon Road #01-02, Singapore 1232
☎(+65) 297 66 78 ⊖ (+65) 297 66 82 [98]

90-00335 CMI Cockerill Mechanical Industries Far East Office
1 Tanglin Road Hex 04-18, Orchard Parade, Podium Block, Singapore 1024
☎(+65) 732 35 00 ⊠ 36983 ⊖ (+65) 732 05 38
Regional Director: Alec Cadi
Deputy Regional Director: Eric Franssen
® Belgium: CMI Mechanical Industries [98]

90-00350 Crompton Instruments (S.E. Asia) Pty Ltd
Block 5000, 02-08 Ang Mo Kio Ave 5, 02-08 Techplace II, Singapore 2056
☎(+65) 481 88 66 ⊖ (+65) 481 82 54
♣ £ 1 000 000 ⚉ 10
General Mgr: D. Wee
® UK: Crompton Instruments Ltd [98]

90-00370 Cylink Corporation
6001 Beach Road, 22-03 Golden Mile Tower, Singapore 0719
☎(+65) 297 61 96 ⊖ (+65) 297 61 95
Manager: James Wong
■ Supplier of commercial encryption devices.
® USA: Cylink Corp. [97]

90-00380 Daewoo Singapore Pte Ltd
7 Temasek Boulevard Hex 26-03, Suntec Tower One, Singapore 038987
☎(+65) 332 37 00 ⊠ 21786 daewoo rs
⊖ (+65) 336 69 60 [98]

90-00400 Daicel Chemical (Asia) Pte Ltd
65 Chulia St, 146-08 OCBC Centre, Singapore 0104
☎(+65) 533 25 11 ⊖ (+65) 532 64 54
® Japan: Daicel Chemical Industries Ltd
Asia Research Centre
15a Senoko Way, Singapore 758037
☎(+65) 756 28 11 ⊖ (+65) 752 37 80
⚉ 5 (1997) ⌂ 1,800 m2
General Manager: Michio Kobayashi [97]

90-00430 Detector Electronics Far East
143 Cecil Street, 15-01 G B Building, Singapore 0106
☎(+65) 229 13 55 ⊖ (+65) 226 63 05
Contact: Cliff Anderson [98]

90-00450 DEUZT Asia-Pacific (Pte) Ltd
PO Box 98 Jurong, Singapore 916404
11 Kian Teck Road, Singapore 628768
☎(+65) 268 53 11 ✉ DEUTZMOTOR ⊠ 23881 dfe rs
⊖ (+65) 264 17 79; 265 30 07; 265 29 73
Regional Financial Controller: Sham Kit Kiong
® Germany: KHD [98]

90-00710 GEC of Singapore Pte Ltd
No.3 Tai Seng Drive, Singapore 535216
☎(+65) 382 82 33 ⊖ (+65) 382 82 00
Regional Managing Dir: R. Ashman
® UK: The General Electric Co. [97]

90-00755 Glamco Aviation Pte Ltd
35 Tannery Rd #10-07 Tannery Block, Ruby Industrial Complex, Singapore 347740
☎(+65) 749 10 77 ⊖ (+65) 749 10 63
General Mgr: James Chia
■ Trading in aviation & military supplies. [97]

90-00900 Hy-Q International (Singapore)
63 Hillview Avenue, 08-22 Lam Soon Industrial Building, Singapore 2366
☎(+65) 762 18 82 ✉ HYQUESING ⊠ 21427 hyqsing
⊖ (+65) 762 39 83 [97]

90-00910 Imrex Aviation Co. Inc.
Block 4010, Ang Mokio Avenue 10, #07-06 TTech Place 1, Singapore 569626
☎(+65) 552 85 88 ⊖ (+65) 552 16 67
Sales & Marketing Dir: Spencer Ong [98]

90-00940 Infotel Technologies (Pte) Ltd
19 Tai Seng Drive #06-00, ST Building, Singapore 1953
☎(+65) 287 68 22 [E] general@infotel.com.sg
⊖ (+65) 287 65 77
Managing Director: S P Leong
Director of Sales and Marketing: James Han
® Germany: Rohde & Schwarz GmbH & Kg
UK: Cossor Electronics Ltd [98]

90-01010 Jardine Matheson (S) Ltd Aviation Division
4 Shenton Way, 17-00 Shing Kwan House, Singapore 0106
☎(+65) 220 51 11 ⊠ 24485 jardines ⊖ (+65) 224 63 70
♣ US$ 12,000,000,000 (1997) ⚉ 5,000
Managing Director: Boon Yoon Chiang
Business Development Manager: Michael Liew
® UK: Westland Helicopters
USA: Hughes Training Inc Link Division [98]

90-01120 Karcher Asia-Pacific Pte Ltd
(† Karcher Pte Ltd)
34 Penjuru Lane, Singapore 609201
☎(+65) 264 23 20 ⊖ (+65) 264 23 21
Managing Director: Kenneth Eng
® Germany: Alfred Karcher GmbH & Co [98]

90-01130 Kelvin Hughes (Singapore) Pte Ltd
(a subsidiary of Kelvin Hughes Ltd, UK)
N°4 Loyang Drive, Singapore 508965
☎(+65) 545 98 80 ⊠ 23482 rs ⊖ (+65) 545 88 92
General Manager: M R Dickinson
® Sweden: Consilium Marine AB
UK: Kelvin Hughes Ltd, Marine Data [98]

90-01180 Kidde Aerospace & Defence Far East
143 GB Building, #15-01, Cecil Street, Singapore 069542
☎(+65) 220 13 55 ⊖ (+65) 226 63 05
Vice President: Tan Hillman
Regional Dir: Patrick Neo
■ Regional office for divisional products. [98]

90-01238 KNT Engineering & Trading Pte Ltd
Block 2 #04-08, Kaki Bukit Avenue 1, Singapore 1441
☎(+65) 841 79 98 ⊖ (+65) 841 79 60
♁ S$ 600 000 ♣ S$ 2 000 000 ⚉ 13 ⌂ 491 sq m
Regional Sales Mgr: Ms Susan Tan
■ Microwave tubes for radar systems. Night vision devices.
Representing:
Varian Associates Inc.
ITT Defense
EEV Limited
Pilkington Optronics [97]

90-01245 Krauss-Maffei Asia Pacific Pte Ltd
539 Yishun Industrial Park A, Mannesmann Centre 2776
☎(+65) 755 66 36 ⊖ (+65) 755 31 33
Chairman: Dipl.-Ing Hinrich J Woebucken
® Germany: Krauss-Maffei AG [98]

90-01250 Lloyd Instruments Ltd Malaysia
26 Jalan SS 3/78, University Gardens, Petaling Jaya, Selangor 47300
☎(+60 3) 736 96 85 ⊖ (+60 3) 736 96 75
® UK: Lloyd Instruments Ltd [98]

90-01290 Lockheed Martin Global Inc
(† Lockheed Martin International Ltd)
(a subsidiary of Lockheed Martin Corporation, USA)
501 Orchard Rd - #21-02, Wheelock Place, Singapore 238880
☎(+65) 738 75 33 ⊖ (+65) 738 88 61
♣ US$ 27,000,000,000 (1997) ⚉ 4 (1997)
⌂ 3,606 sq ft
President South Asia Region: Paul C Johnson
Vice President Business Development: Bob Hammond
■ Resarch, design, development, manufacture and integration of advanced-technology products and services. Core business: aeronautics; electronis; energy; environment; information services; space; strategic missiles and systems integration. [98]

90-01300 Manitou Asia PTE Ltd
(a subsidiary of Manitou BF, France)
25 Tuas Avenue 4, Singapore 639375
☎(+65) 862 07 71 ⊖ (+65) 862 02 16
® France: Manitou BF [98]

90-01330 Miltope Asia-Pacific
331 North Bridge Road, 03-01 Odean Towers, Singapore 0718
☎(+65) 339 6209 ⊖ (+65) 338 5053
Manager: David Soh [97]

90-01350 Moog Singapore Pte Ltd
(a subs. of Moog Inc., USA)
14 Science Park Dr, # 04-06A The Maxwell, Singapore Science Park 0511
☎(+65) 773 62 38 ⊖ (+65) 777 76 27
♣ US$ 1 200 000 (1994) ⚉ 5 ⌂ 200 sq m
Managing Dir: Richard Stamper
Sales Mgr: Shashi Nadesu
Representing:
Moog International [97]

90-01370 MTU Asia Pte Ltd
Jurong Town, PO Box 228, Singapore 9161
☎(+65) 861 59 22 ⊖ (+65) 861 15 50
Managing Director: Mr Beyer
® Germany: MTU Friedrichschafen GmbH [98]

90-01390 Munters Pte Ltd
(a subsidiary of AB Carl Munters, Sweden)
No.6 Genting Rd, #04-00, Jubilee Industrial Building, Singapore 1334
☎(+65) 744 68 28 ⊖ (+65) 744 95 85
♁ US$ 100,000 (1996) ♣ US$ 5,000,000 (1995) ⚉ 16
⌂ 4,600 sq m
Managing Dir: Lim Khay Teg [97]

90-01430 NKI Asia Pacific Pte Ltd
(a subsidiary of NKI BV, Netherlands)
77c Duxton Road, Singapore 089536
☎(+65) 227 78 22 ⊖ (+65) 224 78 22
Chairman: G Aerts
Managing Director: J P G Prins
® Netherlands: NKI BV [98]

90-01450 Noble Denton Singapore Pte Ltd
51 Goldhill Plaza 19-02, Singapore 308900
☎(+65) 356 68 18 ⊖ (+65) 356 70 86
® UK: Noble Denton [98]

Contact Reference

90-01460 Noricum Far East Division
(a division of T&T Technology Trading Ltd, Switzerland)
101 Cecil Street #10-08/09, Tong Eng Building, Singapore 069533
☎(+65) 225 36 11; 225 36 12 ⊖ (+65) 225 58 72
5 118 m2
Managing Director: Peter K Maier
■ Manufacture of artillery systems including Noricum 155 GH N-45. [98]

90-01480 Northrop Grumman (Singapore) Pte Ltd
250 North Bridge Road 15-04, Raffles City Tower, Singapore 0617
☎(+65) 339 07 22 ⊖ (+65) 338 30 42
Director: William Ennis
® USA: Grumman Corp. [97]

90-01670 Precision Technologies Pte Ltd
(† Precision Component & Equipment Pte Ltd)
211 Henderson Road #13-02, Singapore 159552
☎(+65) 273 45 73 [E] pretech@singnet.com.sg
20201 prewee ⊖ (+65) 273 88 98
US$ 3,000,000 (1996) ♣ US$ 11,500,000 (1995)
25 (1997) 5,000 sq ft
Managing Director: B C Wee
General Manager: Richard Wong
■ Defence and aerospace hardware, electrical/electronic components. Equipment distributor and representative. [98]

90-01700 Racal Electronics Singapore Pte Ltd
(a subsidiary of Racal Electronics plc, UK)
26 Ayer Rajah Crescent, No. 04-06/07, Singapore 0513
☎(+65) 779 22 00 ⊖ (+65) 778 54 00
S$ 1,000,000 (1995) ♣ S$ 11,000,000 (1995) 31
6,800sq ft
Director and General Manager: Duncan Fisken
Sales Manager: Paul Ooi
Technical Manager: Lim Poh Chuan
® UK: Racal Electronics [98]

90-01840 Ring Sights (Far East) Ltd
No.1425 Tai-Pan Ramada Hotel, Singapore 0718
☎(+65) 338 60 44 21586 ⊖ (+65) 339 04 53
Regional Dir: Leopold de Coutere
® UK: Ring Sights Holdings Ltd, Ring Sights Ltd
USA: Ring Sights Inc. [98]

90-01850 Rockwell Sales and Support Services
(a division of Rockwell International Corp, USA)
Avionics and Communications Service Centre
18 Loyang Lane, Loyang Industrial, Estate, Singapore 508918
☎(+65) 542 20 78 ⊖ (+65) 542 20 79
Director, Asia Pacific - Collins Air Transport Division: William A Burnore [E] waburnor@collins.rockwell. com
Manager (Area and Singapore): Ben J Thurlow
® USA: Rockwell International Corp. [98]

90-01900 SAMAERO Company Pte Ltd
(a subs of Eurocopter SA France & Singapore Technologies Aerospace)
Tampines Central, PO Box 0326, Singapore 915211
48 Loyang Way, Singapore 508740
☎(+65) 543 41 01 ✉ ECUREUIL 25974 samaro rs
⊖ (+65) 542 87 97
President: Gerald Negrel
VP Commercial: Henri Stell
Vice President Operations: Patrick Collignon
Finance & Admin Mgr: Ivan Ho
Logistics Mgr: Tian Toh Kian
Technical Mgr: Michael Chin Chun Leong
New Business Mgr: Paul A. Elgee
Quality Control Mgr: Kuldeep Agrawal
■ Sales & service of Eurocopter helicopters & spares, SOCATA light aircraft & Turbomeca engines. → 80
® France: Eurocopter, SOCATA, Turbomeca
Germany: Eurocopter [97]

90-01910 Schottel Far East Ltd
23 Gul Drive, Singapore 629471
☎(+65) 861 09 55 ✉ SCHOTTEL 24715 schotfe rs
⊖ (+65) 861 23 01
Managing Dir: Rinus Buijs
® Germany: Schottel-Werft [97]

90-01915 Sema Group Asia Pacific Pte Ltd
111 Somerset Road # 09-01, Singapore 238164
☎(+65) 339 89 11 ⊖ (+65) 339 68 87
® UK: Sema Group plc [98]

90-01928 Siemens Showa Solar Pte Ltd
Blk 164 Kallang Way 05-14/15, Kolam Ayer Industrial Park, Singapore 349248
Managing Dir: Udo Gogol
→ 74
® Japan: Showa Solar Energy K.K.
USA: Siemens Solar Industries [97]

90-01935 Simrad (Asia) Pte. Ltd
73 Science Park Drive, #B1-03/04 Cintech 1, Singapore Science Park 118254
☎(+65) 774 46 67 ⊖ (+65) 774 56 73
S$ 1 701 890 ♣ S$ 11 000 000 (1995) 13
Managing Dir: Robert Manson
Sales & Marketing Dir: Arthur Gothard
Operations & Technical Dir: Alex Campbell
■ A high-tech company active in marine electronics. Sales of vessel steering, positioning, monitoring & navigation systems, subsea systems, process control & data acquisition systems. [97]

90-01960 Sperry Marine Asia
213 Henderson Road, #03-05, Henderson Industrial Park, Singapore 0315
☎(+65) 274 33 32 ⊖ (+65) 271 33 39
Manager: Eo Sing Kons
® USA: Sperry Marine Inc. [98]

90-02018 Teledyne Industries International, Inc.
Singapore Branch Office
2 Handy Rd #09-606, Cathay Building, Singapore 229233
☎(+65) 338 80 77/78/79 26510 tdy s
⊖ (+65) 339 33 16
Vice President: Robert T. Ford
General Mgr: Patrick Foo
® USA: Teledyne Inc. [97]

90-02025 Textron Marine & Land Systems
(† Cadillac Gage Textron)
Regional Representative Office
55 Cairnhill Road 15-06, Cairnhill Plaza, Singapore 0922
☎(+65) 733 06 54 ⊖ (+65) 732 38 70
Managing Director: J L Broome
® USA: Textron Marine and Land Systems [98]

90-02070 UBE Industries Ltd Singapore Representative Office
150 Beach Road, #20-05 Gateway West, Singapore 189720
☎(+65) 291 93 63 34651 ⊖ (+65) 291 90 39 [98]

90-02090 Ultramare Singapore Pty Ltd
79 Loyang Way, Singapore 508766
☎(+65) 545 66 63 [E] ultsin@mbox2.singnet.com.sg
⊖ (+65) 543 08 01
US$ 1,000,000 (1997) ♣ US$ 5,000,000 (1996)
20 (1997) 4,500 m2
Managing Director: Vincent Teo
Sales and Marketing Director: Richard Yow
General Manager: Albert Chin
Purchasing Manager: Charles Lim
■ Wholesale supplier of aviation spares and equipment, ground support equipment. → 85
Affiliated Companies:
Ultramare Australia Pty Ltd
Ultramare Malaysia Sdn Bhd [98]

90-02210 Vetech Pte Ltd
PSA Bldg n°24-04a, 460 Alexandra Road, Singapore
☎(+65) 276 90 88 ⊖ (+65) 276 91 88
Managing Dir: Arne Berglund
■ Products represented include the BV206 All Terrain Carrier, Combat Vehicle 90 (CV90) and recovery units.
® Sweden: Hägglunds Vehicle AB [98]

Slovakia

Head of State:
Michal Kovac

10 Head of State

10-10000 Office of the President
Bratislava Castle, Zamocka Ulitsa, Bratislava
President: Michal Kovac [98]

11 Prime Minister

11-10000 Office of the Prime Minister
Nam Slobody 1, 813 70 Bratislava
☎(+421 7) 379 51 11 ⊖ (+421 7) 39 75 95
Prime Minister: Vladimir Meciar ☎(+421 7) 359 52 85; 31 54 87 (fax)
Vice Prime Minister: Jozef Kalman
☎(+421 7) 359 51 84; 39 67 59 (fax)
Vice Prime Minister and Minister of Finance: Sergej Koslik
☎(+421 7) 39 74 42; 39 75 30 (fax)
Vice Prime Minister: Ms Katarína Tóthová
☎(+421 7) 359 51 85; 39 71 21 (fax)
Head Secretary: Ms Anna Nagyová
☎(+421 7) 39 55 84; 39 76 25 (fax)
Spokeswoman of the Slovak Government: Ms L'udmila Buláková
☎(+421 7) 39 26 42; 359 52 48; 39 76 25 (fax)
Spokeswoman of the Prime Minister: Ms Magda Pospísilova ☎(+420 7) 39 55 14; 35 95 25 54; 39 76 25 (fax)
Press Spokeswoman: Ms Marta Tomasovicová
☎(+420 7) 359 51 48; 39 71 21 (fax)
Press Secretary: Ms Eva Jonáková
☎(+420 7) 359 51 49; 39 67 59 (fax) [98]

12 Ministries

12-10000 Ministry of Defence
Kutuzovova 8, 832 47 Bratislava
☎(+421 7) 25 88 61 ⊖ (+421 7) 25 89 04
Minister of Defence: Jan Sitek
Secretary of State: Jozef Gajdos
Press Officer: Frantisek Kasicky
Defence Policy & Development Section
Defence Policy and Development Section
Kutuzovova 8, 832 47 Bratislava
☎(+421 7) 25 87 74 ⊖ (+421 7) 25 87 74
Director: Col Milan Stránava
Human Resources Development Section
Kutuzovova 8, 832 47 Bratislava
☎(+421 7) 25 87 83 ⊖ (+421 7) 25 87 83
Director: Col Milan Stegena
Logistics Section
Kutuzovova 8, 832 47 Bratislava
☎(+421 7) 25 87 88 ⊖ (+421 7) 25 87 88
Director: Maj Gen Leopold Bilcík
Foreign Affairs Section
Kutuzovova 8, 832 47 Bratislava
☎(+421 7) 25 87 81 ⊖ (+421 7) 25 87 81
Director: Col Ludovít Gál
Economic Policy Section
Kutuzovova 8, 832 47 Bratislava
☎(+421 7) 25 62 43 ⊖ (+421 7) 25 62 43
Director: Col Jozef Zadzora [98]

12-20000 Ministry of Foreign Affairs
Stromová 1, 833 36 Bratislava
☎(+421 7) 370 41 11 ⊖ (+421 7) 370 44 91
Foreign Minister: Zdenka Kramplova
Press Department Director: Juraj Matejovsky [98]

12-21000 Ministry of External Relations
Stromova No 1, 833 36 Bratislava
☎(+421 17) 80 21 91 ⊖ (+421 17) 80 22 39
Director: Dusan Matulay [98]

12-40000 Ministry of the Interior
Martanoviciva Ulitsa c 2, 812 72 Bratislava
Minister of the Interior: Gustav Krajci [98]

12-50000 Ministry of Economy
Mierova 19, 827 15 Bratislava
☎(+421 7) 23 20 41 ⊖ (+421 7) 23 01 22
Minister of Economy: Karol Cesnek ☎(+421 7) 299 82 52
State Secretary: Josef Brhel
☎(+421 7) 299 87 08; 299 87 09
Chief of Office: Juraj Görner
☎(+421 7) 299 87 44; 23 80 29; 299 82 28 (fax)
Cabinet of Minister Director: Ivan Misík
☎(+421 7) 299 87 00; 23 08 18;
299 87 01; 23 01 22 (fax)
Economic Strategy Division: Milos Lelovsky
☎(+421 7) 299 87 48; 299 87 49; 23 19 40 (fax)
Commercial Policy Division: Mrs Anna Jostiaková
☎(+421 7) 299 83 10; 23 23 28; 299 82 37 (fax)
Basic Industries Division: Inrich Sitás
☎(+421 7) 299 87 32; 23 20 49;
299 87 33; 23 92 87 (fax)
General Electric Engineers Division: Ivan Kovácik
☎(+421 7) 299 87 34; 23 19 42;
299 87 35; 23 35 95 (fax)
Chemical and Consumer Goods Industry: Stofan Kasicka
☎(+42 7) 299 87 40; 23 00 61;
299 87 41; 23 14 85 (fax)
Tourism Division: Mrs Milana Jankovicová
☎(+42 7) 299 87 34; 23 19 42;
299 87 35; 23 49 84 (fax) [98]

12-70000 Ministry of Transportation, Post and Telecommunications
(Civil Aviation Administration)
6 Nam Slobody, Post Office 15, PO Box 100,
810 05 Bratislava
☎(+421 7) 25 62 48 ⊖ (+421 7) 25 40 20
Minister: Jan Yasovsky ☎(+421 7) 25 47 98
Deputy Minister: Prof Dezider Szabo
Director Foreign Relations: Mrs Z Dianova
Director General Civil Aviation: Peter Matejcik [98]

25 Defense Forces

25-10000 General Staff of the Armed Forces of the Slovak Republic
Rázusova 7, 911 00 Trencín
☎(+421 831) 53 70 16 ⊖ (+421 831) 53 76 25
Chief of General Staff: Col Gen Jozef Tuchyna [97]

29 Customs & Excise

29-00500 Customs Directorate
Mierová 23, 81511 Bratislava [98]

31 Embassies & Attachés

31-01890 Embassy of the United Kingdom
Panska 16, 81101 Bratislava
☎(+421 7) 531 96 31; 531 96 33
⊖ (+421 7) 531 00 02
Ambassador: H E Mr Peter Harborne
Deputy Chief of Mission and Press and Information Officer: Ms R E Cooper
Commercial Counsellor: P Repka
Defence Attaché: Wg Cdr P M Leadbetter (RAF) [98]

55 Government-owned Industry & Production Facilities

55-01000 Kerametal Co. Ltd
Jasikova ulitsa 2, 826 05 Bratislava
☎(+421 7) 579 11 11 [E] kerametal@netlab.sk
☒ 93233 kemet sk
⊖ (+421 7) 572 17 83 ; 522 58 04; 522 58 96
♧ US$ 12,000,000 (1996) ♣ US$ 1,150,000 (1996)
👥 237 (1996)
Chairman and Managing Director: Igor Junas
Director Sales: Stefan Ziak ☎(+42 7) 522 09 32
Director Commercial Relations: Vladimir Tikl
■ Import and export of metals, ores and defence equipment. [98]

55-01500 Petrimex
Ulitsa Dr V Clementisa 10, 826 02 Bratislava
☎(+421 7) 290 21 11 ☒ 92525 ⊖ (+421 7) 23 81 48
♣ US$ 600,802,000 (1993) 👥 220
Director General: Eng Marian Mojzis
■ Import and export of chemical products, pharmaceuticals and raw materials. Transport and barter operations. [97]

60 Research Organizations, Institutes & Test Establishments

60-01000 Vyskumny ústav zváracsky (VÚZ)
Welding Research Institute
Ulitsa Racianska 71, 832 59 Bratislava
☎(+421 7) 504 61 11; 25 59 66; 525 71 87
[E] rozvoj@vuz.sanelisk ☒ 93384 zvar c
⊖ (+421 7) 25 48 67; 504 65 70
♣ Kcs 204,970,000 (1996) 👥 356 (1996) 👥 29,000 m2
Director: Prof Ing Peter Polak ☎(+421 7) 25 35 00
Operations Director: Ing Lubos Mraz
Director Quality Control, System of Quality: Ing Vladimir Uher ☎(+421 7) 504 65 90
Director of Welding Personnel Traning and Education: Ing Michal Kopcak ☎(+421 7) 504 66 70
Director of Certifiction of Welding Personnel: Ing Ondrej Papajcik ☎(+421 7) 525 35 85; 504 67 39
Director of Certifiction of NDT Personnel: Ing Pavel Gabriska ☎(+421 7) 525 35 85; 504 65 56
Director of Certifiction of Products SKTC 115 State Test Board: Ing Viliam Sloboda
☎(+421 7) 25 70 41; 504 68 04
(+421 7) 25 59 66; 504 64 11: Ing Jana Rychtarikova
Deputy Director for Technical Affairs: Ing Ivan Sujan
☎(+421 7) 25 70 01; 504 65 95
Head of Information Dept: Ing Alojz Jajcay
■ A state institution of science and research. Research and development in the field of materials, weldability, welding technologies, surfacing, brazing/soldering, deposition, spraying, cutting and heat [98]

75 Manufacturing Companies

75-01000 Závody Tazkého Strojárstva (ZTS)
01841 Dubnica nad Váhom
☎(+421 827) 227 73; 262 47 ☒ 75318
⊖ (+421 827) 220 52
♣ CSk 1,500,000,000 👥 5,600
Managing Director: Jozef Barancok
Commercial Director: Juraj Stefanec
Financial Director: Emil Potocek
Production Director: Marian Podpera
Director for Strategy: Bohumil Zlocha
Public Relations:
■ Military technology division produces military armoured vehicles.
Metallurgy division produces steel castings and die forgings. [98]

75-01500 ZTS TEES Martin s.p.
Ceskoslovenskej armády 3/1697, 03657 Martin
☎(+421 842) 314 53; 348 92 ☒ 75235
⊖ (+421 842) 310 36; 328 22
♧ CSk 10,009,000,000 ♣ CSk 1,560,000,000 (1994)
👥 5,000 (1994) 🏭 240,000 sq m
Managing Director: Ján Segla
Sales and Marketing Director: Ján Kubacka
Operations Director: Ján Supala
Strategy and Development Director: Ján Hrivnák
Finance Director: Jan Lajac
Marketing Manager: Pindroch Vladimír
■ Manufacture of diesel piston motors, special machines, technological equipment and gear systems. Builds the T-72 MBT under Russian licence - involved with SABCA and SFIM in an upgrade of the tank, T-72 M2 Moderna. [97]

75-02000 ZVS Dubnica Jas
(a subsidiary of ZVS as)
Stúrova ulitsa, 01841 Dubnica nad Váhom
☎(+42 827) 221 88; 220 33
⊖ (+42 827) 225 29; 227 05
♧ Kcs 550,000,000 (1997) ♣ Kcs 200,000,000 (1997)
👥 850
General Director: Ing Miroslav Solava
Technical Director: Ing Frantisek Tomsa
Purchasing Manager: Ing Milan Grencik
Sales Manager: Ing Vladimir Duris
■ Manufacture and supply of large calibre ammunition and respective parts. Artillery rockets, rocket fuzes, tank and anti-tank guns. [98]

Slovenia

Head of State:
Milan Kucan

10 Head of State

10-10000 Office of the President
Erjavceva 17, 1000 Ljubljana
☎(+386 61) 178 12 22
President: Milan Kucan
Chief of Cabinet: Peter Tos ☎(+386 61) 178 12 06;
Public Relations: Spela Furlan ☎(+386 61) 178 13 05
178 13 52 (fax) [98]

11 Prime Minister

11-10000 Office of the Prime Minister
Gregorciceva 20, 61000 Ljubljana
☎(+386 61) 178 11 00 ⊖ (+386 61) 178 17 21
Prime Minister: Dr Janez Drnovsek
☎(+386 61) 125 91 48; 22 42 40 (fax)
Deputy Prime Minister: Marjan Podobnik
Chief of Cabinet: Mojca Osolnik ☎(+386 61) 178 11 17
Public Relations: Ms Veronika Stabej
☎(+386 61) 178 17 20 [98]

12 Ministries

12-10000 Ministry of Defence
Kardeljeva ploscad 2, 1000 Ljubljana
☎(+386 61) 133 11 11 ⊖ (+386 61) 131 81 64
Minister: Tit Turnsek
Chief of Staff: Zoran Klemencic
Director Foreign Relations: Maja Stonic
State Secretary: M A Boris Znidaric
State Secretary: Teodor Gesak
State Secretary: Marija Ribic
Public Relations: Tanja Pandev ☎(+386 61) 171 25 58;
131 12 42 (fax)

Minister's Office
Kardeljeva ploscad 25, 61000 Ljubljana
☎(+386 61) 11 91 00; 11 92 00; 11 91 95
⊖ (+386 61) 11 83 00; 11 91 45
Head of Minister's Office: Zoran Klemencic

Development and Training Department
Kardeljeva ploscad 25, 1000 Ljubljana

☎(+386 1) 133 11 11 ⊖ (+386 1) 131 81 64
Director: Uros Krek

Logistics Department
Kardeljeva ploscad 26, 1000 Ljubljana
☎(+386 1) 131 92 20 ⊖ (+386 1) 131 92 82
Director: Borut Farcnik

Military Affairs Department
Kardeljeva ploscad 25, 1000 Ljubljana
☎(+386 1) 133 11 11 ⊖ (+386 1) 131 80 67
Director of Military Affairs: Viktor Krajnc (acting)

Civil Defence Department
Kardeljeva ploscad 26, 1000 Ljubljana
☎(+386 1) 131 80 11 ⊖ (+386 1) 131 80 06
Director: Igor Nered

Telecommunications Department
Kardeljeva ploscad 26, 1000 Ljubljana
☎(+386 1) 131 71 18 ⊖ (+386 1) 131 01 83
Director of Telecommunications: Tine Brajnik

Security Affairs Department
Kardeljeva ploscad 26, 1000 Ljubljana
☎(+386 1) 133 11 11 ⊖ (+386 1) 131 81 64
Director: Joze Secnik

General Affairs Department
Kardeljeva ploscad 26, 1000 Ljubljana
☎(+386 1) 133 11 11 ⊖ (+386 1) 131 81 64
Director: Joze Pesak

Center for Strategic Studies
Kardeljeva ploscad 25, 1000 Ljubljana
☎(+386 1) 131 80 50 ⊖ (+386 1) 131 80 51
Director: Mag. Jernej Kozar

Informatics Service
Kardeljeva ploscad 26, 1000 Ljubljana
☎(+386 1) 133 11 11 ⊖ (+386 1) 131 81 64
Director: Viktor Strle

Educational Center
Kardeljeva ploscad 25, 1000 Ljubljana
☎(+386 1) 133 21 70 ⊖ (+386 1) 132 70 36
Director: Milan Gorjanc

Defence Inspectorate
Kardeljeva ploscad 25, 1000 Ljubljana
☎(+386 1) 131 80 27 ⊖ (+386 1) 131 80 32
Director: Antun Sostaric

Fire Protection Inspectorate
Kardeljeva ploscad 26, 1000 Ljubljana
☎(+386 1) 131 00 91 ⊖ (+386 1) 30 13 68
Director: Bogomir Zupancic

Administration for Civil Defence Protection and Disaster Relief
Kardeljeva ploscad 26, 1000 Ljubljana
☎(+386 1) 131 81 84 ⊖ (+386 1) 131 81 17
Director: Bojan Usenicnik [98]

12-20000 Ministry of Foreign Affairs
Gregorciceva 25, 1000 Ljubljana
☎(+386 61) 178 20 00 ⊖ (+386 61) 178 23 74
Minister: Zoran Thaler
Public Relations: Alfonz Naberznik ☎(+38 61) 178 23 50; 178 23 55 (fax) [98]

12-40000 Ministry of Internal Affairs
Stefanova 2, 1000 Ljubljana
☎(+386 61) 125 14 00 ⊖ (+386 61) 21 43 30
Minister of Internal Affairs: Mirko Bandelj
☎(+386 1) 31 83 86
Chief of Public Relations Office: Muzic Janez
☎(+386 61) 21 67 30; 21 74 95 (fax) [98]

12-50000 Ministry of Economic Affairs
Kolnikova 5, 1000 Ljubljana
☎(+386 61) 178 33 11 ⊖ (+386 61) 133 10 31
Minister of Economic Affairs: Metod Dragonja
Public Relations: Verica Vogrincic
☎(+386 61) 178 32 09; 178 33 03 [98]

12-60000 Ministry of Economic Relations and Development
Kolnkova 5, 1000 Ljubljana
☎(+386 61) 178 33 11 ⊖ (+386 61) 178 35 44
Minister: Marjan Senjur
Head of Public Relations: Ms Mariana Jancar
☎(+386 61) 178 35 85; 178 35 44
■ Performs tasks related to trade, planning and economic relations with foreign countries. [98]

12-70000 Ministry of Transport and Communications
Presernova 12, 1000 Ljubljana
☎(+386 61) 125 62 56 ▣ 39160 slruzp si
⊖ (+386 61) 21 87 07
Minister of Transport and Communications: Anton Bergauer ☎(+386 61) 125 62 56
Director: Ludvik Bokal
Public Relations: Valentin Lapanja Furlan
☎(+386 61) 21 20 15; 21 27 76 (fax) [98]

12-80000 Ministry of Science and Technology
Slovenska 50, 1000 Ljubljana
☎(+386 61) 131 11 07 [E] vlado.valda@mzt.si
⊖ (+386 61) 132 41 40
Minister: Dr Lojze Marincek
Head of Public Relations Unit: Vladimir Vaida [98]

12-91000 Government Public Relations and Media Office
Levstikova 10, 1000 Ljubljana
☎(+386 61) 125 01 11 ⊖ (+386 61) 21 23 12 [98]

15 Parliament

15-10000 National Assembly
Subiceva Street 4, 1000 Ljubljana
☎(+386 61) 126 12 22 ⊖ (+386 61) 125 81 60
President of the Assembly: Mag. Herman Rigelnik
☎(+38 61) 15 80 22
Chief of Cabinet: Matjaz Jugovic ☎(+386 61) 125 80 02; 126 22 27 (fax)
Head of Information Service: Janez Pezelj
☎(+38 61) 125 80 71; 125 81 73 (fax) [98]

15-20000 National Council
Subiceva 4, 61000 Ljubljana
☎(+386 61) 125 73 11 ⊖ (+386 61) 21 22 51 [98]

20 Defense Administrations/Services/Agencies

20-01000 Legal Service
(Ministry of Defence)
Kardeljeva ploscad 26, 1000 Ljubljana
☎(+386 1) 133 11 11 ⊖ (+386 1) 11 90 51
Director: Igor Ivancic [97]

25 Defense Forces

25-10000 Slovenian Territorial Defence Staff
Kardeljeva ploscad 24-26, 1000 Ljubljana
☎(+386 1) 21 97 50 ⊖ (+386 1) 21 97 64
Chief of General Staff: Col Gen. Albin Gutman [97]

29 Customs & Excise

29-00500 National Customs Department
Smartinska 130, 61000 Ljubljana
☎(+386 61) 140 41 77 ⊖ (+386 61) 140 52 23 [98]

31 Embassies & Attachés

31-00020 Embassy of the Republic of Albania
Ob Ljubljanici 12, 61000 Ljubljana
☎(+386 61) 132 23 24 ⊖ (+386 61) 132 31 29
Ambassador: H E Mr Napoleon Dhimiter Roshi [98]

31-00100 Embassy of the Republic of Austria
Strekljeva 5, 61000 Ljubljana
Ambassador: H.E. Dr Jutta Stefan-Bastl [98]

31-00210 Embassy of the Republic of Bosnia and Herzegovina
Likozarjeva 6, 61000 Ljubljana
☎(+386 61) 132 22 14 ⊖ (+386 61) 132 22 30
Ambassador: H E Mr Ugljesa Uzelac [98]

31-00250 Embassy of Bulgaria
Ziherlova 6, 61000 Ljubljana
☎(+386 61) 21 36 43 ⊖ (+386 61) 21 36 43
Chargé d'Affaires: Gantcho Ganev [98]

31-00257 Embassy of the Delegation of the Commission of the European Community
Trg Republike 3/XI, 61000 Ljubljana
☎(+386 61) 125 13 03 ⊖ (+386 61) 125 20 85
Chargé d'Affaires: Pierre André Borgoltz [98]

31-00360 Embassy of the People's Republic of China
Stara Slovenska 1, 61000 Ljubljana
☎(+386 61) 125 15 50 51; (+386 61) 140 23 83
⊖ (+386 61) 140 40 07
Chargé d'Affaires: Lu Peixin [98]

31-00425 Embassy of the Republic of Croatia
Grubarjevo nabrezje 6, 1000 Ljubljana
☎(+386 61) 125 62 20; 125 62 45
⊖ (+386 61) 125 81 06
Ambassador: H E Dr Ivica Mastruko
Minister: Ivana Moric
Minister and Counsellor: Zdravko Gavran
Counsellor of the Counsular Section: Mrs Katija Milicic [98]

31-00450 Embassy of the Czech Republic
Kolarjeva 30, 61000 Ljubljana
☎(+3861) 132 80 35 ⊖ (+3861) 31 84 23
Ambassador: H.E. Mr Petr Kypr
Defence Attaché: Col Pavel Vanicky (resides in Vienna)
☎(+43 1) 894 21 25-6
Commercial Attaché: Pavel Svitil [98]

31-00460 General Consulate of the Kingdom of Denmark (honorary)
Resljeva 24/1, 61000 Ljubljana
☎(+386 61) 31 12 36; 31 37 40
⊖ (+386 61) 133 70 98
Honorary Consul General: Rudi Selih [98]

31-00590 Embassy of France
Robbova 18/VI, 61009 Ljubljana
☎(+386 61) 173 44 31 ⊖ (+386 61) 173 44 42
Ambassador: H E Mr Bernard Poncet
Commercial Counsellor: Pierre Compagnon
Defence Attaché: Lt Cdr Loíc Tacher
Cultural Counsellor: Jean Baundry [98]

31-00660 Embassy of Federal Republic of Germany
Presernova 27, 61000 Ljubljana
☎(+386 1) 21 61 66; 22 20 93 ⊖ (+386 1) 125 42 10
Ambassador: H E Mr Günther Seibert [98]

31-00690 Embassy of Greece
Dalmatinova 2, 61000 Ljubljana
☎(+386 61) 31 43 76; 313 82 ⊖ (+386 61) 31 31 82 [98]

31-00810 Embassy of the Republic of Hungary
Dunajska 22, 61000 Ljubljana
☎(+386 1) 131 51 68; 131 60 63
⊖ (+386 1) 131 71 43
Ambassador: H E Dr Istvan Balogh [98]

31-00890 Embassy of Italy
Snezniska 8, 1000 Ljubljana
☎(+386 61) 126 21 94; 126 23 20
⊖ (+386 61) 125 33 02
Ambassador: H E Dr Massimo Spinetti
☎(+386 61) 30 29 77; 125 33 02 (fax)
Deputy Chief of Mission: Dr Manuel Jacoangeli
☎(+386 61) 131 52 95
Defence Attaché: Col Giovanni Nitti
(Resides in Vienna, Italy) ☎(+43 1) 713 36 41
Commercial Attaché: Dr Serafino Placere
General Consulate of Italy
Belvedere 2, 66000 Koper
☎(+386 61) 384 11; 384 12 ⊖ (+386 61) 384 16
General Consul: Dr Michele Esposito [98]

31-01080 Embassy of the Republic of Macedonia
Kardeljeva Pl 1/I, 61000 Ljubljana
☎(+386 61) 168 44 54; 168 54 54
⊖ (+386 61) 168 51 81
Ambassador: H E Dr Dimitar Mircev
Chargé d'Affaires: Stevan Nikolovski [98]

31-01270 General Consulate of the Kingdom of the Netherlands (honorary)
Dunajska 22, 61000 Ljubljana
☎(+386 61) 32 57 96; 32 89 78
⊖ (+386 61) 32 61 58; 32 47 49
Honorary Consul General: Dr Matija Skof [98]

31-01340 General Consulate of the Kingdom of Norway
Dalmatinova 10, 61000 Ljubljana
☎(+386 61) 32 74 10 ⊖ (+386 61) 30 16 21
Consul General: Kuhelj [98]

31-01480 Embassy of the Republic of Romania
Nanoska 8, 61000 Ljubljana
☎(+386 61) 26 87 02 ⊖ (+386 61) 26 87 02
Chargé d'Affaires: Ms Marian Pavelescu [98]

31-01485 Embassy of the Russian Federation
Rozna Dolina Cesta II N° 7, 61000 Ljubljana
☎(+386 61) 26 22 50; 123 12 36
⊖ (+386 61) 15 41 41; 125 41 41
Chargé d'Affaires: Jurij Stepanovic Girenko [98]

31-01840 Embassy of the Republic of Turkey
Livarska 4, 1000 Ljubljana
☎(+386 61) 132 20 12; 132 41 15
⊖ (+386 61) 132 31 58
Ambassador: H E Mr Ilhan Yigitbasioglu
Counsellor: Ms Deniz Kilicer [98]

31-01890 Embassy of the United Kingdom of Great Britain and Northern Ireland
Trg Republike 3/IV, 1000 Ljubljana
☎(+386 61) 125 71 91; 125 31 66; 125 32 66 (commercial section)
⊖ (+386 61) 125 01 74; 125 90 80 (commercial section)
Ambassador: H E Mr David Lloyd
Deputy Head of Mission, Commercial: Ms S Parnell
Defence Attaché: Lieut Col A R Manton
(Resident in Vienna)
British Consul and Management Officer: Miss C Inglehearn
Executive Assistant: Mrs M Horskey [98]

31-01900 Embassy of the United States of America
Prazakova 4, 1000 Ljubljana
☎(+386 61) 30 14 27; 30 14 72 ⊖ (+386 61) 30 14 01
Ambassador: H E Mr Victor Jackovich
☎(+386 61) 30 14 27
Deputy Chief of Mission: J Paul Reid
Consular/Administrative: Danny Hall
Defense Attaché: Col Michael Vrosh
☎(+386 61) 210 190
Public Affairs Officer: Domenick DiPasquale [98]

75 Manufacturing Companies

75-00400 Fotona D.D.
Stegne 7, 1210 Ljubljana
☎(+386 61) 57 69 82 ☒ 39518 ⊖ (+386 61) 57 59 85
👥 480
Marketing Dir, Electro-optics Div.: B. Vukas
■ Design, development & manufacture of electro-optical instruments & systems. [97]

Solomon Islands

Head of State:
H.M. Queen Elizabeth II

Governor-General:
Sir Moses Pitakaka

10 Head of State

10-10000 Office of the Governor-General
Government House, Honiara
Governor-General: Sir Moses Pitakaka [98]

11 Prime Minister

11-10000 Office of the Prime Minister
Government House, PO Box 61, Honiara
Prime Minister: Solomon Mamaloni
Deputy Prime Minister: Francis Saemala [98]

12 Ministries

12-20000 Ministry of Foreign Affairs
PO Box G 10, Honiara
☎(+677) 222 23; 212 50 ✉ FAHON Honiara
☒ 66311 primus ho
Minister for Foreign Affairs: David Sitai [98]

12-40000 Ministry of Home Affairs
Government House, Honiara
Minister: Nathaniel Supa [98]

15 Parliament

15-10000 National Parliament
PO Box G 19, Honiara
☎(+677) 217 51 ⊖ (+677) 238 66
Speaker: Paul Joseph Tovua [97]

31 Embassies & Attachés

31-01890 United Kingdom High Commission
Telekon House, Mendana Avenue, PO Box 676, Honiara
☎(+677) 217 05; 217 06 ⊖ (+677) 215 49
High Commissioner: H E Mr B N Connelly
Deputy High Commissioner: Mrs D M Fitch
Aid Attaché: T G Whiting [98]

South Africa

Head of State:
Nelson Mandela

10 Head of State

10-10000 Office of the President
Bryntirion, PO Box X80, 0002 Pretoria
☎(+27 12) 74 31 31 ☒ 320433 sa
President of the Republic: Nelson Mandela [98]

12 Ministries

12-10000 Department of Defence
Private Bag X427, 0001 Pretoria
☎(+27 12) 26 67 18 ☒ 322113 ⊖ (+27 12) 347 01 18
Minister of Defence: Joe Modise [98]

12-20000 Department of Foreign Affairs
Private Bag X152, 0001 Pretoria
☎(+27 12) 351 10 00; 323 05 27
☒ 320291 sa; 320695 sa; 320696 sa
⊖ (+27 12) 351 02 55; 351 02 60; 323 81 51
Minister of Foreign Affairs: Alfred Nzo
Deputy Minister: Aziz Pahad
Director General: Leo Henry Evans
Head Office
Union Buildings, Private Bag X152, 0001 Pretoria
☎(+27 12) 323 05 27 ⊖ (+27 12) 323 10 11
Director General: Leo Henry Evans [98]

12-40000 Department of Home Affairs and Communications
Private Bag X114, 0001 Pretoria
☎(+27 12) 26 80 81 ☒ 321358 sa, 321353 sa
Minister: Mangosuthu Buthelezi [98]

20 Defense Administrations/Services/Agencies

20-10000 National Intelligence Agency
239 Skinner Street, Consilium Building - PO Box X87, 0001 Pretoria
☎(+27 12) 317 59 11 ✉ NASVEIL pretoria ☒ 30121
⊖ (+27 12) 320 02 19
Minister: D Omar
Deputy Minister: J Nhlanhla
Director General: Dr L D Barnard
Media and Communications Director: W A Theron [97]

25 Defense Forces

25-20000 South African Air Force Headquarters (SAAF)
PO Box X199, 0001 Pretoria
☎(+27 12) 312 27 16; 312 29 11; 312 12 89
⊖ (+27 12) 312 28 03; 312 27 03
Chief of Air Force Staff: Maj Gen A Burger (MMM)
Chief of Air Force: Maj Gen W H Hechter (SD SM MMM)
Chief of the South Africa Air Force: Lieut Gen J Kriel (SSAS SD SM MMM)
Chief of Air Staff Operations: Maj Gen C B Lombard
Chief of Air Staff Logistics: Maj Gen G C Nel (SD SM MMM)
Chief of Air Staff Finance: Brig P Du Plessis (SM MMM)
Chief of Air Staff Intelligence: Brig D Badenhorst (SM MMM)
Chief of Air Staff Plans: Brig J J Botha
Inspector General: Maj Gen F Du Toit

Contact Reference

Liaison Services Director: Brig A J Grobbelaar
Director Media Liaison: Col T Michell
Enquiries: Maj W P Dormehl [97]

25-29100 SAAF Flying School
(† Air Force Ysterplaat)
Langebaanweg
☎ (+27 12) 529 11 ⊖ (+27 12) 521 68
Commanding Officer: Col C Gagiano
Technical Officer: Maj S Van Noordwyk [97]

25-30000 South African Army
SA Army Headquarters, Private Bag X172, 0001 Pretoria
☎ (+27 12) 355 91 11 ⊖ (+27 12) 355 14 14
Chief of the Army: Lt Gen. R. Otto
Deputy Chief of Staff: Maj.Gen. F.A. Botha (SD, SM, MMM)
Inspector General: Maj.Gen. A.L. van Graan (SM, MMM)
Chief of Army Staff Personnel: Maj.Gen. J.M. Dippenaar (SD, SM, MMM)
Chief of Army Staff Intelligence: Maj.Gen. W.G. Lombard (SM, MMM)
Chief of Army Staff Operations: Maj.Gen. C.J. Serfontein (SD, SM, MMM)
Chief of Army Staff Logistics: Maj.Gen. P.P.J. Ooetser (SD, SM, MMM)
Chief of Army Staff Planning: Maj.Gen. A.K. de Jager (SD, SM, MMM)
Chief of Army Staff Finance: Brig. A.J. Wilke (SM, MMM)
Corporate Communication Dir: Brig. C.H. van Zyl (SM, MMM)
SO1 Media Liaison: Lt Col C.F. de Klerk [97]

25-31400 Western Province Command, South African Army
PO Box 1, 8000 Cape Town
☎ (+27 21) 469 11 11; 469 10 00
⊖ (+27 21) 469 10 76; 469 12 99
Officer Commanding: Maj Gen C H Van Zyl
Senior Staff Officer Operations: Col J J Liebenbers
Chief of Staff: Brig D M Tshike
Senior Staff Officer Logistics: Col P J Swanepoel
Senior Staff Officer Communications: Col A H Louw [98]

25-31600 North Western Command Headquarters
South African Army, Military Base, 2521 Potchefstroom
☎ (+27 148) 289 33 63 ⊖ (+27 148) 289 33 66
Officer Commanding: Maj Gen G N Ngwenya
Deputy Commander: Brig R Coetzee
Director of Operations: Col Potgieter
Director of Procurement: Col J van Rendourg [98]

25-31700 Natal Command
Marine Parade, PO Box 10004, 4056 Durban
☎ (+27 31) 369 19 11 ⊖ (+27 31) 369 12 61
Officer Commanding: Maj.Gen. C.E. Le Roux
Deputy Commander: Brig. D.M. Mofokeng
Operations Dir: Col P.W. Rose
Procurement/Logistics Dir: Col J. ud Lingen [97]

31 Embassies & Attachés

31-00090 Australian High Commission
Private Bag X150, 0001 Pretoria
292 Orient Street, 0083 Arcadia, Pretoria
☎ (+27 12) 342 37 40 ⊖ (+27 12) 342 42 22
High Commissioner: H.E. Mr I.W. Porter
Deputy Chief of Mission: P. Lawless
Admin Section: E.J. Oakes [97]

31-00310 Canadian High Commission
Private Bag X13, 0083 Hatfield, Pretoria
☎ (+27 12) 422 30 00 ✉ CANDOM Pretoria
⊖ (+27 12) 422 30 71
High Commissioner: H E Mr Arthur C Perron
Minister Deputy Chief of Mission: Ms Carolyn McMaster
● PO Box 683, 8000 Cape Town
16th Floor, Reserve Bank Bldg, 30 Hout St, 8001 Cape Town
☎ (+27 21) 23 52 40 ◙ 527060 canct sa
⊖ (+27 21) 23 48 93 [98]

31-00350 Embassy of Chile
PO Box 12672, 0028 Hatfield, Pretoria
5th Floor, Campus Centre, 1102 Burnett Street, 0083 Hatfield, Pretoria
☎ (+27 12) 342 15 11; 342 16 21
⊖ (+27 12) 342 16 58
Ambassador: H E Mr Jorge Heine
Counsellor: Patricio Damm
Counsellor: Ramon Rada
Air Attaché: Col Javier Anabalon
Army and Naval Attaché: Col Patricio Cartoni
Second Secretary: Patricio Becker
Third Secretary: Andres Landerretche
● PO Box 2067, 8000 Cape Town
1918 Main Tower, 19th Floor Standard Bank Center, Foreshore, Cape Town.
☎ (+27 21) 21 23 44; 21 23 46 ⊖ (+27 21) 25 30 34 [98]

31-00535 Honoury Consulate of the Republic of Estonia
(† Embassy of Estonia)
16 Hofmeyr Street, 7530 Welgemoed
☎ (+27 21) 913 38 50 ⊖ (+27 21) 913 25 79
Honorary Consul: Riina Knipe
Secretary: Tiina Saadoja ☎ (+27 21) 913 64 54 [98]

31-00580 Embassy of Finland
(† Legation of Finland)
628 Leyos Street, Mulkleneuk, PO Box 443, 0001 Pretoria
☎ (+27 12) 343 02 75 ✉ FINLANDIA Pretoria
Ⓔ finland.embassy@finland.caz ⊖ (+27 12) 343 30 95
Ambassador: H E Mr Tapani Brotherus
Counsellor: Pekka Oprana
Secretary: L M Riitaoja [98]

31-00590 Embassy of France
(June 16 to December 31)
807 George Avenue, Arcadia, 0083 Pretoria
☎ (+27 12) 43 55 51 ◙ 3-20 706 sa
⊖ (+27 12) 43 55 52
Head of Chancery: Ms Nicolas Normand
Armed Forces Attaché: Capt P Dupont
☎ (+27 12) 43 55 06
(January 1 to June 15)
● 1009 Cape Town Centre, Heerengracht Tower, 8001 Cape Town
☎ (+27 21) 22 13 38 ◙ 5-20 687 sa
⊖ (+27 21) 24 84 70
Office of the Commercial Counsellor
PO Box 1427, 2000 Johannesburg
☎ (11) 331 71 36 ✉ COMATTA Johannesburg
◙ 4-85 974 comatta sa
Commercial Counsellor: Jean Philippe Guiltat [98]

31-00660 Embassy of the Federal Republic of Germany
(July 1 to January 14)
PO Box 2023, 0001 Pretoria
180 Blackwood Street, Arcadia, 0083 Pretoria
☎ (+27 12) 427 89 36 ✉ DIPLOGERMA Pretoria
◙ 321386 aapre sa, 322153 aapre sa
⊖ (+27 12) 343 94 01
Ambassador: H E Dr Uwe Kaestner
Deputy Chief of Mission: Dr Peter Schola
Political Secretary: Anke Schlimm
Commercial Counsellor: Karl Wokalek
Press and Information Counsellor: Dr Peter Woeste
(January 15 to June 30)
● 825 Martini Gard. Queen Victoria St., PO Box 4273, 8000 Cape Town
825 Martini Gard. Queen Victoria Street
☎ (+27 21) 24 24 10 ◙ 527799 sa
⊖ (+27 21) 23 76 21; 24 94 03 [98]

31-00690 Embassy of Greece
(June 16 to December 31)
995 Pretorius Street, Arcadia, 0083 Pretoria
☎ (+27 12) 43 73 51; 43 73 53
✉ GREEK EMBASSY Pretoria ◙ 320520 sa
⊖ (+27 12) 43 43 13
Ambassador: H E Mr Pericles A Tsamoulis
First Secretary: Nicolas Protonotarios [98]

31-00890 Embassy of Italy
(June 16 to December 31)
796 George Avenue, Arcadia, 0083 Pretoria
☎ (+27 12) 43 55 41/4 ✉ ITALDIPL ◙ 321397
⊖ (+27 12) 43 55 47
Ambassador: H.E. Dr Mario Piersigilli
Deputy Chief of Mission: Dr Alessandro Cevese
Commercial Secretary: Dr Tosca Barucco [98]

31-01100 High Commission for the Republic of Malawi
(July to December)
770 Government Ave, PO Box 11172, 0028 Hatfield, Pretoria
☎ (+27 12) 342 01 46/7 ✉ MALAWIAN Pretoria
◙ 322017 ⊖ (+27 12) 342 01 46/7
High Commissioner: H.E. Mr W.S. Khoza
Defence Attaché: Col M.C. Chirwa
Second Secretary: A. Kambwiri
Third Secretary: P. Dawa
Third Secretary General Duties: Mrs F.A. Teleka
(January to June)
● 4 Serpentine Road, Orangezicht, Cape Town
☎ (+27 21) 45 61 47 ⊖ (+27 21) 45 61 47 [97]

31-01680 Embassy of Spain
PO Box 1633, 0001 Pretoria
37 Shortmarket Street, 8001 Cape Town
☎ (+27 12) 344 38 75 Pretoria
(+27 21) 22 23 26 Cape Town
⊖ (+27 12) 343 48 91 Pretoria
(+27 21) 22 23 28 Cape Town
Ambassador: H E Mr Miguel Angel Carriedo
Counsellor: Jose Pascual Marco
Defence Attaché: Capt Luis Inclàn (NAVY)
■ January 1st to June 30 in Cape Town, 1st July to 31st December in Pretoria.
Commercial Office
PO Box 781050, 2146 Sandton, Johannesburg
☎ (+27 11) 883 21 02 ⊖ (+27 11) 883 26 24
Commercial and Economic Counsellor: Victor Audera [98]

31-01890 United Kingdom High Commission
255 Hill Street, Accardia, 0083 Pretoria
☎ (+27 12) 43 31 21 ◙ 321323 bremb sa
⊖ (+27 12) 43 32 07
High Commissioner: H E Miss Maeve Fort
Counsellor and Deputy Head of Mission: Dr M J Lyall-Grant
Counsellor, Political: D Woods
Defence and Military Attaché: Brig M L Wildman
United Kingdom High Commission
91 Parliament Street, 8001 Cape Town
☎ (+27 21) 461 72 20 ◙ 527301 ct
⊖ (+27 21) 461 00 17 [98]

40 Government Defense Sales/Procurement Organizations

40-01000 ARMSCOR
Private Bag X337, 0001 Pretoria
370 Nossob Street - Corner Nossob Street and Delmas Drive, Erasmuskloof Extension 4, Pretoria
☎ (+27 12) 428 19 11 ✉ ARMSCOR
◙ 320217; 320567; 321190 ⊖ (+27 12) 428 56 35
♧ R 75,000,000 (1995-96) 👥 1,064
Executive Chairman: Ron Haywood
General Manager Aeronautics and Maritime: Erich H de W Esterhuyse
General Manager, Combat Systems: Jaco S de Jager
General Manager, Foreign Trade: Raymond P Pretorius
General Manager, Finance and Administration: C J Hoffman
General Manager, Corporative Communications: Y Abba Omar
Senior Manager, Planning: André J Buys
Senior Manager, Quality: Jan C Oosthuizen
Senior Manager, Human Resources: Jackie T Kgare
Corporate Communications Manager: Don Henning

■ Aquisition function involved in undertaking product specifications, research and development. Establishment of appropriate technologies and commissioning or contracting suitable suppliers to further develop and manufacture required products and systems including naval, air force and army weapons, ammunition in various calibres, conventional and mine-protected vehicles including troop carriers and gun tractors, armoured fighting vehicles, combat support helicopters, advanced electronic equipment, including communications, surveillance systems and components, mine detection and clearing equipment.
Promotes the international marketing of the above products and systems on behalf of the South African armaments industry. [97]

60 Research Organizations, Institutes & Test Establishments

60-00500 OTB
(† Overberg Test Range (OTR))
(a division of Denel (Pty) Ltd)
Private Bag X12, 7280 Bredasdorp
☎ (+27 2847) 590 10 [E] otbinfor@atb.co.za
[E] wilhwlmw@otb.co.za [E] ferdiep@otb.co.za
⊖ (+27 2847) 592 56
♣ R 33,000,000 (1997) 👥 182 (1997)
General Manager: Dr J G Malan ☎ (+27 2847) 590 10
Marketing Manager: P G W Wessels
Communications Manager: F Pieterse
Operations Manager: J Venter ☎ (+27 2847) 590 10
■ Specialise in flight tests on sophisticated weapon and aviation systems for the international aerospace industry. In addition it can provide a launch service for LEO-satellites and consultation to clients on design and development of test ranges. Products include trajectories, telemetry reception, command & control, meteorological data, photographic documentation and data reduction. High speed target drone is available. [98]

70 Financial Holding Groups, Parent Companies

70-00100
Denel (Pty) Ltd
PO Box 8322, 0046 Hennopsmeer
☎ (+27 12) 428 09 11 ⊖ (+27 12) 347 03 00
♣ US$ 771,000,000
👥 120 (corporate office); 14,000 (group)
Managing Director: Johan P L Alberts
Communications Director: George M Masoek
Communication Manager: Eben Mouton
Promotion Manager: Ms Corrie Strydom
■ Manufacture and supply of high technology industrial and military products, systems and services. [98]

75 Manufacturing Companies

75-00500 Advanced Technologies and Engineering (ATE)
PO Box 632, 1685 Halfway House
998 Sixteenth Rd, Halfway House
☎ (+27 11) 314 21 70 [E] ategen@iafrica.com
⊖ (+27 11) 314 21 51
👥 200 (1997) 🏭 6,000 m2
Managing Director: Jean-Marc Pizano
Technical Director: Patrick Barrier
Sales and Marketing Manager: Frank Archibell
General Manager: Carel de Beer
■ Integrated glass cockpit avionics systems for helicopter and fixed wing aircraft. Main contractor for the Rooivalk combat support helicopter and the SAAF Pilatus PC-7 MK II. New and mid-life upgrades of aircraft avionic systems. Develop and manufacture UAV's for a range of mission applications including artillery fire correction. [98]

75-00520 Aeroflo
(a CSIR-AEC Joint Venture)
PO Box 395, 0001 Pretoria
☎ (+27 12) 841 41 59 ⊖ (+27 12) 841 41 96
Managing Dir: I.J. Rademeyer ☎ (+27 12) 841 41 57
■ Manufacture of Aeroflo particle separators. Design, development and manufacture of air intake particle separators. [97]

75-00760 Alpine Aluminium Cases
(a division of Alpine Gulf Equipment)
PO Box 414, 2157 Sunninghill
Cnr Innes Road & Bismut Street, Jet Park, Boksburg
☎ (+27 11) 397 40 95; 397 40 98
[E] alpine@hixnet.co.zo ⊖ (+27 11) 397 35 15
♣ R 3,300,000 (1996) 👥 28 (1997) 🏭 800 sq m
Managing Director: Peter K Brooks
Operations Director: Ian Cervo
Sales and Marketing Director: Stanley Roup
Purchasing and Administration Manager: Peter Brookes
■ Manufacture superior rigidized aluminium cases. [98]

75-00765 Altech Defence Systems (ADS)
(† UEC Projects)
Mount Edgecombe, PO Box 432, 4300 KwaZulu, Natal
☎ (+27 31) 508 11 11 ⊖ (+27 31) 59 53 60
♧ R 73 000 000 (1996) ♣ R 156 000 000 👥 420
🏭 10 114 sq m
Managing Dir: F. Stark
Sales & Marketing Dir: I. Randeree
Operations & Technical Dir: D.N. Hiles
Purchasing Mgr: D. Devandran
Marketing Mgr: L.A.C. Pretorius
■ Design, manufacture & support of electronic defence products & systems. Fields of activity: Air defence systems, avionics, command, control & intelligence, sonar sensors & training simulators. [97]

75-00770 Analysis Management Systems (AMS)
PO Box 1980, 1685 Halfway House, Midrand
247 Old Pretoria Road, Randjespark, Midrand
☎ (+27 11) 315 10 02 [E] hums@ams.co.za
[W] http://www.ams.co.za ⊖ (+27 11) 315 16 45
♧ R 2,516,521 (1996) ♣ R 59,000,000 (1997) 👥 140
🏭 4,500 sq m2
Managing Director: B J Cilliers
Operations Director: E G de Beer
Marketing Manager: Brig P D de Wet (Rtd)
Purchasing Manager: T Blunden
■ Niche capability areas compose of: Health and Usage Monitoring Systems (HUMS) for aircraft and vehicles. Harsh environment and high performance computers. Guided weapons and guidance systems. Meteorological sensors for combat vehicle fire control systems. Rugged power supplies and servo amplifiers. Precision mechanical sub-systems including gimbals and antennae positioners. Stealth and radar absorbing materials, and structural analysis. [98]

75-00795 ARAM
PO Box 1950, 0001 Pretoria
☎ (+27 12) 21 70 83 ⊖ (+27 12) 325 76 12
Managing Dir: R.A. Williams
Operations & Technical Dir: A.J. Brown
Director: S. Allen
Purchasing Mgr: J.M. Thomas
■ Small-arms repair, refurbishing and modification. Manufacture of spare parts for small-arms and light support weapons. [97]

75-01000 Atomic Energy Corp
PO Box 582, 0001 Pretoria
☎ (+27 12) 316 32 70 [E] aec.marketing@pixie.co.z
[W] http://www.aec.co.za ⊖ (+27 12) 316 21 02
♣ US$ 185,000,000 👥 2,200 🏭 5,500 m2
Chief Executive Officer: Dr W E Stumpf
Vice Chairman: Dr M T de Waal
Head of Pelindaba Technology Products (PTP): L S Hyslop
Head of Technology and Development: K F Fouché
Head of Nuclear Fuel and Production: L S Snyders
Head of Corporate Relations and Development: P H S vd Merwe
Head of Finance and Commercial Services: S M J van Vuuren
Head of Corporate Business Development: A G M Jackson
Purchasing Manager: H R Van Der Walt
Export Marketing Manager: Dr F J Wessels
■ Design, development and manufacture of components, sub-systems and systems. Self-luminous components for gunsights and markers, helicopter sand filters, vehicle air-filters, shelf-stable food products, plating and coating of gun bores, special alloys, laser equipment, precision mechanical components, special valves, pressure vessels, decompression chambers, freon compressors, gas turbine components, fluorination of polymers, including fuel tanks. [98]

75-01100 Aztec Electronics (Pty) Ltd
PO Box 794, 2008 Bedfordview
☎ (+27 11) 828 08 00 [E] aztec@mail.global.co.za
⊖ (+27 11) 828 02 46
♣ R 20,000,000 (1996) 👥 60 (1997)
Chairman: D W Atkins
Managing Director: Stan Marcus
Operations and Technical Director: Paul Sabatier
Sales and Marketing Director: Dave Atkins
■ Design and manufacture of aircraft batteries, battery chargers and analysers. Ground power units and ground support equipment. Controllers and chargers for underground battery locomotives. [98]

75-02220 Bestobell Electronic Technologies
PO Box 8035, 1406 Elandsfontein, Johannsberg
☎ (+27 11) 397 33 77 ⊖ (+27 11) 397 31 00
Chief Executive: A Vermaak
Managing Director: N Berry
Sales Manager: T Kotze
Representing:
EEV Ltd, UK [98]

75-02400 Booyco Engineering
PO Box 15258, 1611 Hurlyvale
☎ (+27 11) 609 86 30 ⊖ (+27 11) 452 24 88
👥 23 (1997) 🏭 1,200 m2
General Manager: J C Pougnet
■ Environmental control systems for armoured vehicles and other applications. Microclimate personal systems. Dust filters, MIL-SPEC fans. Power-pack cooling systems. Environmental life support systems. [98]

75-03000 Cochrane Steel Products (PTY) Ltd
(a subsidiary of LTA Ltd)
18 Botha Street, PO Box 123957, 1451 Alrode, Alberton
☎ (+27 11) 908 12 78 ⊖ (+27 11) 908 12 79
♣ US$ 12 000 000 👥 280
Managing Director: R B Cochrane
Marketing and Sales Director: G J Mitchell
■ Manufacture fixed portable razor wire barriers. Armour for vehicle modification. Bullet-proof vests. [98]

75-03100 Consultancy in Advanced Technology
(a subsidiary of Persetel Holdings)
PO Box 650, 3630 Westville, Kwa-Zulu Natal
☎ (+27 31) 266 94 22 [E] info@cat.co.za
[W] http://www.cat.co.za ⊖ (+27 31) 266 94 74
♣ R 15,500,000 (1997) 🏭 900 m2
Managing Director: Dr Clive Putman
Operations Director: Dave Ritten
Marketing Manager: Louise Tromp
Purchasing Manager: Shane Nicholas
■ Security systems. Image capture, management and transmission systems, data encryption products. [98]

75-04030 Da Gama Textile Company Ltd
PO Box 753, 5200 East London
☎ (+27 431) 945 11 24 ⊖ (+27 431) 945 14 56
♣ R 306 000 000 (1995) 👥 3 300
Managing Dir: T.H. Pearce
Marketing Dir: G.E. Schroeder
Operations/Technical Dir: R. Funnell
Sales Mgr: P.D. Campbell
Purchasing Mgr: G. Gibson
■ Manufacture yarn, fabric & textiles. [97]

75-04200 Delcon Control Systems (Pty) Ltd
PO Box 39, 7856 Diep River, Cape Town
☎ (+27 21) 705 62 95 ⊖ (+27 21) 705 62 94
👥 50 (1996) 🏭 1 120 m2 (1996)
Managing Dir: L. Downes
Sales & Marketing Mgr: Tony Richardson

Marketing Co-Ordinator: Ms Lynda MacKenzie
■ Integration, design, installation & support of naval EW systems. Development of advanced ECM systems. Shipboard control & monitoring systems. [97]

75-05680 Eloptro
(a division of Denel (Pty) Ltd)
De Havilland Rd, PO Box 869, 1620 Kempton Park
☎ (+27 11) 921 40 00 [E] jans@eloptro.co.za ⊠ 742237
⊖ (+27 11) 973 16 27
♧ R 9,000,000 (1997) ♣ R 79,300,000 (1997)
👥 400 (1997) 🏭 28,000 m2
Managing Director: J J Swanepoel
Marketing Director: P J Viljoen
Operations Director: T Van Zyl
Operations Director: H B Meyer
Purchasing Manager: C Viljoen
■ Engineering and manufacturing concern specialising in handheld and high repetition laser rangefinders, thermal imaging systems, night vision equipment, armoured vehicle optronics, infra red detectors, optical components, thin film coatings and submarine periscopes. [98]

75-06000 FUCHS Electronics
(a member of the Reutech Group)
PO Box 57, 1450 Alberton
☎ (+27 11) 864 18 60 ⊖ (+27 11) 908 22 99
👥 300 🏭 10 000 sq m
Managing Dir: J. Budricks
Sales & Marketing Dir: C. Bezidenhout
Operations & Technical Dir: A. Theodouro
Purchasing Mgr: Clive Gerrer
■ Manufactures a wide range of fuzes, naval, mortar, artillery & rocket systems. [97]

75-06100 Gelvenor Textiles
(a division of Avtex Ltd)
PO Box 656, 4000 Durban Natal
☎ (+27 325) 611 16 [E] gelvenor@sprintlink.co.za
⊖ (+27 325) 616 25
♣ US$ 30,000,000 (1996) 👥 532
Managing Director: W R Burger
Technical Director: F J Coetzee
Sales Manager: C I Brookes
Export Manager: F Schirmer
■ Manufacture fabric for sleeping bags, rainwear, groundsheets, webbing, packs and ammunition pouches, parachutes, retardation parachutes, extraction parachutes. Parapack materials, medical fabrics (antistatic barrier fabrics), antistatic fabrics with carbon fibre, kevlar and kermel fire resistant fabrics. [98]

75-06400 Grinel
PO Box 1463, 0001 Pretoria
☎ (+27 12) 803 50 00 ⊖ (+27 12) 803 60 39
♣ US$ 50 000 000 (1996) 👥 500 🏭 3 000 sq m
Managing Dir: V.A. De Jager
Sales & Marketing Dir: C.R. Labuschagne
Operations Dir: K. Bell
Purchasing Mgr: L. Du Toit
■ Design and manufacture of battle proven military communication products. [97]

75-06800 Hausler Scientific Instruments (Pty) Ltd
PO Box 33346, 2043 Jeppestown
☎ (+27 11) 614 16 10; 614 16 19
⊖ (+27 11) 614 81 37
♧ R 10,000 (1997) ♣ R 20,000,000 (1996)
👥 160 (1997) 🏭 6,700 m2
Managing Director: R P Hausler
Sales and Marketing Director: J G Smale
Administration Manager: R Hay
■ Precision manufacturing engineers. Aerial survey instrumentation. Aero engine and helicopter control components. Oleo legs for light aircraft. Precision goniometers, machine-gun mountings, small arms and mortar components and borescopes. Missile and submunition fuzes, fire-control equipment and instrumentation, hydraulic and pneumatic rammer and recoil systems. Electrical and mechanical primers. Sights naval fuzes and primers submarine valves and precision components. Petrol and diesel camshafts. Gun drilling and rifling, riot shotguns, handcuffs, 81mm and 60mm mortars, 9mm PB pistols. [98]

75-07000 Hazmat
(a subsidiary of Armscar)
PO Box 2177, 0127 Pretoria
☎ (+27 12) 804 51 73; 804 51 74; 804 22 92
⊖ (+27 12) 804 51 75
Managing Director: H de Bruin
Operations Manager: Ms Linda van Niekerk
■ Develop and manufacture equipment for human protection in chemically and biologically contaminated environments. [98]

75-08000 Hunting Aviation (Pty) Ltd
(† Field Aviation SA (Pty) Ltd)
PO Box 1910, 1610 Gauteng
☎ (+27 11) 397 14 02; 397 14 03 ⊠ 746830 sa
⊖ (+27 11) 397 14 73
♣ US$ 10,000,000 👥 210 🏭 37,000 m2
Director: A K Stewart
Chairman and Chief Executive Officer: Richard Charter
Financial Director: L Busschau
Operations Director: L S L Wybenga
Marketing Director: Steve Duley
Air Motive Division Manager: Jan Viljoen
Fixed-Wing Division: Carlos Barata
Helicopter Division: Pottie Potgieter
■ Aircraft, helicopter and engine maintenance, overhaul and repair, propellors, fuel components and painting. Manufacture of aircraft components. Fixed-wing FBO facilities. PT6 turboprop and Rolls-Royce Dart overhaul facility. Helicopter charter. [98]

75-08150 Interceptor SA (Pty) Ltd
(a subs. of Interceptor Holdings)
P.O. Box 261, 0026 Laezonia
☎ (+27 12) 669 06 33 ⊖ (+27 12) 669 06 36
👥 30 🏭 2 500 sq m
Managing Dir: Jacques van den Berg
☎ (+27 82) 659 1050
Marketing Dir: Sean McGlynn
Director: Justin van den Berg
■ Manufacture of steel palisade fencing & bullet resistant panelling systems. [97]

75-08300 Irenco
(a fully owned subsidiary of Denel Pty Ltd)
PO Box 520, 0062 Irene
☎ (+27 12) 661 71 90; 671 16 71
⊖ (+27 12) 671 18 70
Managing Director: J H Mulder
Sales and Marketing Director: A P H Phillips
■ Manufacture of gyro-stabilised airborne observation systems for helicopters. Day/night observation turret systems for helicopters and aircraft. [98]

75-08900 Kal-Gard
PO Box 1359, 3631 Wandsbeck
113 Premier Place, Phoenix Industrial Park, Durban
☎ (+27 31) 507 57 00; 507 57 01
⊖ (+27 31) 507 92 77
⊡ 0315079277 ♣ R 40,000,000 (1997) 👥 25
🏭 2,400 m2
Managing Director: Rory Nesbitt
Sales and Marketing Director: Ashley Nesbitt
■ Specialised coatings and lubricants for weapons and ammunition. Specialised automotive lubricants. Preservation kits, weapon oil. [98]

75-09000 Kentron
(a division of Denel (Pty) Ltd)
PO Box 7412, 0046 Hennopsmeer
Nellmapius Drive, Irene, Centurion
☎ (+27 12) 671 12 39 [E] besp@kidd.co.za
⊖ (+27 12) 671 17 79
♣ US$ 80,000,000 (1997) 👥 1,200 (1997)
🏭 60,000 m2
General Manager: J Hans Pretorius ☎ (+27 12) 671 11 00
International Marketing Manager: Paul M Bester
☎ (+27 12) 671 12 39
Public Relations Manager: S Walebye
☎ (+27 12) 67 16 81
International Business Development: A J R Oates
☎ (+27 12) 671 17 76
■ Design, development and manufacture of systems, subsystems and components for military and civil applications.
Missile and weapon systems: integration and support for air-to-ground; air-to-air; anti-aircraft and anti-armour missile systems and subsystems.
Unmanned aerial vehicles: remotely piloted aircraft, integration and support of high and low speed target drone systems, subsystems and aerial observation systems.
Fire control systems for air defence (helicopters) and anti-armour (tank) systems.
Inertial navigation systems: inertial sensors and avionics subsystems; maintenance. Airborne observation platforms and stabilised sighting systems. [98]

75-09500 LIW
(† Littleton Engineering Works)
(a division of Denel (Pty) Ltd)
PO Box 7710, 0001 Pretoria
☎ (+27 12) 620 91 11 ⊖ (+27 12) 620 24 31
♣ R 200,000,000 (1996) 👥 1,750
General Manager: J C S Barnard
Marketing Manager Artillery: B Verhoek
Marketing Manager Armour: P L Eksteen
Marketing Manager Small and Medium Calibre Weapons: F du Rand
Public Relations Manager:
■ Heavy engineering. [98]

75-09800 Logistic Projects
Hydraulic Ground Equipment and Systems
Hydraulic and Electrical
PO Box 4114, 2128 Rivonia, Transvaal
Unit 10, Mifa Park, George Road, Midrand, Transvaal
☎ (+27 11) 314 28 00 ⊖ (+27 11) 314 28 27
👥 12
Chief Executive Officer: M R Holland
Operations and Technical Director: D Robertson
■ Supply of ground equipment packages to the military and civil aviation industries. Comprises a range of equipment from ground power units, hydraulic test benches, bomb lifters, airfield lighting, jacks, air conditioning rigs, refuelling units and other associated equipment. → 80 → 85 [98]

75-11000 Mechem
(a div. of Denel (Pty) Ltd)
329b Mundt St., PO Box 912454, 0127 Silverton
☎ (+27 12) 803 72 90 ⊖ (+27 12) 803 71 89
Sr Marketing Mgr: Steve Britz
Communication Services/Marketing: R.H. Walrond
■ Vehicle systems development: mine-protected & or high mobility & or air-droppable. Warhead systems development: Prefragmented warheads for 68mm, 107mm & 122mm rockets. Anti-tank warheads & miscellaneous explosive devices. Special forces equipment, electronic & mechanical. Bomb disposal equipment, bombsuits & disruptors. [97]

75-11500 MOH-9
(a division of Multotec Wear Linings)
PO Box 16992, 0116 Pretoria North
☎ (+27 12) 546 04 61 ⊖ (+27 12) 546 04 60
👥 130
Managing Director: E Holtz
Marketing and Operations Director: D W Skawran
■ Ceramic armour (3mm - 50mm) for body armour, light vehicle systems and tank composite armour packs. [98]

75-15000 Naschem
(a division of Denel (Pty) Ltd)
Private Bag X1254, 2520 Potchefstroom
☎ (+27 148) 297 71 51 ⊖ (+27 148) 298 11 48
♣ R 230,000,000 👥 593
General Manager: Simon Aing
Marketing: R Heyster ☎ (+27 11) 880 35 58
Contract Manager: J N Wagner
■ Develops and manufactures medium and heavy calibre ammunition products. Provides products to the commercial sector, primarily to the mining industry. Provides manufacturing plants to the industrial sector. [98]

75-15100 National Braiding Industries
(a subsidiary of SA Bias Industries Ltd)
PO Box 1700, 3310 Estcourt
☎ (+27 36) 488 15 05 [E] sabas@co.za
⊖ (+27 36) 488 10 60
👥 115 (1997) 🏭 2,200 m2
Managing Director: Cautts Trotter
Marketing Director: M 1Zaroozny
Operations Director: P J 1Roche
Purchasing Manager: S 1Nandcoomar

Factory Manager: W D 1Smith
■ Parachute cords, elasticised cords, braided threads. [98]

75-15200 New Generation Ammunition
PO Box 9207, 0046 Hennopsmeer
☎(+27 12) 666 81 80 [E] mga@tims.co.za
⊜ (+27 12) 666 89 83
♧ US$ 300,000 (1997) ♣ US$ 2,000,000 (1997)
25 (1997) 500 m2
Managing Director: Ivan R Monsieur
Sales and Marketing Director: Peet Grobler
Operations Director: Andre V Dyk
■ Develop and manufacture handgun and rifle cartridges. [98]

75-15500 Optima Hydraulics
Paarden Eiland, PO Box 278, 7420 Cape Town
☎(+27 21) 510 49 60 ⊜ (+27 21) 510 21 14
♣ R 16 000 000 65
Managing Dir: Helmut Herbst
Sales & Marketing Dir: Heinz Von Fintel
■ Hydraulic masts for radar systems. Heavy duty wheel handling systems. Naval cranes & handling equipment. Advanced system engineering. [97]

75-15850 Parnis Engineering Group (Pty) Ltd
PO Box 83511, 2136 South Hills
☎(+27 11) 613 38 01 ⊜ (+27 11) 613 86 60
♣ R 41,000,000 (1996) 174 14,200 m2 (1997)
Managing Director: Mario Guerini
Sales and Business Manager: Michez Raffanti
Manufacturing Manager: Chris Yan Zyl
■ Aircraft ground support equipment, lapes systems, aircraft arresting gear, refuelling system (hydrants), project management, precision & heavy engineering, fabrication. [98]

75-16300 PMP
(a division of Denel (Pty) Ltd)
Private Bag X334, 0001 Pretoria
☎(+27 12) 318 15 45; 318 14 72
⊜ (+27 12) 386 41 32; 318 14 63
♣ R 190,000 (1996) 1,200
General Manager: C P Wolhuter ☎(+27 12) 318 12 16
Marketing Manager: E Burger ☎(+27 12) 318 14 72
■ Small and medium calibre ammunition. [98]

75-17000 Protechnik Laboratories
PO Box 8854, 0001 Randburg
☎(+27 12) 669 07 30 ⊜ (+27 12) 669 06 35
♣ R 11,000,000 35
Managing Director: H D de Bruin
Technical Director: Dr P C Coleman
■ Manufacture military grade impregnated carbons. Evaluation and advisory services. → 80 [98]

75-18210 Republic Arms (Pty) Ltd
PO Box 33346, 2043 Jepperstown
75 Auret Street, Jeppe, Johannesburg
☎(+27 11) 614 16 19 ⊜ (+27 11) 614 81 37
♧ R 10,000,000 ♣ R 17,000,000 (1997) 165
Managing Director: Roderick Peter Hausler
Sales Director: Jeff Smale
Advertising Manager: Bob Hay
■ Manufacture musler riot shotgun model RAP-401 9mm PB Pistol (compact), model 65 handcuffs, 60mm and 80mm morter sights, and weapon components ammunition primers and fuzers. [98]

75-18230 Reumech Austral
PO Box 1004, 1540 Brakpan
☎(+27 11) 813 10 20 ⊜ (+27 11) 817 15 17
Marketing Manager: G van Zyal
■ Fabrication of armoured vehicle hulls and turrets (Rooikat, G6 and others). [98]

75-18265 Reumech-OMC
(a div. of Reunert Mechanical Systems Ltd)
Private Bag X-049, 1500 Benoni
☎(+27 11) 845 14 40 ⊜ (+27 11) 845 13 79
♣ US$ 80 000 000 (1995) 750 14 000 sq m
Managing Dir: A.P.C. Joubert
Operations/Technical Dir: M.H. Meyer
Marketing Mgr: R.J. Craig
Public Relations Mgr: Ms Josy Sanders
Purchasing Mgr: J.J. Snyman
■ Development, manufacture, upgrade & logistic support of armoured vehicles for military, paramilitary & security applications. Manufacturer of the Mamba mine protected APC, Ratel 6x6 ICV, Eland Armoured Car (including diesel turbo conversion), Rooikat 8x8 AFV, weapon platform for G6 6x6 SP howitzer, modernisation of Olifant MBT & support variants, new generation MBT (Tank Technology Demonstrator). [97]

75-18300 Reunert Technology Systems (Pty) Ltd (Reutech)
(a subs. of Reunert Ltd)
42 James Crescent, Halfway House, PO Box 1750, 2128 Gauteng, Rivonia
☎(+27 11) 315 17 00 ☒ 451003 reutech
⊜ (+27 11) 315 21 03
♣ US$ 100 000 000 1 300
Managing Dir: L.R. Swan
Group Marketing Dir: Gordon Blackbeard
Reutech International Mgr: John Gallon
■ Design & manufacture of tactical communication equipment, mine detection systems. Production of prefragmentation air-to-ground weapon systems. A leader in electronic fuzes for bombs, howitzers, mortars & artillery rockets. Radars, IFF, MSSR & avionic systems, fire control systems for tanks, gun control for AFV, tanks, anti-aircraft guns & naval guns. [97]

75-18340 Reutech
PO Box 1750, 2128 Rivonia
☎(+27 11) 315 17 00 [E] info@reutech.co.za
⊜ (+27 11) 315 21 03
Director: Van Verwesthiuzen
Business Development Director: Graham Field
■ Research, development and manufacture of defence systems and equipment including radar systems, aircraft air-to-ground air delivered weapon systems, ordinance electronic fuzes, mine detection equipment, gun and fire control systems, avionic systems, tactical and airborne communication systems and logistics systems. [98]

75-18400 Reutech Defence Industries (RDI)
(† Aserma Manufacturing)
PO Box 118, 3620 Natal
10 Pineside Road, 3600 Natal
☎(+27 31) 719 57 11 ⊜ (+27 31) 72 35 69; 72 36 43
♣ US$ 25,000,000 600 14,000 sq m
Managing Director: C D Makins
Operations Manager: P Farquarson
Marketing and Sales Manager: P Kamerman
Purchasing Manager: R Williams
■ Specialise in air to ground weapon systems. Prefragmented bombs equivalent to MK, French, English and Eastern Europe configurations. Fuzing (electronic) covering all attack profiles to optimise bomb terminal effect and performance. VHF tactical communications systems. Hand-held, manpack and vehicle radios. Artillery communication systems. V/UHF aircraft communication systems. GPS systems, manpack, vehicle and airborne. Mine detectors, hand-held and vehicular. Air weapon systems (fragmentation bombs, fuzes), custom design solutions, firearms (9mm pistols and shotguns). [97]

75-18500 Reutech Radar Systems (Pty) Ltd Radar Division
(a division of Reunert Defence & Allied Reutech Systems)
Technopark, PO Box 686, 7599 Stellenbosch
35 Elektron Street, 7600 Stellenbosch
☎(+27 21) 880 11 50 ⊜ (+27 21) 880 18 42
♣ R 3,500,000,000 600
Managing Director: Derril Burchell
Operations and Technical Director: Bill Reeler
Sales and Marketing Manager: Ken Smith
General Manager: D Botha
Purchasing Manager: Piet Smit
Marketing Manager: Errol Howard
Business Development Executive: B E Blackbeard
■ Design and manufacture of air surveillance systems. Long and medium range search for early detection of low flying aircraft, helicopters and missiles. Short range anti-aircraft systems, 2D and 3D mobile radars, tracking radars. [97]

75-18510 Reutech Systems
PO Box 35, 1685 Halfway House
☎(+27 11) 652 55 55 [E] info@reutech.co.za
⊜ (+27 11) 815 31 90
Managing Director: Derrl Burchell
Marketing Director: Dave Skuy
■ Manufacture and develop integrated gun and fire directing systems; tactical and airborne communiction systems; avionic systems and products; logistics support and maintenance; early warning acquisition and tracing systems. [98]

75-18700 Rotrax Cars International
PO Box 1488, 2153 Jukskei Park
☎(+27 11) 462 57 00 ⊜ (+27 11) 462 14 33
Managing Dir: Enrico Bernert
Sales & Marketing Dir: Bobby Thornton
Operations & Technical Dir: Ken Gillibrand
Purchasing Mgr: Ralf Stranofsky
■ Developed & trailed the El Macho light scout vehicle & displayed at FIDAE 94, DEXA 94 & IDEX 95. → 85 97

75-19160 S.A.P.H.I. 2000
(† S.A.P.H.I. Group)
PO Box 911-547, 0200 Rosslyn
☎(+27 12) 541 60 03 [E] saphi@iafrica.com
⊜ (+27 12) 541 90 93
♧ R 2,500,000 ♣ R 36,000,000 (1997) 220 (1997)
2,000 m2
Managing Director: P B Brown
General Manager: W Van Ree
Sales Manager: Lennox Lekgetha
■ Manufacturer and designer of military headware including ballistic helmets, backpacks and webbing equipment, hand embroidered military badges, and formal head dress. [98]

Contact Reference

75-19180 Silvetech Batteries
(a division of the Institute for Maritime Technology (Pty Ltd)
PO Box 1178, 7599 Stellenbosch
☎(+27 21) 883 88 00 ⊜ (+27 21) 883 88 02
Managing and Marketing Director: P W Stark
Operations Director: D Botha
Purchasing and Finance Manager: W de Villers
■ One-shot and rechargeable batteries designed to need. 98

75-19200 Somchem
(a division of Denel (Pty) Ltd)
PO Box 187, 7129 Somerset-West
☎(+27 24) 850 29 11 ⊜ (+27 24) 850 20 11
1,000
General Manager: Ferdie Hergst ☎(+27 24) 850 20 00
Marketing Manager: Frank Jenkins ☎(+27 24) 850 20 08
Purchasing Manager: André Engelbrecht ☎(+27 24) 850 20 20
■ Supplier of raw materials, propellants, artillery charges and base bleeds, artillery and anti-armour rocket systems, plants and technology. Specialists in the development and/or refurbishment of weapon systems. [98]

75-19390 Sparks & Ellis Uniforms
(a division of Cape Union Mart International Pty Ltd)
PO Box 2545, 8000 Cape Town
34-40 Barrack Street, 8001 Cape Town
☎(+27 21) 45 76 11 ⊜ (+27 21) 461 77 67
Director: Leonard Jocum
■ Manufacture uniforms and foul weather gear including fire fighting uniforms. → 80 [98]

75-19500 Swartklip Products
(a division of Denel (Pty) Ltd)
PO Box 977, 8000 Cape Town
☎(+27 21) 378 11 00 ⊜ (+27 21) 376 11 08
535 120,000 sq m
Managing Director: J J Botha
Operations and Technical Director: F Laubscher
Marketing Manager: Hugo Bester
Export Marketing Manager: André Wagener
Purchasing Manager: J Bouwer

■ Develop and manufacture pyrotechnics and ammunition including 155mm carrier shells, 40mm rounds, rifle and hand grenades, illuminating flares and signals, anti-riot munitions, smoke generators, EW decoys, marine distress signals, 12 gauge shotgun cartridges, .22 rimfire ammunition and powertool cartridges, practice ammunition and components. [98]

75-21500 Uniross Batteries Pty Ltd
PO Box 28861, 0132 Sunnyside
☎ (+27 12) 327 32 66 ⊖ (+27 12) 327 32 94
♧ US$ 1,000,000 (1997) ♣ US$ 4,000,000 (1997)
👥 82 (1997) 🏠 2,700 m2
Managing Director: Dennis A Polack
Technical Director: M S Southern
Sales and Marketing Manager: R Nyman
Purchasing Manager: H Ismail
■ Design, manufacture and supply small rechargeable nickel-caduim, lithium and lead acid batteries. Supply various types of battery chargers.
● 341 Carl Street, PO Box 28861,
0183 Pretoria West Sunnyside
☎ (+27 12) 327 32 66 ⊖ (+27 12) 327 32 94 [98]

80 Service Companies & Organizations

80-00003 Altech Defence Systems
(† Teklogic (Pty) Ltd)
(a subsidiary of the Altech Group)
PO Box 420, 1685 Halfway House
410 Richards Drive, Halfway House, Midrand
☎ (+27 11) 313 90 00 [E] zarinas@teklogic.co.za
[Tx] 427859 ⊖ (+27 11) 313 91 00
👥 240
Managing Director: J F Stark
Marketing Manager: Jack Erasmus
Manager Operational Systems: Charles Strachan
Manager Simulators and Avionics:
Divisional Marketing Manager: Danie van der Walt
Purchasing Manager: Pieter Benkes
■ Systems and Software Developers: Artillery, command and control systems, air defence, systems upgrade and refurbishment, image processing systems, procedural simulators, aircraft simulators, gunnery simulators and intelligence systems. [98]

80-00100 Avex Air Training Pty Ltd
Private Bag X102, 1685 Halfway House
☎ (+27 11) 315 00 03; 315 00 04 [E] avex@iafrica.com
⊖ (+27 11) 805 36 49
♧ US$ 100,000 ♣ US$ 750,000 👥 8 🏠 200 m2
Director: C E Chandler
Chief Flying Instructor: S D Barr
Accountant: I J Steer
■ Flight training, flight simulator training, ground school.
® UK: Airtour International Pilot Equipment, British Airways Aerad Flight Guides, Civil Aviation Authority - Aeronautical, Jane's Information Group
USA: Aircraft Technical Publishers, Frasca Flight Simulators, Information Handling Services (IHS), National Ocean Service- Aero Charts
Representing:
Air Transport Association (ATA)
British Defence Film Library
Jeppesen & Company [98]

80-01990 C2I2 Systems
PO Box 171, 7701 Rondebosch
☎ (+27 21) 683 54 90 [E] rmyoung@ccii.co.za
[W] http://www.ccii.co.za ⊖ (+27 21) 683 54 35
♣ R 10,000,000 (1997) 👥 20 (1997) 🏠 480 sq m
Managing Director: R M Young
Technical Director: H G Krüger
Operations Director: S van der Walt
Marketing Support Officer: B Meyburgh
■ Manufacture naval combat and platform management systems, employing computer networks and optical fibre media. Naval radar consoles, armoured vehicles and control networks. [98]

80-03000 Chase Management Systems (Pty) Ltd
PO Box 76020, 2144 Wendywood
☎ (+27 11) 884 09 87 ⊖ (+27 11) 884 27 92
Contact: John Miller
® UK: Artemis Management Systems [98]

80-04000 Denel Dendex
(a division of Denel (Pty) Ltd
PO Box 7063, 0001 Pretoria
☎ (+27 12) 428 07 74; 06 66
⊖ (+27 12) 45 39 14; 347 11 32
♣ R 30,000,000 👥 120
General Manager: Dr J F Waldeck
Marketing Manager: P Diepenbroek
■ Engineering and management consultants. Simulation and wargaming, multi-criteria decision support, operations management information systems, configuration management, ergonomics, weapon system analysis, ballistics modelling and analysis, product/system integrity test and evaluation, strategic management services, logistics, project management/risk management, systems engineering. [98]

80-05000 Denel Informatics
PO Box 25293, 0105 Monument Park
☎ (+27 12) 450 91 18 ⊖ (+27 12) 450 28 79
👥 1,500
■ Computer related technology to develop and maintain information related systems which improve effectiveness and productivity in the clients environment. [98]

80-06500 Grinaker System Technologies (GST)
(a div. of Grinaker Electronics Ltd)
PO Box 912-561, 0127 Silverton
292 Mundt St, Waltloo, Pretoria
☎ (+27 12) 421 62 00 [Tx] 3491309 sa
⊖ (+27 12) 349 13 08
👥 250 🏠 8 040 sq m
Managing Dir: C.P. van der Merwe
Operations & Technical Dir: J.B. Grobbelaar
Marketing Mgr: Tinus van Wyk
■ Static, mobile & tactical, monitoring, direction finding & jamming systems for land based, airborne & naval installations covering the frequency range 1MHz to 1Ghz as well as electrified security fence systems. [97]

80-09500 Integrators of Systems Technology Pty (Ltd) (IST)
221 Garstfontein Rd, Menlyn, Pretoria
☎ (+27 12) 348 87 05 ⊖ (+27 12) 348 94 72
♣ R 68 000 000 👥 168
Managing Dir: Derrick du Toit
Group Mgr Military Products: G. Shepperson
Contact: Ms S. Vermaak
■ Systems engineering. Electro-mechanical systems, automatic testing systems, low-speed aerial target, training simulators, aircraft test equipment, shooting training systems, air combat training system testers, aircraft arresting gear, avionics fire-control units, avionics clearance equipment, heavy duty mobile air conditioners, helicopter gun turrets & aerial delivery systems. [97]

80-09950 Logtek (Logistic Technologies)
(a subsidiary of Log-Tek Ltd)
PO Box 1300, 1685 Halfway House
309 15th Road, Randjespark, Midrand, Gauteng.
☎ (+27 11) 314 22 70 ⊖ (+27 11) 314 15 95
Managing and Marketing Director: M N Moolman
Operations and Technical Director: W E Delport
Programme Manager Army Systems: C S Wilsworth
Marketing Coordinator: Ms Ellen Kruijer
[E] ellen@logtek.co.za
■ Logistic engineering, logistic support analysis, integrated logistic support, stock management systems, product support and documentation, computer-aided training programmes. [98]

80-10000 Megkon
PO Box 20143, 0005 Alkantrant
☎ (+27 12) 348 76 10 [E] hennie_f@megkom.com
⊖ (+27 12) 47 59 73
♧ R 3,000,000 (1997) ♣ R 30,000,000 (1996)
👥 80 (1996) 🏠 6,000
Managing Director: Dr Piet Stoker
Operations and Technical Director: Dr Harry Wichers
Sales and Marketing Director: Hennie Fourie
Director: T P Roberts
■ Integrated logistic support, systems engineering, mechanical design and analysis. Specialised automated plant and equipment. Computerised integrated logistic's support systems and pharmaceutical plant. [98]

80-13000 Plessey Communication Systems (Pty) Ltd
Head Office
Private Bag X95, 1685 Halfway House, Midrand
Plessey Office Park, Waterfall Park,
Cnr Alandale & Bekker Rd, 1685 Midrand
☎ (+27 11) 266 00 00 ⊖ (+27 11) 266 01 11
Director: W.A. Gibbens
■ Research, development, manufacture, supply and maintenance of high technology telecommunication equipment.
Cape Town
PO Box 3972, 8000 Cape Town
☎ (+27 21) 419 80 60 ⊖ (+27 21) 419 80 71
Durban
PO Box 18145, 4014 Dalbridge
☎ (+27 31) 305 53 91 ⊖ (+27 31) 305 53 94
Port Elizabeth
PO Box 680, 6000 Port Elizabeth
☎ (+27 41) 38 41 44 ⊖ (+27 41) 38 41 63 [97]

80-25000 Reutech Systems
(† Lasso)
PO Box 35, 1685 Halfway House
☎ (+27 11) 652 55 55 ⊖ (+27 11) 805 31 90
♣ R 180 000 000 (1995) 👥 700 🏠 60 000 sq m
Managing Dir: D. Burchell
Sales & Marketing Dir: M. Kitshoff
Operations & Technical Dir: W. Reeler
Purchasing Mgr: M. Lynch
PR Mgr: D.M. Skuy
■ Provision of turnkey solutions for avionic systems, radar systems, airport ground systems, logistic support systems, gun & fire control systems & communication network systems. [97]

80-30000 Roux Systems Analysis and Test
PO Box 3543, 1620 Kempton Park
☎ (+27 11) 396 19 39 ⊖ (+27 11) 979 14 14
👥 10 🏠 120 m2
Managing Director: I J Roux [98]

80-41000 Southern Oceanics (Pty) Ltd
PO Box 36541, 7442 Chempet
☎ (+27 21) 551 22 33 ⊖ (+27 21) 551 22 75
♣ R 9 000 000 (1995) 👥 25 🏠 1 660 m2
Managing Dir: D. Park-Ross
Sales & Marketing Dir: VAdm L.J. Woodburne
Operations Dir: F. Burman
Purchasing Mgr: Ms Ann Schwaeble
■ Life support engineering, diving bells. Decompression chambers. Mission support. [97]

80-42000 Spesaero Engineering
PO Box 74555, 0040 Lynwood Ridge
☎ (+27 12) 348 68 86 ⊖ (+27 12) 348 52 71
👥 4 🏠 300 m2
Managing Dir: Dr R.S. van Rooyen
■ Aerodynamic design & analysis for fixed-wing aircraft, UAVs, RPVs & helicopters. Software development for flight planning. [97]

80-60000 Thales Advanced Engineering
PO Box 243, 2062 Lonehill
☎ (+27 11) 465 43 12 ⊖ (+27 11) 465 43 15
Director: B Coetzer
Director: P Handley
Personal Assistant: P Bott
■ Development, design and implementation of RF data communications systems, image transmission systems and communications planning systems. [98]

85 Stockists, Distributors, Spare Parts Suppliers

Rotrax Cars International 75-18700

85-00150 Advanced Materials Technology (Pty) Ltd

Garden View, PO Box 751485, 2047 Johannesburg
☎ (+27 11) 392 42 32 ⊖ (+27 11) 974 41 98
👥 25 (1997) 🏭 860 m2
Managing Director: G E Bargehr
Operations and Technical Director: J R Jacinto
Sales and Marketing Director: D J Stevenson
Purchasing Manager: Mrs N Wahlstrand
■ Importer and distributor of advanced composite materials for the manufacture and repair of aerospace composite parts. Resins, film adhesives, honeycomb, prepregs, sandwich panels and ferrules, woven fabrics carbon and aramid fabrics.
® UK: Aerovac Systems, Dexter Adhesives, Dexter Frekote, Hexcel Composites, Hexcel Fabrics
USA: Baltek Balsa USA

AMT Natal
PO Box 659, 7435 Milnerton
☎ (+27 31) 21 13 19 ⊖ (+27 31) 21 28 03

AMT Natal
PO Box 649, 7435 Milnerton
☎ (+27 21) 511 26 69 ⊖ (+27 21) 510 83 89 [98]

85-00300 Astra Aircraft Corp (Pty) Ltd

Astra Helicopter (Pty) Ltd
PO Box 786392, 2146 Sandton
☎ (+27 11) 884 87 80 ⊖ (+27 11) 884 68 01
♣ R 50 000 000
Managing Director: B A Nash
General Manager: G G West
■ Distributor of aircraft and aeronautical equipment. Engineering consultants. → 90
® France: Thomson Training and Simulation
UK: Dzus Fastener Europe, Rediffusion Simulation, Smiths Industries Avitron
USA: Lockheed [97]

85-00700 Avitronics

PO Box 8492, 0046 Centurion
☎ (+27 12) 672 60 00 [E] defence@avitronics.co.za
⊖ (+27 12) 672 60 20; 672 62 22 (marketing)
👥 300 (1997)
Managing Director: Alan Holloway ☎ (+27 12) 672 60 01
Operations Director: Ben Ash ☎ (+27 12) 672 60 02
Sales and Marketing Director: Garrie Radloff
☎ (+27 12) 672 60 29
Purchasing Manager: Sakkie Allers ☎ (+27 12) 672 60 39
Public Relations Executive: Ms Linda Weaver
■ Specialise in electronic warfare systems, avionics, radar and associated equipment. The in-house capabilities in microwave, antenna, electro-optic, digital, software and logistic support technologies provide customised and innovative solutions that satisfy specific and unique requirements. [98]

85-01300 Heliquip Pty Ltd

PO Box 5526, 1685 Halfway House, Gauteng
☎ (+27 11) 315 00 01; 315 00 02 ⊠ ROTORGEN
⊖ (+27 11) 805 34 09
👥 22 (1997) 🏭 1,200 m2
Managing Director: P Piggott ☎ (+27 11) 805 46 47
Sales and Marketing Director: Christopher Piggott
☎ (+27 11) 315 00 01; 315 00 02
Operations and Technical Director: Keith Piggott
Purchasing Manager: Eston Manda
■ Sales and service Eurocopter helicopters and Aerospatiale (Socata) aircraft.

Operations
1685 Halfway House, Grand Central Airport
☎ (+27 11) 315 00 01; 315 00 02
⊖ (+27 11) 805 34 09 [98]

85-01900 Somchem

(a div. of Denel (Pty) Ltd)
PO Box 187, 7129 Somerset-West
☎ (+27 24) 850 20 08; 29 11
⊖ (+27 24) 850 20 12; 21 11
♣ R 185 000 000 (1995/96) 👥 1 000
Managing Dir: F.P. Herbst
Sales & Marketing Dir: J.B. Bestbier
Operations & Technical Dir: N. Leicher
Purchasing Mgr: A.J. Engelbrecht
Public Relations Mgr: F. Gerber
■ Supplier of raw materials, propellants, artillery charges & base bleeds, artillery & anti-armour rocket systems, plants & technology. Also specialists in the development and/or refurbishment of weapon systems. → 80 [97]

85-08000 Trimos-Sylvac SA (Pty) Ltd

PO Box 95672, 0145 Waterkloof
☎ (+27 12) 660 22 96/7 ⊖ (+27 12) 660 22 98
Managing Dir: Fritz Kosa
Marketing Dir: Ms Edith Kosa
■ Suppliers of precision measuring equipment, fixtures, gauges, machine tools & accessories. [97]

90 Sales Agents, Sales & Liaison Offices

Companies represented:

Aerovac Systems, UK	85-00150
Aim Fliteform, UK	90-00800
Aim Henshalls, UK	90-00800
Aircraft Technical Publishers, USA	80-00100
Airtour International Pilot Equipment, UK	80-00100
Alfred Karcher GmbH & Co, Germany	90-10000
AlliedSignal Aerospace, USA	90-16900
Artemis Management Systems, UK	80-03000
Arvico, USA	90-00800
Aviobridge, Netherlands	90-00800
Baltek Balsa USA, USA	85-00150
Barigo, Germany	90-16900
Bendix/King, USA	90-16900
BP Solar Limited, UK	90-01000
Bridport Aviation/Nets/Straps, UK	90-16900
British Airways Aerad Flight Guides, UK	80-00100
Cessna Aircraft, USA	90-03750
Civil Aviation Authority - Aeronautical, UK	80-00100
Cullem Detuners, UK	90-00800
Davis Instruments, USA	90-16900
Davis Weather Systems, USA	90-16900
Dexter Adhesives, UK	85-00150
Dexter Frekote, UK	85-00150
Dzus Fastener Europe, UK	85-00300
Frasca Flight Simulators, USA	80-00100
Helmets Integrated Systems, UK	90-00800
Hexcel Composites, UK	85-00150
Hexcel Fabrics, UK	85-00150
Hobart GPU, USA	90-16900
Information Handling Services (IHS), USA	80-00100
Jane's Information Group, UK	80-00100
Lockheed, USA	85-00300
McDonnell Douglas Helicopter Co., USA	90-03750
Munro Meteo Equipment, UK	90-16900
National Ocean Service- Aero Charts, USA	80-00100
OFEMA, France	90-14950
Penny & Giles Aerospace, UK	90-00800
Precision Air Motive Corp, USA	90-16900
Rediffusion Simulation, UK	85-00300
RLG Docking Guidance Systems, USA	90-00800
Safety Training Systems Inc, USA	90-00800
Smiths Industries Avitron, UK	85-00300
Teledyne Continental Motors, USA	90-16900
Thomson Training and Simulation, France	85-00300
Totex, Japan	90-16900
TRO Learning CBT Products, UK	90-00800
UFA Systems, USA	90-00800
UKW Technik, Germany	90-16900
Ultra Command & Control, UK	90-00800
Ultra Electronics Weapons Systems, UK	90-00800
Unison Industries, USA	90-16900

Sales agents/representatives listed elsewhere:

Astra Aircraft Corp (Pty) Ltd	85-00300

90-00550 AlliedSignal Aerospace Marketing Sales & Service

PO Box 26215 (East Rand), 1627 Kempton Park
c/o SAA Permit Office,
☎ (+27 11) 978 69 36 ⊖ (+27 11) 978 69 35
Field Service Engineer: Joe McQueen
☎ (+27 11) 390 17 56
Field Service Engineer: Ms Nicole Marias
☎ (+27 11) 390 54 05
Field Service Engineer: Joe Parreirinha [98]

90-00800 Atalanta (Pty) Ltd Aviation Division

(a subsidiary of Astra Group Holdings Pty Ltd)
PO Box 786392, 2146 Sandton
☎ (+27 11) 884 87 80 ⊖ (+27 11) 884 68 01
Managing Director: B A Nash
General Manager: Gerald G West
® Netherlands: Aviobridge
UK: Aim Fliteform, Aim Henshalls, Cullem Detuners, Helmets Integrated Systems, Penny & Giles Aerospace, TRO Learning CBT Products, Ultra Command & Control, Ultra Electronics Weapons Systems
USA: Arvico, RLG Docking Guidance Systems, Safety Training Systems Inc, UFA Systems [98]

90-01000 BP Solar Southern Africa

2nd Floor, BP House, 10 Junction Road, 2193 Parktown
☎ (+27 11) 488 53 32 ⊖ (+27 11) 488 51 31
® UK: BP Solar Limited [98]

90-03750 Comair Sales (Pty) Ltd

(a subsidiary of Commercial Air Services Pty Ltd)
PO Box 8050, 1622 Bonaero Park
☎ (+27 11) 921 01 11; 659 24 10 [E] comair@pixie.co.za
⊠ COMAIR johannesburg
⊖ (+27 11) 395 25 46; 701 32 05
👥 10
Managing Director: D Novick
General Manager and Director: M D Moritz
Sales Director: A H Novick
■ Sales and support for Cessna and McDonnell Douglas.
® USA: Cessna Aircraft, McDonnell Douglas Helicopter Co. 98

90-03800 Daewoo Johannesburg Branch

1st Floor Norwich Life Towers, 13 Fredman Drive, 2196 Sandown, Sandton
☎ (+27 11) 784 13 26 ⊠ 427870 sa
⊖ (+27 11) 784 13 01 [98]

90-04100 Day Technical Products (Pty) Ltd

393 Jules St Malvern, PO Box 40476, 2022 Cleveland Johannesburg 2094
☎ (+27 11) 616 51 87 ⊖ (+27 11) 616 58 03
👥 17 (1997) 🏭 1,000 m2
Managing Director: R S Day
Operations Director: R R Day
Marketing and Purchasing Director: C K Bate [98]

90-05000 Fairmont Electonics (Pty) Ltd

4th Floor, 312 Kent Avenue, 2194, Ferndale, Randurg
☎ (+27 11) 886 29 20 ⊠ 424842 ⊖ (+27 11) 886 29 29
👥 38 (1997)
General Manager: R Taitz [98]

90-10000 Karcher (Pty) Ltd

PO Box 11818, 1686 Vorna Valley
☎ (+27 11) 466 24 34 ⊖ (+27 11) 466 21 85
Managing Director: Brendan P Kelly
® Germany: Alfred Karcher GmbH & Co [98]

90-14000 Munters Pty

(a division of AB Carl Munters)
PO Box 21817, 6000 Port Elizabeth
☎ (+27 41) 34 58 32 ⊖ (+27 41) 34 58 38
♣ R 7,000,000 - 8,000,000 👥 9 🏭 200 m2
Divisional Manager: Mike Spearpoint [98]

90-14950 OFEMA South Africa

(a subsidiary of OFEMA, France)
3rd Floor - United Building, PO Box 10347, 0046 Centurion
☎ (+27 12) 663 68 37 [E] ofema@iafrica.com
⊖ (+27 12) 663 68 41
Technical Delegate: K G Smith
® France: OFEMA [98]

Contact Reference

90-16900 **C W Price & Co. (Pty) Ltd**
68 4th Street, PO Box 5, 2063 Marlboro
☎(+27 11) 444 70 56 ⊜ (+27 11) 444 78 48
♣ R 4,000,000 (1995) 5
Director: B I Price
Director: Kelvin L Price
→ 85
® Germany: Barigo, UKW Technik
Japan: Totex
UK: Bridport Aviation/Nets/Straps, Munro Meteo Equipment
USA: AlliedSignal Aerospace, Bendix/King, Davis Instruments, Davis Weather Systems, Hobart GPU, Precision Air Motive Corp, Teledyne Continental Motors, Unison Industries 98

Spain

Head of State:
H M King Juan Carlos I

10 Head of State

10-10000 **Office of H M the King**
Palacio de la Zarzuela, El Prado, Madrid
King of Spain: H M King Juan Carlos I 98

11 Prime Minister

11-10000 **Office of the President of the Government**
Presidente del Gobierno
Complejo de la Moncloa, Edificio Semillas, 28071 Madrid 15
☎(+34 1) 449 72 52
President: Jose Maria Aznar 98

11-20000 **Office of the Vice-President of the Government**
Vicepresidente del Gobierno
Complejo de la Moncloa, Edificio Semillas, 28071 Madrid
Vice President, Minister of the Presidency: Francisco Alvarez-Cascos Fernandez
Vice president, Minister of Economy and Finance: Rodrigo Rato Figaredo 98

12 Ministries

12-10000 **Ministry of Defence**
Paseo de la Castellana 109, 28071 Madrid
☎(+34 1) 555 26 00; 555 46 00 ⊠ 47737 ebci e
⊜ (+34 1) 556 55 72
Minister of Defence: Eduardo Serra Rexach 98

12-20000 **Ministry of Foreign Affairs**
Ministerio de Asuntos Exteriores
Plaza de la Provincia 1, Palacio de Santa Cruz, Madrid 12
☎(+34 1) 366 48 00; 366 50 00 ⊜ (+34 1) 366 50 32
Minister of Foreign Affairs: Abel Juan Matutes
Deputy Director General Scientific and Technical: Vicente Gonzalez Loscertales
Deputy Director General, Land Sea, Air Co-operation: Alfonso Muqoz-Seca Y Fernandez Cuesta
Director General International Economical Relations: Ricardo Zalacain 98

12-40000 **Ministry of the Interior**
Castellana 5, 28071 Madrid
☎(+34 1) 537 10 00 ⊠ 46092 ⊜ (+34 1) 310 27 30
Minister of the Interior: Jamie Mayor Oreja
Directorate of State Security
Calle Fernando do Santo 23, 28010 Madrid
International Relations: Benito Gonzalo
International Relations: Mr Piquares 98

12-50000 **Ministry of Finance and Economy**
Ministerio de Economia y Hacienda
Alcalá 7-9, 28014 Madrid
☎(+34 1) 468 20 00; 458 86 64
Minister of Economy and Finance: Rodrigo Rato Figaredo 98

12-50500 **Ministry of Industry & Energy**
Ministerio de Industria y Energia
Paseo de la Castellana 160, 28046 Madrid
☎(+34 1) 458 80 10 ⊠ 44204
Minister: Josep Pique Camps 98

12-60000 **Ministry of Public Works & Transport**
Ministerio de Obras Publicas, Transportes Y Medio Ambiente
Nuevos Ministerios, 28071 Madrid
☎(+34 1) 553 16 00 ⊠ 22325 minop e
⊜ (+34 1) 554 63 65; 533 22 71
Minister: Rafael Arias-Salgado Montalvo 98

15 Parliament

15-20000 **Congress of Deputies**
Congreso de los Diputados
Floridablanca s/n, 28014 Madrid
☎(+34 1) 429 51 93; 429 51 56; 429 51 76
⊠ 48433 edgp e
President: F T F Martinez-Conde
First Vice-President: Fernandez-Miranda Lozana
Second Vice-President: Joan Marcet i Morera
Third Vice President: Josep Lopez de Lerma i Lopez
Fourth Vice President: Bevia Pastor José Vicente
First Secretary: Maria Bernarda Barrios Curbelo
Second Secretary: Pedro Antonio Rios Martinez
Third Secretary: Carmen del Campo Casasus
Fourth Secretary: José Juan Gonzalez de Txabarri Miranda
Defence Commission
President: Alejandro Muñoz-Alonso Y Ledo
First Vice President: Joan Miquel Nadal I Male
Second Vice President: Jeronimo Nieto Gonzalez
First Secretary: Jordi Marsal I Muntala
Second Secretary: Luis de Torres Gomez
Foreign Affairs Commision
President: Javier Ruperez Rubio
First Vice-President: Rafael Estrella Pedrola
Second Vice-President: Manuel Milian Mestre
First Secretary: Tomas Rodriguez Bolanos
Second Secretary: Ricardo Burballa I Campabadal 98

25 Defense Forces

25-22200 **Strait Air Command / 2nd Air Region**
Mando Aereo del Estrecho (MAEST)
Sevilla
☎(+34 5) 445 20 11 ⊜ (+34 5) 445 46 45
Commander: Lt Gen. Fernando Ostos Gonzales
Deputy Commander & Chief of Staff: Maj.Gen. Enrique Richard Marin
Public Information Officer: Lt. Juan Piquero Cordero 97

25-22400 **Canaries Aviation Command**
Mando Aéreo de Canarias (MACAN)
Las Palmas de Gran Canaria
☎(+34 28) 24 39 03 ⊜ (+34 28) 24 56 96
Commander: Maj Gen Sebastiaan Rodriguez-Barrueco Salvador
Chief of Staff: Maj Gen Antonio Garcia Lozano 98

25-31000 **Army Helicopter Forces Command**
Fuerzas Aeromoviles del Ejército de Tierra (FAMET)
Colmenar Viejo, 28770 Madrid
☎(+34 1) 845 27 50 ⊜ (+34 1) 845 03 40
Commander: Gen Gonez Vilaplana
Chief of Staff: Lieut Jose Monasterio Renteri
Director of Operations: Maj Julio Alondo
Chief Transmissions: Maj José Salas Salvatierra
Director of Logistics: Maj Carbunell
Deputy Chief of Staff Logistics: Maj Carlos De Salas
■ Missions of FAMET include army support and civilian emergency medical services and relief flights. 98

25-40000 **Navy Headquarters**
Cuartel General de la Armada
Montalbán 2, 28014 Madrid
☎(+34 1) 379 50 00 ⊜ (+34 1) 379 51 74
Chief of Naval Staff: Adm. Juan José Romero
Deputy Chief of Staff: VAdm. Francisco Rapallo
Chief of Cabinet: Cdr Carlos Tortosa
Chief Logistics: RAdm. Angel González
Chief Naval Operational Command: RAdm. Juan J. González-Irún
Chief Naval Air Dept: Capt. José Rosano
Chief Plans: RAdm. José Antonio Balbas
Chief Tactics: RAdm. Fernando Riaño
Chief Personnel: VAdm. Marcelino García
Admiral Chief Cantabrian Zone: Adm. Gabriel Portal
☎(+34 81) 33 60 00
Admiral Chief Straits Zone: VAdm. Carlos J. Sánchez de Toca
☎(+34 56) 59 90 00
Admiral Chief Mediterranean Zone: Adm. Justino Antón
☎(+34 68) 12 70 00
Admiral Chief Canarian Zone: VAdm. Adolfo Baturone
Naval Logistic & Support Command
Avenida Pio XII 83, 28016 Madrid
☎(+34 1) 379 55 00
Director: Adm. Eduardo Liberal
Asst Director: Capt. Antonio González-Aller
Dir Construction: VAdm. Nicolás Lapique
Deputy Dir Construction: RAdm. Pedro M. de Benito
Deputy Dir Maintenance: RAdm. Emilio Erades
Chief Technical Services Weapons: VAdm. Carmelo Sánchez
Dir of Supplies & Transports: Gen. José Romero 97

25-41000 **Spanish Fleet Commands**
Montalbán 2, 28014 Madrid
Admiral of the Fleet (ALFLOT): Adm. Manuel Acedo
Chief of Staff: RAdm. Tomás García-Figueras
☎(+34 56) 82 70 00; 82 76 96
Commander Amphibious Command: RAdm. Alfonso Mosquera
Commander Aeronautical Group: RAdm. Lorenzo Montero
Chief of Staff Aeronautical Group: Cdr José Sagrera
Commander Escort Command: Capt. Fernando Armada
Naval Air Arm
Flotilla de Aeronaves del Armada
Naval Base, Rota-Naval, 11530 Cadiz
☎(+34 56) 81 23 50
Chief Naval Air Arm: RAdm. Joachim Pita de Veiga
Commander: Capt. Francisco Cantó
Comandante Director EDAN: Capt. Miguel López 97

25-42000 **Marines General Command**
Commandancia General de la Infanteria de Marina
Montálban 2, 28014 Madrid
☎(+34 1) 379 50 00
Commandant General: Lt Gen. Abelardo Vázquez
Deputy Commander: Maj.Gen. Francisco González 97

25-49100 **Naval War School**
Ave de Burgos 6, 28036 Madrid
☎(+34 1) 379 55 00
Director: VAdm. Francisco Nuñez
Deputy Director: Capt.(N) Angel Abad 97

29 Customs & Excise

29-00500 Directorate General of Customs
Departamento de Aduanas e Impuestos Especiales
Agencia Estatal de Administracion Tributaria
Guzmán el Bueno 137, 28003 Madrid [98]

30 Civil Defense

30-00500 General Directorate of Civil Protection
Dirección General de Protección Ciudadana
Carretera de la Coruña - Km 22, Pinar de las Rozas, 28250 Madrid
Contact: Mrs Almudena González Fernández [98]

31 Embassies & Attachés

31-00230 Brazilian Embassy
Calle Fernando El Santo, 6-28010 Madrid
☎(+34 1) 308 04 59
[E] embajadadebrasil@mad.servicom ☒ 23 291 brmd e
⊖ (+34 1) 308 04 65
Ambassador: H E Mr Luiz Felipa de Seixas-Correa
Counsellor: Eduardo Da Costas farias
Military Naval and Air Attaché: Col Hideo Sato
Counsellor, Political: Carlos Roberto Bevilaqua Penna
First Secretary: Fernando Jablonski
First Secretary: Arnaldo Caiche D'Oliveira
First Secretary: Bruno Luiz Dos Santos Cobuccio
Second Secretary: Fabio Mendes Marzano
Commercial Office
Jacometrezo 4-6, Madrid 13
☎(+34 1) 221 83 68; 231 03 72 [98]

31-00310 Embassy of Canada
PO Box 587, 28080 Madrid
Edificio Goya, Calle Nunez de Balboa 35, Madrid
☎(+34 1) 431 43 00 ✉ DOMCAN Madrid
☒ 27347 domcan e ⊖ (+34 1) 431 23 67
Ambassador: H.E. Mr David Wright
Minister Counsellor & Consul: J. Kneale
Canadian Forces Attaché: Col T.J. Pile
Public Information Officer: Paul Willox
Commercial Office
PO Box 117, 28080 Madrid
☎(+34 1) 431 43 00 ✉ CANADIAN Madrid
⊖ (+34 1) 431 23 67
Commercial Counsellor: G. Bruneau [97]

31-00460 Royal Danish Embassy
Claudio Coello 91, 28006 Madrid
☎(+34 1) 431 84 45 [E] dkembassy@genio.infor.es
☒ 27265 amdane ⊖ (+34 1) 431 91 68
Ambassador: H E Mr John Bernhard
Consul: Esper Larsen Anne-Marie ☎(+34 1) 650 17 61
Agronomy Counsellor: Ib Byrge Sørensen
Commercial Consular: Christian F L d'Auchamp
First Secretary: Nis Lauge Gellert
Secretary of the Embassy: Birgit S Kondrup-Palmqvist [98]

31-00660 Embassy of the Federal Republic of Germany
Calle de Fortuny 8, 28010 Madrid
☎(+34 1) 319 91 00 ☒ 27768 aamdr e
⊖ (+34 1) 310 21 04
Ambassador: H.E. Dr Henning Wegener
Minister Deputy Chief of Mission: Dr Reinhold Huber
Defence Attaché: Col Ulrich Twrsnick
Naval Attaché: Cdr Dieter Krause
First Counsellor, Economics: Dr Bernd Sproedt
First Councellor, Politics: Dr Christian Stocks [97]

31-00690 Embassy of Greece
Avda del Dr Arce 26, 28002 Madrid
☎(+34 1) 564 46 53 ☒ 44135 ⊖ (+34 1) 564 46 68 [98]

31-00850 Embassy of the Islamic Republic of Iran
Jerez 5, Madrid
☎(+34 1) 345 01 12 ☒ 22322 irana e
⊖ (+34 1) 345 11 90
Ambassador: H.E. Mr Bahram Ghassemi
Deputy Chief of Mission: Hamid Reza Nafez Arefi
☎(+34 1) 345 01 16 [97]

31-00870 Embassy of Ireland
Paseo de la Castellana 46-40, 28046 Madrid
☎(+34 1) 576 35 00 ☒ (+34 1) 435 16 77
Ambassador: H E Mr Richard Ryan
First Secretary: Rúairi De Búrca
Third Secretary: Declan Smyth [98]

31-00890 Embassy of Italy
Calle de Lagasca 98, 28006 Madrid
☎(+34 1) 577 65 29 ⊖ (+34 1) 575 77 76
Ambassador: H E Dr Raniero Vanni d'Archirafi
Minister Counsellor and Deputy Chief of Mission: Dr Emilio Barbarani
Naval Attaché: Capt de Navio Claudio Pini
Commercial Counsellor: Marco Baccin
☎(+34 1) 431 35 23
Counsellor: Ms Cristina Ravaglia
Airforce Attaché: Col Alberto Battaglini
Military Attaché: Col Nicola De Santis
First Secretary: Roberto Vellano
First Sectretary: Tosca Barucco
Administrative Secretary: Donatella Fattori
Attaché, Scientific: Angelo Guerrini
Attaché, Cultural: Vito Grasso
Attaché: Italo Fiamingo
Agustin de Bethancourt n.3
28006 Madrid
☎(+34 1) 534 53 93 ⊖ (+34 1) 534 22 04
Defence & Naval Attaché: Capt.Vaasc. Cesare Braccio
☎(+34 1) 534 18 00
Military Attaché: Col. Nicola De Santis
☎(+34 1) 534 22 04
Air Attaché: Col. Alberto Battaglini ☎(+34 1) 534 54 45 [98]

31-00920 Embassy of Japan
Embajada del Japón
Calle Joaquín Costa 29, 28002 Madrid
☎(+34 1) 562 55 46 ✉ TAISHI Madrid ☒ 27652 taisi e
⊖ (+34 1) 562 78 68
Ambassador: H.E. Mr Tatsuo Yamaguchi
Minister: Masami Takemoto
General Consulate of Japan
Consulado General de Japón
Edificio Banca Catalana, Planta 3,
Avenida Diagonal 662-664, 08034 Barcelona
☎(+34 3) 280 34 33 ⊖ (+34 3) 280 44 96
General Consulate of Japan
Consulado General del Japón
Calle Santiago Rusiñol n°12, PO Box 733,
35005 Las Palmas de Gran Canaria
☎(+34 28) 24 40 12 ⊖ (+34 28) 23 23 48 [97]

31-01000 Embassy of the State of Kuwait
Paseo de la Castellana 141, 28046 Madrid 1
☎(+34 1) 572 01 62/4 ☒ 23753 kwit e
⊖ (+34 1) 579 63 50
Ambassador: H.E. Mr Mohammad Al-Sallal
Deputy Chief of Mission: Muhyee Aldeen R. Al-Felaij
Third Secretary: Sami Abdul-Aziz A. Al-Hamad
Attaché: Jazz'aa J.H. Al-Otaibi [97]

31-01270 Royal Netherlands Embassy
Avda. Comandante Franco 32, 28016 Madrid 16
☎(+34 1) 359 09 14 [E] nlgovmad@ctv.es
☒ 27316 holmd e ⊖ (+34 1) 359 21 50
Ambassador:
H E Mr W O Baron Bentinck van Schoonheten
Commercial Counsellor: J A v d Kleÿ
Defence Attaché: Capt J M van der Ham (Navy)
Press and Public Affairs Attaché: E H van As [98]

31-01340 Royal Norwegian Embassy
Apartado 6132, 28080 Madrid
☎(+34 1) 310 31 16 ☒ 23331 embn e
⊖ (+34 1) 310 43 26
Ambassador: H E Mr Jan E Nyheim
Deputy Chief of Mission: Håvard Austad
Military Attaché: Lieut Col Gunner Mjell
(Resident in Paris from August) (1997)
☎(+33 1) 53 67 04 17
Commercial Counsellor: Knut Reintz ☎(+34 1) 344 17 88 [98]

31-01480 Embassy of Romania
Alfonso XIII 157, 28016 Madrid
☎(+34 1) 350 44 36 ☒ 22433 rocom e
⊖ (+34 1) 345 29 17
Ambassador: H.E. Mr Darie Aurel Novaceanu
Counsellor: Vasile Macovei
First Secretary: Dumitru Aldea
Second Secretary: Angelo Florin Florian
Second Secretary: Nicusor Daniel Tanase
Economic Counsellor: Ionel Marin [97]

31-01655 Embassy of the Republic of Slovenia
Hermanos Bécquer, 7-20, 28006 Madrid
☎(+34 1) 411 68 93; 411 67 24 ⊖ (+34 1) 564 60 57
Cultural Attaché: Lidija Sircelj [98]

31-01780 Royal Thai Embassy
Calle del Segre 29-2A, 28002 Madrid
☎(+34 1) 563 29 03; 563 79 59
[E] thai-mdd@asupaa.medusa.es ☒ 49044 siam e
⊖ (+34 1) 564 00 33; 562 41 82
Ambassador: H E Mr Prida Boonyobhas
Minister (Deputy Chief of Mission): Suphat Chitiamukroh
Defence and Naval Attaché:
Gp Capt Amomtep Nabangchang
Commercial Minister: Wiuichai Chemchaeng
Commercial Counsellor: Nalin Kraiviksh
First Secretary: Chaypan Bamrungphong
Attaché: Pottama Saichasassi
Office of the Naval Attache
C /Islas Kuriles, 4-A, 28035 Madrid
☎(+34 1) 373 71 63 ⊖ (+34 1) 373 71 22
Naval Attaché: Gp Capt Amorntep Nabangchang [98]

31-01890 Embassy of the United Kingdom
Calle de Fernando el Santo 16, 28010 Madrid
☎(+34 1) 319 02 00 ☒ 27656 ingla e
⊖ (+34 1) 319 04 23
Ambassador: H E Mr A D Brighty (Accredited to Andorra)
Minister and Deputy Chief of Mission: J A Dew
Counsellor (Commercial): M C Ramscar
Counsellor (European Community, Economic): C J Ingham
Defence and Naval Attaché: Capt P J Pacey (RN)
Military and Air Attaché: Col R J Lawson [98]

31-01900 Embassy of the United States of America
Serrano 75, 28006 Madrid
☎(+34 1) 587 22 00 ☒ 27763 amemb e
⊖ (+34 1) 587 23 03
Ambassador: H E Mr Richard N Gardner
Deputy Chief of Mission: Lawrence Rossin
Commercial Counsellor: Rafael Fermoselle
Scientific Attaché: Marshell Carter-Tripp
Defence Attaché: Capt James Tinsley (USAF)
Head of Defense Co-operation: Capt L E Sheehan (USN)
Public Affairs Officer: Brian E Carlson [98]

31-02010 Embassy of the Federal Republic of Yugoslavia
Calle de Velázquez 162, 28002 Madrid 2
☎(+34 1) 536 50 45 ☒ 44208 emyu e
⊖ (+34 1) 563 04 40
First Secretary: Gordana Vidovic
Counsellor, Political: Zoran Kazazovic
Administration: Dojcin Djujic [98]

32 Military Liaison Offices

32-10000 Office of Defense Co-operation
Security Assistance Office
Romero Robledo 8, 28008 Madrid
☎(+34 1) 543 28 02 ⊖ (+34 1) 549 01 45
Security Assistant Chief: Lieut Col Cliff Graham [98]

Contact Reference

34 Multinational Intergovernmental Organizations

34-25000 WEU Satellite Centre
Apdo de Corroes 511, Base aérea de Torrejón,
28850 Torrejón de Ardoz, Madrid
☎(+34 1) 677 79 99 ⊖ (+34 1) 677 72 28
Director: Bernard Molard
Director of Operations: Geoffrey Quick
Deputy Commander: Frank Asbeck
Director of Procurement: Robert du Meyere [98]

55 Government-owned Industry & Production Facilities

55-00120 Construcciones Aeronáuticas SA (CASA)
Avenue de Aragon, 404, 28022 Madrid
☎(+34 1) 585 72 67; 585 70 00 ⊠ 41726; 41748
⊖ (+34 1) 585 72 74; 585 74 57; 585 76 66;
585 76 67; 585 73 66
♧ Pta 46,425,000,000 (1996)
♣ Pta 110,034,000,000 (1996) 👥 7,931
🏭 3,046,000 sq ft
Chairman and Chief Executive Officer: Raul Herranz
Vice President and General Manager Aircraft Division: Carlos Navarro
Vice President Commercial: Juan A Alonso
Vice President Programmes: Luis Muñoz
Maintenance Division Director: Juan T Ruiz Sequera
Space Division Director: Antonio Fuentes
Marketing Research Manager: Mario Lopez
Marketing Manager: Mariano Barrena
Public Relations and Press Manager: Jose Maria Sanmillan
■ Manufacture C-212 and CN 235 civil and military light transport aircraft, C-101 jet trainer with ground attack capability, major components for several European and American programs (Saab 2000, Boeing 727, 757, McDonnell Douglas MD-11, Harrier II Plus, overhaul Harrier Matador, F-4, C-130 Hercules, components for F-18). Member Airbus consortium and EFA (European Fighter Aircraft), Space programs (ESA and Ariane). Involvement in FLA program. → 75

Barajas Factory (Space Division)
Avenue Aragon 404, 28022 Madrid
☎(+34 1) 586 37 00 ⊠ 48540 casa e
⊖ (+34 1) 747 47 99
Director: Antonio Fuentes

Cadiz Factory
Avenue Marconi 33, Aptdo 65, 11011 Cadiz
☎(+34 56) 24 47 00 ⊠ 76123 casa e
⊖ (+34 56) 26 49 85
Director: Alberto Peces

Getafe Factory
Avenue John Lennon, s/n, 28906 Getafe-Madrid
☎(+34 1) 624 23 22 ⊠ 44382 casa e
⊖ (+33 1) 624 26 77
Director: Jose Julian Fernandez Amigo

San Pablo Factory
Avenue Aeropuerto s/n, Aptdo 715, 41080 Sevilla
☎(+34 5) 451 75 11 ⊠ 72355 casa e
⊖ (+34 5) 467 40 93
Director: Rafael Gonzalez Ripoll

Tablada Factory
Avenue Garcià Morato s/n, Aptdo 429, 41011 Sevilla
☎(+34 5) 445 26 55 ⊠ 72119 casa e
⊖ (+34 5) 445 88 15
Director: Luis E Arizon Del Prado

Puerto Real Factory
Poligono Industrial del Trocadero, Puerto Real, Cadiz
☎(+34 56) 24 47 00 ⊖ (+34 56) 83 48 04
Director: Alberto Peces

Illescas Factory
Carretera Nacional Madrid-Toledo, 45200 Illescas, Toledo
☎(+34 25) 51 34 26 ⊖ (+34 25) 51 34 06
Director: Manuel Martin [98]

55-00140 Brazan SA
(† Empresa Nacional Bazan)
Paseo de la Castellana 55, PO Box 90, 28046 Madrid
☎(+34 1) 335 84 00 ✉ BAZAN ⊠ 27480 bazan e
⊖ (+34 1) 441 50 90
♧ Pta 3,850,000,000 (1996)
♣ Pta 60,000,000,000 (1996) 👥 7,400 (1997)
🏭 700,000 m2
President: Juan Alsina Torrente
Secretary General: Santiago Galindo
Technical Director: Jose Ramon Lopez
Business Development Director: Manuel Lopez
Export Director: Ignacio Gil
Director Procurement: Jacinta Pineda
Financial Director: Hipolito Alvarez
Communication Director: Miguel A Martinez
■ Naval construction, shipbuilding, ship repairs and conversions, turbines and boilers, diesel engines and reduction gears, ordnance and ammunition, manufacture of various parts and components and technological development, design and integration of combat systems, weapon systems and C2 systems (Naval). → 75

El Ferrol Works
PO Box 1, 15400 Ferrol, La Coruña
☎(+34 981) 32 90 11, 32 92 11 ✉ BAZAN
⊠ 85535; 86589 enbf e ⊖ (+34 981) 35 29 50
Shipyard Director: Natalio Rodriguez
Director Turbine Plant: Emilio Segui
Technical Manager: Luis Luengo

Cartagena Works
PO Box 47, 30280 Cartagena, Murcia
☎(+34 968) 50 37 00, 50 63 00 ✉ BAZAN
⊠ 67360; 67460 gale e ⊖ (+34 968) 50 41 78
Shipyard Director: Enrique Llorens
Director Diesel Plant: Jose Del Pino
Commercial Manager Diesel Plant: Salvador Cal
Technical Manager: Pedro Morales

San Fernando Works
PO Box 18, 11100 San Fernando, Cádiz
☎(+34 956) 89 54 11; 89 71 11 ✉ BAZAN
⊠ 76117 bzsf e
⊖ (+34 956) 89 91 11; 89 91 61 (Ord/Ammo)
Shipyard Director: Rodolfo Aller
Director Weapons Systems Plant: Luis Abad
Commercial Manager Weapon Systems: Andres Gandia
Technical Manager: Juan Antonio Moret [98]

55-00250 Iveco Pegaso SA
(† Empresa Nacional de Autocamiones SA)
(a subsidiary of Iveco, Italy)
Avda de Aragón 402, 28022 Madrid
☎(+34 1) 750 10 00; 325 10 00
✉ AUTOCAMIONES Madrid ⊠ 46972 maba e
⊖ (+34 1) 747 47 32
👥 3,000
■ Military all-terrain trucks, armoured wheeled vehicles both for personnel carrier and combat versions in all kinds of configurations and versions, diesel engines for defence products. [98]

55-01000 Santa Barbara SA
Manuel Cortina 2, 28010 Madrid
☎(+34 1) 585 03 90 ⊠ 444466 ensb e
⊖ (+34 1) 585 02 68; 585 02 44
♧ Pta 2,150,000 ♣ Pta 5,160,000,000 👥 2,200
President: Jose Antonio Casanova Gayoso
Vice President Sales and Marketing: Manuel Arroyo
Export Manager: Angel de las Cuevas
■ Tanks, CETME assault rifles, AMELI machine guns, howitzers, guns, small and medium calibre cartridges, artillery ammunition and various types of turrets and mounts. Automatic grenade launcher AGL 40mm. Armoured Personnel Carrier (wheels and tracks). → 75

Granada
El Fargue, 18182 Granada
☎(+34 958) 29 01 00 ⊠ 78408 ⊖ (+34 958) 29 04 09

La Coruna
Palavea s/n, 15170 La Coruna
☎(+34 981) 28 10 99 ⊠ 82210 ⊖ (+34 981) 28 21 81

Oviedo
Marcelino Fernandez 2-1, 33010 Oviedo
☎(+34 98) 528 20 00 ⊖ (+34 98) 529 73 60

Palencia
Plaza Héroes del Alcazar s/n, 34004 Palencia
☎(+34 988) 72 15 00 ⊠ 26235 ⊖ (+34 988) 72 50 82

Murcia
Jabali Viejo s/n, Ctra de Alcantarilla, 30831 Murcia
☎(+34 968) 80 01 00 ⊖ (+34 968) 80 32 39

Vallodolid
Ctra Cabezón del Pisuerga Km 7, 47260 Valladolid
☎(+34 983) 50 00 11 ⊖ (+34 983) 50 04 80

Trubia
Ctra. de la Estación, s/n, 33100 Trubia (Oviedo)
☎(+34 98) 578 41 00 ⊖ (+34 98) 578 48 24

Toledo
Avenda Carlos III, s/n, 45004 Toledo
☎(+34 925) 22 01 42 ⊖ (+34 925) 26 95 03 [98]

60 Research Organizations, Institutes & Test Establishments

60-00250 Asociación de Diplomados y Estudiantes de Criminologia
Association of Graduates and Students of Criminology
Mallorca 165, 08036 Barcelona
☎(+34 3) 589 31 79
President, Criminologist and Private Investigator: Josep Guillot i Colls
Vice President, Criminologist and Psychologist: Jaime Inglauaga i Lopez
Secretary, Criminologist: Prof. Carmina Dieste i Estany
Treasurer, Scientific Criminology: Ms Sandra Tena i Siller
Criminologist and Psychologist: Xavier Perez i Perez
■ Research and training in criminology.
Established in 1989. [98]

60-01900 Instituto de Criminologia de Barcelona
Criminology Institut of Barcelona
Avenida Diagonal 684, 08034 Barcelona
☎(+34 3) 402 43 40; 402 45 77 ⊖ (+34 3) 402 45 78
Director: Prof. José Maria Escriba Gregori
Deputy Director, Penitentiary Law: Ms Esther Gimenez-Salinas
Personality and Anti-social Behaviour: A A Pueyo
Mercantile Law: Vicente Baldo del Castagño
Penal Law Sociology: Roberto Bergalli
Forensic Psychiatry: Luis Borras Roca
Criminological Thought: Ms Encarna Bodelón
Criminalistics: Pedro Cabezas Esteban
Criminal Policy: Luis Maria Díaz Valcarel
Labour Law and Social Security: Ricard Esteban Legarreta
Investigation Methods: Jaime Farras Farras
Victimology: Ms Elena Farre Trepat
Legal Medicine and Toxicology: Gabriel Font Riera
Penal Process Law: Angel García Fontanet
Criminalistics: Maria José Gonzalez Gonzalez
Drug Addiction: Carlos Gonzalez Zorrila
Clinical Criminology and Criminology: Araceli Hernández Hernández
Constitutional Law: Antonio Monreal Ferrer
Penal Law: Fermín Morales Prats
Criminal Psychopathology: Leopoldo Ortega-Monasterio
State, Control, Police, Justice: José Ignacio Rivera Beiras
Police Law: Amadeo Recasens Brunet
■ Research and teaching on criminalistic techniques.
Established in 1955. [98]

60-01950 Instituto de Criminologia de Santiago de Compostela
Institut of Criminology of Santiago de Compostela
Facultad de Derecho - Universidad,
de Santiago de Compostela,
15706 Santiago de Compostela
👥 12 (professsional); 1 (support)
Director: Luis González Guitián
■ Research and teaching on Criminology and Penal Law.
Established in 1978. [98]

60-02000 Instituto de Criminologia de Sevilla
Criminology Institut
Facultad de Derecho, Avenida del Cid s/n, 41004 Seville
Contact: Mapelli Caffarena Borja [98]

60-02050 Instituto de Estudios Penales
Institut of Penal Studies
Facultad de Derecho, Departamento de Derecho Penal, 47002 Valladolid
☎(+34 83) 42 32 16 ⊖ (+34 83) 42 30 12
👥 7
Director: Angel Torio Lopez
■ Research and teaching. Established in 1977. [98]

60-02100 Instituto de Estudios de Policia
Institut of Police Studies
Avenida Pio XII 50, 28016 Madrid
Director: Manuel Alvarez Sobredo
Inspector of Documentation Service: J C López Baeza [98]

60-02200 Instituto de Seguridad Pública de Valencia

Institut of Public Security of Valencia
Avinguda Camp del Túria s/n, 46183 L'Eliana, Valencia
Director: Fernando Salom Herrero [98]

60-02500 Instituto Vasco de Criminologia

Institut Basque of Criminology
Villa Soroa, Avenida Ategorrieta 22, 20013 San Sebastián
☎ (+34 43) 32 14 11; 32 14 12 🄴 szoivac@sz.ehu.es
🅆 http://www.sc.ehu.es/scrwwwiv/ivac.htm l
⊖ (+34 43) 32 12 72
👥 35 (professional); 5 (support)
Director: Dr Antonio Beristain
Research: I Dendaluze
Forensic Medecine: F Etxeberria
Forensic Medecine: G Tamayo
Contact: Ms Inmaculada Iraola
■ Research, teaching and training on criminology, marginality, sociology, victimology, penitentiary problematics, juvenile delinquency, terrorism and torture. Established in 1976. [98]

60-03000 Instituto Universitario de Criminologia

University Institut of Criminologia
Universidad Complutense, Ciudad Universitaria, 28040 Madrid
☎ (+34 1) 394 54 13
Director: Antonio García Pablos de Molina
■ Research and teaching. Established in 1964. [98]

65 Industry, Trade & Armed Forces Associations

65-00100 Agrupación Tecnica Española de Construccion de Material Aerospacial (ATECMA)

Arcipreste de Hita 14, 28015 Madrid
☎ (+34 1) 543 76 51 ⊖ (+34 1) 543 76 51
President: Adolfo Revuelta
General Secretary: Carlos Moro ☎ (+34 1) 543 76 51 [98]

65-00120 Asociación Española de Fabricantes de Armamento y Material de Defensa (AFARMADE)

Spanish Association of Armament and Defence Material Manufacturers
Maria de Molina 3-3°, 28006 Madrid
☎ (+34 1) 561 79 88; 562 77 84; 561 78 01; 561 69 05; 562 92 18; 563 28 23 ⊖ (+34 1) 561 24 51
President: Humberto Figarola Plaja
Vice President: Gabriel Peña Aranda
Director General: Felix Alonso Majagranzas y Acha
Secretary: Ms Carmen García Valdés Yrizar
■ Non-profit making trade association representing Spanish manufacturers of defence and security equipment.
Member Companies:
Aeronáutica Industrial SA, Alcatel Standard Eléctrica SA, Amper Programas de Electrónica y Communicaciones SA, Aries Data Defensa SA, Base Documental de Empresa SA, Ceselsa SA, Computadoras, Redes e Ingeniera SA, Construcciones Aeronáuticas SA, Contrucciones y Auxiliar de Ferrocaril SA, EDB SA, Elt SA, EN Bazan de Construcciones Navales Militares SA, Empresa Nacional de Electrónica y Sistemas SA, Empresa Nacional de Optica SA, EN Santa Barbara de Industrias Militares SA, ENA Telecommunicaciones SA, Ericsson Sistemas Avanzados SA, Esperanza y Compañía SA, Eritel SA, Eurotrónica SA, Explosivos Alaveses SA, Forjas y Aceros de Reinosa SA, Grupo Auxiliar Metalúrgico SA, Industria de Turbo Propulsores SA, Intelligent Decision Systems SA, Instalaza SA, International de Composites SA, International Technology SA, Iveco Pegaso SA, Nissan Motor Iberica SA, Nueva Bressel SA, Peugeot-Talbot SA, Page Iberica SA, Sainsel Sistemas Navales SA, Santana Motor SA, Sener Ingeniería y Sistemas SA, Siemens SA, Simulación Mando y Control SA, Sistemas e Instrumentación SA, Sociedad Anónima de Electrónica Submarina, Sociedad Anónima de Placencia de las Armas, Technologías Reunidas par la Defensa SA, Telefónica Sistemas SA, Unión Española de Explosivos SA. [98]

75 Manufacturing Companies

Brazan SA 55-00140
Construcciones Aeronáuticas SA . . . 55-00120
Santa Barbara SA 55-01000

75-00200 Aeronautica Industrial SA (AISA)

Carretera del Barrio de la Fortuna, 28044 Madrid
Carretera del Barrio de la Fortuna, 10
☎ (+34 1) 379 72 00 ⊖ (+34 1) 508 39 58
♲ Pta 110,000,000 ♣ Pta 1,600,000,000 (1996)
👥 105 (1997) 🏭 300,000 sq ft
President: Juan I Ruiz Sequera ☎ (+34 1) 379 72 03
General Manager: Julio de los Rios ☎ (+34 1) 379 72 00
Marketing Director: Carlos Herraiz
Commercial Manager: S Domingo Fernandez
Purchasing Manager: Rafael Tomas
■ Aircraft hydraulic components, aircraft and helicopter structures. Aircraft and helicopter maintenance, repair, overhaul, upgrading and modernisation. → 80 [98]

75-00440 Akzo Nobel Coatings SA

AD Aerospace Finishes
(a subsidiary of Akzo Nobel Coatings BV)
Feixa Llarga, 14-20 (Zona Franca), 08040 Barcelona
☎ (+34 3) 484 28 91 ⊖ (+34 3) 484 29 00
Technical Sales Manager: P Barreda
■ Primers, topcoats and all specialities for aerospace coating purposes, maskants for chemical milling, epoxies and polyurethane paint systems approved to American and European specifications. [98]

75-00470 Alcatel España SA Radio, Space & Defence Division

(† Alcatel Standard Eléctrica SA Radio, Space & Defence Division)
(a subsidiary of Alcatel Alsthom, France)
Einstein 7 - P.T.M., 28760 Tres Cantos, Madrid
☎ (+34 1) 807 79 00; 803 47 10
⊖ (+34 1) 807 79 93; 803 96 15
♲ Pta 135,000,000,000 ♣ Pta 52,000,000,000
👥 5,485
General Manager, Radio, Space and Defence: L G Echegoyen
Commercial Manager: José L Sanchez Julian
Marketing and Engineering Manager: Carlos Cuellas
Information Contact: Baudilio Sanmartin
■ Defence activities: Infrastructure networks, tactical networks, C3I systems, navaids. [98]

75-00480 Amper Programas SA

Calle Isabel Colbrand N.4, 28050 Madrid
☎ (+34 1) 358 93 91; 358 93 70 ⊖ (+34 1) 358 99 82
♲ US$ 30,000,000 (1994) ♣ US$ 31,000,000 (1995)
👥 220 🏭 10,000 sq m
Managing Director: R Posada
Assistant Manager: R Filias
Commercial Director: F Soler
Quality Control: J C Bordas
Operations Manager: A Perucha
Purchasing Manager: F Gomez Mesh
Finance and Control: F Saiz de Ibarra
Human Resources: A Escavias
General Secretary: L Gomez Luengo
■ Military communications, A/G transceivers, avionics, navigation equipment, automatic pilots, altimeters and attitude indicators, anemometers, accelerometers and maintenance. [98]

75-01400 Astilleros Espanoles SA (AESA)

(Head Office)
Ochandiano 14-16, El Plantio, 28023 Madrid
☎ (+34 1) 387 81 00 ⊖ (+34 1) 387 81 14; 387 81 15
Chairman: Carlos M. de Albornoz
Vice President Commercial: Pedro Morenés
■ Ship construction: newbuilding, repair, conversion, offshore units and engines.

Group Members:
Astano, Ferrol
Astander, Santander
Barreras, Vigo
Juliana, Gijon
Manises Diesel Engine Co., Valencia (joint venture with MAN B&W)

Cadiz Yard
PO Box 39, 11080 Cadiz
☎ (+34 56) 27 33 50; 27 33 54 ⊞ 76021
⊖ (+34 56) 27 83 62; 25 68 28
🏭 304,000 m2
■ Ship repairs & ship conversions, facilities includes 2 dry-dock of 385m x 66.6m & 234m x 42m accepting up to 80 000 DWT, 5 outfitting piers 3 025m & 25 cranes from 4.5 to 400 tons. Tank cleaning & gas-freeing, waste disposal, dirty ballast separation. (DELTA SA).

Puerto Real Yard
PO Box 226, 11080 Cadiz
☎ (+34 56) 47 88 00 ⊞ 76110
⊖ (+34 56) 47 88 37; 47 88 83; 47 88 73
🏭 1,000,000 m2
■ Shipbuilding facilities include 1 dry-dock of 525m x 100m, 1 outfitting pier 500m, 9 cranes from 10 to 600 tons. Maximum vessel size DWT= 1 000 000.

Sestao Yard
PO Box 231, 48080 Bilbao
☎ (+34 4) 495 80 11 ⊖ (+34 4) 496 49 76
🏭 291,000 m2
■ Shipbuilding facilities including 2 new-building berths of 270m x 43.16m and 2 repair dry docks of 155m x 22.36m and 146.6m X 21.8m., outfitting piers 870m and 16 cranes from 7.5 to 100 tons. Maximum vessel size 277 x 42.5 DWT= 138 000.

Seville Yard
PO Box 89, 41080 Seville
☎ (+34 54) 45 10 11 ⊖ (+34 54) 45 76 59
🏭 390,000 m2
■ Shipbuilding facilities including 2 covered berths of 201m x 35m, 1 dry dock of 150m x 22.5m, 1 slipway of 60m x 16m, 13 cranes from 1.2 to 90 tons. Maximum vessel size 187 x 24.2 DWT= 35 000.

Astillero Juliana
Gijon, PO Box 49, 33380 Asturias
☎ (+34 8) 532 12 50 ⊖ (+34 8) 532 59 16
🏭 147,000 m2
■ 2 Berths of 175 x 23.5 each, 2 Dry docks of 170 x 25 mts and 125 x 17 mts. Outfitting Piers 325 mts, 9 cranes from 12 to 40 tons. Maximum vessel size from 1 (m) x b (m) = 175 x 23.5. DWT = 23,000

Astillero Astander
El Astillero, 39680 Cantabria
☎ (+34 42) 54 00 50 ⊖ (+34 42) 54 00 26
🏭 172,000 m2
■ Repairs and conversions of ships. Facilities include:
2 Dry-docks of 231 x 32 mts and 160 x 23.8 mts, 2 slipways of 69.69 x 13.7 mts and 57 x 9.2 mts, outfitting piers 758 mts and 11 cranes from 10 to 40 tons.

Astano
Ferrol, 15480 La Coruna
☎ (+34 81) 34 07 00 ⊖ (+34 81) 34 09 54
🏭 755,000 m2
■ Ship Repairs and Conversions. Facilities include: 2 Berths of 337x 58 mts and 337 x 49 mts, 2 repair dry-docks of 160 x 26 mts and 260 x 36 mts, 11 outfitting piers 2,700 mts, 19 cranes from 10 to 800 tons, maximum vessel size 1(m) x b(m) = 346 x 60, DWT = 365,000. [98]

75-01500 ASTRA Sport SA

(† ASTRA-Unceta y Cia SA)
Ciutat de Berga s/n, PO Box 3, 48300 Guernica, Vizcaya
☎ (+34 4) 625 01 00 ⊠ ASTRA ⊞ 34626 astra e
⊖ (+34 4) 625 51 86
👥 96 (1995)
Managing Director: Suan Mari Brawnbrri
Sales and Marketing Director: Unai Cortabarria
Operations Director: Aitor Brawnbarri
Export Manager: Undi Cortbarria
Purchasing Manager: Mr Braceli
■ Firearms (pistols and revolvers). [98]

75-01660 BP Solar España SA

Poligono Industrial Valportillo, Calle Primera No. 5, 28100 Alcobendas, Madrid
☎ (+34 1) 661 16 14 ⊞ 48496 bpse e
⊖ (+34 1) 661 27 03
👥 110
Commercial Manager: Luis Gordo
■ Manufacture photovoltaic modules (solar energy) used to power airfield telecommunication & navigational aid equipment. Suppliers of complete solar-powered obstruction lighting systems.

Sister companies:
BP Solar Australia Pty Ltd → 75
BP Solar Ltd (UK) → 75
BP Thai Solar Corp.Ltd → 75 [98]

75-01670 **Bressel SA**
Carretera de Madrid-Algete km.5, 28110 Algete (Madrid)
☎ (+34 1) 722 02 35 ⊜ (+34 1) 722 02 95
⁂ 50
Managing Dir: Jaime Barbeito
■ Fuzes for aircraft bombs, naval & field artillery ammunition & rockets. Electronic proximity fuzes. Missile components. [97]

75-01700 **Bretta Spain**
Apartado, 619-01080 Vitiria-Gasteiz
☎ (+34 945) 36 41 30 ⊜ (+34 945) 36 40 70
General Manager: Antonio Rilo [98]

75-01900 **CompAir Holman Iberica SA**
Apartado de Correos N 14-785, 28080 Madrid
☎ (+34 1) 649 92 00 [E] compair@mail.alhsys.es
⊜ (+34 1) 679 92 49
⁂ 40 (1997)
Managing Director: J Merchan ☎ (+34 1) 649 92 33
Finance Director: J Merchan
Marketing Director: F Martin Bermejo
■ Shelters and compressors for the Spanish Army. [98]

75-02850 **Equipos Industriales de Manutención SA (EINSA)**
28812 Alcalá de Henares, Madrid
Antigua Corretera de Barcelona, km 27,300
☎ (+34 1) 880 90 00 ⊜ (+34 1) 880 90 86
♢ Pta 100,000,000 (1995) ♣ US$ 7,000,000 (1994)
⁂ 47 ⌂ 24,000 m2
Managing Director: Pedro Rodriguez Pia
Marketing Manager: Chris A Markham
Plant Manager: Lorenzo Rodriguez Osuna
■ Ground support equipment for military aviation, subdivided into six main categories: Aircraft cargo handling equipment, including mobile, self-propelled, air-transportable hi-lift loaders of up to 15 tonnes capacity and general purpose carts/dollies. Aircraft servicing and test equipment, including mobile hydraulic test trolleys, mobile hydraulic decontamination trolleys, mobile/self-propelled ground power units and compressed air charging carts; mobile hydraulic test stand; autonomous filling unit for sidewinder launcher bottles. Aircraft weapon loading equipment, including self-propelled external weapon loader of up to 6,000 lbs capacity for use on land or on board aircraft carrier. Towing equipment comprising aircraft and cargo tow tractors up to 16,000 lbs draw-bar pull. Maintenance access equipment from fixed and hydraulically operated platforms to aircraft wing stands. Ancillary equipment including but not limited to the following: mobile autonomous floodlight stands; parachute jump towers; specific and universal adapters for weapon loaders; multi-purpose aircraft lifting slings; aircraft tow-bars; aircraft hydraulic jacks. [98]

75-02860 **Equipos Móviles de Campana Arpa SA**
Avda Constitución s/n, Polígono Los Huertos Naves 16-17, 50410 Cuarte de Huerva, Zaragoza
☎ (+34 76) 50 41 64 ⊜ (+34 76) 50 45 90
⁂ 52 ⌂ 3 800 m2
Managing Dir: Antonio Arpa Azofra
Sales Dir: Joaquin Arpa Azofra
Operations & Purchasing Dir: Ricardo Used [97]

75-03050 **Explosivos Alaveses SA (EXPAL)**
(part of Defence Business Unit of the Unit Española de Explosivos SA group)
Avenida del Partenón 16 - 5a Planta,
Campo de las Naciones, 28042 Madrid
☎ (+34 1) 722 02 35 ⊜ (+34 1) 722 01 95
♣ US$ 40,000,000 ⁂ 250
President: Jose Fernando S Junco
Managing Director: J Barbeito
Area Manager: Modesto Martinez
■ Research, development and manufacture of aviaton bombs, ammunitions, mortars, grenades, fuzes, mine systems, submunitions, demolition material and pyrotechnical devices. 76/52 Naval ammunition pre-fragmented with proximity fuze, 5"/54 HE-VT and HE-MT ammunition. Military products meet the Nato specification AQAP-120 or AQAP-110 if it is required.
Factory
Paraje de Ollavarre S/N, Iruña de la Oca, Alava
☎ (+34 45) 36 14 70 ⊜ (+34 45) 36 15 43
● Avenida del Partenón 15 - 5a Planta, 28024 Madrid
☎ (+34 1) 722 02 35 ⊜ (+34 1) 722 02 95 [98]

75-03300 **Fabrica Española de Cartucheria SL (FEC SL)**
Tesifonte Gallego 10, 02002 Albacete
☎ (+34 67) 52 40 07; 50 62 27
⊜ (+34 67) 52 45 53; 50 63 62
♢ US$ 750 000 ♣ US$ 2 000 000 ⁂ 30 ⌂ 2 000 m2
Managing Dir: Rafael Galante Terrer
Commercial Dir: Ms Elena Belmonte Garcia
■ Manufacture of aircraft armament: bombs Mk81, 32, 83, 84; rocket motor 2.75in FFAR mods. Mk 4 & Mk 40; warhead M151 for 2.75in rocket motor; signalling head M156 for 2.75in rocket motor; training bombs MK106, BDU 33, Mk 76 all models; 30mm DEFA ammunition; 20mm Vulcan ammunition; other ammunition. [97]

75-03650 **Grupo Auxiliar Metalurgico SA (GAMESA)**
(a div. of Corporación IBV, Spain)
Portal de Gamarra 40, 01013 Vitoria-Alava
☎ (+34 45) 28 07 00 ▣ 35455 gam e
⊜ (+34 45) 25 16 42
♢ US$ 65 000 000 (1994) ♣ US$ 322 000 000 (1995)
⁂ 2 300 ⌂ 35 000 sq m
Managing Dir: Ahmed Moussa
Sales & Marketing Dir: Javier L. Górriz
Technical Dir: Pedro L. Górriz
Purchasing Mgr: Pedro Etxegaray [97]

75-03781 **Hijos de J. Barreras SA (BARRERAS)**
(a subsidiary of Astilleros Espanoles)
PO Box 35 + 1535, 36280 Vigo, Pontevedra
☎ (+34 86) 23 14 00 ⊜ (+34 86) 20 44 15
♣ Pta 12,000,000,000 ⁂ 350 ⌂ 10,000 m2
Chairman: J G Vinas
Operations Director: F Fernandez
Sales Director: F Portanet
Purchasing Manager: S Vazquez
■ Shipbuilding, repair and conversion, facilities include: 2 berths of 145m x 23m each, 2 slipways of 67m X 14m & 80m x 16m, outfitting piers 425m, 10 cranes from 12 to 45 tons. Maximum vessel size 143 x 24 DWT = 17 000. [98]

75-03790 **Hughes Microelectronics Europe España SA**
Parque Tecnologico Andalucia, Parc 1,
Carretera MA-401, KM 11-Campanillas, 295590 Malaga
Parque Tecnologico de Andalucia,
Parcela 11-Carretera MA-401, KM 11,
Apartado d Campanillas, Malaga
☎ (+34 5) 224 92 00 ⊜ (+34 5) 224 92 47 [98]

75-03800 **INDRA**
(† CESELSA)
C/ Mar Egeo n°4/Polígono Ind'l n°1,
San Fernando de Henares, 28850 Madrid
☎ (+34 1) 396 29 00 ⊜ (+34 1) 396 38 16
♢ US$ 7,439,000 (1996) ♣ US$ 415,752,000 (1996)
⁂ 3,439
President: Javier Monzon
General Manager: Humberto Figarola
Export Manager: Antonio Santamaría
Marketing Communication: Ms Ana Llaguno
■ Missiles, radars, automatic test systems, air defence, simulation, avionics, radar display consoles, electronic warfare, command and control, fire control systems.
Member companies:
Empresa Nacional de Óptica SA, ENOSA 51%
Radar Command & Control Division
Carreiera de Loeches 9, Torrejon de Ardoz (Madrid)
☎ (+34 1) 396 81 97 ⊜ (+34 1) 656 58 87; 396 80 02
Simulation and Automatic Test System
Simulation and Automatic Test Systems
Mar Egeo, 4-Pol Ind 1, 28850 San Fernando De Henares
☎ (+34 1) 396 29 00; 396 38 74 ⊜ (+34 1) 396 38 50 [98]

75-03870 **Industria de Turbo Propulsores SA**
Head Office and Manufacturing Plant (ITP)
Edificio n°300 - Parque Tecnológico,
48170 Zamudio, Vizcaya
☎ (+34 4) 489 21 00 ⊜ (+34 4) 489 21 93
♢ Pta 11,675,000,000 (1995)
♣ Pta 16,051,000,000 (1995) ⁂ 694
Chief Executive Officer and Managing Director: Joaquin Coello
Commercial Director: Ignacio García Beavides
Public Relations Manager: Miss Iraxte de Madariaga
■ Design, research, development, engineering, maintenance, repair, overhaul, assembly and testing of complete engines and components for military, civil, naval and industrial applications.
Commercial Office
Travesía Costa Brava 6-7, (Carretara Colmenar Viejo), 28034 Madrid
☎ (+34 91) 384 80 00 ⊜ (+34 91) 384 80 01
Maintenance and Overhaul Plant
Carretera Torrejón Ajalvir km 3.5,
28850 Torrejón de Ardoz, Madrid
☎ (+34 1) 384 80 00 ⊜ (+34 1) 384 80 01 [97]

75-03900 **Instalaza SA**
Monreal 27, 50002 Zaragoza
☎ (+34 76) 29 34 23 ▣ 58952 intze
⊜ (+34 76) 29 93 31
♢ Pta 931,500,000 ♣ Pta 2,000,000,000 (1995) ⁂ 180
⌂ 16,000 m2
Managing Director: Leoncio Muñoz Bueno
Commercial Manager: Miguel Serrano
■ LC65, 88.9mm rocket launcher (with: CH65 anti-tank ammo, MB66 anti-vehicle and anti- personnel ammo, FIM66 smoke and incendiary ammunition). Rifle Grenades (FTIM61 anti-tank, FTIIM63 anti-tank and anti-personnel, FTV bullet trap). MAT 120 (cargo shell containing 21 submunitions with self destruct function, and self sterilisation function for 120mm mortars.
Madrid Office
Nuñez de Balboa 103, 28006 Madrid
☎ (+34 1) 561 88 35 ⊜ (+34 1) 562 63 50
Commercial Mgr: Miguel Serrano [97]

75-03950 **International Technology SA (ITSA)**
Poligono de Bayas, Parcela N°12, PO Box 283,
09200 Miranda de Ebro (Burgos)
☎ (+34 47) 31 45 00 ⊜ (+34 47) 31 45 71
⁂ 85
Managing Director: Javier Rodriguez Venero
■ Aviation bombs, free fall, retarded and cluster. Missiles - warheads, dummy, practice, HE and other metal parts.
Marketing & Sales Office
calle Emilio Vargas 20 - 6°, 28043 Madrid
☎ (+34 1) 413 58 95; 413 59 63 ⊜ (+34 1) 519 30 74 [98]

75-04200 **Llama Gabilondo y Cia SA**
Portal de Gamarra 50, PO Box 290, Vitoria
☎ (+34 45) 926 24 00 ▣ 35517 llama e
⊜ (+34 45) 26 24 44
Managing Director: Ignacio Quintana
Marketing Director: Miguel Angel Pérez
■ Small arms and pistols [98]

75-04500 **Manises Diesel Engine Co. (MDE)**
(a subsidiary of Astilleros Espanoles)
PO Box 1, 46930 Cuat de Poblet, Vaencia
☎ (+34 6) 154 64 00 ⊜ (+34 6) 154 64 15
⌂ 188,735 m2
■ Manufacture of two stroke diesel engines for vessels and onshore/floating electricity generation plants.
The factory works under licence from Sulzer and MAN B&W. [98]

75-04930 **Proquimed SA**
PO Box 118, 12080 Castellon
☎ (+34 964) 28 00 00 ⊜ (+34 964) 28 20 48 [98]

75-04950 **Roal Saddlery**
Arganzuela 10, 28005 Madrid
☎ (+34 1) 366 96 11; 365 38 83 ⊜ (+34 1) 365 38 83
Managing Director: Jose Carlos Rodiguez
■ Manufacture of holsters. [98]

75-04960 Rockwell Automation SA Regional Headquarters
(a subsidiary of Rockwell International, USA)
Avenida Gran Via 8-10, 08902 l'Hospitalet de Llobregat
☎ (+34 3) 331 70 04 ⊜ (+34 3) 331 79 62
Sales Director, Iberia: Urs T Marti
E umarti@eu1.ra.rockwell.com
® USA: Rockwell International Corp. 98

75-04980 Rockwell Light Vehicle Systems - España SA
Dr Robert No 13, Santa Maria de Palautordera,
08460 Barcelona
☎ (+34 3) 848 23 37 ▣ 50155 rabs e
⊜ (+34 3) 848 07 33
Plant Manager and LVS Country Manager:
Guadalupe Tamayo E gutamayo@auto.rockwell.com
® USA: Rockwell International Corp. 98

75-05000 Rodman Polyships SA
(† Polyships SA)
PO Box 501, Rios Teis, 36200 Vigo
☎ (+34 86) 81 18 10/11 ▣ 83667 plsp e
⊜ (+34 86) 45 29 61
■ Glass reinforced plastic ships for patrol, survey, oceanography, hydrographic vessels, school vessels, fishing, pilot, mooring, rescue and refrigeration barges. Design and construction of hydrodynamics technologies: speed, capacity, consumption, stability and lower maintenance. 98

75-05030 Ros Roca SA
Avda. Cervera s/n, PO Box 31, 25300 Tarrega (Lérida)
☎ (+34 73) 50 81 08 ✉ ROSMOLQUES
E rosrocaexp@redestb.es ▣ 52162 rrsa e
⊜ (+34 73) 50 81 32
♲ Pts 925,000,000 (1996) ♣ Pts 10,195,000 (1996)
👥 415 🏭 250,000 m2
President: Ramón Roca Boncompte
Managing Director: Ramon Roca Enrich
Export Director: José Luis Martinez
Area Manager: Paul Langenkamp
■ Aircraft refuellers (on & underwing syst.), dispensers, catering loaders, invalid passenger movers, water and toilet servicing vehicles. 98

75-05100 Saft Nife Iberica SL
(a subsidiary of Saft SA, France)
Artapadura 11, 01080 Vitoria
☎ (+34 45) 25 99 00 ⊜ (+34 45) 27 57 49
General Manager: Pedro Eguidazu
Marketing Director: Tomas Barrutia
Purchasing Manager: Jose Luis Sarralde
■ Nickel-cadmium batteries for aircraft and for engine starting or APU. Emergency power units. Emergency lighting units on aircraft. Power supply for navigation aids. Maintenance chargers. Special batteries for satellites, missiles, meteorology. Silver-zinc batteries and accumulators. 98

75-05200 Sener Ingenieria y Sistemas SA
Severo Ochoa s/n, Parque Tecnológico Madrileno,
28760 Tres Canto, Madrid
☎ (+34 1) 807 70 00 ⊜ (+34 1) 807 72 01
♲ Pta 4,010,000,000 ♣ Pta 10,000,000,000 (1996)
👥 842 🏭 22,000 sq m
President: Dr Ing Enrique de Sendagorta
Managing Director: Dr Ing José Maria Menéndez Rubio
Chief Executive Officer: Jorge de Sendagorta Gomendio
Foreign Relations Director: Gabriel Vilallonga
Space Department Director: Dr Ing Alvaro Azcárraga
Aeronautics Department Director: Dr Ing Manuel Valls
■ Naval systems technology, air launched weapons systems, ground support and test facilities. Spacecraft mechanical subsystems and associated electronics. Control electronics, software and systems definition studies. Rocket and satellite launching systems.

● Avda de Zugazarte, 56, 48930 Las Arenas, Vizcaya
☎ (+34 4) 481 75 81 ▣ 32745 ⊜ (+34 4) 464 12 60
Information and Press Officer:
Ms Maria Carmen Sendagorta McDonnell
■ Rocket and satellite launching systems, air launched weapon systems, space mechanisms, booms, antennas, control electronics, military software systems analysis, ship design, ground support and test facilities, jet engine design and development. 98

75-05650 TMR-Tecnologías Metalúrgicas Reunidas SA
Nuñez de Balboa 30, 28001 Madrid
☎ (+34 1) 578 02 59 ⊜ (+34 1) 431 23 98
Managing Director: Gregorio Cuenca
Director: Jose L Cano
Operations Director: Antonio Duran
■ Manufacturing company specialising in micro-mechanics, numerical machine control, light alloy foundry, forging, plastic injection, professional printed circuits, hybrid circuits and SMD electronic component insertion. 98

75-05660 Gontrailer SL
Zona Industrial Sector Autopista, Passeig Fluvial,
08150 Parets Del Valles, Barcelona
☎ (+34 3) 562 34 11 ⊜ (+34 3) 562 32 55
♲ Pta 10,000,000 (1996) ♣ Pta 1,200,000,000
👥 75 (1996) 🏭 11,000 sq m
Managing Director: Eduardo Balada
Technical Director: Juan Vales
■ Trailers, semi-trailers, containers, metal bodywork. 98

75-06750 Vibrachoc SA
Poligono Ind da Valdonaire-Nave 207,
Ctra de Moraleja de Enmedio, Km 1,64,
Fuenlabrada, Madrid
☎ (+34 1) 690 16 73 ▣ 45034 vibr e
⊜ (+34 1) 690 40 63
♣ Pta 360,000,000 👥 18
■ Shock absorbing mounts, recoil mounts for machine-guns, panels, electroluminescent synoptic panels, jet engine and APU mountings, anti-crash mounts for helicopter seats, rocket and missile suspensions.
Representing:
Vibrachoc SA, France 98

75-07000 Vorkauf SA
Comandante Franco 3, 28016 Madrid
☎ (+34 1) 359 17 12 ⊜ (+34 1) 350 04 31
♲ Pta 13,000,000 ♣ Pta 400,000,000 (1994) 👥 12
🏭 250 m2
Managing Director: Karl Vorkauf
Marketing Director: Fernández Ramos
Purchasing Manager: Nicolas Alcaide
Vice Manager: D Francisco Fernan Dez Ramos
■ Special materials for armour-plating, tank sprockets, sliprings for turrets, NBC-equipment. Slip rings for tracked armoured vehicles and tracked armoured vehicle turrets. Tracks for armoured vehicles. Slip rings for wheeled armoured vehicles and turrets of wheeled armoured vehicles. NBC protection and prevention equipment, ground based. Special roller bearings. Electric linear telescopic actuators according to MIL-SPEC RF rotary joints. Relays and commuters according to MIL-SPEC, linear and circular ball bearing. 98

75-18000 THORN Europhane SA
Calle Orence 26-2A, 28020 Madrid
☎ (+34 1) 556 84 69 ⊜ (+34 1) 556 62 66
Chief Executive: Guillermo Pereira 98

80 Service Companies & Organizations

Aeronautica Industrial SA 75-00200
Comercial Donald Grant SL 90-00370

80-00060 Centro De Paracaidismo
(† Aero Adventures)
(A division of Jip Aviacio)
Costa Brava, Skydive Spain, PO Box 194,
Ampuriabrava, Gerona
☎ (+34 72) 45 01 11 E cpcb@gma.es
W http://fvest.gma.es/dlleure/cpcb ⊜ (+34 72) 45 07 49
🏭 500 sq m
Managing Director: Peter Jones
Operations Director: Ivan Coufal
Marketing Director: Ms Anna Dalmau
Purchasing Manager: Jaume Comas 98

80-00200 Aries Ingenieria y Sistemas SA
Castellana, 163, 28046 Madrid
☎ (+34 1) 570 27 37 E ariesing@arrakis.es
⊜ (+34 1) 570 27 66
♲ Pta 50,000,000 (1997) ♣ Pta 627,000,000 (1997)
👥 32 (1997) 🏭 400 m2
Managing Director: Javier Rojas
Sales and Marketing Director: Jesus Martinez
Operations and Technical Director: Carlos Arroyo
Purchasing and Public Relations Manager:
Ms Cristina Perez
■ Engineering and design services; systems for aviation, space and defence.
® France: PRECILEC SA 98

80-00700 Grupo 29 SL
(† Elkron Seguridad SA)
(a subsidiary of Elkron SpA)
Avda Alfonso XIII 72, 08208 Sadabell, Barcelona
☎ (+34 3) 723 15 50 ▣ 48975 ⊜ (+34 3) 716 51 99
Managing Director: Joaovin Tejedor
Operations Director: Albertu Sanjaume
Sales Director: Ramun 1Avella
Purchasing Manager: Ms Teresa Casellas
■ Importers and distributors of security equipment.
® Italy: Elkron SpA 97

80-00800 GMV SA
(† Grupo de Mecánica del Vuelo SA)
(a subsidiary of Grupo de Mecanica del Vuelo, SA)
c/o Isaac Newton, 11 Parque,
Technolsgico de Madrid-Tres Cantos, 28760 Madrid
☎ (+34 1) 807 21 00 E webmaster@gmv.es
W http://www.gmv.es ⊜ (+34 1) 807 21 99
♲ Pta 75,000,000 ♣ Pta 1,200,000,000 (1997) 👥 115
🏭 3,000 sq m
President: Prof. J J Martínez-García ☎ (+34 1) 807 21 01
Operations Director: L A Mayo Muñiz
☎ (+34 1) 807 21 59
General Secretary: Dr Enrique Revilla
☎ (+34 1) 807 21 04
Cost Control and Administration Department: F J Martinez
Internet Global Solutions: C Bachmaier
Logistics Systems: M A Martinez
Marketing Manager: Luis F Alvarez ☎ (+34 1) 807 21 59
Quality Assurance: M Diaz
Mission Analysis: M Bello
Data Processing Industrial Applications: P Lopez
Transportation and Positioning Systems: M Perez
Space Systems and Engineering: J Serrano
Ground Facilities for Operations: J Potti 98

80-00840 Hispano Radio Maritima SA (RADIOMAR)
(a subsidiary of Racal Group, UK)
28050 Madrid
Calle San Severo 26.28 Barajas, 10-12, Local 97
☎ (+34 1) 358 97 27 ▣ 42727 ⊜ (+34 1) 358 97 42
♲ Pta 60,000,000 ♣ Pta 400,000,000 👥 20 🏭 800 sq ft
Managing Director: Francisco Barcelo
Financial Director: Leonor Barcelo
Naval Director: F Barcelo
Fishing Director: Pedro Cano
Service Director: Jose Maria Roca
■ Manufacturing, import, sales and service organization of: Communication systems (by satellite), fishing aids, navigation aids, engineering systems. 97

80-00900 Ingenieria de Sistemas (ISDEFE)
Edificio Edison, Calle Edison 4, 28006 Madrid
☎ (+34 1) 411 50 11 E general@isdefe.es
W http://www.isdefe.es ⊜ (+34 1) 411 47 03
♲ Pta 325,000,000 (1996) ♣ Pta 2,733,481 (1996)
👥 242 🏭 4,500 m2
President: Pedro Morenés Eulate ☎ (+34 1) 555 50 00
Managing Director: José V Cebrian ☎ (+34 1) 411 33 99
Public Relations and Image Corporative:
Sra Fernanda Ruiz de Azcarate
■ Technical and managerial support to various defence and civilian government agencies in project systems acquisition in both national and multinational programs. Creates, designs, develops and implements Command, Control and Communication. Air Transport, logistics, security and information technology systems. 98

80-00930 International Business Consulting SA (IBC)
Castellana 123, 28046 Madrid
☎ (+34 1) 388 56 14 ⊜ (+34 1) 759 53 77

Contact Reference

⊕ Pta 10,000,000
President: Jacques Lasry
■ Consultants to Electronic and Aerospace Industry. [98]

80-01000 Libreria Nautica Robinson
Barbara de Braganza 10, 28004 Madrid
☎ (+34 1) 308 48 72 [E] robinson@expocenter.com
[W] http://www.expocenter.com/robinson
⊖ (+34 1) 319 92 55
☗ 5 (1997) ⌂ 120 m2
Managing Director: Ignacio Uranga
Purchasing Manager: Marta Marinero
■ Books and magazines about yachting and professional navigation, and nautic charts of the Spanish coast. [98]

80-01200 Maritima del Mediterraneo
Via Layetana 28, Barcelona
☎ (+34 3) 268 35 03 [W] http://www.marmedsa.es
☒ 5009 ship e ⊖ (+34 3) 268 33 59
♣ Pta 1,980,000,000 (1995) ☗ 250
President: Josep Trius
Public Relations Manager: Christopher Wood
■ Shipping company offering services to ship owners, importers and exporters including chartering, marketing, customs brokers, ship brokers, land transport, towage, stevedoring, depot and repair.

Universal Maritime:
Maritima Eurogulf
Keller Maritima
Ketrans
Jotabe Maritima
Viajes Cosmos
Incentice & Marketing
Blunt Lines
Lluch Transportir [98]

85 Stockists, Distributors, Spare Parts Suppliers

Anatronic SA 90-00115

85-00050 Alava Ingenieros SA
Estébanez Calderón 5, 4A, 28020 Madrid
☎ (+34 1) 572 04 40 ⊖ (+34 1) 570 26 61
⊕ Pta 10,000,000 (1995) ♣ US$ 12,000,000 (1996)
☗ 63 (1997) ⌂ 900 sq m
General Manager: Jaime Alava
Sales Manager: Eduardo Marquina
■ Distributor incorporating two divisions. Test Division: noise and vibration; environmental. Instrumentation Division: transducers, amplifiers and data acquisition systems; recording and systems analysis equipment; ballistic; high vacuum. [98]

85-00300 Commerce International Group Ltd (CIG)
Military Ordnance Defence Ltd
Main European Sales Office
Edificio Atlantico, 5 Ave Marques Del Duero s/n, 29670 S.P.de Alcantara, Marbella
☎ (+34 5) 279 90 00 ☒ 79254 cis
⊖ (+34 5) 278 67 16; 57 29
Director: Paul R. Cradock
Secretary to Managing Dir: Mrs Eva Comninos
■ Defence systems & equipment sales. [97]

85-00400 Compañia Aeronàutica Española SA (CAESA)
Antonio Maura 8, 28014 Madrid
☎ (+34 1) 531 39 07 ✉ CARIBOU [E] caesa@landeres
⊖ (+34 1) 521 53 88
⊕ Pta 10,000,000 (1997) ♣ Pta 150,000,000 (1997)
☗ 15 (1997) ⌂ 400 m2
President: Ricardo Füster (Jr)
Vice President: Arturo Rainer
Interntional Director: Ms Susan Brewer
® Canada: CAE Electronics Ltd
UK: British Aerospace, Vickers Shipbuilding & Engineering., Vosper Thornycroft Ltd.
USA: AlliedSignal Aerospace, Flir Inc., McDonnell Douglas, Schweizer Aircraft Co., Weber Aircraft [98]

85-00950 ROR Rockwell España SA
Ctra d'Granollers a Sabadell km 13 3,
Poligono Argelagues, 08185 Llissa de Vall
☎ (+34 3) 843 95 68; 843 96 93 ⊖ (+34 3) 843 83 59
⊕ Pta 100,000,000 (1997) ♣ Pta 1,600,000,000
☗ 12 (1997) ⌂ 1,500 m2
Managing Director: Albert Juanpere
Operations Director: Daniel Millan
Marketing Director: Mercedes Anduezo
Purchasing Manager: Ms Isabel Hernandez
® USA: Rockwell International Corp. [98]

85-01000 Tecomar S.L.
c/ San Pedro 12, 20110 Pasajes San Pedro
☎ (+34 43) 39 19 44 [E] tecomar@ibm.net
[E] tecomar@arrakis.es [W] http://www.arrakis.es/~tecomar
⊖ (+34 43) 39 19 47
♣ Pta 300,000,000 (1995) ☗ 6 ⌂ 450 m2
Managing and Marketing Director: Jose Antonio Ortega
Marketing Director: Andries Goicoechea
Purchasing Manager: Pedro Posadas
■ Trading and supply of spare parts to the Spanish Armed Forces. [98]

85-05000 Winter Aircraft Products SA
Aristoteles n°5, 28027 Madrid
☎ (+34 1) 880 66 96 ⊖ (+34 1) 880 66 96
President: Ruf S Lopez
General Manager: José D Fernandez
Purchasing Manager: Alberto Diaz
■ Distributor, maintenance, overhaul and support of: Military aircraft spare parts - CASA (C 101, C 212); F5; C-130 Hercules; P3 Orion. Helicopters - Hughes; Alouette; Puma; Superpuma.
Military spare parts: M48; M60; M108; M109; M110; M113.

Representing:
Kollsman, OGMA, Aeroquip, Parker Hannifin Corp., Airwork, 3M. [98]

90 Sales Agents, Sales & Liaison Offices

Companies represented:

Adams Rite Products Inc., USA . . . 90-00920
Aero Adventures, USA . . . 80-00060
Aero-Union, USA . . . 90-00920
AKG, Austria . . . 90-01430
Alfred Karcher GmbH & Co, Germany . . . 90-01148
Allied International Corporation, USA . . . 90-00920
Allied-Signal Aerospace, USA . . . 90-00100
AlliedSignal, USA . . . 90-00920
AlliedSignal Aerospace, USA . . . 85-00400
Alvis plc (Light Tanks), UK . . . 90-02340
Ampex, USA . . . 90-01430
ATMEL, USA . . . 90-00115
Barrett Firearms Manufacturing, USA . . . 90-00400
Bell Textron (Helicopters), USA . . . 90-02340
Bodenseewerk (Missiles), Germany . . . 90-02340
British Aerospace, UK . . . 85-00400
British Rail Research, UK . . . 90-00370
Brunswick Corporation, USA . . . 90-00920
CAE Electronics Ltd, Canada . . . 85-00400
Champion Spark Plug, USA . . . 90-00920
Crompton Eternacell, UK . . . 90-00370
DDC, USA . . . 90-00115
Dornier GmbH (Tow Targets), Germany . . . 90-02340
Dowty Seals, UK . . . 90-00370
Dowty Woodville Polymer, UK . . . 90-00370
EG&G Potron, UK . . . 90-00115
Elkron SpA, Italy . . . 80-00700
Encore Computer Corp., USA . . . 90-00555
Evergreen Int. AV, USA . . . 90-00920
Flir Inc., USA . . . 85-00400
Gayrobot, Switzerland . . . 90-00400
GEC Marconi Avionics, UK . . . 90-00370
GEC Marconi Materials, UK . . . 90-00115
GEC-Marconi (Consultants), UK . . . 90-00400
GEC Plessey, UK . . . 90-00115
General Atronics Corporation, USA . . . 90-00920
Grupo Agusta, Italy . . . 90-00920
H&S Aviation Limited, UK . . . 90-00920
Hants & Sussex (Allison Gas Turbines), UK . . . 90-02340
Heli-Plex, USA . . . 90-00920
Helicopter Welders of Canada, Canada . . . 90-00920
IDT, USA . . . 90-00115
Imo (Launcher), USA . . . 90-02340
Insertia Switch, USA . . . 90-00920
International Aerospace, USA . . . 90-00920
ITT Neo-Dyn, USA . . . 90-01500
Ivie, USA . . . 90-01430
Kyoei, Japan . . . 90-00920
Leach Relais Elektronik GmbH, Germany . . . 90-00920
Litton Industries Inc, USA . . . 90-01200
Lockheed Martin, USA . . . 90-00400
Magnavox, USA . . . 90-00920
Magnavox (Sonobuoys), USA . . . 90-02340
Man-GHH (Military Bridges), Germany . . . 90-02340
Mann Aircraft Co., USA . . . 90-00920
Marotta Scientific Controls Co., USA . . . 90-00920
McDonnell Douglas, USA . . . 85-00400
Micro Precision Textron, USA . . . 90-01500
MII, USA . . . 90-01430
Muirhead Vactric, UK . . . 90-00115
Nagra, Switzerland . . . 90-01430
NEI/APE Allen Gears, UK . . . 90-00370
Neutrik, Liechtenstein . . . 90-01430
Northern Aero Industries Ltd, USA . . . 90-00920
Olaer Industries, France . . . 90-01500
Paravant, USA . . . 90-00115
PRECILEC SA, France . . . 80-00200
Radstone, UK . . . 90-00115
Raytheon Equipment Division (Sonar), USA . . . 90-02340
RFD Ltd, UK . . . 90-00920
Rockwell International Corp., USA . . . 75-04960
Rockwell International Corp., USA . . . 75-04980
Rockwell International Corp., USA . . . 85-00950
Rockwell-Collins (Telecommunications), USA . . . 90-02340
Sabre Industries, USA . . . 90-00920
Schottel-Werft (Special Marine Prop.), Germany . . . 90-02340
Schweizer Aircraft Co., USA . . . 85-00400
Semtech, UK . . . 90-00115
Simplex, USA . . . 90-00920
Sothwest Products Co, USA . . . 90-00920
Soundtech, USA . . . 90-01430
SPAR, Canada . . . 90-00920
Staco Energy Products Co., USA . . . 90-00920
Sunair, USA . . . 90-00920
Talley Defense Systems, USA . . . 90-00920
Technetics Corporation, USA . . . 90-00920
Teledyne Inc., USA . . . 90-02035
Texas Instruments (Defence Systems), USA . . . 90-02340
Tracor Aerospace Austin, USA . . . 90-02340
Universal Propulsion Company, USA . . . 90-00920
Vickers Shipbuilding & Engineering., UK . . . 85-00400
Voith-Schneider, Germany . . . 90-00400
Vosper Thornycroft Ltd., UK . . . 85-00400
Wagner Electric Corporation, USA . . . 90-00920
Wallace & Tiernan Ltd, UK . . . 90-00920
Waterman Hydraulics, USA . . . 90-00920
Weber Aircraft, USA . . . 85-00400
Werclyfe, UK . . . 90-01430
Western Avionic, UK . . . 90-00115
Woerner Engineering Inc., USA . . . 90-00920

90-00020 Acer Computer Iberica SA
Frederic Mompou 5, 3, 2B, Sant Just Desvern, 08960 Barcelona
☎ (+34 3) 499 03 03 ⊖ (+34 3) 499 04 83 [97]

90-00095 AlliedSignal Aerospace Marketing Sales & Services
28042 Madrid
☎ (+34 1) 747 80 42; 587 59 36 ☒ k69500
⊖ (+34 1) 308 34 85; 747 88 73
Field Service Engineer: Fernando Castrodeza [98]

90-00100 AlliedSignal Aerospace Marketing Sales and Service
(a subsidiary of AlliedSignal Aerospace, USA)
Carretera Alcala de Henares, 28800 Madrid
☎ (+34 1) 882 95 00 ⊖ (+34 1) 882 58 61
Field Service Engineer: Willi Ortmann
☎ (+34 1) 677 71 98
® USA: Allied-Signal Aerospace [98]

90-00115 Anatronic SA
Avda de Valladolid 27, 28008 Madrid
☎ (+34 1) 542 44 55; 542 55 66 ⊖ (+34 1) 559 69 75

♲ Pta 20,000,000 (1995) ♣ Pta 1,200,000,000 (1995)
👥 30 🏠 5,000 sq m
Managing Dir: John Pater
Dir Commercial: Javier Cruzado
Advertising Mgr: Jesus Garcia Manes
Public Relations Mgr: Javier Cruzado
■ Distribution of electronics components in Spain & Portugal. → 85
® UK: EG&G Potron, GEC Marconi Materials, GEC Plessey, Muirhead Vactric, Radstone, Semtech, Western Avionic
USA: ATMEL, DDC, IDT, Paravant [97]

90-00200 British Steel España SA
(a subsidiary of British Steel Engineering Steels, UK)
Paseo de la Castellana 135, 28046 Madrid
☎ (+34 1) 572 03 00 ⊜ (+34 1) 572 12 95 [98]

90-00370 Comercial Donald Grant SL
Nunez De Balboa 49, 28001 Madrid
☎ (+34 1) 435 60 19 ⊜ (+34 1) 575 41 66
👥 3 🏠 110 sq m
Marketing Dir: Robert Grant
→ 80
® UK: British Rail Research, Crompton Eternacell, Dowty Seals, Dowty Woodville Polymer, GEC Marconi Avionics, NEI/APE Allen Gears [97]

90-00380 Daewoo Corp Oficina en Madrid
Paseo de la Castellana 135 2b, 28046 Madrid
☎ (+34 1) 571 05 63 ⊠ 45499 dwmd e
⊜ (+34 1) 571 68 27 [98]

90-00400 Defence Development SL
Orense 58, 28020 Madrid
☎ (+34 1) 556 65 63; 556 66 05 ⊜ (+34 1) 597 18 08
♣ US$ 11,000,000 (1996) 👥 3 🏠 140 m2
General Manager: Alvaro Cervera
Marketing Support Manager: Ms Verónica Blanco-Traba
■ Consultant/sales representation of several foreign companies in Spain. Exports Spanish defence products to Middle East countries, mainly Saudi Arabia and Tunisia.
® Germany: Voith-Schneider
Switzerland: Gayrobot
UK: GEC-Marconi (Consultants)
USA: Barrett Firearms Manufacturing, Lockheed Martin [98]

90-00450 Dow Corning Iberica SA
Avda Diagonal 613, 5°, 08028 Barcelona
☎ (+34 3) 405 12 15 ⊠ 50425 ⊜ (+34 3) 405 38 94 [97]

90-00555 Encore Computer Espana SA
Agustin De Foxa, 25-2nd Planta, 28036 Madrid
☎ (+34 1) 733 89 14; 88 93 ⊜ (+34 1) 733 89 71
Contact: Esteban Grijaldo
® USA: Encore Computer Corp. [97]

90-00880 Hughes España SA
Paseo de la Castellana 41, 2° Planta, 28046 Madrid
☎ (+34 5) 319 42 00 ⊜ (+34 5) 210 23 72
Managing Director: Robert E Apple [98]

90-00920 Iberavia Internacional SA (IBERISA)
Hermanos Becquer 5, 28006 Madrid
☎ (+34 1) 563 96 82 ⊠ 23488 ⊜ (+34 1) 563 13 11
♲ Pta 10,000,000 (1997) 👥 16 🏠 400 m2
President: Luis Saenz de Pazos
Director: Joaquin Franco
Sales and Marketing Director: J Manuel Hernandez
® Canada: Helicopter Welders of Canada, SPAR
Germany: Leach Relais Elektronik GmbH
Italy: Grupo Agusta
Japan: Kyoei
UK: H&S Aviation Limited, RFD Ltd, Wallace & Tiernan Ltd
USA: Adams Rite Products Inc., Aero-Union, Allied International Corporation, AlliedSignal, Brunswick Corporation, Champion Spark Plug, Evergreen Int. AV, General Atronics Corporation, Heli-Plex, Insertia Switch, International Aerospace, Magnavox, Mann Aircraft Co., Marotta Scientific Controls Co., Northern Aero Industries Ltd, Sabre Industries, Simplex, Sothwest Products Co, Staco Energy Products Co., Sunair, Talley Defense Systems, Technetics Corporation, Universal Propulsion Company, Wagner Electric Corporation, Waterman Hydraulics, Woerner Engineering Inc. [98]

90-01000 Idecanarias SA
Leon y Castillo 42, Las Palmas 6c, 35002 Canary Islands
☎ (+34 28) 38 01 92 ⊜ (+34 28) 38 00 32 [97]

90-01148 Karcher SA
Poligono Industrial Font del Radium,
Calle Doctor Trueta 6-7, 08400 Granollers, Barcelona
☎ (+34 3) 846 44 47 ⊜ (+34 3) 846 55 05
♲ Pta 100,000 (1995) ♣ Pta 1,600,000 (1995) 👥 64
🏠 2,000 m2
Managing Director: H Joaquin Carrera
Sales and Marketing Director: Jorge Marin
Operations and Technical Director: Francesc Escude
Purchasing Manager: Diego Gallardo
® Germany: Alfred Karcher GmbH & Co [98]

90-01200 Litton Precision Products International
8 Condes del Val, 28016 Madrid
☎ (+34 1) 344 01 44 ⊜ (+34 1) 344 01 02
General Manager: José Luis Jimenez
® USA: Litton Industries Inc [98]

90-01420 Moog Sarl Sucursal en España
Pabellon Altxerri, Poligono Ubegunorio, Guipuzera
☎ (+34 43) 13 32 40 ⊜ (+34 43) 13 31 80
♣ US$ 800 000 👥 5
Branch Mgr: Pierre Gondard ☎ (+34) 43 13 32 40 [97]

90-01430 Neotecnica SAE
(a partner of AKG Austria)
Marques de Urquijo 44, 28008 Madrid
☎ (+34 1) 542 09 00 ⊜ (+34 1) 542 77 42
♲ Pta 20,000,000 👥 20 🏠 700 m2
President: Juan Carlos Tabellini Robbi
Managing Director: Rafael Bonastre Peiro
® Austria: AKG
Liechtenstein: Neutrik
Switzerland: Nagra
UK: Werclyfe
USA: Ampex, Ivie, MII, Soundtech [98]

90-01500 Olaer Oiltech Iberica SA
(† Olaer España SA)
Traversia Industrial No 29, 08907 Hospitalet de Llobregat
☎ (+34 3) 336 14 12 ⊜ (+34 3) 335 71 86
General Manager: Francisco R Samper
® France: Olaer Industries
USA: ITT Neo-Dyn, Micro Precision Textron [98]

90-01600 Peugeot España SA
(† Peugeot Talbot)
7500-28041 Madrid
route de Madrid à Getafe par Villaverde Km
☎ (+34 91) 347 23 70 ⊠ 227590
⊜ (+34 91) 347 23 43; 347 23 44
General Manager: T Armenguad
President: Eduardo Serra Rexach
General Director Defence: Antonio Paris Garrido
Commercial Director Defence: Julian Arribas Mocoroa [98]

90-02035 Teledyne International Office Spain
Pedro de Valdivia #10, Bajo DCHA, 28006 Madrid
☎ (+34 1) 563 60 25 ⊜ (+34 1) 411 71 56
® USA: Teledyne Inc. [97]

90-02100 EBE Europe (España) SA
c/o Proquimed SA, PO Box 118, 12080 Castellon
☎ (+34 964) 73 80 00 ⊜ (+34 964) 73 80 74 [98]

90-02340 Wiresa
(† H.& O. Wilmer SA)
(H.& O. Wilmer Group)
Calle Pinar 6 bis, 28006 Madrid
☎ (+34 1) 411 02 85 ⊠ 27448 ⊜ (+34 1) 562 77 62
♲ US$ 600 000 (1994) ♣ US$ 240 000 000 (1994)
👥 13 🏠 460 sq m
President: Javier Rey
Aeronautical Engineer: Javier Alonso Gil
® Germany: Bodenseewerk (Missiles), Dornier GmbH (Tow Targets), Man-GHH (Military Bridges), Schottel-Werft (Special Marine Prop.)
UK: Alvis plc (Light Tanks), Hants & Sussex (Allison Gas Turbines)
USA: Bell Textron (Helicopters), Imo (Launcher), Magnavox (Sonobuoys), Raytheon Equipment Division (Sonar), Rockwell-Collins (Telecommunications), Texas Instruments (Defence Systems), Tracor Aerospace Austin [97]

Sri Lanka

Head of State:
Chandrika Kumaratunga

10 Head of State

10-10000 Presidential Secretariat
Republic Square, Old Parliament Building, Colombo 1
☎ (+94 1) 248 01; 290 10
President: Mrs Chandrika B Kumaratunga [98]

11 Prime Minister

11-10000 Office of the Prime Minister
58 Sir Ernest de Silva Mawatha, Colombo 7
☎ (+94 1) 57 53 17
Prime Minister: Mrs Sirimavo Bandaranaike [98]

12 Ministries

12-20000 Ministry of Foreign Affairs
Republic Building, Colombo 1
☎ (+94 1) 253 71 ✉ FORINSEC Colombo
⊠ 21139, 21291 forsec ce ⊜ (+94 1) 54 60 91
Minister: Lakshman Kadirgamar
Deputy Minister: D P Wickremasinghe [98]

12-40000 Ministry of Home Affairs
Independence Square, Colombo 7
☎ (+94 1) 59 62 11
Minister: Amarasiri Dodangoda
Deputy Minister: Lackshman Kiriella [98]

31 Embassies & Attachés

31-00510 Embassy of the Arab Republic of Egypt
39 Dickmans Road, Colombo 5
☎ (+94 1) 58 36 21; 50 87 52; 58 52 92
⊠ 21127 boustane ce ⊜ (+94 1) 58 52 92
Ambassador: H E Mr Hani Abdel Hamid El-Abbadi
Chargé d'Affaires a.i.: Hany M M Mostafa
Attaché Administration Affairs: Magdy M E A Nada
Attaché Administration Affairs: Reda R M El Sakt [98]

31-00660 **Embassy of the Federal Republic of Germany**
40 Alfred House Avenue, PO Box 658, Colombo 3
☎(+94 1) 58 04 31; 58 04 34 ⊠ 21119 aaccolo ce
⊖ (+94 1) 58 04 40
Ambassador: H E Dr Michael Schmidt
Deputy Chief of Mission and Commercial Counsellor: Dr Joachim Heidorn
Second Secretary: Peter Kuder
Commercial Secretary: Thomas Heilmaier
Public Information Secretary: Rainer Schmeidchen [98]

31-00920 **Embassy of Japan**
20 Gregory's Road, PO Box 822, Colombo 7
☎(+94 1) 69 38 31; 69 38 33 ⊠ 21148 taishi cce
⊖ (+94 1) 69 86 29
Ambassador: H E Mr Yoji Sugiyama
Deputy Chief of Mission and Minister: Shoji Ogawa
Third Secretary: Kenji Miyata [98]

31-01050 **People's Bureau of the Great Socialist People's Libyan Arab Jamahiriya**
N°120 Horton Place, Colombo 7
☎(+94) 69 73 11; 69 48 74 ⊠ 21246 ashaxbi ce
⊖ (+94) 69 56 71
Chargé d'Affaires and Acting ead of Mission: Abdulhamed M Alwandi
Financial Counsellor: Salem Saleh Elmehdwi
Administrative Attaché: A Hadi M Almuamri [98]

31-01890 **United Kingdom High Commission**
PO Box 1433, Colombo 3
190 Galle Road, Kollupitiya
☎(+94 1) 43 73 36; 43 73 43 ⊠ 21101 ukrep ce
⊖ (+94 1) 43 03 08
High Commissioner: H E Mr D E Tatham
Deputy High Commissioner: P C F Gregory Hood
Defence Adviser: Lieut Col T J O'Donnell
First Secretary Commercial and Economic: C P Haslam [98]

80 Service Companies & Organizations

80-00100 **A&P Consultants (Pvt) Ltd**
85 Jawatte Road, Colombo 5
☎(+94 1) 58 42 78 ⊖ (+94 1) 58 20 80
Chairman and Managing Director: G W Quintus De Silva
Director: Mrs Irene De Silva
■ Manufacturers' representatives for defence products. [98]

90 Sales Agents, Sales & Liaison Offices

90-00300 **CATIC**
29 Layards Rd, Colombo 5
☎(+94 1) 50 08 89 ⊖ (+94 1) 50 84 04 [97]

Sudan

Head of State:
Lieut Gen Umar Hasan Ahmad al-Bashir

10 Head of State

10-10000 **Office of the President**
Koubbeh Palace, Khartoum
☎(+249 11) 753 00
President: Lieut Gen Umar Hassan Ahmad al-Bashir (Chairman of Supreme Council) [98]

10-20000 **Office of the First Vice President**
Khartoum
First Vice President: Maj Gen al-Zubayr Muhammad Salih [98]

12 Ministries

12-10000 **Ministry of Defence**
c/o Army Headquarters, PO Box 371, Khartoum
☎(+249 11) 780 55; 791 59; 758 36; 703 38; 724 60
⊠ 22411 prokaid sd ⊖ (+249) 777 05
Minister: Gen Hassanl Abdul-Rahman [98]

12-20000 **Ministry of Foreign Affairs**
PO Box 873, Khartoum
☎(+249 11) 731 01 ✉ KHARIJIA
⊠ 22459 kharita km; 22460 kharita km; 22461 kharita km
Minister: Ali Osman Mohamaed Taha [98]

12-40000 **Ministry of Internal Affairs**
Khartoum
Minister: Brig Bakeri Hassan Salih [98]

31 Embassies & Attachés

31-00660 **Embassy of the Federal Republic of Germany**
53 Baladia Street, Block 8 D, Plot 2, PO Box 970, Khartoum
☎(+249 11) 777 79; 777 90;
777 91; 777 92; 777 93; 777 94; 777 95
⊠ 22211 aaktm sd ⊖ (+249 11) 77 76 22
Ambassador: H E Dr Werner Daum
Counsellor Deputy Chief of Mission: Robert Klinke
Defence Attaché: Lieut Col B F F von Butler-Brandenfels
Commercial Attaché: Gotthard Freick [98]

31-00690 **Embassy of Greece**
Sharia el Gamhouria, Block 5 No 30, PO Box 1182, Khartoum
☎(+249 11) 734 69 ✉ GREEK EMBASSY Khartoum
⊠ 22951 gremb sd ⊖ (+249 11) 731 55 [98]

31-00890 **Embassy of Italy**
Street 39, Khartoum 2
☎(+249 11) 45 16 14/7 ⊠ 24 034
⊖ (+249 11) 45 12 17
Ambassador: H.E. Dr Francesco Lo Prinzi [98]

31-00920 **Embassy of Japan**
Street No 3, House 24, Block 10 AE, PO Box 1649, Khartoum
☎(+249 11) 47 16 01/02 ✉ TAISHI Khartoum
⊠ 24019 tais sd ⊖ (+249 11) 45 16 00
Ambassador: H.E. Mr Renzo Izawa
Counsellor Deputy Chief of Mission: Sakae Nakajima
First Secretary: Yoshiko Watanabe
Second Secretary: Masaki Yokoi
Second Secretary Political & Economic: Kenjiro Mori
Admin Attaché: Masashi Kasahara [97]

31-01740 **Embassy of Switzerland**
Street 15, Amaraat, PO Box 1707, Khartoum
☎(+249 11) 47 10 10; 47 11 15 ⊖ (249 11) 47 24 04
Chargé d'Affaires a i: Friedrich Bollinger
Head of Chancery: Margrit Letsch [98]

31-01890 **Embassy of the United Kingdom**
Off Sharia Al Baladia, PO Box 801, Khartoum East
☎(+249 11) 77 07 66; 77 07 69 ⊠ 22189 prdrm sd
⊖ (+249 11) 77 64 57
Ambassador: H E Mr Alan F Goulty
Deputy Head of Mission: Bill Rident
Second Secretary Political/Information: Julian Reilly
Third Secretary and Vice Consul: Neal Carlin
Third Secretary, Commercial: Ms Louise Burrett [98]

75 Manufacturing Companies

75-10000 **Nile Corp Ltd Construction Corp**
(a subsidiary of Daewoo Corporation, South Korea)
PO Box 1873, Khartoum
Hara 5, Ruva 2 Sinnat Street, Industrial Area Omdruman, Khartoum
☎(+249 11) 564 68; 547 78 ⊠ 28058 dwskc sd [98]

75-15000 **Palace International Corp Ltd (PIC)**
(a subsidiary of Daewoo Corporation, South Korea)
PO Box 148, Baladia Park, Khartoum North
☎(+249 11) 33 02 04 ⊖ (+249 11) 33 62 30 [98]

Suriname

Head of State:
Jules Wijdenbosch

10 Head of State

10-10000 **Office of the President**
Kleine Combeweg 1, Paramaribo
☎(+597) 47 28 41 ⊖ (+597) 47 52 66
President of the Republic: Jules Wijdenbosch
Vice President: Pretaapnarian Radhakishun [98]

12 Ministries

12-10000 **Ministry of Defence**
Kwattaweg No 29, Paramaribo
☎(+597 4) 742 44 ⊖ (+597 4) 200 55
Minister: Ramon Dwarka Panday [98]

12-15000 **Ministry of Finance**
Onafhankelijkheidsplein 3, Paramaribo
☎(+597) 47 58 50 ⊖ (+597) 47 63 14
Minister: Motilal Mungra [98]

12-20000 **Ministry of Foreign Affairs**
Gravenstraat 6, PO Box 1816, Paramaribo
☎(+597) 47 12 09 ⊖ (+597) 41 08 51
Minister of Foreign Affairs: Faried Pierkhan [98]

12-30000 Ministry of Home Affairs
Onafhankelijkheidsplein 2, Paramaribo
☎(+597) 47 64 61
Minister: Sonny Kertowidjojo 98

12-35000 Ministry of Justice and Police
Gravenstraat 1, Paramaribo
☎(+597) 47 30 33 ⊜ (+597) 41 21 09
Minister: Paul Sjak Shie 98

12-40000 Ministry of Trade and Industry
Havenlaan - Nieuwe Haven, PO Box 557, Paramaribo
☎(+597) 47 50 80 ⊜ (+597) 47 76 02
Minister: Robby Dragman 98

26 Police (Civil)

26-00000 Police Headquarters
Korps Politie Suriname
Gravenstraat 52-54, Paramaribo
☎(+597) 46 21 36 ⊜ (+597) 49 46 04
Minister of Justice and Police: Soeshiel Girjasing
Chief of the Police: C F Hunsel
Deputy Commissioner: D W D Braam 97

31 Embassies & Attachés

31-00230 Brazilian Embassy
Maratakastraat 2, Zorg en Hoop, PO Box 925, Paramaribo
☎(+597) 40 02 00; 40 02 02; 40 02 06
185 brasemb sn ⊜ (+597) 40 02 05
Ambassador: H E Mr Jorge Saltarelli
Military Attaché: Lt Col Vicente Goncalves de Magalhaes
Third Secretary: Flavio Cardone
Vice Consul: Mrs Janete Antonio Batuli 98

Swaziland

Head of State:
H.M. King Mswati III

10 Head of State

10-10000 Office of H M the King
PO Box 1, Lobamba
☎(+268) 531 61, 612 71, 610 80 2134 wd
King of Swaziland: H M King Mswati III 98

11 Prime Minister

11-10000 Office of the Prime Minister
PO Box 395, Mbabane
☎(+268) 422 51, 437 65
Prime Minister: Dr Barnabas Sibusiso Dlamini 98

12 Ministries

12-10000 Ministry of Defense
Kwaluseni, PO Box 1928, Mbabane
☎(+268) 850 00 2189 ⊜ (+268) 841 46
Principal Secretary: Mrs Lucy Dlamini 98

12-20000 Ministry of Foreign Affairs
PO Box 518, Mbabane
☎(+268) 426 61 ✉ DEFA 2036 wd
Minister: Arthur Khoza 98

25 Defense Forces

25-10000 Swaziland Defense Force
c/o St George's Barracks, Manzini
☎(+268) 844 33 2189 ⊜ (+268) 841 46
Army Commander: Brig. G.R. Dube
Deputy Commander: Col Ndambi D. Dlamini
Procurement & Logistics Dir: Lt Col Z.M. Shongwe
Operations Dir: Lt Col C.S. Mhlengethusa 97

31 Embassies & Attachés

31-01890 United Kingdom High Commission
Allister Miller Street, Private Bag, Mbabane
☎(+268) 425 81; 425 82; 425 83 ✉ UKREP MBABANE
2079 wd ⊜ (+268) 425 85
High Commissioner: H E Mr J F Doble
Deputy High Commissioner: D Morley
Director of British Council: Ms F T Mahundla
Third Secretary Management and Consular:
Ms H Thompson 98

31-01900 Embassy of the United States of America
Warner Street, Central Bank Building, PO Box 199, Mbabane
☎(+268) 464 41; 464 45 2016 usaid wd
⊜ (+268) 459 59
Ambassador: H E Mr Allen McKee
Deputy Chief of Mission: Michael Hoze 98

Sweden

Head of State:
H.M. King Carl XVI Gustaf

10 Head of State

10-10000 Office of H.M. the King
Kungl. Slottet, S-11130 Stockholm
☎(+46 8) 402 60 00
King of Sweden: H.M. King Carl XVI Gustaf 98

11 Prime Minister

11-10000 Office of the Prime Minister
Statsrådsberedningen, Rosenbad 4, S-10333 Stockholm
☎(+46 8) 405 10 00 17820 premier s
Prime Minister: Goran Persson 98

12 Ministries

12-10000 Ministry of Defence
Försvarsdepartementet
Vasagatan 8-10, S-103 33 Stockholm
☎(+46 8) 405 10 00 E registrator@defence.ministry.se
W http://www.regeringen.se ⊜ (+46 8) 723 11 89
Minister for Defence: Björn von Sydow
State Secretary: Ms Lena Jönsson
Political Adviser: Ms Ann Linde
Political Adviser: Jonas Hjelm
Press Secretary: Ms Judith Jäderberg
Information Officer: Ms Anna Liedbergius
☎(+46 8) 405 25 24; 21 15 77 (fax)
E anna.liedbergius@defence.m inistry.se 98

12-20000 Ministry of Foreign Affairs & Foreign Trade
Utrikesdepartementet
PO Box 16121, S-103 23 Stockholm 16
☎(+46 8) 405 10 00 ✉ CABINET 10590 minfor s
⊜ (+46 8) 723 11 76
Minister for Foreign Affairs: Ms Lena Hjelm-Wallen
Under Secretary of State, Foreign Affairs: Jan Eliasson
Under Secretary of State, Foreign Trade: Gunnar Lund
War Materiel Inspector: Ambassador Staffan Sohlman
Press & Information Department Ministry for Foreign Affairs
PO Box 16121, S-103 23 Stockholm
☎(+46 8) 405 10 00 ⊜ (+46 8) 723 11 76
Chief of Press & Information Department:
Ambassador Annika Söder
Head of Press Section: Staffan Björck 98

12-30000 Ministry of Transport & Communication
Jakobsgatan 26, S-103 33 Stockholm
☎(+46 8) 405 10 00 17328 mincom s
⊜ (+46 8) 411 89 43
Minister: Ines Uusmann 98

20 Defense Administrations/Services/Agencies

20-02000 Board of Accident Investigation
PO Box 12538, S-102 29 Stockholm
☎(+46 8) 441 38 20 E info@havkom.se
⊜ (+46 8) 441 38 21
General Director: Olof Forssberg ☎(+46 8) 441 28 22
Deputy General Director: Sven-Erik Sigfridsson
☎(+46 8) 441 38 23 98

20-04300 Defence Materiel Administration (FMV)
Försvarets Materielverk
S-115 88 Stockholm
Banérgatan 62
☎(+46 8) 782 40 00 19610 fmv s
⊜ (+46 8) 667 57 99
Director General: Mrs Birgitta Böhlin
Deputy Director General and Chief of Joint Staffs:
Brig Jan Edelswärd
Chief of Army Materiel Command: Maj Gen Kurt Blixt
Chief of Army Materiel Command Staff: Brig Lars Olsson
Head of Weapons Directorate: Brig Hans Ramström
Head of Combat Vehicles and Mobility/Quatermaster Matériel Directorate: Brig Sture Ärlebäck
Head of Army Materiel Logistics Directorate:
Brig Per-Arne Öhman
Chief of Naval Materiel Command:
Rear Adm Bertil Björkman
Chief of Naval Materiel Command Staff: Col Mats Westin
Head of Ships Directorate: Cdre Inge Smedensjö
Head of Underwater Weapons Directorate:
Cdre Emil Svensson
Chief of Air Force Matériel Command:
Maj Gen Staffan Näsström
Vice Chief of Air Force Matériel Command:
Brig Peter Lundberg
Chief of Air Force Matériel Command Staff:
Brig Lennart Källqvist
Head of Aeronautical Systems Directorate:
Brig Göran Langemar
Head of Testing Directorate: Brig Kjell Holmström

Contact Reference

Head of Air Force Matériel Logistics Directorate: Col Bengt Hörnsten
Chief of Joint Systems and Support Command: Örjan Eriksson
Chief of Joint Systems and Support Command Staff: Col Lars Hellgren
Head of Information Systems Directorate: Kenth Carlsson
Head of Electronic Systems Directorate: Kenneth Monthan
Head of Joint Support Directorate: Brig Håkan Willenfors
Head of Spares Supply Command: Brig Jan Sandin
Chief of Joint Matériel Command Staff: Col Urban Jonsson
Chief of Personnel Staff: Bo Svanlind
Chief of Finance Staff: Berit Kjellser
Chief of Commercial Staff: Stephan Källmén
Chief of Information Staff: Lars Wigert
☎ (+46 8) 782 63 85; 782 63 95 (fax)
Chief of Internal Services Directorate: Bertil Nyrén [98]

25 Defense Forces

25-15000 National Home Guard
Rikshemvárnscentrum
S-10787 Stockholm
☎ (+46 8) 788 75 00 [E] rikshvc@hkv.mil.se
⊠ 19633 hkvsbc s ⊖ (+46 8) 664 57 90
Commander: Maj Gen Alf Sandqvist
Chief of Staff: Col Sven Hallberg
Aide de Campe: Lt Bo Wallin [98]

25-20000 Swedish Air Force Headquarters
S-107 85 Stockholm
☎ (+46 8) 788 75 00
Commander-in-Chief: Lt Gen. Keut Harrskog
Chief of Training & Procurement Division: Brig.Gen. Curt Westberg
Chief of Air Force Systems Division: Brig.Gen. Mats Hellstrand
Chief of Flight Safety Division: Col Göte Pudas
Chief of Weather Service Division: Brig.Gen. Anders Söderberg [97]

25-29400 Swedish Air Force Flying Training Wing
Krigsflygskolan, F5
PO Box 501, S-26070 Ljungbyhed
☎ (+46 435) 450 00 ⊖ (+46 435) 450 19
Commander: Col Sven Sjöling
Chief of Staff: Col Bengt Brodin [98]

25-29550 Swedish Air Force Uppsala Schools (F 20)
(† SAF Military Academy)
PO Box 16042, S-75016 Uppsala
☎ (+46 18) 28 10 00 ⊖ (+46 18) 28 19 09
Commander: Col Hans Hagberg ☎ (+46 18) 28 19 00
Chief of Staff: Maj Gunde Borén ☎ (+46 18) 28 19 02
■ Unit for tactical evaluation within early warning and air defence command field.
Schools:
Air Force Academy
Armed Forces Language School
Air Force Flying Commanders School
Air Force Intelligence School
Air Force Air Control and Air Surveillance School
Air Control and Surveillance Operational Test and Evaluation [98]

25-31000 Arméns Tekniska Centrum (ATC)
(† Arméns Materielunderliallscentrum)
Swedish Army Technical Center, S-83185 Ostersund
☎ (+46 63) 15 80 00 ⊖ (+46 63) 15 89 84
Commander: Brig. Gustaf Ankancvona
Deputy Commander & Chief of Staff: Lt Col Ronnie Åbrandt
Chief of the Army Technical School: Lt Col Jan-Erik Lougren
Dir of Procurement: Lt Col Bo Lagerfalk [97]

25-32100 Headquarters, Southern Joint Command
Sődra Militärområdet (Milo S)
(† South Military Region Command)
Box 514, S-291 25 Kristianstad
☎ (+46 44) 10 50 00 ⊖ (+46 44) 10 52 26
Commander: Lieut Gen Sven-Åke Jansson
Chief of Staff: Maj.Gen. Ulf Rubarth
Chief Operations: Col Swen Persson
Information Officer: Lieut Col Jan Thunström [98]

25-39500 Swedish Armed Forces Headquarters
Lidingöven 24, S-107 85 Stockholm
☎ (+46 8) 788 75 00 [W] http://www.mil.se
⊠ 19633 hkvsbc s
⊖ (+46 8) 788 77 88; 788 88 66 (Information Department)
Supreme Commander: Gen Owe Wiktorin
Chief of Joint Operations Staff: Lieut Gen Percurt Green
Vice Chief of Joint Operations Staff and Director of Plans and Policy: Rear Adm Torsten Lindh
Secretary to the Supreme Commander: Mrs Inga-Lisa Karlsson
Aide-de-Camp to the Supreme Commander: Lieut Col Mats Alnevik
Chief of Surgeon General's Department: Rear Adm Björn Zetterström
Chief of Training and Procurement Department: Col Per-Olof Ahlqvist
Chief of Short-Term Planning Department: Col Sten Edholm
Operations Division
Deputy Chief of Joint Operations Staff: Maj Gen Tomas Warming
Chief of Strategy Department: Brig Håkan Syrén
Chief of Operations Department: Col Svante Menckel
Chief of International Department: Col Peter Grunditz
Chief of Logistics Department: Brig Bjorn Rosen [98]

25-40000 Swedish Armed Forces Headquarters Naval Staff
S-10785 Stockholm
☎ (+46 8) 788 75 00 [W] http://www.mil.se
⊖ (+46 8) 788 82 24
Commander-in-Chief and Chief of Naval Staff: Vice Adm Peter Nordbeck
Chief of Training and Procurement Department: Col Goran Gunnarsson
Chief of Naval Systems Department: Cdre Rolf Brehmer
Chief of Information Royal Swedish Navy: Cdr Lars Wallgren [98]

25-50000 Swedish Army Intelligence School
S-546 23 Karlsborg
☎ (+46 505) 850 00 ⊖ (+46 505) 120 84
Commander: Lieut Col Mats Goransson
Chief of Staff: Jens Ramhöj [98]

29 Customs & Excise

29-00500 National Board of Customs
PO Box 2267, S-103 17 Stockholm
☎ (+46 8) 789 73 00
Director General: Ulf Larsson
Deputy Director General: Göran Ekström [98]

31 Embassies & Attachés

31-00050 Embassy of the Republic of Angola
Roskullsvägen 2-18162 Lidingö, Box 3199, S-103 64 Stockholm
☎ (+46 8) 766 03 72; 24 28 90 ⊠ 12449 gasmin s
⊖ (+46 8) 766 12 43; 34 31 27
Ambassador: H E Mrs Josefina Pitra Diakite
First Secretary: Ms Maria Ivone Henrique Tuta
Second Secretary: Ms Maria Helena Couto de Assis Boavida
Attaché: Luís S José Rodrigues [98]

31-00080 Embassy of the Argentine Republic
Grevgatan 5, PO Box 14039, S-104 40 Stockholm
☎ (+46 8) 663 19 65 ⊠ 10029 embarge s
⊖ (+46 8) 661 00 09
Ambassador: H.E. Mr Federico Erhart del Campo
First Secretary: Luis Claudio Hashimoto
Commercial Secretary: Edgardo Víctor Galotti
☎ (+46 8) 661 61 97 [98]

31-00090 Australian Embassy
12 Sergels Torg, PO Box 7003, S-103 86 Stockholm
☎ (+46 8) 613 29 00 ⊠ AUSTEMBA Stockholm
⊠ 10382 austemb s ⊖ (+46 8) 24 74 14
Ambassador: H.E. Ms Judith Pead
First Secretary: Rick Nimmo
First Secretary: Amanda Gorely
First Secretary, Consular & Administrative Affairs: Trevor Spring
First Secretary, Immigration Affairs: Ms Marie Kabala [98]

31-00100 Embassy of Austria
Kommendörsgatan 35, S-114 58 Stockholm
☎ (+46 8) 665 17 70 ⊠ AUSTROAMB Stockholm
[E] austria@algonet.se [W] http://www.algonet.se/~austria
⊖ (+46 8) 662 69 28
Ambassador: H.E. Dr Franz Parak
Minister-Counsellor, Deputy Chief of Mission: Dr Johann Spitzer
Defence Attaché: Brig Peter Hans Grabner
Second Secretary Administration: Martin Krämer
Cultural and Press Counsellor: Dr Liselotte Kitzwegerer
Office Commercial Counsellor
Karlaplan 12, S-115 20 Stockholm
☎ (+46 8) 667 01 30 ⊖ (+46 8) 660 83 78
Commercial Counsellor: Dr Heinz Wimpissinger
Commercial Affairs Attaché: Hans-Christian Kügerl [98]

31-00220 Embassy of the Republic of Botswana
Tyrgatan 11, PO Box 26024, S-100 41 Stockholm
☎ (+46 8) 723 00 35 ⊠ 16848 ⊖ (+46 8) 723 00 87
Ambassador: H.E. Mr Sekgoma Tshekedi Khama
Counsellor, Deputy Chief of Mission: Miss Bernadette S Rathedi
First Secretary: Simon Phalaagae Tau
Commercial Attaché: Ezekiel Raisaka Thekiso
Defence Attaché: Lieut Col Zeke Molefe (based in London) [98]

31-00230 Brazilian Embassy
11-2 Sturegátan, S-114 36 Stockholm
☎ (+46 8) 23 40 10 ⊠ 19 132 suebrem s
⊖ (+46 8) 23 40 18
Ambassador: H.E. Mr Luiz Felipe de Teixeira Soares
Minister Counsellor: Cézar Augusto de Souza Lima Amaral
Naval Attaché: Capt. Marcus Vinicius Oliveira dos Santos
☎ (+44 171) 499 08 77
Air Attaché: Col Cesar Costa ☎ (+44 171) 499 08 77
Second Secretary: Luis Ivaldo Villafane Gomes Santos
Second Secretary: Joao Marcelo de Aguiar Teixeira [97]

31-00360 Embassy of the People's Republic of China
Lidovägen 8, S-115 25 Stockholm
☎ (+46 8) 783 67 30 ⊠ 19608 hoatine s
Ambassador: H E Mr Qlao Zonghuai ☎ (+46 8) 662 32 70
Counsellor (Political Affairs): Gao Feng
Military, Naval and Air Attaché: Senior Col Yu Wanlong
☎ (+46 8) 662 71 55; 662 59 83 (fax)
Assistant Military Naval and Air Attaché: Lieut Col Chen Haoping
Counsellor (Scientific and Technological Affairs): Wu Zhenxiang ☎ (+46 8) 767 75 26
Counsellor (Economic and Commercial Affairs): Xu Yanchun ☎ (+46 8) 767 40 83
Office of Commercial Affairs
Ringvägen 56, S-181 34 Lidingö
☎ (+46 8) 767 40 83 ⊖ (+46 8) 731 84 04
Scientific and Technological Affairs
Herserudsvägen 51, S-181 35 Lidingö
☎ (+46 8) 767 75 26 ⊖ (+46 8) 731 07 40 [98]

31-00450 Embassy of the Czech Republic
Floragatan 13, Box 26156, S-100 41 Stockholm
☎ (+46 8) 440 42 10; 440 42 25
[E] stockholm@embassy.mzv.cz ⊠ 11141 obzamin s
⊖ (+46 8) 440 42 11
Ambassador: H E Mr Petr Kolár
Deputy Chief of Mission: Ivan Jancárek
☎ (+46 8) 440 42 34

Military and Air Attaché: Col Milos Krampera
☎ (+46 8) 440 42 26
Commercial Attaché: Tomás Vacek ☎ (+46 8) 440 42 32
98

31-00500 Embassy of Ecuador
13 Engelbrektsgatan, PO Box 26095, S-100 41 Stockholm
☎ (+46 8) 679 60 43; 679 60 70
E suecia@embajada.ecuador.se
W http://www.embajada-ecuador.se
⊖ (+46 8) 611 55 93
Ambassador: H E Mr Alfonso López
Counsellor: Ms Fanny de Lourdes Puma Puma 98

31-00535 Embassy of Estonia
Storgatan 38, 1 tr, Box 14069, S-10440 Stockholm
☎ (+46 8) 665 65 50; 661 58 10 (consular department)
E sekretar@estemb.se ⊖ (+46 8) 662 99 80
Ambassador: H E Mr Andres Unga
First Secretary (Political Affairs): Jürgo Loo
Second Secretary (Commercial Affairs): Allan Luik
Third Secretary (Consular Affairs): Mrs Reet Ollo 98

31-00580 Embassy of Finland
Jakobsgatan 6, PO Box 7423, S-103 91 Stockholm
☎ (+46 8) 676 67 00 ⊠ 19692 finland s
⊖ (+46 8) 20 74 97
Ambassador: H.E. Mr Heikki Talvitie
Minister & Counsellor: Harry Helenius
Military, Naval & Air Attaché: Col Heikki Hietanen
Secretary: Ms Susan Kosk 98

31-00660 Embassy of the Federal Republic of Germany
9 Skarpögatan, S-11527 Stockholm
☎ (+46 8) 670 15 00 ⊠ 10049 aasv s
⊖ (+46 8) 661 52 94
Deputy Chief of Mission: Eberhard Schuppius
Defence Attaché: Joachim Scholtz
Commercial Counsellor: Gerd Benke 98

31-00810 Embassy of the Republic of Hungary
Strandvägen 74, PO Box 24125, S-104 52 Stockholm
☎ (+46 8) 661 67 62; 662 56 75 ⊖ (+46 8) 660 29 59
Ambassador: H E Dr István Mohácsi
Counsellor (Political Affairs): Dr Erzsébet Nagy
Military and Air Attaché: Col István Horváth (resident in Helsinki)
Third Secretary (Consular Affairs): Péter Győri

Office of the Commercial Counsellor
37 Birger Jarlsgatan, S-111 45 Stockholm
☎ (+46 8) 20 86 05; 20 86 06 ⊠ 19427 ungex s
⊖ (+46 8) 21 86 58
Commercial Counsellor: Zsolt Monszpart
Counsellor (Commercial Affairs): Zoltán Pap
Counsellor (Commercial Affairs): Miklós Csuhay 98

31-00820 Embassy of Iceland
Kommendörsgatan 35, S-114 58 Stockholm
☎ (+46 8) 662 40 16 E icemb.stock@utn.stjr.is
⊖ (+46 8) 660 74 23
Ambassador: H E Mr Hödur H Bjarnason
First Secretary, Deputy Chief of Mission: Bjarni Vestmann
98

31-00850 Embassy of the Islamic Republic of Iran
Västra Yttrige Gärd, Elfviksvägen, PO Box 6031, S-181 06 Lidingö
☎ (+46 8) 765 08 19 ⊠ 17791 iransw s
Ambassador: H E Mr Abdollah Norouzi
Chargé d'Affaires a.i.: Abdolreza R Hanjani 98

31-00860 Embassy of the Republic of Iraq
Baldersgatan 6A, PO Box 26 031, S-100 41 Stockholm
☎ (+46 8) 411 75 02 ⊠ 11 199 iraswed s
⊖ (+46 8) 796 83 66
Chargé d'Affaires en pied: Abdul Wahab Al-Jawary 98

31-00870 Embassy of Ireland
PO Box 10326, S-100 55 Stockholm
☎ (+46 8) 661 80 05 ⊖ (+46 8) 660 13 53
Ambassador: H.E. Mr Martin Burke
Second Secretary: Brian McElduff 98

31-00890 Embassy of Italy
Oakhill, Djurgården, S-115 21 Stockholm
☎ (+46 8) 24 58 05 ⊠ 10453 italdip s
⊖ (+46 8) 660 05 05
Ambassador: H E Mr Franco Ferretti ☎ (+46 8) 24 58 06
First Counsellor: Agostino Chiesa Alicator
☎ (+46 8) 24 58 08
First Secretary: Luigi Ferrari ☎ (+46 8) 24 58 09
Military Naval and Air Attaché: Col Francesco Gianesin
☎ (+46 8) 661 12 04
Commercial Attaché: Lelizia Leonardi ☎ (+46 8) 24 95 77
98

31-00920 Embassy of Japan
Gärdesgatan 10, S-115 27 Stockholm
☎ (+46 8) 663 04 40 ✉ TAISHI Stockholm
⊠ 19948 taisi s ⊖ (+46 8) 661 88 20 98

31-00960 Embassy of Kenya
37 Birger Jarlsgatan, PO Box 7694, S-103 95 Stockholm
☎ (+46 8) 21 83 00/04/09 ⊠ 17811 kenamb s
⊖ (+46 8) 20 92 61
Ambassador: H.E. Prof. Ahmed Idha Salim
Second Secretary: Francis Kiriinya Mwongo
Third Secretary: Johnson Kiplang'at Soi
Attaché: Peter Mungai Wakahia 98

31-01010 Embassy of the Lao People's Democratic Republic
Badstransvägen 11, PO Box 34050, S-112 65 Stockholm
☎ (+46 8) 618 20 10 ⊖ (+46 8) 618 20 01
Ambassador: H.E. Mr Thongsavath Praseuth
First Secretary: Souvanna Phouyabong
Third Secretary: Mrs Chanthalangsy Nalisouphan
Third Secretary: Chansamone Thongpaseuth 98

31-01050 People's Bureau of the Socialist People's Libyan Arab Jamahiriya
74 Valhallavägen, PO Box 10133, S-100 55 Stockholm
☎ (+46 8) 14 34 35 ⊠ 17422 allibya s
Ambassador: H.E. Mr Abuagela M. El Hweg
☎ (+46 8) 85 98 06
Counsellor: Mostafa M. Khaled
Financial Affairs Attaché: Farag K. Saleh
Administrative Affairs Attaché: Haggi M. Halb 97

31-01190 Embassy of Mexico
Grevgatan 3, S-114 53 Stockholm
☎ (+46 8) 663 51 70 E embamex@algonet.se
⊖ (+46 8) 663 24 20
Ambassador: H.E. Mr Lorenzo Vignal
Deputy Chief of Mission: Francisco Alonso
First Secretary: Saul Zambrano
Third Secretary: Mrs Claudia Keller-Lapayre
Attaché Consular Affairs:
Evangelina Olvia G. Moreno de Padilla 98

31-01270 Royal Netherlands Embassy
Götgatan 16a, PO Box 15048, S-104 65 Stockholm
☎ (+46 8) 24 71 80 ⊖ (+46 8) 702 96 83
Ambassador: H E Mr J C Th Bast
Deputy Chief of Mission: J R Wilton
Defence Attaché: Col F W Hoogeland
Commercial Secretary: H E Hageman 98

31-01330 Embassy of the Federal Republic of Nigeria
8 Tyrgatan, PO Box 628, S-101 32 Stockholm
☎ (+46 8) 24 63 90 to 95 ⊠ 17649 nigemb s
⊖ (+46 8) 24 63 98
Minister Chargé d'Affaires a.i.: Joseph C. Okeke
Sr Counsellor Economic: Mohammed K. Ibrahim
Counsellor Political: Gordon H Bristol
Information, Education and Culture: Obinna E Nnama
First Secretary Consular Affairs: Shina F. Alege 98

31-01340 Royal Norwegian Embassy
113 Strandvägen, S-115 27 Stockholm
☎ (+46 8) 665 63 40 ⊠ 10 112 noramb s
⊖ (+46 8) 782 98 99
Ambassador: H.E. Mr Ketil Børde
Deputy Chief of Mission: Anders Helseth
Military, Naval & Air Attaché: Capt. Eivand Indersand (N)

Office of Commercial Affairs
Birger Jarlsgatan 26, PO Box 7224, S-114 34 Stockholm
☎ (+46 8) 440 50 80 ⊖ (+46 8) 440 50 90
Commercial Counsellor: Eivind Nyhuus 97

31-01430 Embassy of the Republic of Poland
Karlavägen 35, S-114 31 Stockholm
☎ (+46 8) 411 41 32; 411 81 66 ⊖ (+46 8) 10 10 69
Ambassador: H.E. Ms Barbara Tuge-Erecinska
Counsellor: Ryszard Liczmanski
Defence Attaché: Col Artur Mazurkiewicz

Office of the Commercial Counsellor
Friggagatan 4, S-114 27 Stockholm
☎ (+46 8) 11 76 37/38; 11 76 67; 20 43 47
⊠ 19847 morhan s ⊖ (+46 8) 21 61 88
Commercial Counsellor: Boleslaw Kosciukiewicz
Commercial Attaché: Jaroslaw Starzyk 98

31-01480 Embassy of Romania
36 Östermalmsgatan, PO Box 26043, S-100 41 Stockholm
☎ (+46 8) 10 86 03/07 ⊠ 10166 agerom s
⊖ (+46 8) 10 28 52
Ambassador: H E Dr Nicolae I Ionescu
☎ (+46 8) 10 28 58
First Secretary, Economic and Commercial Affairs:
Ilie Stanica ☎ (+46 8) 20 56 74
Second Secretary, Consular Affairs: Virgil Goaga
☎ (+46 8) 10 86 07
Second Secretary, Press & Cultural Affairs: Gogu Radu
Counsellor, Political Affairs: Florica Vasiliu
☎ (+46 8) 10 86 03 98

31-01485 Embassy of the Russian Federation
31 Gjörwellsgatan, S-112 60 Stockholm
☎ (+46 8) 13 04 40/42 ⊖ (+46 8) 618 27 03
Minister-Counsellor and Chargé d'Affaires:
Victor Tatarintsev
Counsellor (Economic Affairs): Andrey Lopukhov
Counsellor (Scientific Affairs): Dmitry Pankratov
Defence Attaché: Maj Gen Nikolai Lychak
Assistant Military, Naval and Air Attaché:
Capt N Stanislav Astakhov
Assistant Military, Naval and Air Attaché:
Lieut Cdr Vladimir Muzyka
Assistant Military, Naval and Air Attaché:
Maj Andrey Potaenko
Attaché: Andrey Rusakov

Office of Commercial Affairs
1 Ringvägen, S-181 34 Lidingö
☎ (+46 8) 765 11 80 ⊠ 19 326 vneshto s
Commercial Representative: G. Rakhimbaev
Deputy Commercial Representative: N. Tsoutskov 98

31-01600 Embassy of Saudi Arabia
5 Sköldungagatan, PO Box 26073, S-100 41 Stockholm
☎ (+46 8) 23 88 00 ⊠ 14330 najdiah s
⊖ (+46 8) 796 99 56
Ambassador: Marwan B. Al-Roume 97

31-01655 Embassy of the Republic of Slovenia
Klarabergsgatan 33, 3 tr, S-111 21 Stockholm
☎ (+46 8) 21 79 80; 21 89 21 ⊖ (+46 8) 723 01 75
Ambassador: H E Mr Ivo Vajgl 98

31-01670 South African Embassy
76 Linnégatan, S-115 23 Stockholm
☎ (+46 8) 24 39 50 ✉ LEGSA ⊖ (+46 8) 660 71 36
Ambassador: I.G. de Swardt
Commercial Officer: A.L. Meyer
Public Information Officer: Dr J. Olivier 97

31-01680 Embassy of Spain
14 Hazeliusbacken, Djurgården, S-115 21 Stockholm
☎ (+46 8) 667 94 30 ⊖ (+46 8) 663 79 65
Ambassador: H E Mr Camilo Barcia ☎ (+46 8) 662 54 94
Deputy Chief of Mission: Emilio Lorenzo

Office of the Commercial Counsellor
12 Sergels Torg, S-111 57 Stockholm

☎(+46 8) 24 66 10 ⊠ 10790 ofcomes s
⊖ (+46 8) 20 88 92
Commercial Counsellor: Ramón Chaves

Office of the Defence Attaché
Djurgårdsvägen 21, S-115 21 Stockholm
☎(+46 8) 667 94 48 ⊠ 19053 est s
⊖ (+46 8) 667 94 49
Defence, Army, Navy and Air Attaché:
Lieut Col Julio Garulo [98]

31-01690 Embassy of the Democratic Socialist Republic of Sri Lanka
Strandvägen 39, PO Box 24055, S-104 50 Stockholm
☎(+46 8) 663 65 23/25 ⊠ 12338 lankemb s
⊖ (+46 8) 660 00 89
Ambassador: H.E. Mr Alfred K. David
Commercial Affairs: T. Dharmasena ☎(+46 8) 783 00 28
Second Secretary: R.S.J. Rambukkana
Admin Attaché: M. Rajagopal ☎(+46 8) 667 68 19 [97]

31-01740 Embassy of Switzerland
PO Box 26143, S-100 41 Stockholm
Birger Jarlsgatan 64, S-114 29 Stockholm
☎(+46 8) 676 79 00 ✉ AMBASUISSE Stockholm
⊖ (+46 8) 21 15 04
Ambassador: H.E. Dr Paul André Ramseyer
Deputy Chief of Mission: Georg Zubler
Defence Attaché: Georg von Erlach
Asst Defence Attaché: Niels Fischer [97]

31-01780 Royal Thai Embassy
3 Floragatan, PO Box 26220, S-100 40 Stockholm
☎(+46 8) 791 73 40 ⊠ 13 769 thaidut s
⊖ (+46 8) 791 73 51
Ambassador: H.E. Mr Suchinda Yougsunthon
Deputy Chief of Mission: Kamthorn Sithichoti
First Secretary: Wijak Chittarat ☎(+46 8) 10 22 53
First Secretary: Phipob Boonrod ☎(+46 8) 662 02 76
Third Secretary: Ms Luckana Suankeaw
☎(+46 8) 662 45 73
Third Secretary: Mrs Luckana Sgoontanag
☎(46 8) 665 33 68
Attaché: Mrs Leena Bordeenithikasen
☎(+46 8) 612 68 39
Naval Attaché: Capt. Prajit Pityapat (RTN)
(resident in London)

Office of the Commercial Counsellor
Engelbrektsgatan 19-1st Floor, PO Box 26082,
S-100 41 Stockholm
☎(+46 8) 679 96 72; 679 96 73 ⊖ (+46 8) 679 96 74
Minister & Counsellor, Commercial Affairs:
Inthorn Wankeao ☎(+46 8) 761 64 19
First Secretary, Commercial Affairs: Pansakdi Musiklad
☎(+46 8) 16 64 27 [97]

31-01890 Embassy of the United Kingdom
Skarpögatan 6-8, Box 27819, S-115 93 Stockholm
☎(+46 8) 671 90 00; 671 91 06; 671 91 07
⊖ (+46 8) 662 99 89 (commerce);
(+46 8) 671 91 04 (defence)
Ambassador: H.E. Mr R.B. Bone
Counsellor, Consul-General and Deputy Head of Mission:
Miss E C Robson
Counsellor, Economics & Commerce: A.R. Murray
Defence & Air Attaché: Wg Cdr P.J. Cullum
First Secretary, Commercial: M R Cumming
Naval & Military Attaché: Lt Col A. Mackenzie [98]

31-02010 Embassy of the Federal Republic of Yugoslavia
Valhallavägen 70, PO Box 26209, S-100 41 Stockholm
☎(+46 8) 21 84 36; 21 84 82; 21 84 85
⊠ 11836 jugamba s ⊖ (+46 8) 21 84 95
Ambassador: H E Dr Aleksandar Prlja
Counsellor: Branko Marjanac
First Secretary: Ms Nada Pejovic [98]

60 Research Organizations, Institutes & Test Establishments

60-00100 The Aeronautical Research Institute of Sweden (FFA)
Flygtekniska Försöksanstalten
Box 11021, S-161 11 Bromma
Ranhammarsvägen 12-14
☎(+46 8) 634 10 00 [E] ffa@ffa.se [W] http://www.ffa.se
⊖ (+46 8) 25 34 81
♣ SKr 104,600,000 (1994/95) 👥 200
Director General: Hans Dellner
Business Area Manager, Defence Customers:
Martin Svenzon
Business Area Manager, Public Research:
Anders Gustafsson
Business Area Manager, Other Customers: Ingvar Roos
Head of Experimental Aerodynamics: Magnus Linde
Head of Computational Aerodynamics: Ingemar Lindblad
Head of Flight Systems: Peter Caap
Head of Structures: Ragnar Gillberg
Head of Acoustics: Peter Göransson
Head of Wind Energy: Sven-Erik Thor
Head of Model Manufacturing: Curt Ohlsson
Head of the National Laboratory for Pressure:
Carin Bergström
Librarian: A M Lindholm
Information Officer: Ms Anna Walter
■ FFA is the central government agency for aeronautical research in Sweden. Its principal role is to provide scientific support and technical assistance to Swedish government authorities, domestic and foreign industries and to other organisations. The main areas of research are: aerodynamics, structures, acoustics, flight systems and simulations, wind energy. FFA is the national standards Laboratory for Pressure.
FFA has a full range of wind tunnel facilities from low speed to hypersonic speed. The newest wind tunnel T1500 is a high performance transonic facility. In addition, FFA has an extensive structures laboratory, a flight research simulator and a vibro-acoustic laboratory with an anechoic chamber. [98]

60-01400 National Defence Research Establishment
Försvarets Forskningsanstalt (FOA)
(a Government Agency under the Department of Defence)
S-17290 Stockholm
☎(+46 8) 706 30 00 ✉ FOAGENERAL
⊖ (+46 8) 706 31 00
♣ SKr 600 000 000 👥 1 000
Director-General: Bengt Anderberg
Deputy Dir General: Ola Dahlman
Research Dir: Birger Schantz
Information Dir: Hans Rehnvall
Administrative Dir: Lars von Malmborg
Defence Analysis Dir: Jan Foghelin
Weapons & Protection Dir: Bo Janzon
Human Sciences Dir: Berndt Brehmer
Guidance & Control, Materials, & Underwater Sensors Dir:
Magnus Hagwall
International Secretariat: Jan Rylander

Linköping Research Department (FOA 3)
PO Box 1165, S-S-581 11 Linköping
☎(+46 13) 11 80 00 ⊖ (+46 13) 13 16 65
Sensor Technology Dir: Svante Ödman
Warfare Technology Dir: Lennart Nyström
■ Area of main interest:
Technology of information handling.

Umeå Research Department (FOA 4)
S-S-901 80 Umeå
☎(+46 90) 18 92 30 ⊖ (+46 90) 10 68 00
NBC Defence Dir: Åke Bovallius
■ Areas of main interest:
Biological & chemical warfare agents, their effects & protective measures; radiobiology, radioecology. [97]

60-01900 Stockholm International Peace Research Institute (SIPRI)
Frösunda, S-169 70 Solna
☎(+46 8) 655 97 00 [E] sipri@sipri.se
[W] http://www.sipri.se ⊖ (+46 8) 655 97 33
👥 55
Director: Dr Adam Daniel Rotfeld
Press & Information Officer: Richard Nyström
■ Independent institute for research into problems of peace & conflict, particularly those of disarmament & arms control. [98]

65 Industry, Trade & Armed Forces Associations

65-00350 Association of Swedish Defense Industries
Box 5501, S-114 85 Stockholm
☎(+46 8) 783 80 78 ⊖ (+46 8) 667 88 18
👥 2
Director: Brig Dag Törnblom (retd) ☎(+46 8) 783 80 79
■ Trade association representing Swedish manufacturers of defence equipment.

Member companies:
Barracuda Technologies AB
Bofors AB
Bofors Underwater Systems AB
Ericsson Microwave Systems AB
Ericsson Saab Avionics AB
FFV Aerotech AB
Hägglunds Vehicle AB
Karlskronavarvet AB
Kockums AB
CelsiusTech Electronics AB
CelsiusTech Systems AB
Saab AB Gripen
Saab Dynamics AB
Saab Training and Systems AB
Volvo Aero Corp [98]

65-01900 Swedish Aerospace Industries Association (SAI)
(a member of AECMA)
c/o Saab AB, S-581 88 Linköping
☎(+46 13) 18 00 00; 43 00 ⊠ 50040
⊖ (+46 13) 18 34 20
President: Lars Gissler ☎(+46 13) 18 40 50
■ Aerospace industry association.

Member companies:
Saab AB
Volvo Aero Corporation AB [98]

70 Financial Holding Groups, Parent Companies

Saab Combitech AB 75-09050

70-00190 Allgon AB (publ)
Box 500, S-184 25 Åkersberga
☎(+46 8) 54 08 22 00 [E] info@allgon.se
[W] http://www.allgon.se ⊖ (+46 8) 54 08 24 91
♣ SKr 1,369,100,000 (Group) (1996)
Chairman: Gunnar Bark
President: Per Wejke
Technical Director: Ulf Saldell
Production Director: Torsten Körsell
Chief Financial Officer: Claes Silfverstolpe
Head of Business Area System: Ulf Wahlund
■ Wave propagation products for mobile telephony. Products for digital and analogue systems including repeater systems, base antennas, filters and hybrid combiners.
Group companies:
Allgon Mobile Communications AB, Sweden
Allgon System AB, Sweden
Allgon Innovation AB, Sweden
Allgon WestCom AB, Sweden
Allgon GmbH, Austria
Allgon SA, France
Allgon System Handels GmbH, Germany
Allgon System, Hong Kong
Allgon Enterprises Inc, USA → 75 [98]

70-00200 Asea Brown Boveri AB (ABB)
Kopparbergsv 2, S-721 83 Västerås
☎(+46 21) 32 50 00 [W] http://www.abb.se
⊠ 40918 abb s ⊖ (+46 21) 13 41 12
👥 27,000 (1996)
President: Anders Narvinger
Sales and Marketing Director: Örjan Larsson
Purchasing Manager: Thommy Bähr
Public Affairs Office: Roger Johansson
Press Office: Eric Österberg
Marketing Communications: Ms Margaretha Jakobssen
■ Asea Brown Boveri AB is the parent company of the Swedish part of the ABB Group, with operations in over 200 locations.
Global engineering activities, serving customers in electric power generation, transmission and distribution, industrial and building systems, and rail transportation.

Subsidiaries:
ABB Pressure Systems AB [98]

70-00250 Celsius AB
Celsius Industrier AB
Celsius Industries Corp.
Head Office
Box 7214, S-103 88 Stockholm
☎ (+46 8) 463 00 00 E info@celsius.se
W http://www.celsius.se ⊖ (+46 8) 611 68 94
President and Chief Executive Officer: Egon Linderoth
Vice President: Tomy Hjorth
Vice President Communication and Public Affairs: Kjell Göthe
Media Contact: Lars O Karlsson
■ Air defence, anti-armour, artillery systems, smart ammunition/munitions, maintenance and ILS, combat control systems and fire control systems. Propellants and explosives, programmable ammunition, naval gun systems, stealthy vessels, surface vessels, submarines, torpedoes, remotely operated underwater vehicles, mine hunting and clearance equipment, controllable, smart, and coastal defence mines. Coastal defence missile and artillery systems.
Shareholders:
Swedish Investors (45%)
Overseas Investors (30%)
Swedish Government (25%)
Gothenburg Office
PO Box 8954, S-402 74 Gothenburg
☎ (+46 31) 65 86 00 ⊖ (+46 31) 51 61 05 [97]

70-00450 Kockums Holdings
S-205 55 Malmö
☎ (+46) 40 34 80 00 ⊖ (+46) 40 97 32 81 [97]

70-00500 LM Ericsson
Telefonaktiebolaget LM Ericsson
Head Office
S-126 25 Stockholm
☎ (+46 8) 719 00 00 ☒ 14910 ⊖ (+46 8) 719 19 76
♣ SKr 98 780 000 000 (1995) 👥 85 000 (worldwide)
Chairman: Björn Svedberg
President & CEO: Lars Ramqvist
Exec. VP & CFO: Carl Wilhelm Ros
Sr VP Corporate Treasury: Stephan Almqvist
Sr VP & General Counsel: Erling Blommé
Sr VP Corp. Markets: B. Forssberg
Sr VP Corp. Relations: Lars A. Stalberg
Market Communications Mgr: M. Rönne [97]

70-00600 NCC AB
Head Office
Vallgatan 3, S-170 80 Solna
☎ (+46 8) 655 20 00 E info@ncc.se ⊖ (+46 8) 85 77 75
♣ SKr 340,000,000 (1996)
Chairman: Bernt Magnusson
President and Chief Executive Officer: Jan Sjöqvist
Financial Control and IT: Bo Friberg
Finance: Staffan Bennerdt
NCC Technology and R&D: Lennart Apleberger
Corporate Communications: Hans-Olof Karlsson
■ Civil engineering, construction and real estate. → 80 [98]

70-09000 Saab Group Head Office
(a subsidiary of Investor AB)
S-581 88 Linköping
☎ (+46 13) 18 70 00 ⊖ (+46 13) 18 18 02; 18 71 11 (communications)
♧ SKr 1,703,355,000 (1996)
♣ SKr 8,249,000,000 (1996) 👥 8,400
Chairman: Anders Scharp
President and Chief Executive Officer: Bengt Halse
Vice President: Marcus Wallenberg
Vice President: Lars Westerberg
Executive Vice President and Chief Financial Officer: Goran Sjoblom
Executive Vice President Defence Business Development: Hans Ahlinder
President Saab Dynamics: Anders A Andersson
President Saab Training Systems: Hans Robertson
President Saab Aircraft: Gert Schyborger
President Saab Combitech: Per Risberg
Senior Vice President Corporate Legal Staff: Per Erlandsson
Director of Corporate Personnel: Mats Lindman
Director of Corporate Communications and Public Affairs: Lars Jagerfelt
Representative of Industrial Salaried Employees Board Association: Lars Olsson
Representative of the Board of the Association of University Graduates: Lennart Hedlund
Representative of the Board of the Engineering Workers Union: Ragnar Ludvigsson
→ 75 → 80
Group companies:
Saab AB (parent company)
Saab AB Gripen
Saab Training Systems AB
Saab Combitech AB
Saab Aircraft AB
Saab Dynamics AB
Ericsson Saab Avionics AB (joint owned by Saab and Ericsson)
Stockholm Office
PO Box 70373, S-107 24 Stockholm
☎ (+46 8) 700 67 77 ⊖ (+46 8) 700 67 79 [98]

74 Consortia, Joint Venture Companies

74-00020 Bofors AB
(† Bofors Industries AB)
(a subsidiary of Celsius AB)
S-691 80 Karlskoga
☎ (+46 586) 810 00 W http://www.bofors.se ☒ 73210
⊖ (+46 586) 857 00
♣ SKr 4,218,000,000 (1996) 👥 4,530 (1996)
President: Per Erik Forsmark
Executive Vice President: Uno Carlsson
Executive Vice President: Ingemar Andersson
Director Communications and Public Affairs: Anders Florenius
Public Relations Manager: Kjell Mohlin
■ Anti-armour, anti-aircraft and artillery systems, missiles, guns, small arms, ammunition, explosives and propellants. → 75 [98]

74-00100 Industrigruppen JAS AB
JAS Industry Group
S-581 88 Linköping
☎ (+46 13) 18 39 00 ⊖ (+46 13) 18 39 19
President: Hans Ahlinder
VP, Communications & Public Affairs: Jan Ahlgren
■ Development & manufacture of JAS 39 Gripen multirole combat aircraft & automatic testing equipment for the Swedish Air Force.
Member companies:
Saab AB
Volvo Aero Corporation
FFV Aerotech
Ericsson Microwave Systems AB [97]

74-00400 Volvo Aero Support AB
(a subsidiary of Volvo Aero Corporation)
S-732 82 Arboga
☎ (+46 589) 810 00 ⊖ (+46 589) 198 63; 161 84
♣ SKr 545,000,000 👥 550
President: Erling Vister
■ Maintenance of Swedish Air Force jet engines. Commercial operations. Performs maintenance on engines for small passenger aircraft & helicopters, maintenance of gas turbines. → 80 [97]

75 Manufacturing Companies

75-00800 Aimpoint AB
Jägershillgatan 15, S-213 75 Malmö
☎ (+46 40) 21 03 90 ☒ 32153 ⊖ (+46 40) 21 92 38
President: Nicklas Gerhardsson
Managing Director: K Gernnarsson
■ Manufacture of electronic red-dot sightings for pistols, rifles, handguns and bows. HUD-style, MPS for machine gun and open door mount ACS models for helicopters. [97]

75-00850 Air Target Sweden AB
(† Swedair Target AB)
PO Box 3006, S-171 03 Solna
☎ (+46 8) 730 22 33 ☒ 15025 airtgt s
⊖ (+46 8) 730 34 24
♣ Skr 20,000,000 👥 20 🏭 800 m2
Managing Director: Bo Eklund
Sales and Marketing Director: Christer Nyberg
Operations Dir: Robert Stake
■ Airborne and inboard towing winches, aerial targets. Vector, ATA, ATG and GTA scoring systems, MARQUE scoring station. [97]

75-00900 Airsafe Sweden AB
S-194 82 Upplands Väsby
☎ (+46 8590) 930 70 ⊖ (+46 8590) 715 49
♧ SKr 650 000 ♣ SKr 10 000 000 👥 18 🏭 700 sq m
Managing Dir: Olle Norrvi
Sales & Marketing Dir: Claes Warbrandt
■ Design & manufacture of parachutes & parachute systems. [97]

75-01100 Åkers Krutbruk Protection AB
S-640 60 Åkers Styckebruk
☎ (+46 159) 366 00 E info@akerskrutbruk.se
W http://www.akerskrutbruk.se ⊖ (+46 159) 307 28
♣ SKr 107,000,000 (1996) 👥 70 (1996)
Managing Dir: Jan Lindberg
Sales & Marketing Dir: Bo Klaeppe
Technical Director: Lennart Sjöö
Purchasing Mgr: Kristina Sandberg
■ Development, production and marketing for:
Vehicle protection:
KE and CE non-reactive solutions, signature reduction, shelters.
Personal protection:
Body armour, helmets, shields, CW detectors and water purification.
Ammunition components:
Parachutes, propelling charges and gun grease. Propellant: composite propellant. [98]

75-01300 Allgon Mobile Communications AB
(† Allgon Mobile AB)
(a subsidiary of Allgon AB)
PO Box 500, S-184 25 Åkersberga
☎ (+46 8) 54 08 22 00 E amc@allgon.se
W http://www.allgon.se ⊖ (+46 8) 54 08 24 87
Managing Director: Torsten Körsell
Technical Director: Håkan Bergqvist
After Sales Product Director: Magnus Agild
Sales and Marketing Director: Gunnar Forsblad (OEM)
Purchasing Manager: Peter Lindquist
Marketing Communications Manager: Ms Teija Marie Pierre
■ Antennas for marine, vehicle, portable use stressing mobile telephony.
Subsidiaries in England, France, Turkey, Singapore, Germany & Austria. [98]

75-02500 Atlas Copco Berema AB
(a division of Atlas Copco AB)
Uddvägen 7, Box 767, S-131 24 Nacka
☎ (+46 8) 743 96 00 ⊖ (+46 8) 743 96 50
♣ SKr 875,000,000 (1997) 👥 600 (1997)
Managing Dir: Lars Larsson
Sales & Marketing Dir: Peo Sollerud
Public Relations Mgr: Andres Toom
■ Hydraulic rig mounted breakers, crushers and pulverizers, motor drills and breakers, pneumatic breakers, hand held surface rock drills, kango electric tools, air powered pumps, air line accessories. [98]

75-02700 Barracuda Technologies AB
(a subsidiary of Hägglunds Vehicle AB)
PO Box 160, S-594 23 Gamleby
☎ (+46) 49 31 48 00 E cami@barracuda.se
W http://www.barracuda.se ⊖ (+46) 49 31 20 64
♧ SKr 20,000,000 (1994) ♣ SKr 100,000,000 (1996)
👥 100 🏭 12,000 m2
President: Göte Nilsson
Technical Director: Lars Karlsson
Purchasing Manager: Claes-Göran Karlsson
Public Relation Manager: Ms Bibbi Johansson

■ Camouflage systems and nets, decoy systems. Near infra-red net (VIS/NIR) to terrain requirements. Thermal camouflage (TIR) providing protection against thermal sensors and heat-seeking missiles. Broadband radar nets against long and short wave radar detection. Analyses of local terrain and climactic conditions, investigations of protection of equipment, measurements, project planning and solutions. TOPCAM lightweight camouflage umbrella for rapid deployment. Decoys for vehicles, aircraft and bridges. [98]

75-02900 **BOA Defence AB**
Oxelgrenswagen 34, S-152 42 Södertalje
☎ (+46 85) 508 08 55 ⊖ (+46 85) 508 08 60
Chairman of the Board: Arne Bergsman
Managing Director: Lars Nylin [98]

75-02950 **Bofors Carl Gustaf AB**
(a subsidiary of Bofors AB)
PO Box 613, S-631 08 Eskilstuna
☎ (+46 16) 16 90 00 ⊖ (+46 16) 13 12 24
Managing Dir: Anders Olsson
Sales and Marketing Director: Leif Larsson
Marketing Administration Mgr: Göte Johansson
■ Manufacture of small arms. [98]

75-02975 **Bofors Ant Armour Systems**
S-691 80 Karlskoga
☎ (+46 586) 810 00 ⊖ (+46 586) 857 00
President: Toruald Wästbjörk [98]

75-03000 **Bofors Explosives AB**
(a company of Bofors AB Group)
S-691 86 Karlskoga
☎ (+46 586) 830 50 ⊖ (+46 586) 853 10
President: Sören Gindahl
Finance & Admin Dir: Björn Börjesson
☎ (+46 586) 834 57
Marketing Dir: Ronny Lindström
Technical R&D Dir: Per Lagerkvist ☎ (+46 586) 832 46
Purchasing Mgr: Håkan Karlsson ☎ (146 586) 830 77
■ Manufacture explosives, powders & propellants. [98]

75-03025 **Bofors LIAB AB**
Kåfallavägen, PO Box 154, S-711 23 Lindesberg
☎ (+46 581) 871 00 ⊖ (+46 581) 61 11 96
👥 250
Managing Dir: Karl-Ake Gronlund
Export Sales Dir: Stig Wretling
Sales & Marketing Dir: Peter Bring
Dir Civil Products: Gehard Nilsson
■ Ammunition in calibre ranges from 20mm to 40mm; fuzes, primers, tracers; detonators, pyrotechnics; explosive charges, modernisation and conversion of ammunition, destruction.

Subs:
Bofors Carl Gustav AB → 75
Bofors Explosives AB → 75
Bofors LIAB → 75
Bofors Underwater Systems AB → 75 [98]

75-03030 **Bofors Missiles**
S-691 80 Karlskoga
☎ (+46 586) 810 00 ⊖ (+46 586) 853 90 [98]

75-03040 **Bofors SA Marine AB**
(† SA Marine AB)
(a subsidiary of Bofors Underwater Systems AB, Sweden)
Hjalmar Brantings väg 5, PO Box 627,
S-26126 Landskrona
☎ (+46 418) 516 00 ⊖ (+46 418) 229 52
♧ SKr 3,500,000 ♣ SKr 90,000,000 👥 60
Managing Director: Per-Inge Birgersson
Advertising Manager, Secretary to the Managing Director: Ms Monica Arvidsson
Marketing Manager: Joakim Natt Och Dag
■ Underwater defence products. Sea mines, MCM equipment, ASW equipment. [98]

75-03050 **Bofors Underwater Systems AB**
PO Box 910, S-591 29 Motala
☎ (+46 141) 245 00 ⊖ (+46 141) 111 79
President: Bertil Malmberg
Managing Dir: Jan Erik Tornqvist
Marketing Director: Magnus Karlsson

■ Torpedoes, mine countermeasure equipment, anti-submarine weapons, underwater sensors and submersible vehicles. [98]

75-03052 **Bofors Underwater Systems AB**
SUTEC
(† Underwater Systems AB)
(a division of Bofors Underwater Systems)
PO Box 7073, S-580 07 Linköping
☎ (+46 13) 15 80 60 ⊖ (+46 13) 15 13 20
👥 40 (1997) 🏭 1,400 m2
Managing Dir: Jan-Erik Törnqvist
Sales & Marketing Dir: Bengt Larsson
Operations Dir: Håkan Klevebrant
Purchasing Mgr: Peter Carsk
■ Leading supplier of ROV's, submersibles for salvage, inspection & a wide range of seabed activities, in addition to vehicles for mine identification & disposal. [98]

75-03075 **Bofors Weapon Systems**
S-691 80 Karlskoga
☎ (+46 586) 810 00 ⊖ (+46 586) 857 00
President: Magnus Ingesson [98]

75-03600 **Bülow Air Target Systems AB**
(† Olle Bülow AB)
Hilleshög, S-179 96 Stenhamra
☎ (+46 8) 56 04 21 70 ⊖ (+46 8) 56 04 21 11
♣ SKr 9,000,000 👥 11
Managing Director: Kjell Bülow
■ Design and manufacture of aerial targets, air gunnery training equipment, radar reflectors, miss-distance indicators, tow systems and target reels. Airborne in/outboard mounted tow target winches, helicopter target towing winches. Tow targets for air-to-air, ground-to-air, ship-to-air and air-to-ground gunnery practice. [98]

75-03710 **CelsiusTech AB**
(a subsidiary of Celsius AB)
Head Office
Nettovägen 6, Jakobsberg, S-175 88 Järfälla
☎ (+46 8) 58 08 40 00 [W] http://www.celsiustech.se
⊖ (+46 8) 58 03 22 44
♣ SKr 1,900,000,000 👥 1,800 (1996)
Managing Director: Lars Gahnström
■ The CelsiusTech Group develops, manufactures and markets defence equipment in the fields of electronic warfare (including radar warners and airborne countermeasures), radar, avionics including radio communications, optronic sights, fire control for anti-armour and air defence systems, munition electronics and microwave products, Combat Management Systems for the Navy, the Army and the Air Force and Tactical MCM Data systems. The Air Defence Ground Environment systems include Multi Sensor Fusion systems.

Operationally, CelsiusTech is organised into five subsidiary companies: CelsiusTech Electronics, CelsiusTech Systems, CelsiusTech Naval Systems, YDAB and CelsiusTech Australia. [98]

75-03720 **CelsiusTech Electronics AB**
(a subsidiary of Celsius AB)
Head Office
S-175 88 Järfälla
☎ (+46 8) 58 08 40 00 [W] http://www.celsiustech.se
⊖ (+46 8) 58 03 22 44; 58 08 72 38
♧ SKr 15,000,000 ♣ SKr 1,075,000,000 (1996)
👥 900 (1996)
President: Lars Gahnström
Vice President Marketing: Jan G Linder
Manager Marketing Communications: Ms Berit Ljung
Production Manager: Michael Lövgren
Production Manager - Technology: Kurt Nordfors
Production Manager - Sensors: Gunnar Tellas
Production Manager - Electronic Warfare: Chrisler Ros
Production Manager - Optronics: Ulf Dahlberg
Production Manager - Microwave: Pontus de Laval
Production Manager - Communications: Goran Bergling
Army Electronics: Bjorn Hammarberg

■ Development, manufacture and marketing of sensors, countermeasures, optronics and fire control systems. Electronic warfare, ESM and ECM systems, chaff and flare dispenser systems, surveillance and fire control radars, munition electronics, missile seekers, microwave products, fire control systems, optronics, night vision equipment, precision optics, sound ranging systems, data communication equipment, radio communication systems for airborne, shipborne and ground-based applications.

Branch Office
Agavägen, S-181 84 Lidingö
☎ (+46 8) 731 60 00 ⊠ nobelba s ⊖ (+46 8) 765 92 65
President: Goran Widen
Vice President Marketing: Jan Linder
Marketing and Communication Director: Ms Berit Ljung
Production Manager: Michael Lövgren
Production Manager - Technology: Kurt Nordfors [98]

75-03730
CelsiusTech Systems AB
(a subsidiary of CelsiusTech AB)

CelsiusTech
CELSIUS GROUP

Nettovägen 6, S-175 88 Järfälla
☎ (+46 8) 58 08 40 00 [W] http://www.celsiustech.se
⊖ (+46 8) 58 03 22 44
Managing Director: Ingvar Johansson [98]

75-03740 **CelsiusTech Naval Systems AB**
(a subsidiary of CelsiusTech AB)
Nettovägen 6, S-175 88 Järfälla
☎ (+46 8) 58 08 40 00 [W] http://www.celsiustech.se
⊖ (+46 8) 58 03 22 44
Managing Director: Karl Lyth [98]

75-04000 **Daloc Sheltec AB**
(† JP Sheltec Skyddsrum AB)
(a subs. of Daloc AB)
PO Box 1057, S-141 21 Huddinge
☎ (+46 8) 774 03 20 ⊖ (+46 8) 711 77 15
Managing Dir: Bengt Jakstad
■ Manufacture of NBC shelter equipment. [97]

75-04100 **Dantherm AB**
(a subs. of Dantherm A/S, Denmark)
Kabelgaten 12, PO Box 10163, S-43422 Kungsbacka 1
☎ (+46 300) 166 20 ⊖ (+46 300) 162 16 [97]

75-04820 **Eltex of Sweden AB**
PO Box 608, S-343 24 Elmhult
☎ (+46 476) 488 00 [E] public@eltex.se ⊠ 52016 eltex s
⊖ (+46 476) 134 00
♣ SKr 45,000,000 (1996) 👥 55 (1997)
Managing Director: Olle Rydborn
Sales and Marketing Director: Bengt-Erik Karlberg
Information Manager: Håkan Löfgren
■ Microprocessor-controlled system for charging sealed NiCd batteries. [98]

75-04890
Ericsson Microwave Systems AB
(† Ericsson Radar Electronics AB)
(a subsidiary of Telefonaktiebolaget LM Ericsson)

ERICSSON

S-431 84 Mölndal
☎ (+46 31) 747 00 00; 747 17 03
W http://www.ericsson.se/hepi/bd ⊠ 20905 ericmw s
⊜ (+46 31) 27 78 91; 67 17 03
♧ SKr 30,000,000 (1996) ♣ SKr 4,330,000 (1996)
👥 4,000 (1997) 🏭 165,000 m2
President: Jan-Åke Kark ☎ (+46 31) 747 14 69
Executive Vice President Microwave Communiations: Sivert Bergman ☎ (+46 31) 747 17 04
Vice President Defence Electronics: Svante Bergh ☎ (+46 31) 747 67 67
Finance Director: Hakan Rusch ☎ (+46 31) 747 30 70
Vice President Human Resources and External Relations: Lars Harenstam
Vice President and Marketing Director: Bertil Hellström ☎ (+46 31) 747 16 23 E bertil.hellstrom@emw.ericsson.se
Purchasing Manager: Michal Cadier
Press Officer and Advertising Manager: Lars Köhler ☎ (+46 31) 747 18 41
■ Airborne multimode doppler radars for combat aircraft, AEW&C radar and mission systems (Erieye), radars for tactical air defence (Giraffe, Pointer, HARD), fire control (Eagle), tracking, naval/air surface surveillance (Sea Giraffe), artillery and weapon location (Arthur).
Defence communication systems and computers airborne EW systems (Erijammer), military telecommunications, digital microwave radio links (Mini-Link), radio base station. [98]

75-04950 Ericsson Saab Avionics AB
S-164 84 Stockholm
☎ (+46) 757 30 00 ⊜ (+46) 757 31 15
👥 600
Managing Director: Björn Erman
■ Development and production of EW systems, cockpit display systems and various avionics. [98]

75-05000 FFV Testsystems AB
PO Box 47125, S-10074 Stockholm
☎ (+46 8) 775 54 00 ⊜ (+46 8) 19 00 75
♣ SKr 100 000 000 👥 110
President: Kurt Höglund
■ Development & manufacture of automatic test systems for test of electronic equipment. [97]

75-05200 Fjellman Press i Mariestad AB
Norra Vägen 11, PO Box 7, S-542 21 Mariestad
☎ (+46 501) 138 30 ⊜ (+46 501) 189 72
♧ SKr 2,500,000 ♣ SKr 60,000,000 👥 65
Managing Dir: Tarmo Kittelfors
Sales and Marketing Director: Jan Carlstein
Operations & Purchasing Dir: Bror Gunnarsson
Export Mgr: Ms Christina Fjellman
■ Manufacture of hydraulic presses. [98]

75-05300 Flodins Filter AB
(a subsidiary of Flodins Filters Föruattnings AB)
PO Box 100 11, S-453 29 Lysekil
☎ (+46 523) 150 20 ⊜ (+46 523) 136 58
♧ SKr 2,652,000 (1997) ♣ SKr 37,000,000 (1997)
👥 40 (1997) 🏭 6,000 m2
Managing Director: Bengt Dahlgren
Operations Director: Inger Grunden
Sales and Marketing Director: Roger Karlsson
Purchasing Manager: Reinhold Färg
■ NBC filters for gas masks. NBC shelter filters. [98]

75-05700 Germa AB
(a subsidiary of Ambu International AS, Denmark)
PO Box 2127, S-291 02 Kristianstad
☎ (+46 44) 12 30 30 ⊠ 72615 ⊜ (+46 44) 10 31 79
♧ Skr 7,500,000 ♣ Skr 20,000,000 👥 30 🏭 1,000 m2
Managing Director: Bengt Sjöberg
Operations Director: Dan Romerfeldt
Purchasing Manager: Ms Cecilia Lindberg
■ Immersion, flying and anti-G suits. Rescue equipment. [98]

75-05800 Hägglunds Vehicle AB Head Office
(a subsidiary of Incentive AB)
S-891 82 Örnsköldsvik
☎ (+46) 66 08 00 00 E stmp.bandvagn@haggve.se
⊜ (+46) 66 01 31 18; 66 08 26 12
♧ SKr 150,000,000 (1996)
♣ SKr 1,306,500,000 (1996) 👥 1,018 (1996)
🏭 65,000 m2
President: Sven Kagevall
Vice President and General Manager, Light Vehicle Division: Thomas Edblad
Vice President and General Manager, Combat Vehicle Division: Tommy Widlund
Vice President and General Manager After Sales Division: Lars Bergström
Product Manager: Lars Erik Lindgren
Marketing Manager: Stefan Nordström
Information Officer: Barbro Wikman-Sundsten
■ Manufacture of all-terrain, crash, fire, and rescue vehicles. Light vehicles, combat vehicles, camouflage and auxiliary equipment.
Bv 206 all-terrain carrier amphibian tracked vehicle for military and civilian applications, troop and weapon carrying, command, communications, radar and fire control, first aid stations. Combat Vehicle 90, CV9025 IFV Infantry Fighting Vehicle, CV 9030 IFV Infantry Fighting Vehicle with 30mm cannon, CV 9040 IFV Infantry Fighting Vehicle with 40mm gun, CV 9040 AAV Anti-Aircraft Vehicle with 40mm gun, CV 90105 TML Anti-Tank Vehicle with 105mm gun, CV 90 FOV Forward Observer Vehicle, CV 90 FCV Forward Command Vehicle, CV 90 ARV Armoured Recovery Vehicle. [98]

75-06030 Hedemora Diesel AB
(a subsidiary of NCC AB)
Box 206, S-776 28 Hedemora
☎ (+46 225) 155 40 ⊜ (+46 225) 154 40
♧ SKr 3,000,000 (1996) ♣ SKr 90,000,000 (1996)
👥 38
Managing Dir: Lars Nordström
Technical Dir: Jan Wester
■ Turbo-charged diesel engines in the output range from 1,000kW to 2,500kW. Submarines and off shore installations are a speciality. [98]

75-06500 Imo Pump AB
(a subsidiary of Imo Industries Inc, USA)
Vastberga Alle 50, PO Box 42090, S-12614 Stockholm
☎ (+46 8) 19 01 60 ⊜ (+46 8) 19 97 49
♧ SKr 50,000,000 (1996) ♣ SKr 192,000,000 (1996)
👥 177 (1996)
Managing Director Europe: Anders Sjogren
Operations Director: Hans-Olof Högman
Sales and Marketing Director: Lars Helin
Logistics Manager: Leif Öhman
■ Manufacture a wide range of high-performance rotary, positive displacement pumps, including two-screw, three screw & gear pumps. [98]

75-06950 Interspiro AB
(† AGA Spiro AB)
PO Box 10060, S-181 10 Lidingö
☎ (+46 8) 636 51 00 E info@interspiro.se
⊜ (+46 8) 765 48 53
👥 95
Managing and Sales Director: Hans O Almqvist
Financial Manager: Kjell Jansson
Sales Manager: Lars Naucker [98]

75-07300 Kamewa Group Kamewa AB
(a subsidiary of Vickers plc, UK)
PO Box 1010, S-681 29 Kristinehamn
☎ (+46 550) 840 00 ⊠ 66050 kamewa s
⊜ (+46 550) 181 90
Managing Dir: I. Jensen
Sales Promotion Manager: Anders Pettersson
■ Manufacture naval propellers, rotable thrusters & water jet propulsion systems. Cpp's up to 40.000 hp. Quiet running, supercavitating, shock resistant & ice-strenghtened to Arctic standards. [98]

75-07400 Karlskronavarvet AB
(a subs. of Kockums AB)
S-371 82 Karlskrona
☎ (+46 455) 33 41 00 ⊠ 8395018; 8395028 kkrv
⊜ (+46 455) 179 34
♧ SKr 32 000 000 ♣ SKr 450 100 000 👥 845
Managing Dir: Hans Hedman
Marketing Dir: Bengt Odgren
Technical Dir: Börje Appelquist
Purchasing Mgr: Bo Lindahl
■ Production, maintenance & modernization of ships for the Swedish Navy & Coast Guard, & production of naval & surveillance ships for export. The production programme includes fast missile craft, mine countermeasures vessels of GRP, offshore patrol vessels & surveillance craft. Heavy dry support bridges & multi space bridges. [97]

75-07500 Kockums Industries AB
S-205 55 Malmö
Seostora Vrvfgatan
☎ (+46 40) 34 80 00 ⊠ 33190 kockum s
⊜ (+46 40) 97 32 81 [98]

75-07505 Kockums Submarine Systems AB
S-205 55 Malmö
☎ (+46 40) 34 80 00 ⊜ (+46 40) 97 32 81
■ Submarines, AIP-systems. [98]

75-07520 Kocoverk International AB
Industrivägen 4, S-19162 Sollentuna
☎ (+46 8) 35 26 60 ⊜ (+46 8) 35 34 63
Managing Dir: Max Ternheim
Public Relations Mgr: Ms Carina Ternheim
■ Manufacture of Freon free air conditioning systems for aircraft, radars & electronics. Major projects include delivery of ground support Freon free A/C system to JAS 39 Gripen combat aircraft. Manufacture of aircraft heaters & dehumidifiers. Supplier of special Freon free A/C systems aimed at qualified electronic industries & military end users. [97]

75-08230 AB Carl Munters
PO Box 430, S-191 24 Sollentuna
☎ (+46 8) 626 63 00 ✉ MUNTERSLAB
E camilla@algonet.se W http://www.munters.com
⊠ 17887 munters s; 15921
⊜ (+46 8) 626 63 65; 626 68 96; 754 68 96
♧ SKr 60,000,000 ♣ SKr 1,782,000,000 (1994)
👥 1,571 (1994)
President: Lennart Evrell
Vice President: Henrik Hegardt
Division Manager DH: Anders Fredriksen
Segment Manager Preservation: Anders Bergstrom
Defence Manager: Gerard Mojet (Netherlands Office) ☎ (+31 1720) 332 31; 43 32 31
America Region: Sven Lundin (US Office)
Asia Region: Erik Williamsson
■ Manufacture dehumidifiers and dehumidification systems used for dry-air storage and in-use protection of all military equipment and installations including tanks, howitzers, anti-aircraft systems, small arms, ammunition, ships during lay-up, aircraft and rock-caverns.
Works
PO Box 29, S-748 03 Tobo
☎ (+46 295) 342 60; 348 00
⊜ (+46 295) 344 33; 344 39
General Manager: Håkan Andersson
■ Manufacture of dehumidification systems for dry-air storage of aircraft and protection of active storage aircraft. [97]

75-08234 Munters Dry Air Sverige AB
(† Munters Dry Air AB)
(a subsidiary of AB Carl Munters)
Kung Hans väg 8, PO Box 450, S-191 24 Sollentuna
☎ (+46 8) 626 63 00 ⊠ 17887 munters s
⊜ (+46 8) 754 85 94
Managing Dir: Owe Löfqvist
Defence Sales Mgr, South Region: Andreas Svensson
Defence Sales Mgr, North Region: Hans Arnevall
■ Dehumidification of all kinds of equipment, either operational or stored. [98]

75-08250 NEFAB AB
Östra Storgatan 20, PO Box 2184, S-55002 Jönköpink
[98]

75-08260 New Pac Safety AB
PO Box 174, S-566 23 Habo
☎ (+46 36) 411 39 ⊖ (+46 36) 410 31
♣ SKr 15 000 000 (1992/93) 👥 8
Managing & Sales & Marketing Dir: Jan Bergling
Technical Mgr: Ralph Widenbäck
■ Supply of disposable NBC survival aids. C-Cover dress "Military version", dress for rescue operations, ponchos, gloves & survival kits. → 90 [97]

75-08275 Nitro Nobel AB
(a subs. of Dyno Industrier)
Gyttorp, S-713 82 Nora
☎ (+46 587) 850 00 ☒ 73260 nobele s
⊖ (+46 587) 255 35
♣ SKr 500 000 000 (1995) 👥 450
Product Sales Mgr: Mats Börjesson
■ ISO propyl nitrate (starting fuel for jet engines) - AVPIN. [97]

75-09000 Saab Aircraft AB
S-581 88 Linköping
☎ (+46 13) 18 20 00 [W] http://www.saab.se/aircraft
⊖ (+46 13) 18 27 22
President: Gert Schyborger
VP, Public Relations: Anders Annerfalk
■ Development of Saab 2000 twin turboprop regional airliner & production of Saab 340 twin turboprop regional airliner. [98]

75-09050 Saab Combitech AB
(part of the Saab Group)
Box 1017, S-551 11 Jönköping
☎ (+46 36) 19 40 00 [W] http://www.combitech.se
⊖ (+46 36) 19 45 10
Chairman: Bengt Halse
President: Per Risberg
■ High-tech niches in the fields of space, industry, traffic systems and information systems. → 80 → 70

Subsidiaries:
Combitech Electronics AB
Combitech Innovation AB
Combitech Network AB
Combitech Software AB
Combitech Traffic Systems AB
DocEye AB
Innovativ Vision AB
IV Image Systems AB
Preseco Informationssystem AB
Pronesto AB
Saab Ericsson Space AB
Saab Marine Electronics AB
Saab Survey Systems AB
Telelogic AB

Combitech Units outside Sweden:
Austrian Aerospace GmbH, Austria
Saab Tank Control, Bahrain
MC2 SA, France
Saab Tank Control Deutschland Vertriebs GmbH, Germany
Combitech Traffic Systems (H.K.) Ltd, Hong Kong
Saab Combitech Far East Sdn Bhd, Malaysia
DocEye AS, Norway
Saab Marine Liaison Office, Russia
Telelogic UK Ltd, UK
Saab Marine (UK) Ltd, UK
Saab Systems Inc, USA
Saab Ericsson Space Inc, USA [98]

75-09100 Saab Dynamics AB
(a subsidiary of Saab AB)
S-581 88 Linköping
☎ (+46 13) 18 60 00 ⊖ (+46 13) 18 60 06
♣ SKr 829,000,000 (1996) 👥 700
Chairman: Bengt Halse
President: Anders A. Andersson
Marketing Dir: Gunnar Sprang
Publicity Mgr: Hans G. Wiklund
■ Guided weapons, electronic warfare systems, optronic systems, airborne equipment. Advanced missile systems and subsystems, military electronics, such as optronics and fire control systems. Sighting systems for aircraft and helicopters, steering control systems for submarines, gyroscopes, aircraft equipment.
Main products: Saab RBS15 anti-ship missile system; Saab IRS-700 IR surveillance system; Saab EOS-450 fire control system; Saab ASW-601 anti-submarine warfare system; IR-OTIS (airborne) Infra-Red Search and Track (IRST) system; STRIX passive IR terminally guided 120mm mortar projectile; Saab LVS AA gun fire control system; HeliTOW sighting, launching and guidance system for the TOW missile.
Saab Dynamics was created by the merger of Saab Missiles and Saab Instruments. [98]

75-09150 Saab Ericsson Space AB
(a member of Saab Combitech AB)
S-405 15 Gothenburg
☎ (+46 31) 335 00 00 ⊖ (+46 31) 335 95 20
♣ SKr 467,000,000 (1996) 👥 500
President: Ivan Öfverholm
Exec. Vice President: Ulf Berg
Technical Dir: Peter Möller
Sales Dir: Per-Gunnar Lundh
Manager Communication and Public Affairs: Ms Iréne Svensson ☎ (+46 31) 335 44 63
■ On-board data handling systems and computers for satellites and launchers. Antennas, microwave electronics, satellite separation mechanisms, sounding rocket payloads and guidance. Satellite systems, subsystems and equipment. Guidance and separation systems.
Owned by Saab Combitech 60% Ericsson Microwave Systems 40% [98]

75-09200 Saab AB Gripen
S-581 88 Linköping
☎ (+46 13) 18 00 00 ⊖ (+46 13) 18 18 02
General Manager, Senior Vice President: Hans Krüger
Vice President Communications and Public Affairs, Saab-BAe Gripen: Jan Ahlgren ☎ (+46) 13 18 39 07
Communications and Public Affairs, Saab-BAe Gripen: John Neilson (based in UK) ☎ (+44 1252) 38 47 95
■ Development of JAS 39 Gripen multirole fighter/attack/reconnaisance aircraft, a joint venture of Saab AB and British Aerospace plc, UK. [98]

75-09250 Saab Training Systems AB
(a subsidiary of Saab AB)
S-561 85 Huskvarna
☎ (+46 36) 19 46 00; 38 80 00
⊖ (+46 36) 14 32 85; 38 80 00; 38 80 80
◇ Skr 15,000,000 ♣ SKr 559,000,000 (1996)
👥 248 (1996)
Chairman: Bengt Halse
Managing Director: Hans Robertson
Finance Director: Lars-Eke Jonsson
Marketing Director: Jörgen Lindgren
Advertising Manager: Ms Lena Hult
Publicity Manager: Christer Gynne
■ Laser simulators for army and airforce applications. Live fire range equipment, visual simulators for gunnery training (CGI-based), mobile transfer systems and target equipment. [98]

75-09850 Saft Nife AB
(a subsidiary of Saft SA, France)
Factory
Jungnergatan, PO Box 709, S-572 28 Oskarshamn
☎ (+46 491) 680 00 ☒ 43982 ⊖ (+46 491) 681 80
◇ SKr 150,000,000 (1996) ♣ SKr 505,000,000 (1996)
👥 415 (1996) 🏭 40,000 sq m
Managing Director: Torbjörn Anulf
Operations Director: Willy Wigothsson
Sales and Marketing Director: Håkan Hammarström
Purchasing Manager: Jerry Engström
■ Design and manufacture of an extensive range of advanced batteries for defence.
Representing Saft SA, France. → 90

Sales Office Sweden
Tingsvägen 17, S-191 61 Sollentuna
☎ (+46 8) 625 14 50 [E] sales@saftnife.alcatel.se
⊖ (+46 8) 96 36 42 [98]

75-10150 Sandvik AB
S-811 81 Sandviken
☎ (+46 26) 26 00 00 ✉ SANDVIK ☒ 47000
⊖ (+46 26) 26 10 22
◇ SKr 1,340,000,000 ♣ SKr 28,300,000,000 👥 30,000
President and Chief Executive Officer: Clas Ake Hedstrom
Executive Vice President: Leif Sunnermalm
Executive Vice President: Gunnar Björklund
Public Relations Manager: Bjorn Jonson ✉ Istyle touch
■ Steels, nickel alloys, titanium and zirconium alloys in bar products, tubes, wire and strip. Cemented carbide products, tools and process systems. [98]

75-10200 Scania AB Head Office and Technical Centre
S-151 87 Södertälje
☎ (+46 8) 55 38 10 00 [W] http://www.scania.com
⊖ (+46 8) 55 38 10 37
Chairman: Anders Scharp
Vice Chairman: Lars V Kylberg
President and Chief Executive Officer: Leif Östling
Executive Vice President and Chief Administrative Officer: Göran Löfgren
Executive Vice President and Chief Financial Officer: Lars Ohlsson-Leijon
Executive Vice President and Chief Technical Officer: Håkan Samuelsson
Senior Vice President, Industrial and Marine Engines: Alf Bexell
Senior Vice President, Communications and Public Affairs: Lars Christiansson
Senior Vice President, Sales and Marketing Europe: Urban Erdtman
Senior Vice President, Sales and Marketing Overseas: Åke Norrman
■ Heavy trucks and buses for the transport of goods and passengers. Industrial and marine engines for use as a power source in generator sets, earth-moving and agricultural machinery, and in ships and pleasure craft.

Falun Production Plant
Box 1906, S-791 19 Falun
☎ (+46 23) 477 00 ⊖ (+46 23) 71 13 79

Buses & Coaches
S-641 81 Katrineholm
☎ (+46 150) 585 00 ⊖ (+46 150) 532 30
👥 785
Vice President: Åke Brännström
Manager, Buses & Coaches, Communications: Gunnar Boman ☎ (+46 150) 585 99

Laxå Production Plant
Box 77, S-695 22 Laxå
☎ (+46 584) 108 20 ⊖ (+46 584) 102 01

Luleå Production Plant
Box 815, S-971 25 Luleå
☎ (+46 920) 766 00 ⊖ (+46 920) 896 10

Oskarshamn Production Plant
Box 903, S-572 29 Oskarshamn
☎ (+46 491) 76 50 00 ⊖ (+46 491) 76 54 30

Sibbhult Production Plant
S-280 63 Sibbhult
☎ (+46 44) 495 00 ⊖ (+46 44) 481 08 [98]

75-10300 Siemens AB
Johanneslurdsvägen 12-14, S-19487 Upplands Väsby
☎ (+46 8) 728 10 00 ✉ SIEMENS STOCKHOLM
☒ 19880 siemens s ⊖ (+46 8) 728 11 28
Managing Dir: L.O. Hjalmar
Mgr Special Electronics: Torsten Imme
Information Mgr: Roland Lindén
® Germany: Siemens
USA: Link Flight Simulation, Singer Kearfott [97]

75-10340 SKF
Aktiebolaget SKF
Hornsgt 1, S-415 50 Göteborg
☎ (+46 31) 337 10 00 ✉ KULLAGER
[W] http://www.skf.com ☒ 2350 ⊖ (+46 31) 337 28 32
♣ SKr 33,589,000,000 (1996) 👥 43,123
Chairman: Anders Scharp
Group Chief Executive Officer: Peter Augustsson
Group Communication Director: Lars G Malmer
■ Manufacture of aeroengine and airframe instrument bearings, bearing maintenance tools and services, bearing monitoring systems, actuators, ball and roller screws, mechanical and electromechanical flight control equipment and components. Rolling bearings (aeroengine/airframe and instruments), linear and plain bearings, seals and special steels. [98]

75-10600 Storebro Bruks AB
(† Storebro Royal Cruiser AB)
S-590 83 Storebro
☎ (+46) 49 21 95 00 Ⓔ info@storebro.se
Ⓦ http://www.storebro.se ⊖ (+46) 49 23 03 00
♣ SKr 120,000,000 ♟ 130
■ Special task vessel (SRC 90 E) built for demanding sea duties which combines very high speed, acceleration and braking capability with an extremely tight turning radius. The craft can be equipped for a wide range of marine tasks, for naval forces, police, ambulance, coast guard, search and rescue, customs or commercial requirements. [98]

75-11260 Telelogic AB
(a subsidiary of Saab Combitech AB)
PO Box 4128, S-203 12 Malmo
☎ (+46 40) 17 47 00 Ⓔ info@telelogic.se
Ⓦ http://www.telelogic.se ⊖ (+46 40) 17 47 47
♣ SKr 67,000,000 (1996) ♟ 77
Managing Director: Ingemar Ljungdahl
Marketing Manager: Carl-Eric Mols
■ Develops & promotes computer aided software engineering & testing tools for real-time & distributed systems. → 80 [98]

75-11280 THORN Lighting AB
PO Box 305, S-261-23 Landskrona
☎ (+46 418) 520 00 ⊖ (+46 418) 265 74
♣ SKr 370,000,000 ♟ 280
Chief Executive: Per-Olof Nilsson [98]

75-11800 Volvo Truck Corporation
Volvo Truck International Division
S-405 08 Gothenburg
☎ (+46 31) 59 00 00; 66 60 00 ⊠ 27000
⊖ (+46 31) 66 46 60
Marketing Mgr Government Sales: Uno Hjelmström
■ Fire crash tenders & rescue vehicles. [97]

75-11810 Volvo Aero Corporation
S-461 81 Trollhättan
☎ (+46 520) 940 00 Ⓦ http://www.aero.volvo.se
⊠ 42040 ⊖ (+46 520) 340 10
♧ Skr 790,000,000 (1996)
♣ SKr 4,100,000,000 (1996) ♟ 3,700
President and Chief Executive Officer: Arne Wittlöv
Executive Vice President Economy and Administration: Leif Gustafsson
Vice President Corporate Communications: Mauritz Ljungman
Commercial Engines: Staffan Zackrisson
Military Engines Division: Roland Sundén
Commercial Overhauls: Ragnar Norbäck
Space Propulsion: Bjorn Sjoblom
Purchasing Department: G Bengtsson
■ Development, manufacture and overhaul of military and commercial aircraft engines. Space rocket propulsion systems and products. Heaters, transmission systems and diesel engines. → 80 [97]

75-11820 Volvo Aero Engine Services AB
(a subs. of Volvo Aero Corporation)
PO Box 218, S-161 26 Bromma, Stockholm
☎ (+46 8) 799 21 00 ⊖ (+46 8) 29 89 81 [97]

75-11830 Volvo Aero Turbines AB
(a subs. of Volvo Aero Corporation)
PO Box 9508, S-200 39 Malmö
☎ (+46 40) 31 16 00 ⊖ (+46 40) 22 48 58 [97]

75-11850 AB Volvo Penta
S-405 08 Gothenburg
☎ (+46 31) 23 54 60; 51 03 48 ⊠ 20755
⊖ (+46 31) 51 23 22
♣ SKr 3 878 000 000 (1995) ♟ 1 570
President & CEO: Jan Walldorf
Public Relations: A.C. Emegård
■ Marine & industrial engines, transmissions & accessories. [97]

75-11950 Wärtsilä NSD Sweden AB
(† Wärtsilä Diesel AB)
PO Box 920, S-461 29 Trollhättan
☎ (+46 520) 226 00 ⊠ 42141 diesel s
⊖ (+46 520) 228 50
Managing Director: Lars-Henrik Backas
Vice President, Marine Engines: Ms Kristina Furubacke
■ Medium speed diesel engines in the power range 1 110kw to 3 680kW. [98]

75-12100 J L Weibull AB
PO Box 43, S-232 02 Åkarp
☎ (+46 40) 46 50 80 Ⓔ jlw@jlw.se ⊖ (+46 40) 46 16 77
♟ 15
Managing Director: Johnny Petersson
■ Manufacture of naval simulators, driving simulators, mortar simulators, loading simulators, crew simulators, team trainers, tactical team trainers, video systems, sight systems, upgrading of AA guns. [98]

75-12150 Weinberger Svenska AB
(a subsidiary of Weinberger AG, Switzerland)
Enbärsvägen 6, S-76163 Norrtälje
☎ (+46 176) 167 00; 167 77 ⊖ (+46 176) 110 00
Managing Director: Lars Einemar
Representing:
Weinberger, Switzerland [98]

75-13000 AB Wibe
Head Office and Factory/Tower Systems
PO Box 401, S-792 27 Mora
Wibevägen 1
☎ (+46 250) 280 00 ⊖ (+46 250) 188 55
♣ SKr 652,960,000 ♟ 400
Dir Business Area Mast & Tower Syst. & Mktg Mgr: Mats Karlsson
Advertising Manager: Inger Dahlman
■ Manufacture of guyed masts up to 335m, self-supported towers up to 150m and telescopic masts up to an extended height of 30m for radio and telecommunication. [98]

75-15000 YDAB
(a subsidiary of CelsiusTech AB)
Varvsallén 13, S-871 45 Härnosänd
☎ (+46 611) 55 77 00 Ⓔ info@ydab.se
⊖ (+46 611) 55 77 19
Managing Director: Ingvar Johansson [98]

80 Service Companies & Organizations

80-00500 Enator Telub AB Head Office
(a subsidiary of Enator AB)
2 Lungadalsgatan, S-35180 Växjö
☎ (+46 470) 420 00 ⊖ (+46 470) 420 43
♣ SKr 240,000,000 ♟ 285
Managing Director: Sune Ekfedt
Business Information Manager: Lars Larsson
■ Consultancy, technical and maintenance services including system solutions for customers in the defence market and transport sector.

Enator Communications AB
Ljungadalsgatan 2, S-35180 Växjö
☎ (+46 470) 420 00 ⊖ (+46 470) 194 05
♣ SKr 315,000,000 ♟ 350
Managing Director: Göran Danielsson
Sales and Marketing Director: Bertil Ullergren
Business Information Manager: Lars Larsson
■ Consultancy, technical and maintenance services within the fields of telecommunications, data and radio communications, technical supervision, power supply, electrical and signal environment, IT safety and logistics.

Enator Driftentreprenader AB
Ljungadalsgatan 2, S-35180 Växjö
☎ (+46 470) 420 00 ⊖ (+46 470) 194 05
♣ SKr 100,000,000 (1995) ♟ 175
Managing Director: Gunnnar Eklund
Business Information Manager: Lars Larsson
■ Operation and maintenance in the fields of information processing and telecommunications. Testing of air and ground based missile systems. [98]

80-00600 FFV Aerotech AB Head Office
(a subsidiary of Celsius Industries)
S-732 81 Arboga
☎ (+46 589) 800 00 ⊠ 73246 ffva s
⊖ (+46 589) 61 16 52
♣ SKr 1,300,000 (1995) ♟ 1,800
President: Jan Eiborn
Vice President Marketing, Strategy and Public Relations: Ingemar Hansson
■ Aircraft maintenance and engineering services mainly for military customers. Major projects include participation in the development of the JAS-39 Gripen multirole combat aircraft (special-to-type equipment and test systems, complete maintenance programme). Swedish Defence Forces are the main customers. [97]

80-00620 FFV Aviocomp AB
(a subsidiary of FFV Aerotech AB)
PO Box 13100, S-580 13 Linköping
☎ (+46 13) 23 19 50 ⊖ (+46 13) 29 78 73
President: Jan-Ingvar Johansson
Vice President Marketing: Jan Rosmark
■ Offers complete maintenance programmes for components from regional airline operated aircraft such as Fokker F28, Fokker 50 and 100, Saab 340, Boeing 737-300/400/500 and ATR 42/72. [98]

80-00625 FFV Avionics
S-580 13 Linköping [97]

80-00650 FFV Mätteknik AB Measuring Technology
(a subs. of FFV Aerotech AB)
PO Box 1015, S-73226 Arboga
☎ (+46) 58 98 31 00 ⊖ (+46) 58 91 59 10
♣ SKr 70 000 000 ♟ 115
President: Gilbert Wahlén
■ Consultancy services related to measuring equipment & techniques. Maintenance of measuring equipment & computer hardware. Sole Swedish laboratory for recovery of information from destroyed Winchester-memories. [97]

80-00800 Heliflyg AB
Dala Airport, PO Box 8302, S-781 08 Borlänge
☎ (+46 243) 333 20 ⊠ 74033 ⊖ (+46 243) 396 20
♟ 65
Managing Director: Henry Hansen
Contact: Olav Ondahl
■ Forest fertilising, ambulance service, heavy lifts, inspections, tourist and incentive flights. [97]

80-01200 Logica Svenska AB
(a subsidiary of Logica plc, UK)
Solna Torg 19, S-171 45 Solna
☎ (+46 8) 705 77 00 ⊖ (+46 8) 730 38 55
Managing Director: Soren Jansson
Finance Director: Staffen Askholt
Marketing Director: Ms Monica Villot-Berling [98]

80-01280 Prof Sten Luthander Ingenjörsbyrå AB
LUTAB
Björnnäsvägen 12, S-11342 Stockholm
☎ (+46 8) 674 12 00 Ⓔ info@lutab.se
⊖ (+46 8) 674 12 01
♣ SKr 11,000,000 ♟ 20 (1997) 🏠 1,000 m2
Chairman: Göran Kaijser
Managing Director: Thor Malmberg ☎ (+46 8) 674 12 17
Manager New Projects: Claes Malmberg
■ Consultants, scientific & technical projects. [98]

80-01490 Nyge Aero AB
PO Box 321, S-611 27 Nyköping
☎ (+46 155) 771 00 Ⓔ carl-gustaf.werner@nygeaero.se
Ⓦ http://www.nygeaero.com ⊖ (+46 155) 26 72 52
♧ SKr 10,700,000 (1996) ♣ SKr 130,000,000 (1996)
♟ 110 🏠 6,800 sq m
Managing Director: Björn Sundén
Operations Director: Bo Söderling
Operations Director: Pekka Havbrandt

Contact Reference

Sales Manager: Carl-Gustaf Werner
Purchasing Manager: Roger Bladh
■ Technical & service centre, engine workshop. Aerial targets & qualified flight services. Target towing, target course flights, radar interference from aircraft, flight formation. Thermo-photography, radar calibration & measurement flight services. Sales services. → 85 → 90 [98]

80-01850 **Rollab Aktiebolag**

Järvstigen 5, PO Box 7073, S-171 07 Solna 7
☎ (+46 8) 755 66 99; 85 03 15 ✉ ROLLABOR
⊖ (+46 8) 755 66 99; 85 36 32
♧ SKr 100,000 (1995) ♣ SKr 3,000,000 (1995) 👥 6
🏠 300 m2
Managing Director: John Rosén
Operations and Technical Director: Lasse Olofsson
Technical Director: Knut Fristedt
■ Specialist consultants in the aeronautical and mechanical engineering fields. Rollab subsonic, supersonic and trisonic continuously variable mach number wind tunnels. Tailor-made ancillary equipment. Internal and external strain gauge balances. Calibration rigs and instrumentation. "Schlieren" optical system. [97]

80-01860 **ScanAvia, Scandinavian Aviation Consultants Sweden AB**

Baltzar von Platens Gata 9, S-112 42 Stockholm
☎ (+46 8) 653 01 75 📠 909 teleopr
⊖ (+46 8) 653 14 90
♧ SKr 180,000 ♣ SKr 2,000,000 👥 38
Chairman & Sr Partner: Björn Törnblom
Sr Partner: Sven-Erik Svanberg
Sr Partner: Gunnar Sandberg
Sr Consultant & Deputy Managing Dir: Gösta Almqvist
Sr Consultant: Karl-Erik Ternhem
Sr Consultant: Kjell Svensson
■ Consulting services to: airlines, airport operators, government agencies, ground equipment manufacturers, aircraft and avionics industry related organisations. All facets in analysing, programming, designing and mangement projects. [97]

80-01960 **SSPA Maritime Consulting AB**

PO Box 24001, S-400 22 Gothenburg
☎ (+46 31) 772 90 00 [E] postmaster@sspa.se
⊖ (+46 31) 772 91 24
♣ SKr 52,200,000 (1996) 👥 75 (1997)
Managing Dir: Hans Broberg
Sr Vice President: Willem B. van Berlekom
Information Officer: Li Fredrikson
■ Research & development directed towards the engineering of practical solutions to vital problems of maritime applications. Marine environment services. Risk analysis, F.S.A.. Customized systems design. [98]

80-02000 **Systecon AB**

(a subs. of Enator-Telub AB)
PO Box 5205, S-102 45 Stockholm
☎ (+46 8) 459 07 50 ⊖ (+46 8) 459 07 80
♣ SKr 23 000 000 (1994) 👥 25
Chairman: Sune Ekfelt
Managing Dir: Sven-Erik Westlund
Marketing Dir: Olof Waak
Military Activities: Bengt Thunstedt
Civil Aviation & Space Activities: Lennart Nyh
■ Systems & logistics engineering consulting in the areas of aerospace, military, civil & space. Operations research model development. Computer programs for analysis & optimisation of logistic support systems. Professional training. [97]

80-30000 **Volvo Aero Engine Services Arboga AB**

(a subsidiary of Volvo Aero Corporation)
PO Box 1005, S-732 26 Arboga
☎ (+46 589) 850 00 ⊖ (+46 589) 104 23
■ Engine maintenance for helicopters and commuter aircraft. [98]

85 Stockists, Distributors, Spare Parts Suppliers

Nyge Aero AB	80-01490

85-03920 **Conplan AB**

(† Conplan International AB)
Nämndemansvägen 6, S-178 37 Ekero
☎ (+46 8) 56 03 49 00 [E] conplan@bigfoot.com
[E] 101477.672@compuserve.com
⊖ (+46 8) 56 03 30 08
♧ SKr 100,000 ♣ SKr 5,000,000 👥 2
President: Bo Skough
■ Products & services within the areas of military, police, civil defence, rescue, safety & security equipment. Quartermaster & NBC-protection equipment, camouflage, decoys, laser warning equipment, HF-antenna systems. Spare parts & maintenance systems for military equipment of Swedish & foreign origin. → 80 → 90
® Belgium: NVSeyntex SA
Israel: Aryt/Amcoram Ltd, Biomed Instruments Ltd, Dafna Ltd, Gumsur Ltd, Orlite Ltd
Sweden: TWC Scandinavia AB [98]

90 Sales Agents, Sales & Liaison Offices

Companies represented:

A/S Hydrema, Denmark	90-00885
AEL Defence Corp, USA	90-00050
Alfred Karcher GmbH & Co, Germany	90-01122
AlliedSignal Aerospace Co, USA	90-00350
AP Labs, USA	90-01000
Armtech Industries, USA	90-00180
Aryt/Amcoram Ltd, Israel	85-03920
Auxilec, France	90-00350
Avalon, UK	90-00050
Avionic Instruments Inc, USA	90-01000
Ballard, USA	90-00050
Bancomm, USA	90-00050
Biomed Instruments Ltd, Israel	85-03920
British Steel Engineering Steels, UK	90-00360
Cartwright Electronics, USA	90-01000
Chomerics Inc, USA	90-01660
Communications & Power Industries Inc., USA	90-02200
Cubic Communications, USA	90-00350
Dafna Ltd, Israel	85-03920
Datum Inc, USA	90-00050
Dayton & Granger Inc, USA	90-00350
Detector Electronics Corp, USA	90-00430
Dupree, USA	90-01230
EEV Ltd, UK	90-01660
Efratom GmbH, Germany	90-00350
Electro-Magnetic Processes Inc (EMP), USA	90-01000
Electro Optical Industries, USA	90-01000
Enerdyne Technologies, USA	90-00050
English Electric Valve Co, UK	90-01660
Fela Mikrotechnik AG, Switzerland	90-01230
GEC Avionics, UK	90-00050
Gumsur Ltd, Israel	85-03920
Haigh-Farr, USA	90-00050
Hamilton Standard Division of, USA	90-01000
Hayes Int'l Targets, USA→ UK	80-07600
Herley-Vega Systems, USA	90-00050
Humphrey Inc, USA	90-01940
IDC Intercont Dynamics, USA	90-00180
ILC Data Device Corp., USA	90-00400
Interstate Electronics Corp, USA	90-00050
Jaivision A/S, Denmark	90-01000
Japan Aviation Electronics, Japan	90-01660
Jet Electronics, USA	90-00350
KGS Electronics Inc, USA	90-01000
KMS Advanced Products, USA	90-00050
Kollsman, USA	90-00180
Kollsman Inc, USA	90-00180
Leach International, USA	90-01230
Leach International (GEPE, LMF, LRE), France	90-01230
LEM, France	90-00350
Link Flight Simulation, USA	75-10300
Litton Applied Technology Div, USA	90-01000
Litton-Clifton Precision, USA	90-00350
Litton Industries Inc, USA	90-01240
LMB, France	90-01230
Loral Microcom, USA	90-00050
Loral Test and Information Systems, USA	90-00050
Lucas Deeco, USA	90-00050
Lucas Deeco Corp, USA	90-00050
Lynwood, UK	90-00050
Marconi, UK	90-01660
Marconi Italiana, Italy	90-01660
Marlow Industries, USA	90-01000
Mekel Engineering, USA	90-01000
Microdyne, USA	90-00050
Microdyne Corp, USA	90-00050
Moog Inc, USA	90-01300
Mountain Optec, USA	90-00050
Narda Microwave Corp, USA	90-00050
Negretti Aviation, UK	90-00180
North Atlantic Ind, USA	90-00050
NVSeyntex SA, Belgium	85-03920
Orlite Ltd, Israel	85-03920
OSPL, UK	90-00050
Oxley Developments Co Ltd, UK	90-01660
Park Air Electronics Ltd, UK	90-00500
Penny + Gilles, UK	90-00050
Plath C. GmbH, Germany	90-00350
Portescap, Switzerland	90-01630
Prestolite, UK	90-01230
Prism Instruments Ltd, UK	90-01660
Radiall, France	90-01810
Raytheon Aircraft Company, USA	90-01000
Saunders-Roe Ltd, UK	90-00180
Shaw Aero, USA	90-00180
Siemens, Germany	75-10300
Singer Kearfott, USA	75-10300
Southwest Aerospace, USA	90-01000
TEAC Corp, Japan	90-01000
Thomson-CSF, France	90-02050
Transaero Inc, USA	90-02075
Trivec Avent Antenna Systems, USA	90-00350
Trompeter, USA	90-00050
TT Group Products, UK	90-01913
TWC Scandinavia AB, Sweden	85-03920
West Electronics, USA	90-01000

Sales agents/representatives listed elsewhere:

Conplan AB	85-03920
New Pac Safety AB	75-08260
Nyge Aero AB	80-01490
Saft Nife AB	75-09850

90-00050 **Acal Auriema AB**

Korta Gatan 9, S-171 54 Solna
☎ (+46 8) 764 54 00 [E] instrument.sales@acal.se
⊖ (+46 8) 764 64 40
♧ SKr 150,100 (1997) ♣ SKr 45,000,000 (1997)
👥 17 (1997) 🏠 610 sq m
Managing Director: Hans Ove Pettersson
Sales Manager: Ake Hansson
■ Telemetry, timing, ELINT, antennas, encryption, radar transponders, 1553 test equipment, Tempest, rugged computers.
® UK: Avalon, GEC Avionics, Lynwood, OSPL, Penny + Gilles
USA: AEL Defence Corp, Ballard, Bancomm, Datum Inc, Enerdyne Technologies, Haigh-Farr, Herley-Vega Systems, Interstate Electronics Corp, KMS Advanced Products, Loral Microcom, Loral Test and Information Systems, Lucas Deeco, Lucas Deeco Corp, Microdyne, Microdyne Corp, Mountain Optec, Narda Microwave Corp, North Atlantic Ind, Trompeter [98]

90-00180 **Avicraft AB**

Björnnasvagen 12, S-113 47 Stockholm
☎ (+46 8) 612 52 01 [E] clma@lutab.se
⊖ (+46 8) 674 12 01; 731 84 03
👥 4
President: Claes Malmberg
Director: Claes Wrenninge
® UK: Negretti Aviation, Saunders-Roe Ltd
USA: Armtech Industries, IDC Intercont Dynamics, Kollsman, Kollsman Inc, Shaw Aero [98]

90-00350 **Benavia AB**

Lundagatan 13, PO Box 1122, S-171 22 Solna
☎ (+46 8) 735 04 50 ⊖ (+46 8) 734 03 03

♧ SKr 300,000 ♣ SKr 7,000,000 ጰ 4 ⌂ 250 sq m
Managing Dir: Bengt Olsson
Project Manager: Ms Margareta Bönström
■ Agents and representatives for avionics, electronics, optronics. Aircraft and defence equipment manufacturers. → 75
® France: Auxilec, LEM
Germany: Efratom GmbH, Plath C. GmbH
USA: AlliedSignal Aerospace Co, Cubic Communications, Dayton & Granger Inc, Jet Electronics, Litton-Clifton Precision, Trivec Avent Antenna Systems [98]

90-00360 British Steel Svenska AB
(a subsidiary of British Steel Engineering Steels, UK)
Sjoporten 4, S-417 64 Gothenburg
☎ (+46 31) 779 32 00
⊖ (+46 31) 779 32 27; 779 32 38
® UK: British Steel Engineering Steels [98]

90-00400 DDC Elektronik AB
(a subsidiary of ILC Data Device Corp., USA)
PO Box 18, S-191 21 Sollentuna
☎ (+46 8) 92 06 35 ⊖ (+46 8) 35 31 81
Managing Director: Goran Astrand
Marketing Communications Manager: Lawrence M Margulies
■ Micro electronic hybrids. Test equipment in data bus (Mil-Std-1553), data conversion (synchro, A/D) & power hybrid areas.
® USA: ILC Data Device Corp. [98]

90-00430 Detronics Scandinavia AB
(† Detector Electronics Scandinavia)
PO Box 81, S-260 83 Vejbystrand
☎ (+46 431) 530 02; 532 40 ⊠ 72008
⊖ (+46 431) 522 36
Contact: Sven Frejvall
® USA: Detector Electronics Corp [98]

90-00500 Dynamics Consulting AB
(† British Aerospace (Sweden) AB)
Propellervägen 8b, S-183 62 Täby
☎ (+46 8) 630 11 35 ⊖ (+46 8) 630 11 45
Managing Dir: Ulf Berg
Marketing Exec.: Jan Ahlén
■ Commercial advisers for British Aerospace Defence Ltd.
® UK: Park Air Electronics Ltd [97]

90-00885 Hydrema AB
Hjalmar Petris Väg 44, S-35246 Växjö
☎ (+46 47) 02 00 70 ⊖ (+46 47) 02 83 26
General Mgr: Leif Fasth
® Denmark: A/S Hydrema [97]

90-01000 Kaitech Trade AB
(† Aerotech Trade AB)
(a subsidiary of Kaitec AB, Sweden)
Ekovägen 12, S-172 37 Sundbyberg
☎ (+46 8) 764 54 55 [E] kaitech@mbox300.swipnet.se
⊖ (+46 8) 98 81 30
♧ SKr 300,000 (1997) ♣ SKr 6,000,000 (1996)
ጰ 4 (1997) ⌂ 75 sq m
Managing Director: Göran Kaijser ☎ (+46 8) 28 94 06
Sales and Marketing: Roland Ericsson ☎ (+46 8) 28 94 07
■ Customer oriented marketing and agencies of aviation and military systems and products. Marketing of these products to other customers when appropriate. Modification of products to improve competitiveness when possible.
® Denmark: Jaivision A/S
Japan: TEAC Corp
USA: AP Labs, Avionic Instruments Inc, Cartwright Electronics, Electro Optical Industries, Electro-Magnetic Processes Inc (EMP), Hamilton Standard Division of, KGS Electronics Inc, Litton Applied Technology Div, Marlow Industries, Mekel Engineering, Raytheon Aircraft Company, Southwest Aerospace, West Electronics [98]

90-01122 Karcher AB
PO Box 24, S-42502 Hisings-Karra
☎ (+46 31) 57 03 80 ⊖ (+46 31) 57 51 49
Managing Director: H Soren Hornell
® Germany: Alfred Karcher GmbH & Co [98]

90-01230 Linreco AB
(† LRE Relä + Elektronik AB)
PO Box 2137, S-176 02 Järfälla
☎ (+46 8) 58 35 31 25 [E] linreco@linreco.se
⊖ (+46 8) 58 35 27 25
♧ SKr 100,000 (1996) ♣ SKr 20,000,000 (1996) ጰ 3
Managing Director: Bengt Lindberg
® France: LMB, Leach International (GEPE, LMF, LRE)
Switzerland: Fela Mikrotechnik AG
UK: Prestolite
USA: Dupree, Leach International [98]

90-01240 Litton Precision Products International Inc
Heliosvägen 1c, PO Box 92146, S-120 08 Stockholm
☎ (+46 8) 743 05 85 ⊠ 17922 litton s
⊖ (+46 8) 643 52 20
Managing Director: Anders Ollén
® USA: Litton Industries Inc [98]

90-01300 Moog Norden AB
(a subsidiary of Moog Inc, USA)
Datavagen 18, S-436 32 Askim
☎ (+46 31) 68 00 60 ⊠ 211776 moog s
⊖ (+46 31) 28 75 32
♣ US$ 7,000,000 (1997) ጰ 10 (1997) ⌂ 500 sq m
Managing Director: Martyn Waddington
Sales and Marketing Director: Klas Lindberg
® USA: Moog Inc [98]

90-01630 Portescap Scandinavia AB
Grev Turagatan 20, S-11446 Stockholm
☎ (+46 8) 661 05 05 ⊖ (+46 8) 661 40 12
ጰ 3
Managing Dir: Per Svanberg
® Switzerland: Portescap [97]

90-01660 Pronest AB
Finlandsgaten 18, S-16475 Kista
☎ (+46 8) 632 65 70 ⊖ (+46 8) 751 41 11
♣ SKr 78 000 000 ጰ 21
Chief Exec.: Hans Edlund
■ Marketing of high technology electronics, advanced electronic components and instruments/equipment on the Nordic market.
® Italy: Marconi Italiana
Japan: Japan Aviation Electronics
UK: EEV Ltd, English Electric Valve Co, Marconi, Oxley Developments Co Ltd, Prism Instruments Ltd
USA: Chomerics Inc [98]

90-01810 Radiall AB
(a subsidiary of Radiall, France)
PO Box 6061, S-19106 Sollentuna
☎ (+46 8) 754 73 15 ⊖ (+46 8) 754 49 16
Managing Director: B Stenerfelt
® France: Radiall [98]

90-01913 Salchow Elektronik AB
(a subsidiary of TT Group plc, UK)
PO Box 190, S-16212 Vallingby, Stockholm
☎ (+46 8) 38 01 20 ⊖ (+46 8) 739 97 84
Managing Dir: Lars O. Jaensson
■ Electronic components and products. Sales and distribution.
® UK: TT Group Products
Divisions:
Strato Connectors AB
Fiber Optic Connectors [98]

90-01940 Scandinavian Aerospace Industry
PO Box 5067, S-123 05 Farsta
☎ (+46 8) 93 08 10 ⊖ (+46 8) 94 33 87
Contact: Hans Sandstrom
® USA: Humphrey Inc [97]

90-02050 Thomson-CSF Delegation for Northern Europe
(a division of Thomson-CSF, France)
Grev Turegatan 73, S-114 38 Stockholm
☎ (+46 8) 660 35 39 ⊖ (+46 8) 662 23 71
Delegate for Northern Europe: Mark Knight
® France: Thomson-CSF [98]

90-02075 Transaero Sweden
Granitvagen 4, S-732 00 Arboga
☎ (+46 589) 61 11 12; 111 12
⊖ (+46 589) 178 28; 178 28
Regional Mgr: Stig Eriksson
® USA: Transaero Inc [97]

90-02200 CPI Europe Ltd
(† Varian AB)
(a subs. of Communications & Power Industries Inc., USA)
Englundavagen 9, PO Box 1099, S-171 22 Solna
☎ (+46 8) 445 19 90 ⊖ (+46 8) 82 76 81
Managing Dir: S. Myrberger
® USA: Communications & Power Industries Inc. [97]

90-04000 B Zinn Trading AB
Modeshögsvägen 68, S-231 92 Trelleborg
☎ (+46) 41 03 51 66 ⊖ (+46) 41 03 51 70
ጰ 3
President: Börje Zinn [98]

Switzerland

Head of State:
Arnold Koller

10 Head of State

10-10000 Federal Council
Bundesrat
Office of the President of the Swiss Confederation
Bundeshaus West, CH-3003 Bern
☎ (+41 31) 61 21 11 ⊠ 911340 helv ch
Federal President: Arnold Koller
Vice President: Flavio Cotti
Member of the Federal Council: Ms Ruth Dreifuss
Member of the Federal Council: Adolf Ogi [98]

12 Ministries

12-10000 Federal Military Department
Eidgenössisches Militärdepartement
Bundeshaus Ost, CH-3003 Bern
☎ (+41 31) 32 22 11; 67 12 11 ⊖ (+41 31) 67 51 04
Chief of Military Department: Adolf Ogi [98]

12-20000 Federal Department of Foreign Affairs
Eidgenössisches Departement für Auswärtige Angelegenheiten (EDA)
Département fédéral des affaires étrangères
Bundeshaus West, CH-3003 Berne
☎ (+41 31) 322 21 11 ⊠ 911440 eda ch
⊖ (+41 31) 324 90 47
Head of Department: Flavio Cotti
Chief of Political Affairs: Jakob Kellenberger
Advisor for European Space Co-operation: Dr Peter Creola
Head of Information and Press: Franz Egle
Secretary of Press Office: Ms Manuela Schafer [98]

12-40000 Federal Department of the Interior
Eidgenossisches Departement des Innern
CH-3003 Bern
☎ (+41 31) 61 91 11 ⊠ 912890 edi ch
⊖ (+41 31) 61 80 32
Chief of Department of the Interior: Ms Ruth Dreifuss [98]

Switzerland

12-50000 Federal Department of Justice and Police
Bundeshaus West, CH-3003 Bern
☎(+41 31) 322 41 11 ⊖ (+41 31) 322 78 32
Chief of Justice and Police Department: Arnold Koller [98]

20 Defense Administrations/Services/Agencies

20-10100 Central Office for General Defence
Zentralstelle für Gesamtverteidigung
CH-3003 Bern
☎(+41 31) 324 40 02; 324 40 15; 324 40 07
⊖ (+41 31) 324 40 44
Director: Roland R Favre ☎(+41 31) 324 40 01
Information Officer: Dr Felix Christ ☎(+41 31) 324 40 15 [98]

25 Defense Forces

25-10000 General Staff Headquarters
Gruppe für Generalstabsdienste
Bundeshaus Ost, PO Box 52, CH-3003 Bern
☎(+41 31) 67 52 76; 67 53 72 ⊖ (+41 31) 67 64 26
Chief of Staff and Head of Battle Front Group: Div Carlo Vinzenz ☎(+41 31) 67 51 92
Chief of Staff and Head of Logistics Group: Div Markus Rusch ☎(+41 31) 67 52 91; 67 51 91
Chief of Staff and Head of Planning Group: Div Paul Múller ☎(+41 31) 67 53 56; 67 51 93
Chief of Material Division: Col Andreas Búrgi ☎(+41 31) 67 53 20
Chief of Armament and Equipment: Alfred Zedi ☎(+41 31) 67 53 20; 67 52 81
Chief of Electronics: Urs Waber ☎(+41 31) 67 53 20; 67 52 51
Chief of Air War Resources: Hans Rudolf Gutjahr ☎(+41 31) 67 53 20; 67 52 30
Chief of Ammunition and Explosives: H Habegger
Head Libarian: Mrs Doris Antener

Federal Office of Military Engineering and Fortifications
Rodtmattstrasse 110, CH-3003 Bern
☎(+41 31) 67 31 25 ▣ 912939 bagf ch
Director: Ulrich Jeanloz ☎(+41 31) 67 30 20
Deputy Director: Dipl.-Ing. Andreas Lamparter ☎(+41 31) 67 30 21
Chief of Engineering Division: Col Peter Schaeublin ☎(+41 31) 67 30 07; 67 30 25
Chief of Fortifications Division: Col Ulrich Schletti ☎(+41 31) 67 31 39; 67 30 23
Chief of Information Services: Martin Aeberhard
Chief of Construction Planning Division: Júrg Trick

Federal Office of Signal Troops
Papiermühlestrasse 14, CH-3014 Bern
☎(+41 31) 67 36 28 ▣ 912671 uemch
Director: Div Hansruedi Fehrlin ☎(+41 31) 67 35 01
Deputy Director: Col Hanspeter Alioth ☎(+41 31) 67 35 07
Chief of Training: Maj Hansjürg Nussbaumer

Federal Office for Army Medical Services
Kasernenstrasse 7, CH-3000 Bern 22
☎(+41 31) 324 27 57 ⊖ (+41 31) 324 27 17
Director and Chief Surgeon: Maj Gen Peter Eichenberger (MD)
Assistant Director: Col Jacques-Henri Piguet
Assistant Director: Col GS Roland Wyler (MD)

Federal Division of Military Veterinary Services (AMVET)
Blumenbergstrasse 39, CH-3003 Bern
☎(+41 31) 67 34 46; 67 34 43
Chief Veterinary Surgeon: Brig Hermann Keller ☎(+41 31) 67 34 40
Assistant Director: Dr Jürg Eberle ☎(+41 31) 67 34 41
Head of Administration: Hanns Stauffer

Central Defence Commissariat
Oberkriegskommissariat
Wylerstrasse 52, CH-3003 Bern
☎(+41 31) 67 42 52 ▣ 32451 okkch
Director: Brig Even Gollut ☎(+41 31) 67 42 51
Chief of Commissariat Division: Stephan Jeitziner ☎(+41 31) 67 42 66
Chief of Ammunition Division: Roland Burri ☎(+41 31) 67 42 68

Federal Office of Transport Troops (OFTT)
Blumenbergstrasse 39, CH-3003 Bern
☎(+41 31) 324 29 88 ⊖ (+41 31) 324 49 15
Chief of Finance and Controlling: Lieut Col Mark Rufer ☎(+41 31) 324 28 66

Federal Office of Air Defence Troops
Wylerstrasse 52, CH-3003 Bern
☎(+41 31) 67 33 37 ▣ 912970 alstch
Director: Brig Peter Bieder ☎(+41 31) 67 33 12
Deputy Director: Brig Jean Langenberger ☎(+41 31) 67 33 18
Head of Administration: André Wiedmer

Commissariat for Defence Materiel
Viktoriastrasse 85, CH-3003 Bern
☎(+41 31) 67 20 30; 67 20 93 ▣ 911390 kmvch
Director: Brig Peter Mülheim ☎(+41 31) 67 20 10
Assistant Director: Walter Spillmann ☎(+41 31) 67 20 30 [97]

25-29100 Swiss Air Force Aeromedical Institute
CH-8600 Dübendorf
☎(+41 1) 823 25 19 ⊖ (+41 1) 823 25 20
Director: Dr Jost Suter ☎(+41 1) 823 25 24
Chief of Air Medicine Section: Dr Kurt Hauser ☎(+41 1) 823 25 32
Chief of Medicine Section: Dr H Schnüriger ☎(+41 1) 823 25 25
Chief of Psychological Section: Dr A Salzberg
→ 60 [98]

26 Police (Civil)

26-10000 Swiss Federal Police
(† Federal Police Headquarters)
Taubenstr. 16, CH-3003 Bern
☎(+41 31) 322 45 11 ⊖ (+41 31) 322 98 76
Head of Swiss Federal Police: Urs von Daeniken [97]

29 Customs & Excise

29-10000 Federal Directorate of Customs
Direction Générale des Douanes
Monbijoustrasse 40, CH-3003 Bern
☎(+41 31) 322 65 11; 322 66 07
⊖ (+41 31) 322 78 72
Director General: Rudolf Dietrich ☎(+41 31) 322 65 01
Chief of Staff Division: Samuel Moser ☎(+41 31) 322 65 02
Chief of Frontier Guards Corps Division: Hanspeter Wüthrich ☎(+41 31) 322 65 35
Chief of Operations Division: Reto Coray ☎(+41 31) 322 65 72
Chief of Law and Collection of Duties Division: Christoph Häni ☎(+41 31) 322 65 03
Chief of Customs Tariffs and Foreign Trade Statistics: Roman Bisaz ☎(+41 31) 322 65 09
Chief of Personnel Division: Erwin Mock ☎(+41 31) 322 67 27
Principal Assistant to the Director: Ms M A Bosson [98]

31 Embassies & Attachés

31-00085 Permanant Mission of the Republic of Armenia to the United Nations Office
64 Route de Troinex, CH-1256 Troinex, Geneva
☎(+41 22) 300 45 00 ⊖ (+41 22) 300 45 15
General Consul: Karen Nazarian
Assistant to the Representative: Mr Lazavian [98]

31-00100 Embassy of Austria
Kirchenfeldstrasse 28, CH-3005 Bern
☎(+41 31) 351 01 11-15 ✉ AUSTROAMB Bern
⊖ (+41 31) 351 56 64
Ambassador: H.E. Mr Markus Lutterotti
Minister & Deputy Chief of Mission: Anton Kozusnik
Defence Attaché: Brig. Rüdiger Stillfried ☎(+41 31) 352 24 35
Counsellor (Press and Cultural Affairs): Otto Schüngel [98]

31-00150 Embassy of Belgium
41 Jubiläummstrasse, CH-3005 Bern
☎(+41 31) 351 04 62; 351 04 65
✉ AMBASSADE BELGIQUE Berne ▣ 912896
⊖ (+41 31) 352 59 61
Ambassador: H E Mr Hugo Callebaut
Head of Chancery: Adriaan Catteeuw
Economic and Commercial Affairs Attaché: Ms Christine Vloeberghs
Counsellor: Jana Zikmundova
Military and Air Attaché: Col Willy Etienne Cornellis Matte (resident in Vienna)
Agriculture Counsellor: Pierre Lambotte (resident in Paris) [97]

31-00230 Embassy of Brazil
PO Box 30, CH-3000 Bern 23
Monbijoustrasse 68, CH-3007 Bern
☎(+41 31) 371 85 15 ⊖ (+41 31) 371 05 25
Ambassador: H E Mr C E de Affonseca Alves de Souza
Minister and Counsellor: Paulo Fernando Telles Ribeiro
Counsellor: Sérgio Luis Gomes
Second Secretary: Aurélio Afrânio Garcia Avelino [97]

31-00250 Embassy of the Republic of Bulgaria
Bernastrasse 2, CH-3005 Bern
☎(+41 31) 351 14 55; 351 14 56 ▣ 911786 bubech
⊖ (+41 31) 351 00 64
Ambassador: H E Dr jur Elena Kirtcheva
Counsellor: Roussi Ivanov ☎(+41 31) 352 06 75
Defence Attaché: Col Michail Michailov (resident in Bonn) ☎(+49 228) 36 30 61
Head of the Consular Section: Radi Naydenov ☎(+41 31) 351 13 67
Commercial Attaché: Silvester Kovatchev ☎(+41 31) 351 13 66
Cultural Attaché: Dr Bogdan Mirtchev ☎(+41 31) 351 74 23 [97]

31-00360 Embassy of the People's Republic of China
Kalcheggweg 10, CH-3006 Bern
☎(+41 31) 352 73 33; 352 73 34 ▣ 32203; 33564
⊖ (+41 31) 351 45 73
Ambassador: H E Mr Zhou Zizhong
Chancellor: Tang Zhonglin
Military and Air Attaché: Lin Mingole

Office of the Commercial Counsellor
J V Widmannstrasse 7, CH-3074 Muri
☎(+41 31) 52 14 01; 52 14 03 ▣ 32265 [98]

31-00380 Embassy of Colombia
Dufourstrasse 47, CH-3005 Bern
☎(+41 31) 351 17 00; 351 17 10
⊖ (+41 31) 352 70 72
Ambassador: H E Mr Manuel Jose Cepeda
Deputy Chief of Mission: Ms Anamaria Prieto
Second Secretary: Ricardo Fernandez [97]

31-00450 Embassy of the Czech Republic
Muristrasse 53, CH-3000 Bern 16
☎(+41 31) 352 36 45; 352 11 34
⊖ (+41 31) 352 75 02
Ambassador: H E Mr Richard Belcredi
Deputy Chief of Mission: Milan Kovac ☎(+41 31) 352 39 25
Defence Attaché: Col Josef Monik ☎(+41 1) 894 21 25
Attaché, Defence Procurement: Lieut Col Jiri Horak ☎(+41 1) 894 21 25
Commercial Officer: Ivo Petrík

Office of the Economic Counsellor
Muristrasse 53, CH-3000 Bern 16
☎(+41 31) 352 38 40 ⊖ (+41 31) 352 75 02
Economic Counsellor: Ivo Petrik
Consul: Roman Prosa
Cultural Attaché: Ms Alice Mzyková [98]

31-00500 Embassy of Ecuador
19a Helvetiastrasse, CH-3005 Bern
☎(+41 31) 351 17 55 ⊖ (+41 31) 351 27 71
Ambassador: H E Mr D Cordovez ☎(+41 31) 351 62 54
Counsellor: F Meneses [97]

31-00535 Embassy of Estonia
8 Chemin des Aulx, CH-1228 Plan-les-Ouates, Geneva
☎ (+41 22) 706 11 11 ⊠ 429422 lem ch
⊖ (+41 22) 794 94 78
Honorary Consul: Enn Vallak 98

31-00580 Embassy of Finland
PO Box 11, CH-3000 Bern 15
Weltpoststrasse 4, CH-3015 Bern
☎ (+41 31) 43 30 31; 43 30 32; 43 30 33
⊠ 912023 amfi ch ⊖ (+41 31) 43 30 01
Ambassador: H E Mr Olli Mennander
Deputy Chief of Mission: Nyyrikki Kurkivuori
Military and Air Attaché: Col Finn-Göran Wennström
Secretary: Mr Bärtschi
Office of the Commercial Secretary
Dufourstrasse 24, PO Box 1724, CH-8032 Zurich
☎ (+41 1) 383 41 71 ⊖ (+41 1) 383 44 84
Consul, Commercial: Pekka O Kauranen 98

31-00810 Embassy of the Republic of Hungary
Muristrasse 31, CH-3006 Bern
☎ (+41 31) 352 85 72; 352 85 73; 352 86 03
⊖ (+41 31) 351 20 01
Ambassador: H E Mr Pál Gresznáryk
First Counsellor: Csaba Mohi
Counsellor, Consular Affairs: Jozsef Hercegfalvi
Defence Attaché: Lieut Col Antal Seres
Third Secretary: Károly Sárdi
Office of the Commercial Counsellor
Sulgeneckstrasse 64, CH-3005 Bern
☎ (+41 31) 371 45 96; 371 45 97; 371 88 95
⊖ (+41 31) 371 77 28
Counsellor, Commercial Affairs: Vilmos Kozàry 98

31-00840 Embassy of the Republic of Indonesia
Elfenauweg 51, CH-3006 Bern
☎ (+41 31) 352 09 83; 352 09 85 ⊠ 911525 indon ch
⊖ (+41 31) 352 67 65
Ambassador: H E Mr Machmud Subarkah
Minister and Counsellor: Resi Gunarto Prasodjo
Counsellor: Saut Maruli Siregar 98

31-00850 Embassy of the Islamic Republic of Iran
Thunstrasse 68, CH-3006 Bern
☎ (+41 31) 351 08 01; 351 08 02 ⊠ 912779
⊖ (+41 31) 351 56 52
Ambassador: H E Mr Mohammad R Alborzi
Military Attaché: Col Naeimi Jamshid
☎ (+41 31) 352 73 16
Commercial Attaché: Morteza Khonssari 98

31-00890 Embassy of Italy
Elfenstrasse 14, CH-3006 Bern
☎ (+41 31) 44 41 51 ⊠ 912615
Ambassador: H E Mr Arduino Fornara
Deputy Chief of Mission: Aldo Mantovani
Office of the Military and Air Attaché
Elfenstrasse 3, CH-3000 Bern 16
☎ (+41 31) 352 35 08 ⊠ 912878 itmich
⊖ (+41 31) 352 41 81
Military and Air Attaché: Gen Carlo Libriz
Office of the Economic and Commercial Counsellor
Elfenstrasse 14, CH-3000 Bern 14
☎ (+41 31) 352 11 61; 352 31 34
Economic and Commercial Counsellor:
Carlo Albeerto Barachi Tua 97

31-00920 Embassy of Japan
Ambassade du Japon
PO Box 51, CH-3000 Bern 26
Engestrasse 43, CH-3012 Bern
☎ (+41 31) 300 22 22 ⊠ 912653 ⊖ (+41 31) 300 22 55
Ambassador: H E Mr Mitsuhei Murata
Minister and Deputy Chief of Mission: Haruo Okamoto
Counsellor: Ryozo Kanno
First Secretary: Shinji Denishi
First Secretary: Takashi Goto
Second Secretary, Press and Cultural Affairs:
Kazuhiro Hashimoto
Second Secretary: Masayoshi Hisano
Second Secretary: Masaki Sugamiya
Second Secretary: Yoshiyuki Kurashima
Attaché: Hajime Furukawa
Permanent Mission of Japan to the United Nations Office
Mission Permanente du Japon auprès de l'Office des Nations Unies et des autres Organisations Internationales
3 Chemin des Fins, CH-1218 Grand-Saconnex, Geneva
☎ (+41 22) 717 31 11 ⊖ (+41 22) 788 38 11
Consulate General and Information Service
16 Place Longemalle, CH-1204 Geneva
☎ (+41 22) 311 46 88 ⊖ (+41 22) 311 82 03 98

31-01020 Embassy of Lebanon
Thunstrasse 10, CH-3074 Muri
☎ (+41 31) 951 29 71; 951 29 73 ⊠ 912974 liba ch
⊖ (+41 31) 951 81 19
Deputy Chief of Mission: Samira El-Daher 98

31-01110 Permanent Mission of Malasia
ICC Block H, 20 route de Pré-Bois, CH-1215 Geneva 15
☎ (+41 22) 788 15 05; 788 15 09
E mission.malasia@itu.ch ⊠ 415650 mala ch
⊖ (+41 22) 788 04 92
Third Secretary: Mohd Kamil Ab Ghani 98

31-01190 Embassy of Mexico
Bernastrasse 57, CH-3005 Bern
☎ (+41 31) 351 18 75
E 100753.151@compuserve.com
⊖ (+41 31) 351 34 92
Ambassador: H E Mr Ezequiel Padilla
Deputy Chief of Mission: Sergio Gomar Rocha
Attaché, Commercial and Economic:
Carlos Quesnel Melendez 98

31-01220 Embassy of the Kingdom of Morocco
Helvetiastrasse 42, CH-3005 Bern
☎ (+41 31) 351 03 62; 351 03 63; 351 26 73
E sifamaberne@pingnet.ch ⊠ 912997
⊖ (+41 31) 351 03 64
Ambassador: H E Mr Tahar Nejjar
Premier Counsellor, Economic Affairs:
Abdelouahad Hassouni
First Secretary, Consular Affairs: Said Naitbachir
First Secretary, Political Affairs: Mohamed Mouadi
Second Secretary, Cultural Affairs: Ms Laila Meziani
Attaché, Consular Affairs: Belkheir El Ouardi
Attaché, Administrative Affairs:
Mohammed Ait El Ouidane 98

31-01270 Royal Netherlands Embassy
PO Box, CH-3000 Bern 6
Kollerweg 11, CH-3006 Bern
☎ (+41 31) 352 70 63 ✉ HOLLANDIA BERN
⊠ 32656 nlambch ⊖ (+41 31) 352 87 35
Ambassador: H E Mr Willem Sinninghe Damsté
Deputy Chief of Mission: Saskia N Bakker
Defence Attaché: Col Wietse F Bijlsma 97

31-01330 Embassy of the Federal Republic of Nigeria
Zieglerstrasse 45, CH-3007 Bern
☎ (+41 31) 382 07 26; 382 07 27 ⊠ 912995 ngbech
⊖ (+41 31) 382 16 02
Ambassador: H E Mr Olayemi Mudazir Adisa Abiola
First Counsellor: Joseph C Ezema
Counsellor: Emmanuel Kayode Ojo 98

31-01420 Philippine Embassy
Brunnmattstrasse 24, CH-3007 Bern
☎ (+41 31) 398 22 66 ✉ PHILEMBASSY Bern
⊖ (+41 31) 382 40 47
Ambassador: H E Mr Tomas T Syquia
Minister and Deputy Chief of Mission:
Mrs Phoebe A Gomez 97

31-01430 Embassy of the Republic of Poland
Elfenstrasse 20a, CH-3006 Bern
☎ (+41 31) 352 04 52 ⊠ 912841 plch
⊖ (+41 31) 352 35 16
Ambassador: H E Mr Marek Jedrys
Deputy Chief of Mission: Mirostaw Wojtkowski
Counselllor: Jan Zielinski
Counsellor: Marek Zielinski
First Secretary: Edward Bialkowski
Defence, Military and Air Attaché: Col Maciej Bekier (resident in Vienna)
First Secretary: Dariusz Dobrowolski
First Secretary: Wojciech Jakubowski
First Secretary: Dariusz Torun
Third Secretary: Mieczyslaw Pyzel
Attaché, Financial Affairs: Wojcieck Jakubowski
Press Officer: Edward Biatkowski
Office of the Commercial Counsellor
Elfenstrasse 9, CH-3006 Bern
☎ (+41 31) 43 19 35; 43 19 36 ⊠ 32189
⊖ (+41 31) 43 34 57
Commercial Counsellor: Adam Leszczynski
Commercial Attaché: Wtodzimierz Kotowiecki
Commercial Attaché: Bronislaw Jaworski 98

31-01440 Embassy of Portugal
PO Box 5224, CH-3001 Bern
Jungfraustrasse 1, CH-3005 Bern
☎ (+41 31) 352 83 29 ⊠ 912837 ⊖ (+41 31) 351 44 32
Ambassador: H E Mr Francisco Grainha do Vale
☎ (+41 31) 352 86 02
Deputy Chief of Mission: Mrs Filipa Ponces 97

31-01655 Embassy of the Republic of Slovenia
Schwanengasse 9, CH-3011 Bern
☎ (+41 22) 312 44 18 ⊖ (+41 22) 312 44 14
Chargé d'Affaires: Zeljko Jeglic 98

31-01670 Embassy of South Africa
Jungfraustrasse 1, CH-3005 Bern
☎ (+41 31) 352 20 11 ✉ SALEG Bern
⊖ (+41 34) 352 11 16
Ambassador: H E Mr R S Mompati
Deputy Chief of Mission: E L Mienie
Military and Air Attaché: Col J P Nel
Commercial Officer: D Beznidenhout 98

31-01730 Embassy of Sweden
Bundesgasse 26, CH-3001 Bern
☎ (+41 31) 312 05 63 ✉ SVENSK BERN
⊠ 911770 svskch ⊖ (+41 31) 312 16 92
Ambassador: H E Mr Folkehöfgren
Counsellor: Gunnar Bloom
Military and Air Attaché: Col Claes-Göran Blomkvist
First Secretary: Mrs Kerstin Olsson 97

31-01890 Embassy of the United Kingdom
Thunstrasse 50, CH-3005 Bern
☎ (+41 31) 352 50 21 ⊠ 911929 ukbe ch
⊖ (+41 31) 352 05 83
Ambassador: H E Mr Christopher Hulse (CMG, OBE)
Deputy Head of Mission: I Knight Smith
Counsellor, Science and Technology: D E Lyscom (resident in Bonn)
First Secretary, Politics: C P Greenwood
Defence Attaché: Lieut Col The Lord Crofton
Commercial Attaché: B Hässig 98

31-01900 Embassy of the United States of America
Jubilaeumstrasse 93, CH-3005 Bern
☎ (+41 31) 357 70 11 ⊖ (+41 31) 357 73 44
Ambassador: H E Madeleine M Kunin
Deputy Chief of Mission: Carey Cavanaugh
Head of Commercial Section: James Joy
Defence Attaché: Col Robert McBride (US Army)
Public Affairs Officer: Ms Mary Ellen Koenig 98

31-02010 Embassy of the Federal Republic of Yugoslavia
Seminarstrasse 5, CH-3006 Bern
☎ (+41 31) 352 63 53-55 ⊠ 912848 yu emb ch
⊖ (+41 31) 351 44 74
Minister, Chargé d'Affaires a.i.: M Dragoslav Ignjatijevic
Minister: Vlado Nadazdin
First Counsellor (Consular Affairs): Mme Milka Burkic
Attaché: Veljko Mirkovic 98

Contact Reference

34 Multinational Intergovernmental Organizations

34-10000 European Laboratory for Particle Physics (CERN)
Laboratoire Européen pour la Physique des Particules
(† European Organisation for Nuclear Research)
CH-1211 Geneva 23
☎ (+41 22) 767 61 11 W http://www.cern.ch/
Telex 419000 cer ch Fax (+41 22) 767 65 55
President of the Council: L Maiani
Chairman of the Scientific Policy Committee: J Lefrançois
Chairman of the Finance Committee: M Gigliarelli Fiumi
Director-General: C H Llewellyn-Smith
Director of Research: L Foà
Head of Press and Publications: Neil Calder
☎ (+41 22) 767 41 01; 785 02 47 (fax)
E neil.calder@cern.ch W http://www.cern.ch/press
■ European collaboration in high energy physics research.
Member States:
Austria
Belgium
Czech Republic
Denmark
Finland
France
Germany
Greece
Hungary
Italy
Netherlands
Norway
Poland
Portugal
Slovak Republic
Spain
Sweden
Switzerland
United Kingdom [98]

34-10500 International Civil Defence Organisation (ICDO)
Organisation Internationale de Protection Civile (OIPC)
10/12 Chemin de Surville, CH-1213 Petit Lancy, Geneva
☎ (+41 22) 793 44 33 Fax (+41 22) 793 44 28
President: H E Sergei Shoigu (Russian Federation)
Director of the Civil Directorate of Senegal: Abdoulaye Ndiaye
Director of the Prevention and Civil Protection Directorate of Benin: Innocent Bré Abalo
Secretary General: Sadok Znaidi
Information Officer: Pascal Gondrand
Assistant to the Secretary General: Vladimir Kakusha
■ Promotion of protection and safety of people and property facing all kinds of disaster and includes in its organisation the national authorities of member states which are responsible for civil defence.
Member States:
Algeria, Bahrain, Benin, Burkina Faso, China, Egypt, Indonesia, Jordan, Mali, Pakistan, Panama, Russian Federation, Saudi Arabia, Senegal, Sri Lanka, Tunisia and Zimbabwe. [98]

34-11000 International Committee of the Red Cross
19 Avenue de la Paix, CH-1202 Geneva
☎ (+41 22) 734 60 01 ✉ INTERCROIXROUGE
W http:www.icrc.ch Telex 414226 Fax (+41 22) 733 20 57
President: Cornelio Sommaruga
Vice President: Pierre Keller
Vice President: Eric Roethlisberger
Operations Director: Jean de Courten
Director of Principles, Law and Relations with the Movement: Yves Sandoz
Director General: Peter Fuchs
■ Founder of the Red Cross and Red Crescent Movements. A Swiss, independent and neutral organisation with the role internationaly recognised of humanitarian intermediary between belligerents during armed conflicts.
Promotes international humanitarian law, protecting and assisting the victims of international and civil wars, internal disturbances and tensions by providing medical aid, relief supplies, tracing, information services for prisoners of war and missing persons, services for transmission of family messages, visits to civilian internees, prisoners of war and political detainees.
Has the observer status to the UN General Assembly. [97]

40 Government Defense Sales/Procurement Organizations

40-10000 Defence Technology & Procurement Agency
Gruppe Rüstung
Kasernenstrasse 19, CH-3003 Bern
☎ (+41 31) 324 58 77 Telex 912619 grdch
Fax (+41 31) 324 59 76
Director: Dipl.-Ing Ulrich Lanz ☎ (+43 31) 67 55 52
Chief of Equipment Division: Rudolf Loss
☎ (+41 31) 67 62 52
Chief of Vehicles and Engineering Division: Dipl.-Ing Heinz Gisiger ☎ (+41 31) 67 57 56; 67 55 56
■ Transport vehicles. Engineering, medical and catering equipment. Clothing and personal equipment. [98]

55 Government-owned Industry & Production Facilities

55-00100 Swiss Aircraft & Systems Co
Entreprise Suisse d'Avions et des Systèmes
(† Swiss Federal Aircraft Factory Emmen)
CH-6032 Emmen
☎ (+41 41) 268 41 11 W http://www.sfaerospace.ch
Telex 41412602588 Fax (+41 41) 260 25 88
♣ SFr 280,000,000 (1996) 👥 1,800 (1997)
🏭 14,400 m2
Managing Director: Dr W Glanzmann
Sales and Marketing Director: Bruno Raschle
Purchasing Manager: H R Furrer
SMDN: Marius Greber
■ Research, development and production in the aerospace field. Specialised in structures, electronics and missiles, wind tunnels and engine test facilities, fatigue testing. Assembly, overhaul, maintenance of military aircraft. Assembly and manufacture under licence of aircraft, aircraft structures and missiles. Development and production of unmanned air vehicles and hydraulic launchers (Ranger). → 75 → 80 [98]

55-00250 Swiss Federal Armament Works
Schweizerische Unternehmung für Waffensysteme
Allmendstrasse 86, CH-3602 Thun
☎ (+41 33) 228 11 00 Fax (+41 33) 228 20 47
👥 700 (1997)
Managing Director: Albert Schöllkopf
Commercial Director: Dr R Hochuli
Sales and Marketing Director: Erich Beutter
Sales Manager: Peter Kissling
Marketing Manager: Christian Immer
Purchasing Manager: H Bohren
■ Production of armoured vehicles. → 75 [98]

60 Research Organizations, Institutes & Test Establishments

Swiss Air Force Aeromedical Inst. . . . 25-29100

60-00020 Aerospace Project Development Group
Arbeitsgruppe für Luft-und Raumfahrt (ALR)
Gotthardstrasse 52, CH-8002 Zurich
☎ (+41 1) 202 93 88 Fax (+41 1) 202 24 00
👥 30
President: Georges Bridel
Managing Director: Dipl Ing Valentin Buergel
■ Study and development of light fighter, training and utility aircraft, concept development and consulting in aerospace. Aircraft performance analysis. [97]

65 Industry, Trade & Armed Forces Associations

65-01500 Swiss Association of Machinery Manufacturers (ASIA)
Verein Schweizer Maschinen-Industrieller
Société Suisse des Constructeurs de Machines.
Aeronautical Industry Division
Kirchenweg 4, CH-8032 Zurich
☎ (+41 1) 384 48 44 E gls3@vsm.ch
W http://www.vsm.ch Fax (+41 1) 384 48 48
President: H Kobelt
Secretary: F Akeret
Members companies:
Acutronic (Schweiz) AG
Aerospace AG
Ambrosetti Technologies SA
Alusuisse Schweiz AG
Alcatel Schweiz AG
Condor SA
Derendinger & CIE SA
Farner Air Service AG
FFA Flugzeugwerke Alternrhein AG
Fonderie Andri Neeser SA
Jean Galley SA
Huber + Suhner SA
Ingenierarbro Isler AG
Leica AG
Mecanex AG
Mecaplex AG
Meylan Frères SA
W Moser-Baer AG
Oerlikon-Contraves AG
Okey AG Zurich
Pilatus Flugzeugwerke AG
Precicast AG
Revue Thommen AG
Sauter Bachmann AG
SIP Société Genevoise d'Instruments de Physique
Siemens Schweiz AG
Société Genevoise d'Instruments de Physique
SF Schweiz Unterhehmung for Flugzeuge und Systeme
Siemens Schweiz AG
Steiger SA
Stoppani SA
Sulzer Triebwerke AG [98]

70 Financial Holding Groups, Parent Companies

ABB Asea Brown Boveri Ltd 74-00100

70-00100 Ascom Holding AG
Belpstrasse 37, CH-3000 Bern 14
☎ (+41 31) 999 11 11; 999 27 73 Telex 911119 hagch
Fax (+41 31) 999 27 00; 999 21 17
♣ SFr 2,992,100,000 (1996) 👥 11,739 (1996)
President: Dr Leonardo E Vannotti
Head of Corporate Communications: Peter Jenni
☎ (+41 31) 999 27 73
■ Multinational group specialised in communications and automation of services.
Subsidiaries:
Ascom Hasler AG
Ascom Autophon AG
Ascom Autelca AG
Ascom Telematic AG [98]

70-00700 Liebherr-International AG
CH-1630 Bulle
☎ (+41 29) 331 11 Fax (+31 29) 212 40
♧ SFr 500,000,000 ♣ SFr 3,828,000,000 (1995)
👥 15,159 (1995)
Chairman of the Supervisory Board: Dipl.-W.-Ing Hans Liebherr
Vice Chairman: Dipl.-Ing Willi Liebherr
Vice Chairman: Dipl.-Kfm Isolde Wagishauser
Member of the Supervisory Board: Dipl.-Ing Hubert Liebherr
Member of the Supervisory Board: Dipl.-Ing Markus Liebherr
Member of the Supervisory Board: Heinz Müller
Member of the Supervisory Board: Dr-Ing Karl Schwiegelshohn
Member of the Supervisory Board: Pierre-Alain Thiébaud
Managing Director: Dipl.-Kfm Winfried Böhm
Managing Director: Alfred Schuler

Managing Director: Dipl.-Volksw Paul Wagishauser
Managing Director: Lic-jur Denis Zosso [98]

70-01530 Oerlikon-Bührle Holding AG
Hofwiesenstrasse 135, CH-8021 Zurich
☎ (+41 1) 360 96 96 ⊜ (+41 1) 363 72 60
SFr 243,000,000 ♣ SFr 3,613,000,000 (1996)
15,543 (1996)
Chairman and Chief Operating Officer: Dr Hans Widmer
Finance Director: Dr Beat Baumgartner
Corporate Communications: Ms Yvonne Büchel
Subsidiaries with defence business:
Oerlikon-Contraves → 75 Switzerland
Pilatus Aircraft → 75 Switzerland
Oerlikon-Contraves SpA → 75 Italy [98]

74 Consortia, Joint Venture Companies

74-00100 ABB Asea Brown Boveri Ltd
(jointly owned by ABB AB and ABB AG)
PO Box 8131, CH-8050 Zurich
☎ (+41 1) 317 71 11 W http://www.abb.com ⊠ 823979
⊜ (+41 1) 317 73 21; 311 98 17
♣ US$ 35,574,000 (1996) 214,894 (1996)
President and Chief Executive Officer: Göran Lindahl
Executive Vice President, Industrial and Building Systems: Sune Carlsson
Executive Vice President and Chief Financial Officer: Renato Fassbind
Executive Vice President, Asia Pacific, South Asia: Alexis Fries
Executive Vice President, Power Transmission and Distribution: Sune Karlsson
Executive Vice President, Europe, Middle East and Africa: Eberhard von Koerber
Executive Vice President, Power Generation: Armin Meyer
Executive Vice President, Americas: Howard Pierce
Assistant Vice President: John Fox
■ Electric power generation, transmission and distribution. Industrial and building systems. Rail transportation. → 70 → 75
ABB Corporate Communications
PO Box 8131, CH-8050 Zurich
☎ (+41 1) 317 72 40 ⊜ (+41 1) 311 98 18
Senior Vice President: Volker Leichsering [98]

75 Manufacturing Companies

ABB Asea Brown Boveri Ltd . . . 74-00100
Interspiro AG . . . 90-00950
Swiss Aircraft & Systems Co . . . 55-00100
Swiss Federal Armament Works . . . 55-00250

75-00050 Acutronic (Switzerland) AG
(a subsidiary of Jung Technologies Holding AG)
Kreuzstrasse 100, CH-8645 Jona-Rapperswil
☎ (+41 55) 220 40 40 ✉ ACUTRONIC
E acutronic@active.ch W http://www.acutronic.com
⊜ (+41 55) 220 40 50
SFr 250,000 (1995) 15 (1996) 1,200 m2
Chairman and Chief Executive Officer: Thomas W Jung
■ Manufacture of servo-controlled precision motion systems for test and calibration of navigation instruments and sensors. These products include rate tables, 2-axis and 3-axis motion simulators.
Sister Companies:
Acutronic Ltd (UK)
Acutronic Deutschland GmbH
Acutronic USA Inc
Acutronic Control Sarl [98]

75-00070 Robert Aebi AG
Baumaschinen und Spezialfahrzeuge
Riedhofstrasse 100, CH-8105 Regensdorf
☎ (+41 1) 842 51 11 ⊠ 55557 ⊜ (+41 1) 842 51 20
Sales Promotion Manager: W Bircher
■ Snow removing machines. Fire fighting vehicles and rescue units.
® Germany: Alfred Schmidt GmbH, Mercedes-Benz Unimog, Metz Feuerwehrgerdte [97]

75-00400 Ametsa
7 rue Saint-Pierre, CH-1470 Estavayer le Lac
☎ (+41 37) 63 23 16 ⊠ 942399 prit ch
⊜ (+41 37) 63 19 93
♣ US$ 6,000,000 200
Manager: P Pritchett
■ Tools and tool kits for maintenance and installation of aerospace and military equipment. [97]

75-00500 Andair AG
Schaubenstrasse 4, CH-8450 Andelfingen
☎ (+41 52) 317 18 36 ⊜ (+41 52) 317 21 72
SFr 600,000 (1996) 40 8,000 sq m
Sales Manager: Jurg Jent
■ Development, manufacture and sale of ventilation sanitary equipment for NBC, air raid shelters including explosion protection valves. Sound and vibration damping. → 90
Nyon Office
240 Chemin Valmont, CH-1260 Nyon
☎ (+41 22) 361 46 76 ⊜ (+41 22) 361 87 45
Manager: Jurg Jent [97]

75-00540 Ascom Systec Ltd
(a subsidiary of Ascom Holding AG)
Gewerbepark, PO Box, CH-5506 Mägenwil
☎ (+41 62) 889 59 63 ⊜ (+41 62) 889 59 90
♣ SFr 108,000,000 210
Managing Director: U Ramseier
Sales and Marketing Director: E Werren
Sales Manager: E Widmer
Public Relations Manager: C Blöchliger
■ Expertise and overall solutions for defence communications. Main activities include: radio communications, networks and transmission, C3I systems and special systems.
Bern Plant
Belpstrasse 37, CH-3000 Bern 14
☎ (+41 31) 999 21 11 ⊜ (+41 31) 999 20 79
Hombrechtikon Plant
Eichtal, CH-8634 Hombrechtikon
☎ (+41 55) 254 61 11 ⊜ (+41 55) 254 62 11 [98]

75-00950 ASAI AG
Advanced Small Arms Industries
(† Industrial Technology & Machines AG)
PO Box 260, CH-4503 Solothurn
☎ (+41 32) 622 86 18 ⊜ (+41 32) 622 83 17
SFr 850,000,000 (1997)
Managing Director: Martin Tuma
Operations Director: Paul Molnar
Sales and Marketing Director: Josef Holenf
■ Manufacture of pistols (9mm, 0,38, 0,41AE, 0,40 and 0,45 S&W). Laser designators and mini laser designators, infra-red for hand guns, machine and anti-tank missiles. Test equipment for gun barrels and ammunition. [98]

75-01400 Breitling Ltd
(† Breitling Watch Co Ltd)
Schlachthausstrasse 2, PO Box 1132, CH-2540 Grenchen, SO
☎ (+41 32) 654 54 54 ⊜ (+41 32) 654 54 00
Chairman: E Schneider
Head of Advertising: Ms Valerie Birgat
■ Navitimer special watches for pilots. Chronographs for civil and military applications. [98]

75-01450 Brügger & Thomet Manufacturing Co
Oberlandstrasse 10, CH-3700 Spiez
☎ (+41 33) 54 97 57 ⊜ (+41 33) 54 97 58
♣ SFr 2,500,000 6
Managing Director: Heinrich Thomet
Technical Director: Karl Brügger
■ Development and manufacture of silencers. Authorised to sell and mediate all types of firearms and military equipment. [97]

75-01580 Bucher-Guyer Ltd, Vehicles
(a subsidiary of Bucher Holding AG)
CH-8166 Niederweningen ZH
☎ (+41 1) 857 22 11 ✉ BUCHERGUYER
E bucherguyer@bluewin.ch ⊠ 53199 bgnwg ch
⊜ (+41 1) 857 22 49
SFr 15,000,000 ♣ SFr 220,000,000 (1996) 600
50,000 m2
Chief Executive Officer: Ivan Feigel
Sales and Marketing Director: Manfred Sturm
Operations Director: Alfred Sasse
Purchasing Manager: Emil Schiess
■ Cross-country tactical and logistics vehicles (1,5 to 2 tonnes). Aircraft towing tractors. [98]

75-01700 Cerberus AG
CH-8708 Männedorf
☎ (+41 1) 922 61 11 W http://www.cerberus.ch
♣ SFr 960,000,000 (1995) 5,560
Managing Director: Luzius Hug
Production Director: Franz Reutsch
Finance Manager: Karl Bollmann
Publicity: H Wigger
■ Security systems against fire and crime. Electronics. Smoke detection systems for aircraft. [98]

75-01800 Compagnie Industrielle Radioélectrique SA (CIR)
(a subsidiary of Sextant-Avionique SA, France)
Nevenburgstrasse 7, CH-2076 Gals/BE
☎ (+41 32) 338 98 00
E cir-gals (cir-gals@mail.menet.ch)
⊜ (+41 32) 338 18 92
Director of Finance and Administration: Bernard Kistier [98]

75-01900 Communications & Power Industries International Inc
Zug Branch Office, Chollerstrasse 38, Postfach 3363, CH-6303 Zug
☎ (+41 42) 749 87 32 ⊠ 868 841
⊜ (+41 42) 740 33 40
■ Microwave, power grid and related equipment such as amplifiers and power supplies.
® USA: Communications & Power Industries [97]

75-02400 Crypto AG
CH-6301 Zug
☎ (+41 41) 749 77 22 E crypto@crypto.ch
W http://www.crypto.ch ⊠ 868702 cry ch
⊜ (+41 41) 741 22 72
Chief Executive Officer: Armin Hubber
Advertising Manager: Marco Weber
■ Manufacture and marketing of a comprehensive range of high-security ciphering systems. The product range covers voice, facsimile, text, data and multichannel ciphering devices, as well as integrated networks systems. On-line, off-line, analogues, digital. HF/VHF/UHF radio, telephone and satellite communications can all be securely ciphered. [98]

75-02760 Decolletage SA Saint-Maurice
(a subsidiary of AMP Inc, USA)
Vers-Pré 10, CH-1890 Saint Maurice
☎ (+41 24) 486 24 86 W http://www.amp.com/
⊜ (+41 24) 486 24 00
145 (1997)
Managing Director: Peter Loertscher
Operations Director: Gilbert Urlacher
Sales and Marketing Director: Fredy Bosshart
Purchasing Manager: Simon Bracey [98]

75-03200 EDAK AG
Rheinauerweg 17, CH-8447 Dachsen
☎ (+41 52) 647 21 11 ⊜ (+41 52) 659 63 70
120
Managing Director: K Müller
Sales and Export Manager: Max Reichlin
■ Vibration resistant transport housings for electrical and electronic equipment. [97]

75-03260 EF International SA
36 Avenue Krieg, CH-1208 Geneva
☎ (+41 22) 347 66 55 ✉ ENERGIE Geneva ⊠ 429653
⊜ (+44 22) 347 66 56
SFr 100,000 ♣ SFr 9,000,000 30

Contact Reference

President: C G Invernizzi
Export Manager: Dr C A Ziegler
■ Shielded lightning protection systems electromagnetic, without radioactivity, worldwide patented, for airport telecommunications, aerial power lines, marine, military, oil industry and buildings. [97]

75-03348 Egli Fischer & Co AG
Gotthardstrasse 6, CH-8022 Zurich
☎ (+41 1) 209 81 11 ⊖ (+41 1) 201 22 75
👥 160 (1997)
President: H E Fischer
Managing Director: M Allemann
Sales and Marketing Director: S Ballerini
Purchasing Manager: P Thommen
® Canada: Dalsa, EDAC, Mitel, Silonex
Sweden: Mitel
UK: CML, Colvern, Electrolube, Multicore Solders, Wolfsen
USA: Electro Corp, Laser Diode, Micropac, Supertex, Syntronic, Thomas & Betts [98]

75-03610 EMS-Dottikon AG
(† Swiss Explosives Works Ltd)
(a subsidiary of EMS-Chemie Holding AG)
Hembrunnstraße 17, CH-5605 Dottikon
☎ (+41 56) 616 81 11 ⊠ 827923 dott ch
⊖ (+41 56) 616 81 20
♣ SFr 137,000,00 (1996) 👥 430
Managing Director: Dr F Wittwer
Operations and Technical Director: Dr E Pietrzik
Sales and Marketing Director: F Schwarz
Vice President Marketing, Fine Chemicals: Dr A Hochrainer
Sales Manager: B Martinie
Purchasing Manager: R Blickenstorfer
Sales Manager: U Loewenthal
Sales Manager: O Bachmann
■ Manufacture of RDX and PETN based speciality explosives for military applications. [98]

75-03650 EMS-PATVAG AG
(† Patvag Technik AG)
(a subsidiary of EMS-CHEMIE Holding AG)
Rusna da Furns, CH-7013 Domat, Ems
☎ (+41 81) 632 75 45 ⊖ (+41 81) 632 74 36
♣ SFr 31,000,000 (1996) 👥 75 (1996) 🏠 2,750 sq m
Director: Dr Albert Sommerauer ☎ (+41 81) 632 73 24
Head of Defence Products: Peter Hartmann
☎ (+41 81) 632 63 72
■ Electric fuse systems, electric inductive generators, electric primers, mechanical safety and arming devices. Electronics-assisted multipurpose fuze systems responding to current safety requirements. Generators also designed for superfast reaction hollow charge ammunition. Advanced thin-film primers of diameters down to 4 mm. Primary explosives and pyrotechnical charges. Hermetic seals: vaccum sealed fusions of glass and metal. various applications in the field of electro-technical electronic engineering. [98]

75-03900 ERNI Licht-Technik AG Airport Equipment Division
Stationstrasse 31, CH-8306 Brüttisellen, Zurich
☎ (+41 1) 835 33 43 [E] erni_licht@bluewin.ch
⊖ (+41 1) 833 07 92
Managing Director: Alfred Gall
Sales and Marketing Director: Ralf Zepter
☎ (+41 1) 835 33 44
■ Airport lighting equipment for approach, runways, taxiways and aprons, elevated and flush or semi-flush installation. [98]

75-04350 FFA Flugzeugwerke Altenrhein AG
Saint Gallen, Altenrhein International Airport,
CH-9423 Altenrhein
☎ (+41 71) 858 51 11 ⊠ 882906 ffach
⊖ (+41 71) 858 53 30
♧ SFr 10,000,000 (1996) ♣ SFr 18,000,000 (1996)
👥 80 (1996) 🏠 9,500 m2
President and Chief Executive Officer: Dr Ch Brönimann
Manufacturing Manager: F Rohrbach
Maintenance Manager: J Galliker
■ Design and manufacture of primary/basic training aircraft. Maintenance and overhaul of civil and military prop and jet aircraft. Subcontractor to the aerospace industry. → 80 [98]

75-04700 Gayrobot Undersea Technology
Via Magazzini Generali 131a, CH-6828 Balerna, Chiasso
☎ (+41 91) 43 92 84 ⊠ 842670 ggay ch
⊖ (+41 91) 43 92 85
Managing Director: Guido Gay
Marketing Support Manager: Walter Nardini
■ Underwater remote operated vehicles for mine identification and destruction (PLUTO and PLUTO PLUS models). [98]

75-05850 Hämmerli AG Target Arms Factory
(a subsidiary of SIG Swiss Industrial Company)
Seonerstrasse 37, CH-5600 Lenzburg 1
☎ (+41 62) 892 22 44 ⊖ (+41 62) 891 38 27
♧ SFr 3,000,000 👥 104
Managing Dir: R. Froidevaux
Sales & Marketing Dir: G. Geissmann
Operations Dir: M. Wetter
Purchasing Mgr: Th. Rätzer
■ Manufacture of high performance target rifles and pistols; barrels and parts for the Swiss assault rifle; assembly of the Swiss Army pistol. [97]

75-06300 ILEE AG
Industrial Laser & Electronic Engineering
Schützenstrasse 29, CH-8902 Urdorf
☎ (+41 1) 734 27 77 ⊖ (+41 1) 734 27 22
♧ SFr 700,000 ♣ US$ 6,000,000 👥 25 🏠 2,500 sq m
Joint Owner/President: Karl Staub
Operations Director: Peter Gerber
Purchasing Manager: Bruno Codello
■ Visible and invisible lasers for targeting at night and surveillance. Simlas combat simulation system. [97]

75-07000 Katadyn Products Inc
Birkenweg 4, CH-8304 Wallisellen
☎ (+41 1) 839 21 11 ⊖ (+41 1) 830 79 42
♧ SFr 6,000,000 ♣ SFr 21,415,000 (1996)
👥 80 (1996)
President and Chief Executive Officer: A Schmassmann
☎ (+41 1) 839 21 36
Purchasing Manager: John Fineron ☎ (+41 1) 839 22 25
Finance and Management Services: H J Huettner
☎ (+41 1) 839 21 19
Assistant to the Board: U Feiss
■ Water treatment.
Subsidiaries:
Katadyn Deutschland GmbH
Katadyn USA Inc
Katadyn France SA [98]

75-07100 Knabenhans
La Pineta, CH-6574 Vira
☎ (+41 91) 795 29 52; (+41 79) 400 48 79
⊖ (+41 91) 795 29 52
♧ US$ 25,000 (1996) ♣ US$ 4,000,000 (1996)
👥 4 (1997) 🏠 600 m2
Managing Director: Balz Knabenhans
Sales and Marketing Director: L Jirasek
■ Glasses, safety-glasses, sunglasses with special treatment for military use. Lightweight helmet systems. → 80 → 90 [98]

75-07200 G Kull AG, Lunor
(† G Kull AG)
Aemtlerstraße 96a, CH-8003 Zurich
☎ (+41 1) 451 10 20; 455 50 70 ⊖ (+41 1) 451 16 26
♧ US$ 3,000,000 ♣ US$ 13,000,000 👥 40
🏠 4,000 m2
Managing Director: Rudolf Kull
Operations Director: Ernest Kull
■ Manufacture of shelter components including doors, ventilation and filter units, valves and auxiliary parts under trademark LUNOR. [98]

75-07310 Leica AG Defence & Special Projects Division
(a division of Leica Group)
CH-9435 Heerbrugg
☎ (+41 71) 727 31 31; 727 37 97 (sales)
[W] http://www.leica.com/optronics ⊠ 881222 wi ch
⊖ (+41 71) 727 46 79
👥 1,600
President and Chief Executive Officer:
Dr Joerg Wullschleger
General Manager, Defence and Special Projects:
Dr Linus Zoller
International Sales Manager: Steve White
Communications Manager: Kurt Hegetschweiler
☎ (+41 71) 727 34 64
■ High-precision optronic instruments for military survey and artillery applications. Day and night observation. 98

75-07330 LEISTER Elektro-Gerätebau
(† Karl Leister Elektro-Gerätebau)
CH-6056 Kägiswil (OW)
☎ (+41 41) 66 00 77 [E] leister@access.ch
[W] http://www.leister.com ⊠ 866404 leist ch
⊖ (+41 41) 66 78 16
👥 250
Managing Director: Ms Christine Leister
Sales and Marketing Director: Walter Zurmuehle
Advertising Manager: J P Bachmann
■ Hot air blowers for industrial application. Hot air tools for contactless de-soldering, soldering, shrinking and welding thermoplastics. [97]

75-07335 Lenoir Aviation AG Headquarters
Eichwaldstraße 9, CH-6300 Zug
☎ (+41 41) 741 55 05 ⊖ (+41 41) 741 55 51
President: Daniel B Bourret
Operations Director: Markus Hanser
Marketing Director: Peter Lesch
■ Air-portable, mobile and autonomous hospital for the C130 and C160. Contained mobile and permanent field hospitals. Prefabricated hospitals in various sizes. [98]

75-07350 Lista Ltd
(a division of Lista Holding AG)
CH-8586 Erlen
☎ (+41 71) 649 21 11 [W] http://www.lista.ch
⊖ (+41 71) 649 22 33
👥 365 (1996) 🏠 56,000 m2
President: Fredy A Lienhard
Vice President, Marketing: Ernst Züblin
■ Drawer cabinets for storage of tools, parts, components, instruments, test equipment, sensitive records. Storage and transport systems for pre-set tooling. Workshop cabinets and individual workstations, tools and maintenance trolleys. Mobile and conventional shelving systems. Vertical carousels.
Subsidiary company:
Lista International Corp, USA → 75 [98]

75-08000 Mäder Lacke AG Lach and Farbenfabrik
(a partner of Walter Märder AG)
Industriestrasse 1, CH-8956 Killwangen
☎ (+41 569 417 81 11 ⊠ 826014 dwm ch
⊖ (+41 56) 401 20 16
Marketing Services: Beinhard Drack
■ Manufacture paints and protective coatings and synthetic resins. [98]

75-08300 Mécanex SA
Zone Industrielle Vuarpillière 29, CH-1260 Nyon
☎ (+41 22) 361 77 33 [E] mecanexsa@bluewin.ch
⊖ (+41 22) 361 67 52
♧ SFr 700,000 (1996) ♣ SFr 8,000,000 (1996)
👥 39 (1996) 🏠 2,100 m2
President: Mrs Nicola Thibaudeau
Vice President and Operations Director: Mr Volker Gass
Technical Director: Miss M T Ivora
Marketing Manager: Jean-Frédéric Berthoud
■ Research, development and production of slipring-brush assemblies and high precision mechanisms. [98]

75-08485 MB-Microtec AG
Freiburgstrasse 624, CH-3172 Niederwangen
☎ (+41 31) 981 11 25 ⊖ (+41 31) 981 11 05
Managing Director: O W Thüler
Operations Director: H Etter
Sales and Marketing Director: Jakob Bänziger
■ Self-activated light sources based on radioactive gas. Military applications, including fire control systems, magnifying telescopes, sights, watches and emergency signs. [97]

75-08490 Micronel AG
CH-8317 Tagelswangen
☎(+41 52) 355 16 16 ⊖ (+41 52) 355 16 20
50
General Manager: Hans Hauser
Sales and Marketing Director: Georges Schaerer
Operations Director: Otto Gossweiler
Purchasing Manager: Roland Bösiger
■ Compact-Air-Supply-Units (CASU's) for personal protection and support against chemical warfare. 1-Filter-Unit 55 1/min for gas masks; 4-Filter-Unit 200 1/min for casualty bags, 2-Filter-Unit 95-120 1/min for gas masks and casualty bags. 97

75-08498 Miltec SA Administration and Marketing Office
Tavaro Technologies
(† Tavaro SA)
1-5 Avenue de Châtelaine, CH-1211 Geneva 13
☎(+41 22) 949 09 99 ⊖ (+41 22) 344 22 15
♣ US$ 20,000,000 (1995) 120 17,000 m2
Managing Director: J F Couturier
Research, Development and Marketing Director: J P Golay
Purchasing Manager: P Bonvin ☎(+41 27) 205 61 11
■ Design, manufacture and assembly of clockwork mechanisms (safety and arming devices), fuzes of all types for mortar, artillery ammunition and missiles. 98

75-08500 Mitralux International AG
Frobenstraße 75, CH-4008 Basle
☎(+41 61) 271 72 22 ⊖ (+41 61) 271 78 88
♧ SFr 100,000 (1995) 6
Managing Director: Hanns Meister
■ Portable high intensity area lights. Also usable from helicopters. 98

75-08900 MOWAG
Motorwagenfabrik AG
Unterseestrasse 65, CH-8280 Kreuzlingen
☎(+41 71) 677 55 00 882211 moagch
⊖ (+41 71) 672 28 86
510 (1997)
Director: Dr E H Strasser
Sales and Marketing Director: Ulrich Wenger
Regional Sales Manager: Heinz Koenig
☎(+41 71) 677 55 22
■ Design, development and manufacture of all types of armoured wheeled vehicles in the weight class ranging from 5 to 22 tonnes. 98

75-09250 Oerlikon-Contraves AG
(a division of Oerlikon-Bührle Holding AG)
Birchstrasse 155, CH-8050 Zurich
☎(+41 1) 316 22 11 823402 cozch
⊖ (+41 1) 311 31 54; 316 42 02
♣ SFr 702,000,000 (1996) 2,815 (1996)
President and Chief Executive Officer: E Odermatt
Managing Director, Pyrotec AG: H Meier
Managing Director, Space: Dr M Somaini
Information Officer: Ernst Jäeggi
Managing Director, Defence: J P Chassot
■ Main products and activities:
Defence: Skyshield 35-AHEAD gun and missile; Skyguard twin gun and missile air defence systems; AHEAD system; Gun-King multi-divergence laser sight system; Shorar/Pagoda search and acquisition radar; Gun-Star compact fire control system; Seaguard CIWS system; ADATS missile systems; Ranger UAV-system; training systems and simulators. Automatic cannon for air defence, vehicle and naval armament.
Pyrotec: ammunition in the calibre range 12,7 to 35 mm. Production machinery and ballistic measuring systems. Facilities and equipment for disposal of ammunition.
Space: spacecraft structures, fairings for launchers, inflatable space rigidized structures, deplorment mechanisms, scientific instruments. 98

75-09330 Omnisec AG
Trockenloostrasse 91, CH-8105 Regensdorf
☎(+41 1) 870 95 42 E marketing@omnisec.ch
⊖ (+41 1) 841 03 15
♧ SFr 10,000,000 65 1,400 m2
Chief Executive Officer: Dr J D Lindecker
■ Communication security equipment for voice, fax, data, mesages and multi-channel. 98

75-10100 Pilatus Aircraft Ltd
Pilatus Flugzeugwerke AG
Pilatus Constructions Aironautiques SA
(a division of Oerlikon-Buehrle Holding Ltd)
CH-6370 Stans, NV
☎(+41 41) 619 61 11 ✉PILATUSAIR 866202 pil ch
⊖ (+41 41) 610 33 51
♣ SFr 187,000,000 832 (1996)
Chairman: Dr Ernst Thomke
President and Chief Executive Officer: Oscar J Schwenk
Senior Vice President, Finance, Control and Administration: Oscar Bründler
Vice President, Logistics: Erich Bodenmann
Vice President, Engineering: Dr Oliver Masefield
Vice President, Production: Anton Waldispuehl
Vice President, Sales and Marketing: Dr Philippe Hertig
Manager of Marketing Support Services: W J Cato
■ Design, manufacture and sale of PC-7 and PC-7 Mk II turbo trainers, PC-9 advanced turbo trainer, PC-6 Turbo Porter super STOL aircraft, PC-12 executive/corporate commuter aircraft and the PC-12 Eagle (surveillance aircraft). Maintenance and overhaul of civil and military aircraft and helicopters. Sub-contracting for international programmes including Airbus, F/A 18, Mirage, Hawk F5. Design and manufacture facility in the UK producing Islander and Defender aircraft. 98

75-10250 Polytronic AG
Polytronic International Ltd
Pilatusstrasse 12, CH-5630 Muri
☎(+41 56) 664 64 64 ⊖ (+41 56) 664 16 55
♣ SFr 40,000,000 (1997) 50 (1997)
President: C Thalmann
Managing and Operations Director: Arnold Jost
Marketing Director: Henry Schweizer
Purchasing Manager: U Rüttimann
Marketing Associate: Ms Diana Archer
■ Manufacture and marketing of shooting training equipment. → 90 98

75-10320 Portescap
(a subsidiary of Inter Scan Group Ltd)
157 rue Jardinière, CH-2301 La Chaux de Fonds 1
☎(+41 39) 25 61 11 ✉ESCAP ⊖ (+41 39) 25 65 90
650
Managing Director: Philippe Maquelin
Operations Director: M Gilliéron
Sales and Marketing Director: C Baur
Sales and Marketing Director: N Veignat
Sales and Marketing Director: A Heinkel
Communication Officer: M L Heinkel
Publicity Administrator: Mrs M Heinkel
■ DC micromotors, tachos, optical encoders, disc magnet stepper motors and corresponding drive circuits, reduction gears, gearmotors.
Subsidiary Companies:
Portescap (UK) Ltd → 75
Portescap Deutschland GmbH → 90
Portescap France SA → 90
Portescap Japan Ltd → 90
Portescap US Inc → 90
Portescap Sverige AB → 90 97

75-10900 RECTA SA
(a subsidiary of Suunto Oy, Finland)
3 rue du Viaduc, CH-2502 Bienne
☎(+41 32) 328 40 60 ⊖ (+41 32) 328 40 69
40 (1997) 1,000 m2
Managing Director: Hans Gloor
Sales and Marketing Director: Herbert Steiner
Operations Director: Gilbert Vorpe
Secretary: Ms Tania Ruegg
■ Army field and sighting compasses, wrist compasses, diving compasses, artillery sitometers. 98

75-11100 Revue THOMMEN AG
Hauptstrasse 85, CH-4437 Waldenburg
☎(+41 61) 965 22 22 966099 rtw ch
⊖ (+41 61) 965 81 71
♧ SFr 1,250,000 (1996) ♣ US$ 12,000,000 (1996)
100
Managing Director: R Moser
Marketing Director: F Kaiser
Aircraft Business: H Kalt
■ Altimeters, encoding altimeters, airspeed indicators, mach airspeed, indicators, vertical speed indicators, clocks and chronographs, cabin pressure measuring instruments. Environmental indicators. All instruments are available in NVG compatible lighting. Pitot static test sets. 97

75-11980 Siemens Switzerland Ltd
(a subsidiary of Siemens AG, Germany)
Freilagerstrasse 38, CH-8047 Zurich
☎(+41 1) 495 55 20 ⊖ (+41 1) 495 38 16
♣ SFr 1,600,000,000 3,100
Managing Director: Jörg Bischof
Sales and Marketing Director: Heinz Zürcher
Marketing Manager: Thomas Hedrich
Purchasing Manager: Berd Böhmichen
Advertising Manager: Mrs V Martignier
■ Communications, digital (private networks, switching and transmission equipment, terminals, engineering); radar (tracking, subsystems for signal processing); optronics (thermal imagers); software (C3I, network management). 97

75-13500 Société Suisse des Explosifs
CH-3900 Brigue Glis
☎(+41 27) 922 71 11 473339 ⊖ (+41 27) 922 72 00
100
Managing Director: D Antille
Sales Manager: M P Chiabotti
■ Manufacture of explosives. 98

75-13550 SpanSet Inter AG
(Holding Company of SpanSet Group)
Eichbüelstraße 31, CH-8618 Oetwil am See
☎(+41 1) 929 20 15 ⊖ (+41 1) 929 19 40
General Manager: David Ehnimb
Sales and Marketing Director: H M G Braddell
■ Manufacture and distibution of lifting and lashing equipment approved to UK Defence standards 05-21/1 worldwide, including polyester roundslings, webbing lifting slings and cargo lashings, lifting and securing systems for ammunition, bombs and torpedoes, tank towing and lashing systems, tactical aircraft arrester gear, non-magnetic lifting and securing systems, aircraft tie-down systems and personal safety harness fall arrest equipment. Railway lashing systems. 98

75-13590 Rockwell Automation AG
(† Sprecher + Schuh, Industrial Control Group)
(a subsidiary of Rockwell Automation Ltd, USA)
Buchserstrasse 7, CH-5001 Aarau
☎(+41 62) 837 21 21 ⊖ (+41 62) 837 29 10
750 (1997)
Managing Director: Herbert Jost
Operations Director: Beat Kaenel
Sales and Marketing Director: Werner Oesch
Purchasing Manager: Klaus Amand
Assistant to Managing Director: Ms Monika Mattes
® USA: Rockwell International Corp 98

75-14410 TIRAX AG
(a subsidiary of Swarovski International Holding AG)
Bernstrasse 390, CH-8953 Dietikon 3
☎(+41 1) 742 17 17 ⊖ (+41 1) 742 17 18
♧ SFr 200,000 (1995) ♣ SFr 3,500,000 (1995) 21
950 sq m
President and Managing Director: Carl Leutwyler
Sales and Marketing Director: H R Holzer
Operations Director: R Frey
Purchasing Manager: A Utzinger
Training and Simulation Division: G Leutwyler
■ Combat simulators, shooting installations for the training of armed forces and police forces (day and night). Perimeter security systems. 97

75-14500 Trimos-Sylvac Task Force (TSTF)
Route de St Legier 60, CH-1800 Vevey
☎(+41 21) 944 68 27 ⊖ (+41 21) 944 68 33
Managing Director: David Rotely-Clark
■ Measuring equipment (Metrology), gauge calibration with 100% technical back-up. Electronic height and length measuring with computer link-up facility instruments. High precison multi-purpose linear measuring units. Dial gauge and lever indicator calibration unit with British Standards printout format. Logistic support in operations worldwide. ISO 9000 accredited. → 80 → 90
Trimos Factory
5 Avenue Longmalle, CH-1020 Renens
☎(+41 21) 636 06 05 454054 jeca
⊖ (+41 21) 635 93 57
Chairman of Trimos Division: J Cao
General Manager of Trimos Division: M Meylan

Sylvac Factory
Chemin du Chosalet 16, CH-1023 Crissier
☎(+41 21) 637 67 57 ⊠ 450196 sylch
⊖ (+41 21) 637 67 40
Managing Director of Sylvac Division: D Liechti
Director of Sylvac Division: A Künzi [98]

75-14600 T&T Technology Trading Ltd Noricum Division
Splügenstrasse 12, CH-8002 Zurich
☎(+41 1) 281 02 81 ⊖ (+41 1) 281 02 82
Contact: Heinz Stierli
■ Manufacture of heavy artillery systems. [97]

75-14700 Verres Industriels SA (VIM)
Quartier de la Verrerie 25, CH-2740 Moutier
☎(+41 32) 494 51 11 ⊖ (+41 32) 494 51 91
♧ SFr 3,000,000 ♣ SFr 30,000,000 👥 170
Managing Director: A Seravalli
■ High impact heated glasses and other special security transparencies for armoured, special vehicles and aircraft. [98]

75-14710 Vibro-Meter SA
(a subsidiary of Electrowatt Ltd, Zurich)
Route de Moncor 4, PO Box 1071, CH-1701 Fribourg
☎(+41 26) 407 11 11 [E] vmsa@vibro-meter.ch
⊠ 942292 ⊖ (+41 26) 402 36 62
♣ SFr 70,000,000 (1994) 👥 450 🏠 10,382 m2
Chief Executive Officer: Dr R W Greaves
Purchasing Manager: M Thiébaud
■ Transducers and associated electronic equipment for measurement, monitoring and control of physical parameters on aircraft, gas turbine engines and rocket engines, vibration, speed, ice, displacement, pressure. Helicopter rotor trim and balance systems. Central maintenance computers and engine and propellor interfaces.
Associated companies:
Vibro-Meter Corp (USA)
Vibro-Meter Ltd (UK)
Vibro-Meter GmbH (Germany)
Vibro-Meter France SA (France)
Vibro-Meter Skandinavia A/S (Norway)
Vibrolot Ltd (Poland)
Vibro-Meter SA (Singapore) [98]

75-14750 Von Arx Ltd
Gelterkindenstrasse 24, CH-4450 Sissach
☎(+41 61) 976 44 44 ⊖ (+41 61) 976 44 55
Managing Director: Thierry Kenel
Operations Director: Pascal Kenel
Marketing Director: Heinz Kobi
■ Special machines for treating large surfaces, with interchangeable tools, up to 180 cm width (removal of rubber deposit on runways and grooving). Surface and deck cleaning and scarifying machines. Machines for traffic line removal (100/200/300 mm width). Power tools for surface preparation. [98]

75-15150 Warner Electric SA
Glattalstrasse 525, PO Box, CH-8153 Rümlang
☎(+41 1) 817 16 66 ⊠ WARNERSWISS
⊖ (+41 1) 817 16 77
Marketing Manager: R Koch
■ Electric clutches and brakes, solenoids, overrunning clutches (freewheels) bi-directional and no-backs and ball screws, special actuators. → 90
® USA: Formsprag-Warren, Warner Electric [98]

75-15300 Wicker-Bürki AG
Wipic-Antennen
Riedackerstrasse 17, CH-8153 Rümlang
☎(+41 1) 817 12 22 [E] remo.staubli@spectraweb.ch
⊖ (+41 1) 817 11 30
♣ US$ 3,000,000 (1994) 👥 18
Managing Director: Peter Schütz
Sales and Marketing Director: Raimund Staubli
Operations Director: Klaus Wicker
Purchasing Manager: Peter Wicker
■ Antennas and towers for radio communication, air traffic control, military police use. [98]

80 Service Companies & Organizations

80-00050 Aero Consultants Ltd AG
Grabenstrasse, Postfach 336,
CH-8606 Nänikon Uster, Zurich
☎(+41 1) 941 18 14 ⊠ AERCO
⊠ 826660; 826696 aec ch ⊖ (+41 1) 941 31 37
♧ SFr 400,000
President: Bernhard Arnold
Vice President Marketing: Ms Cristina Müller
Sales and Marketing Manager: A Staub
Sales and Marketing Manager: Beat Barmettler
■ Distributors of composite related materials and consultation for manufacture and repair of highly stressed lightweight structures. Lightweight ballistic armour.
Boronfibres and boron/epoxy prepregs, structural adhesives, foam and primers, modelling boards, silicon rubbers. Vacuum-bag and autoclave materials, tooling resins and prepegs. Sound and heat insulation materials and parts. Automated ultrasonic NDT-scanning systems. Hot bond repair consoles. Chemical milling maskants and 3-D laser projection systems.
® USA: AC Products, Airtech/IPP International, Dexter Aerospace Materials, General Scanning Inc, Heaton Systems, Hi-Temp Insulation, Hysol Adhesives, McDonnell Aircraft, Mosites Rubber Co, Textron Speciality Materials [97]

80-00060 AlliedSignal Aerospace Marketing Sales and Service
(a subsidiary of AlliedSignal Aerospace, USA)
c/o Crossair Ltd, Basel Airport, CH-4030 Basel
☎(+41 61) 325 36 93 ⊖ (+41 61) 325 34 05
Field Service Engineer: Paul Eaton
→ 90
● PO Box 272, CH-CH-2016 Cortlaillod
☎(+41 32) 842 31 31 ⊖ (+41 32) 842 31 87
Director, Central European Sales: Dan Wildi
■ Marketing office for avionics equipment.
● Geneva Airport, PO Box 173,
CH-CH-1215 Geneva Airport 15
☎(+41 22) 798 32 65 ⊖ (+41 22) 788 19 77
Field Service Engineer: Richard Cia
● Swissair Technik (YG), PO Box 1397, CH-8058 Zurich
☎(+41 1) 812 78 30 (recording) ⊖ (+41 1) 810 65 25
Field Service Engineer: Herb Merker
■ On-site products services for AlliedSignal Aerospace customers. [98]

80-00430 Defence Data Consulting Services
10 Boulevard de la Tour, CH-1205 Geneva
☎(+41 22) 320 47 48 ⊠ 428750 ptel ch
⊖ (+41 22) 320 47 48
Services Director: Alain Drouillard
■ Technical research, database on defence related topics. Market research reporting on defence industries. Aerospace editorial, public relations and sales promotions. Value engineering DGCAO. [97]

80-00800 Ingenieurbüro Flugwesen & Biomechanik IFB AG
Bündackerstrasse 67, CH-3047 Bremgarten
☎(+41 31) 301 38 60 ⊖ (+41 31) 302 73 72
President: Dipl.-Ing ETH Stefan Freudiger
■ Design, manufacture, certification and testing of small aircraft prototypes; design of civil and military aircraft structural modifications, ambulance outfits for fixed and rotary-wing aircraft, and helicopter skis. Fatigue analysis. [97]

80-00955 Inter Acoustic AG
International Technical Acoustic Company
Seestraße 241, CH-8804 Au Wädenswil
☎(+41 1) 781 41 30 ⊖ (+41 1) 781 41 17
Managing Director: Dipl.-Ing Wilhelm Körner
■ Acoustic engineering for jet and propeller engine test cells, run-up test facilities (fixed and mobile), hush houses (civil and military) and turnkey projects as general contactor). Consultancy services. [98]

80-01030 Jet Aviation Management AG
Headquarters for Europe/Middle and Far East
PO Box 586, CH-8034 Zurich
☎(+41 1) 382 22 02 [E] jmgt@jetaviation.com
[W] http://www.jetaviation.com ⊖ (+41 1) 382 10 60
♣ SFr 387,300,000 (1996) 👥 2,600 (1997)
President and Chief Operating Officer, Jet Aviation Europe, Middle and Far East: Heinz Köhli
Chief Financial Officer, Jet Aviation Europe, Middle East and Far East: Edgar Schlatter
Public Relations and Advertising Co-ordinator for Europe, Middle and Far East: Lorenz Köhler
Jet Aviation Zürich AG
PO Box 1524, CH-8058 Zurich Airport
☎(+41 1) 816 48 00 [E] jzrh@jetaviation.com ⊠ 825688
⊖ (+41 1) 816 38 63
President and General Manager: Werner Aerne
■ Maintenance of jet, turboprop engines; interior refurbishment and outfitting; avionic retrofit; design and engineering; airline line maintenance.
Jet Aviation Basel
EuroAirport, PO Box 214, CH-4030 Basel-Mulhouse
☎(+41 61) 325 20 07 [E] jbsl@jetaviation.com
⊠ 963133 ⊖ (+41 61) 325 20 90
Vice President and General Manager: Elie Zelouf
Deputy General Manager: Willi Gastner
■ Maintenance of jet engines; interior refurbishment and outfitting; avionic retrofit; engineering and design; airline line maintenance; aircraft sales.
Jet Aviation Geneva
Airport, PO Box 456, CH-1215 Geneva-Cointrin
☎(+41 22) 929 11 11 [E] jgva@jetaviation.com
⊠ 415594 ⊖ (+43 22) 791 01 90
Vice President and General Manager: Günter Huf
■ Maintenance of jet, turboprop engines; interior refurbishment and outfitting; avionic retrofit; engineering and design; airline line maintenance; aircraft sales. [98]

80-01400 Lockheed Martin International SA (LMISA)
(† Lockheed Corporation (International) SA)
(a subsidiary of Lockheed Martin Corporation, USA)
1 Place Longemalle, CH-1204 Geneva
☎(+41 22) 318 40 40 ⊠ LOCKHEED
⊖ (+41 22) 318 40 55
Chairman: M K Som
President, Central and Eastern Europe: R R Clifford
Vice President, Central Europe: E G Jordan
Vice President, Space Imaging EOSAT: M C Wasielewski
Vice President, Eastern Europe: R A Watters (Jr)
■ Business development, finance and administration. → 90
Representing:
Lockheed Martin Corporation, USA [98]

80-01500 Logica Informatik AG
(a subsidiary of Logica plc, UK)
Waaggasse 5, CH-8050 Zurich
☎(+41 1) 313 10 10 [W] http://www.logica.ch
⊖ (+41 1) 313 10 40
👥 40 (1997)
Managing Director: Urs Haintheld
Operations Director: Walter Lohrer
Personnel Department: Ms Evelyne Dingelschweiler
[E] dingetsche@logica.com
■ Services in consultancy, project management, operational analysis, custom built computer and communication systems, including software and hardware implementation and products. [98]

80-03000 Securité Civile-Genève Service Hélicoptère
(† Service Cantonal de la Protection Civile)
Route H C Forestier 32, CH-1217 Meyrin
☎(+41 22) 798 61 31; 798 00 00 (emergency)
⊖ (+41 22) 782 40 20
👥 6
Managing Director: Philippe Wassmer
Operations Manager: François Gindre
Chief Pilot: Jean-Bernard Schmid
■ Ambulance and rescue flights. [98]

80-03500 Technischer Fachbuch Vertrieb AG

Spitalstrasse 12, PO Box 119, CH-2501 Biel
☎(+41 32) 322 61 41 Ⓔ tvf-bielaswissonline.ch
Ⓦ http://www.tvf.ch ⊖ (+41 32) 322 61 30
♣ SFr 5,000,000 ⚭ 10
Managing Director: Prof. Manfred Studer
Sales and Marketing Director: Nick Lambing
® UK: Jane's Information Group [98]

85 Stockists, Distributors, Spare Parts Suppliers

Interspiro AG 90-00950

90 Sales Agents, Sales & Liaison Offices

Companies represented:

AC Products, USA 80-00050
Advanced Control Components Inc, USA 90-02020
Advanced Products NV, Belgium 90-01100
Airtech/IPP International, USA 80-00050
Alfred Karcher GmbH & Co, Germany . . 90-01085
Alfred Schmidt GmbH, Germany 75-00070
AlliedSignal Aerospace, USA 90-00090
Aluminium Co of America, USA 90-00100
B&H Electronics Inc, USA 90-02020
Bendix King, USA 90-00090
BFGoodrich Aerospace, USA 90-00200
Cannon Electric GmbH, Germany 90-00310
Cincinnati Milacron, UK 90-00340
Clarke Hess Communication Res. Corp, USA 90-02020
CML, UK 75-03348
Colvern, UK 75-03348
Communications & Power Industries, USA 75-01900
CPI, USA 90-00365
Dalsa, Canada 75-03348
Data Delay Devices, USA 90-02020
DeVore Aviation, USA 90-00640
Dexter Aerospace Materials, USA 80-00050
Drägerwerk AG, Germany 90-00495
EDAC, Canada 75-03348
EEV Ltd, UK 90-00570
Electro Corp, USA 75-03348
Electrolube, UK 75-03348
Elisra Electronic Systems, Israel 90-02020
Enstrom Helicopter, USA 90-00520
Formsprag-Warren, USA 75-15150
General Microwave, USA 90-02020
General Scanning Inc, USA 80-00050
Hayes Int'l Targets, USA→ UK 80-07600
Heaton Systems, USA 80-00050
Heckler & Koch GmbH, Germany 90-00830
Hi-Temp Insulation, USA 80-00050
Hysol Adhesives, USA 80-00050
Israel Military Industries Ltd, Israel 90-00990
Jane's Information Group, UK 80-03500
JCA Technology Inc, USA 90-02020
Jensen Tool Company, USA 90-00070
Kalmus Engineering International, USA 90-02020
Krauss-Maffei Wehrtechnik, Germany . 90-01480
L-W International, USA 90-02335
Laser Diode, USA 75-03348
Litton Industries Inc, USA 90-01260
McDonnell Aircraft, USA 80-00050
Mercedes-Benz Unimog, Germany 75-00070
Metz Feuerwehrgerdte, Germany 75-00070
Micropac, USA 75-03348
Mitel, Sweden 75-03348
Mitel, Canada 75-03348
Mosites Rubber Co, USA 80-00050
MTS Microtechnik GmbH, Germany 90-02020
Multicore Solders, UK 75-03348
Orbit Advanced Technologies, Israel . . 90-02020
Pantron, Switzerland 90-00360
Pearson Electronics, USA 90-02020
Raychem, USA 90-01820
Redlake Corp, USA 90-02335
RF Monolithics, USA 90-02020
RLC Electronics, USA 90-02020
Rockwell International Corp, USA 75-13590
Sensor Systems Inc., USA 90-00640
Servo Corp of America, USA 90-01930
Silonex, Canada 75-03348
Sonoma Scientific, USA 90-02020
Supertex, USA 75-03348
Syntronic, USA 75-03348
Textron Speciality Materials, USA 80-00050
Thomas & Betts, USA 75-03348
Wainwright Instruments GmbH, Germany 90-02020
Warner Electric, USA 75-15150
Wolfsen, UK 75-03348

Sales agents/representatives listed elsewhere:

AlliedSignal Aerospace Marketing . . 80-00060
Andair AG 75-00500
Knabenhans 75-07100
Lockheed Martin International SA . . . 80-01400
Polytronic AG 75-10250
Trimos-Sylvac Task Force 75-14500
Warner Electric SA 75-15150

90-00070 AMETSA

Aerospace & Military Electronics Technology SA
(† Motorola Military & Aerospace Electronics Inc.)
7 rue Saint Pierre, CH-1470 Estavayer le Lac
☎(+41 37) 63 23 16; 63 23 17 ⊠ 942399 pritch
⊖ (+41 37) 63 19 93
Marketing Director: P Pritchett
■ Rugged duty Mil-Spec tool kits in olive drab with quarter turn spring latches and complete with inch or metric tools.
® USA: Jensen Tool Company [97]

90-00090 AlliedSignal Aerospace King Radio SA

(a subsidiary of AlliedSignal Aerospace, USA)
Route de Prévessin 6, PO Box 100, CH-1217 Meyrin 1
☎(+41 22) 798 58 80 ⊖ (+41 22) 798 58 83
♣ SFr 100,000 (1997) ⚭ 5 (1997)
Sales Director, Central Europe: Daniel Wildi
Field Service Engineer: Gilbert Bergès
Secretary: Mrs Denyse Annaheim
■ Product suport of Bendix/King avionics equipment.
® USA: AlliedSignal Aerospace, Bendix King

Marketing Office
PO Box 272, CH-2016 Cortaillod, NE
☎(+41 32) 842 31 31 ⊖ (+41 32) 842 31 87
Sales Director, Central Europe: Daniel Wildi [98]

90-00100 ALCOA International SA Sales Division

Aluminum Company of America
Avenue d'Ouchy 61, PO Box 249, CH-1000 Lausanne 6
☎(+41 21) 619 20 00 ⊠ 454201 ais ch
⊖ (+41 21) 619 22 64
⚭ 25
Aerospace Sales Manager: R Mocchi
Technical Service Manager: C E Brichet
Communications Manager: Ms G Raeber
■ Supply a full range of aluminium products: plate, sheet, forgings, extrusions, tubes and castings.
® USA: Aluminium Co of America [97]

90-00200 BFGoodrich Aerospace

Via Visconti 1, CH-6500 Bellinzona
☎(+41 918) 25 97 66 ⊖ (+41 918) 25 97 70
Area Manager: Kurt Arner
® USA: BFGoodrich Aerospace [98]

90-00310 Cannon Electric GmbH

Herzogenmühle 18, CH-8304 Wallisellen
☎(+41 1) 830 38 88 ⊖ (+41 1) 830 31 04
Area Manager: U Canova
® Germany: Cannon Electric GmbH [98]

90-00340 Cimcool Europe BV Swiss Branch

(a subsidiary of Cimcool Industrial Products BV, The Netherlands)
Vogelsangstrasse 11, CH-8307 Effretikon
☎(+41 52) 343 68 21 ⊖ (+41 52) 343 56 48
Managing Director: S Wohnlich
® UK: Cincinnati Milacron [98]

90-00360 Corina Corporation AG

Unter Allmend 16, CH-8702 Zollikon
☎(+41 1) 391 96 30 ⊖ (+41 1) 391 96 26
⚭ 5 (1997)
Sales and Marketing Director: Marc Gisler
■ Product development, sale and distribution of Pantron night vision equipment.
® Switzerland: Pantron [98]

90-00365 CPI International Inc

(† Varian International AG)
(a subsidiary of Communications & Power Industries Inc)
Chollerstrasse 38, CH-6303 Zug
☎(+41 41) 749 85 13 ⊖ (+41 41) 749 87 32
European Operations Director: Franco Sibille
■ Microwave, power grid and related equipment such as amplifiers and power supplies.
® USA: CPI [97]

90-00495 Dräger (Schweiz) AG

(a subsidiary of Drägerwerk AG, Germany)
Aegertweg 7, CH-8305 Dietlikon
☎(+41 1) 805 82 82 ⊖ (+41 1) 805 82 80
♣ SFr 10,000,000 (1994) ⚭ 30
Managing Director: Urs Weder
Sales and Marketing Director: Cadurisch Ebneter
Sales and Marketing Director: H J Ebneter
Purchasing Manager: D Willi
Marketing Manager: B Gloor
■ NBC protection filters for shelters and vehicles. Breathing protection equipment.
® Germany: Drägerwerk AG [97]

Contact Reference

90-00520 Eichenberger Aviation AG

Flugplatz, CH-5632 Buttwil
☎(+41 56) 675 50 60 ⊠ 827964 lszu
⊖ (+41 56) 675 50 65
General Manager: Werner Eichenberger
® USA: Enstrom Helicopter
European agent for:
Allison
Cessna
Enstrom Helicopter Corp
O & N Aircraft Modifications Inc
Soloy Corp [98]

90-00570 Erno International AG Electronic

(† Erno Electronic AG)
(a subsidiary of Erno Holding AG)
Niederhaslistrasse 12, CH-8157 Dielsdorf, Zurich
☎(+41 1) 855 53 30 Ⓔ erno@compuserve.com
⊖ (+41 1) 855 53 50
♣ SFr 500,000 ♣ SFr 1,600,000 ⚭ 2 ⌂ 200 m2
Division Manager: Daniel Hoegger
® UK: EEV Ltd [98]

90-00640 Flight Components AG

Bitzibergstrasse 5, PO Box 57, CH-8184 Bachenbülach
☎(+41 1) 861 12 00 ⊖ (+41 1) 861 12 00; 861 17 15
⚭ 20
President: Paul E Brenner
Senior Vice President: Ms Renate I Schlapfer
Sales and Marketing Director: Arthur E Müller
International Sales Manager: Thomas C Brenner
■ Manufacture of ACL 1000 anti-collision light systems; Timearc 6 GPS airborne navigation management systems. Timarc Visulizer: electronic LCD moving map display. → 80
® USA: DeVore Aviation, Sensor Systems Inc. [98]

90-00830 Heckler & Koch SA

8 rue des Beaux-Arts, PO Box 635, CH-2000 Neuchâtel
☎(+41 32) 724 52 88 ⊠ 41952706 hksach
⊖ (+41 32) 724 70 79
Director: Wolfram Mackrodt
■ Quartermaster material, pyrotechnics and raw material, machine tools, electronics.
® Germany: Heckler & Koch GmbH [98]

90-00900 Inter-CB

Postfach 47, CH-3000 Bern 15
☎(+41 31) 952 76 86 ⊖ (+41 31) 952 61 52

👥 2 🏭 30 m2
Managing Director: F Schneider
Operations Director: U Schnneider
■ Military and civil NBC-defence consulting and sales of protective equipment. Decoapparatus E-85 / Mast Leakage Test Kit.

Huber & Suhner AG
CH-8330 Pfäffikon
☎(+41 1) 952 22 11 ⊖ (+41 1) 952 24 24
Information Contact: P Moser
■ Manufacture NBC protective masks.

Aerochem H Lettko GmbH & Co KG
Am Becherweg 2, CH-6501 Ober-Olm
☎(+41 61) 36 90 81 ⊖ (+41 61) 36 97 63
Information Contact: H Lettko
■ Decontamination agents annd equipment; CW detection and simulation agents. Atropine nasal spray.

Korr. Dätwylet AG
CH-6460 Altdorf
☎(+41 4) 44 11 22 ⊖ (+41 4) 44 15 46
Information Contact: M Büeler
■ NBC protective gloves and boots.

Pavag AG
CH-6244 Nebikon
☎(+41 62) 86 81 81 ⊖ (+41 62) 86 35 66
Information Contact: G Waibel
■ Chemical protective clothing, casualty bags, tarpaulins, gloves, overboots, ponchos and first aid bandages.

Katadyn Products Inc
Birkenweg 4, CH-8304 Wallisellen
☎(+41 1) 839 21 11 ⊖ (+41 1) 830 79 42
Information Contact: R Boesch

Micronel AG
Zürcherstrasse 51, CH-8307 Tagelswangen
☎(+41 52) 355 16 16 ⊖ (+41 52) 355 16 20
Information Contact: H Hauser [98]

90-00950 Interspiro AG
(a subs. of Interspiro AB, Sweden)
Export Sales Office, Near- & Middle East
Güterstr. 47, CH-4133 Pratteln
☎(+41 61) 821 82 82 ▣ 968077 ⊖ (+41 61) 821 17 86
👥 16
Regional Export Manager, Near and Middle East: Erwin Hunziker
■ Breathing apparatus. Diving sets. Chemical protective suits, filters & filtermasks, high pressure filling stations. → 75 → 80 → 85 [98]

90-00990 Israel Military Industries Ltd Swiss Liaison Office
(† TAAS Israel Industries Ltd)
PO Box 3000, Bern 6
☎(+41 31) 352 36 77 ⊖ (+41 31) 352 36 18
Director: Z U Urbach
® Israel: Israel Military Industries Ltd [97]

90-01085 Karcher AG
(† Karcher Vaporapid AG)
(a subsidiary of Alfred Karcher GmbH & Co, Germany)
Industriestrasse 16, CH-8108 Dallikon
☎(+41 1) 846 67 11 ⊖ (+41 1) 846 67 12
Managing Director: H R Rutschmann
® Germany: Alfred Karcher GmbH & Co [98]

90-01100 Kubo-Tech AG
Dichtungen und Industriebedarf, Im Langhang 5, CH-8307 Effretikon
☎(+41 52) 354 18 18 E kubotech@bluewin.ch
⊖ (+41 52) 354 18 18
♧ SFr 200,000 👥 50 🏭 2,200 m2
Managing Director: Kurt Bosshard
Technical and Sales Manager: Roger Bartholdi
Sales and Marketing Manager: M Benaiges
Purchasing Manager: H Burkhatter
■ Distributor of seals and gaskets.
® Belgium: Advanced Products NV [98]

90-01260 Litton Precision Products International Inc
(a division of Litton Industries, Inc, USA)
Gubelstrasse 28, CH-8050 Zurich
☎(+41 1) 313 10 01 ⊖ (+41 1) 313 12 55
Managing Director: Peter Windler
Information Contact: Louise E Füglister
® USA: Litton Industries Inc [98]

90-01300 Logica Informatique SA
(a subsidiary of Logica plc, UK)
Rue du Tir 3, CH-1204 Geneva
☎(+41 22) 329 03 29 ⊖ (+41 22) 329 03 45 [98]

90-01480 Netstal-Maschinen AG
(a subsidiary of Krauss-Maffei AG, Germany)
CH-8752 Näfels
☎(+41 58) 36 61 11 ⊖ (+41 58) 34 35 41
Director: Dipl.-Ing Dieter Klug
® Germany: Krauss-Maffei Wehrtechnik [98]

90-01820 Raychem AG
(a subsidiary of Raychem Corp, USA)
Oberneuhofstrasse 8, PO Box 229, CH-6341 Baar
☎(+41 41) 768 65 00 ⊖ (+41 41) 786 65 01
Managing Director: Willy Huber
Sales and Marketing Director: Urs Beer
■ Wire and cable products to MIL/UL/CSA/VG specifications. Thermofit heat shrink tubings and moulded parts and other harness components.
® USA: Raychem [98]

90-01930 Servo Corp
Chemin du Cap 3, CH-1006 Lausanne
☎(+41 21) 29 98 73 ▣ 24420 servo ch
⊖ (+41 21) 29 98 74
Managing Director: V Liardet
® USA: Servo Corp of America [97]

90-01995 Swiss Firearms Trade
Schwamendingerstrasse 10, PO Box 9013, CH-8050 Zürich
☎(+41 1) 311 51 33 ✉ SWISSFIREARMS ▣ 822882
⊖ (+41 1) 311 51 06
Managing Director: H R Hofmann
■ Weapons, ammunition and aircraft accessories. Lasersight/Paralaser 2000, single-point sights. Military booksellers. [97]

90-02020 Telemeter Electronic AG
Im Gree 79, CH-8566 Ellighausen
☎(+41 71) 699 20 20
E m.schroettle@compuserve.com ⊖ (+41 71) 699 20 24
President: Michael Schröttle
® Germany: MTS Microtechnik GmbH, Wainwright Instruments GmbH
Israel: Elisra Electronic Systems, Orbit Advanced Technologies
USA: Advanced Control Components Inc, B&H Electronics Inc, Clarke Hess Communication Res. Corp, Data Delay Devices, General Microwave, JCA Technology Inc, Kalmus Engineering International, Pearson Electronics, RF Monolithics, RLC Electronics, Sonoma Scientific [98]

90-02300 Warner Electric
La Pierreire, CH-1029 Villars Sainte Croix
☎(+41 21) 631 33 55 ⊖ (+41 21) 636 07 04 [97]

90-02335 Weinberger AG
Bernstrasse 88, CH-8953 Dietikon
☎(+41 1) 741 44 66 W http://www.weinberger.ch
⊖ (+41 1) 741 00 33
♧ SFr 800,000 (1997) 👥 30 (1997)
President: Alexander Weinberger
Sales and Marketing Director: F Schreppers
Operations and Purchasing Director: H J Keller
Sales Manager: S Trost
Information Contact: M Gabathuler
■ High speed photo and video instrumentation. Specialised cameras for automotive engineering, research ballistics and other industrial applications.
® USA: L-W International, Redlake Corp
Affiliates:
Weinberger Deutscland GmbH
Weinberger Svenska AB [98]

Syria

Head of State:
Gen. Hafiz al-Asad

10 Head of State

10-10000 Office of the President
Abu Rumaneh, Al-Rashid Street, Presidential Palace, Damascus
☎(+963 11) 33 11 12 ▣ 419160 mumjed sy
President of the Syrian Arab Republic: Gen Hafiz al-Asad
Vice President: Rifaat Al-Asad
Vice President: Abd al-Halim ibn Said Khaddam
Vice President: Muhammad Zuhayr Mashariqa
Minister for Presidential Affairs: Wahib Fadil [98]

11 Prime Minister

11-10000 Office of the Prime Minister
Shahbandar Street, Damascus
☎(+963 11) 22 60 00
Prime Minister: Mahmud Zubi
Deputy Prime Minister and Minister of Defence: Lieut Gen Mustafa Talas
Deputy Prime Minister (Economic Affairs): Salim Yasin
Deputy Prime Minister (Services Affairs): Rashid Akhtarini [98]

12 Ministries

12-10000 Ministry of Defence
Place des Omayades, Damascus
☎(+963 11) 11 21 01 ▣ 411500 padar sy
Deputy Prime Minister and Minister of Defence: Lieut Gen Mustafa Talas [98]

12-20000 Ministry of Foreign Affairs
Al Rashid Street, Damascus
☎(+963 11) 33 12 00; ▣ 419123 karji
Minister: Farouk al-Shara
Minister of State: Nasir Quaddur ☎(+963 11) 33 72 00 [98]

12-30000 Ministry of Economy and Foreign Trade
Salhieh, Ma'mun Bitar Street, Damascus
☎(+963 11) 11 35 13; 11 35 15 ▣ 411982 econom sy
Minister: Muhammad al-Imadi [98]

12-40000 Ministry of Interior
Merjeh Circle, PO Box 6351, Damascus
☎(+963 11) 22 01 00; 22 75 41
Minister: Muhammad Harba [98]

12-60000 Ministry of Finance
Jule Jamal Street, PO Box 13136, Damascus
☎(+963 11) 23 96 24 ✉ FINANCE Damascus
⊖ (+963 11) 22 47 01
Minister: Khaled Al-Mahayni [98]

25 Defense Forces

25-10000 Armed Forces General Headquarters
Omayad Square, Damascus
☎ (+963 11) 22 91 00; 331 42 00
⊖ (+963 11) 223 08 97
Chief of Staff Army & Armed Forces:
Lt Gen. Hikmat Shehabi
Deputy Chief of Staff: Lt Gen. Hasan Tourkmani
☎ (+963 11) 372 08 10; 333 37 05 (fax)
Industrial Defence Establishment Dir:
Maj.Gen. M. Khaled Ahmar
Productive Production Admin Dir:
Maj.Gen. M. Ismael Jahjah ☎ (+963 11) 21 49 14; (+963 11) 22 15 21 (fax)

Army Supply Bureau (ASB)
Arsenal, PO Box 3361, Damascus
☎ (+963 11) 212 53 86 ✉ ARSENAL Damascus
⊠ 411371 fermod sy ⊖ (+963 11) 212 29 39
Director: Maj.Gen. Kamel Eid

Industrial Defence Establishment
PO Box 2330, Damascus
☎ (+963 11) 445 33 13 ⊠ 412909; 411029
⊖ (+963 11) 45 67 49
Director: Maj.Gen. M. Khaled Ahmar

Military Construction Establishment
PO Box 3108, Damascus
☎ (+963 11) 211 12 16 ⊠ 411376
⊖ (+963 11) 223 99 32
Director: Brig.Gen. Riad Salman Issa 97

25-20000 HQ Air Force & Air Defence
Square des Omayades, PO Box 90, Damascus
☎ (+963 11) 333 55 05 ⊠ 411900 safcony sy
⊖ (+963 11) 331 04 40
Commander of Air Forces and Air Defence:
Maj Gen Mohammed Khouli
Chef d'Etat-Major: Gen. DA Hikmat Chehabi 98

25-40000 HQ Naval Forces
Square des Omayades, Damascus
☎ (+963 11) 23 92 48; 42 29 22 ⊠ 451141; 451077
⊖ (+963 11) 41 04 51
Commander of Naval Forces: Maj.Gen. Wael Naser 97

31 Embassies & Attachés

31-00090 Australian Embassy
127A Farabi Street, East Mezzeh, Damascus
☎ (+963 11) 613 24 24 ⊠ 419132 austem sy
⊖ (+963 11) 662 11 95
Ambassador: H.E. Mr Jim Dollimore
Second Secretary Political & Economic: Ms Jane Urquhart 97

31-00350 Embassy of Chile
43 Al-Rachid Street, Charkasieh-Chaura, Damascus
☎ (+963 11) 331 18 91 ⊖ (+963 11) 333 15 63
Ambassador: H E Mr Julio Lagarini
Second Secretary: Gonzalo Fernandez 98

31-00590 Embassy of France
Ata Ayoubi Street, PO Box 769, Damascus
☎ (+963 11) 332 79 93 [E] jeromecauchard@.diplomtic.fr
⊖ (+963 11) 333 86 32
Ambassador: H E Mr C H d'Aragon
Deputy Chief of Mission: Jean-Jacques Beaussow
First Counsellor: Jean-Pierre Filiu
Armed Forces Attaché: Col A Roignan
Assistant Army Attaché: Maj Jean-Claudia Serot
Press Attaché: Jerome Cauchard
Commercial Officer: Jean-Loius Fontenille

Commercial Attaché's Office
Maysaloun Street, Cham Hotel, PO Box 2629, Damascus
☎ (+963 11) 221 34 14; 221 39 87
⊠ 411367 comata sy ⊖ (+963 11) 221 34 14
Commercial Counsellor: Jean Louis Fontenille 98

31-00690 Embassy of Greece
1 rue Farabi Imm Telle, Mezzeh, Damascus
☎ (+963 11) 223 30 09; 223 29 20 ⊠ 411045
⊖ (+963 11) 223 29 20 98

31-00810 Embassy of the Republic of Hungary
Mezzah, East Villas, PO Box 2607, Damascus
Mezzeh, East Villas,
Al Fursan Street, No 102, PO Box 2607
☎ (+963 11) 611 07 87; 611 79 66
⊠ 419151 hungex sy ⊖ (+963 11) 611 79 17
Ambassador: H E Mr Zoltan Pereszlenyi
☎ (+963 11) 613 22 89
Counsellor: Miklos Karoati ☎ (+963 11) 613 25 33
Attaché: Gabor Jurcza
Commercial Counsellor: Zsolt Sajti
☎ (+963 11) 613 25 16
Attaché: Peter Kadaas 98

31-00890 Embassy of Italy
Al Mansour Street 82, PO Box 2216, Damascus
☎ (+963 11) 333 83 38; 333 26 21 ⊠ 411180 itadip sy
⊖ (+963 11) 332 03 25
Ambassador: H E Mr Antonio Napolitano
Defence Attaché: Col Giacomo Lanza
First Secretary: Lorenzo Galanti
Counsellor: Mario Bologna 98

31-01370 Embassy of Pakistan
PO Box 9284, Damascus
☎ (+963 11) 613 26 94; 613 26 95
⊖ (+963 11) 613 26 62
Ambassador: H E Mr Afzal Akbar Khan
☎ (+963 11) 613 26 96
Deputy Chief of Mission and Counsellor:
Shahzada Ziauddin Ali ☎ (+963 11) 613 26 93
Third Secretary, Commercial: Shujjat Ali Rathore 98

31-01730 Embassy of Sweden
Chakib Arslan Street, Abou Roumaneh, PO Box 4266, Damascus
☎ (+963 11) 332 72 61; 332 72 62
✉ SVENSK Damascus ⊠ 411339 svensk sy
⊖ (+963 11) 332 77 49
Ambassador: H.E. Mr Jan Nordlander
Deputy Chief of Mission: Niklas Kebbon
Commercial Officer: Mrs Hannan Ghannoum 97

31-01740 Embassy of Switzerland
PO Box 234, Damascus
☎ (+963 11) 331 18 70/1 ⊖ (+963 11) 332 11 37 97

31-01890 Embassy of the United Kingdom
PO Box 37, Imm Kotob, Damascus
Quarter Malki, 11 Mohammad Kurd Ali Street,
Imm Kotob, Damascus
☎ (+963 11) 371 25 61; 371 25 63
⊖ (+963 11) 373 16 00
Ambassador: H E Mr Basil Eastwood
Counsellor and Deputy Chief of Mission: P Whiteway
Defence, Naval, Military and Air Attaché: Col R Clarke
First Secretary Chancery: James Watson
Commercial Officer: Richard Salt
Information Officer: Ms Rosemary Davis 98

31-02010 Embassy of the Federal Republic of Yugoslavia
Chare Al Jalaa 18, PO Box 739, Damascus
☎ (+963 11) 333 62 22; 333 30 69
⊠ 412646 yuamba sy ⊖ (+963 11) 333 36 90
Ambassador: H E Mr Zoron S Popovich 98

Taiwan

Head of State:
Lee Teng-hui

10 Head of State

10-10000 Office of the President
Chiehshou Hall, 122 Chungking South Road, Section 1, 100 Taipei
☎ (+886 2) 311 37 31 ⊖ (+886 2) 314 07 46
President: Lee Teng-hui
Vice President: Lien Chan 98

10-50000 National Security Council
Chiehshou Hall, 122 Chungking S. Rd, Sec.1, 100 Taipei
☎ (+886 2) 371 85 78 ⊖ (+886 2) 371 85 99
Chairman: Lee Teng-Hui
Secretary General: Liu Chao-shiuan 97

11 Prime Minister

11-10000 Office of the Premier Executive Yuan
1 Chunghsiao East Road, Section 1, 10023 Taipei
☎ (+886 2) 356 15 00 ⊖ (+886 2) 394 87 27
President, Executive Yuan (Premier): Lien Chan
Vice President, Executive Yuan (Vice Premier): Hsu Li-teh
Secretary, Executive Yuan: Chao Shou-po 98

12 Ministries

12-10000 Ministry of National Defence
Chiehshou Hall, Chungking South Road, 100 Taipei
☎ (+886 2) 311 61 17
Minister of National Defence: Gen Chiang Chung-ling (retd) 98

12-20000 Ministry of Foreign Affairs
2 Kaitakelan Boulevard, 100 Taipei
☎ (+886 2) 311 92 92 ⊠ 11299 mofaroc
⊖ (+886 2) 314 49 72
Minister of Foreign Affairs: John Chang 98

12-30000 Ministry of Economic Affairs
15 Foochow Street, 100 Taipei
☎ (+886 2) 321 22 00 ⊖ (+886 2) 391 93 98
Minister: Chih-Kang Wang 98

12-40000 Ministry of the Interior
5-9th Floor - 5 Hsuchow Road, 100 Taipei
☎ (+886 2) 356 50 00 ⊖ (+886 2) 356 62 01
Minister: Feng-cheng Lin

National Police Administration
7 Chunghsiao E. Rd. Sec. 1, 100 Taipei
☎ (+886 2) 321 90 11 ⊖ (+886 2) 394 05 84
Director General: Shih-si Yen ☎ (+886 2) 321 47 43
Deputy Director General: Shu-yu Lee
☎ (+886 2) 392 87 75
Deputy Director General: Yu-tang Yu
☎ (+886 2) 321 54 23
Deputy Director General: Yuan-chin Ting
☎ (+886 2) 392 87 55 98

12-50000 Ministry of Finance
2 Aikuo West Road, 100 Taipei
☎ (+886 2) 322 80 00 ⊖ (+886 2) 396 58 29
Minister: Paul Chiu

Directorate General of Customs
13 Ta Cheng Street, 103 Taipei

☎(+886 2) 550 81 66 ⊖ (+886 2) 550 81 24
Director General: De-Ho Jan
Deputy Director General: Mao Chi-wei
☎(+886 2) 741 28 92
Deputy Director General: Hou Teh-chi
☎(+886 2) 751 69 71 [98]

15 Parliament

15-10000 National Assembly
53 Chunghwan Road, Sec. 1, 100 Taipei
☎(+886 2) 331 13 12 ⊖ (+886 2) 314 20 56
Secretary General: Chen Chin-jang ☎(+886 2) 331 19 86 [97]

15-20000 Control Yuan
2 Chunghsiao East Road, Section 1, 100 Taipei
☎(+886 2) 341 31 83
⊖ (+886 2) 394 09 10; 351 50 85
President, Control Yuan: Tso-yung Wang [98]

15-30000 Legislative Yuan
1 Chungshan South Road, 10040 Taipei
☎(+886 2) 321 15 31 ⊖ (+886 2) 322 35 57
President, Legislative Yuan: Sung-fan Liu [98]

15-40000 Judicial Yuan
124 Chungking South Road, Section 1, 10036 Taipei
☎(+886 2) 371 32 60
President, Judicial Yuan: Chi-yang Shih [98]

20 Defense Administrations/Services/Agencies

20-10000 Government Information Office
2 Tientsin Street, 100 Taipei
☎(+886 2) 322 88 88 ✉ INFORMTION Taipei
⊠ 11636 inform ⊖ (+886 2) 356 87 33
Director General: Jason C. Hu ☎(+886 2) 321 43 25
Deputy Director General: Albert T.H. Yeh
☎(+886 2) 321 36 92
Deputy Director General: Wu Chung-lih
☎(+886 2) 321 12 16
Dept of Compilation & Translation Dir: Lee Shou-hua [97]

25 Defense Forces

25-10000 General Headquarters Combined Service Forces
PO Box 90482, Nankang, Taipei
☎(+886 2) 785 41 21
Commander in Chief: Gen. Wang Wen-hsieh
Inspector General Dir Inspection Dept: Gen. Ting Chih-fa
Deputy Chief of General Staff for Personnel: Lt Gen. Ma Chung-wu
Deputy Chief of General Staff for Intelligence: Lt Gen. Huang Yun-dea
Deputy Chief of General Staff for Operations: Lt Gen. Sung Chuan-chiang
Deputy Chief of the General Staff for Logistics: Lt Gen. Yang Te-chih
Deputy Chief of General Staff for Planning: VAdm. Ho Chao-pin
Military Intelligence Bureau Dir: Lt Gen. Hu Chia-chi [97]

25-20000 General Headquarters Air Force
Air Force Building, Jen-Ai Rd, Sec. 3, 10600 Taipei
☎(+886 2) 711 17 44 ⊖ (+886 2) 731 87 17
Commander in Chief: Gen Huang Hsien-yung [97]

25-30000 General Headquarters Army
PO Box 90601, Lungtan, Taoyuan
☎(+886 3) 479 21 11
Commander-in-Chief: Gen. Lee Chen-lin
Bureau of Military Procurement Dir: Lt Gen. Sun Tao-yu [97]

25-31000 Armed Forces Police Command
Linden N. Road, Taipei
Commanding General: Lt Gen. Tsao Wen-sheng [97]

25-40000 General Headquarters Navy
3015 Pei-An Road, Tachih, Taipei
☎(+886 2) 505 38 11 ⊖ (+886 2) 502 46 98
Commander-in-Chief: Adm. Ku Chung-lien [97]

28 Coastguard, Border Protection Forces

28-10000 Armed Forces Reserve & Coast Guard Command
172 Po-Ai Road, 100 Taipei
☎(+886 2) 311 15 01
Commander: Gen. Wang Juo-yu [97]

29 Customs & Excise

29-10000 Directorate General of Customs
13 Ta Cheng Street, 103 Taipei
☎(+886 2) 550 81 66 ⊖ (+886 2) 550 81 24
Director General: Jan De-ho
Deputy Director General: Mao Chi-wei
☎(+886 2) 554 65 88
Deputy Director General: Hou Teh-chi
☎(+886 2) 554 66 99
Secretary to Director General: Ms Mindy Wei [98]

31 Embassies & Attachés

31-01400 Embassy of Paraguay
1 Alley 52, Lane 117, Tien-Mou West Road, Taipei
☎(+886 2) 873 63 10/1 ⊠ 13744 embapar
⊖ (+886 2) 873 63 12
Ambassador: H.E. Mr Miguel A. Solano Lopez
First Secretary: Bernardo J. Balbuena
Military Attaché: Col Carlos A. Baruja Arbues [97]

60 Research Organizations, Institutes & Test Establishments

60-00500 Aero Industry Development Center (AIDC)
PO Box 90008-10, 407 Taichung
☎(+886 4) 256 23 38 ⊖ (+886 4) 256 22 82
👥 6 000
Dir General: Gen. Wen-Li Lin
Vice Dir General: Dr Shih-Sen Wang
Strategy & Planning Dir: Tzeng-Chy Mao [97]

60-01000 Chungshan Institute of Science and Technology (CSIST)
Lung Tan, PO Box 90008-1, Lung Tan, Taoyuan
☎(+886 3) 471 22 01 ⊖ (+886 3) 471 10 57
President: Vice Adm F P Shen
Vice President: Dr Jackson Chung
Deputy Director Planning Division: Chung-Mei Ho
■ Design, development and manufacture of Hsiung Feng II surface-to-surface (anti-ship) missile, Tien-Kung I and II surface-to-air missiles, Tien-Chien I and II air-to-air missiles, Hsiung Feng II surface-to-surface missile, Sky Bow I surface-to-air missile, Sky Sword I air-to-air missile. Diversification into civil sector by promoting dual-use technologies and spin-off military technologies. [98]

75 Manufacturing Companies

75-00500 Aero Industry Development Center (AIDC)
PO Box 90008-10-14, 40722 Taichung
111 Lane 68 - Fu-Hsing N Road, 407 Taichung
☎(+886 4) 256 21 39; 256 22 26 (public relations); 256 23 27; 256 23 36 (business)
⊖ (+886 4) 256 22 82; 256 23 25 (business); 256 23 70 (public relations)
👥 4,400 (1997)
Chairman: Chuen-Huei Tsai
President: Shih-Sen Wang
Vice President Operations: Yuan-Shi Peng
Vice President Business: Yuean-Yuan Chou
Vice President Engineering: An-Dong Han
Vice President Resources and Services: Tai-Nan Chao
Specialist, Business: Hua-Lang Chang
■ Capabilities: design and manufacture of military aircraft including PL-1 trainer, T-CH-1 trainer, AT-3 trainer, IDF fighter. Manufacture under licence UH-1H helicopter and F-5E/F fighter. Manufacture of aircraft and engine components including F-16 airframe, Falon 900/2000 rudder, B-757 support fitting and GE engine parts. [98]

75-04700 China Steel Corporation
Lin Hai Industrial District, PO Box 47-29, Hsiao Kang, Kaohsiung
☎(+886 7) 802 11 11 W http://www.csc.com.tw
⊖ (+886 7) 802 25 11; 803 32 24
♧ NT$ 85,000,000,000 ♣ US$ 2 230 000 000 (1993)
👥 9,050 (1997)
Chairman of the Board: C Y Wang
President: J Y Chen
Executive Vice President: Y C Chen
Vice President Administration Division: R S Jong
Vice President Commercial Division: J C Tsou
Vice President Finance Division: L M Chung
Vice President Technology Division: G H Cheng
Vice President Engineering Division: K L Du
Vice President Production Division: T T Cheng
Manager Control and Evaluation Section - Corporate Planning Department: C F Tsai
■ Carbon steel plates, wire rods, bars, hot and cold rolled products, electro-galvanised and electrical sheets. Stainless steel hot rolled bands.

Affiliated Companies:
CS Aluminium Corporation
China Steel Express Corporation
China Steel Global Trading Corporation
Gains Investment Corporation
China Steel Structure Corporation
China Steel Chemical Corporation
China Hi-Ment Corporation
China Ecotek Corporation
Taisil Electronic Materials Corporation
Kuei Yi Industrial Co. Ltd
Himag Magnetic Corporation

Taipei Office
5th Floor - 25 Jen-Ai Road, Section 4, 106 Taipei
☎(+886 2) 721 64 11 ⊖ (+886 2) 721 03 93 [98]

75-05850 Ever Bright Precision Ltd
6F-1 n°148 Tung-An Street, 106 Taipei
☎(+886 2) 736 47 09 ⊖ (+886 2) 736 90 84
Managing Director: Larry Chen
® Switzerland: Trimos-Sylvac Task Force [98]

75-08000 Getac Corp
Hsinchu
4th Floor, 1 R&D Road,
Hsinchu Science Based Industrial Park
☎(+886 35) 77 82 75; 79 44 32 ⊖ (+886 35) 77 82 70
Programmes Manager: Kun-Ming Wu
■ Ruggedized portable computers qualified to MIL-Std 810D, 461-C, 167C. [98]

75-10000 Hwan Chong Enterprize Co. Ltd
N°201 Wan Dan Road, Yung Fan Village, Ta-Liao Shing, Kaohsiung Hsien
☎(+886 7) 781 11 31; 781 11 32; 781 11 33
⊖ (+886 7) 782 11 25
👥 306
Manager: M Oka [98]

75-14000 Masbe Corporation Ltd
162 Kwei Kang Village, Lio Chia Hsiang, Tainan
☎(+886 6) 698 22 56 ⊖ (+886 6) 698 92 38
♧ US$ 1 200 000 (1996) 👥 66 🏭 6 000 sq m
Managing Dir: David C. Chen
Operations & Technical Dir: L.P. Chen
Sales & Marketing Dir: Anthony Su
■ Machining, stamping, forming, welding, chemical cleaning & brazing. Testing & inspection, tooling design, H.T. & Plasma coating. [97]

75-18000 Radstone Technology
Fl 4 No 568, Chung Cheng Road, Hsintien, Taipei
☎ (+886 2) 218 58 19 ⊖ (+886 2) 218 62 48
Managing Director: Peter Leggett
General Mgr: Louis Lin
Operations & Technical Dir: Albert Chen
Purchasing Mgr: Shelly Chen
■ Supplier of standard and custom board level, systems and software products for military and industrial applications. Products support open bus architectures (VME) including processors, memories, I/O and enclosures. [98]

80 Service Companies & Organizations

80-00128 AlliedSignal Aerospace Marketing Sales and Service
Kaohsiung Hsien
N°4 Land 15 (A-3), Ho-Sheng West Street,
Ta-Hwa Village, Niao-Sung, 833 Kaohsiung Hsien
☎ (+886 7) 625 33 64 ⊖ (+886 7) 392 78 53
Field Service Engineer: Paul Limsirichai
☎ (+886 9) 273 13 59 (mobile)
■ On-site customer support for AlliedSignal Aerospace. [98]

80-00130 AlliedSignal Aerospace Marketing Sales and Service
2F#12 Da-De 1st Road, 101 Lane, Kangshang, Kaohsiung
☎ (+886 7) 621 96 44 ⊖ (+886 7) 621 96 44
Field Service Engineer: George (Yun-Tien) Tai
■ On-site customer support for AlliedSignal Aerospace. [98]

80-00132 AlliedSignal Aerospace Marketing Sales and Service
5F #70 1-Tung Road, Tainan
Kaohsiung International Airport, UNI Airways, Kaohsiung
☎ (+886) 78 03 69 46 ⊖ (+886) 22 75 93 46
Field Service Engineer: Ching-Chung Lai
■ On-site customer support for AlliedSignal Aerospace. [98]

80-00134 AlliedSignal Aerospace Marketing Sales and Service
205 Tun Hwa North Road, Suite 1205 Bank Tower, Taipei
☎ (+886 2) 719 57 96 ⊖ (+886 2) 719 57 95
Regional Director, Taiwan: Magnus Lou
☎ (+886 2) 719 57 96
Manager, Marketing and Sales: Jay Min [98]

80-00136 AlliedSignal Aerospace Marketing Sales and Service
Taoyuan
c/o China Airlines Maintenance and Engineering Division, CKS International Airport, Taoyuan
☎ (+886 3) 383 40 86 ⊖ (+886 3) 383 40 93
Field Service Engineer: Sam L Lee
■ On-site customer support for AlliedSignal Aerospace. → 90 [98]

80-00140 AlliedSignal Engines (ITEC)
International Turbine Engine Corporation
PO Box 90008-12-3, 40722 Taichung
☎ (+886 4) 254 72 32 ⊖ (+886 4) 254 04 70
Director, ROC Office: Hsin-Chien Liu
Manager, Marketing and Sales: Jay Min
■ On-site customer support for AlliedSignal Aerospace. → 90 [98]

80-00650 Hughes International Corporation
Room 701, Bank Tower, 205 Tun Hwa North Road, 10592 Taipei
☎ (+886 2) 715 12 92 ⊖ (+886 2) 712 48 15 [98]

90 Sales Agents, Sales & Liaison Offices

Companies represented:

AAR Brooks & Perkins, USA . . . 90-01290
AEG Aktiengesellschaft, Germany . . . 90-00050
Amherst Systems Inc, USA . . . 90-02005
AS&E, USA . . . 90-01480
Carco Electronics, USA . . . 90-02005
Chemring Ltd, UK . . . 90-01480
Comptek Federal Systems Inc, USA . . . 90-02005
Comsat-RSI Mark Antenna, USA . . . 90-01480
Conax Florida Corp, USA . . . 90-01480
Daicel Chemical Industries, Japan . . . 90-00440
Danish Camp Supply, Denmark . . . 90-01290
Dantec Measurement Technology AS, Denmark . . . 90-02005
Firearms Training Systems, USA . . . 90-01290
FL Jennings, USA . . . 90-01470
Garrett Security Systems, USA . . . 90-01480
Grumman Corp., USA . . . 90-01340
GTE Government Systems Corp, USA . . . 90-02005
Kamewa AB, Sweden . . . 90-02005
Kollmorgen Corp (Electro-Optics Div), USA . . . 90-02005
Marine Data Ltd, UK . . . 90-01480
Minebea Co. Ltd, Japan . . . 90-01325
Motorla GSTG, USA . . . 90-02005
Nico-Pyrotechnik Hanns-Jürgen Diederichs, Germany . . . 90-01290
NMP Corp, USA . . . 90-01480
OFEMA, France . . . 90-01475
PK Electric, Germany . . . 90-01290
Poongsan Corporation, Korea (ROK) . . . 90-01290
Rockwell International Corp., USA . . . 90-01830
Saint John Shipbuilding Ltd, Canada . . . 90-02005
Sperry Marine Inc, USA . . . 90-01480
Steyr-Daimler-Puch Spezialfahrzeug AG, Austria . . . 90-01290
STN Systemtechnik Nord., Germany . . . 90-00050
Teledyne Inc., USA . . . 90-02020
Trimos-Sylvac Task Force, Switzerland . . . 75-05850
Varian/Electron Device Grp, USA . . . 90-01470
Watking-Johnson International, USA . . . 90-02005

Sales agents/representatives listed elsewhere:

AlliedSignal Aerospace Marketing . . . 80-00136
AlliedSignal Engines . . . 80-00140

90-00020 Acer Incorporated
156 Min Sheng E. Rd. Sec.3, 6 Flr, 105 Taipei, Taiwan
♣ US$ 1 898 000 000 (Group) [97]

90-00050 AEG Taiwan Co. Ltd
Room 305, Worldwide House, 129 Min Sheng E.Rd Sec 3, 105 Taipei
☎ (+886 2) 545 60 50 ⊖ (+886 2) 546 08 86
♧ US$ 800 000 ♣ US$ 10 000 000 (1995) 👥 12
🏭 3 405 sq m
Chairperson: F.D. Maier
Managing Dir: Heinz Rabe
Sales & Marketing Dir: Alex Huang
Operations & Technical Dir: Horst Stöckermann
® Germany: AEG Aktiengesellschaft, STN Systemtechnik Nord. [97]

90-00350 CIC International Ltd
(a subs. of CIC Int'l Ltd, USA)
12F, No. 74, Sung-Teh Rd, Taipei
☎ (+886 2) 758 72 57 ⊖ (+886 2) 758 21 37
♣ US$ 34 260 392 👥 10
President: Henry Liu
■ Traditional weapon systems with their spare parts, such as artillery, motor, AA guns, small arms and its ammunition, missiles, rockets and thermal-imaging equipment. Tanks' AMMOS and its fire-control system. fuzes, mines, grenades. Advanced technology transfer. Thermal and military battery, radios. [97]

90-00400 Daewoo Corp Taipei Branch
N°134 Cosmos Building 3rd Floor, Min Sheng East Road 3, Taipei
☎ (+886 2) 716 00 02 ⊠ 20469 daewoo tw
⊖ (+886 2) 514 81 74 [98]

90-00440 Daicel Kisco Co. Ltd
Asia Enterprise Center Building,
Room 1421, 602 Min Chuan East Road, Taipei
☎ (+886 2) 712 47 49 ⊖ (+886 2) 712 89 21
® Japan: Daicel Chemical Industries [97]

90-00450 Dana Industrial, Asia Pacific
3 Fl No. 35, Lane 32, Kwang-Fu S. Rd, Taipei
☎ (+886 2) 577 81 56 ⊖ (+886 2) 570 63 58
President: George C.T. Lee [97]

90-00950 Kawasaki Heavy Industries Ltd Taipei Office
15th Floor Fu-Key Building, 99 Jen-Ai Road - Section 2, Taipei
☎ (+886 2) 322 17 52 ⊖ (+886 2) 322 50 09 [98]

90-01150 Kyokuto Boeki Kaisha Ltd Taipei Branch Office
10th Floor Capital Building, 158 Sung Chiang Road, Taipei
☎ (+886 2) 551 41 59 ⊖ (+886 2) 561 60 85 [98]

90-01290 Luen Hing Taiwan Co. Ltd
Floor 17-18 - 77, Hsin Tai 5th Road, Section 1, Hsi Chih, Taipei
☎ (+886 2) 698 81 61 ⊠ LHTAIWAN
Ⓔ lhtgroup@ms2.hinet.net ⊖ (+886 2) 698 81 66
♧ US$ 500,000 (1997) ♣ US$ 5,000,000 (1996)
👥 30 (1997) 🏭 3,000 sq ft
Managing Director: Alan C Y Huang
Deputy Managing Director Import Division: Andy Huang
■ Night vision and surveillance equipment. Firearms simulators, pistols, rifles, machine guns, shelters, mine buriers, detectors and ammunition. Mortar training systems.
® Austria: Steyr-Daimler-Puch Spezialfahrzeug AG
Denmark: Danish Camp Supply
Germany: Nico-Pyrotechnik Hanns-Jürgen Diederichs, PK Electric
Korea (ROK): Poongsan Corporation
USA: AAR Brooks & Perkins, Firearms Training Systems [98]

90-01325 Minebea Tehnologies (Pte) Ltd Taiwan Branch
Room C, 13F, Sanhsin Building,
N°167 Fu-Hsing North Road, Taipei
☎ (+886 2) 718 79 86 ⊖ (+886 2) 718 58 83
👥 175
General Manager: T Tsurutu
® Japan: Minebea Co. Ltd [98]

90-01340 Northrop Grumman International Inc.
7th Fl. Cheng Tung Bldg, 212 Cheng Hsiao E. Rd, Sec 4, Taipei
☎ (+886 2) 752 74 37/9 ⊖ (+886 2) 777 30 86
Director: Joe L. Tieng
® USA: Grumman Corp. [97]

90-01470 Oceanic Trading Corp.
10th Floor, Rose Mansion, PO Box 891, Taipei
☎ (+886 2) 707 71 11-4; 703 09 90 ⊠ 11458 oceanic
⊖ (+886 2) 703 07 70
♧ NT$ 18 000 000 (1996) ♣ US$ 5 000 000 (1995)
👥 20 (1995)
President: Sam C.S. Hsu
® USA: FL Jennings, Varian/Electron Device Grp [97]

90-01475 OFEMA Taipei Branch
8th Floor - n°128, Section 3 - Ming Sheng E Road, Taipei
☎ (+886 2) 719 76 17 ⊖ (+886 2) 719 76 18
Technical Delegate: M Baratte
® France: OFEMA [98]

90-01480 Orlando Engineering Co. Ltd
PO Box 101-79, Taipei
2nd Floor - N°121, Min-Sheng East Road, Section 3, 105 Taipei

☎(+886 2) 719 53 54 Ⓔ orlando@ms7.hinet.net
⊖ (+886 2) 715 38 01
NT$ 12,000,000 (1997) ♣ US$ 4,000,000 (1997)
20 (1997) 200 m2
Managing Director: Jesse Lo
General Manager: Dick Lane
■ Marine electronics, weapon systems/control, military sub-systems and components.
® UK: Chemring Ltd, Marine Data Ltd
USA: AS&E, Comsat-RSI Mark Antenna, Conax Florida Corp, Garrett Security Systems, NMP Corp, Sperry Marine Inc [98]

90-01490 Parsons Overseas Company
(† Parsons Corp.)
6/F, 118-1 Chung Cheng Rd, Shihlin, Taipei
☎(+886 2) 838 41 12 ⊖ (+886 2) 836 84 57
Vice President: J.G. Motheral [97]

90-01800 Rafael Armament Development Authority
5FL No 28 Lane 19, Chung-Shan N Rd S., Tien Mu, Taipei
☎(+886 2) 873 23 16 ⊖ (+886 2) 873 23 17
General Mgr: Samuel Hadar [97]

90-01830 Rockwell International Taiwan Company Ltd
Room 2808, International Trade Bldg, 333 Keelung Road, Section 1, 10548 Taipei
☎(+886 2) 720 02 82 ⊠ 28101 ⊖ (+886 2) 757 67 60
Director - Taiwan Operations: C S Lai
Ⓔ cs.lai@nb.rockwell.com
Manager - Asia Pacific (Rockwell Semiconductor Systems): Ms Rita Ger
® USA: Rockwell International Corp. [98]

90-02005 Tai-E Trading Co Ltd
PO Box 81-265, Taipei
Floor 4-1, N°130 - NanKing East Road Section 4
☎(+886 2) 577 33 25 Ⓔ taietrad@msl.hinet.net
⊠ 22328 taicotrn ⊖ (+886 2) 577 99 51
US$ 1,460,000 (1996) ♣ US$ 15,000,000 (1995)
23 (1996) 5,544 sq ft
President: Frederic C Fey
Vice President: Ming-Chao Hsu
Advertising Manager: William Ai
® Canada: Saint John Shipbuilding Ltd
Denmark: Dantec Measurement Technology AS
Sweden: Kamewa AB
USA: Amherst Systems Inc, Carco Electronics, Comptek Federal Systems Inc, GTE Government Systems Corp, Kollmorgen Corp (Electro-Optics Div), Motorola GSTG, Watking-Johnson International [98]

90-02020 Teledyne Industries International Inc.
201 Tun Hwa North Rd, 10th Fl, Bldg B, PO Box 81-665, 105 Taipei
☎(+886 2) 713 81 01
⊠ TWX 910 240 7684 tdy tio taiwan
⊖ (+886 2) 713 81 08
® USA: Teledyne Inc. [97]

90-02100 Texas Instruments Taiwan Ltd
Defence Systems & Electronics Group
23/F, Taipei Internatinal Bldg A, #216, Sec. 2, Tun Hua S. Road, Taipei
☎(+886 2) 377 27 15 ⊖ (+886 2) 377 27 16
Manager: Peter Chang
■ TI products & services include semi- conductors; defence electronics systems; software productivity tools; printers, notebook computers & consumer electronics products; custom engineering & manufacturing services; electrical controls & metallurgical materials. [97]

Tajikistan

President:
Emomali Rahmonov

10 Head of State

10-10000 Office of the President
Presidential Palace, Dushanbe
☎(+7 3772) 23 29 81
President: Emomali Rahmonov [98]

11 Prime Minister

11-10000 Office of the Chairman of the Cabinet of Ministers
Dushanbe
☎(+7 3772) 23 28 04
Prime Minister: Yahyo Azimov
First Deputy Prime Minister: Yuriy Ponosov
Deputy Prime Minister: Bozqul Dodkhudoyeva
Deputy Prime Minister: Sadriddin Ghiyasov
Deputy Prime Minister: Jalol Mansurov
Deputy Prime Minister: Ramazan Mirzoyev
Deputy Prime Minister: Abdurakhman Nazimov
Deputy Prime Minister: Kholis Timurjonov [98]

12 Ministries

12-10000 Ministry of Defence
Dushanbe
Minister: Maj Gen Sherali Khayrulloyev [98]

12-20000 Ministry of Foreign Affairs
Dushanbe
Minister: Talbak Nazarov [98]

12-30000 Ministry of Foreign Economic Relations
Dushanbe
Minister: Tukhtaboy Ghafforov [98]

12-40000 Ministry of Internal Affairs
Dushanbe
Minister: Maj Gen Khomiddin Sharipov [98]

Tanzania

Head of State:
William Benjamin Mkapa

10 Head of State

10-10000 Office of the President
State House, PO Box 9120, Dar es Salaam
☎(+255 51) 232 61, 210 85
President: William Benjamin Mkapa
Vice President: Omar Ali Juma [98]

11 Prime Minister

11-20000 President's Office
The State House, PO Box 776, Zanzibar
☎(+255 51) 308 14; 308 15
President of Zanzibar and Chairman of Revolutionary Government: Salmin Armour [98]

12 Ministries

12-10000 Ministry of Defence and National Service
PO Box 9544, Dar-es-Salaam
☎(+255 51) 388 80; 388 88 ✉ ULINZI ⊠ 41721
⊖ (+255 51) 290 63
Minister of Defence: Edgar Maokola Majogo [98]

12-20000 Ministry of Foreign Affairs
Kivukoni Front, PO Box 9000, Dar-es-Salaam
☎(+255 51) 212 34 ⊠ 41086 foreign tz
Minister of Foreign Affairs: Jakaya Kikwete [98]

12-40000 Ministry of Home Affairs
Ghana Avenue, PO Box 9223, Dar es Salaam
☎(+255 51) 272 91 ⊠ 41231 uhamiaji tz
Minister of Home Affairs: Ali Amer Mohammed [98]

31 Embassies & Attachés

31-00890 Embassy of Italy
Upanga Road 316, PO Box 2106, Dar es Salaam
☎(+255 51) 463 55 ⊠ 41062 ⊖ (+255 51) 463 54
Ambassador: H.E. Dr Torquato Cardilli
First Secretary: Ms Laura Botta
Commercial Officer: Alessandro Falchetto [97]

31-01270 Royal Netherlands Embassy
Cnr Ohio St Garden Ave New ATC Term., PO Box 9534, Dar es Salaam
☎(+255 51) 11 85 66 Ⓔ nlgovdar@intafrica
⊠ 41 050 hollandia ⊖ (+255 51) 11 28 28
Ambassador: H E Mr Sjoerd Leenstra
Deputy Chief of Mission, Head Development Cooperation: Theo van Banning
Political Officer: Cees Roels
Trade Officer: Frans van Ryan
Head of General Affairs: Henk Nouwens [98]

31-01370 Embassy of Pakistan
149 Malik Road, Upanga, PO Box 2925, Dar es Salaam
☎(+255 51) 11 76 30; 11 66 34
✉ PAREP Dar es Salaam ⊖ (+255 51) 11 32 05
High Commisioner: H E Mr Sajjad Haider
☎(+255 51) 11 59 41 [98]

31-01485 Embassy of the Russian Federation
Plot 73 Ali M Mwinyi Road, PO Box 1905, Dar es Salaam
☎(+255 51) 66 60 05, 66 60 06 ⊠ 41747
⊖ (+255 51) 66 68 18
Ambassador: H E Mr Doku Zavgaev
Deputy Chief of Mission: Yuri L Vassiliev
Military Naval and Air Attaché: Guennad V Bazhanov [98]

31-01680 Embassy of Spain
99 B Kinondoni Road, PO Box 842, Dar es Salaam
☎(+255 51) 66 60 18; 66 69 36 ⊠ 41589 embastan
⊖ (+255 51) 66 69 38
Ambassador: H E Mr Luis Gopmez de Aranda
Deputy Chief of Mission: Daniel Soto
Administrative Attaché: Xavier Pejuán [98]

31-01740 Embassy of Switzerland
Kinondoni Road, Plot 79, PO Box 2454, Dar es Salaam
☎(+255 51) 66 60 08; 66 60 09 ⊠ 41322 suisse tz
⊖ (+255 51)

Ambassador: H E Ms Lise Favre
Deputy Chief of Mission: A Burdet
Administration Attaché: G Déléze
Swiss Agency for Development and Cooperation
Plot 581, Msasani Peninsula
☎(+255 51) 60 00 12; 60 01 33 ⊖ (+255 51) 60 00 22
Head of Swiss Agency for Development and Cooperation: Niklaus Zinng
Attaché: Gerhard Siegfried
Attaché: Alessandro Fischer [98]

31-01890 United Kingdom High Commission
Hifadhi House, Samora Ave, PO Box 9200, Dar es Salaam
☎(+255 51) 11 76 60; 11 76 04 ⊖ (+255 51) 11 29 51
High Commissioner: H E Dr Alan Montgomery
Deputy High Commissioner: D A Jones
Defence Adviser: Lieut Col N P Lewis (Resident in Kampala)
First Secretary Development: David Bell
Second Secretary Commercial: D A Paginton
Second Secretary Management: L V Croucher
Second Secretary Chancery/Information: Mrs V Wallace
Second Secretary Consular/Immigration: Peter Palmer [98]

31-01990 Embassy of the People's Democratic Republic of Yemen
353 United Nations Road, PO Box 349, Dar es Salaam
☎(+255 51) 217 22; 295 96 ✉ CADFAN TZ
☒ 41901 cadfan tz ⊖ (+255 51) 468 34
Ambassador: H.E. Mr Mohamed Abdulla Almas ☎(+255 51) 667 91
Minister: Abdulkadir Mohamed Hadi ☎(+255 51) 385 63
Second Secretary: Mohamed Yaha S. Al-Katia'a ☎(+255 51) 333 48
Attaché: Haitham Ali Thabet ☎(+255 51) 447 92 [97]

31-02010 Embassy of the Federal Republic of Yugoslavia
35-36 Ali Hassan Mwinyi Rd, PO Box 2838, Dar es Salaam
☎(+255 51) 463 77; 79 ⊖ (+255 51) 463 78
Chargé d'Affaires a.i.: Djordje Lukic
Attaché: Sasa Rodic [97]

31-02030 High Commission for the Republic of Zambia
Ohio Street/City Drive Junction, PO Box 2525, Dar es Salaam
☎(+255 51) 11 84 81; 11 84 82 ⊖ (+255 51) 11 29 77
High Commissioner: H E Mr Joel Chitafu
Deputy High Commissioner: Eva K Fundafunda
Defence Attaché: Col A S K Nyirongo
First Secretary, Political: Daniel F Y Banda
Second Secretary, Administration: Ms Mwansa G Kabalika
Second Secretary, Finance: Mrs T S M Zunduna
Third Secretary, Personal: Ms Rose Musa
Trade Commissioner: Evaristo Banda [98]

90 Sales Agents, Sales & Liaison Offices

90-00200 CATIC
Representative Office
6 Upanga Road, PO Box 182, Dar es Salaam
☎(+255) 514 60 75 ☒ 41036 catiz tz
⊖ (+255) 514 60 75 [97]

Thailand

Head of State:
H.M. King Phumiphon Adunyadet

10 Head of State

10-10000 Office of HM the King
Grand Palace, Thanon Na Phra Lan, Bangkok 10200
King of Thailand: H M Phumiphon Adunyadet [98]

11 Prime Minister

11-10000 Office of the Prime Minister
Government House, Thanon Nakhon Pathom, Bangkok 10300
☎(+66 2) 282 88 74; 280 37 77; 282 55 15
⊖ (+66 2) 280 36 93
Prime Minister: Gen Chawalit Yongchaiyut (retd)
Deputy Prime Minister: Rangsitphon Sukhawit
Deputy Prime Minister: Thappharangsi Kon
Deputy Prime Minister: Phongphanit Montri
Deputy Prime Minister: Sunthorawet Samak
Minister Attached to Office of Prime Minister: Phonlakun Phonkin
Minister Attached to Office of Prime Minister: Mongkhontham Chingchai
Minister Attached to Office of Prime Minister: Wathacharaphon Piyanat
Minister Attached to Office of Prime Minister: Khamprakop Wirakon
Minister Attached to Office of Prime Minister: Amonwiwat Somphong
Minister Attached to Office of Prime Minister: Sukthana Rakkiat [98]

12 Ministries

12-10000 Ministry of Defence
Thanon Sanam Chai Road, Bangkok 10200
☎(+66 2) 222 11 21; 225 00 98
Defence Minister: Gen Chawalit Yongchaiyut [98]

12-20000 Ministry of Foreign Affairs
Saranrom Palace, Bangkok 10200
☎(+66 2) 225 79 00; 225 79 43; 225 00 96
☒ 82698 saranrom ⊖ (+66 2) 226 13 74
Minister of Foreign Affairs: Prachuap Chaiyasan
Deputy Foreign Minister: Intharawitthayanan Phithak [98]

12-30000 Ministry of Commerce
Thanon Sanam chai, Bangkok 10200
☎(+66 2) 225 84 11; 225 84 21 ☒ 82 389 minomer th
Minister of Commerce: Akkaraserani Narongchai
Deputy Minister: Sanitwongchai Chaloemphon
Deputy Minister: Atsawahem Somphon [98]

12-40000 Ministry of Interior
Thanon Atsadang, Bangkok 10200
☎(+66 2) 222 11 41; 222 11 55; 222 88 66
⊖ (+66 2) 224 61 88
Minister of the Interior: Thianthong Sano
Deputy Minister: Lik Ruangwit
Deputy Minister: Chaichowarat Kiatchai
Deputy Minister: Losunthon Phairot
Deputy Minister: Siriwat Chaipak
Deputy Minister: Capt Yubamrung Chaloem [98]

12-50000 Ministry of Finance
Rama VI Road, Bangkok 10400
☎(+66 2) 273 90 21 ⊖ (+66 2) 273 94 08
Minister of Finance: Tanong Bidaya
Deputy Minister: Chaisaeng Chaturon
Deputy Minister: Chawarat Chanwirakun [98]

31 Embassies & Attachés

31-00090 Australian Embassy
37 South Sathorn Road, Bangkok 10120
☎(+66 2) 287 26 80 ⊖ (+66 2) 287 11 08
Ambassador: H.E. Mr Cavan Hogue
Deputy Head of Mission: James Wise
Commercial Counsellor: Greg Carmody
Defence Attaché: Col Richard D. Warren (AM)
Asst Defence Attaché: Sqn Ldr Micheal Spencer
Counsellor: Peter Stanford
Counsellor Admin: Mal Ferguson
Public Information Officer: Richard Palk [97]

31-00690 Embassy of Greece
President Tor Building 3rd Floor, 412/8-9 Siam Square, Rama 1 Road, Bangkok
☎(+66 2) 251 51 11/7 ⊖ (+66 2) 246 79 73
Chargé d'Affaires a.i.: Adam G. Adamides
Attaché: Panayotis Pappas [98]

31-00830 Embassy of India
46 Soi Prasarnmitr, Sukhumvit Road Soi 23, Bangkok 10110
☎(+66 2) 258 03 00/6 ☒ 82793 indbasy th
⊖ (+66 2) 258 46 27
Ambassador: H.E. Mr Ranjit Gupta
Counsellor Economics & DPR to ESCAP: Dr V.S. Seshadri
Counsellor Consular: K.C. Mahajan
Defence Attaché: Gp Capt. M. John (Designate) ☎(+66 2) 258 47 99
First Secretary Culture & Education: Vinod Kumar
First Secretary Policital & Information: Mrs Narinder Chauhan ☎(+66 2) 258 44 91
First Secretary Commerce & Head of Chancery: A.K. Sinha ☎(+66 2) 259 49 98
First Secretary Consular: S. Mukhopadhyay ☎(+66 2) 261 69 45 [97]

31-01260 Royal Nepalese Embassy
189 Soi Sukhumvit Road, Bangkok 10110
☎(+66 2) 390 22 80; 391 72 40; 392 47 42
✉ NEPEMBASSY BANGKOK ☒ 20216 ronepem th
⊖ (+66 2) 381 24 06
Ambassador: H.E. Mr Sundar Nath Bhattarai
Counsellor & Deputy Chief of Mission: Pradhumna Bikram Shah ☎(+66 2) 390 29 85
Second Secretary: Tara P. Pakharel ☎(+66 2) 390 27 79
Third Secretary: Krishna Govind Dangol ☎(+66 2) 391 72 40
Attaché: Madhusadan Adhikari ☎(+66 2) 391 72 40 [97]

31-01270 Royal Netherlands Embassy
PO Box 404, Bangkok 10501
106 Wireless Road, Bangkok 10330
☎(+66 2) 254 77 01/5 ✉ HOLLANDIA Bangkok
☒ 82691 nedba th ⊖ (+66 2) 254 55 79
Ambassador: H.E. Ms M.L.A.C. van den Assum
Permanent Representative to ESCAP: M.W.H. Crom
Technical Attaché: H. Jeronimus
Attaché: W. van Adrichem [97]

31-01485 Embassy of the Russian Federation
108 Sathorn Nua Road, Bangkok 10500
☎(+66 2) 234 98 24; 234 39 39 ☒ 21947 russemb th
⊖ (+66 2) 237 84 88
Ambassador & Permanent Representative to ESCAP: H.E. Mr Oleg V. Bostorin ☎(+62 2) 234 98 24
Minister Counsellor & Deputy Head of Mission: Andrei A. Tatarinov ☎(+66 2) 234 98 24
Minister Counsellor & Deputy Permanent Rep. to ESCAP: Alexei N. Borodavkin
Military, Air & Naval Attaché: Col Victor M. Nikiforov ☎(+66 2) 235 55 99
Councellor: Alexandre Y. Mitiaguine
First Secretary Political, Press & Information: Igor Akimov
Trade Representation
11th Floor, Charn Issara Tower II,
2922/1 New Petchburi Rd, Bangkok 10310
☎(+66 2) 308 27 51/4 ☒ 87306 vnstorg th
⊖ (+66 2) 308 27 55

Contact Reference

Trade Representative: Serguei A. Briguinevitch
Deputy Trade Representative: Alexandre Mistrov
☎ (+66 2) 308 27 51/4
Deputy Trade Representative: Alexandre V. Loukianov
Deputy Trade Representative: Igor Smirnov 97

31-01600 Embassy of Saudi Arabia

10th Floor, Sathorn Thani Building, 90 Sathorn Nua Road, Bangkok 10500
☎ (+66 2) 237 19 38/41 ⊠ 87648 najdiah th
⊖ (+66 2) 236 64 22
Charge d'Affaires: Mohamed Said Khoja
Second Secretary: Fahad A. Al-Gutaini Al Alamri
Third Secretary: Abdullah A. Al-Dughither 97

31-01655 Consulate of the Republic of Slovenia

298/2 Silom Road, Soi 28, Bangkok
☎ (+66 2) 234 76 37, 234 84 85; 234 24 81
⊖ (+66 2) 237 13 75
Honorary Consul: Phatra Putipanpong 98

31-01730 Swedish Embassy

Nana Post Office, PO Box 1324, Bangkok 10112
20th Fl. First Pacific Place 140 Sukhumvit Road, Bangkok 10110
☎ (+66 2) 254 49 55 ✉ SVENSK Bangkok
⊠ 81174 svensk th ⊖ (+66 2) 254 49 14
Ambassador: H.E. Miss Eva Heckscher
Counsellor: Sven Malmberg
Second Secretary: Ms Anna-Lisa Trulsson Evidon
Second Secretary: Miss Lena Kjellström 97

31-01890 Embassy of the United Kingdom

Wireless Road, Bangkok 10300
☎ (+66 2) 253 01 91; 253 01 99
⊖ (+66 2) 254 95 78 (chancery); 255 95 78 (commercial/information); 255 92 78 (management); 255 60 51 (consular); 254 95 79 (immigration)
Ambassador: H E Sir James Hodge
Counsellor and Deputy Head of Mission: P R Sizeland
Counsellor, Commercial: M Greenstreet
Defence Attaché: Col J D Fielden
Counsellor and Permanent Representative in ESCAP: M Regan
First Secretary, Management: R H Tonkin
First Secretary, Political: Dr N A D Stokes
First Secretary, Defence Supply: P Russell-Jones
First Secretary and Consul: B P Kelly
First Secretary, Commercial: P J W Hardman
First Secretary: G Atkinson 98

75 Manufacturing Companies

75-00250 Daewoo Corp Bangkok Branch

Room 1702 SERM-MIT Tower 17th Floor, 150 Sukhumvit 21 (ASOKE) Road, Long-Toey, Bangkok 10110
☎ (+66 2) 260 84 09; 260 84 10; 260 84 11; 260 84 12; 260 84 13; 260 84 15; 260 84 16
⊖ (+66 2) 260 84 14 98

75-00700 Glory Kawasaki Motors Co Ltd

(joint venture with Kawasaki Heavy Industries Ltd, Japan)
222/1 Rama 9 Road, Huay Kwang, Bangkok 10320
☎ (+66 2) 246 15 10 ⊖ (+66 2) 246 15 17 98

75-01300 Minebea Electronics (Thailand) Ltd

5/2 Moo 8 - Phaholyothin Road km 149, Lopburi Province 15000
Tambol Nikom Srang-Toan-Ang, Amphur Muang, Lopburi Provice 15000
☎ (+66 36) 41 38 11 ⊠ 73304 th ⊖ (+66 36) 41 39 50
👥 6,416
■ Manufacture of PM stepping motor, floppy disk drives, magnetic heads, spindle motors and HI-Brid IC's. 98

75-01400 Minebea Thai Ltd

1 Moo 7, Phaholyothin Road, km 51, Ayutthaya Province 13180
Amphur Bang Pa-in, Ayutthaya
☎ (+66 35) 36 14 29 ⊠ 21473 th ⊖ (+66 35) 36 14 77
👥 12,888
■ Manufacture of electronic equipment. 98

75-01500 NMB Thai Ltd

18 Moo 3, Asia Road, Km 72, Tambol Thanoo, Amphur Uthai, Ayutthaya Province 13210
☎ (+66 35) 33 53 09 ⊠ 73207 th ⊖ (+66 35) 33 53 11
👥 2,338
■ Manufacture of M&I buildings.

NMB Precision Balls Ltd
1 Moo 7, Phaholyothin Road, km 51, Ayutthaya Province 13180
☎ (+66 35) 36 15 07 ⊖ (+66 35) 36 14 78

NMB Hi-Tech Bearings Ltd
1 Moo-7 Phaholyothin Road km 51, Ayutthaya Province 13180
☎ (+66 35) 36 15 07 ⊖ (+66 35) 36 14 78 98

75-01700 Pelmec Thai Ltd

1 Moo 7, Phaholyothin Road, km 51, Ayutthaya Province 13180
Tambol Chiang Rak-Nol, Amphur Bang PA-IN, Ayutthaya Province 13180
☎ (+66 35) 36 11 52 ⊠ 21242 th ⊖ (+66 35) 36 11 58
👥 1,077
■ Manufacture of small-sized ball bearings. 98

75-02035 Thai Caprolactam Ltd

17F LPN Tower, 216/73-75 Nanglinchee Road, Chongnonsi, Yannawa
☎ (+66 2) 285 44 98 ⊖ (+66 2) 285 45 03 98

75-02040 Thai Daewoo Development Co Ltd (TDDC)

(a subsidiary of Daewoo Corporation, South Korea)
Room 1702 SERM-MIT Tower 17th Floor, 150 Sukhumvit, Klong-Toey, Bangkok 10110
☎ (+66 2) 260 84 60; 260 84 61; 260 84 62; 260 85 45
⊖ (+66 2) 260 86 39 98

75-02050 Thai Ferrite Co. Ltd

1 Moo 7, Phaholyothin Road, km 51, Tambol Chiang Rak-Nol, Ayutthaya 13180
Amphur Bang Pa-in, Ayutthaya Province
☎ (+66 35) 36 11 52 ⊠ 21473 th ⊖ (+66 35) 36 11 77
👥 300
■ Manufacture of magnets, soft and hard ferrites. 98

75-02100 THORN Lighting Thailand

21st Floor Times Square Building, Sukhumvit Road, Bangkok 10110
☎ (+66 2) 267 84 35 ⊖ (+66 2) 267 84 37
Chief Executive: K Adisorn 98

75-02200 Transfield Pty Ltd

6th Floor - Olympia Thai Tower, 444 Ratcchadapsiek Road, Bangkok 10320
☎ (+66 2) 512 14 24 ⊖ (+66 2) 512 14 57 98

80 Service Companies & Organizations

80-00050 AlliedSignall Aerospace Marketing and Sales

c/o Thai Airways International, Don Muang Airport, Bangkok 10210
c/o Thai Airways International, Technical Department BKKZZTG, Don Muang Airport, Bangkok 10210
☎ (+66 2) 532 10 43; 563 81 67 ⊖ (+66 2) 533 62 92
Field Service Engineer: Martin Gilbert
☎ (+66 1) 815 19 72
Field Service Engineer: Gary Perritt ☎ (+66 1) 431 11 51
Field Service Engineer: Ed Reppel ☎ (+66 1) 815 63 15 98

90 Sales Agents, Sales & Liaison Offices

Companies represented:

ADAMS, UK	90-01910
Airborne Rotables (UK) Ltd, UK	90-01910
Allied International Corp, USA	90-02000
Bell Helicopter Textron, USA→ Singapore	90-00220
BP Solar Limited, UK	90-00250
Celsius Tech, Sweden	90-01910
Ferranti International, UK	90-01910
Ghisalba SpA, Italy	90-01910
Nida Corporation, USA	90-01910
Officine Galileo SpA, Italy	90-01910
OTE, Italy	90-01910
Pompe Garbarino SpA, Italy	90-01910
ShinMaywa Industries Ltd, Japan	90-02030
Transaero Inc, USA	90-02050

90-00250 BP Thai Solar Corporation Ltd

4th Floor Kian Gwan House 1, 140 Wireless Road, Bangkok 10330
☎ (+66 2) 255 79 45 ⊖ (+66 2) 255 79 45
® UK: BP Solar Limited 98

90-00300 CATIC

Representative Office
Taiping Towers, Condominium 1142/582, Ekamai Road, Bangkok 10110
☎ (+66 2) 391 12 83 ⊖ (+66 2) 391 12 83 97

90-00400 Daewoo Corp

1901-1902 19th Floor M Thai Tower, All Seasons Place 87 Wireless Road, Lumpini, Pathumwan, Bangkok 10330
☎ (+66 2) 645 01 24 ⊠ 82229 daewoo th
⊖ (+66 2) 645 01 29 98

90-00850 The Goodyear Tire & Rubber Company

(t Air Treads)
(a subs. of Aviation Products Div. D782)
Q-House Building, 66 Sukhumuit 21 (Asoke Road), Bangkok 10110
☎ (+66 2) 264 27 16 ⊖ (+66 2) 264 27 09
♤ Baht 74 000 000 ♣ Baht 2 985 917 261 👥 634
Director Aviation Products Asia: H P Sack
Sales Manager Asia East: T Visudhi
Sales Manager Asia West: C Navapol
Operations Manager: Y Nareumol
Product Support Manager Asia: R Preechaga
■ Provides sales, product support & distribution services to the Asia region with offices in Bangkok, Singapore, Malaysia & Tokyo. 98

90-01100 Kawasaki Heavy Industries Ltd Bangkok Office

20th Floor Thaniya Plaza Business, Complex, N°52 Silom Road, Bangkok 10500
☎ (+66 2) 231 23 60 ⊖ (+66 2) 231 23 63 98

90-01150 Kyokuto Boeki Kaisha Ltd Bangkok Liaison Representative (KBK)

6th Floor Chongkolnee Building, 56 Suriwongse Road, Bangkok 10500
☎ (+66 2) 266 59 20 ⊖ (+66 2) 266 59 21 98

90-01695 Siam Aviation

(a subs. of Recon/Optical Inc.)
PO Box 730, General Post Office, Bangkok
☎ (+66 2) 248 76 30/3 ⊖ (+66 2) 248 76 34
President: Vithan Makarabhirom 97

90-01910 Sapac International Co. Ltd

3/1 Soi Chetha, Sanpavuth Rd, Bangna, Bangkok 10260
☎ (+66 2) 250 11 85 ⊖ (+66 2) 252 75 71
Managing Dir: LCdr Narongsak Supaphongs
® Italy: Ghisalba SpA, OTE, Officine Galileo SpA, Pompe Garbarino SpA

Sweden: Celsius Tech
UK: ADAMS, Airborne Rotables (UK) Ltd, Ferranti International
USA: Nida Corporation [97]

90-02000 Suphasuk Co Ltd
(† Allied International Corp)
Bangkok 10210
31/112 Suphasuk Building, 2nd Floor, Soi Chinnakhet 2/16, Neamwongwan Road Don Muang
☎ (+66 2) 589 33 20 ⊖ (+66 2) 580 68 30
Manager: Suphant Komalarajun
® USA: Allied International Corp [98]

90-02030 Thai ShinMaywa Co. Ltd
199 Moo 12 Petchakasem Road, Semutsakorn 74130
☎ (+66 2) 420 00 89 ⊖ (+66 2) 420 47 13
® Japan: ShinMaywa Industries Ltd [97]

90-02050 Transaero Thailand
17/234-235 Soi Jaisamarn, Klongtoey, Bangkok 10110
Sokhumvit Soi No.6 Prakhanong, Klongtoey, Bangkok
☎ (+66 2) 251 11 87 ⊖ (+66 2) 251 11 87
Manager: Terence Heffernon
® USA: Transaero Inc [97]

90-10000 UBE (Thailand) Co Ltd
18F Thaniya Plaza Building, 52 Silom Road, Bangkok 10500
☎ (+66 2) 23 12 41 02 ⊖ (+66 2) 231 24 13 [98]

Togo

Head of State:
Gen. Gnassingbé Eyadéma

10 Head of State

10-10000 Office of the President
Avenue de la Marina, Palais Présidentiel, Lomé
☎ (+228) 21 27 01, 21 31 85 ▣ 5201 prerep to
President of the Republic: Gen Gnassingbé Eyadéma [98]

11 Prime Minister

11-10000 Office of the Prime Minister
Lomé
Prime Minister: Kwasse Klutse
Minister of State in Charge of Industry and Commerce: Elom Koumi Dazie
Minister of State in Charge of Economy and Finance: Barry Moussa Barque [98]

12 Ministries

12-10000 Ministry of National Defense
Route d'Atakpame, Lomé
☎ (+228) 21 28 91; 21 28 92 ▣ 5321 mindef to
Minister of National Defense: Bitokotipou Yagninim [98]

12-20000 Ministry of Foreign Affairs and Co-operation
Avenue Sarakawa, Lomé
☎ (+228) 21 29 10; 21 29 13; 21 34 84
▣ 5239 mineac to
Minister of Foreign Affairs and Co-operation: Koffi Panou [98]

12-40000 Ministry of Interior
Rue Albert Sarrault, Lomé
☎ (+228) 21 45 49; 21 54 16
Minister of Interior and Security: Col Seyi Memene [98]

31 Embassies & Attachés

31-00660 Embassy of the Federal Republic of Germany
Marina, Route d'Aflao, PO Box 1175, Lomé
☎ (+228) 21 23 70; 21 23 38 ▣ 5204 aalome to
⊖ (+228) 22 18 88
Ambassador: H E Dr Dieter Simon
Deputy Chief of Mission: Dr Jürgen Morhard [98]

31-01890 Consulate of the United Kingdom
British School of Lomé, BP 20050, Lomé
☎ (+228) 26 46 06 ⊖ (+228) 26 49 89
Honorary Consul: Mrs J A Sayer [98]

31-01900 Embassy of the United States of America
Rue Pelletier Caventou et Rue Vauban, PO Box 852, Lome
☎ (+228) 21 29 92; 21 29 94; 21 77 17
⊖ (+228) 21 79 52
Ambassador: H E Mr Johnny Young
Deputy Chief of Mission: Terence P McCulley
Commercial Counsellor: Ms Mary Beth Leonard
Public Affairs Officer: George P Newman
General Services Officer: Ms Martina Aintrop [98]

Tonga

Head of State:
H.M. King Taufa'ahau Tupou IV

10 Head of State

10-10000 Office of His Majesty the King
PO Box 6, Nuku'alofa
King of Tonga: HM King Taufa'ahau Tupou IV [98]

11 Prime Minister

11-10000 Office of the Prime Minister
PO Box 62, Nuku'alofa
☎ (+676) 213 00 ▣ 66269 primo ts
Prime Minister: Baron Vaea
Deputy Prime Minister: S Langi Kavaliku [98]

12 Ministries

12-10000 Ministry of Defence
PO Box 72, Nuku'alofa
☎ (+676) 210 99
Minister: H R H Crown Prince Tupouto'a [98]

12-20000 Ministry for Foreign Affairs
PO Box 821, Nuku'alofa
☎ (+676) 226 00 ✉ MINOFA Nuk ▣ 66235 minofa ts
Minister: H R H Crown Prince Tupouto'a [98]

12-50000 Ministry of Police
PO Box 8, Nuku'alofa
☎ (+676) 212 33 ▣ 66224 topolhq ts
Minister: Clive Edwards [98]

31 Embassies & Attachés

31-00090 Australian High Commission
Salote Road, Nuku'alofa
☎ (+676) 232 44; 232 45 ⊖ (+676) 232 43
High Commissioner: H E Mr Andrew Steuart Mullin
First Secretary and Consular Administration: Rob McClory
First Secretary Development Assistance: Bill Pennington
Defence Attaché: Lieut Col D Phillips (Resident in Suva)
Trade Commissioner: R Dravers (Resident in Suva)
Attaché and Vice Consul: Eva Rucinska [98]

31-01890 United Kingdom High Commission
PO Box 56, Nuku'alofa
☎ (+676) 210 20; 210 21 ✉ UKREP NUKU'ALOFA
▣ 66226 ukrep ts ⊖ (+676) 241 09
High Commissioner: H E Mr A J Morris
Management Officer: Mrs Ann C Morris [98]

Trinidad & Tobago

Head of State:
Arthur Napolean Raymond Robinson

10 Head of State

10-10000 Office of the President
Circular Road, St Ann's, Port of Spain
☎ (+1 809) 624 12 61
President of the Republic: Arthur Napoleon Raymond Robinson [98]

11 Prime Minister

11-10000 Office of the Prime Minister
Central Bank Building, Level 19, Corner Independence Square, Port of Spain
☎ (+1 809) 623 36 53 ▣ (+1 809) 627 34 44
Prime Minister: Basdeo Panday
Minister Extraordinaire, Special Responsibility for Tobago, Adviser to the Prime Minister: A N R Robinson [98]

12 Ministries

12-03000 Ministry of External Affairs
Ministry of Foreign Affairs and International Trade
"Knowsley", Queen's Park West, Port of Spain
☎ (+1 809) 623 41 16 ▣ 3321 trinex wg
⊖ (+1 809) 627 05 71
Minister of External Affairs: Ralph Maraj [98]

12-05000 Ministry of Finance
Eric Williams Financial Building, 8th Floor, Independence Square South, Port of Spain
☎ (+1 809) 627 97 00; 627 96 92
⊖ (+1 809) 627 61 08
Minister of Finance: Brian Anthony Kuei Tung [98]

Contact Reference

12-10000 Ministry of Justice & National Security
Government Building, Knox Street, Port of Spain
☎(+1 809) 623 24 41; 623 24 61; 623 24 66
⊖ (+1 809) 627 80 44
Minister of National Security:
Senator Brig Joseph Theodore (retd)
Permanent Secretary: Hant Edwards
Head of the Trinidad and Tobago Defence Force:
Brig Carlton A Alfonso
Staff Officer 1, Administration: Dave L Williams [98]

25 Defense Forces

25-10000 Trinidad and Tobago Defence Force
Defence Force Headquarters, Port of Spain
☎(+1 809) 627 3456 ⊖ (+1 809) 625 7502
Chief of Defence Staff: Brig. Carl Alfonso
VCDS Administration: Capt Anthony Franklin (Navy)
VCDS Operations: Col John Sandy
Trinidad and Tobago Regiment
Teteron Barracks, Port of Spain
☎(+1 809) 625 2211
Commanding Officer 1st Battalion: Lt Col D. Williams
Commanding Officer 2nd Battalion: Lt Col A. Antoine
Commanding Officer Support & Services Battalion:
Lt Col E. Dillon
Trinidad and Tobago Coast Guard
Staubles Bay, Port of Spain
☎(+1 809) 634 4439 ⊖ (+1 809) 637 2678
Commanding Officer: Kayam Mohammed [97]

31 Embassies & Attachés

31-00590 Embassy of France
11 Maraval Road, Tatil Bldg 6th Fl, PO Box 1242,
Port of Spain
☎(+1 809) 622 7446; 622 7447 ▣ 22258 fambat wg
⊖ (+1 809) 628 2632
Ambassador: H.E. Mr Pierre Ariola
Deputy Chief of Mission: Dominique D'Hermies
Public Information Officer: Mrs Monique Tosello [97]

31-00660 Embassy of the Federal Republic of Germany
7-9 Marli Street, PO Box 828, Port of Spain
☎(+1 809) 628 1630/2 ▣ 22316 aaposp wg
⊖ (+1 809) 628 5278
Ambassador: H.E. Mr Holger Eberle
Head of Administration: R. Kummer [97]

31-01270 Royal Netherlands Embassy
69-71 Edward Street, Cnr Park Street, PO Box 870,
Port of Spain
☎(+1 809) 625 1722; 625 1210; 625 2532
✉NEERLANDIA Port of Spain ▣ 22322 nethemb wg
⊖ (+1 809) 625 1704
Ambassador: H.E. Mr W.Th. Six
First Secretary: E. Huguenin [97]

31-01890 British High Commission
PO Box 778, Port of Spain
19 St Clair Avenue, St Clair
☎(+1 809) 622 27 48 ⊖ (+1 809) 622 45 55
High Commissioner: H E Mr Gregory Faulkner
Deputy High Commissioner: R J Lynne
Defence Adviser: Col A Moorby (Resides at Kingston)
Second Secretary, Commercial: N Collyer
Third Secretary: M Louth [98]

Tunisia

Head of State:
Zine el Abidine Ben Ali

10 Head of State

10-10000 Office of the President
Palais de Carthage, 2016 Carthage
☎(+216 1) 74 29 11; 73 05 66 ▣ 15475
⊖ (+216 1) 73 19 99
President: H E Zine el Abidine Ben Ali [98]

11 Prime Minister

11-10000 Office of the Prime Minister
Place du Gouvernement, 1001 Tunis
☎(+216 1) 26 31 23 ▣ 14530 tn ⊖ (+216 1) 56 23 78
Prime Minister: Dr Hamed Karoui [98]

12 Ministries

12-10000 Ministry of National Defence
Boulevard Bab Menara, 1030 Tunis
☎(+216 1) 26 02 44; 26 02 45; 26 00 88
▣ 12580 defnat tn; 13570 defnat tn
⊖ (+216 1) 26 18 04
Minister for National Defence: Habib Ben Yahia [98]

12-20000 Ministry of Foreign Affairs
Place du Gouvernement, La Kasbah, 1006 Tunis
☎(+216 1) 66 00 88 ▣ 13470 mafta tn
⊖ (+216 1) 56 73 88
Minister: Abderrahim Zouari [98]

12-40000 Ministry of Interior
Avenue Habib Bourguiba, 1001 Tunis
☎(+216 1) 24 30 00; 34 31 99 ▣ 13994
⊖ (+216 1) 34 08 88; 34 62 99
Minister: Mohamed Ben Rajeb [98]

12-60000 Ministry of Communication
3 bis, rue d'Angleterre, 1000 Tunis
☎(+216 1) 24 51 36 ▣ 13900 tn ⊖ (+216 1) 35 23 53
Minister: Ahmed Friaa [98]

31 Embassies & Attachés

31-00080 Embassy of the Argentine Republic
10 rue al-Hassan et Houssaine, al-Menzhah V, 1004 Tunis
☎(+216 1) 23 12 22; 76 74 30
🄴 etune@emb_argentina.intl.ar ⊖ (+216 1) 75 00 58
Ambassador: H E Mr Marcelo Eduardo Huergo
Conseiller d'Ambassade, Commererales: Gabriel Volpi
Consciller Adjoint des Affairs Acientfiques et Technologiques: Alberto Angel Julio Prieto [98]

31-00100 Embassy of Austria
16 Rue Ibn Hamdiss, PO Box 23, 1004 Tunis-El Menzah
☎(+216 1) 75 10 91; 75 10 94 ⊖ (+216 1) 76 78 24
Ambassador: H.E. Dr Karl Diem
Commercial Section
21 Avenue Charles Nicolle, Tunis
☎(+216 1) 79 78 75; 79 99 30; 79 98 48 ▣ 15246
⊖ (+216 1) 78 83 68
Commercial Counsellor: Manfred Banholzer [97]

31-00150 Embassy of Belgium
47 rue du 1er Juin, PO Box 24, 1002 Tunis
☎(+216 1) 78 16 55 ▣ 14342 ⊖ (+216 1) 79 27 97
Ambassador: H E Mr Daniel Lerory ☎(+216 1) 78 10 09
First Secretary: Jean-Cedric Janssens de Bisthoven
☎(+216 1) 74 23 94
Attaché, Administrative Affairs: Jean-Pascal Dricot
☎(+216 1) 74 48 98
Counsellor: Jean-Marie Urbain
Attaché, Agricultural: Bruno Dineur
Counsellor: Marcel LaFarge
Delegate of Begium People: Daniel Sotiaux
Economic and Commercial Representati (Economic and Commercial Representative)
Tunis
☎(+216 1) 79 60 80; 79 52 71
⊖ (+216 1) 80 27 84; 80 18 25
Economic and Commercial Attaché: Dirk Verlee
Economic and Commercial Attaaché: Didier Denayer [98]

31-00250 Embassy of the Republic of Bulgaria
5 Rue Ryhane, Cité Mahragéne, 1002 Tunis
☎(+216 1) 28 32 44 ▣ 12289
Ambassador: H E Mr Tchavdar Tchervenkov
Commercial Section
3 Rue de l'Argentine, Tunis
☎(+216 1) 28 21 38; 28 09 95 ▣ 12 289
Commercial Counsellor: Dimiter Stoytchev [98]

31-00660 Embassy of the Federal Republic of Germany
1 Rue el Hamra, Mutuelleville, PO Box 35,
1002 Tunis Belvédère
☎(+216 1) 78 64 55 ⊖ (+216 1) 78 82 42
Contact: G Chaneelie [98]

31-00690 Embassy of Greece
9 Impasse Antelas, Nord Hilton, 1002 Tunis
☎(+216 1) 28 84 11 ✉LEGOGRECE Tunis
▣ 13742 ambgr tn ⊖ (+216 1) 78 95 18
Ambassador: H E Mr Constantine Prevedourakis
Defense Attaché: Lieut Col E Tagalakis
Commercial Attaché: Miss K Daskalogianni [98]

31-00810 Embassy of the Republic of Hungary
12 Rue Ashart-Nord Hilton, 1004 Tunis
☎(+216 1) 78 05 44; 78 48 49 ⊖ (+216 1) 78 12 64
Ambassador: H E Mr László Nikicser
Deputy Chief of Mission: György Pantos
Commercial Counsellor: Albert Czili [98]

31-00890 Embassy of Italy
3 Rue de Russie, 1002 Tunis
☎(+216 1) 34 18 11 ▣ 13501 ⊖ (+216 1) 35 41 55
Ambassador: H.E. Mr Francesco Caruso
First Counsellor: Gergio Busetto
Commercial Counsellor: Giorgio Visetti
Defence Attaché: Col Mario Fatone
Public Information Officer: Dr. Vincenzo De Luca [98]

31-01220 Embassy of the Kingdom of Morocco
39 Avenue du 1er Juin, 1002 Tunis
☎(+216 1) 78 27 75; 78 38 01
🄴 sifmatunis@emb_maroc.intl.tn ▣ 14460
⊖ (+216 1) 78 71 03
Ambassador: H E Mr Abdelkader Benslimane
First Secretary: Mohammed Lhamzaaoui
Military Naval and Air Attaché: Col Mohamed Kamir
Economic Counsellor: Chorkry Mekouar
Consulat
16 Rue Ibn Maudhom, 1082 Tunis
☎(+216 1) 78 36 27; 78 44 42 [98]

31-01430 Embassy of the Republic of Poland
4 Rue Sophonisbe, PO Box 171, Notre Dame, Tunis
☎(+216 1) 28 62 37; 28 71 73 ▣ 14024
⊖ (+216 1) 79 51 18
Ambassador: H E Mr Tadeusz Mulicki
Deputy Chief of Mission: Janusz Czaplarski
Commercial Department
12 Rue Didon, PO Box 171, Notre Dame, Tunis

☎(+216 1) 28 26 21; 28 26 22 ▣ 15374
⊖ (+216 1) 79 09 61
Commercial Counsellor: Roman Adamczyk [98]

31-01485 Embassy of the Russian Federation
2092 El Manar I, 4 Rue Bergamotes, BP 48, Tunis
☎(+216 1) 88 24 46 ⊖ (+216 1) 88 24 78
Ambassador: H.E. Mr B.A. Schiborin
Deputy Chief of Mission: S.V. Verchinine
Commercial Officer: A.M. Laiko
Defence Attaché: N.D. Katchan
Public Information Officer: A.I. Kopylov [97]

31-01610 Embassy of the Republic of Senegal
122 Avenue de la Liberté, Tunis
☎(+216 1) 80 23 97 ▣ 12477 ⊖ (+216 1) 48 07 70
Ambassador: H E Mr Adama A Lo
Deputy Chief of Mission: Mrs Ba [98]

31-01730 Swedish Embassy
87 Avenue Taieb Mehiri, PO Box 112,
1012 Tunis Belvedere
☎(+216 1) 79 54 33; 79 61 22 ✉ SVENSK Tunis
▣ 15258 svensk tn ⊖ (+216 1) 78 88 94
Ambassador: H E Ms Catherine von Heidenstam
Deputy Chief of Mission: Björn-Gosta Sporrong [98]

31-01890 Embassy of the United Kingdom
5 Place de la Victoire, Tunis
☎(+216 1) 79 33 22 ▣ 14007 prdrom tn
⊖ (+216 1) 79 26 44
Ambassador and Consul General: H E Mr R J S Edis
First Secretary and Deputy Head of Mission: B M Bennett
Defence Attaché: Cdr M R B Wallace
Cultural Attaché: J McKenzie
Second Secretary, Commercial: Mrs J M Quinn
Second Secretary, Management: D L Walker
Vice Consul: Mrs V Zaoui [98]

31-02010 Embassy of the Federal Republic of Yugoslavia
4 Rue de Libéria, BP 93, 1002 Tunis-Belvedere
☎(+216 1) 78 30 57; 79 46 26 ▣ 18399
⊖ (+216 1) 79 64 82
Chargé d'Affaires: H E Mr Bratislav Krstic
Attaché: Zoran Toplicanin [98]

Turkey

Head of State:
Suleyman Demirel

10 Head of State

10-10000 Office of the President
Cumhurbaskanligi, Ankara
☎(+90 4) 427 43 30; 440 72 12 ▣ 42303 kosk tr
President of the Republic: Suleyman Demirel [98]

11 Prime Minister

11-10000 Office of the Prime Minister
Basbakanlik, Ankara
☎(+90 4) 118 57 76
Prime Minister: Mesut Yilmaz
Deputy Prime Minister and Minister of National Defence: Ismet Sezgin
Deputy Prime Minister: Mesut Yilmaz [98]

12 Ministries

12-10000 Ministry of National Defence
Milli Savunma Bakanligi, Ankara
Deputy Prime Minister and Minister of Defence: Ismet Sezgin [98]

12-20000 Ministry of Foreign Affairs
Disisleri Bakanligi-Balgat, 06100 Ankara
☎(+90 312) 287 25 55; 212 51 25 ▣ 42203 sfa tr
⊖ (+90 312) 287 18 86
Minister: Ismail Cem [98]

12-40000 Ministry of Interior
Içisleri Bakanligi
Ankara
☎(+90 4) 134 20 50; 118 76 96
Minister of Interior: Murat Basesgioglu [98]

12-60000 Ministry of Finance
Maliye ve Gümrük Bakanligi
Bakanlil, Ankara
☎(+90 4) 324 25 76 ▣ 43260 gtb tr
Minister: Zekeriya Temizel [98]

31 Embassies & Attachés

31-00010 Embassy of the Islamic State of Afghanistan
Cinnah Cad. 88, 06551 Cankaya, Ankara
☎(+90 312) 438 11 21 ▣ 46769
⊖ (+90 312) 438 77 45
Chargé d'Affaires: Sayed Zaher Shah
Second Secretary: Mohammed Qaseem
Third Secretary: Rooz Mohammed Noor [98]

31-00020 Embassy of the Republic of Albania
Ebuziya Tevfik Sok. No: 17, Cankaya, Ankara
☎(+90 312) 441 61 03; 05-10 ⊖ (+90 312) 441 61 04
Ambassador: H E Mr Skender Shkupi
Counsellor: Safet Stafa
Defence Attaché: Col Hayro Limay
First Secretary: Skender Drini
Second Secretary: Nexhat Pilku [98]

31-00030 Embassy of the Democratic People's Republic of Algeria
Sehit Ersan Cad. No. 42, 06680 Cankaya, Ankara
☎(+90 312) 427 87 00; 83 85 ▣ 4268959 amal tr
⊖ (+90 312) 426 89 59
Ambassador: H E Mr Rachid Haddad
Counsellor: Belkacem Smaili
First Secretary: Mechri Gouni
Second Secretary: M'hamed Kadi
Second Secretary: Mébarek Zellagui [98]

31-00080 Embassy of the Republic of Argentina
Ugur Mumcu Cad. 60/3, 06700 Gaziomanpasa, Ankara
☎(+90 312) 446 20 61/62 ▣ 42373 barg tr
⊖ (+90 312) 446 20 63
Ambassador: H E Mr José Pedro Pico
Counsellor: Juan Martin Sabatini
First Secretary: Guillermo Luis Nicolas [98]

31-00090 Embassy of Australia
Nenehatun Cad. No 83, 06700 Gaziomanpasa, Ankara
☎(+90 312) 446 11 80-87 ⊖ (+90 312) 446 11 88
Ambassador: H E Mr David Wyke Evans
First Secretary and Deputy Head of Mission: Geoffrey Leach
Defence Attaché: Maj Gen Geoffrey Carter
First Secretary (Consular and Administration): Andre Jedreski
First Secretary (Immigration): Peter Templeton [98]

31-00100 Embassy of Austria
Atatürk Bulvari 189, PO Box 131,
06680 Küçükesat-Ankara
☎(+90 312) 419 04 31-33 ▣ 46432 oean tr
⊖ (+94 312) 418 94 54
Ambassador: H E Dr Johann Plattner
First Secretary: Dr Walter Gehr
Defence, Military, Air and Naval Attaché: Brig Gen Helge Lerider
Second Secretary and Vice Consul: Josef Saiger
Commercial Counsellor: Peter Daniel Sedlmayer [98]

31-00105 Embassy of the Republic of Azerbaijan
Cemal Nadir Sok. No 20, Cankaya, Ankara
☎(+90 312) 441 26 20-22 ⊖ (+90 312) 441 26 00
Ambassador: H E Mr Hemmed Aliyev
First Secretary: Zakir Hashimov
First Secretary: Tural Rzayev
First Secretary (Consul): Faik Bagirov [98]

31-00130 Embassy of the People's Republic of Bangladesh
Cinnah Cad. No. 78/7-9, Cankaya, Ankara
☎(+90 312) 439 27 50/51 ▣ 46068 bdt tr
⊖ (+90 312) 439 24 08
Ambassador: H E Mr Mahboob Alam
Minister: Syed Maudud Ali
Defence Attaché: Brig Mirza Tabarak Hossain Beg [98]

31-00150 Embassy of Belgium
Mahatma Gandi Cad. No. 55,
06700 Gaziosmanpasa, Ankara
☎(+90 312) 446 82 47-50 ⊖ (+90 312) 446 82 51
Ambassador: H E Mr Eric Kobia
Counsellor: Johan Indekeu
Military, Naval and Attaché: Col F. De Smedt
☎(+90 312) 446 76 86 ▣ 46580
Commercial Attaché: E Schmitz [98]

31-00210 Embassy of the Republic of Bosnia and Herzegovina
Mahatma Gandi Cad. 91/8-9, Gaziosmanpasa, Ankara
☎(+90 312) 446 40 90 ⊖ (+90 312) 446 62 28
Ambassador: H E Mr Hajrudin Somun
Counsellor: Nihad San
Military Attaché: Brig Gen Hazim Sadic [98]

31-00230 Embassy of the Federative Republic of Brazil
Iran Cad. 47/1-3, 06700 Gaziosmanpasa, Ankara
☎(+90 312) 468 53 20-22 ⊖ (+90 312) 468 53 24
Ambassador: H E Mr Luiz Antonio Jardin Gagliardi
Counsellor: Mrs Ivone Vieira F Wester
Second Secretary: Ms Ana Suza Cartaxo de Sa [98]

31-00250 Embassy of the Republic of Bulgaria
Atatürk Bulvari 124, Kavaklidere, Ankara
☎(+90 312) 426 74 55/56 ⊖ (+90 312) 427 31 78
Ambassador: H E Mr Victor Gueorguiv Valkov
☎(+90 312) 427 38 99
Counsellor: Branimir Mladenov
Counsellor: Ivan Ivanov
Counsellor: Nikolay Kondev
Military Attaché: Col Hristo Bonev
First Secretary (Consular Affairs): Tzvetan Tomov
First Secretary (Press): Hristo Popov
Air and Navy Attaché: Lt Col Ivan Karpuzanova [98]

31-00310 Embassy of Canada
Nenehatun Caddesi 75, 06700 Gaziosmanpasa, Ankara
☎(+90 312) 436 12 75 ⊖ (+90 312) 446 44 37
Ambassador: H E Mr Peter J A Hancock
Counsellor (Political): Ron Willson
Counsellor (Commercial): Bertrand Desjardins
Canadian Force Attaché: Col Art Armstrong
First Secretary (Immigration): Mrs Erica Usher [98]

31-00350 Embassy of Chile
Cinnah Cad. No. 45/2, 06690 Cankaya, Ankara
☎(+90 312) 438 94 44; 440 76 31 ▣ 46673 bchi tr
⊖ (+90 312) 438 61 45

Contact Reference

Ambassador: H E Mr Fernando Cistern
Second Secretary: Juan Luis Bianchi [98]

31-00360 Embassy of the People's Republic of China
Gólgeli Sok. 34, 06700 Gaziosmanpasa, Ankara
☎ (+90 312) 436 06 28 ▣ 44532 cham tr
Ambassador: H E Mr Keming Wu
Counsellor: Aiguo Song
Military, Naval and Air Attaché: Col Fei Yu
Commercial Counsellor: Yujun Feng
Cultural Counsellor: Zhongdi Jia
First Secretary: Yooujian Guo
First Secretary: Zhengdon Gao
First Secretary (Press): Xinmin Zhang
First Secretary: Yongsheng Han [98]

31-00425 Embassy of the Republic of Croatia
Kelebek Sok. No. 15/A, Gaziosmanpasa, Ankara
☎ (+90 312) 446 94 60; 08 31 ⊖ (+90 312) 436 62 12
Ambassador: H E Dr Ivica Tomic
Minister Counsellor: Renato Cobal [98]

31-00430 Embassy of the Republic of Cuba
Komsu Sok. 7/2, Cankaya, Ankara
☎ (+90 312) 439 41 10; 53 62 ▣ 46140 cuba tr
⊖ (+90 312) 441 40 07
Ambassador: H E Mr Jorge Castro Benitez
Counsellor: Claudio Garcia Alvarez [98]

31-00450 Embassy of the Czech Republic
Ugur Mumcu Cad. 100/3, Gaziosmanpasa, Ankara
☎ (+90 312) 446 12 44; 48 04 ⊖ (+90 312) 446 12 45
Ambassador: H E Mr Thomas Lane
Minister Counsellor and Deputy Chief of Mission: Ilja Ulrich
Military and Air Attaché: Col Jozef Havran
Counsellor: Vladimir Nemec [98]

31-00460 Royal Danish Embassy
Kirlangic Sokak 42, 06700 Gaziosmanpasa, Ankara
☎ (+90 212) 467 77 60 ▣ 46385 dane tr
⊖ (+90 212) 468 45 59
Ambassador: H E Mr Niels Peter Georg Helskov
Minister Counsellor: Jesper Knudsen
Attaché and Consul: Ms Vibeke Andreasen
Consul (Commercial): Ms Vibeke Andreasen [98]

31-00510 Embassy of the Arab Republic of Egypt
Ataturk Bulvari 126, 06680 Kavaklidere, Ankara
☎ (+90 312) 426 10 26; 64 78 ⊖ (+90 312) 427 00 99
Ambassador: H E Mr Mohammed Elmahdy Fathalla
Counsellor: Sherin Kaher
Counsellor: Elsayed Fouad Khalil
Counsellor: Moustafa A I Sadek
Press and Information Counsellor: Mrs Soheir Elkerdawy
Defence Attaché: Col Alaa Eldin A Fahmy
Assistant Defence Attaché: Lt Col Nabil Zaki Abdelrahman
Second Secretary: Mohammed Fathi Edrees [98]

31-00535 Embassy of Estonia
Caddesi N° 39, 80300 Zincirlikuyu, Istanbul
☎ (+90 212) 211 34 80 ▣ 27397
⊖ (+90 212) 211 34 67
Honorary Consul: Can Paker [98]

31-00580 Embassy of Finland
Galip Dede Sok. 1/19-20 Farabi,
06680 Kavaklidere, Ankara
☎ (+90 312) 426 49 64; 56 85 ▣ 42856 fine tr
⊖ (+90 312) 468 00 72
Ambassador: H E Mr Jan Hendrik Groop
Commercial Counsellor: Timo Hanninen
Second Secretary: Pertti Attinen [98]

31-00590 Embassy of France
Paris Cad. 70, 06540 Kavaklidere, Ankara
☎ (+90 312) 468 11 54-59 ▣ 44553
⊖ (+90 312) 467 94 34
Ambassador: H E Mr Francois Dopffer
First Counsellor: Alain Rouillard
First Counsellor: Mrs Elisabeth Beton Delegue
First Secretary: Laurent Bili
First Secretary (Press): Alain Fort
Defence Attaché: Cl Pierre-Hervé Godard
Commercial Counsellor: Benoit Battistelli [98]

31-00660 Embassy of the Federal Republic of Germany
PO Box 54 Cankaya, 06552 Ankara
114 Atatürk Bulvari, 06540 Kavaklidere, Ankara
☎ (+90 312) 426 54 65/67; 426 54 51 ▣ 44394 aank tr
⊖ (+90 312) 426 69 59
Ambassador: Dr Hans-Joachim Vergau
Deputy Chief of Mission: Dr Werner Zimprich
Defense and Military Attaché: Col (GS) Heinrich Quaden
Air Attaché: Lt Col Jurgen Köpke
Naval Attaché: Cdr Günther Dachs
Commercial Counsellor: Reiner Möckelmann [97]

31-00680 Embassy of the Republic of Georgia
Ulubey Sok. No. 28, Gaziosmanpasa, Ankara
☎ (+90 312) 447 17 20-23 ⊖ (+90 312) 447 17 24
Ambassador: H E Mr Nodar Komahidze
Counsellor and Deputy Chief of Mission: Victor Chikaidze
Counsellor: Vladimer Pangani
Counsellor: Mrs Meri Djangveladze
Counsellor (Press Attaché): Ramin Chelidze
First Secretary: Revaz Bakradze
First Secretary: Badri Vadachkoria
First Secretary: Shalva Gabeskiria [98]

31-00690 Embassy of Greece
Ziya Ur-Rahman Cad 9-11, 06610 Gaziosmanpasa, Ankara
☎ (+90 312) 436 88 60; 436 88 61 ▣ 42334 gr tr
⊖ (+90 312) 446 31 91
Ambassador: H E Mr Dimitrios Nezeritis
Minister Plenipotentiary: Panayotis Vlassopoulos
First Counsellor: Alexandre Couyou
Defence Attaché: Col Constantinos Balanos
Press Counsellor: Stavros Stathoulopoulos
First Secretary: Mrs Cleopatra Liouni
First Secretary: Theoharis Lalacos
Second Secretary: Ioannis Papameletiou
Military Attaché: Lieut Col Emmanouil Xideris
Naval Attaché: Cdr Andonios Sourvinos
Air Attaché: Lieut Col Dimitrios Zois [98]

31-00830 Embassy of India
Cinnah Cad. 77 A, 06680 Canakaya, Ankara
☎ (+90 312) 438 21 95-98 ▣ 46541 inde tr
⊖ (+90 312) 440 34 29
Counsellor: Arvind Gupta
Defence Attaché: Col S K Singh
First Secretary: Mrs Shamma Jain
First Secretary: Madan Kumar Pyasi [98]

31-00840 Embassy of the Republic of Indonesia
Abdullah Cevdet Sok. 10, 06552 Canakaya, Ankara
☎ (+90 312) 438 21 90-92 ▣ 43250 indo tr
⊖ (+90 312) 438 21 93
Ambassador: H E Mr Soelarso Haknyosoebroto
First Secretary: Mr Budiarman
Defence Attaché: Col Inf. Iwan Ridman Sulandjana
Second Secretary (Administrative): Arief Budiman [98]

31-00850 Embassy of the Islamic Republic of Iran
Tahran Cad. No. 10, Kavaklidere, Ankara
☎ (+90 312) 427 43 20-22 ⊖ (+90 312) 468 28 23
Ambassador: H E Mr Mohammad Bagheri
Counsellor: Djalaleddin Namini Mianaji
Military Attaché: Col Reza Aramvash
Counsellor (Economic Affairs): Bahman Jahangir
First Secretary (Cultural Affairs): Zakarya Tarzani
Air Attaché: Lt Col Mohsen Lotfipour [98]

31-00860 Embassy of the Republic of Iraq
Turan Emeksiz Sok. 11, 06700 Gaziosmanpasa, Ankara
☎ (+90 312) 468 74 21-22 ▣ 46562
⊖ (+90 312) 468 48 32
Ambassador: H E Mr Rafi D M El-Tikriti
Counsellor: Dr Saad Al-Samarrai
Military Attaché: Brig Saleh A K Khalaf
First Secretary: Sameer N Ahmad [98]

31-00880 Embassy of the State of Israel
Mahatma Gandi Cad. No. 85, Gaziosmanpasa, Ankara
☎ (+90 312) 446 36 05; 45 84 ▣ 42560
⊖ (+90 312) 426 15 33
Ambassador: H E Mrs Zvi Elpeleg
Counsellor: Benny Omer
Defence Attaché: Col Beny Shefer [98]

31-00890 Embassy of Italy
Atatürk Bulvari 11, 06640 Ankara
☎ (+90 312) 426 54 60 ▣ 46603 ital tr
⊖ (+90 312) 426 58 00
Ambassador: H.E. Dr Michelangelo P. Massamormile
Deputy Chief of Mission: Giovanni Manfredi
Military, Naval & Air Attaché: Capt. Alfredo Saitto
Commercial Counsellor: Dr Roberto Martin
☎ (+90 312) 426 65 13 [97]

31-00920 Embassy of Japan
Resit Galip Cad. 81, Gazi Osman Pasa,
PO Box 31 Kavaklidere, Ankara
☎ (+90 312) 446 05 00 ✉ TAISHI Ankara
▣ 46438 japn tr ⊖ (+90 312) 437 18 12
Ambassador: H.E. Mr Takehiro Togo
Deputy Chief of Mission: Masmitsu Oki
Defense Attaché: Capt. Teruo Tao (Navy)
Commercial Officer: Kenji Yamaga
Public Information Officer: Yoshifumi Sasatani [97]

31-00930 Embassy of the Hashemite Kingdom of Jordan
Mesnevi, Dedekorkut Sokak. No. 18,
06690 Cankaya, Ankara
☎ (+90 312) 440 20 54; 45 94 ▣ 46954 urdn tr
⊖ (+90 312) 440 43 27
Ambassador: H E Mr Saleh KKabariti
Third Secretary: Saqer Abu Shattal [98]

31-00940 Embassy of the Republic of Kazakhstan
Ebuziya Tevfik Sok. No. 6, Cankaya, Ankara
☎ (+90 312) 441 23 01-02 ▣ 461193 sirt tr
⊖ (+90 312) 441 23 03
Ambassador: H E Mr Kanat Saudabayev
Counsellor: Abutalip Ahmetov
Counsellor: Amanjol Jankuliev
First Secretary: Bekjasar Narbayev
First Secretary: Serik Sarsembayev
Second Secretary: Janseit Tuimebayev [98]

31-00970 Embassy of the Kyrgyz Republic
Boyabat Sok. No. 11 Eren Apt., Gaziosmanpasa, Ankara
☎ (+90 312) 446 84 08-11 [E] kirgiz-o@servis2.net.tr
▣ 46405 kijg tr ⊖ (+90 312) 446 84 13
Ambassador: H E Mr Tolomush Okeev
Counsellor-Minister: Talant Koushchoubekov
Second Secretary: Batyr Syrgabaev [98]

31-00990 Embassy of the Republic of Korea
Alaçam Sokak 5, Cankaya, 06690 Ankara
☎ (+90 312) 468 48 22; 468 48 23
⊖ (+90 312) 468 22 79
Ambassador: H.E. Mr Byung Woo Yu
Counsellor: Young Jun Kim
Armed Forces Attaché: Lt Col Song Won Kim
Second Secretary: Hee Chul Lee [98]

31-01000 Embassy of the State of Kuwait
Resit Galip Cad. Kelabek Sok. 110, Gaziosmanpasa, Ankara
☎ (+90 312) 445 05 76-80
⊖ (+90 312) 446 68 39; 437 97 07
Charge d'Affaires: Bader Al-Tunaib
First Secretary: Nasser Al-Muzayyan
Military Liaison Officer: Col Ali Yousuf Malallah [98]

31-01020 Embassy of the Republic of Lebanon
Kizkulesi Sok. 44, Gaziosmanpasa, Ankara
☎ (+90 312) 446 74 85-88 ⊖ (+90 312) 10 23
Ambassador: H E Mr Jaafar Moawi
First Secretary: Khalil El-Habre [98]

31-01050 People's Bureau of the Socialist People's Libyan Arab Jamahiriya
Cinnah Cad. 60, 06690 Cankaya, Ankara
☎ (+90 312) 438 11 10-14 ☒ 42205 alib tr
⊖ (+90 312) 440 38 62
Secretary of the Bureau: Ahmed El-Atrash
First Counsellor (Economic Affairs): Mohammed Bahi
Counsellor (Information and Cultural Affairs): Mohammed Sharshari
Counsellor: Mohammed Khalifa Omar
First Secretary (Administrative): Wanis Agnaiber 98

31-01080 Embassy of the Republic of Macedonia
Filistin Sok. 30/2, Gaziosmanpasa, Ankara
☎ (+90 312) 446 92 04-05 ⊖ (+90 312) 446 92 06
Ambassador: H E Mr Trajan Petrovski
Counsellor: Danco Markovski 98

31-01110 Embassy of Malaysia
Ugur Mumcu Sokak 6, 06700 Gaziosmanpasa, Ankara
☎ (+90 312) 446 35 47/48 ☒ 43616 mal tr
⊖ (+90 312) 446 41 30
Counsellor and Charge d'Affaires: Ahmad Rasidi Hazizi (acting)
First Secretary: Misri Barham
Defence Attaché: Col Hashim Husainay
Assistant Defence Attaché (Navy): Lt Cdr Ahmad Pharodz
Assistant Defence Attaché (Air Force): Maj Adam Sabri 98

31-01190 Embassy of Mexico
Iran Cad. No 45/2, Kavaklidere, 06552 Ankara
☎ (+90 312) 441 32 04 ☒ 42278 mxco tr
⊖ (+90 312) 441 32 03
Ambassador: H.E. Mr Enrique Buj Flores
Commercial Counsellor: Juan José Campuzano Lopez 97

31-01220 Embassy of the Kingdom of Morocco
Resit Galip Cad. Rabat Sokak No. 11, 06700 Gaziosmanpasa, Ankara
☎ (+90 312) 437 60 20/21 ☒ 42869
⊖ (+90 312) 446 84 30
Ambassador: H E Mr Abdellatif Aqallal
First Counsellor: Haddou Esaadi
First Secretary: Mohammed Benjilany 98

31-01270 Royal Netherlands Embassy
Ugur Mumcu Caddesi 16, Gaziosmanpasa, Ankara
☎ (+90 312) 446 15 61 ✉ HOLLANDIA Ankara
☒ 46594 pbas tr ⊖ (+90 312) 446 14 72
Ambassador: H.E. Dr Jenda N.J.B. Horak
Counsellor & Deputy Head of Mission: Rob Bosscher
First Secretary Commercial Affairs: Ms Rini Bus
Defence Attaché: Capt. Robin A. Snouck Hurgronje (RNLN)
First Secretary: Gerrit Kool 97

31-01301 Embassy of New Zealand
Iran Cad. 13/4, 06692 Kavaklidere, Ankara
☎ (+30 312) 467 90 54; 90 56; 90 58
⊖ (+90 312) 467 90 13
Ambassador: H E Mr Clive Wallace Pearson
Deputy Head of Mission: Ms Katie Kerr
First Secretary: Hamis MacMaster 98

31-01340 Royal Norwegian Embassy
Kelebek Sok. 18, 06692 Gaziosmanpasa, Ankara
☎ (+90 312) 437 99 50; 64 30; 446 31 96 ☒ 42244x
⊖ (+90 312) 437 64 30
Ambassador: H E Mr Svein Ole Saether
Second Secretary: Arne Björnstad 98

31-01370 Embassy of Pakistan
Iran Cad. 37, 06700 Gaziosmanpasa, Ankara
☎ (+90 312) 427 14 10-14 ⊖ (+90 312) 467 10 23
Ambassador: H E Mr Inam ul Haque
Counsellor: Ms Seema Naqvi
First Secretary: Farukh A Amil
Defence Attaché: Col Mir Haider Ali Khan 98

31-01420 Embassy of the Philipines
Cayhane Sok. 24, Gaziosmanpasa, Ankara
☎ (+90 312) 446 58 31; 447 03 50
⊖ (+90 312) 446 57 33
Ambassador: H E Ms Minerva Jean A Falcon
Third Secretary: Antonio A Morales 98

31-01430 Embassy of the Republic of Poland
Atatürk Bulvari 241, 06692 Kavaklidere, Ankara
☎ (+90 4) 312 16 94-98 ⊖ (+90 4) 312 39 87
Ambassador: H E Mr Wojciech Hensel
Counsellor: Mrs Irena Tatarzynska
Counsellor: Wladyslaw Gondzik
Defence, Military, Naval & Air Attaché: Maj Gen Andrzej Tyszkiewicz
Commercial Counsellor: Ms Hanna Cieplucha
First Secretary: Stanislew Twardy 98

31-01440 Embassy of Portugal
Kuleli sok. 26, Gaziosmanpasa, 06700 Ankara
☎ (+90 312) 446 18 90; 446 18 91 ☒ 42771 port tr
⊖ (+90 312) 446 18 92
Ambassador: H.E. Mr Duarte Vaz Pinto
First Secretary Deputy Chief of Mission: António Chrystello Tavares
Military Attaché: José Inacio De Sousa (Resident in Rome)
Commercial Counsellor: Carlos Carvalho Vaz 97

31-01460 Embassy of the State of Qatar
Karaca Sok. 19, 06610 Gaziosmanpasa, Ankara
☎ (+90 312) 441 13 64-66 ⊖ (+90 312) 441 15 44
Ambassador: H E Mr Saad Mohamed Al-Kobaisi
First Secretary: Hamad Mohammed Al-Khalifa
Second Secretary: Ahmed Irbrahim Al-Abdulla 98

31-01480 Embassy of Romania
Bükres Sok. No. 4, 06680 Cankaya, Ankara
☎ (+90 312) 427 12 41; 12 43 ☒ 42760
⊖ (+90 312) 427 15 30
Ambassador: H E Mr Alexandru
Counsellor: Nicolae Dascalu
First Secretary: Aurelian Paraipan
First Secretary: Victor Oprea
First Secretary (Consular Affairs): Nicolae Moraru
Defence, Military, Air and Naval Attaché: Lt Col Marian Ilie 98

31-01485 Embassy of the Russian Federation
Karyagdi Sok. No. 5, PO Box 35, 06692 Cankaya, Ankara
☎ (+90 312) 439 21 22; 440 82 17
Ambassador: H E Mr Vadmin Kuznetsov
Minister-Counsellor: Alexandre A Tchitchkanov
Counsellor: Victor I Zagrekov
Counsellor: Alexangre L Prichtchhepov
Counsellor: Nikolay I Belovol
Counsellor: Vasili V Fedotov
Counsellor: Serguei Tamaev
Military and Air Attaché: Col Alexandre Deriglazov
Naval Attaché: Capt Mikhail M Popov
First Secretary: Vladimir M Barinov
First Secretary: Valeri I Golovkin
First Secretary: Mirdjalol M Khushanov
First Secretary: Victor V Melnik
Commercial Representative: Victor N Koptevski 98

31-01650 Embassy of the Republic of Slovakia
Atatürk Bulvari 245, Kavaklidere, Ankara
☎ (+90 4) 426 58 87 ⊖ (+90 4) 446 65 95
Deputy Chief of Mission: Dr Vladimir Lomen
Military Attaché: Col Milan Compel 97

31-01670 Embassy of the Republic of South Africa
Filistin Cad. No. 27, Ankara
☎ (+90 312) 446 40 56-61 ⊖ (+90 312) 446 64 34
Ambassador: H E Mr J Adriaan Eksteen
First Secretary: Andre Stemmet
Second Secretary: Andre de Munnink 98

31-01680 Embassy of Spain
Abdullah Cevdet Sok. No. 8, Cankaya, Ankara
☎ (+90 312) 438 03 92; 440 21 69 ☒ 46528 espa tr
⊖ (+90 312) 439 51 70
Ambassador: H E Mr Carlos Carderera
Defence Attaché: Lt Col Jose M Bohorquez
Commercial Counsellor: Jose Pascual Marco 98

31-01700 Embassy of the Republic of Sudan
Koza Sok. No. 51 G.O.P., 06700 Ankara
☎ (+90 312) 441 38 84-85 ☒ 46719 sudan tr
⊖ (+90 312) 441 38 86
Ambassador: H E Mr Hassan Adam Omer
Minister: Ibrahim Fadlalla El-Hadi
First Secretary: Azhari Abdelrahman
Second Secretary: Abdelazim Alsadig 98

31-01730 Swedish Embassy
Katip Celebi Sokak 7, PO Box 3 Kavaklidere, Cankaya, Ankara
☎ (+90 312) 428 67 35-37 ☒ 42230 swed tr
⊖ (+90 312) 468 50 20
Ambassador: H E Mr Michael Sahlin
Counsellor (Economic, Political, Commercial & Press): Ulf Walden
Second Secretary (Administration and Consular): Ms Karin Hedgards 98

31-01740 Embassy of Switzerland
Ataturk Bulv. 247, 06692 Kavaklidere, Ankara
Atatürk Bulvari 247, 06692 Ankara
☎ (+90 312) 467 55 55-56; 11 98 ⊖ (+90 4) 467 11 99
Ambassador: H E Mr Andre Faivet
Counsellor: Urs Breiter
Defense Attaché: Col Markus Orgis 98

Contact Reference

31-01750 Embassy of the Syrian Arab Republic
Sedat Simavi Sok. No. 40, Cankaya, Ankara
☎ (+90 312) 440 96 57-58 ☒ 46688
⊖ (+90 312) 438 56 09
Ambassador: H E Mr Abdul Aziz Al-Rifai
First Secretary: Dr Nadim Hussein
Third Secretary: Mohammed S Haj Bakri
Third Secretary: Osman Saab 98

31-01780 Royal Thai Embassy
Cinnah Cad. Kader Sok. No. 45/3-4, 06700 Ankara
☎ (+90 312) 467 34 09; 30 59 ☒ 46096 tayl tr
⊖ (+90 312) 438 64 74
Ambassador: H E Mr Chalerm Acharry
Minister-Counsellor: Wibunsawat Watanawarangkun
Minister-Counsellor: Pandit Srinurak
First Secretary: Miss Patranan Pattiya 98

31-01830 Embassy of Tunisia
Kuleli Sok. No. 12 G.O.P., Ankara
☎ (+90 312) 437 77 20; 78 12 ☒ 42215
⊖ (+90 312) 437 71 00
Ambassador: H E Mr Mohammed Megdiche
First Secretary: Mohammed Bel Kefi
Military, Naval and Air Attaché: Mohamed Rached Bargaoui
Counsellor: Mourad Belhassen 98

31-01875 Embassy of Ukraine
Cemal Nadir Sok. No. 9, Cankaya, Ankara
☎ (+90 312) 439 99 73; 440 53 89
⊖ (+90 312) 440 68 15
Ambassador: H E Mr Igor M Turyanskiy
First Secretary: Volodimir I Poberejniy
First Secretary: Adrian M Topale
First Secretary: Valeriy M Mayorov
Military Attaché: Lt Col Oleg I Dudnik 98

31-01880 Embassy of the United Arab Emirates
Mahmud Yesari Sok. No. 10, 06680 Cankaya, Ankara
☎ (+90 312) 440 84 10-13 ☒ 46044
⊖ (+90 312) 438 98 54
Ambassador: H E Mr Yousuf Abdelkhaliq Al-Ansari
First Secretary: Ahmed Mohammed Al-Taneigi
Third Secretary: Abdul Aziz Yousuf Al-Mahmood 98

31-01890 Embassy of the United Kingdom
Sehit Ersan Caddesi 46a, Cankaya, Ankara
☎(+90 312) 468 62 30-42 ⊖ (+90 312) 468 32 14
Ambassador: H E Sir David Logan
Deputy Head of Mission: Norman A Ling
Defence Attaché: Brig Anthony V Twiss
Naval and Air Attaché: Cdr Davis M H Lingard (RN)
Counsellor: Clive D Newell
Counsellor (Cultural Affairs): David Marler
First Secretary: Dr Peter H Tibber
First Secretary (Commercial): Alasdair MacDermott
First Secretary (Consul/Management): John G Fox
First Secretary (Defence Equipment): Hugh Thomas
First Secretary (Cultural Affairs): Ken Fitzsimons
Second Secretary (Political): Frank Baker
Second Secretary (Commercial): W Mike Stevenson
Second Secretary (Economic/Consular): Simon Shellly
Second Secretary (Political/Information): Owen Jenkins
Second Secretary (Cultural Affairs): Richard M Weyers
98

31-01900 Embassy of the United States of America
Atatürk Bulvari 110, Kavaklidere, Ankara
☎(+90 312) 468 61 10 ⊖ (+90 312) 467 00 19
Ambassador: H E Mr Marc J Grossman
Deputy Chief of Mission and Minister-Counsellor: Francis J Ricciardone (Jr)
Counsellor (Political Affairs): Richard K McKee
Counsellor (Regional Affairs): Hugh J Turner (III)
Counsellor (Political and Military Affairs): James L Clunan
Counsellor (Public Affairs): David I Arnett
Counsellor (Economic Affairs): C Eugene Dorris
Counsellor (Administrative Affairs): William A Eaton
Counsellor (Commercial Affairs): James M Wilson
Defence and Air Attaché: Col James Michael Carlin
Naval Attaché: Capt Jeffry D Keho
Army Attaché: Lt Col Robert R Pistana
98

31-01950 Embassy of Venezuela
Cinnah cad 78/2, PK 239 Kavaklidere, Ankara, Cankaya
☎(+90 312) 439 31 98; 438 71 35 ⊠ 46460 lara tr
⊖ (+90 312) 440 66 19
Ambassador: H.E. Mr Ramón Delgado
First Secretary: Mrs Sonia Pittol
97

31-01990 Embassy of the Republic of Yemen
Fethiiye Sok. No. 2, 06700 G.O.P. Ankara
☎(+90 312) 446 26 37; 31 78 ⊠ 46290 yemn tr
⊖ (+90 312) 446 17 78
Ambassador: H E Mr Mohamed Abdulla Al-Gaifi
Minister: Dr Ibrahim Aseed Al-Adoofi
First Secretary: Abdulrab Abubaker Bin Gaber
First Secretary: Mohamed Saleh Salooh
Third Secretary: Khaled Ali A Al-Batool
98

31-02010 Embassy of the Federal Republic of Yugoslavia
Paris Cad. No. 47, 06540 Kavaklidere, Ankara
☎(+90 312) 426 02 36; 24 32; 03 54
⊠ 46680 yuamb tr ⊖ (+90 312) 427 83 45
Charge d'Affaires: Dr Darko Tanaskovic
Minister-Counsellor: Bojan Bugarcic
Military, Air and Naval Attaché: Lt Col Milojko Radosavljevic
First Secretary: Zivorad Simic
98

34 Multinational Intergovernmental Organizations

34-22100 Sixth Allied Tactical Air Force (SIXATAF)
PK 527, Izmir
☎(+90 232) 448 48 48 ⊖ (+90 232) 487 65 82
Commander SIXATAF: Lt Gen. Erol Olcay (Tu AF)
Deputy Commander: Maj.Gen. William Hinton (USAF)
Chief of Staff: Maj.Gen. Ersan Inan (Tu AF)
Chief Logistics Department: Col Mehmet Unlu (Tu AF)
Exec. Officer to COMSIXATAF: Maj. Robert Moore (USAF)
Chief Public Information Officer: Capt. Mehmet Akdeniz ☎(+90 232) 448 48 48 ext 2305
97

55 Government-owned Industry & Production Facilities

55-00050 Aselsan Microwave and Systems Technologies Division (MST)
(† ASELSAN Military Electronics Industries Inc.)
Macunköy Facility
PO Box 101, 06172 Yenimahalle, Ankara
☎(+90 312) 385 19 00 ⊠ ASELSAN Ankara
⊠ 44518; 44422 asel tr ⊖ (+90 312) 354 13 02
♧ TL 968 000 000 000
Vice President and Chief Executive Officer: Mahmut Karadeniz
■ Manufacture C2 information systems, field artillery battery fire control systems, pedestal mounted air defence system, ground surveillance and artillery fire adjustment radar, SIGNIT and ELINT systems, and communication and radar jamming systems.

Microelectronics, Guidance and Electrooptics Division
(† Ayurt Facilities)
PO Box 30, 06011 Etlik, Ankara
☎(+90 312) 847 53 00 (11 lines)
⊖ (+90 312) 847 53 20
♧ TL 484,000,000,000 ♣ TL 3,860,000,000,000
👥 2,300 🏭 81 700 sq m
Chairman: Cengız Ergenemaan
Marketing Director: Melth Candan
■ High technology multi-product electronics. Military VHF/FM tactical man-pack, vehicular, tank radio sets, trunk radio systems, VHF/FM frequency hopping radio, data terminal, field telephone, EW equipment, laser range finder, avionic systems, electronic guidance systems of the STRINGER Missiles, night vision goggles, monoculars and binoculars, portable thermal camera, ACV/Day night thermal sight system, aviator night vision system, night weapon sights, laser infra-red aiming devices, laser designators, mine detector, artillery fire command control communication system. VP-Proximity fuses and radar. HF SARP Secure Adaptive Radio Processor, data crypto equipment, Network Management System, squad radio, HF receivers, hybrid microelectronics, electro-optics surveillance and target acquisition.

Aselsan Electronics Industries, Inc
PO Box 101, 06172 Yenimahalle, Ankara
☎(+90 312) 385 19 00 ⊖ (+90 312) 354 13 02
Chairman: Lieut Gen Gürhan Parker (retd)
Vice President and Division Chief Executive Officer of Communications: Ayhan Gerceker
Marketing Director of the Communications Division: Ugur Karavelioglu
General Manager: M Hacim Kamoy
■ Manufacture tactical radios and data terminals, data encryption equipment, field telephones and data communication systems.
98

55-01310 Makina ve Kimya Endüstrisi Kurumu (MKE)
Genel Müdürlügü, Tandogan Medyani, 06330 Ankara
☎(+90 312) 213 39 35; 223 20 11; 213 41 71 (marketing); (+90 312) 223 13 46; 223 13 78 (procurement) ⊠ 42223 mkga tr; 44363 mkek tr
⊖ (+90 312) 222 22 41 (marketing); (+90 312) 223 01 40 (procurement)
♧ US$ 100,000,000 ♣ US$ 700,000,000 👥 10,700
Chairman: Recai Önder
Deputy General Director: Semra Kaya
Deputy General Director: Dogan Mahmutoglu
Deputy General Director: Sedat Gezen
Deputy General Director: Ahmet Sari
Treasury Director: Ismail Irkin
Marketing and Export Director: Ismet Sökmensüer
Marketing and Export Assistant Director: Alinur Araz
Export Manager: Suphan Gaffaroglu
Advertising Manager: Mrs Canan Ilgun
■ Ordnance manufacturer of ammunition, rockets and explosives for Turkish Forces. Also export and civil sector. Detonators, small arms ammmunition, gas masks factory in Ankara. Artillery ammunition, propellant, rifle factory and weapon industry establishment in Kirikkale. Powder plant and rocket factory in Elmadag. Aircraft gun and machinery factory in Cankiri. → 75
98

65 Industry, Trade & Armed Forces Associations

65-00400 Defence Manufacturers Association
Katip Celebi Sok 2/11, 06680 Ankara
☎(+90 4) 440 55 66 ⊖ (+90 4) 440 55 67
General Secretary: Emin Gök
General Secretary: Mustafa Tasan
■ The association's main target is to develop and direct Turkish defence industry's interests in international markets, and enable the associated companies to take part in a more comprehensive and active way in international co-operation projects. The members of the association are public sector and private companies which actually manufacture for the domestic, NATO and other international markets.
98

74 Consortia, Joint Venture Companies

74-00680 FMC-Nurol Savunma Sanayii (FNSS)
PK 37, 06830 Gölbasi, Ankara
☎(+90 312) 484 00 77 ⊖ (+90 312) 484 00 73
General Mgr: Robert Cunningham
97

74-02130 TUSAS Engine Industries Inc. (TEI)
Muttalip Mevkii Mrk PK 610, Eskisehir
☎(+90 22) 322 20 30; 322 20 40 ⊠ 35356 tmco tr
⊖ (+90 22) 322 20 57
♣ US$ 34,000,000 (1996) 👥 443 🏭 45,000 m2
Managing Director: Tayfun Mutlu (acting)
Purchasing Manager: Mehmet Alemliogu
Sales Leader: Ali Üye
■ Joint venture by General Electric and Turkish partners. Manufactures complex fabrications, rotating and stationary machining for aircraft engines and gas turbines.
98

75 Manufacturing Companies

Makina ve Kimya Endüstrisi 55-01310

75-00010 ABIM Anadolv Bilig islem Merkezi ve Genel Danmismanlik Ltd
Istanbul
Bagoat Cad, Kördere Sok, No 10/3 Kiziltoprak
☎(+90 216) 347 28 30; 414 31 16
🄴 zeki.yvksebilgili@usa.net ⊖ (+90 216) 414 07 28
♧ US$ 1,500,000 ♣ US$6,000,000 (1996) 👥 37
🏭 1,800 m2
Managing Director: Ulug Turan
Assistant General Manager: Zeki Yüksekbilgili
■ Manufacturer of video surveillance system for the civilian and military market.
98

75-00150 BMC Sanayi ve Ticaret AS
Plant
Kemalpasa Asfalti Pinarbasi Kavsagi, 35060 Izmir
☎(+90 232) 479 20 00 ⊠ 53383 bmc tr
⊖ (+90 232) 479 12 88; 479 13 64
♧ TL 1 875 000 000 ♣ TL 9 200 000 000 000 👥 2 200
🏭 75 000 sq m
Vice Chairman: Mehmet Demirpence
Deputy General Mgr Marketing: Turgut Cankiliç
Advertising & Research: Nefi Farman
■ Largest commercial vehicle manufacturer in Turkey, producing a range of trucks, light commercial vehicles & engines for automotive, industrial & marine applications. It has production capacity of 9000 trucks, 6000 light commercial vehicles & 22 000 engines per year.
97

75-00530 EFE Endüstri ve Ticaret A.S.
Ankara Asfalti 282/1 No: 14, 35100 Bornova-Izmir
☎(+90 232) 484 04 98 ⊠ 51801 etm tr
⊖ (+90 232) 878 16 24
♧ TL 10 000 000 000 ♣ TL 135 000 000 000 👥 350
🏭 20 000 m2

General Mgr: Canerhan Tipi
■ Manufacture military trailers & semi-trailers, refuse compactors, cement mixers, batching plants, high loaders for catering, aircraft sewage & watering equipment, hydraulic components & axles. [97]

75-01350 Marconi Kominikasyon A.S.
(a subs of GEC Marconi Communications Ltd)
Muharebe ve Elektronik Cihazlar Tes., Konya Yolu 25 km, 06830 Golbasi, Ankara
☎ (+90 312) 484 51 81; 484 43 30
⊖ (+90 312) 484 43 32
⌂ 4 100 sq m
Managing Dir: George G. Hill
Operations Exec.: Brian C. Sidwell
Sales Exec.: A.Unal Solay
Commercial Exec.: K. Ekrem Kadioglu
■ Design, development & manufacture of electronic communication equipment. [97]

75-01400 Mercedes-Benz Türk A.S.
Eski Londra Asfalt, Kavsagi, 34530 Yenibosna, Istanbul
☎ (+90 212) 559 10 67; 559 17 82
⊖ (+90 212) 559 04 40
♧ TL 45,000,000 (1997) ♣ TL 1,070,000
ᚔ 3,000 (1997)
Managing Director: Eike Lippold
Sales and Marketing Director: Wolfgang Presinger
Government and Military Sales Group Chief: Haluk Altunbas
Department Chief Market Planning and Management: Ozlem Basara
■ Produce city and intercity buses, commercial and military vehicles including, trucks, Unimogs and cross-country vehicles.
® Germany: Mercedes-Benz AG
● Londra Asfalti, Kuleli Kavsagi, 34530 Bakirköy, Istanbul
☎ (+90 212) 661 55 02; 559 10 67; 661 55 05
▣ 28749 otpa ⊖ (+90 1) 559 04 40
Marketing Director: Wolfgang Presinger

Aksaray Plant
Aksaray
☎ (+90 382) 215 35 03; 20 23 ▣ 44 742 samo tr
⊖ (+90 382) 215 13 55
⌂ 33,646 sq m

Hosdere Bus Factory
Hosdere Mevkii, Sivatderesi, Büyükcekmece, Istanbul
☎ (+90 212) 623 19 00; 623 19 09
⊖ (+90 212) 623 19 18
■ Manufacture of city and intercity buses. [98]

75-01450 Mikes Microwave Electronic Systems Inc.
Cankiri Yolu 5 km, 06750 Akyurt, Ankara
☎ (+90 312) 847 51 00 ⊖ (+90 312) 847 51 14
♧ US$ 11 000 000 (1995) ♣ US$ 50 000 000 (1995)
ᚔ 115 ⌂ 6 500 sq m
General Mgr: Pertev Cinalioglu
Operations Dir: Rusen Yazgan
Engineering Dir: Dr M. Ali Tugay
Program Mgr: Atilla Ozkefeli
■ Design, manufacturing & integration of airborne EW Systems. Manufacturing of self protection EW systems for F-16 (AN-ALQ-178 (V) 3). Software & hardware logistics support for EW systems. [97]

75-02080 Transvaro A.S.
Eski Papaz Koprusü Mevkii, Dereboyu Cad. 12, 34660 Halkali, Istanbul
☎ (+90 212) 548 95 90 ⊖ (+90 212) 548 43 86
♧ US$ 5 000 000 (1995) ♣ US$ 5 040 000 (1995) ᚔ 36
⌂ 5 000 sq m
President: Sedat Uyar
Managing Dir: O. Berkan (Acting)
Operations Dir: Nur Debre
Technical Dir: Osman Damarli
Sales & Marketing Dir: Turhan Agikagn
■ Production & logistic service of night vision systems (goggles, binoculars, light & heavy weapon sights, observing devices, vehicle driver sights & periscopes, pilot goggles), laser range finders (eye safe), target pointers & system accessories. Metal mine detectors, hand held metal detectors, training simulation systems, target designators, IR identification devices, cable harnesses, electronic fuses. [97]

75-02120 TUSAS Aerospace Industries Inc. (TAI)
TUSAS Havacilik ve Uzay Sanayii A.S.
PO Box 18, 06692 Kavaklidere/Ankara
☎ (+90 312) 811 18 00
⊖ (+90 312) 811 14 08; 811 14 25
♧ US$ 292,000,000 ᚔ 2,273 ⌂ 130,000 m2
General Manager: Jerry R Jones ☎ (+90 312) 811 14 23
Deputy Managing Director: Dr Birol Altan ☎ (+90 312) 811 28 75
Programs Director: Kaya Ergenç ☎ (+90 312) 811 14 09
Sales and Marketing Manager: Pinar Ozerman
Purchasing Manager: Cüneyt Maltepeli
Public Relations Co-ordinator: Nursel Kōran
■ Manufacture and assembly of fuselage, wing and other components for all types of aircraft and helicopters. Modification and modernisation of fighter and fighter bomber aircraft. Design and develop UAVs, target drone, fighter surveillance aircraft and 20 seater passenger/cargo regional airliner. [98]

80 Service Companies & Organizations

80-00950 Hisas AS
Baagdat Cad 190, 81030 Selamicesme, Istanbul
☎ (+90 216) 363 01 35 Ⓔ hisas@doruk.net.tr
⊖ (+90 216) 386 33 95
♣ US$ 2,000,000 ᚔ 25
Owner: M Burcak Akiska
Managing Director: Ekin Akiska
Marketing Director: Hakan Icen
■ Consulting, representation and documentation services for foreign companies which want to expand their coverage of the Turkish defence market. [98]

80-01200 Proje Yonetim AS
80700 Istanbul
Murbasin Sok, Kaza is Merkezi, B Block K:2 D:7, Besiktap
☎ (+90 212) 275 00 08 ⊖ (+90 212) 266 63 06
® UK: Artemis Management Systems [98]

80-01330 Metis Construction & Trade Co. Inc.
Kas-met Constr., Installation & Trade Co.
Ecs Electronic and Electrical Equipment Industry and Trade Co.
Head Office
Abidin Saver Sokak No.33, 06550 Cankaya, Ankara
☎ (+90 312) 438 11 70/72 ✉ METIS Ankara
▣ 46749 kas tr; 46567 ecs tr ⊖ (+90 312) 439 16 13
♧ US$ 8,403,220 (1995) ♣ US$ 43 871 870 000
ᚔ 1,055
General Mgr: Timur Oztekin
General Co-ordinator: Veysel Sever
General Mgr Marketing: Cengiz Ergin
Finance & Budget Co-ordinator: Mrs F. Nese Özdemir
Planning Mgr: F. Resan Ergul
Bidding Co-ordinator: E. Lutfi Ozcan
■ Industrial, commercial & military construction projects, machinery, electronics, installation, general merchandise & marketing. → 90 [97]

85 Stockists, Distributors, Spare Parts Suppliers

85-02000 Pilot Aviation Services & Tourizm Ltd
Selvili Sokak 15/3, 80620 4.Levent, Istanbul
☎ (+90 212) 281 03 00 ⊖ (+90 212) 281 03 01
ᚔ 8
General Mgr: Mehmet Aksu
■ Commercial representation for aviation spare parts. [97]

90 Sales Agents, Sales & Liaison Offices

Companies represented:

Agusta SpA (Commercial Sales only), Italy . . . 90-00095
Airtractor Inc., USA . . . 90-00095
Artemis Management Systems, UK . . . 80-01200
Blohm & Voss, Germany . . . 90-00810
Boeing Aerospace & Elect., USA . . . 90-00510
CAE Electronics Ltd, Canada . . . 90-00095
Canadair Amphibious Aircraft Div., Canada . . . 90-00095
CMI Mechanical Industries, Belgium . . . 90-00300
Cossor Electronics Ltd, UK . . . 90-00510
EEV Ltd, UK . . . 90-02070
Hollandse Signaalapparaten b.v., Netherlands . . . 90-00810
IAI/Elta, Israel . . . 90-00095
IAI/Mata, Israel . . . 90-00095
IAI/Tamam, Israel . . . 90-00095
Intermarine SpA, Italy . . . 90-00810
Lockheed Corp., USA . . . 90-01200
Manitou BF, France . . . 90-01300
Marconi Company, St. Lucia . . . 90-00095
Mercedes-Benz AG, Germany . . . 75-01400
MTU Friedrichschafen GmbH, Germany 90-01350
NavCom Defense Electronics, USA . . . 90-00510
OTO Melara, Italy . . . 90-00810
Pilatus, Switzerland . . . 90-00510
Pratt & Whitney/UTC, USA . . . 90-00510
Radamec Group, UK . . . 90-00510
Telephonics Inc., USA . . . 90-00095
Thyssen R.T., Germany . . . 90-00810
Thyssen Rheinstahl Tec., Germany . . . 90-00810
Trimos-Sylvac Task Force, Switzerland . 90-01995

Sales agents/representatives listed elsewhere:

Metis Construction & Trade Co. Inc. . 80-01330

90-00087 AlliedSignal Aerospace Marketing Sales & Service
Turkish Airlines, 34830 Instanbul
Turkish Airlines Inc. Technical Department
34830 Ataturk Airport
☎ (+90 212) 663 08 72 ⊖ (+90 212) 573 76 04
Service Engineer: Louis Pacini [98]

90-00095 Astek Ltd
(† Namat Ltd)
Gunes Sokak 6/6, Kavaklidere, 06690 Ankara
☎ (+90 312) 426 79 66; 467 48 58
⊖ (+90 312) 428 11 06
♧ US$ 20 000 ♣ US$ 600 000 ᚔ 8 ⌂ 2 000 sq ft
President: Dr A. Müren Gökeri
Vice President: Bülent Acur (MSME)
■ Aviation related consulting & representation.
® Canada: CAE Electronics Ltd, Canadair Amphibious Aircraft Div.
Israel: IAI/Elta, IAI/Mata, IAI/Tamam
Italy: Agusta SpA (Commercial Sales only)
St. Lucia: Marconi Company
USA: Airtractor Inc., Telephonics Inc. [97]

90-00300 CMI Cockerill Mechanical Industries
Koybasi Caddesi 267, Yenikoy, Istanbul
☎ (+90 212) 277 69 37 ⊖ (+90 212) 277 69 37
Deputy Director: Stephan Mahieu
® Belgium: CMI Mechanical Industries [98]

90-00350 Daewoo Corp
Buyukedere Cad N°59/3, 80670 Maslak, Istanbul
☎ (+90 212) 285 41 20 ⊖ (+90 212) 285 41 25 [98]

90-00510 Elektronik Mekanik Cihaziar Ticaret Ltd (EMC)
Sirketi Cinnah Caddesi, Yasilyurt Sokak 14/5, Ankara
☎ (+90 312) 428 68 18; 428 68 28
⊖ (+90 312) 468 48 25
ᚔ 8
General Mgr: M. Ayhan Izgi
Sales & Marketing Dir: Bedirhan Izgi
Public Relations Mgr: Mrs Hulya Yigit
■ Communication sets, equipment. IFF, several types of radar, aircraft, helicopter, jet engine, missile.
® Switzerland: Pilatus
UK: Cossor Electronics Ltd, Radamec Group
USA: Boeing Aerospace & Elect., NavCom Defense Electronics, Pratt & Whitney/UTC [97]

90-00810 Hatco A.S.
Technical Equipment Representation & Trading Inc.
Acisu Sok No.19/C Macka, 80680 Besiktas, Istanbul
☎(+90 212) 258 99 83; 258 79 91/3
⊠ 27818 htko tr; 27652 erha tr ⊖ (+90 212) 258 79 94
👥 50
Managing Dir: M.E. Hatayli
General Mgr Technical: Cdr Oner Tasdelen (Rtd)
■ Representing defence field companies trading steel, coal, equipment & machinery.
® Germany: Blohm & Voss, Thyssen R.T.
Italy: Intermarine SpA, OTO Melara
Netherlands: Hollandse Signaalapparaten b.v.
Hatco-Ankara
PO Box 236, Küçükesat, Ankara
☎(+90 313) 418 48 48; 418 75 55 ⊠ 42776 ertu tr
⊖ (+90 4) 425 47 40
Director: Capt. Iya Karal
® Germany: Thyssen Rheinstahl Tec. [97]

90-01130 KEP Beynelmilel AS
Valikonagi Cad. Sonu 5-1, Nisantasi, 80220 Istanbul
☎(+90 212) 248 85 11 ⊖ (+90 212) 241 31 30
👥 8 🏠 200 sq m
President: Ahmet Neyzi
Vice President: Halim Neyzi
■ Radio communications, radar, etc. Sales & service & defense consultants.
Representing:
Condor
Southwest Research Institute
Simula [97]

90-01200 Lockheed Corp.(International) SA
Cinnah Caddesi 1/2-B, PO Box 144,
00692 Kavalidere, Ankara
☎(+90 4) 221 21 28
Director: R.G. Todd
® USA: Lockheed Corp. [97]

90-01300 Manitou HMK
(a subsidiary of Manitou BF, France)
Derliger Sokak 29/1, Gazeteciler Mahallesi,
Esentepe, Istanbul
☎(+90 212) 266 80 30 ⊖ (+90 212) 274 17 91
® France: Manitou BF [98]

90-01350 MTU Motor Türbin San. ve Tic. AS
MTU Center, Mürbasan Sok, Barbaros Bulvari, Balmumcu,
80700 Besiktas, Istanbul
☎(+90 212) 288 05 03; 27 59
⊖ (+90 212) 266 77 43; 266 35 34
♣ DM 2,000,000 👥 15 🏠 1,020 sq m
Managing Director: Muammer Iyi
Sales and Marketing Director: Salih Turna
Operations and Technical Director: Ozer Akarslan
Purchasing Manager: Sabri Tuncer
® Germany: MTU Friedrichschafen GmbH [98]

90-01990 Sikorsky United Technologies Int'l Ops
27/2 Ugur Mumcu Cad, GOP 06700, Ankara
☎(+90 312) 447 41 77 ⊖ (+90 312) 447 41 70
♣ US$ 2 000 000 000 👥 8,000 [97]

90-01995 Tekno 2000
(† Takimsas)
Izzetin Calislar Cad n°12/3 B-Block,
34600 Bahcelievler/Istanbul
☎(+90 212) 557 38 30 ⊖ (+90 212) 507 64 43
Managing Director: Hamdi Gulen
Technical Advisor: Ayhan Etyemez
® Switzerland: Trimos-Sylvac Task Force [98]

90-02070 Tork Elektronic Valf Ltd
Rihtim Cad. Funda Han No. 221/509,
80030 Karakoy, Istanbul
☎(+90 212) 251 08 68 ⊖ (+90 212) 252 59 95
Managing Director: S Bayramoglu
® UK: EEV Ltd [98]

Uganda

Head of State:
Yoweri Kaguta Museveni

10 Head of State

10-10000 Office of the President
Parliamentary Buildings, PO Box 7168, Kampala
☎(+256 41) 25 48 81 to 88 ⊠ ADMINISTER
⊠ 61048 resist; 61389 kapital
President: Yoweri Kaguta Museveni
☎(+254 41) 23 16 66
Minister of State in Charge of Karamoja:
Peter Aparite Lokirit ☎(+254 41) 23 61 18; 23 68 21
Minister of State in Charge of Luwero:
Edward Tuddu Makubya
Minister of State (Gender and Community Development):
Janat B Hajati Mukwaya ☎(254 41) 25 63 74 [98]

10-10010 Office of the Vice President
5 Katonga Road, PO Box 7359, Kampala
☎(+254 41) 23 45 90
Vice President: Dr Specioza W Kazibwe
☎(+254 41) 23 65 63 [98]

11 Prime Minister

11-10000 Office of the Prime Minister
Kitante Road, Post Office Building, PO Box 341, Kampala
☎(+256 41) 25 87 41; 25 86 94; 25 43 07 ⊠ 61378
⊖ (+256 41) 24 23 41
Prime Minister: Kintu Musoke ☎(+254 41) 25 95 18
First Deputy Prime Minister: Eriya Kategaya
☎(+254 41) 25 94 98
Second Deputy Prime Minister: Brig Gen Mose Ali
☎(+254 41) 23 39 22
Third Deputy Prime Minister: Paul Orono Etiang
☎(+254 41) 24 16 65 [98]

12 Ministries

12-20000 Ministry of Foreign Affairs
PO Box 7648 and 4411, Kampala
☎(+256 41) 25 75 25; 25 82 51; 25 82 52 ⊠ EXTERIOR
⊠ 61228 exterior, 61007 adminster
⊖ (+256 41) 25 87 22
Minister: Eriya Kategaya ☎(+254 41) 23 39 22 [98]

12-40000 Ministry of Internal Affairs
Crested Towers, PO Box 7191, Kampala
☎(+256 41) 25 83 55 ⊠ 61331 intaff, 61194 affairs
⊖ (+256 41) 23 11 88
Minister for Internal Affairs: Maj Tom Butime
☎(+254 41) 23 65 63; 24 30 88
Permanent Secretary: Norman Isingoma
☎(+256 41) 23 11 03 [98]

31 Embassies & Attachés

31-00660 Embassy of the Federal Republic of Germany
15 Philip Road Kololo, PO Box 7016, Kampala
☎(+256 41) 25 67 67/68; 23 64 21/22
⊠ 61005 aa kamp ⊖ (+256 41) 24 31 36
Ambassador: H.E. Mr Christian Nakonz
Deputy Chief of Mission: Dr Thomas Kurz
Economic Attaché: Mr Henning [97]

31-00890 Embassy of Italy
PO Box 4646, Kampala
11 Lourdel Road, Nakasero-Kampala
☎(+256 41) 24 17 86; 25 64 16; 25 04 42; 23 34 43
⊠ 61261 italdipl ⊖ (+256 41) 25 04 48
Ambassador: H.E. Dr Marcello Ricoveri
First Secretary Deputy Chief of Mission:
Dr Mainardo Benardelli
Commercial Attaché: Dr Saverio Rosini [97]

31-01890 United Kingdom High Commission
10/12 Parliament, PO Box 7070, Kampala
☎(+256 41) 25 70 54; 25 70 59; 23 73 01; 23 73 03
⊠ UKREP KAMPALA ⊠ 61202 ukrep kla
⊖ (+256 41) 25 73 04
High Commissioner: H E Mr Michael Cook
Deputy High Commissioner: M E Rouse
Defence Adviser: Lieut Col N P C Lewis
First Secretary, Aid and Commercial: D Seddon [98]

Ukraine

Head of State:
Leonid Kuchma

10 Head of State

10-10000 Office of the President
Bankivska ulitsa 11, 252009 Kiev
☎(+38 044) 291 53 33 ⊖ (+38 044) 293 73 64
President: Leonid Kuchma [98]

11 Prime Minister

11-10000 Office of the Prime Minister
Gruschevskogo ulitsa 12/2, 252008 Kiev
☎(+38 044) 293 52 27 ⊖ (+38 044) 293 20 93
Prime Minister: Ms Valeriy Pustovoytenko
Deputy Prime Minister for Humanitarian Policy: Ivan Kuras
Deputy Prime Minister for Economic Reform:
Serhiy Tygypko [98]

12 Ministries

12-10000 Ministry of Defence
Bankivska ulitsa 6, 252009 Kiev
Minister: Lieut Gen Oleksandr Kuzmuk [98]

12-20000 Ministry of Foreign Affairs
Mychaylivska Square 1, 252018 Kiev
Minister: Hennadiy Udovenko
National Committee of Ukraine on Disarmament
Mykhaylivska Sq. 1, 252018 Kiev
☎(+38 044) 212 82 65 ⊖ (+38 044) 212 02 14
Deputy Foreign Minister & Head of Committee:
K. Gryshchenko [98]

12-30000 Ministry of Machine Building Military Industrial Complex and Conversion
Surikova Straat, 352035 Kiev
☎(+38 044) 246 32 08 ⊖ (+38 044) 245 62 09
Minister: Vasil Hureyev ☎(+38 044) 245 47 78
Deputy Minister: Ms Valery Kazakov
☎(+38 044) 246 31 71
External Economic Relations Director: Victor Shulyepov [98]

12-40000 Ministry of Internal Affairs
Bogomoljtsa ulitsa 10, 242024 Kiev
☎ (+38 044) 226 20 04 ⊖ (+38 044) 220 19 05
Minister: Yuriy Kravchenko
Head of State Security Service: Volodymyr Radchenko
Public Relations Director: Ivan Levchenko [98]

12-50000 Ministry of Finance
ulitsa Grushevskogo 12/2, 252008 Kiev
☎ (+38 044) 226 20 44 ⊖ (+38 044) 226 25 17
Minister of Finance: Ihor Mityukov [98]

12-60000 Ministry of Engineering, Military-Industrial Complex and Conversion of Ukraine
3 Surykova Straat, 252035 Kiev
☎ (+38 44) 245 47 78 ⊖ (+38 44) 246 32 14
Minister: Mr Guureyev
Deputy Minister: Valery P Kazakov ☎ (+38 44) 245 62 01
Foreign Relations Director: Victor M Shulyepov
☎ (+38 44) 246 32 13; 246 32 55 (fax) [98]

12-70000 Ministry of Communications
vul Kreshchatik 22, Kiev
Minister: Dmytro Khudoliy [98]

15 Parliament

15-10000 Supreme Council (Parliament) of Ukraine
Grushevskogo ulitsa 5, 252019 Kiev
☎ (+38 044) 293 49 92 ⊖ (+38 044) 293 73 64
Chairman of the Parliament: Oleksandr Moroz [98]

31 Embassies & Attachés

31-00100 Embassy of Austria
Ivana Franko 33, 252030 Kiev
☎ (+38 044) 220 57 59; 244 39 43
⊖ (+38 044) 227 54 65
Ambassador: H E Mr Klaus Fabjan
Deputy Head of Mission: Werner Winkler
Attaché: Ms Johanna Hausensteiner
Attaché: Margit Schwaner
Defence Attaché: Maj Simon Palmisano
Cultural Service: Harald Fleischmaann
Secretary-Archivist: Ms Monika Junker-Guldin
Secretary-Archivist: Peter Klaus
Secretary-Archivist: Anton Pfeiffer
Secretary-Archivist: Ms Barbara Steinböck
Office of the Commercial Counsellor
Holoposhtamt /c 62, 252001 Kyiv
☎ (+38 044) 225 13 41; 225 03 18
⊖ (+38 044) 230 25 37
Commercial Counsellor: Dr Rudolf Thaler
Secretary, Archivist: Ms Monika Huber [98]

31-00310 Embassy of Canada
Yaroslaviv val 31, 252034 Kiev
☎ (+38 044) 212 02 12; 212 28 64
⊖ (+38 044) 225 13 05; 212 23 39
Ambassador: H.E. Mr Christopher Westdal
Political Counsellor: Roman Waschuk
Commercial Officer: Larry Duffield [97]

31-00450 Embassy of the Czech Republic
Yaroslaviv val 34A, 252034 Kiev
☎ (+38 044) 212 04 31; 212 21 10
⊖ (+38 044) 224 61 80
Ambassador: H E Mr Pavel Masa
Defence Attaché: Stanislav Hajek
Commercial Officer: Juraj Melioris
Secetary: Ms Paula Hoffmamor [98]

31-00460 Royal Danish Embassy
Vul. Volodymyrska 45, 252034 Kiev
☎ (+38 044) 229 45 37 ⊠ 131457 ambdk ux
⊖ (+38 044) 229 18 31
Ambassador: H.E. Mr Christian Faber-Rod
Minister Counsellor: Peter Melchior
Consul: Kurt Moeller Olesen
Defence Attaché: Col Hans Joergen Foss (resident in Moscow)
Assistant Defence Attaché: Lieut Col Carsten Barloese (resident in Moscow) [97]

31-00535 Embassy of Estonia
Kutuzovi Prospekt 8, Kiev
☎ (+380 44) 294 80 55; 295 87 72 (consular department) 🄴 saatkond@estemb.kiev.ua
⊖ (+380 44) 294 80 55; 295 87 72 (consular department) [98]

31-00580 Embassy of Finland
vul. Striletska 14, 252034 Kiev
☎ (+38 044) 228 70 49; 228 70 47
228 75 51; 212 04 49 ⊖ (+38 044) 228 20 32
Ambassador: H.E. Mr Erik Ulfstedt
First Secretary: Ms Tuula Yrjölä
Second Secretary: Ms Helena Alaruikka
Attaché: Lauri Veijalainen
Defence Attaché: Lt Col Antti Iivonen (resident in Moscow)
Commercial Secretary: Olavi Filppula (resident in Moscow) [97]

31-00660 Embassy of the Federal Republic of Germany
wul. Hontschara 84, 252054 Kiev
☎ (+38 044) 216 71 56; 216 36 09;
219 12 92 (press/culture department) ⊠ 131122 ux
⊖ (+38 044) 216 92 33
Ambassador: H E Mr Eberhard Heyken
Minister Counsellor and Deputy Head of Mission, Political Affairs: Bernd Braun
Counsellor, Financial Assistance: Claus Brümmer
Defence Attaché: Col Jürgen Frick
Naval Attaché: Cdr Hans-Peter Scholz
First Secretary, Cultural and Minority Affairs: Christian Reißmüller
First Secretary, Economic Affairs: Klemens Mömkes
First Secretary, Economic Affairs: Michael Keßler
First Secretary, Scientific Affairs: Wolfgang Böhm
First Secretary and Consul: Oliver Schnakenberg
Second Secretary, Political Affairs: Dr Gernot Biehler
Second Secretary, Press and Protocol: Ms Anka Feldhusen
Second Secretary, Interpreter: Michael Josy
Second Secretary, Head of Administration: Günther Hübner
Information Officer: A Feldhusen [98]

31-00890 Embassy of Italy
Vulitsa Sichnevoho Povstanniya 25, 252015 Kiev
☎ (+38 044) 24 42 94; 29 44 42 42; 294 42 43
⊠ 64131462 itdki su ⊖ (+38 044) 290 51 62
Ambassador: H.E. Mr Vittorio C. Surdo
Defence Attaché: Gen. Paolo Carosella
Consular Office
Pereulok Dzerzhinskogo 5, Kiev
☎ (+38 044) 293 05 69 ⊖ (+38 044) 293 93 62 [97]

31-00920 Embassy of Japan
Regina Business Centre - 7th/8th, Floor, 4 Muzeyny Lane, 252001 Kiev
☎ (+49 30) 57 91 63 00; (+7 044) 462 00 19;
462 00 20 ⊖ (+49 30) 57 91 63 02
Ambassador: H.E. Mr Shoji Suezawa [97]

31-01015 Embassy of Latvia
Desiatinna Straat, 4/6, Kiev
☎ (+38 044) 462 07 08 🄴 latemb@public.ua.net
⊖ (+38 044) 229 27 45
Charge d'Affairs ai/Second Secretary: Ms Anda Abele
Second Secretary: Gunnars Everts
Third Secretary: Dainis Garancs
Commercial Attaché: Mihails Popkovs
Commercial Attaché: Andris Dzenis [98]

31-01650 Embassy of the Republic of Slovakia
Jaroslavov Val 34, 252034 Kiev
☎ (+38 044) 229 79 22; 212 03 10; 212 00 08
⊖ (+38 044) 212 32 71; 212 09 08
Deputy Chief of Mission: Ms Olga Mihalikova
Defence Attaché: Gen Jan Pancik
Commercial Officer: Stefan Kuriscak
Public Information Officer: Marek Estok [98]

31-01730 Embassy of Sweden
PO Box 122, 252030 Kiev
34/33 Ivana Franka Street 3rd Floor
☎ (+38 044) 462 05 80
🄴 ambassaden.kiev@foreign.ministry.se
⊠ 131479 uyut su ⊖ (+38 044) 462 05 81
Ambassador: H E Mr Göran Jacobsson
Defence Attaché: Hans Bjernby (resides in Warsaw)
Counsellor: Ulf Sör
Second Secretary: Mrs Irene Katzam
Third Secretary: Ms Susanne Sjöström [98]

31-01890 Embassy of the United Kingdom
Desyatinna 9, 252025 Kiev
☎ (+38 044) 462 00 11 ⊠ 131429 prodr su
⊖ (+38 044) 462 00 13
Ambassador: H E Mr R Reeve
Deputy Head of Mission and Consul-General: S J Butt
Defence Attaché: Capt L C P Merrick (RN)
Counsellor: The Viscount Asquith
First Secretary, Commercial Know-How Fund: Ms T M V Abbott-Watt
Second Secretary, Information Officer: Mrs J M Gardiner [98]

31-01960 Embassy of the Socialist Republic of Vietnam
Ivashkevych Street 6/33, Lokora Straat 5, Kiev
☎ (+38 044) 295 28 37
🄴 embassy@vietnam.fuenet.kiev.ua
⊖ (+38 044) 295 28 37
Chargé d'Affairs: Dusan Rozbora
Deputy Chief of Mission: Eiji Dinh Dinh
Third Secretary: Ivan Skorupa [98]

40 Government Defense Sales/Procurement Organizations

40-03000 State Scientific and Technical Centre of Artillery and Rifle Arms
PO Box 882/1, 151057 Kiev
☎ (+380 44) 446 24 59 🄴 kba@kba.kiev.ua
⊖ (+380 44) 446 28 34
Managing Director: L Bondarenko
Marketing Director: V Elensky
■ Manufacture tank barrels; 30 mm automatic and universal artillery complexes; anti aircraftself propelled installations with 23 mm guns; mobile missile artillery complexes; mobile morter complexes; mobile missile artillery complexes for extinguishing; gun barrels: howitzers, morters and rifle arms and revolvers. [98]

55 Government-owned Industry & Production Facilities

55-00100 Antonov
Aeronautical Scientific and Technical Complex
1 Tupoleva Street, 252062 Kiev
☎ (+38 44) 442 70 90; 442 70 98; 442 73 04
⊠ 131309 ozon su ⊖ (+38 044) 449 99 96; 442 70 98
General Designer: Piotr V Balabuyev
First Deputy General Designer: Dmitry S Kiva
Deputy General Designer: Oleg K Bogdanov
Deputy General Designer Sales/Marketing: Victor I Bolgak
Director: Vadim N Panin
Commercial Director: Orest G Uskov
Chief Flight Test Centre: Anatoly G Bulanenko
■ Design of new aircraft and performance of a wide range of scientific and research work including wind-tunnel, structures, laboratory and flight tests. Construction of prototypes and promotion of their certification. Provides after sales and engineering support to increase operated aircraft service life. Performs flying and maintenance personnel training and retraining. Carries out international cargo transportation (including outsize cargoes) on a charter basis. Also designs and builds prototypes of surface transports including low-floor tram, trolleybus, RADAN and light railway rolling stock. → 75 [98]

Ukraine

55-01500 Kiev Aircraft Repair Plant
(a subsidiary of the Ministry of Machinery)
94 Vozdukhoflotsky Avenue, Kiev 151
☎ (+38 044) 245 99 26; 245 99 58, 245 98 69
✉ KIEV KRYLO ⊠ 631041 wing su ⊠ 631041 wing ux
⊖ (+38 044) 245 98 06; 245 99 83; 245 98 69
⇔ US$ 750,000,000 (1995)
♣ US$ 320,000,000 (1995) ⁂ 2,200 (1995)
⌂ 400,000 m2
General Director: Anatoly P Kudrin
Technical Director: Gregory R Danyelian
Financial Director: Juri E Medakov
Sales and Marketing Director: Sergey M Podreza
Public Relations Director: E Titlianov
■ Repair, overhaul and modification of AN-24, AN-26 and AN-32 aircraft. Engines for D 36, AL 9 and AL 25. Testing, demagnetisation, leasing and technical training. Export of CIS produced aircraft. [97]

55-01950 Musson Open Stock Company
ul. Vakulenchuka 29, 335053 Sevastopol
☎ (+38 0692) 24 72 00; 23 80 86
⊠ (+38 0692) 23 80 79 ⊠ 197304
⁂ 2,000 ⌂ 44,800
Chairman of the Board - OSC MUSSON President: Nikolay I Krutov
Marketing Director: Gennady I Stepanov
Purchasing Manager: Victor A Podolsky
■ Manufacturer of communication electronics and equipment. Mobile and marine communication equipment.

Musson-Laspee
(Daughter company of Musson Open Stock Company)
ulitsa Vakulenchuka 29, 335053 Sevastopol
☎ (+38 0692) 24 31 84; 24 42 60 ⊠ 197304
⊖ (+38 0692) 24 40 74
⁂ 180 ⌂ 2,800
Director: Vladimir G Semenov
Marketing and Sales Director: Sergey A Linchevsky
Deputy Technical Director: Oleg A Lutov
☎ (+38 0692) 24 05 54
■ Design and manufacturer of mobile communication equipment. [98]

55-02500 SDO Yuzhnoye
3 Krivorozhskaya Street, 320059 Dnepropetrovsk 59
☎ (+380 562) 70 04 47 ✆ (+380 0562) 92 49 79
[E] kbu@public.ua-net
⊠ 127404 orekh su; 131223 orekh su
⊖ (+380 562) 70 01 25
General Director and General Designer: S N Konyukhov
First Deputy General Designer: Alexander N Mashchenko
First Deputy Designer of Research: Professor Yuri Alekseyevich Smetanin
Deputy General Designer of Co-ordination and Management: Evgeny V Kouryachiy
Chief Engineer: Vladimir G Vasilina
Head of Marketing and Foreign Relations Department: Alexander V Degtyarev
■ Design and preparation of rocket and space complexes, launch vehicles ("Cyclone", "Zenit", Dnepr"), spacecrafts ("Sich", "Ocean") and production of their components. Provision of launch services to inject paylod into space; perform tests as follows: strength, pneumohydraulic, electrical, exposure to various environment conditions (temperature, humidity), radio frequency and engine ignition tests. [98]

60 Research Organizations, Institutes & Test Establishments

Erotron 75-01500
Hartron Corporation 75-03000

60-00880 Kvant Scientific Research Institute
5 Dimitrov St, 252006 Kiev
☎ (+38 044) 227 31 50 ⊖ (+38 044) 220 74 68
Director: L.I. Lisitsa
International Relations Dir: Prof. V.V. Molebny
■ Detection (including over-horizon); recognition; fire control; ship navigation; GPS; electro-optics (including fire control & countermeasures); medical instrumentation. [97]

60-00940 "Orion" Scientific Industriel Enterprises
(† Orion Research Institute)
252057 Kiev - 57
☎ (+38 044) 446 93 82, 446 91 64
⊖ (+38 044) 446 52 91
⇔ US$ 15,500,000 ⌂ 34,000 (1996)
Managing Director: Lev G Gassanov
Marketing Director: Mykhylo I Yatsyuk
Operations Director: Semen P Rakitin
Purchasing Manager: Sergey I Kulik
■ Production of components, assemblies, instruments and devices of SHF and EHF equipment for communication, telecommunication systems, transportation, installations for EHF therapy, automobile electronics. [98]

60-00990 Research Institute for Aeroelastic Systems (RIAS)
85 Garnaev St, Crimea, 334812 Feodosiya
☎ (+38 06562) 426 46 ✆ (+38 06562) 426 55
⊖ (+38 06562) 426 55
Managing Director: P M Nikolaev ☎ (+38 06562) 426 46
Marketing Director: Y M Nikolaev ☎ (+38 06562) 426 55
Operations Director: R A Sharafutdinov
☎ (+38 06562) 426 55
Purchasing Manager: V A Yelsakov
☎ (+38 06562) 426 47
■ Research and development of cargo delivery parachute systems both automatic and manual with ram-air or round canopies for loads of 50 kg to 2 tons, paragliders, hot-air balloons, drift anchors of different modifications, wind-protective systems, submarine emergency personal escape floation bags, parachute systems for meteorological probes and rockets. Tents, sleeping bags for northern climatic conditions. [98]

65 Industry, Trade & Armed Forces Associations

65-50000 Ukrinmash, State Scientific and Production Foreign Trade Association
(a joint venture with the Ministry of Engineering, Military-Industrial Complex and Conversion)
30 Akademika Beletskoco Straat, 25156 Kiev
☎ (+38 044) 483 80 54 ⊖ (+38 044) 483 30 82
Director General: Boris Marusich
Commercial Director: Yuriy Syemoushkin
☎ (+38 044) 488 28 77
■ State owned association promoting exportation of Ukrainian products and intermediary in foreign trade. Activities include the attraction of foreign investment and expertise, arrangment of exhibitions, provision of goods handling and storage facilities and organisation of clearance for commercial shipments. [98]

75 Manufacturing Companies

Antonov 55-00100

75-00800 Central Design Bureau Arsenal (CDO)
8 Moskovska Straat, 252601 Kiev 10 GSP
☎ (+38 044) 293 00 62 ✆ (+38 044) 291 49 78
⊠ 131016 prizma ⊖ (+38 044) 293 15 09
Director and Chief Designer: V Busanov
☎ (+38 044) 293 00 62
■ Manufacture and development of: optical and optoelectronics; optical seekers for surface-to-air and air-to-air missiles; gyroscopic systems for independent orientation and navigation; laser measuring systems including laser gyros; optical devices for space applications; optical radars utilizing TV, thermal vision and laser technologies; helmet-mounted target designation and indication systems. [98]

75-01500 Erotron
(a subsidiary of R I Erotron)
90 Antonovich Straat, 290044 Lvov
☎ (+38 322) 34 11 19 ⊠ 734100 ifc su
⊖ (+38 322) 35 41 05
♣ US$ 65,000,000 ⌂ 9,230 m2
Managing Director: Ms Vera P Martynova
Operations Dir: S.P. Kinakh
Sales and Marketing Director: Jacob S Kozlenko
Advertising Manager: Lev R Lib ☎ (+38 322) 39 31 97
Public Relations Mgr: V.P. Dmitruk
☎ (+38 0322) 39 31 16
Purchasing Mgr: S.M. Shatny
■ Monochrome airborne display cathode ray tubes (CRTs), colour airborne display CRTs, monochrome kinescopes, colour kinescopes, photoregistering & flying-spot scanning CRTs, oscilloscope CRTs, projection CRTs. → 60 [97]

75-03000 Hartron Corporation
(† Hartron Research & Production Association)
1 Academician Proskura Straat, 310070 Kharkov
☎ (+38 0572) 43 31 31 ✆ (+38 0572) 44 50 90
[E] root@khartron.kharkov.ua ⊠ 311043 glava su
⊖ (+38 0572) 43 16 77
⇔ US$ 11,400,000 (1996) ♣ US$ 2,200,000 (1996)
⁂ 7,000 (approx) (1996) ⌂ 170,400 sq m
General Manager: Dr Yakiv Eisenberg
Marketing Director: Dr Alexander Bek
☎ (+38 0572) 44 51 80
Operations Director: Oleg Luchenko
Purchasing Manager: Vladimir Sosunov
■ Control systems, computers and electronic equipment for boosters and space vehicles.

The Hartron Corp consists of four design divisions and two factories located in Kharkov and Zaporizhzhia.

Hartron-Arkos Division (Kharkov) designs, tests, supplies and services control systems for complex dynamic objects: rocket equipment, space equipment and aviation equipment.

Hartron-Delt Division (Kharkov) designs, manufactures, tests and services during operation, unified base functional units for digital computing complexes. Computers for radar information and traffic control processing systems, programmable controllers for frequency control of asynchronous drives and basic processes, sensors, parts and units out of technical ceramics.

Hartron-Enkos Division (Kharkov) designs, manufactures, tests, supplies and puts into operation, with subsequent servicing, computerised process control systems and automatic control systems for complex dynamic objects including control systems for spacecraft of various applications.

Electroavtomatika Division (Zaporizhzhia) develops fibre-optical devices and telecommunication systems. The division manufactures, tests and services during operation: unified control systems, including mathematical software, telemeasuring systems for spacecraft. Aircraft engine automatic control systems. Electroavtomatika Plant (Zaporizhzhia) is the pilot production plant of the research and production association Hartron. The plant has been developing automatic control systems for plane engines ND onboard automation devices, systems of guaranteed electric power.

Electropribor Plant (Kharkov) pilot production Hartron, manufactures experimental batches of space objects control systems designed by Hartron. → 60 [98]

75-05600 Neptun Odesa Production Association
28 Promyslovsa Straat, 270031 Odesa
☎ (+38 0482) 32 18 64; 32 35 23
⊖ (+38 0482) 22 51 89; 32 58 23
⇔ US$ 25,000,000 (1997) ♣ US$ 7,000,000 (1997)
⁂ 1,700 (1997) ⌂ 69,000 m2
Managing Director: Alexander I Strashko
Operations Director: Dmitry K Menshchilkov
Marketing Director: Yuri A Kotov
General Director: Vladimir B Tkachenku
Purchasing Manager: Yuri E Mnrtynov
■ Sets of high-frequency communication, protection and alarm protection equipment, tele-automatics, remote disconnecting equipment for 35-1.150 kV and 0.38-35 kV power supply lines. Processing and HV connection equipment. Data transmission equipment. Automated noise measuring units with improved reconnaissance efficiency. [98]

75-05800 G.I. Petrovsky Chemical Association
1 Lenin Street, 349333 Petrovskoye, Lugansk Region
☎ (+380 64) 70 512; 737 86; 710 24
(telephone with fax) ⊠ 115718 ⊖ (+380 64) 225 00 34
Managing Director: Victor Shevtsov
Commercial Director: Andrew Piuk
Operations Director: Vladimir Filatov
Purchasing Manager: Michail Povarachuk

■ Manufacture of nitroether containing and ammonium nitrate industrial explosives. Dry-rolled pastes, pigment concentrates for nitrocellular enamels and leather paints, nitrocellular enamels NC-25, NC-1125, leather paints NC-583, oil paints MA-15, poilvinyl choride flooring with a stamping draft on the textile base. Pentaftale enamels PF-115, PF-167 varnish for furniture, cellular rubber products. [98]

75-06500 Sevastopol JSC Marine Plant
(† Sevastopol S. Ordzhonikidze Marine Plant)
13 Geroiv Sevastopolya St, 335011 Sevastopol, Crimea
☎ (+38 0690) 52 43 20; 36 30 91; 36 30 67
✆ (+38 0690) 36 30 72; 36 30 48 ✉ BOT SEVASTOPOL
Telex 197315 brco su ⊖ (+38 0690) 36 64 81
♣ US$ 18 000 000 (1995) 👥 5 400 (1996)
🏭 1 060 000 sq m
Managing Dir: A. Tcherevaty ☎ (+38 0692) 52 41 50
Head of Marketing & Foreign Relations Dept: K. Tchernyavsky ☎ (+38 0692) 36 36 86
Technical Dir: M. Sorokin
Purchasing Mgr: N. Marderos
■ Repair & reconstruction of ships, construction of marine self-propelled floating cranes, manufacture of ship machinery equipment. [97]

75-10000 Znamya Production Group
2-9 Avtobazovskaya Street, 314008 Poltava
☎ (+38 05322) 384 18 ⊖ (+38 05322) 310 88
♤ US$ 248 000 (1997) ♣ US$ 110,000 (1997)
👥 3,000 (1997) 🏭 150,000 sq m
Director General: Yu A Chelembiy ☎ (+38 05322) 301 24
Operations Director: Yu D Soloshenko ☎ (+38 05322) 314 50
Purchasing Manager: A A Lebedinskiy ☎ (+38 05322) 384 04
■ Manufacture of microwave equipment for countermeasures, radio-location, navigation and communication systems. [98]

90 Sales Agents, Sales & Liaison Offices

90-05000 Daewoo Corp Kiev Branch
Room 403, 404, 405 Ukrimpex, Vorovsky Street 22, Kiev
☎ (+380 44) 216 47 52 ⊖ (+380 44) 216 48 23 [98]

United Arab Emirates

Head of the Union:
Sheikh Sultan bin Zayid Al-Nuhayyan

10 Head of State

10-10000 Office of the President of the UAE and Ruler of Abu Dhabi
Manhal Palace, PO Box 280, Abu Dhabi
☎ (+971 2) 65 20 00 Telex 22220 palace em
President of UAE and Ruler of Abu Dhabi: Sheikh Sultan bin Zayid Al- Nuhayyan [98]

10-20000 Office of the Vice-President of UAE
Za'beel Palace, PO Box 12848, Dubai
☎ (+971 4) 53 45 50 Telex 45688 baldya em
Vice President of the UAE and Prime Minister: Maktum Bin Rashid al-Maktum [98]

10-30000 Supreme Council of the Union
PO Box 545, Abu Dhabi
☎ (+971 2) 34 39 21
Minister of State for Supreme Council Affairs: Majid bin Said al Nuaymi [98]

11 Prime Minister

11-10000 Office of the Prime Minister
PO Box 899, Abu Dhabi
☎ (+971 2) 36 19 75; 43 11 10 Telex 23245 amana em
Prime Minister: Maktum bin Rashid al-Maktum
Minister of State for Cabinet Affairs: Khalfan Sa'id al-Ghayth [98]

11-20000 Deputy Prime Minister's Office
PO Box 831, Abu Dhabi
☎ (+971 2) 65 18 81
Deputy Prime Minister: Sheikh Sultan bin Zayid Al-Nuhayyan [98]

12 Ministries

12-10000 Ministry of Defense
PO Box 2838, Dubai
☎ (+971 4) 53 23 30 Telex 45554 mod em
⊖ (+971 4) 53 19 74
Minister: Muhammad bin Rashid al-Maktum [98]

12-20000 Ministry of Foreign Affairs
PO Box 1, Abu Dhabi
☎ (+971 2) 65 22 00 Telex 22217
Minister of Foreign Affairs: Rashid bin Abdallah al-Nu'aymi
Minister of State: Hamdan bin Zayid Al- Nuhayyan [98]

12-40000 Ministry of the Interior
PO Box 398, Abu Dhabi
☎ (+971 2) 44 76 66 Telex 22398 uaepol em
Minister of Interior: Lieut Gen Muhammad Said al Badi ☎ (+971 2) 44 66 88 (+971 2) 44 89 11 (fax) [98]

12-60000 Ministry of Communications
PO Box 900, Abu Dhabi
☎ (+971 2) 65 19 00
Minister: Ahmad bin Humayd al-Tayir [98]

31 Embassies & Attachés

31-00090 Austrade
Australian Consulate General, PO Box 9303, Dubai
6th Floor Dubai World Trade Centre, Dubai
☎ (+971 4) 31 34 44 Telex 49393 austc em
⊖ (+971 4) 31 48 12
Consul General: Peter Deacon
Office Mgr: Ms Sue-Sharyn Ward [97]

31-00660 Embassy of the Federal Republic of Germany
Al Nahyan Street, PO Box 2591, Abu Dhabi
☎ (+971 2) 43 56 30 ⊖ (+971 2) 43 56 25
Ambassador: H E Dr Helmut P Arndt ☎ (+971 2) 46 31 16
Counsellor/Deputy Chief of Mission: Peter H Muerz
Defence Attaché: Cdr Rinnert
Second Secretary: Cornel Barth
Third Secretary/Vice Consul: Mrs Silke Hahn [97]

31-00690 Embassy of Greece
PO Box 5483, Abu Dhabi
☎ (+971 2) 65 48 47; 65 07 87 Telex 24383
⊖ (+971 2) 65 60 08 [98]

31-00890 Embassy of Italy
PO Box 45092, Abu Dhabi
Al Nhayan Street, Corner Dalma Street, Plot 350W, 14-02 Batine, Abu Dhabi
☎ (+971 2) 43 56 22; 43 35 48 Telex 23023 navit
⊖ (+971 2) 43 66 97
Ambassador: H.E. Mr Pietro Cordone
Deputy Chief of Mission: Dr Enrico de Agostini
Naval, Military & Air Attaché: Capt. Rodolfo Orsini
Trade Promotion Section
PO Box 46752, Abu Dhabi
☎ (+971 2) 32 18 08 Telex 23487 ⊖ (+971 2) 43 43 37 [97]

31-00930 Embassy of the Hashemite Kingdom of Jordan
PO Box 4024, Abu Dhabi
☎ (+971 2) 44 71 00 ⊖ (+971 2) 44 91 57
Ambassador: H.E. Mr Nabil T. Talhouni [97]

31-01270 Royal Netherlands Embassy
PO Box 46560, Abu Dhabi
☎ (+971 2) 32 19 20 Telex 23610 ⊖ (+971 2) 31 31 58
Ambassador: H.E. Mr Hein de Vries
Deputy Head of Mission & First Secretary Commercial: Jeroen Steeghs
Defence Attaché: Cdr Michiel Wouters (RNLN) [97]

31-01650 Embassy of the Republic of Slovakia
PO Box 3382, Abu Dhabi
Medinat Zayed, Villa No. 134/1 Baniyas St., Abu Dhabi
☎ (+971 2) 32 16 74 ⊖ (+971 2) 31 58 39
Chargé d'Affaires a.i. and Commercial Counsellor: Dusan Horniak
Third Secretary: Alexander Micovcin
Admin Officer: Lolita Chebat [97]

31-01690 Embassy of the Democratic Socialist Republic of Sri Lanka
Villa 108, bes. Khalidiya Park Rest., PO Box 46534, Abu Dhabi
☎ (+971 2) 78 88 81 Telex 23333 lakemb em
⊖ (+971 2) 76 65 09
Ambassador: H.E. Mr Aliuthumalebbe M. Samsudeen ☎ (+971 2) 76 21 87
Second Secretary: Padmasiri Jayaweera ☎ (+971 2) 34 97 67
Second Secretary: N.H.Nanasiri Perera ☎ (+971 2) 76 51 22
Attaché: K.P.J. Patrick Silva ☎ (+971 2) 76 45 28 [97]

31-01890 Embassy of the United Kingdom
PO Box 248, Abu Dhabi
☎ (+971 2) 32 66 00 ✉ PRODROME ABU DHABI
Telex 22234 prodro em ⊖ (+971 2) 34 17 44; 31 81 38
Ambassador: H E Mr A D Harris
Counsellor and Deputy Head of Mission: P Morgen
First Secretary, Commercial: G Deane
Second Secretary, Commercial: M Pakes
Defence Attaché: Wg Cdr M Allport
● PO Box 65, Dubai
☎ (+971 4) 52 10 70 Telex 45426 prodr em
⊖ (+971 4) 52 57 50
Consul General and Counsellor: C Wilton
First Secretary, Commercial: H Dunnachie [98]

75 Manufacturing Companies

75-00300 Costain Middle East
Holiday Inn Commercial Tower, PO Box 12118, Dubai
Crown Plaza, Suite 1606
☎ (+97 14) 31 02 58 ⊖ (+97 14) 31 38 43
👥 1 200
General Mgr: S.C. Marshall
■ Design & construct airfields, aircraft shelters, harbours, army & other military camps, battles ranges (urban) & buildings generally. [97]

80 Service Companies & Organizations

Radio-Holland Group Middle East . . 90-01700

80-01350 Emitac
(† Metier Management Systems)
PO Box 2711, Abu Dhabi

☎(+971 2) 77 04 19 ⊖ (+971 2) 72 30 58
® UK: Artemis Management Systems [98]

85 Stockists, Distributors, Spare Parts Suppliers

Colonel Military Supplies 90-00400
Radio-Holland Group Middle East . . . 90-01700

90 Sales Agents, Sales & Liaison Offices

Companies represented:

Artemis Management Systems, UK . . 80-01350
CATIC, China PR 90-00300
Nobel Denton, UK 90-01200
Radio-Holland Group, Netherlands . . . 90-01700
Transaero Inc, USA. 90-03000

90-00100 AlliedSignal Aerospace Marketing Sales & Service (GAMCO)
c/o Gulf Air Maintenance Company, PO Box 4650, Dubai
☎(+971 2) 75 74 59 ⊖ (+971 2) 75 78 30
Field Service Engineer: Gerrit De Vries [98]

90-00110 AlliedSignal Aerospace Marketing Sales & Service
Jebel Ali Free Zone, Lease Office, PO Box 17061, Dubai
Jebel Ali Free Zone,
☎(+971 4) 81 68 56 ⊖ (+971 4) 81 67 30
Sales Director: Malek Mashtoub [98]

90-00300 CATIC
PO Box 3851, Abu Dhabi
☎(+971 2) 72 76 98 ⊖ (+971 2) 77 94 83
® China PR: CATIC [97]

90-00400 Colonel Military Supplies
(† Emirates Impex)
PO Box 6636, Dubai
Defence Roundabout, Shaikh Rashid Bldg, Shop No. 10
☎(+971 4) 44 50 50; 44 62 62 (Dubai)
(+971 6) 59 00 77; 59 50 80 (Sharjah)
(+971 2) 563 43 73 (Abu Dhabi) ⊠ 49111 impex em
⊖ (+971 4) 44 94 49 (Dubai)
(+971 6) 59 00 09 (Sharjah)
(+971 2) 563 43 74 (Abu Dhabi)
♧ $ 2 000 000 ♣ $ 1 500 000 ⁂ 8 ⌂ 250 sq m
Owner: Abdul Karim Abdul Kadir
■ Specialists in defence, police & security equipment. → 85 [97]

90-00450 Daewoo (Gulf)
PO Box 8090, Dubai
Flat N°1701 Al Moosa Tower, Dubai-Abudhabi Road, Dubai
☎(+971 4) 31 33 84 ⊖ (+971 4) 31 36 09 [98]

90-00900 Logica Middle East
(a subsidiary of Logica plc, UK)
City Tower 1, PO Box 26379, Dubai
☎(+971 4) 31 32 20 ⊖ (+971 4) 31 20 03 [98]

90-01200 Noble Denton & Associates (WLL)
PO Box 27510, Abu Dhabi
☎(+971 2) 72 57 39 Ⓔ nodent@ns2.emirates.net.ae
⊠ 24282 nodent ⊖ (+971 2) 77 83 58
® UK: Nobel Denton
● PO Box 12541, Dubai
☎(+971 4) 31 79 52 ⊠ 46983 nodem em
⊖ (+971 4) 31 76 07 [98]

90-01700 Radio-Holland Group Middle East
PO Box 4592, Sharjah
☎(+971 6) 54 30 02 ⊠ 68302 ⊖ (+971 6) 35 60 83
⁂ 35
Managing Dir: G. Borst
Sales & Marketing Mgr: R.C. Joustra
■ Communication equipment, naval electronic equipment, security equipment. → 80 → 85
® Netherlands: Radio-Holland Group [97]

90-03000 Transaero Dubai
PO Box 23252, Dubai
☎(+971 4) 49 45 79 ⊖ (+971 4) 49 45 79
® USA: Transaero Inc [97]

United Kingdom

Head of State:
H.M. Queen Elizabeth II

10 Head of State

10-10000 Office of HM The Queen
Buckingham Palace, London SW1A 1AA
Queen of the United Kingdom: HM Queen Elizabeth II [98]

11 Prime Minister

11-10000 Prime Minister's Office
10 Downing Street, London SW1A 2AA
Prime Minister, First Lord of the Treasury and Minister for the Civil Service: Tony Blair
Deputy Prime Minister and Secretary of State for the Environment, Transport and the Regions: John Prescott
Chancellor of the Exchequer: Gordon Brown
Secretary of State for Foreign and Commonwealth Affairs: Robin Cook
Secretary of State for Defence: George Robertson
President of the Board of Trade and Secretary of State for Trade and Industry: Ms Margaret Beckett [98]

12 Ministries

12-10000 Ministry of Defence
Main Building, Whitehall, London SW1A 2HB
☎(+44 171) 218 90 00 Ⓦ http://www.mod.uk/
⊠ 825911 ⊖ (+44 171) 218 12 89
Secretary of State for Defence: George Robertson ☎(+44 171) 218 21 11; 218 21 13
Parliamentary Private Secretary to the Secretary of State for Defence: David Ames ☎(+44 171) 219 40 25
Minister of State for the Armed Forces: Dr John Reid ☎(+44 171) 218 63 85
Minister of State for Defence Procurement: The Rt Hon Dr the Lord Gilbert ☎(+44 171) 218 66 21
Parliamentary Under Secretary of State: John Spellar ☎(+44 171) 218 24 52
Permanent Under Secretary of State: Richard Mottram ☎(+44 171) 218 28 39
Deputy Under Secretary of State (Policy): R P Hatfield ☎(+44 171) 218 38 30
Assistant Under Secretary of State (Programmes): D J Seammen ☎(+44 171) 22 92
Assistant Under Secretary of State (Systems): D Fisher ☎(+44 171) 218 28 38
Chief of the Defence Staff: Gen Sir Charles Guthrie ☎(+44 171) 218 33 53
Vice Chief of the Defence Staff: ACM Sir John Willis ☎(+44 171) 218 76 57
Deputy Chief of the Defence Staff (Commitments): AM J R Day ☎(+44 171) 218 62 97
Deputy Chief of the Defence Staff (Programmes & Personnel): AM P Squire ☎(+44 171) 218 40 26
Deputy Chief of the Defence Staff (Systems): Vice Adm J H Dunt ☎(+44 171) 218 25 31
Assistant Chief of the Defence Staff (Operations): AVM A J Harrison ☎(+44 171) 218 13 89
Assistant Chief of the Defence Staff (Logistics): Maj Gen G A Ewer ☎(+44 171) 218 36 34
Assistant Chief of the Defence Staff (Policy): AVM J C French ☎(+44 171) 218 23 14
Assistant Chief of The Defence Staff Operational Requirements (Sea Systems): Rear Adm R T R Phillips ☎(+44 171) 218 60 70
Assistant Chief of the Defence Staff Operational Requirements (Land Systems): Maj Gen E F G Burton ☎(+44 171) 218 27 06
Assistant Chief of the Defence Staff Operational Requirements (Air Systems): AVM C C C Coville ☎(+44 171) 218 65 96
Assistant Chief of the Defence Staff (Programmes): Rear Adm N R Essenhigh ☎(+44 171) 218 21 88
Directorate of Defence Programmes: Brig F R Dannatt ☎(+44 171) 218 77 89
Chief Scientific Adviser: Prof Sir David Davies ☎(+44 171) 218 35 60
Chief Scientist: P D Ewins ☎(+44 171) 218 24 68
Deputy Chief Scientist (Scrutiny and Analysis): M J Earwicker ☎(+44 171) 218 20 34
Deputy Chief Scientist (Research and Technology): P M Sutcliffe ☎(+44 171) 218 62 37
Director of Medical Policy: Surgeon Cdre C W Evans ☎(+44 171) 218 90 00
Defence Medical Training Organisation: Maj Gen C G Callow ☎(+44 171) 218 90 00
Resources and Programmes (Navy): I R Lee ☎(+44 171) 218 26 03
Resources and Programmes (Army): T A Woolley ☎(+44 171) 218 61 78
Resources and Programmes (Air): T C McKane ☎(+44 171) 218 60 69
Defence Services Secretary: AVM P J Harding ☎(+44 171) 218 73 58
Directorate of Reserve Forces and Cadets: Brig D A K Biggert ☎(+44 171) 218 57 12
Directorate of Operational Capability: Brig P A Rigdeway ☎(+44 171) 218 64 93
Directorate of Defence Commitments (Joint Warfare): Cdre R P Stevens ☎(+44 171) 218 74 12
Directorate of Naval Operations: Cdre M D McPherson ☎(+44 171) 218 00 14
Directorate of Military Operations: Brig J R Reith ☎(+44 171) 218 25 61
Directorate of Air Operations: Air Cdre S M Nicholl ☎(+44 171) 218 62 21
Directorate of Overseas Military Activity: Cdre J M De Halpert ☎(+44 171) 218 28 55
Overseas Secretariat: D J Bonner ☎(+44 171) 218 61 44
Proliferation and Arms Control Secretariat: P W D Hatt ☎(+44 171) 218 28 78
Directorate of International Organisation: Brig P J Russell-Jones ☎(+44 171) 218 24 92
Director of Defence Policy: B R Mann ☎(+44 171) 218 61 09
Director of Nuclear Policy: Cdre J M Parkinson ☎(+44 171) 218 67 06
Director of Defence Health and Safety: Dr J Connor ☎(+44 171) 305 73 56
Director General Information and Communications Services: Maj Gen W J P Robins ☎(+44 171) 218 74 45
Directorate Defence Communications Service: Air Cdre A M Ferguson ☎(+44 1225) 81 42 79
Defence Housing Executive: C J I James ☎(+44 171) 921 26 84
Press Secretary: Ms G Samuel ☎(+44 171) 218 79 00 [98]

12-10100 Defence Council
Main Building, Whitehall, London SW1A 2HB
☎(+44 171) 218 90 00 ⊠ 825911
Chairman: George Robertson (Secretary of State for Defence) ☎(+44 171) 218 21 11; 218 21 13
Members
Minister of State for the Armed Forces: Dr John Reid ☎(+44 171) 218 63 85
Minister of State for Defence Procurement: The Rt Hon Dr the Lord Gilbert ☎(+44 171) 218 66 21
Parliamentary Under Secretary of State: John Spellar ☎(+44 171) 218 24 52
Chief of the Defence Staff: Gen Sir Charles Guthrie ☎(+44 171) 218 33 53
Vice Chief of the Defence Staff: ACM Sir John Willis

Deputy Chief of the Defence Staff (Commitments): AM J R Day
Chief of General Staff: Gen Sir Roger Wheeler
Assistant Chief of the General Staff: Maj Gen M A Willcocks ☎(+44 171) 218 91 52
Chief of the Air Staff: ACM Sir Richard Johns ☎(+44 171) 218 63 13
First Sea Lord, Chief of Naval Staff: Adm Sir Jock Slater ☎(+44 171) 218 61 93
Chief Scientific Adviser: Prof Sir David Davies ☎(+44 171) 218 35 60 98

12-10200 Defence Council Air Force Board

Main Building, Whitehall, London SW1A 2HB
☎(+44 171) 218 90 00 ✉ 825911
Chairman: George Robertson (Secretary of State for Defence) ☎(+44 171) 218 21 11; 218 21 13

Members

Minister of State for the Armed Forces: Dr John Reid ☎(+44 171) 218 63 85
Minister of State for Defence Procurement: The Rt Hon Dr the Lord Gilbert ☎(+44 171) 218 66 21
Parliamentary Under Secretary of State: John Spellar ☎(+44 171) 218 24 52
Chief of the Air Staff: ACM Sir Richard Johns ☎(+44 171) 218 63 13
Assistant Chief of the Air Staff: AVM T I Jenner ☎(+44 181) 218 69 39
Directorate Air Staff: Air Cdre P A Crawford ☎(+44 171) 218 62 59 98

12-10300 Defence Council Army Board

Main Building, Whitehall, London SW1A 2HB
☎(+44 171) 218 90 00 ✉ 825911
Chairman: George Robertson (Secretary of State for Defence) ☎(+44 171) 218 21 11; 218 21 13

Members

Minister of State for the Armed Forces: Dr John Reid ☎(+44 171) 218 63 85
Minister of State for Defence Procurement: The Rt Hon Dr the Lord Gilbert ☎(+44 171) 218 66 21
Chief of the General Staff: Gen Sir Roger Wheeler ☎(+44 171) 218 78 73
Assistant Chief of the General Staff: Maj Gen M A Willcocks ☎(+44 171) 218 91 52
Adjutant General: Gen Sir Alex Harley
Quartermaster General: Lieut Gen Sir Sam Cowan 98

12-10400 Defence Council Admiralty Board

Main Building, Whitehall, London SW1A 2HB
☎(+44 171) 218 90 00 ✉ 825911
Chairman: George Robertson (Secretary of State for Defence) ☎(+44 171) 218 21 11; 218 21 13

Members

Minister of State for the Armed Forces: Dr John Reid ☎(+44 171) 218 63 85
Minister of State for Defence Procurement: The Rt Hon Dr the Lord Gilbert ☎(+44 171) 218 66 21
First Sea Lord, Chief of Naval Staff: Adm Sir Jock Slater ☎(+44 171) 218 61 93
Assistant Chief of the Naval Staff: Rear Adm J J Blackham ☎(+44 171) 218 66 17 98

12-11000 Ministry of Defence Procurement Executive

MoD (PE) Abbey Wood, Stoke Gifford, Bristol BS12 7DU
☎(+44 117) 913 00 00
Chief of Defence Procurement: Sir Robert Walmsley ☎(+44 117) 913 00 02
Deputy Chief of Defence Procurement (Support): J Howe ☎(+44 117) 913 00 10
Deputy Chief of Defence Procurement (Operations): Lieut Gen Sir Robert Haymon-Joyce ☎(+44 117) 913 00 05 98

12-11500 Ministry of Defence Defence Information Division (D INFO D)

Main Building, Whitehall, London SW1A 2HB
☎(+44 171) 218 66 45 (Public Enquiries)
⊜ (+44 171) 218 12 89
Press Secretary and Chief of Information: Gill Samuel ☎(+44 171) 218 79 00; 218 73 76 (fax)
Deputy Press Secretary and Director of Information (Policy and Procurement): Tony Boardman ☎(+44 171) 218 22 15
Chief Press Officer: Tim Downes ☎(+44 171) 218 29 06; 218 79 07 (out of hours); 218 64 60 fax
Defence Press Officer (Procurement): Andrew Willis ☎(+44 171) 218 77 14
Defence Press Officer (Policy): Ian Jones ☎(+44 171) 218 79 31
Defence Press Officer (Navy): Ms Ailsa McIntyre ☎(+44 171) 218 32 57
Defence Public Relations (Navy): Cdre Barry Leighton ☎(+44 171) 218 79 03; 218 79 18 (fax)
Defence Press Officer (Army): David Davies ☎(+44 171) 218 32 55
Defence Public Relations (Army): Brig Robert Gordon ☎(+44 171) 218 79 04; 218 49 20 (fax)
Defence Press Officer (Air Force): Alan Patterson ☎(+44 171) 218 32 53
Defence Public Relations (Air Force): Air Cdre Gordon McRobbie ☎(+44 171) 218 79 05; 218 76 62 (fax)
Head of Media Planning and Support: Colin Mayhew ☎(+44 171) 218 76 70; 218 12 89 (fax)
Head of Publicity: Henry Bojdys ☎(+44 171) 218 35 38; 218 12 89 (fax)
Exhibitions and Publications: Brian Reed ☎(+44 171) 218 39 49; 218 38 05 (fax)

■ As part of the Defence Information Division the Public Enquiry Office acts as a first point of contact for the public. 98

12-12000 Commercial Contracts Division

(part of the Ministry of Defence)
MoD Abbey Wood, Stoke Gifford, Bristol BS12 7DU
☎(+44 117) 913 00 00 (switchboard)
Director General of Commercial Contracts (DGCC): A Phipps 98

12-13000 Directorate of Intellectual Property Rights

(part of the Ministry of Defence)
MoD Abbey Wood, Stoke Gifford, Bristol BS12 7DU
☎(+44 117) 913 00 00 (switchboard)
Director of Intellectual Property Rights: Mr Beckham 98

12-14300 Defence Housing Executive

(part of the Ministry of Defence)
St Christopher House, Southwark Street, London SE1 0TE
☎(+44 171) 921 10 31 (secretariat)
⊜ (+44 171) 921 27 91
Chief Executive: Colin James

■ Tri-service housing organization established to manage the Defence married quarters estate. 98

12-20000 Foreign & Commonwealth Office

King Charles Street, London SW1A 2AH
☎(+44 171) 270 15 00 (exchange)
W http://www.fco.gov.uk/ ✉ 297711 prdrme g
⊜ (+44 171) 270 28 81
Secretary of State For Foreign & Commonwealth Affairs: Robin Cook
Minister of State: Derek Fatchett
Minister of State: Doug Henderson
Minister of State: Tony Lloyd
Parliamentary Under-Secretary of State: Baroness Symons of Vernham Dean

Non-Proliferation Department
King Charles Street, London SW1A 2AH
☎(+44 171) 270 22 61 ⊜ (+44 171) 270 28 60
Superintending Under Secretary: R H Smith (CMG)
Head of Department: B Cleghorn

Joint Export Promotion Directorate (Joint FCO/DTi Dept)
Kingsgate House, 66-74 Victoria Street, London SW1E 6SW
☎(+44 171) 215 50 00 ⊜ (+44 171) 215 46 44
Superintending Under Secretary (Deputy Director General): Alan Hunt
Director JEPD 1: Denis Walker ☎(+44 171) 215 49 37
Director JEPD 2: Peter Beckingham ☎(+44 171) 215 47 02

Security Policy Department
King Charles Street, London SW1A 2AH
☎(+44 171) 270 31 32 ⊜ (+44 171) 270 28 46
Superintending Under Secretary: R H Smith (CMG)
Head of Department: Richard Gozney ☎(+44 171) 270 31 31 98

12-30000 Department of Trade and Industry (DTI)

1 Victoria Street, London SW1H 0ET
☎(+44 171) 215 50 00 (General Enquiries); 215 15 17 (Innovation Enquiries); (+44 171) 944 48 88 (Business in Europe);
✆ (800) 50 02 00 (Business Link) ✉ 8813148 dthq g
⊜ (+44 171) 222 26 29
President of the Board of Trade and Secretary of State for Trade and Industry: Ms Margaret Beckett
Minister for Trade and Competitiveness in Europe: Lord Simon of Highbury
Minister for Trade: Lord Clinton-Davis
Minister of State: Ian McCartney
Minister for Science, Energy and Industry: John Battle
Parliamentary Under Secretary of State for Competition and Consumer Affairs: Nigel Griffiths
Parliamentary Under Secretary of State for Small Firms, Trade and Industry: Ms Barbara Roche
Permanent Secretary: Michael Scholar (CB)
Deputy Secretary: Alistair MacDonald (CB)

Aerospace and Defence Industries Directorate († Aerospace Division)
151 Buckingham Palace Road, London SW1W 9SS
☎(+44 171) 215 50 00 (switchboard); 215 11 66 (European Commission/World Trade Organisation enquiries); 215 11 25 (ordnance/competition/National Defence Industries Council/military aircraft)
☎(+44 171) 215 58 54 (Missiles/MoD Liaison/ Defence Procurement); 215 11 73 (Market assessment (defence)/statistics); 215 11 65 (Aviation Equipment Sector/Flight simulators/helicopters/naval systems)
☎(+44 171) 215 11 28 (Export Promotion/ trade policy); 215 11 27 (defence technology/MoD liaison (aeronautics)); 215 11 14 (worldwide market trends/export licensing)
☎(+44 171) 215 11 57 (Engineering and Physical Sciences Research Council); 215 11 17 (environmental issues); 215 11 49 (technology transfer) ✉ 8813148 dtihq g
⊜ (+44 171) 215 11 64; 215 13 19
Secretary to the Defence and Aerospace Panel: Dr Martin Ridge
Director: Robert Foster ☎(+44 171) 215 11 59
Director, Aero-engines, Equipment, Helicopters, Naval Defence and Competitiveness: Ms Judy Britton ☎(+44 171) 215 16 84
Director, Aerospace and Defence Industries Policy: John Neilson ☎(+44 171) 215 11 39
Technical Director, Project Assessment and Environment: Malcolm Ralph
Director, Technology Partnership and CARAD Programme: Shelley Ian Charik ☎(+44 171) 215 11 49
CARAD Programme Director, Aerodynamics and Propulsion Systems: Ray Kingcombe ☎(+44 171) 215 11 15
Economic Adviser: Michael Hodson ☎(+44 171) 215 18 48
Sponsorship of Aero-engines and Aerostructures: Peter Lambert ☎(+44 171) 215 11 28
Sponsorship of Aerospace Equipment, Helicopters and Naval Defence Industries: Bob Insley ☎(+44 171) 215 11 65
Promotion of Competitiveness: Kevin Moore ☎(+44 171) 215 12 28
Civil Aircraft and Trade Policy: Matthew Cocks ☎(+44 171) 215 11 66
Aerospace and Defence Strategy and Sponsorship: Matthew Hilton ☎(+44 171) 215 11 25
UK Representation to Airbus Agency: Roly Piggott (based in Paris) ☎(+33 1) 45 52 43 17
Large Commercial Projects: Philip Davies ☎(+44 171) 215 11 14
Small Commercial and Military Projects: Bob Waters ☎(+44 171) 215 11 73
Defence Industries and Procurement, General Advice: Tony Dewick ☎(+44 171) 215 58 54
Environmental Issues: Peter Newton ☎(+44 171) 215 11 17
Materials and Structures: Tony Thorpe ☎(+44 171) 215 11 57
Cryotechnology and ETW, CARAD Budget and Policy: Bob Collier ☎(+44 171) 215 11 56
Helicopters, Equipment and Avionics: Richard Harrison ☎(+44 171) 215 11 27

■ The overall aim of the Directorate is to assist the UK's aerospace and defence industries to compete internationally.

British National Space Centre (BNSC)
151 Buckingham Palace Road, London SW1W 9SS
☎(+44 171) 215 08 31; 215 07 05 (defence space programmes); 215 50 00
⊜ (+44 171) 821 53 87; 215 09 36

Contact Reference

Director General: Derek Davis ☎ (+44 171) 215 08 77
Deputy Director General and Director Space Technology, Chairman Space Technology Advice: Dr David Leadbeater
Associate Director Space Science: Dr Eric Dunford
Director Space Applications and Transportation: Dr David Lumley ☎ (+44 171) 215 08 81
Director Space Science and Microgravity: Dr Paul Murdin ☎ (+44 171) 215 08 05
Director ESA Policy: Hugh Evans ☎ (+44 171) 215 08 94
Press Office: M Blackwell
Liaison Office: Ms Sandra Clifton
■ Co-ordinates the UK's space activities. Technology Centres at DRA Farnborough and Rutherford Appleton Laboratory, Oxford.

Communications and Information Industries Directorate
2nd Floor (green), 151 Buckingham Palace Road, London SW1W 9SS
☎ (+44 171) 215 17 33 (defence telecommunications and radiocommunications hardware and software); (+44 171) 215 19 02 (defence IT hardware, software and services)
Projects Director: David Baxter
Electronics and Engineering Contact: Harley Berry ☎ (+44 171) 215 58 32
■ Industrial sponsorship.

Engineering, Automotive and Metals Directorate
151 Buckingham Palace Road, London SW1W 9SS
Director: Mike O'Shea
Commercial and Specialist Vehicles: Paul Mullins ☎ (+44 171) 215 11 83
Shipbuilding: Ms Ann Wilks
Automotive Directorate: Anthony Vinall
■ Sponsorship of the UK military vehicles industry and liaison with UK military vehicle manufacturers.

Export Market Information Centre
Kingsgate House, 66-74 Victoria Street, London SW1E 6SW
☎ (+44 171) 215 54 44; 215 54 45
⊖ (+44 171) 215 42 31

Export Control Organisation
Export Licensing Unit
6th Floor, Kingsgate House, 66-74 Victoria Street, London SW1E 6SW
☎ (+44 171) 215 50 00 ⊖ (+44 171) 215 85 64
■ The DTI's Export Control Organisation is responsible for administering controls in respect of the export of certain goods from the UK and for the processing of export licences.

Innovation Unit
151 Buckingham Palace Road, London SW1W 9SS
Director: Dr Alistair Keddie ☎ (+44 171) 215 18 28
■ The Unit is responsible for a range of networking and facilitating activities to encourage innovation as a key component of industrial competitiveness.

Technology and Standards Directorate
151 Buckingham Palace Road, London SW1W 9SS
Director: Dr David Evans
Director Technology and Innovation (Policy and Budget): Geoffrey Riggs
Director, Research and Technology Co-ordination: Ian Downing
Director, Private Finance Initiative: Gregor McGregor
■ The work of the Directorate falls into two areas:
Helping to manage DTI's overall policies and activities aimed at helping industry in the areas of innovation, science, engineering and technology;
Managing DTI's work in the area of standardisation. [98]

12-40000 Home Office
50 Queen Anne's Gate, London SW1H 9AT
☎ (+44 171) 273 40 00
Secretary of State for the Home Department: Jack Straw
Minister of State (Criminal Policy): Alun Michael
Minister of State (Prisons, Probation and Europe): Joyce Quin
Parliamentary Under Secretary of State (Deregulation, Drugs and Elections): George Howarth
Parliamentary Under Secretary of State (Immigration and Nationality): Mike O'Brien
Parliamentary Under Secretary of State (Constitutional Issues): The Lord Williams of Mostyn [98]

12-41000 Scottish Office
St Andrew's House, Edinburgh EH1 3DE
☎ (+44 131) 556 84 00 ▣ 72202 scotof g
⊖ (+44 131) 244 35 63
Secretary of State for Scotland: Donald Dewar
Minister for Home Affairs and Devolution: Henry McLeish
Minister for Education and Industry: Brian Wilson
Parliamentary Under Secretary of State (Agriculture, Environment and Fisheries): The Lord Sewell (CBE)
Parliamentary Under Secretary of State (Health and Arts): Sam Galbraith
Parliamentary Under Secretary of State (Local Government and Transport): Malcolm Chisholm

London Office
Dover House, Whitehall, London SW1A 2AU
☎ (+44 171) 270 30 00 ▣ 22890 scotol g [97]

12-50000 Department of the Environment, Transport and the Regions
Elend House, Bressenden Place, London SW1E 5DU
☎ (+44 171) 890 30 00; 276 09 00 ▣ 22221 doemar g
Deputy Prime Minister and Secretary of State for the Environment, Transport and the Regions: John Prescott
Minister for Transport: Gavin Strang
Minister for the Environment: Michael Meacher
Minister of State (Local Government and Housing): Ms Hilary Armstrong [98]

15 Parliament

15-10000 House of Lords
Parliament Office, Westminster, London SW1A 0PW
☎ (+44 171) 219 32 32 ⊖ (+44 171) 219 47 11
Lord Chancellor: The Rt Hon The Lord Irvine of Lairg
Lord Privy Seal and Leader of the House of Lords: The Rt Hon The Lord Richard
Captain of the Honourable Corps of the Gentlemen-at-Arms (Chief Whip): The Lord Carter
Parliamentary Secretary to the Lord Chancellor: Geoff Hoon [98]

15-20000 House of Commons
Westminster, London SW1A 0AA
☎ (+44 171) 219 30 00
W http://www.parliament.uk/commons/ hsecom.htm
Speaker of the House of Commons: Ms Betty Boothroyd
President of the Council and Leader of the House of Commons: Ms Ann Taylor ☎ (+44 171) 219 40 40
Chief Whip and Parliamentary Secretary to the Treasury: Nick Brown
Deputy Chief Whip and Treasurer of Her Majesty's Household: George Mudie

Defence Committee
House of Commons, London SW1A 0AA
☎ (+44 171) 219 57 45 E defcom@parliament.uk
W http://www.parliament.uk/commons/ selcom/defhome.htm ⊖ (+44 171) 219 69 52
Chairman: Bruce George
Committee Clerk: A Kennon

Foreign Affairs Committee
Committee Office, House of Commons, London SW1A 0AA
☎ (+44 171) 219 61 06 E foraffcom@parliament.uk
W http:// www.parliament.uk/commons/selcom/ fachome.htm ⊖ (+44 171) 219 68 64
Senior Committee Member: The Rt Hon Sir Peter Emery
Committee Clerk: Dr C Ward

Home Affairs Committee
House of Commons, London SW1A 0AA
☎ (+44 171) 219 32 76 E homeaffcom@parliament.uk
W http://www.parliament.uk/commons/ selcom/hmafhome.htm ⊖ (+44 171) 219 66 06
Chairman: Chris Mullin
Committee Clerk: C J Poyser

Research Services
House of Commons, London SW1A 0AA
Head of International Affairs and Defence Section: Dr R J Ware ☎ (+44 171) 219 39 78
Head of Public Information: Dr C C Pond ☎ (+44 171) 219 42 72 [98]

20 Defense Administrations/Services/Agencies

20-00200 British Underwater Test and Evaluation Centre (BUTEC)
(part of the Defence Evaluation and Research Agency/DTEO)
Kyle of Lochalsh, Ross-shire IV40 8AJ
☎ (+44 1599) 53 42 62 Ext. 219
⊖ (+44 1599) 53 42 62 Ext. 242 [98]

20-00650 Defence Clothing and Textiles Agency
(Defence Agency)
Skimmingdish Lane, Caversfield, Oxfordshire OX6 9TS
☎ (+44 1869) 87 55 01 ⊖ (+44 1869) 87 55 09
Director: Brig M J Roycroft
COSSEC DCTA: I Wood
■ The DCTA was established as an agency in 1994 to provide a range of clothing, textile products and services and some general stores to meet the needs of the UK Armed Forces and other customers. [98]

20-00700 Defence Codification Agency (DCA)
Ministry of Defence
Kentigern House, 65 Brown Street, Glasgow G2 8EX
☎ (+44 141) 224 20 66 E kareng@dcancb.demon.co.uk
⊖ (+44 141) 224 21 48
Chief Executive: Ken Bradshaw
Director of Operations: John Clark
Director of Business Management: Bob Trainer
Public Relations Officer: Ms K L Gallagher
■ National Codification Bureau. National authority for the implementation of the NATO Codification System. [98]

20-01000
Defence Evaluation and Research Agency (DERA)
(DERA is an Agency of the UK Ministry of Defence)

DERA

Room 2024, Cody Building (A1), DERA Farnborough, Hampshire GU14 0LX
☎ (+44 1252) 39 33 00 W http://www.dera.gov.uk
⊖ (+44 1252) 39 33 99
■ The Defence Evaluation and Research Agency (DERA) brings together virtually all the non-nuclear science and technology resources formerly operated by the UK Ministry of Defence. Under the DERA umbrella these facilities and their services are now available to clients throughout the world, instead of being restricted to the Ministry of Defence. DERA provides scientific advice, innovative engineering and a broad range of technical services to governments and industries throughout the world.
The role of DERA is to ensure that the highest standards of quality and service are maintained in support of UK defence. It fulfils this role by managing the world-class science and technology assets under its control to produce the best solutions to defence problems at the lowest cost. Operational studies and analyses are carried out in various categories of basic and applied research to assist in procurement decisions; it also carries out testing and evaluation of specific equipment and systems, often under operational conditions. Many discoveries and inventions find important civil applications, with rewarding commercial results.
The Defence Evaluation and Research Agency is one of the largest of its kind in the world. It employs nearly 11,000 highly qualified specialists and has an annual budget exceeding £ 1,000,000,000. It now carries responsibility for most of the testing, trialling, research and development facilities and study centres developed by the British armed services and the UK Ministry of Defence over many years. By virtue of its status as an agency, DERA can now for the first time make these facilities and services available to other customers in all parts of the world. → 60 → 80 [98]

20-01015 DERA Centre for Marine Technology

(† Haslar Hydrodynamic Test Centre)
(a division of DERA)
Haslar Road, Gosport, Hampshire PO12 2AG
☎ (+44 1705) 33 55 00 W http://www.dera.gov.uk
⊖ (+44 1705) 33 54 14
♣ £ 120,000,000 (1997) 1,300+ (1997)
Director: Dr John Dering
Marketing Manager: Rob Webber
■ Hydrodynamic model testing & consultancy advice for designers & naval architects covering resistance, propulsion, manoeuvring, seakeeping, ship handling & trials for marine & other vehicles, fixed structures & special equipment. 98

20-01100 Meteorological Office

(Defence Agency)
Bracknell, Berkshire RG12 2SZ
☎ (+44 1344) 42 02 42 W http://www.meto.gov.uk/
⊖ (+44 1344) 85 44 11
Chief Executive Officer: Prof J C R Hunt
Senior Press Officer: Andy Yeatman
☎ (+44 1344) 85 66 55
■ The National Meteorological Service provides meteorological services for civil and military use. Conducts research and is an executive agency of the MoD. Maintains the National Meteorological Library.
→ 60 98

20-01125 Military Survey

(Defence Agency)
Elmwood Avenue, Feltham, Middlesex TW13 7AH
☎ (+44 181) 818 21 81 ⊖ (+44 181) 818 21 48
Military Survey Dir/Chief Exec.: Brig. P.R. Wildman (OBE)
Deputy Chief Exec./Geographic Resources Dir: M. Sands
Geographic Field Support Dir: Col A.P. Walker (OBE)
Geographic Information Dir: Mrs M. Jacobs
Geographic Information Services & Development Dir: B. Thomas
Geographic Production Dir: A. Painter
Colonel Geographic Commitments: Col I. Whittington
Finance Director: S Hall
Head of Corporate Planning: Mrs M. Sinfield
■ To ensure the provision of geographic support to defence operations, planning & training. 98

20-01140 Naval Aircraft Repair Organisation Fleetlands Division (NARO)

Fareham Road, Gosport, Hampshire PO13 0AW
☎ (+44 1705) 54 467 76 ⊖ (+44 1705) 82 30 43
Chief Executive: Steve Hill ☎ (+44 1705) 54 49 10
Corporate Development Director: A Mangan
Finance Director: John Alderton ☎ (+44 1705) 54 42 12
Superintendent Almond Bank: Robert Scott
☎ (+44 1705) 58 33 01
Director of Sales and Customer Service:
Raymond Sparshott ☎ (+44 1705) 54 32 75
■ Repair and overhaul of helicopters, engines and components for army, navy and air force. Servicing for marine gas turbines for ships of the Royal Navy and other NATO countries. 98

20-01145 Pay & Personnel Agency (PPA)

(† Defence Accounts Agency)
Warminster Road, Bath BA1 5AA
☎ (+44 1225) 82 81 25
Chief Exec.: M.A. Rowe
Corporate Services Group: Mr Schiller
■ The aim of the PPA is to provide pay services and personnel information support services in respect of civilian staff of the MoD. 97

20-01500 UK Hydrographic Office (UKHO)

(Trading Fund within the Ministry of Defence)
Admiralty Way, Taunton, Somerset TA1 2DN
☎ (+44 1823) 33 79 00 W http://www.hydro.gov.uk
⊖ (+44 1823) 35 19 45
Chief Executive: RAdm. J.P. Clarke
Public Relations Manager: Ms C Faun
Librarian: P. Woods
■ The UKHO produce hydrographic, oceanographic and other such information for the Royal Navy, sale of unclassified charts and publications to the Merchant Navy. 98

25 Defense Forces

25-10000 Central Defence Staff

Main Building, Whitehall, London SW1A 2HB
☎ (+44 171) 218 90 00 ⊠ 22241 moddcc g
⊖ (+44 171) 218 67 99 (Chief Defence Staff)
Chief of the Defence Staff: Gen Sir Charles Guthrie
☎ (+44 171) 218 33 53
Vice Chief of the Defence Staff: ACM Sir John Willis
☎ (+44 171) 218 76 57
Deputy Chief of the Defence Staff (Commitments): AM J R Day
Deputy Chief of the Defence Staff (Systems): Vice Adm J H Dunt ☎ (+44 171) 218 25 31
Deputy Chief of the Defence Staff (Programmes & Personnel): AM P Squire ☎ (+44 171) 218 40 26 98

25-10100 Chief of the General Staff

Main Building, Whitehall, London SW1A 2HB
☎ (+44 171) 218 25 00 ⊖ (+44 171) 218 49 20
Chief of the General Staff: Gen Sir Roger Wheeler
Assistant Chief of the General Staff: Maj Gen M A Wilcox
Public Relations Director: Brig P C C Trousdell
Military Assistant to the Chief of the General Staff: Maj J T Jackson ☎ (+44 171) 218 71 91
Director of Military Operations: Brig D J M Jenkins
☎ (+44 171) 218 25 61
Director General Territorial Army and Organisation: Maj Gen M D Jackson ☎ (+44 171) 218 24 09
Director General Army Training: Maj Gen S C Grant
☎ (+44 171) 218 23 62
Assistant Chief Defence Staff Operational Requirements Land Systems: Maj Gen E F G Burton
☎ (+44 171) 218 27 06
Director Operational Requirements Land: Brig J W M Kincaid ☎ (+44 171) 218 30 01
Director General Command, Control, Communications and and Information Systems (Army): Maj Gen A H Boyle
☎ (+44 171) 218 61 59
Deputy Signal Officer in Chief Army: Brig S G M Gordon
☎ (+44 171) 218 66 56 97

25-20000 Office of Chief of the Air Staff

Main Building, Whitehall, London SW1A 2HB
☎ (+44 171) 218 63 14 ⊠ 22611 modpr g
Chief of Air Staff: ACM Sir Richard Johns
Anglo American Community Relations: AM Sir Alan Davies (retd)
Chief Executive National Air Traffic Services: D J McLaughian ☎ (+44 171) 832 57 71
Director General Policy and Plans: AVM J D L Feesey
☎ (+44 171) 832 57 81
Director General Projects and Engineering: C G Davies
☎ (+44 171) 832 57 75
Director of RAF Legal Services: AVM J Weeden (LLB)
Head of Secretariat Air Staff: J R G Clark
☎ (+44 171) 218 69 30
Director Public Relations: Air Cdre Gordon McRobbie
☎ (+44 171) 218 35 59

Office of Assistant Chief of Air Staff
Main Building, Whitehall, London SW1A 2HB
☎ (+44 171) 218 90 00
Assistant Chief of the Air Staff: AVM T I Jenner
☎ (+44 171) 218 69 39
Director of Air Force Staff Duties: Air Cdre J C French
☎ (+44 171) 218 62 59
Director of Air Offensive: Air Cdre Brian Pegnall
☎ (+44 171) 218 62 21
Director of Air Defence: Air Cdre M J Battell
☎ (+44 171) 218 34 72
Inspector of Flight Safety: Air Cdre M J Abbott
☎ (+44 171) 430 75 39 97

25-20050 Air Warfare Centre

RAF Bentley Priory, Stanmore, Middlesex HA7 3HH
☎ (+44 181) 838 78 00 ⊖ (+44 181) 838 78 38
Commandant: Air Cdre R V Morris
Chief Research Officer Operational Analysis: J Roe
☎ (+44 181) 838 78 20
Gp Capt Doctrine and Training: Gp Capt R W Gault
☎ (+44 181) 838 78 10

Tactics and Electronic Warfare (Tactics & EW)
RAF Waddington, Lincoln LN5 9NB
☎ (+44 1522) 72 63 00 ⊖ (+44 1522) 72 61 94
Gp Capt Tactics and Electronic Warfare: Gp Capt M Gleave 98

25-20100 RAF Personnel & Training Command HQ

Computer Block, RAF Innsworth, Gloucester GL3 1EZ
☎ (+44 1452) 71 26 12
Chief of Staff: AVM M D Smart

Air Officer Training Group
HQ RAF Personnel & Training Command, Gloucester GL3 1EZ
☎ (+44 1452) 71 26 12
Chief Executive Training Group Defence Agency: AVM Stables
■ The role of the agency is to provide trained servicemen and civilians to meet the operational requirement of the RAF in peace crisis, transition to war and war. 98

25-20500 Aircraft Controllerate

Abbey Wood, Bristol, Avon 97

25-22100 Headquarters Strike Command, Royal Air Force

RAF High Wycombe, High Wycombe, Bucks HP14 4UE
☎ (+44 1494) 46 14 61 ⊖ (+44 1494) 49 60 77
Air Officer Commanding in Chief & C-in-C NATO UK RAF: ACM Sir William Wratten
Deputy Personal Staff Officer to the Commander-in-Chief: Sqn Ldr P A Clay
Deputy Commander-in-Chief and Chief of Staff: AM G A Robertson
Personal Staff Officer to Chief of Staff: Sqn Ldr D L Whittingham
Senior Air Staff Officer: AVM D A Hurrel
Air Commodore Operations: Air Cdre R A Wright
Air Officer Administration: AVM T B Sherrington
Air Officer Engineering and Supply: AVM D J Saunders
Air Officer Training: AVM J A G May
Air Officer Plans: Air Cdre M D Pledger
Directorate of RAF Regiment, Security and Defence Fire Services: Air Cdre I W P NcNeil
Command Mechanical Engineers: Air Cdre K E V Monkhouse
Command Public Relations Officer: C Shepherd
Deputy Command Public Relations Officer: F E T Lewis

No.1 Group Headquarters
RAF Benson, Wallingford, Oxfordshire SN9 6BE
☎ (+44 1491) 83 77 66
Air Officer Commanding: AVM J.R. Day
Sr Air Staff Officer: Air Cdre P.V. Harris

No.11/18 Group Headquarters
RAF Bentley Priory, Stanmore, Middlesex HA7 3HH
☎ (+44 181) 950 40 00
Air Officer Commanding: AVM C.R. Spink
Sr Air Staff Officer: Air Cdre M.K. Widdowson

AO Marine Air Headquarters
RAF Northwood, Northwood, Middlesex HA6 3EP
☎ (+44 1923) 82 61 61
Air Officer Commanding: Air Cdre A.E. Neal
Maritime Air Operations: Wg Cdre A. Ware

HQ Military Air Traffic Operations
Hillingdon House, Uxbridge, Middlesex UB10 0RU
☎ (+44 1895) 25 73 00
Air Officer Commanding: Air Cdre J.R.D. Arscott

No.38 Group Headquarters
RAF High Wycombe, High Wycombe, Bucks HP14 4UE
☎ (+44 1494) 46 14 61
Air Officer Commanding: AVM D.A. Hurrell
☎ (+44 1494) 46 68 11
Sr Air Staff officer: Air Cdre M.J. Gardiner
☎ (+44 1494) 46 63 32 98

25-29000 Headquarters Air Cadets

RAF Cranwell, Sleaford, Lincolnshire NG34 8HB
☎ (+44 1400) 26 12 01 ext. 7630/1
W http://www.garrick.co.uk/aircorps.html
; http://www.open.gov.uk/atc/atc.htm
⊖ (+44 1400) 26 12 01 ext. 7642
Commandant & Air Officer Commanding Air Cadets: Air Cdre J. Ford (RAF)
Chief of Staff: Gp Capt. W.N. Cross (RAF)
Sr Officer Combined Cadet Force: Wg Cdr F. West
Sr Officer Training Ground Staff: Wg Cdr R. Crowder
Public Relations Officer: Mrs D. Housby

Contact Reference

■ The Air Cadet Organisation, integrating both the Air Training Corps & CCF(RAF) sections is a national voluntary youth organisation established to: Promote a practical interest in aviation & the Royal Air Force, provide training valuable in any walk of life (civil or military) & to develop the qualities of leadership. Activities include flying, gliding, annual attachment camps to RAF Stations in the UK & abroad, adventurous training & exchange visits with foreign cadet organisations. Membership is open to both male and female cadets. [98]

25-32100 Headquarters Land Command

Erskine Barracks, Wilton, Salisbury, Wiltshire SP1 0AG
☎ (+44 1722) 33 62 22
Commander in Chief: Gen. Sir Roger Wheeler (KCB CBE)
☎ (+44 1722) 43 32 08
Military Assistant: Lt Col C.M de W. Roskelly
Deputy Commander in Chief: Lt Gen. H.W.R. Pike (DSO MBE) ☎ (+44 1722) 43 32 35
Chief of Staff: Maj.Gen. C.G.C. Vyvyan (CBE)
☎ (+44 1722) 43 32 14
Deputy Chief of Staff: Maj.Gen. R.A. Cordy-Simpson (CB OBE) ☎ (+44 1722) 43 39 49
Command Secretary: J.S. Pitt-Brooke
☎ (+44 1722) 43 32 30
Brigadier General Staff: Brig. A.E.G. Truluck
DACOS Information: Col Jonathan Seddon-Brown
☎ (+44 1722) 43 32 46; 43 36 77 (fax)
SIO: Jim Campbell ☎ (+44 1722) 43 33 45

Headquarters London District
Horseguard's, London SW1A 2AX
☎ (+44 171) 414 23 96
⊖ (+44 171) 414 23 52 (press office)
General Officer Commanding: Maj.Gen. I.C. Mackay-Dick (MBE)
Senior Press and Information Officer:
Col Jonathan Trelawney (retd)
Press and Information Officer: Bob Higson
☎ (+44 171) 414 23 55

Army Headquarters Scotland
Annandale Block, Craigiehall, South Queensferry, West Lothian EH30 9TN
☎ (+44 131) 336 17 61
⊖ (+44 131) 310 20 58 (press office)
General Officer Commanding: Maj.Gen. J.M.F.C. Hall
Senior Press and Information Officer: Lesley Edgar
☎ (+44 131) 310 20 91
Press and Information Officer: Graham Bardgett
☎ (+44 131) 310 20 92

Headquarters 2 Division
Imphal Barracks, Fulford Road, York YO1 4AU
☎ (+44 1904) 65 98 11
⊖ (+44 1904) 66 27 05 (press office)
General Officer Commanding:
Maj.Gen. C.D. Farrar-Hockley (MC)
Senior Press and Information Officer: Ms Sarah Bilton
☎ (+44 1904) 66 20 20
Press and Information Officer: John Davies
☎ (+44 1904) 66 25 33

Headquarters 3(UK) Division
Bulford Camp, Salisbury, Wiltshire SP4 9NY
☎ (+44 1980) 67 29 63 ⊖ (+44 1980) 67 28 55
General Officer Commanding: Maj Gen C N G Delves (DSO)
Senior Press and Information Officer: Marcia Nash
☎ (+44 1980) 67 33 29

Headquarters 4 Division
Steeles Road, Aldershot, Hampshire GU11 2DP
☎ (+44 1252) 34 70 50
⊖ (+44 1252) 34 70 13 (press office)
General Officer Commanding: Lt Gen. Sir N.W.F. Richards (OBE)
Senior Press and Information Officer: Paul Sykes
☎ (+44 1252) 34 70 11
Press and Information Officer: Gordon Skilling
☎ (+44 1252) 34 70 12

Headquarters 5 Division
Copthorne Barracks, Shrewsbury, Shropshire SY3 8LZ
☎ (+44 1743) 23 60 60 ⊖ (+44 1743) 26 20 13
General Officer Commanding: Maj.Gen. R.V. Searby
Senior Press and Information Officer: James Gee
☎ (+44 1743) 26 22 52
Press and Information Officer:
Ms Fleur Cavanagh-Mainwaring
☎ (+44 1743) 26 20 12 [98]

25-33000 Brigade and Battle Group Trainer (North) (BBGT(N))

Catterick Garrison, North Yorkshire DL9 3JS
☎ (+44 1748) 87 21 29 ⊖ (+44 1748) 87 23 04 [98]

25-33750 HQ Quartermaster General

Portway, Monxton Road, Andover, Hampshire SP11 8HT
☎ (+44 1264) 38 21 11 [97]

25-33950 Royal Logistic Corps Headquarters

The Princess Royal Barracks, Blackdown, Surrey
Director General Logistic Support: Maj.Gen. M.S. White
Colonel Commandant: Maj.Gen. D.F.E. Botting
Colonel Commandant: Maj.Gen. G.W. Field [97]

25-39150 Royal Army Medical Corps (RAMC)

Keogh Barracks, Ash Vale,
Aldershot, Hampshire GU12 5RR
☎ (+44 1252) 34 03 25 ⊖ (+44 1252) 34 03 48
Media Ops: Lt Col Gerald Dineley [98]

25-42100 Headquarters Commander-in-Chief Fleet

HMS Warrior, Sandy Lane, Eastbury Park,
Northwood, Middlesex HA6 3HP
☎ (+44 1923) 82 61 61 ⊖ (+44 1923) 83 71 31
Commander-in-Chief Fleet: Adm Sir Michael Boyce
☎ (+44 1923) 83 72 65
FPRO: Michael Hill ☎ (+44 1923) 83 73 36; 83 72 83 (fax)
DFPRO: David Healy ☎ (+44 1923) 83 76 35 [98]

25-42300 Headquarters Flag Officer Naval Aviation

(† Naval Air Command Headquarters)
Royal Naval Air Station Yeovilton,
Yeovil, Somerset BA22 8HL
☎ (+44 1935) 45 55 48 ⊖ (+44 1935) 45 59 29
Flag Officer Naval Aviation: RAdm. T.W. Loughran
Chief Staff Officer Naval Aviation: Cdre P.A. Fish
Staff Public Relations Officer: A R Worner (MVO)
■ Flag Officer Naval Aviation (FONA) is subordinate to CINCFLEET, and is responsible for meeting naval aviation requirements in peace & war. FONA manages all four Naval Air Stations & all Naval Air Squadrons. [98]

25-42320 Headquarters Flag Officer Surface Flotilla

3 The Parade, HM Naval Base,
Portsmouth, Hampshire PO1 3NA
☎ (+44 1705) 72 23 51
Public Relations Officer: Paul Parrack
☎ (+44 1705) 29 16 51; 72 23 91 (fax) [98]

25-42400 Headquarters Naval Home Command

Room 253, Victory Building, HM Naval Base,
Portsmouth, Hampshire
☎ (+44 1705) 72 70 01
Second Sea Lord and Commander-in-Chief of the Naval Home Command: Adm Sir Michael Boyce
Chief Public Relations Officer: Leslie Burgess
☎ (+44 1705) 72 71 15; 72 74 90 (fax)
Information Officer: Ms Fiona Devine
☎ (+44 1705) 72 71 16 [98]

25-42450 Hydrographic Office

Taunton, Somerset TA1 2DN
☎ (+44 1823) 33 79 00 ⊖ (+44 1823) 28 40 77
Hydrographer of the Navy: RAdm. J.P. Clarke
☎ (+44 171) 430 63 71 [98]

25-42500 Flag Officer Sea Training

Grenville Block, HMS Drake, HM Naval Base,
Plymouth, Devon PL2 2BG
☎ (+44 1752) 55 78 22 ⊖ (+44 1752) 55 78 06
Flag Officer: RAdm P M Franklyn (MVO)
Chief Staff Officer Operations and Training:
Capt M W G Kerr [98]

25-45300 Amphibious Trials and Training Unit, Royal Marines

Instow, Bideford, Devon EX39 4JH
☎ (+44 1271) 86 04 84 ⊖ (+44 1271) 86 04 46
Commander: Maj R N Eales [98]

26 Police (Civil)

26-10000 Metropolitan Police Service

New Scotland Yard, 10 Broadway, London SW1H 0BG
☎ (+44 171) 230 12 12 (switchboard)
⊖ (+44 171) 230 42 76
👥 27,864 (police); 16,860 (civilian)
Commissioner: Sir Paul Condon (QPM)
Deputy Commissioner: Brian Hayes (CBE, QPM)
Receiver: Philip Fletcher
Assistant Commissioner, Specialist Operations:
David Veness (QPM)
Director of Personnel: Ms Helen Maslen
Director of Property Services: Trevor Lawrence
Director of Technology: Nicholas Boothman
Director of Public Affairs: Ms Sarah Cullum
Assistant Commissioner (2nd Area - North West):
Baden Skitt (BEM, QPM)
Assistant Commissioner (3rd Area - North East):
Anderson Dunn (QPM)
Assistant Commissioner (4th Area - South East):
Ian Johnston (QPM)
Assistant Commissioner (5th Area - South West):
Paul Manning
Information Contact: Ms Jane Dolby
☎ (+44 171) 230 32 34; 230 35 93 (fax) [97]

26-30000 Cheshire Constabulary Headquarters

Castle Esplanade - Nuns Road, Chester CH1 2PP
☎ (+44 1244) 35 00 00; 61 21 15
E 100624.2204@compuserve.com ⊠ 61261
⊖ (+44 1244) 61 21 87; 61 22 69
£ 98,000,000 (Annual Budget)
👥 2,050 (uniformed); 637 (civilian)
Chief Constable: J M Jones (QPM, MSc)
☎ (+44 1244) 61 20 90
Deputy Chief Constable: I Holding (MA (Oxon)
☎ (+44 1244) 61 20 91
Assistant Chief Constable and Director of Crime Operations: J V Dwyer (BA, MIMgt)
☎ (+44 1244) 61 20 92
Assistant Chief Constable and Director of Operations (Uniform): I R Moody ☎ (+44 1244) 61 20 93
Director of Finance and Planning: M Flanagan (IPFA, DMS)
☎ (+44 1244) 61 29 69
Operations Policy: Superintendent C Unwin (BA (Hons)
North Area Co-ordinator: Chief Superintendent B Ingman
South Area Co-ordinator:
Chief Superintendent G E Gerrard
Crime Policy Co-ordinator:
Detective Chief Superintendent L Lee
Information Systems Co-ordinator:
Chief Superintendent R G Marsden
Quality of Service: Superintendent A Barclay (BSc)
Head of Personnel: G Keenan (BA (Hons)
Head of Press and Public Relations: Brenda Cowling
Public Relations Officer: Inspector Richard Hollinshead
Administration: B Edge (LLB)
■ Provides policing services covering Chester, Ellesmere Port and Neston, Vale Royal, Halton, Warrington, Macclesfield, Congleton, Crewe and Nantwich. [97]

29 Customs & Excise

29-00500 H.M. Customs & Excise

New King's Beam House, 22 Upper Ground,
London SE1 9PJ
☎ (+44 171) 620 13 13 ⊠ 886231; 917905 cehqln g
⊖ (+44 171) 865 50 05
Chairman of the Board: Mrs V P M Strachan (CB)
Deputy Chairman: A W Russell (CB)
☎ (+44 171) 865 50 12
Private Secretary to the Chairman: S Timewell
☎ (+44 171) 865 50 01
Commissioner, VAT Policy: M Brown
Commissioner, Customs Policy: M Eland
Commissioner, Excise and Central Policy: D Howard
Commissioner, Operations Compliance: Mrs E Woods
Commissioner, Operations: A Sawyer
Commissioner, Personnel and Finance: Dennis Battle
Commissioner, Central Operations: Ray McAfee
Commissioner, Information Systems: Alan Paynter

Customs Directorate
New King's Beam House, 22 Upper Ground,
London SE1 9PJ

☎(+44 171) 865 54 05 ⊖ (+44 171) 865 48 67
Commissioner and Director: M J Eland

Press and Information
New King's Beam House, 22 Upper Ground, London SE1 9PJ
☎(+44 171) 865 56 65 ⊖ (+44 171) 865 56 25
Principal Information Officer: Ms Lorna Sinclair

Personnel and Finance Directorate
New King's Beam House, 22 Upper Ground, London SE1 9PJ
☎(+44 171) 865 59 33 ⊖ (+44 171) 865 56 22 [98]

30 Civil Defense

30-00100 Scottish Fire Service and Emergency Planning Division
Saughton House, Broomhouse Drive, Edinburgh EH11 3XD
☎(+44 131) 244 21 70 ⊖ (+44 131) 244 28 19
Head of Fire Service and Emergency Planning Division, The Scottish Office Home Department: Mrs M Gunn
■ Civil defence & emergency planning. [98]

31 Embassies & Attachés

31-00020 Embassy of Albania
4th Floor, 38 Grosvenor Gdns, London SW1W 0EB
☎(+44 171) 730 57 09 ⊖ (+44 171) 730 57 47
Ambassador: H.E. Mr Pavli Mihal Qesku
Counsellor: Vangjel Dheri [97]

31-00030 Embassy of Algeria
54 Holland Park, London W11 3RS
☎(+44 171) 221 78 00 ⊖ (+44 171) 221 04 48
Ambassador: H.E. Mr Ahmed Benjamina
First Counsellor Political: Abdelghani Cheriaf
Defence Attaché: Col Mohamed Tabat
Financial Attaché: Semaine Mouna
Commercial Officer: Chakib Kaid
Public Information Officer: Abderahmane Benguerrah [97]

31-00060 High Commission for Antigua and Barbuda
15 Thayer Street, London W1M 5LD
☎(+44 171) 486 70 73
[E] ronald@antiguahc.sonnet.co.uk
[W] http://www.antigua-barbuda.com/
⊖ (+44 171) 486 99 70
High Commissioner: H E Mr Ronald M Sanders
Counsellor (Administration and Consular): Mrs Althea Banahene
Attaché (Information): Miss Sandra Scotland
Administrative Attaché: Miss Curliss Bart [98]

31-00080 Embassy of the Argentine Republic
65 Brook Street, London W1Y 1YE
☎(+44 171) 318 13 00; 13 30 (Trade Section)
[telex] 913348 argsia g
⊖ (+44 171) 318 13 01; 13 31 (Trade Section)
Ambassador: H.E. Mr Rogelio Pfirter
First Secretary Economic & Commercial Affairs: Alejandro Aragones
Political Councellor: Ms Monica E. Deregibus
Counsellor: Jorge L. Gomez
Press Officer: Ms Bibiana Jones

Office of Commercial Affairs
6th Floor, 100 Brompton Road, London SW3 1ER
☎(+44 171) 589 63 37; 78 57 ⊖ (+44 171) 589 65 50
Economic & Commercial Affairs Councellor: Alejandro Aragones
Secretary for Commercial Affairs: Armando Becher

Office of the Defence Attaché
Suite 2, 60 Cheval Place, London SW7 1EN
☎(+44 171) 584 57 53 ⊖ (+44 171) 584 63 54
Defence Attaché: Col Hector Gallardo
Assistant to the Defence Attaché: Earle H. Anselmo [97]

31-00085 Embassy of Armenia
25a Cheniston Gardens, London W8 6TG
☎(+44 171) 938 54 35; 938 45 14
⊖ (+44 171) 938 25 95
Ambassador: H.E. Mr Nounoufar Sarkissian
First Secretary: Zorab Mnatsakanian
Third Secretary: Edgar Grigorian [98]

31-00090 Australian High Commission
Australia House, Strand, London WC2B 4LA
☎(+44 171) 379 43 34 ✉ AUSTCOM LONDON
[telex] 27565 austlh g ⊖ (+44 171) 240 53 33
High Commissioner: H.E. The Hon. Neal Blewett
☎(+44 171) 887 52 20
Deputy High Commissioner: Ms Rosaleen M McGovern
☎(+44 171) 887 52 11
Defence Advisor and Head of Defence Staff: Brig P L McGuinness ☎(+44 171) 887 56 29
Staff Officer Coordination: Lt Col L Hollingsworth
☎(+44 171) 887 56 27
Army Adviser: Col R L Sayce ☎(+44 171) 887 52 62
Staff Officer Material - Army: Lt Col K L Powell
☎(+44 171) 887 55 09
Naval Adviser: Capt T E Lewis ☎(+44 171) 887 53 79
Staff Officer Engineering - Navy: Cdr K S McBryde
☎(+44 171) 887 54 16
Air Adviser: Gp Capt P B Layton ☎(+44 171) 887 52 64
Staff Officer Engineering - Air Force: Wg Cdr S C Sheedy
☎(+44 171) 887 51 09
Counsellor Defence Science: W R Dickson
☎(+44 171) 887 52 65
Counsellor Defence Logistics and Acquisitions: John Burke
☎(+44 171) 887 52 28
Counsellor Industry, Science and Technology: Ms Carolyn Walsh ☎(+44 171) 887 54 40
Minister (Marketing): Peter Lorschy
☎(+44 171) 887 52 05
Counsellor Defence Industry: David J McCue
☎(+44 171) 887 55 05 [98]

31-00100 Embassy of Austria
18 Belgrave Mews West, London SW1X 8HU
☎(+44 171) 235 37 31 ✉ AUSTROAMB LONDON
[telex] 28327 oebolo g ⊖ (+44 171) 235 80 25
Ambassador: H.E. Dr Eva Nowotny
Minister Deputy Chief of Mission: Dr Herbert Krauss
Defence Attaché: Brig.Gen. Hans-Rüdiger Sulzgruber
Administration Counsellor: Gerhard Gerzla

Office of the Trade Commissioner
45 Princes Gate, London SW7 2QA
☎(+44 171) 584 44 11 [telex] 25668 acomld g
⊖ (+44 171) 584 25 65
Trade Commissioner: Dr Rudolf Engel
Public Relations Officer: Ms Rosemary Hall [98]

31-00105 Embassy of Azerbaijan
4 Kensington Court, London W8 5DL
☎(+44 171) 938 34 12 ⊖ (+44 171) 937 17 83
Ambassador: H.E. Mr Mahmud Mamed Kuliyev
☎(+44 171) 938 34 12
Commercial Counsellor: Altai Efendiev
☎(+44 171) 937 20 97
First Secretary Information and Political Affairs: Heydar Efendiyev ☎(+44 171) 937 64 63
Second Secretary: Rafael Ibraguimov
☎(+44 171) 938 54 82
Second Secretary: Namik Kamranov
☎(+44 171) 937 64 63 [98]

31-00110 High Commission for the Commonwealth of the Bahamas
10 Chesterfield Street, London W1X 8AH
☎(+44 171) 408 44 88 [telex] 892617 bahreg g
⊖ (+44 171) 499 99 37
High Commissioner: H.E. Mr Arthur Foulkes
First Secretary/Consul: Mrs Donna Knowles-Lowe
Second Secretary/Vice Consul: Ms Rhoda H. Jackson
Second Secretary/Vice Consul: Melvin V. Claridge
Attaché (Maritime): Miss Judith Francis
Attaché: Raymond L. Dames [97]

31-00145 Embassy of the Republic of Belarus
6 Kensington Court, London W8 5DL
☎(+44 171) 937 32 88 ⊖ (+44 171) 361 00 05
Ambassador: H E Mr Uladzimir G Shchasny
☎(+44 171) 938 32 23
Counsellor (Press, Culture and Humanitarian Co-operation): Mrs Volga M Galubovich
Counsellor (Economic and Trade): Valery A Dougan
☎(+44 171) 938 16 33
First Secretary (Consular and Visa): Uladzimir Matusevich
☎(+44 171) 938 36 77
Second Secretary (Finance and Administration): Syargei U Yakushevich
Second Secretary (Information): Mikalai Barysevich [98]

31-00150 Embassy of Belgium
103-105 Eaton Square, London SW1W 9AB
☎(+44 171) 470 37 00; 235 40 12
✉ AMBABELGE LONDON [telex] 22823 belam g
⊖ (+44 171) 259 62 13; 823 20 83
Ambassador: H E Mr Lode Willems
Chargé d'Affaires: Jean-Michel Veranneman de Watervliet
Minister Counsellor: Ms Marie-Louise Vanherk
Defence Attaché: Col José Bouzette
☎(+44 171) 235 40 12
Agricultural Counsellor: Jean Marie Chislain Lekeu
First Secretary (Political and Economic Affairs): Hans Timbremont
First Secretary (Press, Information and Culture): Mrs Cathy Buggenhout
First Secretary (Head of Consular Section): Edgard de Tremmerie [98]

31-00180 Benin Consulate
Dolphin House, 16 The Broadway, Stanmore, Middlesex HA7 4DW
☎(+44 181) 954 88 00 [telex] 24620
⊖ (+44 181) 954 88 44 [98]

31-00200 Embassy of Bolivia
106 Eaton Square, London SW1W 9AD
☎(+44 171) 235 42 48; 22 57 ⊖ (+44 171) 235 12 86
Ambassador: H.E. Carlos Morales-Landívar
Counsellor: Gonzalo Chacón
Counsellor: Ms Marta Bosacoma Bonel
First Secretary: Ms Sara Compton-Galindo [98]

31-00220 Botswana High Commission
6 Stratford Place, London W1N 9AE
☎(+44 171) 499 00 31 [telex] 262897
⊖ (+44 171) 495 85 95
Defence Adviser: Lt Col W.M. Matshwa [98]

31-00230 Brazilian Embassy
32 Green St, London W1Y 4AT
☎(+44 171) 499 08 77 [telex] 261157 braslo g
⊖ (+44 171) 493 51 05
Ambassador: H.E. Mr Rubens Antonio Barbosa
Minister Counsellor: Sergio Eduardo Moreira Lima
Minister Counsellor: Eduardo dos Santos
Naval Attaché: Capt. Marcus Vinicius Oliveira dos Santos
Air Attaché: Gp Capt. Cesar Costa
Adviser for Maritime Affairs: Capt. Julio Expedito Curcio Aveline
First Secretary Environment & Human Rights: Bruno de Risios Bath
Press Officer: Ms Ivanise de Melo Maciel

Brazilian Naval Commission for Europe
170 Upper Richmond Rd, Putney, London SW15 2SH
☎(+44 181) 788 81 11/4 [telex] 918851 branav g
⊖ (+44 181) 780 07 30
Naval Adviser: Capt. Carlos Henrique Miranda
Naval Adviser: Cdr Marin Rodrigues

Brazilian Air Commission for Europe
16 Great James St, London WC1N 3DP
☎(+44 171) 405 50 62 [telex] 5121196 cablon g
⊖ (+44 171) 831 81 29
Air Adviser: Wg Cdr Ivo de Almeida Prado Xavier
Air Adviser: Wg Cdr Sergio Lozano da Silva
Air Adviser: Wg Cdr Luiz Carlos Barreto

Commercial Section
32 Green St, Mayfair, London W1Y 4AT
☎(+44 171) 499 08 77 [telex] 5125814 brastc g
⊖ (+44 171) 493 46 21
Counsellor Commercial Affairs: Marcos Bezerra Abbott Galvao [97]

31-00240 Brunei Darussalam High Commission
20 Belgrave Square, London SW1X 8PG
☎(+44 171) 581 05 21 ✉ BRULONGOV London
[telex] 888369 brulon g ⊖ (+44 171) 235 97 17
High Commissioner: H E Pehin Dato Haji Jaya Abdul Latif
First Secretary: Haji Alaihuddin Taha
Defence Adviser: Lt Col Ismail Haji Sarbini
Second Secretary: Mrs Masurai Haji Masri
Second Secretary: Aonan Jaafar [98]

Contact Reference

31-00250 Embassy of the Republic of Bulgaria
186-188 Queen's Gate, London SW7 5HL
☎(+44 171) 584 94 00; 584 94 33; 581 31 44
▣ 25465 bultra g ⊖ (+44 171) 584 49 48
Ambassador: H.E. Mr Stefan L Tafrov
Minister Plenipotentiary: Gueorgui Dagaradin
Military, Naval and Air Attaché: Capt. Ivan Yordanov
☎(+44 171) 584 94 00 X 18
Counsellor (Commercial/Economic): Dimitar Georgiev
First Secretary (Consul): Valeri Hristov
Counsellor (Political Affairs): Andrey Jichev
Counsellor (Political Affairs): Christo Kojouharov
First Secretary (Commercial/Economic): Peter Petrov
Third Secretary (Political and Protocol): Rossen Roufev
Attaché (Administration): Latchezar Tzenov
Attaché (Administration): Tzauko Yortchev [98]

31-00260 Honorary Consulate of Burkina Faso
5 Cinnamon Row, Plantation Wharf, London SW11 3TW
☎(+44 171) 738 18 00 ✉AFROTRADE London SW11
▣ 296420 afro g ⊖ (+44 171) 738 28 20
Ambassador: H E Mr Youssouf Ouedraogo
(based in Brussels)
Honorary Consul-General: Stuart G Singer
Second Counsellor: Raymond Balima (based in Brussels)
Economic Counsellor: Ambrose Balima (based in Brussels)
Financial Attaché: Ouanza Maurice Domboue
(based in Brussels)
Attaché: Mme Catherine Legani (based in Brussels) [98]

31-00290 Embassy of the Republic of Cameroon
84 Holland Park, London W11 3SB
☎(+44 171) 727 07 71/3 ▣ 25176 camlon g
⊖ (+44 171) 792 93 53
Ambassador: H.E. Mr Samuel Libock Nbei
Defence Attaché: Capt Pierre Ndangang
☎(+44 171) 727 07 71 ext. 220
First Counsellor: N Lloyd Forbang
☎(+44 171) 727 07 71 ext. 219
Cultural Counsellor: Samson Negbo Abangma
☎(+44 171) 727 07 71 ext. 209

Economic & Commercial Office
84 Holland Park, London W11 3SB
☎(+44 171) 727 07 71 ▣ 8954 557 meclon g
First Secretary Finance: Seini Katcalla
☎(+44 171) 727 07 71 ext. 231 [98]

31-00310 Canadian High Commission
Macdonald House, 1 Grosvenor Square, London W1X 0AB
☎(+44 171) 258 66 00 ✉DOMINION LONDON
▣ 261592 cdaldn g ⊖ (+44 171) 258 63 33
High Commissioner: H.E. Mr Royce Herbert Frith (QC)
☎(+44 171) 258 63 28
Deputy High Commissioner: Serge April
☎(+44 171) 258 63 29
Commander CDLS & Defence Adviser: Cdre D.E. Miller
☎(+44 171) 258 64 55
Air Adviser: Col P.S. Tanton ☎(+44 171) 258 64 09
Asst Air Adviser: Lt Col J.P. Doherty
☎(+44 171) 258 64 10
Army Adviser: Col J.B. Boileau ☎(+44 171) 258 64 08
Asst Army Adviser: Lt Col R.K. McDiarmid
☎(+44 171) 258 64 06
Naval Adviser: Capt. D.V. Jacobson
☎(+44 171) 258 64 02/3
Asst Naval Adviser: Cdr K.B. Lait ☎(+44 171) 258 64 04
Counsellor Public Affairs: Harry Adams
☎(+44 171) 258 63 71

Commercial/Economic Division
Macdonald House, 1 Grosvenor Square, London W1X 0AB
☎(+44 171) 258 66 00 ⊖ (+44 171) 258 63 33
Minister Commercial/Economic: Robert Burchill
Counsellor Commercial: Christopher MacClean
☎(+44 171) 258 66 10; 258 63 84 (fax)
Counsellor Commercial/Economic: Ms Cécile Latour
☎(+44 171) 258 66 50
Counsellor Commercial/Economic: John Sloan
☎(+44 171) 258 64 63 [97]

31-00350 Embassy of Chile
12 Devonshire Street, London W1N 2DS
☎(+44 171) 580 63 96
[E] echileuk@echileuk.demon.co.uk ▣ 25970 embchi g
⊖ (+44 171) 436 52 04
Ambassador: H E Mr Mario Artaza
Minister (Deputy Chief of Mission): Jose Luis Morales
First Secretary: Christian Rehren
Air Attaché: Col Ricardo Gutierrez
Military Attaché: Col Fernan Gonzalez
Naval Attaché: RAdm Jorge Arancibia
☎(+44 171) 734 41 76
Assistant Naval Attaché: Capt Reginald Thornton
☎(+44 171) 734 08 02
Assistant Naval Attaché: Capt Fernando Mandiola
☎(+44 171) 734 08 03
Commercial Attaché: Homero Sotomayor
Minister Counsellor: Jose Luis Morales
Cultural and Press Attaché: Ms Jacqueline Tichauer

Office of the Military Attaché
12 Devonshire Street, London W1N 2FS
☎(+44 171) 580 63 92; 63 97 ⊖ (+44 171) 436 52 04
Military Attaché: Col Fernan Gonzalez

Office of the Air Attaché
Samuel House, 6 St Albans Street, London SW1Y 4SQ
☎(+44 171) 930 00 28; 41 60; 57 98 ▣ 25758 mal g
⊖ (+44 171) 930 04 99
Air Attaché: Col Ricardo Gutierrez
Deputy Air Attaché: Gp Capt Rolando Hunt

Office of the Naval Attaché
Piccadilly House, 33 Regent Street, London SW1Y 4NE
☎(+44 171) 734 41 76; 08 02 ⊖ (+44 171) 434 07 93
Naval Attaché: RAdm Jorge Aranciba
Assistant Naval Attaché: Capt Reginald Thornton
Assistant Naval Attaché: Capt Fernando Mandiola [98]

31-00360 Embassy of the People's Republic of China
49-51 Portland Place, London W1N 4JL
☎(+44 171) 636 93 75 ▣ 23851 china g
⊖ (+44 171) 636 29 81
Ambassador: H.E. Mr Jiang Enzhu
Minister Counsellor: Wang Qiliang
Private Secretary: Jin Rong

Defence Section
26 Lyndhurst Road, Hampstead, London NW3 5PA
☎(+44 171) 794 75 95 ▣ 4333653
⊖ (+44 171) 433 36 53
Defence Attaché: Maj.Gen. Han Kaihe
Military Attaché: Sr Col Wang Suobao
Air Attaché: Sr Col Li Kewen
Naval Attaché: Sr Capt. Yin Xianyin
Deputy Defence Attaché Technical: Col Shen Lei
Deputy Defence Attaché: Lt Col Mu Changlin

Science & Technology Section
42 Maida Vale, London W9 1RP
☎(+44 171) 286 12 97 ⊖ (+44 171) 286 68 33
Counsellor Science & Technology: Zhao Yingfu
First Secretary: Zhou Dingheng

Commercial Section
1-3 Leinster Gardens, London W2 6DP
☎(+44 171) 723 89 23 ⊖ (+44 171) 706 27 77
Minister-Counsellor: Lu Kexing [97]

31-00380 Embassy of Colombia
3 Hans Crescent, London SW1X 0LN
☎(+44 171) 589 91 77 ⊖ (+44 171) 581 18 29
Ambassador Extraordinay & Plenipotentiary:
H.E. Dr Carlos Lemos-Simmonds
Minister Plenipotentiary: Eduardo Muñoz-Gomez
Minister/Counsellor Commercial Attaché:
Dr Juan Esteban Restrepo-Santamaria
Minister/Counsellor: Hernando Correa-Peraza
Defence Attaché: Brig. Gen. Jose E. Reyes-Lopez
Military Attaché: Col Arcesio Barrero-Aguirre
Naval Attaché: Capt. Victor Manuel Villalba
First Secretary Human Rights: Dr Anibal Palacio
First Secretary: Miss Patricia Acosta-Trujillo
Third Secretary: Victor Raul Martinez
Consul General: Oscar Lopey-Pulecio
First Deputy Consul: Mrs Belen Cardenas De Cordoba
Second Deputy Consul: Mrs Patricia Castaño [97]

31-00410 Embassy of Costa Rica
14 Lancaster Gate, London W2 3LH
☎(+44 171) 706 88 44 [E] info@embcrlon.demon.co.uk
[W] http:// www.embcrlon.demon.co.uk
⊖ (+44 171) 706 86 55
Ambassador: H.E. Jorge Borbón
Minister Counsellor & Consul General: Walter Fonseca
Press Attaché: Mrs Elke Ruge-Moya [98]

31-00420 Embassy of the Republic of Côte d'Ivoire
2 Upper Belgrave Street, London SW1X 8BJ
☎(+44 171) 235 69 91 ▣ 23906 ivory g
⊖ (+44 171) 259 53 20
Ambassador: H E Mr Kouadio Adjoumani
First Counsellor: Florent Kouassi Ekra
Counsellor (Consul and Cultural):
Miss Kinanwonaman Toure
First Secretary (Economic, Commercial and Tourism):
Gaëtan Irie Fouabi
Second Secretary (Finance): Lazare Aman Oi Aman
Second Secretary (Protocol): Alphonse Guei [98]

31-00425 Embassy of the Republic of Croatia
21 Conway Street, London W1P 5HL
☎(+44 171) 387 17 90; 20 22 ⊖ (+44 171) 387 32 89
Ambassador: H.E. Dr Ante Cicin-Sain
Minister Counsellor: Darko Javorski
Counsellor: Damir Kusen
Military Attaché: Lieut Col Branco Gacak
Counsellor (Consular): Mrs Orhideja Sokac
Third Secretary: Ms Maja Erceg
Third Secretary: Mrs Katharin Diane Milas-Rezo
Information Officer: Ms Loreta Bertosa [98]

31-00430 Embassy of the Republic of Cuba
167 High Holborn, London WC1V 6PA
☎(+44 171) 240 24 88 ▣ 261094 cubaex g
⊖ (+44 171) 836 26 02
Ambassador: H E Mr Rodney Lopez
Minister Counsellor: Carlos Acevedo
Commercial Counsellor: Miss Elena Blanco
First Secretary (Political and Press): Manuel Ayuto
Second Secretary (Administration): Anselmo Montiel
Commercial Attaché: Cristobal Lupo [98]

31-00440 Cyprus High Commission
93 Park Street, London W1Y 4ET
☎(+44 171) 499 82 72; 629 53 50 (consular);
491 85 68 (press); 499 28 43 (welfare)
⊖ (+44 171) 491 06 91; 499 13 46 (press)
High Commissioner: H E Mr Vanias Markides
☎(+44 171) 499 28 10
Consul General: George Vyrides
First Secretary: Marios Kountourides
Counsellor Cultural Affairs: Klitos Avgoustinos
Press Counsellor: Soteris Georgallis
Maritime Attaché: Andreas Chrysostomou
Commercial Counsellor: Andreas Georgiades
☎(+44 171) 629 62 88; 52 44 (fa x)

Cyprus Tourism Organisation
211-213 Regent Street, London W1R 8DA
☎(+44 171) 734 89 22 ⊖ (+44 171) 287 65 34
Director, London Office: Orestis Rossides

Cyprus Trade Centre
3rd Floor, 29 Princes Street, London W1R 7RG
☎(+44 171) 629 62 88 ⊖ (+44 171) 629 52 44
Commercial Counsellor: Andreas Georgiades [98]

31-00450 Embassy of the Czech Republic
26 Kensington Palace Gardens, London W8 4QX
☎(+44 171) 243 11 15 ⊖ (+44 171) 727 96 54
Ambassador: H E Mr Karel Kühnl
Minister-Counsellor: Milan Jakobec
Defence Attaché: Col Milan Skalicky
Assistant Defence Attaché: Lt Col Jirí Bednar
Counsellor: Jirí Havlík
Counsellor: Ms Helena Bauschneiderova
First Secretary: Jan Valek
First Secretary: Mrs Jitka Martinova
Public Information Officer: Miss Daniela Moravcikova

Commercial Section
26 Kensington Palace Gardens, London W8 4QY
☎(+44 171) 243 11 15 ▣ 28 276 obzalb g
⊖ (+44 171) 727 96 54
Commercial Counsellor: Karel Antropius [98]

31-00460 Royal Danish Embassy
55 Sloane Street, London SW1X 9SR
☎(+44 171) 235 12 55; 333 02 00
✉AMBADANE London SW1 ⊖ (+44 171) 333 02 31
Ambassador: H E Mr Ole Lønsmann Poulsen
Minister Deputy Chief of Mission: Berno Kjeldsen
Counsellor (Economic Affairs): Peter Biering
Defence Attaché: Capt Poul Groosss
☎(+44 171) 333 02 28 (army/air)
(+44 171) 333 02 27 (navy) ✉DANDEFENCE London
Commercial Counsellor: Gunnar Tetler
Commercial Attaché: Mrs Cecilie Dahl
Counsellor Press and Cultural Affairs: Knud Plougmann [98]

31-00480 High Commission for Dominica
1 Collingham Gardens, South Kensington,
London SW5 0HW
☎(+44 171) 370 51 94/5; 373 87 51
881391 dacom g ⊖ (+44 171) 373 87 43
High Commissioner: H E George E Williams
Second Secretary: Miss Agnes Adonis 98

31-00500 Embassy of Ecuador
50 Montagu Square, London W1H 1TH
☎(+44 171) 224 93 07 ⊖ (+44 171) 258 30 04
Ambassador: H.E. Mr Patricio Maldonado
Minister: Rodrigo Riofrio
Defence and Naval Attaché: Capt Francisco Espinosa
☎(+44 171) 224 93 07
Air Attaché: Col Guillermo Argüello (Based in Paris)
☎(+33 1) 45 63 57 63
Second Secretary:
Sra Marie Soledad Cordova de Valencia
Second Secretary: Rolando Suarez
Defence Attaché's Office
50 Montagu Square, London W1H 1TH
☎(+44 171) 224 93 07 ⊖ (+44 171) 258 30 04
Defence Attaché: Capt Francisco Espinosa 98

31-00510 Embassy of the Arab Republic of Egypt
24 South St, London W1Y 5PJ
☎(+44 171) 493 26 49; 499 79 47
⊖ (+44 171) 499 79 48
Ambassador: H.E. Mr Mohamed I. Shaker
Counsellor: Gehad Refaat Madi
Counsellor Consular Affairs: Zeinab Hassan Choukry
First Secretary: Mrs Fatma Fathelbab Galal
Defence Attaché's Office
24 South St, London W1Y 5PJ
☎(+44 171) 499 79 47; 493 26 49
⊖ (+44 171) 499 79 48
Defence Attaché: Brig. Ibrahim El Shaib
Asst Defence Attaché: Lt Col Mohammed El Keshky
Asst Defence Attaché: Col Tarek Mustaffa 97

31-00520 Embassy of El Salvador
Tennyson House, 159 Great Portland St,
London W1N 5FD
☎(+44 171) 436 82 82
✉EMail: 75047,3102@compuserve.com
⊖ (+44 171) 436 81 81
Ambassador: H.E. Sr Don Manuel Gutierrez Ruiz
First Secretary Consul General:
Ms Violeta Salazar de Corich 97

31-00535 Embassy of Estonia
16 Hyde Park Gate, London SW7 5DG
☎(+44 171) 589 34 28 E sekretar@estonia.gov.uk
⊖ (+44 171) 589 34 30
Chargé d'Affaires: Peep Jahilo
Counsellor: Peeter Kapten
Press & Information Officer: Mrs Kadri Kadakas
Commercial Officer: Andrew Inkley-Leitch 97

31-00570 Embassy of the Republic of Fiji
34 Hyde Park Gate, London SW7 5DN
☎(+44 171) 584 36 61; 36 62 ⊖ (+44 171) 584 28 38
Ambassador: H.E. Mr Filimone Jitoko
Commercial Counsellor: Sung Kangwai
Second Secretary: Sukulu Maraiwai
Third Secretary: Mrs Vuki Vakatawa 98

31-00580 Embassy of Finland
38 Chesham Place, London SW1X 8HW
☎(+44 171) 838 62 00 24786 finamb g
⊖ (+44 171) 235 36 80
Ambassador: HE Mr Pertti Salolainen
Minister (Deputy Chief of Mission): Mrs Kirsti Eskelinen
Counsellor (Economic): Mr Harri Sipri Mikael Salmi
Counsellor (Political): Miss Liisa Anneli Valjento
Defence Attaché: Col Hanno Strang
Counsellor (Industry and Technology):
Mr Markku Antero Sjöstedt
Finland Trade Centre
3rd Floor, 30-35 Pall Mall, London SW1Y 5LP
☎(+44 171) 747 30 00 262360 finntr g
⊖ (+44 171) 747 30 07
Trade Commissioner: Mr Marcus Moberg
Counsellor (Industrial and Scientific):
Mr Markku Antero Sjöstedt 98

31-00590 Embassy of France
58 Knightsbridge, London SW1X 7JT
☎(+44 171) 201 10 00 261905 fralon g
⊖ (+44 171) 201 10 99
Ambassador: H.E. Mr J. Gueguinou
Minister Deputy Chief of Mission: J.P. Taix
Defence & Naval Attaché: RAdm. Y. de Kersauson
☎(+44 171) 201 10 91; 201 10 99 (fax)
Asst Naval Attaché: Cdr D. Bablon
Air Attaché: Col M. Bottine ☎(+44 171) 201 11 08
Asst Air Attaché: Lt Col D. Mounier
Military Attaché: Col P. Bonnet ☎(+44 171) 201 11 03
Attaché for Defence Equipment: L. Barthelemy
☎(+44 171) 201 11 21
Asst Attaché for Defence Equipment Naval: O. Prats
Asst Attaché for Defence Equipment Land/Research:
F. Laure
Asst Attaché for Defence Equipment Procurement:
M. Dumas
Duty Press Officer: V. Fevrier
Commercial Section
21-24 Grosvenor Place, London SW1X 7HU
☎(+44 171) 235 70 80 ✉COMATTA London
263093 comatl g ⊖ (+44 171) 235 85 98
Minister/Counsellor Commercial: O. Louis
Science & Technology Section
6 Cromwell Place, London SW7 2JN
☎(+44 171) 581 07 11 ⊖ (+44 171) 584 79 91
Scientific Counsellor: H. Gibert 97

31-00630 Embassy of the Republic of Gabon
27 Elvaston Place, London SW7 5NL
☎(+44 171) 823 99 86 919418 ambgab g
⊖ (+44 171) 584 00 47
Ambassador: H.E. Mr Vincent Boulé
Chargé d'Affaires a.i: Michel Madoungou
First Counsellor: Paul Patrick Biffot 98

31-00640 The Gambia High Commission
57 Kensington Court, Kensington, London W8 5DG
☎(+44 171) 937 63 16/18 911857 gamext g
⊖ (+44 171) 937 90 95
High Commissioner: H E John P Bojang
First Secretary: Baba S Saho
Finance Attaché: Mrs Olivia A Phillott
Information Attaché: Mrs Sainabou L Ndure
☎(+44 171) 376 00 93 98

31-00660 Embassy of the Federal Republic of Germany
23 Belgrave Square, Chesham Place, London SW1X 8PZ
☎(+44 171) 824 13 00 28191/25634 aaldn g
⊖ (+44 171) 824 13 90
Ambassador: H.E. Dr Oesterhelt
Deputy Chief of Mission: von Butler
Defence Attaché: Brig Gen Fischer
☎(+44 171) 824 13 85
Air Attaché: Col Volker Zimmer ☎(+44 171) 824 13 87
Military Attaché: Col Uchtmann ☎(+44 171) 824 13 91
Naval Attaché: Cdr Kupferschmidt (Navy)
☎(+44 171) 824 13 93
First Counsellor Defence Research: Pipperger
Secretary Defence R&T: H. Lohde
First Counsellor Scientific Affairs: Dr Klaus Wild
First Counsellor Head of Economic Dept:
Dr Bernd Westphal 98

31-00690 Embassy of Greece
1A Holland Park, London W11 3TP
☎(+44 171) 229 38 50 ✉GREEK EMBASSY London
266751 greemb g ⊖ (+44 171) 229 72 21; 221 28 18
Ambassador: H E Mr Vassilis S Zafiropoulos
Minister-Counsellor: Michael Spinellis
First Counsellor and Consul-General: Evangelos Carokis
Defence Attaché: Capt A Economou
☎(+44 171) 727 37 85
Assistant Defence Attaché: Lieut Cdr D Balopoulos
Shipping Attaché: Capt G Papachristodoulou
Counsellor (Economic and Commercial):
Mrs A Missa-Kerkentze
Counsellor: Nikolaos Garilidis 98

31-00710 High Commission for Grenada
1 Collingham Gardens, Earls Court, London SW5 0HW
☎(+44 171) 373 78 09 ⊖ (+44 171) 370 70 40
High Commissioner: H.E. Miss June Lendore
Third Secretary: Samuel Sandy
Attaché: Miss Wendy Grenade 97

31-00770 High Commission for Guyana
3 Palace Court, Bayswater Road, London W2 4LP
☎(+44 171) 229 76 84 ⊖ (+44 171) 727 98 09
High Commissioner: H.E. Mr Laleshwar K.N. Singh
Second Secretary: Ms Sonia Elliott
Second Secretary: Orin Alexander
Admin Attaché: Mrs Carol Barker
Admin Attaché: Cecil Singh-Ramotar 97

31-00820 Embassy of Iceland
1 Eaton Terrace, London SW1W 8EY
☎(+44 171) 730 51 31 ⊖ (+44 171) 730 16 83
Ambassador: H E Mr Benedikt Asgeirsson
Minister Counsellor: Thordur B Gudjonsson
Attaché: Ms Dora Asgeirsdottir 98

31-00830 High Commission of India
India House, Aldwych, London WC2B 4NA
☎(+44 171) 836 84 84 267166 hcildin g
⊖ (+44 171) 836 43 31
High Commissioner: H E Dr L M Singhvi
☎(+44 171) 836 25 56
Deputy High Commissioner: Shri P K Singh
☎(+44 171) 240 01 58
Military Adviser: Brig Rakesh Dhir ☎(+44 171) 240 08 70
Air Adviser: Air Cdre Ak Trikha ☎(+44 171) 240 19 48
Naval Adviser: Cdre P R Franklin ☎(+44 171) 240 23 17
Technical Adviser (Defence): Rajat Pal ☎(+44 171) 18 16
Second Secretary (Trade): Shri Kamal Kumar Ajwaui
☎(+44 171) 836 62 54
Attaché (Commerce): Shri Vikram Singh Rana
☎(+44 171) 836 84 84 ext. 191 & 163 98

31-00850 Embassy of the Islamic Republic of Iran
16 Prince's Gate, London SW7 1PT
☎(+44 171) 225 30 00 E eirilond@eirilond.demon.co.uk
23998; 918648 irangb g ⊖ (+44 171) 589 44 40
Deputy Head of Mission: Muhammad Safaei
Chargé d'Affaires: H.E. Mr Gholamreza Ahsani
☎(+44 171) 584 92 13
Commercial Officer: A Sajjadhan
Public Information Officer: S Haidarpur 98

31-00920 Embassy of Japan
101-104 Piccadilly, London W1V 9FN
☎(+44 171) 465 65 00 ✉NIHONTAISI London W1
25630 taisic g ⊖ (+44 171) 491 93 48; 409 06 55
Ambassador: H E Mr Hiroaki Fujil
Minister Deputy Chief of Mission: Sadaaki Numata
Defence Attaché: Capt Masato Shimada
Finance Minister: Yoshihiko Matsuo
Commercial Minister: Masayoshi Amano
Information and Culture Minster: Masatoshi Muto
Consulate-General of Japan
2 Melville Crescent, Edinburgh EH3 7HW
☎(+44 131) 225 47 77 ⊖ (+44 131) 225 48 28 98

31-00960 High Commission for Kenya
45 Portland Place, London W1N 4AS
☎(+44 171) 636 23 71 262551 kenrln g
⊖ (+44 171) 323 67 17
High Commissioner: H.E. Mr Mwanyengela Ngaili
First Secretary: J K Ariithi
Defence Adviser: Col G L Okanga
Asst Defence Adviser: P. Nyamai
Commercial Attaché: D.N. Mbugua
Financial Attaché: P.A.O. Olando
Public Information Officer: L N Ngaithe 98

31-00990 Embassy of the Republic of Korea
60 Buckingham Gate, London SW1E 6AJ
☎(+44 171) 227 55 00; 55 02
⊖ (+44 171) 227 55 03; 55 23
Ambassador: H E Mr Dong-Jin Choi
Minister: Seung-Hoh Choi
Defence Attaché: Capt. Jang Kil Joo
Defence Logistics Attaché: Maj. Hyun Ju Lee
Assistant Defence Attaché: Maj Il Jung
Counsellor: Seeya Ryoo
First Secretary (Information): Yongman Park
Second Secretary, Political Office: Heung Kyeong Park
Minister Counsellor (Information): Sung Soo Park 98

Contact Reference

31-01000 Kuwait Embassy
2 Albert Gate, Knightsbridge, London SW1X 7JU
☎(+44 171) 590 34 00 ⊠ 893407 kuagen g
⊖ (+44 171) 823 17 12
Ambassador: H E Mr Khaled Al-Duwaisan
First Secretary: Ali Khaled Al-Sabah
First Secretary: Salem Al-Zamanan
Cultural Attaché: Ibrahim Al-Refaie
Military Affairs Office
46 Queen's Gate, London SW7 5HR
☎(+44 71) 589 45 33; 584 15 56; 589 13 21
⊠ 893407 kuagen g
[98]

31-01015 Embassy of Latvia
45 Nottingham Place, London W1M 3FE
☎(+44 171) 312 00 40 ⊖ (+44 171) 312 00 42
Ambassador: H.E. Jànis Lūsis
Counsellor: Ms Marie-Anne Zarina
☎(+44 171) 312 00 40
First Secretary Deputy Chief of Mission: Kàrlis Eihenbaums
First Secretary Consular: Eriks Flaumanis
Second Secretary & Commercial Officer:
Mrs Teréze Svilàne
[97]

31-01020 Embassy of Lebanon
21 Kensington Palace Gardens, London W8 4QM
☎(+44 171) 229 72 65 ⊠ 262048 aliban g
⊖ (+44 171) 243 16 99
Ambassador: H E Dr Mahmoud Hammoud
Secretary: Ms A Dob
Counsellor (Commercial and Public Information):
Dr Bassam Naamani
[98]

31-01030 High Commission for the Kingdom of Lesotho
7 Chesham Place, Belgravia, London SW1 8HN
☎(+44 171) 235 56 86
E 106065.1632@compuserve.com ⊠ 262995 leqosa g
⊖ (+44 171) 235 50 23
High Commissioner: H E Mr Benjamin Masil
Counsellor: Tsotang Donald Moeketsi
First Secretary: Moeketsi Tsiboho
First Secretary: Alina Nzeku
Third Secretary: Mohanuoa Tsotetsi
Admin Attaché: Mpati Mphatsōe
[98]

31-01050 Libyan Interests Section
c/o Royal Saudi Arabian Embassy, 119 Harley Street, London W1N 1DH
☎(+44 171) 486 83 87 ⊠ 266787 msaleh g
⊖ (+44 171) 224 09 07
Head of Libyan Interests Section: Isa B Edaeki
[98]

31-01060 Embassy of Lithuania
84 Gloucester Place, London W1H 3HN
☎(+44 171) 486 64 01 ⊖ (+44 171) 486 64 03
Chargé d'Affaires: Dr Nijole Zambaite
Political Counsellor: Alvydas Medalinskas
Head of Chancery: Imsre Sabalinnas
First Secretary: Algimantas Misevicius
Admin Attaché: Erikas Petrikas
[97]

31-01070 Embassy of Luxembourg
27 Wilton Crescent, London SW1X 8SD
☎(+44 171) 235 69 61 ⊠ 28120 amblux g
⊖ (+44 171) 235 97 34
Ambassador: H.E. Mr Joseph Weyland
Counsellor: Paul Steinmetz
Chancellor: Ms A M Zurn
[98]

31-01100 High Commission for the Republic of Malawi
33 Grosvenor St, London W1X 0DE
☎(+44 171) 491 41 72 ✉ KWACHACOM London
⊠ 263308 g ⊖ (+44 171) 491 99 16
High Commissioner: H.E. Mr Jack Muwamba
Defence Adviser: Col E.M. Kaputolo
[97]

31-01110 Malaysian High Commission
45 Belgrave Square, London SW1X 8QT
☎(+44 171) 235 80 33 ⊠ 262550
⊖ (+44 171) 235 51 61
High Commissioner: H E Dato Kamarudin Abu
Deputy High Commissioner: Jasmi Md. Yusoff
Head of Chancery: Husni Zai Yaacob
Defence Adviser: Col Kamaruddin Mattan
Assistant Defence Adviser: Maj Mohd Ariffin Musa
Minister Counsellor: Muhamad Mustapha
Trade Commission
17 Curzon Street, London W1Y 7FE
☎(+44 171) 499 73 88 ⊖ (+44 171) 493 88 04
[98]

31-01140 Malta High Commission
Malta House, 36-38 Piccadilly, London W1V 0PQ
☎(+44 171) 292 48 00 ⊠ 261102 mltldn g
⊖ (+44 171) 734 18 31
High Commissioner: H E Mr Richard A Matrenza
First Secretary: Joseph Izzo Clarke
First Secretary: Stephen Borg
[98]

31-01190 Embassy of Mexico
42 Hertford Street, Mayfair, London W1Y 7TF
☎(+44 171) 499 85 86 ⊖ (+44 171) 495 40 35
Ambassador: H E Santiago Oñate
Minister and Deputy Head of Mission:
Ramón Brito Moncada
Consul General: José Héctor Ibarra Morales
Minister Economic Affairs: Ms Cecilia Ramos
Minister Cultural Affairs: Juan Manuel Santin
Second Secretary Multilateral Affairs: Jorge Lomónaco
Second Secretary Political Affairs: Ms Alicia Buenrostro
Counsellor Commercial Affairs: Fernando Estandia
Naval Attaché: Vice Adm Mariano Saynes
Military Attaché: Gen Francisco Meza Castro
Consular Section
8 Halkin Street, London SW1X 7DW
☎(+44 171) 235 63 93
Military Affairs Section
8 Halkin Street, London SW1X 7DW
☎(+44 171) 245 93 69
Naval Affairs Section
8 Halkin Street, London SW1X 7DW
☎(+44 171) 235 62 11
Tourism
60/61 Trafalgar Square, London WC2N 5DS
☎(+44 171) 734 10 58
[98]

31-01210 Embassy of Mongolia
7 Kensington Court, London W8 5DL
☎(+44 171) 937 01 50; 937 52 38
⊠ 128849 monemb g ⊖ (+44 171) 937 11 17
Ambassador: H.E. Mr C. Baatar
3rd Secretary: L. Davaadelger
Commercial Attaché: Jantsangiin Gulgou
[98]

31-01220 Embassy of the Kingdom of Morocco
49 Queen's Gate Gardens, London SW7 5NE
☎(+44 171) 581 50 01/04 ⊠ 28389 morosy g
⊖ (+44 171) 225 38 62
Ambassador: H.E. Mr Khalil Haddaoui
Defence Attaché: Col Maj. M. Assari
Counsellor Economic: Ahmed Faouzi
[97]

31-01250 High Commission of the Republic of Namibia
6 Chandos Street, London W1M 0LQ
☎(+44 171) 636 62 44 ⊖ (+44 171) 637 56 94
High Commissioner: H.E. Mr Ulenga Ben Ulenga
Minister Counsellor: Herbert Charles Schultz
Counsellor: Jeroboam Shaanika
[98]

31-01260 Royal Nepalese Embassy
12A Kensington Palace Gardens, London W8 4QU
☎(+44 171) 229 15 94; 62 31 ⊠ 261072 yeti g
⊖ (+44 171) 792 98 61
Ambassador: H.E. Mr Surya Prasad Shrestha
Deputy Head of Mission: P.K. Prasai
Military Attaché: Lt Col Indra B. Karki
Attaché: Chandra B. Maharjan
[97]

31-01300 New Zealand High Commission
New Zealand House, Haymarket, London SW1Y 4TQ
☎(+44 171) 930 84 22; 930 84 00; 973 03 66
✉ DEPUTY London SW1 ⊠ 24368 deputy g
⊖ (+44 171) 839 45 80; 930 84 01; 973 03 70
High Commissioner: H.E. The Hon. John Collinge
Deputy High Commissioner: Michael Chilton
Defence Adviser: Cdre J.G. Peddie (RNZN)
Asst Defence Adviser: Cdr A.L. Peppercorn (RNZN)
Asst Defence Adviser Air: Wg Cdr G.J. Sibbald (RNZAF)
Asst Defence Adviser Army: Lt Col I.M. Gordon (RNZA)
Defence Purchasing Mgr: Lt Cdr M.I. Ternent
[97]

31-01340 Royal Norwegian Embassy
25 Belgrave Square, London SW1X 8QD
☎(+44 171) 591 55 00 ⊖ (+44 171) 245 69 93
Ambassador: H E Mr Kjell Colding (CMG)
Minister Deputy Chief of Mission: Rolf Baltzersen
Defence Attaché: Capt Torstein Seim (RNoN)
Naval and Assistant Defence Attaché: Cdr Svein Buvik
First Secretary: Sigvald Hauge
Minister Press, Information and Cultural Affairs:
Oystein Braathen (CVO)
Private Secretary: W Hanssen
Commercial Division
5th Floor, Charles House, 5-11 Lower Regent Street, London SW1Y 4LR
☎(+44 171) 973 01 88 ⊖ (+44 171) 973 01 89
Commercial Counsellor: Sverre Lindtvedt
[98]

31-01380 Embassy of the Republic of Panama
48 Park Street, London W1Y 3PD
☎(+44 171) 493 46 46 E panama@mail.bogo.co.uk
⊖ (+44 171) 493 43 33
Ambassador: H.E. Mr Aquilino Boyd
Counsellor: Alberto Watson Fabrega
Financial and Commercial Counsellor: Salim Kheireddine
Attaché: Miss Ismenia Velasquez
Cultural Attaché: Miss Anabel Quijada
Cultural Attaché: Miss Itzel Valle
Maritime Attaché: Miss Argelis Cabrera
[98]

31-01400 Embassy of Paraguay
Braemar Lodge, Cornwall Gardens, London SW7 4AQ
☎(+44 171) 937 12 53; 66 29 ⊖ (+44 171) 937 56 87
Ambassador: H E Mr Washington Ashwell
First Secretary: Sr Igor Alberto Pangrazio
Second Secretary: Miss Elisa Victoria Aquino Laterza
Second Secretary: Miss Maria Ester Villar
[98]

31-01410 Embassy of Peru
52 Sloane Street, London SW1X 9SP
☎(+44 171) 235 19 17; 25 45 ✉ LEPRU London
E 100575.761@compuserve.com
⊖ (+44 171) 235 44 63
Ambassador: H E Mr J Eduardo Ponce-Vivanco
Minister Counsellor: Ms Maria S Landaveri
Counsellor for Maritime and Defence Affairs and Alternate IMO Permanent Representative:
Vice Adm Fernando D'Alessio ☎(+44 171) 235 49 36
Counsellor and Alternate IMO Permanent Representative:
Sr George Nicholson
Counsellor and Consul General: Sr Luis Chuquihuara
Public Information Officer: Ms Carmen Azurin
[98]

31-01430 Embassy of the Republic of Poland
47 Portland Place, London W1N 4JH
☎(+44 171) 580 43 24
W http://www.poland-embassy.org.uk ⊠ 265691
⊖ (+44 171) 323 40 18
Ambassador: H.E. Mr Ryszard Stemplowski
Deputy Head of Mission, Counsellor-Minister Plenipotentiary (Political): Witold Sobków
Defence, Military, Naval and Air Attaché:
Cdr Ireneusz Goreczny
Counsellor Maritime Affairs: Jerzy Vonau
☎(+44 171) 253 89 98
Counsellor (Press): Andrzej Kolczynski
Assistant Defence, Military, Naval and Air Attaché:
Lt Col J Czyzyk
Assistant Defence, Military, Naval and Air Attaché:
Lt Col Andrzej Jedrzejewski
Office of the Commercial Counsellor
15 Devonshire Street, London W1N 2AR
☎(+44 171) 580 54 81 ⊖ (+44 171) 323 01 95
Commercial Counsellor-Minister Plenipotentiary:
Piotr Kozerski
Commercial Counsellor: Dr Cezariusz Kwaterski
Commercial Attaché: Dr Elzbieta Raciniewska
Commercial Attaché: Eugeniusz Swiatkiewicz
Commercial Attaché: Marek Jedruszak Jarosla Grabowski
[98]

31-01480 Embassy of Romania
4 Palace Green, London W8 4QD
☎ (+44 171) 937 96 66 [E] romania@roemb.demon.co.uk
[TX] 22232 ⊖ (+44 171) 937 80 69
Minister, Counsellor: Ion Pascu ☎ (+44 171) 937 39 02
[TX] Minister, Counsellor (Deputy Chief of Mission)
Military, Naval and Air Attaché: Col Vasile Huicá
Assistant Military, Naval and Air Attaché:
Maj Gheorghe Savu
Counsellor (Economic): Constantin Soare
☎ (+44 171) 937 96 68
Counsellor (Cultural): Mrs Sînziana Dragos
☎ (+44 171) 937 81 25
Third Secretary (Press and Information): Dragos Tigáu
☎ (+44 171) 937 96 69 [98]

31-01485 Embassy of the Russian Federation
13 Kensington Palace Gardens, London W8 4QX
☎ (+44 171) 229 36 20 ⊖ (+44 171) 727 86 25
Ambassador: H E Mr Yuri Fokin
Commercial Officer: N. Telyatnikov
Defence & Military Attaché: Lt Gen. Vyacheslav N. Pronin
Asst Military Attaché: Col Pavel Yedygin
Air Attaché: Col Anotoliy Pavlovich Belyaev
Naval Attaché: Capt. 1st Rank Aleksandr I. Antropov
Public Information Officer: B. Malakhov
Office of the Military Attachés
16 Kensington Palace Gardens, London W8 4QH
☎ (+44 171) 229 64 51 ⊖ (+44 171) 221 20 38
Defence & Military Attaché: Lt Gen. V.N. Pronin
Asst Defence & Military Attaché: Col P. Yedygin
Naval Attaché: Capt. 1st Rank A. Antropov
Asst Naval Attaché: Capt. 1st Rank P. Tsvetkov
Asst Naval Attaché: Capt. 2nd Rank M. Mytarev
Air Attaché: Col A. Belyaev
Asst Military Attaché: Maj. N. Kudinov
Office of the Trade Delegation
32 Highgate West Hill, London N6 6NL
☎ (+44 181) 340 32 72 [TX] 28577 tradel g
Trade Representative: N. Telyatnikov
Scientific/Technical Attaché's Off.
33 Highgate West Hill, London N6 6NL
☎ (+44 181) 340 32 72 [97]

31-01550 High Commission for Eastern Carribean States
10 Kensington Court, London W8 5DL
☎ (+44 171) 937 95 22 [TX] 913047 ecacom g
⊖ (+44 171) 937 55 14
High Commissioner: H.E. Mr Aubrey E. Hart
Minister Counsellor: Cenio Elwin Lewis
First Secretary: Mrs Coletta President-Aruoma
Second Secretary: Mrs Shirley V. Pemberton
Trade Attaché: Bernard Cornibert
Third Secretary: Ms Jasmine M.L. Baksh
Third Secretary: Mrs Lucy Cornibert
■ Represents:
Saint Christopher & Nevis,
Saint Lucia,
Saint Vincent & the Grenadines [97]

31-01610 Embassy of the Republic of Senegal
21/24 Cockspur Street, London SW1Y 5BN
☎ (+44 171) 930 76 06 [E] g.sar@senegal.btinternet.com
[TX] 917119 ambsen g ⊖ (+44 171) 930 76 07
Ambassador: H.E. Mr Gabriel Alexandre Sar
Counsellor: Mamadou Diarra
First Secretary: Mouhamadou Lamine Diagne
Attaché: Mrs Assietou Ndiaye Diarra [98]

31-01620 The Seychelles High Commission
2nd Floor, Eros House, 111 Baker Street,
London W1M 1FE
☎ (+44 171) 224 16 60 [E] soozyw@aol.com
⊖ (+44 171) 487 57 56
High Commissioner: H E Mr J P Mascarenhas [98]

31-01630 Sierra Leone High Commission
33 Portland Place, London W1N 3AG
☎ (+44 171) 636 64 83/6 ⊖ (+44 171) 323 31 59
High Commissioner: H.E. Mr Alhaji Haroun Buhari
Counsellor: Rupert S.D. Davies
First Secretary: E.M. Sankoh
Financial Attaché: Y.D. Fasuluku
Information Attaché: Gibril Samura [97]

31-01640 High Commission for the Republic of Singapore
9 Wilton Crescent, London SW1X 8RW
☎ (+44 171) 235 83 15/18 [TX] 262564 shciuk g
⊖ (+44 171) 245 65 83
High Commissioner: H.E. Mr J.Y. Pillay
First Secretary: Thigarajah Thirunagaran
First Secretary Commercial: Francis Kwang Yong Tay
☎ (+44 171) 222 07 70
Defence Procurement Office Europe
5 Chesham Street, London SW1X 8RN
☎ (+44 171) 245 61 01 ⊖ (+44 171) 245 19 75
Counsellor: Col Puah Hung Khiang [97]

31-01650 Embassy of the Slovak Republic
25 Kensington Palace Gardens, London W8 4QY
☎ (+44 171) 243 08 03
⊖ (+44 171) 727 58 24; 792 02 15
Chargé d'Affaires: Ján Dömök ☎ (+44 171) 727 94 32
Commercial Counsellor: Jozef Chovan
☎ (+44 171) 727 36 67
Military and Air Attaché: Col Jan Sabo
☎ (+44 71) 792 02 15
Public Information Officer: Viliam Eliás [98]

31-01655 Embassy of the Republic of Slovenia
Suite 1 - Cavendish Court, 11-15 Wigmore Street,
London W1H 9LA
☎ (+44 171) 495 77 75 ⊖ (+44 171) 495 77 76
Ambassador: H.E. Mr Matjaz Sinkovec
Commercial Counsellor: Ciril Kosak
First Secretary: Ms Alenya Sz. Suhadolnik
Attaché: Ms Jana Erzetic [97]

31-01660 Somali Community Information Centre
490 Harrow Road, London W9 4QA
☎ (+44 181) 964 45 40 ⊖ (+44 181) 964 45 40
Director: Mrs Abshir Musse
Commercial Officer: Mr Hussein
Public Information Officer: Mr Sed [97]

31-01670 High Commission for the Republic of South Africa
South Africa House, Trafalgar Square, London WC2N 5DP
☎ (+44 171) 451 71 86 [E] sandf-uk@msu.com
⊖ (+44 171) 839 22 87
High Commissioner: H.E. Mr Mendi Msimang
Defence & Air Adviser: Brig. G.A. Hallowes
Naval Adviser: Capt J A J B Vorster
Army Adviser: Col A.H. Leijenaar [98]

31-01720 High Commission for the Kingdom of Swaziland
20 Buckingham Gate, London SW1E 6LB
☎ (+44 171) 630 66 11 [TX] 28853 swazi g
⊖ (+44 171) 630 65 64
High Commissioner: H.E. Rev. Percy S. Mngomezulu
Counsellor, Deputy High Commissioner: Khizitho Simelane
First Secretary Information: Micah Mgundu Dube [98]

31-01730 Swedish Embassy
11 Montagu Place, London W1H 2AL
☎ (+44 171) 724 21 01; 917 64 00 ✉ SVENSK London
[TX] 28249 svensk g ⊖ (+44 171) 724 41 74
Ambassador: H E Mr Mats Bergquist
Minister Deputy Chief of Mission: Marten Grunditz
Defence and Army Attaché: Col Gustaf Diurlin
☎ (+44 171) 917 64 38; 723 78 81 (fax)
Naval and Air Attaché: Col Nils Eklund
☎ (+44 171) 917 64 37; 723 78 81 (fax)
Counsellor Science and Technology: Rolf Carlström
☎ (+44 171) 724 67 85
Counsellor Press Affairs: Gösta Grassman
☎ (+44 71) 724 67 86
Swedish Trade Office
73 Welbeck Street, 4th floor, London W1M 8AN
☎ (+44 171) 935 96 01 [TX] 22620 swetra g
⊖ (+44 171) 935 41 30 [98]

31-01760 Taipei Representative Office in the UK
50 Grosvenor Gardens, London SW1W 0EB
☎ (+44 171) 396 91 52 ⊖ (+44 171) 396 91 45
Representative: H E Dr Eugene Chien
Director of Public Affairs Division: Mark Liao
Director of Press and Information Division: Calvin Chen
Director of Public Services Division: Paul Wang [98]

31-01780 Royal Thai Embassy
29-30 Queen's Gate, London SW7 5JB
☎ (+44 171) 259 50 05; 50 15 [E] thai@cityscape.co.uk
[TX] 27661 thai ld g ⊖ (+44 171) 235 98 08
Ambassador: H.E. Mr Vidhya Rayananonda
Minister Deputy Chief of Mission: Asipol Chabchitrchaidol
☎ (+44 171) 589 01 73
Defence and Air Attaché: Gp Capt Arkom Ganchanahirun
☎ (+44 171) 938 24 33
Military Attaché: Col Tanasak Patimapragorn
☎ (+44 171) 589 02 09
Assistant Military Attaché: Lt Col Surasak Prachumyaat
☎ (+44 171) 589 02 09
First Secretary Information: Mrs Sirinthip Petchpon
☎ (+44 171) 584 23 84
Minister Counsellor Commercial: Suraphol Jaovisidha
☎ (+44 171) 493 57 49
Office of Air Attaché
2 Victoria Road, London W8 5RD
☎ (+44 171) 938 24 33 ⊖ (+44 171) 937 79 18
Air Attaché: Gp Capt. Arkom Ganchanahirun
☎ (+44 171) 938 24 33
Commercial Counsellor's Office
9 Stafford Street, London W1X 4RT
☎ (+44 171) 493 57 49 [TX] 298706
⊖ (+44 171) 493 47 16
Counsellor Commercial: Mrs Pilai Siripanich
☎ (+44 171) 493 57 49
Ecomonic and Financial Minister:
Chakrabhand Chandanasiri ☎ (+44 171) 589 72 66
Office of Naval Attaché
1/3 Yorkshire House, Grosvenor Crescent,
London SW1X 7EP
☎ (+44 171) 245 11 90 ⊖ (+44 171) 245 11 70
Naval Attaché: Capt. Prajit Pityapat
Assistant Naval Attaché: Capt Anothai Rattarangsi [98]

31-01800 Tonga High Commission
36 Molyneaux Street, London W1H 6AB
☎ (+44 171) 724 58 28 ⊖ (+44 171) 723 90 74
High Commissioner: H E Mrs A Fineanganofo
Counsellor: Miss Kaimana H Aleamotua
First Secretary: Tevita Tupou [98]

31-01820 The High Commission of the Republic of Trinidad and Tobago
42 Belgrave Square, London SW1X 8NT
☎ (+44 171) 245 93 51 [E] trintagov@tthc.demon.co.uk
⊖ (+44 171) 823 10 65
High Commissioner: H E Mrs Sheelagh de Osuna
Deputy High Commissioner: Mr Patrick Edwards
Consular Department: Clive Pegus
Consular Department: Martin Baptiste
Chief of Defence Staff: Brig C A Alfonso
Chief of Staff - Navy: Capt A Franklin
Chief of Staff - Army: Col J Sandy
Commercial Officer: G G A Lowe
Public Information Officer: G Matthews [98]

31-01830 Embassy of Tunisia
29 Prince's Gate, London SW7 1QG
☎ (+44 171) 584 81 17 [TX] 23736
⊖ (+44 171) 225 28 84
Ambassador: H E Mrs Saida Chtioui
Minister Counsellor: Mrs Fatma Houidi
Counsellor: Habib Miled
Counsellor: Kais Darragi
Secretary Social and Culture: Hechmi Dhaoui
First Secretary Finance and Administration:
Mohamed Ali Mansouri
Attaché: Tahar Kricha
Attaché: Khaled Trabelsi
Attaché: Hedi Azaiez [98]

31-01840 Embassy of the Republic of Turkey
43 Belgrave Square, London SW1X 8PA
☎ (+44 171) 393 02 02 ⊖ (+44 171) 393 00 66
Ambassador: H.E. Mr Ozdem Sanberk
☎ (+44 171) 393 02 22
Minister Counsellor: Cinar Aldemir
☎ (+44 171) 393 02 22
Air Attaché: Col Altay Baklacioglu
Naval Attaché: Capt. Ahmet Erdogan

Military Attaché: Lt Col Umit Dundar
☎ (+44 171) 235 68 62
Asst Military Attaché Medical: Maj. Nadir Yildirim
☎ (+44 171) 235 19 59
Economic & Commercial Counsellor: Okan Elgin
☎ (+44 171) 235 49 91 ☎ (+44 171) 235 22 07 (fax)
Financial & Customs Counsellor: Necdet Gokmen
☎ (+44 171) 245 63 18
Third Secretary: Murat Ersoy [98]

31-01860 Uganda High Commission
Uganda House, 58-59 Trafalgar Square,
London WC2N 5DX
☎ (+44 171) 839 57 83 ⊠ 915141 uganda g
⊖ (+44 171) 839 89 25
High Commissioner: H.E. Prof. George Kirya
Deputy High Commissioner: D. Ssozi
Minister Counsellor: A. Nyerwanire
First Secretary: B. Ndinvenka
Second Secretary: Donald Christian Nyakairu
Second Secretary: C. Ssentongo
Financial Attaché: A. Bamweyana
Attaché: Mrs Beatrace Mutabazi [97]

31-01875 Embassy of Ukraine
78 Kensington Park Road, London W11 2PL
☎ (+44 171) 727 63 12; 243 89 23 Visa Section
⊖ (+44 171) 792 17 08
Ambassador: H.E. Prof. Sergui Komissarenko
Minister Plenipotentiary: Dr. Yuriy Sergeyev
Counsellor (Scientific): Prof. Volodymyr Koval
Counsellor (Consular): Olexander Kulakov
Defence Attaché: Col Mykola Natyaga
Counsellor (Economic and Commercial Affairs):
Dr. Boris Savchuk
First Secretary Information & Culture Affairs:
Mykola Kravchenko [98]

31-01900 Embassy of the United States of America
24/31 Grosvenor Square, London W1A 1AE
☎ (+44 171) 499 90 00 ⊖ (+44 171) 491 24 85
Ambassador: H E William Crowe
Deputy Chief of Mission: Robert Bradtke
Commercial Counsellor: Charles Ford
Political Counsellor: Michael Habib
Defence and Naval Attaché: James Mader (USN)
Army Attaché: Col Harry P Allen (USA)
Air Attaché: Col Frederick E Bush (USAF)
Political Military Officer: Charles Peacock
Public Affairs Officer: Michael O'Brien
Press Attaché: John Burgess ☎ (+44 171) 499 52 61 [98]

31-01910 Embassy of Uruguay
140 Brompton Road, 2nd Floor, London SW3 1HY
☎ (+44 171) 584 81 92 ✉ URUBRI London
⊖ (+44 171) 581 95 85
Ambassador: H E Sr Juan Enrique Fischer
Military and Air Attaché: Col Alejandro Bremermann
Naval Attaché: Capt Carlos Martino (Navy)
First Secretary and Consul: Dr Oscar Carbajal
Second Secretary: Dr Fernando Marr Merello
Commercial Officer: W Aldemar Tarigo [98]

31-01950 Embassy of Venezuela
1 Cromwell Road, London SW7 2HW
☎ (+44 171) 584 42 06/7 ✉ EMBAVENEZ London
[E] venezlon@venezlon.demon.co.uk ⊠ 264186 embven g
⊖ (+44 171) 589 88 87
Ambassador: H E Mr Roy Chaderton-Matos
☎ (+44 171) 584 53 75
Minister Counsellor: José Eugenio Lopez-Contreras
First Secretary: Ms Nancy Meza
First Secretary: Luis Carpio
Second Secretary Commercial Section: Carlos Espert
Third Secretary: Andrés Abreu
Cultural Attaché: Ms Gloria Carnevali
Press Attaché: Ms Mariela Briceño de Branger
Attaché: Ms Marie Navarro

Office of Military Attaché
54 Grafton Way, London W1P 5LB
☎ (+44 171) 387 06 95 ⊖ (+44 171) 916 11 55
Military, Naval and Air Attaché:
Capt Alfredo Sanchez-Vegas
Attaché Maritime Section:
Capt Felix Francisco Bravo-Mayol [98]

31-01960 Embassy of the Socialist Republic of Vietnam
12-14 Victoria Road, Kensington, London W8 5RD
☎ (+44 171) 937 19 12; 31 74 (commerce dept)
⊠ 887361 vietem g ⊖ (+44 171) 937 61 08
Ambassador: H E Mr Huynh Ngoc An
Commercial Counsellor: Thai Huy Duong
☎ (+44 71) 938 46 25 (Fax)
Minister Counsellor: Lam Ngoc Bao
First Secretary (Press, Info., Culture and Education):
Ngo Tien Long
First Secretary (Consular Affairs): Nguyen Thi Nhu Hoa [98]

31-02010 Embassy of the Federal Republic of Yugoslavia
5 Lexham Gardens, London W8 5JJ
☎ (+44 171) 370 61 05 ⊠ 928542
⊖ (+44 171) 370 38 38
Ambassador: H E Dr Milos Radulovic
Minister-Counsellor, Deputy Head of Mission:
Radojko Bogojevic
First Counsellor, Economic Affairs: Dr Janko Gogic
Defence Attaché: Capt Milos Ladicorbic
Assistant Defence Attaché: Lieut Col Cedomir Truta
Counsellor, Political Affairs: Mladen Mandic
Counsellor Press and Cultural Affairs: Djordje Mijalkovic
Administrative Attaché: Marko Miljus [98]

31-02030 High Commission for the Republic of Zambia
2 Palace Gate, London W8 5NG
☎ (+44 171) 589 66 55 ✉ ZAMBIANS London
⊠ 263544 zambia g ⊖ (+44 171) 581 13 53
High Commissioner: H.E. Mr Love Mtesa
First Secretary Finance: Lu Poleka
First Secretary Information & Tourism: J. Chikwenda
First Secretary Political: Mr Mkanga
First Secretary Immigration: Mr Mulila
Defence Attaché: Brig.Gen. M.G. Lisita
Second Secretary Political: Mr Mwanambale
Second Secretary Protocol: Mr Nkumbula [97]

31-02040 High Commission for the Republic of Zimbabwe
Zimbabwe House, 429 Strand, London WC2R 0SA
☎ (+44 171) 836 77 55 ⊠ 262115; 262014 zimho g
⊖ (+44 171) 379 11 67
His Excellency the High Commissioner:
H.E. Dr Ngoni Togarepi Chideya
Minister Counsellor: Ms Alice Nyazika
First Secretary: Taonga Mushayavanku
Defence Adviser: Lt Col E. Zabanyana
Defence Admin Assistant: WO1 E. Madzanira
Commercial Counsellor: Samuel Undenge [97]

33 Foreign Forces

33-31530 US Army Research, Development and Standardization Group (UK) (USARDSG-UK)
223 Old Marylebone Road, London NW1 5TH
☎ (+44 171) 514 49 35 [W] http://www.ehis.navy.mil
⊖ (+44 171) 724 14 33
Commander: Col Henry J Atwood (Jr)
☎ (+44 171) 514 49 11
Chief, Standardization Division: Lieut Col Thomas J Kindel
[E] kindel@army.ehis.navy.mil
Chief, European Research Office: Dr Karl Steinbach
■ Liaises with UK Government Agencies in the furtherment of American, British, Canadian and Australian (ABCA) materiel standardization and interoperability. Responds to initiatives and developments from NATO countries and processes requests from US and UK agencies for reciprocal use of equipment for test and evaluation. Coordinates research activities of the US Army in Europe, Africa, Middle East, Former Soviet Union and southwest Asia. Gives financial support to relevant European research efforts. [98]

33-31540 US Defense Contract Management Command-Northern Europe-United Kingdom
(† US Defense Contract Management Area Operations (UK))
1st Floor Hampden Court,
Kingsmead Business Pk, 190 London Rd,
Loudwater, High Wycombe HP11 1JU
☎ (+44 1494) 45 93 75 ⊖ (+44 1494) 45 94 77
Commander: Capt R L Ketts (USN)
Technical Director: R H Johnson
[E] hjohnson@europe.dla.mil [98]

33-45000 US Naval Activities, United Kingdom (COMNAVACTUK)
RAF West Ruislip, Box 60, Ruislip, Middlesex HA4 7UL
☎ (+44 1895) 61 61 03 [E] cnaal@post.nctsl.navy.mil
⊖ (+44 1895) 61 61 77
Commander: Capt Gerald J Corcoran
Chief Staff Officer: Cdr William F Slagle
Operations Officer: Lieut Sharon L Tate
☎ (+44 1895) 61 61 51
Supply Officer: Lieut Cdr William C Lyon
☎ (+44 1895) 61 61 73
Public Affairs Officer: Michael C Sturges-Yebernetsky
☎ (+44 1895) 61 61 03 [98]

34 Multinational Intergovernmental Organizations

34-10150 Allied Naval Communications Agency (ANCA)
Main Building, Whitehall, London SW1A 2HB
☎ (+44 171) 218 21 47 Chairman; 824 40 99 Secretary
⊖ (+44 171) 218 68 53 Chairman; 824 29 78 Secretary
Chairman: Capt. A.C. Lyddon (RN)
Secretary: Cdr J. Wolbaum [97]

34-20050 Commander Standing Naval Force Atlantic (STANAVFORLANT)
c/o HQ Allied Command Eastlant,
Northwood, Middlesex HA6 3HP
☎ (+44 1923) 84 37 63 ⊖ (+44 1923) 84 37 62
Commander: Cdre Peter C Van der Graaf (RNLN)
Chief Staff Officer: Cdr Michael Hellemans (BEN)
Staff Operations Officer: Cdr S Willassen
Deputy Chief of Staff Logistics: Lieut Cdr Joep Westerhof (RNLN)
Staff Public Affairs Officer: Lieut Cdr John M Van Zee (RNLN)
■ On board HNLMS Tromp, September 1997 to April 1998. [98]

34-25000 International Maritime Organization
4 Albert Embankment, London SE1 7SR
☎ (+44 171) 735 76 11 ⊠ 23588
⊖ (+44 171) 735 32 10
Secretary General: William A O'Neil
Head, Information Office: Roger Kohn [E] rkohn@imo.org
■ Provides the machinery for co-operation among Government in the field of governmental regulation and practices relating to technical matters of all kinds affecting shipping engaged in international trade. Encourages and facilitates the adoption of the highest standards in maritime safety, navigation and pollution control and deals with legal matters related to the above. [98]

40 Government Defense Sales/Procurement Organizations

40-10000 Defence Export Services Organisation (DESO)
(Ministry of Defence)
Room 0204, Main Building, Whitehall, London SW1A 2HB
☎ (+44 171) 218 90 00 (switchboard)
⊠ 22241 mod pe g ⊖ (+44 171) 218 70 82
Head Defence Export Services: Charles Masefield
☎ (+44 171) 218 30 42
Director General Marketing: D Bowen
☎ (+44 171) 218 68 28

Military Deputy to Head of DESO: RAdm J F T G Salt
☎(+44 171) 218 30 60
Assistant Under Secretary Export Policy and Finance: Dr A Fox ☎(+44 171) 218 67 48
Private Secretary to Head of DESO: S Meighan
☎(+44 171) 218 30 42
Defence Export Service Secretariat 3: A Greenway
■ Responsible for promoting the export of defence equipment and management contracts; Supports British manufacturers in firm-to-government sales and in negotiating government-to-government sales.

Marketing Services
Metropole Building, Northumberland Avenue, London WC2N 5BL
☎(+44 171) 218 90 00 (switchboard)
Telex 22241 mod pe g ⊖ (+44 171) 807 81 75
Marketing Service Director: M Rickson
☎(+44 171) 807 81 12
Regional Marketing Director 1: C Bentley
☎(+44 171) 807 82 05
Regional Marketing Director 2: J Morrison
☎(+44 171) 807 81 25
Regional Marketing Director 3: Dr D Hughes
☎(+44 171) 807 81 41
Regional Marketing Director 4: J Broadhead
☎(+44 171) 807 81 69
■ Region 1: Kuwait, Eqypt, Qatar, UAE, Oman, Yemen, Bahrain, Jordan, Lebanon, Syria.
Region 2: Indian sub-continent (including Pakistan), Burma, Iran, China, Japan, Vietnam, Cambodia, Korea, Taiwan, Hong Kong, Latin America, Caribbean, Mexico.
Region 3: South East Asia and Australasia including Malaysian Project Office.
Region 4: North America, Europe, Africa, Israel, NATO, UN.

Disposal Sales Agency
Hercules House, 6 Hercules Road, London SE1 7DJ
☎(+44 171) 928 23 45 (switchboard)
Telex 22241 mod pe g
⊖ (+44 171) 928 07 90 (ships); 928 79 76 (technical); 928 50 46 (armaments and aircraft)
⊖ (+44 171) 928 39 74 (contract services); 928 75 19 (industry and recycling)
Chief Executive: M Westgate
[98]

40-10500 Infantry Export Support Team

(part of the Defence Export Services Organisation - DESO)
Warminster Training Centre, Warminster, Wiltshire BA12 0DJ
☎(+44 1985) 22 21 33 ⊖ (+44 1985) 22 22 96
Employees 9 (1997)
Managing Director: Charles Masefield
Operations Director: RAdm Sam Salt
Sales and Marketing Director: Malcolm Rickson
■ The Infantry Export Support Team has access to extensive training and equipment testing facilities.
Companies wishing to obtain the services of the team should contact MDS1 (Infantry) at DESO in London. Demonstrations of equipment using military personnel and training facilities including live firing ranges. Production of still photography and video footage, and presentations as required.
[98]

60 Research Organizations, Institutes & Test Establishments

British Sarozal Ltd 75-09570
Defence Evaluation and Research . . . 20-01000
Meteorological Office 20-01100
RAF School of Aviation Medicine . . . 60-01600
Sonar Research & Development Ltd . . 75-55170

60-00250 British American Security Information Council (BASIC)

Carrara House, 20 Embankment Place, London WC2N 6NN
☎(+44 171) 925 08 62 ⊖ (+44 171) 925 08 61
Employees 11
Director: Daniel Plesch
Asst Dir: Ms Bronwyn Brady
Sr Analyst: Stephen W. Young
Programme Co-ordinator: Ms Nicola Calvert
Admin Asst: Ms Valerie Kemball
■ Information/research charity. Specialist research areas include NATO, OSCE & arms control negotiations, European Security Structures, European Nuclear Forces, US Defence Budget, US Bases/Forces in Europe & the arms trade. Reports are produced regularly aimed mainly at politicians & the media.
[97]

60-00300 Centre for Criminal Justice Studies

University of Leeds, Leeds LS2 9JT
☎(+44 113) 233 50 33 [E] law6cw@leeds.ac.uk
⊖ (+44 113) 233 50 56
Employees 15
Director: Prof. Clive Walker ☎(+44 113) 233 50 22
Deputy Director: Adam Crawford
Deputy Director: David Wall
Member: Ian Brownlee
Member: Dr Josephine Goodey
Member: Ben Fitzpatrick
Member: Peter Seago
Member: Paul Eden
Member: Ms Clare Zurniss
Member: Ms Rachel Edgar
■ Research and study into all aspects of criminal justice systems. Undertakes research projects and arranges conferences and seminars.
Established in 1987.
[97]

60-00350 Cranfield University

Cranfield, Bedfordshire MK43 0AL
☎(+44 1234) 75 01 11 [W] http://www.cranfield.ac.uk
Telex 825072 citech g ⊖ (+44 1234) 75 16 40
♣ £ 85,000,000 Employees 1,200
Chancellor: Lord Kings Norton
Vice Chancellor: Frank R. Hartley
Chairman Helicopter Damage Tolerance: Prof. Philip Irving
Commercial Accountant: Philip Aspinall
☎(+44 1908) 67 29 75
College Administrator: Ron Wingrove
■ Institute of applied research including aeronautical engineering & research, defence systems, propulsion, electronics, communications, data processing, weapon systems & explosives, materials and manufacturing science, management, chemical systems.

College of Aeronautics
Cranfield, Bedfordshire MK43 0AL
☎(+44 1234) 75 01 11 Telex 825072 citech g
⊖ (+44 1234) 75 11 81
♣ £ 2,500,000 Employees 100 Floor area 150,000 sq ft
Chief Executive Officer and Director of Projects: Prof Ian Poll
Chief Designer: Prof. J. Fielding
Head of Cranfield Aerospace: P Lawson
Chief Test Pilot: R. Bailey
■ Aeronautical engineering & research.

School of Mechanical Engineering
Cranfield, Bedfordshire MK43 0AL
☎(+44 1234) 75 00 11
Head: Prof. P. Hutchinson
[98]

60-00400 Engineering and Physical Sciences Research Council (EPSRC)

Polaris House, North Star Ave, Swindon, Wiltshire SN2 1ET
☎(+44 1793) 44 40 00 [E] infoline@epsrc.ac.uk
[W] http://www.epsrc.ac.uk Telex 449466
⊖ (+44 1793) 44 40 10
Chairman: Dr Alan Rudge
Chief Executive: Professor Richard Brook
LS Publications Officer: Ms Juliet Russell
LS Publications Officer: Julian Richards
Media Relations Mgr: Geoff Heaford
■ Research & training in the fields of engineering & physical sciences.
[98]

60-00550 GEC-Marconi Materials Technology Ltd Hirst Division

Elstree Way, Borehamwood, Herts WD6 1RX
☎(+44 181) 732 03 32 Telex 22777 gecbw g
⊖ (+44 181) 732 02 00
Divisional Mgr: Dr D.A. Wilby
Business Development Mgr: C.M. Lennon
Information Services Division: Mrs Angela Tidmarsh
■ Research, development & technical support services. Areas of activity include lasers, image processing, microengineering, stable oscillators, networks, data systems, manufacturing methods & technologies.
[97]

60-00600 GEC-Marconi Materials Technology Ltd

(a division of GEC-Marconi Limited)
Caswell, Towcester, Northamptonshire NN12 8EQ
☎(+44 1327) 35 05 81 Telex 31572 plesca g
⊖ (+44 1327) 35 67 75
♣ £ 20,000,000 (1997) Employees 500 (1997)
Managing Director: Dr D Richardson
Commercial Mgr: J.B. Swift
■ Research, development & production of electronic, opto-electronic & pyroelectric devices & subsystems. GaAs custom MMICs, phased-array radar TX/RX modules. Optical delay lines for microwave signals. OEICs. Solid-state lasers, broadband LEDs, modulators. Optical interconnects. Synthetic diamond windows. Pyroelectric, passive infra-red sensors and arrays.
[98]

60-00700 GEC-Marconi Research Centre

West Hanningfield Road, Great Baddow, Chelmsford CM2 8HN
☎(+44 1245) 733 31 [E] gmrc.enquiries@gecm.com
⊖ (+44 1245) 752 44
Managing Director: M G Ginn
Purchasing Manager: D P Salmond
■ Central research facility for all GEC-Marconi companies. Areas of activity include computer applications, sensors & avionics, radar, space & defence, microwave components, materials, communications, remote sensing.
[98]

60-00870 Institute of Sound & Vibration Research (ISVR)

University of Southampton, Highfield, Southampton SO17 1BJ
☎(+44 1703) 59 22 94 [E] mzs@isvr.soton.ac.uk
Telex 47661 soton un g ⊖ (+44 1703) 59 31 90
♣ £ 4,000,000 (1993) Employees 200
Director: Prof. J.K. Hammond
Public Relations Officer: Mrs M.Z. Strickland
■ Teaching, research & consulting centre for the study of all aspects of sound & vibration. Defence & aerospace activities include work on spacecraft, satellites, aircraft, data processing recording systems, active control, automotive engineering, subjective acoustics & environmental noise.
[98]

60-00900 International Institute for Strategic Studies (IISS)

23 Tavistock St, London WC2E 7NQ
☎(+44 171) 379 76 76; 872 07 70 [E] iiss@iiss.org.uk
[W] http://www.isn.ethz.ch/iiss/ Telex 94081492 mlink g
⊖ (+44 171) 836 31 08
Employees 35
Director: John Chipman
Publications Mgr: Ms Rachel Neaman
Journal Marketing Mgr: Ms Jane Martin
■ An independent centre for research, information & debates on the problems of security, conflict, arms transfers & arms control in the contemporary world.

Oxford University Press
Journals Customer Services
Walton Street, Oxford OX2 6DP
☎(+44 1865) 26 79 17 ⊖ (+44 1865) 26 74 85
Information Contact: Ms Jane Martin
[97]

60-01100 Joint Services Command and Staff College (JSCSC)

Bracknell, Berkshire RG12 3DD
☎(+44 1344) 545 93 ⊖ (+44 1344) 86 16 82
[98]

60-01400 NEL

(† National Engineering Laboratory)
(a subsidiary of NEAG - National Engineering & Assessment Group Ltd)
East Kilbride, Glasgow G75 0QU
☎(+44 1355) 22 02 22 [E] info@nel.co.uk
⊖ (+44 1355) 23 69 30
♣ £ 12,000,000 (1996) Employees 260 (1997)
Director and General Manager: W Paton
Sales and Marketing Manager: Graeme Boyle
Marketing Dept: Marietta Hughes
Manager Defence: A Winski
Pruchasing Manager: Miss F McDonald
■ Independent technology organisation providing research, development, consultance, testing services to the defence, transportation, energy and process industries.
[98]

Contact Reference

60-01600 RAF School of Aviation Medicine
Farnborough, Hampshire GU14 6SZ
☎(+44 1252) 39 38 59 ⊜ (+44 1252) 39 34 69
👥 70
Commandant: Air Cdre A.N. Nicolson
☎(+44 252) 244 61 Ext 2279
Officer Commanding PSF: D P Garland
→ 60
97

60-01800 Roke Manor Research Ltd (RMR)
(a subs. of Siemens Plessey Electronic Systems Ltd)
Roke Manor, Romsey, Hants SO51 0ZN
☎(+44 1794) 83 34 55 ⊜ (+44 1794) 83 34 32
♣ £ 16 000 000 👥 300
Managing Dir: Dr A. Low
Financial Exec.: P. West
PR & Marketing Communications Mgr: Mrs Trudi Morgan
■ A wholly owned Siemens subsidiary specialising in advanced electronics applied research & development covering sensors, communications, simulation & modelling. Range of services include feasibility studies, technical consultancy, concept demonstrators, design for production & evaluation.
97

60-01810 Royal College of Defence Studies
Seaford House, 37 Belgrave Square, London SW1X 8NS
☎(+44 171) 915 48 00 Ⓔ rcds@seaford.demon.co.uk
⊜ (+44 171) 915 49 99
Commander: Lt Gen S C Grant
Secretary: Brig R Tarsnane
Senior Directing Staff Navy: RAdm J H A J Armstrong
Senior Directing Staff Army: Maj Gen P A Chambers
Senior Directing Staff Air Force: AVM P J Goddard
Senior Directing Staff Civilian: AUS N J Thorpe
Librarian: A Brady
■ The college prepares selected senior officers & officials of the UK & certain other countries for high responsibility in the direction & management of defence & security or other related areas of public policy. The course lasts for one year. 75-80 students, of Brigadier & equivalent level, attend each course. About one fifth of the students are civilians.
98

60-01820 Royal Military College of Science (RMCS)
(a faculty of Cranfield University)
Shrivenham, Swindon, Wiltshire SN6 8LA
☎(+44 1793) 78 54 35
Ⓦ http://www.rmcs.cranfield.ac.uk ⊠ 265871
⊜ (+44 1793) 78 39 66
Commandant: Maj Gen A S H Irwin
Deputy Commandant: Brig P Rawlins
Principal: Prof P Hutchinson
Director Administration, Resources and Marketing: Rama Nand-Lal
Faculty Information Officer: Mrs Wendy Randles
Ⓔ randles@rmcs.cranfield.ac. uk
■ A university faculty offering undergraduate, postgraduate and short courses in Science, Engineering, Technology & Management to military as well as civilian students. Research and consultancy facilities are also available on request.
98

60-01830 Royal School of Military Engineering
Headquarters, Chatham, Kent ME4 4UG
☎(+44 1634) 82 22 33 ⊜ (+44 1634) 82 24 19
SO2 G2/G3: G.R.N. Holland
97

60-01850 Rutherford Appleton Laboratory (RAL)
Chilton, Didcot, Oxon OX11 0QX
☎(+44 1235) 82 19 00
Ⓦ http://star-www.rl.ac.uk/ssd.htm1
⊜ (+44 1235) 44 58 48
♣ £ 101,000,000 (1994/95) 👥 1,840
Chairman and Chief Executive: Dr Paul Williams
Commercial Director: Ms Alyson Reed
Space Science Director: Eric Dunford
Research Director: Gordon Walker
Head of Public Relations Space Science Department: W J Curtis
■ Part of the Central Laboratory of the Research Councils (CLRC). Manages and provides technical support for CLRC's space hardware programmes and carries out research in Astronomy, Space Science and Earth Observation. Development of models for radio propagation; high power lasers, (dye, excimer and Ti sapphire) for studying high temperature, high density, plasma; state of the art sensors (neutron, X-Ray, visible and microwave); advanced database techniques, expert systems and data visualisation; a powerful neutron source for determining the structure of materials; microelectronics design (including rad-hard components); instrumentation for Space including environmental test facilities), and the design and fabrication of advanced microstructures.
97

60-02000 Society of Flight Test Engineers (SFTE)
(a division of the Society of Flight Engineers, CA, USA)
European Chapter
c/o B A Wood, DTEO Boscombe Down,
Salisbury, Wiltshire SP4 0JF
☎(+44 1980) 66 25 63
Ⓦ http://www.hughes-ec.com/org/sfte
⊜ (+44 1980) 66 24 94
President European Chapter: Franz Enzinger
☎(+49 8459) 816 46 42
Vice President European Chapter: Ms B A Wood
■ An organisation dedicated to the promotion of common interest and exchange of information in the world of flight test, both civil and military.
98

60-02030 Sowerby Research Centre British Aerospace (Operations) Ltd
PO Box 5, Filton, Bristol BS12 7QW
☎(+44 117) 969 38 31 ⊜ (+44 117) 936 37 33
■ Corporate scientific research centre of British Aerospace (BAe).
98

60-02050 Strategic and Combat Studies Institute (SCSI)
Staff College, Camberley, Surrey GU15 4NP
☎(+44 1276) 41 26 53
⊜ (+44 1276) 41 27 18; 41 26 98
Colonel Defence Studies: Col R A M S Melvin
98

60-02100 University of Essex Human Rights Centre
Wivenhoe Park, Colchester CO4 3SQ
☎(+44 1206) 87 25 58 ⊜ (+44 1206) 87 36 27
Director: Prof. K Boyle
Deputy Director: Dr M Freeman
■ Teaching and research activities in the field of international human rights and humanitarian law.
Established in 1989.
97

60-02200 Vaisala TMI Ltd
(a subsidiary of Vaisala Oy, Finland)
Vaisala House, 349 Bristol Road, Birmingham B5 7SW
☎(+44 121) 683 12 00 ⊜ (+44 121) 683 12 99
Managing Dir: Jonathan Lister
98

65 Industry, Trade & Armed Forces Associations

65-00070 The Association of Consulting Engineers
Alliance House, 12 Caxton Street,
Westminster, London SW1H 0QL
☎(+44 171) 222 65 57 Ⓔ consult@acenet.co.uk
Ⓦ http://www.acenet.co.uk ⊜ (+44 171) 222 07 50
👥 12 🏠 3,500 sq ft
Chief Executive Officer: Hugh Woodrow
Communications Manager: Andy Walker
■ The Association of Consulting Engineers acts in the professional & business interests of its member firms. Members are independent consultants advising their clients & managing on their behalf projects in concept, planning, design, construction supervision & maintenance. Main engineering disciplines are civil, electrical, mechanical, structural & environmental. Member firms provide services related to defence infrastructure.
Affiliated to:
EPCA (for Europe)
FIDIC (for World)
98

65-00100 The Association of Police and Public Security Suppliers (APPSS)
(a division of DMA)
Marlborough House, Headley Road,
Grayshott, Surrey GU26 6LG
☎(+44 1428) 60 77 88 ⊜ (+44 1428) 60 45 67
♣ £ 258,000 (1995) 👥 14
Director General: Alan Sharman ☎(+44 1428) 60 41 41
Director: Jim Gibson
Assistant Director: Peter May
Exhibitions Director: Miss Stella Webb
☎(+44 1428) 60 78 99
Company Secretary: Mrs Elaine A Luck
Marketing Information Officer: Ms Sandra Collett
■ APPSS provides support & advice for all UK companies concerned with selling products & services to Police, Fire & Prison services, Customs & Excise as well as air & seaport security authorities in the UK or overseas. APPSS also provides a free sourcing information service to customers trying to identify suppliers of products & services.
98

65-00200 British Association of Women Police (BAWP)
PO Box 911, Purfleet, Essex RM16 1SZ
Information Contact: Ms Tina Martin
☎(+44 1708) 86 96 29
Information Contact: Ms Carolyn Williamson
☎(+44 121) 236 04 31
■ It aims to increase professionalism in police work; further the utilisation of women in law enforcement and to provide a forum for sharing developments in police administration. Established in 1987.
97

65-00250 British Marine Industries Federation
Meadlake Place, Thorpe Lea Rd,
Egham, Surrey TW20 8HE
☎(+44 1784) 47 33 77 ⊜ (+44 1784) 43 96 78
👥 40
Exec. Chairman: Tony Beechey
Marketing Dir: Denzil Lee
Public Relations & Media Mgr: Catherine Hands-Davies
■ Organisers of the London & Southampton International Boat Shows & Nottingham Boat Show, also European Workboat Exhibition (trade) & the British Marine Trade Show.
97

65-00280 British Naval Equipment Association (BNEA)
4th Floor, 30 Great Guildford St, London SE1 0HS
☎(+44 171) 928 91 99 ⊜ (+44 171) 928 65 99
President: The Right Hon. Lord Younger of Prestwick (KCVO, TD, DL)
Chairman: Capt. J.A.B. Thomas (RN)
Association Dir: Cdr M.F. Whelan (OBE, RN)
■ Defence-related trade association dedicated solely to providing marketing support services to the British naval industry.
BNEA Members:
3M United Kingdom plc
Advanced Military Audio Systems Ltd
Advanced Projects Group
Aeronautical & General Instruments Ltd
Alfa Laval Separation Ltd
Ansell Jones
Babcock Facilities Management Ltd
BAeSEMA Ltd
Bridon Marine Ltd
British Aerospace (Systems & Equipment) Ltd
British Aerospace Defence Ltd
British Maritime Charitable Foundation
British Maritime Technology Limited
Brown Brothers & Co.Ltd
Caird & Rayner Clark
Chernikeeff Instruments Limited
Chesterfield Cylinders Ltd
Clarke Chapman Marine
CompAir Reavell Ltd
Corrintec Ltd
David Brown Special Products Ltd
Devonport Management Limited
Easat Antennas Ltd
EDS Limited
G L Communications Limited
GEC ALSTHOM Engineering Systems Ltd
GEC ALSTHOM Paxman Diesels Ltd
GEC ALSTHOM Ruston Diesels Ltd
GEC-Marconi Limited
Griffin-Woodhouse Ltd

Hall Thermotank
Hamworthy Marine Limited
Insumat Ltd
Kelvin Hughes
Kempsafe Ltd
Kidde Fire Protection Ltd
LEC Marine (Lowestoft) Ltd
Legelec Projects Ltd
Lloyds (Brierley Hill) Ltd
Losalt International Ltd
MacKenzie Tribbeck Associates
MacTaggart, Scott & Co.Ltd
Marine Data Limited
Mirrlees Blackstone Ltd
MTD Ltd
Ocean Fleets Ltd
OMI Logistics Ltd
Orkot Ltd
Power Magnetic & Electronic Systems Ltd
PPI Engineering Group Ltd
Racal Radar Defence Systems Ltd
Racal Thorn Wells Ltd
Railko Ltd
Rolls-Royce Industrial & Marine Gas
Senior Thermal Engineering Ltd
Serck Heat Transfer
Siemens Plessey Systems
Simrad Ltd
Stone Vickers Ltd
Thompson Valves Ltd
Three Quays Marine Services Ltd
UK Hydrographic Office
Varivane Industries Ltd
Vickers PLC - Michell Bearings
Vickers Shipbuilding & Engineering Ltd
Vosper Thornycroft (UK) Ltd
Weir Pumps Ltd
Welin Lambie Ltd
Wichita Co Limited
Widney Aish Ltd
Worchester Controls (UK) Ltd
Wynn Marine Limited
York International Ltd [97]

65-00285 British Security Industry Association (BSIA)

Security House, Barbourne Road, Worcester WR1 1RS
☎ (+44 1905) 214 64 [E] info@bsia.co.uk
⊜ (+44 1905) 61 36 25
👥 14
Chief Executive: David Fletcher
Policy and Public Relations Manager: Tony Makosinski
■ Member companies involved in all aspects of the private security industry: manned guarding; secure transport; safes & locks; CCTV; access control; intruder alarms, physical security equipment & security equipment manufacture. [98]

65-00290 Defence Manufacturers Association of Great Britain Ltd (DMA)

Marlborough House, Headley Road, Grayshott, Hindhead, Surrey GU26 6LG
☎ (+44 1428) 60 77 88 ⊜ (+44 1428) 60 45 67
♧ £ 388,000 (1996) ♣ £ 1,419,000 (1996)
👥 24 (1997)
Dir General: Alan Sharman ☎ (+44 1428) 60 41 41
Deputy Dir General: Mike Evans
Exhibitions Dir: Miss Stella Webb ☎ (+44 1428) 60 78 99
Procurement Information Sea, Air & IT Dir: Robin Kennedy
Procurement Information Commercial Director: George Eynon
Company Secretary: Mrs Elaine Luck ☎ (+44 1428) 60 41 41
Accountant: Mrs Christine Lewry
Market Information Mgr: Brinley Salzmann
Membership Exec.: Miss Julie Southcott ☎ (+44 1428) 60 41 41
Exhibitions Co-ordinator: Miss Lorraine Coonan ☎ (+44 1428) 60 78 99
■ Equipment & services for defence forces throughout the world. DMA provides a free service to government procurement agencies, main contractors & other potential customers for identifying manufacturers/suppliers of a particular product or service within the UK defence industry. [98]

65-00400 Federation of the Electronics Industry (FEI)

Russell Square House, 10-12 Russell Square, London WC1B 5EC
☎ (+44 171) 331 20 00 [E] raxford@fei.org.uk
⊜ (+44 171) 331 20 40
♣ £ 3,900,000 (1996) 👥 35 (1997)
Director General: Anthony Parish
Defence and Capital Systems Director: G Attrill
Aerospace Director: R Axford
Company Secretary: G L J Gill
■ Trade association representing companies in design, development, manufacture & supply of electronic systems, equipment & office products. [98]

65-00500 FEI

(† EEA)
Russell Square House, 10-12 Russell Square, London WC1B 5EE
☎ (+44 171) 331 20 00 [E] feedback@fei.org.uk
[W] http://www.fei.org.uk/eri ⊜ (+44 171) 331 20 40
👥 35+
President: Keith Chapple
Director General: Anthony Parish ☎ (+44 171) 331 20 01
Secretary to the Board: Gordon Gill ☎ (+44 171) 331 20 06
■ Trade association for the information technology, electronics, communications, defence, office equipment & office furniture industries. [98]

65-00600 International Police Association British Section

1 Fox Road, West Bridgford, Nottingham NG2 6AJ
☎ (+44 115) 981 36 38 ⊜ (+44 115) 981 04 89
Chief Executive Officer: A F Carter (BA, MIL)
■ Composed by 91 branches in 11 regions throughout the UK. The International Police Association is an independent body made up of members of the police service. Main purpose is to create bonds of friendship and to promote international co-operation. It also aims to include the development of cultural relations amongst its members, a broadening of their general knowledge and an exchange of professional experience. [97]

65-00700 International Professional Security Association (IPSA)

IPSA House, 3 Dendy Road, Paignton, Devon TQ4 5DB
☎ (+44 1803) 55 48 49
[E] ipsa.secretary@dial.pipex.com
[W] http://ourworld.compuserve.com/ homepages/ipsa
⊜ (+44 1803) 52 92 03
♣ £ 1,000,000 (1996) 👥 8 (1997)
Association Secretary: Mrs C A L Hill
Sales and Marketing Director: P Coker
■ Supports 13 regional councils throughout the UK and members in over 55 countries. [98]

65-01100 The Royal Air Force Benevolent Fund

67 Portland Place, London W1N 4AR
☎ (+44 171) 580 83 43 ⊜ (+44 171) 636 70 05
Chairman: Sir Adrian Squire
Controller: ACM Sir Roger Palin
Administration Director: Gp Capt G A Lucas
Finance Director: Wg Cdr D A F Bailey
Public Relations: Air Cdre C H Reineck
■ The relief of distress or need, actual or potential amongst past & present members of all ranks of the RAF (including the Women's Services), The Royal Auxiliary Air Force, the RAF Reserves & their widows, children & other dependants. [98]

65-01900 Society for Underwater Technology

Head Office
76 Mark Lane, London EC3R 7JN
☎ (+44 171) 481 07 50 ⊠ 886841
⊜ (+44 171) 481 40 01
♧ £ 287 156 (1995) ♣ £ 183 454 (1995) 👥 5
President: Sir Anthony Laughton
Exec. Secretary: I.N.L. Gallett
Conference Organiser: Ms Jean Pritchard ☎ (+44 1224) 82 36 37
Journal Editor: Ms Rowena Morton
■ Multidisciplinary learned society, international membership of individual scientists, engineers & companies with a common interest in underwater technology, ocean science & offshore engineering. Activities include specialist committees; organisation of conferences, courses & seminars; publication of proceedings, quarterly journal & newsletter. Sponsors of Oceanology International since 1969.
Aberdeen Office
Innovation House, Exploration Drive, Bridge of Don, Aberdeen AB23 8GX
☎ (+44 1224) 82 36 37 ⊜ (+44 1224) 82 02 36 [98]

65-01950 The Society of British Aerospace Companies Ltd (SBAC)

60 Petty France, Victoria, London SW1H 9EU
☎ (+44 171) 227 10 00 ✉ THESBAC LONDON SW1
⊠ 262274 ⊜ (+44 171) 227 10 67
Director General: David Marshall
President: M.J. Turner
Vice President: G.F. Page
Deputy President: N.V. Barber
Treasurer: R W R McNulty
Secretary: T G Sidebottom
Head Public Relations & Marketing: Mrs P.H. Cox
■ Trade Association of the United Kingdom's Aerospace Industry & organiser of the Farnborough International Exhibition & Flying Display. [97]

70 Financial Holding Groups, Parent Companies

70-00030 Aardvark Holdings Ltd

Shevock Farm, Insch, Aberdeenshire
☎ (+44 1464) 82 01 22 ⊠ 73509 aard g
⊜ (+44 1464) 82 09 85
Managing Director: David Sadler
Marketing Director: John Belchamber
■ Manufacture and supply of mine clearance and detection equipment, mine training simulators, mobile runway division barriers, multi-terrain vehicles, trailers and mine countermeasure related products. Advice and training services.
Member Companies:
Aardvark Active Barriers Ltd
Aardvark Clear Mine Ltd
Aardvark Commercial Contracts Ltd
Aardvark Training Schools Ltd [98]

70-00120 Adwest Group plc

Power Systems (Defence) Division
Woodley, Reading, Berks RG5 4SN
☎ (+44 1734) 69 71 71 ⊠ 848405 adgrp g
⊜ (+44 1734) 69 01 21
♧ £ 14 852 000 (1995) ♣ £ 36 655 000 (1995) 👥 714
Group Managing Dir: G.R. Menzies
Chief Exec. Defence Div.: H.R.S. Truscott
Subs. with defence business:
MPE Ltd
Air-Log Ltd
Abbott Electronics Ltd
Abbott Electronics Inc., USA
Field & Grant Ltd
Mawdsleys Ltd [97]

70-00130 Alvis plc

215 Vauxhall Bridge Road, London SW1V 1EN
☎ (+44 171) 821 80 80 ⊜ (+44 171) 931 74 33
♧ £ 19,000,000 ♣ £ 101,400,000 (1995) 👥 582
🏭 1,000,000 sq m
Chairman & CEO: Nicholas Prest
Finance Dir: Stuart Mitchell
Group Company Secretary: Tony Pearson
Publicity/Marketing Services Mgr: N.C. Turner
■ Design, manufacture and sale of electro-optical equipment, tracked fighting vehicles, transmissions, specialist wheeled vehicles, ordnance disposal equipment, simulators and other equipment and components for the defence and aerospace industries.
Group members & subsidiaries:
Alvis Industries Ltd
Alvis Vehicles Ltd
Alvis Transmissions Ltd
Alvis Logistics Ltd
Alvis Unipower Ltd
Alvis De-Mil Systems Ltd
Automotive Export Supplies Ltd
Self-Changing Gears Ltd
United Scientific Instruments Ltd
Associate Companies:

ACP Metal Finishing Pte Ltd
Avimo Singapore Ltd
Avimo Precision Optics Pte Ltd
Avimo Alvis Aerospace Ltd [98]

70-00230 BICC Group
Group Head Office
Devonshire House, Mayfair Place, London W1X 5FH
☎ (+44 171) 629 66 22 ⊜ (+44 171) 409 00 70
♧ £ 498,000,000 (1996) ♣ £ 4,668,000,000 (1996)
👥 37,500 (1996)
Chairman: The Viscount Weir
Chief Executive: Alan Jones
Head Corporate Communications: Tim Sharp
Corporate Communications Manager: Ms Nina Daniels
Subsidiaries with defence business:
Balfour Beatty Limited
BICC Brand-Rex Limited [98]

70-00247 T.T. Boughton & Sons Ltd
(† The Boughton Group)
Bell Lane, Amersham, Bucks HP6 6PE
☎ (+44 1494) 76 44 11 [Tx] 83132 ttbam g
⊜ (+44 1494) 76 52 18
♣ £ 30 000 000 👥 390 (group)
Chairman: Trafford Boughton
Managing Dir: Michael Robertson-Smith
Director: Stephen Boughton
Technical Dir: James Hopkinson
Dir & Company Secretary: Richard Boughton
Non-Exec. Dir: Ms Pamela Coles
Subs. with defence businesses:
Reynolds Boughton (Devon) Ltd
Reynolds Boughton Ltd [97]

70-00270 Bridport-Gundry plc
The Court, West Street, Bridport, Dorset DT6 3QU
☎ (+44 1308) 45 66 66 ⊜ (+44 1308) 45 66 05
Managing Director: J Bowden
Sales: J R Laurence
Subsidiary companies:
Bridport Aviation Products
Britannic Aviation [98]

70-00275 British Aerospace plc (BAe)
Corporate Headquarters
Warwick House, PO Box 87,
Farnborough Aerospace Centre,
Farnborough, Hampshire GU14 6YU
☎ (+44 1252) 37 32 32 [Tx] 859598 baefbr g
⊜ (+44 1252) 38 30 00
♣ £ 6,464,000,000 (1996) 👥 44,000 (1995)
Chairman: Bob Bauman
Chief Executive: Sir Richard H Evans
Group Managing Director (Commercial Aerospace): Michael Turner
Group Managing Director (Defence): John Weston
Purchasing Executive: Alastair Henderson-Begg
Finance Director: Richard Lapthorne
Director of Communications: Locksley Ryan
Corporate Communications: Ms Vanessa Lugger
■ Contractor for major defence and aerospace projects. Manufacture and supply of military aircraft, guided weapon systems, ordnance systems and support services. The majority of the Group's UK operations in both defence and commercial aerospace are conducted by British Aerospace (Operations) Ltd.
Subsidiaries and investments:
BAeSEMA Ltd (50%)
Eurofighter Jagdflugzeug GmbH, Germany (33%)
Heckler & Koch GmbH, Germany
Matra BAe Dynamics SAS, France (50%)
Saab-BAe Gripen AB, Sweden (50%)
Reflectone Inc, USA (48%)
Royal Ordnance plc [98]

70-00290 BTR plc
Registered Office
Silvertown House, Vincent Square, London SW1P 2PL
☎ (+44 171) 834 38 48 [W] http://www.btrplc.com
[Tx] 22524 btrldn g ⊜ (+44 171) 834 38 79
♣ £ 8,413,000,000 (continuing sales) (1996)
Chairman: E O M Eilledge
Chief Executive: I C Strachan
Executive Director: P S Aiken
Executive Director: C R Burns
Executive Director: P H M Buysse
Executive Director, Finance: K A O'Donovan
Executive Director: J S Thompson
■ Global manufacturing and engineering group. The Automotive group manufactures and supplies automotive components which are mainly engineered polymers, including sealing systems, anti-vibration systems and drivetrain systems. The Power Drives group manufactures and supplies high performance electric motors, industrial gear units, couplings, roller and conveying chain. The Process Control Group produces valves, actuators, metering systems and industrial batteries. → 75 [98]

70-00320 Chemring Group plc
1480 Parkway, Whiteley, Fareham, Hampshire PO15 7AF
☎ (+44 1489) 88 18 80
[W] http://www.marinedata.co.uk/chemring
⊜ (+44 1489) 88 11 23
♧ £ 1,239,000 (1995) ♣ £ 71,112,000 (1995) 👥 1,341
Chairman: Sir William Barlow
Life President: Ian M. Fairfield (CBE)
Group Managing Dir: David R. Evans
Countermeasure Programmes Dir: Victor A. Pheasant
Non-Exec. Dir: Sir William Barlow (FEng)
Non-Exec. Dir: Gen.Sir John Stibbon (KCB, OBE)
Non-Exec. Dir: M.S.D. Hanlon
Marketing Director: Martin Walmsley
Financial Dir: R.J. Gibbs
Business Development Dir: D.F. Whyte
Company Secretary: Iain Leighton
Publicity Mgr: Ms M.D. Key
Subsidiaries:
Chemring Ltd → 75
Pains-Wessex Ltd → 75
Vacuum Reflex Ltd → 75
Octavius Hunt Ltd → 75
Haley & Weller Ltd → 75
Nova Marine Systems Ltd → 75
Kembrey Engineering Ltd
Kembrey Wiring Systems Ltd
McMurdo Ltd
Pains Wessex (Australia) Pty Ltd
Ronstan International Pty Ltd
Cirra S.A.R.L.
Hutchwilco Ltd
Alloy Surfaces Company Inc
Microsulis Ltd [98]

70-00345 J.S. Chinn Holdings Ltd
Coventry Road, Exhall, Coventry, West Midlands CV7 9FT
☎ (+44 1203) 36 30 80 ⊠ CHINDIT [Tx] 31303 chinn g
⊜ (+44 1203) 64 49 33
♧ £ 6,000,000 ♣ £ 13,000,000 (1995) 👥 333 (1995)
⌂ 171,000 sq ft
Chairman: J S Chinn
Director: J K Atkinson
Chief Exec. & Financial Dir: G Gascoigne
Company Secretary: R P Hipkins
■ Subcontractors in precision fabrication & machining of components & complete assemblies for aero engines, airframe, satellites, missiles & underwater weapons & a wide range of aerospace applications.
Subsidiaries:
J.S. Chinn & Co. Ltd
J.S. Chinn Project Engineering Ltd
A.O. Henton Engineering Co. Ltd
J.S. Chinn Engineering Co. Ltd
Colledge & Morley (Gears) Ltd [98]

70-00350 Cobham plc
(† FR Group)
Brook Road, Wimborne, Dorset BH21 2BJ
☎ (+44 1202) 88 20 20 [E] smartr@cobham.com
[Tx] 41247 ⊜ (+44 1202) 84 05 23; 94 01 PR/Exhibitions
♣ £ 227,700,000 (1995) 👥 3,457 (1995)
Chairman: Sir Michael Knight
Chief Executive: G F Page
Finance Director: A G Irwin
Executive Director/MD (FRA): C F Jones
Executive Director/MD (FRL): R H L Clark
Executive Director/MD (Chelton): G C Cooper
Non Executive Director: Sir Michael Cobham
Non Executive Director: M J Smith
Non Executive Director: The Lord Rockley
Non Executive Director: J W Edington
Company Secretary: J M Pope
Group Public Relations/ Exhibitions Executive: R Smart
■ Cobham plc designs and manufactures equipment, specialised systems and components supplied primarily to the aerospace, defence, industrial and communications markets - and operates, maintains and modifies aircraft, particularly in relation to special mission flight operations. → 75 [98]

70-00360 Costain Group PLC
111 Westminster Bridge Rd, London SE1 7UE
☎ (+44 171) 705 84 44 ⊠ COSDON London SE1
[Tx] 8811804 ⊜ (+44 171) 705 85 99
♣ $ 1 200 000 000 👥 11 900
Chairman: Sir Christopher Benson
Group Chief Exec.: A.C. Lovell
Head of Corporate Communications: T.N.C. Westman
Press Officer: B. Davis
■ International engineering & construction, mining & property group.
Subs. with defence related business::
Costain Building & Civil Engineering Ltd [97]

70-00400 Daimler-Benz UK plc
(a subsidiary of Daimler-Benz AG, Germany)
25 St James's Street, London SW1A 1HA
☎ (+44 171) 839 89 98 ⊜ (+44 171) 839 92 79
Head of Media Relations: Eckhard Zanger
■ Aerospace, propulsion systems, defence systems, automation energy systems technology and information technology services. [98]

70-00550 The EIS Group plc
6 Sloane Square, London SW1W 8EE
☎ (+44 171) 730 91 87 [Tx] 266717
⊜ (+44 171) 730 22 71
♣ £ 540,000,000 (1996) 👥 6,500
Group Sales & Marketing Dir: J.G. Masters
Subsidiary companies:
Horstman Defence Systems Ltd → 75
Mollart Universal Joints Ltd
Flexibox Ltd → 75
Beagle Aircraft Co. Ltd
Zwicky Engineering Ltd → 75
C.F. Taylor (Metalworkers) Ltd
AB Precision (Poole) Ltd
Kontak Manufacturing Co. Ltd
AEM (Stanstead)(Ramsgate) Ltd
Davall Gears Ltd
C&F Millier Ltd
Bradley Electronics Ltd
Aerocontracts Ltd
Aeronautical & General Instruments Ltd (AGI)
Flight Spares Ltd
Compstock Ltd
Stuart Electronics Ltd [98]

70-00610 Fairey Group plc
Station Road, Egham, Surrey TW20 9NP
☎ (+44 1784) 47 04 70 ⊜ (+44 1784) 47 08 48
♣ £ 246,000,000 (1996) 👥 3,000 (1996)
Chairman: Sir Robin Biggam
Vice Chairman: Eric Swainson
Chief Executive: John Poulter
Commercial Dir & Group Secretary: Paul Boughton
Group Financial Director: J G Zacharias
Operations Director: J C Webster
Director: R Williams
Director: Keith Mackrell
Public Relations Mgr: Derek Thurgood
Group companies with defence business:
Arcom Control Systems Ltd
Fairey Hydraulics Ltd
Fairey Industrial Ceramics Ltd
Fairey Arlon Ltd
Fairey Microfiltrex Ltd [98]

70-00735 GEC-Marconi Ltd
(† The Marconi Co. Ltd)
(a subs. of The General Electric Co. plc)
Administrative Headquarters
The Grove, Warren Lane, Stanmore, Middx HA7 4LY
☎ (+44 181) 954 23 11 [Tx] 22616 mardef g
⊜ (+44 181) 954 78 08
👥 40,000 (worldwide)
Joint Chairman: Rt Hon. Sir Geoffrey Pattie (MP)
Joint Chairman: Dr I.G. MacBean
Managing Dir: Peter Gershon
Public Relations Dir: Alan Tull
■ Research, design, development, manufacture & in-service support (including prime contractorship, project management & complete integrated logistic support capabilities) relating to advanced materials, air defence systems, avionics, combat systems, command & control systems, crypto, electronic warfare, electro-optics, fire control systems, military communications, missiles & precision-guided munitions, naval shipbuilding, power systems, radar, remotely piloted vehicles, simulation, sonar, space systems & thermal imaging.

UK Subsidiaries:
GEC-Marconi Avionics
GEC-Marconi Radar & Defence Systems Ltd
GEC-Marconi Communications
GEC-Marconi S3I (Simulation & Software Integration)
GEC-Marconi Infra-Red
GEC-Marconi Materials Technology
GEC-Marconi Secure Systems
GEC-Marconi Underwater Weapons
Matra Marconi Space (UK)
EASAMS
European Subsidiaries:
Marconi SpA, Italy
Marconi Kominikayson AS, Turkey
North American Subsidiaries:
GEC-Marconi Avionics Inc.
GEC-Marconi Dynamics Inc.
GEC-Marconi Electronic Systems Corp.
GEC-Marconi Materials Corp.
Marconi Communications Inc.
Canadian Marconi Company, Canada
Cincinnatti Electronics Corp.
Lear Astronics Corp.
Australasian Subsidiary:
GEC-Marconi Systems Pty
97

70-00740 The General Electric Company plc (GEC)
1 Stanhope Gate, London W1A 1EH
☎(+44 171) 493 84 84 ✉ POLYPHASE London W1
Telex 22451 ⊜ (+44 171) 493 19 74
♣ £ 10,990,000,000 (1996)
Employees 127,000 (worldwide 1996)
Managing Director and Chief Executive Officer: George Simpson
Community Relations Director: Ms Sara Morrison
Marketing Director: Sir Geoffrey Pattie
Strategic Planning: Jack Fryer
Personnel Director: Rob Meakin
Technical Director: Dr David Grant
Publicity and Public Relations Manager: Philip Treloar
Information Officer: Miss Y Clarke
■ Defence systems, avionics, warships, sensors, radio, radar, command and control systems, crypto, materials, satellites, torpedoes and sonar.
Subsidiaries with defence business:
GEC-Marconi Ltd (and related subs.) → 70
GPT Ltd → 74
GEC Plessey Semiconductors → 75
Marconi Electronic Devices Ltd → 75
EEV Ltd → 75
Marconi Instruments Ltd → 75
98

70-00750 GKN plc
Ipsley House, Ipsley Church Lane, PO Box 55, Redditch, Worcester B98 0TL
☎(+44 1527) 51 77 15 ⊜ (+44 1527) 51 77 00
♧ £ 349,000,000 (1995) ♣ £ 3,304,700,000 (1995)
Employees 31,100 (worldwide 1995)
Chairman (non-executive): Sir David Lees
Chief Executive: C K Chow
Human Resources Director: B D Insch
Managing Director GKN Automotive and Agritechnical Products: T C Bonner
Managing Director GKN Aerospace and Special Vehicles: D J Wright
Managing Director GKN Industrial Services: M Beresford
Finance Director: David J Turner
Company Secretary: G Denham
■ A global group of companies engaged in the design, development and manufacture of automotive and agritechnical components and of aerospace and defence products, also the provision of a range of industrial and distribution services. Core businesses are in automotive driveline systems, in the rapid reaction segment of the defence industry through helicopters and light armoured vehicles and in pallet and container pooling.
Subsidiaries with defence business:
GKN Defence Ltd → UK 75
Glover Webb Ltd → UK 75
GKN Westland Aerospace Division → UK 75
GKN Westland Helicopters Division → UK 75
GKN Westland Technologies Division → UK 75
London Office
7 Cleveland Row, London SW1A 1DB
☎(+44 171) 930 24 24 ⊜ (+44 171) 930 32 55
Corporate Dir Public Affairs: N. de Jongh
☎(+44 171) 839 36 38 (fax)
98

70-00760 GKN Westland
(a subsidiary of GKN plc)
Yeovil, Somerset BA20 2YB
☎(+44 1935) 47 52 22 ✉ AIRCRAFT Yeovil
Telex 46277 whyleo g ⊜ (+44 1935) 70 21 31
♧ £ 44,200,000 Employees 8,500
Chairman: Sir David Lees
Chief Exec. GKN Westland Group: D.J. Wright
Financial Dir: Christopher J. Bunker
Commercial Director: A D P Milne
Ext'l Affairs Dir: G.N. Cole
Public Affairs Dir: Chris Loney
Strategic Planning Dir: J. McGowan
Subsidiaries:
GKN Westland Aerospace Ltd
GKN Westland Helicopters Ltd
GKN Westland Technologies Ltd
97

70-00780 Graseby plc
Lynton House, 7/12 Tavistock Square, London WC1H 9LT
☎(+44 171) 383 30 60 ⊜ (+44 171) 383 24 95
♣ US$ 200,000,000 (1996) Employees 1,270 (1996)
Chief Executive: Paul J. Lester
Financial Dir: Lawrence Irving
Human Resources Dir: John Cooper
Secretary to Group Chief Exec.: Mrs Lu Firth
Subsidiary companies:
Graseby Analytical Ltd
Graseby Dynamics Ltd
Graseby Ionics Ltd
Graseby Microsystems Ltd
Graseby Infrared Ltd
Graseby Security Ltd
Graseby Specac Ltd
98

70-00850 Howden Airdynamics Group Ltd
(a Howden Group company)
111 Windmill Rd, Sunbury-on-Thames, Middx TW16 7EF
☎(+44 1932) 76 58 22 Telex 929515 airsrw g
⊜ (+44 1932) 76 10 98
Managing Dir: M.J. Cooper
Contact: Gwyn Jones
Group companies:
Airscrew Howden → 75
Howden Aircontrol → 75
Howden Wade → 75
97

70-00900 Hunting plc
3 Cockspur Street, London SW1Y 5BQ
☎(+44 171) 321 01 23 E PR@hunting.plc.uk
W http://www.hunting.plc.uk ⊜ (+44 171) 839 20 72
♣ £ 1,127,900,000 (1995) Employees 12,680
Chairman: R H Hunting
Managing Dir: K W Miller
Head of Public Relations: Christopher R Simmonds
Subsidiaries and associates with defence: business
Hunting Aviation Ltd → 80
Hunting Defence Ltd → 70
Hunting Engineering Ltd → 75
Hunting-BRAE Ltd → 80
AWE plc
MLRS Europaische Produktions GmbH, Germany
Irvin Group Ltd → 70
Halmatic Ltd → 75
98

70-00910 Hunting Defence Ltd
(a subsidiary of Hunting plc)
Reddings Wood, Ampthill, Bedfordshire MK45 2HD
☎(+44 1525) 84 10 00 ⊜ (+44 1525) 40 58 61
♣ £ 414,000,000 (1996) Employees 7,150 (1996)
Chairman: Field Marshal the Lord Vincent (GBE, KCB, DSO)
Chief Executive Officer: T L Grievson (CBE)
Company Secretary: J W J Jewell
Manager, Marketing and Public Relations: C I P Martin
■ Holding company for Defence Division subsidiaries of Hunting plc. → 80
98

70-00940 Hyder plc
(† Welsh Water plc)
Alexander Gate, PO Box 295, Rover Way, Cardiff CF2 2UE
☎(+44 1222) 50 06 00 ⊜ (+44 1222) 58 56 00
Contact: Brian Roussel
Contact: Hugh Thomas
■ Infrastructure services worldwide. → 80
98

70-00950 IMI plc
Headquarters
PO Box 216, Witton, Birmingham B6 7BA
☎(+44 121) 356 48 48 Telex 336771 imikyn g
⊜ (+44 121) 356 35 26
♣ £ 1 322 000 (1995) Employees 17 031
Chairman: Sir Eric Pountain
Managing Dir: G.J. Allen
Exec. Dir: R. Amos
Finance Dir: A.L. Emson
Exec. Dir: R.B. Pointon
Exec. Dir: N.C. Paul
Public Relations Mgr: G.F. Whelan
■ Main product areas - building products, drinks dispensers, fluid power, special engineering.
IMI companies with defence production:
Eley Ltd → 75
IMI Components Ltd
IMI Yorkshire Alloys Ltd → 75
IMI Marston Ltd → 75
IMI Norgren Ltd → 75
Timet Ltd → 75
97

70-00975 Irvin Group Ltd
(† Irvin Aerospace Ltd)
(a subsidiary of Hunting Defence Ltd)
Reddings Wood, Ampthill, Bedfordshire MK45 2HD
☎(+44 1525) 84 11 48 ⊜ (+44 1525) 84 03 43
♣ £ 31,000,000 (Group) Employees 780
Chairman: T.L. Grievson (CBE)
Chief Exec.: R.J. Whysall
Public Relations Co-ordinator: Mrs J. Adams
Subs. with defence business:
Irvin Aerospace Ltd
Irvin Industrie Spa, Italy
Irvin Industries Canada Ltd
Irvin Industries Inc., USA
98

70-00980 Kvaerner plc
International Headquarters
St James's House, 23 King Street, London SW1Y 6QY
☎(+44 171) 766 20 00 ⊜ (+44 171) 766 23 00
■ Head Office for Kvaerner ASA, Norway
98

70-00990 Leafield Holdings Ltd
Leafield Industrial Estate, Corsham, Wiltshire SN13 9SS
☎(+44 1225) 81 07 71 ⊜ (+44 1225) 81 06 14
♣ £ 9,000,000 (1996) Employees 120 (1996)
Managing Director: R L Tonge
Finance Director: S A Whadcock
Manufacturing Director: P A D Munro
Engineering Director: R A Dyke
Subsidiaries::
Leafield Engineering Ltd
Leafield Logistics & Technical Services Ltd
98

70-01100 Logica plc
Stephenson House, 75 Hampstead Rd, London NW1 2PL
☎(+44 171) 637 91 11 W http://www.logica.com
⊜ (+44 1932) 86 91 04
♣ £ 261 576 000 (1995) Employees 3 900
Managing Dir: Dr Martom Read
Corporate Business Development Dir: Mario Anid
UK Sales & Marketing Dir: Gerry O'Keefe
Operations & Technical Dir: Laurence Julien
Public Relations Admin: Ms Anna Bailey
■ Independent computer software, systems, services & consultancy company.
London Office
51-53 Great Marlborough Street, London W1V 1HL
☎(+44 171) 637 91 11
London Office
84-86 Great Portland Street, London W1N 5PA
☎(+44 171) 637 91 11
Leatherhead Office
Medina House and Byron House, Business Park n°4, Randalls Way, Leatherhead, Surrey KT22 7TW
☎(+44 171) 637 91 11
Cobham Office
Cobham Park - Downside Road, Cobham, Surrey KT11 3LG
☎(+44 171) 637 91 11
Brentwood Office
Regent House - Hubert Road, Brentwood, Essex CM14 4JE
☎(+44 171) 637 91 11
Cambridge Office
Betjeman House, 104 Hills Road, Cambridge CB2 1LQ
☎(+44 171) 637 91 11

Contact Reference

Bristol Office
North Quay - Temple Back, Bristol BS1 6FL
☎(+44 171) 637 91 11

Stockport Office
Hilton House - Lord Street, Stockport SK1 3NA
☎(+44 171) 637 91 11

Aberdeen Office
2 Queen's Gardens, Aberdeen AB1 6YD
☎(+44 171) 637 91 11 [97]

70-01280 LucasVarity plc
(† Lucas Industries plc)
Brueton House, New Road,
Solihull, West Midlands B91 3TX
☎(+44 121) 704 51 71 ⊠ 335334
⊖ (+44 121) 711 27 36
♣ £ 4,600,000,000 (1996) 👥 57,059 (worldwide)
Chairman: Sir Brian Pearse
Chief Executive Officer: Victor Rice
Managing Director Aerospace: Ken Maciver
Managing Director Light Vehicle Braking Systems: Steve Lunn
Managing Director Heavy Vehicle Braking Systems: Leighton Quinn
Managing Director Diesel Systems: Dominique Chauvin
Managing Director Electrical and Electronic Systems: John Plant
Managing Director Aftermarket Operations: John Anthony
Finance Director: Neil Arnold
Personnel Director: Howard Chandler
Director Customer Satisfaction & Quality: Kelvin Kellaway
Director Corporate Communication: Nick Jones
Manager Corporate Communication: Martin Butler
☎(+44 121) 704 51 71
→ 75 [98]

70-01340 Meggitt plc
(Holding Company)
Farrs House, Cowgrove, Wimborne, Dorset BH21 4EL
☎(+44 1202) 84 78 47 ⊖ (+44 1202) 84 24 78
♧ £ 10,852,000 (1995) ♣ £ 235,000,000 (1995)
👥 4,000 (1995)
Chairman (Non-Executive): K H Coates
Chief Executive: M A Stacey
Non-Executive Director: Sir Alan Cox
Non-Executive Director: Sir Donald Spiers
Marketing Director: M D Farge
Group Finance Director: Terry Twigger
Group Company Secretary: Philip Green
■ Holding company for a group of companies whose activities include controls, aerospace and electronics.

Subsidiaries & affiliates:
Avica Equipment → 75
Endevco Corporation → 70
Meggitt Avionics Inc. → USA 75
Target Technology Ltd → 75
Avica US → 75
Bestobell Aviation → 75
Meggitt Avionics → UK 75
Cartwright Electronics Inc, USA
Southwest Aerospace → USA 75
Silicone Engineering UK [98]

70-01345 Messier-Dowty International
(Joint Venture TI Group UK, SNECMA France)
Lambourn Court, Abingdon, Oxfordshire OX14 1UH
☎(+44 1235) 55 06 77 ⊠ 83371
⊖ (+44 1235) 55 31 57
♣ US$ 400 000 000 👥 2 300
Chairman & Chief Exec. Officer: L.A. Edwards
Chief Operating Officer: D. Paris
Chief Financial Officer: B Wasson
Commercial/Legal Dir: M. Steel
Marketing/Strategy Dir: P. Couteaux
Communications & Admin Dir: T. Davis
Engineering Director: J-L Enqerand
Product Support Dir: M. Hurst
Production Director: P Martin
Deputy Finance Director: G de Chaisemartin
■ Design, manufacture & test equipment for a wide range of new aircraft programmes including the Airbus A330/A340, Canadair Challenger, Falcon 2000, IPTN N250, McDonnell Douglas F/A-18, Rafale & Eurofighter 2000. [98]

70-01350 ML Holdings plc
644 Ajax Avenue, Slough, Berkshire SL1 4BQ
☎(+44 1753) 52 38 38 ⊖ (+44 1753) 53 23 21
♣ £ 101,800,000 (1997) 👥 947 (1997)
Chairman: T W B Sallitt (CBE)
Group Chief Executive Officer: D F Crook
Non-Executive Director: J D W Goodman (CB)
Non-Executive Director: Joe Dowling
Non-Executive Director: D L Tucker
Finance Director: J M Bryson (MBA, FCA)
Group Secretary: J D Loveridge
■ Activities include composite structures, safety & survival, defence & component distribution.

Aerospace & defence companies:
ML Aerospace Division
Slingsby Aviation Ltd → 75
Slingsby Engineering Ltd → 75
ML Lifeguard Equipment Ltd → 75
ML Aviation Ltd → 75
ML Wallop Defence Systems Ltd → 75 [98]

70-01370 MS International plc
Carr Hill, Balby, Doncaster, S.Yorks DN4 8DH
☎(+44 1302) 32 21 33 ⊖ (+44 1302) 36 93 29
♣ £ 28 800 000 👥 617
Chairman: Michael Bell

Division with defence business:
MSI Defence Systems Ltd [97]

70-01400 The Nitor Group Ltd
Nitor House, Great George Street,
Godalming, Surrey GU7 1EE
☎(+44 483) 42 62 99 [E] nitor@compuserve.com
⊖ (+44 483) 41 59 53
♧ £ 181,000 (1996) ♣ £ 2,200,000 (1996)
👥 20 (1996) 🏭 250 sq m
Joint Managing Director: Martin E Read
Joint Managing Director: M B Pegrum
Financial Director: R J W Rickman
■ Group of companies specialising in the establishment and support of specialist training and range complexes.

Subsidiaries:
OTEC Special Products Ltd
OTEC Support Services Ltd
Sector Ltd
UK Training Ltd
Nitor Far East Sdn Bhd
Quest Research Limited [98]

70-01700 Pilkington plc
Group Corporate Affairs
Prescot Road, St Helens WA10 3TT
☎(+44 1744) 288 82 ⊖ (+44 1744) 69 26 60
♣ £ 2,872,000,000 (1996)
👥 39,100 (worldwide) (1996)
Chairman: Nigel Rudd
Group Chief Executive: Roger Leverton
Executive Director: Peter Grunwell
Executive Director: Glen Nightingale
Executive Director: Andrew Robb
■ Flat and safety glass products. Defence related activities include optronics, aircraft transparencies and ultra-thin glass for the electronics industry. → 75

Subsidiaries/associated undertakings include:
Pilkington Optronics Ltd
Pilkington Aerospace Inc, USA
Pilkington Aerospace Ltd
Pilkington Micronics Ltd
Pilkington Special Glass Ltd [98]

70-01805 Racal Electronics plc
Western Road, Bracknell, Berkshire RG12 1RG
☎(+44 1344) 38 80 00 [E] racal_pr@compuserve.com
[W] http://www.racal.com ⊖ (+44 1344) 38 80 61
♣ £ 938,000,000 (1994/1995) 👥 15,000
Chairman: Sir Ernest T Harrison
Chief Executive: David C Elsbury (OBE)
Senior Press Relations Officer: C B Talago
■ Principal activities: radio communications, data communications, radar defence systems, command, control, communications and intelligence systems, avionics, airborne navigation, marine electronics, precise positioning and energy technology. Other important activities: acoustics, antennas, automatic and diagnostic test systems, data and communications recording, electronic instrumentation, health and safety, intruder detection, logistic support, microwave components and systems.

Racal Group Companies:
Racal Avionics Ltd → 75
Racal-Datacom Ltd → 80
Racal Radar Defence Systems Ltd → 75
Racal Radio Ltd → 75
Racal-Comsec Ltd → 75
Racal Positioning Systems Ltd
Racal Acoustics Ltd → 75
Racal Antennas Ltd → 75
Racal Safety Ltd → 75
Racal Recorders Ltd → 75
Racal Research Ltd → 80

Corporate Communications Centre
Central Publicity and Information Services
Western Road, PO Box 3620,
Bracknell, Berkshire RG12 1WL
☎(+44 1344) 38 80 00 ⊖ (+44 1344) 38 80 61
Director Corporate Relations: Richard Poston
Group Public Relations Manager: David Nicholas
Press Office Manager: Nigel Edwards [98]

70-01815 Ricardo Group plc
(† Ricardo International plc)
1 Watling Drive, Sketchley Business Park,
Hinkley, Leicestershire LE10 3EY
☎(+44 1455) 63 63 44 ⊖ (+44 1455) 23 06 85
♧ £ 15,000,000 (1996) ♣ £ 95,000,000 (1996)
👥 1,400 (1996)
Chairman: Sir Philip Foreman
Chief Executive Officer: Rodney Westhead
Company Secretary: Ronald Severn
■ Automotive and engines.

Subsidiaries:
Ricardo Test Automation
Ricardo Consulting Engineers
FFD-Ricardo [98]

70-01930 Sema Group plc
Registered Office
233 High Holborn, London WC1V 7DJ
☎(+44 171) 830 44 44 ⊖ (+44 171) 830 18 30
♣ £ 927,000,000 (1996) 👥 14,000 (1996)
Chairman: A. Barrera de Irimo
Managing Dir: P.S.E. Bonelli
Exec. Vice President: Frank Jones
■ Systems integration outsourcing & consulting, servicing all major markets including defence, public sector, telecommunications, transport & finance. [98]

70-01940 Siebe plc
Saxon Hse, 2-4 Victoria St, Windsor, Berkshire SL4 1EN
☎(+44 1753) 85 54 11 ⊠ 848005
⊖ (+44 1753) 84 06 38
♧ £ 120 000 000 ♣ £ 2 117 200 000 👥 32 309
Chairman: Barrie Stephens
Managing Dir & CEO: Allen M. Yurko
Company Secretary: R.P.A. Coles
Deputy Company Secretary: J.G. Latimer
■ The principal activities of the Group are the design & manufacture of control devices & process control systems high technology compressed air & pneumatic equipment, electronics, transfer & filtration equipment for gases & liquids, valves, automotive equipment & lubricators, specialised mechanical engineering, safety & life support products.

Subs. with defence business:
James North & Sons Ltd → 75
Tecalemit Garage Equipment Co. Ltd → 75
James North Australia Pty Ltd → Australia 75
Tecalemit (Australasia) Pty → Australia 75
CompAir Reavell/Broomwade → UK 75
CompAir Australia Pty Ltd → Australia 75
CompAir Luchard SA → France 75
CompAir Holman Iberica SA → Spain 75
Robertshaw Controls Company → USA 75
Barber-Colman Aerospace & Power Controls Div. → USA 75
Siebe North Inc. → USA 75 [97]

70-02100 TI Group plc
Lambourn Court, Abingdon, Oxfordshire OX14 1UH
☎(+44 1235) 55 55 70 [W] http://www.tigroup.com/
⊖ (+44 1235) 55 58 18
♧ £ 596,600,000 (1996) ♣ £ 1,756,600,000 (1996)
👥 25,000 (worldwide)
Chairman: Sir Christopher Lewington
Deputy Chairman (Non-Executive): John M Hignett
Chief Executive, Dowty Group and Chairman, Messier-Dowty: L Anthony Edwards
Chief Executive, John Crane Group: John Potter
Chief Executive, Bundy Group: William J Laule
Group Finance Director: Martin D Angle
Director of Strategic Development: James L Roe
Non-Executive Director: Sir Colin Chandler
President, Asia-Pacific: Robert J M Fisher
Non-Executive Director: Lord Fanshawe of Richmond (KCMG)
Non-Executive Director: John M Harris
Non-Executive Director: Rudolf G Mueller

■ Engineering group concentrating on specialised engineering businesses, operating in selected niches on a global basis.

TI Group's three core businesses:
John Crane International
Bundy International
Dowty Aerospace

TI Reynolds Rings Ltd
(a member of the TI Group)
Hay Hall, Redfern Rd, PO Box 763,
Tyseley, Birmingham B11 2BN
☎ (+44 21) 706 33 33 ☒ 338181 tirey g
⊖ (+44 21) 707 01 27
♧ £ 4,934,000 (shareholders fund)
♣ £ 11,244,000 (1995) 👥 80 🏭 115,000 sq ft
Managing Director: A Cramphorn
Finance Dir: R. Thompson
Commercial Dir: D. Anscombe
■ Manufacture and machining of flash welded, cold form rolled and Tru-Form gas turbine engine rings for use as: structural members, flanges, casings, seals, segments & combustors in aero and marine gas turbine engines, static gas generators and other aerospace, defence and commercial applications. 98

70-02230 Vickers plc

Millbank Tower, Millbank, London SW1P 4RA
☎ (+44 171) 828 77 77 ☒ 27921 viclon g
⊖ (+44 171) 828 65 85
♣ £ 1,197,600,000 (1996) 👥 10,193 (1996)
Chairman: Sir Richard Lloyd
Deputy Chairman and Chief Executive: Sir Colin Chandler
Managing Director Finance and Planning: David Essex
Executive Director: Chris Woodwark
Commercial Director: Andrew John
■ International company with activities in the field of automotive engineering, manufacturing and precision castings, land defence systems and equipment, marine propulsion systems and motion control equipment, superalloys and components for the gas turbine and automotive industries, medical equipment. → 75 → 80

Defence systems/propulsion technology:
divisions and subsidiaries include
Vickers Defence Systems
Ross & Catherall Ltd
Ross Catherall Ceramics Ltd
Vickers Armstrongs Ltd
Trucast Ltd
Certified Alloy Products Inc, USA
Vickers Aerospace Components
Vickers Airmotive
Vickers Precision Machining
Kamewa AB, Sweden
Brown Brothers & Company Ltd
Michell Bearings
Stone Vickers Ltd 98

70-02300 The Weir Group plc

149 Newlands Road, Cathcart, Glasgow G44 4EX
☎ (+44 141) 637 71 11 ☒ 77161 wplcrt g
⊖ (+44 141) 637 22 21
♣ £ 619,900,000 (1996)
Chairman: The Rt Hon The Viscount Weir
Managing Director and Chief Executive: Sir Ronald Garrick (CBE)
Public Relations Manager: Emrys Inker
■ Specialist engineering products and engineering services.

Subsidiaries with defence business:
Hopkinsons Ltd → 75
Peacock Inc. → Canada 75
Strachan & Henshaw Ltd → 75
Weir Engineering Pty Ltd → Australia 75
Weir Materials Ltd → 80
Weir Pumps Ltd → 75
Weir Westgarth Ltd → 75 98

70-02500 Westinghouse Brake & Signal

(a subs. of Hawker Siddeley Group plc)
Foundry Lane, PO Box 85,
Chippenham, Wiltshire SN15 1RT
☎ (+44 1249) 65 41 41 ⊖ (+44 1249) 44 10 26 98

70-02800 Williams Holdings plc

Pentagon House, Sir Frank Whittle Road, Derby DE21 4XA
☎ (+44 1332) 20 20 20 ☒ 37685 wwhder g
⊖ (+44 1332) 38 44 02
♣ £ 1,598,500,000 (1995) 👥 16,899 (1995)
Chairman: Nigel Rudd
Chief Exectutive: Roger Carr
Corporate Finance Director: T Park
Managing Director Kidde International Subsidiaries Division: M Harper
Marketing Director Kidde International Subsidiaries Division: R Farquhar
Finance Director Kidde International Subsidiaries Division: E Nicholas

Subsidiaries with defence business:
Detector Electronics Corp. → USA 75
Kidde-Graviner Ltd → 75
Williams Fairey Engineering Ltd → 75
Walter Kidde Aerospace → 75
Deugra 98

74 Consortia, Joint Venture Companies

BAeSEMA Ltd 80-03070
Shorts Missile Systems Ltd 75-53550

74-05350 EH Industries Ltd (EHI)

(joint subsidiary of GKN Westland Ltd and Agusta, Italy)
Pyramid House, Solartron House,
Farnborough, Hampshire GU14 7QL
☎ (+44 1252) 37 21 21 ⊖ (+44 1252) 38 64 80
♣ £ 200,000,000 (1996) 👥 75 (1996)
Joint Chairman: D. Wright
Joint Chairman: Amedeo Caporaletti
Managing Dir: Dr E.M. Striano
Deputy Managing Dir: A. Moorhead
Business Dir: G. Saponaro
Operations Dir: M. Corsi
Finance & Admin Dir: N. Griffiths
Proposals Mgr: Clive Kellaway
Purchasing Mgr: Alistair Gibson
■ International management company jointly owned by GKN Westland & Agusta to manage the EH 101 helicopter development, production & support programmes. 98

74-06550 GEC ALSTHOM UK Head Office

(jointly owned by GEC plc & Alcatel Alsthom)
Mill Road, PO Box 70, Rugby, Warwickshire CV21 1TB
☎ (+44 1788) 54 66 00 ☒ 312626 gecal g
⊖ (+44 1788) 54 64 40
♣ £ 7,618,000,000 (1995-96 - worldwide) 👥 83,000
Communications: Hazel Nicholas
■ Group activities: power generation, power transmission and distribution, transport, marine and industrial equipment. → 70 → 75 98

74-06600 GPT Limited

(† GEC-Plessey Telecommunications Ltd)
(jointly owned by GEC 60% & Siemens 40%)
New Century Park, PO Box 53, Coventry, W.Mids CV3 1HJ
☎ (+44 1203) 56 20 00 ☒ 31361 gectel g
⊖ (+44 1203) 56 22 47
♣ £ 1 000 000 000 👥 12,000
Managing Dir: A.R. Cobbe
Communications Dir: C.J. Whatley
■ Supplier of telecommunications equipment & systems to world markets to meet a wide range of applications, both civil & military. → 75

Strategic Communications Systems
PO Box 146, Coventry CV3 1LQ
☎ (+44 1203) 56 20 00 ☒ 31361
⊖ (+44 1203) 56 59 41
Director: P.N. Glanfield
Business Development Dir: G. Outterside
Business Development Dir: K. Kerton
■ Design & implementation of strategic defence telecommunication projects in the UK & overseas, from project definition to completion. Current major projects for the Royal Air Force, Royal Netherlands Air Force & the British Army including special military developments of commercial telecommunications projects. 97

74-06850 Matra BAe Dynamics

(† British Aerospace Dynamics)
(a 50:50 partnership of British Aerospace and Lagardère, France)
Six Hills Way, PO Box 19,
Stevenage, Hertfordshire SG1 2DA
☎ (+44 1252) 38 45 82 ☒ 825125; 825126
⊖ (+44 1252) 38 45 82
♣ £ 1,000,000,000 👥 6,100 (1996)
Chairman and Managing Director (France): Pierre Bourgeon
Managing Director (UK): Dave Hewitson
Chief Executive Officer: Noel Forgeard
Senior Executive Vice President: Jean-Paul Gut
Finance Director: Roger Curtis
Group Financial Controller: Colette Ormezzano
Strategy and Planning Director: Pierre-Henri Ricaud
Technical and Operations Director: Bob Harësceugh
Head of Purchasing (UK): David Arnold
Human Resources and Quality Director: John Graham
Head Public Affairs (UK): Charles Carr
Communications Director (France): Mark Vellay
Public Relations Manager: H J Borrington
■ Design, development, manufacture and supply of guided weapons systems and equipment. Matra BAe Dynamics is a joint venture formed by British Aerospace Dynamics and Matra Defence. → 75

Filton
PO Box 5, Filton, Bristol BS12 7QW
☎ (+44 117) 931 86 84 ☒ 449452
⊖ (+44 117) 969 86 75
👥 500

Lostock
Lostock Lane, Bolton, Lancashire BL6 4BR
☎ (+44 1204) 69 65 51 ☒ 63134
⊖ (+44 1204) 69 39 08
👥 300 98

74-06900 Matra Marconi Space UK, Ltd

(a joint venture of GEC plc and Lagardére, France)
Gunnels Wood Road, Stevenage, Hertfordshire SG1 2AS
☎ (+44 1438) 31 34 56
W http://www.matra-marconi-space.com
⊖ (+44 1438) 77 36 37
♣ US$ 1,600,000,000 (1996) 👥 4,760 (1996)
Chairman: Armand Carlier
Managing Director and Deputy Chief Executive Officer: Nick Franks
Financial Director: Jonathan Harwood
Director UK Military Space: Paul Down
Marketing Director: Andrew Roberts
Public Relations Manager: Alistair Scott
■ Design and manufacture of satellites, spacecraft subsystems and equipment for communications, remote sensing and scientific applications together with earth stations, civil and military, ground support and control facilities and launcher systems. Product inventory encompasses the following range of space engineering activities: prime contractorship; system studies; assembly; integration and test; design, development and manufacture of TT&C systems; civil and military earth stations (fixed, transportable and mobile) and image processing systems. → 75

Ground Systems Directorate
Gunnels Wood Road, Stevenage, Hertfordshire SG1 2AS
☎ (+44 1438) 31 34 56 ⊖ (+44 1438) 77 36 37
Director: Jon Sayer

Directorate of Science and Earth Observation (Radar)
PO Box 16, Filton, Bristol BS12 7YB
☎ (+44 117) 900 69 00 ⊖ (+44 117) 900 68 49/50
Director: Dave Link 98

74-07000 Normalair-Garrett Ltd (NGL)

(a subsidiary of GKN Westland Technologies Ltd)
Yeovil, Somerset BA20 2YD
☎ (+44 1935) 47 51 81 ✉ NORMALAIR
☒ 46132 nglyeo g ⊖ (+44 1935) 42 76 00
♣ £ 76,000,000 (1995) 👥 1,400
Chief Executive: Peter Burrows
Executive Director Sales, Marketing and Customer Support: Dr M Tutcher
Finance Director: D A J McKechnie
Operations Director: M Wills
Engineering Director: G R Giles
Head of Corporate Marketing: M P Bednall
Purchasing Manager: E W Staines
■ Design and manufacture of environmental control systems, oxygen systems, hydraulic equipment, data recorders, investment castings, heat transfer and weapon launch and actuation systems.
Environmental and air systems: Cooling and pneumatic equipment for aerospace applications.
Aerospace breathing systems: Life support anti-G protection and oxygen generating systems.
Hydraulics: Aerospace and defence hydraulics and components.
Mechanical control: Actuation and weapon control systems.

Contact Reference

Electronics: Aerospace and industrial electronics, severe environment data acquisition and recording systems, ground replay equipment, analogue and digital recorders. → 75

Customer Support Group
Bournemouth International Airport,
Christchurch, Dorset BH23 6NW
☎ (+44 1202) 58 18 18 ⊠ 41418
⊖ (+44 1202) 58 19 19
♣ £ 37,000,000 (1995) 👥 220 🏠 75,000 sq ft
Exec. Dir: K. O'Donnell
■ Repair, overhaul and technical support of environmental control, pneumatic, heat transfer, actuator, compressor, starter, lighting and electronic systems and equipment. [97]

74-10000 Pilkington Optronics Ltd
(Joint Venture with Pilkington plc)
Glascoed Road, St Asaph, Clwyd LL17 0LL
☎ (+44 1745) 58 80 00 ⊖ (+44 1745) 58 42 58
Chief Executive: T G Jones
Finance Director: Guy Blanguernon
Technology Director: Gordon Smith
Publicity Manager: Tim Orr
Public Relations: M Currie
Business units:
Pilkington Space Technology
Pilkington P.E. Ltd
Overseas Offices:
Pilkington Optronics Inc. → USA 75

Pilkington Optronics Ltd Barr & Stroud
1 Linthouse Road, Govan, Glasgow G51 4BZ
☎ (+44 141) 440 40 00 ⊠ 778114 bsglw g
⊖ (+44 141) 440 40 01
👥 750
Deputy Chief Exec. Officer: Tom O'Neill
Sales & Marketing Dir: Graham Smart
Technology Dir: Gordon Smith
■ Thermal imaging and surveillance systems, laser rangefinders, tank laser sights and thermal fire control systems, submarine periscopes and masts and non-hull penetrating optronic pods, visible & IR optical systems, hard coatings.

Pilkington Thorn Optronics
120 Blyth Road, Hayes, Middlesex UB3 1DLL
☎ (+44 181) 910 24 00 ⊖ (+44 181) 910 83 61
Sales & Marketing Dir: Graham Smart
General Mgr: Tom O'Neill
Technology Dir: Gordon Smith
Sr Contracts Mgr: S.R. Hunt
■ Diffractive & refractive avionic display systems. Infantry sighting & surveillance equipment.

Pilkington Space Technology
Kinmel Park Industrial Estate, Bodelwyddan,
Rhyl, Clwyd LL18 5TY
☎ (+44 1745) 58 45 84 ⊠ 61532
⊖ (+44 1745) 58 42 08
General Mgr: C.A. Kitchen
■ Cover slips for solar cells as used on satellites. Product range also includes second surface mirrors used for the thermal control of spacecraft. [98]

74-18000 Rolls-Royce Turbomeca Ltd
(joint subsidiary of Rolls-Royce plc UK and Turbomeca France)
4/5 Grosvenor Place, London SW1X 7HH
☎ (+44 171) 235 36 41 ✉ ROLLTURB
⊖ (+44 171) 245 63 85
♣ £ 35,000,000 (1996) 👥 5
Company Secretary: A West
General Manager Marketing: A S Williamson
■ Adour turbofan aircraft engines & RTM 322 turboshaft helicopter engines. Can also arrange supplies of marine gas turbine engines, gas turbine engines for tracked armoured vehicles, railway diesel engines, missile propulsion systems & power generation turbines.
® France: Turbomeca SA [98]

74-20000 Thomson Marconi Sonar Limited
(† Ferranti-Thomson Sonar Systems UK Ltd)
(jointly owned by GEC Marconi and Thomson CSF)
Headquarters
Wilkinthroop House, Templecombe, Somerset BA8 0DH
☎ (+44 1963) 37 05 51 🅆 http://www.tmsluk.co.uk
⊠ 46268 tmart g ⊖ (+44 1935) 44 22 00
♣ £ 150,000,000 (1996) 👥 1,750 (1997)
Managing Director: J W Bentley
Purchasing Director: Derek Moore
Operations Director: Mike J Shaw
Sales and Marketing Director: G P D Lock
Publicity Manager: Charles W Braine
Programme Manager for Sonar 2076: Paul Willmott
Sales Manager ASW: N Buckley
Company Lawyer: Christopher Rathbone
■ Sonar systems and suites for surface ships and submarines. Mine-countermeasures systems, acoustic and magnetic ranges, towed arrays, transducers, acoustic modules, systems design and management, airborne sonars, ROV based systems for obstacle identification and power survey. → 75

● Dolphin House, Ashurst Drive,
Bird Hall Lane, Cheadle Heath,
Stockport, Cheshire SK3 0XB
☎ (+44 161) 491 40 01 ⊖ (+44 161) 491 17 96
Manager: J M Richardson

● Lancaster Road, Carnaby Industrial Estate,
Bridlington, Yorkshire YO15 3QY
☎ (+44 1262) 60 28 81 ⊠ 52343
⊖ (+44 1262) 40 00 74
Manager: D. Calvert
■ Acoustic systems manufacture. Polymer research development.

● Redfields Industrial Park, Church Crookham,
Fleet, Hampshire GU13 0XX
☎ (+44 1252) 85 14 85 ⊠ 858945
⊖ (+44 1252) 85 14 49
Manager: H J Allen
■ Systems analysis & engineering capability in acoustics, electronics & sonar systems.

● Surrey Close, Granby Industrial Estate,
Weymouth, Dorset DT4 9TY
☎ (+44 1305) 76 13 33 ⊠ 41167
⊖ (+44 1305) 76 13 33 ext.102
Manager: Dr R.D. Orme
■ HF jammers, sonar intercept, underwater intruder detection.

● Airport Works, Rochester, Kent ME1 2XX
☎ (+44 1634) 84 44 00 ⊠ 96333 gecro c
⊖ (+44 1634) 81 65 79
Manager: R Tester [98]

74-20700 Turbo-Union Ltd
Head Office
PO Box 3, Filton, Bristol BS12 7QE
☎ (+44 117) 979 12 34 ⊖ (+44 117) 979 75 75
Managing Dir: Karlheinz Koch
■ Joint company formed by Rolls-Royce, MTU Munich and FiatAvio to co-ordinate the design, development, manufacture and product support of the RB199 military turbofan engine. [98]

75 Manufacturing Companies

75-00025 A1 Marketing
Suite 8, Grove House, Headley Road, Grayshott,
Hindhead, Surrey GU26 6LE
☎ (+44 1428) 60 46 66 ⊖ (+44 1428) 60 51 50
Proprietor: J W Wylie
■ Manufacture of oceanographic, hydrographic & environmental equipment. → 85 [98]

75-00032 Aardvark Active Barriers Ltd
Shevock Farm, Insch, Aberdeenshire AB5 6XQ
☎ (+44 1464) 82 01 22 ⊠ 73509 aard g
⊖ (+44 1464) 82 09 85
Managing Dir: David Sadler
■ Manufacture of mobile runway division barriers (MRDB). [98]

75-00034 Aardvark Clear Mine Ltd
(a member of the Aardvark Group)
Shevock Farm, Insch, Aberdeenshire AB5 6XQ
☎ (+44 1464) 82 01 22 ⊠ 73509 aard g
⊖ (+44 1464) 82 09 85
Managing Dir: David Sadler
■ Manufacture of mine clearance (flail type) & detection equipment. Also provide comprehensive mine clearance training. [98]

75-00040 AB Connectors Ltd
(a subs. of TT Group plc)
Abercynon, Mountain Ash,
Mid Glamorgan, Wales CF45 4SF
☎ (+44 1443) 74 03 31 ⊖ (+44 1443) 74 16 76
♣ £ 12 000 000 (1995) 👥 258 (1995) 🏠 10 600 sq m
Managing Dir: E.S. Smith
Sales & Marketing Dir: R.A. Crook
Operations Dir: Paul Lewis
■ Design & manufacture of connectors for aerospace, ground-based, military & commercial markets. [98]

75-00050 AB Electronic Assemblies Ltd
(a subs. of TT Group plc)
Tregwilym Industrial Estate, Rogerstone,
Newport, Gwent NP1 9YA
☎ (+44 1633) 89 23 45 ⊖ (+44 1633) 89 57 55
♣ £ 80 000 000 👥 650 🏠 12 000 sq m
Managing Dir: Tony Newton
Commercial Dir: T.D. Hunt
Operations Dir: J. Reardon-Smith
Purchasing Mgr: Nigel Coghlan
■ Printed circuit boards & boxed units approved to ISO 9000 & BS 5750 part 2. Contract assembly, materials management. [98]

75-00080 AB Electronics (Rogerstone) Ltd
(a subsidiary of TT Group plc)
Rogerstone, Newport, Gwent NP1 9YA
☎ (+44 1633) 89 23 45 ⊠ 498968
⊖ (+44 1633) 89 57 55
👥 1,000 (1996) 🏠 20,000 m2
Managing Director: A Newton
Sales and Marketing Director: T D Hunt
■ Mfr microprocessor controlled systems for all applications. Large SMT capability. [98]

75-00140 Abbey Electronics
Lyme Green Business Park, London Road,
Macclesfield, Cheshire SK11 0LR
☎ (+44 1625) 50 37 03 ⊖ (+44 1625) 50 36 76
♣ £ 5,146,000 (1996) 👥 70 (1997) 🏠 3,000 sq m
General Manager: P Clowes
Sales and Marketing Manager: B J McHugh
Purchasing Manager: G J Curzon
Quality Assurance Manager: M Harvey
■ Procurement and manufacture of electrical and electronic systems and cabling/looms. Manufacture of products to customer specifications. ISO 9002 accreditation Reg No. 0697. [98]

75-00180 Abbott Electronics Ltd
(† MPE Ltd)
(part of the Adwest Group)
Brunswick Rd, Cobbs Wood, Ashford, Kent TN23 1EB
☎ (+44 1233) 62 34 04 ⊖ (+44 1233) 64 17 77
♣ £ 2 044 000 👥 54 🏠 20 000 sq m

Dir & General Mgr: Ian Parrott
Sales & Marketing: David Wilks
Sales & Marketing: Ms Yvonne Spiers
Purchasing Mgr: D. Stringer
■ Designers & manufacturers of customised power supply units, wound components, electronic/electrical assembly & wiring for defence & aerospace applications. [97]

75-00280 **Abrasive Developments Ltd**
(a subsidiary of Unicorn International plc)
Norman House, Henley-in-Arden,
Solihull, West Midlands B95 5AH
☎ (+44 1564) 79 22 31 [E] vaqua@waverider.co.uk
[W] http://www.vaqua.com ⊖ (+44 1564) 79 30 79
♣ £ 5,400,000 (1996) 👥 65 (1997)
Managing Director: Robert K Brook
Sales and Marketing Director: Colin Denman
Purchasing Manager: Geoff Gee
■ Abrasive blast machines, wet & dry, for cleaning & finishing with combined degreasing. Aluminium oxide grits, glass beads, vegetable grits, plastic blasting media, silicon carbide grit & metallic abrasives. [98]

75-00290 **ACAL Auriema Ltd Top Optic Group**
(† Top Optic (RH) Ltd)
(a subsidiary of ACAL plc)
442 Bath Road, Slough, Berkshire SL3 9HN
☎ (+44 1628) 60 43 53 ⊖ (+44 1628) 60 37 30
Managing Director: M J Ridgewell
Product Group Manager: R Heaseman
Divisional Manager: D Wickens
Publicity Manager: C W Culpin
■ Design and sale of rigid and flexible endoscopes, light sources and related accessories. CCTV video system - standard and special purpose equipment. [98]

75-00300 **Access International Ltd**
(a division of SGB Services plc)
The Causeway, Maldon, Essex CM9 7XA
☎ (+44 1621) 85 45 11 ⊖ (+44 1621) 85 02 19
Managing Director: Ken Hansel
Sales and Marketing Director: David Graham
Purchasing Manager: Nigel Newton
■ Designers & manufacturers of aircraft maintenance docks, stagings & workstands. Hydraulically controlled passenger steps & undercarriage servicing transporters.

Andover Division
South Way, Walworth Industrial Estate,
Andover, Hampshire SP10 5AD
☎ (+44 264) 32 40 14 ⊠ 47277 ⊖ (+44 264) 35 87 30
Managing Dir: J.E. Green
Production Dir: R. Phillips
Sales Mgr: Brendan Rose
■ Aircraft docks and workstands for aircraft maintenance, production and repair. Tail servicing towers and platforms. Fly-away maintenance kits, access steps. Aluminium and steel fabrications. Air transportable equipment, hydraulically adjustable workstands, steps and stagings. Wheel change transporter. APU access stands. Towable passenger steps. [97]

75-00320 **Accles & Pollock Ltd**
(† TI Accles & Pollock Ltd)
(a division of The Hay Hall Group Ltd)
Brades Road, Oldbury, Warley, West Midlands B69 2DF
☎ (+44 121) 552 15 00 ⊖ (+44 121) 511 15 00
👥 100 (1996) ⌂ 16,000 m2
Managing Director: B Parsons
Operations Director: B Adkins
Marketing Manager: Ms W Moran
Sales Manager: Mrs T S Godwin
Buyer: A Timmington
■ Manufacture of stainless steel nickel alloy & carbon steel tubes for nuclear, aerospace & defence industries. Also manufacture of tubular manipulated assemblies. [98]

75-00360 **Accuracy International Ltd**
PO Box 81, Portsmouth, Hampshire PO3 5SJ
☎ (+44 1705) 67 12 25 ⊖ (+44 1705) 69 18 52
♣ £ 3,400,000 (1997) 👥 40
Managing Director: M D Cooper (MBE)
Operations Director: M A K Habershon
Technical Director: D B Walls
Sales and Marketing Director: M S Kay
Sales Manager: W H Welch
Purchasing Manager: P Turner
■ Design, manufacture & supply of small arms & ancillary equipment for sporting & military use. Infantry, long-range & moderated sniper rifles, CISM & UIT sporting rifles. [98]

75-00410 **Acorn Hardas Ltd**
Clover Street, Kirkby-in-Ashfield, Notts NG17 7LJ
☎ (+44 1623) 75 31 07 ⊖ (+44 1623) 75 45 38
♣ £ 1,300,000 👥 46
Managing Director: A Morrell
General Manager: D J Cox
Technical Director: T Hitchcock
■ CAA & 05-24 appr. finisher for hard & decorative anodising & conversion coating of aluminium, including: NDT, Spray Paint & Silver Plating. Hard Anodising, Chromic Anodising.

Works & Sales Office
Clover Street, Kirkby-in-Ashfield, Notts NG17 7LJ
☎ (+44 1623) 75 31 07 ⊖ (+44 1623) 75 45 38
General Mgr: D.J. Cox [98]

75-00520 **Acutronic Ltd**
(a sister company of Acutronic AG, Switzerland)
Forsyth Road, Sheerwater, Woking, Surrey GU21 5SB
☎ (+44 1483) 72 22 12; 72 22 13 [E] acutronic@aol.com
⊖ (+44 1483) 72 43 41
👥 2
Managing Director: Colin Stevens
■ Inertial guidance test instruments, centrifuges, servo systems and slip ring assemblies. [98]

75-00580 **Adams Electronics**
Unit 10, Forest Row Business Park,
Forest Row, East Sussex RH18 5DW
☎ (+44 1342) 82 38 56
[E] sales@adamselec.demon.co.uk
[W] http://www.adamselec.demon.co.uk
⊖ (+44 1342) 82 61 00
♣ £ 2,800,000 👥 17 ⌂ 2,400 sq m
Managing Director: J D Adams
Operations Manager: Miss Wendy Phillips
Sales and Marketing Director: Robert E Adams
Marketing Manager: Dean S Mann
■ Hand held metal detectors and plastic mine detectors. High sensitivity metal detection in general. [98]

75-00710 **ADE-HML Limited**
(a subsidiary of EIS Group plc)
12 Rigby Lane, Hayes, Middlesex UB3 1EY
☎ (+44 181) 569 11 20 ⊠ 28847 adehml g
⊖ (+44 181) 569 14 47
♣ £ 12,000,000 (1996) 👥 150 ⌂ 11,000 m2
Managing Director: Mike Cardy
Customer Services Director: Craig Morris
Operations Manager: Steve Vickery
Purchasing Manager: R Lewis
■ Manufacturer of hydraulic testing equipment for aircraft, from small trolley for helicopters through to HTD220/HTE221 units for fixed wing aircraft. Hydraulic components test benches. Also complete hangar ground support systems, including ground power units; low pressure air starters; frequency converters; passenger, cargo, access & handling equipment; maintenance units for ramp & hangar; hydrant & tanker refuellers; passenger stairs & towbars. [98]

75-00845 **Advanced Composite Group Ltd**
Composites House, Adams Close,
Heanor Gate Industrial Estate,
Heanor, Derbyshire DE7 7SW
☎ (+44 1773) 76 34 41 [E] advanced@inotts.co.uk
⊠ 377165 adcomp g ⊖ (+44 1773) 53 02 45
♣ £ 17,500,000 (1997) 👥 195 (1997) ⌂ 6,000 m2
Group Chairman: R M Sloman
Group Financial Director: A Moore
Operations Director: S R King
Group Market Development Manager: Dr A Potts
Purchasing Manager: A Wood
Sales Manager: J P Mabbitt
■ Manufacture of low temperature cure prepregs for prototypes, low quantity production and tooling, especially for high temperature or high accuracy applications. Manufacture of components and structures made from low temperature cure prepregs.

Advanced Composite Materials Ltd
Composites House, Adams Close,
Heanor Gate Industrial Estate, Heanor, Derby DE7 7SW
☎ (+44 1773) 53 45 99 ⊖ (+44 1773) 71 92 89
Director and General Manager: S King
Chief Executive: R M Sloman
■ 70°c/150°f cure prepregs with 350°c end use 50°c/90°f cure prepregs for prototypes, High accuracy materials for moulds and components.

Advanced Composite Components Ltd
Composites House, Adams Close,
Heanor Gate Industrial Estate, Heanor, Derby DE7 7SW
☎ (+44 1773) 76 90 55 ⊠ 377165 adcomp g
⊖ (+44 1773) 53 13 71
General Manager: S Cope
Chief Executive: R M Sloman
■ Components from materials produced by Advanced Composite Materials Ltd. Composite moulds, communications reflectors. [98]

75-00900 **Advanced Vision Systems**
Unit 2B Raceview Business Centre, Hambridge Road,
Newbury, Berkshire RG14 5SA
☎ (+44 1635) 314 66 ⊖ (+44 1635) 439 96
♣ £ 500,000 (1996) 👥 6 ⌂ 400 sq m
Partner: M. Woodruff
Partner: R. Daymond
Partner: G. Perry
■ Suppliers of equipment & systems for both overt and covert video surveillance applications. Specialist products include the Viewdapter. ISO 9002 Quality Assurance approved. [98]

75-00903 **Advance Hivolt**
(† Farnell Hivolt)
(a division of Advance International Group)
Durban Road, Bognor Regis, West Sussex PO22 9RL
☎ (+44 1243) 84 18 88 [E] sales@advance-hivolt.com
[W] http://www.advancehivolt.com
⊖ (+44 1243) 82 05 55
👥 100
Managing Director: C J J Austin
Sales and Marketing Director: P C Revell
Technical Dir: M.R. Barnett
Purchasing Mgr: Brian Pellett [98]

75-00904 **Advance Power**
(a division of Advance International Group Ltd)
Wetherby Road, Boroughbridge, York YO5 9UY
☎ (+44 1423) 32 00 00 [E] advancepower.com [W] http://www.advanceint.com/advance-int/power.htm
⊖ (+44 1423) 32 00 01
Managing Director (Acting): M Jayes
Sales and Marketing Director: Tim Joynson
Advertising Manager: Mark Oakley
■ Supplier of switch mode power supplies. Products include ac-dc and dc-dc converters, telecom rectifiers and dc-ac inverters. The company is ISO9001 registered and most products are CE marked and approved to UL, CSA, VDE, BABT and EN safety standards. → 85 [98]

75-00905 **Advance Power Systems**
(† Farnell Advanced Power)
(a subsidiary of Advance International Group Ltd)
Raynham Road, Bishop's Stortford, Herts CM23 5PF
☎ (+44 1279) 65 51 55
[E] 106625.2441@compuserve.com
[W] http://www.advanceint.com/advance-int/ cayson.htm
⊖ (+44 1279) 46 55 67
♣ £ 11,800,000 (1995) 👥 150 ⌂ 70,000 sq ft
General Manager: Jim Henderson
Marketing Manager: Neil Whiting
■ Standby power systems for telecommunications. [98]

75-00930 **AEI Cables Limited**
(a subsidiary of TT Group plc, UK)
Birtley, Chester le Street, Co Durham DH3 2RA
☎ (+44 191) 410 31 11
[E] marketing@aeic.compulink.co.uk ⊠ 53621
⊖ (+44 191) 410 83 12
Managing Director: Alan Gidney
Purchasing Director: J G Murray
Commercial Manager: K R Scott
Technical Manager: G Stainthorpe
Quality Manager: Michael Baker
Publicity Officer: M Woodley
Assistant Sales Manager: D Poole
■ Power cables, general wiring and industrial cables, Firetec - fire performance cables, PTFE and ETFE Cable. Approved to national and international standards.

Merseyside Office
Hawthorne Road, Bootle, Merseyside L20 9PN
☎ (+44 151) 933 95 55 ⊖ (+44 151) 933 10 10

Contact Reference

Gravesend Office
Crete Hall Road, Gravesend, Kent DA11 9AF
☎ (+44 1474) 56 44 66 ⊖ (+44 1474) 32 50 32
National Distribution Centre
Unit 18 Norquest Industrial Park, Pennine View, Birstall, Batley, West Yorkshire WF17 9NF
☎ (+44 345) 22 25 37 ⊖ (+44 345) 22 25 33 [98]

75-01020 **Aero Consultants (UK) Ltd**
(a subs. of Aero Consultants Ltd AG)
13/14 Clifton Road, Huntingdon, Cambs PE18 7EJ
☎ (+44 1480) 43 21 11 ⊖ (+44 1480) 41 29 10
President: B Arnold
Technical Dir: Wilf Bishop
Sales and Marketing Director: A Staub
Purchasing Dir: L. Adamson
→ 85 → 90
® Switzerland: Aero Consultants [97]

75-01040 **Aeromet International plc**
(† Kent Aerospace)
Eurolink Way, Sittingbourne, Kent ME10 3RN
☎ (+44 1795) 41 50 00 ⊖ (+44 1795) 41 50 15
♣ £ 30,000,000 👥 550+
Managing Director: Tom Hogg
Contact: Andrew King
■ Formed from the merger of Kent Aerospace and TKR International, the company provides integrated shaped metal component facilities for the aerospace and defence industries. Development of high strength alloys and materials processing. [98]

75-01050 **Aeronautical & General Instruments Ltd (AGI)**
(a subsidiary of EIS Group plc)
Fleets Point, Willis Way, Poole, Dorset BH15 3SS
☎ (+44 1202) 68 56 61 [E] sales@agi.co.uk
⊖ (+44 1202) 68 56 70
♣ £ 4,000,000 (1997) 👥 60 (1997) 🏭 1,800 sq m
Managing Director: A C Blyth
Marketing Director: V G Welham
Technical Dir: B. Brooks
Purchasing Mgr: J. Wright
Technical Sales Exec.: Mark R. Beisley
Technical Sales Exec.: B. Hunt
■ Electromagnetic & doppler ships log systems; underwater sensors; repeaters. Airfield fog monitoring & display systems, shipborne helicopter landing aids, ILS calibration equipment, meteorological systems, solid-state wind measuring systems & gun barrel bore measuring systems. Ship's helicopter glide path indicator. [98]

75-01101 **Aeroquip Aerospace Marine Group (Trinova Ltd)**
(a division of the Trinova Corporation)
Broadground Rd, Lakeside Industrial Estate, Redditch, Worcestershire B98 8YS
☎ (+44 1527) 51 75 55 ⊠ 338660 clamps g
⊖ (+44 1527) 51 75 56
👥 90 🏭 30 000 sq ft
General Mgr: Mark Day
Sales Manager: Paul Smith
Quality Mgr: B. Ward
■ Manufacture of hoses and fittings, quick disconnect couplings, swivels, V-band clamps and flanges, fuelling systems and associated fluid handling products. [97]

75-01160 **Aerospace Forgings Ltd**
(a member of Cortworth plc)
Churchbridge, Oldbury, Warley, West Midlands B69 2AU
☎ (+44 121) 552 29 21 ⊖ (+44 121) 544 57 31
♣ £ 12,500,000 (1996) 👥 200
Managing Director: R Dixon
Sales and Marketing Director: Karl Bishop
Finance Director: W M Sutton
Operations Director: Ben Harris
Customer Liaison Manager: R M Bourne
Purchasing Manager: Keith White
■ Ferrous, aluminium and titanium alloy forgings from 0-2kg to 100 kg for aerospace and defence related industries. MOD and CAA approved. [98]

75-01180 **Aerospace Metal Composites Limited**
RAE Rd, Farnborough, Hampshire GU14 6XE
☎ (+44 1252) 37 50 01 ⊖ (+44 1252) 37 50 02
♣ £ 800,000 (1996) 👥 8 (1996)
Sales & Marketing Dir: David J. Griffiths
Technical Dir: Andrew D. Tarrant
■ Manufacture of Metal Matrix composites by powder metallurgy. Products include silicon carbide particle reinforced aluminium in the form of billet, plate, sheet, extrusion & forgings. Manufacture of dispersion hardened aluminium-lithium and elevated temperature aluminium. [98]

75-01225 **Aerotech A.T.E. Ltd**
Unit 23, Ghyll Road Industrial Est., Heathfield, East Sussex TN21 8AW
☎ (+44 1435) 86 52 45 ⊖ (+44 1435) 86 55 88
♣ £ 1 000 000 👥 10
Managing Dir: Mervyn Cullimore
Director: Mrs S. Cullimore
■ Multi-axis force measurement systems. Wind tunnel balance design & manufacture marine model test rigs (tow tanks). Precision custom designed & manufactured load cells. [97]

75-01250 **Aerotron Ltd**
24 & 25 Cobham Way, Gatwick Road, Crawley, West Sussex RH6
☎ (+44 1293) 51 66 51; 55 31 01 ⊠ 877142
⊖ (+44 1293) 52 27 59; 56 21 30
♣ £ 25 000 000 👥 18
Managing Dir: A.M. Westley
® USA: Aerotron AirPower Inc, McDonnell Douglas Aircraft [97]

75-01350 **AETC Ltd**
(† AE Turbine Components Ltd)
(a UK Division of PCC Airfoils Inc)
Victoria Avenue, Yeadon, Leeds, West Yorkshire LS19 7AY
☎ (+44 113) 250 51 51 ⊖ (+44 113) 238 60 06/7
👥 800 🏭 40,000 sq m
Managing Dir: P.J. Worsley
Operations Dir Casting: David M. Pawlowski
Operations Dir Machining: A.G. McMahon
Sales and Marketing Manager: Roger H Thornton
☎ (+44 113) 238 60 10
Purchasing Mgr: Martin I. Archer
■ A specialist in precision investment casting, machining & processing of turbine engine airfoils. [98]

75-01400 **A.F. Aerospace Ltd**
(a subsidiary of Lentern (Aircraft) Ltd)
2 Chariot Way, Glebe Farm Industrial Estate, Rugby, Warwickshire CV12 1DA
☎ (+44 1788) 57 84 31 ⊖ (+44 1788) 54 02 68
♣ £ 3,700,000 👥 75 🏭 30,000 sq ft
Managing Director: P A Dean
Director and General Manager: R T Orams [98]

75-01475 **Agfa-Gevaert Ltd**
(a subsidiary of Agfa-Gevaert, Germany)
27 Great West Road, Brentford, Middlesex TW8 9AX
☎ (+44 181) 560 21 31 ⊠ 28154 aguk g
⊖ (+44 181) 231 49 51
♣ £ 175,000,000 (1996) 👥 600 (1996)
Managing Director: H Koskimies
Sales and Marketing Director: L Roberts
Manager Corporate Business: Michael J Morrison
Corporate Services Manager: P I Miller
■ Photographic materials for industrial and aerial applications. Extensive range of equipment and materials. [98]

75-01530 **AIM Aviation (Jecco) Ltd**
(† Jecco Aviation Ltd)
(a subsidiary of AIM Group plc)
Jecco House, Boscombe Grove Road, Bournemouth, Dorset BH1 4PD
☎ (+44 1202) 39 73 88 [E] e-mail@aim.aviation.co.uk
⊠ 418274 ⊖ (+44 1202) 39 87 06
♣ £ 12,000,000 (1997) 👥 250 (1997) 🏭 102,000 sq ft
Managing Dir: Peter S. Forbes
Commercial Dir: Shaun O. McDonald
Technical Dir: Bryan Reade
Operations Dir: Tony Roberts
■ Design, manufacture and installation of aircraft interior furnishing, especially civil and military helicopters. Aircraft sidewall/roof panel trimming. Royal, VIP and executive quality aircraft interiors. Pitot covers and warning pennants. Equipment holdalls and carrying cases. Parachute and survival packs, webbing equipment and straps. Repair to composite panels. Vacuum covering and interior decorative fabrics. Lightweight composite panels. Aircraft interior completion and refurbishing. Maintenance and modifications, aircraft and component paint spraying and avionic repairs. [98]

75-01550 **Airborne Industries**
(a division of Valentec International Ltd)
Penyfan Industrial Estate, Crumlin, Gwent NP1 4EF
☎ (+44 1495) 24 91 61 ⊖ (+44 1495) 24 46 21
♣ £ 5,000,000 (1997) 👥 150 (1997) 🏭 60,000 sq ft
Managing Director: J Fitzgerald
Operations Director: P Thomas
Sales Director: P Hayes
Purchasing Manager: G Main
■ Parachutist training balloons, camouflage nets, water storage tanks, inflatable dummy targets, solar stills, marker balloons, thermal image missile targets. Protective clothing and webbing. High visibility jackets. [98]

75-01555 **Air Cargo Equipment (UK) Ltd**
Unit 12 Space Waye, Feltham, Middx TW14 0TH
☎ (+44 181) 890 07 88
Managing Dir: Graham Walters [97]

75-01560 **Aircraft Equipment International Ltd**
1 Kings Ride Park, Kings Ride, Ascot, Berkshire SL5 8AR
☎ (+44 1344) 63 62 00 ⊠ 848412 airqip g
⊖ (+44 1344) 63 62 05
♣ £ 4,000,000 (1997) 👥 25 (1997) 🏭 5,000 sq m
Chairman: Norman W Angel
Managing Director: Simon Angel
Purchasing Manager: Malcolm Breach
Sales and Marketing: Garry Leach
■ Design, manufacture, stock and distribution of aircraft, military and naval gun systems including 25mm and 30mm Aden and 20mm Hispano-Suiza. Research and development facilities. Sole licenced manufacturer on behalf of Royal Ordnance plc for 30mm Aden guns and spares. Design authority for 25mm Aden gun. → 85 → 90
® UK: Royal Ordnance plc [98]

75-01580 **Aircraft Maintenance Support Services Ltd**
Eagle House, Village Farm Industrial Estate, Pyle, Mid Glamorgan CF33 6NU
☎ (+44 1656) 74 37 00 [E] amss@.celtic.co.uk
⊖ (+44 1656) 74 42 65
♧ £ 100,000 (1997) ♣ £ 6,000,000 (1997)
👥 90 (1997) 🏭 6,000 sq m
Managing Director: D R Barber
Technical Director: P L Summers
Management Accountant: Terri Dodd
■ Designers & manufacturers of ground servicing equipment: aircraft heater trolleys, airconditioning trolleys, airconditioned docking, cargo handling equipment & servicing trolleys. Approved to ISO 9001. → 80 [98]

75-01650 **Aircraft Materials Ltd**
(a subs. of Wardle Storeys plc)
Brunel Road, Newton Abbot, Devon TQ12 4PE
☎ (+44 1626) 32 22 00 ⊖ (+44 1626) 616 54
♣ £ 8 000 000 👥 150 🏭 4 000 sq m
Managing Dir: D.R. Nuwar
Commercial & Engineering Dir: F.W. Gould
Technical Sales Dir: R.H. Edwards
Sales & Marketing Dir: I.H. Bell
Purchasing Mgr: S. Sanders
Sales Mgr: R. Pantry
■ Aerial delivery systems, air transportation and ground support equipment. Aircrew safety restraint systems and parachute harness fittings. Helicopter deplaning systems by abseiling and fast roping. Parachute control equipment. Helicopter underslinging equipment. Applications for both civil and military fixed wing and rotary wing aircraft. Aerospace fastening products, adhesives and tools. [98]

75-01730 Aircraft Tanks Ltd
(a subs. of MSM Group of companies)
Spring Vale Works, Middleton, Manchester M24 2HS
☎ (+44 161) 643 24 62/4 ⊖ (+44 161) 643 34 90
♣ £ 900 000 (1996) 80 100 000 sq ft
Chairman: F. Pedley
Managing Dir: John D. Pedley
Sales & Marketing Dir: Michael Pedley
Financial Dir: I. Warrington
Purchasing Mgr: Ms Sandra Cannon
■ Maintenance, repair & overhaul of aircraft & aircraft powerplants. Manufacture of aircraft fuel tanks & aircrew seats. Sheet metal work & component manufacture. Subcontract work on missile programmes & Ariane space launcher. 97

75-01750 Air Industrial Developments Ltd
Empire Close, Empire Industrial Park, Aldridge, Walsall, West Midlands WS9 8UQ
☎ (+44 1922) 554 16 ⊖ (+44 1922) 578 89
♣ £ 600,000 6 500 sq ft
Managing and Marketing Director: R M Hughes
Operations Director: R Glenn
Purchasing Manager: Graham Patten
■ Pneumatic and hydraulic cylinders and rotary actuators. 98

75-01800 Air-Log Ltd
(a subsidiary of EIS Group plc)
North Lane, Aldershot, Hampshire GU12 4QH
☎ (+44 1252) 244 11 ⊖ (+44 1252) 33 37 90
♣ £ 11,700,000 (1997) 135 (1997) 9,000 sq m
Managing Dir: A.C. Yates
Product Dir: A.W. Mackie
Sales and Marketing Manager: D P Mercer
Sales and Marketing Manager: M W Ambidge
Operations Mgr: J.L. Mullard
Purchasing Manager: M Pegg
■ Electronic test & control equipment, aircraft & missile ground support equipment, portable electric generating sets, weapon & missile handling systems; hydropneumatic suspension systems for fighting vehicles. Mechanical handling equipment & conveyor systems. 98

75-01850 Airmatic Engineering (UK) Ltd
(a subsidiary of Graystone plc)
The Maltings Industrial Estate, Brassmill Lane, Bath BA1 3JL
☎ (+44 1225) 42 62 06 ⊖ (+44 1225) 42 62 07
£ 2,000,000 ♣ £ 1,090,000 (1996/7) 19 (1997) 765 sq m
Dir & General Mgr: R.J. Butler
Managing Director: S Richardson
Operations & Technical Mgr: Ian Whitehead
Purchasing Mgr: G. Tenniswood
■ Manufacture of an extensive range of high precision rotary & inclining tables for use in test, inspection & production departments. Special equipment designed and produced to customers requirements. Expertise in very high accuracy angular positioning, fluid bearings (air & oil) and field service & calibration. 98

75-02050 Airscrew Howden Ltd
(a subsidiary of Howden Group plc)
111 Windmill Road, Sunbury-on-Thames, Middlesex TW16 7EF
☎ (+44 1932) 76 58 22
E 100540.2120@compuserve.com ⊠ 929515 airscw g
⊖ (+44 1932) 76 10 98
£ 480,000 (1997) ♣ £ 10,500,000 (1997) 107 (1997) 5,600 sq m
Managing Director: Bryan Hiscock
Head of Sales: G Jones
Sales Manager: S Galloway
Purchasing Manager: Geoff Withers
■ Design, development & manufacture of air handling devices & specialised components for applications in aerospace, electronic & military vehicle equipment (fans, electric motors & specialist components). Overhaul & repair facility for aerospace components. Weapons carriage and release equipment. 98

75-02150 Air Vehicles Ltd
(† Tiger Hovercraft Air Vehicles Ltd)
(a division of Southern Vectis plc)
Three Gates Road, Cowes, Isle of Wight PO31 7UT
☎ (+44 1983) 29 31 94 ⊠ 86513
⊖ (+44 1983) 29 19 87
♣ £ 800,000 (1997) 12 (1997) 5,000 m2
Managing Director: Charles Eden
Sales and Marketing Manager: Duncan Bray
Administration Manager: Ms Brenda Janeway
Purchasing Manager: Stuart Edwards
■ Design and manufacture of amphibious air cushion vehicles and associated lift and propulsion systems. Design and manufacture of lightweight, high performance marine and transport seating and interior systems. CAA approved. 98

75-02200 AIT Recorders Ltd
(† Walton Radar Systems Ltd)
Electron House, Redfields Bus. Park, Church Crookham, Fleet, Hants GU13 0RD
☎ (+44 1252) 85 18 55 ⊖ (+44 1252) 85 18 08
£ 100 000 ♣ £ 1 500 000 25 6 500 sq ft
Managing Dir: Michael Jones
Operations Director: M King
Purchasing Manager: J Parker
Marketing Mgr: Ms Maureen McMahon
■ Recording systems for surveillance, archival & training use, datalogging & real-time data processing systems & graphics. Design, development & production of specialist electronic systems. → 80 97

75-02370 AD Aerospace Finishes Ltd
(a subsidiary of Akzo Nobel Coatings BV)
136 Milton Park, Abingdon, Oxon OX14 4SB
☎ (+44 1235) 86 22 26 ⊖ (+44 1235) 86 12 88
General Manager: J Rowbotham
■ Primers, topcoats and all specialities for aerospace coating purposes, maskants for chemical milling, epoxies and polyurethane paint systems approved to American and European specifications. 98

75-02372 Albacom Ltd
(previously part of Ferranti Industrial Electronics)
Dunsinane Avenue, Dundee, Scotland DD2 3PN
☎ (+44 1382) 88 93 11 ⊖ (+44 1382) 83 21 48
♣ £ 4,000,000 (1996) 75 13,500 m2
General Mgr: Martin Mackin
Technical Dir: R. Tait
Operations Director: George Robertson
Purchasing Manager: T Knight
■ Intelligent microwave receiver protectors, advanced microwave travelling wave tubes & TWT amplifiers, high & low voltage custom designed power supplies, press fit backplanes & connectors. 98

75-02390 Albright & Wilson Ltd
210-222 Hagley Road West, PO Box 3, Warley, West Midlands B68 0NN
☎ (+44 121) 420 51 13
W http://www.albright-wilson.com ⊠ 336291 albriw g
⊖ (+44 121) 420 55 45
♣ £ 583,400,000 5,000 (1996)
Business Sales Manager (AMGARD): Nick Westop
Sales and Marketing Assistant (AMGARD): Ms Sarah Buff
Business Sales Manager (PROBAN): Gerard Lelotte
Sales and Marketing Assistant (PROBAN): Ms Alison Mills
■ Manufacturer & supplier of PROBAN & AMGARD chemicals & technology for the flame retardant treatment of textiles, polyurethane foams, wood products & plastics. 98

75-02455 British Aluminium Speciality and Aerospace
(† Alcan Speciality & Aerospace)
PO Box 383, Kitts Green, Birmingham B33 0QR
☎ (+44 121) 252 80 00 ⊖ (+44 121) 252 80 01
Managing Director: M Wallis
Sales Director: B Hopkins
Business Development Manager: G Bevan
Marketing Manager: D J Holmes
■ Aircraft plate and extrusions manufacturer. 98

75-02560 ALCOA Manufacturing (GB) Ltd
(a subs. of ALCOA, USA)
PO Box 6, Worcester WR1 2EL
Marmion Hse, Copenhagen St
☎ (+44 1905) 61 35 51 ⊠ 338518
⊖ (+44 1905) 251 61
Sales Mgr Aerospace: S.W. Boney 97

75-02580 S.M. Alexander (Plastics) Ltd
(inc. Protection Safety Div. SMA Wescare, a member of the Impak Group of Companies)
Levellers Lane off Cromwell Road, Eynesbury, St. Neots, Cambridgeshire PE19 2JU
☎ (+44 1480) 47 31 40 ⊠ 82392 robins g
⊖ (+44 1480) 40 69 68
♣ £ 750,000 (1997) 35 (1997) 600 m2
Director: J. McIlroy
■ Lightweight NBC suits, size & design to customer requirements. Protective NBC covering, i.e weapons, transport, medical. Facelet masks & smoke hoods. 98

75-02625 Algram Group Ltd
(† Algram Engineering Co Ltd)
(a subsidiary of Keymed (Medical and Industrial Equipment) Co Ltd)
Eastern Wood Road, Langage Industrial Estate, Plympton, Plymouth PL7 5ET
☎ (+44 1752) 34 23 88 E algram@demon.co.uk
⊖ (+44 1752) 34 24 82
♣ £ 3,400,000 (1997) 105 (1997) 43,000 sq ft
Managing Director: Stephen Brown
Sales and Marketing Manager: J M Dryburgh
Purchasing Manager: Richard Bishop
■ Specialist low volume injection moulders of PPS, PES, PEEK, PEI mouldings. In-house tool making and pattern making, vacuum forming and hand lay-up GRP components. Ground support equipment, ordnance, aircraft components. Approved to BS EN ISO 9002 1994. → 80 98

75-02650 Aljac Engineering Ltd
31 Queen Street, Emsworth, Hampshire PO10 7BJ
☎ (+44 1243) 37 14 41 ⊖ (+44 1243) 37 38 66
4 1,500 sq ft
Managing Director: J R Horne
Operations Director: P S Churchill
Sales and Marketing Director, Purchasing Manager: C A Sellen
■ Design & supply of aviation fuelling systems & aviation fuel quality control equipment including differential pressure gauges, sampling units, bonding reels & deadman handles. 98

75-02680 R. Alkan (UK) Ltd
(a subsidiary of Alkan SA, France)
Robert House, 2A Shute End, Wokingham, Berkshire RG40 1BJ
☎ (+44 1189) 978 55 88; 978 23 95
⊖ (+44 1189) 977 49 39
Chairman: Jean-Paul Blass
Managing Dir: E.D. Beston
Sales and Marketing Director: X Bruneau
Operations Dir: Ph Guitaut
■ Missile ejectors; multiple weapon carriers and composite aircraft pylons; modular passive countermeasure systems; ground attack cartridge launchers; ground support equipment. → 90
® France: Alkan et Cie 98

75-02740 P W Allen & Co Ltd
Internal Security Division
25 Swan Lane, Evesham, Worcester WR11 4PE
☎ (+44 1386) 401 48 ⊠ 335202 fosfre g
⊖ (+44 1386) 76 53 51
♣ £ 2,320,000 (1996) 25 (1996) 2,500 m2
Managing Director: R M Freeman
General Manager: Roy Smith
Purchasing Manager: P Benwell
■ Sophisticated visual inspection equipment such as fibrescopes, conventional and fibrelit endescopes and light probes, also fluorescent and ultraviolet inspection lamps for the internal examination of aircraft, in particular engines and control mechanisms. Also manufacture of security and surveillance equipment such as search mirrors, search endescopes, search inspection kits, fluorescent and ultraviolet search lamps. EOD hook and line sets, EOD lifting tripod and non-magnetic tool kits.
Approved by Bureau Veritas Quality International (BVQI) to quality standard BS 5750/EN 9002/ISO 9002 (certificate No. 2204). 98

75-02745 Allibert Handling
Sommer Allibert (UK) Ltd
(a subsidiary of Allibert Equipment (Sommer Allibert SA), France)
Berry Hill Industrial Estate, Droitwich, Worcester WR9 9AB

☎ (+44 1905) 79 57 96 [W] http://www.allibert.com
⊠ 339844 ⊖ (+44 1905) 79 44 54
♣ £ 1,100,000,000 (Group 1996)
👥 13,000 (Group 1996)
Marketing Mgr: Simon Mendes
National Sales Mgr: Ed Pickin
■ Manufacturer of injection moulded high density polyethylene containers & polypropylene containers. Plastic boxes, trays, crates, bulk bins, pallets, insulated containers, plastic based roll cage pallets, dollies & trolleys, pallet boxes, plastic shelving, bottle crates, pallet tanks, safety & fatigue relief matting, plastic bulk liquid storage tanks, intermediate bulk containers, glass reinforced plastic structures, panels, sections & walkway flooring. [98]

75-02747 AlliedSignal Ltd Ocean Systems Division
(† Bendix Ocean Systems Division)
(a div. of AlliedSignal Aerospace Co.,USA)
Collier House, 163-169 Brompton Road, London SW3 1PY
☎ (+44 171) 581 16 91 ⊖ (+44 171) 584 67 44
♣ US$ 150 000 000 👥 830
Director: John A. Devine
Marketing Dir: C.J. Caughey
■ Design & manufacture helo dipped sonars, underwater mine hunting systems, mine neutralization systems; ship & submarine towed arrays; acoustic mines; torpedoes; surveillance systems, acoustic transducers; acoustic countermeasures; & sonobuoys. [97]

75-02762 The ALPOCO Group
Aluminium Powder Company
Forge Lane, Minworth,
Sutton Coldfield, W.Midlands B76 1AF
☎ (+44 121) 351 46 86 ⊖ (+44 121) 351 76 04
♣ £ 7 135 914 👥 55
Managing Dir: John Harris
Sales & Marketing Dir: J.D. Tait
Operations & Technical Dir: J.W. Wright
Sales Admin: S.E. Dennis
■ Light metal powders in aluminium & magnesium & their alloys including propellant & explosive grades. [97]

75-02770 A.L.S.
Rock Channel Shipyard, Rye, East Sussex TN31 7HL
☎ (+44 1797) 22 71 85 [E] sales@scubaals.demon.co.uk
⊖ (+44 1797) 22 73 50
■ Manufacture and distribution of protective suits, lighting systems, harnesses, navigation beacons, ironmongery and specialist diving equipment. [98]

75-02850 Alvis Industries Ltd
(a subsidiary of Alvis plc)
The Triangle, Walsgrave,
Coventry, West Midlands CV2 2SP
☎ (+44 1203) 53 54 55 ⊠ 31459
⊖ (+44 1203) 53 92 87; 53 92 88
Chairman: N M Prest
Managing Director: C G Rowe
Sales and Marketing Director: R Jones
Publicity/Marketing Services Manager: N C Turner
■ Alvis Industries Ltd acts as a holding company for the individual based operating subsidiaries.
Operating Subsidiaries:
Alvis Vehicles Ltd
Alvis Transmissions Ltd
Alvis Logistics Ltd
Alvis Unipower Ltd [98]

75-02860 Alvis Logistics Ltd
(a subsidiary of Alvis plc)
Unit 5, Quinn Close, Manor Park,
Coventry, West Midlands CV3 4LH
☎ (+44 1203) 53 92 99 ⊠ 31332
⊖ (+44 1203) 53 92 81; 53 92 90
♣ £ 15,200,000 (1995) 👥 58 🏭 3,100 sq m
Managing Director: M I Cooper
Director: C G Rowe
General Manager Export Sales: C Byrne
Commercial Manager: A Pavey
■ Incorporating United Scientific Instruments, company provides spares and support for armoured vehicles and military hardware (particularly instrumentation). Supplies the Wheelbarrow Explosive Ordnance Disposal (EOD) robot and associated equipment. Incorporating Alvis De-Mil Systems Ltd. Representation and marketing services for other defence companies, UK & worldwide. → 80 [98]

75-02862 Alvis Transmission Ltd
(† Alvis Aerospace & Transmissions Ltd)
(a subsidiary of Alvis Industries Ltd)
The Triangle, Walsgrave,
Coventry, West Midlands CV2 2SP
☎ (+44 1203) 53 54 55
⊖ (+44 1203) 53 93 01; 53 93 02
♣ £ 8,000,000 (1997) 👥 130 (1997) 🏭 7,000 sq m
Chairman: N M Prest
Managing Director: Stuart Wilkins
Sales and Marketing Director: Herve Jardon
Technical Director: David Hoyle
Operations Director: Lou O'Toole
Purchasing Manager: Christine Carver
Publicity/Marketing Services Manager: N C Turner
■ Design, development & manufacture mechanical transmission systems & other precision mechanical components & assemblies, for AFVs, aircraft, railcars, buses & specialist industrial applications. [98]

75-02863 Alvis Unipower Ltd
(a subsidiary of Alvis Industries Ltd)
Watford Business Park, Watford, Hertfordshire WD1 8QU
☎ (+44 1923) 81 65 55 ⊖ (+44 1923) 22 86 21
♣ £ 15,000,000 (1996) 👥 50 (1996)
Managing Director: I D Clayton
Business Development Director: H G Barnes
Engineering Director: D Picton
Head of Sales and Marketing: R J Hamilton-McLeod
Operations Manager: J L Miller
■ Military vehicles, wheeled heavy duty, high mobility including; tank transporters, bridging vehicles, tactical refuellers, load carriers and recovery vehicles.
Airfield crash tenders, rear engine, 4x4, 6x6 and 8x8.
Other services include: service, training, spare parts, post design services and in-depth repairs. → 80 → 85 [98]

75-02865 Alvis Vehicles Ltd
(a subsidiary of Alvis Industries Ltd)
The Triangle, Walsgrave,
Coventry, West Midlands CV2 2SP
☎ (+44 1203) 53 54 55 ⊠ 31459
⊖ (+44 1203) 53 92 86; 53 92 87
♣ £ 25,000,000 👥 280
Managing Director: C G Rowe
Sales Dir: I. Mitchell
Operations Dir: R. Garner
Commercial Dir: G.C. Abel
Engineering Dir: R.S. Wigley
Logistics Director: B E Eardley
Publicity & Marketing Services Mgr: N. Turner
■ Design, development & manufacture of AFVs for governments worldwide. Repair, overhaul, maintenance & upgrade of armoured vehicles. High integrity aluminium welding & fabrication for aerospace & other applications. → 80 [98]

75-02950 Ampair
PO Box 416, Poole, Dorset BH12 3LZ
☎ (+44 1202) 74 99 94 ⊖ (+44 1202) 73 66 53
Director: H G Durrant
Marketing: I W Marriott
■ Generators driven by submerged or trailing turbines, proven range of wind powered battery charging equipment. Flexible solar panels. High output alternators. All products are manufactured from top quality materials to withstand the rigours of marine and hostile environments. UK distributor for Canon Solar Panels. → 85 [98]

75-02970 Ampep plc
(a subsidiary of Akteboliget SKF, Sweden)
Strode Road, Clevedon, North Somerset BS21 6QQ
☎ (+44 1275) 87 60 21 ⊖ (+44 1275) 87 84 80
♧ £ 1,250,000 (1995) ♣ £ 7,200,000 (1995) 👥 147
🏭 70,000 sq ft
Managing Director: A J Grogono
Manufacturing Director: B C Herbison
Sales and Marketing Director: B O Jenkins
Supplies Manager: D F Bradbury
Quality Assurance Manager: P D Butler
■ Self lubricating plain spherical, rod end and journal bearings for aerospace and industrial applications. [98]

75-03000 Amphenol Ltd
Thanet Way, Whitstable, Kent CT5 3JF
☎ (+44 1227) 77 32 00 ⊠ 96157
⊖ (+44 1227) 27 65 71
♣ £ 25 000 000 👥 450
Managing Dir: J.B. Backhouse
Financial Dir: Ian Scott
Sales & Marketing Dir: D. Redfern
Sales Promotion Mgr: Norman Evans
Aerospace Marketing Mgr: R. Graves
■ Design, development & manufacture of interconnection systems to national & international specifications. Specialists in data highways, fibre optics, EMC/EMP protection, weapons release. Application specific power.
Industrial Tech Div. Spectra-Strip
Romsey Industrial Estate, Greatbridge Road,
Romsey, Hants S05 8XA
☎ (+44 1794) 51 75 75 ⊠ 47467
⊖ (+44 1794) 51 62 46
Marketing Mgr: S. Pritchard
Pyle National
Sherbrook Rd, Daybrook, Nottingham NG5 6AT
☎ (+44 115) 920 81 16 ⊖ (+44 115) 920 88 96
Scotland
Unit 29 Lynedoch St Ind Estate,
Greenock, Scotland PA15 4AY
☎ (+44 1475) 88 88 98 ⊖ (+44 1475) 72 91 09 [97]

75-03050 AMP of Great Britain Ltd
(a subs. of AMP Inc.)
Merrion Ave, Stanmore, Middx HA7 4RS
☎ (+44 181) 954 23 56 ⊠ 929547
⊖ (+44 181) 954 62 34
👥 1 500
Managing Dir: J. Keith Drysdale
Publicity Mgr: John Dean ☎ (+44 181) 420 82 51
Purchasing Mgr: Phil Grayer
■ Precision crimped terminals & splices, connectors, programming systems & switches for the electronic & electrical industries.
® USA: AMP [97]

75-03156 Andrew Ltd
(a subsidiary of Andrew Corp., USA)
Mulberry Business Park, Fishponds Road, PO Box 134,
Wokingham, Berkshire RG11 2FJ
☎ (+44 1189) 77 68 86 ✆ (800) 25 00 55
⊖ (+44 1189) 79 40 29
Corporate Marketing Mgr: Steve Gibson
Govt Sales Mgr: Simon Lawrence
Marketing Communications Mgr: Ms Michelle Spencer
■ Antenna systems for military microwave & HF radio links. Coaxial cable systems, fibre-optic gyros. EW SIGINT equipment, tracking systems. Electro-optical communications systems equipment.
® USA: Andrew Corp. [98]

75-03330 ANI Aurora plc
(† Aurora plc)
(a subsidiary of Australian National Industries Ltd, Australia)
Aurora House, Meadow Hall Road, PO Box 644,
Sheffield S9 1JD
☎ (+44 114) 261 00 11 ⊖ (+44 114) 256 11 46
♣ £ 120,000,000 (1997) 👥 1,300 (1996)
Group Managing Director: Alan Sheehan
Executive Director: Peter Wilson
■ Manufacture and distribution of a wide range of hot forged, extruded and rolled, near-to-net complex shaped components in nickel and titanium based alloys for aeroengines, airframes, fighting vehicles, submarines, guns and missiles, etc. → 85 [98]

75-03355 Antex (Electronics) Ltd
2 Westbridge Industrial Estate,
Tavistock, Devon PL19 8DE
☎ (+44 1822) 61 35 65 ⊖ (+44 1822) 61 75 98
♣ £ 2,000,000 👥 40
Managing Dir: I.J. Lunt
Director: C.S. Dicker
Sales Admin Mgr: J. Tomkies
■ Soldering equipment. [98]

75-03410 A.P. Metalcraft Ltd
Henley Road, Walsgrave,
Coventry, West Midlands CV2 1AD
☎ (+44 1203) 61 34 41 ⊖ (+44 1203) 60 21 17
♣ £ 1,000,000 (1997) 👥 33 (1997) 🏭 7,400 sq m
Chairman and Managing Director: K J Sheen
Purchasing Manager: Mrs M Bradshaw
■ Adjustable storage cabinets. Cabin furniture, storage and security equipment. Sheet metal fabrication service. [98]

75-03415 APPH Ltd
(† AP Precision Hydraulics Ltd)
(a div. of the BBA Group)
8 Pembroke Court, Chancellor Rd, Manor Park,
Runcorn, Cheshire WA7 1TG
☎ (+44 1928) 57 93 66 ⊖ (+44 1928) 57 94 54
♧ £ 17 000 000 (1994) ♣ £ 28 000 000 (1994) 👥 180
🏭 34,000 m2 (Aircraft Division)
Managing Dir: T.P. Whitehead
Operations Dir: Brian Jones
Finance Dir: K. Pearson
Purchasing Dir: Clive Dixon
Sales & Marketing Dir: Miriam McKee
Sales Mgr Aerospace Defence Projects: D. Gostridge
Chief Engineer: M.J. Hall
■ Design and production of precision hydraulic units and complete remote control systems for military and civil aircraft. Aircraft landing gear, flying controls and systems. Pumps, valves, self sealing couplings and filters.

APPH (Bolton) Ltd
(† AP Precision Hydraulics (Bolton) Ltd)
Great Bank Road, Wingates Industrial Park,
Westhoughton, Bolton BL5 3XN
☎ (+44 1942) 85 19 00 ⊖ (+44 1942) 85 19 01
👥 40
Chairman: T P Whitehead
Managing Dir: David Haslam
Technical Mgr: J. Sharples
Sales Mgr: B.M. Rowe
Marketing Manager: A Jones
■ Design & production of precision hydraulic units & systems for military vehicle & engineering applications. Power braking systems for tracked & wheeled fighting vehicles, steering & gearbox controls. Military bridge articulators. Micronic filters, strainers & self-sealing couplings. [97]

75-03420 Apex Storage Systems Ltd
(† Apex Storage Systems Ltd)
Apex House, Garamonde Drive,
Wymbush, Milton Keynes MK8 8ND
☎ (+44 1908) 56 12 22 ⊠ 826945
⊖ (+44 1908) 56 74 55
♣ £ 14,000,000 (1997) 👥 160 (1997)
Operations Dir: Mike Tucker
Sales & Marketing Dir: Edward Hutchison
Purchasing Mgr: Barry Wood
Marketing Executive: K A Brookbank
Marketing Assistant: Craig Triggs
■ Products include: Rack clad building systems pallet racking, cantilever racking, longspan & adjustable shelving, mobile & static document shelving & modular shelving for use in vehicles, ships & semi-permanent static installations; louvre panels for high density small parts storage. [98]

75-03520 Applied Holographics plc
40 Phoenix Road, Crowther, District 3,
Washington, Tyne and Wear NE38 0AD
☎ (+44 191) 417 54 34
Ⓔ 100407.1630@compuserve.com
Ⓦ http://www.applied-holographics.com
⊖ (+44 191) 416 32 92
♣ £ 8,000,000 (1995-96) 👥 140 (1996)
Chief Executive: Dr D J Tidmarsh
Technical Director: Dr T Hammond
Sales Director: R Cartwright
Marketing Manager: Mrs Astrid Mitchell
Purchasing Manager: Dennis Meadows
■ Hologram security devices used in such areas as product integrity, personnel security and document security. [98]

75-03530 Applied Microengineering Ltd
The Innovation Centre, 68 Milton Park,
Abingdon, Oxon OX14 4RX
☎ (+44 1235) 83 39 34 ⊖ (+44 1235) 83 39 35
♣ £ 250 000 (approx.) 👥 5
Sales & Marketing Dir: R. Santilli
Technical Dir: Tony Rogers
■ AML design, develop & manufacture micro-engineered sensors & other devices for industry on a contract basis. Extensive expertise exists in capacitive sensor technology, including for pressure sensors and ultrasonic (PVdF) transducer usage, design & manufacture. [97]

75-03560 APT Radar Systems Ltd
Unit B, Sprint Ind. Estate, Chertsey Rd,
Byfleet, Surrey KT14 7LA
☎ (+44 1932) 35 17 11/2 ⊖ (+44 1932) 35 40 80
♣ £ 1 000 000 👥 18
Managing Dir: David Baggaley
Sales & Marketing Dir: Ed Crosby
Purchasing Mgr: John Butt
■ Design & manufacture of tracking radar systems. Repair, overhaul, refurbishment of radar systems. Radar spares, logistic support, post design services, radar sub-systems & enhancement packages. Avionic/radar test equipment. Meteorological systems. Language training systems. [97]

75-03590 Aquabeam Ltd
Willow Lane, PO Box 3, Grantham, Lincolnshire NG31 6AA
☎ (+44 476) 56 69 98 ⊖ (+44 476) 59 13 39
Managing Director: M Sewell
■ Designers & manufacturers of a comprehensive range of underwater & hyperbaric lighting equipt & systems including NATO Approved & Coded luminaries & accessories. [98]

75-03600 Aqua Bell Ltd
Waterside, Brundall, Norfolk NR13 5PY
☎ (+44 1603) 71 30 13 ⊖ (+44 1603) 71 41 41
Managing Director: V J Bell
■ Boats from 28ft to 36ft built to customer specifications. [98]

75-03700 Arcom Control Systems Ltd
(a subsidiary of the Fairey Group plc)
Unit 8-10 Clifton Road, Cambridge CB1 4BW
☎ (+44 1223) 41 12 00 ⊖ (+44 1223) 41 04 57
👥 108
Managing Dir: Andrew Walters
■ Data logging systems for aircraft & helicopters, trials visual display units & power supplies, computer equipment. MIL-STD-1553B interfaces. Test equipment. RTU & scada products. [98]

75-03703 Arkonia Systems Ltd
(† Watkins-Johnson (UK) Ltd)
35 Woolmer Way, Bordon, Hampshire GU35 9QE
☎ (+44 1420) 48 86 46 ⊖ (+44 1420) 47 78 45
👥 35
Managing Director: Len Robinson
Commercial Executive: Len Murrell
Production Manager: Colin Arbon
Customer Services Administrator: Mrs Marjorie Howell
■ Electronics manufacturer. Designs, develops and manufactures a wide range of RF based products from paging systems to radar jammers. [98]

75-03705 Aremco Products
(a Joint Venture company)
Foxoak Street, Cradley Heath, West Midlands B64 5DQ
☎ (+44 1384) 56 85 66; 45 26
Ⓔ aremco@btinternet.com ⊖ (+44 1384) 63 89 19
Managing Director: Miss C B Case
Commercial Director: M P Kennard
Sales Office Manager: Doug Beasley
Purchasing Manager: Ms Sandra Hackett
■ Security and access control products, barriers and gates. [98]

75-03715 Arjo Ltd.
St. Catherine Street, Gloucester GL1 2SL
☎ (+44 1452) 50 02 00 ⊠ 43461
⊖ (+44 1452) 52 52 07
Managing Dir: A. Wozencroft
Sales & Marketing Dir: M. Murray
Marketing Communications Mgr: Mrs Clift
Key Accounts Mgr: F.W. Dummett
Events Co-ordinator: Mrs L. Seager
■ Special baths for burns treatment. Mecabed - net suspension & turning bed. Shower trolley. Bath hoists for general ward use. [97]

75-03729 Armalon Ltd
44 Harrowby Street, London W1H 5HX
☎ (+44 171) 262 18 81 ⊖ (+44 171) 402 09 59
♧ £ 10,000 (1997) ♣ £ 112,000 (1995/96)
👥 3+ (1997) 🏭 160 sq m
Managing Director: Peter P Sarony
Marketing/Sales Director: Mrs J Anne Sarony
■ Design, research & development, manufacture and sale of small arms systems, improved components, accessories, sighting systems, accoutrements, ammunition, range equipment and installations including target systems. Also other defence and law enforcement products and equipment, training and consultancy. Sales agent, distributor and representative. → 80 → 90 → 85 [98]

75-03775 Armourshield Limited
PO Box 456, Manchester M60 2LL
☎ (+44 161) 834 25 50 ⊖ (+44 161) 832 35 57
Managing Dir: Michael Sacks
Operations & Technical Dir: Glynn Jones
■ Manufacture of specialised armour systems. Suppliers of combat tested equipment to many of the World's most elite special forces teams. [97]

75-03790 Armour Technology International Ltd
PO Box 99, Leamington Spa, Warwickshire CV31 3RH
☎ (+44 1926) 45 03 93 ⊠ 94070449 arma g
⊖ (+44 1926) 45 00 55
Director: N.J. Smith
■ Modular bullet & fragment-resistant body armour. Ceramic, laminated & composite armour; stab-resistant & riot control vests; lightweight armour for equipment, vehicles, vessels & aircraft; structural & architectural armour; parasitic & retrofit armour. [98]

75-03800 Armstrong Fastening Systems
(a division of Caparo Group Ltd)
Clough Road, Hull HU6 7PR
☎ (+44 1482) 34 14 01 ⊖ (+44 1482) 34 83 55
👥 41
Managing Director: E B Barmby
Production Manager: L A Burgoyne
Sales and Marketing Manager: Miss K Robinson
■ Fastening Systems. [98]

75-04000 Arromet International plc
(† Kent Aerospace plc)
(a subsidiary of Charles Baynes plc)
Eurolink Way, Sittingbourne, Kent ME10 3RN
☎ (+44 1795) 41 50 00 ⊖ (+44 1795) 41 50 31
♣ £ 16,000,000 (1996) 👥 350 🏭 100,000 m2
Managing Director: Duncan Crighton
Sales Director: Andrew King
Sales and Marketing Director: Colin Cropley
Operations Director: Tom Hogg
Technical Director: Peter Thompson
Purchasing Manager: Chris Coombes
■ Manufacture fully machined precision sand & investment castings in magnesium & aluminium alloys for the aerospace, defence & motorsport industries to aluminium & magnesium specifications. [98]

75-04050 Ascot Systems
(a subs. of Ascot Circuits Ltd)
Woods Way, Goring-by-Sea,
Worthing, West Sussex BN12 4QY
☎ (+44 1903) 50 30 41 ⊖ (+44 1903) 50 72 50
♣ £ 2 000 000 👥 20
Managing Dir: Phil North
Marketing Mgr: Neil Kimmance
■ Design & manufacture computer training equipment under the brand name "Course Master". Military applications are for classroom training, where computers are applied to either scientific or information technology training, as at RMCS Shrivenham, or to defence and communications equipment simulation training. [97]

75-04100 Ashcroft Ceramic Capacitors
(a division of FG Taylor Industrial Holdings Ltd)
Fulledge Works, Holmes Street,
Burnley, Lancashire BB11 3BG
☎ (+44 1282) 83 15 40 ⊖ (+44 1282) 43 00 04
♣ £ 100,000 (1997) 👥 5 (1997) 🏭 3,000 sq ft
Managing Dir: Simon Taylor
Sales Manager: Ms Karina Taylor
Quality Manager: Philip Darwood
■ Manufacture of multi layer leaded cased ceramic capacitors. BS 9075, N0023, N0024, F0002, CECC 30701-007, 30601-009, and ISO 9002. [98]

75-04235 Astrosyn International Technology plc
The Old Courthouse, New Road Avenue,
Chatham, Kent ME4 4QJ
☎ (+44 1634) 81 51 75; 81 55 11 ⊠ 965033
⊖ (+44 1634) 82 65 52
♣ £ 3,500,000 (1997) 👥 12 (1997)
Managing Dir: John M. Melville
Commercial Sales Mgr: Mrs Marion Bovington
■ Design & manufacture of rotating electrical machines. [98]

75-04295 Atlantic Precision Engineering Ltd
(† Coin Plastic Industries Ltd)
Aragon Raod, Blackbush Business Park,
Yateley, Hampshire GU46 6ET
☎ (+44 1252) 87 02 25 ⊖ (+44 1252) 87 10 82
♧ £ 250 000 (1995) ♣ £ 900 000 (1995) 👥 22
Chairman: D.R. Hudson
Managing Dir: R.E. Partridge
General Mgr: C.C. Redrup
Purchasing Mgr: S. Denyer
Sales Mgr: Ivan Jones
■ Aerospace approved precision machining in metallics & engineering plastics, including asbestos. Aerospace approved moulding facility (Atlantic Plastics Ltd). [98]

75-04300 Atlas Hydraulic Loaders Ltd
(a subsidiary of Atlas Weyhausen GmbH)
Vere Road, Blackwood, Kirkmuirhill, Lanark ML11 9RR
☎ (+44 1555) 89 32 81 [E] atlas@atloi.demon.co.uk
⊖ (+44 1555) 89 48 58
♧ £ 8,000,000 (1996) ♣ £ 32,000,000 (1995)
👥 312 (1996) 🏭 14,864 m2
Managing Dir: Bryan Findlay
Sales Mgr, Special Projects: B.D. McCulloch
Purchasing Mgr: Alan Clark
■ Manufacture and supply lorry loaders, cranes and excavators to Defence Forces. [98]

75-04400 Audiotel International Ltd
Corby Road, Weldon, Corby, Northamptonshire NN17 3AR
☎ (+44 1536) 26 66 77
[E] sales@audiotel-international.com ⊠ 341775 audint g
⊖ (+44 1536) 26 67 11
👥 35 (1997)
Managing Director: Andrew Martin
Sales Director: Adrian Hickey
Marketing Manager: Ms Julie Walker
■ Design and manufacture a range of electronic countermeasure equipment including a multiple response scanning receiver and a non-linear junction detector. Systems used for internal security and EOD/IEDD applications. [98]

75-04440 Autocar Electrical Equipment Ltd
Lumenition Division
49-51 Tiverton St, London SE1 6NZ
☎ (+44 171) 403 43 34 ⊖ (+44 171) 378 12 70
♣ £ 500 000 👥 13
Managing Dir: M.A. Ford
General Mgr & Sales: C.R. Bailey
■ Electronic ignition systems & electronic engine speed limiters. [97]

75-04470 Automatic Systems Laboratories Ltd
(a division of Quadramatic plc)
28 Blundells Rd, Bradville,
Milton Keynes, Bucks MK13 7HE
☎ (+44 1908) 32 06 66 [E] aslco.com
⊖ (+44 1908) 32 25 64
♣ £ 2,900,000 (1995) 👥 50
Managing Director: Bruce Wilson
Operations Director: Ms Judy Densham
General Sales Manager: John McKay
■ Manufacture of precision thermometry and calibration equipment. [97]

75-04485 Avalon Electronics Ltd
Langhorne Park House, High Street,
Shepton Mallet, Somerset BA4 5AQ
☎ (+44 1749) 34 52 66
[W] http://www.aval.demon.co.uk/avalon
⊖ (+44 1749) 34 52 67
♣ £ 1,400,000 👥 12 🏭 500 sq m
Director: Terry Mason
Operations Dir: J. Knott
■ Manufacture wide band analogue, digital & video tape recorders for use with S-VHS video cassettes. Suitable for aerospace, satellite & general communications applications. [98]

75-04550 Aveling Barford (Machines) plc
Invicta Works, Grantham, Lincs. NG31 6JE
☎ (+44 1476) 56 55 51 ⊠ 377861
⊖ (+44 1476) 57 58 27
♣ £ 65,000,000 (1995) 👥 700
Managing Director: Peter Youlton
Sales Director: Nick Hodges
Purchasing Manager: John Pritchard
■ Rapid runway repair equipment. Off-highway construction equipment, articulated dumptrucks, 6/6, non-armoured. 6/6, all wheel steer motor graders. Rough terrain telescopic forklifts. Drops/ammunition pallets, dry freight trailers, fifth wheel couplers. [97]

75-04560 Avery Berkel
(† GEC Avery Ltd)
(a subs. of The General Electric Co. plc)
Smethwick, Warley, W. Midlands B66 2LP
☎ (+44 121) 558 11 12 ⊠ 336490
⊖ (+44 121) 555 60 62
♣ £ 200 000 000 👥 6 000
Group Managing Dir: Michel Steyaert
UK Managing Dir: Rupert Soames ☎ (+44 1922) 43 43 43
UK Sales Dir: Stuart Roberts ☎ (+44 1922) 43 43 43
Public Relations Officer: Ms Ingrid Maxam
☎ (+44 1922) 43 43 43
■ Electronic & intrinsically safe weighing systems widely used in munitions, medicine, research & specialist aircraft or tank weighing. High accuracy counting & checkweighing systems control stocks from minute components to bulk supplies. [97]

75-04570 Avesta Sheffield Ltd
Shepcote Lane, PO Box 161, Sheffield S9 1TR
☎ (+44 114) 244 33 11
[W] http://www.avestasheffield.com ⊠ 547025
⊖ (+44 114) 244 82 80
♣ £ 2,100,000,000 (1995) 👥 2,100
■ Armour steel plates. [98]

75-04870 Avica Equipment
(a Meggitt Aerospace Company)
Boundary Way, Hemel Hempstead, Herts HP2 7SL
☎ (+44 1442) 647 11 ⊖ (+44 1442) 23 00 35
♣ £ 20 000 000 👥 150
Managing Director: Eric Lewis
Sales & Commercial Dir: Roger Tyson
Sales Mgr: Mark Bolton
Sr Sales Engineer: Barry Parker
■ Designs, develops & manufactures a wide range of products & systems for aircraft, space vehicles, nuclear, cryogenic & marine applications e.g. metal bellows, expansion joints, hoses, ducting, gimbals, hose reels, seals, clamps, rigid & flexible metallic ducting systems. [98]

75-04950 Avimo Ltd
(a subsidiary of Avimo Group Ltd)
Lisieux Way, Taunton, Somerset TA1 2JZ
☎ (+44 1823) 33 10 71
[E] 101317.1734@compuserve.com ⊠ 46126
⊖ (+44 1823) 27 44 13
♧ £ 4,000,000 (1997) ♣ £ 11,000,000 (1997)
👥 218 (1997) 🏭 6,805 m2
Managing Director: Dr A D Wilson
Commercial Director: R Hartnell
Business Development Director: S B Hillier
Technical Director: K Venables
Finance Director: G Francis
Operations Director: N Runkee
Sales Manager Aerospace: S Russell
Purchasing Manager: W Vaughan
Publicity and Exhibitions Manager: Chris Graves
■ Design, development and manufacture of electro-optical instrumentation for military and civil applications. Equipment includes day surveillance systems, image intensified night observation and weapon aiming devices, thermal imaging, vehicle sights, laser rangefinders and HUDs. Provides sighting systems and equipment for Challenger 2, LeClerc, Starstreak, Scimitar Warrior, Scorpion, AS90 and TADS/PNVS system for the UK Apache. Pitot and pitot static heads and probes, electrically de-iced static ports, rate gyro gunsights, attitude monitors, pipe couplings, anti-icing control units, instrument panels and threat warning receivers.
Project management, environmental testing, logistic and product support. [98]

75-04970 Avionic Components Ltd
(a subsidiary of Hampson Industries plc)
Rowleys Green Industrial Estate, Longford,
Coventry, West Midlands CV6 6AN
☎ (+44 1203) 66 10 47 ⊖ (+44 1203) 63 71 29
♣ £ 3,000,000 (1997) 👥 60 (1997) 🏭 2,000 m2
Managing Director: M Sharrocks
Operations Director: E Adamsonis
Purchasing Manager: L Peffers
■ Precision engineering of complex and critical components for the world's aerospace, defence and high technology industries. Aerospace and industrial gas turbine components. Mechanical components for aircraft, fighting vehicles, surface vessels, missiles and ground support equipment. [98]

75-05100 Avon Inflatables Ltd
(† Avon Marine)
Military/Commercial Dept
Dafen, Llanelli, Carmarthanshire, S. Wales SA14 8NA
☎ (+44 1554) 74 11 55 ⊖ (+44 1554) 74 15 00
♣ £ 12 500 000 👥 250
Chairman: David Powell
Managing Dir: M. Sheil
Technical Dir: R. Harding
Commercial Sales Dir: A. Morgan
Military Sales Exec.: G.A. Tetley
Marketing Mgr: Ken Watson
Purchasing Mgr: C. Dornan
■ Wide range of craft including rigid hull craft (RIBs) high speed assault craft, rescue boats, work boats & clearance diving craft. [97]

75-05130 Avon Rubber plc
Corporate Headquarters, Manvers House, Kingston Road,
Bradford-on-Avon, Wiltshire BA15 1AA
☎ (+44 1225) 86 11 00 ⊠ 44142
⊖ (+44 1225) 86 11 99
♧ £ 180,000,000 (1996) ♣ £ 350,000,000 (1996)
👥 6,100 (1997)
Chairman: Christopher King
Chief Executive: Steve Willcox
Director (Global Automotive): Steve Stone
Director (Technical Products, N America): Don Samardich
Director (Technical Products, Europe): Roger Hunt
Sales/Marketing Director (Automotive): Mike James
Sales/Marketing Director (Technical Products):
Keith Butler
Executive Director: Hamish McPhie
Company Secretary: C L Martin
Group Publicity Manager: Gordon Morris
■ Manufacture of products using elastomeric, thermoplastic and composite materials for industrial, commercial, marine, defence, leisure and other markets. Manufacture of automotive components and related products.
Subsidiaries with defence business:
Avon Technical Products
CQC Ltd
Bell Avon Inc., USA
Avon Spencer-Moulton SA, France [98]

75-05150 Avon Technical Products
(a division of Avon Rubber plc)
Bath Road, Melksham, Wiltshire SN12 8AA
☎ (+44 1225) 70 63 38 ⊠ 44142
⊖ (+44 1225) 70 08 28
♣ £ 327,000,000 (1996) 👥 5,900 (1997)
🏭 150,000 m2
Managing Director: Roger Hunt ☎ (+44 1225) 79 18 23
Operations Director: R Head
Sales and Marketing Director: Gary Dunn
Purchasing Manager: Ian Muspratt
Sales Manager Fabrications: Paul Smart
Sales Manager AFV Tracks/Wheels: Pat Grace

Advertising Manager: Gordon Morris
☎(+44 1225) 86 39 11
■ Design, development & manufacture of polymer components for defence forces; NBC suits & respirators; flexitanks for transportation & storage of bulk liquids; running gear components for armoured fighting vehicles; acoustic/signature reduction; clothing & personal equipment. Hovercraft components & dry diving suits. 98

75-05190 AVX Ltd
(a subsidiary of AVX Corp, USA)
Admiral House, Harlington Way, Fleet, Hants GU13 8BB
☎(+44 1252) 77 00 00 W http://www.avxcorp.com
858473 ⊜ (+44 1252) 77 00 01
♣ £ 100 000 000 ⁂ 2,000
Corp. Vice President AVX Corporation: Alan Gordon
Marketing Dir: Craig Hunter
Business Sector Mgr: Mick Pye
Communications Mgr: Ian Ford
■ Multilayer ceramic capacitors and tantalum capacitors. Trimmers, potentiometers, resonators, resistors, clock oscillators, buzzers, resistor networks, filters, transient voltage suppressors, fuses, high voltage capacitors, high temperature capacitors, RF microwave capacitors, glass dielectric capacitors. EMI filters. 97

75-05420 Balmoral Group Ltd
Balmoral Park, Loirston, Aberdeen AB9 2BY
☎(+44 1224) 85 90 00 ⊜ (+44 1224) 85 90 59
♣ £ 40,000,000 (1995) ⁂ 550
Chairman & Managing Dir: James S. Milne
Group Public Affairs Mgr: Stephen Gibb
■ High tech engineering specialist in the design & manufacture of glass fibre/elastomer/syntactic foam products. Marine equipment, modular building suppliers and manufacturers of tanks.
Associate companies & divisions:
Balmoral Norge A/S (Norway)
Balmoral Group Houston Inc. (USA)
Balmoral Hitec Buoyancy
Balmoral Buoyancy Hire
Balmoral Hydrospace Test Centre
Balmoral Injection Mouldings
Balmoral Nav-Aids
Balmoral Marine
Balmoral Marine Equipment Hire
Balmoral Lloyds Testing
Balmoral Research
Balmoral Glassfibre
Balmoral Plastic Engineering
Balmoral Wire Rope Services
Balmoral Tanks & Pipes
Balmoral Webco Pipeline Systems
Balmoral Modular Building Systems.
Balmoral Environmental Engineering 98

75-05430 Baltex Fabrics
Burr Lane, Ilkeston, Derbyshire DE7 5JD
☎(+44 115) 932 24 03 W http://www.vsl.co.uk/baltex
377106 baltex g ⊜ (+44 115) 944 06 30
♣ £ 1,000,000 ⁂ 25
Managing Dir: A. Marley
Marketing Dir: Charles Wood
Operations Dir: D. Housley
■ Manufacture of specialist knitted fabrics including NOMEX & KEVLAR for reinforcement & fire protection. Also aircraft transparency polishing cloths. BSF137 1989. 98

75-05450 F. Bamford & Co. Ltd
Ajax Works, Whitehill, Stockport, Cheshire SK4 1NT
☎(+44 161) 480 65 07 ⊜ (+44 161) 480 76 93
♧ £ 100,000 (1996) ♣ £ 4,000,000 (1996)
⁂ 39 (1996) 4,000 sq m
Chairman & Managing Dir: A.W. Smith
■ Propellers & sterngear with specialised experience in FP propulsion systems including fully feathering CPPs. F. Bamford (Instruments) Ltd: Flowmeters for fuel, chilled water and hydraulics, diesel engine monitoring systems. 98

75-05630 The Barden Corporation (UK) Ltd
(a division of FAG Kugelfischer George Schaefer AG)
Plymbridge Road, Estover, Plymouth, Devon PL6 7LH
☎(+44 1752) 73 55 55 E sales@barden.co.uk
⊜ (+44 1752) 73 34 81
♣ £ 20,000,000 (1996) ⁂ 350 (1996) c. 500 m2
Managing Director: Dr Graham S Stirling
Business Development Executive: M J Allen
■ Manufacture of high precision ball bearings & bearing assemblies including custom made designs for specific applications. End uses include gyros, servos, generators, fans, cold air units & fuel control systems for military & civil applications. Supplier of Kamatics products including plain and spherical bearings. → 85 98

75-05830 Barrow Hepburn Sala Ltd
4 Old Mill Road, Portishead, Bristol BS20 9BX
☎(+44 1275) 84 61 19
E 101635.547@compuserve.com
⊜ (+44 1275) 84 91 14
⁂ 100
Managing Director: David Eaton
Export Director: Mike Baldwin
Operations Director: Doug Biddulph
Sales and Marketing Manager: John Evans
Purchasing Manager: Keith Pointon
■ Manufacture of specialist harnesses and fall arrest devices to Ministry of Defence standards. 98

75-05900 E.P. Barrus Ltd
Launton Road, Bicester, Oxfordshire OX6 0UR
☎(+44 1869) 36 36 36 ⊜ (+44 1869) 36 36 00
⁂ 120 10,900 sq m
Chairman: Andrew Feilden
Managing Dir: Robert D. Glen
Publicity Mgr: G. Lambert
■ Custom-built Mariner outboard engines. Diesel and petrol Alterrain Vehicles. 98

75-05930 Barry Controls Ltd
(† Muffelite Ltd)
(a subsidiary of Applied Power Inc)
Molesey Road, Hersham,
Walton-on-Thames, Surrey KT12 3PQ
☎(+44 1932) 25 52 55 ⊜ (+44 1932) 25 52 56
♣ £ 5,500,000 ⁂ 80
Managing Dir: D. Golby
Technical Dir: Edward Quinton
European Marketing Mgr: Ian Slater
ID Mgr: D. Mead
Purchasing Mgr: Mrs G. Pearson
■ Vibration, shock and noise control engineers. Design and manufacture of vibration/shock isolators, avionics trays and cases. → 80 98

75-05940 Bartington Instruments Ltd
10 Thorney Leys Business Park,
Witney, Oxfordshire OX8 7GE
☎(+44 1993) 70 65 66 E sales@bartington.com
W http://www.bartington.com/ ⊜ (+44 1993) 77 48 13
♣ £ 645,487 (1994) ⁂ 12 325 sq m
Managing Director: Mrs Tessa Bartington
Technical Director: Geoffrey Bartington
Technical Sales Manager: Cyril Chapman
■ Mag-03 Three Axis Sensors:
These provide precision measurements of static and alternating magnetic fields in three axes. They are available in cylindrical, square or submersible enclosures with measuring ranges from +/-70 to +/-1,000 microtesla.
During 1996 the following products were developed for the US Navy: Mag-03RC Three Axis Range Magnetometer, designed for use in degaussing ranges, and Grad-03-12 Three Axis Flugate Gradiometer, designed for closed loop degaussing applications. 98

75-06000 BAS Components Ltd
(† Bintcliffe Turner Ltd)
(a subsidiary of TT Group plc)
Cramptons Road, Sevenoaks, Kent TN14 5DS
☎(+44 1732) 45 00 11; 74 18 81
⊜ (+44 1732) 45 58 84; 74 31 66
♧ £ 1,188,000 (1996) ♣ £ 10,500,000 (1996)
⁂ 170 (1996) 10,200 m2
Managing Director: Peter Kemsley
Sales and Marketing Director: Ken Hastings
Operations Director: John Fuller
Purchasing Manager: John Perry
■ Fasteners & components, cold forged & CNC precision machined for aerospace & defence applications. Components supplied for missiles & satellite systems. → 85 98

75-06010 Baugh & Weedon Ltd
Beech Business Park, Tillington Road, Hereford HR4 9QJ
☎(+44 1432) 26 76 71 ⊜ (+44 1432) 35 90 17
♣ £ 750,000 (1996) ⁂ 35 (1996)
Managing Director: Mrs P Baugh
Sales Dir: John Bone
■ Ultrasonic non-destructive testing equipment, subsea & topside. Equipment for materials testing, inspection & quality control. Magnetic particle inspection equipment & accessories. Ultrasonic thickness gauges. Quality approval to BS EN ISO9001: 1994. → 80 98

75-06030 Baxter Woodhouse & Taylor Ltd
Group Headquarters
London Road, Adlington, Macclesfield, Cheshire SK10 4NJ
☎(+44 1625) 87 22 61 ⊜ (+44 1625) 87 94 72
♣ £ 12,000,000 (1997) ⁂ 300 (1997) 80,000 sq ft
Chief Executive Officer, Sales and Marketing Director: George Sampson
Financial Director: R V Garlick
Purchasing Manager: Vernon Hopkinson
■ Manufacture ultra-lightweight, high strength conditioned air systems for aircraft and fighting vehicles. Super acoustic ducting (silencers) and fitted thermal and acoustic panel (bags) for aircraft and fighting vehicles and railroad rolling stock. → 70 98

75-06045 BBH Windings Ltd
(a subsidiary of Advance International Group Ltd)
South Church Road, Bishop Auckland, Co Durham
☎(+44 1388) 60 22 65
E 101546.3334@compuserve.com W http://www.advanceint.com/advance-int/bbh.htm
⊜ (+44 1388) 66 31 42
♣ £ 6,150,000 (1995) ⁂ 230 72,000 sq ft
Managing Director: M Jayes (acting)
Marketing Manager: Garry Wetherill
■ High frequency ferrite cored inverters and chokes, laminated transformers, toroidal transformers, lighting transformers and portable tool/isolation transformers. 98

75-06075 BCB International Ltd
Clydesmuir Road Industrial Estate, Cardiff CF2 2QS
☎(+44 1222) 46 44 63 E bcb@celtic.co.uk 497789
⊜ (+44 1222) 49 11 66
⁂ 68 5,500 sq m
Managing Dir: Andrew R. Howell
Vice President & Medical Sales: Hugh Field
Export Sales Mgr: Paul Ames
Purchasing Mgr: Gayle Bentley
■ Manufacture of rations, survival equipment & personal equipment i.e. webbing, rucksacks, first aid & medical equipment, clothing, body armour. 98

75-06080 BDL Systems Ltd
Suite 29, Westmead House, 123 Westmead Road,
Sutton, Surrey SM1 4JH
☎(+44 181) 288 81 12 ⊜ (+44 181) 288 81 13
■ Registered as compliant with ISO9001 Part 1 (BS5750 Pt1) & ISO9000 Parts 3 (Software TickIT scheme) certification is via second party certification (UK MoD) to the following scope: electronic, mechanical & software design, development, production & related consultancy. Cable connected, timed & remotely controlled exploders/devices. Microprocessor based control & measurement systems. Target systems. High intensity lamps & torches, battery chargers. Test equipment & telemetry transceivers.
Under Vehicle Surveillance System, Mallett broad band radio frequency jammer, range of target mechanisms for indoor or outdoor use. 98

75-06100 Beaufort Air-Sea Equipment Ltd
(a subsidiary of BTR plc)
Beaufort Road, Birkenhead, Merseyside L41 1HQ
☎(+44 151) 652 91 51 629478 basebd g
⊜ (+44 151) 653 66 39
Managing Dir: D. Bexon
Sales & Marketing Dir: Alan Southall
Export Sales Mgr: Mike Peddie
■ Manufacture, design & supply of survival equipment including inflatable liferafts, life vests, pressure clothing, immersion & survival suits. NBC personal & collective protection. Fabric structures, e.g. helicopter landing pads and inflatable domes. Pressure and anti-G suits. 98

Contact Reference

75-06125 Beck Electronics Ltd
(a subsidiary of Oxley Group plc)
Main Cross Road, South Denes,
Great Yarmouth, Norfolk NR30 3PX
☎ (+44 1493) 85 62 82 ⊖ (+44 1493) 85 01 69
♣ £ 3,200,000 (1996) ⚉ 110 ⌂ 5,000 sq m
Managing Director, Operations: R L Brook
Sales and Marketing Director: S E Williams
Purchasing Manager: J Brook
■ Manufacturer of EMI/RFI filters, STD ranges approved to BS9000, custom designs & arrays; single layer ceramic & mica capacitors; mains filters. [98]

75-06130 Becker Industrial Coatings
(a subsidiary of A.B. Wilhelm Becker)
Goodlass Rd, Speke, Liverpool L24 9HJ
☎ (+44 151) 448 10 10 ⊖ (+44 151) 448 25 00
♣ £ 20,000,000 (1996) ⚉ 110 (1996)
Managing Director: J R Lyon
Quality Director: H H Bryan
Sales and Marketing Director: N E Rawlinson
Operations Director: P J Kilgallon
Purchasing Manager: T Hughes
■ Special paints for defence equipment. Approved supplier of primers, undercoating & finishing paints against MoD specification type materials. [98]

75-06135 Bedford Trucks
(† AWD (Bedford) Ltd)
(a division of Marshall SPV Ltd; a subsidiary of Marshall of Cambridge (Holdings))
The Airport, Cambridge CB5 8RX
☎ (+44 1223) 37 30 89 ▣ 81208 marfly
⊖ (+44 1223) 37 30 83
Managing Director: A Lines
Marketing and Communications Manager: T Scown-Geary
Operations Director: D Revell
Purchasing Manager: D Waldron
■ All-terrain 4x4 & 6x6 trucks in the range 12 tons to 26 tons gross vehicle weight plus 4x2 & 6x4 on/off road trucks suitable for a wide variety of applications, in particular, logistic support for the modern fighting force. [98]

75-06140 Bedwas Bodyworks
(a division of U.V.E.)
Pantglas Industrial Estate, Bedwas,
Newport, Gwent NP1 8XA
☎ (+44 1222) 88 57 81 ⊖ (+44 1222) 88 73 13
♣ £ 11,000,000 (1997) ⚉ 240 (1997)
Managing Director: Geraint Jones David
Sales and Marketing Director: Derrick G Painter
Sales Manager: Mervyn G Jenkins
Technical Estimator: Lee Perry
■ Manufacturers of specialist bodywork. Police vehicles, custodial and security vehicles. [98]

75-06170 Beldam Crossley Ltd
(† Henry Crossley (Packing) Ltd)
(a subsidiary of Pexion Ltd)
Hill Mill, Temple Rd, Astley Bridge, PO Box 7,
Bolton, Lancashire BL1 6PB
☎ (+44 1204) 49 47 11 ▣ 63149
⊖ (+44 1204) 49 32 03
♣ £ 10,795,000 (1996) ⚉ 222 (1996) ⌂ 7,000 sq m
Managing Dir: C.J. Millington
Sales & Marketing Dir: P. Childs
Operations & Technical Dir: D. Lee
Purchasing Mgr: D. Wareham
■ Crossley Obturator - a one-piece frusto-conical seal of composite construction specifically designed & developed to seal gun breech mechanisms. [98]

75-06190 Belix Co. Ltd
Techview Hse, Cox Lane, Chessington, Surrey KT9 1SB
☎ (+44 181) 397 09 21 ▣ 8951747
⊖ (+44 181) 391 59 69
♣ £ 3,500,000 ⚉ 70
Managing Dir: Brian A. Read
Sales Dir: Stephen L. Pellett
■ Specialist airborne military power supplies. [97]

75-06195 Benchmark Physical Laboratories Ltd
(† BPL Healey Instruments Ltd)
61 Union Street, Dunstable, Bedfordshire LU6 1EX
☎ (+44 1582) 66 83 05 ⊖ (+44 1582) 47 28 63
♣ £ 1,500,000 ⚉ 26 ⌂ 3,000 sq m
Managing Director: B E Cook
General Manager, Sales and Marketing: M J Wall
Operations and Technical Manager: J Swinbourne
Purchasing: S Gould
Sales Administrator: S Angel
■ Design, development & manufacture of instrumentation to Defence Spec. 66.9 part 2. Hermetically sealed for tracked & wheeled vehicles. Also electrical equipment.

● Sheet Road Industrial Estate,
Ludlow, Shropshire SY8 1NE
☎ (+44 1584) 87 24 04/5; 54 18 ▣ 35138 healey g
⊖ (+44 1584) 87 72 74 [98]

75-06200 Bennett Safetywear Limited
Mersey Road, Crosby, Liverpool L23 3AF
☎ (+44 151) 924 39 96 ⊖ (+44 151) 924 65 48
♣ £ 2,800,000 (1996) ⚉ 64 (1996) ⌂ 1,675 sq m
Managing Director: Dr B G Bennett
Sales and Marketing Director: D C Bennett
Purchasing Manager: P F Spooner
■ Manufacture of personal protective clothing & gloves for military, security & industrial applications. [98]

75-06525 Berghaus Ltd
Military Sales Dept
1 Stephenson Rd, District 12,
Stephenson Industrial Estate,
Washington, Tyne and Wear NE37 3HR
☎ (+44 191) 415 02 00 ⊖ (+44 191) 416 98 86
♣ £ 25 000 000 ⚉ 500 ⌂ 85 750 sq ft
Managing Director: Andy Lewis
Sales and Marketing Director: Mike Littlewood
Operations Director: Rob Blundell
Purchasing Manager: Ms Hilary Pearce
Marketing Services Mgr: J. Hardman
■ Design & manufacture of specialist footwear, clothing & load carrying systems. [98]

75-06600 Bestobell Aviation
(a Meggitt Aerospace Company)
Farnham Road, Slough, Berkshire SL1 4UY
☎ (+44 1753) 53 20 31 ▣ 847291 avia g
⊖ (+44 1753) 53 91 26
♣ £ 9,500,000 (1996) ⚉ 160 (1996) ⌂ 11,000 sq m
Managing Dir: E.J. Lewis
Financial Director: M Bodeker
Business Development Director: D J Norman
Operations & Technical Dir: C. Wills
Sales and Marketing Director: R Tyson
Purchasing Manager: T Angio
■ Design & manufacture specialist components for aerospace & defence applications; flexible sleeves, connectors & bellows; heat & fire-resistant hoses, engine & bulkhead fireseals & gaskets, passsenger cabin window seals; thermal & acoustic insulation; coupling systems; periseals & vibreaker units. Aerodynamic moulded seals. [98]

75-06630 B.I. Engineering
Crane Close, Denington Industrial Estate,
Wellingborough, Northants NN8 2QG
☎ (+44 1933) 22 80 12 ⊖ (+44 1933) 44 19 35
♣ £ 4,900,000 (1996) ⚉ 105 (1996) ⌂ 5,600 m2
Managing Dir: J.V. Matcham
■ Design, manufacture, repair & packaging of engine cooling systems, air filtration, exhaust systems. Sheet metal fabrication comprising of stowage bins, fuel & oil tanks for military vehicles. Fabricators in non-ferrous, ferrous & armour materials. Processes, including hot tin dipping, phosphating & painting to full military specifications. [98]

75-06640 BICC Brand-Rex Limited
(an agency company of BICC plc)
Viewfield Industrial Estate, Glenrothes, Fife KY6 2RS
☎ (+44 1592) 77 21 24 ⊖ (+44 1592) 77 53 14
Managing Dir: I. Mack
Commercial Dir: W. McLeod
Export Dir: Alberto Alfina
Sr Product Engineer: Ken Hodge
Product Engineering Mgr: Iain Ballingall
Marketing: Ms Lyn Oswald
■ Cable & cable systems, power communications & fibre optics for defence applications. Complete range of airframe & shipwiring cables. [98]

75-06680 Bigg & Robertson Plastics Ltd
Tribune Drive, Trinity Trading Estate,
Sittingbourne, Kent ME10 2PG
☎ (+44 1795) 47 73 41 ⊖ (+44 1795) 47 65 44
♣ £ 350,000 (1995) ⚉ 14 ⌂ 1,000 sq m
Managing Director: L A Saunders
Purchasing Manager: Ms Valerie Harman
■ Manufacture of vacuum formed plastic components in conductive & conventional materials primarily for electronic equipment housings, covers & storage boxes. [98]

75-06700 M.A. Bigland (Preparations) Ltd
Ludlow Road, Knighton, Powys LD7 1LP
☎ (+44 1547) 52 82 16; 52 82 37 ▣ 35684 talus g
⊖ (+44 1547) 52 05 07
⚉ 42 (1996)
Managing Director: Michael A Bigland
■ Semi-submersible recovery vehicles. Aircraft engine test rig & ground support equipment, boat launching equipment. [98]

75-06720 Birkbys Plastics Ltd
(a subsidiary of the Marubeni Corp.)
Headlands Road, PO Box 2,
Liversedge, West Yorkshire WF15 6QA
☎ (+44 1924) 40 37 21 ⊖ (+44 1924) 40 00 51
♣ £ 50,000,000 (1996) ⚉ 670 ⌂ 223,000 sq ft
Managing Dir: M.J. Hands
Operations/Technical Dir: I. Hunter
Manufacturing Director: Paul Jenkins
Sales & Marketing Dir: I. Rendell
Marketing Manager: V Allport
Marketing Manager: Ms Isla McGukin
Operations Manager: Paul Wilson
■ Full engineering & design service including material evaluation, environmental testing, CAD/CAM design engineeering, specialist post process applications & assembly, injection, structural foam & gas counter pressure moulding capabilities. → 80 [98]

75-07015 Blakley Electrics Ltd
Head Office
Conington Road, Lewisham, London SE13 7LJ
☎ (+44 181) 852 43 83 ⊖ (+44 181) 318 52 84
♣ £ 7 400 000 (1995/6) ⚉ 125+
Chairman: D.W. Blakeley
Managing Dir: P.J. Blakley
Technical Dir: J.M. Lay
Finance & Purchasing Dir: J.O. Church
Sales & Marketing Dir: M.S. Cole
Sales Office Manager Lewisham: J N Robinson
■ Manufacture of outdoor switchboards, residal current devices, transformers, power & lighting systems.

● Marsh Lane, Ware, Hertfordshire SG12 9QQ
☎ (+44 1920) 46 39 61 ⊖ (+44 1920) 46 46 82
Production Operations Dir: B.J. Cobb
■ Depots at Edinburgh, Wakefield, Lewisham/London.

● Unit 26, Stenhouse Ind. Estate, Stenhouse Mill Wynd,
Edinburgh, Scotland EH11 3XX
☎ (+44 131) 443 80 05 ⊖ (+44 131) 443 35 14
Dir Scotland: G. Rason
General Mgr: K.I. Rason

● Unit 55 Monckton Rd Ind. Estate,
Wakefield, West Yorkshire WF2 7AL
☎ (+44 1924) 29 00 66 ⊖ (+44 1924) 29 04 55
Dir North/Midlands: M.S. Cole
Sales Depot Mgr: B.J. Miles [97]

75-07240 BMS Trading Ltd
BCM BMS Arms, London WC1N 3XX
☎ (+44 1747) 87 10 91 ⊖ (+44 1747) 87 14 27
Managing Dir: Derek Bremner
Marketing & Sales Dir: Robert Partridge
■ Manually operated 5.56mm military rifles. [98]

75-07500 Bodycote Metallurgical Coatings Ltd
(† Zinc Alloy Ltd)
(a division of Bodycote International plc)
Shakespeare Street,
Wolverhampton, West Midlands WV1 3LR
☎ (+44 1902) 45 29 15 ⊖ (+44 1902) 35 29 17
♧ £ 2,300,000 (1995) ♣ £ 4,400,000 (1995)
⚉ 102 (1996) ⌂ 5,000 m2
Managing Director: Carl Savage
Sales Director: R D Blakemore

■ Sherardizing - a zinc iron alloy coating for steel components, used for its wear-resistant, corrosion protection properties → 80 [98]

75-07600 Bondina NBC Materials
(part of Freudenberg Nonwovens L.P.)
PO Box 3, Greetland, Halifax, W.Yorks HX4 8NJ
☎ (+44 1422) 31 30 00 ⊠ 51198
⊖ (+44 1422) 31 32 32
♧ £ 8 000 000 (1995) ♣ £ 30 000 000 (1995)
👥 400 (1995) 🏠 36 000 sq m
Managing Dir: Dr D. Ireland
NBC Business Mgr: Dr C. Barson
Technical & Marketing Assistant: Miss Karen Smith
■ All NBC fabric concepts are based in part upon non-woven fabrics with additional carbon powder, granules or spherical absorbers to provide exceptional chemical protection. The additional properties of flame retardancy, air permeability & oil & water repellency can all be added without compromising the chemical protection properties. [97]

75-07670 Booth Industries Limited
(a member of Booth Industries Group plc)
Nelson Street, PO Box 50, Bolton BL3 2AP
☎ (+44 1204) 36 63 33 ⊖ (+44 1204) 38 08 88
Managing Director: Keith Chilton
Operations Director: John Hamer
Sales and Marketing Director: Michael Stott
Purchasing Manager: Mrs Elaine Whittle
■ Hardened shelters and doors to withstand shock, blast, fire, gas and bacteriological attack. Specialist constructors of arch liner packages for hardened shelters. Water storage tanks. [98]

75-07740 Border Holdings (UK) Ltd
The Grove, Craven Arms, Shropshire SY7 8DB
☎ (+44 1588) 67 27 11 ⊠ 35587
⊖ (+44 1588) 67 27 27
♣ £ 10,000,000 👥 45 🏠 10,000 sq m
Managing Director: David P Beddow
Operations Director: Trevor Jordan
Purchasing Manager: E W Kok
Company Secretary: A J Overs
■ Replacement parts for motor vehicles (seats, hoods, engines, gear boxes, axles). Spare parts for Landrover vehicles. Approved to BS EN ISO 9002 1994. [98]

75-07750 Boss Group Limited
Grovebury Rd, Leighton Buzzard, Beds. LU7 8SR
☎ (+44 1525) 21 50 00 ⊠ 825781
⊖ (+44 1525) 21 51 05
👥 700 🏠 45 acres
Chairman: R.A. Bischof
Managing Dir: J. Porter
Publicity Mgr: Ms Louise England
■ Lift trucks (front-lift, side-lift & container-handling trucks). Integrated warehouse systems. [97]

75-07770 Bowers Group
32 Leeds Old Road, Bradford BD3 8HU
☎ (+44 1274) 66 07 09 ⊖ (+44 1274) 66 38 96
♣ £ 7 000 000 👥 100
Sales Mgr: Lauri Wilson
General Sales Mgr: Richard Sharp
■ Internal micrometers (mechanical, electronic or digital), thread gauges & hardness testers. Dial calipers. Cylinder bore gauges. [97]

75-07780 Bowthorpe plc
Gatwick Road, Crawley, West Sussex RH10 2RZ
☎ (+44 1293) 52 88 88
Ⓦ http://www.bowthorpe-plc.com
⊖ (+44 1293) 54 19 05
♧ £ 22,500,000 ♣ £ 470,400,000 (1995)
👥 7,000 (1995)
Chairman: H A Vice (Non-executive)
Chief Executive: N K Brookes
Financial Director: C M McCarthy
Treasurer: Miss Helen Ralston
■ International group of companies engaged in the design and manufacture of electronic and electrical components used by defence, aerospace and avionics industries. Products cover aircraft instrumentation, flight data recorders, data acquisition, data recording, position sensors, securing and fixing devices, power supplies, air data computers, engine monitors, ice and snow detectors, heat sinks, rotary switches.

Subsidiary companies and divisions:
Bowthorpe Thermometrics
Hellermann Insuloid
Hellermann Electric
Hellerman Do Brasil Ind & Comercio Ltda
Hellerman France SA, France
Redpoint Thermalloy Ltd
Wessex Advanced Switching Products Ltd
B&D Instruments & Avionics Inc, USA
El.Bo.Mec Srl, Italy
Paul Hellermann GmbH, Germany
Autronics Corp., USA
Monitor Products Corp, USA
Optim Electronics Corp, USA
The Flight Data Company Ltd
Thermometrics Inc, USA
Thermalloy Inc, USA
Penny & Giles Aerospace Ltd
Penny & Giles Electronic Components Ltd
Penny & Giles Position Sensors Ltd
Penny & Giles Data Systems Ltd
Tempo Instruments Inc, USA
Tyton Hellerman Corp., USA [98]

75-07785 Bowthorpe Thermometrics
(a subsidiary of Bowthorpe plc)
Crown Industrial Estate, Priorswood Road, Taunton, Somerset TA2 8QY
☎ (+44 1823) 33 52 00
Ⓦ http://www.thermometrics.com
⊖ (+44 1823) 33 26 37
Managing Director: D Kennedy
Sales and Marketing Director: K Wilson
Operations Director/Purchasing: P A Chalker
Marketing Analyst: C Westhead
■ Design and manufacture of high performance NTC and PTC thermistors for temperature compensation, measurement, control requirements plus self-regulating PTC thermistor heaters. [98]

75-07790 BP Chemicals Ltd Advanced Materials Division
Aerospace Composites
Avonmouth Rd, Avonmouth, Bristol BS11 9DU
☎ (+44 117) 937 93 45 ⊖ (+44 117) 982 46 40
👥 320 🏠 22,300 sq m
Managing Director: John L Williamson
Operations Director: Kevin L Payne
Business Development Manager: John M Janke
Business Development: David J Hendy
■ Design, development, manufacture and repair of advanced composite structures and components for aerospace industry applications. Current products include aircraft composite structures and nacelle components, Extensive resources available for customer directed development programmes. [98]

75-07800 BP Solar Limited
(a subsidiary of British Petroleum plc)
Chertsey Road, PO Box 191, Sunbury on Thames, Middlesex TW16 7XA
☎ (+44 1932) 77 95 43 Ⓔ hammonm@bp.com
⊠ 296041 bp sunag ⊖ (+44 1932) 76 26 86
♣ US$ 60,000,000 (1996) 👥 500 (1996) 🏠 6,400 sq ft
Managing Director: D J Shah
Operations Director: J Driver
Sales and Marketing Director: R Evans
Marketing Executive: Mark Hammonds
Global Sciences and Technology Manager: A K Turner
Purchasing Manager: Mrs I Wright
■ Manufacture of solar electric power systems for remote applications including telecommunications, navigational aids, telemetry, lighting and cathodic protection. Solar battery chargers for vehicles, tactical radios. Complete solar powered obstruction light, water pumping and purification systems.
Sister Companies:
BE Solar East Asia, Malaysia
BP Solar Inc, USA
BP Solar Australia Pty Ltd
BP Solar España SA
BP Thai Solar Corp. Ltd
BP Solar Arabia Ltd
Tata BP Solar India Ltd [98]

75-07850 Bradford Cylinders Ltd
Soho Works, Allerton Rd, Bradford, West Yorkshire BD8 0BA
☎ (+44 1274) 49 56 11 ⊖ (+44 1274) 54 71 19
♣ £ 5,000,000 (1996) 👥 75 (1996)
Managing Dir: D.G. Stocks
Sales Dir: A.V. Attwood
■ Hydraulic & pneumatic actuators, piston accumulators brake/balance systems, propeller shaft sealing systems. [98]

75-07855 Bradley Doublelock Ltd
Victoria Works, Bingley, West Yorkshire BD16 2NH
☎ (+44 1274) 56 04 14 Ⓦ http://www.syltone.co.uk
⊖ (+44 1274) 55 11 14
♣ £ 6,000,000 (1996) 👥 120 (1996)
Managing Dir: P.A. Hanson
Operations & Technical Dir: W.P. Austin
Sales & Marketing Dir: L.I.J. Lambert
Purchasing Manager: P O'Hagen
■ Towing equipment for military vehicles & trailers. Over-run coupling devices & towing hooks. [98]

75-07860 Bradley Electronics Ltd
(a member of the E.I.S Group)
Head Office
Neasden Lane, London NW10 1RR
☎ (+44 181) 450 78 11
Ⓔ 101336.624@compuserve.com ⊠ 25583
⊖ (+44 181) 450 83 57
♧ £ 800,000 (1996) ♣ £ 5,000,000 (1997)
👥 90 (1997) 🏠 2,601 sq m
Managing Director: D Ray
Director of Marketing: B R Epton
Purchasing Manager: J Herzog
Technical Librarian: David Hoffman
■ Design, development, manufacture, installation and servicing of electronic equipment for ground, sea and airborne use. Aircraft test equipment; Ground navigational aids; Ships polemast antenna systems with separate transmit & receive UHF antennas; VLF receiving equipment; Auxiliary power units; ground telemetary monitoring equipment; meteorological systems. Repair and recalibration of electrical and electronic test equipment.
● Maylands Avenue, Hemel Hempstead, Herts HP2 4SP
☎ (+44 1442) 24 22 33 ⊠ 82110
⊖ (+44 1442) 27 88 40
Systems Mgr: R. Peters
■ Complete electrical systems for armoured & soft skinned vehicles, vehicle suppression equipment. [98]

75-08000 Brett Harris Ltd
Waterloo Mills, Howden Road, Silsden, West Yorkshire BD20 0HA
☎ (+44 1535) 65 44 79 ⊖ (+44 1535) 65 25 03
■ Manufacture of synthetic sleeping bags, windshield and fibre filled jackets. [98]

75-08350 Bridon International Ltd
Excelsior Works, Western Ave, Cardiff CF4 3XW
☎ (+44 1222) 61 93 42 ⊖ (+44 1222) 61 42 64
Chief Inspector: B.R. Nunn
■ Wire rope slings, strops & bridles.
● Carr Hill, Balby, Doncaster, S.Yorks DN4 8DG
☎ (+44 1302) 34 40 10 ⊠ 547981 ropes g
⊖ (+44 1302) 640 48
Technical Dir: Dr A.G. Stacey
Sales Mgr: J. Tandy
Sales Mgr Wire: A.G. MacGregor
■ Underwater towing cables, aircraft carrier arrester & bridle cable, barrier packs, mooring & salvage ropes, target towing & glider launch, balloon cables, tank & vehicle towing. Armoured electromechanical cables/stainless steel wire for coils, springs, whip aerials/UHT wire. [97]

75-08400 Bridport Aviation Products
Britannic Aviation
(part of the Bridport-Gundry (UK) Ltd Group)
The Court, West Street, Bridport, Dorset DT6 3QU
☎ (+44 1308) 45 66 66 ✉ NETTING Bridport
⊖ (+44 1308) 45 66 05
♣ £ 7,000,000 👥 170+
Managing Director: J Bowden
Production Director: C Dowle
Sales Manager: Alan C Newman
Defence Sales Manager: J R Laurence
Commercial Sales Manager: S R Broom
Purchasing Manager: I Kentfield

Contact Reference

■ Camouflage & concealment systems. Decoys. Helicopter slings, underslung & internal restraint nets. Container nets for ship-to-shore resupply & aerial delivery. Air cargo restraint systems. Lightweight torpedo recovery systems. Sports nets. Design & development facilities. Aircraft spares. → 85 [98]

75-08550 Brighton Sheet Metal Ltd

The Hyde, Auckland Drive,
Lower Bevendean, Brighton BN2 4JW
☎ (+44 1273) 60 22 16 ⊖ (+44 1273) 67 41 53
♣ £ 4,500,000 (1995/96) 👥 80 ⌂ 2,900 sq m
Managing Dir: Philip N. Clark
Technical Sales Dir: Malcolm P. Holman
Operations: B. Jonen
Purchasing Mgr: M. Cooper
■ Specialists in sub-contract, precision CNC fabrications & presswork with phosphating, plating & comprehensive paint finishing facilities. Approved to ISO9002. [97]

75-08600 Brimotor Ltd

Unit 10, Orchard Centre, North Farm Road,
Tunbridge Wells, Kent TN2 3DR
☎ (+44 1892) 53 75 88 ⊖ (+44 1892) 52 77 24
Director: Guy Mossery
■ Portable, mobile & fixed generators & telescopic floodlighting. [98]

75-08760 Bristol Uniforms Ltd

Wathen Street, Staple Hill, Bristol BS16 5LL
☎ (+44 117) 956 31 01
[E] 101500.1114@compuserve.com
⊖ (+44 117) 956 59 27
♣ £ 7,000,000 👥 120
Managing Director: Ian Hill
Commercial Director: Roger Startin
■ Manufacture of fire protective clothing, fireman helmets, safety & rescue equipment, designed specifically for civil & military airports. [98]

75-08882 British Aerospace Airbus Limited

(a subsidiary of British Aerospace plc)
PO Box 77, Bristol BS99 7AR
☎ (+44 117) 969 38 31 ⊖ (+44 117) 936 28 28
👥 4,100 (1997)
Chairman: Mike Turner
Managing Dir: Chris Geoghegan
Product Operations Dir: Ray Wilson
Head of Public Affairs: Howard Berry
Public Relations Officer: Trevor Mason
Head of Airbus Procurement: Alan Wakeham
■ Design, manufacture & supply of components & major structural assemblies for Airbus & other programmes from its Filton & Chester sites. Aviation Services subsidiary company which overhauls, maintains & converts civil & military aircraft. [98]

75-08886 British Aerospace Military Aircraft

Warton Aerodrome, Preston, Lancashire PR4 1AX
☎ (+44 1772) 63 33 33 ⊠ 67627/8 g
⊖ (+44 1772) 85 47 41
👥 15,000 (1997)
Managing Director: Mike Rouse
Public Relations Manager: Andrew Bunce
Public Relations Officer: Ms Julia Sevin
■ Main activities include development of Eurofighter with Germany, Italy and Spain. To market, adapt, manufacture and support export variants of the JAS-39 Gripen with SAAB. Production and support of the Panavia Tornado with Dasa and Alenia. Production and support of the VSTOL Harrier and Sea Harrier, joint production and support of the AV8B/GR7 Harrier I and Harrier II plus for the US Marine Corps, Spanish and Italian Navies and T45A Goshawk for the US Navy with McDonnell Douglas; development, production and support of the Hawk advanced jet trainer/light fighter. Complete systems integration capability. BAe Military Aircraft has been involved in BMD studies for many years. It currently provides the UK focal point for NIAG SG-37 on extended air defence.

Brough
Brough, North Humberside HU15 1EQ
☎ (+44 1482) 66 71 21 ⊠ 592735
⊖ (+44 1482) 66 39 25
Public Relations: Robert Gillies ☎ (+44 1482) 66 32 70

Dunsfold
Dunsfold Aerodrome, Godalming, Surrey GU8 4BS
☎ (+44 1483) 27 21 21 ⊠ 859475
⊖ (+44 1483) 20 03 41

Samlesbury
Samlesbury Aerodrome, Balderstone, Lancashire BB2 7LF
☎ (+44 125) 81 23 71 ⊠ 63435 ⊖ (+44 125) 76 80 00 [98]

75-08890 British Aerospace (Systems & Equipment) Ltd (BASE)

(a subsidiary of British Aerospace plc)
Clittaford Road, Southway, Plymouth, Devon PL6 6DE
☎ (+44 1752) 69 56 95 ⊠ 45708
⊖ (+44 1752) 69 55 00
👥 1,000 ⌂ 35,000 sq m
Managing Dir: Nigel Randall
Deputy Managing Dir: Terry Bleakman
Finance Director: Anthony Pettit
Engineering Dir: R.J. Minion
Purchasing Manage: David Jeremiah
Head of Public Affairs: Chris Tear [97]

75-09000 British Aluminium Plate

(† Alcan Plate)
(Part of Alcan Aluminium UK Limited)
Kitts Green Rd, PO Box 383, Birmingham B33 9QR
☎ (+44 121) 252 80 00 ✉ ALCANPLATE Bham
⊠ 337417 ⊖ (+44 121) 252 80 01
♣ £ 60 000 000 👥 430
Managing Dir: M.G. Wallis
Sales & Marketing Dir: Bernard Hopkins
Technical Support Manager: D J Holmes
Business Development Mgr: G. Bevan
■ Manufacturer & distributor of plate in high-technology aluminium alloys including aluminium lithium alloys, semi-fabricated products & other advanced materials in the aerospace & defence markets. [97]

75-09020 British Aluminium Tubes

(a subsidiary of British Aluminium Ltd)
Studley Road, Redditch, Worcestershire B98 7HN
☎ (+44 1527) 48 45 00 ⊖ (+44 1527) 48 45 01
♣ £ 23,000,000 (1995) 👥 156 (1997)
General Mgr: John Orwin
Sales Mgr: John E. Blackham
Business & Technical Development Mgr: Dr C. Jackson
Commercial Mgr: Steve Rowbotham
■ Aluminium tubes seamless extruded, seamless drawn & porthole drawn tubes. [98]

75-09030 British Filters Ltd

(a division of Quinton Hazell Automotive Ltd)
St. Modwen Road, Plymouth, Devon PL6 8LH
☎ (+44 1752) 66 76 75 ⊖ (+44 1752) 67 34 15
♣ £ 10,000,000+ (1997) 👥 197 (1997)
Managing Director: R Morgan
Product Manager Aerospace/Defence: A P H Bamford
Purchasing Manager: P Turner
■ Aerospace fuel, oil, hydraulic, air filters, bulk fluid filters and shock absorbing units. MOD, CAA and Rolls-Royce approved. [98]

75-09150 British International Industries Ltd (BII)

(a subsidiary of Diesel Power Ltd, UK)
12 Mitcham Industrial Estate, Mitcham CR4 2AP
☎ (+44 181) 646 33 75 ⊠ 929475
⊖ (+44 181) 640 84 71
Managing Director: V von Dewitz
Sales Director: C R Cradock
Purchasing Manager: Ms Sarah Mills
■ Suppliers of military & commercial aircraft spare parts, ground support equipment, diesel generating sets & diesel engine spares. [97]

75-09500 British Rollmakers Corporation

(† Midland Rollmakers Ltd)
(a subs. of Sheffield Forgemasters Ltd)
Weston Road, Crewe, Cheshire CW1 1DB
☎ (+44 1270) 21 34 12 ⊠ 36169
⊖ (+44 1270) 21 23 51
♣ £ 27 000 000
Managing Dir: Haydn Beresford
■ Production of cast iron & steel alloy rolls.
(† R.B. Tennent Ltd)
(a subs. of Sheffield Forgemasters Ltd)
● Whifflet Foundry, Coatbridge, Scotland ML5 4RR
☎ (+44 1236) 42 25 61 ⊠ 77484 tennebt g
⊖ (+44 1236) 42 90 48
Works Mgr: A.F. Robertson
■ Specialists in cast steel roll manufacture. [97]

75-09570 British Sarozal Ltd

Handrail House, Maygrove Rd, London NW6 2EG
☎ (+44 171) 328 21 11 ✉ SAROZAL ⊠ 298655
⊖ (+44 171) 624 51 44
⌂ 7 500 sq ft
Managing Dir: J.M. Treister
Sales & Marketing Dir: A. Baptista
Operations & Technical Dir: Leonard Simmonds
■ Manufacture of solid state direct replacements for thermionic valves. Design service for special solid-state equivalents of valves and RF power transistors to meet a wide range of defence requirements. Sole suppliers of Rank Brimar range of radar cathode ray tubes for marine and airport radars including Siemens/Plessey Watchman Radar. Supply a full range of electronic & electrical maintenance spare parts for equipment & installations used on British ships sold to foreign Navies. → 60 [97]

75-09600 British Steel Engineering Steels

Aldwarke Lane, PO Box 50, Rotherham S60 1DW
☎ (+44 1709) 37 12 34 ⊖ (+44 1709) 82 62 33
♣ £ 500,000,000 (1996) 👥 3,800 (1996)
Managing Director: Don Turner
Commercial Director: Martin Howell
Technical Director: Gwyn Davis
Personnel Director: Garry Carter
Finance Controller: Ian Cooper
Director Manufacturing: Clive Jackson
Managing Director, Bright Bar: Bob Millard
Managing Director, Stockholders: Graham Stringer
Managing Director, Metal Recycling: Malcolm Armson
Logistics Manager: Bill Ford
Manager Information Technology: Mark Pope
Manager Business Development: Richard Welburn
General Sales Manager: David Jackson
■ Manufacture & supply of stainless, alloy & carbon steel billets & bars to aerospace/defence engineering customers or through forgers, tubemakers, rerollers & stockists. Remelted steels including (Vacuum Arc Remelted and Electroslag Refined). Products cover the full range of International & UK steel specifications.

Aldwarke - Rotherham Works
Aldwarke Lane, PO Box 50, Rotherham S60 1DW
☎ (+44 1709) 38 21 41 ⊖ (+44 1709) 82 62 33
■ Freecutting, machining, forging, general engineering and bearing steels for automotive and related markets.

Roundwood Mill
Rotherham Works, Aldwarke Lane, PO Box 50,
Rotherham S60 1DW
☎ (+44 1709) 38 21 41 ⊖ (+44 1709) 82 62 33
■ Specialist producer of bar in coil form.

Stocksbridge Works
Stocksbridge, Sheffield S30 5JA
☎ (+44 114) 288 23 61 ⊖ (+44 114) 288 50 33
■ Specialist steels for aerospace, power generation, oil and gas extraction, general engineering and bearing industries. Supplier of steel to the worldwide automotive industry.

Thrybergh Bar Mill
Rotherham Works, Aldwarke Lane, PO Box 50,
Rotherham S60 1DW
☎ (+44 1709) 38 21 41 ⊖ (+44 1709) 82 62 33

Tinsley Park Mill
Shepcote Lane, Sheffield S9 1US
☎ (+44 114) 243 43 53 ⊖ (+44 114) 243 16 40
■ Production of spring steel flats to BS 970 Part 2 1988, DIN 4620, DIN 59145, DIN 59146 and DIN 1017 for automotive, railway and general engineering applications.

Bright Bar
Bloomfield Road, Tipton, West Midlands DY4 9EP
☎ (+44 121) 521 22 22 ⊖ (+44 121) 520 89 29
Managing Director: Bob Millard
General Sales Manager: Ian Clarke
■ Bright drawn, turned and precision ground steel bars

Lee Bright Bar
PO Box 54, Meadowhall, Sheffield S9 1HU
☎ (+44 114) 242 23 11 ⊖ (+44 114) 242 59 69

Roundwood Bright Bar
Aldwarke Lane, PO Box 73, Sheffield S60 1RS
☎ (+44 1709) 71 01 01 ⊖ (+44 1709) 71 94 85
■ Bright turned and precision ground bar and heat treated products.

Stockholders
Coleford Road, Sheffield S9 3QE
☎ (+44 114) 244 72 64 ⊖ (+44 114) 243 09 41

Managing Director: Graham Stringer
General Sales Manager: Peter Gill
■ Stockholding, processing and distribution of black/bright steel bars. Carbon, freecutting and alloy steel specifications, tool and high speed steels.

Sheffield Depot
Coleford Road, Sheffield S9 3QE
☎(+44 114) 244 72 64 ⊜ (+44 114) 243 09 61

West Bromich Depot
Oldbury Road, West Bromich, West Midlands B70 9DX
☎(+44 121) 553 49 51 ⊜ (+44 121) 500 53 84

Bolton Depot
Unit 2, Pullman Court, Lever Street, Manchester Road, Bolton BL2 1HL
☎(+44 1204) 37 09 99 ⊜ (+44 1204) 39 66 84

Hetton Depot
Hetton Lyons Industrial Estate,
Hetton-le-Hole, Houghton-le-Spring,
Tyne and Wear DH5 0RD
☎(+44 191) 526 32 88 ⊜ (+44 191) 517 01 38

Slough Depot
Oxford Avenue, Slough, Berkshire SL1 4LQ
☎(+44 1753) 57 01 71 ⊜ (+44 1753) 69 49 85

AMS Metal Stockholders
Unit A Lecturers Close, Bolton, Lancashire BL3 6DG
☎(+44 1204) 39 44 74 ⊜ (+44 1204) 53 17 62
General Manager: John Burgess
■ Metal service centre, stocking, cutting, boring and machining. Aerospace and speciality materials.

Midland Service Centre
Horsley Fields, PO Box 25, Wolverhampton WV1 3DY
☎(+44 1902) 35 17 73 ⊜ (+44 1902) 35 17 41
Director and General Manager: Ian Maw
■ Specialist distributor of bars and billets. Cutting and processing facilities including reeling, shotblasting, billet burning, shearing, cropping and sawing.

Metal Recycling
Station Works, Warwick Road,
Tyseley, Birmingham B11 2HL
☎(+44 121) 706 11 10 ⊜ (+44 121) 706 84 59
Managing Director: Malcolm Armson

Allen Rowland
Station Works, Warwick Road,
Tyseley, Birmingham B11 2HL
☎(+44 121) 706 11 10 ⊜ (+44 121) 706 84 59
■ Metal reclamation and merchanting of ferrous scrap, low residual scrap for steelworks and iron foundries.

Parkfield Iron and Steel
Cable Street, Bliston Road, Wolverhampton WV2 2JE
☎(+44 1902) 45 15 87 ⊜ (+44 1902) 35 10 67
■ Ferrous scrap collection and processing.

Church and Bramhall Metals
Ashburton Road East, Trafford Park, Manchester M17 1JL
☎(+44 161) 872 57 15 ⊜ (+44 161) 872 83 51
■ Ferrous and non-ferrous scrap collection and processing. [98]

75-09750 British Vita plc
Vita Salford Division
Seaford Road, Salford M6 6NB
☎(+44 61) 736 53 43 ⊜ (+44 61) 745 75 21
♣ £ 16,000,000 (1996) ⅲ 140 (1996)
Managing Director: T Hardiment
Business Manager: David Dykes
Purchasing Manager: Alan Egerton
■ Manufacturer of polymeric coated fabrics, closed cell PVC foams, adhesives, sealants, coatings & calendered compounds. [98]

75-09900 G.B. Britton
(a subs. of UK Safety Ltd)
Lodge Rd, Kingswood, Bristol BS15 1JB
☎(+44 117) 967 40 14 ⊜ (+44 117) 967 44 74
♣ £ 27 000 000 (1995) ⅲ 500 ⌂ 125 504 sq ft
Managing Dir: John Newman ☎(+44 117) 935 33 36
Financial Dir: Ian Smith ☎(+44 117) 935 33 36
Sales Dir: Paul Cockburn ☎(+44 117) 935 27 77
Admin. Exec.: Tony Bateman
Advertising Mgr: Ms Laura Carter ☎(+44 117) 935 27 77
■ Defence, military, combat & protective footwear manufacturer. [97]

75-09980 Brock's Explosives Ltd
(a subs. of Explosive Developments Ltd)
Gateside, Sanquhar, Scotland DG4 6JP
☎(+44 1659) 505 31 ⊜ (+44 1659) 505 26
♣ £ 2 000 000 (1995) ⅲ 12
Factory Mgr: James Young
■ Military and commercial pyrotechnics including bird scaring cartridges, buoyant smock grenades, arctic/snow and smoke grenades for safety purposes. Practice bomb cartridges. High/plastic explosive filling capacity. Comprehensive research and development, test facilities. [97]

75-10100 Brookes & Gatehouse Ltd
Premier Way, Abbey Park, Romsey SO51 9AQ
☎(+44 1794) 51 84 48 [W] http://www.marinedata.co.uk/b&g/b&ghome.html ⊜ (+44 1794) 51 80 77
♣ £ 5,600,000 (1996) ⅲ 70 (1996) ⌂ 30,000 sq ft
Managing Director: Robert Hill
Sales Manager: Simon Pettit
Purchasing Manager: Tim Coletta
■ Design and manufacture of autopilot's wind and water speed direction sensors, heading reference units, logs, ultrasonic echo sounders and weather monitoring equipment. GPS, electronic charting and head up displays. [98]

75-10103 Brookhouse Paxford Ltd
(a subsidiary of Brookhouse Patterns Ltd)
Redwongs Way, Huntingdon, Cambridgeshire PE18 7HB
☎(+44 1480) 45 35 37 ⊜ (+44 1480) 41 31 25
♣ £ 3,000,000 (1995) ⅲ 70 (1997) ⌂ 3,500 sq m
Managing Dir: Bernard Garner
Operations Director: Tony Harrison
Purchasing Manager: Neil Johnson
Technical Sales Representative: Peter Ketteridge
■ Composites & reinforced plastics engineering. [98]

75-10150 David Brown Defence Equipment Ltd
(a subsidiary of David Brown Group plc)
Park Road, Lockwood,
Huddersfield, West Yorkshire HD4 5DD
☎(+44 1484) 42 21 80 ⊜ (+44 1484) 43 52 92
♣ £ 26,000,000 (1996) ⅲ 270 (1996)
Managing Director: R S Tadman
Sales and Marketing General Manager: G Vere-Laurie
Purchasing Manager: V Hebda
■ Transmissions for tracked armoured vehicles. Final drives, test drives. Gears & gear related products. Test rigs & test equipment. Overhaul & upgrade of transmissions & drive systems.

Subsidiaries:
David Brown Heatech Ltd
David Brown Trans Axle Ltd
Ganz David Brown [98]

75-10250 John Brown Engineering Ltd
Kvaerner Energy Limited, Thermal Power Division
(a subsidiary of Kvaerener plc)
Glasgow Road, Clydebank, Dunbartonshire G81 1YA
☎(+44 141) 952 20 30 ✉ JAYBENG ⊠ 778395 jabeng g
⊜ (+44 141) 952 53 27
ⅲ 850 (199)
Managing Director: C J Packard
Sales and Marketing Director: C Shields
Marketing Manager: S J Rodgers
■ Industrial gas turbines and associated equipment for power generation (10.4 MW to 224 MW) and mechanical drive applications (14,000 HP to 38,000 HP). Turbines can be integrated into total energy systems providing power, and steam for district heating, desalination and repowering applications. Combined cycle plants for fuel economy. Projects for complete power stations undertaken with full turnkey scope. [98]

75-10330 S.G. Brown
(a division of Voster Thornycroft (UK) Ltd)
1 Garnett Close, Greycaine Industrial Estate,
Watford WD2 4JL
☎(+44 1923) 47 08 00 ⊜ (+44 1923) 47 08 38
Managing Director: J Blott
Military Sales Manager: J Harrison
■ Warship steering control & automatic pilot systems, navigation systems & equipment. [98]

75-10385 Brownell Ltd
(† Brownell Climatic Controls Ltd)
Helena Works, Mordaunt Rd, London NW10 8PX
☎(+44 181) 965 92 81
[E] 100655.3520@compuserve.com ⊠ 8952723 brnell g
⊜ (+44 181) 965 32 39
♣ £ 2 100 000 ⅲ 70 ⌂ 44 000 sq ft
Chairman & Managing Dir: G.G. Hooper
Director: N.G. Hooper
Technical Dir: Martin K. Partridge
■ Control & measurement equipment for the prevention of problems caused by moisture & other contaminents. This includes mould growth, condensation, corrosion & electrical short circuits. Rapid decompression valves for airborne & air transportable equipment. [97]

75-10400 Brunton's Propellers Ltd
384 Feraday Close, Gorse Lane Industrial Estate,
Clacton-on-Sea, Essex CO15 4TR
☎(+44 1255) 42 00 05
[E] 101713.3563@compuserve.com
⊜ (+44 1255) 42 77 75
♣ £ 850,000 (1996) ⅲ 10 (1996) ⌂ 300 m2
General Manager: Adrian Miles
Technical Sales: Sean Bianchi
Purchasing Manager: Phil Gant
■ Designers & manufacturers of marine propellers, sterngear & associated castings. Accredited by all leading classification societies. AQAP 4 ED2. Approved by MoD, BS5750/ISO9000 (September 94). The marine propellers of our supply are manufactured to ISO R484 class 'S' I or II to suit patrol craft, supply vessels, tugs, fishing boats etc, as appropriate. [98]

75-10450 Brush Electrical Machines Ltd (BEM)
(† British Electrical Machines)
Head Office
PO Box 18, Loughborough, Leics LE11 1HJ
☎(+44 1509) 61 15 11 ⊠ 341091 brulob g
⊜ (+44 1509) 61 04 40
♣ £ 100 000 000 ⅲ 1 600
Managing Dir: D.N. Wells
Operations & Technical Dir: J.S. Tyrell
Sales & Marketing Dir: G.A. Bazire
Purchasing Mgr: N. Buxton
Business Development Mgr: B.L. Woodward
■ Manufacturers of large electric motors, generators & associated control equipment including purpose built machines. [97]

75-10830 BTR Silvertown Ltd
(a subs. of BTR plc)
HQ and Precision Mouldings Division
Horninglow Road, Burton-on-Trent, Staffs DE13 0SN
☎(+44 1283) 51 05 10 ⊠ 34419 btrbtn g
⊜ (+44 1283) 51 00 52
♣ £ 20,000,000 (1997) ⅲ 340
Managing Director: Ian Dugan
Operations Director: Paul Myerscough
Divisional Dir: Ewart James
Business Development Mgr: John Neave
Purchasing Manager: Ken Markham
Sales Manager: G Shaw
■ Manufacture of high performance elastomeric mouldings for industrial, aerospace & military applications. Specialists in rubber to metal bonding, anti-vibration & shock control. Fighting vehicle running gear such as wheels, bushes and pads.

Andre Silentbloc Division
(† Andre Ltd/Silentbloc Ltd)
Horninglow Road, Burton-on-Trent, Staffs DE13 0SN
☎(+44 1283) 51 05 10 ⊜ (+44 1283) 51 00 52
Business Manager: Andy Toft
■ Anti-vibration mountings, AFV running gear (bushes, links, top rollers, track pads, wheels), rubber injection mouldings, rubber bonded components.

Engineering Division
(Cutless Bearings & Expansion Joints)
Horninglow Road, Burton-on-Trent, Staffs DE13 0SN
☎(+44 1283) 51 05 10 ⊜ (+44 1283) 51 00 52
Sales Mgr: Ms Isabelle Lloyd
Business Mgr: Mark Goldby
Technical Mgr: Martin Stockwell
■ Rubber lined water lubricated bearings for marine, dredging & industrial applications. Flexible rubber expansion joints for pipework applications. Shock & anti-vibration mounts for naval applications.

Technical Division (Precision Mouldings)
Horninglow Road, Burton-on-Trent, Staffs DE13 0SN
☎(+44 1283) 51 05 10 ⊜ (+44 1283) 51 00 52
Business Manager: Mark Golby
■ Precision components manufactured in advanced polymeric materials for airframe, electrical generation, & computer applications. [98]

75-10870 Bunce (Ashbury) Ltd
Ashbury, Swindon, Wilts SN6 8LW
☎ (+44 1793) 71 02 12 ⊖ (+44 1793) 71 04 37
Joint Managing Director: Philip Bunce
Joint Managing Director: Stephen Bunce
Joint Managing Director and Sales Director: Andrew C Bunce
Purchasing Manager: Paul Newman
■ Snow clearing equipment for runways, airport areas and roads. [98]

75-10900 Burfield & Co. (Gloves) Ltd
Manor Road, Martock, Somerset TA12 6JH
☎ (+44 1935) 82 32 78 ⊠ BURFIELD MARTOCK
⊖ (+44 1935) 82 60 75
♧ £ 35,000 (1996) ♣ £ 2,000,000 (1996) 👥 90 (1996)
🏭 16,000 sq ft
Chairman: S N Burfield
Managing Director, Operations: C R Burfield
Sales and Marketing Director, Purchasing Manager: M N Burfield
■ Comprehensive range of leather dress gloves & gauntlets including water-resistant infantry combat, pilots sweat resistant, helicopter winchmen, anti-flash & service dress gloves for all ranks. [98]

75-11032 Burr-Brown International Ltd
(a subs. of Burr-Brown Corp., USA)
1 Millfield House, Woodshots Meadow, Croxley Centre, Watford, Herts WD1 8YX
☎ (+44 1923) 23 38 37 ⊖ (+44 1923) 23 39 79
♧ £ 2 200 000 (1994) ♣ £ 13 000 000 👥 12
Managing Dir: A. Collister
Applications Engineer: Bob Lee
■ Technical support for the company's range of analog & mixed signal integrated circuits. [97]

75-11050 Busak + Shamban Ltd
(† Shamban Europa (UK) Ltd)
Focus House, Ashborne Way, Solihul, West Midlands B90 4QU
☎ (+44 121) 744 12 21 ⊖ (+44 121) 733 24 42
♣ £ 6,000,000 (1996) 👥 35 🏭 1,500 m2
Managing Dir: M.D. Harris
■ Manufacturer & supplier of high performance bearings & seals for the defence & aerospace industry. Advanced fluid sealing systems & bearings for high temperatures, pressures & aggressive media.
® Denmark: W.S. Shamban Europa A/S
UK: Willis Engineered Polymers Ltd
USA: American Variseal, Quadion Corp., Shamban [98]

75-11090 Butyl Products Ltd
11 Radford Crescent, Billericay, Essex CM12 0DW
☎ (+44 1277) 65 32 81
[E] enquiries@butylproducts.co.uk
[W] http://www.butylproducts.co.uk
⊖ (+44 1277) 65 79 21
👥 80
Managing Director: R E Knights
General Manager: R Martin
■ NBC overboots. Storage tanks, E.G. Silos, transport bags, storage bags, water linings of reservoirs, canals and lakes. [98]

75-11125 CAE
Sedgewall Communications Group Ltd
(a subsidiary of Sedgewall Ltd)
1 Apex Business Centre, Boscombe Road, Dunstable, Bedfordshire LU5 4SB
☎ (+44 1582) 47 55 55 ⊠ 82362 bateco g
⊖ (+44 1582) 47 55 53
♣ £ 2,000,000 👥 30
Chairman: R.F. Aird
Managing Director: Alex Becker
Central Manager: Peter Johnson
Purchasing Manager: Ms Rita Alan
■ Telecommunications equipment, (Test access/monitoring equipment), Fibre optic modem accessories; Matrix switches-mics., headphones, handsets, speakers, power supplies, cabinets, jackfields, cables, data patching, custom design products, etc.
See also:
Morse Equipment (71-40950) [98]

75-11145 CAE Invertron Ltd
(a division of CAE Electronics plc)
Albert Drive, Burgess Hill, West Sussex RH15 9TN
☎ (+44 1444) 24 75 35
✆ (0800) 14 60 69 (customer services)
[E] caei@pavilion.co.uk [W] http://www.cae.ca
⊖ (+44 1444) 24 48 95
♣ £ 12,000,000 (1996) 👥 200 (1997) 🏭 3,000 m2
Chairman: N M Prest
Managing Director: Denis Sauzier
Operations Director: Vince Margetts
Sales and Marketing Director: Paul Matthews
Purchasing Mgr: Nick Brockes
Marketing Exec.: Simon Kelly
Marketing Exec.: Alan Foster
Marketing and Public Relations Officer: Ms Sara Gunning
■ Produces land-based simulators for military training, including:
Direct Fire Trainers: Tank Gunnery Trainers, Anti-Tank Trainers, INGUARD - Invertron New Generation Air Defence Trainers;
Indirect Fire Trainers: INFRONT - Invertron Forward Observation Artillery Trainer, Integrated Trainers for example OPV and A-class, Systems Trainers such as MSTAR and AS90;
Battle and Command & Control Systems Trainers: INCOMBAT - Invertron Command and Battlegroup Trainers;
Recognition Trainers: INTREST - Invertron Target Recognition System Trainer, Single Student System or Integrated Classroom Trainer. [98]

75-11250 Caledonian Mining Co. Ltd
Carlton Works, Carlton-on-Trent, Newark, Nottinghamshire NG23 6NT
☎ (+44 1636) 82 16 45 ⊖ (+44 1636) 82 12 61
👥 280
Chairman: C J MacLeod
Managing Director: P Hinton
Director Sales: K Mitchell
Company Secretary: K Wheat
■ Consultants in international construction methods. Manufacturers of steel containers and buildings. Civil contracting. → 80 [98]

75-11260 Caley Ocean Systems
(† Anderson Caley)
(a division of Dresser UK)
Mavor Avenue, East Kilbride, Glasgow G74 4PU
☎ (+44 13552) 466 26 ⊠ 776653
⊖ (+44 13552) 293 59
♣ £ 4,000,000 👥 30
Director and General Manager: David Cooper
Sales and Marketing Manager: Gregor McPherson
Operations Director: D Mitchell
Business Manager North America: Paul Kjolseth
Purchasing Assistant: J M Lee
■ Design, manufacture, installation and commissioning hydraulic systems for naval, oceanographic, communications and offshore systems. Main products are shipborne launch and recovery systems for manned & unmanned submersibles, R.I.B. inflatables and towed bodies. Also produce offshore cable handling machinery and heave compensation units, together with their relevant control systems.
Overseas representatives:
KBK → Japan
Kung Hee → Korea
Tripulsor → Sweden
Global Teamwork → Malaysia
IL-Jin Trading → South Korea
Silver Streak International → India [98]

75-11350 Camlock Engineering Ltd
Unit 3, Tarsmill Court, Rotherwas Industrial Estate, Hereford HR2 6JZ
☎ (+44 1432) 27 95 53 ⊖ (+44 1432) 26 60 10
Managing Director: S Altham
Director: D Altham
■ Personnel deployment equipment. [98]

75-11400 Cameron Balloons Ltd
St John's Street, Bedminster, Bristol BS3 4NH
☎ (+44 117) 963 72 16 ⊖ (+44 117) 966 11 68
♣ £ 4,500,000 👥 96 🏭 42,000 sq ft
Managing Director: D.A. Cameron
Group Sales and Marketing Director: Philip Dunnington
Technical Director: T R Sage
Sales Director: J N H Purvis
Production Director: J Howard
Marketing Manager: Alan Noble
■ Design and manufacture hot air balloons, hot air ships and helium airships. [98]

75-11480 W. Canning plc
Canning House, 133 Great Hampton Street, Birmingham B18 6AS
☎ (+44 121) 236 86 21
[E] 101776.3431@compuserve.com [W] http://www.canning.com ⊖ (+44 121) 236 33 20
♣ £ 117,000,000 👥 1,200
Managing Director: G J Green
Operations Director: G Shard
Sales and Marketing Director: P Hannam
Purchasing Director: D Brown
Marketing Manager: W A Dawson
■ Surface finishing chemicals, vibratory finishing and polishing aeronautical adhesive types and mask sealants. Adhesive fuel storage biocides and water treatment biochemicals for odour and grease control. [98]

75-11500 Caparo Group Ltd
(† Caparo Industries plc)
Caparo House, Birchills, Walsall, West Midlands WS2 8QE
☎ (+44 1922) 43 00 43 ⊖ (+44 1922) 43 00 18
♣ £ 430 000 000 👥 4 000
Chief Exec.: Ambar Paul
Purchasing Dir: C.M. Shaw
Financial Dir: John Smith
■ Industrial group with main business activities in steel tubes, steel sections & wire, fasteners & other industrial products.
Subs. with defence business:
Armstrong Fastenings Ltd
CMT Engineering Ltd
Clydesdale Engineering Ltd [97]

75-11620 Cargo Aids
(Division of Drallim Industries Ltd)
Brett Drive, De La Warr Road, Bexhill-on-Sea, East Sussex TN40 2JP
☎ (+44 1424) 21 66 11/6 ⊠ DRALLIMIND Bexhill-on-Sea
⊖ (+44 1424) 21 66 36
Managing Director: Robert McBrien
Contracts Mgr: Richard Bradford
■ Design and manufacture helicopter underslung cargo hooks and associated load weighing systems, aircraft tie-downs (webbing and chain), cargo and weapon restraint systems, airborne floor and seat tube fitting. ISO 9001 and CAA (A1 and JAR 145) approved. [97]

75-11700 Carmichael International Ltd
(a subsidiary Trinity Holdings Plc)
Weir Lane, Worcester WR2 4AY
☎ (+44 1905) 42 00 44
[E] 101522.2726@compuserve.com
⊖ (+44 1905) 42 01 20
♧ £ 5,000,000 (1996) ♣ £ 16,000,000 (1996)
👥 130 (1996) 🏭 3,700 m2 (1996)
Chairman: S L Burton
Managing Director: N J Crespi
Sales & Marketing Dir: R. Briscoe
Operations Director: M J Kingston
Engineering Dir: M. Day
Purchasing Mgr: M. Kingston
■ Design and manufacture of an extensive range of fire fighting and specialist emergency vehicles. Riv-Hydraulic platforms. [98]

75-11750 Carr & Day & Martin Ltd
Lloyds House, Alderley Road, Wilmslow, Cheshire SK9 1QT
☎ (+44 1625) 53 91 35 ⊠ 669737
⊖ (+44 1625) 54 84 88
♣ £ 4,000,000 👥 120
General Mgr/Dir: Peter Wilkes
■ Solid fuel cooker for instant boiling & cooking. Floor & furniture polishes; shoe polishes; metal polish; household polishes & cleaners; industrial polishes & cleaners; leather oil & dressings; saddle soaps; hoof oil; dubbin; floor dressings. [98]

75-11800 Carrington
(a division of IPT Group Ltd)
Calder Works, Thornhill Road, Dewsbury, West Yorkshire WF12 9QQ
☎ (+44 1924) 46 51 61
[W] http://www.carrington.fp.co.uk ⊠ 55449
⊖ (+44 1924) 45 75 96
♣ £ 40,000,000 (1996) 👥 700 (1996)
Managing Director: Jim Newman
Operations Dir: P.D. Morse
Sales Mgr: Simon Tanner

■ Manufacture of specialist technical fabrics. 98

75-11900 Cartographic Engineering Ltd
(a subs. of Cartographical Services Ltd)
Landford Manor, Salisbury, Wiltshire SP5 2EW
☎(+44 1794) 39 03 92 ⊠ 477575 sharet g
⊖ (+44 1794) 39 08 67
👥 10
Managing Dir: W.B. Bucknall
General Mgr: Mrs R.J. Payne
■ Manufacture & supply of aerial photo viewing devices, photogrammetric & photo interpretation equipment. 98

75-12000 Carville Ltd
Station Rd, Dorking, Surrey RH14 1HQ
☎(+44 1306) 88 16 81 ⊖ (+44 1306) 87 62 65
♲ £ 750 000 (1996) ♣ £ 2 140 000 (1996) 👥 55
🏭 2 400 sq m
Managing Dir: Gordon Campbell
Sales & Marketing Dir: Gary Willmott
Operations & Technical Dir: Kevin Gibson
■ Precision machinist in plastic materials specialising in acrylics (including full heat-treatment). Range covers: optical, scientific, military, medical & avionic applications. → 80 97

75-12050 Case (United Kingdom) Ltd
(a subsidiary of Case Corp. Ltd, USA)
Marketing Division
Wheatley Hall Road, PO Box 121,
Doncaster, South Yorkshire DN2 4PN
☎(+44 1302) 73 31 05 ⊖ (+44 1302) 76 03 14
♣ £ 300,000,000 👥 2,600
Sales/Marketing Director: Bob Kelsall
MoD Contracts Manager: David Sunshine
■ Tractors/construction equipment. 98

75-12070 Castrol (UK) Ltd
(a subsidiary of Castrol International)
Industrial Div.
Burmah Castrol House, Pipers Way,
Swindon, Wiltshire SN3 1RE
☎(+44 1793) 51 27 12 ✉ CASTROL Swindon
⊠ 449221 burcas g ⊖ (+44 1793) 51 35 06
♣ £ 135,000,000 (1996) 👥 700 (1996)
Managing Director: Rob Griffin
Industrial Dir: John Carey
Communications Executive: Ms Nanda Spiniello
☎(+44 1793) 45 23 19
Purchasing Manager: Iain Fleming
Product Group Manager: R J F Tweedale
■ Manufacturers of lubricants & allied products to the aviation & aerospace industries. 98

75-12100 Cathelco Limited
Marine House, Hipper Street South, Chesterfield S40 1SS
☎(+44 1246) 20 77 02 E sales@cathelco.co.uk
W http://www.cathelco.co.uk ⊠ 547534 mussel g
⊖ (+44 1246) 20 65 19
♣ £ 2,500,000 (1996) 👥 22 (1996)
Managing Director: J Salisbury
Sales and Marketing Director: A MacDonald
Purchasing Manager: R Oliver
Publicity Manager: R Woolley
■ Designers & manufacturers of anti-fouling systems to protect seawater pipes, pumps & condensers on naval surface craft & submarines. Currently specified by 16 navies around the world. Cathelco Ltd also produce C-Shield, an impressed current cathodic protection system which is designed to prevent hull corrosion on naval vessels. 98

75-12250 Celab Ltd
(a subsidiary of Celab Power Management Ltd)
Woolmer Way, Bordon, Hampshire GU35 9QE
☎(+44 1420) 47 70 11 ⊖ (+44 1420) 47 20 34
♲ £ 500,000 ♣ £ 2,300,000 (1996) 👥 70 (1996)
🏭 1,700 sq m
Chairman & Managing Dir: B.C.D. Wood
Operations Director: C J Wood
Company Secretary: R. Mier
UK Sales Mgr: Christopher Edmonds
Purchasing Mgr: Emer Larkin
Sales Exec.: Andrew Crawford
■ Design & manufacture of high density military power supplies for rigorous applications in airborne, ship & ground environments.

Multiple rail modular systems with full BITE facilities. Comprehensive sub-contract manufacturing facility including post design if required. 98

75-12270 CEM Systems Ltd
Unit 4 Ravenhill Business Park, Ravenhill Road,
Belfast BT6 8AW
☎(+44 1232) 45 67 67 E sales@devel.cemsys.com
⊖ (+44 1232) 45 45 35
Contact: Ms Donna McDonnell
■ Integrated security management system "Access 1000" for small to medium-sized installations. The Access 2000 system is designed for use on medium to large-scale facilities including office towers, multi-building complexes and airports. Visual Imaging Pass Production System (VIPPS) for card production, ensuring that only authorised personnel and visitors can gain access to secure areas, where their location can be pinpointed at any time. 98

75-12280 Centaur Precision Limited
(† Precision Castparts Corp. (UK) Ltd)
(a subsidiary of Triplex Lloyd plc)
Parkway Close, Parkway Industrial Estate,
Sheffield S94 WH
☎(+44 114) 273 87 64 ⊖ (+44 114) 275 45 26
♣ £ 10,500,000 (1997) 👥 208 (1997) 🏭 60,000 sq ft
Managing Dir: John Harris
Sales Dir: Robert Bruce
Operations Director, Purchasing Manager: John Harwood
Finance Director: John Middleton
■ High technology precision castings in steel, vacuum & air melt nickel alloys, cobalt alloys. Specialists in complex shapes for aero-engine, airframe & military applications to a maximum dimension of 1,000mm. 98

75-12410 Ceramicast (Precision Investment Castings) Ltd
Castings House, Boundary Road,
Woking, Surrey GU21 5BX
☎(+44 1483) 75 16 66 ⊖ (+44 1483) 75 18 88
♲ £ 30,000 (1996) ♣ £ 1,450,000 (1996) 👥 50 (1996)
🏭 18,000 sq ft
Managing Dir: R. Gordon
Purchasing Dir: T. Bowen
Operations Mgr: S. Wherlock
Sales & Marketing Mgr: S. Davis
■ Suppliers of high quality lost wax precision investment castings, in a wide range of alloys including low carbonsteel, high carbon steel, stainless steel, cobalt & nickel base superalloys, copper base & aluminium alloys. Fully approved to BS EN ISO9002 1994, QAS 3640/65. CAA A8-2. → 85 98

75-12440 WB Chain Co Ltd
(an FKI plc company)
Worcester Road, PO Box 5,
Stourport-on-Severn DY13 9BP
Worcestershire
☎(+44 1299) 82 83 84 ⊖ (+44 1299) 82 83 85
Managing Director: C Burdon
Operations Director: G T Baldwin
Sales and Marketing Director: J D Coupe
Commercial Manager: G A Deakin
Purchasing Manager: J D Dalton
■ Chains for lifting and mechanical handling. Carbon and alloy steel chains, armoured face conveyor chains for mining, drag alloy chains for trawling, chain slings, haulage chains for mining. 98

75-12500 Chapman & Smith Ltd
Safir Works, East Hoathly, Lewes, E. Sussex BN8 6EW
☎(+44 1825) 84 03 23 ⊖ (+44 1825) 84 08 27
♣ £ 16 000 000 👥 64
Chairman: M.A. Smith
Managing Dir: G.H. Mason
Operations/Technical Dir: Dr D. Harris
Purchasing Mgr: P.J. Deadman
■ Ear defenders, industrial respirators & respiratory protective equipment; a range of hearing protection; two person communication systems. 97

75-12550 Charcoal Cloth (International) Ltd
(a subsidiary of Chemviron Carbon Ltd)
High Tech House, Commerce Way,
Arena Business Park, Rainton Bridge,
Houghton Le Springs DH4 5PP

☎(+44 191) 584 69 62 ⊖ (+44 191) 584 67 93
♲ £ 2,000,000 (1996) ♣ £ 2,000,000 (1996)
👥 30 (1996) 🏭 25,000 sq m
Managing Director: Aled Davies
Sales/Marketing Director: Dr Alan Grint
Consultant Military Sales: Richard Langston
■ NBC personal protection equipment & materials. British Armed Forces NBC facelet mask for respiratory protection against chemical warfare attack. Activated charcoal cloth materials as high efficiency filter media in NBC protection devices, including masks, air & water purification & clothing. Respirator canisters, collective NBC protective systems & NBC field dressings. Lightweight high protection NBC suit. Anti-riot & crowd control equipment. British police riot facelet mask for eye & respiratory protection against tear gases. 98

75-12600 Chelsea Instruments Ltd
2/3 Central Ave, West Molesey, Surrey KT8 2QZ
☎(+44 181) 941 00 44 E chelsea@compuserve.com
W http://www.chelsea.co.uk ⊖ (+44 181) 941 93 19
♣ £ 2,500,000 (1996) 👥 45 (1997) 🏭 18,000 sq ft
Managing Dir: David Chatterton
Marketing Dir: Richard J. Burt
Operations Dir: Brian Wardhaugh
Purchasing Mgr: Ms A. Crisp
■ Environmental sensor systems, submarine sensors, data handling, towed bodies, oceanographic equipment. 98

75-12700 Chelton (Electrostatics) Limited
(a Cobham plc company)
Fieldhouse Lane, Marlow, Buckinghamshire SL7 1LR
☎(+44 1628) 47 20 72 E mkt@chelton.co.uk
⊠ 849363 ⊖ (+44 1628) 48 22 55
👥 275 (1997) 🏭 13,580 sq m (1997)
Chairman and Managing Director: G C Cooper
Financial Director: P D Long
Technical Director: M P Brierley
Operations Director: D B Ireson
Engineering Director: D A Griffiths
Marketing Director: C Esplin-Jones
■ Supplier of airborne antannas and associated RF and microwave equipment. It is also an advisory, research and development organisation concerned with antenna and static discharger technology. Chelton engages in the development of antenna systems jointly with defence and civil aviation authorities, aircraft contractors, airline operators and system manufacturers. The company has recently extended its product base to include a range of receivers and U/VHF transceivers. Its advisory services include antenna siting assessment based on full scale polar diagram measurement. Chelton maintains two free space radiation pattern test ranges, anechoic chambers capable of measurement at 40 GHZ and a full range of environmental test facilities.
Associated companies:
Atlantic Microwave, USA
Chelton Communication Systems Inc., USA
Chelton, France
Chelton Inc., USA
Comant, USA
Kevlin Corp., USA
Northern Airborne Technology, USA
Chelton, France
Rayan, France
Northern Airborne Technology, Canada
Cobham Composites, UK
Complas, UK
Continental Microwave and Tool Co Inc, USA 98

75-12800 Chemring Countermeasures
(a division of Pains-Wessex Limited)
High Post, Salisbury, Wiltshire SP4 6AS
☎(+44 1722) 41 16 11 ⊖ (+44 1722) 78 27 66
Marketing Director: Martin Walmsley
Sales Manager: D G Radford
■ Electronic countermeasures (chaff) and decoys, EMI & RFI screening fabrics, SOLAS liferaft radar reflectors, radar reflective materials, project management and systems evaluation, military training targets incorporating thermal signature, automated machined parts in brass & pulse plating equipment for PCB manufacture. 98

75-13045 J.S. Chinn & Co. Ltd
(a subsidiary of J.S. Chinn Holdings Ltd)
Coventry Road, Exhall, Coventry, West Midlands CV7 9FT
☎(+44 1203) 36 30 80 ⊖ (+44 1203) 36 80 00
♣ £ 11,000,000 👥 164 🏭 219,000 sq ft
Director: J S Chinn
Director: G Gascoigne
Managing Director: J K Atkinson

Contact Reference

Business Manager: Mark Cunnington
Works Manager: Stefan Wiezak
■ Intricate fabrications for airframes and aero-engines in stainless steel, hi-nickel alloys, aluminium and titanium. Vacuum brazing. CNC milling, turning and boring for six axis control and pipe bending. Repair and overhaul. Prototype development production programmes. Precision CNC CAD/CAM machining and tool making. [98]

75-13050 J.S. Chinn Engineering Co. Ltd
(a subsidiary of J.S. Chinn Holdings Ltd)
Faraday Road, Harrowbrook Industrial Estate,
Hinckley, Leicester LE10 3DE
☎ (+44 1455) 61 63 52 ▣ 31303
⊖ (+44 1455) 25 12 83
👥 45 🏭 289,000 sq ft
Managing Dir: T. Latimer
Director: J.S. Chinn
■ Precision aerospace fabrications to 20 tonnes, metal spinning and flow forming, assembly and handling jigs, ground equipment and test rigs, satellite packaging and handling equipment. Special purpose test equipment aerospace. [98]

75-13060 J.S. Chinn Project Engineering Ltd
(a subsidiary of J.S. Chinn Holdings Ltd)
Deep Pit Road, Speedwell, Bristol BS5 7UZ
☎ (+44 117) 935 50 67 🄴 jschinnpe@aol.com
⊖ (+44 117) 935 59 87
♢ £ 200,000 (1997) ♣ £ 1,400,000 (1997)
👥 31 (1997) 🏭 1,000 m2
General Manager: R Haydon
Sales and Marketing Director: John Nutton
■ Manufacture of handling equipment for aircraft & gas turbines, Cargo X-ray systems. Mechanical, electronic & electrical design engineering. System software design. EMC & environmental assessment/test services and engineering project control. → 80 [98]

75-13150 Chubb Alarms Ltd
(a division of Chubb Security plc)
Marketing Deptartment
42-50 Hersham Road,
Walton-on-Thames, Surrey KT12 1RY
☎ (+44 1932) 24 61 41 ✆ (800) 28 24 94
⊖ (+44 1932) 22 36 14
♣ £ 80,000,000 (1997) 👥 1,300 (1997)
Managing Director: D Hughes
Sales and Marketing Manager: Chris Booth
Purchasing Manager: George Cassie
■ Design, supply, installation and maintenance of security systems. Products include: intruder alarms, CCTV, access control, fire detection systems and manned guarding services. 31 branches across the UK. → 80
Divisions:
CCTV & Access Control
Integrated Security Systems
Fire Detection [98]

75-13200 Chubb Fire Products & Systems Ltd
(a subsidiary of Chubb Security plc)
Lancaster Road, Cressex Industrial Estate,
High Wycombe, Bucks HP12 3QF
☎ (+44 1494) 44 11 14 ⊖ (+44 1494) 53 17 10
👥 1,250
Managing Dir: R.R. Key
Publicity Mgr: A. Rice
■ Manufacture & servicing of fire protection equipment including specialist military portable extinguishers, extinguishing systems, fire fighting foams & dry powders. Fire detection equipment installations & contract maintenance service. Chubb Fire decommissions & refills exsisting Halon 1301 systems & portable 1211 extinguishers using their registered refilling plant & Halan bank. Also complete refurbishment of extinguishers & trolley units.
Subsidiary:
Chubb Fire Products & Systems [98]

75-13205 Chubb Safe Equipment Co.
(a subs. of Chubb Security plc)
Wednesfield Road, PO Box 61,
Wolverhampton WV10 0EW
☎ (+44 1902) 45 51 11 ▣ 334604 chuwtn g
⊖ (+44 1902) 35 10 77
♣ £ 23 000 000 👥 600
Managing Dir: Rodney J. Bunyan
Export Dir: Frank Morris
■ Safes, vaults & strong rooms. Fire protection cabinets for computer & paper media. Cell doors, steel cabinets, spiral reinforcing, tangbar pre-cast blocks. Handcuffs. [98]

75-13207 Chubb Security plc
Chubb House, Staines Road West,
Sunbury-on-Thames, Middlesex TW16 7AR
☎ (+44 1932) 78 55 88
⊖ (+44 1932) 78 77 29; 78 26 54
👥 18,000
Chairman: Sir Ernest Harrison (OBE)
CEO: D.J. Peacock
Exec. Dir: D.F. Langley
Operations Director: Ian Thomas
Financial Dir: J.A. Biles
Non-Exec. Dir: M.J. Blackburn
Non-Exec. Dir: Sir Colin Corness
Publicity Mgr: D.C. Hall
■ Manufacture and installation of locks, safes, intruder and fire detection systems, security communications and central station equipment. [98]

75-13300 Circuitwise Electronics Ltd
Carlton Commerce Centre, 4 Dukes Road,
Southampton, Hampshire S14 0SQ
☎ (+44 1703) 67 19 67 ⊖ (+44 1703) 67 18 92
♢ £ 500,000 (1996) ♣ £ 1,200,000 (1996) 👥 9
🏭 1,000 sq m
Managing Dir: P. McMurray
Operations/Technical Dir: A. Williamson
■ Manufacture of electronic countermeasure (ECM) systems; transmitting, jamming and deception equipment. GSM monitoring systems and equipment. [98]

75-13350 Civil Defence Supply
Hunt House, Welbourn, Lincolnshire LN5 0JF
☎ (+44 1400) 27 38 51-3 ⊖ (+44 1400) 27 38 54
Managing Director: E N Bauer
Operations Director: Mrs J Bauer
Sales and Marketing Director: Mrs G Reed
Purchasing Manager: Mrs E Burgess
■ Manufacture and supply of specialised police and military projects including hand-held cordless searchlight, special duty chargers, tactical riot shields. [98]

75-13430 CJB Developments Ltd
Airport Service Rd, Portsmouth, Hants PO3 5PG
☎ (+44 1705) 66 49 11 ▣ 86351 cjbdev g
⊖ (+44 1705) 69 78 64
♣ £ 5 000 000 👥 85
Managing Dir: Bob Worlidge
Operations & Technical Dir: R.G. Couzens
Commercial Mgr: T. Martin
■ Air purification equipment for submarines & other sealed environments. [97]

75-13460 Clansman Armour Systems
(a subs. of Clansman Lightweight Armour Systems (pty) Ltd)
Ashling, The Croft, Bures, Suffolk CO8 5JB
☎ (+44 1787) 22 82 13 ⊖ (+44 1787) 22 84 62
♢ US$ 20 000 000 ♣ US$ 30 000 000 👥 120
🏭 1 500 m2
Managing Dir: Charles Hutchinson
Director: Nicholas Ferguson
Purchasing Mgr: John Watton
■ Manufacture military & civilian protective clothing & bomb disposal suits. [97]

75-13500 Clarke & Smith Mfg Co Ltd
(a subsidiary of Clarke & Smith Industries Ltd)
Melbourne House, Melbourne Road,
Wallington, Surrey SM6 8SD
☎ (+44 181) 669 44 11 ⊖ (+44 181) 669 40 90
Sales Director: John Simpson
Sales Manager: Mark Walford
■ Manufacture of portable sound reproducing equipment, ground station public address and amplifier and induction loop systems. [98]

75-13530 Clarke Chapman Marine
(a subsidiary of Rolls-Royce plc)
Saltmeadows Road, PO Box 9,
Gateshead, Tyne and Wear NE8 1SW
☎ (+44 191) 477 22 71 ▣ 53239 ccltd g
⊖ (+44 191) 477 10 09
Director and General Manager: J Allen
Sales Manager: J B Patton
■ Ship's deck machinery, including mooring, cargo, scientific, towing, traction and specialist winches, anchor windlasses and capstans. Jib-type deck cranes for cargo grabbing and container handling. Replenishment-at-sea system for naval vessels. Towed array sonar handling systems. Specialist naval cranes for high integrity missile, ammunition, RAS fuelling and helicopter handling duties. Truck-mounted balloon handling systems for RAF parachute training. [98]

75-13550 Clark Masts Teksam Ltd
Binstead, Isle of Wight PO33 3PA
☎ (+44 1983) 56 36 91 ⊖ (+44 1983) 56 66 43
♣ £ 2,500,000 (1996) 👥 65 (1996)
General Mgr: Colin G. Thornton
Sales Mgr: C.P. Harris
Publication Department Manager: C Littlejohn
■ Fast erecting air operated telescopic & mechanically extended mast systems. Models available for field erection or vehicle mounting. Used in telecommunications, broadcasting, surveillance, intelligence gathering & transmission.
Overseas offices/agents:
Clark Masts Teksam NV, Belgium. [98]

75-13750 Cleco Limited
(† Cleco Electric Industries Ltd)
Riverside, Market Harborough, Leics LE16 7PZ
☎ (+44 1858) 41 07 10 🄴 cleco.mis@dial.pipex.com
⊖ (+44 1858) 41 00 54
♣ £ 25,000,000 👥 300 (1997) 🏭 50,000 sq ft
Chairman: D M Jinks
Managing Director: Trevor A Pemberton
Director: P M Binnie
Sales Director: Andrew J Smith
Operations General Manager: Philip Wise
Commercial Manager: M C Jinks
Finance Authority: Peter Mortimer
Senior Sales Secretary: Mrs S L Muggleton
■ Manufacture and design of specialist narrow aisle materials handling and equipment. [98]

75-13771 Clement Clarke International
Airmed House, Edinburgh Way, Harlow, Essex CM20 2ED
☎ (+44 1279) 41 49 69 ✉ CLEMCLARKE
▣ 329118 execom g ⊖ (+44 1279) 63 52 32
♣ £ 14,600,000 (1997) 👥 180 🏭 125,000 sq ft
Managing Director: Brian Woodcock
Sales Director: Paul Burchett
Quality Manager: Mike Wilkinson
International Sales Manager: Dennis O'Leary
International Sales: Franca Cognata
■ Aviation, marine and telecommunications headsets.
Overseas agents:
Miles Air Products, Canada
Tronic Bits Melbourne, Australia
Hunting Aviation SA, Australia
Airadio Technology, Singapore [98]

75-13810 Cleveland-Guest (Engineering) Ltd
North Valley Road, Colne, Lancashire BB8 9RQ
☎ (+44 1282) 86 42 84 🄴 info@cleveland-guest.co.uk
🅆 http:// www.subnet.co.uk/cleveland-guest
⊖ (+44 1282) 86 17 71
♣ £ 4,000,000 (1997) 👥 103 (1997)
Managing Dir: Steve Hollis
Operations & Technical Dir: Stuart Grant
Purchasing Mgr: Bruce Lamb
Sales Engineer: Alec Bolton
■ Manufacture of airframe parts and engine parts/components, aircraft ground equipment, test rigs, all types of military spares and ancillary equipment. Registered for precision machined components, raw materials procurement, fabrications and assemblies to BS 5750 pt 2, ISO 9002, EN 29002.
Approvals include:
Aerostructures Hamble
Airbus Industries
AIT Ltd
APP Hydraulics
British Aerospace

Dowty Aerospace
GEC
Hurel Dubois
Lucas Aerospace
Rolls-Royce
Shorts
Westlands 98

75-13860 C.L.S. Group
256 Moseley Road, Birmingham B12 0DL
☎(+44 121) 440 19 77 ⊖ (+44 121) 440 17 67
♣ £ 17,000,000 (1996/97) 350 (1996/97)
Managing Director: Neil Dolley
Director: James Allen
Sales and Marketing Director: Paul Berny
Purchasing Manager: Ms Sally Mobley
■ Remanufacturing all types of vehicle electrical, diesel fuel injection & turbo charger equipment for road vehicles, off highway vehicles & fighting vehicles. 98

75-13865 W.J. Clow (Midlands) Ltd
Units 3/4 Meadow Street, Walsall, W.Midlands WS1 3QP
☎(+44 1922) 325 25 ⊖ (+44 1922) 72 20 89
Managing Dir: E. Barker
Sales & Marketing Dir: A.W. Keates
■ Design, develop & manufacture of a range of assault ladders. 97

75-13868 Clyde Sturtevant Ltd
Clyde Pneumatic Conveying Ltd
(† Sturtevant Engineering Co. Ltd)
(a subsidiary of Clyde Blowers plc)
Shaw Lane Industrial Estate,
Doncaster, South Yorkshire DN2 4SE
☎(+44 1302) 32 13 13 ☒ 547273
⊖ (+44 1302) 73 97 21
♣ £ 28,000,000 (1995) 50 93,475 m2
Managing Dir: J.M. Blower
Sales & Marketing Dir: R.C. Sims
Operations & Technical Dir: J.P. Quail
Purchasing Mgr: D. Merills
■ Mobile vacuum cleaning equipment for servicing military aircraft & helicopters; power operated inflation/deflation machines for liferafts & air-filled equipment, portable flameproof equipment for OTTO fuel spillage collection in submarines. 98

75-13870 Clydesdale Engineering Ltd
(a subsidiary of Caparo Industries plc)
Marriott Road, Netherton,
Dudley, West Midlands DY2 0LA
☎(+44 1384) 25 25 87 ☒ 336532
⊖ (+44 1384) 23 10 05
♣ £ 21,000,000 (1996) 300 (1996)
♣ £ 8,000,000 (1996)
Chief Exec.: John Wood
Special Products Sales Mgr: Steve Earle
■ Manufacture of drop forgings & conveyor equipment, conveyor systems, bearings & handtools. 98

75-13880 Clydesdale Forge Co.
Brenda Rd, Hartlepool, Cleveland TS25 2BP
☎(+44 1429) 26 81 01 ⊖ (+44 1429) 23 33 46
♣ £ 12 000 000 200
Managing Dir: J Wood
Sales Director: K Newton
Sales Manager: S Earle
■ Press & hammer forgings for applications in defence equipment, aerospace, nuclear power generation & petrochemical industries. 98

75-13890 Clydesdale Whitehouse Co.
(a div. of Clydesdale Engineering Ltd)
Marriott Rd, Netherton, Dudley, West Midlands DY2 0LA
☎(+44 1384) 25 25 87 ⊖ (+44 1384) 45 99 52
♣ £ 2 000 000 40
Managing Dir: Neil Dainty
Sales Dir: Alan Middleton
Sales Mgr: R. Siviter
■ Tool manufacturer. 98

75-13940 CMT Dynamics
(† CMT Engineering Ltd)
(a division of Caparo Group)
Corngreaves Rd, PO Box 36,
Cradley Heath, West Midlands B64 7DQ
☎(+44 1384) 56 32 20 ⊖ (+44 1384) 56 32 25
♣ £ 5 000 000+ 100 55 000 sq ft
Operations Director: R D Powell
Purchasing Manager: M D Raybould
■ Manufacture anti-vibration and shock mounts for vibration shock and noise control. 98

75-13950 Cobham Composites Limited
(a Chelton Group Company)
Devey House, Gelders Hall Road, Shepshed,
Loughborough, Leicestershire LE12 9NH
☎(+44 1509) 50 45 41 ☒ 341252
⊖ (+44 1509) 50 75 63
92 36,000 sq ft
Managing Director: Laurence Marshall
Finance Dir: G. Masterton
Operations Director: Anthony Pedley
Sales and Marketing Director: I Richardson
Sales and Marketing Manager: Roger Manwaring
Purchasing Manager: Peter Bruce
■ Composite components for aerospace and military applications. 98

75-14030 Alan Cobham Engineering Ltd (ACEL)
(a Cobham plc company)
Holland Way, Blandford, Dorset DT11 7BJ
☎(+44 1258) 45 14 41 ☒ 417254
⊖ (+44 1258) 45 14 14
80 5,000 sq m
Chairman: G F Page
Managing Director: P J Hebard
Financial Director, Company Secretary: J F Annis
Sales and Marketing Director: I Marcham
■ Supplier of ground based turnkey installations which control the flow, filtration, separation and conditioning of both liquid and gaseous hydrocarbons. Also supplies turnkey facilities for the simulation of gas turbine bleed air at high temperatures as well as fuel for the testing of aircraft systems and components. 98

75-14070 COES (Derby) Ltd
Thirsk Place, Ascot Drive, Derby DE24 8JL
☎(+44 1332) 34 09 80; 34 63 01
⊖ (+44 1332) 34 07 74
♧ £ 10,000 ♣ £ 1,000,000 45 2,000 sq m
Managing, Sales & Marketing Dir: R.C. Harrison
Operations & Technical Dir: J. McDermott
Company Secretary: Mrs J.C. Harrison
Purchasing Mgr: V.A. Heath
■ Design & manufacture of fibre reinforced plastics, ABS vacuum forming & welvic moulding. Specialising in:
Tool control systems: A method & container, designed to meet & hold a specific tooling requirement for designated work in many fields including ship board, aircraft & vehicles.
Transportation protection: engine blanking covers, plugs, caps & sleeves for the aerospace industry. 98

75-14090 Coherent Optics (Europe) Ltd
(† Vinten Electro-Optics Ltd)
(a subsidiary of Coherent Inc, USA)
Unit 28, Ashville Way, Whetstone, Leicestershire LE8 3NU
☎(+44 116) 286 71 10 ⊖ (+44 116) 286 73 59
♧ £ 300,000 (issued share capital) (1997)
♣ £ 5,300,000 (1996) 65 (1997) 35,000 sq ft
Managing Director: Robin Henderson
Technical Director: W H Gray
Sales and Marketing Director: John Henson
Purchasing Manager: Lisa Betts
■ Precision optical manufacturing & thin film coating company. Areas of specialist expertise are optical components for infra-red imaging & tactical sighting equipment. Complex optical thin film coating design & manufacture is offered throughout the visible & far infra-red spectral regions. 98

75-14100 Cold Formed Products Ltd
24 Saint Mary's Road, Plaistow, London E13 9AD
☎(+44 181) 471 27 27 ⊖ (+44 181) 470 17 06
♣ £ 2,000,000 + (1997) 50 (1997) 2,000 sq m
Managing Dir: A.G. Palmer
Operations Director: Ron Cox
Sales and Marketing Director: David Binks
Purchasing Manager: Peter Breame
■ Specialist impact instrusion & forging of components in aluminium & its alloys for aerospace & defence markets. 98

75-14120 Colebrand Ltd
Colebrand House, 20 Warwick Street, London W1R 6BE
☎(+44 171) 439 10 00 ⊖ (+44 171) 734 33 58
♣ £ 10,000,000 90
Operations Manager: D F Brown
Defence Sales: C M Stubbs
■ Design, manufacture, supply and application of radar and thermal signature control materials to UK and NATO armed forces. BS5750 Parts 1 and 2. 98

75-14150 Colledge & Morley (Gears) Ltd
(a subsidiary of J.S. Chinn Holdings Ltd)
Curriers Close, Charter Avenue, Canley,
Coventry, West Midlands CV4 8AW
☎(+44 1203) 46 23 28 ☒ 31303
⊖ (+44 1203) 69 40 08
♣ £ 1,700,000 48
Managing Director: M Millward
Director: J S Chinn
■ Manufacture precision gearing to customer specifications. Spur & helical gears, splines, racks, wormshafts, wheels & bevel gears. 98

75-14200 Colvic Craft plc
Earls Colne Industrial Park, Earls Colne,
Colchester, Essex CO6 2NS
☎(+44 1787) 23 39 93 ⊖ (+44 1787) 22 43 30
♣ £ 6 000 000 100 3 300 sq m
Managing Dir: C.C. Burns
General Mgr: V.J. Pascoe
Sales & Marketing Dir: Tony Morgan
Purchasing Mgr: Dean Fuller
■ Construction of IGM Patrol boats for the Greek coastguard. 97

75-14300 Com Dev Europe
Triangle Business Park, Stoke Mandeville,
Aylesbury, Bucks HP22 5SX
☎(+44 1296) 61 44 44 ⊖ (+44 1296) 61 44 00
Managing Dir: B.N.F. Eddleston
Sales & Marketing Dir: C. Mclaren
Technical Dir: R. Cameron
Purchasing Mgr: G. Brown
■ Design, manufacture & test microwave equipment for spacecraft & defence. 97

75-14410 Commercial Hydraulics Keelavite Ltd
(a unit of Commercial Intertech Corp, USA)
Tachbrook Park Drive, Tachbrook Park, Warwick CV34 6TU
☎(+44 1926) 88 90 00 ⊖ (+44 1926) 88 91 72
♣ £ 13 000 000 111
Managing Dir: Peter J. Cleife
Sales Mgr: Alan Dobson
■ Hydraulic test rigs, hydraulic and pneumatic aircraft ground support equipment, hydraulic power packs, control and actuation systems, bomb trepanner, wavemaker, engine driven electric generator sets.
Associated Group companies:
Commercial Hydraulics Liverpool Ltd
Commercial Hydraulics Gloucester Ltd
Commercial Hydraulics Kontak Ltd 97

75-14415 Commercial Hydraulics Kontak Ltd
(a unit of Commercial Intertech Corp, USA)
Belton Park, Londonthorpe Rd, Grantham, Lincs NG31 9SJ
☎(+44 1476) 641 05 ⊖ (+44 1476) 664 62
♣ £ 11 000 000 120 97

75-14460 Communications & Measurement Technologies Ltd
12 Howard Way, Cromwell Business Park,
Newport Pagnell, Bucks MK16 9QR
☎(+44 1908) 21 82 00 W http://www.cmt.co.uk
☒ 826097 aplusb g ⊖ (+44 1908) 21 73 27
♣ £ 1,500,000 15
Managing Director: Danny Woolard
Advertising Manager: S Thomas
Marketing: B Whitford
■ A full range of data communications products covering VHF, UHF and HF frequencies with particular emphasis on GPS (Global Positioning Systems) related systems for applications including off/onshore survey, navigation & positioning for civil & defence applications. 98

75-14520 CompAir BroomWade Ltd
BroomWade Works, Hughenden Ave, PO Box 7, High Wycombe, Bucks. HP13 5SF
☎(+44 1494) 60 53 00 ☒ 837371
⊖ (+44 1494) 46 26 24
Managing Director: Paul Jarvis
Finance Director: S C Haigh
Operations Director: Tim Hair
Sales Director: Brian Jeffers
■ Compressed air equipment manufacturer. [97]

75-14530 CompAir Reavell Ltd
Ranelagh Rd, PO Box 44, Ipswich, Suffolk IP2 0AE
☎(+44 1473) 60 22 22 ☒ 98254 reavel g
⊖ (+44 1473) 60 17 04
Finance Dir: N.J. Lomax
■ Reciprocating air compressors. Water cooled & air cooled high pressure compressors. Breathing air systems. [97]

75-14543 Complas Limited
(a Chelton Group Company)
Avenue 1, Station Lane, Witney, Oxon OX8 6XS
☎(+44 1933) 77 88 81 ⊖ (+44 1993) 77 60 89
Managing Director: G C Cooper
Deputy Managing Director: L Marshall
Finance Director: G Masterton
Sales Manager: I Richardson
Operations Manager: D Seymour
■ Manufacture of radomes. [98]

75-14545 Compounding Ingredients Ltd
(a subsidiary of Synair Corporation, USA)
Tyre Related Products Department
217 Walton Summit Center, Bamber Bridge, Preston, Lancashire PR5 8AQ
☎(+44 1772) 32 28 88 ⊖ (+44 1772) 31 58 53
♣ £ 5,000,000 (1996) 👥 46 (1997)
Managing Director: P B Astles
Product Manager: D Sherwell
■ TYRFIL puncture-proofing systems for low speed & off the road use. TYRLYNER system self-healing tyres for highway & agricultural use. MONOTHANE one component polyurethane elastomer. DUOTHANE two component high performance polyurethane elastomer. CILBOND rubber to metal bonding agents. [98]

75-14550 Compton Webb Ltd
(a division of Vermilion Corp)
Gosforth Road, Derby DE24 8HU
☎(+44 1332) 34 26 16 ⊖ (+44 1332) 37 23 60
Managing Director: K A Hunt
Sales Director: K F Birch
Export Sales Director: R G Body
Export Sales Manager: N J Waller
■ Uniforms: military combat, dress & ceremonial. Police, fire service etc. uniforms. Protective clothing & outerwear (waterproofs, overcoats). Shirts, pullovers, caps, berets. Underwear. Insignia & accessories. Design & cloth consultancy service. [98]

75-14600 L.N. Compton Ltd
90-100 Glover Street, Birmingham B9 4EN
☎(+44 121) 772 04 71 ⊖ (+44 121) 766 50 99
♧ £ 10,000 ♣ £ 900,000 (1995/96) 👥 22 🏭 8,000 sq ft
Managing Dir: John Pagan
Works Mgr: Steve Hennessey
■ Sewn canvas & PVC Nylon for military vehicle covers & penthouses. Tents, curtains, tarpaulins & gun covers. [98]

75-14620 Computing Devices Canada Ltd
(a division of Computing Devices Canada Ltd)
PO Box 1, Newquay, Cornwall TR7 3AP
☎(+44 1637) 87 43 04 ⊖ (+44 1637) 85 01 02
👥 1,100 (1997)
President: Dave Scott
Sales & Marketing Dir: Ernie Trischuk
Operations & Technical Dir: John Watts
International Marketing Manager: John Bussey [98]

75-14650 Computerised Training Systems Ltd (CTS)
Duke House, Queensway Court, Arkwright Way, Scunthorpe, South Humberside DN16 1AL
☎(+44 1724) 28 10 37 ⊖ (+44 1724) 27 17 69
♧ £ 450,000 (1995/96) ♣ £ 16,000,000 (1996) 👥 70
🏭 10,000 sq m
Managing Dir: Bryan Hogg
■ Design and development of a wide range of civil and military related computer based trainers for instructors, students and desk top simulators including hardware systems, software and training software (courseware). Courseware includes basic aircraft engineers licencing, Tornado GR1, F3, integrated avionics, weapons and escape systems, cabin crew and aircrew training and the CTS speciality maintenance training for civil and military aircraft systems. [97]

75-14700 Computing Devices Company Ltd
(a subsidiary of Ceridian Corp, USA)
Castleham Rd, St Leonards-on-Sea, E.Sussex TN38 9NJ
☎(+44 1424) 85 34 81 ☒ 95568 cdcltd g
⊖ (+44 1424) 85 15 20
♧ £ 2 000 000 (1995) ♣ £ 35 000 000 (1995) 👥 450
🏭 120 000 sq ft
Chairman & Managing Dir: W.J. Uttley-Moore
Operations & Technical Dir: S. Knowles
Purchasing Mgr: S. Horne
■ Specialises in systems for high performance aircraft & military vehicles. Current programmes include: airborne & ground based sensor image processing & recording for reconnaissance systems, airborne computing & weapon management. Aircraft such as the Tornado, Hawk, Harrier, EFA, Jaguar, F18 & EH101 helicopter rely on systems designed & produced by the company. [97]

75-14780 Concoat Ltd Materials and Machinery Division
Alasan House, 2c Albany Park, Frimley Road, Camberley, Surrey GU15 2PL
☎(+44 1276) 69 11 00 🄴 concoat@ax.compulink.co.uk
⊖ (+44 1276) 69 12 27
♧ £ 30,000 (1997) ♣ £ 2,600,000 (1996/97)
👥 30 (1997) 🏭 1,000 m2
Managing Director: Graham K Naisbitt
Technical Director: David Greenman
Purchasing Manager: A Hamilton
■ Application engineers in chemical compounds for the electronics industry. Conformal coatings and application machinery for coatings. Non CFC cleaning materials, automatic cleaning and testing cleanliness equipment. Conformal coating application services. [98]

75-14810 Concurrent Computer Corporation Ltd
(a subsidiary of Concurrent Computer Corporation, USA)
227 Bath Road, Slough, Berkshire SL1 4AX
☎(+44 753) 53 45 11 🅆 http://www.ccur.com
☒ 847376 ⊖ (+44 753) 51 32 68
♣ £ 20,000,000 👥 240
UK Managing Director: Michael Baker
European Marketing Director: Philip Burgin
Regional Business Manager Northern Europe: Patrick Quinn
European Marketing Communications Manager: Claude Leduc
■ International developer & manufacturer of event-driven, high-performance computer systems for a variety of time-critical applications. [97]

75-14855 Conjay Arms Co. Ltd
PO Box 74, Woking, Surrey GU21 5EU
☎(+44 181) 960 68 79 🄴 cbap@netcomuk.co.uk
⊖ (+44 181) 903 56 42; 960 24 11
Managing Dir: Gordon D. Conway
Sales & Marketing Mgr: Daren Conway
■ Experts in S.A.A. development of improved types. Defence contractors. Importers & distributors of defence, anti-terrorist & anti-riot equipment. 9mm CBAP armour piercing ammunition. → 85
® UK: American Derringer, Glaser Safety Slug, Grifon Global, North Amercian Arms, Olympic Arms, Safari Arms USA: Knights' Manufacturing Co., Pachmayr, Versapod [98]

75-14860 Contemporary Systems Design Ltd (CSD)
(a div. of Molynx Ltd)
Albany St, Newport, Gwent NP9 5XW
☎(+44 1633) 82 19 74 ⊖ (+44 1633) 85 08 93
♣ £ 12 000 000 (1996) 👥 132 🏭 7 000 m2
Managing Dir: Dr O. Ta'eed
Operations Dir: Richard Langdon
Operations/Technical Dir: G. Stevens
Sales & Marketing Dir: Robert van der Mey
Purchasing Mgr: Ken Bryl
■ Electronic surveillance equipment for use in ISA COIN & by the Police, customs & drug enforcement agencies. The company also produces EOD search equipment & radio control systems for EOD vehicles such as Wheelbarrow & Hobo. [97]

75-14875 Control Techniques Dynamics Ltd
(† Moore Reed & Co. Ltd.)
(a div. of Emerson Electric Co., USA)
South Way, Walworth Industrial Estate, Andover, Hants SP10 5AB
☎(+44 1264) 38 76 00
🄴 ctdynam2!mailpo!ctd@ctdynam.attmail.co m ☒ 47654
⊖ (+44 1264) 35 65 61
♣ £ 40,000,000 👥 180 🏭 40,000 sq ft
Managing Dir: Martin Boughtwood
Finance Director: J Owen
Operations/Technical Dir: Paul Watson
Purchasing Mgr: Richard Turpin
Sales Mgr: Colin J. Arch
Production Manager: B Smith
Technical Manager: D Driscoll
Marketing Manager: D Elliott
■ Components, syschros, resolvers, AC servo motors, servo motor generators, DC motors and generators, magnetic pick-offs, contact, optical and solid state encoders, stepper motors, packaged servo systems, absolute and incremental digital displays. Torque motors and brushless DC motors. [97]

75-14900 Henry Cooch & Son Ltd
Unit 2 Platt Industrial Estate, Boroughgreen, Sevenoaks, Kent TN15 8LN
☎(+44 1732) 88 44 84 ✉ COOCH Sevenoaks
☒ 957486 cooch g ⊖ (+44 1732) 88 26 81
♧ £ 5,000 (1996) ♣ £ 1,250,000 👥 20 🏭 1,000 sq m
Managing Director: S H Cooch
Sales Director: B A Chantler
■ Mobile floodlighting units (trailer-mounted equipment for rapid deployment). Diesel/hydraulic mobile cranes. [98]

75-14920 William Cook George Blair Ltd
(a subsidiary of William Cook plc)
Stanhope, Bishop Auckland, County Durham DL13 2YR
☎(+44 1388) 52 82 48 ⊖ (+44 1388) 52 88 79
👥 350 (1997)
Managing Dir: Philip John Kite
Sales & Marketing Dir: David Palframan
Operations & Technical Dir: Paul Hagger
Purchasing Mgr: Michael Orchard
■ Design and manufacture of tracks and components for fighting vehicles. [98]

75-14950 Coopers Filters Ltd
(a subs. of FIAAM Filter SpA, Italy)
Llanfoist Works, Abergavenny, Gwent NP7 9LW
☎(+44 1873) 85 20 41 ⊖ (+44 1873) 85 80 46
♣ £ 29 000 000 (1995) 👥 350 🏭 22 128 sq m
Managing Dir: R. Martinez
Sales Dir: D.A. Cunningham
■ Manufacture automotive related filtration systems & filters. [97]

75-15120 Corrintec Ltd
(† Morgan Berkeley & Co. Ltd)
(a subsidiary of Cathelco Ltd)
Moorside Road, Winnall, Winchester, Hampshire SO23 7SF
☎(+44 1962) 86 96 22 🄴 enq@corrintec.co.uk
⊖ (+44 1962) 86 66 28
👥 23
Managing Director: T M Reeve-Tucker
Marketing Co-ordinator: Ms Corrie Johnstone
Marine Manager/Sales: Mike Wallbridge
Industrial/Offshore Manager/Sales: Pete Woodland
■ Design, manufacture, install and maintain cathodic protection systems (hull corrosion prevention equipment) for the complete spectrum of military vessels. Expert engineers travel worldwide to service any make of cathodic protection system. → 80 [98]

75-15140 Cosens Engineering Ltd
(† Cosens & Co Ltd)
Perryfield House, Pennsylvania Road,
Portland, Dorset DT5 1HF
☎(+44 1305) 82 29 22 ⊖ (+44 1305) 82 68 05
♣ £ 1,219,000 (1996) 👥 35 (1996) 🏠 4,000 sq m
Managing Director: Trevor Primett
Operations Director: David Willshire
■ Design, manufacture and refurbishment (in-house, alongside and slipway), of medium and heavy metal fabrications including pontoons, small ships and boats and sub-contracted G.R.P. repairs. Provision of shot blasting, painting and anti-corrosive zinc and aluminium metal spray treatments. Approved scope to BS EN ISO 9001. [98]

75-15213 Costain Civil Engineering Ltd
(† Costain Marine International Ltd)
(a subsidiary of Costain Group plc)
Costain House, Nicholson Walk,
Maidenhead, Berkshire SL6 1LN
☎(+44 1628) 84 24 44 ⊖ (+44 1628) 84 21 37
♣ £ 170,000,000 (1996) 👥 2,600 (1996)
Managing Director: G V May
Operations Director: R H Walker
Marketing and Business Development Manager: J W Rumboll ☎(+44 1628) 84 22 49
■ Worldwide design and construction of ports, harbours, airports, dockyards, and defence establishments. Training ranges (CQBs). → 80 [98]

75-15235 Cotag International
(a division of Amtech Europe Ltd)
Mercers Row, Cambridge CB5 8EX
☎(+44 1223) 32 15 35
[E] username@cotag.mhs.compuserve.com
⊖ (+44 1223) 36 67 99
♣ £ 7,000,000 (1997) 👥 140 (1997) 🏠 18,000 sq ft
Chairman & CEO: Stuart M. Evans
Development Director: Nigel Rix
Finance Director: Ann Utley
Technical Dir: J.B. Falck
Sales Director: Spencer Hall
Director of Marketing: Simon Houghton
Manufacturing Director: David Ireland
Marketing Manager: J Sparkes
Purchasing Manager: Ian Hamilton
■ Electronic access control systems. Hands free, proximity cards and tags. Automatic vehicle identification. [98]

75-15250 Cougar Limited
Defence Division
Stone Pier Yard, Shore Road,
Warsash, Hampshire SO31 9FR
☎(+44 1489) 58 33 32 ⊖ (+44 1489) 57 55 35
♣ £ 3,500,000 👥 50
Managing Director: Paul Brett
Operations Director: J Salmon
Sales and Marketing Director: Peter Currington
Sales and Marketing Director: Steve Curtis
Purchasing Manager: Sam Gamble
■ Cougar Ltd manufacture the Ultra-Fast Patrol Boat (UFPB) which has speeds in excess of 55kt. Cougar supply to both defence and commercial customers. [98]

75-15330 Courtaulds Aerospace
(a subsidiary of Courtaulds plc)
473 Foleshill Road, Coventry, West Midlands CV6 5AQ
☎(+44 1203) 68 87 44 ⊠ 311895
⊖ (+44 1203) 68 73 13
♧ £ 3,000,000 (1997) ♣ £ 12,000,000 (1997)
👥 180 (1997) 🏠 9,000 m2
General Manager, Sales and Marketing: Roger Medwell
Operations Director: Robert Turner
Purchasing Manager: Jim Cantas
■ Leading producer of combat helmets in ballistic nylon, aramid & hybrid combinations. Body armour, aircrew helmets. Lightweight composite vehicle armour & armoured vehicles. Specialise in high performance injection, compression & autoclave moulding.
Courtaulds Aerospace launched its new Integrated Protection range of products at DSA in April. The Integrated Protection packages provide for a complete system designed to meet users specific needs, including vehicles. [98]

75-15390 Coutant-Lambda Ltd
(† Coutant Electronics Ltd)
Kingsley Avenue, Ilfracombe, Devon EX34 8ES
☎(+44 1271) 86 37 81 ⊖ (+44 1271) 86 48 94
♣ £ 20 000 000 (1995) 👥 350
Managing Dir: Sandy Anderson
Sales & Marketing Dir: Martin Southam
Technical Dir: Adam Rawicz
Purchasing Mgr: Ms Janet Sanders
■ High stability military & commercial DC power supplies, DC regulators, electronic power supplies. AC-DC switchmode. [97]

75-15650 Cow Industrial Polymers Ltd
(a subs. of BTR plc)
41 Imperial Way, Croydon, Surrey CR0 4RR
☎(+44 181) 686 68 33 ⊖ (+44 181) 681 19 22
👥 150
Managing Dir: E.A. Thompson
■ Manufacturers of compression & injection mouldings in a wide range of natural & synthetic rubbers. [97]

75-15680 CP Cases
(† BDS Cases)
(a subsidiary of C.P. Mega)
Unit 18 Westbrook Road, Trafford Park,
Manchester M17 1AY
☎(+44 161) 873 81 81 [E] cpcases@dial.pipex.com
[W] http://www.dialspace.dial ⊖ (+44 161) 876 65 99
👥 50 🏠 2 000 sq m
Director: Hans Van Ek
Commercial Design Engineer: David Hindle
Operations Mgr: Noel Cahill
Marketing: Karl Haw
■ Special to content flight/transit cases in aluminium, wood, plastic, vulcanised fibre & padded bags. All available with CNC engineered foam inserts. Typical applications are for transit or operational protection of electronics, communications, munitions & optical devices. All cases are reusable. BS-5750 Part 1-ISO 9001-EN29002.
● Worton Hall Industrial Estate, Worton Rd,
Isleworth, Middlesex TW7 6EW
☎(+44,181) 568 18 81 ⊖ (+44 181) 568 11 41 [98]

75-15700 CQC Plc
(a subsidiary of Avon Rubber plc, UK)
Riverside Road, Barnstaple, Devon EX31 1NB
☎(+44 1271) 723 45 ⊖ (+44 1271) 720 90
♣ £ 5,000,000 👥 150
Managing Dir: Tony Lloyd
Sales Dir: Peter Gulliford
Works Dir: Michael Rennie
■ NBC protective suits; NBC casualty bags; sleeping bags; rucksacks; tents; personal load carriage equipment and combat body armour. [98]

75-15750 Crane Fruehauf Ltd
Toftwood, Dereham, Norfolk NR19 1JF
☎(+44 1362) 69 53 53 ⊖ (+44 1362) 65 43 54
♣ £ 100 000 000 (1995) 👥 1 000 (1995)
Managing Dir: T.H. Lynchy
Communications Dir: G.S. Thomson
Sales Admin Mgr: R. Fear
■ Military trailers and semi-trailers from 16 tonnes to 100 tonnes capacity. [97]

75-15772 Cranfield Precision
(† Cranfield Precision Engineering Ltd)
(a division of Landis Lund)
90 Central Avenue, Wharley End,
Cranfield, Bedfordshire MK43 0JR
☎(+44 1234) 75 42 22 [E] cpecran@cix.compulink.co.uk
⊠ 825072 ⊖ (+44 1234) 75 42 23
👥 66
Division Managing Director: G N Portas ☎(+44 1234) 75 40 20
Division Director of Sales and Marketing: Dr K G Dobson
Operations: R G Bent
Purchasing Manager: N Whiting
■ Ultra precision single point diamond turning & grinding machines; Precision metrology machines; precision positioning & indexing table; submicron & sub-arc second tranducers: customised machine tools for ultra-precision manufacture. [98]

75-15870 Crawford Collets
(a subsidiary of The 600 Group)
Tower Hill Works, Witney, Oxfordshire OX8 5DS
☎(+44 1993) 70 39 31 ⊠ 83244
⊖ (+44 1993) 77 83 10
♣ £ 5,500,000 👥 135
Director and General Manager: G Wilkin
Sales Director: V J Marston
Sales Manager: Phil J Waller
Purchasing: Mrs Susan Panons
■ Collets, feed fingers, tool holders, Collet chucks, toolposts, VSI tooling & Ainjest/metradial. [98]

75-15915 Cray Marine
Riverside Road, Barnstaple, Devon EX31 1LY
☎(+44 1271) 721 41 ⊠ 46324 ⊖ (+44 1271) 461 49
♣ £ 12 000 000 👥 190
Managing Dir: Alan Hayman
Sales & Marketing Mgr: Duncan Nicholson
■ Systems engineering. Repair & refurbishment of weapon systems & sonar systems. Transducer design & manufacture. [97]

75-16030 Crestbury Ltd
J.W. Automarine
Hempstead Road, Holt, Norfolk NR25 6DL
☎(+44 1263) 71 35 22; 71 26 68; 71 38 93
[E] sales.crestbury@paston.co.uk
[W] http://www.jwautomarine.co.uk
⊖ (+44 1263) 71 11 52
♣ £ 3,000,000 (1996) 👥 65
Managing Dir: John Wise
Sales Dir: Chris Henry
Sales Mgr: Malcolm Felmingham
Production: Andrew Turner
■ Design & manufacture of buoyancy & underwater lifting equipment for all military, salvage & marine civil engineering applications. Flotation systems, flexible liquid containers & weapon bags & covers for numerous service customers including PR1VM mine & submarine ordnance recovery bag. → 80 [98]

75-16050 Crewsaver
Mumby Road, Gosport, Hants PO12 1AQ
☎(+44 1705) 52 86 21 ⊠ 86607
⊖ (+44 1705) 51 09 05
■ Manufacture a wide range of industrial lifejackets, buoyancy aids and related accessories. [97]

75-16070 Crompton Instruments Ltd
(a division of BTR Control Systems, USA)
Freebournes Road, Witham, Essex CM8 3AH
☎(+44 1376) 51 26 01 ⊖ (+44 1376) 51 83 20
♣ £ 7,000,000 (UK); £ 13,000,000 (export) 👥 500
Managing Director: R S Stuart
General Manager: G A Wignell
Sales Director: J R Harkus
Marketing Director: M T Ashton
Marketing Manager: B E Harker
Purchasing Manager: I Ayres
■ Electronic & electro-mechanical indicating instruments for measurement & control, including sealed & non-sealed analog & digital meters, 'PROTECTOR' relays 'PALADIN' transducers, shunts & CT's all of which conform to most international standards.
International Sales Offices:
Australia, Canada, Japan, Netherlands,
Singapore, USA, Germany,
Hong Kong → 90 [98]

75-16136 Cronite Precision Castings Ltd
(a division of AFE Metal, France)
Blacknell Lane, Crewkerne, Somerset TA18 7YA
☎(+44 1460) 733 94 ⊖ (+44 1460) 720 79
♧ £ 52,000 (1996) ♣ £ 3,800,000 (1996) 👥 76 (1996)
🏠 3,000 m2
Managing Dir: D.A. Bond ☎(+44 1460) 733 94 Ext. 302
Sales Dir: C.A. Anderson ☎(+44 1460) 733 94 Ext. 303
Quality Assurance/Technical Mgr: R.I. Lindsay
Production Mgr: G. Lomas
■ Precision investment castings in all ferrous alloys, carbon steel, alloy steel, stainless & nickel chrome. [98]

75-16140 CRP Marine
Stanley Way, Stanley, Skelmersdale, Lancs WN8 8EA
☎(+44 1695) 72 43 43 ⊖ (+44 1695) 72 14 44
♣ £ 10 000 000 (1995/96) 👥 95
Managing Dir: Andrew Myrer
Marine Dir: Geoff Shelton
Sales & Marketing Dir: Des Hatfield
Operations & Technical Dir: David Lace
Purchasing Mgr: Peter Wright

Contact Reference

■ ROV buoyancy blocks for the US Navy, Super Scorpio, fenders for Royal Navy, floats for MoD, Allied Navies & US Air Force. Bulkheads for towed arrays & other specialised mouldings for the defence industry. Surface & sub-sea buoys/buoyancy. [97]

75-16150 Crosfield Group
(a subsidiary of Unilever)
PO Box 26, Warrington, Cheshire WA5 1AB
☎ (+44 1925) 41 61 00 ☒ 627067
⊜ (+44 1925) 65 98 28
👥 1,300 (1997)
Chairman: J D Hinnigan
Sales Manager Desiccants: K French
Assistant Marketing Services Manager: L M Hubbard
■ Desiccant silica gel for the protection of goods from moisture corrosion during storage and transit. ISO 9002 accredited. [98]

75-16220 CSC (Specialised Vehicles) Ltd
Somerden Road, Kingston-Upon-Hull HU9 5PE
☎ (+44 1482) 79 84 00 ⊜ (+44 1482) 78 72 38
♣ £ 5 000 000 (1996) 👥 97 🏭 75 000 sq ft
Managing Dir: C.T. Berridge
Operations & Technical Dir: P. Poulson
Sales & Marketing Mgr: J.J. Middlebrook
■ Specialised transportable containers. Products include shelters for communications systems, mobile workshops, binned storage units, instrumentation vehicles, various forms of mobile laboratories & medical clinics, classrooms, accommodation units & enclosures for simulators. [97]

75-16250 CSIP
(a subs. of ECA SA, France)
69 St. Thomas St, Weymouth, Dorset DT4 8EL
☎ (+44 1305) 77 90 20 ⊜ (+44 1305) 77 80 95
General Mgr: R.W. Parkes
■ Design, supply & support of equipment for defence, oceanology & pollution control applications. [97]

75-16430 John Curran Ltd
(a subsidiary of Rumney Crane & Eng Ltd)
Curran Road, PO Box 72,
Cardiff, South Glamorgan CF1 1TE
☎ (+44 1222) 22 06 41 ☒ 497492 chacom g
⊜ (+44 1222) 37 78 41
♣ £ 4,000,000 👥 60
Managing Director: Dr P Sandom
Sales Director: G Nurton
■ Design, manufacture & install aero engine test facilities for military aviation. Radar turntable manufacture & installation. Satellite communication antenna manufacture & installation. [98]

75-16440 Curtis Instruments (UK) Ltd
51 Grafton St, Northampton NN1 2NT
☎ (+44 1604) 297 55 ⊜ (+44 1604) 298 76
♣ £ 4,200,000 (1996) 👥 40 🏭 9,000 sq ft
Managing Director: R W Sadler
Operations Director: Rob Francis
Sales Manager: David Wilkes
Purchasing Manager: Barry Langsford
Marketing Services Co-ordinator: Mrs Tanya Stock
■ Elapsed time indicators. Battery monitoring instrumentation. DC motor speed controllers. Care, safety & weighing equipment. [97]

75-16444 Custom Cameras Ltd
27 St Cuthberts Street, Wells, Somerset BA5 2AW
☎ (+44 1749) 67 58 34 ⊜ (+44 1749) 67 57 84
♧ £ 10,000 (1997) ♣ £ 300,000 (1997) 👥 8 (1997)
🏭 300 m2
Managing Director: R S Harris
Director: A E Moon
■ Special-purpose low-light television cameras. Digital TV frame stores, miniature TV cameras, radiation (gamma) resistant cameras, waterproof cameras. [98]

75-16495 Cyclops Engineering
(a division of Avesta Sheffield Ltd)
Cyclops Works, Carlisle Street, Sheffield S4 7LJ
☎ (+44 114) 243 05 13 ☒ 547025
⊜ (+44 114) 272 84 45
Sales Manager: T J Connolly
■ Supplies three grades of wrought steel armour plate to meet the requirements of the defence industry. Also supplies wear resistant plate. → 85 [98]

75-16500 Cygnus Marine Ltd
Britannia Yard, Penryn, Cornwall TR10 9ER
☎ (+44 1326) 37 29 70 ⊜ (+44 1326) 37 45 85
♧ £ 100 000 ♣ £ 2 500 000 👥 65 🏭 50 000 sq ft
Production Director: John Peters
Sales Director: Ernie Cooke
Commercial Director: Andrew Clynick
■ Work boats & patrol vessels. Glassfibre-reinforced plastic mouldings for any purpose. [98]

75-16525 D&C Cases Ltd
(a subsidiary of Polyformes Ltd)
Unit 4, Porters Wood, Valley Road Industrial Estate,
St Albans, Hertfordshire AL3 6PQ
☎ (+44 1727) 84 44 94 [E] polyformes@compuserve.com
⊜ (+44 1727) 84 02 51
♣ £ 600,000 👥 12 🏭 1,600 m2
Managing Director: Simon Dabson
■ Design and manufacture bespoke transportation containers for the aviation industry. [98]

75-16550 Daco Scientific Ltd
Vulcan House, Calleva Industrial Park,
Aldermaston, Berkshire RG7 8PB
☎ (+44 118) 981 73 11 [E] sales@daco.co.uk
⊜ (+44 118) 981 99 63
♣ £ 1,500,000 👥 50 🏭 7,500 sq ft
Managing Dir: P.M. Kerrison
Sales Director: Chris D Jones
Sales Director: Ed Davidson
Purchasing Manager: Peter Davis
Sales Manager: Rob Stevens
Office Supervisor: Ms Debbie Simmons
■ Design & manufacture of trackballs, joysticks, hand controllers for defence & aerospace related applications. Market proximity & pressure sensors for hostile environments. Precision pressure calibration equipment & primary standards, air data pressure test sets, pressure transducers. Accredited by NAMAS for calibration of pressure-sensitive devices. → 80 → 90
® USA: Kaman Instrumentation Corp, Ruska Instruments, Teledyne Industries [98]

75-16700 Dantherm Ltd
(a subs. of Dantherm A/S, Denmark)
Hither Green, Clevedon, Avon BS21 6XT
☎ (+44 1275) 87 68 51 ⊜ (+44 1275) 34 30 86 [97]

75-16750 Darchem Engineering Ltd
(a subs. of The Weir Group plc)
Stillington, Stockton-on-Tees, Cleveland TS21 1LB
☎ (+44 1740) 63 04 61 ☒ 58549 dareng g
⊜ (+44 1740) 63 05 29
👥 450
Chairman: K. Gamble
Managing Director: J W Menzies
Sales Director: J R Padbury
Operations Director: J Gagg
Business Development Manager: K J Wood
■ High temperature insulation materials and systems. Precision fabrications and honeycomb core. Marine gas turbine intake and exhaust systems. Sheet metal profiling. Specialist forming techniques. [98]

75-16780 Dasic International Ltd
Winchester Hill, Romsey, Hampshire SO51 7YD
☎ (+44 1794) 51 24 19 ⊜ (+44 1794) 52 23 46
♣ £ 2 000 000 👥 22
Sales & Marketing Dir: R. Billingham
Technical Dir: J.L. Belk
■ Production & supply of approved maintenance chemicals for the aviation & transport industries & oil dispersants. [98]

75-16800 Data Acquisition Ltd
Electron House, Higher Hillgate,
Stockport, Cheshire SK1 3QD
☎ (+44 161) 477 38 88 ⊜ (+44 161) 480 51 42
♧ £ 120,000 (1996) ♣ £ 1,481,000 (1996)
👥 35 (1997) 🏭 1,300 sq m
Managing Director: Kourosh Mehrabani
Sales and Marketing Manager: Ian Drage
Purchasing Manager: Des Ratcliffe
■ Manufacture of DATA Acquisition Multi-Analyser systems. Suppliers of vibration analysis systems. Sub-contract facility for design and build of electronic equipment and systems.
® USA: Wilcoxon Research Inc. [98]

75-16820 Data Disposal & Destruction
Security Centre, Osier Way, London E10 5SB
☎ (+44 181) 556 56 08 ⊜ (+44 181) 558 92 30
♣ £ 1,400,000 👥 33 (1997)
Managing Director: Mike J Bowles
Sales and Marketing Director: Ms Josephine Bowles
Operations and Technical Director: Daniel Sheehan
Purchasing Manager: Ms Jayne Golborn
■ Supply of service and disintegration equipment for the security destruction of all types of confidential data. ISO9002 registered company. [98]

75-16885 Datong Electronics Ltd
(a subsidiary of Network Security Holdings Ltd)
Clayton Wood Close, West Park, Leeds LS16 6QE
☎ (+44 113) 274 48 22 ⊜ (+44 113) 274 28 72
♣ £ 1,500,000 (1993/94) 👥 40 (1997) 🏭 15,000 sq ft
Managing Dir: B. Hall
Sales Dir: Mrs G.K. Tong
Operations Dir: Ms Teresa Jones
Sales Mgr: Dar Shivtiel
Company Administrator: Ms Sandra Wilshaw
■ Design & manufacture of anti-bugging & radio direction finder systems. [98]

75-17010 Davies Industrial Communications Ltd
Sutton Road, Wiggington, Yorks YO3 3RB
☎ (+44 1904) 76 01 11 ⊜ (+44 1904) 76 03 71
♣ $ 1 500 000 👥 30
Managing Dir: J. Davies
Sales Dir: M.G. Winter
■ Electronic hearing protection, audio ancillaries, communications headsets, intercom equipment & tactical microphones. [97]

75-17100 Davin Optical Ltd
(a subsidiary of David Optical Holdings Ltd)
9A Chester Rd, Borehamwood, Herts WD6 1LT
☎ (+44 181) 905 14 14
⊜ (+44 181) 207 65 81 (Design/Sales)
207 48 53 (Administration/Production)
♧ £ 380 000 ♣ £ 3,000,000 👥 70
Chairman: D.J.J. Holland
Managing Dir: R.G. Chilton
Technical Dir: E.J. Judd
Marketing Mgr: A. Hedges
Purchasing Manager: Mark Britton
■ Design and manufacture of night vision sighting equipment for small weapons, night vision photographic equipment and observation equipment for military and security use. Design and manufacture of optical components and sub-systems for use in visible and infra-red wavebands. Optics to customer specifications. Optical design service. [97]

75-17270 Deben Group Industries Ltd
Deben Way, Melton, Woodbridge, Suffolk IP12 1RB
☎ (+44 1394) 38 77 62 ⊜ (+44 1394) 38 03 73
♧ £ 500 000 (1996) ♣ £ 2 000 000 (1996) 👥 30
🏭 7 500 sq m
Managing Dir: P.H. Walker
Sales & Marketing Dir: A.R. Gibson
■ Electronic active hearing protection. Passive hearing protection. Sniper scopes MIL-SPEC. TASCO. Military & civilian binoculars TASCO. Camouflage nets. [97]

75-17300 DM Aerospace Ltd
(† Dedicated Microcomputers Group Ltd)
(a subsidiary of the Dedicated Microcomputers Group Ltd, UK)
16 Hilton Square, Pendlebury,
Swinton, Manchester M27 1DB
☎ (+44 161) 727 33 83 [E] mhorne@dmicros.com
⊜ (+44 161) 727 33 79
♣ £ 35,000,000 (Group) (1996) 👥 350 (Group) (1996)
Managing Director: Mike Newton
Operations Director: D Sprott
Sales and Marketing Director: G Oliver
Technical Manager: M A Horne
Purchasing Manager: Ms Sharon Plant

■ Specialising in the manufacture of control equipment for security applications, with the uniplex video multiplexer. Available in colour and monochrome, the system allows multiple images to be displayed on screen and recorded to tape. 98

75-17350 Deenside Ltd
Meadows House, Henry Street, Northampton NN1 4JE
☎(+44 1604) 332 60 ⊠ 31612 otss g
⊖ (+44 1604) 60 43 98
Managing Director: Alan Hickling
■ Protective clothing incl. knife resistant vests, forearm and elbow guards, shin and knee guards, and groin guards. Protective equipment for horses and mounted police officers. 98

75-17400 Deep Sea Seals Ltd
(a subs. of John Crane Marine Int'l)
4 Marples Way, Havant, Hants PO9 1NX
☎(+44 1705) 49 21 23 ⊠ 86122 dss g
⊖ (+44 1705) 49 24 70
👥 1 000
Managing Dir: J.F. Cousins
Sales Dir: K.W. Harris
Naval Sales Mgr: C.L. Hyde
Marketing Mgr: Mrs P.E.L. Salter
■ Specialists in design and manufacture of seals for propeller shafts, rudderstocks, stabilisers, thrusters, bulkheads, pumps, and waterjets for surface and submersible craft. Cable handling systems. → 80 → 90 98

75-17460 Deloro Stellite Ltd
(a subsidiary of Thermadyne, USA)
Swindon Road, Stratton St Margarets,
Swindon, Wiltshire SN3 4QA
☎(+44 793) 82 24 51 ⊠ 44215 ⊖ (+44 793) 82 38 14
♣ £ 16,000,000 👥 100+
Managing Director: K A Pearce
Sales and Marketing Director: M J Thompson
■ Wear resistant alloy products in the form of hardfacing rods, electrodes, powders, wires, cast components, powder metal parts and specialised hardfacing equipment. 98

75-17470 Delta (Manganese Bronze) Ltd
(a subs. of Delta plc)
Handford Works, Hadleigh Road, PO Box 22,
Ipswich, Suffolk IP2 0EG
☎(+44 1473) 25 21 27 ⊖ (+44 1473) 21 82 29
👥 200+
Managing Dir: K. Benson
Sales Dir: Alan E. Eklid
Export Sales Dir: I. Richardson
Works Mgr: C. Joynson
UK Sales Mgr: S. Kidd
■ Manufacturer of semi-finished wrought & continuous cast copper alloys which include aluminium bronzes, manganese & naval brasses & bronzes, phosphor bronzes, gunmetals & beryllium free & high conductivity alloys. These are in the wrought form in rod, bar, plate, forgings, billet & wire. Continuous cast products in rod, hollow, rectangle & profiles. 98

75-17500 Denbridge Digital
(incorporating dB Electronics and Datawave Technology)
62-68 Strand Road, Bootle, Liverpool L20 4BG
☎(+44 151) 933 01 18 ⊖ (+44 151) 933 67 07
General Manager: Harry Limming
Sales and Marketing Manager: Malcolm Lidsay
■ Design and manufacture of a wide range of radar based surveillance systems for coastal, port, harbour and EEZ zones with navy, coastguard, customs, VTS fishery protection. 98

75-17530 Denian Electronic Ltd
The Beeches, Old Castle Road,
Salisbury, Wiltshire SP1 3SF
☎(+44 1722) 32 01 50 ⊖ (+44 1722) 33 87 82
♣ £ 200,000 + 👥 3
Director: I R Douglas
Sales and Marketing Director: D C Adams
■ Communications security products. 98

75-17550 H.R. Denne Ltd
(a subsidiary of R. Elwood & Co Ltd)
50 Stuart Road, London SE15 3BE
☎(+44 171) 639 88 55 ⊖ (+44 171) 358 10 90
♧ £ 221,700 (1996) ♣ £ 3,000,000 (1996) 👥 18
Director: A C Samways
Sales and Marketing Director: J Murrell
Purchasing Manager: Mrs M Fryer
■ Uniform, especially ancillary items. Corporate clothing suppliers and manufacturers. Personnel care, security and survival. 98

75-17570 Dennis Specialist Vehicles
Dennis Way, Guildford, Surrey GU1 1AF
☎(+44 1483) 57 12 71 ✉ TRINITY DENNIS Guildford
⊠ 859211 ⊖ (+44 1483) 30 68 36
♣ £ 48,700,000 👥 413
Chairman: V. Treves
Managing Dir: John I. Smith
Sales & Marketing Dir: R. Heard
Operations & Technical Dir: Robert Coombes
General Mgr Fire Sales: Roger Paice
Purchasing Mgr: Tony Barnett
■ Major fire appliance including water, foam, dry powder tenders, hydraulic platform, emergency tenders.

Dennis Fire
Opus Park, Moorfield Road, Guildford, Surrey GU1 1SZ
☎(+44 1483) 82 50 00 ⊖ (+44 1483) 82 50 03 98

75-17705 Deritend Precision Castings Ltd
(a subsidiary of Triplex Lloyd plc)
Ten Acres, Berry Hill Industrial Estate,
Droitwich Spa, Worcester WR9 9BP
☎(+44 1905) 77 25 34 ⊖ (+44 1905) 79 57 72
♧ £ 7,700,000 (1995) ♣ £ 20,000,000 (1996)
👥 260 (1996) 🏭 11,000 sq m
Managing Director: P C Farrow
Sales and Marketing Director: D I Poole
Finance Director: Mrs H Gilhooly
Operations Director: P Postans
Technical Director: P Atkinson
Senior Marketing Executive: T J Noyes
■ Manufacture of precision investment castings, armour quality steel, ferrous, vacuum and superalloys. Melting capacity: 500 Kg air; 350 Kgs vacuum. 98

75-17800 Desoutter UK Sales Ltd
Eaton Road, Hemel Hempstead, Herts HP2 7DR
☎(+44 1442) 34 43 00 ⊠ 21392
⊖ (+44 1442) 34 46 00
👥 46 (1997)
President: Jack Demao
General Mgr: D. Wish
Marketing: David Johnston
Marketing Co-ordinator: Ms Lucy Morgan
■ Manufacturers & suppliers to NATO forces of pneumatic & electric power tools & systems for maintenance & manufacture of defence equipment. Subsidiaries or main agents in all major world markets. 98

75-17950 Dexine Rubber Co Ltd
Spotland Bridge, Rochdale, Lancashire OL11 6AU
☎(+44 1706) 400 11 ⊖ (+44 1706) 52 77 14
♧ £ 600 000 ♣ £ 3 300 000 👥 145 🏭 12 500 sq m
Sales Manager: A H Hollinshead
■ Mouldings, bonded components, extrusions, sheeting in all natural & synthetic rubbers. 98

75-18025 Diamond H Controls Limited
(a subsidiary of Cortworth plc, UK)
Vulcan Road North, Norwich, Norfolk NR6 6AH
☎(+44 1603) 42 52 91 ⊖ (+44 1603) 42 49 07
👥 350 (1997)
Managing Dir: A.C. Hobson
Sales & Marketing Dir: P. Hodge
Technical Mgr: A.E. Clark
Sales Mgr: M.G. Temple
Sr Buyer: J. Ingham
■ Manufacture of electrical, electronic & gas controls. 98

75-18040 Dicoll Electronics Limited
Unit 7, Bordon Trading Estate, Oakhanger Road,
Bordon, Hampshire GU35 9HH
☎(+44 1420) 47 37 73 E sales@dicoll.co.uk W http://www.dicoll.co.uk ⊖ (+44 1420) 47 86 64
♣ £ 2,500,000 (1997) 👥 30 (1997)
Managing Director: Brian Collier
Administration Manager: Ms Jeraldine Kennedy
Sales Manager: David Thorpe
■ Videographic processors for PC/AT, VME, S BUS.
® USA: Parallax Graphics Inc. 98

75-18050 Didsbury Engineering Co. Ltd
(a division of Didsbury Holdings)
Clifton Works, Manor Road, Manchester M19 3EJ
☎(+44 161) 224 62 24 ⊖ (+44 161) 224 20 98
♣ £ 3,000,000 (1997) 👥 35 (1997) 🏭 6,000 m2
Group Managing Director: Rod Daniels
Managing Director: Keith Rogerson
Operations Director: C Beddall
Sales and Contracts Manager: J G Hegg
■ Mechanical handling equipment primarily involved with defence, civil air and companies associated with industrial and marine safety.
Design and manufacture of specialised portable lifting equipment. Low permeability manual winches, overhead runway weapon handling, naval systems, solving lifting/handling problems. Lifeboat release gear. 98

75-18220 Divex Ltd
Westhill Industrial Estate, Wetshill, Aberdeen AB32 6TQ
☎(+44 1224) 74 01 45 ⊖ (+44 1224) 74 01 72
General Mgr: B. Bigg
■ Diving equipment, air compressor sets, associated plant & tools. Underwater cutting & welding systems, underwater video systems.
● Pressure Products House, Westhill Ind. Est., Aberdeen
☎(+44 1224) 74 01 45 ⊠ 739762 gaservg
⊖ (+44 1224) 74 01 72
Manager: D. Smith 98

75-18250 B. Dixon-Bate Ltd
Unit 45, First Avenue, Deeside Industrial Park,
Deeside, Flintshire CH5 2LG
☎(+44 1244) 28 89 25 ⊖ (+44 1244) 28 84 62
♧ £ 85,000 (1994/95) ♣ £ 1,700,000 (1994/95) 👥 46
🏭 8,000 sq m
Managing Dir: T.F. Barnes
Operations Dir: P. Ashley
Sales Mgr: P.L. Miles
■ Manufacture of a range of towing pintles, jaws & hooks for all types of military vehicles. 98

75-18270 DMS Technologies
Unit A7, The Premier Centre, Abbey Park,
Romsey, Hampshire SO51 9AQ
☎(+44 1794) 83 01 11 E info@dmstech.co.uk
W http://www.dmstech.co.uk ⊖ (+44 1794) 83 01 03
♣ £ 3,000,000 (1996) 👥 25 (1996) 🏭 6,000 sq ft
Managing Dir: M. Winter
Sales Dir: P. Driver
Production Dir: A. Fenn
Purchasing Mgr: Miss L. Puttock
Public Relations Mgr: Miss K. Dugard
■ Manufacture of a range of battery power supplies for various applications & industry sectors, including portable instrumentation, engine starting, telecommunications, military applications, medical equipment, switch tripping & closing, uninterruptible power supplies & standy by power. 98

75-18320 domnick hunter Limited
Dukesway, Team Valley Trading Estate,
Gateshead, Tyne & Wear NE11 0PZ
☎(+44 191) 402 90 00 E domnickh.co.uk
⊖ (+44 191) 482 62 96
👥 750 (1996)
Managing Director: K Murtagh
Marketing Director: D Wilde
Operations Director: D Swan
Marketing Services Manager: R Moore
Purchasing Manager: S Carter

domnick hunter ltd
Industrial Division
Dukesway, Team Valley Trading Est,
Gateshead, Tyne & Wear NE11 0PZ
☎(+44 191) 402 90 00 ⊖ (+44 191) 482 62 96
Managing Dir: C.T. Billiet
Operations Dir: R.M. Fielding
Sales & Marketing Dir: D. Gallant
Marketing Services Mgr: R. Moore
Purchasing Mgr: S. Scott
■ High efficiency purification & drying of compressed air & other gases. Specialists in the membrane filtration field. 98

75-18400 Dopra Systems Integration Ltd
(† Dowty Systems Integrations Ltd)
(a member of the Dopra Group)
Marshfield, Chippenham, Wiltshire SN14 8NN
☎ (+44 1225) 89 18 81 ⊠ 449520
⊖ (+44 1225) 89 14 40
♣ £ 30 000 000 👥 230 ⌂ 16 000 sq m
Managing Dir: Bruce Brain
Sales & Marketing Dir: Kevin Chevis
Operations & Technical Dir: Robin Chillman
Purchasing Mgr: Ms J. Rose
Emergency Systems Sales Mgr: F. Wells
■ Specialises in providing systems solutions including centralised command and control, administration, digital switching, mobile data and telemetry, as well as specialised consultancy. → 80

● Langley Road, PO Box 451,
Chippenham, Wiltshire SN15 1LZ
☎ (+44 1249) 44 37 77 ⊠ 449318
⊖ (+44 1249) 44 35 98
Managing Dir: Bruce Brain [97]

75-18410 Doncasters plc
(† Inco Engineered Products Ltd)
Head Office
28-30 Derby Road, Melbourne, Derbyshire DE73 1FE
☎ (+44 1332) 86 49 00 ⊖ (+44 1332) 86 48 88
👥 1,500 (1996)
Chairman: Dr I L Dillamore
Commercial Director: J Stevenson
Business Development Manager: Peter Bridges
Purchasing Manager: J Laud
Marketing Services Manager: Miss P Innes
■ A group of companies engaged in the manipulation and shaping of metals for a variety of industries, with particular emphasis on the production of aerospace components. Including forging, fabricating, forming, casting and machining of titanium, nickel and aluminium alloys, and high strength and corrosion-resistant steels. → 80
Subsidiaries with defence business:
Doncasters Monk Bridge, UK 75
Doncasters Structures, UK 75
Doncasters Sheffield, UK 75
Doncasters Inc, USA 75 [98]

75-18420 Doncasters AMTech
(† IEP Airfoils AMTech Division)
(a Doncasters plc company)
Weycroft Avenue, Axminster, Devon EX13 5HH
☎ (+44 1297) 345 67 ⊖ (+44 1297) 345 67
General Manager: D Gallier
■ Sub-contract electrochemical machining of compressor and turbine blades, nozzle guide vanes, disks and rings. [98]

75-18430 Doncasters Blaenavon
(† IEP Doncasters Blaenavon Division)
(a Doncasters plc company)
Forge Side, Blaenavon, Gwent NT4 9XG
☎ (+44 1495) 79 03 45 ⊖ (+44 1495) 79 15 65
■ Forged & rolled rings & castings, flash butt-welded rings & seamless precision rolled rings in nickel, titanium, copper, aluminium alloys & stainless steels. Screw press & isothermal forged turbine & compressor disc in steel, titanium, nickel alloys & powder alloys. [98]

75-18450 Doncasters Monk Bridge
(† IEP Airfoils Monk Bridge Division)
(a Doncasters plc company)
Whitehall Road, PO Box 108, Leeds LS1 1PE
☎ (+44 113) 244 62 62 ⊖ (+44 113) 243 51 91
♣ £ 20,000,000 (1997) 👥 300 (1997)
Commercial Director: J Stevenson
Commercial Manager: David Rooke
Director of European Operations: Kevin Boyce
Purchasing Manager: J Laud
International Sales Manager: B Biddle
■ Conventional forged turbine and compressor blades, vanes partially or fully machined, in nickel alloys, titanium, powder and oxide dispersion strengthened alloys and stainless steels.
Isothermal, hammer, press and extrusion forgings in alloy, stainless and maraging steels. For defence applications such as rocket motor preforms, nozzles and hangers. [98]

75-18455 Doncasters Moorside
(† IEP Airfoils Moorside Division)
(a Doncasters plc company)
Ripponden Road, Oldham OL1 4BB
☎ (+44 161) 624 74 17 ⊖ (+44 161) 624 74 17
General Mgr: T. Connolly
■ Precision machined & coined hollow blades. [98]

75-18460 Doncasters Renston
(† IEP Airfoils Renston Division)
(a Doncasters plc company)
Whitehall Road, Leeds, West Yorkshire LS1 IPE
☎ (+44 113) 243 12 22 ⊖ (+44 113) 246 56 81
■ Design & manufacture of dies & tooling. [98]

75-18465 Doncasters Structures Bramah Division
(† IEP Structures Bramah Division)
(a Doncasters plc company)
Holbrook Works, Halfway, Sheffield S20 3GB
☎ (+44 114) 248 39 81 ⊖ (+44 114) 247 41 05
Commercial Mgr: N. Reece
Commercial Mgr: N. Middleton
■ Fabrications, pressings & machined components in titanium, nickel, aluminium & stainless steel. Superplastic & hot forming, diffusion bonding & electroforming for engines & airframes.
IVC Division
Unit 14, Bromford Road Ind Estate, Credenda Road, West Bromwich, West Midlands B70 7JE
☎ (+44 121) 511 11 15 ⊖ (+44 121) 544 52 53
Commercial Manager: Y Zaka
■ RFI shielding of plastic enclosures in a range of metal coatings including copper, nickel, aluminium, silver & stainless steel. Coatings for optical, microwave & infrared reflectance. [98]

75-18573 Dowty Aerospace
(an operating division of TI Group plc)
Lambourn Court, Abingdon, Oxfordshire OX14 1UH
☎ (+44 1235) 55 99 99 ⊖ (+44 1235) 55 31 57
♣ US$ 400,000,000 👥 3,000
Chief Executive Officer: L A Edwards
Managing Director: Peter Wright
Financial Director: C J Taft
HR Director: R Jennings
→ 70
Operating Companies:
Dowty Aerospace Aviation Services (Gloucester)
Dowty Aerospace Aviation Services (Sterling)
Dowty Aerospace Aviation Services (Singapore)
Thermal Processing Group
Dowty Aerospace (Wolverhampton)
Dowty Aerospace (Los Angeles)
Dowty Aerospace Propellers
Dowty Aerospace Hydraulics
Dowty Aerospace (Isle of Man)
Dowty Aerospace (Yakima)
TI Reynolds Rings Ltd
KFW Canada
EBTEC Corp
King Fifth Wheel Co.
NCI Inc.
Valley Manufacturing
Tru Form Inc.
Lewis & Saunders [98]

75-18580 Dowty Aerospace Hydraulics
(a member of the TI Group plc)
Arle Court, Cheltenham, Gloucester GL51 0TP
☎ (+44 1242) 22 11 55 ⊖ (+44 1242) 27 75 77
♣ £ 20,000,000 👥 150
Sales & Marketing Dir: S.M. Ridley
☎ (+44 1242) 27 76 05
■ Design & manufacture of hydraulic systems & components. [98]

75-18581 Dowty Aerospace Isle of Man
School Road, Onchan, Isle of Man
☎ (+44 1624) 67 33 17 ⊖ (+44 1624) 67 24 72
Dir & General Mgr: M. McDonald
■ Manufacture of hydraulic components. [98]

75-18587 Dowty Aerospace Propellers
(a member of the TI Group plc)
Anson Business Park, Cheltenham Road East, Gloucester GL2 9QN
☎ (+44 1452) 71 48 88 ⊠ 43246
⊖ (+44 1452) 71 13 33
👥 300 ⌂ 127,000 sq ft
Managing Director: Kim Ward
Sales and Product Support Director: Geoff Hanson
Engineering Director: M H Burden
Finance Director: R Lathan
Operations Director: S Hawkins
■ Design & manufacture of civil & military propeller systems. [98]

75-18589 Dowty Aerospace Wolverhampton
(a division of the TI Group plc)
Wobaston Road, Wolverhampton, Staffordshire WV9 5EW
☎ (+44 1902) 39 77 00 ⊠ 338220 dowty g
⊖ (+44 1902) 39 43 94
♣ £ 34,000,000 (1995) 👥 475 ⌂ 40,000 m2
Director and General Manager: Paul Strothers
Financial Director: R J Wall ☎ (+44 1902) 39 44 00
Manufacturing Director: Cliff Johnson
☎ (+44 1902) 39 43 17
Commercial Director: R J Willcox ☎ (+44 1902) 39 42 08
Sales and Marketing Manager: G Holden
☎ (+44 1902) 39 42 56
Purchasing Manager: David Storey
■ Design, development & manufacture of primary & secondary flight control actuation systems, airframe accessory drive gearboxes & servo-valves. Supply of high performance propellant tank systems & components for spacecraft programmes. Provision of submarine cable handling equipment & shipboard mine counter-measures winching systems. [98]

75-18592 Dowty Controls
(a division of Dowty Electronic Systems)
Knaves Beech Business Centre, Loudwater, High Wycombe, Bucks HP10 9UT
☎ (+44 628) 81 08 10 ⊖ (+44 28) 81 08 13
Dir & General Mgr: Andy N. Hamment
Operations/Technical Director: G Hatton
Sales and Marketing Director: A J Fisher
Purchasing Manager: P Sherry
■ Design, development, manufacture & support of avionic equipment, engine controls, utility & monitoring systems for aerospace applications. Advanced engine & flight controls. [97]

75-18620 Dowty Seals Ltd
(a subsidiary of the TI Group plc)
Ashchurch, Nr.Tewkesbury, Gloucester GL20 8JS
☎ (+44 1684) 29 91 11 ✉ DOWTYSEALS
[E] sales@dowtyseals.co.uk
[W] http://www.dowtyseals.co.uk ⊠ 43163 dowty g
⊖ (+44 1684) 85 22 92
♣ £ 44,000,000 (1995) 👥 997 (1995)
Managing Dir: D. Richardson
Finance Dir: D.H. Brown
General Mgr Aerospace: R. Davis
Technical Mgr Aerospace & Defence: R.M. Smith
Technical Sales Manager: Keith Tripp
■ A range of elastomers from Nitriles to Perflouroelastomers formulated and mixed on site. A range of international O-Rings including military specification, square section seals, CX seals (PTZFE/elastomer composite seals), T-seals, PTEF and elastomer rotary shaft seals, wiper and scraper seals, U-rings, J-rings, accumulator seals, bearing rings, anti-extrusion rings, elastomeric moulded gaskets, custom designed technical mouldings, conductive elastomers and products for EMI/RFI shielding, Dowty bonded seals and plates, rubber to metal bonded components, Dowprint gaskets, optimising of designs for specific applications using design techniques, FEA, testing and in-house rubber formulating.

● Thornton Trading Estate,
Milford Haven, Dyfed SA73 2RS
☎ (+44 1646) 69 03 95 ⊠ 48134
⊖ (+44 1646) 69 72 77

● Unit 6, Dyffryn Industrial Estate, Pool Road,
Newtown, Powys SY16 3RD
☎ (+44 1686) 62 92 22 ⊠ 35434
⊖ (+44 1686) 62 48 75

● Alton Road, Ross on Wye, Hertfordshire HR9 5NF
☎ (+44 1989) 76 55 55 ⊠ 35154
⊖ (+44 1989) 76 32 05 [98]

75-18730 D.S.G. Lathes Ltd
(† Dean Smith & Grace Ltd)
Pitt Street, PO Box 15, Keighley, West Yorkshire BD21 4PG
☎ (+44 1535) 60 52 61 ⊜ (+44 1535) 68 09 21
♧ £ 700,000 (1997) ♣ £ 2,000,000 (1997)
👥 30 (1997) 🏠 2,000 m2
Managing Director: R W Hole
Sales Manager: W Pell
■ Manufacture of precision lathes (numerically controlled lathes & centre lathes). [98]

75-18750 DSB Special Batteries Limited
Ruben House, Crompton Way,
Crawley, West Sussex RH10 2QR
☎ (+44 1293) 61 19 30; 61 22 53; 44 90 00 (sales)
⊜ (+44 1293) 55 23 16; 40 33 96
♧ £ 50,000 (1997) ♣ £ 10,000,000 (1997)
👥 75 (1997) 🏠 1,700 sq m
Managing Director: S Boyes
Sales and Marketing Director: J Merrell
Operations Director: S Bowler
UK Sales Manager: E Pearce
■ Manufacture & distribution of custom built primary & secondary batteries, battery chargers & battery management systems. Certified to ISO 9001. [98]

75-18850 Dudleys Screenwipers
Hepworth House, Brook Street,
Redditch, Worcester B98 8NF
☎ (+44 1527) 630 57 ⊜ (+44 1527) 668 36
♣ £ 1,600,000 👥 47
Managing Dir: J.P. Eddy
Marketing/Sales Dir: K. Rogers
■ Designers and manufacturers of military and heavy duty windscreen wiper systems, straightline, pantograph, pendulum and clear view screens, electrically, hydraulically and pneumatically driven and associated products. [98]

75-18900 Dunlop Ltd
Metalastik Vibration Control Systems
(a subs. of BTR plc)
Evington Valley Rd, Leicester LE5 5LY
☎ (+44 116) 273 02 81 ⊠ 34397 mvcsdl g
⊜ (+44 116) 273 56 98
♣ £ 35 000 000 👥 675
Sr Sales Engineer Marine & Ind Projects: W. Wreglesworth
■ Anti-vibration/shock mountings & other rubber-bonded-to-metal components. [97]

75-18990 Dunlop Adhesives
(a subs. of BTR plc)
Chester Rd, Birmingham B35 7AL
☎ (+44 121) 373 81 01 ⊜ (+44 121) 384 28 26
Managing Dir: D. Holdham
Sales & Marketing Dir: D.M. Waterson
Operations & Technical Dir: J. Kennedy
Sales Mgr Specialist Adhesives: D. Robinson
Purchasing Mgr: P. Vickers
■ Manufacture and supply of adhesives and sealants, including NASA/ESA approved space grade materials. Outgassing test services. [97]

75-19000 Dunlop Aviation Division
(a member of the BTR Group)
Holbrook Lane, Coventry CV6 4AA
☎ (+44 1203) 66 66 55 ⊠ 31677 dlpcov g
⊜ (+44 1203) 66 22 94
Managing Director: B P Walton-Knight
Director and General Manager Operations: M Attwood
Director and General Manager Programs: J T Whelan
Director and General Manager Dunlop Aviation Services: R A Davies
Sales and Customer Support: S Smith
Public Relations Officer: Ms Dianne Plunkett
■ Design and manufacture of aircraft wheels, steel and carbon brakes. Aircraft brake management systems, digital electronic anti-skid and computerised landing gear control systems. [98]

75-19005 Dunlop Aircraft Tyres Limited
Fort Dunlop, PO Box 504, Birmingham B24 9QH
☎ (+44 121) 384 88 00 ⊜ (+44 121) 377 71 50
👥 200 (1997) 🏠 24,200 sq m
Managing Director: A R Milner
Sales and Marketing Director: M C Simons
General Sales Manager: D R Jones
Purchasing Manager: R Cross
■ Design and manufacture of aircraft tyres for civil/military aircraft applications. [98]

75-19020 Dunlop Hiflex Ltd
(a subs. of BTR plc)
Telford Rd, Churchfields Ind Est, PO Box 2,
Salisbury, Wiltshire SP2 7PQ
☎ (+44 1722) 33 62 31 ✉ HI-FLEX Salisbury
⊠ 47560 hiflex g ⊜ (+44 1722) 32 39 14
♣ £ 12 000 000 (1994) 👥 220
Managing Dir: M.J. Morgan
Marketing Mgr: S.E. Allen
MoD Co-ordinator: I.J. Hooper
Advertising Co-ordinator: Mrs Claire Taylor
Technical Mgr: D.W. Wilson
■ Manufacture & develop flexible hose assembly fittings & associated hydraulic power products.

Manipulated Tube Division
Pennywell Industrial Estate, Sunderland, Tyne & Wear
☎ (+44 191) 534 51 51 ⊠ 537715 dunlop g
⊜ (+44 191) 534 54 42
Commercial Mgr: P. Keast
■ Rigid tube manipulation & design. [97]

75-19035 Dunlop Marine Safety Ltd
(a div. of Dunlop Beaufort Ltd)
Beaufort Road, Birkenhead, Merseyside L41 1HQ
☎ (+44 151) 652 82 76 ⊠ 629797
⊜ (+44 151) 653 64 64
Managing Dir: D. Bexon
Sales & Marketing Dir: A. Southall
Technical Dir: M. Martin
Business Mgr: R.N. Lester
■ Design manufacture & supply of commercial marine liferafts aviation & offshore safety & survival clothing & equipment. [98]

75-19050 Dunlop Oil & Marine
(a subs. of BTR plc)
Moody Lane, Pyewipe,
Grimsby, North East Lincs DN31 2SP
☎ (+44 1472) 35 92 81 ⊠ 52184 dlpgy g
⊜ (+44 1472) 36 29 48
👥 250
Managing Dir: Simon Marsden
Sales & Marketing Dir: Bryan Allison
Sales Mgr: Don Cunningham
■ Comprehensive range of specialised hoses. [97]

75-19055 Dunlop Precision Rubber Division
(a member of the BTR Group)
Ashby Road, Shepshed,
Loughborough, Leicestershire LE12 9EQ
☎ (+44 1509) 50 21 51 ⊜ (+44 1509) 50 54 35
Managing Dir: G.J. Midgley
Financial Controller: P. Gunn
Business Mgr Aerospace Products: T. Hunt
Purchasing Mgr: R. Stacey
Sales Mgr: I. Wood
Personnel Mgr: B. Wilson
Technical Mgr: B.P. Lee
■ Rubber moulders, specialising in airframe & engine seals. [98]

75-19100 Du Pont (UK) Limited
Maylands Avenue, Hemel Hempstead, Herts HP2 7DP
☎ (+44 1442) 21 85 00 ⊠ 825591
⊜ (+44 1442) 24 94 63
♣ £ 39 333 000 (1994)
UK Sales Mgr: Dr D. Mays
Telemarketing Representative: T. Prince
■ Manufacture a wide range of synthetic rubbers, plastics and textiles supplied to defence contractors as raw materials. [98]

75-19140 Dynamic Controls Ltd
Dynamic Works, Union Street, Royton, Oldham OL2 5JD
☎ (+44 161) 633 39 33 ⊜ (+44 161) 633 41 13
♣ £ 1,500,000 (1997) 👥 36 (1997)
Managing Director: Arthur Heap
Operations Director: Andrew Hill
Operations and Technical Director: Ben Ruscillo
■ Design & manufacture of high & low pressure valves, pressure reducing stations & distribution blocks, control valves, pressure reducing regulators, back pressure regulators & pressure actuated systems for submarines, surface ships & industrial applications. [98]

75-19205 Ealing Electro-Optics
(a div. of 600 UK Ltd)
Greycaine Road, Watford, Herts WD2 4PW
☎ (+44 1923) 24 22 61 ⊜ (+44 1923) 23 42 20
👥 90
Managing Dir: John G. Cleather
Sales Mgr: D. Ashby
■ Design & manufacture of optical components & opto-mechanical assemblies for use in the visible & infrared wavebands & optical systems. Specialist products include Thermal Imager & Optical Transfer Function test equipment. [97]

75-19280 ECC Simulation Ltd
(† ECC (Simulation) Ltd)
(a subsidiary of ECC International Corp, USA)
Unit 2 Home Farm Business Centre, Home Farm Road,
Brighton, East Sussex BN1 9HU
☎ (+44 1273) 67 16 70
[E] 100442.3327@compuserve.com
⊜ (+44 1273) 60 01 77
♣ £ 10,000,000 (1997) 👥 160 (1997) 🏠 4,600 m2
Managing Director: Roger McIntosh
Technical Director: I Lamont
Marketing Director: Philip Sercombe
Programmes Director: C Neal
Purchasing Manager: M Catt
Marketing Manager: N Guard
■ Simulators (crew, operator, maintenance) for the military, commercial and industrial sectors. Computer-based instruction. → 80 [98]

75-19300 Edbro plc
Nelson Street, Bolton, Lancashire BL3 2JJ
☎ (+44 1204) 52 88 88 ⊠ 63196 edbroh g
⊜ (+44 1204) 53 19 57
♣ £ 28,500,000 (1996) 👥 616 (1996) 🏠 44,000 m2
Chief Exec. & Managing Dir: Dr J.F.W. Griffiths
Finance Dir & Military Bodywork Sales: S. Matthews
Sales and Marketing Director: R Stewart
Fabrications Mgr: L. Williamson
Marketing Mgr: C.K. Nowell
Purchasing Manager: A Pope
■ Vehicle bodywork. Hydraulic systems; cylinders, PTO/pumps, skiploaders. Special cylinders for earthmoving equipment & vehicle mounted hydraulic cranes. [98]

75-19370 Eddystone Radio Ltd
(a subs. of GEC-Marconi Communication Ltd)
8/9 Birkdale Road, Selly Oak, Birmingham B29 6UB
☎ (+44 121) 471 26 00 ⊠ 337081 eddybm g
⊜ (+44 121) 471 42 00
👥 500
Managing Dir: Christopher Pettitt
General Mgr: Dale Harvey
Business Mgr Communications: David Lakin
Business Mgr Broadcast: Simon Mason
Contracts Mgr: Ted Stacey
■ Solid state communications receivers, HF SSB transceivers, low power broadcast transmitters & die-cast aluminium boxes. [97]

75-19410 Edgcumbe Instruments Ltd
Main Street, Bothwell, Glasgow, Scotland G71 8EZ
☎ (+44 1698) 85 25 74 [E] sales@edgcombe.co.uk
⊜ (+44 1698) 85 44 42
♣ £ 6,600,000 👥 75
Managing Director: Tom J Buckle
Operations Director: David Harrison
Sales Manager: John Armstrong
Sales Office Manager: John A Powell
Commercial: Bill Austin
■ Insulation testers, phase/earth loop testers, high voltage indicators and detectors. Live line testers up to 33 kV, test equipment for telecommunications including pulse-echo cable fault locators, level meters and level generators, earth and contact locators. Portable appliance testers. Business equipment safety testers. [98]

75-19500 EES Technology Ltd
25 Eastways, Witham, Essex CM8 3AL
☎ (+44 1376) 51 41 14 [E] sales@eestech.com
[W] http://www.eestech.com ⊜ (+44 1376) 51 61 16
♧ £ 1,000 (1995) ♣ £ 637,201 (1995) 👥 16 (1997)
🏠 400 m2
Managing Director: Neil Pitman
Purchasing Manager: Trevor Rogers

■ Design, development & manufacture of precision time & frequency standards for both military & commercial use. Applications include radar systems & global positioning systems (GPS). [98]

75-19590 EEV Ltd
(† English Electric Valve Co Ltd)
(a subsidiary of GEC plc)
106 Waterhouse Lane, Chelmsford, Essex CM1 2QU
☎(+44 1245) 49 34 93 ⊠ENELECTICO Chelmsford
Ⓦ http://www.eev.com Ⓔ info@eev.com
⊡ 99103 eevcmd g ⊖ (+44 1245) 49 24 92
♣ £ 80,000,000 (1996/97) ⅆ 1,900 (1997)
⌂ 564,456 sq yd
Managing Dir: J.C. Spinks
Marketing Dir: V. Leverett
Director: G.J. Rowlands
Director: C.V. Neale
Director: D.R. Boylan
Director: D.S. Lunn
Director: R.S. Best
Publicity Mgr: A.D. Christelow
■ Electron tubes for radar, communications and EW, IR devices; night vision, TV camera tubes, image intensifiers; display devices including LCD's; hydrogen thyratrons; microwave tubes; travelling wave tubes; frequency agile magnetrons; receiver protection. CCD sensors and cameras for industrial, scientific and security purposes. Slapper detonators for electronic safety and arming circuitry. UV detectors. Caesium arc lamps. [98]

75-19660 Effer (UK) Ltd
16d Hollingworth Court, Ashford Road,
Maidstone, Kent ME14 5PP
☎(+44 1622) 69 03 92 ⊖ (+44 1622) 67 00 66
♣ £ 1,250,000 ⅆ 6
Managing Director: Robb Gilbert
■ Construction of stiff boom, knuckle boom & telescopic hydraulic cranes for 1-400 tonnes. ISO 9001. → 90
UK Agents for:
Davit International, Effer SPA, Fassmer GmbH [98]

75-19670 EG&G Astrophysics Research Ltd
(† Astrophysics Research Ltd)
Coronation Road, Cressex Business Park,
High Wycombe, Bucks HP12 3TP
☎(+44 1494) 45 16 61 ⊡ 849338
⊖ (+44 1494) 47 13 17
ⅆ 15 000 (worldwide)
Managing and Operations Director: P Mcguire
Sales and Marketing Director: A Aouajane
Sales Office Manager: A Parker
■ Manufacturers of X-Ray equipment for security systems. (Baggage & freight X-ray inspection system.) [98]

75-19675 EG&G Sealol Ltd
(a div.of EG&G Corp., USA)
Coronation Road, Cressex Business Park,
High Wycombe, Bucks HP12 3TP
☎(+44 1494) 45 16 61 ⊡ 838982
⊖ (+44 1494) 45 24 25
♣ £ 10 000 000 ⅆ 110 ⌂ 52 000 sq m
Commercial Manager: Ted Broadfield
■ Develop & manufacture a range of products for the defence industry including: mechanical seals, bellows accumulators, solenoid valves, metal bellows, complete systems & assemblies. [98]

75-19810 EKA Limited
(a subsidiary of Oxley Forster Holdings Ltd)
Valkyrie House, 38 Packhorse Road,
Gerrards Cross, Bucks SL9 8EB
☎(+44 1753) 88 98 18 ⊖ (+44 1753) 88 00 04
ⅆ 10
Chairman & Managing Dir: William O. Forster
Liaison Mgr: John E. Fadelle
Company Secretary: Ms Kathryn A. Clark
■ Supply of specialised hydraulic recovery & lifting equipment (& winches through an affiliate) for incorporation into recovery vehicles. Complete recovery vehicles, rail transfer & side loading transport equipment & the provision of technical training & back-up services. Design & development services for truck-based loading systems. → 85
Production Facility
Grovebury Road, Leighton Buzzard, Beds LU7 8SL
☎(+44 1525) 85 18 57 ⊖ (+44 1525) 37 01 27
■ Assembly & supply of products & spares. [98]

75-19870 Electricars Ltd
Unit 15 Carlyon Industrial Estate,
Atherstone, Warwicks. CV9 1LQ
☎(+44 1827) 71 68 88 ⊡ 342759
⊖ (+44 1827) 71 78 41
♣ £ 5 000 000 ⅆ 46
Managing Dir: R.H. Mason
UK Sales Mgr: R. Welch
■ Electric battery operated industrial & road vehicles, tow tractors & platform vehicles. Diesel operated tow tractors a variety of ground service support vehicles.
Associated companies:
M&M Electric Vehicles Ltd
Electricars Ltd
Eaton Axles Ltd
Chatelec Vehicles Ltd - India
Harbilt Ltd
FMW Ltd [97]

75-20200 A.V. Electronics Ltd
Unit E, Goodwood Airfield,
Chichester, West Sussex PO18 0PM
☎(+44 1243) 53 22 50 ⊖ (+44 1243) 53 65 09
♣ £ 800,000 ⅆ 10 ⌂ 700 sq ft
Managing Dir: J.R. Myers
Operations Director: K Uren
Purchasing Manager: Ms C Destin
■ Headsets, helmets, intercoms, both passive and active noise. Headsets with built in radio units for rotor trimming. Full headset repair service. [98]

75-20250 Electron Tubes Limited
(† THORN EMI Electron Tubes)
(a subsidiary of Electron Technologies Limited)
Bury Street, Ruislip, Middlesex HA4 7TA
☎(+44 1895) 63 07 71 Ⓔ rmca@electron-tubes.co.uk
Ⓦ http://www.electech.demon.co.uk
⊖ (+44 1895) 63 59 53
♣ £ 8,000,000 (1997) ⅆ 200 (1997)
Managing Director: J P Frederiksen
Marketing Director: Dr A G Wright
Technical Director: R McAlpine
■ High technology, low light level detectors. Photomultipliers & related accessories. Design, develop and manufacture silicon detectors to defence standards. BS/EN/ISO 9001 registered company. [98]

75-20400 Electrothermal Engineering Ltd
417/419 Sutton Road, Southend-on-Sea, Essex SS2 5PH
☎(+44 1702) 61 22 11 ⊠ELECTROTOP
Ⓦ http://www.electrothermal.c om ⊡ 995387 thermo g
⊖ (+44 1702) 61 98 88
♧ £ 100,000 ♣ £ 3,833,153 ⅆ 80 ⌂ 30,000 sq ft
Chairman: David N Whittaker
Managing Director: James Morris
Technical Director: D Overfield
■ Feasibility studies, project definition, development, design, prototype manufacture, installation, inspection, testing, post design services, repair, modification kits & spares. Surface and stress relieving equipment. Cooking & boiling vessels; vehicle heaters; de-frosting trolleys; light projector starting units; electrically heated ovens; environmental chambers. BSEN9001 FM28083. → 80 [98]

75-20450 Elequip Projects Ltd
(† Elequip Ltd)
(a subs. of Silvermines plc)
Gloucester Crescent, Wigston, Leics LE18 4YN
☎(+44 116) 277 27 11 ⊡ 341263
⊖ (+44 116) 277 93 69
♧ £ 132 000 ♣ £ 25 000 000 ⅆ 230 ⌂ 7 424 m2
Chairman & CEO: C. Jansen
Managing Dir Projects: F.W. Hull
Field Sales Dir: Peter G. Alvey
Operations & Technical Dir: J. Gemmell
Purchasing Mgr: D. Kirby
■ Electrical & mechanical contracting. Building management systems. Control systems. Diesel generating sets. Low & high voltage switchgear & motor control gear. Project management. [97]

75-20475 Elettronica (UK) Ltd
(a subs. of Elettronica SpA, Italy)
Park View House, 100 Wickam Rd,
Fareham, Hants PO16 7HT
☎(+44 1329) 28 28 32 ⊖ (+44 1329) 22 03 51
ⅆ 60
Chairman: Sir Frank Holroyd
Technical Dir: E.T. Hill
Marketing Dir: J.T. Burrows
■ Microwave division:
Microwave components (0.5GHz to 18GHz); amplifiers; TX/RX modules; attenuators; switches; regulators; mixers; oscillators; filters.
Systems division:
EW systems research & development: passive & active; coherent jamming. Civil systems: IR surveillance & warning; UHF radar systems; sensor fusion.
® Italy: Elettronica Spa [97]

75-20550 Eley Ltd
PO Box 705, Witton, Birmingham B6 7UT
☎(+44 121) 356 88 99 ⊠KYNOCH Birmingham
Ⓔ 101622.1253@compuserve.com ⊡ 336826 eley g
⊖ (+44 121) 331 41 73
ⅆ 140
Managing Director: K J Irving
Marketing Director: P R Thomas
Commercial Manager: A Ward
■ Shotgun and rimfire cartridges. Primer caps and signal flares. Licencing of Eleyprime process for priming. [98]

75-20600 Elga Ltd
(a subsidiary of Protean plc)
High Street, Lane End, High Wycombe, Bucks HP14 3JH
☎(+44 1494) 88 77 00 Ⓔ elga@elga.demon.co.uk
Ⓦ http://www.elga.co.uk ⊖ (+44 1494) 88 10 07
♣ £ 15,488,000 (1995) ⅆ 250
Managing Director: K Rippington
Operations Director: J Pethers
Technical Director: A Mortimer
Production Director: A Wardrop
Technical Manager: P Darnes
Sales Manager: S Weston
■ Water purification equipment for metal finishing & thrust augmentation. [98]

75-20610 Elite Manufacturing Co Ltd
(a division of Frank Wilson (Filing) Ltd)
Elite Works, Station Road, Manningtree, Essex CO11 1DZ
☎(+44 1206) 39 21 71; 39 32 52 ⊡ 987675 elite g
⊖ (+44 1206) 39 14 65
♣ £ 2,500,000 ⅆ 45 ⌂ 43,000 sq ft
Director: M A Prager-Macholl
Commercial Manager: M J Wright
Purchasing Manager: H Thomas
■ Mobile tool control systems for air and land based defence systems. Suspended plan files, rotary files, magnetic planning systems. Universal storage cabinets. Recognised BS EN ISO9002, Reg. No. Q5790 and MoD Defence contractor. [98]

75-20730 Emrad Ltd
Langham Park, Catteshall Lane,
Godalming, Surrey GU7 1NG
☎(+44 1483) 41 86 00 Ⓔ enquiries@emrad.com
Ⓦ http:// www.emrad.com ⊖ (+44 1483) 41 86 18
♧ £ 175,000 (1996) ♣ £ 470,000 (1996) ⅆ 12 (1996)
⌂ 1,000 sq m
Director: Dr Richard J Chignell
Sales and Marketing Director: M J Bushell
Design and Production Manager: C Harrison
■ Concept demonstrator, man portable plastic and metallic detectors. Manufacturers of PipeHawk Ground Probing Radar System. Service providers of PipeHawk Surveys, underground location of utilities. → 80 [98]

75-20750 Emsworth Shipyard
Thorney Road, Emsworth, Hampshire PO10 8BP
☎(+44 1243) 37 52 11 ⊖ (+44 1243) 37 72 59
♧ £ 100,000 (1996) ♣ £ 4,000,000 (1996)
ⅆ 20 (1997) ⌂ 1,200 sq m
Managing Dir: A.W. Chappell
Operations Mgr: T. Smith
Yard Mgr: Tim Smith
Contracts Mgr: Mike Andrews
■ Builders of boats for use by pilots, customs & excise, police, naval patrols & for general harbour duties; assault craft. → 80 [98]

75-20770 Encrypta Electronics Ltd
Waterside Court, Albany Street,
Newport, South Wales NP9 5NT
☎(+44 1633) 85 98 59 Ⓔ mark@encrypta.com
⊖ (+44 1633) 85 97 55
♣ £ 1,500,000 ⅆ 18 (1997)
Operations Director: Simon Fiera

Marketing Director: Mark Hayward
Purchasing Manager: Mrs Nicky Smith
Marketing Assistant: Ms Sarah MacCallum
■ Electronic security sealing products and systems. Electronic seals replace disposable mechanical seals and record seal numbers with times and dates of opening and closing. Can be permanently mounted on trucks, containers, doors, diplomatic bags or equipment. 98

75-20960 Engineering Dynamics (Southern) Ltd (ED)
Watt Close, East Portway, Andover, Hampshire SP10 3LU
☎ (+44 1264) 35 07 78/9 ☒ 477407 zenmon g
⊖ (+44 1264) 33 42 15
♣ £ 1 000 000 ⛬ 12 ⌂ 670 sq m
Managing Dir: J.K. Baker
Sales & Marketing Dir: T. Cowie
Operations & Technical Dir: A. Hutchings
Purchasing Mgr: D. Weeks
■ Anti-vibration mountings and systems. Noise control equipment, environmental testing. Fabrications of steel and aluminium components and structures. Vinration and structural testing. 98

75-20980 Enkev (UK) Ltd
Station Road - Unit 3, Silloth, Cumbria CA5 4AG
☎ (+44 16973) 326 00 ⊖ (+44 16973) 329 94
♣ £ 1,000,000 ⛬ 8
Managing Director: R Tomlinson
■ Packaging material made of rubberised hair (Hairlok), rubberised coir (Cocolok). Specially suitable for ammunition and sensitive electronic equipment. 98

75-21060 Enterprise Control Systems Ltd
31 High Street,
Wappenham, Northamptonshire NN12 8SN
☎ (+44 1327) 86 00 50 ⊖ (+44 1327) 86 00 58
Managing Director: Colin Bullock
Office Manager: Mrs Haysey
■ Design and manufacture of microwave video transmission and telemetry equipment. Professional equipment for Police, MoD and outside broadcasters. 98

75-21090 Envirodoor Markus Ltd
(a subsidiary of the Envirohold Group)
Great Gutter Lane, Willerby, East Yorkshire HU10 6BS
☎ (+44 1482) 65 93 75 Ⓔ sales@envirodoor.com
⊖ (+44 1482) 65 51 31
♣ £ 4,000,000 ⛬ 35 ⌂ 10,000 sq ft
Managing Dir: M De V Roberts
Operations Director: R Newsam
Marketing Assistant: Andrew Acum
■ Shiplift doors, X-ray doors, blastproof, airtight doors, specialist industrial doors, energy saving doors. 98

75-21150 EPC Group Ltd
PO Box 22, Bordon, Hampshire GU35 9PD
☎ (+44 1420) 47 83 72 ☒ 859535
⊖ (+44 1420) 47 83 59
♣ £ 6 000 000 ⛬ 9
Chairman: Ian R. Flack
Marketing Mgr: Dick Williams
Administrator: A.M. Blanford
■ Sub-miniature illuminated dot sight for pistols, SMGs, assault rifles, hunting rifles & shotguns.
Subsidiary:
EPC Worldwide Ltd 98

75-21470 Erskine Systems Ltd
(† Erskine-Dale Power Systems plc)
(a subsidiary of TT Group plc)
Lee De Forest Hse, Eastfield,
Scarborough, North Yorkshire YO11 3DU
☎ (+44 1723) 58 35 11 ☒ 52562
⊖ (+44 1723) 58 12 31
♣ £ 10,000,000 (1996) ⛬ 140 (1996)
⌂ 55,000 sq ft (1996)
Managing Dir: C. Wootton
Sales & Marketing Dir: A. Burnley
Operations Director: Tony Graves
UK Sales Mgr: Roger L. Milner
Purchasing Mgr: John Bowman
■ Manufacturer of battery chargers, DC power supply equipment, emergency lighting, UPS and frequency converters (400Hz). 98

75-21500 Eschmann Bros. & Walsh Ltd
Equipment Division
Peter Rd, Lancing, West Sussex BN15 8TJ
☎ (+44 1903) 75 33 22 ☒ 877075
⊖ (+44 1903) 76 67 93
♣ £ 26 000 000 (1996) ⛬ 270
Managing Dir: P. Battersby
Operations Dir: J.D. Russell
■ Operating tables, portable suction & electrosurgery products, bench top sterilizers. 97

75-21565 E-T-A Circuit Breakers Ltd
(a subsidiary of E-T-A Elektrotechnishe Apparate GmbH, Germany)
Telford Close, Aylesbury, Buckinghamshire HP19 3DG
☎ (+44 1296) 203 36 Ⓔ info@etacbe.co.uk
Ⓦ http://www.etacbe.co.uk ⊖ (+44 1296) 884 97
♣ £ 3,800,000 (1996) ⛬ 20 (1997) ⌂ 1,500 m2
Managing Dir: J.D. Adams
Sales & Marketing Mgr: A.G. Pipe
Purchasing Mgr: S.A. Linnett
■ Manufactures a wide range of circuit breakers for components, equipment & systems protection. 98

75-21567 ETC Ltd
Warrington Road, Manor Park Industrial Estate,
Runcorn, Cheshire WA7 1SB
☎ (+44 928) 57 94 76 ⊖ (+44 928) 57 94 75 98

75-21575 Euravia Engineering & Supply Co. Ltd
Euravia House, Colne Road, Kelbrook, Lancs BB8 6TE
☎ (+44 1282) 84 44 80 ⊖ (+44 1282) 84 42 74
⌂ 1 300 sq m
Managing Dir: D.G. Mendoros
Purchasing Mgr: J. Sutcliffe
■ Supply & technical support in engineering & fabrication for the aviation market. Aero-engine repair centre. Prime contractor to UK MoD. Repair, overhaul & modification for jet engines & APU's. ISO 9001, JAR 145, CAA, NATO & NAMSA Approved. 97

75-21580 Eurochart
Efficienta Magnetic Planning Systems
Old Bell Works, School Lane, Pilling, Lancashire PR3 6HB
☎ (+44 1253) 79 01 11; 09 22 ⊖ (+44 1253) 79 01 22
Partner: P.C. McCaffrey
Partner: C. McCaffrey
■ Planning systems & information boards. 98

75-21583 European Data Systems
Cornbrash Park, Chippenham, Wiltshire SN14 6RA
☎ (+44 1249) 44 66 57 Ⓦ http://www.rugged-pc.com
⊖ (+44 1249) 44 88 47
Director: Geoff Shingles
Contact: Robb Lawton
■ Rugged military and industrial computer hardware manufacturer. 98

75-21701 Eve Trakway Ltd
(a division of the Eve Group)
Coxmoor Road, Sutton-in-Ashfield, Notts NG17 5LA
☎ (+44 1623) 51 53 33
Ⓔ 106103.2724@compuserve.com
⊖ (+44 1623) 44 01 54
Managing Director: D J Robinson
Administration Director: P A Murdoch
Sales and Marketing Manager: Shaun Bennett
■ Aircraft recovery equipment, including sledges & portable access roads. 98

75-21800 Evered Tube Division
(† Evered & Co. (Metals) Ltd)
(a division of Cole Tubes Limited)
Heath Street South, Spring Hill, Birmingham B18 7PZ
☎ (+44 121) 456 59 59 Ⓔ evered@atwall.com
⊖ (+44 121) 456 52 11
♧ £ 900,000 ♣ £ 1,500,000 (1996) ⛬ 17 (1997)
⌂ 3,000 m2
Managing Director: A W Cole
Operations Director: M A Wilkinson
Sales and Marketing Director: M A Thomson-Tur
Purchasing Manager: L Treagust

■ Raw precision waveguide tubing in all standard sizes, sections & alloys. Services on non standard sizes including thin wall for low weight applications: satellite communications/surveillance systems. Thick wall, for high powered ground based radar systems. Waveguide supplied in either straight lengths or complete with flanges & fully manipulated to meet product application. 98

75-21970 Evotec Ltd
27 Furham Field, Hatch End, Pinner, Middx HA5 4DX
☎ (+44 181) 428 47 23 ⊖ (+44 181) 420 17 64
♣ US$ 15 000 000 ⛬ 12
Managing Dir: Cyril J.F. Townson
Operations Dir: Christopher Shillinglaw
Supplies Mgr: Ms Diane Douglass
■ Production of food ration packs, lightweight survival packs & supplementary rations. First Aid Kits. Comprehensive field catering equipment, mobile field kitchens & insulated food containers. Gelled safety fuel & associated cooking equipment. Advisory & consultancy service. 97

75-21990 Exchem Explosives
(a division of Exchem plc)
Rough Close Works, PO Box 4,
Alfreton, Derbyshire DE55 7AB
☎ (+44 1733) 83 22 53 ⊖ (+44 1733) 52 01 38
Managing Director: R R Keen
Sales and Marketing Director: G R Temple 98

75-22000 Exchem plc
Commonwealth House, 1/19 New Oxford Street,
London WC1A 1NF
☎ (+44 171) 405 60 65 ⊖ (+44 171) 405 54 94
Managing Director: Alain Thibaudeau
Sales and Marketing Director: J Murphy
■ Explosives casting & filling of ordnance stores. Manufacture of speciality chemicals. 98

75-22050 Exide Batteries Ltd
Gate No.3, Pontypool Ind. Estate, New Road,
Pontypool, Gwent NP4 5DG
☎ (+44 1495) 75 00 75 ☒ 261186
⊖ (+44 1495) 75 02 22
Managing Dir: G. Sykes
Sales & Marketing Dir: A.A.F. Summers
Marketing Services Mgr: D. Rudd
■ Extensive range of automotive batteries suitable for military vehicles. 97

75-22150 EW Simulation Technology Ltd
(† Anaren Microwave Ltd)
Unit 1 The Royston Centre, Lynchford Road, Ash Vale,
Aldershot, Hampshire GU12 5PQ
☎ (+44 1252) 51 29 51 Ⓔ ewst@compuserve.com
⊖ (+44 1252) 51 24 28
♣ £ 1,000,000 (1997) ⛬ 15 (1997) ⌂ 600 sq m
Managing Dir: Dr R.S. Andrews
■ EW simulators, on-board trainers, receiver support & manufacture. 98

75-22160 A.R. Fabb Bros. Ltd
29/31 Risborough Road, Maidenhead, Berkshire SL6 7BJ
☎ (+44 1628) 235 33 ⊖ (+44 1628) 227 05
Director: J.R. Fabb
■ Gold wire embroiderers & accoutrement manufacturers. Ceremonial helmets. Flags & banners. Military band regalia. 98

75-22170 Facet Industrial UK Ltd
(a subsidiary of Mark VI Industries Inc, USA)
Treforest Industrial Estate,
Pontypridd, Mid Glamorgan CF37 5YL
☎ (+44 1443) 84 41 41
⊖ (+44 1443) 84 20 44; 84 42 82
♣ £ 6,772,000,000 (1996) ⛬ 65 (1997) ⌂ 46,860 sq ft
Managing Director: Michael F Smart
UK Director: Frank Ennis
Export Sales Director: Andras Juhasz
Purchasing Manager: Wayne Gibbs
Marketing Co-ordinator: Ms Melanie J Smart
☎ (+44 1443) 84 42 82 ext. 238
■ Filtration and separation equipment. Specialists in fuel handling, oil recovery and environmental protection. Approved to ISO 9001. 98

Contact Reference

United Kingdom

75-22190 W.H.J. Fagg & Son Ltd
Driberg Way, Braintree, Essex CM7 1NB
☎(+44 1376) 55 07 77 ⊖ (+44 1376) 55 29 55
♣ £ 1 500 000 (1995) 39 1 400 sq m
Managing Dir: W.T. Fagg
Sales & Marketing Dir: I. Brown
Marketing & Sales Mgr: J.E. Fagg
Production Mgr: M. Fagg
■ Manufacture small arms cleaning & maintenance kits. Also component manufacturing & tooling packages. [97]

75-22200 Fairey Arlon Ltd
(a subsidiary of the Fairey Group plc)
Fareham Industrial Park, Fareham, Hampshire PO16 8XG
☎(+44 1329) 82 61 61 ⊖ (+44 1329) 82 57 58
♣ £ 10,000,000 150 (worldwide)
Marketing Dir: Ged Palmer
■ Hydraulic filtration equipment fitted as proprietary items on various systems e.g. Ground support equipment, ship stabilizer equipment, hydraulic control equipment of variable pitch propellers, hydraulic lifting equipment. [98]

75-22255 Fairey Hydraulics Ltd
(a subsidiary of Fairey Group plc)
Claverham, Bristol BS19 4NF
☎(+44 1934) 83 52 24 ☒ 444518 FAIRCL G
⊖ (+44 1934) 83 53 37
♣ £ 27,000,000 (1996) 320 (1996)
Chairman: J.W. Poulter
Managing Director: S R Frost
Commercial Director: C Boorman
Financial Director: P Butterworth
Engineering Director: M J Anthony
Production Director: T A Brown
Marketing Director: N J Guttridge
Publicity Manager: R J Harris
Purchasing Manager: M J Robinson
■ Design and manufacture of specialised actuation systems and associated equipment for the aerospace, defence and industrial sectors. Business areas include aircraft flight controls, hydraulic systems and landing gear, guided weapon and combat vehicle actuation systems, subsea and rail track control equipment, and repair and overhaul services. → 80 [98]

75-22260 Fairey Industrial Ceramics Ltd
(† Doulton Industrial Products)
(a Fairey Group plc company)
Filleybrooks, Stone, Staffordshire ST15 0PU
☎(+44 1785) 81 32 41 ⊖ (+44 1785) 81 87 33
♣ £ 8,500,000 220
Managing Dir: Colin Tiernan
■ High pressure air filters. Breathing air filters. Potable water filters. High temperature precision ceramics. [98]

75-22300 Fairey Microfiltrex Ltd
(a subsidiary of Fairey Group plc)
Fareham Industrial Park, Fareham, Hampshire PO16 8XG
☎(+44 1329) 28 56 16
[E] 100434.2340@compuserve.com
⊖ (+44 1329) 82 24 42
♣ £ 8,000,000 (1996) 110 3,000 sq m
Managing Dir: Adrian Ryder
Operations/Technical Dir: T.J. Cardy
Financial Dir: R. Everitt
Sales & Marketing Dir: Terry Stott
International Sales Mgr: L. Lee
Purchasing Mgr: Les White
■ Superfine filtration for the aerospace, chemical, nuclear and process industries. Suppliers of cleanable and disposable filter elements, differential pressure indicators, last chance discs and assemblies. Quality approved to DEFSTAN 05-91, BCAR AS-1 and ISO 9001. [98]

75-22350 Fairless Engineering Supplies Co. Ltd
56 Skinner Street, Stockton-on-Tees, Cleveland TS18 1EG
☎(+44 1642) 67 60 70 ⊖ (+44 1642) 60 64 01
Director: J R Nelson
■ Bulkhead glands, deck glands, attachable equipment glands, threaded and welded types and heatshrink moulded alternatives. To MoD (navy) specifications and NATO service drawing numbers for use in ships. EMP/RF1 glands. [98]

75-22370 Fairmile Fencing Ltd
Gower Street, St Georges, Telford, Shropshire TF2 9DB
☎(+44 1952) 62 00 62
⊖ (+44 1952) 61 72 55; 61 00 11
♣ £ 9,000,000 (1996) 110 (1996) 7,000 m2
Managing Dir: I.D. Ripley
Sales & Marketing Dir: K.R. Greenhalgh
Sales Mgr: D. Sneath
Purchasing Mgr: George Bailey
■ Manufacture fencing, perimeter fences, vehicle control barriers & pedestrian & crowd control turnstiles. [98]

75-22400 Leonard Farnell & Co. Ltd
North Mymms, Hatfield, Herts AL9 7SR
☎(+44 1707) 26 44 88 ☒ 22149
⊖ (+44 1707) 26 83 47
♣ £ 1 000 000 19
Director: S. Robertson
■ Designers & manufacturers of civil engineering materials test & quality control instruments & apparatus. [97]

75-22410 Farr Europe
272 Kings Road, Tyseley, Birmingham B11 2AB
☎(+44 121) 707 82 11 ☒ 337065 farrco g
⊖ (+44 121) 706 99 86
♣ £ 5 800 000 145
Managing Director: Don A Parker
Operations Director: Jamie B Kav
Sales and Marketing Director: Richard D Brennan
■ Supply a wide range of engine air cleaners and filtration products for military equipment. [98]

75-22420 Farsound Engineering Ltd
Westminster Instrument Development Co. Ltd
(a subs. of Farsound Holdings plc)
Unit 3, Highams Park Industrial Est., Jubilee Avenue, London E4 9JD
☎(+44 181) 523 33 77 ⊖ (+44 181) 523 12 49
♧ £ 500 000 (1996) ♣ £ 3 500 000 (1996) 50
10 000 sq m
Managing Dir: R.B. Gulperin
Production Dir: Melvin R. Atkinson
Purchasing Mgr: J. Glynn
■ Design & manufacture of aircraft bonding leads, jumper assemblies. Manufacture of support clips, clamps, clip & cleat bushes, lanyard assemblies, turned parts & aircraft standard parts. Official distributor North Bridge Engine Bolts. Westminster - design & manufacture of aircraft turn & slip gauges. [97]

75-22450 Thomas Fattorini Ltd
Regent Street Works, Birmingham B1 3HQ
☎(+44 121) 236 13 07 ⊖ (+44 121) 200 15 68
♣ £ 4 000 000
Managing Dir: G.T.B. Fattorini
Director: T.R. Fattorini
■ National honours & awards, military medals, badges, belt buckles & other uniform accoutrements, buttons, ceremonial swords, insignia & regalia.

London Office
189 Regent St, London W1R 7WF
☎(+44 171) 734 02 68 ⊖ (+44 171) 734 22 01
Director: T.R. Fattorini
■ Badges, medals, uniform buttons, swords. National & military honours. [97]

75-22460 Faure Herman Industrial Ltd
(a subsidiary of Faure Herman SA, France)
Unit 7 Wirral Business Center, Gorsey Lane/Dock Road, Birkenhead, Wirral L41 1JW
☎(+44 51) 638 45 09 ⊖ (+44 51) 639 61 64 [98]

75-22485 FBM Marine Limited
(a member of the FBM Marine Group)
Cowes Shipyard, Pelham Rd, Cowes, IoW PO31 7DL
☎(+44 1983) 29 71 11 ☒ 86466 fambro g
⊖ (+44 1983) 29 96 42
260
Managing Dir: Malcolm Keith
Commercial Dir: Simon O'Neill
Operations & Technical Dir: D. Border
Marketing Mgr: Mrs Heather Cox
■ Design & manufacture of boats & ships up to 120m for military, para-military & commercial operators, also floating pontoon systems. The company incorporates Brooke Marine Shipbuilders Ltd, Fairey Marine Ltd, Fairey Marinteknik Ltd, Cheverton Workboats. [97]

75-22490 FDB Electrical Ltd
(† Findlay Durham & Brodie Electrical)
(a subs. of Findlay, Durham & Brodie Ltd)
Reynard Mills Trading Estate, Windmill Road, Brentford, Middlesex TW8 9NZ
☎(+44 181) 568 46 21 ⊖ (+44 181) 569 78 99
♧ £ 3,000 ♣ £ 1,500,000
Chairman: C.J.W. Durham
Managing Dir: J.C. Allen
Sales Mgr: G.R. Miles
■ Manufacture & supply of electrical earth monitoring & earth leakage protection devices; energy management systems & circuit breakers. → 85 [98]

75-22520 Feedback Instruments Ltd
(a subsidiary of Feedback plc)
Park Road, Crowborough, East Sussex TN6 2QR
☎(+44 1892) 65 33 22 [E] feedback@fdbk.demon.co.uk
[W] http://www.fbk.com ⊖ (+44 1892) 66 37 19
Managing Director: A J Whiteley
Operations Director: D J Marks
Sales and Marketing Director: K J Hamilton
Marketing Manager: C E Tucker
Purchasing Manager: R J Pope
■ Weather satellite receiving systems, electronic test instruments, CBL & 'hands on' equipment for technical training. [98]

75-22550 Fenner Fluid Power
Ashton Road, Harold Hill, Romford, Essex RM3 8UA
☎(+44 1708) 34 38 51 ⊖ (+44 1708) 37 39 93
♣ £ 10 000 000 130 5 200 sq m
Managing Dir: N. Hobson
Operations/Technical Dir: S. Usher
Sales & Marketing Dir: G. Ward
■ Manufacture of air operated pressure pumps, hydraulic power packs, piston air motors, vane air motors, vacuum pumps hydraulic pumps & motors. Manufacture of hydralic pumps & test rigs for systems, missiles, vessels, hoses, valves, mortar bombs & steering assemblies. Specialist rigs for gas & diving equipment. [97]

75-22600 Fernau Avionics Ltd
(a subsidiary of Firan Corp, Canada)
Airport Executive Park, President Way, Luton, Bedfordshire LU2 9NY
☎(+44 1582) 48 31 11 ⊖ (+44 1582) 48 44 04
♣ £ 10,000,000 (1996) 74 12,000 sq ft
Managing Director: Dr G Morris
Financial Director: Michael Craggs
Sales and Marketing Director: Ian Souter
Purchasing Manager: Tony Sullivan
■ Manufacture ground based navigation systems for air traffic management including: DME, TACAN, VORTAC, DF, ILS & DVOR. [98]

75-22650 Ferno (UK) Ltd
Stubs Beck Lane, Cleckheaton BD19 4TZ
☎(+44 1274) 85 19 99 ⊖ (+44 1274) 85 11 11
♣ £ 5,000,000 38
Managing Dir: John Wilby
Sales & Marketing Mgr: Margaret Slater
Purchasing Mgr: Bruce Burns
Asst Product Mgr: Jennifer Jackson
■ Design & manufacture of ambulance & emergency systems & equipment, stretchers, stretcher trolleys & chairs, splints & backboards, rescue equipment, resuscitators, defibrillators & first-aid kits. [98]

75-22850 Denis Ferranti Meters Ltd
(a member of Denis Ferranti Group)
Caernarfon Road, Bangor, Gwynedd LL57 4SP
☎(+44 1248) 37 03 70 ☒ 61188 dfm g
⊖ (+44 1248) 35 38 65
♣ £ 14,000,000 400 261,360 sq ft
Business Development Dir: D.J. Dibble
General Mgr: K. Chandler
Sales Mgr: Brian J. Snape
■ Manufacture to print. Electronic, electrical/mechanical engineers approved to ISO9001. [98]

75-22925 Fibreforce Composites Ltd
Fairoak Lane, Whitehouse Ind.Est., Runcorn, Cheshire WA7 3DU
☎(+44 1928) 70 15 15 ⊖ (+44 1928) 71 35 72
Managing Director: S J Webb

■ Design & production service for composite materials. Ceramic fibre fabrics for high-temperature applications. 98

75-23000 Field & Grant Ltd
(a member of the Adwest Group)
Narrowboat Way, Blackbrook Business Park,
Dudley, West Midlands DY2 0XQ
☎(+44 1384) 45 97 77 ⊠ 336718 chacoh g
⊖ (+44 1384) 45 62 07
♣ £ 9 122 000 ⁂ 205
Managing Dir: G. Lovett
Sales Dir: P. Bartlam
Technical Mgr: A. Ansari
■ Manufacture of AC & DC motors 0.5 - 2 kW (esp. flameproof), switchgear & motor controls for power applications 10 - 2000 kW (ground support systems & marine). 97

75-23210 H. Fine & Son Ltd
Victoria Hse, 93 Manor Farm Road,
Wembley, Middlesex HA0 1XB
☎(+44 181) 997 50 55 W http://www.hfine.co.uk
⊖ (+44 181) 997 84 10
⁂ 70 (1996) ⌂ 2,500 m2
Managing Director: I Fine
Operations Director: P Hollingworth
Sales and Marketing Director: J Garner
Purchasing Manager: L Collin
■ Manufacture & design of protective covers, penthouses, tilts & curtains, personnel equipment, webbing equipment, clothing including immediate response jackets, radio back packs, holdalls, kitbags, desert water bags, strap assemblies, gun muzzle covers. → 80 98

75-23215 Fine Tubes Ltd
(a subsidiary of Superior Group Inc, USA)
Estover Works, Plymouth, Devon PL6 7LG
☎(+44 1752) 73 58 51 E sales@finetubes.co.uk
W http://www.finetubes.co.uk ⊖ (+44 1752) 73 33 01
♧ £ 115,000 ♣ £ 18,000,000 (1997) ⁂ 300 (1997)
⌂ 20,000 sq m
Managing Director: R W Johnson
Sales and Marketing Manager: H Marshall
Manager Marketing Services: F P Massy-Dawson
Technical Manager: D J Rippon
Purchasing Officer: R Watkins
■ Manufacture of stainless steel, nickel and titanium tubing for engineering, electronics, aerospace, nuclear and petrochemical industries. Tubular components and fabrications in a similar range of materials. 98

75-23220 Finecast (Maidenhead) Ltd
Clivemont Road, Cordwallis Estate,
Maidenhead, Berkshire SL6 7BU
☎(+44 1628) 728 14 ⊖ (+44 1628) 78 32 54
♧ £ 100,000 (1997) ♣ £ 2,400,000 (1996)
⁂ 95 (1997) ⌂ 22,500 sq ft
Chairman and Joint Managing Director: D F Mills
Joint Managing Director: A C Everill
Sales Mgr: C.G. Harding
Operations Mgr: P.M. Wynn
Purchasing Mgr: K. Wall
■ Lost wax investment castings in aluminium, copper-based and steel alloys for aerospace, military and commercial applications. 98

75-23600 Firth Rixson plc
Firth Rixson Hse, Carbrook Hall Rd, Sheffield S9 2EJ
☎(+44 114) 261 22 22 ⊖ (+44 114) 261 22 23
♣ £ 130,000,000 (1996) ⁂ 1,100 (1996)
CEO: David J. Hall
Finance Dir: N. MacDonald
Marketing Mgr: M. Wolstenholme
■ Forgings & rolled rings for the aerospace & other defence industries in steel, nickel & titanium alloys, which include turbine & compressor discs, gears, shafts, bearing rings, undercarriage components, missile & rocket engine forgings & engine parts. Cast & wrought superalloys. Steel & alloy iron castings.

Darley Dale Plant
Darley Dale, Nr Matlock, Derbyshire DE4 2JB
☎(+44 1629) 73 36 21 ⊖ (+44 1629) 73 42 73
Sales Dir: P.J. Shaw
■ Closed die forged & railway products.

Rotherham Plant
Sheffield Road, PO Box 22, Rotherham S60 1AD
☎(+44 1709) 37 12 71 ⊖ (+44 1709) 27 89 87
Sales Dir: A.C. Inglis
■ Ring rolled & open die forged products.

Glossop Plant
Shepley St, Glossop, Derbyshire SK13 9SA
☎(+44 1457) 85 43 51 ⊖ (+44 1457) 85 55 29
Sales Dir: P.N. Harris
■ Cast & wrought superalloy products.

Scunthorpe Plant
Dawes Lane, Scunthorpe DN16 1DN
☎(+44 1724) 27 21 00 ⊠ 527242
⊖ (+44 1724) 27 21 01
Sales Director: D Griffin
■ Steel & alloy iron castings. 98

75-23670 Fisco Products Ltd
Brook Road Industrial Estate, Rayleigh, Essex SS6 7XD
☎(+44 268) 74 70 74 ⊖ (+44 268) 78 28 01
♣ £ 6,000,000 (1996) ⁂ 135 (1996)
Managing Director: Peter Fisher
Sales and Marketing Director: John Fisher
Marketing Manager: M Bryce
■ Measuring tools for precision engineering and general workshop use, spirit levels, long and short measuring tapes, engineers rules, surveyors rods and precision squares. Electronic measuring and leveling instruments. 98

75-23800 Flexibox Ltd
(a subs. of EIS Group plc)
Nash Rd, Trafford Park, Manchester M17 1SS
☎(+44 161) 872 24 84 ⊠ 667281 flxman g
⊖ (+44 161) 872 16 54
♣ £ 60 000 000 ⁂ 1 000
Managing Dir: F.K. Wright ☎(+44 161) 886 63 60
Dir & General Mgr: D.C. Mitchell ☎(+44 161) 886 63 50
Sales Dir: J.I. Proudfoot ☎(+44 161) 886 63 10
Technical Dir: N.M. Wallace ☎(+44 161) 886 62 50
Purchasing Mgr: M. McCartney ☎(+44 161) 886 63 53
Publicity Mgr: D.A. Tricker ☎(+44 161) 886 62 61
■ Design & manufacture of mechanical face seals & ancillary products. Design & manufacture of power transmission couplings marketed under metastream/powerstream. Mechanical shaft seals & flexible power transmission couplings. Shaft alignment & torque monitoring equipment.

Associated companies in:
Argentina, Australia, Benelux, Brazil, Canada, France, Germany, India, Italy, Japan, Mexico, Norway, Singapore, S.Africa, Spain, Sweden, USA, Venezuela, CIS & Thailand. 97

75-23850 Flight Refuelling Limited FR-HiTEMP Division
(a Cobham plc company)
Brook Road, Wimborne, Dorset BH21 2BJ
☎(+44 1202) 88 21 21 ⊠ 41247
⊖ (+44 1202) 88 00 96
⁂ 840 ⌂ 25 acre
Chairman: G F Page
Director: A G Irwin
Managing Director: R H L Clark
Personnel Director: Simon Clifton
Financial Director: W French
Director: Manufacturing: J S Adamson
Director/General Manager: FR-HiTEMP: R J Annett
Director/General Manager: Military Systems Division: P S Turner
Head of Sales and Marketing: FR-HiTEMP: P E Jones
■ Flight Refuelling is known for the "Probe and Drogue" method of refuelling. It also has experience in the designs, development and operation unmanned air vehicles and airborne target systems including the associated ground control, lanuch and recovery equipment.
The company's FR-HiTEMP division provides the civil and military aircraft industries with an extensive range of components for fluid and air systems including jet pumps, control valves, couplings, level sensors and drain valves.
Flight Refuelling is able to undertake large scale collaborative or subcontracted manufacturing programmes.
Associated companies:
FR-HiTEMP, UK
FR-HiTEMP Inc, USA
Sargent Fletcher, USA → 80 98

75-23900 FLS Aerospace (Light Aircraft) Ltd
(a division of FLS Aerospace Ltd)
Bournemouth International Airport, NW Industrial Area,
Christchurch, Dorset BH23 6NW
☎(+44 1202) 50 02 00 ⊖ (+44 1202) 59 32 71
⁂ 10 (1997) ⌂ 6,000 m2
Chairman: Stephen Harpøth
Quality Manager: B T Pearce
Purchasing Manager: E Hunt
■ Aircraft spares overhaul and manufacture of Optica OA7 observation aircraft and Sprint training aircraft. → 80 98

75-24020 Fluorocarbon Company Limited
Caxton Hill, Hertford, Herts SG13 7NH
☎(+44 1992) 55 07 31 ⊖ (+44 1992) 58 46 97
♧ £ 550,000 (1997) ♣ £ 12,000,000 (1997)
⁂ 310 (1997)
Managing Director: B J McAllister
Operations Director: A P Jones
Sales Manager: R Nichols ☎(+44 161) 777 63 00
Marketing Representative: Ms Julie Harrison
Purchasing Manager: R Cash
■ Machined & moulded components in PTFE & high-performance engineering plastics. Precision metal engineering, elastomeric mouldings. Approved applicators of chemical/corrosion-resistant coatings. AQAP-4, BS5750 pt2, CAA. → 80

Fluorocarbon Seals Ltd
Mitchell Close, West Portway Industrial Estate,
Andover, Hampshire SP10 3TJ
☎(+44 1264) 35 16 44 ⊖ (+44 1264) 33 40 13 98

75-24040 FMI Composites Ltd
(a subsidiary of Fiber Materials Inc)
3 Queen St, Galashiels, Selkirkshire, Scotland TD1 1QQ
☎(+44 1896) 75 77 11 ⊠ 727548
⊖ (+44 1896) 75 89 62
Director: Duncan Cunningham
■ Composite materials & components.
® USA: Fiber Materials Inc. 98

75-24095 Focal Signs Ltd
(† Focal Displays Ltd)
12 Wandle Way, Mitcham, Surrey CR4 4NB
☎(+44 181) 640 68 21 ⊖ (+44 181) 646 41 15
♣ £ 10,000,000 ⁂ 60 ⌂ 11 000 sq m
Managing Director: M.W. Padberg
Operations Director: Dave Morgan
Sales Office Mgr: John Pharoah
Public Relations/Marketing Asst: Paul Lovelock
■ Safety warning signs to IATA and IMO specifications. Also interior and exterior direction signs. 97

75-24100 Foden Trucks
(a division of Paccar (UK) Ltd)
Moss Lane, Sandbach, Cheshire CW11 3YW
☎(+44 1270) 75 84 00 ⊖ (+44 1270) 76 27 58
♣ £ 85,000,000 ⁂ 600 (1997)
Managing Dir: Alan Mayne
Sales Dir: John Pigott
Sales Dir: Paul Brunning
General Marketing Mgr: Gary Rowlands
Mgr Military & Special Products Dept: Ian M. Gallimore
Purchasing Mgr: David Storer
Advertising and PR Mgr: Spencer Harrop
■ Development, testing & production of premium military & commercial vehicles for worldwide distribution. 98

75-24150 Foerster UK Ltd
Unit 9 Wrens Court, 50 Victoria Road,
Sutton Coldfield B72 1SY
☎(+44 121) 321 11 66 ⊖ (+44 121) 321 10 66
Managing Director: John Flynn
■ Mine detectors, scanning equipment. 98

75-24155 Force Computers (UK) Ltd
(a subsidiary of Solectron Corp., USA)
Alton House Office Park, Gatehouse Way,
Aylesbury, Buckinghamshire HP19 3XU
☎(+44 1296) 31 04 00
W http://www.forcecomputers.com
⊖ (+44 1296) 31 04 20
⁂ 450 (1996)
Managing Director: Ian Denchfield
Manager Administration: F Khamlichi

■ Provides embedded, open standard real-time and UNIX computer platforms to the industrial control, telecommunications, medical, aerospace and government sectors worldwide. ISO 9001 qualified supplier. [98]

75-24160 L.P. Foreman & Sons Ltd
(a subsidiary of GEC-Marconi)
Willow Works, Roxwell Road, Chelmsford, Essex CM1 2LZ
☎ (+44 245) 26 45 21 ⊖ (+44 245) 49 52 32
♣ £ 2,000,000 (1997) 👥 40 (1997) 🏭 20,000 sq m
Managing Director: H B Palmer
Operations Director: P Watts
Sales/Marketing Director: A Joyce
Purchasing Manager: G Watts
■ Designers & manufacturers of packing cases, crates, fitted carrying cases, special cartons & multi-wall containers for the transportation of delicate, high value equipment. [98]

75-24165 Forged Rolls (UK) Ltd
(a subs. of Sheffield Forgemasters Ltd)
Hawke Street, Sheffield S9 7SU
☎ (+44 114) 261 15 51 ⊠ 547734 fgroll g
⊖ (+44 114) 261 14 96
♣ £ 13 000 000
Managing Dir: Peter Carless
■ Manufacture high performance steel forged work rolls. [97]

75-24168 Forgemasters Steel & Engineering Ltd
(a subs. of Sheffield Forgemasters Ltd)
Brightside Lane, PO Box 286, Sheffield S9 2RW
☎ (+44 114) 244 90 71 ⊠ 54279 sfmg g
⊖ (+44 114) 242 21 03
♣ £ 52 000 000
Managing Dir: Peter Birtles
Operations Dir: Peter Marsh
Sales Dir: Peter Fretwell
Sales Dir: John Almond
Purchasing Mgr: John Pritchard
■ Manufacture of steel & forgings including forged back-up rolls & section rolls from 15 to 150 tonnes. [97]

75-24200 Forward Industries Ltd
(a subsidiary of OMI International plc)
South Marston Industrial Estate, PO Box 8, Swindon, Wiltshire SN3 4RA
☎ (+44 1793) 82 32 41 ⊠ 44317 vfpben g
⊖ (+44 1793) 82 84 74
♣ £ 7,000,000 (1997) 👥 125 (1997) 🏭 100,000 sq ft
Managing Dir: J.J. Forrest
Sales and Marketing Director Hydraulics: D P Drudy
Sales and Marketing Director Nuclear: N Brencaley
Purchasing Manager: John Dodgson
■ Hydraulic pumps, valves, actuators & systems for on-board or support functions, weapons loading, general services & simulation. Hangar ring main & mobile systems for servicing aircraft hydraulics, pneumatics & fuel systems. [98]

75-24380 Fothergill Engineered Fabrics Ltd
(† Performance Fabrics)
PO Box 1, Summit, Littleborough, Lancashire OL15 0LU
☎ (+44 1706) 37 24 14
E 106127.647@compuserve.com
⊖ (+44 1706) 37 64 22
♣ £ 11,000,000 (1996) 👥 130 (1996) 🏭 10,776 m2
Managing Director: D Nichols
☎ (+44 1706) 37 24 14 ext. 6049
Operations and Technical Director: J Mootrie
☎ (+44 1706) 37 24 14 ext. 6011
Business Manager Aerospace/Defence: P W Bradshaw
☎ (+44 1706) 37 24 14 ext. 6041
Purchasing Manager: D Eastwood
☎ (+44 1706) 37 24 14 ext. 6023
■ Woven fibre assemblies for textile applications in the armouring of vehicles, fixed installations & for use in personal protection (Flack jackets, bullet resistant vests, EOD suits & helmets). Textiles for composite use in combat aircraft & minehunters. Fabrics for use in counter surveillance systems. Cloths to be incorporated into NBC assemblies for personal protection. [98]

75-24400 FPT Industries Ltd
(a division of GKN Westland Aerospace Ltd)
The Airport, Portsmouth, Hampshire PO3 5PE
☎ (+44 1705) 66 23 91 ✉ FIREPROOF Portsmouth
⊠ 86106 ⊖ (+44 1705) 67 08 99
♣ £ 10,000,000 👥 220 (1997) 🏭 150,000 sq ft
Managing Dir: I.P. Smith
Operations Director: Stewart Smith
General Mgr Engineering: T.L. Woodcock
Head of Sales & Marketing: S.J. Tutton
Purchasing Mgr: M. Spooner
■ Flexible, crash-resistant and self-sealed fuel cells and assemblies, emergency flotation equipment for aircraft, helicopters and vehicles, flotation collars, salvage bags, hyclad flexible gaskets, rubber proofed fabrics, rubber mouldings, extrusions, adhesives and lacquers. Rigid plastic mouldings for fuel and hydraulic reservoirs. Explosion suppressants. Air portable fuel containers. [98]

75-24480 F.R. Ordnance International Ltd
(a division of Ranger Arms Co. Ltd)
Frogs Hall, Hertfordshire CM22 6PE
☎ (+44 1279) 87 00 66
E 106427.1673@compuserve.com
⊖ (+44 1279) 87 00 45
👥 5 (1997) 🏭 500 sq m
Managing Dir: M.G.A. Ranger
Operations Dir: G. Felton
■ Manufacturers of MC51 sub-machine gun. Sales and procurement related services on all military equipment worldwide. → 90 [98]

75-24500 Francis & Lewis International Ltd
11 Barnett Way, Barnwood, Gloucester GL4 3RT
☎ (+44 1452) 61 65 00 ⊖ (+44 1452) 61 69 00
♣ US$ 5,000,000 (1996) 👥 42 (1996) 🏭 3,700 m2
Managing Director: W J Haley
General Manager: J R Garbutt
Sales Manager: John E Sturman
Project Manager: J Morse
Purchasing Manager: P Verso
■ Wide product range of towers and masts for telecommunications, broadcasting, and cellular applications. Support structures for radar equipment, flood lighting and meteorological instruments. BS 5750 part 1 approved. BS/EN/ISO 9001. [98]

75-24660 Frazer-Nash (Midhurst) Ltd
(† Frazer-Nash Engineering Ltd)
Bepton Road, Midhurst, West Sussex GU29 9LZ
☎ (+44 1730) 81 61 41 ⊖ (+44 1730) 81 56 06
♧ £ 660,000 (1995) ♣ £ 2,400,000 (1995)
👥 27 (1997) 🏭 880 sq m
Managing Director: G T Hoy
Operations Director: J Marshall
Sales and Marketing Director: N Cox
Purchasing Manager: P May
■ Component manufacture. [98]

75-24700 Freeman & Proctor Ltd
(a subsidiary of Fairmead Group Ltd)
PO Box 22, Nuneaton, Warwicks CV11 4UU
Unit 3/5 Centrovell Ind. Estate Caldwell Road, Nuneaton
☎ (+44 1203) 38 20 32 ⊠ 312348 fmpr g
⊖ (+44 1203) 37 43 53
♣ £ 4 000 000 👥 100
Managing Dir: R.A. Collett
■ Specialist services in the following areas:- Welding & brazing of most types of ferrous & non-ferrous materials. Manufacture and manipulation of complex tube assemblies. Machining of precision components and fittings in small or large batch quantities. Repair and salvage welding of high value cast or machined parts. Comprehensive testing including:- Dye penetrant & fluorescent flaw detection, magnetic flaw detection, air and hydraulic pressure testing, radiological examination, aqua and vapour blasting, preening for fatigue resistance. Approved to BS5750 PT2 and ISO9002 standards. [97]

75-24730 Freudenberg Technical Products LP
(† George Angus & Co Ltd)
(a partner of Freudenberg & Co, Germany)
Coast Road, Wallsend, Tyne and Wear NE28 9NR
☎ (+44 191) 202 22 00
E 106027.1374@compuserve.com ⊠ 53628 gaco g
⊖ (+44 191) 202 23 61
♣ £ 60,000,000 👥 500 🏭 20,000 m2
Managing Director: C M Lloyd
Purchasing Manager: G Walton
Operations: C Wade
Operations: R Evans
Operations: J Quieruga
Sales and Marketing: D Brown
■ Shock absorber seals, vibration dampers, auto seals. [98]

75-24800 Froude Consine
(a division of FKI Engineering plc)
Headquarters
Blackpole Road, Worcester WR3 8YB
☎ (+44 1905) 85 68 10 E sales@froudeconsine.co.uk
W http://www.froudeconsine.co.uk/fcsales
⊖ (+44 1905) 85 68 11
♣ £ 14,000,000 (1995/96) 👥 154 🏭 15,000 sq m
Managing Director: R L Gott
Sales and Marketing Director: B Hemstock
Financial Director: P Smith
Technical Director: S Sharpe
Production Director: A Turner
Manager of Marketing: A W Nash
■ Design, manufacture of engine test equipment & instrumentation for the marine, aero & automotive industries. Manufacturers of both fixed site & demountable facilities. [98]

75-24870 FTL Seals Technology Limited
(† FTL Fluid Seals Ltd)
Bruntcliffe Avenue, Leeds 27 Business Park, Morley, Leeds LS27 0TG
☎ (+44 113) 252 10 61 ⊠ 55821
⊖ (+44 113) 252 26 27
👥 30 🏭 930 m2
Managing Director: Malcolm J Harwood
Operations Director and Purchasing Manager: Phil Todd
Sales Director: Tony Snowdon
Sales Engineer Defence: Andrew Dean
■ Mechanical face seals for tracked vehicles and complete track roller assemblies. [98]

75-24880 FT Technologies Ltd
Church Lane, Teddington, Middlesex TW11 8PA
☎ (+44 181) 943 08 01 E sales@fttech.co.uk
W http://www.fttech.co.uk ⊠ 24667
⊖ (+44 181) 943 32 83
👥 12 (1997)
Managing Dir: P.F. Elgar
Sales Dir: Dr S.J. Norton
■ Design, development & production of electronic systems for naval & aerospace applications. Product range includes synchro & resolver custom interfaces, angle position indicators, digital directors & synchro/resolver test sets. ISO9001 & ISO9000-3 (software) approved. [98]

75-24885 Fugro-UDI Ltd
(† UDI-Wimbol Ltd)
(a subsidiary of Fugro NV, Netherlands)
Denmore Road, Bridge of Don, Aberdeen AB23 8JW
☎ (+44 1224) 25 75 00 E info@fugro-udi.co.uk
⊠ 73361 udi g ⊖ (+44 1224) 25 75 01
♣ £ 22,000,000 👥 150+ 🏭 3,781 sq m
Managing Director: Graeme Kidd
General Manager: John Meaden
Technical Sales Manager: Roy Clarke
☎ (+44 1224) 25 75 11
Engineering Manager: Mark Walsh
☎ (+44 1224) 25 75 18
Technical Support Mgr: Dr Colin MacLean
Purchasing Manager: Warroy Dow
■ Scanning sonar for remotely operated underwater vehicles, diver guidance, debris location, underwater survey. Underwater black/white, colour & low light TV & lighting systems. Emergency wireless underwater communications system. Underwater connectors. Design & manufacture of subsea engineering projects. Offshore survey & positioning, (DGPS) & ROV operations. Small high quality (low volume) engineering workshops for electrical & metalwork services. Machine shop machines in titanium, steels, alloys, nylons & plastics. → 80 → 90 [98]

75-24890 Furneaux Riddall & Co. Ltd
Alchorne Place, Portsmouth, Hampshire PO3 5PA
☎ (+44 1705) 66 86 21 ⊖ (+44 1705) 69 05 21
👥 28
Managing Dir: G. Weston

■ Manufacture of rechargeable Halogen Handlamps and a range of safety torches. [98]

75-24900 Furon Limited

Crusader House, High Street Maxey,
Peterborough PE6 9HQ
☎(+44 1778) 34 81 11 ⊖ (+44 1778) 34 29 34
■ Control systems for vehicles, plant & equipment. Thermoplastic hoses (SYNFLEX). [97]

75-25100 Gallay Ltd

Paterson Road, Wellingborough, Northants NN8 4BZ
☎(+44 1933) 22 48 01 ⊖ (+44 1933) 27 99 02
Managing Director: A T Bryant-Fenn
■ Air conditioning systems for armoured vehicles and shelters; electronics cooling; engine cooling systems; air filtration systems. Engine cooling systems - radiators, oil coolers, charge coolers & header tanks. [98]

75-25150 Garmin (Europe) Ltd

European Office
Unit 5 The Quadrangle, Abbey Park Industrial Estate,
Romsey, Hampshire SO51 9AQ
☎(+44 1794) 51 99 44 ⊖ (+44 1794) 51 92 22
Director International Marketing: Dan J Bartel
Sales Manager: Steve Gubbins
■ Design, manufacture and marketing of navigation and communication electronics equipment, particularly GPS products. [98]

75-25205 Garrard & Co Ltd

(part of the Asprey Group plc)
112 Regent Street, London W1A 2JJ
☎(+44 171) 734 70 20 ⊖ (+44 171) 734 07 11
Corporate Division, Military: C F A Gough
■ Military statuettes and insignia. Specialist replica models in precious and non-precious metals, end-of-contract presentations. [98]

75-25210 Gates Hydraulics Ltd

(a subs. of Gates Rubber Company Ltd, USA)
Station Road, St Neots, Cambs PE19 1QF
☎(+44 1480) 47 53 33 ⊖ (+44 1480) 21 50 48
♣ £ 35 000 000 (1996) 👥 400
Managing Dir: A.J. Roberts
Operations Dir: J. Boyd
Technical Dir: G.R. Keast
Marketing Dir: David Howse
Purchasing Mgr: M. Jones
■ Hydraulic hose, couplings, assemblies & assembly machinery. Hydraulic hose adaptors, nylon air brake tubing fittings & assembly tools, hydraulic hose test equipment. [97]

75-25220 GEC ALSTHOM Diesels Ltd

(part of GEC ALSTHOM, France)
Paxman Works, Hythe Hill, Colchester, Essex CO1 2HW
☎(+44 1206) 79 51 51 [W] http://www.gecalsthom.com
⊠ 98152 ⊖ (+44 1206) 79 78 69
👥 921 (1997)
Managing Director: Dr Harvey Perkins
Sales and Marketing Director: W L Pavry
Purchasing Manager: J Gant
Group Public Relations Manager: Simon Shand-Brown
■ High and medium speed diesel engines in the power range 500 KWb to 7500 KWb for marine propulsion, auxiliary power and land based power use. Control packages for marine and land based power plants.
Subsidiary companies:
GEC ALSTHOM Ruston Diesels Ltd
GEC ALSTHOM Paxman Diesels Ltd
GEC ALSTHOM Regulateurs Europa Ltd
GEC ALSTHOM Diesels Inc. → USA
GEC ALSTHOM Regulateurs Europa BV → Holland [98]

75-25225 GEC ALSTHOM Engineering Systems Ltd

Nuclear, Marine & Defence Div.
Cambridge Road, Whetstone, Leicestershire LE8 6LH
☎(+44 116) 275 07 50 ⊠ 34611
⊖ (+44 116) 275 01 32
♣ £ 80 000 000 👥 1 800
Director: Chris J. Ealing
Group Mgr Marine & Defence Div.: Alan Gardiner
Sales Mgr: Keith Costall
■ Specialise in design, manufacture & installation of mechanical handling equipment. The company provides integrated replenishment at sea systems for use in conditions up to seastate 6; quick release devices & fluid cargo handling including NATO standard probes & receivers for refuelling at sea plus a wide range of elevators for aircraft, stores, weapons & personnel transfer, also aircraft & torpedo handling systems. [98]

75-25230 GEC ALSTHOM Gears Ltd

Leicester Road, Rugby, Warwickshire CV21 1BD
☎(+44 1788) 54 21 21 ⊠ GEC GEAR Rugby
⊖ (+44 1788) 56 08 76
♣ £ 8,000,000 (1996) 👥 125 (1996)
Managing Director: K M Batchelor
Marine Sales Manager: J R Evans
■ Specialise in design and manufacture of high accuracy gearing for marine, industrial and traction applications. The company supplies naval main propulsion gearing, from simple plain reduction gears for offshore patrol vessels to complex combining gears for CODOG, CODAD, COLAG and other propulsion systems for frigates and corvettes.
Representing:
Allied Marine Services Inc, USA [98]

75-25240 GEC ALSTHOM Large Machines Ltd

(a subsidiary of GEC ALSTHOM Ltd)
Mill Road, Rugby, Warwickshire CV21 1TB
☎(+44 1788) 54 21 21 ⊖ (+44 1788) 54 12 80
♧ £ 25,000,000 (1996/97) ♣ £ 60,000,000 (1996/97)
👥 750 (1997)
Managing Dir: P.T.J. Featherstone
Sales Dir: M.A. Lovie
Operations Director: G B Durand
Engineering Dir: D.W. Grieve
Purchasing Dir: C.C. Nelson
■ Design & manufacture of high voltage AC electric motors & generators for industrial & marine applications including naval propulsion for surface ships & submarines. Supply of specialist electric motors & generators, for basic research (eg: pulse generator for the JET nuclear fusion project). [98]

75-25242 GEC ALSTHOM Large Steam Turbines

(a subsidiary of GEC ALSTHOM)
Newbold Road, Rugby, Warwickshire CV21 2NH
☎(+44 1788) 57 71 11 ⊠ 94023430 GATE G
⊖ (+44 1788) 53 19 81
Operations Director: A Melsom
Commercial General Sales Manager: Mike Wilcox
☎(+44 1788) 53 17 49
■ Steam turbines for all power generation applications. [98]

75-25243 GEC ALSTHOM Medium Turbo Machines

(† GEC ALSTHOM Turbine Generators Ltd)
(a subsidiary of GEC ALSTHOM)
Newbold Road, Rugby, Warwickshire CV21 2NH
☎(+44 1788) 57 71 11 ⊠ 94023430 GATE G
⊖ (+44 1788) 53 17 33
Contracts Manager: R Cocksedge
Commercial Manager: P Critchley
■ Naval main propulsion turbines, steam turbines for power generation. [98]

75-25245 GEC ALSTHOM Paxman Diesels Ltd

(† Paxman Diesels Ltd)
(a management company of GEC ALSTHOM Diesels Ltd)
Paxman Works, Hythe Hill, Colchester, Essex CO1 2HW
☎(+44 1206) 79 51 51 ⊠ PAXMAN Colchester
⊠ 98152 gecpdc g ⊖ (+44 1206) 79 78 69
👥 411 🏭 10 hectares
Managing Director: J Branscombe
Operations and Technical Director: J N Ramsden
Sales and Marketing Director: R Guest
Purchasing Mgr: J.P. Gant
General Sales Mgr: J. Hill
■ Manufacture high speed diesel engines with high power to weight ratio. Powers 610-3917 KWb for main propulsion and onboard power generation for naval craft, fast ferries, luxury yachts, industrial land-based power generation, off-highway vehicles and rail locomotives. [98]

75-25250 GEC ALSTHOM Regulateurs Europa Ltd

(a management company of GEC ALSTHOM Diesels Ltd)
Port Lane, Colchester, Essex CO1 2NX
☎(+44 1206) 79 95 56 ⊠ 98438
⊖ (+44 1206) 79 26 85
👥 60
General Manager: T Richardson
Sales and Marketing Director: Derek J Rigby
■ Design and supply of electronic, hydraulic and mechanical governors, monitoring and control equipment, and power plant control systems. [98]

75-25255 GEC ALSTHOM Ruston Diesels Ltd

(† Ruston Diesels Ltd)
(a management company of GEC ALSTHOM Diesels Ltd)
Vulcan Works, Newton-le-Willows WA12 8RU
☎(+44 1925) 22 51 51 [E] ruston@compuserve.com
⊠ 627131 ⊖ (+44 1925) 22 20 55
♧ £ 11,612,000 (GEC Alsthom Diesels Ltd)
♣ £ 40,630,000 (1996) 👥 463 🏭 36,838 2m
Managing Director: J M MacKinnon
Commercial Director: C B Foulkes
☎(+44 1925) 22 51 51
Operations and Technical Director: R Hunt
■ Manufacturer of medium speed diesel, gas, heavy fuel oil and dual fuel engines. Powers 500-7500KW (670-10000bhp) for main propulsion and on board power generation for naval and commercial applications including tugs and fast ferries as well as land based power plants and freight locomotives. [98]

75-25260 GEC ALSTHOM T&D Power Electronic Systems Ltd

Lichfield Rd, PO Box 27, Stafford ST17 4LN
☎(+44 1785) 25 71 11 ⊖ (+44 1785) 25 25 40
♣ £ 50 000 000 👥 400
Sales Mgr: J.R. Bottom
Business Development Mgr: Michael H. Baker
■ Static Frequency Convertor (using pulse width modulation for control) developed for 400Hz power supplies for sonar & weapon systems. High performance fixed & variable frequency invertors using switched-mode techniques. High stability & high voltage AC/DC power sources, using diode, thyristor, GTO thyristor & power MOSFET devices & including custom built convertor equipment. Harmonic filters including active filters from 1 Mvar up & from 5kV up. [97]

75-25300 GEC-Marconi Aerospace Ltd

(† GEC Aerospace Ltd)
(a subs. of General Electric Company plc)
Abbey Works, Titchfield, Fareham, Hants PO14 4QA
☎(+44 1329) 85 30 00 ⊖ (+44 1329) 85 37 97
♣ £ 50,000,000 (1995/96) 👥 704
Managing Dir: Nick Franks
Manufacturing Dir: Barry Capp
Marketing Exec.: Peter Bradshaw
Purchasing Mgr: Norman Austen
Marketing Admin Mgr: C. Howting
■ Aircraft fuel systems including engine driven & electrically driven fuel pumps. Radar coolant pumps. High temperature pneumatic actuators, valves & systems, generation systems for aircraft & fighting vehicles. Intelligent power control systems.
Avery Hardoll
Abbey Works, Titchfield, Fareham, Hants. PO14 4QA
☎(+44 1329) 84 42 02 ⊠ 86214
⊖ (+44 1329) 85 38 04
Marketing Mgr: David Irving
General Mgr: Steve Woolcock
■ Aviation fuelling equipment: hydrant pit boxes & valves, sampling vent & drain valves, API couplers with pressure control & deadman, in-line pressure controllers, bulkmeters, hose units, aircraft & ground adaptors. [97]

75-25310 GEC-Marconi Aerospace Systems

Airport Works, Rochester, Kent ME1 2XX
☎(+44 1634) 84 44 00 ⊖ (+44 1634) 82 73 32
Managing Director: M H R Peto
■ System integration and co-ordination. [98]

75-25320 GEC-Marconi Avionics Ltd
(a management company for GEC-Marconi Ltd)
Head Office
Airport Works, Rochester, Kent ME1 2XX
☎(+44 1634) 84 44 00 ⊠ 96333 gecroc g
⊖ (+44 1634) 82 73 32
♣ £ 537,000,000 (1997) ⚉ 8,000 (1997)
Managing Director: Saul Lanyado
Technical Director: J F Roulston
Financial Director: D M Frost
Commercial Director: A J Cowdery
Personnel Director: E J Bradley
■ Avionics manufacturer: flight controls and actuators, navigation and electro-optics, display systems, airborne radar, communications, fuel systems and engine management. Product support facility.
Associated Company:
GEC-Marconi Avionics Inc. → USA 75
Flight Systems Division
Airport Works, Rochester, Kent ME1 2XX
☎(+44 1634) 84 44 00 ⊖ (+44 1634) 82 73 32
Divisional Director: R W Dennis
Mission Avionics Division
Airport Works, Rochester, Kent ME1 2XX
☎(+44 1634) 84 44 00 ⊖ (+44 1634) 82 73 32
Divisional Managing Director: M T Sweeney
Power Systems Division
Abbey Works, Titchfield, Fareham, Hampshire PO14 4QA
☎(+44 1329) 85 30 00 ⊖ (+44 1329) 85 37 97
Divisional Managing Director: C West
Radar Systems Division
Crewe Toll, Ferry Road, Edinburgh EH5 2XS
☎(+44 131) 332 24 11 ⊖ (+44 131) 332 06 90
Divisional Managing Director: A E Cook
Support Division
6 South Gyle Crescent, Edinburgh EH12 9EA
☎(+44 131) 337 24 42 ⊖ (+44 131) 334 88 36
Divisional Managing Director: W D Hamilton 98

75-25335 GEC-Marconi Communications Ltd
(† Marconi Communication Systems Ltd)
(a subs. of GEC-Marconi Ltd)
Marconi House, New St, Chelmsford, Essex CM1 1PL
☎(+44 1245) 35 32 21 ⊠ 99201 marcon g
⊖ (+44 1245) 28 71 25 (Group 2/3)
⚉ 1 800 (1996)
Managing Dir: Ian McNamee
Engineering Dir: Dr Alan Martin
Military Communications Dir: John Dale
Publicity Mgr: Peter A.T. Turrall
■ Mobile communications & satellite communications equipment. HF & VHF military combat net radio systems, tactical strategic & naval communications, HF high power transmitting & receiving stations.
Military Communication Division
Marconi Hse, New St, Chelmsford, Essex CM1 1PL
☎(+44 1245) 35 32 21 ⊠ 99201 marcon g
⊖ (+44 1245) 28 71 25
Divisional Dir: John Dale
Business Mgr Tactical: Tony Barber
Business Mgr Strategic: Mike Steel
Business Mgr Naval: John Rosie
■ Research, design, development, production, installation & commissioning of naval & military communications equipment & systems of all types, civil land based HF communication systems, land & maritime. Military radio relay, man pack, vehicle, containerised & fixed radio equipment & systems. Full support services including studies, documentation, training, repairs & spares.
Contractor for the Royal Navy & overseas for ship, submarine & shore station communication systems. → 80
Telecommunications & Control Division
Marconi House, New St, Chelmsford, Essex CM1 1PL
☎(+44 1245) 35 32 21 ⊠ 99201 marcon g
⊖ (+44 1245) 28 71 25
Divisional General Mgr: Gary Dawson
■ Design & manufacture of long-range tropospheric scatter communication links which can be rapidly deployed.
Mobile Communications Division
Marconi House, New Street, Chelmsford, Essex CM1 1PL
☎(+44 1245) 35 32 21 ⊠ THULIUM Chelmsford
⊖ (+44 1245) 35 87 76
Divisional General Mgr: Phil Robinson
■ Design & manufacture mobile radios, PMR, base stations & systems. 97

75-25337 GEC-Marconi Dynamics Ltd (GMDL)
The Grove, Warren Lane, Stanmore, Middx HA7 4LY
☎(+44 181) 954 23 11; 420 31 82 ⊠ 22616 mardef g
⊖ (+44 181) 954 19 05; 954 78 08
Managing Dir: Chris Chant
Press Officer: Richard Coltart
■ Weapons system contractor, high technology capability in the field of missile seeker technology, including radar, IR & electro-optics seekers. Partner in international programmes with strong links to North American & European manufacturers. Products cover surface-to-air, air-to-air, air-to-ground & anti-ship including Principal air-to-air missile (PAAMS) for ship defence & Brimstone an autonomous anti-tank missile for the Harrier & Attack Helicopter. 97

75-25338 GEC-Marconi Infra-Red Limited
Millbrook Industrial Estate, PO Box 217,
Southampton, Hampshire SO15 0EG
☎(+44 1703) 70 23 00 ⊠ 47640
⊖ (+44 1703) 31 67 77
⚉ 250 (1996)
Managing Director: Dr G Read
Operations and Technical Director: R P Sutton
Business Development Manager: Dr G K Hall
Purchasing Manager: I Whitehead
■ Photoconductive and photovoltaic infrared detectors of mercury cadmium telluride, including focal plane multiplexed linear and staring arrays in packages/dewars for Joule Thomson or engine cooling. Thermoelectrically cooled infrared MCT detectors. Pyroelectric detectors. 98

75-25340 GEC-Marconi Materials Technology Ltd
(a subsidiary of GEC-Marconi Ltd)
Materials & Sensors Division
Wood Burcote Way, Towcester, Northants NN12 6JW
☎(+44 1327) 35 05 81 ⊖ (+44 1327) 35 69 96
♣ £ 8,000,000 (1995/96) ⚉ 80
Managing Dir: Dr R. Humphryes
General Mgr: Martin Woolley
Sales Mgr: Philip Poole
Contracts Mgr: P.J. Driver
Technical Exec.: Steve Morris
Sales & Marketing Engineer: Mark Kelk
■ Design, development & manufacture of radar absorbent & radar transparent materials & components. Radar camouflage, EMI suppression, testing facilities, radomes. Research & development department carries out contract work for specific applications. 97

75-25363 GEC Plessey Semiconductors
(a subsidiary of GEC plc)
Cheney Manor, Swindon, Wiltshire SN2 2QW
☎(+44 1793) 51 80 00 Ⓦ http://www.gpsemi.com
⊖ (+44 1793) 51 84 11
♣ US$ 350,000,000 (1996) ⚉ 3,000 (1996)
Managing Dir: Tom Urwin
Finance Director: Tony Gallagher
Marketing Dir: Ernie Pusey
Military Marketing Mgr: Ken Robertson
Marketing Communications Mgr: Jeff Salter
■ Bipolar and CMOS integrated circuits. Applications specific standard ICs. Semi-custom and custom designed include both cell based and gate array techniques. Also Power Semiconductors. Markets served include communications, comsumer, and computing.
● Unit 1, Crompton Road, Groundwell Ind. Estate, Swindon, Wiltshire SN2 5AY
☎(+44 1793) 72 66 66 ⊖ (+44 1793) 51 85 50
● Tamerton Road, Roborough, Plymouth, Devon PL6 7BQ
☎(+44 1752) 69 39 99 ⊖ (+44 1752) 69 32 00
● Tweedale Avenue, Hollinwood, Oldham, Lancashire OL9 7LB
☎(+44 161) 682 68 44 ⊖ (+44 161) 668 78 98
● Landsdowne Road, Chadderton, Oldham, Lancashire OL9 9TY
☎(+44 161) 634 05 15 ⊖ (+44 161) 678 71 61
● Doddington Road, Lincoln LN6 3LF
☎(+44 1522) 50 05 00 ⊖ (+44 1522) 50 05 50 98

75-25405 GeoAcoustics Ltd
Shuttleworth Close, Gapton Hall Industrial Estate,
Great Yarmouth, Norfolk NR31 0NQ
☎(+44 1493) 60 06 66 Ⓔ sales@geoacoustics.co.uk
Ⓦ http:// www.geoacoustics.co.uk
⊖ (+44 1493) 65 11 00
⚉ 31 (1997) ⌂ 2,000 sq m
Managing Dir: D.R. Stone
Operations Mgr: W. Hone
Production Mgr: P. Roling
■ Acoustic telemetry, oceanographic products, seabed survey systems. 98

75-25415 Geoquip Ltd
Kingsfield Industrial Estate, Derby Road, Wirksworth, Matlock, Derby DE4 4BG
☎(+44 1629) 82 48 91/5 ⊖ (+44 1629) 82 48 96
♣ £ 2,000,000 ⚉ 20
Managing Director: Peter Elliott
Technical Director: Ian Macalindin
Marketing Director: Brian Perrins
■ Provides a range of perimeter intrusion detection sensors including fence-mounted & underground detection systems. 98

75-25430 Gilbert Gilkes & Gordon Ltd
Canal Iron Works, Kendal, Cumbria LA9 7BZ
☎(+44 1539) 72 00 28 Ⓔ chris@gilkes.com
Ⓦ http://www. gilkes.u-net.com/ ⊖ (+44 1539) 73 21 10
♣ £ 500,000 (1996) ♣ £ 10,000,000 (1996)
⚉ 165 (1996)
Managing Director: P R W Hensmann
Manager: C B Harrison Beck
Sales and Marketing Manager: R J Pengelly
■ Pumps for cooling marine diesel engines & other marine applications. 98

75-25450 Gill Instruments Ltd
Solent House, Cannon Street,
Lymington, Hampshire SO41 9BR
☎(+44 1590) 67 17 54 Ⓔ gill@gill.co.uk
⊖ (+44 1590) 67 64 09
♣ £ 3,200,000 (1996) ⚉ 50 ⌂ 1,000 m2
Managing Dir: M.J. Gill
Operations Dir: R.S. Duff
Sales Manager: G Westearth
Purchasing Manager: Alison James
■ Manufacture anemometers for shipboard & weapons firing use. 98

75-25500 GEL Engineering Ltd
(† Gilpack Ltd)
(a subsidiary of Ross Group plc)
Nutsey Lane, Salisbury Road,
Totton, Southampton SO40 3ZY
☎(+44 1703) 66 30 30 ⊠ 47526 giltpak g
⊖ (+44 1703) 66 31 55
♣ £ 6,000,000 ⚉ 80 ⌂ 319,000 sq ft
Managing Director: Adrian Norman
Programme Director: Peter Manville
Purchasing Manager: W Drake
■ Weapon/engine containers. Aircraft component test rigs. Helicopter gearbox test rigs. Modular containers, relocatable plant and equipment rooms, buildings and technical shelters. Specialist vehicle bodies and specialist high speed gearboxes. 98

75-25600

GKN Defence Ltd

Hadley Castle Works, PO Box 106,
Telford, Shropshire TF1 4QW
☎(+44 1952) 22 45 00 ⊖ (+44 1952) 24 49 47
Chief Executive: D J Wright
Deputy Chief Executive: J W Rosenthal
Sales and Marketing Director: T A Woodhouse
Operations Director: J Horton
Director of Purchasing and Logistics: C C Wade
Head of Public Affairs and Communications: M W Docherty
■ Tracked & wheeled armoured personnel carriers, fighting vehicles, trailers & other defence components. 98

75-25625 GKN Westland Aerospace Ltd
(a division of GKN plc)
Castle Street, East Cowes, Isle of Wight PO32 6RH
☎(+44 1983) 29 41 01 ⊠ 86761 wad g
⊖ (+44 1983) 29 10 06
♣ £ 200,000,000 (1997) ⚉ 3,535 (1997)
Chief Executive: Chris C Gustar
Finance Dir: C. Rushton
Commercial Dir: J.M. George

Managing Dir Structures Div.: B. Hinton
Dir & General Mgr Transmissions: M. Askew
Public Affairs Manager: A J Roden
Business Mgr: N.A. May
■ Design, manufacture, qualification & testing of flight critical structures & components from advanced composite & metal for civil aircraft. Design & manufacture of transmissions & gears for helicopter & aeroengine applications. Also repair & overhaul of complete gearboxes. 98

75-25630 GKN Westland Aerospace Transmissions

(a subsidiary of GKN Westland Aerospace)
Box 212, Yeovil, Somerset BA20 2YB
☎ (+44 1935) 70 40 50 ▣ 46277 whl yeo g
⊖ (+44 1935) 70 26 74
♣ £ 50,000,000 (1995) ♛ 676 (1995)
⌂ 28,697 sq m (1995)
Director and General Manager Transmissions: Michael J Askew ☎ (+44 1935) 70 34 17
Head of Sales and Marketing: Richard E Jones ☎ (+44 1935) 70 40 50
Head of Operations: Mrs Nuala Griffin ☎ (+44 1935) 70 23 39
Head of Contracts: Kevin Pointon ☎ (+44 1935) 70 43 08
Head of Engineering: Philip Peters ☎ (+44 1935) 70 35 55
Head of Finance: James Schofield ☎ (+44 1935) 70 21 53
Head of Purchasing: Steve Parry ☎ (+44 1935) 70 27 20
■ Design, manufacture & supply of gearboxes & transmission systems for helicopters & aeroengine applications. Also repair & overhaul of complete gearboxes. 98

75-25650 GKN Westland Helicopters Ltd

(† Westland Helicopters Ltd)
(a subsidiary of GKN plc)
Westland Works, Yeovil, Somerset BA20 2YB
☎ (+44 1935) 47 52 22 ✉ AIRCRAFT Telex Yeovil
▣ 46277 whlyeo g ⊖ (+44 1935) 70 21 31
Chief Executive: Richard I Case
Export Sales & Marketing Dir: Phillip Burton
Public Affairs Dir GKN Westland: Chris Loney
Chief Test Pilot: Colin Hague
■ Production of a complete range of helicopter systems. Sea King & Commando; Super Lynx & Battlefield Lynx; EH101 naval ASW, military utility & civil helicopter, WAH-64D Longbow Apache attack helicopter. 98

75-25675 GKN Westland Industrial Products Ltd

(a subs. of GKN Westland Helicopters Ltd)
Winterstoke Rd, Weston Super Mare, Avon BS24 9AB
☎ (+44 1934) 63 55 55 ▣ 444136
⊖ (+44 1934) 63 61 11
♛ 375
Dir & General Mgr: Clive Sibley
Financial Controller: John Barker
■ Manufacture/support for military/civil aircraft & aerospace industry. Fabrication machining treatments. Repair & overhaul facilities for aerostructures, transmission, rotor blades, electrical & hydraulic systems. → 80 97

75-25690 GKN Westland Technologies Ltd

(a subsidiary of GKN plc)
PO Box 27, Yeovil, Somerset BA20 2YD
☎ (+44 1935) 47 51 81 ▣ 46132 nglyeo g
⊖ (+44 1935) 42 76 00
Managing Dir: Peter Burrows
PR Mgr: G.D. Carley
■ Design, development & manufacture of high technology systems & components for the aerospace defence industries together with full customer support services.
Subsidiary companies:
Normalair-Garrett Ltd
Hermetic Aircraft International Corp., USA
Westland-Sitec GmbH, Germany 98

75-25700 Glover Webb Ltd

(a subsidiary of GKN plc)
Hamble Lane, Hamble,
Southampton, Hampshire SO31 4NY
☎ (+44 1703) 45 28 11 ▣ 477527 glover g
⊖ (+44 1703) 45 29 64
Managing Director: P J Bruce
■ Design & manufacture of armoured personnel carriers on 4x4, 6x6 & 8x8 vehicles, scout cars, armoured command vehicles, armoured buses & ambulances, mine clearance vehicles, mobile workshops & communication vehicles, radar relay units, air transportable mobile air traffic control cabins, plot control vehicles, prison vehicles & ambulances, long range desert patrol vehicles, gun mount pedestals, protected & armoured cash/bullion carrying vehicles, amphibious vehicles, outside broadcast vehicles, outdoor video screen walls & mobile video screen walls. 98

75-25900 Godiva Ltd

Charles Street, Warwick CV34 5LR
☎ (+44 1926) 82 50 00 Ⓔ godivagroup@ukbusiness.com
▣ 311761 ⊖ (+44 1926) 82 50 66
♣ £ 12,000,000 (1996) ♛ 120
Managing Director: M Partridge
Operations Director: S Coyne
Sales and Marketing Director: D Elwkk
Sales Manager: David Ross
Purchasing Manager: J Higgs
■ Full range of firefighting pumps, vehicle mounted, trailer or fixed & portable. 98

75-26135 Goodyear Great Britain Ltd

(a subsidiary of Goodyear Tire & Rubber Co., USA)
Stafford Rd, Bushbury, Wolverhampton WV10 6DH
☎ (+44 1902) 223 21 ▣ 33889 gtrwol g
⊖ (+44 1902) 71 03 96
♣ £ 430 000 000 (1995) ♛ 4 850
Managing Dir: M. Molinari
Sales & Marketing Dir: W. Hall
Mgr Government Contracts: R. Walters
■ Aviation products worldwide, sales & service of aircraft tyres & tubes; repair of aircraft wheels & brakes. Manufacture of car & truck tyres.
Goodyear Aviation Products
Suite 10-11, Watermans Court,
Kingsbury Crescent, The Causeway,
Staines, Middx TW18 3BA
☎ (+44 1784) 46 28 83 ⊖ (+44 1784) 46 28 97
Sales Mgr: J.W. Bell
■ Aviation products. 97

75-26300 Goyen Controls (UK) Ltd

Unit 3B, Beechwood, Chineham Business Park,
Basingstoke, Hampshire RG24 8WA
☎ (+44 1256) 81 78 00 ⊖ (+44 1256) 84 31 64
♛ 13
Managing Director: Keith Salmon
■ Manufacture of solenoid valves. 98

75-26450 Graff Electronic Machines Ltd (GEM)

Woodhill Road, Collingham,
Newark, Nottinghamshire NG23 7NR
☎ (+44 1636) 89 30 36 ⊖ (+44 1636) 89 33 17
♣ £ 1,000,000 (1996) ♛ 19 (1996)
Chairman: A.V. Tushingham
Managing Dir: Alan Leonard
Sales & Marketing Dir: Roger Platts
Purchasing Manager: Alan Leonard
■ High speed audio copiers for duplicating audio tapes at 16 times normal speed. Available as single copiers or models with unlimited expansion, stereo or mono & capable of copying from both open reel & cassette masters. Specified by the British home office for use in police stations & similar environments for the high speed copying of interview tapes. Also used in training personnel. Bulk tape erasers & custom wound blank audio cassette tapes also manufactured & supplied. 98

75-26500 Graham Engineering Ltd

Whitewalls Industrial Estate, Edward St,
Nelson, Lancashire BB9 8SY
☎ (+44 1282) 69 51 21 ⊖ (+44 1282) 69 84 98
♛ 110
Managing Dir: P. Manley
Sales & Marketing Mgr: Terry P.E. Ward
■ Precision machinists, fabricators & welding specialist engineers with CNC laser cutting, Xray, deep draw presswork & vacuum heat treatment facilities for manufacturing products for the aerospace, defence, nuclear, pharmaceutical & power industries. 97

75-26590 Graseby Dynamics Ltd

(a subsidiary of Graseby plc)
459 Park Avenue, Bushey,
Watford, Hertfordshire WD2 2BW
☎ (+44 1923) 22 85 66 ⊖ (+44 1923) 24 02 85
♣ £ 14,000,000 (1996) ♛ 160 (1996)
Managing Director: C P Fowler ☎ (+44 1923) 23 84 83
Financial Director: A Dunn
Technical Director: Dr R B Turner ☎ (+44 1923) 24 44 64
Manufacturing Manager: J Spensley ☎ (+44 1923) 23 84 83
Personnel Manager: D M Strong
Marketing Manager: B R Langford
Purchasing Manager: F Merridew
Sales and Marketing Executive: N P Bloomfield
Ordnance & Marine Division
459 Park Avenue, Bushey,
Watford, Hertfordshire WD2 2BW
☎ (+44 1923) 22 85 66 ⊖ (+44 1923) 21 15 91
Commercial Manager: N Day
Business Development Manager: R Mawkin
■ Specialist in the field of electronic/electrical & mechanical engineering fulfilling UK & overseas governments requirements in areas relating to: (a) Electronic electro-mechanical fuzing safety and arming. (b) ASW sonar systems, torpedo, decoy systems, transducers & sonar calibration equipment. Secure transponder systems. (c) Target miss-distance indicator.
Security Division
459 Park Avenue, Bushey,
Watford, Hertfordshire WD2 2BW
☎ (+44 1923) 81 67 66 ⊖ (+44 1923) 81 62 83
Sales Manager: G Greenhough
■ Explosive vapour detectors & metal detectors, mailroom x-ray screening systems.
Ionics Division
459 Park Avenue, Bushey,
Watford, Hertfordshire WD2 2BW
☎ (+44 1923) 23 84 83 ⊖ (+44 1923) 22 13 61
Senior Sales and Marketing Executive: T P G Otter
■ Designer and manufacturer of toxic/hazardous vapour detection and monitoring equipment for military, commercial and environmental applications. 98

75-26630 Graseby Microsystems Ltd

(† Newmarket Microsystems Ltd)
(a division of Graseby plc)
Exning Road, Newmarket, Suffolk CB8 0AU
☎ (+44 1638) 66 33 81 ⊖ (+44 1638) 66 06 28
♣ £ 6,000,000 (1997) ♛ 100 (1997) ⌂ 60,000 sq m
General Manager: Miss Helen McMenamin
Business Development Manager: Andy Whitaker
Technical Manager: John Rogers
Operations Manager: T Bache
Purchasing Manager: Mark Neeves
■ Design & manufacture of custom microcircuits using Copper Fineline technology & Multi-Chip Module (MCM) technology. 98

75-26640 Graseby Specac Ltd

(† SPECAC Ltd)
(a subsidiary of Graseby plc)
River House, 97 Cray Avenue, St Mary Cray,
Orpington, Kent BR5 4HE
☎ (+44 1689) 87 31 34 ▣ 898188 specac g
⊖ (+44 1689) 87 85 27
♣ £ 4,600,000 (1997) ♛ 85 (1997) ⌂ 22,000 sq ft
Managing Dir: Andy Little
Production Mgr: Paul Smith
Technical Support Mgr: Dr Alfred Afran
Engineering Mgr: Dr G. Poulter
Sales & Marketing Mgr: Nigel Woodhead
■ Infra-red & laser optical components including polarisers, systems & thermal imaging optics, FIR interferometers & laser beam expanders, optical filters & accessories for FT-IR spectroscopy.
Overseas agents:
Graseby Specac Inc, USA 98

75-26750 Greene, Tweed & Co Ltd (GTC)

(a subsidiary of Greene, Tweed & Co, USA)
Finch Close, Nottingham NG7 2NN
☎ (+44 115) 986 65 55 Ⓔ mktng@gtweed.com
⊖ (+44 115) 986 62 35
♛ 60 (1997) ⌂ 14,000 sq m
Managing Dir: Roger De Silva
Sales Mgr Aerospace & Defence: Gary Appleby
Marketing Mgr: Chris Walker
■ Manufacture high performance extended life fluid sealing components & advanced thermoplastic materials. → 85 98

75-26950 Grenson Electronics Ltd
(a subsidiary of Eldec Corp., USA)
High March, Daventry, Northamptonshire NN11 4HQ
☎(+44 1327) 70 55 21 ⊖ (+44 1327) 87 92 21
♣ £ 1,700,000 (1997) ⛬ 44 (1997) 🏭 650 sq m
Managing Director: D Guppy
Operations Director: Stuart Taylor
Purchasing Manager: Jeremy Carless
Sales Manager: D C Wheeler
■ Design & manufacture of AC/DC & DC/DC low voltage convertors. Full military specifications satisfied, custom built or modified standard units. Existing designs used in aircraft, naval & tank applications. [98]

75-27025 Griffon Hovercraft Ltd
PO Box 7, Sarisbury Green, Southampton SO31 8YS
☎(+44 1703) 40 35 47 ⊖ (+44 1703) 40 67 47
♣ £ 2,000,000 ⛬ 20
Managing Director: J H Gifford
Sales Manager: G A Gifford
■ Manufacture long range diesel-engined amphibious hovercraft for the Army, Navy & Coastguards worldwide including UK Royal Marines. [98]

75-27028 Ground Clearance International
Unit 13, Reach Road, Burwell, Cambridge CB5 0AH
☎(+44 1638) 74 19 70 ⊖ (+44 1638) 74 14 33
■ Manufacture Uni-Sift 3MU anti-personnel and anti-tank mechanical mine clearance machine. [98]

75-27030
Grove Europe Ltd International Sales & Marketing
(† Grove Worldwide)

Crown Works, Pallion, Sunderland, Tyne & Wear SR4 6TT
☎(+44 191) 565 62 81 ⊠ 53484 cranes g
⊖ (+44 191) 564 04 42
⛬ 1 200
Operations Director: T Haver
Senior Vice President Sales: W R Lawson
Vice President Manlift Sales and Marketing: G Bradshaw
General Manager, International Marketing: N E L Day
General Manager, Military Sales and UK Military Liaison: A W Mathias
■ Manufacturer of mobile hydraulic cranes including all terrain, rough terrain, truck & materials handling cranes, plus aerial work platforms of the scissor, telescopic & articulating telescope types. Extensive parts & service support, full refurbishment & special build capability. [98]

75-27100 Guardall Ltd
(a subs. of Chubb Security plc)
Lochend Industrial Estate, Newbridge,
Edinburgh, Scotland EH28 8PL
☎(+44 131) 333 29 00 ⊖ (+44 131) 333 49 19
Director and General Manager: Andrew Smith
Managing Dir: Richard Fernie
■ Perimeter protection "Fenceguard" microwave perimeter fence. [98]

75-27130 Guartel Ltd
16 Alliance Court, Alliance Road, London W3 0RB
☎(+44 181) 896 02 22 ⊖ (+44 181) 896 03 33
♣ £ 2 000 000 ⛬ 6
Managing Dir: M.F. Arnold
Operations/Technical Dir: P.E.V. Arnold
■ Manufacture & development of mine & metal detection equipment. [97]

75-27170 Gunpell & Co. Ltd
Wildacres, Reading Road North, Fleet, Hants GU13 8HS
☎(+44 1252) 61 75 90 ⊖ (+44 1252) 81 55 60
♣ £ 100 000 - 200 000 ⛬ 5
Managing Dir: J.A. Vertanness (C.Eng.)
■ Firearms & small arms ranges design, supply & construction. [97]

75-27200 Hadland Photonics Ltd
(a division of Photonics Holdings Ltd)
Harrow Yard, Akeman Street,
Tring, Hertfordshire HP23 6AA
☎(+44 1442) 82 15 00 Ⓔ post@hadland.co.uk
⊖ (+44 1442) 82 15 99
♣ £ 6,000,000 (1997) ⛬ 40 (1997) 🏭 1,000 sq m
Director: M Hadland
Managing Director: H Davies
■ Manufacture and market high speed and ultra high speed camera systems for evaluating events down to 1 picosecond time scale. Total instrumentation systems for ballistic ranges. Airborne video tape recorders, monitors and head-up display cameras, night vision goggles and aircraft crack detection equipment. → 90
® Denmark: Terma
Germany: BKT Germany
Japan: TEAC Corp.
Sweden: Scandiflash
UK: MSI
USA: Physical Research Inc., SRL Inc., Videospection
Subsidiaries:
Hadland Photonics GmbH → Germany 75
Hadland Photonics Inc. → USA 90
Hadland Photonics Pty Ltd → Australia 90 [98]

75-27220 Hagen
(† Hagen Morfax Ltd)
23 Arrow Business Park, Shawbank Road, Lakeside,
Redditch, Worcester B98 7YN
☎(+44 1527) 51 48 88
Ⓔ arrowvale@ax.compulink.co.uk
⊖ (+44 1527) 51 43 21
♣ £ 4,000,000 (1997) ⛬ 80 (1997) 🏭 1,500 m2
Managing Director: M P Thompson
Manager: Colin Wood
■ Covert surveillance equipment and radio communications accessories for police, special forces & customs authorities. Battery chargers & electronic subcontract design and manufacture. BSEN ISO9001. [98]

75-27300 A.W. Hainsworth & Sons Ltd
Spring Valley Mills, Stanningley,
Pudsey, W. Yorkshire LS28 6DW
☎(+44 113) 257 03 91 ⊖ (+44 113) 255 81 87
⛬ 250
Marketing Mgr: Ms Gillian Rushton
Worsted Mgr: T.C. Hainsworth
■ All wool cloths for cermonial uniforms. Nomex cloths for fire resistant coveralls & protective clothing. Tapes webbings for personal equipment. Felted woollen, cotton or synthetic fibre sleeving in lengths & rolls. Industrial cloths for all purposes. [97]

75-27350 Hale Hamilton (Valves) Ltd
Frays Mill Works, Cowley Rd,
Uxbridge, Middlesex UB8 2AF
☎(+44 1895) 23 65 25 ⊖ (+44 1895) 23 14 07
♣ £ 10,500,000 (1996) ⛬ 200 (1997)
Managing Dir: R.W. Bartlett
Technical Director: G Pearson
Sales & Marketing Mgr Defence Div: W.R. Dormer
Purchasing Manager: R Partridge
■ Air pressure reducing stations for gas turbine & diesel engine starting systems; general purpose reducing stations; constant-flow valves for trim adjustments in submarines; stop valves, filters & pipe couplings; valves for pure-air thermal imaging applications. All onboard HP air systems for surface & sub-surface vessels. Valves for missile launch systems. [98]

75-27400 Haley & Weller Ltd
(a subsidiary of Chemring Group plc)
Wilne Mill, Draycott, Derby DE72 3QJ
☎(+44 1332) 87 24 75 ⊖ (+44 1332) 87 30 46
♣ £ 5,228,000 (1995) ⛬ 109 (1997) 🏭 28 acres
Chairman: P G Billington
Managing Director: D R Evans
Director Finance: P Simpson
General Manager: T P E Potter
Technical Manager: D Kinnear
Technical Marketing Manager: R Harvey
Sales Support/Marketing: Ms J Rice
■ Pyrotechnic & specialist explosive products for defence. Linear cutting charge, mild detonating cord, canopy severance systems. Delay elements. Flares, signalling, ground marking etc. Simulators for all training purposes. Specialised defence obscurants. → 85 [98]

75-27420 J&E Hall Limited
(† APV Baker Ltd)
Regd Office
6 Prospect Place, Dartford, Kent DA1 1BU
☎(+44 1322) 22 34 56 ⊖ (+44 1322) 29 14 58
Marketing Dir: J.F. Nason
■ Design, installation, commission manufacture of freezing equipment (industrial). Land & offshore aftersales. Naval refrigeration & air conditioning.
Hall Termmotank
Ste 6, Cameron Ct, Cameron St,
Hillington Industrial Estate, Glasgow G52 4JG
☎(+44 141) 882 62 34 ⊠ 778065
⊖ (+44 141) 882 77 92
General Mgr: John Coatsworth [97]

75-27430 E C Hallam Engineering (Leicester) Ltd
Smeeton Road, Kibworth, Leicester LE8 0LG
☎(+44 116) 279 23 30 ⊖ (+44 116) 279 22 62
Chairman and Managing Director: E Carey Hallam
■ Aircraft towing and recovery tractors. Shunting tractors, motor graders, towed and coupled graders. Bulldog Mk II range of four-wheel drive runway tractors; NA-11, TC-13, TC-14 and TC-18. ISO 9001 accredited. [98]

75-27650
Hall & Watts (Defence Optics) Group
(a subsidiary of Hall & Watts Ltd)
266 Hatfield Road, St Albans, Hertfordshire AL1 4UN
☎(+44 1727) 85 92 88 ⊖ (+44 1727) 83 56 83
♧ £ 1,000,000 ♣ £ 4,500,000 (1996) ⛬ 38 (1996)
🏭 2,200 m2
Group Managing Director: John Hoskins
Sales and Marketing Manager: Alan Mitchell
Purchasing Clerk: Mark Spratt
■ Production of a range of military optical equipment & provision of a full technical training, repair, recalibration & certification service to ISO 9000 standard. [98]

75-27700 Halmatic Ltd
(a subsidiary of Hunting Defence Ltd)
Saxon Wharf, Lower York Street,
Noatham, Southampton SO14 5QF
☎(+44 1703) 33 74 77 ⊠ 47646 halmat g
⊖ (+44 1703) 33 74 78
♧ £ 100,000 ♣ £ 17,000,000 (1996) ⛬ 150
🏭 60,000 sq ft
Managing Dir: Peter G. Jones
Sales Dir: Michael Burnham
Quality Assurance & Special Projects: Colin G. Dove
Financial Dir: Peter C.W. Lunt
■ Glass fibre, aluminium boats. Range of complete specialised military & combat craft upto 30m based on over 20 different hull designs. Hulls up to 30m are completed under Lloyds supervision. Glass fibre hulls, decks & superstructures for fast patrol/attack craft up to 33m. → 80 [98]

75-27730 Hamech Ltd
Head Office
Shirling Rd, South Manton Park,
Swindon, Wiltshire SN3 4TS
☎(+44 1793) 83 11 41 ⊖ (+44 1793) 83 11 94
♣ £ 25 000 000 (1995/96) ⛬ 240 🏭 100 000 sq ft
Managing Dir: V. Parry
Sales & Marketing Dir: A.C. Nash
Marketing Services: Ms Rita Cox
General Mgr Special Accounts: R.H. James
Exec. Special Accounts: William B. Jeram
Promotions Controller: John Knowles
■ Manufacture & supply of mechanical handling equipment. Fork trucks, reach trucks & electric tow tractors (6-20 ton). Electric & internal combustion engine products.
Manufacturing Plant
Bramley, Basingstoke, Hants RG26 5DW
☎(+44 1256) 88 14 44 ⊖ (+44 1256) 88 11 29 [97]

75-27850 Hamworthy Marine
(† Hamworthy Pumps & Compressors Ltd)
Fleets Corner, Poole, Dorset BH17 0JT
☎(+44 1202) 66 55 66 ☒ 41348 hampac g
⊖ (+44 1202) 66 63 63
480
Managing Dir: Kelvyn Derrick
Sales & Marketing Dir: Ian Ballard
Marketing Mgr: N. White
Sales Co-ordinator: W.G. Adams
■ Manufacture & supply of horizontal & vertical centrifugal pumps for marine duties, heavy duty & positive displacement pumps, air compressors for shipboard uses. Naval dockyard services & packaged sewage treatment plant & incinerators for liquid & solid waste disposal. Oily water separators. Cargo pumps. High lift rudders. [97]

75-27970 The Harboro Rubber Co. Ltd
St Mary's Road, Market Harborough, Leics LE16 7EE
☎(+44 1858) 41 06 10 ⊖ (+44 1858) 41 00 06
♣ £ 4 000 000 147
Managing Dir: J.E. Briggs
Marketing Dir: N. Tulloch
Operations/Technical Dir: G. Ruffell
Sales Mgr: M. Rose
Purchasing Mgr: J. Winter
■ Rubber moulding service. Footwear components, keypads. [97]

75-28050 Harkness Hall Ltd
The Gate Studios, Station Road,
Borehamwood, Hertfordshire WD6 1DQ
☎(+44 181) 953 36 11 [E] harknesshall.com
⊖ (+44 181) 207 36 57
♣ £ 6,000,000 (1996) 125 (1997) 2,800 m2
Managing Director: John Lawton
Sales and Marketing Director: I D Sim
■ Fixed and portable screens for front and rear projection and simulators. Divider curtains for aircraft hangers and space management applications. Theatre rigging equipment. [98]

75-28140 Harris Brett Ltd
Waterloo Mills, Howden Road, Silsden BD20 0HA
☎(+44 1535) 65 44 79 [E] snugpak@legend.co.uk
[W] http://www.westyorks.com/snugpak
⊖ (+44 1535) 65 25 03
■ Manufacture sleeping bags, accessories and clothing. [97]

75-28150 T S Harrison & Sons
(a subsidiary of 600 Group plc)
Union Street, PO Box 20,
Heckmondwike, West Yorkshire WF16 0HN
☎(+44 1924) 40 37 51 [E] mail@harrison.co.uk
[W] http://www.harrison.co.uk ☒ 55217
⊖ (+44 1924) 40 72 85
♧ £ 100,000 (1997) ♣ £ 30,000,000 (1997)
520 (1997) 20,000 m2
Managing Director: Anthony Sweeten
Operations Director: Dr Stuart Cessford
Sales Director: D H Smith
Marketing Director: Dr Stephen Lebeau
Marketing Manager: Stephen Lawless
Purchasing Manager: Ms Sophie Hinley
■ Manufacture of geared head and variable speed centre lathes from 280mm to 760mm swing capacity and 635 to 1000mm centre centre distances. Alpha electronic centre lathe from 400 to 554 swing capacity and 650 to 3000 centre distance. [98]

75-28180 Arthur Hart Webbing
(† Arthur Hart & Son Ltd)
(a division of Bridport Gundry UK plc)
Viney Bridge Mills, South Street,
Crewkerne, Somerset TA18 8AE
☎(+44 1460) 733 11 [E] sales.harts@ndirect.co.uk
[W] http://www.ndirect.co.uk/~harts ☒ 46543 a hart g
⊖ (+44 1460) 742 23
♣ £ 1,800,000 (1996) 43 (1996)
General Manager: J A Yuill
Finance Manager: Mrs M Barnett
Technical Mgr: E.D. Gaskell
Plant Mgr: R. Bidgood
Sales Mgr: Alan Bate
■ Manufacture of webbing & narrow fabrics for aerospace, defence & industrial markets. Specialised webbings & narrow fabrics for aerospace use, including manufacture to US specifications. [98]

75-28300 Hatcham Rubber Ltd
(a subsidiary of Wellington Holdings plc)
Mead Lane, Hertford, Hertfordshire SG13 7AU
☎(+44 1992) 55 43 21 ⊖ (+44 1992) 55 42 52
103 (1996)
Managing Director: I M K Hurst
Financial Director: Mike Kennedy
Operations Director: A Din
Technical Director: R Morgan
Commercial Manager: D Mottershead
Marketing Manager: Ms Karen Brewster
■ Manufacturers of unvulcanised rubber compounds and vulcanised rubber sheeting. [98]

75-28475 Hawke Cable Glands Ltd
Oxford St West, Ashton-under-Lyne, Lancs OL7 0NA
☎(+44 161) 308 36 11 ⊖ (+44 161) 308 44 86
♣ £ 13 500 000 277
Divisional Sales Dir: Dave Curtis
■ Underwater & severe environment electrical connectors & penetrator systems. Suitable for use in subsea electrical technology and space conditions. → 85 [97]

75-28480 Hawke Connection Systems
Oxford Street West, Ashton under Lyne, Lancs OL7 7NA
☎(+44 161) 308 36 11 ⊖ (+44 161) 339 96 56
20
Marketing Dir: David Curtis
■ Full range of high integrity cable terminations for severe environments. BASEEFA approved EExd flameproof connectors. European manufacturer of lockheed challenger marine ROV mateable subsea connectors. Manufacturer of FITA products for subsea. UK distributor for E.O. connector range. Defence contractors listing approval (No. 46FH01). BS5750 part 1 approved and certified (FM10608). [97]

75-28490 Hawker Batteries Ltd
(a div. of BTR Plc)
Challenge Court, Barnett Wood Lane,
Leatherhead, Surrey KT22 7LW
☎(+44 1372) 36 05 20 ⊖ (+44 1372) 36 20 82
Chairman & CEO: W.J. Richardson
Research & Development Dir: Dr Geoffrey May
Marketing Dir: Mike Dunckley

Chloride Industrial Batteries Ltd
Clifton Junction, PO Box 5, Swinton, Manchester M27 2LR
☎(+44 161) 794 46 11 ☒ 669087 chlor g
⊖ (+44 161) 793 66 06
500
Chairman: John Richardson
Managing Dir: Ian Wells
Operations Dir: Ian Murdoch
Marketing & Defence Sales Dir: A.G. Smith
Technical Sales Mgr: Lian Delaney
Commercial Mgr: Phil Yates
Product Planning Mgr: Ian Harrison
■ Batteries for submarine & torpedo propulsion, fighting vehicles, degaussing & sonobuoy operation, standby & other applications.

Hawker Energy Products Ltd
Stephenson St, Newport, Gwent NP9 0XJ
☎(+44 1633) 27 76 73 ☒ 498431
⊖ (+44 1633) 28 17 87
110
Managing Dir: Ian Wells
General Mgr: Paul Rendell
Marketing Dir: N. Finn
Sales Mgr: B. Sergeant
Technical Mgr: D. Rice
■ Manufacture of sealed, pure lead, low maintenance batteries with MoD, CAA & military approvals for standby aircraft, marine use & in commercial grades for GSE. [97]

75-28500 Hawker Eternacell Ltd
(† Crompton Eternacell Ltd)
(a subsidiary of BTR plc)
River Drive, South Shields, Tyne and Wear NE33 2TR
☎(+44 191) 456 14 51 [E] sales.eternacell@btr.inc.com
☒ 53210 ⊖ (+44 191) 456 63 83
150 7,500 sq m
Chairman: W.J. Richardson
Managing Director: David Edwards
Finance Director: Tim Perrins
Operations Director: B Gibson
UK Sales Director: Dr Anne Brown
Purchasing Manager: Deborah Charlton
European Sales Manager: Ms Lisa Charlton
■ Designs & manufactures a range of lithium & traditional primary batteries for industrial & defence applications. Major suppliers of Lithium-SDX batteries to armed forces. [98]

75-28750 Haynes & Cann Ltd
1-9 Overstone Road, Northampton NN1 3JL
☎(+44 1604) 261 43 ⊖ (+44 1604) 261 43
♧ £ 278,880 (1995) ♣ £ 973,177 (1995) 20 (1997)
650 sq m
Managing Dir: Brian Keech
Purchasing Mgr: Peter A. Shears
■ Specialist military footwear for the flying services including boots for aircrews, fire crash crews & parachute jumping instructors, fabric flying helmets, associated ear capsules & aircrew survival hoods. Wide range of post design service & development work on the above & other items of aircrew equipment. [98]

75-28760 Haynes International Ltd
Parkhouse St, Openshaw, Manchester M11 2ER
☎(+44 161) 230 77 77 ⊖ (+44 161) 223 24 12
♣ £ 10 000 000 65
Managing Dir: Phillip Crawshaw
Sales Mgr: Michael Marycz
■ High performance nickel & cobalt based alloys. Nickel & cobalt alloy wrought materials in sheet, plate, billet, bar, wire, tubular pipe fittings, welding consumables. Titanium Alloy Tubing. [97]

75-28770 HDA Forgings Ltd
(a subsidiary of High Duty Alloys Ltd)
Windsor Road, Redditch, Worcestershire B97 6EF
☎(+44 1527) 642 11 ⊖ (+44 1527) 59 17 60
375 (1997)
Managing Director: Dr A J Scanlon
Finance Director & Company Secretary: G A Berry
Sales and Marketing Director: P J R Masters
Commercial Director: D W Smith
Operations Director: P R Langston
Sales Administration: D A Pinfield
■ Manufacture of hand, die, close-to-form & precise-to-form forgings & precision forged gas & steam turbine & compressor blades in aluminium, nickel & titanium-based alloys & alloy steels. Manufacture of cast aluminium alloy billet for in-house use as forging stock. [98]

75-28860 Hebron & Medlock Engineering Ltd
821 Bath Road, Brislington, Bristol BS4 5NT
☎(+44 117) 977 53 23 ⊖ (+44 117) 971 07 92
♣ £ 2,500,000 30
Managing Dir: Ken Barnes
General Mgr: Mike Ferris
■ Design & develop sight mountings, sight hoods, linkages, gearboxes, clinometers for fighting vehicles (armoured & unarmoured); gunners training systems & crew seats. [98]

75-28880 HEC (Aerospace) Ltd
Cranleigh Gardens Industrial Estate,
Southall, Middlesex UB1 2BZ
☎(+44 181) 574 59 58 ⊖ (+44 181) 571 48 40
♧ £ 1 000 000 ♣ £ 1 750 000 40 10 000 sq m
Managing Dir: J.J. Corcoran
Sales & Marketing Dir: J.J. Corcoran
General Mgr: A.J. Thomas
■ Precision mechanical component manufacturers for the aerospace, defence, automotive, process, telecommunications and associated industries. [98]

75-28900 Helio Mirror Co. Ltd
(a subs. of Avimo Europe Ltd)
Crabtree Manorway South, Belvedere, Kent D17 6AY
☎(+44 181) 311 41 40 ⊖ (+44 181) 311 10 04
♧ £ 500 000 (1995) ♣ £ 12 000 000 (1995) 160
4 400 sq m
Managing Dir: Alan J. Spencer
Business Development Dir: Marcus A. Bartlett-Rawlings
Operations Dir: D. Reeve
Purchasing Mgr: K. Sleafer

Contact Reference

■ Machine Gun turrets & small calibre weapon turrets for light armoured vehicles & APCs, cupolas, hatches, gun mounts, gearboxes, power traverse and elevation systems multi-purpose grenade dischargers, vision blocks, drivers' periscopes & a large range of direct vision periscopes. Sight protection & cleaning, electronic tracking mounts, encoder gearbox systems, laser detection - defensive aid systems. [97]

75-28970 Hellermann Electric
Pennycross Close, Plymouth, Devon PL2 3NX
☎(+44 1752) 70 12 61 ⊖ (+44 1752) 79 00 58
Managing Director: D Elliott
■ Provides cable care services to the defence market. Identification systems, heat shrinkable moulded parts, insulation sleeving-heatshrink and non-heatshrink systems, cable screening systems, cable harness systems and cable stripping and cutting tools. [98]

75-28990 Hellermann Insuloid
Sharston Works, Leestone Road,
Wythenshawe, Manchester M22 4RH
☎(+44 161) 945 41 81 ⊖ (+44 161) 945 37 08
Managing Director: Tony Cunliffe
Marketing Assistant: Ms Debra Carter
■ Cable and wire fixing systems, including INSULOK cable-tie range. Design and application advice. [98]

75-29000 Helmet Integrated Systems Ltd
(a subsidiary of Middlemace Ltd)
Moat Factory, Wheathampstead,
St Albans, Hertfordshire AL4 8QT
☎(+44 1582) 83 42 11 ▣ 825587 helmts g
⊖ (+44 1582) 83 42 10
👥 215
Managing Dir: I. Hoyle
Sales & Marketing Dir: K.J. Korny
Marketing Services Mgr: Ms Yvonne Peat
Military Sales Mgr: S. Skeggs
■ Approved contractor under British MoD Quality Assurance Standard BS5750 Part 1, ISO 9001 for design, development, PDS & manufacture of helmets:- Flying protective, flight deck, combat, firemen, parachutist, public order, motorcyclist, visors, headsets, boom microphones accessories, electronically enhanced & associated equipment. [98]

75-29010 Helston Gunsmiths
Water Ma Trout, Helston, Cornwall TR13 0LW
☎(+44 1326) 56 14 40; 57 33 85
[E] helston-gun@demon.co.uk
⊖ (+44 1326) 57 32 21; 57 47 47
🏭 4,000 sq m
General Mgr: C.D. Price
■ Suppliers of military small arms to the police and military, spares for light weapons and ordnance, forensic science services. → 80 → 85 [98]

75-29040 A.O. Henton Engineering Co Ltd
(a subsidiary of J.S. Chinn Holdings Ltd)
Cotes Road, Burbage, Hinckley, Leicestershire LE10 2HJ
☎(+44 1455) 23 83 31 ▣ 31303
⊖ (+44 1455) 25 10 23
♣ £ 4,500,000 (1996) 👥 75 (1996) 🏭 2,000 m2
Managing Director: D J English
Director: J S Chinn
Works Director: C W Greenaway
Purchasing Manager: Edwin Manley
■ Components for aerospace applications, form grinding, spark erosion, CNC turning, milling & boring of light alloys, titanium, stainless & hi-nickel alloys. [98]

75-29045 Hepworth Marine International
(† B. Hepworth & Co Ltd)
Hepworth House, Brook Street,
Redditch, Worcestershire B98 8NF
☎(+44 1527) 612 43 ⊖ (+44 1527) 668 36
♣ £ 3,800,000 👥 47
Managing Director: J P Eddy
Marketing and Sales Director: K Rogers
■ Design and manufacture of marine windscreen wiper systems, straightline, pantograph, pendulum and clear view screens; electrically, hydraulically and pneumatically driven and associated products. Anti-glare sun blinds. [98]

75-29150 Hexcel Composites Ltd
(a division of Hexcel Corporation, USA)
Duxford, Cambridge CB2 4QD
☎(+44 1223) 83 31 41 ▣ 81101
⊖ (+44 1223) 83 88 08
👥 500 🏭 35,000 m2
President & Managing Dir EuroMaterials: W. Hunt
Head of Sales & Marketing EuroMaterials: Mrs S. Preston
Head of Research & Development: G. Green
Communications Manager: Mrs Rachel A Atkinson
■ Manufacture of a complete range of carbon fibre, glass & aramid advanced composite structural materials including adhesives, honeycombs, laminates for makers of civil & military aircraft, tanks, lorries, boats & radars.
® USA: Hexcel Corporation [98]

75-29180 Hiatt & Company Ltd
111-115 Baltimore Rd, Great Barr, Birmingham B42 1DN
☎(+44 121) 357 43 47 ⊖ (+44 121) 358 77 68
♲ £ 595 000 (1996) ♣ £ 2,000,000 (1996)
👥 40 (1996) 🏭 20,000 sq ft (1996)
Chairman: Geoffrey W Cross
Operations Director: C J Dipple
Purchasing Manager: Mrs C M Ebanks
■ Manufacture high quality arrest and escort handcuffs, truncheons and batons. Suppliers of riot shields, helmets, whistles and other police equipment. → 85 [98]

75-29250 High Integrity Systems Ltd (HIS)
Astra Centre, Edinburgh Way, Harlow, Essex CM20 2BE
☎(+44 1279) 45 00 00 ⊖ (+44 1279) 42 91 49
Managing Dir: Ms Daphne Gordon
Dir Marketing: Dr Richard Emmerson
■ Computing solutions for defence applications. Systems development contractor & supplier of training & consultancy services. → 80 [97]

75-29430 Hippo Marine Products Ltd
Administration
Little Green Court, Staplehurst, Kent TN12 0BJ
☎(+44 1580) 89 13 08 ⊖ (+44 1580) 89 34 85
♣ £ 1 000 000 👥 20 🏭 10 000
Chairman: John Wickham
■ Surface/sub-surface marker buoys. Flotation units, floating fenders. Manufacture of navigation & mooring buoys in polyurethane elastomer, GRP & steel. Manufacture of items in GRP. Manufacture of fendering & floating fenders. Suppliers of rubber fendering.
Design Office & Factory
Churchill Industrial Estate, Marlborough Rd,
Lancing, W. Sussex BN15 8UF
☎(+44 1903) 75 25 35 ⊖ (+44 1903) 85 12 66
Managing Dir: P.J. Wickham [97]

75-29470 HMD Seal/Less Pumps Ltd
(a division of Sundstrand Fluid Handling Corporation)
Hampden Park Industrial Estate,
Eastbourne, East Sussex BN22 9AN
☎(+44 1323) 50 12 41 [E] pumps@hmd.co.uk
[W] http://www3.mistral.co.uk/pumps ▣ 87375 noseal g
⊖ (+44 1323) 50 33 69
♣ £ 12,000,000 (1996) 👥 143 (1996) 🏭 3,742 m2
Managing Director: J Veness
Senior Marketing Manager: D Twyford
UK Sales Manager: C Guppy
Purchasing Manager: R Verhoevan
■ Seal/Less zero leakage magnet drive pumps for naval duties. Repair & modification services, including pumps of other makes. Design, research & development facilities for toxic radioactive transfer duties. [98]

75-29550 Hobson & Sons (London) Ltd
Kenneth Road, Thundersley, Essex SS7 3AF
☎(+44 1268) 79 30 97
⊖ (+44 1268) 79 48 10; 56 60 24
♣ £ 6,250,000 (1996) 👥 160
Chairman: C D Brown
Managing Director: C J Lane
Operations and Technical Director: C A Meads
☎(+44 181) 692 34 67
Sales and Marketing Director: K P Palmer
■ Uniforms and equipment for all services and occasions. [98]

75-29580 Hocking NDT Ltd
Inspec House, 129-135 Camp Rd,
St Albans, Herts AL1 5HP
☎(+44 1727) 84 03 21 ▣ 825562 (for Hocking NDT)
⊖ (+44 1727) 84 50 58
👥 48
Managing Dir: D.H. Hocking
Sales Co-ordinator: R. Lake
Marketing Mgr: Tony Hocking
■ Design & manufacture of a wide range of portable eddy current non-destructive testing instruments & systems. Crack detectors, corrosion detectors, electrical conductivity meters. Provides equipment for numerous world customers including air forces, government agencies & public corporations within the aerospace/defence industry. [97]

75-29590 Hodge Clemco Ltd Aerolyte Systems Division
Orgreave Drive, Sheffield S13 9NR
☎(+44 114) 254 06 00 ⊖ (+44 114) 254 02 50
♣ £ 4,000,000 👥 55
Managing Director: M A Eagle
■ Aerolyte plastic media stripping systems, dry stripping cabinets, plastic media, media recovery systems, walk-in booths, approved operator safety equipment and clothing, spray coating equipment and internationally recognised training courses.
® USA: Aerolyte Systems, Clemco Industries [98]

75-29650 Holliday Dyes & Chemicals Ltd
(† L.B. Holliday & Co. Ltd)
(a div. of Holliday Chemical Holdings)
Leeds Road, PO Box B22, Huddersfield, W.Yorks HD2 1UH
☎(+44 1484) 42 18 41 ▣ 51433
⊖ (+44 1484) 51 53 28
Managing Dir: B. Haith
Operations/Technical Dir: Dr S. Partington
Sales & Marketing Dir: M.V. Russell
Purchasing Mgr: A. Haigh
General Sales Mgr: G.A. Pearson
■ Smoke Dyestuffs. [97]

75-29652 Hollow Extrusions Ltd
(a subsidiary of Hay Hall Group)
Redfern Road, PO Box 764,
Tyseley, Birmingham B11 2BW
☎(+44 121) 706 66 77 [E] sales@hollow-extrusions.com
⊖ (+44 121) 765 41 71
♣ £ 6,000,000 (1995) 👥 120
Managing Dir: C.T. Powell
Operations Director: R J Edwards
Sales Exec.: F.C. Hopkins
■ Hollow extruded defence components for missile bodies, munitions, pyrotechnics & hydraulic applications. [98]

75-29653 Holmes Halls Tanners Ltd
(a subs. of Profitwell Ltd)
Sculcoates, Hull, N. Humberside HU5 1RL
☎(+44 1482) 44 94 49 ▣ 597237 tanner g
⊖ (+44 1482) 44 53 63; 34 23 42
♣ £ 25 000 000 👥 250 (1996)
Managing Dir: T.W. Ulliott
Sales & Marketing Dir: Ms Jean McDonald
Operations Dir: Stuart Parker
Purchasing Mgr: G.K. Mallinson
■ Suppliers of sole and upper leather for the manufacture of boots and shoes to Ministry of Defence standard. Registered with the MoD for AQAP 4 Edition 2. [97]

75-29655 Holscot Industrial Linings Ltd
Alma Park Rd, Alma Park Industrial Estate,
Grantham, Lincs NG31 9SE
☎(+44 1476) 747 71 ▣ 37641 holsct g
⊖ (+44 1476) 635 42
♣ £ 2 000 000 (1993) 👥 40
Managing Dir: David Joyce
Sales Dir: Martin Daff
■ Specialists in melt processible fluoroplastics, FEP, PFA, PVDF or ECTFE. Production of fluoroplastic items for a wide range of industries & applications.
Northern Division
Unit 9 Burnmill Industrial Estate, Burnmill Rd,
Leven, Fife KY8 4RA
☎(+44 1333) 275 55 ⊖ (+44 1333) 42 75 30 [97]

75-29730 Honeywell Control Systems Ltd
(a subsidiary of Honeywell Inc, USA)
Honeywell House, Bracknell, Berkshire RG12 1EB

☎ (+44 1344) 65 60 00
⊘ (800) 52 11 21 (UK); (800) 328 51 11 (USA)
Ⓔ info.centre@uk.honeywell.com
Ⓦ http://www.sensing.honeywell.com
Ⓦ http://www.honeywell.com ⊖ (+44 1344) 65 62 40
♣ US$ 7,300,000,000 (1996) 👥 53,000 (worldwide)
Chairman and Managing Director: Dennis Kennedy
Quality Director: Martin Kruger
Divisional Sales Manager: Ray Ventura
Communications Manager: R N Slater
■ Push button assemblies, gear selectors for armoured personnel carriers, proximity devices and sub-systems for AOM market. Secure data communications products, opto-electronic components. Solid state sensors for pressure, flow, current and temperature.
® USA: Honeywell Inc.

Micro Switch Division
Honeywell House, Bracknell, Berkshire RG12 1EB
☎ (+44 1344) 65 60 00 ▣ 847064
⊖ (+44 1344) 65 60 15
Divisional Director: Alan Wright
Divisional Sales Mgr: Ray Ventura 98

75-29800 Hopkinsons Ltd
(a subs. of The Weir Group PLC)
PO Box B27, Birkby, Huddersfield, W.Yorks. HD2 2UR
☎ (+44 1484) 42 21 71 ▣ 51682
⊖ (+44 1484) 51 80 92
♣ £ 40 000 000 👥 830
Managing Dir: John Steele
Sales Dir: D. Perkin
Publicity Mgr: R.H. Lodge
■ Valves, boiler mountings, electric actuators, sootblowers & steel castings. Suitable for fighting ships & naval, army & air force establishments with steam raising plants. Repair facilities & mobile services available worldwide. 97

75-29920 Horstman Defence Systems Ltd
(a member of EIS Group plc)
Locksbrook Rd, Bath, Avon BA1 3EX
☎ (+44 1225) 42 31 11 ▣ 444363
⊖ (+44 1225) 44 73 57
♣ £ 8 000 000 👥 120
Managing Dir: Alan J. Brooks
Sales & Marketing Dir: Paul G. Briggs
Technical Dir: Tim Holman
Purchasing Mgr: Ms Susie Bates
■ Naval instrumentation, retransmission units & electro-mechanical drives. Suspension systems & gearboxes for armoured fighting vehicles. Precision gears sub-contract components. Auxiliary power units & spares. 97

75-30000 Hoskins Medical Equipment Ltd
(a subsidiary of Huntleigh Technology plc)
Admail 1001, Birmingham B1 1HJ
☎ (+44 121) 707 66 00 ⊖ (+44 121) 707 66 88
Director and General Manager: G Cox
Sales Director: P Beresford
Customer Services Manager: L Riley
Export Sales Manager: B Zealey
■ Hospital beds, lockers, bed tables, trolleys, obstetric equipment and wooden furniture. 98

75-30100 Houchin Aerospace Ltd Factory and Administration Centre
(a subsidiary of TT Group plc)
Hilton Road, Ashford, Kent TN23 1DZ
☎ (+44 1233) 62 32 11 ⊠ GARFODIA
▣ 96178 houchin g ⊖ (+44 1233) 63 84 03
♣ £ 11,000,000 (1996) 👥 125 (1997) ⌂ 130,000 sq ft
Managing Director: Brian Groves
Financial Director: Steve Drain
Sales and Marketing Director: Martin Hoare
Sales Manager: Merrick Chrusciel
Purchasing Manager: Morph Parsons
■ Civil and miliitary ground power units and fixed 400Hz distribution systems. 98

75-30180 House of Hardy
Fibatube Division
Willowburn, Alnwick, Northumberland NE66 2PF
☎ (+44 1665) 60 25 50 ▣ 537500 g
⊖ (+44 1665) 60 23 89
Managing Dir: Chris Bond
Technical Dir: George Minto
Customer Services: Simon Harbord
■ Manufacture bespoke tubular components using composite materials to engineering standards for squareness, concentricity, pressure integrity & surface finish. 97

75-30190 Hovair Systems
North Lane, Aldershot, Hants GU12 4QH
☎ (+44 1252) 31 99 22 ⊖ (+44 1252) 33 43 87
♣ £ 12 000 000 👥 120
Managing Dir: A.C. Yates
Operations & Technical Dir: Q. Simpson
Sales & Marketing Dir: K.A. Tracey
Divisional Mgr: A.W. Machie
■ Air film bearings/heavy load handling & manoeuvring equipment. Low friction & very low effort equipment to move heavy loads including aircraft, vehicle & marine engines, A/C sub-assemblies, wheeled & tracked vehicles. Vehicle turntables materials handling equipment. 97

75-30200 Hovermarine International Ltd
Spitfire Quay, Hazel Rd, Woolston,
Southampton, Hants SO2 7GB
☎ (+44 1703) 44 31 22 ▣ 47141 whl g
⊖ (+44 1703) 44 44 29
♣ £ 2 000 000 (1996) 👥 50 ⌂ 15 000 m2
Managing Dir: P.J. Hill
Sales Mgr: D. Williams
Technical Consultant: E.G. Tattersall
■ Manufacture high speed surface effect marine craft, glassfibre weapon shields, plastic piping, chemical tanks. Craft for police, coast guard & civilian applications. Research & development undertaken for advanced naval vessels. 97

75-30240 Howden Aircontrol Ltd
(a subsidiary of Howden Airdynamics Group)
Hawthorne Road, Staines, Middlesex TW18 3AY
☎ (+44 1784) 46 61 66 ⊖ (+44 1784) 46 58 94
♣ £ 10,000,000 👥 70
Managing Dir: M. Cooper
Marketing Dir: J.R. Baxter
Operations Dir: N. Iles
■ Design & manufacture military & aerospace air control systems including vapour cycle or air cycle cooling, NBC filtration equipment, engine cooling systems, ground support trolleys & environmental control systems. 98

75-30250 Howden Compressors Ltd
(a subsidiary of Howden Group plc)
133 Barfillan Drive, Glasgow G52 1BE
☎ (+44 141) 882 33 46 ⊠ ROTACOMP Glasgow
▣ 778711 rotacomp g ⊖ (+44 141) 882 86 48
👥 150 ⌂ 13,517 m2
Managing Director: I McCreery
Marketing Co-ordinator: Ms Susan Laidlaw
■ Manufacture of compressors for air conditioning of naval vessels and special air compressors for frigates. 98

75-30260 Howden Wade Ltd
(a subsidiary of the Howden Group plc)
Crowhurst Road, Brighton, East Sussex BN1 8AJ
☎ (+44 1273) 50 63 11 ▣ 87511 wade g
⊖ (+44 1273) 55 71 23; 55 65 05 (sales)
♣ £ 5,000,000 (1995-96) 👥 106
Managing Director: J Lambert
Marketing Manager: Mark Nash
Ⓔ mark.nash@howden-wade.co. uk
Operations Manager: Caspar Ingrams
Purchasing Manager: Ms Christina Rice
■ OEM, repair and overhaul of superchargers, roots and screw type blowers and air motors. Design, development and production of and temperature switches, fire suppression suppression systems and hydraulic rigs. OEM, authority for fluid condition monitoring systems. Applications: military and civil aircraft, military vehicles, naval vessels and general hydraulics. 98

75-30280 Howmet Limited
Exeter Casting
Kestrel Way, Exeter, Devon EX2 7LG
☎ (+44 1392) 707 31 ▣ 42863 ⊖ (+44 1392) 41 31 47
♣ £ 40 000 000 (1991) 👥 790
General Manager: S Sanders

Howmet Exeter Alloys
Heron Road, Sowton Industrial Estate,
Exeter, Devon EX2 7LL
☎ (+44 1392) 43 27 51 ▣ 42524
⊖ (+44 1392) 21 17 74
General Manager: Scott Sanders
■ Manufacture of superalloys for the investment casting industry. 98

75-30350 HPC Engineering plc
Victoria Gardens, Victoria Industrial Estate,
Burgess Hill, West Sussex RH15 9RQ
☎ (+44 1444) 24 16 71 ⊖ (+44 1444) 24 75 87
♧ £ 7,300,000 (1995/96) ♣ £ 22,900,000 (1995/96)
👥 220 (1997) ⌂ 10,000 sq m (1996)
Managing Director: Sir Gawaine Baillie
Sales Director Precision Engineering and Hydraulic Divisions: Jim Hunter
Sales Director Compressor Division: Geoff Houlgate
Technical Sales Engineer: Dennis Taylor
Technical Manager Compressor Division: David Lawrence
Chief Engineer Precision Engineering and Hydraulic Divisions: Ken Holt
Secretary and Personal Assistant: Miss Carole Taylor
■ Subcontracting precision engineers (CNC machining). Manufacturer of hydraulic valves. Supply and servicing of compressed air systems. 98

75-30360 H.P.P. (UK) Ltd
82 Cliveland St, Birmingham B19 3SN
☎ (+44 121) 359 64 65 ⊖ (+44 121) 359 07 46
♣ £ 1 000 000 👥 13
Director: M. Barford
■ Helmets, shields, leg & arm guards, handcuffs, batons, body armour, high-power torches. Leather pouches, belts. 97

75-30480 Hughes Microelectronics Europe Ltd
(a subsidiary of Hughes Aircraft Co, USA)
Queensway Industrial Estate, Fullerton Road,
Glenrothes, Fife KY7 5PY
☎ (+44 1592) 75 43 11 ▣ 72148
⊖ (+44 1592) 61 01 86
♣ £38,000,000 👥 630
Managing Director: Dr Paul Davies
Chief Executive Officer: A Cook
Marketing Manager: Stephen Wood
Sales Manager: Kevin Swann
■ Design and manufacture of electronics, semiconductors, interconnectors and microcircuits. 97

75-30580 Humber Fabrications (Hull) Ltd
(† Humber Inflatables Ltd)
246 Wincolmlee, Hull HU2 0PZ
☎ (+44 1482) 22 61 00 ⊖ (+44 1482) 21 58 84
♣ £ 300 000 👥 10
Director: F. Roffee
■ Inflatable & rigid hull inflatable boats for military use. 98

75-30700 Hunting Engineering Ltd
(a subsidiary of Hunting Defence Ltd)
Reddings Wood, Ampthill, Bedfordshire MK45 2HD
☎ (+44 1525) 84 10 00 ▣ 82105 huneng g
⊖ (+44 1525) 40 58 61
♣ £ 61,000,000 (1996) ⌂ 25,000 sq m
Chairman: T.L. Grievson (CBE)
Managing Dir: B.E. Hibbert
Commercial Dir: G.A. Wells
Financial Dir: J.W.J. Jewell
Manager, Marketing and Public Relations: C I P Martin
■ Advanced engineering techniques & systems management of defence projects from conception, through design & development, to production & customer support for the UK & overseas. Engaged on future advanced guided & unguided weapons, launchers, warheads & submunitions including runway cratering munition, hard target penetrators specialists in current & future threat analysis including delivery systems, countermeasures to defence systems & provision of light trial payloads. → 80

Sister companies:
Hunting-BRAE Ltd
Halmatic Ltd
Irvin Group Ltd

Communications Division
Reddings Wood, Ampthill, Bedfordshire MK45 2HD
☎ (+44 1525) 84 10 00 ⊖ (+44 1525) 40 58 61
⌂ 500 sq m

Contact Reference

■ Integrated information distribution systems, installations for military tactical CIS, staff & battle trainers, mobile power systems, aircraft support equipment & electromagnetic assessment. [98]

75-30770 Husky Computers Limited
(a subs. of Peek plc)
Eden Rd, Walsgrave Triangle Business Park,
Coventry, Warwicks CV2 2TB
☎ (+44 1203) 60 40 40 ⊖ (+44 1203) 60 30 60
👥 300+
Managing Dir: A.J. Faulkner
Marketing Mgr: Keith Gallagher
Marketing Services Co-ordinator: Ms Christine Smith
■ Microprocessor technology applications. → 80 [97]

75-30785 Hydro-Bond Engineering Ltd
Hydro House, Woodside Road,
Bridge of Don, Aberdeen AB23 8EF
☎ (+44 1224) 82 29 96 ⊠ 739454
⊖ (+44 1224) 82 51 42
♧ £ 100,000 ♣ £ 2,200,000 (1996) 👥 51
🏭 12,000 sq ft
Managing Director: Douglas J Whyte
Sales Manager: Ian Temple
Chief Engineer: Bob Holmes
Production Manager: Mark Hewdey
■ Manufacture of underwater electrical connectors. Pressure resistant underwater electrical connectors & penetrators. [98]

75-30788 Hydro Marine Systems
(a div. of Pressure Products Group Ltd.)
Westhill Industrial Estate, Pressure Products House,
Westhill, Aberdeen AB32 6TQ
☎ (+44 1224) 74 01 45; 84 18 41
⊖ (+44 1224) 74 01 72
👥 6
Divisional Mgr: Gordon Blakemore
☎ (+44 1224) 84 17 65
Welding Supervisor: Martin Liddle
Marketing/Operations: Kevin Walker
☎ (+44 1224) 84 17 36
■ Design & manufacture portable friction stud welding machines, capable of use underwater & in explosive atmospheres. [97]

75-30790 Hydrocable Systems
Unit 2A, Woodside Road,
Bridge of Don, Aberdeen AB23 8EF
☎ (+44 1224) 70 66 11 ⊖ (+44 1224) 70 60 06
👥 26 (1997) 🏭 500 m2
Chairman: D J Whyte
Managing Director: Ms Irene Buxton
Technical Sales Manager: Ian Temple
■ Engineers specialising in design and manufacture of underwater and harsh environment cables. [98]

75-30795 Hydrovision Ltd
Howemoss Avenue, Kirkhill Industrial Estate,
Dyce, Aberdeen AB2 0GP
☎ (+44 1224) 77 21 50 [E] rovs@hydrovision.co.uk
[W] http://hydrovis.demon.co.uk ⊖ (+44 1224) 77 21 66
Managing Director: C J Tarmey
■ Manufacturer of underwater remotely operated vehicles. [98]

75-30900 The Hymatic Engineering Company Ltd
(a Cobham plc company)
Burnt Meadow Road, North Moons Moat,
Redditch, Worcestershire B98 9HJ
☎ (+44 1527) 649 31 ⊖ (+44 1527) 59 91 17
👥 375 🏭 15 acre
Chairman: G F Page
Managing Director: A J Jan-Janin
Director: A G Irwin
Financial Director, Company Secretary: W J Weaver
Manufacturing Director: R E Twine
■ Designs, develops and manufactures cryogenic cooling systems and products for a range of applications including infrared detection, thermal imaging, superconductivity and millimetric radar for the defence, industrial and scientific markets. Provider of high pressure stored gas systems for mechanical actuation, inflation and underwater depth keeping devices. The company's fluid control technologies address a range of airborne and ground vehicle requirements associated with cabin and avionic conditioning, pressurisation and cooling of electrical equipment, fuel pressurisation, air control valves and solenoids. [98]

75-30920 Hypertac Ltd
(a Smiths Industries Ltd Company)
36/38 Waterloo Road, London NW2 7UH
☎ (+44 181) 450 80 33 ⊠ 928071
⊖ (+44 181) 208 34 55
👥 150
Managing Director: Mrs T Fagnant
Operations Director: P Oakland
Sales and Marketing Manager: R Muir
■ Design & manufacture of high integrity electronic connectors: two-part PCB, rack & panel, circular, microminiature. Qualified to BS 9000 & 9525, ESA/SCC 3401, Boeing ARINC 600, Mil-C-5015, 26482, 26500, 55302. Unique "Hypertac" low insertion force socket.
Subsidiary Companies:
SI-TAC Connectors, USA
The Aero-Electric Connector, USA [98]

75-30925 Hy-Q International (UK) Ltd
(a subs of Hy-Q International, Australia)
Station Road East, Whittlesford, Cambridge CB2 4NL
☎ (+44 1223) 83 44 44 ⊠ 817196 hyqeng
⊖ (+44 1223) 83 45 89
♧ £ 3 000 000 ♣ £ 4 000 000 (1995) 👥 85
🏭 5 000 sq m
Managing Dir: K. Jackson
Operations/Technical Dir: D. Ford
Marketing Mgr: P.J. Ince
Purchasing Mgr: G. Reeves
■ Manufacturer of quartz frequency control products, crystals & filter oscillators. [97]

75-30950 Hyster Europe Ltd
(a subsidiary of Nacco Materials Handling, Inc)
Berk House, Basing View,
Basingstoke, Hampshire RG21 1HQ
☎ (+44 256) 46 11 71 [W] http://www.hyster.co.uk
⊖ (+44 256) 567 33
Managing Dir: Paolo de Chiara
Sales/Marketing Dir: Terry Foreman
■ Manufacture of fork lift trucks.
Irvine, Scotland
Portland Road, Irvine, Ayrshire KA12 8JG
☎ (+44 1294) 31 56 00 ⊠ 77485 [98]

75-31060 ICL Technical Plastics Ltd
(a subsidiary of ICL Plastics Limited)
Hopehill Road, Glasgow G20 7NF
☎ (+44 141) 332 13 31 ⊖ (+44 141) 332 31 29
♣ £ 2,000,000 👥 40 🏭 70,000 sq m
Managing Director: F F D Stott
Operations Director: C Brown
Sales and Marketing Director: P Neil-McLachlan
Purchasing Manager: Ms Tracey McCafferty
■ Shields, body protection: limb protectors, helmets, batons, overalls & holdalls. [98]

75-31070 Icom (UK) Ltd
Unit 9, Sea Street, Herne Bay, Kent CT6 8LD
☎ (+44 1227) 74 17 41 [E] icomsales@icomuk.co.uk
[W] http:// www.icomuk.co.uk ⊖ (+44 1227) 74 17 42
♣ £ 9,500,000 👥 45 🏭 20,000 sq ft
Managing Director: Phil Hadler
Sales and Marketing Director: Robert Stockley
Marketing Manager: Dale Blackman
■ Hand-portable, mobile, base station VHF/UHF/HF communications equipment. Monitor & surveillance receivers. Secure communications (radio) VHF/UHF (digital encryption & analogue). → 80 → 85 [98]

75-31080 Icore International Ltd
(a subsidiary of Smiths Industries plc)
Leigh Road, Slough, Berkshire SL1 4BB
☎ (+44 1753) 57 41 34 [E] icore@icore.co.uk
⊖ (+44 1753) 82 36 74
👥 150
Managing Dir: B.A. Henderson
Sales & Marketing Dir: Mrs C. Parkes
Technical Dir: Dr G.N. Burland
UK Sales Mgr: D.A. MacLean
Hose Product Mgr: B.C. Miller
■ Manufacture electrical interface & interconnection systems for defence, aerospace, power generation & communications. Re-enterable flexible harness systems for all forms of cable protection, EMI/EMC shielding & screening. Fliteline flexible pressure hoses.
Associated companies:
Icore International Inc → USA 75
Icore International GmbH → Germany 75
Icore International → France 75 [98]

75-31100 IDM Electronics Ltd
(a Kaydon Company)
30 Suttons Park Avenue, Suttons Business Park,
Reading, Berkshire RG6 1AW
☎ (+44 118) 966 60 44 [E] idmelect@msn.com
⊖ (+44 118) 966 65 24
♧ £ 135,000 ♣ £ 4,000,000 👥 75 🏭 26,000 sq ft
Managing Dir: C. Haggett
Sales Mgr: S. Hilton
■ Design & manufacture of precision slip ring assemblies and fibre optic rotary joints. Miniature slip ring assemblies for inertial guidance systems. [98]

75-31155 IMI Marston Ltd
(† Marston Palmer Ltd)
(a subsidiary of IMI plc)
Wobaston Road, Fordhouses, Wolverhampton WV10 6QJ
☎ (+44 1902) 39 77 77 ✉ MAREX Wolverhampton
[E] aht@attmail.com ⊠ 337411 marstn g
⊖ (+44 1902) 39 89 79
♣ £ 30,000,000 (1997) 👥 500 (1997)
Managing Director: K R Collie
Aerospace Director: S L Hiscox
Sales and Marketing Manager: David N Gosling
Purchasing Manager: R C Cogzell
■ Heat exchangers for engine and aircraft systems, including cooling devices for electronics equipment. Flexible hoses, pipe couplings, electrical harnesses and conduits. Pneumatic de-icers for aircraft leading edges. Metallic ducts and ducting systems. Non-metallic flexible joints and couplings. [98]

75-31158 IMI Norgren Ltd
(† Norgren Martonair Ltd)
(a subsidiary of IMI plc)
Eastern Avenue, PO Box 22,
Lichfield, Staffordshire WS13 6SB
☎ (+44 1543) 41 43 33 [W] http://www.norgren.com
⊠ 338555 normar g ⊖ (+44 1543) 26 80 52
♣ £ 64,950,000 👥 1,400
Managing Director: M I A.N. Welburn
Sales Director: Keith Stockdale
Communications Manager: R J Lawson
■ Development and manufacture of pneumatic control and automation components. [98]

75-31260 Inco Alloys Ltd
(a subsidiary of Inco Limited, Canada)
Wiggin Works, Holmer Road, Hereford HR4 9SL
☎ (+44 1432) 38 22 00 ⊠ 35101 wiggfd g
⊖ (+44 1432) 26 40 30
♣ £ 118,000,000 (1996) 👥 1,050 (1996)
Managing Director: R D Causon
Sales and Marketing Director: G B Nairn
Marketing Manager: I Hedley
Purchasing Manager: D C Mellor
Sales Promotion and Advertising Manager: K R Owens
■ Manufacturer of high-performance heat & corrosion-resisting nickel based superalloys (Nimonic, Inconel, Incoloy & Inco) for gas turbine engineering. Products include billet, bar, sheet, strip, tube, wire & extrusions. New products include nickel & aluminium alloys made by mechanical alloying.
Subsidiaries:
Inco Alloys International Ltd, France → 75
Inco Alloys International Ltd, Germany FR [98]

75-31300 Industrial Acoustics Company Ltd (IAC)
(a subsidiary of Industrial Acoustics Company Inc, USA)
IAC House, Moorside Road,
Winchester, Hampshire SO23 7US

☎(+44 1962) 87 30 00 ✉ INDCOUSTIC ⊠ 47117
⊖ (+44 1962) 87 31 11
♣ £ 20,000,000 (1996) ₦ 200 (1997) ⌂ 10,000 m2
Managing Director: George Dawson
Operations Director: Dave Grange
Pdurchasing Manager: Gerald Keenan
Marketing Dir: Simon L. White
Press Officer: Simon White
■ Aviation division: design and manufacture of fully automated air-cooled engine test facilities and ground run-up systems. Project management turnkey capability from initial specifications to completion.
Gas turbine and special product division: design and manufacture of land-based and marine gas turbine noise control systems.
ISO9001 accredited. [98]

75-31430 Information Security Systems Ltd
Windsor House, Abbey Barn Road,
High Wycombe, Bucks HP11 1XG
☎(+44 1494) 46 59 58 [E] infosecsys@btinternet.com
⊖ (+44 1494) 46 48 38
♣ £ 900,000 (1997) ₦ 9 (1997) ⌂ 450 m2
Managing Dir: B.E. Dunk
Director: C.J. Yorke
Director: A.S. Miles
Director: A.C. Twitchen
■ ISS is approved to ISO9001 Quality standards. Design, development & manufacture of Tempest PC & peripheral products. Systems integration. Military message handling systems. → 80 [98]

75-31450 Ingersoll-Dresser Pumps (UK) Ltd
(† Worthington Simpson Operations, Dresser Pump Division)
(a division of Ingersoll-Dresser Pump Co, USA)
PO Box 17, Newark, Nottinghamshire NG24 3EN
☎(+44 1636) 49 46 00 ⊠ 37451
⊖ (+44 1636) 70 59 91
₦ 440 (1996) ⌂ 25,000 sq m
Managing Director, UK Operations: John Edmunds
Sales/Marketing Director: A Williams
MoD Department Manager: Ernest J White
Marketing Manager: Ian Peel
Purchasing Manager: S Etheridge
Advertising Manager: Michael Ball
Personnel Manager: Richard Palmer
■ Pumps for naval vessels; services for which these are supplied include bilge/ballast, sea water circulating, fire & cargo transfer. → 80 [98]

75-31500 J T Inglis & Sons Ltd
Riverside Works, Dundee DD1 3LU
☎(+44 1382) 46 21 31 [E] inglis@taynet.co.uk ⊠ 76439
⊖ (+44 1382) 46 28 46
♣ £ 2,500,000 (1996) ₦ 35 (1996)
Managing Director: K G Andrew
Operations Director: G Carroll
■ Weather protective canvas. Waterbags, tarpaulins, vehicle covers and shelters. Ranger lightweight rapidly deployable multi-purpose military shelter, transparent to radar. Military tents. [98]

75-31530 Insight Vision Systems Ltd
Unit 6, Merebrook Ind.Units, Hanley Rd,
Malvern, Worcs WR13 6NP
☎(+44 1684) 31 00 01 ⊠ 334480 insigt g
⊖ (+44 1684) 31 05 10
♣ £ 2 000 000 (1996) ₦ 20 ⌂ 9 000 sq ft
Managing Dir: R.M. Harper
Operations & Technical Dir: Tim Norman
Sales Mgr: R.C. Saunders
Purchasing Mgr: David Loader
■ Design, manufacture and supply of specialist CCTV and thermal imaging equipment. [97]

75-31605 Installocean Ltd
Rack Close Road, Alton, Hants GU34 1JJ
☎(+44 1420) 54 14 48 ⊖ (+44 1420) 54 14 06
Managing Dir: Alan Greig
Operations & Technical Dir: Ian Bellamy
Sales Mgr: M. Martin
Survey Mgr: Nicholas Philp
■ Provision of scientific expertise & monitoring systems in the fields of oceanographic & environmental data acquisition for offshore, inshore & estuarine applications. [97]

75-31630 Instro Precision Ltd (IPL)
Hornet Close, Pysons Road Industrial Estate,
Broadstairs, Kent CT10 2YD
☎(+44 1843) 60 44 55 ⊖ (+44 1843) 86 10 32
♣ US$ 5 000 000 ₦ 60
Managing Director: John Morcom
Marketing Manager: Jason Morley-Smith
■ Design & manufacture tripods, goniometers & tracking heads to support night vision systems, laser rangefinders, laser target markers & thermal imagers, radar, CCTV for artillery & mortar fire control surveillance systems for high risk areas. Platforms for autotracking high precision systems with multi payloads, control consoles. [98]

75-31660 Insumat Ltd
(a subsidiary of Darchem Engineering Ltd)
Eastbrook Road, Eastern Avenue, Gloucester GL4 3DB
☎(+44 1452) 37 72 01 ⊖ (+44 1452) 61 13 33
♣ £ 5,000,000 (1997) ₦ 120 (1997)
General Manager: Kevin Cleaver
Operations Manager: Roger Sage
Sales and Marketing Director: John Padbury
Quotations Manager: Kay Hicks ☎(+44 1452) 31 72 00
■ Thermal insulation & fire protection. Sheet metal fabrication in stainless steel, titanium, nimonic and composites. Air intake and exhaust silencers and exhaust systems for naval vessels. Gun barrel insulation for MBT's. [98]

75-31800 International Combustion Ltd
(a subsidiary of Rolls-Royce Power Engineering Ltd)
Sinfin Lane, Derby DE2 9GJ
☎(+44 1332) 27 11 11 ⊠ 37581 energy g
⊖ (+44 1332) 27 12 34
♣ £ 80,000,000 (1997) ₦ 800 (1997)
Managing Director: Brian Roulston
Marketing Manager: John R Dickson
■ Subcontracting work & nuclear component manufacture. Boilers for direct fired or heat recovery applications. Mechanical erection, repair & maintenance. Metallic vessels & general fabrications. [98]

75-31820 International Procurement Services Ltd
39 Star Street, London W2 1QB
☎(+44 171) 258 37 71
[E] 100145.2747@compuserve.com
⊖ (+44 171) 724 79 25
Managing Dir: G.R. Hall
Sales & Marketing Dir: D.W. Judd
■ Manufacture & supply of surveillance & counter-surveillance equipment. [98]

75-31840 International Radiator Services Ltd
Serck Marston
Serck Services International
(† Serck Services)
(a subs. of BTR plc)
Washford Industrial Estate, Claybrook Drive,
Redditch, Worcs B98 0DZ
☎(+44 1527) 51 01 11 ⊠ 312242 midtlx g
⊖ (+44 1527) 51 02 01
♣ £ 60 000 000 ₦ 950
Managing Dir: C. Prichard
Financial Dir: M.J. Weetman
Sales & Marketing Dir: D.M.A. Birtles
Operations Dir: B. Waters
Advertising & Promotions Mgr: A. Camwell
■ Supply of radiators for all types of vehicle & engine cooling systems, industrial heat exchangers, heater radiators, oil coolers, number plates, clutches, brake shoes, air brakes, power steering, water pumps, turbochargers & road springs.
Factory
Sandhill Drive, Narborough, Leics LE9 5DF
☎(+44 116) 275 07 77 ⊖ (+44 116) 275 08 01 [97]

75-31860 International Rectifier Co.(GB) Ltd
(a subs. of International Rectifier Corp. USA)
Holland Rd, Hurst Green, Oxted, Surrey RH8 9BB
☎(+44 1883) 73 20 20 ⊠ 95219
⊖ (+44 1883) 73 34 10
♤ £ 8 400 000 (1995) ♣ £ 105 000 000 (1995) ₦ 500
Managing Dir: R. Mueller
Sales & Marketing Dir: N. Thornburn
Technical Mgr: George Pearson
Purchasing Mgr: R. Baychuk
Publicity Co-ordinator: B. Donegan
■ Production of power MOSFETs and IGBTs in discrete format. Schottky diodes in discrete format. Co-packaged IGBT's with ultra fast diodes in discrete format. → 90 [97]

75-31945 Interspiro Ltd
(part of Comasec Group)
Halesfield 19, Telford, Shropshire TF7 4QT
☎(+44 1952) 58 28 22 ⊖ (+44 1952) 58 18 89
♤ £ 150 000 ♣ £ 2 300 000 ₦ 20 ⌂ 12 000 sq ft
Managing Director: J Thorp
Military Business Manager: Anthony A Bianco
Commercial Office Manager: Mrs Kay Billinge
Company Secretary: Mrs Tracey Evans
Purchasing Manager: Wayne Shillam
■ Sale and manufacture of respiratory protective equipment, breathing apparatus, diving sets and training systems, BA, CBWD, and diving systems and HAZMAT supplier to Ministry of Defence (UK), US Air Force, US Navy, other NATO and WEU members and many other armed forces. [98]

75-31946 Interstate Technical Products
Great Western Rd, Martock Industrial Estate,
Martock, Somerset TA12 6DT
☎(+44 1935) 82 32 01 ⊖ (+44 1935) 82 62 08
♣ £ 200 000 ₦ 4
Manager: R.S. Lawson
■ Custom loading of ammunition for pistols, rifles, shotguns and weapons to 40mm. Shotgun cartridges for police & military use including riot control. Advisory service for shotgun cartridge production plant. Manufacture of ceremonial blanks, 3PDR, 25PDR, 105m pack. → 80 [97]

75-31948 Intertechnique (Aerospace) Ltd
(a subsidiary of Intertechnique SA, France)
Fishponds Road, Wokingham, Berkshire RG11 2QJ
☎(+44 1734) 78 97 89 ⊠ 848225 inaero g
⊖ (+44 1734) 79 45 11
♤ £ 485,000 ♣ £ 5,540,000 (1996) ₦ 15 ⌂ 4,000 sq ft
Managing Dir: C.I. Campbell
Commercial Mgr: D. Brewin
Purchasing Mgr: D. Smith
■ Fuel gauging, booster & transfer pumps, valves, flowmeters, level sensors, pipe couplings & connectors, AC & DC current sensors, fuel pump assembly, test & overhaul, oxygen systems, electrical power supply switchgear, cockpit switches & indicators & hydraulic systems. → 80
® France: ECE, EROS, Faure Herman, Intertechnique SA [98]

75-32000 ION Track Instruments
(† Ai Cambridge Ltd)
London Road, Pampisford, Cambridgeshire CB2 4EF
☎(+44 1223) 49 55 10 ⊖ (+44 1223) 83 01 33
♣ £ 5 000 000 ₦ 90
Managing Dir Security Div.: A. Jenkins
Sales & Marketing Mgr: S. Williams
Operations Mgr: Richard Drage
Sales & Admin Mgr: V. Linsdell
■ Explosive detectors & explosive ordnance disposal equipment. Drug detectors. [98]

75-32007 Ionix High Performance Wiring Systems
(a div. of Volex Group plc)
Festival Way, Stoke-on-Trent, Staffs. ST1 5TF
☎(+44 1782) 27 45 45 ⊖ (+44 1782) 21 29 92
♣ £ 106 097 000 ₦ 3 629
General Mgr: M.R. Mitchell
Operations/Technical Mgr: R.A. Davies
Purchasing Mgr: M. Flavell
■ Design & manufacture of harnesses & woven ribbon cablelooms. Encapsulation & small moulding facilities. Fully automatic test equipment for cable & electronic wiring systems. [97]

75-32040 Ipeco Europe
(a division of Ipeco Aerospace Ltd)
(a subsidiary of Ipeco Holdings plc)
Aviation Way, Southend Airport,
Southend, Essex SS2 6UN
☎(+44 1702) 20 93 02 ⊠ 9312111463 pa
⊖ (+44 1702) 54 07 82

Contact Reference

👥 95
Managing Director: I Campbell
Financial Director: C Backhouse
Technical Director: D Scott
Purchasing Manager: D Marsh
Personal Assistant to Managing Director: Ms Linda Negrone
■ Design and manufacture of aircrew seats. [98]

75-32050 Irvin Aerospace Limited
(a subs. of Irvin Group Ltd)
Icknield Way, Letchworth, Hertfordshire SG6 1EU
☎(+44 1462) 48 20 00 ✉IRVIN ⊖ (+44 1462) 48 20 07
♣ £ 11 000 000 👥 300 🏭 150 000 sq ft
Chairman: Roger J. Whysall
Managing Dir: M.A. Warland
Engineering Dir: Dr M.G. Woollard
Marketing Mgr: G.A. Linaker
Public Relations: Mrs. J.C. Adams
■ Passive decoy & target systems. Parachutes for personnel, cargo, weapon & aircraft retardation. Recovery systems for missiles, space vehicles & cargo delivery systems. Safety harnesses & air drop sequence control devices. Parachute packing & servicing. [98]

75-32060 Iskra Agency Ltd
(a subs. of Fotona, Slovenia)
Unit A6, Redlands, Coulsdon, Surrey CR5 2HT
☎(+44 181) 668 71 41 ⊖ (+44 181) 668 31 08
♧ £ 50 000 ♣ £ 1 084 000 👥 6 🏭 10 000 sq ft
Managing Dir: V. Sodin
■ Laser range finders & thermal imaging sights for infantry, artillery & EFCS. [97]

75-32170 ITT Cannon RF Products
(† ITT Cannon/Sealectro)
(a division of ITT Industries Inc, USA)
Jays Close, Viables Estate,
Basingstoke, Hampshire RG22 4BW
☎(+44 1256) 31 12 00 ⊖ (+44 1256) 32 33 56
♣ £ 14,000,000 (1995) 👥 300 (1997) 🏭 53,000 sq ft
Managing Director: K Wheelhouse
Commercial Director: Jim Bellis
Publicity Manager: Shane Dennis
Product Manager: Graham Cunningham
■ Manufacture and sale of electronic components including RF coaxial connectors SMA, SMB, SMC, SSMB, SSMC and K series, connector/cable assemblies, microwave components, flexible and semi-rigid cables. SSIS micro miniature microwave connectors. → 90
® France: ITT Cannon
Germany: ITT Cannon GmbH
Japan: ITT Cannon Ltd
USA: ITT Cannon RF Products [98]

75-32230 Iver Precision Engineers Ltd
Ridgeway Trading Estate, Thorney Lane,
Iver, Buckinghamshire SL0 9HZ
☎(+44 1753) 65 18 18 ⊖ (+44 1753) 63 00 12
♣ £ 2,500,000 (1996) 👥 60 (1996) 🏭 1,260 m2
Managing Director: Kevin Smith
Technical Director: Simon Waterhouse
Operations Director: Martyn Snoad
Sales and Marketing Director: Alan Cook
Purchasing Manager: Bradley Smith
■ Precision machinists & assemblers for the aircraft, aerospace, automotive, defence & computer industries. [98]

75-32296 Jason's Cradle
Land and Marine Products Ltd
Kingsley House, Old Park Farm, Kingsley,
Bordon, Hampshire GU35 9LU
☎(+44 1420) 47 44 84 ⊖ (+44 1420) 48 90 02
Sales and Marketing Director: Jason Watling
Company Secretary: S L Matthews
■ Manufacture 'Jason's Cradle', a man overboard life saving appliance. Suitable for use on all craft including lifeboats and rescue boats. [98]

75-32297 JC Bamford Excavators Ltd (JCB)
Rocester, Staffordshire ST14 5JP
☎(+44 1889) 59 03 12 Ⓦ http://www.jcb.co.uk
▣ 36372 ⊖ (+44 1889) 59 05 88
♣ £ 704,600,000 (1995) 👥 3,000 (1997)
Chairman: Sir Anthony Bamford
Chief Executive Officer: M Coyne
Corporate Marketing Director: Neil Allen
Purchasing Director: Les Green
Press Officer: Nigel Chell
■ Military engineering equipment: backhoe loaders, crawler & wheeled excavators, articulated wheeled tractors. Military materials handling equipment: rough terrain forklifts, loaded forward reach handlers.
Excavating machines for a wide range of combat engineer duties. Hydraulic power packs. Materials handling machines for all military re-handling applications. Supplier to the Royal Navy, Royal Air Force, British Army & over 30 defence forces overseas. [98]

75-32300 Mediworld Ltd
(† Pan Servico (Surgical) Ltd)
436 Streatham High Rd, London SW16 1DA
☎(+44 181) 764 18 06 ▣ 917134 pansco g
⊖ (+44 181) 679 24 89
👥 15 🏭 6 000 sq ft
Marketing Dir: M. Shez Khan
Sales: M.M. Quinn
Admin Mgr: F. Williams
Advertising Mgr: V. Mardy
■ Manufacture & supply of medical & surgical equipment supplies. → 90 [97]

75-32330 Jarvis Manufacturing Co.
116 Seaside, Eastbourne, East Sussex BN22 7QP
☎(+44 1323) 41 19 93 ⊖ (+44 1323) 64 96 62
♣ £ 1,400,000 (1996) 👥 60 🏭 12,000 sq ft
Partner: R I Hawgood
Partner: M J Harris
Technical Administrator: W G Cobbold
■ Manufacture of ancillary rescue & life-saving equipment. Specialists in all leather, nylon & PVC goods. Tank crew helmets. Anti-static bags, webbing, pouches & covers. [98]

75-32370 Jasmin Simtec Ltd
(† Jasmin plc)
Sellers Wood Drive, Bulwell, Nottingham NG6 8UX
☎(+44 115) 927 37 41
Ⓔ jasmin@jasminuk.demon.co.uk
⊖ (+44 115) 927 86 14
♧ £ 5,000,000 ♣ £ 10,030,000 (1995) 👥 120 (1995)
Managing Director: R A J Plant
Operations Director: D H Andrews
Sales/Marketing Director: J R Michaelis
Purchasing Manager: C Morris
■ Specialists in explosive, chemical and radiation detection systems, part task trainers, emulation and simulation systems, build to print manufacture and design of control and information systems. → 80 [98]

75-32400 Jaybeam Ltd
(a subsidiary of Davey Bickford Smith & Cie, France)
Rutherford Drive, Park Farm South,
Wellingborough, Northants NN8 6AX
☎(+44 1933) 40 84 08 ⊖ (+44 1933) 40 84 04
♧ £ 1,000,000 (1996) ♣ £ 6,263,000 (1996)
👥 120 (1996) 🏭 4,500 sq m
Managing Dir: M.J. Wilkins
Technical Dir: Dr K.A.K. Hanna
Sales & Marketing Dir: J. Southwood
Company Secretary: C. Clarke
Purchasing Mgr: P. Bray
■ Manufacture & supply of antenna systems in the range 150 KHz to 2 GHz. Customer advice & installation service. [98]

75-32480 JDR Cable Systems Ltd
(† Jacques Cable Systems Ltd)
(a subs. of Bromsgrove Industries plc)
11 Padnal, Littleport, Ely, Cambs CB6 1NS
☎(+44 1353) 86 00 22 ⊖ (+44 1353) 86 13 88
👥 150 (1995) 🏭 10 000 sq m (1995)
Managing Dir: Neil McAdam
Operation & Technical Dir: Nick O'Hear
Sales & Marketing Dir: Glen Bloomfield
Purchasing Mgr: Barry Hills
Defence Sales Mgr: Ray Ward
■ Design & manufacture of custom built underwater cables, terminations & associated equipment. UK agents for the Jupiter & Souriau range of marine connectors. Military products include both dynamic strain cables for a variety of tow applications (satellite tethers/aerial target) & static cables for use in ranges and manufacture an extensive range of diving umbilicals. Cables can include optical fibres, electrical conductors, steel or non-metallic strength members such as Kevlar or Spectra. Complete cable systems including shipboard handling, winch & inter-connects & cable harnesses incorporating polyethylene moulding. [97]

75-32490 Jeckells Sails Ltd
(an associate company of Plain Sailing Ltd)
Station Road, Wroxham, Norfolk NR12 8UT
☎(+44 1603) 78 22 23 ⊖ (+44 1603) 78 40 23
♣ £ 600,000 👥 9
Managing Director: Chris Jeckells
Sales and Marketing Director: Will Knights
Purchasing Manager: Mel Jones [98]

75-32500 Jessop Saville Ltd
(a subs. of Sheffield Forgemasters Ltd)
Hawke Street, Sheffield S9 2SU
☎(+44 114) 256 05 12 ▣ 547899 jessav
⊖ (+44 114) 243 72 67
♣ £ 8 000 000
Managing Dir: Keith Comaz
■ Machined & forged steel components. [97]

75-32520 Jetway Associates Ltd
(† Berwyn Engineering Ltd)
Euridge Works, Thickwood, Colerne,
Nr Chippenham, Wiltshire SN14 8BG
☎(+44 1225) 74 23 01 ⊖ (+44 1225) 74 34 57
♣ £ 500,000 (1996) 👥 20 (1997) 🏭 15,000 m2
Managing Dir: Mark Read
Sales & Marketing: David Brierley
Purchasing Mgr: Chris Phillips
■ Marine safety equipment, rocket launchers, machine gun mounts, ovens, furniture, lockers. Suppliers of HRUs to NATO. Decoy launchers. → 80 → 85 [98]

75-32815 Johnson Matthey Plc Noble Metals Business Unit
Orchard Rd, Royston, Herts SG8 5HE
☎(+44 1763) 25 30 00 ▣ 817351 jmchq g
⊖ (+44 1763) 25 33 13
♣ £ 20 000 000 (1996) 👥 170 🏭 4 000 sq m
Sales & Marketing Dir: T.K. Gillinder
Manufacturing Dir: W.G. Hall
General Mgr: W.F. Sandford
Sales & Marketing Mgr: P.N. Duncan
Purchasing Mgr: H.C. Jarman
■ Manufacture of precious metal thermocouples, alpha platinum for PRTs & wire scrim for use as radar absorbent materials. Potential future involvement in thrust chambers for satellite propulsion. [97]

75-32895 Jotron (UK) Ltd
(a subsidiary of Jotron Electronics AS)
Enterprise Court, Crowhall Road,
Cramlington, Northumberland NE23 9LZ
Nelson Industrial Estate
☎(+44 1670) 71 20 00 ▣ 53637 radsch g
⊖ (+44 1670) 59 02 65
♧ £ 100,000 (1997) ♣ £ 2,000,000 (1997) 👥 10
🏭 4,000 m2 (1997)
Managing Director: I A Ross
Operations Manager: I A Forbes
Sales Manager: J Dixon
■ Manufacture of ground-to-air communications equipment in the VHF & UHF radio bands. Also manufacture marine emergency beacons & lights. [98]

75-33210 Kango Ltd
12 Flag Business Exchange, Vicarage Farm Rd,
Peterborough PE1 5TX
☎(+44 1733) 86 62 00 ⊖ (+44 1733) 86 62 66
Managing Dir: Gösta Torssell
Operations Mgr: Victor Lilly
UK Sales & Marketing Mgr: Bernard Smith
Purchasing Mgr: George Cameron
Marketing Mgr: Conor Riordan
Marketing Co-ordinator: Miss Julie Jarvis [97]

75-33220 Karcher (UK) Ltd
(a subsidiary of Alfred Karcher GmbH & Co, Germany)
Karcher House, Beaumont Road,
Banbury, Oxon OX16 7TB
☎(+44 1295) 26 75 11 ⊠ 051837783
⊖ (+44 1295) 26 64 36
👥 90
Managing Director: G J Jones
Operations Manager: R Higgs
■ Manufacture of field kitchens and decontamination equipment. Aircraft and vehicle cleaning equipment. High pressure washing systems.
Representing:
Alfred Karcher GmbH & Co [98]

75-33240 Karl International Ltd
(a division of Karl Holdings Ltd)
Springfarm Industrial Estate,
Antrim, Northern Ireland BT41 4NZ
☎(+44 1849) 46 42 31 ⊖ (+44 1849) 46 71 09
♣ £ 32,000,000 (1996) 👥 515 (1996)
Managing Director: C G Blackbourne
Technical Director: Ronnie Stephenson
Marketing Manager: Ms Gayle Davidson
Purchasing Manager: Keith Wood
■ Defensive building systems. Hardened structures, border security complexes. [98]

75-33245 Kawasaki Heavy Industries (UK) Ltd
3rd Floor Bracton House, 34-36 High Holborn,
London WC1V 6AA
☎(+44 171) 404 45 55 ⊖ (+44 171) 404 45 56 [98]

75-33250 Kawasaki Motors (UK) Limited
748-749 Deal Avenue, Trading Estate, Slough, Berkshire [98]

75-33252 Kawasaki Motors (UK) Ltd
(a subsidiary of Kawasaki Heavy Industries Ltd, Japan)
1 Dukes Meadow, Millboard Road,
Bourne End, Buckinghamshire SL8 5XF
☎(+44 1628) 85 10 00 ⊖ (+44 1628) 85 09 60 [98]

75-33255 Kawasaki Precision Machinery (UK) Ltd
(a subsidiary of Kawasaki Heavy Industries Ltd, Japan)
Ernesettle Lane, Ernesettle, Plymouth, Devon PL5 2SA
☎(+44 1752) 36 43 94 ⊖ (+44 1752) 36 48 16 [98]

75-33260 Kawasaki Robotics (UK) Ltd
(a subsidiary of Kawasaki Heavy Industries Ltd, Japan)
Greengate, Middleton, Manchester M24 1SA
☎(+44 161) 956 54 00 ⊖ (+44 161) 956 54 04 [98]

75-33300 KCI Carruthers Ltd
(† Kone Carruthers Ltd)
(a subsidiary of KCI Konecrantes International Corp, Finland)
Peel Park Place, East Kilbride, Glasgow G74 5LR
☎(+44 1355) 22 05 91 ⊖ (+44 1355) 26 36 54
♣ £ 10,000,000 👥 100 🏭 20,000 m2
Managing Director: G W Adie
Operations Manager: W Cameron
Sales and Marketing Manager: H Abbott
Purchasing Manager: R W Wallace
■ Manufacture overhead travelling cranes. [98]

75-33370 Kearsley Airways Ltd
Romeera House, Stansted Airport,
Stansted, Essex CM24 1QL
☎(+44 1279) 87 10 00 ⊠ 81465 kalair g
⊖ (+44 1279) 87 11 87
♧ £ 100,000 ♣ £ 6,000,000 (1996) 👥 100 (1997)
🏭 45,000 sq ft
Chairman: Madhoo Mehta
General Manager: S N Gurney
Sales Manager: B Rowell
Quality Manager: M R V Bowman
Instrument Manager: M N Harris
■ Repair & overhaul aircraft rotable components, undercarriages, wheels, brakes, hydraulic pumps & motors, electrical equipment, fuel booster pumps, landing lights, beacons/strobe lights, gearboxes, driveshafts, instruments, non-destructive testing, flying control, boosters, oxygen regulators, ballscrews & actuators.
Manufacture Aflo fuel systems. → 80 [98]

75-33400 Kelan Circuits Ltd
(a subsidiary of Advance International Group Ltd)
Wetherby Road, Boroughbridge, North Yorkshire YO5 9UY
☎(+44 1423) 32 11 00 [E] kelan@compuserve.com
⊖ (+44 1423) 32 11 07
♣ £ 3,500,000 (1995) 👥 100 🏭 36,000 sq ft
Managing Director: Mike Regan
Sales and Marketing Director: Ken Byers
■ Printed circuit boards - PTH, non-PTH, multilayer. [98]

75-33453 Kelvin Hughes Ltd
(† QUBIT Kelvin Hughes Ltd)
(a subsidiary of Smiths Industries)
New North Road, Hainault, Ilford, Essex IG6 2UR
☎(+44 181) 500 10 20 ⊠ 896401 kelhue g
⊖ (+44 181) 559 85 22
👥 450
Managing Dir: D. Sheppard
Sales & Marketing Dir: D. Pickles
Operations & Technical Dir: A. Norris
Purchasing Mgr: C. Corris
Marketing Services Mgr: B. Morgan
■ Manufacturers of commercial & naval marine electronic equipment, including high definition surveillance radar, navigation & plotting aids, echo sounders. Suppliers of Admiralty charts, nautical publications & instruments. [98]

75-33455 Kelsey Instruments Ltd
University of Warwick Science Park, Coventry CV4 7EZ
☎(+44 1203) 41 17 88 ⊖ (+44 1203) 41 64 23
♣ £ 2,500,500 (1997) 👥 30 (1997) 🏭 700 m2
Managing Director: D R Smith
Operations Director: Bibb Stuart
■ Computer-based destructive testing systems for static & fatigue test applications in the aerospace, automotive & civil engineering industries. [98]

75-33525 Kembrey Wiring Systems Ltd
(a subsidiary of Chemring Group plc)
Kembrey Park, Swindon, Wiltshire SN2 6BJ
☎(+44 1793) 69 33 61 ⊖ (+44 1793) 61 42 98
♣ £ 5,332,381 (1996) 👥 102 (1997) 🏭 110,000 m2
Managing Director: John J Lynn
Financial Director: P A Rayner
Sales and Marketing Director: Robert Hart
Operations Director: S P Hardy
Commercial Director: R Martin
Purchasing Manager: Eddie Gilmore
■ Wiring assemblies & panels for aerospace & harsh environment applications. Design & distribution of connector & cable accesories. Components for the telecommunications industry. [98]

75-33600 T.W.Kempton (Corporate Clothing) Ltd
47 Cyprus Road, Leicester LE2 8QT
☎(+44 116) 283 21 92 ⊖ (+44 116) 244 06 77
♣ £ 8,350,000 (1996) 👥 265 (1996)
Managing Director: D S Kenyon
■ Uniform knitwear (woollen pullovers). [98]

75-33700 KEC Limited
(† Kern Electrical Components Ltd)
Orpheus House, Calleva Park, Aldermaston,
Reading, Berkshire RG7 8TA
☎(+44 1734) 81 15 71 ⊖ (+44 1734) 81 15 70
♧ £ 100,000 (1996) ♣ £ 630,000 (1996) 👥 9 (1997)
🏭 275 m2 (1997)
Managing Director: David Y Dyson
Administration Director: Cheryl Watson
Purchasing Manager: Tracey Lane
Advertising Manager: David Dyson
■ Electrical connector accessories for EMC applications. Through bulkhead fittings for EMC applications. Cable & enclosure systems specialising in EMI/RFI/EMP & TEMPEST areas. [98]

75-33880 Kidde-Graviner Ltd
(† Graviner Ltd)
(a Williams Holdings Co)
Mathisen Way, Poyle Road, Colnbrook,
Slough, Berkshire SL3 0HB
☎(+44 1753) 68 32 45 ⊖ (+44 1753) 68 51 26
♣ £ 22,000,000 (1996) 👥 200 (1996) 🏭 6,000 sq m
Managing Director: Hugh Carpenter
Sales and Marketing Director: Chris Boyd
Operations/Techical Director: Chris Milburn
Product Assurance Manager: Peter Mannifield
Marketing Analyst: Ms Jean Hood
■ Supplies a full range of fire detection & suppression systems & safety breathing equipment for the protection of military & civil, fixed & rotary wing aircraft & military vehicles.
Fire detection devices range from continuous heat detectors, through point type heat & smoke detectors, to ultra-fast UV & IR optical detectors. Suppression equipment utilises a wide range of environmentally acceptable gaseous & powder-based extinguishants.
Also a specialist in safety & arming technology & high pressure stored energy systems for applications such as guided weapon control surface actuation, infra-red seeker cooling & thermal imagers. The product range includes detection & reporting systems for aircraft. Systems can either be permanently installed on the aircraft or be packaged as portable units.
Registered Office
Pentagon House, Sir Frank Whittle Road, Derby DE2 4XA
☎(+44 1332) 20 20 20 [98]

75-34050 W&J Knox Ltd
(a subsidiary of Cosalt plc)
PO Box 1, Kilbirnie, Ayrshire KA25 7DY
☎(+44 1505) 68 25 11 ⊖ (+44 1505) 68 29 80
👥 250 🏭 120,000 sq ft
Managing Director: J Traynor
Production Director: J Templeton
Marketing Manager: Mrs Julie Macdonald
■ Garnished camouflage nets, twines, cords and knitted fabric. [98]

75-34100 Kongsberg Simrad Ltd
(† Simrad Ltd)
(a subsidiary of Kongsberg Group ASA, Norway)
Headquarters
Campus 1, Aberdeen Science and,
Technology Park, Balgownie Road,
Bridge of Don, Aberdeen AB22 8GT
☎(+44 1224) 22 65 00 ⊖ (+44 1224) 22 65 01
♣ £ 13,000,000 👥 106 (1997) 🏭 30,000 sq ft
Managing Director: David E Ireland
Technical Director: D J Stewart
Quality Assurance Manager: Dave Morland
■ Develop, manufacture and supply ship dynamic positioning systems, seabed mapping echo sounders, sonar systems, advanced acoustic tracking systems, marine instrumentation and underwater television cameras.
Wick
Airport Industrial Estate, Wick, Caithness KW1 4QR
☎(+44 1955) 36 06 ⊠ 75565 ⊖ (+44 1955) 55 11
👥 60 🏭 20 000 sq ft
Manager: Bill Baxter
■ Develop, manufacture and supply: Harsh environment television equipment, electro-optic sensors for remote surveillance, inspection, target tracking and aquisition.
Customer Support Department
Woolmer Way, Bordon, Hampshire GU35 9QE
☎(+44 1420) 48 32 06
Manager: Mike Topp
■ In-service support of propietary products for UK Ministry of Defence, including Sonar 2059 ultra-short baseline acoustic traking system. CDS/PDS management, installation and trials support. [98]

75-34150 Kopex International Ltd
(† Uni-Tubes Ltd)
(a subs. of Smiths Industries plc)
675 Ajax Avenue, Slough, Berkshire SL1 4BG
☎(+44 1753) 53 49 31 ⊠ 848347
⊖ (+44 1753) 69 35 21
♣ £ 8 000 000 👥 25
Managing Director: K Cooper
Technical Director: Ian Gibson
Operations Director: G Richardson
Purchasing Manager: D Hillier
Marketing Manager: Ms Amanda Wood

■ Manufacture of electrical conduits for EMC/EMP protection. [98]

75-34160 K S Paul Products Ltd (KSP)
Eley Estate, London N18 3DB
☎ (+44 181) 345 55 66 ✉ ANTIWEAR ▣ 23362
⊖ (+44 181) 884 32 55
♣ £ 6,000,000 (Group) 👥 60
Chief Executive: M S G Day
Sales Admin Manager: Mrs L G Warman
Business Manager: J Connal
Export Sales Manager: C Pulleu
Purchasing Manager: P McGeough
■ Solid lubricants, molybdenum disulphide, bonded coating, pastes, colloidal dispersions, materials for the lubrication of aircraft mechanisms and instruments. MQAD approved. ISO 9001. 1987 NATO specification. [98]

75-34180 Kulite Sensors Ltd
Kulite House, Stroudley Road, Kingsland Business Park, Basingstoke, Hampshire RG24 8UG
☎ (+44 1256) 46 16 46 ▣ 858434 kulit g
⊖ (+44 1256) 47 95 10
♣ £ 2,166,000 (1996) 👥 15 (1997)
Managing Director: Geoff Bancroft
Company Secretary: Mrs Kim Walters
■ Pressure transducers/transmitters for use on civil and military aircraft. [98]

75-34280 Labtech Ltd
Broadiaxe Industrial Park, Presteigne, Powys LD8 2UF
☎ (+44 544) 26 00 93 ▣ 35203 ⊖ (+44 544) 26 03 10
♣ £ 15 000 000 (1995) 👥 50 🏭 32 000 sq ft
Managing Dir: David Latham
Sales & Marketing Dir: Ms Karen Brassington
Operations & Technical Dir: T. Bateman
Purchasing Mgr: Eric Latham
■ Manufacture circuit boards, microwave PTFE circuits and antennas. [97]

75-34430 Lanarealm Ltd
Unit Kingfisher Place, Clarendon Road, Wood Green, London NV22 6XF
☎ (+44 181) 888 12 22
[E] 101356.3667@compuserve.com
⊖ (+44 181) 888 08 70
👥 50 (1996) 🏭 990 m2
Managing Director; Sales and Marketing: J H Hartley
Managing Director: A M Plight
Operations Director: F R Packer
Purchasing Manager: B Hall
■ Manaufacture of electro-optical systems including night vision goggles, helmet mounted displays, bore stable CCD zoom lens operating in visual and infrared. [98]

75-34665 Lantor (UK) Ltd
(a division of the IITP Group Ltd)
St Helens Road, Bolton, Lancashire BL3 3PR
☎ (+44 1204) 614 26 ⊖ (+44 1204) 617 22
♣ £ 5,000,000 (1997) 👥 65 (1997)
Managing Dir: Roger Tattersall
Sales & Marketing Dir: Richard Keidish
Purchasing Mgr/Technical Mgr NBC: John Retford
Industrial Sales Mgr NBC: John Parker
■ Manufacture activated charcoal non-woven anti-gas fabrics for use in personal protective equipment - NBC suits, casualty bags. Development work with in-house facilities for future end user needs. Manufacture breather felts for carbon composites in the aerospace industry. [98]

75-34870 Law Enforcement International Ltd
PO Box 162, Wembley, Middlesex HA9 0AI
☎ (+44 181) 903 83 05 ⊖ (+44 181) 903 83 02
Director: G S Felton
■ Manufacture silencers, silenced weapon systems and laser gun sights. Design, development and manufacture of "special weapon" systems. Procurement services. → 80 [98]

75-34875 John Lawson Display & Technical Models
(† John Lawson Modelmakers)
Viking House, 21 Viking Way, Metheringham, Lincolnshire LN4 3DW
☎ (+44 1526) 32 16 28 ⊖ (+44 1526) 32 21 40
Managing Director: John R Lawson
Operations and Technical Director: B G Cross
■ Manufacture of engineering, architectural & topographical models for major defence companies. [98]

75-34910 Leafield Engineering Ltd
(A subsidiary of Leafield Holdings Ltd)
Leafield Industrial Estate, Corsham, Wilts SN13 9SS
☎ (+44 1225) 81 07 71 ⊖ (+44 1225) 81 06 14
♣ £ 4,000,000 👥 90
Managing Director: R A Dyke
Purchasing Manager: Mathew Wells
■ Design, development, manufacture and through life support of mechanical, electro-mechanical, electronic, explosive and pyromechanical equipment and devices. [98]

75-35150 Lentern (Aircraft) Ltd
Main Road, Hawkwell, Nr Hockley, Essex SS5 4JJ
☎ (+44 1702) 20 25 53 [E] lenternuk@aol.com
▣ 99427 lenhoc g ⊖ (+44 1702) 20 62 48
♣ £ 21,500,000 (1996) 👥 235 (1996) 🏭 51,000 sq ft
Managing Director: P A Dean
■ Aerospace fasteners & components. Manufacture & stockists British, American, French & German standards & to customers own specifications. BSI & CAA registered & approved. → 85 → 70

Subsidiaries:
Lentern International Inc. → USA 90
A.F. Aerospace Ltd. → UK 75

Heathrow Division
Branch Sales Offices & Stores
1 Viscount Ind. Est., Horton Rd, Colnbrook, Slough SL3 0DF
☎ (+44 1753) 68 61 61 ▣ 847325
⊖ (+44 1753) 68 36 15 [98]

75-35160 Leo Electron Microscopy Ltd
(† Leica Cambridge Instruments Ltd)
Clifton Road, Cambridge CB1 3QH
☎ (+44 1223) 41 41 66 ▣ 81162
⊖ (+44 1223) 41 27 76
👥 120 🏭 79 959 sq ft
Managing Dir: R.B. Kalbag
Operations & Technical Dir: Nicholas Brodrick
Operations & Technical Dir: Dr Henry Waters
Sales & Marketing Dir: David Gower
Sales & Marketing Dir: Tony Edwards
Purchasing Mgr: Tony Evans
Marketing Support Officer: J. Williamson
■ Scientific instrument manufacturers. [97]

75-35350 Leyland & Birmingham Rubber Co. Ltd
(a subs. of BTR plc)
Golden Hill Lane, PO Box 2, Leyland, Lancs PR5 1UB
☎ (+44 1772) 42 14 34 ⊖ (+44 1772) 43 64 01
♣ £ 12 000 000 (1996) 👥 270
Managing Dir: Norman Barker
Business Mgr: Simon Gledhill
■ Manufacture & market a range of NBC (nuclear, biological & chemical) defence equipment & a variety of rubber-based products for military applications. Rubber based products include extruded & moulded components for fighting vehicles and a large selection of refuelling & connecting hoses, inflatable fabrications & spread fabric. High voltage & anti-static rubber matting is produced for a variety of military applications. Natural rubber & silicone detonator plugs. A range of products can be manufactured in low smoke/low toxicity materials. [97]

75-35400 Leyland Trucks Ltd
(† Leyland DAF Ltd)
Headquarters
Croston Road, Leyland, Preston, Lancashire PR5 3LZ
☎ (+44 1772) 62 58 02 ⊖ (+44 1772) 62 58 00
♣ £ 200,000,000 (1994/95) 👥 750 (1997)
Executive Chairman: J P Gilchrist
Sales and Marketing Director: Stuart Heys
☎ (+44 1772) 62 58 01 (+44 1772) 62 58 00 (fax)
International Military Sales Dir: George Maze
☎ (+44 772) 62 58 04 (+44 772) 62 58 00 (fax)
Operations & Technical Dir: K. Howard
Purchasing Mgr: J. Lynch
■ Light, medium and heavy commercial vehicles. Military trucks. [98]

75-35450 LFD Limited
Victory Business Centre, Somers Road North, Portsmouth PO1 1PJ
☎ (+44 1705) 83 82 73 ⊖ (+44 1705) 87 18 54
Managing Dir: Len Fleck
Marketing Dir: Maj. Mick Jarvill
■ Manufacture of covert night vision lighting systems and searchlights. [98]

75-35560 Light Alloy Ltd
Dales Road, Ipswich, Suffolk IP1 4JR
☎ (+44 1473) 74 04 45 ⊖ (+44 1473) 24 00 02
♣ £ 2,000,000 👥 15
Managing Director: W G Eckford
■ High duty aluminium alloy transport containers & shock protected equipment housings for military applications. [98]

75-35650 Lightweight Body Armour Ltd
2 Macadam Close, Drayton Fields, Daventry, Northamptonshire NN11 5BT
☎ (+44 1327) 26 12 82 ✆ (800) 71 61 40
[E] 100704.2211@compuserve.com
⊖ (+44 1327) 26 13 17
♣ £ 2,500,000 👥 24 (1997)
Chairman: A Macgill
Managing Director: R Perks
Sales and Marketing Director: J M Hughes
Ballistics: Tony Dedman
UK and European Sales Manager: C Oliver
■ Design & manufacture of advanced armour products. Protection includes anti-ballistic personal armour, flak, EOD suits, ballistic helmets, riot control equipment, armour for vehicles, helicopters, light vessels & sensitive electronic equipment. Confidential ballistic testing service available to third parties on a contractual basis. → 80 [98]

75-35660 Lightwood & Son Ltd
4/7 Highgate St, Birmingham B12 0XS
☎ (+44 121) 440 30 60 ⊖ (+44 121) 446 41 65
♧ £ 30 000 (1995) ♣ £ 200 000 (1995) 👥 11
🏭 10 000 sq m
Managing Dir: H.M. Johnson
Operations & Technical Dir: G. Turpin
Purchasing Mgr: James Gleeson
Sales Mgr: P. Johnson
■ Cleaning equipment. [97]

75-35800 Lines Instrument Engineers Ltd
43 Honey Hill Road, Kingswood, Bristol BS15 4JB
☎ (+44 117) 967 52 85 ▣ 449752
⊖ (+44 117) 961 18 34
Managing Director: A L Dimond
Sales Director/Secretary: Mrs J M Le Bas
■ Manufacturers of hydrostatic load cells, crane scales. Platform scales, beam weighers, vertical bottle weighers and chlorine drum weighing systems with digital read-out. [98]

75-36065 Linread Northbridge
(† North Bridge Fasteners)
(a division of McKechnie plc)
Viking Road, Aylestone Lane, Wigston, Leicester LE18 2BL
☎ (+44 116) 288 11 92 [E] ln-leicester@compuserve.com
▣ 342471 norfas g ⊖ (+44 116) 281 26 94
♧ £ 5,251,000 (1997) ♣ £ 11,000,000 (1997) 👥 170
🏭 8,200 m2
Managing Director: Edward Brooks
Sales and Marketing Director: A J Baker
Operations Director: Paul Brown
Operations Manager: Manesh Pandya
Commercial Manager: Peter Palij
Sales Manager: Mark Doherty
■ Precision forged fasteners. [98]

75-36068 Lion Profiles
New Road, Sandy, Bedfordshire SG19 1NY
☎ (+44 1767) 69 17 65 ⊖ (+44 1767) 68 33 54
♣ £ 6,000,000 👥 11 🏭 28,000 sq ft
Managing Dir: Charles Fletcher
General Office Administration: Mrs Julie Smith
■ Fully finished precision aluminium extrusions, CNC machining, rolled aluminium. Metal and alloy sections and plates.

RKB Precision Products Ltd
New Road, Sandy, Bedfordshire SG19 1NY

☎ (+44 1767) 68 07 31 ▣ 825417
⊖ (+44 1767) 69 15 75
♣ £ 4,000,000 ⇮ 160
Sales: Paul Rawson
■ Aluminium precision sand casting, high pressure die castings, gravity casting, CNC machining, paint finishing. [98]

75-36070 **Lister-Petter Ltd**
(a member of the BTR Group)
Dursley, Gloucestershire GL11 4HS
☎ (+44 1453) 54 41 41 [E] sales@lister-petter.co.uk
▣ 43261 lpltd g ⊖ (+44 1453) 54 67 32
⇮ 900 (1997)
Managing Dir: Terry Sharpe
Production Dir: Hugh Tipper
Sales Manager: A Hurley
Marketing Mgr: P.R. Starr
■ Manufacturer of a wide range of air & water cooled diesel engines with power outputs upto 112kW (150bhp). Supplies military customers with diesel engines that have proved their worth in combat and other vital roles. [98]

75-36080 **D.J.Litt (Firearms) Ltd**
Unit 3, Maesglas Industrial Estate,
Newport, Gwent NP9 2XF
☎ (+44 1633) 25 00 25; 84 32 52
⊖ (+44 1633) 24 49 78
♣ £ 4 000 000 ⇮ 20
Managing Dir: D.J. Litt
Technical Sales Dir: C.G. Eriksson
Director: A.D. Litt
Public Relations Mgr: M.C. Fewings
■ Small arms of all description, including presentation grades. Range installations. Specialist advice for bodyguard training.
Pyrotechnics. Specialist uniform requirements. Sporting firearms. Optics for small arms. UK distributor of McMillan sniper system, Heym rifles, Burris telescopic sights plus others. → 80 → 85 → 90 [97]

75-36150 **Lloyd Instruments Ltd**
Forum House, 12 Barnes Wallis Road, Segensworth East,
Fareham, Hampshire PO15 5TT
☎ (+44 1489) 57 42 21 ▣ 477042 jayjay g
⊖ (+44 1489) 88 51 18
⇮ 80 (1997)
Managing Director: John Richings
Sales and Marketing Manager: Robin Bramble
Operations Director: Brian Reed
Purchasing Manager: Chris Jones
Marketing Executive: Ms Suzanne Sharpe
■ Manufacture materials testing equipment (2.5kN-300kN max. strength) for determining the physical characteristics of a wide range of materials including destructive testing of components and assemblies by tension, compression and flexural testing methods. Also manufacturer of the Davenport polymer test instrumentation range.
Associated companies/sales offices:
Lloyd Instruments SA, France
John Chatillon & Sons Inc, USA
Lloyd Instruments Ltd, Malaysia
Erichsen Prüftechnik Wuppertal GmbH, Germany [98]

75-36270 **Logan Fenamec (UK) Ltd**
(a subs. of Fabricom S.A.)
UK Head Office
Sutton Rd, Hull HU7 0DR
☎ (+44 1482) 78 56 00 ⊖ (+44 1482) 78 56 99
♧ £ 2 900 000 (1995) ♣ £ 15 400 000 (1994) ⇮ 180
⌂ 6 500 sq m
Managing Dir: Edward Ferris ☎ (+44 1482) 78 56 06
Sales & Marketing Dir: T. Railton ☎ (+44 1482) 78 56 35
Operations Dir: M. Jeffery ☎ (+44 1482) 78 56 36
Purchasing Mgr: B. Treloar ☎ (+44 1482) 78 56 08
■ Warehousing & distribution system design & implementation. Automated package & parcel sorting systems. Unit load conveyors & systems. Airport baggage handling systems. [97]

75-36300 **Londesborough Security Equipment plc**
(a member of the Londesborough plc Group of Companies)
8 Gordon Court, Victoria Road,
Stechford, Birmingham B33 8AN
☎ (+44 121) 786 10 04 ⊖ (+44 121) 786 10 04
♣ £ 2,000,000 (1996) ⇮ 36 (1996) ⌂ 11,000 sq m
Managing Director: Lieut Cdr R B W Lowndes
Marketing and Sales Director: Maj J Hutchinson
Advertising Manager: J Sinclair
■ Vehicle access and entrapment systems; portable road barriers & blocks. Underground road barrier systems. Vehicle halting devices; rapid response road and lightweight portable road barriers. Anti-riot weapon systems. Helicopter landing systems; handlamps. NATO codified military hand searchlights. Airfield runway denial weapon systems; tracked vehicle destructors. NBC overboots; military footwear. [98]

75-36500 **Lorraine Electronics Head Office**
(a subsidiary of Dataglen Ltd)
716 Lea Bridge Road, Leyton, London E10 6AW
☎ (+44 181) 558 42 26 ⊖ (+44 181) 558 13 38
♣ £ 4,000,000 (1996) ⇮ 20
Director: David Benn
Public Relations Manager: Richard Ingham
■ Audio surveillance products and services. Subminiature transmitters and receivers, covert recording devices.
Showroom
59 South Audley Street, Mayfair, London W1Y 5FB
☎ (+44 171) 493 40 07 ⊖ (+44 171) 493 43 37
Contact: Ms Jo Ann O'Neill [98]

75-36530 **LRC Products Ltd**
(a subsidiary of London International Group)
London International House, Turnford Place,
Broxbourne, Hertfordshire EN10 6LH
☎ (+44 1992) 45 11 11 ⊠ LONRUBMANF London
▣ 21644 ⊖ (+44 1992) 47 15 36
♣ £ 200,000,000 (1996) ⇮ 4,000 (worldwide 1997)
Managing Director - Marigold Industrial: Shaun Davies
Sales and Marketing Manager - Marigold International: Martin Ireland
National Account Manager: S J Langstaff
■ Manufacture and marketing of consumable products for hospital and industrial markets.
Products include "Marigold" range of protective gloves for industrial, chemical resistant, laboratory and specialist applications. [98]

75-36650 **Lucas Aerospace**
Brueton House, New Road,
Solihull, West Midlands B91 3TX
☎ (+44 121) 627 59 08 ⊖ (+44 121) 500 64 05
Marketing Services Manager: Ms Julie Hewitt
☎ (+44 121) 627 59 76 (fax) [98]

75-36655 **Lucas Aftermarket Operations**
(† Lucas Assembly & Test Systems)
Test Equipment Division
Unit 9, Monkspath Business Park, Highlands Rd,
Solihull, Birmingham B90 4NY
☎ (+44 121) 697 59 99 ⊖ (+44 121) 697 59 50
♣ £ 12 000 000 ⇮ 30 ⌂ 750 sq m
Managing Dir: Ian Joesbury
Operations Dir: Gary Christian
Sales & Marketing Dir: Richard Green
Product Mgr: R.J. Savory
■ Marketing, sales & service of a range of products for servicing diesel fuel injection pumps. Products include test & calibration equipment, attachment kits, reconditioning machines, inspection aids, cleaning equipment & specialised overhaul tools. [97]

75-36675 **Lucas Control Systems Products**
(† Lucas NSF Ltd)
(a subsidiary of Lucas Varity)
Ingrow Bridge Wks, Ingrow Lane,
Keighley, West Yorkshire BD21 5EF
☎ (+44 1535) 66 11 44 ⊖ (+44 1535) 66 14 74
♣ £ 7,000,000 (1995/96)
⇮ 130 (European Sales 1995/96)
Operations Manager: L Bennett
Market Communications Mgr Europe: Roy Whitaker
European Sales Manager: D R Priestley
■ Manufacturing BS9000/CECC approved lever and rotary switches as well as rotary and linear solenoids. Typical applications include military radio and communications, fighting vehicles, radar, shipboard and airborne equipment.
® USA: Eaton Corporation, Lucas Ledex Inc. [98]

75-36700 **Lynwood**
(† Lynwood Scientific Developments Ltd)
(a subsidiary of NAI Technologies Inc, USA)
Lynwood House, The Trading Estate,
Farnham, Surrey GU9 9NN
☎ (+44 1252) 73 44 88 [E] sales@lynwood.com
[W] http://www.lynwood.com ⊖ (+44 1252) 73 44 66
Managing Director: Paul Rossiter
Director of Business Development: Peter Hurst
Director of Engineering: Richard Young
Business Manager, Air Systems: Tony Woollaston
Business Manager, Sea Systems: Tim Giles
Business Manager, Land Systems: Colin Furniss
Sales Manager Europe: David Cockarill
Customer Service Manager: Pat Parker
Sales Executive: Ms Samantha Smith
Marketing Co-ordinator: Ms June Yates
■ Design, manufacture, integration and support of a range of rugged PCs, workstations and flat panel displays based on COTS technology for land, sea and airborne forces. TEMPEST products, network design, cabling, installation and commissioning. → 80 [98]

75-36710 **Lynx Engineering UK Ltd**
Lower Clough Mill, Pendle Street, Barrowford,
Nelson, Lancashire BB9 8PH
☎ (+44 1282) 69 42 90 ⊖ (+44 1282) 61 33 90
♧ £ 150,000 (1997) ♣ £ 1,400,000 (1997)
⇮ 38 (1997)
Managing Director: Les Nuttall
■ Prototype development, small batchwork, production machining, toolmaking. Manufacturing facilities: CNC turning, CNC milling, grinding, spark eroding, precision pressing, CMM inspection, waterjet cutting. [98]

75-36720 **Lyon & Brandfield Ltd**
4/5 Inverness Mews, London W2 3JQ
☎ (+44 171) 229 91 25 ⊖ (+44 171) 229 92 17
♧ £ 1,000 ♣ £ 3,300,000 (1995) ⇮ 30 ⌂ 3,000 sq m
Director: Howard Freckleton
Sales/Marketing Director: David R Sofaer
Operations and Technical Director: John Bird
Purchasing Manager: Philip Wade
■ Manufacturers of night vision photographic equipment, surface & underwater night vision equipment, hostile fire warning indicators, biocular red dot weapon sights. British & American military aircraft & helicopter spares. British armoured fighting vehicle & tank parts. Protective clothing for police & military. [98]

75-36830 **Mabey & Johnson Ltd**
(a subs. of Mabey Holdings Ltd)
Floral Mile, Twyford, Reading, Berks RG10 9SQ
☎ (+44 1734) 40 39 21 ⊠ MEBEYJON Reading
▣ 848649 mabeyt g ⊖ (+44 1734) 40 39 41
♧ £ 15 000 000 (1995) ♣ £ 31 000 000 (1995) ⇮ 210
⌂ 4 500 sq m
Managing Dir: R.C.E. Forsyth
Operations/Technical Dir: C.J. Carter
Export Dir: R.J. Glover
Purchasing Dir: K. Jones
■ Rapidly erected military bridges to 60m span & MLC100 in single or multiple carriageways for applications throughout the support area or in civilian emergency. [97]

75-36850 **Maccaferri Ltd**
(† River & Sea Gabions Ltd)
(a subsidiary of Officine Maccaferri SpA)
4B The Quorum, Oxford Business Park, Garsington Road,
Oxford OX4 2JY
☎ (+44 865) 77 05 55 ⊖ (+44 865) 77 45 50
⇮ 16
Managing Director: R C Greenford
Sales/Marketing Director: Colin Evans
Consultant: A D Crowhurst
Technical Manager: P Staten
■ Products are used by the military for applications such as retaining walls, piers for 'Bailey' bridges together with more specialised military uses including free standing walls around bunkers & munitions stores. [98]

75-36870 **Machine Technology Ltd**
(† Arcadia Engineering Co.)
Arcadia Works, 22/23 Arcadia Ave, London N3 2JU
☎ (+44 181) 349 48 14 ⊠ ELMACH
⊖ (+44 181) 346 62 51
♧ £ 10 000 ♣ £ 1 400 000 ⇮ 11 ⌂ 8 000 sq ft
Managing Dir: Mrs Christa T. Dimont

Contact Reference

Technical Dir: Simon Dimont
Sales & Marketing Dir: John Hunter
Operations Dir: Peter Forman
Purchasing Mgr: Kirk Doglas
■ Assemblers of opto-electronic proximity fuzes. Telerobotics systems integrators. [97]

75-36900 MacTaggart Scott & Co. Ltd
Hunter Ave, PO Box 1, Loanhead, Midlothian EH20 9SP
☎ (+44 131) 440 03 11 ⊠ 72478 mascot g
⊖ (+44 131) 440 44 93
♣ £ 15 000 000 (1994) 👥 320 🏠 6 997 sq m
Managing Dir: P.R. Prenter
Technical Sales Dir: T.M. Craig
Publicity & Advertising (Graphics): R.R. Mitchell
Business Development Mgr: J. Partridge
■ Aircraft lifts, FW & RW handling systems, aircraft arresting gear, missile & weapons hoists, stores/provision hoists; missile & weapons handling systems, low magnetic handling equipment & davits. Submersible launch recovery systems. Diving bell/sonar deployment systems. Submarine hydraulic pumps & vent valves & other products. [97]

75-37000 Magnesium Elektron
(part of British Aluminium Limited)
PO Box 23, Manchester M27 8DD
☎ (+44 161) 911 10 00 ⊠ 667817 melman g
⊖ (+44 161) 911 10 10
👥 250
Managing Dir: D.R. Hirons
Sales & Marketing Dir: K. White
Market Development Mgr: S.R. Harris
☎ (+44 161) 911 12 20
Purchasing Mgr: D. Bird
■ Extruded magnesium alloy bars & sections, machined shapes. Cast alloys for aerospace, lightweight & man portable equipment. Particulate magnesium products (powders, raspings & granules). [98]

75-37050 Maine Group Ltd
Maine Engineering
Home Park, Kings Langley, Hertfordshire WD4 8LZ
☎ (+44 1923) 26 04 11 ⊖ (+44 1923) 26 71 36
♧ £ 380,000 (1996) ♣ £ 4,500,000 (1996)
👥 91 (1996) 🏠 35,000 sq ft
Chief Executive Officer: R T Klottrup
Operations Director: R J Davis
Finance Director: K C Huxham
Engineering Sales Manager: M L Buck
Sales Administration: Mrs S Goksu
■ Prime contractor to the UK MoD Navy for design & manufacture warship accommodation furniture, bunk systems, kit lockers. Specialist in aluminium furniture for fast patrol craft, submarines and barrack room accommodation. [98]

75-37065 Malakoff & Wm. Moore Ltd
(a subsidiary of Lithgows Ltd)
North Ness, Lerwick, Shetland ZE1 0LZ
☎ (+44 1595) 69 55 44 ⊠ 75434
⊖ (+44 1595) 69 57 20
♧ £ 100,000 (1997) ♣ £ 5,378,000 (1996)
👥 105 (1997) 🏠 10,000 sq m
Managing Director: Allan M Gould
Operations and Technical Director: Angus Graham
Sales and Marketing Director: Mike Comerford
Purchasing Manager: Dick Page
■ Engineering, fabrication and construction of aluminium hulled workboats. [98]

75-37100 Mannesmann Rexroth Ltd
(† G L Rexroth Ltd)
(a subsidiary of Mannesmann-Rexroth GmbH)
Cromwell Road, St Neots, Huntingdon, Cambs PE19 2ES
☎ (+44 1480) 760 41 ⊠ 32161 ⊖ (+44 1480) 22 32 00
♧ £ 1,700,000 ♣ £ 35,000,000 👥 235
Managing Director: J Willmott
Sales Director: R G D Wight
Financial Director: J Willmott
Sales and Marketing Manager: B C Bonham
Technical Services Manager: G G D French
■ Oil hydraulic control systems and equipment. Complete hydraulic motion platforms for simulators. [98]

75-37165
Manroy Engineering Ltd
Hobbs Lane, Beckley, East Sussex TN31 6TS
☎ (+44 1797) 26 05 53 ⊖ (+44 1797) 26 03 74
Managing Director: Mrs Madeleine Swainbank
Sales Director: Roy Swainbank
Technical Director: Michael Wreford
■ Manufacturer of .50 M2HB, QCB & 7.62mm GPMG machine guns. Supplier of ordnance & small arms spares including machine guns, rifles, pistols, kits, gauges, mounts and canvas. Weapon refurbishment & conversion. Manufacture M3, M63, M31 boat & soft mount for .50 M2HB machine guns. [98]

75-37270 Marconi Instruments Ltd (MI)
(a subsidiary of The General Electric Co plc)
Longacres House, Norton Green Road,
Stevenage, Hertfordshire SG1 2BA
☎ (+44 1438) 74 22 00
🄴 webmaster@marconi-instruments.co.uk
🅆 http://www.marconi-instruments.com ⊠ 23350
⊖ (+44 1438) 72 76 01
👥 1,021
Managing Director: Peter J Smith
Sales Director: Roy Lester
Commercial Manager (Contracts): C W Thurston
Manager Military Sales: T E Tinsley
Senior Press Relations Officer: Ms Emma Cookson
■ Design and production of electronic test instrumentation and ATE, with worldwide military and commercial customer base. Incorporation of test instrumentation into specialised test systems.

Headquarters
Longacres, St Albans, Hertfordshire AL4 0JN
☎ (+44 1727) 85 92 92 ⊠ 23350
⊖ (+44 1727) 85 74 81
Business Manager, Products: Tim Pegg
Business Manager, Distributed Products: Steve Gledhill
Design and Manufacture Product Group Manager: Kevin Ingrams
Installlation and Maintenance Product Group Manager: Jon Stafford
■ Design and production of electronic test instrumentation. Incorporation of test instruments into specialised test systems.

Special Test Systems Business Unit
Marconi Instruments Telford Building,
Donibristle Industrial Park,
Near Dunfermline, Fife KY11 5JZ
☎ (+44 1383) 64 64 64 ⊠ 727779
⊖ (+44 1383) 64 64 68
■ Design and production of in-circuit and functional ATEs, including turnkey software packages and test management systems.

Customised Applications & Systems
Dunfermline, Scotland
MICAS Business Manager: Peter Hand
■ Project management support for electronic production and service environments. Electronic hardware and software design, test fixturing, programming, in-circuit and funtinal test systems, test area management software. [98]

75-37286 Marine Acoustics Ltd
Unit 3, Hogwood Lane Ind Est,
Finchampstead, Berks RG11 4QN
☎ (+44 1734) 73 49 00 ⊖ (+44 1734) 32 80 96
Director: Dr Brian Phillips
Sales & Marketing Mgr: Bruce Swale
■ Design & manufacture of both research & production towed array systems including new generation, digitally multiplexed streamers; programmable low frequency acoustic sources; navigation transponders & high integrity acoustic command & data telemetry units; miniature echo sounders & submarine mounted sonars; wide range of OEM acoustic transducer products encompassing calibrated hydrophones & low noise preamplifiers. [97]

75-37288 Marine Data Ltd
Queensway, New Milton, Hants BH25 5NN
☎ (+44 1425) 61 94 29
🅆 http://www.tcp.co.uk/~/acadia/ ⊠ 47674
⊖ (+44 1425) 61 21 25
♣ £ 700 000 👥 27
Managing Dir: K.R. Simmonds
Sales and Marketing Manager: Charles Livingstone
■ Development & supply of magnetometer & gyro compass & re-transmission units of all types. Supply of 400Hz static inverters up to 1.5 kVA rating. [97]

75-37290 Marine Electronics Ltd
10 Barras Lane Industrial Estate,
Vale, Guernsey, CI GY6 8EQ
☎ (+44 1481) 531 81 ⊖ (+44 1481) 531 82
♧ £ 55,000 (1997) ♣ £ 450,000 (1996) 👥 8 (1997)
🏠 1,600 sq m
Managing Director: Alan Green
Operations/Technical Director: Gordon Collier
■ Underwater acoustic scanning sonars. Transponder beacons. Acoustic positioning systems. Acoustic releases. [98]

75-37530 Marshall of Cambridge Aerospace Ltd
(† Marshall of Cambridge (Engineering) Ltd)
(a division of Marshall of Cambridge (Holdings) Ltd)
The Airport, Newmarket Road, Cambridge CB5 8RX
☎ (+44 1223) 37 32 63 🄴 01354.66@compuserve.com
🅆 http://www.marshall-group.co.uk ⊠ 81208
⊖ (+44 1223) 37 33 73
👥 1,800 (1997) 🏠 1,700,000 sq ft
Chairman & CEO: Michael Marshall
Managing Dir Aircraft Div.: Martin Broadhurst
Engineering Director: R Ward
Production Director: Grahame Nix
Commercial Dir: Chris Bunney
Dir Marketing & Business Development: M.E. Milne
General Mgr Marketing: Keith Hogan
■ Aircraft & space systems components design, manufacture & assembly. Fixed & rotary wing modification, refurbishment, flight test & certification. Avionics, electronics & weapons system design, embodiment & certification. Fixed & rotary wing repair & maintenance. Design & production of specialist military vehicle bodies & transportable containers. → 80 [98]

75-37560 Marshall SPV Ltd
Marshall Specialist Vehicles
The Airport, Cambridge CB5 8RX
☎ (+44 1223) 37 37 37 ⊠ 81208
⊖ (+44 1223) 32 10 32
♣ £ 50,000,000 👥 500+
Marketing and Communications Manager: Terence Scown-Geary
■ Design, manufacture and assembly of Bedford Trucks, worldwide supply and support of Bedford genuine parts, design, sales and manufacture of bus bodies and specialist commercial vehicles, military engineering and low bake paint spraying facilities for the largest commercial and military vehicles. → 85 [98]

75-37900 Mawdsley's Ltd
(a member of the Adwest Group)
Zone Works, Dursley, Glos. GL11 5AE
☎ (+44 1453) 54 41 31 ⊖ (+44 1453) 54 65 71

♣ £ 3 951 000 👥 85
Managing Dir: G. Lovett
Technical Mgr: P. Allen
Sales & Marketing Dir: R. Hinder
■ Electric motors & power generating sets for fighting ships. Mobile Ground Power Units. [97]

75-38080 McLennan Marine Ltd
24 Doman Road, Camberley, Surrey GU15 3DF
☎ (+44 1276) 265 06 ⊖ (+44 1276) 265 05
♣ £ 1,000,000 (1997) 👥 15 (1997) 🏠 750 m2
Managing Director: G A Ewer
General Manager: Hywel J Thomas
Marketing Manager: Bill Sheraton
■ Night vision systems & equipment. [98]

75-38170 McMurdo Ltd
(a subsidiary of Chemring Group plc)
Rodney Road, Fratton Estate,
Portsmouth, Hampshire PO4 8SG
☎ (+44 1705) 73 53 61 ⊖ (+44 1705) 75 50 20
♣ £ 11,500,000 (1997) 👥 200 (1997)
Managing Director: J J Lynn
Sales and Marketing Director: Bob Hart
Purchasing Manager: M Lee
■ Electronic connectors, cable assemblies and marine location safety equipment. [98]

75-38200 McQueen Ltd
Target Division
Nether Road, Galashiels, Scotland TD1 3HE
☎ (+44 1896) 75 48 66 ⊖ (+44 1896) 75 07 74
♧ £ 5,000,000 (1997) ♣ £ 40,000,000 (1997)
👥 600 (1997) 🏠 102,000 sq m
Managing Director: Michael M Gray (OBE)
Target Division General Mgr: Gerry Moncur
■ Specialist in manufacture of targets for military, police & sports shooting. Specifically anti-terrorist training targets. [98]

75-38230 McTay Marine Limited
The Magazines, Port Causeway,
Bromborough, Merseyside L62 4YP
☎ (+44 151) 346 13 19 ⊖ (+44 151) 334 00 41
♣ £ 10,000,000 (1996) 👥 135 (1996)
Managing Director: M R Brodie
Sales/Marketing Director: R McBurney
Purchasing Manager: Mrs Sandra Parish
■ Manufacture of landing craft, assault boats, military patrol boats, torpedo recovery vessels, minesweepers, boom defence vessel tugs, high speed survey vessels. [98]

75-38270 MDH Limited
(a subsidiary of Kynoch Group plc)
34A Walworth Road, Andover, Hampshire SP10 5AA
☎ (+44 1264) 36 21 11
[E] 101376.1130@compuserve.com
⊖ (+44 1264) 35 64 52
♧ £ 2,000,000 (1997) ♣ £ 10,000,000 (1997)
👥 140 (1996) 🏠 160,000 sq ft
Managing Director: Clive Woods
Business Manager - Defence: C T Cheshire
Operations Director: P Cornielle
General Sales Manager: J R Heather
Purchasing Manager: D Shelley
■ NBC protection, heating & cooling systems for vehicles, ships, tents & hardened shelters. Electronic equipment cooling systems. Approved to AQAP-1, BS5750-pt1, ISO9001, EN29001. [98]

75-38275 Measurement Devices Ltd
(† Measurement Devices Engineering Ltd)
Silverburn Crescent, Bridge of Don Ind. Estate,
Aberdeen AB23 8EW
☎ (+44 1224) 82 41 41 [E] mdl.uk.@zetnet.co.uk
⊖ (+44 1224) 82 49 87
♧ £ 1 000 000 (1996) ♣ £ 2 500 000 👥 20 🏠 600 sq m
Managing Dir: Stephen L. Ball
■ Manufacture of navigation systems & equipment. Communications systems. [97]

75-38330 C.P. Mega
Worton Hall Ind. Est., Worton Rd,
Isleworth, Middx TW7 6ER
☎ (+44 181) 568 18 81 ⊖ (+44 181) 568 11 41
👥 40
Managing Dir: Peter Ross
Commercial Design Engineer: Steve Jenkins
Commercial Design Engineer: Chris Gers
Marketing Co-ordinator: Andrew Smith
■ Protective, re-usable equipment cases & padded bags for military & commercial applications.
● Unit 18 Westbrook Road, Trafford Park, Manchester M17 1AY
☎ (+44 161) 873 81 31 ⊖ (+44 161) 876 65 99 [97]

75-38335 MEO Products Ltd
Unit 8, Arrow Industrial Estate, Eelmoor Road,
Farnborough, Hampshire GU14 7QN
☎ (+44 1252) 52 43 77 ⊖ (+44 1252) 52 43 80
Managing Director: Mr Verden [98]

75-38340 Meggitt Avionics (UK)
(a Meggitt Aerospace Company)
6 Manor Court, Barnes Wallis Road,
Fareham, Hampshire PO15 5TN
☎ (+44 1489) 57 90 90 ☒ 47671 nzstn g
⊖ (+44 1489) 57 90 13
👥 229
Managing Dir: D. Gemmell
Operations Director: Vic Blackwell
Sales & Marketing Dir: Peter O'Sullivan
Marketing Mgr: Chris Butler
■ Instruments, sensors & controls for flight, engines & system applications. Engine & Airframe fatigue life monitors, (EMS & EHM). Gas turbine IR Pyrometry. "Bryans" ground & test bench equipment. Battery charger & analyser. Aerial reconnaissance & hand held camera systems. Ships data distribution systems. Engine telegraph & display systems. Wind speed & direction systems. [98]

75-38345 Meggitt Composites
(a subsidiary of Meggitt plc)
New Road, Netherton, Dudley DY2 9AF
☎ (+44 1384) 23 02 02 ⊖ (+44 1384) 23 97 22
👥 105 (1997) 🏠 6,000 sq m
Managing Director: E J Lewis
Business Development Director: D J Norman
Operations and Technical Director: G Lacey
Operations Director: R Gibbs
Purchasing Manager: J Billingham
■ Aircraft seat backs. Design & manufacture of composite structures & ballistic protection products. [98]

75-38360 Meggitt Oxygen Systems
(a Meggitt Aerospace Company)
Edinburgh Place, Edinburgh Way, Templefields,
Harlow, Essex CM20 2DJ
☎ (+44 1279) 44 20 81 ⊖ (+44 1279) 43 77 94
Business Development Mgr: Colin Faulkner
■ Design, development & manufacture of life support & oxygen breathing systems. Capabilities also include manufacture of pressure vessels & control valves for pressurised gas stored energy systems. [98]

75-38550 Mermaid Marine Engines Ltd
Ferndown Industrial Estate, 70-72 Cobham Road,
Wimborne, Dorset BH21 7RN
☎ (+44 1202) 89 18 24 ⊖ (+44 1202) 89 58 82
♣ £ 2,000,000 👥 20
Managing Director: C R Snare
Operations Director: N C Saw
■ Suppliers of diesel & gasoline engines up to 500hp for small craft equipped with weapon systems. [98]

75-38700 Messier-Dowty Limited
(† Dowty Aerospace Landing Gear Ltd)
(a member of the TI Group)
Cheltenham Road, Gloucester GL2 9QH
☎ (+44 1452) 71 24 24 ☒ 43246
⊖ (+44 1452) 71 38 21
👥 750 🏠 44,000 m2
Managing Director: A J Stevens
Industrial Director: Les George
Finance Dir: G. Capps
Engineering Dir: D.J. Wood
Production Dir: G. Skinner
Sales & Marketing Dir: S. Luxmoore
Dir Material: P.M. Booth
Marketing Services Mgr: P.T. Hall
■ Design, development, manufacture & support of fully integrated landing gear systems for commercial & military aircraft. [98]

75-38750 Metalite Aviation Lighting
(a division of Metalline International Ltd)
Winster Grove, Great Barr, Birmingham B44 9EJ
☎ (+44 121) 360 22 22 ☒ 339410 metlin g
⊖ (+44 121) 366 60 03
♣ £ 2,000,000 + (1996) 👥 35 (1997)
Managing Director: Glyn T Davies
Marketing Manager: Paul Hartshorn
Sales Co-ordinator: Ms Lisa Horrocks
■ Manufacturer of portable airfield and helipad lighting equipment for military and civil applications. NATO approved Minimum Operating Strip Lighting Kit (MOSKIT) and civil derivative Civil Airfield Lighting Kit (CALKIT). Other portable items include Combined Omnidirectional Runway and Taxiway Edge Light (COREL), Unidirectional Battery Powered Approach Light (UAL), Visual Spectrum Precision Approach Path Indicator (VIS PAPI), Night Vision Goggle compatible PAPI (NVG PAPI), Metalite Electroluminescent Helipad Lighting System (MELHS) - also available as fixed system.
Manual and remotely controlled systems, MAPAL (LE/58) ground light, MERE-2, 2/T and METE-2, 2/T runway and taxiway edge lights. Ancillary items including battery chargers, generators, master switching units and purpose-built trailers for MOSKIT and CALKIT. Research, design, development after sales support and parts availability. [98]

75-38760 Metallisation Ltd
(a subs. of Metallisation Industries Ltd)
Pear Tree Lane, Dudley, W. Midlands DY2 0XH
☎ (+44 1384) 25 24 64 ⊖ (+44 1384) 23 71 96
♧ £ 2 450 000 (1996) ♣ £ 6 000 000 👥 63 🏠 2 000 m2
Managing Dir: Peter Hall
Technical Dir: Dr Terry Lester
Commercial Dir: C.S. Hewitt
■ Manufacture metal spraying equipment, EMI/RFI shielding, moulding, lightning strike protection. Corrosion protection & engineering reclamation. [97]

75-38800 Metheus Corporation
First Base, Beacontree Plaza, Gillette Way,
Reading RG2 0BP
☎ (+44 118) 931 21 12 [E] europe@metheus.com
[W] http://www.metheus.com ⊖ (+44 118) 875 61 14
👥 40 (1996)
Managing Director: Angus Mitchell
Applications Engineer: K Starling
■ Design, manufacture and marketing of high resolution display controllers. Products designed to industry standards in platforms (DEC, HP, IBM, SUN, INTEL) form factor (6U VME & PCI) and software (X-Window System) for compatibility and ease of integration. Recent product additions allow live video and radar information to be displayed under full X Window control. [98]

75-38875 Metool Products Ltd
(a subsidiary of Kabelschlepp GmbH, Germany)
Lilac Grove, Beeston, Nottingham NG9 1PG
☎ (+44 115) 922 59 31 ✆ (0500) 63 86 65 (METOOL)
[E] postmaster@metool.demon.co.uk
⊖ (+44 115) 922 45 78
♣ £ 5,000,000 👥 65 🏠 8,000 sq m
Managing Director: Steven Rynton
Projects Director: Malcolm Dade
Purchasing Manager: Nick Hague
Marketing Assistant: Ms Maria Narducci
■ Cable reels, hose reels and earth bonding reels. Cable/hose drag chain systems, conductor and festoon systems. [98]

75-38890 Metrum Information Storage Ltd
2 Weller Drive, Finchampstead,
Wokingham, Berkshire RG40 4QZ
☎ (+44 118) 973 30 00 ⊖ (+44 118) 973 43 63
♣ £ 2,560,000 (1996) 👥 18 (1997) 🏠 550 sq m
Managing Dir: C.G. Beeton
Operations Dir: M.D. De Roux
■ Military magnetic tape data recorders. [98]

75-38900 MFC Survival Ltd
Naval Yard, Tonypandy, Rhondda, S.Wales CF40 1JS
☎ (+44 1443) 43 30 75 ✉ AIRLIFT, TONYPANDY
☒ 49722 ⊖ (+44 1443) 43 78 46
♣ £ 1 250 000 (1994/95) 👥 58 🏠 12 000 sq ft
Managing Dir: D.B. Jones
Sales & Marketing: Mike Sollis
Operations & Technical: D.J. Sheppard
Purchasing Mgr: B.J. Lea

Company Secretary: Mrs B. Richards
■ Inflatable emergency response shelters, single & modular systems. Air lift cushions, recovery low & high pressure, inflatable water storage reservoirs. [97]

75-38905 MFD Capacitors Ltd
Penley, Clwyd, N.Wales LL13 0LY
☎(+44 1978) 71 05 51 ⊖ (+44 1978) 71 05 01
21 250 sq m
Director: G.G. Watson
Sales & Marketing Dir: Bernard Morgan
■ Manufacture of electrical capacitors to approved standards. [97]

75-38912 M.G. Electric (Colchester) Ltd
(part of the Martin Group of companies)
Wyncolls Road, Colchester, Essex C04 4HT
☎(+44 1206) 84 22 44 sales@mgelectric.co.uk
⊖ (+44 1206) 84 58 49
♣ £ 4,000,000 50 35,000 sq ft
Managing Director: M J Martin
Commercial Director: G M Martin
Engineering Director: A Clark
Sales Director: T Woods
Purchasing Manager: J Brown
Marketing: M Morris
■ Manufacture of a range of medical suction equipment. Marketed under the name SAM suction. May be specified for emergency field use or complete hospital systems, both central piped, electric mobile & portable units. [98]

75-38915 MHH Engineering Co. Ltd
(a subs. of Bullough plc)
Tannery Lane, Bramley, Guildford, Surrey GU5 0AJ
☎(+44 1483) 89 27 72 ⊖ (+44 1483) 89 85 55
♣ £ 3 900 000 (1996) 80 1 720 m2
Managing Dir: D.G. Allen
Production Dir: G. Ellera
Purchasing Mgr: D.B. Giddins
■ Torque wrenches & other torque tools. Air pressure regulators, push-pull gauges & special-purpose gauges. [97]

75-38920 Mibert Precision Ltd
(a subsidiary of Hampson Industries plc)
129 Scudamore Road, Leicester LE3 1UQ
☎(+44 116) 232 22 33 ⊖ (+44 116) 232 23 11
£ 2,000,000 (1997) ♣ £ 3,600,000 (1997) 80
1,800 sq m
Managing Director: Chris Machin
Operations Director: Mark Bertolini
Purchasing Manager: Vic Ironmonger
■ Precision machining of components for the world's aerospace, defence & high technology industries. Aerospace & industrial gas turbine components, mechanical components for aircraft, fighting vehicles, surface vessels, missiles & ground support equipment. [98]

75-38935 Michell Bearings
(a division of Vickers plc)
Scotswood Road, Newcastle upon Tyne NE15 6LL
☎(+44 191) 273 02 91 ⊖ (+44 191) 272 27 87
262 20,000 m2
Managing Director: Mike Conway
Operations Director: B Hurley
Sales Director: Neal Henderson
Engineering Director: Bob Knox
Purchasing Manager: L M Richardson
■ Propeller lineshaft bearings, thrust bearings, bulkhead sealing glands, for all classes of surface warships & submarines. [98]

75-39010 Micro Metalsmiths Ltd
(† M.M. Microwave Ltd)
Thornton Rd, Pickering, North Yorkshire YO18 7JB
☎(+44 1751) 47 50 58 57676
⊖ (+44 1751) 47 63 64
♣ £ 4 500 000 (1995) 170 20 000 sq ft
Managing Dir: Christopher Shaw
Marketing Dir: Paul Birnie
Exec. Sales Mgr: M.T. Stevens
Field Sales Mgr: Paul Horsley
Purchasing Mgr: Neil Lishman
■ Design, development & manufacture of waveguide components & sub-assemblies for the UK, European & American radar & communications industries. Specialists in the design of waveguide systems & antenna feeds. Manufacture of investment castings. Fully equipped machine shop.
® France: MBElectronique
Germany: Municom GmbH
India: New World Technologies
Israel: Racom
Spain: BFI
Sweden: Pronesto AB
Switzerland: Schiesser Electronc Ag
USA: Fusion Inc, RF Microwave Limited
UK Agents:
Raysheel Limited
Castech Associates
Hoskins & Munson
G.T. Products
Electromat Ireland Limited [97]

75-39160 Microturbo Ltd
(a division of Labinal, France)
Fort Wallington, Fareham, Hampshire PO16 8TT
☎(+44 1329) 28 36 11 ⊖ (+44 1329) 22 06 31
75
Chairman: Sir Colin Fielding
Managing Director: Michael J Foote
Finance Director: Michael E Hansford
Operations Director: M L Chalkey
Head of Sales and Marketing: R B Harrison
Purchasing Manager: Ms D Frankpitt
■ Design, development, manufacture and repair of small gas turbine engines and their associated equipment. Turbojet engines for missiles and RPV's. Gas turbine starting systems for aircraft. Gas turbine driven auxiliary power units. Electronic and electrical products. [98]

75-39330 Middlesex Group Limited
Weyside Division
Telford Rd, Basingstoke, Hants RG21 6YU
☎(+44 1256) 537 11 ⊖ (+44 1256) 84 26 13
£ 250 000 (1996) ♣ £ 12 000 000 130
5 500 sq m
Chairman & Managing Dir: P.S. Foulds
Operations & Technical Dir: L.E. Foulds
Purchasing Mgr: S. McNair
■ Middlesex/Weyside Division specialises in the precision machining of components & high-quality assemblies for the defence & aerospace industries. [97]

75-39500 Milldale Limited Pulvermatic Sales
(a subsidiary of Ropner plc)
Excalibur House, Firth Street,
Huddersfield, West Yorkshire HD1 3AH
☎(+44 1484) 53 42 14 ⊖ (+44 1484) 53 44 04
£ 350,000 ♣ £ 3,000,000 50 2,500 sq m
Managing Director: Crowther Laycock
Technical Manager: David Pharaoh
Sales Administration: Miss Dawn A Rogers
■ Manufacturers and servicers of baling, shredding and business machines. [98]

75-39700 Minebea Electronics (UK) Ltd
Kelburn Business Park, Port Glasgow, Scotland PA14 6PD
☎(+44 1475) 74 52 31 ⊖ (+44 1475) 74 40 46
71
Managing Director: J A Morrison
■ Manufacture of power supply equipment. [98]

75-39800 Mineral Derivatives Ltd
Dalton Way, Middlewich Business Park,
Middlewich, Cheshire CW10 0HU
☎(+44 1606) 83 63 46 ⊖ (+44 1606) 83 64 08
£ 10 000 ♣ £ 2 200 000 25 22 000 ft sq
Managing Dir: M.D. Markman
Business Mgr: S.P. Jolly
■ Breather valves & humidity indicators for use on transit & air-portable containers. Desiccators for use on equipment. Container hardware. Drisorb desiccant bags conforming to UK, French, German & American specifications. → 90 [97]

75-40050 Mirrlees Blackstone Limited
(a BTR Group Company)
Hazel Grove, Stockport, Cheshire SK7 5AH
☎(+44 161) 483 10 00 667314; 667315
⊖ (+44 161) 487 14 65
♣ £ 65,000,000 (1996) 800 (1997)
46,500 m2 (1997)
Managing Dir: Alan Marsh
Sales Dir: C.L. Smith
Marketing Manager: R Buckingham
Publicity Manager: John D Vesey
■ Diesel and dual fuel engines and generating sets from 500bhp (400kw) to 18000 bhp (13000kw). Pumping and compressor sets. Rail traction and marine engines. [98]

75-40100 Andrew Mitchell & Co. Ltd
Bishop Works, Amulree Street, Glasgow G32 7SL
☎(+44 141) 778 54 61 MONARCH Glasgow
⊖ (+44 141) 778 81 44
£ 1,300,000 (1996) ♣ £ 6,000,000 (1996)
130 (1996) 10,000 m2
Managing Dir: Duncan Kirk
Sales Manager: Brian Kier
Export Mgr: T. Thomson
■ Field accommodation & equipment including Mitco, Coverguard, flexible alarmed sheeting to cover sensitive loads or items eg: main battle tanks in transit & static storage.
● Hainault Road, Little Heath, Romford, Essex RM6 5ST
☎(+44 181) 590 60 70 23207
⊖ (+44 181) 599 46 13
Sales Mgr: N. Gillies
■ Canvas & PVC products, tilts & tarpaulins. Portable shelters for storage, workshop or medical facilities in the field. Specialists in army tents, marquees. Kit & security bags & camp beds. [98]

75-40150 Mitel Telecom Ltd
(a subs. of Mitel Corp.)
Headquarters
Mitel Business Park, Portskewett, Gwent NP6 4YR
☎(+44 1291) 43 00 00 497360
⊖ (+44 1291) 43 04 00
♣ £ 60 000 000 850 300 000 sq ft
Managing Dir: Alan Kirkham
Sales Dir: Cliff Tomaszewski
Head of Marketing: Tim Redpath
MoD Sales: Ian Crombie
Business Admin: Ms Ros Chapman
■ Manufacture PABX's up to 3500 lines, call cost accounting systems, feature telephones and a range of semiconductors for telecommunictions and computer use. → 80 [97]

75-40160 MJ Technologies
Whitwick Business Park, Coalville, Leicestershire LE67 4JP
☎(+44 1530) 81 55 01 ⊖ (+44 1530) 81 15 19
♣ £ 3,000,000 30 10,000 sq ft
Joint Managing Director: M G Whitmore
Joint Managing Director: J E Duffin
Marketing Manager: Mrs H V Hockedy
■ Manufacturers of advanced process systems for aerospace and automotive industries including laser machining centres, EDM, ECM, stem and capillary systems for drilling, cutting and welding of complex components. Total solution for manufacturing, including inbuilt inspection, auto load/unload, full process performance monitoring and control. Products include Delta and Alpha multi-axis CNC laser systems (compact, high capacity, high speed performance); Robo Tech EDM systems with vision inspection facility and refeed electrode; Trio and Duo Tech EDM systems for flexible multi-hole drilling applications. Maxi/Midi/Mini Tech EDM machines. [98]

75-40170 MK Services Ltd
24 Eden Way, Pages Ind Est, Billington Rd,
Leighton Buzzard, Beds. LU7 8TZ
☎(+44 1525) 38 23 33 ⊖ (+44 1525) 85 00 73
♣ £ 1 100 000 11
Managing Dir: M.P. Kirby
Sales & Marketing Mgr: A.D. Lawton
■ Suppliers of marine geophysical & hydrographic survey equipment & consumables, ROVs, seabed mapping sonars, towed array positioning systems, enhanced sonar & seismic processors & an extensive range of acoustic transponders, flotation & imaging systems. [97]

75-40200 ML Aviation Ltd

(a subsidiary of ML Holdings plc)
Arkay House, Weyhill Road,
Andover, Hampshire SP10 3NR
☎ (+44 1264) 33 33 22 ⊠ 47692 walind g
⊖ (+44 1264) 33 21 86; 33 56 13
♣ c. £ 20,000,000 (1997) 👥 204
Managing Director: R C Bolter
Sales Director: C T Cornish
Purchasing Manager: Angela Woolford
■ Airborne weapon carriage & release systems, including ejector release units, rail launchers, multi-store bomb carriers, practice bomb carriers, chaff & flare decoy dispensers. Complete aircraft weapon systems. Ordnance safety devices including electronic fuses & radio frequency attenuating connectors. Incorporating ML Wallop Defence Systems.

Branch Office
ML Building, 644 Ajax Avenue, Slough, Berkshire SL1 4BQ
☎ (+44 1753) 52 38 38 ⊠ 847633 mla slg
⊖ (+44 1753) 53 23 21

● Craydown Lane, Middle Wallop,
Stockbridge, Hampshire SO20 8DX
☎ (+44 1264) 78 14 56 ⊖ (+44 1264) 78 20 84
Business Mgr Pyrotechnics: J. Alexander [98]

75-40230 ML Lifeguard Equipment Ltd

(a subsidiary of ML Holdings plc)
Canol-Y-Dre, Ruthin, Denbighshire LL15 1YU
☎ (+44 1824) 70 43 14 ⊠ 61687
⊖ (+44 1824) 70 57 01
👥 185
Managing Director: A Baxendine
Head of Marketing: G Billington
Technical Director: C A Edmond
■ Military aircrew/naval protective clothing & equipment. Life saving equipment, life rafts & inflatable boats for defence commercial marine purposes. Anti-riot, anti-terrorist & bomb disposal equipment. Life support division assesses, tests & qualifies aircrew personal equipment.

Branch Office
Vauxhall Industrial Estate, Johnstown Road,
Ruabon, Wrexham LL14 6HA
☎ (+44 1978) 81 00 07; 81 00 46
⊖ (+44 1978) 81 00 48 [98]

75-40240 ML Wallop Defence Systems Ltd

(† Wallop Industries Ltd)
(a division of ML Aviation Ltd)
Craydown Lane, Middle Wallop,
Stockbridge, Hampshire SO20 8DX
☎ (+44 1264) 78 14 56 ⊠ 47692 walind g
⊖ (+44 1264) 78 20 84
♣ £ 12,500,000 (1995/96) 👥 45
General Manager: J G Alexander
Sales Manager, Pyro: A N Gibson
Technical Manager: J E Parsons
■ Pyrotechnics, infrared and chaff decoys, airborne, shipborne and ground-based passive defence systems. Controlled aerial delivery systems (remotely guided payload parachute). [98]

75-40540 Molecular Products Ltd

Mill End, Thaxted, Essex CM6 2LT
☎ (+44 1371) 83 06 76
[E] sales@molecularproducts.co.uk
⊖ (+44 1371) 83 09 98
👥 62
Managing Dir: Ian McKernan
Sales & Marketing Dir: Mick Mahon
Technical Sales Exec.: Darryl Mansfield
■ Chemical absorbents & catalysts for purification of breathable gases. Oxygen generating chemicals for life support in closed environments: submarines, diving equipment, closed circuit breathing apparatus, shelters & medical equipment. [98]

75-40548 Mondiale Security Ltd

(a partner of Capital Group)
2 D'Eynsford Road, London SE
☎ (+44 171) 252 55 01 ⊖ (+44 171) 277 40 50
👥 12 ⌂ 4 500 sq ft
Managing Dir: Arnold Harrison
■ Specialist vehicles, discreetly armoured limousines & associated products. [97]

75-40600 Montford Instruments Ltd

4 McNicol Drive, Park Royal, London NW10 7AW
☎ (+44 181) 965 03 61 ⊖ (+44 181) 965 36 21
Managing Director: G Martell
■ Design, development and manufacture of environmental test chambers and stress screening equipment. [98]

75-40670 Moog Controls Ltd

(a subs. of Moog Inc., USA)
Ashchurch, Tewkesbury, Glos GL20 8NA
☎ (+44 1684) 29 66 00 ⊠ 43229
⊖ (+44 1684) 29 67 60
♣ US$ 20 285 000 (1995) 👥 168 ⌂ 6 000 sq m
Managing Dir: Ron Livecchi
Chief Aerospace Engineer: Brian Ballard
Engineering Dir: Nigel Cottell
■ Electro-hydraulic/pneumatic and electro-mechanical servo control systems and components, valves, actuators, motors and gassing systems. [97]

75-40860 The Morgan Crucible Co. plc

Morgan House, Madeira Walk, Windsor, Berks SL4 1EP
☎ (+44 1753) 83 70 00 [E] mail@morgancrucible.com
[W] http:// www.morgancrucible.com ⊠ 849025
⊖ (+44 1753) 85 08 72
♣ £ 847,800,000 (1995) 👥 13,007 (1995)
Group Managing Director: Dr Bruce Farmer
Finance Director: Graham Swetman
Subsidiary company:
MBM Technology Ltd → 75 [98]

75-40865 Morgan Marine Ltd

Llandybie, Ammanford, Dyfed SA18 3JG
☎ (+44 1269) 85 04 37 ⊖ (+44 1269) 85 06 56
♣ £ 5,000,000 (1997) 👥 100 ⌂ 7,000 sq m
Managing Director: R H C Morgan
Operations: Martin W Ingram
Sales and Marketing: R I Young
Purchasing: T Mathews
■ Transportable prefabricated buildings (TPBs) constructed from glass reinforced polyester (GRP) composite materials with options for traditional and environmentally acceptable appearances (simulated brick or stone). Buildings can incorporate RF protection, EMI screening and NEMP shielding. [98]

75-40945 Morse Controls Ltd

IMO Industries Inc. USA
Christopher Martin Rd, Basildon, Essex SS14 3ES
☎ (+44 1268) 52 28 61 ⊠ 99237
⊖ (+44 1268) 28 29 94
♣ £ 12 000 000 👥 226
Managing Dir: D.H. Owen
Sales & Marketing Dir: Douglas J. Taylor
Chief Designer: N. Sadeghi
■ Aircrew seats & harnesses, high/low temperature remote control systems, cable tension regulators. Mechanical & electromechanical actuators. [97]

75-40950 Morse Equipment

(a subs. of Sedgewall Ltd)
1 Apex Business Centre, Boscombe Road,
Dunstable, Bedfordshire LU5 4SB
☎ (+44 1582) 47 55 55 ⊠ 82362 bateco g
⊖ (+44 1582) 47 55 53
♣ £ 2 000 000 👥 35 ⌂ 650 sq m
Managing Director: A Becker
■ Morse training equipment, morse keys, tactical keys; field equipment, handsets, headphones & microphones.
See also:
CAE (75-11125) [98]

75-41000 G.E. Mortley, Sprague & Co. Ltd

Lyons Crescent, Tonbridge, Kent TN9 1ET
☎ (+44 1732) 77 03 58 ⊖ (+44 1732) 77 06 50
♧ £ 50 000 ♣ £ 1 000 000 👥 25
Managing Dir: G.E. Mortley
General Mgr: R.F. Mungeam
Company Secretary: M. Freestone
■ Manufacturer of hand-driven generators for military man-pack transceivers & printed circuit boards & electronic assemblies. [97]

75-41050 Motorola Ltd Space and Systems Technology Group

(a subsidiary of Motorola Inc, USA)
Intech House, Wilbury Way,
Hitchin, Hertfordshire SG4 0AP
☎ (+44 1462) 42 14 40 ⊖ (+44 1462) 42 13 48
Chairman: Mike Alderson
General Manager: Robert Richards [98]

75-41100 Mountain Range

PO Box 1, Alston, Cumbria CA9 3TS
☎ (+44 1434) 38 17 42 ⊖ (+44 1434) 38 18 05
♣ £ 2 000 000 👥 73
Managing Dir: D.C. Udberg
Operations & Technical Dir: Ms May Blackstock
Financial Dir: Ken Sayer
■ Manufacturers of waterproof clothing & equipment. [97]

75-41210 MPE Ltd

(† Wego Condenser Co. Ltd)
(a member of Adwest Group)
PO Box 11, Liverpool L33 7UL
Hammond Road, Kirby Industrial Estate
☎ (+44 151) 547 26 00 ⊖ (+44 151) 546 46 42
♣ £ 4 765 000 👥 133
Managing Dir: Peter Cotterill
Technical Dir: Jan Nalborzyk
Sales & Marketing: Michael Johnson
■ EMC/RFI electrical filters & shields, specialist plastic film, ceramic & oil impregnated capacitors. EMC/RFI system analysis, test & solutions. [97]

75-41260 MPL Powerware Systems Ltd

Prescott Road, Poyle, Colnbrook, Berkshire SL3 0BE
☎ (+44 1753) 68 62 00 [W] http://www.exide.com
⊖ (+44 1753) 68 68 27
👥 50 (1997)
Chairman: S Smith
General Manager: R Potts
Sales and Marketing Manager: C Chilton
Service Manager: Robert Miller
■ Uninterruptible power supplies (UPS). [98]

75-41265 MS Instruments plc

Electron House, 33-49 Farwig Lane,
Bromley, Kent BR1 3RE
☎ (+44 181) 290 02 00
[E] firstname_surname@msinstruments.co.uk
[W] http://www.msinstruments.co.uk/
⊖ (+44 181) 464 65 96
♣ £ 1,500,000 👥 20
Director/General Manager: P A H Everington
Managing Director: Dr B D McCarthy
Operations and Technical Director: A H Philbrook
■ Instrumentation for the measurement of exact position, velocity, rate of fire & direction of origin of ballistic ammunition, for use in the development of weapons and ammunition, training of marksmen & helicopter hostile fire indication. [98]

75-41270 MSA (Britain) Limited

(a subsidiary of Mine Safety Appliances Company, USA)
East Shawhead, Coatbridge ML5 4TD
☎ (+44 1236) 42 49 66 [E] info@msabritain.co.uk
[W] http:// www.msabritain.co.uk/msa ⊠ 778396 minsaf g
⊖ (+44 1236) 44 08 81
♧ £ 160,000 (1997) ♣ £ 10,000,000 (1997)
👥 150 (1997) ⌂ 13,000 m2
Managing Director: Nigel J B Trodd
Sales and Marketing Director: Douglas H C Cunningham
Purchasing Manager: Gerry Sweeney
■ Breathing apparatus, compressed air, full facepiece masks, gas and particulate filters, chemical suits, full enclosure, air regeneration systems, chemical oxygen breathing apparatus, self-rescue devices, safety helmets. [98]

75-41280 MSI-Defence Systems Ltd

(a div. of MS International plc)
Salhouse Rd, Norwich, Norfolk NR7 9AY
☎ (+44 1603) 48 40 65 ⊠ 975067 desnor g
⊖ (+44 1603) 41 56 49
♧ £ 3 500 000 ♣ £ 9 000 000 👥 180 ⌂ 7 000 sq m
Managing Dir: G.P. Hough
Marketing Dir: Barry Larke
Engineering Dir: Rodney F. Smith

Contact Reference

■ Design, manufacture & support of naval defence equipment & tactical shelters/cabins including power driven stabilised weapon & sensor platforms, target designation sights, navigational & tactical plotting tables, submarine masts & deployment systems. Full service, spares, training & post design services.

Support Services
10 Cambridge Road, Granby Ind.Est.,
Weymouth, Dorset DT4 9XA
☎(+44 1305) 76 01 11 ⊠ 418251 desway g
⊖ (+44 1305) 76 02 22
■ Provides complete in-service support to defence industries & related government establishments. Project management, upkeep & post-design services, support & introduction into service, documentation services. Specialist life support & military diving equipment. Sonar directing gear, explosive ordnance recognition & disposal equipment. 97

75-41330 MTH Ltd
42 Queens Road, Farnborough, Hampshire GU14 6DT
☎(+44 1252) 51 92 51 ⊖ (+44 1252) 52 44 94
♣ £ 250,000 (1997) 👥 5 (1997)
Managing Dir: J. Green
Technical Dir: B. Beesley
■ Design and manufacture of data acquisition, control systems and software for MoD repair and maintenance of in service equipment. Manufacture of "build-to-print" electro-mechanical assemblies. → 80 98

75-41404 Muirhead Vactric Components Ltd
(a subsidiary of Silvermines Group plc)
Oakfield Road, Penge, London SE20 8EW
☎(+44 181) 659 90 90 [E] sales@muirheadvactric
⊖ (+44 181) 659 99 06
♧ £ 7,000,000 (1995) ♣ £ 10,000,000 (1997)
👥 200 (1997) 🏭 72,000 sq ft
Managing Director: Douglas Dunbar
Sales Director: Mrs Jean Glasspool
Technical Marketing Director: Dr Brian Bakewell
Director of Quality and Engineering Operations: David Francis
Purchasing Manager: Peter Simpson
■ Manufacturer of servo components and control systems to ISO 9001. Products include synchros/resolvers, AC & DC motors, engine control components, LVDT's/RVDT's, torque motors, shaft encoders, gearheads and stepping motors.
® USA: Performance Control Inc, Pickering Controls Inc 98

75-41550 Muller England Limited
(† Muller & Co. (England) Ltd)
Cleobury Mortimer, Kidderminster, Worchester DY14 8DT
☎(+44 1299) 27 02 71 ⊖ (+44 1299) 27 08 77
♧ £ 3,000,000 ♣ £ 12,000,000 (1997) 👥 260 (1997)
🏭 12,00 m2 (1997)
Managing Director: G K Farr
Sales Director: P R Bethell
Purchasing Director: G C Broome
Operations/Technical Manager: D Bamford
■ Manufacture components for aerospace, arms & weapons industries & complete weapon systems. 98

75-41600 Mullion Manufacturing Ltd
44 Northfarm Road, South Park Industrial Estate,
Scunthorpe, South Humberside DN17 2AY
☎(+44 1724) 28 00 77 ⊖ (+44 1724) 28 01 46
👥 30
Managing Dir: Philip Crossley
Purchasing Mgr: J. Goodwin
Sales Coordinator: A.L. Boulton
Sales Manager: I Crampton
■ Manufacture a wide range of buoyancy & thermal protective safety clothing and equipment. 98

75-41690 Multipoint Communications Ltd
Satellite House, Eastways Industrial Park,
Witham, Essex CM8 3YQ
☎(+44 1376) 51 08 81 ⊠ 987313 mpoint g
⊖ (+44 1376) 50 22 33
♣ £ 5 000 000 👥 35 🏭 15 000 sq ft
Chairman & Marketing Dir: J.D. Rogers
Managing Dir: S.J. Rogers
Technical Dir: J. Kennett
Technical Dir: D. Atkinson
Operations Dir: D. Stringer
Purchasing Mgr: M. Williams
■ Design, manufacture & installation of fixed satellite earth stations & mobile earth stations & VSAT networks. 97

75-41695 IBV Multipower Ltd
(a subs. of I.B.V. Ltd)
Unit 8, Harmill Industrial Estate, Grovebury Rd,
Leighton Buzzard, Beds. LU7 8SQ
☎(+44 1525) 37 17 72 ⊖ (+44 1525) 37 07 66
👥 40
Managing Dir: L. Hamilton
Operations/Technical Dir: N. Beavan
Advertising Mgr: Mrs Wendy Kay
Sales & Marketing: Barry Norton
■ Batteries and chargers for handheld & military radios. 97

75-41700 Multitone Electronics plc
Kimbell Road, Basingstoke, Hampshire RG22 4AD
☎(+44 1256) 32 02 92 [E] info@multitone.co.uk
⊖ (+44 1256) 46 26 43
👥 390
Managing Director: Michael Walker
Technical Director: Terry Seddon
Sales and Marketing Director: Richard Jeffries
Marketing Communications Manager: Ms Anna Sutherland
Marketing Co-ordinator: Ms Sarah Chukoury
■ Designs, manufactures and markets a wide range of radio paging systems covering a variety of applications. ISO9001 and ISO9002 approved. 98

75-41780 Murata Electronics (UK) Ltd
Oak House, Ancells Road, Ancells Business Park,
Fleet, Hampshire GU13 8UN
☎(+44 252) 81 16 66 ⊖ (+44 252) 81 17 77
♣ £ 30,000,000 (UK only) 👥 280
Marketing Dir: Mike Vrettos
Business Dev. Mgr, Military Sector: Bill Chisholm
■ Electronic & electro-mechanical components.
® Japan: Murata Mfg Co. Ltd
USA: Murata Erie N.America 98

75-41870 Musto Ltd
Christy Way, Laindon, Essex SS15 6TR
☎(+44 1268) 49 15 55 ⊠ 995466 mustac g
⊖ (+44 1268) 49 14 40
♣ £ 5,000,000 👥 130
Managing Director: F Keith Musto
Assistant Managing Director: Nigel Musto
Sales Director: Paul McKee
■ Manfacture protective clothing. 98

75-41890 Nanoquest Defence Products Ltd
(† Vickers Instruments)
(a division of British Aerospace plc)
Green Park Business Centre,
Sutton-on-the-Forest, York YO6 1ET
☎(+44 1347) 81 12 34 ⊖ (+44 1347) 81 12 13
♣ £ 7,000,000 (1996) 👥 50 (1996) 🏭 15,000 sq ft
Site Director: Paul Johnson
Company Secretary: Paul Sapsford
UK Sales Mgr: K.A. Lansley
Sales & Marketing Mgr: Ken Jarvis
Operations Mgr: Martin Worrall
■ Design, development & manufacture optical, opto-mechanical & opto-electronic sighting, alignment & harmonisation & simulation systems for military applications. 98

75-41900 Nautronix Ltd
(† Nautronix Helle Ltd)
(a subsidiary of Nautronix Ltd, Australia)
Nautronix House, Howe Moss Avenue, Kirkhill,
Dyce, Aberdeen AB21 0GP
☎(+44 1224) 77 57 00 [E] info@nautronix.com.au
[W] http://www.nautronix.com.au ⊖ (+44 1224) 77 58 00
♧ £ 1,000 (1996) ♣ £ 2,315,615 (1996) 👥 26 (1997)
🏭 10,000 m2
Managing Director: Mark Patterson
Technical Manager: David Maxwell
Commercial/Internal Sales Manager: Donald Thomson
Manufacturing Manager: Lindsay MacDonald
Quality/IT Manager: Lynda Ritchie
Accountant and Company Secretary: Neil Smith
Materials/Purchasing Supervisor: Mike Laughlin
■ Dynamic positioning systems, acoustic positioning & tracking systems. Design & manufacture quality underwater products & systems, including underwater communication systems, acoustic tracking & relocation systems, pingers, sonars, wire diver phones & helium unscramblers. 98

75-41955 Navstar Systems Ltd
Mansard Close, Westgate, Northampton NN5 5DL
☎(+44 1604) 58 55 88
[W] http://www.telecom.com/navstar
⊖ (+44 1604) 58 55 99
♣ £ 4,000,000 👥 55
Managing Director: R Duckman
Technical Director: Tony Haddrell
Accounting Manager: D Jenkinson
Business Development Manager: Clive De La Fuente
Sales Manager: M Samson
■ XR5.M Military GPS Receiver, 6 and 12 channel versions. XR5 Kernel-OEM GPS boardsets. UHF datalink. Centimetre accuracy systems. 98

75-42270 Neal Recording Systems Ltd
(† Lee James Electronics Ltd)
(a division of Canford Audio plc)
1 Hanlon Court, Royal Industrial Estate,
Jarrow, Tyne and Wear NE32 3HR
☎(+44 191) 489 93 79 ⊖ (+44 191) 483 35 95
♣ £ 1,000,000 (1996) 👥 25 (1996) 🏭 1,100 sq m
Managing Director: Dr D L D Mitchell
Operations Director: M Hindmarsh
Technical Sales Manager: Michael Sylph
Purchasing Manager: H Mann
■ Fixed location & portable communications recording systems. Cassette recorders, cassette transcribers. Audio & video interview recorders. Digital voice recorders. 98

75-42310 Neotronics Ltd
Parsonage Road, Takeley,
Bishops Stortford, Herts CM22 6PU
☎(+44 1279) 87 01 82 ⊠ 817126 neotro g
⊖ (+44 1279) 87 03 77
♣ £ 20,114,000 (1995)
■ Electronic gas detectors, monitors and analysers. 98

75-42318 Nera Telecommunications
(† Continental Microwave Technology Ltd)
171 Camford Way, Sundon Park,
Luton, Bedfordshire LU3 3AN
☎(+44 1582) 49 11 49 ⊠ 826877 cmlg
⊖ (+44 1582) 58 18 73
👥 140
Managing Dir: A. Lockwood
Operations & Technical Dir: D. Cowan
European Sales & Marketing Mgr: Peter Moss
■ Development & manufacture of oscillators, transponders, amplifiers & systems for military satcoms, radar & defence applications. Oscillators, amplifiers, converters, receivers, SNG transportable. Fixed ground stations for satellite communications. 97

75-42320 New Chapel Electronics Ltd
London Road, Fairford, Glos GL7 4DS
☎(+44 1285) 71 21 44 ⊖ (+44 1285) 71 31 43
👥 100 🏭 200 sq m
Managing Dir: Piers Hubbard-Miles
Technical Dir: M. May
Purchasing Mgr: R. Williams
■ Design & manufacture cable & wiring assemblies & connectors. 97

75-42330 Newage International Ltd
Barnack Road, PO Box 17, Stamford, Lincolnshire PE9 2NB
☎(+44 1780) 48 40 00 ✉ NEWAGE G ⊠ 32268 g
⊖ (+44 1780) 48 41 00
♣ £ 65,000,000 👥 950
Managing Dir: S.L. Zeller
Marketing Services Mgr: N. Sellwood
Sales Mgr Europe: Rod Logan
■ Design, manufacture & service of ground support AC generators from 0.5 to 3375kva. 98

75-42520 Newton Derby Ltd
Alfreton Road, Derby DE21 4AG
☎(+44 1332) 34 76 76 ⊖ (+44 1332) 29 45 29
👥 60 🏭 112,000 sq ft

Managing Director: P D Keely
Operations and Technical: A Palmer
Sales and Marketing: D Butcher
Purchasing Manager: Mrs M K Reed
■ Design, develop & manufacture (with post design services) electrical power supply & generating systems, static & mobile, for ground, aircraft & shipboard use. 98

75-42555 NIS Ltd
Ackhurst Road, Chorley, Lancashire PR7 1NH
☎(+44 1257) 26 56 56 ⊜ (+44 1257) 27 55 01
250 (1997) 25,000 m2
Managing Director: Liam Ferguson
Operations Director: Pat Jackson
Sales and Marketing Director: Phil Robson
Sales Manager: I L Dickson
Purchasing Manager: Ken Booth
Group Marketing Assistant: J Justin Shaw
■ Engineering design, development, manufacturing, testing, installation and commissioning of electro-mechanical systems and equipment. 98

75-42570 Nitech Ltd
4-6 Highfield Business Park, Churchfields Trading Estate, St Leonards-on-Sea TN38 9UB
East Sussex
☎(+44 1424) 85 27 88 ⊜ (+44 1424) 85 10 08
18
Managing Director: Peter F Barker
Financial Director: Laurie Matthews
MoD Sales Co-ordinator: Mrs Maggie Wakeline
Sales and Marketing Co-ordinator: Mrs Julie Hyland
■ Manufacture and distribution of X-cell range of rechargeable handlamps and batteries, portable lighting, floodlighting and power systems for defence, security and emergency use. 98

75-42580 Nitram Vacuum Heat Treatment Co. Ltd
Silverdale Rd, Hayes, Middx UB3 3BN
☎(+44 181) 573 51 11 ⊜ (+44 181) 756 10 23
4
Marketing & Sales Dir: D.R. King
■ Vacuum heat treatment services. 97

75-42645 Nobo (UK) Limited
Alder Close, Compton Industrial Estate, Eastbourne, East Sussex BN23 6QB
☎(+44 1323) 64 15 21 E info@nobo.co.uk
⊜ (+44 1323) 41 03 28
250 (1997) 105,000 m2
Managing Director: D Anderson
Operations Director: John Britton
Marketing Administration Manager: Mrs R J Taylor-Byrne
Export Projects Manager: J Dellar
Purchasing Manager: Mrs S Richardson
■ Flipcharts & other training aids. Notice boards, whiteboards, lecterns, letter boards, magnetic maps & display systems. Overhead projectors & accessories, conference furniture. 98

75-42700 Normalair-Garrett Limited
(a subsidiary of AlliedSignal Aerospace, USA)
Yeovil, Somerset BA20 2YD
☎(+44 1935) 47 51 81 ⊜ (+44 1935) 42 76 00
Managing Director: Dr Peter J Burrows
☎(+44 1935) 44 62 55; 44 65 71 (fax)
Quality Director: Peter F Bayliss ☎(+44 1935) 44 62 51
Personnel Director: Roger A Gard
☎(+44 1935) 44 62 10; 44 65 83 (fax)
Engineering Director: George R Giles
☎(+44 1935) 44 62 57
Finance Director: Duncan A McKechnie
☎(+44 1935) 44 64 43
Director: Michael A O'Loughlin ☎(+44 1935) 44 66 66; 44 65 77 (fax)
Director, Sales, Marketing and Customer Support: Mike G Tutcher ☎(+44 1935) 44 64 55
Operations Director: Mike K Wills ☎(+44 1935) 44 61 22
■ Manufacture and sale of electromechanical, electronic and environmental control systems and hydraulic equipment. Administrative office. → 90 98

75-42727 The North Face (Europe) Ltd
(a subsidiary of The North Face Inc, USA)
Industrial Estate, PO Box 16,
Port Glasgow, Renfrewshire PA14 5XL
☎(+44 1475) 74 13 44 ⊜ (+44 1475) 74 45 72
♧ £ 5,275,000 (1996) ♣ £ 18,227,000 (1996)
145 (1996) 10,000 sq m
Chief Executive Officer: Karl Heinz Salzburger
Director of European Development: Jean Pascal Papin
General Manager: Duncan J McCallum
Director of Marketing: Stephen L Jessey
Director of Product Development: Charlie Campbell
Factory Manager: Ian T Spence
Purchasing Manager: Margaret Ludgate
Financial Controller: Ms Una L MacDonald
■ Designs and distributes technically sophisticated outerwear, skiwear, functional sportswear, tents, sleeping bags and back packs. → 85 98

75-42730 James North & Sons Ltd
(a subs. of Siebe plc, UK)
Market Street, PO Box 3, Hyde, Cheshire SK14 1RL
☎(+44 161) 368 58 11 ☒ 669332
⊜ (+44 161) 367 85 71
♣ £ 12 000 000 200
Managing Director: R A Dickens
Sales and Marketing Director: J C Thorne
Electrical Equipment Supply Manager: Mrs Sarah Thompson
Product Manager: Ms Helen Bond
Product Manager: Sam Glover
Purchasing Manager: Ms Margaret Hanson
■ Specialists in protective clothing and safety equipment including head protection, eye protection, ear protection, respiratory chemical clothing and hand protection. 98

75-42740 The Northern Rubber Company Ltd
(a subsidiary of Tomkins plc)
Retford, Nottinghamshire DN22 6HH
☎(+44 1777) 70 67 31 ☒ RUBBER
⊜ (+44 1777) 70 97 39
♣ £ 10,000,000 - 15,000,000 300 25,000 sq m
Managing Dir: D.E.P. Owen
Sales & Marketing Dir: M. Thompson
Operations Dir: M. Bowskill
Purchasing Mgr: T. Grant
■ Lightweight silicone/glass & neoprene/nylon shapes. Wire supported ducts, flanges, bellows, sleeves, mouldings, sheeting, extrusions, fluorosilicone clamp loops & coated fabrics. Silicone/Viton composite ablative tile & lining mouldings. All polymers including silicone & viton. Rubber flashings & expansion joints. Rocket motor linings. 98

75-42760 Nova Marine Systems Ltd
Rodney Rd, Portsmouth, Hampshire PO4 8SG
☎(+44 1705) 77 50 00 ⊜ (+44 1705) 81 90 87
♣ £ 1 273 672 22
Dir & General Mgr: A.N. Higgins
Director: D.R. Evans
Director: J.J Lynn
■ Electronic marine search & rescue products. Radio control centre equipment. 97

75-42770 NSW Technology Limited
The Innovation Centre, Aberdeen,
Offshore Tech. Pk, Exploration Drive,
Bridge of Don, Aberdeen AB23 8GX
☎(+44 1224) 70 35 95 ⊜ (+44 1224) 70 35 96
♣ £ 500,000 (approx.) 6
Managing Director: Angus Wing
Sales/Marketing Director: Ken Mitchell
■ Supplier of NSW's range of custom-designed electrical, FO and electro-hydraulic cables and umbilicals for subsea control systems. Applications include diving systems, subsea trenching machines, ROVs, control, instrumentation, TV, telecomms and power supplies; mechanical and electrical terminations. UK agents for Subconn and GISMA U/W electrical & FO connectors, PMI cable accessories and POLYNET surface protection nets.

Subconn Electrical Connectors c/o NSW Technology Ltd
The Innovation Centre, Aberdeen,
Offshore Tech. Pk, Exploration Drive,
Bridge of Don, Aberdeen AB23 8GX
☎(+44 1224) 70 35 95 ⊜ (+44 1224) 70 35 96
■ A range of underwater-pluggable electrical and FO connectors including low-cost circular and low profile series, field-installable and military application specials.

Poly-Net UK
(a div. of NSW Technology Ltd)
14 C.R.Bates Industrial Est, Wycombe Road, Stokenchurch, Bucks HP14 3RJ
☎(+44 1494) 48 55 09 ⊜ (+44 1494) 48 21 61
Divisional Mgr: Ms Barbara Potts
■ Poly-Net is a protective netting made from Polyethylene, supplied in tubular and flat-sheet form, for protection of high finished surfaces. 98

75-42850 Nullifire Ltd
(a div. of Dore Holdings)
Torrington Ave, Coventry CV4 9TJ
☎(+44 1203) 85 50 00 ⊜ (+44 1203) 46 95 47
♣ £ 5 000 000 (1995) 100
Managing Dir: Nigel Maris
Sales & Marketing Mgr: Geoff Pimm
Purchasing Mgr: Soolie Lowe
■ Supply & coating of fire protective intumescent coatings & seals to munitions, canistors, actuators & valves. 97

75-42875 N.V. Tools Ltd
St James's Rd, Brentwood, Essex CM14 4LL
☎(+44 1277) 21 44 55 ☒ 99432
⊜ (+44 1277) 22 73 41
♧ £ 30 000 ♣ £ 246 431 863 108 40 000 sq ft
Chairman: C.F. Bennett
Managing Dir: Peter R. Bennett
General Mgr: T. Mullen
■ Defence-related contracts for MoD & overseas governments involving precision engineering and/or electronics. 97

75-42895 D.G. O'Brien UK Ltd
(a division of D.G. O'Brien Inc, USA)
Leigh Court, Suite 3, Pill Road,
Abbots Leigh, Avon BS8 3RA
☎(+44 1275) 37 51 51 ⊜ (+44 1275) 37 51 56
General Mgr: Michael H. Pachul
■ Design & manufacture of electrical & optical connectors, cable assemblies & pressure hull fittings for use on submarines, research vehicles & in hostile environments. 98

75-42900 Ocean Dynamics
Cross Square, St Davids, Pembrokeshire SA62 6SL
☎(+44 1437) 72 13 90; 72 16 86
⊜ (+44 1437) 72 07 47
♧ £ 750 000 12
Managing Dir: S.L. White
Sales & Marketing Dir: J. Cowling
■ Consultants & manufacturers of specialised craft & equipment, high-speed, all weather, jet powered rigid-hulled inflatable boats, for attack, rescue, high surf & covert operations. → 80 98

75-42920 Octavius Hunt Ltd
(a subs. of Chemring Group PLC)
Dove Lane, Redfield, Bristol BS5 9NQ
☎(+44 117) 955 53 04 ⊜ (+44 117) 955 78 75
♣ £ 1 850 000 50 30 000 sq ft
Managing Dir: Ms Jan Reynolds
Director: D.R. Evans
Director: P.S. Maton
Director: M. Monks
■ Windproof/waterproof matches, fusée matches & smoke grenades, supplied in waterproof packaging or boxes for inclusion in any size of field rations or survival kits. 97

75-42950 Octec Ltd
The Western Centre, Western Road,
Bracknell, Berkshire RG12 1RW
☎(+44 1344) 86 10 51 E octec@octec.co.uk
W http://www. octec.co.uk ⊜ (+44 1344) 86 09 83
♣ £ 2,000,000 25
Managing Director: Gordon A Cain
■ Automatic video target tracking and position analysis systems. 98

75-42967 Oki Systems (UK) Ltd
(a subsidiary of Oki Electric Industry Co Ltd, Japan)
550 Dundee Road, Slough Trading Estate,
Slough, Berkshire SL1 4LE
☎(+44 1753) 81 98 19 ☒ 848716
⊜ (+44 1753) 81 98 99
♣ £ 27,000,000 100
Managing Director: H C C Jones
Marketing and Sales Director: Stephen Martin
■ Manufactures dot matrix and LED/laser printers. 98

Contact Reference

75-42985 Oldbury Components Ltd
(† Oldbury Trailers Ltd)
Bull Lane, Wednesbury, West Midlands WS10 8RW
☎ (+44 121) 502 14 11 ⊖ (+44 121) 556 70 25
♣ £ 2,000,000 (1997) 👥 20 (1997)
🏭 1,000 sq m (1997)
Managing Director: B J Lang
Technical Director: Clifford Large
Financial Director: Peter J Wood
Operations Manager: Douglas J Pearson
Purchasing Manager: Robert Cregeen
■ Special-purpose trailers for military requirements. Trailer axles & suspension systems, miscellaneous trailer components. [98]

75-43020 Olympus Industrial
KeyMed (Medical & Industrial Equipment) Ltd
(† Olympus KMI)
(a subsidiary of Olympus Optical Co Ltd, Japan)
KeyMed House, Stock Road,
Southend-on-Sea, Essex SS2 5QH
☎ (+44 1702) 61 63 33
[E] 106160.1063@compuserve.com ☒ 995283
⊖ (+44 1702) 46 56 77
♧ £ 1,010,000 (1996) ♣ £ 70,000,000 (1996)
👥 560 (1996) 🏭 12,380 sq m
Head International Sales: Mike Batty
Sales and Marketing Director: P A Hillman
International Marketing Manager: Ms J Walker
Division Manager, Industrial Sales: M A E Johnson
European Sales Manager: Richard Armstrong
Middle East Sales Mgr: Gary Woodside
UK Sales Manager: G D Lewis
■ Supply remote visual inspection equipment: fibrescopes, borescopes, videoscopes, video-imaging systems for internal inspection of engines, gas turbines etc, and for search and surveillance applications. Full advisory, repair and back-up services available worldwide. → 80 [98]

75-43050 Omitec Electro-Optics Ltd
(a subsidiary of O.M.I. International)
Parc Dyffyn Industrial Estate, Prestatyn, Clwyd
☎ (+44 1745) 88 62 22 ⊖ (+44 1745) 68 65 50
♣ £ 6,500,000 (1997) 👥 95 (1997)
Managing Director: David Marks
Operations Director: Mac Picken
Purchasing Manager: Simon Birchenough
■ Design and manufacture of advanced optical systems for military or aerospace applications. [98]

75-43130 Optex
22-26 Victoria Road, New Barnet, Hertfordshire EN4 9PF
☎ (+44 181) 441 21 99 [E] info@optexint.com
[W] http://www.optexint.com ⊖ (+44 181) 449 36 46
👥 60 (1997)
Managing Director: R E Collins
Operations Director: J Maciejewki
Sales and Marketing Director: G P Chappell
Senior Sales Executive: Neil D Davies
Purchasing Manager: A Khan
■ Communications equipment, microwave links for broadcast TV and covert work. Also miniature TV, TV cameras, data links etc. → 85 [98]

75-43135 Optima Enclosures Ltd
Macmerry Industrial Estate,
Tranent, East Lothian EH33 1EX
☎ (+44 1875) 61 07 47 ⊖ (+44 1875) 61 24 86
♣ £ 7,000,000 (1996) 👥 120 (1997)
Managing Director: David Powell
Financial Controller: Scott Lindsay
Field Sales Manager: Pat Sherry
Operations Manager: Tommy Smith
Sales Office Supervisor: Evan Bell
■ Manufacture of standard & customised electronic enclosures, including tempest. RFI shielded 19" racks. [98]

75-43140 Ormiston Wire Ltd
(† P. Ormiston & Sons Ltd)
1 Fleming Way, Worton Road,
Isleworth, Middlesex TW7 6EU
☎ (+44 181) 569 72 87 ✉ ORMISTON
⊖ (+44 181) 569 86 01
♧ £ 300,000 (1995) ♣ £ 1,400,000 (1995) 👥 27
🏭 10,000 sq ft
Managing Dir: M.V. Ormiston
Technical Dir: K.J. Barker
Purchasing Manager: Frances Delaney
■ Non-ferrous & stainless steel, fine wires, braided wires, stranded wires & aerial wires. [98]

75-43160 Orseal Ltd
Orseal Valves
(† Richard Threlfall Ltd)
(a subsidiary of Richard Threlfall Group Ltd)
Bridgeman Place Works, Salop Street,
Bolton, Lancashire BL2 1DD
☎ (+44 1204) 274 31 (6 lines) ✉ THRELFALL Bolton
[E] richard@threlfall. telme.com ⊖ (+44 1204) 38 13 31
♧ £ 50,000 (1997) ♣ £ 250,000 (1997) 👥 7 (1997)
🏭 5,000 sq ft
Managing Director: D Speak
Purchasing Manager: C Fakira
■ Manufacturers of valves in Aluminuim, bronze & gunmetal. Range covers ball, plug, swing check, globe, needle & 3 way valves for use in submarines, ships & offshore structures. [98]

75-43290 OTEC Special Products Ltd
(a member of The Nitor Group)
Nitor House, Great George Street,
Godalming, Surrey GU7 1EE
☎ (+44 483) 42 62 99 ☒ 94011723 nitr g
⊖ (+44 483) 41 59 53
Marketing Director: M E Read
■ Supply and installation of specialist products for training and operations including targetry, projected interactive digital video training systems and other range material and equipment. [98]

75-43400 Oxley Developments Co. Ltd
Priory Park, Ulverston, Cumbria LA12 9QG
☎ (+44 1229) 58 26 21 ☒ 65141
⊖ (+44 1229) 58 50 90
♣ £ 8,500,000 (1994/95) 👥 300
Managing Dir: Dr G.S. Edwards
Operations Dir: D.R. Norrie
Sales Dir: E.D. Maguire
Works Dir: Trevor Armistead
Marketing Mgr: Stephen J. North
Purchasing Mgr: M. Fieldhouse
■ Design consultants & manufacture of special passive components including RFI suppression filters & custom built arrays & filtered connectors; sealed solid state indicator lamps including NVG/ANVIS compatible & IR free lighting; trimmer capacitors; printed circuit connectors including surface mounting test points; terminals, plugs & sockets; non-volatile elapsed time indicators & event counters; Microwave ceramic packages & fibre optic products. → 80 [97]

75-43450 Package Control (UK) Ltd
(a subs. of Package Control Nederland BV)
Unit 6, Station Approach Industrial Estate,
Fleet, Hampshire GU13 8QY
☎ (+44 1252) 81 55 65 ⊖ (+44 1252) 81 61 80
♣ £ 1 200 000 👥 10
Managing Dir: James Laitt
Marketing: Mrs Linda Insley
■ Load restraint & lashing straps using polyester or nylon webbing. Variety of tensioning buckles. Heavy duty roller conveyer systems. [98]

75-43470 P.A.D. International
National Trading Estate, Bramhall Moor Lane, Hazel Grove, Stockport, Cheshire SK7 5AA
☎ (+44 161) 456 56 60 ⊖ (+44 161) 456 25 66
♣ £ 750,000 (1997) 👥 9 (1997)
Managing Dir: Harry Lee ☎ (+44 161) 456 56 60
Sales & Marketing Dir: Keith Taylor
■ Manufacture of anti-stab & ballistic vests for male & female officers. [98]

75-43505 Page Aerospace Ltd
(† Page Engineering Company Ltd)
Forge Lane, Sunbury-on-Thames, Middx TW16 6EQ
☎ (+44 1932) 78 76 61 ☒ 27520 page g
⊖ (+44 1932) 78 03 49
👥 200
Managing Dir: P.R. Gibbs
Associate Dir Commercial: M.A. Lipyeat
Sr Sales Engineer: N. Thomas
Product Support Mgr: D.G. Barrett
■ Central warning systems; audio tone generators; dimmer units. Central maintenance panels; light & switch light modules; pilot & voice alerting systems; engine fire shut-off switches; annunciators. Control grips, wheels, ministicks & throttle tops; night vision goggle compatible components, fluorescent ballast units. Emergency battery packs. [97]

75-43510 Pains-Wessex Ltd
Pains-Wessex Schermuly
(a subsidiary of Chemring Group plc)
High Post, Salisbury, Wilts SP4 6AS
☎ (+44 1722) 41 16 11 ⊖ (+44 1722) 41 21 21
♣ £ 33,500,000 (1996) 👥 350 (1997) 🏭 100 acres
Managing Director: David Evans
Military Business Director: P A Greenwood
Technical Director: Peter Cutler
Purchasing Director: David G Hyland
Sales Support Manager: Peter Davies
■ A complete range of service pyrotechnics including products for illumination, signalling, search & rescue, training, law enforcement, countermeasures & minefield breaching. [98]

75-43517 Pall Aerospace
(† Aircraft Porous Media Europe)
(a division of Pall Europe Ltd; a subsidiary of Pall Corporation, USA)
Europa House, Havant Street,
Portsmouth, Hampshire PO1 3PD
☎ (+44 1705) 30 33 03 [W] http://www.pall.com
☒ 86251 palleu g ⊖ (+44 1705) 30 25 06
♣ US$ 960,376,000 (1996) 👥 270
Managing Dir: C.S. Hutchings
Commercial Dir: Kenneth P. Smith
Technical Dir: P. Stallard
Advertising Mgr: S. Osborn
■ Filtration products for aircraft hydraulic, fuel & lube oil systems & air intake protection systems for gas turbines & air fed auxiliaries.
® USA: Aircraft Porous Media, Mectron Industries, Pall Corp., Pall Land & Marine [98]

75-43520 Pall Land & Marine
(a Division of Pall Europe Ltd)
Europa House, Havant Street,
Portsmouth, Hampshire PO1 3PD
☎ (+44 1705) 30 33 03 [W] http://www.pal.com ☒ 86251
⊖ (+44 1705) 30 25 06
♣ US$ 1,000,000,000 (1996) 👥 9,000 (worldwide)
Managing Dir: C.S. Hutchings
Marketing Dir: R. Collins
Commercial Dir: Kenneth P. Smith
Sr Advertising Mgr: S.R. Osborn
Sales Mgr: T. Cottrell
■ Design & manufacture of a range of filtration products for naval, military, aerospace & ground support systems including fuel, hydraulic, lubrication, transmission, pneumatics & engine air intake. [98]

75-43650 Pandect Precision Components Ltd
Wellington Road, High Wycombe, Bucks HP12 3PX
☎ (+44 1494) 52 63 03 ⊖ (+44 1494) 46 55 57
♣ £ 3,000,000 (1996) 👥 120 (1996)
Technical Director: W Palmer
Technical Sales Manager: Michael J Ward
Quality Director: D J Murphy
■ Design, manufacture and development of precision electrical and electro-optic slip-ring assemblies, simulator instruments and precision wire-wound potentiometers. Complete instrument repair service. [98]

75-43660 Panorama Antennas Ltd
Frogmore, London SW18 1HF
☎ (+44 181) 874 53 00 ☒ 885886
⊖ (+44 181) 870 76 41
♣ £ 3,200,000 (1995/96) 👥 50
Managing Dir: A.P. Jesman
Sales & Marketing Dir: C.E. Jesman
Sales Mgr: John Dudley
■ Manufacturers of a comprehensive range of antennas for defence & security applications. [97]

75-43750 Park Air Electronics Ltd (PAE)
(a subsidiary of Northrop Grumman Corp, USA)
Northfields, Market Deeping, Peterborough PE6 8LG

☎ (+44 1778) 34 54 34 [E] sales@parkair.co.uk
⊖ (+44 1778) 34 28 77; 34 12 86
♣ £ 15,000,000 (1996) 👥 125 (1997) 🏢 48,000 sq ft
Chief Executive Officer: Paul Barry
Operations Manager: Chris Tarrant
Commercial Manager: Nicola Lehair
Technical Director: R W Allis
Sales: David Bartlett
Marketing: Charles Houseago
■ Suppliers of ground-to-air, ship-to-shore & ship-to-ship radio products & air traffic control systems in the VHF/UHF frequency band. [98]

75-43850 Parker-Hale Ltd
Bisley Works, Golden Hillock Rd, Birmingham B11 2PZ
☎ (+44 121) 766 69 96 ⊖ (+44 121) 772 01 29
Managing Dir: R.C. Hale
■ Manufacturer & distributor of small arms components & accessories, target shooting equipment & accessories. [97]

75-43880 Parker Hannifin plc Chomerics Division
(† Chomerics Europe Ltd)
Parkway, Globe Park, Marlow, Buckinghamshire SL7 1YB
☎ (+44 1628) 48 60 30 ⊖ (+44 1923) 47 63 03
👥 97 (1997) 🏢 17,000 m2
General Mgr: D.K. Jones
Marketing Mgr: G.D. Evan-Hart
Sales Mgr Europe: M van den Brul
■ Electromagnetic shielding materials; test & rectification services. Electrically conductive gaskets, windows, ventilation panels, caulks, adhesive tapes & cable shields for electromagnetic compatibility. Also thermally conductive materials. Conductive airframe, firewall & lighting management seals. [98]

75-43900 Parker Plant Limited
Canon Street, PO Box 146, Leicester LE4 6HD
☎ (+44 116) 266 59 99 ⊠ 34542
⊖ (+44 116) 268 12 54
♧ £ 4,000 (1996) ♣ £ 30,000,000 (1996)
👥 400 (1997) 🏢 18 acres
Managing Dir: Simon Wilkinson
Operations/Technical Dir: Trevor Shaw
Sales & Marketing Dir: Gino Ciuffardi
Purchasing Mgr: Ray Love
■ Fixed & mobile asphalt mixing plant, including batch & continuous process types. Fixed & mobile crushing & screening plant, including feeders, compression crushers, impact breakers, screens, conveyors, washers, dewaterers & storage hoppers. [98]

75-43950 L.B. Parkes & Co Ltd
Station Street, Walsall, West Midlands WS2 9JZ
☎ (+44 1922) 72 07 20 [E] sales@lbparkes.co.uk
[W] http://www.lbparkes.co.uk ⊖ (+44 1922) 72 34 00
♣ £ 1,800,000 (1997) 👥 82 (1997) 🏢 2,000 m2
Managing Dir: Stephen L. Parkes
Sales Manager: Ian Parsons
Operations: Peter Lester
■ Anodizing by sulphuric and chromic methods. Polishing. Chromating powder coating, finishing of aluminium for the aviation and defence industries. Approved by BS to BSEN ISO9000, EN 29002-1987, OAS 3138/20, OAS 3640-65 and UK CAA. [98]

75-44000 Park Products Ltd
(a member of the Fife Indmar Group)
Ordnance Street, Blackburn, Lancashire BB1 3BN
☎ (+44 1254) 67 11 21/2/3 [E] sue@park.telme.com
⊠ 635274 ppl g ⊖ (+44 1254) 69 83 24
♣ £ 4,500,000 👥 66
Managing Director: Edwin N Hunter
Technical Director: K H Hayes
Operations Director: R Sabey
Sales and Marketing Manager: G Garen
Purchasing Manager: C McCormeick
■ Manufacture of marine motors to DGS 2088, waterproof plugs, sockets & switchgear. Searchlights, Hylite mobile lighting towers. BS5750 part 2 approved, MoD AQAP 4. [98]

75-44035 Parsons Chain Co. Ltd
(an FKI plc company)
Worcester Rd, Stourport-on-Severn, Worcs DY13 9AT
☎ (+44 299) 82 77 00 ⊠ 339711 parcon g
⊖ (+44 299) 82 76 59
♣ £ 21 000 000 👥 295
Managing Dir: R.A. Rawlings
Sales Dir: C.D. Flatman
Marketing Mgr: R.A. Kington
Area Mgr: Jeremy Lenox
■ Chains for lifting & mechanical handling. Mining chain & components for haulage & armoured flexible conveyors, security chain, mine clearance chains, fascine chains & a wide range of chain for other specialist applications. [97]

75-44055 Partek Cargotec Ltd
(† Cargotec (UK) Ltd)
(a division of Partek Cargotec AB, Sweden)
Government Business Operations
Harlescott Lane, Shrewsbury, Shropshire SY1 3AG
☎ (+44 1743) 44 22 32 ⊖ (+44 1743) 46 99 35
♧ £ 4,360,000 (1995) ♣ £ 35,000,000 (1995)
👥 301 (1995)
Managing Dir: N.J. Eden (OBE)
Operations Dir: A.C. Richardson
Engineering & Quality Dir: S.W. Locker
Purchasing Mgr: M.M. Lloyd
Sales & Marketing Mgr: M.E. Pugh
■ Design & manufacture of MULTILIFT, HIAB & LEEBUR hydraulic load handling equipment for military logistic vehicles. Specific applications for ammunition transport & resupply, ISO container handling, ribbon bridge transportation, launch & retrieval & engineer applications. DROPS/PLS systems supplied to Canada, Czech Republic, Italy, Malaysia, Norway, Romania, UK, USA & UN forces. → 90 → 80 [98]

75-44065 Pathfinder Fox Ltd
Clifton Technology Park, Wynne Avenue,
Clifton, Manchester M27 2HB
☎ (+44 61) 794 81 37 [E] pathfinder.fox@gconnect.com
⊖ (+44 61) 794 01 57
Director: Robert Atkins
■ Manufacture of Kinetics equipment protector cases.
Gyro Stablilised Day/Night Binocular (M-25) with night vision capability. Miniature Integrated Camera Eye (MICE) system which allows images seen by the observer to be transmitted by microwave to a remote control centre located kilometers away. [98]

75-44080 Paul Evers Holsters
Unit 7, Bear Brand Complex, Woolton, Liverpool L25 7SF
☎ (+44 151) 428 18 00 ⊠ 669755 office g
⊖ (+44 151) 421 01 15
👥 10 (1995) 🏢 4 210 sq ft
Propietor/Chief Exec.: A.W.P. Evers
Sales: S.E. Johns
Purchasing Mgr: Mrs Enid Jones
■ Purpose built Military and Law Enforcement Leather/Nylon Holsters, Belts & Pouches. [98]

75-44200 PD Technical Mouldings plc (PDTM)
(† PD (Technical Mouldings) Ltd)
Rotation House, 20 Mayday Road,
Thornton Heath, Surrey CR7 7HL
☎ (+44 181) 689 43 36 ⊖ (+44 181) 683 24 91
♧ £ 50,000 (1996) ♣ £ 3,000,000 (1996) 👥 40 (1996)
🏢 50,000 sq ft
Chairman: R.G. Seddon
Managing Dir: S.J. Smith
Advertising Mgr: John Bryant
■ Pipe Fascine narrow gap bridging used by British & overseas armies. Water carriage pack for carrying drinking water to British Army in the field. Other defence related products include pumping assemblies, manifold equipment, bridge pads, steel fence posts & insulate food containers. Trailers, runway repair screed boards, armour plate targets. Demountable fire-fighting tenders. Fuel pump sets. [98]

75-44270 Pearpoint Ltd
Pearpoint House, 47 Woolmer Way,
Woolmer Trading Estate, Bordon, Hampshire GU35 9QE
☎ (+44 1420) 48 99 01 [E] tomm@pearpoint.co.uk
⊖ (+44 1420) 47 75 97
♣ £ 8,000,000 (1996) 👥 120 (1997) 🏢 3,500 (1997)
Managing Director: Tom MacMorran
Operations Operations: Shaun Lynes
Marketing Director: Ms Mary Perry
Purchasing Manager: Mike Easton
■ Miniature, monochrome, high sensitivity, high resolution surveillance cameras. Advanced gen 2 ICCD night vision cameras. X-Y colour scanning cameras which dispense with external pan/tilt. 24 hour surveillance cameras, colour and ICCD (night) sections. Ruggedized, environmentally sealed units. ISO9001. [98]

75-44330 Pearson Engineering Ltd
(a subsidiary of Bywell Holdings Ltd)
Wincomblee Road, Walker,
Newcastle-upon-Tyne NE6 3QS
☎ (+44 191) 234 00 01 ⊖ (+44 191) 262 04 02
♧ £ 500,000 (1997) ♣ £ 15,000,000 (1997)
👥 120 (1997) 🏢 80,000 sq ft
Managing Dir: Simon R. Reece
Operations/Technical Dir: Randall Flack
Sales Dir: Dennis Doble
Purchasing Mgr: Tom Clark
■ Design & manufacture a range of mine countermeasure equipment including mineploughs & scatterable mine clearance systems. Combat dozer blades, earth anchor blades & clear lane marking systems for minefields & similar dangerous areas. [98]

75-44400 Penman Engineering Ltd
Heathhall, Dumfries DG1 3NY
☎ (+44 1387) 25 27 84 ⊖ (+44 1387) 26 73 32
♧ £ 1,400,000 (1996) ♣ £ 10,200,000 (1995)
👥 160 (1996) 🏢 15,000 sq m
Managing Director: Edwin N Hunter
Technical Director: K H Hayes
Operations Director: R Sabey
Sales and Marketing Manager: G Garen
Purchasing Manager: C McCormick
■ Design & manufacture range of Hotspur light armoured vehicles based on Land Rover chassis, and armoured vehicles. [98]

Contact Reference

75-44450 Pennant International Limited
Pennant Court, Staverton Technology Park, Staverton,
Cheltenham, Gloucester GL51 6TL
☎ (+44 1452) 71 49 14 ⊖ (+44 1452) 71 49 20
♧ £ 930,000 (1996) ♣ £ 3,700,000 (1996)
👥 70 (1996) 🏢 2,000 m2
Managing Director: J J J Thompson
Operations Director: S J Phillips
Sales and Marketing Director: R J Lewis
Commercial Director: P G Barwell
Sales and Marketing Manager: P A Jones
■ Design and manufacture of training equipment for high technology applications. Computer based training courseware, electronic classrooms through to cockpit procedure trainers and maintenance trainers. → 80 [98]

75-44495 Penny & Giles Blackwood Group
(a subsidiary of Bowthorpe plc)
Unit 35/36 Nine Mile Point, Industrial Estate,
Cwmfelinfach, Gwent NP1 7HZ
☎ (+44 1495) 20 20 00 ⊖ (+44 1495) 20 20 06
♣ £ 10,000,000 (1996) 👥 220 (1996) 🏢 5,500 m2
Managing Director: T R Evans
Operations Director: T M S Jones
Sales and Marketing Director: N D Parke
Purchasing Manager: T Collins
■ Measurement and control instrumentation for the avionics industry: conductive plastic angular transducers, magneto resistive encoders, inductive transducers, solenoids and solenoid valves. [98]

75-44550 Penny & Giles Aerospace Ltd
6 Airfield Way, Christchurch, Dorset BH23 3TT
☎ (+44 1202) 48 17 71 ⊖ (+44 1202) 48 48 46
Managing Director: S Boulton
Marketing Manager: P E Walker
■ Design and manufacture of digital air data systems, computer-controlled air data test sets, primary sensors and transducers. Ice and snow detection systems. Accident data recorders and ground replay equipment. [98]

75-44560 Penny & Giles Data Systems Ltd
(a Bowthorpe plc company)
The Mill, Wookey Hole, Somerset BA5 1BB
☎ (+44 1749) 67 54 54
[E] 100750.641@compuserve.com [W] http://www.penny-giles.co.uk ⊖ (+44 1749) 67 10 92

United Kingdom

Managing Director: Adrian Girling
Sales and Marketing: Edwin Kayes
■ High performance data recording systems used in weapons system tests; telemetry, ASW; ballistics, communications, remote sensing satellite data acquisition, processing and archiving; digital data mass storage. Multi-channel analogue recorders. Digital D-1 cassette and optical tape recorders. Data acquisition systems. [98]

75-44563 Penny & Giles Drives Technology Ltd
(a subsidiary of Bowthorpe plc)
1 Airspeed Road, Christchurch, Dorset BH23 4HD
☎ (+44 1425) 27 14 44 ⊖ (+44 1425) 27 26 55
Managing Director: Hal Chenhall
■ Design and manufacture of controllers for wheelchairs. [98]

75-44565 Penny & Giles Electronic Components Ltd
(a subsidiary of Bowthorpe plc)
Unit 35-36, Nine Mile Point Industrial Estate, Cwmfelinfach, Gwent NP1 7HZ
☎ (+44 1495) 20 20 00 ⊖ (+44 1495) 20 20 06
Managing Director: T R Evans
Sales and Marketing Director: N D Parke
Operations Director: T M S Jones
Purchasing Manager: T Collins
■ Manufacture of joystick controllers, displacement transducers, precision potentiometers, inductive position transducers, aerospace solenoids, windspeed and direction indicators and transmitters. CAA, BS5750 and ISO9000 approved. [98]

75-44570 Penny & Giles Position Sensors Ltd
(a subsidiary of Bowthorpe plc)
15 Airfield Road, Christchurch, Dorset BH23 3TJ
☎ (+44 1202) 40 94 09
E sales@pgsensors.demon.co.uk
W http://www.penny-giles-controls.co.uk
⊖ (+44 1202) 40 94 10
👥 9
Managing Director: Roy Dewey
Operations Director: David Taylor
Sales and Marketing Director: David Gardiner
Business Development Manager: M Iles
Purchasing Manager: Richard Terndrup
■ Linear, rotary & inclination displacement transducers. [98]

75-44580 Penny & Giles Instrumentation Ltd
(a subsidiary of Bowthorpe plc)
4 Airfield Way, Christchurch, Dorset BH23 3TS
☎ (+44 1202) 47 66 21 ⊖ (+44 1202) 47 65 01
Managing Director: Charles White
■ Design and manufacture of telemetry instrumentation. [98]

75-44670 Percival Associates Ltd
Knowle Sidings, Knowle, Fareham, Hampshire PO17 5LZ
☎ (+44 1329) 83 38 14 ⊖ (+44 1329) 83 40 13
👥 3,000
Managing Director: Neil Percival
Operations and Technical Director: David Headley
Sales and Marketing Director: Mark Radford
Purchasing Manager: Ms Irene Percival
■ Manufacture airborne fire containment covers, cargo and passenger restraint systems. Cargo floor and cabin hardware. CAA AI and JAR145 approved. Sales and service for cabin fire extinguisher systems. → 90
® France: Industria
UK: First Tech.Fire & Safety
USA: Skylock Industries Inc. [98]

75-44680 Perex Ltd
Ellinor Touch Technology Division
7 Arkwright Road, Reading, Berkshire RG2 0EA
☎ (+44 118) 931 10 66 ⊖ (+44 118) 931 11 69
♣ £ 8,000,000 (1995) 👥 65 🏭 2,500 sq m
Managing Dir: Mike Clarke
Sales & Marketing Dir: John Rodger
Operations & Technical Dir: Jim Carter
Purchasing Mgr: David Crowther
■ Touch screen manufacturer. Ellinor Technology Div: Manufacture of force sensing platform and integration of touch sensitive surfaces to CRT & flat panel displays.
Drive Technology Div: Design and manufacture of cartridge tape drive mechanisms including a range of products with NATO numbers and conforming to ECMA 46/79. [98]

75-44750 Varity Perkins
(a division of LucasVarity plc)
Frank Perkins Way, Eastfield, Peterborough PE1 5NA
☎ (+44 1733) 674 74 ✉ PERKOIL Peterborough
▣ 32501 ⊖ (+44 1733) 58 22 40
♣ US$ 1,000,000 (1995) 👥 4,500
Group Managing Director: Michael Baunton
Corporate Press Relations Manager: Kevin Horsley
Group Communications Manager: Jon Chandler
■ Diesel engines of two, three, four, six, eight, twelve and sixteen cylinder configurations covering power outputs from 3.5 to 2,237 kW (5 to 3,000 bhp).

Perkins Engines (Shrewsbury) Ltd Military Projects Division
Lancaster Road, Shrewsbury, Shropshire SY1 3NX
☎ (+44 1743) 21 20 00 ▣ 35171/2 pesl g
⊖ (+44 1743) 21 27 01
👥 900
Managing Director: Brian Amey
Sales and Marketing Director: Richard Halahan
Business Development Manager: David Elmes
■ Supply comprehensive range of diesel engines for military and naval applications in a power band stretching from 5 to 3,000 bhp. The range includes the high output CV Series, the SE series of six, eight, twelve, and sixteen cylinder engines, the Eagle and phaser in-line automotive engines and the 100 series generator engines. Engines tailored to military applications. Manufacture, assembly and testing of the X300 transmission for armoured infantry combat vehicles. Facilities for design, development, manufacture and test of complete power packs available. [98]

75-44800 Permabond
(a division of National Starch & Chemical)
Woodside Road, Eastleigh, Hampshire SO50 4EX
☎ (+44 1703) 62 96 28 ⊖ (+44 1703) 62 96 29
♣ £ 20,000,000 👥 110
Operations Manager: Nick Siegler
Sales Manager: R Milne
Business Development Manager: Jeff Wigham
Marketing Manager: Paul Sennett
Quality Assurance Manager: David Wade
■ Develop and manufacture of adhesives, sealants, potting compounds and encapsulants required by structural and engineering industries, composite fabrications and manufactures of electrical, electronic and optical devices. Approved to BS5750 Pt 1 (EN29001, ISO9001). [98]

75-44850 Permali Gloucester Ltd
(† Permali Ltd)
(a subsidiary of BTR plc)
Bristol Road, Gloucester GL1 5TT
☎ (+44 1452) 52 82 82
E sales@permali-gloucester.ltd.uk
⊖ (+44 1452) 50 74 09
♣ £ 10,000,000 (1995) 👥 100 (1995) 🏭 12,000 m2
Managing Director: A J T King
Operations Director: G Dunn
Sales and Marketing Director: P R Sprason
Purchasing Manager: S J Lear [98]

75-45055 PGM Ballscrews Ltd
(a subsidiary of THK Japan)
Bodmin Road, Wyken, Coventry CV2 5DZ
☎ (+44 1203) 61 62 31 E chrisw.pgm@pipemedia.co.uk
⊖ (+44 1203) 61 10 32
♧ £ 5,000,000 ♣ £ 4,200,000 👥 100 🏭 4,500,sq m
Managing Dir: I. Kitamura
Dir & General Mgr: C.J. Wolfe
Financial Mgr: Neil Southwell
■ Precision ballscrews, ballsplines & special form lead screws, used in control surface actuators, test equipment, gearbox drive shafts, armament, missile launchers, autoland systems, flight refuelling systems. [98]

75-45065 PG Products Ltd
Unit 6, Midfield Road, Mitchelston Industrial Estate, Kirkcaldy, Fife, Scotland KY1 3NL
☎ (+44 1592) 65 31 48 E pgproducts@dial.pipex.com
⊖ (+44 1592) 65 30 79
♧ £ 70,000 (1997) ♣ £ 470,000 (1997) 👥 15 (1997)
Managing Director: D O Simpson
■ Design and manufacture of protective equipment. Riot helmets, shields and protective visors for military helmets. Assault and fire clothing. [98]

75-45260 Photon Beard Ltd
Unit K3, Cherry Court Way, Stanbridge Road, Leighton Buzzard LU7 8UH
☎ (+44 1525) 85 09 22 ⊖ (+44 1525) 85 09 22
♣ £ 600,000 👥 6
Managing Director: Graham Wainwright
Operations and Technical Director: R J Beard
Sales Director: Alan V Gooch
■ Manufacture of portable tungsten and digital imaging flourescent lighting equipment, photographic darkroom equipment, security cameras, equipment cases and viewing lightboxes. Suppliers of portable tungsten kits and digital imaging flourescent lighting equipment (for video and film recording), security camera and equipment cases, darkroom easels (for specialised photographic work). Camera battery equipment, battery lighting equipment and lighting filters to the MoD and security forces. Camera battery equipment, battery lighting, equipment and lighting filters. [98]

75-45300 PI Castings Ltd
Davenport Lane, Altringham, Cheshire WA14 5DS
☎ (+44 161) 928 58 11 ⊖ (+44 161) 927 70 23
♣ £ 5 000 000 👥 120 🏭 90 000 sq ft
Managing Dir: Dr K.J. Albutt
Sales Director: Brian Mills
Operations Director: C J Batty
Sales Manager: Ian Cookson
Purchasing Manager: E E Hewitt
■ Precision investment castings (lost wax process) in most types of steels, stainless steels, aluminium alloys, copper based materials & tool steels. [98]

75-45400 Pilatus Britten-Norman Ltd (PBN)
(a subsidiary of Pilatus Aircraft, Switzerland)
Bembridge, Isle of Wight PO35 5PR
☎ (+44 1983) 87 25 11 ▣ 86866 pbnbem g
⊖ (+44 1983) 87 32 46
♣ £ 20,000,000 (1996) 👥 165 (1997) 🏭 93,000 sq ft
CEO & Marketing Dir: Anthony Stansfeld
Technical Dir: Robert Wilson
Sales & Marketing Dir: Walter Stark
Marketing Communications Mgr: Ms Sheila Dewart
■ Manufacture, support, repair & overhaul of BN2 Islander & Defender series of twin engined utility aircraft. [98]

75-45450 Pilcher-Greene Ltd
Consort Way, Burgess Hill, W.Sussex RH15 9NA
☎ (+44 1444) 23 57 07 ⊖ (+44 1444) 24 17 30
♣ £ 1 000 000 👥 30 🏭 40 000 sq ft
Managing Dir: M.L. Pilcher
Sales & Marketing Dir UK: B. Reynolds
Operations & Technical Dir: J. Gunn
Purchasing Mgr: G. Muir
■ Specialist vehicle builders, ambulances & fire engines. [97]

75-45500 Pilkington Aerospace Ltd
(† Triplex Aircraft & Special Products Ltd)
(a member of the Pilkington Group)
Eckersall Road, Kings Norton, Birmingham B38 8SR
☎ (+44 121) 606 41 00
E 106321.1047@compuserve.com
⊖ (+44 121) 606 41 91
♣ £ 14,000,000 (1997) 👥 240
Managing Dir: P. Molineux
Technical Dir: R.S. Bruce
Marketing Dir: E.S. Hinkley
Manufacturing Mgr: C.J. Woodhouse
Publicity Co-ordinator: Ms S.C. Smith
Purchasing Manager: A D Tither
■ Windows & transparencies for all aircraft, helicopters, hydrofoils, ships, tanks, fighting vehicles & space applications. De-icing/de-misting systems, anti-static coatings & bird/hail impact protection. [98]

75-45570 Pilkington's Tiles Ltd
(a subs. of BTR plc)
Clifton Junction, PO Box 4, Swinton, Manchester M27 8LP

☎(+44 161) 727 10 00/11 11/11 33
⊖ (+44 161) 727 11 22
♣ £ 25 000 000 👥 500
Managing Dir: B.H. Thompson
Sales Dir: Ian Crowther
Contracts Sales Mgr: Phil Taylor
Product Mgr: Simon James
■ Supplies tiles to military bases, glazed wall tiles & unglazed floor tiles. [97]

75-45700 Pittards plc
(† Pittard Garnar plc)
Sherborne Road, Yeovil, Somerset BA21 5BA
☎(+44 1935) 47 43 21
[W] http://www.pittardsleather.com ⊠ 46141
⊖ (+44 1935) 42 71 45
♧ £ 23,300,000 (1996) ♣ £ 109,000,000 (1996)
👥 1,092 (1996)
Group Managing Director: J W W Pittard
Technical Director (Gloving Div.): Reg Hankey
Company Secretary: Mrs J Williams
■ High specification performance leather for boots, gloves and shoes. [98]

75-45720 PK Electronic International Ltd
Neville House, 55 Eden Street,
Kingston-upon-Thames, Surrey KT1 1BW
☎(+44 181) 541 52 11 [E] pkelectronic@msn.com
⊖ (+44 181) 541 56 24
👥 7 🏠 200 sq m
Managing Director: Derek Myers
■ Design, development & manufacture of professional electronic intelligence gathering & security equipment. Procurement services & specialist training courses. Representing PK Electronic International group of companies. → 80 [98]

75-45730 Plalite Ltd
Design Division
Star House, 8 Star Hill, Rochester, Kent
☎(+44 1634) 84 49 44 ⊠ 965430 plalit g
⊖ (+44 1634) 84 49 46
♣ £ 3 000 000 (1994) 👥 48 🏠 2 100 sq m
Managing Dir: A.P. Hardman
Project Dir: A.G. Smith
Purchasing Mgr: M. Foord
■ Provides a complete engineering, design, development & manufacturing service to government departments & industry. Quality standards: ISO 9001, EN 29001 & BS 5750 part 1.
Works and Administration
Unit 18, Sittingbourne Ind Park, Crown Quay Lane,
Sittingbourne, Kent ME10 3JZ
☎(+44 1795) 47 63 67 ⊖ (+44 1795) 47 63 69 [98]

75-45735 Planar Advance Inc.
Finchley House Business Centre, 707 High Rd,
London N12 0BT
☎(+44 181) 445 44 16 ⊖ (+44 181) 343 92 62
Consultant Display Products: Charles Sebesta
■ Manufacture of taut shadow mask colour CRT's for avionics applications. Field sequential colour monitors. A.M.L.C.D. displays (Active Matrix). E.L. display (Electroluminescent). A.M.E.L. displays (Active Mantrix E.L. Products are primarily aimed at the military market. [97]

75-45740 Pland
(† TRF Pland Ltd)
(† Plant Stainless Ltd)
(a member of Spring Ram Corp plc group of companies)
Lower Wortley Ring Road, Leeds LS12 6AA
☎(+44 113) 263 41 84
[E] 106333.2100@compuserve.com
⊖ (+44 113) 231 05 60
♣ £ 10,000,000 (1997) 👥 145 (1997)
Manufacturing Director: Jim Cochrane
Industrial Sales Executive: R D Poole
Marketing Officer: J P Selwood
■ Field kitchen equipment. Six gallon stainless steel field cooking containers complete with insulated outer container. Fixed base stainless steel catering sinks & tables. [98]

75-45950 Plastico Ltd
(a subs. of D.Green & Co. (SH) Ltd)
100 Morden Rd, Mitcham, Surrey CR4 4DA

☎(+44 181) 646 04 56 ⊖ (+44 181) 646 54 40
♧ £ 45 000 (1996) ♣ £ 7 800 000 (1996) 🏠 2 700 sq m
Managing Dir: D.T. Wiggins
Operations/Technical Dir: B. Hash
Sales & Marketing Dir: C.A. Wiggins
Purchasing Mgr: N. Trim
■ Meal service packs to order which include: Cutlery, food sachets, fresheners hygienically machine wrapped & sealed. [97]

75-46050 Plenty Ltd
(a subsidiary of EIS plc)
Hambridge Road, Newbury, Berkshire RG14 5TR
☎(+44 1635) 423 63 [E] pumps@plenty.co.uk
[E] filters@plenty.co.uk ⊠ 848023 ⊖ (+44 1635) 497 58
♣ £ 27,000,000 (1995) 👥 462 🏠 40,500 m2
Chairman: S C M Hunt
Financial Director: S G Garrett
General Manager - Filter Products: R Broad
General Manager - Mixer Side Entry Products: B R Rogers
General Manager - Mixers Top Entry Products: P Knights
General Manager - Pump Products: J Fisher
General Manager - Business Development: A E Walton
General Manager - Plenty Steam Traps: M Ashworth
General Manager - GD Engineering: P Phelan
General Manager - Manufacturing: B Goodall
■ Design and manufacture of pumps, filters, mixers, steam traps and pipeline equipment. Positive displacement of oil and oil/water pumps for ship and shore installation (ship bunkering and bilge, hydraulic, oil firing, helicopter refuelling). Pipeline filters for both ship and shore installation with any liquid. Approved suppliers to MoD standard AQAP-4 ED2. Approved to BS5750 Part 1 and ISO 9001. Filters for oil production; processing and distribution, gas processing and distribution, power generation, marine and defence, petrochemical, water and waste, food, drink and steel production.
Plenty Limited Manufacturing Plant
Hambridge Road, Reading, Berkshire RG14 5TR
☎(+44 1635) 423 63 ⊠ 848023 plenty g
⊖ (+44 1635) 497 58
👥 257 🏠 11,150 sq m
Plenty Mirrlees
Earl Haig Road, Hillington, Glasgow G52 4JN
☎(+44 141) 882 03 14 ⊠ 778368 triro g
⊖ (+44 141) 882 27 52
🏠 2,415 sq m
■ Design, manufacture and after sales support of positive displacement triple screw pumps, industrial mixers and machining services.
Plenty Steam Traps
Unit 1 Lakeside Business Park,
Thurmaston, Leicester LE4 8AS
☎(+44 1162) 69 51 72 ⊠ 347323
⊖ (+44 1162) 69 36 95
🏠 1,140 sq m
■ Design, manufacture and repair of steam traps and pipe assemblies.
Plenty Fine Filtration
Eastgate Works, Marine Road, Southampton SO1 1JP
☎(+44 1703) 22 45 75 ⊖ (+44 1703) 33 07 31
🏠 880 sq m
■ Manufacture of sheet metal fabrications and metallic/non-metallic filters.
GD Engineering
Retford Road, Worksop, Nottingham S80 2PY
☎(+44 1909) 48 23 23 ⊠ 54149 gdc g
⊖ (+44 1909) 47 79 02
🏠 1,420 sq m
■ Design and manufacture of pig launchers, receivers, closures, pigging and pipeline cleaning equipment. Design and management of sphere tees. [98]

75-46330 Plint & Partners Ltd
Oaklands Park, Wokingham, Berkshire RG41 2FD
☎(+44 118) 978 18 98 [E] sales@plint.co.uk
[W] http://www.plint.co.uk ⊖ (+44 118) 977 22 81
♧ £ 410,000 (1997) ♣ £ 3,000,000 (1997)
👥 40 (1997) 🏠 1,480 m2
Managing Director: George Plint
Marketing Director: Dr Alex F Alliston-Greiner
■ Educational equipment for scientific and technical training in aeronautics, fluid mechanics, machines, thermodynamics and tribology. Industrial equipment for engine and transmission testing. Tribology equipment for materials and lubrication testing. Process control training equipment. [98]

75-46340 Plumett Ltd
Victoria Road, Burgess Hill, Sussex RH15 9LD
☎(+44 1444) 24 15 71 ⊖ (+44 1444) 23 23 38

♣ £ 2 000 000 👥 19 🏠 12 000 sq m
Managing Dir: James A. Brome
Sales & Marketing Mgr: W.J.M. Scott
Purchasing Mgr: R. Wells
Commercial Mgr: F.B. Langworthy
■ Specialised winch systems for cable laying & vehicle recovery. Air launchers for use with grapnels or line throwing. [97]

75-46348 Polyformes Ltd
Cherrycourt Way, Stanbridge Road,
Leighton Buzzard, Beds LU7 8UH
☎(+44 1525) 85 24 44 [E] polyformes@compuserve.com
⊖ (+44 1525) 85 04 84
♣ £ 2,200,000 👥 52 🏠 6,000 m2
Managing Director: Simon Dabson
Marketing: John Henrick
■ Design and manufacture a wide range of standard and special purpose transit and storage containers to meet specific customer needs worldwide. [98]

75-46350 Polymathic Engineering Co. Ltd
Fazeley Rd, Tamworth, Staffs B78 3JR
☎(+44 1827) 634 41/5 ⊖ (+44 1827) 31 07 65
♣ £ 1 000 000 👥 24 🏠 2 500 sq m
Managing Dir: Michael R. Pickering
Sales & Marketing Dir: T. Shaw
■ Manufacturer of the "Coolie" range of lift trucks, hand/hydraulic, battery/electric or pneumatic operation. [97]

75-46356 Polywarm Products Ltd
Cambuslang Rd, Farme Cross,
Rutherglen, Glasgow G73 1RS
☎(+44 141) 647 23 92/4; 647 82 48 ⊠ 779968
⊖ (+44 141) 613 15 69
♧ £ 5 300 000 (1995) ♣ £ 11 000 000 (1995) 👥 192
🏠 14 200 sq ft
Managing Dir: Ian Jackson
Sales & Marketing Dir: Andrew Dunlop
Operations & Technical Dir: Jim Friel
Purchasing Mgr: Ms Jane Watts
■ Sleeping bags for military and civil forces. Fire-retardant quilts and bedding accessories. [97]

Contact Reference

75-46420 Portescap (UK) Ltd
(a subsidiary of Portescap, Switzerland)
Headlands Business Park, Salisbury Road,
Ringwood, Hampshire BH24 3PB
☎(+44 1425) 47 11 00 ⊖ (+44 1425) 47 77 55
♣ £ 7,000,000 👥 70
Managing Director: Brian C Kieran
Field Sales Manager: Jonathan Rumbol
Publicity Administration: Ms Deborah Simon
■ Manufacture of precision gearheads and linear actuators.
® Switzerland: Portescap [98]

75-46500 Powersport International Ltd
(a subsidiary of Hawtin plc)
Queens Road, Bridgend Industrial Estate,
Bridgend, Mid Glamorgan CF31 3UE
☎(+44 1656) 66 11 64 ⊖ (+44 1656) 66 13 82
♣ £ 6,900,000 (1996) 👥 78 🏠 60,000 sq ft (1997)
Managing Director: P T Cravos
Operations Director: R Davies
Sales Director: P Hanson
Purchasing Manager: D Jenkins
Marketing: N Canning
■ Multigym military fitness units. [98]

75-46600 Premaberg Manufacturing Ltd
(a subsidiary of Premaberg Holdings Ltd)
Premaberg House, 22-24 High Street,
Halstead, Essex CO9 2AP
☎(+44 1787) 47 56 51 ✉ PREMABERG Halstead
⊠ 987806 prehal g ⊖ (+44 1787) 47 50 46
👥 60 🏠 25,000 sq ft
Managing Director, Marine Division: Graham J Wadley
Sales Department: Julian A Durrant
■ Weather louvres. Air intake filtration and filters for gas turbine and diesel engines. Stainless steel grilles and diffusers. [98]

75-46655 Premier Fuel Systems Limited
Willow Road, Trent Lane Industrial Estate,
Castle Donington, Derby DE7 2NP
☎(+44 332) 85 05 15 ⊖ (+44 332) 85 07 49

♣ £ 2,000,000 ⚉ 40
Managing Director: R A Fletcher
Operations Director: R E Fletcher
Purchasing Manager: P Bates
■ Premier manufacture flexible safety fuel and auxiliary equipment for the aircraft and military markets. 98

75-46660 Pre-Mac (Kent) Ltd
(a subsidiary of Global Environmental Management Ltd)
Unit 5, Morewood Close, Sevenoaks, Kent TN13 2HU
☎ (+44 1732) 46 03 33 ⊖ (+44 1732) 46 02 22
♧ £ 100,000 (1997) ♣ £ 1,000,000 (1997)
⚉ 10 (1997) ⌂ 5,500 sq ft
Sales and Marketing Director, Acting Chairman: Colin A Bearsby
Managing Director: Colin M Knox
Operations and Technical Director: R W Higham
Sales Executive: Cameron Kiggell
■ Designers & manufacturers of water purification equipment. 98

75-46690 Premier Hazard
(† Premier Hazard Systems (UK) Ltd)
Moorfield Estate, Yeadon, Leeds LS19 7BN
☎ (+44 113) 239 11 11 ⊖ (+44 113) 239 11 31
♣ £ 5,000,000 (1997) ⚉ 55 (1997) ⌂ 2,052 sq m
Managing Director: Bryn J Tennant
Sales Director: Peter Ballard
■ Manufacture of a complete line of emergency vehicle warning equipment including strobe & halogen lightbars, beacons, sirens, speakers, switch control & auxiliary lighting products. 98

75-46755 Price Western Leather Co. Ltd
Vennland Way, Ponsford Road,
Minehead, Somerset TA24 5DX
☎ (+44 1643) 70 50 71 ⊖ (+44 1643) 70 21 27
♣ £ 1,600,000 (1995) ⚉ 42
Managing Director: John A Littlewood
Sales/Marketing: Gary Parsons
Purchasing Manager: June Phillingham
■ Holsters, leather accessories. Other police equipment. Procurement services. 98

75-46790 Primary Image Ltd
Millbank House, 171-185 Ewell Road,
Surbiton, Surrey KT6 6AP
☎ (+44 181) 339 96 69 ⊠ 928893
⊖ (+44 181) 339 90 91
Director: A. Davenport
Sales Mgr: Stephen Bennet
Technical Dir: Paul Pontin
■ High-speed image processing for automatic recognition, computer graphics for simulation & training, computer hardware & software design for advanced real time applications. 98

75-46865 Prosig Computer Consultants Ltd
Group Headquarters
Link House, High Street, Fareham, Hampshire PO16 7BQ
☎ (+44 329) 23 99 25 W http://www.prosig.co.uk
⊖ (+44 329) 23 91 59
♧ £ 100,000 (1996) ♣ £ 2,200,000 (1996)
⚉ 45 (1997)
Chairman: Dr Colin Mercer
Managing Dir: Chris Evans
Purchasing Manager: Ms P Hay
■ High performance data capture. Analysis & display of data through a range of technology software & hardware products. Software & turnkey systems for all aspects of scientific & engineering data collection & analysis. Application areas in defence range from sound & mechanical vibration to processing of signals from imaging & radar systems. → 80 98

75-46870 Prostab International Ltd
Unit 10, Boyn Valley Trading Est., Boyn Valley Rd,
Maidenhead, Berkshire SL6 4EJ
☎ (+44 1628) 78 49 31 ⊖ (+44 1628) 77 35 19
♣ £ 2 300 000 ⚉ 23
Managing Dir: Simon J.C. Hill
Commercial Mgr: J.G. Lee
Sales Mgr: John McNicoll
Purchasing Mgr: David Carter
■ Television-related specialist systems & products for direct military use & other defence applications. 97

75-46890 Protector Technologies Ltd
(† Protector Safety Limited)
Pimbo Road, West Pimbo,
Skelmersdale, Lancashire WN8 9RA
☎ (+44 1695) 271 71 ⊖ (+44 1695) 505 73
♣ £ 12,000,000 (1997) ⚉ 155 (1997) ⌂ 80,000 sq ft
Managing Director: Maurice C Vallois
Operations Director: Alan Morris
United Kingdom Sales Manager: P A Upcott
Purchasing Manager: S Rodick
■ Manufacturer of industrial safety equipment for protection of head, eyes, face, hearing and respiration. 98

75-47000 Protovale (Oxford) Ltd
(† Protovale Electronic Research Instruments)
Rectory Lane Ind.Estate, Kingston Bagpuize,
Abingdon, Oxon OX13 5AS
☎ (+44 1865) 82 09 45; 82 12 77
⊖ (+44 1865) 82 05 73
♣ £ 500 000 ⌂ 5 000 sq ft
Director: D.C. Farrow
Marketing Mgr: F. Dolby
■ Mine detection equipment: design & manufacture of metal detectors, based on the pulse induction principle, used for deeply sited metallic mines. 97

75-47030 Pulnix Europe Ltd
(a subsidiary of Takenaka Group, Japan)
Pulnix House, Aviary Court, Wade Road,
Basingstoke, Hampshire RG24 8PE
☎ (+44 1256) 47 55 55 E office@pulnix.co.uk
E market@pulnix.co.uk E sales@pulnix.co.uk
W http://www.pulnix.co.uk ⊖ (+44 1256) 46 62 68
♧ £ 250,000 (1994) ♣ £ 5,500,000 (1996)
⚉ 28 (1996)
Managing Director: Kenneth Ono
Technical Manager: Barrie Henwood
UK Sales Manager: Stuart Mann
European Sales Manager: Mike Huddison
■ Security detection and video devices. → 85 → 90 98

75-47100 Puma Manufacturing Ltd
(a subsidiary of NRG Generating Limited)
37 Sandwich Road, Ash, Canterbury, Kent CT3 2AJ
☎ (+44 1304) 81 28 18 ⊠ 96366 puma g
⊖ (+44 1304) 81 33 31
♧ £ 1,000,000 (1997) ♣ £ 9,000,000 (1996)
⚉ 60 (1997) ⌂ 35,000 sq ft
Chairman: Brian M Tatton
Managing Director: John Inglis
Operations Director: Paul Willett
Financial Director: P W Croney
Director of Marketing: Andrew Inglis
Purchasing Manager: Ivan Beavis
■ Design, manufacture, installation & maintenance of diesel generating equipment & control panels. → 80
Associate companies:
O'Brien Energy (Europe) Ltd
Powerent Ltd 98

75-47120 Purification Products Ltd
Reliance Works, Saltaire Road,
Shipley, West Yorkshire BD18 3HL
☎ (+44 1274) 53 01 55 ⊖ (+44 1274) 58 04 53
Managing Director: W Q C MacKenzie
Sales Director: Mike Johnson
Operations and Technical Director: David Bellerby
■ Protective textiles. 98

75-47290 Pyle-National
(a division of Amphenol Ltd)
100 Sherbrook Road,
Daybrook, Nottinghamshire NG5 6AT
☎ (+44 115) 920 81 16 ⊖ (+44 115) 920 88 96
♣ £ 4,000,000 (1997) ⚉ 35 (1997)
Managing Dir: J.B. Backhouse
Sales & Marketing Dir: D. Redfern
Operations Director: Chris Parsons
Operations Mgr: A. Pleass
Sales & Marketing Mgr: P. Saunders
Product Marketing Mgr: R.G. Woodroffe
Purchasing Manager: Andy Cable
■ Design, development & manufacture connector systems for military & civil environments, MIL-C-26500, MIL-C-83723, MIL-C38999. Cable harnessing & filtered connectors are also a speciality. 98

75-47300 Pyser-SGI Ltd Francis Barker Division
Fircroft Way, Edenbridge, Kent TN8 6HA
☎ (+44 1732) 86 41 11 ⊖ (+44 1732) 86 55 44
Managing Director: M J Wimsey
Sales Manager: Ian Searle
■ Manufacturer of hand-held military products including the Small Arms Collimator (SAC) and laser collimator for weapon zeroing, prismatic compasses for marching, gun laying and artillery purposes, a full range of MIL-spec binoculars, PYSIS, commercial low-cost Night Vision Equipment, together with navigation & orienteering compasses. Francis Barker also offer prototype, development, repair and recalibration facilities under the Ministry of Defence. The company is accredited to ISO9001 Quality Management System. 98

75-47350 QED Design and Manufacture Ltd
Fairbourne Manor, Harrietsham, Kent ME17 1LH
☎ (+44 1622) 85 99 88 E gordon@tigra.demon.co.uk
⊖ (+44 1622) 85 91 19
♣ £ 2,000,000 (1996) ⚉ 40 (1996) ⌂ 2,000 sq m
Managing Dir: Gordon Harrold
Sales & Marketing Dir: Maurice Morsia
■ Design, development, testing & manufacture missile & equipment transit & storage containers, soft bags & covers, specialist containers & ground handling equipment. Advanced composites & structural foam moulding. 98

75-47450 Quasar Microwave Technology Ltd
Battle Road, Heathfield, Newton Abbot, Devon TQ12 6XU
☎ (+44 1626) 83 42 22 ⊠ 42616 gamsby g
⊖ (+44 1626) 83 29 94
♣ £ 4,800,000 (1997) ⚉ 95 (1997) ⌂ 40,000 sq ft
Managing Dir: Dr Chris Hughes
Sales & Marketing Dir: R.A. Shrubsole
Operations & Technical Dir: M. Jackson
■ Flexible/rigid waveguide transmission line runs, waveguides for satellite applications, nickel electroformed waveguides, flexible & flex/twist rectangular & double ridge waveguides, waveguide components.
Subsidiary:
Gabriel Microwave Ltd 98

75-47653 Racal Acoustics Ltd
(a division of Racal Electronics plc)
Waverley Industrial Park, Hailsham Drive,
Harrow, Middlesex HA1 4TR
☎ (+44 181) 427 77 27 ✉ ACOUSTICS
⊠ 926288 racal g ⊖ (+44 181) 427 03 50
♣ £ 21,000,000 ⚉ 360
Managing Director: Lewis Macdonald
Technical Director: Peter Smith
Commercial Director: Anthony Dent
Publicity Manager: A J Wilson
■ Electro-acoustic audio communications products. 98

75-47655 Racal-Airtech Ltd
(† Airtech Computer Security Ltd)
Meadow View Hse, Crendon Ind Est, Long Crendon,
Aylesbury, Bucks HP18 9EQ
☎ (+44 1844) 20 18 00 ⊖ (+44 1844) 20 85 50
Managing Dir: Alan Lamb
Marketing Mgr: Terry Plume
Advertising Mgr: Mrs Sarah Gorton
■ Medium & high grade communications & computer security equipment, software & systems. Security for wide area networks, remote LAN access & electronic commerce (payment systems). Personal authentication systems. Design & systems integration for custom data security projects. → 80 97

75-47656 Racal Antennas Ltd
(a div. of Racal Electronics Plc)
First Avenue, Millbrook Trading Estate,
Southampton SO15 0LJ
☎ (+44 1703) 70 57 05 ✉ ANTENNAS, Southampton
⊖ (+44 1703) 70 11 22
Managing Director: Paddy Graham
Technical Director: John Rowntree
Commercial Director: Derek C Henderson
Publicity Manager: John Horner

■ Manufacture of antennas, masts and support systems for tactical and strategic applications from MF to microwave frequencies. Cellular mobile and base station antennas. [98]

75-47660 **Racal Avionics Ltd**
(a division of Racal Electronics plc)
88 Bushey Road, London SW20 0JH
☎ (+44 181) 946 80 11 ⊠ 928592
⊖ (+44 181) 946 75 30
👥 400 (1997)
Managing Director: Sean O'Malley
Operations Director: M Miller
Marketing Director: Leo Gallagher
Purchasing Manager: J Mansi
Senior Publicity Manager: Trevor Rodgers
■ Civil and military airborne navigation, flight management and instrument landing systems. Ground navigation aids. Aeronautical satellite communications systems. Area navigation computers and mission management systems.
Manufacturing
88 Bushey Road, London SW20 0JW
☎ (+44 181) 946 80 11 ⊖ (+44 181) 946 75 30
General Manager: John Mountifield
■ Civil and military airborne navigation, flight management and instrument landing systems. Ground navigation aids. Aeronautical satellite communications systems. Area navigation computers, mission management systems. [98]

75-47665 **Racal-Comsec Ltd**
(a division of Racal Electronics plc)
Milford Industrial Estate, Tollgate Road,
Salisbury, Wiltshire SP1 2JG
☎ (+44 1722) 32 39 11 ⊖ (+44 1722) 33 00 16
♣ £ 5,000,000 👥 50
Director and General Manager: Trefor Watkins
Commercial Director: John Hall
■ Manufacture of a range of encryption devices to protect voice and data transmissions over radio, telephone, satellite and other communication systems; remote ground sensing systems. Facilities for RFI/Tempest/EMC testing and consultancy. [98]

75-47670 **Racal-Datacom Ltd**
(a division of Racal Electronics plc)
Landata House, Station Road, Hook, Hampshire RG27 9PE
☎ (+44 1256) 76 39 11 ⊠ 858294 racmil g
⊖ (+44 1256) 76 47 17
Managing Director: M Lea
Director and General Manager Network Systems: C Musgrave
Publicity Manager: Gordon Brake
Press Officer: David Morrow
Press Officer: Ms Carol Moore
■ Local area data networks, wide area data networks (national & international), modems, data multiplexers, communications resource management systems, X.25 circuit & packet switched networks, ISDN remote access, PC connectivity, IT & DATA security. [98]

75-47675 **Racal Health & Safety Group**
12-16 Bristol Road, Greenford,
Greenford, Middlesex UB6 8XT
☎ (+44 990) 16 81 18 ✆ 0800 72 22 57 ⊠ 922101
⊖ (+44 181) 832 87 99
♣ £ 60,000,000 👥 200
Managing Director: John Gill
Operations Director: Geoff Sadler
Publicity Manager: Ms Audrey Cooper
Purchasing Manager: Jonathan Haywood
■ Design and production of respiratory protective devices for commercial, industrial and military environments. [98]

75-47676 **Racal Instruments Ltd**
(a subsidiary of Racal Electronics plc)
480 Bath Road, Slough, Berkshire SL1 6BE
☎ (+44 1628) 60 44 55 ⊖ (+44 1628) 66 20 17
👥 300 (1997)
Managing Director: Colin Robertson
Head of Defence Marketing: K K Vartanian
Operations Executive: Paul Canning
Sales and Marketing Director: Greg Maton
Financial Director: John Taylor
Commercial Director: Tony Smith
Purchasing Manager: Robert Wilson
Publicity Manager: Mike Janicki
■ Design and manufacture of measuring instruments, VXI modules, first line, mobile and base automatic test equipment, application test packages and support services for maintenance of electronic systems. [98]

75-47682 **Racal-MESL Ltd Head Office**
(a subsidiary of Racal Electronics plc)
Lochend Industrial Estate,
Newbridge, Edinburgh EH28 8PL
☎ (+44 131) 333 20 00 ⊖ (+44 131) 333 36 07
♣ £ 10,000,000 (1995) 👥 196 🏠 5,000 sq m
Managing Director: M McSwan
Sales Director: Dr Eddie Patterson
Director, Microwave Products: Dr W T Nisbet
Director, Electronic Products: Dr John Arthur
■ Microwave components and subsystems. Surface Acoustic Wave (SAW) components and subsystems. LCD and EL display modules. [98]

75-47688 **Racal Radar Defence Systems Ltd**
(a subsidiary of Racal Electronics plc)
Manor Royal, Crawley, West Sussex RH10 2PZ
☎ (+44 1293) 52 87 87 ⊠ 27720 rdds g
⊖ (+44 1293) 64 46 44
👥 2,000
Managing Director: Barton J Clarke
Technical Director: Peter East
Sales and Marketing Director: John Wilson
Senior Publicity Manager: Trevor J Rodgers
☎ (+44 1293) 64 41 25; 61 17 70 (fax)
Publicity Manager: Ms Sarah Turley
☎ (+44 1293) 64 41 25
■ Defence radar systems. ELINT, ESM and ECM systems. CCIS and C3I systems. Miss distance indicators; surveillance and target cueing radar. Non-communications electronic warfare systems and radar displays for fixed and rotary wing aircraft, surface ships and submarines. High resolution colour display systems, colour map display systems, ground-based ELINT systems, battlefield ESM and ELINT collection systems. Action Information Systems, SWSE/SWSA, high resolution display systems. [98]

75-47689 **Racal Recorders Ltd**
(a division of Racal Electronics plc)
Hardley Industrial Estate, Hythe,
Southampton, Hampshire SO45 3ZH
☎ (+44 1703) 84 32 65 E recorders@racalrec.co.uk
⊖ (+44 1703) 84 89 19
👥 500 🏠 7,722 m2
Managing Director: David Kempson
Operations and Technical Director: A Miles
Marketing Director: Bruce Titheridge
Senior Marketing Manager: G Burn
General Sales Manager: G Clark
Purchasing Manager: S Newton
Publicity Manager: Ms Georgina Froud
UK Sales Manager: K Ingram
■ Communications and instrumentation recorder manufactures. Design, manufacture and supply of voice recorders and advanced digital systems. [98]

75-47698 **Racal Radio Ltd**
(a division of Racal Electronics plc)
Western Road, PO Box 3621,
Bracknell, Berkshire RG12 1WJ
☎ (+44 1344) 38 70 00 ⊖ (+44 1344) 38 74 03
♣ £ 152,000,000 (1996) 👥 1,600 (1996)
Managing Director: R J Moon
Deputy Managing Director: S S Willder
Deputy Managing Director: J F L Pidgeon
Sales and Marketing Director: F Mulholland
Senior Publicity Manager: H S Kernohan
UK Marketing Manager: A W L Vosper
■ UHF/VHF/HF transceivers for portable and vehicle mobile systems. Supplier of HF/VHF/UHF frequency hopping systems (JAGUAR and Panther), tactical trunk communications systems, COUGARNET and COUGAR 2000 area communications systems. Point to point fixed frequency and adaptive communications systems. HF/VHF/UHF receivers. Communications electronic warfare systems. Military ground mobile TACSATCOM terminals. [98]

75-47750 **Radamec Defence Systems Ltd**
Bridge Road, Chertsey, Surrey KT16 8LJ
☎ (+44 1932) 56 11 81 ⊠ 929945 radepo g
⊖ (+44 1932) 56 87 75
♣ £ 5 600 000 👥 100 🏠 20 000 sq ft
Managing Director: John S Dobson
Finance Dir & Company Secretary: Lionel Moore
Sales Mgr: Roger C. Joel
■ Electro-optical systems for surveillance, tracking and weapon fire control. Design and production of precision servo driven platforms for all military environment applications, video processing and auto-tracking sub-systems and military CCD TV cameras. [98]

75-47755 **Radamec Control Systems Ltd**
(† Newbury Diesel Controls Ltd)
(a subsidiary of Radamec Group plc)
Unit A2, Cyril Vokins Road, Hambridge Lane,
Newbury, Berkshire RG14 5XB
☎ (+44 1635) 405 28 ⊖ (+44 1635) 474 53
♣ £ 350,000 👥 10 🏠 500 sq m
Director and General Manager: T N Barnard
Technical Director: K J Hall
■ Bridge control systems: type MICROSHIFT and SERVOSTEP for the control of engine speed, CPP propellers and reversing gearboxes. Monitoring systems: type DARAD for alarms data logging and event recording. Supporting equipment: includes telegraphs, tachometry systems and engine safety systems. [98]

75-47760 **Radar**
95 Lord Street, Redcar, Cleveland TS10 3HR
☎ (+44 1642) 49 06 52 ⊖ (+44 1642) 49 42 14
♣ £ 500,000 👥 10 🏠 200 sq m
Managing Dir: L. Williams
■ Manufacture of uniforms. [98]

75-47810 **Radstone Technology plc**
Water Lane, Towcester, Northamptonshire NN12 6JN
☎ (+44 1327) 35 94 44 E sales@radstone.co.uk
W http://www.radstone.co.uk ⊖ (+44 1327) 35 81 12
✧ £ 5,670,000 (1994/95) ♣ £ 19,320,000 (1995/96)
👥 320 🏠 90,000 sq ft
Managing Director: Dr Charles H Paterson
Sales Dir: John S. Umney
Operations Dir: Peter J. Cavill
■ Supplier of standard and custom board level, systems and software products for military and industrial applications. Products support open bus architectures (VME) including processors, memories, I/O and enclosures. [98]

75-47850 **Railko Ltd**
(a division of United Industries)
Loudwater, High Wycombe Buckinghamshire HP10 9QU
☎ (+44 16285) 249 01 ⊠ 848406
⊖ (+44 1628) 81 07 61
👥 120
Managing Director: Dr G D Wells
Operations and Technical Director: T Biswell
Sales and Engineering Director: P Cumberlidge
Publicity Manager: Mrs B A Rosser
Purchasing Manager: S Devlin
■ Design and manufacture of industrial thermoplastic and thermosetting bearings for marine, railway, automotive and general engineering applications. [98]

75-48305 **Rapiscan Security Products Ltd**
(† IAL Security Products)
Unit B1, The Fleming Centre, Fleming Way,
Crawley, West Sussex RH10 2NN
☎ (+44 1293) 54 06 61 ⊠ ADDON
⊖ (+44 1293) 54 20 19
♣ £ 6,500,000 👥 50
Managing Director: Anthony S Crane
Operations Director: Roger Davies
Purchasing Manager: Rob Jones
Sales Department: F Rogers
Sales Department: Angus De Villiers
Sales Department: Nick Downer
■ X-ray detection equipment of weapons and bombs. [98]

75-48340 **Raydex/C.D.T. Ltd**
(a division of Cable Design Technologies Corporation, USA)
Church Street, PO Box 3,
Littleborough, Lancashire OL15 8HG
☎ (+44 1706) 37 40 15 ⊖ (+44 1706) 37 05 76

Contact Reference

♣ 270
Operations & Technical Dir: S.J. Prescott
Sales & Marketing Dir: D.T. Bolton
Sales Mgr: M.B. Chew
Purchasing Mgr: L. Fitzsimmons
■ High performance electrical wires & cables including coaxial & thermocouple cables plus optical fibre cables. Insulations include PTFE, Kapton, polyester, ETFE, PFA, FEP, PDF & low toxic zero halogen materials. 98

75-48350 The Ray Engineering Company (RENCOL)
(a division of Lilleshall Plastics and Engineering)
216-218 Southmead Road, Southmead, Bristol BS10 5EB
☎ (+44 117) 950 66 66 E sales@rencol.com
W http://www.rencol.com ☒ 44763 rayeng g
⊖ (+44 117) 950 45 50
♣ £ 5,000,000 ♣ 140
Managing Director: B J McKenna
Operations Director: N Murphy
Group Sales Manager: D Shawe
Marketing Manager: I van Duijneveldt
■ Large range of plastic knobs, handles and handwheels. Plastic injection & compression mouldings using thermoset and thermoplastic materials. Custom moulding to customer specifications.
Overseas Agents:
DFS International (USA)
DNL Kunststoffen BV (Netherlands)
Nortel Manufacturing Ltd (Canada)
DZUS Fastener GmbH (Germany)
Arthur Spence (Sales) Ltd (Ireland)
BRD Klee (Denmark)
Société TDI (France)
OPAC SL (Spain) 98

75-48367 Raytheon Cossor Electronics Ltd
(† Cossor Electronics Ltd)
(a subsidiary of Raytheon Co, USA)
The Pinnacles, Elizabeth Way, Harlow, Essex CM19 5BB
☎ (+44 1279) 42 68 62 ☒ COSSOR HARLOW ☒ 81228
⊖ (+44 1279) 41 04 13
♣ £ 5,250,000 (1996) ♣ £ 35,000,000 (1996)
♣ 550 (1996) ♣ 15,000 sq m (1996)
Chairman: Sir Kenneth MacDonald
Managing Director: R W Barry
Business Development Director: I H Bainbridge
Purchasing Manager: M A Mutimer
Manufacturing Executive: J S Warren
Technical Executive: P Fraser
■ Secondary surveillance rad, IFF radar interrogators and airborne transponders. VHF/UHF ground to air communications, modems and GPS systems. 98

75-48370 RBR Armour Ltd
88-94 Old Kent Road, London SE1 4NU
☎ (+44 171) 703 10 05 ⊖ (+44 171) 703 55 05
♣ £ 10,000 (1997) ♣ £ 15,000,000 (1997)
♣ 130 (1997) ♣ 43,000 sq ft
Managing Director: Douglas Garland
Technical Director: James F Barron
Marketing Director: Terry F Hand
Regional Sales Director: Ian Dunbavard
Purchasing Manager: David Pope
■ Riot jackets, soft and hard body armour. Ballistic helmets and shields. 98

75-48380 Reflectone UK Ltd
(† British Aerospace Simulation Ltd)
Building 20A-1, PO Box 77, Filton, Bristol BS99 7AR
☎ (+44 117) 936 44 38 ⊖ (+44 117) 936 44 62
♣ £ 35,000,000 (1996) ♣ 110 ♣ 60,000 sq ft
Managing Director: J G Hancock
Sales and Marketing Director: M R Wain
Operations Manager: A J Warrell
Commercial Manager: C K Blackburn
■ Design and manufacture of civil and military aircraft flight simulators, tactical air defence weapons trainers, EW training systems, air traffic control trainers. Radar air defence trainers. Naval and maritime trainers. Ships bridge simulators. 98

75-48400 RE Components Ltd
(a subsidiary of BTR plc, part of Dunlop Aviation Division)
Station Road, Bagworth,
Coalville, Leicestershire LE67 1BF
☎ (+44 1530) 23 03 61 ⊖ (+44 1530) 23 01 13
♣ £ 3,000,000 ♣ 150
Managing Director: Gordon Midgley
Business Manager: K D Ambrose
Sales Manager: Tony Carter
Purchasing Manager: Ron Stacey
■ Precision rubber mouldings and diaphragms for defence equipment and other applications. 98

75-48530 Redifon MEL Limited
(subsidiary of Thomson-CSF Communications)
Head Office & Factory
Newton Road, Crawley, West Sussex RH10 2TU
☎ (+44 1293) 51 88 55 ⊖ (+44 1293) 54 00 45
♣ £ 25,000,000 (1996) ♣ 350 (1997)
Managing Dir: M. Purvis
Sales & Marketing Dir: W. Mawer
Financial Dir: Martin Parmenter
Publicity Mgr: Mark Rouson
■ Communications and security equipment, computerised systems and data management.
Communications Division
Newton Road, Crawley, West Sussex RH10 2TU
☎ (+44 1293) 51 33 55 ⊖ (+44 1293) 56 10 96
Programmes Dir: Colin Shaw
Engineering Dir: Peter Harrop
Purchasing Mgr: Carolyne Trew
Sales & Marketing Support Mgr: Simon Kane
■ HF, VHF & VLF receivers & transmitters, radio communications systems for naval vessels. Portable & fixed air traffic control equipment for military airfields. Loop antennas. BS EN ISO 9001 approved.
SPT Division
Woodrolfe Rd, Tollesbury, Maldon, Essex CM9 8XD
☎ (+44 1621) 86 95 58 ☒ 995133 spt g
⊖ (+44 1621) 86 88 98
♣ 35
General Mgr: Tim Thomas
Sales Mgr: Keith Payne
■ Design & manufacture of main broadcast, entertainment, telephone, communal aerial & intercom systems for civil & military marine, offshore or land-based applications.
Technology Division
Newton Road, Crawley, West Sussex RH10 2TU
☎ (+44 1293) 51 88 55 ⊖ (+44 1293) 53 01 01
General Mgr: John Higgins
Sales Mgr: Bob Garlick
■ Electronic design & development services. MoD quality approvals. TickIT quality approvals. BS EN ISO 9001 approved. 98

75-48960 Redpoint Thermalloy Ltd
(a subsidiary of Bowthorpe plc)
Cheney Manor, Swindon, Wiltshire SN2 2QN
☎ (+44 1793) 53 78 61 ⊖ (+44 1793) 61 53 96
Managing Director: R Johnson
■ Manufacture of extruded and pressed aluminium heat-sinks and heat transfer devices for power semiconductors and other electronic equipment. 98

75-49170 Reliant Design Development Co Ltd
60 Woolmer Way, Bordon, Hampshire GU35 9QF
☎ (+44 1420) 47 83 41 ⊖ (+44 1420) 48 93 22
♣ £ 30,000 (1997) ♣ £ 1,095,000 (1996) ♣ 15 (1997)
♣ 8,500 sq ft
Managing Director: S W Edwards
Operations Director: D H Cox
Sales and Marketing Manager: J B Hodgson
Purchasing Administration: D Fleming
■ Design, development & manufacture, installation, servicing, post design services for test rig systems & associated control gear - ground support equipment. Mobile & fixed electronic control boxes. Access equipment. Repair, service & refurbishing facility. Fluid dispensing equipment for the automotive industry. 98

75-49210 Remsdaq Ltd
(a subsidiary of DAQ Electronics Inc, USA)
Parkway, Deeside Industrial Park,
Deeside, Flintshire CH5 2NN
☎ (+44 1244) 28 64 95 ⊖ (+44 1244) 28 64 96
♣ £ 10,334,000 (1996) ♣ 107 (1996) ♣ 5,000 sq m
Managing Dir: T.J. Breen
Operations Dir: R.J. Colston
Marketing & Sales Mgr: John P. Warner
Purchasing Mgr: B. Bythell
■ Supplies integrated security systems; mobilising systems; supervisory control (SCADA); building management; communication & network management systems worldwide to governments, international, military & key commercial organisations. 98

75-49225 Renair Antenna
(† Bantex Ltd)
11/15 Chase Road, London NW10 6PT
☎ (+44 181) 965 30 01 ⊖ (+44 181) 965 57 73
♣ £ 750 000 ♣ 20
Managing Dir: J. Futter
Operations Dir: N. Levin
Sales & Marketing Dir: S. Pandya
■ Communications antennae, ground-based & shipborne. 97

75-49226 Rendar Ltd
(a subsidiary of Advance International Group Ltd)
Durban Road, Bognor Regis, West Sussex PO22 9RL
☎ (+44 1243) 86 67 41 ✆ (888) 221 96 71 (USA only)
E sales@rendar.co.uk W http://www.rendar.com
⊖ (+44 1243) 84 14 86
♣ £ 3,100,000 (1995) ♣ 65 ♣ 20,000 sq ft
Managing Director: David Fairlamb
Marketing and Sales Manager: Duncan Palmer
■ Manufacture and distribution of jack plugs and sockets, low power mains connectors, cordsets. → 85 98

75-49227 Renown Engineering Ltd
Dudley Lane, Dudley,
Cramlington, Northumberland NE23 7RA
☎ (+44 191) 250 01 13 ⊖ (+44 191) 250 19 80
♣ £ 2 500 000 (1995) ♣ 55 ♣ 35 000 sq ft
Managing Dir: Paul Brown
Director: Peter Renney
Purchasing Mgr: David Youhill
Admin: J. Boon
■ Manufacture & repair of vehicle assemblies, sub-assemblies & components, roadwheels, bridging structures, weapon mountings, fabrication of units in aluminium armour & maraging steels, metal arc spraying & painting. 97

75-49250 Research Engineers (RE)
11-31 Orsman Road, London N1 5RD
☎ (+44 171) 739 78 11 ⊖ (+44 171) 739 66 15
♣ £ 100,000 ♣ £ 1,900,000 (1995/96) ♣ 25
Managing Director: A N D Goodhart
Sales Office Manager: D Garnsworthy
Sales Manager: T Collins
■ Airfield lighting equipment including Precision Approach Path Indicators (PAPI) for tactical and fixed applications. → 80 98

75-49260 Reson Offshore Ltd
(† Reson UK)
(a subsidiary of Reson A/S, Denmark)
Bowtech Building, Howemoss Crescent,
Kirkhill Industrial Estate, Dyce, Aberdeen AB21 0GN
☎ (+44 1224) 72 74 27 ⊖ (+44 1224) 72 74 28
♣ 4 (1997)
Managing Director: Claus R Steenstrup
Sales Manager: Allan Jack
■ SeaBat 6012 - minehunting real-time, multibeam sonar system. Update rate - up to 30 times per second. ROV or surface vessel mounted, 200m range, frequency: 455 Khz, use: minehunting and surveillance. SeaBat 8100 real-time forward looking sonar, frequency: 240 kHz, 40 times a second update. Mine detection, obstacle avoidance, deep water operations; range 300 m. In use with the Danish, US, French and Swedish Navies.
SeaBat 9001 - real-time multibeam bathymetric sonar. Update rate: up to 900 data points per second. ROV, ROTV or surface vessel mounted, 100m range, frequency: 455 Khz, use: detailed seafloor mapping, ground training, speed: up to 12 knots.
SeaBat 8100 - real-time multibeam bathymetric sonar. Frequency: 240 Khz. Range: 300m. ROV or surface vessel mounted. Depth resolution: 1cm. 98

75-49280 Rest Assured Limited
Pontygwaith, Ferndale, Rhondda, Mid. Glamorgan
☎ (+44 1443) 73 05 41 ⊖ (+44 1443) 75 69 00
♣ £ 7 000 000 ♣ 200
Managing Dir: I.P. Smith
Sales Dir: Brian Wright
Purchasing Mgr: Michael Dickins
■ Manufacture of beds & mattresses for various conditions. 97

75-49340 Reynolds Boughton Ltd
(a member of the Boughton Group)
Engineering Division
Bell Lane, Amersham, Buckinghamshire HP6 6PE
☎(+44 1494) 76 44 11 ✆ (800) 59 22 45
[E] rb@boughton.co.uk ⊠ 83132 ⊖ (+44 1494) 76 52 18
♧ £ 500,000 (1997) ♣ £ 18,000,000 (1997)
👥 180 (1997) 🏭 120,000 sq ft
Managing Dir: Stephen R. Boughton
Technical Dir: James Hopkinson
Sales Dir: V.R. French
Technical Sales Mgr: Edward Cox
■ Manufacture of winches, forks & buckets for mounting on wheeled & tracked vehicles.
Vehicle Division
Bell Lane, Amersham, Buckinghamshire HP6 6PE
☎(+44 1494) 76 44 11 ⊠ 83132
⊖ (+44 1494) 76 52 18
Sales Dir: V.R. French
■ Special vehicles, vehicle bodies, winches & cranes. Missile launching & logistic support equipment.
Reynolds Boughton (Devon) Ltd
Winkleigh Airfield, Winkleigh, Devon EX19 8DR
☎(+44 183) 78 35 55 ⊠ 42741 ttbfer g
⊖ (+44 183) 78 37 68
Chairman: T.Trafford Boughton
Managing Dir: Stephen R. Boughton
Technical Works Dir: Roger Hearn
■ Airport equipment, firefighting machines, MoD approved, ground support vehicles. [98]

75-49345 Reynolds Industries Ltd
(a subsidiary of Reynolds Industries Inc)
Navigation House, Canal View Road,
Newbury, Berkshire RG14 5UR
☎(+44 1635) 311 37 [E] rey@globalnet.co.uk
⊖ (+44 1635) 52 19 36
♣ £ 3,200,000 (1996/97) 👥 38 (1997) 🏭 2,700 m2
Managing Dir: Nick S. Bance
Operations & Technical Dir: Paul Belsey
Mgr Microwave Div.: Brian H. Huntsman
Sales Mgr: Neil C.M. Smithers
Purchasing Mgr: Malcolm Horne
■ High voltage connectors, high & low voltage cable assemblies for the aerospace industry. Microwave cable assemblies flexible & semi-rigid to 40 GHz. High voltage capacitors, transient protection devices used in military & airborne electronic equipment. Custom built electro-mechanical & electro optical interconnect products for a wide range of space & aerospace applications. [98]

75-49500 RFI Shielding Ltd
Warner Drive, Rayne Road, Springwood Industrial Estate,
Braintree, Essex CM7 7YW
☎(+44 1376) 34 26 26 ⊖ (+44 1376) 34 64 42
♣ £ 4,300,000 (1996) 👥 70 (1996)
Managing Director: Barry Comley
Sales and Marketing Manager: Philip Mizen
Purchasing Manager: Paul Campen
■ Screening products, conductive gasketing, shielded windows. Honeycomb or one piece air vents for EMC, Tempest FCC & VDE screening applications. DEF STAN AQAP-4 approved, BS5750 part 2, ISO 9002. Full free applications advice service. → 80 [98]

75-49550 Rheometric Scientific Ltd
Surrey Business Park, Kiln Lane, Epsom, Surrey KT17 1JF
☎(+44 372) 74 33 86 [W] http://www.rheosc.com
⊠ 8811417 ⊖ (+44 372) 72 71 02
♣ £ 5,000,000 👥 48
Managing Director: Paul Newbatt
Sales and Marketing Director: Paul Larcey
Sales and Marketing Director: Rob Bell
■ Manufacturers of flammability testing equipment - LOI (Limiting Oxygen Index) BS 2782, ISO 4589, ASTMD 2863, LOI to + 400°C. NBS smoke chamber BS 6401, ESTM E-662. Cone calorimeter ISO 5660, ASTM E-1354.
Manufacturers of thermal analysis equipment - DSC Differential Scanning Calorimeter. TGA Thermogranimetric analyser. STA Simultaneous (TGA/DSC) Thermal Analyser, DMTA Dynamic Mechanical Thermal Analyser, DETA Dielectric Thermal Analyser.
Manufacturers of Rheological Testing Equipment for both solid and liquid material. [98]

75-49565 RHP Aerospace
(a subsidiary of N.S.K., Japan)
Oldends Lane, Stonehouse, Gloucestershire GL10 3RH
☎(+44 1453) 82 23 33 ⊖ (+44 1453) 82 59 45
♣ £ 15,000,000 (1997) 👥 350 (1997)
Managing Director: C Jackson
Divisional Sales and Marketing Manager: C Follows
European Sales Manager: K Burrows
Engineering Manager: M Thwaite
Purchasing Manager: J Panes
Design Manager: L Banstead
■ Precision bearings - aeroengine, gearbox & accessory, airframe. [98]

75-49570 D&BA Richards & Son
Valley Mills, Southfield St, Nelson, Lancashire BB9 0LD
☎(+44 1282) 69 10 68 ⊖ (+44 1282) 69 48 56
♣ £ 250 000 (1995) 👥 10 🏭 1 000 sq m
Works Mgr: Adrian Richards
■ Metal workers & fabricators, CNC plasma & oxygen fuel profiling, guillotining, bending, MIG, TIG & PVC welding. Suppliers to the aerospace & defence industries. [97]

75-49630 Richmond Packaging (UK) Ltd
New Rd, Winsford, Cheshire CW7 2NY
☎(+44 1606) 55 74 22 ⊖ (+44 1606) 86 10 63
♧ £ 1,700,000 (1996) ♣ £ 2,400,000 (1996)
👥 100 8996) 🏭 3,600 sq m (1996)
Managing Director: I S Harty
Commercial Director: J J England
■ Decontamination kits. Residual vapour detector kits. Water testing kits, small liquid & powder packs. [98]

75-49700 Ring Sights Holdings Co. Ltd
Headquarters
PO Box 22, Bordon, Hampshire GU35 9PD
☎(+44 1420) 47 22 60 [E] ringsights@easynet.uk.co
⊖ (+44 1420) 47 83 59
♣ £ 10,000,000 (1997) 👥 12 (1997)
Managing Dir: Ian R. Flack
Research & Development Dir: Brig. Fraser Scott
■ Design, manufacture & sale of patented unit power collimator sights & weapon interfaces for pistols, rifles, cannons, SMG & machine guns.
Subsidiaries:
Ring Sights Optronics Ltd →
Ring Sights Ltd → 75
Ring Sights (India) Pvt Ltd → India 75
Ring Sights Far East Ltd → Singapore 90
Ring Sights Defence Ltd [98]

75-49820 Ivar Rivenaes Ltd
9 Argent Centre, Silverdale Road,
Hayes, Middlesex UB3 3BL
☎(+44 181) 561 05 40 ⊠ 9419375 rmcldn g
⊖ (+44 181) 848 87 89
Director: J Hayward
Sales Manager: Gordon C Winson
■ "R-MC", a water based compressor cleaning liquid for diesel, gas turbine & aviation jet engines. [98]

75-49830 River Don Castings Ltd
(a subs. of Sheffield Forgemasters Ltd)
Brightside Lane, PO Box 99, Sheffield S9 2RX
☎(+44 114) 256 00 08 ⊠ 54432
⊖ (+44 114) 243 43 61
♣ £ 24,000,000 👥 230
Managing Director: J H Beresford
Commercial Director: G J Marston
■ Carbon, alloy and stainless steel castings. [98]

75-49860 RM Trailers Ltd
New Farm Industrial Estate,
New Alresford, Hampshire SO24 9QF
☎(+44 1962) 73 25 60 ⊖ (+44 1962) 73 40 27
♣ £ 750,000 👥 16 🏭 5,000 sq ft
Managing Director: R Gifford-Hull
Director: M Gifford-Hull
■ Manufacture military trailers of all types for equipment up to 15 tonnes gross weight for on/off highway use. Any design to order. [98]

75-49880 Roband Electronics plc
Charlwood Works, Charlwood, Horley, Surrey RH6 0BU
☎(+44 1293) 84 30 00 [E] sales@roband.co.uk
⊖ (+44 1293) 84 30 01
👥 100+ (1996) 🏭 3,500 sq m
Managing Director: A L Gold
Technical Director: K V Sharman
■ Design and manufacture of electronic power supply systems both LV & HV. Registered to BS5750 part 1. 98

75-50000 Rockwell Light Vehicle Systems (UK) Ltd
(† Rockwell Automotive Body Systems (UK) Ltd)
Fordhouse Lane, Birmingham B30 3BW
☎(+44 121) 459 11 66 ⊠ 337297
⊖ (+44 121) 459 98 08
Managing Director: Robert S Marty
☎(+44 121) 625 34 71
■ Automotive access control systems.
® USA: Rockwell International Corp. [98]

75-50050 Rockwell Vehicle Drivetrain Systems
Rockwell Automotive (UK) Ltd
(a division of Rockwell International)
Alcester, Warwickshire B49 6HT
☎(+44 1789) 76 41 23 ⊖ (+44 1789) 48 85 39
♧ £ 6,500,000 (UK only) (1997) ♣ £ 26,000,000 (1997)
👥 70 (1997) 🏭 4,000 m2
Managing Director: Brad Arnold
Sales and Marketing Director: Clive Harrup
Operations Director: P Britt
UK Sales Manager: Neil Winsper
Purchasing Manager: Neil Kirby
■ Axles, brakes.
® USA: Rockwell Int'l
● Rochwell Maudslay, Great Alne,
Alcester, Warkwickshire B49 6HT
☎(+44 81) 577 28 00 ⊖ (+44 81) 577 23 56
European Sales Dir: Clive Harrup
® USA: Rockwell International [98]

75-50060 Rocol Site Safety Systems Ltd
(† Rocol Ltd)
(a division of Morgan Crucible plc)
Rocol House, Swillington, Leeds LS26 8BS
☎(+44 113) 232 28 26 ⊠ 55249
⊖ (+44 113) 232 28 50
♣ £ 25,000,000 👥 300
Managing Director: Stephen King
Operations Director: Mike Walker
Sales Director: Andrew McNeill
Advertising Manager: Ms Sandra Kay
■ Lubrication - Aerospace greases, corrosion preventatives, MoD approved lubricants for ships and vehicles, metal forming products (MoD and aerospace).
Safety - Anti-slip and anti-corrosion deck paints (flight and weather) and related products. Specialist anti-slip coatings for vehicles and ground installations, safety consumables. [98]

75-50180 Rollalong
(a subsidiary of The Energy Group plc)
Woolsbridge Industrial Park, Three Legged Cross,
Wimborne, Dorset BH21 6SF
☎(+44 1202) 82 45 41 ⊖ (+44 1202) 82 65 25
♧ £ 8,200,000 (1997) ♣ £ 22,000,000 (1996)
👥 250 (1997) 🏭 200,000 sq ft
Managing Director: J R Day
Sales and Marketing Director: John S Jackson
Operations Director: David Turnbull
Purchasing Manager: Tom Smith
Sales Administrator: P S Brown
■ Design and manufacture of relocatable buildings including camps, clinics, laboratories. Mobile laboratory trailers, mobile workshops and trailers. [98]

75-50208 Rollet H. & Co. Ltd
Unit 10 Cotswold Mews, 30 Battersea Square,
London SW11 3RA
☎(+44 171) 228 64 94 ⊖ (+44 171) 350 28 72
♣ £ 1,000,000 👥 30
Sales Director: J H Medlicott
■ High precision waveguide & microwave components; protective plastic coating & light engineering service. Non-ferrous metal stockists & producers of precision tubes, copper sections, copper, brass & aluminium tubes. [98]

75-50250 Rolls-Royce plc
Head Office
65 Buckingham Gate, London SW1E 6AT
☎(+44 171) 222 90 20 ⊠ 918091 rrldn g
⊖ (+44 171) 233 17 33

Contact Reference

♣ £ 4,291,000,000 (1996) ⁂ 42,900 (1996)
Chairman: Sir Ralph H Robins
Chief Executive: John Rose
Group Marketing Director: Richard Turner
Managing Director Aerospace Group: Colin Green
Managing Director Industrial Power Group: Paul Heiden
Engineering and Technology Director: Stewart C Miller
Financial Director: Michael Townsend
Director of Human Resources: John Rivers
Director of Corporate Communications:
Peter Barnes-Wallis ☎ (+44 171) 222 90 20
Company Secretary and General Counsel: Charles Blundell
■ Design, development & manufacture of gas turbine engines for civil & military aircraft. Designs, constructs & installs complete power generation, transmission & distribution systems & major equipment, marine propulsion (including nuclear) oil & gas pumping, offshore & defence markets.

Principal Subsidiary Undertakings:
Registered in UK
Aerospace Group
Rolls E.L. Turbofans Ltd
Rolls-Royce Aero Engine Sesrvices Ltd
Rolls-Royce Commercial Aero Engines Ltd
Rolls-Royce Engine Controls Ltd
Rolls-Royce Int'l Support Services Ltd
Rolls-Royce Military Aero Engines Ltd
Rolls-Royce Inc.
Rolls-Royce Capital Inc.
Rolls-Royce Turbomeca Ltd (England & France)
Rolls-Royce & Partners Finance Ltd
Rolls-Royce Overseas Holdings Ltd
Rolls-Royce Power Ventures Ltd
Rolls-Royce Materials Handling Ltd
Rolls-Royce and Associates Ltd
Rolls-Royce Ind. & Marine Gas Turbines Ltd
Rolls-Royce Leasing Ltd
Rolls-Royce Industrial Power (India) Ltd
Rolls-Royce Nuclear Engineering Ltd
Rolls-Royce Nuclear Engineering Services Ltd
Rolls-Royce Power Engineering plc
Rolls-Royce International Ltd
Rolls-Royce Canada Ltd
Rolls-Royce Gas Turbine Engines (Canada) Inc
Rolls-Royce Holdings Canada Inc.
Rolls-Royce Industries Canada Inc.
Mexico: Ferranti-Packard de Mexico SA De CV
Parsons Power Generation Systems Ltd
Peebles Electric Ltd
Reyrolle Ltd
Corporate
Middle East Equity Partners Ltd
Thompson Kennicott Ltd
Sawley Packaging Co. Ltd
Industrial Power Group
Allen Power Engineering Ltd
Clarke Chapman Ltd
International Combustion Ltd
NEI Brantford International Ltd (51%)
NEI Overseas Holdings Ltd
Incorporated overseas
Aerospace Group
Brazil: Motores Rolls-Royce Ltda
France: Rolls-Royce Technical Support SARL
Allison Engine Company Inc.
Industrial Power Group
Australia: Rolls-Royce Ind. Power (Pacific)
Canada: Bristol Aerospace Ltd
Ferranti-Packard Transformers Ltd
Parsons Turbine Generators Canada Ltd
New Zealand: Rolls-Royce Ind. Power Ltd
South Africa: NEI Africa Holdings Ltd (60%)
Northern Engineering Ind. Africa Ltd (56%)
Zambia: Cutler Hammer Zambia Ltd
NEI Zambia Ltd
Zimbabwe: NEI Holdings Zimbabwe (Private) Ltd
Corporate:
Australia: Rolls-Royce of Australia Pty Ltd
Guernsey: Nightingale Insurance Ltd
Netherlands: Rolls-Royce Partner Finance NV
USA: Rolls-Royce North America Inc.
Principal Associated Undertakings
Registered in England or Scotland
Aerospace Group
Rolls Smiths Engine Controls Ltd
Turbo-Union Ltd (England, Germany & Italy)
Industrial Power Group
Rolls Wood Group (Repair & Overhauls) Ltd
Cooper Rolls Ltd
Corporate
Incorporated overseas
Aerospace Group
Germany: BMW Rolls-Royce GmbH (England & Germany)
Eurojet Turbo GmbH (England, Germany, Italy & Spain)
MTU, Turbomeca, Rolls-Royce GmbH (England, France & Germany)
Spain: Industria de Turbo Propulsores SA
Switzerland: IAE International Aero Engines AG (England, Germany, Italy, Japan & USA)
USA: William-Rolls Inc. (Europe & North America)
Industrial Power Group
Canada: Cooper, Rolls Corp.
India: Easun Reyolle Relays & Devices Ltd
USA: Cooper Rolls Incorporated (Europe & North America)
Corporate
Saudi Arabia: Middle East Propulsion Company Ltd
USA: R-H Component Technologies L.C.

Industrial Power Group
Registered Office
NEI Hse, Regent Centre, Newcastle-upon-Tyne NE3 3SB
☎ (+44 191) 401 41 00 ☒ 537900 neincl g
⊖ (+44 191) 285 66 54
♣ £ 1 400 000 000 ⁂ 22 000
Public Affairs Mgr: Martin Nield
☎ (+44 191) 401 4167 (Fax)

London Office
65 Buckingham Gate, London SW1E 6AT
☎ (+44 171) 222 90 20 ☒ 918091
⊖ (+44 171) 233 17 33

Subsidiary companies:
Rolls-Royce Power Engineering plc
Rolls-Royce Ind. & Marine Gas Turbines Ltd
Rolls-Royce Nuclear Engineering Ltd
Rolls-Royce Industries Canada Inc

Industrial & Marine Gas Turbines
Ansty, Coventry CV7 9JR
☎ (+44 1203) 62 40 00 ☒ 31636 rrimd g
⊖ (+44 1203) 62 46 66
Managing Dir: R.W. Sunerton
Head of Marketing: J.A. Roberts
■ Marine Spey gas turbines. Industrial Avon & RB211 gas turbines.

Power Engineering plc
NEI House, Regent Centre,
Newcastle-upon-Tyne NE3 3SB
☎ (+44 191) 401 41 00 ⊖ (+44 191) 285 66 54
Managing Dir: A.D.J. Perkins

Subsidiary companies:
Allen Power Engineering Ltd
Clarke Chapman Ltd
NEI Control Systems Ltd
International Combustion Ltd
Parsons Turbine Generators Ltd
Peebles Electric Ltd
Reyrolle Ltd [98]

75-50255 Rolls-Royce Commercial Aero Engines Ltd

PO Box 31, Derby DE24 8B5
☎ (+44 1332) 24 24 24 ☒ 37645
⊖ (+44 1332) 24 99 36
♣ £ 2 139 000
Managing Dir: J.P. Cheffins
Management: Chris Winfield
Marketing Dir: D.J. Wicks
■ Full range of aero engines for commercial airlines, regional, business & executive aircraft. Products: Trent family, RB211 series, Tay engine collaborative, V2500, FJ44. BMW Rolls-Royce BR700 series. [97]

75-50258 Rolls-Royce Aero Engine Services Ltd

Mavor Ave, Nerston Industrial Est., Nerston, East Kilbride, Glasgow G74 4PY
☎ (+44 13552) 202 00 ✉ ROLLS ROYCE
☒ 778432 rr glw g ⊖ (+44 13552) 304 11
Managing Dir: I. Lloyd
■ Repair, overhaul & rework of Rolls-Royce & other manufacturers' engines. → 80 [97]

75-50260 Rolls-Royce MatEval

Unit 7, Raynesway Park, Raynesway, Derby DE21 7BH
☎ (+44 1332) 75 19 00 ⊖ (+44 1332) 75 17 51
♣ £ 2 000 000 ⁂ 25
General Mgr: Barry Dabell
■ Design & manufacture of custom-made NDT inspection systems & equipment including Ultrasonics & X-ray equipment for inspection of rockets, aircraft structures & components, munitions, nuclear reactor pressure vessels & primary components. Portable inspection devices. Feasibility studies, consultancy & customer support. [97]

75-50265 Rolls-Royce Military Aero Engines Ltd

(† Military Engine Business)
(a div. of Rolls-Royce plc)
PO Box 3, Filton, Bristol BS12 7QE
☎ (+44 117) 979 12 34 ☒ 44185 rr bsl g
⊖ (+44 117) 979 75 75
⁂ 2 600 ⌂ 239 000 sq m
Managing Dir: J. Ferrie
Finance Dir: R. Tomlinson
Production Dir: L. Belliew
Customer Operations Dir: S.M. Lyons
Personnel Dir: R. Edwards
Fleet Engines Dir: C.B. Fairhead
EJ200/FLA Dir: P.F. Wilkins
Information Technology Dir: R.G. Taylor
Quality Assurance Dir: R.G. Mair
Pegasus & Adour F405 Dir: C.M. Pratt
Commercial Dir: S.A. Smith
Future Combat Engines Dir: C.A. Hughes
RTM322 & MTR390 Dir: G.K. Waddington
Dir Business Development: G.H. Scott
Engineering Dir: H.C. Hillier
■ Development, manufacture & support of engines for fixed wing application: EJ200 reheated turbofan (as partner in Eurojet Turbo GmbH); RB199 reheated turbofan (as partner in TurboUnion); Pegasus vectored thrust turbofan; Adour turbofan (in collaboration with Turbomeca); RB168-807 turbofan (in collaboration with FiatAvio and CELMA of Brazil); Viper turbojet. FJ44 turbofan (with Williams International) & Olympus 593 supersonic reheated turbojet (in collaboration with SNECMA). Engines for rotary wing applications: RTM 322 turboshaft (in collaboration with Turbomeca & Piaggio); MTR 390 turboshaft (in collaboration with MTU & Turbomeca); Gem turboshaft & Gnome turboshaft. Airbreathing missile propulsion systems including Odin ramjet for Sea Dart missile. [97]

75-50270 Rolls-Royce Motor Cars Ltd Specialist Engines

(a subsidiary of Rolls-Royce Motor Cars Ltd)
Pyms Lane, Crewe, Cheshire CW1 3PL
☎ (+44 1270) 25 51 55 ☒ 36121 rrmtrs g
⊖ (+44 1270) 53 48 16
♣ £ 8,000,000 (1996) ⁂ 100 (1996) ⌂ 7,400 m2
Director and General Manager: J E Vickerman
☎ (+44 1270) 25 51 55 ext. 3425
Sales and Business Development Manager:
Hugh A J McDonnell ☎ (+44 1270) 53 53 02
MoD Account Manager: Paul Kettle
Manufacturing and Quality Manager: Graham Ridsdale
Engineering Project Manager: Graham Smith
■ Manufacturer of marine diesel generators 30kW-950kW, including low magnetic signature applications. Steering units for military vehicles. Overhaul and refurbishment of diesel and petrol engines for marine and military use. Specialists in military applications. Approved to ISO 9001 Standard. Manufacture and re-manufacture. → 80 [98]

75-50400 Rose Bearings Ltd

(a subsidiary of Minebea Group, Japan)
Doddington Road Industrial Park, Doddington Road, Lincoln LN6 3RA
☎ (+44 1522) 50 09 33 ✉ ROSE ⊖ (+44 1522) 50 09 75
⁂ 588
President: R P Chapman
Company Secretary: M N Stansfield
Sales Manager: D M Shore
Operations Manager: A C Willerton
Purchasing: L Turner
■ Manufacturers of rod end and plain spherical bearings and journal bushes.

Skegness Office
Church Road, Skegness, Lincolnshire PE25 3RS
☎ (+44 1754) 76 71 11 ⊖ (+44 1754) 61 03 71
Operations Dir: K. Kojima
■ Commercial ball bearings. [98]

75-50555 Ross & Catherall Ltd

(a subsidiary of Vickers plc)
Forge Lane, Killamarsh, Sheffield S31 8BA
☎ (+44 114) 248 64 04 ⊖ (+44 114) 247 59 99
♣ £ 27,658,000 (1996) ⁂ 180 (1996) ⌂ 15,201 sq m
Chairman: T.M. Mackie
Managing Dir: Dr C.R. Thomas
Financial Dir: A. Barnett
Technical Sales Dir: Dr P.E. Waudby
Exec. Dir: D.J. Cowen

■ Manufacture of vacuum melted high performance alloys for directionally solidified & single crystal turbine blades & nozzle guide vanes, permanent cast magnets, personal body protection equipment.
Associated companies:
Certified Alloy Products Inc.
Ross Catherall Ceramics Ltd
Trucast Ltd
Vickers Aerospace Components
Vickers Airmotive
Vickers Precision Machining [98]

75-50570 Rotadata Ltd
Bateman Street, Derby DE23 8JQ
☎(+44 1332) 34 80 08 [E] sales@rotadata.co.uk
⊜ (+44 1332) 33 10 23
£ 100,000 (1995) ♣ £ 5,500,000 (1995)
80 (1995) 5,500 m2
Managing Director: John Taylor
Sales Director: Derek J Taylor
Operations Director: Graham Barsby
Administrator: Ms Jackie Milburn
■ Fixed & traversing aerodynamic probes for aircraft & aircraft engines. Tip clearance measurement equipment. Application of sensors to engine test components. Design & manufacture of turbo-machinery test rigs & wind tunnel models. High speed sliprings & telemetry. → 80 [98]

75-50575 Rotech Components Ltd
(a subs. of Schleifring und Appaaratebau)
Abex Rd, Newbury, Berks RG14 5EL
☎(+44 1635) 363 63 ⊜ (+44 1635) 383 34
♣ £ 2 550 000 (1995) 52 2 000 sq m
Managing Dir: J. Considine
Sales & Marketing Dir: A.J. Manahan
Technical Mgr: G. Knapp
Production Mgr: D. Chappell
■ High-precision electrical and optical rotary-joint devices and slip-ring assemblies for the military and professional engineering markets.
Associated company:
Schleifring u.Apparatebau GmbH, Germany → 75 [97]

75-50590 Rotherham Precision Engineers Ltd
Unit 3A, Bedworth Road, Longford, Coventry CV6 6BP
☎(+44 1203) 36 36 80 ⊜ (+44 1203) 36 35 00
♣ £ 2,500,000 (1996) 80 1,500 m2
Managing Dir: S. Fish
Commercial Dir: C.R. Peters
■ Machining, assembly & testing of fuzes, timing & retarder devices. → 80 [98]

75-50600 Rowlinson Packaging Ltd
(† Rowlinson (Wirebound Containers) Ltd)
Unit 1, Green Lane, Wardle, Nantwich, Cheshire CW5 6BN
☎(+44 1829) 26 05 71 ⊜ (+44 1829) 26 07 18
£ 3 200 000 (1995/96) ♣ £ 7 700 000 120
Joint Managing Dir: David Rowlinson
Joint Managing Dir: John Williams
Purchasing Mgr: John Carter
■ Manufacturer of protective cases & containers. [97]

75-50620 Royal Ordnance
(a subsidiary of British Aerospace plc)
Headquarters
Euxton Lane, Chorley, Lancashire PR7 6AD
☎(+44 1257) 26 55 11 ⊜ (+44 1257) 24 26 09
♣ £ 500,000,000 5,000
Managing Director: Stephen Henwood
☎(+44 1257) 24 29 00
Sales and Commercial Director: Barry Flower
☎(+44 1257) 24 22 12
Financial Director: Alan Fraser
Director, International Business Africa and Asia Pacific: David Sewell ☎(+44 1257) 24 26 46
Director, International Business Central and South America, Middle East, North Africa and India: Tony Dodd ☎(+44 1257) 24 27 01
International Business Manager, Eastern Europe and Former Soviet Union: Neil McCaughan
☎(+44 1257) 24 20 06
Business Unit Director, Ammunition: Adrian Watts
☎(+44 1257) 24 23 80
Business Manager, Ammunition, UK/MoD: Alan Weeden
☎(+44 1257) 24 20 17
Business Manager, Ammunition NATO/Europe: Paul Hough ☎(+44 1257) 24 26 81
Business Unit Director, Small Arms Ammunition: Brian Fleet ☎(+44 1270) 88 66 00
Business Unit Director, Explosives and Propellants: Bob Smith ☎(+44 1257) 24 26 50
Business Unit Director, Electronics and Fuzes: Alan Humphreys ☎(+44 1254) 29 20 01
Business Unit Director, Environmental Services Group: Dr Malcolm Green ☎(+44 1296) 65 21 00
Business Director, Explosive Ordnance Disposal: John Smith ☎(+44 1257) 24 24 85
Media and Advertising Manager: Andrew Jeacock
☎(+44 1252) 38 47 98
Personnel: Jonathan Crookall ☎(+44 1257) 24 26 25
■ The manufacturing sites at Chorley, Birtley, Bishopton, Radway Green, Glascoed Bridgwater produce ammunition for small arms, cannon, mortars, tank guns & artillery; gun propellants & combustible charge containers; shell & cartridge cases; rocket motor bodies, high explosive & rocket propellant manufacture; assembly & filling of warheads, bombs & shells. The manufacturing site at Blackburn designs, develops & manufactures mechanical & electronic devices for safe functioning of ammunition, guided weapons & barrier warfare systems. Includes fuzes, safety & arming units, electronic subsystems; thick film technology; vehicle intercom systems. The sites at Summerfield & Westcott design & manufacture solid & liquid rocket motors for military & civil use. The site at Nottingham designs & manufactures rifles (H&K UK), cannon, mortars, tank guns, artillery systems & armoured vehicles.
Overseas subsidiary:
Heckler & Koch Group, Germany
Muiden Chemic BV, Holland
Large Guns & Vehicles
Kings Meadow Road, Nottingham NG2 1EQ
☎(+44 115) 968 23 00 ⊠ 37531 g
⊜ (+44 115) 968 21 18
Business Unit Director: Tony Buxton
■ Manufacture of small arms, cannon, mortars, tank guns & artillery; fighting vehicle production; AFV upgrading, including the areas of firepower, mobility & armour; internal security systems; engineering equipment.
Product Support Group
186 Beardall Street, Hucknall, Nottinghamshire NG15 7JU
☎(+44 115) 968 11 93 ⊜ (+44 115) 968 19 37
Rocket Motors Division
Summerfield, Kidderminster, Worcester DY11 7RZ
☎(+44 1562) 82 40 61 ⊠ 337727 imisrs g
⊜ (+44 1562) 82 81 26
430
Business Unit Dir: David Singleton
☎(+44 1562) 82 80 00
Commercial Dir: Eddie Crompton ☎(+44 1562) 82 81 09
■ Research, design, develop & manufacture & in-service support of rocket motors fuelled by all types of solid & liquid propellant for use in land, sea & air-launched guided missiles, underwater & in space. Associated products include gas generators, ammunition protection & vertical launch systems, thrust vector control systems. Environmental test facilities. [98]

75-50630 Royal Ordnance Speciality Metals (ROSM)
(a subs. of Royal Ordnance plc)
PO Box 27, Wolverhampton, West Midlands WV10 7NX
☎(+44 1902) 78 39 39 ⊜ (+44 1902) 78 33 52
♣ £ 3,600,000 (1996) 56 (1997)
Sales Mgr: C.P. Smith
Manufacturing Manager: C Cooling
■ Manufacturer of the full range of small to large calibre tungsten alloy kinetic energy penetrators, having the expertise & knowledge necessary to meet the exacting requirements of British service specifications & international requirements. ISO 9001 approved. [98]

75-50650 RTK Marine
(a subs. of Moseley (Holdings) Ltd)
446 Blandford Rd, Hamworthy, Poole, Dorset BH16 5BL
☎(+44 1202) 68 55 81 ⊜ (+44 1202) 68 33 47
♣ £ 2 500 000 (1995) 25
Managing Dir: John G. Street
Sales & Marketing Dir: David Woods
■ Range of amphibious patrol & assault craft up to 14 metres long, Rigid Raiding Craft of 5 & 6.5 metres for ship to shore movement of troops & equipment. [97]

75-50750 Rubb Buildings Ltd
(a subsidiary of Rubb Holdings A/S)
Dukesway, TVTE, Gateshead, Tyne and Wear NE11 0QE
☎(+44 191) 482 22 11 ⊜ (+44 191) 482 25 16
♣ £ 4,500,000 60
Managing Director: Finn Haldorsen
Operations and Purchasing Manager: Bill Wood
■ Manufacturer of demountable buildings & shelters suitable for all types of military purposes including aircraft hangars, warehouses, workshops & field support buildings. 98

75-51020 R.X. Products Ltd
Pontygwindy Industrial Estate,
Caerphilly, Mid.Glamorgan CF83 3HU
☎(+44 1222) 88 21 26 ⊜ (+44 1222) 88 32 26
♣ £ 750 000 10
Managing Dir: R.J. Lewis
Sales & Marketing Dir: P. Balmer
Director: R.W. Cross
■ Target equipment. [97]

75-51150 Sabre Computers International Ltd
Defence Products Div.
Process Hse, 43 Selsdon Road,
South Croydon, Surrey CR2 6PY
☎(+44 181) 681 82 41 ⊜ (+44 181) 688 18 48
♣ US$ 2,000,000 18
Managing Dir: David P. Balfour
Technical Dir: Dr Martyn J. Eccles
Marketing Mgr: Major (Rtd) John C. Paterson
■ Manufacture ballistics measurement equipment (including Milligan Products) and training aids used on land, sea and air. Ballistics instrumentation (including pressure measurement projectile velocity to NATO STANAG 4114 accuracy etc.) Sabre Advanced Miss Distance Indicator (SAMDI) for airborne, marine and ground target measurements. Telescope precision IR tracker for missiles and ILS. Subsidiary HPD Ballistics Ltd manufactures test guns, weapon rests and associated range equipment.
USA Office:
Sabre Computers International Ltd → USA 90 [98]

75-51300 Safe Training Systems Ltd
Holly House, Maidenhead Road,
Wokingham, Berkshire RG40 5RR
☎(+44 1344) 48 35 63 ⊜ (+44 1344) 48 51 75
♣ £ 300 000 (1995) 5
Managing Dir: D.R. Ward
■ Design, manufacture & sales of simulators for hazardous materials monitors - radiation, chemical warfare agents & toxic chemicals. [98]

75-51470 Saft Nife Ltd
(a subsidiary of Saft SA, France)
Station Road, Hampton, Middlesex TW12 2BY
☎(+44 181) 979 77 55 ⊠ 23572 saft ukg
⊜ (+44 181) 783 04 94
♣ £ 31,000,000 220
Managing Director: Ken Klinker
Commercial Director: Dr J Thompson
Director of Portable and Advanced Batteries: R Boyens
Technical Director: R N Thomas
Marketing Executive: Ms Antoinette Leach [98]

75-51650 Samas Roneo Ltd
(† Samas Vickers Ltd)
Hawley Road, Dartford, Kent DA1 1NY
☎(+44 322) 22 34 77 ⊠ 9413372 vfm g
⊜ (+44 322) 55 54 02; 55 63 57; 52 16 14
♣ £ 32,000,000 720
Export Manager: Paul Moss
Marketing Manager: W Whitnell
■ Steel & wood office furniture, military lockers, files, cupboards, seating, file cabinets, security filing cabinets, security cupboards. [98]

75-51750 Sandhurst Mfg Co Ltd
Cherry Holt Road, Bourne, Lincolnshire PE10 9LQ
☎(+44 1778) 42 64 26 [E] info@sandhurst.co.uk
⊜ (+44 1778) 39 30 07
£ 100,000 (1997) ♣ £ 6,000,000 (1997)
30 (1997) 10,000 sq ft
Managing Director: Tim Dean
Sales and Marketing Director: Chris Archer
Operations Director: Robin Brand
Purchasing Manager: Brian Kelham
Public Relations, Promotion and Marketing: I S McConnachie
■ Design, assembly, sale and service of earthmoving plant and equipment. [98]

Contact Reference

75-51800 SAS Communications Ltd
(a subsidiary of SAS Eurotech Group Ltd)
Kings Hill, West Malling, Kent ME19 4QQ
☎(+44 1732) 84 94 44 ⊜ (+44 1732) 87 04 44
♣ £ 1,000,000 ♚ 15 ♛ 12,000 sqft
Managing Dir: Steve Snook
Technical Dir: Phil Greenhalgh
Marketing Dir: Richard Romain
Purchasing Mgr: John Norburn
■ Professional communications equipment & systems. Fibre-optic communications specialists. MoD defence contractor. 97

75-51820 Sascal Displays
(a division of Sascal Ltd)
Head Office and Sales
Unit 1, Hayes Metro Centre, Springfield Road, Hayes, Middlesex UB4 OLE
☎(+44 181) 573 03 03 W http://www.sascal.com
⊜ (+44 181) 569 15 15
♧ £ 25,000 (1997) ♣ £ 1,500,000 (1997) ♚ 25 (1997)
♛ 800 m2
Managing Director: S Caldecourt
Purchasing Manager: R Kashap
■ Development and design of colour and monochrome CRT monitors and computerised control systems. Used in many industrial and ATC military applications where environmental requirements can be demanding. Certified ISO 9001.

Service & Repair
Unit 1 Bridge Works, Kingston Road, Leatherhead, Surrey KT22 7SU
☎(+44 1372) 37 80 80 ⊜ (+44 1372) 37 86 86 98

75-51830 SAS Limited
(† Safety Air Services Ltd)
North Florida Road, Haycock Industrial Estate, St Helens, Merseyside WA11 9TN
☎(+44 1942) 72 42 48 ⊜ (+44 1942) 27 07 71
♧ £ 2,000,000 (1996) ♣ £ 2,200,000 (1996)
♚ 30 (1996)
Managing Director: B O'Neill
Sales Manager: Russell Chaplin
Administrator: Pauline Gregson
■ Design & manufacture high & low pressure air & gas compressor systems, panels, recompression chambers, systems & allied diving equipment, training & installations. 98

75-51850 Saunders-Roe Ltd (SRL)
Millington Road, Hayes, Middlesex UB3 4AZ
☎(+44 181) 573 38 00 ⊜ (+44 181) 561 34 36
♣ £ 5 000 000 (1994/95) ♚ 100
Sales Dir: T.C. Bartlett
■ Self-powered light sources, Trilux rifle sights, Betalight route markers; aircraft panel illumination. 98

75-51970 Schaffner EMC Ltd
(a subsidiary of Schaffner Elektronik AG)
5 Ashville Way, Molly Millers Lane, Wokingham, Berkshire RG41 2PL
☎(+44 118) 977 00 70 W http://www.schaffner.com
⊜ (+44 118) 979 29 69
♚ 20
Managing Director: Michael C Lowe
Operations Director: A L Cogswell
Sales and Marketing Director: M C Lowe
Purchasing Manager: M de Brion
■ Components, electrical, electronic, electromechanical & electromagnetic. 98

75-51990 Schering-Plough Ltd
(a subs. of Schering-Plough Corp., USA)
Mildenhall, Bury St Edmunds, Suffolk IP28 7AX
☎(+44 1638) 71 63 21 ⊜ (+44 1638) 71 78 58
Contract Sales Director: J P Drinkhill
Managing Director: Peter Martin
Purchasing Manager: Steve Dewar
■ Water purification tablets. Pharmaceuticals. 98

75-52070 Schmidt UK Limited
(† Schmidt Manufacturing & Equipment)
(a subsidiary of Schmidt Holding Europe GmbH, Germany)
Southgate Way, Orton Southgate, Peterborough PE2 6GP
☎(+44 1733) 36 33 00 ⊜ (+44 1733) 36 33 33
♣ £ 22,000,000 (1996) ♚ 240 (1997) ♛ 12,000 m2
Managing Director: J Atkinson
Sales and Marketing Director: A F Jenkinson
Operations Director: I R McAllister
■ Suction sweepers for airfields, stores, roadways, gangways & areas. Snow clearing equipment. 98

75-52100 Sciaky Electric Welding Machines Ltd
212 Bedford Avenue, Trading Estate, Slough, Berkshire SL1 4RH
☎(+44 1753) 52 55 51 ⊜ (+44 1753) 82 14 16
Chairman: Mario Sciaky
Managing Dir: A.P. West
Sales Mgr: P. Morrison
■ Resistance, electron beam fusion & friction welding machines. Laser welding machines & electron beam & laser heat treatment equipment. → 80 98

75-52160 Scitec (EMA) Limited
(a subsidiary of Scitec Limited, Australia)
Apex House, London Road, Bracknell, Berkshire RG1 2XH
☎(+44 1344) 30 30 37 ⊜ (+44 1344) 30 30 57
Regional General Manager: Mike Mayhew
Company Secretary: Ms Ethi G Oepen
■ Research, development, manufacture and sales of network products for enterprise-wide networks. Technology to provide local and wide area networks that handle voice, fax, data and LAN traffic including advanced systems for audio and video transmission via satellite. → 80 98

75-52250 Sea Connections Systems Ltd
Unit 8, Bessemer Way, Harfreys Industrial Estate, Great Yarmouth, Norfolk NR31 0LX
☎(+44 1493) 65 27 33 ⊜ (+44 1493) 65 28 40
♣ £ 1 126 000 ♚ 18
Sales Dir: Chris Knight
Technical Dir: David Parker
Marketing Co-ordinator: Miss M. Harrison
■ Manufacturers & suppliers of underwater electrical connectors. Able to offer a variety of moulding tasks combining almost any range of underwater electrical connector to the majority of sub-sea cable jacket materials including polyurethane to polyethlene. A BS-EN-ISO 9002 accredited company. 97

75-52390 Securicor Datatrak Ltd
(† Datatrak Ltd)
(a subs. of the Securicor Group)
Wiltshire House, Country Park, Swindon, Wilts SN1 2NR
☎(+44 1793) 50 01 00 ⊜ (+44 1793) 50 01 01
♣ £ 4 175 000 ♚ 80
Managing Dir: Andy Bird
International Business Dir: D. Palmer
Technical Dir: Mike Wade
Sales & Marketing Mgr: Ken Hills
■ Automatic vehicle location & monitoring system (Datatrak). 97

75-52450 Security Research Ltd
(a member of the Audiotel Group)
Corby Road, Weldon, Corby, Northamptonshire NN17 1DZ
☎(+44 1536) 40 09 88 E sales@security-research.com
⊜ (+44 1536) 26 67 11
♧ £ 70,000 (1996) ♣ £ 2,000,000 (1996) ♚ 35 (1997)
♛ 1,700 m2
Managing Director: Andrew Martin
Operations Director: Vincent Lynch
Sales Manager: Adrian Hickey
Marketing Manager: Ms Julie Walker
Purchasing Manager: Ms Ann Kay
■ Design, manufacture and supply of communication and information security systems, video and counter-surveillance systems, anti-terrorist, technical surveillance equipment. 98

75-52600 Seetru Ltd
Albion Dockside Works, Bristol BS1 6UT
☎(+44 117) 927 92 04 E enquires@seetru.co.uk
W http://www.seetru.co.uk ⊜ (+44 117) 929 81 93
♣ £ 4,500,000 (1996/97) ♚ 125 (1996/97)
Managing Dir: O.H. Varga
Sales Dir: Richard T. Boswell
UK Sales Mgr: D.J. Leech
Export Sales Mgr: T.F. Riley
■ Design & manufacture of liquid level indicators & pressure relief valves & coded pipe couplings. 98

75-52700 Self-Changing Gears Ltd
(a subsidiary of Alvis Transmissions Ltd)
The Triangle, Walsgrave, Coventry, West Midlands CV2 2SP
☎(+44 1203) 53 54 55 ⊠ 31459
⊜ (+44 1203) 53 92 05
♣ £ 8 722 000 ♚ 175
Managing Dir: S. Wilkins
Operations Dir: D. Hoyle
Sales & Marketing: H. Jardon
Publicity & Marketing Services Mgr: N.C. Turner
■ Crossdrive transmission systems for tracked fighting vehicles & MBTs. Design & development of advanced mechanical transmission equipment. 98

75-52710 Selwood Ltd
(a subsidiary of Pascon Group Ltd)
Bournemouth Road, Chandlers Ford, Eastleigh, Hants SO5 3ZL
☎(+44 1703) 26 63 11 ⊠ 477481 pascon g
⊜ (+44 1703) 27 10 12
Managing Director: Kevin P Bright
Operations Director: John W Hill
Sales/Marketing Director: Chris Graham
Purchasing Manager: John R Howarth
■ Manufacture of pumps. 98

75-52730 Sencea Ltd
Unit 17c, Wincombe Business Park, Shaftesbury, Dorset SP7 9QJ
☎(+44 1747) 85 54 00 E sales@sencea.co.uk
W http://www.sencea.co.uk ⊜ (+44 1747) 85 29 00
♚ 4
Managing Director: Daniel Lewis
■ Manufacture of compact lightweight non-directional beacons, antennas, remote controls, monitors and solar power. 98

75-52800 Sensitised Coatings Ltd
Bergen Way, North Lynn Ind.Est., King's Lynn, Norfolk PE30 2JL
☎(+44 1553) 76 48 36 ⊜ (+44 1553) 76 03 77
♣ £ 2 000 000 (1995) ♚ 25
Managing Dir: David Eatwell
Marketing Mgr: Tony Morgan
■ Manufacture recording paper rolls for sonar, echosounder, maritime surveillance, hydrographic, oceanographic, seismic & satellite signal monitoring systems. Thermal paper rolls for facsimile equipment, non-impact printers & data terminals. 97

75-52820 Sensor Electronics Limited
BWB House, Flemingate, Beverley, East Yorkshire HU17 0NG
☎(+44 1482) 86 12 69
■ Portable security systems. 98

75-52830 SEOS Displays Ltd
(† Specialised Electro-Optical Services Ltd)
Marchants Way, Burgess Hill, West Sussex RH15 8QY
☎(+44 1444) 87 08 88 E info@seos.co.uk
⊜ (+44 1444) 87 07 77
♧ £ 96,300 (1997) ♣ £ 2 400 000
♣ £ 6,974,000 (1996) ♚ 35 ♛ 18,000 (1997)
Managing Director: O J Wynn
Operations Dir: S J Elmer
Head of Sales and Marketing: Mrs Sarah Minett
Purchasing Manager: Robin Lilley
Marketing: J B Swainston
■ Design & development organisation specialising in opto-electronics & the manufacture of a range of high quality, cost effective colour display systems for commercial & military simulation. 98

75-52860 Serck Aviation Div.
(a member of the BTR Group)
Warwick Road, Birmingham B11 2QY
☎(+44 121) 693 13 00
⊠ 337341 serckb g; 339805 serctc g
⊜ (+44 121) 766 60 13
♣ £ 10 000 000 ♚ 100 (1996)
General Sales Mgr: Steven Salt
Finance Controller: David Sellers
Operations Mgr: Innes Mitchell
Sales & Marketing Mgr: J.B. Payne
Commercial Mgr: John Felton
■ Design & manufacture of heat exchange equipment for aerospace applications. 97

75-52880 **Serck Heat Transfer**
(a member of the BTR Group)
Warwick Rd, Birmingham B11 2QY
☎(+44 121) 766 66 66 ⊠ 337341 serckb g
⊖ (+44 121) 766 60 14
♣ £ 23,000,000 (1996) 👥 300 (1997)
Managing Dir: Chris Teed
Operations Director: Richard Cartwright
General Mgr: Andrew Davis
General Mgr: Andrew King
Marketing Co-ordinator: Ms Ellen Gifford
■ Heat exchange equipment for military vehicles. Heat exchange equipment for naval applications. 98

75-53000 **Seton Healthcare Group plc**
Tubiton House, Medlock Street, Oldham OL1 3HS
☎(+44 161) 652 22 22 ✉ TUBITON ⊠ 669956
⊖ (+44 161) 626 90 90
♣ £ 40,000,000 👥 600
Managing Director: Iain Cater
Operations Director: Bernard Humphreys
Sales and Marketing Director: Ian Adamson
Public Relations Co-ordinator: Ms Brenda Alexander
■ Medical equipment and supplies: pressure therapy products, support bandages and stockings, wound care and leg ulcer treatment, infection control, dressing retention bandages and stockinettes, orthapaedic and chiropody products, medical sundries. 98

75-53050 **Seward Medical Limited**
131 Great Suffolk St, London SE1 1PP
☎(+44 171) 357 68 17 (UK Sales);
357 67 27 (Export Sales) ⊠ 914702 sewmed g
⊖ (+44 171) 357 65 63
👥 60 🏭 18 000 sq ft
Director: C.J. Measures
■ Operating tables for field & hospital use, mobile medical units, surgical instruments, infant incubators, suction pumps, lights, patient monitoring equipment & a range of disposable hospital protective clothing/masks. 97

75-53100 **SEXTANT Avionique UK Branch Office**
(a subsidiary of SEXTANT Avionique SA, France)
108 Hawley Lane, Farnborough, Hampshire GU14 8HY
☎(+44 1252) 37 00 11 ⊖ (+44 1252) 37 00 44
General Representative: Roy Tingay 98

75-53200 **Sheffield Forgemasters Ltd**
The Old Rectory, School Hill, PO Box 108,
Whiston, Rotherham S60 1QR
☎(+44 1709) 82 82 33 ⊠ 547039 sfhog
⊖ (+44 1709) 36 77 65
♣ £ 150 000 000 👥 1 700+
Chief Exec.: Phillip M. Wright
Divisional Managing Dir: D. Fletcher
Divisional Managing Dir: D.J. Kavanagh
Sales Dir Primary Products: Peter Fretwell
■ Suppliers of high technology steels, forgings, castings & rerolled products for aerospace, defence, nuclear & conventional power generation, oil & gas sectors. Materials include low, medium & high alloy & stainless steels, semi-nickel & nickel alloys & superalloys. 97

75-53350 **Shipham & Co. Ltd**
(a subs. of Tomkins plc)
Hawthorn Ave, Hull HU3 5JX
☎(+44 1482) 32 31 63 ⊖ (+44 1482) 22 40 57
♣ £ 6 000 000 👥 150 🏭 58 000 sq ft
Managing Dir: D. Gibson
Sales & Marketing Dir: N. Ellwood
Operations & Technical Dir: J. Litherland
Sales & Marketing Mgr: S. Flannagan
Purchasing Mgr: S. Walker
■ Manufacture & supply of valves, components & castings including gate, globe, check, ball & relief valves. Special valve design facility. 97

75-53450 **Shorrock Integrated Systems**
(† Shorrock Major Systems Ltd)
Shadsworth Road, Blackburn, Lancashire BB1 2PR
☎(+44 1254) 68 87 40 E sis@shorrock.co.uk
W http://www/shorrock.co.uk ⊠ 635151 shorok g
⊖ (+44 1254) 69 53 06
♧ £ 16,111,000 (1996) ♣ £ 88,130,000 (1996)
👥 1,226 (1996)
General Manager and Director: M Deasey
Operations Director: R Demain
Systems Engineering Director: J Paton
Sales and Marketing Director: D J Beatty
Purchasing Manager: E O'Hara
Technical Executive: F Hoyle
Export and Sales Administration: A Settle
■ Design, manufacture and installation of security systems, microwave and other perimeter intruder detection devices, large scale access control and event monitoring systems. Complete service for security and fire applications. Consultancy, design, manufacture, installation, maintenance and training. 98

75-53550
Shorts Missile Systems Ltd (SMS)
Alanbrooke Road, Castlereagh, Belfast BT6 9HB
☎(+44 1232) 45 84 44 ⊖ (+44 1232) 73 38 72
♣ c. £ 70,000,000 👥 500+
Chief Executive Officer: Terry Stone
Contact: Denise Clarke
Project Mgr, Business Development: Ronnie Harrison
Information Contact: Andy Lyttle
■ A 50:50 joint venture company of Short Brothers plc of Belfast and Thomson-CSF of France, SMS is in the position of being able to offer customer proven integrated systems combining command to line of sight (CLOS) missiles, command and control, and automatic firing posts in one package. → 74 98

75-53600 **Siebe Gorman & Co. Ltd**
(a subsidiary of Siebe plc)
Avondale Way, Cwmbran, Gwent NP44 1TS
☎(+44 1633) 86 12 11 ⊠ 498108
⊖ (+44 1633) 87 27 59
♧ £ 400,000 ♣ £ 6,500,000 👥 100 🏭 370,000 sq ft
Director of Business Development: Nino M Granatiero
Product/Marketing Co-ordinator: Andrew Rayson
Product Manager: Kevin White
Purchasing Manager: Graham Thomas
■ Compressed air breathing apparatus. Airline breathing equipment, escape sets, air compressors. Activated charcoal cloth (C-Tex). Resuscitators. Air-supplied hoods, masks and suits. 98

75-53700 **Siemens Environmental Systems Ltd Radiation Monitoring**
(† Siemens Plessey Controls Ltd)
(a management subsidiary of Siemens plc)
Sopers Lane, Poole, Dorset BH17 7ER
☎(+44 1202) 78 20 78 E radmon@poole.siemens.co.uk
⊠ 41272 plespl g ⊖ (+44 1202) 78 21 57
👥 150 (1997)
Managing Director: Dr A Low ☎(+44 1202) 78 27 48
General Manager: Robin G Carter
Sales Manager: B Acton ☎(+44 1202) 78 23 74
■ Design, development, manufacture and support for nuclear radiation detection and monitoring equipment for seaborne and land force applications. Ship installed radiac systems for surface ships and submarines. Portable dose rate meters, decontamination meters and personnel dosimetry systems. Electronic personal dose and dose rate meters for individual use. Decontamination simulator and training aids. Radioaudible detection and spectral analysis. 98

75-53750 **Siemens plc**
(a subsidiary of Siemens AG, Germany)
Siemens House, Windmill Road, Bracknell RG12 8FZ
☎(+44 1344) 39 60 00 ⊖ (+44 1344) 39 61 33
Chief Executive Officer: J Gehrels
Managing Director: A J Wood
Finance and Commercial Director: G Steinhardt
Design and Development Manager: R Owen
Business Manager, Special Projects: Robert Miller
Press and Public Relations Manager: Rodney M Tucker
Marketing Communications: David K Close
■ Radio and landline communications equipment.
Manufacturing Facility
Eaton Bank Trading Estate, Varey Road,
Congleton, Cheshire CW12 1PH
☎(+44 1260) 28 30 00 ⊖ (+44 1206) 27 10 27
General Manager: Richard Mellor 97

75-53770 **Siemens Plessey Systems**
(a business within Siemens Plessey Electronic Systems Ltd)
Grange Road, Christchurch, Dorset BH23 4JE
☎(+44 1202) 40 85 12 ⊠ 418417
⊖ (+44 1202) 40 48 03
👥 2,500
Managing Dir: Clive Dolan
Finance Dir: Nigel Billingham
Technical Dir: T.J. Hewson
Personnel Dir: P. Bird
Publicity Mgr: D. Burton
■ Prime contractor management, system design, development & integration & turnkey project management for air defence radar systems, C3I systems and tactical communications systems. Logistic support, training and all other services in support of these requirements. → 80
Air Defence Div.
Newport Road, Somerton, Cowes, Isle of Wight PO31 8PF
☎(+44 1983) 29 41 41 ⊖ (+44 1983) 20 23 26
Managing Director: R D Barnes
Marketing Executive: A Ford
■ Strategic & tactical radar systems for air defence. Systems analysis & modelling. Air defence display, data handling & communication systems. Naval surveillance, target indication & missile detection radars, fixed & mobile coastal defence radars. Long, medium & short range tactical radars for theatre, corporations & divisional level air defence. Multi-function radars for naval & battlefield defence. SAM system radars. Integrated sensor & control systems, naval combat systems, refit & modernisation. Modular radar enhancement packages.
Defence Systems Div.
Grange Road, Somerford, Christchurch, Dorset BH23 4JE
☎(+44 1202) 48 63 44 ⊖ (+44 1202) 40 42 21
Managing Dir: D.H. Barnes
Marketing Exec.: D A Phillips
■ Command, control communications & information (C3I) systems, satellite ground stations, integrated C3I & EW systems for all armed services & theatres. Systems analysis & modelling. Electronic surveillance & countermeasure systems. Adaptive HF strategic communications systems, communications ESM receiver, analysis & DF systems. Secure relational database management systems & applications software for CCIS. Satellite communications ground terminals & ECM-resistant modems.
Tactical Communications Systems Div.
Grange Rd, Somerford, Christchurch, Dorset BH23 4JE
☎(+44 1202) 48 63 44 ⊖ (+44 1202) 40 88 88
Managing Dir: R.F. Clarke
Marketing Exec.: C.M.G. Grafton
■ Trunk & radio communications systems for all armed services & theatres. Systems analysis & modelling. Secure digital circuit, packet & message switches, terminal equipment & ancillaries for area communications networks. HF, VHF & UHF combat radio systems for manpack, vehicle & multi-role requirements. Advanced ECM-resistant tactical HF & VHF radio systems. VHF & UHF radio systems for shipborne & ground station applications. Gun sound ranging systems.
Production Plant
Newport Road, Cowes, Isle of Wight PO31 8PF
☎(+44 1983) 29 41 41 ⊖ (+44 1983) 20 23 26
Operations Director: R F Wyatt
■ Manufacture, assembly, test and systems integration for products listed under business divisions.
Production Plant
Vicarage Lane, Ilford, Essex IG1 4AQ
☎(+44 181) 478 30 40 ⊖ (+44 181) 553 88 99
Operations Director: R Wyatt
■ Manufacture, assembly, test and systems integration for products listed under business divisions. 98

75-53778 **Sigma Security Devices Ltd**
(† Sigma Security Services Ltd)
(subs. of Nightingale Holdings Dover Ltd)
134-136 Snargate Street, Dover, Kent CT17 9DA
☎(+44 1304) 20 50 50 E sigma@nightingale.co.uk
W http:// www.nightingale.co.uk/sigma
⊖ (+44 1304) 21 55 63
♣ £ 80,000 👥 3
Managing Director: D H Nightingale
Financial Director: Phillip R Henley
Operations Director: J Potter
■ Manufacture of rapid entry equipment. 98

75-53780 **Signal Processors Ltd**
Cambridge Science Park, Milton Road,
Cambridge CB4 4GJ

Contact Reference

☎(+44 1223) 42 03 57 [E] sales@spl.co.uk
[W] http://www.spl.co.uk/satcomms/index.html
⊖ (+44 1223) 42 08 83
£ 20,000 ⌂ 10,000 sq ft
Managing Director: Dr Robin J Smith-Saville
Financial Director: Geoffrey J Walker
Intrac Operations Director: Dr Martin Smith
Commercial Director: Robert B Michaelson
DBR Operations Director: Dr Kenneth Gardner
Marketing and Sales: Ms Ann Luxon
■ Satellite communications subsystems, antenna controllers, DSP systems, ruggedised modems, high accuracy beacon receivers, concatenated FEC systems. [98]

75-53785 Signature Industries Ltd

Tom Cribb Road, Thamesmead, London SE28 0BH
☎(+44 181) 316 44 77 ⊖ (+44 181) 854 51 49
♣ £ 14,000,000 (1997) 260 (1997)
Managing Dir: D.M. Cairnie
Operations/Technical Dir: F. Bassett
General Mgr: K. Kearns
Purchasing Mgr: T. Gage
■ Personal Locator Beacon, CAA & FAA approved. [98]

75-53790 Signature Management

(† Dowty Signature Management)
(a member of the TI Group plc)
Alton Road, Ross-on-Wye, Herefordshire
☎(+44 1989) 76 55 55 ⊠ 35708
⊖ (+44 1989) 76 35 02
General Mgr: Keith Webster
Sales & Marketing Mgr: Bob Ibell
■ Specialises in the management & design of infra-red & acoustic signatures. Products & services include acoustic & anechoic coatings, radar reflective decoys & radar absorbent materials. [97]

75-53830 Silverstone Electronics Ltd

12 Vincent Avenue, Crownhill, Milton Keynes MKB OAB
☎(+44 1908) 26 52 23 ⊖ (+44 1908) 26 51 43 [98]

75-53908 Simon Access (UK) Ltd

(† Simon Gloster Saro Ltd)
(a subs. of Simon Engineering plc)
Gloucester Trading Est., Hucclecote, Glos GL3 4AD
☎(+44 1452) 37 13 21 ⊖ (+44 1452) 37 10 24
♣ £ 20 000 000 (1995) 275 ⌂ 45 000 sq m
Managing Dir: B. Woodward
Sales & Marketing Dir: V.M. Edwards
Opertions/Technical Dir: D. Phyall
Purchasing Mgr: N. Gregory
■ Design & manufacture of aircraft refuelling vehicles, airfield fire-fighting vehicles, aerial ladder platforms, mobile access equipment. [97]

75-54250 Sira Ltd
Electro-Optics Division

South Hill, Chislehurst, Kent BR7 5EH
☎(+44 181) 467 26 36 [W] http://www.sira.co.uk
⊠ 896649 ⊖ (+44 181) 467 65 15
♣ £ 10,000,000 (1996) 160 (1997)
Managing Director: Prof. Richard A Brook
Marketing Manager: John Hother
■ Independent research and development organisation specialising in the design and manufacture of electro-optical systems. Reflectometers, thermal imager test equipment, high-speed IR spectro radiometers, laser TV projectors, laser scanners, special purpose CCD cameras, scanning sub-systems for laser radars, special purpose microscopes and telescopes and robust sensors. [98]

75-54300 SIRS Navigation Ltd

186A Milton Road, Swanscombe, Kent DA10 0LX
☎(+44 1322) 38 36 72 ⊖ (+44 1322) 38 24 18
♣ £ 800,000 (1996) 17 (1995) ⌂ 3,500 sq ft
Managing Director: R W Eady
■ Manufacture and repair of aircraft, army and naval compasses, sextants, clocks, azimuth circles and general light engineering. [98]

75-54400 Sleeman Engineering

(a subs. of Firth Rixson plc)
Heath Rd, Wednesbury, W.Midlands WS10 8XF
☎(+44 121) 568 62 22 ⊖ (+44 121) 568 62 44
♣ £ 2 200 000 10

Managing Dir: C. Fox
Operations & Technical Dir: D. Griffin
Sales & Marketing Dir: Ian Shiach
■ High hardness steel armour. Cut profiled armour. Ballistic testing. Armoured kits for vehicles. → 80 [97]

75-54410 Slingsby Aviation Ltd

(a subsidiary of ML Holdings plc)
Ings Lane, Kirkbymoorside, York YO6 6EZ
☎(+44 1751) 43 24 74 ⊖ (+44 1751) 43 11 73
183
Managing Dir: R.D. Haworth
Finance Controller: C Wilson
Contracts Dir: S.A. Cooper
Aircraft Manufacturing Mgr: J. Howard
Purchasing Mgr: J.N. Dowkes
Operations: J. Bevan
■ Advanced composite structures for aerospace & marine applications, Firefly training aircraft & hovercraft. [98]

75-54420 Slingsby Engineering Ltd

(a subsidiary of ML Holdings plc)
Ings Lane, Kirkbymoorside, York YO6 6EZ
☎(+44 1751) 43 17 51 ⊠ 57911 sel g
⊖ (+44 1751) 43 13 88
98
General Manager: D W Berry
Finance Dir: Dave Berry
Quality Exec.: P.A. Ambury
Sales/Marketing Mgr: Keith Pennington
Operations Manager: K Asken
Technical Mgr: D.W. Hartley
Purchasing Mgr: Dave Daniel
■ Subsea system engineers. Underwater vehicles, specialist electronic/acoustic equipment, subsea robotics, underwater electrical penetrators & connectors. [98]

75-54430 H.C. Slingsby plc

Preston St, Bradford BD7 1JF
☎(+44 1274) 72 15 91 ⊠ 517450
⊖ (+44 1274) 72 30 44
♣ £ 11 770 350 187
Sales Office Mgr: J.D. Oman
Sales: P. Hardy
Advertising Dept: T.L. Grunwell
Advertising Asst: David Scarlett
■ Materials handling & access equipment. Machine & load-mover skates, drum & cylinder trolleys, wheelbarrows & other handling & access aids.

London Office
Unit 8 Delta Park, Smugglers Way,
Wandsworth SW18 1EG
☎(+44 181) 877 07 78 ⊠ 517450
⊖ (+44 181) 877 30 63
Branch Mgr: S. Curtis [97]

75-54540 A. Smith Gt Bentley Ltd

Centre Park, Clacton Road, Frating,
Colchester, Essex CO7 7DL
☎(+44 1206) 25 03 80 ⊖ (+44 1206) 25 05 09
♣ £ 2,810,000 68 ⌂ 3,000 sq m
Managing Director: Ian C Smith
Technical Sales Manager: J M Evans
■ Silenced static and mobile power generators, mobile communication vehicles and shelters. Outside broadcast units, specialist containers and mobile units, sound-proofed power systems. [98]

75-54570 Smith-Clayton Forge (Aerospace)

(a div. of British Steel Forgings)
Tower Works, PO Box 22, Lincoln LN2 5DT
☎(+44 1522) 52 54 92 ⊖ (+44 1522) 53 02 76
Managing Dir: D.J. Jones
Commercial Mgr: P J Williams ☎(+44 1522) 56 50 44
■ Large critical die forgings for the aerospace industry, including turbine discs, compressor discs, airframe & undercarriage parts, produced in a wide range of steel, titanium & nickel-based superalloys. [97]

75-54590 H.R. Smith Group

H.R. Smith (Technical Developments) Ltd
Street Court, Kingsland,
Leominster, Hereford & Worcs HR6 9QA
☎(+44 1568) 70 87 44 [E] hrsmith@demon.co.uk
⊠ 837264 smith g ⊖ (+44 1568) 70 87 13
90 ⌂ 30 000 sq m (1995)
Managing Director, Operations Director: R E L Smith
Group Marketing Manager: R Smith
UK Sales Manager: Frank Webb
Purchasing Manager: D Downey
■ Technical developments for advanced airborne antenna systems. Land & marine antenna systems for civil & defence industries. Electrostatic discharges (aircraft).

H.R. Smith (Microwaves) Ltd
Street Court, Kingsland,
Leominster, Hereford & Worcs HR6 9QA
☎(+44 1568) 70 87 44 ⊠ 837264
⊖ (+44 1568) 70 87 13
Managing Dir: R.E.L. Smith
Marketing Mgr: Richard Smith
UK Sales Mgr: F.R. Webb
■ Design & manufacture of microwave antenna (1-94 GHz) for aerospace & defence industries; frequency selective surfaces for radomes.

Specmat Ltd
Street Court, Kingsland,
Leominster, Hereford & Worcs HR6 9QA
☎(+44 1568) 70 87 44 ⊠ 35445 smith g
⊖ (+44 1568) 70 87 13
Marketing Mgr: Richard Smith
UK Sales Mgr: F.R. Webb
■ Advanced thermoplastic composites for aerospace & defence industries, incl. radomes; primary, secondary aircraft structures.

Techtest Ltd
Street Court, Kingsland,
Leominster, Hereford & Worcs HR6 9QA
☎(+44 1568) 70 87 44 ⊠ 837264
⊖ (+44 1568) 70 87 13
Managing Dir: R.E.L. Smith
Sr Sales Exec.: L. Crompton
Sales Mgr: Frank Webb
Production Mgr: G. Laing
Marketing Mgr: R Smith
■ Manufacture precision portable and battery operated antenna test sets from 27 MHz to 4-4 GHz, AM/FM signal sources, primary standard WBCO attenuator and frequency linear up-convertors. SAR homing systems, R&D capability. Personal Locator Beams, Emergency Locator Transmitters. [98]

75-54620 Smith & Nephew plc

2 Temple Place, Victoria Embankment, London WC2R 3BP
☎(+44 171) 836 79 22 ⊖ (+44 171) 240 70 88
♣ £ 1,026,000,000 (1995) 12,000
Chief Executive: John Robinson
Finance Director: Peter Hooley
Company Secretary: Michael Parson
Director of Business Development: Crispin Simon
Group Director of Information Technology: John Coleman
Group Communications Manager: Peter Nind
Group Communications Executive: Davina Newsham
■ Development, manufacture and marketing of tissue repair products, primarily for bone, joints, skin and other soft tissue.

Smith & Nephew Healthcare Limited Medical Products
Healthcare House, Goulton Street, Hull HU3 4DJ
☎(+44 1482) 22 22 00 ⊖ (+44 1482) 22 22 11
400
Managing Director: Dr David Webster
Business Director, Medical Products: Ed Holmes
Business Director, Surgical Products: Ian Cutts
Technical Services Manager: Mike Eyre
Customer Care Manager: Paul Ormston
Medical Information Officer: Colin Knight

Smith & Nephew Healthcare Surgical Products
The Technopark, Newmarket Road, Cambridge CB5 8PB
☎(+44 1223) 56 81 00 ⊖ (+44 1223) 56 80 98

Smith & Nephew Homecraft
Sidlings Road, Lowmoor Road Industrial Estate,
Kirkby-in-Ashfield NG17 7JZ
Nottingham
☎(+44 1623) 75 40 47 ⊖ (+44 1623) 75 55 85
■ Wound management, casting and support, rehabilitation, trauma, orthopaedic implants, endoscopy and ENT microsurgery.

Smith & Nephew Medical Limited
Hessle Road, PO Box 81, Hull HU3 2BN
☎(+44 1482) 32 51 81 ⊖ (+44 1482) 32 83 26
1,100
Managing Director: Nick Hildyard
■ Woundcare, casting, bandaging and first aid products.

Smith & Nephew Consumer Products Limited
Alum Rock Road, Birmingham B8 3DZ
☎(+44 121) 327 47 50 ⊖ (+44 121) 327 61 72

Smith & Nephew Group Research Centre
York Science Park, Heslington, York YO1 5DF

☎(+44 1904) 45 10 00 ⊖ (+44 1904) 45 10 30
Group Director, Research and Development: Dr Alan Suggett
Director of Group Research: Dr Gareth Lloyd-Jones

Smith & Nephew Medical Fabrics
Brierfield Mills, Brierfield, Nelson, Lancashire BB9 5NJ
Managing Director: Darrell Jenkins
■ Manufacture of fibre-based materials for healthcare applications including gauze, non-woven swabs, cotton wool, procedure packs, bandages and advanced base fabrics for adhesive bandages, casting products and wound dressings. [98]

75-54630 Smith & Prince Ltd
Unit 1 Boundary Business Court, 92-94 Church Road, Mitcham, Surrey CR4 3TD
☎(+44 181) 687 00 22 ⊖ (+44 181) 687 83 48
♣ £ 400,000 (1996) 👥 12 (1997) 🏭 4,200 m2
Director: M A Mumford
Director: D M Burrell
Company Secretary: A J Smith
■ Power distribution system for military and civil 40 kW field power networks. Hand-held explosion-proof fluorescent lights and torches, lighting equipment for aircraft servicing, power distribution equipment for aircraft servicing. [98]

75-54650 Smith's Electric Vehicles Ltd
Marquisway, Team Valley Trading Estate, Gateshead, Tyne and Wear NE11 0RX
☎(+44 191) 487 13 11 ☒ 53120 rintea g
⊖ (+44 191) 482 02 43
♣ £ 5,000,000 👥 110
Managing Director: J D Smith
Sales Director: S L Jollie
Commercial Manager: L A Dodd
■ Battery powered commercial vehicles for horizontal movement of personnel and stores. Four wheel trucks and vans with maximum weight up to 5,080 kg including batteries and charging equipment. [98]

75-54700 Smiths Industries Aerospace
(a subsidiary of Smiths Industries plc)

Headquarters
765 Finchley Road, Childs Hill, London NW11 8DS
☎(+44 181) 458 32 32 [E] plc@smithsind.co.uk
[W] http://www.smithsind.com ☒ 928761 esseye g
⊖ (+44 181) 209 05 26
♣ £ 377,100,000 (1996) 👥 4,755 (1996)
Chairman: Sir Roger Hurn
Chief Executive: Keith Butler-Wheelhouse
Corporate Director Human Resources: David Spencer
Financial Director: Alan Thompson
Director Strategy and Business Development: Dr G K Thornton
Director Press and Public Relations: R Plumley
Publicity Manager: N H Fox
■ Supplier of civil and military aerospace and defence systems for aircraft, helicopters, RPVs, ships and fighting vehicles. → 80 → 70

Cheltenham
Bishops Cleeve, PO Box 188, Nr Cheltenham, Gloucester GL52 4ZA
☎(+44 1242) 67 33 33 ⊖ (+44 1242) 66 17 98
Managing Director Defence Systems Cheltenham: N J Wilton

Cheltenham
Bishops Cleeve, PO Box 244, Nr Cheltenham, Gloucester GL52 4ZB
☎(+44 1242) 67 85 55 ⊖ (+44 1242) 66 17 97
Managing Director Civil Systems United Kingdom: M D Jones

Product Support
Bishops Cleeve, PO Box 162, Nr Cheltenham, Gloucester GL52 4YB
☎(+44 1242) 67 33 55 ⊖ (+44 1242) 67 65 25
Managing Dir: J. Legg
■ Worldwide support to operators of SI equipment. [98]

75-54870 Soil Machine Dynamics Ltd (SMD)
Wincomblee Road, Walker, Newcastle Upon Tyne NE6 3QS
☎(+44 191) 234 22 22 ⊖ (+44 191) 234 04 44
♣ £ 14 000 000 (1995) 👥 70 🏭 11 700 sq m
Managing Dir: John Reece
Marketing Dir: Dr A.D. Trapp
Operations Dir: Dr Tim Grinsted
■ Seabed ROV development for military applications. [97]

75-55000 Solomat Ltd
(a division of Neotronics Ltd)
Parsonage Road, Takeley, Bishop's Stortford, Herts CM22 6PU
☎(+44 1279) 87 12 52 ☒ 817126
⊖ (+44 1279) 87 03 77
■ Manufacture of environmental air & water quality monitoring instrumentation for spot checks & continuous data acquisition. [98]

75-55150 Somers Forge Ltd
(† Walter Somers Ltd)
(a division of Folkes Group plc)
Haywood Forge, Prospect Road, Halesowen, West Midlands B62 8DZ
☎(+44 121) 585 59 59 [E] somers@folkes.forgings.ltd.uk
⊖ (+44 121) 585 66 99
♣ £ 15,000,000 (1996) 👥 110 (1996)
🏭 24,000 m2 (1996)
Managing Dir: R.C. Fradgley
Works Dir: K. Jasper
Sales Director: P V Mitchell
Sales Manager: J Warr
■ Manufacturers of forgings for use in ships, aircraft, missiles, guns etc. Carbon, alloy & stainless steels, die & mould steels. Approved to British MoD 05-24 & similar requirements. [98]

75-55170 Sonar Research & Development Ltd
Unit 1B, Grovehill Industrial Estate, Beverley, North Humberside HU17 0JW
☎(+44 1482) 86 95 59 [E] sales@srd-ltd.demon.co.uk
⊖ (+44 1482) 87 21 84
♣ £ 750,000 (1996) 👥 20
Managing Director: Dr Peter Stevens
Business Development Manager: Bill Austin
Sales Manager: Brian Walker
■ Sonar acoustics, transducers, hydrophones, ships correlation speed logs and electronic scanning sonar. Doppler current profilers, acoustic tide monitors and telemetry. → 60 → 80 [98]

75-55202 Sonic Communications (Int.) Ltd
Communications Centre
The Green, Castle Bromwich, Birmingham B36 9AJ
☎(+44 121) 749 49 00 ☒ 7495244
⊖ (+44 121) 749 52 44
♧ £ 25 000 ♣ £ 5 000 000 (1994) 👥 70+
🏭 15 000 (1996)
Managing Dir: David John Bryan
Sales & Marketing Dir: Neil Barker
Operations Dir: Adrian Wright
Purchasing Mgr: Simon Jinks
Advertising Mgr: Geoff Platten
■ The company has developed world leading voice activated transmission (VOX) techniques, rejecting background noise & only transmitting speech, for many applications e.g. Police firearms teams, SAR winchmen, armoured vehicles etc. [97]

75-55208 Sound Attenuators Ltd
(a member of The Salex Group)
Eastgates, Colchester, Essex CO1 2TW
☎(+44 1206) 86 69 11 ⊖ (+44 1206) 86 59 87
♧ £ 20 000 ♣ £ 16 000 000 👥 200
Managing Director: C M McKay
Sales and Marketing Director: R Stubbings
Operations and Technical Director: Ian Evans
Director: B E Barnet
Industrial Sales Manager: Bill Hood
☎(+44 1455) 63 68 00
Purchasing Manager: M Penrose
■ Manufacture, supplies and installs a wide range of noise control equipment. Fan attenuators, sound-resistant doors, acoustic enclosures, vibration isolation equipment. Gas turbine attenuating packages. Sound damping, absorbent and barrier materials. Diesel engine silencers using active noise control. [97]

75-55212 Souter Shipyard Ltd
(subs. of Edward Williams (Holdings) Ltd)
Admiralty Gate, Thetis Road, Cowes, Isle of Wight PO31 7DJ
☎(+44 1983) 29 47 11 ✉ SOUTER
⊖ (+44 1983) 29 98 42
♧ £ 1 425 000 (1996) ♣ £ 2 000 000 (1995) 👥 80
Director: Clive Benton
Sales & Marketing Dir: Bob Milner
Operations & Technical Dir: John Tucker
Purchasing Mgr: Peter Willis
■ Range of workboats, personnel support vessels & patrol craft available up to 35m in length, in aluminium alloy, GRP & steel. [97]

75-55213 South Midlands Communications Ltd (SMC)
S.M. House, School Close, Chandlers Ford Ind. Est., Eastleigh, Hampshire SO53 4BY
☎(+44 1703) 25 51 11 ☒ 477351 smcomm g
⊖ (+44 1703) 26 35 07
♣ £ 6 000 000 👥 60
Managing Dir: J. Lightfoot
Sales & Marketing Dir: D.W. Peaty
Antenna Sales Mgr: M.A. Bennett [97]

75-55214 Southcombe Brothers Ltd
Stoke-sub-Hamdon, Somerset TA14 6QD
☎(+44 1935) 82 35 67 ⊖ (+44 1935) 82 29 18
♧ £ 863,000 (1994) ♣ £ 7,000,000 (1996)
👥 225 (1996) 🏭 100,000 sq ft
Managing Director: R Michael Southcombe
Managing Director: T A Southcombe
Purchasing Manager: E J Carver
■ Cut and sew production of fabric and leather gloves using leather from it's own tannery. [98]

75-55230 SpanSet Ltd
(a member of SpanSet Group)
Telford Way, Middlewich Business Park, Middlewich, Cheshire CW10 0HX
☎(+44 1606) 73 74 94
[W] http://www.offshoretechnology.com/contractors/safety/spanset/index.html
⊖ (+44 1606) 73 75 02
♣ £ 5,800,000 👥 96
Managing Director: A J Eaden
Sales Director: C Hamilton
Military Sales Manager: M A Taylor
Marketing Manager: Mrs K Moore
Purchasing Manager: Glyn Williams
Marketing Assistant: E Smith
■ Manufacture and distribution of lifting and lashing equipment approved to BS 5750 Pt1 including polyester roundslings, webbing lifting slings and cargo lashings, lifting and securing systems for ammunition, bombs and torpedoes, tank towing and lashing systems, tactical aircraft arrester gear, non-magnetic lifting and securing systems, aircraft tie-down systems and personal safety harnesses. Agent for the 'Stinger-Spike Systems' - tyre deflating device. Range of fall arrest equipment and associated products. Cargo protection systems. Personnel safety training. [98]

75-55240 Special Melted Products Ltd (SMP)
(a subsidiary of Sheffield Forgemasters Ltd)
Atlas House, Attercliffe Road, Sheffield S4 7UY
☎(+44 114) 272 00 81 [E] flilley@specmp.co.uk
[W] http://www.specmp.co.uk ⊖ (+44 114) 276 15 56
♣ £ 36,000,000 (1996) 👥 350 (1996)
Managing Director: K B Lomax
Operations Director: Richard Wright
Sales Director: J C Gerrard
Senior Sales Manager: D Whiteley
Purchasing Manager: Mrs H Walker
Sales Manager: Fred Lilley
■ Vacuum induction melted, vacuum arc & electroslag remelted steel & nickel based materials in the form of ingots, billets, bars, rolled rings, shaped forgings & many other forms for the aerospace, nuclear energy, medical, defence industries & offshore. Current suppliers to the UK & US defence departments. [98]

75-55555 Spectrol Reliance Ltd
(a Kearney-National company)
Garrard Way, Swindon, Wiltshire SN3 3HY
☎(+44 1793) 52 13 51 [E] spectroluk@aol.com ☒ 44692
⊖ (+44 1793) 53 92 55
♣ £ 10,000,000 (1996) 👥 270 (1997) 🏭 60,000 sq m
Managing Director: V Smith
Manufacturing Director: C Harris
Sales/Marketing Director: V Slauvinski
General Sales Manager: Graham Flynn
Purchasing Manager: R Cooper
Export Sales Manager: G Hoey
Internal Sales Engineer: Mike Abrams

■ Included in Spectrol's manufacturing activities are a wide range of potentiometers, including BS approved trimmers and military approved precision potentiometers; custom designed sensors. [98]

75-55620 Spectrum Technologies Ltd
Mid Glamorgan Science Park, Technology Drive,
Bridgend, Mid Glamorgan CF31 3NA
☎ (+44 1656) 65 54 37 ✆ (500) 74 93 11
[E] 106266.2507@compuserve.com
⊖ (+44 1656) 65 59 20
♣ £ 4,000,000 (1996) 👥 40 (1997) 🏭 1,000 m2
Managing Director: Dr P H Dickinson
Operations Director: A Jones
Sales Admin and Marketing Manager: Ms Elaine Hardy
■ Design and manufacture of ultra-violet laser wire marking and processing equipment. "Capris" cable printing and identification systems: a range of modular wire marking and processing equipment for use in electrical wire harness manufacture, suitable for applications from maintenance and low volume production to high volume production. [98]

75-55700 Spencer Clark Metal Industries plc
(a subsidiary of Aurora plc)
Greasborough Street, Rotherham, Yorkshire S60 1RG
☎ (+44 1709) 82 86 00
[E] aurora.met@spclark.clemon.co.uk ⊠ 54131 scmi g
⊖ (+44 1709) 82 85 07
♣ £ 12,000,000 (1997) 👥 160 (1997)
Managing Director: Ian Gale
Sales Director: D M Bilston
Finance Director: M Phillips
Operations Director: Alex McDonald
■ Quality steels & alloys for aerospace & advanced technologies. Specialist converter of steel, nickel & titanium alloys into rolled bars, sections & profiles, rolled sheet. Distributor of bar & billet to gas turbine manufacturers. [98]

75-56050 SPM Instrument UK Ltd.
Bolholt Works, Walshaw Rd, PO Box 14,
Bury, Lancashire BL8 1PY
☎ (+44 161) 761 48 37 ⊖ (+44 161) 797 52 09
👥 16
Managing Dir: P.J. Brown
■ Condition monitoring equipment, including shock pulse method for monitoring rolling element bearings, shaft alignment, vibration & maintenance tools. [97]

75-56220 SSS Gears Ltd
Park Road, Sunbury-on-Thames, Middlesex TW16 5BL
☎ (+44 1932) 78 06 44 ⊖ (+44 1932) 78 00 18
Chairman and Managing Director: H A Clements
Director: J M Neeves
Director: I G Fairhead
Purchasing Manager: P J Shaw
■ Gear-tooth-type SSS automatic over-running clutches in sizes from 3 to over 400,000 hp, torque capacities from 1,000 to 3,000,000 lb ft for marine and industrial applications which require a smooth and rapid connect/disconnect system. Used by 27 world navies for the main propulsion drive of over 500 surface warships. [98]

75-56222 St Bernard Composites Ltd
(† St Bernard Plastics Ltd)
Saberhouse, Lynchford Road,
Farnborough, Hampshire GU14 6JE
☎ (+44 1252) 30 40 00 ⊠ CARBOLAC
⊖ (+44 1252) 30 40 01
♣ £ 9,000,000 (1997) 👥 200 (1997) 🏭 10,000 sq m
Chairman: C.J.B. Wood
Managing Dir: E. Tindall
Marketing Dir: J.R.G. Wood
Purchasing Mgr: Nigel Wood
■ Engineers in advanced composite products, structures include engine, wing and weapon components. [98]

75-56350 W.F. Stanley & Co. Ltd
33 Avery Hill Rd, New Eltham, London SE9 2BW
☎ (+44 181) 850 55 51 ⊖ (+44 181) 294 13 33
♧ £ 200,000 ♣ £ 1,000,000 👥 19 (1997) 🏭 8,000 sq m
Managing Director: W A Mears
Sales Office Manager: Ms Janice Carter
■ Optical/mechanical instruments, prismatic compasses, optical sighting systems for man-portable and artillery weapons. → 80 [98]

75-56355 Stanley Plastics Ltd
Holmbush Industrial Estate,
Midhurst, West Sussex GU29 9HX
☎ (+44 1730) 81 62 21 ⊖ (+44 1730) 81 28 77
Managing Dir: P.W. Stanley
■ Acrylic engineering: conning towers, viewports, hemispheres. [98]

75-56450 Steatite Insulations Ltd
(a subsidiary of Steatite Holdings Ltd)
2 The Square, Broad Street, Birmingham B15 1AP
☎ (+44 121) 643 68 88
[E] 101465.3353@compuserve.com
⊖ (+44 121) 643 20 11
♣ £ 13,000,000 (1996) 👥 80 🏭 5,000 sq m
Group Director: Richard L. Placito
Director: Frank A. Green
■ Precise time & frequency systems, VME computers & telemetry military & aerospace. [98]

75-56780 Stella-Meta
(† Stella-Meta Filters Ltd)
(a subsidiary of Thames Water plc)
Laverstoke Mill, Whitchurch, Hampshire RG28 7NR
☎ (+44 1256) 89 59 59 ⊠ 859605 stella g
⊖ (+44 1256) 89 20 74
Managing Director: David Carslaw
Operations and Purchasing Director: Kevin Rowe
Sales and Marketing: Richard Simcox
■ Production of mobile potable water supplies to WHO standard from freshwater otherwise unfit for drinking, brackish & seawater. NBC contaminated supplies are also treatable. Flexible water storage tanks & chlorination equipment. [98]

75-57100 STI International Ltd
(a subsidiary of Meridian Medical Technologies Inc)
34 Riverside, Sir Thomas Longley Road,
Findsbury, Rochester, Kent ME2 4DP
☎ (+44 1634) 71 57 85 ⊖ (+44 1634) 72 10 82
♣ £ 25,000,000 (Group) 👥 15 🏭 750 sq m
Managing Director: James H Miller
Marketing Director: Geoffrey Ritson
Operations Director: John Wilmot
■ Design & manufacture a family of automatic pre-filled self-injecting syringes for battlefield & emergency service use by medically untrained personnel. [98]

75-57325 Stone Foundries Ltd
(a subs. of Langham Industries Ltd)
Woolwich Rd, London SE7 8SL
☎ (+44 181) 853 46 48 ⊠ 23395 stones g
⊖ (+44 181) 305 19 34
♣ £ 7 000 000 👥 175
Chairman: J.C. Langham
Managing Dir: Alan Alderton
■ High grade aerospace castings in magnesium, aluminium, steel & copper based alloys up to Class 1 radiography standards. [97]

75-57335 Stone Manganese Marine Ltd
(a subsidiary of Langham Industries Ltd)
Dock Road, Birkenhead, Merseyside L41 1DT
☎ (+44 151) 652 23 72 ⊖ (+44 151) 652 23 77
♧ £ 10,000,000 ♣ £ 5,000,000 (1995) 👥 99
Chairman: J M Langham
General Manager: D R Quillam
Managing Director: Dr G Patience
Technical Manager: L Bodger
■ Marine propellers from 1 ton to over 100 tons in weight with diameters up to 12 m. Naval propellers are manufactured for eight navies for submarines, frigates, aircraft carriers or through deck cruisers.
A special alloy called "Sonoston" has been developed & sold throughout the world for use mainly in submarines. A special silicon bronze has also been developed for minesweepers. [98]

75-57345 Stop-Choc Ltd
Banbury Avenue, Slough, Berkshire SL1 4LR
☎ (+44 1753) 53 32 23 ⊖ (+44 1753) 69 37 24
♣ £ 3,500,000 👥 45
Managing Dir: J.M.H. Dunn
Engineering Dir: L.C. May
Chief Exec. Sales: R.W. Cooper
Sales Mgr: R.W. Steer
■ Design & manufacture of all metal & elastomeric anti-vibration mountings, helicopter blade lead/lag dampers, avionics & military racking & accessories, illuminated instrument cockpit panels with integrated electronics, pushbutton controls & alpha numeric displays. ISO 9001. → 80 → 85 [98]

75-57347 Storage Concepts (Europe) Ltd
(a subsidiary of Storage Concepts Inc, USA)
9 Lancaster Court, Coronation Road,
High Wycombe, Bucks HP12 3TD
☎ (+44 1494) 44 22 02 [E] chris@storcon.co.uk
[W] http://www.storageconcepts.com
⊖ (+44 1494) 45 02 81
Managing Director: Chris Cuthbert
■ Software, hardware and technical support for guaranteed-bandwidth data storage and retrieval. Applications include airborne and satellite ground station image capture, motion analysis and telemetry. → 80 [98]

75-57350 Thos. Storey (Engineers) Ltd
61 High Street, Chobham, Surrey GU24 8AF
☎ (+44 1276) 85 69 85 ⊠ 21868
⊖ (+44 1276) 85 52 68
Managing Dir: Peter Dawson
Managing Dir: K.J. Churchill
Sales Dir: L. Hodgson
■ Unit construction bridging equipment, pontoons, push-tugs & propping systems marketed internationally & with a wide variety of military & civil applications. [98]

75-57460 Stott Benham
(a member of the Electrolux Group)
Bordesley Green Road, Birmingham B9 4UA
☎ (+44 121) 623 38 00 ⊖ (+44 121) 623 38 10
Managing Director: Å Stråberg
Operations Director: J Booth
Sales/Marketing Director: Ms Jayne Hall
Purchasing Manager: Naresh Surati
Publicity Manager: T J Stack
■ Food preparation, cooking & serving equipment, beverage preparation & service & related products, including advanced high-capacity dishwashing systems. Also a complete design, planning & installation service for integrated catering units is provided for military units & prisons. [98]

75-57515 Strategic Imaging Systems
8 St Peter's Road, Parkstone, Poole, Dorset BH14 0PQ
☎ (+44 1202) 71 61 00 [E] tony@sisys.co.uk
⊖ (+44 1202) 71 61 40
♣ £ 1,500,000 (1996) 👥 25 (1997) 🏭 4,000 sq m
Managing Director: G A Douglas
■ Temporary ID systems for visitors and contractors at high security sites. Low cost, high speed ID passes with image for all people on site. Storage of image and all details for investigation. High quality staff badging and access control, and automatic number plate recognition. Site access control - visitors, contractors, staff and vehicles. [98]

75-57525 Structure-flex Ltd
24 Grove Lane, Holt, Norfolk NR25 6EG
☎ (+44 1263) 71 29 11 ⊠ 975431
⊖ (+44 1263) 71 11 50
♣ £ 9,000,000 (1997) 👥 150 (1997) 🏭 4,000 sq m
Managing Dir: Ian Doughty
Sales & Marketing Dir: David Dawber
■ Manufacture flexible containers for water, fuel & non hazardous chemicals, intermediate bulk containers for powered or granular commodities, vehicle covers & tarpaulins, tentage, camouflage systems, load restraints & inflatable lifting equipment. Registered to AQAP-4, ISO 9002, BS5750 Pt 2, contractor & sub-contractor to MoD (UK). [98]

75-57530 Strumech Versatower Ltd
(a subsidiary of Strumech Holdings)
Portland House, Coppice Side, Brownhills,
Walsall, West Midlands WS8 7EX
☎ (+44 1543) 45 23 21 ⊠ STRUMECH Walsall
⊖ (+44 1543) 36 10 50
♣ £ 5,000,000 (1997) 👥 89 (1997)
🏭 7,000 sq m (1997)
Managing Director: Frank V Weetman

Operations Director: P C Colley
Sales and Marketing Executive: M Stringer
Purchasing Manager: Wayne Drewett
■ Manufacturer of VERSATOWER range of telescopic and tilt-over towers for communications, lighting, CCTV security and instrumentation. Trailer-mounted towers for field deployment, including emergencies. [98]

75-57820 Supacat Ltd
The Airfield, Dunkeswell, Honiton, Devon EX14 0RA
☎(+44 1404) 89 17 77 ⊖ (+44 1404) 89 17 76
Director: Nick Jones
Director: David Clayton
Marketing Manager: Mike Gormley
■ Design & manufacture of Supacat high mobility 6x6 vehicles for civil & military use & associated equipment. Supacat glider launching winch & specialist trailers. Vehicle development projects for other manufacturers. [98]

75-57910 Supercraft Ltd
Canada Rd, Oyster Lane, Byfleet, Surrey KT14 7JS
☎(+44 1932) 35 19 41 ⊖ (+44 1932) 34 08 07
♤ £ 100 000 (1996) ♣ £ 4 700 000 (1995/96)
👥 100 (1996) 🏭 4 000 sq m
Joint Managing Dir: D.P. Vincent
Joint Managing Dir: D.R. Baggott
Technical Dir: Geoff Green
Commercial Dir: Ted Johnson
■ Design & manufacturing services for Aerospace & Defence Industries specialising in precision fabrication welding & assembly. Airframe assemblies, avionics casing & consoles. Naval equipment racking & consoles. Facilities include CAD, CAM & DNC machinery. [97]

75-57920 Superform Aluminium
(a subsidiary of British Aluminium Ltd)
Cosgrove Close, Worcester WR3 8UA
☎(+44 1905) 87 43 00 [E] superform@enterprise.net
⊖ (+44 1905) 87 43 01
♣ US$ 5,000,000 (1996) 👥 40 (1996) 🏭 30,000 sq ft
General Mgr: F.K. Holliday
Sales & Marketing Mgr: A.M. Lowerson
■ Manufacturer & distributor of superplastic components in aluminium alloys. [98]

75-57950 Superwinch Ltd
Abbey Rise, Whitchurch Road, Tavistock, Devon PL19 3DR
☎(+44 1822) 61 41 01 ⊖ (+44 1822) 61 52 04
♣ £ 5 000 000 (1996) 👥 90
Managing Dir: F. Tolsdorf
Sales & Marketing Dir: C.K. Farnworth
Operations Dir: David Anderson
■ Electric, hydraulic & mechanical winches. [98]

75-57955 Surveillance Technology Group
62 South Audley Street, London W1Y 5FB
☎(+44 171) 408 02 87; 629 02 23
⊖ (+44 171) 629 95 38; 499 78 46
♣ US$ 30,000,000 (1997) 👥 45 (1997)
Managing Director: H Silver
Sales and Marketing Director: David Ross
Operations Director: Mark Johnston
Product Manager: Mark Stevens
Purchasing Manager: Ms Johanna Brook
■ Covert surveillance equipment. [98]

75-58000 Sutcliffe Speakman Carbons Ltd
(a subs. of Sutcliffe Speakman plc)
Lockett Road, Ashton-in-Makerfield, Lancashire WN4 8DE
☎(+44 1942) 27 54 00 ⊖ (+44 1942) 27 56 03
♣ £ 8 000 000 👥 100
Managing Dir: Ralph Lowery
Sales & Marketing Dir: Allan Singleton
Operations Dir: Dr Steve Regan
Purchasing Mgr: Paul Mallard
■ MoD approved activated carbon for use in NBC filters, respirators, NBC suits. Air conditioning filtration systems. [98]

75-58080 Swiss Net UK plc
(† John C. Small & Tidmas plc)
(a subsidiary of Swiss Net Company)
Hartley House Industrial Estate, Hucknall Road, Sherwood, Nottingham NG5 1FD
☎(+44 115) 969 25 00 ⊖ (+44 115) 969 32 70
♣ £ 2,000,000 (1996) 👥 80 (1996)
Managing Dir/Sales Mgr: Mike Mitchell
Production Co-ordinator: Mark Tacey
■ Fine netting fabrics, mosquito nets, parachute netting & anti infra-red military face nets. [98]

75-58325 Synrein Plastics Ltd
Grosvenor Works, Grosvenor Street, Ashton-under-Lyne Lancashire OL7 0RG
☎(+44 161) 330 85 73 ⊖ (+44 161) 343 21 95
♣ £ 300,000 👥 11 🏭 10,000 sq m
Managing Director: S Wrigley
Purchasing Manager: D Gaick
Sales and Marketing: J Conlen
■ Body bags, PVC mattress covers, pillow slips, PVC aprons, gowns and curtains. [98]

75-58400 TA Mfg Ltd
(a subsidiary of TA Mfg Co.)
39/41 St Catherine Street, Gloucester GL1 2BS
☎(+44 1452) 30 76 78 ⊖ (+44 1452) 30 88 18
Regional Sales Mgr: Adrian R. Brewster [98]

75-58500 Tamares Ltd
Prince's House, 36-40 Jermyn Street, London SW1Y 6DN
☎(+44 171) 629 66 36 ⊖ (+44 171) 629 66 76
Managing Dir: P.C. Zabhidowicz
Director: Lt Col. K.V. Stewart
■ Tampella type weapons & ammunition including mortars & 155mm towed & independent gun howitzer. Mortar training ammunition (full bore, realistic & reusable).
® USA: Pocal Industries Inc. [97]

75-58550 Tanner Brothers (Greenfield) Ltd
Waterside Mills, Greenfield, Oldham, Lancashire OL3 7NH
☎(+44 1457) 87 22 73 ⊖ (+44 1457) 87 01 33
👥 180
Managing Dir: J. Tanner
Sales Mgr: J.W. Caveney
■ Manufacturer of cotton & synthetic fabrics, MoD ISO 9002 (pending). [98]

75-58650 Tapley Instrumentation
(a division of D. Evans Electrical Ltd)
42 Fontenoy Street, Liverpool L3 2DR
☎(+44 151) 231 14 42 ⊖ (+44 151) 236 17 31
♣ £ 300,000 👥 15
General Mgr & Dir: David Smith
Product Mgr: Jim Atherton
Admin Mgr: P.J. Jones
■ Design & production of runway friction metering equipment for civil & military applications. [98]

75-58660 Target Technology Ltd (TTL)
(a Meggitt Target Systems Company)
Brunswick Ind. Centre, Ashford, Kent TN23 1ED
☎(+44 1233) 63 97 62/64 ⊖ (+44 1233) 62 48 83
♣ £ 6,000,000 👥 63
Managing Dir: Terence Timms
Technical Dir: Philip Newman
Head of Sales and Marketing: Brian Tyekiff
Operations Manager: Jason Fenn
■ Design, manufacture & operation of aerial targets, remotely piloted vehicles (RPVs), launchers, RPV powerplants & ground support equipment. Manufacture of radio telecommand, telemetry, tracking & other associated electronic equipment for unmanned target & surveillance aircraft. Transponders. [98]

75-58740 TBA Textiles
(† TBA Armour Textiles)
(a subsidiary of T+N plc)
PO Box 40, Rochdale, Lancashire OL12 7EQ
☎(+44 1706) 474 22 ⊖ (+44 1706) 35 42 95
♣ £ 32,000,000 👥 250 🏭 14,000 sq m
Chief Executive: Dr A Ruddy
General Manager: Ian Hedley
Purchasing Manager: David Prytharch
Production Manager: Norman Barker
Business Development and Sales Manager: Colin Mills
Technical Manager: Phil Ingham
Marketing Manager: Mrs Emma Palmer
■ Textiles for thermal & personal protection, fireblocker fabrics, ballistic body armour, armoured ballistic protection & associated products. Firefly cavity barrier system. Woven material in fire blocking situation in rail & air transport. [98]

75-58750 TDS Circuits Plc
(a div. of Zincocelere SpA, Italy)
Whitebirk Estate, Blackburn, Lancs. BB1 5TD
☎(+44 1254) 66 13 21 ⊖ (+44 1254) 58 33 73
♣ £ 10 200 000 (1995) 👥 200 🏭 10 000 m2
President: A. Bellazini
Managing Dir: Graham Partridge
Sales Dir: G. Silva
■ Manufacture high technology printed wiring. Approved to BS9761, 9762, 9763 & CECC. Carries USA underwriters laboratories recognition. Professional printed circuit boards for space & avionics, data-processing & telecommunications. [97]

75-58800 Tebbutt & Hall Ltd
(† Tebbutt & Hall Bros Ltd)
Military Footwear Div.
Coleman St, Raunds, Northants NN9 6NL
☎(+44 1933) 62 33 23 ⊖ (+44 1933) 46 02 23
♣ £ 5 000 000 👥 130 🏭 25 000 sq m
Managing Dir: D.R.G. Burman
Marketing Dir: B.D. Perkins
Manufacturing Dir: M.J. Turner
■ Military footwear to UK MoD specification. Military footwear for export. [97]

75-58840 TEC Electrical Components Ltd
Progress Way, Enfield, Middx EN1 1UZ
☎(+44 181) 366 12 91 ⊖ (+44 181) 367 77 48
♣ £ 6 500 000 (1996) 👥 130 🏭 6 500 sq m
Managing Dir: S. Hunt
Technical Dir: J.A. Barker
Commercial Mgr: D. Mercer
Purchasing Mgr: B. Scott
■ Design, development & manufacture of complete systems, or sub-system elements & components for a wide range of defence, industrial & commercial applications. Products include electrical connectors & switches, lamps, illuminated information display panels & special purpose electrical components.
Subsidiary:
TEC Panelgraphic [97]

75-58850 Tecalemit Garage Equipment Co Ltd
(a subsidiary of Siebe plc)
Belliver Industrial Estate, Roborough, Plymouth, Devon PL6 7BW
☎(+44 1752) 70 12 12 [telex] 45124
⊖ (+44 1752) 79 63 29
Managing Dir: P.A. Crowder
Finance Director: J Simms
Operations/Technical Dir: Trevor Potter
Marketing Services Manager: M I Rhodes
Export Mgr: F. Howarth
Purchasing Manager: J Beesley
■ Manufacture a wide range of vehicle servicing equipment including lifts, brake testers, MOT test equipment & a full range of lubrication equipment. Approved supplier to NATO Forces. [98]

75-58860 Tecalemit Systems Ltd
(† Interlube Systems Ltd)
(a division of Siebe plc)
Estover Road, Plymouth, Devon PL6 7PS
☎(+44 1752) 77 57 81 ⊖ (+44 1752) 77 73 16
♣ £ 25,000,000 (1997) 👥 330 (1997) 🏭 35,000 sq m
Managing Director/Operations Director: P Moate
Purchasing Manager: M Radbourne
PR Consultant: M Merrifield
■ Vehicle lubrication systems (automatic). Weapon & radar lubrication systems, lubrication systems for steering, stabilisers & weapon handling equipment. High pressure hydraulic hose & fittings. [98]

75-58900 Technix Rubber and Plastics Limited
(a member of the British Marine Industrial Federation)
Unit 13, City Commerce Centre, Marsh Lane, Southampton, Hampshire SO14 3EW
☎(+44 1703) 63 55 23 ⊖ (+44 1703) 63 56 89
Managing Director: Paul Collins
■ Rubber engineers, manufacturers and distributors of industrial rubber and plastic products extrusions, mouldings, gaskets, sheeting, hose, ducting, matting sponge, adhesives and sealants. → 85 [98]

United Kingdom

75-59020 Tector Visual Systems
(a division of Graff Electronic Machines Ltd)
Woodhill Road, Collingham,
Newark, Nottinghamshire NG23 7NR
☎ (+44 1636) 89 22 46 ⊖ (+44 1636) 89 33 17
♣ £ 1,000,000 🏠 10,000 sq ft
Chairman: A V Tushingham
General Manager/Marketing Director: A L Leonard
■ Low-cost computer generated image (CGI) systems & target image generators for flight simulators & air combat trainers. Total weapon aiming training systems for anti-aircraft & SAM trainers. Air traffic control visual simulation. Optical display systems. Medium & high resolution colour monitors. [98]

75-59200 Telektron Ltd
Control & Systems Engineers
Armfield Close, West Molesey, Surrey KT8 2RJ
☎ (+44 181) 979 12 34 ⊖ (+44 181) 941 41 11
♧ £ 500 000 (1995) ♣ £ 1 800 000 👥 40 🏠 1 200
Managing Dir: P.D. MacAulay
Operations & Technical Dir: M.J. Francis
Sales & Marketing Mgr: P. Varnom
■ Warship main machinery control systems & equipment : pneumatic & electronic control systems design & manufacture. Electro pneumatic equipment, solenoid valves, rotary pnematic actuators & hazardous area equipment. [97]

75-59250 Television Installation Services (Mansfield) Ltd (TIS)
Millway, Mansfield Woodhouse, Notts NG19 9BG
☎ (+44 1623) 42 58 00 ⊖ (+44 1623) 65 07 67
♣ £ 4,000,000 (1996) 👥 85 (1997) 🏠 800 m2
Technical Director: C Lewis
■ Design, manufacture & installation of a wide range of specialist communications & security systems including CCTV surveillance. [98]

75-59305 Telxon Ltd
(a subsidiary of Telxon Corp, USA)
Old Orchard, High Street, Poole, Dorset BH15 1AE
☎ (+44 1202) 78 53 00 E userid@telxon.co.uk
⊖ (+44 1202) 66 52 70
👥 80 🏠 15,000 sq ft
Managing Director: W Collins
Sales Dir: Carl Graham
Dir Software Group: K. Minns
Dir Technical Resources: K. Williams
Sales Exec: Peter Turner
Marketing Manager: David Spokes
■ Data collection & transmission systems & terminals. [98]

75-59318 Teradyne Limited
(a subsidiary fo Teradyne Inc)
Units 5, The Western Center,
Bracknell, Berkshire RG12 1RW
☎ (+44 1344) 42 68 99 ⊖ (+44 1344) 72 58 83
♣ US$ 1 200 000 (1995) 👥 5 000 (1995)
Sales Mgr: Andrew Parkinson
■ Automatic test equipment for commercial & military applications used in production, maintenance or repair of electronics. → 80 [98]

75-59320 Terex Equipment Ltd
(a subsidiary of Terex Corp., USA)
Newhouse Industrial Estate,
Motherwell, Lanarkshire ML1 5RY
☎ (+44 1698) 73 21 21 ⊠ 77141 terex g
⊖ (+44 1698) 73 40 46
👥 550 (1997)
Managing Director: W S Buchan
Operations Director: C Robertson
Sales/Marketing Director: J Rotherford
Purchasing Manager: T Gardner
■ Earthmoving equipment for use in military engineering applications and for logistic support of combat arms over difficult terrain. [98]

75-59525 Thompson Carmichael Ltd
Great Bridge Road, Bilston, West Midlands WV14 8NP
☎ (+44 1902) 35 31 41 ⊠ 337329 tanker g
⊖ (+44 1902) 40 55 09
Managing Dir: J.F. Davies
Sales Mgr: P.A. Gardner
■ Produce transport tankers & aircraft refuellers for civil & military applications. Aircraft de-icer vehicles & bulk liquid road tankers. [98]

75-59540 Thomson-CSF UK Ltd
(a subsidiary of Thomson-CSF S.A., France)
UK Corporate Office
81 Piccadilly, London W1V 0HL
☎ (+44 171) 355 45 43 ⊖ (+44 171) 629 18 76
♣ £ 500,000,000 (UK Group) (1995)
👥 6,000 (UK) (1996)
Managing Director: B C Asbee
Director Business Development: P Hitchcock
Corporate Development Director: P Kahn
☎ (+44 171) 355 45 43
® France: Thomson-CSF S.A. [98]

75-59567 Thomson Training & Simulation Ltd
(a subsidiary of Thomson-CSF)
(merger of Link-Miles & Rediffusion)
Gatwick Road, Crawley, Sussex RH10 2RL
☎ (+44 1293) 56 28 22 W http://www.ttsl.co.uk
⊖ (+44 1293) 56 33 66
♣ £ 173,000,000 (1996) 👥 2,000
Chairman & CEO: Louis Le Portz
Civil Programs Dir: Denis Chedeville
Military Programs Director UK: Steve Grinham
Military Sales & Marketing Dir: Michael Turley
Civil Sales & Marketing Director: Michel Orman
Communications Mgr: Miss Lisa Bottle
■ Simulators & training systems for civil & military aircraft, land forces, tanks, naval ships, submarines, power stations (nuclear & conventional), road driving & entertainment. Computer generated visual systems. Computer based desktop training packages for any application. Computer based flight management systems trainer. [98]

75-59750 THORN Lighting Ltd
(a division of TLG plc)
Elstree Way, Borehamwood WD6 1HZ
☎ (+44 181) 905 13 13 ⊖ (+44 181) 967 63 43
♣ £ 380,000,000 (1995/96) 👥 4,200 (1995/96)
Chief Executive: Hamish Bryce
Government Department Manager: Clive Goodier
Publicity Manager: Hugh King
■ Comprehensive range of lighting equipment including airfield lighting.
Subsidiaries in:
Australia, Austria, Belgium, Czechoslovakia, Denmark, France, Finland, Germany, Holland, Hong Kong, Ireland, Italy, Malaysia, N.Zealand, Norway, Portugal, Singapore, Sweden, Taiwan, Spain, Thailand.
UK Sales
3 King George Close, Eastern Avenue West,
Romford, Essex RM7 7PP
☎ (+44 1708) 76 60 33 ⊖ (+44 1708) 77 62 69
Airfield Manager: Tony Hayter
Government Contracts: Clive Goodier
■ Lighting design, manufacture & distribution. National procurement contract administration. Consultancy.
International Sales
3 King George Close, Eastern Avenue West,
Romford, Essex RM7 7PP
☎ (+44 1708) 660 33 ✉ THORNLITE RFORD
⊠ 897759 thlite g ⊖ (+44 1708) 74 18 27
General Manager: David Hazell
Regional Manager: Ian Killick [98]

75-59760 THORN Microwave Devices Ltd
Swallowfield Way, Hayes, Middlesex UB3 1DQ
☎ (+44 181) 573 55 55 W http://www.tmd.co.uk/
⊖ (+44 181) 569 18 39
♣ £ 10,000,000 (1997) 👥 160 (1997)
Managing Director: P J Butcher
Marketing Director: P Plurien
Technical Director: I G Haig
Business Development Director: V H Smith
Publicity Manager: Ms Heather Skinner
■ A supplier of microwave tubes & microwave receivers to the international radar, EW & communications markets. Products include: magnetrons (airborne & surface applications); high-power klystrons; travelling wave tubes; sub-systems & surveillance microwave receivers for ELINT/ ESM. High voltage power supplies; instrumentation TWT amplifiers. [98]

75-59770 THORN Security Ltd
(† AFA Minerva)
International Division
Security House, The Summit, Hanworth Road,
Sunbury-on-Thames, Middlesex TW16 5DB
☎ (+44 1932) 74 33 33 ⊠ 8814916
⊖ (+44 1932) 74 31 55
♣ £ 200,000,000 (1996) 👥 3,400 (worldwide) (1996)
Managing Director: Joe Matheson
Commercial Manager Marine Department: Brian Rodricks
General Manager International: Shaun Bond
Marketing Services/PR Manager: Richard Nelson
■ Design, supply, installation & servicing of fire detection & extinguishing systems, fire equipment, security systems & sensors, CCTV & perimeter security systems & building management systems. [98]

75-59800 Thwaites Ltd
Welsh Rd, Cubbington,
Leamington Spa, Warwicks. CV32 7NQ
☎ (+44 1926) 42 24 71 ⊖ (+44 1926) 33 71 55
👥 250 🏠 26 462 sq. m
Managing Dir: D.N. Whittaker
Export Sales Mgr: R.J. Adams
■ Specialist manufacturer of 2 & 4 wheel drive dumpers with payload capacities from 750 kg to 16 000 kg. [97]

75-59930 Tilt Measurement Ltd
Horizon Hse, London Road,
Baldock, Hertfordshire SG7 6NG
☎ (+44 1462) 89 45 66 ⊖ (+44 1462) 89 59 90
♣ £ 500,000 (1996) 👥 10 (1997) 🏠 1,000 m2
Managing Dir: A.M.B. Shaw
Director: N.J. Granger
■ Alignment system for naval weapons & navigation equipment. Accelerometer, instrumentation & control systems. Tilt transducers & systems. [98]

75-59940 Time Electronics Ltd
Botany Industrial Estate, Tonbridge, Kent TN9 1RH
☎ (+44 1732) 35 59 93 ⊖ (+44 1732) 77 03 12
👥 35
Office Mgr: Mrs C. Hall
Purchasing Mgr: P. Hickey
■ Portable & programmable calibration equipment. [97]

75-59950 Timet UK Ltd
(a division of Timet, USA)
PO Box 704, Witton, Birmingham B6 7UR
☎ (+44 121) 356 11 55 ⊖ (+44 121) 356 54 13
♣ £ 100,000,000 (1996) 👥 800 (1997)
Managing Director: Christian Leonhard
Operations Director: Brian Hadley
Sales/Marketing Director: Ian Hodges
Sales Manager: Martin Hudson
Marketing Manager: Chris Hall
■ Titanium and titanium alloys, ingots, bars, billet, sheet, rod, tube, extrusions, wire and castings. Superconducting composites wires and cables. → 85 [98]

75-60150 Tirfor Ltd
Old Lane, Holbrook Ind.Estate, Halfway, Sheffield S19 5GZ
☎ (+44 114) 248 22 66 ⊖ (+44 114) 247 56 49
Sales & Marketing Mgr: T. Fennell
Marketing Services Mgr: Mrs Marion Lee
■ Tirfor hand-operated recovery equipment, load measuring equipment, access platforms, materials handling equipment. [97]

75-60360 Todd Research Ltd
Robjohns Rd, Chelmsford, Essex CM1 3DP
☎ (+44 1245) 26 22 33 ⊠ 99327
⊖ (+44 1245) 26 94 09
♣ £ 2 000 000 👥 50
Sales Exec.: Ms Sandra Downs
■ Manufacturers of a comprehensive range of security X-ray apparatus designed to help identify mail bombs & other illegal objects. [97]

75-60363 Tomarc Limited
Unit 3 Elmdon Trading Estate, Bickenhill Lane,
Birmingham B37 7HE
☎ (+44 121) 782 38 88 ⊠ 337856
⊖ (+44 121) 782 34 22
♣ £ 2,000,000 👥 25
Chairman: T Smeaton
National Sales Manager: G Roberts
Purchasing Manager: Stan Steanes
Senior Sales Co-ordinator: Mrs Jane O'Rourke
■ Workshop and vehicle maintenance tools. [98]

75-60365 F.H. Tomkins Buckle Co. Ltd
Brockhurst Crescent, Walsall, West Midlands WS5 4QG
☎ (+44 1922) 72 30 03 ⊖ (+44 1922) 72 31 49
Managing Dir: P. Graham
Sales Dir: A. Fleming
■ Buckles & metal fastenings. Buckles & metal fittings for shoes, clothing, luggage, harness & saddlery. [98]

75-60400 The Torrington Company Ltd
(a div. of Ingersoll-Rand Co., USA)
Fafnir Bearing Division
Upper Villiers Street, PO Box 18,
Wolverhampton WV2 4NT
☎ (+44 1902) 261 01 ⊖ (+44 1902) 71 19 03
Operations Mgr: Gary Kenmir
Production Planning Mgr: Phil Rowley
■ Manufacture precision ball and parallel roller bearings. [98]

75-60460 Touringsport (Rosker Ltd)
(a subsidiary of Allcord Ltd)
Unit 13, Quay Lane Ind. Est.,
Gosport, Hampshire PO12 4LJ
☎ (+44 1705) 52 87 11 ⊖ (+44 1705) 52 91 08
♧ £ 20,000 ♣ £ 1,300,000 (1996) 👥 9 (1997)
🏭 40 sq m
Managing Director: P Glossop
Sales Director: R D White
Purchasing Manager: J Moore
■ Manufacture specialist tents, outdoor pursuits equipment & specialist survival gear. [98]

75-60600 Toye, Kenning & Spencer Ltd
(a subsidiary of Toye & Co. plc)
Regalia House, Newtown Road,
Bedworth, Warwickshire CV12 8QR
☎ (+44 1203) 31 56 34 ⊠ 312235
⊖ (+44 1203) 64 30 18
♧ £ 4,000,000 👥 314
Chairman and Managing Director: B E Toye
Group Sales Dir: P.R.S. Frost
General Sales Mgr: S.M. Mills
Purchasing Manager: D Fox
■ Uniform accoutrements, regalia & accessories covering ceremonial, service & combat dress. [98]

75-60960 TransInstruments Recording Systems Div.
Imo Industries Ltd
(a div. of Imo Industries Inc., USA)
Block A. The Green Buildings, Kempshott Hill,
Basingstoke, Hants RG22 7LL
☎ (+44 1256) 571 71 ⊠ 858103 ⊖ (+44 1256) 575 88
♣ £ 7 500 000 👥 170 🏭 55 000 sq ft
Managing Dir: Ron Pabers
Sales & Marketing Dir: John Goodson
Operations & Technical Dir: Jerry Moreland
Purchasing Mgr: Brian Davies
Sales Support Engineer: John Symonds
Recording Systems Mgr: David Underhill
■ Instrumentation rotary tape recorders, data acquisition, computer compatible signal conditioning, recording oscillographs, transducers for the measurement of pressure, vibration, acceleration & liquid density. [97]

75-60970 Transradio Ltd
(a subsidiary of Radiall, France)
10 Perivale Industrial Park, Horsenden Lane South,
Perivale, Middx UB6 7RL
☎ (+44 1819) 97 88 80 ⊠ 92300
⊖ (+44 1819) 97 01 16
General Manager: Stephen Phipson
■ Coaxial connectors 0-26 GHz; coaxial cable assemblies, flexible and semi-rigid for use from 0-26 GHz; multipin connectors; fibre optic connectors; coaxial and waveguide microwave devices up to 26 GHz; 2 mm and 4 mm banana plugs and patch leads. Sub D connectors. [98]

75-61000 Trelawny Pneumatic Tools
(† John Trelawny Ltd)
(a division of Fulton Group Ltd)
12-14 Longfield Road,
Leamington Spa, Warwickshire CV31 1XB
☎ (+44 1926) 88 37 81 ⊖ (+44 1926) 45 03 52
♣ £ 4,000,000 (1995) 👥 50
Managing Dir: R. Stuart Harrigan
Sales Manager: Rob Chapman
Operations & Technical Mgr: Chris Dee
Purchasing Mgr: John Butler
Customer Service Mgr: R.G. Barton
■ Pneumatic needle scaling guns, scaling hammers, long reach scalers & other large area ship deck scaling machines. Electric, pneumatic, petrol, diesel & gas powered hand tools. [98]

75-61075 Tremlett Boat Sales Ltd
Topsham, Exeter, Devon EX3 0PD
☎ (+44 1392) 87 36 80; 87 45 80
⊖ (+44 1392) 87 62 77
♣ £ 400 000 👥 10
Managing Dir: Christopher Tremlett
Director: Mrs D. Tremlett
■ Design & building fast offshore powerboats up to 100 ft, used as patrol boats, police/custom launches & assault craft. Boats can be designed to customer's specifications. → 90

Tremlett Powerboats Ltd
Topsham, Exeter, Devon EX3 0PD
☎ (+44 1392) 87 36 80; 87 45 80
⊖ (+44 1392) 87 62 77 [97]

75-61090 A.F. Trenchers Ltd
Edgworth Road, Sudbury, Suffolk CO10 6TG
☎ (+44 1787) 31 18 11 [E] trenchers.co.uk
[W] http://www.trenchers.co.uk ⊖ (+44 1787) 31 08 88
♧ £ 10,000 (1997) ♣ £ 1,000,000 (1997) 👥 10 (1997)
🏭 16,000 sq ft
Managing Director: H M Jurgens
Operations Director: H B Fulford
Purchasing Mgr: R. Turner
■ Chain type trenching machines designed to cut narrow trenches for cable & pipe laying. [98]

75-61110 Trenton Technical Services Ltd
Unit 6, 28 Botley Road,
Hedge End, Southampton SO3 3HE
☎ (+44 1489) 78 11 82 ⊠ 85546 vyner g
⊖ (+44 1489) 78 11 82
Managing Director: D J Hillier
Director: A R Docherty
■ Maintenance and logistical support services, training systems and equipment. Indoor live fire systems including video target trainer, multi-media training system, panoramic range target system, firearms incident real situation trainer and live fire target screen unit. [98]

75-61130 Trilux Night Sights
Millington Road, Hayes, Middlesex UB3 4AZ
☎ (+44 181) 573 38 00 ⊖ (+44 181) 561 34 36
■ Manufacture self-illuminated gunsights. In low light conditions where traditional sights are ineffective Trilux gives you the power to sight your target with confidence. [98]

75-61140 Trimble Navigation Europe Ltd
(a subs. of Trimble Navigation Ltd)
Trimble Hse, Meridian Office Park, Osborn Way,
Hook, Hampshire RG27 9HX
☎ (+44 1256) 76 01 50 ⊖ (+44 1256) 76 01 48
♣ US$ 250 000 000 👥 80
Managing Dir: Mike Coles
Technical Sales: W.J. Thornton
Military Sales Mgr: R.D. Stephenson
■ High accuracy navigation position finding, surveying & mapping systems based upon the GPS satellite constellation. Receivers, sensors, software suitable for all applications from guidance systems for drones through to troop movement monitoring & tracking. → 80
® USA: Trimble Navigation Ltd [97]

75-61150 Trimite Ltd
Arundel Road, Uxbridge, Middlesex UB8 2SD
☎ (+44 1895) 25 12 34 ⊠ 934444 trimit g
⊖ (+44 1895) 25 67 89
♣ £ 50,000,000 👥 500
Commercial Executive: N F Holmes
Marketing Manager: A P Lance
Laboratory/Technical Manager: K A Davis
Advertising & Public Relations Officer: J R Brooks
■ Manufacture of surface coatings for military & civil aircraft, helicopters, missiles & ancillary equipment. [98]

75-61160 Tritech International Ltd
Peregrine Road, Westhill Business Park,
Westhill, Aberdeenshire AB32 6JL
☎ (+44 1224) 74 41 11 [E] sales@tritech.co.uk
⊖ (+44 1224) 74 17 71
♣ £ 2,000,000 (1997) 👥 12 (1997) 🏭 1,000 sq m
Managing Dir: Richard Marsh
Operations Dir: Dick Wright
Sales Manager: Kevin A Parker
Marketing: Ms Anne Findlay [98]

75-61330 Tronic Electronic Services Ltd
Sandside Road, Ulverston, Cumbria LA12 9EF
☎ (+44 1229) 58 05 00 [E] tronic@co.uk
⊖ (+44 1229) 58 66 04
Chairman: J M Alcock
Managing Director: Harold Dearden
Sales and Marketing Director: Paul Alcock
Operations Manager: Steve Walker
Marketing Manager: Phil Logan
Sales Manager: Adam Kendall
Marketing Analyst: Ms Nicola Noss
■ Design and manufacture of severe environment electrical connectors and subsea cables/hoses for total package provision. Products include subsea mateable (CE, MINICE, ROV); downhole (packer and wellhead penetrators, potheads, splices) and topside (EExD) connectors, for instrumentation and power applications. Connectors; junction boxes & specialised pressure housings. [98]

75-61360 Trucast Ltd
(a member of the Propulsion Technology Group of Vickers plc)
Marlborough Road, Ryde, Isle of Wight PO33 1AD
☎ (+44 1983) 56 76 11 ⊠ 86180
⊖ (+44 1983) 56 76 18
♣ £ 21 700 000 (1995) 👥 350
Managing Dir: I.J. Bulwer
Marketing & Technical Support Dir: R. Palmer (MBE)
Technical Dir: S. Leyland
Purchasing Mgr: D. Lankshear
■ Air melt and high vacuum casting for the aerospace, defence & other industries. [97]

75-61380 Truflo Marine Valves Ltd
(a subsidiary of Charles Baynes plc)
Westwood Road, Birmingham B6 7JF
☎ (+44 121) 327 47 89 [W] http://www.truflo.co.uk
⊖ (+44 121) 327 41 32
♣ £ 7,000,000 👥 70
Dir & General Mgr: Robert Bowser
Finance & Commercial Dir: John Lokier
Technical Mgr: Robert Watson
Operations Manager: Steve Chapman
Purchasing Mgr: Robert Beathe
Sales and Marketing Manager: David Bowen
■ Supplier of marine valves including shock proof hull penetration systems & special valves for weapon discharge & AIP systems to shipbuilders worldwide for both submarines & surface fighting ships. [98]

75-61395 TSS (UK) Ltd
New Mill, New Mill Lane, Witney, Oxfordshire OX8 5TF
☎ (+44 1993) 77 77 00 [E] tssmail@tssuk.co.uk
⊖ (+44 1993) 77 77 01
♣ £ 3,800,000 (1996) 👥 85 (1997) 🏭 22,000 sq ft
Managing Director: Robert A Manson
Sales and Marketing Manager: Rob Balloch
Buyer: Tania Bailey
Marketing Coordinator: D M Hopkins
■ Marine motion monitors: position, heading, heave, roll and pitch data for multibeam echosounders and positioning systems compensation. Sub-sea metal detection systems. Marine technology projects and consultancy. Deck motion monitors. → 80 [98]

75-61410 Tufnol Ltd
Wellhead Lane, PO Box 376,
Perry Barr, Birmingham B42 2TB
☎ (+44 121) 356 93 51 [E] sales@tufnol.co.uk
[W] http://www.tufnol.co.uk ⊖ (+44 121) 331 42 35
♣ £ 10,000,000 👥 210 (1997)
Managing Director: J R Trotman
Operations Director: K J Aston
Sales Manager, SDI Contact: R Thomason
Marketing Manager: R P Godwin
Purchasing Manager: P Goodwin
Marketing Officer: R Starkey

■ Manufacturer of high-pressure laminates, filament wound tube & pultruded composite profiles. A full machining service is available for these materials. → 85 [98]

75-61500 Tungum Hydraulics Ltd
(a subsidiary of Tungum Holdings Ltd)
The White House, Arle,
Cheltenham, Gloucester GL51 0AD
☎(+44 1242) 24 55 21 ⊜ (+44 1242) 22 26 56
£ 100,000 ♣ £ 2,500,000 ⅲ 100
Managing Director: G W Burndred
Marketing Director: T Clark
■ Supply of Tungum alloy tube & bar. Supply of "PG Series" high-pressure tube couplings in tungum alloy. Manufacture of pipework assemblies.
Subsidiary:
Dudley Engineering Services Ltd [98]

75-61510 TTL Dynamics
(† Turbine Technik Ltd)
PO Box 1642, Salisbury, Wiltshire SP3 5TJ
☎(+44 1722) 33 93 39 ⊜ (+44 1722) 33 96 69
Chairman: Richard J. Biddle
Purchasing Mgr: M. Simcock
■ High speed marine interceptor craft. Supplier of Marine gas turbines. [97]

75-61520 Turbotech Precision Products Ltd
Atlantic St, Broadheath, Altrincham, Cheshire WA14 5DF
☎(+44 161) 927 78 73 ▣ 668606
⊜ (+44 161) 927 70 23
Managing Dir: G.D. Richards
■ Manufacture of precision cast aluminium compressor wheels. [97]

75-61650 Typhoon International Ltd
Mortlake Court, 28 Sheen Lane,
East Sheen, London SW14 8LW
☎(+44 181) 876 98 18 ⊜ (+44 181) 876 92 23
£ 650 000 ♣ £ 3 000 000 ⅲ 80 ⌂ 2 230 sq m
Chairman & Joint Managing Dir: C. van der Lande
Joint Managing Dir: R. Gugen
Purchasing Mgr: S. Ward
■ Design & manufacture of drysuits for marine, military & aviation environments. [97]

75-61700 UCC International Limited
(a part of UCC International Group Ltd)
PO Box 3, Thetford, Norfolk IP24 3RT
☎(+44 1842) 75 42 51 [E] fluidpower@ucc.co.uk
[W] http://www.ucc.co.uk ⊜ (+44 1842) 75 37 02
ⅲ 134
Managing Dir: David Button
Sales Dir: Brian Jenkins
Public Relations Mgr: John Botly
■ Condition & contamination monitoring equipment. Filters & fluid power products for hydraulic systems. Flowmeters. [98]

75-61780 U E F Chesterfield Cylinders
(† Chesterfeild Cylinders Ltd)
(a division of United Engineering Forgings Ltd)
Derby Road, Chesterfield, Derbyshire S40 2EA
☎(+44 1246) 27 72 11 [E] post@chestcyl.demon.co.uk
▣ 54231 cyls g ⊜ (+44 1246) 21 01 01
♣ £ 25,000,000 ⅲ 370
Managing Director: J Winchester
Operations Director: J Kelly
Sales and Marketing Director: G S B Larcombe
Sales Manager Special Products: R G Atkinson
■ Design and manufacture of seamless steel high pressure gas cylinders and other cylinders. [98]

75-61830 Ultrafine Technology Ltd
Unit 14, Brook Lane Business Centre, Brook Lane North,
Brentford, Middlesex TW8 0PP
☎(+44 181) 569 99 20 ⊜ (+44 181) 569 96 49
♣ £ 850,000 ⅲ 10
Managing Director: John Patterson
Sales and Marketing Director: David Robertson
Operations and Purchasing Director: Ms Carole Ferris
Technical Sales Support: Chris Hill-Kelly
■ Quartz fibrescopes/endoscopes & UV light sources & liquid light guides. Endoscope servicing & repairs (all makes). [98]

75-61840 Ultralife Batteries (UK) Ltd
(† Dowty Batteries)
(a subsidiary of Ultralife Batteries Inc, USA)
18 Nuffield Way, Abingdon, Oxon OX14 1TG
☎(+44 1235) 52 05 02 ⊜ (+44 1235) 53 57 66
ⅲ 85
Managing Dir: John Welsh
Technical Dir: Paul Bennett
Sales and Marketing Director: Graham Archdale
Purchasing Mgr: Mike Stevens
Technical Sales Engineer: Gregor Ritchie
■ Manufacture of lithium & sea water activated batteries. Specialists in meeting rigorous defence specifications. [98]

75-61900 Underwater & Environmental Technology Ltd
8 Colebrook Way, Weyhill Road, Andover, Hants SP10 3BB
☎(+44 1264) 35 65 66 ⊜ (+44 1264) 33 21 43
Managing Dir: Keith Swinden
Sales & Marketing Dir: Frank Ashley
Operations/Technical Dir: Dave Rundle
■ Specialists in the design & manufacture of the highest integrity electrical penetrators, connectors & cable harnesses for submarine use. Design and manufacture of hydrophones for the offshore & exploration industries. [98]

75-62050 Uniross Batteries Ltd
Unit 4, Blackfriards Rd, West End Trading Estate,
Nailsea, Avon BS19 2DJ
☎(+44 127) 585 81 01 ⊜ (+44 127) 585 72 02
Managing Dir: John Searle [97]

75-62130 United Moulders Ltd
(a subsidiary of International Process Technology Ltd)
Caker Stream Road, Alton, Hampshire GU34 2PL
☎(+44 1420) 866 16 [E] sales@uml.co.uk
[W] http://www.uml.co.uk ⊜ (+44 1420) 859 41
♣ £ 5,000 (1996) ⅲ 85 (1997) ⌂ 1,500 m2
Managing Director: John Perrins
Operations Director: Daryl Barker
Sales Manager: Steve Woods
Purchasing Manager: Steve Stewart
■ Automatic inflation valves, plastic moulded items, life-saving accessories, prototyping, toolmaking for plastic moulding. [98]

75-62300 Universal Carbon Fibres
(a member of IPT Group)
Gomersal Mills, Cleckheaton, West Yorkshire BD19 4LU
☎(+44 1274) 85 32 20 ⊜ (+44 1274) 85 33 06
General Manager: Peter L Cole
Sales Executive: Peter D Burnell
Research and Development Director: N Saville
■ Panotex fire resistant fabric, based on carbon fibres. Versions coated in steel powder or aluminium include protection against burning petrol, napalm & nuclear flash - ideal for protecting personnel, petrol dumps, vehicles or equipment. Finished cloth or products such as uniforms, tents etc. [98]

75-62550 Vacu-Blast Ltd
(a subs. of BTR plc)
Woodson House, Ajax Ave, Slough, Berkshire SL1 4DJ
☎(+44 1753) 52 65 11 ✉ VACUBLAST
▣ 848183 vblast g ⊜ (+44 1753) 53 80 93
♣ £ 10 000 000 (1995) ⅲ 140 ⌂ 7 000 sq m
Managing Dir: M. Riley
Sales & Marketing Dir: R.E. Brickwood
Operations & Technical Dir: Mark Torres
Finance Dir: Brian Brook
Purchasing Mgr: Peter Pritchard
■ Shot peening, abrasive blasting & plastic media blasting equipment & service.
Subsidiaries:
Vacu-Blast → USA 75
Vacu-Blast International → France 90 [97]

75-62600 Vacuum Reflex Ltd
(a subsidiary of Chemring Group plc)
Head Office & Works
Barrack Square, Martlesham Ind.Est.,
Ipswich, Suffolk IP5 7RF
☎(+44 1473) 62 40 80 ⊜ (+44 1473) 62 23 71
♣ £ 1,666,000 (1997) ⅲ 43 (1997)
⌂ 10,000 sq ft (1997)
Executive Director: B C Leveratt
Director: J.A. Armstrong
Manufacturing Dir: J G Eaton
Director: B.P. Berlanny
Director: D.R. Evans
Purchasing Mgr: M.N. Wincote
■ Thermal & weather protective garments for military applications. Producers of survival suits for cold water immersion & military pattern lifejackets. Electrically heated clothing for military & civil applications. Producers of radar reflective balloon covers in various sizes for use in calibration of radar installations. [98]

75-62627 Valeport Ltd
(† Valport Marine Scientific Ltd)
Unit 7, Townstal Industrial Estate,
Dartmouth, Devon TQ6 9LX
☎(+44 1803) 83 40 31 [E] sales@valeport.co.uk
[W] http://www.valeport.co.uk ⊜ (+44 1803) 83 43 20
ⅲ 28
Managing Director: Charles Quarterly
Operations Director: Peter Adams
Sales and Marketing Manager: Kevin R Edwards
Purchasing Manager: John Davis
■ Range of current meters (conventional and electromagnetic) CTDS probes, winches and accessories. Specific application packages catered for, including manufacture, research and development. [98]

75-62630 Vapormatt Ltd
Rue a Chiens, St Sampson's, Guernsey, CI GY2 4AG
☎(+44 1481) 531 95 ⊜ (+44 1481) 523 69
Managing Director: S I Ashworth
Sales and Marketing Director: R S Ashworth
Operations and Technical Director: T I Ashworth
Purchasing Manager: M P Ashworth
Sales Admin: J. Le Prevost
■ Paint stripping, cleaning techniques. Environmentally friendly paint & conformable coating, cleaning & stripping. Concurrent degreasing & cleaning of precision mechanical & electrical components. Decontamination of radioactive components.
Sales Offices:
Hydratec Systems Ltd
Wings Pte Ltd, Singapore [98]

75-62660 Vascon
25g Lowndes Street, London SW1X 9JF
☎(+44 171) 235 77 50 ⊜ (+44 171) 245 61 94
♣ £ 1,000,000+ ⅲ 7 (1997) ⌂ 200 sq m
Chairman: Peter Cadbury
Managing Director: Peter Harris
Sales/Marketing Director: M Philips
Purchasing Manager: S Burbridge
■ Manufacture of Digital Global Listening Post. Supplier of surveillance and ECM equipment. → 85 [98]

75-62720 VCH International
(† General Instruments Lamps Ltd)
Beetons Way, Bury St Edmunds, Suffolk IP32 6RA
☎(+44 284) 76 24 11 [E] admin@vch.ltd.uk
⊜ (+44 284) 75 44 06
♣ £ 15,000,000 (1996) ⅲ 250 (1996) ⌂ 8,000 m2
Managing Director: M B Hare
Technical Director: J R Cole
Operations Director: M A Hemingway
Sales Director: R S Clarke
Purchasing Manager: D J Unwin
■ Indicators, miniature, sub-miniature, incandescent neon and LED lamps. High energy switches. [98]

75-62740 Vega-Cantley Instrument Company Limited
Unit J, Eskdale Rd, Uxbridge, Middx UB8 2RT
☎(+44 1895) 25 35 58 ▣ 917870 vega g
⊜ (+44 1895) 25 95 40
♣ £ 1 500 000 (1995) ⅲ 30 ⌂ 1 200 sq m
Managing Dir: L.M. Rodriguez
Financial Dir: B.R. Jones
Secretary: S. Elliot
■ Radar transponders, radar tracking systems, telemetry systems & associated test equipment, broadening into a range of transponders in the bands 800 MHz to 30 GHz, data encoders & decoders, antenna switching units, with a range antenna complimentary to their group. Quality registration to BS & ISO9001. [97]

75-62743 Veribest International Ltd
(† Integraph (UK) Limited)
(a subsidiary of Intergraph Corp, USA)
Delta Business Park, Great Western Way,
Swindon, Wiltshire SN5 7XP
☎(+44 1793) 51 11 99 W http://www.veribest.com
Telex 444164 intgb Fax (+44 1793) 49 29 40
Employees 25
Managing Director: Mark Turner
Operations Director: Paul Smith
Sales and Marketing Director: David Doyle
Purchasing Manager: Neil Cushing
■ Suppliers for electronic design automation (EDA) products used in the creation of electronic systems & components. [98]

75-62750 Vernon Gauging Systems Ltd
Wilbury Way, Hitchin, Herts SG4 0TS
☎(+44 1462) 42 21 11 Fax (+44 1462) 45 36 67
♣ £ 4 000 000 (1996) Employees 85 Area 22 082 sq m
Managing Dir: J.S. Thackray
Sales Dir: T.A. Jellis
Publicity Mgr: Ian Evans
■ Electronic & pneumatic multi-dimensional inspection equipment designed for quality control of turbine engine components & finely toleranced parts for the defence industry. [97]

75-63010 Vessa Limited
(a subs. of BTR plc)
Mill Lane, Alton, Hants GU34 2PX
☎(+44 1420) 832 94 Fax (+44 1420) 54 15 63
Managing Dir: Keith Bell
Operations/Technical Dir: James Parker
Production Services Mgr: Tom Cornfield
Marketing Services Co-ordinator: Mrs Judy Brown
■ Manufacture & supply of artificial limbs & components (including Quantum modular limb system), prosthetic & orthotic workshop equipment & post-amputation mobility aids (PPAM). [97]

75-63020 Vickers Aerospace Components
(† Vickers Precision Components)
(a Vickers plc company)
Whitchurch Rd, Shrewsbury, Shropshire SY1 4DP
☎(+44 1743) 23 51 81 Telex 35623
Fax (+44 1743) 46 22 03
Managing Dir: D. Murray
Operations & Technical Dir: D. Marsh
Finance Dir: A. Deacon
Purchasing Mgr: A. Harper
■ Manufacture of aero engine components: turbine blades & aerofoil products; precision machining, metal forming & fabrication of nozzles, blades, combustors & other components; servicing & repairs. [97]

75-63030 Vickers Bridging
(a division of Vickers plc)
Spring Road, Ettingshall, PO Box 37,
Wolverhampton WV4 6YN
☎(+44 1902) 40 50 50 Telex 337412
Fax (+44 1902) 49 14 59
♧ £ 35,000,000 (1997) ♣ £ 50,000,000 (1997)
Employees 183 (1997) Area 7,206 sq m
Managing Director: W J McGawley
Operations Director: J A Eden
Finance Director: C S Oliver
Head of Business Development: K D Sargent
Sales and Marketing Director: B Truman
Supply Manager: M Zouch
Technical Manager: T J Davis
Works Manager: M Brassington
■ Design & manufacture of mechanical defence systems, including lightweight MLC 70 modular bridges. Prime contractor for production of the UK MoD bridging for 1990's systems. [98]

75-63040 Vickers Defence Systems
(a div. of Vickers plc)
Armstrong Works, Scotswood Road,
Newcastle-upon-Tyne NE99 1BX
☎(+44 191) 273 88 88 Telex 53104 vicarm g
Fax (+44 191) 273 06 98
Chief Executive: Colin Clark
Finance Dir: John Hammerton
Commercial Director: I S McNeeney
Operations Dir: Bill McGawley
Sales and Marketing Director: Brian Trueman (acting)
☎(+44 191) 226 70 33
General Manager Sales and Marketing: John Slade
☎(+44 191) 226 70 25
Public Affairs: Peter McKenzie ☎(+44 191) 226 70 26
Marketing Manager: Ray Smith ☎(+44 191) 226 70 27
■ Design & manufacture of armoured fighting vehicles including the Challenger Main Battle Tank. Specialist in weapon & vehicle systems integration & design of armoured vehicle derivatives. Turn-key training packages for military & civil applications.

Leeds Facility
Manston Lane, Crossgates, Leeds LS15 8ST
☎(+44 113) 264 81 23 Telex 55314
Fax (+44 113) 260 76 74
Public Affairs: Peter McKenzie [98]

75-63100 Vickers Shipbuilding & Engineering Ltd (VSEL)
VSEL
(a subs. of The General Electric Company plc)
Barrow-in-Furness, Cumbria LA14 1AF
☎(+44 1229) 82 33 66 Fax (+44 1229) 87 40 00
♧ £ 1 000 000 ♣ £ 428 800 000 (1994/95) Employees 5 446
CEO: B.V. George
Overseas Sales Dir: N.C. Forsyth
Finance Dir: R.D. Holden
Projects Dir: Dr D.E. Gilbert
Operations Dir: Dr R.H. Rigby
Commercial Dir: P.A. Martin
Personnel & Quality Assurance Dir: R.J.G. Davies
Managing Dir Armaments: P. Kenyon
Product Support Dir: D.E. Shepherd
Purchasing Dir: G.A. Hayes
Company Secretary: J. Bruell
Marketing Mgr Central: G.R. Sharpe
■ Project management, design, development & build of nuclear & conventional submarines, large surface ships (naval & commercial) & naval & land based weapon systems. VSEL also provides a wide range of products & services for through-life support.
VSEL, Yarrow Shipbuilders & NNC Ltd are wholly owned subsidiaries of GEC plc & together they form GEC Marine, a new organisation dedicated to world-class shipbuilding, armaments, nuclear engineering & project management.

Surface Ships
Barrow-in-Furness, Cumbria LA14 1AF
☎(+44 1229) 82 33 66 Fax (+44 1229) 83 23 24
Project: G.D. Thompson ☎(+44 1229) 87 35 54
Sales: R. Mawson ☎(+44 1229) 87 33 57
■ Builder of medium & large surface warships in the UK. In-house design capability in Naval Architecture, Marine & technical systems & combat systems.

Armaments
Barrow-in-Furness, Cumbria LA14 1AF
☎(+44 1229) 87 59 82 Fax (+44 1229) 87 44 02
Managing Dir: P. Kenyon
Land Systems Sales Mgr: Col R.G. Webster
☎(+44 1229) 87 42 84
■ Artillery systems house serving the needs of land & naval defence markets world-wide. Specialist in 155mm artillery, towed & self-propelled howitzers & MCNG's. Developing business areas include armoured fighting vehicles, naval ammunition & missile systems.

Submarines
Barrow-in-Furness, Cumbria LA14 1AF
☎(+44 1229) 82 33 66 Fax (+44 1229) 83 23 24
Sales & Projects: B. Everard ☎(+44 1229) 87 33 73
■ Submarine specialists. System design & construction of submarines including combat systems.

Product Support
Barrow-in-Furness, Cumbria LA14 1AF
☎(+44 1229) 87 36 75 Fax (+44 1229) 87 30 48
Sales UK: D.E. Shepherd ☎(+44 1229) 87 37 19
International Sales: B. Boden ☎(+44 1229) 87 41 97

General Engineering
Barrow-in-Furness, Cumbria LA14 1AF
☎(+44 1229) 87 34 02 Fax (+44 1229) 87 35 99
Sales: D. Wilton ☎(+44 1229) 87 43 01
■ Heavy fabrication services. Heavy, medium & light machining services. [97]

75-63150 Vickers Systems Division TRINOVA Ltd
(† Vickers Systems Ltd)
(a division of Vickers Inc., USA)
Aerosapce Marine Defence
Larchwood Avenue, Bedhampton, Havant, Hants PO9 3QN
☎(+44 1705) 48 72 70 Telex 86749 vsport g
Fax (+44 1705) 49 24 00
Employees 70 Area 3,500 m2
Operations Director: Colin Hunt ☎(+44 1705) 48 72 04
European Engineering Manager: Graham Barrow
☎(+44 1705) 48 72 66
Sales Manager: Jeremy Smithson
☎(+44 1705) 48 72 70
Purchasing Manager: Keith Watts ☎(+44 1705) 48 72 03
Sales Application Engineer: Keith Austin
☎(+44 1705) 48 72 13
■ Design and production of fluid power and fuel control systems and components including hydraulic, electrohydraulic and fuel pump equipment; pumps, motors, electric motor pumps, integrated packages, test stands and ground kart support equipment. [98]

75-63205 Vidisco Ltd
Kinetic Centre, Theobald Street,
Borehamwood, Hertfordshire WD6 4SE
☎(+44 181) 386 66 13; (+44 956) 52 40 88 (mobile)
Fax (+44 181) 386 66 13
Managing Director: M Hilsenrad
Financial Director: S K Freedman
Marketing Director: L Newmark
■ X-ray inspection equipment. A-600 and Micro-Ray 150 portable video-based x-ray inspection systems. FoXray laptop PC based portable x-ray inspection system. MIC80A mail inspection cabinets. X-ray inspection systems. EOD search equipment, mail inspection cabinets. [98]

75-63230 Viking Ltd.
Chatsworth Hse, Portland Close, Houghton Regis,
Dunstable, Beds. LU5 4AW
☎(+44 1582) 60 36 00 Fax (+44 1582) 47 11 14
♣ £ 2 950 000 (1995) Employees 55 Area 2 000 sq m
Managing Dir: F.W. Nicholls
Sales & Marketing Dir: John Parrott
Technical Dir: A. Whitney
Operations Dir: J. Logsdail
■ Connectors, interconnection systems and components. [97]

75-63353 W. Vinten Ltd
(† Vinten Military Systems Ltd)
Vicon House, Western Way,
Bury St Edmunds, Suffolk IP33 3SP
☎(+44 1284) 75 05 99 Telex 818315 vicon g
Fax (+44 1284) 75 05 98
Employees 200
Chief Executive: B J Hughes
Engineering Director: A C Leacy
Sales and Marketing Director: J D Rust
Commercial Director: S G Theunissen
Financial Director: P N Hockett
Marketing Executive: Mrs S Grainger
■ Design, development & manufacture of low, medium & high altitude reconnaissance pod systems; airborne countermeasures system (chaff/flare); airborne video HUD & gunsight recording systems; framing & panoramic reconnaissance cameras; electro-optical (EO) sensors; infra-red linescan systems; modular imagery evaluation systems; engineering ground support systems; technical training facilities. [98]

75-63355 VIP-Heinke
(† Vietaulie Industrial Polymers)
(a Glynwed International Business)
St Peter's Road, Huntingdon, Cambridgeshire PE18 7DJ
☎(+44 1480) 41 13 33 Telex 32221
Fax (+44 1480) 41 39 91
Employees 230
Managing Dir: Dr J.R. Hirst
Sales Dir: J. Amphlet
Sales Exec.: R. Veazey
Operations Dir: M. Douglas
Purchasing Mgr: M. Martin
■ Manufacture of natural and synthetic rubber seals and gaskets for the pipeline industry. Approved for use with potable water, drainage and gas applications. Moulders of most rubber products. Non-pipeline products-tunnel segment gaskets and rail pads. [98]

75-63360 The Vision Research Co. Ltd
2 Marples Way, Havant, Hampshire PO9 1UH
☎(+44 1705) 49 23 35
W http://www.visionresearch.com
Fax (+44 1705) 49 21 62
♣ £ 4,500,000 (1995/6) Employees 53 (1996) Area 35,000 sq ft
Managing Director: Dr Victor Chapman
Finance Director: Stephen Emery
Sales, Deputy Director: Tony Lyon-Lee
Operations Director: Alvern Ember

Manufacturing Director: Tony Wahlberg
Director of Sales: David C Walmsley
Technical Director: Warwick Silber
Purchasing Manager: Debbie Clark
■ Design & manufacture video processing equipment for CCTV security, including Video Movement Detection equipment, multi-camera recording and frame store devices, as well as video transmission products using telephone lines for remote site monitoring.
● "Parklands", Forest Road,
Denmead, Hampshire PO7 6XP
Promotions Manager: Pieter Ter-Borg 98

75-63390 Vitafoam
(a member of The British Vita Group)
Middleton, Manchester M24 2DB
☎(+44 161) 643 11 33 ☒ 667722
⊖ (+44 161) 655 31 98
♣ £ 65,000,000 (1995) ⅏ 791
Managing Director: David Campbell
Operations Director: Douglas Clayton
Marketing Manager: John Liddie
Purchasing Manager: Martin Cosgrove
■ Barrier foams for mattresses, pillows, heat and sound insulation. 98

75-63395 Vitalograph Ltd
Maids Moreton House, Buckingham MK18 1SW
☎(+44 1280) 82 28 11
E 10433.1244@compuserve.com
⊖ (+44 1280) 82 33 02
♣ £ 7,000,000 (1996) ⅏ 150 (1996)
Managing Director: Bernhardt Garbe
Sales and Marketing Director: Raymond Griffiths
International Sales and Marketing Manager: Lewis Weidman
■ Design and manufacture of medical diagnostic equipment including peakflow meters, lung function test equipment and cardio-pulmonary resuscitation equipment. 98

75-63450 Vokes Ltd
(a subsidiary of BTR plc)
Henley Park, Guildford, Surrey GU3 2AF
☎(+44 1483) 56 99 71 E environment@btrinc.com
☒ 859235 vokes g ⊖ (+44 1483) 23 53 84
♣ £ 25,000,000 ⅏ 350
Managing Director: P King
Sales/Marketing Director: P Baker
Marketing Services Manager: J J Griffiths
Product Manager: B Lodge
■ Manufacturers of fuel & lubricating oil filters for gear boxes, diesel engines, gas & steam turbines. Air filters for air intake & extract systems. Bellows expansion joints for exhaust systems & pipework installations. Oil treatment plants for transformer insulating oil & other fluids including hydraulic oils. Microfilters for the sub-micron filtration of liquids & vent gases, including pure water supply systems. Pressure filters for process gases including air & liquids. 98

75-63600 Volvo Penta UK Ltd
(a subsidiary of AB Volvo Penta, Sweden)
Otterspool Way, Watford, Hertfordshire WD2 8HW
☎(+44 923) 22 85 44 ⊖ (+44 923) 24 67 00
Sales Manager-Marine Engineering: Clive Dunsford
■ Marine propulsion, auxiliary and industrial engines 0-612 hp 0-450 kw, gearboxes, outdrives & associated equipment, steering and controls transmission systems, sterngear, propellers, exhaust, cooling & fuel systems, PTO drives instrumentation, alarms for workboats and patrol craft, mobile and stationary plant. 98

75-63610 Volvo Truck and Bus Ltd
(a subsidiary of AB Volvo, Sweden)
Wedgnock Lane, Warwick CV34 5YA
☎(+44 1926) 40 17 77 ☒ 31444
⊖ (+44 1926) 49 09 91
Managing Director: Stephen Johnson
Sales and Marketing Director: Norman McIlwraith
Manager, Government Sales: Christopher Dove
■ Wheeled tactical and logistics vehicles. 98

75-63800 Vosper Thornycroft (UK) Ltd
Head Office
Woolston, Victoria Road,
Woolston, Southampton SO19 9RR
☎(+44 1703) 44 51 44 ☒ 47682 vtwool g
⊖ (+44 1703) 42 15 39
♣ £ 238,000,000 (1995/96) ⅏ 3,000 (1996)
Chairman: Lord Wakeham
Managing Dir: Martin Jay
Managing Dir Systems Group: George W. Cameron
Managing Director Enterprises Group: Terry J Madden
Financial Dir: Chris F. Girling
Operations Dir: B.E. Wrankmore
Commercial Director: Christopher J Bentley
Managing Dir Shipbuilding Group: Dick W. Taylor
Non-Exec. Dir: Sir Robert Hill
Non-Executive Director: Norman F Baldock
Non-Executive Director: Anthony L Dorey
Non-Exec Dir: Graham Raper
Public Relations Consultant: P. Rood
Purchasing Manager: M Mardell
■ Design and construction of frigates, corvettes, strike craft, glass reinforced plastic minehunters, patrol craft, machinery controls, power distribution systems, simulators, stabilisers, bow thrusters, water jet propulsors and steering gear. Minesweeping equipment. The company also provides facilities management, maintenance management systems, spares, training, computer based training and naval base consultancy services. → 80 → 90

Support Projects Division
223 Southampton Road, Paulsgrove,
Portsmouth, Hampshire PO6 4QA
☎(+44 1705) 37 94 81 ☒ 86115 vtpsth g
⊖ (+44 1705) 38 11 24
General Manager: G Middleton
■ Comprehensive logistic support services including ILS, maintenance management systems, spares provisioning, configuration control & technical documentation.

Hydraulic Power Unit
Northarbour Road, Cosham,
Portsmouth, Hampshire PO6 3TL
☎(+44 1705) 38 33 11 ☒ 86869 vthypo g
⊖ (+44 1705) 32 51 33
Dir & General Mgr, HPU: J. Blott
■ Comprehensive range of non-retractable fin stabilisers for all naval surface ships; steering gears for OPVs, corvettes, frigates and destroyers, retractable and fixed bow thrusters, water jets for workboats and marine ride control systems. Water jet units.

VT Controls
Cow Lane, Wymering, Portsmouth, Hampshire PO6 3TR
☎(+44 1705) 37 35 11 ☒ 86528 vtprod g
⊖ (+44 1705) 32 68 85
Director and General Manager: R Truscott
Sales Manager: N Hogg
■ Advanced digital machinery control & surveillance systems for naval ships. Fully interactive real time simulators for machinery control, operator & maintainer training, power distribution systems. Versatile console systems. Design studies & ship performance simulations. Ship control & manoeuvring systems for minehunter. Integrated bridge systems & electrical propulsion packages. Electrical power management, systems for industry & aviation. Industrial electronic & PLCs. Customers include power generation & offshore industries. 98

75-64000 Wabco Automotive UK Ltd
(a division of American Standard Companies Inc, USA)
Group Sales Dept.
PO Box 6, Morley, Leeds, West Yorkshire LS27 0HQ
☎(+44 113) 251 25 10 ⊖ (+44 113) 238 16 33
♣ £ 120,000,000 (1996) ⅏ 890 (1996)
Managing Director: M C R Broughton
Operations Director: P H Scholey
Sales Director: R Thomas
Purchasing Manager: M Taylor
Process Owner Marketing: R A Waddell
■ Manufacturers of full air, air/hydraulic & full power hydraulic braking systems for all types of vehicles including light, medium & heavy CV's, bus & coach, agricultural tractors, off-road & 4 x 4's. Specialists in electronically controlled automotive systems including ABS, ASR, ECAS (suspension), speed limitation/cruise control & cabin temperature control (ATC). 98

75-64120 James Walker & Co. Ltd
Lion House, Woking, Surrey GU22 8AP
☎(+44 1483) 75 75 75 ☒ 859221 lionwk g
⊖ (+44 1483) 74 61 19; 74 61 03
♣ £ 33 000 000 ⅏ 900
Managing Dir: G. Bullock
Advertising Office: P.D. Halligan
Marketing Co-ordination Mgr: Ian Anderson
National Sales Mgr: A.H. Wootton
Asst National Sales Mgr: A.J. Cowell
■ Manufacturers of precision-moulded seals, profiled-extrusions & press-cured sheets in a variety of elastomers, conductive elastomers, reinforced elastomers and elastomer proofed fabrics servicing the requirements of a multitude of industries including defence & aerospace to marine, steel & electronics. 97

75-64150 Wallis Autogyros
Reymerston Hall, Norfolk NR9 4QY
☎(+44 1362) 85 04 18 ⊖ (+44 1362) 85 04 18
545 sq m
Proprietor: Wg Cdr K H Wallis
■ Consultancy, design, construction, test & operation of light autogyros for military, naval, police & other working roles. Manufacture of all components for the Wallis Aircraft. → 80 98

75-64450 Wardle Storeys plc
Brantham Works, Brantham,
Nr Manningtree, Essex CO11 1NJ
☎(+44 1206) 39 24 01 ☒ 98201 storbm g
⊖ (+44 1206) 39 62 31
♧ £ 43,175,000 (1996) ♣ £ 110,387,000 (1996)
⅏ 1,829 (1996)
Managing Dir & CEO: Brian R. Taylor
Finance Director, Company Secretary: D J Wilman
Assistant Company Secretary: G D Hoad
■ Designs & manufactures marine life saving equipment for shipping, aviation & offshore industries. Supplies a wide range of emergency escape, special forces & weapon delivery parachutes. Manufacture aircraft recovery equipment, specialised plastic sheet and related products in the industrial, commercial and automotive fields. Plastic sheet coatings for metalwork.

Subs. with defence business:
GQ Parachutes Ltd → 75
RFD Ltd → 75

Wardle Storeys (Safety & Survival Equipment) Ltd GQ Parachutes Ltd
Isfryn Industrial Estate, Blackmill,
Bridgend, Mid Glamorgan CF35 6EQ
☎(+44 1656) 84 03 00 ☒ 498667
⊖ (+44 1656) 84 03 96
♣ £ 8,099,756 ⅏ 131
Managing Dir: D. Nuwar
■ A complete range of parachutes from man & supply drop to weapon delivery chutes & associated equipment, ejection seats & aircraft braking. Towed targets, aircrew restraint & survival equipment.

Wardle Storeys (Safety & Survival Equipment) Ltd RFD
Kingsway, Dunmurry BT17 9AF
☎(+44 1232) 30 15 31 ☒ 747264
⊖ (+44 1232) 62 17 65
♣ £ 16,651,000 ⅏ 287
Managing Dir: W.S. McChesney
Sales Director: A C Harris
Technical Dir: R.E. Bell
■ Designs & manufactures marine life-saving equipment for shipping, aviation & offshore industries. 98

75-64850 Wavetek Ltd
(† Datron Instruments Ltd)
(a subsidiary of Wavetek Corp, USA)
Calibration Division
52 Hurricane Way, Norwich NR6 6JB
☎(+44 1603) 40 48 24 ⊖ (+44 1603) 48 36 70
♣ £ 50,000,000 (Datron £ 7,000,000)
⅏ 600 (Datron 130)
Managing Dir: Paul Stevenson
Sales Mgr: D.M. Walker
Sales Support Mgr: B.E. Foster
Purchasing Mgr: Ian McLachlan
■ Wavetek can provide a full range of test & measurement equipment from DC to 1.1 GHz. Calibration Div. manufactures precision bench & systems digital multimeters, multifunction calibrators, DC & LF standards. 97

75-64860 Wayne Kerr Electronics
(a subsidiary of Advance International Group Ltd)
Durban Road, Bognor Regis, West Sussex PO22 9RL
☎(+44 1243) 82 58 11 E sales@wayne-kerr.co.uk
W http://www.wayne-kerr.co.uk ⊖ (+44 1243) 82 46 98
♣ £ 10,000,000 (1996) ⅏ 160 (1997)
Managing Director: Paul Parshley
General Manager: Rachel Maulin
Marketing Director: Ian Hunt
Product Specialist: A Furr

Marketing Manager: C M Sefton
■ NATO codified test & measuring instruments & stabilised power supplies. High voltage power supplies, ATE systems, fiber optics. Suppliers to Ministry of Defence (UK) and to Department of Defense (USA). 98

75-64880 Weald Electronics Ltd (WE)
Slinfold Lodge, Horsham, West Sussex RH13 7RN
☎ (+44 1403) 79 07 15 ⊖ (+44 1403) 79 07 34
♧ £ 2,000 ♣ £ 1,500,000 ꗳ 40 ⌂ 10,266 sq ft
Director: Ms J.A. Lane
Director: Mrs R.J. Lane
Director: F.C. Lane ☎ (+44 1403) 79 06 61
■ Design and manufacture LF electrical connectors suitable for commercial and military applications. UK MoD, British Standard and CECC approved. All UK sales are via F.C. Lane Electronics Ltd.
® UK: F.C. Lane Electronics Ltd
Agents:
Infratron (Germany) 98

75-64950 Webley & Scott Ltd
(a subs. of Harris & Sheldon Group Ltd)
Frankley Industrial Park, Tay Rd, Rubery, Rednal, Birmingham B45 OPA
☎ (+44 121) 453 18 64 ⊖ (+44 121) 457 78 46
ꗳ 50 ⌂ 2 800 sq m
Managing Dir: A.L. Hall
■ Manufacture of anti-riot guns & signal pistols. 97

75-65100 Weir Pumps Ltd
(a subsidiary of The Weir Group plc)
149 Newlands Road, Cathcart, Glasgow G44 4EX
☎ (+44 141) 637 71 41 ⊠ 77161
⊖ (+44 141) 637 73 58
♧ £ 7,500,000 (1995) ♣ £ 150,000,000 (1995)
ꗳ 1,900 (1996) ⌂ 216,000 m2
Managing Director: John F Hood
Sales Dir: Graham Bibby
Operations Director: J K Frew
Sales Manager Naval Equipment: I M Topping
Purchasing Manager: M B Leiper
Commercial Manager: John W Anderson
■ Centrifugal pumps & associated equipment for ships (including warships) & other industrial applications. Steam turbines & electric motors, including submersible motors. 98

75-65120 Wellman Furnaces Ltd
Cornwall Rd, Smethwick, W.Mids. B66 2LB
☎ (+44 121) 558 31 51 ⊠ 337598
⊖ (+44 121) 558 17 03
♣ £ 9 000 000 ꗳ 90
Managing Dir: R. Hancox
Sales & Marketing Dir: Michael D. Bullen
Market Development Mgr: D. Lupton
Purchasing Mgr: M. Hill
Project Mgr: C.L. Winters
■ Heat treatment furnaces for aerospace industries, dockyards (naval) and ammunition factories. 97

75-65130 Wellman Garage Equipment
(† FKN Bradbury)
Bradbury Drive, Springwood Industrial Estate, Braintree, Essex CM7 2ET
☎ (+44 1376) 55 52 00 ⊠ 98361 fkibrd g
⊖ (+44 1376) 55 00 90
♣ £ 150,000,000 (1997) ꗳ 310 (1997) ⌂ 15,000 m2
Director and General Manager: Chris Yates
Managing Director: Barry Jameson
Operations Director: Derrick Wood
Purchasing Manager: David Stroud
■ Manufacture hydraulic jacks and vehicle lifts (Approved by UK Ministry of Defence to AQAP 4: Edition 2). 98

75-65135 Weltonhurst Ltd
32-38 Blackburn Road, Darwen, Lancashire BB3 1QJ
☎ (+44 1254) 70 21 24 ⊖ (+44 1254) 87 33 54
♧ £ 2,100,000 (1997) ♣ £ 8,000,000 (1997)
ꗳ 140 (1997) ⌂ 5,000 sq m
Managing Director: Paul A Burrows
Operations Director: P Wareing
Purchasing Manager: S Wood
■ Plastics blow-moulding specialists: jerrycans (plastic containers for water, petrol & diesel oil.) 98

75-65145 Welwyn Components Ltd
(a subs. of TT Group plc)
Bedlington, Northumberland NE22 7AA
☎ (+44 1670) 82 21 81
⊖ (+44 1670) 82 94 65; 82 74 34
♣ £ 35 000 000 (1995) ꗳ 670
CEO: T. Blackshaw
Managing Dir: John Tallack
Sales & Marketing Dir: A. Jude
Operations/Technical Dir: Brian Rutledge
Purchasing Mgr: Steven Ward
Sales Mgr: K. Adamson
■ Manufacture of electronic components - resistors, microcircuits & hybrids. 97

75-65300 Wesley Group Ltd
Head Office
Wesley House, Wortley Moor Lane, Leeds LS12 4HT
☎ (+44 113) 279 00 15 ⊠ 556386
⊖ (+44 113) 263 65 61
♧ £ 500,000 (1997) ♣ £ 3,000,000 (1997)
ꗳ 50 (1997) ⌂ 20,000 sq ft
Sales and Marketing Dir.: M.R. West
Production Dir.: E. Bulley
■ High pressure jet washing equipment for cleaning of plant and equipment and for de-icing & decontamination of equipment. 98

75-65330 Wessex Advanced Switching Products Ltd
(a subsidiary of Bowthorpe plc)
Southmoor Lane, Havant, Hampshire PO9 1JW
☎ (+44 1705) 45 37 11 ⊖ (+44 1705) 47 39 18
Managing Director: R Middleton
■ Design and manufacture of rotary switches, keypads, membrane switches and touch screens. 98

75-65333 Wessex Electronics Ltd
114-116 North Street, Downend, Bristol BS16 5SE
☎ (+44 117) 957 14 04 ⊠ 449160 wessex g
⊖ (+44 117) 957 38 43
♣ £ 2,800,000 (1994) ꗳ 51 (1995) ⌂ 6,500 m2
Managing Dir: Philip Westwood
Sales Mgr: M. Butt
■ Design & manufacture of RF & microwave amplifiers (small signal & power, narrow & broadband). Also sub-systems (design capability includes related components parts). BS EN ISO 9001. Company has the expertise to supply RF & microwave amplifiers & sub-systems for most ground-based communications, radar & other related items. Also has the expertise to design sub-systems for spaceborne applications (but not to test) to meet space qualified requirements.
Product line includes:-
RF braodband power amplifiers, microwave amplifiers LNA to HPA, high power pulse amplifiers, up/down converters, broadband microwave phase shifters, passive components, TWT dirvers with muting & switching, "active" combiners & splitters, microwave swept/CW sources. → 90
® France: Socitec
USA: California Instruments 98

75-65370 Western Trailers
22 The Green, Stoke Gifford, Bristol BS12 6PD
☎ (+44 117) 969 42 73 ⊖ (+44 117) 931 18 28
♧ £ 400 000 ♣ £ 800 000 (1995) ꗳ 12 ⌂ 20 000 sq m
Managing Director: Roger Davies
Operations Director: Neal Davies
Sales and Marketing Director: Ralph Davies
Purchasing Manager: John Davies
■ Design and manufacture of tanker trailers. 98

75-65410 Westair Ltd
(a subsidiary of Westair (Holdings) Ltd)
Central Ave, W. Molesey, Surrey KT8 2QZ
☎ (+44 181) 941 41 84 ⊠ 263943
⊖ (+44 181) 979 58 82
♧ £ 1,000,000 (1994/95) ♣ £ 3,000,000 (1994/95)
ꗳ 43
Managing Dir: J.A. Luck
Technical Dir: Brian Barton
Commercial Mgr: C. Stemp
■ Low & medium voltage power distribution systems for naval dockyards, airfields & hangars. Portable field distribution equipment. Lighting for military vehicles and ships. Flameproof lights for use in hazardous areas. Electrical & electro-mechanical equipment for fighting vehicles. → 80 97

75-65450 Westford Plastics & Engineering Ltd
Westford, Wellington, Somerset TA21 0DU
☎ (+44 1823) 66 23 77 ⊖ (+44 1823) 66 32 38
♣ £ 500,000 (1995) ꗳ 18
Director: R.A. Selby
Sales Mgr: A.C. Morris
Purchasing Mgr: A. Gordon
■ Machining, moulding & fabricating in plastics. 98

75-65680 Westley Ltd
(a subsidiary of Westley Engineering Group plc)
Doulton Road, Cradley Heath, PO Box 1, Warley, West Midlands B64 5QS
☎ (+44 1384) 41 01 11 🄴 bronze@westley.co.uk
⊠ 338548 wesfab g ⊖ (+44 1384) 41 01 16
♣ £ 6,250,000 (1995/96) ꗳ 125
Managing Director: T P Westley
Sales Director: A M Steventon
Sales Engineer: W Miles
Telephone Sales: R Myles
■ Copper based castings, pattern making and machining, marine components, sonar equipment housings, propeller shaft bearings. Casting capacity up to 10 tons, with specialisation in aluminium bronze. All manufacturing facilities approved by LRQA to BS5750 (1987) part 2, and MoD approved to AQAP-4. 98

75-65695 Westwind Air Bearings Limited
(a Cobham plc company)
Holton Road, Holton Heath, Poole, Dorset BH16 6LN
☎ (+44 1202) 62 72 00 ⊖ (+44 1202) 62 72 02
ꗳ 450 ⌂ 10 acre
Chairman: G F Page
Managing Director: K S Surbuts
Director: A G Irwin
Technical Director: C Pavey
Financial Director, Company Secretary: Mrs M J Vincent
■ Designs and manufactures air bearing spindle systems over 90% of which are exported. Much of the company's product is customer- specific and major industries supplied are: Electronics (printed circuit board drilling, routing and semi-conductor production), Automotive (coating systems) and Pre-press (image setting equipment). 98

75-65700 Westwire Harnessing Limited
Headlands Trading Estate, Swindon, Wiltshire SN2 6JQ
☎ (+44 1793) 53 72 17 ⊖ (+44 1793) 42 10 39
♣ £ 1 000 000 ꗳ 20
Managing Dir: Geoff Kennington
Sales Dir: Trevor Humphrey
■ Manufacture of defence standard electrical harnesses, looms, cable assemblies & electrical boxes. All supplied tested on automatic test equipment. 97

75-65750 W.H. Barker & Sons Engineers Ltd
Etna Works, Duke Street, Fenton, Stoke-on-Trent ST4 3NS
☎ (+44 1782) 31 92 64 ⊖ (+44 1782) 59 97 24
Fencing Sales Mgr: M. Tonks
■ Manufacture & supply of high security palisade systems. 97

75-65798 Whipp & Bourne
(a div. of FKI Engineering plc)
Switchgear Works, Castleton, Rochdale OL11 2SS
☎ (+44 1706) 320 51 ⊖ (+44 1706) 34 58 96
ꗳ 269 ⌂ 12 474 m2
Managing Dir: Brian Bullock
Engineering Dir: S.E. Lane
Service Division Dir: M.E. Leach
Manufacturing Dir: M. Paylor
Financial Dir: T.J. Smith
Sales & Marketing Dir: L.C. Toma
Military Sales Mgr: S.J. Short
■ Design & manufacture switchgear. 97

75-65800 Whippendell Electrical Ltd
Whippendell Road, Watford, Hertfordshire WD1 7PU
☎ (+44 1923) 22 82 01 ⊖ (+44 1923) 22 80 07
♧ £ 4,140 (1996) ♣ £ 758,000 (1996) ꗳ 20 (1996)
⌂ 10,000 m2 (1996)
Chairman: B.W.H. Swabey
Managing Director: D G Swabey
Divisional Manager: B R Human
Sales Mgr Defence: A. Hopkins
Production Mgr: B. Spreadbury

Contact Reference

■ Automatic motor starters, starter components, switchboards and ancillary apparatus as supplied to the Royal Navy. 98

75-65857 White Cross Rubber Products Ltd
White Cross, Lancaster LA1 4XQ
☎ (+44 1524) 625 55 ⊜ (+44 1524) 84 32 66
♣ £ 2,500,000 (1995/96) 👥 34 (1996) 🏭 55,000 sq ft
Chief Exec.: V. Williams
Managing Dir: R. Harper
Sales & Marketing Exec.: R. Mackrell
■ Sheeting & coated fabrics for defence use in NBC field shelters, NBC protective clothing, tentage, protective covers & packs; flexible tanks for military vehicles & aircraft. 98

75-65890 C.Whitehead & Co. Ltd
8 Barnfield Road, Park Farm Trading Estate, Folkestone, Kent CT19 5DT
☎ (+44 1303) 85 13 11 ⊜ (+44 1303) 85 09 07
♣ £ 3,000,000 (1996) 👥 125 (1996)
Managing Director: I G Meredith
Sales/Marketing Director: Ms A M Meredith
Purchasing Manager: R L Thomason
■ All types of personal load-carrying equipment in nylon webbing. 98

75-65950 White's Electronics (UK) Ltd
(a subsidiary of White's Electronics Inc, USA)
13 Harbour Road, Inverness IV1 1SY
☎ (+44 1463) 22 34 56 E sales@whelects.demon.co.uk
W http://www.whites.co.uk ⊜ (+44 1463) 22 40 48
⬦ £ 150,000 (1996) ♣ £ 1,750,000 (1996)
👥 20 (1996) 🏭 1,700 m2
General Manager: G A McRae
■ Military mine detectors. 98

75-66240 J&D Wilkie Ltd
Gairie Works, Kirriemuir, Angus DD8 4BL
☎ (+44 1575) 57 25 02 ⊜ (+44 1575) 57 45 64
⬦ £ 402 000 ♣ £ 18 000 000 (1995) 👥 200
🏭 250 000 sq ft
Managing Dir: R.M. Rowan
Sales Dir: R.W. Low
■ Industrial textile manufacturers supplying thermal, IRR & urban camouflage to MoD. Also CARM fabric & a range of packaging materials plus hardened collective protection fabrics. 97

75-66250 Wilkinson Sword Ltd
Sword Division
19-21 Brunel Road, London W3 7UH
☎ (+44 181) 749 10 61 ⊠ 897067 swords g
⊜ (+44 181) 749 23 04
♣ £ 1,320,000 (1997) 👥 20 (1997) 🏭 3,350 m2
General Manager: Malcolm Ordever
Purchasing Manager: Mike Medlin
■ Swords, scabbards, fighting knives, survival knives & associated products. 98

75-66285 Williams Fairey Engineering Ltd
(† Fairey Engineering Ltd)
(a subs. of Williams Holdings plc)
Crossley Road, Heaton Chapel, PO Box 41, Stockport, Cheshire SK4 5BD
☎ (+44 161) 432 02 81 ⊠ 667866 fairen g
⊜ (+44 161) 431 35 75
♣ £ 36 000 000 👥 120 🏭 13 acres
Managing Dir: Lindsey Thomas
■ Tactical modular military bridging system (Medium Girder Bridge); manportable components, rapidly built by hand. Single span, multispan, floating, ferry. High integrity aluminium alloy fabrications. 98

75-66310 Francis B. Willmott Ltd
Forward Works, River Street, Birmingham B5 5SB
☎ (+44 121) 772 09 07 ⊜ (+44 121) 773 36 21
⬦ £ 1,600,000 (1997) ♣ £ 1,500,000 (1997)
👥 30 (1997)
Managing Director: J G Andrews
Production Manager: Graham R Rowley
Purchasing Manager: D Booth
■ Suspension torsion bars, torsion springs and starter gear rings, flywheels and flexplates. 98

75-66450 Wilson & Garden Ltd
17-21 Newtown Street, Kilsyth, Glasgow G65 0JX
☎ (+44 1236) 82 32 91 ✉ UNIQUE Kilsyth
E wilsongar@aol.com W http://www.besanet.org.uk
⊜ (+44 1236) 82 56 83
⬦ £ 6,900 (1996) ♣ £ 3,744,000 (1996) 👥 120
🏭 100,000 sq ft
Managing Dir: J.M. Cooper
Sales Mgr: R.A. Cooper
Operation/Technical Mgr: N.A. McConnachie
Purchasing Mgr: M. McGuire
■ Educational equipment, revolving-surface blackboards, white boards, notice boards, pin boards, cine screens & accessories. Information boards & flipcharts. 98

75-66600 Winkelmann (UK) Ltd
(† Winkelmann Security Systems Ltd)
39 Star Street, London W2 1QB
☎ (+44 171) 706 79 60 ⊜ (+44 171) 402 70 04
♣ £ 1,000,000 👥 5
Managing Director: Stephen A Read
Purchasing Manager: Ms Claire Walker
■ Manufacturer of offensive & defensive countersurveillance equipment for police & military operations. 98

75-66750 Wolsey Electronics Ltd
(a subsidiary of TT Group plc)
c/o AB Microelectronics Ltd, Dinas Isaf Estate, Edmondstown, Tonypandy, Mid-Glam CF40 1NX
☎ (+44 443) 43 07 17 ⊠ 498405 ⊜ (+44 433) 43 13 83
Managing Director: M Eccleshall
■ Signal distribution equipment, alarm systems. 98

75-66760 Woodbridge Engineering Co Ltd
Red House Works, Red House Farm, Ashbocking, Suffolk IP6 9LD
☎ (+44 1473) 78 57 24 ⊜ (+44 1473) 78 58 07
♣ £ 2 000 000 👥 50 🏭 10 000 sq ft
Managing Dir: J. Bloomfield
Operations & Technical Dir: M. Bloomfield
Sales & Marketing Dir: S. Elliot
Purchasing Mgr: P. Davis
■ Produce a range of special & standard tools for fighting vehicles, also supply parts for the audio industry for use on public address & loudspeaker systems for the military, oil industry & domestic markets.

Leiston Office
Masterlord Ind Estate, Leiston, Suffolk
☎ (+44 728) 83 19 23 ⊠ 987139 mtsg 97

75-66790 Wood & Douglas
Lattice House, Baughurst, Basingstoke, Hants RG26 5LL
☎ (+44 118) 981 14 44 ⊜ (+44 118) 981 15 67
♣ £ 2 500 000 👥 26
Managing Dir: Alan Wood
Operations & Technical Dir: Andy Stickland
Sales & Marketing Mgr: Chris Young
Purchasing Mgr: Shaun Dainton
■ Design & manufacture of video scramblers & miniature radio links & video links. 97

75-66870 Woods of Colchester Ltd
(incorporating Keith Blackman; a subsidiary of The General Electric Co plc)
Tufnell Way, Colchester, Essex CO4 5AR
☎ (+44 1206) 54 41 22 ⊠ 98422 woods g
⊜ (+44 1206) 57 44 34
♣ £ 50,000,000 (1996) 👥 750 (1996)
Managing Director: D J Priest
Sales Director: H E Nicoll
Technical Dir: W.T.W. Cory
Works Dir: W. Dearnley
Purchasing Mgr: N. Kidman
Sales Mgr MoD Systems: P.W. Beston
Business Manager MoD: P M Greenwood
■ Fans for fighting ships (centrifugal & axial for surface ships, nuclear-powered & conventional submarines). Mixed flow fans in aluminium & stainless steel for low magnetic signature vessels - ie SRMH's. Air treatment units, UC range of coolers, NBC filtration fans & battery agitation blowers. Approved MoD testing house. BS 5750 PT1 (150-9001) - 1987. EN 29001. DEF STAN 0591. (Formaly AQAP-1). Full maintenance & repair facility available. 98

75-66900 Woodville Polymer
(† Dowty Woodville Polymer)
(a subs. of the TI Group plc)
Hearthcote Road, Swadlincote, Derbyshire DE11 9DX
☎ (+44 1283) 22 11 22 ⊜ (+44 1283) 22 29 11
Managing Dir: Mick J. Davies
Commercial Dir: Richard Walker
■ Design & manufacture of elastomeric based components & systems. Diverse range of capabilities stealth materials, thermal & acoustic insulation materials. Design & manufacture customised inflatable & non-inflatable polymer-based fabrications for sealing & non-sealing applications. Products supplied include Slot Seals for swing wing aircraft, railway carriage gangway diaphragms, air bags for car driven protection & conformable packers. Also supplied are spread fabrics, calendered sheet for specialist applications, a range of respiratory & filter equipment & materials. The company is ISO 9001 approved.

Dowty Woodville Polymer Fabrications
Hearthcote Road, Swadlincote, Derbyshire DE11 9DX
☎ (+44 1283) 22 11 22 ⊜ (+44 1283) 22 29 11
Dir & General Mgr: Kevin Fisher
■ Designs & manufactures customised inflatable & non-inflatable fabrications for sealing & non-sealing applications. Products supplied include slot seals for swing wing aircraft, railway carriage gangway diaphragms, air bags for car driver protection & conformable packers. Also supplied are spread fabrics, calendered sheet for specialist applications, a range of respiratory & filter equipment.

Woodville Environmental & Safety Products
(† Dowty Environmental & Safety Products)
(a member of the TI Group plc)
Hearthcote Rd, Swadlincote, Derbyshire DE11 9DX
☎ (+44 1283) 22 11 22 ⊜ (+44 1283) 22 29 11
Sales Mgr: Julian Jordan
■ Supplies fire smoke escape hoods & activated carbon fabrics - treated. These fabrics can be used in IR decoy devices & for NBC protection, casualty evacuation bags & hoods. 97

75-67000 Worcester Controls (UK) Ltd
(a subsidiary of BTR plc)
Military Sales Division
Burrell Road, Haywards Heath, West Sussex RH16 1TL
☎ (+44 1444) 41 41 33
E sales@worcestercontrols.co.uk
W http://www.worcestercontrols.co.uk ⊠ 87189 wovco g
⊜ (+44 1444) 45 94 68
👥 300 (1997)
Marketing Mgr: R.A. Bennett
MoD/Contractor Sales Mgr: Thomas Bourn
■ A wide range of ball valves & actuators 98

75-67080 A.J. Worthington (Leek)
Portland Mills, Queen Street, Leek, Staffordshire ST13 6LW
☎ (+44 1538) 39 96 00 ✉ WORTHINGTON Leek
⊜ (+44 1538) 37 19 68
👥 140 (1997)
Managing Dir: B.J. Chesworth
Director: A. Bourne
■ Braided cordage for parachute rigging lines used in the production of parachute escape systems and supply dropping assemblies; braided cords, tubular tapes and sleeving for safety equipment, protective clothing and tents made from nylon, polyester, polypropylene, Kevlar, Twaron, Nomex Dyneema and Spectra fibres. 98

75-67150 Robert Wright & Son (Coachworks) Ltd
Head Office
Calgorm Industrial Estate, Ballymena, N. Ireland BT42 1PY
☎ (+44 266) 412 12 ⊜ (+44 266) 497 03
♣ £ 11 000 000 👥 300 (1996) 🏭 8 489 m2 (1996)
Managing Dir: Jeff Wright
Sales Dir: Jack Kernohan
Design Dir: Trevor Erskine
Material Dir: Ian Kerr
Marketing Mgr: Aveyline Finlay
■ Aluminium bodies from 7-12 metres suitable for PSV. Manufacture of passenger carrying buses & coaches. 97

75-67280 Wyman-Gordon Ltd
(† Cameron Forged Products)
Houstoun Road, Livingston, West Lothian EH54 5BZ
☎ (+44 1506) 44 62 00 ⊜ (+44 1506) 44 63 00

♣ £ 35,000,000 (1996) 👥 290 (1996)
Managing Director: D G Pattle ☎ (+44 1506) 44 64 11
Operations Director: R Meikle
Plant Director: C R Lake ☎ (+44 1506) 44 64 62
Technical Director: F J Logan ☎ (+44 1506) 44 64 51
Purchasing Manager: D Weir
Account Representative: A S Turner
☎ (+44 1506) 44 62 64 [98]

75-67300 Wynn Marine Ltd
(a subsidiary of Hoplite Ltd)
Wynn House Lansdown Estate,
Cheltenham, Gloucestershire GL51 8PL
☎ (+44 1242) 23 22 66 ✉ WYNN
⊖ (+44 1242) 23 11 31
♧ £ 120,000 (1996) ♣ £ 3,500,000 (1996)
👥 47 (1996) ⌂ 2,500 sq m
Managing Director: Derek Poulton
Sales and Marketing Director: Tony Parker
Purchasing Manager: Ms Vanda Elderidge
■ Manufacture marine windscreen wiper equipment, temperature controllers for heater bridge windows. [98]

75-67400 Xelflex
Euxton Mill, Dawbers Lane, Euxton, Lancashire PR7 6EB
☎ (+44 1257) 26 99 33 ⊖ (+44 1257) 24 13 61
♣ £ 2,500,000 (1994/95) 👥 65 ⌂ 3,000 sq m
Managing Director: W Trendle
Commercial Director: J A Marson
Operations Manager: G Dyer
Technical Manager: R C Martland
■ Rubber, full compounding, moulding, tool manufacture. Calendering in all types of polymers including silicones & flurocarbon rubbers. Approved by CAA, MoD & BS EN ISO9002. [98]

75-67500 Yarrow Shipbuilders Ltd
(a subsidiary of GEC plc)
South Street, Scotstoun, Glasgow G14 0XN
☎ (+44 141) 959 12 07 ⊖ (+44 141) 958 06 42
♣ £ 200,000,000 (1995/96) 👥 2,000 (1997)
Director General Manager: W Semple
Technical Director: N Grant
Commercial Manager: J Feely
Purchasing Manager: I Atkinson
Marketing Manager: D Holdsworth
■ Total surface ship capability including design, build, lead yard & support services, which can meet the requirements of any modern navy. The company employs a strong multidisiplinary team in ship design & weapon system integration, making extensive use of CAD technology. The combination of modular building techniques & the application of advanced technology in ship design & signature reduction, results in high performance warships.
Yarrow Shipbuilders, NNC & VSEL are wholly owned subsidiaries of GEC plc & together they form GEC Marine, a new organisation dedicated to world-class shipbuilding, armaments, nuclear engineering & project management. [98]

75-67520 Yewlands Engineering Co. Ltd
(a subs. of Quadrant Group plc)
Fowler Rd, Hainault Ind. Estate, Hainault, Essex IG6 3XB
☎ (+44 181) 500 19 33 ⊖ (+44 181) 500 95 53
♣ £ 6 000 000 (1995) 👥 144 ⌂ 8 500 sq m
Managing Dir: Ken Reed
Operations Dir: Ronnie Young
Sales & Marketing Dir: Roy Fletcher
Purchasing Mgr: Mrs J. Page
Overseas Business Representative: L.R. Fletcher
■ Close tolerance manufacture of sheet metal fabrications & assemblies in many specialised materials as required by the MoD & aerospace industry, machined components & high pressure pipes. → 80 [97]

75-67560 Yuasa Automotive Batteries (Europe) Limited
(† Lucas Yuasa Batteries Ltd)
(a subsidiary of Yuasa Corporation, Japan)
Formans Road, Sparkhill, Birmingham B11 3DA
☎ (+44 121) 777 32 92 ⊖ (+44 121) 777 77 87
♣ £ 40,000,000 (1996) 👥 575 (1996)
Managing Director: C J Williamson
Sales and Marketing Director: G Trevitt
Engineering and OE Sales Manager: K J B McEwan
Purchasing Manager: S Stone
■ Lead acid automotive batteries. [98]

75-67600 Zellweger Analytics Ltd
(† Sieger Ltd)
Hatch Pond Hse, 4 Stinsford Rd, Nuffield Estate,
Poole, Dorset BH17 0RZ
☎ (+44 1202) 67 61 61 ✉ GASALARM ℡ 41138
⊖ (+44 1202) 67 80 11
👥 360 ⌂ 105 000 sq ft
Managing Dir: E. Bradley
Operations Dir: R. Van-Ewyk
Sales & Marketing Dir: J. Trautwein
Communications Mgr: J.I. Markley
Purchasing Mgr: Alan Trickett
■ Flammable & toxic gas detection equipment. [97]

75-67680 Zephyr Flags & Banners
Midland Road, Thrapston, Northamptonshire NN14 4LX
☎ (+44 1832) 73 44 84
E 106016.3121@compuserve.com
⊖ (+44 1832) 73 30 64
♧ £ 28,800 (1996) ♣ £ 2,000,000 (1995) 👥 45 (1997)
⌂ 5,500 m2
Managing Director: Christopher Skey
Operations Director: Bryan Skey
Sales and Marketing Manager: Simeon Skey
Purchasing Manager: Mrs S Garner
■ National flags, government & military flags and standards, for external or internal display. Flagstaffs in glass fibre (GRP) & aluminium. [98]

75-67700 Zero Cases (Europe)
Samuel Groves & Co. Ltd
(a subsidiary of Zero Corp. USA)
Norton St, Hockley, Birmingham B18 5RQ
☎ (+44 121) 554 20 01 ⊖ (+44 121) 523 29 24
⌂ 10,000 m2
Managing Director: Tony Joynson
Marketing Manager: Peter Whitbread
■ Manufacturer of cases and enclosures from deep drawn aluminium and fabricated aluminium to military specification. [98]

75-67720 ZF Great Britain Ltd
(† ZF Gears (GB) Ltd)
(a subsidiary of ZF Friedrichshafen AG, Germany)
Abbeyfield Road, Lenton, Nottinghamshire NG7 2SX
☎ (+44 115) 986 92 11 ⊖ (+44 115) 986 92 61
♧ £ 4,000,000 (1995) ♣ £ 17,429,596 (1996)
👥 130 (1996) ⌂ 4,753 m2
Managing Dir: J.G. Poore
Operations & Technical Dir: M. Brierley
☎ (+44 115) 935 54 07
Financial Controller: J.P. Lawrence
☎ (+44 115) 935 54 36
Account Mgr: G.J. Truckel ☎ (+44 115) 935 54 55
■ Manufacturer of gearboxes & transmission systems for high speed marine craft & for military vehicles.
® Germany: ZF Friedrichshafen AG [98]

75-67800 Zuken-Redac Ltd
Green Lane, Newtown Industrial Estate,
Tewkesbury, Glos. GL20 8HE
☎ (+44 1684) 29 41 61 ℡ 43108
⊖ (+44 1684) 29 97 54
♣ £ 8 600 000 (1994) 👥 150
Managing Dir: K. Yamamoto
Dir Corporate Marketing Div.: G. Gruel
Sr Marketing Communications Officer: Miss S. Graves
Publicity Mgr: Mrs. C. Hage
Marketing Communications Co-ordinator: Mrs B. Rowe
■ Specialists in EDA systems for the electronics industry. Advanced software systems, CADSTAR & VISULA, for use on industry standard computers ranging from IBM compatible machines to sophisticated engineering workstations. [97]

75-67900 Zwicky Engineering Ltd
(a subs. of EIS Group plc)
Molly Millars Lane, Wokingham, Berks RG41 2RY
☎ (+44 1734) 77 13 31 ℡ 848852 zwicky g
⊖ (+44 1734) 77 48 56
♣ £ 5 500 000 (1995) 👥 60
Managing Dir: Antony Wise
Sales Dir: Graham Cornwall
Manufacturing Mgr: Bob Wicks
Sales Mgr: Ray Smith
Sales Engineer: Bryan Lawrence
■ Skyhi & Mingays hydraulic tripod/pillar jacks for all types of civil & military aircraft including helicopters. Aircraft towbars, washing/de-icing vehicles/trailers & access booms/platforms. Equipment & components tailored to suit customer requirements. Ancilliary products include Zwicky filters & Viking pumps. 97

80 Service Companies & Organizations

Contact Reference

United Kingdom

80-00200 Aardvark Training Schools Ltd
Shevock Farm, Insch, Aberdeenshire
☎ (+44 1464) 82 01 22 ✉ 73509 aard g
⊖ (+44 1464) 82 09 85
Managing Dir: David Sadler
■ Courses in minefield countermeasures. [98]

80-00420 Advanced System Architectures Ltd (ASA)
North Block, Bentley Hall, Blacknest,
Alton, Hampshire GU34 4PU
☎ (+44 1420) 238 15 [E] sales@asa.co.uk
⊖ (+44 1420) 237 41
♣ £ 500,000 (1997) ♣ £ 2,500,000 (1997)
👥 50 (1997) 🏭 7,000 m2
Managing Director: Christopher Williams
Technical Director: Stephen Rhodes
Projects Director: David James
■ Multi-sensor, multi-target tracking systems. Radio coverage planning tools for Ground Forces. Secure Voice Gateways and multi-party teleconferencing systems. → 75 [98]

80-00450 Advisory, Consulting & Technical Services
2 High Chimneys, Windlesham, Surrey GU20 6LT
☎ (+44 1344) 224 02 ⊖ (+44 1483) 72 83 03
👥 3
Director: Alan Garratt
■ Engineering consultancy for international aerospace & defence contracts. Specialise in overhaul, repair & maintenance for commercial and military customers in the UK, Europe and USA. Advisory services for airline, airport development for EC, Tacis & Phare project consultancy. [98]

80-00480 AEA Technology plc
A1 Harwell, Didcot, Oxon OX11 0RA
☎ (+44 1235) 43 55 55 [E] enquiry@aeat.co.uk
[W] http://www. aeat.co.uk ⊖ (+44 1235) 43 28 59
♣ £ 257,300,000 (1994/95) 👥 3,380
Chairman: Sir Anthony Cleaver
Chief Exec.: Dr Peter Watson
Marketing & Sales Dir: Mike Watson
Business Development Dir Defence: David Harris
☎ (+44 1235) 43 30 95 (+44 1235) 43 20 64 (fax)
Business Development Mgr Defence Sea: Tim Langdon
☎ (+44 1235) 43 30 76 (+44 1235) 43 20 64 (fax)
Business Development Mgr Defence Land: Michael Gill
☎ (+44 1235) 43 37 90 (+44 1235) 43 20 64 (fax)
Business Development Mgr Defence Aviation: Al Hayes
☎ (+44 1235) 43 28 46
Business Development Mgr Defence Nuclear:
Glyn Stanton ☎ (+44 1235) 43 22 91
(+44 1235) 43 20 64 (fax)
■ An international science & engineering services business which solves technical, safety & environmental problems for industries & governments around the world. Offering solutions based on know-how built up over 40 years of leading edge science & engineering, first in the nuclear industry, but increasingly for a wide range of other industries such as manufacturing, defence, transport, oil & gas, power generation, generation, chemicals, water & healthcare.
UK sites: Harwell, Culham, Risley, Winfrith, Dorset, Dounreay, Caithness.
14 overseas offices. [97]

80-00520 Aerobel Defence Technology Ltd
356 West Barnes Lane, PO Box 90D,
New Malden, Surrey KT3 6JG
☎ (+44 181) 336 17 33 [E] nicholasw@aerobel.co.uk
[W] http://www.aerobel.co.uk/aerobel
⊖ (+44 181) 942 89 09
♣ US$ 2,500,000 (1995/96) 👥 9 (1997) 🏭 250 sq m
Managing Director: Nicholas Wrobel
Operations Manager: Alan Dickerson
Purchasing Manager: Janis Wrobel
■ Aerobel is a UK Government-approved defence company with quality assurance certification equivalent to ISO 9001-3. It provides realtime visual systems for design, prototyping/MMI & training. Products include: DIS-compliant reconfigurable crewstations, driving simulators, ship bridge simulators, tactics trainers, naval charting aids. [98]

80-00650 Aerospace Inspection Training
Hangar A, Norwich Airport, Norwich, Norfolk NR6 6EG
☎ (+44 1603) 42 61 65
[E] 101630.1667@compuserve.com
✉ 975384 aafnwig g ⊖ (+44 1603) 42 40 19
👥 7
Managing Dir: George V. Crowe
Operations/Technical Dir: John Manson
Centre Mgr: Ms Danielle S Crowe
■ Training in UK and on-site for aerospace NDT to worldwide recognised standards, Boeing approved to MIL-STD 410 and BAC 5953/BSS 7600 BINDT accredited. PCN test centre. [98]

80-00700 Airlie Associates
80 Roewood Lane, Macclesfield, Cheshire SK10 2PQ
☎ (+44 1625) 42 50 65 ✉ 665080
⊖ (+44 1625) 42 50 65
👥 1
Managing Director: Michael John Blamey
■ Analytical services for chemical effects. Ballistic protection fabrics. NBC protection fabrics. Functional fabric, development. [98]

80-00830 AirteQ Ltd
(† Airtech-SAV Ltd)
Media House, Presley Way, Crown Hill,
Milton Keynes MK8 0ES
☎ (+44 1908) 57 01 19 ⊖ (+44 1908) 57 01 18
♣ £ 4 000 000 (1996) 👥 52 🏭 3 750 sq m
Managing Dir: Phil Colley
Operations Dir: Dave Fairhurst
Operations Dir: K.P. Ottaway
Sales & Marketing Dir: Ken Dowling
Purchasing Mgr: Ms F. Sale
■ Provision of multi-media training & simulation systems for real-time applications such as firearms training, identification & recognition training & driver instruction. Emergency procedure & C3 training systems. Development of associated software. Development of training & marketing strategies. [97]

80-01000 Alifabs Design & Construction Ltd
(† Alifabs Ltd)
Fairoaks Airport, Hangar B1, Chobham, Surrey GU24 8HU
☎ (+44 1276) 85 65 85 [E] alifabs@btinternet.com
[W] http:// www.btinternet.com/~alifabs/alifabs.htm
⊖ (+44 1276) 85 71 77
♣ £ 1,500,000 (1997) 👥 15 (1997) 🏭 18,000 sq ft
Director: A J Pactat
Sales and Marketing Director: R Smith
■ Manufacture airport visual control rooms & air traffic control cabs. Sonarview anti-condensation ATC glazing systems. Air traffic control consoles. [98]

80-01030 AlliedSignal Aerospace Marketing Sales and Service
(a subsidiary of AlliedSignal Aerospace, USA)
Southall Lane, Unit 3A Harlequin Centre,
Southall, Middlesex UB2 5NH
☎ (+44 181) 574 10 10 ⊖ (+44 81) 571 22 34
Vice President Military Market, Europe, Middle East and Africa: Serge Buchakjian ☎ (+44 181) 843 53 05
Executive Director, Regional Airline Propulsion, United Kingdom and Europe: William Brenner
☎ (+44 181) 843 53 06
Director, Europe, Africa and Middle East:
William (Bill) Kirkham
Field Service Engineer: Richard Brand
☎ (+44 181) 571 72 82; 571 23 36 (fax)
Field Service Engineer: Ian Gilbert
Field Service Engineer: Melvyn W Goodbun
Field Service Engineer: Mike Nealon
☎ (+44 181) 571 10 10
■ Marketing and support for AlliedSignal Aerospace products. → 90
Representing:
AlliedSignal Aerospace, USA

● c/o British Airways Heathrow Airport,
North BLK, TB "D" (E-77), Middlesex, London TW6 2JA
☎ (+44 181) 562 40 08 (recording); 562 57 14
⊖ (+44 181) 759 72 03
Field Service Engineer: Adrian Thomas

● Stockport, Cheshire SK7 1QR
c/o British Aerospace PLC-AVRO
International, Post Station 40, Woodford Aerodrome
☎ (+44 161) 955 41 97 ⊖ (+44 161) 955 41 98
Field Service Engineer: Chris Carlson

● Stockport, Cheshire SK7 1QR
BAeRJ Technical Support, c/o British Aerospace (PS33)
Chester Road, Woodford
☎ (+44 161) 955 31 39 ⊖ (+44 161) 955 31 84
Field Service Engineer: Tom Tompkins
Field Service Engineer: Richard Malins

● 18 Llangorse Close,
Stourport-On-Severn, Worcs DY13 8LJ
☎ (+44 181) 574 10 10 (London)
⊖ (+44 181) 571 22 34 (London)
Area Director: Mike Nealon

● 4 Rowans Close, Farnborough, Hampshire GU14 9EJ
☎ (+44 1276) 367 87 ⊖ (+44 1276) 367 84
Field Service Engineer: John Parson

● PO Box 747, Gatwick, West Sussex RH6 0FH
c/o British Airways, 4th Floor, Hangar 6
☎ (+44 1293) 56 70 75 (recording)
⊖ (+44 1293) 56 79 90
Field Service Engineer: Jim Cowler
Field Service Engineer: Anthony Shaw
Administrator: Ms Gill Cox

● 6 Hilltop Farm, Harrowby Lane,
Grantham, Lincolnshire NG31 9HA
☎ (+44 1476) 59 28 38 ⊖ (+44 1476) 57 28 28
Field Service Engineer: Reg Marratt

● c/o Raytheon Aviation Company, Hawarden Airport,
Broughton, (North) Wales CH4 0BA
☎ (+44 1244) 52 37 66 ⊖ (+44 1244) 52 30 30
Field Service Engineer: Philip J Thacker

● Jetstream Aircraft,
Prestwick, Ayshire, Scotland KA9 2RW
☎ (+44 1292) 67 22 78 ⊖ (+44 1292) 47 42 24
Field Service Engineer: Willie Jones
Field Service Engineer: Kirk Knisley [98]

80-01040 AlliedSignal Commercial Avionics Systems
(a subsidiary of AlliedSignal Aerospace, USA)
Unit 3A Harlequin Centre, Southall Lane,
Southall, Middlesex UB2 5NH
☎ (+44 181) 571 34 22 ✉ 912071 asilon g
⊖ (+44 181) 571 23 36
Service Centre Manager: Alan Howell [98]

80-01045 AlliedSignal Ltd
(a subsidiary of AlliedSignal Aerospace, USA)
AlliedSignal Aerospace Marketing Sales and Service: Engine Repair and Overhaul
65 President Way, London Luton Airport,
Luton, Bedfordshire LU2 9NB
☎ (+44 158) 239 38 00 ⊖ (+44 158) 243 50 40
Site Leader: J S Choulet
Marketing and Sales: Jim Bryden
LF500 Turbofan Product Support: Peter Cleasby
TPE 331 Turboprop Product Support: Brady McNamara
Product Support, APU: Peter Soar

■ Repair and overhaul of LF502/507 and TPE 331 turbine aero-engines. Support for AlliedSignal APUs including test cells. → 90 [98]

80-01050 Allott & Lomax
Fairbairn House, Ashton Lane,
Sale, Manchester M33 6WP
☎ (+44 161) 962 12 14 ▣ 667859 alomax g
⊖ (+44 161) 969 51 31
♣ £ 20 000 000 (1996) 👥 700
Director: E.J. Hill
Director: J.H. Mills
Associate Dir: N.P. Hinings
Marketing Mgr: S.R. Crow
■ Consulting engineers. Advanced structural analysis including the non-linear behaviour of structures under impact & seismic loadings. Project management and advisory services. [97]

80-01060 Amtac Laboratories Ltd
(part of the Amtac Group)
Norman Road, Altrincham, Cheshire WA14 4EP
☎ (+44 161) 928 89 24 [E] admin@amtac.co.uk
⊖ (+44 161) 927 73 59
♣ £ 2,000,000 (1997) 👥 70 (1997) 🏭 30,000 sq ft
Managing Director: Dr K Albutt
Operations Director: Ian Dalgetty
Sales & Marketing Dir: Alan Kirwilliam
■ Consultants & test house (metallurgical & other materials). Quality assurance services. [98]

80-01200 Andrich International Ltd
51 Market Place, Warminster, Wiltshire BA12 9AZ
☎ (+44 1985) 84 61 81 [E] andrich@andrich.com
⊖ (+44 1985) 84 61 63
Managing Dir: Richard Curtis
Director: Peter Le Marchand
■ Conference and exhibition organisers. Marketing, business development and training services. Research, reports and studies. Disaster and emergency planning and response. High technology leisure. [98]

80-01230 Anglo American Airmotive Ltd
(a subsidiary of Agricultural & General Aviation Ltd)
Building 602, Bournemouth International Airport,
Christchurch, Dorset BH23 6NE
☎ (+44 1202) 57 97 56; 57 00 23 ▣ 41537 agatas g
⊖ (+44 1202) 57 91 65; 57 02 03
♣ £ 5,000,000 (1997) 👥 40 (1997) 🏭 2,500 m2
Managing Director: K W Norris
Commercial Director: W R Norris
Marketing Director: L J Norris
Finance Director: M Cornelius
Chief Engineer: D Munday
■ Sole distrubutors for Nw Piper Inc (UK and Ireland) aircraft sales and spare parts. Aircraft engineering, avionics (distributors for Allied Signal and dealers for Narco). Pilot training and social club (proprietors of Bournemouth Flying Club). [98]

80-01240 Anite Systems Ltd
(† Cray Systems Ltd)
Head Office
5 Genesis Business Park, Albert Drive,
Woking, Surrey GU21 5RW
☎ (+44 1483) 74 00 14 [W] http://www.anitesystems.com
⊖ (+44 1483) 74 02 31
♣ £ 80,000,000 (1996) 👥 1,200 (1996)
Chairman and Chief Executive: Michael Shone
Technical Director: Dr David Sloggett
Marketing Manager: Ms Andrea Newill
Marketing Executive: Ms Elizabeth Cooper
■ Software & consultancy for avionics & defence systems.
Registered Office
The Chess Building, Caxton Way,
Watford, Hertfordshire WD3 8XH
Space & Defence Division
Das House, Quayside, Temple Back, Bristol
☎ (+44 117) 927 78 54 ⊖ (+44 117) 929 09 17
♣ £ 10,000,000 (1994/95) 👥 110
Managing Dir: Bill Black
Sales & Marketing Mgr: P.J. Bostock
■ Software systems design & development for command & control, simulation & training. Earth observation data processing & communications. [98]

80-01251 Aonix Europe Ltd
(† Alsys Ltd)
(a subsidiary of Thomson-CSF UK)
Partridge House, Newtown Road,
Henley-on-Thames, Oxon RG9 1EN
☎ (+44 1491) 57 90 90 [W] http://www.aonix.co.uk
⊖ (+44 1491) 57 18 66
♣ £ 3,000,000 (1996) 👥 26 (1997) 🏭 8,000 sq m
Managing Director: M L Kent
Marketing Manager: B C Dodwell
Operations and Technical Director: A Larkham
■ Consultancy, special engineering, related to software intensive projects & special expertise in ADA development tools. Training in ADA & teleUSE. [98]

80-01253 A.P. Publications Ltd
(† IUSD Publishing Company)
377 St John Street, London EC1V 4LD
☎ (+44 171) 837 59 21
[E] 106142.1713@compuserve.com
[W] http://www.ap.appublications.co.uk
⊖ (+44 171) 837 11 97
Director: E. Patterson
Director: S Blackman
■ Publication devoted to the design of underwater systems. [98]

80-01254 APTEC Surface Engineering Ltd
Technology Transfer and Development Division
Southbrook Road, Gloucester GL4 3DN
☎ (+44 1452) 30 08 00 ⊖ (+44 1452) 30 00 50
Director: Colin John [98]

80-01255 Arab Defence Journal
3 Park Place, 12 Lawn Lane, Vauxhall, London SW8 1UA
☎ (+44 171) 582 22 20 ▣ 918046 contat g
⊖ (+44 171) 820 01 95
♣ US$ 900 000 👥 22
Advertising Dir: Salim Zreik
Advertising Mgr: Simon Taylor
■ Arabic subscription monthly magazine for defence technology to government & military personnel in the Middle East. A monthly Arabic defence magazine audited by the ABC in London. A source on defence matters for officers in the army, navy & airforce in each of the 16 countries of the Arab World. [98]

80-01257 Archer Mullins Limited
12 Acorn Business Centre, Northarbour Road, Cosham, Portsmouth, Hampshire PO6 3TH
☎ (+44 1705) 38 05 87 ⊖ (+44 1705) 38 08 17
👥 36
Managing Director: Roger C Dimmock
Manager (Training): John Tumilty
■ Defence Consultants and Corporate Training Practitioners. Consultancy services encompass Policy, Operations, Equipment and Bid Support for Maritime, Ground and Air Forces. Training products and services cover all aspects of training: TNA, training objectives, course design, delivery of training, customer/ field support, competence assessment systems and instruments, computer based training. ISO 9001 assured. [98]

80-01259 Artemis Management Systems Ltd
CSC Artemis Limited
(† Lucas Management Systems Ltd)
(a division of Gores Technology Group, Gores International, USA)
Slough Office
Artemis House, 219 Bath Road,
Slough, Berkshire SL1 4AA
☎ (+44 1753) 60 00 00 [E] info@artemispm.com
[W] http://www.artemispm.com ⊖ (+44 1753) 60 62 99
♣ £ 50,000,000 (1997) 👥 600 (worldwide 1997)
Managing Director: Steve Yager
Vice President, European Operations: Howard Magee
Sales Director (UK General Manager): Robert Legge
UK Marketing Manager: Ms Anne Jensen
■ Supplier of the Artemis range of project management, resource management, cost management and time tracking systems featuring relational database planning and scheduling systems, flexible report writer and sophisticated graphics capabilities. [98]

80-01260 Arthur D. Little Ltd (ADL UK)
(a subsidiary of Arthur D. Little Inc, USA)
Berkeley Square House, Berkeley Square,
London W1X 6EY
☎ (+44 171) 409 22 77 ▣ 261109
⊖ (+44 171) 491 89 83
♣ US$ 370,000,000 (Worldwide) 👥 150
Managing Director: Stephen Lawrence
Defence and Aerospace Manager: Mike Price
Manager: M J S Price
Cambridge Consultants Ltd
Science Park, Milton Road, Cambridge CB4 4DW
☎ (+44 1223) 42 00 24 ▣ 81481 ccl g
⊖ (+44 1223) 42 33 73
Managing Dir: Howard Biddle
Environmental Mgmt Unit (Europe)
Science Park, Milton Road, Cambridge CB4 4DW
☎ (+44 1223) 42 00 24 ⊖ (+44 1223) 42 00 21
Managing Dir: Roger Hill
■ Management & technology consulting. [98]

80-01500 Assessment Services
(† Siemens-Plessey Assessment Services Ltd)
(a subsidiary of Siemens Plessey Systems)
Segensworth Road, Titchfield,
Fareham, Hampshire PO15 5RH
☎ (+44 1329) 44 33 22 [E] as@as.neag.co.uk
[W] http://www.neag.co.uk ▣ 869489 plesas g
⊖ (+44 1329) 44 34 21
👥 800 (1997)
Managing Director: R W Heighington
Financial Director: R Trollope
Marketing Executive: C J Brett
Marketing Manager: Ms Kate Ryder
■ Independent, confidential test & consultancy service for industry & government. Environmental, reliability, EMC & TEMPEST testing of systems, equipment & components. Telecommunications testing to BABT. Consultancy for design & specifications of environmental, EMC, EMP & TEMPEST aspects. Test programme planning & management. Test facility specification & installation. Off-site test & trials management. NAMAS reg. No.0141, No.1377, No.0141 Si, TickIT & CAA approved, BABT accredited & BS EN ISO 9001 registered.
● Snitterfield Road, Bearley,
Stratford-upon-Avon CV37 0EX
☎ (+44 1789) 73 11 55 ⊖ (+44 1789) 73 12 64
Marketing Exec.: C.J. Brett
■ EMC test site.
● 12 Marshall Road, Hillmead,
Swindon, Wiltshire SN5 9FZ
☎ (+44 1793) 87 19 90 ⊖ (+44 1793) 87 29 90
Marketing Exec.: C.J. Brett
■ Telecommunications test site. [98]

80-01520 AST Group Ltd
(† International Dry Stripping Ltd)
(a subs. of AST Group Ltd)
Alexandra Way, Ashchurch Business Centre,
Tewkesbury, Gloucestershire GL20 8NB
☎ (+44 1684) 29 17 77 ⊖ (+44 1684) 29 87 77
♣ £ 1 200 000 (1995) 👥 24 🏭 1 115 sq m
Managing Dir: James Ryan
Operations & Technical Dir: R.E.A. Cook
Commercial Production Dir: Malcolm Noad
Sales Mgr: M.J. Shipman
■ Surface coating removal/corrosion control & re-finishing. [97]

80-01550 WS Atkins plc
Woodcote Grove, Ashley Road, Epsom, Surrey KT18 5BW
☎ (+44 1372) 72 61 40 [E] wsatkinsinfo@wsatkins.co.uk
▣ 266701 atkins g ⊖ (+44 1372) 74 00 55
♣ £ 235,800,000 (1996) 👥 7,500
Chairman: C.T. Wyatt
Technical Dir: Richard A. Kidd
Sales and Marketing Director: Martyn Fox
Head of Public Relations: Paul Jowers
Defence Services: K. Vickers
■ Management and technical support to equipment procurement projects including: Command and control ARM and ILS project management technical, project risk management scientific, technological advice planning, investment, efficiency and logistical advice, telecommunications and data technology. ITT preparation and tender assessment, management information systems, design of hardened support facilities and ranges. [98]

Contact Reference

80-01600 Atlantic Commercial (UK) Ltd
Hilltop House, Cuckney, Nottinghamshire NG20 9NS
☎(+44 1909) 72 26 22 ⊖ (+44 1909) 72 04 84
♧ £ 50 000 👪 8 🏠 5 000 sq ft
Chairman: R W Dunk
Sales and Marketing Director: P D Morris
Operations & Technical Dir: A.R. Schlesinger
■ Brokers in the procurement of all types of defence equipment but mainly in the field of infantry weapons & their associated ammunition, from former Eastern Bloc in addition to conventional Western producers.
London Area Office
44 Summerlands, Cranleigh, Surrey GU6 7BP
☎(+44 1483) 27 66 07 ⊖ (+44 1483) 27 18 67 98

80-01650 Auspex Systems Limited
(a subsidiary of Auspex Systems Inc.)
670 Eskdale Road, Winnersh Triangle, Wokingham, Reading RG41 5TS
☎(+44 118) 969 09 00 E chardy@auspex.co.uk
W http:// www.auspex.com ⊖ (+44 118) 969 22 66
♣ $ 170,000,000 (1995) 👪 700
Managing Director: Mike Smith
Marketing/Sales Director: Chris Hardy
Advertising Manager: Brendan Moran
■ Provider of high performance optimised network file servers for client-server environments with requirements for high reliability, scaleable architecture, fast response times and large storage capacities. 98

80-01653 The Austin Company of UK Ltd
(a subsidiary of The Austin Company, USA)
35 Ballards Lane, London N3 1XW
☎(+44 181) 346 93 51 ⊖ (+44 181) 349 91 16
♣ £ 20,000,000 (1996) 👪 70 (1996)
Managing Director: N H Vincent
Operations Director: M McCarthy
Sales and Marketing Director: A Hildyard
Purchasing Manager: I F N Sainthouse
■ Design and construction of industrial, commercial and governmental facilities. 98

80-01850 Aviation Spares Ltd
Unit D5, 53 Haviland Rd, Ferndown Ind.Estate, Wimborne, Dorset BH21 7RY
☎(+44 1202) 87 53 36 ⊠ 41468 aviairg
⊖ (+44 1202) 89 42 90
🏠 10 000 sq ft
Chief Executive: B.J. Adams
Chief Buyer: Bill Stanford
■ Stockist & distributor of European & American aircraft parts & spares, including airframe & engine accessories, avionics, instruments & hardware. Also maintenance, repair & overhaul. ISO9002 Lot traceability approved. → 85 97

80-01900 Avimed Ltd
19 Cliveden Place, London SW1W 8HD
☎(+44 171) 730 80 45 E ianperry@pop3.hiay.co.uk
⊖ (+44 171) 730 19 85
👪 6
Senior Consultant: Dr I C Perry
■ Worldwide consultant services: aviation medicine training and pilot selection. Aviation medicine equipment. Complete systems environmental and occupational health services for armed forces and airlines. Provision of national air ambulance schemes and equipment on a worldwide basis. Accident investigation. Expert witness. Fires and explosions. 98

80-03050 Babtie Group Ltd
95 Bothwell Street, Glasgow G2 7HX
☎(+44 141) 204 25 11 ⊖ (+44 141) 226 31 09
♣ £ 47,000,000 (1995) 👪 1,500 (1997)
Chairman: H Perfect
Chief Executive Officer: W Mitchell
Marketing Manager: Alan Waugh
■ Consulting engineers and planners covering the complete range of civil and structural engineering work. 98

80-03060
BACTEC International Ltd

Alpha House V, Laser Quay, Culpeper Close, Rochester, Kent ME2 4HU
☎(+44 1634) 29 67 57 E bactecint@aol.com
⊖ (+44 1634) 29 67 79
♣ £ 1,500,000 👪 8 🏠 1,200 sq m
Managing Director: Guy Lucas
Operations Manager: Mark Hurlston
Purchasing Manager: Ms Christine Bridges
Business Manager: Nick Brewer
Company Secretary: J M Lucas
■ Provides a wide range of Explosive Ordnance Disposal (EOD) services including: Battle Area Clearance, De-Mining, Range Clearance, Emergency Response, Mines Awareness, Land & Underwater survey & clearance. EOD, mine clearance & search training. Provision of EOD & associated equipment & general engineers equipment.
Consultancy.
Accredited to ISO9002, DCL Reg No 543B02.
United Nations & EC registered. 98

80-03070 BAeSEMA Ltd
(a joint venture of British Aerospace plc and Sema Group plc)
Apex Tower, 7 High Street, New Malden, Surrey KT3 4LH
☎(+44 181) 942 96 61 W http://www.baesema.co.uk
⊖ (+44 181) 942 97 71
👪 1,700
Managing Director: Alex Dorrian
Marketing Director: Alan Nicholson-Florence
Technical Director: Dr Mike Christie
■ International supplier and systems integrator in naval, military and selected industrial markets with expertise in ship design, supply and support, ship platform and propulsion systems, naval combat and combat management systems, mine warfare systems, active towed array sonars and countermeasures systems, communications management systems, strategic consultancy, logistics services, airborne software and systems, road, rail and air transport management systems and services, energy control systems and services.
BAeSEMA incorporates YARD (naval engineering expertise) and CORDA (defence operational analysis consultancy) as well as 50% of Aerosystems International (Ael) - a joint venture with GKN-Westland Helicopters.
Specialises in serving highly demanding customers operating in defence and civil environments requiring public accountability and adherence to strict quality and safety standards and technical experience.
Much of the company's current business is in the naval area making use of expertise gained from the exploitation of information technology and engineering skills at sea. → 74
● Scientific House, 40-44 Coombe Road, New Malden, Surrey KT3 4QF
☎(+44 181) 942 96 61 ⊖ (+44 181) 949 80 67 98

80-03073 Balfour Beatty Ltd
(a subsidiary of BICC plc)
One Angel Square, Torrens Street, London EC1V 1SX
☎(+44 171) 216 68 00 ⊖ (+44 171) 216 69 50
♣ £ 1,700,000,000 (1995) 👪 16,000
Chief Exec.: Mike Welton
Public Relations: Ms Marjorie Hooper
■ Power construction, building, engineering & general contracting company. 98

80-03350 Sir J.H. Biles (Naval Services) Ltd
Hamilton House, Blackhall Lane, Paisley, Scotland PA1 1TA
☎(+44 141) 848 17 71 ⊖ (+44 141) 889 54 79
♣ £ 1,250,000 👪 50
Managing Dir: Hugh C. McLachlan
Operations & Technical Dir: Peter McArthur
■ Marine consultants. Comprehensive technical service to world navies together with their ancillary vessels & facilities necessary to support their active roles. Specialisations relate to ship weapons platforms, both surface & sub-sea. 98

80-03570 BMT Defence Services Ltd
(a subsidiary of British Maritime Technology Ltd)
210 Lower Bristol Road, Bath, Avon BA2 3DQ
☎(+44 1225) 47 36 00
E 100631.2573@compuserve.com
⊖ (+44 1225) 44 87 14
♣ £ 10,000,000 (1997) 👪 170 (1997) 🏠 1,700 m2
Managing Director BMT Group and Chief Executive: John Gallagher
Managing Director BMT Defence: Roger Swann
Operations Director: Roger Cooper
Commercial Director: Lance Holyoake
Technical Director: Les Sheppard
Head Naval Architect: Paul Buckton
Head of Engineering: Roy Quilliam
Head of Combat Systems: Geoff Brooks
Head of Project Services: Ms Jenny Clarke
Projects Mgr Submarines: Paul Tasker
■ BMT Defence Services comprises of a team of naval architects, engineers, combat system engineers, project managers and consultants to advise and monitor on all stages of major naval build and modernisation programmes for surface ships, submarines and auxiliary vessels of all types. Services include concept design, project definition, production oversight, quality assurance, risk, safety, ship management, in-service support life support through to decommissioning and disposal. The company is also able to call on the resources of BMT Group with over 80 years experience in maritime technology. 98

80-03573 BMT Marine Information Systems Ltd
(a subsidiary of BMT Group Ltd)
Grove House, 7 Ocean Way, Ocean Village, Southampton SO14 3TJ
☎(+44 1703) 23 22 22 E mis@bmtmis.demon.co.uk
⊖ (+44 1703) 23 28 91
♧ £ 50,000 (1996) ♣ £ 500,000 (1996) 👪 10 (1996)
Managing Dir: Andrew Tyler
Sales and Marketing Director: Peter Batt
Financial Controller: Andrew Swales
■ Environmental consultancy, shoreline management, applied oceanography & marine information systems. 98

80-03579 BMT Quality Assessors Limited
(a division of TTL Chiltern Ltd)
Stirling Enterprise Park, Stirling Road, Stirling, Scotland FK7 7RP
☎(+44 1786) 45 08 91 E stirlinro@hlchiltern.co.uk
⊖ (+44 1786) 45 10 87
👪 9
Managing Director: R Foster
Sales and Marketing Director: J Hooper
Operations Director: S Barclay
■ Third party quality assurance and certification to appropriate British, European and International standards ISO 9000 and TickIT scheme. Particular speciality is QA assessment and certification for those providing products, services and equipment to the Ministry of Defence and their sub-contractors. 98

80-03595 BMT Seatech Ltd
(a subs. of British Maritime Technology Group Ltd UK)
Grove House, 5 Ocean Way, Ocean Village, Southampton SO14 3TJ
☎(+44 1703) 23 22 22; 63 51 22
⊖ (+44 1703) 23 28 91; 63 51 44
♣ £ 1 000 000+ 👪 22
Managing Dir: Dr Philip Thompson
Sales & Marketing Dir: Gwynne Lewis
Operations Dir: Dr Ian Dand
■ Specialist marine technical consultancy providing structural, hydrodynamic & marine operational services & products. 97

80-03650 Bodycote H.I.P. Ltd
(† H.I.P. Ltd)
(a subsidiary of Bodycote International plc)
Carlisle Close, Sheffield Road,
Sheepbridge, Chesterfield S41 9ED
☎ (+44 1246) 26 08 88 ⊖ (+44 1246) 26 08 89
👥 28 🏭 13 000 sq ft
Managing Director: B Birch
General Mgr: D.J. Roberts
Technical Mgr: Dr L.E. Tidbury
Marketing & PR Officer: Ms S. Andrews
■ Sub-contract hot isostatic processing service for defect healing and consolidation of cast and powder nickel based superalloy turbine components (blades/discs), titanium alloys (engine and airframe), steel & alumimium components. Also for diffusion bonding & cladding using dissimilar materials. [98]

80-03800 Brassey's (UK) Ltd
33 John Street, London WC1N 2AT
☎ (+44 171) 753 77 77 [E] brasseys@dial.pipex.com
[W] http://www.brasseys.com ⊖ (+44 171) 753 77 94
♣ £ 1,000,000 👥 20
Managing Director: Miss Jenny Shaw
Operations & Technical Dir: James Cook
Purchasing Manager: Philip Hardy
Marketing Manager: Warren Prentice
■ Specialist publisher of book and journals on defence and international affairs. Extensive list of titles covering all aspects of military technology. [98]

80-03950 Bristow Helicopters Ltd
Redhill Aerodrome, Redhill, Surrey RH1 5JZ
☎ (+44 1737) 82 23 53 ⊠ BRISTOW Redhill
⊠ 21913 brstow g ⊖ (+44 1737) 82 28 84
👥 2 000 (worldwide)
Managing Dir & Chief Exec: Bryan Collins
Operations Dir: Capt. Mike Norris
Technical Dir: Dennis Russell
Commercial & Technical Services Dir: Allan C. Brown
■ Services include basic & advanced pilot & engineer training, repair & overhaul, consultancy & logistics support. Design & installation of major equipment modifications include complete helicopter SAR packages & helicopter integrated flight data recorders. Health & usage monitoring systems. [97]

80-03990 British Aerospace (Consultancy Services) Ltd
620-622 Birchwood Boulevard, Birchwood,
Warrington, Cheshire WA3 7QW
☎ (+44 1925) 82 23 37 ⊖ (+44 1925) 82 24 83
■ Integrated project management cost engineering and design consultancy services for the construction requirements of the manufacturing, aerospace and defence industries. [98]

80-03995 British Aerospace Flight Training (UK) Ltd
Prestwick Airport, Ayrshire KA9 2RW
☎ (+44 1292) 67 10 22 ⊖ (+44 1292) 67 10 10 [98]

80-04000 British Aerospace Systems and Services
Warton Aerodrome, Mill Lane, Warton,
Preston, Lancashire PR4 1AX
☎ (+44 1772) 63 33 33 ⊠ 67642
⊖ (+44 1772) 85 52 86
👥 5,808
Managing Director: Steve Mogford
Head of Public Affairs and Communications:
Ms Penny Studholme
☎ (+44 1772) 85 62 49; 85 81 07 (fax)
■ Comprehensive prime contract services for major defence programmes, including specialist training, civil engineering, maintenance and support and total flying training services in military and commercial sectors. [98]

80-04090 British International Helicopters
Aberdeen Airport, Dyce, Aberdeen AB2 0DT
☎ (+44 1224) 77 13 53 ⊖ (+44 1224) 77 04 91
Managing Dir: Paul Conway [97]

80-04175
British Nuclear Fuels plc (BNFL)

Risley, Warrington, Cheshire WA3 6AS
☎ (+44 1925) 83 20 00 ⊖ (+44 1925) 83 43 72
♣ £ 1,304,000,000 (1994/95) 👥 14,000 (1996)
Chief Executive Officer: J Taylor
Head of Commercial (Waste Management and Decommissioning): M Simpson
Business Relationship Manager: P D Mitchell
Senior Account Manager UK Group Commercial:
A S Daniel
■ Provision of a comprehensive range of services relevant to users of radioactive material including:
Waste management;
Effluent treatment;
Decommissioning;
Material handling, storage & processing;
Instrumentation;
Comprehensive engineering services;
Engineering & project services. [98]

80-04550 BTR Technology Services Ltd
(a subs. of BTR plc)
Horninglow Rd, Burton-on-Trent, Staffs DE13 0SN
☎ (+44 1283) 51 05 10 ⊖ (+44 1283) 51 00 78
Business Development Mgr: J. Marshall
■ Factorial experiment design, statistical process control, historical data analysis operational techniques, software & training consultancy. [97]

80-05280 Cable and Wireless (Marine) Ltd
(† BT (Marine) Ltd)
Berth 203, Western Docks, Southampton, Hants SO1 0HH
☎ (+44 1703) 77 55 77 ⊖ (+44 1703) 35 87 43
Marketing Communications Mgr: K. Taylor
■ Development & installation of subsea communication systems. [97]

80-05300 Cambridge Consultants Ltd
(a subsidiary of Arthur D Little Ltd)
Science Park, Milton Road, Cambridge CB4 4DW
☎ (+44 1223) 42 00 24 ⊠ 81481
⊖ (+44 1223) 42 33 73
♣ £ 24,000,000 (1997) 👥 330 (1997) 🏭 18,000 sq m
Managing Director: Howard Biddle
Sales and Marketing Director: Alan D Murphy
Marketing Communications Manager:
Ms Diane Aston-James
Purchasing Manager: Francis Pullen
Defence Business Development: Gordon Oswald
■ Contract design and development for electronic technology, computing and information processing, physics, mechanical engineering, product design, production and process engineering. [98]

80-05320 Cambridge Neurodynamics Ltd Head Office
St John's Innovation Centre, Cowley Road,
Cambridge CB4 4WS
☎ (+44 1223) 42 11 07 [E] info_uk@neurodynamics.com
[W] http://www.neurodynamics.com
⊖ (+44 1223) 42 10 96
♣ £ 3,000,000 (1996) 👥 40 (worldwide)
Managing Director: Dr M Lynch
Technical Director: R Gaunt
Sales Director: Chris Hartley
Marketing Manager: Ms Emma Barnes
■ Advanced recognition and signal processing systems. Automatic classification, tracing and analysis of targets. [98]

80-05500 Carmichael & Sweet Consultancy Ltd
Rodney House, 44 Craneswater Ave,
Southsea, Hants PO4 0PB
☎ (+44 1705) 73 91 01 ⊠ 86626 mynews g
⊖ (+44 1705) 82 17 78
♣ £ 100 000 👥 24
Director: R.H. Gaskin
General Mgr: J.R. Soper
■ Professional & cost-effective consultancy covering defence & security in UK. Contractors to the UK MoD. Publishers of professional reports on the British worldwide defence markets for industry & commerce. [97]

80-05550 Cartographical Services Ltd
Landford Manor, Landford, Salisbury, Wiltshire SP5 2EW
☎ (+44 1794) 39 03 21 ⊖ (+44 1794) 39 08 67
♧ £ 19,500 (1997) ♣ £ 1,400,000 (1997) 👥 44 (1997)
🏭 53,000 sq ft
Chairman: W.B. Bucknall
Managing Dir: J.B. Waterman
Operations/Technical Dir: R.M. Whitfield
Company Secretary: R. Henry
■ Aerial & ground surveys including the preparation of digital mapping & photogrammetric mapping.
Survey Centre
Waterworks Road, Worcester WR1 3EZ
☎ (+44 1905) 290 85 ⊖ (+44 1905) 61 77 71
■ Ground surveys. [98]

80-05800 CD-ROM Systems (Europe) Ltd
(† CD-FICHE Distributors Ltd)
4 Lloyds Court, Manor Royal,
Crawley, West Sussex RH10 2XT
☎ (+44 1293) 52 52 71
[W] http://www.cd-rom-systems.co.uk ⊠ 878647
⊖ (+44 1293) 56 20 66
♧ £ 5,000 (1997) ♣ £ 450,000 (1997) 👥 6 🏭 400 m2
Managing Director: James Phillips
Commercial Director: Pauline Moore
General Manager: Robin Fitton
Sales Executive: Gary Wilkins
Purchasing: Fran Harman
■ Full sales, service & support of CD-ROM hardware & software. Suppliers of optical storage products including CD-ROM, CD-Writers, CD-Networking, CD-Towers & document management with optical character recognition. Full consultancy, installation & post sales support available on our complete range. [98]

80-05895 Central Research Laboratories Ltd (CRL)
(a subsidiary of Scipher Ltd)
Dawley Road, Hayes, Middlesex UB3 1HH
☎ (+44 181) 848 64 44 [E] charmson@crl.co.uk
[W] http://www.crl.co.uk ⊖ (+44 181) 848 66 82
♣ £ 13,000,000 (1997) 👥 125 (1997) 🏭 10,000 m2
Managing Director: Dr John White
Marketing Department: Ms Claire Harrison
■ British company providing innovation and advanced technology products to companies worldwide in the following areas: electro optics, instrumentation products, electronics, displays, sensors, micro-engineering, audio technology, computing, transaction security and industrial design. [98]

80-05900 CHAM (Concentration, Heat & Momentum) Ltd
Bakery House, 40 High Street, Wimbledon Village,
London SW19 5AU
☎ (+44 181) 947 76 51 [E] phoenics@cham.co.uk
[W] http:// www.cham.co.uk ⊖ (+44 181) 879 34 97
♣ £ 4,000,000 (1997) 👥 50 (1997)
Managing Director: Prof D B Spalding
Business Development Manager: Dr Ulysses Cartalos
Sales Manager: Mrs Sylvie Stevens
Marketing Executive: Ms Sophie Moeng
■ Software house & consultancy organisation specialising in the computer modelling of processes involving fluid flow, heat transfer, chemical reaction & combustion. Applications include aerodynamics, hydrodynamics, compressor/turbine design & propulsion system modelling. CFD software products include PHOENICS-2. [98]

80-06000 Chilport (U.K) Limited
Wide Street, Portland, Dorset DT5 2JP
☎ (+44 1305) 86 03 00 ⊖ (+44 1305) 86 13 43
■ Specialist security services. [97]

80-06150 Clearway Explosives Services Ltd
Willowbank House, Willowbank Road, Aberdeen AB11 6YG
☎(+44 1224) 33 23 02 ⊖ (+44 1224) 58 14 72
♣ £ 150,000 (1996) ꔮ 3 (1997)
Managing Director: John Belchamber
☎(+44 1224) 33 23 02
Training Manager: Robb Cumming
☎(+44 1224) 33 23 03
■ Specialists in the use of explosives on land & underwater for: platform removal, obstruction clearance, trenching, structural cutting, salvage etc.. Clearance of mines & unexploded ordnance from war zones. Military firing range clearance. Inspection & repackaging or disposal of explosives & ordnance. Diving courses for coastguard, military & commercial personnel. Provision of low cost mine disposal equipment & MCMVs, throughwater acoustically actuated explosive firing units & mine lifting bags. [98]

80-06250 Cogent
(a Nortel Professional Services Business)
Waterview House, 1 Roundwood Avenue, Stockley Park, Uxbridge, Middlesex UB11 1AP
☎(+44 181) 867 56 00 ⊖ (+44 181) 867 56 01
♣ US$ 125,000,000 ꔮ 600 ⌂ 20,000 sq m
Managing Director: Brian Gowers
Marketing Director: Mike Greenway
■ Consultancy, design, implementation and support of telecommunication systems. → 75

Cogent Defence
Meadows Road, Queensway Meadows Industrial Estate, Newport, Gwent NP9 0SS
☎(+44 1633) 29 22 96 ⊖ (+44 1633) 29 22 20
Managing Director: M H Roberts
Marketing Director: Mike Greenway
Business Manager Secure Systems: Steve Gibson
Business Manager Naval Systems: Bob Waters
Business Manager Tactical Systems: Len Tyler
Business Manager Special Installations: Ken Ackland
Contact: Brian Chubb
■ Secure and tactical communication systems for Army, Air Force and Navy. → 75

Cogent Integrated Networks
Great Eastern House, Edinburgh Way, Harlow, Essex CM20 2NQ
☎(+44 181) 945 54 00 ⊖ (+44 181) 945 58 08
Managing Director: Bob Swindle
Director Data Networks: Richard de Fraine
Business Manager Network Services: Andy Thomas
Business Manager Transport: Bob Deverall (Hong Kong office)
Business Manager Strategic Communications: Malcolm Nicholls (Hong Kong Office)
Contact - External Relations: Brian Skingley
■ Provides telecommunication network services to the UK MoD. Provides data and voice networks for government and military services within Europe as well as security products for the financial, law enforcement and other communities.
Industrial and Transport divisions provide consulting, design and implementation services for data and telecommunication networks and services within their business sectors.
Strategic communications provide a voice and data network offering for defence forces worldwide. → 75 [98]

80-06400 Commslogic Ltd
(a subsidiary of Nisaba Group Ltd)
Nisaba House, Waterfront Business Park, Fleet Road, Fleet GU13 8QT
☎(+44 1252) 77 67 76 ⊖ (+44 1252) 77 67 78
♣ £ 3,500,000 (1996/97) ꔮ 65 (1997) ⌂ 11,000 sq ft
Managing Director: G Wokes
Operations Director: E Dennehy
Marketing Director: S Hunt
Technical Director: A Brown
■ Communications consultants: Specification writing; tender evaluation; skill hire; project management; equipment evaluation and testing; Tempest skills; prototype development; software development, evaluation and testing system design. [98]

80-07200 Cranfield Aerospace Ltd
Cranfield University, Cranfield, Bedfordshire MK43 0AL
☎(+44 1234) 75 40 46 ✉ 825072
⊖ (+44 1234) 75 11 81
♧ £ 100 000 (1994/95) ♣ £ 3 250 000 (1994/95)
ꔮ 50 (1994/95) ⌂ 3 000 (1994/95)
Chairman/Head of College of Aeronautics: Prof. D.I.A. Poll
Engineering Dir: P. Lawson ☎(+44 1234) 75 40 46
■ All aspects of aeronautical engineering, including the concept, design & implementation of major modifications to civil & military aircraft, for role changes, modernisation, flight trials or research. [98]

80-07370 Crew Services Ltd
Fairfield House, Kingston Crescent, Portsmouth PO2 8AA
☎(+44 1705) 65 11 56
[E] crew_services@compuserve.com
⊖ (+44 1705) 65 54 33
♣ £ 6,500,000 (1996-97) ꔮ 120 (1997) ⌂ 700 sq m
Managing Director: N D Wood
Operations Director: J R S Dewar
Commercial Manager: M D Day
Business Manager: M T Hutton
Business Manager: H Slinger
Business Manager: R W Needham
Business Manager: R G Shore
Business Manager: D G Thomsett
Marketing Co-ordinator: R D Pound
■ Independent consultancy providing support to UK Government, MoD and NATO covering all aspects of C2/C3I system procurement, combat system integration and engineering, conceptual studies, IS/IT support, training and marketing. [98]

80-07500 CSC Computer Sciences Ltd
(† Computer Sciences Corp)
279 Farnborough Road, Farnborough, Hampshire GU14 7LS
☎(+44 1252) 36 30 00 [W] http://www.csc.com
⊖ (+44 1252) 36 44 85
♧ US$ 383,000,000 (1996)
♣ US$ 4,242,000,000 (1996) ꔮ 33,850 (1996)
Marketing Director: Mojtaba Ghassemian
Operations Director: Guy Hains
Sales and Marketing Director: Steve Thomas
☎(+44 1252) 36 33 55
Purchasing Manager: Kathleen Ramey
Business Development: Adam Wainwright
☎(+44 1252) 36 39 79
Business Consultant: Craig Stevens
☎(+44 1252) 36 33 90
Business Consultant: Nigel Wylde
Sales Support: Mike Ford ☎(+44 1252) 36 40 35
Defence Practice: David Hunt ☎(+44 1252) 36 35 36
■ Professional and IT services and consultancy for defence, government and commercial sectors. Systems integration. [98]

80-07600 CSE Aviation Ltd
Oxford Airport, Kidlington, Oxon OX5 1RA
☎(+44 1865) 84 12 34 ✉ 83204 aztec g
⊖ (+44 1865) 84 12 07
♣ £ 22 000 000 ꔮ 320 ⌂ 200 000 sq ft
Managing Dir: R. Crooke
Commercial Dir: T. Trivett
Operations & Technical Dir: S. Lewis
Marketing Mgr: Peter D. Moxham ☎(+44 1865) 84 42 76
■ Operators of aircraft & systems for defence related training including target towing. Agents for a wide range of defence related equipment. Consultancy services for Tow Target operations & equipment. Pilot & engineer training. Light aircraft including business jets, piston engine, propeller, electrics, hydraulics, instruments & avionic maintenance & overhaul. Avionic installation design. → 85 → 90
® Brazil: Embraer do Brazil
UK: Sabre Computers Int'l
USA: Hayes Targets, Teledyne Continental Rotors (Engines), Textron Lycoming (Engines) [97]

80-07650 CTP Logistics Ltd
(† Cheltenham Technical Publications Ltd)
Cheltenham House, Bath Road, Cheltenham, Gloucestershire GL53 7LR
☎(+44 1242) 51 91 19 [E] sales@ctp.co.uk
⊖ (+44 1242) 22 60 06
♧ £ 1,000,000 (1997) ♣ £ 3,500,000 (1997)
ꔮ 110 (1997) ⌂ 1,500 m2
Managing Director: Simon Wade
Operations Director: Gordon O'Brien
Finance Director: David Exton
Sales Manager: Steve Herbert
■ Logistic support, authoring to military and commercial specifications in all disciplines. Illustrated parts catalogues, spares provisioning and related logistics data handling. NATO codification and database management. Publishing with SGML, internet and CD-Rom. Quality system approved to BS EN 9001. [98]

80-07720 Data Cell Ltd
(a subsidiary of TVX Incorporated, USA)
Hattori House, Vanwall Business Park, Maidenhead, Berkshire SL6 4UB
☎(+44 1628) 41 54 15 [E] info@datacell.co.uk [W] http://www.datacell.co.uk ⊖ (+44 1628) 41 54 00
♣ £ 6,000,000 (1997) ꔮ 30 (1997)
Sales and Marketing Director: Ian Young
Operations Director: Jo Rogers
Sales Manager: Mark Norris
Purchasing Manager: Lin Jones
■ Design and develop image-processing and multi-sensor tracking systems and components for very high speed tracking applications. Very experienced in advanced image analysis. → 75 [98]

80-07740 Data Sciences Ltd
(a subsidiary of IBM)
Meudon Avenue, Farnborough, Hampshire GU14 7NB
☎(+44 1252) 80 55 55 ✉ SOFTWARES Farnborough
[W] http://www.datasci.co.uk ⊖ (+44 1252) 80 55 56
♣ £ 106,000,000 (1995/96) ꔮ 1,800
Managing Director: G Windett
Business Development Director: K Newman
Production and Technology Director: J Palfreyman
Information Officer: S M Farrell
■ Software and systems integration specialist. Electronic warfare, command and control software technology. Operational requirements analysis, feasibility studies, project definition studies, total system design, software implementation, operator and maintenance training, post-design support. Tailored IT Facilities Management (FM). [98]

80-07750 Datel Technology Ltd
(a division of Datel Ferranti Group)
Old Docks House, 90 Watery Lane, Preston, Lancs PR2 1AU
☎(+44 1772) 76 06 00 ⊖ (+44 1772) 76 65 05
♣ £ 15,000,000 (1995/96) ꔮ 200
Managing Dir: G.J. Pugh
Technical Dir: S. Williamson
Commercial Dir: K. Morris
Business Development: B. MacPherson
■ Software & systems engineering covering; navigation, flight control, EW, weapons control, sonar, interfaces, communication, digital databus & simulation systems on aircraft, ships, submarine & land equipment. Accredited to ISO 9001/9000-3 software (TICKIT Scheme). Requirements: Analysis, system studies, design, implementation, commissioning & post design service support. Specialists in real-time safety-critical software/computer systems. Also provides software engineers on contract basis through in-house staff agency.

London Office
Country House, 190 Great Dover St, London SE1 4YB
☎(+44 171) 403 55 55 ⊖ (+44 171) 403 55 53 [97]

80-08180 Dee Organics Laboratory
(a subsidiary of Hyder Environmental)
Huntington Water Treatment Works, Huntington, Cheshire CH3 6EA
☎(+44 1244) 31 12 42 [W] http://www.hyder.com
⊖ (+44 1244) 34 86 83
ꔮ 300+ (1997)
Managing Director: Dr Mark Kibblewhite
Operations Director: Peter Dignum
Operations Manager: Richard Whiteman [98]

80-08190 Defence Public Affairs Consultants Ltd (DPAC)
11 Bouverie Square, Folkestone, Kent CT20 1BD
☎(+44 1303) 22 04 93 ⊖ (+44 1303) 22 10 95
Managing Director: Peter Bishop
Operations Director: David McDine
Operations Manager: Roger Tutt
■ Provides professional services to major companies employing their own in-house PR & marketing teams & also smaller companies who have no dedicated specialists. [98]

80-08200 Defence Research & Analysis
461 Harrow Road, London W10 4RG
☎(+44 181) 960 44 88 ⊖ (+44 181) 960 68 86
ꔮ 2 (+ consultants)
Partner: Lt Col Robert Jarman
Partner: Miss Yvonne Headington

■ Market Research - Corporate Strategy; bespoke analysis of defence issues for the financial, industrial & media communities. [98]

80-08300 Deloitte & Touche Consulting Group

(† Touche Ross Management Consultants)
(a member firm of Deloitte Touche Tohnatsu International)
Stonecutter Court, 1 Stonecutter St, London EC4A 4TR
☎ (+44 171) 936 30 00 ☒ 884739 trlndn g
⊖ (+44 171) 583 11 98
♣ £ 85 500 000 (1995) 👥 700
Partner in Charge: John Everett
Sales & Marketing Dir: David Evans
■ Accountants & management consultants. Services offered include feasibility studies, project evaluation, financial appraisals, cost analysis & management of systems projects. [97]

80-08375 Dennis & Gemmill International Ltd

Headquarters
The Courtyard, Smith Street, London SW3 4EE
☎ (+44 171) 730 71 38 ⊖ (+44 171) 730 92 27
♣ £ 500 000 (approx.) 👥 5
Chairman: John M. Dennis
■ Executive search services internationally to a broad spectrum of clients in electronics, computing, telecommunications, simulation, aerospace and other related high-tech industries in accordance with the standards of AESC & BAESC.
US Affiliate:
William C. Houze & Co. → USA 80 [97]

80-08400 ICL DESC

(† DESC Limited)
DESC House, Suttons Business Park,
Early, Berkshire RG6 1AZ
☎ (+44 1189) 35 23 53 ☒ 847557
⊖ (+44 1189) 35 99 00
♣ £ 110,000,000 (1996) 👥 500
Chairman: T.K. Todd
Managing Dir: M.S. Coote
Director Business Development: G. Brodie
■ Specialists in systems integration, project management of CCIS in the operational and military intelligence market places. Prime contractor for NSTN, UITAIR, CHOTS, CCIS, AMRAD (R), ALEBI and other projects in the operational CIS, messaging and intelligence areas. Secure IT infrastructures. [98]

80-08450 Designability Ltd

56 High St, Dilton Marsh, Westbury, Wilts BA13 4DY
☎ (+44 1373) 82 20 37 ⊖ (+44 1373) 86 53 21
♣ £ 100 000 👥 3 🏭 1 000 sq m
Director: Brian Cunnington
■ Design & build of composite light aircraft. Road vehicles & mechanical & composite components. [97]

80-08500 Devonport Royal Dockyard Ltd (DML)

(† Devonport Management Ltd)
(owned by consortium led by Brown & Root)
Devonport Royal Dockyard, Plymouth, Devon PL1 4SG
☎ (+44 1752) 60 56 65 ⊖ (+44 1752) 55 28 87
♣ £ 250,000,000 👥 4,000 🏭 300 acres
Chairman: Tony Pryor
Chief Executive: Dennis Gilbert
Finance Dir: Henry Warren
Contracts Dir: Gareth Unsworth
Surface Ships Business Unit Director: David Kilborn
Business Development Dir: Peter Whitehouse
Media and Corporate Affairs Manager: Reg Shield
Marketing Services Mgr: Mrs Wendy Smith
Personnel Director: Terrance Slater
Submarine Director: Ewen MacLachlan
■ Refitting & modification of surface warships & submarines. Overhaul of mechanical, electrical & electronic components with refurbishment & manufacture of spares. Weapon systems overhaul & integration. Design support services.
Consortium members:
Brown & Root (UK) Ltd
BICC (Balfour Beatty)
Weir Group plc [98]

80-08600 Dowty Aerospace Aviation Services

Cheltenham Road, Gloucester GL2 9QH
☎ (+44 1452) 71 31 11 ☒ 43246/7
⊖ (+44 1452) 71 19 54; 71 16 79
👥 160 🏭 100,000 sq ft
Managing Dir: Dr S.J. Lee ☎ (+44 1452) 71 18 08
Sales & Marketing Mgr: A.R. Rowe
☎ (+44 1452) 71 13 85
■ Overhaul & repair of aircraft components including landing gear, actuators, hydraulics, composite & metal bladed propellers, gearboxes & ram air turbines. Advance exchange & loan equipment for BAe 146, HS 748, ATP & Jetstream 31, Dash 8, Fokker F27, F28, F50 & F100, Metro III, Saab 340 & Shorts SD 360.
Associated overseas companies:
Dowty Aerospace Aviation Services USA
Dowty Aerospace Aviation Services Singapore [98]

80-08700 DSD Ltd

3 Highview, High Street, Bordon, Hampshire GU35 0AX
☎ (+44 1420) 47 88 88 ⊖ (+44 1420) 47 79 26
♲ £ 82 380 (1995) ♣ £ 600 000 (1994) 👥 20
🏭 300 sq m
Marketing Director: John Harris
Operations Director: J P Whitehead
Purchasing Manager: A K Robb
■ Technical assessment & analysis for vehicles, weapons, communications, computer systems, simulators & remote control. Analysis of operational data, testing & trials. Design, development, manufacture & modification of special electrical, electronic, hydraulic & mechanical equipment for defence purposes. Approved to ISO 9001. [98]

80-08750 DSSD plc

Q101 Building, DRA Farnborough, Hampshire GU14 6TD
☎ (+44 1276) 67 55 88; 67 55 66 (media enquiries)
♣ £ 150,000,000+ 👥 1,300+
Chief Executive: Graham Love
■ Formerly the support services division of the Defence Evaluation and Research Agency, DSSD specialises in the facilities management of outsourced business processes in sensitive, hazardous or difficult environments. Its core areas of service include: total managed services, site services, financial services, information services and support services such as logistics management, reprographics and imaging. Provision of security for financial, physical, computer, electronic and personnel assets. [98]

80-08800 Dun & Bradstreet UK

Holmers, Farm Way, High Wycombe, Bucks HP12 4UL
☎ (+44 1494) 42 20 00 ☒ 83128
⊖ (+44 1494) 42 22 60
♣ £ 70 000 000 👥 1 000
Marketing Mgr: Philip Mellor
■ Dun & Bradstreet has the most comprehensive international database providing commercial information & services worldwide. [97]

80-08830 Dundridge College

(part of GEC-Marconi Radar & Defence Systems Limited)
Harberton, Totnes, Devon TQ9 7PP
☎ (+44 1803) 86 51 18
🄴 training.dundridge@dial.pipex.com
⊖ (+44 1803) 86 31 95
■ Courses in air defence, electronic warfare, joint warfare, air traffic control and general military training. Customised courses as required. [98]

80-08850 Dynamic Graphics Ltd

1 Forest Court, Oaklands Park, Fishponds Rd,
Wokingham, Berkshire RG41 2QJ
☎ (+44 1734) 77 47 55 🄴 dgi@dgl.co.uk
⊖ (+44 1734) 77 47 21
Managing Dir: Peter Irwin
Business Development Mgr/ Exhibition Co-ordinator:
A.D. Kittridge
■ EarthVision. Surface & volume modelling. Bathymetric mapping/visualization, oceanographic property modelling, temperature, salinity & velocity. Pollution modelling & military site clean-ups. [97]

80-08900 Dytecna Ltd

Roebuck Rd, Chessington, Surrey KT9 1LN
☎ (+44 181) 397 33 55 ☒ 333095
⊖ (+44 181) 391 53 49
♣ £ 10 000 000 👥 160 🏭 8 000 sq m
Chairman: J.A. Fulford
Managing Dir: J.J.C. Fulford
Sales & Marketing Dir: P.A. Bates
Operations & Technical Dir: D.B.L. Durham
Purchasing Mgr: R. Sankey
■ Engineering Services: Specification, design, manufacture, installation & trialling of integrated communications systems for ground mobile & transportable applications. Independent consultancy, EMC & TEMPEST engineering, standards interpretation. Independent technical support services. Antenna design, development & manufacture.
Logistics Services: Integrated Logistic Support ILS services. ILS consultancy, feasibility studies & system assessments. Electronic technical publications, parts catalogues & handbooks. Spares procurement & logistic database management. [97]

80-09000 EASAMS Ltd

Lyon Way, Frimley Rd, Camberley, Surrey GU16 5EX
☎ (+44 1276) 645 66 ☒ 858115 easams g
⊖ (+44 1276) 69 66 82
♣ £ 100 000 000 + (1995) 👥 1 800
Managing Dir: Dr T. Maini
Defence Division Dir: P. Kenyon
Technical Dir: G. Oddy
Business Development Dir: M. Ferranti
Defence Business Development: K.G.T. Robinson
Information Officer: D. Picken
■ Objective system engineering capability covering all stages of system procurement, concept/feasibility studies, operational analysis, mathematical modelling, cost/risk/performance assessment, software development, overall systems design, implementation, integration, project management, training & in-service support. Experienced both in providing specialist consultancy services (operational, technical, system assessment, project management) & undertaking major prime/sub-contractor roles (including leading multi-national consortia to deliver complex, £100M+, aerospace & defence systems). [97]

80-09050 Eaton Limited

(† Arrow-Hart (Europe) Ltd)
Plymbridge Road, Estover, Plymouth, Devon PL6 7PN
☎ (+44 1752) 73 11 55 ⊖ (+44 1752) 73 73 72
👥 9
Senior Sales Co-ordinator: S A Moorman
■ Electromechanical switches. Customer service centre. [98]

80-09100 E.C. Transport (Wimborne) Ltd

PO Box 209, Poole, Dorset BH15 4AH
☎ (+44 1202) 66 66 00 ☒ 418173
⊖ (+44 1202) 66 55 55
♣ £ 2,900,000 👥 12 (1997)
Managing Director: Charles Ian Rycroft
Transport Manager: André Paul Linford
■ European road transport & international shipping of explosives, ammunition & defence material (own fleet of ADR vehicles). Transport of hazardous waste. Abnormal load movement in UK & abroad. [98]

80-09200 EDS Defence Ltd

(a division of EDS Ltd)
1-3 Bartley Wood Business Park, Bartley Way,
Hook, Hampshire RG27 9XA
☎ (+44 1256) 74 20 00 ⊖ (+44 1256) 74 23 49
♣ £ 70,000,000 👥 1,000 (1997)
Managing Director: Fred Bullock
Sales and Marketing Director: Peter Appleby
Marketing and Communications Manager: Tim Magness
Purchasing Manager: Dave Mears
Marketing Communications Executive:
Ms Anna Silverwood ☎ (+44 1256) 74 23 45
■ Advanced computer technology & systems integration. Software systems, consultancy products & services, systems management for land, sea & air forces.
Tewksbury Office
Unit 30, Alexandra Way, Ashchurch Industrial Estate,
Tewkesbury, Gloucestershire GL20 8NB
☎ (+44 1684) 29 21 01 ⊖ (+44 1684) 29 52 02
Business Support Manager: Chris Hunt

Contact Reference

■ Advanced computer technologies, S/W development, systems integration & consultancy services for communications & surveillance systems for land & air forces.

Camberley Office
Pembroke House, Pembroke Broadway,
Camberley, Surrey GU15 3XD
☎ (+44 1276) 68 62 00 ⊜ (+44 1276) 68 35 11
Manager: Libby Ralph
■ Advanced image processing technologies, including neural network & factal processing. SAR processing, software engineering products, including PCTE.

Milton Keynes Office
Wavendon Tower, Wavendon, Milton Keynes MK17 5LX
☎ (+44 1908) 58 58 58 ⊠ 827622 scicon g
⊜ (+44 1908) 58 46 20
Manager: Jonathan Tinkler
■ Licenced evaluation facilities (CLEF). Systems assurance & evaluation for high integrity systems.

Uxbridge Office
1 Roundwood Avenue, Stockley Park,
Uxbridge, Middlesex UB11 1BQ
☎ (+44 181) 848 89 89 ⊜ (+44 181) 756 01 30
Manager (Support Operations): Neil Hancock
■ Digital terrain mapping. Logistics IT systems. Operational analysis, simulation & modelling, human factors engineering including training needs analysis.

Fleet Office
Centrum House, 101-103 Fleet Road,
Fleet, Hampshire GU13 8NZ
☎ (+44 1252) 62 21 61 ⊜ (+44 1252) 61 64 80
General Manager, Research and Studies Group: Lawrence Collingbourne
■ Research & studies. 98

80-09300 Electron Beam Processes Ltd
Bridge Wharf, Chertsey, Surrey KT16 8LJ
☎ (+44 1932) 56 65 21 ⊜ (+44 1932) 56 54 34
♧ £ 10,000 (1997) ♣ £ 1,000,000 (1997) 👥 32 (1997)
🏠 11,000 sq ft
Managing Director: D P Vincent
Director: F J Becket
Production Director: R T Short
Sales and Marketing Director: S B Nightingale
Administration Manager: J M Harris
■ Electron beam, micro-plasma and precision pulsed TIG welding and non-destructive testing services to Ministry of Defence, ISO9002, CAA standards. 98

80-09380 Electro Mechanical Systems Ltd (EMS)
Eros House, Calleva Industrial Park,
Reading, Berkshire RG7 8LN
☎ (+44 1189) 81 73 91 ⊜ (+44 1189) 81 76 13
♣ £ 6,000,000 (1996) 👥 36
Chairman: G.R. Swainston
Managing Dir: W.S. Oliver
Dir & Product Mgr: D.W. Allen
Sales and Marketing Director: J S Mead
Manufacturing Dir: D.H.F. Davies
■ Standard & custom designed electric linear actuators, AC & DC motors, gearheads, tachogenerators, encoders, coupling components, electronic drive systems, up to 500 Watts. Manipulator heads, tools & accessories for remote handling of radioactive isotopes, chemicals, inflamable or explosive substances. 98

80-09550 EPS Logistics Technology Ltd
Staplehurst Road, Sittingbourne, Kent ME10 1XS
☎ (+44 1795) 42 44 33 ⊜ (+44 1795) 42 60 35
♧ £ 1,000,000 (1996/97) ♣ £ 3,000,000 (1997)
👥 85 (1997) 🏠 65,000 sq ft
Chairman and Chief Executive Officer: H C Boucher
Business Development Manager: S Reed
☎ (+44 1795) 41 52 30
■ Design, development and manufacture of environmental protection systems, transport and storage containers and ground handling equipment for defence and aerospace (missiles, satellite sub-assemblies, aero-engines).
DRIPAK: Flexibly-joined container for transport of electronic assemblies.
DRICLAD and TEXIKOON covers.
DRICASE: Long-term storage system for high-value equipment.
Support services, project feasibility, project management. Logistics management services and CAD. → 75 98

80-09600 ERA Technology Ltd
Cleeve Road, Leatherhead, Surrey KT22 7SA
☎ (+44 1372) 36 70 00 Ⓔ info@dera.co.uk
Ⓦ http://www.era.co.uk ⊠ 264045 weahld g
⊜ (+44 1372) 36 70 99
♣ £ 19,500,000 (1995) 👥 460 🏠 3,350 sq m
Chairman: Admiral Sir Lindsay Bryson
Managing Director: M J Withers
Technology Director: Dr Shiraz Adatia
RF Technology Division Manager: Dr N Williams
Operations Support Division Manager: Mrs P A Shepherd
Purchasing Manager: Mrs J Wright
■ Research, development, design & testing to international standards in electrical & electronic fields including EMC, EMP, computers, antennas, microwaves, motors, electronic & mechanical properties of materials. EMC/EMP design, testing & consultancy, vehicle electrical systems assessment, data transmission system assessment, testing, radar signal processing consultancy, control system design & consultancy, software assessment (including safety criteria & high integrity). 98

80-09700 ESDU International plc
27 Corsham Street, London N1 6UA
☎ (+44 171) 490 51 51 Ⓔ esdu@esdu.com ⊠ 266168
⊜ (+44 171) 490 27 01
♧ £ 100,000 (1996) ♣ £ 2,300,000 (1996)
👥 40 (1996) 🏠 570 m2
Managing Dir: John Castle
Operations/Technical Dir: John Shaw
Customer Services Mgr: Ms Joyce Culling
■ Engineering design & analysis data, including computer software for aerospace & defence industries. 98

80-09730 ESL Services (GB) Ltd
(a subsidiary of ESL Ltd)
Berkeley Square House, London W1X 5LA
☎ (+44 171) 493 40 80 ⊜ (+44 171) 499 04 91
Administration Director: Peter Bradshaw 98

80-09850 European Helicopters Ltd (EHL)
(a subs. of Lynton Group Inc., USA)
Denham Airfield, Uxbridge, Middlesex UB9 5DF
☎ (+44 1895) 83 47 77 ⊠ 8953801 lynton g
⊜ (+44 1895) 83 41 94
Managing Director: Christopher Tennant
General Manager UK: G R J Holder
Sales and Marketing Manager: Janey Ellis
Technical Manager: Steve Borrowdale
■ Civilian to military conversions, upgrading of helicopters and systems. 98

80-09855 European Security Operatives Head Office
1 Queens Road, Westgate-on-sea, Thanet, Kent CT8 8AW
☎ (+44 1843) 83 60 17 ⊜ (+44 1843) 83 60 17
Managing Director, Finance and Sales: C Champ
☎ (+44 411) 05 66 15 (mobile)
Ⓔ cchamp2795@aol.com
Director and Secretary: J Morly
☎ (+44 839) 46 98 02 (pager)
■ Specialist high risk management, personal bodyguarding, VIP protection and all aspects of high risk security. 98

80-09870 Explorocean Technology Ltd
Unit 1A/B Thorpe Industrial Estate, Crabtree Road,
Thorpe, Nr Egham, Surrey TW20 8RW
☎ (+44 1784) 47 21 30
Ⓔ explorocian-technology@compuserve.com
⊜ (+44 1784) 47 30 32
Director: D. Noble
Purchasing Mgr: K. Lennan
■ Oceanographic, hydrographic, geophysical and meteorological equipment sale and hire. 98

80-09890 Expotel Hotel Reservations
(a division of The Modern Group of Companies, Jersey)
Kingsgate House, Kingsgate Place, London NW6 4HG
☎ (+44 171) 328 98 41 Ⓔ lores@expotel.atfmail.com
Ⓦ http://www.expotel.co.uk ⊠ 2620599
⊜ (+44 171) 328 80 21
♣ £ 80,000,000 + (1996) 👥 300 (1997)
Chairman: Maurice Segal
Managing Director: J Segal
Sales Office Director: Kevin Wayt
Operations Director: Stephen Kaye
General Manager - Marketing: Andrew Parsons
Marketing Services Manager: Chris Robinson
Office Manager: M O'Neill
■ The company offers an accommodation service for events in the UK and abroad. 98

80-10300 FHP International plc
178-202 Great Portland Street, London W1N 6JD
☎ (+44 171) 631 01 19 ⊜ (+44 171) 580 58 40
♣ £ 5 000 000 👥 130
Partner/Director: A E Fenn
Director: J P Doyle
■ All types of security services for military & civil installations. 98

80-10450 Fireseal International Ltd
19 Cliveden Place, London SW1 W8HD
☎ (+44 171) 730 80 45
Ⓔ firesealperry@pop3.hiway.co.uk
⊜ (+44 171) 730 19 85
👥 20
Managing Director: Dr Ian Perry
Operations Director: Denis Bloomfield
Sales and Marketing Director: Ray Averre
■ Research & development in fire retardant & fire resistant products. 98

80-10500 Flagship Training
(Vosper Thornycroft is a shareholder with GEC & Procord)
223 Southampton Road, Paulsgrove,
Portsmouth, Hampshire PO6 4QA
☎ (+44 1705) 32 13 13 ⊜ (+44 1705) 32 74 69
Managing Director: Simon Withey
Marketing Director: Phillip Knights 98

80-10600 The Flight Data Company Ltd
(a subsidiary of Bowthorpe plc)
The Lodge, Harmondsworth Lane,
West Drayton, Middlesex UB7 OLQ
☎ (+44 181) 759 34 55 Ⓔ marketing@flightdata.co.uk
Ⓦ http://www.fdata.demon.co.uk
⊜ (+44 181) 564 90 64
♣ £ 4,000,000 (1997) 👥 25 (1997) 🏠 11,000 sq ft
Chairman: R B Potter
Managing Director: Ms F Pearson
Sales and Marketing Director: Geoff Hughes
Technical Director: P Waller
Financial Controller: Keith M Beggs
Product Development Director: P T A Roberts
■ Supply of software, consultancy and design services and computer systems, for all aspects of recording, replay and analysis of flight data. 98

80-10650 Flight Operations Ltd
(a subs. of Northern Exec. Aviation Ltd)
Hangar 522, Manchester Airport,
Wilmslow, Cheshire SK9 4LL
☎ (+44 161) 436 66 66 ⊠ 668777 neamcr g
⊜ (+44 161) 436 34 50
♧ £ 400 000 (1995) ♣ £ 4 250 000 (1995) 👥 35
🏠 3 300
Managing Dir: D.L. Antrobus
Engineering Mgr: P. Whitham
Operations Mgr: G. O'Neill
■ Civil aircraft operations for the military. Target towing, EW operations, V.I.P & air ambulance flying. 98

80-10660 Fluid Gravity Engineering Ltd
Chitlee Manor, Haslemere Road,
Liphook, Hants GU30 7AZ
☎ (+44 1428) 72 77 77 ⊜ (+44 1428) 72 77 72
♣ £ 1 200 000 (1995) 👥 18 🏠 1 380 sq m
Managing Dir: R.M. Smith
Director: Dr P.G. Marriott
Sales & Marketing Dir: Dr R. Smith
Purchasing Mgr: W. Carnegie
■ Provides scientific and engineering research and development support. Specialist areas of expertise include: Computational fluid dynamics, aerodynamic Computional fluid dynamics, aerodynamic flow fields and aerothermochemistry phenomena. Thermal and structural response. Electromagnetic emission, plasma and magnetohydrodynamic phenomena. Systems performance assessment. Scientific software development and application. Scientific data assessment and presentation. Project management. 97

80-10850 FR Aviation Limited
(a Cobham plc company)
Bournemouth International Airport,
Christchurch, Dorset BH23 6NE
☎ (+44 1202) 40 90 00 ⊠ 41147 frhurn g
⊖ (+44 1202) 58 09 36
♧ £ 10,000,000 (1996) ♣ £ 48,000,000 (1996)
👥 1,800 (1997) 🏭 21,216 m2
Chairman: G F Page
Director: A G Irwin
Managing Director: C F Jones
Finance Director: P Richardson
Director of Marketing: R Rhodes
Administration Director: M Leaker
Engineering Director: M Cogan
Director: Flight Operations: J R Myers
Director: Special Projects: D E Lewis
Director: B Wallis
Purchasing Manager: Peter Browne
■ FR Aviation's fleet of Dassault Falcon 20 aircraft is used to provide contractorised air defence training for the Royal Navy, Royal Air Force and other NATO countries, including threat simulation, electronic warfare and target towing. Additional aircraft include three Dornier 228s which undertake European Economic Zone surveillance around the shores of the UK. Comprehensive flight inspection and engineering support facilities which provide services ranging from component repair and overhaul to full scale aircraft maintenance and role conversions.
Associated companies:
FRA Services, UK
FRA Serco, UK (Joint Venture)
Flight Precision, UK (JV)
Corporate Air Spares, UK
Sigma Aviation, Malaysia (JV)
Air Précision, France
AVDEF/Précision Air Services, France (JV)
Aeroflight Services Gesellschaft, Germany (JV) [98]

80-10950 Frazer-Nash Consultancy Limited
Stonebridge House, Dorking Business Park,
Dorking, Surrey RH4 1HJ
☎ (+44 1306) 88 50 50 [E] fnc@frazer.demon.co.uk
⊖ (+44 1306) 88 64 64
👥 65 (1997) 🏭 18,500 sq ft
Managing Director: Andrew G Milton
Engineering Director: William T Chester
Director of Business Management: Giles A Beattie
Technical Director: Dr Christopher C H Guyott
Marketing Co-ordinator: Miss Tracey A Ives
■ Provision of technical consultancy services and software simulation products.
Naval Systems: Design and development of systems involving acoustic and non-acoustic signature management, weapon handling and discharge, towed arrays, propulsion, advanced battery technology, ventilation, shock, blast, human protection and safety management.
Land Systems: Design and development of platforms and gun systems involving concept generation, vulnerability analysis, signature management, ride, handling and stability, structures, composite materials, occupant protection, combustion, ballistics and ammunition design.
Air Systems: Design of flight structures, including composite materials and crashworthiness, design of flight equipment including weapon release systems, gas turbine development, missile systems and noise reduction. Development and use of both infra-red signature and radar cross section prediction techniques. [98]

80-10970 Frost & Sullivan Ltd
4 Grosvenor Gardens, London SW1W 0DH
☎ (+44 171) 730 34 38 ⊖ (+44 171) 730 33 43
Managing Director: Andrew Small
Marketing Manager: Ms Lisa Shorey
Direct Marketing-Europe: Ms Mary Birabwa
■ The Company produces European, American and worldwide market research reports on the defence, aerospace and other high technology industries. Training seminars. [98]

80-11000 GECI Ltd
(a subs of GECI International, France)
Hanover House, 76 Coombe Rd,
Kingston-upon-Thames, Surrey KT2 7AH
☎ (+44 181) 541 18 77 ⊖ (+44 181) 541 46 97
♧ £ 1 000 000 ♣ £ 3 500 000
Sales Dir: John Greening [97]

80-11650 GED (Southampton) Ltd
(a subsidiary of Hebron & Medlock (Holdings) Ltd)
Omega Enterprise Park, Electron Way, School Lane,
Chandlers Ford, Eastleigh SO53 4SE
☎ (+44 1703) 27 33 22 [E] ged@hmsoton.demon.co.uk
⊖ (+44 1703) 27 33 00
♣ £ 2,200,000 (1997) 👥 85 (1997) 🏭 1,000 sq m
General Manager: L Evans
Sales and Marketing: B. Cox
Operations and Technical: D Sell
■ Engineering design, development and support services. [98]

80-11670 General Offshore (UK) Ltd
(a subsidiary of the General Offshore Corporation, USA)
Park Place, 12 Lawn Lane, London SW8 1UD
☎ (+44 171) 793 77 43 ⊖ (+44 171) 735 63 23
♣ £ 5,000,000 (1997) 👥 12
Managing Director: A J Smith
Marine Projects Manager: Nigel Shaw
Operations Manager: P F Williams
■ Marine engineering, submarine cable installation & burial, beach civil works, underwater range maintenance & installation, vessel operation & maintenance. [98]

80-11775 GKN Westland Design Services
(a subs. of GKN Westland Aerospace Ltd)
Winterstoke Rd, Weston-Super-Mare, Avon BS24 9AB
☎ (+44 1934) 63 55 55 ⊠ 4441360
⊖ (+44 1934) 41 51 88
👥 300 🏭 33 800 sq ft
Dir & General Mgr: Mike Breward
Business Development Mgr: Tony Gray
■ Comprehensive design & engineering service to aerospace & defence related industries in aerostructures, mechanical systems, electrical installation, avionics, weapon systems, powerplant installation, hydraulics, ground support equipment. Finite element stress analysis, weights engineering & technical publications. Supply of agency staff for long or short term contracts. [97]

80-11800 GKN Westland Systems
(† Westland System Assessment Ltd)
(a subs. of GKN Westland Aerospace)
91 Preston Road, Yeovil, Somerset BA20 2DN
☎ (+44 1935) 40 11 00 ⊖ (+44 1935) 41 10 51
♣ £ 5 000 000 (1995) 👥 120 🏭 20 000 sq m
General Mgr: David J. Wright
Training Systems Mgr: J.A. Davidson
Operational Studies Mgr: P.K. Hulett
■ Defence operational analysis, training systems & logistic support analysis. [98]

80-11952 Gracemoor Consultants (UK) Ltd
25 Gilbert Street, London W1Y 1RJ
☎ (+44 171) 629 48 82 [E] gracemoor@dial.pipex.com
⊖ (+44 171) 493 01 76
👥 6 + consultants
Managing Director: Dr B L G Carre
Programmes Director: G G Birchal
Defence Adviser: Maj Gen A C P Stone
■ Consultants in strategy, market research, project and bid support, lobbying, representation and marketing. Formed in 1986 Gracemoor is a specialist in Anglo-French defence relationships. Its client base extends throughout the European Union, NATO and the Middle East. [98]

80-12050 H+S Aviation Ltd
(a subsidiary of Vector Industries Ltd)
Airport Service Road, Portsmouth, Hampshire PO3 5PJ
☎ (+44 1705) 30 40 00 ⊠ 86403 hxs g
⊖ (+44 1705) 30 40 20
♧ £ 7,500,000 (1996) ♣ £ 34,300,000 (1996) 👥 357
🏭 200,000 sq ft
Managing Director: Alan C Robinson
☎ (+44 1705) 30 40 25
Finance Director: John E R Barker ☎ (+44 1705) 30 40 21
Marketing and Business Development Manager:
David M McRobert ☎ (+44 1705) 30 40 22
■ Overhaul and repair:
Engines: RR Nimbus, RR Dart, P&W Canada PT6 T-series, PW100 and PW901 APU. Overhaul and repair of Allison model 250, General Electric CT7/T700 and CT58.
Propellers: Hamilton Standard, Hartzell, Dowty, McCauley, BAe Dynamics Group. Mechanical Components for Bendix, Scintilla and Woodward.
Fuel systems: Bendix, Chandler Evans, Hamilton Standard, TRW.
Training: Engineering courses for P&W Canada PW100, PT6, general electric CT7/T700 Allison Model 250.
Robinson Way Office
Robinson Way, Portsmouth, Hampshire PO3 5TH
☎ (+44 1705) 30 40 20 ⊠ 86343 hxs g
⊖ (+44 1705) 30 40 20
■ Overhaul and repair: Allison Model 250, General Electric CT7/T700 & CT 58, Rolls-Royce Dart and Nimbus engines and associated components. [98]

80-12220 H. Hartley Engineering Design & Recruitment Consultants
The Design Centre, 25 Wellington St,
Kirkham, Lancashire PR4 2JN
☎ (+44 1772) 68 52 23
Managing Director: Harry Hartley
■ Engineering design consultants: project engineers, designers, draughters, jig and tool designers; systems design (eg. flying/parachute); recruitment consultants for technical, managerial, secretarial and manual staffing requirements. [98]

80-12240 Hawkeye Surveillance Systems Ltd
Brigg Road, Scunthorpe, South Humberside DN16 1AU
☎ (+44 1724) 28 94 11 ⊖ (+44 1724) 27 13 49
Managing Director: David G King
Sales and Marketing: Mrs H Hind
■ Surveillance systems utilising low-light intensified cameras (24 hour surveillance capability in daylight, dusk and total darkness) without illumination. [98]

80-12250 R. Hazel & Co
(a subsidiary of Court & Judicial Publishing Company Ltd)
PO Box 39, Henley-on-Thames, Oxon RG9 5UA
☎ (+44 1491) 64 10 18
Managing Dir: C.G.A. Parker
Publisher: C. George
■ Publisher of the Police and Constabulary Almanac, The Annual Official Register of UK Police Forces and Organisations, International Security Directory, Hazell's Guide and the Bar List. [98]

80-12300 HeavyLift Cargo Airlines Ltd
(a member of the Trafalgar House Group)
London-Stansted Airport, Stansted, Essex CM24 1QW
☎ (+44 1279) 68 06 11 ⊠ 817618
⊖ (+44 1279) 68 06 15
♣ £ 25 000 000 👥 120
Managing Dir: Michael Hayles
Commercial Dir: Graham Pearce
■ Specialists in the transportation of outsize loads particularly for the aerospace & defence industries.
HeavyLift Aircraft Engineering Ltd
Southend Airport, Southend-on-Sea, Essex SS2 6YU
☎ (+44 1702) 54 16 16 ⊠ 995522 hleng g
⊖ (+44 1702) 53 03 00
■ Maintenance & refurbishment of civil transport aircraft particularly B707/727/737 & BA 146 [97]

80-12745 Hunterskil Howard plc
(a subs. of Wolseley plc)
Technical Information Division
Mitchell House, Southampton Rd,
Eastleigh, Hants SO50 5PA
☎ (+44 1703) 64 45 54 ⊖ (+44 1703) 64 45 02
♧ £ 10 360 000 (1996) ♣ £ 5 000 000 (1996) 👥 110
🏭 1 200
Managing Dir: Alan McBride
Operations/Technical Dir: Tony Watkins
Sales & Marketing Dir: Calvin Barnett
Purchasing Mgr: Tony Abbott
■ Technical documentation: User, maintenance & software manuals, illustrated parts catalogues & information systems building. Experience includes: Communications, radar, weapons systems, avionics, simulators, test equipment, rail plant & propulsion, computer & software systems. Training: Course design, development & presentation, training manuals & CBT. Information retrieval systems, multi-media, CD-ROM, SGML, CALS, Quality Assurance to BS5750 Part 1, ISO9001. [97]

80-12760 Hughes International Corporation
140 Park Lane - Suite 4, London W1Y 3AA
☎(+44 171) 491 07 29 ⊜ (+44 171) 493 29 53
Vice President: W W Bender
® Belgium: Hughes Europe SA/NV [98]

80-12785 Hunting Aviation Ltd
(a subsidiary of Hunting plc)
Pembroke House, East Midlands Airport,
Castle Donington, Derby DE74 2SL
☎(+44 1332) 81 37 47 [E] huntavia@compuserve.com
⊜ (+44 1332) 85 36 18
♣ £ 220,000,000 (1996) 👥 4,270
Chief Executive: Tom L Grievson
Finance Director: Gavin H Gracie
Strategic Planning Manager: Stephen J Pain
Marketing Services Manager: Niall D Buntain
■ Engineering, repair, overhaul, maintenance, modification and completion of airframes, engines, avionics and aircraft accessories. Aviation contracting services. Fixed base and engineering operations, air cargo charter and ad hoc operations. [98]

80-12786 Hunting-BRAE
(a subsidiary of Hunting Defence Ltd)
AWE, Aldermaston, Reading, Berkshire RG7 4PR
☎(+44 1189) 82 41 11 ⊜ (+44 1189) 82 53 20
♧ £ 11,000,000 (1996) ♣ £ 300,000,000 (1996)
👥 5,200 (1996)
Chairman: T L Grievson
Managing Director: R A Bradley
Director Finance: S Ash
Commercial Director: P J Clifton [98]

80-12790 HVR Consulting Services Ltd (HVR-CSL)
Selborne House, Mill Lane, Alton, Hampshire GU34 2QJ
☎(+44 1420) 879 77 ⊜ (+44 1420) 898 19
♣ £ 3,250,000 (1997) 👥 58 🏭 7,500 sq ft
Managing Director: Peter D Cook
Director: J Coyle
Director: C A Dalton
General Manager: R Atkinson
■ Risk assessment and equipment cost analysis for UK MoD and defence related industry, operational analysis and modelling studies software development and management. Trials design and management including the analysis of trials data from military equipment trials concerned with system effectiveness, safety and fitness for role. Support to MoD in the assessment of tenders from industry, including the use of decision analysis software and to industry in the preparation of bids to MoD. Database application development for MoD project offices and technical establishments. [98]

80-12900 Hyder Consulting Limited
(† Acer Consultants Limited)
(a subsidiary of Hyder plc)
2-3 Cornwall Terrace, Regent's Park,
Regents Park, London NW1 4QP
☎(+44 171) 544 66 00
[E] 100351.1331@compuserve.com
[W] http://www.hyder.com ⊜ (+44 171) 470 00 19
♧ £ 65,012,000 ♣ £ 133,942,000 (1996)
👥 3,206 (1997)
Executive Director: Stuart Doughty
Sales and Marketing Director: Ian Watson
Advertising and Public Relations Manager: Mary Melvin
■ Provision of technical, environmental and management services worldwide. Range of planning, design, project management and management contracting services for defence sector work. Projects have been undertaken for military and defence authorities in the UK, the USA, Cyprus, Malaysia, Oman, Saudi Arabia, the UAE and elsewhere. Work has been undertaken for army, navy and air force bases and installations on many kinds of technical, staorage and administration buildings, hardened control centres, mechanical handling systems and berthing facilities, associated building services, housing, leisure and medical facilities, runway, roads, drainage, security fencing, CCTV and communications systems.

Birmingham Office
Aston Cross, Rocky Lane, Aston, Birmingham B6 5RQ
☎(+44 121) 333 44 66 ⊜ (+44 121) 333 42 75
Contact: Stewart Tyler

Penarth Office
Plymouth House, Plymouth Road,
Penarth, Cardiff CF64 3YF
☎(+44 1222) 70 43 21 ⊜ (+44 1222) 70 97 93

Hyder Environmental
Beacon House, William Brown Close,
Llantarnum Park, Cwmbran NP44 3AB
☎(+44 1633) 87 65 11 ⊜ (+44 1633) 87 08 99
Contact: Ms Clare Dugdale

Southern Laboratories
Mid Glamorgan Science Park, 2 Technology Drive,
Bridgend CF31 3NA
☎(+44 1656) 64 75 57 ⊜ (+44 1656) 64 65 25
Contact: Geraint Venn

Northern Laboratories
Howard Court, Manor Park, Runcorn, Cheshire WA7 1SJ
☎(+44 1928) 57 99 55 ⊜ (+44 1928) 57 99 70
Contact: Ms Yvonne Bussell
Contact: Peter Dignum

Runcorn Office
Howard Court, Manor Park, Runcorn, Cheshire WA7 1SJ
☎(+44 1928) 57 99 55 ⊜ (+44 1928) 57 92 06
Contact: Simon Binyon

Belfast Office
Rivers Edge, 13-15 Raven Hill Road, Belfast BT6 8DN
☎(+44 1232) 46 06 57 ⊜ (+44 1232) 46 15 34
Contact: Roy Deddis

Civil Engineering Laboratory
Griffin Lane, Aylesbury, Buckinghamshire HP19 3BP
☎(+44 1296) 865 00 ⊜ (+44 1296) 865 88
Contact: David Frankland [98]

80-13100 ICL Enterprises
International Computers Ltd
Observatory House, Windsor Rd, Slough, Berks SL1 2EY
☎(+44 1753) 60 44 05 ⊜ (+44 1753) 60 48 00
♣ £ 250 000 000 👥 6 000
Managing Dir: Tim Gibson
Marketing Mgr (Defence): Graham Cooper
■ Provides integrated systems solutions & managed services to meet user requirements. Supplies & manufactures a full range of PC, midrange & mainframe computer systems. Provides security, FM, training systems design & integration. Supplies CHOTS infrastructure, UKAIR CIS, FAAIP. Supplies UNIX systems through DGCIS catalogue. → 75 [97]

80-13310 Inovex Technology Ltd
33 St. Thomas Street, Winchester, Hants SO23 9HJ
☎(+44 1962) 86 65 86 ⊜ (+44 1962) 87 78 14
♣ £ 1 300 000 (1994) 👥 19
Managing Dir: Dr M.G. Notley
Commercial Dir: J.R. D'Silva
Marketing Mgr: C.J. Wood
■ Specialist in the design of leading edge, complex analogue & digital micro-electronics & sophisticated software systems. Hardware expertise includes designs using ASICS, thick films, EPLDS, analogue & digital VLSI, fibre optics & power electronics spanning DC to microwave bandwidths. Software skills include low & high level languages & operating systems on micro, mini & mainframe platforms. ADA, Fortran C, C, Assembler, Pascal & Basic are some of the languages utilized under UNIX, OS/9, MS-DOS, OS/2, VMS & VM. ISO 9001 (TICK IT). Supplier of Part Task Trainers & operational analysis & planning tools for short range air defence (SHORAD) applications. [97]

80-13320 INRI UK Ltd
(a subsidiary of Inter-National Research Institute Inc, USA)
Alpha House, Chilworth Research Centre,
Southampton SO16 7NS
☎(+44 1703) 76 04 84 [E] sales@inri.co.uk
[W] http://www.inri.com ⊜ (+44 1703) 76 04 83
♧ £ 907,579 (1996) ♣ £ 1,522,641 (1996)
👥 45 (1997) 🏭 1,100 sq m
Managing Director: Brian E Luff
Purchasing Manager: Miles Bulpitt
■ Computer software company specialising in command & control systems for the military, developing the architectural framework of command, control, communications & intelligence (C3I) systems.
® USA: Inter-National Research Inst. [98]

80-13328 Instrument Services Ltd
(a subsidiary of Dowding & Mills plc)
60 Wilbury Way, Hitchin, Hertfordshire SG4 0TA
☎(+44 1462) 42 12 34 ⊜ (+44 1462) 42 00 12
♧ £ 6,000 (1996) ♣ £ 10,000,000 (1996)
👥 150 (1997) 🏭 27,000 sq ft
Chairman: J.H.R. Cole
Managing Dir: Kenneth H. Holden
Director: Mike L. Hume
Sales Dir: Brian E. Willats
■ Maintenance, repair and calibration of electronic equipment. Supply of products including Tempest and RFI/EMC test equipment, fibre optic modems, multiplexers, LAN's, EMP/P pulse products and Tempest approved computers and peripherals.

● 71b Whitecraigs Road, Glenrothes, Fife KY6 2RX
☎(+44 1592) 77 30 08 ⊠ 727795
⊜ (+44 1592) 77 28 77
Managing Dir: Kenneth H. Holden
General Mgr: Bill J. Findlay
■ Maintenance repair & calibration of electronic equipment.

● Fulwood Road South, Fulwood Industrial Estate,
Sutton-in-Ashfield, Notts NG17 2JZ
☎(+44 1623) 55 51 10 ⊜ (+44 1623) 55 50 22
Managing Dir: Kenneth H. Holden
■ Maintenance, repair & calibration of electronic equipment.

● 60 Wilbury Way, Hitchin, Herts SG4 0TA
☎(+44 1462) 42 12 34 ⊜ (+44 1462) 42 00 12
Managing Dir: Kenneth H. Holden
Operations Dir: Michael L. Hume
Sales Engineer: Stewart Baker
■ Supply of products including Tempest and RFI/EMC test equipment, fibre optic modems, multiplexers, LAN's, EMP/P pulse products and Tempest approved computers and peripherals. → 85 → 90
® USA: Antenna Research Associates, Cominvest System AB, Dynamic Sciences Inc., Instruments for Ind., PRANA Research & Development, RFI Corporation, Systemware Inc., Thomson-CSF, Varian Associates, Versitron

● 11 Pit Road, Bellshill Industrial Estate,
Bellshill, Lanarkshire ML4 3JA
☎(+44 1698) 84 48 44 ⊜ (+44 1698) 74 08 61
Branch Manager: Ken W Baxter

● Gooch Avenue, Barrington Industrial Estate,
Bedlington, Northumberland NE22 7DQ
☎(+44 1670) 82 51 60 ⊜ (+44 1670) 82 44 59
Branch Manager: Paul McCarron

● Park Gate Close, Bredbury Park Way,
Bredbury, Stockport SK6 2SL
☎(+44 161) 406 78 78 ⊜ (+44 161) 406 79 79
Branch Manager: Paul Jennings [98]

80-13330 Interactive Development Environments Ltd (IDE)
1 Stirling House, Stirling Road, The Surrey Research Park,
Guildford GU2 5RF
☎(+44 1483) 57 90 00 ⊜ (+44 1483) 312 72
👥 7
Managing Dir: C. Kenber
Sales Mgr: Adrian Jones
■ Provides interactive software development environments. [97]

80-13340 Intereurope Technology Ltd
(a subsidiary of Intereurope Technology Services plc)
Head Office
21/23 East Street, Fareham, Hampshire PO16 0BZ
☎(+44 1329) 82 30 47 ⊜ (+44 1329) 82 20 58
♣ £ 2,000,000 (1997) 👥 120 (1997)
Chairman and Managing Director: D H Immanuel
Financial Dir & Company Secretary: A.R. Horn
Group Coordinator, MoD: T. O'Riordan
■ Preparation of engineering documentation, operating & maintenance documentation, software design & documentation, engineering design & prototype manufacture, drawing office services, training, translation services, project management, recruitment of contract & permanent technical staff. [98]

80-13345 Intermec UK Ltd
(a subsidiary of Intermec Corporation USA)
Attenborough House, 15 Bennet Road,
Reading, Berkshire RG2 0QX
☎(+44 734) 87 65 94 [W] http://www.intermec.com
⊠ 847244 intermec g ⊜ (+44 734) 87 63 05
♣ £ 7,000,000 (1997) 👥 44 (1997)
Managing Director: John Gover
UK Sales Manager: J A Jackson
■ Intermec UK provide a complete range of equipment and services for automated data capture solutions using bar coding, RF networks, smart cards and RF tags. [98]

80-13410 International Hospitals Group
Hertford Place, Denham Way,
Rickmansworth, Hertshire WO3 2XB
☎ (+44 1923) 71 01 71 ⊖ (+44 1923) 71 01 10
♣ £ 100 000 000 👥 4 000+
Group Managing Dir: Hertford M. King
Purchasing Mgr: Stuart Hembury
Business Mgr: Ms Margaret Jennings
■ Healthcare services. Hospital management - Middle East Military Hospitals. Hospital Planning. Healthcare consultancy. [97]

80-13500 Istec Consultants
Spinney Estate, Hoddesdon Road, St Margarets,
Ware, Hertfordshire SG12 8EJ
☎ (+44 1920) 87 17 34 ⊖ (+44 1920) 87 13 57
■ Designs and supplies specialised military equipment, primarily related to weapon systems technology. Small arms, automatic cannons, weapon ancillaries, anti-riot systems, gun platforms and EOD. [98]

80-13520 ISVR Consultancy Services
(part of the Institute of Sound & Vibration Research)
Acoustics Laboratory
University of Southampton, Highfield,
Southampton, Hampshire SO17 1BJ
☎ (+44 1703) 59 21 62 ⊖ (+44 1703) 59 27 28
♣ £ 1 000 000 👥 12
Manager: R.A. Davis
■ The laboratories of the Institute of Sound & Vibration Research (ISVR) are available for use by external organisations. Acoustic facilities include anechoic & reverberation chambers & the A.B.Wood laboratory with an 8m x 8m x 5m tank for underwater work, supported by a wide range of expertise in the analytical, experimental, technical & instrumentation aspects of acoustics. [98]

80-13550 ITEC Ltd
51 Market Place, Warminster, Wiltshire BA12 9AZ
☎ (+44 1985) 84 61 81 [E] itec@andrich.com
⊖ (+44 1985) 84 61 63
Director: Richard Curtis
■ Conference & exhibitions organisers for the training & simulation industry. [98]

80-13570 ITT Defense Ltd
(† ITT Defense International UK)
Jays Close, Viables Estate,
Basingstoke, Hampshire RG22 4BW
☎ (+44 1256) 31 16 00 ⊖ (+44 1256) 84 05 56
Managing Dir: J.D.B. Craen
Sales & Marketing Dir: P.G. Bedwin
Operations & Technical Dir: G. Young
Purchasing Mgr: P. Revill
® USA: ITT Defense [97]

80-13700 Jane's Information Group Ltd (JIG)

Jane's INFORMATION GROUP

Sentinel House, 163 Brighton Road,
Coulsdon, Surrey CR5 2NH
☎ (+44 181) 700 37 00 [E] info@janes.co.uk
[W] http://www.janes.com ⊠ 916907 janes g
⊖ (+44 181) 763 10 05
Managing Director: Alfred Rolington
Publishing Director: Bob Hutchinson
Advertisement Sales Director: Simon Kay
Production Director: David Moyes
Financial and Commercial Director: Mike Staton
Director: Ms Gina Dawkins
Director: Joe Bria
Publisher: Ian Kay
Publisher: Ms Fabiana Angelini
Publisher: Alan Condron
Publisher: Ms Janine Boxall
Company Secretary: Bernard Laverick
■ Jane's Information Group is the leading supplier of professional information to the defence and aerospace communities, and a major information provider to selected areas of transportation worldwide.
Sub:
Jane's Information Group (USA) → 80
Defence Publications:
Jane's Aero-Engines (binder)
Jane's Armour & Artillery Upgrades (yearly)
Jane's Air-Launched Weapons (binder)
Jane's All The World's Aircraft (yearly)
Jane's Ammunition Handbook (yearly)
Jane's Armour & Artillery (yearly)
Jane's Avionics (yearly)
Jane's C4I Systems (yearly)
Jane's Aircraft Upgrades (yearly)
Jane's Defence Systems Modernisation (monthly magazine)
Jane's Defence Upgrades (newsletter)
Jane's Defence Spending Worldwide (binder)
Jane's Defence Weekly
Jane's Electro-Optic Systems (yearly)
Jane's Electro-Optic Markets (binder)
Jane's Fighting Ships (yearly)
Jane's Helicopter Markets and Systems (binder)
Jane's Infantry Weapons (yearly)
Jane's Intelligence Review (monthly magazine)
Jane's International Defence Directory (yearly)
Jane's International Defense Review (monthly magazine)
Jane's Land-Based Air Defence (yearly)
Jane's Major Warships (yearly)
Jane's Marine Propulsion (binder)
Jane's Military Communications (yearly)
Jane's Military Vehicles & Logistics (yearly)
Jane's Mines and Mineclearance (yearly)
Jane's Missiles & Rockets (newsletter)
Jane's Naval Construction and Retrofit Markets (binder)
Jane's Naval Weapon Systems (binder)
Jane's Navy International (monthly magazine)
Jane's NBC Protection Equipment (yearly)
Jane's Radar & EW Systems (yearly)
Jane's Simulation & Training Systems (yearly)
Jane's Strategic Weapon Systems (binder)
Jane's Underwater Technology (yearly)
Jane's Underwater Warfare Systems (yearly)
Jane's Unmanned Aerial Vehicles & Targets (binder)
Jane's US Chemical/Biological Defense Guidebook (yearly)
Jane's World Air Forces (binder)
Jane's World Defence Industries (binder)
Jane's World Armies (binder)
Aerospace Publications:
Jane's International ABC Aerospace Directory (yearly)
Jane's Space Directory (yearly)
Jane's Aircraft Component Manufacturers (binder)
Jane's Air Traffic Control (yearly)
Jane's Airports, Equipment & Services (yearly)
Jane's Airports & Handling Agents (yearly)
Jane's Airport Review (magazine)
Jane's High-Speed Transportation (yearly)
Jane's Urban Transport Systems (yearly)
Jane's World Airlines (binder)
Jane's World Railways (yearly)
Electronic Publications:
Most publications, including magazines, are available on CD-ROM for Personal Computers and many UNIX platforms. Jane's Defence Magazines are available ONLINE.
Jane's Sentinel (Regional security, economic & commercial assessments):
Central Europe and the Baltic (regional security assessment)
Balkans (regional security assessment)
Eastern Mediterranean (regional security assessment)
South China Sea (regional security assessment)
South China Sea (regional economic & commercial assessment)
Southern Africa (regional security assessment)
Commonwealth of Independent States (regional security assessment)
China & North East Asia (regional security assessment)
South Asia and Indian Ocean (regional security assessment)
South America (regional security assessment)
Gulf States (regional security assessment)
Central Africa (regional security assessment)
Central America and the Caribbean (regional security assessment)
Gulf States (regional economic and commercial assessment)
North Africa (regional security assessment)
Special Reports:
Published market special reports
Commissioned special reports & studies
Police and Security Publications:
International Police Review (magazine)
Jane's Police and Security Directory (Europe) (yearly)
Jane's Police and Security Equipment (yearly)
Police Review (weekly magazine)
Police Science and Technology Review (quarterly magazine)
Policing Today (quarterly magazine)
Public Safety Professional (report in 3 volumes) [98]

80-13850 Jeppesen UK Ltd
(a subsidiary of Jeppesen Sanderson)
Kingston House, Stephenson Way, Three Bridges,
Crawley, West Sussex RH10 1TN
☎ (+44 1293) 54 77 81 ⊠ 877881 fltpln g
⊖ (+44 1293) 55 25 46
♣ £ 1,800,000 👥 24 (1997)
Director: Edward Hazelwood
Manager, Sales and Service: Derek Towns
■ Distributor of Jeppesen Dataplan computerised flight advisory systems. Jetplan: Computerised flight planning service. Metplan: Computerised graphic & non-graphic weather service. 24 hour operational flight support. Aeronautical charting service. [98]

80-13870 John Laing International Ltd
Paramount House, Maylands Avenue,
Hemel Hempstead, Herts HP2 4XH
☎ (+44 1442) 27 55 00 ✉ ROTCARTNOC London
⊖ (+44 1442) 24 30 61
♧ £ 198,000,000 (1996) ♣ £ 1,254,000,000 (1996)
👥 8,600 (1996)
Managing Dir: J. Cockburn
Sales & Marketing Mgr: K. Dawson
Estimating Mgr: V.W. Allen
Purchasing Mgr: G. Baker
■ Company involved in the design, management, construction & equipping of all types of defence related projects. [98]

80-13900 John Reed (Specialist Market Research & Analysis)
(† Defence Industry)
15 Tremena Gardens, St Austell, Cornwall PL25 5QH
☎ (+44 1726) 612 25 ⊖ (+44 1726) 612 25
👥 4
Managing Dir: John Reed
■ Understakes specialist research & analysis in areas of particular interest to the defence industry.
● 15 Truro Road, St Austell, Cornwall PL25 5JE
☎ (+44 1726) 612 25 ⊖ (+44 1726) 612 25 [98]

80-13950 Kalamos International Ltd
(a member of the Polaris Group)
North Star Business Park,
Woodrow Way, Gloucester GL2 6DX
☎ (+44 1452) 38 00 00 [E] stevekalamos@polaris.co.uk
⊖ (+44 1452) 38 44 38
■ Geophysical technical marketing & services. [98]

80-14000 Robert Keen Ltd
Anchor Street, Chelmsford, Essex CM2 0LB
☎ (+44 1245) 35 55 41 ⊖ (+44 1245) 49 24 03
♣ £ 2 000 000 👥 45 🏭 12 000 sq ft
Joint Managing Dir: V.S. Trew
Joint Managing Dir: R.D. Nicks
■ Precision sub-contract machining, specialising in CNC milling & turning, gun drilling & EDM. BS 5750 & AQAP-4 approved. [98]

80-14180 Kockums Computer Systems (UK) Ltd
(a subsidiary of Kockums Computer Systems AB, Sweden)
Armstrong Technology Centre, Davy Bank,
Wallsend, Tyne and Wear NE28 6UY
☎ (+44 191) 263 60 00 ⊠ 53476
⊖ (+44 191) 262 64 00
♣ $ 1,500,000 (1997) 👥 35 (1997)
President and Sales/Marketing Director: Roy Metcalfe
Operations/Technical Director: A Drozdowicz
Purchasing Manager: P Gilroy
■ Writing, selling and supporting software for the design and building of ships. [98]

80-14200 Landair International Ltd
Brickworth Lane, Whiteparish,
Salisbury, Wiltshire SP5 2QE
☎ (+44 1794) 88 42 29 ⊖ (+44 1794) 88 45 81
♣ US$ 2,500,000 (1996/97) 👥 20
Chairman: Brian R Barry
Technical Director: Ken B Brown
Sales and Marketing Director: Gerry Dodd

Project Manager: Ms Rowena Padel
■ Analysis & solution of complex defence & security problems through the medium of studies & where appropriate, mathematical modelling. Balanced military, scientific & technical expertise of operations, equipment & markets worldwide. [98]

80-14210 Laser-Scan Ltd
(a subs. of Laser-Scan Holding plc)
Cambridge Science Park, Milton Rd, Cambridge CB4 4FY
☎(+44 1223) 42 04 14 ☒ 817346
⊜ (+44 1223) 42 00 44
£ 100 000 (1996) ♣ £ 6 172 000 (1995) 112
34 000 sq ft
President: W. Coleman
Deputy Chairman: Peter A. Woodsford
Managing Dir & CEO: Dr M.J. Jackson
Finance Dir: Roger J. Winter
Government & Defence Sales Mgr: Mike J. Stanbridge
Government Projects Mgr: Bill Savage
■ Specialists in digital map production, the exploitation of digital map data for defence purposes & the integration of remotely sensed data. It supplies GIS (Geographic Information Systems) software specifically designed for integration into wider complex applications eg: C3I, mission planning.
Object oriented database & applications development environment software. [97]

80-14215 La-Vey Security International Ltd
PO Box 393, Alton, Hampshire GU34 5YL
☎(+44 1420) 56 14 42 ⊜ (+44 1420) 56 14 52
Managing Dir: Dean La-Vey
Sales & Marketing Dir: Richard Burletson
■ Specialist in electronic counter-measures. Specialised products & techniques. [97]

80-14218 Leafield Logistics & Technical Services Ltd
(a subsidiary of Leafield Holdings Ltd)
Leafield Industrial Estate, Corsham, Wiltshire SN13 9SS
☎(+44 1225) 81 07 71 ⊜ (+44 1225) 81 20 96
♣ £ 4,000,000 (1996) 25 (1996)
Managing Director: R L Tonge
Operations Manager: R Shoemark
Purchasing Manager: N Coleman
■ Logistic and technical support of ex-UK Royal Navy vessels purchased and operated by other Navies. Inventories are held for propulsion, weapons and general spares. [98]

80-14300 Lockheed Martin Tactical Technologies
71 Buckingham Avenue, Slough, Berkshire SL1 4PN
☎(+44 1753) 69 64 88 ⊜ (+44 1753) 69 62 18
Programme Director: David J Peck

Electronic Systems
71 Buckingham Avenue, Slough, Berkshire SL1 4PN
☎(+44 1753) 69 64 88 ⊜ (+44 1753) 69 62 18
Programme Director: David J Peck
■ Technical support point of contact & programme management for military electronics systems. Electronic warfare, missile warning systems, range support systems, antennas, missile systems & fuzes, related simulators, EO sensor/cameras, display systems & space information technology.

Test & Information Systems
71 Buckingham Ave, Slough, Berkshire SL1 4PN
☎(+44 1753) 69 64 88 ⊜ (+44 1753) 69 62 18
Sales Manager Europe: David Heyes
■ Data acquisition and analysis systems for real-time telemetry and space TT&C applications. Telemetry tracking, reception and video compression for aircraft, missile and space applications. MIL-STD-1553 A/B bus products for avionics testing, simulation and monitoring. Telemetry receivers. Virtual instrument product line for 1553/PCM telemetry. Full product/hardware maintenance support in country.

Advanced Distributed Simulation Division
71 Buckingham Avenue, Slough, Berkshire SL1 4PN
☎(+44 1753) 69 64 88 ⊜ (+44 1753) 69 62 18
Director European Operations: Steven Fleming
■ Marketing & support point of contact for core technologies related to Synthetic Environments, specifically computer image generators, vehicle simulations, computer generated forces, systems integration & networking products. [98]

80-14350 Logica UK Limited
(† Logica Defence & Civil Government Ltd)
(a subs of Logica plc)
Cobham Park, Downside Rd, Cobham, Surrey KT11 3LG
☎(+44 171) 637 91 11
⊜ (+44 1932) 86 91 02; 86 91 04
♣ £ 228 000 000 3 400
Chairman: Cliff Preddy
Managing Dir acting: Laurence Julien
Sales & Marketing Dir: Gerry O'Keefe
Technical Dir: John Millar
Defence Systems Division Dir: Anthony Cole
Civil Government & Consultancy Division Dir: Ms Louise Bennett
Civil Government Business Development Mgr: Ms Sandy Hopkinson
Sales & Marketing Mgr Defence: Denis Harris
■ Services in consultancy, project management, operational analysis, custom built computer & communications systems, including software & hardware implementation & products.
● 68 Newman Street, London W1A 4SE
☎(+44 171) 637 91 11 ⊜ (+44 171) 637 82 29 [97]

80-15100 MacDonald Dettwiler Ltd
Broadmede Farnham Business Park, Weydon Lane, Farnham, Surrey GU9 8QL
☎(+44 1252) 71 71 12 ⊜ (+44 1252) 71 04 45
Business Development Mgr: P. Walton [97]

80-15150 Mackenzie Tribbeck Associates (MTA)
(a division of MTA Holdings Ltd)
Springfield House, 23 Oatlands Drive, Weybridge, Surrey KT13 9LZ
☎(+44 1932) 23 12 77 [E] mta@mta.co.uk
[W] http://www.mta.co.uk ⊜ (+44 1932) 25 42 07
£ 374,000 (1997) ♣ £ 1,300,000 (1997)
30 (1997) 557 sq m
Managing Consultant: H R Cairns
Marketing Director: R Thomson
Financial Controller: D P Arter
Purchasing Manager: D P Arter
■ Production of computer systems across a varied spectrum of applications in government, health, aerospace, industry, finance and communications sectors. Provides systems management support in all sectors including requirements capture and specifications, concept studies, feasibility and project definition studies. Research and development in science and technology in such fields as surveillance, image processing, data compression, decision support, simulation, modelling and gaming. These separate operating companies are wholly owned by MTA Holdings Ltd. [98]

80-15180 Mantix Systems Ltd
Mantix House, London Road, Bracknell, Berks RG2 2XH
☎(+44 1344) 30 15 15 ⊜ (+44 1344) 30 10 83
♣ £ 3 500 000 50
Managing Dir: Tom Ulrich
Operations/Technical Dir: Dick Eve
Sales Dir: Martin Hess
Marketing Mgr: Ms Mary Anne Trotman
■ Provides a range of products, support & consultancy services, including in-house & on-site training, software & technical support, business analysis, consultancy & product customisation, based around the "Cascade" programme management system. [97]

80-15620 Marr Technical Services Ltd
(a subsidiary of J Marr Ltd)
St Andrew's Dock, Hull, East Yorkshire HU3 4PN
☎(+44 1482) 32 78 73 ✉ MARR HULL ☒ 592214
⊜ (+44 1482) 32 11 87
5 (1996)
Managing Director: J G Marr
Director: David B Gallacher
Commercial Director: B J Hind
■ Naval architects and project managers. Experienced in conversion, refit and modernisation of defence related research survey and trials ships. [98]

80-15650 MASS Consultants Limited
(a subs. of Silicon Valley Group plc)
Grove House, Great North Rd, Little Paxton, Cambs PE19 4EL
☎(+44 1480) 47 22 11 ⊜ (+44 1480) 40 73 66
♣ £ 2 800 000 55 750 sq m
Managing Dir: Dr Ray Ayres
Business Operations Dir: Malcolm Lowes
■ MASS is a Technical Consultancy specialising in Systems, Software & Electronic Engineering. Particular expertise exists in guided weapons, radar, electro-optics, C3I, electronic warfare, aerodynamics, mathematical modelling & simulation. In addition, the Company has established a capability in the development of embedded software & the design & build of special purpose test, monitoring & hardware-in-the-loop simulation systems related to the above. [97]

80-16300 Metra Martech Limited
Glenthorne House, Hammersmith Grove, London W6 0LG
☎(+44 181) 563 06 66 [E] research@metra.demon.co.uk
[W] http://www.metra-martech.com
⊜ (+44 181) 563 00 40
♣ £ 1,400,000 21
Managing Director: Peter Gorle
Senior Consultant: Andrew Clive
Senior Consultant: John White
■ Specialist consultancy based on technically knowledgeable staff and wide international experience. Market analysis and business development special expertise. Aerospace background, knowledge on lasers, sensors, advanced materials, air & water pollution. [98]

80-16450 MFD International Ltd
(† MFD Consultancy Ltd)
Bell Hse, 32 Bell St, Romsey, Hants SO51 8GW
☎(+44 1794) 51 61 71 ⊜ (+44 1794) 52 44 60
£ 300 000 ♣ £ 1 230 000 30 1 360 sq m
Managing Dir: R. Dixon
Director: H.A. Bays
Engineering Mgr: J. Hitchings
Engineering Mgr: R. Edwards
Engineering Mgr: J. Sercombe
Engineering Mgr: D. Cooley
■ Consulting engineers & designers with qualified staff in mechanical, electrical, civil & structural, environmental, computing disciplines. Services offered: consultancy, design & development, project management, management, stress analysis. Areas of expertise include building hardened structures & telecommunication structures. [97]

80-16500 Microinfo Ltd
Omega Park, PO Box 3, Alton, Hampshire GU34 2PG
☎(+44 1420) 868 48 [E] ops@ukminfodemon.uk
[W] http://www.microinfo.co.uk ⊜ (+44 1420) 898 89
28
Director: R.B. Selwyn
Director: C.R. Fish
Marketing Mgr: Mrs N. Temple
■ A UK company offering a unique combination of information resources. A worldwide service is provided to government, commerce, industry, educational, scientific & research organisations. Distributor of CD-ROM database products including: Global Defence Information, Defence Library & TAURUS. [98]

80-16560 Microlease plc
Forbes House, Whitefriars Estate, Tudor Road, Harrow, Middlesex HA3 5SS
☎(+44 181) 420 02 00 ☒ 8953165 microl g
⊜ (+44 181) 420 02 99
♣ £ 8,300,000 100 (1996)
Managing Dir: P.D. Rennie
Commercial Dir: Peter Spillman
Sales Dir: Michael Bannon
Marketing Mgr: Nigel Rennie
■ Electronic & optical test & measurement equipment hire company for aviation & defence applications. [98]

80-16780 Military Aircraft Spares Ltd
Unit 100, Taunton Trading Est, Norton Fitzwarren, Taunton, Somerset TA2 6RX
☎(+44 1823) 35 10 02 ⊜ (+44 1823) 35 10 06
Managing Dir: E.G. Stickley
Sales & Marketing Dir: Ashley Mills
Business Mgr: D.P. Hooper
■ Contracted supplier for MOD RAF surplus aircraft spares. [97]

80-16850 Milo Holdings Ltd
Field House Barn, Chineham Lane, Sherborne St John, Basingstoke, Hants RG24 9LR
☎(+44 1256) 533 55 ⊜ (+44 1256) 84 24 47

👥 8 🏭 1 500 sq ft
Managing Dir: R.M.W. Sitton [97]

80-16880 M.J. Electronic Services (International) Limited
(† MJ Electronics)
Unit 8, Axe Road, Colley Lane Industrial Estate,
Bridgwater, Somerset TA6 5LJ
☎ (+44 1278) 42 28 82
[E] 101756.501@compuserve.com
⊖ (+44 1278) 45 33 31
✤ £ 300,000 (1994) ♣ £ 250,000 (1994) 👥 6 (1994)
🏭 1,500 sq m (1994)
Managing Dir: M. Cornish
Sales & Marketing Dir: T. Burgin
Operations & Technical Dir: S. Williams
Purchasing Mgr: E. Matthews
■ Supply and service UPS and D.C. power systems. [98]

80-16895 MMB International Ltd
(a subsidiary of Macclesfield Motor Bodies (UK) Ltd)
Unit A, Calamine Street, Macclesfield, Cheshire SK11 7HU
☎ (+44 1625) 61 50 25 ⊖ (+44 1625) 51 15 13
♣ £ 1,500,000 (1996) 👥 25
Managing Director: I G Woolhouse
General Manager: G G Hibbert
Purchasing Manager: Ms Mary O'Toole
■ Civil, military and airport ambulances. Cross country (4x4) and domestic (4x2) models. [98]

80-17000 MMM Consultancy Group Ltd
(a member of MMM Holdings Ltd)
Norden House, Basing View, Basingstoke, Hampshire
☎ (+44 1256) 46 42 26
[E] basingstoke@mmmconsult.com
[W] http://www.mmmconsult.com ⊖ (+44 1256) 84 20 54
♣ £ 1,600,000 (1996) 👥 30
Managing Director: N A Lee ☎ (+44 121) 455 87 44
Director Consultancy Services: A Wright
■ Independent consultancy with specialist services covering: logistics and distribution layout, handling systems, manufacturing services, information technology, project management. [98]

80-17200 Modemetric Limited
Modemetric House, Chosen View Road,
Cheltenham, Gloucestershire GL51 9LT
☎ (+44 1242) 25 20 00 [E] modemet.demon.co.uk
⊖ (+44 1242) 25 24 00
✤ £ 223,000 (1997) ♣ £ 1,250,000 (1996/97)
👥 15 (1996/97) 🏭 3,000 sq ft
Managing Director: Terry Allen
Marketing Quality Manager: Adrian Leatherland
Operations Manager: Ms Liz Carpenter
Engineering Manager: Neil Allen
■ Systems design and analysis. Software design and development. Software test and integration. [98]

80-17400 Mott MacDonald
St Anne House, 20-26 Wellesley Road,
Croydon, Surrey CR9 2UL
☎ (+44 181) 686 50 41 [E] marketing@mottmac.com
[W] http://www.mottmac.com ⊠ 917241 mottay g
⊖ (+44 181) 681 57 06
♣ £ 210,000,000 (1996) 👥 4,398 (1997)
Chairman: R Beresford
Marketing Director: A G Powlesland
Defence point of contact: Roger Hewings
■ Defence engineering services. [98]

80-17500 MSL Secure Systems Instrument Services (Hitchin) Ltd
(a division of Dowding & Mills plc)
60 Wilbury Way, Hitchin, Hertfordshire SG4 0TA
☎ (+44 1462) 42 12 34 ⊖ (+44 1462) 42 00 12
■ Secure communications and specialist engineering services. Fully automated Tempest test system. MSL Secure Systems is registered as an Industrial Tempest Scheme (ITS) company and is fully approved by Government Communications Headquarters (CESG/GCHQ). System design and configuration, installation, commissioning and onsite training. [98]

80-17900 Mycon Marketing Services Ltd (MMS)
Ionix House, Festival Way, Stoke-on-Trent,
Staffordshire ST1 5SQ
☎ (+44 1782) 26 51 77 [E] pconnor@mycon.co.uk
[W] http:// www.mycon.co.uk ⊖ (+44 1782) 26 11 32
✤ £ 28,500 (1996) ♣ £ 200,000 (1996) 👥 5 (1996)
🏭 1,500 sq ft
Managing Dir: Paul B. Connor
Director: George G. Connor (OBE)
Marketing Services Mgr: Miss R.E. Smith
■ A full range of marketing services for defence business both in the UK & worldwide. Marketing consultancy, market research, advertising, public relations, marketing & sales personnel training, exhibition, seminar & conference planning. → 90 [98]

80-17950 National Technical Information Service (NTIS)
Omega Park, PO Box 3, Alton, Hants GU34 2PG
☎ (+44 1420) 868 48 [E] ops@ukminfo.demon.uk
[W] http://www.microinfo.co.uk ⊖ (+44 1420) 898 89
Director: R.B. Selwyn
Director: C.R. Fish
Associate Dir: Ms Maureen L. Main
Associate Dir: Ms Nicola S. Temple
■ All US government publications including reprint service for MIL-SPEC standards & national aerospace standards. Distributor of CD-ROM database products including: The Jane's series plus Global Defence Information on CD ROM & the US Governments Defense Library on DISC. Also represents the CALS information center. [98]

80-18000 Navtech Systems Limited
(† MetRoaTech Limited)
The Hollies, Naseby Road, Clipston, Market Harborough, Leicester LE16 9RZ
☎ (+44 1858) 52 54 54 [E] sales@navtech.demon.uk
⊖ (+44 1858) 52 54 62
♣ £ 360,000 (1996) 👥 6 🏭 200 sq m
Technical Director: Roger D Cooper
Operations Director: Terry Hoyle
Product Manager: Phil Charles
■ Time, date and position tagging onto video film for airborne surveillance and reconnaissance. Applications of GPS for specific user requirements, provision of position intelligence for steerable microwave TV transmitting and receiving antennae for long range television links, including directing of receiving/transmitting dishes to synchronous satellites. GIS presentation of simultaneous video and mapping from GPS encoded video. [98]

80-18200 Newey & Eyre Group Ltd
(a subsidiary of Hagemeyer)
Donne House, Calthorpe Road,
Edgbaston, Birmingham B15 1QX
☎ (+44 121) 455 97 27 ⊖ (+44 121) 454 09 48
♣ £ 400,000,000 👥 2,500
Managing Director: John Hudson
Commercial Director: Stewart Hodges
Marketing Manager: Ian Lund
■ Distributor of electrical equipment including switch & fuse gear, & cables. [98]

80-18250 New Markets Ltd
Theocsbury House, 18-20 Barton Street,
Tewkesbury, Gloucestershire GL20 5PP
☎ (+44 1684) 29 15 44 ⊖ (+44 1684) 29 15 45
♣ £ 400,000 👥 6
Managing Director: Ted Elwes
Sales and Marketing Director: David Richman
Purchasing Manager: Andrew Turner
■ Sales & marketing abroad for defence manufacturers worldwide. Market research, market consultancy, countertrade & offset assignments for defence manufacturers worldwide. Supply of components for non-combatant vehicles. → 85 → 90 [98]

80-18300 Noble Denton Europe Ltd
(a subsidiary of Noble Denton Holdings Ltd)
Head Office
Noble House, 131 Aldersgate Street, London EC1A 4EB
☎ (+44 171) 606 49 61 [E] marketing@nodent.co.uk
⊠ 296928 ⊖ (+44 171) 606 50 35
♣ £ 11,000,000 👥 200 🏭 18,500 sq ft
Chairman: Dr A.A. Denton
Managing Director: John Trickey
Interim Financial Director: Brendan Connolly
Marketing Co-ordinator: Ms Ruth O'Sullivan
■ Leading international company of engineering, marine and oceanographic consultants. This multi-disciplinary company provides expert practical and technical advice on marine related projects including transportation and installation of large or unusual structures, mooring system design and analysis, risk, reliability and safety studies and the analysis/interpretation of metocean data.

Marine Services
42 Victoria Street, Aberdeen AB1 1XA
☎ (+44 1224) 64 22 55 [E] ndeaberdeen@nodent.co.uk
⊖ (+44 1224) 64 70 46 [98]

80-18400 Norsk Data Ltd
(a subsidiary of Telenor Bedrift, Norway)
Benham Valence, Newbury, Berkshire RG20 8LU
☎ (+44 1635) 355 44 [E] kayr@serviceteam.co.uk
[W] http:// www.serviceteam.co.uk ⊠ 849819 norskd g
⊖ (+44 1635) 51 10 52; 340 05
♣ £ 36,500,000 (1996) 👥 714 (1997) 🏭 107,000 m2
Managing Director: Clive Higgins
Operations Director: Linda Grange
Defence Sales Manager: John Parker
Purchasing Manager: Dorothy Phippard
Marketing Manager: Ms Kay Binns
■ Multi-vendor IT support services: hardware maintenance, software support, help desk, network integration, network management. Consultancy services. Information management systems: Document Image Processing, workflow, computer output to laser disk, archiving. Managed on-line services: electronic picture libraries, Internet/ Intranet services. [98]

80-18500 NNC Ltd
(a member of the GEC Group of Companies)
Booths Hall, Chelford Road,
Knutsford, Cheshire WA16 8QZ
☎ (+44 1565) 63 38 00 ⊖ (+44 1565) 63 36 59
♣ £ 47 300 000 (1994/95) 👥 1 000
Managing Director: J Durston
Business Director: Fraser Ashman
Consultancy Director: Kevin Routledge
Purchasing Manager: Tim Hardman
Corporate Communications Manager: Ms Sharon Lindsay
Business Development Manager: A F Curzon
■ Project management, engineering & technical consultancy. Service range from facilities project management, through turnkey design & build contracts, nuclear engineering services to a wide range of safety & technical consultancy. NNC's manufacturing & technology centre houses a full suite of workshops & laboratories & facilities for manufacture of special process plant & equipment.
NNC, VSEL & Yarrow Shipbuilders are wholly owned subsidiaries of GEC plc & together they form GEC Marine, a new organisation dedicated to world-class shipbuilding, armaments, nuclear engineering & project management. [98]

80-18550 NPA Group Ltd
1 Fircroft Way, Edenbridge, Kent TN8 6HS
☎ (+44 1732) 86 50 23 ⊠ 95442
⊖ (+44 1732) 86 65 21
♣ £ 1 000 000 👥 20 🏭 375 sq m
Managing Dir: Nigel Press
Remote Sensing Mgr: Richard Chiles
Purchasing Mgr: David Morten
■ Digital & photographic enhancement of satellite imagery. Official SPOT image distributor in UK. Production & interpretative facilities. [97]

80-18800 The Numerical Algorithms Group Ltd (NAG)
Wilkinson House, Jordan Hill Road, Oxford OX2 8DR
☎ (+44 1865) 51 12 45; 31 17 44 (response centre)
[E] infodesk@nag.co.uk. [W] http://www.nag.co.uk/
⊖ (+44 1865) 31 01 39; 31 17 55 (response group)
♣ £ 5,500,000 (1996) 👥 85
Chairman: Dr D F Hartley
Director: Dr Brian Ford
Operations Director: Gareth Hudson
Sales and Marketing Director: John Piper
Marketing Manager: Terry Burgess
Senior Sales Executive, Defence: Andrew Giles
■ Production and distribution of numerical, symbolic, statistical and visualisation software for scientific, engineering, financial analysis and research applications. Sub-programme categories in a range of computer languages including Ada, C, Fortran, Pascal and Turbo Pascal.

Iris Explorer Centre (Europe)
Wilkinson House, Jordan Hill Road, Oxford OX2 8DR
☎ (+44 1865) 51 63 77 E helpdesk@iec.co.uk
W http://www.nag.co.uk/welcome_iec.html
Fax (+44 1865) 51 63 88
■ Visual programming system for data visualisation, animation, manipulation and analysis for scientific and engineering use.
Applications include: mechanical and electrical engineering, aerospace and automotive engineering, geology, physics, chemistry, computational fluid dynamics, medical imaging and research, multimedia and financial modelling. 98

80-19050 Oceano Instruments (UK) Ltd
(a subsidiary of Mors SA)
9 New Broom Park, Granton Park Ave Industrial Estate, Edinburgh, Scotland EH5 1RS
☎ (+44 131) 552 03 03 E sales@oceano.demon.co.uk
W http://www.oceano.demon.co.uk
Fax (+44 131) 552 66 19
♣ £ 1,000,000 (1996) 5 (1997) 200 m2
Managing Dir: R.K. Riddell
Sales & Marketing Mgr: Steve Palmer
■ Sales & service of underwater acoustic & oceanographic instrumentation. Long & short baseline acoustic positioning. Acoustic telemetry. Acoustic command (releases, valves, explosives, pyrotechnics). Acoustic control of mine disposal devices. Acoustic beacons & pingers for torpedo/ missile trajectography & relocation. Sound velocimeters. Currentmeters, tidegauges, metocean buoys, etc. → 75 98

80-19400 Ogle Design Ltd
Birds Hill, Letchworth, Hertfordshire SG6 1JA
☎ (+44 1462) 68 26 61 Fax (+44 1462) 67 48 14
♣ £ 2,800,000 55 25,000 sq ft
Marketing and Sales Manager: A Barlow
■ Full product development service including design, human factors, engineering, full size/scale exterior/interior modelling and prototyping with over 40 years experience in the aerospace and automotive industries. Projects include the design and build of trainers/simulators for military and civil markets. Running ProEngineer, Alias, Unigraphics, Varimetrix and AutoCAD. Accredited to ISO 9001. 98

80-19450 OMI Logistics Limited
(a subs. of OMI International plc)
2-10 Cawte Road, Southampton, Hants SO15 3TD
☎ (+44 1703) 90 82 00 Fax (+44 1703) 33 61 17
♣ £ 12 000 000 450
Managing Dir: G. Mullis
Commercial Dir: J. Cornish
Operations/Technical Dir: Andrew Mair
■ Acting in support of major asset makers & users in the defence industry. Services include Integrated Logistic Support (ILS) to MIL-STD-1388, Software tools for reliability engineering (AR&M products), technical publications, technical documentation, spares provisioning, NATO codification, electronic component consultancy, obsolete component research, technical graphics, computer services, database design & management, micrographics. Services to ISO9000 within tight security regime. CALS Bureau Service, Data Conversion Bureau Services, Standard General Mark-Up Language (SGML) consultancy, Interactive Electronic Technical Manuals, A, R & M software, spares management and rationalisation, micrographic bureau services, computer generated illustration, NATO codification. 97

80-19600 OTEC Support Services Ltd
(a member of The Nitor Group)
Nitor House, Great George Street, Godalming, Surrey GU7 1EE
☎ (+44 483) 42 62 99 Telex 94011723 nitr g
Fax (+44 483) 41 59 53
Managing Director: Martin Read
■ Project managers and contractors undertaking turnkey requirements for all range, training complexes and barrack constructions. 98

80-19700 O.T.P. Ballistics
(Assoc. of PLS Ballistic Service, USA)
59 Correnden Road, Tonbridge, Kent TN10 3AX
☎ (+44 1732) 35 69 26 Fax (+44 1732) 35 69 26
♣ £ 50 000 3
Principle Consultant: Thomas W.C. Page
■ Ballistics consultancy service covering all aspects from gun to target. Wound ballistics. NATO standards to safe and suitable for service status. Qualification testing to NATO 4224. 98

80-19900 Oxford Air Training School (OATS)
(a division of CSE Aviation Ltd)
Oxford Airport, Kidlington, Oxford OX5 1RA
☎ (+44 1865) 84 12 34 Telex 83204 aztec oxford
Fax (+44 1865) 37 87 97
♣ £ 5,000,000 500
Vice President: Peter Moxham ☎ (+44 1865) 84 42 76
Managing Dir CSE Aviation: R. Crooke
Head of Oxford Air Training School: K.B. Latton
■ Military training for fixed wing & helicopter students. Engineering training provided for fixed wing & helicopter mechanics in the full range of skills. Courses can be either to British or foreign licence standards or specifically tailored to customer's requirements. English & Foundation courses provided where required. 98

80-20250 Paint Services Group
Weydown Road, Haslemere, Surrey GU27 1BT
☎ (+44 1428) 65 12 46 Fax (+44 1428) 66 14 71
♣ £ 3,600,000 24
Managing Director: A K Wood
Marketing and Sales Director: G R Yates
Technical Manager: C Bolton
■ PSG has an extensive knowledge of both UK and US Defence Coatings and a range of coatings approved by the Ministry of Defence and the US military. 97

80-20300 Parity Solutions Ltd
(a division of Parity plc)
Wimbledon Bridgehouse, 1 Hartfield Road, Wimbledon, London SW19 3RU
☎ (+44 181) 543 53 53 W http://www.parity.co.uk
Fax (+44 181) 545 64 56
♣ £ 35,000,000 (1996) 480 (1997)
Managing Director: Keith Jennings
Operations Director: Ken Richardson
Sales and Marketing Director: Richard Hawthorn
Purchasing Manager: Ms Linda Clarke
Marketing Manager: Ms Sharon Gore
■ Project management; consultancy; system design & implementation; feasibility and full study; procurement support. 98

80-20450 Parmatic Betex Limited
(t BT Sales & Services Ltd)
(a division of Paramatic Filter Corporation, USA)
69 Carwood Road, Sheffield S4 7SA
☎ (+44 114) 256 00 66 Telex 547676 chamcom g
Fax (+44 114) 242 36 15
24
Managing Director: C Lucarelli
Operations Manager: D Taylor
Area Sales Manager, Aerospace and MoD: Paul Deakin
■ Design and installation of all types of fume and dust extraction systems for automotive, industrial and special systems for use in aerospace markets and MoD. 98

80-21000 Penter Associations Limited
Unit 1 Grange Farm Business Park, Grange Road, Hugglescote, Leicestershire LE67 2BT
☎ (+44 1530) 83 32 32 Fax (+44 1530) 83 32 03
Managing Dir: T. Carpenter
Operations/Technical Dir: F. Clarke
Sales & Marketing Dir: Miss J. Wright
Purchasing Mgr: Miss D. Meadows
■ Risk management services. Surveillance, security personnel, control & restraint training & defensive driving techniques. 97

80-21700 Photo-Scan Systems Ltd
(a subs. of Relyon Group plc)
Dolphin Estate, Windmill Rd, Sunbury-on-Thames, Middx TW16 7HG
☎ (+44 1932) 78 97 41 ✉ PHOTO-SCAN Sunbury Middx
Fax (+44 1932) 78 70 67
♣ £ 11 286 000 (1995) 110
Managing Dir: Peter Hanksworth
Technical Dir: P. Fletcher
Marketing Mgr: Ms Katy Sunter
Purchasing Mgr: Andrew Duffus
■ Specialists in the design, installation & maintenance of closed circuit television systems. CCTV surveillance systems for interior or exterior use, covering security (site areas, access points, perimeters, buildings etc.) plus safety monitoring & process control applications.
● Unit 4 Stafford Park 15, Telford, Shropshire TF3 3BB
☎ (+44 1952) 29 02 33 Fax (+44 1952) 29 09 65
Manager: M. Clayton
● Unit 4, Mosshall Ind.Est., W.Main St, Blackburn W. Lothian EH47 7LY
☎ (+44 1506) 63 34 40 Fax (+44 1506) 558 63
Manager: D. Houston 97

80-22350 Plasfin Ltd
Unit 262 Colthrop Lane, Thatcham, Berks RG19 4NT
☎ (+44 1635) 87 37 30 Fax (+44 1635) 87 36 16
♣ £ 800 000 11
President: Mrs G.I. Mallaburn
General Mgr: R.R. Henson
Works Mgr: Ian Passey
■ Applies a wide range of thermoplastic & thermoset materials to customers own metalwork. Processes components for Westland Helicopters & our approval is Q827. Specialises in the electrostatic application of Rilsan Nylon II. 97

80-22500 Plastechnol Ltd
3 Eelmoor Road, Farnborough, Hampshire GU14 7QN
☎ (+44 1252) 51 30 01 E info@plastechnol.co.uk
Fax (+44 1252) 54 97 49
♣ £ 2,100,000 (1997) 45 (1997) 2,500 sq m
Managing Director: John Whittaker
Sales and Marketing Director: James Purves
■ Precision machining; moulding composite materials & structural forms; atmospheric deep-sea diving suits. Spall lines for military vehicles plus other composite mouldings incorporating fire retardancy, bullet and blast resistance for applications such as vehicles, doors and anti-bandit counters. 98

80-23500 A.T. Poeton (Gloucester Plating) Ltd
Eastern Avenue, Gloucester GL4 3DN
☎ (+44 1452) 30 05 00 E atp@poetong.demon.co.uk
W http://www.poeton.co.uk Fax (+44 1452) 30 00 50
♣ £ 4,000,000 125
Managing Dir: A.R. Poeton
Sales & Marketing Administrator: Gordon Gisborne
■ Precision hard chrome, hard anodizing, chromic & sulpheric anodizing, electroless nickel, industrial silver, plasma & flame spraying. New generation composite coatings.

A.T. Poeton (Cardiff) Ltd
Penarth Road, Cardiff CF1 7UL
☎ (+44 1222) 38 81 82 Fax (+44 1222) 38 81 85
Director: B.N. Poeton
■ Precision hard chrome, hard anodising, chromic & sulpheric anodising, electroless nickel, industrial silver, plasma/flame spraying. New generation composite coatings.

Poeton Coating Technology Ltd
Brokenford Lane, Totton, Southampton, Hampshire SO40 9DY
☎ (+44 1703) 86 59 55 Fax (+44 1703) 87 30 22
General Mgr: Trevor Amos
■ Precision hard chrome, hard anodising, chromic & sulpheric anodising, electroless nickel, industrial silver, plasma/flame spraying. New generation composite coatings. 98

80-23700 Polaris International Ltd
(a member of the Polaris Group)
North Star Business Park, Woodrow Way, Gloucester GL2 6DX
☎ (+44 1452) 38 00 00 E mil@polaris.co.uk
Fax (+44 1452) 38 44 38
10 (1997)
Polaris Group Director: David F Chapman
Director: Martin J Smith
Technology Manager: C Somerville
■ Geophysical technical marketing & services. Oil exploration. Integrated navigation systems. 98

80-23800 Police Aviation Services Ltd (PAS)
Gloucestershire Airport, Staverton, Cheltenham, Gloucestershire GL51 6SS
☎ (+44 1452) 85 79 99 Fax (+44 1452) 85 78 88

♣ £ 8,000,000 (1997) 👥 120 (1997)
Managing Director: Capt M R S Trumble
Executive Director: Capt S R Bidmead
Operations Director: Capt J E Ball
Sales and Marketing Manager: Capt I Payne
Maintenance Manager: J Cook
■ Provision of helicopters, fixed wing, pilots specialist role equipment & engineering support to law enforcement & emergency services. Engineering/integration of role equipment/communications, training for above especially visual/electronic search techniques. 98

80-23830 Poligrat (UK) Ltd
(a subsidiary of Poligrat GmbH, Germany)
2 Holder Road, Aldershot, Hampshire GU21 4RH
☎ (+44 1252) 33 63 37 ⊜ (+44 1252) 227 91
♣ £ 900,000 (1996) 👥 18 (1996)
Managing Director: R Hall
Joint Managing Director: A D Gosling
■ Pickling and electropolishing of stainless steel and high nickel alloys. → 75 98

80-23850 Powerent Ltd
(Joint Venture between Puma Manufacturing Ltd and The Tatton Group of Companies)
Sandwich Industrial Estate, Sandwich, Kent CT13 9LY
☎ (+44 1304) 62 00 62 ⊜ (+44 1304) 62 00 72
♧ £ 50,000 (1997) ♣ £ 2,000,000 (1997) 👥 6 (1997)
🏭 8,000 m2
Managing Director: B M Tatton
Operations Director: M E Tatton
Purchasing Manager: Mrs Jatta Young
■ Hire of generating sets & auxiliary equipment, cables, transformers & switchgear. 98

80-23925 Praxair Surface Technologies Ltd
(a subs. of Praxair Inc., USA)
Drakes Way, Greenbridge Industrial Estate, Swindon, Wilts SN3 3HX
☎ (+44 1793) 51 25 55 ⊠ 44828
⊜ (+44 1793) 41 02 89
♣ £ 16 527 000 👥 240
Managing Dir: John E. Land
Sales & Marketing Dir: Paul Gosling
Operations Dir: Paul Kent
Sales Mgr Aerospace Business: P. Umbrich
Purchasing Mgr: David Wells
■ Provision of heat, wear & corrosion resistant coatings to metal substrates. Coatings are both ceramic & metallic & are applied to aerospace & other defence components.
● Southam Industrial Est, Westfield Rd, Southam, Warwickshire CV33 0JH
☎ (+44 1926) 81 23 48 ⊜ (+44 1926) 81 77 75 97

80-23930 Praxis Critical Systems Ltd
(† Praxis plc)
20 Manvers Street, Bath BA1 1PX
☎ (+44 1225) 46 69 91 E info@praxis-cs.co.uk
W http://www.praxis-cs.co.uk ⊜ (+44 1225) 46 90 06
♣ £ 3,500,000 (1996) 👥 60 (1996)
Managing Director: Dr David Allen
Aerospace Account Manager: D Clutterbuck
■ Specialists in engineering complex software intensive systems with particular expertise in techniques for the management, design and verification of safety-related software. 98

80-23995 Premier Product Support Services Ltd
Monkswell House, Manse Lane, Knaresborough, North Yorkshire HG5 8NQ
☎ (+44 1423) 86 78 67 ⊜ (+44 1423) 86 88 44
Managing Dir: Barry C. Stonard
■ A comprehensive range of technical services to government establishments & industry. 97

80-24250 Price & Orphin Limited
(a subsidiary of Hampson Industries plc)
Lower Canal Road, Newtown, Powys SY16 2JQ
☎ (+44 1686) 62 55 17 E sales@pontown.demon.co.uk
⊜ (+44 1686) 62 58 78
♣ £ 2,200,000 (1996/97) 👥 68 (1997) 🏭 20,000 sq ft
Managing Dir: Tony Wildish
Commercial Mgr: Andrew Smart
Purchasing Mgr: Stephen Moore
■ Specialising in the manufacture of high-precision components for aerospace & defence applications. In addition to machining standard & exotic metals the company is well-known for its precision presswork, grinding, honing, drilling & tapping. Prototype work & short runs are acceptable. Machining standards & inspection carry the approval of BS5750 part 2 & ISO9002. 98

80-24300 Prior Data Sciences
Hankley Wood House, Thursley Rd, Elstead, Surrey GU8 6LW
☎ (+44 1252) 70 24 49 ⊜ (+33 1252) 70 29 90 97

80-24400 Product Risk Associates
(† Willis Corroon Product Safety Ltd)
2 Queens Court, Queens Road, Richmond, Surrey TW10 6LA
☎ (+44 181) 940 14 24 ⊜ (+44 181) 940 14 24
Principal: Howard Abbott
■ Management of product risks including: hazard analysis & risk assessment; product liability loss control; product recall planning & management; education & training with realistic simulation exercises. 98

80-24500 Profile Security Services Ltd
(a subsidiary of Profile Holdings Ltd)
374 Wandsworth Road, London SW8 4TD
☎ (+44 171) 498 35 11 E profilesecurity@compuserve
⊜ (+44 171) 498 82 46
♣ £ 6,000,000 (1997) 👥 500 (1997)
Managing Director: J A Forbes ☎ (+44 1224) 21 19 92
Sales and Marketing Director: A M D Prendergast
Operations and Technical Director: S Morley
Chief Accountant: R Heath
■ Consultancy, training, close protection & surveillance. 98

80-24700 Programming Research Ltd (PRL)
Glenbrook House, 1-11 Molesey Road, Hersham, Surrey KT12 4RH
☎ (+44 1932) 88 80 80 E info@prqa.co.uk
W http://www.prqa.co.uk ⊜ (+44 1932) 88 80 81
👥 18 (1997) 🏭 780 m2
Managing Director: Paul Blundell
Technical Director: Fergus Bolger
Business Development Manager: John Heathcote
Purchasing Manager: Mrs Bhav Chauhan
Marketing Executive: Robert J Kellam
■ Software quality assurance. Static fault screening and audits of software written in C, C++ and Fortran software for safety and non safety-related environments. 98

80-25000 QES Ltd
Quality Electronic Services
Business & Technology Centre, Bessemer Drive, Stevenage, Hertfordshire SG1 2DX
☎ (+44 1438) 74 98 49 ⊜ (+44 1438) 31 84 20
♣ £ 500,000 (1996) 👥 15 🏭 250 m2
Managing Dir: P.E. Talbot
■ Service, support & repair of military electronics (in particular, power conversion equipment to approx 5KVA, ie linear & switched mode power supplies as originally manufactured by most UK & US sources). Design & manufacture of electronic equipment including power supplies, test equipment & wound components. Build-to-Print manufacture & modification programmes. Service, repair & manufacture to BSENISO 9002. → 80 98

80-25200 Qualitair Engineering Services Limited (QESL)
(a subsidiary of Qualitair Holdings Ltd)
Qualitair House, London Road, Six Mile Bottom, Newmarket CB8 0UL
☎ (+44 1638) 57 04 56 W http://www.qualitair.co.uk
⊜ (+44 1638) 57 04 57
♣ £ 7,000,000 (1996/97) 👥 150 (1996/97)
Chairman and Managing Director: Peter Bottomley
General Manager: Michael Donohoe
■ Supplier of aircraft maintenance personnel with fixed wing or rotary expertise. Work is performed at customers facilities on line base maintenance, specialist modifications or emergency repair projects. 98

80-26300 Racal Research Ltd
(a subsidiary of Racal Electronics plc)
Worton Drive, Worton Grange Industrial Estate, Reading, Berkshire RG2 0SB
☎ (+44 1734) 86 86 01 ⊜ (+44 1734) 23 83 99
👥 200 (1997)
Managing Director: P K Blair
Commercial Manager: T J Wiseman
■ Research and development, study services. 98

80-26320 Racal Survey Ltd-Products
(† Racal Positioning Systems Ltd)
(a subsidiary of Racal Electronics plc)
Burlington House, 118 Burlington Rd, New Malden, Surrey KT3 4NR
☎ (+44 181) 942 24 64; 942 24 88
⊜ (+44 181) 942 08 35
♣ £ 6,000,000 👥 900 (worldwide)
Chairman: M R Richardson
Managing Director: B R Wood
Publicity Manager: Mrs Jane Holden
New Business Executive: S Clark
■ ROV's and surveys for oil and gas industries for land and near-shore applications. Associated operation, maintenance and training, vehicle and vessel tracking systems. 98

80-27820 Ricardo Aerospace Ltd
(† SAC Technology Ltd)
(a subs. of Ricardo Group plc)
Brunswick House, Upper York Street, Bristol BS2 8QN
☎ (+44 117) 924 00 88 ⊜ (+44 117) 924 61 81
♣ £ 18 000 000 👥 350
Managing Dir: John Baker
■ Independent consulting & design engineering company with activities in the field of aircraft & gas turbines. 97

80-27840 Ricardo Hitec Ltd
(† SAC Hitec Ltd)
(a subs. of Ricardo Group plc)
Club Street, Bamber Bridge, Preston, Lancashire PR5 6FN
☎ (+44 1772) 340 51 ⊜ (+44 1772) 62 92 61
♧ £ 10 000 ♣ £ 13 700 000 👥 300
Managing Dir: A. Bodnar
Sales Dir: D.R. Bradley
Project Mgr Proposals: G.K. Poole
■ Provides a full design, construct & project management service to the defence, nuclear & manufacturing industries. These services cover a broad range of technologies/skills including mechanical, electrical, software computing & control robotics & special purpose machine developments. 97

80-28000 Ridgeway International Ltd
69 High Street, Wallingford, Oxon OX10 0BX
☎ (+44 1491) 83 97 80
E ridgeway_uk@compuserve.com ⊠ 847994
⊜ (+44 1491) 83 97 65
♧ £ 5,110 ♣ £ 6,000,000 👥 22
Joint Managing Director: Christopher Hawker
Joint Managing Director: Robert Williamson
Director: E J Evelegh
■ Transport system managers for the international movement, storage & insurance of explosives, ammunition & missiles, worldwide by land, sea & air for governments & commerce. 98

80-28100 Ridsdale Ranges
(† Vickers Proof Range)
(a subs. of Vickers Defence Systems)
The Steel, Ridsdale, Northumberland NE48 2EZ
☎ (+44 1434) 22 02 08 ⊜ (+44 1434) 22 02 08
General Manager: Barry Torpy
Public Relations Mgr: Peter McKenzie
■ 860 acres of remote research facilities which is able to offer customers private secluded cross-country vehicle test tracks, explosive testing areas & gun ranges up to 200m for weapon systems up to 155mm calibre. 98

80-28200 Ring Sights Ltd
(a subsidiary of Ring Sights Holdings Co Ltd)
PO Box 22, Bordon, Hampshire GU35 9PD
☎ (+44 1420) 47 22 60 E ringsights@easynet.co.uk
⊠ 859535 ⊜ (+44 1420) 47 83 59
♣ £ 150,000 👥 5
Managing Director: I R Flack
Research and Development Manager: Brig Fraser Scott

■ Research & development of simple sights for rifles, MGs, pistols, 40 AGL, LAW & cannon. [98]

80-28600 RO Future Systems and Concepts Group

(† Future Systems Group)
(a div. of British Aerospace Defence Co., Royal Ordnance Division)
PO Box 243, Shrivenham, Swindon, Wiltshire SN6 8QD
☎(+44 1793) 78 64 23 ⊠ 444112
⊖ (+44 1793) 78 64 40
20
Business Manager: R W Ince
■ Security consulting services. [98]

80-29100 Rolls-Royce & Associates Ltd (RRA)

(a subs. of Rolls-Royce plc)
PO Box 31, Derby DE24 8BJ
☎(+44 1332) 66 14 61 ⊠ 37616
⊖ (+44 1332) 62 29 29
1 800
Managing Dir: J.M. Glanville
Operations Dir: D.E. Gorham
Commercial Dir: C. Mead
Technical Dir: Dr S. Garwood
Public Relations/Affairs: A. Rice
■ Responsible for the design, development, supply & in-service support of the nuclear steam raising plant for the Royal Navy. Also undertakes project & facility management, turnkey contracts, safety studies, high integrity system/software validation, inspection & consultancy. → 75 [97]

80-29200 Rood Technology UK Ltd

(a subs. of Rood Testhouse Int'l NV, Netherlands)
Test House, Mill Lane, Alton, Hants GU34 2QG
☎(+44 1420) 880 22 ⊖ (+44 1420) 872 59
♣ £ 34 000 000 (1995) 160
Managing Dir: Ian Reid
Sales & Marketing Dir: Alistair Hoy
Operations & Technical Dir: Patrick Faulkener
Purchasing Mgr: Roger Gough
■ Incoming/outgoing IC inspection, programming PLD quality & reliability investigations, characterisation failure analysis, construction analysis, die processing, custom packaging, memory modules testing BS & CECC quality approved. → 75

● Mill Lane, Alton, Hants GU34 2QG
☎(+44 1420) 880 22 ⊖ (+44 1420) 872 59
Production Mgr: Patrick Faulkener
Sales & Marketing Mgr: Alistair Hoy
■ Has the capability to package discretes & ICs (monolithic & multichip) in all standard ceramic package outlines. This includes a high percentage of work to space level standards (ESA/SCC-9000). ISO9002, BS9000 & CECC approved. [97]

80-29325 Rotortech Ltd

Howemoss Drive, Aberdeen Airport,
Dyce, Aberdeen AB2 0GL
☎(+44 1224) 72 27 22 ⊖ (+44 1224) 77 59 00
60
Managing Director: John Cloggie
Sales and Marketing Executive: C S Forsyth
Purchasing Manager: Stuart Park
■ The UK Eurocopter customer service centre for repair & overhaul along with the Sikorsky European Blade repair facility. A comprehensive helicopter repair, refurbishment, up-rating & modification service in addition to component repair & overhaul. Design & manufacture composites. → 75 [98]

80-29340 R-R Industrial Controls Ltd

(a subs. of Rolls-Royce Industrial Power Group)
Kingsway, Team Valley,
Gateshead, Tyne & Wear NE11 0QJ
☎(+44 191) 487 08 11 ⊖ (+44 191) 482 00 06
♣ £ 27 000 000 (1995) 300+
Managing Dir: Peter Conway
Commercial Dir: Philip Derges
Chief Engineer: Clive Ross
Sales & Marketing Mgr: Stewart Gibbins
■ Control & instrumentation including SCADA, telemetry, PLC's & high integrity software systems. Power electronics including AC/DC drives, automatic voltage regulators (AVR's) & turbine supervisory equipment. Nuclear plant including neutron flux monitoring, health physics, reactor core monitors & gas & liquid monitoring. Quality Assurance approved EAQA BS5750 Pt1, AQAP-1 & 13, BS5882, ISO9001. [97]

80-29350 Rubery Owen Group Services Ltd

(a subs. of Rubery Owen Holdings Ltd)
Defence Equipment Div.
Darlaston, PO Box 10, Wednesbury, W. Mids WS10 8JD
☎(+44 121) 526 31 31 ⊖ (+44 121) 526 28 69
♣ £ 940 000 (1995) 35
Managing Dir: J.E. Owen
Exec. Dir: R.J.C. Oakes
■ Design & development of special purpose trailers & ancillary equipment for MoD & the provision of associated military spares. BS EN1SO 9001 1994. [97]

80-30100 SAC Technographic

Centre Court, 1301 Stratford Road,
Hall Green, Birmingham B28 9HH
☎(+44 121) 702 22 02 ⊖ (+44 121) 702 29 44
♣ £ 10 500 000 (1994) 350
Managing Dir: Roy Pierson
Sales & Marketing Dir: Andrew King
Account Mgr: D. Stevens
■ Technical documentation, description, operation & maintenance manuals. Clients in the aerospace, defence, power generation, oil, gas & manufacturing industries. Multi-media communications. Computer based training.

MSC Division
Cow Lane, Wymering, Portsmouth, Hampshire PO6 3TR
☎(+44 1705) 37 35 11 ⊖ (+44 1705) 32 68 85
40
General Mgr: C. McNab
■ Computer systems, trainers and simulators, computer based training. Naval base project management and consultancy services. In addition the Division operates the Marine Training Centre which provides technician courses. [98]

80-30220 SAIC Ltd

Poseidon House, Castle Park, Cambridge CB3 0RD
☎(+44 1223) 46 21 21 ⊖ (+44 1223) 31 50 57
Managing Dir: John J. Yale
Operations & Technical Dir: J.F. Gillam
Business Development Manager: Guy Bewsher
Business Management Manager: Andrew Seale
Purchasing Mgr: D. Denton
■ Defence systems integrators. → 75
® USA: SAIC Science Applications International, SAIT SAI Technology [98]

80-30300 Saladin Security Ltd

(† Saladin Holdings Ltd)
7 Abingdon Rd, London W8 6AH
☎(+44 171) 376 26 55 ⊠ 89675 saladn g
⊖ (+44 171) 937 58 05
♣ £ 2 000 000 70
Managing Dir: Bruce W. Norris
Office Mgr: Ms J. Mitchell
■ Corporate defence consultants & specialists in Special Force training & equipment worldwide. Wide experience in fields of counter-terrorism, surveillance & counter-insurgency.

Saladin Trading Ltd
7 Abingdon Road, London W8 6AH
☎(+44 171) 376 26 55 ⊠ 89675
⊖ (+44 171) 937 58 05
Manager: Ms Debbie Richards [97]

80-30800 Samitac

6 Westfield Road, Garlinge, Margate CT9 5NZ
☎(+44 1843) 23 14 61
♣ US$ 180,000 (1995) 50
Managing Director: Steven Hughes
Marketing Manager: Michael Stalley
■ Specialists in mine clearing, mine awareness, training in mines and unexploded ordnance. [98]

80-31550 Scientific Computers Ltd

Premiere House, Betts Way, London Road,
Crawley, West Sussex RH10 2GB
☎(+44 1293) 40 36 36 Ⓔ info@scl.com
⊖ (+44 1293) 40 36 41
15
Managing Dir: Dr Alan Hall
Sales & Marketing Mgr: Ms Alison Bolton
■ Responds to specialised needs in the scientific areas of computation, software development & simulation. Knowledge resource centre: full hardware/software support & training provided in area of artificial intelligence, image processing, solid modelling, real-time simulation & monitoring & process control systems. → 90 [98]

80-31600 Science Systems (Industrial) Ltd

(a subsidiary of Science Systems (Holdings) Ltd)
23 Clothier Road, Bristol BS4 5SS
☎(+44 117) 971 72 51 ⊠ 44772
⊖ (+44 117) 972 18 46
♣ £ 16,000,000 260
Managing Director: Michael Love
Space Systems Chief Executive: John Haynes
Director: Peter Turner
■ Consultancy, software design, development and project management for aerospace, avionics and defence applications. Kernel control product - integrated suite of TT&C programs providing real-time monitoring and control of space satellites. [98]

80-31800 S.Com Computer Systems Engineering Ltd

Buckingham House, Buckingham Street,
Aylesbury, Buckinghamshire HP20 2LA
☎(+44 1296) 31 14 11 Ⓔ response@scom.co.uk
Ⓦ http://www.scom.com ⊖ (+44 1296) 43 68 95
♣ £ 35,000,000 70
Managing Director: Paul Roebuck
Marketing Manager: Mike Sullivan
■ Specialists in supplying engineering and IT contract staff to the defence and aerospace industries. [98]

80-31900 Scorpio Navigation Services

Bennett House, 1 High Street,
Edgware, Middlesex HA8 7DF
☎(+44 181) 951 44 46 ⊖ (+44 181) 951 56 50
Managing Director: D L G Edwards
Operations Director: S R Gale
Marketing Director: D T Wilson
■ Differential GPS systems, automated hydrographic survey system, electronic chart & planning system, precision Loran receiver, navigation data integration systems. [98]

80-32000 S&D Group

(† S&D Security (Equipment) Ltd & STI Ltd)
45 Lower Belgrave Street, London SW1W 0LS
☎(+44 171) 730 81 05 ⊖ (+44 171) 730 47 47
Director: Lt Col John M. Gaff (GM)
■ Search & disposal equipment. Anti-terrorist advisory service. Drug detection equipment. [98]

80-32400 SDMS Security Products Ltd

(† Security & Defence Marketing Services Ltd)
(a subsidiary of Complete Security Ltd)
365 Fulham Road, Chelsea, London SW10 9TJ
☎(+44 171) 376 40 91 ⊠ 8954111 reply
⊖ (+44 171) 351 44 68
♣ £ 1,400,000 (1996) 3
Managing Director: D Knights
■ Supply of anti-terrorist equipment & services & also a wide range of specialist security equipment. [98]

80-33300 AV Seawork Limited

(a subs. of Vosper Thornycroft Holdings plc)
AVS House, 16 Little Park Farm Road, Segensworth,
Fareham, Hampshire PO15 5TD
☎(+44 1489) 88 51 33 ⊠ 86115 vtpsth g
⊖ (+44 1489) 88 55 29
♣ £ 6 000 000 350
Managing Dir: Jonathan Beadon
Sales & Marketing Mgr: Ian Strugnell

■ Facilities management of defence-related & non-defence establishments & service organisations. The provision of craftsmen, technicians & management to marine & defence organisations. Personnel are supplied on a recruitment basis, on contract hire or the company can take on a full project management contract. Areas of work include surveys, design, planning, operations, administration, maintenance, training & management. [98]

80-33400 Sector Ltd
(a member of The Nitor Group)
PO Box 1, Godalming, Surrey GU7 1EE
☎ (+44 483) 42 62 99 ☒ 261376 nitr g
⊜ (+44 483) 41 59 53
♧ £ 10,000
Managing Director: M B Pegrum
Director: M E Read
Company Secretary: R J W Rickman
■ International consulting engineers, design and commissioning of shooting ranges and special training installations for police, internal security and counter-terrorist forces. Supervision for construction, equipment installation and commissioning of custom-built and other training facilities. [98]

80-33445 Secure Consulting International Registered Office
33-35 Bell Street, Reigate, Surrey RH2 7AW
Secure Consulting International Ltd
Administration Office
Meridian House, 18 Pollards Drive,
Horsham, West Sussex RH13 5HH
☎ (+44 1403) 25 84 68 [E] simulation@mcmail.com
⊜ (+44 1403) 25 85 68
Managing Director: John Wingate
Company Secretary: Trevor Munday
■ Consultancy, marketing and advice on all aspects of procurement and liaison with UK and foreign defence ministries regarding military aviation training. [98]

80-33500 Security Industry Training Organisation Ltd (SITO)
Security House, Barbourne Road, Worcester WR1 1RS
☎ (+44 1905) 200 04 ⊜ (+44 1905) 61 36 25
♣ £ 900 000 👥 10
General Mgr: Raymond Clarke
■ Training provider to security industry. [98]

80-34000 SEED Technical Services Ltd
North Site, Leigh Sinton Road,
Malvern, Worcestershire WR14 1LL
☎ (+44 1684) 56 80 80 ⊜ (+44 1684) 89 29 19
♣ £ 450 000 👥 18
Managing Dir: Raymond May
Operations & Technical Dir: Andrew Crowther
Drawing Office Managing Director: Andrew Griffin
☎ (+44 1684) 58 30 47
■ Provides a document/manual publishing service. Operator handbooks, technical descriptions and maintenance manuals for complex hardware/software systems. Produces interactive electronic technical manuals on CD. [98]

80-34150 Serco Aviation Services
(† Serco-IAL Limited)
(a subsidiary of Serco Ltd)
Bailbrook College, London Road West,
Bath, Avon BA1 7JD
☎ (+44 1225) 85 51 00 ⊠ INTAERIO Bath
☒ 444122 ial g ⊜ (+44 1225) 85 52 53
Managing Director: David Buckley
Operations Director: Alan Shaw
Sales and Marketing Director: Rick Sturge
Purchasing Manager: Peter Sands
Marketing Manager: Ms Elaine Markey
■ Provision of technical services in support of aviation including airport management, ATC, rescue and firefighting services, operations, maintenance and integration of communications and security systems, consultancy studies and related training services. [98]

80-34200 Services Sound & Vision Corp. (SSVC)
Chalfont Grove, Narcot Lane,
Gerrards Cross, Bucks SL9 8TN
☎ (+44 1494) 87 44 61 [W] http://www.ssvc.com
☒ 817693 ⊜ (+44 1494) 87 29 82
♣ £ 49,000,000 (1995) 👥 1,000 (1995)
Managing Dir: David Crwys-Williams
☎ (+44 1494) 87 82 39
Operations Director: Mike Weston
Director of Broadcasting: Peter McDonagh
☎ (+44 1494) 87 85 62
Head of Corporate Marketing: Ian Hyde
☎ (+44 1494) 87 85 10
■ Audio-visual services for training and security. Videos and film production and facilities house. Interactive video based training and simulations systems. Satellite and broadcast transmission services. [98]

80-34550 SGS Yarsley International Certification Services Ltd
(† SGS Yarsley Technical Services)
(division of the Société Générale de Surveillance, Switzerland)
SGS House, Portland Road,
East Grinstead, West Sussex RH19 4ET
☎ (+44 1342) 41 00 88 ✆ (800) 90 00 94 ☒ 8951511
⊜ (+44 1342) 30 53 05
♣ £ 10,000,000 👥 180 (1997)
Managing Dir: Reg Easy
Operations Dir: Peter Marriott
Marketing Coordinator: K Glazier
■ Independent certification body accredited by the National Accreditation Council for Certification Bodies (NACCB) for its quality & environmental management systems certification. Also offer a wide range of ISO 9000 & BS 7750 training courses. [98]

80-34570 Shade (Computer Services) Ltd
1/3 Patford St, Calne, Wiltshire SN11 0EF
☎ (+44 1249) 81 57 57 ☒ 445982 shade g
⊜ (+44 1249) 81 62 07
♣ £ 700 000 👥 16
Managing & Sales Dir: Graeme Robinson
Purchasing Mgr: Andy Fox
■ Systems house employing both hardware & software engineers. Supplies turnkey computer systems for a wide range of scientific & technical applications for both commercial & military customers. [97]

80-34700 Shaw Packing Ltd
Export House, Meltham,
Huddersfield, West Yorkshire HD7 3AP
☎ (+44 1484) 85 06 78 ⊜ (+44 1484) 85 08 28
🏠 100 000 sq ft
Personnel Officer: Mrs J.M. Stephenson
■ PHST, ILS. Design, post design services, testing & production of packaging to all military specifications.
Accrington Office
Bridge Street, Church, Accrington, Lancs BB5 4HU
☎ (+44 1254) 87 11 55 ⊜ (+44 1254) 87 13 77
🏠 75 000 sq ft
Bilston Office
Meadow Lane, Bilston, West Midlands WV14 9NQ
☎ (+44 1902) 88 07 21 ⊜ (+44 1902) 88 04 30
🏠 68 000 sq ft
Managing Dir: M.J. Butterworth
Sales Dir: M.L. Parker
Hartlepool Office
Brenda Road, Hartlepool, Cleveland TS25 2BW
☎ (+44 1429) 27 65 61 ⊜ (+44 1429) 27 32 24
👥 15 🏠 15 000 sq ft
Managing Dir: M.J. Butterworth [97]

80-35500 Shield International Ltd
13-15 Moor Road, Bardstone, Dorset BH18 8AZ
☎ (+44 1202) 65 95 50 ⊜ (+44 1202) 65 94 40
👥 9
Managing Dir: Robert A. Beers (BEM)
Company Secretary: Mrs M. Beers
■ Target equipment. Portable/self contained units, controlled by radio, microprocessor, & line driver. Audio Visual Systems. Firearms range design, construction, manufacture & provision. → 75 [98]

80-35600 Shorts Support Services Division (Airwork Ltd)
(† Airwork Ltd)
(a division of Bombadier Inc, Canada)
Bournemouth International Airport,
Christchurch, Dorset BH23 6NW
☎ (+44 1202) 36 52 00 ⊜ (+44 1202) 57 36 92
♣ £ 200,000,000 (1996) 👥 3,500+ (1997) 🏠 1,038 m2
Managing Director: Richard Beasley
Operations Director: Philip McBride
Purchasing Manager: Mel Stringer
Business Development Manager: Ian Smith
Marketing Co-ordinator: Roger Marsh
■ Operation and support of military and civil aircraft, and related engineering. Operation and maintenance of airfield aids and installations, surveillance, radar and communication systems. Aviation consultants. Provision of training services.
Supplies Division
15 Cobham Road, Ferndown Ind. Est.,
Wimborne Road, West Staplehill,
Wimborne, Dorset BH21 7PE
☎ (+44 1202) 89 16 66 ⊜ (+44 1202) 89 61 76
Manager: P.L. Groombridge [98]

80-35700 Shredaway Ltd
Park House, Green Hill Crescent, Watford, Herts WD1 8QU
☎ (+44 1923) 24 60 60 ⊜ (+44 1923) 24 61 61
♧ £ 100 000 ♣ £ 160 000 👥 6
Managing Dir: Jonathan Sandford
Sales & Marketing Dir: Anthony Pearlgood
■ Provide an on-site security shredding system for the destruction of confidential data. Widely used by the MoD, the company's mobile shredder can destroy 500kgs per hour at sites throughout the UK. [97]

80-35850 Sidem (UK) Ltd
Old Customs Warehouse, 4 Huntspill Road,
Highbridge, Somerset TA9 3DD
☎ (+44 1278) 78 65 42 ⊜ (+44 1278) 78 06 02
♣ £ 1 000 000 👥 5
Managing Dir: David L. Jenkins
■ International & UK land defence procurement consultants including counter-trade. Manufacture weapon mounts & softmounts. M3, M63, M31 for light & heavy machine guns & automatic 40mm grenade launchers. Small arms dealers & spare parts suppliers. Sales of SACO CAL.50M2HB, M60, M60E3 & MK19 40mm ME's. → 75
® USA: Saco Defense Inc. [98]

80-36050 Signal Computing Ltd
20 Nugent Rd, The Surrey Research Park,
Guildford, Surrey GU2 5AF
☎ (+44 1483) 57 99 00 ⊜ (+44 1483) 56 28 36
♧ £ 242 000 (1996) ♣ £ 3 200 000 (1995) 👥 60
🏠 1 000 sq m
Managing Dir: Peter Middleton
Operations Dir: John Hogg
Sales & Marketing Dir: Peter Upton
■ Computer systems integration, software & hardware design & development. Real time systems including simulation, modelling, data acquisition, recording, analysis & signal processing. Information systems. Navigation, particularly GPS. Intelligence applications. Equipment control & monitoring applications. [97]

80-36070 Silicone Engineering
(a Meggitt Aerospace Company)
Greenbank Business Park, Blakewater Road,
Blackburn, Lancashire BB1 3HU
☎ (+44 1254) 26 13 21 ⊜ (+44 1254) 58 35 19
👥 121
Managing Director: Eric Lewis
Operations Manager: Dennis Beadman
Sales and Marketing Manager: Trevor Galling [98]

80-36500 S.I.P. Services International Ltd
52 Upper Brook Street, Mayfair, London W1Y 1PG
☎ (+44 171) 355 19 11
[E] sipinternational@compuserve.com
⊜ (+44 171) 355 18 64
👥 20
Manager: Ben Fawcett
■ Specialist in counter electronic surveillance (debugging), close protection, investigations and covert surveillance. [98]

80-37000 Skytech Aviation Services Ltd
Hurst Farm, Winchfield, Hampshire RG27 8SL
☎ (+44 252) 84 47 75 ✆ SKYTECH ☒ 858728 skytec
⊜ (+44 252) 84 51 30
Chairman: Richard F H Noble
Managing Director: Julian C D Balaam
Company Secretary: John Taylor
■ Aircraft leasing, finance and marketing. [98]

80-39500 Smith System Engineering Ltd
Surrey Research Park, Guildford, Surrey GU2 5YP
☎ (+44 1483) 44 20 00 [W] http://www.smithsys.co.uk
⊜ (+44 1483) 44 21 44
♣ £ 13,000,000 (1996) ⁂ 140 (1997)
Managing Dir: Dr T.J. Black
Dir/Head of Defence Business: Dr H.J. Pearson
Marketing Dir: R.J.M. Sweet
Purchasing Mgr: M.A. Bennett
Public Relations Administration: Ms T Farbrother
■ Independent system engineering consultants providing complete system engineering services in the defence sector, including technology, research, procurement support, strategy studies, policy advice & the development & implementation of demonstration hardware & software systems. [98]

80-40000 Sonda Aviation Enterprises Limited
Unit 5, Riverside Business Centre, Brighton Road, Shoreham-by-Sea, West Sussex BN43 6RE
☎ (+44 1273) 44 14 77 [E] office@sonda_sim.co.uk
⊜ (+44 1273) 44 14 88
♧ £ 46,000 (1996) ♣ £ 400,000 (1996) ⁂ 16 (1996)
⌂ 200 m2
Managing Director: John E Zacharakis
Operations Director: D Hazzelby
Purchasing Manager: G Fry
■ Military and commercial flight simulation and training systems consultancy/products. Supply of software solutions and products designed to improve the performance of simulators and related training devices. [98]

80-41000 Sonsub International Ltd
(† Sonsub North Sea Ltd)
(a subsidiary of Saipem, Italy)
Tern Place, Denmore Road, Bridge of Don, Aberdeen AB23 8JX
☎ (+44 224) 84 34 34 [E] it@sonsub.co.uk
[W] http://www.sonsub.co.uk ⊜ (+44 224) 84 34 35
♣ £ 15,700,000 (1996) ⁂ 140 (1997) ⌂ 2,400 m2
Managing Director: Robert Addis
Operations: Keith Robathon
Sales and Marketing Manager: Derek A Cruickshank
Purchasing Manager: Bill McGougan [98]

80-45100 Spearhead Exhibitions Ltd
Ocean House, 50 Kingston Road, New Malden, Surrey KT3 3LZ
☎ (+44 181) 949 92 22 [E] bob@spearhead.co.uk
[W] http://www.spearhead.co.uk
℡ 8954102 attn spearhead ⊜ (+44 181) 949 81 68
⁂ 30 ⌂ 625 sq m
Managing Dir: Bob Munton
■ Specialist organisers of international exhibitions & conferences. [98]

80-45700 Specialist Training & Technical Services Ltd (STATS)
Gresham House, 2 Union St, Bingham, Notts NG13 8AD
☎ (+44 1949) 83 86 63 ⊜ (+44 1949) 83 19 93
♣ £ 6 346 819 (1994) ⁂ 110
Managing Dir: Peter Broekhuizen
Manager: Brian D'Souza
■ STATS can provide: Specialist training including the specialist personnel who carry out a training needs analysis, prepare all courseware and implement the course or programme. Also provides all specialist training equipment for the classroom or training environment including computer based trainers, simulators, models, films and student/instructor courseware. [97]

80-45800 Spice Communications
140 Barnhorn Road, Little Common, Bexhill, East Sussex TN39 4QG
☎ (+44 1424) 84 28 92 [E] sue.spice@virgin.net
⊜ (+44 1424) 84 82 00
⁂ 3
Proprietor: Ms Sue Spice
■ Provides specialised public relations/marketing service for the defence industry. Full trade show support, media liaison, offers expertise in all aspects of defence marketing. [98]

80-46000 SR Comms Management Ltd
(† Systems Reliability plc)
(a subs. of TSB International)
The Herculean, Porz Avenue, Houghton Hall Office Park, Dunstable, Beds LU6 3DH
☎ (+44 1582) 86 08 60 ℡ 825541
⊜ (+44 1582) 86 08 00
♣ £ 24 000 000 ⁂ 220
Managing Dir: John Woolford
Marketing Dir: Peter Middleton
Sales Dir: S. Robins
Principal Accounts Div.: B. Crawshaw
Contact: Jacquie Cox
Head of Product Mktg: Russel Pedder
Sales Mgr: David Heseltine
■ Offers a complete range of Telecom voice network management systems for all sizes of company. Our systems are capable of assimilating into one reporting system information from a variety of communications systems. Systems reliability products can provide concise information of traffic flows, communication costs and alarms. [97]

80-46200 SRC (Consultants) Ltd
(† Safety & Reliability Consultants Ltd)
Oceanic House, 89 High Street, Alton, Hampshire GU34 1LG
☎ (+44 1420) 54 16 36 ⊜ (+44 1420) 54 16 34
♣ £ 5 000 000 ⁂ 80
Operations Director: Mrs Karen Roberts
Head of Software Integrity: Anthony Darlison
Sales and Marketing Manager: Antony Dawe
■ Safety, reliability, maintainability, availability, environmental risk assessment, occupational health & safety, project risk analysis & bid support. Cost analysis, Manprint CFQ management & support system & software integrity assessment, software product assurance, quality assurance, software verification & validation. [98]

80-46500 SRI Consulting
(a subsidiary of SRI International, USA)
London Office
Stanford House, 2 Manchester Square, London W1M 5RF
☎ (+44 171) 446 61 66 ✉ SRI INTL EUR ℡ 946125
⊜ (+44 171) 446 61 99
♣ £ 5,000,000 (1995) ⁂ 45 ⌂ 1,000 sq m
Managing Director: Roger Camrass
Technical Director: Charles Stancomb
■ Management consultancy specialising in information security, IT strategy, business process redesign, lean distribution and various other activities. [98]

80-46700 STASYS Ltd
The Granary, 1 Waverley Lane, Farnham, Surrey GU9 8BB
☎ (+44 1252) 73 25 00 [E] julian.ranger@stasys.co.uk
⊜ (+44 1252) 73 25 01
♣ £ 2,500,000 (1997) ⁂ 40 (1997)
Operations & Technical Dir: Julian F.O. Ranger
Sales & Marketing Dir: John D. Curtis
SDI Contact: Julian F O Ranger
■ Independent consultants, particularly for military and civil Command and Control applications. Extensive experience in the integration of ground and airborne C2 systems, including Multi-function Information Distribution Systems (MIDS), joint Tactical Information Distribution System (JTIDS)/Link 16, IJMS, Link 11, Link 1 and Link 14. Specialists in air traffic control activities and systems and personnel security issues. [98]

80-47000 Stirling Dynamics Ltd
2 Tyndall's Park Road, Clifton, Bristol BS8 1PG
☎ (+44 117) 974 11 95; 15 44
[E] sdl@online.rednet.co.uk ⊜ (+44 117) 923 86 36
♣ £ 1,200,000 (1996) ⁂ 28 ⌂ 2,340 sq m
Managing Dir: Dr Robert Stirling
Operations & Technical Dir: Dr David Cowling
■ Mathematical modelling & computer simulation of dynamic systems, including feedback control, actuation & structural dynamics. Control systems analysis & design, including advanced multivariable methods. Flight dynamics & control, engine dynamics & control, aeroelasticity, ground vibration testing, flight flutter testing. Structural modal analysis & testing. Static & dynamic finite element analysis. Submarine & underwater vehicle platform control. Precision servo system design & supply. specialised software for dynamic analysis & real time control. [97]

80-47200 Stothert & Pitt
Clake Chapman Ltd
(a subsidiary of Rolls Royce IPG)
Lower Bristol Road, PO Box 25, Bath, Avon BA2 3DJ
☎ (+44 1225) 31 44 00 ℡ 44311 spale g
⊜ (+44 1225) 33 25 29
⁂ 80
Managing Director: F G J Gilroy
Sales and Marketing Director: Colin Dowding
■ Design & contracting for a wide range of specialist cranes, container handling cranes & shipyard cranes, including cantilever types. Rope luffing jib & cantilever cranes for defence application are designed per BS5882 & the necessary seismic studies, safety cases & other analyses included to ensure the integrity & reliability of the product. Services available are: design, contract & project management, erection, commissioning after sales service, provision of spares & rehabilitation. [98]

80-47400 Strachan & Henshaw Ltd
(a subsidiary of The Weir Group PLC)
Ashton House, Ashton Vale Road, PO Box 103, Bristol BS99 7TJ
☎ (+44 117) 966 46 77 ℡ 44170 sandh g
⊜ (+44 117) 963 42 59
♣ £ 75,000,000 ⁂ 750 ⌂ 45,000 m2
Chairman: Kevin Gamble
Managing Dir: Ken Grove
Engineering Director: David Smillie
Finance Dir: Mark Hardy
Defence & Aerospace Dir: Richard Craddy
Defence & Aerospace Marketing Mgr: Mike Bravery
Marketing Exec.: Cdr Brian King (RN retd)
■ Design, development, manufacture, erection and commissioning of purpose built equipment including underwater weapons handling and discharge in submarines, weapon handling in surface ships, submarine signal ejectors, towed sonar array handling, nuclear fuel, bulk materials handling, marine and aircraft lifts. Vehicle lifts/ramps, ships lifts for deck to deck transfer of vehicles, aircraft maintenance platforms, air cargo handling facilities, personnel and stores lifts. Aero engine test rigs. → 75 [98]

80-47500 Strat*X Ltd
(a subsidiary of Strat*X SA, France)
Windsor Court, 11 Windsor End, Beaconsfield, Bucks HP9 2JJ
☎ (+44 1494) 68 07 00 ⊜ (+44 1494) 68 08 80
♣ US$ 1,000,000 ⁂ 30
Managing Director: Niall McDonagh
Marketing Manager: Michael Loveless
■ Specializes in the development and implementation of tailored management development programs in business and marketing strategy. [98]

80-47800 STS Recruitment
Radley House, 8 St Cross Road, Winchester, Hampshire SO23 9HX
☎ (+44 1962) 86 94 78 [E] sts@tcp.co.uk
⊜ (+44 1962) 84 19 82
Director: John W. Spencer
■ Specialist recruitment consultancy for aerospace, defence & communications industries worldwide. [98]

80-49300 Symonds Group Ltd
(† Travers Morgan Limited)
(a subsidiary of Energy & Technical Services Group)
24/30 Holborn, London EC1N 2LX
☎ (+44 171) 421 20 00 ⊜ (+44 171) 421 22 22
♣ £ 70,000,000 (1996) ⁂ 1,300 (1997)
Chief Executive: Christopher Booy
Sales and Marketing Director: Tony Ferrett
Marketing Officer: R. Parkinson
■ Consultants in: Engineering, transport, environment, project management, cost management & facilites management.
Overseas locations:
Antigua, Australia, Barbados, Malaysia, New Zealand, Oman, Saudi Arabia, USA, Ireland, Bahrain, Abu Dhabi, Netherlands Antilles.
UK locations:
Altrincham, Belfast, Birmingham, Bristol, Cardiff, Carmarthen, Colwyn Bay, East Grinstead, Edinburgh, Leeds, Lincoln, London, Reading and York.

● Symonds House, Wood Street, East Grinstead, West Sussex RH19 1UU
☎ (+44 1342) 32 71 61 ⊜ (+44 1342) 31 34 55
Projects Dir: R.W. Flack [98]

80-49480 Systematic Software Engineering Ltd
(a subsidiary of Systematic Software Engineering AS, Denmark)
The Coliseum, Riverside Way,
Camberley, Surrey GU15 3YL
☎ (+44 1276) 67 55 33 [E] ssel@systematic.demon.co.uk
[W] http://www.systematic.dk ⊖ (+44 1276) 67 55 44
Managing Director: A D Patterson
Sales and Marketing Director: W Peach
Marketing Co-ordinator: Ms Jacqueline Snowdon
■ Software systems house specialising in formatted messages to the AdatP3, USMTF, OTH-Gold and ADFORMS standards for the Military and EDIfact standard for electronic trading between businesses and organisations. Modular software packages for EW systems providing facilities for operational support with threat data, parameter information and analysis to enable automated identification and support to countermeasures. [98]

80-49500 Systems Consultants Services Ltd
Grove House, Fair Mile,
Henley-on-Thames, Oxon RG9 6AD
☎ (+44 1491) 41 21 02 [E] info@scs-ltd.co.uk
⊖ (+44 1491) 41 20 82
👥 30
Managing Dir: A.E.S. Carter
Marketing Dir: J.W. Lyde
Finance Dir: J.D. Tydeman
Business Mgr: M.J. Riddick
■ Defence consultancy providing consultants who combine military & specialist technical expertise with an in-depth knowledge of the MoD & Armed Forces environment. Accredited to ISO9001 standard. [98]

80-49600 Systems Engineering & Assessment Ltd
Beckington Castle, PO Box 800, Bath BA3 6TB
☎ (+44 1373) 83 18 00 ⊖ (+44 1373) 83 11 33
♣ £ 3,250,000 (1996/97) 👥 85 (1997) 🏭 2,500 m2
Managing Director: Ian Dale-Staples
Defence Systems Director: Dave Young
Space Systems Director: Paul Phillips
Technical Director: Ron Barnes
Engineering Director: Allan Phillips
■ Provides project management & technical specialist support to defence procurement agencies. Support ranges from research & feasibility studies, preparation of tender documentation, tender evaluation, risk management, contract monitoring through to trials & contract acceptance. Remote sensing modelling systems & workstations, for both civil & military purposes. Hardware & software is developed & supplied. [98]

80-50100 TADS Video Production
(† TADS Television & Training)
(part of Training Advisory Development Services Ltd Group of Companies)
37 High Street, Shaftesbury, Dorset SP7 8JE
☎ (+44 1747) 85 58 88 [E] tads@shaftesbury.telme.co.uk
⊖ (+44 1747) 85 58 89
👥 6 (+ 27 assocs.)
Managing Dir: M.J. Wills
■ Specialise in the production of films & videos for the defence industry & Armed Forces for sales, marketing, training & trial coverage applications. Tender application support. [97]

80-50110 TA Group Ltd
(† TA Consultancy Services Ltd)
(incorporating Rex Thompson & Ptnrs Ltd)
(a subsidiary of Fluor Daniel Ltd)
Newnhams, West Street, Farnham, Surrey GU9 7EQ
☎ (+44 252) 71 14 14 ⊖ (+44 252) 73 51 55
♣ £ 20,000,000 (1996) 👥 500 (1996) 🏭 2,000 sq m
Managing Director: Brian Thomas
Director: Dr G A Robins
■ Engineer & software consultants providing technical & project management services in operational analysis & simulation, reliability & safety technology, software engineering & project management. Safety critical software IV & V. [98]

80-50150 T&N Technology Ltd
(a subs. of T&N plc)
Cawston House, Cawston, Rugby, Warwicks CV22 7SA
☎ (+44 1788) 81 25 55 ⊖ (+44 1788) 81 69 52
♣ £ 10 000 000 👥 250
Managing Dir: Alan Begg
Information Officer: R.L. Wagstaffe
■ Automotive/Aerospace components, research & development. [97]

80-50200 Tarmac Professional Services
(† TBV Consult)
(a div. of Tarmac plc)
The Lansdowne Building, Lansdowne Rd,
Croydon, Surrey CR0 2BX
☎ (+44 181) 256 40 00 ⊖ (+44 181) 256 45 61
♣ £ 107 000 000 (1994) 👥 1 893
Managing Dir: Steve Reding
Operations Dir: Martin Kennard
Marketing Dir: Shonagh Hay
Press Officer: Paul Wilkinson
■ Multi-disciplinary design, project management surveying services & technical advisory consultancy. Has experience in building & civil engineering works including airport, ports defence facilities. Building design to resist explosions, NBC, EMP & electronic eavesdropping. Monitoring & testing of ventilation systems for NBC protection.
Explosives: effects on structures; storage of explosives.
Maritime & Coastal Engineering: dredging surveys & management.
Surveys: topographic, hydrographic, airfield obstruction surveys.
Airfields: ground lighting & control systems, navaids - design, testing, commissioning & inspection. Pavements - non destructive testing & evaluation, friction testing, maintenance advice & training.
Masts & Towers: design, inspection & maintenance.
Water Supplies: design, appraisal & testing, including boreholes, reservoirs & treatment works.
Fuel/POL: bulk storage & distribution systems - advice on design, operation & maintenance.
Contaminated Land: investigations & advice on remedial work. [97]

80-50300 Taylor Woodrow Construction Ltd
(† Taylor Woodrow International Ltd)
(a subsidiary of Taylor Woodrow plc)
345 Ruislip Road, Southall, Middlesex UB1 2QP
☎ (+44 181) 578 23 66
[E] cirwinchilds@taywoodint.co.uk
[W] http://www.taywood.co.uk ☒ 23502 tayint g
⊖ (+44 181) 231 11 18
♧ £ 540,200,000 (1996) ♣ £ 489,000,000 (1996)
👥 10,000 (staff) 11,000 (operatives) (1996)
Managing Dir: B.C. Russell
Marketing Manager: Nigel R Marks
Business Development Mgr: Christopher Irwin-Childs
■ Construction of all types of multi-disciplined defence facilities to serve land, sea or airborne forces & national security. Services available include design, research, procurement, resourcing, construction management, offset projects, operational support, maintenance & quality assurance.
Offices in:
Australia, Botswana, Canada, China, Denmark, Germany, Ghana, Gibraltar, Hong Kong, India, Indonesia, Kuwait, Korea, Malaysia, Malta, Oman, Saudi Arabia, Singapore, Spain, South Africa, USA, Vietnam & Zimbabwe. [98]

80-50400 Taylor Woodrow Management & Engineering Ltd
(a subs. of Taylor Woodrow Construction Holdings Ltd)
Taywood House, 345 Ruislip Road,
Southhall, Middlesex UB1 2QX
☎ (+44 181) 575 48 27 ☒ 24428
⊖ (+44 181) 575 46 78
♣ £ 25 000 000 (1996) 👥 100
Managing Dir: S.A. Hardiman ☎ (+44 181) 575 48 27
Operations/Technical Dir: J.B. Sargent
Marketing Mgr: Chris Buxton ☎ (+44 181) 575 44 76
Purchasing Mgr: Peter East
■ Provide an integrated engineering & project management service for multi-disciplined projects at home & overseas. [97]

80-51000 TDSi Micro Applications Ltd
(† HPA Micro Applications Ltd)
(a subsidiary of Time and Data Systems International Ltd)
Quay House, The Quay, Poole, Dorset BH15 1HA
☎ (+44 1202) 66 99 10
[E] 100442.422@compuserve.com
[W] http://www.bucc.co.uk/vpp/tdsi.html
⊖ (+44 1202) 66 99 12
♣ £ 600,000 (1997) 👥 12 (1997) 🏭 3,000 m2
Managing Director: Philip M Gater
Director and General Manager: Keith Whittington
■ Systems house offering feasibility studies, consultancy, research and development, production and installation of defence and space systems using C++, ADA, Centura and object design methods. ISO 9001 and ISO 9003 certified. [98]

80-52400 Technical Indexes Ltd
(a subsidiary of Information Handling Services, USA)
Willoughby Road, Bracknell, Berkshire RG12 8DW
☎ (+44 1344) 42 63 11 [E] systems@techni-dex.co.uk
[E] http://www.technidex.co.uk ⊖ (+44 1344) 42 49 71
♣ £ 10 000 000 👥 150
Managing Director: Philip Stow
Sales and Marketing Director: Andrew Loch
Product Manager: Ian Johns
Marketing Manager: Mrs J Westgate
Marketing Executive: Chris Copland
Marketing Assistant: Miss Sophie Johnson
■ Defence standards, specifications, Defcons, quality assurance documents, british standards, BS9000, CECC and IECQ detail specifications, American MIL-Spec standards, handbooks and other military documents. UK and US NATO stock numbers. [98]

80-52800 Technology Project Services (International) Ltd (TPS)
8-10 Grosvenor Gardens, London SW1W 0DH
☎ (+44 171) 730 92 65 [E] mail@tps.co.uk
[W] http://www.tps.co.uk ⊖ (+44 171) 730 47 30
♧ £ 5,256,263 (1996) ♣ £ 39,892,655 (1996)
👥 550 (1996)
Managing Director: Michael G Doubleday
Finance Director: Roger S Cox
Quality and Training Manager: P G Winny
Marketing Manager: M Ellingham
■ Supply of qualified & experienced staff on contract basis to the defence, aerospace, telecommunications & electronics industries. Skills include project/programme management, systems/hardware/software & engineering, technical authorship, CAD & board draughting. Technology project services to ISO9002 accredited & an approved MoD & NATO supplier.
Scottish Division
31/33 Hopetoun Road, South Queensferry,
West Lothian EH30 9RB
☎ (+44 131) 331 30 30 ⊖ (+44 131) 331 25 18
Dir Scottish Div.: Joseph M. Forbes
■ Software development services. [98]

80-52900 TEDCO International Ltd
Kings Scholars House, 230 Vauxhall Bridge Road,
London SW1V 1AU
☎ (+44 171) 828 73 02 ☒ 8811310 tedco g
⊖ (+44 171) 828 26 28
♣ £ 15,000,000 👥 10
Director: A K Misra
Company Secretary: Mrs H Hobsbaum
■ Specialist in assisting companies find & develop export opportunities to India, having extensive coverage in India with a professional team of military advisers in main offices in Delhi, Bombay & Calcutta. [98]

80-54500 Thermal Processing Group Ltd
(a member of the TI Group plc)
Colliery Lane, Exhall, Coventry CV7 9NW
☎ (+44 1203) 36 77 86 ⊖ (+44 1203) 64 53 25
♣ £ 6,300,000 👥 125 🏭 8,000 m2
Managing Director: Trevor H Sturt
☎ (+44 585) 56 06 33 (mobile)
Technical and Sales Director: Derek Alty
☎ (+44 585) 32 80 30 (mobile)
Financial Director: Peter Hodgson
☎ (+44 585) 56 06 32 (mobile)
■ Specialists in subcontract metal joining & heat treatment.
Other Divisions at:
Cambridge, Sheffield & Skelmersdale. [98]

80-55000 Thompson Kennicott Limited
(a subs. of Rolls-Royce Industrial Power Group)
PO Box 100, Wolverhampton WV4 6JY
☎ (+44 1902) 35 33 53 ⊖ (+44 1902) 35 46 19
Managing Dir: A. Macdonald-Hardie

Sales & Marketing Dir: C.A. Marson
Sales Mgr: P.M. Granger
■ Water & effluent treatment plant design & manufacture. Containerised water treatment. [97]

80-55300 Three Bridges Publishing Ltd
Intersec
Bridge House, Aviary Road,
Pyrford, Woking, Surrey GU22 8TH
☎(+44 1932) 34 04 18 ⊜ (+44 1932) 34 04 19
Publishing Dir: Miss Susan Ayre
■ Publishers of a monthly journal covering high level international security topics. [98]

80-55400 Three Quays Marine Services Ltd (TQMS)
12-20 Camomile Street, London EC3A 7AS
☎(+44 171) 929 22 99 ▣ 8813407 marsaf g
⊜ (+44 171) 929 16 50
♣ £ 2 000 000 ₶ 85
Managing Dir: Patrick Mellis
Project Mgr: M.H. Peacock
Project Co-ordinator: W.B. Davies
■ Naval architects & consulting engineers for ship design & specifications, commissioning trials, fleet support, spares procurement & distribution, project management, construction, conversion, refit & repair oversight. [97]

80-56600 TNT Express (UK) Ltd
(a division of KPN Postal & Telecommunications, Netherlands)
TNT House, Holly Lane,
Atherstone, Warwickshire CV9 2RY
☎(+44 1827) 30 30 30 ✆ (800) 10 06 00
▣ 347371 tnt ath ⊜ (+44 1827) 30 13 01
♣ £ 352,000,000 (1996) ₶ 8,500 (1997)
Managing Director: Tom Bell
Operations Director: Nigel Barton
Sales and Marketing Director: Chris Atkinson
Purchasing Manager: Norman Duckworth
Group National Accounts Manager: Neville W Cooper
■ Delivery services for defence manufacturers, suppliers, national and private organisations. Same day and next day deliveries, nationwide. Extensive ground and air network for transportation of all sizes of freight, documents, parcels and pallets for rapid and secure movement throughout the UK. [98]

80-58500 TRH Pictures
Bradley's Close, 74-77 White Lion Street, London N1 9PF
☎(+44 171) 520 76 47 ⊜ (+44 171) 520 76 06
₶ 2 (1997)
Director: E Nevill
■ Specialist picture library covering defence, aviation & space for advertising marketing & training publications. Photographic services. [98]

80-58600 Triad Group plc
Weyside Park, Catteshall Lane,
Godalming, Surrey GU7 1XE
☎(+44 1483) 86 02 22 ▣ 859391
⊜ (+44 1483) 86 01 98
♣ £ 11 700 000 (1995/96) ₶ 160
Managing Dir: A.M. Fulton
Operations/Technical Dir: T. Eglen
Marketing Mgr: Miss A.S.J. Jones
■ Supplier of computer consultancy, software development, systems integration & software support. Deliver business solutions involving the most appropriate technology & using highly trained & experienced consultants, analysts, designers & programmers. Triad is certificated to BS EN ISO 9001:1994 against TickIT for all its services. [97]

80-58700 TRL Technology Ltd
Shannon Way, Tewkesbury, Gloucestershire GL20 8ND
☎(+44 1684) 27 87 00 ▣ 437444
⊜ (+44 1684) 85 04 06
♣ £ 6,000,000 (1996/97) ₶ 76 ⌂ 190 sq m
Managing Director: Roger Allan
Marketing Director: Peter Chisholm
Technical Director: Hugh Jarman
Marketing Manager: Dave Hall
Purchasing Manager: Malcolm Cubitt
■ Development and manufacture of high technology electronic systems:-
Versatile satellite communications modems and ancillary equipment specialising in fast acquisition in low signal to noise ratio systems.
SATCOM link test equipment and automatic bit error rate test sets and mobile earth station test units.
Satellite communications monitoring and test systems. High speed data manipulation equipment/encryption systems.
Radio electronic warfare systems, radio receivers, jammers, signal analysers and robust data links.
Radio surveillance systems and radio direction finding systems.
Bespoke DSP hardware and software design.
Real time/time critical software systems.
Cabin and tower installations for turnkey video and communications applications. [98]

80-58770 TSB Metrology Ltd
Unit 7 - Crown Business Centre, Horton Road,
West Drayton, Middlesex UB7 8HZ
☎(+44 1895) 43 00 01 ⊜ (+44 1895) 43 00 02
♣ £ 3,000,000 (1996) ₶ 15 (1996) ⌂ 550 sq m
Managing Director: Simon Lucas
Sales Director: Steve White
® Switzerland: Trimos-Sylvac Task Force [98]

80-58800 Turbine Support Ltd
5A Daish Way, Dodnor Industrial Estate,
Newport, Isle of Wight PO30 5XB
☎(+44 1983) 82 62 50 [E] tur@supp.demon.co.uk
⊜ (+44 1983) 82 62 53
Managing Director: Peter O Collins
Technical Director: Cliff Camm
■ Rolls-Royce Gnome hovercraft and helicopter gas turbine repair and overhaul. Dynamic balancing services, CAA and ISO 9002 approved. [98]

80-60000 UK Training Ltd
(a member of The Nitor Group)
Nitor House, Great George Street,
Godalming, Surrey GU7 1EE
☎(+44 483) 42 62 99 ▣ 94011723 nitr g
⊜ (+44 483) 41 59 53
Director: M B Pegrum
Director: M E Read
■ Provides complete training services for police and military organisations. Training equipment support including maintenance, servicing and operator training. [98]

80-61000 UMC International plc
(t Underwater Maintenance Co Ltd)
Mayflower Close, Chandlers Ford Ind. Est.,
Eastleigh, Hants SO53 4AR
☎(+44 1703) 26 98 66 ⊜ (+44 1703) 25 31 98
♧ £ 250 000 ♣ £ 2 500 000 (1995) ₶ 50 ⌂ 880 m2
Managing Dir: D.F. Jones
Director: S.R. Moore
Technical Mgr: J. Jackson
Operations Mgr: P. Jupp
■ Diving contractors specialising in underwater ship maintenance. Also markets equipment & services in the re-application of acoustic cladding, underwater hull cleaning & propeller polishing. Permanent operating bases in the UK (Southampton, Plymouth & Glasgow), Netherlands, Malta, Dubai & Fujairah. [97]

80-62000 Uniplex Ltd
(a member of CP Software Group)
Redwood House, Maylands Ave,
Hemel Hempstead, Herts HP2 7DE
☎(+44 1442) 23 03 30 ✆ (800) 58 16 43
▣ 826194 uniplx g ⊜ (+44 1442) 23 02 22
♣ £ 40 000 000 ₶ 140
UK General Mgr: Guy Norgrove
[E] guy.norgrove@uniplex.co.uk
Sales & Marketing Dir: Andrew Forster
Operations & Technical Dir: Ms Alison Coates
Purchasing Mgr: Steve Bucknall
■ Designs & builds workgroup & document management software. [97]

80-65500 Vickers Medical Ltd
(a division of Vickers plc)
Ruxley Corner, Sidcup, Kent DA14 5BL
☎(+44 181) 309 04 33 ⊜ (+44 181) 309 09 19
♧ £ 10,000,000 (1997) ♣ £ 22,000,000 (1997)
₶ 160 (1997)
Managing Director: Michael McAdam
Operations Director: Martin Clements
Sales and Marketing Manager: Robin Thuiller
Purchasing Manager: Les Smith
Marketing Co-ordinator: Ms Eileen Restell
Marketing Copywriter: Ms Natalie Spencer
[E] nspencer@vickersmedical.co .uk
■ Medical electronics for critical care. Disposable face masks and electrodes. [98]

80-67000 Vosper International Ltd
Kelvin Campus, West of Scotland Science Park,
Glasgow G20 0SP
☎(+44 141) 945 55 29 [E] vosper@valtos.com
[W] http://www.vosper.co.uk ⊜ (+44 141) 945 53 56
♣ £ 1,000,000 (1995) ⌂ 2,000 m2
Managing Director: Brian F M Morrison
Company Secretary: Mrs Margaret Lamb (Administration)
Senior Consultant Ship Design: Anthony J Gould
Senior Consultant Support Systems: Paul Gillon
Naval Architect: Dr Bärbel Carstens
■ Ship design, marine engineering consultancy providing technical & support services to defence authorities & shipyards engaged in the operation & construction of light frigates, patrol craft & high performance vessels. Supplies computerised maintenance management systems (VIMMS) for marine & land based operations. [98]

80-70500 The Walters Research Company
9 Kittles Yard, Edinburgh, Scotland EH9 1PJ
☎(+44 131) 553 55 34 ⊜ (+44 131) 553 60 33
₶ 4
Managing Dir: Stewart Walters
■ Provides management consultancy research markets, diversification market development, marketing & sales IT systems. [98]

80-71500 Watson & Lewis Ltd
354 High Road, Willesden, London NW10 2EG
☎(+44 181) 459 62 92; 451 12 12
⊜ (+44 181) 451 19 90
♣ £ 300,000 ₶ 10
General Manager: S Ash
■ Precision plating of components. [98]

80-72000 Weir Materials Ltd
(a subsidiary of The Weir Group plc)
Park Works, Newton Heath, Manchester M40 2BA
☎(+44 161) 203 42 62 ▣ 667193
⊜ (+44 161) 205 89 66
Dir & General Mgr: Glenn Byrne
■ Supply of stainless steel & other high alloyed materials to the marine/defence industry, platform mooring systems/materials for underwater cameras, pipe work fittings and flanges for the offshore industry petrochem, nuclear, FGD, chemical, oil & gas industries.
Corrosion resistant alloy components for: civil structures, heat exchangers, long term storage systems (fuel, active waste etc.). Suppliers of corrosion resistant fasteners for marine, FGD & other industries. [98]

80-72250 Weir Westgarth Ltd
(a subs. of The Weir Group PLC)
149 Newlands Road, Cathcart, Glasgow G44 4EX
☎(+44 141) 633 13 36 ▣ 77165
⊜ (+44 141) 633 32 89
♣ £ 20 000 000 ₶ 85
Chairman: Sir Ron Garrick
Managing Dir: Paul Capell
Sales Dir: Bill Stewart
Operations/Technical Dir: Clyne Gall
■ Undertakes design, supply & contract management for engineering projects in the field of water treatment & desalination. → 75 [97]

80-72500 Welcom Software Technology International (WSTI)
(a subsidiary of Welcom Software Technology, USA)
South Bank Technopark, 90 London Road,
London SE1 6LN
☎(+44 171) 401 26 26 [E] info@wsti.co.uk
[W] http://www.wst.com ⊜ (+44 171) 922 88 65
₶ 100+ (Group)
Managing Dir: Tony Welsh
Operations/Technical Dir: John Owen
Marketing Manager: Ms Sara Robinson
Marketing Co-ordinator: V Balfour

■ Supply a range of project management software products & services. The software can be used for a diverse range of projects within the defence industry, covered by the standing arrangement scheme of authorised demanders by the CCTA (SA273). Products include: Open Plan Professional and Open Plan Desktop, and the Cobra performance measurement system, CSPEC compliant. 98

80-73000 Western Tactical Systems
3 Meadow Rise, Bewdley, Worcestershire DY12 1JP
☎ (+44 1299) 40 43 76 ⊖ (+44 1299) 40 43 76
Chief Executive: Robert C Waddington
■ Military consultants & independent agents for the supply, procurement & leasing of defence equipment. 98

80-76000 Wilmont Marine Services
Imperial House, 15 19 Kingsway, London WC2B 6UN
☎ (+44 171) 379 55 59 ⊖ (+44 171) 836 99 47 97

80-76200 Woldtare Export Limited
(† Woldtare Ltd)
The Mill, Northiam, Rye, Sussex TN31 6QT
☎ (+44 1797) 25 29 50 ⊖ (+44 1797) 25 29 60
♣ £ 2 000 000 (1995) 7
Managing Dir: P.M. Mott
Sales Dir: T. Angus
■ Supplies weapons, ammunition, parts & all related services to defence ministries worldwide. 97

80-76250 Wood Group Aero Ltd
(† GTC Gas Turbine Ltd)
(a subsidiary of Wood Group Gas Turbines)
Baldovie Industrial Estate, Drumgeith Road, Dundee DD4 0NZ
☎ (+44 1382) 73 93 11 76234 gtc uk g
⊖ (+44 1382) 77 61 67
♣ £ 15,000,000 (1995) 210 7,500 sq m
Operations Dir Aero Division: F.P. Hall
■ Overhaul, repair & servicing of industrial, civil & military gas turbine engines. 98

85 Stockists, Distributors, Spare Parts Suppliers

85-00020 Aerco Aviation Electrical & Radio Co. Ltd
Aviation Electrical & Radio Co Ltd
16-17 Lawson Hunt Industrial Park, Broadbridge Heath, Horsham, West Sussex RH12 3JR
☎ (+44 1403) 26 02 06 ⊠ AERCO Horsham
⊖ (+44 1403) 25 97 60
£ 1,000,000 ♣ £ 5,000,000 (1997) 50
10,000 sq ft
Managing Director: M J Griffiths
Financial Director: P D Ward
Sales and Marketing Manager: S Taylor
■ Distributer of electrical connectors, lamps, indicators, lighting equipment, fuzes and other circuit protection items. Control knobs; relays; terminal blocks; cross referencing of defence/NATO part numbers. Approved supplier to MoD. 98

85-00070 Allmakes Ltd
(† Allmakes Projects Ltd)
176 Milton Park, Abingdon, Oxon OX14 4SW
☎ (+44 1235) 82 11 22 ⊠ QUICKPARTS Oxford
⊖ (+44 1235) 82 11 33
58
Director: Hugh Smith
■ Distributors & stockists of spare parts for British support vehicles. Carries out in-country refurbishment of vehicles on a contract basis. 97

85-00085 Apollo Metals plc
Apollo House, Bordesley Green, Birmingham B9 4SJ
☎ (+44 121) 773 25 26 ⊖ (+44 121) 359 37 14
♣ £ 71,400,000 (1996)
Managing Director Apollo Metals plc, Executive Chairman Aviation Metals Ltd: Derek Webb
Executive Chairman Apollo Metals UK, Deputy Managing Director Apollo Metals plc: Clive Orford
Joint Managing Director Apollo Metals UK: Keith Ingle
Joint Managing Director Apollo Metals UK: Bruce Maggs
Managing Director Apollo Metals International Holdings Ltd: John Ferguson
Sales Director Aviation Metals Ltd: Colin Hicks
Operations Director Apollo Metals UK: Will Clare
Marketing Manager Apollo Metals UK: John Oldfield
Purchasing Manager Apollo Metals UK: Mark Jackson
General Manager Apollo Metals London and Oxford (Southern Region): Cliff Campbell
General Manager Apollo Metals Birmingham and Manchester (Northern Region): Malcolm Mitson
■ Aluminium plate & bar stockholder & processor, offers plate in 18 alloys, 6 tempers, 10 widths, 12 lengths & 65 thicknesses up to 15ins, bar in diameters up to 16ins & squares up to 8ins.

Apollo Metals (International) Ltd
Apollo House, Bordesley Green, Birmingham B9 4SJ
☎ (+44 121) 773 25 26 ⊖ (+44 121) 359 37 12

Apollo Metals (UK) Ltd
Apollo House, Bordesley Green, Birmingham B9 4SJ
☎ (+44 121) 773 25 26 E apollometals.co.uk
⊖ (+44 121) 772 23 56

Apollo Metals (Birmingham) Ltd
Bangor Road, Bordesley Green, Birmingham B9 4TX
☎ (+44 121) 773 36 61 ⊖ (+44 121) 773 39 60

Apollo Metals (London) Ltd
Lockfield Avenue, Enfield, Middlesex EN3 7PX
☎ (+44 181) 804 40 32 ⊖ (+44 181) 443 03 88

Apollo Metals (Manchester) Ltd
Unit 20 Newby Road Industrial Estate, Hazel Grove, Stockport, Cheshire SK7 5DA
☎ (+44 161) 483 96 62 ⊖ (+44 161) 483 96 68

Apollo Metals (Oxford) Ltd
Old Kinlock Warehouse, Stanton Harcourt Road, Eynsham, Oxfordshire OX18 1JT
☎ (+44 1865) 88 27 47 ⊖ (+44 1865) 88 38 16

Aviation Metals Ltd
Michigan Drive, Tongwell, Milton Keynes MK15 8JE
☎ (+44 1908) 21 00 12 ⊖ (+44 1908) 21 00 66 98

85-00195 Aviation Jersey Ltd
Goose Green, Beaumont, St Peter, Jersey, CI JE3 7BR
☎ (+44 1534) 253 01 ⊠ AVIO Jersey 4190161 aviojy
⊖ (+44 1534) 594 49
Director: Hedley Griffiths
■ Specialise in the supply of spares and technical services for Rolls-Royce non-current aero engines, as well as military, commercial and industrial engines. 98

85-00200 BACO Metal Centres
(† Alcan Metal Centres Ltd)
(a subsidiary of British Aluminium)
Kiltonga Industrial Estate, Newtownards, Co. Down BT23 4TJ
☎ (+44 247) 84 41 00 ⊖ (+44 247) 84 41 99
Managing Director: Derek Maddox
Sales and Marketing Director: David Howell
Service Centre Manager: Ms Lynda Burns
■ Stockist of semi-fabricated products in aluminium & stainless steel for aerospace & defence. Titanium, brass, bronze & alloy steels. 98

85-00235 Britorion Ltd
PO Box 98, Alton, Hampshire GU34 4YL
☎ (+44 1420) 221 34 ⊖ (+44 1420) 221 34
♣ £ 250,000 4 500 sq ft
Managing Director: Col J M Evans (retd)
■ Uniforms & webbing, footwear, ceremonial & other equipment. Tentage, bedding, underwear, tropical & weatherproof wear, ration packs, first-aid kits & other personal equipment. 98

85-00240 BSD Steel Service Centre
(† BSD Tubes)
(a subsidiary of British Steel plc)
Walker Road, Walker Industrial Estate, Guide, Blackburn BB1 2LJ
☎ (+44 1254) 551 61 ⊖ (+44 1254) 67 08 36
Sales Mgr: S. McMahon
■ Supply of tubes & tubular fittings in carbon, alloy & stainless steels. 98

85-00300 Carel Components Ltd
(a subsidiary of Caynton Ltd)
24 Endeavour Way, Wimbledon Park, London SW19 8UH
☎ (+44 181) 946 98 82 ⊖ (+44 181) 946 62 59
Managing Dir: A.C. Janse
Dir & Secretary: J.C. Knight
Group Marketing Mgr: J. Finch-Taylor
Customer Service Mgr: Mrs J. Gilleland
■ Sensors & heaters for military & aerospace.
® USA: Minco Products 98

85-00320 Certacom Ltd
(† Wescom (UK) Ltd)
1 Holtspur Lane, Wooburn Green, High Wycombe, Bucks HP10 0AA
☎ (+44 1628) 52 79 72 ⊖ (+44 1628) 81 00 06
♣ £ 5,000,000 (1996) 18
Managing Dir: Brian Cook
Sales Dir: D.J. Walton
Operations & Technical Dir: John Bennett
Sales Mgr: Terry Field
Purchasing Mgr: Toby Howick
■ Telephone signalling equipment & systems. Conversion products, echo cancellers, channel banks. → 90 98

85-00350 Compstock International
(a subsidiary of EIS Ltd)
Eagle Trading Estate, Brookers Road, Billingshurst, West Sussex RH14 9YZ
☎ (+44 1403) 78 52 75 ⊖ (+44 1403) 78 23 10
♣ £ 8,000,000 (1996) 25 (1996) 2,800 sq m
Managing Director: W McKnight
Senior Sales Executive: Paul Bailey
Export Sales Manager: Lionel E Lloyd
■ Specialises in the supply of electronic components, particularly as spares for communications & radar equipment. Wide range of electronic tubes, semi-conductors & passive components. 98

85-00370 Crayford Special Equipment Co. Ltd
Berkeley House, Wood Lane End, Hemel Hempstead, Herts HP2 4RF
☎ (+44 1442) 21 93 22 ⊖ (+44 1442) 21 93 23
♣ £ 1,000,000 (1996) 3 (1997)
Managing Director: A M Brooks
■ Import and distribution of specialist all terrain vehicles and associated mobile field equipment. 98

Contact Reference

85-00375 Crellon Microsystems
(part of Gothic Crellon Ltd, a subsidiary of ML Holdings)
3 The Business Centre, Molly Millars Lane,
Wokingham, Berkshire RG11 2EY
☎ (+44 734) 77 61 61 [E] crellon@gcrellon.co.uk
[Telex] 847571 ⊜ (+44 734) 77 60 95
♣ £ 20,000,000 (Gothic Crellon 1995/96)
[Employees] 90 (1995/96)
Managing Director: John Walker
Sales and Marketing Manager: Steve Sydes
Business Manager: Shaun Froome
■ Computer solutions for industrial and other harsh environments using range of rugged IBM-compatible PCs and rackmount PC workstations. COTS and commercial VMEbus boards. Embedded PCs and compact PCI. → 90
[98]

85-00380 CryoService Ltd
Blackpole Trading Estate East, Blackpole Road,
Worcester WR3 8SG
☎ (+44 1905) 75 45 00 [Telex] 335092
⊜ (+44 1905) 75 40 60
Chairman: Colin Kinnear
Technical Director: John Carver
Sales Manager: R Meeks
■ UK suppliers of argon, nitrogen, CO2 special gases and mixtures for MoD, laboratory and research institutions. Supplier of cryogenic equipment and cylinders. [98]

85-00390 Cylink Ltd
(a subsidiary of Cylink Corp., USA)
Unit 3, Intec 2, Wade Road,
Basingstoke, Hampshire RG24 8NE
☎ (+44 1256) 84 19 19 [E] info@cylink.co.uk
⊜ (+44 1256) 32 41 56
[Employees] 19 (1997)
Operations & Technical Dir: K. Hardy
General Mgr: David J. Moseley
Purchasing Mgr: Mrs Sally Brown
■ Worldwide supplier of data communications encryptors, remote access control systems & spread spectrum wireless communications products.
® USA: Cylink Corp. [98]

85-00425 Diamond Point International Ltd
(a subsidiary of Diamond Point Inc, USA)
9 Enterprise Close, Rochester, Kent ME2 4LY
☎ (+44 1634) 72 23 90 [W] http://www.dpie.com
⊜ (+44 1634) 72 23 98
♣ £ 1,500,000 (1996) [Employees] 15 (1996)
Chairman/General Manager: Sherman Kemmer
Sales and Marketing Director: John Vaines
Public Relations Manager: Sherman Kemmer
Support Manager: Mark Jeffrey
■ Provide a range of I/O solutions to the simulation & training industry incl. digital I/O, analogue I/O, synchro I/O, resolver I/O, repeater links & intelligent I/O controllers. [98]

85-00450 Dowty Energy Control Products
(a member of the TI Group plc)
Alton Road, Ross-on-Wye, Herefordshire
☎ (+44 1989) 56 56 36 ⊜ (+44 1989) 654 10
Dir & General Mgr: Michael A. Icke
■ Supplies non-burn foams for acoustic & thermal insulation & sound attenuation materials. Also supplies fabric reinforced diaphragm materials, silicone sheet & silicone sponge. Provides a specialist polymer conversion service including mouldings, extrusions & die cut parts. → 80 [97]

85-00475 Henry Du Bora
38 Station Road, Aldershot GU11 1HT
☎ (+44 1252) 22 668 ⊜ (+44 1252) 319 963
Contact: Ms. J. Nicholls
■ Specialists in military clothing, equipment, badges & insignia. [97]

85-00500 Edgar Brothers
Catherine Street, Macclesfield, Cheshire SK11 6SG
☎ (+44 1625) 61 31 77 ⊜ (+44 1625) 61 52 76
Director: A. Edgar
■ Importers & wholesalers of arms & ammunition. Tactical lights & laser sights. [97]

85-00560 European Market Liaison (UK) Ltd (EML)
Unit 11C, Weldale Street, Reading, Berkshire RG1 7BX
☎ (+44 1734) 56 63 45 ⊜ (+44 1734) 59 21 78
♣ £ 2,200,000 (1996) [Employees] 22 (1997) [Floor area] 6,000 sq ft
Managing Director: Thomas Lee
Sales and Marketing Director: Alex Longson
Operations Director: R Bone
Purchasing Manager: Ms Maureen Reid
■ Company provides specialized line test equipment & test sales used by OUTSIDE PLANT technicians in both maintenance & construction of cable plant for PTT/Telecom operators such as BT, British Rail, Mercury, London Underground, Military etc. [98]

85-00610 The Fasteners Centre plc
Hale House, Ghyll Industrial Estate,
Heathfield, East Sussex TN21 8AW
☎ (+44 1435) 86 60 11 ⊜ (+44 1435) 86 66 20
♣ £ 3,500,000 (1997) [Employees] 32 (1995) [Floor area] 1,100 sq m
Managing Director: D G R Hale
Technical Director: B R Goode
Sales Director: D P Webb
Marketing and Public Relations Administrator: Ms Dawn Hayward
■ Circlips and retaining rings. Smalley spiral retaining rings and wave springs, special designs and design information. [98]

85-00620 C.J. Fox & Sons Ltd
Head Office
Shoreham Airport,
Shoreham-By-Sea, West Sussex BN43 5FN
☎ (+44 1273) 46 40 46 ⊜ (+44 1273) 46 45 77
♣ £ 11,000,000 (1996) [Employees] 78 (1997)
Managing Director: Sean O'Connor
Operations/Technical Director: R M Scott
Sales and Marketing Director: John Williams
Purchasing Manager: S Sewell
■ Specialist fasteners for defence & aerospace sales/design.
® USA: ATI, Bandy, Briles, Cherry Textron, FIT, Hartwell Corp., Jonathan Manufacturing, Lokfast, Nylon Molding Corp., Omega Technologies, RFM (Republic Fasteners), Rivnut Engineered Products, USI
Northern Office
222 Europa Boulevard, Gemini Business Park,
Warrington, Cheshire WA5 5TN
☎ (+44 1925) 41 71 17 ⊜ (+44 1925) 44 46 46 [98]

85-00630 FP Diesel Ltd
(† Power Components Ltd)
(a subsidiary of Fel-Pro Inc., USA)
Unit A, Hanworth Lane Business Park, Hanworth Lane,
Chertsey, Surrey KT16 9JX
☎ (+44 1932) 57 15 54 [E] fpdiesel@aol.com
⊜ (+44 1932) 57 15 85
[Employees] 30
Managing Director: Tony Lax
Operations Director/Purchasing Manager: Dennis Lax
Sales Manager: Peter Harvey
Purchasing Manager: Phillip Foschi
Office Manager: K F Matthews
■ Export of diesel engines and components for vehicle and marine propulsion and power generation. Export replacement parts for diesel engines - Caterpillar Detroit and Cummins, Mack, Navistar and Komatsu. [98]

85-00660 Garphyttan Industrier
Hilton Road, Aycliffe Industrial Estate,
Newton Aycliffe, Durham DL5 6SX
☎ (+44 1325) 31 01 10 ⊜ (+44 1325) 31 18 34
♣ £ 8,000,000 (1996) [Employees] 38
General Manager: G Balfour
Sales Manager: B Clancy
Sales Manager: D Mitchell
■ Automatic adjusters & drain valves for pneumatic brake systems. → 80 → 90 [98]

85-00690 GE Silicones
(a division of General Electric Silicones Europe, Netherlands)
Old Hall Road, Sale, Cheshire M33 2HG
☎ (+44 161) 905 50 20
✆ (800) 96 61 65; (800) 96 61 14 (fax)
⊜ (+44 161) 905 50 22
General Manager and Director: John Krenicki
☎ (+31 164) 29 22 93
Sales Director: M Masten ☎ (+31 164) 29 24 58
Sales Manager Northern Europe: B G Warren
☎ (+44 161) 905 50 21
■ Supplier of silicone, elastomers, resins, ablative materials, greases & fluorosilicones. Approved by MoD. [98]

85-00700 Gould Alloys Ltd
Carrwood Industrial Park, Carrwood Road,
Chesterfield S41 9QB
☎ (+44 1246) 26 33 00 [E] gouldalloys@mail.att.net
⊜ (+44 1246) 26 09 99
♧ £ 20,000 (1996) ♣ £ 8,400,000 (1996) [Employees] 48 (1996)
[Floor area] 2,600 sq m
Managing Director: P C L Gould ☎ (+44 1246) 26 33 20
Operations Manager: Neil Murdoch
☎ (+44 1246) 26 33 01
Sales and Marketing Manager: Gerry Clark
Export Sales Manager: Andrew Vintin
■ Stockists of carbon alloy stainless high temp. high-grade ferrous & non-ferrous materials. Bars, sheets & tubes. [98]

85-00760 HCC Industries International
49-53 Hazelwood Road, Northampton NN1 1LG
☎ (+44 1604) 23 31 01 ⊜ (+44 1604) 23 29 07
Manager: Jim Sherlock
® USA: HCC Industries Inc. [98]

85-00764 HOPPECKE Batteries UK Ltd
Unit 2, The Glades, Festival Park, Stoke-on-Trent ST1 5SQ
☎ (+44 1782) 27 13 77 ⊜ (+44 1782) 27 32 02
Managing Director: C Barrett
Sales and Marketing Director: Julian Revington
Operations Director: David Redican
® Germany: HOPPECKE [98]

85-00769 ImageBase Technology Ltd
Research House, Fraser Road,
Perivale, Middlesex UB6 7AQ
☎ (+44 181) 991 13 96 [E] imagebas@aol.com
⊜ (+44 181) 991 26 98
Marketing Manager: M Penwarden
Marketing Manager: Ms Mary Musker
■ Distributor of the 'Photophone' range of products, a still image storage & transmission system. Primary users include law enforcement agencies, aerospace industries & the armed forces. [98]

85-00775 Interpoint (UK) Ltd
(a subsidiary of Interpoint Corp, USA)
Victoria House, 18-22 Albert Street,
Fleet, Hampshire GU13 9RL
☎ (+44 1252) 81 55 11 [W] http://www.interpoint.com
⊜ (+44 1252) 81 55 77
♣ US$ 6,000,000 (1996) [Employees] 4 (1996)
Vice President International: Simon Abel
■ Full military and space specification power converters from 1 to 1000 watts. EMI filters and transient protection circuits. → 90 [98]

85-00900 Jet Support Ltd
14 Dalebury Road, London SW17 7HH
☎ (+44 181) 767 73 31 [W] http://www.jetsupport.co.uk
[Telex] 916428 jet sup g ⊜ (+44 181) 672 50 40
♧ £ 10,000 (1996) ♣ £ 3,000,000 (1996) [Employees] 8 (1996)
[Floor area] 700 sq m
Managing Dir: Mark H.B. Blomfield
Operations Dir: Richard Worby
Sales & Marketing Dir: Ms Alison Noble
Purchasing Mgr: Dennis Theobald
■ Suppliers of defence/aviation products. [98]

85-00950 JVC Professional Products (UK) Ltd
(a subs. of Victor Company of Japan)
Ullswater House, Kendal Avenue, London W3 0XA
☎ (+44 181) 896 60 00 ⊜ (+44 181) 896 60 60
[Employees] 65
Managing Dir: S.W. Heap
Operations Dir: Don Cameron
Financial Dir: B. Jackson
Sales Dir: D.M. Coupe
Marketing Dir: J.R. Carpenter
Purchasing Manager: John Waters [98]

85-01160 Knight Strip Metals Ltd
Knuway House, Cranborne Road,
Potters Bar, Hertfordshire EN6 3JL
☎(+44 1707) 65 02 51
E 101375.2531@compuserve.com
W http://www.knight-group.co.uk
⊖ (+44 1707) 65 12 38
♣ £ 15,000,000 (1996) 98
Chairman: B.W. Knight
Managing Dir: D.J. Pearson
Financial Controller: J.A. Proudfoot
Sales Dir: B.J. Morley
Operations & Technical Dir: A. Woodhouse
Business Development Manager: J E Turner
Purchasing Mgr: B. Bundey
■ Stockholder and processor of all types of metal strip and wire. Mild, carbon and stainless steels, brass, beryllium copper, phosphor bronze, nickel and nickel alloys. Titanium and titanium alloys. CAA approved. BS EN ISO 9002: 1994. QAS 3289/17. 98

85-01210 F.C. Lane Electronics Ltd
Slinfold Lodge, Horsham, West Sussex RH13 7RN
☎(+44 1403) 79 06 61 (sales); 79 02 67 (admin)
⊖ (+44 1403) 79 08 49
♧ £ 4,000 ♣ £ 4,200,000 35 8,000 sq ft
Managing Dir: Frederick C. Lane
Head Sales & Marketing: Mrs Rosemary Lane
■ Distributor holding extensive stocks of a wide range of electrical connectors, cable ties from UK, European and USA manufacturers. Family related connector designer manufacturer on same site. Ministry of Defence, BSI, CECC approved. Vast library NATO reference numbers. → 90
Associated Company:
Weald Electronics → 75 98

85-01230 Leach International UK Ltd
(† LRE Relays & Electronics Ltd)
(a subsidiary of Leach International Europe SA, France)
Signal House, Jacklyns Lane,
Alresford, Hampshire SO24 9JJ
☎(+44 1962) 273 44 33 ☒ 477516 lre uk g
⊖ (+44 1962) 273 46 85
Managing Director: Keith Joy
Sales Manager: P W Searle
■ Supplier of thin integrated keyboard systems, advisory & central warning panels, illuminated pushbutton switches, anti-icing controllers, electronics. Supplier of relays, contactors, time delay relays, solid state intelligent switching devices, knobs & fasteners. → 90
® France: Gepe, Leach International Europe SA, Les Modeles Francais
USA: Dupree, Leach International 98

85-01240 Light Aero Spares Ltd (LAS)
Shebbear, Beaworthy, Devon EX21 5RQ
☎(+44 1409) 28 15 78; 28 16 10
E aero@mail.zynet.co.uk ⊖ (+44 1409) 28 18 13
8 (1997)
Managing Director: Stephen Saywell
Sales Manager: R Howe
Sales Manager: S Powell
■ All types of aircraft spares. 98

85-01320 MT Aerospace Ltd
(† Marvin Tomkins Ltd)
Claverton House, Love Lane,
Cirencester, Gloucestershire GL7 1YG
☎(+44 1285) 88 52 01 ⊖ (+44 1285) 65 88 59
♧ £ 750,000 (1997) ♣ £ 925,000 (1996/97)
5 (1997) 285 sq m
Chairman: Mrs Patricia M. Tomkins
Managing Dir: Paul R. Tomkins
■ Aerospace stockists & agents. → 90 98

85-01450 Messier-Dowty Customer Support Centre - Europe
Gloucester Office, Cheltenham Road, Gloucester GL2 9QH
☎(+44 1452) 71 24 24 ☒ 43246
⊖ (+44 1452) 71 17 68
Customer Support Director: Martin Akerman
Commercial Manager: Emma Tew
Technical Manager: Mike Trendell
Customer Support Executive: Ms Tessa Baxter
☎(+44 1452) 71 18 16
■ Provide spare parts, AOG service & technical support for Messier-Dowty & Dowty Aerospace equipment in Europe, Middle East & Africa. 98

85-01560 Neatworld Ltd
(a subsidiary of Manor Projects Ltd)
Collingwood House, Allington Road Industrial Estate,
St Neots, Cambridgeshire PE19 2UX
☎(+44 1480) 47 07 00 ⊖ (+44 1480) 47 07 01
♧ £ 1,000,000 ♣ £ 1,600,000 (1995) 15
10,000 sq ft
Managing Director: Anthony Pearlgood
Sales and Marketing Director: Colin Ashton
Operations and Purchasing Manager: Ian Richardson
■ Supply industrial shredding, compacting & baling equipment. Applications include security shredding, product recovery, waste minimisation & recycling. 98

85-01570 Nera Ltd
(† ABB Nera Ltd)
(a division of Nera A.S.A., Norway)
20 Imperial Way, Croydon, Surrey CR0 4RR
☎(+44 181) 686 57 01 ☒ 946384 eb ume g
⊖ (+44 181) 686 68 11
♣ £ 45,000,000 (1997) 140 (1997) 30,000 sq m
Chief Executive Officer: M L Braime
Divisional Managing Director, Mobile Products and Systems: R J Chewter
Government and Defence Sales Manager: J A Boys
Purchasing Manager: D Silcock
→ 75 → 80
® Denmark: Skanti Radio A/S
Norway: Comrod Antennas, Nera AS 98

85-01600 Park Cameras Ltd
115 Lower Church Road,
Burgess Hill, West Sussex RH15 9AA
☎(+44 1444) 24 53 16 ⊖ (+44 1444) 24 53 19
♧ £ 52 586 (1996) ♣ £ 1 150 000 (1995) 19
350 sq m
Managing Dir: Reg Atkins
Director: Denis Atkins
General Mgr: Ms Angela Mitchell
Sales Mgr: Chris Reskze
■ Stockists & suppliers of leading photographic & binocular equipment. 97

85-01650 Pattonair International Ltd
(a subs. of Umeco plc)
Kingsway Business Park, Forsyth Road, Sheerwater,
Woking, Surrey GU12 5SA
☎(+44 1483) 77 46 00 ⊖ (+44 1483) 77 46 19
♣ £ 7 000 000 (1996) 50 3 000 sq m
Managing Dir: J.R. Pound
Sales Dir: J. Crabtree
■ Stockist of seals, clamps & fasteners to British, American & European Standards. Stockist of aviation lamps. 97

85-01660 Penrith Survival Equipment
The Square, Morland, Penrith, Cumbria CA10 3AZ
☎(+44 1931) 71 44 44 ⊖ (+44 1931) 71 44 50
♣ £ 500 000 (1995) 6
Partner: Colin Westgarth
Partner: Mrs Sheila Westgarth
■ Complete range of survival & alternative military equipment. Personal sleeping systems & personal protection clothing. Thermal garments. Survival kits, food rations, personal navigation equipment. → 80 97

85-01677 Positioning Resources Ltd
64 Commerce St, Aberdeen AB11 5FP
☎(+44 1224) 58 15 02 ⊖ (+44 1224) 57 43 54
Managing Dir: George F. Ritchie
Sales & Marketing Dir: Ms Judith Collier
Technical Sales: Neil Sutherland
Marketing: Miss Lorraine Mitchell
■ Specialise in providing a complete geodetic surveying service using the global positioning system - GPS. 97

85-01690 R.A. Poole and Co (Stainless) Ltd
(† I.M.S. Stainless Fasteners Ltd)
3/5 Chequers Lane, Walton-on-the-Hill,
Tadworth, Surrey KT20 7SF
☎(+44 1737) 81 41 23
E r.a.poole@st.st.fasteners.demon.co.uk
⊖ (+44 1737) 81 42 56
♧ £ 100,000 (1997) ♣ £ 3,000,000 (1997)
18 (1997) 5,000 sq ft
Managing Director, Advertising Manager: R A Poole
Sales Director: Tony Gray
Purchasing Manager: Tony Wigg
■ All types of stainless steel & non-corrosive screws & allied fasteners. Crinkle washers. 98

85-01700 Repaircraft plc
The Common, Cranleigh, Surrey GU6 8LU
☎(+44 1483) 27 35 36 E hg@repaircraft.co.uk
W http://www.repaircraft.co.uk ⊖ (+44 1483) 27 80 78
♧ £ 100,000 (1997) 2,000 m2
Managing Dir: Adrian N. Stainton
Operations Mgr: M. Cooper
Sales & Marketing Mgr: J. Tallon
Purchasing Mgr: M. Adams
■ Supply Division:
Supply of spare parts & systems for defence & aerospace equipment.
Manufacturing Division:
Design, manufacture, overhaul & repair of defence & aerospace equipment.
Representation Division:
Representation & sale of overseas defence equipment in the United Kingdom. → 75 → 80 → 90 98

85-01850 Samson Distraco U.K. Ltd
PO Box 13, Bacup, Lancashire OL13 9FB
☎(+44 1706) 87 07 33 ⊖ (+44 1706) 87 06 99
UK Director: P Waite
■ Stockist/distributor/supplier of Israel Military Industries (I.M.I.) products. 98

85-01870 Security Defence Technologies Ltd
Kinetic Centre, Theobald Street,
Borehamwood, Hertfordshire WD6 4SE
☎(+44 181) 386 36 17 ⊖ (+44 181) 986 66 13
Financial Director: S K Freedman
Sales and Marketing Director: L Newmark
■ Import and export of security equipment. Distributor of Videsco X-ray inspection systems. 98

85-01910 Simpson Lawrence
Head Office
218-228 Edmiston Drive, Glasgow G51 2YT
☎(+44 141) 427 53 31/8 ⊖ (+44 141) 427 54 19
Managing Dir: Tony Kent
Sales & Marketing Dir: David Guthrie
Purchasing Logistics Mgr: E. Vallance
■ Worldwide marine equipment distributors. → 75
Branch Offices
Station Road, Lawford, Manningtree, Essex CO11 2LL
☎(+44 1206) 39 34 74/5 ⊖ (+44 1206) 39 61 76
● Unit G, Fort Wallington, Fareham, Hampshire PO16 8TT
☎(+44 1329) 82 33 00 ☒ 86645 afloat g
⊖ (+44 1329) 82 32 82
● Unit 8, Tamar View Industrial Estate,
Saltash, Cornwall PL12 6LD
☎(+44 1752) 84 51 51 ⊖ (+44 1752) 84 87 87 98

85-01920 Stanley Productions
147 Wardour St, London W1V 3TB
☎(+44 171) 494 45 45; 439 03 11
⊖ (+44 171) 437 21 26; 434 05 92
Managing Dir: S. Aarons
Sales Mgr: Steve Langston
■ Supply of video cameras, video & audio recorders, monitors, television, tripods camcorders & specialised items involved in surveillance & CCTV. 97

85-01936 Stedall (Vehicle Fittings) Ltd
(a subsidiary of Vector Industries)
Badminton Road Trading Estate, Yate, Bristol BS17 5JS
☎(+44 1454) 31 95 24 ⊖ (+44 1454) 31 20 77
♣ £ 5,000,000 (1997) 25 (1997) 7,000 m2
Managing Dir: R. Miller
Operations & Technical Dir: P. Ryemill
Sales & Marketing Dir: A. Leahy
Purchasing Manager: A Leahy
■ Supplier of vehicle body building products and custom fittings for specialist commercial vehicles including coaches, ambulances, fire tender, horse boxes, rigids and tippers. 98

85-01940 Subroc Ltd
PO Box 3, Cranleigh, Surrey GU6 7NF
☎ (+44 1483) 27 31 49 ⊠ JOLONCAPE
⊜ (+44 1483) 27 40 61
Managing Director: Mark H Yeats
■ Arms, ammunition & related equipment. [98]

85-01990 Swindens Patents Ltd
115 Northgate St, Bury St Edmunds, Suffolk IP33 1HQ
☎ (+44 1284) 76 66 52 ⊜ (+44 1284) 76 83 81
♧ £ 3 000 ♣ £ 80 000 👥 4
Managing Dir: R.S. Worthington
Sales & Marketing Dir: A. Marangos
■ Supplier company for "Swindens" revolving head vice. Tested & approved by the British MoD. [97]

85-02010 Tiernay Metals Ltd
(a subs. of Tiernay Metals, USA)
Unit 8, Gunnels Wood Park, Stevenage, Herts SG1 2BH
☎ (+44 1438) 31 55 77 📠 825894
⊜ (+44 1438) 31 43 85
♧ £ 3 000 000 (1996) ♣ £ 5 000 000 (1996) 👥 18
🏭 14 700 sq ft
Managing Dir: Brian Austin
Operations Dir: David Dainton
Commercial Mgr: Pierre Sheppard
■ Aircraft quality aluminium extruded & roll formed sections. [97]

85-02040 Titanium International Ltd
Key House, Granby Ave,
Garretts Green, Birmingham B33 0SP
☎ (+44 121) 789 80 30 (admin); 789 80 88 (aerospace); 789 79 81 (manufacture/tech)
⊜ (+44 121) 784 80 54
♣ £ 9 000 000 👥 40
Chairman: J.S. Paddock
Managing Dir: A.E. Lawrence
Sales & Marketing Mgr: N.J. Aston
■ Stockholders of special metals: titanium, zirconium & alloys as billet, bar, wire plate, sheet, tube, pipe & fittings. Forgings to order. [97]

85-02060 T.M.S. International
24 Diamond Ridge, Camberley, Surrey GU15 4LD
☎ (+44 1276) 68 29 60 ⊜ (+44 1276) 68 29 60
♧ £ 50 000 ♣ £ 200 000 👥 4 🏭 300 sq m
Managing Dir & Proprietor: A. Akeroyd
Operations & Technical Dir: Ian Davidson
■ Aerospace, light aircraft sales. Spares sourcing for civil & military. Brokering & consultancy service. Security, surveillance services. Marine, custom built for special applications. [97]

85-02080 TRACE International Ltd
Barn End House, Rose Estate,
Marlow Bottom, Bucks SL7 3ND
☎ (+44 1491) 57 14 27 ⊜ (+44 1491) 48 46 46
Managing Director: Brian Hollis [98]

85-02120 Tynemount Ltd
Unit 5 Northfield Farm, Great Shefford,
Hungerford, Berkshire RG17 7DQ
☎ (+44 1488) 64 88 65 📠 849676 tynmnt g
⊜ (+44 1488) 64 88 16
Managing Dir: John K. Watson
Sales Dir: R.D. MacGregor
■ Supply & logistics support of electronic spare parts & components for communications radar & other electronic equipment used by navy, army, police & air force. [98]

85-02450 Viking Arms Ltd
New York Mills, Summerbridge,
Harrogate, North Yorkshire HG3 4BW
☎ (+44 1423) 78 08 10 ⊜ (+44 1423) 78 15 00
♣ £ 1,700,000 👥 10
Managing Dir: M.H.F. Brown
Director: Edward C Horton
Accountant: A Groves
■ Importer, exporter and wholesaler. → 90 [98]

85-02500 Viking Life Saving Equipment Ltd
(a subs. of Viking Life-saving Equipment, Denmark)
Ocean House, Caxton Street North, London E16 1JL
☎ (+44 171) 474 85 00 📠 896920
⊜ (+44 171) 511 63 35
Managing Dir: M. Harriman
■ Sales & service of Viking liferafts & marine safety/survival products. [98]

85-02770 Wessex Marine Equipment Ltd
Logistics House, 2nd Ave Business Park,
Millbrook, Southampton SO15 0LP
☎ (+44 1703) 51 05 70 ⊜ (+44 1703) 51 05 60
👥 11 (1997) 🏭 722 sq m
Managing Director: A J Hall
■ Suppliers of marine & defence equipment (naval stores, supplies & equipment). Design & manufacture of maritime training aids. → 80 [98]

85-02900 Wood Harris Ltd
1 Cleeve Court, Cleeve Road,
Leatherhead, Surrey KT22 7NN
☎ (+44 1372) 36 29 90 [E] Dennys@btinternet.com
⊜ (+44 1372) 36 29 20
♣ £ 2,000,000 👥 11
Director: Paul Jubert
Director: John Jubert
■ Uniform & camouflage combat clothing, fire retardant uniform clothing. NBC protective clothing. Anti-microwave clothing. [98]

85-03000 W S Ocean Systems
Unit 4, Omni Business Centre, Omega Park,
Alton, Hampshire GU34 2QD
☎ (+44 1420) 54 15 55 [E] sales@wsocean.com
⊜ (+44 1420) 54 14 99
♧ £ 10,000 (1997) ♣ £ 1,800,000 (1996) 👥 19 (1997)
🏭 300 m2
Managing Director: R F Scrivens
Operations and Technical Director: D Menzies
Purchasing Manager: M Gill
■ Provides oceanographic and environmental instruments to ocean research, survey and associated markets. [98]

90 Sales Agents, Sales & Liaison Offices

Companies represented:

A/S Hydrema, Denmark . . . 90-00915
A.T.I., France . . . 90-00018
Abex Aerohydraul, Germany . . . 90-00015
Acer Inc., Taiwan . . . 90-00019
ACSL, USA . . . 90-01810
Adams Rite Sabre International, USA . . . 90-00042
Aero Consultants, Switzerland . . . 75-01020
Aerolyte Systems, USA . . . 75-29590
Aerospatiale, France . . . 90-00037
Aerotron AirPower Inc, USA . . . 75-01250
AID/RC Allen, USA . . . 90-00042
Aircraft Porous Media, USA . . . 75-43517
Airflyte electronics Company, USA . . . 90-01162
Airpax Corp., USA . . . 90-00930
Airtech International, USA . . . 90-00040
Alkan et Cie, France . . . 75-02680
Allen Aircraft, USA . . . 90-00011
Allen Aircraft Products, USA . . . 90-00042
Allied International Corp., USA . . . 90-00092
AlliedSignal Aerospace, USA . . . 90-00042
Altoflex, France . . . 90-00018
American Derringer, UK . . . 75-14855
American Variseal, USA . . . 75-11050
AMP, USA . . . 75-03050
Andrew Corp., USA . . . 75-03156
Antenna Research Associates, USA . . . 80-13328
Appleton Pohl GmbH & Co., Germany . . . 90-02260
Applied Dynamics Int'l, USA . . . 90-01810
Arthur Maurer GmbH, Germany . . . 90-02260
Astro Instrument Corporation, USA . . . 90-01162
Astronautics Corp of America, USA . . . 90-00042
Astronautics Corporation of America, USA . . . 90-00160
Atacs Products Inc., USA . . . 90-00040
Aten Srl, Italy . . . 90-00155
ATI, USA . . . 85-00620
Aubert & Duval SA, France . . . 90-00163
Bandy, USA . . . 85-00620
Barco NV, Belgium . . . 90-00200
Bell Helicopter Textron, USA . . . 90-00220
Bergman Forge, USA . . . 90-01895
Beta Transformer Technology Corp., USA . . . 90-00425
BFI, Spain . . . 75-39010
Bird Electronic, USA . . . 90-00155
BKT Germany, Germany . . . 75-27200
Bloksma BV, Netherlands . . . 90-01920
Boeing Defense & Space Group, USA . . . 90-00240
Bofors AB, Sweden . . . 90-00255
Boonton Electronics Corp, USA . . . 90-00155
BP Metal Composites Ltd, UK . . . 90-01520
Brent Europe Ltd, UK . . . 90-00280
Briles, USA . . . 85-00620
Bryce Manufacturing, USA . . . 90-00042
CAE Electronics GmbH, Germany . . . 90-00855
CAE Electronics Ltd, Canada . . . 90-00855
California Instruments, USA . . . 75-65333
CATIC, China PR . . . 90-00324
CCS International Ltd, USA . . . 90-00325
CE Precision Assemblies Inc., USA . . . 90-00965
CEF Calco, USA . . . 90-00042
CEISTM, France . . . 90-00042
CelsiusTech AB, Sweden . . . 90-01930
Chase Walton Elastomers, USA . . . 90-01895
Chelton (Electrostatics) Ltd, UK . . . 90-01520
Cherry Aerospace, USA . . . 90-00335
Cherry Textron, USA . . . 85-00620
Clemco Industries, USA . . . 75-29590
Cole, USA . . . 90-00018
Cominvest System AB, USA . . . 80-13328
Communication Techniques, USA . . . 90-00155
Computing Devices Canada Ltd, Canada . . . 90-00364
Comrod Antennas, Norway . . . 85-01570
Comtronic, Germany . . . 90-00018
Concurrent Computer Corp., USA . . . 75-14810
Courtaulds Aerospace, UK . . . 90-00280
Cristek Interconnects Inc., USA . . . 90-00965
Custom Electronics Inc, USA . . . 90-00028
Cylink Corp., USA . . . 85-00390
David Clark Corp, USA . . . 90-00042
DeSoto, Inc., USA . . . 90-00040
Detector Electronics Corp., USA . . . 90-00430
Detroit Diesel Corp., USA . . . 90-00435
Dexter - Hysol Adhesives, USA . . . 90-00040
Dowty Defence & Air Systems Ltd, UK . . . 90-01520
Dukane Corp, USA . . . 90-00042
Dupree, USA . . . 85-01230
Dynamic Sciences Inc., USA . . . 80-13328
E.G.G. Birtcher, USA . . . 90-00018
E.H. Industries Ltd, UK . . . 90-01520
Eaton Corporation, USA . . . 75-36675
EBG, Austria . . . 90-00028
ECE, France . . . 75-31948
EDO Corp., USA . . . 90-00110
Elcom Systems, USA . . . 90-00155
Elettronica Spa, Italy . . . 75-20475
Elin Energieanwendung GmbH, Austria . . . 90-02260
Embraer do Brazil, Brazil . . . 80-07600
EMS Inc, USA . . . 90-00155
Enviro Systems Inc, USA . . . 90-00042
EROS, France . . . 75-31948
ETEL S.A., Switzerland . . . 90-01162
Eurocopter, France . . . 90-01480
F.C. Lane Electronics Ltd, UK . . . 75-64880
Faure Herman, France . . . 75-31948
FCC Systems Ltd, UK . . . 90-00042
FCX Systems, USA . . . 90-00042
Fiber Materials Inc., USA . . . 75-24040
Fiber Resin Corp., USA . . . 90-00040
First Tech.Fire & Safety, UK . . . 75-44670
FIT, USA . . . 85-00620
Flightline Electronics Inc, USA . . . 90-00042
Fredericks Co., USA . . . 90-00930
Froude Consine, UK . . . 90-01480
FSE GmbH, Germany . . . 90-02260
Fuji Industries Co. Ltd, Japan . . . 90-00700
Fusion Inc, USA . . . 75-39010
GEC-Marconi Avionics Ltd, Sensors Div., UK . . . 90-01520
General Pneumatics Inc, USA . . . 90-00042
Gepe, France . . . 85-01230
Glaser Safety Slug, UK . . . 75-14855
Grace NV, Belgium . . . 90-00530
Greaves Cotton & Co.Ltd, India . . . 90-00780
Grifon Global, UK . . . 75-14855
Grimes Aerospace Company, USA . . . 90-00790
H.R.Textron, USA . . . 90-00042
Harper Engineering, USA . . . 90-00042
Hartman, USA . . . 90-00042
Hartwell Corp., USA . . . 85-00620

Haydon Switch & Instr. Inc., USA	90-01826
Hayes Targets, USA	80-07600
HCC Industries Inc., USA	85-00760
Heinrich Stamm GmbH, Germany	90-02260
Hexcel Corporation, USA	75-29150
Hindustan Aeronautics, India	90-00850
Honeywell Inc., USA	75-29730
Hoover Industry, USA	90-00042
HOPPECKE, Germany	85-00764
Hughes Europe SA/NV, Belgium	80-12760
Humphrey Inc., USA	90-00930
I.E.C. Corporation, USA	90-00270
I.G.G. Component Technology Ltd, UK	90-01520
ICI - Fiberite, USA	90-00040
ILC Data Device Corp., USA	90-00425
Inductosyn International, USA	90-01162
Industria, France	75-44670
Instruments for Ind., USA	80-13328
Inter-National Research Inst., USA	80-13320
Intertechnique SA, France	75-31948
Itochu Corporation, Japan	90-00980
ITT Cannon, France	75-32170
ITT Cannon GmbH, Germany	75-32170
ITT Cannon Ltd, Japan	75-32170
ITT Cannon RF Products, USA	75-32170
ITT Defense, USA	80-13570
Ivaltel, USA	90-00042
J C Carter, USA	90-00042
Janco Corp., USA	90-00965
Jay-El-Products, USA	90-00018
JCA Technology, USA	90-00155
Jerrik Connecting Devices Inc., USA	90-00965
Jonathan Manufacturing, USA	85-00620
K&L Microwave, USA	90-00155
Kahn Industries, USA	90-00042
Kaman Instrumentation Corp, USA	75-16550
Kavlico Corporation, USA	90-01100
Kearfott Guidance & Nav., USA	90-00930
Kearfott Guidance & Navigation, USA	90-00042
KHD AG, Germany	90-01110
Knights' Manufacturing Co., USA	75-14855
Kollmorgen Artus, France	90-01162
Kollmorgen Industrial Drives, USA	90-01162
Kollmorgen Inland Motor, USA	90-01162
Kollmorgen PMI, USA	90-01162
Kongsberg Gruppen A/S, Norway	90-01165
Labinal PM, France	90-01480
Lambda Novatronics, USA	90-00930
Latecoere, France	90-00855
Leach International, USA	85-01230
Leach International Europe SA, France	85-01230
Lee Co., USA	90-01235
Lemo, Switzerland	90-01238
Les Modeles Francais, France	85-01230
Lewis & Saunders, USA	90-00042
Litton Industries Inc, USA	90-01240
LJF SNC Hutchinson, France	90-00047
LMS International, Belgium	90-01260
Lockheed Martin Corp., USA	90-01280
Lohmann & Stolterfoht GmbH, Germany	90-01920
Lokfast, USA	85-00620
Lucas Aerospace Ltd, Engine Systems Div., UK	90-01520
Lucas Ledex Inc., USA	75-36675
Lucas Western Inc, USA	90-00042
M.C. Gill Corp., USA	90-00040
Magnetic Metal Corp., USA	90-01895
Malwin Electronics Corp, USA	90-00042
Manitou BF, France	90-01313
Martin Lohse GmbH, Germany	90-02260
Marubeni Corp., Japan	90-01325
Mason Electric Co. Inc., USA	90-01323
Mathematica, USA	90-01810
MBElectronique, France	75-39010
MCB SA, France	90-01826
McDonnell Douglas Aircraft, USA	75-01250
McDonnell Douglas Corp., USA	90-01330
Mectron Industries, USA	75-43517
Messier Fonderie, France	90-01895
Micro Analysis & Design, USA	90-01810
Micro Saint, USA	90-01810
Microphase Corp., USA	90-01346
Minco Products, USA	85-00300
Mitchell & Gauthier, USA	90-01810
Mitsubishi Heavy Industries, Japan	90-01400
MJP Corporation Ltd, Finland	90-01410
MSI, UK	75-27200
Municom GmbH, Germany	75-39010
Murata Erie N.America, USA	75-41780
Murata Mfg Co. Ltd, Japan	75-41780
Nera AS, Norway	85-01570
New Hampshire Ball Bearings Inc, USA	90-01435
New World Technologies, India	75-39010
Normalair-Garrett Ltd, Customer Support, UK	90-01520
North Amercian Arms, UK	75-14855
Northhills Electronics, USA	90-00042
Nylon Molding Corp., USA	85-00620
OFEMA, France	90-01480
Olektron, USA	90-00155
Olympic Arms, UK	75-14855
Omega Technologies, USA	85-00620
Omnirec, USA	90-00028
Orbit Advanced Technology, Israel	90-00155
Pacer Systems Inc., USA	90-01555
Pachmayr, USA	75-14855
Pall Corp., USA	75-43517
Pall Land & Marine, USA	75-43517
Pamtech, USA	90-00155
Parallax Graphics Inc., USA	75-18040
Parsons Corp., USA	90-01595
Passive Power, USA	90-00155
Paulstra SNC Hutchinson, France	90-00047
Penny & Gile, France	90-01480
Penny & Giles Data Recorders Ltd, UK	90-01520
Performance Control Inc, USA	75-41404
Photronics Inc., USA	90-00965
Physical Research Inc., USA	75-27200
Pickering Controls Inc, USA	75-41404
Pocal Industries Inc., USA	75-58500
Polyphaser Inc, USA	90-00155
Portescap, Switzerland	75-46420
PRANA Research & Development, USA	80-13328
PRECILEC SA, France	90-01826
Primex Aerospace Corp, USA	90-00042
Pronesto AB, Sweden	75-39010
PTI Technologies, USA	90-00042
Quadion Corp., USA	75-11050
Quantic Holex, USA	90-00042
Quartztek, USA	90-00155
Racom, Israel	75-39010
Raytheon, USA	90-01825
Remote Ocean Systems, USA	90-00270
Res-Net, USA	90-00155
RF Microwave Limited, USA	75-39010
RFI Corporation, USA	80-13328
RFM (Republic Fasteners), USA	85-00620
Rivnut Engineered Products, USA	85-00620
Rockwell Int'l, USA	75-50050
Rockwell International, USA	75-50050
Rockwell International Corp., USA	75-50000
Rockwell International Corp., USA	90-01830
Rolls-Royce plc, UK	90-01520
Rose Bearings Ltd, USA	90-01435
Rotodata, UK	90-01480
Royal Ordnance plc, UK	75-01560
Ruska Instruments, USA	75-16550
Sabre Computers Int'l, UK	80-07600
Saco Defense Inc., USA	80-35850
Safari Arms, UK	75-14855
Safe Flight Instrument Corp., USA	90-01885
SAIC Science Applications International, USA	80-30220
Saint Switch Inc., USA	90-00965
SAIT SAI Technology, USA	80-30220
Scandiflash, Sweden	75-27200
Schiesser Electronc Ag, Switzerland	75-39010
Schorch, Germany	90-02260
Schottel-Werft, Germany	90-01920
Scott Aviation, USA	90-00042
Seacon/Brantner Associates, USA	90-00270
Security Systems Int'l, Switzerland	90-01935
Sekisui, Japan	90-00155
Semtech, USA	90-00028
Sensonics Inc., USA	90-00930
Senstar Corp., Canada	90-01937
SFIM Industries, France	90-01945
Shamban, USA	75-11050
Shell Aerospace Product, USA	90-00040
Singapore Aerospace, Singapore	90-01959
Skanti Radio A/S, Denmark	85-01570
Skylock Industries Inc., USA	75-44670
SL Auburn, USA	90-00042
Socitec, France	75-65333
Spectral Dynamics Inc, USA	90-01973
Spectrolab, USA	90-00042
Sperry Marine Inc, USA	90-01975
SRL Inc., USA	75-27200
Stewart & Stevenson Services Inc, USA	90-00042
Sumitomo Precision Products Co Ltd, Japan	90-01985
Sun Microsystems Inc., USA	90-01987
Systemware Inc., USA	80-13328
TCI Inc., USA	90-02040
TEAC Corp., Japan	75-27200
Teledyne Allvac, USA	90-02044
Teledyne Continental Rotors (Engines), USA	80-07600
Teledyne Fluid Systems, USA	90-00042
Teledyne Industries, USA	75-16550
Terma, Denmark	75-27200
Testek Inc, USA	90-00042
Textile Technologies, USA	90-00040
Textron Lycoming (Engines), USA	80-07600
Thermion Aerospace, USA	90-00040
Thomson-CSF, France	90-02060
Thomson-CSF, USA	80-13328
Thomson-CSF S.A., France	75-59540
Transicoil Inc., USA	90-00965
Trimble Navigation Ltd, USA	75-61140
Trimos-Sylvac Task Force, Switzerland	80-58770
Trumag, Switzerland	90-02260
TRW Inc, USA	90-02100
Turbomeca SA, France	74-18000
Turboméca, France	90-01480
USI, USA	85-00620
UTI Corp., USA	90-00040
Vaisala Oy, Finland	90-02140
Vanguard, USA	90-00028
Vapormatt Ltd, UK	90-00912
Varian Associates, USA	80-13328
Vectron Inc, USA	90-00155
Versapod, USA	75-14855
Versitron, USA	80-13328
Vibro-Meter SA, Switzerland	90-02238
Videospection, USA	75-27200
Visual Numerics, USA	90-01810
Voith Group, Germany	90-02260
Voith Safeset AB, Sweden	90-02260
W.S. Shamban Europa A/S, Denmark	75-11050
Weldon Pump Division, USA	90-00042
Westland Helicopters Ltd, UK	90-01520
Wilcoxon Research Inc, USA	90-01973
Wilcoxon Research Inc., USA	75-16800
Willis Engineered Polymers Ltd, UK	75-11050
Wolfram Research, USA	90-01810
XL Microwave, USA	90-00155
Zeihl Abegg, Germany	90-02260
ZF Friedrichshafen AG, Germany	75-67720

Sales agents/representatives listed elsewhere:

Aero Consultants (UK) Ltd	75-01020
Aircraft Equipment International Ltd	75-01560
AlliedSignal Ltd	80-01045
AlliedSignal Aerospace	80-01030
Armalon Ltd	75-03729
Certacom Ltd	85-00320
Crellon Microsystems	85-00375
CSE Aviation Ltd	80-07600
D.J.Litt (Firearms) Ltd	75-36080
Daco Scientific Ltd	75-16550
Deep Sea Seals Ltd	75-17400
Effer (UK) Ltd	75-19660
F.C. Lane Electronics Ltd	85-01210
F.R. Ordnance International Ltd	75-24480
Fugro-UDI Ltd	75-24885
Garphyttan Industrier	85-00660
Hadland Photonics Ltd	75-27200
Instrument Services Ltd	80-13328
International Rectifier Co.(GB) Ltd	75-31860
Interpoint (UK) Ltd	85-00775
ITT Cannon RF Products	75-32170
Leach International UK Ltd	85-01230
Mediworld Ltd	75-32300
Mineral Derivatives Ltd	75-39800
MT Aerospace Ltd	85-01320
Mycon Marketing Services Ltd	80-17900
New Markets Ltd	80-18250
Normalair-Garrett Limited	75-42700
Partek Cargotec Ltd	75-44055
Percival Associates Ltd	75-44670
Pulnix Europe Ltd	75-47030
R. Alkan (UK) Ltd	75-02680
Repaircraft plc	85-01700
Scientific Computers Ltd	80-31550
Tremlett Boat Sales Ltd	75-61075
Viking Arms Ltd	85-02450
Vosper Thornycroft	75-63800

Wessex Electronics Ltd 75-65333

90-00011 AAR International Inc
(† Allen Aircraft International Inc)
(a division of AAR Corp, USA)
Cardinal Point, Newall Road, London Heathrow, Hounslow, Middlesex TW6 2BP
☎ (+44 81) 759 40 22 ⊠ 263321 aaint g
⊖ (+44 81) 897 90 05
👥 15
Vice President: Pat Aherne
Purchasing Manager: Guy De Rooy
Assistant Controller: Daniel Wyatt
■ Aircraft engines, spare parts and components. → 85
® USA: Allen Aircraft [98]

90-00012 ABB Seatec Ltd
2 High Street, Nailsea, Bristol BS19 1BS
☎ (+44 1275) 81 01 00 ⊖ (+44 1275) 85 14 67
♣ US$ 50,000,000 👥 258 [98]

90-00015 Abex Aerohydraul UK Office
48 Victoria Road, Burgess Hill, West Sussex RH15 9LR
☎ (+44 1444) 87 00 40 ⊖ (+44 1444) 87 00 95
Managing Director: G Schrank
Operations Director: M Fleischer
Sales and Marketing Director: W Jung
Purchasing Manager: J Passera
Sales Manager: M J Harvey
® Germany: Abex Aerohydraul [98]

90-00018 Acal Electronics Ltd
(a subs. of Acal plc)
Wokingham Road, Bracknell, Berkshire RG12 1ND
☎ (+44 1344) 72 33 44 ⊖ (+44 1344) 72 33 45
♧ £ 3 000 000 ♣ £ 18 000 000 👥 80 🏭 26 000 sq ft
Managing Dir: A.J. Laughton
Sales Dir: D.L. Hooper
Technical Dir: A. Moore
Office Mgr: Paul Gregory
■ Data acquisition & telemetry equipment. Avionic electronic components & equipment. Distributors of aerospace & defence equipment. MIL STD 1553 test equipment.
® France: A.T.I., Altoflex
Germany: Comtronic
USA: Cole, E.G.G. Birtcher, Jay-El-Products [97]

90-00019 Acer UK Ltd
Maddison House, Thomas Rd, Woodford Green HP10 0PE
☎ (+44 1628) 53 34 22 ⊖ (+44 1628) 52 40 71
® Taiwan: Acer Inc. [97]

90-00026 ADEC Propulsion Ltd
ADEC House, Fitzherbert Rd, Farlington, Portsmouth PO6 1RU
☎ (+44 1705) 21 51 38 ⊖ (+44 1705) 37 67 44
Managing Dir: Jeremy Hunt
■ Sales of propellers & waterjets. [97]

90-00027 Advance Visual Optics Ltd
Hawk House, Mustang Road, Earl's, Colne Industrial Park, Earl's Colne, Colchester, Essex CO6 2NS
☎ (+44 1787) 22 36 88 Ⓔ avoltd@easynet.co.uk
Ⓦ http://www.avoltd.co.uk ⊖ (+44 1787) 22 35 01
👥 9 (1997) 🏭 4 000 sq m
Managing Director, Marketing Manager: John Tuffen
Operations/Technical Director: Andy Hibbert
Purchasing Manager: Dave Hawes
■ Advance Visual Optics Security Sales Division supply for all security applications incorporating EOD search & surveillance. "Roofeye". TVS (3D imaging). → 75 [98]

90-00028 Advanced Power Components Ltd
47 Riverside, Medway City Estate, Rochester, Kent ME2 4DP
☎ (+44 1634) 29 05 88 ⊖ (+44 1634) 29 05 91
👥 20 (1995) 🏭 1 200 sq m (1995)
Managing Dir: R. Robinson
Sales & Marketing Dir: J. Mitchell
→ 75
® Austria: EBG

USA: Custom Electronics Inc, Omnirec, Semtech, Vanguard [97]

90-00037 Aerospatiale UK Limited
(a subsidiary of Aerospatiale, France)
16 Old Bond Street, 5th Floor, London W1X 3DB
☎ (+44 171) 499 50 94 ⊖ (+44 171) 495 00 64
Managing Director: Paul Busquet de Caumont
® France: Aerospatiale [98]

90-00040 Aerospheres (UK), Ltd
Trafalgar House, Grenville Place, London NW7 3SA
☎ (+44 181) 959 36 11 ✆ 802 77 65 99
⊖ (+44 181) 959 81 99; 906 17 00
♧ US$ 250 000 (1995) ♣ US$ 2 000 000 (1994) 👥 10
🏭 500 sq m
Managing Dir: B. Nemenyi
Sales & Marketing Dir: Mrs Dianne Jacobs
Operations & Technical Dir: C. Angus
Purchasing Mgr: Rodger Holton
■ Products: Aerospace chemicals (coatings, sealants, adhesives & composite materials).
® USA: Airtech International, Atacs Products Inc., DeSoto, Inc., Dexter - Hysol Adhesives, Fiber Resin Corp., ICI - Fiberite, M.C. Gill Corp., Shell Aerospace Product, Textile Technologies, Thermion Aerospace, UTI Corp. [97]

90-00042 Aerotech World Trade Ltd
(a subsidiary of Aerotech World Trade Corp)
St Peters Road, Maidenhead, Berkshire SL6 7QU
☎ (+44 1628) 345 55 Ⓔ ukhq@aerotech.uk.com
⊖ (+44 1628) 78 10 70
♧ £ 10,000 (1997) ♣ £ 5,000,000 (1997) 👥 23 (1997)
🏭 10,000 sq ft
Managing Dir: K.A. Mines
Sales Dir: Aleem H. Shaikh
Business Development Dir: Robert Mosedale
Product Mgr Aerospace: Alan Lawrence
Sales Office Mgr: Ms Sharon Smith
Product Mgr Aerospace: John Bates
■ Aerospace and defence equipment distribution.
® France: CEISTM
UK: FCC Systems Ltd
USA: AID/RC Allen, Adams Rite Sabre International, Allen Aircraft Products, AlliedSignal Aerospace, Astronautics Corp of America, Bryce Manufacturing, CEF Calco, David Clark Corp, Dukane Corp, Enviro Systems Inc, FCX Systems, Flightline Electronics Inc, General Pneumatics Inc, H.R.Textron, Harper Engineering, Hartman, Hoover Industry, Ivaltel, J C Carter, Kahn Industries, Kearfott Guidance & Navigation, Lewis & Saunders, Lucas Western Inc, Malwin Electronics Corp, Northhills Electronics, PTI Technologies, Primex Aerospace Corp, Quantic Holex, SL Auburn, Scott Aviation, Spectrolab, Stewart & Stevenson Services Inc, Teledyne Fluid Systems, Testek Inc, Weldon Pump Division [98]

90-00044 A G Electro-Optics Ltd
Tarporley Business Centre, Tarporley, Cheshire CW6 9UY
☎ (+44 1829) 73 33 05 ⊖ (+44 1829) 73 36 79
♣ £ 3,500,000 (1994) 👥 13 (1995) 🏭 550 sq m (1995) [98]

90-00045 AGEMA Infrared Systems Ltd
Arden House, West Street, Leighton Buzzard, Beds LU7 7DD
☎ (+44 1525) 37 56 60
Ⓔ 101473.3271@compuserve.com
Ⓦ http://www.agema.com ⊠ 826586 agema g
⊖ (+44 1525) 37 92 71
Managing Dir: Brian McQueen Smith
Sales and Marketing Director: Ms Jenny Jeffreys
Surveillance Business Mgr: Peter Jenkins
■ Infra-red thermal measurement systems for the 2-5μ and 8-14μ wave bands. Non-classified equipment available for FLIR simulation target signature analysis & IR countermeasure research. [98]

90-00047 Aircotech Ltd
Ashmansworth, The Old Cottage, Newbury, Berks RG15 9SG
☎ (+44 1635) 25 40 08 ⊖ (+44 1635) 25 49 88
👥 1
Managing Dir: Derek Lessware
® France: LJF SNC Hutchinson, Paulstra SNC Hutchinson [97]

90-00090 Allgon Ltd
(a subs. of Allgon Antenna AB, Sweden)
Unit 11 The Courtyard, Whitwick Business Park, Stenson Rd, Coalville, Leics. LE67 4JP
☎ (+44 1530) 51 00 13 ⊖ (+44 1530) 51 03 73
♣ £ 2 000 000 👥 12 🏭 300 sq m
Managing Dir: John Lovelock
■ Mobile communications antennas; base station antennas, combiners, multicouplers, filters & hybrids for mobile telephones. [97]

90-00092 Allied International Corp.
11 Bonded Road, Upton St. Leonards, Glous GL4 8AG
☎ (+44 1452) 52 67 59 ⊖ (+44 1452) 33 05 39
👥 1 🏭 500 sq m
Manager: Nigel Abell
® USA: Allied International Corp. [98]

90-00103 Alpha Networks Ltd
Station House, Station Rd, Liphook, Hants GU30 7DR
☎ (+44 428) 72 32 74 ⊖ (+44 428) 72 54 11
Managing Dir: Ken Tankard
Sales & Marketing Dir: Garry Collings
Technical Mgr: Dennis Guy
Purchasing Mgr: Sheelagh Hoblyn
Representing:
3-Com
ACC
Accton
Acer
Andest
Andrews
APC
Ascom
Bay Networks
Blue Lynx
Cabletron
Chase
Cisco
Controlware
Cray
Dataflex
D-Link
Fibernet
Gandalf
Hayes
H.P. Micom
Microsoft
Multitech
Newbridge
OST
Penril
Psion
Racal
Retix
Shiva
SMC
Sonix Specialist Spider
Synoptics
Telebit
Teleglobe
Tricom
Unipalm
U.S. Robotics
Wisdm
Wyse [97]

90-00105 Ambu International (UK) Ltd
Charlton Rd, Midsomer Norton, Bath, Avon BA3 4DR
☎ (+44 1761) 41 68 68 ⊖ (+44 1761) 41 94 29
Chief Exec.: C. Marshall
■ Resuscitation equipment & life saving training aids. [97]

90-00110 AMI GFV Marine Ltd
(† GFV Marine Ltd)
18 Canute Road, Ocean Village, Southampton, Hants SO14 3FT
☎ (+44 1703) 49 90 50 ⊖ (+44 1703) 49 90 49
♣ £ 1 500 000 (1995) 👥 8
Managing Dir: R.A. Linse
Operations & Technical Dir: Mike Woods
Sales & Marketing Dir: John Cox
■ Sales & service of EDO Corp. Acoustic Div. underwater acoustic products. U/W telephones, harbour intruder sonars, Doppler sonar speed logs, obstacle avoidance sonars, correlation velocity logs, a complete range of projectors, hydrophones & transducers. Bathymetric echo sounder & new ECDIS system. → 80
® USA: EDO Corp.

Service Depot
18 Canute Road, Ocean Village,
Southampton, Hants SO14 3FS
☎(+44 1703) 49 90 50; 39 98 60
⊖ (+44 1703) 49 90 49; 33 66 74 [97]

90-00120 Ampex Great Britain Ltd
(a subsidiary of Ampex Corporation)
Ampex House, Beechwood,
Chineham Business Park, Chineham,
Basingstoke, Hampshire RG24 8WA
☎(+44 1256) 81 44 10 ⊖ (+44 1256) 81 44 74
Managing Dir: Willy Björklund
National Sales Mgr: Paul Drew
Operations Mgr: Allen Parrish
Purchasing Mgr: Ms Maureen Brake
■ Airborne & laboratory Digital Cassette Recorders, mass storage devices & magnetic tape for instrumentation, professional video & mass storage applications. [98]

90-00140 Anglo Swiss Optical Co. Ltd
(a subsidiary of Hall & Watts Ltd)
266 Hatfield Road, St Albans, Hertfordshire AL4 1UN
☎(+44 1727) 85 92 88 ⊖ (+44 1727) 83 56 83
♧ £ 50,000 (1997) ♣ £ 1,500,000 (1994) 👥 5
Managing Dir: John Hoskins
■ Sales consultants for the special products (military) of the LEICA Co. plc, for the Middle East & Gulf, Africa & the Far East. [98]

90-00145 Anritsu-Wiltron European Centre
Capability Green, Luton, Beds LU1 3LU
☎(+44 1582) 41 88 53 ⊠ 826750 anrseu g
⊖ (+44 1582) 313 03 [97]

90-00150 Applied Dynamics International Ltd (ADI)
Oxford House, Oxford Street,
Wellingborough, Northants NN8 4HG
☎(+44 1933) 726 66 ⊠ 317100 adiltd g
⊖ (+44 1933) 730 24
Contact: Victor Strachan [98]

90-00155 Aspen Electronics Ltd
1-3 Kildare Close, Eastcote, Ruislip, Middlesex HA4 9UR
☎(+44 181) 868 13 11
W http://www.aspelect.demon.co.uk
⊖ (+44 181) 866 65 96
👥 26 🏠 7,500 sq ft
Managing Director: P A Johnson
Technical Director: R J Smith
General Manager: Howard Venning
Sales and Marketing Director: Martin Evans
■ Design, manufacture, procurement, configuration and supply of RF/microwave components, systems and measuring instruments, fibre optic components and measuring instruments, special and general purpose test equipment, security and test equipment including installation, maintenance, repair and associated recalibration. → 80
® Israel: Orbit Advanced Technology
Italy: Aten Srl
Japan: Sekisui
USA: Bird Electronic, Boonton Electronics Corp, Communication Techniques, EMS Inc, Elcom Systems, JCA Technology, K&L Microwave, Olektron, Pamtech, Passive Power, Polyphaser Inc, Quartztek, Res-Net, Vectron Inc, XL Microwave [98]

90-00160 Astronautics Corporation of America
European Operations
28 Tekels Avenue, Camberley, Surrey GU15 2LR
☎(+44 1276) 67 73 66
General Mgr: David J. Chalklin
■ Marketing office. → 80
® USA: Astronautics Corporation of America [98]

90-00163 Aubert & Duval
5 Clifton Court, Corner Hall,
Hemel Hempstead, Herts HP3 9XY
☎(+44 1442) 21 66 55 ⊠ 825816 audval g
⊖ (+44 1442) 25 97 29
Office Mgr: P. Ellis
® France: Aubert & Duval SA [97]

90-00168 Austernetics Pty Ltd
PO Box 61, St Albans, Herts AL1 2XA
☎(+44 1727) 85 90 00 ⊠ 299840 aaiuk g
⊖ (+44 1727) 83 17 43
Chairman: P.C.F. Morgan
Sales Dir: T.I. Morgan [97]

90-00180 Avio-Diepen BV
(a subsidiary of Avio-Diepen BV, Netherlands)
2 Kennet House, 19 High Street,
Hungerford, Berks RG17 0LG
☎(+44 1488) 68 46 25; 68 11 22
⊖ (+44 1488) 68 31 84
Manager, UK: Clive Ealing [98]

90-00182 Avionics Specialities
Wayside House, Church Street, Upton Grey,
Basingstoke, Hampshire RG25 2RA
☎(+44 1256) 86 28 19 ⊖ (+44 1256) 86 27 95
European Mgr: Brian Thrussell [97]

90-00187 AWA New Zealand Ltd
Hodford House, 17-27 High Street,
Hounslow, Middlesex TW3 1TA
☎(+44 181) 577 13 31 ⊠ 927662
⊖ (+44 181) 570 94 30 [98]

90-00200 Barco Ltd
(a subsidiary of Barco NV, Belgium)
50 Suttons Park Avenue, Reading, Berkshire RG6 1AZ
☎(+44 118) 966 46 11 W http://www.barco.com
W http://www.lyceum.com/~barco
⊖ (+44 118) 926 77 16
👥 50 🏠 2,500 m2
Managing Dir: John Bullivant
Market Development Manager C3I: Tony Stanford-Beale
Sales & Marketing Mgr Projection: Martin Howe
Sales & Marketing Mgr Information Displays: Mike Coleman
Sales & Marketing Mgr Graphics: Ms Marilyn Bowman
■ Large screen industrial monitors, video, data & graphics projection systems, rugged displays, avionics, display generators & air traffic control displays systems. High end colour critical printing systems-maps & security printing. → 80
® Belgium: Barco NV [98]

90-00220 Bell Helicopter - Textron
(a div. of Textron Limited)
8 Chesterfield Hill, London W1 7RG
☎(+44 171) 647 30 00
👥 5
Regional Mgr Europe: Terry D. Burnal
® USA: Bell Helicopter Textron [97]

90-00235 Berendsen Fluid Power
Sandy Way, Amington Industrial Estate,
Tamworth, Staffordshire B77 4DS
☎(+44 1827) 693 69 ⊖ (+44 1827) 651 65
👥 180 (1997)
Managing Dir: R. Knott
Communications Mgr: Paul Hensman
■ Hydraulic components for:- aircraft fuel test systems, armoured vehicles, ships & submarines, hydraulic pumps, including those for low viscosity fluids, motors, linear & semi-rotary actuators & valves.
Representing:
Vickers
Salami
Marzocchi
Auburn Gear
JSB
Fairey Arlon
Ausco
Eaton
Enerpac
Hydr-App
Leister
Micro Precision Textron
Bosch (Pneumatics)
Honeywell
Hydrokraft
Tangye [98]

90-00240 Boeing Defense & Space Group
17 Julius Court, Brentford Dock, Justin Close,
Brentford, Middx. TW8 8QH
☎(+44 181) 568 23 61 ⊖ (+44 181) 568 23 61
UK Consultant: Air Cdre O.J. Truelove (RAF Retd)
® USA: Boeing Defense & Space Group [97]

90-00255 Bofors (UK) Representative Office
Quayside House, Navigation Way, Preston PR2 2YP
☎(+44 1772) 76 94 94 ⊖ (+44 1772) 72 11 38
UK Representative: Roy Williams
® Sweden: Bofors AB [98]

90-00270 Bowtech Products Ltd
Howe Moss Crescent, Kirkhill Industrial Estate,
Dyce, Aberdeen AB21 0GN
☎(+44 1224) 77 23 45 E bowtech@bowtech.co.uk
W http://www.bowtech.co.uk ⊖ (+44 1224) 77 29 00
♧ £ 10,000 (shares) (1997) ♣ £ 1,100,000 (1997)
👥 9 (1997) 🏠 600 m2
Managing Director: Stephen Bowring
Sales and Marketing Director: Mike Winstanley
Sales Administrator: Ms Diane Farquhar
■ Represents three manufacturers who supply underwater lamps, underwater cameras, underwater electrical connectors, slip rings for power, signal & fibre optic coupling & positioning systems for cameras & lights.
® USA: I.E.C. Corporation, Remote Ocean Systems, Seacon/Brantner Associates [98]

90-00280 Brittain Engineering Ltd
(† Brittain Engineering (Guildford) Ltd)
509 Footscray Road, PO Box 367, London SE9 3UJ
☎(+44 181) 302 03 10 ⊖ (+44 181) 309 13 21
Chairman and Managing Director: Paul M Beza
Operations Director: Mrs Anne-Marie N Martin
■ Exporters, export agents, processes and plant engineering, aircraft and electronics industries. Non-destructive testing, speciality chemicals for maintenance and overhaul of engines and airframes, sealants, corrosion protection, organic coatings, PCB manufacturing processes. Industrial furnaces.
® UK: Brent Europe Ltd, Courtaulds Aerospace [98]

90-00290 Burmarc Ltd
Unit 12, Beaver Industrial Estate,
Liphook, Hampshire GU30 7EU
☎(+44 1428) 72 47 77 E burmarc@dircon.uk
⊖ (+44 1428) 72 46 52
♣ £ 1,000,000 (1997) 👥 6 (1997) 🏠 200 m2
Managing Director: H R J Bush
Director: S J Bush
■ Supply of digital electronic fluxgate compass systems for vessels & vehicles. Specialist fluxgate heading sensors & military systems for a wide variety of applications.
Representing::
KVH Industries Inc, USA [98]

90-00323 Caterpillar Overseas SA
Caterpillar (UK) Ltd
(a subsidiary of Caterpillar Inc, USA)
Regal Court 42/44 High St, Slough, Berks SL1 1EL
☎(+44 1753) 388 35 ✉ CATSALES Slough ⊠ 849397
⊖ (+44 1753) 69 44 25
👥 15
Area Mgr: D.R. Noiles
■ Manufacture of heavy construction and mining equipment, engines, power plants, industrial trucks, articulated and rigid frame dump trucks.
Subs. with defence business:
Artix Ltd
Caterpillar Belgium
Caterpillar France
Caterpillar (UK) Ltd [97]

90-00324 CATIC Trading Development (UK) Ltd
(a subsidiary of the China National Aero- Technology Import and Export Corporation)
Unit 9, Iron Bridge House, Bridge Approach,
London NW1 8BD
☎(+44 171) 586 38 54 E catic@msn.com
⊠ 266865 catic g ⊖ (+44 171) 586 44 46
♧ £ 150,000 (1997) ♣ £ 2,000,000 (1997) 👥 4 (1997)
🏠 150 m2
Director: Hongkui Cai
® China PR: CATIC
Representative Office
12 Lower Merton Rise, Swiss Cottage, London NW3 3RA

Contact Reference

☎(+44 171) 586 83 49 ▣ 266865 catic g
⊖ (+44 171) 586 44 46 [98]

90-00325 CCS Communication Control Systems

62 South Audley Street, London W1
☎(+44 171) 629 02 23 ▣ 8814709
⊖ (+44 171) 629 95 38
Manager: M Miguel
■ Anti-bugging, anti-wiretapping, bullet-proof, armouring cars, countersurveillance, bomb and lie detection equipment.
® USA: CCS International Ltd [98]

90-00335 Cherry Division of Textron Inc.

c/o 10 The Beeches, Harbury, Warwickshire CV3 39LW
☎(+44 1926) 61 22 18; 0806 72 55 59
⊖ (+44 1926) 61 30 95
President: George Andrews
Sales and Marketing Director: Ron Manley
Operations and Technical Director: Michael Harhan
European Product Development Director: C P Houghton
® USA: Cherry Aerospace
UK Distributor
c/o C.J. Fox & Sons Ltd, Shoreham Airport,
Shoreham by Sea, W. Sussex BN43 5FN
☎(+44 1273) 46 40 46 ▣ 87642 fox g
⊖ (+44 1273) 46 45 77
Chairman: Paddy O'Connor
■ Automatic blind riveting, nut-plate rivets, lockbolts, blindbolts & fasteners for composites. [98]

90-00357 CMP Batteries Ltd

14 Gunnels Wood Industrial Estate, Gunnels Wood Road,
Stevenage, Herts SG1 2BH
☎(+44 1438) 35 90 90 ⊖ (+44 1438) 72 76 84
Co-ordinator: P Meeds [98]

90-00364 Computing Devices Hastings (UK) Ltd UK Office

(a division of Computing Devices Canada)
PO Box 1, Newquay, Cornwall TR7 3AP
☎(+44 1637) 87 43 04 ⊖ (+44 1637) 85 01 02
Senior Vice President and Chief Executive Officer: John Watts
Sales and Marketing Director: Ernest Trischuk
International Marketing Manager: John Bussey
® Canada: Computing Devices Canada Ltd [98]

90-00367 ComStream Corporation

Range House, Walnut Tree Close,
Guildford, Surrey GU2 4US
☎(+44 1483) 44 09 33 ⊖ (+44 1483) 44 08 76 [97]

90-00372 Conrac UK Limited

(a subsidiary of Conrac Elektron GmbH, Germany)
177 The Avenue, Aylesford, Kent ME20 7RF
☎(+44 1622) 71 80 90 ⊖ (+44 1622) 71 80 94
♣ DM 1,000,000 (1996) ⛹ 2
Manager: Adrian Bailey [98]

90-00403 Daewoo UK Ltd

Templar House, 82 Northolt Road,
South Harrow, Middlesex HA2 0YL
☎(+44 181) 423 72 00 ▣ 932241 daewoo g
⊖ (+44 181) 864 60 70 [98]

90-00405 Daicel Polymers Ltd

5-6 Shenley Pavilions, Chalkdell Drive,
Shenley Wood, Milton Keynes MK4 6LB
☎(+44 1908) 50 67 77 ⊖ (+44 1908) 50 67 78 [97]

90-00425 DDC (UK) Ltd

(† ILC Data Device Corp.)
(a subsidiary of ILC Data Device Corporation, USA)
European Operations
Mill Reef House, 9-14 Cheap Street,
Newbury, Berkshire RG14 5DD
☎(+44 1635) 81 11 40 Ⓔ jdg@ilc-ddc.demon.co.uk
▣ 848826 nab g ⊖ (+44 1635) 322 64
⛹ 10
Director of Operations Europe: John Gasking
Marketing Communications Manager: Lawrence M Margulies
Sales Manager: John Northfield
■ Synchro/resolver to digital converters, angle indicators & simulators, MIL-STD-1553 data bus products, custom hybrid micro circuits.
® USA: Beta Transformer Technology Corp., ILC Data Device Corp. [98]

90-00430 Detector Electronics (UK) Ltd

(a subsidiary of Detector Electronics Corporation, USA)
Poyle Road, Colnbrook, Slough SL3 0HB
☎(+44 1753) 68 30 59 ▣ 848124 gravin
⊖ (+44 1753) 68 45 40
Managing Director: Garth Watkins
® USA: Detector Electronics Corp. [98]

90-00435 Detroit Diesel (Overseas) Corp.

Riverside House, Riverside Way, PO Box 63,
Northampton NN1 5NX
☎(+44 1604) 246 17 ▣ 31329
Sr Vice President International: Paul Moreton
® USA: Detroit Diesel Corp. [97]

90-00480 Draeger Ltd

(a subs. of Drägerwerk AG, Germany FR)
Kitty Brewster Industrial Estate, Blyth,
Northumberland NE24 4RG
☎(+44 670) 35 28 91 ⊖ (+44 670) 35 62 66
⛹ 310
Chairman: C Dräger
Operations/Technical Director: Vince Smith
Purchasing Manager: Fenwick Yeouart
Marketing Communications Manager: Richard Beckwith
■ Manufacturers of gas detection instruments, respiratory protection equipment and diving equipment. → 75 [98]

90-00510 Eaton Ltd

Administrative Center
Eaton House, Staines Road, Hounslow, Middx TW4 5DX
☎(+44 181) 572 73 13 ✉ EATONYALE Hounslow
▣ 27798 ⊖ (+44 181) 572 70 02
Int'l Co-ordinator Northern Europe: Dr Gunther Vieten [98]

90-00520 ED Electronics Ltd

(a subsidiary of EDI Inc., USA)
Shelley House, The Avenue, Lightwater, Surrey GU18 5RF
☎(+44 1276) 47 26 37 Ⓔ electronic-disigns.com
⊖ (+44 1276) 47 37 48
♣ US$ 60,000,000 (1996) ⛹ 135 (1997)
Director: Richard Sawers
■ High density semiconductor memories for military applications including SRAM, DRAM, Flash & EEPROM. [98]

90-00530 Emerson and Cuming Microwave Products Ltd

(† Emerson and Cuming (UK) Ltd)
30 Catherine Gardens, Hounslow, London TW3 2PR
☎(+44 181) 560 64 84 ⊖ (+44 181) 569 89 80
⛹ 1
Sales Manager, United Kingdom: M J Abrey
® Belgium: Grace NV [98]

90-00535 Encore Computer (UK) Ltd

(a subsidiary of Encore Computer Corp, USA)
Marlborough House, Mole Business Park, Randalls Road,
Leatherhead, Surrey KT22 7BA
☎(+44 1372) 81 42 00 Ⓔ encore.com ▣ 947707
⊖ (+44 1372) 81 42 01
⛹ 120 ⌂ 35,000 sq ft
International Operations Director: Ed Aston
General Manager: Michael Edgley-Pyshorn
Marketing Manager: Chris Atkins
Sales Manager: Michael Edgley Pyshorn
Sales Executive: Glen Robinson
Sales Executive: Jeremy Smith
Senior European Material Planner: Evan Sepion
Business Development Manager: Chris Atkins
Technical Services Analyst: Andy Salter
Marketing Communications Executive: Angus Reid
Customer Support Manager: Tom Higgins
■ Manufacture of storage & real-time computers based on UNIX & real-time operating systems for commercial, scientific & research applications.

Northern Office
Regent House, Heaton Lane, Stockport, Cheshire SK4 1BS
☎(+44 161) 480 09 07 ▣ 665574 selcom g
⊖ (+44 161) 480 17 88
Contact: Neil Chapman [98]

90-00700 Fuji Industries Co Ltd

197 Knightsbridge, London SW7 1RB
☎(+44 171) 225 29 06 ▣ 911556
⊖ (+44 171) 225 29 10
Representative: S Kohda
® Japan: Fuji Industries Co. Ltd [98]

90-00760 Gordon Audio Visual Ltd

Symes Mews, 37 Camden High St, London NW1 7JE
☎(+44 71) 387 33 99 ✉ GORCAM London
▣ 264413 greshm g
Managing Dir: Martin Ellwood
Director: R.E. Johnson
■ High speed photography & film analysis projection systems. [97]

90-00780 James Greaves & Co.

(a division of Greaves Limited, India)
Brazennose House, Brazennose Street, PO Box 406,
Manchester M60 2JA
☎(+44 161) 834 09 91 ⊖ (+44 161) 832 07 53
♣ £ 5,125,000 (1996/97) ⛹ 12 (1996/97)
Accountant: K Valentine
® India: Greaves Cotton & Co.Ltd [98]

90-00790 Grimes Aerospace (UK) Ltd

Bourne House, Cores End Road,
Bourne End, Buckinghamshire SL8 5AR
☎(+44 1628) 85 09 85 ⊖ (+44 1628) 85 08 90
Managing Director: Mike Dubose
Vive President Product Support: Mark Fiedler
European Regional Manager: Patrick Douneen
® USA: Grimes Aerospace Company [98]

90-00795 Gulf Helicopters Company

(a division of Gulf Air GSC, Bahrain)
Unit 1 Stockwell Works, Stephenson Way,
Crawley RH10 1TN
☎(+44 1293) 40 13 33 ⊖ (+44 1293) 61 15 66
Managing Dir Doha: M. Evans
UK Office Mgr: C. Perceval
■ Purchasing agents for Arabian Gulf. Operates helicopter charters from its base at Doha Qatar for Arabian region. [98]

90-00820 Hanjin Heavy Industries Co Ltd London Office

Alhambra House, 27-31 Charing Cross Road,
London WC2H 0AU
☎(+44 171) 839 47 00 ⊖ (+44 171) 839 54 00 [98]

90-00830 Heckler & Koch UK

(a division of Royal Ordnance plc)
Kings Meadow Road, Nottingham NG2 1EQ
☎(+44 115) 968 23 00 ⊖ (+44 115) 968 21 18
General Manager: Nick Hutt
Sales and Marketing Director: Armin Bähr
Operations Manager: Mike Thornton
Purchasing Manager: Tim Lynch [98]

90-00845 HIMA Sella Ltd

Carrington Field Street, Stockport, Cheshire SK1 3JN
☎(+44 161) 477 83 32 Ⓔ sales@himsela.demon.co.uk
Ⓦ http://www.himasela.demon.co.uk
⊖ (+44 161) 480 43 76; 476 30 95 (Sales)
♣ £ 4,000,000 (1995) ⛹ 60
Managing Dir: P. Garland
Marketing & Sales Dir: J. Dolan
Engineering Dir: M. Hibert
■ Systems integration & installation of Theissen training systems complete range of targetry equipment for infantry & artillery use. Radio remote control & data transmission systems. Control systems. → 75 → 80 [98]

90-00850 Hindustan Aeronautics Ltd

Room N° 602 - India House, High Commission of India,
Aldwych, London WC2B 4NA
☎(+44 171) 497 23 60 ▣ 264616 halldn g
⊖ (+44 171) 497 83 98

Resident Manager: Mr Kanakappa
® India: Hindustan Aeronautics [98]

90-00855 William Hinks & Partners
Suite 25A, 78 Buckingham Gate, London SW1P 6PE
☎ (+44 171) 222 03 90 ⊖ (+44 171) 222 48 98
Managing Dir: William W. Hinks
® Canada: CAE Electronics Ltd
France: Latecoere
Germany: CAE Electronics GmbH [97]

90-00912 Hydratec Systems Ltd
(a daughter company of Vapormatt Ltd)
Queen Street, Taunton, Somerset TA1 3AX
☎ (+44 1823) 25 79 76 ⊖ (+44 1823) 33 64 46
Sales and Marketing Director: A R Loweth
Sales and Marketing Director: R S Ashworth
Sales Administration: J Le Prevost
® UK: Vapormatt Ltd [98]

90-00915 Hydrema (UK) Ltd
Camp Hill Close, Dallamires Lane,
Ripon, N. Yorkshire HG4 9XX
☎ (+44 1765) 60 15 71 ⊖ (+44 1765) 60 79 06
Sales & Marketing Dir: T. Wadsworth
■ Manufacture & supply of high quality construction equipment. Range comprises articulated loader excavators & articulated rough terrain dump trucks.
® Denmark: A/S Hydrema [97]

90-00930 IFO International (UK) Ltd
3 Beechwood Drive, Keston, Kent BR2 6HN
☎ (+44 1689) 85 67 68 ⊖ (+44 1689) 85 59 33
♣ £ 500,000 (1996)
Director: Peter A. Leach
■ Suppliers of gyroscopes, accelerometers, servo components, elapsed time indicators, events counters & electro levels.
® USA: Airpax Corp., Fredericks Co., Humphrey Inc., Kearfott Guidance & Nav., Lambda Novatronics, Sensonics Inc. [98]

90-00948 Intergraph (UK) Ltd
(a subsidiary of Intergraph Corp., USA)
Delta Business Park, Great Western Way,
Swindon, Wiltshire SN5 7XP
☎ (+44 1793) 61 99 99 E bfrench@ingr.com
W http://www.intergraph.com/uk ⊖ (+44 1793) 61 85 08
👥 250 (UK based)
Managing Director: Roger Hill
Operations and Technical Director: Dennis Goodyear
Product Marketing Manager, GIS/Mapping: Peter Cook
Marketing Services Manager: Ms Kate Mills
■ CAD, CAD/CAM, CAE and digital mapping systems, GIS and decision planning software. [98]

90-00965 International Precision Products BV
London End, PO Box 3, Beaconsfield, Bucks HP9 2HU
☎ (+44 1494) 67 17 47 ⊖ (+44 1494) 67 18 25
Director: J.R. Wolff
Sales Mgr: J. Oakenfull
® USA: CE Precision Assemblies Inc., Cristek Interconnects Inc., Janco Corp., Jerrik Connecting Devices Inc., Photronics Inc., Saint Switch Inc., Transicoil Inc. [97]

90-00980 Itochu (Europe) plc
The International Press Centre, 76 Shoe Lane,
London EC4A 3JB
☎ (+44 171) 827 08 22 ✉ CITOH ⊠ 264931 itochu g
⊖ (+44 171) 583 36 18
Aerospace Dept Mgr: H. Sato
® Japan: Itochu Corporation [97]

90-00995 Jaivision Ltd UK
(† Jørgen Anderson Ingeniorfima)
Audley House, North Bridge Rd,
Berk Hampstead, Herts HP4 1EH
☎ (+44 1442) 87 96 69 ⊖ (+44 1442) 87 92 81
Managing Dir: Jørgen Andersen [97]

90-01050 JCB Sales Ltd
(a div. of J.C.Bamford Excavators Ltd)
3 Deanery Street, Mayfair, London W1Y 5LH
☎ (+44 171) 491 25 76 ⊠ 36372
⊖ (+44 171) 491 10 28
♣ £ 650 000 000 👥 3 000
Managing Dir: J. Patterson
Marketing Dir: N. Allan
Military Sales Mgr: D. Hall [98]

90-01100 Kavlico UK
(a division of Kavlico Corporation, USA)
Unit 1 Drake Walk, Atlantic Wharf,
Cardiff, South Glamorgan CF1 5AN
☎ (+44 1222) 46 34 49 ⊖ (+44 1222) 45 08 52
👥 3 (1997)
UK Technical Sales Mgr: Paul Chilcott
■ Position and pressure transducers for the aerospace and defence business.
® USA: Kavlico Corporation [98]

90-01110 KHD Great Britain Ltd
(inc. KHD Deutz & Deutz MWM Div.)
2 St. Martins Way, London SW17 0UT
☎ (+44 181) 781 72 00 ✉ EMail: khdgb@dial.pipex.com
⊠ 8954136 khdlon g ⊖ (+44 181) 947 63 80
👥 100
Managing Dir: A. McQuillan
Sales Mgr Deutz MWM Division: D. Emmas
Marketing Mgr: Heinz Kessel
■ Sales, marketing & service in the UK for Deutz air and liquid cooled engines and Deutz MWM water cooled engines.
® Germany: KHD AG [97]

90-01162 Kollmorgen Hightech Limited
(† Hightech Components Ltd)
(a subs. of Kollmorgen Corp., USA)
Servo House, 14-16 Mulfords Hill,
Tadley, Hants RG26 3JA
☎ (+44 1189) 81 39 22 ⊖ (+44 1189) 81 15 32
♣ £ 5 000 000 (1994) 👥 11
Managing Dir: Graham R. Horner
■ Servo components for high-technology users.
® France: Kollmorgen Artus
Switzerland: ETEL S.A.
USA: Airflyte electronics Company, Astro Instrument Corporation, Inductosyn International, Kollmorgen Industrial Drives, Kollmorgen Inland Motor, Kollmorgen PMI [97]

90-01165 Kongsberg Gruppen ASA
(a subsidiary of Kongsberg Gruppen ASA, Norway)
4 Stanstead Close, Rowlands Castle PO9 6AN
☎ (+44 1705) 41 28 09 ⊠ 847847
⊖ (+44 1705) 41 38 09
UK Representative: Capt. M.J.M. Wilkin
■ Main areas of activity include anti-ship missiles for fighter aircraft, helicopters & naval vessels; missile launchers; missile control & test systems; ground based systems; naval defence systems; tactical communications; trainers & simulators; satellite-based surveillance & special special electronics.
® Norway: Kongsberg Gruppen A/S [98]

90-01170 Krauss-Maffei (UK) Ltd
(a division of Krauss-Maffei Wehrtechnik GmbH)
Europa Boulevard, Gemini Business Park,
Warrington WA5 5TR
☎ (+44 1925) 64 41 00; 65 15 55
⊖ (+44 1925) 23 42 84
♧ £ 150,000 ♣ £ 2,200,000 👥 50 🏭 6,000 sq m
Managing Director: N A Wilkins
Operations Director: V Kelly
Sales Director: Tony Slim [98]

90-01180 Kyokuto Boeki Kaisha Ltd London Branch Office (KBK)
5th Floor St Alphage House, 2 Fore Street,
London EC2Y 5DA
☎ (+44 171) 628 66 16 ⊖ (+44 171) 628 61 98 [98]

90-01235 Lee Products Ltd
(a subs. of the Lee Company, USA)
3 High Street, Chalfont St Peter,
Gerrards Cross, Bucks SL9 9SU
☎ (+44 1753) 88 66 64 ⊖ (+44 1753) 88 95 88
General Manager: Michael S Stott
European Sales Director: Peter A Roberts
Sales Engineer: Roger J. Brooks
Sales Engineer: Lloyd S Bailey
Sales Office Manager: Ms Kate J Paton
■ Suppliers of miniature hydraulic control components for aerospace & military vehicle applications.
® USA: Lee Co. [98]

90-01238 Lemo UK Ltd
(a subsidiary of Lemo SA, Switzerland)
12 North Street, Worthing, West Sussex BN11 1DU
☎ (+44 1903) 23 45 43 ⊖ (+44 1903) 20 62 31
♣ £ 4,000,000 👥 34 (1997)
Managing Director: Richard R Thomas
Technical Dir: G.P. McFarlane
Sales & Marketing Dir: B. Bostock
Purchasing Mgr: P. Hardey
■ Swiss precision self-latching electrical connectors, coax, multi-pin, high density, high voltage, environmentally sealed fibre optic.
® Switzerland: Lemo [98]

90-01240 Litton Precision Products International Inc
6 First Avenue - Globe Park, Marlow, Bucks SL7 1YA
☎ (+44 1628) 48 60 60 ⊖ (+44 1628) 47 24 38
General Manager: Clive A McGovern
® USA: Litton Industries Inc [98]

90-01260 LMS UK
Westwood House - Unit 10,
Business Park, Westwood Way, Coventry CV4 8HS
☎ (+44 1203) 474 700 ⊖ (+44 1203) 856 310
→ 80
® Belgium: LMS International [98]

90-01280 Lockheed Martin International Inc.
(† Martin Marietta International Inc.)
Shortlands, Hammersmith, London W6 8BX
☎ (+44 181) 846 87 58 ⊖ (+44 181) 846 87 75
President European Region: Ian R Stopps
Vice President of Business Development: Gary W Mowery
Secretary: Mrs Maguy Webster
® USA: Lockheed Martin Corp. [98]

90-01290 M/A COM Limited
Centennial Court, Easthampstead Road,
Bracknell, Berkshire RG12 1YQ
☎ (+44 1344) 86 95 95 ⊖ (+44 1344) 30 00 20
♣ £ 350,000,000 (1996) 👥 3,900 (1997)
Vice President Sales and Marketing: Roger Tarrant
Global Marketing Manager: Alistair Blaxill
Area Manager: Malcolm Weatherhead
■ Components and sub-systems for avionics, defence, radar, surveillance, Navaid and telecommunication applications. Products range from semiconductor and ferrite based devices and components; connectors and cable assemblies; antennas; through to advanced microwave multifunction assemblies, synthesizers and subassemblies. [98]

90-01295 MacArtney AS
(a div. of MacArtney A/S)
10 Hilltop, Windermere, Cumbria LA23 2H9
☎ (+44 15394) 422 50 ⊖ (+44 15394) 423 99
♣ £ 7 000 000 (1995) 👥 30
Managing Dir: Martin MacArtney ☎ (+45 75) 11 66 77
Technical Dir: Lars Hansen
Sales & Marketing Dir: Anders Andersen
Area Mgr: Doug Huntington ☎ (+44 1539) 44 22 50
■ Supply a range of products related to the underwater technology market.
Representing:
Focal Technologies Inc.
Deep Ocean Engineering
Seatec B.V.
Navylle bv
GMI [97]

90-01300 Magellan Systems Corporation
European Marketing and Sales Office
Suite 4, 9 Dragoon House, Hussar Court,
Waterlooville, Hampshire PO7 7SF
☎ (+44 1705) 79 11 88 ⊖ (+44 1705) 79 21 88
European Sales Manager: Mark Luffingham
☎ (+44 1705) 79 21 88 [98]

Contact Reference

90-01313 Manitou Site Lift Ltd
(a subsidiary of Manitou BF, France)
Ebblake Industrial Estate - Verwood,
Wimborne, Dorset BH31 6BB
☎(+44 1201) 82 53 31 ⊖ (+44 1201) 81 30 27
® France: Manitou BF
98

90-01323 Mason Electric Co. Inc.
(a subsidiary of Mason Electric Co. Inc., USA)
International Sales Office
Gables Court, Main Street, Frolesworth,
Leicestershire LE17 5EG
☎(+44 1455) 20 20 64 ⊖ (+44 1455) 20 20 99
155 50,000 sq ft
President: William E Southern
Technical Director: Terry Allen
Sales Manager: Michael Nystrom
Purchasing Manager: Ms Charlotte Gesualdi
International Sales and Marketing Manager: Michael P Frost
■ Design, development and manufacture controls and switching for all types of fixed and rotary wing aircraft, land based vehicles, tanks and for missile firing.
® USA: Mason Electric Co. Inc.
98

90-01325 Marubeni UK plc
(a subs. of Marubeni Corp., Japan)
120 Moorgate, London EC2M 6SS
☎(+44 171) 826 86 00; 88 32 ⊖ (+44 171) 826 86 19
Managing Dir: H. Takahata
Aerospace & Defence Dept Mgr: G. Toya
Aerospace Dept: Miss C. Baughan
® Japan: Marubeni Corp.
97

90-01330 McDonnell Douglas Ltd
(a subs. of McDonnell Douglas Corp., USA)
European Operations
102 Jermyn St, St James's, London SW1Y 6EE
☎(+44 171) 930 50 00 ⊖ (+44 171) 930 33 00
Managing Dir: Robert J. Madden
Helicopters Dir: Stu Park
® USA: McDonnell Douglas Corp.
97

90-01340 Metrix Instrument Company
European Office
Western Way, Bury St Edmunds, Suffolk IP33 3SZ
☎(+44 1284) 76 59 77
E bob.matrix.europe@dial.pipex.com
W http://www.industry.net/metrix
⊖ (+44 1284) 76 02 56
Managing Director: Ted Wrona
Sales and Marketing Director: Dean P Wichert
European Manager: Bob Lorenzetto
■ Sales of a wide range of high quality vibration sensing, monitoring and recording instruments for a wide range of applications.
98

90-01346 Microphase International
(a subsidiary of Microphase Corp, USA)
40 The Findings, Hawley,
Nr. Farnborough, Hampshire GU14 9EG
☎(+44 1276) 351 07 ⊖ (+44 1276) 363 88
Marketing Dir: Roger Burfoot
® USA: Microphase Corp.
98

90-01400 Mitsubishi Heavy Industries Europe Ltd
Head Office
Bow Bells Hse, Bread St (Cheapside), London EC4M 9BQ
☎(+44 171) 634 51 11 ✉MITSUBISHI ⊠ 888994
⊖ (+44 171) 248 13 29
President: Yoshihumi Sawamoto
® Japan: Mitsubishi Heavy Industries
97

90-01410 MJP Corporation Ltd Sales Office
17 Park Close, Kingston-upon-Thames KT2 6DW
☎(+44 802) 62 53 79
Sales Director: Jari Kuusisto
® Finland: MJP Corporation Ltd
98

90-01425 Multronics Ltd
41 West Street, Marlow, Buckinghamshire SL7 2LS
☎(+44 1628) 47 46 41 ⊖ (+44 1628) 89 01 38
3

Managing Director: B J Mulady
■ Represents manufacturers of aircraft equipment, de-icing controllers, generators and static inverters, pressure switches, flow control metering, transformer rectifier units, power supplies and engine instruments.
Representing:
Aurora Bearing Company
DCHS - Dynamic Controls
Fluid Components International
Howell Exports
ITT Aerospace Controls Division
Leland Electrosystems Inc
Power Paragon Inc
Shallco
98

90-01429 Nikon Ltd
Nikon House, 380 Richmond Road,
Kingston-upon-Thames, Surrey KT2 5PR
☎(+44 181) 541 44 40 ✆ (800) 23 02 20
⊖ (+44 181) 541 45 84
General Mgr Consumer & EI: Simon Coleman
Professional Products Mgr: Jeremy Gilbert
■ Cameras, lenses & accessories. Digital Camera Systems. 35mm & flat bed scanners, binoculars & large format lenses.
98

90-01435 NMB (UK) Ltd
(a subsidiary of Minebea Co, Tokyo)
1 Sterling Centre - Eastern Road,
Bracknell, Berkshire RG12 2PW
☎(+44 1344) 42 66 11 ⊠ 51883578
⊖ (+44 1344) 48 55 22
♣ £ 1,500,000,000 (1996) 3,500 (worldwide - 1996)
Managing Director: M Wetson
■ Sales and distribution in the UK and Ireland of all Group products. These include bearings (ball and roller).
® USA: New Hampshire Ball Bearings Inc, Rose Bearings Ltd
98

90-01480 Nozaki Europe, Ltd London Office
21-26 Garlick Hill, London EC4 2AU
☎(+44 171) 248 43 55 ⊠ 886247
⊖ (+44 171) 248 20 45
President: K Maeda
® France: Eurocopter, Labinal PM, OFEMA, Penny & Gile, Turboméca
UK: Froude Consine, Rotodata
98

90-01505 Ocean Acoustic Developments Ltd
Pendower Leverlake Road, Widemouth Bay,
Bude, Cornwall EX23 0AF
☎(+44 1288) 36 16 52 ⊖ (+44 1288) 36 10 97
2
Director: John Hodgson
Marketing and Sales Director: Diane Hodgson
■ Underwater acoustics research & development & software sales.
98

90-01510 Oki Electric Industry London Office
Central House, Balfour Road,
Hounslow, Middlesex TW3 1HY
☎(+44 181) 577 90 00 ⊖ (+44 181) 572 50 66
General Manager: Naotake Nishihama
98

90-01520 Okura & Co (UK) Ltd
(a subs. of Okura & Co.Ltd, Japan)
Albangate, 125 London Wall, London EC2Y 5AS
☎(+44 171) 600 48 24 ⊠ 883798 okura g
⊖ (+44 171) 283 23 28; 22 42
Managing Director: T Kitamura ☎(+44 171) 600 60 01
Sales and Marketing Director: R Hough ☎(+44 171) 600 48 24
® UK: BP Metal Composites Ltd, Chelton (Electrostatics) Ltd, Dowty Defence & Air Systems Ltd, E.H. Industries Ltd, GEC-Marconi Avionics Ltd, Sensors Div., I.G.G. Component Technology Ltd, Lucas Aerospace Ltd, Engine Systems Div., Normalair-Garrett Ltd, Customer Support, Penny & Giles Data Recorders Ltd, Rolls-Royce plc, Westland Helicopters Ltd
98

90-01555 Pacer Systems Inc.
c/o City Marketing Ltd, 6 Long Lane, London EC1A 9HA
Investor Relations: Richard Bernstein
® USA: Pacer Systems Inc.
97

90-01575 Paint Services Group Limited (PSG)
Weydown Road, Haslemere, Surrey GU27 1BT
☎(+44 1428) 65 12 46 ⊖ (+44 1428) 66 14 71
24
Managing Director: A K Wood
Deputy Managing/Sales Director: G R Yates
Operations Director: A E T Swayne
■ Suppliers of aerospace and defence coatings and environmentally friendly cleaners to the aviation market.
Representing::
Courtaulds Aerospace
AD Aerospace
Sherwin Williams
Pratt & Lambert "Jet Glo"
98

90-01595 Parsons Group International
(† The Ralph M Parsons Co Ltd)
(a subsidiary of Parsons Corporation, US)
Parsons House, Kew Bridge Road,
Brentford, Middlesex TW8 0EH
☎(+44 181) 742 12 00 W http://www.parsons.com
⊖ (+44 181) 995 17 60
1,000
Managing Director: Andor Richter
Sales and Marketing Vice President: James R Shappell
Marketing Manager: Stephen Aldridge
® USA: Parsons Corp.
98

90-01630 Perry Tritech
(a member of the Coflexip Stena Offshore Group)
Greenbank Crescent, East Tullos, Aberdeen AB12 3BG
☎(+44 1224) 87 71 11 ⊖ (+44 1224) 89 88 11
♧ £ 510 000 (1995) ♣ £ 3 000 000 (1995) 19
930 sq m
Managing Dir: John Jacobson
Operations/Technical Dir: David Troup
Purchasing Mgr: Andrew Smith
Sales Engineer: Peter MacInnes
Representing::
Alcatel Kabel
Burton
Insite Systems
AK Industries
Deep Sea Power & Light
Sunstar Electric
Immetrix
Stereographic
Atchley
97

90-01800 Radiometer Limited
(a subsidiary of Radiometer Medical A/S)
Manor Court, Manor Royal,
Crawley, West Sussex RH10 2PY
☎(+44 1293) 51 75 99 ✆ (800) 22 03 99 ⊠ 878718 g
⊖ (+44 1293) 53 15 97
Managing Dir: Peter Lovett
Product Mgr: Kevin M. Bage
Sales & Marketing Mgr: Alan Jarman
Service Mgr: Roger Wynveldt
■ Sales & service of blood gas, co-oximetry & electrolyte analysers; transcutaneous analytical & testing instruments. Also arterial samplers, associated reagents, accessories & pH buffers electrodes. → 80
98

90-01810 Rapid Data Ltd
Amelia House, Crescent Road,
Worthing, West Sussex BN11 1RL
☎(+44 1903) 82 12 66 ⊖ (+44 1903) 82 07 62
♣ £ 100,000 10
Managing Director: William A Havranek
Marketing Director: Ms Celia Lovering
Operations Director: Ole Sorensen
Purchasing Manager: Ms Marie Jaine
■ Microsaint/animation (interactive, easy-to- use, discrete simulation system with graphics - wide field of use, for PC and MAC). Also interactive program for numerical linear algebra and matrix computation, for PC to Cray; easily interfaced with ACSL.
® USA: ACSL, Applied Dynamics Int'l, Mathematica, Micro Analysis & Design, Micro Saint, Mitchell & Gauthier, Visual Numerics, Wolfram Research
98

90-01825 Raytheon International Inc
7 Old Park Lane, London W1Y 3LJ
☎(+44 171) 493 71 72 W http://www.raytheon.com
⊖ (+44 171) 629 08 57
72,000 (1996)
Vice President, UK Country Manager: Alan W Jones

■ International business development office of the Raytheon Company.
® USA: Raytheon [98]

90-01826 RDS Servo Components
(a subsidiary of Review Display Systems Ltd)
Horton Place, Horton's Way, Westerham, Kent TN16 1BT
☎ (+44 1959) 56 33 45 ⊖ (+44 1959) 56 44 52
♣ £ 10,000,000 (1996) 25 (1996) 500 m2
Managing Director: R D Smith
Sales and Marketing Manager: P Blake
Sales Manager (Servo Components Division): R C Nash
■ Agents/distributors for overseas companies manufacturing ranges of equipment including: DC torque motors, synchros, resolvers, RVDTS, LVDTS, hysteresis motors, rotary & linear actuators including custom assemblies optical shaft encoders incremental and absolute, miniature stepping motors rotary and linear. → 85
® France: MCB SA, PRECILEC SA
USA: Haydon Switch & Instr. Inc. [98]

90-01829 The Rochester Corp.
2nd Floor, Taylor Building, 62-64 Bromham Road, Bedford MK40 2QG
☎ (+44 1234) 32 70 13 826818
⊖ (+44 1234) 32 70 62
♣ US$ 40 000 000 (worldwide) 300 (worldwide)
32 500 (worldwide)
Int'l Sales & Marketing Dir: David J. Harris
Int'l Technical Sales Mgr: David I. Buchan
■ Wide range of electro-mechanical & electro-optic cable systems to the highest military quality & inspection standards. Supplier to the military & oceanographic markets. [97]

90-01830 Rockwell-Collins (UK) Ltd Avionics and Communications
(a subsidiary of Rockwell International Corp, USA)
Suttons Business Park, Earley, Reading, Berkshire RG6 1LA
☎ (+44 118) 926 11 11 847286 colrad g
⊖ (+44 118) 966 40 16
Managing Director: Peter J Humphrey
Director of Marketing: Steve J Banham
■ Avionics and telecommunications systems.
® USA: Rockwell International Corp. [98]

90-01885 Safe Flight Instrument Ltd
PO Box 188, Hook, Hants RG27 8XE
☎ (+44 1252) 84 54 64 ⊖ (+44 1252) 84 54 65
Managing Dir: Derek Holden
® USA: Safe Flight Instrument Corp. [97]

90-01895 Samet Consultants
10 Blythwood Road, Pinner, Middx. HA5 3QB
☎ (+44 181) 868 90 77 ⊖ (+44 181) 868 84 99
3
Owner: A.E. Pryce
Owner: M.A. Pryce
■ Transformer laminations, sand castings, silicon rubber products & forgings.
® France: Messier Fonderie
USA: Bergman Forge, Chase Walton Elastomers, Magnetic Metal Corp. [97]

90-01900 SAN Elektroheat
PO Box 106, Oldham OL2 7DU
☎ (+44 1706) 29 95 00 ⊖ (+44 1706) 29 98 50
5
Sales Mgr: Harold Comerford
Sole Agent for:
Svend A. Nielsen A/S
Manufacture & design company. [98]

90-01920 Schottel Representation (UK) Ltd and Ireland
Twistwood, Hurst Road, Walton-on-the-Hill, Tadworth, Surrey KT20 5BN
☎ (+44 1737) 81 41 25 ⊖ (+44 1737) 81 40 42
Managing Director: Geoffrey Hewitt
☎ (+44 1737) 81 41 25
■ Rudder propellors (azimuthing etc.). Gear- boxes, clutches and couplings, bearings, box coolers, tubular heat exchanges.
® Germany: Lohmann & Stolterfoht GmbH, Schottel-Werft
Netherlands: Bloksma BV [98]

90-01930 SEAB Ltd
(a subsidiary of CelsiusTech AB, Sweden)
146 High Street, Billericay, Essex CM12 9DF
☎ (+44 1277) 62 53 23
E 101601.740@compuserve.com
⊖ (+44 1277) 63 05 90
Managing Director: Ms Margaret A Robins
Technical Dir: Frank Sumner
® Sweden: CelsiusTech AB [98]

90-01932 Seatronics Ltd
(a division of Seatronics Ltd, Norfolk UK)
Unit 23, Denmore Industrial Estate, Bridge of Don, Aberdeen AB2 8JW
☎ (+44 1224) 70 48 44
E 101233.3177@compuserve.com
⊖ (+44 1224) 82 26 14
21 (1997)
Managing Director: M A Stephens
Base Manager: Ross MacLeod
Sales Manager: M Tranah
■ Navigation systems, scanning sonars, sidescan sonars (single and duel frequency), profilers, pipeliners, echosounders, seismic and sonar processors, magnetometers, gyrocompasses, current meters, tide gauges, underwater positioning systems, motion compensators, graphic recorders, winches, electronic engineers. → 80 [98]

90-01935 Security Systems International Corp
UK Rep.
BCM Box 646, London WC1N 3XX
☎ (+44 171) 584 57 16 262433 mono (Ref.No 2602)
⊖ (+44 171) 831 94 89 (Ref.No.2602)
Marketing Dir: Joseph C. Sands
■ Electronic security systems for governments or very large industrial projects. → 80
® Switzerland: Security Systems Int'l [98]

90-01937 Senstar (UK) Ltd
5 King Charles Court, Vine Street, Evesham, Worcestershire WR11 4RF
☎ (+44 1386) 76 58 65 ⊖ (+44 1386) 76 57 09
Operations Dir: Martin Kelly
® Canada: Senstar Corp. [98]

90-01945 SFIM (UK)
(† Sarum Systems)
(a subsidiary of SFIM Industries, France)
94 London Road, Amesbury, Wilts SP4 7DY
☎ (+44 1980) 62 41 84 477019
⊖ (+44 1980) 62 41 84
Business Development Director: Derek Riley
® France: SFIM Industries [98]

90-01957 Simrad Optronics Ltd
(a subs. of Simrad Optronics A/S, Norway)
3 Oakwood Industrial Park, Gatwick Road, Crawley, West Sussex RH10 2AZ
☎ (+44 1293) 56 04 13 ⊖ (+44 1293) 56 04 18
♣ £ 6,914,000 (1996) 11 (1997)
Managing Director: Robert A Adams
Marketing Manager: Brian K D Allman
Marketing Manager: Gerry Smart
Marketing Manager: Roger Cross
Office Manager: Ms Tricia Inker
■ Laser rangefinders. Night vision equipment. [98]

90-01959 Singapore Aerospace (UK) Pte Ltd
(a subsidiary of Singapore Technologies Aerospace, Singapore)
Building 6002, Taylors End, Stansted Airport, Stansted, Essex CM24 1RL
☎ (+44 1279) 68 17 70 E bdbc@st.com.sg
⊖ (+44 1279) 68 05 98
Regional Director: Bernard Cheong
® Singapore: Singapore Aerospace [98]

90-01963 Somer Electronics Ltd
169 Junction Road, Burgess Hill, West Sussex RH15 0JW
☎ (+44 1444) 87 10 73 ⊖ (+44 1444) 24 88 72
Managing Director: Ken Somerville
Sales and Marketing Director: Ilmo Pietila
Marketing Manager: Maggie Baldwin

■ Design, sales and marketing of laser rangefinders. Consultancy service advising and assisting other companies on their marketing/technical problems in electro-optics and other defence related fields. → 80 [98]

90-01970 Spar Aerospace (UK) Ltd
Springfield Road, Hayes, Middlesex UB4 0TY
☎ (+44 181) 813 54 44 ⊖ (+44 181) 813 63 30
£ 200 000 ♣ £ 1 600 000 17 5 000 sq ft
Managing Dir: Douglas C. Stuart
Operations Dir: Roy Robinson
Purchasing Mgr: Richard Covill [97]

90-01973 Spectral Dynamics (UK) Ltd
(a subsidiary of Spectral Dynamics Inc, USA)
Fulling Mill, Fulling Mill Lane, Welwyn, Hertfordshire AL6 9NP
☎ (+44 1438) 71 66 26 E spectral@sd-star.co.uk
⊖ (+44 1438) 71 66 28
11
Operations Director: Ian Trafford
Sales and Marketing Director: Philip G Bell
■ Sales and service of dynamic test and analysis products.
® USA: Spectral Dynamics Inc, Wilcoxon Research Inc [98]

90-01975 Sperry Marine Limited
(a subsidiary of Litton Sperry Marine Inc, USA)
Molly Millars Lane, Wokingham, Berkshire RG41 2SH
☎ (+44 1189) 78 94 44
W http://www.sperry-marine.com 848563
⊖ (+44 1189) 77 63 06
30
Managing Director: Adm P D Miller
Sales and Marketing Director: J Nolasco da Cunha
Military Sales Manager: Keith A Mortimer
■ Sales and service support of military systems within an area encompassing the Indian Sub-Continent, the Gulf and Europe Products include ring laser based SINS, heading and attitude systems, Azimuth gyro compasses, navigation radars, doppler sonar speed logs, ESM systems and Mil-Spec integrated bridges.
® USA: Sperry Marine Inc [98]

90-01985 Sumitomo Precision Products
(a division of Sumitomo Precision Products Co Ltd, Japan)
33 Cavendish Square, London W1M 9HF
☎ (+44 171) 409 02 37
E 106001.2562@compuserve.com
⊖ (+44 171) 409 00 42
Yen 10,284,000 (1996) ♣ Yen 33,374,000 (1996)
1,100 (1996)
General Manager: Ikuzo Maeno
■ Supplier of aircraft propellers, landing gear, pneumatics, hydraulics and heat exchangers.
® Japan: Sumitomo Precision Products Co Ltd [98]

90-01986 Sundstrand Aerospace Europe
Kingfisher House, 160-162 High Street, Egham Surrey TW20 9HP
☎ (+44 1784) 43 80 86 ⊖ (+44 1784) 43 80 92
Marketing: Martin Ryan [98]

90-01987 Sun Microsystems Ltd
(a subsidiary of Sun Microsystems Inc, USA)
Farnborough Aerospace Centre, 2 Woodside Road, Farnborough, Hampshire GU14 6XR
☎ (+44 1252) 37 22 72 W http://www.sun.com
859017 ⊖ (+44 1252) 51 10 43
■ Manufacturer of systems for client/server computing: work stations, servers, net working products & related products that use the UNIX operating system.
® USA: Sun Microsystems Inc. [98]

90-02030 Technical Sales International
PO Box 17, Lewes, East Sussex BN7 3LS
☎ (+44 1273) 47 43 66 ⊖ (+44 1273) 47 77 87
♣ £ 1,000,000 5
Director: B.C. Clarke
■ Equipment for mixing explosives & propellants & complete manufacturing plants. [98]

90-02032 Technical Services International
(a subsidiary of TSI Group Inc, USA)
42 Rushington Busines Park, Totton, Hampshire SO40 9LA

☎(+44 1703) 86 12 06 [E] salek@tsi-group.com
⊖ (+44 1703) 86 85 31
♧ US$ 600,000 (1997) ♣ US$ 2,200,000 (1997) 👥 22
🏠 1,000 m2
Managing Director: John Earley
Sales Director: Nick Kantharia
Purchasing Manager: Paul Clifton
■ Technical supply services. Machinery, equipment and supplies for military and governmental use. Specialists in radio telecommunications and electronic/electrical related projects. General or broad range suppliers. [98]

90-02040 Technology for Communications International Ltd (TCI)
(a subs. of TCI, USA)
Sterling House, 27 Hatchlands Road,
Redhill, Surrey RH1 1RH
☎(+44 1737) 76 33 77 ⊖ (+44 1737) 77 90 37
Managing Dir: C.S. Metcalfe
Sales & Marketing Dir: M.T. Richards
■ Communication & broadcasting antennas. Emitter locating & direction finding systems. Spectrum search, signal classification & monitoring systems. ESM & ECM systems.
® USA: TCI Inc. [97]

90-02044 Teledyne Allvac
Unit 18 (Freeport Zone), The Gateway Industrial Estate,
Birmingham B26 3QD
☎(+44 121) 782 98 88 ⊠ 935008 tdy uk g
⊖ (+44 121) 782 98 00
Sales & Marketing Dir: Simon G.S. Marshall
Operations Mgr: Brian Slater
® USA: Teledyne Allvac [97]

90-02060 Thomson Electronic Components Ltd
(a subsidiary of Thomson-CSF, France)
Unit 4 Cartel Business Centre, Stroudley Road,
Basingstoke, Hampshire RG24 8FW
☎(+44 1256) 84 33 23 ⊠ 858121
⊖ (+44 1256) 32 31 72
♧ £ 500,000 (1996) ♣ £ 14,701,000 (1996)
👥 34 (1996) 🏠 750 sq m
Managing Director: S P Waldron
Sales Manager Electron Tubes: R Ferman
Sales Manager Semiconductors: D Taylor
Sales Manager Passive Components: J Marcais
■ Sales outlet for products manufactured by Thomson-CSF in France: professional electron tubes & electronic components for the defence industry including image & transmitting tubes, memories, bipolar, MOS & GaAs integrated circuits, ASICs, discrete semiconductors, capacitors, ferrites, quartz products & microwave components.
® France: Thomson-CSF [98]

90-02073 Transaero UK
The Nook, Duns Tew, Oxford OX6 4JR
☎(+44 1869) 34 71 74 ⊖ (+44 1869) 34 75 23
Regional Mgr: John C. Thornton [97]

90-02100 TRW Systems Overseas Inc
(a subsidiary of TRW Inc, USA)
Third Floor, 16 Upper Grosvenor Street, London W1X 9PB
☎(+44 171) 629 60 09 ⊖ (+44 171) 629 60 79
Managing Director - Europe: Frank R Wiemann
■ Command, control, communications, space satellites.
® USA: TRW Inc [98]

90-02140 Vaisala (UK) Ltd
(a subsidiary of Vaisala Oy, Finland)
Suffolk House, Fordham Road,
Newmarket, Suffolk CB8 7AA
☎(+44 1638) 67 44 00 ⊖ (+44 1638) 67 44 11
👥 14
Managing Director: R E W Pettifer
Marketing Manager: Ms Helen Carson
® Finland: Vaisala Oy [98]

90-02238 Vibro-Meter Ltd
(a subsidiary of Vibro-Meter SA)
Aviation Division
Bramhall Technology Park, Pepper Road, Hazel Grove,
Stockport, Cheshire SK7 5BW
☎(+44 161) 483 08 14 ⊖ (+44 161) 483 28 50
👥 15
Managing Director: Geoff W Thomas
Project Manager: Robert Cant
■ Transducers & associated electronic equipment for measurement of physical parameters e.g. vibration monitoring, ice detectors, speed probes, helicopter rotor trim & balance.
® Switzerland: Vibro-Meter SA [98]

90-02260 Voith Engineering Ltd
(a subsidiary of Voith Turbo GmbH & Co KG, Germany)
6 Beddington Farm Road, Croydon, Surrey CR0 4XB
☎(+44 181) 667 03 33 ⊠ VOITH Croydon
⊖ (+44 181) 667 04 03
👥 70 🏠 3,000 sq m
Managing Director: M Lornie
General Manager Hydrostatics: David Gilbert
Manager, Rail Products: Alan Morris
General Mgr Stat. Drives: David Rees
Public Relations and Marketing Executive: Ms V Binoeder
■ Suppliers of Voith hydraulic internal gear pumps for tank recovery vehicles, hydrodynamic transmissions, retarders and fans; Hawe hydraulic valves for battlefield pallet controls; hydraulic systems for naval sonar controls; torpedo test stands and equipment. Overhaul and repair of above in own workshop.
® Austria: Elin Energieanwendung GmbH
Germany: Appleton Pohl GmbH & Co., Arthur Maurer GmbH, FSE GmbH, Heinrich Stamm GmbH, Martin Lohse GmbH, Schorch, Voith Group, Zeihl Abegg
Sweden: Voith Safeset AB
Switzerland: Trumag [98]

90-02280 Volda Bamford Ltd
Ajax Works, Whitehill, Stockport, Cheshire SK4 1NT
☎(+44 161) 480 04 43 ⊠ 665432
⊖ (+44 161) 480 76 93
👥 4
Managing Director: Mrs K Smith
Operations and Technical Director: A W Smith
Sales Mgr: D.J. Love
■ Sales, marketing & service of Bamford fixed pitch propulsion system, Volda C.P. propulsion systems & engineering & service of Volda/Bamford combined C.P. propulsion systems. [98]

United States of America

Head of State:
William Jefferson Clinton

10 Head of State

10-10000 Executive Office of the President
White House (US Presidency)
1600 Pennyslvania Avenue NW, Washington, DC 20500
☎(+1 202) 456 14 14 ⊠ 440074
President: William Jefferson Clinton
Vice President: Albert Gore
Chief of Staff: Erskine Bowles
Deputy Chief of Staff: John Podesta
National Security Adviser: Sandy Berger
Assistant Secretary of State (Political): Thomas McNamara
National Security Adviser: Anthony Lake
Asst to the Pesident & Dir Office of Communications: George Stephanopoulos
Deputy Director Office of Communications: Rick Seidman
Director White House Military Office: John A. Gaughan
Deputy Assistant to the President & Press Secretary: Arthur Jones
Press Secretary: Dee Dee Myers [98]

12 Ministries

12-10000 Department of Defense
(Office of the Secretary of Defense)
The Pentagon, Washington DC 20301-3110
☎(+1 703) 545 6700
Secretary of Defense: William Cohen
☎(+1 703) 695 52 61
Deputy Defense Secretary: John White
☎(+1 703) 695 63 52
Under Sec. of Defense (Acquisition)
Under Secretary of Defense Acquisition and Technology
3010 Defense Pentagon, Washington, DC 20301-3010
Under Secretary for Defense Acquisition & Technology: Paul G. Kaminski ☎(+1 202) 695 2381
Under Secretary of Defense (Policy)
2000 Defense Pentagon, Washington, DC 20301-2000
Under Secretary of Defense (Policy): Walter B Slocombe
☎(+1 703) 697 7200
Asst Sec. (Personnel & Readiness)
Under Secretary of Defense (Personnel and Readiness)
The Pentagon, Washington, DC 20301-3110
Under Secretary of Defense: Edwin Dorn [98]

12-20000 Department of State
2201 C Street NW, Washington, DC 20520
☎(+1 202) 647 5291 ⊠ 440080 secstate
Secretary of State: Mrs Madeleine Albright [98]

12-60200 United States Customs Service
1301 Constitution Avenue NW, Washington, DC 20229
☎(+1 202) 927 55 80
[W] http://www.customs.ustreas.gov
Commissioner of Customs: George J Weise
Deputy Commissioner: Samuel H Banks
Chief of Staff: Holm Kappler
Assistant Commissioner Congressional and Public Affairs: Jose Padilla
Assistant Commissioner International Affairs: Douglas M Browning [98]

20 Defense Administrations/Services/Agencies

20-01000 Defense Advanced Research Projects Agency (DARPA)
3701 North Fairfax Drive, Arlington, VA 22203-1714
☎(+1 703) 696 00 90 ⊖ (+1 703) 696 97 81
Director: V Larry Lynn ☎(+1 703) 696 24 00
Deputy Director and Director of Joint Dual-Use Program: Dr H Lee Buchanan (III) ☎(+1 703) 696 24 02
Deputy Director for Management: Ron H Register
☎(+1 703) 696 24 03
General Counsel: Richard Dunn ☎(+1 703) 696 24 07
Comptroller: Greg Colocotronis ☎(+1 703) 696 23 90
Administration and Small Business Director: Dr Bud Durand ☎(+1 703) 696 23 79
Contract Management Director: R Timothy Arnold
☎(+1 703) 696 23 81
Defense Sciences Director: Dr Lawrence H Dubois
☎(+1 703) 696 22 83
Electronics Technology Director: Dr Kaigham J Gabriel
☎(+1 703) 696 22 52
Information Systems Director: Thomas Swartz
☎(+1 703) 696 74 38
Information Technology Director: Dr Howard Frank
☎(+1 703) 696 22 28
Sensor Technology Director: Dr Matthew Ganz
☎(+ 703) 516 74 13
Tactical Technology Director: Dr David A Whelan
☎(+1 703) 696 23 07
DARPA/DISA Joint Programme Director: Don C Eddington
☎(+1 703) 284 88 90
HAE UAV Program Director: Charles Heber
☎(+1 703) 524 51 99
→ 60 [98]

20-03220 Center for Defense Information
1500 Massachusetts Ave N.W., Washington, DC 20005
☎(+1 202) 862 0700 [W] http://www.cdi.org
⊠ 904059 wsh (CDI) ⊖ (+1 202) 862 0708
👥 25
Director: VAdm. John J. Shanahan (USN retd)

Director of Research: David Johnson
Research Librarian: Jeffrey Mason
■ Provide information on military issues to Congress, the media and the public. [97]

20-04350 Defense Intelligence Agency (DIA)

The Pentagon, Washington, DC 20301
☎ (+1 703) 695 73 53 ⊖ (+1 703) 695 36 87
Director: Lieut Gen Patrick Hughes
Deputy Director: Michael F Munson
☎ (+1 703) 697 51 28
Chief of Staff: John T. Berbrich ☎ (+1 703) 695 7646
Public Affairs Officer: Martin Culp [97]

20-04360 Defense Logistics Agency Headquarters (DLA)

8725 John J. Kingman Road, Ste 2533,
Fort Belvoir, VA 22060-6221
☎ (+1 703) 767 7100 (civilian locator)
Director: VAdm. Edward M. Straw (SC, USN)
Principal Deputy Dir: R.E. McCoy ☎ (+1 703) 767 5222
Dpty Dir Aquisition Commander Defense Contract Mgmt Command: Maj.Gen. R.W. Drewes (USAF)
☎ (+1 703) 767 2403
Deputy Dir Material Management: Maj.Gen. W.P. Hallin (USAF) ☎ (+1 703) 767 1600
Comptroller: RAdm. J.D. McCarthy (SC, USN)
☎ (+1 703) 767 7300
General Counsel: B. Baird ☎ (+1 703) 767 6060
Deputy Dir Corporate Administration: A.C. Ressler
☎ (+1 703) 767 5208
Exec.Dir Contract Management: R.P. Scott
Exec.Dir Procurement: R.L. Molino ☎ (+1 703) 767 1400
Exec.Dir Supply Management: J. Jones
☎ (+1 703) 767 2600
Deputy Exec.Dir Distribution: F. Baillie
☎ (+1 703) 767 3600
Exec.Dir Strategic Programming & Contingency Ops: Ms C. Gallo
Exec.Dir Human Resources: Mrs J.C. Smith
☎ (+1 703) 767 6445
Exec.Dir/Chief Information Officer: T.J. Knapp
☎ (+1 703) 767 3100
Staff Dir Business Management: L. Reno
☎ (+1 703) 767 5305
Administrator DLA Support Center: R.A. Martínez
☎ (+1 703) 767 5550
■ The DLA is a combat support agency. It provides supplies to the military services & supports service acquisition of weapons & other materials. It is involved with joint planning with the services for parts for new weapons systems, through to production & disposal of material that is obsolete or no longer required.
The Agency also provides supply support, contract administration services & technical & logistics services to all branches of the military.

Defense Electronics Supply Center
1507 Wilmington Pike, Dayton, OH 45444-5000
☎ (+1 513) 296 6111
Commander: Col J.W. House (USA)
☎ (+1 513) 296 6841
Deputy Commander: Capt. D.R. Haven (USN)
☎ (+1 513) 296 6145

Defense Industrial Supply Center (DISC)
700 Robbins Ave, Philadelphia, PA 19111-5096
Commander: Brig.Gen. G.B. Higginbotham (USMC)
Deputy Commander: N. Ranalli

Defense General Supply Center
Richmond, VA 23297-5000
Commander: Capt. M.E. Finley (SC, USN)
☎ (+1 804) 279 3801
Deputy Commander: F. Lotts ☎ (+1 804) 279 3803

Defense Fuel Supply Center (DFSC)
8725 John J.Kingman Rd, Suite 2941,
Ft Belvoir, VA 22060-6222
☎ (+1 703) 427 9700
Commander: Brig.Gen. L. Wilson (USAF)
Deputy Commander: Capt. F.C. Chitty (USN)
■ Contracting activity for fuel delivery & related services

Defense Personnel Support Center
2800 S.20th St, Philadelphia, PA 19101-8419
Commander: Brig.Gen. C Freeman (USN)
☎ (+1 215) 737 2300
Deputy Commander: G. Allen ☎ (+1 215) 737 2308

Defense Construction Supply Center
PO Box 3990, Columbus, OH 43216-5000
Commander: RAdm. E.A. Elliot (USN)
Deputy Commander: Ms M. Barnett
☎ (+1 614) 692 2167

Defense Distribution Region East
New Cumberland, PA 17070-5001
Commander: Col J.L. Vick (USAF)

Defense Distribution Region West
Stockton, CA 95296-0002
Commander: Capt. M. Casey (USN)
☎ (+1 209) 982 2001
Deputy Commander: Col G. Melton (USA)
☎ (+1 209) 982 9001

Defense Reutilzation and Marketing Serivce
74 Washington Avenue North,
Battle Creek, MI 49017-3092
Commander: Capt D.A. Hempson (USN)
☎ (+1 616) 961 5972
Deputy Commander: Col S. Hockett (USAF)
☎ (+1 616) 961 5978

Defense National Stockpile Center
8725 John J.Klingman Road, Ste 3339,
Ft Belvoir, VA 22060-6223
Administrator: R.J. Connelly
☎ (+1 703) 767 5500 ext 4528
Deputy Administrator: R.M. O'Brien
☎ (+1 703) 767 5500 ext 4528

Defense Logistics Services Center (DLSC)
Federal Center, Battle Creek,
74 Washington Avenue North, MI 49017-3084
☎ (+1 616) 961 4286 ⊖ (+1 616) 961 4265
Commander: Col R. Haglund (USMC)
☎ (+1 616) 961 4989
Deputy: W.Y.T. Takasaki ☎ (+1 616) 961 4990 [97]

20-04370 Defense Mapping Agency (DMA)

8613 Lee Highway, Fairfax, VA 22031-2137
✆ (800) 455 08 99 (NIMA help desk)
⊖ (+1 703) 275 85 61
Acting Director: Rear Adm Joseph J J Dantone
☎ (+1 703) 275 59 00
Deputy Director for Operations: Leo Hazelwood
☎ (+1 703) 275 85 08
Deputy Director for Systems and Technology: William Mularie ☎ (+1 703) 275 59 10
Deputy Director for Corporate Affairs: W Douglas Smith
☎ (+1 703) 275 59 05
General Counsel: Ed Obloy ☎ (+1 703) 275 85 86
Inspector General: Col Leonard Maggio (USAF)
☎ (+1 703) 275 84 36
Commercial Office: Clay Ancell (acting)
☎ (+1 703) 275 85 14
Office of Congressional and Public Liaison: Terence Meehan (Acting) ☎ (+1 703) 275 84 09
■ The agency is responsible for providing global geospatial information and services to support and advance national security objectives.

Installation Management Group
4600 Sangamore Road, Bethesda, MD 20816-5003
☎ (+1 301) 227 20 98

Acquisition & Technology Group
4600 Sangamore Road, Bethesda, MD 20816-5003
☎ (+1 301) 227 20 93

Operations Group
12310 Sunrise Valley Drive, Reston, VA 22091-3414
☎ (+1 703) 264 73 00

DMA Hydrographic/Topographic Center
4600 Sangamore Road, Bethesda, MD 20816-5003
⊖ (+1 301) 227 46 75
Director: William N Hogan ☎ (+1 301) 227 20 00

DMA Combat Support Center
6001 MacArthur Boulevard,
Washington, DC 20816-5001
⊖ (+1 703) 227 33 65
Director: Col Robert F Kirby (US Army)
☎ (+1 301) 227 31 11

DMA Systems Center
4600 Sangamore Road, Bethesda, MD 20816
⊖ (+1 301) 227 15 56
Director: Earl Phillips ☎ (+1 301) 227 20 93

DMA Reston Center
12310 Sunrise Valley Drive, Reston, VA 22091-3414
⊖ (+1 703) 264 34 05
Director: Paul L Peeler (Jr) ☎ (+1 703) 264 73 00

DMA Aerospace Center
3200 South Second Street, St Louis, MO
⊖ (+1 314) 263 45 79
Director: William J Brown ☎ (+1 314) 263 43 07 [98]

20-04372 Armed Forces Institute of Pathology

(Walter Reed Army Medical Center)
6825 16th Street NW, Bldg 54,
Washington, DC 20306-6000
☎ (+1 202) 782 2100 ⊖ (+1 202) 782 7164
Director: Col M.J. Dickerson (USA) ☎ (+1 202) 782 2111
Deputy Dir: Col J.C. Durham (US Army)
☎ (+1 202) 782 2104
Deputy Dir: Capt. Glenn Wagner (USN)
☎ (+1 202) 782 2103
Public Affairs Officer: C. Kelly ☎ (+1 202) 782 2115
■ One of the world's leading pathology institutes, AFIP is involved with consultation, education & research for civilian & military pathology. → 60 [97]

20-04383 Defense Reutilization & Marketing Office - Tucson

(Defense Logistics Agency)
Davis-Monthan AFB, Tucson, AZ 85707-4373
☎ (+1 520) 228 88 19 Ⓦ http://www.drms.dla.mil
⊖ (+1 520) 228 81 01
Manager: H Ross
■ The DRMO disposes of surplus Government (DoD) property. [98]

20-04500 Defence Systems Management College

9820 Belvoir Road, Fort Belvoir, VA 22060-5565
☎ (+1 703) 805 3360 ⊖ (+1 703) 805 2639
Commander: Brig.Gen. Richard Black (USA)
☎ (+1 703) 805 33 60
Deputy Commander: Edward Hirsch
☎ (+1 703) 805 21 55
Operations Director: Col Charles W Westrip (USA)
☎ (+1 703) 805 21 40
Managing Editor: Collie Johnson ☎ (+1 703) 805 28 92 [97]

20-11000 Inter-American Defense Board

2600 16th Street NW, Washington, DC 20441
☎ (+1 202) 939 6600 ⊖ (+1 202) 939 6620
Chairman: Maj.Gen. James R. Harding (US Army)
Secretary IADB: Col Ronald J. Cruz (USMC) [97]

20-12000 Federal Emergency Management Agency (FEMA)

Room 806, Federal Center Plaza, 500 C Street SW,
Washington, DC 20472
☎ (+1 202) 646 4600 Ⓔ eipa@fema.gov
⊖ (+1 202) 646 4086
Director: James Lee Witt
Chief of Staff: Ms Jane Bullock
Operations Dir: Bruce Baughman
Sr Procurement Officer: Robert Boyer
Deputy Public Affairs Dir: Phillip S. Cogan [97]

20-15000 National Aeronautics & Space Administration (NASA)

300 "E" Street SW, Washington, DC 20546
☎ (+1 202) 358 16 00
⊖ (+1 202) 358 42 10 (Public Affairs)
Administrator: Daniel S. Goldin
Deputy Administrator (Acting): John R. Dailey
Inspector General: Ms Roberta Gross
Comptroller: Malcolm Peterson
General Counsel: Edward A. Frankle
Associate Administrator for External Affairs: John Schumacher
Associate Administrator for Procurement: Ms Deidre Lee
Associate Administrator for Public Affairs: Peggy Wilhide
Associate Administrator for Safety & Mission Assurance: Fred Gregory
Associate Administrator for Aeronautics: Dr Robert E. Whitehead
Associate Administrator for Space Flight: Wilbur C. Trafton
Associate Administrator for Space Science: Wesley Huntress
Sr Press & Information Officer: Brian Welch

NASA Inspector General
NASA HQ, 300 "E" Street, Washington, DC 20546
☎ (+1 202) 358 12 20
Inspector General: Ms Roberta Gross

Ames Research Center (ARC)
Moffett Field, CA 94035
☎ (+1 415) 694 50 00
⊖ (+1 415) 604 39 53 (Public Affairs)
Director: Dr Henry MacDonald

Contact Reference

Dir NASA Office of Inspector General: David L. Gandrud
Aerospace Systems Dir: Thomas Snyder

Dryden Flight Research Facility (DFRF)
PO Box 273, Edward AFB, CA 93523
☎(+1 805) 258 33 11 ⊖ (+1 805) 258 35 66
Manager: Martin A. Knutson
Public Affairs Officer: Cam Martin

Goddard Space Flight Center (GSFC)
Greenbelt, MD 20771
☎(+1 301) 286 20 00
⊖ (+1 301) 286 81 42 (Public Affairs)

Goddard Institute for Space Studies (GISS)
2880 Broadway, New York, NY 10025
☎(+1 212) 678 55 00
⊖ (+1 212) 678 55 52 (Public Affairs)
Director: Dr James E. Hansen

John F Kennedy Space Center (KSC)
Kennedy Space Center, FL 32899
☎(+1 407) 867 71 10 ⊖ (+1 407) 867 26 92
Director: Roy Bridges
Small Business Specialist: Norman Perry

Lyndon B Johnson Space Center (JSC)
Houston, TX 77058
☎(+1 713) 483 31 11
⊖ (+1 713) 483 20 00 (Public Affairs)
Director: George W.S. Abbey
Deputy Director: James Wethenbee
Public Affairs Director: Douglas K Ward

Langley Research Center (LaRC)
Mail Stop 115, Hampton, VA 23665-5225
☎(+1 804) 864 10 00
⊖ (+1 804) 864 63 33 (Public Affairs)
Director: Jeremiah Creedon
Office of External Affairs Dir: Gary A. Price

Lewis Research Center (LeRC)
2100 Brookpark Road, Cleveland, OH 44135
☎(+1 216) 433 40 00
⊖ (+1 216) 433 81 43 (Public Affairs)
Director NASA Office of Inspector General: Ralph D. Rhodes
Director: Don Campbell
Public Relations Officer: John Hairston

Marshall Space Flight Center (MSFC)
Huntsville, AL 35821
☎(+1 205) 544 21 21
⊖ (+1 205) 544 58 52 (Public Affairs)
Director: Dr J. Wayne Littles
Public Affairs: John Taylor
Payload Projects Office Mgr: Harry G. Craft (Jr)
Mission Mgr: Anthony O'Neil
Mission Scientist: Dr Marsha Torr
Chief Engineer: Robert Beaman
Science & Engineering Dir: Dr George McDonough
Shuttle Project Office & Acting Space Shuttle Main Engine Project Mgr: Alexander A. McCool
Redesigned Solid Rocket Motor Proj. Mgr: Victor Keith Henson
External Project Mgr: Gerald C. Ladner

Michoud Assembly Facility
PO Box 29300, New Orleans, LA 70189
☎(+1 504) 255 33 11 ⊖ (+1 504) 255 21 52
Mgr & Public Affairs Liason: Dr Mathias P. Siebel

Stennis Space Centre (SSC)
(† National Space Technology (NSTL)) Laboratories
Stennis Space Centre Station, Bay St, St Louis, MS 39529
☎(+1 601) 688 22 11
⊖ (+1 601) 688 10 94 (Public Affairs)
Manager: Roy Estess
Public Affairs Officer: Ms Myron Webb

Wallops Flight Facility
Wallops Island, VA 23337
⊖ (+1 804) 824 19 71
Public Affairs Specialist: Keith Koehler
☎(+1 804) 824 1579

White Sands Test Facility
PO Drawer MM, Las Cruces, NM 88004
☎(+1 505) 524 50 11
Manager: Rob R. Tillett [98]

20-18000 Naval Criminal Investigative Service

Washington Navy Yard, Bldg #111,
Washington, DC 20388-5380
Director: SA Roy D. Nedrow ☎(+1 202) 433 8800
Asst Dir for Investigations: SA John F. McEleny
☎(+1 202) 433 8800
Asst Dir for Counter Intelligence: SA David L. Brant
☎(+1 202) 433 9578
Comptroller: Ken Burns ☎(+1 202) 433 9620
Asst Dir for Government Liaison & Public Affairs: SA Robert J. Finan (II) ☎(+1 202) 433 6790
Asst Dir for Inspection & Planning: SA Thomas Fischer
☎(+1 202) 433 8830 [97]

20-20000 Strategic Defense Initiative Organisation

The Pentagon, Washington, DC 20301-7100
☎(+1 703) 695 87 43 ⊖ (+1 703) 697 40 27
Director: Amb Henry F Cooper
Deputy Director: Maj Gen O'Neill
Director International and External Programmes: Dr Martin
Deputy for Strategic Defense: Dr Carlson
Director Contracts: Col Stewart [98]

20-25000 US Air Force Institute of Technology

2950 "P" Street, Wright Patterson AFB,
Dayton, OH 45433-7765
☎(+1 937) 255 22 16 ⊖ (+1 937) 255 21 35
Commandant: Col John H Russell
AFIT/PA: Maj Roger Overturf
NCOIC, AFIT/PA: MSgt Jay E Harris
■ Graduate & professional education & research. → 60 [98]

25 Defense Forces

25-10000 Joint Chiefs of Staff

9999 Joint Staff Pentagon, Washington, DC 20301
Chairman: Gen. John M Shalikashvili (US Army)
☎(+1 703) 697 9121
Vice-Chairman: Gen Joseph W Ralston (USAF)
☎(+1 703) 614 8948 [98]

25-12300 United States Strategic Command HQ

(† Strategic Air Command HQ)
Offutt AFB, Omaha, NE 68113
☎(+1 402) 294 41 30 ⊖ (+1 402) 294 48 92
Commander-in-Chief: Gen Eugene E Habiger
Deputy Commander-in-Chief: VAdm Dennis A Jones
Deputy Chief of Staff: Col Marion E Callender (Jr)
Director Manpower and Personnel (J1): Col C M (Mike) Robards
Director Intelligence (J2): Brig. Gen. Tiiu Kera
Director Operations and Logistics (J3/4): RAdm Richard A Buchanan
Deputy Director Operations and Logistics (J3/4): Brig. Gen. Timothy J McMahon
Director Plans and Policy (J5): Maj.Gen. Ronald C Marcotte
Director C42 (J6): Brig.Gen. Robert F Behier
Chief Public Affairs (J020): Capt. Robert S Pritchard [98]

25-12500 US Special Operations Command (SOCOM)

7701 Tampa Point Boulevard,
MacDill AFB, FL 33621-5323
☎(+1 813) 840 51 00 ⊖ (+1 813) 828 40 35
47,000 (active, reserve and National Guard)
Commander in Chief: Gen Henry H Shelton (USA)
☎(+1 813) 840 51 03
Deputy Commander in Chief and Chief of Staff: Rear Adm Raymond Smith (USN) (Jr)
☎(+1 813) 840 51 10
Operations Director (J3): Maj Gen Maxwell C Bailey (USAF) ☎(+1 813) 828 43 33
J5 Plans, Policy and Strategic Assessments Director: Brig Gen Douglas Brown
Washington Operations Director: Col Albert R Brownfield (USA) ☎(+1 703) 697 26 02
Senior Enlisted Advisor: CMSgt Michael I Lampe (USAF)
Press and Public Affairs Officer: Col Barry E Willey (USA)
☎(+1 813) 828 46 00
Media Relations Officer: George Grimes
Chief Community Relations: Chet Justice
☎(+1 813) 828 45 57; 828 40 35 (fax) [98]

25-12550 Special Operations Acquisition Center (SOAC)

(US Special Operations Command)
MacDill AFB, FL 33621-5323
☎(+1 813) 828 94 82 [E] bowlers@soac.hqscom.mil
Acquisition Executive: Gary L Smith
■ The Centre serves as the centralised management agency for the research, development and acquisition of special operations-peculiar equipment, supplies and services. [98]

25-12600 US Transportation Command

Scott AFB, IL 62225-5357
☎(+1 618) 256 48 28 ⊖ (+1 618) 256 87 76
Commander in Chief: Gen. Walter Kross (USAF)
Deputy Commander in Chief: Lt Gen. Hubert G Smith (USA)
Operations and Logistics Director: Maj.Gen. William J Begert (USAF)
Plans and Policy Director: RAdm. Richard Naughton (USN)
Director Command, Control, Communication & Computer Systs: Brig.Gen. Stephen E Kelley (USAF)
Public Affairs Officer: Capt. Alan Dooley (USN)
☎(+1 618) 256 48 28
Deputy Public Affairs Officer: Maj. Steve Christy (USA)
☎(+1 618) 256 61 68 [98]

25-15000 Headquarters US Pacific Fleet

Pearl Harbor 96860-7000
☎(+1 808) 471 37 69 ⊖ (+1 808) 422 07 71
Commander-in-Chief, US Pacific Fleet: Adm. Archie Clemins ☎(+1 808) 471 97 27
Deputy Commander in Chief and Chief of Staff: RAdm. Tim Beard ☎(+1 808) 471 98 00
Deputy Chief of Staff for Operations, Plans and Communications: RAdm. David Polatty
☎(+1 808) 471 06 12
Assistant Chief of Staff Fleet Supply: RAdm. Leonard Vincent ☎(+1 808) 471 39 83
Deputy Chief of Staff for Fleet Maintenance Officer: RAdm. James Taylor ☎(+1 808) 471 04 01
Deputy Chief of Staff for Shore Installation MGMT: RAdm. A T (Tom) Church ☎(+1 808) 471 49 42
Deputy Chief of Staff for Training (COMTRAPAC): Capt. Robert Danberg (Acting) ☎(+1 619) 524 14 52
Deputy Chief of Staff Resources, Requirements and Assessments: RAdm. Thomas Elliott
☎(+1 808) 471 98 27
Fleet Civil Engineer: RAdm. Thomas Dames
☎(+1 808) 474 11 41
Public Affairs Officer: Capt. Keith Arterburn
☎(+1 808) 471 37 69
Deputy Fleet Public Affairs Officer: Cdr Kevin Wensing
☎(+1 808) 471 37 69
Commander Naval Air Force: VAdm. Brent M Bennitt
☎(+1 619) 545 11 37
Commander Naval Surface Force: VAdm. Alexander J Krekich ☎(+1 619) 437 27 35
Commander Submarine Force: RAdm. W G Ellis
☎(+1 808) 471 09 11
Commander 3rd Fleet: VAdm. H Browne
Commander 7th Fleet: VAdm. R Natter [98]

25-19100 National Defense University

Building 62, 300 5th Avenue, Fort McNair,
Washington, DC 20319-5066
☎(+1 202) 287 39 28; 685 39 38 (Public Affairs)
⊖ (+1 202) 287 39 35
President: Lieut Gen Richard A Chilcoar
Vice President: Ambassador William G Walker
☎(+1 202) 475 08 12

Industrial College of the Armed Forces
Building 59, 408 4th Avenue, Fort McNair,
Washington, DC 20319-5062
☎(+1 202) 685 43 33

National War College
Building 61, 300 D Street, Fort McNair,
Washington, DC 20319-4344
☎(+1 202) 685 43 44

Armed Forces Staff College
Normandy Hall, 7800 Hampton Boulevard,
Norfolk, VA 23511-6097
☎(+1 804) 444 56 76

Information Resources Management College
Building 62, 300 5th Avenue, Fort McNair,
Washington, DC
☎(+1 202) 685 38 37
Dean: John M. Carabello

Institute for National Strategic Studies
Building 62, 300 5th Avenue, Fort McNair,
Washington, DC 20319-5066
☎(+1 202) 685 38 37 [98]

25-19600 Joint Military Intelligence College
(† Defense Intelligence College)
Washington, DC 20340-5100
☎(+1 202) 231 33 44 ⊜ (+1 202) 231 49 77
President: A. Denis Clift
Deputy: Col Lewis Wallace ☎(+1 202) 231 33 51
Provost: Dr Ronald Garst ☎(+1 202) 231 3322
Academic Dean: Dr Max L. Gross ☎(+1 202) 231 3335
Resource Dean: John R. Robinson ☎(+1 202) 231 3373
98

25-19900 John C Stennis Space Center
Hancock County, MS 39529-6000
☎(+1 601) 688 33 41 ⊜ (+1 601) 688 10 94
Centre Director: Roy S Estess
Special Assistant to the Director: Jon Roth
Deputy Director: Mark K Craig
Procurement Officer: Ms Kim Stone
Public Affairs Officer: Ms Myron Webb
Public Affairs Secretary: Ms Melanie B DiAntonio
■ NASA's center for large propulsion systems testing. All main engines which power the Space Shuttle into orbit are tested at Stennis Space Center prior to flight. At Stennis, NASA is the host agency to 18 other federal and state agencies including the US Navy, the Environmental Protection Agency, the National Marine Fisheries Service, the National Data Buoy Center and others. 98

25-19920 Joint Services Survival, Evasion, Resistance and Escape Agency
Fort Belvoir, VA
☎(+1 703) 806 54 18; 806 64 19 (training division); 806 66 07 E jssa@ix.netcom.com
⊜ (+1 703) 806 62 75
👥 105
Commander: Col John C Chapman
Deputy Commander: Robert Dussault
■ Develop area SERE contingency guides, evasion charts, blood chits, training programs and films and aids for use by joint commands in regional and counterdrug operations. Operating locations in Virginia, Washington and Florida. 98

25-21200 Air Mobility Command (AMC)
Scott Air Force Base, IL 62225-5001
☎(+1 618) 256 1110
👥 118 718 (53 825 active duty)
Commander: Gen. Walter Kross ☎(+1 618) 256 3205
Vice Commander: Lt Gen. Charles T. Robertson (Jr)
☎(+1 618) 256 3293
Logistics Dir: Brig.Gen. Michael A. Moffitt
☎(+1 618) 256 3300
Commander Tanker Airlift Control Centre:
Brig.Gen. William Welser (III) ☎(+1 618) 256 8209
Operations Dir: Maj.Gen. Edward F. Grillo (Jr)
☎(+1 618) 256 3315
Plans Dir: Maj.Gen. Walter S. Hogle (Jr)
☎(+1 618) 256 3311

Direct Reporting Unit of AMC
Air Mobility Warfare Center, Bldg 5656, Texas Avenue, Ft Dix, NJ 08641

HQ AMC/PA
503 J Street, Room 214, Scott AFB, IL 62225-5335
☎(+1 618) 256 4502 ⊜ (+1 618) 256 5592
HQ AMC/PAM: Dave Patterson ☎(+1 618) 256 5003
E pattersd@hqamc.sarb.af.mil

15th Air Force
Travis AFB, CA 94535-2171
☎(+1 714) 655 11 10
■ 15th Air Force units
43rd Air Refueling Group (KC-135), Malmstrom AFB, MT
22 Air Refueling Wing (KC-135), McConnell AFB, KS
60th Air Mobility Wing (C-5, C-41, KC-10), Travis AFB, CA
62nd Airlift Wing (C-14), McChord AFB, WA
92nd Air Refueling Wing (KC-135), Fairchild AFB, WA
375th Airlift Wing (C-9, C-21) Scott AFB, WA
615th Air Mobility Operations Group, Travis AFB, CA
615th Air Mobility Support Group, Hickham AFB

21st Air Force
McGuire AFB, NJ 08641-5002
☎(+1 609) 724 2104 ⊜ (+1 609) 724 6999
■ 21st Air Force units
19th Air Refueling Wing (KC-135), Robins AFB, GA
89th Airlift Wing (C-9, C-21, VC-25, C-135, C-137, UH-1), Andrews AFB, MI
319th Air Refueling Wing (KC-135), Grand Forks AFB, ND
436th Airlift Wing (C-5), Dover AFB, DE
437th Airlift Wing (C-17, C-141) Charleston AFB, SC
621st Air Mobility Operations Group, McGuire AFB, NJ
621st Air Mobility Support Group, Ramstein AB, GE
624th Air Mobility Support Group, Pope AFB, NC
640th Air Mobility Support Squadron, Howard AFB 97

25-21400 Air Intelligence Agency
(† Air Force Intelligence Command)
Kelly AFB, San Antonio, TX 78243-5000
☎(+1 512) 977 2166 ⊜ (+1 512) 977 4948
👥 16 000 (Worldwide)
Commanding General: Maj.Gen. Michael V. Hayden
Vice Commander: Brig.Gen. Jeffrey S. Pilkington
Public Affairs Dir: Lt Col Martha J. Cenkci
☎(+1 512) 977 2166
Operations Dir: Lt Col Neal T. Robinson
Logistics Dir: Col Phillip L. Jackson
Deputy Director of Public Affairs: Capt. Amber L. Cargile
■ Deliver flexible collection, tailored air & space intelligence, weapons monitoring & information warfare products & services.

Major Subordinate Units::
67th Intelligence Wing Kelly Air Force Base, TX
Air Force Informaton Warfare Center Kelly Air Force Base, TX
National Air Intelligence Center Wright-Patterson Air Force Base, OH
544th Intelligence Group Peterson Air Force Base, CO
497th Intelligence Group Fort Meade, MD 97

25-21500 Air Force Space Command (HQ AFSPC)
Peterson AFB, Colorado Springs, CO 80914-5000
☎(+1 719) 554 37 31 ⊜ (+1 719) 554 60 13
👥 26,681
Commander: Gen Howell M Estes
Vice Commander: Lt Gen. Patrick Caruana
Comptroller: Brig.Gen. Cynthia A. Deese
Operations: Maj.Gen. Donald G. Cook
Personnel: Col Raymond G. Chapman
Information Management Dir: Col Robert F. Koenig
Communication Computer Systems:
Maj.Gen. William J. Donahue
Logistics: Col Jerry M. Drennan
Plans: Maj.Gen. Lance W. Lord
Requirements: Brig.Gen. H. Marshall Ward

Responsibilities:
Supports terrestrial forces & civil & commercial space activities.

Major Units:
21st Space Wing - Ballistic Missile Warning, Peterson AFB.
30th Space Wing - DoD, civil & commercial spacelift, Vandenburg AFB.
45th Space Wing - DoD, civil & commercial spacelift, Patrick AFB & Cape Canaveral.
50th Space Wing - Satellite Operations, Falcon AFB.
90th Missile Wing - Peacekeeper & Minuteman III, FE Warren AFB
91st Missile Wing - Minuteman III, Minot AFB.
321st Missile Group - Minute III, Grand Forks AFB.
341st Missile Wing - Minuteman III, Malmstrom AFB.
721st Space Group - Mission Support, Cheyenne Mountain AS.
750th Space Group - Satellite Control, Onizuka AS. 97

25-21600 Air Force Special Operations Command (AFSOC)
100 Bartley Street, Hurlburt Field, FL 32544-5273
☎(+1 850) 884 55 15; 23 23 ⊜ (+1 850) 884 72 49
👥 12,187 (9,244 active duty)
Commander: Maj Gen Charles R Holland
Public Affairs Officer: Ms Shirley A Sikes
E sikess@hqfsoc.hurlburt.af.mil
■ Serves as the Air Force component of US Special Operations Command.
Provides unconventional warfare, direct action, special reconnaisance, counterterrorism and foreign internal defence support to the unified commands.
Also operates units from RAF Mildenhall, UK and Kadena AB, Japan.

16th Special Operations Wing
Hurlburt Field, FL 32544-5000
☎(+1 850) 884 74 64

720th Special Tactics Group
Hurlburt Field, FL 32544-5000
☎(+1 904) 884 22 88
■ Also: 352nd Special Operations Group at RAF Alconbury, UK; 353rd Special Operations Group at Kadena AB, Japan.

USAF Special Operations School
Hurlburt Field, FL 32544-5000
☎(+1 904) 884 62 22

18th Flight Test Squadron
Hurlburt Field, FL 32544-5000
☎(+1 904) 884 69 24

AFSOC Air Support Operations Squadron
Fort Bragg, NC
☎(+1 910) 396 02 30 98

25-21700 US Air Force Air Education & Training Command HQ (AETC)
100 H Street, Suite 3, Randolph AFB, TX 78150-4330
☎(+1 210) 652 11 10 (operator assistance)
Commander AETC: Gen Lloyd W Newton
☎(+1 210) 652 55 12
Vice Commander: Lieut Gen John C Griffith
Director Comptroller: Brig Gen Rodney W Wood
Director of Logistics: Col Sharla J Cook
Civil Engineer: Col David M Cannan
Director of Personnel: Col Michael J Duffee
Commander Recruiting Service:
Brig Gen Walter E Buchanan (III)
Director of Medical Services and Training:
Col Jacqueline Morgan
Director of Plans and Programs: Brig Gen Karen S Rankin
Director of Operations: Maj Gen Charles H Coolidge
Commander AFSAT (Air Force Security Assistance Training Squadron): Col Jan R Heining
Inspector General: Col William L Reynolds
Director of Public Affairs: Col Valerie Elbow
☎(+1 210) 652 68 24; 652 20 27 (fax)
Chief of News Division: Lieut Col Suzanne M Randle

Mission:
Recruits, accesses, commissions, trains and educates Air Force enlisted and officer personnel. Provides basic military training, initial and advanced technical training and professional military and degree-granting education. Conducts joint medical service, readiness and Air Force security assistance training.

Responsibilities:
Recall of Individual Ready Reservists. Mobility and contingency tasking support to combatant commanders.

Air University
55 LeMay Plaza South, Suite 120,
Maxwell AFB, AL 36112-6335
☎(+1 334) 953 11 10; 953 20 14
⊜ (+1 334) 953 33 79
Commander: Lieut Gen Joseph J Redden
Public Affairs Director: Capt Robert J Gonzales
■ Conducts professional military, graduate and professional continuing education for officers, enlisted personnel and civilians to prepare them for command, staff, leadership and management responsibilities. Air University units include:
Air War College (Maxwell AFB)
Air Command and Staff College (Maxwell AFB)
Squadron Officer School (Maxwell AFB)
USAF Senior NCO Academy (Gunter Annex, AL)
Air Force Quality Inst.(Maxwell AFB)
College for Aerospace Doctrine,
Research & Education (Maxwell AFB)
Ira C. Eaker College for Professional Development (Maxwell AFB)
Air Force Institue of Technology (Wright-Patterson AFB, OH)
Extension Course Institute (Gunter Annex AL)
Air University Library (Maxwell AFB)
HQ Civil Air Patrol-USAF (Maxwell AFB)
42nd Air Base Wing (Maxwell AFB)
Community College of the Air Force (Maxwell AFB)
Reserve Officer Training Corps (Maxwell AFB)
Officer Training School (Maxwell AFB)
First Sergeant Academy (Maxwell AFB)

37th Training Wing
Lackland AFB, TX 78236-5109
Commander: Brig Gen Robert J Courter (Jr)
■ Controls 737th Military Training Group, 37th Training Group and Defense Language Institute and Inter-American Air Forces Academy.

81st Training Wing
Keesler AFB, MS 39534-2604
Commander: Brig Gen Andrew J Pelak (Jr)
■ Controls 81st Training Group, 81st Logistics Group, 81st Support Group and 81st Medical Group.

17th Training Wing
Goodfellow AFB, TX 76908-4410
Commander: Col Kelvin R Coppock
■ Controls 17th Training Group.

82nd Training Wing
Sheppard AFB, TX 76311-2941
Commander: Brig Gen Scott C Bergren
■ Controls 82nd, 782nd, 882nd, 982nd Training Groups, 82nd Logistics, 82nd Medical and 82nd Support Group.

Air Force Recruiting Service
Randolph AFB, TX 78150-4527

Contact Reference

Commander: Brig Gen Walter E Buchanan (III)
■ Recruiting Groups: 360th (Hanscom AFB), 367th (Robins AFB), 369th (Lackland AFB) and 372nd (Hill AFB).

US Air Force Reserve Office
Maxwell AFB, AL 36112-6106
Commander: Brig Gen Brian A Arnold
■ Controls: 330th Technical Training Wing.

59th Medical Wing
Lackland AFB, TX 78236-5300
☎ (+1 210) 670 73 51
Commander: Maj Gen Paul K Carlton (Jr)
■ Controls: 3400th Technical Training Group and the 3320th Correction and Rehabilitation Squadron.

12th Flying Training Wing
Randolph AFB, TX 78150-4559
Commander: Brig Gen Garry R Trexler
■ Conducts pilot, navigator training and flight screening.

336th Training Group
Fairchild AFB, WA 99011-8628
Commander: Col Kenneth M Page
■ Conducts survival training. Sub-units at: Eielson AFB (Alaska) and NAS Pensacola.

14th Flying Training Wing
Columbus AFB, MS 39710-1000
Commander: Col Michael G Lee

47th Flying Training Wing
Laughlin AFB, TX 78843-5230
Commander: Col Gary A Winterberger

56th Fighter Wing
Luke AFB, AZ 85309-1420
Commander: Brig Gen Carrol H Chandler

58th Special Operations Wing
Kirtland AFB, NM 87117-5861
Commander: Col John H Folkerts

64th Flying Training Wing
Reese AFB, TX 79489-5000
Commander: Col Henry W Horton

71st Flying Training Wing
Vance AFB, OK 73705-5015
Commander: Col Daniel J Mumaugh

80th Flying Training Wing
Sheppard AFB, TX 76311-2051
Commander: Col Kenneth M Decuir

97th Air Mobility Wing
Altus AFB, OK 73523-5047
Commander: Brig Gen David R Love

173rd Fighter Wing
Klamath Falls, OR 97603-1925
Commander: Col Billy J Cox

162nd Fighter Wing
Tucson International Airport, Tucson, AZ 85706-6052
Commander: Brig Gen Joseph I Mensching

325th Fighter Wing
Tyndall AFB, FL 32403-5541
Commander: Brig Gen John H Campbell

360th Recruiting Group
Hanscom AFB, MA 01731-2300
Commander: Col Claude N Richards (Jr)

369th USAF Recruiting Group
Lackland AFB, TX 78236-5122
Commander: Col Darrel R Greer

59th Medical Wing
Wilford Hall Medical Center,
Lackland AFB, TX 78236-5000
Commander: Brig Gen Paul K Carlton (Jr)

81st Medical Group
Keesler AFB, MS 39534-2519
Commander: Brig Gen Pedro N Rivera

372nd Recruiting Group
Hill AFB, UT 84056-5838
Commander: Lieut Col Ronald A Seyle

619th Training Support Group
Randolph AFB, TX 78150-4404
Commander: Lieut Col Michael B Hainsey [98]

25-22000 US Air Force Materiel Command (AFMC)

4375 Chidlaw Rd, Suite 6, Wright-Patterson AFB,
OH 45433-5006
☎ (+1 513) 257 6308 (public affairs)
⊖ (+1 513) 257 2558
👥 120 763 (38 561 active duty)
Commander: Gen. Henry Viccellio (Jr)
Vice Commander: Lt Gen. Lawrence P. Farrell (Jr)
☎ (+1 513) 257 6888
Command Civil Engineer: Brig.Gen. Todd Stewart
☎ (+1 513) 257 6576
Directorate, Test & Operations: Maj.Gen. F. Gideon
☎ (+1 513) 257 2109
Directorate, Personnel: Steve N. Smith
☎ (+1 513) 257 2912
Directorate of Engineering & Technical Management: J. Bair ☎ (+1 513) 257 0066
Directorate of Financial Management: Brig.Gen. D. Samic
☎ (+1 513) 257 6781
Directorate of International Affairs: Col W.F. Havelka
☎ (+1 513) 257 9845
Inspector General: Col K. Sullivan ☎ (+1 513) 257 6331
Public Affairs Dir: Col Michael Warden
☎ (+1 513) 257 6306
Directorate of Logistics: Brig.Gen. Dennis Haines
☎ (+1 513) 257 2635
Office of Communications: O. Howard
☎ (+1 513) 257 3626
Office of the Staff Judge Advocate: Brig.Gen. O. Waldrop
☎ (+1 513) 257 6513
Directorate of Contracting: Brig.Gen. R. Roellig
☎ (+1 513) 257 7000
Directorate of Science & Technology: Maj.Gen. R. Paul
☎ (+1 513) 257 6977
Directorate of Plans & Programs: Maj.Gen. S. Kostelnik
☎ (+1 513) 257 7100
Directorate of Requirements: Maj.Gen. R. Bridges
☎ (+1 513) 257 3024
Chief News Division: Ron Fry
■ Cradle to grave management of research development, test, acquisition & support for Air Force systems.

Major AFMC Installations:
Arnold AFB: (Arnold Engineering Development Center) Flight simulation test complex.
Brooks AFB: (Armstrong Laboratory & Human Systems Center) Concentrates on human aspects of USAF weapon systems & facilities.
Edwards AFB: (Air Force Flight Test Center) Flight test centre for all USAF & NASA aircraft.
Eglin AFB: (Air Force Developement Test Center) Tests & evaluates munitions (non-nuclear), electronic combat systems & navigation/guidance systems.
Griffiss AFB: (Rome Laboratory) Research & development for CSI.
Hanscom AFB: (Electronic Systems Center) Development & acquisition of C3I systems.
Hill AFB: (Ogden Air Logistics Center) Maintenance for ICBM, F-16, F-4, Maverick, GBU-15 & laser-guided bombs.
Kelly AFB: (San Antonio Air Logistics Center) Maintenance for C-5, C-17, T-37 & T-38 aircraft.
Kirtland AFB: (Phillips Laboratory) Research & development for space & missile related technology.
Los Angeles AFB: (Space & Missile Systems Center) Development & acquisition of launch & space systems for military communications, navigation, weather & surveillance applications.
McClellan AFB: (Sacramento Air Logistics Center) Maintenance for all F-111, A-10, F-117A & F-22 aircraft.
Newark AFB: (Aerospace Guidance & Metrology Center) Repair of inertial guidance & navigation systems.
Robins AFB: (Warner-Robins Air Logistics Center) Maintenance for F-15, C-141 & C-130 aircraft.
Tinker AFB: (Oklahoma City Air Logistics Center) Maintenance for B1-B, B-2, B-52, E-3, E-4 & 135 series aircraft.
Wright-Patterson AFB: (Aeronautical Systems Center & Wright Laboratory) Research, development & acquisition of aerospace systems, electronics, materials & manufacturing technologies.

Ogden Air Logistics Center
Hill Air Force Base, 7981 Georgia St, UT 84056-5990
☎ (+1 801) 777 5201 ⊖ (+1 801) 777 3721
Commander: Maj.Gen. S. Condon
Director of Public Affairs: Maj. Bob Ballew

Oklahoma City Air Logistics Center
Tinker Air Force Base, 3001 Staff Drive, Ste 1AG78A,
OK 73145-5000
☎ (+1 405) 739 2026 ⊖ (+1 405) 739 2882
Commander: Maj.Gen. C. Perez
Director of Public Affairs: Gene Pickett

San Antonio Air Logistics Center
807 Buchner, San Antonio, TX 78241-5842
☎ (+1 210) 925 7951 ⊖ (+1 210) 925 9739
Commander: Maj.Gen. J. Childress
Director of Public Affairs: Maj. Dewey Ford

Sacramento Air Logistics Center
McClellan Air Force Base, 5241 Arnold Ave, Suite 3,
CA 95652-1089
☎ (+1 916) 643 1341 ⊖ (+1 916) 643 0172
Commander: Maj.Gen. E. Tattini
Director Public Affairs (SM-ALC/PA): Lt Col Dennis Linn

Warner Robins Air Logistics Center
Robins Air Force Base, 215 Page Rd, Ste 106,
GA 31098-1662
☎ (+1 912) 926 2137 ⊖ (+1 912) 926 9597
Commander: Maj.Gen. R. Smith
Public Affairs Officer: Lt Col Dian Lawhon

Space & Missile Systems Center
2400 East El Segundo Blvd,
Los Angeles, CA 90009-2960
☎ (+1 310) 363 0030 ⊖ (+1 310) 363 2549
Commander: Lt Gen. Lester Lyles

Phillips Laboratory
Kirtland AFB, 3550 Aberdeen Ave SE, NM 87117-5776
☎ (+1 505) 846 1911 ⊖ (+1 505) 846 0423
Commander: Col Michael Heil
Public Affairs Dir: Juventino R. Garcia
→ 60

Electronic Systems Center
Hanscom AFB, Nine Eglin St, MA 01731-2118
☎ (+1 617) 377 5191 ⊖ (+1 617) 377 5077
Commander: Lt Gen. C. Franklin
Director of Public Affairs (ESC/PA): Lt Col Charles Merlo

Rome Laboratory
Griffiss AFB, 26 Electronic Parkway, NY 13441-4514
☎ (+1 315) 330 3053 ⊖ (+1 315) 330 3909
Commander: Col F. Boulds
Public Affairs Officer: F.L. Crumb

Aeronautical Systems Center
Wright-Patterson Air Force Base, 1865 4th St, Suite 15,
OH 45433-7129
☎ (+1 513) 257 3334 ⊖ (+1 513) 476 4022
Commander: Maj.Gen. K. Eickmann
Public Affairs Officer: Lt Col Marc Martens

Wright Laboratory
Wright Patterson AFB, OH 45433-6523
☎ (+1 513) 255 3334 ⊖ (+1 513) 476 4022
Commander: Col Richard W. Davis
Public Affairs Officer: Lt Col Marc Martens
■ The Wright Laboratory comes under the jurisdiction of the Aeronautical Systems Center, USAF Material Command.

Human Systems Center
Brooks Air Force Base, 2510 Kennedy Circle, Suite 1,
Brooks AFB, TX 78235-5120
☎ (+1 210) 536 32 34 ⊖ (+1 210) 536 32 35
Commander: Brig Gen R Belihar
Chief of Public Affairs (HSD/PA): Maj Peter Kirk

Air Force Flight Test Center (AFFTC)
Edwards AFB, 5 Pollux Drive, CA 93524-1225
☎ (+1 805) 277 3510 ⊖ (+1 805) 277 2732
Commander: Brig.Gen. R. Engel
Director Public Affairs (AFFTC/PA): Lt Col Janet Reese

Arnold Engineering Development Ctr (AEDC)
Arnold AFB, 100 Kindel Drive, Suite B-213,
TN 37389-2213
☎ (+1 615) 454 42 04; 455 52 01
⊖ (+1 615) 454 67 20
Commander: Col M. Wiedemer
Public Affairs Officer (AEDC/PA): Capt. Steve Doub

Air Force Development & Test Center (AFDTC)
Eglin AFB, 101 West D Ave, Suite 129, FL 32542-5498
☎ (+1 904) 882 3931 ⊖ (+1 904) 882 4894
Commander: Brig.Gen. S. Cranston
Public Affairs Officer: Maj. Jerry Renne

US Air Force Office of Scientific Research (AFOSR)
Bolling AFB, 110 Duncan Ave, Suite B115,
DC 20332-6448
☎ (+1 202) 767 6010 ⊖ (+1 202) 767 4961
Deputy Assistant Secretary: Dr Hellmut Hellwig
Public Affairs Officer: Ms Jane Knowlton [97]

25-22100 Pacific Air Forces Headquarters

25 East Street, Suite I-106, Hickam AFB, HI 96853-5496
☎ (+1 808) 449 93 60 ⊖ (+1 808) 449 86 01
Commander, Pacific Air Forces: Gen Richard B Myers
☎ (+1 809) 449 45 00
Vice Commander: Lieut Gen Eugene D Santarelli
Director of Staff: Col George Wagaski (III)
☎ (+1 808) 449 45 15
Director of Safety: Col Larry Andrew Reseter
☎ (+1 808) 449 69 32
Inspector General: Col David T Hanson
☎ (+1 808) 449 94 20
Director of Operations: Maj Gen Steven R Polk
☎ (+1 808) 449 63 21
Director of Logistics: Brig Gen Richard E Brown (III)
☎ (+1 808) 449 90 46
Director of Plans: Brig Michael M Dunn
☎ (+1 808) 449 12 93
Director of Financial Management and Comptroller: Col Josephine A Kumer ☎ (+1 808) 449 98 50
Director of Civil Engineering: Col Frank J Destadio
☎ (+1 808) 449 52 54
Director of Personnel: Col James G Ferguson
☎ (+1 808) 449 29 40

Director of Communications and Information:
Col Bernard K Skoch ☎(+1 808) 449 63 38
Director of Security Forces: Col Doss Von Brandenstein
☎(+1 808) 449 59 65
Director of Public Affairs: Col Donald L Black
☎(+1 808) 449 93 60
Director of Services: Col Dale H Collins
☎(+1 808) 449 29 63
Staff Judge Advocate: Col Thomas J Fiscus
☎(+1 808) 449 95 61
Command Surgeon: Col James G Roudebush
☎(+1 808) 449 23 32 ext. 224
Command Chaplin: Col Richard B Higgins
☎(+1 808) 449 28 40
Command Historian: Dr Timothy Keck
☎(+1 808) 449 56 35
■ PACAF operates four air forces:
5th Air Force, Yokota AB, Japan
7th Air Force, Osan AB, South Korea
11th Air Force, Elemendorf AFB, Alaska
13th Air Force, Andersen AFB, Guam

11th Air Force (PACAF)
6920 12th Street, Suite 120,
Elmendorf AFB, AK 99506-2530
☎(+1 907) 552 81 51
Commander: Lieut Gen Patrick K Gamble
☎(+1 907) 552 21 00
■ Includes: 354th Wing at Eielson AFB, AK and 3rd Wing at Elmendorf AFB.

13th Air Force (PACAF)
Building 23003, Andersen AFB, Guam 96543-4033
☎(+1 671) 366 13 13 ⊖ (+1 671) 366 13 22
Commander: Maj Gen John R Dallager
☎(+1 671) 366 13 17; 366 13 18
■ Includes: 497th Fighter Training Squadron (Singapore) and 36th Air Base Wing (Guam).

15th Air Base Wing
800 Scott Circle, Hickham AFB, HI 96853-5328
☎(+1 808) 449 24 90
Commander: Col Ann M Testa ☎(+1 808) 449 63 41 [98]

25-22500 Air Force Comunications Agency

(† Air Force C4 Agency)
203 West Losey Street,
Scott Air Force Base, IL 62225-5219
☎(+1 618) 256 25 71 ⊠ 62290
⊖ (+1 618) 256 22 90; 256 61 27
Commander: Col Gilbert Hawk
Vice Commander: Col Stephen E Bovich
☎(+1 618) 256 25 73
Executive Officer: Maj. Barb Peters
Resources Director: Alvin P Jones ☎(+1 618) 256 59 89
Systems and Procedures Director: Col Bill Crawford
☎(+1 618) 256 31 89
Interoperability and Technical Director:
Col David L Rakestraw ☎(+1 618) 256 50 83
Plans and Analysis Director: Lt Col James Shaw
☎(+1 618) 256 39 85
Public Affairs Chief: Lori L Manske ☎(+1 618) 256 43 96 [98]

25-22560 US 750th Space Group

(† Air Force Satellite Test Center) (AFSPACECOM)
Onizuka Air Force Base, Onizuka, CA 94089-1236
☎(+1 408) 752 4026 ⊖ (+1 408) 751 4024
Commander Space Group 750 SG/CC:
Col Timothy A. Roberts (USAF)
Deputy Commander 750 SG/CD: Lt Col James E. Porter (III, USAF)
Mission Support Squadron 750 MSS/CC:
Lt Col John W. Passey (USAF)
Commander Logistics Support Squadron 750 LSS/CC:
Maj. Susanne L. Holcomb (USAF)
Commander Operation Support Squadron 750 OSS/CC:
Lt Col Dwight J. Miller (USAF)
Commander Space Operations Squadron 21 SOPS/CC:
Lt Col Ricky B. Kelly (USAF)
Commander Communications Squadron 750 CS/CC:
Lt Col Michael S. Kidd (USAF)
Commander Medical Squadron 750 MDS/CC:
Col William S. Swindling (USAF)
Commander Contracting Flight 750 CONF/CC:
Maj. Arthur J. Brown (USAF)
Commander Space & Missile Center Det 2, SMC/CC:
Col Daryl J. Joseph (USAF)
Commander Operations Group OD-4/CC:
Col Ronald K. Henley (USAF)
Commander Space & Missile Center SMC/CWO:
Lt Col Ronald S. Lassonde (USAF)
Commander Space Operations Squadron 5 SOPS/CC:
Lt Col Michael L. Wolfert (USAF)
Commander Communications Squadron DET 7/CC:
Capt. Wayne Kodama (USAF)
Commander Satellite Operations Squadron OD-4/DC:
Robert Brandon
Commander Saatellite Operations Squadron OD-4 DH:
Ms Irene Mills
Commander Satellite Operations Squadron OD-4/DX:
Edward T. Cope
Commander Mission Support Squadron OD-4/DM:
Ms Janet I. Cohoon [97]

25-22700 Alaskan Command (Joint Forces) (ALCOM)

Elmendorf AFB, Anchorage, AK 99506-5001
☎(+1 907) 552 23 41 ⊖ (+1 907) 552 54 11
Commander: Lt Gen Patrick K Gamble (USAF)
Public Affairs Officer: Maj. Joseph Mecadon (USAF) [98]

25-23050 Air Force Operational Test & Evaluation Center

Kirtland AFB, NM 87117-7001
☎(+1 505) 846 45 33 ⊖ (+1 505) 846 97 26
👥 855
Commander: Maj Gen George B Harrison
[E] harrisog@afotec.af.mil
Public Affairs Officer: Ms Therese Longenfelder
■ Plans and conducts realistic, objective and impartial operational tests and evaluations to determine the operational effectiveness and suitability of Air Force systems and their capacity to meet mission needs. → 60 [98]

25-25000 US Air Force Security Forces Control

8601 F Avenue, Kirkland AFB, NM 87117-5516
👥 141
Commander: Col Lawrence R Mayes
Operations Director: Col James A Jones
Public Affairs Officer: Harold L Washington (Jr)
■ Provides expertise for the security of nuclear weapons and weapon systems.
Prepare guidance on air base defense operations and continuation training and guidance for law enforcement, resources protection and antiterrorism USAF-wide.
Develop and implement base-level training and combat arms training and maintenance programs.
Assist in planning, allocating and evaluating Security Police resources, equipment and future technology requirements.
Develop and maintain tables of allowance identifying Security Police equipment requirements.
Manage Air Force corrections activities.
Facilities at US Disciplinary Barracks Fort Leavenworth; Naval Consolidated Brig NAS Miramar; Naval Consolidated Brig Charleston Naval Weapons Center, SC. [98]

25-25100 Air Force Civil Engineering Support Agency

139 Barnes Drive, Suite 1, Tyndall AFB, FL 32403-5319
☎(+1 904) 283 61 44 ⊖ (+1 904) 283 64 99
👥 225
Commander: Col H Dean Bartel
Public Affairs Officer: MSgt Joseph Gonzales
■ Provides tools, practices and professional support to maximize Air Force civil engineer capabilities in base and contingency operations. [98]

25-25150 Air Force Military Personnel Center (AFMPC)

550 C Street west, Suite 01,
Randolph AFB, TX 78150-4703
☎(+1 210) 652 42 52 ⊖ (+1 210) 652 23 91
Commander: Maj Gen Susan L Pamerleau
Public Affairs Officer: Capt Michael J Rein [98]

25-25300 US Air Force Safety Center

9700 Avenue G, South East, Kirtland AFB,
NM 87117-5670
☎(+1 505) 846 23 72
⊖ (+1 505) 846 27 21; 853 05 65
Commander: Maj Gen Francis C Gideon (Jr)
☎(+1 505) 846 23 72
Commander of Headquarters Section and Executive Officer: Maj Keneth E Stanfill ☎(+1 505) 846 23 72
Vice Commander: Col Bernard B Burklund
Chief, Ground Safety: Col Robert Scott
☎(+1 505) 846 07 80
Chief, Flight Safety: Col Robert Osterholt
☎(+1 505) 846 06 44
Chief, Policy, Plans and Programs: Col Roger C Craig
Chief, Data Operations and Analysis: Todd G Fattor
Weapons, Space and Nuclear Safety Director:
Col Harold J Camacho (Jr) ☎(+1 505) 846 13 61 [98]

25-25400 Air Force Management Engineering Agency

Randolph AFB, TX 78150-5001
Commander: Col Charles F. Pibrell
Operations Dir: Lt Col Rudy Bruback
Public Affairs Officer: Greg Titus [97]

25-29180 Air War College USAF

Maxwell AFB, Montgomery, AL 36112-6427
☎(+1 334) 953 72 88 ⊖ (+1 334) 953 40 28
Commandant: Maj Gen Bruce D Smith
Vice Commandant: Col Donald D Karle (Jr)
Dean of Academic Affairs: Ronald J Kurth
Media Advisor: Lieut Col David L Simmms
■ Educates senior officers to lead at strategic level. [98]

25-29200 US Air Force Academy

2304 Cadet Drive, Suite 320,
Colorado Springs, CO 80840-5016
☎(+1 719) 333 29 90; 333 18 18 (Operator Asst)
⊖ (+1 719) 472 40 94
👥 2,121 (active duty exluding cadets)
Superintendent: Lt Gen Paul E Stein
☎(+1 719) 333 41 40 ✉ steinpe.supt@usafa.af.mil
Commandant of Cadets: Brig Gen Steven R Lorenz
☎(+1 719) 333 42 90
Dean of the Faculty: Brig Gen Ruben A Cubero
☎(+1 719) 333 42 70
Director of Public Affairs: Lieut Col Doug McCoy
[E] mccoydd.pa@usafa.af.mil [98]

25-30000 Department of the Army

Pentagon, Washington, DC 20310-1502
☎(+1 703) 697 43 14 ⊖ (+1 703) 697 21 59
Deputy Inspector General: Maj.Gen. L.R. Jordan
☎(+1 703) 695 1500
Chief of Public Affairs: Maj.Gen. F.A. Gorden
☎(+1 703) 695 5135
Deputy Chief Public Affairs: Col Robert E. Gaylord
☎(+1 1 703) 697 4482
Command Information & Products Division:
William Drobnick ☎(+1 703) 697 4640
Public Communications Division: Ms Judy Johnston (Acting) ☎(+1 703) 697 2564

Office of the Secretary of the Army
Washington, DC 20310-0100
☎(+1 703) 695 1717
Secretary of the Army: Hon. Togo D. West (Jr)
☎(+1 703) 695 3211
Under Secretary of the Army: Hon. Joe R. Reeder
☎(+1 703) 695 4311
Comptroller of the Army: Lt Gen. Merle Freitag
Asst. Secretary (Research & Development):
Gilbert F. Decker (Hon.) ☎(+1 703) 695 6153
Military Dpty of Asst Secretary R&D & Acquisition:
Lt Gen. Ronald V. Hite ☎(+1 703) 697 0356
Asst Secretary (Installations, Logistics & Environment):
Hon. Robert M. Walker ☎(+1 202) 695 6527
Asst Secretary (Manpower & Reserve Aff):
Hon. Ms Sara E. Lister ☎(+1 703) 697 9253
Asst Secretary (Financial Management):
Hon. Ms Helen T. McCoy ☎(+1 202) 695 22 54
Asst Secretary (Civil Works): Hon. H. Martin Lancaster
☎(+1 202) 697 8986
Dir Information Systems for C4: Lt Gen. Otto J. Guenther
☎(+1 703) 697 7494

Office of the Chief of Staff
Washington, DC 20310-0200
☎(+1 703) 695 2077
Chief of Army Staff: Gen. Dennis J. Reimer
Exec.Officer to the Chief of Staff: Col James L. Campbell
☎(+1 703) 695 2077
Sergeant Major of the Army: Sgt Maj. Gene C. McKinney
☎(+1 703) 695 2150
Vice Chief of Staff: Gen. Ronald H. Griffith
☎(+1 703) 695 4371
Exec.Vice Chief of Staff: Col Roy Whitcomb
☎(+1 703) 695 7372
Army Staff Dir: Lt Gen. John A. Dubia
☎(+1 703) 695 3542

Deputy CoS (Logistics)
Washington, DC 20310-0500
☎ (+1 703) 695 4102
Dpty Chief of Staff Logistics: Lt Gen. John G. Coburn
☎ (+1 703) 695 4102
Asst Dpty Chief of Staff (Logistics):
Maj.Gen. William N. Farmen ☎ (+1 703) 697 5301
Plans & Operations Dir: Col Hawthorne L. Proctor
☎ (+1 703) 695 3282
Supply & Maintenance Dir: Brig.Gen. Charles S. Mahan
☎ (+1 703) 697 8401
Resources Management Dir: Stephen R. Burdt
☎ (+1 703) 614 4391
Transport, Energy & Troops Support Dir:
Brig.Gen. Boyd E. King ☎ (+1 703) 695 0955

Deputy CoS (Personnel)
Washington, DC 20310-0500
☎ (+1 703) 695 6003
Deputy Chief of Staff Personnel:
Lt Gen Frederick E. Vollrath ☎ (+1 703) 695 2250
Military Personnel Management Dir:
Maj.Gen. Thomas Sikora ☎ (+1 703) 695 2497

National Guard Bureau
Pentagon, Room 2E394, Washington, DC 20310-2500
☎ (+1 703) 697 6987 ⊖ (+1 703) 693 3422
Chief National Guard: Lt Gen. Edward D. Baca
Inspector General: H. Cronin Byrd ☎ (+1 703) 693 4125
Joint Staff Dir: Thomas L. Link ☎ (+1 703) 697 7449
Air National Guard Dir: Maj.Gen. Donald Shepperd
☎ (+1 703) 693 4750
Army National Guard Dir: Maj.Gen. William Navas
☎ (+1 703) 697 5559
Public Affairs Dir: Dan Donahue
The Pentagon, Room 2D419 ☎ (+1 703) 695 0421

Office of the Chief of Engineers
20 Massachusetts Avenue NW, Pulaski Bldg,
Washington, DC 20314-1000
☎ (+1 202) 272 0001/0660
Chief of Engineers: Lt Gen. Arthur E. William
☎ (+1 202) 761 0001
Deputy Chief of Engineers: Maj.Gen. Pat M. Stevens
☎ (+1 202) 761 0002
Civil Works Dir: Maj.Gen. Stanley G. Genega
☎ (+1 202) 761 0099
Military Programs Dir: Maj.Gen. Albert J. Genetti (Jr)
☎ (+1 202) 761 0379
Director of Military Engineering & Topographic Div.:
Col T.J. Sheehy ☎ (+1 202) 695 2782 [97]

25-32100 US Army Space & Strategic Defense Command (USASSDC)
PO Box 15280, Arlington, VA 22215-0280
☎ (+1 703) 607 19 38 ⊖ (+1 703) 607 14 91
Commanding General: Lieut Gen Edward G Anderson (USA) ☎ (+1 703) 607 18 73
Deputy Commanding General: Brig Gen Willie B Nance (Located in Huntsville, AL) ☎ (+1 205) 955 35 01
Assistant Chief of Staff, Operations: Col Robert Billings
☎ (+1 703) 607 19 36
Assistant Chief of Staff, Personnel, Administration and Logistics: Lieut Col David Stoddard
☎ (+1 703) 602 53 69
US Army Space Command: Col Otis B Ferguson (located in Colorado Springs) ☎ (+1 719) 554 88 88
Commander US Army Kwajalein Atoll: Col Scott B Cottrell
Inspector General: Col Mike Putnam
Chief Scientist: Dr Darrell Collier
Chief of Resource Management: Col Garth T Bloxham
☎ (+1 205) 955 31 00
High Energy Laser System Test Facility Directorate Director: Col Larry Anderson
Space and Missile Defence Battle Laboratory:
Laurence H Burger ☎ (+1 205) 955 30 70
Missile Defence and Space Technology Centre:
Dr J Richard Fisher ☎ (+1 205) 955 35 20
Principle Assistant for Contracting: Mark J Lumer
Public Affairs Officer: William M Congo (located in Huntsville)
Public Affairs Officer: Michael C Biddle (Located in Washington) ☎ (+1 703) 607 20 39
Public Affairs Officer: John Cummings
Command Historian: James A Walker
☎ (+1 205) 955 47 78 [98]

25-32200 US Army Pacific (USARPAC)
(† US Army Western Command (WESCOM))
Fort Shafter, HI 96858
☎ (+1 808) 438 22 06; 471 71 10
⊖ (+1 808) 438 33 90
Commander in Chief: Lieut Gen Robert L Ord (III)
Deputy Commander, Chief of Staff:
Maj Gen Stephen Silvasy ☎ (+1 808) 438 24 22
Deputy Commander for Army National Guard:
Maj Gen Jan P Wepster ☎ (+1 808) 438 38 21
Deputy Commander for US Army Reserve:
Maj Gen Calvin Lau ☎ (+1 808) 438 60 29
Commander Sergeant Major: CSM C Lopez
☎ (+1 808) 438 97 24
Liaison Officer: Lieut Col Beghtol
E beghtol@hqda.army.mil
Public Affairs Officer: Col H T Linke
☎ (+1 808) 438 93 75; 438 91 60 (fax)
E appaadm@shafter-emh3.army. mil
Chief Media Relations Division: J Fairbank
Science Adviser: J Davis
☎ (+1 808) 438 14 80; 438 62 42 (fax)
E apsaadm@shafter-emh3.army. mil
Command Surgeon: Col F R Smith
☎ (+1 808) 433 57 16; 438 60 35 (fax)
E apmdadm@shafter-emh3.army. mil
Provost Marshal: Col J Saunders
☎ (+1 808) 438 97 17; 438 80 22 (fax)
Inspector General: Col J Cray
☎ (+1 808) 438 16 92; 438 60 07 (fax)
E apigadm@shafter-emh3.army. mil
Deputy Chief of Staff for Personnel: Col A Arnold
☎ (+1 808) 438 93 45; 438 86 62 (fax)
E appeadm@shafter-emh3.army. mil
Deputy Chief of Staff for Intelligence: Col M Pernell
☎ (+1 808) 438 92 14; 438 25 66 (fax)
E apidadm@shafter-emh3.army. mil
Deputy Chief of Staff Operations & Plans: Col T Scott
☎ (+1 808) 438 6071; 438 6293 (fax)
E apopadm@shafter-emh3.army. mil
Deputy Chief of Staff for Logistics: Col L Collignon
☎ (+1 808) 438 22 65; 438 34 60 (fax)
E aplgadm@shafter-emh3.army. mil
Deputy Chief of Staff, Resource Manager: Col C Holland
☎ (+1 808) 438 22 81; 438 89 54 (fax)
E aprmadm@shafter-emh3.army. mil
Deputy Chief of Staff for Engineering: Col D Kershaw
☎ (+1 808) 438 10 25; 438 86 90 (fax)
E apenadm@shafter-emh3.army. mil
Deputy Chief of Staff, Info Manager: Col A Franks
☎ (+1 808) 438 10 10; 438 75 81 (fax)
E apimadm@shafter-emh3.army. mil
Deputy Chief of Staff Reserve Component Integration:
Col W Carwille ☎ (+1 808) 438 12 20; 438 36 61 (fax)
Assistant Chief of Staff for Reserves: Col A Sepe
☎ (+1 808) 438 92 23; 438 27 76 (fax)
E aprcadm@shafter-emh3.army. mil
Chief Special Operations Theatre Support: Col Griffin
☎ (+1 808) 438 30 50; 438 67 60 (fax)
E sotse@shafter-emh3.army. mil
Commander US Army Chemical Activity Pacific: Col Kilgore
☎ (+1 808) 441 39 49 E kilgore@shafter-emh3.army. mil [98]

25-32210 US Army Alaska
AK 96508-7850
☎ (+1 907) 384 2180
E fs2-amh%cdrusarak@sf-emh2.army.mil
Commanding General & Deputy Commander US Alaskan Command: Maj Gen K.W. Simpson
Deputy Commander: Col Brophy ☎ (+1 907) 353 6680
Public Affairs Officer: Maj. McGuire ☎ (+1 907) 384 1542 [97]

25-32400 US Army Training & Doctrine Command (TRADOC)
Fort Monroe, VA 23651-5000
☎ (+1 804) 727 3333 ⊖ (+1 804) 727 3368
Commanding General: Gen. W. Hartzog
Deputy Commanding General: Lt Gen. J. Miller
Exec. Officer: Col E. Buckley ☎ (+1 804) 680 2922
Dpty Commanding General & Commander Combined Arms Command: Lt Gen. L. Holder
Bldg 52, Fort Leavenworth, KS
Dpty Commanding General & Commander Combined Arms Support Command: Maj.Gen. T. Robinson
Bldg 10500, Fort Lee, VA
Dpty Commanding General for US Army Reserve:
Maj.Gen. Evo Riguzzi (Jr)
Dpty Chief of Staff Base Operations Support:
Maj.Gen. N. Snyder
Dpty Chief of Staff Combat Development:
Maj.Gen. J. Cravens
Dpty Chief of Staff Doctrine: Maj. Gen. R. Scales
Dpty Chief of Staff Training: Maj.Gen. C. Ernst
Dpty Chief of Staff Resource Management:
Brig.Gen. J. Hennessee
Dpty Chief of Staff Analysis: M. Bauman
Bldg 52, Fort Leavenworth, KS
Dpty Chief of Staff Information Management: H. Markey (Acting)
Dpty Chief of Staff for Intelligence: Col N. O'Dawe
Assistant Deputy Chief of Staff: Col Bob Killebrew
Director Future Battles Directorate: Col Mike Starry
Post Commander: Col M. Rochelle
Chief of Public Affairs: Col G. Stinnett
☎ (+1 804) 727 3333
Public Information Branch: H. Perritt
☎ (+1 804) 727 4465
Command Information Branch: P.B. Wallace
☎ (+1 804) 727 3464 [97]

25-32430 US Army Transportation Center
Fort Eustis, VA 23604-5015
☎ (+1 757) 878 49 20 ⊖ (+1 757) 878 15 02
Commanding General: Maj Gen Daniel G Brown
☎ (+1 757) 878 48 01
Chief of Staff: Col Thomas C Brown
☎ (+1 757) 878 48 03
Garrison Commander: Col Scott M Bergeron (Jr)
☎ (+1 757) 878 29 08
Public Affairs Officer: Ronald A Johnson [98]

25-32432 Deployment/Deployment Systems Department (JSDTC)
Fort Eustis, VA 23604-5015
☎ (+1 757) 878 60 69 ⊖ (+1 757) 878 49 00
Chief of Department: Lieut Col Stephen S Slyfield
Director of Operations: Jesse Hamilton
■ Focal point for strategic and army related deployment doctrine issues. [98]

25-32434 US Army NCO Academy
Fort Eustis, VA 23604
☎ (+1 757) 878 44 40 ⊖ (+1 757) 878 34 54
Commandant CSM: Harold Cameron
Public Affairs Officer: Ms Diane Jennings [98]

25-32436 US Army Training Support Center (ATSC)
Fort Enstis, VA 23604
☎ (+1 757) 878 57 08 ⊖ (+1 757) 878 26 74
Director: Col Lauren Mullendore
Deputy Director: Dr Charles S Duncan
Public Affairs Officer: Ms Diane Jennings
■ ATSC manages the standards in the Armywide Range Standardisation/Modernisation Weapons Training Programme. [98]

25-32440 US Army Combined Arms Support Command
Fort Lee, VA 23801-5000
☎ (+1 804) 734 45 58 ⊖ (+1 804) 734 46 59
Commanding General: Maj Gen Daniel G Brown
Chief of Staff: Col Howard W McMillan
Garrison Commander: Col Robert W Ralston
Media Relations Officer: Ms Joy Whitmore
Public Affairs Officer: Ms M K Eder [98]

25-32445 US Army Combined Arms Center
Fort Leavenworth, KS 66027-1399
☎ (+1 913) 684 56 04 ⊖ (+1 913) 684 36 24
Commanding General: Lieut Gen L D Holder
Chief of Staff: Col Stephen H Robinette
Garrison Commander: Col Rolland A Dessert
Deputy Commanding General Training:
Brig Stanley F Cherrie (III)
Inspector General: Lieut Col G Bakian
Public Affairs Officer: Lt Col Jim Gleisberg
Deputy Chief of Staff for Resource Management:
Col Albert Fleumer
Commander of Battle Command Training Programmes:
Col John Kain
Procurement Director: Rob Bristow
Logistics Director: Benny Doyal
Public Information Officer: Ms Janet Wray

US Army TRADOC Analysis Center (TRAC)
Fort Leavenworth, KS 66027-2345
☎ (+1 913) 684 51 32 ⊖ (+1 913) 684 43 68
Director: Michael Bauman [98]

25-32450 US Army Intelligence & Security Command (INSCOM)
8825 Beulah Street, Fort Belvoir, VA 22060-5246
Commanding General: Brig Gen Thomas
☎ (+1 703) 706 1603
Deputy Commander: Col Varnado ☎ (+1 703) 706 12 32

Chief of Staff: Mr Hollingsworth ☎ (+1 703) 706 12 32
Deputy Chief of Staff, Personnel: Col Henning ☎ (+1 703) 706 12 07
Deputy Chief of Staff, Security: Mr Xiques ☎ (+1 703) 706 27 36
Deputy Chief of Staff, Operations: Col Demoney ☎ (+1 703) 706 29 75
Deputy Chief of Staff, Logistics: Col Wright ☎ (+1 703) 706 13 48
Deputy Chief of Staff, Information Management: Col Lucia ☎ (+1 703) 706 16 33
Deputy Chief of Staff, Resource Management: Col Wharton ☎ (+1 703) 706 13 75
Inspector General: Col Mastrangelo ☎ (+1 703) 706 11 58
Chief of Public Affairs: J Lau ☎ (+1 703) 806 53 26 [98]

25-32550 US Army Simulation, Training & Instrumentation Command (STRICOM)

12350 Research Parkway, Orlando, FL 32826-3276
☎ (+1 407) 380 81 16 ⊜ (+1 407) 380 42 58
Commanding General: Brig.Gen. John Geis
Deputy Commanding General: James M Skurka
Chief of Staff: Col Jim Godwin
Technical Director: Jerry Stahl
Chief Plan and Operations Office: Ralph Nelson
Total Quality Management: Ms Elaine Perrin
Project Manager for Combined Arms Tactical Trainer: Col Hammond
Project Manager for Training Devices: Col Noble Johnson
Deputy Project Manager for Training Devices: Phillipe Sprinkle
Project Manager for Instrumentation Targets and Threat simulators: Col Mark Russell
Deputy Project Manager for Instrumentation Targets and Threat Simulators: Henry Jehan
Project Manager for Distributed Interactive Simulation: Col Michael Rodgers
Project Manager for Distributed Interactive Simulation: Stan Goodman
Product Manager for Combined Arms Assessment Network: Lt Col Greek
Project Manager Combat Support Training Systems: Lt Col Steven Kessinger
Deputy Project Manager Combat Support Training System: Steven G Milburn
Project Manager Close Combat Training Systems: Lt Col Fair
Deputy Project Manager Close Combat Training System: Bill Rucker
Project Manager Air Combat Training Systems: Lt Col Hallagan
Deputy Project Manager Air Combat Systems: Donald Jones
Deputy Project Manager for Family of Simulations: Pat Spangler
Director Instrumentation Management Office: J Russell Longenbach
Director Resources Management: Mack Perry
Director Research and Engineering Management: Ed Trier
Director Acquisition: Dave Creech
Director Logistics: Dave Manning [98]

25-32600 US Army Medical Command Headquarters

2050 Worth Rd, Bldg 2792,
Fort Sam Houston, TX 78234-6000
☎ (+1 210) 221 6213 ⊜ (+1 210) 221 8483
Commanding General: Lt Gen. Alcide M. LaNoue
Deputy Commanding General: Maj.Gen. James B. Peake
Deputy Commander Health Care Operations: Brig.Gen. Robert F. Griffin
Deputy Commander Training: Maj.Gen. James B. Peake
Chief of Staff/Deputy Commander Admin.: Brig.Gen. Jerome V. Foust
Directorate Clinical Operations: Col A. Zaloznik
Directorate Logistics: Col J. Canella
Operations Dir: Col B. Furbish
Commander US Army Dental Command: Col L. Rouse
Commander US Army Veterinary Command: Col P. Barrows
Inspector General: Col A. Rhodes
Chief Information Officer: Col S. Baker (II)
Command Sergeant Major: CSM R. Adams [97]

25-32700 US Army Security Assistance Command (USASAC)

AMC Building, 5001 Eisenhower Ave.,
Alexandria, VA 22333-0001
⊜ (+1 703) 274 4338
Commanding General: Maj.Gen. W.A. Fitzgerald
Deputy Commanding General: Col D. Caton
Chief of Staff: Col T. Chase
Administration Security Division: D. Uber [97]

25-32750 US Army Aviation Systems Command (AVSCOM)

4300 Goodfellow Blvd, St Louis, MO 63120-1798
☎ (+1 314) 263 1002; 3161
Commanding General: Maj.Gen. Richard E. Stephenson ☎ (+1 314) 263 1002
Deputy Commanding General: Maj.Gen. Donald R. Williamson ☎ (+1 314) 263 1004
Chief of Staff: Col Robert S. Young ☎ (+1 314) 263 1007
Technical Dir: Thomas L. House ☎ (+1 314) 263 1010
Public Affairs Officer: Steve Wright ☎ (+1 314) 263 1160 [97]

25-32760 US Army Special Operations Command (USASOC)

Fort Bragg, NC
☎ (+1 910) 432 60 05 ⊜ (+1 910) 432 10 46
Commanding General: Lieut Gen Peter J Shoomaker
Command Sergeant Major: Cmd Sgt Maj William H Rambo
Public Affairs Officer: Lieut Col Walter E Pierce (III)

US Army Special Forces Command (Airborne)
Fort Bragg, NC
☎ (+1 910) 432 60 05 ⊜ (+1 910) 432 10 46
Commander: Maj Gen Kenneth R Bowra
Command Sergeant Major: Cmd Sgt Maj Richard Efird
Public Affairs Officer: Lieut Col Walter E Pierce (III)
USASOC Public Affairs Liaison to SFC: Maj Thomas McCollum
■ Includes:
1st Special Forces Group (Airborne) Fort Lewis, WA
3rd Special Forces Group (Airborne) Fort Bragg, NC
5th Special Forces Group (Airborne) Fort Campbell, KY
7th Special Forces Group (Airborne) Fort Bragg, NC
10th Special Forces Group (Airborne) Fort Devens, MA
19th Special Forces Group (Airborne), (ARNG Unit), Draper, UT
20th Special Forces Group (Airborne), (ARNG Unit), Birmingham, AL
112th Special Operations Signal Battalion (Airborne), Fort Bragg, NC
528th Special Operations Support Battalion (Airborne), Fort Bragg, NC

US Army John F Kennedy Special Warfare Center and School
Fort Bragg, NC
☎ (+1 910) 432 69 39 ⊜ (+1 910) 432 10 46
Commanding General: Maj Gen William P Tangney
Command Sergeant Major: Cmd Sgt Maj Paul J Shedlock
Public Affairs Officer: Lieut Col Walter E Pierce (III) ☎ (+1 910) 432 60 05
USASOC Public Affairs Liaison to SWCS: Paul Nelson ☎ (+1 910) 432 69 39
■ Includes:-
1st Special Warfare Training Group (Airborne) Fort Bragg, NC

US Army Civil Affairs and Psychological Operations Command (Airborne)
Psychological Operations Command, Fort Bragg, NC
☎ (+1 910) 432 60 05 ⊜ (+1 910) 432 10 46
Commander: Brig Gen Bruce B Bingham
Command Sergeant Major: Cmd Sgt Maj Thomas Jackson
Public Affairs Officer: Lieut Col Walter E Pierce (III) ☎ (+1 910) 432 60 05
USASOC Public Affairs Liaison to USACAPOC: John Thompson ☎ (+1 910) 432 60 05
■ Includes:
351st Civil Affairs Command, Reserve, Mountain View, CA
352nd Civil Affairs Command, Reserve, Riverdale, MD
353rd Civil Affairs Command, Reserve, Bronx, NY
358th Civil Affairs Brigade, Reserve, King of Prussia, PA
361st Civil Affairs Brigade, Reserve, Pensacola, FL
96th Civil Affairs Battalion (Airborne), Fort Bragg, NC
2nd Psychological Operations Group, Reserve, Parma, OH
4th Psychological Operations Group, (Airborne), Fort Bragg, NC
7th Psychological Operations Group, Reserve, San Francisco, CA

75th Ranger Regiment
Fort Benning, GA
Commander: Col Stanley A McChrystal
Command Sergeant Major: Cmd Sgt Maj Michael T Hall

160th Special Operations Aviation Regiment
Fort Campbell, KY
Commander: Col Richard A Cody
Command Sergeant Major: Cmd Sgt Maj Robert J Bishop [98]

25-32800 US Army Communications & Electronics Command (CECOM)

Fort Monmouth, NJ 07703-5000
Commanding General: Maj.Gen. Gerard P. Brohm
Deputy to Commanding General: Victor J. Ferlise
Deputy Business Management: Frank Fiorilli
Chief of Staff: Col Patrick T. Thornton
Command Sergeant Major: Cmd Sgt Maj. Timothy McCarthy
Dir Int'l Programs & Logistics: Eugene Bennett
Dir C4I Acquisition Center: Edward Elgart
Dir Material Management: William Richl ☎ (+1 908) 532 47 55
Commander Fort Monmouth Garrison: Lt Col Joel Nimsl
Commander Vint Hill Garrison (Warrenton): Col Mark Cornwell
Chief Public Affairs: Henry Kearney ☎ (+1 908) 53 22 534

Research, Dev. & Engineering Center
Myer Center, NJ 07703-5201
⊜ (+1 908) 532 0749
Director: Robert Giordano

Night Vision Electro-Optics Dir.
Fort Belvoir, VA 22060-5677
⊜ (+1 703) 704 1100
Director: Dr R. Buser

C4I Logistics & Readiness Center
CECOM Office Building, NY 07703-5008
Director: Col Andrew Follmer (Acting) [97]

25-32900 US Army Tank-Automotive and Armament Command (TACOM)

Detroit Arsenal, Warren, MI 48397-5000
☎ (+1 810) 574 50 00
Commanding General (TACOM): Maj Gen Edward L Andrews (Jr) ☎ (+1 810) 574 51 31
Chief of Staff: Col Hank Pankey ☎ (+1 810) 574 62 97
Dir of the Research & Dev. & Engineering Center (TARDEC): J Chapin ☎ (+1 810) 574 61 44
Command Sergeant Major: CSM H Nicholson ☎ (+1 810) 574 64 63
Acquisition Center Director: D Mehney ☎ (+1 810) 574 70 25
Programme Executive Officer, Ground Combat Support Systems: Maj Gen J Michitsch ☎ (+1 810) 574 66 62
Programme Executive Officer Tactical Wheeled Vehicles: W Wynbelt ☎ (+1 810) 574 66 01 [98]

25-32905 US Army Soldier Systems Command

Natick, MA 01760-5000
Commander: Brig.Gen Henry T. Glisson
Deputy to the Commander: Dr Robert W. Lewis
Chief of Staff: Col Richard Ross
Logistics Dir: Stephen De Beradinis
Command Sergeant Major: CSM Warren Lamonte Williams
Chief Public Affairs: Harvey Keeve

US Army Natick Research, Development & Engineering Center
Natick, MA 01760-5012
☎ (+1 508) 651 4300; 651 5340 ⊜ (+1 508) 651 5390
Commander: Col Morris E. Price (Jr) ☎ (+1 508) 651 4206
Technical Dir: Philip Brandier (acting)
Deputy Commander: Lt Col David F. Faulkner ☎ (+1 508) 651 4214
Deputy for Installation and Support Services: Thomas A. Rhodes ☎ (+1 508) 651 4205
Associate Technical Dir for Engineering & Acquisition: Lt Col David F. Faulkner ☎ (+1 508) 651 4612
Associate Technical Dir for Technology: Ms Colleen Cathcart ☎ (+1 508) 651 4243
Mobility Directorate: Edward J. Doucette ☎ (+1 508) 651 4636
Survivability Directorate: Ronald Granchelli (ACTG) ☎ (+1 508) 651 4308
Science and Technology Directorate: Dr Abner S. Salant ☎ (+1 508) 651 4577
Sustainability Directorate: Gerald A. Darsch (ACTG) ☎ (+1 508) 651 4402
Advanced Systems Concept Directorate: Dr Matthew L. Herz ☎ (+1 508) 651 4793
Natick Acquisition Center Dir: Ms Cheryl A. Deluca ☎ (+1 508) 651 4514

US Army Research Institute of Environmental Medicine
Natick, MA 01760-5000
☎ (+1 508) 651 4811 ⊜ (+1 508) 651 5298
Commander: Col Joel T. Hiatt ☎ (+1 508) 651 4811
Deputy Commander: Lt Col David Penetar ☎ (+1 508) 651 4811

Contact Reference

US Navy Clothing & Textile Research Facility
Natick, MA 01760-5000
☎(+1 508) 651 4172 ⊜ (+1 508) 651 4783
Officer in Charge: Cdr Keith T. Adams
☎(+1 508) 651 4172
Technical Dir: Dr Barbara Avellini ☎(+1 508) 651 4172
97

25-32920 US Army Test & Evaluation Command (TECOM)

Aberdeen Proving Ground, Ryan Bldg,
Aberdeen, MD 21005-5055
☎(+1 410) 278 1003 ⊜ (+1 410) 278 9029
Commanding General: Maj.Gen. Richard W. Tragemann
☎(+1 410) 278 1003
Deputy Commander: Col James Kriebel
☎(+1 410) 278 1009
Chief of Staff: Col Roger F. Brown ☎(+1 410) 278 1011
Technical Dir: Raymond G. Pollard ☎(+1 410) 278 1016
Procurement Dir: Saul S. Goldberg ☎(+1 410) 278 1191
Test & Assessment Dir: Gary L. Holloway
☎(+1 410) 278 1315
Corporate Information & Technology Dir:
Edward G. Stauch ☎(+1 410) 278 1465
Personnel Dir: Col Robert Patrick ☎(+1 410) 278 1155
Engineering Dir: Thomas L. Roller ☎(+1 410) 278 1118
Security & Intelligence Dir: James F. Stewart
☎(+1 410) 278 1279
Chief Public Affairs Officer: Gary A. Holloway
☎(+1 410) 278 1141
→ 60

US Army Garrison
Aberdeen Proving Ground, Maryland 21005-5001
☎(+1 410) 278 4006 ⊜ (+1 410) 278 2001
Commander: Col Jame M. Bosley

US Army Aberdeen Test Center
Aberdeen Proving Ground, Maryland 21005-5059
☎(+1 410) 278 2740 ⊜ (+1 410) 278 2283
Commander: Col Richard O. Bailer
Technical Dir: James W. Fasig

US Army Cold Regions Test Center
APO AP, Alaska 96508-7850
☎(+1 907) 873 4215 ⊜ (+1 907) 873 1990
Commander: Lt Col Gary E. Payne
Technical Dir: Jerold Barger

US Army Electronic Proving Ground
Ft Huachuca, Arizona 85613-7110
☎(+1 602) 538 6891 ⊜ (+1 602) 538 6361
Commander: Col Wayne L. Sittler

US Army Aviation Technical Test Ctr
Ft Rucker, AL 36362-5276
☎(+1 205) 255 8000 ⊜ (+1 205) 255 8232
Commander: Col Jack O. Shafer
Technical Dir: Jim McCrory

US Army Dugway Proving Ground
Dugway, UT 84022-5000
☎(+1 801) 831 3314 ⊜ (+1 801) 831 3096
Commander: Col James R. King
Technical Dir: William J. Haslem

US Army White Sands Missile Range (WSMR)
White Sands, NM 88002-5031
☎(+1 505) 678 1101 ⊜ (+1 505) 678 5950
Commander: Brig.Gen. Jerry L. Laws
Technical Dir: George Orlicki

US Army Yuma Proving Ground
Yuma, AZ 85365-9124
☎(+1 602) 328 2163 ⊜ (+1 602) 328 6249
Commander: Col Richard R. Walker
Technical Dir: James L. Wymer

US Army Redstone Technical Test Ctr
Redstone Arsenal, Redstone, AL 35898-8052
☎(+1 205) 876 3552 ⊜ (+1 205) 876 9866
Director: Larry H. Johnson
Deputy Dir: Carl E. Roberts
97

25-32940 US Army Troop Support Command (TROSCOM)

4300 Goodfellow Boulevard, St Louis, MO 63120-1798
☎(+1 314) 263 2247
Commanding General: Maj.Gen. Thomas L. Prather
Aide-de-Camp: Capt. Olger D. Burch
Deputy to the Commander: James B. Emahiser
Chief of Staff: Col David Archer ☎(+1 314) 263 2208
Material Readiness Dir: Col John J. Sosnowski
☎(+1 314) 263 3176
Troop Support Acquisition Center Dir: Harold Mabrey
☎(+1 314) 263 23 06
Public Affairs Officer: Paul E. Powell (Jr)
☎(+1 314) 263 9075
97

25-33000 US Army Industrial Operations Command (IOC)

Rock Island, IL 61299-6000
Commanding General: Maj.Gen. James W. Monroe
☎(+1 309) 782 6111
Deputy Commander & Executice Dir: Col David M. Emling
☎(+1 309) 782 5944; 782 5140
Chief of Staff: Col Jerald E. Kleager ☎(+1 309) 782 6511
Secretary of the General Staff: Maj. J. Eric Dietz
☎(+1 309) 782 5621
Business Relations: Craig Rel ☎(+1 309) 782 5880
Small Business: Bruce Myers ☎(+1 309) 782 7302
Public Affairs Officer: Stephen D. Abney
☎(+1 309) 782 5421 ☎(+1 309) 782 5011 (fax)

Rock Island Arsenal
Rock Island, IL 61299-5000
☎(+1 309) 782 6035
Commander: Col John L. Storm
Public Affairs Officer Rock Island: Ms Vicki M. Stapes
☎(+1 309) 782 4786
■ HQ Army armament, munitions & chemical command.
97

25-33015 US Army Defense Ammunition Center

3700 Army Depot Road, Savanna, IL 61074-9639
☎(+1 815) 272 89 01 W http://www.dac.mil
⊜ (+1 815) 273 89 85
Director: Larry V Gulledge ☎(+1 815) 273 89 01
Training Directorate: C P Stroo ☎(+1 815) 273 89 39
Technology Directorate: J Q Wheeler
☎(+1 815) 273 80 84
Operations Directorate: A J Podsiadlo
☎(+1 815) 273 80 49
Engineering Directorate: W F Ernst ☎(+1 815) 273 89 22
Public Affairs Officer: Mrs L K Theis ☎(+1 815) 273 89 50
98

25-33020 US Army Missile Command (USAMICOM)

Redstone Arsenal, Alabama 35898-5000
☎(+1 205) 876 41 61; 21 51
W http://www.redstone.army.mil
Commanding General: Maj Gen James M Link
☎(+1 205) 876 21 01
Deputy to the Commanding General: Ernest A. Young
☎(+1 205) 876 2875
Chief of Staff: Col Bruce A. Block ☎(+1 205) 876 1120
Command Sergeant Major: CSM B.F. Sundey
☎(+1 205) 876 1874
Deputy Dir Weapons Systems Management:
Ted W. Gandy
Public Affairs Officer: David G. Harris
☎(+1 205) 876 4161

Research Dev. & Engineering Center (AMSMI-RD)
Redstone Arsenal, Bldg 5400, Huntsville, AL 35898-5240
Director: Dr William C. McCorkle ☎(+1 205) 876 3322
Deputy Dir: Col Donald H. Watt ☎(+1 205) 876 3382
Chief of Procurement Office: Sharon Zirbel (acting)
☎(+1 205) 876 7307

Integrated Materiel Mgmt Center (AMSMI-MMC)
Redstone Arsenal Bldg 5681, Huntsville, AL 35898-5230
Director: James L. Flinn (III) ☎(+1 205) 876 3106
Deputy Dir: John R. Chapman ☎(+1 205) 876 3108

Acquisition Center (AMSMI-AC)
Redstone Arsenal, Bldg 4488, Huntsville, AL 35898-5280
Director: Marlene Cruze ☎(+1 205) 876 7161
Contract Operations Dir: James Brannon
☎(+1 205) 876 1161

Defense Megacenter-Huntsville
Redstone Arsenal, Bldg 5201, Huntsville, AL 35898-5170
Director: James C. Musch ☎(+1 205) 876 0048
Technical Dir: Charles R. Raley ☎(+1 205) 876 4822

Program Executive Office for Missile Defense
Redstone Arsenal, Bldg 5250, Huntsville, AL
Program Executive Officer: Brig.Gen. Richard Black

Program Executive Office Tactical Missiles (SFAE-MSL)
Redstone Arsenal, Huntsville, AL 35898-8000
Program Exec. Officer (SFAE-MSL): George G. Williams
☎(+1 205) 876 0714
Exec. OFCR/R&D Co-ordination: Capt. Maxwell G. Carroll
☎(+1 205) 876 0714

Redstone Arsenal
Huntsville AL 35898-5010
☎(+1 205) 876 2151

■ HQ Army Missile Command, Strategic Defense Command, Ordnance, Missile & Munitions Center & School, US Army Test Measurement/Diagnostic Equipment Activity.
97

25-33050 US Army Armor Center & School

Fort Knox, KY 40121-5000
☎(+1 502) 624 1101/1102/2121
⊜ (+1 502) 624 6434
Commanding General & Cdt US Army Armor School:
Maj.Gen. L. Maggart
Chief of Staff: Maj.Gen. C. Melton
Garrison Commander: Col D. Harris
Chief of Staff: Col J. Veach
97

25-33100 US Army Air Defense Artillery Center and Fort Bliss HQ

Fort Bliss, TX 79916
☎(+1 915) 568 38 98; 568 34 01
⊜ (+1 915) 568 56 15
Commanding General: Maj Gen John Costello
Deputy Commanding General: Brig Gen Dennis D Cavin
Public Affairs Officer: Mrs Jean Offutt
Chief of Staff: Col Jeffrey W Gault
98

25-33150 US Army Operational Test & Evaluation Command

Park Center IV, 4501 Ford Avenue,
Alexandria, VA 22302-1458
☎(+1 703) 681 93 62; 681 93 61; 681 93 63
W http://www.optec.army.mil ⊜ (+1 703) 681 37 79
Commander: Maj Gen Larry G Lehowicz
Chief of Staff: Col David B Kent
Public Affairs Officer: Ms Cynthia McGovern
→ 60
98

25-33200 US Army Aviation Branch & Center

Fort Rucker, AL 36362-5000
☎(+1 334) 255 26 06; 255 26 00
Commanding General: Maj Gen Daniel J Petrosky
☎(+1 334) 255 26 06
Deputy Commander: Brig Gen Burt S Tackaberry
☎(+1 334) 255 28 08
Deputy Assistant Commandant of US Army Reserve:
Col Hubert Reed ☎(+1 334) 255 11 30
Deputy Assistant Commandant of National Guard:
Lieut Col Robert Godwin ☎(+1 334) 255 35 31
Chief of Staff: Col Jan E Callen ☎(+1 334) 255 2500
Deputy Chief of Staff: Lieut Col Donald Isbell
☎(+1 334) 255 90 26
Combat Developments Director: Col Jesse Danielson
☎(+1 334) 255 2703
Evaluation and Standardisation:
Lieut Col Michael Bozeman ☎(+1 334) 255 26 03
Training, Doctrine and Simulation: Col William Powell
☎(+1 334) 255 33 20
Commander 1st Aviation Brigade (Air Assault):
Col Mark Wentlent ☎(+1 334) 255 90 83
Commander of Aviation Training Brigade:
Col Michael Mulvenon ☎(+1 334) 255 26 07
Command Sergeant Major: CSM Marvin Horne
☎(+1 334) 255 35 49
Army Air Traffic Control Activity Director:
Lieut Col Donald Stuck ☎(+1 334) 255 32 19
Inspector General: Lieut Col Curtis Herr Boldt
☎(+1 334) 255 21 18
Resource Management Director: Lieut Col Karen Lloyd
☎(+1 334) 255 31 58
Plans, Training, Mobilisation and Security Director:
Lieut Col Richard Johnson ☎(+1 334) 255 23 00
Information Management Director: Floyd Leighton
☎(+1 334) 255 21 74
Public Works Director: Col Kenneth Clow
☎(+1 334) 255 21 13
Logistics Director: Ms Jane Gillis ☎(+1 334) 255 24 00
Contracts Director: Ms Gloria Wheeler
☎(+1 334) 255 34 04
Reserve Component Support: Col James Mowery
☎(+1 334) 255 11 30
Adjutant General: Milton Wesley ☎(+1 334) 255 35 25
Public Affairs Officer: Maj James R Yonts

US Army Aviation Technical Test Center
Edward Air Force Base, CA 93523
Commander: Col Shafer ☎(+1 334) 255 80 00
Test Support Director: Lieut Col Cole
☎(+1 334) 255 80 92
Flight Systems Test Director: Maj Waller
☎(+1 334) 255 81 49

Airworthiness Qualification Test Director: Lieut Col Walsh
☎(+1 805) 527 39 01

US Army Aeromedical Center
Fort Rucker, Building 301, AL 36362-5000
☎(+1 334) 255 73 59 ⊖ (+1 334) 255 79 90
Commander: Col Craig Urbauer

US Army Aeromed Research Laboratory
Fort Rucker, Building 6901, Al 36362-5000
☎(+1 334) 255 69 17
Commander: Col Dennis Shanahan
☎(+1 334) 255 69 17; 255 69 37 (fax)
Deputy Commmander: Lieut Col Collins
☎(+1 334) 255 69 14
Programs and Plans Director: Dr Kimball
☎(+1 334) 255 69 08 [98]

25-33210 US Army Aviation Logistics School
Fort Eustis, VA 23604-5414
☎(+1 804) 878 68 47
Assistant Commandant: Col Hoppes
Aviation Trades Training Director: Lieut Col Mathewson
☎(+1 804) 878 54 05
Department of Aviation Systems Training: Major Enderk
☎(+1 804) 878 50 39
Attack Helicopter Training Director: Lieut Col Meyers
☎(+1 804) 878 45 64 [98]

25-33400 US Army Signal Center & Fort Gordon
Fort Gordon, GA 30905-5000
☎(+1 706) 791 7003/6001 ⊖ (+1 706) 793 6095
Commanding General: Maj.Gen. Douglas D. Buchholz
☎(+1 706) 791 4588
Deputy Commanding General:
Brig.Gen. Jerry W. McElwee ☎(+1 706) 791 3185
Chief of Staff: Col Randy C. Hinds ☎(+1 706) 791 2114
Garrison Commander: Col Larry C. Crockett
☎(+1 706) 791 6300
Contracting Dir: Ms Patricia D. Wallace
☎(+1 706) 791 1800
Public Works Dir: Lt Col David P. Plank
☎(+1 706) 791 3225
Information Management Dir: Jose C. Miller
☎(+1 706) 791 5381
Community Activities Dir: Charles H. Large
☎(+1 706) 791 4140
Plans, Mobilisation & Security Dir:
Lt Col Christopher D. Feeney ☎(+1 706) 791 2634
Resource Management Dir: Col Laszok
☎(+1 706) 791 3117
Regimental Training Dir: Col Richard Lee
☎(+1 706) 791 6206
Human Resourses Dir: W.H. Sharpe
☎(+1 706) 791 3840
Combat Developments Dir: Col Robert L. Forrester
☎(+1 706) 791 3985
Public Affairs Officer: James L. Judgings
☎(+1 706) 791 7003 [97]

25-33500 US Army Infantry Headquarters Fort Benning
Fort Benning, GA 31905-5000
☎(+1 404) 545 2211; 545 5111
👥 18 385 (active duty)
Commanding General/Commandant:
Maj.Gen. John W. Hendrix
Deputy Commanding General, Assistant Commandant:
Brig.Gen. Robert T. Clark
Chief of Staff US Army Infantry Center:
Col Gregory C. Camp
Chief of Staff USAIS: Col Johnny Retzell
☎(+1 404) 545 5231
Garrison Commander: Col Richard L. Rutledge
Commander Infantry Training Center: Col Johnny Brooks
☎(+1 404) 544 8216
Command Sergeant Major US Army Center:
CSM George Conrad
Command Sergeant Major US Army Infantry School:
CSM Sam Spears (III)
Engineering Dir: Col W.D. Brown ☎(+1 404) 545 5515
Logistics Dir: Dale Ellis ☎(+1 404) 545 5151
Public Affairs Officer: Rich McDowell
☎(+1 404) 545 2211 [97]

25-36500 Milan Army Ammunition Plant
General Dynamics Ordnance Systems
Milan, TN 38358
☎(+1 901) 686 69 90 ⊖ (+1 901) 686 69 30
Commander: Lieut Col Billy J Dowdy
☎(+1 901) 686 60 87
Executive Officer: Capt Thomas J Langowski
☎(+1 901) 686 67 84
Vice President and Plant Manager: Ms Janet M Kendrick
Chief of Contract Management Division: Lawrence E Smith
☎(+1 901) 686 62 44
Chief of Safety and Security Office: Ronald A Neff
☎(+1 901) 686 62 47
Chief of Industrial Operations Division: Patrick H Brew
☎(+1 901) 686 62 51
Industrial Relations Director: Ms Debra Lee
Engineering Director: M Harris
Quality Director: R Adcock
Administration Co-ordinator and Plant Manager: L Buckner
■ Ammunition plant operated by Lockheed Martin. Sole US manufacture of the M433 40mm HEDP grenade and M430 long-range version. 105mm Tank Training Rounds, M490 and M724, 105mm Tank D.U. Rounds, M833, M900, 81mm Mortars HE M889 and M821, 60 mm Mortars HE M888 and M720, 120mm Mortars HE M933 and M934, M74 Grenade, Mortar Propelling Charges and Ignition Cartridges. [98]

25-37200 US Army CECOM Research, Development & Engineering Center
(Night Vision & Electronic Sensors Directorate)
10221 Burbeck Rd, Suite 430,
Fort Belvoir, VA 22060-5806
☎(+1 703) 704 1158 ⊖ (+1 703) 704 1215
Director: Dr Rudolf G. Buser ☎(+1 703) 704 1172
Ground/Air Sensor Systems Div. Dir: Donald Jenkins
☎(+1 703) 704 1129
Science & Technology Dir: Dr John Pollard
☎(+1 703) 704 1199
Countermine/CCD Div.: Dr David Heberlein
☎(+1 703) 704 1040
Survivability/Radar Div. Dir: Joseph Swartz
☎(+1 908) 544 3123
Tech. Support & Operations Div. Dir: Larry Fillan
☎(+1 703) 704 1168
Chief Productability/ Affordability Branch: James Schaffer
☎(+1 703) 704 1319
Chief Programme Support Branch: Al Wedemeyer
☎(+1 703) 704 1433
Chief Visionics Modeling & Simulation Branch:
Ms Luanne Obert ☎(+1 703) 704 1754
A/Chief IR Technology Branch: Stu Horn
☎(+1 703) 704 2025
Chief Advanced Laser Branch: Ward Trussel
☎(+1 703) 704 1355
Chief Soldier Systems Branch: Martin Weaver
☎(+1 703) 704 1297
Chief Ground Vehicle Branch: Dennis Van Derlaske
☎(+1 703) 704 1258
Chief TECRAS/Physical Security Branch: Fred Petito
☎(+1 703) 704 1734
Chief Mine Detection Branch: Robert Barnard
☎(+1 703) 704 1066
Chief Environmental Branch: Ms Beverly Briggs
☎(+1 703) 704 1073
Chief Mine Neutralisation Branch: Richard Weaver
☎(+1 703) 704 1090
Chief CCD Branch: Grayson Walker ☎(+1 703) 704 2594
Chief Radar CI Branch: Allen Tarbell ☎(+1 908) 544 3103
Chief Survivability Branch: Allen Lisuzzo
☎(+1 908) 544 3106
Chief Resource & Planning Branch: Ms Kathy Yamarik
☎(+1 703) 704 1144
Chief Facilities Support Branch: Ronald Petrie
☎(+1 703) 704 1136
A/Chief Airborne Applications Branch: Don Reago
☎(+1 703) 704 1301
Program Information Officer: Ms Martha McCaslin
☎(+1 703) 704 1158
■ Develops equipment to support the Combat Support (combat engineer equipment) and the Combat Service Support (logistics equipment) elements of the US Army.
→ 60 [97]

25-38980 US Army Tank-Automotive and Armaments Command (ARDEC)
(part of the US Army Materiel Command)
Picatinny Arsenal, NJ 07806-5000
Armament Research, Development and Engineering Centre
☎(+1 201) 724 40 21
Commander: Brig.Gen. James W Boddie (Jr)
Deputy Commander: Col Jay D Johnson
Chief of Staff: Col Brian F Lee
Technical Director: Carmine Spinelli
Public Affairs Officer: Peter J Rowland
☎(+1 201) 724 63 65; 724 63 64
Project Manager for Tank Main Armament Systems:
Col Raymond Pawlicki
Project Manager for SADARM: Col James E Unterseher
Project Manager for Paladin: E C Gagnor
Project Manager for 9mm Pistol Small Arms: Lt Col Irish
Project Manager AFAS, Crusader: Col William Sheaves (III)
Project Manager Mines, Countermines and Demolitions:
Col Thomas Dresen
Product Mortar Systems Manager: Lt Col Steve Davis
Army Fuze Management Office: L J McConnell
→ 60 [98]

25-39200 US Army War College
Carlisle Barracks, Carlisle, PA 17013-5050
☎(+1 717) 245 41 01; 44 00
⊖ (+1 717) 245 42 24; 36 03
Commandant: Maj.Gen. Richard A. Chilcoat
Deputy Commandant: Col E.L. Roper
Public Affairs Officer: Lt Col John W. Falkenbury
☎(+1 717) 245 4101

Strategic Studies Institute
Carlisle Barracks, Carlisle, PA 17013-5050
☎(+1 717) 245 4212 ⊖ (+1 717) 245 3820
Director: Col R.H. Witherspoon
Research Dir: Dr E.H. Tilford

US Army Military History Institute
Upton Hall, Carlisle Barracks, Carlisle, PA 17013-5050
☎(+1 717) 245 4134 ⊖ (+1 717) 245 4370
Director: Col S.L. Bowman
Academic Affairs Dir: Col R.A. Brace
Dept of Command, Leadership & Management Dir:
Col H.F. Harback
Department of National Security & Strategy Dir:
Dr G.L. Guertner
Dept of Military Strategy, Planning & Operations Dir:
Col G.F. Crider

US Army Physical Fitness Research Institute
Carlisle Barracks, Carlisle, PA 17013-5050
☎(+1 717) 245 4511 ⊖ (+1 717) 245 4558
Director: Col P.T. Harig
Center Dir for Strategic Leadership: Prof. D.B. Campbell
Operations Group Dir: Col R. Smith [97]

25-39300 US Army Military Police School
Fort McClellan, AL
☎(+1 205) 848 4334
Commandant: Brig.Gen. D. Foley
Assistant Commandant: Col M. Sudnik
☎(+1 205) 848 3626
Command Sergeant Major: CSM Malavet
☎(+1 205) 848 3603 [97]

25-39400 US Army Command and General Staff College
Fort Leavenworth, KS 66027-1352
☎(+1 913) 684 24 80 ⊖ (+1 913) 684 37 49
Deputy Commandant: Brig Gen Joseph R Inge [98]

25-39450 US Army Adjutant General School
(Sub. to US Army Training & Doctrine Command - TRADOC)
Fort Jackson, SC
☎(+1 803) 751 8440
Commandant: Col G. Gresh
Assistant Commandant: Lt Col T. Wikstrom
Sergeant Major: SGM Nicholson [97]

25-41300 Space & Naval Warfare Systems Command Headquarters (SPAWAR)
2451 Crystal Drive, 5 Crystal Park,
Arlington, VA 22245-5200
☎(+1 703) 602 9831 ⊖ (+1 703) 602 1659
Commander: RAdm. George F.A. Wagner
☎(+1 703) 602 3006
Deputy Commander: Dr Dale G. Uhler
☎(+1 703) 602 4967
DEO for Space Comms & Sensors: John W. Desalme
☎(+1 202) 602 11897
Chief Engineer: RAdm. Osie V. Combs
☎(+1 703) 602 8873
Deputy Chief Engineer for Logistics:
Capt. Gerard A. Van Houweling ☎(+1 202) 602 8874

Advanced Technology & Prototype Systems: RAdm. L. Ed Allen ☎(+1 703) 602 9223
Space Technology Systems Directorate: Capt. Dwight Denson ☎(+1 703) 767 9600
Information Support Systems Directorate: Capt. Michael C. Bachmann ☎(+1 703) 602 2747
Dir for Corporate Operations Organization Staffing: Ronald Knudsen ☎(+1 703) 602 1006
Dir Information & Electronic Warfare Systems Program: Capt. Charles T. Ristoercelli ☎(+1 703) 602 4812
Command & Control & Comms Systems: Robert J. Martin ☎(+1 703) 602 8874
Intelligence, Surveillance & Recconaisance Systems: Capt. Gerald K. Nifintoff ☎(+1 703) 602 1120
Marine Corps Liaison: Col Steven W. Mock ☎(+1 703) 602 9800
Inspector General: Capt. Robert R. Rossi ☎(+1 703) 602 1554
Comptroller: Capt. Michael T. Rader ☎(+1 703) 602 3260
Contract Dir: Capt. Jay M. Cohen ☎(+1 703) 602 7777
Congressional & Public Affairs: Lt Cdr Rebecca Brenton ☎(+1 703) 602 9831
Director for Small & Disadvantaged Business: Ms Betty S. Geesey ☎(+1 703) 602 9844
Scientific & Technical Intelligence Liaison Officer: Ms Gloria J. Whitley ☎(+1 703) 602 7606
Command Master Chief: CTRCM James A. Harmon ☎(+1 703) 602 7550

US Naval Command & Control and Ocean Surveillance Centre San Diego (NCCOSC)
53660 Oceanview Drive, Suite 317, San Diego, CA 92147-5088
☎(+1 619) 553 9396 ⊜ (+1 619) 553 0152
Commander: Capt. A.W. Lengerich ☎(+1 619) 553 0170

Washington Office (NCCOSC)
2451 Crystal Drive, Room 1100, Arlington, VA 22245-5200
☎(+1 703) 602 2766 ⊜ (+1 703) 602 1071
Commander: Capt. A.W. Lengerich ☎(+1 703) 602 2766

Research, Development, Testing & Evaluation Division (RDTE)
53560 Hull Street, San Diego, CA 92152-5001
☎(+1 619) 553 3000 ⊜ (+1 619) 553 5188
Commander: Capt. Kirk Evans ☎(+1 619) 553 3000

RDTE Division Detachment Warminster
PO Box 5152, Warminster, PA 18974-0591
☎(+1 215) 441 2350 ⊜ (+1 215) 441 1998
Director: Walter J. Schoppe

RDTE Activity Pacific
675 Lehua Avenue, Pearl City, HI 96782-3356
☎(+1 808) 474 3166 ⊜ (+1 808) 474 7685
Officer in Command: Cdr Sherman Metcalf

RDTE Facility
PSC 488 Box 128, Guam
☎(+671 355) 515 951 54
Officer in Command: Cdr Gary G. Durante

NISEEAST Charleston
4600 Marriott Drive, N.Charleston, SC 29406-6504
☎(+1 803) 974 5000 ⊜ (+1 803) 743 1866
Commander: Capt. Ronald Polkowsky

NISEEAST Detachment Norfolk
Box 1376, Norfolk, VA 23501-1376
☎(+1 804) 558 6601 ⊜ (+1 804) 396 1957
Officer in Command: Cdr William Rodriguez

NISEEAST Detachment, St Ingoes
Villa Road, St Inigoes, MD 20684-0010
☎(+1 301) 862 8721 ⊜ (+1 301) 862 8743
Resident in Charge: Rey Bald

Navy Management Systems Support Office (NAMASSO)
1441 Crossway Blvd, Cheapeake, VA 23320-2842
☎(+1 804) 523 8000 ⊜ (+1 804) 523 8035
Commanding Officer: Capt. Dana Fuller 97

25-41450 Naval Air Warfare Center Weapons Division

China Lake, CA
☎(+1 619) 939 9011 (operator)
☎(+1 619) 939 0050 (information line)
⊜ (+1 619) 939 2056
Commander: RAdm Dana B. McKinney ☎(+1 619) 939 2201
Vice Commander: Capt. Jack Dodd ☎(+1 805) 989 7113
Deputy Commander for R&D: Sterling Haaland ☎(+1 619) 939 3409
Executive Assistant: Cdr L. Langford ☎(+1 619) 939 2290
Public Affairs Officer: Steven F. Boster ☎(+1 619) 939 3511

Naval Air Weapons Station
Nr Ridgecrest, China Lake, CA 93555-6001
Commanding Officer: Capt. Stanley Douglas ☎(+1 619) 939 2211
Executive Officer: Cdr Brian Yri ☎(+1 619) 939 3511 97

25-41455 Naval Air Warfare Center Weapons Division

Point Mugu, CA 93042-5001
☎(+1 805) 989 8094 (Public Affairs)
⊜ (+1 805) 989 1785
Commander: RAdm. Dana B. McKinney
Public Affairs Officer: Alan Alpers

Naval Air Weapons Station
Point Mugu, CA 93042-5001
Commander: Capt. Stephen D. Beal 97

25-41500 Naval Space Command HQ

Dahlgren, VA 22448-5300
☎(+1 703) 663 6100
Commander: RAdm. Katharine L. Laughton (USN)
Deputy Commander: Col Victor Simpson (USMC)
Public Affairs Officer: Gary R. Wagner 97

25-41630 Naval Atlantic Undersea Test & Evaluation Center (AUTEC)

Naval Undersea Warfare Center, PO Box 24619, West Palm Beach, FL 33416-4619
☎(+1 407) 832 8590; 8566 (West Palm Beach)
⊜ (+1 407) 832 8566
Technical Dir: Dr Sirmalis
Program Mgr: Thomas Zeh 97

25-41660 Navy Amphibious Base Coronado

San Diego, CA 92155-5000
☎(+1 619) 437 3024 ⊜ (+1 619) 437 0970
Commanding Officer: Capt. Edward W. Kelly ☎(+1 619) 437 2078
Exec. Officer: Cdr Michael Meyer
Public Affairs Officer: William C. Bartkus ☎(+1 619) 437 3024 97

25-41710 US Naval Diving & Salvage Training Center

Panama City, FL 32407-5002
☎(+1 904) 234 4651 ⊜ (+1 904) 235 5253
Commander: Cdr Michael C. Herb
Procurement/Logistics Dir: Lt Robert L. Williams
Public Affairs (HTC/SW/DV): Marc A. Tochterman 97

25-41800 Naval Research Laboratory (NRL)

Code 1230, Washington, DC 20375-5320
☎(+1 202) 767 2541 ⊜ (+1 202) 767 6991
Commanding Officer: Capt. B.W. Buckley ☎(+1 202) 767 3403
Director Naval Center for Space Technology: P.G. Wilhelm ☎(+1 202) 767 6547
Associate Dir of Research (Directorate Systems): R. LeFande ☎(+1 202) 767 3324
Associate Director of Research (Material Science & Tech.): Dr B.B. Rath ☎(+1 202) 767 3566
Assoc. Dir of Research Ocean & Atmosphere Science & Tech Dir: Dr E.O. Hartwig ☎(+1 202) 404 8690
Associate Director of Research: R. Doak ☎(+1 202) 767 2371
Director of Research: Dr T. Coffey ☎(+1 202) 767 3301
Contracting Manager: J.C. Ely ☎(+1 202) 767 5227
Financial Management: D.T. Green ☎(+1 202) 767 3405
Inspector General: Capt. R.E. Leonard ☎(+1 202) 767 3621
Supply Officer: Ms C. Hartman ☎(+1 202) 767 3446
Chesapeake Bay Manager: M. Kosky ☎(+1 410) 257 4002
Research & Development Support: S. Harrison (Acting) ☎(+1 202) 767 3371
Human Resources: Ms B. Duffield ☎(+1 202) 767 3421
Technical Information: P.H. Imhof ☎(+1 202) 767 2187
Superintendent Radar: Dr G.V. Trunk ☎(+1 202) 767 2936
Superintendent Information Technology: Dr R.P. Shumaker ☎(+1 202) 767 2903
Superintendent Optical Sciences: Dr T.G. Giallorenzi ☎(+1 202) 767 3171
Tactical Electronic Warfare: Dr J.A. Montgomery ☎(+1 202) 767 6278
Superintendent Chemistry: Dr J.S. Murday ☎(+1 202) 767 3026
Superintendent Material Science & Technology: Dr D.U. Gubser ☎(+1 202) 767 2926
Chief Scientist & Dir Fluid Dynamics/Comp. Physics Lab.: Dr Jay P. Boris ☎(+1 202) 767 3055
Superintendent Condensed Matter & Radiation Sciences: Dr D.J. Nagel ☎(+1 202) 767 2931
Superintendent Plasma Physics: Dr S. Ossakow ☎(+1 202) 767 2723
Superintendent Electronics Science & Technology: Dr G.M. Borsuk ☎(+1 202) 767 3525
Deputy for Space Systems: Dr R. LeFande ☎(+1 202) 767 3324
Superintendent Space Systems Development: R.E. Eisenhauer ☎(+1 202) 767 0410
Superintendent Spacecraft Engineering: H.E. Senasack (Acting) ☎(+1 202) 767 6407
Superintendent Acoustics: Dr E.R. Franchi (Acting) ☎(+1 202) 767 3482
Superintendent Remote Sensing: Dr P. Schwartz (Acting) ☎(+1 202) 767 3421
Superintendent Oceanography Div. Stennis Space Center: Dr W.B. Moseley ☎(+1 601) 688 4670
Superintendent Marine Geosciences Div.: Dr H.C. Eppert ☎(+1 601) 688 4650
Superintendent Marine Meteorological Div. Monterey: Dr J.B. Hovermale ☎(+1 408) 656 4721
Superintendent Space Science Division: Dr H. Gursky ☎(+1 202) 767 6343
Public Affairs Officer: Richard H. Baturin ☎(+1 202) 767 2541
→ 60

● 4555 Overlook Ave SW, Code 1230, Washington, DC 20375-5320
☎(+1 202) 767 2541 ⊜ (+1 202) 767 6991
Commanding Officer: Capt. B.W. Buckley 97

25-44210 Naval Surface Warfare Center Carderock Div. Detachment, SFTF

8010 N. Ocean Drive, Dania, FL 33004-3303
☎(+1 954) 926 4044; 4000 ⊜ (+1 954) 927 7898
Officer in Charge: Lt Cdr Robert C. Thorne ☎(+1 954) 926 4004
Detachment Dir: Robert J. King ☎(+1 954) 926 4000
Range Operations Mgr: Bill Paulson ☎(+1 954) 926 4003
Military Personnel Officer: YNC (SW) Raimundo Rivera ☎(+1 954) 926 4044

■ Supports the Naval Surface Warfare Center, Carderock Division in conducting air, surface & underwater systems testing under real life environmental conditions.

Provision of shipping & receiving, storage & consulting services. 97

25-44270 US Naval Pacific Meteorology and Oceanography Center West

(Joint Typhoon Warning Center)
PSC 489, Box 12, Guam 96536-0051
☎(+1 671) 349 4268 ⊜ (+1 671) 349 4032
Commanding Officer: Capt. James F. Etro
Exec. Officer: Cdr Charles M. White 97

25-44410 New Hampshire Satellite Tracking Station

Manchester, NH
☎(+1 603) 471 2000 ⊜ (+1 603) 471 2213
Commander: Lt Col Louis E. Christensen
Technical Dir: Maj. Louis E. Richardson
Public Affairs Representative: Sgt Lew D. Barron

23rd Space Operations Squadron
Amherst, NH 03031-1514
New Boston Air Station, 317 Chestnut Hill Road 97

25-44900 US Naval Academy

121 Blake Rd, Annapolis, MD 21402-5000
☎(+1 410) 293 2291 ⊜ (+1 410) 293 3133
Superintendent: Adm. Charles R. Larson ☎(+1 410) 293 2402
Public Affairs Officer: Capt. Tom Jurkowsky ☎(+1 410) 293 2291 97

25-44920 Naval Education and Training Center (NETC)

Newport, RI 02841
☎(+1 401) 841 3538 ⊜ (+1 401) 841 2655
Commander NETC: Capt. Ronald C. Bogle (USN) ☎(+1 401) 841 3715
Deputy Chief: Capt. Billie E. Crawford (USN) ☎(+1 401) 841 3932
Dir Operations: Cdr Michael J. Waite (USN)

Comptroller/Dir for Supply: Cdr Aubrey Lane (SC, USN)
Public Affairs Officer: David Sanders
☎ (+1 401) 841 3538 [97]

25-45350 US Marine Landing Force Training Command Atlantic

Amphibious Reconnaisance School
Fort Story, VA 23459
☎ (+1 757) 422 77 67 ⊖ (+1 757) 438 75 94
Commander: Maj Chris Kazmierczak
Deputy Commander: Capt Douglas Clark [98]

25-45400 Marine Corps Systems Command HQ

Command Support Directorate,
Quantico, VA 22134 5010
☎ (+1 703) 784 3636 ⊖ (+1 703) 784 3636
Commander: Maj.Gen. Carol A. Mutter
Command, Control, Communications Dir: Col J. D'Lugos
☎ (+1 703) 784 2820 ☎ (+1 703) 784 3212 (fax)
Amphibious Warfare Technology Dir: David A. Vaughn
☎ (+1 703) 784 2761 ☎ (+1 703) 784 2764 (fax)
Marine Corp. Tactical Systems Support Activity Dir:
Col David L. Chadwick ☎ (+1 619) 725 2617
☎ (+1 619) 725 2812 (fax)
Intelligence Systems Programme Mgr:
James R. McGillicudy ☎ (+1 703) 784 2581
☎ (+1 703) 784 3561 (fax)
Communications Systems Programme Mgr:
Ronald G. Shelkey ☎ (+1 703) 784 3162
☎ (+1 703) 784 3561 (fax)
Air Defense Systems Programs Mgr:
Col John R. Quattlebaum ☎ (+1 703) 784 2645
☎ (+1 703) 784 2655 (fax)
Command Information Systems Program Mgr:
Col James E. Vesely ☎ (+1 703) 784 2645
☎ (+1 703) 784 2655 (fax)
Common Computer Resources: Lt Col Christopher C. Platt
☎ (+1 703) 784 5610
☎ (+1 703) 784 5609; 5618 (fax)
Ammunition Programme Mgr: George J. Phillips
3033 Wilson Blvd Arlington, VA 22201-001
☎ (+1 703) 696 1163 ☎ (+1 703) 696 1171 (fax)
Combat Service Support Systems Mgr: Douglas A. Davis
☎ (+1 703) 784 4278 ☎ (+1 703) 784 4287 (fax)
Training Systems Programme Mgr:
Lt Col Michael G. Przepiora ☎ (+1 703) 784 3310
☎ (+1 703) 784 3241 (fax)
Ground Weapons Programme Mgr: Col Gary Dockendorff
☎ (+1 703) 784 2006 ☎ (+1 703) 784 5642 (fax)
Amphibious Assault Vehicle Program Mgr:
Edward M. Lerner ☎ (+1 703) 784 4507
☎ (+1 703) 784 4436 (fax)
Light Armored Vehicles Program Mgr: Col Richard L. Owen
US Army Tacom Warren, MI 48397 5000
☎ (+1 313) 786 8338 ☎ (+1 313) 574 5472 (fax)
Engineer System Mgr: Col William F. Thompson
☎ (+1 703) 784 2242 ☎ (+1 703) 784 3292 (fax)
Advanced Amphibious Assault Vehicle Mgr:
Col James M. Feigley ☎ (+1 703) 696 1105
☎ (+1 703) 696 1117 (fax)
→ 60 [97]

28 Coastguard, Border Protection Forces

28-40000 US Coast Guard Headquarters (USCG)

2100 Second St SW, Washington, DC 20593-0001
☎ (+1 202) 267 1587 ⊖ (+1 202) 267 4307
Commandant US Coast Guard: Adm. Robert E. Kramek
Vice Commandant US Coast Guard:
VAdm. Richard D. Herr
Chief of Staff: VAdm. James M. Loy
Chief Operations Directorate: RAdm. Norman T. Saunders
Chief Office of Acquisition: RAdm. Robert C. North
☎ (+1 202) 267 2007
Head Public Affairs: Capt. W. "Bud" Schneeweis
☎ (+1 202) 267 1587
Public Affairs Officer (G-CP-2): Lt Cdr Pat Philbin
☎ (+1 202) 267 1933 ☎ Fax (+1 202) 267 4307

US Coast Guard Pacific Area
Coast Guard Island, Alameda, CA 94501-5100
☎ (+1 510) 437 3324 ⊖ (+1 510) 437 5918
Commander: VAdm. Roger T. Rufe
Commander Pacific Maintenance & Logistic Command:
RAdm. Fred L. Ames

US Coast Guard Atlantic Area
Governors Island, New York, NY 10004-5099
☎ (+1 212) 668 7114
Commander: VAdm. Kent H. Williams
Commander Atlantic Maintenance & Logistic Command:
RAdm. Douglas H. Teeson (II)

First Coast Guard District
Coast Guard Bldg, 408 Atlantic Ave,
Boston, MA 2210-22091
☎ (+1 617) 223 8515 ⊖ (+1 617) 223 8523
Commander: RAdm. John L. Linnon (Jr)

Fifth Coast Guard District
431 Crawford St, Portsmouth, VA 23705
☎ (+1 804) 398 6272 ⊖ (+1 804) 398 6238
Commander: RAdm. Kent H. Williams

Seventh Coast Guard District
51 SW First Ave, Federal Building, Miami, FL 33130
☎ (+1 305) 356 5641 ⊖ (+1 305) 356 7402
Commander: RAdm. John W. Lockwood

Eighth Coast Guard District
500 Camp St, New Orleans, LA 70130
☎ (+1 504) 589 6198 ⊖ (+1 504) 589 2142
Commander: RAdm. Timothy W. Josiah

Ninth Coast Guard District
1240 East 9th St, Cleveland, OH 44199
☎ (+1 216) 522 3900 ⊖ (+1 216) 522 4447
Commander: RAdm. Gerald F. Woolever

Eleventh Coast Guard District
CG Island, Bldg 42, Alameda, CA 94501-5100
Commander: VAdm Roger T. Rufe

Thirteenth Coast Guard District
915 Second Ave, Seattle, WA 98174
☎ (+1 206) 220 7237 ⊖ (+1 206) 220 7245
Commander: RAdm. David J. Spade

Fourteenth Coast Guard District
300 Ala Moana Blvd, Honolulu, HI 96850
☎ (+1 808) 541 2051; 541 2123 ⊖ (+1 808) 541 2175
Commander: RAdm. Thomas H. Collins

Seventeenth Coast Guard District
PO Box 3-5000, Juneau, AK 99802
☎ (+1 907) 463 2065 ⊖ (+1 907) 463 2072
Commander: RAdm. Ernest R. Riutta

US Coast Guard Academy
New London, CT 06320
☎ (+1 203) 444 8270 ⊖ (+1 203) 444 8289
Superintendent: RAdm. Paul E. Versaw [97]

28-50000 US Coast Guard Research & Development Center

1082 Shennecossett Rd, Groton, CT 06340-6096
☎ (+1 203) 441 2600 ⊖ (+1 203) 441 2792
Commanding Officer: Capt. J.R. Taylor
Resource Dir: James Green
■ Marine safety, marine environmental protection, aids to navigation, marine fire research, buoy technology, search and rescue, marine engineering & human factors engineering. → 60 [97]

31 Embassies & Attachés

31-00080 Embassy of the Argentine Republic

1600 New Hampshire Avenue NW,
Washington, DC 20009
☎ (+1 202) 939 64 00 [E] feeeuu@atina.com
⊠ 197805 embargwash ⊖ (+1 202) 332 31 71
Ambassador: H E Dr Raúl Granillo Ocampo
Minister (Deputy Chief of Mission): Santos Goni-Marenco

Office of the Military Attaché
1810 Connecticut Avenue NW, Washington DC 20009
☎ (+1 202) 667 49 00/02; 316 85 86 ⊠ 440131 agmaui
⊖ (+1 202) 332 57 13
Military Attaché: Gen. Julio A Hang
Assistant Military Attaché: Col Gonzalo Gomez
Assistant Military Attaché: Lt Col Jorge E Beauchamp
Assistant Military attaché: Lt Col Alberto P Holgado

Office of the Naval Attaché
630 Indiana Avenue NW, Washington, DC 20004
☎ (+1 202) 626 21 00 ⊠ 89500 Navalgar
⊖ (+1 202) 626 21 80
Naval Attaché: RAdm. Norberto Pereiro
Assistant Naval Attaché: Capt. Jorge F Peyon

Office of the Air Attaché
2405 I Street NW, Washington, DC 20037
☎ (+1 202) 452 85 00
⊠ 197724 agraerusawsh; 892606 agrerusawsh
⊖ (+1 202) 822 80 51
Air Attaché: Brig. Ricardo J Ciaschini
Assistant Air Attaché: Com. Huberto Tulian
Assistant Air Attaché: Com. Ernestp Cuello

Office of the Economic Counsellor
1901 L Street, NW Suite 640, Washington, DC 20036
☎ (+1 202) 331 37 23 ⊖ (+1 202) 857 52 11
Minister (Economic & Commercial): Carlos A Magarinos

Office of the Financial Attaché
1600 New Hampshire Avenue NW,
Washington, DC 20009
☎ (+1 202) 939 64 57 ⊖ (+1 202) 588 06 50
Minister: Joaquin A Cottani [98]

31-00110 Embassy of the Commonwealth of the Bahamas

2220 Massachusetts Ave, NW, Washington, DC 20008
☎ (+1 202) 319 2660 ⊖ (+1 202) 319 2668
Counsellor Deputy Chief of Mission: Ms Sheila Carey
First Secretary & Consul: Mrs Edda Dumont-Adolph
Second Secretary & Consul: Eugene Newry
Attaché: Ms Monique Major [97]

31-00140 Embassy of Barbados

2144 Wyoming Avenue NW, Washington, DC 20008
☎ (+1 202) 939 92 00/02; 939 92 18/19
[E] barbados@oas.org ⊠ 64343 barwash
⊖ (+1 202) 332 74 67
Ambassador: H E Dr Courtney N Blackman
Minister/Counsellor: Ricardo R Browne
Commercial Counsellor: Ms Myrtle D Bishop
Press/Information Officer: Ms Annalee Babb [98]

31-00150 Embassy of Belgium

3330 Garfield St NW, Washington, DC 20008
☎ (+1 202) 333 6900 ✉ AMBABELGE Washington, DC
⊠ 892409 ambel wsh, 440139 ambel ui
⊖ (+1 202) 333 3079
Ambassador: H.E. Mr André Adam
Minister Deputy Chief of Mission: Patrick Van Haute
Defense & Armed Forces Attaché:
Brig.Gen. Jozef De Heyn
Asst Defense & Armed Forces Attaché:
Lt Col Jean-Pierre Bastin
Counsellor (Scientific): Peter Grognard
Minister-Counsellor (Economic): Patrick van Haute
Attaché (Commercial): Patrick Everarts de Velp
Information Officer: Ms Catherine Anderson

Military Supply Office
8810 Spectrum Drive, Landover, MD 20785
☎ (+1 301) 350 3900 ⊖ (+1 301) 350 5892
Chief Supply Office: Lt Col René Hancisse [97]

31-00240 Embassy of the State of Brunei Darussalam

Suite 300, 3rd Floor, Watergate,
2600 Virginia Avenue NW, Washington, DC 20037
☎ (+1 202) 342 01 59 ⊖ (+1 202) 342 01 58
Ambassador: H.E. Pengiran Anak Dato Puteh
Deputy Chief of Mission: Shofry Abdul Ghafor (Acting)
Second Secretary: Mrs Hasney Mat Yassin
Third Secretary (Consular): Buyong Arshad
Third Secretary (Finance and Administration):
Haji Morshidi Hj. Mohd. Yusof
Attache Communication: Hasnan Hj. Seruddin
Attache: Ms Zubaidah Daud [98]

31-00250 Embassy of the Republic of Bulgaria

1621 22nd St NW, Washington, DC 20008
☎ (+1 202) 387 7969 ⊖ (+1 202) 234 7973
Ambassador: H.E. Ms Snezhana Botusharova
Deputy Chief of Mission: Todor Staykov
Military, Naval & Air Attaché: Col Petar V. Petrov
First Secretary (Cultural): Boyan Papazov
First Secretary (Consul): Tsanko Bogdanov
First Secretary (Economic): Ivan Gospodov
Second Secretary Political: Valentin Popov
Second Secretary Trade, Science: Ms Nikola Mishev
Second Secretary Trade: Panteley Penev
Third Secretary Political: Vlado Lukanov
Third Secretary Press: Mihael Christov
Attaché Political: Kalin Tinchev [97]

31-00260 Embassy of Burkina Faso

2340 Massachusetts Ave NW, Washington, DC 20008
☎ (+1 202) 332 5577; 332 6895 ⊠ 440399 ambvol
⊖ (+1 202) 462 2865
Ambassador: H.E. Mr Gaetan R. Ouedraogo
Second Counsellor: Simplice H. Guibila

Contact Reference

Financial Attaché: Boureima Ouedraogo
Economic Counsellor: Souleymane Ouedraogo
Public Information Officer: Cécile P. Toé [97]

31-00310 Embassy of Canada

501 Pennsylvania Ave NW, Washington, DC 20001
☎ (+1 202) 682 1740 ⊠ BEAVER Washington
⊠ 89664 domcan a wsh ⊖ (+1 202) 682 7726
Ambassador: H.E. Raymond Chrétien
Deputy Chief of Mission: Douglas Waddell
Commercial Counsellor: Mark Bailey
Defense Counsellor: David Buxton
Counsellor (Defense Procurement): E. Lewicki
Commander Canadian Defence Liaison Staff: Maj.Gen. Donald Williams
Counsellor (Science & Technology): Dr D. Strange
Public Affairs Director: T.R. Colli

Office of Canadian Forces Attachés & Defence Liaison Staff (DLS)
501 Pennsylvania Ave NW, Washington, DC 20001
☎ (+1 202) 682 1740 ⊠ 89532 candssitc wsh
Military Attaché: Col W.M. Holmes
Naval Attaché: Lt Col J.B. Dick
Asst Military Attaché: Col K.R. Betts
Asst Naval Attaché: Cdr J.T. Legaaden
Asst Air Attaché: Lt Col J.M. Westrop
Counsellor (Defence Research): R Thomas
Attaché (Defence R&D): Dr B.P. Luker [97]

31-00350 Embassy of Chile

1732 Massachusetts Ave NW, Washington, DC 20036
☎ (+1 202) 785 1746 ⊠ 247003 echi ur
⊖ (+1 202) 887 5579
Ambassador: H.E. Mr John Bieh
Deputy Chief of Mission: Roberto Alonso Budge
Counsellor: Manuel Hinojosa
Counselor: Rodrigo Gaete
Counselor: William Patrickson
Second Secretary: Eduardo Escobar
Second Secretary: Antonio Pena
Second Secretary: Rodrigo Fernández
Second Secretary: Francisco Gormaz
Second Secretary: Ricardo Rojas
Third Secretary: Mathias Francke
Third Secretary: Ms María del Carmen Domínguez
Economic & Commercial Attaché: Mario Matheus

Office of Air Attaché
1920 L St NW, Suite 201, Washington, DC 20036
☎ (+1 202) 872 1334/35 ⊖ (+1 202) 872 1361
Air Attaché: Gen. Jaime Estay Vivero
Asst Air Attaché: Col José Ignacio Concha

Office of Military Attaché
2174 Wisconsin Ave NW, Washington, DC 20007
☎ (+1 202) 965 9661/62 ⊖ (+1 202) 965 9318
Military Attaché: Brig.Gen. Ricardo Izurieta Caffarena
Asst Military Attaché: Lt Col Jaime Brahm

Office of Naval Attaché
1736 Massachusetts Ave NW, Washington, DC 20036
☎ (+1 202) 466 2500 ⊖ (+1 202) 223 9733
Naval Attaché: RAdm. Onofre Torres Colvin
Asst Naval Attaché: Capt. Tomás Schlack [97]

31-00420 Embassy of the Republic of Côte d'Ivoire

2424 Massachusetts Avenue NW, Washington DC 20008
☎ (+1 202) 797 03 00 ⊖ (+1 202) 483 84 82
Ambassador: H E Mr Kofi Moise Koumoue
Counsellor: Fry Kouadio
Counsellor (Economic): Bah Jeannot Zoro Bi
Counsellor (Economic): Djelhi Y Adamoh
First Secretary (Consular and Cultural Affairs): Severin Mathias Akeo
First Secretary (Financial): Vincent N'Zi Yao [98]

31-00440 Embassy of the Republic of Cyprus

2211 R St NW, Washington, DC 20008
☎ (+1 202) 462 5772 ⊖ (+1 202) 483 6710
Ambassador: H.E. Mr Andreas J. Jacovides
Deputy Chief of Mission & Counsellor: Leonidas Markides
First Secondary & Consul: Pavlos Anastasiades
Counsellor (Press): Miltos Miltiadou
Press Attaché: Dr Lilian Tsappa ☎ (+1 202) 232 8993-4 [97]

31-00535 Embassy of Estonia

2131 Massachusetts Avenue NW,
Washington, DC 20008
☎ (+1 202) 588 01 01 [E] loa@estemb.org
⊖ (+1 202) 588 01 08
Commercial Counsellor: Okke Metsmaa

Estonian Mission to the United States
Suite 2415, 630 Fifth Avenue, New York, NY 10111
☎ (+1 212) 247 76 34 (consular department);
247 14 50 (visas) ⊖ (+1 212) 262 08 93
Ambassador Consul General: H E Mr Ersnt Jaakson

Honorary Consulate of Estonia
21515 Vanowen Street, Suite 211,
Canoga Park, Los Angeles CA 91303
☎ (+1 818) 884 58 50
[E] 73671.2064@compuserve.com
⊖ (+1 818) 593 29 73
Honorary Consul: Jaak Treiman

Honorary Consulate of Estonia
500 Union Street, Suite 930, Seattle 98101
☎ (+1 206) 467 14 44 ⊖ (+1 206) 467 81 29
Honorary Consul: Mart Kask

Estonian Embassy to the United Nations
630 Fifth Avenue, Suite 2415, New York NY 10111
☎ (+1 212) 247 04 99 [E] estun@undp.org
⊖ (+1 212) 262 08 93
Ambassador: H E Mr Trivimi Velliste [98]

31-00550 Delegation of the European Commission

2300 M St NW, 3rd floor, Washington, DC 20037
☎ (+1 202) 862 9500 ⊖ (+1 202) 429 1766
Ambassador: H.E. Mr Hugo Paemen
☎ (+1 202) 862 9510
Deputy Head of Delegation: John Richardson
☎ (+1 202) 862 9520
Counsellor (Science & Technology): José Costa
☎ (+1 202) 862 9575
Counsellor (Political Affairs): Lodewijk Briët
☎ (+1 202) 862 9595
Counsellor (Development): Aslam Aziz
☎ (+1 202) 862 9582
First Secretary (Economic Affairs): Werner Schüle
☎ (+1 202) 862 9580
First Secretary (Commercial Affairs): Bernd Langeheine
☎ (+1 202) 862 9525
First Secretary (Agriculture): Tasos Haniotis
☎ (+1 202) 862 9560
First Secretary/Transportation Energy & Environment: Yves Devellennes ☎ (+1 202) 862 9570
Press & Public Affairs: Theo Hustinx
☎ (+1 202) 862 9530 [97]

31-00570 Embassy of Fiji

Suite 240, 2233 Wisconsin Ave NW,
Washington, DC 20007
☎ (+1 202) 337 83 20 ⊖ (+1 202) 337 19 66
Chargé d'Affaires: Ratu Finau Mara [98]

31-00660 Embassy of the Federal Republic of Germany

4645 Reservoir Rd NW, Washington, DC 20007
☎ (+1 202) 298 4295
⊠ 248321 aawn ur, 440234 aawn ui
⊖ (+1 202) 298 4321
Ambassador: H.E. Mr Juergen Chrobog
Deputy Chief of Mission: Thomas Matussek
Counsellor (Economic & Commercial): Volker Schlegel
Counsellor (Science & Technology): Wolfram Schött
Defence Attaché: RAdm. Peter Schwitz
Naval Attaché: Capt. Friedrich Jacobi
Air Attaché: Col Michael Padberg
Military Attaché: Col Gerhard Stelz
Counsellor (Defense Research & Engineering): Diethelm Stephan

FRG Liaison Office of Armaments Sector for Defense Materiel
11150 Sunrise Valley Drive, Reston, VA 22091
☎ (+1 703) 715 8260 ⊖ (+1 703) 715 8240
Chief: Mr Gewehr [97]

31-00740 Embassy of Guatemala

2220 R St NW, Washington, DC 20008
☎ (+1 202) 745 4952/4 ⊖ (+1 202) 745 1908
Ambassador: H.E. Mr Pedro Miguel Lamport
Deputy Chief of Mission: Alfonso Quinonez
Defense Military Naval & Air Attaché: Col Roberto Letona
Commercial Attaché: Roberto Rosenberg
First Secretary: P. Mendoza [97]

31-00890 Embassy of Italy

1601 Fuller St NW, Washington, DC 20009
☎ (+1 202) 328 5500
Ambassador: H.E. Mr Ferdinando Salleo
☎ (+1 202) 328 5570; 5571
Deputy Chief of Mission: Antonio Puri Purini
☎ (+1 202) 328 5539; 5540
Counsellor Economic & Commercial Affairs: Mrs Anna Blefari Schneider
Scientific Attaché: Dr Emanuele Mannarino

Office of the Armed Forces Attachés
5301 Wisconsin Avenue, NW, Washington, DC 20015
☎ (+1 202) 966 3754 ⊖ (+1 202) 573 1109
Defence Technological Research Attaché: Col Ubaldo Serino ☎ (+1 202) 966 3809
Defence, Air & Defence Co-operation Attaché: Brig.Gen. Augusto Bertinaria ☎ (+1 202) 966 3809
Naval Attaché: Brig.Gen. Francesco Castagnetti
☎ (+1 202) 966 4069; 4071
Naval Attaché: Cdr Pietro Francini
☎ (+1 202) 966 4165; 4184
Asst Military Attaché: Col Claudio Periotto
☎ (+1 202) 966 4069; 4071
Asst Naval Attaché: Cdr Antonio Cervone
☎ (+1 202) 966 4165; 4184
Asst Air Attaché: Col Gianni Ammoniaci
☎ (+1 202) 966 3809 [97]

31-00910 Embassy of Jamaica

1520 New Hampshire Ave, NW, Washington, DC 20036
☎ (+1 202) 452 0660 ⊠ 64352 emjam uw
⊖ (+1 202) 452 0081
Ambassador: H.E. Dr Richard L. Bernal
Minister Counsellor: Ms Pamela Coke
Commercial Officer: Dr Winsome Leslie
Public Information Officer: Ms Hillary Brown [97]

31-00970 Embassy of Kyrgyzstan

Suite 705, 1511 K St.NW, Washington, DC 20005
☎ (+1 202) 347 3732; 3733 ⊖ (+1 202) 347 3718
Chargé d'Affairs: Almas Chukin
Second Secretary (Consul): Aibek Tilevalier
Press Attaché: Asan Usupov [97]

31-01015 Embassy of Latvia

4325 17th St, NW, Washington, DC 20011
☎ (+1 202) 726 8213 ⊖ (+1 202) 726 6785
Ambassador: H.E. Mr Ojars Kalnins
Deputy Chief of Mission: Einars Semanis
Second Secretary Commercial: N. Popens
Information Officer: Ms Liana Eglitis
Consular Office: Edgars Bondars [97]

31-01060 Embassy of Lithuania

2622 16th St NW, Washington, DC 20009
☎ (+1 202) 234 5860 ⊖ (+1 202) 328 0466
Ambassador: H.E.Mr Alfonsas Eidintas
Minister: Ms Dalia Grybauskaite
Counsellor: Stasys Sakalauskas
Counsellor: Vytautas Zalys
First Secretary: Darius Pranckevicius [97]

31-01070 Embassy of Luxembourg

2200 Massachusetts Ave NW, Washington, DC 20008
☎ (+1 202) 265 4171 ⊠ 64130 ⊖ (+1 202) 328 8270
Ambassador: H.E. Mr Alphonse Berns
Deputy Chief of Mission: Carlo Krieger
Admin Attaché: Ms Colette Kinnen [97]

31-01090 Embassy of the Republic of Madagascar

2374 Massachusetts Avenue NW, Washington DC 20008
☎ (+1 202) 265 55 25/26 [E] malagasy@embassy.org
⊠ 440167 ambamad ⊖ (+1 202) 265 30 34
Charge d'Affaires ad Interim: Biclair Andrianantoandro
Cultural Attaché: Voahirana Ralay [98]

31-01140 Embassy of Malta

2017 Connecticut Ave NW, Washington, DC 20008
☎ (+1 202) 462 3611-12 ⊖ (+1 202) 387 5470
Ambassador: H.E. Dr Albert Borg Olivier de Puget
First Secretary & Consul: Ms Vanessa Grima Baldacchino [97]

31-01170 **Embassy of the Islamic Republic of Mauritania**
2129 Leroy Place NW, Washington DC 20008
☎ (+1 202) 232 57 00 [W] http://www.emb.org/mauritania
⊠ TWX 710 8221 151, 248 532 abar ur
⊜ (+1 202) 319 26 23
Ambassador: H E Mr Bilal Ould Werzeg
[E] werzeg@embofirm.org
Deputy Chief of Mission: Abdallahi Ould Kebd
First Secretary: Mohamed Ould Hamadi [98]

31-01210 **Embassy of Mongolia**
10201 Iron Gate Road, Potomac, Washington, DC 20854
☎ (+1 202) 333 71 17 ⊠ 3738761
⊜ (+1 202) 298 92 27 [98]

31-01235 **Embassy of the Union of Myanmar**
2300 South Street NW, Washington DC 20008
☎ (+1 202) 332 90 44/45 ⊠ 248 310 burm ur
⊜ (+1 202) 332 90 46
Ambassador: H E Mr U Tin Winn
Defence Attaché: Col Thet Win
Third Secretary: Hmway Hmway Khyne [98]

31-01270 **Royal Netherlands Embassy**
4200 Linnean Ave NW, Washington, DC 20008
☎ (+1 202) 244 5300 ✉ HOLLANDIA Washington
⊜ (+1 202) 362 3430
Ambassador: H.E. Mr A.P.R. Jacobovits de Szeged
Economic Minister: A.C.M. Hamer
Counsellor (Science & Technology):
Paul M. Op Den Brouw
Defense & Air Attaché: Maj.Gen. C.J. Vanden Burg
Asst Naval Attaché: Cdr R. Bloemendael
Military Attaché: Col G. Van Oppen
Asst Military Attaché: Lt Col R. Harting
Naval Attaché: Capt. W.I. Wilms
Asst Air Attaché: Lt Col J. Smit
Defense Co-operation Attaché: Col J.G.C. Kiemeney
Asst. Defense Co-op. Attaché: F.S.M. Van Spaendonck [97]

31-01300 **Embassy of New Zealand**
37 Observatory Circle NW, Washington, DC 20008
☎ (+1 202) 328 4800 ✉ TOTARA Washington
⊜ (+1 202) 265 9238
Ambassador: H.E. John Wood
Deputy Chief of Mission: George Troup
Minister (Commercial): Dell Clark-Higgie
Defence, Naval & Air Attaché (Head of NZDS):
Air Cdre Maurice Manttan
Asst Naval Attaché: Cdr P.C. Redman
Military Attaché: Lt Col M.J. Pearce
Asst Air Attaché: Wg Cdr T.M. Greening [97]

31-01340 **Royal Norwegian Embassy**
2720 34th St NW, Washington, DC 20008
☎ (+1 202) 333 6000 ⊠ 892374 noramb wsh
⊜ (+1 202) 337 0437
Ambassador: H.E. Mr Kjeld Vibe
Deputy Chief of Mission: Karsten Kepsvik
Counsellor Security & Defense Policy:
Are-Fastein Norheim
First Secretary Economic Affairs: Aasmund Rygnestad
Defense Attaché: Maj.Gen Bjorn Nygard
Defense Co-operation & Air Attaché: Lt Col Roay A. Teien
Military, Naval & Asst Defense Attaché: Lt Col Vidar Falck
Asst Military, Naval, Air & Defense Co-operation Attaché:
Lt Cdr Steinar Torjusen
Counsellor Scientific Affairs:
Gunnar Kristian Fuhr Wilhelmsen
Counsellor Press & Public Affairs: Tore Tanum [97]

31-01420 **Embassy of the Philippines**
1600 Massachusetts Avenue NW, Washington DC 20036
☎ (+1 202) 467 94 07 ⊜ (+1 202) 467 94 37
Ambassador: H E Mr Raul Ch. Rabe
Deputy Chief of Mission: Antonio Basilio
Defence Attaché: Col Danilo Augusto B Francia
Military Attaché and Assistant Defence Attaché:
Col Pedro Cesar C Ramboanga
Trade/Commercial Counsellor: Josefa Dizon [98]

31-01655 **Embassy of the Republic of Slovenia**
1525 New Hampshire Avenue NW,
Washington DC 20036
☎ (+1 202) 667 53 63 ⊜ (+1 202) 667 45 63
Ambassador: H E Dr Ernest Petric

Permanent Mission of the Republic of Slovenia to the United Nations
600 Third Avenue, 24th Floor, New York NY 10016
☎ (+1 212) 370 30 07 ⊜ (+1 212) 370 18 24
Ambassador and Head of the Permanent Mission:
H E Dr Danilo Türk

Consulate General of Slovenia
600 Third Avenue, 24th Floor, New York NY 10016
☎ (+1 212) 370 30 06 ⊜ (+1 212) 370 35 81
Consul General: Matjaz Kovacic

Honorary Consulate of the Republic of Slovenia
1111 Chester Avenue, Suite 520, Cleveland, Ohio 44114
☎ (+1 216) 589 92 20 ⊜ (+1 216) 589 92 10
Honorary Consul: Dr Karl Bonutti [98]

31-01690 **Embassy of the Democratic Socialist Republic of Sri Lanka**
2148 Wyoming Ave NW, Washington, DC 20008
☎ (+1 202) 483 4025 ⊠ 49616790 sri lanka
⊜ (+1 202) 232 7181
Ambassador: H.E. Mr Jayantha Dhanapala
Minister: Prasad Kariyawasam
Commercial Counsellor: Asoka Dharmawardhane
First Secretary: Sumanasena Abeywarna [97]

31-01830 **Embassy of Tunisia**
1515 Massachusetts Ave NW, Washington, DC 20005
☎ (+1 202) 862 1850 ⊠ 248377 tnis ur
⊜ (+1 202) 862 1858
Ambassador: H.E. Mr Azouz Ennifar
Counsellor (Commercial Affairs):
Slaheddine Ben Mahmoud

Office of the Armed Forces Attaché
1515 Massachusetts Ave NW, Washington, DC 20005
☎ (+1 202) 862 1850; 223 5837
Defense, Military, Naval & Air Attaché:
Col Mohamed Bezzez
Asst Defense Attaché: Lt Col Abdesselem Zoubeidi [97]

31-01875 **Embassy of Ukraine**
Washington, DC
☎ (+1 202) 342 6031 ⊜ (+1 202) 333 0698
Ambassador: Dr Yuri Scherbak
Military & Air Attaché: Lt Col V. Havrylov
Commercial Officer: M. Reznik
Public Information Officer: N. Zarudnaya [97]

31-01890 **Embassy of the United Kingdom**
3100 Massachusetts Avenue NW,
Washington, DC 20008
☎ (+1 202) 588 65 00 ⊠ 216760 britem wsh
⊜ (+1 202) 588 78 70 (Chancery); 588 78 66 (Management); 588 79 01 (Trade); 588 78 50 (Consular)
Ambassador: H E Christopher Meyer
Minister: S J L Wright
Minister Economic: H P Evans
Minister Hong Kong Economic and Trade Affairs: H P Pang
Minister Hong Kong Economic and Trade Affairs:
C I C Jackson
Defence Attaché & Head of British Defence Staff:
Maj.Gen. C G C Vyvyan
Minister of Defence Material: J C Taylor
Naval Attaché and Commander British Naval Staff:
Cdre D J Anthony
Military Attaché and Commander British Army Staff:
Brig. R F Baly
Air Attaché and Commander: Air Cdre David Norriss
Counsellor External Affairs: J C R Gray
Counsellor Science and Technology: C Whaley
Counsellor Defence, Science and Equipment:
Dr P A Kennedy
Counsellor Defence Supply: J Scopes
Attaché Defence Equipment-Air: R A Stern
Attaché Defence Equipment-Maritime: R Richards
Attaché Defence Equipment-Land: A Dixon
Attaché Defence Equipment Intellectual Property Rights:
R Luetchford
Attaché Defence Supply: T Kennedy
First Secretary (Press and Public Affairs):
R M F Chatterton-Dickson
First Secretary (Politico/Military): W J Shapcott
First Secretary (Trade Promotion): F A Blogg [98]

31-01910 **Embassy of Uruguay**
1918 F St NW, Washington, DC 20006
☎ (+1 202) 331 1313/16 ✉ URUWASHI Washington
⊠ 904 158 uruguay wsh; 64 310 uruwash
⊜ (+1 202) 331 8142
Ambassador: H.E. Dr Alvaro Diez de Medina
Deputy Chief of Mission: Alvaro Moerzinger

Office of Military, Naval & Air Attachés
Suite 701, 1825 K St NW, Washington, DC 20006
☎ (+1 202) 466 3164; 3175; 3177; 3179
⊜ (+1 202) 835 9099
Defence & Military Attaché: Gen. Yelston Bagnasco
☎ (+1 202) 466 3177 ⊠ 248278 fauw ur
Asst Military Attaché: Col Washington Garcia
Naval Attaché: Capt. Juan Carlos Dogliotti
☎ (+1 202) 466 3179
Air Attaché: Col Dr Roberto Amorin ☎ (+1 202) 466 3177

Office of the Commercial Counsellor
747 3rd Avenue 21st Floor, New York, NY 10017
☎ (+1 212) 751 7137 ⊠ 621134 dec ny
⊜ (+1 212) 758 4126
Minister-Counsellor (Commercial Affairs): Eduardo Casabo

Financial Office
2021 L Street NW #201, Washington, DC 20036
☎ (+1 202) 223 9833 ⊜ (+1 202) 223 2119
Minister: Carlos Steneri [97]

31-01950 **Embassy of Venezuela**
1099 30th Street NW, Washington, DC 20007
☎ (+1 202) 342 22 14 ✉ EMBAVENEZ Washington
⊠ 440071 vla ui; 248350 venz ur ⊜ (+1 202) 342 68 20
Ambassador: H.E. Mr Pedro Luis Echeverria
Minister Deputy Chief of Mission: Francisco Velez
Minister Counsellor Cultural Affairs: Gonzalo Palacios
Counsellor: Douglas Goussot
Counsellor: Ms Luisa Perez Contreras
Counsellor for Legal and Financial: Ms Mary Dager
First Secretary: Ms Rebaca Guevara
First Secretary: Edgar Moros [98]

31-02030 **Embassy of the Republic of Zambia**
2419 Massachusetts Ave NW, Washington, DC 20008
☎ (+1 202) 265 9717 to 19 ⊠ 440 247 zambian
⊜ (+1 202) 332 0826
Ambassador: H.E. Mr Dunstan W. Kamana
Deputy Chief of Mission: George R. Nzala
First Secretary (Economics): Benny Chundu
First Secretary (Political): Mike K. Chargwe
Defence, Military & Air Attaché:
Brig.Gen. Gilbert C. Chileshe
Second Secretary (Accounts): Paul Mulenga
Third Secretary (Personal Secretary): Ms Teresa Tembo [97]

34 Multinational Intergovernmental Organizations

34-20000 **Supreme Allied Command Atlantic (SACLANT)**
Forces Alliées de l'Atlantique (GQG)
Headquarters
Norfolk, VA 23551-2490
☎ (+1 804) 445 3600 ⊜ (+1 804) 445 3234
Supreme Allied Commander Atlantic:
Gen. John J. Sheehan (USMC)
Deputy Commander: VAdm. Sir Ian D.G. Garnett (UK Navy)
Chief of Staff: RAdm. Merrill W. Ruck (USN)
ACOS for Operations: RAdm. Russell D. Moore (CAN)
ACOS for Plans & Policy: RAdm. Robert G. Sprigg (USN)
ACOS Communications Electronics:
RAdm Jacques Waltmann (RNN)
Chief Public Information: Capt. Craig R. Quigley (USN)
☎ (+1 804) 445 3227

Striking Fleet Atlantic (STRIKFLTLANT)
Norfolk, VA 23551-2490
☎ (+1 804) 444 2422
Commander: VAdm. Jay L. Johnson (USN)
Public Affairs Officer: Cdr Paul Weishaupt (USN)

Striking Fleet Atlantic Carrier Striking Force (CARSTRIKFOR)
Norfolk, VA 23551-2490

Striking Fleet Atlantic ASW Striking Force (ASWSTRIKFOR)
Norfolk, VA 23551-2490

Commander-in-Chief Western Atlantic (CINCWESTLANT)
Norfolk, VA 23551-2490
☎(+1 804) 444 6294
Public Affairs Officer: Capt. George Farrar (US Navy) [97]

34-60000 NATO Medium Extended Air Defence System Design and Development, Production and Logistics Management Agency (NAMEADSMA)
Building 1, 620 Discovery Drive, Suite 300,
Huntsville AL 35806
☎(+1 205) 922 39 01 ⊜ (+1 205) 922 39 00
General Manager: Brig Gen H Meunier [97]

60 Research Organizations, Institutes & Test Establishments

Armed Forces Institute of Pathology	20-04372
ASTA Corp.	80-03355
Astron Corp.	75-09980
Defense Advanced Research Projects	20-01000
Marine Corps Systems Command HQ	25-45400
Naval Research Laboratory	25-41800
Phillips Laboratory	25-22000
UBC Inc.	75-91383
US Air Force Operational Test &	25-23050
US Air Force Institute of Technology	20-25000
US Army CECOM Research,	25-37200
US Army Operational Test &	25-33150
US Army Test & Evaluation Command	25-32920
US Army Tank-Automotive & Armaments	25-38980
US Coast Guard R&D Center	28-50000

60-02500 Bend Research
64550 Research Rd, Bend, OR 97701-8599
☎(+1 541) 382 4100 ⊜ (+1 541) 382 2713
♣ $ 6 000 000 ⚉ 57 ⌂ 18 000 sq ft
Chief Exec. Officer: Walter C. Babcock
Marketing Dir: Mark H. Henry
■ Research & development in the areas of separations, controlled-release & bio-processing applications of membranes & thin-films, including systems for desalting water; purifying wastes & concentrating food; extracting & concentrating metals; separating gases; biotechnological separations & purification. [97]

60-03000 Calspan SRL Technology
800 Connecticut Ave NW, Ste 111,
Washington, DC 20006
☎(+1 202) 887 4700 ⊜ (+1 202) 861 0321
Chairman: Joseph P. Allen (Ph.D)
President: David H. Langstaff
VP Corporate Communications: Al Salandra
■ Operates technical facilities & provides high-technology services, systems & products, including professional & technical research & development, testing & evaluation. Simulation & training in aeronautics, transportation sciences, space, human system technology, information systems & chemical surety for US Government, industry & international customers.

Calspan Operations
PO Box 400, Buffalo, NY 14225
☎(+1 716) 632 7500 ⊜ (+1 716) 631 4131
⚉ 500 ⌂ 600 000 sq ft
General Mgr & Sr Vice President: Jack E. Wagner
VP Operations: Larry S. Horwitz
■ R&D, test & evaluation facilities for conducting programmes for Government & commercial customers. [97]

60-03300 Center for Remote Sensing
Boston University, 725 Commonwealth Ave,
Boston, MA 02215
☎(+1 617) 353 97 09 Ⓦ http://www.bv.edu
⊠ 286995 univ bsn ⊜ (+1 617) 353 32 00
⚉ 18
Director: Dr Farouk El-Baz
Admin Asst: C. Smith [97]

60-03500 Cortana Corp.
520 N. Washington St, Ste 200,
Falls Church, VA 22046-3538
☎(+1 703) 534 8000 Ⓔ corporate@cortana.com
⊜ (+1 703) 534 8005
⚉ 18
President: K.J. Moore
■ Maritime analysis & research with emphasis on submarine design. [97]

60-09100 Institute for Defense Analyses (IDA)
1801 N.Beauregard St, Alexandria, VA 22311-1772
☎(+1 703) 845 2000 ⊜ (+1 703) 845 2569
♣ $ 107 264 165 (1994) ⚉ 925
Chairman: W. Jarvis Moody
President: Gen. Larry Welch
VP Administration & Finance: Ms Ruth L. Greenstein
Treasurer: Craig A. Webster
■ Federally funded research & development centre with primary function of assisting the office of the Secretary of Defense, the Joint Staff, the Unified & Specified Commands & Defense Agencies, in the areas of systems evaluation, technology assessment as well as resource & acquisition analysis, information systems, strategy & force assessments. [97]

60-10250 Jet Propulsion Laboratory, NASA (JPL)
4800 Oak Grove Drive, Pasadena, CA 91109
☎(+1 818) 354 43 21; 354 50 11 (public information)
Ⓦ http://www.jpl.nasa.gov
⊠ 675421; TWX 910 588 3294
⊜ (+1 626) 354 22 04; 354 34 37; 354 65 01
♣ $ 1,115,000,000 (projected 1997) ⚉ 5,100
Director: Dr Edward C Stone
Deputy Director: Larry N Dumas
Associate Director: Kirk M Dawson
Chief Engineer: John R Casani
Chief Scientist: Dr Moustafa T Chanine
Director of Telecommunications and Mission Operations Directorate: Ms Gael Squibb
Director of Space and Earth Science Programmes Directorate: Dr Charles Elachi
Director of Technology and Applications Programs Directorate: Dr R Rhoads Stephenson (acting)
Director of Business Operations Directorate: Daryl T Gant
Director, Engineering and Science Directorate: William J Weber
Director, Mars Exploration Directorate: Norman R Haynes
Controller: William H Harrison
General Counsel: Harry M Yohalem
Manager of Public Information Office: Franklin O'Donnell
■ Operated by the California Institute of Technology for NASA, with planetary exploration as its primary mission. The laboratory also carries out research for the US DoD and other Federal Agencies. Working on the Mars Pathfinder mission.

East Coast Technology & Applications Programs Office
Mail Stop 300 C, 1500 Planning Research Drive,
McLean, VA 22101
☎(+1 703) 556 29 01 ⊜ (+1 818) 354 76 94
Director: Dr Edward C Stone
Marketing Director: Dr Philip Eckman [98]

60-11111 Kennedy Aerospace Corp.
Long Beach Airport, PO Box 667, Long Beach, CA 90801
☎(+1 562) 599 66 95 ⊜ (+1 562) 427 99 33
President: T W Kennedy
■ Unmanned aerial vehicles, multi-role flight simulators and electric vehicles. [98]

60-13150 Mark Three
1410 Central Avenue South West #23,
Albuquerque, NM 87104
☎(+1 505) 243 42 57 ⊜ (+1 505) 768 23 10
⚉ 6
General Operations Director: John A Anderson
Project Co-ordinator and Government Liaison Manager: Ms Donna D Kind Anderson
■ Research and development of small arms and special operations weapon systems. [98]

60-14500 Nevada Automotive Test Center (NATC)
PO Box 234, Carson City, NV 89702
☎(+1 702) 629 20 00 ⊜ (+1 702) 629 20 29
⚉ 120 (1996)
President: Henry C Hodges (Jr)
Ⓔ hhodgesjr@nate.nate-ht.com
■ An independent research, development, test, evaluation and certification facility for vehicle, vehicle component systems, transport materials and weapons and ordnance. 3,000 miles of test courses, railroad transportability area, EMI/EMC, 2 large environmental chambers (56'LX 16'HX16'W and 32'LX20'WX 21'H-88' combinable length) and the full range of support facilities and service. performance and durability tests are conducted in accordance with military and commercial standards. Testing environment is also supported with in-house instrumentation and computer aided engineering tools including CATIA, SDRC I-DEAS, TNSC/NASTRAN and DADS. Capability to quantify domestic, international and military user duty cycles to accurately predict warranty and mission performance. [98]

60-16000 Orlando Technology Inc.
PO Box 1218, Shalimar, FL 32579
☎(+1 904) 651 50 97 ⊜ (+1 904) 651 14 91
⚉ 50
President: Dan Matuska
Sales and Marketing Director: R Wallace
Program Manager: Rick Snow
■ Research and development of conventional ordnance. → 75 → 80 [98]

60-17000 RAND
1700 Main Street, PO Box 2138,
Santa Monica, CA 90407-2138
☎(+1 310) 393 04 11 Ⓦ http://www.rand.org
⊜ (+1 310) 393 48 18
⚉ 1,000
President: James A Thomson
Executive Vice President: Michael D Rich
Vice President and Director of Army Research Division: James T Quinlivan
Vice President and Director of Project Air Force Division: Brent D Bradley
Public Information Director: Jess Cook
Acting Director of National Security Research Department: Eugene Gritton
Treasurer: Kennith M Krug
■ Private, not for profit research institution.

Washington Office
2100 High Street North West,
Washington, DC 2005-4707
☎(+1 202) 296 50 00 ⊜ (+1 202) 296 79 60
Associate Chair of Research Staff Management Department: David Chu
● 342 Maddison Avenue, Suite 1532,
New York, NY 10173
☎(+1 212) 661 58 00 (Council for aid to education); 661 31 66 (New York external affairs) [98]

60-17500 R&B Enterprises
20 Clipper Road, W. Conshohocken, PA 19428-2721
☎(+1 610) 825 1960 ⊜ (+1 610) 825 1684
♣ $ 4 500 000 (1995) ⚉ 60 (1995)
President: Robert D. Goldblum
Engineering Services Mgr: James L. Press
■ EMI testing. [97]

60-18000 Research Triangle Institute (RTI)
PO Box 12194, Research Triangle Park, NC 27709-2194
☎(+1 919) 541 60 00 Ⓔ listen@rti.org
Ⓦ http://www.rti.org ⊜ (+1 919) 541 58 59
♧ $ 50,000,000 (1996) ♣ $ 143,000,000 (1996)
⚉ 1,450 (1996) ⌂ 60,000 m2
Director: F Thomas Wooten
Director of Research: Alvin M Cruze
Sales and Marketing Director: Ms Karen Mead
☎(+1 919) 541 67 68
Communications Manager: Reid Maness
☎(+1 919) 541 70 44
■ Provides research and development services for the military in semiconductors, avionics, computer graphics, electronic systems antenna technology, aerosol dispersal, medicine, human services and environmental protection. [98]

60-19150 David Sarnoff Research Center
(a subsidiary of SRI International)
201 Washington Road, PO Box CN 5300,
Princeton, NJ 08543-5300

☎ (+1 609) 734 20 00; 734 25 07
⊠ 843497 rca labs prin
⊖ (+1 609) 734 22 21; 734 28 70
👥 725
President and Chief Operating Officer: James E Carnes
Vice President of Business Development:
Dr Miles A Libbey
Director of Communications Services: Robert Harris
Manager of Media Relations: W Shumann [98]

60-19160 Scaled Composites Inc.

(a subsidiary of Wyman-Gordon Co.)
1624 Flight Line, Mojave, CA 93501-1663
☎ (+1 805) 824 45 41 ⊖ (+1 805) 824 41 74
👥 100 🏭 66,600 sq ft
President and Chief Executive Officer: Burt Rutan
Vice President and General Manager: Michael W Melvill
Vice President Business Development: Doug Shane
■ Triumph Business Jet, ARES, composite prototype design, fabrication & flight test. [98]

60-19200 Southwest Research Institute (SRI)

6220 Culebra Road, PO Drawer 28510,
San Antonio, TX 78228-0510
☎ (+1 210) 684 51 11
[W] http://www.swri.org/indexpix.htm ⊠ 244846
⊖ (+1 210) 522 35 47
♣ $ 259,000,000 👥 2,500 🏭 1,200 sq m
President: Martin Goland
Executive Vice President: Richard B Curtin
Vice President of Planning and Program Deveolpment:
Dr Jay M Lewallen
Vice President of Engine and Vehicle Research: Nigel Gale (III)
Purchasing Director: Paul Easley
Communications Director: Craig Witherow
■ Applied research and development in engineering and applied sciences. Consultation, prototype hardware and field services, NDT evaluation, electromagnetics, electronics, digital systems, robotics, automation, materials, chemistry, environmental, space, ballistics, military safety.

Washington DC Office
Crystal Gateway One, Suite 1104,
1235 Jefferson Davis Highway, Arlington, VA 22202
☎ (+1 703) 416 05 00 ⊖ (+1 703) 416 21 89
Director: W E Woolam

Houston Office
7500 San Felipe, Suite 825, Houston, TX 77063
☎ (+1 713) 977 13 77 ⊖ (+1 713) 977 12 89
Director: Dr Noel C Willis

Southfield Office
26100 American Drive, Suite 603, Southfield, MI 48034
☎ (+1 810) 353 25 50 ⊖ (+1 810) 353 26 66
Director: Dr Leonard F Herk (Jr) [98]

60-20000 United Technologies Research Center (UTRC)

Silver Lane, East Hartford, CT 06108
☎ (+1 860) 727 7000 ⊖ (+1 860) 727 7310
♣ $ 95 000 000 (1994) 👥 800
Vice President UTRC: John F. Cassidy (Jr)
Vice President Military After Sales Service: James D Taiclet
Manager Business Development: Karl F Winckel [97]

60-25000 Vista Control Corporation

27825 Fremont Court, Santa Clarita, CA 91355
20746 Tulsa Street
☎ (+1 805) 297 44 30 [E] mssfv@aol.com
[W] http://www.vistaoc.com ⊖ (+1 805) 257 47 82
♣ $ 7,500,000 (1996) ♣ Chatsworth, CA 👥 40
🏭 1,800 sq m
Managing Dir: Ron Rambin
Public Relations/Marketing: Ms Marcia Steil
■ Design & manufacture rugged 6U VME boards & systems for military & industrial applications. New products include a Power PC 603 SBC with PCI mezzanine, a 6U VME navigator with on-board GPS & inertial aiding & a family of rugged I/O boards. Products fit applications including flight control & navigation, stabilisation, weapon control & C3I systems. [97]

65 Industry, Trade & Armed Forces Associations

65-00080 Aerospace Industries Association of America (AIA)

1250 Eye Street, NW, Washington, DC 20005-3924
☎ (+1 202) 371 84 00 ⊠ TWX 710 8220 134
⊖ (+1 202) 371 84 70
👥 42
President & CEO McDonnell Douglas Corp.:
Harry C. Stonecipher
President: Don Fuqua
Director of Communications: Ms Alexis Allen
☎ (+1 212) 371 85 44 [E] alexis@aia-aerospace.org
Secretary Treasurer: George F. Copsey
President & CEO Lockheed Martin Corp.:
Norman R. Augustine
Chairman, President & CEO Harris Corp.:
James F. Hardymon
Sr VP Government & Commercial Marketing, Raytheon Co.: Renso L. Caporali
President, Aerospace Industries Assoc.: Don Fuqua
Exec. VP Texas Instruments Incorporated:
William F. Hayes
Chairman, President & CEO Northrop Corporation:
Ken Kresa
Vice Chairman, Hughes Electronic Corp.: Michael T. Smith
Vice President Americas Ind. Expertise:
Ms Diane A. Albano
VP & General Mgr DuPont Company: J. Michael Bowman
Sr Vice President Teledyne Inc.: Hudson B. Drake
President & COO Gulfstream Aerospace:
Fred A. Breidenbach
President Loral Missiles Group: Felix W Fenter
President Aerospace Division The BFGoodrich Company:
David L. Burner
President AlliedSignal Aerospace: Daniel P. Burnham
President Parker Bertea Aerospace: Stephen L. Hayes
Chairman BH Aircraft Company Inc.: Vincent E. Kearns
President Kaman Aerospace Corporation:
Walter R. Kozlow
President FMC Corporation: Larry D Brady
Vice President & COO Sundstrand Corp.:
Ronald F. McKenna
President & COO Honeywell Inc.: D. Larry Moore
President & CEO GE Aircraft Engines: Eugene F. Murphy
Sr VP & COO Rockwell International Corporation:
Kent M Black
Chairman & CEO General Dynamics Corp.:
James R. Mellor
President & CEO Rohr Inc.: Robert H. Rau
Chairman, CEO The Boeing Company: Frank A. Shrontz
Exec. Vice President United Technologies:
Roger I. Ramseier
President Computing Devices Int'l: Ronald L. Turner
President & COO Thiokol Corp.: James R. Wilson
President Aerojet: Roger I. Ramseier
President & CEO Alliant Techsystems Inc.:
Richard Schwartz
President Aerospace Systems & Components:
Terry D. Stinson
President & Chairman Williams Int'l: Sam B. Williams
Vice President International Programmes: Joel Johnson
Executive VP & General Manager Space:
Timothy W Hannemann
Chairman Electronic Systems: Richard A Linder
President & CEO GE Aircraft Engines: Eugene F Murphy
Executive VP United Technologies Corp.: William F Paul
Executive VP & COO Aerospace Sunderland Corp.:
Robert J Smuland
Chairman Chrysler Technologies Corp.: Paul E Wright
Chief Executive Officer Hexcel Corp.: John J Lee
■ Trade association representing US manufacturers of commercial, military & business aircraft, helicopters, aircraft engines, missiles, spacecraft & related components & equipment.

Represent:
AAI Corporation
Aerojet, A Segment of GenCorp
Allian Techsystems Inc.
AlliedSignal Aerospace
American Pacific Corporation
Argo-Tech Corporation
BH Aircraft Company Inc.
The Boeing Company
Chrysler Technologies Corporation
Coltec Industries Inc.
Computing Devices International
Digital Equipment Corporation
Dowty Aerospace
DuPont Company
FMC Corporation
GEC-Marconi Electronic Systems Corporation
General Dynamic Corporation
General Electric Company
The BFGoodrich Company
Gulfstream Aerospace Corporation
Harris Corporation
Heath Tecna Aerospace Company
HEICO Corporation
Hexcel Corporation
Honeywell Inc.
Hughes Electronics Corporation
ITT Defense and Electronics Inc.
Interturbine Corp.
Kaman Aerospace Corporation
Lockheed Martin Corporation
Loral Vought Systems Corporation
Lord Corporation
Lucas Aerospace Inc.
McDonnell Douglas Corporation
Northrop Grumman Corporation
Parker Hannifin Corporation
Raytheon Company
Rockwell International Corporation
Rohr Inc.
Rolls- Royce North America Inc.
Sundstrand Corporation
Teledyne Inc.
Teleflex Inc./TFX Aerospace Group
Texas Instruments Incorporated
Textron Inc.
Thiokol Corporation
TRW Inc.
UNC Inc.
United Technologies Corporation
Williams International
Westinghouse Electric Corporation [98]

65-00100 Air Force Association (AFA)

1501 Lee Highway, Arlington, VA 22209-1198
☎ (+1 703) 247 58 00 [E] info@afa.org
[W] http://www.afa.org ⊖ (+1 703) 247 58 53
Executive Director: John A Shaud
President: Doyle E Larson
Chairman of the Board: R E Smith
National Secretary: Ms Mary Anne Thompson
National Treasurer: Charles H Church (Jr)
■ An independent non-profit aerospace organisation. [98]

65-00110 American Defense Preparedness Association (ADPA)

2101 Wilson Blvd, Ste 400, Arlington, VA 22201
☎ (+1 703) 522 1820 ⊖ (+1 703) 522 1885
President: Lt Gen. Lawrence F. Skibbie (retd)
VP Administration & Finance: Col B.P. Prokuski (Retd)
■ Organisation dedicated to providing information & education to support the understanding of the peacetime needs of defence preparedness.
ADP has 24 000 individual & over 700 corporate members. It presents more than 55 technical meetings annually. [97]

65-00120 American Society of Naval Engineers (ASNE)

1452 Duke Street, Alexandria, VA 22314-3458
☎ (+1 703) 836 67 27 [W] http://www.jhuapl.edu/ASNE
⊖ (+1 703) 836 74 91
👥 12
Executive Director: Capt Dennis Kruse (US Navy, retd)
Technical Director: Ms Lee Dona
Membership Manager: Ms Judy Owen
Meetings Manager: Ms Sally Cook
Operations Manager: Capt Dennis Pignotti (retd)
Staff Accountant: Bob White
Publications Coordinator: Ms Sara Keeney
■ A professional forum for the discussion and exchange of scientific, technical and operational developments in the field of naval engineering. [98]

65-00130 Armed Forces Communications & Electronics Association (AFCEA)

4400 Fair Lakes Court, Fairfax, VA 22033-3899
☎ (+1 703) 631 61 00 [E] admin@afcea.org
[E] afcea@infoboard.be ⊖ (+1 703) 631 46 93
President and Chief Executive Officer:
Lieut Gen Norman Wood (USAF, retd)
☎ (+1 703) 631 62 12

Contact Reference

■ Member companies involved in communications, electronics, command and control, computer sciences, teleprocessing, intelligence systems and related technologies for use by the military, government and industry. 98

65-00140 Association of Old Crows (AOC)

The AOC Building, 1000 North Payne Street,
Alexandria, VA 22314-1696
☎ (+1 703) 549 16 00 ⊜ (+1 703) 549 25 89
$ 2,600,000 (1997) ♣ $ 1,750,000 (1997)
9 (1997) 4,600 sq ft
Executive Director: Vern Luke
Conventions Director: Ms Carolyn Wood-Holmes
■ Association for members with interests in the specialist field of electronic warfare, information operations and defense electronics. Publication of the Journal of Electronic Defense magazine. 98

65-00150 Association of the US Army (AUSA)

2425 Wilson Boulevard, Arlington, VA 22201
☎ (+1 703) 841 43 00 ✆ (800) 336 45 70
W http://www.ausa.org E ausahq@aol.com
⊜ (+1 703) 243 25 89; 525 90 39; 841 35 05 (public affairs)
♣ $ 12,000,000 (budget 1996) 70
President: Gen Jack Merritt (retd)
Vice President of Finance and Administration: Arthur S Welch
Industry Affairs Director: Joseph P Hollis (Jr)
■ A private non-profit organisation which represents the US Army. Conducts classified and unclassified symposia, regional security seminars and local meetings designed to enhance the Army's readiness and welfare. 98

65-00155 Association of Naval Aviation Inc. (ANA)

5205 Leesburg Pike, Ste 200, Falls Church, VA 22041
☎ (+1 703) 998 7733 ✆ (800) 666 9262
⊜ (+1 703) 671 6052
6
Chairman: Adm. Wesley L. McDonald (retd)
President: VAdm. Jerry L. Unruh (retd)
Dir Administration: Mrs Sue B. Henderson
Editor: Capt. Rosario Rausa (USN retd)
Secretary: Cdr Charles Sammons (USN rtd)
■ 53 Chapters located in areas of large naval aviation strength. 97

65-00200 British American Security Information Council

Washington Office
1900 L Street NW, Suite 401-2, Washington, DC 20036
☎ (+1 202) 785 1266 ⊜ (+1 202) 387 6298
Director: Daniel T. Plesch
Deputy Dir: Dr Natalie J. Goldring
Analyst: Ms Nicola Butler
Financial & Administrative Co-ordinator: Ms Nicole Hecker
Analyst: Ms Susannah L. Dyer
Program Asst: Joel Johnston
Research & Planning Asst: Ms Sami Fournier
■ Information/research charity. Specialist research areas include NATO, OSCE & arms control negotiations. European Security Structures, European Nuclear Forces, US Defense Budget, US Bases/Forces in Europe & the arms trade. Reports are produced regularly, aimed mainly at politicians & the media. 97

65-00300 Council of Defense & Space Industry Associations (CODSIA)

1250 Eye Street NW, Suite 1200, Washington, DC 20005
☎ (+1 202) 371 84 14 ⊜ (+1 202) 371 84 70
Chairman: Ms Meredith Murphy
Chairman Operating Committee: Peter C Scrivner
Administration Officer: Ms Ruth W Franklin
Member Associations:
Aerospace Industries Association
American Defense Preparedness Association
American Electronics Association
American Shipbuilding Association
Contract Services Association
Electronic Industries Association
Mfr Alliance for Productivity & Inovation
National Security Industrial Association
Professional Services Council
Shipbuilders Council of America 98

65-05620 Electronic Industries Association (EIA)

2500 Wilson Boulevard, Arlington, VA 22201
☎ (+1 703) 907 75 00 W http://www.eia.org
E publicaffairs@eia.org ⊜ (+1 703) 907 75 01
President: Peter F McCloskey E peterm@eia.org
Vice President, Secretary and General Counsel: John J Kelly
Vice President of Public Affairs: Mark V Rosenker
Vice President of Government Division: Dan Heinemeier
Vice President, Standards and Technology: Dan Bart
Vice President, Market Research Department: T Glyn Finley
Vice President, Components Group: Peter J Walsh
Staff Vice President, Government Relations Department: Ms Donna Gleason
Operations Director: Peter Moore
Consumer Electronic Shows Director: Ms Wendy Jones
■ Full-service national trade organisation representing the entire spectrum of companies involved in the manufacture of electronic components, parts, systems and equipment for communications, industrial, government and consumer end uses.
EIA government division advances the interests of the US electronic communication and information technology industries with regard to government market requirements, technical standards at federal, state and local levels. 98

65-10000 National Security Industrial Association (NSIA)

1025 Connecticut Ave NW, Ste 300,
Washington, DC 20036-5405
☎ (+1 202) 775 1440 ⊜ (+1 202) 775 1309
President: Gen. Thomas C. Richards (USAF (Rtd)
VP Membership & Public Relations: David H. Burpee
■ Association representing about 300 industrial, research, legal & educational organisations with interests related to national security. 97

65-10100 Navy League of the United States (NLUS)

2300 Wilson Boulevard, Arlington, VA 22201-3308
☎ (+1 703) 528 17 75 E mail@navyleague.org
W http://www.navyleague.org ⊜ (+1 703) 528 23 33
♣ US$ 7,470,999 (1995) 38 (1997) 14,000 sq ft
National President: Jack Kennedy
Executive Director: Charles L Robinson
■ A civilian organization dedicated to supporting America's maritime services, including the US Navy, Marine Corps, Coast Guard and merchant marine. A non-profit organization with a membership of approx. 70,000 individuals. The main objective of the NLUS is to motivate and educate Americans to the importance of sea power. Non-profit educational association. Sponsors Sea-Air-Space Expo, the world's largest maritime trade show. Publishes Sea Power magazine and an annual Almanac of Seapower. 98

65-18250 Refractory Metals Association

Metal Powder Industries Federation
105 College Road East, Princeton, NJ 08540
☎ (+1 609) 452 77 00 W http://www.mpif.org
⊜ (+1 609) 987 85 23
Administration Director: Peter K Johnson
■ Provides information on uses of tungsten, cobalt, tantalum, niobium, molybdenum and other refractory metals. 98

65-20000 Society of Automotive Engineers (SAE)

400 Commonwealth Drive, Warrendale, PA 15096-0001
☎ (+1 412) 776 48 41 ✉ 866355 saeinwnde
⊜ (+1 412) 776 00 02
$ 30,000,000 350 (1997)
Marketing Promotions Coordinator: Ms Helen Rehshaw
■ Develops engineering standards and material specifications. Organises technical meetings and expositions for aerospace and ground vehicles industries.
Aerospace Related Meetings:
Aerotech (annual)
Environmental Systems
Airline Plating and Metal Finishing Forum (annual)
General Aviation Aircraft Meeting and Exposition (bi-annual)
Joint Propulsion Conference Aerospace Atlantic (annual) 98

65-30000 Washington Aerospace Alliance

1904 Third Avenue, Suite 405, Seattle, WA 98101
☎ (+1 206) 224 43 44 E tradec@aol.com
⊜ (+1 206) 224 43 40
International Coordinator: Raymond Ma 98

70 Financial Holding Groups, Parent Companies

Trinity Industries, Inc. 75-90950

70-00390 Amcast Precision

11000 Jersey Boulevard, Rancho Cucamonga, CA 91730
☎ (+1 909) 987 47 21 ⊜ (+1 909) 980 79 58
Rapid Prototype Development: Winston J. Harris
■ Produces rapid prototype investment castings in ferrous and non ferrous alloys. 98

70-01000 AlliedSignal Inc.

101 Columbia Rd, PO Box 2245,
Morristown, NJ 07962-2245
☎ (+1 201) 455 2000
✉ 136410 alcmtna mwn b; TWX 710 986 7482
⊜ (+1 201) 455 4807
♣ $ 11 827 000 000 (1993) 86 400
Chairman and Chief Executive Officer: Lawrence A Bossidy
Executive Vice President and President Engineered Products: Frederic M. Poses
Executive Vice President and President AlliedSignal Aerospace: Daniel P Burnham
Exectutive Vice President and President AlliedSignal Automotive: John W. Barter
Senior Vice President, General Counsel and Secretary: Peter M Kreindler
Executive Vice President Human Resources and Communications: Donald J Redlinger
Senior Vice President Quality and Productivity: James E Sierk
Senior Vice President International: Paul R Schindler 98

70-02600 British Aerospace Holdings Inc

Suite 200 Washington Technology Park,
15000 Conference Center Drive,
Chantilly, VA 20151-3819
☎ (+1 703) 802 00 80 ⊜ (+1 703) 227 16 10
■ Operating and holding company for British Aerospace (BAe) subsidiaries operating in North America, Central America, South America and the Caribbean. Management of commercial investments in US companies. 98

70-02650 Ceridian Corporation

8800 34th Ave South, Minneapolis, MN 55440-1640
☎ (+1 612) 853 8100
♣ $ 1 333 000 000 (1995) ♣ $ 916 300 000 (1994)
10 200
Chairman, President & CEO: Lawrence Perlman
President & CEO Computing Devices Int'l: Ronald L. Turner
President & CEO The Abitron Company: Stephen B. Morris
President & CEO Ceridan Human Resources Group: Ron Somers
VP Communications: Ms Linda J. Jadwin
Vice President: Michael Kotten
Vice President: Ms Loren Gross
Director: Ms Ruth Margaret Davis
Public Relations Dir: Ms Nancy Foltz
☎ (+1 612) 853 5229
Associated Companies & Subsidiaries:
Computing Devices International → USA 75
Computing Devices Company Ltd → UK 75
Computing Devices Canada → Canada 75 97

70-02700 Chrysler Corp.

Corporate Offices
1000 Chrysler Drive, Auburn Hills, MI 48326-2766
☎ (+1 810) 576 5741
♣ $ 53 200 000 000 (1995) 126 000 (worldwide)
Chairman & CEO: Robert J. Eaton
Vice Chairman & CAO: Thomas G. Denomme
President & COO: Robert A. Lutz
Exec. Vice President & CFO: Gary C. Valade 97

70-03000 ESCO Electronics Corp.

8888 Ladue Road, Suite 200, St Louis, MO 63124
☎ (+1 314) 213 72 00 ⊖ (+1 314) 213 72 50
♣ $ 438,543,000 (1996) 👥 2,400 (1996)
🏠 1,482,000 sq ft
Chairman and Chief Executive Officer: D J Moore
Senior Vice President and Chief Financial Officer: P M Ford
Senior Vice President, General Counsel and Secretary: W Stark
Vice President of Planning Development: D W Snoke
Vice President of Planning: H E Johnson
Vice President and Treasurer: D H Nonnenkamp
Senior Vice President of Human Resources and Administration: P A Hutchinson
Investor Relations Director: Ms P K Moore

■ Through its operating subsidiaries ESCO Electronics produces a wide range of electronic products (including radar systems and electronic warfare equipment); armament products (including sighting, fire control and integration systems for helicopter and mobile ground-based anti-armour missile systems); automatic test equipment for high performance aircraft; mobile tactical systems; and a range of fluid flow valves, manifolds and filter products, utility load management systems, and anechoic materials.

Operating subsidiaries:
Rantec Microwave & Electronics Inc. → 75
Systems & Electronics Inc. (SEI) → 75
PTI Technologies including PTI Ltd
Filtertek, Inc.
EMC Test Systems, LP
Distribution Control Systems Inc. (DCSI)

Washington Office
1235 Jefferson Davis Highway, Suite 305,
Arlington, VA 22202-3284
☎ (+1 703) 920 76 00 ⊖ (+1 703) 920 14 06
Vice President of Washington Operations: L K Brewer 98

70-04000 Figgie International Inc.

4420 Sherwin Rd, Willoughby, OH 44094
☎ (+1 216) 953 2700 ⊖ (+1 216) 951 1724
♣ $ 359 900 000 (1995) 👥 2 500
Chairman & CEO: John P. Reilly
Corp. Communications Mgr: Ira Gamm

■ Figgie International is comprised of four principal divisions and subsidiaries serving consumer, technical & industrial markets.

Subsidiaries (Defence):
Interstate Electronics Corp → 75
Scott Aviation → 75 97

70-05050 GenCorp

175 Ghent Rd, Fairlawn, OH 44333-3300
☎ (+1 330) 869 42 00 ⊖ (+1 330) 869 42 11
👥 10,000
Chairman & CEO: John B. Yasinsky
Sr Vice President & CFO: D. Michael Steuert
Exec. Vice President: Roger Ramseier
VP Communications: Ms Rosemary Younts
VP Human Resources: Sam Harmon
VP Plastic Films Div.: David Mittiga
VP Vehicle Sealing Div.: Wayne Smith
VP Wallcovering Div.: James Ward
VP Penn Racquet Sports: Gregg Weida
VP Designed Plastics Div.: Dalton Windham
Vice President Specialty Polymers: Marvin Zima
Media Relations: Ms Rosmary Younts

■ Technology-based manufacturing company with positions in numerous polymer product markets as well the automotive, aerospace and defence industries. Rocket propulsion, defence electronics, ordnance, automative and polymer products. Designs and produces original equipment components and systems for domestic and transplant automotive companies. Manufacture specialty polymers and engineered plastics for consumers and industry.

Business Units:
Aerojet → 75 97

70-05150 Gores International

6260 Lookout Road, Boulder, CO 80301
☎ (+1 303) 581 31 00 ⊖ (+1 303) 581 06 49
President - Artemis Management Systems: Steven Yager 98

70-05200 GTE Corp.

One Stamford Forum, Stamford, CT 06904
☎ (+1 203) 965 20 00 Ⓦ http://www.gte.com
📠 4750071; TWX 710 4743 042 ⊖ (+1 203) 965 29 36
♣ $ 19,994 000 000 (1994) 👥 111,000
Chairman and Chief Executive Officer: Charles R Lee
Vice Chairman & President - GTE Telephone Operations: Kent B Foster
Vice Chairman and President International: Michael T Masin

Business Units:
GTE Telephone Operations
GTE Service Corporation
GTE Data Services
GTE Supply
GTE Main Street
GTE Entertainment
GTE Mobilenet
GTE Telecommunications Services
GTE Information Services
GTE Card Services
GTE Airfone
GTE Technology & Systems
GTE Government Systems
GTE Laboratories
GTE Worldwide Telecommunications Services
GTE Intelligent Network Services 98

70-06000 ITT Industries Inc.

(† ITT Corp.)

World Headquarters
Corporate Headquarters
4 West Red Oak Lane, White Plains, NY 10604
☎ (+1 914) 641 20 00 ⊖ (+1 914) 696 29 50
♧ $ 627,000,000 (1995) ♣ $ 8,884,000,000 (1995)
👥 56,000
Chairman, President & CEO: Travis Engen
Senior Vice President & CFO: Ms Heidi Kunz
Senior Vice President, Corporate Development Director: Martin Kamber
Senior Vice President, General Counsel: Vincent Maffeo
Senior Vice President, Controller: Richard J Townsend
Vice President, Investor Relations Dir: Ralph D. Allen
Vice President and Director Corporate Relations: Thomas Martin

■ Design, manufacture & support of high technology electronic systems & components for defence & commercial markets.
Manufacture products, systems & services for the movement, measurement & control of fluids.
Supply systems & components to the automotive industry.

Primary Business Segments:
Automotive
Defence & Electronics
Fluid Technology 98

70-11000 Kaman Corp.

Corporate Office
Blue Hills Avenue, Bloomfield, CT 06002
☎ (+1 860) 243 7100 Ⓦ http://www.kaman.com
📠 TWX 710 4253 411 ⊖ (+1 860) 243 6365
♣ $ 899 476 000 (1995) 👥 5 400
Chairman, CEO & President: Charles H. Kaman
President Kaman Aerospace Corp.: Walter R. Kozlow
President Kaman Diversified Tech. Corp.: Richard E.W. Smith
President Kaman Industrial Tech. Corp: T.Jack Cahill
Exec. VP & CFO: Robert M. Garneau
Exec. VP & President Kaman Music Corp.: C. William Kaman (II)
Sr VP, Chief Legal Officer & Secretary: Candace A. Clark
Sr VP & Chief Information Officer: Ronald M. Galla
Sr VP Finance: Robert H. Saunders (Jr)
VP Government Relations: Patrick L. Reneham
VP Secretary & General Counsel: Glenn M. Messemer
VP Public Relations: J. Kenneth Nasshan
Vice President: Russell H. Jones
Asst Controller: Michael J. Morneau
Asst Controller: Raymond H. Picard

■ Widely diversified company providing high-technology products and services to industrial, commercial and defence markets.

Subsidiary companies:
Kaman Aerospace Corp. → 75
Kaman Sciences Corp. → 80
Kaman Electromagnetics Corporation
Kaman Instrumentation Corp. → 75
Kaman Industrial Technologies Corp
Kamatics Corp.
Kaman Diversified Technologies Corp
Kaman Music Corp
Advanced Energetic Materials Corp (AEMC) 97

70-12000 McDonnell Douglas Corp. (MDC)

PO Box 516, St Louis, MO 63166
☎ (+1 314) 232 02 32 ⊖ (+1 314) 777 10 96
♣ $ 13,834,000,000 (1996) 👥 63,900
Chairman: John F McDonnell
President and Chief Executive Officer: Harry C Stonecipher
President, Douglas Aircraft Company: Walter J Orlowski
President, McDonnell Douglas Aerospace: Michael M Sears
Executive Vice President and Chief Financial Officer: James F Palmer
Senior Vice President and General Counsel: F Mark Kuhlman
Senior Vice President, Communications and Community Relations: Frederick W Hill
Senior Vice President and General Manager, McDonnell Douglas Helicopter Systems: Dean C Borgman
Senior Vice President, Human Resources and Quality: Laurie Broedling
Senior Vice President, Washington Operations: Stanley Ebner
Vice President, Corporate Communications: Larry McCracken
Vice President and Chief Information Officer: Baxter D Tate
Vice President and Controller: Mark Schroeder
Vice President and Treasurer: Roger A Krone
Vice President, Corporate Ethics and Ombudsman: Harold S Coyle

Subsidiaries:
Douglas Aircraft Co.
McDonnell Douglas Aerospace
McDonnell Douglas Canada Ltd
McDonnell Douglas Finance Corporation
McDonnell Douglas Technical Services Co.

Washington Office
1300 Wilson Boulevard, Suite 800, Arlington, VA 22209
☎ (+1 703) 526 25 00
Senior Vice President, Washington Operations: Stanley Ebner
Washington DC Communications Director: Bob O'Brien

McDonnell Douglas Helicopter Systems Headquarters (MDHS)
5000 East McDowell Road, Mesa, AZ 85205-9797
☎ (+1 602) 891 30 00 📠 3719337 md hc c mesa
⊖ (+1 602) 891 45 59
👥 3,800 🏠 1,300,000 sq ft
Senior Vice President, McDonnell Douglas Helicopter Systems: Dean C Borgman
Vice President, Production, Engineering and Operations: Al Winn
Vice President, Finance and Supplier Management: C A York
Vice President, Apache Programs: Charles A Vehlow
Vice President, Commercial Light Helicopters: Ervin Hunter
Director, Marketing: Terry Higginbotham
General Manager, Human Resources and Communications: Ms Catherine Loucks

■ Attack helicopter (AH-64A and AH-64D Longbow Apache); MD 500 series commercial and light attack helicopters; MD 600N; MD Explorer helicopters, research and development on composite and IR suppression materials; conventional armament including M242 and M230 chain gun; M242 and M230 chain gun; automatic cannon feed, mechanical and auxiliary components.

McDonnell Douglas Finance Corp.
4060 Lakewood Boulevard, 6th Floor,
Long Beach, CA 90808
☎ (+1 310) 627 30 00 ⊖ (+1 310) 627 65 74
President, McDonnell Douglas Finance and MDC Realty Co.: Tom Motherway 98

70-16100 The Parsons Corp.

100 West Walnut Street, Pasadena, CA 91124
☎ (+1 818) 440 20 00 📠 675336 ⊖ (+1 818) 440 26 30
👥 9,500
Chairman: Ray W Judson
President and Chief Executive Officer: James F McNulty
Senior Vice President and Chief Financial Officer: Curtis A Bower
Media Relations Manager: Ms Jamie Brown

Subsidiaries with defence business:
Parsons Transport Group
Parsons Power Group
Parsons Infrastructure and Technology Group

Washington, DC Office
1133 15th Street North West, Washington, DC 20005
☎ (+1 202) 775 60 00 📠 892695 parsons wash
⊖ (+1 202) 775 60 05
Senior Vice President and Manager: W J Hargett

Houston Office
PO Box 34311, Houston, TX 77234-4311
☎ (+1 713) 943 58 43 ⊖ (+1 713) 947 58 42
Manager: Mark H Griebe

Atlanta Office
57 Executive Park South, North East, Suite 590,
Atlanta, GA 30329
☎ (+1 404) 325 07 70 ⊖ (+1 404) 325 83 69
Vice President and Manager: D Burstein

Hawaii Office
567 South King Street, Honolulu, HI 96813
☎ (+1 808) 523 54 64 ⊖ (+1 808) 545 78 37
Manager: G Watada

Fairfax Office
Two Flint Hill, 10521 Rosehaven Street, Fairfax, VA 22030
☎ (+1 703) 591 75 75 ⊖ (+1 703) 591 13 05
Vice President: M R Hockenbury

Cleveland Office
19101 Villaview Road, Suite 301, Cleveland, OH 44119
☎ (+1 216) 486 90 05 ⊖ (+1 216) 486 61 19
Vice President: W H Rownd

Charlotte Office
4701 Hedgemore Drive, Charlotte, NC 28209
☎ (+1 704) 529 62 46 ⊖ (+1 704) 529 03 74
Manager: K M Barry

St. Louis Office
425 Woodmill Road South, Suite 150,
St Louis, MO 63017
☎ (+1 314) 434 29 00 ⊖ (+1 314) 576 27 02
Vice President: R B Bax [98]

70-17000 Primex Technologies, Inc

(† Olin Corporation)
10101 9th Street North, St Petersburg, FL 33716
☎ (+1 813) 578 81 13 ⊖ (+1 813) 578 81 19
♣ $ 500,000,000 (1997) 👥 2,600 (1997)
President and Chief Executive Officer: James G Hascall
Communications Director: Edwin E Alber

Business units:
Ordnance and Tactical Systems Division
Aerospace and Electronics Division

Washington, DC Office
1730 K Street, North West, Washington, DC 20006
☎ (+1 202) 331 74 00 ⊖ (+1 202) 331 07 33
General Manager: Al Calabnese [98]

70-18000 The Racal Corporation Corporate Headquarters

(a subsidiary of Racal Electronics plc, UK)
11 Penn Plaza, New York, NY 10001
1601 North Harrison Parkway, Sunrise, FL
☎ (+1 212) 268 09 18 [W] http://www.racal.com
⊖ (+1 212) 563 90 87
Vice President, Corporate Communications:
Edward W Cheatham
■ Defence-related products and services including data and radio communications, electronic warfare, test and recording instrumentation, switching systems, marine electronics, navigation and positioning systems, defence radar and avionics. Provides financial and management services to subsidiaries.

Subsidiaries:
Racal Data Group/Racal-Datacom Inc
Racal Communications Inc
Racal Avionics Inc
Racal Canada Inc
Racal Instruments Inc
Racal NCS Inc
Racal Pelagos Inc
Racal Antennas Inc
Racal Filter Technologies Ltd
Racal Protection
Racal Recorders Inc
Racal Health & Safety Inc
Racal Pelagos Canada Inc
EFI Corp
Racal-Datacom Canada Inc

Corporate Communications Center
(a subsidiary of Racal Electronics plc, UK)
11 Penn Plaza, New York, NY 10001
☎ (+1 212) 268 09 18 ⊖ (+1 212) 563 90 87
Vice President, Corporate Communications:
Edward W Cheatham [98]

70-18500 SEXTANT Holdings Inc

(a subsidiary of Sextant Avionic SA, France)
Beacon Center 1924 NW, 84th Avenue, Miami, FL 33126
☎ (+1 305) 597 63 21; 597 63 00
⊖ (+1 305) 597 63 63
Vice President and General Manager: Ms Priscilla Moore
☎ (+1 305) 597 63 14
Executive Assistant to the Chief Executive Officer:
Ms Annie Pou-Garcia [98]

70-19100 Sundstrand Corp.

Corporate Offices
4949 Harrison Avenue, PO Box 7003,
Rockford, IL 61125-7003
☎ (+1 815) 226 60 00 ⊠ 257440; TWX 910 631 4255
⊖ (+1 815) 226 26 99
♣ $ 1,372,700,000 (1994) 👥 9,200
Chairman and Chief Executive Offiver: Robert H Jenkins
Corporate Vice President and Controller:
DeWayne Fellows
Executive Vice President, Chief Financial Officer and Treasurer: Paul Donovan
Executive Vice President and Chief Operating Officer Aerospace: Ronald F McKenna
Executive Vice President and Chief Operating Officer Industrial: Patrick L Thomas
Vice President, General Counsel and Secretary:
Richard M Schilling
Director License and Internal Agreements: James Wanner
Director: J P Bolduc
Director: Gerald Grinstein
Director: Charles Marshall
Director: Klaus Murmann
Director: Richard A Abdoo
Director: Ward Smith
Director: Berger G Wallin
Corporate Manager Public Affairs: Bradley J Considine
■ Components and subsystems for aerospace and industrial markets.

Business units:
Sundstrand Aerospace
The Falk Corp.
Milton Roy Co.
Sullair Corp. [98]

70-19500 Talley Industries Inc.

2702 North 44th Street, Suite 100-A, Phoenix, AZ 85008
☎ (+1 602) 957 77 11 ✆ (800) 321 86 23
⊖ (+1 602) 852 69 72
👥 3,507
Chairman and Chief Executive Officer: Paul L Foster
President and Chief Operating Officer: Jack C Crim
Vice President, General General Counsel and Secretary:
Mark S Dickerson
Vice President and Treasurer: Daniel R Mullen
Treasurer: Donald L Corey
Treasurer: Thomas A Dickey
■ Operates in four main business segments; government products and services, industrial products, specialty products, realty.

Subsidiaries with defense business:
Talley Defense Systems Inc.
Electrodynamics Inc.
Rowe Industries Inc.
John J. McMullen Associates Inc.
Universal Propulsion Co. Inc.

Washington, DC Office
1215 Jefferson Davis Highway, Suite 1105,
Arlington, VA 22202
☎ (+1 703) 414 50 70 ⊖ (+1 703) 414 50 80
Vice President of Government Relations: George W Poole [98]

70-20250 Textron Inc.

40 Westminster Street, Providence, RI 02903-2525
☎ (+1 401) 421 2800 [W] http://www.textron.com
⊖ (+1 401) 421 2878
👥 55,000
Chairman & CEO: James F. Hardymon
President & COO: Lewis B. Campbell
Exec. VP Corporate Development: Thomas P Hollowell
Exec. VP & CFO: Dennis G. Little
Exec. VP Government & International: Ms Mary L. Howell
Exec. VP & General Counsel: Wayne W. Juchatz
Exec. VP Admin & Human Resources Officer:
William F. Wayland

Defence related divisions:
Textron Marine & Land Systems → 75
Textron Systems Division → 75

Washington Office
1101 Pennsylvania Avenue, NW, Suite 400,
Washington, DC 20004
☎ (+1 202) 637 3800 ⊖ (+1 202) 637 3860
Exec. Dir Public Affairs: Ms Eugene Kozicharow [97]

70-31000 United Technologies Corp. (UTC)

1 Financial Plaza, Hartford, CT 06101
☎ (+1 860) 728 70 00 [W] http://www.utc.com
⊠ 99212 utc hfd a; TWX 710 4255 642
⊖ (+1 860) 728 79 44; 728 79 79
♣ $ 23,500,000,000 (1996) 👥 173,800 (1996)
Chairman, President and Chief Executive Officer:
George David
Executive Vice President Government Affairs/Chair International Operations: Ms Ruth Harkin
Senior Vice President Human Resources and Organisation:
William L Bucknall (Jr)
Senior Vice President Science and Technology:
Robert J Hermann
Vice President Strategic Planning: Ari Bousbib
Vice President Controller: Jay Haberland
Vice President Industrial Relations: James F Perretta
Senior Manager, Public Relations: Peter Murphy
■ Diversified manufacture with products for aerospace, building and automotive industries.

Divisions & Subsidiaries:
Carrier Corp.
Hamilton Standard Div. → 75
International Fuel Cells Corp.
Sikorsky Aircraft → 75
Pratt & Whitney Canada Inc. → 75
Otis Elevator Co. → 75
Pratt & Whitney → 75
United Technologies Research Center
UT Automotive

United Technologies International Affairs
1401 Eye Street, NW, Suite 600, Washington, DC 20005
☎ (+1 202) 336 74 00 ⊖ (+1 202) 336 75 30
Chair, International Relations: Ms Ruth Harkin [98]

74 Consortia, Joint Venture Companies

74-00050 Advanced Energetic Materials Corp. (AEMC)

1111 Jefferson Davis Highway,
Arlington, VA 22202-3225
☎ (+1 703) 553 0380
■ Joint venture company set up by Kaman Corp. & French company SNPE. Products & systems for defense & industrial applications based on insensitive explosive materials technology. → 75 [97]

74-00300 Bell-Boeing V-22 Program Office

Suite 400, Crystal Gateway, No 1,
1235 Jefferson Davis Highway, Arlington, VA 22202
☎ (+1 703) 414 33 50 [W] http://www.boeing.com
[W] http://www.bellhelicopter.textron.com
⊖ (+1 703) 414 33 78
Manager of Tiltrotor Comms, Bell Helicopter Textron:
Terry Arnold ☎ (+1 817) 280 64 40
Manager of Media Relations, Boeing Helicopters:
Nick Kernstock ☎ (+1 610) 591 28 22
■ Co-producers of the V-22 Osprey Tiltrotor Aircraft for the US Joint Military Services. [98]

74-00400 CFM International Inc. (CFMI)

(Joint venture between General Electric, USA and SNECMA, France)
PO Box 15514, Cincinnati, OH 45215
☎ (+1 513) 552 33 00 [W] http://www.cfmsl.com
⊖ (+1 513) 552 33 06
♣ $ 2,800,000,000 (1996)
President and Chief Executive Officer: Gerard Laviec
Executive Vice President: William S Clapper
Executive Vice President: Pierre Fabre
Finance Director: William T Fitzgerald
■ Jointly owned company formed by General Electric Co.(USA) & SNECMA (France) to provide overall program management & a single customer interface for the CFM56 engine. [98]

74-09500 International Aero Engines AG (IAE)

400 Main Street, M/S 121-10, East Hartford, CT 06108
☎ (+1 860) 565 55 15 [W] http://www.v2500.com
⊠ 436031 intlaero ⊖ (+1 860) 565 55 10
Chairman: Karl Krapek
President: Barry Eccleston
■ A consortium company responsible for the direction of the V2500 turbofan programme, for sales of powerplants and providing worldwide spares and in-service support to operators.

Shareholders in IAE:
Pratt & Whitney (through United Technologies Corp.), Rolls Royce Plc, MTU and Japanese Aero Engines Corporation. [97]

75 Manufacturing Companies

Advanced Energetic Materials 74-00050
AIM Enterprises Inc. 80-01350
Atlantic Research Marketing Systems 80-03365
Global Security Complex 80-12950
Lockridge Mapstrap Co. 80-18612
Manufacturers Export Organization . . 85-01250
Orlando Technology Inc. 60-16000
Prosig USA Inc 90-16727
Q-Systems International Inc. 75-71960
Tracor Information Systems Company 80-33154
US Airmotive Inc. 85-02600

75-00010 AACO Inc.
9626 Rainier Avenue South, Seattle, WA 98118-5935
☎(+1 206) 722 15 71 ⊞ TWX 650 249 65 88
♣ $ 5,000,000 ☗ 6
President: H. Davis Crone
Manager: Ms Leslee Lemka
■ Manufacture avionic test equipment. Perform read-out services for flight data recorders. [98]

75-00015 AAF International
PO Box 35690, Louisville, KY 40232
215 Central Avenue, Louisville, KY 40208
☎(+1 502) 637 00 11 ✆ (800) 501 3146
⊖ (+1 502) 637 03 21
☗ 6,100
President and Chief Operating Officer: Gerald Boehrs
President - Air Filtration: Robert Brymer
President - Environmental: Dennis Marcott
Marketing Communications Manager: Dennis Russ
■ Air cleaning and filtration equipment, noise suppression equipment. Dust collection equipment, air pollution control equipment. [98]

75-00020 AAI Corp.
(a subsidiary of United Industrial Corp.)
PO Box 126, Hunt Valley, MD 21030-0126
☎(+1 410) 666 14 00 ⊠ AAICKYVL Baltimore MD
Ⓦ http://www.aaicorp.com ⊖ (+1 410) 628 32 15
♣ $ 206,000,000 (1995) ☗ 1,500 ⌂ 1,000,000 sq ft
President and Chief Executive Officer: Richard R Erkeneff
Vice President and Chief Financial Officer: Paul J Michaud
Vice President, General Counsel and Secretary: Howard M Bloch
Vice President and General Manager of Defence Systems: Maurice P Ranc
Vice President and General Manager of Weather Systems: G Russell Zink
Vice President and General Manager of Transportation Systems: Jackson R Bell
Vice President and General Manager of Fluid Test Systems: Thomas E Wurzel
Vice President and General Manager of Operations: Joseph F Burger
Total Quality Director: Howard E Butz
Information Technology Director: David A Powell
Human Resources Director: Ms Suzan J Feild
Corporate Communications Manager: Ms Anne Stone
■ Development and production of high-technology mechanical and electronic products serving military, government and commercial customers.
Prime contractor and subcontractor specialising in systems analysis, design, development, manufacture, installation and contractor logistics support.
The Corporation is organised around its core competencies: Defence Systems; Fluid Test Systems; Weather Systems and Transportation Systems.
ISO 9001 certificated.
Subsidiaries:
AAI/Engineering Support Inc
AAI/Systems Management Inc
AAI/ACL Technologies [98]

75-00120 AAR Advanced Structures
12633 Inkster Rd, Livonia, MI 48150
☎(+1 313) 522 2000 ⊞ TWX 810 242 2979
⊖ (+1 313) 522 2240
VP & General Mgr: Edmund Swain
Dir Sales: W.D. Compton
■ Design, engineering & manufacture military & commercial air cargo restraint systems, nuclear shielding & specialised storage containers, rubber shelters & special projects. [97]

75-00122 Aardvark Tactical Inc.
132 1/2 La Porte Street, Arcadia, CA 91006
☎(+1 818) 821 34 50 Ⓔ info@nonlethal.com
Ⓦ http://www.nonlethal.com ⊖ (+1 818) 821 04 44
Presdient and Chief Executive Officer: Jon Becker
■ Provider of training and equipment to military units and police organisations for non-lethal weapons. [98]

75-00150 Abbott Electronics Inc.
(† Abbott Transistor Laboratories Inc.)
(a subs. of the Adwest Group, UK)
2727 S. La Cienega Blvd, Los Angeles, CA 90034
☎(+1 310) 202 8820 ⊖ (+1 310) 836 1027
♣ $ 10 000 000 ☗ 100 ⌂ 22 000 sq ft
President: H. Emami
Sales & Marketing Mgr: Bill Standen
■ Power Modules mainly sine-wave inverters, DC to DC converters, AC to DC switches, power transformers, overvoltage protectors & input filters. [97]

75-00180 Abbott Interfast Corp.
190 Abbott Drive, Wheeling, IL 60090
☎(+1 708) 459 6200 ⊞ 254591 ⊖ (+1 708) 459 4076
☗ 180 ⌂ 100 000 sq ft
Chairman: M.R. Binder
President: R.M. Binder
Sr Vice President Sales: J.A. Calabaese
Controller: S.G. Breitsenstein
■ Special, standard & miniature nylon insert locknuts & all metal locknuts, captive nuts, threaded fasteners, automatic screw machine parts, sealing locknuts for integral fuel tanks per MIL 85615, special metal parts, cold formed & machined, subassemblies; windshield nuts; switch-lok (anti-tamper, miniature dip switches). [97]

75-00260 ABLE Corp.
23695 Via Del Rio, Yorba Linda, CA 92887-2715
☎(+1 714) 692 02 00 ⊖ (+1 714) 692 02 59
♣ $ 12,200,000 (1996) ☗ 107 (1997)
President and Chief Executive Officer: Stephen D Ehrick
Sales and Marketing Director: Ed Palmer
Purchasing Manager: Steve Breunle
Contract Administrator: Ms Nancy Dobar
■ M1 tank oil cooling and scavenge fans, C13 suction boost pumps, static line retrieval winch, flap asymmetry brakes, CH53 oil cooling blower, LCAC oil cooling and scavenge fans, C-17 ramp heater blower. [98]

75-00280 Abrams Instrument Corp.
1322 Rensen St, Lansing, MI 48910
☎(+1 517) 882 7211 ⊖ (+1 517) 882 3136
President: Merlyn Gates
■ Counters, timers, stepping motors, photogrammetric instruments, electrical & electronic controls. Manufacturing & assembly services. [97]

75-00300 AC Inc.
1085 Jordan Road, Huntsville, AL 35811-9376
☎(+1 205) 851 90 20 ⊖ (+1 205) 851 90 25
♧ $ 1,500,000 (1997) ♣ $ 20,000,000 (1996)
☗ 130 (1997) ⌂ 120,000 sq ft
President: John R Riche
Operations Director: Bobby Bayless
Sales and Marketing Manager: John 'Butch' Cryder
Purchasing Manager: Dwayne Griffin
■ High volume guided missile container and component part manufacturing, with capabilities to fabricate fibreglass, aluminium or steel products. Welding, sheetmetal and machine shop facilities. Manufacture Harpoon, Phoenix, Maverick, Chaparral and Stinger Missile Containers. [98]

75-00330 Accurate Energetic Systems, LLC
5891 Highway 230 West, McEwen, TN 37101-9801
☎(+1 615) 729 42 07 ⊖ (+1 615) 729 42 17
♣ $ 10,000,000 (1996) ☗ 53 (1996)
President: John Sonday
Operations Director: Mickey Hambright
Sales and Marketing Director: Marty Liggins
Explosives Sales Manager: Bill Robertson
Powder Sales Manager: Marty Liggins
Purchasing Manager: Dave Haley
Personnel Manager: Ms Paula Pittman
■ High explosives. Propellant for small arms ammunition. [98]

75-00360 AAI/ACL Technologies Inc.
(a subsidiary of AAI Corporation)
1505 East Warner Avenue, Santa Ana, CA 92705
☎(+1 714) 979 55 00 ✆ (800) 232 5784
Ⓦ http://www.acitech.com
⊖ (+1 714) 979 63 53; 979 39 05
☗ 100 ⌂ 200,000 sq ft
President: Thomas Wurzel
Vice President of Research and Development and Technical Director: Jim Bybee
Operations and Technical Director: Ted Marwitz
Sales and Marketing Director: Robert V Dodson
International Technical Sales Director: Mark Krusemark
Manager of Government and Military Programs: Richard Orton
Administration Manager, Business Development: Ms Judy Gamble
Purchasing Manager: Don Harris
■ Design and manufacture of computer-driven and dedicated hydraulic, pneumatic and fuel test equipment for servo valves, actuators, pumps, motors, CSD/IDGs, engine accessories and other aircraft components; hydraulic and pneumatic power supplies and fluid cooling systems. Carrier-based aircraft dollies. [98]

75-00380 Acousticom Corp.
28180 Clay Street, Elkhart, IN 46517-1070
☎(+1 219) 293 05 34 ✆ (800) 664 05 34
Ⓔ acousticom@skyenet.net
Ⓦ http://www.acousticom.com ⊖ (+1 219) 294 72 50
♣ $ 1,650,000 (1996) ☗ 35 (1996) ⌂ 10,000 sq ft
President and Chief Exectutive Officer: Ms Jean A Cochran
Vice President and Chief Engineer: Mark Cochran
Marketing Manager: Terry Fisher
Purchasing Manager: Douglas Cochran
■ Manufacturer of dynamic noise cancelling microphones, dynamic earphones, audio communication cables, cable assemblies and related spare parts. [98]

75-00460 AC Tridon - North America
8100 Williams Drive, PO Box 65, Smyrna, TN 37167
President & COO: M. Hottinger
VP Sales & Marketing: J. Bray
■ Manufacturer specialising in wiper & wiper systems, engineered clamps & electronic modules, serving both the automotive & industrial products industries. [97]

75-00560 Actronics Inc.
166 Bear Hill Road, Waltham, MA 02154-1002
☎(+1 617) 890 70 30 ⊖ (+1 617) 890 75 40
☗ 20 ⌂ 10,000 m2
President: David C Ewing
Operations Director: Joseph Heine
Director of Advertising: Edward Beucker
■ Temperature & pressure valves & actuators applicable to fuel & oil heating & cooling systems for aircraft. [98]

75-00590 Acutronic USA Inc
(a subsidiary of Acutronic AG, Switzerland)
139 Delta Drive, Pittsburgh, PA 15238
☎(+1 412) 963 94 00 Ⓔ aus@acutroni.com
⊖ (+1 412) 963 08 16
☗ 20
Managing Director: Louis A de More
■ Intertial guidance test instruments, motion simulators, centrifuges, servo systems, controllers. [98]

75-00630 Adams Rite Sabre International (ARPI)
540 W. Chevy Chase Dr., Glendale, CA 91204
☎(+1 818) 241 5121
⊞ TWX 910 4972 276 adms rte gla
⊖ (+1 818) 500 7239
☗ 240 ⌂ 60 000 sq ft
President: Charles Collins
Technical Dir: Jim Slate
Sales & Mktg Dir: Jonas Williams
Regional Sales Mgr: Daniel J. Scandalito
■ Electromechanical flight controls, computer cabinetry latches, bellows reservoirs, door bolting systems, pressure regulating/relief valves, pilot wheels & grips, quadrants, electronic faucets. [97]

75-00730 ADC Telecommunications Inc.
4900 West 78th Street, Minneapolis, MN 55435-5480
☎(+1 612) 938 80 80 ✆ (800) 366 38 91
Ⓦ http://www.adc.com ⊖ (+1 612) 946 32 92

♧ $ 366,500,000 (1993) ♣ $ 828,000,000 (1997)
👥 2,900 (1997) 🏭 400,000 sq ft
Chairman: W J Cadogan
Public Relations Director: Ms Lynne M High
Director of Marketing Communications: Kevin Johnson
Account Manager Government Sales: Jerry Kerr
■ Leading supplier of transmission and networking systems for voice, data & video services & of physical connectivity products for fibre optic, twisted-pair, coaxial & wireless networks. [98]

75-00790 Advanced Aerodynamics & Structures Inc. (AASI)
3501 Lakewood Boulevard, Long Beach, CA 90808
☎ (+1 562) 938 86 18 [W] http://www.aasiaircraft.com
⊖ (+1 562) 938 86 20
👥 130
Chairman, President and Chief Executive Officer:
Dr Carl L Chen
Executive Vice President: Gene Comfort
■ Turboprop single-engined and jet aircraft. Unmanned aircraft, UAV's, RPV's. Advanced graphite compositer. [98]

75-00795 Advanced Circuitry
4811 West Kearney, Springfield, MO 65803-9579
☎ (+1 417) 862 07 51 ⊖ (+1 417) 862 07 34
President: Robert Schutz
■ Printed circuit boards, double-sided, multi-layer, multi-layer backpanel (commercial and military). [98]

75-00820 Advanced Composites Group Inc.
(a subsidiary of Advanced Composites Group Ltd, UK)
5350 South 129th East Avenue, Tulsa, OK 74134
☎ (+1 918) 252 39 22 ⊖ (+1 918) 252 73 71
♣ $ 15,000,000 (1997) 👥 48 (1997) 🏭 2500 m2
President: C Ridgard
Operations Director: Steve Larby
Sales and Marketing Manager: John Powers
Purchasing Manager: Dave Collins
■ Manufacture speciality prepreg material for tooling prototype and production composite mouldings, components and assemblies. [98]

75-00830 Advanced Composites Inc.
1154 South 300 West, Salt Lake City, UT 84101
☎ (+1 801) 467 12 04 ⊖ (+1 801) 467 43 67
👥 30 🏭 30,000 sq ft
President: Randall J Philpot
Salesand Marketing Director: David W Erickson
Purchasing Director: Jeff Sleight
Production Director: Daniel K Buckmiller
■ Design and manufactures filament wound composite structures for the defence, aerospace, commercial and recreational industries. [98]

75-00895 Advanced Electromagnetics
PO Box 711719, Santee, CA 92072
☎ (+1 619) 449 94 92 ⊖ (+1 619) 449 15 53
♣ $ 3,700,000 (1996) 👥 21 🏭 10,000 sq m
President and Chief Executive Officer: Gabriel A Sánchez
General Manager: Ms Barbara Sanchez
Sales and Marketing Director: Peter Deal
Purchasing Manager: Ms Michele Cochran
■ Manufacture of RF microwave absorber, design and construction of EMC and anechoic chambers. Consultancy. [98]

75-00900 Advanced Friction Technology
12410 N.E. Whitaker Way, Portland, OR 97230
☎ (+1 503) 256 5641 ⊖ (+1 503) 256 5671
👥 10
Managing Dir: Jim McMahen
■ Brake lining & friction material manufacturer, Kevlar reinforced composite sheet stock. [97]

75-00905 Advanced Ground Systems Engineering Corporation (AGSE)
(a subsidiary of K B Industries, USA)
1265 North Kraemer Boulevard, Anaheim, CA 92806
☎ (+1 714) 632 82 00 [E] agse@juno.com
📠 TWX 910 346 7745 ⊖ (+1 714) 632 90 95
♣ US$ 12,000,000 (1997) 👥 80 (1997) 🏭 70,000 sq ft
President: John E Marshman
Vice President and Operations: Joel R Nunn
Vice President of Sales and Marketing: Bobby D Jackson
Inside Sales Manager: Renée Green
Quality Control Manager: Ed Callahan
Purchasing Agent: Ms Maryan Kosareff
Chief Engineer: Joel Morey
Plant Manager: Wayne Ordway
■ Aircraft work docks, tail stands, fixed and portable access stands, split engine shipping systems, engine transport stands, bootstrap engine change equipment, module containers for jet engines, special tools. [98]

75-00930 Advanced Processing Labs Inc.
5871 Oberlin Drive, San Diego, CA 92121
☎ (+1 619) 546 86 26 ✆ (800) 822 75 22
[W] http://www.aplabs.com ⊖ (+1 619) 546 02 78
♣ US$ 14,000,000 (1997) 👥 80 (1997) 🏭 40,000 sq ft
President: Dr Doug Giese
Chief Operating Officer: Richard Wank
Vice President of Sales: Steve Gills
■ Real time systems integrator, manufacture of COTS, VME based test and processing systems, simulation, telemetry data processing and satellite ground use. [98]

75-00940 Advanced Research & Applications (Aracor)
425 Lakeside Drive, Sunnyvale, CA 94086
☎ (+1 408) 733 77 80 ⊖ (+1 408) 732 19 96
♣ $ 10,000,000 👥 57 🏭 27,000 sq ft
President: Dr Robert A Armistead
Vice President: Dr J H Stanley
Vice President: H E Peck
Vice President: B H LeBalcer (III)
■ Manufacture a range of industrial x-ray imaging systems used for non-destructive inspection (NDI) and cargo screening. Principal products include digital radiography and computed tomography systems. Manufacture model 4100 line of automatic semiconductor irradiation systems for radiation testing semiconductor electronics at the wafer level of processing. Custom systems and research services are also provided under research and development contracts. [98]

75-00950 Advanced Semiconductor Inc.
7525 Ethel Avenue, North Hollywood, CA 91605-1912
☎ (+1 818) 982 12 02
[E] asiusa@asi.ccmail.compuserve.com
⊖ (+1 818) 765 30 04
👥 100 (1997)
President: Fred Golob
Vice President, Sales: Allen Klein
Marketing Manager: Bill Meyer
Purchasing Manager: Ms Dolly Barrera
■ Manufacture electronic components and supplies for use in communication and other electronic military equipment. [98]

75-01000 Advanced Technical Corp.
(† Alvaradio Industries Inc.)
1915 Armacost Avenue, Los Angeles, CA 90025
☎ (+1 310) 478 35 61 ⊖ (+1 310) 826 40 58
♣ $ 2,000,000 👥 20 🏭 20,000 sq ft
President: A T Bigman
Marketing Director: Thomas E Abel [98]

75-01030 Advanced Training Systems Inc. (ATS)
4524 Highway 61 North, St Paul, MN 55110
☎ (+1 612) 429 8091 ⊖ (+1 612) 429 8702
♣ $ 1 000 000 (1996) 👥 12 (1996) 🏭 3 500 sq m
President & Marketing Mgr: Edwin J. Taylor
Operations/Purchasing Mgr: E.J. Schneider
■ Portable military & police electronic shooting ranges, featuring computerised target controls & hit response targets. [97]

75-01105 Aerial Machine & Tool Corp.
US 58, PO Box 222, Vesta, VA 24177-0222
☎ (+1 540) 952 20 06
[E] aerialm-tcorporation@postoffice
[W] http://www.worldnet.att.net ⊖ (+1 540) 952 22 31
👥 81 🏭 23,000 sq ft
President and Chief Executive Officer: Peter T Coe
Executive Vice President: John Riccitello
Sales Director: Ms Gail V Clark
Purchasing Manager: J Terrance Quinn
■ Aircraft seat belts, shoulder and specialised harnesses, SAR equipment (helicopter rescue hooks, rescue swimmers harnesses, survivors slings, hoisting vests). Parachute accessories. Air deployment equipment. Machined parts. MIL-I-45208A. [98]

75-01108 Aero Innovations Inc.
2800 East 12th Street, Little Rock, AR 72202-4300
☎ (+1 501) 374 23 76 [W] http://www.aeromed.com
⊖ (+1 501) 374 56 03
👥 9 (1997)
President: Kevin G Saunders
Vice President of Marketing: Severyn Ciosmak
Marketing and Communications Manager: Ms Kelly Smith
■ Design, manufacture and certification of air medical and medevac interiors for aircraft and helicopters. Specialise in head of state and military applications. [98]

75-01113 Aero Products Research Inc. (APR)
11201 Hindry Avenue, Los Angeles, CA 90045
☎ (+1 310) 641 72 42 ⊖ (+1 310) 649 17 03
👥 100 🏭 30,000 sq ft
President: Leonard Wilstein
Vice President: Ms Joyce Wilstein
Sales And Marketing Director: Craig Spencer
Operations Director: Roger Schamp
Purchasing Manager: Edwin Rusk
■ Manufacture pilot's equipment and pilot training materials for general aviation and military use. Products include plotters and plotting boards, navigation computers, kneeboards, related pilot aids and custom OEM products printed on metal or plastic. [98]

75-01116 Aero Systems Engineering Inc. (ASE)
(a subsidiary of Celcius Inc., USA)
358 East Fillmore Avenue, St Paul, MN 55107
☎ (+1 612) 227 75 15 📠 297043 aerosystem stp
⊖ (+1 612) 227 05 19
♣ $ 20,000,000 (1996) 👥 150 (1996) 🏭 150,000 sq ft
Chairman: Christer Persson
President: Leon Ring
Director of Engineering: Grant Radinzel
Director of Test Cell Marketing: Dick Thomalla
Director of Test Cell Operations: Michael Browne
Director of Wind Tunnel Marketing and Operations:
Don Kamis
Director Fluidyne Aero Test Laboratory: Dick Brasket
■ Design and manufacture gas turbine engine test facilities, test stands and test benches, including turnkey test cells, thrust stands, adaptors, full instrumentation and computer systems, wind tunnels and vehicle test systems, wind tunnel aero/accoustic testing services. [98]

75-01117 Aero Tec Laboratories Inc. (ATL)
Spear Rd Industrial Park, Ramsey, NJ 07446
☎ (+1 201) 825 1400 ✆ (800) 526 5330
⊖ (+1 201) 825 1962
👥 50 🏭 70 000 sq ft
President: Peter J. Regna
VP Operations: Louis D'Amico
VP R&D: Albert S. Baris
VP Sales: David H. Dack
■ Fuel cell bladder tanks in accordance with MIL-T-6396-D & FAA TSO C80, collapsible pillow-type fuel storage containers to MIL-T-52983 & 100 to 100,000 gallons & air portable fuel drums. [97]

75-01118 Aero Union Corp.
100 Lockheed Avenue, Chico, CA 95973-9024
☎ (+1 916) 896 30 00 ⊖ (+1 916) 896 30 99
♣ $ 14,000,000 (1996) 👥 150 (1996)
President and Chief Operating Officer: V Alvistur
Senior Vice President of Sales and Marketing:
John M Oswald ☎ (+1 916) 896 30 14
Director of Public Relations: Chris Berge
☎ (+1 916) 896 30 08
Purchasing Manager: Vince Decco
■ In flight refuelling pods. Fuel bulk tanks 1800 gals for C130. On board cargo powered roller systems. Aerial firefighting, conversions and operations. [98]

75-01150 Aerocom/Squitter
792 South Military Trail, Deerfield Beach, FL 33442
☎ (+1 954) 428 79 55 [E] squitter@squitter.com
[W] http:www.squitter.com ⊖ (+1 954) 420 08 68

👥 16 (1997) 🏭 360 m2
President: Ajax Melo
Operations Director: John Gonzales
Purchasing Manager: Bill Clark
Marketing Manager: Louis Oricchio
■ Measurement and control systems with applications in meteorology, agriculture, industry and cargo tracking. Data retrieval. [98]

75-01500 Aerojet
(a segment of GenCorp.)
Headquarters
Highway 50 & Aerojet Road, PO Box 13222,
Sacramento, CA 95813-6000
☎ (+1 916) 351 8501 ✉ AEROGENCOR
⊖ (+1 916) 351 8667
♣ $ 520 000 000 (1995) 👥 3 300 (1996)
President: Roger I. Ramseier ☎ (+1 916) 351 8616; 351 8668 (fax)
VP Tactical, Defense & Armaments: Joe Carleone ☎ (+1 916) 355 2187; 355 5187 (fax)
VP Strategic & Space Propulsion: Bob Davis ☎ (+1 916) 355 3755; 355 3743 (fax)
VP Space Surveillance Systems: Carl Fischer ☎ (+1 818) 812 2001; 812 2408 (fax)
VP Earth Sensing Systems: George 1Padgett ☎ (+1 818) 812 8110; 969 7750 (fax)
VP SADARM: Ron Abbott ☎ (+1 818) 812 2927; 969 4801 (fax)
VP Sacramento Operations: Bob Harris ☎ (+1 916) 355 2721; 355 2448 (fax)
VP Azusa Operations: Walt Chin ☎ (+1 818) 812 2234; 812 2792 (fax)
VP Tennessee Operations: Curtis Burnette ☎ (+1 615) 753 1216; 753 8645 (fax)
VP Washington Operations: Hal Howes ☎ (+1 202) 828 6810; 828 6849 (fax)
VP & Dir Engineering: Sabby Sabnis ☎ (+1 818) 812 2110; 812 2792 (fax)
VP Strategic Business Dev.: Tom Peoples ☎ (+1 916) 351 8551; 351 8668 (fax)
VP Human Resources: Terry Cochran ☎ (+1 916) 351 8511; 351 8614 (fax)
VP Contracts Proposals & Pricing: Stan Barnett ☎ (+1 916) 351 8601; 351 8664 (fax)
VP Communications: Edie Cartwright ☎ (+1 916) 351 8501; 351 8667 (fax)
VP Law: Bill Phillips ☎ (+1 916) 351 8604; 351 8665 (fax)
VP/Controller: Mike Martin ☎ (+1 916) 351 8555; 351 8668 (fax)
VP Environmental Safety & Health: Suzanne Phinney ☎ (+1 916) 351 8572; 351 8664 (fax)
Procurement Dir: Bill Ashbaugh ☎ (+1 916) 355 3447; 355 2448 (fax)
■ Major aerospace/defence contractor specialising in rocket propulsion systems, space sensor systems & smart munitions. Infrared development, production & testing facilities.
The company builds the sensor for the Defense Support Program's family of surveillance satellites and design & production of highly advanced environmental sensor systems.
Aerojet is developing the SADARM smart munition & warheads & propulsion systems for numerous missiles including the US Army's Nike-Hercules ground-to-air missile, Sidewinder, Hawk, Sparrow III, Tow 2B & Hellfire.
Azusa Operations
1100 West Hollyvale Street, PO Box 296,
Azusa, CA 91702
☎ (+1 818) 812 1000
Tennessee Operations
Old Highway 11 East, PO Box 399,
Jonesborough, TN 37659
☎ (+1 615) 753 1200
Washington Operations
1025 Connecticut Ave N.W., Ste 1107,
Washington, DC 20036
☎ (+1 202) 828 6800 ⊖ (+1 202) 828 6849 [97]

75-01800 Aerolyte Systems Division
(a division of Clemco Industries Corp.)
One Cable Car Drive, Washington, MO 63090
☎ (+1 314) 239 03 00 [E] clemco@mail.usmo.com
⊖ (+1 314) 239 07 88
👥 160 (1997)
President: Arnie Sallaberry
Vice President of Marketing: Ms Patti Roman
Market Development Specialist: Jeffrey F Cleveland
■ Manufacture dry paint stripping equipment for paint removal without the use of chemicals.
Subsidiaries:
Clemco Internatinal GmbH → Germany FR 90
Hodge Clemco → UK 90 [98]

75-01920 Aeromax Industries Inc.
9546 Topanga Canyon Boulevard, Chatsworth, CA 91311
☎ (+1 818) 701 95 00 [E] pilot@aeromax.com
[W] http://www.aeromax.com ✉ 283162 amax ur
⊖ (+1 818) 886 28 45
👥 8 (1997) 🏭 3,000 sq ft
President: Richard D Berube
■ Manufacture airframe and engine parts for military FW and RW aircraft (A-4, C-130, F-4, F-5). Specialising in F-5 parts; RKO kits, distribution boxes, special tools, circuit cards, power supplies. MFG J85 engine parts. [98]

75-02200 Aeronca Inc.
(a subs. of Fleet Aerospace Inc., Canada)
1712 Germantown Rd, Middletown, OH 45042-9983
☎ (+1 513) 422 2751 ✆ (800) 881 9017
⊖ (+1 513) 422 0812
♣ $ 32 000 000 (1994) 👥 220 (1996) 🏭 42 014 sq m
Executive VP & General Mgr: James O. Stine
Dir Marketing: Thomas P. Kueterman
Controller: John Furbay
Dir MIS/MRP: Keith B. Wyman
■ Airframe & jet engine components, assemblies & parts, power plant packages or engine pods, brazed structures, bonded structures, stretch forming, honeycomb, materials and structures, missile assembly. [97]

75-02350 Aeroquip Corp.
(a subs. of Trinova Corp.)
1695 Indian Wood Circle, PO Box 631,
Maumee, OH 43537-0631
☎ (+1 419) 891 7622 ✉ 223412; TWX 810 253 1947
⊖ (+1 419) 891 7614
President & CEO: H.M. Selland
Group VP Aeroquip Industrial Americas Group: E.T. Smith
Group VP Aeroquip Aerospace Group: Stephen C. Schwab
Marketing Communications Mgr: Lee Brower
■ Hose & fittings for military use on aircraft & vehicles, moulded rubber goods, clamps & couplings, sound isolation couplings, moulded plastic products. [97]

75-02450 Furon Forming Center of Excellence
3711 South Hudson Street, PO Box 18319,
Seattle, WA 98118
☎ (+1 206) 760 22 01 ⊖ (+1 206) 723 68 69
♣ $ 12,000,000 👥 121 (1997) 🏭 132,000 sq ft
Director: Mike Craddock
Purchasing Manager: Lance Korb
Product Leader: Robert Bogue ✉ robert.bogue@usa.furon.com
■ Dryliner, manufacture & assembly vacuum & pressure formed enclosures, composite components, foam ducts, self skinning foam components & gaskets. [98]

75-02480 Aerospace Systems
(† Teledyne Aerospace Systems)
(a div. of Nortech Systems Inc.)
1007 E. 10th St, PO Box 998, Fairmont, MN 56031-0998
☎ (+1 507) 235 3355 ⊖ (+1 507) 235 6088
♣ $ 10 000 000 (1995) 👥 105 🏭 50 000 sq ft
Sales Mgr: Kim Rock
Technical Dir: Mike Wasmund
Sales Mgr: Ms Kim Rock
Purchasing Mgr: Avis Wagner
■ Custom moulded shielded cable assemblies. [97]

75-02501 Aerospace Systems & Components Inc.
1620 South Broadmoor, Wichita, KS 67207
☎ (+1 316) 686 7392 ⊖ (+1 316) 686 8542
👥 35
President: Ken W. Seaman
Marketing Mgr: Paul T. Hetherington
Operations & Technical Dir: Pat Smith
■ Design and manufacture of aerospace-quality pneumatic valves, plus thermal actuation devices for use with gas turbines and air cycle machines. [97]

75-02750 AGEMA Infrared Systems
(a Spectra Physics Company)
550 County Ave, Secaucus, NJ 07094
☎ (+1 201) 867 5390 ✆ (800) 969 5390
⊖ (+1 201) 867 2191
👥 20
President: David J. Smith
VP Finance: Mark Neilsen
VP Surveillance Sales: Benkt Linnander
Communications Mgr: Jay levy
Applications Mgr IR Surveillance Prod.: Greg McIntosh
■ Manufacture infrared imaging & data acquisition systems for use in target signature measurement & analysis & other military & defence applications, surveillance & materials & composite evaluation. [97]

75-02870 AGM Container Controls Inc.
3526 East Fort Lowell Road, PO Box 40020,
Tucson, AZ 85717-0020
☎ (+1 520) 881 21 30 ✆ (800) 995 55 90
[E] agm@azstarnet.com ⊖ (+1 520) 881 49 83
👥 55 (1997)
President: Ms Joyce C Stewart
Sales and Marketing Director: Howard N Stewart
Sales Manager: Ms JoAnne M Quirion
Purchasing Manager: Jim McCorry
Marketing: Ms Lee Ann Allin
■ Manufacture environmental control hardware for moisture-sensitive electronic and optical equipment, mobile electronic shelters, re-usable shipping and storage containers: breather valves, desiccators, desiccant, humidity indicators, record holders, shock overload indicators, tie down straps and tie down shelving.
Overseas agents/reps:
Adtech Inc./Aero-Spec Products Inc. (Japan)
Dydra (France)
Mineral Derivatives Ltd (UK)
Sued-Chemie AG (Germany FR)
Yail-Noa Ltd Agencies (Israel)
Glenn Drew Australia (Australia/New Zealand) [98]

75-02950 AIL Systems Inc.
(a subsidiary of the Eaton Corp.)
455 Commack Road, Deer Park, NY 11729
☎ (+1 516) 595 50 00 ✆ (800) 944 11 80
⊖ (+1 516) 595 57 54
Chairman of the Board: Neil A. Armstrong
President & CEO: James M. Smith
VP Programs & Technology: Edmond Bresnihan
VP Chief Finance Officer: Darrell Reed
VP Eningeering & Manufacturing: Gordon Corlew
Dir Plans & Business Dev.: Jack Migliaccio
Dir Corporate Development: Harvey Kreisberg
Special Assistant to the President: Ms Patricia A. Esposito
■ Design, develop & manufacture electronic warfare & radar systems for US military services. Products include the ALQ-99 Tactical Jamming System for EA-6B & EF-111A aircraft, ALQ-161 Defensive Avionics System for B-1B bomber, TPS-74 Ground Surveillance Radar & Airborne Modular Radar for the US Army.
Technical Services Operations (TSO)
254 East Avenue, K-4, Lancaster, CA 93535
☎ (+1 805) 723 50 63 ⊖ (+1 805) 948 70 03
Manager: Lawrence Schwartz [98]

75-03000 Thermax/CDT, Inc
(a division of Cable Design Technologies)
8946 Winnetka Avenue, Northridge, CA 91324
☎ (+1 818) 701 05 65 ✆ (800) 423 58 73
⊖ (+1 818) 701 79 64
Vice President: R Canny
Sales Manager: Ted Lind
■ High temp wire and cable for defence, electronics and aerospace applications, coaxial cable for microwave, radar and radio communication. [98]

75-03020 AIM Inc.
19200 Middletown Road, Parkton, MD 21120-9693
☎ (+1 410) 329 68 01 ⊖ (+1 410) 357 59 16
♣ $ 5,000,000 👥 55 🏭 31,000 sq m
Vice President of Sales and Marketing: LeRoy Phillips
Public Relations Director: Ms Carole Currey
Sales Manager: Bert Ochterbeck
■ MK 22 MoD 4 rocket motor. [98]

75-03050 Aimsco Inc.
5707 Airport Way South, PO Box 80304,
Seattle, WA 98108-0304
☎ (+1 206) 762 80 14 ✆ (800) 426 02 44
✉ TWX 910 444 13 33 ⊖ (+1 206) 762 83 94
👥 50 (1995) 🏭 25,000 sq m
President: C. Ernst

Vice President: Carl Teichman
Dir Sales & Marketing: Butch Davis
Operations/Technical Dir: M.L. Petersen
Purchasing Mgr: Don Ellsworth
■ Aerospace fasteners & military fastening devices.
Associate:
C.W. Warren Co. [98]

75-03100 Airborne Power
(† MagneTek Tempe)
(a division of Power Paragon Inc.)
2002 East Watkins Street, Suite 1, Phoenix, AZ 85034
☎ (+1 602) 253 46 63 ⊖ (+1 602) 253 47 73
♣ $ 4,000,000 (1997) 👥 35 (1997) 🏠 17,000 sq m
General Manager: Alan Ball
Operations Director: Steve DiDomenico
Marketing Manager: Ms Loren K Watson
Purchasing Manager: Ms Betty Ueltzen
■ Complete line of AC/DC converters, DC/AC inverters, converters and power supplies for aviation, industry and military applications. [98]

75-03200 Air Cargo Equipment Corp.
(a div. of Zero Corporation)
2930 E.Maria Street, Rancho Dominguez, CA 90221
☎ (+1 310) 898 2200 ⊖ (+1 310) 764 1498
♣ $ 30 000 000 👥 175
President: William G. Dye
Marketing Dir: Craig Reisacker
Sales Mgr Composites: Chris Lorenson
Purchasing: Tirzah James
■ Intermediate size containers for US Marine Corps. Galley partitions & equipment. Air cargo baggage containers. In-plane conveyor systems. [97]

75-03280 Air Command International Inc.
PO Box 1345, Wylie, TX 75098
☎ (+1 972) 442 66 94 [E] aircommand@aircommand; aircmd@biztod.com [W] http://www.aircommand.com
⊖ (+1 972) 442 34 45
👥 13 🏠 14,700 sq ft
President: Harold F Smith
Managing Director: Douglas H Smith
Operations Director: John Olsen
Sales and Marketing Director: Ms Sharon Caldwell
■ Manufacture range of Commander single and two-seat aircraft and fully enclosed two-seat side-by-side Commander 582 autogyro sport. [98]

75-03400 Aircraft Missile Parts Mfg. Co. (AMPM)
255 Easy St, Unit F, Simi Valley, CA 93065-1798
☎ (+1 805) 522 39 11 ⊖ (+1 805) 522 18 14
♧ $ 500 000 👥 5 🏠 2 000 sq ft
President: Pat Panza
VP Government Sales: Don Rosskof
■ Bolts, spacers, screws, nuts, knobs, standoffs, keys, bushings, pins, washers. [97]

75-03440 Aircraft Parts Corp. (APC)
160 Finn Court, Farmingdale, NY 11735
☎ (+1 516) 249 30 53 ✉ AIRCOR Fdle ▣ 961341 aircor
⊖ (+1 516) 249 25 77
👥 108 🏠 37,800 sq ft
General Manager: S Sherman
Engineering Director: Ray Denniston
Manager Electronic Controls: Peter Matamala
Sales Manager: Joe Galan
Military Sales: Ms Eileen Bergold
Domestic Sales: Ms Rosemary Pavlicek
Export Sales: Ms Maureen Schnaue
■ Manufacture starter-generators, generator control units, piston engine starters, carbon pile voltage regulators. [98]

75-03600 Air Cruisers Co
(a subsidiary of Zodiac SA Holding, France)
Monmouth County Airport Route 34, PO Box 180, Belmar, NJ 07719
☎ (+1 908) 681 35 27 ▣ 132410 aircruise
⊖ (+1 908) 681 91 63
President and Chief Executive Officer: José Redento
Sales Director: Lou Perdoni [98]

75-03645 Turbine Engine Components Textron
(a subsidiary of Textron Inc.)
23555 Euclid Avenue, Cleveland, OH 44117-1798
☎ (+1 216) 692 52 00 ⊖ (+1 216) 692 56 65
👥 133 🏠 70,000 m2
Plant Manager: James Pinto
Manufacturing Manager: Herb Searson
Sales Manager: William Scully
Purchasing Manager: Eric Larsen [98]

75-03650 Airflow Co.
295 Bailes Lane, Frederick, MD 21701
☎ (+1 301) 695 6500 ▣ 160998 airflow ut
⊖ (+1 301) 695 4057
👥 140 🏠 120 000 sq ft
President: Robert W. Hunt
Sales Mgr: Philip Whitaker
Controller: David Teach
■ Manufacturer of purpose-built environmental control systems for temperature, humidity & particulate filtration & heat exchangers & purification loops for de-ionized water & special fluids.
Divisions:
Impac Div.
Dryomatic Div.
Datac Div. [97]

75-03690 Air International Inc
(† Sigma International Inc)
321 S. Fairbank, Addison, IL 60101
☎ (+1 708) 628 32 44 ⊖ (+1 708) 628 66 23
President: R. Darling [97]

75-03700 Air Logistics Corp.
3600 East Foothill Boulevard, Pasadena, CA 91107-3120
☎ (+1 818) 795 99 71 ⊖ (+1 818) 795 25 28
President and Chairman: G H Schirtzinger
Vice President: D L Buckley
■ Aerospace, missile, vehicle and jet aircraft GSE. Fibre optic cable components. [98]

75-04120 Aitech Defense Systems Inc.
3080 Olcott Street, Suite 105A, Santa Clara, CA 95054
☎ (+1 408) 980 62 00 [W] http://www.rugged.com
⊖ (+1 408) 980 61 61
♣ 15,000,000 + (1996) 👥 100 + (1996)
President: Tsvi Lavi
Marketing Manager: Peter J Sasure
■ Complete line of 100% compatible full conduction-cooled Mil-Spec rugged and space qualified VMEbus systems and boards for off-the-shelf solutions. [98]

75-04205 Alan Industries Inc.
745 Greenway Drive, PO Box 1203, Columbus, IN 47201-2210
☎ (+1 812) 372 8869 ⊖ (+1 812) 372 5909
♣ $ 2 000 000 (1995) 👥 35 (1995) 🏠 9 600 sq ft
President: William D. Kennedy
VP Sales & Marketing: Blayde A. Kennedy
Public Relations Dir: Ms Susan J. Kennedy
■ RF & microwave passive components. [97]

75-04350 Alden Electronics Inc.
40 Washington St, Westborough, MA 01581
☎ (+1 508) 366 8851 ✆ (800) 225 9492 ▣ 948404
⊖ (+1 508) 898 2427
♣ $ 10 000 000 👥 105 🏠 51 000 sq ft
President & CEO: Arnold Kraft
■ Manufacture computer-based, colour weather display systems for the processing, manipulation, storage & display of weather graphics; ground stations for the receipt of weather satellite pictures, 256 gray scale printers for satellite & intelligence imaginery. [97]

75-04400 Alden Electronics Incorporated
(† Alden International Inc.)
40 Washington St, Westborough, MA 01581
☎ (+1 508) 366 8851 [E] info@alden.com
[W] http://www.alden.com ✆ (800) 225 9492
⊖ (+1 508) 898 2427
President: Arnold A. Kraft
Vice President: John Alexanderson
International Marketing: Michael J. Connelley
Government Marketing: Kevin P. O'Reilly
Contact: J. Michael Connelley
■ Weather satellite receiving, weather radar display & weather map facsimile systems. 97

75-04410 Alden Sales Co.
PO Box 7062, Richmond, VA 23221
☎ (+1 804) 282 7246 ⊖ (+1 804) 282 7512
President: Mark Sternheimer
VP Engineering: Zach Sternheimer
■ Military aircraft instruments, autopilots, drones, inverters, electronics. [97]

75-04430 Algiers Iron Works & Dry Dock Co. Inc.
(sub of Bollinger Machine Shop & Shipyard)
434 Powder St, PO Box 6068, New Orleans, LA 70174
☎ (+1 504) 362 7960 ⊖ (+1 504) 361 1679
♣ $ 3 500 000 (1990) 👥 35
President: Richard "Dick" Bollinger
VP & Co-ordinator Repair Yards: Tillman Esteve [97]

75-04460 Alinabal Inc.
28 Woodmont Road, Milford, CT 06460-2832
☎ (+1 203) 877 3241 ⊖ (+1 203) 874 5063
👥 250
Managing Dir: Phil Brodt
Marketing Dir: Gaston (Gus) Levesque
■ Manufacture stampings, rod-ends, bearings, linkage assemblies & electromechanical assemblies. [97]

75-04580 Allard Industries Inc. Aerospace Display Systems
(† BFGoodrich Aerospace Display Systems Inc.)
2321 Topaz Drive, Hatfield, PA 19440
☎ (+1 215) 822 60 90 ⊖ (+1 215) 822 79 74
♣ US$ 9,000,000 (1993) 👥 100
Managing Dir: Robert G. Martin
Operations & Technical Dir: Andrew Orr
Sales & Marketing Dir: Larry C. Green
Purchasing Mgr: David Paul
Sales Mgr European Region: Denise Dietrich-Bokum
■ Custom design & manufacture dichroic liquid crystal displays & modules for military & commercial aerospace applications. [98]

75-04610 Allen Osborne Assoc., Inc.
756 Lakefield Road, Westlake Village, CA 91361-2624
☎ (+1 805) 495 84 20 [E] aoa@netcom.com
⊖ (+1 805) 373 60 67
♣ $ 19,000,000 👥 60 🏠 2,555 sq m
President: Allen W Osborne
Vice President of Business Development: Robert W Snow
■ GPS systems, telescopic masts. [98]

75-04630 Allen Telecom Inc.
(† The Antenna Specialists Co.)
30500 Bruce Industrial Parkway, Cleveland, OH 44139-3996
☎ (+1 216) 349 84 00 [W] http://www.allentele.com
⊖ (+1 216) 349 84 07
♣ $ 169 000 000 (1993) 👥 1,200
President: Erik van der Kaay
VP Marketing: Dale Bastian
Managing Dir: John Kepple
Marketing Dir: Martha Levasseur
Int'l Operations Dir: Jerry Stupka
Technical Dir: Tom Riesterer
Purchasing Mgr: Tony de Tota
Marketing Administrator: Ms Stacy Pierce
■ Manufacture mobile and base station antennas for bands between 25 MHz - 960 MHz. Manufacture radio site equipment: duplexers, combiners, multicouplers, isolators, loads and filters. [98]

75-04635 Allfast Fastening Systems Inc.
PO Box 3166, City of Industry, CA 91744-1098
15200 E. Don Julian Rd, City of Industry, CA 91745
☎ (+1 818) 968 9388 ▣ TWX 910 5841 815
⊖ (+1 818) 968 9393
♧ $ 11 000 000 ♣ $ 16 000 000 (1995) 👥 180
🏠 90 000 sq ft
President & CEO: James H. Randall
VP Sales & Marketing: Glenn R. Willey
VP Engineering: Ralph Luhm
Manufacturing Dir: Alan Reoch
Chief Financial Officer: Don E. Hawes
UK Sales Mgr: Chris Osborne

■ Manufacture aerospace & communication solid & blind rivets, AN, NAS, MS solid rivets & Mil Spec R-007885, NAS 1738/39 & 1398/99, MS 20600 and nut plate type blind rivets, materials and specials. Blind rivet installation tools. [97]

75-04660 Alliant Techsystems Inc.

Corporate Offices
600 Second Street NE, Hopkins, MN 55343
☎ (+1 612) 931 60 00 ⊖ (+1 612) 931 54 23
Chairman, President and Chief Executive Officer: Richard Schwart
Group Vice President, Defence Systems: Hugo Fruehauf
Group Vice President Aerospace Systems: Dr Peter Bukowick
Group Vice President, Space and Strategic Systems: Paul A Ross
Group Vice President Emerging Business: Don Willis
Vice President Human Resources: Robert E Gustafson
Vice President Engineering and Technology: Roger P Heinisch
Vice President and General Counsel: Daryl L Zimmer
Vice President ICBM Prime Integration Program: Arlen D Jameson
Vice President and Treasurer: Galen K Johnson
Vice President and Chief Financial Officer: Scott S Meyers
Vice President and Controller: Ms Paula J Patineau
Vice President Washington DC Operations: William Martin
Vice President Public Affairs: Ms Kristi Rollag Wangstad
Secretary and Associate General Counsel: Charles H Gauck
■ Operations are conducted within four business groups: Aerospace, Defence & Marine Systems & Emerging Business.
Aerospace comprises four business groups: Space/Strategic, Ordanance, Tactical & Defence Electronics.
Defence Systems comprises four business areas: Tank Ammunition, Close Combat Systems, Artillery Systems and Air-Delivered Systems. Marine Systems comprises two business areas: Naval Systems & Engineering Services.
Emerging Business operates through four business units: Global Environmental Solutions Inc., Power Sources Center, Signal Analysis Center & Advanced Technology Applications. [98]

75-04670 Allied Research Corp. (ARC)

8000 Towers Crescent Drive, Vienna, VA 22182
☎ (+1 703) 847 52 68 ✆ (800) 847 53 22
⊖ (+1 703) 847 53 34
♣ $ 103,660,137 (1996) 👥 507
Chief Executive Officer: Dr Jay R Sculley
Chief Administrative Officer: Ms Karen M Bauer
Corporate Vice President: William Glenn Yarborough (Jr)
■ ARC/Mecar/Barnes and Reinecke is a diversified armaments designer and manufacturer of ammunition and gun systems. Development, design and systems technical support for US 155 and 8" howitzers M109/110. Manufacturer of equipment for update of various military vehicles. Mecar rifle grenades and KE ammunition are standard in thirty-five countries.

Subsidiary companies:
Mecar SA → Belgium 75
Barnes & Reinecke → 80
Allied Research Corp. International (ARCI) → UK 90

Public Relations Office
8000 Towers Crescent Drive, Suite 750, Vienna, VA 22102
☎ (+1 703) 847 52 68 ⊖ (+1 703) 847 53 34
Public Relations: Ms Susan Urbaniak [98]

75-04690 Allied Screw Products Inc.

815 East Lowell Avenue, PO Box 543, Mishawaka, IN 46546-0543
☎ (+1 219) 255 4718 ⊖ (+1 219) 255 4173
♣ $ 5,000,000 (1997) 👥 62 (1997) 🏭 20,000 sq ft
Chairman: Neil Silver
President: K.E. Holderman
Vice President: Ms Pam Rubenstein
Operations & Technical Dir: Patrick A. Szucs
■ Manufacture parts for actuators, wheel & brake assemblies, fuel controls & auxiliary pumps on all military aircraft. Also manufacture internal fine pitch gears, pinions, splines & serrations; fine pitch gear racks, worm gears, anti-backlash gears & sector gears. [98]

75-04700 AlliedSignal Aerospace

2525 West 190th Street, Torrance, CA 90504-6099
☎ (+1 310) 323 95 00 ⊖ (+1 310) 512 22 21
President: Daniel P Burnham ☎ (+1 310) 512 36 36; 512 11 97 (fax)
President, Government Services: Ms Karen K Clegg (Kansas City) ☎ (+1 816) 997 32 12; 977 70 16 (fax)
President Commercial Avionics Systems: David Dittemore (Olathe, KS) ☎ (+1 913) 768 30 95; 791 13 07 (fax)
President Engines: Greg L Summe (Phoenix, AZ) ☎ (+1 602) 231 27 75 231 21 04 (fax)
President, Electronic Systems: Raymond C Stark (Teterboro, NJ) ☎ (+1 201) 393 22 00 393 66 77 (fax)
President, Aerospace Equipment Systems: John R Tucker ☎ (+1 310) 512 55 19 512 40 30 (fax)
Senior Vice President Marketing, Sales and Service: Joseph B Leonard ☎ (+1 310) 512 37 10 512 30 40 (fax)
Senior Vice President, Business Strategy: Robert A Choulet ☎ (+1 310) 512 11 60
Senior Vice President, Total Quality: Barry Holmes ☎ (+1 310) 512 11 27; 512 23 05
Senior Vice President, Engineering and Technology: Kenneth A Pickar ☎ (+1 310) 512 31 31; 512 12 47 (fax)
Vice President and General Counsel: Terrance L Carlson ☎ (+1 310) 512 37 27; 512 56 46 (fax)
Vice President, Information Systems and Chief Information Officer: Paul J Hoedeman ☎ (+1 310) 512 23 24; 512 19 17 (fax)
Vice President, Human Resoureces: R Peter Mercer ☎ (+1 310) 512 50 50; 512 19 73 (fax)
Vice President, Finance: J Thomas Zusi ☎ (+1 310) 512 47 74; 412 56 47 (fax)
Vice President, Public Affairs: Dennis J Signorovitch ☎ (+1 310) 512 40 49; 512 24 90 (fax)
■ Manufacture aircraft guidance, navigation, communication and identification systems. Flight controls and cockpit instrumentation: weather radar, fuel controls and ignition systems. Commercial and military aircraft wheels, brakes and struts. Gas turbine engines, APUs, turboprop and turbofan engines. Environmental systems, cabin pressurisation systems. Systems and controls for aircraft, missiles and undersea applications. Provide technical services and spacecraft and satellite tracking support to DoD, DoE and NASA.

Microelectronics and Technology Center
9140 Old Annapolis Road, Columbia, MD 21045
☎ (+1 301) 964 40 00 ⊖ (+1 301) 992 58 13
Director and General Manager: Dr Patrick N Keating
■ Design and manufacture of microelectronics and other electronics systems: VLSI, military microprocessors and semiconductor devices. Aerospace research and development.

Center for Process Improvement
699 Route 46 East, Plant 1, MC 1/W18, Teterboro, NJ 07608-1173
☎ (+1 201) 393 31 83 ⊖ (+1 201) 393 66 46
Director: Charles Schmidt
■ Internal project management and technical training, application of CAD/CAM/CAE tools.

West Coast Support Operations
117 East Providencia Avenue, Burbank, CA 91502-1996
☎ (+1 213) 849 39 61 ✆ (800) 348 26 57
⊖ (+1 818) 845 60 68
Site Leader: John Collins ☎ (+1 818) 556 53 06; 840 83 09 (fax)
■ Repair and overhaul Bendix fuel controls and other systems.

Engineering and Technology
2525 West 190th Street, Torrance, CA 90504-6099
☎ (+1 310) 512 31 31; 512 12 47
⊖ (+1 310) 512 12 47
Senior Vice President: Kenneth A Pickar
Vice President, Technology: Ms Linda Capuano
■ Administrative Headquarters for Engineering and Technology.

Ceramic Components
2525 West 190th Street, Torrance, CA 90504-6099
☎ (+1 310) 512 59 16 ⊖ (+1 310) 512 59 01
General Manager: Ms Maxine Savitz ☎ (+1 310) 512 56 78
Director, Engineering/Manufacturing: Mike Meiser ☎ (+1 310) 512 56 76
Business Development Manager: Ron Walecki
■ Administration and manufacturing of ceramic components for aerospace and other high technology applications.

AlliedSignal Aerospace Equipment Systems
2525 West 190th Street, Torrance, CA 90504-6099
☎ (+1 310) 323 95 00 ⊖ (+1 310) 512 22 21
President: John R Tucker ☎ (+1 310) 512 55 19; 512 40 30 (fax)

Business Units:
Aircraft Landing Systems
AlliedSignal Aerospace Canada
Engine Systems and Accessories
Environmental Control Systems
Power Management and Generation Systems
TurboGenerator Systems
● 11100 North Oracle Road, Tucson, AZ 85737-9588
☎ (+1 602) 469 10 00; 469 65 15 (sales) ⊠ 187181
⊖ (+1 602) 469 66 00; 469 59 76 (sales)
Site General Manager: Ms Vicki E Panhuise ☎ (+1 520) 469 66 04
Director, Strategic Planning: Michael Anderson ☎ (+1 520) 469 66 21; 469 66 06 (fax)
■ Research, development and maufacture of monitoring and diagnostic systems; cabin pressure controls, electronic engine controls, air data sensors, smoke and flame detectors.

(† Garrett Fluid Systems Division)
● PO Box 22200, Tempe, AZ 85285-2200
1300 West Warner Road, Tempe, AZ 85284-2896
☎ (+1 602) 893 50 00; 893 44 06 (public affairs); 893 54 00 (service, repair and overhaul)
⊖ (+1 602) 893 44 06 (public affairs)
893 54 00 (service, repair and overhaul)
Site General Maager: Tim McLung ☎ (+1 602) 893 47 78; 893 43 71 (fax)
■ Design, development and manufacture of actuators, starters and control systems. Undersea vehicle propulsion systems.

Engine Systems and Accessories
(† Bendix Engine Controls, Allied-Signal)
717 North Bendix Drive, South Bend, IN 46620
☎ (+1 219) 231 30 00; 231 33 01 (marketing); 231 42 83 (engineering) ⊖ (+1 219) 231 33 35
Vice President and General Manager: David Berges ☎ (+1 219) 231 31 01; 231 31 05 (fax)
Director, Engineering: Dan Jones ☎ (+1 219) 231 42 57; 231 43 30 (fax)
■ Administrative headquarters.

Rocky Mount Operations
3475 Wesleyan Boulevard North, Rocky Mount, NC 27804
☎ (+1 919) 977 21 00 ⊖ (+1 919) 972 62 22
Site Manager: Thomas J LaMantia ☎ (+1 919) 972 666 69; 985 71 00 (fax)
■ Manufacture engine controls, subsystems and hydromechanical and pneumatic actuation devices.

Aircraft Landing Systems
(† Bendix Wheels & Brakes Division)
3520 Westmoor Street, South Bend, IN 46628-1373
☎ (+1 219) 231 20 00 ⊖ (+1 219) 231 20 20
Vice President and General Manager: Carl Montalbine
Director, Friction Materials: Rod Mamula ☎ (+1 219) 231 23 91; 231 32 98 (fax)
Director, Wheels and Brakes: Brad Morton ☎ (+1 219) 231 20 40; 231 20 75 (fax)
■ Landing gear systems, wheels, brakes, hydraulic components and brake control for commercial and military aircraft.

San Tan Remote Test Facility
PO Box 22200, Tempe, AZ 85285-2200
☎ (+1 602) 496 79 00 ⊖ (+1 602) 496 79 56
■ Support for AlliedSignal Fluid Systems products.

Aircraft Landing Systems
253 East Harris Avenue, South San Francisco, CA 94080
☎ (+1 415) 588 56 19; 558 56 20; 558 56 21
⊖ (+1 415) 558 90 17
Manager: Luther Jones
Assistant Manager: Ms Sandra Schier
Product Support Administrator: Ms Beatrice Tao-Yu
■ Sales and Servicing of aircraft landing systems.

Environmental Control Systems
2525 West 190th Street, Torrance, CA 90504-6099
☎ (+1 310) 323 95 00
⊖ (+1 310) 512 22 21; 532 71 31
Vice President and General Manager: John Hamilton ☎ (+1 310) 512 21 63; 512 39 81 (fax)
Site General Manager: Bruce Bringleson ☎ (+1 310) 512 35 61; 512 50 40 (fax)
■ Supplier of electrical power systems, environmental control systems and space and heat transfer systems

TurboGenerator Systems
2525 West 190th Street, Torrance, CA 90504-6099
☎ (+1 310) 323 95 00 ⊖ (+1 310) 512 15 61
Vice President and General Manager: Tony Prophet ☎ (+1 310) 512 37 46
■ Power generation systems.

Aircraft Landing Systems, Kansas City Support Center
1730 North Topping, Kansas City, MO 64120
☎ (+1 816) 231 84 00 ⊠ 42394 bendixabs nksc
⊖ (+1 816) 231 27 75
Manager: John C Foster
Assistant Manager: Brian Little
Assistant Manager: Greg Nachbar
■ Aircraft wheels and brakes, sales and service.

Power Management and Generation Systems

118 Highway 35, Eatontown, NJ 07724
☎ (+1 908) 542 20 00 ⊖ (+1 908) 389 78 51
Site Leader: Dennis Strouse ☎ (+1 908) 389 77 08; 389 78 52 (fax)

AlliedSignal Commercial Avionics Systems
(† Bendix/King General Aviation Avionics Division)
400 North Rogers Road, Olathe, KS 66062-1294
☎ (+1 913) 782 04 00 ⊖ (+1 913) 791 13 02
President: David Dittemore ☎ (+1 913) 768 30 95; 791 13 07 (fax)

(† Bendix/King Air Transport Avionics)
● PO Box 9327, Fort Lauderdale, FL 33309
2100 North West 62nd Street, Fort Lauderdale, FL 33309
☎ (+1 305) 928 21 00 ⊖ (+1 305) 928 30 00
Plant Manager: Ms Jeanine McLaughlin ☎ (+1 954) 928 22 50; 928 30 92 (fax)
■ Manufacture of a range of air transport products including communication and navigation avionics, traffic alert and avoidance systems (TCAS) and weather radar.

● 400 North Rogers Road, Olathe, KS 66062-1294
☎ (+1 913) 768 04 00 ⊠ 669916 kingrad
⊖ (+1 913) 791 13 02
Vice President, Customer Operations: Gary C Glynn
Vice President, Sales and Marketing: Greg Shuk ☎ (+1 913) 768 54 54; 791 13 04 (fax)
■ Avionics, communication, navigation, weather radar, EFIS and TCAS equipment for military and commercial applications.

c/o Douglas Aircraft Company
3855 Lakewood Boulevard M/S D15-0019, Long Beach, CA 90846
☎ (+1 310) 593 15 22 ⊖ (+1 310) 497 66 78
Field Service Engineer: Phil Collins ☎ (+1 310) 890 21 44; 987 70 22 (fax)
■ On-site customer support services.

● 348 Deodora Street, Vacaville, CA 95688
☎ (+1 707) 446 69 01 ⊖ (+1 707) 446 88 18
Manager, Regional Sales Southwestern US: Chuck Freeland
■ Sales Office.

● 1850 22nd Street, Suite 6, Boulder, CO 80302
☎ (+1 303) 473 06 23 ⊖ (+1 303) 473 06 25
Manager, Regional Sales Central US: Dan Prosser
■ Sales Office.

● 1 Aviation Way, Hangar 3, DeKalb-Peachtree Airport, Atlanta, GA 30341
☎ (+1 770) 455 84 61 ⊖ (+1 770) 455 84 53
Manager, Regional Sales Southeastern US: Ray Bennet

● 2920 Haskell Avenue, Lawrence, KS 66046
☎ (+1 913) 842 04 08 ⊖ (+1 913) 842 31 25
Plant Manager: Steve Spademan ☎ (+1 913) 842 04 06 ext. 2177; 842 52 95 (fax)
Boards Focus Factory Manager: Leo Ruhlman ☎ (+1 913) 824 04 06 ext.2249
Units Focus Factory Manager: Bill Trivette ☎ (+1 913) 842 04 06 ext. 2231

● 101 North New Century Parkway, New Century Airport, KS 66031
☎ (+1 913) 791 60 17 ⊖ (+1 913) 782 03 45
Facility Manager: Dennis A Yarber ☎ (+1 913) 791 60 91
■ Aviation research, development and engineering.

● 1935 Midfield Road, Wichita, KS 67209
☎ (+1 316) 943 93 00 ⊖ (+1 316) 943 20 53
Supervisor: John O'Gorman
■ Service centre.

● 19 Gary Circle, Guilford, NH 03246
☎ (+1 603) 267 17 00 ⊖ (+1 603) 267 17 01
Senior Field Service Engineer: Herbert J Furnee

● 4389 International Gateway, Suite 221/223, Columbus, OH 43219
☎ (+1 614) 231 30 33 ⊖ (+1 614) 231 31 48
Area Manager, Regional Sales: Rick Brainard
Field Service Engineer: Jim Mills ☎ (+1 614) 231 31 03

● 445 Flagstaff Drive, Greenburg, PA 15601
☎ (+1 412) 837 42 42 ⊖ (+1 412) 837 48 98
Manager, Regional Sales: Robert L Raube

● 6406 Ira Ingram Drive, Austin, TX 78749
☎ (+1 512) 301 23 40 ⊖ (+1 512) 301 34 91
Manager, Regional Sales South and Central America: Steve Brown

● PO Box 97001, Redmond, WA 98073-9701
15001 North East 36th Street, Redmond, WA 98073
☎ (+1 206) 885 37 11 ⊖ (+1 206) 885 20 61
Account Manager, Western Americas, Douglas Aircraft Company and Canadian Airlines International: Ms Susan Thim ☎ (+1 206) 885 82 27
Manager, Airline Marketing, Western Americas, American Airlines and TransWorld Airlines: Matt Marinos ☎ (+1 206) 885 86 79
Manager, Government Marketing US Air Force/US Army: Harvey Pekich ☎ (+1 206) 885 83 54; 885 88 55 (fax)
Manager, International Sales Asia/Pacific: Richard Philips ☎ (+1 206) 885 83 28
Customer Advocate, Boeing Programs: Ethan Powell ☎ (+1 206) 885 86 64
■ Sales support.

Redmond Support Center
PO Box 97001, Redmond, WA 98073-9701
8747 148th Avenue North East Building C, M/S 36
☎ (+1 206) 885 86 81 ⊖ (+1 206) 885 41 14
Manager, Seattle Support Facility: John Edgar ☎ (+1 206) 885 86 09
Technician: Boyce Choate ☎ (+1 206) 885 86 15
■ Maintenance centre for avionics equipment.

AlliedSignal Electronic Systems
(† Allied-Signal/Bendix Guidance & Control)
699 Route 46 East, Teterboro, NJ 07608-1173
☎ (+1 201) 288 20 00 ⊖ (+1 201) 393 66 10
President: Raymond C Stark ☎ (+1 201) 393 22 00; 393 66 77 (fax)

Business Units:
Communications Systems
Guidance and Control Systems
Inertial and Sensor Products
Ocean Systems

Communications Systems
(† Bendix Communications Division)
1300 East Joppa Road, Baltimore (Townsend), MD 21286-5999
☎ (+1 410) 583 40 00 ⊖ (+1 410) 337 74 85
Vice President and General Manager: Alan Dietrich ☎ (+1 410) 583 40 01; 583 59 73 (fax)
■ Develop and manufacture electronic information systems including: security systems, tactical systems, combat identification systems and radar systems.

Guidance and Control Systems, Target Systems
400 South Beiger Street, Mishawaka, IN 46544
☎ (+1 219) 255 21 11 ⊠ 810 297 2664 bendix mistl
⊖ (+1 219) 259 91 06
Controller: John Graf ☎ (+1 219) 254 42 50; 259 18 84 (fax)
General Manager: Tom Reiter ☎ (+1 219) 254 42 14;
Engineering Manager: Tom Collins ☎ (+1 219) 254 42 16; 254 42 29 (fax)
Operations Manager: Dick Przybelski ☎ (+1 219) 254 42 00; 254 42 12 (fax)
Manager, HS&E: John Sauter ☎ (+1 219) 254 42 89
■ Supersonic target manufacture.

Inertial and Sensor Products, Courter Operations
375 North Lake Street, Boyne City, MI 49712
☎ (+1 616) 582 65 26 ⊖ (+1 616) 582 72 57
General Manager: David Bennet ☎ (+1 616) 582 38 01; 582 34 57 (fax)
■ Manufacture precision control, indicator and reference instrumentation and gyroscopic devices for aircraft and missile systems.

Inertial and Sensor Products, Cheshire Operations
250 Knotter Road, Cheshire, CT 06410
☎ (+1 203) 250 35 00 ⊖ (+1 203) 250 36 26
General Manager: Richard Razza ☎ (+1 203) 250 37 01; 250 37 69 (fax)
Plant Controller: John M Rodhouse ☎ (+1 203) 250 37 50
Manager, Human Resources: Bernard Davis ☎ (+1 203) 250 36 90; 250 36 91
■ Manufacture: accelerometers, gyro precision instruments for aerospace.

Ocean Systems
15825 Roxford Street, Sylmar, CA 91342-3597
☎ (+1 818) 367 01 11 ⊠ 662900 bx sylm
⊖ (+1 818) 367 04 03
Vice President and General Manager: Steven Schroer ☎ (+1 818) 833 25 31; 367 69 99 (fax)
■ Research and development. Airborne, and shipborne electroacoustic systems. Avionic guidance and control repair.

Guidance and Control Systems
1525 Perimeter Parkway, Suite 125, Huntsville, AL 35806
☎ (+1 205) 837 58 30; 837 58 31; 837 58 32
⊖ (+1 205) 837 25 21
Senior Staff Engineer: Jaime B Fernandez
Senior Staff Engineer: Peter Van Hoff
Senior Staff Engineer: Bernard P Woods

Guidance and Control Systems
12540 Corellian Court, Cerritos, CA 90703
☎ (+1 310) 860 33 75 ⊖ (+1 310) 860 33 75
Engineer: Clifford Miyaoi ☎ (+1 310) 721 06 05
■ Design and development of avionics systems.

● 699 Route 46 East, Teterboro, NJ 07608-1173
☎ (+1 201) 288 20 00 ⊖ (+1 201) 393 66 10
Vice President and General Manager, Inertial and Sensor Products: W James Close ☎ (+1 201) 393 25 56; 288 82 70 (fax)
Vice President and General Manager, Guidance and Control Systems: Dean M Flatt ☎ (+1 201) 393 22 75; 393 65 04 (fax)
Vice President, Business Development, Electronic Systems: Raymond Wallace ☎ (+1 201) 393 23 33; 393 30 43 (fax)
■ Supplier of systems in the areas of avionics, guidance and control, oceanics and communications and sensor products.

Guidance and Control Systems
699 Route 46 East, Teterboro, NJ 07608-1173
☎ (+1 201) 288 20 00 ⊖ (+1 201) 393 66 10
Vice President and General Manager: Dean M Flatt
Vice President, Engineering: Michael Dunphy ☎ (+1 201) 393 24 00; 393 67 45 (fax)
Vice President, Advanced Programs and New Business Development: Joel Levinthal ☎ (+1 201) 393 32 99
Director, Space Systems: Rick Abramowitz ☎ (+1 201) 393 29 59
Director, Cockpit Management: John Borghese ☎ (+1 201) 393 36 96; 393 66 89 (fax)
Director, Automated Test Systems: Rakesh Khanna ☎ (+1 201) 393 25 99; 393 66 05 (fax)
Director, GCS Business Development: Paul Mangano ☎ (+1 201) 393 31 32; 393 65 04 (fax)
■ Supplier of avionics, space, target and automatic test systems.

Inertial and Sensor Systems
699 Route 46 East, Teterboro, NJ 07608-1173
☎ (+1 201) 288 20 00 ⊖ (+1 201) 393 66 10
Vice President and General Manager: W James Close
■ Administrative Office.

Inertial and Sensor Products, Instrument Systems
PO Box 97001, Redmond, WA 98073-9701
15001 North East 36th Street, Redmond, WA 98073
☎ (+1 206) 885 82 94 ⊖ (+1 206) 883 21 04
General Manager: Norman Klein ☎ (+1 206) 885 83 00
Marketing, New Business Development: Brian Derrick ☎ (+1 206) 885 82 94
■ Design and manufacture of precision instruments.

AlliedSignal Engines
(† Garrett Engine Division, AlliedSignal)
PO Box 52181, Phoenix, AZ 85034-2892
111 South 34th Street
☎ (+1 602) 231 10 00 ⊖ (+1 602) 231 57 13
President: Gregory L Summe ☎ (+1 602) 231 27 75; 231 21 04 (fax)

Business Units:
CFE Company
Commercial APUs
Commercial Propulsion
International Turbine Engine Corporation (ITEC)
Light Helicopter Turbine Engine Company (LHTEC)
Military/Helicopter Engines

● PO Box 52181, Phoenix, AZ 85072-2181
111 South 34th Street, Phoenix, AZ 85304-2892
☎ (+1 602) 231 10 00
⊖ (+1 602) 356 20 29 (public affairs)
President: Gregory L Summe ☎ (+1 602) 231 27 75; 231 21 04 (fax)
Vice President, Military/Helicopter: Carl F Baerst
Vice President, Human Resources: Dick Fincher
Vice President, Commercial Propulsion: Ms Dominique Hedon
Vice President, Manufacturing Operations: Marc A Hoffman
Vice President, Commercial APUs: Thomas A Johnson
Vice President and General Counsel: Gjon Nivica
Vice President, Quality and Continuous Improvement: Ms Mari Pitcher
Vice President, Stratford Integration: Mike Propen
Vice President, Strategic Planning and Strategic Business Development: Carl Schopfer
Director, Public Affairs: Kyle Hultquist ☎ (+1 602) 356 20 50
■ Gas turbine engines for aircraft and Helicopters including: turboprop, turbofan turboshaft and auxiliary power units.

● PO Box 52187, Phoenix, AZ 85034
402 South 36th Street, Phoenix, AZ 85072-2187
☎ (+1 602) 231 10 00 ⊖ (+1 601) 231 39 82
Director, Operations, Gears: Richard Barlow
Director, Operations, Case/Wheel/Disk: Tom Flessor
Director, Materials: Tom Gottlieb
Director, Procurement: Dave Terruciano
Director, Operations, Static Components: Mark Thurman

Commercial APUs
PO Box 52181, Phoenix, AZ 85072-2181
111 South 34th Street, Phoenix, AZ 85034-2892
☎ (+1 602) 231 75 00 ⊖ (+1 602) 231 75 05
Vice President, Commercial APUs: Thomas A Johnson
■ Commercial air transport gas turbine auxiliary power units.

Commercial Propulsion
PO Box 52181, Phoenix, AZ 85072-2181

111 South 34th Street, 67-00/MS 503-314
☎ (+1 602) 231 10 00
Vice President, Commercial Propulsion:
Ms Dominique Hedon ☎ (+1 602) 231 40 45; 231 52 89 (fax)

International Turbine Engine Corporation
PO Box 52181, Phoenix, AZ 85072-2181
111 South 34th Street
☎ (+1 602) 231 12 29
President: Paul J Hurley (Jr)
Director, Finance and Contracts: William R Clement
Director, Manufacturing Operations: Mark A Guenther
Director, Integrated Logistics: Jimmy B Smith
Director, Engineering: Ms Loren E Wagner
■ Turbo engines, turbo machinery and aircraft equipment. Joint venture between AlliedSignal Engines and Aerospace Industrial, Taiwan.

Miliary/Helicopter Engines
PO Box 52181, Phoenix, AZ 85072-2181
149 South 27th Street
☎ (+1 602) 231 10 00 ⊖ (+1 602) 231 30 48
Vice President, Military/Helecopter Engines: Carl F Bearst
■ Airborne APUs.

Military/Helicopter Engines
PO Box 52181, Phoenix, AZ 85072-2181
111 South 34th Street
☎ (+1 602) 231 10 00 ⊖ (+1 602) 231 30 48
Vice president, Military/Helicopter Engines: Carl A Baerst
☎ (+1 602) 231 23 61
■ Military fixed wing, marine and industrial aircraft and helicopters.

San Tan Remote Test Facility
PO Box 52181, Phoenix, AZ 85072-2181
111 South 34th Street M/S 302-207
☎ (+1 602) 496 79 00 ⊖ (+1 602) 496 79 56
Site Supervisor: John Hertz ☎ (+1 602) 496 79 34; 499 14 06 (fax)

CFE Company
(a joint venture between AlliedSignal Engines and GE Aircraft Engines)
PO Box 62332, Phoenix, AZ 85082-2332
111 South 34th Street, Department 67-60, MS 503-3J
☎ (+1 602) 231 45 70 ☒ 187113 ⊖ (+1 602) 231 50 30
President: Walter J Daub
Director, Program Operations, Marketing and Sales:
Mike Bush ☎ (+1 602) 231 48 38; 231 21 23 (fax)
Director, Product Support: Jim Davis
☎ (+1 602) 231 50 09
■ Turbofan, high-bypass engines.

● 6201 West Imperial Highway, Los Angeles, CA 90045
☎ (+1 310) 568 37 19 ⊖ (+1 310) 568 37 15
Director, Retrofit Programs: Conrad Daigle
☎ (+1 310) 568 37 07
Sales Manager: Don Sterling ☎ (+1 310) 568 37 28
■ Sales and marketing for Falcon 20 aircraft.

(† Textron Lycoming Turbine Engine Division)

● 550 Main Street, Stratford, CT 06497
☎ (+1 203) 385 20 00 ☒ 964242 ⊖ (+1 203) 385 31 22
Site Director: Michael Meshey
☎ (+1 203) 385 35 90; 385 25 00; 285 35 34 (fax)
Director, Materials Management: Michael Dunleavy
☎ (+1 203) 385 29 06; 385 5142 (fax)
Director, Human Resources: Brian McMenamin
☎ (+1 203) 385 15 94; 385 21 60 (fax)
Director, Manufacturing Assembly and Test, Stratford Repair and Overhaul: Ms Kerry O'Sullivan
☎ (+1 203) 385 30 79; 385 35 91 (fax)
■ Manufacture of Engines.

Light Helicopter Turbine Engine Company (LHTEC)
(a partnership between AlliedSignal Engines and Allison Engine Company)
12400 Olive Boulevard, Suite 400, St Louis, MO 63141
☎ (+1 314) 576 54 19 ⊖ (+1 314) 576 59 73
Executive Vice President: Ms Lynn E Snyder
(based in Indianapolis, IN) ☎ (+1 317) 230 53 51
Executive Vice President: Don J Christiansen
(based in Phoenix, AZ) ☎ (+1 602) 231 47 60
Manager, Business Development: Fred Dickens
(based in Indianapolis, IN) ☎ (+1 317) 231 65 15
Manager, St Louis Office Government Operations:
Herbert D Jones
■ Manufacture GTS turboshaft systems for civil and military helicopter engines.

(† Lycoming Engines)

● 400 South Buncombe Road, PO Box 1404, Greer, SC 29652-1404
☎ (+1 864) 989 37 00 ⊖ (+1 864) 989 37 46
Site Leader, Parts and Component Repair: John Conover
☎ (+1 864) 989 37 05; 989 38 59

AlliedSignal Government Services
PO Box 419159, Kansas City, MO 64141-6159
2000 East 95th Street, Kansas City, MO 64131-0395
☎ (+1 816) 997 20 00 ⊖ (+1 816) 997 33 31
President: Ms Karen K Clegg ☎ (+1 816) 997 32 12; 997 70 16 (fax)
Vice President and Director, Electronic Products:
Dale W Clements ☎ (+1 816) 997 22 86; 997 70 16 (fax)
■ Government owned headquarters for non-nuclear components for nuclear weapons.

Business Units:
AlliedSignal Technical Services Corporation
Federal Manufacturing and Technology

● One Bendix Road, Columbia, MD 21045-1897
☎ (+1 301) 964 70 00 ☒ 087860 benfld colb
⊖ (+1 301) 730 67 75
President and Chairman: Ivan M Stern
☎ (+1 401) 964 70 01; 964 77 24 (fax)
■ Integrated management and field engineering services, systems engineering, logistics planning and support, data acquisition and presentation, computer programming, systems analysis and design, RFI/ECM measurement analysis, repair and overhaul of electronic and electromechanical equipment. Operation, maintenance and support of traffic control facilities and equipment, including operation and maintenance of aircraft navigational and communication equipment. Design and modification of airport surveillance radar systems, planetary and interplanetary space mission tracking, laser satellite tracking.

Technical Services Corporation
AGRC Building 24309, PO Box 620550, Fort Rucker, AL 36362
AGRC Building 24309, Fort Rucker, AL 36362-0550
☎ (+1 334) 598 88 58; 503 11 07
⊖ (+1 334) 598 50 37
Program Manager: Ms Gail M Riley ☎ (+1 334) 598 50 31
Quality Assurance Specialist: Ken Cross
Deputy Program Manager: James Hodges
■ Operate a range complex for the US Army.

Technical Services Corporation, Goldstone Deep Space Communications Complex
850 East Main Street, PO Box 997, Barstow, CA 92311
☎ (+1 619) 255 80 00 ⊖ (+1 619) 255 80 43
Program Manager: Robert Frazier ☎ (+1 619) 255 83 22
■ Data acquisition and maintenance services for deep space tracking.

Technical Services Corporation ITIMS
1616 Industrial Boulevard, Suite A, Chula Vista, CA 91911
☎ (+1 619) 424 42 00 ✆ (800) 892 46 23
⊖ (+1 619) 423 74 16
Program Manager: Greg Bitel ☎ (+1 619) 424 42 99
■ Installation and maintenance of shipboard electronic systems.

Technical Services Corporation, Livermore Operations
2021 Las Positas Court, Suite 157, Livermore, CA 94550
☎ (+1 510) 455 78 70 ⊖ (+1 510) 373 06 08
Manager: Ronald E Burch ☎ (+1 510) 455 78 73
Manager, Laser Support: Charles Ellerbee

Technical Services Corporation
2500 East Foothill Boulevard, Suite 400, Pasedena, CA 91107-3464
☎ (+1 818) 584 46 00 ⊖ (+1 818) 793 88 20
Director: William D Edmonson ☎ (+1 818) 584 46 01

Technical Services Corporation, FCIP
3065 Beyer Boulevard, Suite 106, San Diego, CA 92154
☎ (+1 619) 690 60 92 ⊖ (+1 619) 690 60 98
Program Manager, FCIP: Charles R Walker
■ Installation and repair of electronic systems.

Technical Services Corporation
1150 Academy Park Loop, Suite 100, Colorado Springs, CO 80910-3715
☎ (+1 719) 596 74 43; 596 74 44
⊖ (+1 719) 596 15 09
Director, Business Development: Ted D Kerr
■ Computer systems and government programs maintenance.

Technical Services Corporation
4060 East Bijou, Suite 100, Colorado Springs, CO 80909-6823
☎ (+1 719) 637 67 00 ⊖ (+1 719) 597 66 49
Program Manager: Gerald T Hager
■ Defence satellite maintenance.

Government Operations and Marketing
1001 Pennsylvania Avenue, Suite 700, Washington, DC 20004
☎ (+1 202) 662 26 50 ⊖ (+1 202) 662 26 74
Vice President, Government Relations:
Martin L (Buzz) Hefti ☎ (+1 202) 662 26 83; 662 26 24 (fax)
Director, International Government Relations:
David T Rogers ☎ (+1 202) 662 26 67; 662 26 65 (fax)
■ Government Operations and Marketing Headquarters.

Technical Services Corporation, MPF
6200 Flagship Circle, Blount Island, Jacksonville, FL 32226
☎ (+1 904) 696 52 22 ⊖ (+1 904) 696 53 85
Director, Prepositioning Programs: Donald W Hill (Jr)
Manager, Contracts/Purchasing: Harris Bischoff
☎ (+1 904) 696 52 20
Manager, Finance/Accounting: Ms Sara Hansen
☎ (+1 904) 696 52 13
Manager, Human Resources: Gary Parkinson
☎ (+1 904) 696 52 21; 696 51 94 (fax)
■ Logistics and maintenance for maritime prepositioning force (MPF) program.

Technical Services Corporation, Mila Spaceflight Tracking and Data Network
Kennedy Space Center, MILA, FL 32899
☎ (+1 407) 867 35 12
⊖ (+1 407) 867 44 16; 867 17 66
Program Manager: Glenn D Smith
■ Operation and maintenance of satellite tracking station.

Technical Services Corporation
5801 Allentown Road, Suite 400, Camp Springs, MD 20746
☎ (+1 301) 702 41 00
⊖ (+1 301) 702 41 12, 702 41 16
Director, Systems Engineering and Integration Government Tactical Systems: Ms Sharon Devlin

Technical Services Corporation
Fort Detrick, College Estate Station, PO Box G, Frederick, MD 21702
☎ (+1 301) 619 74 11 ⊖ (+1 301) 619 28 18
Program Manager: Thomas G Brothers
■ Satellite communications station operator.

Technical Services Corporation, Goddard Corporate Park
7515 Mission Drive, Lanham-Seabrook, MD 20706
☎ (+1 301) 805 30 00 ⊖ (+1 301) 805 30 32
Vice President, Civilian Space: Philip H Johnson
☎ (+1 301) 870 35 82; 870 20 65 (fax)
■ Provide support for space flight mission operations.

Technical Services Corporation
Blossom Point Road, PO Box 578, LaPlata, MD 20646
☎ (+1 301) 870 35 82 ⊖ (+1 301) 870 20 65
Senior Manager: R A Allison
■ Tracking and data acquisition for spaceflight.

Technical Services Corporation, NESAE Headquarters
21515 Great Mills Road, Lexington Mills, MD 20653
☎ (+1 301) 737 10 00 ⊖ (+1 301) 737 10 28
Program Manager: Paul R Schenel
☎ (+1 301) 737 10 20 ext. 335; 737 10 32 (fax)
Program Manager: Ms Carole B Woodburn
☎ (+1 301) 737 89 16

Technical Services Corporation
PO Box 5370, Alberquerque, NM 87185-5370
☎ (+1 505) 768 57 52; 284 38 26
Manager, Alberquerque Operations: Jimmy D Wilson
■ Administration and support; seismic and instrumentation systems.

Technical Services Corporation, NASA/NMOS White Sands Complex
PO Box 9000, Las Cruces, NM 88004-9000
☎ (+1 505) 527 70 00 ⊖ (+1 505) 527 73 02
Program Manager: Gardiner W Rhoderick
☎ (+1 505) 572 70 04
■ Operation of ground terminal for space tracking systems.

Technical Services Corporation, White Sands Test Facility
PO Drawer MM, Las Cruces, NM 88004
☎ (+1 505) 524 52 77 ⊖ (+1 505) 524 51 06
Program Manager: Ben Boykin
■ Research and development.

Technical Services Corporation
NASA Lewis Research Center, 2100 Apollo Drive, Brook Park, OH 44142
☎ (+1 216) 997 09 30 ⊖ (+1 216) 997 09 41
Regional Manager: Richard R McCoy
☎ (+1 216) 997 09 31

Technical Services Corporation
1010 Bankton Drive, Hanahan, SC 29406
☎ (+1 803) 774 12 21 ⊖ (+1 803) 744 10 71
Program Manager: Richard G Strickland
☎ (+1 803) 744 08 48
■ Installation and integration of electronic equipment.

Technical Services Corporation
4925 LaCross Road, Suite 205, North Charleston, SC 29418
☎ (+1 803) 308 93 91 ⊖ (+1 803) 308 93 03
Manager, Business Development: Lloyd Vick
■ Provider of business develpment services for AlliedSignal Technical Services Corporation.

Technical Services Corporation
Northpark, Suite 120, 3125 Ashley Phosphate Road, North Charleston, SC 29418
☎ (+1 803) 552 31 21 ⊖ (+1 803) 552 31 29
Project Manager: John S Fisher
■ Installation and repair of electronic and telecommunications systems.

Technical Services Corporation
501 Gemini Avenue, Houston, TX 77058-2753
☎ (+1 713) 283 73 00 ⊖ (+1 713) 283 73 99
Director, Houston Operations: Edward M Briggs
☎ (+1 713) 283 73 11
■ Space programs ground support systems.

Technical Services Corporation, Goddard Space Flight Center Resident Office
16821 Buccaneer, Suite 206, Houston, TX 77058-2323
☎ (+1 713) 280 82 68; 483 58 52
⊖ (+1 713) 483 58 53
Office Manager: Jim Martin ☎ (+1 713) 483 58 53
■ Provider of communications and data networks for the Space Shuttle, the International Space Station and NASA.

Technical Services Corporation, Electronic Systems Test Laboratory
Mail Code EV4, Building 44, Room 115, NASA Johnson Space Center, Houston, TX 77058
☎ (+1 713) 483 00 82 ⊖ (+1 713) 483 53 82
Project Manager: John Ross
■ Technical analysis of the performance of space-to-space and space-to-ground communication systems.

Technical Services Corporation, NESSEC
4011A-1 Seaboard Court, Portsmouth, VA 23701
☎ (+1 804) 465 13 98 ⊖ (+1 804) 465 38 14
Project Manager: Gary H Scott
■ Design, Manufacture and servicing of voice, video and data telecommunications systems.

Technical Services Corporation, GSFC/Wallops Flight Facility
Building E106, Room 211, Wallops Island, VA 23337
☎ (+1 804) 824 19 05 ⊖ (+1 804) 824 25 29
Program Manager: Ken Griffin
■ Support in areas including radar and telecommunications.

AlliedSignal Public Affairs
2525 West 190th Street, Torrence, CA 90504-6099
Vice President, Public Affairs: Dennis J Signorovitch
☎ (+1 310) 512 40 49; 512 24 90 (fax)
Director, Marketing Communications: Ms Jan Angus
☎ (+1 310) 512 36 59; 512 22 00 (fax)
Director, Media Relations: Ron Iori ☎ (+1 310) 512 16 56
Director, International Communications: Donald Wilt
☎ (+1 310) 512 51 41

AlliedSignal Aerospace Marketing Sales and Service
2525 West 190th Street, Torrance, CA 90504-6099
☎ (+1 310) 323 95 00 ⊖ (+1 310) 512 22 21
Senior Vice President: Joseph P Leonard
☎ (+1 310) 512 37 10; 512 30 40 (fax)
Business Units:
Airline Services
Airsupply
Military Customer Support
Repair and Overhaul

● 2690 Cumberland Parkway, Suite 450, Atlanta, GA 30339
☎ (+1 770) 433 12 10 ⊖ (+1 770) 436 04 34
National Account Director: John Gentilella

● 3000 Presidential Drive, Suite 300, Fairborn, OH 45324-6271
☎ (+1 513) 427 27 23 ⊖ (+1 513) 427 29 40
Account Team Leader: John Mitchem ☎ (+1 427 28 80
Senior Government Account Manager: Brian E Trhlin
Account Manager, Air Transport Markets: Russel G Brax
☎ (+1 513) 427 29 15
Account Manager, Commercial Air Transport Markets: Gary S Dixon
Account Manager, Commercial Air Transport Markets: David S Evans
Account Manager, Commercial Air Transport Manrkets: Aubyn E Thomas

● 1525 Perimeter Parkway, Suite 115, Huntsville, AL 35806
☎ (+1 205) 837 22 27 ⊖ (+1 205) 837 25 21
Manager: Hector Negroni

● Centerpoint Three, 600 Six Flags Drive, Suite 500, Arlington, TX 76011
☎ (+1 817) 640 61 52 ⊖ (+1 817) 649 52 34
Account Team Leader and North American Government Sales: George Gabela
Account Team Manager: Brad Deacon
☎ (+1 817) 999 49 29
■ Field service engineering office.

Garrett Aviation Services
6201 West Imperial Highway, Los Angeles, CA 90045
☎ (+1 310) 568 38 23 ⊖ (+1 310) 568 37 90
Field Service Engineer: Homer Shiroma

● 3221 McKelvey Road, Suite 204, Bridgeton, MO 63044
☎ (+1 314) 739 06 06 ⊖ (+1 314) 739 56 42
Director: Kenneth Thompson
Manager: Herbert D Jones
Manager: Steven Privett
Manager: David Ryan

● 40 North East Loop 410, Suite 420, San Antonio, TX 78216
☎ (+1 512) 366 06 03 ⊖ (+1 512) 366 45 82
National Account Manager, Military Customer Support: Carl (Denny) Portz
98

75-04800 Allvac
(† Teledyne Allvac)
(a division of Allegheny Teledyne)
2020 Ashcraft Avenue, PO Box 5030, Monroe, NC 28111-5030
☎ (+1 704) 289 45 11 ✆ (800) 537 55 51 ▣ 866653
⊖ (+1 704) 289 40 18
President: John V Andrews
Senior Vice President of Operations: C Frank Elliot
Senior Vice President of Sales and Marketing: Henry S Ashbaugh
Senior Vice President of Administration: Rick Harshman
Vice President of Marketing: Tom E Williams
Manager, Marketing Communications: Howard L Freese
■ International supplier of high performance alloys for critical environments: nickel-base, titanium-base alloys and specialty steels (stainless, high speed, tool, high strength maraging and ultra-high alloy steels) produced in a broad range of alloys and grades.
Product forms: Ingot, billet, straight bar, hot rolled shaped bar, coiled bar and rod, hot rolled and cold drawn wire, coiled or straight length rod and wire.
98

75-04950 Thomson Software Products
(† Alsys Inc.)
Headquarters
10251 Vista Sorrento Parkway, St 300, San Diego, CA 92121
☎ (+1 619) 457 2700 ✆ (800) 833 0085 ▣ 855300
⊖ (+1 619) 452 2117
♣ $ 75 000 000 ⅏ 280 ⌂ 40 000 sq ft
President - N.America & Pacific Rim: Ben Goodwin
VP Sales: Steve Phillips
Marketing Dir: Abby Pinard
Marketing Communications Mgr: Ms Andrea Williams
■ Provider of software products & services for mission-critical application development, enterprise wide access & reporting. Main product lines include: Ada development environments; Teleuse user interface management system; Nomad HGL & Middlewall for information management systems.
(a subs. of Thomson-CSF, France)
● 101 Merrit Seven, Norwalk, CT 06856
☎ (+1 203) 845 5000 ⊖ (+1 203) 845 5252
President & CEO: Jean-Luc Badault
■ Ada Software development environments, real-time systems, programming tools, training products & services. Fourth generation software packages for complex databases. Multiplatform 6UI development tools.
MA Sales Office
67 South Bedford St, Burlington, MA 01803-5152
☎ (+1 617) 270 0030 ⊖ (+1 617) 270 6882
→ 80
97

75-05050 Aluminum Co. of America (ALCOA)
PO Box 8025, Bettendorf, IA 52722
☎ (+1 319) 344 3000 ▣ 812469; 866470
⊖ (+1 319) 344 3021
♣ $ 12 499 700 000 (1995) ⅏ 72 000
Chairman: Paul H. O'Neill
Dir Marketing & Sales East: J.W. Harrington
Admin Mgr: Ms B.A. Grimm
■ Aluminium sheet, plate, foil, extrusions, tube, wire, rod, bar. Aluminium lithium products. Forgings & castings. Steel & titanium forgings. Arall laminates.
Forging Div.
1600 Harvard Ave, Cleveland, OH 44105
☎ (+1 216) 641 3600 ▣ 980638 ⊖ (+1 216) 641 4032
Marketing Mgr: R.R. DeLay
Extrusion/Tube Div.
PO Box 7500, Lafayette, IN 47903
☎ (+1 317) 448 3600 ⊖ (+1 317) 448 3577
Marketing Mgr: M.A. Gregory
97

75-05110 AM General Corporation
(a subsidiary of The Renco Group Inc)
105 North Niles Avenue, South Bend, IN 46617
☎ (+1 219) 284 29 29 ✆ (800) 732 5493
⊖ (+1 219) 284 29 10
President and Chief Executive Officer: James A Armour
Public Relations Director: Craig C MacNab
E cmacnab@am-general.com
■ Produces the Hummer M998 series 1 1/4-ton 4x4 High Mobility Multipurpose Wheeled Vehicle (HMMWV), also offers a complete Integrated Logistics Support (ILS) program including field service, training, parts and publications.
Remanufacture M44 2 1/2-ton trucks into M35A3. Systems technical support and engineering services.
98

75-05200 AMCO Engineering
3801 Rose St, Schiller Park, IL 60176-2190
☎ (+1 847) 671 6670 ⊖ (+1 847) 671 9469
⅏ 360
President: Thomas Anderson
VP Sales: Tom Ligman
Dir Marketing: J.R. Walenda
■ Manufacturer of modular enclosures for electronic/electrical applications, including commercial & shielded cooling devices. Full line of accessories available for both. Also available - heavy duty structural aluminium castings & extrusions for enclosure or design applications.
97

75-05450 American Aerospace Controls Inc. (AAC)
570 Smith Street, Farmingdale, NY 11735
☎ (+1 516) 694 51 00 ⊖ (+1 516) 694 67 39
President: Ms Ruth Roberts
Sales Manager: Greg D'Abramo
■ Current, voltage & power sensing instruments for industrial, military & space applications.
98

75-05530 American Aluminum Co.
230 Sheffield Street, Mountainside, NJ 07092-2303
☎ (+1 908) 233 35 00 ⊖ (+1 908) 233 32 41
⅏ 120 ⌂ 50,000 sq ft
President: Robert J. Brucker
VP Sales & Marketing: Ronald R. King
Purchasing Mgr: Al Schaefer
■ Custom deep drawn components from aluminium, brass, steel, copper & stainless steel. Secondary operations including piercing, spinning, spotwelding, welding, machining, dip brazing & paint finishing. Nose cones, ogives, carrying cases & night vision equipment cases.
98

75-05540 American Blimp Corporation Airborne Surveillance Group
(a division of American Blimp Corporation)
302 Ritchie Highway, Severna Park, MD 21146
☎ (+1 410) 544 65 07 W http://www.americanblimp.com
⊖ (+1 410) 544 65 09
⅏ 40 (1997) ⌂ 1,900 m2
President: James Thiele
Vice President of Marketing: E J Brandreth
■ Design and manufacture airships for civil and surveillance purposes.
98

75-05790 American Eurocopter
(a subsidiary of Eurocopter)
2701 Forum Drive, Grand Prairie, TX 75052
☎ (+1 972) 641 00 00 W http://www.eurocopterusa.com
⊖ (+1 972) 641 36 71
♣ $ 1,118,000 (1995) ⅏ 233 (1996) ⌂ 33,000 sq m
Chairman: Gerard Chauvallon
President and Chief Executive Officer: Francois Guisolphe
☎ (+1 972) 641 34 00
Director of Marketing and Communications: Ms Brenda Reuland ☎ (+1 972) 641 34 92
■ Helicopter customisation, sales and services. → 80
® France: Eurocopter
Germany: Eurocopter
98

75-05920 American Induction HTG Corp.
(a subsidiary of the Ajax Magnethermic Corporation, USA)
33842 James J Pompo Drive, PO Box 248,
Franser, MI 48026-3468
☎(+1 810) 294 17 00 [E] aih@rust.net
[W] http://www.americaninduction.com
⊖ (+1 810) 294 22 93
♣ $ 12,000,000 (1996) 85 3,437 m2
President: Paul N Lavins
Vice President Sales and Marketing: Douglas R Brown
Director of Marketing: Eugene H Diggs
■ Design, develop, build-induction heating systems including all ancillary equipment such as work handling, conveyors, robotics, loaders, feeders & cooling recirculators. [98]

75-05940 American Cleaning Manufacturing Co
(† American Kleaner Manufacturing Co Inc)
(a subsidiary of Alfred Karcher GmbH & Co, Germany)
9000 Rochester Avenue, Rancho Cucamonga, CA 91730
☎(+1 909) 481 43 33 ⊖ (+1 909) 481 43 30
♣ $ 25,000,000 (1995) 140 130,000 sq m
Vice President and Chief Executive Officer: Steve Wishek
■ Manufacture of maintenance cleaning equipment, steam cleaners, pressure washers, NBC decontamination systems and field feeding systems.
® Germany: Alfred Karcher GmbH & Co [98]

75-06000 American Laser Systems Inc. (ALS)
PO Box 6944, Santa Barbara, CA 93160
☎(+1 805) 967 04 23 ⊖ (+1 805) 683 43 82
♣ $ 100,000 (1997) 3 (1997) 400 sq ft
President: Jackie Parker
■ Current Probes. [98]

75-06100 American Nucleonics Corp.
(a subsidiary of AIL Systems Inc.)
696 Hampshire Rd, Westlake Village, CA 91359
☎(+1 805) 496 2405 ⊖ (+1 805) 379 2392
President: Barry Fitzgerald [97]

75-06180 American Oil & Supply Co. (AOS)
238 Wilson Ave, Newark, NJ 07105
☎(+1 201) 589 0250 ✆ (800) 582 3267
⊖ (+1 201) 465 5083
President: Stanley J. Ziemski
General Mgr: John S. Ziemski
■ Manufacture & distribute petroleum and synthetic lubricants to military and civil aerospace. Also manufacture thermal greases or (heat sink compounds) used in electronics (computer) assembly. [97]

75-06455 American Scientific Corp.
3250 Holly Way, Chula Vista, CA 91910-3217
☎(+1 619) 426 12 80; 422 17 54
⊖ (+1 619) 426 12 80
20 4,000 sq ft
President and General Manager: Louis C Gerken
Public Relations Director: Ms Sherry Gerken
Marketing Manager: John Ortega
Production Engineer: Donald Woolstenhulme [98]

75-06500 Ameriel Inc.
4110 Shawnee Lane, Atlanta, GA 30319
☎(+1 770) 457 67 60 ⊖ (+1 770) 986 82 22
President: William C Beggs
■ Strobe lighting systems 0 to 8 km long for commercial and military airports and other applications. Modular maintenance. FAA approved. [98]

75-06510 Ametek Inc.
Station Square, Paoli, PA 19301
☎(+1 610) 647 21 21 ✆ (800) 473 12 86
[W] http://www.ametek.com ⊖ (+1 610) 296 34 12
♣ $ 868,661,000 (1996) 6,500
Chairman: Walter E Blankley
Vice President: Robert W Harris
Communications Manager: James P McKinley
■ Fractional horsepower (vacuum & blower) electric motors, mass fuel-flow meters and engine monitoring systems for jet aircraft, instruments to measure, control and record pressure/temperature flow and level for process, aircraft and heavy vehicles, water filtration products, high-purity metal powders & strip and clad metal products.

Ametek Aerospace Products Inc.
50 Fordham Road, Wilmington, MA 01887
☎(+1 508) 988 41 01 ⊖ (+1 508) 988 49 44
■ Flightdeck instruments, display systems, engine monitoring systems, fuel flow, temperature & speed sensors, electrical & thermocouple cable assemblies, liquid quantity systems for commercial & military aircraft engines & land & marine gas turbines.

Aerospace Products
900 Clymer Avenue, Sellersville, PA 18960
☎(+1 215) 257 65 31 ⊖ (+1 215) 257 18 88
■ Aircraft display systems, fluid gauging systems, cockpit indicators, fuel hydraulic/water quantity measuring systems, data converters & recorders, maintenance computers, pressure transmitters, active matrix indicating systems.

Seattle Support Center
4560 150th Avenue NE, Redmond, WA 98052
☎(+1 206) 881 11 72 ⊖ (+1 206) 881 72 92
■ OEM repair of indicators & sensors manufactured by Ametek Aerospace Products & Aerospace Products Div. FAA Repair Station No. HU6RR600N

Corporate Office
Station Square, Paoli, PA 19301
☎(+1 610) 647 21 21 ⊠ 834637 ⊖ (+1 610) 296 34 12

Mansfield & Green Division
8600 Somerset Drive, Largo, Fl 34643-2713
☎(+1 813) 536 78 31 ⊖ (+1 813) 539 68 82
Sales and Marketing Manager: John Gregor
■ Noise monitoring instruments, pressure and temperature calibrators, electronic force gauges, tachometers & stroboscopes.

US Gauge Division
(† Aerospace Products)
900 Clymer Avenue, Sellersville, PA 18960
☎(+1 215) 257 65 31 ⊠ 847445 ⊖ (+1 215) 257 47 11
General Mgr: Vito Parato
■ Pressure and temperature gauges for general industrial use & automotive instrumentation.

PMT Pressure Measurement Products
820 Pennsylvania Boulevard, Feasterville, PA 19047
☎(+1 215) 355 69 00 ⊠ 834799 ⊖ (+1 215) 355 29 37
Operations Manager: James P Mellon
■ Pressure measurement products including transducers, pressure transmitters, pressure calibrators, level systems and pneumatic controllers. [98]

75-06600 Amherst Systems Inc.
30 Wilson Road, Buffalo, NY 14221
☎(+1 716) 631 06 10 [W] http://www.amherst.com
⊖ (+1 716) 631 06 29
♣ $ 19,900,000 (1997) 220 (1997) 50,000 sq ft
Director Business Development: Robert L Cockrell
[E] cockrell@amherst.com
■ Signal environment simulators for testing of EW systems or operator training. [98]

75-06620 AMI Instruments Inc.
(a subsidiary of Huges Training Inc.)
PO Box 471800, Tulsa, OK 74147-1800
1914 West Reno, Building A, Broken Arrow, OK 74012
☎(+1 918) 258 07 07 ⊖ (+1 918) 258 00 51
♣ $ 8,000,000 (1996) 81 (1996) 22,000 sq ft
President: Maurice Arnott [E] marnott@ccgate.hac.com
Operations Director: Rick Franke
Marketing Director: Graham Coe
Engineering Manager: Keith Eggimann
Purchasing Manager: Gary Poulsen
■ Design and manufacture of simulated instruments, avionics and cockpit displays for military and commercial training equipment. Design and develop instrumentation for aircraft, ship and power station engineering research simulators. [98]

75-06640 A-Mold Corp
4600 Mason-Montgomery Road, Mason, OH 45040
☎(+1 513) 459 17 60 ⊖ (+1 513) 459 70 60 [98]

75-06750 Ampex Data Systems Corporation
(† Ampex Corporation)
(a subsidiary of Ampex Corporation)
500 Broadway, MS 4205, Redwood City, CA 94063-3199
☎(+1 415) 367 38 88 [W] http://www.ampex.com
⊠ 348464; TWX 910 378 5920 ⊖ (+1 415) 367 46 69
♣ US$ 95,700,000 500
Sales Director: Thomas Klang
Director of Corporate Communications:
Ms Karen Schweikher
■ Manufacture data/instrumentation tape recording equipment, magnetic tape, high data rate solid state buffer, computer interfaces, data multiplexing, demultiplexing equipment and professional video equipment. [97]

75-06800 Amphenol Corp. Amphenol Aerospace Operations
(a div. of Amphenol Corporation)
40-60 Delaware St, Sidney, NY 13838-1395
☎(+1 607) 563 5011 ⊠ 932456; TWX 510 261 7550
⊖ (+1 607) 563 5118 (G2+3); 563 5157 (G2)
♣ $ 500 000 000 1 300 750 000 sq ft
VP & General Mgr: G. Anderson
Marketing Dir: R. Aiken
National Sales Mgr: Mark Wycallis
Int'l Marketing Mgr: Charles J. Gill
Purchasing Mgr: Keith Young
■ High performance cylindrical & rectangular fiber optic & electrical connectors for the military/aerospace industry.

Amphenol Corporation
Headquarters
358 Hall Ave, Wallingford, CT
☎(+1 203) 265 8600 ⊖ (+1 203) 265 8628
President: Martin Loeffler [97]

75-06880 Amptek Inc.
6 De Angelo Drive, Bedford, MA 01730-2204
☎(+1 617) 275 22 42 [E] sales@amptek.com
[W] http://www.amptek.com ⊖ (+1 617) 275 34 70
25
President: John A Pantazis
Vice President: Alan C Huber
Purchasing Director: N Stavros
Advertising Director: Ms Linda Hantzis
■ Manufacturer of state-of-the-art hybrid preamplifiers, shapers, discriminators & counters to be used with solid state detectors, proportional counters, photodiodes, PM tubes, CEMs MCPs & others. [98]

75-06883 Amron L.L.C.
(a subsidiary of Pohlman, Inc., USA)
525 Progress Avenue, Waukesha, WI 53186-5948
☎(+1 414) 547 16 61 ⊖ (+1 414) 544 84 81
♣ $ 20,000,000 200 (1997) 50,000 sq m
President: Philip F Lange
Sales Director: Daniel F Borta
Operations Director: Richard R Peters
Purchasing Manager: John C Kennedy
■ Ammunition components including steel and aluminium cartridge cases, projectiles, grenade and submunition parts. Automotive airbag components. [98]

75-07140 Analog Modules Inc.
126 Baywood Avenue, Longwood, FL 32750-3426
☎(+1 407) 339 43 55 [E] analog@magicnet.net
⊖ (+1 407) 834 38 06
♣ $ 5,670,000 55 (1995) 12,000 sq ft
President and Chief Executive Officer: Ian D Crawford
Managing Director: William C Hudson
Director of Public Relations and Sales Support Manager:
Ms Kelly L Parker
Purchasing Manager: Ms Lea Ann Pelling
■ Specialists in analogue electronics and electro-optics, high technology modules and designs. [98]

75-07300 AN/COM Industries Inc.
5521 Cleon Avenue, PO Box 530,
North Hollywood, CA 91603
☎(+1 818) 769 55 18 [E] ancom@ancom.com
⊠ 4998094 ⊖ (+1 818) 769 55 44
1,200 m2
President: E D Wubben
Vice President of Marketing: Edward J Wubben
■ Manufacturing ground tactical communications equipment and spare parts. [98]

Contact Reference

75-07452 Andrew Sensor Products
(† Andrew Kintec Inc.)
(a division of Andrew Corp.)
10901 Roosevelt Boulevard, Ste 1200,
St Petersburg, FL 33716-2305
10901 Roosevelt Boulevard, Suite C1200
☎ (+1 813) 573 78 66; 577 36 11
W http://www.andrew.com
⊖ (+1 813) 573 13 45; 577 78 73
♣ $ 2,000,000 ﾎ 10 (1997) ⌂ 2,900 m2
Managing Director: Sid Bennet
Engineering Director: Allen Donnie
Sales Manager: Scot Thompson 98

75-07630 Angus Electronics Co.
(† Esterline Angus Instrument Corp.)
(a subsidiary of Esterline Technologies)
PO Box 24000, Indianapolis, IN 46224
☎ (+1 317) 244 76 11 ✆ (800) 899 76 11 ✉ ESTANGO
E 102476.3667@compuserve.com
W http://www.industry.net/angus ⊠ 27419
⊖ (+1 317) 247 47 49
ﾎ 52 ⌂ 63,550 sq m
President: William A. Hedges
Sales and Marketing Director: Tom Pilcher
Advertising Mgr: Ms Christie Sandi
Purchasing Manager: Ms Arlene Pozek
■ Manufacture electronic instrumentation for test, analysis & recording. 98

75-07800 Anritsu Wiltron
(a subs. of Anritsu)
685 Jarvis Dr., Morgan Hill, CA 95037-2809
☎ (+1 408) 776 8300 ⊖ (+1 408) 776 1744
♣ $ 100 000 000 ﾎ 800 ⌂ 275 000 sq ft
President: M. Evans
Mgr Marketing Comm.: John Hamilton
Communications Co-ordinator: Ms Maureen Lohe
■ Electronic test and measuring instruments including sweep generators, RF detectors, telecommunications test instruments, network analyzers, frequency synthesizers, microwave components, microwave connectors, transmission line test and fault location systems. Counters, microwave frequency opticaltime-domain reflectometers, fibre optic spectrometers, infra-red spectrum analyzers, low-frequency spectrum analysers, microwave spectrum analyzers, optical. 97

75-07810 Ansul Incorporated
(a subs. of Tyco International Ltd)
One Stanton Street, Marinette, WI 54143
☎ (+1 715) 735 7411 ✆ (800) 346 3626
⊖ (+1 715) 732 3479
ﾎ 560
Vice President: Karl Kinkead
Dir Marketing: Gary A. Nadolny
Mgr Marine Govt Business: John P. Goudreau
Advertising Mgr: Jim Cox
■ MIL-SPEC fire suppression systems, MIL-SPEC hand portable fire extinguishers, servicing of shipboard fire suppression products. 97

75-07940 AOtec Inc.
(a subsidiary of American Optical Corp.)
14 Mechanic Street, Southbridge, MA 01550-2555
☎ (+1 508) 765 21 71 W http://www.aotec.com
⊖ (+1 508) 764 85 06
♣ $ 15,000,000 (1994) ﾎ 120 (1995)
Managing Director: William Dhooghe
Director Government Business: Ronald J Tillen
Operations Director: Joe Arnone
■ Develops and manufactures: ballistic and laser protection systems & devices including aviator visors, gas mask lenses, spectacles, sun, wind & dust goggles and combat eyeware; plastic optics including HMD components, laser warning systems. 98

75-07980 APD
3336 East La Palma Ave, Anaheim, CA 92806
☎ (+1 714) 630 6068 ⊖ (+1 714) 630 6640
Director: John Leary
■ Fluoropolymer fluid handling components. 97

75-08000 Apex Broach & Machine Co.
6401 East Seven Mile Road, Detroit, MI 48234-2874
☎ (+1 313) 891 86 00 E apexl@apexbroach.com
W http://www.apexbroach.com ⊖ (+1 313) 891 50 83
ﾎ 35 (1997) ⌂ 5,220 sq m
President: Gust Headbloom (Jr)
Vice President and General Manager: Thomas P Heinrich
Operations and Technical Director: George Iapalucci
■ Design & manufacture of standard & special broaching machines & hydraulic presses, broach tooling/fixtures/holders/pullers/sharpeners. 98

75-08015 API-Harowe
(† Harowe Servo Controls)
(American Precision Industries)
110 West Town Road, West Chester, PA 19380
☎ (+1 610) 692 27 00 ⊖ (+1 610) 696 22 64
Managing Dir: Pat Dulin
Sales Dir: Bob Hyatt
Engineering Dir: Dave Hawes
Operations Dir: Dave Janelle
Purchasing Mgr: Bill Rutherford
Sales and Marketing Manager: Mike Preitice
■ Brushless resolvers, AC and Brushless DC motors, stepper motors, gearheads, encoders. 98

75-08030 APM Hexseal
44 Honeck Street, Englewood, NJ 07631-4134
☎ (+1 201) 569 57 00 W http://www.hexseal.com
⊖ (+1 201) 569 41 06
ﾎ 29 (1996)
Chief Executive Officer: Milton Morse
Vice President Sales and Marketing: Peter Engstrom
Vice President Production: Kenneth Schwinn
■ Switch seals, sealing fasteners, o-rings, coloured lamp filters. 98

75-08050 Applied Data Sciences Inc.
PO Box 814209, Dallas, TX 75381-4209
☎ (+1 972) 243 01 13 ✆ (800) 240 0113
E sysadm@appdatsci.com W http://www.appdatsci.com
⊖ (+1 972) 243 02 17
♣ $ 1,500,000 ﾎ 8 ⌂ 6,000 m2
President: Wayne Penner
Vice President: James Goode
Sales Manager: William A Helmke
Purchasing Manager: Ms Mayla Piering
■ Engineering design house. Custom designs (hardware and software) for interfaces, interconnections, communications, data links and test systems. These designs are typically used data acquisition systems, nuclear reactor simulators or flight simulators. Also, manufacturing and assembly. 98

75-08100 Applied Dynamics International (ADI)
3800 Stone School Road, Ann Arbor, MI 48108
☎ (+1 313) 973 13 00 ⊠ 230238 appdynint arb
⊖ (+1 313) 668 00 12
ﾎ 72
President and Chief Executive Officer: J D McIntosh
Senior Vice President: C R Moores
Vice President Sales and Marketing: W O Grierson
Asia Sales Manager: Maurice F Snyder
■ Manufacturer of digital computer systems for simulating models of dynamic, continuous systems. The SIMsystem, an integrated system, provides an ideal environment for real-time simulation of technological advanced equipment such as aircraft, turbines, missiles & satellites.
Overseas Representatives:
Hinditron Computers Pvt. Ltd (India)
Omikron Delta (1927) Ltd. (Israel)
Mecs Srl (Italy)
Kyokuto Boeki Kaisha Ltd. (Japan)
Dewell Industrial Co.Ltd. (Korea)
Sayyadain (Private) Ltd. (Pakistan)
Alpha Precision Instrument Corp. (China)
Applied Dynamics In'tl (UK) 98

75-08130 Applied Electro Mechanics Inc. (AEM)
4016 Rock Hall Road, Pt of Rocks, MD 21777
☎ (+1 301) 874 51 66 ⊖ (+1 301) 874 52 82
ﾎ 20 (1997) ⌂ 30,000 sq ft
President: Mrs P Banes
Marketing and Public Relations Director: Edward C Jenkins
E tjenkgrp@access.digex.net
■ High powered public address systems, fixed and portable, airborne/land/sea. 98

75-08165 Applied Microwave Energy Inc.
31127 Via Colinas 805,
Westlake Village, CA 91362-3921
☎ (+1 818) 991 46 24 ⊖ (+1 818) 991 38 01
♣ $ 1,800,000 (1993) ﾎ 10
President: Helmut Boehm
Vice President of Sales and Marketing: M Tanigawa
■ EW equipment, ship and airborne. EW antennas. Test and evaluation systems for electronic countermeasures. 98

75-08200 Applied Science Laboratories (ASL)
(a div. of Applied Science Group Inc.)
175 Middlesex Tpk, Bedford, MA 01730
☎ (+1 617) 275 4000 ⊖ (+1 617) 275 3388
ﾎ 15
Dir Marketing: Jose F. Velez
VP Marketing: John Flannery
■ Develop and manufacture a full range of eye trackers for simulators, weapon aiming, human factors, etc. Also conducts basic research in visual sciences. 97

75-08210 Applied Signal Technology
400 West California Avenue, Sunnyvale, CA 94086
☎ (+1 408) 522 33 88 ✆ (800) 374 35 60
E customer@appsig.com ⊖ (+1 408) 522 29 60
♣ US$ 77,400 000 (1996) ﾎ 596 (1997)
⌂ 264,000 m2
President and Chief Executive Officer: Gary L Yancey
☎ (+1 408) 522 3300
Vice President of Operations: Kenneth D Snow
☎ (+1 408) 522 3497
Vice President of Engineering: Ben Scribner
☎ (+1 408) 522 33 09
Sales and Marketing Director: Brian G Bedrosian
☎ (+1 408) 522 33 88
Materials Manager: Ms Cynthia O'Connell
☎ (+1 408) 522 34 22
Operations Administrator: Sandy Marian P White
■ Communications intelligence processors, PCM equalizers/demodulators, mobile radio processor, digital FDM demultiplexers, flexible modem demodulators, automatic multichannel processors. Signal analysis software packages. 98

75-08370 ARA Inc. (ARA)
1123 South Hatcher Street, Unit C, PO Box 3824,
City of Industry, CA 91744-0824
☎ (+1 818) 330 16 03 ⊖ (+1 818) 333 92 50
♣ $ 623,100 (1997) ﾎ 8 (1997) ⌂ 10,000 sq ft
President: Bernard Mazelsky
■ Crashworthy armoured and unarmoured seats for helicopters and aircraft. 98

75-08380 ARC Industries Inc.
PO Box 033498, Indialantic, FL 32903-0498
325 Poinciana Drive, Melbourne, FL 32935-6814
☎ (+1 407) 254 19 17 ⊖ (+1 407) 254 85 44
⌂ 3,000 sq ft
President: Frank T De Angelo
Technical Director: F Matthews
Sales Manager: Brad Fletcher
■ Worldguide, navigation magnetic digital compass. Electronic compass and sensors with displays and BCD, RS422. 98

75-08400 Arcata Associates Inc.
4220 Arcata Way, N. Las Vegas, NV 89030
☎ (+1 702) 642 9500 ⊖ (+1 702) 642 6589
♣ $ 8 500 000 (1995) ﾎ 160 (1995)
Vice President: Nancy Wong
■ Diversified electronics firm with an outstanding record of performance in manufacturing electronic systems & providing technical support services. → 80 97

75-08500 Arde Inc.
Head Office and Plant
500 Walnut Street, Norwood, NJ 07648
☎ (+1 201) 784 98 80 E ardeinc@aol.com
⊖ (+1 201) 784 97 10
♣ $ 7,000,000 (1996) ﾎ 40 (1996) ⌂ 58,000 sq ft
President: Reginald Wood
Vice President Finance: Mel Goldstein
Production Director: Jeff Ehalt
Business Development Manager: Robert E Holland

■ Research, development and manufacturing of lightweight metal and metal/fibre composite pressurant and propellant tanks, with and without propellant management system and ASME code. 98

75-08600 ARES Inc.
Building 818, Front Street, Erie Industrial Park,
Port Clinton, OH 43452-9399
☎ (+1 419) 635 21 75 ⊖ (+1 419) 635 26 40
25 (1997) 5,000 m2
President: Nelson Wall (Marketing)
Director Manufacturing: Herbert Roder
Purchasing Manager: Ms Ann Paulsen
■ Weapons, hand-held & vehicle mounted; machine guns, rifles & automatics. 98

75-08620 Arga Controls Inc.
35 E. Glenarm St, Pasadena, CA 91105-3418
☎ (+1 213) 682 3314 ⊖ (+1 818) 799 9407
♣ $ 2 000 000 35
CEO & President: H. Galman
VP Sales & Marketing: Ms Linda Galman
■ Meter & relays for power systems. 97

75-08630 Argo-Tech Corp.
23555 Euclid Avenue, Cleveland, OH 44117
☎ (+1 216) 692 60 00 ⊖ (+1 216) 692 52 93
550 (1997)
President and Chief Executive Officer: Michael Lipscomb
☎ (+1 216) 692 62 27
Vice President Sales and Marketing: Ken Hicks
☎ (+1 216) 692 66 07
Vice President Operations: David L Chrencik
Customer Support Manager: Dan Caschera
☎ (+1 216) 692 68 82
Purchasing Manager: Richard Doganiero
Contracts Manager: David Loney
■ Systems, subsystems & components for aerospace, military, commercial, domestic & foreign markets for engine fuel, lubrication pumps & airframe boost pumps. 98

75-08850 Armour of America Inc.
PO Box 1405, Beverly Hills, CA 90213
☎ (+1 310) 532 06 90 E armourinc@aol.com
W http://www.armourofamerica.com
⊖ (+1 310) 532 86 96
50
President: Arthur Schreiber
General Manager: John Nehmens
Sales Manager: Jeff Thompson
■ Helicopter armour and soft body armour. Patrol boat armour 50 calibre AP. C-130 transport armour for flight deck, paratrooper doors and lox bottle. 98

75-08950 Armtec Defense Products Co.
(a subsidiary of Esterline Technologies)
85-901 Avenue 53, PO Box 848, Coachella, CA 92236
☎ (+1 619) 398 01 43 TWX 910 332 1585
⊖ (+1 619) 398 46 09
210
President: Robert Harris
VP Marketing: Jack J. Alt
VP Operations: Nick Bielik
Dir Marketing Europe: Mike J. Kelly
Dir Marketing Asia Pacific: Paul H. Bassin
■ Manufacture combustible ordnance products, including 60mm, 81mm and 120mm mortar propellant containers, 155mm combustible cases, combustible cases for 120mm tank rounds and other products. 98

75-08995 The Arnold Engineering Company
(a subsidiary of SPS Technologies Inc.)
300 North West Street, Marengo, IL 60152
☎ (+1 815) 568 20 00 257448 ⊖ (+1 815) 568 22 29
508
President: Walter T Benecki
Sales and Marketing Director: Shuk Rashiki
■ Magnetic materials for electric motors, power supplies, aircraft instrumentation and telecommunications. 98

75-09020 Array Connector Corp.
12555 SW 130th Street, Miami, FL 33186-6226
☎ (+1 305) 234 1000 ✆ (800) 585 9250
⊖ (+1 305) 234 6601
♣ $ 6 500 000 (1996) 65 (1996) 20 700 sq ft
Chairman & CEO: William C. McPherson
Federal Procurement Liason: Ms Sharon Reisenfeld
Quality Assurance Mgr: Joe Cooner
Manufacturing Mgr: Mark Howerton
■ Manufactures Mil C-26482 I, Mil C-5015 and rack & panel electrical connectors. Also manufactures applications specific to customer specifications. Distributes QPL/Mil-Spec connectors. 97

75-09150 Arrowprop Co. Inc.
PO Box 610, Meeker, OK 74855
☎ (+1 405) 279 3833 3791754
⊖ (+1 405) 279 2377
15 16 000 sq ft
President: Don L. Miller
■ Manufacturer of test propellers (test clubs), air cushion vehicle fans, mini RPV propellers, ultra-light aircraft propellers, experimental aircraft propellers. 97

75-09240 Arvico Corporation
8283 Currency Drive, Riviera Beach, FL 33404
☎ (+1 561) 863 15 50 ⊖ (+1 561) 863 84 63
♣ $ 4,000,000 - 5,000,000 (1997) 36 (1997)
1.25 acres
President, Owner: Arthur F Brown
Executive Vice President: Clark Bentley
Vice President of Engineering and Production: John Marsden
Sales Manager: Ms Karen Kirk
■ Jet engine starting systems for military and commercial aircraft, portable airstart trailers, airstart hose assemblies. Combination Jetstart trailer with up to 60 kVA, 400 Hz diesel generator installed (sold to US and foreign Air Forces). 400HZ diesel powered GPUs from 60-140 kVA, trailer truck or self propelled. 28.5 VDC GPUs, aircraft plugs and cables, load banks. 98

75-09450 Associated Aerospace Activities Inc.
2502 Williams Street, San Leandro, CA 94577
☎ (+1 510) 483 90 20 ✉ AEROACT 176507
⊖ (+1 510) 357 77 40
♣ $ 3,000,000 25
President: D E Johnson
Corporate Secretary: Ms Rebecca MacDougall 98

75-09500 Associated Aircraft Manufacturing & Sales Inc.
784 NW 57th Court, Fort Lauderdale, FL 33309-2092
☎ (+1 954) 772 6606 514304 ⊖ (+1 954) 772 1633
♣ $ 10 000 000 (1996) 60 (1996) 90 000 sq ft
President: Germain J. Seguin
Sales & Marketing Dir: Jack Seguin
Public Relations Mgr: Germain J. Seguin
■ Manufacture of airframe parts for aircraft such as F5, F18, C130, CH47, P3. Manufacture of aircraft electronics, avionics & related spares. → 80 97

75-09700 Astech Manufacturing Inc.
(† Astech/MCI)
3030 Red Hill Avenue, Santa Ana, CA 92705
☎ (+1 714) 250 1000 ⊖ (+1 714) 250 4982
President & CEO: Ross K. Anderson
Technical Dir: Steve Hjelmstrom
Marketing Mgr: Thomas W. Brizes
Purchasing Mgr: Pat Fleming
■ Manufacture lightweight, high-strength, heat-resistant, noise-suppression structures for military & commercial aircraft, missiles and the space shuttle including jet engine exhaust nozzles, nacelle structures, turbine frames & ducts, missile wings and vanes. 97

75-09955 Astro-Med Inc.
600 East Greenwich Avenue,
West Warwick, RI 02893-2321
☎ (+1 401) 828 40 00 ✆ (800) 343 40 39
E astro-med@astro-med.com
W http://www.astro-med.com ⊖ (+1 401) 822 24 30
♣ $ 31,333,000 (1993) 220
President: Everett Pizzuti
Chief Exec. Officer: Albert Ondis
Advertising & Marketing Mgr: Ms Donna Ray
■ Manufacture a complete line of chart recorders & high speed specialty printers designed to meet the requirements of the defence & general military industries. 98

75-09980 Astron Corp.
22560 Glenn Drive, Sterling, VA 20164-4440
☎ (+1 703) 450 55 30 ⊖ (+1 703) 450 97 53
♧ $ 2,000,000 (1997) ♣ $ 3,000,000 (1996) 30
3,000 sq m
President: Joseph R Jahoda
Executive Vice President: Joseph McElroy
Sales and Marketing Director: Jim Talbert
Purchasing Manager: Ms Joanne McElroy
■ Research, development and manufacture of advanced antennas for transmitting and receiving broadband for HF/VHF/UHF. → 60 98

75-10000 Astronautics Corp. of America
4115 North Teutonia Avenue, Milwaukee, WI 53209
☎ (+1 414) 447 82 00 ⊖ (+1 414) 447 82 31
2,300 900,000 sq ft
President: Dr R.E. Zelazo ☎ (+1 201) 785 6000
CEO: N.K. Zelazo ☎ (+1 414) 447 8200
Exec. Vice President: N.Z. Paige ☎ (+1 201) 785 6000
Senior Vice President and General Manager: Michael Russek ☎ (+1 414) 447 82 00 Ext. 302
VP Business Development: R.D. Seinfeld
☎ (+1 414) 447 8200 Ext. 306
VP Engineering: Robert Lessin
☎ (+1 414) 447 8200 Ext. 400
VP Material: Joe Potts ☎ (+1 414) 447 8200 Ext. 126
Business Development Mgr: Paul Zahn
☎ (+1 414) 447 8200 Ext. 343
■ CRT displays & aircraft instruments including HSI, ADI, BDHI & engine-performance indicators. Air data computers, multi-functions display systems, cockpit integration. Mechanical & ring laser gyros, spaceborne, shipborne, airborne & ground-based inertial navigation systems. Synchros, resolvers, brushless motors, actuators & cryogenic refrigeration.

Subsidiary:
Kearfott Guidance & Navigation Corp. → 75
Wayne, N.J.
Astronautics C.A. Ltd
Bnei-Brak, Israel
Joint Venture:
Astronautics Kearfott Electroautomatica (AKE), St Petersburg, Russia 98

75-10090 ATC Inc
3050 Sidco Drive, Nashville, TN 37204
☎ (+1 615) 244 89 94 ⊖ (+1 615) 244 89 97 98

75-10145 Atlantic Microwave Corporation
(a Chelton Group Company)
58 Main Street, Route 117, Bolton, MA 01740
☎ (+1 508) 779 6963 ⊖ (+1 508) 779 2906
60
Chief Exec. Officer: D. Gaggin
Engineering Mgr: D. Tombs
Sales Manager: F Cavallaro
■ Manufacture radar antenna feeds & waveguide components. 98

75-10150 Atlantic Research Corp. (ARC)
(a subsidiary of Sequa Corp.)
Corporate Headquarters
5945 Wellington Road, Gainsville, VA 20155/1699
☎ (+1 703) 754 50 00 ⊖ (+1 703) 754 53 16
867 (1997)
Chairman and Chief Executive Officer: James R Sides
Vice President and Chief Financial Officer: Patrick J Jenkins
Executive Assistant to the Chief Executive Officer: Ms Judy Briggs
Vice President Human Resources and Administration: Patrick F Davis
Vice President and General Counsel: Carlyle C Ring (Jr)
Corporate Secretary: Ms Barbara F Biendl
■ High technology company specialising in rocket propulsion for departments and agencies of the US Government and private industry clients. → 80

ARC Aerospace Group
Group HQ: 5945 Wellinton Road, Gainesville, VA 22065
☎ (+1 703) 754 50 00

Solid Propulsion Division
5945 Wellington Road, Gainesville, VA 22065
☎ (+1 703) 754 50 00 ⊖ (+1 703) 754 52 81
Vice President and General Manager: B Frank Rohrback
■ Design, development and production of small tactical motors and gas generators. Advanced tactical and strategic propulsion systems.

Arkansas Operations
PO Box 1036, Camden, AR 71701

Contact Reference

☎(+1 501) 574 31 24 ⊜ (+1 501) 574 25 09
Vice President: Robert E Sherton
■ High volume production of solid rocket propellant. Large rocket motors. State-of-the-art technology in analysis, testing, QA and production.

ARC Liquid Propulsion
PO Box LPO, Niagara Falls, NY 14304
☎(+1 716) 731 60 00 ⊜ (+1 716) 731 62 81
Vice President and General Manager: Gerald L Greene
■ Liquid rocket engines for space propulsion.

Advanced Materials
5945 Wellington Road, Gainesville, VA 22065
☎(+1 703) 754 57 40 ⊜ (+1 703) 754 56 38
Vice President and General Manager: Jerry L Fields
■ High-temperature, lightweight advanced composite materials structures. [98]

75-10175 Atlas Copco Berema Inc
(a div. of Atlas Copco North America Inc.)
161 Lower Westfield Road, Holyoke, MA 01060
☎(+1 413) 746 0020 ⊜ (+1 413) 746 5383
300 30 000 m2
President: Peter Bigwood [98]

75-10240 AudioPack Sound Systems
10011 Walford Avenue, Cleveland, OH 44102
☎(+1 216) 651 00 66 E audiopack@audiopack.com
3710290 ⊜ (+1 216) 651 23 88
25 24,000 sq ft
President: Jonathan D Adams
Operations Manager: Ms Dianne Boduszek
■ Clear voice communications for NBC environments. Protective mask-compatible speech amplifiers, portable & self-contained with 100m range. Portable sound systems for tactical, training & garrison needs. Miniature radios & wireless communicators. [98]

75-10360 Auto Crane Co.
(a subsidiary of Ramsey Ind.)
PO Box 581510, Tulsa, OK 74158
☎(+1 918) 438 27 60 ⊜ (+1 918) 438 66 88
Marketing Manager: Mike McGuire
Advertising Co-ordinator: Ms Paula Kaufman
■ Remote-controlled cranes, truck-mounted or stationary, for field service, maintenance and marine applications. Capacities from 1,000 - 14,000 lbs (454 - 6,350 kg). Hydraulically or electrically powered. Also articulated loader cranes from 7 ton/meter to 17 ton/meter. [98]

75-10380 Automated Production Equipment Corp. (APE)
142 Peconic Ave, Medford, NY 11763-3260
☎(+1 516) 654 1197 ✆ (888) DSOLDER
⊜ (+1 516) 289 4735
25
President: William Scheu
VP Sales & Marketing: Edward Lyons
■ PCB repair/rework equipment, digital soldering/desoldering stations. SMT reflow/desolder infra-red hot-air systems. Circuit repair kits. [97]

75-10580 Autronics Corp
(a subsidiary of Bowthorp plc, UK)
312-314 East Live Oak Avenue, Arcadia, CA 91006
☎(+1 818) 445 54 70 W http://www.autronics.com
⊜ (+1 818) 446 00 14
President: John Kuperhand
Director of Marketing and Sales: James H Ward (Jr)
■ Design and manufacture of power supplies for sensing and control products and power conversion systems for military and commercial aircraft applications. High reliability sensing and control devices for a range of US military/aerospace programs. [98]

75-10800 Aviation Fluids Service Inc. (AFS)
PO Box 2489, St Louis, MO 63114
950 Kingsland, St Louis, MO
☎(+1 314) 721 2910 ✆ (800) 325 4270
⊜ (+1 314) 721 4519
Vice President: Mike West
■ Assembly lube for Skydrol MCS352; assembly lube for MIL-SPEC fluids, AFS682; Skydrol WC393 thrust reverser lube, turbine engine lubes, sealants, paint kits; deicer, speciality packaging & blending; greases. [97]

75-11000 Aviation Simulation Technology Inc. (AST)
2285 Old Hwy 50, Ottawa, KS 66067
☎(+1 913) 242 0952 ✆ (800) 423 3950
⊜ (+1 913) 242 4719
Chairman: Charles Mann
■ Manufacture low to mid cost simulators for piston, turboprop & turbojet aircraft, with CGI colour visual display & worldwide navigation. [97]

75-11050 Aviation Support Technology International Inc. (ASTI)
320 W. Lone Cactus Dr, Ste 10, Phoenix, AZ 85027
☎(+1 602) 582 1707 ⊜ (+1 602) 582 1841
$ 1 600 000 (1996) ♣ $ 288 000 (1995) 4 (1996)
6 800 sq ft
President: David W. Hill
Dir Aerospace Systems: Gary M. Gould
■ Specialists in GSE & spares for all types of equipment, engine components & military hardware. Manufacture (US Government) code OAH-69 for airframe and structural components for A-37, T-37, F-5E/F & T-33. [97]

75-11100 Avica US
(† General Connectors Corp.)
(a Meggitt Aerospace Co.)
1175 Aviation Place, San Fernando, CA 91340
☎(+1 818) 361 0900 67735 genconnco
⊜ (+1 818) 361 0518
138
Operations Director: Steve Lueneburg
Dir Sales & Marketing: Roger Tyson
General Sales Mgr: Ms Katherine Griffin
■ Pneumatic system components for aircraft. Bellows Ball Joints. Peri-Seal connectors. Thermal insulation. Pneumatic ducting systems components. [98]

75-11200 Avionics Specialities
(† Teledyne Avionics)
(a subs. of Aerosonic Corporation)
Charlottesville-Albemarle Airport, PO Box 6400, Charlottesville, VA 22906
☎(+1 804) 973 3311 510 587 5469 tdyavio cvl
⊜ (+1 804) 973 2976
President: David A. Baldini
Technical Dir: David Ashworth
Mgr Marketing: John Rafferty
■ Aircraft products; vertical speed indicators, angle of attack systems for military & commercial aircraft, stall warning & prevention systems, engine performance monitoring systems. [97]

75-11250 Avirex Ltd
33-00 47th Ave, Long Island, NY 10012
☎(+1 718) 482 1860 ⊜ (+1 718) 482 1881
70 30 000 sq ft
President: Jeff Clyman
Exec. VP: Ms Jacky Clyman
■ Clothing & other accessories. → 90 [98]

75-11300 AVO International
(† Biddle Instruments)
510 Township Line Rd, Blue Bell, PA 19422
☎(+1 215) 646 9200 6851045
⊜ (+1 215) 643 2670
Marketing Communication Mgr: Peg O'Neil
■ Manufacture electrical testing instruments. [97]

75-11400 Avondale Industries Inc. Shipyards Division
(† Avondale Shipyards Inc.)
PO Box 50280, New Orleans, LA 70150-0280
☎(+1 504) 436 21 21; 436 53 93
⊜ (+1 504) 436 53 04; 436 57 81; 436 53 03
♣ $ 576,308,000 (1995) 5,300 (1995)
55,184 sq m
Chairman, President and Chief Executive Officer: Albert L Bossier (Jr) ☎(+1 504) 436 50 02
Vice President of Advanced Program and Marketing: Ron McAlear ☎(+1 504) 436 53 93
Corporate Vice President of Modular Construction: R P Meric (Jr) ☎(+1 504) 436 50 60
Office Manager: Ralph Scheeler ☎(+1 504) 436 53 79
■ Naval shipbuilding, Landing Ship Docks (LSD), auxiliary oilers (T-AO) and minehunters (MHC), strategic sealift ships (T-AKR) for the US Navy. [98]

75-11405 AV Technology LLC
(† Verne Corp.)
(an afiliate of General Dynamics)
50405 Patricia Drive, Chesterfield, MI 48051
☎(+1 810) 949 58 50 ⊜ (+1 810) 949 87 60
110 50,000 sq ft
President & COO: James C. Leblanc
Exec. Vice President: Richard S. LeBlanc
VP Operations: James E. Jambriska
VP Business Development International: David M. Merry
VP Business Development: Thomas S. Jambriska
Engineering Dir: Douglas J Kouri
Purchasing Manager: Ms Cindy A. Berlin
■ Design, develop, fabricate, test, support & manufacture weapon systems, light armoured combat vehicles & armoured security vehicles. Also provide repair, rebuild or modernisation of existing in-service combat/security vehicles & weapon systems. [98]

75-11450 Aydin Corp. Corporate Office
700 Dresher Road, Horsham, PA 19044
☎(+1 215) 657 75 10 W http://www.aydin.com
⊜ (+1 215) 657 38 30
$ 90,327,000 ♣ $ 116,578,000 (1996) 1,200
600,000 sq ft
President and Chairman: I Gary Band
Executive Vice President: John F Vanderslice
Marketing Contact: Ms Carole Eagles
■ Airborne and ground telemetry, including transmitters, mux and demux, colour display terminals and computer systems. Telecommunication equipment and systems for satellite, troposcatter and line of sight networks; digital communications and digital radios. Radars and EW.

Aydin Communications Systems Group
700 Dresher Road, Horsham, PA 19044
☎(+1 215) 657 86 00 E govtsystems@aydin.com
⊜ (+1 215) 657 54 70
President: Klaus D Oebel
Sales: Dick Burk
■ Design and manufacture command and control equipment and systems, air traffic control systems and air defence consoles.

Aydin Products Group
Newtown Industrial Commons, PO Box 328, Newtown, PA 18940-0328
47 Friends Lane, Newton, PA 18940-0328
☎(+1 215) 968 42 71 E telemetry@aydin.com
⊜ (+1 215) 968 32 14
400
President: John Vanderslice
Telemetry Sales: Ms Linda Tyre
Radio Sales: Bob Davis ☎(+1 715) 392 47 47
Monitor/Workstation Sales: Larry Radbill ☎(+1 215) 658 46 28
■ Airborne and ground based telemetry and data acquisition equipment and systems, microwave and digital radios, ruggedised workstations, flat panel and CRT monitors.

US Government Programs
5205 Leesburg Pike, Suite 103, Falls Church, VA 22041
☎(+1 703) 998 11 15 E aydin@001@crols.com
⊜ (+1 703) 820 48 04 [98]

75-11500 B&D Instruments & Avionics Inc.
(a subsidiary of Bowthorpe plc, UK)
209 West Main, Valley Center, KS 67174
☎(+1 316) 755 12 23 ✆ (800) 444 9173
⊜ (+1 316) 755 10 33
65 (1997) 1,860 sq m
President: Howard P Botts
Vice President of Marketing and Sales: Gary D Richards
Sales and Marketing Director: Bob Trimmer
Purchasing Manager: Bill R Botts
■ Design, manufacture and service of aircraft instrumentation and microprocessor-based aviation components. [98]

75-11600 Babcock Inc.
14930 East Alondra Boulevard, La Miranda, CA 90638-5752
☎(+1 714) 994 65 00 E babcockinc@msm.com
W http://www.babcockinc.com ⊜ (+1 714) 994 30 13
♣ $ 25,000,000 + 250 135,000 sq ft
President: R S Dixon
Chief Financial Officer: Bob Kulow
Operations Director: Tom Baranowski
Sales ansd Marketing Director: Eric Bagdasarian
Purchasing Manager: Ms Barbara Enke

■ Custom power supplies and high reliability relays for military and aerospace environment. Commercial flat panel displays. [98]

75-11645 Badger Fire Protection Inc.
(† Figgie Fire Protection Systems)
4251 Seminole Trail, Charlottesville, VA 22911
☎(+1 804) 973 4361 ✆ (800) 446 3857 ⊠ 893444
⊜ (+1 804) 973 1589
Operations Dir: Ken Novak
Sales Dir: Ron Baellow
Marketing Mgr: Tod Cohen
Purchasing Mgr: Jeff Bruek
■ Badger & American LaFrance portable dry chemical, halon, CO2, water & wheeled extinguishers, Powhatan brass products for fire protection & industrial castings. Automatic sprinkler devices, special hazard fire systems. [97]

75-11660 Baganoff Engineering Inc.
400 Brookes Drive, Suite 103, Hazelwood, MO 63042
12 4,000 sq ft
President: Fred Baganoff
■ Aircraft ground support equipment. [98]

75-11800 Ballard Technology Inc.
3229A Pine Street, Everett, WA 98201
☎(+1 206) 339 02 81 ✆ (800) 829 15 53
[E] sales@ballardtech.com [W] http://www.ballardtech.com
⊜ (+1 206) 339 09 15
Customer Support Engineer: Kevin S. Christian
■ Manufacture test equipment to MIL-STD-1553, space shuttle, ARINC 629 & ARINC 429 avionics data buses. Boards & software for PC/XT/AT, VME computers, industry pack & PCMCIA [98]

75-11850 Ball Aerospace and Technologies Corporation (BATC)
10 Longs Peak Drive, PO Box 1235, Broomfield, CO 80038-1235
☎(+1 303) 469 55 11
[W] http://www.ball.com/aerospace/hplist. html
⊜ (+1 303) 460 23 15
♣ $ 362,000,000 (1996) 2,300 1,008,285 sq ft
President and Chief Executive Officer: D W Van Landingham
Senior Vice President, Administration: R W Dean
Senior Vice President, Telecommunications Products Division: J P Dummigan
Senior Vice President, Aerospace Systems Division: E L Vande Noord
Vice President, Finance and Accounting: E P Morgan
Vice President, Public Affairs: A E Bracken
■ Provide systems engineering services. Design and manufacture complete spacecraft and space systems, space and scientific sensors, cryogenic subsystems, antenna systems and video products.

Aerospace Systems Division
1600 Commerce Street, Boulder, CO 80301
☎(+1 303) 939 40 00 ⊠ 45605 TWX 9109403241
⊜ (+1 303) 442 48 12
Senior Vice President, Aerospace Systems Division: E L Vande Noord
■ Aerospace systems, spacecraft and platforms, sensors and instruments, stored cryogenics, systems engineering.

Telecommunication Products Division
10 Longs Peak Drive, Broomfield, CO 80021-2510
☎(+1 303) 469 55 11 ⊠ 204875 ballcec
⊜ (+1 303) 460 23 15
Senior Vice President, Telecommunication Products Division: J P Dummigan
■ Antenna subsystems, RF systems, video products, tactical subsystems, satellite communications systems and secure communications systems. [98]

75-11860 Efratom, Time and Frequency Products, Inc
(a subsidiary of Datum Inc.)
3 Parker, Irvine, CA 92618-1696
☎(+1 714) 770 50 00 ✆ (800) EFRATOM
[W] http://www.datum.com ⊜ (+1 714) 770 24 63
375 (approx. 1997)
President: Heinz Badura
Operations Director: Al Wilson
Sales and Marketing Director: Ed St Criox
Director of Marketing Planning: Ms Alison Whisler
Purchasing Manager: Steve Breit
■ Research, develop and manufacture time and frequency devices. [98]

75-11920 Barco Chromatics
(a division of Barco NV, Belgium)
2558 Mountain Industrial Boulevard, Tucker, GA 30084-3893
☎(+1 770) 493 70 00 ✆ (800) 743 3576
[E] products@bcx.barco.com
[W] http://www.lyceum.com/~barco ⊜ (+1 770) 492 51 01
♣ $ 500,000,000 (1995) 3,200 (1995)
Executive Vice President and General Manager: H David Scott
Vice President of Finance: Steven Ames
Sales and Marketing Director: Robert Creasy
Technical Director: Tom Burdette
Public Relations Manager: Ms Peggy Grimm
Purchasing Manager: Gerarl Melton
■ Design and manufacture commercial and industrial rugged workstations, display controllers, high resolution monitors, high brightness monitors and rugged flat planel/avionics displays. [98]

75-11930 Barfield Inc (BIC)
(a subsidiary of SOGERMA, France)
4101 North West 29th Street, PO Box 025367, Miami, FL 33142
☎(+1 305) 871 39 00 ✆ (800) 321 10 39
⊜ (+1 305) 871 56 29
♣ US$ 150,000,000 300
President and Chief Executive Officer: Marc Paganini
Senior Vice President of Sales and Marketing: Patrick J Harris
Operations Director: Sylvan Babon
■ Test equipment, instrument repair/overhaul, engineering services, distribution of over 100 OEMs. → 80 [98]

75-11960 Barber-Colman Co. Aerospace & Power Controls Division
(a division of Siebe plc, UK)
1354 Clifford Avenue, PO Box 2940, Loves Park, IL 61132-2940
☎(+1 815) 637 30 00 ⊜ (+1 815) 877 50 60
360
Vice President and General Manager: C D Spitler
Financial Controller: Mrs M A Streid
Sales and Marketing Director: Steve Rusk
Purchasing Manager: Don Pearson
■ Aircraft environmental control systems. Air cycle machines. Aircraft pneumatic pressure regulators; butterfly, check, and shut-off air valves; electromechanical rotary & linear actuators; solid-state electronic controllers for aircraft cabin and windshield temperature control. Bleed air leak detection systems. [98]

75-12000 Barnes Aerospace Group
Group Headquarters
169 Old Kennedy Rd, Windsor, CT 06095
☎(+1 860) 298 7740
⊠ 62032359; TWX 910 240 3706
⊜ (+1 860) 298 7738
650
Group President & COO: Theodore E. Martin
Group Controller: Daniel Bishop
Vice President Sales & Marketing: Paul R. Jorgensen
■ Manufacture jet engine components & overhaul of jet engine parts. Titanium super plastic forming & diffusion bonding.

Europe/Middle East Sales Office:
France → 90

Jet Die
5300 Aurelius Rd, PO Box 25066, Lansing, MI 48909
☎(+1 517) 393 5110 ⊜ (+1 517) 393 2243
President: Ronald T. Colligan
VP Advanced Technology: Marvin M. Sutton
■ Manufacture hot formed precision sheet metal shapes; design, development and production of lightweight components and structures, specialising in super plastic forming & diffusion bonding of titanium alloys.

Windsor Airmotive
7 Connecticut South Dr., East Granby, CT 06026
☎(+1 203) 653 5531
⊠ TWX 910 240 3706 windsor airmot
⊜ (+1 203) 653 0397
110 5 000 sq m
President: Harry G. Saddock
VP Sales & Marketing: Clayton Moore
Operations Mgr: Michael Esposito
US Sales Rep.: Kendall A. McArthur
Customer Service Mgr: Samuel S. Thresher
■ Aircraft engine component repairer, specialising in brazing, plasma coating, electron beam & TIG welding, restoration of large complex structural components such as turbine frames, cases, knife edge airseals, bearing housings, shafts & discs. → 80

Windsor Manufacturing Co.
169 Kennedy Rd, Windsor, CT 06095
☎(+1 203) 688 6411
⊠ TWX 710 447 5241; 62032359
⊜ (+1 203) 688 7088
President: Cedric Beckett
VP Sales & Marketing: Paul R. Jorgensen
VP AVCO & Military Sales: William Gonet
■ Manufacture critical aerospace engine components, specialising in providing cost effective, process solutions to the production of complex components. [97]

75-12060 Bartlett Lifting Devices Inc.
6830 West 63rd Street, Chicago, IL 60638
☎(+1 312) 586 82 00 ⊜ (+1 312) 586 82 05
President: Frederick P Bartlett
■ Lifting fifth wheels mounted on truck tractors. Converts any truck tractor into a trailer and container yard tractor. [98]

75-12200 Bath Iron Works Corp.
700 Washington St, Bath, ME 04530
☎(+1 207) 442 15 93 ⊜ (+1 207) 442 15 67
♣ $ 500 000 000 (1990) 11 000
CEO: William E. Haggett
President: Allan C. Cameron
VP Sales & Marketing: William D. Potter
Dir Public Relations: James A. McGregor
Manager: R.C. Larson
■ Construction of Naval surface combatant ships. DDG-51 Class AEGIS destroyers. CG-47 Class AEGIS cruisers. Shipbuilding & program related consulting services.

Portland Facility
40 Commercial Street, Portland, ME 04101
☎(+1 207) 761 48 00 [98]

75-12250 Bauer Aerospace
50 Tower Lane, PO Box 1470, Avon, CT 06001
☎(+1 203) 677 9707 ⊜ (+1 203) 674 9395
♣ $ 10 000 000 (1996) 65 72 000 sq ft
General Mgr: Louis J. Auletta (Jr)
■ Design & manufacture aircraft component test equipment. Commercial & military applications. Automated & manual test stands. Fuel, oil, hydraulic, pneumatic, electrical. [97]

75-12540 Beaird Industries Inc.
(a subsidiary of Trinity Industries Inc.)
601 Benton Kelly Street, Shreveport, LA 71106-7198
☎(+1 318) 865 63 51 ⊜ (+1 318) 868 17 01
650 750,000 sq ft
President and General Manager: John Wawrzenaik
Marketing Services Manager: Robert W Rademacher
■ Desalinators and silencers for the entire marine industry. Heavy contract steel fabrication and machining. [98]

75-12600 Becker Avionics Inc
(an affiliate of Becker Flugfunkwerk GmbH, Germany)
912 Airport Road, Roanoke, TX 76262
☎(+1 817) 491 0080 ⊜ (+1 817) 491 0081
10 (1996) 710 sq m
President: F.Patrick Carr (Jr)
Sales Dir: Bill Inman [97]

75-12740 BEI Sensors & Systems Co.
(† BEI Sensors & Motion Systems Co.)
(a subsidiary of BEI Electronics Inc.)
13100 Telfair Avenue, Sylmar, CA 91342
☎(+1 818) 362 71 51 ✆ (800) 959 05 06
⊜ (+1 818) 362 18 36
500
President and Chief Executive Officer: Dr Asad Madni
Vice President of Commercial Marketing: Gregg Stokely
■ High technology proprietary design and manufacturing company. Main activities are development and manufacture of precision sensors, electro-mechanical motion control components and systems for industrial and government customers.

Encoder Systems Division
13100 Telfair Avenue, Sylmar, CA 91342

☎(+1 818) 341 61 61 ⊜ (+1 818) 882 45 53
👥 67
General Manager: Ramesh Pavasia
Sales and Marketing: Al Trendacosta
■ Builds specially packaged absolute and incremental encoders for defence/military applications. Other product specialties include multi-turn absolute encoders, instrument-grade trackballs, rotating polygon scanners and fibre optic encoders.

Precision Systems and Space Division
1100 Murphy Drive, Maumelle, AR 72113-6139
☎(+1 501) 851 40 00 ⊜ (+1 501) 851 54 76
👥 89
Vice President and General Manager: Ron Roberts
Assistant General Manager: Charles Hudson
■ Design and produce pointing and positioning components and servo controls used in space, tracking and precision instrumentation systems. Standard optical encoders give angular resolution of over 33 million absolute positions per turn with accuracy to 0.3 arcseconds.

Duncan Electronics Div.
15771 Red Hill Avenue, Tustin, CA 92680
☎(+1 714) 256 75 00 ⊜ (+1 714) 258 81 20
Vice President and General Manager: Dan Hardwick
Marketing Manager: Ms Barbara Wiesinger
■ Precision position sensing products for instrumentation and control applications. Manufacture of stock and custom, small modular and enclosed incremental optical encoders. Units typically used on light machinery in protected environments including clean factories, offices and laboratories.

Edcliff Instruments Division
13100 Telfair Avenue, Sylmar, CA 91342
☎(+1 818) 362 17 51 ⊜ (+1 818) 362 18 36
Manager: Ramesh Pavasia
■ Flight hardware to monitor engine oil, fuel, cabin pressure, ground support equipment.

Industrial Encoder Division
7230 Hollister Avenue, Goleta, CA 93117
☎(+1 805) 968 07 82 ⊜ (+1 805) 968 31 54
👥 60
General Manager: Glenn Avolio
Sales Manager: Scott Orlosky
■ Manufacture incremental optical encoders for industrial applications demanding reliability under severe operating conditions. Resolutions from 1 to 200,000 counts per turn available. Ruggedised housings and heavy-duty bearings withstand up to 300lbs and shock up to 200 g.

Kimco Magnetics Division
804-A Rancheros Drive, San Marcos, CA 92069
☎(+1 619) 744 56 71 ⊜ (+1 619) 744 88 15
👥 75
General Manager: Anthony Morreale
Sales Manager: Mike Lopez
■ Design and manufacture fractional horsepower brushless DC motors, linear and rotary voice coil actuators and specialty electromagnetic devices. Commercial, military and aviation products available.

Systron Donner Inertial Division
2700 Systron Drive, Concord, CA 94518
☎(+1 510) 671 66 01 ⊜ (+1 510) 671 66 47
Sales Director: Jeff Lamparter
General Manager: Gerard Brasuel
■ Design and manufacture accelerometers and rate sensors for commercial, military and aerospace applications. [98]

75-12800 **Belfort Instrument Co.**
727 South Wolfe Street, Baltimore, MD 21231-3513
☎(+1 410) 342 26 26 [W] http://www.belfort-inst.com
BELF ⊜ (+1 410) 342 70 28
♣ $ 6,000,000 👥 45 🏭 80,000 sq ft
President: John Hoover
Sales Director: Dennis Kast
■ Design and manufacture of meteorological instruments and systems; laser beam ceilometer. [98]

75-12905 **Bell Avon Inc.**
(a joint venture with Avon Rubber plc)
1200 Martin Luther King Jnr. Blvd, Picayune, MS 39466
1200 Martin Luther King Junior Boulevard
☎(+1 601) 799 12 17 ⊜ (+1 601) 799 13 60
♣ $ 22,000,000 (1998) 👥 120 (1997) 🏭 40,000 sq ft
President: Larry K Sipe
Sales Manager: Donald Lefebre
Project Manager: David Poole
■ Hovercraft skirts, collapsible pillow tanks (fuel or water). Inflatable buildings. [98]

75-13100 **Bell Helicopter Textron Inc.**
(a subsidiary of Textron)
PO Box 482, Fort Worth, TX 76101
☎(+1 817) 280 20 11; 280 82 00 ✉ BELLCRAFT
[W] http://www.bellhelicopter.textron.com
☒ 6733926 bell uw ⊜ (+1 817) 280 23 21
👥 8,000 (1996)
Chairman: Webb F Joiner
Vice Chairman: Lloyd Shoppa
President: Terry D Stinson
Senior Vice President of Customer and Marketing Support: P H Parsinen ☎(+1 817) 280 20 70
Senior Vice President of Research and Engineering: Troy M Gaffey ☎(+1 817) 280 43 79
Senior Vice President of Commercial and Foreign Military Business: Fred N Hubbard (III) ☎(+1 817) 280 46 86
Vice President of International Business Development: Rex A Marion
Vice President of Government Contracts: M Dan McCrary
Vice President of USG Military Business: Jim P Schwalbe ☎(+1 817) 280 89 21
Vice President of Material: L G Gleim
Public Affairs and Advertising Director: Carl Harris ☎(+1 817) 280 24 25
Manager of Used Helicopters Department: Darrell Wilson ☎(+1 817) 280 28 65
Advertising Manager: Dick Davies ☎(+1 817) 280 81 50
Media Relations Manager: Bob Leder ☎(+1 817) 280 84 16
Tiltrotor Communications Manager: Terry Arnold ☎(+1 817) 280 24 25
Promotions and Trade Shows Manager: Ms Patricia Butler ☎(+1 817) 280 81 45
■ Light and heavy-lift helicopters and VTOL aircraft for military and commercial use. Service, overhaul and maintenance, including components and accessories.

Textron Inc. Washington, DC Office
1101 Pennsylvania Avenue North West, Suite 400, Washington, DC 20004-2504
☎(+1 202) 637 38 00 ⊜ (+1 202) 637 38 60
Vice President of Government Marketing: John Zugschwert
International Vice President: Y Meyersick ☎(+1 401) 45 72 54
Executive Director of International Business Development: Ms Audrey Deegan ☎(+1 202) 637 38 25 [98]

75-13300 **Bender Shipbuilding & Repair Co.**
265 South Water Street, PO Box 42, Mobile, AL 36603-2305
☎(+1 334) 431 80 00
[E] corporate@mobile.bendership.com
[W] http://www.bendership.com ⊜ (+1 334) 432 22 60
👥 500
President: T B Bender (Jr)
Ship Repair Director: R A (Bob) Beckman
Sales and Marketing Director: Frank G Terrell (Jr)
Purchasing Manager: Jon Harrison
Marketing Assistant: Ms Linda Lewis
■ Construction, conversion and repair of military vessels including patrol boats and support craft ranging in size from less than 40' up to at least 150'. [98]

75-13540 **Benthos Inc.**
49 Edgerton Drive, North Falmouth, MA 02556
☎(+1 508) 563 10 00 ✆ (800) 446 12 22
[E] info@benthos.com [W] http://www.benthos.com
☒ 820673 ⊜ (+1 508) 563 64 44
♣ $ 11,700,000 (1996) 👥 75 (1997) 🏭 11,000 m2
Director: Lawrance W Gray
Sales and Marketing Director: Kevin McCarthy
Communications Manager: Peter Zentz
Purchasing Manager: Ms Kathie Martin
■ Specialist in undersea systems technology. Supplies equipment including imaging systems, deepsea photographic systems and underwater video robotic systems. [98]

75-13605 **Beretta USA Corp**
(a subsidiary of Beretta SpA, Italy)
17601 Beretta Drive, Accokeek, MD 20607
☎(+1 301) 283 2191 ⊜ (+1 301) 375 7677
🏭 86,000 sq ft
Executive Vice President: Robert Bonaventure
Product Support Manager: Wesley Lang
■ Pistols, submachine guns for military and police forces.
® Italy: Beretta SpA [98]

75-13700 **Bering Technology Inc.**
(† Ocean Microsystems)
1357 Dell Ave, Campbell, CA 95008-6629
☎(+1 408) 364 6500 ✆ (800) 237 4641
⊜ (+1 408) 374 8309
President: Leung Lok
Sales & Marketing Dir: George Wissler
Engineering Dir: Steve Sun
Purchasing Dir: Steve McDonald
Marketing Associate: Ms Juliette Katz
■ Computer peripheral & mass storage products. Specifically, hard disk, MO disk drive subsystems, Bermoulli disk subsystems, SyQuest removable hard disk subsystems, RAID systems for Networking & CD-Recordable drives. [97]

75-13720 **Bertan High Voltage**
(† Bertan Associates Inc.)
(a subs. of DEL Electronics)
121 New South Rd, Hicksville, NY 11801
☎(+1 516) 433 3110 ⊜ (+1 516) 935 1766
👥 110 🏭 40 000 sq ft
Sales Mgr: Paul Oranges
■ Manufacture precision high voltage power supplies & test instruments for automatic test equipment, nuclear instrumentation, precision CRT displays, photo-multipliers & other applications for a precise & stable source of high voltage up to 100 000 volts. [97]

75-13750 **Beta Company**
2137B Flintstone Drive, Tucker, GA 30084
☎(+1 770) 270 07 73 ✆ (800) 669 BETA
☒ 154111 ocom ut ⊜ (+1 770) 270 05 99
President: Ms Silvia Kaiser-Block
■ Manufacture C-Mag rifle magazine and storage systems.
International Sales:
Beta AG → Liechenstein 90 [98]

75-13770 **BFGoodrich Aerospace**
Headquarters
250 North Cleveland-Massillon Road, PO Box 5501, Akron, OH 44334-0501
☎(+1 330) 374 22 00 [W] http://www.bfgaerospace.com
⊜ (+1 330) 374 36 39
♣ $ 1,200,000,000 (1996) 👥 9,000
President: Marshall Larsen
Vice President Legal: Kevin J Ryan
Group Vice President Aircraft Integrated Systems: John Grisik
Group Vice President Landing Systems: Ernest Schaub
Group Vice President Maintenance Repair & Overhaul: Robert L Avery
Communications Dir: Brayton Harris
Vice President Business Development: Steve Huggins
■ A diversified manufacturer and supplier of aerospace & defence products and services.

Aircraft Wheels & Brakes
101 Waco Street, PO Box 340, Troy, OH 45373
☎(+1 513) 339 38 11 ⊜ (+1 513) 339 38 13
Commercial General Manager: Mike Piscatella
Military General Manager: Bruce Coggins
■ Brake business units of BFGoodrich Aerospace design, develop, manufacture, market and provide worldwide customer support for a wide variety of aircraft wheels; steel and carbon brakes. Commercial Wheels and Brakes primarily supports Airbus, Boeing & Lockheed commercial aircraft. Regional and Business Wheels & Brakes supports the general aviation & commuter aircraft industries. Military Aircraft Wheels & Brakes works with NASA, the US and allied armed forces and major defense contractors.

Carbon Products
50 William White Boulevard, Pueblo, CO 81001
☎(+1 719) 948 35 00 ⊜ (+1 719) 948 34 22
■ Manufacture carbon-carbon materials in support of BFGoodrich aircraft brake programs.

Landing Gear Services Division
6445 NW 25th Street, Miami, FL 33122
☎(+1 305) 871 34 20 ⊜ (+1 305) 876 95 57
■ Landing gear overhaul and repairs.

Lighting Systems Division
(a subsidiary of BFGoodrich Aerospace)
129 Fairfield Street, Oldsmar, FL 34677
☎(+1 813) 855 44 28 ⊜ (+1 813) 855 55 72
■ Airport and aircraft lighting systems and controls.

Ice Protection Systems Division
1555 Corporate Woods Parkway, Uniontown, OH 44685
☎(+1 216) 374 30 40 ⊜ (+1 216) 374 22 90
General Manager: Dick McMurry

■ Design and manufacture heated products and pneumatic de-icing boots and electrothermal de-icing grids and associated controls for airfoil and structural components.

Evacuation Systems Division
3414 South 5th Street, Phoenix, AZ 85040
☎(+1 602) 232 40 00
⊖ (+1 602) 232 41 00; 232 41 22
General Manager: William Wathall
■ Manufacture inflatable evacuation slides and slide rafts for commercial aircraft of all manufacturers and survival rafts for business jet aircraft.

Avionics Systems Division
5353 52nd Street SE, Grand Rapids, MI 49508
☎(+1 616) 949 66 00 ⊖ (+1 616) 949 93 76
General Mgr: Wes Perry
Sales Dir Military/Commercial: Ms Adrienne Stevens
■ Design & manufacture avionic & navigation equipment, including navigation management systems, DMEs, LORAN, TACAN, TCAS-1 collision warning systems, worldwide navigation databases, stand by altitude indicators. The Stormscope line of weather mapping instruments (former 3M product).

Engineered Polymer Products Division
6061 BFGoodrich Boulevard, Blount Island, Jacksonville, FL 32226
☎(+1 904) 757 36 60 ⊖ (+1 904) 757 71 16
General Mgr: Jerry La Reau
■ Products include sonar domes & acoustic windows for US & Allied Navies, suspension & seating components for military vehicles & water-lubricated marine bearings.

Rosemount Aerospace Division
14300 Judicial Road, Burnsville, MN 55306
☎(+1 612) 892 43 00 ⊖ (+1 612) 692 44 30
General Mgr: Ron Hodges
■ Design & manufacture a wide-range of airdata products & sensors.

Landing Gear Division
8000 Marble Avenue, Cleveland, OH 44105
☎(+1 216) 341 17 00 ⊖ (+1 216) 429 40 38
General Manager: Jack Carmola
■ Manufacture landing gear for commercial & military aircraft (including B747; B1 & B2; C17). [98]

75-13771 BFGoodrich Avionics Systems Inc

(a division of BFGoodrich Aerospace)
5353 52nd Street SE, Grand Rapids, MI 49588-0873
☎(+1 616) 949 66 00 ✆ (800) 253 95 25
[W] http://www.avionics.bfg.com ⊖ (+1 616) 285 42 24
👥 428 (1997) 🏭 110,000 sq ft
Vice President: Wes Perry
Director Engineering: Gary Watson
Director Manufacturing: Jay LaFoy
Director Customer Service: Ray Poyner
Manager Business Development: Joe Ehtier
Manager Air Transport Sales: Tom McDonald
Manager Business and General Aviation Sales: William Gay
Marketing Serices Representative: Ms Carol Comer
■ Manufacture, development and service of aircraft safety systems and components including the Skywatch traffic advisory systems, TCAS I traffic alert annd collision avoidance system, Stormscope weather mapping system, GtI-3000 electronic standby attitude indicator, electromechanical standby attitude indicators, vertical and directional gyros, servo actuators, capstans, emergency power supplies and invertors. Maintains a global support network at a number of factory authorised service centres. [98]

75-13772 BFGoodrich Specialty Chemicals

(a division of The BFGoodrich Company)
9911 Brecksville Road, Brecksville, OH 44141
☎(+1 216) 447 50 00 [W] http://www.bfgoodrich.com
⊖ (+1 216) 447 56 69

Specialty Adhesives Group
9911 Brecksville Road, Brecksville, OH 44141
☎(+1 216) 447 50 00 ⊖ (+1 216) 447 56 69 [98]

75-13774 BHP Hawaii Inc.

(† BHP Petroleum Americas (Hawaii) Inc)
(a subs. of The BrokenHill Propietary Company Ltd)
PO Box 3379, Honolulu 96842
PRI Tower, 733 Bishop St, Suite 2700, Honolulu, Hawaii 96813
☎(+1 808) 547 3111 ⊠ 7430292
⊖ (+1 808) 547 3145
Group General Mgr & President: Henry G. Neal
VP Enrironmental & Government Affairs: George E. Bates
VP Corp. Communications: Ms Andrea L. Simpson
Federal Affairs Mgr: Ms Pamela J. Whitted
■ Manufacture and distribute jet fuel, diesel, gasoline, marine and bunker fuels, residual fuel oils, synthetic natural gas and propane in Hawaii and the Pacific.

Washington, DC Office
1225 Eye Street NW, Ste 500, Washington, DC 20005
☎(+1 202) 682 0611 ⊖ (+1 202) 682 0616

BHP Petroleum Pacific Islands Inc.
(† PRI South Pacific Inc.)
American Samoa Office
PO Box 488, Pago Pago, American Samoa 96799
☎(+1 684) 633 5331 ⊠ 782510 ⊖ (+1 684) 633 5101
Managing Dir: Paul Stevenson [97]

75-13795 Big Sky Racks Inc.

PO Box 729, Bozeman, MT 59771-0729
☎(+1 406) 586 93 93 ⊖ (+1 406) 585 73 78
Managing Director: Ms Jennifer Hefty
Operations Director: Ted Morford
■ Law enforcement vehicle weapon mounts. [98]

75-14083 BIW Cable Systems Inc.

(a subsidiary of Draka USA)
22 Joseph E. Warner Boulevard, North Dighton, MA 02764-1300
☎(+1 508) 822 54 44 ⊖ (+1 508) 822 19 44
👥 190
President and Chief Executive Officer: D Maistrellis
Vice President Sales and Marketing: S M Adams
■ Military specification wire and cable product. [98]

75-14085 BIW Connector Systems Inc.

(a subsidiary of Draka USA)
500 Tesconi Cir., Santa Rosa, CA 95401-4692
☎(+1 707) 523 23 00 ⊖ (+1 707) 523 35 67
♣ $ 11,000,000 (1996) 👥 100 🏭 45,000 sq m
Executive Vice President and General Manager: John Hathaway
Operations Director: Greg Glynn
Marketing Manager: Eckhard Konkel
Purchasing Manager: L MacPhee
■ Underwater electrical connectors. MIL-C-24217, MIL-T-24552, MIL-C-7974. [98]

75-14095 Blast Deflectors Inc. (BDI)

18 C Hangar Way, Watsonville, CA 95076
☎(+1 408) 722 9928 ⊖ (+1 408) 722 9686
🏭 8 500 sq ft
President: Stanley Lynn
Exec. VP & General Mgr: Christopher Lynn
■ Jet blast deflector fences for military & civil aircraft. Noise protection for airport ground run-up facilities. [97]

75-14100 Bliley Electric Co.

2545 West Grandview Boulevard, PO Box 3428, Erie, PA 16508
☎(+1 814) 838 35 71 [E] info@bliley.com
[W] http://www.bliley.com ⊖ (+1 814) 833 27 12
👥 147 (1997) 🏭 62,500 sq ft
President: R D Bliley
Vice President Marketing: D M Bliley
Operations Director: D B Siggins
Purchasing Manager: T J Weunski
■ Manufacture quartz crystals and crystal oscillators. [98]

75-14250 The Boeing Company

PO Box 3707, Seattle, WA 98124
7755 E. Marginal Way S., Seattle, WA 98108
☎(+1 206) 655 21 21 ✉ BOEINGAIR ⊠ 329430
⊖ (+1 206) 655 11 77
♣ $ 22,681,000,000 (1996) 👥 150,000
Chairman and Chief Executive Officer: Philip M Condit
Director: Robert Beck
Director: John E Bryson
Director: John B Fery
Director: Paul E Gray
Director: Harold J Haynes
Director: Stanley Hiller (Jr)
Director: Donald E Petersen
Director: Charles Pigott
Director: Frank Shrontz
Director: Ms Rozanne L Ridgway
Director: George Weyerhauser
Senior Vice President: Lawrence W Clarkson
Senior Vice President - People: Jim Dagnon
Senior Vice President and Chief Financial Officer: Boyd E Givan
Senior Vice President: C Gerald King
Senior Vice President: Alan R Mulally
Senior Vice President: John D Warner
Senior Vice President: Ronald B Woodard
Vice President Communications and Investor Relations: Larry Bishop
Vice President and General Counsel: Theodore J Collins
Vice President US Government Affairs: Christopher W Hansen
Vice President State and Local Government Affairs and Community Relations: John F Hayden
Vice President and Treasurer: David A Jaeger
Vice President and Controller: Gary Beil
Staff Vice President Trust Investments: Gary B Bland
Vice President Employee and Union Relations: Jerry Calhoun
Vice President Facilities and Capital Investments: Arlington W Carter
Vice President Employee Benefits, Insurance and Taxes: Dennis J Crispin
Vice President Engineering and Technology: Robert A Davis
Staff Vice President and General Auditor: Corwin N Lott
Vice President Compensation: Michael Valliere
Vice President International Business: Ray Waldmann
Corporate Secretary and Corporate Counsel: Ms Heather Howard
■ Operates primarily in four market segments: the development, production and marketing of commercial & military aircraft; development & product of missiles, space exploration equipment; development & production of defence electronic systems & computer services & large-scale information networks.

Products:
Commercial jet aircraft
Military aircraft
Helicopters
Guided missiles
Space vehicles
Defence & space systems
Defence electronics

Business Groups:
Boeing Commercial Airplane Group
Boeing Defense & Space Group
Boeing Computer Services

Boeing Defense & Space Group
(a division of The Boeing Co.)
20403 68th Avenue S., PO Box 3999, Seattle, WA 98142
☎(+1 206) 773 21 21
President: Alan Mulally
Executive Vice President Production: John Schmit
Vice President & Chief Counsel: Kenneth J. Brown
VP & General Mgr Product Support Div: Larry L. Debrot
Vice President Information and Communication: Denton R Hanford
VP & General Mgr Missiles & Space Div: James P. Noblitt
VP Information Systems: Terence V. Milholland
VP Planning & Administration: Willis D. Smith
VP Operations: Dennis H. Williams
VP Engineering: Larry Winslow
VP Human Resources: H. William Wolfe
VP Contracts: Matthew E. Brislawn
VP Information & Electronics Systems Div: John B. Dempster
VP Business Development: James W. Evatt
VP Finance: Scott E. Carson

Boeing Aerospace Operations
1355 North Atlantic Avenue, PO Box 320220, Cocoa Beach, FL 32932-0220
☎(+1 407) 783 02 20 ⊖ (+1 407) 783 60 90
Vice President & General Mgr: Frank P. Smith
■ Responsible for providing flight equipment processing contract support to the Johnson Space Center in Houston, TX.

Information and Electronic Systems Division
PO Box 3999, Seattle, WA 98124
20403 68th Avenue S., Kent, 98032
☎(+1 206) 773 21 21 ⊖ (+1 206) 657 30 33
Vice President & General Mgr: John B. Dempster
■ Responsible for a variety of defence systems, including E-3 AWACS, 767 AWACS, Maritime Patrol Avionics, as well as design & manufacture of electronics for all Boeing divisions.

Helicopters Division
PO Box 16858, Philadelphia, PA 19142
Stewart Avenue and Industrial Hwy, Ridley Park, PA 19078
☎(+1 610) 591 21 21 ⊖ (+1 610) 591 48 20
Vice President and Assistant General Manager: James Morris
■ Responsible for a growing line of Boeing rotorcraft, including the CH-47 & CH-46, the V-22 Osprey & the RAH-66 Comanche.

Military Airplanes Division
PO Box 3999, Seattle, WA 98124
9725 E. Marginal Way S., Seattle, WA 98108
☎(+1 206) 655 21 21 ⊖ (+1 206) 662 06 60

Contact Reference

Vice Presidennt and General Manager: Frank Statkus
- Responsible for the company's prime & subcontracting roles in a number of programmes, including the B-2 and F-22, key contracted work for JAST & ASTOVL fighter concept definitions & extensive engineering research efforts on future military aircraft.

Missiles & Space Division
PO Box 240002, Huntsville, AL 35824
499 Boeing Blvd, Huntsville, AL 35806
☎ (+1 205) 461 21 21 ⊖ (+1 205) 561 22 52
Vice President & General Mgr: James Noblitt
- Responsible for Boeing's partcipation in a variety of US DoD & NASA programmes, including the Avenger air defence system, Space Station Freedom & the Inertial Upper Stage.

Product Support Division
PO Box 7730, K12-12, Wichita, KS 67277
3801 South Oliver, Wichita, KS 67210
☎ (+1 316) 526 21 21 ⊖ (+1 316) 526 18 45
VP & General Mgr: Larry L. Debrot
- Provides follow-on support to Boeing customers as well as modification programmes including the KC-135 re-engine programme & upgrading of B-52 Stratofortress bombers.

Boeing Information & Support Services
PO Box 24346, Seattle, WA 98124
2810 160th Avenue, SE, Bellvue, WA 98008
☎ (+1 206) 763 51 66 ⊖ (+1 206) 865 29 58
President: John D. Warner
VP Business Resources: Wally E. Alder
- Provides information management services & computing resources to all Boeing operating divisions & to government customers worldwide. [98]

75-14600 Bollinger Shipyard Inc.
(† Bollinger Machine Shop & Shipyard Inc.)
PO Box 250, Lockport, LA 70374-0250
☎ (+1 504) 532 2554 ✆ (800) 888 3894 ☒ 584127
⊖ (+1 504) 532 5546; 532 7295
👥 460 (1996) 🏭 14 850 sq m
Chairman & CEO: Donald T. Bollinger
Treasurer: Ms Charlotte Bollinger Toups
- Ship construction & repair, from push boats to patrol boats. [97]

75-14700 Bose Corporation
The Mountain, MS271, Framingham, MA 01701
☎ (+1 508) 879 7330 Ext. 4250 ⊖ (+1 508) 879 3049
Marketing Mgr, Government Programmes: Ms Mary Ann West
- Active noise reduction headsets for flight & armoured vehicle applications. Supplied to US Air Force, US Army, US Navy & Belgian Air Force. [97]

75-14750 Bosma Machine and Tool Corporation
4754 East US 40, Tipp City, OH 45371-9400
☎ (+1 513) 667 4488 ✆ (800) 377 4480
[E] ben@bosma.com ⊖ (+1 513) 667 5864
♣ $ 7 000 000 👥 70 🏭 55 000 sq m
Manager Dir: Leendert (Lee) Bosma
- Manufacturing company specialise in build to order & production manufacturing of medium to large parts. [97]

75-14760 Boston Whaler Inc. Commercial Products
4121 South US Highway One,
Edgewater, FL 32141-7221
☎ (+1 904) 428 00 57 ⊖ (+1 904) 423 91 87
♣ US$ 55,000,000 (1996) 👥 500 (1996)
🏭 250,000 sq ft
Manager of Military Accounts: Douglas M Nettles
- Offshore patrol boats from 19ft to 27ft in glass fibre foam cord construction. Designed for a variety of law enforcement applications and special missions such as harbour patrol, drug interdiction and anti-terrorist intervention. [98]

75-14780 Boulder Scientific Co.
PO Box 548, Mead, CO 80542-0548
☎ (+1 970) 535 44 94 [E] mklem@worldnet.att.net
[W] http://www.chem.com/boulder ⊖ (+1 970) 535 45 84
👥 53 (1997)
President: John M. Birmingham
Operations Director: Dr Mel Talbot
Sales Manager: Dave Enders
Purchasing Manager: Ms Susanne Zar
- Curing agents and catalysts for solid fuel systems. [98]

75-14855 BMY-Wheeled Vehicles
(a division of Harsco Corp.)
13311 Industrial Parkway, Marysville, OH 43040
☎ (+1 513) 644 69 75 ⊖ (+1 513) 644 68 87
👥 270
Director: Jerry Dodd
- Manufacture of military trucks. Production includes the M939A2 5-ton truck for the US Army. [98]

75-14860 Bowmar Instrument Corp.
5080 North 40th Street, Suite 475, Phoenix, AZ 85018
☎ (+1 602) 957 02 71 ⊖ (+1 602) 381 13 14
♧ $ 8,502,000 (1996) ♣ $ 25,317,000 (1996)
👥 205 (1997)
President and Chief Executive Officer: Thomas K Lanin
Vice President of Finance: Joseph G Warren (Jr)
Office Manager: Ms Cheryl Curtis
- Manufacture of precision mechanical and electromechanical systems, fuzing, S&A devices, rotating components and joystick control systems, keyboards/data entry systems, solid state memory microcircuits. Multichip modules.

Bowmar Technologies Division
(† Bowmar Aerospace Division)
8000 Bluffton Road, Fort Wayne, IN 46809
☎ (+1 219) 747 31 21 ⊖ (+1 219) 747 96 01
President and General Manager: Mark Morton

White Microelectronics Division
4246 East Wood Street, Phoenix, AZ 85040
☎ (+1 602) 437 15 20 ⊖ (+1 602) 437 91 20
President and General Manager: H Shokrgozar [98]

75-14990 Break-Free Inc.
1035 South Linwood Avenue, PO Box 25020,
Santa Ana, CA 92705-4396
☎ (+1 714) 953 19 00 [E] break-free@worldnet.att.net
⊖ (+1 714) 953 04 02
Vice President and General Manager: Dwight B Woodruff
Vice President of Break-Free Defence Products Division: Mel Davies ☎ (+1 702) 356 07 75
- Cleaner, lubricant and preservative for weapons and weapon systems. Break-Free CLP, Break-Free BOR-CAP system. Bore cleaning and preserving kits for small calibre cannon, mortars, tanks, artillery and navy guns. Pneumatic bore cleaning system for 76mm - 205mm armour and artillary. [98]

75-15110 Breeze Eastern Corp.
(a division of TransTechnology Corp.)
700 Liberty Avenue, Union, NJ 07083
☎ (+1 908) 686 40 00 ✆ (800) 929 19 19
⊖ (+1 908) 686 92 92
President: Bob White
Vice President Marketing: David Orth
Purchasing Director: B Campbell
Marketing Representative: M Thompson
Marketing Representative: O Balten [98]

75-15170 Bren-Tronics Inc.
10 Brayton Court, Commack, NY 11725
☎ (+1 516) 499 51 55 [E] mrbrentron@aol.com
⊖ (+1 516) 499 55 04
👥 95 🏭 32,000 sq ft
President: Bill Enman
Vice President Sales and Marketing: Leo A Brenna
Vice President Operations: Mike Roamer
Purchasing Manager: Ms Maria Swist
- Batteries for military & industrial equipment. [98]

75-15210 Brimar Inc.
(† Rank Brimar Inc.)
(a subsidiary of Brimar Ltd, UK)
25358 Avenue Stanford, Valencia, CA 91355-1214
☎ (+1 805) 295 57 70
[E] 102521.1245@compuserve.com
⊖ (+1 805) 295 50 87
♣ $ 9,000 (1996) 👥 5 (1997)
President and Chief Executive Officer: Glenn Barone
Sales Manager: Robert T Hoffman
Marketing Administration: Ms Susan K Raty
- Manufacture high performance special purpose Cathode Ray Tubes and display equipment for leisure, commercial, industrial and defence applications. [98]

75-15240 Broadcast Electronics Inc.
4100 North 24th Street, Quincy, IL 62301-9632
☎ (+1 217) 224 96 00 [E] bdcast@bdcast.com
[W] http://www.bdcast.com ⊖ (+1 217) 224 96 07
👥 150
CEO/President: Jack Nevin
Operations & Technical Dir: Don Rust
Purchasing Mgr: Dick Boll
Marketing Services Mgr: Kim Winking
Government Sales Mgr: Tom Beck
- Radio broadcast equipment. [98]

75-15250 Broadcast Microwave Services Inc.
(a subsidiary of Cohu Inc.)
5636 Ruffin Road, San Diego, CA 92123
☎ (+1 619) 560 86 01 ✆ (800) 669 96 67
[E] bms@bms-inc.com [W] http://www.bms-inc.com
⊖ (+1 619) 560 16 37
👥 55 🏭 1,400 m2
President and Chief Executive Officer: Graham Bunney
Director of Military Marketing: Robert Jacobs
Operations Director: Jack Simmons
Purchasing Manager: Quinn Bremes
- Microwavé data links for UAV and manned surveillance, aircraft systems. Automated tracking antenna systems. [98]

75-15260 Broco Inc
8690 Red Oak Street, Rancho Cucamonga, CA 91730
☎ (+1 909) 483 32 22 ✆ (800) 845 72 59
[E] broco@worldnet.att.net ⊖ (+1 909) 483 32 33
President: Rick Ferry
Vice President: Gary L Maines
Sales Director: Thomas Joos
- Forced entry/breaching, lightweight portable exothermic torch systems. For Cutting steel, aluminum, titanium and other alloys. Applications include shipboard damage control/rescue, Special Forces engineer's demolition tool, armour vehicle recovery. [97]

75-15270 Bromma Inc.
(a subsidiary of Bromma Conquip AB, Sweden)
PO Box 451, Roxboro, NC 27573
☎ (+1 910) 599 31 41 ✆ (800) 334 01 15
[W] http://www.bromma.com ⊖ (+1 910) 599 44 99
Managing Director: Robert Clark
Vice President of Sales: Rackley Dawson
✉ rackdawson@aol.com
Operations Director: James Malone
Purchasing Manager: Donald W Hubbard
- Manufacture of container handling equipment from manual to telescoping 20' fixed to 20' to 48' telescopic for installation on Boom Cranes, FLT, Overhead Cranes, RTG, Ship to Shore and ship gear. [98]

75-15295 Stewart R. Browne Mfg Co.Inc.
1165 Hightower Trail, PO Box 500008,
Atlanta, GA 31150
☎ (+1 770) 993 96 00 ⊖ (+1 770) 594 77 58
👥 10
President: James S Browne
Vice President Sales and Marketing: Alan R Browne
Sales and Marketing Director: Mike Vacarro
Purchasing Manager: David Prisock
- Explosion-proof flashlights and inspection lights, static ground clamps. Aircraft maintenance lighting. [97]

75-15298 Brownell & Co. Inc.
(† Brownell Defense & Aviation Products)
429 East Haddam-Moodus Road, PO Box 362,
Moodus, CT 06469
☎ (+1 860) 873 86 25 [E] pkb12956@concentric.net
⊖ (+1 860) 873 19 44
President: Anthony A Ferraz
Sales Manager: Ms Pamela K Bronson
Operations Manager: Ms Sharon Rose
- Knotless underslung cargo nets, internal cargo restraint nets. [98]

75-15300 Brownells Inc.
200 South Front St, Montezuma 50171
☎ (+1 515) 623 5401 ⊖ (+1 515) 623 3896
👥 95
President: Frank Brownell
Dir Advertising: Larry Weeks
Government Quotes: Paul Mazan

■ Gunsmithing tools & supplies, M-16 Armourer's Kit, Remington shotgun service kit, small arms cleaning rods & accesories. [97]

75-15360 Bruce Industries Inc.
PO Box 1700, Dayton, NV 89403
☎ (+1 702) 246 01 01 ⊖ (+1 702) 246 04 51
👥 250 (1997) 🏠 160,000 sq ft
Chairman: Tom Bruce
Managing Director: Lou Abbott
Operations Director: Steve Jaffe
Engineering Manager: Sam Howe
Marketing and Customer Services Manager: Carla Durand
Purchasing Manager: Kishan Karia
■ Specialised illumination systems and electrical/lighting equipment for aerospace ground and airborne applications. Power distribution systems for military applications.
Overseas agents:
Abercorn Aero (Canada) [98]

75-15460 Brunton Co.
620 E. Monroe Ave, Riverton, WY 82501-4928
☎ (+1 307) 856 6559 ⊖ (+1 307) 856 1840
♣ $ 4 500 000 (1995) 👥 55 🏠 5 400 sq m
President: H. Herron
■ Compasses, digital direction devices, binoculars & cutlery. [97]

75-15600 Bulova Technologies L.L.C.
101 North Queen Street, PO Box 4787,
Lancaster, PA 17604-4787
☎ (+1 717) 299 25 81 [E] bulova@success.net
[W] http://www.bulovatech.com ⊖ (+1 717) 397 36 08
♣ $ 72,000,000 (1996) 👥 400 🏠 73,000 m2
President and Chief Executive Officer: Stephen L. Gurba
Vice President Marketing: Lee Tannehill
International Marketing Director: Tony Germann
■ Development & manufacture of fuzing systems & safe/arm devices for: missiles, rockets, tanks, artillery & other munitions. [98]

75-15620 Bullet Guard Corp.
3963 Commerce Drive, West Sacramento, CA 95691
☎ (+1 916) 373 04 02 ⊘ (800) 233 56 32
⊖ (+1 916) 373 02 08
👥 14 (1997) 🏠 15,000 sq ft
President: Ms Sharon Durst
Sales Director: Timothy T Klagge
Purchasing Manager: Karlin Lynch
■ Bullet resistant systems including transaction windows, doors, security enclosures and ballistic glass fibre. → 85 [98]

75-15630 Bunnell Plastics
1-295 & Harmony Rd, Mickleton, NJ 08056
☎ (+1 609) 423 6630 ⊖ (+1 609) 423 8182
Director: Gary Garwood
■ Teflon tubing, roll covers, custom extrusions. [97]

75-16005 Burton Electrical Engineering
(a division of WPI)
111 Maryland Street, PO Box 217,
El Segundo, CA 90245-0217
☎ (+1 310) 322 06 15 ⊘ (800) 847 2878
⊖ (+1 310) 322 45 20
👥 60
General Manager: Preston D Schultz
■ Manufacture cables and connectors. [98]

75-16006 Busak + Shamban Inc. Shamban Seals Division
(† W.S. Shamban & Co.)
Aerospace Products Group
2531 Bremer Road, PO Box 176, Fort Wayne, IN 46801
☎ (+1 219) 749 96 31 ▣ 228480
⊖ (+1 219) 749 84 20 (Engineering Dept)
(+1 219) 749 83 56 (Sales offers)
Marketing Manager: Ms Margaret Fish
Aerospace Product Mgr: Ray Gillies
Seals Division
2531 Bremer Road, PO Box 176, Fort Wayne, IN 46801
☎ (+1 219) 749 96 31 ▣ 228480 ⊖ (+1 219) 749 00 66
General Manager: Tom Zobitz
■ Design and manufacture of non-metallic seals for aerospace & defense industries. Broad range of standards plus development and test facilities to support custom design engineering.
Busak + Shamban West Export Group
(† Shamban International)
3904 Del Amo Boulevard, Suite #805,
Torrance, CA 90503
☎ (+1 213) 371 10 25; 371 10 96 (Export) ▣ 4720467
⊖ (+1 213) 371 31 35
General Mgr West: Larry Kaford
General Mgr Export: Larry Neipling
American Variseal
510 Burbank Street, Broomfield, CO 80038
☎ (+1 303) 465 17 27 ▣ 759242 ⊖ (+1 303) 469 48 74
Div. General Mgr: Cliff Goldstein [98]

75-16007 Bushmaster Firearms Inc.
999 Roosevelt Trail, Windham, ME 04062
☎ (+1 207) 892 20 05 ⊘ (800) 998 79 28
[E] bushmaster@maine.com
[W] http://www.bushmaster.com ⊖ (+1 207) 892 80 68
♣ $ 18,000,000 (1995) 👥 62 🏠 60,000 sq ft
President: Richard Dyke
Director Marketing/Sales: Thomas Barr
Advertising/Public Relations Manager: Jonathan Clark
■ Manufacture of AR15/M16 type weapons and M203 grenade launcher. Also specialty weapons parts for national and international military, police and civilian markets. [98]

75-16009 B&W Engineering Corp.
Commerce Park, 3303 Harbor Blvd, E-1,
Costa Mesa, CA 92626
☎ (+1 714) 540 99 75
[W] http://www.b-w-engineering.com
⊖ (+1 714) 540 26 62
👥 7 (1997) 🏠 2,000 m2
President: Garry L Black
■ Manufacturers of Particle Impact Noise Detection (PIND) and mechanical shock test equipment for microelectronics and MIL-STD compliance. [98]

75-16016 C & R Systems Inc.
21 Standen Drive, Hamilton, OH 45015
☎ (+1 513) 860 15 44 ⊖ (+1 513) 860 15 56
Chief Executive Officer and Chief Financial Officer: Darrell Topmiller
Vice President Sales: Ron Nabors [E] ronnabors@juno.com
■ Manufacture electronic enclosure systems for commercial and military applications. Power ready systems built to MIL-SPEC 45208. [98]

75-16020 WPI-Boston Division Inc. Cable Systems/Mutron
(a division of WPI-Salem)
210 Broadway, Everett, MA 02149
☎ (+1 617) 389 70 80 ⊖ (+1 617) 389 47 58
♣ $ 5,000,000 👥 70
Manufacturing Director: Joseph Galli
Vice President and General Manager: E F Vecchia
Sales and Marketing Director: Darrell L Fernald
Material Manager: Eugene F Vicchia (Jr)
■ Manufacturer of engineered and build-to-print cable assemblies. Capabilities include cabling, shielding, jacketing, moulding and assembly. [98]

75-16210 Calgon Carbon Corp.
PO Box 717, Pittsburgh, PA 15230-0717
☎ (+1 412) 787 67 00 ⊘ (800) 4 CARBON
⊖ (+1 412) 787 63 24
♣ $ 291,898,000 👥 1,097
President: Colin Bailey
Senior Technical Sales Representative: Glenn Trump
■ Activated carbons for chemical warfare agents (Whetlerite, cooperite). Systems for air, groundwater & waste water treatment. [98]

75-16230 Calif Gasket & Rubber Corp.
1601 W. 134th St, PO Box 1037,
Gardena, CA 90249-0037
☎ (+1 310) 323 4250 ⊘ (800) 635 7084
⊖ (+1 310) 329 9721
♣ $ 2 800 000 👥 42
President: Scott Franklin
Sales Mgr: Ron Burger
■ Custom die cut gaskets; metal stamping, forming, shims, washers; custom molded rubber products, rubber to metal bonding, custom seals. [97]

75-16260 California Avi-Tron Corp.
(a subsidiary of Singapore Technologies Pte)
1973 Via Arado, Rancho Dominguez, CA 90220
☎ (+1 310) 886 88 00 ⊖ (+1 310) 762 14 93
👥 75 (1997) 🏠 100,000 sq ft
Chief Executive Officer: Ron A Lim
Commercial Director: David K Ruck
Purchasing Manager: Ms Jenny Faber
■ Manufacture precision sheetmetal riveted and welded assemblies for commercial and military aerospace. Major programmes included infrared suppressor housing for Cobra helicopter, cowling and engine mounts for CL-215 water bomber (Bombardier), plenums for a variety of APU engines, heat shields, mufflers and aircraft structures. FAA approved repair station. Approved for Boeing D1-9000 "A" and MIL-I-45208A requirements. Boeing licenced repair and overhaul centre for Boeing Aircraft Pneumatic Ducts. [98]

75-16280 California Instruments Corp.
9025 Balboa Avenue, San Diego, CA 92123-1509
☎ (+1 619) 279 86 20 ⊘ (800) 4AC-POWER
[E] sales@calinst.com [W] http://www.calinst.com
⊖ (+1 619) 279 81 39
👥 80 (1997) 🏠 31,000 sq ft
President: Brian Hull
Controller: R Bladel
Operations Director: G R Riggle
National Sales Manager: Gerald R Holland
Regional Sales Manager: D Shipman
Marketing Manager: J J Kope
Purchasing Manager: Paul Thobois
Product Marketing: H van Eijkelenburg
Marketing Administrator: Ms Janice N Olney
■ Manufacture AC power products. Primarily linear power converters (fixed and IEEE programmable) for single- and poly-phase applications, current sources, GPIB PC controllers and software. Precision products for regulatory testing (IEC 555, IEC1000, IEC1000-4-11, RTCA/DO-160C and MIL-STD-704D). CE marked to meet new European standards. [98]

75-16350 California Microwave Inc.
555 Twin Dolphin Dr, Ste 650, Redwood City, CA 94065
☎ (+1 415) 596 9000 ▣ 352016 ⊖ (+1 415) 596 6600
♣ $ 468 000 000 (1995) 👥 2 605 🏠 556 000 sq ft
Chairman, President & CEO: Philip F. Otto
President, Government Group: Gilbert F. Johnson
Sr VP Business Dev.: Dr Roger Engel
VP International Ops: Lt Gen. Charles Eichelberger (retd)
Staff VP Corp. Communications: Ms Stephanie M. Day
VP Business Development: Roger Engel
■ Supplies electronic surveillance systems and equipment to US Government, to domestic & foreign telephone carriers & to major corporations. Designs and implements microwave-based communications and communications and intelligence systems.
Subsidiary company:
Satellite Transmission Systems Inc.
Government Electronics Div.
PO Box 2800, Woodland Hills, CA 91365
☎ (+1 818) 992 8000 ⊖ (+1 818) 992 5079
President: Donald Kenins
VP Business Development: Lester Baney
Airborne Systems Integration Div.
1362 Brass Mill Road, POBox 358, Belcamp, MD 21017
☎ (+1 410) 272 2228 ⊖ (+1 410) 272 4245
President: William Shaver
VP Marketing/Advanced Dev.: Charles Angelo
■ Electronic and electro-optical surveillance systems.
Government Communications Systems
10820 Guilford Road, Ste 202,
Annapolis Junction, MD 20701
☎ (+1 410) 792 2225
[W] http://www.cmi-ged.com/corp.htm
⊖ (+1 410) 792 2407
VP & General Mgr: Nelson Ward
Business Development Dir: Terrence Hartley
■ Design, manufacture & install satellite communication earth stations & networks. Also manufacture a complete line of mobile ground communications equipment & small product earth stations. [97]

75-16480 Cambridge Parallel Processing (CPP)
(† Active Memory Technology Inc.)
16755 Von Karman Avenue, Suite 120, Irvine, CA 92606
☎(+1 714) 261 89 01 ✆ (800) 995 8879
[E] info@cppus.com [W] http://www.cppus.com
⊜ (+1 714) 261 88 02
👥 40 (worldwide)
President: Philip Buckley
Vice President of Manufacture: Gordon Parey
Vice President of Business Development: Bruce Alper
Vice President of Administration and Marketing:
Ms Michele M Herbel ☎(+1 714) 724 83 05; 724 83 99 (fax)
■ Manufacture of DAP (Distributed Array of Processors) computer systems. [98]

75-16630 Cannatech Inc.
15 Spring Hollow Drive, Erial, NJ 08081
☎(+1 609) 629 74 16 ⊜ (+1 609) 875 05 92
President: Martin F Concannon
Sales and Marketing Director: T Concannon
■ Test, inspection, measurement, calibration and alignment equipment. [98]

75-16642 Cannon-Muskegon Corp.
(a subsidiary of SPS Technologies Inc.)
2875 Lincoln Street, PO Box 506,
Muskegon, MI 49443-0506
☎(+1 616) 755 16 81 ⊜ (+1 616) 755 49 75
👥 130
President: Joseph Snowden
Vice President of Technology: Ken Harris
■ Develop and manufacture vacuum induction and AOD (Argon Oxygen Decarburization) refined nickel, cobalt and iron base alloys for aerospace, medical and other applications. [98]

75-16650 Capco Inc.
PO Box 1028, Grand Junction, CO 81502
☎(+1 970) 243 87 50 ⊜ (+1 970) 243 84 81
👥 100+ (1997) 🏭 48,000 sq ft
President: Stephen K Wood
Vice President and Marketing Manager: John A Younker
Engineering Manager: Eric Goertz
PC and Customer Service Manager: Forrest Stewart
■ Electronic, mechanical components and assemblies for military customer applications. Pyrotechnics. [98]

75-16700 Carco Electronics
195 Constitution Drive, Menlo Park, CA 94025
☎(+1 415) 321 81 74 ⊜ (+1 415) 321 18 90
♧ $ 4,000,000 (1996) ♣ $ 10,000,000 (1996)
👥 102 (1997) 🏭 4,600 sq m
President and Chief Executive Officer: John M Carter
President and Chief Operating Officer: Hugh D Carter
Treasurer: William R Meckfessel
Purchasing Manager: Clay E Falkner
■ 3-axis flight simulators, microwave radar and radome boresight systems, rate tables, air bearings, radar pattern range, test equipment, lab. model 2-axis, linear positioners. [98]

75-16720 Carey Industries Inc.
PO Box 620, Danbury, CT 06813
☎(+1 203) 744 72 80 [E] carey@usa.nai.net ⊠ 969642
⊜ (+1 203) 791 27 62
President: R J Carey
Director of Operations: R A Baker
■ Organic dyes & compositions for coloured pyrotechnic smoke devices including grenades, artillery shells, rockets & aircraft signals. Dyes for military camouflage textiles.
Sales & Distribution Centre:
Charlotte, NC [98]

75-16750 Carleton Technologies Inc.
(a Cobham plc company)
10 Cobham Drive, Orchard Park, NY 14127
☎(+1 716) 622 00 06 ⊠ TWX 710 2641 446
⊜ (+1 716) 662 07 47
♣ $ 45,000,000 (1996)
👥 252 (including subsidiaries 1996) 🏭 130,000 sq ft
President: Paddy Cawdrey
Director: G F Page
Director Finance: D Vogel
Marketing Director: Graham Hodge
Director: J Buck
Director: G Yarborough
■ Design & manufacture of high technology pneumatic pressurisation systems & components, used in a large number of strategic & tactical missiles. Life support systems & equipment for airborne, spaceborne and undersea applications.
Associated companies:
Carleton Technologies Inc, Florida, USA
Carleton Life Support Technologies, Canada [98]

75-16753 Carlin Manufacturing Inc.
(a subs. of TCBY Enterprises Inc.)
3714 N.Valentine, Fresno, CA 93722
☎(+1 209) 276 0123 ⊜ (+1 209) 222 1538
♣ $ 10 000 000 👥 70
President: Ralph H. Goldbeck
■ Mobile kitchens, field feeding systems, communications vehicles, command vehicles, mobile shower units and latrines. [97]

75-16760 Carolina Metals Inc.
(a subs. of Nuclear Metals Inc.)
PO Box 1366, Barnwell, SC 29812
☎(+1 803) 259 2321 ⊜ (+1 803) 259 3622
👥 75 🏭 78 000 sq ft
President: Sherman D. Brady
Sales Mgr: John Nicholson
■ Converts UF6 (gas) UF4 (powder) and reduces UF4 to depleted uranium metal. [97]

75-16800 Carter Manufacturing Corp.
237 Sugar Road, Bolton, MA 01740-1092
☎(+1 508) 779 55 01 ⊜ (+1 508) 779 61 66
♣ $ 2,000,000 (1997) 👥 30 (1997) 🏭 6,000 sq ft
President: Orin Bowlby
VP Sales: Leo Nichols
■ Avionics - precision variable resistors. Missile systems - precision variable resistors and assemblies. Radar - precision variable resistors. [98]

75-16850 Carleton Technologies Inc.
3910 Riga Boulevard, Tampa, FL 33619-1344
☎(+1 813) 623 3711 ⊜ (+1 813) 623 5373
♣ $ 50 000 000 👥 300
President: T.C. Cannon
VP Sales & Marketing: Lee Walters
■ Underwater breathing equipment. [97]

75-16910 Cartwright Electronics Inc.
(a subsidiary of Meggitt PLC, UK)
655 West Valencia Drive, Fullerton, CA 92632-2104
☎(+1 714) 525 23 00 ⊜ (+1 714) 525 94 61
♣ $ 16,000,000 👥 100
President: B L Leathers
Vice President of Business Development: Robert A Bradley
■ Electronic scoring systems and electronic warfare equipment for aircraft protection. Systems are based on pulse doppler radar techniques. Complete systems capability includes expertise in antennas, microwave technology, radar and automatic data processing. [98]

75-16950 Case Corporation
(† J.I. Case Co.)
700 State Street, Racine, WI 53404
☎(+1 414) 636 6011 ⊜ (+1 414) 636 7809
Public Relations Mgr: Rogher A. Zuehl
■ Materials handling equipment. [97]

75-16990 Caswell International Corp.
1221 Marshall Street North East, Minneapolis, MN 55413
☎(+1 612) 379 20 00 [E] caswellintl@internetmci.com
[W] http://www.caswellintl.com ⊜ (+1 612) 379 23 67
👥 55 (1996)
President and Chief Executive Officer: Paul T Faust
Managing Director: Douglas Jones
Operations Director: John Roy
Marketing Director: William C Brogdon
Sales and Marketing Director: Don Addy
Vice President Military Marketing: Jeff Nelson
Purchasing Manager: Gary Hollenbeck
■ Indoor & outdoor shooting range equipment & systems. Target mechanisms for all direct fire weapons from pistols to tank main guns. Mobile shooting ranges, live & laser fire video shooting equipment. [98]

75-17100 Caterpillar Inc. Defense & Federal Products
PO Box 470, Peoria, IL 61629
☎(+1 309) 578 65 02; 578 30 89; 578 28 37
⊠ 404435; 4330283; 210044; 6871139
⊜ (+1 309) 578 37 39
♣ $ 16,522,000 000 (1996) 👥 57,026
Chairman and Chief Executive Officer: Donald V Fites
Manager Engine Marketing Government Sales:
Sib Sciabica
Defense and Federal Products: Frank M Winberg
Foreign Military Sales: Jim McCormack
■ Earthmoving, construction and materials handling machinery including tractors, loaders, lift trucks, excavators and dump trucks. Diesel engines and turbines. Paving products, agricultural tractors and forest machines. [98]

75-17180 CC&L
717 Plantation St, Worcester, MA 01605
☎(+1 508) 852 3072 ⊜ (+1 508) 582 3759
Director: Dave Devine
■ Release films, graphic arts media & custom coating & laminating products & services. [97]

75-17250 CCS International Ltd
360 Madison Avenue, New York, NY 10017
☎(+1 212) 557 30 40 [E] ccsnychq@aol.com
[W] http://www.spyzone.com ⊜ (+1 212) 983 12 78
Advertising Director: Ms Marsha Pearl
■ Manufacture electronic security and counter intelligence equipment: bug detectors, wiretap detectors, telephone scramblers, night vision viewers, lie detection, bulletproof apparel and vehicles. Consulting services and electronic sweep service to ensure privacy on premises. → 80

New York
444 Madison Avenue, New York, NY 10022
☎(+1 212) 688 85 00 [E] ccs444@aol.com
⊜ (+1 212) 888 64 60

Washington, DC
1027 Connecticut Avenue NW, Washington, DC 20036
☎(+1 202) 887 17 17 [E] ccswashdc@aol.com
⊜ (+1 202) 887 18 18

Miami
1435 Brickell Ave, Miami, FL 33131
☎(+1 305) 358 43 36 [E] ccsfla@aol.com
⊜ (+1 305) 358 48 05

Los Angeles
9557 Wilshire Blvd, Los Angeles, CA 90212
☎(+1 310) 274 77 00 [E] ccsbh@aol.com
⊜ (+1 310) 274 02 12 [98]

75-17280 Celesco Transducer Products Inc.
(a subsidiary of Tedea-Huntleigh, Inc.)
7800 Deering Avenue, Canoga Park, CA 91304
☎(+1 818) 884 68 60 ✆ 800 423 54 83
[E] info@celesco.com [W] http://www.celesco.com
⊠ 4998157 celesco cnp ⊜ (+1 818) 340 11 75
♣ $ 5,000,000 (1996) 👥 50+ (1996) 🏭 7,000 m2
Managing Director: Michael Katz
General Manager, Sales and Marketing Director:
Jim Bishop
Operations Director: Pini Wolman
Purchasing Manager: Pat Weber
Sales Supervisor: Ms Cyndi Stevens
Engineering: Quenton Olsen
■ Design and manufacture of differential variable reluctance pressure transducers for missiles and rockets. Position and velocity transducers, linear to rotational measurement modules. [98]

75-17290 Central Engineering Company
2930 Anthony Lane, Minneapolis, MN 55418-3275
☎(+1 612) 781 6557 ⊜ (+1 612) 789 6417
👥 130
President: Loren Swanson
Sr VP Marketing & Sales: Graham Ball
Marketing Mgr: Steve Biessener
■ Aircraft engine test facilities & equipment. [97]

75-17385 Century Electronics
Perillo Industries Inc.
5701 Lindero Canyon Road 1-100,
Westlake Village, CA 91362
☎(+1 818) 706 82 24 ⊜ (+1 818) 706 82 26
👥 86 (1997) 🏭 23,500 sq ft

President: Ms Mary E Perillo
Executive Vice President and General Manager: Joseph J Perillo
Director of Public Relations: Ms Donna Bukowski
Contracts Manager: Lloyd Stephens III
Chief Engineer: David Ahronowitz
■ Design and manufacture high reliability power supplies. [98]

75-17397 Ceramic Devices Inc.
(a subs. of PCT Holding)
434 Olds Station Rd, Wenatchee, WA 98801
☎ (+1 509) 665 6500 ⊖ (+1 509) 664 2373
♣ $ 2 500 000 (1995) 👥 42 (1996) ⌂ 1 500 sq ft
President: Ivan G. Sarda
Sales Mgr: Donna Gardner
■ Small military-qualified business. Manufacture EMI/RFI ceramic filters & capacitors. MIL-F 28861, Class "S" & "B". [97]

75-17400 Cercom Inc.
1960 Watson Way, Vista, CA 92083
☎ (+1 760) 727 62 00 [E] cercom@ix.netcom.com
⊖ (+1 760) 727 62 09
♣ $ 5,000,000 (1996) 👥 54 ⌂ 3,000 sq m
President: Richard Palicka
Operations and Technical Director: Jack Rubin
Purchasing Manager: Barbara Reyna
■ Producer of lightweight high-performance armour ceramic material for use in armour systems. Materials are silicon carbide, boron carbide, titanium diboride and special aluminium oxide. Ceramic bearing blanks and high performance high-temperature structural silicon nitride components are supplied on a custom basis to the aerospace and defence industries. Optical quality silicon carbide. [98]

75-17750 Cessna Aircraft Company
(a subsidiary of Textron Inc.)
PO Box 7704, Wichita, KS 67277-7706
One Cessna Boulevard, Wichita, KS 67215-8710
☎ (+1 316) 941 64 88
[W] http://www.cessna.textron.com/
⊖ (+1 316) 941 78 12
👥 7,500
Chairman: Russell W Meyer (Jr)
President and Chief Operating Officer: David G Assard
Vice Chairman: Gary W Hay
Senior Vice President of Advanced Programs: Bruce E Peterman
Executive Vice President of Operations: Charles B Johnson
Vice President of Finance: Michael Shonka
Vice President of Public Relations: David M Franson
Vice President of Marketing: Phil Michel
Director of Caravan Sales: Robert Conover
Director, Corporate Communications: Ms Jan McIntire
Sales Manager, Western Division: Ernest Edwards
■ Manufacture business jet and utility turboprop aircraft. Manufacture aircraft parts and other aircraft accessories, aviation training systems.
Distributor::
Reims Aviation SA, France
Citation Marketing
Box 7706, Wichita, KS 67277
☎ (+1 316) 941 7000 ⊠ 417342 ⊖ (+1 316) 941 6640
Sr Vice President: Gary W. Hay
Turboprop Aircraft Marketing
(† Propeller Aircraft Marketing)
PO Box 7704, Wichita, KS 67277
☎ (+1 316) 946 6000 ⊠ 4319020
⊖ (+1 316) 941 7250
Director: Robert Conover [98]

75-17950 Chamberlain Manufacturing Corp.
(a subsidiary of Duchossois Industries)
Corporate Headquarters
845 Larch Avenue, Elmhurst, IL 60126-1196
☎ (+1 708) 279 36 00 ⊠ 721476 ⊖ (+1 708) 530 60 57
♣ $ 80,000,000 (1995) 👥 700
President: Merton L. Townsend
Vice President: James R. Vieregg
■ Manufactures projectiles for howitzers, mortars, naval guns. Also produces automatic weapons & commercial small arms.
Subs. with defence business:
Saco Defense Inc. [98]

75-17975 Champion Friction Co.
PO Box 1459, Eugene, OR 97440
☎ (+1 503) 687 8301 ✆ (800) 547 6180 ⊠ 364412 intr
⊖ (+1 503) 344 0104
President: William DeChent
Sales Mgr: Tracy A. Combs
■ Woven brake pads, replacements for all authorized linings as used on M113 armoured personnel carriers & M548 cargo carriers. Caliper brake pads & parts for industrial brakes & clutches, industrial friction material formed, fabricated & manufactured. [97]

75-17980 Champion Spark Plug Co. Aviation Products
(a div. of Cooper Industries)
1230 Old Norris Road, Liberty, SC 29657
☎ (+1 864) 843 5365 ⊠ 415190 champion
⊖ (+1 864) 843 5414
VP & General Mgr: R.W. Matz
VP Sales & Marketing: S.R. Fletcher
Operations/Technical Dir: Steve Straub
Purchasing Mgr: Clyde Heddem
■ Aviation spark plugs, turbine igniters, leads, exciters, oil filters, spark plug cleaning & testing equipment & service tools. Power converters, cockpit displays. [97]

75-18050 Channel Industries Inc. (CII)
839 Ward Drive, Santa Barbara, CA 93111
☎ (+1 805) 967 01 71 ⊖ (+1 805) 683 34 20
President: Robert M Callahan
Marketing Manager: Edward A Bickel
■ Piezoelectric ceramics (lead zirconate/lead titanate and barium titanate) for ultrasonic and underwater applications. [98]

75-18065 Chapy Corporation
Box 601, Sierra Madre, CA 91025-0601
☎ (+1 818) 358 29 29 [E] chappy@earthlink.net
⊖ (+1 818) 358 29 29
♣ $ 470,000 (1995) 👥 9 (1995) ⌂ 2,000 sq m
President: Gordon Chappelle
Chief Executive Officer: Ms Betty Sue Ramsey
Vice President of Sales and Marketing: Ms Priscilla Madox
■ Develop and manufacture robotic aircraft for (RPV-remote piloted vehicle, drones) commercial or military missions.
Distributor: Robotic aircraft systems. Services: Commercial aerial photo and video surveying; low-cost. [98]

75-18100 Checkmate Industries of USA Ltd
17 East 48th St, New York, NY 10017
☎ (+1 212) 750 5177 ⊖ (+1 212) 750 5179
👥 10
President: Vartan Keshishian
Sales & Marketing Dir: M. Haladjian
■ Military & police uniforms, footwear, webbing equipment, protective bulletproof vests, protective bulletproof helmets, transport equipment & vehicles, antiriot equipment, uniforms, military tents & camouflage equipment. → 85 [97]

75-18120 Chelton Inc.
(a subsidiary of Chelton Group Company, UK)
698 Metro Park Circle, PO Box 976, Lewisville, TX 75067
☎ (+1 972) 221 17 83 [E] info@cheltoninc.com
⊖ (+1 972) 436 27 16
President: Michael S Wilkenson
Vice President of Marketing: Jeffery A Salata
■ Aircraft static dischargers. Aircraft antenna systems. Aircraft audio systems. Direction finding and homing equipment.
® Canada: Northern Airborne Technology Ltd
France: Chelton/Rayan
UK: Chelton Electrostatics Ltd [98]

75-18200 Chicago Bullet Proof Systems
(† Chicago Bullet Proof Equipment Co.)
2250 Western Avenue, Park Forest, IL 60466-1298
☎ (+1 708) 481 34 00 ⊖ (+1 708) 481 79 22
President: Francis J. Sazama
Sales & Marketing Mgr: Ms Susan A. Sazama
■ Design and manufacture bullet/attack resisting architectural products. [98]

75-18275 Chomerics
(a division of Parker Hannifin Corp.)
77 Dragon Court, Woburn, MA 01888
☎ (+1 617) 935 48 50 [E] chomerics@aol.com
[W] http://www.chomerics.com ⊖ (+1 617) 933 43 18
👥 500
General Manager: Akbar Naderi
Sales Manager: Joe Travers
Operations Manager: John Pierog
Purchasing Manager: Ms Julie Farrington
Marketing Communications Coordinator: Ms Martha J Curran
■ Develops, manufactures & sells materials formulated or engineered to solve electronic packaging problems. Typical applications are electro-magnetic interference shielding, grounding, insulating, heat dissipating, bonding, sealing, static protection, switching & interconnecting.
Divisions:
Chomerics Europe, UK
Parker Hannifin PLC [98]

75-18360 Chromalloy Gas Turbine Corp.
(a subsidiary of Sequa Corp.)
4430 Director Drive, PO Box 200150, San Antonio, TX 78220
☎ (+1 512) 359 55 70 ⊖ (+1 512) 359 55 70
Chairman and Chief Executive Officer: Dr Martin Weinstein
■ Coating, repair and manufacturing of blades, vanes and other components for jet aircraft engines. [98]

75-18425 Crystallume
(a div. of EDI)
125 Constitution Dr., Menlo Park, CA 94025-1118
☎ (+1 408) 324 9681 ⊖ (+1 408) 653 1710
👥 30
President: Don McGuinness
VP Sales & Marketing: Ken Buckley
Research Dir & General Mgr: Dr John Herb
Advertising Dir: Ms Liz Marsolais
■ Diamond coatings using the chemical vapour deposition process (CVD). Products include diamond coated carbide cutting tools for machining non-ferrous materials. Development of diamond coated ball bearings, diamond radiation detectors and transistors. [97]

75-18480 Churchill Corp. (CC)
344 Franklin Street, Box G, Melrose, MA 02176
☎ (+1 617) 665 4700 ✆ (800) 225 3932
⊖ (+1 617) 662 5291
♧ $ 250 000 000 ♣ $ 2 000 000 (1995) 👥 20 (1995)
President & General Mgr: M.W. Schermerhorn
■ Airborne electronic cases, Mil-C-172C, Arinc 404 and 600. [97]

75-18570 CIC International Ltd
38-01 23rd Avenue, Astoria, NY 11105
☎ (+1 718) 204 09 00 ⊠ COMMINTAL ⊠ 237593
⊖ (+1 718) 728 76 63
♧ $ 31,250,000 ♣ $ 423,677,509 (1996)
👥 252 (1996) 👥 252 ⌂ 118,240 sq ft
Chairman: Satiris G. Fassoulis
President: Michael Kane
Exec. Vice President: Dino A. Seferian
Vice President Purchasing: Robert C. Thompson
Vice President Operations: Robert Volvo
Vice President Sales & Marketing: Robert C. Perry
Vice President: James Chladek
Vice President: W.T.E. Finan
Operations Director: Joseph Ceva
Advertising: J. Gordon
■ Supplier of defence products to various defence establishments & munitions manufacturers. Principal products include a full range of shells and ammunition, howitzers, motor gun carriages & military vehicles. Maintains spare parts inventories for military vehicles, performs tank retrofits & acts as a consultant in modernizing military vehicles. → 80 → 85
Affiliated Companies:
CIC Iberia SA
CIC Task Ltd
CIC Canada (Reg.)
Global Defense Products Inc. → 75
Darien Office
989 Post Road, Darien, CT 06820
☎ (+1 718) 204 10 10 ⊠ TWX 710 5814 409
President: Michael Kane
■ Financial and general administrative operations.
Astoria Office
Warehouse and R&D

38-01 23rd Avenue, Astoria, NY 11105
☎ (+1 718) 204 10 10
Asst Vice President: Lawrence Spitz
■ Research & development of military electronics & military steels. Maintains military vehicular spare parts. [98]

75-18580 Cicoil Corp.
(a subs. of The New Cicoil Corp./DBA)
24960 Ave Tibbitts, Valencia, CA 91355-3428
☎ (+1 805) 295 1295 ⊖ (+1 805) 295 0813
♣ $ 3 500 000 👥 77 🏭 14 500 sq ft
President: Peter G. Nogradi
Vice President: Thomas A. Nogradi
National Sales & Marketing Mgr: James Alderson
Operations Mgr: Ed Haddad
Manufacturing Mgr: Nell Rider
Chief Engineer: Lew Fluhrer
Quality Assurance: Fran Robbins
Admin Mgr: Ms Fran Kirkendoll
■ Extremely flexible silicone cable & harness assemblies. Used in dynamic applications where bending movements & torque are applied. Meets military & NASA outgassing & test requirements. [97]

75-18590 Jonathan Arthur Ciener Inc.
8700 Commerce St, Cape Canaveral, FL 32920
☎ (+1 407) 868 2200 ⊖ (+1 407) 868 2201
♧ $ 500 000 ♣ $ 500 000 👥 8 🏭 30 000 sq ft
Managing Dir: Jonathan Arthur Ciener
Technical Dir: Bill Reynods
Marketing Dir: Ms Emily Heath
Purchasing Mgr: Peter Zemelka
■ 22LR conversion kits for: Colt AR15/M16, Ruger Mini-14 & AC-556 AK Rifles (Kalashnikov design) 7.62x39 and/or 5.56mm (.223 Rem.), Colt M1911A1 .45 automatic pistol, Beretta 92 Pistol, Taurus PT92 pistol.
Firearm noise suppressors & suppressed firearms. Belt fed conversion for Colt conversion for Colt AR15/M16. M203 40mm & 37mm Grenade Launchers. Over/Under Remington 870 shotgun to mount to Colt AR15/M16. Prototype firearm development. [97]

75-18640 Cincinnati Electronics Corp.
(a subsidiary of Canadian Marconi Co.)
7500 Innovation Way, Cincinnati, OH 45040-9699
☎ (+1 513) 573 61 00 ⊖ (+1 513) 573 62 90
Chairman, President and Chief Executive Officer: J Wimmers
■ Designs, manufactures, markets and supports communications systems and equipment; space electronics; infra-red systems & detectors; radar systems; electronic warfare systems & equipment; high power transmitters; and hybrid microcircuits. In addition Cincinnati Electronics Corp. is engaged in integrated logistics support; contract production (built-to-print); and environment and metrology services. [98]

75-18675 The Cincinnati Gear Co. (CINTI)
5657 Wooster Pike, Cincinnati, OH 45227
☎ (+1 513) 271 7700 ⊖ (+1 513) 271 0049
👥 320 🏭 350 472 sq ft
Chairman: Walter L. Rye
President: Frank J. Posinski
Advanced Programme Sales Mgr: Ken M. Kiehl
Media Co-ordinator: Ms Jo Lynn Robinson
■ Marine propulsion systems to handle 50,000 hp. Larger requirements can be accommodated. Epicyclic and parallel shaft high precision gear drive for diesel, gas turbine and electric motor applications. [97]

75-18680 Circle Seal Controls Inc.
(a subsidiary of Watts Industries Inc.)
2301 Wardlow Circle, PO Box 3300, Corona, CA 91718
2301 Wardlow Circle, Corona, CA 91720
☎ (+1 909) 270 62 00 [E] mktgsvcs@circle-seal.com
[W] http://www.circle-seal.com ⊖ (+1 909) 270 62 01
👥 244 🏭 105,000 sq ft
Marketing Services Manager: Ms Ann S Grider
Marketing Services Co-ordinator: Ms Jeanne Scholz
■ Products range from check, relief, shutoff, solenoid, electromechanical valves, gauges and regulators to integrated manifolds and complete fluid control systems. [98]

75-18685 CI Systems Inc
30961 West Agour Road, Suite 109, Westlake Village, CA 91361
☎ (+1 818) 865 04 02 ⊖ (+1 818) 865 04 03
🏭 500 sq m
Chief Executive Officer: E F Tzafrir
■ Manufacture blackbodies, collimators, imaging spectroradiometers, simulators, boresight systems and other turnkey electro-optical test systems. [98]

75-18850 David Clark Co. Inc.
360 Franklin Street, PO Box 15054, Worcester, MA 01615-0054
☎ (+1 508) 756 62 16 [E] sales@davidclark.com
[W] http://www.davidclark.com ⊖ (+1 508) 753 58 27
👥 305 (1997)
Vice President Sales and Marketing: Robert J Murphy
Product Manager Government: Dennis Buzzell
Purchasing Manager: Gerald Leach
■ Communications systems headsets, hearing protectors. [98]

75-19000 Cla-Val Co.
Automatic Control Valves
PO Box 1325, Newport Beach, CA 92659-0325
☎ (+1 714) 722 4826 ✆ (800) 942 6326 ⊠ 678429
⊖ (+1 714) 548 5453
👥 450 🏭 250 000 sq ft
Government Sales Mgr: Mike Uffer [97]

75-19040 Cleveland Crystals Inc.
19306 Redwood Avenue, Cleveland, OH 44110-2738
☎ (+1 216) 486 61 00 [E] sales@clevelandcrystals.com
[W] http://www.clevelandcrystals.com
⊖ (+1 216) 486 61 03
👥 47
President and Chief Executive Officer: Lee R Shiozawa
Sales Manager: Philip D Marlowe
■ Manufacture crystals and devices for optical, electro-optical and solid-state applications including harmonic generators, pockells cells, wave retarders, parametric oscillators, laser attenuators, polarizers. [98]

75-19170 CMS Inc.
(† Conventional Munitions Systems Inc.)
(a subsidiary of Daimler Benz AG, Germany)
4904 Eisenhower Boulevard, Suite 310, Tampa, FL 33634
☎ (+1 813) 882 44 77 [E] cms_inc@msn.com
⊖ (+1 813) 884 18 76
♣ $ 80,000,000 (1996) 👥 600 (1996)
President and Chief Executive Officer: Frank P Ragano
Vice President of Business Development: Edward J White
Director of Marketing and Communication: Bill Lewis
Director of Environmental Business Development: John Chionchio
■ Structured to develop, test and produce munition products as a team member with with major US prime contractors. Specialise in advanced technology for conventional warheads and associated delivery mechanisms. Principal product activity includes; advanced warheads, dragon anti-tank system and dispensers for aircraft weapons delivery. Modelling and simulation. Environmental clean-up of hazardous sites containing explosives. [98]

75-19195 Coastcom
1151 Harbor Bay Parkway, Alameda, CA 94502
☎ (+1 510) 523 60 00 ✆ (800) 433 34 33
[E] info@coastcom.com ⊖ (+1 510) 523 61 50
👥 100
President: E M Buttner
Executive Vice President: Ms P Diane Frank
Sales Director: John K Collins
Marketing Services Manager: Frank Falcon
Marketing and Communications: Ms Joan Black
■ Specialise in TI and fractional TI equipment for small to medium sized TI networks. Develop and manufacture TI multiplexers. [98]

75-19400 Cohu Inc.
Electronics Division
5755 Kearny Villa Road, San Diego, CA 92123-1111
☎ (+1 619) 277 67 00 [W] http://www.cohu.com/cctv
⊖ (+1 619) 277 02 21
♣ $ 25,000,000 (1997) 👥 180 (1997)
President and Chief Executive Officer: C A Schwan
President: J M Brown
International Sales Manager: Vijay Rana
US Sales Manager: Joe Olmstead (Jr)
Marketing Manager: John R Foster
Marketing and Communications Specialist: Ron Rohe
■ Closed-circuit, airborne and surveillance television systems. [98]

75-19420 Cole Instrument Corp.
2650 South Croddy Way, Santa Ana, CA 92704-5238
☎ (+1 714) 556 31 00 [E] switch@earthlink.net
[W] http://www.cole-switches.com ⊖ (+1 714) 241 90 61
♣ $ 3,000,000 (1997) 👥 100 (approx. 1997)
🏭 15,000 sq ft
President: Ricardo L Garcia
Director of Marketing and Sales: James N Batliner
Purchasing Manager: Peter Nagle
■ Manufacture precision rotary switches. [98]

75-19730 Colorado Video Inc.
PO Box 928, Boulder, CO 80306
6235 Lookout Rd., Boulder, CO 80301
☎ (+1 303) 530 9580 ⊖ (+1 303) 530 9569
Chairman: Glen Southworth
President: Robert Cornilsen
■ Video telecommunications equipment used for transmission of still video images over a narrow communications channel. Freeze-frame television systems, use standard or data telephone lines, microwave, FM, VBI, TV or satellite transponder subcarriers. Also manufacture time division video multiplexers and remote monitoring equipment. [97]

75-19850 Columbus Jack Corp.
1000 South Front St, Columbus, OH 43206
☎ (+1 614) 443 7492 ⊠ 245472 ⊖ (+1 614) 444 9337
👥 40 🏭 40 000 sq ft
President: Dennis B. Mellman
Vice President: Ms Karen Hart
VP Sales: Steven H. Stout
■ Aircraft jacks, towbars, tyre bead breakers. [97]

75-19860 Comant Industries Inc
(a Chelton Group Company)
12920 Park Street, Santa Fé Springs, CA 90670
☎ (+1 310) 946 66 90 ⊖ (+1 310) 946 51 50
👥 70
Chairman: Michael S Wilkerson
President: J Stites
Vice President Marketing: J Lerma
■ General aviation antennas & GPS antennas. [98]

75-19893 Combined Systems Inc.
226 Newtown Road, Plainview, NY 11803
☎ (+1 516) 777 7888 ⊖ (+1 516) 777 7898
👥 25
President: Jacob Kravel
VP Sales & Marketing: Mike Brunn
International Marketing Dir: Jack Hananya
■ Manufacture primers, fuzes & components for HEAT & HEPT tank ammunition, M72 LAWs & 2.75in rockets. Also flares, simulators, training services, booby traps propelling charges & CN/CS/smoke ammunition. [97]

75-20050 Commercial Intertech Corp.
(† Commercial Shearing Inc.)
Group Headquarters
1775 Logan Ave, PO Box 239, Youngstown, OH 44501-0239
☎ (+1 330) 746 8011 ⊖ (+1 330) 746 1148
♣ $ 622 000 000 (1995) 👥 4 572
Eastern Sales Mgr Mil. & Ind. Products: Ms M. Sauline
Advertising Mgr: James Cartwright
■ Broad range of hydraulic pumps, motors, control valves & cylinders & hydrostatic steer transmissions. Also medium to heavy custom stampings, & pre-engineered steel buildings.
Subs. with defense business:
Oildyne Inc. [97]

75-20290 Communications-Applied Tech.
11250-14 Roger BaconDrive, Reston, VA 22090-5202
☎ (+1 703) 481 0068 ✆ (800) 229 3925
⊖ (+1 703) 471 4428
♧ $ 750 000 000 (1995) ♣ $ 1 000 000 (1995) 👥 10
🏭 200 sq m
President: G. Seth Leyman
Sales Dir: Seth Leyman
Purchasing Mgr: Mandy Thai
Operation Mgr: Sandra Logan
■ Manufacture very low power, duplex, voice radio systems (wireless ICS for aircraft). [97]

75-20300 Communications Instruments Inc. (CII)
(a division of CII Technologies)
Highway 74 East, PO Box 520, Fairview, NC 28730
☎ (+1 704) 628 17 11 Ⓔ info@ciidiv.com
⊖ (+1 704) 628 14 39
♣ $ 35,000,000 (1997) 450 (1997) 95,000 sq ft
President: Mike Steinback
Vice President of Operations: Ted Anderson
Vice President of Sales and Marketing: Raymond McClinton
Purchasing Manager: Gary Mehling
■ Military grade relays for low level switching to 10 amps; hermetically sealed; MIL-STD-5757, MIL-STD-39016, MIL-STD-6106, MIL-STD-28750 and MIL-STD-28776. 98

75-20393 Communications & Power Industries
(† Varian Associates Inc.)
607 Hansen Way, PO Box 51110, Palo Alto, CA 94303
☎ (+1 415) 846 29 00 ⊖ (+1 415) 846 32 76
♣ $ 120,000,000 1,800
Chief Executive Officer: Al D Wilunowski
Chief Finacial Officer: Ms Lynn E Harvey
Vice President of Western Hemesphere Sales: John R Beighley

Beverly Microwave Products Division
(† Varian Crossed Field)
150 Sohier Road, Beverly, MA 01915-5595
☎ (+1 508) 922 60 00 940795 bluy varas
⊖ (+1 508) 927 80 61
President: Dennis Gleason
■ Magnetrons, crossed field amplifiers, ring-loop TWTs, transmitters and receiver protectors for military radar, EW, data transmission and communication applications.

Satcom Division
811 Hansen Way, Palo Alto, CA 94303
☎ (+1 415) 846 37 00 ⊖ (+1 415) 424 17 44
President: Jim Commendatore
Marketing Manager: Frank Blaha

EIMAC Division
(† Varian Power Grid Tube Products)
301 Industrial Way, San Carlos, CA 94070
☎ (+1 415) 592 12 21 ⊖ (+1 415) 594 41 88
President: Fred Koehler
Marketing Manager: Mark Kushner
■ Power grid tubes and ancillary hardware, klystrode tubes, ceramic and metal products, cavity subsystems.

Microwave Power Tube Division
(† Varian Microwave Power Tube Products)
811 Hansen Way, Palo Alto, CA 94303-1031
☎ (+1 415) 846 28 00 ⊖ (+1 415) 846 30 97
Marketing Manager: Mike Cheny
■ Coupled cavity tubes designed to meet customer specifications for land-based, shipboard and airborne environments. Klystrons, gyrotrons, microwave components and systems, electromagnets.

Travelling Wave Tube Division
607 Hansen Way, Palo Alto, CA 94304-1110
☎ (+1 415) 846 29 00 ⊖ (+1 415) 846 83 47
President: Al Ferreira
Marketing Manager: Tom Webb
■ High and low power pulse and CW broadband helix TWTs and transmitters used in EW, communications, radar and instrumentation systems operating in airborne, shipboard and ground environments. 98

75-20400 Compac
1320-13 Lincoln Avenue, Holbrook, NY 11741
☎ (+1 516) 585 34 00 ⊖ (+1 516) 585 35 34
30 (1997) 8,000 m2
Vice President: Charles F Groves
Operations Director: Peter Rad
Sales and Marketing Director: Ms Sandy Barthman
Purchasing Manager: V Valenti
Corporate Marketing Manager: Wally Valenti
■ RFI/EMI/EMC shielded enclosures, board mounting systems RF transfer switches. 98

75-20650 Comptek Federal Systems Inc.
(a subs. of Comptek Research Inc.)
2732 Transit Rd, Buffalo, NY 14224
☎ (+1 716) 677 4070 ⊖ (+1 716) 677 0014
♣ $ 70 000 000 (1995) 650
President & CEO: John J. Sciuto
Sr Vice President Business Development: Joseph F. Sciuto
■ Design, development, test & manufacture a variety of high technology EW (Signal processing) & C3I systems & products for both the operational & training environments. EW training systems & Signal environment simulators/stimulators. → 80
● Two Crystal Park, Suite 700, Arlington, VA 22202
☎ (+1 703) 979 6200 ⊖ (+1 703) 979 6082
● 5161 Verdugo Way, Camarillo, CA 93012
☎ (+1 805) 987 9739 ⊖ (+1 805) 388 2424
● 1025 Briggs Rd, Suite 148, Mt. Laurel, NJ 08054
☎ (+1 609) 778 1616 ⊖ (+1 609) 778 7305
● Trailer Complex TR-101, PO Box 42-125, Point Mugu, CA 93044
☎ (+1 805) 986 5255 ⊖ (+1 805) 488 1276
● 5327 Windsor Drive, PO Box 150, Dahlgren, VA 22448
☎ (+1 703) 663 0900 ⊖ (+1 703) 663 2289
● 44 Castillian Drive, Goleta, CA 93117-3027
☎ (+1 805) 685 4571 ⊖ (+1 805) 685 7853
● 4849 Viewridge Avenue, San Diego, CA 92123-1667
☎ (+1 619) 560 4767 ⊖ (+1 619) 560 7133
● Park West, 2877 Guardian Lane, Suite 200, Virginia Beach, VA 23452
☎ (+1 804) 463 8500 ⊖ (+1 804) 463 5675 97

75-20680 Computer Optics Inc.
120 Derry Road, PO Box 7, Hudson, NH 03051
☎ (+1 603) 889 21 16 Ⓔ coi@jlc.net
⊖ (+1 603) 889 23 93
20 20,000 sq ft
President: Gordon Kane
Vice President and General Manager Marketing: Edward A Wollensak
Operations Director: Dr Jonathan Kane
Purchasing Manager: J B Smith
■ Precision optics - components to systems. Terminal guidance HUD, cockpit simulator, tank optics, TV & systems, visible through infrared spectrum in both zoom & fixed focal length lenses. 98

75-20720 Computer Sciences Corp. Corporate Headquarters (CSC)
2100 East Grand Avenue, El Segundo, CA 90245
☎ (+1 310) 615 03 11 TWX 910 3486 286
⊖ (+1 310) 322 98 05
♧ $ 525,314,000 (1997)
♣ $ 5,616,048,000 (1996/97) 40,980
Chairman, President and Chief Executive Officer: Van B Honeycutt
Vice President, General Counsel and Secretary: Howard D Fisk
Vice President and Deputy General Counsel: Harvey N Bernstein
Vice President and Chief Financial Officer: Leon J Level
Vice President, Corporate Development: Lawrence Parkus
Vice President, Corporate and Marketing Communications: C Bruce Plowman
Director of Investor Relations: Spencer Davis
■ Design and development of system software and application programs, system engineering, facilities management, systems integration, remote computing services and communication systems.

Systems Group Headquarters
3170 Fairview Park Drive, Falls Church, VA 22042
Group President: Milton E Cooper

Applied Technology Division
6565 Arlington Boulevard, Falls Church, VA 22046
☎ (+1 703) 237 20 00 64671 ⊖ (+1 703) 237 18 59
President: Jerome B Himes

Systems Engineering Division
3160 Fairview Park Drive, Falls Church, VA 22042
☎ (+1 703) 876 10 00 ⊖ (+1 703) 560 87 11
President: Ronald R Ross

System Sciences Division
4061 Powder Mill Road, Calverton, MD 20705
☎ (+1 301) 572 49 00 ⊖ (+1 301) 672 75 02
President: George O Meyerson

Integrated Systems Division
304 West Route 38, PO Box N, Moorestown, NJ 08057
☎ (+1 609) 234 11 00 ⊖ (+1 609) 234 15 83
President: Michael W Laphen

Technical Management Group Headquarters
Virginia Technology Center, 3160 Fairview Park Drive, Falls Church, VA 22042
☎ (+1 703) 237 20 00 ⊖ (+1 703) 849 10 05
Group President: Thomas C Robinson

Consulting Group Headquarters
One University Office Park, 29 Sawyer Road, Waltham, MA 02154
Group President: James P Saviano

CSC Financial Services Headquarters
9500 Arboretum Boulevard, Austin, TX 78759
Group President: Thomas R Madison (Jr) 98

75-20730 Computing Devices International
(a subsidiary of Ceridian Corp.)
8800 Queen Avenue South, Bloomington, MN 55431-1996
☎ (+1 612) 921 60 00 Ⓦ http://www.cdev.com
⊖ (+1 612) 921 68 69
♣ $ 553,000,000 (1996) 3,000 (1997)
President: Ronald L. Turner
Public Relations Manager: Maurice Chenier
☎ (+1 612) 921 61 20
■ Electronic subsystems & information processing products & services for defense & civilian government applications. Militarized avionics computer subsystems, anti-submarine warfare subsystems, communications systems, display subsystems, ground subsystems, tactical reconnaissance systems, shipboard systems, intelligence & surveillance systems, space processing subsystems, and airline operations. 98

75-20780 Comstream Corporation
(a subsidiary of Spar Aerospace Ltd, Canada)
Headquarters
10180 Barnes Canyon Road, San Diego, CA 92121
☎ (+1 619) 458 18 00 ⊖ (+1 619) 657 54 04
♣ $ 203,506,000 (1996) 470 (1996) 165,000 sq ft
Chief Executive Officer: Frederick D Lawrence
Vice President Sales and Marketing: Sal Benti
Vice President Operations/Technical: John Wadsworth
Purchasing Director: Dave Rogers
Marketing Director: Leila Tardif
■ Satellite based communications technology, supplies interactive & broadcast networks & digital componentry. Systems include TDMA & SCPC earth stations, modems, receivers, uplinks, network management systems & custom chips. Point-to-multipoint voice, data, imaging & fax transmission. Point-to-point, full mesh or "star" connectivity. Uplink only for low cost network hubs. Applications: Mesh networking for private business & government services. Long distance telephony trunking terminals & voice backbone networks. LAN interconnection. Video conferencing. Data acquisition by a central site polling many remote locations (SCADA). Digital Data Applications. Real time financial market information. News/weather gathering and distribution. Computer file transfer. Photograph or image transmission. Paging networks. Digital Audio Applications. Radio programme distribution. Remote audio broadcasting. News gathering. Remote studio recording. Business music, store & forward information services. Major systems designs & implementation of large turnkey satellite communication systems. 98

75-20810 Comtech Systems Inc.
(a subs. of Comtech Telecommunications Corp.)
3100 Communications Rd, St Cloud, FL 32769
☎ (+1 407) 892 6111 ⊖ (+1 407) 957 3402
35
Managing Dir: Richard Burt
Marketing Dir: Robert DiCicco
■ Manufacturer of digital tactical troposcatter communication systems, tactical transportable antennas and line of sight microwave systems. 97

75-20825 Conap Inc.
(a subsidiary of Cytel Ind.)
1405 Buffalo Street, Olean, NY 14760-1139
☎ (+1 716) 372 96 50 ✆ (800) 836 36 66
Ⓦ http://www.conap.com ⊖ (+1 716) 372 15 94
136 (1997) 83,000 sq ft
President: Allen M Hodges
Operations Director: Robert Rieck
Vice President Sales and Marketing: D A Williamson
Vice President Research and Development: W J Palladino
Customer Services Manager: John R Brewer
■ Polyurethanes & epoxies. Potting & encapsulating systs, conformal coatings, urethane elastomers & tooling resin systs & adhesives & sealants. 98

75-20833 Conax Florida Corp.
2801-75th Street, North, St. Petersburg, FL 33710
☎ (+1 813) 345 80 00 ⊖ (+1 813) 345 42 17
130 43,000 sq ft
President: Jeffrey K Eckhart
Marketing Manager, Life Support: Tom Hallis

Contact Reference

Marketing Manager, Aerospace: Al Guenthardt
■ Manufacture military life support equipment and electro-explosive devices. [98]

75-20835 CONCO Inc.
4000 Oaklawn Drive, Louisville, KY 40219
☎ (+1 502) 969 13 33 ⊖ (+1 502) 962 21 90
150 (1997) 75,625 sq ft
President: Robert D Pope
Manufacturing Director: Clyde E Mullins
Contract Administrator: Robert C Corcoran
Contract Administration Specialist: Ms Carolyn Carter
Materials Manager: Jerry L Gesenhues
■ Metal containers. [98]

75-20845 Concorde Battery Corp.
2009 San Bernardino Rd, West Covina, CA 91790
☎ (+1 818) 813 1234 ⊖ (+1 818) 338 3549
President: Don W. Godber
VP Marketing: E.F. (Skip) Koss
■ Vented, sealed, valve regulated lead acid batteries for aircraft, ships, motive power, standby power, ground support equipment & vehicles. [97]

75-20950 Condor Systems Inc.
2133 Samaritan Drive, San Jose, CA 95124
☎ (+1 408) 371 9580 ⊖ (+1 408) 371 9589
♣ $ 41 000 000 (1994) 325 110 000 sq ft
President: Robert E. Young II
Chief Technical Officer: John L. Barnum
VP Business Development: Vernon Dale
VP Advanced Program Development: David Klingler
VP Engineering: Drew Downs
Operations Dir: Tom Michalski
Purchasing Mgr: Cy Wheaton
■ Design & manufacture MIL-SPEC microwave receivers (AN/ALR-81), millimeter-wave down converters, elint signal processors, sigint systems, shaped reflector antennas, digitally based direction-finding & positioning systems.

Condor Systems East
45180 Business Ct, Sterling, VA 22170
☎ (+1 703) 709 8300 ⊖ (+1 703) 709 8304 [97]

75-21050 Connecticut Machine Products Inc. (CMP)
(a subs. of Dynatech International Corp.)
1 Rodeo Drive, Edgewood, L.I. NY 11717
☎ (+1 516) 243 1700 ⊠ 144646 ⊖ (+1 516) 243 1022
President: David A. Yarom
Executive VP: Isaac Robinson
Marketing Mgr: George J. Pond
General Mgr: Terry Clark
■ Manufacturer & supplier of military replacement parts, helicopter engine spares, ordnance components. → 85 [97]

75-21520 Conrac Systems Inc
(a subsidiary of Conrac Elektron GmbH, Germany)
730 East Cypress Avenue, Monrovia, CA 91016
☎ (+1 818) 303 00 95 ⊖ (+1 818) 303 54 84
President: Steve Pirrone
Sales Dir: Guy Nitti
Dir Marketing: David J. West
■ Aircraft stall warning systems, helicopter true airspeed systems, pressure transducers, vanes. Monochrome & raster-scan color video display monitors for airborne, shipboard & ground-based applications. Full mil.& ruggedized displays & console work stations. [98]

75-21700 Container Research Corp. (CRC)
Hollow Hill Rd, PO Box 159, Glen Riddel, PA 19037
☎ (+1 215) 459 2160 ⊖ (+1 215) 358 9297
150 (1996) 225 000 sq m
President: Stanley E. Rines
Sales & Marketing Mgr: Girard L. Hemminger
■ Re-usable metal shipping & storage containers for military & commercial specifications. Electronic cabinets & enclosures; aircraft wingstands, maintenance platforms. [97]

75-21800 Continental Electronics Corp.
4212 S. Buckner Blvd, PO Box 270879,
Dallas, TX 75227-0879
☎ (+1 214) 381 7161 ⊖ (+1 214) 381 4949
240 130 000
President: R.M. McDonald
VP Marketing: J. Ross Faulkner
Advertising & Promotions: Steve A. Claterbaugh
■ High-powered RF transmitters. [97]

75-21830 Continental Microwave and Toll Co Inc
(a Chelton Group Company)
10 Merrill Industrial Drive,
Hampton, New Hampshire 03842
☎ (+1 603) 926 35 95 ⊖ (+1 603) 926 98 77
60 40,000 sq ft
President: D V Gaggin
General Manager: D L Wall
Director Programmes: J H Rifel [98]

75-21850 Contraves Inc
(a subsidiary of Oerlikon-Contraves AG)
615 Epsilon Drive, Pittsburgh, PA 15238
☎ (+1 412) 967 77 00 ⊖ (+1 412) 967 75 73
♣ $ 60,500,000 (1996) 414
Managing Director: Mrs N Baldini
■ Design and manufacture of precision instrumentation and systems for electro-optical, electro-mechanical, inertial guidance test and simulation applications. [98]

75-21928 Control Products Corp.
1513 West Jefferson, Grand Prairie, TX 75051-1443
☎ (+1 972) 264 03 68 ⊖ (+1 972) 262 16 53
♧ $ 2,500,000 (1997) ♣ $ 12,000,000 (1997)
120 (1997) 25,000 sq m
President: H Wayne Hanks
Sales and Marketing Director: C Pieroway
Technical Director: Tom Ferguson
Operations Director: Ron Jensen
Purchasing Manager: T Brown
■ Machined parts and sheet metal assemblies. Exterior lights, aircraft cabin lights, aircraft. [98]

75-22010 Coors Ceramics Co. Structural Products Group
(a subsidiary of ACX Technologies Inc.)
600 9th Street, Golden, CO 80401
☎ (+1 303) 277 41 21 ⊖ (+1 303) 277 45 96
1,200 45,000 sq m
Vice President and General Manager: William Howe
Product Director: Gary Klapper
Engineering, Sales and Products: Tom Riley
■ Complete line of ceramic components for armour systems for light through heavy threats. Custom ceramics for structural and electronics application. [98]

75-22135 Cortex Vicsion Systems, Inc
PO Box 14282, Research Triangle Park, NC 27709
☎ (+1 919) 361 96 06 ✆ (800) 766 57 23
[E] cfoley@cortexvision.com ⊖ (+1 919) 572 24 71
16
President: Robin Thompson
Marketing Director: Ms Carolyn Foley
■ E-clips surveillance system which requires security personnel to respond immediately to situations to assess their potential. [98]

75-22200 Cotag International
17304 Preston Road, Building E100, Dallas, TX 75252
☎ (+1 214) 733 6000 ⊖ (+1 214) 733 6091 [98]

75-22300 Cox & Company Inc.
200 Varick Street, New York, NY 10014
☎ (+1 212) 366 02 00 ⊠ TWX 710 5812 839
⊖ (+1 212) 366 02 22
250 110,000 sq ft
President: Warren Achenbaum
VP Marketing: Dennis M. Phillips
VP Engineering: Thomas Ferguson
Operations Mgr: Patrick L. Kelly
Contracts Mgr: Skip Seyer
■ Design, engineer & manufacture sophisticated temperature control systems & advanced de-icing systems. These include electrically heated anti-icing & de-icing inlets for aircraft operated in the Western World. Supplier to the major aerospace companies worldwide. [98]

75-22350 Craig Systems
(a subsidiary of Eastern Technologies)
10 Industrial Way, PO Box 648,
Amesbury, MA 01913-4848
☎ (+1 508) 388 56 62 ✆ (800) 662 72 44
⊠ TWX 710 3470 590 ⊖ (+1 508) 388 08 45
160
President: Lewin Jaffe
Vice President Marketing and Sales: J Roger Chouinard (Jr)
■ Design and manufacture mobile electronic equipment, shelters and transporters and other ancillary items. [98]

75-22400 C-RAN Corp.
699 4th Street North West, Largo FL 33770-2408
☎ (+1 813) 585 38 50 [E] c-rancorp@spintrail.com
⊖ (+1 813) 586 17 77
♣ $ 5,000,000 (1996) 7 (1997) 12,000 sq ft
President: A C Kruger
General Manager: Ms Maddy Nason
Marketing Manager: Mark Rose
Engineer: A C Kruger
■ Military survival radios, RF generators, wire harness assemblies, URM military radio test sets. [98]

75-22452 Crane Co. Hydro-Aire Inc.
(a subsidiary of Crane Co., CT)
PO Box 7722, Burbank, CA 91510-7722
3000 Winona Avenue, Burbank, CA 91504
☎ (+1 818) 526 26 00 ⊖ (+1 818) 842 61 17
500 (1997)
President: R Boushie
Vice President of Sales and Marketing: Vern Sanders
Manager of Advertising and Public Relations: Virgil A Jorgensen
Purchasing Manager: Dave Boileau
■ Electronic, mechanical fluid and hydraulic products. Hardware and software design, development, qualification, manufacturing, quality assurance, customer and product support. [98]

75-22500 John Crane Inc.
(a subs. of TI Group plc, UK)
6400 Oakton St, Morton Grove, IL 60053
☎ (+1 847) 967 2400 ✆ (800) SEALING ⊠ 687 1540
⊖ (+1 847) 967 3915
1 500
Exec. Vice President: Ron Warren
Marketing Communications Dir: Ms Barbara Conti
Marketing Communications Mgr: Ms Carolyn Staven
■ Mechanical seals, fluorocarbon products, mechanical & hydraulic packings, oil seals.

John Crane Marine
1536 Barkclay Drive, Buffalo Grove, IL 60089
☎ (+1 847) 808 9240 ⊖ (+1 847) 808 9295
■ Seals for US navy & military vehicles & tanks. [97]

75-22570 Cray Research Inc.
(a subsidiary of Silicon Graphics Inc)
655A Lone Oak Drive, Eagan, MN 55121
☎ (+1 612) 452 66 50 [W] http://www.cray.com
⊖ (+1 612) 683 71 99
2,300
Chairman & CEO: J. Phillip Samper
President & COO: Robert Ewald
Chief Financial Officer: Larry Betterley
Sr Vice President Supercomputing Systems: Ms Irene Qualters
Vice President Business Systems: Bobbi Hazzard
Vice President Service: Mick Dunsworth
Vice President Business Systems: Hazzard Bobbi
Vice President Marketing: Karl Freund
General Mgr Government Systems: Paul Iwanchuch
■ Manufacture computers & advanced computer systems for civil & defence applications. [98]

75-22590 Criminal Research Products Inc.
206-218 E. Hector St, PO Box 408,
Conshohocken, PA 19428
☎ (+1 610) 828 5326 ✆ (800) 635 5225
⊖ (+1 610) 825 1228
25
VP International Sales: Richard Lawler
■ Manufacturer crime scene investigation/fingerprint products. Electoral marking inks & supplies. [97]

75-22700 CTL-Aerospace Inc.
5616 Spellmire Drive, World Park, Cincinnati, OH 45246
☎ (+1 513) 874 7900 ⊖ (+1 513) 874 2499
♣ $ 18 000 000 (1996) 👥 230 (1996) 🏠 96 000 sq ft
President & CEO: James C. Irwin
General Mgr: James T. Irwin
■ Aircraft-aerospace composite HWD, radomes, seals, H/C panels filament winding. [97]

75-22800 CTS Corporation
905 North West Boulevard, Elkhart, IN 46514
☎ (+1 219) 293 75 11 ⊖ (+1 219) 293 61 46
♣ $ 321,297,000 (1996) 👥 3,815 (1996)
President: Joseph P. Walker
Marketing Communications Supervisor: Robert L. Ramsby
Advertising Mgr: Robert L. Ramsby
■ Variable resistors & associated switches, rotary & DIP switches, trimmers, cermet fixed & variable resistors, crystals, clock oscillators & precision crystal oscillators, thick & thin film hybrid circuits, resistor networks, surge networks, loudspeakers, connectors, backplanes and LED-based optical data link transmitters/receivers.
CTS Corporation Interconnect Products
9210 Science Centre Drive, New Hope, MN 55428-3621
☎ (+1 612) 536 36 02
👥 200
National Sales Mgr: Don DeCosse
■ Manufacturer of electronic interconnects. Products include metal backpanels & headers to MIL-C-28754, printed circuit backpanels to MIL-C-28859 & molded connectors to MIL-C-55302. Also MIL-A-28870, MIL-P-55110, MIL-Q-9858 & MIL-STD-2000. [98]

75-22850 Cubic Communications Inc. OAR DF Products Group
(† OAR Corp.)
9535 Waples Street, San Diego, CA 92121-2953
☎ (+1 619) 643 5800 ⊖ (+1 619) 643 5803
♣ $ 18 000 000 (1995) 👥 124 🏠 141 083 sq m
President: Ron Tosta
Sr Marketing Asst: Ms Denise Léger
■ Multi-mode communications equipment, DF systems and tactical antennas. [97]

75-22900 Cubic Defense Systems Inc.
(a subsidiary of the Cubic Corporation)
PO Box 85587, San Diego, CA 92186-5587
9333 Balboa Avenue, San Diego, CA 92123-1589
☎ (+1 619) 277 67 80 ✉ CUBIC ⊖ (+1 619) 277 93 29
♣ $ 370,000,000 (1995) 👥 2,800 🏠 1,200,000 sq ft
President and Chief Executive Officer: Walter Zable
Vice President, Marketing: Stephen R Sampson
Public Relations Director: Abe Wischnia
Engineering Director: Edwin Salkeld (Jr)
Purchasing Director: Michael Tuffley
Director of Air Ranges, Business Development: Bud R Biloon
Public Relations Manager: Gerry Reeves
☎ (+1 619) 505 23 30
■ Electronic training systems - air and ground - including Tactical Aircrew Combat Training System/Air Combat Manoeuvring Instrumentation (TACTS/ACMI). Miles 2000, JRTC. CMTC, Air Warrior. Design and manufacture of advanced location systems for ASW and rescue operations. Space and communications equipment and systems.
Washington, DC Office
1235 Crystal Gateway One, Ste 1102,
Arlington, VA 22202
☎ (+1 703) 415 1600 ⊖ (+1 703) 415 1608
VP E.Region Office: Alex Balberde [98]

75-22980 Cummins Engine Co. Inc.
PO Box 3003, Columbus, IN 47202-3005
5th & Jackson St, Columbus IN 47201
☎ (+1 812) 377 39 20 [W] http://www.cummins.com
⊖ (+1 812) 377 80 32
♣ $ 4 247 900 000 (1993) 👥 23,600
Director Military Business: Stephen A Hewitt
■ Manufacture of compact, heavy-duty diesel engines suitable for all military power requirements from 76hp to 6,000hp. [98]

75-23050 Curtiss-Wright Corp.
1200 Wall Street West, Lyndhurst, NJ 07071
☎ (+1 201) 896 84 00 ⊖ (+1 201) 438 56 80
👥 1,700
Chairman & President: David Lasky
■ Produces precision components & systems. Provides highly engineered services to the aerospace, industrial, flow control & marine markets. [98]

75-23055 Curtiss-Wright Flight Systems Inc.
(a subsidiary of Curtiss Wright Corp.)
300 Fairfield Road, Fairfield, NJ 07004-1962
☎ (+1 201) 575 22 00 ⊖ (+1 201) 575 22 20
👥 422
President: Robert E Mutch ☎ (+1 201) 575 22 22
Director Marketing: Michael J McGovern
☎ (+1 201) 575 22 65
■ Control & actuation components & systems. [98]

75-23085 Custom Electronics Inc.
Browne Street, Oneonta, NY 13820
☎ (+1 607) 432 38 80 [E] ceisales@otsego.net
[W] http://www.otsego.net/custom ⊖ (+1 607) 432 39 13
♣ $ 5,500,000 (1996) 👥 65 (1996) 🏠 30,000 sq ft
President and Chief Executive Officer: P S Dokuchitz
Managing Director: David Dokuchitz
Vice President Sales and Marketing: Duane Hathaway
Operations Director: Graydon Rogers
Manager Sales: Michael Schulte
Purchasing Manager: Gary Sandike
■ Manufacture high voltage reconstituted mica paper capacitors for use in radar power supplies, ECM power supplies, high-voltage transmitters for missile applications, TWT power supplies for satellites, detonation systems, ignition systems and other high voltage applications. [98]

75-23120 Cyberchron Corp.
US Route 9, PO Box 160, Cold Spring, NY 10516
☎ (+1 914) 265 3700 [TX] 255684 ⊖ (+1 914) 265 4154
♣ $ 15 000 000 (1994) 👥 45 (1996)
President: Chris Fadden
Sales & Marketing Administrator: Chris Smith
■ Ruggedized military computers & defence-related systems integration. Severe environment computer systems. Rugged peripherals, keyboards & monitors. [97]

75-23134 Cylink Corp.
910 Hermosa Court, Sunnyvale, CA 94086
☎ (+1 408) 735 5800 ⊖ (+1 408) 720 8294
👥 300
President & CEO: Lewis C. Morris
VP International: Colin Petty
Chief Technical Officer: J.K. Omura
■ Manufacture advanced communications security equipment and spread spectrum radios for voice, fax and data networks. Bulk digital encryption products operate from 1.2 kbps to 50 mbps. Voice encryption equipment operates with standard telephones and radios. Fax encryption for all CCITT group 3 and 4 fax machines. Spread spectrum radios provide operation for voice circuits and data networks at speeds up to 2.048 mbps. [97]

75-23145 Dabico Inc.
(† Culligan & McCallum)
2995 Airway Avenue, Costa Mesa, CA 92626
☎ (+1 714) 545 79 00 [E] genmail@dabico.com
[W] http://www.dabico.com ⊖ (+1 714) 545 76 76
👥 55
President and Chief Executive Officer: Robert M Dabic
Vice President and General Manager: Bill Fischer
Marketing Director: Jesse Hunter
Purchasing Manager: David Nuccio
■ Manufacturer of prefabricated, aircraft underground utility dispensing pits providing: fuel, engine start air, air conditioning, 400/60/50 Hz and DC electrical power, wash water, potable water, lavatory drainage, hydraulic fluid, compressed shop air, security, static grounding, fire protection and concrete vault accessibility. [98]

75-23360 Daedalus Research Inc.
PO Box 6067, North Logan, UT 84341
☎ (+1 801) 752 10 10 ⊖ (+1 801) 753 19 71
♣ $ 1,100,000 (1997) 👥 12 (1997)
President: Ned Allen [E] nallen@sisna.com
Program Engineer: Steven R Jacobson
[E] sjacobson@sisna.com
■ Manufacture STF unmanned air vehicles, UAV propellers, V/STOL air vehicle hybrid (gas/electric) powerplants. Sub-scale prototype design, fabrication and flight testing services. → 80 [98]

75-23365 Daewoo America Development Inc (DADI)
(a subsidiary of Daewoo Corporation, South Korea)
Unisys Center II 377E, Butterfield Road Suite 425,
Lombard, IL 60148
☎ (+1 708) 810 00 49 ⊖ (+1 708) 810 19 14 [98]

75-23370 Daewoo America Development Inc (DADI)
(a subsidiary of Daewoo Corporation, South Korea)
4135 Providence Point Drive SE, Issaquah, WA 98029
☎ (+1 206) 392 19 22 ⊖ (+1 206) 392 29 28 [98]

75-23570 Danner Shoe Manufacturing Co.
(a division of LaCrosse Footwear Inc., USA)
12722 N.E. Airport Way, Portland, OR 97230
☎ (+1 503) 251 11 00 [E] danner@teleport.com
⊖ (+1 503) 251 11 19
President: Eric E Merk (Sr)
Sales and Marketing Director: Joe Schneider
Purchasing Manager: David Sprecher
■ Footwear for military & other services. [98]

75-23580 Dare Electronics
3245 S. County Road 25A, Troy, OH 45373-0419
☎ (+1 937) 335 00 31 ⊖ (+1 937) 339 69 48
♣ $ 1,000,000 👥 25
President and Chief Executive Officer: Ms Karen Beagle
■ Design & manufacture custom electronic & electromechanical components. [98]

75-23625 Dassault Systèmes
(a subsidiary of Dassault Aviation Group, France)
777 Terrace Avenue, Hasbrouck Heights, NJ 07604
☎ (+1 201) 262 40 40
President: Charles Edelstenne
Vice President: Ben R Brewster [98]

75-23645 Datacom Inc.
651 Anchors Street North West, PO Box 278,
Ft Walton Beach, FL 32548-3803
☎ (+1 904) 244 61 21 ⊖ (+1 904) 244 63 44
👥 50 🏠 32,000 sq ft
President and Chief Executive Officer: Anthony J Skeete
Marketing: Earnie J DeVille
■ Testing range instrumentation control and display equipment and systems, data acquisition, processing and display systems and networks. Training, maintenance and documentation support services, EW/Avionics, equipment and systems for extraordinary environments: shipboard; airborne; ground mobile; (semi-trailer, truck, step-van); hostile fixed site. Proof of design and reverse engineering, manufacturing (including build-to-print) and test. [98]

75-23700 Datametrics Corporation (DmC)
(† Diametrics Corp.)
21135 Erwin Street, Woodland Hills, CA 91367-3713
☎ (+1 818) 598 62 00 ⊖ (+1 818) 598 62 99
♣ $ 25,000,000 (1997) 👥 130 (1997) 🏠 65,000 sq ft
President and Chief Financial Officer: Adrien Maught
Vice President of North American Sales: Jim Foti
Director of International Operations: Dale Cooke
Operations Director: Jim Sturgeon
Purchasing Manager: Larry Marrowe
■ Design, develop and manufacture military and ruggedised printers, black and white and colour printer/plotters, keyboards, rugged computers, workstations and portable computers for a variety of C31 applications. [98]

75-23860 DATATAPE Incorporated
605 East Huntington Drive, Monrovia, CA 91017
☎ (+1 818) 358 95 00 [W] http://www.datatape.com
[E] info@datatape.com [TX] 67-5417 ⊖ (+1 818) 358 91 00
👥 400
President and Chief Executive Officer: Dom Saccacio
Vice President of Business Development: Bruce Peters
Operations Director: Bill Kruger
Sales and Marketing Director: Garth Orgill
Purchasing Manager: Lee Rentfrow

Marketing and Communications: Janet Westergaard
■ Manufactures high peformance data storage systems, products and services for various military/aerospace and commercial applications. [98]

75-24000 Datron/Transco Inc.
(a subs. of Datron Systems Inc.)
200 W. Los Angeles Ave, Simi Valley, CA 93065-1650
☎ (+1 805) 584 1717 ⊖ (+1 805) 526 0885
♧ $ 14 241 000 (1995) ♣ $ 70 033 000 (1995) 👥 183
🏠 140 000 sq ft
President: Donald R. Rondeau
VP Marketing: John Ricardi
Marketing Dir: David Benson
Technical Dir: Regan Walden
Purchasing Mgr: John Sclimenti
■ Manufacture antennas & systems used in satellite communications. Radar & telemetry tracking & microwave antennas & components. Coaxial & wave guide antennas. [97]

75-24020 Datron World Communications Inc.
(† Trans World Communications Inc.)
(a subsidiary of Datron Systems Inc.)
304 Enterprise Street, Escondido, CA 92029
☎ (+1 619) 747 10 79 📠 695433 ⊖ (+1 619) 741 16 58
♣ $ 30,000,000 (1996) 👥 200
President: John DiGioia
Vice President of Sales: Richard Dailey
■ Communications equipment including HF-SSB base and mobile equipment, VHF-FM manpack radios. [98]

75-24290 Davis Aircraft Products Co. Inc.
1150 Walnut Ave, Bohemia, NY 11716
☎ (+1 516) 563 1500 📠 TWX 510 2260 235 davisaft ny
⊖ (+1 516) 563 1117
♣ $ 8 000 000 👥 100 🏠 30 000 sq ft
President: Bruce T. Davis
VP: Francis J. Nester
Marketing Dir: Ms Jill L. Davis
Sales Dir: Ms Sylvia A. Prill
■ Seatbelts, restraint & cargo tiedowns & straps; helicopter blade clamps; floorplates & studs; military hardware; buckles, fittings, clips.

Davis Restraint Systems
Suite 15, 820 South Palm Ave,
Alhambra, CA 91803-1529
☎ (+1 818) 289 3555 ⊖ (+1 818) 289 3587 [97]

75-24700 DEC Technologies Inc.
501 Industrial Park Road, Piney Flats, TN 37686
☎ (+1 423) 538 09 19 ⊖ (+1 423) 538 08 38
♣ $ 3,000,000 (1996) 👥 6 (1996) 🏠 12,000 sq ft
Vice President and Director of Engineering: Rick Rathbun
■ Design, engineering, sales and certification. [98]

75-24820 Deep Ocean Engineering Inc.
1431 Doolittle Drive, San Leandro, CA 94577
☎ (+1 510) 562 9300 ⊖ (+1 510) 430 8249
♣ $ 3 500 000 👥 28
President & CEO: Dick Rosen
Vice Chairman: Graham S. Hawkes
Sr Vice President: Dr Philip Ballou
Sr Vice President: David C. Jeffrey
VP Manufacturing: John Edwards
■ Design & manufacture manned & remotely operated submersibles for observation, intervention & subsea work. Products include remotely operated Phantom vehicles & manned submersibles. [97]

75-24840 DeepSea Power & Light Inc.
4819 Ronson Court, San Diego, CA 92111
☎ (+1 619) 576 12 61 [E] info@deepsea.com
[W] http://www.deepsea.com ⊖ (+1 619) 576 02 19
👥 45 (1997) 🏠 750 m2
President: Mark Olsson
Managing Director: John Chew
Marketing Manager: Bob Novak
Sales: Mike Barbee
■ Manufacturer of underwater imaging systems, video cameras, lights and batteries. [98]

75-24870 Deft Chemical Coatings
17451 Von Karman Ave, PO Box 19507, Irvine, CA 92714
☎ (+1 714) 474 0400 ⊖ (+1 714) 474 7269
👥 100
President: William A. Desmond
Sales Mgr: Ms Tracy Garrett (Jr)
■ Polyurethane coatings for aerospace, water reducible primers, epoxy primers, MIL-SPEC coatings, skydrol resistant epoxy primers. [97]

75-24980 Defense Technology Corporation of America (DTCoA)
National Sales, Training Division & Corporate Offices
2136 Oil Drive, Casper, WY 82604
☎ (+1 307) 235 2136 (Corp. Office)
☎ (+1 307) 235 6900 (National Sales Div.)
☎ (+1 307) 235 2911 (Training Div.)
[W] http://www.defensetechnology.com
⊖ (+1 307) 473 2713 (Corp. Office)
⊖ (+1 307) 473 2605 (National Sales Div.)
⊖ (+1 307) 235 0889 (Training Div.)
♣ $ 25 000 000 👥 115
President & CEO: Robert L. Oliver
National Sales Div.: William E. Moles
Training Div.: Steven D. Mattoon
■ Manufacture less than lethal products for law enforcement, corrections & the military. Products include OC sprays, tear gas, smoke grenades, tear gas & smoke projectiles, Stinger projectiles, distraction devices, flameless expulsion grenades, tear gas guns, helmets, shields, batons & other police & military related items.

International Division
7400 SW 50th Terrace, Suite 101, Miami, FL 33155
☎ (+1 305) 669 3150 [E] dtcoa@ipof.fla.net
⊖ (+1 305) 669 3046
International Sales: James W. Coberly [97]

75-24990 Dekoron
1199 S. Chillicothe Rd, Mantua, OH 44202
☎ (+1 216) 562 3008 ⊖ (+1 216) 562 3631
Director: Chuck Hayes
■ Instrumentation wire & cable, tubing; optical fiber cables, self-regulating heating cable. [97]

75-25360 Delco Systems Operations
(a member of the Aerospace & Defense Business Sector of GM Hughes Electronics Corp.)
6767 Hollister Ave, Goleta, CA 93117
☎ (+1 805) 961 5630 ⊖ (+1 805) 961 6226
👥 600 🏠 438 000 sq ft
President: M.G. Johnston
Business Development Dir: A. Howell
Marketing Coordinator: P. Penniman
■ Conducts R&D design development & manufacture of turreted weapons systems & advanced land combat electronic systems. [97]

75-25362 Del Mar Avionics
(† Del Mar Engineering Laboratories)
1621 Alton Avenue, Irvine, CA 92714-4878
☎ (+1 714) 250 3200 📠 6831110
⊖ (+1 714) 261 0529
President: Bruce Del Mar
■ Acoustic target scoring systems; magnetic & optical recording systems. [97]

75-25365 Del Norte Technology Inc.
1100 Pamela Drive, PO Box 696, Euless, TX 76040-6812
☎ (+1 817) 267 35 41 [E] dnti@delnorte.com
[W] http://www.onramp.net/delnorte
⊖ (+1 817) 354 57 62
👥 50
Chief Executive Officer: Fritz-Alan Korth
Vice President of Operations: M A Lang
Marketing Administrator, Sales Support: Ms Christina D Santos
■ Global positioning systems (GPS). Geodetic, differential - mobile and reference station unit. Digital distant measuring unit. Trisponder Systems. [98]

75-25390 Delfin Systems
(a division of Delfin Systems, California)
12007 Sunrise Valley Drive, Ste 300, Reston, VA 22091
☎ (+1 703) 758 0190 ⊖ (+1 703) 758 09 41
👥 200
President: M C Baird ☎ (+1 408) 748 12 00
Vice President and General Manager of Signal Products Division: Jerry Florence
Director of Business Development: Allen Offutt
■ Analysis, development and integrated systems solutions for information resource managements. Includes experts in hardware and software engineering, artificial intelligence, signal processing and operations analysis. → 80 [98]

75-25550 Delta Electronics Inc.
5730 General Washington Dr., PO Box 11268,
Alexandria, VA 22312
☎ (+1 703) 354 3350 ✆ (800) 833 5828 📠 901963
⊖ (+1 703) 354 0216
♣ $ 5 000 000 (1995) 👥 42 (1995) 🏠 2 200 sq m
President: John Wright
VP & Marketing: Joseph S. Novak
■ Communications systems & equipment. [97]

75-25651 Denro Inc.
(† Denro Laboratories Inc.)
(a subsidiary of Firan Corp., Canada)
9318 Gaither Road, Gaithersburg, MD 20877
☎ (+1 301) 840 15 97 ⊖ (+1 301) 869 3192
♣ US$ 35,000,000 (1996) 👥 302 (1996) 🏠 10,700 m2
President: Edward C Hanna
Vice President and Chief Operating Officer: Jeffrey Clute
Vice President and Chief Financial Officer: Martin A Blade
Vice President of Business Development: Terry Blummer
Vice President International: Irwin D Nathanson
■ Design and manufacture air traffic control voice communications and C3 (VCS) for the Federal Aviation Administration, US Department of Defense and international customers. Also manufacture secure voice and data communication systems. [98]

75-25745 Detector Electronics Corp. (Det-Tronics)
(a subsidiary of Williams Holdings plc, UK)
Corporate Headquarters
6901 West 110th Street, Minneapolis, MN 55438
☎ (+1 612) 941 56 65 ✆ (800) 765 FIRE
[E] detronics@detronics.com [W] http://www.detronics.com
⊖ (+1 612) 829 87 50
👥 250 🏠 80,000 sq ft
President: G F Slocum
Vice President: M Willey
Operations Director: R Bredehoft
Sales Manager: Jim Vogel
Sales Manager: Cliff Anderson
Sales Manager: Garth Watkins
Purchasing Manager: G Vollendorf
■ UV, UV/IR and IR flame detection systems. Protection for aircraft hangars & military installations. Combustion & H25 gas detection systems. Burner management systems.

Houston Office
11210 Steeplecrest Drive, Suite 104, Houston, TX 77065
☎ (+1 281) 970 26 46 ⊖ (+1 281) 970 26 67
Regional Sales Manager: Mark Carpenter

Aston Office
466 Conchester Highway, Aston, PA 19014
☎ (+1 610) 497 55 93 ⊖ (+1 610) 485 20 78
Regional Sales Manager: Bill Rhodes [98]

75-25770 Detoronics Corp.
10660 East Rush Street, PO Box 3805,
South El Monte, CA 91733
☎ (+1 818) 579 71 30 📠 TWX 910 587 3436
⊖ (+1 818) 579 19 36
♣ $ 3,000,000 👥 50 🏠 20,000 sq ft
President: Kenneth S Clark
Operations Director: M Goldman
Customer Services Manager: Ms Lydia Lin
■ Hermetic sealing-glass ceramic to metal, microwave coaxial connectors, headers terminals window, viewports, high temperature and radiation resistant vacuum tight seals. Mil.Spec type hermetic connectors. Pressure and vacuum. [98]

75-25830 Detroit Diesel Corp.
(a subs. of Penske Corporation)
13400 West Outer Drive, Detroit, MI 48239-4001
☎ (+1 313) 592 5000 ⊖ (+1 313) 592 7580
♣ $ 2 100 000 👥 6 200 (worldwide 1995)
🏠 2 932 500 sq ft
Chairman: Roger S. Penske
President: Ludvik F. Koci
VP Construction & Industry Sales: Jeffrey A. Sylvester
VP Defence Programs: Tedd Grulke
Sr VP Purchasing: Walter Ware
Mgr Public Affairs: Beth Socall

■ Diesel engines in 40-2200 HP range for combat vehicles, marine propulsion & ground support equipment. [97]

75-25850 Detoronics Corporation
10660 East Rush Street, PO Box 3805,
SO El Monte, CA 91733-3432
☎ (+1 818) 579 71 30 ⊠ TWX 910 587 3436
⊖ (+1 818) 579 19 36
♣ $ 3,000,000 👥 45 🏭 20,000 sq ft
Chief Executive Officer: Richard J Baroda
President and Marketing Manager: Kenneth S Clark
Vice President and Treasurer: Ms Marcia A Baroda
Operations and Technical Director: S Shah
Operations Director: M Goldman
Purchasing Manager: R Mehta
Manager of Advertising and Public Relations: Ms Dawna Clark
Customer Service Manager: Diana Rhoads
■ Glass to metal hermetically sealed electronic, circular and rectangular connectors, headers, terminals and HI reliability devices. Pressure and vacuum. [98]

75-25930 Deutsch Relays Inc.
65 Daly Road, East Northport, NY 11731-6304
☎ (+1 516) 499 60 00 ⊖ (+1 516) 499 60 86
♣ $ 25,000,000 (1996) 👥 200 (1996) 🏭 130,000 sq ft
President and Chief Executive Officer: Jean-Marie Painvin
Managing Director: Tom Sadusky
Director Marketing: James Bedell
Operations/Technical Director: Ed Cheswick
Purchasing Manager: Jack Dickinson
■ High reliability, hermetically sealed relays. Moulded relay sockets, timing relays. [98]

75-25950 Deutz Corp.
(a subsidiary of Deutz Motor AG, Germany)
3883 Steve Reynolds Boulevard, Norcross, GA 30093
☎ (+1 404) 564 71 00 ⊖ (+1 404) 564 72 22
President: Werner Schmitz ☎ (+1 404) 564 7221
Director of Marketing: Robert Mann
☎ (+1 404) 564 72 41
Marketing Services Manager: Walter Krieger
☎ (+1 404) 564 72 65
■ Diesel engines. [98]

75-26100 The Dewey Electronics Corp.
27 Muller Road, Oakland, NJ 07436
☎ (+1 201) 337 47 00 ⊖ (+1 201) 337 39 76
♣ $ 8,181,000 (1993)
President: Gordon C Dewey
Vice President Operations: Edward L Proskey
Treasurer: Thom A Velto [98]

75-26440 Dialight Corp.
(a subsidiary of Roxboro Group)
1913 Atlantic Avenue, Manasquan, NJ 08736-1095
☎ (+1 908) 223 94 00 [E] dial.dialight@mcimail.com
⊖ (+1 908) 223 87 88
👥 500
President: Graham Ryan
Director Sales and Marketing: Michael Murphy
Marketing Communications Manager: Ms Diane Plasko
■ Manufacturer of Circuit Board Indicators (CBI) including NVIS & secure devices. [98]

75-26450 Diamond Electronics Inc.
(† Arvin/Diamond)
(a division of ULTRAK, USA)
PO Box 200, Lancaster, OH 43130
☎ (+1 614) 756 92 22 ✆ (800) 443 66 80
[W] http://www.diamondelectronics.com
⊖ (+1 614) 756 42 37
♣ $ 19,000,000 👥 135 🏭 72,000 sq ft
President: Richard Tompkins
Director of Marketing: Patrick Kula
Operations Director: Jerry Zielinski
Engineering Manager: Michael A Bolte
Purchasing Manager: Mark Guinan
■ Manufacturer of closed circuit television systems. Specialising in systems for hostile environments & custom designed systems. Typical uses - target location & threat evaluation. [98]

75-26455 Diamonite Products Inc.
(a division of the Ferro Corp.)
453 West McConkey Street, Shreve, OH 44676-9769
☎ (+1 330) 567 21 45 ⊖ (+1 330) 567 22 60
👥 300
General Manager: David Banfield
Sales and Marketing Manager: Ms Carol Bernauer
■ Technical ceramics for electrical, heat & wear applications. Application engineering support available. [98]

75-26650 Digital Technology Inc. (DTI)
(a subs. of Dynatech Corp.)
2300 Edwin C. Moses Blvd, Dayton, OH 45408
☎ (+1 513) 443 0412 ✆ (800) 852 1252
⊠ 753248 dti dtn ⊖ (+1 513) 226 0511
General Mgr: James L. Swyler
Marketing Mgr: Steven L. Guentner
■ Manufactures & promotes computer interface products for MIL-STD-1553 databus communications in simulation, emulation & test markets. [97]

75-26660 Dimension Technologies Inc.
315 Mount Read Boulevard, Rochester, NY 14611-1911
☎ (+1 716) 436 35 30 ⊖ (+1 716) 436 32 80
VP, Sales & Marketing: Thomas Dickenson
■ Autostereoscopic (3D without glasses) LCD based display systems for commercial, industrial & military applications. Firm is also developing technlogies for high definition, high brightness & hologram-like look-around for commercial LCDs. [98]

75-26850 Dixon Industries Corp.
386 Metacom Ave, Bristol, RI 02809
☎ (+1 401) 253 2000 ⊠ 927706 dixoncorp brol
⊖ (+1 401) 253 8211
President: Bill Mansell
Director: Cirillo Lou
■ Rulon, meldin, bearings, custom components. [97]

75-27000 DNE, Inc.
(† DNE Technologies Inc.)
(a subsidiary of The Alpine Group, Inc., USA)
50 Barnes Park North, PO Box 30, Wallingford, CT 06492
☎ (+1 203) 265 71 51 ✆ (800) 370 44 85
[W] http://www.dnetech.com ⊖ (+1 203) 265 78 01
👥 204 (1996)
President: William Gill (Jr)
Vice President Business Development: Robert H Griffith
Director of Operations: Kevin Dooley
Vice President Sales and Marketing: Charles F Reese
Purchasing Manager: Ms Sheri Macri
■ Designs & produces multiplexers, cryptographic support devices, switching & network control systems & Tempest accredited printers & terminals for the secure & nonsecure communications environment. For the aerospace market the company produces ice detection equipment, sensors, & cockpit control systems. [98]

75-27050 Donaldson Co. Inc.
Defense & Aerospace Products
1400 West 94th Street, PO Box 1299,
Minneapolis, MN 55440
☎ (+1 612) 887 35 47 ⊠ 290434 ⊖ (+1 612) 887 32 21
♣ $ 700,000,000 (1997) 👥 5,000
Defence and Aerospace Products Director: Clark Lewis
Sales Mgr: Thomas M. Rice
■ Aircleaners (2 stages and self-cleaning) for vehicles, silencers and hydraulic filters. Replacement filters and parts. [98]

75-27110 Doron Precision Systems Inc.
174 Court Street, PO Box 400,
Binghamton, NY 13902-0400
☎ (+1 607) 772 16 10 ⊖ (+1 607) 772 67 60
👥 100 (1997) 🏭 97,000 sq m
Managing Dir: Karl J. Hirsham
Sales & Marketing Dir: Owen "Bill" Murray
PR & Advertising Mgr: Ms R. Kathy McCain
■ Ground vehicle training: Producer of simulation systems used in ground vehicle training. Standard ground vehicle simulation systems, portable desktop training systems and customised systems available.
Motion system: Hydraulic three degree of motion systems for use in flight, ground vehicle, marine and gunnery training. [98]

75-27120 Dosimeter
(† Dosimeter Corp.)
(a div. of Artisan Electronics)
(a div. of Morganite Industries Inc.)
5 Eastmans Road, Parsippany, NJ 07054
☎ (+1 617) 487 2292 ✆ (800) 322 8258 ⊠ 214648
⊖ (+1 617) 890 4711
President & CEO: Paul A. Roba
Marketing & Sales Mgr: Robert M. Rich
■ Military specification radiacmeters, dosimeters, radiac chargers & radiation detection equipment. [97]

75-27200 Douglas Aircraft Company
3855 Lakewood Boulevard, Long Beach, CA 90846-0001
☎ (+1 562) 593 55 11
👥 14,000
President, Douglas Aircraft Company: Walter J Orlowski
President, MDC China: John L Fugh
Deputy President, Douglas Aircraft Company: Edward C Bavaria
Vice President and General Manager, Sales and Marketing: Donald V Black
Vice President and General Manager, Product Support: George G Field
Vice President and General Manager, Production Programs: Allen C Haggerty
Vice President and General Manager, Manufacturing: Philip M Lowry
Vice President and General Manager, MD-95 Program: James R Phillips
Vice President, Chief Financial Officer and Business Operations: Michael J Cave
Vice President and Associate General Counsel: Eddie R Island
Vice President, Human Resources: Gary J Hedges
Vice President, Supplier Manager: Stephen J Pohl
Director, Communications: Douglas L Jacobsen
General Manager, Quality: Richard P Doherty
■ Manufacture and development of MD-80 and MD-90 series twin jets and MD-11 trijet. [98]

Contact Reference

75-27280 Dowty Aerospace Yakima
(† Dowty Decoto Inc.)
(a subsidiary of the TI Group plc)
2720 West Washington Avenue, PO Box 9907,
Yakima, WA 98909-0907
☎ (+1 509) 248 50 00 [E] dowty@wolfenet.com
[W] http://www.dowty.com ⊠ TWX 510 777 5352
⊖ (+1 509) 452 70 22
♣ $ 26,634,000 (1996) 👥 231 (1997) 🏭 100,000 sq ft
President and Chief Executive Officer: William A Wheeler
☎ (+1 509) 248 50 00 ext. 200
Vice President Finance: R Galloway
☎ (+1 509) 248 50 00 ext. 282
Vice President Operations: Don Johnston
☎ (+1 509) 248 50 00 ext. 295
Engineering Director: Jim Helm
☎ (+1 509) 248 50 00 ext. 250
Sales Director: Mark Simon
☎ (+1 509) 248 50 00 ext. 268
Quality Director: George Tuncil
☎ (+1 509) 248 50 00 ext. 260
Senior Account Manager: Dean Holmes
Purchasing Manager: Steve Miller
■ Design, develop and manufacture hydraulic assemblies for aerospace, naval and land vehicles. Products include electro-hydraulic servoactuators, internal locking actuators, high energy weapons release systems, repeatable release holdback bars, hydraulic fuses and shuttle valves, brake control valves.

Washington, DC Office
PO Box 5000, Sterling, VA 22170
☎ (+1 703) 450 5930 ⊖ (+1 703) 430 9060
General Mgr: Tom White [98]

75-27315 Dowty Turbine Engine Components Group
Crestwood Industrial Park, PO Box 68,
Mountaintop, PA 18707-0068
President: Mark Celasniak
Controller: Lou Ferrello [98]

75-27320 dp Manufacturing Inc.
5647 South 122nd East Avenue, Tulsa, OK 74146
☎ (+1 918) 250 24 50 ✆ (800) dpwinch
[E] sales@dpwinch.com [W] http://www.dpwinch.com
⊖ (+1 918) 250 06 90
♣ $ 10,000,000 👥 100
President: Don R Pruitt
Government and Military Sales Director: Tom Daniel

Advertising: Greg Reigel
■ Hydraulic planetary winches for tank/truck recovery from 3,000lb capacity to 100,000lb capacity. (1360kg to 45360kg). [98]

75-27360 DRS Precision Echo, Inc. Data Systems Group
(† Media Technology Group)
(a subsidiary of DRS Technologies, Inc.)
3105 Patrick Henry Drive, Santa Clara, CA 95054
☎ (+1 408) 988 05 16 [W] http://www.precisionecho.com
⊖ (+1 408) 727 74 91
👥 120-150 (1997) 🏭 2,000 m2
President: RAdm. Stuart F Platt (USN retd)
Vice President Programs and Product Management: James Murphy
Vice President Business Development: James M Dunford (Jr)
Sales and Marketing Director: Rene Chikhani
Purchasing Manager: Ms Page Ligget
Marketing Administrator: Ms Carmen Erwin Moeller
■ Designs and manufactures high-density digital rotary-head and longitudinal recording and playback systems for intelligence, surveillance and other military and industrial applications. [98]

75-27380 DRS Laurel Technologies
(† Laurel Technologies)
423 Walters Avenue, Johnstown, PA 15904
☎ (+1 814) 269 41 41 [W] http://www.lrltech.com
⊖ (+1 814) 269 46 61
♣ $ 40,000,000 👥 265 🏭 77,000 sq m
President: John J Donnelly
General Manager: Vincent Dente
Controller: Pat Kilmartin
Vice President Operations: Bill Kohler
Vice President Sales and Marketing: Russ Clark
Purchasing Manager: Ed Rose
■ Provides high-quality manufacturing, integration and test for military- quality electronic assemblies, cables and harnesses and circuit card assemblies. [98]

75-27400 DRS Technologies, Inc.
(† DRS Inc.)
Corporate Headquarters
5 Sylvan Way, Parsippany, NJ 07054
☎ (+1 201) 898 15 00 [W] http://www.drs.com
⊖ (+1 201) 898 47 30
♧ $ 33,990,000 (1996) ♣ $ 101 454 000 (1996)
👥 1,170 (1997) 🏭 328,900 sq ft
President and Chief Executive Officer: Mark S Newman
Executive Vice President, General Counsel and Secretary: Ms Nina Laserson Dunn
Vice President Finance: Ms Nancy R Pitek
Vice President: RAdm. Stuart F Platt (USN Rtd)
Vice President: Richard Ross
Vice President: Paul G Casner (Jr)
Assistant Vice President Corporate Communications: Ms Patricia M Williamson
■ Design, develop and manufacture leading edge systems used for the processing, display and storage of data for government and commercial markets worldwide. Provide a broad range of electronic sensor, electronic imaging and electro-optical systems. Offers a full complement of technical support services.

Washington, DC Office
1215 S.Jefferson Davis Hwy, Ste 1004,
Arlington, VA 22202
☎ (+1 703) 416 8000 ⊖ (+1 703) 416 8010

DRS Ahead Technology, Inc.
(† Ahead Technology Inc.)
6410 Via del Oro, San Jose, CA 95119
☎ (+1 408) 226 99 91 ⊖ (+1 408) 226 91 95
President: Arthur Honegger

DRS Ahead Technology, Inc.
520 Blanding Woods Road South,
St. Croix Falls, WI 54024
☎ (+1 715) 483 32 33 ⊖ (+1 715) 483 13 23
Senior VP and General Manager: Donald A Puls
Operations Manager: Joel Mitchell

DRS Ahead Technology, Inc.
686 Mendelssohn Avenue, Golden Valley, MN 55427
☎ (+1 612) 545 13 21 ⊖ (+1 612) 545 03 35
Vice President Manufacturing: William Schrepfer

DRS Ahead Technology, Inc.
145 South Third Street, Dassel, MN 55325
☎ (+1 320) 275 33 25 ⊖ (+1 320) 275 20 96
Operations Manager: Robert Liston [98]

75-27620 DTC International
PO Box 24364, New Orleans, LA 70184
☎ (+1 504) 486 4686 ⊖ (+1 504) 486 6329
♧ $ 1 500 000 (1996) ♣ $ 6 000 000 (1995) 👥 8
🏭 6 400 sq ft
Managing Dir: S. Shay
Sales Dir: M.L. Malloy
Purchasing Mgr: M. McIver
■ Hard armour retrofit packages for vehicles & ships, personal body armour, bomb suppression equipment including blankets, sleds & EOD suits. Exports & Imports [97]

75-27650 Dukane Corp. SeaCom Div.
2900 Dukane Dr., St Charles, IL 60174
☎ (+1 708) 584 2300 ⊠ 720426 ⊖ (+1 708) 584 5154
♣ $ 80 000 000 (1995) 👥 725
President: Thomas A. Greenacre
■ Underwater locators for aircraft & flight recorders. Acoustic beacons. [97]

75-27730 Dunlin Mfg Co. Div. Skypen Industries Inc.
(† Dunlin Corporation)
(a div. of Skypen Industries Inc.)
PO Box 353, Plainview, NY 11803-0353
☎ (+1 516) 249 18 71 ⊖ (+1 516) 249 19 32
♣ $ 500,000 (1995) 👥 5 🏭 195 sq m
President: Harvey Pensky
■ Military replacement parts. [98]

75-27750 E.I. Du Pont de Nemours & Co.
Barley Mill Plaza, PO Box 80016,
Wilmington, DE 19880-0016
☎ (+1 302) 892 86 94 ✆ (800) 441 7515
✉ DUPONT Wilmington [W] http://www.dupont.com
⊖ (+1 302) 992 41 33
Chief Executive Officer: John Knol
Aerospace Enterprises Director: William O McCabe
Assistant Director, Aerospace: Arthur R Teffean [98]

75-27900 Dynaco Corp.
(a subsidiary of Palomar Technologies)
1000 South Priest Drive, Tempe, AZ 85281
☎ (+1 602) 968 20 00 ⊖ (+1 602) 921 98 30
♣ $ 18,000,000 (1995) 👥 200 (1995) 🏭 20,000 sq m
President: Lyle Jensen
Vice President of Sales and Marketing: Peter K Royal
Operations and Technical Director: Ms Adeline Boomgaard
Purchasing Manager: Nick Stocke
■ Defence and aerospace flexible circuits (types I-IV, MIL-P-50884C) and rigid boards (MIL-P-55110). Cable harnesses (RF, molded, fibre optic wire). [98]

75-27960 Dynamation Research
2301 Pontius Avenue, Los Angeles, CA 90064
☎ (+1 310) 477 12 24 ⊠ 323604 ⊖ (+1 310) 479 56 56
♣ $ 10,000,000 🏭 29,000 sq ft
President: Gal Lipkin
Manufacturing Manager: Mark Bernstein
Sales Manager: Brian Smith
■ Manufacture Mil-Spec chemicals, wire and cable, webbing cloth. Manufacture and distribution of sealing systems (3000/8000 psi). AN, NS, NAS, BAC, 7M, ST7M, hardware for Mil 2 comm. aircraft, military hardware.

Aerospace Fittings
11150 Tennessee Avenue, Los Angeles, CA 90064
☎ (+1 310) 312 16 91 ⊠ 323604 ⊖ (+1 310) 575 04 70
President: Daniel O'Mally
Sales Manager: Bill Beck
Manufacturing Manager: Mark Bernstein
■ Manufacture of hydraulic/pneumatic fittings for 3000, 4000, 5000 and 8000 psi systems, using all advanced materials; manufacture of fittings seals for leakage connection AN, MS, NAS, Mil-Spec. [98]

75-27980 Dynamic Air Engineering Inc.
620 East Dyer Road, Santa Ana, CA 92705
☎ (+1 714) 540 10 00 ⊖ (+1 714) 545 91 45
♧ $ 1,000,000 (1997) ♣ $ 4,000,000 (1997)
👥 45 (1997) 🏭 5,100 sq m
President: H Morrison
Marketing Manager: Carl A Grisier
■ Design and manufacture centrifugal, tubeaxial, vaneaxial fans and blowers, speciality motors, aircraft and tactical enclosure ECUs. [98]

75-28100 Dynamic Engineering Inc. (DEI)
703 Middle Ground Blvd, Newport News, VA 23606
☎ (+1 804) 873 1344 ⊖ (+1 804) 873 2183
♣ $ 20 000 000 👥 240 🏭 8 900 sq m
President: David H. Mullins
VP, Dir Marketing: Richard L. Best
Operations/Tech Dir: Jim Brunges
Purchasing Mgr: Ms Debbie Ross
■ R&D prototype hardware (MIL-I-45208A, MIL-Q-9858A), precision metals & composite fabrication in support of research in aircraft, rotorcraft, missile systems, test stands, propulsion systems & remotely piloted vehicles. [97]

75-28240 Dynamic Metal Products Co.
967 Parker Street, Manchester, CT 06040-2208
☎ (+1 860) 646 40 48 ⊖ (+1 860) 645 91 70
♣ $ 32,000,000 (1997) 👥 260 (1997)
President: Howard Miller
Chief Operating Officer: Alan J Kick
Operations Director: Richard Wood
Purchasing Manager: Dwight Craig
■ Manufacturer of jet engine parts including sheet metal fabrications, weldments & machined parts. [98]

75-28265 Dynasound Inc.
Sound Masking and Audio Security Products
6439 Atlantic Boulevard, Norcross, GA 30071-1212
☎ (+1 770) 242 81 76 ✆ (800) 989 6275
[W] http://www.dynasound.com ⊖ (+1 770) 242 88 58
President: Thomas A Koenig
[E] ltkoening@dynasound.com
Director of Marketing Services: Ms Libby Thacker
Operations Director: Ms Lisa Minton
■ Manufactures Audio Surveillance Countermeasures Systems that consist of Gaussian random (non-deterministic) noise masking generators and amplifiers that drive devices to protect windows, doors, walls, air ducts, plenums, pipes and speakers from being sources for both human and electronic eavesdropping. [98]

75-28267 Dynatech International Corp.
1 Rodeo Drive, Edgewood, NY 11717
☎ (+1 516) 243 17 00 ⊖ (+1 516) 243 10 22
👥 40
Chairman: David Yarom
Sales and Marketing Director: Karen Simms
■ Manufacturing and engineering sciences with advanced technologies and efficient logistics support programmes. Supplier of spare parts to support T53, T55, T56 and other engines.

Subsideries:
Connecticut Machine Products Inc.
Dynatech Engineering
Air Instro Inc. [98]

75-28325 Dynepco, Inc
(a subsidiary of Daicel Chemical Industries Ltd, Tokyo)
84 North Main Street, Wharton, NJ 07885
☎ (+1 201) 361 89 00 [E] dynepco@nae.com
[W] http://www.dynepco.com ⊖ (+1 201) 361 96 06
♣ $ 5,000,000 👥 58 (1997) 🏭 140,000 sq m
Managing Director: Tom Cunningham
Vice President and General Manager: Ron Kubrak
Operations Director: Ms Lynn Whitney
Sales & Marketing Mgr: Ms Valerie Telford
® Japan: Daicel Chemical Industries [98]

75-28350 Dynex/Rivett Inc.
770 Capitol Drive, Pewaukee, WI 53072-2549
☎ (+1 414) 691 03 00 ⊖ (+1 414) 691 03 12
♣ $ 15,000,000 (1996) 👥 150 (1997)
President and Chief Executive Officer: Raymond M Warell
Advertising Manager: Bruce M Peight
■ Axial piston hydraulic checkball design pumps; compact low-speed, high-torque, hydraulic vane motors; hydraulic valves to 1040 bar; pumps 700 to 1040 bar, split flow hydraulic pumps. [98]

75-28430 Eagle Industrial Truck Mfg. Inc.
(† Eagle Bob Tail Tractors Inc.)
PO Box 8, Grosse Ile, MI 48138

☎(+1 313) 671 81 81 ⊖ (+1 313) 671 62 57
👥 25 (1995) 🏭 28,000 sq ft
President: Dennis Morrow
Marketing Director: Ralph Gillespie
Chief Engineeer: John Perry
■ Aircraft and ground support equipment tractors, warehouse tractors, trailers. 98

75-28500 E-A-R Specialty Composites
(a division of Aearo Company, USA)
7911 Zionsville Road, Indianapolis, IN 46268-1650
☎(+1 317) 692 11 11 ⊖ (+1 317) 692 31 11
👥 240 (1997)
Executive Vice President: M Rand Mallitz
Director Marketing and Sales Service: Richard Holub
Operations/Technical Director: Kevin McCormick
Purchasing Manager: William Bogden
Marketing Assistant: Ms Rose Clouse
■ Noise & vibration control materials, including MIL-SPEC submarine vibration damping tiles & marine barriers. 98

75-28800 Eaton Corporation
Corporate Headquarters
Eaton Center, Cleveland, OH 44114-2584
☎(+1 216) 523 50 00 ✆ (800) 386 70 14
[W] http://www.eaton.com ⊖ (+1 216) 523 47 87
♣ $ 6,961,000,000 (1996 Net) 👥 54,000 (1996)
Chairman and Chief Executive Officer: S R Hardis
President and Chief Operating Officer: A M Cutler
Executive Vice President and General Counsel: G L Cherlein
Vice President Chief Financial and Planning Officer: A T Dillon
Vice President Corporate Affairs: J W Hushen
Senior Public Relations Manager: R M Romain
Secretary and Assistant General Counsel: E R Franklin
Subs. with defence business:
AIL Systems Inc. → 75
Truck Components Operations
PO Box 4013, Kalamazoo, MI 49003
13100 East Michigan Avenue, Galesburg, MI 49053
☎(+1 616) 342 30 00 ⊖ (+1 616) 342 33 12
Senior Vice President Truck Components: T W O'Boyle
■ Medium & heavy-duty truck transmissions, axles & brakes.
Commercial & Military Control Ops.
4201 North 27th Street, Milwaukee, WI 53216
☎(+1 414) 449 60 00 ✉ EATON CUTLER
[TX] 26716 eaton ch mil ⊖ (+1 414) 449 62 21
Senior Vice President Semiconnductor and Specialty Systems: B R Bachman
Cutler-Hammer
5 Parkway Plaza, Pittsburgh, PA 15220
☎(+1 412) 937 61 00 ⊖ (+1 412) 937 67 40
Senior Vice President Cutler-Hammer: J L Becherer
■ Principal products: Cutler-Hammer. Relays, starters, switches, circuit breakers & systems for aerospace, industrial and military applications. 98

75-28980 ECC International Corp
Marketing and Executive Offices
175 Strafford Avenue, Wayne, PA 19087-3377
☎(+1 610) 687 26 00 ⊖ (+1 610) 254 92 68
■ Technology-based training for complex military and industrial systems, including weapons systems and communication networks. Provision of training analysts, human factors specialists full customer support. Simulation hardware and software. → 80
Systems Design and Production Center
2001 W. Oak Ridge Road, Orlando, FL 32809-3803
☎(+1 407) 859 74 10 ⊖ (+1 407) 851 18 71 98

75-29030 Edal Industries Inc.
4 Short Beach Road, East Haven, CT 06512-3519
☎(+1 203) 467 25 91 ⊖ (+1 203) 469 59 28
♣ $ 3,000,000 👥 60 🏭 15,000 m2
President: R Cowles
Vice President Sales: D Vagnini
Operations Director: F Truffa
■ Manufacture of rectifiers. 98

75-29100 Edison Aerospace & Defense
46 Mystic Circle, Bay Shore, NY 11706-7139
☎(+1 516) 968 13 18 ⊖ (+1 516) 968 62 60
👥 3 (1997) 🏭 2,000 sq ft
Sales and Marketing Director: Bill Harris
■ Supply all types of test equipment for testing fuel, hydraulic, pneumatic, electronic, electrical systems & components. 98

75-29250 EDO Corporation
14-04 111th Street, College Point, NY 11356-1434
☎(+1 718) 321 40 00 [W] http://www.edocorp.com
[TX] 423049 ⊖ (+1 718) 321 41 94
♧ $ 14,162,000 (1995) ♣ $ 91,113,000 (1995) 👥 846
🏭 50,000 sq m
Chairman: Frank A Fariello
Executive Vice President and Chief Operating Officer: Ira Kaplan
President & CEO: Frank A. Fariello
VP Finance & Treasurer: Kenneth A. Paladino
VP General Counsel: Marvin D. Genzer
VP Administration: W.J. Frost [E] wfrost@edocorp.com
■ Design & manufacture advanced electronic, mechanical, acoustic & composite products for military space, marine & commercial aviation markets.
Washington Office
2001 Jefferson Davis Highway, Suite 1105, Arlington, VA 22202-3688
☎(+1 703) 415 15 60 ⊖ (+1 703) 415 15 64
Vice President: Edward Zebrowski
Marine and Aircraft Systems
14-04 111th Street, College Point, NY 11356-1434
☎(+1 718) 321 40 00 ⊖ (+1 718) 939 01 19
General Mgr: Frank Otto
Finance Dir: John Scanlan
Research & Development Dir: Jonathan Schere
Marketing Mgr: Carl Fisher
■ Airborne/helicopter towed acoustic and magnetic mine countermeasure systems, high performance aircraft ejection and stores release units, bomb racks and missile launchers.
Combat Systems
1801-E Sara Drive, Chesapeake, VA 23320
☎(+1 804) 424 10 04 ⊖ (+1 804) 424 16 02
General Mgr: Milo Hyde
Controller: Dan Woods
Research & Development Dir: Jim Brunelle
Marketing Mgr: Julius Ceasar
■ Manufacture military and commercial electronic, ASW processors, sonar systems and command and control systems.
Electro-Optics (Barnes Engineering)
88 Long Hill Cross Roads, Shelton, CT 06484-0867
☎(+1 203) 962 17 77 ⊖ (+1 203) 926 92 85
■ Manufacture spacecraft earth sensors using electro-optic technology, thermal and infra-red testing of semiconductors, thermal and infra-red sensor systems.
Acoustic & Ceramic Products
2645 South 300 West, Salt Lake City, UT 84115-2968
☎(+1 801) 486 74 81 ⊖ (+1 801) 484 33 01
Vice President: Douglas Moore
General Mgr Acoustic Products: Joseph F. Luckey
General Mgr Ceramic Products: Gary D. Springfield
■ Manufacture of piezelectric components for various applications.
Fiber Science
506 North Billy Mitchell Road, Salt Lake City, UT 84116
☎(+1 801) 537 18 00 ⊖ (+1 801) 363 95 54
General Mgr: Michael G. Therson
■ Manufacture components for waste/water tanks for commercial aircraft. 98

75-29500 EEV Inc.
(a subsidiary of GEC Inc., USA)
4 Westchester Plaza, Elmsford, NY 10523
☎(+1 914) 592 60 50 [W] http://www.eevinc.com
⊖ (+1 914) 682 89 22
👥 36
President: Jack Collard
Vice President Finance and Administration: Rudy Winter
Sales Director: Stuart Hesselson
Vice President Marketing: Mike Kirk
Vice President Marketing: Howard Paul
Publicity Manager: Ms Ann Sayers
■ Professional electron tubes including travelling wave tubes, magnetrons, integrated microwave packages, klystrons, thyratrons, TR limiters, EMPs, electro-optical devices & LCDs.
® UK: EEV Ltd 98

75-29520 EFW Inc.
4700 Marine Creek Parkway, Fort Worth, TX 76136-6969
☎(+1 817) 234 66 00 [E] efw_inc@efw.aom
⊖ (+1 817) 234 67 00
🏭 155,000 sq ft
General Manager and Chief Executive Officer: T Taylor
Vice President of Programs: Lieut Gen C Pitman
Vice President of Operations: R Horowitz
Marketing Director: S Israel
■ Weapon system upgrades, complete cockpit upgrades, helmet mounted displays, digital map systems, rugged handheld computers, liquid colour displays and mission computers. Systems engineering, integrated logistics support, depot level maintenance, product support and build-to-print services. → 90 → 80
Representing::
US Military
Lockheed Martin
Boeing Helicopters
American Eurocopter
United Defense LP
Bell Helicopter
McDonnel Douglas
Washington Operations
1111 Jefferson Highway, Suite 605, Arlington, VA 22202-4306
☎(+1 703) 416 51 03 ⊖ (+1 703) 416 51 08 98

75-29552 EG&G Electro-Optics
35 Congress Street, Salem, MA 01970
☎(+1 508) 745 32 00 ⊖ (+1 508) 745 08 94
Manager: Paul Beech
■ Flash tubes for airborne laser targeting units. 98

75-29553 EG&G Mechanical Components Group
50 Sharpe Drive, Cranston, RI 02920
☎(+1 401) 463 56 40 ⊖ (+1 401) 463 68 20
President: Luciano Rossi
Aerospace Sales Dir: Peter Bennett
Business Development Dir: James Mellencamp
■ Pneumatic ducting systems & components, exhaust ducts, solenoid valves, dynamic mechanical shaft seals & welded metal bellows seals & devices. 98

75-29555 EG&G Inc.
45 William Street, Wellesley, MA 02181
☎(+1 617) 237 51 00 [W] http://www.egginc.com
⊖ (+1 617) 431 42 55
♧ $ 194,915,000 (1996) ♣ $ 1,427,252,000 (1996)
👥 15,000 (199)
Chairman, President and Chief Executive Officer: John M Kucharski
Senior Vice President and Chief Financial Officer: John F Alexander
Vice President: Angelo Castellana
Vice President: Luciano Rossi
Vice President Investor Relations and Corporate Communications: Ms Deborah S Lorenz
■ Design, develop & manufacture amorphous silicon detectors, micro-machined products, X-ray inspection systems & sensors for the automotive & medical markets, testing operations for light trucks & automobiles. 98

75-29558 EG&G Power Systems
1330 East Cypress Street, Covina, CA 91724
☎(+1 818) 967 95 21 ⊖ (+1 818) 967 31 51
General Mgr: Paul Beech
■ Power supply & switch mode power units for military & industrial use. 98

75-29562 EG&G Reticon
345 Potrero Avenue, Sunnyvale, CA 94086-4197
☎(+1 408) 738 42 66 ⊖ (+1 408) 738 69 79
President: Chris Raanes
Marketing/Communications Supervisor: Jody Small
■ Solid state image sensors and cameras. 98

75-29564 EG&G Rotron
9 Hasbrouck Lane, Woodstock, NY 12498
☎(+1 914) 679 24 01 [TX] TWX 510 247 9033
⊖ (+1 914) 679 70 80
President: Peter Stewart
■ Fans & blowers for cooling electronic equipment - land, sea & air based platforms.
Subsidiary companies:
FHA Inc., USA 98

75-29566 EG&G Sealol Engineered Products Division
15 Pioneer Avenue, Warwick, RI 02888
☎(+1 401) 781 47 00 [TX] TWX 681-4081
⊖ (+1 401) 781 09 30
VP & General Mgr: Edward Broadfield

■ Seals & bellows devices used in systems involving engines, gear-boxes, pumps & compressors. Applications in aerospace, marine, military & industrial markets. Welded-metal bellows seal valves for high-tech fluid containment & control systems. [98]

75-29590 EIP Microwave Inc.
1745 McCandless Drive, Milpitas, CA 95035-8024
☎ (+1 408) 945 14 77 ✆ (800) 232 3471
Ⓦ http://www.eipm.com ⊜ (+1 408) 945 09 77
♣ $ 6,492,000 (1997) 👥 45 (1997)
Chairman of the Board and Chief Executive Officer: J Bradford Bishop
Vice Chairman, Secretary and Treasurer: John F Bishop
President and Chief Operating Officer: Lewis R Foster
Sales and Marketing Director: Ivan Andres
Director of Sales: Bill White
Applications Engineer: Steve Ashby
■ Microwave/millimeter wave test equipment, frequency counters and signal sources. [98]

75-29650 Elcom Systems Inc.
932 Clint Moore Road, Boca Raton, FL 33487-2801
☎ (+1 407) 994 17 74 Ⓔ sales@elcomsystems.com
Ⓦ http://www.elcomsystems.com/home/
⊜ (+1 407) 994 17 70
♣ $ 200,000 👥 5
President: L Pollachek
Vice President of Sales and Marketing: H Pollachek
■ RF coaxial attenuators and terminations DC-4.2 GHz. [98]

75-29660 Eldec Corp.
(a subsidiary of Crane Co.)
16700 13th Avenue West, PO Box 97027,
Lynnwood, WA 98046-9727
☎ (+1 425) 743 13 13 ⊜ (+1 425) 743 82 34
▥ 6289-9566 👥 950 🏭 258 400 sq ft
President: Arian VanKoevering
Vice President General Manager Aerospace and Defence: Roy Robinson
Vice President Finance: Russell Fleetwood
Director Business Development and Management: Bruce Vogele
Director Product Support: John Ardussi
■ Custom designs and manufactures, services and supports electronics and electromechanical products and sub-systems for severe environment aerospace, defence, military ground based and marine applications, such as: proximity switches and sensors, position indication and control systems, true mass flowmeters, fuel gauging systems, liquid level sensing systems, high accuracy pressure sensors, electrical power controllers and systems, battery management systems, transformer-rectifiers and regulated converters, DC-DC switch mode power converters, power conditioning products, CRT power supplies and display backlight drivers.
International Subsidary:
Eldec France [98]

75-29680 Electric Tachometer Corp.
68th and Upland Streets, Philadelphia, PA 19142
☎ (+1 215) 726 77 23 ⊜ (+1 215) 724 36 82
♣ $ 2,300,000 👥 45
President: T J Porter
Vice President Sales and Marketing: F M Porter
■ Speed indicators (propeller shaft engine speed, turbine speed, bow-master). Rudder angle, engine order, log systems. [98]

75-29685 Electro Corporation
1845 57th Street, Sarasota, FL 34243-2296
☎ (+1 941) 355 84 11 Ⓦ http://www.electrocorp.com
⊜ (+1 941) 355 31 20
👥 175
President: Steve Anderson
Marketing and Sales Manager: John Kacinko
Product Manager: Michael Sons
Operations manager: Michael Stapleton
Marketing Services: Ms Lois Y McVay
■ Speed and proximity sensors. [98]

75-29690 Electrodynamics Inc. Minelco Products
(a subsidiary of Talley Industries Inc.)
1200 Hicks Road, Rolling Meadows, IL 60008
☎ (+1 847) 259 07 40 ⊜ (+1 847) 255 38 27
👥 270 🏭 50,000 sq ft
President: John W Kravcik
Operations Director: Don Quant
Sales Manager: Jim Stillwell
Purchasing Manager: Lou Pacini
■ Miniature electronic fault indicators and switches used in the military and commercial avionic and aerospace industries. LED & incandescent lighting indicators & annunciator panels. [98]

75-29692 Electro-Flex Heat Inc.
5 North Wood Drive, PO Box 88,
Bloomfield, CT 06002-1120
☎ (+1 860) 242 62 87 ⊜ (+1 860) 242 72 98
♣ $ 5,000,000 (1997) 👥 56 (1997) 🏭 20,000 sq ft
President and Chief Executive Officer: J R Dabasinskas
Director Advertising: Robert L Murphy
Operations Director: Karl Almquist
■ Silicone & kapton electrical flexible heaters. [98]

75-29695 Electroid Corp.
(a division of Valcor Engineering Corp.)
45 Fadem Road, Springfield, NJ 07081-3189
☎ (+1 201) 467 81 00 ⊜ (+1 201) 467 56 56
👥 100 (1995) 🏭 56,000 sq ft
President: Jack Rosa
Chairman Engineering: Rocco Dragone
Vice President Aerospace Marketing: Edward Lazorchak
Manager Advertising and Public Relations: Ms F S Foster
Product Manager: L Melanson
Purchasing Manager: T Muranski
■ Electromagnetic clutches/brakes, solenoids, power supplies, tension control devices. [98]

75-29700 Electro Impulse Laboratory Inc.
1805 Corlies Avenue, PO Box 278,
Neptune, NJ 07754-0278
☎ (+1 908) 776 58 00 Ⓔ eil@electro-impulse.com
Ⓦ http://www.electro-impulse.com
⊜ (+1 908) 776 67 93
♣ $ 7,400,000 👥 35 (1997)
President: Mark Rubin
■ Design and manufacture custom cooling equipment for military/industry. Also manufacture RF high power dummy loads and RF calorimeters. [98]

75-29760 Electro-Mavin
2985 East Harcourt Street,
Rancho Dominguez, CA 90221-5501
☎ (+1 310) 632 98 67 ✆ (800) 421 2442
⊜ (+1 310) 632 35 57
General Manager, Vice President: Michael Canyon
Purchasing: Robert Averbach
■ Coax.connectors, M39012, M55302, M28748, M81511, M83723, D38999, MIL-C 5015, UG connectors. [98]

75-29800 Electro-Methods Inc.
330 Governors Highway, PO Box 54,
South Windsor, CT 06074
☎ (+1 203) 289 86 61 ▧ TWX 710 425 6016
⊜ (+1 203) 289 18 68
♣ $ 29,000,000 (1996) 👥 175 🏭 120,000 sq ft
President: Randall Fries
Director Finance: Paul Keller
Vice President Operations Support: Dani Stephens
Purchasing Manager: Ms Joyce O'Donnell
Sales Manager: Ralph Foster
■ Aircraft parts manufacturer. [98]

75-29810 Electro-Metrics Inc.
(a Penril Company)
231 Enterprise Road, Johnstown, NY 12095
☎ (+1 518) 762 26 00 ▧ TWX 710 446 4798
⊜ (+1 518) 762 28 12
👥 73 🏭 65,000 sq ft
President: K W Bach
Chief Engineer: Carl Herzog
Manager Sales and Marketing: Michael A DiCaprio
Marketing Services Coordinator: Ms Valerie J Zabo
■ Communications equipment; instrumentation, antennas and sensors for signal intercept, monitoring, distribution, measurement, towers, turntables and amplifiers. [98]

75-29812 Electro-Optical Systems Group DRS Photronics, Inc.
(a subsidiary of DRS Technologies, Inc.)
270 Motor Parkway, PO Box 11368,
Hauppauge, NY 11788
☎ (+1 516) 231 95 00 ⊜ (+1 516) 231 95 01
President: Richard Ross
Director Marketing and Sales: Peter C Connolly
■ Design and manufacture multi-platform boresight equipment for rotary and fixed-winged aircraft. The company also manufactures sighting systems, infrared (IR) components and assemblies utilised in various weapons and targeting systems for defence applications.
DRS Optronics, Inc.
(a subsidiary of DRS Photronics)
2330 Commerce Park Drive, NE, Palm Bay, FL 32905
☎ (+1 407) 984 90 30 Ⓦ http://www.drs.com
⊜ (+1 407) 984 87 96
♣ $ 14,000,000 (1997) 👥 120 (1997) 🏭 54,000 sq ft
President: Richard Ross
Vice President of Operations: Irving J Lowe
Executive Director, Business Development: Edwin T Carlson
Engineering Director: Daniel D Young
Director, Opto Mechanical Business: Stan R Sheridan
Purchasing Manager: Ms Sallye Bates
Administration Assistant: K A Kelly
■ Range of products from infra-red, passive and panoramic sighting devices to sophisticated electro-optical equipment. [98]

75-29850 Electronic Concepts Inc.
526 Industrial Way West, Eatontown, NJ 07724-2212
☎ (+1 908) 542 78 80 Ⓔ sales@eci-capacitors.com
Ⓦ http://www.eci-capacitors.com ⊜ (+1 908) 542 05 24
♣ $ 15,000,000 (1997) 👥 200 (1997) 🏭 5,900 sq m
President: Bernard Lavene
Vice President Sales and Marketing: Philip Lepore
Operations Manager: Robert Landis
Purchasing Manager: Mark Fitzpatrick
■ Film dielectric capacitors. [98]

75-30100 Electronic Space Systems Corp. (ESSCO)
Old Powder Mill Road, Concord, MA 01742-4697
☎ (+1 508) 369 72 00 Ⓔ essco@tiac.net
⊜ (+1 508) 369 76 41
♣ $ 20,000,000 👥 135 🏭 60 000 sq ft
Chairman: Albert Cohen
Director, International Sales: Edwing P Twarog
International Sales Manager: Dale Meyers
Advertising Mgr: Stephen J. Contons
Marketing Administrator: Ms Katie Gallant
■ Radomes (metal space frame, sandwich, solid laminate, dielectric, air supported) for a variety of military & commercial applications. Precision moulded sandwich reflectors & feeds for aircraft antennas; precision antenna systems for satellite communications, radio astronomy and precision pointing and tracking applications. [98]

75-30150 Electronic Specialty Corp. (ESC)
14511 NE 13th Avenue, PO Box 3501,
Vancouver, WA 98668-3501
☎ (+1 360) 573 50 00 ⊜ (+1 360) 573 46 35
♣ $ 5,000,000 (1996) 👥 100 (1996) 🏭 5,500 m2
Chairman: H Zamarano
Vice President of Sales: Leon Fuller
Operations Director: Phil Dryden
Sales Manager: Ms Kim Wilson
Purchasing Manager: Ms Darlene Green
■ Crystal can, hermetically sealed, military grade relays and solenoids, contract assembly, voltage sensors, current and phase sensors.
Displays and Technologies, Inc (D&T)
(a subsidiary of Electronic Speciality Corp.)
14511 North East 13th Avenue, PO Box 3501,
Vancouver, WA
☎ (+1 360) 573 50 00 Ⓔ distech@teleport.com
⊜ (+1 360) 573 46 53
♣ $ 4,000,000 (1996) 👥 45 (1996) 🏭 5,500 m2
Chairman: H Zamarano
Managing Director: Leon LeCave
Operations Director: Frank Prevé
Sales Manager: Ms Kim Wilson
Purchasing Manager: Ms Donna Gerahty
■ Fabricate colour and contrast enhancement filters for flat panel displays. Ruggedization of displays through manufacture and optical bonding of filters, heaters, EMI/RFI shielding, special backlighting and NVIS compatibility. [98]

75-30220 Electronic Systems Group DRS Electronic Systems
(a division of DRS Technologies, Inc.)
138 Bauer Drive, Oakland, NJ 07436
☎ (+1 201) 337 38 00 ⊖ (+1 201) 337 47 75
President: Paul G Casner (Jr)
Senior VP Business Development: Frank Tortorelli
■ Design and manufacture signal processing systems, high-resolution display workstations, sonar trainer systems, radar scan converters and coastal surveillance systems for military and industrial applications.
DRS Medical Systems
31 Industrial Avenue, Mahwah, NJ 07430
☎ (+1 201) 825 84 14 ⊖ (+1 201) 825 85 16
President: Ron Hadani
General Manager: William O Koenig
DRS Technical Services, Inc.
5721 Thurston Avenue, Virginia Beach, VA 23455
☎ (+1 757) 464 40 94 ⊖ (+1 757) 464 17 22
Director: Roy Russell
DRS Technical Services, Inc.
2535 Camino Del Rio South, San Diego, CA 92108
☎ (+1 619) 299 32 10 ⊖ (+1 619) 299 77 59
Director: David Leedom [98]

75-30450 Electro Optical Industries Inc.
859 Ward Drive, Santa Barbara, CA 93111
☎ (+1 805) 964 67 01 [E] eoi@sonatech.com
⊖ (+1 805) 967 85 90
♣ 60 (1997) ⌂ 1,800 m2
President and Chief Operating Officer: Bruce W Malcolm
Sales Engineer: Steve White
■ Black body radiation sources, differential temperature targets, infrared detector test sets, radiometers, spectroradiometers infrared and visible target simulators, collimators, modulators. → 80 [98]

75-30500 Electrophysics Corp.
373 Rt. 46 West, Fairfield, NJ 07004
☎ (+1 201) 882 02 11 ∅ (800) 759 95 77
[E] info@electrophysicscorp.com
[W] http://www.electrophysicscorp.com
⊖ (+1 201) 882 09 97
♣ 25 (1997) ⌂ 2,000 sq m
Manager: Dr Frank J Vallese
Operations Director: Paul Stadelmann
■ Manufacture of infra-red & low-light level night vision viewers and TV cameras. [98]

75-30700 Electro-Tec Corp.
(a division of Kaydon Corp., USA)
1501 North Main Street, Blacksburg, VA 24060
☎ (+1 540) 552 21 11 [W] http://kaydoncorp.com
⊖ (+1 540) 951 38 32
♣ 200 (1997) ⌂ 100,000 sq ft
Vice President and General Manager: Ms JoAnna Sutton
Vice President of Sales: Paul M Schlueter [98]

75-30720 Elgar Corporation
(† Sorensen Company)
9250 Brown Deer Road, San Diego, CA 92121
☎ (+1 619) 450 00 85 ∅ (800) 733 5427 ⊠ 211063
⊖ (+1 619) 458 02 67
♣ 400 (1995) ⌂ 100,000 sq ft
President: Ken Kilpatrick
Vice President of Sales: Samuel A Lewis
Vice President of Operations: Ron Garrett
Vice President of Engineering: Norm Precourt
Marketing and Communications Manager: Ms Maggie Aguilar ☎ (+1 619) 458 02 91
Purchasing Manager: Lee Frabkhauser
■ Power converters. [98]

75-30800 Elgin Industries
1100 Jansen Farm Drive, Elgin, IL 60123-2555
☎ (+1 708) 742 17 20 [E] elginind@worldnet.att.net
⊖ (+1 708) 742 22 20
♣ 150
CEO & President: Martin Skok
VP Sales & Marketing: Bill Skok
■ Manufacture of high quality engine parts & chassis parts. [98]

75-30850 Elmwood Sensors Inc.
500 Narragansett Park Dr., Pawtucket, RI 02861-4325
☎ (+1 401) 727 1300 ∅ (800) 356 9663
⊖ (+1 401) 728 5390
President: T. Berlam
Marketing Dir: Stephen Fox
■ Manufacture thermal sensing & control products. [97]

75-30880 Eltec Instruments Inc.
PO Box 9610, Daytona Beach, FL 32120-9610
☎ (+1 904) 252 04 11 ∅ (800) 874 7780
⊖ (+1 904) 258 37 91
♣ 65
Sales and Marketing Manager: Ms Pamela Corrick
■ Passive infrared detectors and outdoor perimeter security devices. [98]

75-30900 Elts Unlimited Inc.
15 River Street, PO Box 1683, New Rochelle, NY 10801
☎ (+1 914) 576 27 02 ∅ (800) 382 00 79
⊖ (+1 914) 576 70 75
♣ $ 1,000,000 (1996) ♣ 15 (1997) ⌂ 8,000 sq ft
President: Ms Joan Goodman
Vice President Sales and Marketing: Ms Patricia M Wyllie
Purchasing Manager: Ms Judith Weber
Associated companies:
Emergency Beacon Corp. → 75
Glatzer Industries Corp. → 75 [98]

75-30985 Emco Inc.
(a subsidiary of Mid-South Industries Co.)
201 Industrial Parkway, PO Box 2193, East Gadsden, AL 35903
☎ (+1 205) 492 88 90 ⊖ (+1 205) 492 76 57
♣ $ 75,000,000 (1997) ♣ 360 (1997) ⌂ 110,000 sq ft
Managing Director: Jerry Weaver
Operations Director: Al Richards
Product Development Manager: James L Slichenmyer
Purchasing Manager: Michael Cook
■ Manufacturing facilities include large transfer die and high speed progressive die fabrication; precision secondary machining and metal finishing operations and high volume, automated assembly lines to support defense ordnance and munitions contracts. [98]

75-31050 Emergency Beacon Corp. (EBC)
15 River Street, PO Box 178, New Rochelle, NY 10801
☎ (+1 914) 235 94 00 ⊖ (+1 914) 576 70 75
♣ $ 1,000,000 + (1996) ♣ 16 (1997) ⌂ 8,000 m2
President: Ms Joan Goodman
Vice President Marketing: Ms Patricia M Wyllie
Vice President Production: Ms Judith L Weber
■ RF, R&D, DC to 1GHz, flight test. 1 aircraft test bed, multiengine. Expertise in low power, high efficiency transmitters & receivers.
Associated Company:
Glatzer Industries Corp.
ELTS Unlimited Inc. [98]

75-31100 Emergency One Inc.
(a subsidiary of Federal Signal Corp.)
1601 South West 37th Avenue, PO Box 2710, Ocala, FL 34474
☎ (+1 352) 237 11 22 [W] http://www.e-one.com
⊖ (+1 352) 237 11 61
President: Ted Fries
Vice President of Sales and Marketing: Jim Schwaller
Sales and Marketing Director: Jim Weigle
Advertising Director: Ms Diane Mathes
■ Manufacture of Titan HPR (High Performance Rescue) vehicle for commercial airports and P-23 ARFF (Aircraft Rescue Fire Fighting) vehicle for the US Air Force. [98]

75-31150 Emerson and Cumming Microwave Products Inc
(a subsidiary of Emerson & Cuming Microwave Products NV, Belgium)
869 Washington Street - Suite 1, Canton, MA 02021
☎ (+1 617) 821 94 40 ∅ (800) 650 57 40
[E] emicrowav@aol.com ⊖ (+1 617) 828 96 43
♣ BFr 70,000,000 (1996) ♣ 16 (1997)
General Manager: M P Carpenter
Operations Director: Ihor Mykyta
Marketing Manager: J T Gear
■ Manufacture of microwave absorbers, RF attenuation materials, absorbers for antenna pattern control, surface current attenuation materials, RCS reduction materials, shielding materials. [98]

75-31450 Emro Engineering Co. Inc.
PO Box 536, Summit, NJ 07901
☎ (+1 201) 539 2466 ⊖ (+1 201) 539 2466
Chairman & President: Robert W. Ruppert
VP Sales: R.W. Carver
■ Aircraft simulation systems. [97]

75-31540 ENDECO/YSI Inc.
13 Atlantis Dr, Marion, MA 02738-1448
☎ (+1 508) 748 0366 ⊠ 929451 endeco mnms
⊖ (+1 508) 748 2543
President: Malte von Matthiessen
Mgr General Business: Kevin J. McClurg
Sales Mgr: Roger Race
■ Oceanographic, hydrographic, towed underwater systems, wave track directional buoys. Water quality environmental instruments. [97]

75-31545 Endevco Corp.
(a Meggit plc Company)
30700 Rancho Viejo Road, San Juan Capistrano, CA 92675
☎ (+1 714) 493 81 81 [E] communication@endevco.com
[W] http://www.endevco.com ⊠ 685608
⊖ (+1 714) 661 72 31
♣ 440
President: Phil Conrad
Finance Director: Tom Plante
Director: Mike Will
■ Vibration, shock and pressure measurement systems used in ground test and flight test applications. [98]

75-31560 Engineered Air Systems Inc. (EASI)
(a division of Engineered Support Systems, Inc.)
1270 North Price Road, St Louis, MO 63132
☎ (+1 314) 993 58 80 ⊖ (+1 314) 567 40 52
♧ $ 14,096,000 (1996) ♣ $ 60,000,000 (1997)
♣ 220 (1997) ⌂ 400,000 sq ft
Chairman: Michael F Shanahan (Sr)
President: John J Wichlenski
Vice President of Marketing: Ronald W Davis
Purchasing Director: Jerry L Lyles
■ Provide systems integration, management skills and services required by all branches of the military. Design, manufacture and market systems and ground support equipment (including environmental control systems for Minuteman and Peacekeeper), NBC equipment, air conditioners, fuel and water distribution equipment and containers.
® UK: Aardvark Group [98]

75-31565 Engineered Fabrics Corp.
(a division of K&F Industries)
669 Goodyear Street, Rockmart, GA 30153
☎ (+1 770) 684 78 55 ⊖ (+1 770) 684 74 38
♣ 500
President: Roger Martin
Vice President of Marketing: Terry Lindsey
■ Fuel tanks for aircraft, helicopter and ground vehicles. Electrical heated aircraft ice protection equipment. Military and commercial coated fabric products. Elastomer cements and coatings. Air inflatable retarders. [98]

75-31568 Engineered Magnetics Inc. (EM)
18435 Susana Road, Rancho Dominguez, CA 90221
☎ (+1 310) 635 95 55 ⊠ 9103256208 engrd magntics
⊖ (+1 310) 631 03 13
♣ 94 (1997) ⌂ 100,000 sq ft
President: S J Miller
Vice President of Engineering and General Manager: Ernest R Schoop
Vice President, Marketing: Ronald J H Chilcott
Matériel Director: Ms Arleen Bailey
Operations Manager: Anthony Pniewski
■ Electrical power conditioning equipment for military and aerospace including power supplies, DC/DC converters, inverters, battery chargers and systems, missile launcher power units, UPS, electronic ballasts. [98]

75-31570 Engineered Plastics Inc.
211 Chase Street, PO BOX 227, Gibsonville, NC 27249-0227
☎ (+1 919) 449 41 21 [E] epi@netpath.net
[W] http://www.industry.net/epi ⊖ (+1 919) 449 63 52
♣ 30 (1997) ⌂ 7,000 sq m
President: D M Davidson III

National Sales Manager: R C Ratliff
Plant Manager: R D Brashier
Purchasing Manager: B Mattison
■ Custom fabricated components. Precision machining - plastics and non-ferrous metals. Thermoforming/vacuum forming, injection moulding. [98]

75-31590 Engineered Systems Co. (ESCO)
(a division of Datron Inc.)
2550 Market Street, Aston, PA 19014
☎ (+1 610) 494 80 00 ⊜ (+1 610) 49- 89 89
♣ $ 30,000,000 👥 100 🏭 11,625 sq ft
President: T S Melvin
Vice President Marketing: G Greer MacMasters
Director of Engineering: E S Lopez
Operations Director: J Guida
Purchasing Manager: R McHale
■ Aircraft arresting systems, and special material handling systems. [98]

75-31600 Engineering Equipment Co.
1020 West 31st Street, PO Box 1590,
Downers Grove, IL 60515-0790
☎ (+1 630) 963 78 00 [E] engreqmt@aol.com
⊜ (+1 630) 963 71 23
♧ $ 5,000,000 ♣ $ 40,000,000 👥 30
President: Robert J. Peot
VP Sales & Marketing: Frank J. Cullen
■ Design & production of ballistic resistant security vehicles for government, commerce & personal safety. Firefighting vehicles. [98]

75-31660 ENIDINE INCORPORATED
(a subsidiary of International Motion Control Inc.)
7 Centre Drive, Orchard Park, NY 14127-2293
☎ (+1 716) 662 19 00 ✆ (800) 852 85 08
[W] http://www.enidine.com ⊜ (+1 716) 662 19 09
👥 145 🏭 40,000 sq ft
President: Benjamin Houghton
Director New Business Development: Thomas J Miller
Operations Director: Roger Metz
Sales and Marketing Director: Dennis Morrow
Marketing Manager: Kristan Celani
Purchasing Manager: Bob Reault
■ Design & manufacture of energy absorption & vibration isolation products including hydromechanical buffers, dashpots, dampers, isolators, rate controls, time delays, actuators, gas spring replacement units, seat recline devices, wire rope isolators, elastomeric isolation mounts, air springs and shock absorbers. [98]

75-31670 Prosser/Enpo
(† Enpo Pump Co.)
(a subs. of Crane Pumps & Systems, Inc.)
420 Third Street, PO Box 603, Piqua, OH 45356-0603
☎ (+1 397) 773 24 42 [E] cranegov@juno.com
⊜ (+1 937) 773 22 38
President: Paul Baldetti
VP Sales & Marketing: Tom Stone
Dir Government Operations: Wayne R. Marshall
■ Submersible direct current bilge pumps for amphibious military vehicles. Self-priming centrifugal pumps including bulk transfer for water or fuel. End section centrifugal pumps. Macerator centrifugal sewage pumps. Centrifugal sewage pumps. Centrifugal shipboard sewage transfer pumps. [98]

75-31700 The Ensign Bickford Co. Aerospace & Specialty Products Group
(a division of Ensign-Bickford Industries)
660 Hopmeadow Street, Simsbury, CT 06070
☎ (+1 860) 658 44 11 ⊜ (+1 860) 843 26 21
Managing Director: Leonard J Mecca
Marketing Director: John J Walsh
Business Development Manager: Michael K Voisine
☎ (+1 860) 843 23 29
■ Ordnance components and systems for aerospace, military and specialty systems. [98]

75-31750 Enstrom Helicopter Corp.
2209 22nd Street, PO Box 490, Menominee, MI 49858
☎ (+1 906) 863 12 00 [E] enstrom@cybrzn.com
⊜ (+1 906) 863 68 21
🏭 105,000 sq ft
President and Chief Executive Officer: Robert Tuttle
Vice President Marketing and Service: Robert J Cleland
Vice President Engineering: Robert Jenny
Vice President Manufacturing: John Hansen
Vice President Controller: Charles Schumacher
■ F28F & 280FX turbocharged piston helicopters. TH28/480 Turbine powered helicopters for commercial, utility & executive applications. [98]

75-31850 Environmental Elements Corp. (Enelco)
PO Box, Baltimore MD 21203
3700 Koppers Street, Baltimore MD
☎ (+1 410) 368 70 92 [W] http://www.eec1.com
⊜ (+1 410) 368 70 81
👥 135
Chairman: F Bradford Smith
President and Chief Executive Officer: E H Verdery
Vice President and General Manager: S Michael Dunseith
Vice President and Director, International: Michael G Mullendore
Vice President and New Product Development: Gregory Carleton
Project Director, Power Division: Robert W Tisone
Project Director, Industrial Division: Donald R Hug
Controller: Ms Susan L Rosebery
Investor Relations and Corporate Communications: Ms Lisa A Morris
■ Design and manufacture air pollution control systems. Electroplastic precipitator, fabric filter, flue gas desulphurisation and acid gas cleaning systems. Aftersales parts and services. [98]

75-31860 Environmental Technologies Group Inc
1400 Taylor Avenue, PO Box 9840,
Baltimore, MD 21284 9840
☎ (+1 301) 321 5114; 5200 ⊜ (+1 301) 321 5255
♣ $ 42 000 000 👥 335
■ Specialises in the design & manufacture of electro-mechanical products such as chemical agent detectors & meteorological measuring equipment. [97]

75-31920 Envirovac Inc.
(a subsidiary of Evac International OY)
1260 Turret Drive, Rockford, IL 61115-1486
☎ (+1 815) 654 83 00 ⊜ (+1 815) 654 83 06
♣ $ 30,000,000 👥 110 🏭 50,000 sq ft
President: Tom Beatty
Director of Marketing: Walt Peters
Operations Director: Jim Egger
Purchasing Manager: Paul Meehan
■ EVAC Sewage treatment systems for marine-vessels and ORCA marine sanitation device. Vacuum powered toilet systems for commercial building, marine, aviation and trash. [98]

75-32000 Equipment & Supply Inc.
4507 Highway 74 W., Monroe, NC 28110
☎ (+1 704) 289 6565 ⊜ (+1 704) 283 1206
👥 32 🏭 140 000 sq m
President: Andrew A. Adams
Office Mgr: Ms Bobbie K. Rivers
Sales Mgr: L.L. Jester (Jr)
■ Quality aircraft tooling & equipment, aircraft test equipment, maintenance & work stands, hydraulic jacks, aircraft & engine special tools, aircraft towbars, cargo systems. [97]

75-32020 The Eraser Co. Inc.
Oliva Drive, PO Box 4961, Syracuse, NY 13221
☎ (+1 315) 454 32 37 ✆ (800) 724 05 94
[E] eraser@dreamscape.com ⊜ (+1 315) 454 30 90
President: Ralph E Bevard
Operations Director: Roger Falconer
Sales Manager: Ms Linda Cahill
■ Wire stripping, component lead forming, reeling & measuring equipment, infra-red heat tools, industrial fibreglass brushes & wire wheels. [98]

75-32025 ERG Inc.
900 Stanford Avenue, Oakland, CA 94608-2320
☎ (+1 510) 658 97 85 ⊜ (+1 510) 658 74 28
👥 25
President: B O Benson
Vice President of Sales and Marketing: Bryan D Leyda
Director of Public Relations: Ms Melody A Carter
■ Duocel foam metal and components, Trident D-5 foam metal nozzle baffle, Tomahawk shock attenuators. Titan IV fluid cooler interchanger. [98]

75-32250 Esmet Inc.
1406 5th Street SW, PO Box 9238,
Canton, OH 44702-2062
☎ (+1 330) 452 91 32 ✆ (800) 321 08 70
[E] esmet@imperium.net ⊜ (+1 330) 452 25 57
👥 48
President: J S Ripple
■ High security weapon storage equipment, high security locks, fittings for wire, synthetic rope & cable. [98]

75-32360 Esterline Technologies
10800 NE 8th Street, Suite 600, Bellevue, WA 98004
☎ (+1 206) 453 94 00 ⊜ (+1 206) 453 29 16
♣ $ 352,843,000 (1996) 👥 3,100 (1996)
Chairman, President and Chief Executive Officer: Wendell P Hurlbut
Executive Vice President and Chief Financial Officer: Robert W Stevenson
Executive Vice President and Chief Operating Officer: Robert W Cremin
Group Vice President: Larry A Kring
Group Vice President: Stephen R Larson
Group Executive: James J Cich
Vice President Human Resources: J M Greenberg
Director Corporate Development: Rickard Carovano
Manager Corporate Communications: Brian D Keogh
Communications Specialist: Ms Susan M Conn
■ Cable assemblies, combustible ammunition components, information display indicators, specialty elastomeric compounds in custom moulded shapes, clamps and support blocks for wiring installations, control sticks.
Subs. with defense business:
Armtec Defense Products Co.
TA Mfg Co.
Hytek Finishes Co.
Korry Electronics Co.
Mason Electric Co.
Midcon Cables Co.
Auxitrol, SA [98]

75-32550 Eurometaal Inc
PO Box 321, Broad Run, VA 22014
☎ (+1 540) 349 09 21 ⊜ (+1 540) 349 17 20
President: Al Houweling [98]

75-32650 Evans & Sutherland Computer Corp. Simulation Division
600 Komas Drive, Salt Lake City, UT 84108
☎ (+1 801) 688 10 00 ⊜ (+1 801) 588 45 11
♣ $ 131,000,000 (1996) 👥 800
President and Chief Executive Officer: James R Oyler
Vice President of Marketing: Greg Phipps
General Manager of Government Business Sector: Ronald Sutherland
General Manager of Commercial Business Sector: Stuart Anderson
Business Development: Dave Janke
■ Computer-generated visual systems for flight simulation, mission rehearsal, naval simulation and vehicle engineering applications. Image generators, simulator domes, displays, screens and projectors. Night vision, FLIR and sensor simulation.
Other locations::
Orlando, Florida; Horsham, UK; Munich, Ger; Beijing, China. [98]

75-32655 Eventide Inc.
One Alsan Way, Little Ferry, NJ 07643-1001
☎ (+1 201) 641 12 00 [W] http://www.eventide.com
⊜ (+1 201) 641 16 40
Marketing Manager: Ms Lana Steffen
■ Argus moving map displays. The Argus works in conjunction with Loran-C, GPS & other long-range navigation systems. TSO & FAA-approved. UK CAA approval on Argus 7000 model. [98]

75-32657 Everest Imaging Ltd
PO Box 80269, Austin, TX 78708-0269
☎ (+1 512) 833 94 08
[W] http://www.everestimaging.com
⊜ (+1 512) 833 94 10
♣ US$ 18,000,000 - 20,000,000 (1996) 👥 30
🏭 7,000 sq ft
President: Jim Wagner
Sales and Marketing Director: Andrew Wood
Purchasing and Operations Controller: Art Daniels
Marketing Manager: Ms Ann Tyler

■ Manufacture inspection equipment (such as boroscopes) with industrial applications including inspection of aircraft engines. [98]

75-32680 Everpure Inc.
(a subsidiary of Culligan Water Technologies Inc.)
660 Blackhawk Drive, Westmont, IL 60559-9005
☎ (+1 630) 654 40 00 Ⓦ http://www.everpure.com
⊖ (+1 630) 654 11 15
240 110,000 sq ft
Manager Marketing Communications: Richard N Verson
Manager Technical Markets: Greg Johnson
■ Water purification & treatment equipment. [98]

75-32725 Exceltech International Corp.
1110 Industrial Boulevard, Sugarland, TX 77478-2569
☎ (+1 281) 240 67 70 Ⓦ http://www.sanilec.com
⊖ (+1 281) 240 67 62
♣ $ 14,000,000 45
President: A J Vaccaro
Vice President of Sales and Marketing: Dan Shepard
Operations Director: Frank Martin
Purchasing Manager: Robert Skinner
Marketing Services Administrator: Ms Jolene Wolfe
■ Omnipure marine sewage treatment plants to disinfect waste for safe discharge into harbours and ports. [98]

75-33000 Exstar International Corp.
301 E-1 N. Green Meadows Drive, Wilmington, NC 28405
☎ (+1 910) 452 47 37 ✆ (800) 500 44 34
⊖ (+1 910) 452 47 38
♣ US$ 600,000 (1995) 6 2,000 m2
President: Charles E Mattes
General Manager: Cecil Hunt
■ Manufacture and supplies water purification equipment (reverse osmosis) for military vessels including minesweepers, submarines, patrol boats and frigates. Also supplies Reverse Osmosis Water Purification Units (ROWPU) for Army and Marine field use. [98]

75-33140 Ez Form Cable Inc.
315 Peck Suite, Building 24, New Haven, CT 06513-2933
☎ (+1 203) 785 82 15 ⊖ (+1 203) 785 04 66
♣ $ 6,500,000 (1996) 45 (1996) 15,000 sq ft
President: Eduard Parker
Vice President Sales: Frank Volpe
■ Miniature semi-rigid coxial cable (Mil-C-17), coaxial cable assemblies, delay lines. [98]

75-33200 Fairchild Defense
(a division of Orbital Sciences Corporation)
Headquarters
20301 Century Boulevard,
Germantown, MD 20874-1181
☎ (+1 301) 428 64 77 Ⓦ http://www.orbital.com
⊖ (+1 301) 428 64 33
♣ US$ 68,000,000 (1996) 800 (1996)
President and Chief Operating Officer: Jack Frohbieter
Executive Vice President of Engineering: John Bigelow
Executive Vice President of Engineering: Jerry Catbone
Senior Vice President and General Manager of Fairchild Defence: Donald P Ryan
Vice President of Finance, Controller and Treasurer: Seth Brown
Vice President of Marketing: Jim Thomason
Marketing Services Manager: Ms Leslie Fox
■ Design, develop and manufacture digital avionics systems including data management/ mass memory systems, mission support systems, reconnaissance management systems, weapon management systems and armoured data systems. [98]

75-33330 Fairey Microflitrex Inc.
(a subsidiary of Fairey Microfiltrex, UK)
454 South Anderson Road, Rock Hill, SC 29730
☎ (+1 803) 327 50 08 Ⓔ fairey.microfiltrex@infoavc.net
Ⓣ 264436 ⊖ (+1 803) 327 10 52
♣ US$ 15,000,000 (1996) 120 (1997) 35,000 m2
Managing Director: Adrian Ryder
Sales and Marketing Director: Terry Stott
Operations Director: Jon Cardy
Purchasing Manager: Les White
Sales and Aplications Manager: Jeff Mothersbaugh
■ Manufacture fuel, oil, air and hydraulic filters, filter manifolds and differential pressure indicators.
® UK: Fairey Arlon Ltd [98]

75-33400 Fairfield Manufacturing Company Inc
US 52 South, PO Box 7940, Lafayette, IN 47903-7940
☎ (+1 317) 474 34 74 ⊖ (+1 317) 477 73 42
1,100 500,000 sq ft
Vice President Sales and Marketing: Frederick G Sharp
■ Provider of custom-designed gears and gearing systems, proprietary gearing systems and complete drive assemblies to defence manufacturers. [98]

75-33600 Fairview Machine
427 Boston Street, Topsfield, MA 01983-1218
☎ (+1 508) 887 21 41 Ⓔ fairview@netway.com
⊖ (+1 508) 887 36 89
♧ $ 3,000,000 (1997) ♣ $ 5,000,000 (1997)
50 (1997) 30,000 sq m
President: Michael J. Moulton
Managing Dir: David Snow
VP Sales & Marketing: Jim Biron
Sales & Marketing Dir: Ed Pinho
Operations/Technical Dir: Ron Moulton
Purchasing Mgr: Mike Cerat
■ All aerospace, missile & electronics component parts. [98]

75-33800 Farrand Controls
(† Inductosyn International)
(a div. of Ruhle Companies)
99 Wall St, Valhalla, NY 10595-1447
☎ (+1 914) 761 2600 ⊖ (+1 914) 761 0405
55 100 000 sq ft
President & CEO: Frank Ruhle
Exec. VP/Treasurer: D.S. Feron
Sales & Marketing Mgr: G.P. Zalewskli
■ Precision angle transducer for inertial navigation, precision angle readout, precision linear readout, angle encoding, readouts for antenna positioning controls. [97]

75-33870 Fastening Systems International Inc. (FSI)
1206 East MacArthur Street, Sonoma, CA 95476-1372
☎ (+1 707) 935 11 70 Ⓔ sales@fsirivet.com
Ⓦ http://www.fsirivet.com ⊖ (+1 707) 935 18 28
♣ $ 2,700,000 (1996) 7 (1997)
President: Roger M Nikkel
■ Manufacture blind fastening tool kits and fastener kits for repair, modification and maintenance of military aircraft and ground vehicles. [98]

75-33877 FATS, Inc. (FATS)
(† Firearms Training Systems Inc.)
7340 McGinnis Ferry Road, Suwanee, GA 30174
☎ (+1 770) 813 07 55 ⊖ (+1 770) 688 35 30
♣ $ 65,000,000 (1995) 280 2,998 m2
President and Chief Executive Officer: Peter Marino
International Sales Director: J C de Ledebur
International Sales Administrator: Ms Carol Bryant
■ Small arms simulators, including shoot/no-shoot decision-making trainers, marksmanship trainers for individuals and teams. Collective trainers for entire squads. [98]

75-34030 Federal Industries Inc.
645 Hawaii Street, PO Box 999,
El Segundo, CA 90245-0999
☎ (+1 310) 297 40 40 ⊖ (+1 310) 297 40 04
22
President and Chief Executive Officer: Avi Wacht
Vice President of Sales and Marketing: Asher Bartov
■ Kits and spare parts for military equipment. [98]

75-34080 FEI Microwave Inc.
(a subsidiary of Frequency Electronics Inc.)
55 Charles Lindbergh Boulevard, Mitchel Field, NY 11553
☎ (+1 516) 794 45 00 ⊖ (+1 516) 794 43 40
♣ $ 30,000,000 (1997) 300 (1997) 20,830 m2
President: Martin Bloch
Director Marketing: Gene Kushner
■ Microwave subsystems for ground and space satellite systems. [98]

75-34100 Fenwal Safety Systems
700 Nickerson Road, Marlborough, MA 01752
☎ (+1 508) 481 5800 ⊖ (+1 508) 485 1485
200
Vice President and General Manager: John Sullivan
IEP Sales & Marketing: D.A. Garuicci
■ Specialist manufacturer of heat detectors including bleed air leak detection, industrial explosion protection. [98]

75-34330 Fibertek
1180 N. Mountain Springs Pkwy, PO Box 910,
Springville, UT 84663-0910
☎ (+1 801) 489 3672 ⊖ (+1 801) 489 3671
♣ $ 12 800 000 (1995) 150 11 800 sq m
Managing Dir: Gil Apodaca
Business Development Mgr: Samuel L. Myers
Operations Dir: Bob Schick
Purchasing Mgr: Grant Hall
■ Helicopter structure & external fuel tanks, missile structures & launch tubes, marine composites; housings, transducer shells, armour panels & kits for air, sea & ground vehicles, water & waste tanks. [97]

75-34340 FIC Corp.
12216 Parklawn Drive, Rockville, MD 20852
☎ (+1 301) 881 81 24 ✆ (800) 638 6594
⊖ (+1 301) 881 05 30
President and Chief Executive Officer: John C Linton
Vice President: Robert E Frye
Marketing Manager: Arthur E Rosfeld
■ Manufacture military fuzeholders in accordance with MIL-F-19207. Distribute all lines of fuzes, terminal blocks, military models and commercial. → 85 [98]

75-34370 Fike Corp.
(† Fike Metal Products Corp.)
704 South 10th Street, PO Box 610,
Blue Springs, MO 64015
☎ (+1 816) 229 34 05 Ⓔ fmpsales@fike.com
Ⓦ http://www.fike.com Ⓣ 42263 ⊖ (+1 816) 228 92 77
300 (1997) 200,000 m2 (1997)
Chairman: Lester Fike (Jr)
President: Maxwell Jewell
Sales and Marketing Director: Robert DeGood
Advertising Manager: Jeff Salmon
International Operations: Paul Mastilak
■ Pressure relief devices: rupture discs. [98]

75-34480 FJW Optical Systems Inc.
629 South Vermont Street, Palatine, IL 60067-6949
☎ (+1 847) 358 25 00 ⊖ (+1 847) 358 25 33
9
President and Chief Executive Officer: Frank Warzak
Purchasing Manager: Leonard Kern
Sales and Marketing: Lavera Warzak
■ Infra-red viewers. [98]

75-34700 Flight Products Inc.
6060 Nicolle Street, PO Box 5790, Ventura, CA 93003
☎ (+1 805) 644 57 02 Ⓣ 182068 ⊖ (+1 805) 644 43 64
♧ $ 10,000,000 ♣ $ 20,000,000 220 60,000 sq ft
President: Clint Butcher
Vice President of Sales: Peter Brocklesby
Vice President of Manufacturing: Larry Franc
■ Manufacture airframe components and electronics for C130, F4, F5 and F16 aircraft. [98]

75-34800 Flight Suits Ltd
1675 Pioneer Way, El Cajon, CA 92020
☎ (+1 619) 440 27 00 ✆ (800) 748 6693
Ⓔ mailbox@flightsuits.com Ⓦ http://www.flightsuits.com
⊖ (+1 619) 440 46 18
♧ $ 1,500,000 (1996) ♣ $ 6,100,000 (1996) 90
19,000 sq ft
President: James A Wegge
Sales and Marketing Director: Thom Nutter
■ Nomex flight suit, jackets, gloves. Flight helmets. Leather flight jackets. [98]

75-34840 Flightline Aircraft Instruments
(† Aero Mechanism Inc.)
(a subs. of Flightline Electronics)
7500 Main Street, Fishers, NY 14453
☎ (+1 716) 924 4000 ⊖ (+1 716) 924 5732
♣ $ 2 500 000 30
President: William Blossom
VP Sales & Marketing: Timothy McGraw
■ Design, development & manufacture of pressure sensing instruments for commercial & military aircraft, surface ships & submarines. [97]

75-35060 Floturn Inc.
120 Progress Place, Cincinnati, OH 45246-1718
☎ (+1 513) 671 02 10 ⊖ (+1 513) 671 70 33
♣ $ 25,000,000 (1997) 👥 135 (1997)
President: Robert V Glutting
Vice President Sales and Marketing: Tony D Grouse
Director Advertising: Phil Wesley
Vice President Engineering: Mike North
Purchasing Manager: Mike DeRosa
■ Shape charge liners (Copper) for 105 mm, 120 mm, anti-tank. Parts for Dragon, Hellfire, SMAW & many others requiring the use of the "shearforming process". [98]

75-35200 Fluorglas
14 McCaffrey St, Hoosick Falls, NY 12090
☎ (+1 518) 686 7301 ⊖ (+1 518) 686 4840
Director: Ed Scheytt
■ PTFE coated glass tapes, fabrics & fabricated parts; other pressure sensitive adhesive tapes, extruded & skived PTFE films. [97]

75-35320 FN Manufacturing Inc (FNMI)
(a subsidiary of FN Herstal SA, Belgium)
US Highway 1 North, PO Box 24257, Columbia, SC 29224
☎ (+1 803) 736 05 22 ⊖ (+1 803) 699 93 73
♣ US$ 48,000,000 👥 400 🏭 140,000 sq ft
President and Chief Executive Officer: James H Ritter [98]

75-35330 Foerster Instruments Inc.
(a subsidiary of Institut Dr Förster, Germany)
140 Industry Drive, RIDC Park West,
Pittsburgh, PA 15275-1028
☎ (+1 412) 788 89 76 ✆ (+800) 635 06 13
[E] foerster@usaor.net [W] http://www.foerstergroup.com
⊖ (+1 412) 788 89 84
♣ $ 10,000,000 (1996) 👥 62 (1996) 🏭 26,500 sq m
President: R.D. Shaffer
National Sales Manager: Tim Brown
Purchasing Mgr: J. May [98]

75-35650 Formsprag-Warren
Dana Corp, Warner Electric Division
23601 Hoover Road, PO Box 778, Warren, MI 48090
☎ (+1 810) 758 50 00 ✆ (800) 927 32 62
⊖ (+1 810) 758 52 04
Plant Manager: Bill Wilson
Sales and Marketing Manager: Michael P Travis
■ Overrunning clutches, sprag retainer assemblies, form-lock and reverse-lock no-backs to counteract reversibility of gear trains and ball screws. [98]

75-35680 Fotec Inc.
151 Mystic Avenue, Medford, MA 02155-4615
☎ (+1 617) 396 61 55 ✆ (800) 537 82 54
[E] info@fotec.com [W] http://www.std.com/fotec
⊖ (+1 617) 396 63 95
👥 25
Sales Manager: William B Pierce
■ Equipment & instrumentation for fibre optic communications, including test & installation equipment for fibre optic comms networks (LANs and FDDI), telephone systems & tactical communications systems. [98]

75-35752 Foxtronics Inc.
3448 West Mockingbird Lane, Dallas, TX 75235
☎ (+1 214) 358 44 25 ✆ (800) 259 24 93
⊖ (+1 214) 358 15 19
♣ $ 2,200,000 (1996) 👥 14 (1997) 🏭 16,000 sq ft
President and Chief Executive Officer: David W Hill
Director of Manufacturing: Don Lecceadone
Sales and Marketing Director: Bob Underwood
■ Design and manufacture of electrical ground power units, aircraft security systems, aircraft NiCad battery temperature systems, aircraft antenna tuning systems, aircraft audio amplifiers. → 90 [98]

75-35753 FP Diesel LP
(a member of the FEL-PRO corporate family)
5412 McFadden Avenue, Huntington Beach, CA 92649
☎ (+1 714) 898 99 55 ✆ (800) 237 74 25
[E] marketing@fpdiesel.com ▣ 683432
⊖ (+1 714) 894 93 56
👥 200
Marketing: Thuc Le
■ Diesel engines and components for vehicles and small surface ship propulsion and power generation. [98]

75-35755 FPI Inc.
Hydrodyne Div.
3125 Damon Way, Burbank, CA 91505
☎ (+1 818) 841 9667 ⊖ (+1 818) 841 8057
👥 22 🏭 1 700 sq m
President: James Boyd
VP Marketing & Sales: Yvon Pierre Cariou
Technical Dir: W. Darlington
Purchasing Mgr: D. Quesenberry
■ Machined static metal seals, high pressure 20 000 PSI, high & low temperature, cryogenic to 2 000°F, machined metal bellows, inconel & other high strength materials. Filters: Fluid & gasses. Precision burst discs assembly (cryogenic & aerospace), relief valves, solenoid valve regulators. Machining of complex parts, exotic alloys. [97]

75-35760 FPT Inc.
(a subs. of Westland Inc.)
25 Anita Ave, La Selva, CA 95076
☎ (+1 408) 662 0432 ⊖ (+1 408) 662 0807
Contact: John M. Golden
■ Flexible, crash-resistant, self-sealing fuel cells, flotation equipt for helicopters, inflatable seats, self-sealing couplings, mouldings, airport canopies, adhesives, lacquers & coated fabrics. [97]

75-35850 Franz Inc.
1995 University Avenue, Suite 275,
Berkeley, CA 94704-1058
☎ (+1 510) 548 36 00 [W] http://www.franz.com
⊖ (+1 510) 548 82 53
President: Russel Henke
Operations Director: Jack Kemp
Sales and Marketing Director: Jim Veitch
Purchasing Manager: Ms Margaret Schleuter
Marketing Assistant: Ms Cathy Foderaro
■ Provides object-oriented development tools for artificial intelligence, automation, computer-aided design and other fields. Provides Allegro Cl with CLOS, an industry-standard common Lisp and Allegro Composer, an interactive window-based development environment. [98]

75-35900 Frasca International Inc.
906 East Airport Road, Urbana, IL 61801
☎ (+1 217) 344 92 00 [E] marketing@frasca.com
[W] http://wwww.frasca.com ⊖ (+1 217) 344 92 07
👥 200
President: Rudy Frasca
Operations/Technical Dir: Marty Hendrickson
Sales/Marketing Dir: Joel Prichard
Technical Training Specialist: Don Cook
Advertising/Promotions Mgr: Ms Peggy Prichard
Purchasing Mgr: Ray Ronk
■ Aircraft simulators both fixed & rotary wing. [98]

75-35950 Fraser-Volpe Corp.
1025 Thomas Drive, Warminster, PA 18974
☎ (+1 215) 443 52 40 [E] fraservolp@aol.com
⊖ (+1 215) 443 09 66
♣ $ 6,000,000 (1993) 👥 50 (1995) 🏭 20,000 sq ft
President: Joseph B Volpe (Jr)
Managing Director: C R White
Operations Director: A J Aldinger
Sales and Marketing Director: Mike Imhof
Purchasing Manager: H Smith
■ Innovative electro-optical systems for military & law enforcement applications. Systems vary from the M61 Fire Control System to the Stedy-Eye stabilised monoculars, binoculars & platforms. [98]

75-35970 Frazier Aviation Inc.
11311 Hartland Street, North Hollywood, CA 91605
☎ (+1 818) 985 17 11 ⊖ (+1 818) 985 23 68
👥 35
President: L P Frazier
Vice President and General Manager: Charles E Ricard
Director of Manufacture and Overseas Operations: Robert L Frazier III [98]

75-36000 Frederick Electronics SFA/DataComm Inc.
(a division of SFA Inc.)
7450 New Technology Way, PO Box 502,
Frederick, MD 21701-0502
☎ (+1 301) 662 59 01 ✆ (800) 270 2669
[E] sales@sfa.com [W] http://www.sfa.com ▣ 197639
⊖ (+1 301) 694 62 79
♣ $ 7,000,000 (1997) 👥 55 (1997) 🏭 110,000 sq ft
Chairman: Dr William C Moyer
President: Robert Fonow
Controller: Richard Deater
Engineering Director: Courtland Thomas
Administration Director: Ms Lisa Broome
Operations Director: Talbott Arnold
Sales Manager: Joe Gibson
Business Development Manager: Ms Julie Bloecher
Marketing Assistant: Ms Heather Beck
Senior Sales Representative: Phil Conti
Senior Sales Representative: Charlie Miller
■ HF Modems, Automatic Link Establishment (ALE) Radio Controllers. [98]

75-36100 Freeman Marine Equipment Inc.
28336 Hunters Creek Road, PO Box F,
Gold Beach, OR 97444
☎ (+1 541) 247 70 78 ⊖ (+1 541) 247 21 14
👥 52
Managing Director: F A Shindler
■ Manufacture marine hatches, doors, portlights, windows and accessories. [98]

75-36190 Frequency Electronics Inc. (FE)
55 Charles Lindbergh Boulevard, Mitchel Field, NY 11553
☎ (+1 516) 794 45 00 ⊖ (+1 516) 794 43 40
♣ $ 30,000,000 (1997) 👥 300 (1997) 🏭 230,000 sq ft
Managing Director: Joe Franklin
Director of Marketing: Gene Kushner
Purchasing Manager: Peter Klopsis [98]

75-36300 Capewell Aerial Delivery Systems Frost Engineering Division
(† Frost Engineering)
(a division of Capewell Components Co.)
105 Nutmeg Road South, South Windsor, CT 06074
☎ (+1 860) 635 40 45 ⊖ (+1 860) 610 01 20
♣ $ 6,000,000 (1996) 👥 75 (1997) 🏭 50,000 sq ft
Managing Director: Ken MacCormac
President: John F Limbach ☎ (+1 303) 761 1010
[E] johnf@capewell.com
Purchasing Manager: Bill Gorlach
■ Development & production of life support type equipment (personnel lowering devices, parachute releases, restraint harness etc.) & aerial cargo delivery (by parachute) components & systems - principally for military use. Training services for above product lines. [98]

75-36340 Froude Engineering Inc.
39201 Schoolcraft, Ste B16, Livonia, MI 48150
☎ (+1 313) 462 15 66 ▣ 234057 ⊖ (+1 313) 462 12 77
👥 5
President: R Gott
Sales Manager: Raymond Rucky
Manager: Ken Barnes
Office Administrator: Ms Ann Johnson
■ Test equipment & measurement instrumentation including engine & vehicle dynamometers, transmission test rigs, automatic computer control systems & engine & plant services. [98]

75-36380 Furon Company
(† The Fluorocarbon Co.)
29982 Ivy Glenn Dr., Laguna Niguel, CA 92677
☎ (+1 714) 831 5350 ⊖ (+1 714) 643 1548
♣ $ 344 886 000 (1996) 👥 2 400 (1996)
President & COO: T. Noonan
Dir Communications: R. Bissell
■ Specified high performance materials for insulation, shielding, fire blocking, sound and vibration dampening. Special silicone rubber for high/low temp gaskets. Surface protection and masking against chemicals. Low friction surfaces. [97]

75-36400 Furon CHR
(† CHR Industries Inc.)
(a div. of Furon)
CHR Division
407 East St, New Haven, CT 06509-9988
☎ (+1 203) 777 3631 ✆ (800) 777 2647
⊖ (+1 203) 787 1725
👥 126

Managing Dir: Woody Swift
Managing Dir: Ed Scheytt
Product Leader: Yudkin Scott
■ Pressure sensitive tapes, silicone rubber, silicone & teflon coated fabrics.

Laguna Niguel
29982 Ivy Glenn Drive, Laguna Niguel, CA 92667
☎ (+1 714) 831 5350 ⊖ (+1 714) 643 1548

Rolling Meadows
Rolling Meadows, IL 60008
☎ (+1 708) 392 8090 ⊖ (+1 708) 392 0403

Anaheim
3340 East La Palma Avenue, Anaheim, CA 92806
☎ (+1 714) 630 5818 ⊖ (+1 714) 630 6632

Mickleton
I-295 and Harmony Road, Mickleton, NJ 08056
☎ (+1 609) 423 6630 ⊖ (+1 609) 423 8182

Los Alamitos
4412 Corporate Center Drive, Los Alamitos, CA 90720
☎ (+1 714) 995 1818 ⊖ (+1 714) 761 1270

Sunnyvale
1032 Morse Avenue, Sunnyvale, CA 94089
☎ (+1 408) 734 4711 ⊖ (+1 408) 744 1761

Fremont
322 East Warren Avenue, Fremont, CA 94539
☎ (+1 510) 651 2022 ⊖ (+1 510) 651 3440

Cape Coral
1531 Commerce Creek Blvd, Cape Coral, FL 33909
☎ (+1 813) 995 8111 ⊖ (+1 813) 995 8027

Mundelein
1150 East allanson Road, Mundelein, IL 60060-3898
☎ (+1 708) 949 0850 ⊖ (+1 708) 949 0198

Holmseville
8351 County Road 245, Holmseville, OH 44633-8351
☎ (+1 216) 279 3711 ⊖ (+1 216) 279 3791

Aurora
1199 South Chillicothe Road, Aurora, OH 44202
☎ (+1 216) 562 9111 ⊖ (+1 216) 562 5717

Macromeric
1395 Danner Drive, Aurora, OH 44202
☎ (+1 216) 995 1600 ⊖ (+1 216) 995 1699

Kent
642 Middlebury Road, Kent, OH 44240
☎ (+1 216) 678 0046 ⊖ (+1 216) 678 4631

Mantua
10585 Maint Street, Mantua, OH 44255
☎ (+1 216) 274 3171 ⊖ (+1 216) 274 5199

Bristol
386 Metacom Avenue, Bristol, RI 02809-5126
☎ (+1 401) 253 2000 ⊖ (+1 401) 253 1608

Houston
1977 Kindred Street, Houston, TX 77049
☎ (+1 713) 458 5830 ⊖ (+1 713) 458 0502

Mt Pleasant
1300 Industrial Boulevard, Mt Pleasant, TX 75455
☎ (+1 903) 572 3475 ⊖ (+1 903) 572 1331

N. Bennington
Water Street, PO Box 428,
North Bennington, VT 05257-0428
☎ (+1 802) 442 8884 ⊖ (+1 802) 442 8488

Seattle
3711 South Hudson Street, Seattle, WA 98118
☎ (+1 206) 723 5600 ⊖ (+1 206) 723 6869 [97]

75-36500 Furno Co. Inc.
168 San Lorenzo St, Pomona, CA 91766-2391
☎ (+1 909) 629 3041 ⊖ (+1 909) 623 7882
👥 35
President: Jim Smith
Operations Dir: C. Thomas
Purchasing Mgr: L.J. Rozar
■ Job-machine shop specialising in ultra-precision & miniature machining & fabrication; contract manufacturing. [97]

75-36550 Furon Co. Holmesville Site
(† Felsted)
(a subsidiary of Furon Co., USA)
8351 C.R. 245, Holmesville, OH 44633
☎ (+1 330) 279 37 11 ✆ (800) 262 33 57
⊖ (+1 330) 279 37 91
Director: D Klingelhoefer
Product Leader: E B Rucinski
■ Cables, controls, electronic controls, dipsticks, FAST cable assembly programme. [98]

75-36650 Futronix Inc.
612 East Taylor Avenue, Sunnyvale, CA 94086-3814
☎ (+1 408) 735 11 22 [E] futronix@juno.com
⊖ (+1 408) 735 16 16
👥 8
President: A H Erlund
Vice President of Marketing: J Mooney
■ Manufacture 24hr battery clock used in mobile command vehicles. [98]

75-36680 FWD International Inc.
1020 West 31st Street, PO Box 1590,
Downers Grove, IL 60515
☎ (+1 630) 963 78 00 ⊖ (+1 630) 963 71 23
♣ $ 2,000,000 👥 40
Vice President: F J Cullen
■ Manufacture heavy duty all-wheel drive vehicles for military forces. Heavy equipment transporter system. MMV modular military vehicle. [98]

75-36700 FXC Corp.
3410 South Susan Street, Santa Ana, CA 92704-6997
☎ (+1 714) 556 74 00 [E] fxc@pia.com
[TX] TWX 910 5952 801 ⊖ (+1 714) 641 50 93
♣ $ 8,000,000 👥 75 🏭 30,000 sq ft
President: Frank Chevrier
Sales and Marketing Director: Howard Weiske
Purchasing Manager: Hans Wachter
Marketing: Rick Velazquez
Sales Administrator: Ms Cheryl Hicks
■ Life support equipment including: parachute assemblies, personnel parachutes, cargo chutes, recovery chutes and harnesses/containers; automatic sensing devices, barometric timers, ripcord releases, speed sensors, cargo releases, weapons delivery systems; HALO systems: high altitude, free-fall and pre-breathers oxygen; test equipment, altitude test chambers. Cargo Delivery Systems with GPS capability. RPV recovery systems.

Guardian Parachute
3412 South Susan Street, Santa Ana, CA 92704-6997
☎ (+1 714) 557 80 32 [TX] TWX 910 5952 801
⊖ (+1 714) 641 50 93 [98]

75-36715 Gabriel Electronics Inc.
Scarborough Industrial Park, PO Box 70,
Scarborough, ME 04070-0070
☎ (+1 207) 883 5161 ⊖ (+1 207) 883 4469
👥 240
VP Sales & Marketing: Donald Crisman
■ Microwave point-to-point parabolic antennas & transmission line systems. Installation services, program management & tower/site maintenance programs. [97]

75-36755 Galileo Corp.
(† Galileo Electro-Optics Corp.)
Galileo Park, PO Box 550, Sturbridge, MA 01566
☎ (+1 508) 347 91 91 [W] http://www.galileocorp.com
⊖ (+1 508) 347 38 49
♣ $ 42,634,000 (1996) 👥 180 (1997)
President: William T Hanley ☎ (+1 508) 347 42 58
Chief Financial Officer: Gregory Riedel
☎ (+1 508) 347 42 22
Sales Director: Robert White ☎ (+1 508) 347 43 41
Purchasing Manager: Cecile Bernier
Marketing Communications Coordinator:
Ms Melissa Duggan ☎ (+1 508) 347 43 18
■ Manufacture & market fiberoptic & electro-optics components which transmit, sense or intensify light or images. [98]

75-36900 GAP Instrument Corp.
100 Horse Block Road, Yaphank, NY 11980-9504
☎ (+1 516) 924 17 00 [E] ceo@gap.net
⊖ (+1 516) 924 17 99
♧ $ 200,000,000 (1997) ♣ $ 40,000,000 (1997)
👥 150 (1997) 🏭 43,000 sq ft
Chief Executive Officer: James M Edwardson
President: Robert Baer
Operations Director: Lawrence Monahan
Sales and Marketing Director: Barry Green
Chief Engineer: Robert Lafferty
Purchasing Manager: John Stadt
■ Servo packages and electromechanical systems solid state converters, analog and hybrid subsystems. Computer-sequenced power management systems, navigation and fire control systems, special test and monitoring systems.

Also Value Added Network licensed to traffic to and from the Federal Acquisition Network. [98]

75-36920 Garmin International
1200 East 151 Street, Olathe, KS 66062
☎ (+1 913) 397 82 00 ✆ (800) 800 1020
[W] http://www.garmin.doc ⊖ (+1 913) 397 82 82
President: Gary Burrell
Vice President: Min Kao
Marketing Director: Gary Kelley
Media: Jim Stothard
■ Designs, manufactures and markets navigation and communications electronics worldwide for consumer, business and military applications. [98]

75-36950 Garrett Metal Detectors
1881 West State Street, Garland, TX 75042
☎ (+1 972) 494 61 51 ⊖ (+1 972) 494 18 81
👥 75 (1995) 🏭 70,000 sq ft
President and Chief Executve Officer: Charles L Garrett
Managing Director: Bob Podhrasky
Director of Sales and Marketing: Jim Dobrei
Operations and Technical Director: Jeff Flansbaum
■ Design & manufacture metal detectors ranging from units that fit in one's palm to highly sophisticated instruments that utilize computerized microprocessor technology. [98]

75-37412 GEC-Marconi Aerospace Inc.
(† Plessey Dynamics Corp.)
(a subsidiary of GEC-Marconi Avionics, UK
110 Algonquin Parkway, Whippany, NJ 07981
☎ (+1 201) 428 98 98
[E] 76245.1237@compuserve.com
⊖ (+1 201) 884 22 77
♣ $ 40,000,000 (1996/97) 👥 260
President: David A Sapio
Vice President Business Development: David Crook
Vice President Operations: Robert Gill
Director Product Support: Frank Fallon
Purchasing Manager: F Fedorczyk
Sales Manager: Don Powers
■ Electro-mechanical actuators for aircraft, ground defence systems, etc. Electrical motors and starter motors for aircraft. Power transmission equipment. [98]

75-37413 GEC-Marconi Avionics Inc.
2975 Northwoods Parkway, Norcross, PO Box 81999, Atlanta, GA 30366
☎ (+1 770) 448 1947 [TX] 708447 ⊖ (+1 404) 449 6128
👥 240 🏭 84,000 sq ft
Chairman: P D Clark
Vice President, Sales and Marketing Director: M McGlade
Corporate Information Officer: Peter Simmons
Vice President, Operations Director: H Kalbach
Director Programs: Vince Di Loreto
Vice President Administration: Ed Kelso
Vice President Engineering: Alan McEndoo
■ Design, manufacture & integrated logistic support of the avionic & industrial products supplied by GEC-Marconi Avionics to the United States. Also provides support to other associated GEC companies.
® UK: GEC-Marconi Avionics Ltd

Seattle
Westpark Building N, 8510 154th Avenue NE,
Redmond, WA 98052-3500
☎ (+1 206) 883 2525 ⊖ (+1 206) 881 3370
Manager: P. Clarke

Los Angeles
3400 Airport Avenue, Suite A, Santa Monica, CA 90406
☎ (+1 310) 391 63 30 ⊖ (+1 310) 391 68 60
Sales Mgr: W. Gist

Fort Worth
6410 Southwest Boulevard, Suite 128,
Fort Worth, TX 76109
☎ (+1 817) 763 02 81 ⊖ (+1 817) 732 36 81
Manager: J Harlow [98]

75-37414 GEC-Marconi Hazeltine Corp.
(† GEC-Marconi Electronic Systems Corp.)
(a subsidiary of The General Electric Company plc, UK)
164 Totowa Road, PO Box 975, Wayne, NJ 07474-0975
☎ (+1 201) 633 60 00 [W] http://www.systems.gec.com
⊖ (+1 201) 633 61 67
♣ $ 320,000,000 (1997) 👥 2,100 (1997) 🏭 67,200 m2
President and Chief Executive Officer: Mark Ronald
Vice President Operations: G A Mucha
Vice President Business Development: John P Sputz

Director of Purchasing: J S Schroeder
Marketing Support Services Manager: Raymond J Spellman
■ CNI programs (Link 16 Tactical including JTIDS, MIDS, system integration, advanced CNI, Programmable Digital Radio and SATCOM).
Electronic Identification Systems (Mark XII products, IFF systems, Combat ID) Communication Programs (airborne receivers, adaptive wireless network systems, software programmable radio, spread spectrum LPI, Remote Monitoring System).
Avionic Products (Doppler navigation, landing systems, altimeters).
Display systems (airborne C2 display systems, military colour displays, ground-based air defence and data display systems).
Inertial products (multisensor, two-axistrate gyro, Advanced Guidance Packages).
Ocean systems (anti-submarine warfare). [98]

75-37415 GEC Marconi Hazeltine Corp Electronic Systems Division
(† Hazeltine Corp.)
(a subsidiary of GEC Marconi North America Group)
450 Pulaski Road, Greenlawn, NY 11740-1606
☎ (+1 516) 261 70 00 ⊖ (+1 516) 262 80 02
♣ $ 150,000,000 (1997) 👥 850 (1997) 🏭 585,000 sq ft
President: R E Daugherty ☎ (+1 516) 262 81 14
Vice President Engineering & Technology: E D Linville ☎ (+1 516) 262 80 08
Vice President Business Development: J T Flynn ☎ (+1 516) 262 80 30
Director of Business Development: J T McCabe ☎ (+1 516) 262 83 71
Manager Business Development: S A Kinser
■ Information electronics. Identification systems, display systems, navigational aid systems, ASW systems & intelligent antenna systems for cell sites. → 80
Ocean Systems
115 Bay State Drive, Braintree, MA 02184-5203
☎ (+1 617) 848 34 00 ⊖ (+1 617) 843 21 53
General Manager: R Kielmeyer [98]

75-37425 GEC Plessey Semiconductors
Sequoia Research Park, 1500 Green Hills Road, Scotts Valley, CA 95066
☎ (+1 408) 438 29 00 ⊖ (+1 408) 438 70 23
Representing:
GEC Plessey Semiconductors → UK [98]

75-37427 Geier & Bluhm Inc.
594 River Street, Troy, NY 12180-1599
☎ (+1 518) 272 69 51 [W] http://www.geier-bluhm.com
⊖ (+1 518) 272 05 33
♣ $ 1,350,000 (1996) 👥 26 (1996) 🏭 1,000 sq m
President: Russell L. Oster
Vice President: David B. Oster
■ Spirit levels & level assemblies, circular levels, tubular levels, plate levels, machinists' levels, inclinometers. [98]

75-37447 General Atomics
PO Box 85608, San Diego, CA 92186-9784
☎ (+1 619) 455 30 00 ⊘ (800) 854 2233 ⊠ 695065
⊖ (+1 619) 455 36 21
👥 1,200 (1997) 🏭 1,000,000 sq ft
Chairman and Chief Executive Officer: J Neal Blue
Senior Vice President: David I Roberts
Business Development Manager: Marvin J Woodring
Marketing Communications Manager: Charles S Luby
■ Research and development of nuclear spacepower, electromagnetic systems, advanced materials, robotic systems, chemical weapon demilitarization, signature control, automated non-destructive inspection, repair technologies, hazardous waste elimination, unmanned air vehicles, electronics, superconducting magnets/ materials and advanced computing. [98]

75-38400 General Electric Co. (GE)
Corporate Headquarters
3135 Easton Turnpike, Fairfield, CT 06431
☎ (+1 203) 373 22 11 ⊠ 643893; TWX 710 453 1089
⊖ (+1 203) 373 29 92
♣ $ 79,179,000,000 👥 239,000
Chairman & CEO: John F. Welch (Jr)
Vice Chairman: Paolo Fresco
Vice Chairman: John D Opie
Sr Vice President Human Resources: William J. Conaty
Sr Vice President Research & Development: Dr Lewis Edelheit
Sr Vice President General Counsel & Secretary: B.W. Heineman (Jr)
Sr Vice President Finance: Dennis D. Dammerman
Senior Vice President and Chief Information Officer: Gary M Reiner
Vice President Sr Counsel: Ms Pamela Daley
Vice President Public Relations: Ms Joyce Hergenhan
Corporate Media Relations: Bruce Bunch
→ 80
Defence business operations.:
GE Aircraft Engines → 75
Business Information Center
Suite 200, One Winners Circle, Capital Plaza, Albany, NY 12205
☎ (+1 518) 869 55 55 ⊘ (800) 626 20 04 ⊠ 145354
⊖ (+1 518) 869 28 28 [98]

75-38500 General Electrodynamics Corp.
8000 Calendar, PO Box 150089, Arlington, TX 76015-6089
☎ (+1 817) 572 03 66 ⊘ (800) 551 6038
[E] info@gecscales.com [W] http://www.gecscales.com
⊖ (+1 817) 572 03 73
♧ $ 4,200,000 (1997) ♣ $ 8,500,000 (1997)
👥 75 (1997) 🏭 20,000 sq ft
President: Tom Smith
Executive Vice President: Harold O Thomas
Sales and Marketing Director: James Jobe
Operations and Technical Director: Andrew Spear
Purchasing Manager: Sam McQuerns
■ Weight and balance aircraft weighing equipment, ground support test and calibration equipment, high resolution instrumentation and portable truck weighing equipment. [98]

75-38700 General Instrument Corp.
8770 West Bryn Mawr Avenue, Chicago, IL 60631
☎ (+1 773) 695 10 00 ⊖ (+1 773) 695 10 21
♣ $ 2,600,000,000 (1996) 👥 14,000
Chairman and Chief Executive Officer: Richard S Friedland
Vice President Corporate Communications: Richard D Badler
Vice President and Chief Financial Officer: Charles T Dickson
■ Development of technological systems and product solutions for the interactive delivery of video, voice and data.
Field Service Facility
961 Norfolk Square, Norfolk, VA 23502
☎ (+1 804) 455 88 10 ⊖ (+1 804) 455 85 01
Grant Pass Office
PO Box 1209, Grants Pass, OR 97526-0306
☎ (+1 503) 479 3331 ⊠ TWX 510 755 0775
⊖ (+1 503) 479 0011
Warner Robins Office
60 Green St, Suite A-1, Warner Robins, GA 31093
☎ (+1 912) 922 4404 ⊖ (+1 912) 922 3075 [98]

75-38900 General Motors Corp. (GMC)
Corporate Headquarters
3044 West Grand Boulevard, Detroit, MI 48202
☎ (+1 313) 556 50 00 ⊖ (+1 313) 974 58 58
♣ $ 164,000,000,000 (1996) 👥 647,000 (1996)
Chairman, CEO and President: John F Smith (Jr)
Vice Chairman: Harry J Peace
Executive Vice President: Louis R Hughes
Executive Vice President: G Richard Wagoner
Executive Vice President: J T Battenberg III
Chairman Hughes Electronics: C Michael Armstrong
Executive Vice President and Chief Financial Officer: J Michael Loch
Vice President Communications: John C Onoda
Director Media Relations: John F Mueller
■ Supply of tactical & utility vehicles, electronics, missiles, spacecraft, heavy-duty transmissions & computer services to US Armed Forces.
Subs. with defence business:
GM Hughes Electronics Corp. → 75
Hughes Aircraft Co. → 75
Delco Electronics Corp. → 75
Electronic Data Systems Corp. [98]

75-38920 General Oceanics Inc. GO Environmental
1295 NW 163rd Street, Miami, FL 33169
☎ (+1 305) 621 28 82 [E] 73410.515@compuserve.com
[W] http://www.business1.com/genocean
⊖ (+1 305) 621 17 10
Managing Director: Larry Kopel
Director of Sales and Marketing: Dan Schaaf
Technical Director: Rick Wood [98]

75-38950 General Pneumatics
440 Washington Street, Orange, NJ 07050-1932
☎ (+1 201) 678 6555 ⊖ (+1 201) 678 9632
♧ $ 1 000 000 ♣ $ 2 500 000 👥 32 🏭 20 000 sq ft
President: Irving Raskin
Sr Contract Administrator: Pete Napierkowski [97]

75-39010 General Thermodynamics Inc.
(a division of Insilco Corp.)
190 West Marquette, Oak Creek, WI 53154-2033
☎ (+1 414) 764 56 00 ⊖ (+1 414) 764 56 30
👥 220
President: Matthew Connell
VP Sales & Marketing: David Ulmen
Purchasing Mgr: Keith Kasch
■ Radiators & oil coolers for diesel powered equipment. [98]

75-39220 Genicom Corp.
Relay Products Div.
1 Solutions Way, Waynesboro, VA 22980-1999
☎ (+1 540) 949 14 71 ⊠ TWX 710 839 04 07
⊖ (+1 540) 949 14 16
♣ $ 16,000,000 (1995) 👥 250
Vice President and General Manager: Dennis M Harbin
Director of Relay Marketing and Sales: Gary Peltier
Operations Manager: Richard Hopkins
■ MIL-R-39016 qualified hermetically sealed crystal can relays. [98]

75-39300 Gentex Corp.
PO Box 315, Carbondale, PA 18407
☎ (+1 717) 282 3550 ⊖ (+1 717) 282 8555
President: L.P. Frieder (Jr)
Marketing Dir: Charles Rudolf
■ Manufacture military life support systems, oxygen masks, air, ground & space shuttle helmets. Manufacture Electret microphones, optical & opthalmic products, laser protective optics. Aluminized, engineered, ballistic & chemical defence textiles.
Electro-Acoustic Products
Five Tinkham Ave, Derry, NH 03038
☎ (+1 603) 434 0311 ⊖ (+1 603) 434 3002
General Mgr: Marcus Smith
■ Manufacture miniaturised Electret microphones, conversational communications units, custom electronic components.
Western Operations
11525 Sixth St, Rancho Cucamonga, CA 91730
☎ (+1 909) 481 7667 ⊖ (+1 909) 481 7759
General Mgr: Tony VanHaastert
■ Manufacture aviation oriented oxygen masks and related hardware, custom helmets.
Gentex Office Inc.
(† Optical Products, W.Main Branch)
PO Box 307, Dudley, MA 01570
☎ (+1 508) 943 3860 ⊖ (+1 508) 949 0261
General Mgr: Robert Nahmias
■ Manufacture lenses. [97]

75-39530 Gerlinger Carrier Co.
665 Murlark NW, PO Box 5288, Salem, OR 97304
☎ (+1 503) 585 41 91 ⊖ (+1 503) 364 13 98
Sales Manager: Rob Cripe
■ Straddle carriers. [98]

75-39540 GET Engineering Corp.
9350 Bond Avenue, San Diego, CA 92021
☎ (+1 619) 443 82 95 [E] getntds@sierra.net
[W] http://www.getnids.com ⊖ (+1 619) 443 86 13
♣ $ 8,000,000 (1996) 👥 25
President and Chief Executive Officer: Rodney Tuttle (Jr)
Sales and Marketing: Wally Moeller
■ Naval tactical data systems (NTDS) interface adaptors MIL-STD-1397 for VME, PC/AT, SBUS, PCI application firmware, CAB/ES/accessories. [98]

75-39550 G&H Technology Inc.
750 West Ventura Boulevard, Camarillo, CA 93010
☎ (+1 805) 484 05 43 ⊖ (+1 805) 987 50 62
♣ $ 50,000,000 👥 185 🏭 120,000 m2
Chairman and Chief Executive Officer: Thomas C Cleary
President: Carl Weiss

Vice President of Sales and Marketing: Ms Di Ann Norkus
Technical Writer: Dennis Archambault
■ Manufacture connectors for special and MIL-SPEC applications. [98]

75-39570 Gibbs Rifle Co.
RR 2, Box 214, Martinsburg, WV 25401
☎ (+1 304) 274 04 58 [W] http://www.gibbsrifle.com
⊖ (+1 304) 274 00 78
♣ $ 3,000,000 (1995) 👥 31 (1995) 🏢 40,000 sq ft
Chairman: Val J Forgett (Jr)
President: Valmore J Forgett (III)
■ Manufacturer of Parker Hale M85 Sniping rifles, official sniping rifle of Canada, New Zealand, Austrailia and Sri Lanka. Also used by the Brazilian armed forces. Import, export and distribution of surplus arms and ammunition. → 85 [98]

75-39610 Gichner Systems Group
Optima Electronic Packaging Systems
2166 Mt Industrial Boulevard, Tucker, GA 30084-5088
☎ (+1 770) 496 40 00 [E] sales@optimaeps.com
[W] http://www.optimaeps.com ⊖ (+1 770) 496 40 41
👥 150 (1997)
President and Chief Executive Officer: J M O'Connor
Director of Advertising: Ms Valerie Cruz
Sales Director: R F Hall
Purchasing Manager: Danny Ford
■ Instrument packaging systems; electronic cabinets, desk systems, instrument cases and power and cooling accessories. [98]

75-39620 M.C. Gill Corp.
4056 East Street, El Monte, CA 91731-1970
☎ (+1 818) 443 40 22 [W] http://www.mcgillcorp.com
⊖ (+1 818) 350 58 00
👥 235 🏢 231,000 m2
President and Chief Executive Officer: Stephen E Gill
Director of Marketing: Irv J Freund
■ Ballistic laminates, ducting/tubing, contoured shaped composites, honeycomb Sandwich Panels, Composite Laminates. [98]

75-39720 RJR Industries
(† GKS Inc.)
25670 First Street, Westlake, OH 44145-1434
☎ (+1 216) 835 08 10 ⊖ (+1 216) 835 11 60
♣ $ 500,000 (1997) 👥 7 🏢 20,000 sq ft
President: Robert R Rankin [98]

75-39750 Glaser Safety Slug Inc.
PO Box 8223, Foster City, CA 94404
☎ (+1 510) 785 77 54 ✆ (800) 221 34 89
⊖ (+1 510) 221 66 85
President: C Kurt Canon
Administration Vice President: Ms Marilyn A Canon
■ Manufacture and distribution of the Glaser Safety Slug, a proprietary line of personal defence ammunition, Glaser gun accessories, including cheekpieces for the Colt AR-15 and Car-15. [98]

75-39755 Glasseal Products
485 Oberlin Avenue, Lakewood, NY 08701
☎ (+1 908) 370 91 00 ⊖ (+1 908) 370 71 07
👥 150
President: Rick Ferraid [98]

75-39760 Glatzer Industries Corp. (GIC)
15 River Street, PO Box 1291, New Rochelle, NY 10801
☎ (+1 914) 576 27 00 ✆ (800) 382 00 79
⊖ (+1 914) 576 70 75
♣ $ 1,000,000 (1996) 👥 16 (1997) 🏢 8,000 sq ft
President: Ms Joan Goodman
Vice President Sales and Marketing: Ms Patricia M Wyllie
Technical Director: Jorge Ramos
Purchasing Manager: Ms Judith Weber
Associated Company:
Emergency Beacon Corp.
ELTS Unlimited Inc. [98]

75-39769 Global Airtech International
(a subs. of Global Airtech Mfg Inc.)
10312 Norris Ave, Unit A, Pacoima, CA 91331
☎ (+1 818) 897 3317 ⊖ (+1 818) 897 3646
♣ $ 5 000 000 👥 15 🏢 9 885 sq ft
President: Rodney Chiamulon
Marketing Mgr: Steven Fendrik
Office Mgr: Ms Angela Davis
■ Aerospace hardware & fasteners, large turbine & aircraft parts. Over 136 000 part numbers stocked. US overhaul consolidation services for foreign airlines. → 80 → 85 [97]

75-39770 Global Defense Products Inc.
38-01 23rd Avenue, Astoria, NY 11105
☎ (+1 718) 204 1010 ✉ GLOBCOM [T] 237593
⊖ (+1 718) 626 3566
♧ $ 13 255 000 (1994) ♣ $ 162 007 534 (1995) 👥 74
🏢 26 230 sq ft
President: S.G. Fassoulis
Vice President: James Chladek
VP & General Mgr: Robert Spitz
Operations & Technical Dir: Kenneth Lateer
Dir Sales & Marketing: Robert Volvo
Purchasing Mgr: Robert Thompson
■ Supplier of military hardware, weapons and ammunition. → 80
Subsidiary company:
Global Communications Co. → 75 [97]

75-39775 Global Communications Co.
(a div. of Global Defense Products Inc.)
38-01 23rd Avenue, Astoria, NY 11105
☎ (+1 718) 204 10 10 ✉ GLOBCOM [T] 237593
⊖ (+1 718) 626 35 66
♧ US$ 4,250,000 (1995) ♣ US$ 146,674,335 (1996)
👥 96 🏢 14,250 sq ft
VP & General Mgr: James Chaldek ☎ (+1 718) 204 1414
Vice President: John Rosenblum
Sales/Marketing Dir: Joseph Ceva
Purchasing Mgr: Robert Thompson
■ Designs, assembles and supplies military electronics including C3 systems design and development. → 80 [98]

75-39820 Glover Machine Company
210 South Morocco, PO Box 210187,
Dallas, TX 75211-4446
☎ (+1 214) 331 87 73; 331 83 73
⊖ (+1 214) 331 83 77
♣ US$ 8,500,000 (1996) 👥 98
President: John A Glover
Executive Vice President: W H Wagner
■ Machined Aircraft parts and assemblies. Major structural parts including bulkheads, wing panels, longerons, spars, etc. [98]

75-39950 GNY Equipment L.L.C.
(† Garsite Inc.)
71E Industry Court, Deer Park, NY 11729
☎ (+1 516) 667 10 10 ✆ (800) 645 58 02
⊖ (+1 516) 242 34 44
♧ $ 500,000 (1997) ♣ $ 5,000,000 (1996)
👥 16 (1997) 🏢 1,858 sq m
President: Herman Kornahrens
International Sales Manager: William Peil
■ Manufacture and distribution of aviation ground support and liquid handling equipment, including service pits, pumping systems and carts, fuelling, cabinets, floating suctions, pantographs, surge suppressors, fuelling valves, nozzles, hoses and swing joints. → 85 [98]

75-40050 The Goodyear Tire & Rubber Co.
1144 East Market Street, Akron, OH 44316-0001
☎ (+1 330) 796 21 21 ✉ GOODYEAR [T] 640550
⊖ (+1 330) 796 12 37
♣ $ 12,288,000,000 (1994) 👥 90,000
Chairman and Chief Executive Officer: Samir F Gibara
President & COO: Samir F. Gibara
Exec. VP, President & General Mgr Europe: William J. Sharp
VP Public Affairs: John P. Perduyn
VP Business Development: Clark E. Sprang
Military Sales Mgr: Jon P. Dupont
VP Product Supply: H. Clay Orme
Mgr Government Sales & Marketing: John G. Appeldorn
■ Tyres for aircraft and ground vehicles.
Subsidiaries:
Air Treads Inc. → 75
Corporate Government Relations
901 15th Street, NW Suite 350, Washington, DC 20005
☎ (+1 202) 682 92 50
Vice President: Ms Isabel Jasinowski
● 115 South Wayne Street, St Marys, OH 45885
☎ (+1 419) 394 33 11, ext. 337 ⊖ (+1 419) 394 16 19
Marketing Mgr Military & Truck, Tank & Vehicle: Jerry L. Bohaty
■ Tank track shoe assemblies and other general products.
The North American Export Operations
D/737, 1144 East Market Street D/953,
Akron, OH 44316-0001
☎ (+1 216) 796 65 88 ⊖ (+1 216) 796 63 33
Director Export Operations: Eric Brown
■ Ground vehicle tyres. [98]

75-40060 Alan Gordon Enterprises Inc.
1430 Cahuenga Boulevard, PO Box 315,
Hollywood, CA 90078
☎ (+1 213) 466 35 61 [E] info@a-g-e.com
[W] http://www.age.com ⊖ (+1 213) 871 21 93
♣ $ 5,000,000 (1997) 👥 61 (1997) 🏢 2,700 sq m
🏢 1,700 m2
President: Grant Loucks
Vice President, Marketing: Don Sahlein
Vice President, US Government Contracts Administration: Robert Kuhagen
Purchasing Manager: J Lansing
Advertising Manager: G Piotrowsky
■ Airborne photographic reconnaissance and mapping systems, equipment, accessories and ground support equipment. Photogrammetric and stereoplotting equipment. Cine theodolites, photography tracking mounts, photography trailers. [98]

75-40070 W.L. Gore & Associates Inc.
551 Papermill Road, Newark, DE 197
☎ (+1 302) 738 48 80 ⊖ (+1 302) 292 41 59
👥 6,000 (worldwide)
President: Robert Gore
Public Relations Manager: Ms Heidi Cofran
■ Manufacture of a wide range of polytetrafluoroethylene (PTFE) products which are marketed worldwide under the GORE trademark. Products include wires and cables, medical products, textile laminates, microfiltration membranes, gaskets and sealants, etchants and weaving fibres. [98]

75-40310 Graflex Inc
1281 North Ocean Drive, Suite 201,
Singer Island, FL 33404
☎ (+1 561) 842 26 00 [E] graflex@aol.com
⊖ (+1 561) 842 30 20
♣ $ 1,000,000 + 👥 10 🏢 4,500 sq ft
Managing Director: Brad Ganther
Sales and Marketing Director: V Brewer
Production Director: J Henson
Purchsing Director: A Ganther
■ Manufacture small and large calibre precision muzzle boresights. Motorised zoom lenses for military and commercial applications. [98]

75-40355 Graphic Datakits International (GDI)
14500 Trinity Boulevard, Suite 110, PO Box 155157,
Fort Worth, TX 76155-0157
☎ (+1 817) 283 78 75 [T] 732561 ⊖ (+1 817) 354 07 90
👥 7
President: Jim (James) C Neely
Sales Department: Ms JoAnne Michener
Connector Microtooling Systems Inc.
14500 Trinity Blvd, Ste 110, PO Box 155157,
Fort Worth, TX 76155-0157
☎ (+1 817) 283 4882 [T] 732561 ⊖ (+1 817) 354 0790
President: Jim (James) C. Neely [98]

75-40360 The Graphics Co. of America
PO Box 4, San Marcos, TX 78667
☎ (+1 210) 629 1400 ⊖ (+1 512) 353 5333
👥 10 🏢 20 000 sq ft
President: Floyd MacKenzie
Advertising Mgr: Ms Danis Cheryl
■ Digital plotters → 80
Affiliate:
The Recorder Co.
PO Box 8, San Marcos, TX 78667
☎ (+1 210) 629 1400 ⊖ (+1 210) 353 5333
President: Floyd MacKenzie
■ CAD/CAM systems & equipment for the design engineer. Strip chart recorders. [98]

75-40380 Greene, Tweed & Co. (GTC)
Detwiler Road, Kulpsville, PA 19443-0305
☎ (+1 215) 256 95 21 ⊖ (+1 215) 256 01 89

Contact Reference

President: F.A. Paino
Marketing Mgr Aerospace Division: Nick Hartle
■ Seals for hydraulic systems. Non-metallic bearing materials, gasketing materials, rotary exclusion devices, engineering thermoplastics. → 85 [98]

75-40390 Grenadier Engineering Inc.
(† Grenadier Industries Inc.)
3603 East, 8620 South, Salt Lake City, UT 84121
☎ (+1 801) 944 1619 ⊖ (+1 801) 944 5303
♣ $ 3 000 000 ⚉ 6 ⌂ 7 500 sq ft
President: Jon P. Crawford
■ Manufacture M79 & M203 grenade launchers. Has developed M203 style launcher for FAL, G-3 & AK-47, Galil & FN & FNC, Mk 19 automatic grenade launcher. Supply of all types of 40mm ammunition. [97]

75-40400 Grimes Aerospace Company
Product Support Center
240 Twain Avenue, Urbana, OH 43078
☎ (+1 937) 653 52 25 (customer service)
⊖ (+1 937) 652 23 22
Chairman & CEO: Michael Dubose
Sr VP Engineering: Andrew Vesel
Sr VP Operations: Roger Jacobson
Sr VP Marketing & Strategic Business Dev: Dana Skaddan
Sr VP Sales & Support: Harry Henderson
■ Manufacture aerospace systems, subsystems & components including: exterior & interior aircraft lighting systems; signs; crew stations; cockpit info displays; aircraft electrical equipment; aircraft emergency systems; passenger service modules; windshield wiper systems; on-board microprocessors; pneumatic, hydraulic & fuel valves; control systems; air moving devices; heat exchangers; rigid couplings; flex joints; environmental controls; information systems; power management systems; fluid controls.
● 550 Route 55, Urbana, OH 43078
☎ (+1 513) 652 1431 ⊖ (+1 513) 652 1518
● 5001 Airport Plaza Drive, Suite 110,
Long Beach, CA 90815
☎ (+1 310) 982 1530 ⊖ (+1 310) 982 1536
● Plaza at Yarrow Bay, 10220 NE Points Dr., Suite 205, Kirkland, WA 98033
☎ (+1 206) 889 5426 ⊖ (+1 206) 827 4738
● 5 Century Drive, Bristol, CT 06010
☎ (+1 203) 584 3500 ⊖ (+1 203) 589 2621 [98]

75-40715 GSI Corporation
1917 Greenspring Drive, Timonium, MD 21093-4178
☎ (+1 410) 252 17 68
Ⓔ 103354.3170@compuserve.com homepages/gsicable
⊖ (+1 410) 561 17 68
President: Joseph Petrella (Jr)
Vice President of Manufacturing: Bill Rigler
■ Electronic flat woven cable, related assemblies and wire marking systems for use in military aircraft, ASW sonobuoys, towed array sonars, simulators, computers, missiles, satellites, radars, laser target designators, ECM gear. [98]

75-40730 G.T.C. RF Products Group
(a div. of GTC)
216 W.Florence Ave, Inglewood, CA 90301
☎ (+1 310) 673 8422 ⊖ (+1 310) 672 2905
♧ $ 7 000 000 (1995) ♣ $ 10 000 000 (1995) ⚉ 55
⌂ 1 200 sq m
Managing Dir: A. Barrios
Sales & Marketing Dir: Ilan Israely
Operations/Technical Dir: E. Bainvoll
Purchasing Mgr: J. Hobb
■ RF power amplifiers & microwave communications products. [97]

75-40800 Guardian Technologies International
22570 Markey Court, Dulles, VA 20166-6901
☎ (+1 703) 444 79 31 ✆ (800) 462 78 80
Ⓔ gti@interramp.com Ⓦ http://www.guardiantech.com
⊖ (+1 703) 444 98 94
♣ $ 2,000,000 (1997) ⚉ 25 (1997) ⌂ 2,900 m2
President: Joseph F Fernandez
Chief Executive Officer: Oliver L North
Sales and Marketing Director: Robert Nieves
Advertising Manager: Joseph C Fernández
Purchasing manager: John Carman
■ Ballistic protective vests manufactured with Spectra woven fabric and Spectra Shield. [98]

75-40900 Guildline Instruments Inc.
(Affiliate of Guildline Instruments Ltd, Canada)
103 Commerce Street, Suite 160, Lake Mary, FL 32746
☎ (+1 407) 333 3327 ⊖ (+1 407) 333 3309
♣ $ 5 000 000 (1993) ⚉ 50
President: Anthony Anderson
VP Sales: Bob Gangawer
Operations Mgr: Lenhard N. Eifler
■ Electronic test & measurement equipment, oceanographic equipment & calibration standards and oceanographic instrumentation. [97]

75-40950 Gulfstream Aerospace Corporation
PO Box 2206 M/S C-10, Savannah, GA 31402-2206
500 Gulfstream Road, Savannah, GA 31408
☎ (+1 912) 965 30 00 ☒ 546470 gulf aero
⊖ (+1 912) 965 32 58
⚉ 5,000
Chairman Emeritus: Allen E Paulson
Chairman: Theodore Forstmann
Vice Chairman: Bryan Moss
President and Chief Operating Officer: Fred A Breidenbach
Executive Vice President and Chief Financial Officer: Chris A Davis
President, Gulfstream Aircraft Inc.: W W (Bill) Boisture
Senior Vice President and General Counsel: Ira P Berman
Senior Vice President, Government Operations: Ms Helen Newman
Vice President, International Marketing: Gene Rainville
Vice President, Gulfstream Financial Sevices Corp: David Rowe
Vice President, Co-Production: Joseph T Lombardo
Vice President, North American Sales: Frank Tyson
Vice President, North American Sales: Joe Walker
Vice President, Government and Special Missions Marketing: Buddy Sams
Vice President, Gulfstream Shares: Raynor B Reavis
Vice President, Marketing and Customer Services: Ms Marsha W Grovenstein
■ Manufacture Gulfstream IV-SP and GV corporate and business jets, also offered in a variety of special government/military requirements.

Washington DC Office
1000 Wilson Boulevard, Suite 2701, Arlington, VA 22209
☎ (+1 703) 276 95 00 ☒ 804705 gulfjet sav
⊖ (+1 703) 276 95 04
Senior Vice President, Government Operations: Ms Helen Newman
Vice President, Government and Special Missions Marketing: Buddy Sams

Gulfstream Aerospace Technologies
Wiley Post Airport, PO Box 22500,
Oklahoma City, OK 73123
☎ (+1 405) 789 50 00 ☒ TWX 910 830 6870
⊖ (+1 405) 787 52 06
President: John Podger [98]

75-41130 Gull Electronic Systems Div.
(a division of Parker Hannifin Corp.)
300 Marcus Boulevard, PO Box 9400,
Smithtown, NY 11787
☎ (+1 516) 231 37 37 ✉ GULLAIR Smith
Ⓔ markers@gull.com Ⓦ http://www.parker.com
⊖ (+1 516) 434 81 52
⚉ 400 (1997) ⌂ 200,000 sq ft
Vice President and General Manager: John Stadelmann
Director Operations: Thomas Renken
Director Sales and Marketing: P J Geraghty
Marketing Representative: John Bunting
■ Manufacture fluid measurement/management systems, flight deck displays and systems, utility controllers and flight inspection systems. [98]

75-41150 Gurley Precision Instruments
(† Teledyne Gurley)
514 Fulton Street, Troy, NY 12181-0088
☎ (+1 518) 272 63 00 ✆ (800) 759 18 44
Ⓔ encoders@gurley.com Ⓦ http://www.gurley.com
⊖ (+1 518) 274 03 36
♣ $ 10,000,000 (1996) ⚉ 75
Managing Dir: O. Patrick Brady
VP Sales & Marketing: Gerald S. Gordon
Operations Dir: Dr Boris Mintseris
Purchasing Mgr: Ann Hogle
■ Manufacture of optical encoders & reticles. [98]

75-41450 HAL Communications Corp.
1201 West Kenyon Road, PO Box 365,
Urbana, IL 61801-0365
☎ (+1 217) 367 73 73 Ⓔ halcomm@halcomm.com
Ⓦ http://www.halcomm.com ⊖ (+1 217) 367 17 01
⚉ 25
President: George W. Henry (Jr)
Marketing Mgr: Kenneth M. Broga
■ Radio-teleprinter systems including modems, terminals & error correction devices. [98]

75-41630 Halocarbon Products Corp.
887 Kinderkamack Road, PO Box 661,
River Edge, NJ 07661
☎ (+1 201) 262 88 99 Ⓦ http://www.halocarbon.com
☒ 134378 ⊖ (+1 201) 262 00 19
⚉ 100
Manager Sales and Marketing: Ronald Epstein
■ Fluorinated organic chemicals. [98]

75-41700 Halter Marine Group
(† Moss Point Marine Inc./Halter Marine Inc. /Trinity Marine Group)
(a subsidiary of Trinity Industries Inc.)
13085 Industrial Seaway Road, PO Box 3029,
Gulfport, MS 39503
☎ (+1 601) 896 00 29 ✉ HALMAR ☒ 6821246 halmar
⊖ (+1 601) 897 48 28
President: John Dane III
Vice President of International Sales: James G Rivers Jr
Vice President of Domestic Sales: Sid Mizell
Vice President of Government Sales: Anil Raj
Advertising Manager: A J Rizzo (Anderson & Rizzo Inc.)
■ Design and build military patrol vessels of all types: patrol and assault vessels, landing craft, research vessels, lighter barges, pilot boats, tugs, oceanographic vessels, in steel aluminium and GRP (glass fibre). [98]

75-41800 Hammonds Technical Services Inc.
15760 West Hardy, Suite 400, Houston, TX 77060
☎ (+1 713) 999 29 00 ✆ (800) 582 42 24
Ⓔ hammonds@worldnet.att.net ⊖ (+1 713) 847 18 57
⚉ 30 (1997)
President: Carl Hammonds
Vice President of Sales: Walter Chartrand
Production Manager: David Wynn
Purchasing Manager: Ed Sanders
■ Multi-additive injection equipment for jet fuel and diesel. [98]

75-41870 Harco Manufacturing Co. Inc.
7700 SW 69th Avenue, Portland, OR 97223-9548
☎ (+1 503) 244 75 71 ✆ (800) 394 75 71
⊖ (+1 503) 244 75 89
♣ $ 1,500,000 (1997) ⚉ 18 (1997) ⌂ 10,500 sq ft
President and Chief Financial Officer: Rod L Harris
■ Engine exhaust mufflers & related products. Oil/water separators. [98]

75-41900 Hardcote Technologies
(† Chromalloy Precision Products)
1505 Royal Parkway, Euvess, TX 76040
☎ (+1 817) 283 46 81 ⊖ (+1 817) 283 09 12
⚉ 60 (1997)
Vice President and General Manager: James Connelly
Customer Service Manager: Gary Jones
■ Helicopter rotory wing head components, gas turbine engine nozzles and disks. [98]

75-41930 Hardigg Cases
(a division of Hardigg Industries Inc.)
147 North Main Street, PO Box 201,
South Deerfield, MA 01373
☎ (+1 413) 665 21 63
✆ (800) 542 73 44; (800) JHARDIGG
Ⓔ literature@hardigg.com ⊖ (+1 413) 665 80 61
⚉ 240
President: James S Hardigg
Vice President: Jamie M Hardigg
Business Manager, Container Products: Bill Hamer
International Sales Manager: Iain Boyd
■ Mil-Spec shipping cases and rugged 19 inch rack shock-mounted enclosures. Over 275 off-the-shelf sizes. Hardigg specialises in tough cushioned roto-moulded cases for delicate equipment in harsh environments. Program experienced engineering department. [98]

75-42050 Harris Corporation Corporate Office
1025 West NASA Boulevard, Melbourne, FL 32919
☎ (+1 407) 727 91 00
✆ (800) 442 77 47 (800 4 HARRIS) ext. 700
[E] webmaster@harris.com [W] http://www.harris.com
♣ $ 3,621,200,000 (1996) 👥 27,600 (1996)
Chairman, President and Chief Executive Officer: Phillip W Farmer
President, Lanier Worldwide, Inc.: Wesley E Cantrell
President, Semiconductor Sector: John C Garret
President, Communications Sector: Guy W Numann
President, Electronic Systems Sector: Albert E Smith
Senior Vice President and Chief Financial Officer: Bryan R Roub
Vice President, General Counsel and Secretary: Richard L Ballantyne
Vice President, Internal Audit: James L Christie
Vice President and Controller: Robert W Fay
Vice President, Human Resources: Nick E Heldreth
Vice President, Quality and New Processes: John G Johnson
Vice President, Information Management: Herbert N McCauley
Vice President, Corporate Development: Ronald R Spoehel
Vice President and Treasurer: David S Wasserman
■ Structured around four main business areas: Electronic Systems Sector, Semiconductors Sector, Communications Sector and Lanier Worldwide Office Products.

RF Systems Division
(† RF Communications Group)
1680 University Avenue, Rochester, NY 14610
☎ (+1 716) 244 58 30 ⊖ (+1 716) 244 29 17
Media Contact: Ms Susan Giuseppi
☎ (+1 716) 242 39 26
■ HF, VHF & UHF radio equipment, turnkey communication systems, digital video image transmission systems & integrated systems. [98]

75-42450 Robert W. Hart & Son Inc.
401 Montgomery Street, Nescopeck, PA 18635
☎ (+1 717) 752 36 55 ⊖ (+1 717) 752 10 88
President: Wallace E Hart
Export Manager: Erik Latranyi [98]

75-42700 Havis-Shields Equipment Corp.
395 Jacksonville Road, PO Box 2099, Warminster, PA 18974
☎ (+1 215) 957 07 20 ✆ (800) 524 99 00
[W] http://www.havis.com ⊖ (+1 215) 957 07 29
👥 100+ (1997) 🏭 60,000 sq ft
General Mgr: Joseph Bernert
Engineering Dir: David E'nama
Sales & Marketing Dir: Michael Bernert
Dir Advertising: Mel Marmer
Marketing Co-ordinator: Karin Evitts
■ Communications console; spotlight; prisoner transport vehicles; vehicle modifications and installations; high intensity lighting products. [98]

75-42740 Haynes International Inc.
1020 W. Park Ave, Kokomo, IN 46904-9013
☎ (+1 317) 456 6000 ✆ (800) 354 0806
Telex 272280; 272232; TWX 810 2691 1000/920
⊖ (+1 317) 456 6905
👥 800 (1996) 🏭 1 000 000 sq ft
President: Michael D. Austin
VP Sales: Charles J. Sponhaugle
Manager: Edward Bickel
Marketing Mgr: Jim Calird
■ Corrosion resistant wrought alloys. High temperature wrought alloys.

Overseas Affiliates:
Haynes International SRL
Haynes International Sarl
Haynes International Ltd
Nickel Contor AG, Switzerland
Nickel Contor GmbH, Germany FR [97]

75-42770 HCC Industries Inc.
4232 Temple City Boulevard, Rosemead, CA 91770
☎ (+1 626) 443 89 33 ⊖ (+1 626) 575 24 37
President and Chief Executive Officer: Andrew Goldfarb
Vice President: Dennis Moore
Chief Financial Officer: C Baterman
■ Manufacture hermetic connectors, seals and headers. Custom design and MIL-STDS. [98]

75-42780 HDL Research Laboratory Inc.
PO Box 2108, Brenham, TX 77834-2108
406 West Horton Street, Brenham, TX 77833-2348
☎ (+1 409) 836 23 00 [E] hdlpower@aol.com
⊖ (+1 409) 836 10 56
♣ US$ 3,000,000 (1996) 👥 50 🏭 88,000 m2
President: Kai S Lichtenberg
Technical Director: Larry LeDoux
Sales and Marketing Director: Ms Lisa Parker
Purchasing Manager: Dom Newman
Assistant Administrator: Ms Heide M Gross
■ Custom military power supplies used in C4I and other electronic warfare equipment. [98]

75-42860 Heckler & Koch Inc
(a subsidiary of British Aerospace Defence Ltd, UK)
21480 Pacific Boulevard, Sterling, VA 20166-8903
☎ (+1 703) 450 19 00 [E] hkimage@aol.com
⊖ (+1 703) 450 81 60
👥 50 (1997) 🏭 40,000 sq ft
Vice President, Federal Operations: John N Meloy
Director of Creative Services: Steve Galloway
Marketing Manager: Jim Schatz
■ Marketing and sales of samll arms to commercial (USA/Canada), military, federal and law enforcement markets (USA).
® Germany: Heckler & Koch [98]

75-42880 Heflin Steel Co.
PO Box 1249, Phoenix, AZ 85001
1111 West Maricopa Freeway, Phoenix, AZ 85007
☎ (+1 602) 252 80 61 ⊖ (+1 602) 256 95 38
♣ $ 10,000,000 👥 40
President: Ronald G Stewart
Armor Products Manager: Greg Wiggins
■ Distributor and fabricator of military specification and commercial grade armour steel. Large inventory on hand. On site ballistic testing. [98]

75-43000 Heller Truck Body Corp.
138 US Highway Route 22, Hillside, NJ 07205-1837
☎ (+1 201) 923 9200 ⊖ (+1 201) 923 9269
■ Transportation products. [97]

75-43080 Hercules Engines Company
(† Hercules Engines Inc.)
101 11th Street SE, PO Box 24101, Canton, OH 44701-4101
☎ (+1 330) 438 10 28; 454 56 31 ✆ (800) 345 06 62
[E] engines@aol.com ⊖ (+1 330) 438 13 13
♣ $ 20,000,000 (1996) 👥 150 (1996)
President: John R Dienes ☎ (+1 330) 438 13 74
Managing Director: Tim Harmon ☎ (+1 330) 438 13 26
Operations Director: Tom D Huggins
☎ (+1 330) 438 10 29
Sales Director: Michael C Ervin ☎ (+1 330) 438 10 26
Purchasing Manager: Bruce K Weick
☎ (+1 330) 438 13 67
International Military Sales: Paul F Harmon
☎ (+1 330) 438 10 28 [98]

75-43090 Herley-Vega Systems
(† Vega Precision Laboratories Inc.)
(a subsidiary of Herley Industries)
10 Industry Drive, Lancaster, PA 17603
☎ (+1 717) 397 27 77 [W] http://www.herley.com
⊖ (+1 717) 397 70 79
♣ $ 30,000,000 (1997) 👥 130 (1997) 🏭 71,000 sq m
President: Myron Levy
Chief Technical Officer: G. Klein
International Marketing Dir: G. Hopp
Domestic Marketing Director: R Umbarger
■ Design, engineer and manufacture UAV tracking command and control systems, TM transmitters, radar transponders, flight termination receivers. [98]

75-43100 Hermetic Aircraft International Ltd Corp. (HAIC)
(a subs of GKN Westland Technologies Ltd)
100 Corporate Drive, Holtsville LI, NY 11742
☎ (+1 516) 758 4242 Telex 961329 haic fdle
⊖ (+1 516) 758 1428
♣ $ 21 000 000 👥 98
President: R. Fauchon
VP Marketing: John M. Minietta
VP Finance & Commerce: J. Fawsett
■ Sales, overhaul & repair of pneumatic, life support, electronic heat transfer & fuel systems. [97]

75-43120 Hermetic Seal Corporation
(a subsidiary of HCC Industries Inc.)
4232 Temple City Boulevard, Rosemead, CA 91770
☎ (+1 626) 443 89 31 ⊖ (+1 626) 443 66 10
👥 400
General Manager: Andrew Goldfarb
Vice President: Dennis Moore
■ Manufacture hermetically sealed headers, terminals and connectors. [98]

75-43150 Hesco Inc.
2139 Greenville Rd, LaGrange, GA 30240
☎ (+1 706) 884 7967 ⊖ (+1 706) 882 4683
President: W.C. Senn
National Sales Mgr: Kerry R. Kindler
■ Meprolight tritium illuminated sights. Sighting aid for low light & daylight conditions. [97]

75-43200 Hewlett-Packard Company Corporate Offices
PO Box 10301, Palo Alto, CA 94303-0890
3000 Hanover Street, Palo Alto, CA 94304-1185
☎ (+1 415) 857 15 01 ✆ (800) 752 09 00
Telex TWX 910 373 1267 ⊖ (+1 415) 857 55 18
♣ $ 38,420,000,000 👥 112,000 (worldwide)
Chairman, President and Chief Executive Officer: Lewis Platt
Executive Vice President, Finance and Administration and Chief Financial Officer: Robert P Wayman
Executive Vice President and General Manager, Test and Measurement Organisation: Edward W Barnholt
Executive Vice President and General Manager, Computer Organisation: Richard E Belluzzo
Senior Vice President and General Manager, Measurement Systems Organisation: Douglas K Carnahan
Senior Vice President, Corporate Affairs and General Counsel: S T Jack Brigham (III)
Senior Vice President, Research and Development and Director HP Laboratories: Joel S Birnbaum
Senior Vice President, Strategic Initiatives Europe: Franco Mariotti
Senior Vice President, Personnel: F E Peterson
Public Relations Representative: Ms Carol Parcels
■ Key Product and Service Segments:
Computer Products Service and Support; Computer and systems (hardware and software) networking products, printers, plotters, scanners, diska dn tape drives, terminals and handheld calculators; support and maintenance services, parts and supplies. Products are used for business, industrial and scientific applications.
Electronic Test and Measurement Instrumentation; Instruments and measurement systems used for design, production and maintenance of electronic equipment; support and maintenance services.
Medical Electronic Equipment and Service; Instruments and information systems used for patient care monitoring; diagnostic cardiology; computer equipment, systems integration and application software; support and maintenance services; hospital supplies.
Analytical Instrumentation and Service; Gas and liquid chromatographs, mass spectrometers and spectrophotometers used to analyse chemical compounds; laboratory data and information management systems; support, supplies and maintenance services.

Direct Marketing Organization
PO Box 58059, Santa Clara, CA 95051-8059
✆ (800) 452 48 44

Geographic Operations
3000 Hanover Street, Palo Alto, CA 94304-1185
☎ (+1 415) 857 29 50 ⊖ (+1 415) 857 70 45
Senior Vice President and Managing Director: Alan Bickel [98]

75-43280 Hexcel Composites
5794 West Las Positas Boulevard, PO Box 8181, Pleasanton, CA 94588-8781
☎ (+1 510) 847 95 00 ⊖ (+1 510) 734 96 76
President: J Koshak
■ Polymer composites in preimpregnation form, honeycomb, weaving, panels & fabrication. [98]

75-43300 Hexcel Corporation
5794 West Las Positas Boulevard, PO Box 8181, Pleasanton, CA 94588-8781 [98]

75-43320 Hexcel Structures
(† Heath Tecna Aerospace Co.)
(a division of Hexcel Corporation)
PO Box 97004, Kent, WA 98064-9704
19819-84th Avenue South, Kent, WA 98032
☎ (+1 206) 872 75 00 ⊜ (+1 206) 395 12 37
♧ $ 100,000,000 ⚉ 1,000 ⌂ 1,000,000 sq ft
President Structures and Interiors: Justin Taylor
Vice President Structures: Mike Carpenter
Director Business Development: Bob LaFramboise
☎ (+1 253) 395 48 01
Sales Marketing Analyst: Jeff Salee
☎ (+1 253) 395 48 12
■ Comprehensive design, development & fabrication of advanced composite structures for defence & aerospace applications. [98]

75-43465 Hill Industries Inc.
19734 Dearborn Street, Chatsworth, CA 91311
☎ (+1 818) 885 03 00 ⊜ (+1 818) 885 50 91
♣ US$ 10,000,000 (1997) ⚉ 25
Chairman and Chief Executive Officer: Harry Lebovitz
President: Bruce Littell
■ Logistic support for US aircraft sold internationally.
Hill Aviation Logistics
(a division of Hill Industries Inc.)
19734 Dearborn Street, Chatsworth, CA 91311
☎ (+1 818) 885 03 00 ⊜ (+1 818) 885 50 91
Vice President of International Business Development: Roland Scott
Contracts Director: H Nagy
■ Total logistic support for US aircraft sold internationally.
Hill Aerospace & Defense
19734 Dearborn Street, Chatsworth, CA 91311
☎ (+1 818) 885 03 00 ⊜ (+1 818) 885 50 91
Chief Executive Officer: Harry Lebovitz
■ Design, engineering, manufacturing aircraft components and subsystems including actuators, dampers, precision structural and electro-mechanical systems. [98]

75-43500 Hi-Shear Corp.
(a subsidiary of GFI Industries, France)
2600 Skypark Drive, Torrance, CA 90509-2975
☎ (+1 310) 326 81 10 ⊠ HISHEAR
⊜ (+1 310) 784 40 96
President: Patrick Meade
Executive Vice President: F Gill
■ Fasteners and fastening systems installation tooling for the aerospace industry. [98]

75-43550 Hitchiner Manufacturing Co. Inc.
PO Box 2001, Milford, NH 03055-2001
☎ (+1 603) 673 11 00 [W] http://www.hitchner.com
⊜ (+1 603) 673 79 60
♣ $ 138,000,000 (1996) ⚉ 2,500 (1997)
⌂ 70,000 sq m
President: John H Morison III
Vice President Sales and Marketing International: Frederick F Marston (Jr)
Vice President Sales and Marketing North America: James E Pasqualoni
■ Investment cast components in ferrous & non-ferrous alloys for light arms, tanks, aircraft & missile systems. [98]

75-43980 Hoffman Engineering Corp.
22 Omega Drive, PO Box 4430, Stamford, CT 06907-0430
☎ (+1 203) 425 89 00
[E] sales@hoffmanengineering.com
[W] http://www.hoffmanengineering.com
⊜ (+1 203) 425 89 10
⚉ 45 (1997) ⌂ 22,000 sq ft
President: Albert C Schafer
Vice Presidennt Sales and Marketing: Jack De Soye
Operations/Technical Director: Paul De Soye
Purchasing Manager: Ron Curtin
■ Lighting technology laboratory; photometry & spectral measurements, products & services; avionic display hardware. [98]

75-44060 Holt Integrated Circuits
45 Parker, Irvine, CA 92618-1605
☎ (+1 714) 859 8800 ✆ (800) 222 HOLT
⊜ (+1 714) 859 9643
♣ $ 5 000 000 (1996) ⚉ 27
President & CEO: William G. Holt
Marketing Dir: Gerald Donaldson
■ CMOS avionis data bus, display drivers & selected high volume, analogue and/or digital asic products. [97]

75-44300 Hornady Manufacturing Co.
PO Box 1848, Grand Island, NE 68802
☎ (+1 308) 382 1390 ⊜ (+1 308) 382 5761
⚉ 130
President: Steve Hornady
Marketing Mgr: Roger Carvalho
Advertising Dir: Douglas Engh
■ Manufactures a complete line of pistol & rifle bullets & reloading tools & accessories for metallic & shotshell reloading including dies, wads & scales. Also manufactures high-quality rifle & pistol ammunition in over 20 different calibres. [97]

75-44400 Howden Fluid Systems
(a subsidiary of Howden Defense Systems, Inc.)
72 Santa Felicia Drive, Santa Barbara, CA 93117-2893
☎ (+1 805) 968 07 02 ⊜ (+1 805) 968 16 35
♣ $ 13,000,000 (1997) ⚉ 59 (1997) ⌂ 15,000 sq ft
President: Ms Debra Hetherington
Operations Director: Nick Boyer
Sales and Marketing Director: William Godecker
Purchasing Manager: Phil Anderson
Sales Manager: Ms Marge Kozaki
■ Component level and complete integrated cooling systems for ECM, EW, laser and space applications; fluid conditioning systems; fuel and lube pumps for airborne and naval craft. Products are specialised to meet specific customer needs.
Howden Defense Systems Inc.
16952 Millikan Avenue, Irvine, CA 97214-5045
☎ (+1 714) 863 05 60 ⊜ (+1 714) 660 87 46
Chief Executive Officer: Michael S Quinn
■ Fans and compressors, heat transfer, refrigeration, NBC environmental control systems, hydraulic and fuel system components, ammunition.
Subsidiaries:
Howden Airdynamics Inc.
Howden Fluid Systems
Western Design Howden [98]

75-44550 Howmet Corporation
(† Howmet Turbine Components Corp.)
475 Steamboat Road, PO Box 1960, Greenwich, CT 06836-1960
☎ (+1 203) 661 46 00 ⊠ 6730017
⊜ (+1 203) 661 04 22
♣ $ 1,100,000,000 (1996) ⚉ 10,000 (1996)
President & CEO: David L. Squier
Senior Vice President Sales and Marketing: Alan Bergquist
Senior Vice President US Operations: James R Stanley
Sr VP Int'l Operations: Mark Lasker
Vice President Corporate Planning: Paul Wilson
Vice President Finance: John Ritter
VP Technology: Neil E. Paton
VP General Counsel/Secretary: Roland A. Paul
Vice President Human Resources: Mike Malady
Treasurer: Jeffrey A. Jankowski
■ Investment cast components for gas turbine engines superalloy, aluminium and titanium alloys.
Specialty Products
1600 South Warner Road, Whitehall, MI 49461
☎ (+1 616) 894 56 86 ⊜ (+1 616) 894 76 07
General Manager: Cliff Sickles
Dover Alloy
Roy Street, PO Box 371, Dover, NJ 07801
☎ (+1 201) 361 23 10 ⊜ (+1 201) 989 54 05
General Manager: J Bodner
Dover Casting
Roy Street, Dover, NJ 07801
☎ (+1 201) 361 03 00 ⊜ (+1 201) 328 77 01
General Manager: Jim Johnson
Hampton Casting
One Howmet Drive, PO Box 9365, Hampton, VA 23670
☎ (+1 804) 838 46 80 ⊜ (+1 804) 827 19 65
Manager IGT: Larry Corliss
La Porte Casting
1110 East Lincoln Way, La Porte, IN 46350
☎ (+1 219) 326 74 00 ⊜ (+1 219) 324 31 93
General Manager: T Trotter
Morristown Casting Support
5650 Commerce Boulevard, Morristown, TN 37814
☎ (+1 615) 587 49 10 ⊜ (+1 615) 581 18 28
General Manager: David Uram
Howmet Research Center
1500 South Warner Street, Whitehall, MI 49461
☎ (+1 616) 894 56 86 ⊜ (+1 616) 894 78 26
VP Technology: Neil Paton
Dir Quality & Technical Services: Dr Frederick L. Norris
Dir Advanced & Spraycast Technology: Thomas Tom
Dir Foundry & Ceramics Technology: Robert Grunstra
Dir Mfr Technology: Michael McRae
Mgr Surface Technology: Jeffrey S. Smith
Thermatech
555 Benston Road, Whitehall, MI 49461
☎ (+1 616) 894 56 86 ⊜ (+1 616) 894 74 00
General Manager: Cliff Sickles
Ti-Cast
1600 South Warner Road, Whitehall, MI 49461
☎ (+1 616) 894 56 86 ⊜ (+1 616) 894 75 12
General Manager: Eric Foos
Titanium Ingot
555 Benston Road, Whitehall, MI 49461
☎ (+1 616) 894 56 86 ⊜ (+1 616) 894 73 54
General Manager: Stanley Gillish
Whitehall Casting
One Misco Drive, Whitehall, MI 49461
☎ (+1 616) 894 56 86 ⊜ (+1 616) 894 73 47
General Manager: Dennis Albrechtsen
Hot Isostatic Pressing
1600 South Warner Street, Whitehall, MI 49461
☎ (+1 616) 894 56 86 ⊜ (+1 616) 894 74 99
General Manager: Cliff Sickles
Wichita Falls Casting
6200 Central Freeway, PO Box 1616, Wichita Falls, TX 76307
☎ (+1 817) 855 81 00 ⊜ (+1 817) 855 14 79
General Manager: T Rose
Machining
Winsted Industrial Park, Price Road, Winsted, CT 06098
☎ (+1 203) 379 33 14 ⊜ (+1 203) 379 42 39
General Manager: E Miller [98]

75-44580 Howmet Tempcraft Inc.
(† Tempcraft Inc.)
3960 South Marginal Road, Cleveland, OH 44114
☎ (+1 216) 391 38 85 ⊜ (+1 216) 391 71 62
General Manager: T Zuk [98]

75-44670 HSC Controls Inc.
(a subsidiary of Woodward Governor Co.)
390 Youngs Road, Buffalo, NY 14221
☎ (+1 716) 631 90 40 ⊜ (+1 716) 631 82 20
♣ $ 17,000,000 (1997) ⚉ 215 (1997)
General Manager: Clint Hull
Sales Manager: Sam Sanfilippo
■ Design & manufacture of servovalves & servoactuators. [98]

75-44857 Hughes Danbury Optical Systems Inc. (HDOS)
(a subsidiary of Hughes Aircraft Co.)
100 Wooster Heights Road, Danbury, CT 06810-7589
☎ (+1 203) 797 52 66 [E] info@hdos.hac.com
⊜ (+1 203) 797 51 14
⚉ 750 (1997)
President: Dr John C Rich
Operations Director: Frank Krausz
Sales and Marketing Director: C Richard Hale
Manager Public Relations: Thomas J Arconti
Purchasing Manager: Les Jacobs
■ Designs and builds high performance EO/IR sensors and subsystems for tactical ground and airborne vehicles. Products include a unique laser threat detection and warning system for rotary and fixed-wing aircraft. Other tactical systems include airborne reconnaissance cameras, EO/IR countermeasures, new and advanced laser warning sensors, high performance optical windows, window subassemblies and filters. [98]

75-44865 Hughes Electronics Corp.
(a subsidiary of General Motors Corp.)
7200 Hughes Terrace Co, CO1 C118, PO Box 80028, Los Angeles, CA 80080-0028
☎ (+1 310) 568 72 00 [W] http://www.hughes.com
⊜ (+1 310) 568 69 42
♧ $ 2,900,000,000 (1996)
♣ $ 15,900,000,000 (1996) ⚉ 86,000 (1996)
Chairman and Chief Executive Officer: C Michael Armstrong
Vice Chairman: Michael T Smith
Vice Chairman and Chief Financial Officer: C H Noski
Executive Vice President: Steven Dorfman
Executive Vice President: J C Weaver
Senior Vice President of Research and Technology: A N Chester

Senior Vice President of Marketing and International: G C C Chang
Senior Vice President and General Counsel: J J Higgins
Senior Vice President of Human Resources and Administration: T G Westerman
Senior Vice President and Treasurer: Ms Roxanne Austin
Senior Vice President: Jack Shaw
Corporate Communications: Richard Doré
☎(+1 310) 568 63 24
■ Automotive, space and defence electronics. Mission Oriented Product Lines:
Command, control and information distribution systems, electronic warfare and surveillance, space exploration, communication and applications, tactical and strategic weapon systems.
Equipment Product Lines:
Communications equipment, fabrication and assembly products, information processing and display equipment, sensors and illuminators, test equipment and instrumentation, aero and space vehicles.
Component Product Lines:
Electro-optical components, interconnection components and devices, microwave and high frequency devices, microcircuits and electronic modules.
Service Product Lines:
Systems engineering, integration and installation, modification, maintenance and support, testing and analysis.

Subsidiary companies:
Hughes Aircraft Co. → 75
Delco Electronics Corp. → 75
Hughes Network Systems Inc.
Hughes Telecommunications and Space Co
DIRECTV Inc.

Washington, DC Office
1100 Wilson Boulevard, Arlington, VA 22209-3978
☎(+1 703) 525 15 50 ⊠ TWX 710 955 0228
⊖ (+1 703) 525 77 38
Vice President of Government Operations: William D Merritt

Hughes Aircraft Company
1100 Wilson Boulevard, 20th Floor, Arlington, VA 22209-3978
☎(+1 703) 525 15 50
Chairman: M T Smith
President and Chief Operating Officer: J C Weaver
Senior Vice President of Merger and Transition Activities: B L Abrahams
Senior Vice President of Sensors and Communications Systems: K C Dahlberg
Senior Vice President of Weapons Systems: D McPherson
Senior Vice President and Chief Financial Officer: C S Ream
Senior Vice President of Information Systems: C Leader
Vice President of Business Development: B L Abrahams
Vice President of Operations and Materials: R L Horowitz
Vice President of Engineering: S C Iglehart
Vice President: D R Infante
General Counsel and Secretary: J T Kuelbs

Hughes Telecommunications and Space Company (TSC)
7200 Hughes Terrace, PO Box 80028, Los Angeles, CA 90045
☎(+1 310) 568 72 00
👥 7,500
Executive Vice President and Chairman: S D Dorfman
Vice President and President of Hughes Space and Communications Co. (HSC): D I Cromer
HSC Vice President of Government Programs: M J Gianelli
HSC Chief Financial Officer: D A Gonzales
HSC Vice President of Government Marketing: R R Rankine
HSC Vice President of Commercial Programs and Marketing: M J Hovterman
General Counsel and Legal Vice President: J A Smolker

Hughes Network Systems Inc. (HNS)
11717 Exploration Lane, Germantown, MD 20876
☎(+1 301) 428 55 00
Corporate Senior Vice President, Chairman and Chief Executive Officer: J A Shaw
HNS Vice President and Digital Cellular Systems Division: P Kaul
President of HNS International: A M Werth
Chief Financial Officer: J Lucchese
Executive Vice President and Assistant Secretary: J L Peeler
Executive Vice President and General Manager of Satellite Network Division: B Pourmand
Senior Vice President: E V Hoversten
Senior Vice President: J Ferenz
Senior Vice President: R Patel
Senior Vice President: S Revkin
Senior Vice President of Business Development: E R Cacciamani
Senior Vice President of Engineering: D D Fraley
Vice President, Secretary and Corporation Counsel: S P Carrier

Hughes Missile Systems Company (HMSC)
PO Box 11337, Tucson, AZ 85734
☎(+1 520) 794 30 00 ⊖ (+1 520) 794 13 15
👥 7,500
President: L L Francesconi
Advanced Programmes Director: D L Gieseking
Advanced Design and Technology Director: R Sheinberg
Engineering Director: G S Shelton
Operations Director: M C Irvine
Production Line Manager, Precision Strike: B A Ganoe
Production Line Manager, Tactical Systems: W M Leonard
Product Line Manager, Surface- to-Air Defense Systems: J K Lockard
Product Line Manager, Air-to-Air Missiles: C D Anderson
Public Relations: Ms C M Niccum ☎(+1 520) 794 85 65; 794 13 15 (fax)
■ Develops and produces advanced tactical guided missile systems. Associated fields of interest include missile guidance, propulsion systems, launch equipment, ancillary sub-systems and components.

Santa Barbara Research Center (SBRC) (a subsidiary of Hughes Aircraft Co.)
75 Coromar Drive B21/56, Goleta, CA 93117
☎(+1 805) 968 35 11; 562 41 27
⊠ TWX 910 3341 885 ⊖ (+1 805) 685 82 27
President: Michael E Allgeier
Marketing Director: Ms Lucy Thomas-Harrington
■ Rapid response fire and explosion detection and suppression systems for combat vehicles, aircraft, ships and commercial programs through application of advanced dual spectrum infrared technology.

Hughes Communications, Inc.
150 Hughes Way, PO Box 9712, Los Angeles, CA 90810
☎(+1 310) 607 40 00 ⊖ (+1 310) 640 12 65
President: J F Farrell
■ Design and manufacture communications satellite ground stations and provides satellite communications services to domestic and international customers. Also operates a worldwide dedicated communication system for the U.S. Navy.

Electron Dynamics Division
3100 West Lomita Boulevard, PO Box 2999, Torrance, CA 90509-2999
☎(+1 310) 517 60 00 ⊠ TWX 910 347 7321
⊖ (+1 310) 517 67 14
👥 1,271
Division Manager: E Illokken
■ High power travelling-wave tubes, travelling-wave tube amplifiers and cryogenic coolers for ECM, radar and communications.

Technical Products Division
2051 Palomar Airport Road, Carlsbad, CA 92009
☎(+1 619) 931 30 00 ⊖ (+1 619) 438 91 91
Division Manager: Gary E Gist
■ Manufacture image and display products, automatic assembly and test equipment.

Microelectronics Circuits Division
500 Superior Avenue, Newport Beach, CA 92658
☎(+1 714) 759 24 11 ⊠ TWX 910 596 1374
⊖ (+1 714) 759 27 20
👥 1,000
Group Vice President and Divisional Manager: Donald Schrock
■ Advanced microelectronic technology.
Manufacture multiplex and audio switching systems, aircraft cabin entertainment and communications systems.

Hughes Training Inc. (HTI)
(† Hughes Simulation Systems Inc.)
(a GM-Hughes Electronics company)
Arlington Operation
2200 Arlington Downs Road, Arlington, TX 76011
☎(+1 817) 695 2000 ⊖ (+1 817) 695 22 22
President: S I Moore
Vice President and Chief Operating Officer:
Vice President of Engineering and Operations: Dwayne Humphrey
Corporate Communications: Glenn Hillin
■ Manufacture a full range of training equipment, image generation products and simulated instruments. Designs visual system displays and databases. Provides complete training systems services. → 80

Subsidiaries:
AMI Instruments Inc. → 75

Herndon Operation
13775 McLearen Road, Herndon, VA 22071
☎(+1 703) 481 45 14 ⊖ (+1 703) 689 00 18
General Manager: Roger Charbonneau
■ Manufacture of full mission simulators, operational flight trainers and air cushion vehicle simulators.

Hughes Research Laboratories (a subsidiary of Hughes Aircraft Co.)
3011 Malibu Canyon Road, Malibu, CA 90265
☎(+1 310) 317 50 00 ⊠ 652310 hacresl mlbu
⊖ (+1 310) 317 54 83
👥 440
Senior Vice President and Director: Arthur N Chester
■ Long range applied scientific research in physics, chemistry, electronics and information sciences. [98]

75-44900 Hughes Training, Inc
(a division of Hughes Electronics Corp.)
PO Box 6171, Arlington, TX 76005-6171
☎(+1 817) 619 22 00
⊖ (+1 817) 619 35 55; 619 37 77
👥 4,900 (1997) 🏭 400,000 sq ft
President: Stuart I Moore
Executive Vice President and Chief Operating Officer: Richard A Baldridge
Vice President of NASA Operations: Howard B Marshal
Vice President of Training Equipment: Bill Eschenbruechen
Vice President of Training: E R Field
Vice President of Advanced Programs: Gary N Nesta
Vice President of Human Resources: David Letts
Vice President of Marketing: Roland Berry
General Counsel: Chris Puffen
Executive Sales and Marketing Director: Jon K Foster
Director of Operational Strategies: Larry Roush
Purchasing Manager: Roy Fyfe
Business Development Specialist: Richard Oyler
■ Specialises in the design and manufacture of military training simulators for surface, airborne and subsurface ASW applications, surface team tactics training and ground C2 systems. Provides professional classroom and simulator instruction, support of training facilities and devices and total training systems. Design and manufacture commercial training simulators for biotechnology, rail/transit, air traffic control ship control systems and satellite- based cellular-transmitted position reporting systems. [98]

75-45000 Charles A. Hulcher Co. Inc.
909 G Street, Hampton, VA 23661
☎(+1 757) 245 61 90 ⊖ (+1 757) 247 94 31
👥 6
President: Ms Betty H Giles
Corporate Secretary: L D Hulcher
■ Photography instrumentation for the government, including 35mm and 70mm sequence cameras and panoramic cameras. [98]

75-45050 Humphrey Inc. Electro-Mechanical Instruments
9212 Balboa Ave, San Diego, CA 92123
☎(+1 619) 565 6631 ⊖ (+1 619) 565 6873
♣ $ 21 900 000 👥 330 🏭 40 000 sq ft
President & Chairman: Errol Ekaireb
Vice President & General Mgr: H.H. Kries [97]

75-45150 Hunter Manufacturing
30524 Aurora Road, Solon, OH 44139-2740
☎(+1 440) 248 61 11
🄴 75144.2613@compuserve.com
⊖ (+1 440) 248 16 91
Chief Executive Officer: Thomas J Gruber
President: I Eugene Strine
Sales Manager: John S Kennedy
■ Developed heater products for the military. Supplies include hot air, hot water, diesel fuel burning & engine coolant heaters. [98]

75-45270 Hycomp Inc.
165 Cedar Hill St, Marlboro, MA 01752
☎(+1 508) 485 6300 ⊖ (+1 508) 481 1547
♣ $ 4 000 000 👥 43
President: George Riley
VP Sales & Marketing: Sumner J. Eagerman
■ Thin film products, custom hybrids, surface mount hybrids. [97]

75-45280 Hyde Products Inc.
28045 Ranney Parkway, Cleveland, OH 44145-1144
☎(+1 216) 871 4885 ✆ (800) 999 0009
⊖ (+1 216) 871 1143
👥 40
President: Thomas P. Mackey
VP Sales: Jim Mackey
Marketing Mgr: Aaron G. Bresnahan
Purchasing Mgr: Ms Karen Perhacs
■ Marine deck machinery & steering systems. [97]

75-45290 Hydraflow
13259 East 166th Street, Cerritos, CA 90703
☎ (+1 562) 926 95 82 ⊜ (+1 562) 926 86 96
♣ US$ 25,000,000 (1997) 👥 120 (1997)
Chief Executive Officer: L E Ullrich
President: D Ullrich
Senior Vice President: C Ayloush
Vice President and Operations Director: Ken Johnson
Sales and Marketing Director: J Lewis
■ Commercial/military oxygen hose assemblies, flexible fluid couplings for fuel and ECS, quick disconnects for electronic cooling applications, electronic cooling hose assemblies. [98]

75-45550 Hydroscience Technologies Inc.
2659 Nova Drive, Dallas, TX 75229-0304
☎ (+1 972) 247 90 52 ✆ (800) 628 23 02
Ⓦ http://www.saemux.com ⊜ (+1 972) 484 32 81
🏭 8,250 m2
President and Chief Executive Officer: Fred G Woodland
☎ (+1 972) 247 90 52 ext 107
Vice President of Operations: Jerry L Fortenberry
☎ (+1 972) 247 90 52 ext 102
Sales and Marketing Director: Sammy Fox
Director: Bailey
■ Production of ASW shipboard electronics and acoustic towed arrays; logistic support, reliability analysis, personnel training, depot level repairs; Navy array maintenance facilities. → 80 [98]

75-45700 Hytek Finishes Co.
(a division of Esterline Technologies)
8127 S 216th Street, Kent, WA 98032
☎ (+1 206) 872 71 60 Ⓦ http://www.hytekfinishes.com
⊜ (+1 206) 872 72 14
👥 185 🏭 96,000 sq m
President: Jim J Cich (Acting)
Sales Director: Pat McShane
Operations Director: Clif Johnson
Purchasing Manager: Leigh Goodier
■ Specialised metal finishing, anodizing and organic coatings. [98]

75-45780 Ibex Aerospace Technologies Inc
(a subsidiary of ECE, France)
2016 Trade Center Way, PO Box 420149,
Naples, FL 33942
☎ (+1 813) 597 22 99 ⊜ (+1 813) 597 37 05 [98]

75-45800 IBM Corporation
Corporate Headquarters
New Orchard Road, Armonk, NY 10504
☎ (+1 914) 499 19 00 ⊜ (+1 914) 499 50 99
♣ $ 75,947,000,000 (1996) 👥 240,615 (1996)
Chairman and Chief Executive Officer: Louis V Gerstner (Jr)
Chief Financial Officer: Lawrence R Ricciardi (Acting)
Senior Vice Presdient Strategy: J Bruce Harreld
Senior Vice President Sales and Services:
Ned C Lautenbach
■ Create, develop and manufacture of advanced information technologies including computer systems, software, networking systems, storage devices and microelectronics. [98]

75-45890 Icore International Inc.
(a subsidiary of Smiths Industries plc)
180 North Wolfe Road, Sunnyvale, CA 94086
☎ (+1 408) 732 54 00 ⊜ (+1 408) 720 85 07
Executive Vice President: I O Humphries
Technical Operations Manager: Willy Tucker
■ Manufacture high security flexible and repairable harness systems.
Associated companies:
Icore International GmbH, Germany FR
Icore International Ltd, UK [98]

75-45895 ICI Devoe Coatings
4000 DuPont Circle, Louisville, KY 40207
☎ (+1 502) 893 14 20 ⊜ (+1 502) 893 1475
♣ $ 416,242,000
Managing Director: Ron Raley
Business Development Manager: James R Dechant
■ Manufacture and supply of heavy duty protective coatings including tank linings. Specialists in epoxy, urethane and rich zinc coatings. [98]

75-45900 Ideal Aerosmith Inc.
1510 Gateway Drive NE, East Grand Forks, MN 56721
☎ (+1 218) 773 2455 ▣ TWX 910 9494 070 ideal aero
⊜ (+1 218) 773 7665
👥 25 (1995) 🏭 17 000 sq ft
President & CEO: L. C. Rogers
Administration Dir: Ms B. Schultz
■ Flight instrument test equipment & precision metering needle valves. Inertial guidance test equipment & motion simulators. Thermal test equipment. [97]

75-45920 Ideal Electric Co.
330 East 1st St, Mansfield, OH 44903
☎ (+1 419) 522 3611 ▣ 987410 ⊜ (+1 419) 522 9386
♣ $ 50 000 000 👥 310 🏭 250 000 sq ft
President & CEO: Michael M. Vucelic
VP Marketing: Michael Colangero
■ Design, development, manufacturing & service of the electric generators, motors & controls (1 000 to 30 000 kw). [97]

75-46020 IKA Works Inc
(† Propex Research & Engineering)
2635 North Chase Parkway SE,
Wilmington, NC 28405-7499
☎ (+1 910) 452 76 93 ⊜ (+1 910) 452 70 59 [98]

75-46050 ILC Data Device Corp.
Data Device Corp. (DDC)
105 Wilbur Place, Bohemia, NY 11716-2482
☎ (+1 516) 567 56 00 ✆ (800) 332 57 57; 17 72
Ⓦ http://www..ilcdde.com ⊜ (+1 516) 567 73 58
👥 600 🏭 160,000 sq ft
President: Clifford Lane
Vice President Sales and Marketing: Herb Jarecki
Marketing Communications Manager:
Lawrence M Margulies
■ Data converters, MIL-STD-1553 products, hybrids and military sub-systems.

Garden Grove Office
12832 Valley View Street, Suite 213,
Garden Grove, CA 92645
☎ (+1 714) 895 97 77 ⊜ (+1 714) 895 49 88
Sales Mgr: Howard Ross

Loudoun Technical Center
21400 Ridgetop Circle Suite 110, Sterling, VA 20166
☎ (+1 703) 450 79 00 ⊜ (+1 703) 450 66 10
Sales Mgr: Dan Arnhols

Union Office
2092 Morris Avenue, Union, NJ 07083
☎ (+1 908) 687 04 41 ⊜ (+1 908) 687 04 70
Sales Mgr: Bill Welford

ILC Dover Inc.
One Moonwalker Road, Frederica, DE 19946-2080
☎ (+1 302) 335 39 11 ⊜ (+1 302) 355 07 62
President: Homer Reihm
VP/Dir Marketing: J. McMullin
■ Space suits & accessories, aircrew equipment, aerostats, RPVs, sonobuoys & balloons.

BETA Transformer Technology Corp.
40 Orville Drive, Bohemia, NY 11716
☎ (+1 516) 244 73 93 ⊜ (+1 516) 244 88 93
Vice President Engineering: Len Marro
Marketing Communications Manager:
Lawrence M Margulies [98]

75-46070 ILC Technology Inc.
399 Java Drive, Sunnyvale, CA 94089
☎ (+1 408) 745 79 00 Ⓦ http://www.ilct.com
⊜ (+1 408) 744 08 29
♣ $ 54,200,000 (1996)
Chairman of the Board and Chief Executive Officer:
Henry Baumgartner
President and Chief Operating Officer: Richard Capra
Vice President of Marketing and Sales: John Lucero
Marketing and Communications Manager:
Ms Heather Brown
■ Design and manufacture CERMAX xenon arc lamps, light sources and power supplies for medical and industrial fibre optics applications. Other products include pulsed and continuous wave xenon and krypton arc lamps for lasers, DAYMAX metal-halide lamps, CAPMAX mercury capillary lamps, super high pressure mercury xenon short arc lamps, IR sources for military applications and space qualified lighting for spacecraft and satellites. [98]

75-46080 Ilex Systems Inc.
1988 Tarob Court, Milpitas, CA 95035
☎ (+1 408) 945 02 94 ⊜ (+1 408) 945 92 89
♣ US$ 50,000,000 (1996) 👥 500 (1997)
President: Joseph S Lopez
Vice President of Marketing: Robert H. Robinson
✉ brobinson@corp.ilex.com
Government Marketing: Dick Barry
■ Manufacture of TEMPEST & non-TEMPEST secure devices including facsimile, optical character reader & STU-III Digital Gateway products. Also provide supervisory & control systems for electric, water, gas utilities, software and systems consulting services.
Distributors:
Tempest Europe ApS → Denmark 90
Presentey Engineering → Canada 75
● 10400 Baton Place, Suite 101, Fairfax, VA 22030
☎ (+1 703) 691 73 80 ⊜ (+1 703) 591 07 66 [98]

75-46110 Image Interpretation Systems Inc. (API)
(† Amarel Precision Instruments Inc.)
(a subs. of Amarel)
78 Schuyler Baldwin Drive, Fairport, NY 14450
☎ (+1 716) 223 2374 ⊜ (+1 716) 223 3413
♧ $ 2 300 000 (1995) ♣ $ 8 200 000 (1995) 👥 46
🏭 1 789 sq m
President & CEO: Willard W. Hunter
Sales Dir: George Jaynes
Operations Dir: Paul Nothnagle
Product Mgr: Tony Amarel
Purchasing Mgr: Bob Orcutt
■ Design, manufacture & distribution of custom optical microscopes, illumination systems & photo interpretation workstations. [97]

75-46140 IMI-Tech Corp.
(a subsidiary of Inspec Group plc, UK)
538 Haggaro Street, #402, Plano, TX 75074
☎ (+1 214) 516 07 02 ✆ (800) 666 7610
Ⓦ http://www.imi-tech.com ⊜ (+1 214) 516 06 24
👥 49 (1997) 🏭 15,000 sq m
Managing Director: Robert Tait
Operations Director: Howard Ryser
Commercial Development Director: Ms Betty Hartman
Technical Manager: Dr Scott Snider
■ Manufacturer of Solimide polyimide flexible foams used for fire protection, thermal and acoustic insulation in aircraft fuselages and cushioning for both aircraft and space applications.
Other applications include insulation for ducting, pipes, bulkheads, bilge and bleed air. [98]

75-46150 Imo Industries Inc.
1009 Lenox Drive, Building Four West, PO Box 6550,
Lawrenceville, NJ 08648-0550
☎ (+1 609) 896 7600 ▣ 843333 ⊜ (+1 609) 896 7688
♣ $ 400 000 000 👥 3 900 (1996)
Chairman, President & CEO: Donald Farrer
Sr. VP, Gen. Counsel & Secretary: Thomas J. Bird
VP Human Resources: Donald F. Vosburgh
Investor Relations Dir: Robert B. Derr
Marketing Communications Dir: Jerry D. Hickey
■ Diversified manufacture of pumps, fluid sensors, motion control products & remote control products & systems. Worldwide operations.
Subsidiary companies:
Baird Corp.
Varo Inc.
Optic-Electronic Corp.

IMO Pump
Airport Road, PO Box 5020, Monroe, NC 28111-5020
☎ (+1 704) 289 6511 ⊜ (+1 704) 289 8354
General Mgr: Berry F. Major
■ Manufactures a wide range of high-performance rotary, positive displacement pumps, including two-screw, three-screw & gear pumps. [97]

75-46700 Infinetics Inc.
Friskem Division
201 Vandever Avenue, PO Box 2330,
Wilmington, DE 19899-2330
☎ (+1 302) 658 24 71 ✉ FRISKEM WIL
⊜ (+1 302) 658 24 75
General Manager: M Schwartz
■ Weapon search equipment. Audio spot announcers. [98]

75-46701 Infodex Incorporated
10 Venus Drive, Wolcott, CT 06716
☎ (+1 203) 879 49 44 ⊖ (+1 203) 879 34 04
♣ $ 4,300,000 (1996) ⛬ 42 (1997) ⌂ 19,000 sq ft
Chairman: Ernest H Ulm
Chief Executive Officer: Robert M Snow
Sales and Marketing Director: Roger Lemire [98]

75-46703 InfoLink Technologies Inc.
777 North 1890 West, Provo, UT 84601
☎ (+1 801) 375 75 07 [E] infolink@aol.com
⊖ (+1 801) 375 75 37
⛬ 19
President: John L Tolman
Operations Director: Ms Jennye Harding
Sales and Marketing Director: Ms Janine Oddis
Purchasing Manager: Weston Tolman
■ Full-text searchable, document archive and retrieval services. Also offers the Infolink Scan Center. [98]

75-46750 Inframetrics Inc.
16 Esquire Road, North Billerica, MA 01862-2598
☎ (+1 508) 670 55 55 ✆ (800) GO INFRA
[W] http://www.inframetrics.com ⊖ (+1 508) 667 27 02
♣ $ 65,000,000 (1996) ⛬ 250 (1996)
President: Jay Teich
Sales and Marketing Director: Dan Turner
■ Design & manufacture of thermal imaging systems for military & paramilitary applications. Products include short-, medium- & long-range infra-red surveillance systems and an infra-red gimballed aerial sensor. [98]

75-46847 Interface Flooring Systems Inc.
(† Interface Inc.)
Orchard Hill Road, PO Box 1503, LaGrange, GA 30241
☎ (+1 706) 882 18 91 [W] http://www.ifsia.com
⊖ (+1 706) 882 05 00
♣ US$ 625,067,000 (1994) ⛬ 715
President and Chief Executive Officer: John Wells
Senior Vice President of Sales and Marketing: Greg Colando
Vice President of Government Programs and Director of Sales Training: Scott W Landa
Marketing Director: John McIntosh
Communications Manager: Ms Tracy F Cook
■ Supplier of office furnishing for DoD. Floor covering, fabrics, coatings (epoxy floors), upholstery, carpet. [98]

75-46855 InterMed Video Technologies Inc
18 Commerce Road, Newtown, CT 06470
☎ (+1 203) 270 91 00 [E] sales@intermedvideo.com
[W] http://www.intermedvideo.com ⊖ (+1 203) 270 96 19
⛬ 14 ⌂ 10,000 sq ft
President: Gregg M Johnstone
Sales Director: Ms Rita Weller
Public Relations Manager: Jay Mongillo
■ Products: video products for the military including multiscan monitors, cameras and video recorders. Latest products are hard disk recorders. [98]

75-46860 International Machine & Tool Corp.
115 Maple Street, Warwick, RI 02888-2189
☎ (+1 401) 467 69 05 ⊖ (+1 401) 467 58 85
⛬ 30 (1997) ⌂ 2,500 m2
President and Chief Executive Officer: W J Elsdoerfer
Vice President Sales and Marketing: Ronald W Elsdoerfer
■ Manufacture test systems for aerospace applications, particularly automated positioning tables & single & multi-axis rate tables, tracking pedestals & motion simulators. [98]

75-46870 Interpoint A Crane Company
(† Interpoint Corporation)
(a subsidiary of Crane Co., USA)
10301 Willows Road, PO Box 97005,
Redmond, WA 98073-9705
☎ (+1 425) 882 31 00 ✆ (800) 822 87 82
[E] power@intp.com [W] http://www.interpoint.com
⊖ (+1 425) 882 19 90
♣ $ 70,000,000 (1996) ⛬ 600 (1997) ⌂ 5,400 sq m
President: Sal Ronchetti
Division Vice President: Dave Uvelli
Director of North American Sales: Don Bosmann
Marketing Manager: Ms Michelle Manson
■ Fully qualified MIL-STD-1772 company manufacturing military & space quality hybrid electronics. Specialising in power converters from 1 Watt to 1 000 Watt, EMI RFI filters & transient protection circuits, especially developed for distributed power systems. Fully compliant to MIL-STD-883. → 90 [98]

75-46890 Instrument Enclosures
4088 South "K" St, Tulare, CA 93274
☎ (+1 209) 688 1100 ⊖ (+1 209) 688 1193
♣ $ 2 000 000 (1994) ⛬ 45 ⌂ 18 000 sq ft
President: William B. Taylor
Vice President: Ms Sandra J. Taylor
Sales Dir: Dennis Fazekas
Purchasing Mgr: José Huertas
■ Watertight, aluminium instrument cases for transport or field use. [97]

75-47020 Insulgard Corp.
5133 Lawrence Place, Hyattsville, MD 20781-1028
☎ (+1 301) 927 88 55 ✆ (800) 638 67 18
[E] insulgard@aol.com [W] http://www.insulgard.com
⊖ (+1 301) 927 45 31
⛬ 30
President: M R French (Jr)
Managing Director: M Tranby
Marketing Manager: Brian Shelton
■ Bullet resisting architectural security products. [98]

75-47050 Intamar Logistics Inc. (ILI)
620 Sentry Park, Blue Bell, PA 19422
☎ (+1 215) 834 1150 ⊖ (+1 215) 834 8240
President: William T. Hord
Treasurer: Ms Lillian S. Hord
■ Pilots personal equipment such as flight helmets, flight suits, personal survival equipment, air droppable sea rescue kits & assemblies, parachute systems. [97]

75-47100 Integrated Power Corp.
(a subs. of Westinghouse Electric Co.)
7618 Hayward Road, Frederick, MD 21702
☎ (+1 301) 663 8279 ⊖ (+1 301) 631 5199
♣ $ 20 000 000 ⌂ 35 000 sq ft
President: Paul Garvison
VP Sales/Marketing: Craig Shoots
■ Design & manufacture of self-contained, transportable, trailer mobile & fixed, solar-electric power systems including solar-diesel and solar-TEG hybrid systems.
Remote power systems for industrial/military telecommunications. Applications including integrated equipment and power systems shelters. [97]

75-47160 Intelect Inc.
(a subs. of Intelect Communications Systems Limited)
1100 Executive Drive, Richardson, TX 75081
☎ (+1 214) 437 1888 ⊖ (+1 214) 437 9041
⛬ 150 ⌂ 35 500 sq ft
Chairman: Pete Ianace
VP Sales & Marketing: Willard F. Barnett
Regional Sales Mgr, Europe, Africa, M.East: John Garrett
■ Design & manufacture air traffic control & air defence ground environment communications equipment, fiber optic multiplexers & teleconferencing switches. [97]

75-47300 Interface Products Inc.
4630 North Avenue, Oceanside, CA 92056
☎ (+1 760) 945 02 30 ⊖ (+1 760) 945 02 39
⛬ 75 (1997) ⌂ 35,000 sq ft
President: William J Lang
Sales Manager: Bill Lang
Engineering Manager: James Phillips
Manufacturing Manager: Warren Whitehead
Marketing Administrator: B Blakenship
■ Design, engineer and manufacture a broad line of controls, switches and displays including hand held key pads and displays, control display units, annunciators, lighted keyboards, integrated displays, special components and Liquid Crystal Display modules. [98]

75-47400 Intergraph Corp. Federal Systems Division
2015 Mercator Drive, Reston, VA 20190
☎ (+1 703) 264 56 00 ✆ (800) 747 22 32
[W] http://www.intergraph.com/federal
⊖ (+1 703) 264 71 31
♣ $ 1,950,000,000 (1996) ⛬ 8,500 (1996)
President and Chief Executive Officer: Jim Meadlock
Vice President of Marketing: Thomas Baybrook
Advertising Director: Ms Martha McGrath
Public Affairs Manager, Federal Systems: Chris Waager
■ Intergraph develops, manufactures, sells and supports computer systems for the technical desktop, the combination of compatible technical applications and personal productivity tools in a single desktop computer. Hardware products include workstations, servers, scanners and plotters. The company's integrated software applications are used for computer- aided engineering, design, analysis, manufacturing, publishing and earth sciences such as mapping/geographical information systems. [98]

75-47865 International Data Sciences Inc. (IDS)
501 Jefferson Blvd, Warwick, RI 02886-1317
☎ (+1 401) 737 9900 ✆ (800) IDS DATA
⊖ (+1 401) 737 9911
♣ $ 4 200 000 (1995) ⛬ 45 ⌂ 1 858 sq m
President & CEO: Dr Renato D'Antonio
Exec. VP & Treasurer: Edward M. Salzillo
Technical Dir: Howard Rowland
Sales Dir: Jeannette Gouin
Purchasing Mgr: Cindy Stone
■ Data communication equipment, Ethernet/token ring monitors, WAN protocol analyzers, T1 & fractional T1 testers, bit error rate testers, analog test sets, breakout boxes, LAN & voice cable testers. [97]

75-48120 International Logistics Group (ILG)
6145-D Wall Street, Sterling Hights, MI 48312
☎ (+1 810) 979 21 11 ⊖ (+1 810) 979 21 12
♧ $ 8,600,000 (1996) ♣ $ 28,000,000 (1996)
⛬ 77 (1997) ⌂ 33,000 sq ft
President: Theodore J Henke
Managing Director: Dorne R Rigby
Operations Director: Billie Taylor
Sales and Marketing Director: Tony Donagon
Advertising Manager: D Dietz
Purchasing Manager: Ms Kathy Corbiel
■ Manufacture and supply of military spare parts to the US Government and NATO countries. [98]

75-48200 International Polymer Corp.
5602 Armour Drive, Houston, TX 77020-8005
☎ (+1 713) 673 77 00
[E] internationalpolymerhouston@postoffice .att.net
⊖ (+1 713) 673 44 88
⛬ 50 (1997) ⌂ 62,000 m2 (1997)
Manager: Ben Still
Technical Sales: D Reid
■ Formulators and fabricators of high performance engineered thermoplastics, injection and compression molding and machining. Supplier of molded bar, tube, advanced composite parts and coatings. [98]

75-48330 International Technologies Inc.
(a subsidiary of the International Machine & Tool Company, USA)
68 Albany Road, Warwick, RI 02888
☎ (+1 401) 467 69 05 ⊖ (+1 401) 467 58 85
⛬ 15 (1997) ⌂ 1,500 m2
Vice President: R W Elsdoerfer
■ Manufacture computer based and remote data acquisition and control systems and subsystems for distributed control applications and measurement and analysis devices. [98]

75-48650 International Transducer Corp. (ITC)
(a subsidiary of Channel Industries, Inc.)
869 Ward Drive, Santa Barbara, CA 93111-2920
☎ (+1 805) 683 25 75 ⊖ (+1 805) 967 81 99
⛬ 52 (1997) ⌂ 27,000 sq m
President: William A Bunker
Assistant to the President: J Johnstone
International Sales Manager: Ms Christine M Bashforth
■ Design, manufacture and test acoustic transducers (hydrophones and projectors) for underwater use (Sonar). [98]

75-48680 Interspiro
31 Business Park Drive, Branford, CT 06405
☎ (+1 203) 481 38 99 ⊖ (+1 203) 483 18 79
♣ $ 17,000,000 (1996) ⛬ 35 (1997)

Contact Reference

Managing Director: Ken Warner
Operations Director: Jack Alexander
Sales and Markerting Director: Mike Brookham
Purchasing Manager: Dorleen Horn
■ Manufacture, sell/service self contained breathing apparatus & diving equipment. 98

75-48700 Interstate Electronics Corp.
(a subsidiary of Figgie International Inc)
PO Box 3117, Anaheim, CA 92803
1001 East Ball Road, Anaheim
☎(+1 714) 758 05 00 ✆(800) 854 69 79
⊜(+1 714) 758 41 48
👥 850 (1997) 🏠 360,000 sq ft
President: Lawrence A LaCotti
Vice President of Engineering: Joseph Rabick
Vice President of Program Management: John Schwartz
Vice President of Business Development: C Stephen Jungers
Vice President of Operations: Michael Patrick
Manager of Communications: Ms Joan C Hansen
Purchasing Manager: Frank Thompson
Manager of Business Development: Peter Rothenberg ☎(+1 714) 758 40 95 Ⓔ peter_rothenberg@iecorp.com
■ Instrumentation and missile tracking systems. Militarised, ruggedised flat-panel and CRT displays. GPS positioning and timing receivers. GPS-based tracking, navigation and flight management. Data collection, recording and signal processing systems. Fast signal acquisition SATCOM modems. GPS systems, satellite communications, band width, on-demand. 98

75-48960 Irvin Industries Inc.
California Division
3621 So. Harbor Blvd, Ste 100,
Santa Ana, CA 92704-5316
☎(+1 714) 662 1400 ⊜(+1 714) 662 1586
♣ $ 14 000 000 👥 250
Chief Exec. Officer: Thomas Montroy
VP Marketing & Technical Development: Phillip Delurgio
VP Business Development: Miles K. Yano
■ Parachutes for personnel, cargo, weapon & aircraft retardation. Recovery systems for missiles, space vehicles & cargo delivery systems. Safety harnesses & air drop sequence control devices. Bomb retarders & cargo nets. 97

75-49100 Iscan Inc.
89 Cambridge Street, Burlington, MA 01803
☎(+1 617) 273 44 55 Ⓔ info@iscaninc.com
Ⓦ http://www.iscaninc.com ⊜(+1 617) 273 00 76
President: Rikki Razdan
VP Sales: Alan M Kielar
■ Manufacturer of automatic video tracking systems for weapons pointing, test range tracking & advanced seekers; manufacturer of eye movement monitoring systems for advanced man-machine interface, including cockpit fire control. 98

75-49452 ITT Cannon
(a division of ITT Industries Inc, USA)
666 E Dyer Road, PO Box 35000,
Santa Ana, CA 92705-6500
☎(+1 714) 557 47 00 ⊜(+1 714) 754 21 42
President: Iain Duffin
Sales Director of Commercial Connectors: Bruce Blankfield
Director of Marketing Co-ordination and Business Planning: Ms Sandra J Wood
■ Manufacturer of circular, rack and panel, D subminiature, PCB, coaxial microminiature fibre optic connectors and cable assemblies. Also design of IC memory cards and connectors.

Switch Products
8081 Wallace Road, Eden Prairie, MN 55344
☎(+1 612) 934 44 00 ⊜(+1 612) 934 91 21
Site Manager: Rick Haneler
Marketing: Darrell J Wilk

Panoma
(† ITT Industrial Technology Corp)
1500 East 9th Street, Pomona, CA 91766
☎(+1 909) 623 34 63 ⊜(+1 909) 629 33 17
General Manager: Tom Barber

RF Products
585 E Main Street, New Britain, CT 06051
☎(+1 860) 223 27 00 ⊜(+1 860) 225 27 81
Vice President Marketing and Sales: Al Melite
■ RF/coaxial connectors & cable assemblies. 98

75-49455 ITT Defense & Electronics
(a division of ITT Industries)
1650 Tysons Boulevard, Suite 1700, McLean, VA 22102
☎(+1 703) 790 63 00 Ⓦ http://www.ittind.com
⊜(+1 703) 790 63 60
♣ $ 1,500,000,000 (1996) 👥 13,000 (1996)
President and Chief Executive Officer: Louis J Giuliano
Vice President and Controller: Edward W Williams
Vice President, International Marketing: Frank Parker
Vice President and General Counsel: James O Murrell
Vice President, Corporate Development and Operations Support: Ralph Meoni
Marketing Communications Director: Ms Millie Rogers
Public Relations: Ms Rebecca Reed
■ Supplier of high technology command, control and communications, electronic defence, electro-optical and surveillance systems, navigation and meteorological space payloads and related services.

ITT Aerospace/Communications Division
1919 West Cook Road, PO Box 3700,
Fort Wayne, IN 46801
☎(+1 219) 487 60 00 ⊜(+1 219) 487 60 03
President and General Manager: Marvin R Sambur
Vice president and Director of Business Development: Barry Abzug
Vice President and Director, Engineering: Richard Arra
Quality Assurance Director: Mike Hayman
Manufacturing Director: Ed Stuczynski
Public Affairs Manager: John Kirkwood
■ Design, development and production of forward area battlefield communications and signal processing equipment and systems. Voice and data communications receiving and transmitting systems; key management systems; surveillance and reconnaissance receiver systems; voice and data secure communications equipment; meteorological and navigation spaceborne payloads. Qualified for ground, airborne and space applications. Featuring microprocessor control and standard digital interface.

ITT Avionics
100 Kingsland Road, Clifton, NJ 07014
☎(+1 201) 284 01 23 ⊜(+1 201) 284 33 34
President and General Manager: Henry J Driesse
Vice President and Director of Business Development: John Berte
Electronic Warfare Marketing Director: Robert Palazzo
AN/ALQ-165 Programme Director: Ron Rowe
Design Engineering Director: Alan Kaufman
Marketing Communications Manager: Ms Rachelle Tucci
■ Situational awareness and self-protection systems for strategic, tactical and special operations aircraft and airborne communications, navigation and identification systems.

ITT Night Vision
ITT Electro-Optical Products
7635 Plantation Road North West, Roanoke, VA 24019
☎(+1 703) 563 03 71 ⊠ 829458 ⊜(+1 703) 366 90 15
👥 600
President and General Manager: Neil Gallagher
Military Marketing Director: John Dornalski
Marketing Communications Manager: Mrs Laurel Holder ☎(+1 703) 563 03 71 ext. 374
■ Manufacture night vision goggles, image intensifier tubes, pocket scopes, weapon sights and special-purpose tubes.

ITT Federal Services Corp.
1 Gateway Plaza, 1330 Inverness Drive,
Colorado Springs, CO 80910
☎(+1 719) 574 58 50 ⊜(+1 719) 380 83 09
President and General Manager: John R Spearing
■ Provides professional technical and support services for military base/range operations and maintenance; complex electronic systems development and integration; education and training centres. → 80

ITT Gilfillan
7821 Orion Avenue, PO Box 7713,
Van Nuys, CA 91409-7713
☎(+1 818) 988 26 00 ⊜(+1 818) 901 24 38
🏠 650,000 sq ft
President: Victor Rios
Vice President and Controller: Jay Larsen
Vice President of Advanced Engineering: Dr David E Hammers
Division Vice President and Engineering Director: Dr John M Milan
Division Vice President and Contracts Director: Ms Esther L Wilcox
Product Assurance Director: Douglas L Marsh
Vice President of Material Production and Logistics Director: Joe Macias
Senior Scientist: Jeffrey T Nemit
Administration Manager: Julian A Topper
■ Air defence systems, coastal and border surveillance systems, naval air defence systems, air traffic control systems. Radar and C3I subsystems.

ITT Gallium Arsenide Technology Center
7670 Enon Drive, Roanoke, VA 24019
☎(+1 703) 563 86 00 ⊜(+1 703) 563 86 96
Director: Paul Garnick
■ Research and manufacturing facility producing integrated GaAs microwave and digital components. 98

75-49950 IVEX Corp.
4355 International Boulevard, Norcross, GA 30093
☎(+1 770) 564 11 48 Ⓔ marketing@ivex3d.com
Ⓦ http://www.ivex3d.com ⊜(+1 770) 381 06 22
♣ $ 10,000,000 👥 37 (1997) 🏠 20,000 sq ft
President: I John Klipper
Vice President of Systems: William B Martin
Chief Operating Officer: Didier Bredy
Sales and Marketing Director: Ms Diana Rutledge
Purchasing Manager: Curtis Clay
■ Visual systems for simulation, specialising in low cost, visuals for flight, gunnery and target simulators. Interactive 3D visualization systems for virtual reality and training. 98

75-50100 Jay-El Products Inc.
(a subsidiary of Ducommon Incorporated)
23301 South Wilmington Avenue,
Carson, CA 90745-6209
☎(+1 310) 513 72 00 Ⓔ sales@jldmt.com
⊜(+1 310) 522 89 94
♣ $ 27,000,000 (1996) 👥 230 (1997) 🏠 120,000 m2
President and General Manager: Robert A Borlet
Vice President of Sales and Marketing: Anthony P Shadbolt
Vice President of Operations: Darrell Baldwin
Vice President of Microwave Products: Don Collertta
Purchasing Manager: Bill Brown
■ Illuminated pushbutton switches, data-entry keyboards, annunciators, illuminated electromechanical assemblies, RF and microwave switches. 98

75-50200 JDK Controls Inc.
424 Crown Point Circle, Grass Valley, CA 95945
☎(+1 916) 273 46 08 ⊜(+1 916) 273 07 69
👥 59 (1997)
Vice President of Marketing: Reece Kendall
Operations and Technical Director: Glen Darby
■ Precision co-moulded conductive plastic potentiometers and assemblies. 98

75-50350 Jennings Technology Corp.
(† Joslyn Jennings Corporation)
(a div. of the Joslyn Corp.)
970 McLaughlin Ave, San Jose, CA 95122-2696
☎(+1 408) 282 0358 ⊜(+1 408) 286 1789
👥 200 🏠 140 000 sq ft
President: J.E. Berkeland
VP Manufacture: Bob Doerr
Marketing Dir: Dwane Gisi
Purchasing Mgr: Kevin Keyser
Advertising Co-ordinator: Ms Michele Sheets
■ High voltage vacuum capacitors, relays, coax, switches, contactors & interrupters. Miniature relays. 97

75-50380 Jered Brown Brothers Inc.
(a division of Vickers America Holdings)
1608 Newcastle Street, Brunswick, GA 31520-6729
☎(+1 912) 262 20 00 Ⓔ jeredbrown@thebest.net
⊜(+1 912) 262 20 51
♣ $ 40,000,000 (1996) 👥 175 (1996)
Managing Director: C A Hooson
Chief Financial Officer: Robert R Motter
Sales Manager: Dale L Stoddard
Purchasing Manager: Danny F Hill
■ Ships steering gear, retractable bow planes, anchor windlasses, winches, capstans, cranes, elevators plus specialised handling equipment. 98

75-50920 JFS Inc.
(† Johnson Firearm Specialties Inc.)
PO Box 12204, Salem, OR 97309
☎(+1 503) 581 32 44 ✆(800) 477 32 44
⊜(+1 503) 581 31 91
President: David Johnson
■ Weapon-related components, ground-based. 98

75-50940 JML Optical Industries Inc.
690 Portland Ave, Rochester, NY 14621-5196
☎ (+1 716) 342 8900 ⊖ (+1 716) 342 6125
👥 125
President: Joseph M. Lobozzo II
VP Sales: Michael E. McCusker
■ Manufacturer of precision lenses & lens systems for industrial & military applications. [97]

75-51000 The Johnson John Company
1481 14th Street, Detroit, MI 48216-1806
☎ (+1 313) 496 06 00 ✆ (800) 991 13 94
⊖ (+1 313) 496 02 52
♧ $ 4,000,000 (1997) ♣ $ 8,900,000 (1997)
👥 85 (1997) 🏭 4,500 sq m
President: Charles A Keermaekers
Vice President Sales and Marketing: Richard H Dancy (Jr)
Vice President Military Sales: Mark Ficarra
Vice President Commercial Sales: Tim Rose
■ Tarpaulins/related cut and sew fabricating straps, truck frames for tarpaulins. [98]

75-51100 Joslyn Sunbank Corporation
(a subsidiary of Danather Corp.)
1740 Commerce Way, Paso Robles, CA 93446-3620
☎ (+1 805) 238 28 40 Ⓔ sunbank@tcsm.net
Ⓦ http://www.sunbank.com ⊖ (+1 805) 238 02 41
👥 250
President and Chief Executive Officer: Ed Prutzman
Vice President of Marketing: Patrick Banakis
Customer Service Supervisor: Ms Erin Whitener
Marketing Administration Assistant: Ms Connie Dennington
■ Manufacturer of connector accessories including backshells, protective covers, cable clamps and EMI/RFI accessories, and Sunband shield termination system and flexible conduit systems. [98]

75-51130 New Philadelphia Fan Company
(† Joy Technologies Inc.)
(a Howden Group Company)
PO Box 5000, New Philadelphia, OH 44663-5000
☎ (+1 216) 339 11 11 Ext. 262 Ⓔ nppnavy@bright.net
⊖ (+1 216) 339 82 11
👥 190 (1997) 🏭 220,000 m2
President and Chief Executive Officer: Jim Miller
Operations Director: Tom Uhl
Sales and Marketing Director: Stan Harris
Product Manager: Bill Mathias
Purchasing Manager: John Capozella
■ Shipboard ventilation fans. [98]

75-51150 Joyce Telectronics Corp.
2049 Range Road, Clearwater, FL 2124
☎ (+1 813) 461 35 25 ⊖ (+1 813) 461 61 72
♧ $ 500,000 (1996) ♣ $ 750,000 (1996) 👥 15
🏭 8,000 sq ft
President: Peter Joyce
Vice President: Michael Joyce
■ Aircraft and shipboard headset microphones, amplifiers, telephone handsets, headset adapters, ground communication cables and switches used in airlines, helicopters, military and space applications. Telephone, telegraph and aircraft equipment.

Gardena Plant
604 West 182nd Street, Gardena, CA 90248
☎ (+1 213) 321 66 50
Polymer Sales Specialist: Merrill Weidner
■ Microwave absorber, anechoic chambers, dielectric materials. [98]

75-51200 JP Instruments Inc.
3204-I West MacArthur Blvd, Santa Ana, CA 92704
☎ (+1 714) 557 3805 ✆ (800) 345 4574
⊖ (+1 714) 557 9840
👥 5
President: Joseph Polizzotto
Vice President: Larry Elbert
PR Mgr: Chris Sam
■ Design & manufacture aircraft instruments. [97]

75-51220 JPS Communications Inc.
5720M Capital Boulevard, Raleigh, NC 27616
☎ (+1 919) 790 10 11 ⊖ (+1 919) 790 14 56
👥 30 (1996)
President: Peter W Pflasterer
Vice President of Marketing: Donald K Scott (Jr)
Sales Director: Roger D Williams
Purchasing Manager: Bob Mitchell
Marketing Administrator: Ms Julia Atwell
■ Designs, develops, manufactures and sells innovative communications peripheral equipment, primarily for UHF, VHF and HF enhancement applications. [98]

75-51250 J-Tec Associates Inc.
5255 Rockwell Drive NE, Cedar Rapids, IA 52402-2020
☎ (+1 319) 393 52 00 ✆ (800) 959 08 72
Ⓔ j-tec@j-tecassociates.com
Ⓦ http://www.j-tecassociates.com
⊖ (+1 319) 393 52 11
♧ $ 570,000 (1996) ♣ $ 2,200,000 (1996) 👥 35
🏭 2,200 m2
President: Gary J. Roling
Dir Marketing: Carole W. Morio
Sales Dir: Randy Brekke
Operations/Technical Dir: Spike Staebler
Purchasing Mgr: Bonnie Bernhard
■ Vortex sensor flowmeters: tank crosswind sensors, engine airflow measurement sensors & blow-by flowmeters for engine health monitoring systems. [98]

75-51265 Julie Research Laboratories
508 West 26th Street, New York, NY 10001-5504
☎ (+1 212) 633 66 25 Ⓔ julie4@mail.idt.net
⊖ (+1 212) 691 33 20
👥 25
President: Ms Loebe Julie
Sales Manager: Marc Ruocco
Purchasing Manager: George Ott
■ Electronic standards 2 calibrators. Precision, ultra-stable resistors & resistor networks used mainly in navigational units (IMU) of missiles & subs. Precision voltage dividers. [98]

75-51280 Kachina Communications Inc.
30 North Alamos Drive, PO Box 1949,
Cottonwood, AZ 86326
☎ (+1 520) 634 78 28 Ⓔ info@kachina-az.com
Ⓦ http://www.kachina-az.com ⊖ (+1 520) 634 80 53
♣ $ 25,000,000 (1996) 👥 40 (1997) 🏭 1,200 sq m
President: Lester A Earnshaw
Vice President: Cameron Earnshaw
Marketing Director: Alfred W Hugueny
■ Manufacture HF/SSB communications equipment including commercial and military product lines. [98]

75-51350 Kaiser Aerospace & Electronics Corp.
950 Tower Lane, Suite 800, Foster City, CA 94404
☎ (+1 415) 349 74 00 ⊖ (+1 415) 349 82 84
Group President: S J Hill
President: Dr H J Smead
Senior Vice President: E Durbin
Vice President and Controller: J E Chapin

Subsidiaries with defence business:
Kaiser Optical Systems Inc. → 75
Kaiser Electronics → 75
Polhemus Inc → 75

Kaiser Electronics
2701 Orchard Parkway, San José, CA 95134
☎ (+1 408) 432 30 00 ⊠ 4945435 kaiser elctrncs
⊖ (+1 408) 432 84 40
👥 500
President: David Mineck ☎ (+1 408) 432 30 00 ext. 1140
Executive Vice President of Vision Systems International: Jim Stahl ☎ (+1 408) 433 97 20
Business Development Director: Joe Garcia ☎ (+1 408) 432 30 00 ext. 1148
Programmes Director: Paul Reynolds ☎ (+1 408) 432 30 00 ext. 1354
Government Affairs Director: Jim Atkins ☎ (+1 703) 447 23 02
■ Manufacture of advanced head-up displays, liquid crystal multifunction displays, helmet displays and display processors for combat aircraft.

Kaiser Electro-Optics
2752 Loker Avenue, West, Carlsbad, CA 92008
☎ (+1 619) 438 92 55 ⊖ (+1 619) 438 68 75
President: Jerry T Carollo ☎ (+1 619) 438 92 55 ext. 242
Marketing Manager: Ben J Mall ☎ (+1 619) 438 92 55 ext. 207

Kaiser Electroprecision
17000 South Red Hill Avenue, Irvine, CA 92714-5676
☎ (+1 714) 250 10 15 ⊖ (+1 714) 250 04 97
👥 306 🏭 230,000 sq ft
President: Ron McMahon ☎ (+1 714) 250 10 15 ext. 504
Vice President of Engineering: Tony Najjar
Sales and Marketing Director: James Basler ☎ (+1 714) 250 10 15 ext. 545
■ Manufacture pressure reoilers; liquid measurement devices; air start couplings and nipples; speciality quick disconnects; lavatory components; secondary flight control and service actuators; throttle quadrants; electromechanical and hydromechanical geared actuation products. ISO9000 standard applies.

Kaiser Compositek
1095 Columbia Street, Brea, CA 92621-2928
☎ (+1 714) 990 63 00 ⊖ (+1 714) 990 26 18
President: T Fulton
Sales and Marketing Manager: B Jones
Engineering Manager: J Noyes
■ Engineered composites. Manufacture of hydrospace products, (remote operated vehicles, bouyancy products, saturation diving support structures), pressure vessles filament wound aerospace products (missiles, fins/airfoils/frame members/aircraft ducts, engine stator blades, air turbine blades).

Kaiser Marquart, Inc.
16555 Saticoy Street, Van Nuys, CA 91409
President: Dominick Bellezza
Director of Aeropropulsion: W Escher
Sales and Marketing Manager: R Heubner
Engineering Manager: A Lucci
■ Manufacture of rocket engines, ramjet engines and ram-air turbine generators. [98]

75-51355 Kaiser Optical Systems Inc. (KOSI)
(a subsidiary of Kaiser Aerospace & Electronics Corp.)
371 Parkland Plaza, PO Box 983,
Ann Arbor, MI 48106-0983
☎ (+1 313) 665 80 83 Ⓦ http://www.kosi.com
⊖ (+1 313) 665 81 99
👥 80 (1997) 🏭 30,000 sq ft
President and General Manager: John H Ward (III)
Vice President and Controller: David M Brown
Vice President of Spectrographic Products: Harry Owen
Chief Scientist: W S Colburn
Business Development Director: James L McNaughton
■ Design and manufacture advanced holographic instrumentation and systems for avionics display, analytical spectroscopy, industrial process ccontrol, laser eye protection and electro-optical systems. [98]

75-51450 Kalyn-Siebert Inc.
(a subsidiary of Rail America Inc.)
PO Box 487, Gatesville, TX 76528-0487
☎ (+1 254) 865 72 35 ✆ (800) 525 96 89
⊖ (+1 254) 865 72 34
♣ $ 14,596,310 (1996) 👥 161 (1997) 🏭 197,671 m2
Senior Vice President and General Manager: Robert B Coward
Vice President of Engineering: Gerg Jordan
Operations Director: Tom Smith
Assistant Controller: Ms Lynn Gohlke
Marketing Manager: John Cope
Technical Manager: Tom Smith
Marketing Manager: Wes Chandler
Purchasing Manager: Ms Barbara Terrell
Sales Co-ordinator: Ms Melinda Ivy
■ Trailer manufacturing. [98]

75-51500 Kaman Aerospace Corp.
(a subs. of Kaman Corp.)
Old Windsor Rd, PO Box 2, Bloomfield, CT 06002-0002
☎ (+1 203) 242 4461; 243 7100; 243 7551
Ⓦ http://www.kaman.com ⊠ 99326
⊖ (+1 203) 243 7514
♣ $ 794,092,000 👥 1 200
President: Walter R. Kozlow
VP Engineering: David J. White
VP Finance: John R. Peloso
Director: Charles H. Kaman
Director: Harvey S. Levenson
Marketing Representative: William Brown
■ Helicopters (SH-2, K-Max), aircraft components, engine nacelle components, aerospace & energy systems, remotely piloted vehicles, space systems, electro-optical systems. Detail parts offloads. [97]

75-51555 Kaman Diversified Technologies Corp.
Blue Hills Avenue, PO Box 1, Bloomfield, CT 06002-0001
⊖ (+1 203) 243 6365
President: Richard E.W. Smith [97]

75-51560 Kaman Instrumentation Corp.
(a subsidiary of Kaman Corporation)
1500 Garden of the Gods Road, PO Box 7463,
Colorado Springs, CO 80933-7463
☎(+1 719) 599 11 32 ✆ (800) 552 62 67
[E] info-cos2@kaman.com ☒ 452412
⊜ (+1 719) 599 18 23
President: M Grier
Marketing and Technical Development Manager: B Duvall
Product Marketing Manager: John Drollinger
Sales Manager: Terry Dillahunty
Sales Administration: Ms Amanda Vignery
■ Non-contact displacement measuring systems capable of resolution to billionths of an inch. Used in airborne laser weaponry and night vision systems to SPC and SQC applications in factory automation. High performance SiO2 cable assemblies for aerospace, space and EW applications. [98]

75-51640 Kavlico Corp.
14501 Los Angeles Avenue, Moorpark, CA 93021
☎(+1 805) 523 20 00 [E] kavilco@aol.com
[W] http://www.kavilco.thomasregister.com ☒ 651481
⊜ (+1 805) 531 65 30
👥 1,000 🏠 270,000 m2
Marketing Director: Dave Carney
■ Position sensors, pressure sensors, force transducers, LVDT's, RVDT's.
Subsidiary company:
Kavlico International, UK [98]

75-51641 Kawasaki Heavy Industries (USA) Inc
509 Lexington Avenue - Suite 2705, New York, NY 10022
☎(+1 212) 759 49 50 ⊜ (+1 212) 759 64 21
Houston Branch
333 Clay Street - Suite 4480, Houston, TX 77002-4103
☎(+1 713) 654 89 81 ⊜ (+1 713) 654 81 87 [98]

75-51642 Kawasaki Loaders Inc
(a subsidiary of Kawasaki Heavy Industries Ltd, Japan)
2140 Barrett Park Drive, Suite 101, Kennesaw, GA 30144
☎(+1 770) 499 70 00 ⊜ (+1 770) 421 68 42 [98]

75-51643 Kawasaki Loaders Manufacturing Corp USA
(a subsidiary of Kawasaki Heavy Industries Ltd, Japan)
60 Amlajack Boulevard, Newnan, GA 30265
☎(+1 770) 251 70 00 ⊜ (+1 770) 251 70 04 [98]

75-51644 Kawasaki Motors Corp USA
(a subsidiary of Kawasaki Heavy Industries Ltd, Japan)
2009 East Edinger Avenue, Santa Ana, CA 92711 [98]

75-51645 Kawasaki Motors Corp USA
9950 Jeronimo Road, Irvine, CA 92718-2016
☎(+1 714) 770 04 00 ⊜ (+1 714) 460 56 28 [98]

75-51646 Kawasaki Motors Manufacturing Corp USA
(a subsidiary of Kawasaki Heavy Industries Ltd, Japan)
Lincoln Plant
6600 Northwest 27th Street, Lincoln, NE 68524
☎(+1 402) 476 66 00 ⊜ (+1 402) 476 66 72
Maryville Plant
Highway 71, South Maryville, MO 64468
☎(+1 816) 582 58 29 ⊜ (+1 816) 582 58 26 [98]

75-51670 KBI
900 Pingree Rd, Lake in the Hills, IL 60102-9670
☎(+1 847) 658 8561 ✆ (800) 527 8278 ✉ KBI DIESEL
☒ 722 448 ⊜ (+1 847) 658 9280
👥 30
CEO/President: James W. Burke
VP Sales & Marketing: James O. Burke
Sales Co-ordinator: Ms Marlene H. Moser
■ Engine starting fluid (ether) injection systems used for cold start of diesel engines. Maintains a complete cold room facility with digital data acquisition capability for virtually any R&D project. → 80 [97]

75-51705 KDI Precision Products Inc.
3975 McMann Road, Cincinnati, OH 45245
☎(+1 513) 943 20 00 ⊜ (+1 513) 943 22 88
♣ $ 45,000,000 (1997) 👥 325 (1997) 🏠 230,000 sq ft
President & CEO: Salvatore Mira
VP Product Development: Max R. Cone
VP Contracts: James Gilzinger
Marketing Mgr: William Kurtz
■ Ordnance safety, arming & fuzing systems & components; avionics; automated assembly & test equipment & components. Proximity sensors & target detecting devices; signal processing & switching systems. [98]

75-51760 Kearfott Guidance & Navigation Corp.
(a subsidiary of Astronautics Corp. of America)
150 Totowa Road, Wayne, NJ 07474-0946
☎(+1 201) 785 60 00; 785 60 10
⊜ (+1 201) 785 60 25
♣ $ 140,000,000 (1997) 👥 1,700 (1997)
🏠 1,300,000 sq ft
President and Chief Executive Officer: Dr Ronald E Zelazo
Executive Vice President: N Paige
Operations Director: Howard J Mintz
Vice President Business Development and Management: Harvey E Truppi
Vice President Engineering: Thomas Shanahan
Vice President Material and Purchasing: Joseph Potts
Purchasing Manager: Richard Varina
Director Business Development: Stephen Beiter
Manager Marketing/ Administrative Services: Joseph A Smalz
Supervisor, Administration Services: A Bleezarde
■ Advanced inertial and stellar-inertial navigation & guidance using ring laser and tuned rotor gyroscopes. Designs, manufacture and integrates systems for navigation, guidance & control of aircraft, space, land and marine vehicles, inertial components including gyroscopes, platforms, accelerometers, servo mechanisms, pilots displays & control panels, GSE, hydraulic and electro mechanical controls, motors, synchros, resolvers, actuators. [98]

75-51950 STC Keltec Corp.
(† Keltec Florida Inc.)
(a subsidiary of Signal Technology Corporation)
84 Hill Avenue, Fort Walton Beach, FL 32548
☎(+1 904) 244 00 43 [W] http://www.sigtech.com
⊜ (+1 904) 664 05 46
♣ $ 31,000,000 (1996) 👥 300 (1997)
President: Wayne Armstrong
Vice President of Marketing and Contracts: Robert W Bechtold
Operations Director: Dave Cheney
Sales Manager: Al Coronis
Purchasing Manager: Edward Noone
■ Customised high and low voltage power supplies, travelling wave tube amplifiers, standard DC to DC power converters, customised DC to DC power converters, power supplies for space applications. [98]

75-52000 Keltron Power Systems
(† Keltron Corp.)
(a division of Preferred Technologies)
225 Crescent Street, Waltham, MA 02154-3420
☎(+1 617) 894 87 00 ⊜ (+1 617) 899 96 02
👥 55
President: Les Hammer
Operations and Technical Director: Joan Kimberley
Purchasing Manager: Ms Janice Donahue
Sales Coordinator: Ms Donna Smith
■ Manufacture of high voltage power supplies for CRTs. [98]

75-52310 Kenrich Petrochemicals Inc.
140 East 22nd Street, Bayonne, NJ 07002-0032
☎(+1 201) 823 90 00 ✆ (800) LICA KPI
⊜ (+1 201) 823 06 91
♣ $ 4,655,000 (1996) 👥 28 🏠 75,000 sq ft
President: Salvatore J. Monte
Vice President: Erika G. Monte
Operations Director: Charles A Lucania
Sales and Marketing Director: Hartmut H Schlaubitz
■ Titanate, Zirconate & Aluminate coupling agents for enhanced polymeric & energetic composites & solid propellants. Plasticisers for polymeric composites. [98]

75-52330 Kern Engineering & Manufacturing Corp.
1146 East Ash Avenue, Fullerton, CA 92631
☎(+1 714) 992 96 30 [E] sales@kerneng.com
⊜ (+1 714) 447 42 42
♣ US$ 5,000,000 (1995) 👥 45 🏠 14,000 sq ft
President: Roland A Kern
Operations Director: Donald Hamilton
Sales Manager: Ms Carol Sue Kirkland
Engineering Manager: Eric Griswold
■ Manufacture machined products for connector and cable assemblies, RFI/EMI/EMP/ TEMPEST requirements. [98]

75-52350 Kester Solder
(a division of Litton Industries)
515 East Touhy Avenue, Des Plaines, IL 60018-2675
☎(+1 847) 297 16 00 ✆ (800) 2 KESTER
[W] http://www.kester.com ⊜ (+1 847) 390 93 38
♣ $ 110,000,000 (1996) 👥 274 (1997)
President: L D Kramer
Vice President of Marketing: Eric Slezak
Vice President of Sales: Chuck Jirek
Plant Manager: Paul Gassensmith
Purchasing Manager: Ms Anna Marie Micek
Advertising Manager: Dennis Gulino
■ Manufacture, supply and distribution of Solders and related materials for the electronics manufacturing industry. Products include solder pastes, liquid fluxes, flux cored and solid wire solder, solder preforms, bar solder, plating anodes, PC fabrication chemicals, residue removers and cleaners, temporary solder masks, conductive links and advanced polymer products. [98]

75-52450 Kevlin Corporation
5 Chelton Place, Wilmington, MA 01887
☎(+1 508) 657 3900 ⊜ (+1 508) 658 5170
👥 76
President: John J. Moran
VP Sales: John F. Rice
Treasurer: Daniel J. Whelan
Quality Assurance Mgr: John J. Malin
■ RF rotary couplers & microwave components for radar systems. [98]

75-52500 Walter Kidde Aerospace
(a division of Kidde Technologies Inc.)
4200 Airport Drive, Wilson, NC 27896
☎(+1 919) 237 70 04
[E] rowev@wkai.lan.mhs.compuserve.com
☒ 802846; TWX 510 929 1673 ⊜ (+1 919) 237 85 33
👥 250
Vice President and General Manager: J Sosnowski
Vice President of Sales and Marketing: Tom Hillman
Director of Sales and Customer Support: Vincent M Rowe
Sales and Marketing Assistant: Ms Donna Freeman
■ Design and manufacture of continuous fire and overheat detection cable, optical-based fire and explosion detection sensors, conventional and special application high-rate-discharge fire/explosion extinguishing vessels and integrated protection systems for aerospace and defence applications. [98]

75-52510 Kidde-Fenwal
(† Fenwal Inc.)
(a division of Williams Holdings plc, UK)
400 Main Street, Ashland, MA 01721
☎(+1 508) 881 20 00 ✆ (800) 872 65 27
⊜ (+1 508) 881 67 29
👥 450 (1997)
President: R H de Marle
Chief Financial Officer: Paul Cousineau
Vice President Sales and Marketing: Peter G Clark
■ Fire protection for all installations including high sensitivity smoke detection. [98]

75-52660 Kinetic Systems Corp.
900 N State Street, Lockport, IL 60441-2292
☎(+1 815) 838 0005 ✆ (800) DATA NOW
⊜ (+1 815) 838 4424
♣ $ 10 000 000 (1995) 👥 78 (1996) 🏠 60 000 sq ft
President: James M. Stephenson
Dir Sales: Arun Sheth
Dir Marketing & Int'l Sales: Philippe Roy
■ Manufacture high-speed, high-performance data acquisition & control systems conforming to CAMAC (IEEE-583) standard, fastbus (IEEE-960), VME & VXI. [97]

75-52670 King Fifth Wheel Co.
(a subsidiary of TI Group plc)
PO Box 68, Mountain Top, PA 18707
☎ (+1 717) 474 63 71 Ⓦ http://www.microserve.net/
⊖ (+1 717) 474 99 01
👥 125
President: Peter Celusniak
Marketing Manager: Mark Kelly
■ Manufacture of high quality flash welded & cold rolled rings for turbine engine & missile applications.
Associated company:
TI Reynolds Rings Ltd, UK [98]

75-52877 KJM Fabritek Inc.
PO Box 162, Marietta, GA 30061-0162
☎ (+1 770) 426 82 51 ⊖ (+1 770) 426 82 52
👥 5
Managing Director: Klaus Maltenieks
Secretary, Treasurer: C B Maltenieks
■ Manufacture brass cartridge casings for small arms ammunition assembly. Also produce contract ammunition on a contract-contract basis for government or military agencies.
Licenced exporters of ammunition, ammunition components and ammunition manufacturing machinery and ancillary equipment. Consultants for design and start up of foreign small arms ammunition plants.
Aeroballistics Group
PO Box 2163, Marietta, GA 30061
☎ (+1 770) 419 90 04 ⊖ (+1 770) 426 82 52
👥 5
Managing Director: James Maltenieks
■ Manufacturer of long range rifle ammunition (LORRAMM) and military contract ammunition. Engineering consultants in ballistics, technical analysis. Also produce munitions training devices for artillery or tank guns from 76mm up to 125mm for subcalibre use up to 25mm. [98]

75-52880 K&L Microwave Inc.
(a div. of Dover Technologies Co.)
408 Coles Circle, Salisbury, MD 21801
☎ (+1 410) 749 2424 ⊖ (+1 410) 749 5725
👥 520
President: Charles Schaub
VP Marketing & Sales: Jim Price
VP Engineering: David Howett
VP Purchasing: Tom Pusey
Advertising Dir: Ms Peggy W. Morey
■ Design & manufacture of microwave & RF components, from 500 KHz to 40 GHz. This broad range of components includes miniature filters, surface mount filters, GPS filters, cavity & waveguide filters, tunable filters, frequency agile & integrated subsystems, integrated subassemblies as well as coaxial switches to 26.5GHz & RF switching matrices. Suspended substrate. [97]

75-52900 Klein Associates Inc.
(a subs. of Oyo Corporation)
High Resolution Sonar
Klein Drive, Salem, NH 03079-1249
☎ (+1 603) 893 6131 ✉ SONAR ⊠ 947439
⊖ (+1 603) 893 8807
👥 30 🏭 54 000 sq ft
President: William H. Key (Jr)
Marketing Asst: Ms Patricia J. Durgin
Exec. Vice President: George Skuse
■ HYDROSCAN side scan Sonar and sub-bottom profiling systems for minehunting and channel clearance operations. System 2000, a fully integrated digital side-scan and 5000 multi-scan focused sonar system.
Oyo Geospace
9777 W. Gulf Bank Rd Ste 5, Houston, TX 77040
☎ (+1 713) 849 0804 ⊖ (+1 713) 849 4915 [97]

75-53030 H. Koch & Sons Co.
5410 East La Palma Avenue, Anaheim, CA 92807-2099
☎ (+1 714) 779 70 00 Ⓔ koch@ape.net
⊖ (+1 714) 779 71 41
👥 45 (1997) 🏭 50,000 sq ft
President: P M Zatezalo
Vice President of Marketing: J E Masterson
Controller: Charles Fiely
■ Tailored design and manufacturing of emergency egress lighting systems, custom vacuum packaging, restraint systems, inertia RSELS, aircrew safety hardware and survival equipment.
® Australia: Air N Sea Pty Ltd
Belgium: SKB Trading NV
Israel: American Aviation
Japan: Tel America
Singapore: Southwest Automotive Singapore
UK: Airspace Products Ltd [98]

75-53150 Kollsman
Kollsman Inc.
220 Daniel Webster Hwy, Merrimack, NH 03054
☎ (+1 603) 889 2500 ✉ KIC MERR
⊠ 943537; TWX 710 366 6883 kolsm
⊖ (+1 603) 889 7966
♧ $ 115 000 000 (1995) ♣ $ 109 200 000 (1994)
👥 655
President & Corporate Sr VP: Ronald H. Wright
VP Int'l Marketing: Charles F. Richmond
Dir Manufacturing: Joseph Moffatt
VP Avionics Marketing: Charles Torrey
VP Domestic Marketing: John Hamm
Marketing Communication Specialist:
Ms Elizabeth M. Conover ☎ (+1 603) 886 2604
Mgr Market Research: Steve A. Russell
Subcontracts Mgr: Chris Smith
■ EO/IR night vision systems. Aircrew combat manoeuvring instrumentation ranges. Weapon training systems. Cockpit instruments. Medical & analytical instruments. Manufactures fire control system for the M60 A3 tank & thermal night vision equipment. Major producer of aircraft instruments & microprocessor based test equipment for military & commercial customers.
Washington DC Operations
Kollsman Suite 600, 1577 Spring Hill Road, Springfield, VA 22182
☎ (+1 703) 356 6970 ⊖ (+1 703) 356 6973
VP Washington Ops: Al Friedrich
Kollsman
2227 Drake Avenue, Suite 8, Huntsville, AL 35805
☎ (+1 205) 882 0303 ⊖ (+1 205) 883 0399
Regional Sales Mgr: John Langbehn [97]

75-53160 Kongsberg Simrad Inc
(† Simrad Inc)
1225 Stone Drive, San Marcos, CA 92069
☎ (+1 619) 471 22 23 ⊖ (+1 619) 471 11 21
👥 36 (1995) 🏭 25,000 sq ft
Exec. Vice President: Tony Gardiner
Sales/Marketing Dir: James Allan
Admin Mgr: Larry Crews
■ Underwater & harsh environment TV, including diver systems with lights, cameras, communications, surface controls; cable systems with lights, cameras, pan & tilt units, displays; harsh surface environment & custom video systems. Night vision devices & acoustic tracking systems for underwater vehicles. → 90 [98]

75-53250 Korry Electronics Co.
(a subsidiary of Esterline Technologies)
901 Dexter Avenue North, Seattle, WA 98109
☎ (+1 206) 281 13 00 ✆ (800) 257 89 21
Ⓔ info@korry.com Ⓦ http://www.korry.com
⊖ (+1 206) 281 13 65
👥 500 (1997) 🏭 123,000 m2
President: Alan Cornell
Director Engineering: Kent Byington
Product Planning Director: Frank Houston
Director Human Resources: Ms Marlene Winter
Operations Director: Gary Dytrt
Promotional Services Manager: Peggy Keene
■ Manufacture of lighted pushbutton switches, indicators, integrated switchpanels, keyboards and night vision filters. [98]

75-53300 Kratos Pressure Products
(a subsidiary of Rogerson Aircraft Corp.)
1694 Von Karman, Irvine, CA 92606
☎ (+1 714) 422 23 47 ⊖ (+1 714) 422 23 11
Vice President and General Manager: John Schultz
■ Manufacture of standard pressure gauges, pressure switches and transmitters, and temperature compensated pressure gauges. [98]

75-53670 KVH Industries Inc.
110 Enterprise Center, Middletown, RI 02840
☎ (+1 401) 847 33 27 Ⓦ http://www.kvh.com
⊠ 382051 ⊖ (+1 401) 849 00 45
♣ $ 25,600,000 (1996) 👥 150 (1997) 🏭 75,000 sq ft
President: Martin Kits van Heyningen
Managing Director: Alan Sylvestre
Vice President Business Development:
Christopher Burnett
Vice President Manufacturing: Bruce Costa
Vice President Marketing: Jim Dodez
Public Relations Director: Sandy Oxx
Purchasing Manager: Ms Kirsten Quinn
■ Manufacture GPs compatible electronic fluxgate compass systems with automatic compensation for ships & boats, landing craft, reconnaissance vehicles, light armour vehicles, tanks & self propelled guns. Also manufacture handheld electro-optic instruments which integrate electronic compasses & electronic ratiometric rangefinders with sighting scopes for unsurpassed accuracy. [98]

75-53690 L&F Industries
2110 Belgrave Avenue, PO Box 40, Huntington Park, CA 90255-2713
☎ (+1 213) 588 22 31 ⊖ (+1 213) 588 22 37
♣ $ 10,000,000 (1997) 👥 75 (1997)
President and Chief Executive Officer: David Chivens
Vice President of Sales: Robert Jones
■ Document disintegration systems, stretch forming presses, hot forming presses, jet engine test stands, rocket sleds, nuclear shielding doors. [98]

75-53700 L&S Machine Co. Inc.
(a subs. of Precision Aero Tech)
PO Box 12264, Wichita, KS 67277-2264
☎ (+1 316) 942 0181 ⊖ (+1 316) 942 6082
♣ $ 16 000 000 (1996) 👥 120 🏭 140 000 sq ft
President: Don Chmelka
Sales & Marketing: Tom Kennedy
■ Manufacturer precision machined detail & mechanical & electro-mechanical assemblies, precision machined hinges, overhaul & repair. [97]

75-53850 Ladish Co.Inc.
5481 South Packard Avenue, Cudahy, WI 53110
☎ (+1 414) 747 26 11 ✉ LADISHCO
Ⓦ http://www.ladishco.com Ⓦ http://www.forging.com
⊖ (+1 414) 747 29 63
👥 1,060 (1997)
President: Ms Kerry Woody
Vice President Sales and Marketing: G J Vroman
■ High-performance forgings for the aerospace, aircraft & aircraft engine industries. 98

75-53935 Lancaster Machining Inc.
3180 West Fair Avenue, PO Box 641, Lancaster, OH 43130-0641
☎ (+1 614) 687 00 44 ⊖ (+1 614) 687 33 41
♧ $ 1,000,000 (1997) ♣ $ 1,300,000 (1997)
👥 13 (1997) 🏭 16,000 sq ft
President: Ted M Figgins
Vice President Sales: Ms Julie A Figgins
Operations Director: Bill Figgins
■ Aircraft machine parts. [98]

75-54390 Lear Astronics Corp.
(† Astronics Division, Lear Siegler)
(a subsidiary of GEC-Marconi Avionics Ltd, UK)
3400 Airport Avenue, PO Box 442, Santa Monica, CA 90406
☎ (+1 310) 915 68 81 ⊖ (+1 310) 915 83 84
👥 850 (1997)
President: David L Dallob
Vice President of Engineering: Geert Gevaert
Vice President of Contracts: I V Fedoroff
Director of Business Development: James Joyner
Director of Flight Systems: Ty Trbovich
Manager of Unmanned Systems: Einar Johnson
Purchasing Manager: Dennis Bent
Senior Marketing Analyst: Ms Mary Corcoran
■ Automatic flight control systems, integrated avionic systems, muzzle velocity measuring systems, ground-based electronic systems, unmanned air vehicles, stick force sensors, fuel management systems.
Developmental Sciences Centre
Lear Astronics Corp.
(a subsidiary of GEC-Marconi Avionics Ltd, UK)
1930 South Vineyard Avenue, PO Box 50000, Ontario, CA 91761
☎ (+1 909) 947 76 00 ⊖ (+1 909) 947 18 23
Business Unit Manager: Pete Aston
UAV Systems Marketing Manager: Ralf Persson
Ground Power Marketing Manager: John Male
■ UAV systems, aviation ground power. [98]

Contact Reference

75-54430 Learjet Inc.
(a subsidiary of Bombardier Inc., Canada)
PO Box 7707, Wichita, KS 67277-7707
☎(+1 316) 946 20 00 Ⓦ http://www.learjet.com
Ⓦ http://www.aero.bombardier.com ☒ 417441
⊜ (+1 316) 946 22 20
4,597 (1996)
President: Jim Robinson
Vice President and Chief Financial Officer: Colin Fernie
Vice President of Bombardier Aviation Services: James Ziegler
Vice President of Operations: Bill Remy
Vice President of Marketing and Sales: Bill Monroe
Vice President of Used Aircraft Sales: Gordon Wishart
Vice President, Bombardier Test: Peter T Reynolds
Vice President of Human Resources: Jeff Bahr
Aerospace and Government Contracts and Pricing Director: Ms Linda Meyer-Gosling
Marketing Flight Operations Director: Robert W Agostino
Public Relations and Strategic Planning Director: Rolland Vincent
Purchasing Manager: Brad Muir
■ Manufactures Learjet twin-jet aircraft suitable for commercial and some military and para-military roles. Aircraft service and refurbishment; aircraft certification, flight testing and aerospace subcontracting. Military support aircraft. Subcontractor to Lockheed Martin on Space Shuttle program (skin panel assemblies on intertank structure and solid rocket booster beam). [98]

75-54630 Lektro Inc.
1190 SE Flightline Drive, Warrenton, OR 97146
☎(+1 503) 861 22 88 ✆ (800) 535 87 67
Ⓔ sales@lektro.com ⊜ (+1 503) 861 22 83
♣ $ 7,000,000 (1997) 40 (1997) 68,000 sq ft
President: Eric Paulson
Operations Director: Jim Magoulas
Purchasing Manager: Don Cyrus
■ Manufacture battery powered towbarless aircraft handling vehicles capable of handling military & commercial transport aircraft. [98]

75-54670 Lenox Instrument Co.
265 Andrews Rd, Trevose, PA 19053-3422
☎(+1 215) 322 9990 ⊜ (+1 215) 322 6126
♣ $ 1 000 000 (1996) 40 10 000 sq m
President: John W. Lang (Jr)
VP Sales & Marketing: William J. Lang
■ Borescopes for non-destructive testing (NDT), including engine inspection. [97]

75-55060 L'Garde Inc.
15181 Woodlawn Avenue, Tustin, CA 92780-6487
☎(+1 714) 259 07 71 ⊜ (+1 714) 259 78 22
40 (1997) 32,000 sq ft (1997)
Chief Operating Officer: Gayle D Bilyeu
Marketing Director: Dr Costa Cassapakis
Technical Director: Gordon Veal
■ R&D of new concepts for spacecraft and weapon systems. Production of ICBM decoys for test. Inflatable space structures for antennas, solar concentrators, thermal radiators, robotic arms and solar arrays. R&D in controls technology and elastomers. Development and production of military solid-state and tape data recorders/players. [98]

75-55163 Life Support International Inc. (LSI)
200 Rittenhouse Circle, Bldg 4 West, Bristol, PA 19007
☎(+1 215) 785 2870 ☒ TWX 910 250 2972
⊜ (+1 215) 785 2880
25
President: Richard A. Brower
Vice President: Mrs Christina Belmonte
Operations Dir: Gregory L. Yerkes
■ Designs, develops, manufactures and supplies equipment for military aviation and marine applications. Products include parachutes, life rafts, life preservers, survival kits and flight clothing. Also offers materials and technology. [97]

75-55175 Global Atmouspherics, Inc.
(a subsidiary of Sankosha Corp.)
2705 East Medina Road, Suite 111, Tucson, AZ 85706
☎(+1 602) 741 28 38 ⊜ (+1 602) 741 28 48
72 (1997) 27,000 sq ft
President: Patrick Zumbusch
Director: Bruce Thatcher ☎(+1 520) 320 16 61
Product Manager: Lee Lawry ☎(+1 520) 204 11 44
■ Lightning detection equipment and services (data) for all branches of domestic and foreign military organisations. The equipment reduces the hazards of lightning strike damage, increases base uptime (44%) and provides tracking of thunderstorms. [98]

75-55220 Linear Technology Corp.
1630 McCarthy Boulevard, Milpitas, CA 95035-7417
☎(+1 408) 432 19 00 ✆ (800) 454 63 27
Ⓦ http//www.linear-tech.com ⊜ (+1 408) 434 05 07
♣ $ 378,000,000 (1996) 2,000 (1997)
388,600 sq ft
President: Robert H Swanson
Vice President of Marketing: Louis Dihlardo
Vice President of Sales: Tim Cox
Operations Director: Clive Davies
Product/Marketing Director: Robert Scott
Purchasing Manager: Phillip Di Girolamo
Advertising Manager: Douglas Dickinson
Public Relations Manager: Roy L Twitty
■ JAN "B" and JAN "S" 38510 analog ICs. MIL-STD-883C analog ICs, SMD analogue ICs, military temperature range standard ICs. [98]

75-55250 Lips Propellers
(a subs. of Lips BV, Netherlands)
Headquarters/East Coast Ops
3617 Koppens Way, Chesapeake, VA 23323
☎(+1 804) 485 5275 ⊜ (+1 804) 487 3658
President: J.F. Baur [97]

75-55300 G.W. Lisk Co. Inc.
2 South Street, Clifton Spring, NY 14432-1118
☎(+1 315) 462 2611 ⊜ (+1 315) 462 7661
♣ $ 50 000 000 (1996) 425 (1996) 200 000 sq m
President: I.A. Morris
VP Sales & Marketing: D.J. Riggs
■ Designs & manufactures solenoids, solenoid valves & LVDTs. [97]

75-55350 Lista International Corp
(a subsidiary of Lista Ltd, Switzerland)
106 Lowland Street, Holliston, MA 01746-2094
☎(+1 508) 429 13 50 ⊜ (+1 508) 626 03 53
120
President: Donald Brown [98]

75-55360 Lister-Petter Inc.
(a subsidiary of Lister-Petter Ltd, UK)
815 East 56 Highway, Olathe, KS 66061
☎(+1 913) 764 35 12 ⊜ (+1 913) 764 54 93
52 6,975 sq m
President: Phil Cantrill
Marketing Manager: Ms Julie Harrington
New Products Manager: David Stringer
National Sales Manager: Joe Campbell
■ Air and water cooled diesel engines 2.5-480 bhp, diesel powered generating sets 5-480 kW. [98]

75-55370 Litton FiberCom
(† FiberCom Inc.)
(a subsidiary of Litton Industries, Inc.)
PO Box 11966, Roanoke, VA 24022-1966
☎(+1 540) 342 67 00 ✆ (800) 537 68 01
Ⓔ info@fibercom.com Ⓦ http://www.fibercom.com
⊜ (+1 540) 342 59 61
140 (1997) 77,000 sq ft
President: Louis Tabor
Vice President, Channel Management: Jack B Freeman
Vice President, Product Development: Kenneth Ferris
Sales Director: Stu Pflaum
Pricing and Contracts: Chet Kostrzewa
Marketing Communications: Ms Janie Osterhaus
■ Development and manufacture of fibre optic systems for military and defence applications including ruggedised, TEMPEST, fault tolerant fibre optic Ethernet LANs, FDDI and SAFENET. Avionic, shipboard and tactical LANs. TEMPEST data links, fibre optic radio/radar remoting systems and military fibre optic transmission systems for Tri-Tac and Eurocom data types. Manufacturer of ATM access products for broadband applications including video conferencing, telephony consolidation. [98]

75-55400 Litton Industries Inc.
Corporate Offices
21240 Burbank Boulevard,
Woodland Hills, CA 91367-6675
☎(+1 818) 598 50 00 Ⓦ http://www.litton.com
☒ 674991 coraacw ⊜ (+1 818) 598 59 40
♣ $ 3,600,000,000 (1996) 33,500 (1996)
Chairman and Chief Executive Officer: John M Leonis
President and Chief Operating Officer: Michael R Brown
Senior Vice President and Chief Financial Officer: Rudolph E Lang (Jr)
Senior Vice President and Group Executive of Navigation Guidance and Control Systems Division: Larry A Frame
Senior Vice President and Group Executive of Information Systems: Alden V Munson
Senior Vice President and General Counsel: John E Preston
Senior Vice President and President of Ingalls Shipbuilding: Gerald J St Pé
Vice President and Group Executive of Navigation Guidance and Control Systems Division: Harry Halamandaris
Vice President of Corporate Communications and Investor Relations: Ms Kathleen M Wailes
Vice President of Corporate Development: H Thomas Hicks
Vice President and President of Data Systems Division: Allen E Powers
Corporate Vice President and Managing Director of LITEF: Richard T Hopman
Corporate Vice President and Associate General Counsel: Frank C Marshall (Jr)
Corporate Vice President and and Group Executive of Interconnect Products: Donald E Lepore
Corporate Vice President of Washington Office: John E Gordon
Corporate Vice President and Director of Industrial Relations: Mathias J Diederich
Corporate Vice President and Controller: Carol A Wiesner
Corporate Vice President and Treasurer: Timothy G Paulson
Corporate Vice President and President of the Applied Technology Division: Charles B Hofmann
Corporate Vice President, Assistant General Counsel and Secretary: Ms Jeanette M Thomas
Corporate Vice President, and President of PRC Inc.: James H Frey
Corporate Vice President, Guidance and Control Systems Division: Darwin D Beckel
Director, Corporate Communications: Robert Stangarone
■ Supplier of advanced electronic products for many military programs within all branches of the US Armed Forces, NATO and allied countries. Inertial guidance and navigation systems; command, communications and control systems, strategic and tactical electronic warfare systems, information technology systems and services. Components and units, night vision technology, travelling wave tubes, laser materials and servo-components. The Company is also a primary builder of large multimission surface combat ships for the US Navy and a major provider of overhaul, repair, modernisation, ship design and engineering services.

Corporate Washington Office
1725 Jefferson Davis Highway,
Suite 601 Crystal Square 2, Arlington, VA 22202-3585
☎(+1 703) 413 39 00 ⊜ (+1 703) 413 39 70
Corporate Vice President and Head of Washington Office: John E Gordon

Overseas Offices:
Switzerland (European Office); Japan (Far East Office)

Information
(† Command, Control & Communications Systems Group)
5490 Canoga Avenue, PO Box 4040,
Woodland Hills, CA 91365-4040
☎(+1 818) 715 57 45 ⊜ (+1 818) 715 57 37
Corporate Senior Vice President and Group Executive: Alden V Munson

Data Systems Division
29851 Agoura Road, Agoura Hills, CA 91301
☎(+1 818) 991 96 60 ☒ 662643 ⊜ (+1 818) 707 43 11
Corporate Vice President and Division President: Allen E Powers
■ Tactical command, control and communications systems, tactical missile defence systems, hand-held digital communications terminals, air defence systems, data processing and display equipment and systems, tactical data links, integrated ship electronics, optical processors, target detection and tracking equipment.

Overseas Marketing Office:
Belgium → 90

Litton Computer Services
4747 Hellyer Avenue, San José, CA 95138
☎(+1 408) 363 24 00 ⊜ (+1 408) 363 24 01
President: Henry T Bodurka

■ Services and systems; computer programming and software development; data systems design; systems engineering and engineering services; operations analysis; technical communications programs; systems operational test and evaluation; turnkey data systems, logistic systems. Computer processing and associated services, batch precessing, time sharing, remote job entry, on-line data base management, computer programming.

Litton Electro-Optical Systems (LEOS)
3414 Herrman Avenue, Garland, TX 75041
☎ (+1 972) 840 56 00 ⊖ (+1 972) 271 21 95
Division President: Steven Lambert
Operations Director: Urdie Heitmeyer
Director of Marketing: Steve Slusher
Purchasing Manager: Drew Jerima
■ Electro-optical devices and systems; image intensifiers, night vision systems, second and third generation, infra-red imaging products, detectors, focal plane arrays, laser optical components, lens assemblies and optical coatings.

Electronic Warfare Systems Group
2140 Burbank Boulevard,
Woodland Hills, CA 91367-6675
☎ (+1 818) 598 59 55 ⊖ (+1 818) 598 33 13
Senior Corporate Vice President and Group Executive: Harry Halamandaris

Amecom Division
5115 Calvert Road, College Park, MD 20740-3898
☎ (+1 301) 864 56 00 ▣ TWX 710 826 9650
⊖ (+1 301) 864 52 75
President: Charles B Hofmann
Vice President and General Manager: Michael Gering
■ Electronic warfare systems; airborne early warning systems, emitter targeting systems, threat avoidance receivers, passive detection systems; RF communications systems; shipboard monitor and control systems; integrated communication switching systems for air traffic control; defense telecommunication systems. Facsimile equipment and systems for graphic communications.

Applied Technology Division
4747 Hellyer Avenue, San José, CA 95150-7012
☎ (+1 408) 365 40 26 ⊖ (+1 408) 365 40 40
President: Charles B Hofmann
■ Threat warning, electronic support measures, EW collection systems, active decoys, advanced self protection integrated integrated EW suites, military and space computers; simulation, test and training systems, EW subsystem components, installation services. Electronic Warfare Systems including airborne and shipboard surveillance and targeting systems, electronic support measure (ESM) systems and radar warning systems; worldwide field engineering and EW equipment installation services for aircraft and ships.

OM= Laser Systems Division
Litton Laser Systems
(t International Laser Systems)
(a subsidiary of Litton Industries Inc.)
2787 South Orange Blossom Trail, Apopka, FL 32703
☎ (+1 407) 295 40 10 ⊖ (+1 407) 297 48 48
👥 525 (1997)
President: Gregory M Misiak
Vice President of Finance: Don Dimos
Vice President of Technical Operations: Bill Wilson
Vice President of Business Development: Robert Del Boca
Comptroller: Ms Leona Owens
Executive Director of Advanced Programs: Steve Guch
Executive Director of Strategic Plannibg: Dr Doug Stanley
Executive Director of Marketing: A Hamilton Evans
Director of Human Resources: Lee Sumner
Manager of Purchasing: Pete Morris
Senior Marketing Representative: Bob Raulerson
■ Developer and manufacturer of military laser rangefinders and designators. Also develop and manufacture frequency-agile medium energy laser transmitters for space, underwater, and ground-combat applications; carbon dioxide and solid-state coherent lasers and laser radars for applications in infra-red countermeasures, 3D imaging, vibration sensing and battlefield IFF.

Navigation, Guidance & Control Systems Group
5500 Canoga Avenue, Woodland Hills, CA 91367-6698
☎ (+1 818) 715 30 24 ⊖ (+1 818) 715 44 77
Corporate Senior Vice President and Group Executive: Larry A Frame

Aero Products Division
21050 Burbank Boulevard, Woodland Hills, CA 91367
☎ (+1 818) 226 20 00 ▣ 662619 ⊖ (+1 818) 226 21 99
President: Leon A Webrand
■ Designs and provides ARINC regulated navigation and guidance systems for commercial airlines, general aviation and military air transport aircraft. Manufacture Omega navigation systems; global positioning systems; attitude and heading reference systems; and inertial navigation systems (gimballed and laser gyro based), as well as inertial reference systems and specialised units for search, surveillance and photogrammetric precise track guidance technology. Provides flight management and navigation management systems, air data computers, air data modules and radio reference sensors.

Navigation, Guidance & Control Syszems Division
5500 Canoga Avenue, Woodland Hills, CA 91367-6698
☎ (+1 818) 715 40 40 ▣ 698483; TWX 910 494 2091
⊖ (+1 818) 715 44 67; 715 33 68; 715 27 97
Corporate Vice President and President: Darwin D Beckel
■ Inertial navigation systems, attitude and heading reference systems, inertial platforms and inertial measurement systems for all classes of military vehicles and weapons. Special test and other support equipment. Production of advanced inertial systems utilizing laser and fiber optic gyros employing either gimbalisolated instruments or in strapdown application. Advanced research and development in inertial systems and sensors including laser and fiber optic techniques. Fiber optic acoustic arrays for anti-submarine warfare.

Litton Solid State
3251 Olcott Street, Santa Clara, CA 95054-3095
☎ (+1 408) 988 13 31 ⊖ (+1 408) 970 99 50
President: Thomas W Parker
■ Custom multifunction assemblies for radar, electronic countermeasures, communications, missile, and space systems. Engineering and manufacturing capabilities include GaAs devices, monolithic microwave integrated circuits, amplifiers, oscillators and other components.

Electron Devices Division
960 Industrial Road, San Carlos, CA 94070-4194
☎ (+1 415) 591 84 11 ▣ TWX 910 376 4900
⊖ (+1 415) 591 56 23
President: James D Benham amplifiers and microwave tube high voltage power supplies).
Microwave devices and subsystems (Helix travelling wave tubes including mini-tubes, coupled-cavity travelling wave tubes, klystrons, switch tubes, M-type backward wave oscillators, travelling wave tube

Electronic Components and Materials
147 Columbia Turnpike, Suite 306,
Florham Park, NJ 07932-1212
☎ (+1 201) 966 01 60 ⊖ (+1 201) 966 01 58
Corporate Senior Vice President and Group Executive: Donald A Lepore

Winchester Electronics
400 Park Road, Watertown, CT 06795-0500
☎ (+1 203) 945 50 00 ▣ TWX 710 478 3392
⊖ (+1 203) 945 51 91
President: Allen J Bernardini
■ Electronic and electrical connectors for computers and other industrial, commercial and military uses; telecommunications and design; connector accessories and tools; insulated terminals.

Litton Poly-Scientific
1213 North Main Street, Blacksburg, VA 24060
☎ (+1 703) 552 30 11 ✆ (800) 336 59 17
▣ TWX 710 875 3694 ⊖ (+1 703) 953 18 41
President: Lawrence J Ball
■ Rotary components including slip ring assemblies, fiber optic rotary joints, Clifton Precision line of small-to-medium size DC motors, resolvers and synchros for commercial, industrial and military applications. Electronic security systems including perimeter protection devices, audio discriminators and alarm monitoring and access control systems. Fibre optic components and systems including transmitters/receivers, couplers, WDM's, data buses and sensor systems.

Litton Instruments & Life Support
2734 Hickory Grove Road, Davenport, IA 52804-1299
☎ (+1 319) 383 60 00 ▣ 468429; TWX 910 525 1197
⊖ (+1 319) 383 63 23
President: John J Heffernan
■ Developer of instruments and life support equipment, such as liquid oxygen systems, onboard oxygen generating systems, nitrogen inerting units and rebreather apparatus for firefighters. Self-contained systems for producing gaseous oxygen or nitrogen or liquid oxygen, nitrogen for commercial and military applications. Also, miniature cryogenic coolers for applications involving infra-red (IR) detectors.

Ingalls Shipbuilding
1000 Litton Access Road, Pascagoula, MS 39567
☎ (+1 601) 935 11 22 ▣ 589951; TWX 510 990 3051
⊖ (+1 601) 935 11 26; 935 11 27
President: Gerald J St Pe'
■ Design, engineering, construction, repair, overhaul, modernization, and fleet support of advanced surface ships. |98|

75-55635 Lockheed Martin Corp.
6801 Rockledge Drive, Bethesda, MD 20817
☎ (+1 301) 897 60 00 ▣ 898437; TWX 710 8249 050
⊖ (+1 301) 897 60 83
♣ $ 26,875,000,000 (1996) 👥 180,000 (1996)
Chairman and Chief Executive Officer: Norman R Augustine
President and Chief Operating Officer: Vance D Coffman
Executive Vice President and Chief Financial Officer: Marcus C Bennett
Vice President of Washington Operations: David Osterhout
Vice President and Controller: Robert E Rulon
Vice President of Corporate Development: John F Egan
Vice President of International Marketing: Robert Trice
Vice President of Domestic Business Development: John F Manuel
Vice President of Corporate Communications: Ms Susan M Pearce
Vice President of News and Information: Charles P Manor (III) ☎ (+1 301) 897 62 58
■ Core businesses in aeronautics, electronics, energy and environment, information and technology services, materials, space systems and launch vehicles and missiles and tactical systems.

Space & Strategic Missiles Sector
6801 Rockledge Drive, Bethesda, MD 20817
☎ (+1 303) 897 60 00
President: Mel Brashears
Vice President of Business Development and Washington Operations: Brian D Dailey
■ Domestic and international space programs, producing the Titan and Atlas family of space launch vehicles, Centaur high-energy upper stage, satellites and spacecraft systems for military, civil government and commercial customers. Provide spacecraft payload integration services and ground-based support for space systems and Trident II ballistic missiles and the THAAD missile.

Information & Technology Services Sector
6801 Rockledge Drive, Bethesda, MD 20817
☎ (+1 301) 897 60 00
President: Peter B Teets
Vice President of Business Development: Harold Kennedy
■ Designs, develops, integrates and operates information systems and simulation and automated test systems for government and commercial customers.
Provides technical support to US government agencies. Responsible for all ground processing tasks for NASA Space shuttle. Manufacturer of the Space Shuttle external fuel tank.

Electronics Sector
6801 Rockledge Drive, Bethesda, MD 20817
☎ (+1 301) 897 60 00 ⊖ (+1 301) 897 66 54
President: Thomas A Corcoran
Vice President of Business Development: Howard E Wolnowsky
■ Designs, develops and produces high-performance defence and electronic systems for undersea, shipboard, land-based and airborne applications. Major product lines include advanced-technology missiles; night navigation and targeting systems for aircraft; submarine and surface ship combat systems; airborne, ship and land-based radar; control systems and thrust reversers for aircraft; ordnance and armament systems; secure communications systems; countermeasures, surveillance systems and avionics.

Aeronautics Sector
6801 Rockledge Drive, Bethesda, MD 20817
☎ (+1 301) 897 60 00
President: James A Blackwell (Jr)
Vice President of Business Development: James F Madewell
■ Design and produce F-22 air superiority fighter, C-130 transport, P-3 anti-submarine and maritime patrol aircraft and F-16 fighter. The Lockheed Martin 'Skunk Works', part of the Aeronautics Sector, produces jet engine thrust reversers for commercial aircraft, worldwide logistics and management services to US government agencies and systems integration and modification of aircraft. |98|

75-55636 Lockheed Martin Skunk Works
(a division of Lockheed Martin Corp.)
1011 Lockheed Way, Palmdale, CA 93599
☎ (+1 805) 572 29 74 ▣ 67208 ⊖ (+1 805) 572 30 76
👥 5,300 🏭 375,000 sq m
President: Jack S Gordon

Contact Reference

■ Advanced development projects. Programmes include the F-117 stealth fighter, U-2 high altitude reconnaissance aircraft, SR-71 Blackbird technical support, C-130 special mission aircraft, rapid vehicle and system prototyping, classified programmes. 98

75-55637 Lockheed Martin Aeronautical Systems (LMAS)
(a division of Lockheed Martin Corp.)
86 South Cobb Drive, Marietta, GA 30063
☎(+1 404) 494 44 11 ⊖ (+1 404) 494 75 18
President: W Bulloch
Executive Vice President: A G Hansen
Vice President of Engineering: B P Osborne
Vice President of Administration: A Goldfarb
Vice President of Human Resources: R L Whillock
Vice President of Hercules Programmes: H L Burnette
Vice President and General Manager of F-22 Team Office: C T Burbage
Vice President of Business Development and Product Support: T A Graham
Vice President and P-3 Korean Program Manager: B F Bernstein
Vice President of International Marketing: B A Ramsey
Vice President of International Sales for Asia Pacific and the Middle East: J R Hundley
Vice President of Sales for Europe, Africa and the Western Hemisphere: W E Perkins
Vice President of New Business Development: J D Summers
Vice President of US Government Requirements: R G Kirkland
Executive Director of Public Affairs: B Johnstone
Commercial Practices Director: M A Hewson
Public Relations Manager: R Crockett
■ C-130 Hercules, P-3 Orion, F-22 and L-100 Hercules programmes. C-141 Starlifter/C-5 Galaxy support and modifications, Airlifter derivative programmes. 98

75-55640 Lockheed Martin Missiles & Space
(a subsidiary of Lockheed Martin Corp.)
1111 Lockheed Way, PO Box 3504,
Sunnyvale, CA 94089-3504
☎(+1 408) 742 43 21 ▣ 346409 lmsc suvl
⊖ (+1 408) 756 86 54
👥 13,600
President: K M Henshaw
Executive Vice President: L L Vic Victorino
Vice President of Communications: Ms Janet Wrather
Communications News Bureau Manager: Eric DeRitis ☎(+1 408) 742 75 31 E jimgraham@lmsc.lockheed.com
■ Fleet ballistic missiles, systems integration, NASA and US Air Force spacecraft and systems, advanced theatre missile defence systems, data exploitation, space research, commercial remote sensing, research and technology programmes, integrated open architecture data system networks.

Washington Office
1725 Jeff Davis Highway, Crystal Square 2, Suite 300, Arlington, VA 22202
☎(+1 703) 413 56 00
Vice President of Space and Strategic Missile Sector Public Relations: Virnell Bruce ☎(+1 301) 897 69 82 98

75-55650 Lockheed Martin Tactical Aircraft Systems.
(a division of Lockheed Martin Corp.)
Lockheed Boulevard, Fort Worth, TX 76108-0748
☎(+1 817) 777 20 00 ⊖ (+1 817) 777 21 15
♣ $ 3,000,000,000 👥 12,000
President: Dain M Hancock
VP & F22 Programme Dir: Ms Charla K. Wise
VP F-16 Programmes: Robert T. Elrod
VP Special Programmes: Charles A. Anderson
VP & Fast Programme Mgr: David J. Wheaton
■ F-16, F-16 mid-life upgrade, F-22, military electronics, F-16 interface avionics, mission planning equipment, training equipment, special mission aircraft, special programmes, national aerospace plane, FS-X fighter (Japan), F-111 modifications & support, IDF fighter (Taiwan), Joint Advanced Strike Technology (JAST).
Facilities: Fort Worth, TX, headquarters; aircraft design, engineering, fabrication assembly, modification & testing. 98

75-55710 Loc Performance Products Inc.
201 Industrial Drive, Plymouth, MI 48170
☎(+1 313) 453 23 00 ⊖ (+1 313) 453 51 80
♣ $ 18,000,000 (1996) 👥 150 (1996) 🏭 116,000 sq ft
President: Victor Vojcek
■ Roadwheel arms, suspension components & suspension mod kits for M-48, M-60 tanks & M-113APC, M-88, M-1, M-2 Bradley, M-109 Howitzers recovery vehicles. → 80 98

75-56090 LORI Heat Transfer Operations
(a division of Allied Signal Aerospace)
6930 North Lakewood, Tulsa, OK 74117-1807
☎(+1 918) 272 80 00 ⊖ (+1 918) 272 24 62
👥 180 (1996) 🏭 28,000 m2
Managing Director: Tom Wilson
Marketing Director: Rich Parsons
Operations Director: Gary Cook
Purchasing Manager: Ms Sharon Edgar
Marketing Coordinator: Ms Marcia Arnold
Spare Parts Department: Ms Karen Scott
■ Heat transfer manufacturing and repair for heat exchangers, oil coolers and Pre coolers. 98

75-56150 LSI Computer Systems Inc.
1235 Walt Whitman Road, Melville, NY 11747
☎(+1 516) 271 04 00 E alsicsl@aol.com
⊖ (+1 516) 271 04 05
♣ $ 5,000,000 (1997) 👥 30 (1997) 🏭 14,000 sq ft
President: Attila Tetik
Chief Executive Officer: Al Musto
Vice President: Alvin Kaplan
Sales Manager: Ms Sherry Smith
Purchasing Manager: Ms Barbara Kaplan
■ Custom & standard integrated circuits in ceramic packages under USA military specifications. 98

75-56250 Lucas Aerospace Cargo Systems
2604 Highway 20 N., Jamestown, ND 58401
☎(+1 701) 252 6250 ⊖ (+1 701) 251 1046
Controller: Kirk Heim
Engineering Mgr: Mark Krump
Operations Mgr: Keith Larson
Quality Assurance Mgr: A. Pedeberg
Programmes/Contracts Mgr: Mike Dronen
■ Cargo handling, systems weapons launching systems.

Geared Systems
6125 Silver Creek Drive, PO Box 680910,
Park City, UT 84068
☎(+1 801) 649 1900 ⊖ (+1 801) 649 1909
General Mgr: Collin Hull
Dir Engineering: Greg Pérez
Finance Dir: Ken Kreider
Materials Dir: Dan Krynski
Programme Management Dir: Richard Thomas
Human Resources Dir: Ms Donna Block
Quality Dir: William Shoop
Manufacturing Dir: Tim Doohen
Contract, Commercial & Marketing Dir: Richard Quirin
Operations Mgr: Mike McKibbin
■ Aircraft and engine accessory drives, actuation systems (rotary & linear).

Cargo Systems
610 Neptune Ave, Brea, CA 92622
☎(+1 714) 671 4500 ⊖ (+1 714) 671 4503
VP & General Mgr: Richard Pogue
Marketing Mgr: Jack Noe
VP Operations: Tom Schlessinger
Marketing Communications: John Cologne
Hoists Product Mgr: Gary Olson
Human Resources Mgr: Ron Schulfer
■ Aerospace AC & DC motors, brushless DC motors, electromechanical actuators, electronic controls, cargo power drive units, hoists & winches.

Power Transmission
211 Seward Avenue, Utica, NY 13503
☎(+1 315) 793 1200 ⊖ (+1 315) 793 1209
General Mgr: Frank Robilotto
Marketing Communications: Dyann Scharf
Human Resources Mgr: Elaine Holt
Commercial Product Mgr: Bob Findlay
Shafts Product Mgr: Ed Marron
Actuation Product Mgr: Steve Croke
■ Power transmission shafting and couplings, high pressure hydraulic pumps, flex pivots, test equipment, aircraft shafts, integrated actuation package, torquemeters.

Power Systems
777 Lena Drive, Aurora, OH 44202
☎(+1 216) 995 1000 ⊖ (+1 216) 995 1116
General Mgr: Mike Branch
VP Marketing: Bob Gibson
Controller: Mike King
VP Human Resources: Curt Tschantz
VP Operations: David Hammond
■ DC generator systems, AC electrical power systems, electromechanical actuation systems.

Engine Control Systems
99A South St, Hopkinton, MA 01748
☎(+1 508) 435 6630 ⊖ (+1 508) 435 1246
Site Mgr: Brian Klinka
Advanced Systems Mgr: Paul Maker
■ Provide technical support to US customers for engine control systems. 97

75-56260 Lucas Aerospace Power Equipment Corp.
(† Lear Siegler Power Equipment Corp.)
(a subsidiary of Lucas Aerospace Inc.)
777 Lena Drive, Aurora, OH 44202
☎(+1 216) 995 10 00
▣ 985488; TWX 810-422-2213 lsi pedclv mphs
⊖ (+1 216) 995 11 16
General Manager: Stephen Spiller
Director Customer Support: Gregory A Raye
Vice President Human Resources: Curtiss G Tschantz
Programme General Manager Business & Commuter Aviation: Andy Barton
Commercial Manager: Charles E Gebby
Controller: Keith Gainer
Quality Assurance: Ronald Kunes
■ DC generator systems; DC starter generator systems; AC electrical power systems; electromechanical actuation systems. 98

75-56265 Lucas Aerospace Power Transmission (LAPTC)
(† Bendix Fluid Power Division)
(a division of Lucas Aerospace Inc.)
211 Seward Avenue, PO Box 457, Utica, NY 13503-0457
☎(+1 315) 793 12 00 ⊖ (+1 315) 793 12 09
♣ $ 30,000,000 (1997) 👥 240 🏭 280,000 sq ft
Plant Manager: William Reigel
Director Public Relations: Ms Elaine Holt
Director of Marketing: Steve Croke
Purchasing Manager: Don Pachulski
■ Design & mfg: flexible power transmission shafts & couplings for aircraft, industrial & marine applications; integrated actuation packages; aircraft & engine hydraulic pumps; flexural pivot bearings; automated test equipment; torquemeters. 98

75-56300 Lucas Western Inc.
Lucas Aerospace Cargo Systems
2604 Highway 20 N., Jamestown, ND 58401
☎(+1 701) 252 6250 ⊖ (+1 701) 251 1046
VP & General Mgr: Richard Pogue
Marketing Mgr: Jim Boyd
Human Resources Mgr: Bill MacBride
■ Weapons launching systems, cargo handling systems, aerostructures.

Lucas Aerospace Geared Systems
6125 Silver Cr. Drive, PO Box 680910,
Park City, UT 84068
☎(+1 801) 649 1900 ⊖ (+1 801) 649 1909
General Mgr: Colin Hall
■ Gears, aircraft-mounted and engine-mounted accessory drive gearboxes.

Lucas Aerospace Cargo Systems
610 Neptune Ave, Brea, CA 92622
☎(+1 714) 671 4500 ⊖ (+1 714) 671 4503
VP Operations: Tom Schlessinger
Marketing Mgr: Jack Noe
Human Resources Mgr: Ron Schulfer
■ Aerospace AC & DC motors; brushless DC motors; electromechanical actuators; electronic controls; cargo power drive units, hoists & winches. 97

75-56350 Ludlow Corp.
One Minden Rd, Homer, LA 71040
☎(+1 318) 927 2531 ⊖ (+1 318) 927 9520
Marketing Mgr: Robert J. Tippit
■ Mil spec packaging. 97

75-56360 Ludlum Measurements Inc.
501 Oak Street, PO Box 810, Sweetwater, TX 79556
☎(+1 915) 235 54 94 ✆ (800) 622 08 28
E ludlum@camalott.com ⊖ (+1 915) 235 46 72
👥 125
President: Donald G Ludlum
General Manager: Paul Fritz
Sales and Marketing Director: Dwane Stevens

Purchasing Manager: Ms Ann Taylor
Sales Co-ordinator: Mrs Lucy Byrd
■ Design and manufacture instruments to measure and detect nuclear radiation. [98]

75-56450 Luminator Aircraft Products
(† Glar-Ban)
(a subs. of Mark IV Industries Inc.)
1200 E. Plano Parkway, Suite 300,
Plano, TX 75094-0030
☎ (+1 214) 881 LITE ⊖ (+1 214) 881 8900
👥 350
President: John Hartzier
Customer Service: B. Pressnell
■ Night vision goggle compatible filters & lighting products including dual mode (white & IR emitting) search lights, NVG convertible utility cockpit light, bezels, post lights, formation lights, navigation lights, flood lights, flip filters, interior lighting, passenger comfort & safety equipment. [97]

75-56480 Luminescent Systems Inc.
(a subs. of Loctite Corp.)
101 Etna Rd, PO Box 9004, Lebanon, NH 03766-9004
☎ (+1 603) 448 3444 ⊖ (+1 603) 448 3452
👥 200 🏭 80 000 sq ft
President: Peter Gundenmann
Operations Dir: Frank Johns
Purchasing Mgr: Keith Nadeau
Marketing Mgr: Tim Zeigler
■ Electroluminescent lighting products incl. formation lights, primarily for military aircraft. Emergency exit lighting, primarily for commercial transport. Electroluminescent lighting for Night Vision Goggle (NVG) compatible, non-NVG compatible cockpits & other aviation or military crew stations. Avionics panels, keyswitches, ruggedised keyboards.
● 77 Olean Road, East Aurora, NY 14052
☎ (+1 716) 655 0800 ⊖ (+1 716) 655 0309 [97]

75-56700 Luxtron Inc. Emergency Lighting Div.
15 Hale St, Haverhill, MA 01830-4833
☎ (+1 508) 372 5211 ⊖ (+1 508) 372 7051
♣ $ 500 000 👥 30 🏭 12 000 sq ft
President: William E. Slusher
Marketing & Sales Mgr: Bruce F. Morse
■ Uninterruptible power systems for computer back-up & lighting. [97]

75-56705 Lyncole XIT Grounding
3547 Voyager Street, Suiite 104, Torrance, CA 90503
☎ (+1 310) 214 4000 ⊖ (+1 310) 214 1114
♣ $ 1 200 000 (1994) 👥 24
President: Elizabeth Robertson
Sales Dir: John "J.B." Bird
Public Relations Mgr: Ms Amy Sharpe
■ Electrical earth grounding services. [97]

75-56850 Mack Electric Devices Inc.
(a div. of FSCM-82691)
North & Paxson Avenues, Wyncote, PA 19095-1205
☎ (+1 215) 884 8123 ⊖ (+1 215) 886 5030
♧ $ 1 000 000 ♣ $ 1 000 000 👥 25 (1994)
🏭 50 000 sq ft
President: N.K. McKinney
Sales Mgr: Don Butterly
■ Manufacture ice detectors for aircraft MIL-D-8181B, MIL-S-1632M. [97]

75-56930 Macneal-Schwendler Corporation
815 Colorado Boulevard, Los Angeles, CA 90041-1777
☎ (+1 213) 258 91 11 ⊖ (+1 213) 259 38 38
Chairman: Dr Richard H MacNeal
President and Chief Executive Officer: Thomas C Curry
Senior Vice President: Dr Dennis A Nagy
Chief Financial Officer: Louis A Greco
Vice President, Solver Development: Bruce E MacNeal
Vice President and General Manager, Aerospace Business Unit: Ken Blakely
Vice President, Modeling Development: Jim White
■ Product design in industries including; automotive, aerospace, heavy machinery, defence, electromechanical and medical. Computer software, analysis systems and programmes. [98]

75-56953 MA/COM Antenna & Cable Div.
(† Adams-Russell Inc.)
(a subs. of M/A-COM Inc.)
110 Haverhill Road, Amesbury, MA 01913
☎ (+1 508) 834 7100 ⊖ (+1 508) 834 7356
♣ $ 43 000 000 (1996) 👥 360 (1996)
Managing Dir: Kenneth J. Andrews
VP International Ops: Kenneth M.A. Lewandowski
VP Finance: David J. McLachlan
VP Marketing Corp.: Roger Tarant
VP Planning & Acquisitions: Dr Lee J. Cooper
VP Systems & Special Programs: Louis J. Gargasz
Sales Dir: Alan Cook
Marketing: Dave McDonald
■ Manufacture of high technology microwave & digital components, subassemblies & equipment for the military electronics, signal intelligence & aerospace markets.

Antenna & Microwave Div.
110 Haverhill Rd, Rt 110, Amesbury, MA 01913
☎ (+1 508) 834 7100
👥 350 🏭 118 000 sq ft
VP & General Mgr: Kenneth J. Andrews
Dir Sales & Marketing: Alan Cook
Sales Mgr: Fred J. Kaczor
■ Manufacture coaxial cable assemblies, antennas & double-ridged waveguide components for electronic warfare & C3 industries.

Interconnect Products Division
1370 Main St, Waltham, MA 02154
☎ (+1 617) 891 4700
General Mgr: Robert P. Widener
■ Design & development of signal analysis & enhancement equipment.

Micro-Tel Div.
10713 Gilroy Rd, Hunt Valley, MD 21030
☎ (+1 301) 667 0077 ⊠ 87442
♣ $ 10 000 000 🏭 31 000 sq ft
VP & General Mgr: William R. Trussell
■ Microwave surveillance equipment & systems covering 01-40 GHz. Microwave test equipment including signal generators, synthesizers, attenuation measurement receivers, antenna pattern receivers & microwave TDRs.

Microwave Products Div.
9533 Irondale Ave, Chatsworth, CA 91311
☎ (+1 818) 882 3434 ⊠ TWX 910 494 2771
Mgr National Sales: Bill Petry
■ Slotted waveguide planar array antennas & components. Precision mechanical devices.

Semiconductor Center
80 Cambridge St, Burlington, MA 01803
☎ (+1 617) 273 3333 ⊠ 200155
VP & General Mgr: Ron Guimond
■ Design & manufacture of monolithic microwave integrated circuits. Complete foundry services. [97]

75-57000 MacTaggart Scott & Co. Ltd
66 Main Street, PO Box 429, North Easton, MA 02356
☎ (+1 508) 238 9889 ⊖ (+1 508) 238 9889
Sales Mgr: J.C. Urciuoli [97]

75-57020 Mag Instrument Inc.
1635 South Sacramento Avenue,
Ontario, CA 91761-7769
☎ (+1 909) 947 10 06 ⊖ (+1 909) 947 31 16
👥 450 (1995)
President and Chief Executive Officer: Tony Maguca
Vice President of Sales and Marketing: John Wyatt
Engineering and Technical Director: Fred McAlister
Marketing Assistant: Dawn Murschel
■ Manufacture high quality, precision machined anodized aluminium flashlights and accessories. [98]

75-57050 Magellan Systems Corp.
(a subsidiary of Orbital Sciences Corp., USA)
960 Overland Ct., San Dimas, CA 91773
☎ (+1 909) 394 50 00 ✆ (800) 669 44 77
✉ http://www.magellangps.com ⊖ (+1 909) 394 70 50
President and Chief Executive Officer: Randy D Hoffman
Vice President of Marketing and Sales: Richard R Sill
Aviation Marketing Manager: Chris Carver
National Sales Manager Aviation Products: Ralph Fisch
☎ (+1 816) 886 57 06
Public Relations Manager: Don Meyer
■ The NAV 1000M5 and GPS commander and Arabic Hawk synchronize to battlefield in time and space. Their applications range from infantry/special forces to co-ordinating indirect fire weapons with precision accuracy, additionally providing a basis for command and control networks. [98]

75-57120 Magnavox Electronic Systems Company West Coast Division
(† Magnavox Adv. Products & Systems Co.)
2829 Maricopa Street, Torrance, CA 90503
☎ (+1 310) 618 1200 ⊠ 696101 ⊖ (+1 310) 618 7001
Senior VP & General Mgr West Coast Div.: Vito F. Brenna
■ Communications satellite terminal equipment, navigation satellite receivers & data processors for position location & automatic ship navigation systems. Omega navigation receivers. GPS receivers. Differential GPS equipment. [97]

75-57330 Magnesium Elektron Inc.
Reade Manufacturing Div.
100 Ridgeway Blvd, Lakehurst, NJ 08733
☎ (+1 908) 657 6451 ⊖ (+1 908) 657 6628
VP Marketing: B.A. Gwynne [97]

75-57380 Magnetic Seal Corp.
365 Market Street, PO Box 445, Warren, RI 02885
☎ (+1 401) 247 28 00 ⊖ (+1 401) 247 28 05
President: Richard B Colby
Executive Vice President: Bruce S Place
■ Magnetic, face type, mechanical shaft seals utilising magnetic force, eliminating springs. Used extensively in aircraft applications including gearboxes, transmissions, fuel pumps, hydraulic pumps. [98]

75-57400 Magnum Marine Corp.
2900 NE 188th St, N.Miami Beach, FL 33180
☎ (+1 305) 931 4292 ⊖ (+1 305) 931 0088
👥 60 (1994)
President: Katrin Theodoli
Purchasing Dir: Gene Mazarrado
■ High performance, ocean-going, glass fibre pleasure and patrol craft from 27' to 90'. [97]

75-57600 Mamor Works
PO Box 15, Caldwell, NJ 07006
☎ (+1 201) 228 74 02 ⊖ (+1 201) 228 34 80
♧ $ 8,000,000 👥 95 🏭 360 sq m
President: Mann Werber
Sales and Marketing Director: Gail Chill
■ Production of telescopic masts up to 30m. Lifting equipment. Self-braking motors. [98]

75-57650 Mannesmann Tally Corp.
(a subs. of Mannesmann Capital Corp.)
8301 S. 180th St, PO Box 97018, Kent, WA 98064-9718
☎ (+1 206) 251 5500 ✆ (800) 843 1347
⊖ (+1 206) 251 5520
👥 2 400
CEO/President: William Munro
VP Sales & Marketing: Richard Waddell
Dir Public Relations: R. Bruce Ridley
■ Manufacture computer printers in all major technologies, laser, dot matrix, line matrix, inkjet, thermal transfer. [97]

75-57950 Marconi Dynamics Inc.
(a subs. of GEC Marconi Radar and Defense Systems)
5701 Lindero Canyon Road, 4-100,
Westlake Village, CA 91362
☎ (+1 818) 991 0300 ⊖ (+1 818) 879 2522
President: A.R. Connolly
■ Design, flight test & manufacture advanced tactical guided munitions, weapon control systems & support equipment involving system engineering, aerodynamics, propulsion, guidance and control, tracking, payload, airborne avionics & EW. Technologies include EO sensors, laser tracking systems, digital signal processing & microprocessing guidance and control. [97]

75-58000 Marine Electric Systems Inc.
(† Marine Electric Railway)
2 Bridewell Place, PO Box 1135, Clifton, NJ 07014-1135
☎ (+1 201) 471 68 00 ⊖ (+1 201) 471 28 11
👥 100 (1996) 🏭 50,000 sq ft
President: H. Epstein
Vice President Sales and Marketing, Director Public Relations: G Mandell E gmandell@aol.com
Operations & Technical Dir: J. Cardamone

■ Qualified products: salinity indicating, temperature monitoring, signal conditioner & power supply, motor generator, battery charger, DC power supply. Proximity switches (Marine) engine monitoring systems. [98]

75-58280 Marinette Marine Corp. (MMC)
Foot of Ely St, Marinette, WI 54143
☎ (+1 715) 735 9341 ⊖ (+1 715) 735 3516
♣ $ 56 000 000 (1995) 👥 520
President & CEO: Daniel L. Gulling
VP Operations: David Cassidy
VP Development: Bernard Bentgen
Marketing Dir: Peter S. Anderson
Purchasing Dir: Douglas Oitzinger
Contract Administration Dir: Robert Cassibry
Planning Dir: James Wilson
Controller: David Roland
■ Design & manufacture vessels under 150m. [97]

75-58350 Marotta Scientific Controls Inc.
78 Boonton Avenue, PO Box 427,
Montville, NJ 07045-0427
☎ (+1 201) 334 78 00 ✆ (888) MAROTTA
[E] sales@marotta.com [W] http://www.marotta.com
⊖ (+1 201) 334 12 19
👥 250 (1997) 🏭 115,000 sq ft
Chairman and Chief Executive Officer: Thomas S Marotta
Senior Vice President Operations: Ken Smith
Executive Vice President: Mark V Laurenzi
Director Aerospace and Industrial Sales: Steve Bruneau
Director Military and Energy Sales: Bob Nunamann
Purchasing Manager: Ed Nowack
Huntsville Office
6767 Old Madison, PO Box 1248,
Huntsville, AL 35807-1248
☎ (+1 205) 971 12 00 ⊖ (+1 205) 971 12 02
■ High pressure control valves, regulators and systems. Hydraulic-electro pneumatic controls.
Santa Ana Office
1851 East First Street, Suite 1240, Santa Ana, CA 92705
☎ (+1 714) 550 12 63 ⊖ (+1 714) 550 77 08 [98]

75-58510 Marsh McBirney Inc
4539 Metropolitan Court, Frederick, MD 21704
☎ (+1 301) 874 5599 ✆ (800) 368 2723
⊖ (+1 301) 874 2172
President & CEO: L.B. Marsh
Operations Dir: Frank Cauley
Marketing Dir: John Cole
Purchasing Mgr: Dave Hubert
Advertising Mgr: Marcia A. Kinley
■ Manufacturers of electromagnetic water current meters and flowmeters. [97]

75-58540 Martin Electronics Inc.
(a subsidiary of MEI Holdings Inc.)
Route 1, PO Box 700, Perry, FL 32347-9751
☎ (+1 904) 584 26 34 [E] mei@perry.gulfnet.com
⊖ (+1 904) 584 20 44
♣ US$ 18,000,000 (1995) 👥 250 (1996)
🏭 164,000 sq ft
Chief Executive Officer: R B York
President: T B Breslin
Vice President of Sales and Marketing: James Hamilton
Public Relations Director: Ms Ann Nola
■ Grenade fuses, IR decoy flares, IR screening grenades, handheld smoke illumination signals, igniters, distraction devices. [98]

75-58700 Mason Electric Co. Inc.
(a subsidiary of Esterline Technologies)
605 8th Street, PO Box 311,
San Fernando, CA 91341-0311
☎ (+1 818) 361 33 66 ✆ (800) 232 77 00
[E] sales@mason-electric.com
[W] http://www.mason-electric.com
⊖ (+1 818) 365 55 05; 365 68 09 (sales)
👥 155 (1997) 🏭 50,000 m2
President: William E Southern
Design Vice President: John Giors
Sales Manager: Michael J Nystrom
Purchasing Manager: Ms Charlotte Gesualdi
■ Custom controls, grips, switches, transducers for military and commercial applications. [98]

75-58970 Massa Products Corp.
280 Lincoln Street, Hingham, MA 02043
☎ (+1 617) 749 48 00 [E] mpc@massaprod.com
[W] http://www.massaprod.com ☒ TWX 710 3486 932
⊖ (+1 617) 740 20 45
♣ $ 7,500,000 (1997) 👥 80 (1997) 🏭 70,000 sq ft
President: D P Massa
Executive Vice President: Richard M Carpenter
■ Sonar transducers, sonar systems. Underwater cable assemblies. [98]

75-58990 Matrix Systems Corp.
5177 North Douglas Fir Road,
Calabasas, CA 91302-1440
☎ (+1 818) 222 23 01 [E] tech@matrixsystems.com
[W] http://www.matrixsystems.com ⊖ (+1 818) 222 23 04
👥 30 🏭 14,000 sq ft
President and Chief Executive Officer: Jack Singleton
Purchasing Manager: Ms Marilyn DeVestern
■ Coaxial, audio and video switches and systems. RS232 and IEEE-488 interface modules and chassis. VXI bus modules. [98]

75-59200 Maxi-Blast Inc.
630 East Bronson Street, South Bend, IN 46601
☎ (+1 219) 233 11 61 ✆ (800) 535 38 74
[E] info@maxiblast.com [W] http://www.maxiblast.com
⊖ (+1 219) 234 07 92
♧ US$ 1,000,000 👥 21 🏭 29,000 sq ft
President: Gail E Barker
Sales Director: Michael A Golubski
Sales and Marketing Director: Jerry L Rose
■ Manufacture of Mil-Spec Plastic Blast Cleaning media and specialised plastic blast machinery and equipment. The equipment is used for military and aerospace paint and coating removal. → 85 [98]

75-59220 Maxwell Laboratories Inc.
8888 Balboa Avenue, San Diego, CA 92123-1506
Vice President of Sales and Marketing: David Gideon [98]

75-59250 Maxwell Technologies Inc. Federal Division
(a subsidiary of Maxwell Technologies Inc.)
9244 Balboa Avenue, San Diego, CA 92123-1506
☎ (+1 619) 279 51 00 [W] http://www.maxwell.com
⊖ (+1 619) 576 78 83
♣ US$ 100,000,000 (1997) 👥 640 (1997)
🏭 13,850 m2
Executive Vice President of Operations: Sean M Maloy
Business Development Director: Michael Kropf
Vice President Finance: Gary J Davidson
■ Pulse power systems and components, including HV, high energy capacitors and capacitor banks, triggered spark gap switches, trigger and delay generators, pulse generators, EMP simulation systems, power supplies and power conditioning systems. → 80
Overseas Representatives:
Laseroptics SA (Argentina)
Coherent Scientific Pty Ltd (Australia)
Tecelinco (Brazil)
Comet Industrial Co. (Korea)
CH Trading Co. (US Office)
AMS Electronic GmbH (Germany)
AMS Electronic Ltd (England)
KBK (Kyokuto Boeki Kaisha) Ltd (Japan)
CODAI (Isreal)
Optilas Iberica (Spain)
Optilas BV (Netherlands)
Optilas SA (France)
S-Cubed Division
3398 Carmel Mountain Road, San Diego, CA 92121
☎ (+1 619) 450 24 23 ⊖ (+1 619) 450 10 29
■ Performs complex analytical services. Survivability and hardening studies. Nuclear weapons effects phenomenology studies. Materials response studies. [98]

75-59370 M.C. Aerospace Corp.
118 Indianwood Road, PO Box 126, Lake Orion, MI 48362
☎ (+1 248) 693 83 11 ⊖ (+1 248) 693 15 11
♣ $ 10,000,000 (1996) 👥 75 (1996)
VP Sales & Marketing: Fred Stark
General Mgr: Ed Forrest
Chief Engineer: Mark Dallas
■ Specialised hydraulic & pneumatic assemblies for aerospace & military land vehicles. Numerous hydraulic systems for various aircraft applications. [98]

75-59480 McDonnell Douglas Aerospace (MCAIR)
PO Box 516, St Louis, MO 63166-0516
☎ (+1 314) 232 02 32 ⊖ (+1 314) 234 70 77
👥 28,000
President, McDonnell Douglas Aerospace: Mike M Sears
Senior Vice President, Helicopter Systems: Dean C Borgman
Senior Vice President, Business Development: Thomas M Gunn
Vice President and General Manager, Business Management: Robert L Brand
Vice President and General Manager of US Navy/US Marine Corps Programs: Gerald E Daniels
Vice President and General Manager, Human Resources: Vincent T DeBlaze
Vice President and General Manager, Space and Defence Systems: Gale Schulter
Vice President, Associate General Counsel: John G Judy
Vice President and General Manager, Engineering: William A Norman
Vice President and General Manager, US Air Force Programs: James B Peterson
Vice President and General Manager, Aerospace Support: James C Restelli
Vice President and General Manager, Quality: John W Steurer
Vice President and General Manager, Supplier Management and Procurement: William L Stowers
Vice President and General Manager, Advanced Systems and Technology-Phantom Works: David O Swain
Vice President and General Manager, Production Operations and General Services: John Van Gels
Vice President, Special Programs: Chuck Suyo
General Manager, Communications and Community Relations: Tom Downey
■ Manufacture and development of F/A-18 Hornet F/A-18 Super Hornet, F-15 Eagle, AV-8B Harrier II, T-45 Goshawk, Advanced Cruise Missile, Harpoon Missile, AH-64 Apache helicopter, 500 Series commercial and light helicopter.
Huntington Beach
5301 Bolsa Avenue, Huntington Beach, CA 92647
☎ (+1 714) 896 33 11 ⊖ (+1 714) 896 13 13
👥 6,200
Vice President and General Manager, Space and Defence Systems: Gale Schluter
Vice President and General Manager, Advanced Space: Chuck Ordahl
Vice President and General Manager, Space Station: Rodney Linford
Vice President and General Manager, C41: Roger Roberts
Vice President, Huntsville: Charles F Marvin
Vice President, Quality Systems and Services: Stephen D Wallace
Director, Communications: Ms Anne Toulouse
Associate General Counsel: Eric Mulloy
General Manager, Engineering Services Division: Ronald Elkins
■ Space station, space transportation systems, space base technical services, Delta II rocket, DCX Delta Clipper, X33 and Mast Mounted Sight.
● 2401 East Wardlow Road,
Long Beach, CA 90807-4418
☎ (+1 310) 496 51 95 ⊖ (+1 314) 496 51 06
👥 9,340
Senior Vice President, Military Transport Aircraft: Donald Kozlowski
Vice President and General Manager, C-17 Program: David Spong
Vice President and General Manager, C-17 Deputy Program Manager: Bill Gendron
Vice President, Program Integration: Randy Mizer
Vice President, Advanced Transport Aircraft Systems: Ms Grace Robertson
Vice President, MD-17: John Feren
Vice President, Business Development: Stu Thompson
Vice President, Product Assurance and Services: Ms Bonnie Soodick
Vice President, Supplier and Materials Manager: Ms Pam Willis
Director, Communications: Rick Fuller
Director, Human Resources: Mark Keough
General Manager, Business Management: Craig Saddler
General Manager, Product Definition: Ron Suiter
General Manager, Macon Operations: Don Pope
■ C-17 transport aircraft. [98]

75-59550 McFadden Systems Inc.
11835 E. Smith Ave, Santa Fe Springs, CA 90670
☎ (+1 310) 948 4414 ⊖ (+1 310) 948 1291
👥 35
President: Larry Hayashigawa

Admin Asst: Bridgette Wilridge
■ High performance servomechanisms. Control force loading systems & motion bases for flight simulators. [97]

75-59580 MCG Electronics Inc.
12 Burt Drive, Deer Park, NY 11729-5747
☎ (+1 516) 586 51 25 ✆ (800) 851 15 08
Ⓔ mcginfo@specdata.com Ⓦ http://www.mcgsurge.com
⊖ (+1 516) 586 51 20
👥 48
CEO/President: Mike Coyle
Managing Director: H Striegl
VP Sales & Marketing: Don Worden
Dir Advertising: James Lane
Purchasing Manager: K Connors
Media Manager: Ms Christine Jelley
■ Manufacture heavy duty power surge protectors for AC power lines & data/signal/telephone lines. [98]

75-59600 McInnes Steel Co.
441 East Main Street, Corry, PA 16407-2013
☎ (+1 814) 664 96 64 ⊖ (+1 814) 664 94 52
♣ $ 43,000,000 👥 200
President and Chief Operating Officer: Anthony A Montani
Corporate Manager, Sales and Marketing: Walter A Cusick
■ Open die steel forgings & seamless rolled ring components.

Erie Bronze & Aluminium
6300 West Ridge Road, PO Box 8099, Erie, PA 16516
☎ (+1 814) 838 86 02 ⊖ (+1 814) 838 89 17
👥 69
Group General Manager: Kevin O'Connell
■ Bronze & aluminium castings. [98]

75-59700 MCL Inc.
501 South Woodcreek Road, Bolingbrook, IL 60440-4999
☎ (+1 630) 759 95 00 Ⓦ http://www.mcl.com
⊖ (+1 630) 759 50 18
♧ $ 19,000,000 (1996) ♣ $ 32,000,000 (1996)
👥 150 (1996) 🏠 5,400 sq m
President and Chief Executive Officer: Frank P Morgan
Marketing Director: Ron Adamson
Marketing and Sales Director: Tom Dolan
■ High power microwave oscillators & amplifiers & severe environment ECM power supplies. [98]

75-59710 McMellon Bros Inc.
915 Honeyspot Rd, Stratford, CT 06497-7113
☎ (+1 203) 375 5685 ⊖ (+1 203) 381 0414
♣ $ 2 000 000 👥 25 🏠 20 000 sq ft
President: V.A. Zaveruha
Sales Mgr: Thomas Miller
■ Special large nuts, turbine engines main shaft nuts, special nuts in helicopter transmissions & rotor head. [97]

75-59720 McNab Inc.
20 North Mac Questen Parkway,
Mount Vernon, NY 10550-1841
☎ (+1 914) 699 16 16 ⊖ (+1 914) 699 16 71
♣ $ 3,000,000 👥 35
President: Horace A Teass
Sales Manager: John Byrnes
■ Water monitoring systems, salinity, PH, shaft horsepower meters, torsion meters, stack smoke opacity monitors, oil in water monitors, linear window wipers. [98]

75-59800 Measurement Systems Inc. (MS)
777 Commerce Drive, Fairfield, CT 06432
☎ (+1 203) 336 45 90 ⊖ (+1 203) 336 59 45
👥 85
President: Keith Thompson
Vice President Sales and Marketing: Gerry Larocque
■ Hand controls, electrical output joysticks, track balls, controls for air-to-ground guided missiles and fire control systems. Stick grips, control wheels and insulator controls. [97]

75-59815 Mechanical Equipment Co. Inc. (MECO)
861 Carondelet St, New Orleans, LA 70130-3707
☎ (+1 504) 599 4000 ⊖ (+1 504) 599 4100
👥 250
Exec. VP Operations: George V. Gsell
■ Design & manufacture water purification systems for military ships such as DDG-51 class & special package systems (trailer mounted) ROWPU (Reverse Osmosis Water Purification Unit). [97]

75-59823 Mechanical Seal
4412 Corporate Center Dr, Los Alamitos, CA 90720
☎ (+1 714) 220 6542; 995 1818 ✆ (800) 544 0080
⊖ (+1 714) 220 6528
Director: John Leary
Director: Jim Fortini [97]

75-59830 Megapulse Inc.
8 Preston Court, Bedford, MA 01730-2380
☎ (+1 617) 275 20 10 ⊠ 923358 ⊖ (+1 617) 275 41 49
♣ $ 10,000,000 (1997) 👥 90 (1997) 🏠 45,000 sq ft
President: William Roland
Exec. Vice President: Edward L. McGann
Vice President Research & Development: Paul R. Johannessen
Vice President Operations: Patrick Patent
Sales and Marketing Director: Felix Sebastian
Systems Engineering Mgr: Duane Marshall
Purchasing Manager: Bill Daigle
■ Complete Loran systems. Solid state Loran transmitters & timing receivers. Loran simulators. Solid state dynamic LF-VLF antenna tuning systems. High power pulsed energy. Low frequency propagation research. [98]

75-59842 Meggitt Avionics Inc.
(† Armtec/Ragen Inc.)
(a Meggitt Aerospace Company)
10 Ammon Drive, Manchester, NH 03103-7406
☎ (+1 603) 669 0940 ⊠ 710220 1747 edison man
⊖ (+1 603) 669 0931
👥 177
President: Steve Marshall
Aerospace Sales Dir: Ms Roberta Marino
Production Dir: Richard LeCompte
Controller: Leon Demere
■ Aircraft displays & systems: fuel quantity & fuel flow systems and transducers, engine instrument displays, air data instruments, aural tone generators and ECM equipment. [98]

75-59845 Meggitt USA Inc.
540 North Commercial Street, Manchester, NH 03101
☎ (+1 603) 669 9971 ⊠ 710220 1747 edison man
⊖ (+1 603) 669 0930
👥 5
President: Bennett F. Moore [98]

75-59850 Mekel Engineering Inc.
(a subsidiary of Metter Industries Inc.)
2800 Saturn Street, Brea, CA 92821
☎ (+1 714) 996 56 00 Ⓦ http://www.mekel.com
⊖ (+1 714) 996 56 96
♣ $ 6,000,000 (1997) 👥 30 (1997) 🏠 1,000 sq m
Vice President and General Manager: Daniel C Lorti
Engineer Director: Julian E Hansen
Operations Director: Roger Mitch
Controller: Jacques Rubinstein
■ Head-Up Display (HUD) gunsight film and video cameras, periscope cameras (35 mm), photographic projection viewers/printers, hand cranked generators, microfilm/microfiche scanners, film transports. [98]

75-59950 Meller Optics Inc.
120 Corliss Street, Providence, RI 02904-2602
☎ (+1 401) 331 3717 ⊖ (+1 401) 331 0519
♣ $ 4 000 000 👥 30 🏠 920 sq m
President: David Lydon
VP Sales: Thomas F. Wilson
■ Manufacture optical components for offensive & defensive systems. [97]

75-60220 Mercury Computer Systems Inc.
199 Riverneck Road, Chelmsford, MA 01824
☎ (+1 508) 256 13 00 ✆ (800) 229 20 06
Ⓔ info@mc.com Ⓦ http://www.mc.com
⊖ (+1 508) 256 35 99
♣ $ 58,300,000 (1996) 👥 300 (1997) 🏠 9,600 sq ft
President and Chief Executive Officer: James Bertelli
Vice President and Chief Financial Officer: G Mead Wyman
Director of Materials: George Koslosky
Manager of Marketing Communications: Gary Olin
■ A technology company sepcialising in high performance computing solutions for diagnostic medical imaging, commercial and defence image and signal processing and servers for digital video editing. [98]

75-60250 Merex Inc.
Merex Aircraft Co. Inc.
2649 Townsgate Road, Suite 200,
Westlake Village, CA 91361
☎ (+1 805) 446 27 00
⊖ (+1 805) 446 28 00; 446 28 01
♧ $ 1,200,000 (1997) ♣ $ 10,000,000 (1997)
👥 25 (1997) 🏠 20,000 sq ft
President: Ahmad Shams
Operations Director: Richard Rynd
International Sales Director: Ms Rebecca Hill
Purchasing Manager: Sheri Butcher
■ Manufacture/contracting spare parts for military aircraft and helicopters (F-15, F-4, F-5, C-130, Bell, Hughes, Sikorsky), military vehicles, communications, radars. Representing major US aerospace firms. → 80 → 85

Subsidiary companies:
Merex SA (Argentina)
Merex Singapore [98]

75-60385 Metalcraft Inc.
9331-A Philadelphia Road, Baltimore, MD 21237-4140
☎ (+1 401) 687 55 00 ⊖ (+1 401) 687 55 03
♣ $ 5,000,000 👥 25 🏠 15,000 sq ft
President and Chief Executive Officer: Ernest E Ellis (Jr)
■ Military and commercially approved Halon 1301 handheld portable fire extinguishers. CO2 and dry chemical portable extinguishers, hand held are also available. [98]

75-60400 Metal Dynamics Corp.
1145 North Iroquois Street, Tulsa, OK 74106-5901
☎ (+1 918) 582 01 24 Ⓦ http://www.tulsaweb.com/mdc
⊖ (+1 918) 587 04 42
👥 35 🏠 2,350 sq m
President: Don Doss
Vice President of Sales: Steve Pinix
■ Sand castings and investment castings in copper base, nickel base and stainless steel alloys. [98]

75-60600 Meteor Communications Corp. (MCC)
8631 So 212th St, Kent, WA 98031
☎ (+1 206) 872 2521 ⊖ (+1 206) 872 7662
♣ $ 4 000 000
President: Donald Sytsma
Technical Dir: Tom Donich
Purchasing Mgr: Lois Godman
■ Wireless packet-switched data communications networks using meteor burt & extended line-of-sight technologies providing turnkey solutions for mobile & remote locations. → 80 [97]

75-60660 Metheus Corporation
(a subs. of Singapore Technologies)
1600 NW Compton Drive, Beaverton, OR 97006
☎ (+1 503) 690 1550 Ⓔ info@metheus.com
⊖ (+1 503) 690 1525
President: Lee Yip Loi
Exec. Vice President: Chuck Eng
Marketing Mgr: Mark Mehall
■ Design, manufacture & market high resolution display controllers for air traffic control, air defence, C4I & other mission critical applications. The products are designed around industry standards in platforms (DEC, HP, Sun), form factor (6U VME) & software (X-Window System) for compatibility & ease of integration. Recent product additions allow live video & radar information to be displayed under full X Window control.

North America, Western Region
Western US
1600 NW Compton Drive, Beaverton, OR 97006
☎ (+1 503) 690 1550 ⊖ (+1 503) 690 1525

North America, Western Region
190 Littleton Road, Westford, MA 01886
☎ (+1 508) 392 8600 ⊖ (+1 508) 392 8629 [97]

75-60700 Metraplex Corp.
7435 New Technolgy Way, Frederick, MD 21701
☎ (+1 301) 662 8000 ⊖ (+1 301) 695 8099
♣ $ 6 000 000 (1996) 👥 45 (1996) 🏠 15 000 sq ft
President: Joseph Linarducci
VP Sales: Charles B. Weaver

■ Airborne & ground based telemetry systems. 97

75-60800 Metric Systems Corp.
(a subsidiary of Tech Sym Corp.)
Electronics Division
645 Anchors Street, Fort Walton Beach, FL 32548-9990
☎(+1 904) 302 33 22 E mscmrkt@eng-msc.com
⊠ 702447 ⊖ (+1 904) 302 33 71
♣ US$ 50,000,000 ⁂ 800 ⌂ 214,000 sq ft
Managing Director: George Lee
International Marketing Director: Stan Wilusz
Operations Director: Robbie Epstein
■ Design and develop aircraft instrumentation systems to include GPs PODs for air combat training, test and evaluation. Produce coastal border surveillance radar and falcon II radar. Design and manufacture air cargo systems, a variety of aircraft GSE, shipping and storage containers, electronic enclosures and structural components of ground-based satellite communications antennas. Design, develop and manufacture threat radar warning systems and related radar sub-systems. 98

75-60820 Metrosonics Inc
PO Box 23075, Rochester, NY 14692
☎(+1 716) 334 73 00 E salesinfo@metrosonics.com
W http://www.metrosonics.com ⊖ (+1 716) 334 26 35
Managing Director: Al Stolberg
Marketing Co-ordinator: Ms Cindy Kearns
■ Design and manufacture of field measurement instrumentation. 98

75-60830 Metrum
(† Metrum Information Storage)
(a division of Bell Technologies Inc.)
4800 East Dry Creek Road, Littleton, CO 80122-3700
☎(+1 303) 773 47 00 ⊖ (+1 303) 773 49 09
⁂ 150 (1997)
President and Chief Operating Officer: Jack Krauss
Sales Director: Robert Starnes
Purchasing Manager: Jack Sassona
Marketing Administrator: Roschelle Funderburg
■ Manufacture of data storage and graphics instrumentation, test instruments and digital imaging products for industrial and government customers. Products include: digital and analogue instrumentation recorders. 98

75-60930 Michigan Wheel Corp.
(a subsidiary of the Robins Group)
1501 Buchanan Avenue SW,
Grand Rapids, MI 49507-1612
☎(+1 616) 452 69 41 E info@miwheel.com
W http://www.miwheel.com ⊖ (+1 616) 247 02 27
⁂ 250 (1997) ⌂ 150,000 sq m
President: John Panarites
Vice President Sales and Marketing: Martin Ronis
Operations Director: John Orthaber
Director Advertising: Ms Amy Turner
Director Public Relations: John Baar
Materials Manager: Tom Whittaker
■ Marine propellers. 98

75-60950 Micrel Inc.
Corp. Offices & Test Div.
1849 Fortune Drive, San Jose, CA 95131
☎(+1 408) 944 0800 ⊖ (+1 408) 944 0970
♣ $ 45 000 000 ⁂ 365 ⌂ 157 000 sq ft
President: Ray Zinn
Marketing/Sales Dir: Richard Hall
Technical Dir: Warren Muller
■ Silicon foundry services; obsoleted PMOS wafer processing; CMOS si-gate wafer processing; CMOS metal gate, NMOS metal & silicon gate; bipolar wafer processing; Mil-Std-883B class B screening services; discontinued/alternate source IC manufacture; analog, digital high voltage on one IC. CMOS/DMOS smart power custom ICs. → 80
Wafer Fab Div.
1849 Fortune Drive, San Jose, CA 95131
☎(+1 408) 944 0800 ⊖ (+1 408) 944 0970
General Mgr: John Husher
97

75-60980 Micro Communications Inc.
PO Box 4365, Manchester, NH 03108-4365
☎(+1 603) 624 43 51 ✆ (800) 545 06 08
⊖ (+1 603) 624 48 22
♣ $ 2,500,000 (1996) ⁂ 30 (1997) ⌂ 2,100 m3
Managing Director: Al Kula
Chief Operating Officer: Dennis Heymans
E dheymans@sprynet.com
Operations Director: Gary Krochmal
Purchasing Manager: Bob Pichette
Engineering Manager: Paul Smith
■ Waveguide and rigid coax components. 98

75-61090 Microsemi Corp.
2830 South Fairview Street, PO Box 26890,
Santa Ana, CA 92799-6890
☎(+1 714) 979 82 20 E webmaster@microsemi.com
W http://www.microsemi.com ⊖ (+1 714) 557 59 89
♧ $ 29,408,000 (1996) ♣ $ 157,435,000 (1996)
⁂ 2,000 (1997)
President and Chief Executive Officer: Philip Frey (Jr)
Vice President and General Manager: Lane Jorgensen
Purchasing Manager: Ms Myriam Roberts
Marketing Services Co-ordinator: Ms Elaine Romero
■ Discrete semiconductors: rectifiers, zeners.
Colorado
800 Hoyt Street, Broomfield, CO 80020
☎(+1 303) 469 21 61 ⊖ (+1 303) 466 37 75
Scottsdale
8700 East Thomas Road, Scottsdale, AZ 85251
☎(+1 602) 941 63 00 ⊖ (+1 602) 947 15 03
Watertown
580 Pleasant Street, Watertown, MA 02172
☎(+1 617) 926 04 04 ⊖ (+1 617) 924 12 35 98

75-61170 Microturbo Inc
(† Microturbo North America Inc)
(a subsidiary of Labinal, France)
2707 Forum Drive, Grand Prairie, TX 7505-7027
☎(+1 214) 660 55 45 ⊖ (+1 214) 660 71 81
⁂ 13 ⌂ 4,500 sq m
President: Dennis Nichols
Senior Vice President: Chris Christensen
Contracts Manager: Ms Lisa Williams
■ Manufacture of small turbojet engines and specialised turbine powered products for the defence and aerospace markets. These include turbojet engines, engine starting systems, auxiliary power units for airborne and ground applications and auxiliary power and environmental control systems, plus related gas turbine products including smoke generators. 98

75-61300 Midcon Cables Co.
(a subsidiary of Esterline Technologies)
2500 Davis Boulevard, Joplin, MO 64802
☎(+1 417) 781 43 31 ⊖ (+1 417) 781 72 37
⁂ 130
President: Charles D Wheeler
■ Cable & harness assemblies, "Black Boxes", fiber optic assemblies, panel & chassis assemblies & any format of interconnect system. 98

75-61530 MIL-BASE Industries Inc.
197 Old York Road, PO Box 354,
New Cumberland, PA 17070-0354
☎(+1 717) 774 09 68 E 1e0276@sshare.com
⊖ (+1 717) 774 04 07
⁂ 10
President: Paul Sandnes
Vice President: Ms Kathleen Sandnes
Sales Dir: Thomas C. Brucklacher (Jr)
■ Military helicopter slings, tow cables, winch lines, wire rope assemblies, chain assemblies, grounding kits for communication vehicles, engline and transmission lifting slings, boat and aircraft lifting slings, various assemblies for US Naval mine warfare, as well as vitrually rope or wire rope related product. 98

75-61550 Military Logistics MFC
252 East Semoran Boulevard, Casselberry, FL 32707
E sales@millog.com ⊖ (+1 407) 851 20 06
⁂ 250 (1994) ⌂ 25,000 sq m
Chief Executive Officer: Les Kaltenecker
Operations Director: John Tulple
Sales and Marketing Director: Dan Savich
■ Manufacturer and stockist of a wide range of sea, land and air based equipment. M-65 field jackets, battle dress uniforms, canteen covers 2qts, sniper veils, boonie hats, walking shorts, US military sweaters, flight helmet bags, LSA weapons oil, USMC evac blankets, butt pack (3 day), weapons cleaning kit, poncho, poncho liners, MRE's, duffle and medic bags, Fritz helmet covers, ammo pouch .45 cal, LC2 canteen covers, strap M-1972 nylon general purpose, range bag, jungle hammock M-1965, alice pack LC2, swab patch cotton, bore cleaner, cleaner preservant lubricant, insect repellent, cots, folding wooden, Al folding, cot cover canvas, dog tag machine imprinting, dog tag supplies. 98

75-61570 Military Systems Group Inc.
736 Fesslers Lane, Nashville, TN 37210
☎(+1 615) 256 4248 ⊖ (+1 615) 256 8066
President: E.R. Maples
Operations Mgr: Ms K. B. Horton
■ Design & manufacturer of mounting systems in non-corrosive materials for minor calibre weapons & supply of complete systems including weapons, mounts, spare parts, O&M documentation & training for all system components. 97

75-61650 R.A. Miller Industries Inc.
14500 168th Street, PO Box 858,
Grand Haven, MI 49417
☎(+1 616) 842 94 50 W http://www.rami.com
⊖ (+1 616) 842 27 71
⁂ 145 (1997) ⌂ 50,000 sq ft
President: Robert M. Lynas
Operations/Technical Dir: Paul Miller
Contract Administrator: Robert Khodl
Purchasing Mgr: Bruce Bareham
■ Manufacture communications antenna used on ground vehicle, both wheel & track. Aircraft & ship antenna's. 98

75-61655 Miller Industries Towing Equipment Inc.
8503 Hilltop Drive, Ooltewah, TN 37363
☎(+1 615) 238 41 71 ✆ (800) 292 03 30
⊖ (+1 615) 238 53 71
President: Jeff Badgley
Controller: Bob DeVries
Export and Government Sales Director: Tom Griffin
Operations Director: Tom Cox
Sales and Marketing Director: Randy Olsen
Purchasing Manager: Bill Emery 98

75-61665 Millitech Corp.
South Deerfield Research Park, PO Box 109,
South Deerfield, MA 01373
☎(+1 413) 665 85 51 ⊖ (+1 413) 665 25 36
♣ $ 17,000,000 (1997) ⁂ 135 (1997) ⌂ 65,000 sq ft
President and Chief Executive Officer: John Youngblood
Vice President Finance: Willie S Day
Director Manufacturing: Robert Browning
General Manager: Ransom Reynolds
Manager Sales and Marketing: Ms Abbie Mathew
Senior Sales Coordinator: Ms Diane Greenman
E dgreenman@millitech.com
■ Design & manufacture of components, instruments & subsystems in the millimeter & submillimeter wavelength section of the electromagnetic spectrum. Customers in the military, commercial & research markets, both domestic & international. 98

75-61810 Miltope Corp.
500 Richardson Road South, Hope Hull, AL 36830
☎(+1 334) 284 8665 ✆ (800) MILTOPE
⊖ (+1 334) 613 6302
♣ $ 75 569 000 (1994) ⁂ 300 ⌂ 170 000 sq ft
President & CEO: George K. Webster
Vice President Human Resources: Edward Crowell
Marketing Communications Mgr: Greg Myers
■ Tactical PC/AT compatible computers and peripherals, militarized or ruggedized, mass memory devices, optical hard drives and solid state with extensive line of printers.
Alabama Office
Miltope Corporation, 1101 S. Brundidge Street,
Troy, AL 36081
☎(+1 334) 566 7740 ⊖ (+1 334) 566 5754
■ Manufacturing facility for computer products & peripherals.
Vermont Office
Miltope Corporation, 76 Pearl Street,
Springfield, VT 05166

☎ (+1 802) 885 4100 ⊖ (+1 802) 885 4175
■ Manufacture/design facility for mass memory products. [97]

75-62000 MITEQ Inc.
100 Davids Drive, Hauppauge, NY 11788-2034
☎ (+1 516) 436 74 00 [W] http://www.miteq.com
⊖ (+1 516) 436 74 $30
♣ $ 60,000,000 (1996) 👥 430 (1996) 🏭 165,000 sq ft
President: Arthur Faverio
Sales and Marketing Director: Ted Heil
Corporate Communications Manager: Hillar E Kiiss
■ Manufacture active microwave components from DC to 60 GHz, communications & receiving subsystems & systems. [98]

75-62200 MMR Technologies Inc.
1400 North Shoreline Boulevard, Suite A5,
Mountain View, CA 94043-1312
☎ (+1 415) 962 96 20 [W] http://www.mmr.com
⊖ (+1 415) 962 96 47
♣ $ 2,500,000 (1997) 👥 16 (1997) 🏭 700 m2
Chairman and Chief Executive Officer: Dr William A Little
Vice President of Sales: Robert L Paugh
Purchasing Manager: Herb Edman
■ Joule-Thomson coolers for IR detectors. [98]

75-62240 M&N Aerospace
(† M&N Plastics)
(a member of The NORDAM Group)
409 Hwy 199 West, PO Box 1239, Springtown, TX 76082
☎ (+1 817) 523 7245 ⊠ 758516 mnpla
⊖ (+1 817) 523 6236
VP & General Mgr: David C. Whetstone
Marketing Dir: Jerry Massey
■ Manufacture helicopter replacement windows and accessories, radome repair, structural composites repair, thermoformed interior components, glass fibre component manufacture and repair. → 80 [97]

75-62295 Modcomp
(† Modcomp/Cerplex L.P.)
1650 West McNab Road, Ft Lauderdale, FL 33309
☎ (+1 954) 974 13 80 ✆ (800) 322 32 87
[W] http://www.modcomp.com ⊠ TWX 510 956 9414
⊖ (+1 954) 977 19 00
👥 200 (1995)
President and Chief Executive Officer: John Clary
Vice President of Finance, Operations and Administration: Manfred Appel
Sales Director: Joe Rutkowski
Marketing Manager: Ms Ellen Rowley
Purchasing Manager: Ken Innes
Marketing Analyst: Mike Duvall
■ Real-time computer systems, products and services to the industrial automation, energy, transportation, scientific and communications markets. [98]

75-62400 Molding Solutions
(† P&S Products Inc.)
781 Enterprise Drive, Lexington, KY 40510-1015
☎ (+1 606) 231 00 31
[E] moldingsolutions@worldnet.att.net
[W] http://www.thenerve2.com/molding solutions
⊖ (+1 606) 254 88 84
👥 30 🏭 2,323 sq m
Owner: Paul Heflin
Vice President: Ms Sherry Heflin
Operations Director: John Hilbers
Manager: Paul Price
Marketing: Joseph V Murphy
■ Aircraft seals and gaskets, pistol grips (M16 and M249), buffer assemblies, grommets, mounts, bellows, seals, rubber to metal bonding, clean room. [98]

75-62410 Moller International
1222 Research Park Drive, Davis, CA 95616
☎ (+1 916) 756 50 86 [W] http://www.moller.com
⊖ (+1 916) 756 51 79
♣ $ 5,000,000 👥 40 🏭 34,500 sq ft
Vice President: Jack G Allison
Managing Director: Paul Moller
Freedom Engine Project Manager: Ron Tubergen
■ Three product lines: high speed (400 mph) VTOL four passenger aircraft; several models of VTOL Unmanned Aerial Vehicles employing ducted fans and rotary engines; high performance rotary engines. [98]

75-62450 Monitor Products Company Inc.
(a subsidiary of Bowthorpe plc)
502 Via del Monte, Oceanside, CA 92054
☎ (+1 619) 433 4510 ⊖ (+1 619) 434 0255
President: Chuck Inskeep
Director of Sales and Marketing: Donald H McNamara
■ Design and manufacture of crystal and clock oscillators and quartz crystals used in telemetry and command satellite systems, military aircraft, test equipment and computers. [98]

75-62550 Moog Inc.
East Aurora, NY 14052
☎ (+1 716) 652 2000 ⊠ 415239 ⊖ (+1 716) 687 4457
♣ $ 374 284 000 (1995) 👥 3 050 (1996)
Chairman, President & CEO: Robert Brady
Exec. VP & CAO: Joe C. Green ☎ (+1 716) 652 2000
Exec. VP & CFO: Robert Banta ☎ (+1 716) 652 2000
VP & General Mgr Aircraft Group: Robert Maskrey ☎ (+1 716) 652 2000
VP & General Mgr Systems Group: Richard Sherrill ☎ (+1 716) 652 2000
VP International: Stephen Huckvale ☎ (+44 1684) 29 66 00 (UK)
■ Design & manufacture of electrohydraulic, electromechanical & electropneumatic controls, components & systems to the defense & aerospace industries. [97]

75-62625 Morgan Marine
(† Morgan Crane Co. Inc.)
3122 West Adams Street, Santa Ana, CA 92704
☎ (+1 714) 549 46 33 ⊖ (+1 714) 549 46 28
♣ $ 3,200,000 (1996) 👥 16 (1996) 🏭 14,800 sq ft
President and Chief Executive Officer: William R Morgan
Sales and Marketing Director: Richard Rice
■ Torpedo handling cranes, non magnetic cranes - mine warfare, ROV cranes, ammunition handling cranes and stores cranes, knuckleboom and telescopic cranes. [98]

75-63000 Mountain Optech Inc.
(a subsidiary of Phillips Service Industries)
4775 Walnut Street, Suite A, Boulder, CO 80301
☎ (+1 303) 444 28 51 [E] mktg@mt-optec.com
[W] http://www.mt-optec.com ⊖ (+1 303) 444 44 31
👥 16 (1997) 🏭 10,385 m2
Managing Director: Dennis Sauercrey
Sales and Marketing Director: Mason F Fein ☎ (+1 303) 444 28 51
Operations Director: Dayle Johnston
Purchasing Manager: Roy Gawthrop
Marketing: Ms Patricia McIntyre
■ Manufacture rugged mass memory storage devices for severe environment applications (military and space). [98]

75-63125 MPD Technologies Inc. MPD and Republic Products
(† Republic Electronics)
(a division of Microwave Power Devices Inc.)
49 Wireless Boulevard, Hauppauge, NY 11788
☎ (+1 516) 273 98 00 ⊖ (+1 516) 273 98 20
👥 350
President: Edward Shubel
Vice President Finance: Paul Donofrio
Vice President Advanced Technologies: John F Michaels
Director Business Development: Robert J Califra
General Manager: Larry Konopelko
Administration Assistant Business Development: Ms Diane Dickler
■ Radar environmental simulators (RES), threat signal generators, TACAN beacon simulators, IFF test set, MLS/DME N/P test set. [98]

75-63130 MPD Technologies Inc.
(† Microwave Power Devices Inc.)
49 Wireless Boulevard, Hauppauge, NY 11788-3935
☎ (+1 516) 231 1400 ⊖ (+1 516) 231 8081
👥 255
President: Ed Shubel
VP Sales & Marketing: Timothy Ursprung
Dir Business Development: Joseph Frumkin
Marketing & Sales Coordinator: Marge Lisowy
■ Design & manufacture high power solid state RF microwave power amplifiers & related subsystems in the frequency range of 1MHz to 20GHz, with power levels from 1W to 20kW. End products include: solid state amplifiers, drivers, transmitters, jammers, modulators, transceivers & power supplies. Applications include defence electronics, telecommunications, cellular communications, medical, & laboratory instruments. [97]

75-63150 MRC Precision Metal Optics
(† Micro Robotics Center)
6455 Parkland Drive, Sarasota, FL 34243
☎ (+1 941) 753 87 07 ⊖ (+1 941) 751 42 49
President: Frank Montone
■ Precision metal optical components and subsystems. Product variations: aspheres, spheres, flats, rotating polygons, exotics. Materials: Aluminum, beryllium, nickel, steel, copper, customer specified. Optical fabrication services: polishing, diamond turning, assemblies. Quality control to MIL-1-45208A. [98]

75-63255 MTL Systems Inc.
3481 Dayton-Xenia Rd, Dayton, OH 45431-0299
☎ (+1 513) 426 3111 ⊖ (+1 513) 426 8301
👥 125
Dir Business Development: Robert W. Bissell
■ Provider of engineering services & electronic products which have applications for imaging technologies, electronic warfare systems & computer based data acquisition systems. [97]

75-63445 Munters Cargocaire
(† Cargocaire Engineering Corp.)
(a subsidiary of AB Carl Munters, Sweden)
79 Monroe Street, PO Box 640, Amesbury, MA 01913
☎ (+1 508) 388 06 00 ✆ (800) 843 53 60
[E] cargo@munters.com [W] http://www.munters.com
⊠ 947476 ⊖ (+1 508) 388 45 56
👥 189 (1996)
President: Sven Lundin
Operations Director: Paul Dinnage
Sales and Marketing Director: David Simkins
General Manager: Mike McDonald
Purchasing Manager: Ms Gail Cassin
■ Manufacture complete line of dehumidification systems (air dryers) for the military market, used on all types of operational and equipment in storage. [98]

75-63480 Mustang Survival (USA) Corp
(† Mustang Manufacturing Inc)
3870 Mustang Way, PO Box 5844,
Bellingham, WA 98227
☎ (+1 206) 676 17 82 ⊖ (+1 206) 676 50 14
👥 20 🏭 20,000 sq ft
President: Davies Dwight
Director US Operations: Steve Sible
Manager Government Products: Ivan Paulin
■ Manufacture a range of USCG approved flotation/anti-hypothermia suits, coats and vests for law enforcement, military, commercial marine, sporting and aviation employment. Special emphasis is also placed on flame retardant, insulated and inflatable vest technology. GSA contract # GS-07F-8086B. [98]

75-63490 MWG Co Inc.
PO Box 971202, Miami, FL 33197
10655 SW 185 Terrace, Miami, FL
☎ (+1 305) 253 83 93 ✆ (800) 428 93 94
[E] mwg@shadow.net [W] http://www.shadow.net/~mwg
⊖ (+1 305) 232 12 47
President: Ned C Scheer
■ 90 round magazine for the M-16 and flash suppressor for the .50 cal machine gun. Heavy-duty scope mounts and rings. Night vision mounting systems. [98]

75-63495 M W Microwave Corp.
111 Eads Street, West Babylon, NY 11704
☎ (+1 516) 420 96 86 ⊖ (+1 516) 420 06 21
President: Maurice Wiesenfeld
VP Marketing: Ms Miriam Wiesenfeld
■ Waveguide & coax components & assemblies, attenuators, couplers, adapters, horns, terminations, detectors, bends, twists, tuners, shorts, tees & phase shifters. → 80 [98]

75-63550 NABCO Inc.
1501 Reedsdale St Suite 502, Pittsburgh, PA 15233
☎ (+1 412) 231 8558 ⊜ (+1 412) 321 9122
President: Jay Markey
■ Total containment bomb disposal system. [97]

75-63580 NAI Technologies Inc.
(† North Atlantic Industries Inc.)
Corporate Headquarters
2405 Trade Centre Ave, Longmont, CO 80503
☎ (+1 303) 776 5674 ⊜ (+1 303) 776 5481
♣ $ 60 000 000 (1995) 👥 325 (1995)
President & CEO: Robert A. Carlson
CFO: Richard A. Schneider
VP Marketing: Randy Twogood
VP Engineering: Dick Nickles
VP Operations: Kenneth R. Kanter
Dir Western Regional Sales Office: Jim Foti
Westlake Village, CA
■ Designs, manufactures & markets computers, computer peripherals, computer systems, communications & instrumentation products to government & the aerospace industry. Product lines include micro computers, data storage devices, printers, teleprinters, computer systems & precision instrumentation.
Codar Technologies Inc.
2405 Trade Centre Ave, Longmont, CO 80503
☎ (+1 303) 776 0472 ⊜ (+1 303) 776 8532
♣ $ 30 000 000 👥 210
Systems Div.
7125 Riverwood Drive, Colombia, MD 21046-1526
☎ (+1 410) 312 5800 ⊜ (+1 410) 312 5850
♣ $ 15 000 000 👥 45 (1995)
President: Stephen Rice
VP Marketing: Stephen Rice
■ Provides TEMPEST computer products to defense/security communities. TEMPEST products include: Desktop & Rackmount 386/486 PC, Rackmount Multibus II systems, disk subsystems, laser printers, inkjet printers & DOT matrix printers. [97]

75-63590 NanoMaterials, Inc.
9 Preston Drive, Barrington, RI 02806
☎ (+1 401) 433 70 22
[E] 75770.2047@compuserve.com
[W] http://www.nanomaterials.com ⊜ (+1 401) 433 70 01
♣ $ 1,000,000 (1996) 👥 12 (1997) 🏭 50,000 sq m
President and Chairman: Ms Emily S Reade
Operations and Technical Director: David Shepperton
Purchasing Manager: Stephen Munson
Sales Manager: Charles Reade
■ Manufacturer and global distributor of high purity inorganic (metal, ceramic and composite) and organic powders (fine, ultrafine, nanophase, nanoparticles, nanocrystals, fullerenes and nanotubes). → 85 → 90
® Canada: Cabot, Cerac, Dow Chemical, Dupont, Elkem, Engelhard, GAF, Handy & Harmon, INCO
France: Baikowski International, Pechiney
Germany: De Gussa, Metallgesellschaft, Frank & Schulte
Japan: Mitsubishi Steel
Sweden: Avesta Nyby
UK: Johnson Matthey
USA: 3M, Alcan Aluminium, American Minerals, Hart Metals, Hoegannes, La Roche, Pfizer Inc., Reynolds Metals Co., TAM, Trans-Tech Inc., Wall Colmonoy, Washington Mills Abrasive, Zinc Corporation of America [98]

75-63600 Napco International Inc.
(a subs. of Venturian Corp.)
1600 Second St South, Hopkins, MN 55343
☎ (+1 612) 931 2400 ✉ NORAUTO ⊠ 290436
⊜ (+1 612) 931 2402
♧ $ 10 000 000 ♣ $ 24 789 000 (1994) 👥 103 (1994)
🏭 138 000 sq ft
President: Gary B. Rappaport
Exec. Vice President: Charles B. Langevin
Advertising Mgr: Ms Jan Gass
■ Supplier of defence related equipment. Primarily track & wheeled vehicles, Raider 22 ft patrol craft, .50 cal. maching guns & vehicle & manpack VRC/PRC radios & related accessories.
Aerospace Div.
PO Box 570, Minneapolis, MN 55440
☎ (+1 612) 931 2400 ⊜ (+1 612) 931 2402
Defense Electronics Div.
1600 Second Street South, Hopkins, MN 55343
☎ (+1 612) 931 2481 ⊜ (+1 612) 931 2402
Marketing Dir: Paul M. Flemming
■ Manufacture & stock for immediate delivery AN/PRC-77 & AN/VRC-12 series radios & ancillaries including the AN/VIC-1, also enhancements including the KY-189 secure handset & a line of continuous duty FLIR thermal imaging systems. Complete parts service for overseas customers. [97]

75-63800 National Bullet Proof Inc.
9855 S. 78th Avenue, Hickory Hills, IL 60457
☎ (+1 708) 430 3200 ⊜ (+1 708) 430 3406
♣ $ 1 500 000 👥 30
Managing Dir: Fred Biedess
Marketing & Sales Dir: W. Wally Newman
■ Bullet resistant products. Windows, security detention cells, pass through devices, stainless steel door & windows, aluminium doors & windows, steel doors. [97]

75-63850 National Forge Co.
Irvine, PA 16329
☎ (+1 814) 563 75 22 ⊠ 914446 ⊜ (+1 814) 563 92 09
👥 800
President: Roger E Clark
Executive Vice President: William E Erickson
Director Planning and Projects: Richard D'Alessandro
Director Marketing: Ms Dana A Beyeler
Program Manager: Richard Brewster
■ Integrated manufacturer of precision, high specification parts, components and systems for aerospace, aircraft, naval and ordnance applications. [98]

75-63870 National Nuclear Corporation
(a subsidiary of Thermo Instruments,Inc.)
1275 Hammerwood Avenue, Sunnyvale, CA 94089-2231
☎ (+1 408) 745 66 50 [W] http://www.nnc-usa.com
⊜ (+1 408) 745 67 75
♣ $ 3,500,000 (1996)
Marketing Manager: Mark McCarthy
■ Radiation detection and shielding equipment. Contamination monitoring systems, neutron detectors and nuclear fuel rod scanners. Dosimeters, rameters, area monitors and sample counters. Activation foils, flux wires and products for nuclear medicine. [98]

75-63900 National Steel & Shipbuilding Co. (NASSCO)
PO Box 85278, San Diego, CA 92186-5278
Harbour Drive and 28th Street, San Diego 92113
☎ (+1 619) 544 34 00 [W] http://www.nassco.com
⊜ (+1 619) 544 35 41
👥 4,500 (1997)
President and Chief Executive Officer: R H Vortmann
Senior Vice President of Operations: D Spanninga
Senior Vice President of Business Affairs and Marketing: A W Lutter (Jr)
Vice President of Materials: J C Scott
Marketing Manager: Trevor C Lucey
[E] tlucey@nassco.com
■ Naval and commercial shipbuilding and repair.
Washington Office
Nassco, PO Box 16614, Arlington, VA 22215
☎ (+1 703) 418 11 95 ⊜ (+1 703) 418 35 85
Vice President of Washington Operations: J M Temenak [98]

75-64000 National Wire & Cable Corp.
136 San Fernando Road, PO Box 31307,
Los Angeles, CA 90031
☎ (+1 213) 225 56 11 [W] http://www.nationalwire.com
⊜ (+1 213) 225 46 30
♣ $ 14,000,000 (1997) 👥 206 (1997) 🏭 126,000 sq ft
Chairman: Harry Appleton
President: A J Evangelista
Operations Director: Jim Wiley
Sales and Marketing Director: Bill Kisb
Advertising Manager: Ms Nancy L Schmidt
Purchasing Manager: Ms Mary Veron
Contract Administrator: Victor E Alfonsi
■ Electronic wire and cable manufacturer. [98]

75-64025 Naval Electronics Inc.
6717 Benjamin Road, Tampa, FL 33634
☎ (+1 813) 885 60 91 ✉ NAVAL Tampa
[E] navale@aol.com [W] http://www.navale.com
⊠ 289237 navl ur ⊜ (+1 813) 885 37 89
🏭 2,000 sq ft
President: Lyman K Duggan
Purchasing Manager: Thomas A Mailly
Government Sales: Ron Hoel
■ Marine TV antennas and distribution systems. [98]

75-64040 Navcom Defense Electronics Inc.
4323 Arden Drive, El Monte, CA 91731-1997
☎ (+1 818) 442 8689 ⊜ (+1 818) 444 7619
👥 120 🏭 325 000 sq ft
VP Operations: Ken Russo
Sales Dir: James Keller
Purchasing Mgr: Phil Klippel
Marketing Administrator: Ms Vicki L. Simmons [97]

75-64050 Navigation Data Systems Inc. (NDS)
520 Commerce Point, PO Box 23850,
New Orleans, LA 70183
☎ (+1 504) 734 55 66 ⊜ (+1 504) 734 50 81
President: A William Marchal
■ Manufactures automatic vehicle tracking and data link systems for land, air and marine applications. Also sells and supports Loran-C chains on a worldwide basis. [98]

75-64070 Navigation Science Corp.
31127 Via Colinas, Suite 807,
Westlake Village, CA 91362
☎ (+1 818) 991 97 94 [E] nsc@ix.netcom.com
⊠ 757124 ⊜ (+1 818) 991 98 96
👥 20 🏭 4,000 sq ft
President: Fred J Joubert
Advertising Manager: Ms T Harris
Purchasing: Ms Diane Powers
■ Manufacture of electronic ground navigation equipment including VOR and NDB. Specialise in turnkey systems integration in navigation, communications, ATC, meteorological instruments, runway and approach lights. [98]

75-64085 Navtec Inc.
PO Box 10998, West Palm Beach, FL 33419
☎ (+1 561) 96 02 [E] owlhh@aol.com
⊜ (+1 561) 881 91 93
President: Howard Hornsby
Sales and Marketing Director: Bill Willing
Operations Director: Ben Johns
Purchasing Manager: Mike Fenner
■ Advanced marine technologies company specializing in the development and implementation of remotely operated marine surveillance and tracking for government and commercial applications. Computer mapping systems and neural networks. [98]

75-64090 NCI Inc.
(a member of the TI Group plc, UK)
401 Sweeten Creek Industrial Park,
Asheville, NC 28803-1799
☎ (+1 704) 274 45 40 ⊜ (+1 704) 274 93 81
👥 170 (1997) 🏭 85,000 sq ft
President: R J Ballard
Controller: Carlton Debnam
Marketing Manager: David King
■ International supplier of precision machined aero engine components. Machining is performed on high temperature nickel, cobalt and titanium based aerospace materials. [98]

75-64235 Neotronics NA Inc.
(member of Neotronics Technology plc Gp)
4331 Thurmond Tanner Road, Flowery Branch, GA 30542
☎ (+1 404) 967 2196 ✆ (800) 535 0606
⊜ (+1 404) 967 1854 [97]

75-64240 Ness Manufacturing Inc.
3300 S. Platte River Drive, Englewood, CO 80110
☎ (+1 303) 761 3300 ⊜ (+1 303) 761 2728
👥 20
President: John E. Ness (Jr)
■ Manufacture aerial tow targets & honeycomb structures, for gunnery training. [97]

75-64260 Networks Electronic Corp.
9750 DeSoto Avenue, Chatsworth, CA 91311
☎ (+1 818) 341 04 40 ⊜ (+1 818) 718 71 33
♣ $ 4,000,000 (1996) 👥 64 (1996) 🏭 3,150 m2
President and Chief Executive Officer: David Wachtel
Executive Vice President, Technical: Mike Tabassi
Operations Director: Ms Carol Hummel

Sales and Marketing Director: Terri Mayo
Purchasing Manager: Ms Val Edwards
■ Spherical bearings, bushings and rod ends, squib switches, igniters, actuators and miniature ordnance devices. 98

75-64300 Newport News Shipbuilding & Drydock Co.
4101 Washington Avenue, Newport News, VA 23607
☎(+1 757) 380 20 00 W http://www.nns.com
⊠ 823453; TWX 710 8800 007
♣ $ 1,870,000,000 (1996) ⛹ 18,000 (1996)
Chairman and Chief Executive Officer: William Fricks
Senior Vice President and Chief Financial Officer: David J Anderson
Vice President and General Counsel: Stephen B Clarkson
Executive Vice President: Thomas C Schievelbein
Vice President of Operations: John E Shephard
Vice President of Finance: Thomas J Bradburn
Vice President of Human Resources: Alfred Little (Jr)
Vice President of Engineering: Mark Pelaez
Vice President of Corporate Communications: T Michael Hatfield
Vice President of Submarines: Georege A Wade
Vice President of Marketing: W Greg Cridlin (Jr)
Vice President (Nuclear): James A Palmer (Jr)
Vice President of Materials: W G Cooper
Media Relations: Jerri F Dickseski ☎(+1 757) 380 23 41
Marketing and Communications: Ms Karen Kennedy
■ Design, construction, overhaul and repair of nuclear and conventionally powered ships. Construction of Double hull product tankers. Sole manufacturer of Nimitz class aircraft carriers. Design yard for Los Angeles and Seawolf class attack submarines.

Integrated Logistics Support
12129 Jefferson Avenue, Newport News, VA 23602
☎(+1 757) 688 23 12 98

75-64380 Nichols Brothers Boat Builders Inc.
5400 South Cameroon Road, PO Box 580, Freeland, WA 98249
☎(+1 360) 331 55 00 E nichols@whidbey.com
⊖ (+1 360) 331 74 84
⛹ 200 (1997) ⌂ 39,600 sq ft
President: Mathew J Nichols
Vice President/Chief Engineer: Justin Nichols
Vice Presdient Marketing: Bryan O Nichols
■ Locations in Freeland and Langley. Complete design and engineering, manufacturing and heavy material moving capabilities for new vessels and conversion and repair programmes. 98

75-64390 Night Vision Equipment Co. Inc.
PO Box 266, Emmaus, PA 18049-0266
☎(+1 610) 391 91 01 E nvec@ptdprolog.net
⊖ (+1 610) 391 92 20
President: William H Grube
Vice President Business Development: Gene Adcock
Government Marketing Director: J P A Maldonado
Domestic Sales Manager: Richard W Urich
Purchasing Manager: Harrison Grube
■ Designer, manufacturer, distributor of night vision systems for military, security & law enforcement applications. Product: second & third generation goggles; weapon sights; pocketscopes; camera-adaptable night vision surveillance devices; infra-red & tactical illuminator weapon-aiming systems; seismic intrusion detectors. → 85 98

75-64400 NMB (USA) Inc
9730 Independence Avenue, Chatsworth, CA 91311
☎(+1 818) 709 17 70 ⊠ 9104941232
⊖ (+1 818) 709 57 78
⛹ 18
President: M Yamanaka 98

75-64750 Nobles Manufacturing Inc.
(† Nobles Industries Inc.)
1105 East Pine St, St Croix Falls, WI 54024
☎(+1 715) 483 3079 ⊠ 297403 ⊖ (+1 715) 483 1884
⛹ 100 (1996) ⌂ 30 000 sq m
Chairman: Ted L. Priem
President: Ted Priem
Dir Engineering: George Bukovich
■ Manufacture flexible ammunition chutes for land, sea & air in various sizes. 97

75-64760 Noise Com Inc.
(a subsidiary of Wireless Telecom Group Inc.)
East 64 Midland Avenue, Paramus, NJ 07652
☎(+1 201) 261 87 97 E info@noisecom.com
W http://www.noisecom.com ⊖ (+1 201) 261 83 39
♣ $22,500,000 (1996) ⛹ 65 (1996)
President: Dale Sydnor
Vice President Sales: Bent Hessen-Schmidt
Director of Advertising: Alex Kim
Marketing Manager: Ms Sheree Parker
■ Solid state noise sources and jammers. Communication Test Equipment. 98

75-64800 Norca Machinery Company
800 Axinn Avenue, PO Box CS-9401, Garden City, NY 11530
☎(+1 516) 222 70 70 E norca@idt.net ⊠ 229114 nrc ur
⊖ (+1 516) 222 88 11
President: R D Friedlander
Managing Director: Eytan Erez
Operations Director: Gene Foley
Sales and Marketing Director: Allen Rosen
Senior Contract Administrator: E F Rabinowitz
Assistant Sales Manager: Allen Rosen
■ Manufacturer/contractor providing production & maintenance parts for military vehicles, both ground support & airborne. 98

75-64880 The NORDAM Group
510 South Lansing, PO Box 3365, Tulsa, OK 74101
☎(+1 918) 628 7345 ⊖ (+1 918) 438 9543
President: Ray Siegfried
■ Aircraft repair & overhaul, powerplant, flight control surfaces, engine components, ducts, aircycle machines, heat exchangers, windows, lenses, windshields, advanced composites. → 80

Other NORDAM Group Members:
LORI
LORI Asia, Singapore
World Aviation Associates
M&N Aerospace

Programme Development Office
11200 E.Pine, Tulsa, OK 74116
☎(+1 918) 437 8110 ⊖ (+1 918) 234 6758

Manufacturing Division
PO Box 3365, Tulsa, OK 74101
510 South Lansing, Tulsa OK 74120
☎(+1 918) 587 4105 ⊠ 492377 nordam tul
⊖ (+1 918) 583 2604
⛹ 1 100 ⌂ 3 500 m2
President: Charles B. Ryan
Marketing Dir: Andrew Altendorf
■ Manufacture honeycomb, hot bonded panels & structures including radomes, airframe structural components, aircraft interiors, helicopters & military aircraft. Powerplant honeycomb structures; bonded control surfaces; plastics, plexiglass & glass fibre assemblies.

Repair Division
PO Box 3365, Tulsa, OK 74101
☎(+1 918) 628 7345 ⊠ 158105 ⊖ (+1 918) 438 9543
Information Contact: Bob Brooks

Transparency Division
PO Box 3365, Tulsa, OK 74101
☎(+1 918) 587 4105 ⊠ 163469 ⊖ (+1 918) 560 5669
Information Contact: Rick Armstrong

Helicopter Program
PO Box 3365, Tulsa, OK 74101
☎(+1 918) 628 7346 ⊖ (+1 918) 438 9543
Information Contact: Rex A. Smith

Hush Kit Program
PO Box 3365, Tulsa, OK 74101
☎(+1 918) 560 5560 ⊖ (+1 918) 560 5528
Information Contact: Jack Arehart 97

75-64900 Norden Systems Inc.
(a subs. of Northrop Grumman Corporation)
10 Norden Place, PO Box 5300, Norwalk, CT 06856
☎(+1 203) 852 5000 ⊖ (+1 203) 852 7858
♣ $ 200 000 000 (1995) ⛹ 900 ⌂ 630 000 sq m
President: Ken Hutchinson
Director: Mike D'Angelo
VP Finance & Administration: Fred Bassett
Business Development Dir: Robert Whitcher
Programme Mgr: Robert Eisenberg
Export Licensing Mgr: Robert Maslow
Marketing Communications: Ms Diane Janeway
■ Airborne & shipboard radars, command & control systems, military computers & display systems.

Long Island Plant
75 Maxess Rd, Melville, NY 11747-3182
☎(+1 516) 694 0900 ⊖ (+1 516) 694 0938 97

75-65000 Norotos Inc.
201 East Alton Avenue, Santa Ana, CA 92707-4416
☎(+1 714) 662 31 13 ⊖ (+1 714) 662 79 50
♣ $ 3,100,000 (1996) ⛹ 40 (1997) ⌂ 12,200 sq m
President: Ronald R Soto
Operations Director: Brian Crawford
Purchasing Manager: Mrs Linda M Soto
■ Manufacturer of night vision components and precision machined commercial products. 98

75-65200 North American Aerodynamics Inc.
107 Carver Drive, Roxboro, NC 27573
☎(+1 910) 599 92 66 ⊖ (+1 910) 599 94 42
⛹ 100 ⌂ 60,000 sq ft
President: John P Higgins
■ Personnel, cargo, drogue and sport parachutes. 98

75-65210 North American Marine Jet Inc.
PO Box 1232, Benton, AR 72018
☎(+1 501) 778 4151 ⊖ (+1 501) 778 6381
⛹ 25
President: Leonard Hill
Marketing Dir: Glenn Prentice
■ Marine water jet propulsion units. 97

75-65250 North American MORPHO Systems Inc (NAMSI)
(a subsidiary of SAGEM SA, France)
Tacoma Financial Building, 1145 Broadway Plaza, Tacoma, WA 98402
☎(+1 206) 383 36 17 ⊖ (+1 206) 591 88 56
♣ $ 55,000,000 (1996) ⛹ 250
Chairman: D Dupuy
Chief Executive Officer: J M Suchier
■ Sale of the Automated Fingerprint Identification System (AFIS) for criminal and civil applications. Fingerprint database conversion. 98

75-65300 North Hills Signal Processing
(† North Hills Electronics Inc.)
(a subsidiary of Porta Systems Corp.)
575 Underhill Boulevard, Syosset, NY 11791
☎(+1 516) 682 77 40; 682 77 06
⊖ (+1 516) 682 77 04
♣ $ 7,000,000 (1996) ⛹ 50
Managing Director: Prem Chandran
Sales Manager: Steve Muth
Purchasing Manager: Bill Duckham
■ Mil-STD-1553 data bus components and testers and wideband signal transformers. 98

75-65450 Northrop Grumman Corporation
1840 Century Park East, Los Angeles, CA 90067-2199
☎(+1 310) 553 62 62 W http://northgrum.com
⊖ (+1 310) 201 30 23; 553 20 76
♣ $ 8,100,000,000 (1996) ⛹ 46,500
Chairman, President and Chief Executive Officer: Kent Kresa
Corporate Vice President and General Manager - Data Systems and Services Division: Herbert W Anderson
Corporate Vice President and General Manager - Commercial Aircraft Division: Ralph D Crosby (Jr)
Corporate Vice President and Chief Human Resources, Communications and Administrative Officer: Marvin Elkin
Corporate Vice President and General Manager - Electronic Sensors and Systems Division: James G Roche
Corporate Vice President, Government Affairs: Robert W Helm
Corporate Vice President and General Counsel: Richard R Molleur
Corporate Vice President and Secretary: James C Johnson
Corporate Vice President and Treasurer: Albert F Myers
Corporate Vice President and General Manager - Electronics and Systems Integration Division: John E Harrison
Corporate Vice President and Controller: Nelson F Gibbs
Corporate Vice President and General Manager - Military Aircraft Systems Division: William H Lawler
Corporate Vice President and Chief Strategic Planning, Advanced Development and Programs Officer: Charles L Jones (Jr)

Contact Reference

Corporate Vice President and Chief Financial Officer: Richard B Waugh (Jr)
Corp. VP & General Mgr Commercial Aircraft Div.: Gordon L. Williams
VP Business Development: William Carrier
VP Military Aircraft Systems Div.: Gerard Dufresne
■ Design, manufacture and systems integration: military surveillance, combat aircraft, defence electronics and systems, airspace management and information systems, marine propulsion, precision weapons and commercial and military aerostructures.

Divisions:
Commercial Aircraft Division
Data Systems & Services Division
Electronic Sensors & Systems Div.
Electronics & Systems Integration Div.
Military Aircraft Systems Div.

Washington Office
1000 Wilson Blvd, Ste 2300, Arlington, VA 22209
☎(+1 703) 525 6767 ⊖ (+1 703) 276 0711

Eastern Regional Office
1919 Commerce Drive, Suite 455, Hampton, VA 23666
☎(+1 804) 838 3360 ⊖ (+1 804) 827 9379
Corp. Staff: William G. Goodyear

Military Aircraft Systems Division Headquarters
One Hornet Way, El Segundo, CA 90245-2804
☎(+1 310) 332 10 00 ⊖ (+1 310) 332 33 96
Corporate Vice President and General Manager: William H Lawler
■ Major products and activities include the B-2 low-observable strategic penetration bomber, the F/A-18 Hornet strike fighter, EC-2C Hawkeye early warning aircraft, the BQM-74 and Chukar unmanned aerial targets, remanufacture and modification of B707-300 aircraft into E-8C for the Joint STARS targeting and battlefield management system, EA-6B Prowler remanufacturing, reconfiguration of the F-14 Tomcat fleet defence fighter and F-5 and T-38 upgrades.

Grumman St Augustine Corp Military Aircraft Div. Facility
5000 US 1 North, St Augustine, FL 32085-3447
☎(+1 904) 825 3300 ⊖ (+1 904) 824 1743

Grumman Lousianna Corp. Military Aircraft Div. Facility
4400 Legion Street, Lake Charles, LA 70615
☎(+1 318) 493 2100 ⊖ (+1 318) 493 2144

Electronic and Systems Integration Division
South Oyster Bay Road, Bethpage, NY 11714-3582
☎(+1 516) 346 28 12 ⊖ (+1 516) 575 58 36
Corporate Vice President and General Manager: John E Harrison
■ Surveillance and battlefield management systems: Joint STARS and E-2C Hawkeye; precision weapon systems, including the BAT 'Brilliant' anti-armour submunition; electronic warfare systems including airborne radar and infrared countermeasures and electronics and readiness support systems.

Surveillance & Electronic Warfare Systems
South Oyster Bay Road, Bethpage, NY 11714-3582
☎(+1 516) 575 1840 ⊖ (+1 516) 346 2048

Combat Support Systems
2301 West 120th Street, PO Box 5032, Hawthorne, CA 90251-5032
☎(+1 213) 600 3000
⊖ (+1 213) 600 4088; 4099; 3510; 6494

Electronic Warfare Systems
600 Hicks Road, Rolling Meadows, IL 60008-1098
☎(+1 847) 259 9600 ▣ 270566 ⊖ (+1 847) 870 5705

Electronics and Readiness Support Systems
Sunrise Highway, Great River, NY 11739-0544
☎(+1 516) 224 3206 ⊖ (+1 516) 224 3356

Surveillance and Battle Management Systems
2000 NASA Boulevard, PO Box 9650, Melbourne, FL 32902
☎(+1 407) 951 5000 ⊖ (+1 407) 951 5379

Electronic Sensors and Systems Division
1580-A West Nursery Road, Linthicum, MD 21290
☎(+1 410) 993 24 63 ⊖ (+1 410) 993 23 94
Corporate Vice President and General Manager: James G Roche
■ Production of radars and electronics for combat aircraft and battlespace management systems including the F-22 and F-16 fighters, B1-B bomber, Longbow/Apache attack helicopter, C-130 transport, E-3 AWACS and E-8 Joint STARS as well as military space and undersea systems.

Data Systems and Services Division
2411 Dulles Corner Park, Herndon, VA 22071
☎(+1 703) 713 40 00 ⊖ (+1 703) 713 40 67
Corporate Vice President and General Manager: Herbert W Anderson
■ Expertise in software and hardware engineering, computer graphics, networking, supercomputers, high level systems architecture, machine intelligence and correlation. Professional and technical services. Specialise in complex systems integration, information processing and advanced data processing problems for US DoD, NASA and industry.

Systems Support Services
10 Orville Drive, Bohemia, NY 11716
☎(+1 516) 563 6907 ⊖ (+1 516) 563 6881

External Information Systems
2411 Dulles Corner Park, Suite 500, Herndon, VA 22071-3430
☎(+1 703) 713 4100 ⊖ (+1 703) 713 4476

Technical Services
1250 Grumman Place, Titusville, FL 32780
☎(+1 407) 264 4522 ⊖ (+1 407) 268 8357

Technical Services and Northrop Worldwide Aircraft Inc.
Worldwide Aircraft Inc., 21 NW 44th Street, Lawton, OK 73502-0108
☎(+1 405) 353 2733 ⊖ (+1 405) 353 0153

Technical Services and Information Systems
Systems, 16511 Space Center Drive, Houston, TX 77055
☎(+1 713) 280 3801 ⊖ (+1 713) 280 3838

Systems Engineering and Integration
12000 Aerospace Avenue, Houston, TX 77034
☎(+1 713) 929 7025 ⊖ (+1 713) 929 7073

Space Station Hab/Lab Operation
12000 Aerospace Avenue, Houston, TX 77034
☎(+1 713) 929 7001 ⊖ (+1 713) 929 7040

Commercial Aircraft Division
9314 West Jefferson Boulevard, Dallas, TX 75211
☎(+1 214) 266 20 11 ⊖ (+1 214) 266 57 61
Corporate Vice President and General Manager: Ralph D Crosby
■ Structural airframe work including: fuselage and tail sections, control surfaces, engine nacelle systems, thrust reversers for a variety of aircraft. Manufacture General Electric CF6-80 jet engines and major aerostructure components for McDonnell Douglas C-17 military transport.

Hawthorne Facility
One Northrop Avenue, Hawthorne, CA 90250-3277
☎(+1 310) 332 1000 ⊖ (+1 310) 332 3066

Georgia Production Site
One Northrop Place, Perry, GA 31069
☎(+1 912) 988 5000 ⊖ (+1 912) 987 2356

(† Vought Aircraft Company)
(a subsidiary of Northrop Grumman)
Northrop Grumman Commercial Aircraft Division Headquarters
PO Box 655907, Dallas, TX 75265-5907

Stuart Facility
PO Box 9015, Stuart, FL 34995-9015
☎(+1 407) 220 5300 ⊖ (+1 407) 220 5205

Milledgeville Facility
90 Highway 22 West, Milledgeville, GA 31061-9699
☎(+1 912) 454 4200 ⊖ (+1 912) 454 4314

Grumman Olson
Route 1, Box 441, Montgomery, PA 17752
☎(+1 717) 547 1681 ⊖ (+1 717) 547 6180

Advanced Systems and Technology
8900 East Washington Boulevard, Pico Rivera, CA 90660-3783
☎(+1 310) 948 87 06 ⊖ (+1 310) 948 04 09
■ The centre is responsible for developing new military business for the Electronic Sensor & Systems, Electronic & Systems Integration & Military Aircraft Systems divisions as well as supporting the other divisions, fostering non traditional business and maintaining the company's technology base. 98

75-65510 Norton Co. Abrasives Marketing Group
1 New Bond Street, Worcester, MA 01615-0008
☎(+1 508) 795 50 00 ▣ 920428 norton wor b
⊖ (+1 508) 795 41 30
Vice President: Richard B Kennedy
Marketing Manager: Lou Gatti
■ Abrasives, grinding wheels, coated abrasive sheets, discs and belts. Bear-Tex non-woven abrasive blending and finishing products. 98

75-65520 Norton Performance Plastics Corp.
(a division of the Saint Gobain Group and Norton Group)
335 North Diamond Street, Ravenna, OH 44266
☎(+1 330) 296 99 48 ⊖ (+1 330) 296 79 55
♣ $ 10,000,000 (1997) 👥 70 (1997) 🏭 75,000 sq ft
Technical Director: Ben MacKenzie
Marketing Manager: Ronald Bauer
Business Manager: Joseph Hensel
Marketing Communications Manager: Ms Nancy Van Emburg
Purchasing Manager: John Byerly
Operations: Dean Burner
Engineering: Donald Moyer
Controller: Lee Jarvis
Quality: Dan Orwig
■ Design, develop, test and manufacture radomes for aircraft, missiles and helicopters. Repair composite structures and radomes. FAA approved repair station. 98

75-65600 Nova Electric
(a subs. of Technology Dynamics Inc.)
100 School St, Bergenfield, NJ 07621-2915
☎(+1 201) 385 0500 ⊖ (+1 201) 385 0702
♣ $ 5 000 000 👥 80 🏭 20 000 sq ft
President: Aron Levy
VP Sales & Marketing: Howard Schrier
■ Ruggedized, 19" rack mount uninterruptible power systems for communications & electronic warfare systems. Rated 1KVA to 15KVA, meets MIL-STD-461, 810, 1399, 167, 704 & MIL-STD-901. 400Hz frequency converters. 97

75-65730 Nuclear Associates Division Victoreen Inc.
100 Voice Road, Carle Place, NY 11514
☎(+1 516) 741 63 60 ✆ (888) 466 82 57
E sales@nucl.com W http://www.nucl.com
⊖ (+1 516) 741 54 14
👥 30 (1997) 🏭 26,000 sq ft
Vice President and General Manager: Martin Ratner
Marketing Director: Ms Stephanie Felney
Operations Director: Peter Wojick
■ Ionizing radiation detectors & dosimeters. 98

75-65760 Nuclear Metals Inc.
2229 Main Street, Concord, MA 01742
☎(+1 508) 369 54 10 E sales@nucmet.com
W http://www.nucmet.com ▣ TWX 710 3471 056
⊖ (+1 508) 369 40 45
♣ $ 30,000,000 (1997) 👥 190 (1997) 🏭 180,000 sq ft
President and Chief Executive Officer: Robert E Quinn
Marketing Manager: James H Scarboro
Manager Power Products: John D Nicholson 98

75-65770 Nuclear Research Corp.
NRC Industries
125 Titus Avenue, PO Box H, Warrington, PA 18976
☎(+1 215) 343 59 00 E mktg@nuc.com
W http://www.nrc-ind.com ⊖ (+1 215) 343 46 70
👥 200 🏭 64,200 sq ft (2 facilities)
President: Earl M Pollock
Executive Vice President: Jack Cooley
Vice President Engineering: Dr S Pandey
Marketing Manager: Jack A Malz
Technical Marketing Specialist: Al Zirkes
■ Manufacturer of military radiac equipment offering three major nuclear product lines including radiation monitoring systems for nuclear utilities & processing facilities nucleonic gauges & military radiac instruments. 98

75-65807 NU Van Technology Inc.
2155 Hwy 1187, PO Box 759, Mansfield, TX 76063-0759
☎(+1 817) 477 17 34 ✆ (800) 487 1734
⊖ (+1 817) 473 39 42
♣ $ 12 000 000 (1995) 👥 110 (1996)
President: Bill Smith
VP Sales & Marketing: Mike Acker
■ Manufacturer of curtain sided semi trailers & truck bodies. Manufacturer of tank retriever lowboys. 97

75-65810 O.I. Analytical CMS Field Products Group
(† CMS Research Corp.)
(a subsidiary of O.I. Analytical, TX)
200 Chase Park South, Suite 100, Birmingham, AL 35244
☎(+1 205) 733 69 11 E cms6900@aol.com
⊖ (+1 205) 733 69 19
👥 40 🏭 8,000 sq ft
President: Dr Gary D Sides

Operations and Technical Director: Dr Craig Lagrone
Sales Manager: Gregory R Houston
Purchasing Manager: Ms Sarah Sides
■ Development and production of detection and alarm systems for toxic chemical vapours including chemical warfare agents. Product line based on an automated GC air monitoring system known as MINICAMS. → 85 [98]

75-65830 O'Brien Energy Services Inc.
(† O'Brien Energy Systems Inc.)
920 Church Street, Wilmington, DE 19899
☎ (+1 215) 658 7100 ⊜ (+1 215) 654 2133
♣ $ 12 000 000 👥 80
President: Robert J. Smallarombe [97]

75-65850 Obrien D G Inc.
PO Box 159, Seabrook, NH 03874
☎ (+1 603) 474 55 71 ⊜ (+1 603) 474 31 61
♣ $ 9,500,000 👥 100
President: Donald G O'Brien
Vice President Marketing:
Chief Financial Officer and Director of Administration: John J Vendola
■ Design, manufacture electrical & optical connectors, cable assemblies & pressure hull fittings for use on submarines, research vehicles & in hostile environments. [98]

75-65860 Ocean Data Equipment Corp.
141 Washington St, E. Walpole, MA 02032
☎ (+1 508) 660 6010 ✆ (800) 732 2326
[E] mkt@seabeam.com ⊜ (+1 508) 660 6061
👥 20 🏭 60 000 sq ft
International Marketing Mgr: Robert N. Lobecker
■ Design, develop & produce oceanographic instrumentation & related products. Comprehensive line of bathymetric systems consisting of shallow, medium & deep water echosounders, tow fish, parametric narrow beam sonars, bottom & Chip sub-buttom classification software processors, doppler speed log systems & related peripherals. The company specialises in customising standard products to meet customer-specific applications. [97]

75-66020 Ocean Technology Systems
Underseas Systems International
2950 Airway Avenue (D-3), Costa Mesa, CA 92626-6002
☎ (+1 714) 754 78 48 [E] ots4com@aol.com
⊜ (+1 714) 966 16 39
♧ $ 400,000,000 (1996) ♣ $ 3,500,000 (1995)
👥 38 (1996) 🏭 7,000 sq m
President: Michael R Pelissier
Operations Director: Jerry Peck
Purchasing Manager: Ms Teri Halter
■ Manufacture of underwater communication systems for both tactical and working divers: support equipment, full face masks, wireless communication hardware, U/W, PA. [98]

75-66050 Odetics Inc. Space Division
1585 South Manchester Avenue, Anaheim, CA 92802
☎ (+1 714) 780 76 02 [W] http://www.odetics.com
⊜ (+1 714) 780 76 49
♣ $ 140,000,000 (1996) 👥 600 (1996) 🏭 250,000 sq ft
President: Crandall Gudmundson
Vice President of Marketing: Jerry Muench
■ Space recorders. [98]

75-66100 OEA Aerospace Inc.
(a subsidiary of OEA Inc.)
PO Box KK, Fairfield, CA 94533
☎ (+1 707) 422 18 80 [W] http://www.oeaa.com
⊜ (+1 707) 422 32 42
👥 450 (1997)
President: Ben E Paul
Managing Director: John N Zinsmaster
Operations Director: Rick L Todd
Administration Secretary: Ms G Harrison
■ Specializing in design, development and manufacture of explosive-actuated devices, aircraft escape systems, missile staging, destruct, safety and release devices, mild detonating cord, linear shaped charges, explosive bolts. [98]

75-66220 OFC Corporation
2 Mercer Road, Natick, MA 01760
☎ (+1 508) 655 16 50 [E] ofccorp@ofccorp.com
[W] http://www.ofccorp.com ⊜ (+1 508) 653 07 29
♣ $ 13,000,000 (1997) 👥 105 (1997) 🏭 50,000 sq ft
President: John F. Blais (Jr)
Corporation Marketing Dir: E.A. Wollensak
■ Custom filters, lenses, diamond turned optics and electro-optical assemblies for use in the IR, visible and UV spectrum; aspheric, diffractive components, multimirror systems, IR coatings and patterned filters.

Optical Filter Division
2 Mercer Road, Natick, MA 01760
☎ (+1 508) 655 16 50 ⊜ (+1 508) 653 07 29
♣ $ 7,000,000 (1997) 👥 55 (1997)
President: John F. Blais (Jr)
VP & General Mgr: Ray Boulay
Sales Mgr: Ms Jewel Anderson

Diamond Turning Division
69 Island Street, Keene, NH 03431
☎ (+1 603) 357 76 62 ⊜ (+1 603) 357 77 64
♣ $ 6,000,000 (1997) 👥 50 (1997) 🏭 20,000 sq ft
Vice President and General Manager: Jim Briggs
Sales Manager: Bob Clark [98]

75-66400 O'Gara-Hess & Eisenhardt Armoring Co.
(a subsidiary of The O'Gara Company, USA)
Headquarters
9113 LeSaint Drive, Fairfield, OH 45014
☎ (+1 513) 874 21 12 ✆ (800) 697 03 07
[W] http://www.ogara.com ⊜ (+1 513) 874 12 62
♣ $ 82,000,000 (1996) 👥 450 (worldwide 1996)
🏭 100,000 m2
Chairman and Chief Executive Officer: Bill O'Gara
President and Chief Operating Officer: Michael J "Chip" Lennon
■ Commercial manufacture of armoured cars, limousines & ambulances. Blast protection for automobiles or military vehicles for governments & foreign or US domestic corporations. [98]

75-66550 Omi Corp.
(† Opto Mechanik Inc.)
(a subs. of DRS/Photronics Corp.)
2330 Commerce Park Dr, NE Ste 2, Palm Bay, FL 32905
☎ (+1 407) 984 9030 ⊜ (+1 407) 984 8796; 253 4494
♣ $ 28 000 000 (1995) 👥 120 🏭 54 000 sq ft
VP Operations: Irv J. Lowe
Engineering Dir: Daniel D. Young (PhD)
Marketing Dir: Murray W. Boggs
Marketing Mgr: Patrick J. Zajac
■ Range of products from infra-red, passive and panoramic sighting devices to electro-optical equipment. [97]

75-66620 Omniglow Corp
96 Windsor Street, West Springfield, MA 01089
☎ (+1 413) 739 82 52 ⊜ (+1 413) 737 86 36
■ Manufacture chemical light markers. [98]

75-66875 Ontic Engineering & Manufacturing Inc.
(a subsidiary of Askeland Corp.)
7230 Coldwater Canyon Avenue, PO Box 7044, North Hollywood, CA 91609
☎ (+1 818) 982 42 20 [W] http://www.ontic.com
⊜ (+1 818) 982 76 16
♧ $ 1,500,000 (1998) ♣ $ 9,000,000 (1998)
👥 52 (1998)
President: Ted Allred
Sales and Marketing Director: Tola Korsunski
Business Development Director: Frank Costello
■ Fluid pumping and regulating devices, pumps, valves (hyd/fuel), hyd piston and actuators. Electromechanical assemblies, motors AC/DC, actuators, tachometer generators. [98]

75-66880 Optek Technology Inc.
1215 West Crosby Road, Carrollton, TX 75006-6905
☎ (+1 972) 323 22 00 [E] optek1@optekinc.com
[W] http://www.optekinc.com ⊜ (+1 972) 323 23 96
♣ $ 67,300,000 (1996) 👥 1,700 (1996)
🏭 316,000 sq m
President and Chief Executive Officer: Tom Filesi
Vice President of Sales and Marketing: Andrew Kahn
Purchasing Manager: Jim Hurst
■ Opto-electronic components: infra-red LED's, photo detectors, assemblies, optocoupler, custom components. Small signal surface mount components. Hall effect devices. Fiber optic components. High rel and commercial versions. [98]

75-66952 Optical Cable Corp.
PO Box 11967, Roanoke, VA 24022-1967
☎ (+1 540) 265 06 90 ✆ (800) 622 77 11
[W] http://www.occfiber.com ⊜ (+1 540) 265 07 24
♣ $ 46,000,000 (1997) 👥 142 (1997) 🏭 148,000 sq ft
President and Chief Executive Officer: Robert Kopstein
Senior Vice President of Sales: Luke Huybrechts
Vice President of Marketing: Ms Susan S Adams
[E] sadams@occfiber.com
■ Manufacture of fibre optic cables for data communications and local area networks (LANS). [98]

75-66953 Optical Components Inc. (OCI)
1175 East Edna Place, Covina, CA 91724
☎ (+1 626) 967 52 81 [W] http://www.primenet.com/~oci/
[E] daleadv@primenet.com ⊜ (+1 626) 967 25 13
♣ $ 5,500,000 (1997) 👥 60 (1997) 🏭 33,000 sq ft
President: Ms Lynn Montgomery
Vice President of HP/Marketing: Dale Josephson
Vice President of Coating: Steve Brazelton
Managing Director: Ms Judy King
Purchasing Manager: Chuck Holt
■ Precision optical components for aerospace and military purposes (prisms, windows, mirrors, lenses and optical sub-assemblies). IR materials, through visible glasses and fused silicas. In-house coating facilities for reflective, anti-reflective and beamsplitter coatings. [98]

75-66965 Optical Systems Tech., Inc.
(† Star-Tron Technology Corp.)
110 Kountz Lane, Freeport, PA 16229
☎ (+1 412) 295 28 80 ⊜ (+1 412) 295 33 66
♣ $ 1,400,000 (1996) 👥 10 (1996) 🏭 800 m2
President: Paul F Maxim
Operations Director: Dr Eugene Pochapsky
Marketing Manager: Michael R Matzko
Purchasing Agent: Ms Vicki DeThomas
■ Night vision equipment. [98]

75-66980 Optim Electronics Corp.
12401 Middlebrook Road, Germantown, MD 20874
☎ (+1 301) 428 72 00 ⊜ (+1 301) 353 01 29
President: R Moore
■ Manufacture of MEGADAC real-time data acquisition systems and test control software (TCS). [98]

75-67100 OptoElectronics Textron
(† OptoElectronics Inc.)
(a subsidiary of Textron Inc.)
1309 Dynamic Street, PO Box 750039, Petaluma, CA 94975-0039
☎ (+1 707) 763 59 00 ⊠ 340186 ⊜ (+1 707) 762 73 83
👥 100 (approx.)
General Manager: Bill Dohery
Marketing Director: Henry Flesh
Engineering Manager: E Sutcliffe
Operations Manager: Richard F Kingsbury
Purchasing Manager: John Kruse
Quality Assurance Manager: Ms Beverly Bartels
Controller: Victor A Pectol
■ Manufacture PbS and PbSe infra-red detectors and related components. [98]

75-67220 Orbit Engine Company (USA) Inc
5550 Occidental Highway, Tecumseh, Michigan 49286
☎ (+1 517) 423 66 23 ⊜ (+1 517) 423 60 79 [98]

75-67350 Ordex Inc.
5777 Center Rd, Valley City, OH 44280
☎ (+1 330) 225 1390 ✆ (800) 856 1758
⊜ (+1 330) 225 1370
♣ $ 2 377 000 (1995) 👥 15 🏭 3 100 sq m
President: T.F. Muranko
Vice President: P.S. Johnson
Exec. Vice President: D.H. Muranko
Operations & Technical Dir: P. Pasek
■ Large stocks of Continental motors & engine parts for USA tracked vehicles & foreign vehicle retrofits. [97]

75-67500 Orion Aviation
27520 Hawthorne Boulevard, Palos Verdes, CA 90274
☎(+1 310) 544 48 49 🄴 whin@orionav.com
⊖ (+1 310) 544 29 75
President: Thomas B Whinfrey
■ Manufacture of high speed vertical takeoff and landing (VTOL) unmanned aerial vehicles (UAVs). 98

75-67505 Orkal Industries Corp.
333 Westbury Avenue, Carle Place, NY 11514-1658
☎(+1 516) 333 21 21 🄴 orkal@orkal.com
⊖ (+1 516) 333 21 75
👥 17 (1997)
President: Kalman Schmidt
Project Manager: Irwin Seigel
■ Logistic support, engineering products & procurement support. 98

75-67525 Oro Manufacturing Co.
5000 Stitt Street, PO Box 5018,
Monroe, NC 28111-5018
☎(+1 704) 283 21 86 ⊖ (+1 704) 283 02 69
♣ $ 2,500,000 (1996) 👥 39 (1996) 🏭 80,000 sq ft
President and General Manager: R W Engel
☎(+1 704) 283 21 86 ext 316
■ Manufacture military aircraft seats and related items. 98

75-67530 Ortel Corp.
2015 West Chestnut Street, Alhambra, CA 91803-1542
☎(+1 818) 281 36 36 ⊖ (+1 818) 281 82 31
♣ $ 75,000,000 (1996) 👥 600 (1997)
President and Chief Executive Officer: Wim Selders
Vice President of Sales: Bill Moore
Business Manager: John Rinks
■ Manufacturer of high speed (up to 18GHz) analogue fibre optic links for antenna remoting, delay lines, secure broadband communications networks and transient data capture. 98

75-67550 Oshkosh Truck Corp.
2307 Oregon Street, PO Box 2566,
Oshkosh, WI 54903-2566
☎(+1 414) 233 91 50 🅆 http://www.oshkoshtruck.com
⊖ (+1 414) 233 95 06
👥 1,400 (1997)
Chairman & CEO: R.Eugene Goodson
President and Chief Operating Officer: Robert Bohn
Exec. Vice President: Paul Hollowell
Purchasing Mgr: Bruce Herrmann
Marketing Services Mgr: Ms Kirsten Skyba
■ Heavy duty vehicles. US Army palletised load system vehicles, US Army M911 & M1070 heavy equipment transporters, US Air Force multi-purpose snow-removal vehicles, US Navy & Air Force aircraft rescue & fire-fighting equipment, US Army heavy expanded mobility tactical trucks. Truck trailers, electronic van trailers for US Army. 98

75-67750 Overly Manufacturing Co.
574 W. Otterman St, PO Box 70, Greensburg, PA 15601
☎(+1 412) 834 7300 ⊖ (+1 412) 830 2871
♣ $ 10 000 000 👥 130 🏭 70 000 sq ft
Sales Mgr: John T. Brooks
■ A custom manufacturer of security doors, including acoustical doors, attack & bullet resistant units, security vault doors, RFI, EMI & acoustical shielding and EMP/blast resistant doors. 97

75-67870 Oxley Inc.
(assoc.of Oxley Developments Co.Ltd)
25 Business Park Drive, PO Box 814, Branford, CT 06405
☎(+1 203) 488 1033 ⊖ (+1 203) 481 6971
Operations Mgr: Mrs C. Carr
Development Mgr: M. Sloan
■ Design & manufacture of passive electronic components for use in military systems.
® UK: Oxley Developments Co.Ltd 97

75-68000 Pacer Systems Inc.
Corporate Headquarters
900 Technology Park Drive, Billerica, MA 01821
☎(+1 508) 667 8800 ⊠ 948198; TWX 710 3326 400
⊖ (+1 508) 667 8873
♤ $ 4 873 544 (1995) ♣ $ 27 648 179 (1995) 👥 260
Chairman: John C.(Jack) Rennie
President: Sigmund Goldblum
Sr Vice President Admin Finance: Rudolph Koczera
VP Business Development: Paul Serotkin
Vice President Human Resources: Roberto Steinberg
■ Information technology & shipboard navigation systems. Maritime patrol.
Engineering Services. Anti-submarine warfare. Wargaming. Expeditionary warfare. Torpedo counter-measures. Air data systems for vehicles. Global positioning systems navigational and test systems. → 80

Horsham, PA
4 Horsham Business Center, 300 Welsh Rd,
Horsham, PA 19044
☎(+1 215) 657 7800
Vice President: Norman Wolfe
Marketing Mgr: Fred Jaffin

Lexington Park, MD
401 Essex Drive, Lexington Park, MD 20653
☎(+1 301) 862 3509 ⊖ (+1 301) 862 5232
Vice President: Michael Woodbury

Middletown, RI
26 Valley Rd, Unit 101, Middletown, RI 02840
☎(+1 401) 849 0059 ⊖ (+1 401) 849 9868

Panama City, FL
7104 Laird Street, Panama City Beach, FL 32407
☎(+1 904) 234 3940 ⊖ (+1 904) 234 1168
Vice President: Stephen Crooks 97

75-68350 Pacific Scientific Company Corporate Office
620 Newport Center Drive, Suite 700,
Newport Beach, CA 92660
☎(+1 714) 720 17 14 🅆 http://www.pacsci.com
⊖ (+1 714) 720 10 83
♤ $ 82,926,000 (1996) ♣ $ 294,779,000 (1996)
👥 2,375 (1996) 🏭 14,000 sq ft (corporate headquarters)
Chairman, President and Chief Executive Officer: Lester Hill
Executive Vice President of Finance and Administration: Richard V Plat
Senior Vice President: William T Fejes
Vice President: Steven L Breitzka
Vice President: Richard D Knoblock
Vice President: Joseph R Monkowski
Vice President: Ronald B Nelson
Controller: William H Amadon
Treasurer: Peter A Swan
Assistant Corporate Secretary: Ms Ailen Younanpour
■ Manufacture electrical equipment and safety equipment.

HTL/Kin-Tech Division
1800 Highland Avenue, Duarte, CA 91010
☎(+1 818) 359 93 17 ⊖ (+1 818) 358 56 09
President: Richard G Knoblock

Energy Dynamics Divsion
7073 West Willis Drive, Chandler, AZ 85226
☎(+1 602) 796 11 00 ⊖ (+1 602) 796 13 22
President: Thomas L Walsh

Fisher Pierce Division
90 Libbey Parkway, Weymouth, MA 02189
☎(+1 607) 340 07 00 ⊖ (+1 607) 340 02 17
President: Steven L Breitzka

Hiac/Royco Division
11801 Tech Road, Silver Spring, MD 20904
☎(+1 301) 680 70 00 ⊖ (+1 301) 622 07 14
President: Joseph R Monkowski

Motor & Control Division
4301 Kishwaukee Street, Rockford, IL 61109
☎(+1 815) 226 31 00 ⊖ (+1 815) 226 31 22
President: Ronald B Nelson

Electro Kinetics Division
402 East Gutierrez Street, Santa Barbara, CA 93102
☎(+1 805) 963 20 55 ⊖ (+1 805) 962 67 35
President: Richard G Knoblock 98

75-68620 Palmer Manufacturing Co Inc.
243 Medford Street, PO Box K, Malden, MA 02148-7301
☎(+1 617) 321 04 80 ⊖ (+1 617) 321 22 26
♣ $ 32,000,000 (1996) 👥 190 🏭 65,000 m2
President and Chief Executive Officer: Frank Moda (Jr)
Senior Vice President Sales and Marketing: Raymond Gosselin
Quality Director: Paul Morse
Operations Director: John Ladge
Engineering Manager: Steve Warshafky
Materials Manager: Mike Munday
■ Aircraft engine components, rings, spacers, seals, fuel control components, machining services, shroud segments, liners, disks. 98

75-68800 Para-Flite Inc.
(a member of Wardle Storeys plc, UK)
5800-J Magnolia Avenue, Pennsauken, NJ 08109
☎(+1 609) 663 12 75
🄴 103734,3341@compuserve.com
⊖ (+1 609) 663 30 28
👥 125 (1997) 🏭 24,000 ft2
President: Elek Puskas
Vice President Marketing: Joseph A Andrzejewski
Operations Director: William F Bruno
■ Tactical gliding parachute systems, automatic aerial cargo delivery systems. 98

75-68900 Parker Bertea Aerospace
(a subsidiary of Parker Hannifin Corp.)
18321 Jamboree Road, Irvine, CA 92612-1073
☎(+1 714) 833 30 00 ⊠ 678383 ⊖ (+1 714) 851 32 77
👥 4,000
President: Steve Hayes
Group Vice President: James Lowes
Group Vice President: Frank Nichols
■ Design, manufacture & service fluid systems & components & related electronic controls for aerospace & other high technology markets.

Air and Fuel Division
18321 Jamboree Road, Irvine, CA 92612-1073
☎(+1 714) 833 30 00 ⊖ (+1 714) 851 33 41
VP & General Mgr: Jim Sabin
■ Fuel management subsystems including pressurisation & vent systems, temperature control systems, in-flight refuelling receptacles & nozzles, modulating motor-operated valves, level-control High temperature bleed air valves, air turbine starters, aircraft fire suppression systems, vapour-cycle environmental control systems, space & missile propulsion control valves. valves, selector valves, ejector pumps, refuelling adaptors & check valves.

Airborne Division
711 Taylor Street, PO Box 4032, Elyria, OH 44036-6229
☎(+1 216) 284 63 00 ⊖ (+1 216) 322 56 25
👥 175
Business Unit Mgr: John R. Hruska
Marketing Mgr: Paul LeFevre
■ DC fuel pump products, AC fuel pumps, air pumps, bleed-air temperature control systems & components, regulator valves, filters, gauges & fittings.

Aircraft Wheel and Brake Division
1160 Center Road, Avon, OH 44011-1297
☎(+1 216) 937 62 11 ✆ (800) 272 54 64
⊖ (+1 216) 937 54 09
👥 125
Business Unit Mgr: John Hruska
Marketing Mgr: Paul LeFevre
■ Aircraft wheels, brakes, master cylinders, parking valves & hydraulic systems & components.

Control Systems Military Division
14300 Alton Parkway, Irvine, CA 92618-1898
☎(+1 714) 833 30 00 ⊖ (+1 714) 586 84 56
VP & General Mgr: Bob Barker
VP Marketing: Jim Ryder
■ Hydraulic primary flight actuation & engine control systems, integrated fly-by-wire electrohydraulic servomodules, electrically controlled actuation systems, electromechanical actuators, flight control components, ballistic tolerant hydraulic systems & components.
Hydraulic bootstrap reservoirs, pressure operated & servo-controlled actuators, accumulators, selector valves, hydraulic control modules, locking actuation devices, integrated nose-wheel steering systems.

Control Systems Commercial Division
1425 West 2675 North, Ogden, UT 84404-2696
☎(+1 801) 786 30 00 ⊖ (+1 801) 786 30 43
Plant Mgr: Clyde Noorda

Customer Support Commercial Division
16666 Von Karman Avenue, Irvine, CA 92606-4997
☎(+1 714) 660 83 10 ⊖ (+1 714) 660 83 90
Business Ops, Commercial Airframes & Spares Dir: Dick Miller
Business Ops, Commuter, Business Jet Dir: Bob Richardson
■ More than 50 000 parts in 24-hour, 7-day-a-week, on-line inventory, plus overhaul & repair services in support of the world's airlines. FAA Repair Station #AU4R063M, Class I & Class II.
Maintenance, overhaul & repair of hydraulic, pneumatic & fuel components & systems for military fixed & rotary wing aircraft.

Air Fuel Division
8505 West McDowell Road, Tolleson, AZ 85353
☎(+1 602) 231 06 00 ⊖ (+1 602) 936 43 87
Plant Mgr: Jack Wetzel

Gas Turbine Fuel Systems Division
17325 Euclid Avenue, Cleveland, OH 44112-1290
☎ (+1 216) 531 30 00 ✉ 980636 ⊖ (+1 216) 383 92 47
Group VP & General Mgr: Frank Nichols
■ Fuel nozzles, atomisers, spraybars, sprayrings, valves, metering devices, manifolding for gas turbine main combustion & augmentation.
● 101 Parker Drive, Andover, OH 44003
☎ (+1 216) 293 57 56
⊖ (+1 216) 293 56 51; 293 64 06
Plant Mgr: John Mattern
● 124 Columbia Street, Clyde, NY 14433
☎ (+1 315) 923 23 11 ⊖ (+1 315) 923 93 06
Plant Mgr: Michael Roth
US Overhaul and Exchange Facility
777 North Glen Harbor Boulevard, Glendale, AZ 85307
☎ (+1 602) 872 18 00 ⊖ (+1 602) 872 85 95
Plant Mgr: Neil McAllister
Gull Electronic Systems Division
300 Marcus Boulevard, PO Box 9400,
Smithtown, NY 11788-2030
☎ (+1 516) 231 37 37 ⊖ (+1 516) 434 81 52
Group VP & General Mgr: John Stadelmann
VP Marketing: Lawrence Berke
■ Fuel & oil management systems, fuel & oil quantity gauging systems, cockpit instruments, flow measurement systems, engine monitoring systems, fuel tank inerting systems, special & general purpose computer systems for aerospace applications.
● 55 Engineers Road, Smithtown, NY 11787
☎ (+1 516) 231 37 37 ⊖ (+1 516) 231 36 99
Nichols Aircraft Division
48 Woerd Avenue, Waltham, MA 02254-9115
☎ (+1 617) 894 06 50 ⊖ (+1 617) 894 92 35
👥 175
General Mgr: John Meston
Business Unit Mgr Airframes & International: Jack Lyons
Business Unit Mgr Engines: Bill Walker
Customer Support Mgr: Jeff O'Reilly
■ Lub & scavenge pumps for aircraft gas turbine engines, APU's, helicopter transmissions & accessories. FAA overhaul & repair service for lubrication pumps.
United Aircraft Products Division
200 Railroad Street, PO Box 37, Forest, OH 45843-0037
☎ (+1 419) 273 2511 ⊖ (+1 419) 273 3285
👥 175
General Mgr: Greg Durall
Marketing & Contracts Mgr: Chris Schaefer
■ Heat transfer & inerting systems & components including heat exchangers, cooling systems, oil tanks, valves, regulators, pressure vessels, heat pipes & on-board inert gas generating systems. [98]

75-68910 Parker Hannifin Corporation
6035 Parkland Boulevard, Cleveland, OH 44124-4141
☎ (+1 216) 531 30 00 ✆ (800) 272 75 37
W http://www.parker.com ⊖ (+1 216) 486 06 18
♣ $ 3,600,000,000 (1996) 👥 33,000 (1996)
Chairman: Patrick S. Parker
President & CEO: Duane E. Collins
Corporate Communications Dir: Richard Charlton
Aviation Mgr: Allen Maurer
Marketing Research Analyst: Ms Kathleen Sloe
■ Manufacture motion control components & systems for industrial, automotive, aviation, space & marine markets.
Operations with defence business:
Parker Bertea Aerospace → 75
Stratoflex Aerospace/Mil. Con Div.
220 Roberts Cut-Off, PO Box 10398,
Fort Worth, TX 76114-0398
☎ (+1 817) 738 65 43 ⊖ (+1 817) 738 99 20
General Mgr: Steve Hamilton
Marketing Mgr: E.H. Larkin
Sales: Chuck Marshall
■ Manufacture hose, hose assemblies, hose fittings & adapters.
Parker Seal/O-Seal Division
7664 Panasonic Way, San Diego, CA 92173-4206
☎ (+1 619) 661 70 00 ⊖ (+1 619) 671 32 02
👥 210
General Mgr: Steve Barnes
■ Manufacture sealing devices.
Stratoflex Divsion, Symetrics Business Unit († Symetrics Inc.)
3353 Old Conejo Road, Newbury Park, CA 91320-2162
☎ (+1 805) 498 45 86 ⊖ (+1 805) 449 10 42
👥 120 (1995)
Sales Mgr: Stephen Calvin
Operations Mgr: William Cartmill
Sr Buyer Mgr: Ms Yolanda Donato
Staff Assistant: Jean Korngiebel
Sales Contact: Ms Barbara Lehto
Sales Contact: Stephen Calvin
■ Manufacture quick disconnect fluid couplings. [98]

75-68980 Partsource Inc.
(a div. of the Chestnut Group)
515 Bonham Ave, Columbus, OH 43211-2619
☎ (+1 614) 299 5427 ⊖ (+1 614) 299 2576
♣ $ 12 000 000 👥 250
President: John Runnette
General Mgr: Bob Kent
■ Spring, stampings, wire forms. [97]

75-68995 Pathfinder Systems Inc. (PSI)
274 Union Blvd, Suite 430, Lakewood, CO 80228
☎ (+1 303) 763 8660 ⊖ (+1 303) 763 8902
👥 6
President: Sheila L. Jaszlics
■ Computer software & systems development of battle simulations. Artificial neural nets, requirements & design analysis & rapid prototyping. Advanced concepts in long-endurance UAVs & virtual reality. [97]

75-69125 Pelican Products Inc.
23215 Early Avenue, Torrance, CA 90505
☎ (+1 310) 326 4700 ✆ (800) 473 5422
⊖ (+1 310) 326 3311
👥 300
Exec. Vice President: Jerry Linehan
Operations & Technical Dir: John Hoven
Sales & Marketing Dir: Walter Kaihatu
Purchasing Mgr: Diane Rome
Sr Sales Co-ordinator: Ms Sharon Ward
■ Manufacture high intensity, submersible flashlights, watertight and airtight cases. [97]

75-69135 Pemco World Air Services
(a subsidiary of Precision Standard Inc.)
1225 17th Street, Suite 1800, Denver, CO 80202
☎ (+1 303) 292 67 67 W http://www.pemcoair.com
⊖ (+1 303) 292 68 88
President and Chief Executive Officer: Matthew L Gold
Vice President of Business Development: David Mitchell
Vice President of Commercial Sales: Frederik Groth
Sales Director: Martyn Craig
Public Relations Manager: Ms Suzanne Smith
■ Aircraft maintenance, repair and overhaul.
Hayes Targets
PO Box 707, Leeds, AL 35094-0707
☎ (+1 303) 699 61 51 ✉ HAYES TGTS ✉ 782550
President: S W Teague (Jr)
Marketing Manager: David Miller
■ Air targets [98]

75-69200 Penstock Inc.
(a division of Avnet Corp.)
520 Mercury Drive, Sunnyvale, CA 94086-4018
☎ (+1 408) 730 03 00 ✆ (800) PENSTOCK
W http://www.penstock.avnet.com
⊖ (+1 408) 730 47 82
♣ $ 120,000,000 (1997) 👥 212 (1997) 🏭 34,000 m2
President: Bruce White
Public Relations Director: Steven Ulett
■ RF and microwave semiconductors, connectors, cable assemblies, components. Active and passive devices. Frequency ranges from DC-50GHZ. → 85 [98]

75-69260 Pentastar Electronics Inc.
(a div. of Chrysler Technologies Corp.)
110 Wynn Drive, Huntsville, AL 35805
☎ (+1 205) 895 2076; 895 2076
⊖ (+1 205) 895 2356; 895 2356
♣ $ 50 000 000 👥 400
President: Jewell Toney
Dir Public Relations: Bill Young
Business Development Mgr: Bill Young
■ Manufacture of electronic equipment for both military & industrial markets. Automatic Test Systems (ATS), Automatic Test Equipment (ATE) & related Test Program Sets (TPS). Support for military vehicles with emphasis on heavy, medium & light armour. Design, development & manufacture of complete vetronics systems. Electro-luminescent displays, power supplies & other VETRONIC products. Transportation & power control units & systems for industry. Hybrid & electric drive vehicles. [97]

75-69270 Perceptronics Inc.
21010 Erwin Street, Woodland Hills, CA 91367
☎ (+1 818) 884 34 85 ⊖ (+1 818) 340 60 67
♣ $ 6,000,000 (1997) 👥 40 (1997) 🏭 50,000 sq ft
Chief Executive Officer: Dr Gershon Weltman
Chief Operating Officer: Tom Lubaczewski
Chief Financial Officer: Robert Anderson
President: Dr Richard Vestewig
Vice President of Marketing: Chris Harz [98]

75-69280 Peregrine Industries Inc.
PO Box 3429, Ketchmun, ID 83340
☎ (+1 208) 726 9753 ⊖ (+1 208) 726 0957
🏭 29 000 sq ft
President: Richard A. Voit
■ Manufacture small arms & infantry weapons including pistols, rifles & support weapons. [97]

75-69290 Perimeter Products Inc.
(a subsidiary of Perimeter Products Inc.)
1130 Terra Bella Avenue, Mountain View, CA 94043
☎ (+1 415) 966 85 50 E mkt@perimeterproducts.com
W http://www.perimeterproducts.com
⊖ (+1 415) 966 85 17
👥 20 🏭 13,000 sq ft
President and Senior Engineer: Mike E Trimble
Vice President of Finance: Julie Ly
Marketing Director: Ms Martha A Lee
Operations Director: Mike Swanson
Engineering Manager: Robert C William
Operations Manager: William Boyd
Sales Manager, Western Region: Ms Jeanne McGuire
Sales Manager, Eastern Region: Lowell Thompson
Marketing Assistant: Kris Ryan
■ Design, develop and manufacture FPS-2-2 fence protection system - fence mounted sensor. Comgard MX-1000, Helisensor, telegate, MPS-4100 microwave protection system, PAS-120 personal alarm system and PA-RTS 01 personal alarm system. [98]

75-69360 Perry Technologies
(an operating unit of Lockheed Martin Corporation)
100 East 17th Street, Riviera Beach, FL 33404
☎ (+1 407) 842 52 61 W http://www.perrytech.com
⊖ (+1 407) 842 53 03
♣ $ 12,000,000 (1997) 👥 70 (1997) 🏭 50,000 sq ft
Managing Director: William V Hayes
Operations Manager: John Pfeifler
Finance Manager: Michael Parrish
Engineering Manager: Dr Paul Rushfeldt
Purchasing Manager: Ms Grace Sawyer
Marketing Services: Mike Mulleavey
■ Submersibles, diving systems, remotely operated/autonomous underwater vehicles, undersea habitats, engineering services, special fabrication, programme management, support, test and evaluation facilities. [98]

75-69380 Persoft Inc.
465 Science Drive, Madison, WI 53711-1058
☎ (+1 608) 273 60 00 ✆ (800) EMULATE
E sales@persoft.com W http://www.persoft.com
⊖ (+1 608) 273 82 27
👥 125 (1997)
President: Thomas Wolfe
Sales and Marketing Director: Gary Larkins
Vice President Sales and Marketing: Ms Susan Hughes
Director Advertising: Ms Terri Hayes
■ Designs, manufactures & distributes terminal emulation & connectivity software which enables PC LAN connectivity to multiple host systems (VAX, UNIX or Data General mainframes). [98]

75-69500 Phaostron Instrument & Electronic Co
717 North Coney Ave, Azusa, CA 91702-2205
☎ (+1 818) 969 6801 ⊖ (+1 818) 334 8057
👥 200 🏭 67 000 sq ft
President: Paul McGuirk
Vice President: Don Yates
■ Engine & control instruments, sensors & precision mechanisms for prime flight instruments. Panel meters & ruggedised meters to MIL-M-10304. [97]

75-69560 Phase Two Industries
1940 Zanker Road, San José, CA 95112-4216
☎ (+1 408) 441 94 90 E mktg@pti2000.com
⊖ (+1 408) 441 95 26

Contact Reference

♣ $ 4,250,000 ⚉ 35 ⌂ 2,000 sq m
President: Burton W Graves
Executive Vice President: Ms Carol A Graves
Purchasing Manager: Kobi Goessel
■ Systems and software, operational software, memory loader verifiers, EW simulators, test instrumentation, signal processors. 98

75-69800 Photo-Sonics Inc.
820 South Mariposa Street, Burbank, CA 91506
☎ (+1 818) 842 21 41 E info@photosonics.com
⊖ (+1 818) 842 26 10
♧ US$ 9,000,000 (1997) ♣ US$ 15,000,000 (1997)
⚉ 80 (1997) ⌂ 40,000 sq m
President: John Kiel
Vice President of Operations: Philip Kiel
Vice President: Roy Edwards
Sales Video Products: John Nelson
■ High speed film cameras, photo/image interpretation systems, video recorders and equipment, tracking mounts.
Subsidiary:
ARAN Electronics, Israel 98

75-69810 PhotoTelesis
(a subs. of Texas Intruments)
700 Lincoln Center, 7800 IH 10 West,
San Antonio, TX 78230
☎ (+1 210) 349 2020 ✆ (800) 822 0233
E rtindal@photot.com ⊖ (+1 210) 341 1020
♣ $ 15 000 000 (1995) ⚉ 50 ⌂ 2 500 sq m
Managing Dir: David Monroe
Technical Dir: Anthony Garret
Sales/Marketing Dir: Anthony Garrett
International Marketing Dir: Richard Tindal
Purchasing Mgr: Tres Ellis
■ Manufacture imagery transmission systems that operate over dial-up telephone, radio, satcom, & COMSEC devices. 97

75-70000 Piezo Technology Inc.
2525 Shader Rd, PO Box 547859,
Orlando, FL 32854-7859
☎ (+1 407) 298 2000 ⊖ (+1 407) 293 2979
⚉ 170
President: Dr W.H. Horton
Dir Sales & Marketing: Paul A. Dechen
■ Crystal filters, oscillators and custom hybrid circuits. 97

75-70100 Pilkington Optronics Inc.
(a subsidiary of Pilkington Optronics Ltd, UK)
7550 Chapman Avenue, Garden Grove, CA 92641-2108
☎ (+1 714) 373 6061 ⊖ (+1 714) 373 6074
Chief Exec. Officer: T.G. Jones
■ Design, development & manufacture of electro-optical systems, sub-system devices & components for space & defence requirements.
® UK: Pilkington Optronics Ltd, Pilkington P.E. Ltd, Pilkington Thorn Optronics
603 King Street, Suite 300
Alexandria, VA 22314
☎ (+1 703) 836 7474 ⊖ (+1 703) 836 7475
Manager: Tom Schilling 98

75-70200 Pioneer Aerospace Corp
(† Pioneer Parachute Co Inc)
(a subsidiary of Zodiac SA Holding, France)
45 South Satellite Road, PO Box 207,
South Windsor, CT 06074-0207
☎ (+1 203) 528 00 92 ▣ TWX 710 4272 994
⊖ (+1 203) 528 81 22
▥ 62892115
President: Gilles Dobray
■ Engineering and manufacture of parachutes and related items; analysis, design, development, testing manufacturing, integration and support of advanced recovery and retrieval systems for aerospace vehicles and components. 98

75-70250 The New Piper Aircraft Inc.
2926 Piper Drive, Vero Beach, FL 32960
☎ (+1 561) 567 43 61 W http://www.newpiper.com
⊖ (+1 561) 778 21 44
♣ $ 100,000,000 (1997) ⚉ 500 (1997)
⌂ 360,000 sq m
President and Chief Operating Officer: Charles Suma
Vice President of Product Services: Werner Hartieb
Vice President of Manufacturing: Buddy Watson
Director of Customer Services:
Ms Susie Luther Stringfellow
Director of Engineering: Bill Moreu
Director of Quality Assurance: Rick Murdoch
Director of Purchasing: John Mattingly
Director of Marketing and Sales: Larry Bardon
Manager of International Sales: Lee Walther
Manager of Domestic Sales: Rick Zimper
Manager of Orders and Delivery: Steve Thompson
Manager of Marketing Services: Kimberly Wheeler
■ Manufacture of a complete line of piston aircraft: Seminole, Archer III Warrior III, Saratoga II HP, Saratoga II TC, Malibu Mirage, Arrow and Seneca V. 98

75-70300 PK Electronic International
405 Park Avenue, New York, NY 10022
☎ (+1 212) 759 8855 ⊖ (+1 212) 421 5024
⚉ 10
President: Damian Krause
Managing Dir: Ruben Carrica
■ Design, development & manufacture of professional electronic surveillance & counter-surveillance equipment for use by government & law enforcement agencies. 98

75-70320 Planar America Inc.
(† Planar Systems Inc.)
(a subsidiary of Planar Systems Inc.)
1400 NW Compton Drive, Beaverton, OR 97006-1992
☎ (+1 503) 690 11 00 E sales@plannar.com
⊖ (+1 503) 690 14 93
♣ $ 80,000,000 (1995) ⚉ 400
President: Jim Hurd
Technical Director: Curt Stevens
Marketing Director: Richard Phillips
Marketing Manager: Erick C Petersen
E erick_petersen@plannar.com
Federal Systems Mgr: Thomas V. Curran
■ Provides strategic direction, marketing focus, technical co-ordination and financial support to the operating divisions.
● 1400 NW Compton Drive, Beaverton, OR 97006-1992
☎ (+1 503) 690 69 67 ⊖ (+1 503) 690 14 93
♣ $ 80,000,000 (1995) ⚉ 300
President: Jim Hurd
Marketing Mgr: Erick Petersen
Federal Systems Mgr: Thomas V. Curran
■ Electroluminescent flat panel displays, monitors, terminals & accessories.
● 13950 S.W. Karl Braun Drive, PO Box 4001,
Beaverton, OR 97076
☎ (+1 503) 614 41 11 ⊖ (+1 503) 614 41 01
Managing Dir: E.L. Herman
Sales & Marketing Dir: R. Groppi
Technical Dir: T. Long
■ Manufacture military CRTs, LCDs and advanced display products. 98

75-70325 Plane Tooling Inc.
(a subs. of Hydro-Gerätebau GmbH & Co. KG)
1297-120th Ave NE, Bellevue, WA 98005
☎ (+1 206) 641 47 04 ⊖ (+1 206) 641 48 46 97

75-70335 Plastic Fabricating Co. Inc.
1650 South McComas, Wichita, KS 67213
☎ (+1 316) 942 12 41 ▣ TWX 510 100 5836 plastic fab
⊖ (+1 316) 942 06 87
♣ US$ 20,000,000 (1997) ⚉ 200 (1997)
⌂ 165,000 sq ft
President: W C Waskey
Vice President of Sales and Marketing: F E Tipton
Operations Director: D A Steiner
Purchasing Manager: B Miller
Contracts, Administration and Customer Service: M Norris
■ Custom fabrication of detail parts and sub-assemblies for the aircraft and aerospace industries, including composite fabrication, low pressure laminating, thermoplastic sheet forming and compression /transfer moulding. 98

75-70360 Plenty Inc
6630 Roxburgh Drive, Suite 175, Houston, TX 77041
☎ (+1 713) 937 38 63 ▣ 795231 oilplus hou
⊖ (+1 713) 937 09 68 98

75-70400 Plexsys Interface Products, Inc.
PO Box 301459, Portland, OR 97294-9459
☎ (+1 503) 251 04 55 E techsupt@plexsysipi.com
W http://www.plexsysipi.com ⊖ (+1 503) 251 04 49
♣ $ 1,400,000 (1997) ⚉ 13 (1997)
President: John J LeDoux E ledouxjj@plexsysipi.com
Vice President Research and Development:
Walter J LeDoux
Vice President Operations: Jonathan G Stacey
E staceyj@plexsysipi.com
■ Manufacture of Ascot IV, radar simulator, video mapper, display console, other ancillary C2 equipment for search radar and ATC installations. Offers an entry level and advanced Air Weapons Controller school. 98

75-70810 Pointer Inc.
1027 North Stadem Drive, Tempe, AZ 85281
☎ (+1 602) 966 16 74 ⊖ (+1 602) 968 80 20
⚉ 11 ⌂ 3,424 sq ft
President: C A Koster
Secretary: Ms Estella Koster
■ Altimeter encoder (blind); emergency locator transmitter, specialty ELT antennas.
Subsidiary:
Pointer Avionics Canada 98

75-70855 Polhemus Inc. Colchester Operations
(a Kaiser Aerospace & Electronics company)
1 Hercules Drive, PO Box 560, Colchester, VT 05446
☎ (+1 802) 655 31 59
✆ (800) 357 47 77 (USA and Canada)
⊖ (+1 802) 655 14 39
⚉ 44 (1997) ⌂ 17,000 m2
President: Philip G Cooper
Vice President of Engineering: Herschell Murry
Vice President of Sales and Marketing: Bill Panepinto
Sales Director: Tom Knoflick
Purchasing Manager: Ms Marie Bartlett
■ Research, development and production of head tracker systems and helmet mounted sights for target acquisition and helmet mounted displays. Magnetrak Plus - helmet mounted sight. Fastrak, a high resolution, low data latency tracker for simulation and virtual reality applications. Advanced head tracker for US Army Comanche helicopter. Digitrak, a high performance, DSP based head tracking systems for Helmet Mounted Displays (HMD). 98

75-70858 Polycast Technology Corp.
(a division of Uniroyal Technology Corp.)
69 Southfield Avenue, Stamford, CT 06902-7614
✆ (800) 243 90 02 E rfrvooa@prodigy.com
⊖ (+1 203) 323 29 25
♣ $ 45,000,000 ⚉ 300
President: Mike Hess
Operations Director: Mike Ilesco
Director of Marketing: Steven DuPont
■ Aircraft transparencies. 98

75-70870 Polytronic Inc
(a subsidiary of Polytronic AG, Switzerland)
106 Peacock Circle, Safety Harbour, FL 34695
☎ (+1 813) 796 32 65 ⊖ (+1 813) 796 32 65
President: Henry Schweizer
Marketing Manager: Ingrid Liddanc
■ Manufacture and marketing of shooting training equipment. → 90 98

75-71070 Power Conversion Inc.
(a subs. of BTR plc, UK)
495 Boulevard, Elmwood Park, NJ 07407
☎ (+1 201) 796 4800 ✆ (800) 452 1211 ▣ 130292
⊖ (+1 201) 796 6243
⚉ 1 200
President: William D. Vlasich
Vice President: James R. Sullivan
Sales Admin: Lori Geimke
■ Manufactures a range of lithium primary batteries for industrial & defence applications. Major supplier of Lithium-SDX batteries to US armed forces. 97

75-71080 Powerex Inc.
200 Hillis Street, Youngwood, PA 15697-1800
☎ (+1 412) 925 72 72 ✆ (800) 451 14 15
⊖ (+1 412) 925 43 93
♣ $ 75,000,000 (1996) ⚉ 325
President and Vice President Sales: S R Hunt
Operations Director: Larry Eckerman
Sales and Marketing Director: Ron Williams
Communications Manager: Ms Claudia Tapper
Purchasing Manager: Ed Hunt

■ Military JAN & JANTX [98]

75-71085 Power Spectra Inc.
919 Hermosa Court, Sunnyvale, CA 94086
☎(+1 408) 737 79 77 ⊖ (+1 408) 732 18 32
♧ US$ 4,000,000 (1996) ♣ US$ 3,000,000 (1996)
👥 29 (1996) 🏭 2,500 m2
President: Gordon Smith
Executive Vice President and Sales Director: Charles Byer
Director: Michael I Gamble
Operations and Technical Director: Jon Grant
■ Bulk avalanch semiconductor switch (Bass). Pulsed optical sources and impulse generators. [98]

75-71087 PowerTech Inc.
0-02 Fair Lawn Avenue, Fair Lawn, NJ 07410
☎(+1 201) 791 50 50 [W] http://www.power-tech.com
⊖ (+1 201) 791 68 05
♣ $ 5,000,000 (1997) 👥 25 (1997)
President: Alex Polner
Marketing Manager: A Martin [E] alexmart@ix.netcom.com
■ High current transistors & darlingtons. High reliability & military types. [98]

75-71090 Power Technology Inc.
7925 Mabelvale Cutoff, Mabelvale, AR 72103
☎(+1 501) 568 19 95
[W] http://www.powertechnology.com
⊖ (+1 501) 568 19 94
President: Thomas H Burgess
Advertising Manager: Walter Burgess
Sales: Glenn Sullivan
■ Manufacture laser simulators for testing laser rangefinders, laser power supplies, Nd: YAG lasers, laser diode pulsers, fibre-optic fusion splicers. Custom design opto-electronic test and measurement equipment in the fields of ranging, countermeasures, threat detectors, threat simulators and threat assessors. [98]

75-71095 Power Ten ICN
486 Mercury Drive, Sunnyvale, CA 94086-4706
☎(+1 408) 738 59 59 [E] order@powerten.com
⊖ (+1 408) 738 43 26
👥 50
President: Joe Varozza
Vice President of Sales, Marketing and Public Relations: Vince Mutasco
■ AC-DC power supplies. [98]

75-71150 Pratt & Whitney
(a div. of United Technologies Corp.)
400 Main St, East Hartford, CT 06108
☎(+1 203) 565 4321 ✉ AIRCRAFT
▣ TWX 710 425 2991
⊖ (+1 203) 565 8345/77; 565 1174
👥 32 000
President: Karl J. Krapek
President, Large Commercial Engines: Robert A. Wolfe
Exec. VP Operations: Mark S. Coran
Exec. VP Technical: Jon E. Wohler
Exec. VP Engineering & Tech.: Bernard L. Koff
Sr VP & General Counsel: Richard M. Whiston
Sr VP Finance: Jean Mordo
Sr VP International: Robert E. Rosati
VP Advanced Engineering: Dan Webb
VP Human Resources & Organization: John P. Leary
VP Group Product Integrity & Quality Assurance: Roger Chericoni
VP Aftermarket Operations: William A. Kerr
VP Communications: Martin W. Kraegel (Jr)
VP Information Technology: John M. Watkins (Jr)
VP Public Affairs: Lawrence R. Churchill
VP Customer Support: Gary Minor
■ Design, development, support & marketing of propulsion systems installed a wide variety of fighter, transport & surveillance aircraft. Rocket engines & propulsion systems for various space programs.

Subsidiary company:
Pratt & Whitney Canada Inc. → 75

Washington, DC Office
1401 Eye St NW, Washington, DC 20005
☎(+1 202) 336 7400 ⊖ (+1 202) 785 7428; 785 7451
Sr. VP Washington Office: William F. Paul
Public Relations Dir: James P. Linse

Government Engines & Space Propulsion
PO Box 109600, West Palm Beach, FL 33410-9600
☎(+1 407) 796 2000 ▣ 803478 pwa fia wpb
⊖ (+1 407) 796 6514
President: John Balaguer
Sr VP Programs & Business Development: R. Edward Ford (Jr)
VP F119 Program: Walter N. Bylciw
VP Communications: Robert G.H. Carroll (III)
VP Human Resources: William V. Panetta
VP Customer Support & Services: Raymond J. Van Overschelde

Large Commercial Engines (LCE)
400 Main Street, East Hartford, CT 06108
☎(+1 203) 565 4321 ✉ AIRCRAFT
▣ TWX 710 4252 991 ⊖ (+1 203) 565 8345/77
President: Robert Wolfe
Sr VP Propulsion Systems: Robert Leduc
VP Customer Support: Gary Minor
LCE Communications Dir: Mark P. Sullivan
General Mgr Turbo Power & Marine Systems: Randall J. Hogan

Operations
400 Main St, East Hartford, CT 06108
☎(+1 203) 565 4321 ✉ AIRCRAFT
▣ TWX 710 4252 991 ⊖ (+1 203) 565 8345; 565 8377
VP Personnel & Ind. Relations: James McMahon

Purchasing Dept
400 Main St, East Hartford, CT 06108
☎(+1 203) 565 4321 ▣ TWX 710 4251 046
⊖ (+1 203) 565 1286; 565 7807; 565 7817 [97]

75-71200 Precise Flight Inc.
63120 Powell Butte Road, PO Box 7168, Bend, OR 97701
☎(+1 503) 382 86 84 [E] preciseflight@preciseflight.com
[W] http://www.preciseflight.com ⊖ (+1 503) 388 11 05
♧ $ 1,200,000 ♣ $ 1,300,000 (1996) 👥 11 (1996)
🏭 4,000 sq ft
President: Mrs Connie LeHuquet
Vice President: Mike Demith
■ Speedbrakes, standby vacuum systems, instrument source warning indicators, pulselite visual contact light systems. RMD wing tip lights. [98]

75-71330 Precision Filters Inc.
240 Cherry Street, Ithaca, NY 14850-5023
☎(+1 607) 277 35 50 [E] pfinfo@pfinc.com
⊖ (+1 607) 277 44 66
♣ $ 4,000,000 (1996) 👥 47 (1997) 🏭 1,440 m2
CEO/President: Donald F. Chandler
■ Computer-controlled instruments: amplifiers, filters, signal conditioners, frequency-band translators, data acquisition & recording systems. [98]

75-71350 Precision Machining Inc.
3820 Hopkins St., Pensacola, FL 32505-5223
☎(+1 904) 434 5331 ⊖ (+1 904) 438 4947
♣ $ 12 500 000 👥 203
President: Stanley L. Pendergrass
VP Sales: Donald Pendergrass
■ Military containers & equipment. Bridging systems & related launching equipment & boats. Mine clearance equipment & systems. [97]

75-71360 Precision Sensors
(a division of United Electric Controls ...p.)
50 Seeman's Lane, PO Box 509, Milford, ...460
☎(+1 203) 877 27 95 ⊖ (+1 203) 877 ...
♣ $ 6,000,000 👥 45 (997) 🏭 10,000 s...
Vice President and General Manager: T S...
Purchasing Manager: J Palmer
Supervisor Sales and Marketing: S Rolli...
Government Sales: Mrs Rani Martin
■ Pressure & flow switches for aircra... es, ground defence, armoured vehicles & FLIR/r... ems. [98]

75-71365 Precision Standard I... (PCSN)
1225 17th Street, Suite 1800, Denver, ...)202
☎(+1 303) 292 65 65 ✉ http://www.pe... air.com
▣ 810 733 3687 ⊖ (+1 303) 292 66 11
Chairman and Chief Executive Officer: Matthew L Gold
Vice President and Treasurer: Tim Rabon
Group Controller: Ms Shere McBryde
Public Relations Manager: Ms Suzanne Smith
■ Diversified company engaged in the maintenance and modification of aircraft for military and commercial customers. Provision of aircraft support services. Design and manufacture of aircraft cargo handling systems. Manufacture aerial target systems and develop and manufacture rocket launch vehicles and guidance systems. → 80

Subsidiary with defence business:
Pemco Aeroplex Inc. [98]

75-71370 Precision Tube Co. Inc. Coaritube Division
(a division of Meuller Industries Inc.)
620 Naylor Mill Road, Salisbury, MD 21801
☎(+1 410) 546 39 11 ⊖ (+1 410) 546 39 13
👥 25
President: Gene Passmore
General Manager: Charles Bogese
Sales Manager: Douglas McNeil
■ Manufacture of semi-rigid and flexible coaxial cable and cable assemblies, delay lines. [98]

75-71395 Premier Hazard Systems Inc
532 Alpha Drive, R.I.D.C. Industrial Park, Pittsburgh, PA 15238
☎(+1 412) 487 6880 ✆ (800) 394 9276
⊖ (+1 412) 487 6881
♣ $ 600 000 👥 12
Sales Mgr: Jim Fisher
■ Manufacture of a complete line of emergency vehicle warning equipment including strobe & halogen lightbars, beacons, sirens, speakers, switch control & auxiliary lighting products. [98]

75-71400 Premier Microwave Corp.
33 New Broad Street, Port Chester, NY 10573
☎(+1 914) 939 89 00 ⊖ (+1 914) 939 92 39
♣ $ 6,000,000 (1997) 👥 45 (1997) 🏭 2,275 m2
President: Merrill May
Operations Director: Michael Deegan
Sales Manager: Mel Radner
Purchasing Manager: John Barbieri
■ Design & manufacture passive microwave components & sub systems & magnets for high power klystrons. [98]

75-71436 PressCut Industries Inc.
(† Oriental Gasket & Packing)
2828 Nagle Street, Dallas, TX 75220-4595
☎(+1 214) 956 91 00 ✆ (800) 442 49 24
[E] info@presscut.com [W] http://www.presscut.com
⊖ (+1 214) 350 47 13
♧ $ 3,000,000 (1997) ♣ $ 7,000,000 (1997)
👥 50 (1997) 🏭 50,000 sq ft
President and Chief Executive Officer: J L Swanson
☎(+1 214) 956 91 00
Vice President of Sales and Marketing: Greg Watson
☎(+1 214) 956 91 00
Marketing Administrator: Ms Barbara King
■ Gaskets, pre-formed packings, custom fabrication of non-metallic materials including rubber, felt, paper, cork, fabrics. Specialists in EMI-RFI shielding. [98]

75-71438 Pressure Systems Inc.
(† TRW Pressure Systems Inc.)
6033 East Bandini Boulevard, Los Angeles, CA 90040
☎(+1 213) 722 02 22 ⊖ (+1 213) 721 60 02
♣ US$ 25,000,000 (1996) 👥 104 (1997) 🏭 9,662 sq m
President and General Manager: Alex Roberts
Marketing Manager: Michael S Hersh
[E] mikeh@psi-pci.com
■ Manufacture pressure vessels for launch vehicles and spacecraft. [98]

75-71440 Presto Casting Co.
(a subsidiary of Haley Industries Ltd)
5440 West Missouri, Glendale, AZ 85301
☎(+1 602) 939 94 41 ▣ TWX 910 9501 242 presto phx
⊖ (+1 602) 937 56 19
♣ $ 13,000,000 👥 155
VP & General Mgr: Bob Cuff
■ Medium sized aluminium & magnesium castings for the aerospace industry. Provides both green sand & dry sand moulding capabilities. Complete in-house, facilities for Layout, Heat Treat, Destructive & Non-destructive testing. [98]

75-71460 Primex Physics International
(† Physics International Co and Olin Chemicals)
(a subsidiary of Primex Technologies)
2700 Merced Street, San Lieandro, CA 94577-0599
President: William W Smith
Vice President of Electromagnetic Systems: Charles Stallings
■ Military ordnance development, manufacture and test range, high explosives technology, survivability of military weapons systems, computation pulsed high voltage devices, plasma physics research, high powered microwave systems. [98]

Contact Reference

75-71472 Primex Technologies, Inc
(† Olin Ordnance)
1010 9th Street North, St Petersburg, FL 33716
☎(+1 813) 578 81 00 ⊜ (+1 813) 578 81 19
♣ $ 5,000,000 (1996) 👥 2,600 (1997)
Chief Executive Officer: James G Hascall
Executive Vice President: Doug Demaine
Vice Chairman: A A Catani
Communications Director: Edwin E Alber
International Defence Marketing Manager: D Bartles

Ordnance & Tactical Systems Division
(† Olin Ordnance)
1010 9th Street North, St Petersburg, FL 33716
☎(+1 813) 578 81 00 ⊜ (+1 813) 578 81 19
♣ $ 400,000,000 (1996) 👥 1,500 (1995)
President: Mike Wilson
Communications Director: Edwin E Alber
Business Development Director:
Brig Gen William R Holmes (retd)
International Defence Marketing Manager: Dean Bartles
■ Manufacture large and medium calibre ammunition, ball powder propellant. Shaped charge warheads. Advanced solid and liquid gun propulsion systems. Non-lethal and competent weapon systems, 155mm artillery combined effects munitions.

Marion Operations
Highway 148 South, Marion, IL 62959-0176
☎(+1 618) 985 82 11 ⊜ (+1 618) 985 12 66
■ Manufacture medium calibre ammunition and solid propellant products.

Red Lion Operations
200 East High Street, Red Lion, PA 17356
☎(+1 717) 244 45 51
■ Manufacture large calibre ammunition and heavy metal components.

St Marks Operations
Highways 98 and 363, St Marks, FL 32355
☎(+1 904) 925 61 11
■ Manufacture ball powder propellant.

Downey Operations
9236 East Hall Road, Downey, CA 90241
■ Manufacture medium calibre ammunition and air dispensing systems.

Aerospace and Electronics Division
(† Olin Aerospace Company)
(a division of Primex Technologies)
11441 Willows Road, North East, PO Box 97009, Redmond, WA 98073-9709
☎(+1 206) 885 50 00 ⊜ (+1 206) 882 58 04
♣ $ 107,000,000 (1996) 👥 630 (1996) 🏭 273,000 sq ft
President: William W Smith
Vice President of Marketing: J J Galbreath
Vice President of Operations: G D Fisher
Purchasing Director: M J Ciucci
■ Monopropellant hydrazine rocket engines for satellite control and manoeuvring; munition dispensing systems. Electronic power supplies, airborne electronic products and ground support equipment. 98

75-71500 Programmed Composites Inc.
(a subsidiary of Pressure Systems Inc.)
5605 East La Palma Avenue, Anaheim, CA 92807
☎(+1 714) 701 73 00 ⊜ (+1 714) 701 73 73
♣ US$ 15,000,000 👥 102 🏭 5,200 sq m
President and General Manager: Alex Roberts
Marketing Manager: Markham Hacke
■ Manufacture composite spacecraft structures, heatpipe panels, strats, precision reflectors, solar array substrates and structural assemblies. 98

75-71560 Projects Unlimited Inc.
3680 Wyse Road, PO Box 14538, Dayton, OH 45414-2539
☎(+1 513) 890 19 19 Ⓦ http://www.pui.com
⊜ (+1 513) 890 05 69
♣ $ 38,000,000 (1995) 👥 110 (1997) 🏭 42,000 sq ft
President: D L Wyse
Executive Vice President: Robert W Conrad
Ⓔ conrad@pui.com
Senior Vice President: Ivan L Clark
Manager Government Sales: Jeff Seymour
Purchasing Manager: Ms Cindy Rogers
■ Cable assemblies, circuit board assemblies, relays, networks, control assemblies, blackbox, ESS, PTH/SMT. 98

75-71670 Protection Development International Corp. (PDIC)
PO Box 2048, Corona, CA 91718
☎(+1 909) 734 75 31 Ⓔ info@armour-pdi.com
Ⓦ http://www.armour-pdi.com ⊜ (+1 909) 734 75 70
♣ $ 1,500,000 (1996) 👥 16 (1997) 🏭 10,000 sq ft
President: Norman E Smith
Executive President: Ms Mary Ann Smith
Purchasing Manager: Ms Theresa Roszkowicz
■ Manufacturing, engineering and marketing company specialising in the fields of explosives, ballistics and security. 98

75-71800 PTI Technologies Inc.
(a subsidiary of ESCO Electronics Corp)
950 Rancho Conejo Boulevard, PO Box 2000-91319, Newbury Park, CA 91319-1796
☎(+1 805) 499 26 61 Ⓔ filters@ptitechnologies.com
⊜ (+1 805) 375 22 96
♣ $ 47,000,000 (1996) 👥 237 🏭 115,000 sq ft
Chairman and Group Vice President: J Michael Conway
☎(+1 805) 375 22 00
Senior Vice President, Operations: Bob Buchanan
☎(+1 805) 375 22 63
Senior Vice President, Finance: Sam Chapetta
☎(+1 805) 375 22 67
Aerospace Marketing Director: David Atkinson
☎(+1 805) 375 22 32
Purchasing Manager: Ms April Romain
Contracts Manager: Ms Linda Oliart
☎(+1 805) 375 23 04
■ Filtration technology for aerospace, defence, fluid processing and fluid power industries. Products and services include commercial and high performance filter modules, filter cartridges and accessories. 98

75-71815 Pulnix America Inc.
1330 Orleans Drive, Sunnyvale, CA 94089
☎(+1 408) 747 03 00 ✆ (800) 445 54 44
Ⓦ http://www.pulnix.com ⊜ (+1 408) 747 06 60
Marketing Communications Manager: Ms Janet Worthen
■ Manufacture miniature CCD video cameras/lenses and night vision equipment for covert surveillance, high reliability deployment, passive infra-red sensor and photoelectric beams. 98

75-71850 The Purdy Corp.
586 Hillard Street, PO Box 1898, Manchester, CT 06045-1898
☎(+1 860) 649 00 00 Ⓔ sales@purdytransmissions.com
Ⓦ http://www.purdytransmissions.com
▣ TWX 710 4274190 ⊜ (+1 860) 645 62 93
👥 110 (1997) 🏭 150,000 sq ft
President and Chief Executive Officer:
Andrew A Sadanowicz
Exec. Vice President: George Purdy
Vice President Marketing: Samuel L DeLoach
General Manager: John Cicia
Purchasing: Peter Robinson
■ Precision machined aerospace components for fixed & rotary wing aircraft, turbine, missile engine & sensor pointing systems (exotic metals) incl. complex gearing, complete gearboxes, transmissions & critical helicopter rotor components; overhaul. 98

75-71890 Puritan-Bennett Aero Systems Co. (PBASCO)
(a subsidiary of Nellcor Puritan Bennet Inc.)
10800 Pflumm Road, Lenexa, KS 66215
☎(+1 913) 338 98 00 ⊜ (+1 913) 338 72 45
👥 300 (1995) 🏭 1,500 sq m
Vice President and General Manager: Ernest E Ross
■ Aircraft oxygen systems, portable and fixed, gaseous and chemical; low and high pressure oxygen and hose assemblies, cylinders and regulators; crew and passenger masks for all types of aircraft.

Wemac Division
18475 Pacific Street, Fountain Valley, CA 92708
☎(+1 714) 962 88 74 ⊜ (+1 714) 962 09 70
👥 50
Operations Manager: B E Thomas
Inside Sales: Ms Joyce Kirtz
■ Passenger and crew air valves, reading lights, information signs and complete passenger service units.

Kansas Facility
10800 Pflumm Road, Lenexa, KS 66215
☎(+1 913) 338 98 00 ⊜ (+1 913) 469 84 19
■ Chemical generators, PBE, chemical deployment units, manifolds, door latches, crew masks. Airborne closed circuit TV. 98

75-71900 Purolator Products Co.
(a division of Mark IV Industries Inc.)
Facet Filter Products Division
8439 Triad Drive, Greensboro, NC 27409-9621
☎(+1 910) 668 44 44 ⊜ (+1 910) 668 44 52
👥 160 🏭 85,000 sq ft
General Manager: Larry Nicolette
Sales and Marketing Director: Kevin Nelson
Engineering Director: Tony Simone
Purchasing Manager: April Allison
■ Airframe and engine mounted filters and elements. Land vehicle fuel, lube oil, hydraulic filters and elements. 98

75-71960 Q-Systems International Inc.
1148 107th Street, Arlington, TX 76011-3109
☎(+1 817) 649 26 00 Ⓔ qsitexas@aol.com
⊜ (+1 817) 649 26 41
♧ $ 1,000,000 ♣ $ 2,000,000 👥 25 (1995)
🏭 1,500 sq m
President: Edwin C Hodges
■ Manufacturer of personal body armour, soft and hard armour manufacturer, in-house ballistic research and development. Research and development, programme management, test planning and execution, data analysis and report preparation, operations research analysis, engineering services, live fire testing, on site battle damage assessment. Protective personnel systems. Protective curtains and Spall curtains; combat protection systems and equipment; aircraft marine and architectural survivability systems. Consulting, armour, security and counter intelligence. Threat assessment. → 75 98

75-71970 Q-Tech Corp.
10150 Jefferson Boulevard, Culver City, CA 90232-3502
☎(+1 310) 836 79 00 Ⓔ sales@q-tech.com
⊜ (+1 310) 836 21 57
♣ $ 8,000,000 👥 140 🏭 22,000 sq ft
President: Howard Phillips
Vice President: Brian Rose
Purchasing Manager: S Rappaport
Sales Manager: Gary Slacum
■ Crystal controlled clock oscillators, surface mount crystals, VCXOs, VHF and UHF sources, microprocessor drivers, baud rate generators, hybrid devices, sine-wave sources. High stability oscillators. Microprocessor controlled crystal oscillators. 98

75-72250 Quantrad Sensor
2360 Owen Street, Santa Clara, CA 95054
☎(+1 408) 727 78 27 Ⓔ quantrad@quantrad.com
Ⓦ http://www.quadrad.com ⊜ (+1 408) 727 78 28
♣ $ 1,000,000 👥 12
Managing Director: Nick Szluk
Sales and Marketing Director: Ed Browning
Operations Manager: Frank Ziemba
■ Electro optical photodiodes and nuclear radiation detectors. X-ray flourescence analysers for metal and other media. 98

75-72380 Racal Communications Inc.
(a subsidiary of The Racal Corporation)
5 Research Place, Rockville, MD 20850
☎(+1 301) 948 44 20 ⊜ (+1 301) 948 63 71
President and Chief Executive Officer: Mark Lipp
Vice President of Marketing: Mitchell Herbets
Director of Marketing: Theodore Neiman
Marketing Administration: Ms Susanne A Cook
■ Manufacturer of tactical radio communication equipment.
® UK: Racal Electronics Plc 98

75-72400 Racal Health & Safety Inc.
(† Racal Airstream Inc.)
(a subsidiary of The Racal Corporation)
7305 Executive Way, Frederick, MD 21701-8368
☎(+1 301) 695 82 00 ▣ 893450 racair
⊜ (+1 301) 695 44 13
Managing Director: Peter Cooper
■ Manufacturer and supplier of personal respiratory and safety equipment for industry/medical/agriculture and military. 98

75-72402 Racal Instruments Inc.
(† Racal-Dana Instruments Inc.)
(a subsidiary of The Racal Corp.)
PO Box 19541, Irvine, CA 92623
4 Goodyear Street, Irvine, CA 92618-2002

☎(+1 714) 859 89 99 ✆ (800) 722 25 28
W http://www.racalinst.com ⊜ (+1 714) 859 71 39
President: Gordon W Taylor
Vice President of Sales and Marketing: Malcolm Levy
Marketing and Communication: Ms Anja Miller
■ Manufacturers of automatic test instrumentation for military and commercial systems. 98

75-72404 Racal Antennas, Inc.
(a subsidiary of The Racal Corporation)
501 Tradeway, Mineral Wells, TX 76067
☎(+1 817) 325 23 41 ⊜ (+1 817) 325 43 77
Vice President of Marketing: Anthony Martin
■ Design, development and manufacture of commercial base-station antennas for the telecommunications industry. 98

75-72408 Racal Avionics Inc.
(a subsidiary of The Racal Corporation)
8851 Monard Drive, Silver Spring, MD 20910-1878
☎(+1 301) 495 66 95 ⊜ (+1 301) 585 7578
General Manager: Stephen L Kemp
■ Design, development, manufacture and support of airborne avionics systems for military, commercial and general aviation. 98

75-72409 Racal Pelagos Inc.
(a subsidiary of The Racal Corporation)
5434 Ruffin Road, San Diego, CA 92123
☎(+1 619) 292 89 22 ⊜ (+1 619) 292 53 08
Senior Vice President: Randall J Ashley
■ Supplier of high-resolution geophysical surveys, seabed mapping and oceanographic programs. 98

75-72410 Racal Data Group/Racal-Datacom Inc.
(a subsidiary of the Racal Corporation)
PO Box 407044, Fort Lauderdale, FL 33340-7044
1601 North Harrison Parkway, Sunrise, FL 33323-2899
☎(+1 305) 846 16 01 ✆ (800) RACAL 55
✉ RACMILINTL ☒ TWX 510 955 4145
⊜ (+1 305) 846 49 72
👥 2,500
Chairman, Racal Data Group: Paul Kozlowski
President, Racal-Datacom Inc.: Louis P Bender (III)
■ International designer and manufacturer of products and systems providing secure, managed access to multimedia information networks. Maintenance and support services for digital, analogue and fibre optic products including digital service units, modems, multiplexers, security systems and network management systems. 98

75-72430 R&D Aeronautical Engineering Co. Inc. (RDAE)
PO Box 1093, Wylie, TX 75098
☎(+1 972) 442 22 94 ⊜ (+1 972) 442 34 45
♣ $ 1,500,000 (1997) 👥 17 (1996) 🏭 35,800 sq ft
President: Harold F Smith
Sales and Marketing Director: Daniel E Smith
■ OEM, KD2R-5 aerial target drone, OEM, O-100-3 drone engine. Parts, engines, services O-154, O-100, RPV consulting, miniature aerial targets, mini-RPV and mini-RPV aerial reconnaissance packages. 98

75-72460 Radar/Digital Systems
10556 Combie Road, Auburn, CA 95602-8908
☎(+1 916) 269 23 40 ⊜ (+1 916) 269 17 77
President: Richard R Fay E rfay@ports.com
■ Radar surveillance systems for surface search and air traffic movement. Radar extractors and radar display processors to enhance or upgrade any radar. 98

75-72500 Radian Technology
(a subsidiary of REMEC, Inc.)
4211 Burton Drive, Santa Clara, CA 95054-1512
☎(+1 408) 980 98 77 E sales@radiantech.com
⊜ (+1 408) 980 16 14
♣ US$ 9,100,000 (1996) 👥 70 (1997) 🏭 25,000 sq ft
President and Chief Executive Officer: Jim Mongillo
Vice President of Sales and Marketing: Len Johnson
Director of Public Relations: Ms Kim Taverrite
Purchasing Manager: Ken McLaren
■ VCOs, DTOs, synthesizers, filters, multiplexers switch filters, set-on receivers. Microwave components. 98

75-72600 Radio Research Instrument Co. Inc.
584 North Main Street, Waterbury, CT 06704-3506
☎(+1 203) 753 58 40 E kbps32e@prodigy.com
W http://www.techexpo.com/www/radiores
⊜ (+1 203) 754 25 67
♣ $ 2,000,000 (1997) 👥 15 (approximately, 1997)
🏭 27,500 sq m
President: P J Plishner
Executive Vice President: E B Doyle
■ Radar systems and spares, threat emitters, high-power RF sources, antennas, az-el pedestals, microwave tubes. 98

75-72610 Radstone Technology Corp
(a subs. of Radstone Technology plc, UK)
50 Craig Road, Montvale, NJ 07645
☎(+1 201) 391 2700 ⊜ (+1 201) 391 2899
US General Sales Manager: Laurence Abzug
■ High technology, microprocessor-based VMEBUS products at all levels from cards to complete systems with operating systems & firmware, for both civilian and military environments. 98

75-72690 Ralbo Inc.
850 Deming Way, Sparks, NV 89431-6455
☎(+1 702) 359 1818 ⊜ (+1 702) 359 1874
♣ $ 600 000 (1994) 👥 10 🏭 8 000 sq ft
CEO/President: Ralph J. Garton
VP Sales & Marketing: Robert J. Garton
■ Design & manufacture of hydraulic & pneumatic valves & pressure switches. Mainly for self propelled guns, tanks & on board ships. 97

75-72700 RAMFAN Corporation
2786 Via Orange Way, Spring Valley, CA 91978
☎(+1 619) 670 95 90 ✆ (800) 4 RAMFAN
⊜ (+1 619) 670 73 45
♣ $ 6,000,000 👥 25
President: Wayne Allen
Marketing Director: Michael Grimes
■ Manufacture of RAMFAN portable ventilation system, water turbine driven, vane axial fan. Electric, gasoline and air powered fans for confined space ventilation. Supplier of spare parts to naval supply centres for pumps, turbines and blowers. 98

75-72810 RAMO Manufacturing Inc.
412 Space Park South, Nashville, TN 37211
☎(+1 615) 832 67 00 ⊜ (+1 615) 832 90 98
👥 60 🏭 38,000 m2
Managing Director: J Stephen Sarles
Operations Director: Charles Richards
Purchasing Manager: Don Gahagan
■ Complete manufacture of the .50 calibre Mz Browning family of weapons including the M2HB, M2AC and M3R guns and spare parts, accessories and mounts. Design, development and manufacturing for producing related equipment and supplies. 98

75-72830 Ramsey Winch Co.
(a subsidiary of Ramsey Industries)
PO Box 581510, Tulsa, OK 74158
☎(+1 918) 438 27 60 ⊜ (+1 918) 438 66 88
Advertising Co-ordinator: Ms Paula Kaufman
■ Vehicle mounted winches, hydraulically, mechanically or electrically driven, for self-recovery, pulling, towing and retrieval operations. Capacities from 3,000 - 100,000lbs (1,360 - 45,360kg) 98

75-72900 Rand Instrument Corp.
Atlanta, GA 30326
3350 Peachtree Road
☎(+1 404) 261 99 99 ⊜ (+1 404) 261 77 75
President and Chief Executive Officer: Alan E Rand
Vice President: Ray L Bridger
Sales and Marketing Director: Jerry Cieszynski
■ Manufacturer of electro-optical surveillance systems. 98

75-72905 Randolph Products Co.
701-12th Street, PO Box 830, Carlstadt, NJ 07072-0830
☎(+1 201) 438 37 00
E 71273.3465@compuserve.com
⊜ (+1 201) 438 42 31
♣ $ 8,000,000 (1995) 👥 65 (1995) 🏭 12,000 sq m
President: John Randolph
■ Primers; strippers, varnishes, sealers, military specification paints, enamels, paints, lacquers, dopes, thinners. 98

75-72915 Rantec Microwave & Electronics Inc.
(a subsidiary of ESCO Electronics Corp)
24003 Ventura Boulevard, Calabasas, CA 91302-1447
☎(+1 818) 591 81 89 ⊜ (+1 818) 591 84 89
President: G Wilson
Vice President of Anechoic Products: M Mawdsley
Sales Manager: F Peterson
■ Microwave systems, anechoic/shielding systems and power systems. 98

75-72920 Rapid Acquisition Targeting Systems (RATS)
260 Columbia Avenue, Fort Jersey, NJ 07024
☎(+1 201) 224 0332 W http://www.ratslasers.com
⊜ (+1 201) 224 0101
President: Ira Marlowe
■ Rapid acquisition targeting system for use on H&K MP5, MP5K, MP5K-PDW, MP510, MP5-40, HK53, Steyr A.U.G. & M16 weapons. 97

75-72925 Rapid Deployment Towers Inc.
500 Throckmorton Street, Suite 1410,
Fort Worth, TX 76102
☎(+1 817) 332 63 11 ✆ (800) 856 45 44
⊜ (+1 817) 336 89 29
President: John W Hughes E jwhii@aol.com
■ Mobile rapid deployment towers (RapUp) capable of deploying radar, remote cameras, IR sensors or any other communications or intelligence gathering device at heights up to 300 ft. 98

Contact Reference

75-73200 Raytheon Aircraft Company
(† Beech Aircraft Corp.)
(a subsidiary of Raytheon Company)
PO Box 85, Wichita, KS 67201-0085
☎(+1 316) 676 76 03 ✉ BEECHCRAFT
☒ 417422 beechcraft ⊜ (+1 316) 676 56 87
♣ $ 2,000,000,000 (1995) 👥 14,080 (worldwide 1996)
🏭 3,420,528 sq ft
President and Chief Executive Officer: Arthur E Wegner
Vice President and Chief Financial Officer: James Gray
Vice President, Strategic Planning/New Business Development: C Doug Mahin
Vice President and Assistant Controller, Financial Analysis: Robert Welton
Vice President, Engineering: Steve Hanvey
Vice President, Operations: Richard Danforth
Vice President and General Counsel: Wayne W Wallace
Vice President, Human Resources: Richard Griffiths
Vice President, Marketing: Scott Kalister
Vice President, International Marketing: William Newman
Vice President, Material: Gary M Hanssen
Vice President, Programme Management: Robert A Brill
Vice President, Operations: John Booth
Vice President, Quality Assurance: W D Wise
Vice President, Sales and Marketing: Keith Nadolski
Vice President, Special Mission Sales: Tariq (Ted) Farid
Vice President International Sales, Far East: Charles Lloyd
Vice President International Sales, Latin America: Nick Schneider
Vice President, Government Marketing: Richard L Klass
Vice President, JPATS: David H Riemer
Corporate Affairs Director: James M Gregory
Ground Test Laboratories Manager: Robert J Simmons
Advertising Manager: David Pishko
Public Relations Manager: Ms Pat Zerbe
☎(+1 316) 681 76 02
■ Aircraft training systems; utility and special mission aircraft; aircraft subassemblies and subcontracting; logistics support, ground support and servicing equipment; subsonic and supersonic missile target systems and support.

Salina Div.
Airport Industrial Center, B655, Salina, KS 67401-9999
☎(+1 913) 827 9684 ⊜ (+1 913) 681 8286
Manager: John Mercer

Little Rock Division
Adams Field, Little Rock, AR 72203
☎(+1 501) 372 1501 ⊜ (+1 501) 372 5130
Manager: Roy Pickens

Montek Div.
2268 South 3270 West, Salt Lake City, UT 84119-1193
☎(+1 801) 973 4300 ☒ 168201 ⊜ (+1 801) 974 7581
👥 393 (1993)
Vice President and General Manager: David A. Williams

Vice President Finance & Administration: R.F. Nash
Administrative & Technical Services Director: William M. Snarr
■ Design & manufacture ground and ship-based navigational aids for military, commercial and general aviation; precision distance measuring equipment and military/commercial aircraft flight controls including the advanced direct drive servo valve. 98

75-73302 Raytheon E-Systems Corporate Headquarters
(† E-Systems Inc.)
(a subsidiary of Raytheon Company)
Corporate Headquarters
PO Box 6056, Greenville, TX 75403
☎ (+1 972) 661 10 00 ☒ 703365 ⊖ (+1 903) 457 93 91
👥 16,241 (1997)
Chairman: A Lowell Lawson
President and Chief Executive Officer: A Lowell Lawson
Senior Vice President and Corporate Officer: Dr Terry W Heil
Senior Vice President and Corporate Officer: Peter A Marino
Senior Vice President and Corporate Officer: Brian D Cullen
Vice President of Finance and Chief Financial Officer: James W Pope
Vice President of New Business Development: Harry L Thurmon
Vice President of Strategic Planning and Development: Dr James R Collins
Vice President of Government and International Operations: David J Matthews
Vice President of Corporate Information and Administration: Arthur E Hobbs
Vice President, Secretary and General Counsel: Mike C Eberhardt
International Operations Director: Bill K Doran
Corporate Communications Manager: Joe D Curtis
Advertising and Sales Promotion Manager: Ken Casady
Purchasing Manager: Richard A Pemble
■ Development and production of high technology defence electronics systems. Aircraft modification, overhaul and maintenance services primarily for government users. → 80

Huntsville AL Office
4910 University Square, Suite 1A, Huntsville, AL 35816
☎ (+1 205) 895 95 00 ⊖ (+1 205) 895 94 28
General Manager: Bert Turner

Dayton Office
5100 Springfield Pike, Suite 305, Dayton, OH 45431-1263
☎ (+1 513) 254 84 64 ⊖ (+1 513) 254 03 02
General Manager: Ken Sexton

Oklahoma City Office
1380 South Douglas Boulevard, Suite 101, Midwest City, OK 73130
☎ (+1 405) 737 76 10 ⊖ (+1 405) 737 70 45
General Manager: Don Johnston

Ft Monmouth Office
766 Shrewsbury Avenue, Tinton Falls, NJ 07724
☎ (+1 908) 747 31 01 ⊖ (+1 908) 747 27 58
General Manager: Rod Adams

Hampton VA Office
Suite 460, 1919 Commerce Drive, Hampton, VA 23666
☎ (+1 804) 825 19 73 ⊖ (+1 804) 825 19 51
General Manager: Tom Ehrenberger

ECI Division
1501 72nd Street North, PO Box 12248, St Petersburg, FL 33733-2248
☎ (+1 813) 381 20 00 ✉ ECI
☒ 523455; TWX 810 863 0377 ⊖ (+1 813) 343 12 95
Vice President and General Manager: Dr James C Garrett

■ Design, develop and qualify communication products, systems and software to provide critical voice, data, imagery and telemetry communications for ship, aircraft and ground station applications. Capabilities include technologies essential to advanced communications systems and terminals, signal processing and distribution equipment, transmitters, receivers and antennas. Capabilities and products include: Robust, redundant multi-media communications systems including software/hardware design, build, integration and test. Command, control, communications and intelligence systems and products for use with satellite systems. Variety of intelligence terminals incorporating advanced microprocessors, latest generation embedded COMSEC devices, VME architecture, anti-jam techniques and innovative mechanical packaging. Airborne internal communications systems using high-speed, broad bandwidth data bus technology. Variety of radio products including a compact UHF satellite communications radio with DAMA/TDMA add-on compatibility primarily intended for aircraft applications. Fibre optics and advanced fibre optic transmission systems capable of operation at a digital bit rate of 5 gigabits/second and another broadband, flat-response system operating over a wide dynamic range up to 20 gigahertz frequency.

Garland Operations
1200 Jupiter Road, PO Box 660023, Dallas, TX 75266-0023
☎ (+1 214) 272 05 15 ☒ 732306; TWX 910 860 5187
⊖ (+1 214) 272 81 44
Vice President and General Manager: Michael Keebaugh
■ Digital communications systems, data systems, command and control systems and electronic warfare systems. High speed digital data processing, information systems, systems support and services.

Greenville Operations
Majors Field, PO Box 6056, Greenville, TX 75403-6056
☎ (+1 903) 455 34 50 ☒ 6829260; 732392
⊖ (+1 903) 457 44 13; 457 39 49
Vice President and General Manager: A J Doshier
Vice President of Finance: D E Bench
Vice President of Human Resources: R J McNulty
Vice President of Government Systems: D S Cleverly
Vice President of Engineering: H D Smith
■ Mission analysis and design, manufacture, test, installation and integration of mission equipment in airborne, naval and ground based platforms specializing principally in reconnaissance, surveillance, electronic warfare; command, control, communications and intelligence systems. Logistics support, management and equipment operation and utilization training. Complete turnkey systems. Head-of-State and other key individual aircraft conversions and communication systems. Military, commercial and private aircraft maintenance and modification.

Falls Church Operations
7700 Arlington Boulevard, Falls Church, VA 22046-1572
☎ (+1 703) 560 50 00 ☒ 899494; TWX 710 831 9802
⊖ (+1 703) 280 46 27
👥 1,951
Vice President and General Manager: Walt P Havenstein
■ Design and manufacture of airborne remotely controlled reconnaissance systems, airborne and ground-based intelligence collection systems, automated signal processing systems, electronic countermeasure systems, intrusion detection or physical security systems and special purpose receivers.

University Center
44983 Knoll Square, Ashburn, VA 22011
☎ (+1 703) 729 60 00 ⊖ (+1 703) 729 71 02
■ Intelligence processing collection systems and other EW systems. Tactical and security electronics systems, RPVs.

DBA (EMED)
(† E-Systems Medical Electronics, Inc.)
(a subsidiary of E-Systems Inc.)
11550 IH-10 West, San Antonio, TX 78230
☎ (+1 210) 641 83 40 ✆ (800) 270 3633
E info@emed.com W http://www.raytheon.com/esys
⊖ (+1 210) 641 74 28
♣ $ 17,000,000 (1995) 👥 75 (1997)
Managing Director: Jim Karlak
Operations Director: Gary Tibbals
Sales Director: Larry Ware
Purchasing Manager: Ron Carter
Marketing Communications Manager: Ms Noreen Huante
■ Produce systems for real-time transmission of images, particularly for medical purposes. 98

75-73380 RBC Transport Dynamics Corp.
(† BFM Transport Dynamics Corp.)
3131 Segerstrom Ave, Santa Ana, CA 92704
☎ (+1 714) 546 3131 ⊖ (+1 714) 545 9885
♣ $ 35 000 000 👥 130
President: Mike Hartnett
Dir Sales & Marketing: Chris Sommers (Aerospace Bearings)
■ Manufacture self-lubricating bearings for aerospace & industrial applications. 97

75-73860 Reade Advanced Materials
(a division of Reade International)
PO Box 15039, Riverside, RI 02915-0039
☎ (+1 401) 433 70 00 E ramsales@reade.com
W http://www.reade.com ⊖ (+1 401) 433 70 01
♣ $ 10,000,000 👥 45 🏭 50,000 sq m
President and Chairman: Charles Reade (Jr)
Managing Director: S M Munson
Operations and Technical Director: Ms Bethany Cochran
Purchasing Manager: E S Reade
■ Manufacture and distribution of metal, ceramic and composite materials for military applications. Consultants for size reduction technologies and manufacturing safety. Market feasibility and start-up studies. → 80 → 85 → 90
® Canada: Alcan Aluminium, Cabot, Cerac, Dow Chemical, Dupont, Elkem, Engelhard, GAF, Handy & Harmon, INCO
France: Baikowski Int'l, Pechiney
Germany: De Gussa, Metallgesellschaft, Frank & Schulte
Japan: Mitsubishi Steel
Sweden: Avesta Nyby
UK: Johnson Matthey
USA: 3M, Alcan Aluminium, American Minerals, Hart Metals, Hoegannes, La Roche, Pfizer Inc., Reynolds Metals Co., TAM, Trans-Tech Inc., Wall Colmonoy, Washington Mills Abrasive, Zinc Corp. of America

Reade Communications Group
PO Box 15039, Riverside, RI 02915
☎ (+1 401) 433 70 00 ⊖ (+1 401) 433 70 01
General Mgr: S.M. Munson 98

75-73930 Recon/Optical Inc. Corporate Headquarters
550 West Northwest Highway, Barrington, IL 60010-3094
☎ (+1 847) 381 24 00; 381 28 18
W http://www.roi.bourns.com
⊖ (+1 847) 381 49 87 (Contracts); 381 28 18 (Business Development)
President & CEO: William L. Owens
VP Business Development: Noel J. Deitrich
■ Design and manufacture of film and E-O reconnaisance systems, visual and IR sighting systems and E-O advanced technology systems.

District of Columbia (DC)
Sales Office, 1800 Diagonal Road, Suite 410, Alexandria, VA 22314-2840
☎ (+1 703) 683 44 40 ⊖ (+1 703) 683 45 54
Director: Joel T. Coney

Pacific Optical
2660 Columbia Street, Torrance, CA 90503-3802
☎ (+1 310) 328 58 40 ⊖ (+1 310) 320 05 50
Director: James W. Dietz

US Divisions:

CAI
550 West Northwest Highway, Barrington, IL 60010-3094
☎ (+1 847) 381 24 00
⊖ (+1 847) 381 49 87; 381 28 18
President and Chief Executive Officer: William L Owens
Vice President Business Development: Bill Stearns
Vice President Engineering: Ms Heidi Venuti
Vice President and Chief Financial Officer: Stephen M Myskowski
Vice President Advanced Reconnaisance Systems: André G Lareau
Product Assurance Director: John J Stachura
Human Resources Director: David DeRita
Contracts Director: Ronald L Polasek
■ Design & manufacture of film & E-O reconnaissance systems, visual & IR sighting systems & E-O advanced technology systems.

Pacific Optical Division
2660 Columbia Street, Torrance, CA 90503
☎ (+1 310) 328 58 40 ⊖ (+1 310) 320 05 50
🏭 50,000 sq ft
Vice President and General Manager: James Dietz
Vice President Marketing: Andrew R Crucitti
■ Designer & manufacturer of visible & IR optics, diffraction-limited reconnaissance lenses and complex optical subsystems. 98

75-74150 Reflectone Inc
(an affiliate of British Aerospace plc UK)
4908 Tampa West Boulevard, Tampa, FL 33634-2481

☎ (+1 813) 885 74 81 ✆ (800) 237 8414
🅆 http://www.reflectone.com ⊖ (+1 813) 887 12 54
♣ $ 100,269,000 (1996) 👥 890
🏭 19,490 m2 (Tampa); 2,230 m2 (Dulles); 2,044 m2 (UK)
Chairman of the Board: Stella F Thayer
President and Chief Executive Officer, President Reflectone UK (acting): Richard G Snyder
☎ (+1 813) 887 15 00
Vice President - Finance and Treasurer: Edward W Bettke
Finance Director: Dave Fish
Director: Syd Gillibrand
Director: Vice Adm Robert F Schoultz (USN retd)
Director: Dale R States
Vice President Commercial and Entertainment Products: Derek R Alden
Corporate Communications Manager: Ms Cathren M Little
☎ (+1 813) 887 14 41; 887 15 24 (fax)
🄴 little@reflectone.com
■ Design & manufacture complex simulation & training devices for military, commercial aviation & entertainment industry worldwide; contractor simulator instruction, operation & maintenance services; operation of training centers.
Subsidiaries:
Reflectone Training Systems Inc.
Reflectone UK Ltd (RUKL) [98]

75-74200 Regent Mfg Inc.
(a subsidiary of Aviation Methods Inc.)
11905 Regentview Avenue, Downey, CA 90241-5587
☎ (+1 310) 862 11 74 ⊖ (+1 310) 861 96 24
👥 46 🏭 9,290 m2
President: M Walker
Finance Director: Greg Holmes
Marketing and Sales Director: John Brammer
Customer Services Manager: Jerry Jarvis
Purchasing Manager: D Feltnor
■ Lifting equipment and maintenance platforms. [98]

75-74280 Relli Technology Inc.
500 S.W. 12th Avenue, Deerfield Beach, FL 33442
☎ (+1 954) 426 9570 ⊖ (+1 954) 426 9577
♧ $ 3 000 000 (1996) ♣ $ 15 000 000 (1996) 👥 30
🏭 15 000 sq m
President: Reuven Ginton
Managing Dir: Thom Wilker
Technical Dir: Shlomo Jacob
Marketing Dir: Harvey Friedman
Purchasing Mgr: Gideon Klein
■ Defence & aerospace contractor to the US & other Governments. Total capability in parts & components for military applications in Army, Navy & Air Force. [97]

75-74285 RELTEC Corporation
(† Reliance Comm/Tec Corp.)
5875 Landerbrook Dr, Suite 250, Cleveland, OH 44124
☎ (+1 216) 460 3600 ✆ (800) 233 2334
⊖ (+1 216) 460 3690
👥 3 850
President: Dudley Sheffler
VP Sales: Dick Schwob
Public Relations: Toby Hoopes
■ Digital loop carrier; analog carrier; line conditioning equipment; central office hardware such as connectors-protectors, modules & main distribution frames; central office power such as rectifiers & power boards, outside plant enclosures, network services. [97]

75-74400 Remmele Engineering Inc.
10 Old Highway 8 S.W., New Brighton, MN 55112
☎ (+1 612) 635 41 00 ⊖ (+1 612) 635 41 $68
♣ $ 90,000,000 (1996) 👥 480 (1996)
President and Chief Executive Officer: Tom Moore
Vice President Sales & Mktg, Director Public Relations: Bert Casper
■ Design & manufacture of aircraft tooling/fixtures, high speed machining of aircraft parts, machining & fabrication of parts for aerospace/communications/missile applications, design & build of special machines for automated production. [98]

75-74430 REMOTEC Inc.
(a subs. of Northrop Grumman Corp.)
114 Union Valley Road, Oak Ridge, TN 37830
☎ (+1 423) 483 0228 ⊖ (+1 423) 483 1426
♣ $ 6 000 000 👥 60 🏭 27 000 sq ft
President: James Rambo
Marketing/Sales Dir: Shawn G. Farrow
Purchasing Mgr: Howard W. Harvey
■ Hazardous duty robots. Uses include explosives handling, nuclear surveillance & maintenance, HAZMAT response, SWAT operations & airport security, mine field detection. [97]

75-74550 Reproduction Technologies Inc.
110 East Vistula, PO Box 790, Bristol, IN 46507-0790
☎ (+1 219) 848 52 33
🅆 http://www.cyberlink-inc.com/rti
⊖ (+1 219) 848 53 33
♣ $ 1,000,000 (1996) 👥 15 (1996) 🏭 27,000 sq m
President: Michael Stoll 🄴 stoll@cyberlink-inc.com
■ Voice logging tape recorders for slow speed 24 hour logging of telephone conversations & radio transmissions. "911" recording machines. [98]

75-74610 RESON Inc.
(a subsidiary of RESON AS, Denmark)
300 Lopez Road, Goleta, CA 93117
☎ (+1 805) 964 62 60 🄴 sales@reson.com
🅆 http://www.reson.com ⊖ (+1 805) 964 75 37
♣ $ 7,000,000 (1996) 👥 56 (1997) 🏭 11,000 sq ft
Chairman and Chief Executive Officer: Jens R Steenstrup
Vice President of Operations: Tony Parker
Purchasing Manager: Ms Lori Bonneau
Marketing Co-ordinator: Dennis Tivey
■ Definitive sea bed mapping. Multibeam bathymetric sonar system and multibeam forward looking sonar system. [98]

75-74850 Revere Transducers Inc.
14192 Franklin Avenue, Tustin, CA 92780
☎ (+1 714) 731 12 34 ✆ (800) 872 4784
🅆 http://www.reveretransducers.com
⊖ (+1 714) 731 20 19
👥 200 🏭 98,000 sq m
Vice President and General Manager: Alfred Friere
National Sales Manager: Richard Glass
Marketing Manager: Ms Linda Buck
Applications Engineer: Chris Palos
Purchasing Manager: Bill Hardwick
■ Manufacture load cells, pressure sensors, aircraft weighing kits, pressure transducers and force translators for aerospace, military and industrial applications. [98]

75-74970 RF International Inc.
(a subs. of RF Telecommunications Ltd, UK)
PO Box 2127, Apopka, FL 32704-2127
☎ (+1 407) 886 7751 ⊖ (+1 407) 886 7752
President: Roger Capper
■ Military & civil communications systems. [97]

75-75000 RGB Spectrum
950 Marina Village Parkway, Alameda, CA 94501
☎ (+1 510) 814 70 00 🅆 http://www.rgb.com
⊖ (+1 510) 814 70 26
♣ $ 6,200,000 👥 40
President: R P Marcus
Operations Director: Carl Lyle
Sales and Marketing Director: Tony Spica
■ Video windowing systems that integrate video with computer generated graphics, FLIR or radar on high resolution monitor. Applications include training, manufacturing inspection, scientific visualisation, surveillance, C3I, simulation and teleconferencing. [98]

75-75350 The Richards Corp.
44931 Falcon Place, Sterling, VA 20166-9510
☎ (+1 703) 834 55 50 ⊖ (+1 703) 481 50 07
♣ $ 2,300,000 (1996) 👥 28 (1997) 🏭 38,000 sq ft
President: Stanley I Richards
Manager, Sales Support Group: Ms Nina L Cullen
■ Photo interpretation equipment (light tables for viewing of aerial photography). [98]

75-75355 Richard Manufacturing Co.
950 W. Kershaw, Ogden, UT 84401-3404
☎ (+1 801) 627 1224 ⊖ (+1 801) 394 3643
♣ $ 1 800 000 👥 15 (1996) 🏭 1 350 sq m
General Mgr: Brent Call
■ Electronic & cable assembly, test sets. F16 electronic spares. [97]

75-75370 Richmond Technology Inc.
(a division of ITW)
1897 Cotton Avenue, PO Box 1129, Redlands, CA 92373
☎ (+1 909) 794 21 11 ⊖ (+1 909) 794 40 03
👥 200 (1997) 🏭 115,000 sq ft
President: Michael Mersch
Vice President: Ed Weggeland
Operations Director: Doug Pastoor
Purchasing Manager: L Myers
■ Manufacturer of composite or plastic components, anti-static films, laminations for military barriers to prevent electrostatic damage (ESD), personnel grounding wrist straps and static prevention packaging. [98]

75-75470 R/M Equipment Inc.
6975 NW 43rd St, Miami, FL 33166
☎ (+1 305) 477 9312 ⊖ (+1 305) 477 9620
Vice President: Todd Griffin
■ Manufacture the M-203/PI a patented/improved version of the M-203 40mm grenade launcher, adaptable to any combat rifle or sub-machine gun. In service with armed forces & law enforcement agencies in the US & more than a dozen other countries. [97]

75-75480 RMI Titanium Company
1000 Warren Avenue, Niles, OH 44446
☎ (+1 330) 544 76 33 ⊖ (+1 330) 544 77 96
♣ $ 251,400,000 (1996) 👥 902 (1996)
Executive Vice President Commercial and Research: John H Odle
Investor Relations and Marketing Services Director: Ms Crystal L Revak
Operations Director: Rich Vandegrift
Sales and Marketing Director: Fred Janowski
Purchasing Manager: Brad Smith
■ Titanium milled products for use in aerospace & industrial applications, including sheet, strip, plate, bar, billet, ingot, pipe & welded tube.

Micron Metals
7186 West Gates Avenue, Salt Lake City, UT 84120
☎ (+1 801) 250 59 19 ⊖ (+1 801) 250 72 95
President: Vicky Woodland

Tradco Inc.
1701 West Main Street, Washington, MO 63090
☎ (+1 314) 239 78 16 ⊖ (+1 314) 239 32 14
General Mgr: George E. Schrecengost [98]

75-75500 Denbridge Digital
(† RMSL Traffic Systems Inc)
(a subsidiary of Denbridge Corporation, Canada)
4 South Fourth Street, Warrenton, VA 20186-3306
☎ (+1 540) 349 23 00 🄴 denbridge@aol.com
⊖ (+1 540) 349 99 87
♣ $ 15,000,000 (1994) 👥 75 (1995)
Managing Director: Colin B Matthews
Director, OEM Accounts: Peter H Roelofs
Operations: Warren Burks
■ Display compression, recording and transmission products for radar equipment. [98]

75-75600 Roanwell Corp.
180 Varick Street, New York, NY 10014
☎ (+1 212) 989 10 90 ⊖ (+1 212) 989 32 85
♣ $ 4,500,000 (1997) 👥 80 (1997) 🏭 40,000 sq ft
President: Kenneth Dobbins
Operations/Technical Dir: Lech Poradowski
Contract Mgr/Sales: Ray LeFebvre
Purchasing Mgr: Marcelle Wabah
■ Headsets, handsets, microphones, earphones & cable assemblies. [98]

75-75630 The Robar Companies Inc.
21438 North 7th Avenue, Suite B, Phoenix, AZ 85027
☎ (+1 602) 581 26 48 🄴 robarguns@earthlink.com
⊖ (+1 602) 582 00 59
👥 26 🏭 3,000 sq m
President: Robert A Barrkman
Operations Director: A Sheppard
Sales and Marketing Director: D Mehalechko
■ 2.75 FFAR manufacturing and packing. M72E4 LAW/sniper weapons, pistol and carbine manufacture. → 80 → 90
● 21438 N. 7th Ave Ste. B, Phoenix, AZ 85027
☎ (+1 602) 581 2648 ⊖ (+1 602) 582 0059
■ Security training & equipment. [98]

Contact Reference

75-75640 Robertshaw Controls Co.
2809 Emerywood Parkway, PO Box 26544,
Richmond, VA 23261-6544
☎ (+1 804) 756 6500 ▣ 710 956 0121
⊖ (+1 804) 756 6563
President: B. Henderson
VP Finance: M.R. Light
■ Aircraft potentiometers & position sensors. [97]

75-75650 Robinson Helicopter Co.
2901 Airport Drive, Torrance, CA 90505
☎ (+1 310) 539 05 08 ⊖ (+1 310) 539 51 98
500 (1997) 350,000 sq ft
President: Frank D Robinson
Vice President of Manufacturing: Wayne Walden
Vice Product of Product Support: Kurt L Robinson
Marketing Director: Tim A Goetz
Sales Manager: Ms Milly Donahue
■ Manufacture of Robinson R22 and R44 piston helicopters used for law enforcement, primary and instrument flight training, powerline and pipeline patrol, radio traffic watch. [98]

75-75900 Rockford Powertrain Inc.
1200 Windsor Road, Rockford, IL 61132-2908
☎ (+1 815) 633 74 60
Ⓦ http://www.rockfordpowertrain.com
⊖ (+1 815) 633 11 31
600 (1997) 634,000 sq ft
Sales Director: John E Phillips
Marketing Director: W David Wilfong
Operations Director: Don Ferguson
Purchasing Manager: Rick Reiland
Advertising Manager: Ms Trish Maher
■ Power-shift & fully automatic transmissions for military vehicles up to 1500 hp. Drive shafts & U-joints, fandrives for diesel engines and mechanical clutches for trucks. [98]

75-75930 Rockwell International Corp.
World Headquarters
600 Anton Boulevard, Suite 100, PO Box 5090,
Costa Mesa, CA 92628-5090
☎ (+1 714) 424 42 00 Ⓦ http://www.rockwell.com
▣ 664465 rockwell rkwl
Chairman and Chief Executive Officer: Don Davis (Jr)
Senior Vice President and Chief Financial Officer: W Michael Barnes
Senior Vice President, General Counsel and Secretary: William J Calise (Jr)
Vice President and Treasurer: Dennis J Popovec
Senior Vice President, Avionics and Communications: J D Cosgrove
Senior Vice President, Rockwell Semiconductor Systems: Dwight W Decker
Senior Vice President, Technology and Business Development: William D Fletcher
Senior Vice President, President and Chief Operations Officer of Rockwell Automation: Jodie K Glore
Senior Vice President, Organisation and Human Resources: Joel R Stone
Senior Vice President, Automotive and Acting President, Heavy Vehicle Systems: Larry D Yost
Senior Vice President, Communications: Earl S Washington
Vice President, Law: John R Stocker
Vice President, Business Practices: Ms Bette Bardeen
Vice President and Controller and Acting Vice President of Corporate Development: Lawrence J Komatz
General Auditor: William A Sante (II)
General Tax Counsel: Charles C Stoops (Jr)
■ Global electronics company with market positions in industrial automation, semiconductor systems and avionics.

Pittsburgh Office
625 Liberty Avenue, Pittsburgh, PA 15222-3123
☎ (+1 412) 565 20 00 ▣ 866213 rockwell pgh a
⊖ (+1 412) 565 73 89
Vice President and Treasurer: Dennis J Popovec
Vice President and Controller: Lawrence J Komatz
Director Treasury and Investor Relations: David H Brehm
General Auditor: William A Sante (II)
General Tax Counsel: Charles C Stoops (Jr)

Washington Office
1745 Jefferson Davis Highway,
Arlington, VA 22202-3402
☎ (+1 703) 412 69 00 ⊖ (+1 703) 412 68 11
Vice President of Public Policy and Governmental Affairs: Raymond Garcia

Science Center
1049 Camino Dos Rios, PO Box 1085,
Thousand Oaks, CA 91360
☎ (+1 805) 373 45 45 ⊖ (+1 805) 373 47 75
Vice President and General Manager: Joseph T Longo
■ The centre is the central research laboratory for Rockwell International, conducting fundamental and applied research in the physical and information sciences in support of divisions within Rockwell, and for private and public agencies.
Capabilities in high-speed electronics, electro-optics, optics, materials science and information and computational sciences.

Automotive Headquarters
2135 West Maple Road, Troy, MI 48084-7186
☎ (+1 248) 435 10 00 ▣ 49976991
⊖ (+1 248) 435 13 95
President Heavy Vehicle Systems: Larry D Yost (acting)
President Light Vehicle Systems: Robert A Calder
■ Producer of components for heavy- and medium-duty trucks, buses, trailers and off-road vehicles.

Rockwell Clutch Company North American Truck Systems
2135 West Maple Road, Troy, MI 48084-7186
☎ (+1 248) 435 10 00 ▣ 4997691
⊖ (+1 248) 435 19 68; 435 87 80
■ Joint venture between EXEDY Corporation and Rockwell International, to design, develop and manufacture clutches for the North American heavy-duty truck markets.

Rockwell International Suspension Systems Company Inc.
Light Vehicle Systems
2135 West Maple Road, Troy, MI 48084-7186
☎ (+1 810) 435 10 00 ▣ 4997691
⊖ (+1 810) 435 12 44
■ Joint venture between Mitsubishi Steel Manufacturing Company Ltd of Japan and Rockwell International, to develop and produce suspension systems and components.

Rockwell Wabco Vehicle Control Systems
Heavy Vehicle Systems
2135 West Maple Road, Troy, MI 48084-7186
☎ (+1 810) 435 10 00 ▣ 4997691
⊖ (+1 810) 435 13 93; 435 80 02
■ Joint venture with WABCO Automotive Products Group of American Standard Inc., to supply vehicle control systems, including anti-lock braking systems and other components.

Rockwell Heavy Vehicle Suspension Systems, Inc.
Heavy Vehicle Systems
472 Elm Ridge Avenue, Canal Fulton, OH 44614
☎ (+1 330) 854 17 22 ⊖ (+1 330) 854 17 33
■ A joint venture between Rockwell International Corporation and Suspensions Inc., supplies air suspension systems for trailers.

Rockwell Collins, Inc.
400 Collins Road NE, Cedar Rapids, Iowa 52498
☎ (+1 319) 295 10 00 ⊖ (+1 319) 295 54 29
President: J D Cosgrove

Collins Air Transport Division
400 Collins Road NE, Cedar Rapids, IA 52498
☎ (+1 319) 295 10 00 ⊖ (+1 319) 295 54 29
Vice President and General Manager: Neal J Keating
■ Supplier of integrated avionics solutions to airlines. Avionics subsystems include electronic flight instruments, displays, autopilot-flight directors, weather radar, TCAS II, global positioning systems and comm-nav-surveillance sensors.

Collins Avionics and Communications Division
400 Collins Road NE, Cedar Rapids, IA 52498
☎ (+1 319) 295 51 00 ⊖ (+1 319) 295 47 77
Vice President and General Manager: Robert M Chiusano
■ Supplier of avionics, communication and navigation products and systems. Include HV, VHF, UHF and UHF SATCOM communications for air, land and sea; HF voice and data modems; cockpit and flight management systems; colour flat panel displays; data links (JTIDS); Global Positioning System (GPS) for ships, planes, vehicles and people; modular avionics and systems integration design services. Produce the 'TransitMaster' information management system which provides real-time vehicle location, position, passenger information and safety and annunciation; lightweight portable secure and nonsecure satellite telephones and wearable personal computers.

Collins General Aviation Division
400 Collins Road NE, Cedar Rapids, IA 52498
☎ (+1 319) 395 10 00 ⊖ (+1 319) 395 54 29
Vice President and General Manager: Robert J Hirvela
■ Supplier of integrated electronics solutions to regional airlines and business aircraft operators. Avionics subsystems include electronic flight instruments, autopilot-flight directors, weather radar, TCAS II, flight management systzems and com-nav-surveillance sensors.
Sales offices in: Tuscon, AZ; Carson, CA;: Atlanta, GA; Wichita, KS; Waterford, MI; Point Pleasant Beach, NJ; Irving, TX.

Communications Systems Division
3200 East Renner Road, Richardson, TX 75082-2402
☎ (+1 972) 705 39 50 ⊖ (+1 972) 705 33 98
Vice President and General Manager: Glenn C Andrew
■ Design, develop and produce information message handling and communication systems and products that support C3 for land, sea and air applications.

Collins Subsidiary and Service Business
400 Collins Road NE, Cedar Rapids, IA 52498
☎ (+1 319) 295 10 00 ⊖ (+1 319) 259 54 29
Vice President and General Manager: Robert P Marovich

Intelligent Transportation Systems
400 Collins Road NE, Cedar Rapids, IA 52498
☎ (+1 319) 295 10 00 ⊖ (+1 319) 295 54 29
President: John R Girotto
■ Design and manufacture electronic systems for travellers, commercial vehicles and railroads. Systems include turn by turn route guidance, traveller information, communications, vehicle monitoring, on board computing and integrated computing and display systems.

Rockwell Semiconducter Systems, Inc.
PO Box C, Newport Beach, CA 92658-8902
4311 Jamboree Road, Newport Beach, CA 92660-3095
☎ (+1 714) 221 46 00 ⊖ (+1 714) 221 40 78
President: Dwight W Decker
Sales offices in: Newport Beach, Santa: Clara, Thousand Oaks, CA; Clearwater, FL; Atlanta, GA; Itasca, IL; Westford, MA.

Digital Infotainment Division
9868 Scranton Road, San Diego, CA 92121-3707
☎ (+1 619) 452 75 80 ⊖ (+1 619) 452 12 49
Vice President and General Manager: Anthony C D'Augustine

Mechanical Filters
2990 Airway Avenue, Costa Mesa, CA 92626-6018
☎ (+1 714) 641 53 15 ⊖ (+1 714) 641 53 20
Business Director: Lee G Cornett
■ Design, develop and manufacture electrical bandpass filters in the 3 kHz to 600 kHz centre frequency range. Bandwidths range from 20 kHz to 30 kHz. Applications include: HF, UHF and VHF radios: DGPS receivers; cellular base stations and Omega navigation receivers.

Multimedia Communications Division
PO Box C, Newport Beach, CA 92658-8902
4311 Jamboree Road, Newport Beach, CA 92660-3095
☎ (+1 714) 221 46 00 ⊖ (+1 714) 221 40 78
Vice President and General Manager: Vijay C Parikh
■ Facsimile and PC modem devices for personal communications electronics.

Network Access Division
PO Box C, Newport Beach, CA 92658-8902
4311 Jamboree Road, Newport Beach, CA 92660-3095
☎ (+1 714) 221 46 00 ⊖ (+1 714) 221 40 78
Vice President and General Manager: Raouf Y Halim
■ Produce digital and mixed-signal ICs for multimedia, graphics, communications and imaging applications.

Wireless Communications Division
PO Box C, Newport Beach, CA 92658-8902
4311 Jamboree Road, Newport Beach, CA 92660-3095
☎ (+1 714) 221 46 00 ⊖ (+1 714) 221 40 78
Vice President and General Manager: Vijay C Parikh (acting)
■ Total system solutions for advanced cordless telephony and global positioning systems receiver engines.

Rockwell Telecommunications, Inc.
300 Bauman Court, PO Box 8007, Wood Dale, Il 60191
☎ (+1 630) 227 80 00 ⊖ (+1 630) 227 81 86
Vice President and General Manager: Ms Mary I Reilly (acting)
■ Design, manufacture and service of call centre technologies for applications in industries including aerospace, utilities and government institutions.

Rockwell Automation
Headquarters
1201 South Second Street, PO Box 2086,
Milwaukee, Wisconsin 53204
☎ (+1 414) 382 20 00 ▣ 4311016
⊖ (+1 414) 382 44 44
President and Chief Operating Officer: Jodie K Glore

■ This division brings together Allen-Bradley, Reliance Electric, Dodge and Rockwell Software. Manufacture products including: control logic, sensors, human-machine interface, engineered motors, mechanical power transmission equipment and software. Technical resources available worldwide. Applications in the manufacturing and process industries.

Dodge Mechanical Group
PO Box 499, Greenville, SC 29602-0499
6040 Ponders Court, Greenville, SC 29615-4617
☎ (+1 864) 297 48 00 ⊖ (+1 864) 281 24 87
Senior Vice President: Joseph D Swann

Global Sales & Support Group
1201 South Second Street, Milwaukee, WI 53204
☎ (+1 414) 382 20 00 ⊠ 4311016
⊖ (+1 414) 382 44 44
Group Vice President: Richard E Shelton

Control & Information Group
1 Allen-Bradley Drive, Mayfield Heights, OH 44124-6118
☎ (+1 440) 646 50 00 ⊖ (+1 440) 646 30 75
Senior Vice president: Keith D Nosbusch

Drives & Motion Control Group
6400 West Enterprise Drive, Mequon, WI 53092
☎ (+1 414) 242 82 00 ⊖ (+1 414) 242 82 13
Senior Vice President: Timothy J Clark

Industrial Control Group
1201 South Second Street, PO Box 2086,
Milwaukee, WI 53204
☎ (+1 414) 382 20 00 ⊠ 4311016
⊖ (+1 414) 382 44 44
Senior Vice President: David L Pugh

Sprecher + Schuh North America
15910 International Plaza Drive,
Houston, TX 77032-2439
☎ (+1 281) 931 70 00 ⊖ (+1 281) 931 90 18
President: John W Hamm

Electrical Groups
PO Box 248020, Cleveland, OH 44124-8020
6065 Parkland Boulevard,
Mayfield Heights, OH 44124-6106
☎ (+1 216) 266 58 00 ⊖ (+1 216) 266 76 66
Executive Vice President: Randolph W Carson

Packaged Control Products
PO Box 424, Milwaukee, WI 53201-0424
788 North Jefferson Street, Milwaukee, WI 53202-3718
☎ (+1 414) 382 21 00 ⊖ (+1 414) 382 36 81
Vice President: Robert B Anderson 98

75-75960 Rocky Mountain Instrument Co.
1501 South Sunset, Longmont, CO 80501-6750
☎ (+1 303) 651 22 11 E 6322108@mcimail.com
⊖ (+1 303) 651 26 48
♣ US$ 6,500,000 (1995) 80 4,500 m2
President: Dr Yubong Hahn
Sales and Advertising Director: Ms Debbie Hunt
Sales Administration: Ms Julie Case
■ Optical components and subassemblies. Laser Optics. 98

75-75970 Rodale Electronics Inc.
603 Chestnut Street, Garden City, NY 11530
☎ (+1 516) 222 00 50 ⊖ (+1 516) 222 02 23
President: John B Clement
Operations Director: Vince Maida
Advertising Manager: Marguerite Fullman
■ ECM equipment used for training, test & evaluation. Radar simulator support systems used to train ESM system operators, airborne, pod & on-board mounted.
Overseas agents:
Ericsson Radar Electronics AB, Sweden 98

75-75980 Rodenstock Precision Optics Inc.
(a subsidiary of Optische Werke G. Rodenstock, Germany FR)
4845 Colt Road, Rockford, IL 61109
☎ (+1 815) 874 83 00 E rodstk@bossnt.com
W http://www.rodenstockoptics.com
⊖ (+1 815) 874 63 74
Vice President: Bruce E Nielsen
■ Provides aspheric, plano & spherical precision optical components & systs for ultra-violet, visual, near infra-red & infra-red requirements.
® Germany: Optische Werke G.Rodenstock 98

75-76000 Rogers Corp.
1 Technology Drive, PO Box 188, Rogers, CT 06263-9999
☎ (+1 860) 774 9605 ⊖ (+1 860) 779 5509
♧ $ 60 098 000 ♣ $ 140 293 000 900

President: H.H. Birkenruth
VP Sales & Marketing: R.D. Wachob
Mgr Marketing Communications: R.F. White
■ Develops & manufactures specialty polymer composite materials & components for targeted applications in the imaging, communications, computer & peripheral, consumer products & transportation markets. 97

75-76030 Rogerson Aircraft Corp.
2201 Alton Parkway, Irvine, CA 92714
☎ (+1 714) 660 06 66 ⊠ TWX 910 595 2843
⊖ (+1 714) 660 79 65
250 (1997) 225,000 sq ft
Chairman: Michael J Rogerson
Marketing Services Manager: Ms Rowen Pastrana

Rogerson Aircraft Controls
16940 Von Karman, Irvine, CA 92606
☎ (+1 714) 422 23 74 ⊖ (+1 714) 442 23 11
Vice President and General Manager: John Schultz
■ Manufacture fuel and air valves, air regulators, flow sensors and water separators.

Rogerson Aircraft Systems
16940 Von Karman, Irvine, CA 92606
☎ (+1 714) 442 23 53 ⊖ (+1 714) 442 32 11
Vice President and General Manager: John Schulz
■ Manufacture vacuum lavatory systems and de-ice and anti-ice systems.

Rogerson ATS
16940 Von Karman, Irvine, CA 92606
☎ (+1 714) 442 23 77 ⊖ (+1 714) 442 23 11
Vice President and General Manager: John Schultz
■ Manufacture auxiliary fuel systems and advanced tactical in-flight refueling systems. 98

75-76035 Rogerson Kratos
(a subsidiary of Rogerson Aircraft Corporation, USA)
403 South Raymond Avenue, PO Box 91210,
Pasadena, CA
☎ (+1 818) 449 30 90 ⊖ (+1 818) 795 03 22
123 (1997) 55,000 sq ft
Chairman: Michael J Rogerson ☎ (+1 714) 442 23 45
President: Cannon Matthews
Vice President of Operations: Dan Richardson
☎ (+1 818) 449 30 90
Vice President of Engineering: Ronald R Ross
Marketing Manager: J Roy Glass
■ Manufacture of flight deck instrumentation and electronic cockpit information system displays. Provides both original equipment and retrofit electronics for commercial and military applications. 98

75-76150 Rohr Inc.
(† Rohr Industries Inc.)
850 Lagoon Drive, Chula Vista, CA 91910-2098
☎ (+1 619) 691 41 11 W http://www.rohr.com
⊠ 695038 rohr inc chul
♣ $ 805,000,000 (1995) 4,213 (1997)
Chairman and Chief Executive Officer: Robert Rau
Senior Vice President and Chief Financial Officer: Laurence Chapman
Senior Vice President of Programs, Technology and Quality Assurance: John R Johnson
Senior Vice President of Operations: Graydon Wetzler
Senior Vice President of Customer Support: David R Watson
Vice President of Engineering: Richard J Warters
Vice President of Materiel: David W Shaw
Vice President of Human Resources: David Ramsay
Vice President of Government and Space Products Division: Keith D Gentry
Vice President of Manufacturing - Chula Vista: Greg Peters
Manager of Corporate Communications: Ms Valorie McClelland
Human Resources Associate: Ms Laura Hawley
■ Component manufacture and supplier to prime manufacturers in the aerospace industry. Products include aircraft engine nacelles, thrust reversers, pylons and other structural and engine components for commercial, military and business aircraft. → 80
Subsidiary companies:
Rohr Aero Services Inc. → 80
Rohr Europe SA → France 75
Rohr Aero Services Asia

Rohr-Riverside Plant
8200 Arlington Avenue, Riverside, CA 92503-1499
☎ (+1 714) 351 54 00
⊖ (+1 714) 351 52 87; 351 55 56; 351 50 88
646
General Manager: Colin Cramp

Rohr-Foley
1300 West Fern Street, PO Box 280,
Foley, AL 36536-0280
☎ (+1 205) 952 30 00 ⊖ (+1 205) 952 33 29
338 265
General Manager: Paul H Nygaard

Rohr-Hagerstown
Showalter Road, Hagerstown, MD 21740
☎ (+1 301) 790 95 00 ⊖ (+1 301) 790 96 21
142
General Manager: Philip Vacca

Rohr-Sheridan
Highway 270 East, Sheridan, AR 72150
☎ (+1 501) 942 8100 ⊖ (+1 501) 942 8169
144
General Mgr: Donald Hinkle

Rohr-Heber Springs
5201 Highway 210 East, Heber Springs, AR 72543
☎ (+1 501) 362 95 00 ⊖ (+1 501) 362 95 20
141
General Mgr: Van Halfacre 98

75-76200 Rohr - San Marcos
2005 Technology Way, San Marcos, TX 78666
☎ (+1 512) 754 36 00 ⊖ (+1 512) 754 36 01
130
President: George Ryan 98

75-76650 Rosemount Aerospace Inc.
(a subsidiary of The BFGoodrich Company)
14300 Judicial Road, Burnsville, MN 55306
☎ (+1 612) 892 43 00 ⊖ (+1 612) 892 44 30
1,070 31,000 sq m
General Mgr: Ron Hodges
Marketing Communications Mgr: Sol Mirelez
■ Design, development & manufacture distributed & integrated air data systems, air data computers, air data sensing probes, flush-based air data systems, angle of attack systems, ice detectors, total temperature sensors, surface & immersion temperature sensors, structure heaters, cooling effect sensors, optical turbine temperature sensors. 98

75-76750 Rotek Inc.
Aurora, OH 44202
☎ (+1 216) 562 4000 ✆ (800) 221 8043
⊠ 986448; 241559 ⊖ (+1 216) 562 4620
President: Brian Webb
VP Sales & Marketing: Michael A. Drobik
Operations & Technical Dir: Lynn Myers
Purchasing Mgr: David Bolton
Mgr Military & Precision Sales: Mark McInturff
■ Manufacture bearings for military applications. 97

75-76900 Rowe Industries Inc.
(a subs. of Talley Industries)
6225 Benore Rd, PO Box 6877, Toledo, OH 43612
☎ (+1 419) 729 9761 ⊖ (+1 419) 729 1794
100 56 000 sq ft
President: Woody Bower
VP Quality: S. Edelstein
Operations/Tech Dir: Mark Robinson
Purchasing Mgr: Dennis Bullimore
■ Manufacturer of high voltage cable assemblies, connectors and specialised wire & cable, including severe environment systems. 97

75-77015 Ruggedtronics
24711 Redlands Boulevard, Suite G.,
San Bernardino, CA 92408
☎ (+1 909) 796 53 74 ⊖ (+1 909) 796 77 04
♣ US$ 734,000 (1996) 6 283 m2
President: Don Parker
■ Ruggedized peripherals for severe environments. 98

75-77040 Ruska Instrument Corp.
(a subsidiary of Druck Ltd, UK)
3601 Dunvale Road, Houston, TX 77063
☎ (+1 713) 975 05 47 E ruska@ruska.com
⊠ 762404 ruska inst hou ⊖ (+1 713) 975 63 38
85 (1997)
Managing Director: Darrel Chelette
Marketing Director: Ms Amy Flanagan
■ Deadweight gauges, digital transfer standards - fixed and portable, pressure instrumentation testing and calibration. Air data test sets. Instrumentation for phase behaviour studies/engineering reservoir analysis (pressure-volume- temperature). 98

Contact Reference

75-77340 **Saco Defense Inc.**
(a subsidiary of Chamberlain Manufacturing Corp.)
291 North Street, Saco, ME 04072-0890
☎ (+1 207) 283 36 11 ⊠ 944408 ⊖ (+1 207) 283 13 95
👥 500 🏭 425,000 sq ft
President: Bruce Makas
Vice President of Sales and Marketing: Michael C Costello
International Sales Manager: Thomas McNamara
■ MK19 Mod 3 40mm machine gun, M60 machine gun series: M60, M60D, M60E2, M60E3, M60E3 (Enhanced), M60E4 (LMG, mounted and Coax), M2HB machine gun series:
M2HB, M2HB (Quick Change Barrel), barrels 5.56 to 40mm, ILS (including spares, training manuals) (All weapons offered as systems: including ILS, mounts, ammunition, FCs → 80 [98]

75-77345 **Sadler Aircraft Corp.**
8225 East Montebello, Scottsdale, AZ 85250
☎ (+1 602) 994 46 31 ⊖ (+1 602) 481 05 74
♣ $ 1,500,000 (1996) 👥 6 (1997) 🏭 6,000 sq ft
President: William G Sadler [E] wsadler@sadlerair.com
Sales Manager: F Brent Stewart
■ Production of Sadler Piranha light armed surveillance aircraft. Production of Sadler UAV18-50 surveillance and attack drone system. [98]

75-77350 **Safariland Ltd Inc.**
3120 E. Mission Blvd, PO Box 51478, Ontario, CA 91761
☎ (+1 909) 923 7300 ⊖ (+1 909) 923 7400
VP Operations: Scott T. O'Brien
Sales Mgr: Thomas Wylie
Advertising Mgr: Ms Norma E. Osborn
■ Law enforcement equipment, belts, holsters, pouches accessories, body armour. [97]

75-77400 **Safe Flight Instrument Corp.**
20 New King St, White Plains, NY 10604-1206
☎ (+1 914) 946 9500 ⊠ 6818104
⊖ (+1 914) 946 7882
👥 150 🏭 47 000 sq ft
President: Leonard M. Greene
Vice President & General Mgr: Thomas Robertazzi
Vice President & Chief Financial Officer: Donald F. Greene
Dir Sales & Marketing: Paul J. Davidson
Mgr Marketing Communications: Ms Helen Pringle Gibbs
■ Avionics manufacture, providing angle-of-attack based safety and performance equipment for all types of aircraft. Systems include stall warning, speed control, speed command of attitude and thrust (SCAT), and automatic throttles, airborne wind shear warning with recovery guidance. [97]

75-77550 **Saft America Inc.**
(a subsidiary of Saft SA, France)
711 Industrial Boulevard, PO Box 1886,
Valdosta, GA 31601
☎ (+1 912) 247 23 31 [W] http://www.saftamerica.com
⊖ (+1 912) 245 28 27
👥 265 (1997) 🏭 200,000 sq m
President and Chief Executive Officer: F D Westfall (Jr)
Vice President and General Manager: Philippe Boulais
Vice President of Sales and Marketing: Syd Reames
Manager of Marketing Communications: Ms Sarah Smart
■ Manufacture and market nickel cadmium, lithium and thermal batteries for fixed-wing and rotary-wing aircraft, communications equipment, navigational and guidance systems.
® France: SAFT

Lithium Battery Division
313 Crescent Street, Valdese, NC 28690
☎ (+1 704) 874 41 11 ⊖ (+1 704) 874 24 31
Manager: Ric Raines

Research and Development Center
107 Beaver Court, Cockeysville, MD 21030
☎ (+1 410) 771 32 00 ⊖ (+1 410) 771 11 44 [98]

75-77610 **Sage Laboratories Inc.**
11 Huron Drive, Natick, MA 01760
☎ (+1 508) 653 08 44 ⊖ (+1 508) 653 56 71
♧ $ 10,096,000 (1996) ♣ $ 9,769,000 (1996)
👥 51 (1996) 🏭 30,000 sq ft
Chairman and Chief Executive Officer: Carl A Marguerite
President and Chief Operating Officer: Louis J Lanzillo (Jr)
Vice President Sales: Anthony J Cieri
Director of Space and Electromechanical Switches: J Andrew Funk
Sales and Marketing Director: Peter A Alfano
Purchasing Manager: Michael J Cahill
■ Design, manufacture & marketing of advanced microwave components. Electromechanical- switches, rotary joints, phase shifters, connectorised-passive RFD microwave components. Caselles Drop-Ins-Wireline, Wirepac, Hybrids & couplers. [98]

75-77670 **Sageware Inc.**
480 San Antonio Road, Suite 230,
Mountain View, CA 94040
☎ (+1 415) 917 76 60 [E] info@sageware.com
⊖ (+1 415) 917 76 79
👥 25
President: John Lehman
■ KnowledgeSets: develop search knowledgebases for use with any search engine in any application. Updated every 90 days.
Subjects include; industries, geo-politics a and geo-economics, company financial analysis and legislation.
Sageware hold a facility security clearance. [98]

75-77700 **SAI Industries Corp.**
1207 Fulton, San Antonio, TX 78201-9514
☎ (+1 210) 736 26 11 [E] saiindco@flash.net
⊖ (+1 210) 736 26 13
♧ $ 500,000 (1997) ♣ $ 1,000,000 (1997)
👥 25 (1997) 🏭 3,800 sq ft
President and Chief Executive Officer: Natu P Patel
Operations Director: Art Ramirez
■ Aircraft work for USAF. Engine and structural aircraft work for Dee Howard, Fairchild and Mooney aircraft. CNC/Lathe/ Mill screw machine and sheet metal parts. Cable, circuit board and electrical mechanical assembly. [98]

75-77850 **Sanders**
(† Lockheed Sanders Inc.)
(a Lockheed Martin Company)
Headquarters
65 Spit Brook Road, Nashua, NH 03061-0868
☎ (+1 603) 885 43 21 ⊠ 943430; TWX 710 2288 958
⊖ (+1 603) 885 3655
👥 5,000 🏭 2,300,000 sq ft
President: Dr John R Kreick
Vice President Programs and Missions Success: Charles E Franklin
Senior Vice President Finance: Howard L Rhude
Vice President Operations: Steven J Walker
Vice President Engineering: Robert K Henry
Vice President Business Development: Joseph A Giacoponello
Vice President Human Resources: Bradford Loggans
Vice President Contracts: John M Boland (Acting)
Procurement Director: Dave Zalewski
Company Counsel: Norman Marsh (Jr)
Communications Director: Marvin Braman
Public Relations Manager: Joseph Wagovich
■ The company is engaged in the development, manufacture & sale of advanced electronic systems & products in a broad range of defence & commercial markets.
It is a major producer of aircraft electronic combat & self-protection systems & automated mission planning systems. Other major business areas include tactical communications, surveillance & intelligence systems, avionics & space electronics & microwave electronic subsystems & telecommunications systems.

Communications NHQ 1-765
PO Box 868, Nashua, NH 03061-0868
☎ (+1 603) 885 28 17 ⊖ (+1 603) 885 28 13
Communications Director: Marvin Braman

Countermeasures Division
95 Canal Street, Nashua, NH 03061
☎ (+1 603) 885 38 24
⊠ 943430 sanders nasu; TWX 710 2281 894
⊖ (+1 603) 885 61 09
Vice President and General Manager: Howard Arndt (Acting)
Business Development Director: William Jones
■ Self-protection countermeasures systems (radio frequency, infra-red & optical); missile warning systems; integrated electronic combat systems; electronic support measures systems; modular avionics & automated test systems.

Microwave Electronics Division
65 Spit Brook Road, Nashua, NH 03060
☎ (+1 603) 885 22 42 ⊠ 943430; TWX 710 2281 894
⊖ (+1 603) 885 05 20
Vice President and General Manager: Kenneth J Cusack (Acting)
Business Development Manager: John A Windyka
■ Design, development of advanced microwave & millimetre wave components utilising internal GaAs/Inp foundry & advanced multi-chip module packaging.

Surveillance Systems Division
Pope Technical Park, 65 River Road, Hudson, NH 03051
☎ (+1 603) 885 27 18 ⊠ 943430 ⊖ (+1 603) 885 77 01
Vice President and General Manager: Douglas M Barrett
Business Development Director: Ranee Walleston (Acting)
■ Signal acquisition, analysis, location & jamming; electronic support measures (ESM); signal exploitation/automatic threat recognition; surveillance systems equipment.

Information Systems Division
130 Daniel Webster Highway, Merrimack, NH 03054
☎ (+1 603) 885 74 48 ⊖ (+1 603) 885 35 04
Vice President and General Manager: Douglas M Barrett (Acting)
Business Development Director: Ralph San Antonio
■ Mission planning systems; training & simulation systems; electronic combat training, air defense training, on-board training, interactive team training.

Avionics Division
95 Canal Street, Nashua, NH 03060
☎ (+1 603) 885 51 82 ⊖ (+1 603) 885 72 64
Vice President: John M Lydiard III
Business Development Director: James E Tatem
■ Electronic flight instrumentation systems (EFIS); embedded processors; integrated avionics design, development & production; non-intrusive antenna array design, development & production; military specification, data storage/retrieval equipment; miligraphic displays, cockpit display subsystems; space and airborne video recorders.

Advanced Systems Directorate
955 Perimeter Road, Manchester, NH 03103
☎ (+1 603) 645 55 17 ⊖ (+1 603) 645 56 81
Director: Dr Douglas Mook
Business Development Manager: William McNary (Acting)
■ Advanced low frequency active sonar systems & subsystems; non-acoustic systems design & development; advanced signal processing systems design & development; airborne acoustic signal processing.

International Business Development
65 Spit Brook Road, PO Box 0868,
Nashua, NH 03061-0868
☎ (+1 603) 885 33 84 ⊠ 943430 ⊖ (+1 603) 885 38 54
Director: R Wheeler (Acting) [98]

75-78150 **Sargent Fletcher Inc.**
(an Associate of Cobham plc, UK)
9400 Flair Drive, El Monte, CA 91731
☎ (+1 818) 283 7171 ⊠ TWX 910 5873 426
⊖ (+1 818) 579 9183
President: W. Skibbe
VP Marketing: Richard Simpson
VP Commercial: G. Catrambone
Tanks & Pods Dir: Frank Garcia
Purchasing Mgr: Paul Stevens
■ Aerial in-flight refuelling systems, external fuel tanks, special-purpose tanks, fire bombs, training missiles, metal containers. Aircraft parts fabrication. [97]

75-78330 **SBS Technologies Inc.**
(† SBS Engineering Inc.)
2400 Louisiana Boulevard, North East,
Albuquerque, NM 87110
☎ (+1 505) 875 06 00 [E] sbssale@sbse.com
[W] http://www.sbs-avionics.com ⊖ (+1 505) 875 04 00
♣ $ 53,000,000 (1996) 👥 270 (1996)
President: Christopher J Amenson
Vice President of Sales: Derek Head
Operations Director: Leonard Schuler
Flight Simulation Systems Designer: John Reeser
Multimedia Marketing Specialist: Ms Kristine Kessler
■ Provider of MIL-STD-1553 and ARINC 429/575 interfaces. Product backplanes include VME, ISA, PC/104, PCMCIA and PCI/PMC. Over thirteen operating systems are supported. C, Ada and DLL source code libraries are supplied at no charge. Features include: multi-functionality with simultaneous BC, 31 RTs and Dual Function Monitoring operations, Bus analysis, Simulation and Drop-out playback. [98]

75-78356 **Schaeffer Magnetics Inc.**
9175 Deering Avenue, Chatsworth, CA 91311-5802
☎ (+1 818) 341 51 56 [E] smi@schaeffer.com
[W] http://www.schaeffer.com ⊖ (+1 818) 341 38 84
♣ $ 20,000,000 (1997) 👥 150 (1997)
President: Ernest Schaeffer
President: Dr Keith Deconde

Executive Vice President: Max Jacobsen
Vice President of Finance: Ms Margaret Lorenger
Vice President of New Business: Jim Hammond
■ Supplier of spaceflight quality motors and actuators. [98]

75-78410 Schott Glass Technologies Inc.
(a subsidiary of Schott Corp)
400 York Avenue, Duryea, PA 18642-2026
☎(+1 717) 457 74 85 ⊖ (+1 717) 457 69 60
👥 450
President: John G. Gulbin
Sales & Marketing Dir: Allan C. Kreutzer
Vice President Operations: S Krenitsky
Marketing Services Mgr: David Schimmel
Optical Sales Mgr: Josef Nimmerfroh
Purchasing Mgr: Richard Blaum
■ Optical glass, colour filter glass, contrast enhancement filters, laser filters, night vision goggle compatible filter glass. [98]

75-78460 Schramm Inc.
800 E. Virginia Ave, Westchester, PA 19380-4439
☎(+1 610) 696 2500 ⊠ 835455 ⊖ (+1 610) 696 6950
♣ $ 24 000 000 (1995) 👥 100
CEO/President: R. Schramm
VP Sales & Marketing: R. Edwards
■ Rotary drilling rigs for water well, blasthole & environmental drilling. [97]

75-78850 Scott Aviation
(a division of Figgie International Company)
225 Erie Street, Lancaster, NY 14086
☎(+1 716) 683 51 00 ⊖ (+1 716) 681 10 89
👥 717 (1996)
President: G. Lindemann
VP Sales & Mktg Aviation & Govt.: M.R. Kaletta
Manufacturing Dir: Richard Delisle
Mgr Aviation Sales: Thomas B. Harmon
■ Life support equipment for military & government applications. Includes: Aviation oxygen systems (mask, hose, hood, regulator/valves); self contained solid state chemical oxygen breathing equipment for emergency escape/firefighting; air supplied & air purifying respiratory protection equipment; hyperbaric deep dive systems; fuel cell regulators; gas detection monitoring & central control systems; aviation instruments.

Scott Air Purifying Products
1201 Kalamazoo Street, S.Haven, MI 49090
☎(+1 616) 637 21 21 ⊠ 729450

Scott Health and Safety
309 West Crowell Street, Monroe, NC 28110
☎(+1 704) 283 21 25 [98]

75-78900 SeaArk Marine Inc.
(† MonArk Boat Co.)
PO Box 210, Monticello, AR 71655
☎(+1 501) 367 9755 ⊠ 783028 ⊖ (+1 501) 367 2120
♣ $ 6 000 000 (1995) 👥 150 (1996)
President: Zach McClendon (Jr)
VP Finance: James G. Eoff
Sales Mgr: Ken McFalls
VP Manufacturing: Robert Hall
■ Patrol boats, crew boats, utility craft, personnel launches, landing craft, survey boats, research vessels, fireboats, search & rescue craft, training craft, harbour security boats in aluminium. [97]

75-78910 Seaboard Electronics Co.
70 Church Street, New Rochelle, NY 10805-3204
☎(+1 914) 253 80 73 ⊖ (+1 914) 235 83 69
👥 20 (1997) 🏭 8,000 sq ft
President: Jerry Hacker
Sr Sales Engineer: S. Grandhi
■ Manufacture electronic security systems; portable body alarm systems; wireless fire alarm systems; remote control systems. [98]

75-78917 Sealtron Inc.
9705 Reading Road, Cincinnati, OH 45215-3592
☎(+1 513) 733 84 00 ⊖ (+1 513) 733 01 31
👥 125
General Manager: Norm Allard
Sales Manager: Gordon Porter
■ Manufacture Mil-Spec and special purpose connectors. [98]

75-78930 Sea Tel Inc.
1035 Shary Court, Concord, CA 94518-2409
☎(+1 510) 798 79 79 ⌀ (888) 798 79 79
⊖ (+1 510) 798 79 86
♣ $ 15,000,000 (1997) 👥 80 (1997) 🏭 3,000 m2
President and Chief Executive Officer: Patrick Matthews
Marketing Director: Richard D Stallbaum
■ Stabilised antenna platforms, satellite tracking equipment, glass fibre radomes to 14 feet in diameter. L,S,C,X,C Ku, bard applications. [98]

75-78940 Seaward International Inc.
PO Box 98, Clearbrook, VA 22624-0098
☎(+1 540) 667 51 91 ⌀ (800) 828 53 60
[W] http://www.seaward.com ⊖ (+1 540) 667 79 87
President: Frank A March
Vice President: Robert B Taylor
■ Dock fenders; ship-to-ship fenders; marine buoys. [98]

75-79000 Second Chance Body Armor Inc.
7919 Cameron Street, PO Box 578,
Central Lake, MI 49622
☎(+1 616) 544 57 21 ⌀ (800) 253 70 90
[W] http://www.secondchance.com ⊖ (+1 616) 544 98 24
👥 65 (1997) 🏭 10,000 sq ft
President and Chief Executive Officer: Richard C Davis
Senior Vice President: Karen P Davis
Vice President Advertising: David A Kotwick
Group Vice President Advertising and Technology: Ed Bachner
Group Vice President Sales and Marketing: Paul Banducci
Operations and Technology Director: Larry McCraney
Purchasing Manager: Ms Vicki Dawson
■ Manufacture, sales & service of concealable & external personal body armour. → 80 → 90 [98]

75-79010 Sector Microwave Industries
999 Grand Boulevard, Deer Park, NY 11729-5705
☎(+1 516) 242 23 00 ⊖ (+1 516) 242 81 58
♣ $ 7,000,000 (1996) 👥 46 (1997) 🏭 2,800 sq m
President: Victor Nelson
Vice President Sales and Marketing: John Philips
Sales Marketing Mgr: Thomas J. Nelson
■ Manufacture waveguide switches, coaxial switches & dual waveguide & coaxial switches used on ships, missiles, aircraft, spacecraft, ground stations or mobile uplinks. [98]

75-79080 Security Defense Systems Corp.
139 Chestnut St, Nutley, NJ 07110-0243
☎(+1 201) 235 0606 ⌀ (800) 325 6339
⊖ (+1 201) 235 0132
♣ $ 2 000 000 👥 6 🏭 2 500 sq m
General Mgr: Donald M. Airey
VP Engineering: Gary Korkala
Sales Mgr: John R. Levereth
■ Manufacture fixed metal detection systems & EOD/mail X-Ray machines. [97]

75-79100 Security Engineered Machinery (SEM)
5 Walkup Drive, PO Box 1045, Westboro, MA 01581
☎(+1 508) 366 14 88 ⌀ (800) 225 92 93
[E] semco@worldnet.att.com
[W] http://www.semshred.com ⊖ (+1 508) 366 68 14
👥 40 (1997) 🏭 3,000 m2
Chief Executive Officer: Leonard Rosen
International Sales Director: Frank Blanchard
■ Document destruction systems including shredders & disintegrators to destroy paper, microforms, circuit boards, ID badges, unburst printouts. Single models or complete turnkey systems. [98]

75-79225 SE International Inc.
436 Farm Road, PO Box 39, Summertown, TN 38483
☎(+1 615) 964 35 61 ⌀ (800) 293 57 59
[E] seiinc@usit.net ⊖ (+1 615) 964 35 64
♣ $ 1,000,000 (1995) 👥 8 (1995) 🏭 2,750 m2
President: Ms Susan Skinner
Director of Marketing: Corey Walker
Purchasing Manager: Ms Deborah DeVoursney
■ Ionizing radiation detectors. [98]

75-79300 Semi-Alloys Co.
888 South Columbus Avenue,
Mount Vernon, NY 10550-5019
☎(+1 914) 664 28 00 [E] semialoys@pipeline.com
⊖ (+1 914) 664 53 54
👥 350
President: D K Loyd
Operations Director: Al Orzechowski
Sales Manager: Fred Amori
Purchasing Manager: Bob Putkowski
■ Electronic packaging components. [98]

75-79320 Sensidyne Inc.
16333 Bay Vista Drive, Clearwater, FL 34620
☎(+1 813) 530 3602 ⌀ (800) 451 9444
⊖ (+1 813) 539 0558
President: Bryan Trvex
Sales Mgr: Michael Egan
■ Chemical agent detectors. Gas detection & air sampling equipment. [97]

75-79340 Senso-Metrics Inc.
4584 Runway Street, Simi Valley, CA 93063
☎(+1 805) 527 36 40 ⊖ (+1 805) 584 29 60
♣ $ 2,500,000 👥 28 🏭 17,000 sq ft
Chief Executive Officer: Don Evans
President: Gary Johnson
Vice President of Marketing: Stan Rubens
Purchasing Manager: Bryan Pierce
■ Design, develop and manufacture a wide range of semiconductor strain gauge type pressure transducers for military, aerospace, commercial and industrial markets. Standard product line and capability to design and manufacture pressure transducers and systems to suit individual application requirements. [98]

75-79400 Sensor Systems Inc.
8929 Fullbright Avenue, Chatsworth, CA 91311
☎(+1 818) 341 53 66
[E] sensorsystems@worldnet.att.net
⊖ (+1 818) 341 90 59
👥 200 (1997) 🏭 60,000 sq ft
President: Ms Mary E Bazar
Vice President: Si Robin
■ Aircraft antennas. [98]

75-79450 SEOS Displays
(† 3-D Structures)
Manufacturing Division
9020 Gap-Newport Pike, Avondale, PA 19311
☎(+1 610) 869 8500 ⊖ (+1 610) 268 0200
♣ $ 3 500 000 👥 34
President: Owen Wynn
■ Visual display systems for simulation & curved projection screens.

North American Office
PO Box 11526, Daytona Beach, FL 32120 [98]

75-79490 Sequa Corp.
Corporate Headquarters
200 Park Avenue, New York, NY 10166
☎(+1 212) 986 55 00 ⊖ (+1 212) 370 19 69
♣ $ 1,459,000,000 (1996) 👥 9,350 (1996)
Chairman and Chief Executive Officer: Norman E Alexander
Vice President Corporate Communications: Ms Linda G Kyriakou
Vice President Corporate Development and Strategy: Amrit Kanwal
■ Manufacture aircraft engine components, solid rocket propellant motors for missiles & space.

Subs. with defence business:
Atlantic Research Corp. → 75
Chromalloy Gas Turbine Corp. → 75 [98]

75-79530 Sermatech International Inc
(a subsidiary of Teleflex Inc)
155 South Limerick Road, Limerick, PA 19468
☎(+1 610) 948 51 00 [W] http://www.sermatech.com
⊖ (+1 610) 948 08 11
Vice President Marketing: Robert Schubert
Marketing Manager Communications: Mike Stock
■ Protective turbine engine coatings, offers Mil.spec. inorganic, organic & organo-metallic coatings for the protection of aerospace & military ordnance components. → 80 [98]

75-79650 Servo Corp. of America
111 New South Rd, Hicksville, NY 11801-5259
☎(+1 516) 938 9700 ⊖ (+1 516) 938 9644
President: Stephen A. Barre

Contact Reference

Sr Vice President: Kent R. Martin
Operations Dir: Walter Jensen
Programmes Dir: Ron Freedman
Purchasing Mgr: Ben Cipriani
■ VHF/UHF direction finding equipment for ground based aviation & marine (shipboard & shore based) applications. [97]

75-79678 SFIM Inc
(a subsidiary of SFIM Industries, France)
2701 Forum Drive Grand Prairie, Texas 75053-4005
☎ (+1 972) 641 34 61 ⊖ (+1 972) 641 37 11
♣ $ 2,500,000 ⚉ 11
Managing Director: Jean Baudin [98]

75-79800 Shakespeare Products Group Electronics & Fibreglass Division
(† Shakespeare Co.)
(a subsidiary of K2 Inc)
PO Box 733, Newberry, SC 29108
☎ (+1 803) 276 55 04 ✆ (800) 800 90 08
W http://www.shakespeare-ef.com
⊖ (+1 803) 276 89 40
⚉ 600 (1997) ⌂ 7,000 sq m
President: Wayne Merck
Technical Director: Jim Davidson
Operations Director: Ray Jeffords
Sales and Marketing Director: Ms Lynn Derrick
Sales Manager: Denny King
Purchasing Manager: Bill Simpson
■ Manufacture vertical monopole antennas for MF/HF/VHF frequencies. Applications include shipborne, vehicular and ground-based antennas. Broadband technology available. [98]

75-79900 Shaw Aero Development Inc.
4227 Progress Ave, PO Box 8657,
Naples, FL 33941-8657
☎ (+1 941) 643 3310 ✆ (800) 997 7429
⊖ (+1 941) 643 2996
⚉ 27 (1995)
CEO/President: James R. Shaw
Executive VP & General Mgr: F.Grant Westerman
Sales & Marketing Dir: John R. Glebas
Purchasing Mgr: John Gracey
Marketing Mgr: Thomas G. Morgan
■ Liquid level sensors, electrical grounding contacts, expandable rubber plugs, leakproof fuel caps & fillers for trucks & off road vehicles. [97]

75-79950 Shaw Aero Devices Inc.
12291 Towne Lake Drive, Fort Myers, FL 33913
☎ (+1 941) 769 56 44 ⊖ (+1 941) 768 02 32
⚉ 175 (1997) ⌂ 40,000 sq ft
Managing Director: James Shaw
Sales and Marketing Director: James Driscoll
Operations and Technical Director: James Cline
Purchasing Manager: Rich Crouse
■ A wide range of products which include oil reservoirs and accessories, flame arresters, lavatory service panel waste drain valves.
Facilities include an environmental test chamber, flame test stand and lighting strike booth.
Quality systems to Mil-I-45208 and Mil-Q-9858. [98]

75-80150 Sheldahl Inc.
1150 Sheldahl Road, Northfield, MN 55057
☎ (+1 507) 663 80 00 ✆ (800) 874 35 32
W http://www.sheldahl.com ⊖ (+1 507) 663 85 45
♣ $ 114,000,000 (1996) ⚉ 1,100 ⌂ 450,000 sq ft
President and Chief Executive Officer: James E Donaghy
Senior Vice President: Edward L Lundstrom
Vice President Technology: Sydney J Robert
Market Research Manager: Terry F Origer
Product Manager Composite Materials: Dan J Roach
Product Manager Aerospace Materials: David R Keenan
Product Manager Flexible Interconnect: David P Becker
■ Active supplier to the military/aerospace industry. Product line includes flexible circuitry, passive thermal control for spacecraft & satellites, radar absorbing materials, engineered flexible structures. [98]

75-80205 Shenango Steel Buildings Inc.
33 Carbaugh Street, West Middlesex, PA 16159-4103
☎ (+1 412) 528 99 25 ✆ (800) 548 06 25
E shenangotmc@pathway.net
⊖ (+1 412) 528 24 52; 528 26 43
President: James M. Campbell
Manfacturing Dir: Albert Reed
Sales Department: Thomas Clark
Purchasing Mgr: Robert Nelson
■ Metal building design, fabrication & construction. [98]

75-80300 Sherwin Inc.
5530 Borwick Avenue, South Gate, CA 90280
☎ (+1 562) 861 63 24 ⊖ (+1 562) 923 83 70
♧ US$ 800,000 (1995) ♣ US$ 2,000,000 (1995)
⚉ 14 (1995) ⌂ 1,600 sq m
President: A G Sherwin
Operations and Technical Director: Sam Robinson
Purchasing Manager: Dale Winans
■ Inspection penetrants, fluorescent and visible dye (non-destructive testing). [98]

75-80455 Ship Analytics Inc.
North Stonington Professional Center,
183 Providence - New London Turnpike,
North Stonington, CT 06359
☎ (+1 860) 535 3092 ⊖ (+1 860) 535 0560
⚉ 75 (1995)
Chairman & CEO: A.J. Pesch
Operations/Technical Dir: Dr J.A. Hammer
VP Marketing: R.C. Cook
■ Shiphandling simulation, training system development, computer generated image systems, instructional technology & operability engineering. Products include Action Speed Tactical Trainers, Passive Acoustic Analysis Trainers, Bridge/Combat Information Center Simulators, Part-Task Trainers, Shiphandling Simulators (military & commercial). [97]

75-80457 Shock-Tech Inc.
(† Stop Choc Corp.)
55 Whitney Road, Mahwah, NJ 07430
☎ (+1 201) 848 1000 E shocktech@worldnet.att.net
⊖ (+1 201) 848 1355
⚉ 75 ⌂ 5,000 sq m
President: Serge Seguin
Technical Dir: Bert Slater
Sales Dir: Herb Kulik
Purchasing Mgr: Bill de Young
® France: Vibrachoc SA
Germany: Stop-Choc GmbH
Spain: Vibrachoc SA
UK: Stop-Choc Ltd [98]

75-80460 Shooting Systems Group Inc.
(† Assault Systems)
1075 Headquarters Park Dr., Fenton, MO 63026-2478
☎ (+1 314) 343 3575 ✆ (800) 325 3049
⊖ (+1 314) 349 3311
⚉ 50 ⌂ 9 200 sq ft
President: Bruce E. Bogue
Secr. & Treasurer: Christine M. Bogue
■ Manufacture of Nylon holsters, weapons cases & slings. [97]

75-80555 Siebe North Inc.
4090 Azalea Drive, PO Box 70729,
Charleston, SC 29415-7528
☎ (+1 803) 554 0660 ▣ 6843143
⊖ (+1 803) 744 2857
President: W.L. Reed
VP Finance & Admin: M. Wilson
■ Safety clothing & equipment [97]

75-80560 Siecor Corp.
(joint venture Corning Inc. and Siemens Corp.)
PO Box 489, Hickory, NC 28603-0489
☎ (+1 704) 327 50 00 ✆ (800) SIECOR1
⊠ http://www.siecor.com ▣ 800510
⊖ (+1 704) 327 59 73
⚉ 5,000 (1997)
President and Chief Executive Officer: Joseph D Hicks
Senior Vice President of Sales: William G Martin
Sales and Marketing Director: Mike Amicone
Public Relations Specialist: Ms Lisa Crawford
■ Manufacturer of fiber optic cables and accessories for voice, data and video communications applications. Products include fiber optic cable, cable assemblies, connectors, splicing and test equipment, interconnect and splice hardware. Also offer engineering services and hands-on fibre optic training courses. → 80 [98]

75-80565 Siemens Electromechanical Components Inc.
(† Potter & Brumfield)
(a subsidiary of Siemens Corp. USA)
200 South Richland Creek Drive, Princeton, IN 47671
☎ (+1 812) 386 10 00 E info@sec.siemens.com
⊖ (+1 812) 386 20 72
⚉ 3,000 (1997)
Vice President Sales: Richard Didriksen
Marketing Communications Manager: Karl R Grubb
■ Electromechanical, solid state & time delay relays; circuit breakers. [98]

75-80570 Siemens Solar Industries
(† Arco Solar Inc.)
4650 Adohr Lane, PO Box 6032, Camarillo, CA 93010
☎ (+1 805) 482 6800 ▣ 6716260
⊖ (+1 805) 388 6395
President: Oswald Gernot
Marketing Communications Mgr: Brett Tarnet
Dir N. American Sales: Raju Yenamandra
■ Photovoltaic energy systems used as power supplies for: communications, water pumping, cathodic protection, navigational aids, railroads, data transmission, lighting & other remote power applications. Photovoltaics represent a primary or supplementary source of energy in remote locations. [97]

75-80640 Sierra Technologies Inc. Sierra Research Division
(† Sierra Research)
485 Cayuga Rd, PO Box 222, Buffalo, NY 14225-0222
☎ (+1 716) 631 6200 ⊖ (+1 716) 631 7849
♣ $ 60 000 000 ⚉ 344
President: William E. Romans
VP Operations: Eugene McCabe
VP Programmes: Joseph A. Tomasino
■ Flight inspection (automatic, semi- automatic, portable, total) avionics upgrade (F-5); airborne platform telemetry system; unmanned threat emitter system (UMTE); aircraft survivability equipment trainer IV (Aset IV); mobile threat emitter system (MOTES) station keeping equipment (SKE). [97]

75-80650 Sierracin/Sylmar Corporation
(† The Sierracin Corp.)
12780 San Fernando Road, Sylmar, CA 91342
☎ (+1 818) 362 67 11 E sscmperry@aol.com
⊖ (+1 818) 362 78 01
⚉ 425 (1997) ⌂ 310,000 sq ft
Vice President of International Marketing: Lee R Marshall
President of Sylmar Division: Robert Hart
International Contracts Administrator:
Ms Donna McChesney
■ Transparency systems for the aerospace industry, including all-plastic (acrylic and polycarbonate), plastic/glass composite and all-glass transparencies, both heated and unheated, for military and commercial aircraft. [98]

75-80651 Sierracin/Magnedyne
1360 Park Center Dr., Vista, CA 92083
☎ (+1 619) 727 1811 ⊖ (+1 619) 727 7747
President: Keith Bosecker
General Sales Mgr: Alex D. Vass Gal
■ Motion control systems: autofocus drive motors for imaging infra-red seekers on missiles; mirror drives & docking actuators on satellites; surgical tool drive systems. [97]

75-80700 SIGARMS Inc.
(a subsidiary of SIG Swiss Industrial Company, Switzerland)
Corporate Park, Exeter, NH 03833
☎ (+1 603) 772 23 02 E sai@nhultranet.com
⊖ (+1 603) 772 90 82
♧ $ 40,000,000 (1996) ♣ US$ 50,000,000 (1996)
⚉ 100 (1996) ⌂ 50,000 sq ft
President and Chief Executive Officer: George N Schneider
Executive Vice President and Genneral Manager:
Hermann H Kloetzer
Director Law Enforcement and Contract Sales:
David A Flander
Sales and Marketing Director: Alan Newcomb
Director of Purchasing: Robert A Giles
■ Manufacturer, distributor. Training for law enforcement personnel.
® Germany: J.P. Sauer & Sohn GmbH
Switzerland: SIG [98]

75-80790 Sigma Signals Corporation
314 Highland Mall Boulevard, Austin, TX 78752
314 Highland Mall Boulevard, Suite 411
☎ (+1 512) 451 90 88 Ⓔ sigmasignal@ix.netcom.com
Ⓦ http://www.netcom.com/~sigmasig/info. html
▣ (+1 512) 451 90 89
♣ $ 500,000 (1996) 👥 10 (approximately)
President: A Wayne Sefcik
Marketing Director: James L Scogin
■ Radar test and measurement products and services. Radar cross section measurement systems, data aquisition hardware and software, doppler spectrum measurement, chaff electronic countermeasures, radar systems, software engineering. Ultra-portable RCS processors, full feature RCS and doppler data acquisition and processing, dynamic ship and aircraft RCS processing software, full feature chaff signature test and analysis software, RCS data storage and retrieval systems, synthetic PPI - radar target track with map overlays, doppler spectral analysis software. → 80 [98]

75-80795 Signal Technology Corp Olektron Operation
(† St Olektron Corp)
28 Tozer Road, Beverly, MA 01915
☎ (+1 508) 922 00 19 Ⓦ http://www.sigtech.com
⊖ (+1 508) 927 93 28
♣ $ 20,000,000 (1996) 👥 127 🏭 42,000 sq ft
President: Sam Smoorler
Managing Director: J Croke
Operations Director: Tom Casale
Sales and Marketing Director: Tom Zinkowsky
Vice President Sales and Marketing: D P Ritchie
Purchasing Manager: A Rusnock
■ RF and IF/log signal processing components. [98]

75-80880 Sino Swearingen Aircraft Company
(† Swearingen Aircraft Inc.)
1770 Sky Place Boulevard, San Antonio, TX 78216
☎ (+1 210) 258 3900 ⊖ (+1 210) 258 3917
👥 150
Advertising Mgr: Timothy P. Lynch
■ Manufacture Swearingen SJ30 twin turbofan business jet. [97]

75-80900 Sikorsky Aircraft
(a div. of United Technologies Corp.)
6900 Main St, S119A, Stratford, CT 06497-9129
☎ (+1 203) 386 4000 ✉ SIKORSKY
▣ 964372; 494607; TWX 710 453 1330
⊖ (+1 203) 386 7300
👥 8 900 🏭 2 000 000 sq m
President & CEO: Eugene Buckley
Exec. VP Operations & Engineering: Dennis Campbell
Sr VP Administration: George Kay
VP Govt Business Development: Gary F. Rast
VP Int'l Marketing: James J. Saherwhite
VP Int'l Programs/Japan: D.L. Powell
VP H-60 Army & Air Force Product Line: Arthur O'Leary
VP H-60 Navy Product Line: Mike Baxter
VP RAH-66 Comanche: Arthur Linden
VP Worldwide Customer Service: Michael Blake
VP Communications: James Lynch
Planning & Scheduling Dir: Albert J. Schwabenbauer
Dir Advertising: M. Spigener (Jr)
■ Design & manufacture helicopters for military & commercial markets. Product line includes Blackhawk (UH-60/S-70), Seahawk (SH-60/CV HELO), Super Stallion (CH-53E/S-80) & Jayhawk SH-60J & S-76. Development of RAH-66 Comanche for US Army in partnership with Boeing Helicopters.

Customer Service Center
PO Box 109610, West Palm Beach, FL 33410-9610
☎ (+1 407) 775 5200
Mgr Technical Support: Fred Baxter [97]

75-80930 Silent Partner Body Armor Inc. Personal Armour Systems Division (SPBA)
612-618 Third Street, Gretna, LA 70053-5898
☎ (+1 504) 366 48 51 ✆ (800) 321 57 41
⊖ (+1 504) 364 89 06
President and Chief Executive Officer: T Tyler Züfle
Vice President: Dr Hugo Axmann
■ Manufacture ballistic armour in concealable and tactical deployment modes; new armour system allowing a 50% weight reduction for same level of protection now available; AFV up armour kits; high-speed coastal patrol boats; RPV's and architectural armour.

Associated company:
South Central Ordnance Inc. → 75 [98]

75-81020 Silver Eagle Manufacturing Co.
1761 NE Argyle Street, Portland, OR 97211
☎ (+1 503) 283 31 66 ✆ (800) 547 67 92
Ⓦ http://www.silvereaglemfg.com ⊖ (+1 503) 286 75 25
👥 150 (1997) 🏭 80,000 m2
Sales: Mary T Tannler
■ High mobility trailers for use with the High Mobility Multi-purpose Wheeled Vehicle (HMMWV). Truck converter dollies, trailer suspension systems, 5th wheel assemblies.

Trailer Manufacturing Plant
5825 NE Skyport Way, Portland, OR 97218
☎ (+1 503) 335 21 36 ⊖ (+1 503) 335 21 71 [98]

75-81250 Simula Government Products, Inc.
(a subsidiary of Simula Inc.)
7414 South 51st Street, Tempe, AZ 85283-4307
☎ (+1 602) 893 62 45 Ⓔ isdept@simula.com
▣ 6052910031 ⊖ (+1 602) 897 19 06
♣ US$ 31,000,000 (1996) 👥 200 (1997) 🏭 7,900 m2
President: S P Desjardins ☎ (+1 602) 893 75 33
Vice President of Operations: William Rogrers
☎ (+1 602) 893 75 33
Product Manager: Ms Maryann Shryack
Project Coordinator: Ms Theresa L McElwee
■ Aircraft and energy-absorbing seating systems for military and commercial applications, armour emgineered for air/land /sea applications and inflatable restraint systems for aircraft and automobiles. [98]

75-81480 SKC Inc.
863 Valley View Road, Eighty Four, PA 15330-9614
☎ (+1 412) 941 97 01 ✆ (800) 752 84 72
Ⓔ skcinc@skcinc.com Ⓦ http://www.skcinc.com
⊖ (+1 412) 941 13 69
Manager Marketing: Ms Carol J Knouse
Sales Manager: Ms Deborah Dietrich
Purchasing Agent: Ms Deborah Orpen
■ Air sampling equipment and collection media for chemical hazards. [98]

75-81500 SKF USA Inc.
(a subsidiary of ABSKF, Sweden)
1100 First Avenue, King of Prussia, PA 19406-1352
☎ (+1 610) 962 43 00 ⊖ (+1 610) 265 14 57
President: Krister Peil
Dir Communications: W. Gondolf
■ Manufacture and marketing of bearings and bearing-related products to original equipment manufacturers and all markets where replacement bearings are required. Products include ball, cylindrical, spherical, tapered, mounted, plain and Slimline bearings. Maintenance tools and services. → 80

MRC Bearings
(a unit of SKF USA Inc.)
402 Chandler St, PO Box 280,
Jamestown, NY 14702-0280
☎ (+1 716) 661 26 00 ⊖ (+1 716) 661 27 40
President: Bengt Nilsson
VP Aerospace Business Unit: Jim Donohue
■ Custom engineered precision rolling bearings & related components for power generation & transmission equipment in aero-engine & helicopter applications. Manufacture & markets ball bearings, cylinderical roller bearings & spherical roller bearings from an 8mm bore up to a 600 mm OD; Precision: ABEC 1-9; Chrome steel & various special alloy steels; Quality control providing full traceability. [98]

75-81520 Skurka Engineering Company
PO Box 2869, Camarillo, CA 93011-2869
☎ (+1 805) 484 88 84 ⊖ (+1 805) 482 77 71
👥 110 🏭 70,000 sq m
President: Morris Skurka
Executive Vice President: Howard Skurka
Sales and Marketing: Robert Riley
■ Manufacture electric AC and DC motors, fans, blowers, transducers, speed switches. [98]

75-81550 Skyline Industries Inc.
1201 Forum Way South, PO Box 821,
Ft Worth, TX 76101-0821
☎ (+1 817) 551 19 67 ⊖ (+1 817) 551 71 79
♣ $ 8,000,000 (1996) 👥 65 (1996) 🏭 50,000 sq ft
President: Melvin Haas [98]

75-81700 J.T. Slocomb Company
68 Matson Hill Road, S. Glastonbury, CT 06073-9790
☎ (+1 860) 633 94 85 ⊖ (+1 860) 633 23 88
♣ $ 11,750,000 (1997) 👥 96 (1997) 🏭 30,000 m2
President: John G Limato
Vice President Sales and Materials: Joseph Snyder
Operations Director: Paul Baars
■ Manufacture of turbine engine weldments & fabrications, compressor stator assemblies, seal assemblies, bearing housing, duct assemblies. [98]

75-82060 Smith & Wesson
(a subs. of Tomkins plc.)
2100 Roosevelt Ave, PO Box 2208,
Springfield, MA 01102-2208
☎ (+1 413) 781 8300 ▣ 6817064 smiws uw
⊖ (+1 413) 731 8980
President: Don L. Hinmon
VP Sales & Marketing: R.L. Scott
VP Product Engineering: Kevin Foley
Internacional Sales & Marketing Dir: Brian P. Lisankie
■ Manufacturing facility of handguns (pistols/revolvers); handcuffs; law enforcement training academy. [97]

75-82090 Smiths Industries Aerospace & Defense Systems Inc
(a subsidiary of Smiths Industries Aerospace, UK)
Washington Office
Site 1100, 1225 Jefferson Davis Highway,
Arlington, VA 22202
☎ (+1 703) 416 76 40 ⊖ (+1 703) 892 94 04

Grand Rapids
4141 Eastern Ave SE, Grand Rapids, MI 49518-8727
☎ (+1 616) 241 70 00 ⊖ (+1 616) 241 75 33
President Defence Systems North America: R F Ehr
Manager North American Press and Public Relations: Ms Jennifer Villarreal
■ Instruments, systems & components for aircraft reference & navigation, missile guidance, spacecraft control & guidance, anti-submarine warfare control, weapon systems operations, commercial flight management, aerospace ground support, solid-state data recording, self-contained navigation systems, weapon management systems, communication control systems, air data computers & digital magnetic tape recording systems.

Stamford
One Circle West, PO Box 10377,
Stamford, CT 06904-2377
☎ (+1 203) 351 8300 ▣ 4750060 sli ui
⊖ (+1 203) 351 8305
President Defence Systems International: Jerry Hines

Clearwater
14180 Roosevelt Boulevard, Clearwater, FL 34622
☎ (+1 813) 531 7781 ▣ 6815304 smith uw
⊖ (+1 813) 539 0177

Florham Park
7-11 Vreeland Road, Florham Park, NJ 07932
☎ (+1 201) 822 1300 ▣ TWX 710 9868 504
⊖ (+1 201) 822 1625
■ Design of weapon management systems, communication control systems, air data computers & digital magnetic tape recording systems for fixed & rotary wing military aircraft & the Space Shuttle.

Malvern
255 Great Valley Parkway, Malvern, PA 19355
☎ (+1 215) 296 5000 ▣ TWX 510 668 5613 hsigvly
⊖ (+1 215) 296 7805
President Civil Systems Division: Douglas L Clark
■ Avionic products & systems. Design and manufacture displays, flight and engine instruments, fuel management systems and electronic chronometers for commercial aircraft. [98]

75-82100 SMR Technologies Inc.
1420 Wolf Creek Trail, PO Box 326,
Sharon Center, OH 44274-0326
☎ (+1 216) 239 1000 ▣ TWX 710 938 1721
⊖ (+1 216) 239 1352
👥 250
President: Oscar J. Mifsud
VP Sales & Marketing: David B. Smith
VP & CFO: Patrick L. Ryan
■ Engineered rubber products (life rafts, helicopter flotation, seals, hovercraft skirts, speciality inflatable products.) [97]

United States of America

75-82170 Snorkel
(a division of Figgie International Inc.)
PO Box 1160, St Joseph, MO 64502-1160
400 Jules Street, St Joseph 64501
☎ (+1 816) 364 03 17 [E] snorkel@snorkelusa.com
⊖ (+1 816) 364 03 80
♣ $ 160,000,000 👥 850
President: R A Solon
Vice President of Marketing: Don Roach
Vice President of Sales: Daniel Malman
■ Fire products: aerial type water/foam delivery systems for fire fighting apparatus. Aerial work platforms: telescopic and articulating booms and scissorlifts. [98]

75-82200 Soderberg Manufacturing Co. Inc.
20821 Currier Road, Walnut, CA 91789
☎ (+1 909) 595 12 91 ⊖ (+1 909) 595 14 74
👥 65 (1997) 🏢 22,700 sq ft
President: B W Soderberg
Vice President: W G Lee
Marketing Manager: A C Knudsen
Purchasing Manager: Curtis B Postel
■ Aircraft lighting, electronics, land vehicle lighting, lighting related products. [98]

75-82300 Solar Turbines
(a subsidiary of Caterpillar Inc.)
2200 Pacific Highway, PO Box 85376,
San Diego, CA 92186-5376
☎ (+1 619) 544 5000 ☒ 695045 ⊖ (+1 619) 544 24 44
👥 4,400
Sales and Marketing Director: Peter Carroll
Manager of Marketing Communications: Susan Stratton
■ Manufacture and overhaul gas turbines and gas turbine-driven generator sets from 1 to 11 MW for a variety of mobile, stationary, marine and air-transportable military requirements.

Washington DC Office
818 Connecticut Avenue NW, Suite 600,
Washington, DC 20006-2702
☎ (+1 202) 293 43 27 ⊖ (+1 202) 293 43 36 [98]

75-82330 Solomat
(a subsidiary of Neotronics of North America)
The Waterside Building, 26 Pearl Street,
Norwalk, CT 06850
☎ (+1 203) 849 31 11 ∅ (800) 765 66 28
[E] iaq@solomat.com [W] http://www.solomat.com
☒ 661913 wui ⊖ (+1 203) 847 93 20
Sales and Marketing Director: Richard Stonier
Office Manager: Ms Helen Rosselli [98]

75-82360 Sonomed Inc.
3000 Marcus Ave, Lake Success, NY 11042
☎ (+1 516) 354 0900 ∅ (800) 227 1285
⊖ (+1 516) 354 5902
♧ $ 2 000 000 (1995) ♣ $ 4 000 000 (1995) 👥 25
🏢 600 sq m
CEO/President: Louis Katz
VP Sales & Marketing: William C. Eaton
Purchasing Mgr: Jim McShay
■ Manufacturer of diagnostic ultrasound systems for ophthalmic applications. [97]

75-82400 South Central Ordnance Co.
230 LaFayette Street, Gretna, LA 70053
☎ (+1 504) 366 48 51 ⊖ (+1 504) 364 89 06
Owner: T Tyler Züfle
■ Barrelled weapons, weapon-related systems and equipment, ammunition, logistic services. → 80 [98]

75-82440 Southern Ammunition Co. Inc.
4232 Meadow Street, Loris, SC 29569-3124
☎ (+1 803) 756 3262 ⊖ (+1 803) 756 3582
♣ $ 5 000 000 👥 17 🏢 69 000 sq m
President: Patrick C. Henry III
■ Manufacture of small arms ammunition (new). Arms dealer/broker for surplus small arms, ammunition, machinery & aircraft. → 80 [97]

75-82470 Southwest Aerospace
(a Meggitt Target Systems Company)
2672 Dow Avenue, Tustin, CA 92680-7208
☎ (+1 714) 832 13 33 [E] swa@meggitt.com
⊖ (+1 714) 832 60 90
♣ $ 12,000,000 👥 40 (1997) 🏢 50,000 sq ft
President: Dr Roger Brum
Sales and Marketing Director: Ian G Matyear
■ Design, development and manufacture of aerial tow targets, reeling machines, IR augmentors, chaff/flare dispensers, aircraft pods and towed countermeasures. [98]

75-82700 Space Machine & Engineering Corp.
2346 16th Avenue North, St. Petersburg, FL 33713-5626
☎ (+1 813) 323 22 21 [E] spacemen@ix.netcom.com
⊖ (+1 813) 323 23 76
♣ US$ 1,400,000 (1996) 👥 17 (1997) 🏢 1,400 m2
President and Marketing Director: Allen H Loyd
Operations Director: William R LLoyd
Sales Director: Ms Nancy Sardige
Purchasing manager: Michael Carroll
■ Waveguide components and related machined components. [98]

75-82850 Spartanics
3605 Edison Place, Rolling Meadows, IL 60008-1077
☎ (+1 847) 394 57 00 [E] sales@spartanics.com
⊖ (+1 847) 394 04 09
♣ $ 10,000,000 (1997) 👥 85 (1997) 🏢 40,000 sq ft
President: William N Knotts
Marketing Manager: Ron Martere
Research and Development Administration: Steve Willits
■ Designs & produces small arms training simulators. [98]

75-82900 Sparton Corp.
Corporate Office
2400 East Ganson Street, Jackson, MI 49202
☎ (+1 517) 787 86 00 ∅ (800) 248 95 79
⊖ (+1 517) 787 18 22
♧ $ 51,529,879 (1996) ♣ $ 102,824,946 (1996)
👥 1,500 (1996)
Chairman & CEO: John J. Smith
President & COO: David W. Hockenbrocht
Corporate Communications Dir: Thomas M. Sussex (Jr)
■ Activities are focused on electronics contract manufacturing for commercial, governmental customers and telecommunications products.

Sparton Electronics
(a subsidiary of Sparton Corporation)
Johnson Lake Road, DeLeon Springs, FL 32130
☎ (+1 904) 985 46 31 ∅ (800) 824 06 82
⊖ (+1 904) 985 50 36
VP & General Mgr: Douglas E. Johnson
■ Electronic contract manufacturing, sonobuoys and underwater acoustic devices. [98]

75-82930 Sparton Engineered Products Inc. Flora Group
(a subsidiary of Sparton Corp.)
Old Route 50 West, PO Box 399, Flora, IL 62839
☎ (+1 618) 662 21 01 ⊖ (+1 618) 662 45 28
General Sales Mgr: Sherry Hoard
■ Horns, air & electric, low air pressure buzzers, solenoid valves & terminals. [98]

75-82950 Sparton Technology Inc.
(† Sparton Southwest Inc.)
(a subsidiary of Sparton Corp.)
4901 Rockaway Boulevard, SE, Albuquerque, NM 87124
☎ (+1 505) 892 53 00 ∅ (800) 772 78 66
⊖ (+1 505) 892 55 15
VP & General Mgr: Dick Mico
Marketing Mgr: Jim Harper
■ Design, development & manufacturing of electro-mechanical assemblies, CCAs, cable assemblies, switches, encoders, weapons controllers, specialised test equipment, intrusion detection systems & monitoring & control systems. [98]

75-82970 Spaulding Composites
One Monogram Place, Rochester, NH 03867
☎ (+1 613) 332 0555 ⊖ (+1 613) 332 5357
♣ $ 35 000 000 👥 400
■ Filament wound products for launch tubes ballistic protection materials, moulded components for aircraft, vehicles & assemblies. Sound dampening advanced composites electrical & thermal insulation products. [97]

75-83005 Specialty Bar Products
(† Blairsville Machine Products)
200 Martha Street, PO Box 127,
Blairsville, PA 15717-0127
☎ (+1 412) 459 7500 ⊖ (+1 412) 459 0944
♧ $ 10 000 000 ♣ $ 40 000 000 👥 150 (1995)
🏢 120 000 sq m
President and Chief Executive Officer: Jeffrey Tott
Technical Director: Derek Saxby
Operations and Technical Director: Brian Murray
■ Track parts for tanks, howitzers, personnel carriers. [98]

75-83035 Specialty Plastic Products Inc.
(a subs. of The Speciality Group Inc.)
530 Sherwood Ave, Dunmore, PA 18512
☎ (+1 717) 961 2042 ⊖ (+1 717) 961 5176
♣ $ 25 000 000 👥 450
Chairman & CEO: Ronald J. Wanchisen
President: John P. Sweeney
Marketing Mgr: Patrick Murphy
Purchasing Mgr: Janice Warrick
■ Ballistic armour for troop & armoured vehicles including combat helmets & protective vests. Also, personnel accessory items, including field packs shoulder straps, suspenders, ammunition cases & similar products.

Overseas subsidiary:
Specialty Plastic International Inc.
(Christiansted, St.Thomas, US Virgin Islands 08802; contact via US headquarters). [97]

75-83040 Specialty Steel & Forge
26G Law Drive, Fairfield, NJ 07004-3293
☎ (+1 201) 808 8300 [E] info@steelforge.com
[W] http://www.steelforge.com ⊖ (+1 201) 808 4488
♣ $ 12 000 000 👥 32 🏢 35 000 sq ft
President: Lewis A. Weiss
Sales Mgr: Ms Sue Breault
Operations Mgr: Ms Cathie Galli
■ Stockist & manufacturer of specialty steel & open die forging in stainless steel, nickel, titanium alloy steel & carbon. Specialising in emergency deliveries. ISO9002 registered SGS. [97]

75-83050 Spec Tool Co.
PO Box 1056, Pico Rivera, CA 90660-1056
☎ (+1 213) 723 95 33 [E] spectool@sure.net
Vice President and General Manager: A G Fink
Airline Sales Manager: Bernerd E Braak
■ Airframe and powerplant maintenance tools and equipment; AGE/SE. [98]

75-83055 Specmat Technologies Inc.
(a subsidiary of H.R. Smith Group of Companies, UK)
1714 Triangle Park, Maryville, TN 37801
☎ (+1 423) 977 61 92 ⊖ (+1 423) 984 16 53
👥 6
Managing Director: Richard E L Smith
Purchasing Manager: Dennis Downey
Sales: Steve Gibson
■ Thermoplastic composites, metallized components, lightweight thermoplastic foams. SAR, ELT, PLB. Airborne antennas. [98]

75-83075 Spectral Dynamics Inc.
1983 Concourse Drive, San Jose, CA 95131
☎ (+1 408) 474 1700 ⊖ (+1 408) 474 1780
👥 100
President: Stewart Slykhous
Sales & Marketing Dir: Todd Davis
■ Manufacture, sales & service of dynamic test & analysis products. [97]

75-83090 SpecTran Corp.
50 Hall Rd, Sturbridge, MA 01566-1279
☎ (+1 508) 347 2261 ⊖ (+1 508) 347 2747
♣ $ 38 6 000 000 (1995) 👥 300
CEO/President: Glenn E. Moore
VP Sales: Edward J. Connor (Jr)
Marketing Dir: Ms Patricia Forti
■ Develop, manufacture & market a wide range of fiber optic products. These include multimode & single-mode optical fiber & cable for use in data communications & telecommunications applications. Also develop special performance fibers, coatings, cable, cable assemblies & other value-added products for use in a variety of specialty markets. [97]

75-83095 Spectrex Inc.
218 Little Falls Road, Cedar Grove, NJ 07009
☎(+1 201) 239 83 98 ✆ (800) 452 21 07
E spectrex@spectrex-inc.com
W http://www.spectrex-inc.com ⊜ (+1 201) 239 76 14
Managing Director: Eric J Zimm
Sales and Marketing Manager: Christine D Kaye
Purchasing Manager: Ms Michelle Fiducia
■ Automatic Fire & Explosion Suppression System for armoured vehicles. 98

75-83100 Spectrolab Inc.
(a subs. of Hughes Electronics Co.)
12500 Gladstone Ave, Sylmar, CA 91342-5373
☎(+1 818) 365 4611 ⊜ (+1 818) 365 7680
Marketing Mgr: Bryan P. Roland
■ High intensity xenon searchlights for airborne, shipboard & land usage. Infra-red filters are available for each model searchlight for use with night vision goggles. 97

75-83250 Sperry Marine Inc.
(a subsidiary of Litton Industries)
1070 Seminole Trail, Charlottesville, VA 22901
☎(+1 804) 974 20 00 W http://www.sperry-marine.com
Telex 822411; TWX 510 5875 463 ⊜ (+1 804) 974 22 59
Employees 750
Chairman: John F Lehman
Chief Executive Officer: George A Sawyer
President: Paul D Miller
Executive Vice President of Operations: John V DeMaso
Executive Vice President and Chief Financial Officer: Joseph A Stroud
Vice President Strategic Planning: Benjamin Bingham
Vice President of Domestic and International Finance: N E Bickers
Vice President Engineering: Clark Graham
Contracts Administrator: W L Spicuzza
Marketing Communications: Ms Elizabeth Jones
■ Design, develop and manufacture marine electronic products and systems for defence and commercial markets. Products include integrated shipsystems, ship bridge consoles, inertial navigation systems, gyrocompasses, steering systems, gyrofin stabilisers, marine surface search radars, airborne search radars, vessel traffic systems, dopplar speed logs, GMDSS communication systems, heading repeaters and marine navigation software. 98

75-83300 I.Spiewak & Sons Inc.
505 Eighth Ave, New York, NY 10018
☎(+1 212) 695 16 20 ✆ (800) 223 6850
Employees 300
Chairman: Gerald Spiewak
President & CEO: Michael Spiewak
Exec. Vice President & CFO: Roy Spiewak
Uniform Sales Dir: Ed Philips
Export Sales Dir: Lynn Wockenbrod
Controller: Pat Cunninlna
National Sales Mgr: Sol Jacobs
■ Manufacture & sell all types of outerwear, insulated & protective clothing. Specialising in uniforms, industrial jackets & parkas. 97

75-83330 Spiral System Instruments Inc.
7830 Old Georgetown Road, Bethesda, MD 20814
☎(+1 301) 657 16 20 E infosbi@spiralbiotech.com
W http://www.spiralbiotech.com ⊜ (+1 301) 652 50 36
Vice President: Ms Susan Schalkowsky
■ Air sampling equipment. 98

75-83350 Spitz Inc.
(a subsidiary of Transnational Industries Inc.)
Routes 1 and 202, PO Box 198, Chadds Ford, PA 19317
☎(+1 610) 459 52 00 ⊜ (+1 610) 459 38 30
President: Charles H Holmes (Jr)
Sales and Marketing Director: Jon Shaw
Operations Director: John Fogleman
Manager Advertising: Mike Bruno
Purchasing Manager: Rich Hanna
■ Manufacture planetarium/space theatre and simulation dome/projection screen. 98

75-83550 SRL Operations
(a div.of Calspan SRL Corp.)
2800 Indian Ripple Rd, Dayton, OH t5440-3696
☎(+1 513) 426 6000 ⊜ (+1 513) 427 0205
Employees 500 Floor area 180 000 sq ft
Sr Vice President & General Mgr: John W. Pitts
Vice President Marketing: Ron Jones
■ High resolution display systems, EW systems, night vision systems, aerosystems research, instrumentation systems, materials research, aircraft modification/testing, simulation systems, computer systems/services and support, NDT evaluation systems and instruments, human factors engineering and research. → 80 97

75-83570 SSS Clutch Co. Inc.
610 West Basin Rd, New Castle, DE 19720
☎(+1 302) 322 8080 ⊜ (+1 302) 322 8548
Vice President: Morgan L. Hendry
Advertising Co-ordinator: M. Dills
■ Clutches for main naval propulsion systems. SSS-Tosi Reversing System is being supplied for the new US Navy AOE-6 Ship Program.
® Italy: Ansaldo
UK: SSS Gears Ltd 97

75-83580 StacoSwitch
1139 Baker Street, Costa Mesa, CA 92626
☎(+1 714) 549 30 41 W http://www.stacoswitch.com
⊜ (+1 714) 549 09 30
Employees 80 (1997)
President: Jim Gust
Vice President Sales: Dan Sugg
Vice President Manufacturing: Bill Swan
Vice President Engineering: Ted Nichols
Vice President Marketing: Rich Brown
Vice President Admin: Pauline Cook
Controller: Don Dietz
■ Lighted display pushbutton switches and indicators, MIL-SPEC performance, DPDT/MOM and ALT switch action and electronic controllers for lighting and switching functions. 98

75-83650 Standard Armament Inc.
631 Allen Avenue, Glendale, CA 91201
☎(+1 213) 849 13 4 ⊜ (+1 213) 848 69 0
Employees 45 Floor area 20 000 m2
President: Curtis Correll
Managing Director: Gary Correll
■ Flexible feed chuting. 98

75-83700 Standard Steel
(a division of Freedom Forge Corp.)
500 North Walnut Street, Burnham, PA 17009-1644
☎(+1 717) 248 49 11 ⊜ (+1 717) 248 80 50
♣ $ 160,000,000 (1996) Employees 1,050 (1996)
Floor area 2,200,000 sq ft
Chief Exec. Officer: Herbert Graves
President: James Spendiff
Vice President Sales & Marketing: John M. Hilton
General Mgr: J.O. Parke
Planning & Development: Richard C. Rheem
■ Manufacture seamless rolled rings, railway wheels, axles & mounted sets, ingots, billets, springs. 98

75-83750 Stanford Research Systems Inc.
1290 Reamwood Avenue, Building D, Sunnyvale, CA 94089-2233
☎(+1 408) 744 90 40 E info@srsys.com
W http://www.srsys.com ⊜ (+1 408) 744 90 49
President: William Green
Purchasing Director: Ms Susan Gust
Sales Manager: Daniel Melick
■ General purpose high resolution test and measurement instruments including time interval and frequency counters, pulse and delay generator, function generator, FFT spectrum analyzers, LCR meters, thermocouple monitors programmable filters, LORAN-C 10MHz frequency standard, lock-in amplifiers, boxcar averagers, photon counters, multi- channel sealer/averager. Optical choppers, preamplifiers and amplifiers. 98

75-83900 Stanley Aviation Corporation
(a Cobham plc company)
2501 Dallas Street, Aurora, CO 80010
☎(+1 303) 340 52 00 ⊜ (+1 303) 360 89 65
Employees 220 Floor area 15 acre
Chairman: G F Page
President, Chief Executive Officer: K W Greene
Vice President: Marketing: M Kurzorfer
Vice President: Finance: W M Thomas
Vice President: Engineering: R Rossway
Director: A G Irwin
■ Supersonic wing manufacture for US Air Force high-speed aircraft development programme. Manufactures a complete range of aircraft engine handling and maintenance equipment including systems to ship, split and store the large engines in use on the latest long range twin jet aircraft from Boeing and Airbus. Supplies fuel system components based on the GAMAH range of couplings, recently enhanced with the development of composite and zero-leak metal seal products which also have applications in the process and industrial sectors. Manufacturing services for aircraft components and structures backed by CAD and automated manufacturing systems.
Associated companies:
European Sales Office, UK
Thailand Sales Office, Thailand 98

75-83952 Staplex Air Sampler Division
777 Fifth Avenue, Brooklyn, NY 11232-1695
☎(+1 718) 768 33 33 ✆ (800) 221 08 22
E info@staplex.com W http://www.staplex.com
⊜ (+1 718) 965 07 50
Technical Dir: William Flynn
Sales Dir: Doug Butler
Communications Mgr: Phil Reed
■ Air sampling equipment, for airborne particulates, dust, asbestos, radioactive particles, and industrial hygiene. 98

75-83955 Star Case Manufacturing Co. Inc.
648 Superior, Munster, IN 46321
☎(+1 219) 922 4440 ✆ (800) 822 7827
E starcase@pla-net.net ⊜ (+1 219) 922 4442
Employees 30 Floor area 30 000 sq ft
President: Denis R. Toma
VP Sales & Marketing: Ralph G. Hoopes
Technical Dir: Roger Hallberg
Purchasing Mgr: Darren Eason
■ Custom shipping containers. 97

75-83957 H.C. Starck
(† NRC Inc.)
(a subsidiary of Bayer Corp., USA)
45 Industrial Place, Newton, MA 02161-1950
☎(+1 617) 630 58 00 ⊜ (+1 617) 630 58 79
♣ $ 50,000,000 (1997) Employees 340 (1997)
Sales and Marketing Director: Gerard Villani
Operations Director: P L Sallade
Product Marketing Manager: Tom Ramlow
☎(+1 617) 630 58 77
■ Tantalum and niobium mill products and semi-finished form. 98

75-83990 State College Operations
(a subsidiary of E-Systems/Raytheon)
300 North Science Park Road, PO Box 60, State College, PA 16804-0060
☎(+1 814) 238 43 11 ✆ (800) 827 43 11
⊜ (+1 814) 234 77 20
Employees 1,000 Floor area 400,000 sq ft
Vice President and General Manager: David A Woodle
■ Designs and supplies intelligence systems and services. Develops systems from requirements analysis through design, fabrication, integration and testing, documentation, maintenance and integrated logistics support.
Washington Operation
800 International Drive, Linthicum, MD 21090
☎(+1 301) 850 78 90 98

75-84175 Sterling Lacquer Manufacturing Co.
Sterling Chemical Coatings
3150 Brannon Avenue, St Louis, MO 63139
☎(+1 314) 776 44 50 ✉ STERLAC Telex 442385
⊜ (+1 314) 771 18 58
Employees 50 (1997) Floor area 3,000 m2
President: Leo V Mitchell 98

75-84180 Stewart & Stevenson Services Inc.
2707 North Loop West, Houston, TX 77474
☎(+1 281) 856 70 40 ✆ (800) 221 FMTV
⊜ (+1 218) 867 15 18
♣ $ 1,187,161,000 (1996) Employees 4,879 (1996)
Chief Executive Officer: Robert L Hargrave
☎(+1 713) 868 77 00

Contact Reference

Group Vice President and Chief Operating Officer: C LaRoy Hammer ☎ (+1 713) 867 1430
Marketing and Business Development Director: Bill Trader ☎ (+1 713) 867 17 74

Tactical Vehicle Systems Division
(a division of Stewart & Stevenson)
5000 I-10 West, Sealy, TX 77474
☎ (+1 713) 867 70 40 ✆ (800) 221 FMTV
✉ http://www.ssss.com ⊖ (+1 713) 867 15 18
♣ $ 1,187,161,000 (1996) 👥 4,879 (1996)
Marketing Account Representative: Ms Denise Cook ☎ (+1 713) 867 15 76
■ Manufacture the US Army's family of medium tactical vehicles as well as speciality all-terrain vehicles. [98]

75-84200 Steward Davis International Inc
10929 Van Owen, North Hollywood, CA 91605
☎ (+1 818) 985 35 00 ⊠ 4973015 amexpo
⊖ (+1 818) 985 57 71
🏭 2,800 m2
President: Stanley W Epstein
Operations and Technical Director: Raymond Kennedy
■ Manufacture and service auxiliary gas turbine engine interface systems. Aftermarket original equipment and upgrade of existing environmental systems. → 80 [98]

75-84220 Stinar Corp.
3255 Sibley Memorial Highway, St. Paul, MN 55121-1697
☎ (+1 612) 454 51 12 ⊖ (+1 612) 454 51 43
♧ $ 4,000,000 (1996) ♣ $ 9,500,000 (1996)
👥 80 (1996) 🏭 42,000 sq ft
President: Gene G Stinar
Operations Director: R L Stinar
Secretary and Treasurer: G F Stinar
Purchasing Manager: Ms Jane Jacobs
Purchasing Agent: Ken Fitzgerald
■ Ground support equipment. [98]

75-84240 Signal Technology Corp. Systems Operation
(† St Olektron Corp.)
(a division of Signal Technology Corp.)
37 Sutton Road, Webster, MA 01570
☎ (+1 508) 943 74 40 [W] http://www.sigtech.com
⊖ (+1 508) 949 18 04
👥 45 (1997) 🏭 1,000 sq m
President and Chief Executive Officer: George Hreczuck
Business Development Director: Bill Uffelman [E] uffelman@sigtech.com
Operations Director: Edward Mezzapelle
Purchasing Manager: Ms Pamila Brooks
■ Signal distribution & direction-finding sub-systems. [98]

75-84250 Stocker & Yale Inc.
32 Hampshire Road, Salem, NH 03079
☎ (+1 603) 893 87 78 [E] stocker@worldnet.att.net
[W] http://www.stkr.com ⊖ (+1 603) 893 56 04
♧ $ 11,959,999 (1996) ♣ $ 11,655,000 (1996)
👥 115 (1996) 🏭 80,000 sq m
President: James Bickman
Managing Director: Mark Blodgett
Operations Director: Nelson Gauthier
Sales and Marketing Director: William Michaud
Sales Manager: Michael A Lavey
Purchasing Manager: Ms Grace Longo
■ Optical inspection & measuring equipment military compasses; survival compasses military watches; M3 borescope illuminated magnifiers, microscope illuminators; compact work lights. [98]

75-84280 Stow Mfg Co.
Brandywine Highway, PO Box 490, Binghamton, NY 13902
☎ (+1 607) 723 64 11 ⊖ (+1 607) 723 93 50
President: Mark Hotchkiss
Vice President Sales and Marketing: F Hotchkiss
■ Manufacture light construction equipment. Electronic assembly. [98]

75-84290 Strandflex Division of MSW, Inc.
(a division of Maryland Specialty Wire)
6149 Sutliff Road, Oriskany, NY 13424
☎ (+1 315) 736 52 08 ✆ (800) 628 04 48
⊖ (+1 315) 736 52 91
👥 85 (1997) 🏭 100,000 sq ft
General Manager: Dennis Kuhns ☎ (+1 410) 785 25 00
Plant Manager: D Waddell
Sales and Marketing Manager: M Wallace
Purchasing Manager: Douglas Rouse
■ Wire rope to MIL-W-83420 for flight controls, compositions A&B, type I. [98]

75-84300 Stratoflex Aerospace
(a division of Parker Hannifin Corporation)
Military Connectors Division
220 Roberts Cut-Off Road, PO Box 10398, Forth Worth, TX 76114-0398
☎ (+1 817) 738 65 43 ⊖ (+1 817) 738 99 20
👥 300
General Manager: Steve Hamilton
Marketing Manager: Earl H Larkin
Product Manager: Eddie Phillips
Chief Research, Development, and Engineering Manager: Rick Deiss
Materials Manager: P H Cummings
Manufacturing Manager: D Calevich
Administration Assistant: Ms Cheryl Simms
■ Aerospace and military hose, hose fittings, hose assemblies; hydraulic fuses and flow regulators; quick disconnect couplings. [97]

75-84435 Stringer Welding & Machine Inc.
201 South Third Street, Dugger, IN 47848
☎ (+1 812) 648 22 05 ⊖ (+1 812) 648 30 18
👥 35 🏭 22,000 sq m
President: J Stephen Sarles
Operations and Technical Director: Douglas Brochin
■ Manufacture and supply of various ordnance items to the US military. Specialise in gun mounts for weapons up to and including .50 calibre. Complete fabricating and machining capabilities for ferrous and non-ferrous products. [98]

75-84500 Strong Enterprises
11236 Satellite Boulevard, Orlando, FL 32837
☎ (+1 407) 859 93 17 [E] strongen@aol.com
[W] http://www.strongparachutes.com
⊖ (+1 407) 859 69 78
👥 47 (1997) 🏭 20,000 sq ft
President: Edward Strong
General Manager: Gerald Kopp
■ Paratroop, personnel and cargo parachute systems. Capability of building complete parachute systems to any specifications. [98]

75-84800 Sumitomo Electric Lightwave Corp.
(† Sumitomo Electric Fibre Optics Corp.)
(a subsidiary of Sumitmo Electric Fibre Optics Corp., Japan)
78 Alexander Drive, PO Box 13445, Research Triangle Park, NC 27709
☎ (+1 919) 541 81 00 [W] http://www.sel-rtp.com
⊖ (+1 919) 541 82 67
👥 500
President: H Horima
Executive Vice President: Robert Swirbul
Purchasing Manager: John Wright
■ Fibre optic cable, connectors, cable assemblies, fusion splicing systems, electro optic products, VSB/AM and digital video transmission equipment, air blown fibre systems. [98]

75-84900 Summit Machine Builders Corp.
(† Hierath Automated Systems Corp.)
550 W 53rd Place, Denver, CO 80216
☎ (+1 303) 423 6741 ⊖ (+1 303) 423 7405
Chairman: Len Hierath
President & CEO: Scott Harris
Sales Mgr: Peter Thompson
Sales & Marketing: Frank Gacnik
Sales Dir: Gary Barker
■ Design & manufacture equipment for loading explosive, ordnance & pyrotechnic powders by net weight to high precision. [97]

75-84950 Sunair Electronics Inc.
Sunair International Sales Corp.
3101 SW 3rd Ave, Ft Lauderdale, FL 33315
☎ (+1 954) 525 1505 ✉ SUNAIR ⊠ 514443
⊖ (+1 954) 765 1322
♧ $ 13,404,689 (1995) ♣ $ 3,333,268 (1995)
👥 50 (1995) 🏭 6,225 sq m
Chairman, CEO & President: Robert Uricho
VP Operations: Manuel Garrido
VP Marketing: James Laurent
■ High frequency point-to-point radio communications. [97]

75-85105 Sundstrand Aerospace
(a division of Sundstrand Corp.)
4747 Harrison Avenue, PO Box 7002, Rockford, IL 61125-7002
☎ (+1 815) 226 60 00 ⊠ 257440; TWX 901 631 4255
⊖ (+1 815) 226 72 88
Executive Vice President and Chief Operating Officer: Ron McKenna
Vice President Business Development: James Smith
Vice President Aerospace Programs: David Linton
Vice President Aerospace Operations and Engineering: Kenneth Eitenmiller
Vice President Quality: George Sorensen
Vice President Finance: James Gingrich
Electronics Project Management Director: Peter Burkott
Vice President Customer Support: James F Petersen
Group Leader Communications: Daniel Blascoe
■ Aircraft pumps, electrical generating systems, gearboxes, actuation systems & components, generators, electric motors, starting systems, pneumatic conditioning & control systems, fans, refrigeration systems, missile power units, emergency turbine systems.

Washington Office
2101 Wilson Blvd, Suite 1002, Arlington, VA 22201
☎ (+1 703) 276 1626 ⊖ (+1 703) 528 9415 [98]

75-85120 Sundstrand Power Systems
(a division of Sundstrand Aerospace)
4400 Ruffin Road, PO Box 85757, San Diego, CA 92186-5757
☎ (+1 619) 627 69 63
⊠ TWX 910 335 1502; 683 4042 ⊖ (+1 619) 627 61 96
⊞ 62034550
Vice President Power Systems Programs: Tim Morris
Director Engineering: William Treece
Director Operations: Jack Matteson
General Manager Product Support: Jerry Todd
Manager Finance: Bob Ridgeway
Manager Business Development: Keith Topliffe
■ Small gas turbine engines, generator sets and APU's for commercial and military aircraft and vehicles. Fans and vapour cycle systems. Propulsion engines for RPVs. [98]

75-85126 Superform USA Inc
PO Box 5375, Riverside, CA 92517
☎ (+1 909) 351 4100 ⊖ (+1 909) 351 1189
♣ $ 5,000,000 (1996) 👥 20
Contact: Mike Reynolds
■ Sheet aluminium components & assemblies for military & commercial land & air systems. Superform specialises in prototype through production quantities of close tolerance lightweight structures. [98]

75-85200 Survival Products Inc.
5614 South West 25th Street, Hollywood, FL 33023
☎ (+1 305) 966 73 29 ⊖ (+1 305) 966 35 84
👥 12
President: Charles G Rogers (Jr)
Vice President of Marketing: Ms Donna W Rogers
Director of Accounting: Ms Deborah Rogers
■ Manufacture, sales and service emergency equipment rafts, vests and slides. [98]

75-85230 STI
2275 Research Blvd, Rockville, MD 20850
☎ (+1 301) 926 1800 ✆ (800) 638 8093
⊖ (+1 301) 926 6186
♣ $ 30 074 900 (1996) 👥 268
President & CEO: James H. Miller
Technical Dir: John G. Wilmot
Marketing Dir: G. Ritson
■ Develop & manufacture military auto-injectors for self-administration of chemical warfare agent antidotes. [97]

75-85265 Swanson Erie Corp.
(a subsidiary of Swanson Systems Inc.)
814 East 8th Street, PO Box 1217, Erie, PA 16512-1217
☎ (+1 814) 453 58 41 [E] swankm@erie.net
[W] http://www.swanson-erie.com ⊖ (+1 814) 455 51 09
♣ $ 25,000,000 (1996) 👥 150 (1996) 🏭 10,000 m2
CEO/President: Douglas Swanson

VP Sales & Marketing: Kevin Moore
Technical Dir: Ben Green
Purchasing Mgr: Ron Morse
■ Machine chassis for detonator loading (lowa loaders). Fuze making machinery (assembly). Miscellaneous multiple station assembly equipment. 98

75-85350 Swiftships Inc.
1105 Levee Rd, PO Box 1908, Morgan City, LA 70380
☎ (+1 504) 384 1700 ⊖ (+1 504) 384 0914
👥 200
Chairman: Dennis Spurgeon
President: Robert Ness
Exec. Vice President: Calvin J. Leleux
■ Construct high quality vessels for government & industry, using modern materials & technology. Develop high speed offshore vessels constructed of steel & marine aluminium alloys.
Washington DC
10725 River Rd, Potomac, MD 20854 97

75-85400 Switlik Parachute Co. Inc.
PO Box 1328, Trenton, NJ 08607
☎ (+1 609) 587 33 00 ⊖ (+1 609) 586 66 47
$ 2,800,000 (1997) ♣ $ 11,500,000 (1997)
👥 155 (1997) 🏭 85,000 sq ft
President: Richard Switlik (Sr)
Vice President: Michael C. Dilts
Marketing Dir: Richard Switlik (Jr)
Sales: Greg Switlik
■ Inflatable life rafts & vests. Anti-exposure suits, survival equipment. 98

75-85450 Swift Textile Metalizing Corp.
PO Box 66, Bloomfield, CT 06002-0066
☎ (+1 860) 243 11 22 ⊖ (+1 860) 243 08 48
👥 23 (1997) 🏭 2,300 sq m
Vice President: Ernest Smith
Marketing Manager: Steven Smith
■ Electrically conductive, flexible metallized fabrics, filaments, yarns and finished articles (hook and loop fasteners, gloves, tubing) for electromagnetic interference (EMI/RFI) shielding, reflection and static control. Wovens, non-wovens and knits to 180 inches wide. Substrates include nylon, polyester, carbon, glass. Highly durable metal coatings. Just in time delivery. Custom fabrics developed to meet specific needs. 98

75-85530 Symtron Systems Inc.
(† Austin Electronics)
17-01 Pollitt Drive, PO Box 950,
Fair Lawn, NJ 07410-2815
☎ (+1 201) 794 02 00 ✆ (800) 288 39 73
⊖ (+1 201) 703 90 31
♣ $ 10,000,000 (1990) 👥 100 🏭 105,367 sq ft
President and Chief Executive Officer: John J Henning
Managing Director: James Hanson
Sales and Marketing Manager: David Joynt
■ R&D and manufacturing of: Training and simulation systems equipment, including firefighter trainer. 98

75-85535 Synair Corporation
2003 Amnicola Highway, PO Box 4269,
Chattanooga, TN 37406
☎ (+1 423) 698 8801 ⊖ (+1 423) 624 0321
President: E.N. Gomberg
■ Tyrfil puncture proofing system for low speed & off road use. Tyrlyner system, self healing tyres for highway & agricultural use. Monothane, one component polyurethane elastomer. Duothane, two component high performance polyurethane elastomer. Cilbond, rubber to metal bonding agents. 98

75-85540 Syndex Aerospace Inc.
3415 West Lomita Boulevard, Suite B,
Torrance, CA 90505
☎ (+1 310) 326 61 24 ⊖ (+1 310) 326 64 35
President: John D Campbell
■ Aircraft spin recovery parachute systems. Ejection seat subsystems. Missile parachute recovery systems. Aerospace research and development. Arming and safety devices, explosive devices, pyromechanical devices. 98

75-85550 Synergistics Inc.
9 Tech Circle, Natick, MA 01760-1023
☎ (+1 508) 655 13 40 E syn@ticc.net
W http://www.synergisticsinc.com ⊖ (+1 508) 651 29 02
👥 25
President: David Longworth
Chief Executive Officer: William Tetrick
Vice President Sales and Marketing: Robert Pogorelc
Purchasing Manager: Thomas Fennell
■ Manufacture industrial & commercial access control systems & video imaging identification systems. 98

75-85650 Systems & Electronics Inc.
(† Electronics and Space Corp.)
(a division of ESCO Electronics Corp.)
201 Evans Lane, M S 4901, St Louis, MO 63121
☎ (+1 314) 553 43 61 W http://www.seistl.com
⊖ (+1 314) 553 49 49
♣ $ 220,000,000 (1996) 👥 1,200 (1997)
President: G A Potthoff
Senior Vice President and General Manager: D M Purdy
Vice President of Business Development: V Richey
Vice President of Engineering: V Reznack
Operations Director: Ken Bieke
Communications Manager: Ms Ginette Sevier
Purchasing Manager: Bruce Clark
■ Developer and manufacturer of tactical mobile systems. Products include: 60k loader, M1000 semi-trailer HET, Hemat trailer, Patriot missile canister, weapons platforms, tactical bridging systems and aircraft cargo handling systems.
Airborne radar, electronic intelligence, sighting and guidance systems, test and measurement systems. 98

75-85950 TA Mfg. Co.
(a subsidiary of Esterline Technologies)
375 West Arden Avenue, Glendale, CA 91203
☎ (+1 818) 240 46 00 ⊖ (+1 818) 241 39 48
👥 205
President: George M Jones
Vice President Marketing and Sales: Rick Gentle
■ Specialty elastomeric compounds in custom-molded shapes; design, production of clamps & support blocks for wiring & tubing installations. 98

75-86000 Tacoma Boatbuilding Co.
1840 Marine View Drive, Tacoma, WA 98422
☎ (+1 206) 572 3600 ⊖ (+1 206) 572 0548
$ 14 500 000 (1994) ♣ $ 11 500 000 (1994)
🏭 58 000 sq m
Purchasing Mgr: Terry Larsen
VP & General Mgr: Ray Nichols
■ Designs & builds commercial & military ships for the US & foreign markets, specializing in fast multi-mission patrol boats, fully tested & operational upon delivery. Ship repair, full line of deck machinery & manufacture of controllable pitch propellers. 97

75-86020 Taconic Plastics Ltd
136 Coonbrook Rd, PO Box 69,
Petersburgh, NY 12138-0069
☎ (+1 518) 658 3202 ✆ (800) 833 1805; 272 2503
⊖ (+1 518) 658 3204
♣ $ 35 000 000 👥 160
General Mgr: C.B. Smith
Operations/Technical Dir: Bob Nurmi
Purchasing Mgr: Carol Goodermote
■ PTFE coated glass fibre fabrics, used to produce advanced composites. Process materials, copper clad PTFE/wovenglass substrates for microwave printed circuit boards. 97

75-86030 Tactair Fluid Controls Inc.
(a subs. of Young & Franklin Inc.)
942 Old Liverpool Road, Liverpool, NY 13088
☎ (+1 315) 451 3928 ⊖ (+1 315) 451 8919
Vice President: Craig Froelich
Marketing & Development Mgr: Michael Yates
■ Design & manufacture hydraulic components for brake, nose wheel steering, landing gear & utility system control for use on military & civil fixed winged aircraft. 97

75-86080 Talbert Manufacturing Inc.
1628 West State Road 114, Rensselaer, IN 47978-9169
☎ (+1 219) 866 71 41 ✆ (800) 348 52 32
⊖ (+1 219) 866 54 37
👥 250
President: Davies Wakefield
Executive Vice President: Stephen A Kingman
Vice Presidnet Sales and Marketing: Douglas Miller
■ Designs and manufactures a full line of lowbed trailers and specialized heavy equipment moving trailers in a broad range of configurations and capacities. 98

75-86120 Talley Defense Systems Inc.
(a subsidiary of Talley Industries Inc.)
3500 N Greenfield Road, PO Box 849, Mesa, AZ 85211
☎ (+1 602) 898 22 00 ⊖ (+1 602) 898 23 58
♣ $ 31,000,000 (1996) 👥 260 (1996) 🏭 610,000 sq ft
President and General Manager: Edward T Ryan (Jr)
Vice President of Marketing: William W Mogan
■ Manufacture gel propellants, gas generators, inflation systems, rocket motors, artillery range extension devices, submunition expulsion and dispersion systems, weapon systems, aircrew escape rockets. 98

75-86366 Curtiss-Wright Flow Control Corporation
(† Target Rock Corp.)
(a subsidiary of Curtiss Wright Corp.)
1966E Broadhollow Road, Farmingdale, NY 11735
☎ (+1 516) 293 3800 ⊖ (+1 516) 293 6144
👥 168
President and General Manager: Martin R Benante
■ Valve and actuator, controller supplier for nuclear aircraft carriers & submarines. 98

75-86385 Taurus International Manufacturing Inc.
(a subsidiary of Forjas Taurus SA, Brazil)
16175 North West 49th Avenue, Miami, FL 33014-6312
☎ (+1 305) 624 11 15 ⊖ (+1 305) 623 75 06
Marketing Manager: Tom Conrad
■ Manufacture, revolvers semi-automatics in .22, .25, .32, .38, .357, .40, .44, .45 and 9mm calibres. 98

Contact Reference

75-86420 Taylor Devices Inc.
90 Taylor Drive, PO Box 748,
North Tonawanda, NY 14120-0748
☎ (+1 716) 694 08 00
W http://members.aol.com/taydevices
⊖ (+1 716) 695 60 15
♣ $ 8,300,000 (1995) 👥 95 🏭 35,500 sq ft
President: Douglas P. Taylor
Vice President: Richard Hill
Advertising Dir: S. Taylor
Marketing Dir: John Mayfield
Contract Administrator: Sandi Taylor
■ Liquid springs, shock absorbers, shock isolators, energy absorbers, single & double acting dampers & snubbers, hydraulic components produced on a custom, design to order basis. 98

75-86750 Technical Communications Corp.
100 Domino Drive, Concord, MA 01742
☎ (+1 508) 287 51 00 E info@tccsecure.com
W http://www.tccsecure.com ⊖ (+1 508) 371 12 80
♣ $ 14,000,000 (1997) 👥 55 (1997) 🏭 35,000 sq ft
President: Roland Gerard
Vice President Sales: Ty Glasgow
Executive Vice President: John Gill
Director Business Development: Panos Lekkas
Operations Director: Walter Kopek
Purchasing Manager: Bob Berg
Marketing Manager: Ms Janet LeClair
■ Communications security equipment for voice, data and facsimile applications. Equipment provides strategic security for high level government and military customers. Available in MIL-SPEC models. 98

75-86760 The Technical Materiel Corp. (TMC)
700 Fenimore Rd, Mamaroneck, NY 10543
☎ (+1 914) 698 4800 ✆ (800) 742 5862 (PICK-TMC)
⊖ (+1 914) 698 4805
$ 1 000 000 ♣ $ 1 000 000 👥 25
President: Neil H. de Pasquale
■ Communications products, including a computer-controlled synthesized exciter. 97

75-86764 Technical Products Group Inc. Corporate Office
3353 Peach Tree Road, Suite 920, Atlanta, GA 30326

☎(+1 404) 231 7272 ⊜ (+1 404) 231 7277
President: J. Carter

Falls Church
5201 Leesburg Pike, Suite 610, Falls Church, VA 22041
☎(+1 703) 824 0300 ✆ (800) 877 3121
⊜ (+1 703) 824 0308
♣ $ 120 000 000 (1995) 👥 1 200 (1996)
🏭 20 300 sq m
VP Marketing: Thomas Flynn ☎(+1 703) 824 0300
Marketing Services Mgr: Ms Diane Sheerar
☎(+1 540) 783 9622

Marion Composites
(† Brunswick Corp.)
150 Johnston Road, Marion, VA 24354
☎(+1 540) 783 3121 ⊜ (+1 540) 783 9667
President: H.Dwight Byrd ☎(+1 540) 783 3121
■ Product information: Radomes, military and commercial shelters, helicopter structures, turbine engine components, aircraft control surfaces.

Lincoln Composites
4300 Industrial Ave, Lincoln, NE 68504
☎(+1 402) 464 8211 ⊜ (+1 402) 464 2247
President: Jim Fuller
Business Development Dir: Richard J. Rashilla
■ Filament wound composites structures, rocket motor cases, pressure vessels, fuel tanks, tubular products, vehicle and missile structures, NGV fuel tanks, oil and gas tubular, industrial rollers.

Intellitec
2000 Brunswick Lane, Deland, FL 32724
☎(+1 904) 736 1700 ⊜ (+1 904) 736 2250
President: Hank Lattanzi
Marketing Dir: Saul Fein
■ Camouflage netting, chemical warfare detector/alarms, advanced electrical and electro-optical components. [97]

75-86800 Technitrol Inc.
1952 E. Allegheny Ave, Philadelphia, PA 19134
☎(+1 215) 426 9105 ▣ 834245 ⊜ (+1 215) 426 2836
♧ $ 43 242 ♣ $176 419 000 (1996) 🏭 120 000 sq ft
President: Thomas J. Flakoll
Treasurer: Albert Thorp (III)
■ MIL-STD-1553B interface transformed & data bus couplers. Delay lines, power transformers, manchester/encoder recorders.
Electronic components, metallurgical products & test & measurement Products. [97]

75-86890 Technology Applications & Services Co.
(a subs. of Diagnostic/Retrieval Systems)
200 Professional Drive, Gaithersburg, MD 20879
☎(+1 301) 921 8100 ⊜ (+1 301) 977 6158
President: Paul G. Casner (Jr)
VP, General Mgr: Terrence L. DeRosa
Business Development: Timothy A. Wilson
■ Design & manufacture display & emulation systems for military & industrial applications. The company also provides technical services & support. [97]

75-86900 Technology Dynamics Inc.
100 School St, Bergenfield, NJ 07621-2915
☎(+1 201) 385 0500 ⊜ (+1 201) 385 0702
♣ $ 6 000 000 👥 51
President: Aron Levy
VP Sales & Marketing: Mark Jacobus
■ Power supplies & DC/DC converters, ruggedized, rugged and full MIL designs up to 2000 watts. Shock and vibration per MIL-STD-810, MIL-STD-461 complaint intended for shipboard, airborne and ground support applications. [97]

75-87150 Tecknit Inc.
(a subsidiary of the TWP Company)
129 Dermody Street, Cranford, NJ 07016
☎(+1 908) 272 55 00 🄴 tecknit@tecknit.com
🅆 http://www.tecknit.com ⊜ (+1 908) 272 27 41
👥 225 (1997) 🏭 79,000 sq m
President & CEO: Paul D. Beck
Managing Dir: Dan Whitmore
Operations/Technical Dir: Donald Dravis
Sales and Marketing Director: Jim Putvinski
Purchasing Mgr: Richard Panas
Mgr Corporate Communications: James L. Pean
R&D Engineering Mgr: James Sauer
■ EMI shielding products & components. [98]

75-87350 Tektronix Inc.
PO Box 1000, Wilsonville, OR 97070
☎(+1 503) 627 7111 ✉ TEKTRONIK ▣ 151754
⊜ (+1 503) 685 3777
♧ $ 1 408 253 000 👥 10 000
Chairman, President & CEO: Jerome J. Meyer
Group VP Int'l: John Vold
Corp. Marketing Mgr: Carl Schuetze
CMI Mgr: Sandy Thomas [97]

75-87400 Telair International
(† Teleflex Control Systems)
(a subsidiary of Teleflex Inc.)
1950 Williams Drive, Oxnard, CA 93030
☎(+1 805) 988 19 02 ✆ (800) 989 48 27
🅆 http://www.teleflex.com ▣ 6730028
⊜ (+1 805) 988 49 52; 983 24 92
👥 200 🏭 220,000 sq ft
President: Alex Hauner
Sales and Marketing Director: Luke McLaughlin
Site Manager: William Greer
Purchasing Manager: Dave Schafer
■ Electro-mechanical actuators and actuation systems: pneumatic and fuel valves, cargo systems, electric motors, gearboxes. Repairs. → 80 [98]

75-87506 Teledyne Continental Motors Battery Products Operations
(† Teledyne Battery Products)
PO Box 7950, Redlands, CA 92375
☎(+1 909) 793 31 31 ✆ (800) 456 00 70
⊜ (+1 909) 793 58 18
👥 75
Vice President and General Manager: John E James
Marketing Director: Ms Jayne M Fischer
Contracts Director: Bill Turnpaugh
Domestic and International Sales Manager: Daniel Rankin
■ Manufacture lead acid batteries for aircraft and military ground applications. [98]

75-87508 Teledyne Brown Engineering Energy Systems
(† Teledyne Energy Systems)
10707 Gilroy Road, Hunt Valley, MD 21031
☎(+1 410) 771 8600 ⊜ (+1 410) 771 8620
👥 102
General Mgr: William C. Kincaide
■ Radioisotope power for spacecraft, thermoelectric generators, electrolytic gas generators, systems engineering, life support systems. [97]

75-87510 Teledyne Brown Engineering
(a subs. of Teledyne Inc.)
Cummings Research Park, PO Box 070007,
Huntsville, AL 35807-7007
☎(+1 205) 726 1000 ▣ 593077; TWX 810 7262 103
⊜ (+1 205) 726 1033
President: James F. McGovern
VP Business Development: Aaron Jestice
■ Systems engineering, space vehicles & payloads, military optics, radar, data processing, missiles, test evaluation & training equipment, fabrication-services modules.

● 300 Sparkman Dr., Huntsville, AL 35807
☎(+1 205) 726 1000
▣ TWX 810-726-2103 tdybrwn htv
⊜ (+1 205) 726 1033
Government Sales: Charles E. Grainger

● 3700 Pender Dr., West Oaks Ex.Park, Ste 200, Fairfax, VA 22030
☎(+1 703) 352 8500
Army Intelligence Programs Mgr: Glen Stroup [97]

75-87515 Teledyne Casting Service Cast Products Operations
(† Teledyne Aircraft Products)
4200 W. Valley Blvd, PO Box 2348,
Pomona, CA 91769-2348
☎(+1 909) 595 2252 ⊜ (+1 909) 594 2891
Marketing Dir: Robert A. Zugehar
■ Aluminium & magnesium cast products requiring maximum physical properties. Large & complex geometry, exacting quality specifications, at weights ranging from one pound to one ton. [97]

75-87516 Teledyne Continental Motors
Aircraft Products Div.
Brookley Ind. Complex, PO Box 90,
Mobile, AL 36601-0090
☎(+1 334) 438 3411 ✉ CONTENT ▣ 505519
⊜ (+1 334) 432 2922
👥 750 🏭 2 000 000 sq ft
President: Bryan L Lewis
VP Finance: Bob Jacklin
VP Operations: Howard Atwood
Sales & Service Dir: Tim Archer
■ Manufacture air cooled & liquid cooled piston engines, rotary engines for RPVs, auxiliary power units for military vehicles & other applications, turbine engines for aircraft & ground power applications. [97]

75-87518 Teledyne Controls
(† Teledyne Aircraft Products Controls)
(a subs. of Teledyne Inc.)
12333 West Olympic Blvd, Los Angeles, CA 90064-1021
☎(+1 310) 820 4616 ▣ 4720988
⊜ (+1 310) 820 7813
President: Jefferson Z. Amacker
VP Business Dev.: James Land
■ Condition monitoring systems & equipment, flight test instrumentation, flight deck display systems, aircraft subsystems, air data computers, angle of attack transducers, crash survivable flight data recorders, airport runway visual range sensors. [97]

75-87522 Teledyne Electronic Technologies
(† Teledyne Microelectronics)
(a subs. of Teledyne Industries)
12964 Panama St, Los Angeles, CA 90066
☎(+1 310) 822 8229 ⊜ (+1 310) 574 2015
👥 550 🏭 126 000 sq ft
President: Marvin H. Fink
National Sales Mgr: Paul Galletta
Marketing Mgr: Bill Kennedy
Marketing Admin: Ms Dianna German
■ Hybrid microcircuits, converters, advanced microelectronic digital analog circuits, solid state memories, flat panel information displays, custom resistor networks, hybrid RF/microwave integrated circuits. Fiber optic transmitter receiver, power supplies, power hybrids & multichip VHSIC/GaAs or ECL digital packaging. [97]

75-87523 Teledyne Advanced Materials
(† Teledyne Firth Sterling)
1 Teledyne Place, Interchange City Ind. Park,
La Vergne, TN 37086-3590
☎(+1 615) 641 4200 ⊜ (+1 615) 641 4333
President: David R. Gray
■ Tungsten metal alloy & cemented tungsten carbide kinetic energy penetrators including fragmentation. [97]

75-87524 Teledyne Brown Engineering Geotech/Instruments
(† Teledyne Geotech)
3401 Shiloh Rd, Garland, TX 75041-9007
☎(+1 214) 271 2561 ▣ 170584 ⊜ (+1 214) 271 0223
President: O. Charles Williams, Jr.
■ Geophysical instruments & systems, services & systems engineering, field installation, meteorological instrumentation & systems. [97]

75-87526 Teledyne Brown Engineering Hastings Instruments
(† Teledyne Hastings-Raydist)
PO Box 1436, Hampton, VA 23661-0436
☎(+1 804) 723 6531 ▣ TWX 710 8820 085
⊜ (+1 804) 723 3925
■ Precise radiolocation systems & equipment for ships & aircraft: used for hydrography, ship testing, mine countermeasures, anti-submarine warfare, search & rescue. [97]

75-87530 Teledyne Brown Engineering Environmental Services Main Facility
(† Teledyne Isotopes)
50 Van Buren Ave, Westwood, NJ 07675-0306
☎(+1 201) 664 7070 ▣ 134474 ⊜ (+1 201) 664 5586
President: Donald F. Schutz

■ Radialogical safety services. Radioisotope measurement services. [97]

75-87534 Teledyne Electronic Technologies Mountain View Facility

(† Teledyne MEC)
1274 Terra Bella Avenue, PO Box 7127,
Mountain View, CA 94039-7127
☎(+1 415) 691 9800 ⊖ (+1 415) 697 4353
👥 400 (1990)
President: Charles E. Kelley
Product Marketing Mgr: Ms Bridget A. Daniel
■ Travelling wave tubes (TWTS), high power amplifiers, memory sub-systems. Transmitters, power supplies. [97]

75-87540 Teledyne Electronic Technologies Microwave Components Business Unit

(† Teledyne Microwave)
1290 Terra Bella Ave, Mountain View, CA 94043-1885
☎(+1 415) 968 2211 ⊖ (+1 415) 966 1521
♣ $ 20 000 000 👥 175
President: James Ewan
Communications: Melvin Lee
■ Isolators, filters, multiplexers, oscillators, coaxial switches, amplifiers, delay devices, equalizers. Microwave MMIC components & subsystems for commercial, military & space applications. [97]

75-87545 Teledyne Casting Service Picco Plant

(† Teledyne Aircraft Products)
16800 Chestnut St, City of Industry, CA 91745-1820
☎(+1 818) 961 0711 ⊖ (+1 818) 961 0894
Marketing Dir: Robert A. Zugehar
■ Castings produced from steel, stainless steel, nickel or cobalt base to any superalloy. [97]

75-87547 Teledyne Advanced Materials Powder Alloys

350 Allwood Road, Clifton, NJ 07012-1701
☎(+1 201) 773 4850 ⊖ (+1 201) 773 4421
♣ $ 15 000 000 👥 70
■ Tungsten alloy counterweights & ballasts for flight control surfaces. [97]

75-87550 Teledyne Electronic Technologies Relays Facility

(† Teledyne Relays)
12525 Daphne Ave, Hawthorne, CA 90250
☎(+1 213) 777 0077 ⊖ (+1 213) 779 9161
General Mgr: Walt Strack
■ Electro mechanical & hybrid non latching relays, through hole & surface mountable.

Solid State Relays
12964 Panama Street, Los Angeles, CA 90066
☎(+1 310) 822 8229 ⊖ (+1 310) 574 2015
President: Marvin Fink
■ Solid state relays & solid state power controllers for aerospace, military & space equipment. Product features include integrated short-circuit protection, current overload protection & status feedback. [97]

75-87555 Teledyne Ryan Aeronautical McCormick Selph Ordnance & Electronics

(† Teledyne McCormick Selph)
3601 Union Road, PO Box 6, Hollister, CA 95024-0006
☎(+1 408) 637 37 31 (extension 236)
⊖ (+1 408) 637 14 50
👥 240 (1997) 🏭 270,000 sq ft
President: Gerry McCartha
Business Development Director: Ray D Weiss
■ Design, development and manufacture of controlled explosive and electronic products for the aerospace, automotive and petroleum industries. Products include linear explosives, ordinance and electronic systems and components. [98]

75-87560 Teledyne Ryan Aeronautical

2701 Harbor Drive, PO Box 85311,
San Diego, CA 92186-5311
☎(+1 619) 291 7311 ⊖ (+1 619) 260 5400
👥 1 300 🏭 1 000 000 sq ft
President: R.A.K. Mitchell
VP Business Development: A.C. Richards
VP Special Programmes: N.S. Sakamoto
VP Engineering D.L.: D.L. Hirsch
VP Human Resources: K.A. Carson
VP Programmes: R. Scurlock
VP Quality Assurance: D.A. Zalk
VP Operations: R.A.R. Pettit
VP Controller: A. Honrud
VP Contracts: G.M. Soltys
■ Target drones & UAV's - prime contractor. Subcontract aircraft structures & tooling.

TCAE Turbine Engine Unit
(† Teledyne CAE)
1330 West Laskey Rd, PO Box 6971,
Toledo, OH 43612-0971
☎(+1 419) 470 3000 ⊠ TWX 810-4210-086
⊠ 205293 tydn cae tol; 989563 tdyn tol
⊖ (+1 419) 470 3052
♣ $ 50 000 000 👥 200
Business Development: Robert S. Van Huysen
☎(+1 419) 470 3224
■ Design, develop & manufacture gas turbine propulsion engines for training aircraft, RPV's, drones & cruise missiles.

Washington, DC Office
2111 Wilson Blvd, Ste 1100, Arlington, VA 22201-3008
☎(+1 703) 558 0493 [97]

75-87570 Teledyne Fluid Systems Brookpark Facility

(† Teledyne Republic/Sprague)
15655 Brookpark Rd, Cleveland, OH 44142-1670
☎(+1 216) 267 2700 ⊖ (+1 216) 267 1918
President: Michael A. Gaudiani [97]

75-87580 Teledyne Aircraft Products Thermatics

Hwy 301 -A- Business, PO Box 909,
Elm City, NC 27822-0909
☎(+1 919) 236 4311 ⊠ 579441 ⊖ (+1 919) 236 3613
Vice President: Jim Harris
VP Engineering: Bruce Pike
Marketing Dir: Tim Owens
VP Operations: Dave Berkebile
VP Sales: Harry Helmer (Jr)
■ High temperature wire & cable for military and commercial use; insulating materials; teflon, vinyl, nylon, polyethylene. [97]

75-87600 Teleflex Inc.

630 West Germantown Pike,
Plymouth Meeting, PA 19462
☎(+1 610) 834 63 01 Ⓦ http://www.teleflex.com
⊖ (+1 610) 834 82 28
♣ $ 931,000,000 (1997) 👥 9,400 (1997)
Chairman: Lennox K Black
President and Chief Executive Officer: David S Boyer
Senior Vice President: Dr R Carriker
Vice President of Investor Relations: Ms Janine Dusossoit
☎(+1 610) 834 63 62 [98]

75-87610 Telegenix Inc.

(† Telegenix/Grim Corp.)
(a subsidiary of Inductotherm Industries Inc.)
26 Olney Avenue, PO Box 5550,
Cherry Hill, NJ 08034-9736
☎(+1 609) 424 52 20 Ⓦ http://www.telegenix.com
⊠ 6851048 induct ⊖ (+1 609) 424 20 22
♣ $ 10,000,000 👥 100 🏭 45,000 sq ft
President and Chief Executive Officer: John O'Meara
Vice President Finance, Purchasing and Manufacturing: Randall Raymond
Operations Director: Tim Phelps
Sales and Marketing Director: Justin Mortimer
Manager Government and Commercial Marketing: A W Henry
Purchasing Manager: Ms Jean Wivel
■ Manufacture communications and control electronics systems and electronic graphic displays, for use in air traffic control, vessel control and command and control. Programmable communications voice switches for internal and external communications, signalling interface to radio, telephone and intercom channels. Communications consoles for data, voice or telephone communications between central operator and remote operators. Intelligent voice and data acquisition control and transmission systems used for remote radio control using voice and control data on a single telephone line. Contract manufacturing. [98]

75-87860 Tel-Instrument Electronics Corp. (TIC)

728 Garden St, Carlstadt, NJ 07072
☎(+1 201) 933 16 00 ⊖ (+1 201) 933 7340
♣ $ 3 000 000 👥 20 🏭 19 000 sq ft
President: Harold K. Fletcher
Government Projects Mgr: Robert Melnick
Sales Mgr: Jack Nemeth
■ Avionics test gear for general aviation & military; test gear of all kinds; electronics equipment made to customer specifications or "build to print". [97]

75-87865 Teltron Technologies Inc.

(a subsidiary of the Video Display Corporation)
2 Riga Lane, Birdsboro, PA 19508
☎(+1 610) 582 94 50 ⊖ (+1 610) 582 08 51
👥 20
Chief Operating Officer: Arthur H Mengel
Operations Manager: Eugene A Dymacek
Purchasing Manager: Kristin Lodwig
■ Imaging devices for the medical, government, broadcast, security and instrumentation industries.

Camera Tube Manufacturing Facility
2 Riga Lane, Birdsboro, PA 19508
☎(+1 610) 582 94 50 ⊖ (+1 610) 582 08 51 [98]

75-87900 Tempo Instruments Inc

(a subsidiary of Bowthorpe plc, UK)
87 Modular Avenue, Commack, NY 11725
☎(+1 516) 864 37 00 ⊖ (+1 516) 864 21 96
President: R Brown
■ Design and manufacture of electronic controls and power supply systems for military applications (aircraft navigation systems, automatic test equipment, flight control radar and data displays). [98]

75-88400 Texas Optoelectronics Inc.

714 Shepherd, Garland, TX 75042-6833
☎(+1 972) 487 00 85 Ⓦ http://www.texas-opto.com
⊖ (+1 972) 276 80 59
♣ US$ 8,000,000 👥 150
President and Chief Executive Officer: Don Collum
Vice President of Marketing and Business Development: Martin R McCoy
Operations Director: Les Major
Purchasing Manager: James Beaty
■ Photo detectors, LED's, Hi rel displays, military hybrids and sub assemblies. [98]

75-88600 Texstar Inc.

(a subsidiary of BBA US Holdings Inc.)
802 Avenue J East, PO Box 534036,
Grand Prairie, TX 75053-4036
☎(+1 972) 647 13 66 Ⓔ marketing@texstar.com
⊖ (+1 972) 641 28 00
♣ $ 25,000,00 (1995) 👥 280 🏭 380,000 sq ft
President: David Rollings
Senior Vice President of Engineering and Quality: Jim Irion
Vice President of Operations: Carey Glenn
General Manager - Marketing: Gary Goleman
Purchasing Manager: Ms Mary Lee DeFord
■ Specialised plastics and composite fabrication, commercial and military aircraft transparencies, ducting, interiors, wing tip lenses, coatings. [98]

75-88655 Textron Systems Division

(† Textron Defense Systems)
(a div. of Textron Inc.)
201 Lowell St, Wilmington, MA 01887
☎(+1 508) 657 1276 ⊠ TWX 710 347 6930
⊖ (+1 508) 657 6644
President: Richard J. Millman
Exec. Vice President: R.J. Milauskas
VP Business Development: R.P. Buckley
Marketing Communications Mgr: Claire Flynn

Contact Reference

■ Smart weapons development & production based on sensor fused munition technology; aircraft landing systems; surveillance systems: infrared detectors; energy technology; special high-performance materials sold worldwide to aerospace automotive, industrial & sporting goods manufacturer as well as major oil & chemical processing companies. 97

75-88656 Textron Fastening Systems Textron Aerospace Fasteners

(† Cherry Textron)
(a subsidiary of Textron Inc.)
1224 East Warner Avenue, PO Box 2157,
Santa Ana, CA 92707-0157
☎(+1 714) 545 55 11 ⊜ (+1 714) 850 60 93
659 (1996) 362,000 sq ft
President: George A Andrews
Vice President of Aerospace Operations: Michael Harhen
Vice President of Aerospace Sales and Marketing: Ron Manley
Sales Director of Pacific Rim, Central and South America: Steven R Karno
Director of European Sales: Bill Walker
Advertising Manager: Al Wahlberg
Advertising Accounts Manager: Erik Sorensen
■ Cherry aerospace fasteners for metal and composites and shear pins. 98

75-88660 Textron Marine & Land Systems

(† Textron Marine Systems)
(a division of Textron Inc.)
6600 Plaza Drive, New Orleans, LA 70127-2584
☎(+1 504) 245 66 00 6711199 tmsnola
⊜ (+1 504) 245 69 70
♣ $ 200,000,000 646
President: John J Kelly
Executive Vice President: James W Kratzer
Vice President of Manufacturing: Donald W Horew
Vice President of Marketing: Irwin F Edenzon
Vice President of Marketing: Windsor E Ward
Public Relations and Marketing Representative: Ms Paige Eaton
■ Manufacture technologically advanced marine craft, air cushion vehicles, including Landing Craft Air Cushion (LCAC), all purpose surface effect ships and heavy weather patrol and rescue vessels. Also manufacture armoured vehicles including water cannon vehicles, APC's, turrets, suspension systems and artillery systems. 98

75-88667 Thermacore Inc.

(a subs. of DTX Corp.)
780 Eden Rd, Lancaster, PA 17601-4271
☎(+1 717) 569 6551 ⊜ (+1 717) 569 4797
♣ $ 7 500 000 65 5 000 sq m
President: Richard W. Longsderff
VP Sales & Marketing: Donald M. Ernst
■ Avionics cooling equipment, total system thermal management, contract research, development on heat pipes, thermal management systems. 97

75-88675 Thermal Structures Inc.

2362 Railroad St, Corona, CA 91720
☎(+1 909) 736 9911 ⊜ (+1 909) 736 1064
100
Marketing & Sales Dir: John Schwab
Manager: James N. Bevis
■ Provides engineering, design & manufacture advanced, light weight & temperature efficient insulation blankets & heat shields. 97

75-88680 Thermalloy Inc

2021 West Valley View Lane, PO Box 810839,
Dallas, TX 75381-0839
☎(+1 972) 243 43 21 ⊜ (+1 972) 241 46 56
Director: J Skelly
■ Heat-sinks and related assemblies for cooling military and aerospace electronics equipment. 98

75-88690 Thermodyne International Ltd

1841 Business Parkway, Ontario, CA 91761
☎(+1 909) 923 99 45 ⊜ (+1 909) 923 75 05
♣ $ 10,000,000 120 50,000 sq ft
Chairman: Walter C Wolf
President: Frank Bucalo
Senior Vice President: Gary S Ackerman
■ Transit and operating cases for electronic and sensitive equipment. 98

75-88700 Thermo-Jet Standard Inc.

PO Box 55976, Houston, TX 77055
☎(+1 281) 331 44 60 ⊜ (+1 281) 331 44 60
President: John A Melenric
■ Pulse jet engines for RPV, drone propulsion. 98

75-88720 Thermometrics Inc

(a subsidiary of Bowthorpe plc)
808 US Highway 1, Edison, NJ 08817
☎(+1 908) 287 28 70 ⊜ (+1 908) 287 88 47
President: J Carter
■ Manufacture of NTC/PTC thermistors and related products for temperature compensation, measurement and control applications. 98

75-88740 Thiokol Corpoartion

2475 Washington Boulevard, Ogden, UT 84401-2398
☎(+1 801) 629 22 70 W http://www.thoikel.com
⊜ (+1 801) 629 22 51
♣ $ 889,500,000 (1996)
Chief Executive of the Board, President and Chief Executive Officer: James R Wilson
Senior Vice President and Chief Financial Officer: Richard L Corbin
Executive Vice President, Human Resources and Administration: James E Mucnulty
Vice President and General Manager, Defence and Launch Vehicles Division: Winston N Bunbridge
Vice President and General Manager, Space Operations: Joseph A Lombardo
Vice President and Controller: Michael R Ayers
Vice President, Government Relations: Tidal W McCoy
Vice President, Tax and Tax Counsel: Paul Cherecwich
Vice President, Compensation and Benefits: Richard T Smith
Vice President and General Manager, Ordnance Operations: Luther C Johnson
Vice President, Information Systems: Brad S Stout
Treasurer: Nicholas J Iuanow
Secretary: Edwin M North
■ Design and manufacture solid rocket propulsion and ordnance for space, strategic and tactical defence applications. Manufacturer of precision fastening systems for aerospace. Products include Space Shuttle RSRM, Castor Series, STAR Series, ballistic and tactical missile propulsion systems, Demilitarisation technologies, flares and training devices.

Government Relations Office
1735 Jefferson Davis Highway, Suite 1001,
Arlington, VA 22202-3461

Space Operations
PO Box 707, Brigham City, UT 84302-0707
☎(+1 801) 863 35 11 ⊜ (+1 801) 863 23 91

Elkton Defense and Launch Vehicles
PO Box 241, Elkton, MD 21922-0241
☎(+1 410) 392 10 00 ⊜ (+1 410) 392 12 05

Defense and Launch Vehicle Division
PO Box 689, Brigham City, UT 84302-0689

Public Relations
2475 Washington Boulevard, Ogden, UT 84401-2398
☎(+1 801) 629 22 57

Science and Engineering Division
PO Box 707, Brigham City, UT 84302-0707

Fastening Systems, Huck International, Inc
3724 East Columbia Street, Tuscon, AZ 85714

Aerospace Fasteners
PO Box 5268, Carson, CA 90749

Industrial Fasteners
PO Box 8117, Waco, TX 76710 98

75-88850 Thomas Electronics Inc. (TE)

100 Riverview Drive, PO Box 435, Wayne, NJ 07470
☎(+1 201) 696 52 00 ⊜ (+1 201) 696 82 98
385 (1997) 60,000 sq ft
Chairman: H.A. Ketchum
VP Sales: Bruce Piaget
■ Cathode ray tubes and cathode ray tube assemblies for commercial, industrial and military applications. Backlights.

● L. Berkley Davis Industrial Park, PO Box 71,
Clyde, NY 14433
☎(+1 315) 923 20 51 ⊜ (+1 315) 923 44 01 98

75-88860 Thomas Instrument & Machine Co. Inc.

3040 East First Street, Brookshire, TX 77423
☎(+1 281) 375 6300 ⊜ (+1 281) 375 5264
♣ US$ 10,000,000 120
President: Thomas R Blackburn
Vice President: William D Clifford
■ Aircraft parts, munitions parts, missile parts. Contract precision machining, mechanical assembly and testing, turnkey product manufacturing, specialising in CNC machining of titanium, super alloys and other premium quality materials. 98

75-88890 Thompson Aero Products

Thompson Aeronautical Consultants (TAC)
2 Bridgeport, Bella Vista, AR 72715
☎(+1 501) 855 77 07 ✆ (800) 237 01 40
⊜ (+1 501) 855 77 07
2
Owner: William D Thompson
■ Manufacture hinged plate spoilers/dive brakes, auxiliary vacuum systems and position indicating systems for light planes. Aviation books on Cessna Aircraft. → 80 98

75-89400 Timet

1999 Broadway, Denver, CO 08020
☎(+1 303) 296 5600 ⊜ (+1 303) 296 5640
1 300
Managing Dir: Andrew R. Dixey
VP Sales: J.P. Monahan
Technical Dir: Thomas A. Buck
Int'l Sales & Marketing Dir: C.E. Carlson
Purchasing Mgr: Hoy E. Frakes (Jr)
■ Titanium alloys. 97

75-89560 Tiodize Co. Inc.

5858 Engineer Drive, Huntington Beach, CA 92649
☎(+1 714) 898 43 77 ⊜ (+1 714) 891 74 67
55 52,000 sq ft
President: Thomas R Adams
Vice President, Controller: Ms Debbie Harris
Advertising Director: Larry M Krupnak
Research and Development Manager: Gary R Wittman
Project Manager: Wade Friedrichs
■ Hi-tech corrosion coatings, solid film moly disulfide lubricants, teflon coatings, composite fastener products, self-lubricating composite bearings, hard anodize w/teflon impregnation, mold releases, and titanium anodizing with no dimensional change, cleaner/degreasers and rust preventive penetrant/lubricants. → 80

Processing Plant
15701 Industry Lane, Huntington Beach, CA 92649
☎(+1 714) 898 43 77 ⊜ (+1 714) 891 74 67

Detroit Office
28966 Wall Street, Wixom, WI 48393
☎(+1 313) 348 60 50 ⊜ (+1 313) 348 25 20 98

75-89580 Tisdale Co. Inc.

PO Box 2728, Conroe, TX 77301
☎(+1 409) 856 15 00 E tisdale@flex.net
⊜ (+1 409) 856 16 00
50
President and Chief Executive Officer: Lloyd B Tisdale
Operations Director: Dusty Adams
Sales and Marketing Director: Bob Duncan
Purchasing Manager: John Tisdale
■ Air conditioning, heating, ventilation and filtration for facilities, test rooms, aircraft ground support and other special built to order products. Also process refrigeration packages, air handlers, chemical air filters and other related products for temperature and environmental control. 98

75-89620 The Titan Corp

3033 Science Park Road, San Diego, CA 92121
☎(+1 619) 552 95 00 W http://www.titan.com/
E corpcomm@titan.com. ⊜ (+1 619) 552 99 09
♣ $ 133,976,000 (1995) 1,500
President and Chief Executive Officer: Dr Gene W Ray
Senior Vice President and Chief Financial Officer: Eric DeMarco
Senior Vice President, General Counsel and Secretary: Phillip England
Corporate Controller: Ms Deanna Hom Petersen
Business Development Manager: Ms Elizabeth K Butterfield
■ Laser systems and light power, high voltage electronic subsystem for directed energy devices; Mil-Spec and ruggedised computer modules and systems. Secure satellite communication systems, C4I, modems, earth-based terminals, test and evaluation services, software design and support. → 80

Subsidiary companies:
Titan Eldridge
Titan Beta
Titan TRT
Titan Unidyne
Titan Information Systems
Titan Scan
Titan Environmental
Titan Systems
Titan PSI
Titan Linkabit
Division Head Office
9410 Topanga Canyon Boulevard, Suite 104,
Chatsworth, CA 91311-2376
☎ (+1 818) 709 37 05 ⊖ (+1 818) 341 26 63
Division President: Tony Fredrickson [98]

75-89630 Titan Electronics
(† Severe Environment Systems Co.)
(a div. of The Titan Corp.)
Headquarters
4215 Sorrento Valley Blvd, San Diego, CA 92121
☎ (+1 619) 452 5259; 5200 ⊖ (+1 619) 597 1616
♣ $ 149 414 000 👥 150
President: R. Rowe
Public Relations Dir: Ms Michelle Mueller
National Sales Dir: W. Mathews
Operations Dir: L. Ragonsesi
Engineering Dir: G. Pennington
Purchasing Mgr: D. Moore
■ Microcomputer board modules & systems for severe environments.
Western Region
4215 Sorrento Valley Blvd, San Diego, CA 92121
☎ (+1 619) 452 5209 ⊖ (+1 619) 597 1616
Manager: W. Matthews
Northen Region
11305 Cromwell Ct., Woodbridge, VA 22102
☎ (+1 703) 490 5114 ⊖ (+1 703) 490 5123
Manager: William I. Hall [97]

75-89650 Titan Linkabit
(a subsidiary of The Titan Corp.)
3033 Science Park Road, San Diego, CA 92121
☎ (+1 619) 552 95 00 ⊖ (+1 619) 552 99 09
👥 200
President: Ron Gorda
Vice President of Marketing and Sales: Tom Trimble
Public Relations Manager: Ms Michelle Mueller
Business Development Manager:
Ms Elizabeth K Butterfield
■ Modems, decoders, radios, voice compression devices, Demand Assigned Multiple Access (DAMA), small satellite communications terminals (UHF, SHF, EHF ranges). [98]

75-89750 Titeflex Corp.
603 Hendee St, PO Box 90054,
Springfield, MA 01104-3003
☎ (+1 413) 739 5631 ⊖ (+1 413) 781 1207
👥 500 🏭 330 000 sq m
President: James Marlatt
Marketing Dir: Ms Sheila Weaver
■ Teflon & stainless steel flexible hose, fittings & hose assemblies for aircraft military & commercial. [97]

75-89755 TIW Systems Inc.
2211 Lawson Lane, Santa Clara, CA 95054
☎ (+1 408) 654 56 00 Ⓔ info@tiw.com
Ⓣ 346352 tiw sys suvl
⊖ (+1 408) 654 56 13; 654 56 14
♣ US$ 45,000,000 (1995) 👥 200 (1996) 🏭 85,000 sq ft
President: Dr Rein Luik
Executive Vice President: Louis E Becker
Marketing Director: Rob Wellins
Manager of Microwave Engineering: K Goudey
■ Satellite antenna systems and earth stations. Digital communications subsystems. [98]

75-89778 Tomsed Corp.
1412 Stewart Street, Fuquay-Varina, NC 27526
☎ (+1 919) 552 61 00 ✆ (800) 334 55 52
Ⓔ tomsed@bellsouth.net ⊖ (+1 919) 552 71 31
♧ $ 500,000 (1994) ♣ $ 2,500,000 (1994) 👥 17
🏭 2,240 sq m
Managing Director: Thomas C Sedivy
Marketing Manager: Robert T Sedivy
Purchasing Manager: Glenn Huff
■ Manufacture turnstiles for large public events. [98]

75-89800 Torwico Electronics L.P.
(a partner with Magnetika Inc., USA)
410 Oberlin Avenue South, Lakewood, NJ 08701
☎ (+1 908) 364 18 00 ⊖ (+1 908) 363 62 34
👥 200 (1997) 🏭 40,000 sq ft
President: Robert Savino
Sales Mgr: Stanley Karr
■ Custom designed magnetic components intended for use in high-rel industrial & military applications. [98]

75-90000 Tracor Inc
6500 Tracor Lane, Austin, TX 78725-2006
☎ (+1 512) 926 28 00 Ⓔ pr@corp1.tracor.com
Ⓦ http://www.tracor.com ⊖ (+1 512) 929 22 62
♣ $ 1,082,505,000 (1996) 👥 10,515 (1996)
Chairman, President and Chief Executive Officer:
James B Skaggs
Vice President and Chief Financial Officer: Robert K Floyd
Vice President, Treasurer: Woody P Endsley
Vice President, Controller: Thomas V Talbott
Vice President of Business Development: Roger W Sadler
Corporate Vice President of Quality, Safety and Environmental Programs: Leroy J Kana
Vice President of Technology and Information Systems:
John M Rock
Vice President of Human Resources: Samuel Crawford
Public Relations Director: Ms Marian Herbst Kelley
☎ (+1 512) 929 22 73
■ Airborne shipboard and ground countermeasures systems, active expendables, chaff and flare expendables, anti-ballistic missile penetration aids, systems integration.
Subsidiaries with defence business:
GDE Systems Inc.
Tracor Information Systems Company
Tracor Aerospace Inc.
Tracor Applied Sciences Inc.
Tracor Flight Systems Inc.
Vitro Corporation [98]

75-90010 Tracor Aerospace Inc.
(a subsidiary of Tracor Inc.)
6500 Tracor Lane, Austin, TX 78725-2006
☎ (+1 512) 926 28 00 Ⓔ mhk@corpl.tracor.com
Ⓣ 776410 ⊖ (+1 512) 929 23 80
President: George R Melton
Vice President of Marketing: Dave Wallace
Vice President of Operations: Ben Medley
Vice President of Advanced Programs: Ed Katkic
Divisional Vice President and ALE-47 Program Manager:
Robert H Swanson
Vice President and General Manager of Information and Display Systems: Edward C Rizzotti
Vice President and General Manager of Expendables:
Morris Creel
Vice President annd General Manager of Electronic Systems: Larry Wehrheim
Director and General Manager of Mine Countermeasures:
Lex Allen
Marketing Division: Ms Gina Tiller
■ Develops electronic countermeasures dispenser systems for use on board US Air Force, Navy and Army tactical and transport aircraft and helicopters, as well as aircraft of other governments. Other products include chaff, active radio frequency, flare decoys, a minefield breaching system, and other countermeasures, simulation and radar warning receiver systems.
Advanced Programs Division
6500 Tracor Lane, Austin, TX 78725-2070
☎ (+1 512) 929 24 84 ⊖ (+1 512) 929 28 91
Division Vice President: Ed Katkic
Airborne Electronic Countermeasures Division
6500 Tracor Lane, Austin, TX 78725-2070
☎ (+1 512) 929 27 79 ⊖ (+1 512) 929 23 20
Vice President and General Manager: Robert H Swanson
North Carolina Operations
PO Box 1179, Lillington, NC 27546
☎ (+1 919) 893 20 94 ⊖ (+1 919) 893 88 07
General Manager: Bob Martin
Electronic Systems Inc.
(† Tracor-AEL)
305 Richardson Road, Lansdale, PA 19446-1485
☎ (+1 215) 996 29 00 ⊖ (+1 215) 996 20 92
Vice President and General Manager: Larry Wehrheim
Expendales Division
6500 Tracor Lane, Austin, TX 78725-2070
☎ (+1 512) 929 45 72 ⊖ (+1 512) 929 28 91
Vice President and General Manager: Morris Creel
Information and Display Systems
6500 Tracor Lane, Austin, TX 78725-2070
☎ (+1 512) 929 45 72 ⊖ (+1 512) 929 29 93
Divisional Vice President and General Manager: Ed Rizzotti
Mine Countermeasures Division
6500 Tracor Lane, Austin, TX 78725-2070
☎ (+1 512) 929 43 24 ⊖ (+1 512) 929 29 80
Director and General Manager: Lex Allen [98]

75-90200 Trammel Crow Residential
(a subsidiary of Daewoo Corporation South Korea)
6400 Congress Avenue Suite 1000, Boca Rato, FL 33487
☎ (+1 561) 997 97 00 ⊖ (+1 561) 997 86 49 [98]

75-90750 Transicoil Inc.
(a subs. of Eagle Picher Ind. Inc.)
PO Box 9011, Vally Forge, PA 19485-1011
☎ (+1 610) 539 4400 ✆ (800) 323 715
⊖ (+1 610) 539 3400
👥 200 🏭 135 000 sq ft
President & CEO: Robert N. Carlson
VP Marketing: P. Siegrist (Jr)
Operations/Tech. Dir: J. Swiger
Purchasing Mgr: P. Pullin
■ Servo motors, resolvers, synchros, motor generators, tachometers, gearheads, precision electromechanical assemblies, integrated switch panels, control panels, fiber optic displays. Annunciators. Engine instrumentation & liquid oxygen indicators, microprocessor based engine monitoring systems, solid state displays. [97]

75-90920 Trilectron Industries Inc.
(a subsidiary of HEICO)
12297 US Highway 41 North, Palmetto, FL 34220-2504
☎ (+1 941) 723 18 41 Ⓣ TWX 510 823 0091
⊖ (+1 941) 723 31 60
♣ $ 18,000,000 - 20,000,000 (1997) 👥 125 (1997)
🏭 35,000 sq ft
President: Sigmund Borax
Manager, Technical Business Development:
Eric Brocklesby
■ Manufacture GPU's for military and commercial aircraft, air conditioners, load banks, battery chargers. [98]

75-90950 Trinity Industries, Inc.
2525 Stemmons Freeway, PO Box 568887,
Dallas, TX 75356-8887
☎ (+1 214) 631 44 20
♣ $ 2,496,000,000 (1996) 🏭 13,000,000 sq ft
Chairman, Vice President and Chief Executive Officer:
W Ray Wallace
Senior Vice President: Ralph A Banks (Jr)
Senior Vice President: Richard G Brown
Senior Vice President: John T Sanford
Treasurer: Neil O Shoop
■ Manufacturer of railcars, marine products, construction products, pressure and nonpressure containers and metal components. → 70 [98]

75-90970 Triplett Corporation
1 Triplett Drive, Bluffton, OH 45817
☎ (+1 419) 358 50 15 ✆ (800) TRIPLET
Ⓦ http://www.triplett.com ⊖ (+1 419) 358 79 56
👥 150 (1997) 🏭 88,000 sq m
President and Chief Executive Officer: Daniel J DeLong
Operations Director: Todd R Haug
Sales and Marketing Director: Tony L Augsburger
Purchasing Manager: Ms Donna Davenport
■ Analogue & digital test equipment. [98]

75-91050 Triton Technology Inc.
125 Westridge Drive, Watsonville, CA 95076
☎ (+1 408) 722 73 73 Ⓔ sales@tritontech.com
⊖ (+1 408) 722 14 05
President: Frederick L Newton
Marketing Director: John M Tamplin
■ Development of data acquisition and image processing systems for side scan sonar, multi-channel seismic and GPR; also, virtual reality products such as 3D mosaics. [98]

75-91070 Trompeter Electronics Inc. (TEI)
31186 La Baya Drive, PO Box 5069,
Westlake Village, CA 91362-4047
☎ (+1 818) 707 2020 ⊖ (+1 818) 706 1040
♧ $ 19 000 000 👥 170 🏭 37 000 sq ft 🏭 44 000 sq ft
President: Jack Kantola
Sales & Mktg Mgr: Raymond Calvin

Advertising: Ms Rebecca Gillmore
Sales Administration: Ms Phyllis McFarlane
■ Coax, twinax, triax. RF connectors patching connectors & patch fields, cable & cable assemblies. [97]

75-91110 TrueTime
(† Kinemetrics/Truetime)
2835 Duke Court, Santa Rosa, CA 95407
☎ (+1 707) 528 12 30 ✉ TRUETIME Santa Rosa
[W] http://www.truetime.com ⊠ 176687 truetime saro
⊖ (+1 707) 527 66 40
👥 80 🏭 30,000 sq ft
Vice President, Sales Manager: Rick Dielman
Marketing Manager: Michael R Tope
■ Time and frequency receivers. [98]

75-91120 Tru-Form Incorporated
(a member of the TI Group plc, UK)
1141 Highway 315, Wilkes-Barre, PA 18702-6928
☎ (+1 717) 821 2750 ⊖ (+1 717) 826 0347
👥 32
General Mgr: Mitch Handler
■ Manufacture cold rolled rings for the gas turbine industry. Using starting material of either flash welded rings or seamless forgings the company produces net or near net shaped rings for critical operations including combustors & cases in nickel & cobalt superalloys. [98]

75-91300 Turbomeca Engine Corp. (TEC)
(a subsidiary of Turbomeca, France)
2709 Forum Drive, Grand Prairie, TX 75052
☎ (+1 972) 606 76 00 ✆ (800) 662 63 22
⊖ (+1 972) 606 76 92
♣ $ 40,000,000 (1994) 👥 125 (1995)
Managing Director: Dennis B Nichols
Operations Director: Gerard Russo
Sales and Marketing Director: Forrest D Feiock
Purchasing Manager: Rick Rutkauskas
Marketing Communications Manager: Ms Melanie Beames
■ Manufacture and product support of Turbomeca gas turbine engines.
® France: Turbomeca [98]

75-91320 Tuthill Corporation Cablecraft Division
4401 South Orchard Street, PO Box 11372,
Tacoma, WA 98411
☎ (+1 253) 475 10 80 [E] cables@ix.netcom.com
⊖ (+1 253) 474 31 02
♣ $ 20,000,000 (1997) 👥 168 (1997) 🏭 6,700 sq m
President: Keith Clarno
Operations/Technical Director: Dan Miller
Purchasing Manager: Dick Curtis
Aircraft Marketing Manager: Jack L Nelson
Production Manager: Pete Clement
■ Mechanical push-pull cables for air, land and sea applications. [98]

75-91350 Twin Disc Inc.
1328 Racine Street, Racine, WI 53403-1758
☎ (+1 414) 638 40 00 [E] http://www.twindisk.com
⊖ (+1 414) 638 44 82
♣ $ 164,232,000 (1996) 👥 750 (1997)
Chairman & CEO: Michael E. Batten
VP Marketing: John Spano
Mgr Marketing Communications: David Johnson
■ General industrial clutches, reduction gears machine tool clutches, marine transmissions, power shift transmissions for off-highway equipment, torque converters, universal joints. Automatic transmission systems for off-highway vehicles. [98]

75-91355 II-VI Inc.
375 Saxonburg Blvd, Saxonburg, PA 16056-9449
☎ (+1 412) 352 1504 ⊖ (+1 412) 352 4980
♧ $ 16 000 000 (1995) ♣ $ 28 000 000 (1995) 👥 420
CEO: Carl J. Johnson
President: Francis J. Kramer
Sales & Marketing Dir: Bernard Bruns
Public Relations Dir: James Martinelli
Advertising Dir: Bernard E. Bruns
■ Infra-red optical components, assemblies & sub-assemblies for FLIR & other night vision applications. Electro-optic modulators for laser radar. CdTe & CdZnTe single crystal substrates for use in epitaxial growth of HgCdTe detectors for IR systems.
II-IV is currently developing products for the radiation detection market through its recently formed eV PRODUCTS Division. Markets addressed by eV's hybrid pre-amplifiers & radiation detectors for nuclear instruments include industrial gauging, medical imaging, health monitoring & high energy physics research & development. [97]

75-91370 Tyton Hellerman Corp
(a subsidiary of Bowthorpe plc)
7930 North Faulkner Road, PO Box 23055,
Milwaukee, WI 53224
☎ (+1 414) 355 11 30 ⊖ (+1 414) 355 73 41
President: D Plesha
■ Cable and wire fixing systems and cable ties. [98]

75-91380 UB Corp.
9829 Wilsky Boulevard, Tampa, FL 33165-1399
☎ (+1 813) 884 14 63 ⊖ (+1 813) 882 31 92
♣ $ 1,843,668 (1997) 👥 20 (1997) 🏭 2,026 m2
President and Chief Engineer: R J Upcavage
Vice President and Finance Officer: L J Bauer (Jr)
Vice President of Sales and Marketing: C M Fairbanks
■ Aircraft antennas, GPS antennas, RF subsystems, microwave assemblies, waveguide assemblies, molding of radomes. [98]

75-91381 UBE Machinery Inc
5700 South State Street, Ann Arbor, MI 48108
☎ (+1 313) 741 70 00 ⊖ (+1 313) 741 70 17 [98]

75-91382 UFC Co.
(† Raymark Friction Co.)
123 E. Stiegel St, Manheim, PA 17545
☎ (+1 717) 665 2211 ✆ (800) 233 0346
⊖ (+1 717) 665 2794
♣ $ 24 000 000 (1995) 👥 320
President: Bradley C. Smith
Sales & Marketing Dir: Ms Doreen F. Tomao
Public Relations Dir: Roderick E. Laukhuff
■ High strength, high heat resistant & nonasbestos products for lining of missiles, aircraft brake piston insulator, tank braking & steering lining. [97]

75-91383 UBC Inc.
6101 Jones Road, Suite 6, Tampa, FL 33634
☎ (+1 813) 884 60 76 ⊖ (+1 813) 884 83 18
♣ $ 1,500,000 (1997) 👥 9 (1997) 🏭 1,000 sq m
President: Ms Rebecca M Crane
Vice President of Sales and Marketing: Patrick E Crane
[E] crane@gate.net
Operations Director: E C Johnson
Sales and Marketing Director: Robert J Shields
■ Research and development engineering. Basic research in phenomenology associated with electromagnetics, limited production of antennas, trackers and sensors. → 60 [98]

75-91385 Ulbrich Stainless Steels & Special Metals Inc.
57 Dodge Avenue, North Haven, CT 06473
☎ (+1 203) 239 44 81 [W] http://www.ulbrich.com
⊖ (+1 203) 239 00 49
♣ $ 125,000,000 (1997) 👥 450 (1997) 🏭 500,000 sq ft
Chairman: Frederick C. Ulbrich (Jr)
President: Chris Ulbrich
■ Stainless steel, nickel & cobalt alloys, titanium & alloys for aerospace, aircraft & defence applications. High temp applications including jet engine honeycomb, seals, gaskets, spring & instrument components, recorder tapes. Round, flat & specialty shaped stainless steel, nickel alloy & cobalt alloy wire. [98]

75-91505 UMPCO
7100 Lampson Avenue, Garden Grove, CA 92641-3914
☎ (+1 714) 897 35 31 ⊖ (+1 714) 897 26 11
👥 90
President: Dan Miller
Sales and Marketing Director: Stephen M Jennie
■ Clamps, brackets, U-bolts and light stampings. [98]

75-91507 USM, Inc
(† United Scale Models Inc.)
(a subsidiary of Granite Hill Vineyards)
5250 Gulfton, Houston, TX 77081-2935
☎ (+1 713) 661 41 44 [E] usmghv7@aol.com
⊖ (+1 713) 661 54 07
👥 55
President: Joy Johnson
Vice President: Steve Johnson
Managing Director: Scott D Morgan
Sales and Marketing Director: Paul R Johnson
Purchasing Manager: Ms Eve Boring
Office Manager: Ms Susan Fish
■ Training devices, simulators, mockups and prototypes, scale models. → 80 [98]

75-91800 Uniroyal Chemical Company Inc. Specialty Chemicals
World Headquarters, Middlebury, CT 06749
☎ (+1 203) 573 3602 ✆ (800) 322 3243
⊖ (+1 203) 573 3541
Worldwide Marketing Mgr: A. Eisenstein
■ Propellant stabilizers & antioxidants. Epoxy curatives. [97]

75-92050 United Defense L.P.
(a subsidiary of FMC Corporation)
1525 Wilson Boulevard, Suite 700, Arlington, VA 22209
☎ (+1 408) 289 28 82 ⊖ (+1 408) 289 42 94
♣ $ 1,000,000,000 (1996) 👥 5,800 (1996)
President and Chief Executive Officer: Thomas W Rabaut
Vice President and Chief Financial Officer: Francis Raborn
Vice President of Marketing: Dennis A Wagner
Vice President of Government Affairs: John A Mullett
Vice President of Communications: R William Highlander
■ Manufacture chemicals and machinery for industry, agriculture and government. Main markets include industrial chemicals, performance chemicals, precious metals, defence systems, machinery and equipment.
● 1100 Briars Road, PO Box 15512, York, PA 17404
☎ (+1 717) 225 80 00 ⊖ (+1 717) 225 83 00
Vice President and General Manager: Peter C Woglom

Ground Systems Division (DSG)
2830 De La Cruz Boulevard, PO Box 58123,
Santa Clara, CA 95052-8123
☎ (+1 408) 289 28 82 ⊖ (+1 408) 289 42 94
Vice President of Communications: R William Highlander
■ Operates within the US as a major supplier of tracked vehicles to the US Army and Marine Corps and of large calibre gun mounts and missile launching systems to the US Navy. High quality alloy and carbon steel castings and forgings for both military and commercial markets. DSG operates internationally to market vehicles and systems in over 50 countries and works with customers to develop training, maintenance and logistic support programs.
● 2830 De La Cruz Boulevard, PO Box 58123,
Santa Clara, CA 95050
☎ (+1 408) 289 01 11
Vice President and General Manager: Arthur L Roberts
☎ (+1 408) 289 39 86 ☎ (+1 408) 289 09 87 (fax)
International Marketing: Brian E Anderson
☎ (+1 408) 289 40 57 ☎ (+1 408) 289 31 24 (fax)

Armament Systems Division
4800 East River Road, Minneapolis, MN 55421
☎ (+1 612) 571 92 01 ⊖ (+1 612) 572 98 26
Vice President and General Manager: Frederick M Strader
☎ (+1 612) 572 65 04

Paladin Production Operation
56 California Avenue, Chambersburg, PA 17201
☎ (+1 717) 261 59 00 ⊖ (+1 717) 261 59 17
Operations Director: Peter F Scott ☎ (+1 717) 261 59 03

Steel Products Division
2101 West 10th Street, PO Box 1030,
Anniston, AL 36201
☎ (+1 205) 237 28 41 ⊖ (+1 205) 235 96 91
Division Manager: Elmer L Doty ☎ (+1 205) 235 98 30
■ Tracked armoured vehicle components and commercial casting, forgings and fabrications. [98]

75-92100 United Electric Motor Co.
4354 Industrial Center, San Antonio, TX 78217-4812
☎ (+1 210) 655 09 21 ✆ (800) 545 61 05
[E] uemc@aol.com ⊖ (+1 210) 655 64 24
♣ US$ 3,800,000 (1995) 👥 41 🏭 40,000 m2
President: Ms Melanie McKown
Contract Administrator: Melvin McMullen
■ Manufacture fabric items used on military aircraft and warning streamers. Overhaul and repair of ground powered generators. Also packaging for the government. → 80 [98]

75-92650 Universal Avionics Systems Corp
(† Universal Navigation Corp)
3260 East Lerdo Road, Tucson, AZ 85706
☎(+1 520) 295 23 00 ✆ (800) 321 52 53
Ⓦ http://www.uasc.com. ⊖ (+1 520) 295 23 95
Executive Vice President: Charles H Edmondson
Vice President Marketing: Donald D Berlin
Director of Marketing Communications: Stephen B Scolnik
Director, Marketing Administration: David Upchurch
Director of Customer Services: Grady Dees
■ Flight management/navigation management systems; long- and short-range navigation sensors; air-to-ground two way datalink. Cockpit voice recorder. Cabin display system. [98]

75-92700 Universal Propulsion Co. Inc. (UPCo)
(a div. of Talley Industries Inc.)
25401 N. Central Ave, Phoenix, AZ 85027-9801
☎(+1 602) 516 3340 ☒ 668487 upco phx
⊖ (+1 602) 516 3364
72 000 sq ft
President & General Mgr: Harold G. Watson
Exec. VP/VP Operations: L.C. Kempton
VP Marketing: Robert F. Sadler
■ Design, develop & manufacture emergency ballistic devices & systems, ejection seat & egress systems; also produces highly engineered propellants, for rocket catapults, rocket motors, initiators, gas generators, drogue guns, thrusters, ballistic systems. [97]

75-92870 UO Equipment Co.
PO Box 924615, Houston, TX 77292-4615
☎(+1 713) 686 18 69 ✆ (800) 231 6372
⊖ (+1 713) 688 00 01
♣ $ 600,000 (1997) 6 (1997)
President: Robert R Wright (Jr)
■ Resuscitators. Bottled oxygen. [98]

75-92915 UPship
Route 2, Box 53-4, Elba, AL 36323
☎(+1 334) 897 61 32 ⊖ (+1 334) 897 34 34
1 (1996)
President: Jesse Blenn
■ Design and manufacture helium airships. [98]

75-92930 Urdan Industries (USA) Inc.
30161 Southfield Rd, Suite 310, Southfield, MI 48076
☎(+1 810) 642 3882 ⊖ (+1 810) 649 2084
President: Dani Gat [97]

75-92950 Urethane Products Corp.
17007 South Broadway, Gardena, CA 90248-3113
☎(+1 310) 532 36 62 ✆ (800) 913 00 62
Ⓦ http://www.urethaneproducts.com
⊖ (+1 310) 532 98 84
25
General Manager: Gerald Thermos
Ⓔ jthermos@urethaneproducts.com
Operations Director: Alfred Medrand
Purchasing Manager: Ms Joan Walker
■ Marine fender, protector dome, cartridge plugs, urethane sheets. [98]

75-93000 US Lighter Than Air Corporation
750 Commercial Street, Eugene, OR 97402
☎(+1 541) 683 49 83 ⊖ (+1 541) 342 38 06
15 (1997) 1,000 m2
Vice President: Joe Olma
Operations Director: Martin Grant
Purchasing Manager: Ted Kruysman
■ Design and manufacture non-rigid helium airships. [98]

75-93025 US Paint
831 South List Street, Saint Louis, MO 63103
☎(+1 314) 621 05 25 ⊖ (+1 314) 621 07 22
Managing Director: Walt Lindner
Operations Director: Kent Brandt
Technical Director: Chuck Ray
Purchasing Director: Ms Susan Klein
Sales Director: Jeff Roemer
Advertising Assistant: Ms Kristy P Chriscinske
■ Complete coating systems for the aviation and aerospace markets, original contractors specified colours, military specifications. [97]

75-93070 Vacco Industries
(a subsidiary of ESCO Electronics Corp)
10350 Vacco Street, South El Monte, CA 91733
☎(+1 818) 443 71 21 ⊖ (+1 818) 442 69 43
Group Vice President and President: J M Conway
Senior Vice President and General Manager: J E Rub
Vice President of Sales: R J McCredie
■ Specialised flow control valves, filters and filtration systems, photochemical machining. [98]

75-93087 Vaisala Artais Division
(† Artais Weather Check Inc.)
(a division of Vaisala Inc., USA)
7450 Industrial Parkway, Plain City, OH 43064
☎(+1 614) 873 68 80 ✆ (800) 327 AWOS
Ⓦ http://www.vaisala.com/artais.htm
⊖ (+1 614) 873 68 90
38 (1997) 15,000 sq ft
Director: Charles R Shanklin
General Manager: Mike Smyser
Plant Manager: Jamie Robertson
International Marketing: James K Ferguson
Chief Engineer: Roberto Arpino
Applications Engineer: Jake Trimpey
■ Automated weather reporting equipment which collects, analysis and disseminates weather information from strategically located sensors. This info is processed automatically & broadcast via a NAVAID to airborne pilots. Mobile systems available. [98]

75-93165 Valley Manufacturing Corp
(a member of the TI Group plc, UK)
120 Hazle Street, Wilkes Barre, PA 18703
☎(+1 717) 825 4544 ⊖ (+1 717) 825 0170
92
General Mgr: David Bjornberg
Controller: Joseph Niemec
■ Supplier of precision machined gas turbine engine components. In addition to normal turning & milling, the company offers laser drilling, piercing, EDM (electro-discharge machining), room temperature vulcanizing & assembly. Materials include sophisticated high temperature alloys, titanium & stainless steel. [98]

75-93170 Valley Todeco
(a subs. of McKechnie plc)
12975 Bradley Ave, Sylmar, CA 91342-3830
☎(+1 818) 367 2261 ⊖ (+1 818) 364 6036
182
President: William G. White
Sales & Marketing Manager: Robert Bishop
Dir Public Relations: Sam Contino
■ Manufactures fasteners, slotted entry bearings & related products for the aircraft & aerospace industry. All products manufactured share the common characteristics of complex close tolerance, high temperature corrosion resistance. [97]

75-93250 Vantage Associates Inc.
6350 Nancy Ridge Drive, San Diego, CA 92121-2914
☎(+1 619) 453 36 80 ⊖ (+1 619) 453 37 68
♧ $ 4,000,000 (1997) ♣ $ 5,000,000 (1997)
70 (1997) 350 m2
President: Dr Louis J Alpinieri
Vice President: Paul A Roy
Sales and Marketing Director: W Penrose
Purchasing Manager: R Tarabocchia
■ Custom composite structural and RF products, including high performance antennas and radomes, lightweight graphite composite structures and avionic racks.

● 199 West 146th Street, Gardena, CA 90248
☎(+1 310) 329 00 46 ⊖ (+1 310) 323 62 84 [98]

75-93450 Vermont Composites, Inc.
(† Courtalds Aerospace)
6 Shields Drive, Bennington, VT 05201
☎(+1 802) 442 99 64 Ⓔ sales@composites.sover.net
⊖ (+1 802) 447 36 42
♣ $ 8,000,000 (1997) 100 (1997) 30,000 sq ft
Managing Dir: Dan Maneely
Sales & Marketing Dir: Garth Kenyon
Technical Dir: Bill Berg
■ Design & fabrication of a range of composite enclosures for weight sensitive applications in the air & on the ground. [98]

75-93462 Versatron Corporation
(a division of Wescam Inc.)
103 West North Street, Healdsburg, CA 95448
☎(+1 707) 433 30 00 ⊖ (+1 707) 433 71 10
♣ $ 12,000,000 (1996) 80 (1996) 2,800 sq m
Managing Director: Doug Smith
Operations Director: Dominic Houlemard
Sales and Marketing Director: Bill Harwood
Purchasing Manager: Tom Wallace (acting)
Marketing Manager: Steve McCann
■ Development and production of 4 and 5 axis gyrostabilised systems for military and commercial surveillance applications. [98]

75-93470 VEAM
(a division of Litton Systems, Inc., USA)
100 New Wood Road, Watertown, CT 06795-3347
☎(+1 860) 274 96 81 ⊖ (+1 860) 274 49 63
General Manager: Peter Bialobrzeski
Sales and Marketing Director: Keith L Olsen
Purchasing Manager: William Porzio
■ Manufacture multipin, metal shelled cylindrical electrical, fibre optic and pneumatic connectors, and accessories for all other types of connectors for military and commercial applications. VEAM specialises in heavy duty multipin power connectors and custom configurations for hostile environments.

Litton Precision Products International
100 New Wood Road, Watertown, CT 06795-3347
☎(+1 203) 274 96 81 ⊖ (+1 203) 274 49 63
President: Ms Constance Zaqar [98]

75-93490 Vectronics Microwave Corp.
113 Lincoln Blvd, Middlesex, NJ 08846-1045
☎(+1 908) 356 2377 ⊖ (+1 908) 356 6782
♣ $ 1 300 000 (1994) 20 1 250 m2
President & Sales & Marketing Dir: J.G. Vogler
Treasurer/Controller: Ruth K. Vogler
■ Design & manufacture of microwave component, subassemblies & microwave integrated circuits for ECM, radars, communications & other defence-related systems. [97]

75-93520 Veeder-Root Co.
125 Powder Forest Drive, PO Box 2003,
Simsbury, CT 06070-7584
☎(+1 860) 651 27 00 Ⓦ http://www.veeder-root.com
⊖ (+1 860) 651 27 19
President: Steven Sigmon
Vice President Sales: David Dyckman
Director of Marketing: Dean Cheramie
Marketing Communications Specialist: Ms Catherine Rowell [98]

75-93570 Vehicle Systems Development Corp. (VSDC)
1271 West 9th Street, PO Box 356, Upland, CA 91786
☎(+1 909) 981 32 36 ⊖ (+1 909) 981 32 36
♧ US$ 350,000 (1997) ♣ US$ 2,100,000 (1997)
12 (1997) 6,500 sq ft
Vice President of Operations: John P Forsyth
Administration and Purchasing Director: C A Forsyth
Manufacturing Manager: Steven Hecht
■ Special purpose armoured vehicles, armoured vehicle conversions, aircraft ground mobility systems (FloTrak TM). Special C3 armoured vehicles, money/valuables trucks and weapon transporters. → 80 [98]

75-93650 Velcon Filters Inc.
4525 Centennial Boulevard,
Colorado Springs, CO 80919-3350
☎(+1 719) 531 58 55 Ⓔ vfsales@velcon.com
Ⓦ http://www.velcon.com ⊖ (+1 719) 531 56 90
♣ $ 27,790,652 250
President: David Taylor
General Manager: Robin Mason
Senior Manager of Marketing Services: Ms Cheryl Scripter
■ Design and manufacture of aviation fuel filtration equipment for airport and aircraft refuelling vehicles. [98]

75-93760 Vernitron Corp.
645 Madison Ave, New York, NY 90022
☎(+1 212) 593 7900 ⊖ (+1 212) 754 6348
Motion Control Group
1601 Precision Park Lane, San Diego, CA 92073
☎(+1 619) 428 5581 ⊖ (+1 619) 428 5090
♣ $ 30 000 000 305

Contact Reference

General Mgr: Albert A. Ruppert
VP Sales: David Mighdoll
Technical Dir: Charles Nunziatta
Purchasing Mgr: Douglas Crawford
■ Permanent magnet DC torque motors (brush & brushless types), DC tachometers, resolvers (single & multispeed), MIL 45208A synchros, shipboard synchro signal converters & retransmission units, shipboard alarms & annunciators, shipboard heading indicators, actuator systems, optical & thermal scanners-oscillating & rotating, pancare resolvers-brush & brushless.

Motion Control Div.
1601 Precision Park Lane, San Diego, CA 92173
☎(+1 619) 428 5581 ✆ (800) 777 3393
⊖ (+1 619) 428 5090
■ Motors, drivers, resolvers, encoders, tachometers, actuators, scanners & motion systems.

Sensor Systems Division
2800 Anvil Street North, St Petersburg, FL 33710
☎(+1 813) 347 2181 ✆ (800) 688 2181
⊖ (+1 813) 347 7520
■ Potentiometers, pressure transducers & systems. [97]

75-93790 ViaSat Inc.
2290 Cosmos Court, Carlsbad, CA 92009-1585
☎(+1 760) 438 80 99 [W] http://www.viasat.com
⊖ (+1 760) 438 84 89
♣ $ 47,000,000 (1997) 285 (1997) 97,000 sq ft
President: Mark Dankberg
Vice President of Government Sales and Marketing: James Collins
Vice President of Commercial Sales and Marketing: Andy Paul
■ Design and manufacture rugged, high performance communications, networking and signal processing products. DAMA modems and network controllers at UHF, SHF and Ku band. High performance TDMA, burst TDMA, Anti-Jam and LPI modems. Multi-media and multi-level secure networks. Advanced Data Controller (MIL-STD 188-184) for Satcom and terrestrial channels. Embeddable INFOSEC module. JTIDS terminal products. Multiple emitter C3 and EW simulators. [98]

75-93795 Viatran Corp.
300 Industrial Drive, Grand Island, NY 14072
☎(+1 716) 773 17 00 ✆ (800) 688 00 30
⊖ (+1 716) 773 24 88
♣ $ 10,800,000 100 30,000 sq ft
President: Don Joslyn
Vice President Commercial Sales: Neil C Riordan
Operations Director: D Fleishmann
VP Engineering: Bob Moss
Marketing Manager: Dave Curtiss
Purchasing Manager: Mike Hallock
■ Electronic pressure & level instrumentation (transducers/transmitters) 5" WC through 100K PSI ranges, MV/V, 0-5V & 4-20mA outputs. [98]

75-93800 Vibration Mountings & Controls Inc. (VMC)
(a subs. of Aeroflex Inc.)
113 Main Street, PO Box 37, Bloomingdale, NJ 07403
☎(+1 201) 838 1780 ✆ (800) LOW VIBE
⊖ (+1 201) 492 8430
Managing Dir: Keith Larson
Director: Harvey R. Blav
Director: Michael Gorin
Director: Leonard Borow
VP Finance & Treasurer: Michael Gorin
Vice President/Controller: Ms Miriam Nolan
Technical Dir: Frank Andrews
Marketing Dir: Philip Littlewood
Sales Mgr: Greg Sherman [97]

75-93820 Vickers Inc. Sterer Division
(† Sterer Engineering & Manufacturing Co.)
(part of Vickers Aerospace-Marine-Defense Group)
4690 Colorado Boulevard, Los Angeles, CA 90039
☎(+1 818) 409 02 00 ⊖ (+1 818) 241 37 72
Vice President and General Manager: Jim Wehti
Marketing Director: R John Williams
■ Design and production of high quality fluid control and actuator components and systems for hydraulic, pneumatic and fuel engineering disciplines. Products include fly-by-wire electrohydraulic assemblies, flight surface actuators, integrated control modules, fuel pressure controls, pneumatic and fluid system controls. [98]

75-93830 Vickers Inc. Aerospace-Marine-Defense
(a subsidiary of Trinova Corp.)
Vickers Fluid Power Division
5353 Highland Drive, PO Box 10177, Jackson, MS 39206-1177
☎(+1 601) 981 28 11; 987 33 40 ⊠ 585439
⊖ (+1 601) 987 52 61
900 250,000 sq ft
Vice President and Manager of Fluid Power Division: Charles F Lyke
Marketing Director: Robert H Muller
Engineering Director: Dr John A Halat
■ Design and production of a broad range of fluid power and fuel control systems and components including hydraulic, electrohydraulic and fuel pump equipment; pumps, motors, electric motor pumps, integrated packages, test stands and ground kart support equipment.

World HQ
3000 Strayer, PO Box 632, Maumee, OH 43537
☎(+1 419) 867 26 53 ⊖ (+1 419) 867 26 49
9,500
President: John Weber
Vice President of AMD Group: Larry Deininger
■ Manufacture, operations, engineering, sales and service for commercial airline, military, executive and general aviation hydraulic and defence products.
Group Members: Vickers Systems Ltd (UK), Vickers Systems GmbH (Germany), Vickers SA (France), Vickers Systems SpA (Italy), Vickers Tedeco Div. (Glenolden, PA), Vickers Electrtomech Div. (Grand Rapids, MI), Vickers Sterer Div. (Los Angeles, CA).

Vickers TEDECO
24 East Glenolden Avenue, Glenolden, PA 19036
☎(+1 215) 583 94 00 ⊠ 831429 TWX 510 669 1323
⊖ (+1 215) 583 39 85
Vice President and General Manager: Tariq Jesrai
Sales and Marketing Director: Frank Di Pasquale
Engineering Director: Henry M Belman
■ Condition monitoring products and fluid systems devices for FW and RW engines and gear boxes. Chip detectors, full flow and quantitative debris monitoring, sight gauges, breather vents, valves, electro-optic level sensors. [98]

75-93840 Victaulic Co. of America
PO Box 31, Easton, PA 18044-0031
4901 Kesslersville Rd, Easton PA
☎(+1 610) 559 33 00 [E] vic@mail.com
[W] http://www.victaulic.com ⊖ (+1 610) 250 88 17
2,000
Chairman & CEO: Joseph M. Trachtenberg
VP Marketing/Business Dev.: Thomas M. Kiley
Military Product Sales: Frank Azzalina
Military Sales: Michael A. Segala
Marketing Services Mgr: G.K. Greene
■ Pipe couplings, fittings & valves for military structures, vehicles, ships & rapid deployment piping for military defence & related applications. [98]

75-94000 Vidaro Corp.
333 Martinel Drive, PO Box 550, Kent, OH 44240-0550
☎(+1 330) 673 74 13 ⊖ (+1 330) 673 02 28
113
President: Edward B Davis
Operations Director: Tom Rosenblatt
Office Manager: Ms Rosemarie Einholz
Purchasing Manager: Howard Fleischmann
■ Acid resistant and flame retardant clothing, clean room (lint free) and anti-static clothing. [98]

75-94090 Videotek Inc.
243 Shoemaker Rd, Pottstown, PA 19464-6433
☎(+1 610) 327 2292 ✆ (800) 800 5719
⊖ (+1 610) 327 9295
115
VP Marketing: Paul Calkin
■ Manufacture a comprehensive line of products for television broadcasting & teleproduction, including: digital & analog products, digital colour corrector, plus waveform monitors, vectorscopes, combination waveform monitor/vectorscopes, sync generators, timing equipment, test signal generators, production/post production switchers, routing switchers, distribution amplifiers, colour monitors, audio programme monitors, demodulators, in NTSC & PAL. [97]

75-94300 Visual Instrumentation Corp. (VIC)
1110 W. Avenue L-12, No 1-C, Lancaster, CA 93534-7034
☎(+1 805) 945 7999 ⊖ (+1 805) 723 5667
♣ $ 2 000 000 (1995) 12 (1996) 5 000 sq ft
President: Gilbert J. Pendley
Operations & Technical Dir: Jerry Kane
Purchasing Mgr: J.M. Valdez (Jr)
Marketing Administrator: James E. Bauer
■ Photo-optical & video instrumentation equipment including: hi-speed cameras, (Film & video), film readers, analysis equipment & accessory products, including airborne, seaborne & land based systems and components. [97]

75-94320 Visual Methods Inc.
35 Charles Street, Westwood, NJ 07675
☎(+1 201) 666 39 50 [W] http://www.visualmethods.com
⊖ (+1 201) 666 79 31
16 4,000 sq m
President: Herman Kruegle
■ Surveillance systems & equipment, night vision systems & equipment, video access control. [98]

75-94430 Voltage Multipliers Inc. (VMI)
8711 West Roosevelt, Visalia, CA 93291-9458
☎(+1 209) 651 14 02
[W] http://www.voltagemultipliers.com
⊖ (+1 209) 651 07 40
135 (1997)
President: Dennis Kemp
Vice President of Marketing: Ken Haag
Operations Director: John Yakuna
Sales and Administration Manager: Ms Karen Spano
Purchasing Manager: Ms Pat Kelton
■ High voltage and low voltage military rectifiers and rectifier assemblies. [98]

75-94550 Voss Industries Inc.
2168 West 25th Street, Cleveland, OH 44113-4172
☎(+1 216) 771 76 55 ⊖ (+1 216) 771 28 87
♣ $ 15,000,000 (1997) 200 (1997) 200,000 sq ft
President: W J Voss
Vice President Sales: W J Soroczak
Purchasing Manager: Bill Pugh
■ Standard & special T-bolt clamps, V-band couplings & mating flanges, strap assemblies, rigid "V" couplings & machined flanges, CNC machining, stampings, light fabrications & hydraulic bulge-formed shapes for aerospace & defence applications. Also MIL-SPEC painting, spot welding & heliarc welding.

Voss Aerospace
2168 West 25th Sreet, Cleveland, OH 44113
☎(+1 216) 771 76 55 ⊖ (+1 216) 771 28 87

Heat Exchange Applied Technology (HEAT)
2186 West 25th Street, Cleveland, OH 44113
☎(+1 216) 241 43 28 ⊖ (+1 216) 771 28 87 [98]

75-94580 Votaw Precision Technologies Inc.
(a subsidiary of Kaiser Aerospace & Electronics Corp.)
13153 Lakeland Road, Santa Fe Springs, CA 90670
☎(+1 562) 944 06 61 ⊖ (+1 562) 944 06 75
♣ $ 15,000,000 (1995) 140 200,000 sq ft
President and Chief Executive Officer: Stuart Gordon
Vice President of Sales and Marketing: Scott Wallace
Purchasing Manager: Ms Yvonne Hurt
■ Manufacture of fixtures for the aerospace industry. Performs machining of aerospace components. Fabrication and tooling. F-22, Titan, Shuttle; Delta II; Atlas. [98]

75-94900 Wall Colomonoy Corp.
30261 Stephenson Highway, Madison Heights, MI 48071
☎(+1 313) 585 6400 ⊖ (+1 313) 585 7960
480
President & COO: Joseph M. Maria
VP Sales & Marketing: W.P. Clark (Jr)
General Mgr: John R. Sturch
■ Manufacture & overhaul of turbine engine components & aircraft exhaust systems. Manufacture of brazing & hard surfacing materials. Overhaul of aircraft heat exchangers. CNC machining of critical tolerance dimension components, brazing & heat treating employing atmospheric furnaces. [97]

75-95105 Warner Electric
(a subs. of Dana Corp.)
449 Gardner St, South Beloit, IL 61080
☎ (+1 815) 389 3771 ✆ (800) 34 3369
Telex 210189; TWX 910 3505 873 ⊖ (+1 815) 389 2582
$ 200 000 ♣ $ 120 000 000
Advertising Support Mgr: Ray Freiwald
■ Precision ball bearing screws & complete aircraft/defence actuators.
Tipton Branch
222 N. West St, Ste 150, Tipton, IN 46072-1655
☎ (+1 317) 675 8466 Telex (+1 317) 675 8466
Market Representative: Brian A. Taylor 97

75-95150 Warren-Knight Instrument Co.
(a division of Warren Industries Inc.)
2045 Bennett Road, Philadelphia, PA 19116
☎ (+1 215) 464 93 00 E accuracy@warrenind.com
W http://www.warrenind.com
President: J.H. Warren (III)
Sales Mgr: Richard M. Marron
■ Precision alignment, angle & distance measuring instruments. Optical, mechanical & electronic. 98

75-95200 Watkins-Johnson Co.
Headquarters
3333 Hillview Ave, Stanford Ind.Park, Palo Alto, CA 94304
☎ (+1 415) 493 4141 Telex 348415 wj pla
Telex TWX 910 373 1253/1254 wj pla
⊖ (+1 415) 813 2402
♣ $ 387 000 000 ♚ 2 000
President & CEO: Dr W.Keith Kennedy
Corp. Exec. Vice President: K.D. Gilbert
Corp. VP & CFO: Scott G. Buchanan
VP Manufacturing: Allen Ibara
VP Engineering: Patrick Brady
VP Microwave Products: Malcom J. Caraballo
Financial Controller: Darryl T. Quan
Advertising Dir: Stephen Witmer
■ Defence electronic systems, electronic devices & related equipment.
Communication Electronics Technology Div.
700 Quince Orchard Rd, Gaithersburg, MD 20878-1794
☎ (+1 301) 948 7550 Telex 898402; TWX 710 828 0546
⊖ (+1 301) 921 9479
Manager: Marc Elgaway
■ Produces modular receiving & signal-processing equipment for reconnaissance, surveillance & direction finding.
Semiconductor Equipment Group
440 Kings Village Rd, Scotts Valley, CA 95066-4081
☎ (+1 408) 438 2100 Telex 346327 ⊖ (+1 408) 438 2318
Microwave Products Group
3333 Hillview Ave, Stanford Ind.Park, Palo Alto, CA 94304
☎ (+1 415) 813 2402 ⊖ (+1 415) 493 1027
VP & Manager: Malcolm J. Caraballo 97

75-95210 Watson Industries Inc.
3041 Melby Rd, Eau Claire, WI 54703
☎ (+1 715) 839 0628 ✆ (800) 222 4976
⊖ (+1 715) 839 8248
♣ $ 1 000 000 ♚ 14 ⌂ 500 sq m
President: William S. Watson
Marketing Dir: Drew Karnick
■ Development & manufacture angular inertial, linear inertial, magnetometer sensing devices, systems for civilian & military applications. 97

75-95270 Wavetek Corporation Instruments Division
11995 El Camino Real, Ste 301, San Diego, CA 92130
☎ (+1 619) 279 2200 ✆ (800) 874 4835
Telex TWX 910 3351 007 ⊖ (+1 619) 565 9558
♣ $ 180 000 000 ♚ 830
Managing Dir: Terence J. Gooding
General Mgr: Joe Budano
Controller: James Hartman
Technical Dir: Derek Morikawa
Marketing Dir: Mike Sciulli
Corporate Counsel: Kathynn A. Valverd
■ Manufacture electronic test & measurement equipment.
Communications Division
5808 Churchman Bypass, Indianapolis, IN 46203-6109
☎ (+1 317) 787 3332 Telex TWX 810 3413 226
⊖ (+1 317) 788 5999
♚ 120
General Mgr: Joe Budano 97

75-95543 Wegmann USA Inc.
(a subs. of Wegmann & Co GmbH, Germany)
37000 Cohen Place, PO Box 11648,
Lynchburg, VA 24506
☎ (+1 804) 522 0348 ⊖ (+1 804) 948 4415
♚ 3 000
President & General Mgr: Charles T. Warren
VP Finance: Dan Ashwell
Technical Mgr: John W. McGuinn
■ M1A1/A2 ammunition rack, 76 mm vehicular mounted grenade launcher, through sight video, target systems, armoured vehicle upgrades, training systems. 97

75-95544 Weinschel Corporation
(† Lucas Weinschel)
(a subsidiary of Microwave Components Enterprises, Inc.)
5305 Spectrum Drive, Frederick, MD 21703-7362
☎ (+1 301) 831 47 01 ✆ (+1 301) 638 20 48
E sales@weinschel.com W http://www.weinschel.com
⊖ (+1 301) 831 45 70
♚ 108 (1997) ⌂ 34,750 m2
President and Chief Executive Officer: Robert L Stephens
Vice President of Sales and Marketing: Geoffrey D Smith
Vice President of Finance and Administration: Kurt Brown
Vice President of Operations: Frank W MacLean
Materials Manager: Hector Hoffmaister
■ Microwave components, test and calibration equipment. 98

75-95700 Wesco Aircraft Hardware Corp.
27727 Ave Scott, Valencia, CA 91355
☎ (+1 805) 295 1414 ⊖ (+1 805) 295 0695
♚ 150
President: Randy Snyder
VP Sales & Marketing: Hal Weinstein
■ Fasteners, hydraulic fittings. 97

75-95750 West Electronics
5 Terri Lane, Unit 9, PO Box 366, Burlington, NJ 08016
☎ (+1 609) 387 43 00 ⊖ (+1 609) 387 43 04
$ 3,000,000 (1997) ♣ $ 2,000,000 (1997)
♚ 30 (1997) ⌂ 15,000 sq ft
President: Frank S. Brainard (Jr)
Vice President: Bradford P. Brainard
Advertising Mgr: Bradford P. Brainard
■ Power supply for AIM-9 sidewinder missile launcher; AIC-14 & AIC-14A intercommunication equipment. 98

75-95800 Western Design Howden
(† Western Design Corp.)
(a subsidiary of Howden Group plc)
16952 Millikan Avenue, Irvine, CA 92714
☎ (+1 714) 863 05 60 ⊖ (+1 714) 660 87 46
♣ $ 18,000,000 (1997) ♚ 65 (1997) ⌂ 4,500 sq m
President: Michael S Quinn
Vice President and General Manager: Charles J Panasewicz
Marketing Manager: Gregory C Hill
Purchasing Manager: David Engar
■ Linear linkless ammunition storage and handling systems (medium calibre) in airborne (rotary and fixed-wing), naval and ground based applications. Include gun feeder systems, ready magazines, reserve magazines, and ammunition suppport equipment. Robotic autoloader systems for tanks and artillery in 105mm, 120mm, 140mm, and 155mm calibres. 98

75-95850 Western Electrochemical Co. (WECCO)
(a subsidiary of American Pacific Corp.)
PO Box 629, Cedar City, UT 84720
☎ (+1 801) 865 50 00 ⊖ (+1 801) 865 50 05
♣ $ 30,000,000 (1996) ♚ 175
Chief Executive Officer: Fred D Gibson (Jr)
President: James J Peveler E jpevler@apfc.com
Sales and Marketing Director: Finnegan
Purchasing Manager: Doug Christian
■ Manufacturer of ammonium perchlorate, sodium perchlorate, potassium perchlorate. 98

75-95930 Western Filament Inc.
630 Hollingsworth Street,
Grand Junction, CO 81505-1000
☎ (+1 970) 241 87 80 ⊖ (+1 970) 241 86 82
♚ 55
President: Burke Wright
Sales and Marketing Director: Ms Gil Angelotti
Sales Manager: Wayne Wright
Purchasing Manager: Ms Lou Coghill
Sales: Ms Charlotte Vecchia
■ Manufacture electrical/electronic wire harnessing devices and systems; round and flat braids, twisting, dying, coating and overbraiding of high-tech fibers. 98

75-95950 Western Filter Corp.
(a subsidiary of Western Filter)
26235 Technology Drive, Valencia, CA 91355-1139
☎ (+1 805) 295 08 00 ⊖ (+1 805) 295 08 01
♣ US$ 14,000,000 ♚ 130 ⌂ 9,400 m2
President: Z Paul Akian
Vice President of Sales and Marketing: Matt D Porlier
Operations Director: John McCauley
Sales and Marketing Director: Paul McKie
Purchasing Manager: Rick Padilla
■ Filter elements, assemblies and related devices such as relief valves, by-pass valves, differential pressure indicators electrical and mechanical for hydraulic, fuel and lube systems. 98

75-96280 Westport Development & Manufacturing Co. Inc.
34 Prindle Hill Road, Orange, CT 06477-0540
☎ (+1 203) 795 08 71 ⊖ (+1 203) 799 38 57
♚ 40
President: R R Hellman (Sr)
Executive Vice President: Richard B Hudson
■ Pressure, temperature & position sensing instruments, hydroformed bellows & bellow assemblies. 98

75-96350 Wheeled Coach Industries
(a subsidiary of Collins Industries, Inc)
2778 North Forsyth Road, Winter Park, FL 32792-6672
☎ (+1 407) 677 77 77 ✆ (800) 422 82 06
W http://www.wheeledcoach.com ⊖ (+1 407) 679 13 37
♚ 600
President: Gary Goetsch
Vice President of Sales: Bob Otto
Managing Director: Robert L Collins
Operations Director: Len Whalen
Sales and Marketing Director: Wilson Jones
International Sales Manager: Paul F Holzapfel
■ Type I, II and III ambulances. Two wheel drive and four wheel drive. Wheeled Coach currently has two GSA contracts for ambulances. 98

75-96400 Wheeler Systems Inc. (WSI)
PO Box 16688, Arlington, VA 22215
☎ (+1 954) 731 15 64 ⊖ (+1 954) 731 38 60
♣ US$ 300,000 (1997) ♚ 10 (1997) ⌂ 500 m2
President and Treasurer: E J Wheeler (Jr)
■ Marine engineering and technical support to government and private industry. Primary area: naval ship development/acquisition. (Currently: FFG frigate, DDG/AEGIS destroyer, CG AEGIS cruiser, battleship reactivation, MCM ships, PHM hydrofoil). Includes: oceanographic/biomedical research/engineering; electronic C3 and navigation systems. Marine overhaul planning, equipment overhaul and pre-fabrication, alteration installation teams, data systems development/training. → 80 98

75-96730 Wilcox Electric Inc.
(a subsidiary of Thomson Corporation, USA)
2001 NE 46th Street, Kansas City, MO 64116
☎ (+1 816) 453 26 00 Telex 434412 ⊖ (+1 816) 459 43 64
♣ $ 50,000,000 (1997) ♚ 400 (1997) ⌂ 320,000 sq ft
President: William P Marberg
Technical and Marketing Director: Benjamin Sandzer-Bell
Purchasing Manager: Ken Knoll
Marketing Assistant: Ms Donna Shue
■ Manufacture ground-based air navigation and landing systems: ILS, DME kit, VOR/DME, NDB, radar, air traffic control and communication equipment, GPS, LAAS, MSE, MSRT. → 80 98

75-96750 Wilcoxon Research (WR)
21 Firstfield Road, Gaithersburg, MD 20878
☎ (+1 301) 330 88 11 ✆ (800) Wilcoxon
E sensors@wilcoxen.com Telex 898342
⊖ (+1 301) 330 88 73
♣ $ 15,000,000 + (1997) ♚ 125 (1997)
Operations Director: James Michaelis
Sales and Marketing Director: Paul Taylor
Marketing and Communications Manager: Ms Becky Bouwman

Contact Reference

Purchasing Manager: James Kurtz
■ Manufacture of vibration instrumentation, specializing in piezoelectric accelerometer and velocity sensors. Vibration transducers and hydrophones are being used in military applications in the USA and Europe. [98]

75-96800 Wildey Inc.
PO Box 475, Brookfield, CT 06804
☎ (+1 860) 355 90 00 ⊖ (+1 860) 354 77 59
👥 10 (1997) 🏭 10,000 sq ft
Managing Director: Wildey J Moore
Marketing/Sales Director: Mrs Linda A Moore
■ Anti-terrorist & offensive gas-operated autoloading handguns chambered for 45 Winchester Magnum, 300, 357, 10mm & 475 Wildey Magnums. Can be equipped with laser, optic or other sighting systems. Suppressors & bipod mount available. [98]

75-96830 The Will-Burt Co.
169 S.Main St, PO Box 900, Orrville, OH 44667-0900
☎ (+1 330) 682 7015 ⊖ (+1 330) 684 1190
♣ $ 25 000 000 (1996) 👥 300 🏭 43 000 sq m
Managing Dir: Dennis B. Donahue
Operations/Technical Dir: Scott Carter
Mast Sales Dir: D.S. Barlow ☎ (+1 330) 684 5214
■ Design & manufacture a wide range of pneumatic & mechanical telescoping masts.

Factory
401 Collins Blvd, Orrville, OH 44667
☎ (+1 216) 682 7015 ⊖ (+1 216) 684 1190
TMD Sales Mgr: Donald S. Barlow [97]

75-96950 Williams International
2280 West Maple Road, PO Box 200,
Walled Lake, MI 48390
☎ (+1 810) 624 52 00 ⊖ (+1 810) 669 00 40
Chairman and Chief Executive Officer: Sam Williams
President and Chief Operating Officer: Gregg G Williams
■ Gas turbine engines.

Washington DC Office
Suite 809, 1735 Jefferson Davis Highway,
Arlington, VA 22202
☎ (+1 703) 413 22 07 ⊖ (+1 703) 413 22 11
Contact: Ray Preston

Ogden Manufacturing Plant
3450 Sam Williams Drive, PO Box 9785,
Ogden, UT 84409
☎ (+1 801) 627 05 50 ⊖ (+1 801) 621 05 46
Plant Mgr: Lloyd McCaffrey [98]

75-97050 Winco Inc.
225 Cordova Ave S., Le Center, MN 56057-1805
☎ (+1 507) 357 6821 ⊖ (+1 507) 357 4857
👥 110 (1996) 🏭 65 000 sq ft
CEO/President: Ralph Call
Dir Public Relations: Ms Joyce Forcelle
■ Portable generators/stand-by power, mobile lighting systems/mobile power systems. [97]

75-97150 Witters & Co
PO Box 31532, Seattle, WA 98103-1532
☎ (+1 206) 286 09 33 ✆ (800) 321 99 47
⊖ (+1 206) 286 06 14
♣ $ 2,500,000 (1995) 👥 18 🏭 15,000 sq ft
President and Chief Executive Officer: Fred H Witters (Jr)
■ Manufacture hydraulic security gate operators which can swing, slide or vertically lift a gate. Also have hydraulic barrier arm operators & pneumatic barrier gate operators. [98]

75-97278 N.Wood Counter Laboratory Inc.
PO Box 509, Chesterton, IN 46304
☎ (+1 219) 926 3571 ⊖ (+1 219) 926 3571
Managing Dir: Ms Marjory Crawford
■ Manufacture of radiation detectors, components in radiation measuring equipment. [97]

75-97280 Stephan Wood Products Inc.
605 Huron, PO Box 669, Grayling, MI 49738
☎ (+1 517) 348 5496 ⊖ (+1 517) 348 2427
♧ $ 3 000 000 👥 35 🏭 28 000 sq ft
President: Francis G. Tenbusch
■ Truck, bus & van accessories & components specifically for troop & cargo transport, seats, racks, canvas, bows. [97]

75-97290 Woodward Governor Company
5001 North Second Street, PO Box 7001,
Rockford, IL 61125-7001
☎ (+1 815) 877 74 41
[E] woodward@wgc.woodward.com
⊖ (+1 815) 636 60 33
👥 3,211 (1996)
Chairman: John A. Halbrook
Managing Director: Phil Turner
Operations Director: Steve Ebbing
Sales and Marketing Director: Tom Gendron
Purchasing Manager: Dennis Benning
Personnel Manager: Jerry Forberg
■ Airframe/engine monitoring & control systems for large & small aircraft. Aircraft & engine maintenance & repair. Gas turbine components.

Aircraft Controls Group
(† Large Aircraft Controls)
5001 North Second Street, PO Box 7001,
Rockford, IL 61125-7001
☎ (+1 815) 877 74 41 ✉ WOODWARD Rockford
⊖ (+1 815) 636 60 33
VP & General Mgr: C. Phillip Turner
Marketing Manager: Chad Preiss

Aircraft Controls Group
(† Aircraft Product Service Center)
One Woodward Way, PO Box 405,
Rockton, IL 61072-0405
☎ (+1 815) 624 44 00 ✉ WOODWARD Rockford
⊖ (+1 815) 624 46 90

Industrial Controls Group
1000 East Drake Road, PO Box 1519,
Fort Collins, CO 80522-1079
☎ (+1 970) 482 58 11 ⊠ 45691 ⊖ (+1 970) 498 30 58
Group Vice President: Chuck Kovac

Turbomachinery Controls
3800 North Wilson Avenue, PO Box 3800,
Loveland, CO 80539-3800
☎ (+1 970) 663 39 00 ⊖ (+1 970) 962 70 50
General Mgr: Terry Shetler [98]

75-97600 WPI Termiflex Corp.
(a subs. of WPI Group)
316 Daniel Webster Highway, Merrimack, NH 03054
☎ (+1 603) 424 3700 ✆ (800) 832 0512
⊖ (+1 603) 424 0330
👥 64 🏭 38 000 sq ft
President: Vel Cassler
VP Sales & Marketing: Ms Mary Lou Welch
Dir International Sales: Gian Piero Oggero
■ Rugged, handheld, control/display & PC based units which serve as the operator interface to microprocessor-based systems, such as simulators, flight line test systems, mobile battlefield weapons, navigational systems, naval systems, radar. A military qualified model is available.

Overseas Agents:
Marchwood Assoc. (UK)
Amtex Electronics (Australia)
Ineltro AG (Switzerland)
Souriau (France)
NC Scandcomp (Sweden)
Racom Electronics (Israel)
Sunin Trading (Republic of Korea) [97]

75-97650 Wright Industries Inc.
707 Spence Lane, Nashville, TN 37217
☎ (+1 615) 361 66 00 ✆ (800) 782 42 02
⊖ (+1 615) 361 45 49
♣ $ 150,000,000 (1997) 👥 825 (1997) 🏭 400,000 sq ft
President: Bailey Robinson
Vice President: Bob Leas
Vice President of Marketing and Sales: Ken Verble
Purchasing Manager: Bill McMinn
■ Custom automation designer and builder specialising in all types of turnkey assembly and machining systems. Tools, dies and contract machining. [98]

75-97700 Wyman-Gordon Co. (WGC)
244 Worcester Street, PO Box 8001,
North Grafton, MA 01536-8001
☎ (+1 508) 839 44 41 ✆ (800) 343 60 70
✉ WYGORDON [E] marketing@wyman.com
[W] http://www.wyman-gordon.com
⊠ TWX 710 3471 454; 174395 ⊖ (+1 508) 839 75 00
♧ $ 110,000,000 (1996) ♣ $ 500,000,000 (1996)
👥 3,300 (1996)
Chairman: John M Nelson
President and Chief Executive Officer: David P Gruber
President Wyman-Gordon Forgings: J Douglas Whelan
President Wyman-Gordon Investment Casting Inc.: Frank J Zugel
President Scaled Composites Inc.: Burt Rutan
Vice President Business Ventures: Colin Stead
Vice President: James V Lombardi
Vice President Energy Products: Derek Pattle
■ Manufacture high quality, technologically advanced forging, extruded and investment casting components and composite airframe structures for the commercial aviation, power generation/energy and defence industries.

Marketing and Sales Office
244 Worcester Street, PO Box 8001,
North Grafton, MA 01536-8001
☎ (+1 508) 839 44 41 ⊠ 3471454; 174395
⊖ (+1 508) 839 75 00
Vice President Chief Financial Officer and Treasurer: Andrew C Genor
Vice President Production Operations: J Stewart Smith
Vice President Materials: Charles Yarbrough

Wyman-Gordon Forgings Inc.
244 Worcester Street, PO Box 8001,
North Grafton, MA 01536-8001
☎ (+1 508) 839 44 41

Wyman-Gordon Forgings Inc.
10825 Telge Road, PO Box 40456,
Houston, TX 77240-0456
☎ (+1 281) 856 99 00

Wyman-Gordon Investment Castings Inc.
Tilton/Frankling Division, Granite Street, Tilton, NH 03276
☎ (+1 603) 286 43 01 [98]

75-97720 Wyman-Gordon Investment Castings Inc.
(a division of Wyman-Gordon Co.)
Poquonnock Road, Groton, CT 06340
☎ (+1 860) 445 74 21 [E] wgic@aol.com
⊖ (+1 860) 449 68 35
♣ US$ 150,000,000 (1997) 👥 1,500 (1997)
President: Frank J Zugel
Vice President Aerospace: Tom des Enfants [98]

75-97760 Xi CAD North America Inc.
PO Box 2375, Stamford, CT 06906-0375
60 Strawberry Hill Ave, Ste 204, Stamford, CT 06902
☎ (+1 203) 964 9689 ⊖ (+1 203) 323 5971
President: N. Igol Levy
■ Mini-RPV navigation control systems, digital mapping & mission planning. [97]

75-98000 XL Computing Corp.
(† Hetra Computers and Communications Industries Inc.)
10305 102nd Terrace, Sebastian, FL 32958
☎ (+1 561) 589 73 44 ✆ (888) RUGGED1
[E] sales@xlcomputing.com
♣ US$ 10,000,000 (1997) 👥 95 (1997) 🏭 45,000 sq ft
President: Alan Bishop
Vice President of Business Development: Kent Bartlett
Operations Director: Edward Howard
■ A wide array of Tempest and non-Tempest products including desktop, rack-mount and ruggedized computer systems and peripherals. [98]

75-98020 R.M. Young Co.
2801 Aero-Park Drive, Traverse City, MI 49686
☎ (+1 616) 946 39 80 [E] met.sales@youngusa.com
[W] http://www.youngusa.com ⊖ (+1 616) 946 47 72
♣ $ 3,000,000 👥 30 🏭 17,000 sq ft
President and General Manager: Thomas M Young
Sales Director: John Campbell
Purchasing Manager: Richard Dormitzer
Advertising Manager: Mrs Pauline Young
■ Quality meteorological instruments for use in meteorological research, air & water quality, energy studies & other related fields. The Company has three principal product categories: Sensitive wind instruments; functional wind instruments; temperature instruments. [98]

75-98050 ZARC International Inc.
PO Box 5800, Bethesda, MD 20824
☎ (+1 301) 564 37 37 ✆ (800) 882 70 11
⊖ (+1 301) 897 52 20
President: Cameron Logman
International Marketing Manager: Amir Mohammadi
■ Manufacture of CAP-STUN non-lethal weapon system. Supplied to law enforcement and military agencies. [98]

75-98150 Zero Enclosures
(a division of Zero Corp.)
500 West 200 North, PO Box 540310,
North Salt Lake, UT 84054-0310
☎ (+1 801) 298 59 00 ⊖ (+1 801) 292 94 50
👥 350 🏭 300,000 sq ft
President: Jay Manwaring
Vice President of Sales and Marketing: Hyrum Mead
Marketing Manager: Evon Lynch Young
■ Design and manufacture of enclosures and accessories for the electronics industry, including heavy-duty protective housings for the government/military market and carrying cases for military or industrial use. [98]

75-98190 Zeta
(a division of Sierra Networks, Inc.)
2811 Orchard Parkway, San José, CA 95134
☎ (+1 408) 434 36 00 [E] zeta@zeta-sni.com
⊖ (+1 408) 433 02 05
♣ US$ 20,000,000 👥 110 🏭 70,000 sq ft
President: Norman D Stecklain
Business Development and Communications Systems: Joseph Rock
Communications and Supplies Manager: Ms Barbara Dominguez
■ Design and manufacture microwave frequency generation, signal-processing components and subsystems, receivers, communication signal intercept, direction finding and location systems and RF high power amplifier products for the defence and aerospace industries for use in guidance data link, radar, electronic warfare and communication systems. [98]

75-98210 Zeus Industrial Products Inc.
620 Magnolia Street, PO Box 2167,
Orangeburg, SC 29116
☎ (+1 803) 531 21 74 ✆ (800) 526 38 42
[W] http://www.zeusinc.com ⊖ (+1 803) 533 56 94
👥 650 (1997) 🏭 200,000 sq ft
President: John Worley
Vice President: Michael C Tourville
Marketing and Advertising Director: C Eduardo Braga
Sales Manager: Alan Andrews
■ Custom and standard extruded Fluoropolymer and heat shrinkable tubing. Convoluted tubing, PTFE, FEP, PFA and ETFE, PVDF, PEEK, DE. Spiral cut tubing. [98]

75-98250 Zippertubing Co.
13000 South Broadway, PO Box 61129,
Los Angeles, CA 90061-1120
☎ (+1 310) 527 04 88 ✆ (800) 321 81 78
[W] http://www.zippertubing.com ⊖ (+1 310) 767 17 14
👥 46 (1997) 🏭 2,000 sq m
President and Chief Executive Officer: T Plummer
Managing Director: David Kenyon
Sales and Marketing Director: M Bolen
■ "Zip-on" shielding and jacketing for wire, cable and custom applications. [98]

75-98260 Zitel Corporation
47211 Bayside Parkway, Fremont, CA 94538-6517
☎ (+1 703) 847 54 82 ✆ (800) 622 50 20
⊖ (+1 703) 440 96 96
♣ $ 30,800,000 (1997) 👥 160 (1997)
Vice President of Sales and Marketing: Donald Olker
Director of Public Relations: Ted Buxton
Federal Sales Manager: Mark Burg
■ Design, manufacture, test and system integration of high performance disk memory systems for UNIX and PC. [98]

75-98400 Zwick Energy Research Organization Inc.
5471 Argosy Drive, Huntington Beach, CA 92649-1038
☎ (+1 714) 891 16 40 ⊖ (+1 714) 895 41 17
👥 100 🏭 85,000 sq ft
President: Eugene B Zwick
Vice President and General Manager: C David Edwards
Purchasing Manager: Galen Usher
■ Hardware oriented research, product development and manufacturing of components and systems in the areas of combustion, heat transfer, cryogenics, hydraulics and pneumatics. Ground support equipment produced include nitrogen/oxygen servicing carts, mobile air compressors, mobile aircraft air conditioners, burnerless and 1 pass direct-fired heater systems for aircraft de-icers. [98]

80 Service Companies & Organizations

80-00025 A&T International Corporation
(a subsidiary of Analysis & Technology Inc.)
PO Box 220, North Stonington, CT 06359
☎ (+1 860) 599 39 10 ext 2236 [W] http://www.aati.com
⊖ (+1 860) 599 65 05
Executive Vice President and Chief Financial Officer: V Lehman Woods [E] lwoods@atinc.com [98]

80-00035 AAR Corp
One AAR Place, 1100 North Wood Dale Road,
Wood Dale, IL 60191
☎ (+1 630) 227 20 00 ⊖ (+1 630) 227 20 19
♣ $ 504,990,000 (1996) 👥 2,200
Chairman: Ira A Eichner
Chief Executive Officer: David Storch
Vice President Special Projects: Terry Davidson
Vice President Corporate Comunications: Ms Ann T Baldwin
Vice President Controller: Timothy J Romenesko
Public Relations/Advertising Manager: Lee Haines
■ AAR provides services including maintenance, overhaul & repair; sale & lease of aircraft & aircraft engines. Also distributes aviation components. Manufactures navigational instruments & gyroscopes, pallets, rail systems & shelters, proprietary composites. → 85
Subsidiary:
AAR Technical Service Center Corp. → 80 [98]

80-00040 AAR Defense Systems
(† Allen Aircraft)
(a division of AAR Corp.)
1100 North Wood Dale Road, Wood Dale, IL 60191
☎ (+1 630) 227 20 00 ⊖ (+1 630) 227 29 29
Vice President and General Manager: Frank Consdorf
Manager Corporate Communications: Ms Jeannine Fons
■ Supplier of aircraft & aircraft equipment (avionics, rotables, engines & engine parts) for both fixed & rotary wing, as well as technical, maintenance support & contract service support to military & government aviation worldwide. [98]

80-00045 AAR Technical Service Center Corp
747 Zeckendorf Boulevard, Garden City, NY 11530
☎ (+1 516) 222 90 00 ⊖ (+1 516) 222 09 87
President: Ami Zelcer
Vice President Sales and Marketing: M Stone
Vice President Operations: Anthony Imparato
■ Overhaul, repair & exchange of jet aircraft system components including instruments, avionics, gyros & pneumatic, hydraulic, electrical, fuel & electronic accessories. [98]

80-00050 Abacus Technology Corp.
5454 Wisconsin Ave, Suite 1100,
Chevy Chase, MD 20815
☎ (+1 301) 907 8500 ⊖ (+1 301) 907 8508
👥 250 (1995)
President: Dennis J. Yee
Vice President: Bill Magro
Asst Mgr Administrative Services: Lisa Feinberg
Sales Mgr: Marc Zoellner
■ Research & consultancy, specialising in providing information systems, systems engineering & programme management related services. Clients include the US Department of Defense. [97]

80-00080 Absolute Security
516 East Main Street, Riverton, WY 82501
☎ (+1 307) 856 24 49 [E] sharons@trib.com
⊖ (+1 307) 856 70 90
General Manager: Michael Spaulding
■ Consulting for electronic and physical security for corporations and key personnel. [98]

80-00100 ACCO Foreign Shipping Inc.
10880 North West 27th St, Ste 200, Miami, FL 33172
10880 North West 27th Street, Suite 200
☎ (+1 305) 591 83 80 [E] accoter@newworld.bridge.net
⊖ (+1 305) 594 59 60; 591 83 84
♣ US$ 7,900,000 (1996) 👥 22 🏭 30,000 sq ft
President: Larry Germi
Managing Director and Vice President of Sales and Marketing: Ali Germi
Public Relations Director: Ms Pauline Germi
Sales and Marketing Director: Ms Marilin Morales
Purchasing Manager: Ms Ann Amortegui
■ International freight forwarders, customs brokers. Non vessel operating carrier (NVOC). [98]

80-00150 Ada PROS Inc.
12224 Grassy Hill Court, Fairfax, VA 22033-2819
☎ (+1 703) 591 5247 ⊖ (+1 703) 591 5005
President: Dr Lewis Gray

Contact Reference

■ Provide confidential, intensive consulting & training on software process improvement & standards for software development. Consulting services include process improvement coaching, project planning & review of project work products. Training services focus on the SEI's Capability Maturity Model (CMM) & on MIL-STD-498 & also ISO/IEC 12207, "SoftwareLife Cycle Processes". [97]

80-00250 Advanced Applications Consultants Inc. (AACI)
7113 Jayhawk Street, Annandale, VA 22003-5851
☎ (+1 703) 642 22 24 [E] aaci1@aol.com
⊖ (+1 703) 642 55 50
President: Michael A Daniels
Advertising Manager: M Daniels
■ Integrated resource management for quality and productivity. Specialise in CM, DM, ILS and QA systems management. Competency and compliance audits, re-engineering solutions. Develop and publish technical text and handbooks, as well as courses and seminars (US and Overseas) for the above. Scientific and technical data search/retrieval. Buisness contacts in the CIS.
® Australia: Logistics Pty Ltd [98]

80-00440 Advanced Processing Laboratories, Inc.
5871 Oberlin Drive, San Diego, CA 92121
☎ (+1 619) 546 86 26 ✆ (800) 822 75 22
[E] info@sd.aplabs.com [W] http://www.sd.aplabs.com
⊖ (+1 619) 546 02 78
♣ $ 12,000,000 (1997) 👥 85 (1997) 🏭 4,000 m2
President: Dr Douglas Giese
Vice President of Sales and Marketing: Steve Gills
■ Integrates off-the-shelf hardware with existing and custom software for real-time applications in ASW, telemetry, telecommunications, signal processing, avionics test, data aquisition. Products include real-time VME station (TM), a development and run time platform, and high end VME interface cards. [98]

80-00550 Aereon Corp.
20 Nassau Street, Princeton, NJ 08542-0322
☎ (+1 609) 921 21 31 ✉ DYNAIRSHIPS
President and Treasurer: William McE Miller (Jr)
Technical Director: F Putman
■ Research and development of the deltoid Aerobody class of aircraft for military surveillance and/or cargo markets. Research and development also of a low-cost, hybrid, VTOL aerial crane concept. [98]

80-00650 Aero Corporation
(a subs. of Whitehall Corp.)
E.H'Way 90, Lake City Mun A'Port, PO Box 1990, Lake City, FL 32056-1990
☎ (+1 904) 758 3000 ⊠ 49609220 aero corp
⊖ (+1 904) 752 7960
♣ $ 50 000 000 👥 700 🏭 200 000 sq m
President & CEO: Mark Owen
Sr Sales & Marketing Dir: Lou Churchville
Human Resources Mgr: Tom Haden
■ An aircraft maintenance facility. FAA Certified in the maintenance of Lockheed L-382, L-188, Boeing B737, B727 & McDonnell Douglas DC8, DC9 aircraft. [97]

80-00800 Aero International Associates
La Guardia Airport, Marine Terminal Hangar 7, New York, NY 11371
☎ (+1 718) 225 7100 ⊖ (+1 718) 225 3053
President: William Miller
Exec. Vice President: Ms Dorothee C. Miller
Vice President: Gen.(Retd) M. Beyer
Sales Mgr: R.F. Baar
■ Transport & executive aircraft sales; procurement; finance & leasing; aviation consultants & appraisers. → 90 [97]

80-00840 Aero-Mach Labs Inc.
7707 East Funston, Wichita, KS 67207
☎ (+1 316) 682 77 07 ⊖ (+1 316) 682 34 18
👥 90 (1997) 🏭 50,000 sq ft
Manager Marketing and Sales: Ron Rucker
■ Aircraft instrument overhaul and repair service. [98]

80-00845 Aerospace Maintenance Systems Inc. (AMS)
1765 The Exchange, Suite 360, Atlanta, GA 30339
☎ (+1 770) 953 68 00 ⊖ (+1 770) 953 44 77
President: Louis Bertrand
Manager, Representative: John T Lyons
■ Training equipment for military and commercial aircraft.
® France: AMS Ground Support [98]

80-00850 Aero Space Modifications
PO Box 5867, Scottsdale, AZ 85261
☎ (+1 602) 483 87 53 ⊖ (+1 602) 483 71 90
Owner and Director of Research and Development: Ralph L Payne
■ Military and civilian aircraft modifications; flight test and specialized flight vehicle applications. Complete logistical support for all civil and military aircraft and flight vehicles. [98]

80-01050 The Aerospace Corp.
PO Box 92957, Los Angeles, CA 90245-4691
2350 East El Segundo Blvd, El Segundo, CA
☎ (+1 310) 336 5000 ⊠ 664460 ⊖ (+1 310) 336 7055
♣ $ 328 000 000 (1996)
Chairman Board of Trustees: Dr Ruth M. Davis
Vice Chairman: Gen. Lawrence A. Skantze
President & CEO: Edward C. Aldridge (Jr)
Exec. Vice President: Dr George A. Paulikas
Group VP Space Systems Group: Richard M. Allman
Group VP National Systems Group, VP Advanced Progams Ops: Dr Michael Daugherty
Head Public Affairs Office: Earl D. Flick
■ A non profit corporation. Operates a Federally funded research & development center serving as an architect-engineer for advanced space systems. It is the integral engineering arm for US national security space & launch system programmes. The corporation also co-operates with NASA, other civil agencies & international space organisations in order to apply space-related skills to critical issues.

Washington, DC Corporate Office
1000 Wilson Blvd. Suite 2600, Arlington, VA 22209-3988
☎ (+1 703) 812 0600
VP Space Technology Applications: Ronald K. Sable

Eastern Range
Kennedy Space Center, PO Box 21205, Florida 32815-0205
☎ (+1 407) 853 6666
Principal Dir: Timothy H. Hanrahan

Western Range
PO Box 5068, Vandenberg AFB, CA 93437-0068
☎ (+1 805) 853 6666 ⊠ 8662003
⊖ (+1 805) 866 2003
Principal Dir: Donald L. Nichols

Houston Office
Johnson Space Center, Mail Code ZR12, Houston, TX 77058-3696
☎ (+1 713) 483 3540
Project Engineer: Larry J. Dungan

Sunnyvale
1320 Orleans Drive, Sunnyvale, CA 94089-1135
☎ (+1 408) 744 6400
Systems Dir: Richard R. Kistler

Albuquerque Office
Kirtland AFB, PO Box 9045, Albuquerque, NM NM 87119
☎ (+1 505) 846 0222
Principal Dir: Alfred N. Sorensen

Colorado Springs
1150 Academy Park Loop, Suite 227, Colorado Springs, CO 80910
☎ (+1 719) 573 2440 ⊖ (+1 719) 550 2559
General Mgr: James La Frieda

Crystal City
PEO for Space Support Office
1725 Jefferson Davis Highway, Suite 702, Arlington, CA 22202
☎ (+1 703) 413 2360
Principal Dir: John A. Kinsey

Herndon
The Hallmark Building, 13873 Park Center Road, Suite 187, Herndon, VA 22071
☎ (+1 703) 709 9749
Group VP National Systems Grp & VP Advanced Program Ops: Jon H. Bryson [97]

80-01350 AIM Enterprises Inc.
10126 Residency Road, Manassas, VA 20110-2007
☎ (+1 703) 361 71 77 ✉ AIM AERO STN
⊠ 901951, TWX 710 831 0895 ⊖ (+1 703) 361 14 05
♣ $ 7,000,000 (1996) 👥 25
President: Dr Luis A Ambroggio
Vice President and General Manager: Mrs Nancy A Noble
Sales and Marketing Director: José Ibarra
Purchasing Manager: Jeff Putiyan
■ Specialises in total logistical support for many US manufactured airframes. Support services include component overhaul, manufacture of flight critical items and distribution. Systems support for the US Government and overseas customers include A-4, F-4, F-18, F-16, E-3A(AWACS), C-130, F-5 (aircraft) and S-61, SH-60, UH-60, CH-53, UH-1, AH-1, 204/205/212 (helicopters). Missiles: Hawk, Patriot. → 75 → 85 [98]

80-01450 Air Cargo Management Group
1501 4th Avenue, Suite 1620, Seattle, WA 98101
☎ (+1 206) 587 65 37 [E] acmg@wolfnet.com
[W] http://www.wolfenet.com/ ⊖ (+1 206) 587 65 40
👥 6
Managing Director/Owner: Edwin C Laird
Project Director: Robert V Dahl
Associate Director: Lee W Hibbets
Executive Assistant: Ms Jackie Edinger
■ Aviation consultants. [98]

80-01600 Aircraft Instrument & Radio Co. Inc.
1853 S. Eisenhower Court, PO Box 9487, Wichita, KS 67277-0487
☎ (+1 316) 945 0445 ⊖ (+1 316) 945 8014
👥 50
President: Martin Potash
VP Sales & Marketing: Jerry Urcavich
■ Sale, lease, exchange & consignment programs for avionics, instrumentation, rotable, expendable & accessory equipment. Purchase surplus equipment. In-house FAA approved repair facility provides complete service/repair capabilities for avionics & instrumentation with loaner service available. → 90 [97]

80-01900 Albion Group International Inc. Head Office
55 Railroad Avenue, Greenwich, CT 06830
☎ (+1 203) 625 08 00 [E] albiongrp@aol.com
⊖ (+1 203) 625 82 28
👥 50
Managing Director: Douglas L Olson
Administration Director: Lisa Colvin
■ Design, negotiation & fulfillment of offset programs in Europe, Middle East & Asia. Emphasis on investment in high technology projects. Fields of expertise in international investments in offset projects related to software, robotics, telecommunications & aerospace. [98]

80-02000 Allied Defense Industries Inc.
Route 3, Box 94L, Martinsburg, WV 25401
☎ (+1 304) 263 7722 ⊖ (+1 304) 263 3112
♣ $ 1 700 000 👥 10
President: Clifford W. Smith
Land Systems Dir: Jerry A. Brown
Naval & Marine Systems Mgr: Terry Silampa
■ Provides marketing & representation for defence business both in US & International. Provides Washington office/representation, provides spare parts, sales & distribution sourcing & purchasing services. → 85 → 90 [97]

80-02200 AMADEUS Inc
(a subsidiary of Vickers Shipbuilding and Engineering Ltd, UK)
3182 Golansky Boulevard, Suite 201, Dale City, VA 22192
☎ (+1 703) 583 97 02 [E] vseline@erols.com
⊖ (+1 703) 583 96 30
👥 3
President: Ralph J Cook (Jr)
Sales Director: Francis Hoerster
Office Manager: Ms Deborah Harpster
■ Management and marketing assistance to offshore manufacturing companies who wish to offer equipment and services to US Department of Defense as a prime or sub-contractor. [98]

80-02350 American Management Systems (AMS)
4050 Legato Road, Fairfax, VA 22033
☎ (+1 703) 267 8000 ⊖ (+1 703) 267 5067
♣ $ 632 400 000 (1995) 👥 5 750 (1996)
Chairman: Charles Rossotti
Vice Chairman: Patrick W. Gross
President: Philip Giunstini

Exec. Vice President Secretary Treasurer: Frank A. Nicolai
Chief Exec. Officer: Paul Brands
Treasurer: Frank A. Nicolai
Corporate Communications: Ms Pauline Weger
■ AMS supports both the US & international defence community in business process re-engineering, industrial operations, logistics/engineering/maintenance, technical support, strategic systems planning. Life-cycle management documentation, systems design & development prototyping, acquisition & procurement, quality assurance & independent verification & validation of software products, facilities maintenance management & systems implementation. [97]

80-02390 Amer Ind. Inc.
1310 Braddock Place, Alexandria, VA 22314
☎ (+1 703) 836 5900 ⊖ (+1 703) 836 8691
President & CEO: Ms Patricia J. Parson
VP Operation: Charles E. Coflin
VP Technology Transfer: Richard P. Brown
Corporate Communications Dir: Ms Nancy D. Runnels
Proposal Specialist: Ms Emily C. Elks
■ Computer systems/telecommunications support, information resources management, integrated logistics support, program management, systems engineering, training technology. [97]

80-02470 AMI International
(a subsidiary of American Maritime International Inc.)
2510 Perry Avenue, Bremerton, WA 98310
☎ (+1 360) 373 26 86 [E] amiinter@silverlink.net
[W] http://www.silverlink.net/amiinter/
⊖ (+1 360) 478 89 62
President: Guy Ames Stitt
Vice President: Michael J Brown
Vice President: Mark Underwood
■ Diversified naval consulting services to worldwide clients through focused market analyses, threat analyses, naval ship project reports and analyses of naval issues. Publishes worldwide Naval Ship Projections Report and Naval Designs catalogue. [98]

80-02600 Analysis & Technology Inc. (A&T)
Corporate Office
Route 2, PO Box 220, North Stonington, CT 06359
☎ (+1 860) 599 39 10 [W] http://www.aati.com
[Telex] 643329 ⊖ (+1 860) 599 89 07
♣ $ 144,000,000 (1996) 👥 1,558 (1997)
Operations Director: Jay W Ryerson
Purchasing Manager: Ken Levy
Corporate Marketing Office: Ms Suzanne C Cattanach
■ Research and analysis, system engineering, test & evaluation, interactive multimedia training, engineering support, equipment design, develop technical management support, publications and audio-visual media, software engineering.
Subsidiaries:
A & T International Corp
Analysis & Technology (Aust) Pty Ltd
Interactive Media Inc. [98]

80-02620 ANSER
(† Analytic Services Inc.)
1215 Jefferson Davis Hwy, Ste 800, Arlington, VA 22202
1215 Jefferson Davis Highway, Suite 800
☎ (+1 703) 416 20 00 [W] http://www.anser.org
⊖ (+1 703) 416 44 51
♣ $ 49,245,000 (1996) 👥 640 (1997)
President and Chief Executive Officer: J M Fabian
Senior Vice President and Chief Operating Officer: E M Mignogna
Senior Vice President of Finance and Chief Financial Officer: Ms Joy R Murtaugh
Vice President of Corporate Development and Marketing: L A Cox
Vice President of Human Resources and Support Planning: K Stacy
■ Research and systems analysis, technology assessment, cost-effectiveness studies, development planning, resource allocation in weapon, aeronautical and space systems for USAF, DoD and other government agencies, international aerospace systems development. [98]

80-02700 AOG Services Command
(a subsidiary of Charter International SA)
Long Beach Airport, PO Box 667, Long Beach, CA 90801
☎ (+1 562) 599 66 95 ⊖ (+1 562) 427 99 33
President: Dr Ted W Kennedy
■ Worldwide emergency parts service for air transport and military. → 85 [98]

80-03100 Arca Systems
2540 N 1st St # 301, San José, CA 95131-1016
☎ (+1 408) 434 6633 ⊖ (+1 408) 434 0262
♣ $ 3 662 489 (1993) 👥 35
President: Dr. William F. Wilson
VP Operations: R.Kenneth Bauer
■ Security integration services (consulting, engineering, training) to system integrators, application developers & end-users in government & business. Topics include: MLS, trust, assurance, certification & accreditation, internet security, firewalls for worldwide web. [97]

80-03130 Arinc
(† Arinc Research Corp.)
2551 Riva Rd, Annapolis, MD 21401
☎ (+1 410) 266 4650 ✉ ARINC
[Telex] 9650643; TWX 710 8678 553 ⊖ (+1 410) 266 2329
♧ $ 63 160 000 ♣ $ 278 000 000 (1995) 👥 1 162
🏠 111 572 sq ft
Chairman & CEO: J.L. Pierce
VP Marketing: R.F. Jefferson
VP Research Division: F. Jacoby
■ Development, design concepts, performance analysis, interoperability, integration analysis, installation, test & evaluation, planning & implementation, LAN & gateway development, program support, data systems, DBMS installations, electronic warfare & training. [97]

80-03250 Artemis Management Systems Inc
Texas Office
5251 Westheimer, Suite 850, Houston, TX 77056
☎ (+1 713) 626 15 11 ⊖ (+1 713) 626 10 04
Regional Director: Perry Nolen
Virginia Fairfax Office
One Flint Hill, 10530 Rosehaven Street, Suite 600, Fairfax, VA 22030-2842
☎ (+1 713) 293 74 00 ⊖ (+1 713) 277 10 52
Regional Sales Manager: Jon Swain
California Office, Irvine Sales
18200 Von Karman Avenue, Suite 1050, Irvine, CA 92612
☎ (+1 714) 660 71 00 ⊖ (+1 714) 660 70 20
Irvine - Development
California Office, Irvine Development
18200 Von Karman Avenue, Suite 1050, Irvine, CA 92612
☎ (+1 714) 851 19 99 ⊖ (+1 714) 660 70 20
Florida Offices
100 Second Avenue South, Suite 200, St Petersburg, FL 33701
☎ (+1 813) 823 14 01 ⊖ (+1 813) 823 38 15
Massachusetts Office
25 Burlington Mall Road, Suite 300, Burlington, MA 01803
☎ (+1 617) 238 67 67 ⊖ (+1 617) 238 67 72
Michigan Office
39111 West Six Mile, Livonia, MI 48034
☎ (+1 313) 591 18 37 ⊖ (+1 313) 591 65 45
Pennsylvania Office
1060 First Avenue, Suite 202, Maschellmac Office Complex, King of Prussia, PA 19406
☎ (+1 610) 878 54 00 ⊖ (+1 610) 878 54 39 [98]

80-03350 Associated Technical Communicators
23289 Ventura Boulevard, Woodland Hills, CA 91364-5141
☎ (+1 818) 225 8473 ⊖ (+1 818) 225 8477
President: David C. Brower
Manager: Karmen I. Brower
■ Provide services for commercial & government oriented companies. Consulting in aviation communications, command, control & general engineering systems. [97]

80-03352 Bell Technologies Inc. Testing Division
(† Associated Testing Laboratories Inc.)
(a division of Bell Technologies Inc.)
1133 Route 23 South, Wayne, NJ 07470
☎ (+1 201) 628 13 63 [W] http://www.belltechinc.com
⊖ (+1 201) 628 78 84
👥 22 (1997) 🏠 11,000 sq ft
Director of Operations: Edward Kulcsar
Office Manager: J Snider
■ Independent testing laboratory servicing the military, industry & consumers. Offering dynamics, electronics, environmental simulation, reliability, qualification failure analysis & package testing on components & completed systems.
● Northwest Ind'l Park, 53 Second Ave, Burlington, MA 01803
☎ (+1 617) 272 90 50 ⊖ (+1 617) 272 81 84
👥 50 🏠 27,000 sq ft
President: Edward E. Kulcsar
Vice President: Robert H. Croft [98]

80-03355 ASTA Corp.
3250 Holly Way, Chula Vista, CA 91910-3217
☎ (+1 619) 422 17 54 ⊖ (+1 619) 426 12 80
♣ $ 400,000 👥 5 🏠 200 sq m
President: Louis Gerken
Sales and Marketing Director: Jay Ortega
Advertising Manager: Ms Sherry Gerken
Public Relations Manager: Ms Valerie Rios
■ Publication of defence related books:- ASW, submarines, mines, torpedoes, UAVs, scientific research, military and civilian airships. → 60 [98]

80-03358 Astroways/Defense Group
(† Astroways USA Corp.)
(a subsidiary of Charter International SA)
Long Beach Airport, PO Box 667, Long Beach, CA 90801
☎ (+1 562) 599 66 95 ⊖ (+1 562) 427 99 33
President: Dr Ted W Kennedy
■ Commercial and military aircraft and heicopter sales. Organisation of corporate and military flight training programmes for undeveloped nations. [98]

80-03365 Atlantic Research Marketing Systems Inc. (ARMS)
230 West Center Street, West Bridgewater, MA 02379
☎ (+1 508) 584 78 16 ⊖ (+1 508) 588 80 45
President: Richard Swan
■ Design manufacturer specialising in integration of the advanced laser based identification friend or foe, NV and day optics systems onto the M16/M4 carbine, machine guns and other weapon systems to include such types as man portable anti-armour, platform, armour or aircraft. Quick change repeatable throw lever attachments. → 75 [98]

80-03757 Aviation Products Division D782
(† Air Treads Inc.)
(a subsidiary of Goodyear Tire & Rubber Co.)
1144 East Market Street, Akron, OH 44316-0001
☎ (+1 330) 796 63 06 ⊖ (+1 330) 796 65 35
♣ $ 11,000,000,000 (1997) 👥 9,000 (1997)
General Mgr: Tom Walker ☎ (+1 330) 796 6322
Marketing Mgr: Jim Pickering ☎ (+1 330) 796 6323
Sales Mgr: William Rathbun ☎ (+1 330) 796 6325 [98]

80-03758 Avio-Diepen Inc.
(a subsidiary of Avio Diepen Holding BV, NL)
114 Southfield Parkway, Suite 120, Forest Park, GA 30050
☎ (+1 404) 361 20 58 ⊖ (+1 404) 361 23 27
👥 8 (1997)
Vice President and General Manager: Herman Kooijman
■ FAA accredited distributor. Stocking warehouses in Atlanta, The Netherlands, London and Hong Kong. [98]

80-03760 Avionics Southwest
(a subsidiary of Avionics of Southern California Inc.)
2667 East 28th Street, Suite 510, Signal Hill, CA 90806
☎ (+1 562) 595 98 66 [E] avionicssocal@internetmci.com
⊖ (+1 562) 595 98 66
♣ $ 300,000 (1996) 👥 3 (1997) 🏠 366 sq m
President and Owner: Tom Woscek
■ Repair and overhaul of airline and military avionics components. Specialising in airborne radar equipment. [98]

80-04070 Barnes & Reinecke Inc.
(a subs. of Allied Research Corp.)
425 East Algonquin Road, Arlington Hts, IL 60005
☎ (+1 847) 640 7200 ⊖ (+1 847) 640 0354
♧ $ 2 000 000 ♣ $ 13 000 000 👥 117 🏠 122 000 sq ft
President & CEO: B.Herb Ellis ☎ (+1 847) 640 3700

Contact Reference

VP & COO: Richard J. Huss ☎ (+1 847) 640 3730
■ Contract technical services, consulting engineering, GSE & test equipment, prototyping & production, preparation of technical documentation. [97]

80-04100 Barton-Aschman Associates Inc.
(a subsidiary of The Parson Corp)
820 Davis St, Evanston, IL 60204
☎ (+1 708) 491 1000 ⊜ (+1 708) 475 6053
275
President: Barry D. Lundberg
■ Traffic engineering, transportation planning, land development analysis & planning. [97]

80-04140 bd Systems Inc.
385 Van Ness Ave, Ste 200, Torrance, CA 90501-1482
☎ (+1 213) 618 8798 ⊜ (+1 213) 212 0753
♧ $ 3 500 000 (1996) ♣ $ 22 000 000 (1996) 267
President: Ms Clarisa F. Howard
VP Technical Operations: Thomas A. Hohman
Sales & Marketing Dir: Carl F. Asiala
■ Military & space systems engineering requirements analysis, survivability; Integrated logistics; Computer systems software development. Management information systems & product assurance. [97]

80-04150 BDM International Inc.
(a subsidiary of BDM Holdings Inc.)
1501 BDM Way, McLean, VA 22102-3204
☎ (+1 703) 848 50 00 [W] http://www.bdm.com
⊜ (+1 703) 848 53 25
♣ $ 890,000,000 (1995) 8,000
President and Chief Executive Officer: Philip A Odeen
Executive Vice President and Chief Financial Officer: C Thomas Faulders
Executive Vice President: William Hoover
Executive Vice President: Dr W E Sweeney
Corporate Vice President Human Resources: Robert Waters
Corporate Vice President Business Development: Jack Todd
Senior Vice President Corporate Government Affairs Investor Relations: Todd A Stottlemyer
Senior Vice President International Research: Paul Ello
Senior Vice President Integrated Supply Chain: Ron Jepper
Senior Vice President Defense Systems Engineering/Analysis: Ms Pam Sullivan
Senior Vice President Systems Support and Operations: Chuck Zang
Senior Vice President Systems Development/Integration (Defense): Jim Cassidy
Senior Vice President Systems Development/Integration Civil: Paul Leslie
Senior Vice President Systems Engineering Analysis: Bennie DiBona
Senior Vice President Integrated Supply Solutions: George Ashkar
Senior Vice President Business Operations: Bob Fries
Senior Vice President IT Services: Doug Coburn
President BDM Technologies: Ms Helen Seltzer
President Air Safety Management: Susan Coughlin
President Enterprise Management Services: Roy Woodle
■ Professional and technical services in defense, communications, logistics, systems integration, manufacture, information systems, business process re-engineering, enterprise management and testing. [97]

80-04360 Beltramo & Associates
13039 Sky Valley Road, Los Angeles, CA 90049
☎ (+1 310) 472 33 75 [E] beltramo@worldnet.att.net
President: Dr Michael N Beltramo
■ Consulting assistance in developing competitive strategies. Areas of particular expertise include: conversion, proposal preparation, international coproduction programs, offsets, foreign military sales and the estimation of life cycle costs. [98]

80-04500 Betac Corp.
2001 N. Beauregard St, Suite 1100, Alexandria, VA 22311
☎ (+1 703) 824 3100 ⊜ (+1 703) 824 0333
♣ $ 35 000 000 (1996) 350 90 000
President: Earl Lockwood
Chief Financial Officer: Brad Chalk
■ Professional systems engineering support in command, control, communication & intelligence. [97]

80-04505 BGS Systems Inc.
One First Avenue, Waltham, MA 02254-9111
☎ (+1 617) 891 00 00 ⊜ (+1 617) 890 00 00
♣ $ 48,578,000 (1997) 230
President & CEO: Harold S. Schwenk (Jr)
Chief Operating Officer: James McGuire
Vice President Product Marketing: Jay Lipovich
VP Marketing Mgr: Tim Grieser
Marketing Programs Manager: Ms Madine Romm
■ Software developers, performance and capacity planning software. [98]

80-04510 Billfish Inc. San Pedro Boat Works
Berth 44 Outer Harbor, San Pedro, CA 90731-7291
☎ (+1 310) 832 72 03 ⊜ (+1 310) 514 06 86
♣ US$ 6,000,000 (1996) 100 (1996)
President: Andy Wall (Jr)
Vice President: Ronald J Rados
■ Ship and boat repair services. [98]

80-04590 Bodycote Taussig Inc.
(† Taussig Associates Inc.)
(a subsidiary of Bodycote International)
7530 Frontage Road, Skokie, IL 60077
☎ (+1 847) 676 21 00 ✆ (800) 323 36 57
[E] taussigmet@aol.com ⊜ (+1 847) 676 21 32
45
Managing Director: Corwyn M Berger
Operations Director: Lyle R Jacobs
Marketing Manager: Norwin A Merens
Purchasing Manager: Ms Evelyne C Johnson
■ Metallurgical testing & consulting services. [98]

80-04700 Booz, Allen & Hamilton Inc.
Headquarters & Worldwide Technology Business
8283 Greensboro Drive, McLean, VA 22102
☎ (+1 703) 902 5000 ⊜ (+1 703) 902 3333
♣ $ 1,100,000,000 (1995) 6,600 (worldwide 1995)
Chairman & CEO: William F. Stasior
President: Ralph W. Shrader
Vice President: J.M. McConnell
Dir Corporate Communications: Ms Marie Lerch
■ Technology, engineering & management consulting, systems engineering & integration, civil/government consulting.

New York Office
101 Park Ave, New York, NY 10178
☎ (+1 212) 697 1900 ✉ BOOZ ALLEN 620196
⊜ (+1 212) 551 6732
President: Brian Dickie [97]

80-04800 Leo J. Brennan Inc.
2359 N. Livernois, Troy, MI 48083-1692
☎ (+1 810) 362 3131 ⊜ (+1 810) 362 2355
♣ $ 1 000 000 9 3 000 sq ft
President: Leo J. Brennan
■ Technical services relating to all types of communications including: literature, catalogues, manuals, training materials, proposals, technical papers. Full capabilities include: writing, design, art & illustrations, electronic retouching, desk top publishing, slides & video tapes. [97]

80-04920 Dayton T. Brown Inc.
Engineering & Test Division
Church Street, Bohemia, NY 11716
☎ (+1 516) 589 63 00 [E] test@dtbrown.com
[W] http://www.dtbrown.com ⊜ (+1 516) 589 40 46
350
President: Dayton T Brown (Jr)
Senior Vice President Engineering and Test Division: C Kenneth Morrelly
Vice President and Marketing Director: James S Mattimore
Contracts Manager Special and Support Departments: Ms Janice Luscher
Test Systems Strategic Business Unit Manager: John Herlihy
Engineering Testing Strategic Business Unit Manager: Ronald Loveland
Systems Test Group Strategic Business Unit: Michael Avari
■ Testing & engineering services for product quality, reliability, performance evaluation, design optimisation, identify operating & regulatory requirements. Design & manufacture of hydraulic, electromechanical & fuel test systems. [98]

80-04940 BRUHN NewTech Inc
10500 Little Patuxent Parkway, Suite 680, Columbia, MD 21044-3505
☎ (+1 410) 465 13 37 [E] bpwnwt@aol.com
[W] http://www.bruhn-newtech.com
⊜ (+1 410) 750 18 33
■ Production of operational software systems for hazard prediction, warning and reporting of the effects of nuclear, biological and chemical incidents. [98]

80-04960 Burdeshaw Associates, Ltd (BAL)
4701 Sangamore Road, Bethesda, MD 20816-2508
☎ (+1 301) 229 58 00 [E] bal@interramp.com
[W] http://www.burdeshaw.com ⊜ (+1 301) 229 50 45
♣ $ 8,000,000 31 (600 Associates)
President: Brig Gen William B Burdeshaw (retd)
Executive Vice President: John J Welch (Jr)
Senior Vice President of Naval Programs: Vice Adm Joseph B Wilkinson (retd)
Senior Vice President of Air Force Programs: Lieut Gen Stephen B Croker (retd)
Senior Vice President of Army Programs: Lieut Gen Max W Noah (Retd)
Vice President of Business Development: J Michael Stacy
Vice President, Navy: Rear Adm Lowell J Holloway (retd)
Vice President, Air Force: Col James W Connally (retd)
Vice President, Army: Col Milton S Newberry (rtd)
Vice President, International: Lieut Gen Phillip C Gast (retd)
Vice President, Logistics: Maj Gen E B Leedy (retd)
Vice President, Army: Maj Gen Wayne Knudson (retd)
Director, International: William Piekney
Purchasing Director: Joseph R Brown
Manager, Strategic Business Information: Steven S Klement
■ Diversified services company dedicated to assisting clients in matching technology and capabilities to defence and government markets. Services include: concept development, operations analysis, market analysis, strategic planning, acquisition strategy, proposal support, business planning and worldwide marketing. [98]

80-05000 3C Systems, Inc.
620 Argyle Road, Wynnewood, PA 19096
☎ (+1 610) 649 1462 ⊜ (+1 610) 649 3459
President: Murray Kornhauser
Operations Mgr: Thomas Bane
■ Impact/dynamics. Analytical and experimental studies, development/testing prototypes of ordnance components (fuzes, protective systems). Explosive sensitivity modelling/testing, small scale test facility configurations. Sharp Shield prevents sympathetic detonation. 200 000-G shock test facility. [98]

80-05160 Callahan Publications
PO Box 1173, McLean, VA 22101
☎ (+1 703) 356 19 25 ⊜ (+1 703) 356 96 14
Editor and Publisher: Vincent F Callahan (Jr)
■ Publisher, Defence Acquisition Report, The Underwater Letter, The Space Letter, Washington Environmental Protection Report. [98]

80-05165 Cambridge Neurodynamics Ltd
101 Eisenhower Parkway, Rosewood, NJ 07068-1029
☎ (+1 201) 364 00 10 [E] info_usa@neurodynamics.com
[W] http://neurodynamics.com ⊜ (+1 201) 364 05 50 [98]

80-05170 Cametoid Technologies Inc.
(a subsidiary of Newman Aerospace Canada)
150 Colonial Road, Manchester, CT 06040-2306
☎ (+1 203) 646 46 67 [E] coatings@cametoid.com
[W] http://www.cametoid.com ⊜ (+1 203) 649 39 34
♣ $ 2,750,000 (1997) 25 (1997) 24,350 sq ft
President: Desmond G Newman
Managing Dir: Dennis Twiss
Technical Director: David N Palmer
Office Manager: Ms Janet Shelto [98]

80-05200 Carbotek Inc.
16223 Park Row, Suite 100, Houston, TX 77084
☎ (+1 281) 578 88 99 [E] cwk@carbotek.com
⊜ (+1 281) 578 82 65
♧ $ 1,300,000 (1997) ♣ $ 2,200,000 (1997)
16 (1997) 22,000 sq ft
President: Dr C W Knudsen

■ Process (chemical engineering) research for space materials processing. [98]

80-05230 Cas Inc.
650 Discovery Dr., PO Box 11190,
Huntsville, AL 35814-1190
☎ (+1 205) 895 8878 ⊖ (+1 205) 895 7635
♟ 480
President: Fredric Clark
VP Sales & Marketing: Charles G. Vaughn
■ Weapon systems analysis, systems engineering, software engineering, test engineering, logistics analysis, program management, simulation development & operation. [97]

80-05240 Cascade General Portland Shipyard
(† Cascade General)
5555 North Channel Avenue, Building 71,
Portland, OR 97217
☎ (+1 503) 285 11 11 ✆ (800) 505 19 30
[W] http://www.casgen.com ⊖ (+1 503) 285 03 61
♧ $ 30,000,000 (1996) ♣ $ 90,000,000 (1996)
♟ 950 (1997) ⌂ 46,000 sq m
President: Frank Foti
Executive Vice President: Andrew G Rowe
Vice President Production: Surendra Menon
Sales and Marketing Director: Ms Terri Deskins
Marketing and Customer Service Coordinator: Ms Vicki Mallonee
■ Repair & conversion facility. Services include blasting & coating, engine repair, boiler repair, piping & valve repairs, steel renewal, electrical/electronic repairs & installations, shafting work, pump repairs & habitability upgrades. [98]

80-05260 C-CUBED Corp.
6800 Versar Center, Suite 300, Springfield, VA 22151
☎ (+1 703) 658 96 85 [W] http://www.c-cubedcorp.com
⊖ (+1 703) 658 96 89
♣ US$ 17,000,000 (1997) ♟ 225 (1997)
⌂ 70,000 m2 (1997)
President: E J Bednar
Executive Vice President: S Wesley Teague
Chief Financial Officer: Ms Rhonda L Lugar
■ Diversified technical services, particularly in the areas of communications, combat systems, command, computer networks control, integrated logistics, information systems, video teleconfering electronic publishing and other multi-media multimedia applications. [98]

80-05400 Century International Arms Inc.
1161 Holland Drive, Boca Raton, FL 33487
☎ (+1 561) 998 32 00 ✆ (800) 991 48 67
[E] info@centuryarms.com
[W] http://www.centuryarms.com ⊖ (+1 561) 998 32 10
President: Michael Sucher
Exec. Vice President USA: André Blouin
Vice President: Brian Sucher [98]

80-05425 Cexec Inc.
8618 Westwood Center Dr, Ste 100, Vienna, VA 22182
☎ (+1 703) 893 3220 ⊖ (+1 703) 556 0829
♧ $ 6 700 000 ♣ $ 27 900 000 ♟ 320 ⌂ 4 000 sq m
President: Douglas C. Rhodes
Exec. Vice President: James W. Moss
■ Service automated data processing (ADP) consulting & system engineering firm. [97]

80-05450 Chase Initiatives
(† Chase Inc.)
150 Cadish Ave, Hull, MA 02045
☎ (+1 617) 925 0560 ⊖ (+1 617) 925 0560
♟ 1 (1996)
Principal Scientist & Sole Proprietor: David M. Chase
■ Mathematical modelling; analytical research & development in hydroacoustics & fluid dynamics directed mainly at prediction & control of flow-related self noise in sonar systems. [97]

80-05490 Cincinnati Testing Laboratories Inc. (CTL)
417 Northland Blvd, Cincinnati, OH 45240
☎ (+1 513) 851 3313 ⊖ (+1 513) 851 3336
♟ 30 ⌂ 2 325 sq m
President: Glenn A. Huber
Sales/Marketing Dir: Mark A. Huber
Operations Dir: Dr Brian G. Gieseke
Purchasing Mgr: Julie Higgins
■ Materials testing & evaluation services. [97]

80-05740 Colonial Export, Inc.
155 Algonquin Parkway, Whippany, NJ 07981-1601
☎ (+1 201) 503 97 21 [E] colonialexport@msn.com
[W] http://www.colonialexport.com ⊖ (+1 201) 503 97 24
Managing Director: Alan Antaki
Marketing Director: Nick Gileta
Advertising Director: Ms Liliane Antaki
■ PHT Supply Company, Marli Shipping Inc. and Colonial Export Inc. are mutual companies that operate jointly to service unusual and compeling urgent military needs. Procurement, export and delivery needs are authorised under 10 USC 2304(c) (2) and 41 USC 253(c) (2). [98]

80-05855 Commercial Marketing Assistance International (CMA INTL)
7515 Topton Street, Suite 100,
New Carrollton, MD 20784
☎ (+1 301) 577 93 40 ⊠ 532282 dmausa wash
⊖ (+1 301) 577 70 19
Director: John A Schaffer [E] jaschaff@wam-und.edu
■ Export sales & marketing to China & USA of military & non-military items from suppliers worldwide. → 90 [98]

80-05860 Commodore Aviation Inc.
(a subsidiary of Israel Aircraft Industries International, Israel)
4900 North West 36 Street, PO Box 661078,
Miami, FL 33152
☎ (+1 305) 869 19 00 [E] comaviation@juno.com
⊖ (+1 305) 869 19 52
♣ US$ 35,000,000 (1997) ♟ 450 (1997)
⌂ 103,000 m2
President: M Hendel
Vice President of Operations: A Ben Basat
Marketing and Contracts Director: R E Weltmann
Purchasing Director: M De Micco
■ Maintenance and overhaul services. [98]

80-05896 Component Repair Technologies Inc. (CRT)
8507 Tyler Boulevard, Mentor, OH 44060
☎ (+1 216) 255 17 93 ⊖ (+1 216) 255 41 62
President: Thomas E Wheeler
Operations and Technical Director: C R Bart
Market Development and Planning Manager: Jay R Varma [E] crtvarma@aol.com
Purchasing Manager: John Smigelski
■ FAA repair station. Repair/rework P&W GE & CFM engine frames, cases, housings & rotating parts. [98]

80-06200 Condor Enterprises Inc.
Aurora Municipal Airport, 43W776 US30,
Sugar Grove, IL 60554-9617
☎ (+1 630) 466 41 12 ⊖ (+1 630) 466 41 25
♟ 9 ⌂ 50 sq m
President: Dr Marjorie F Stampolis
Vice President: Dr Ernst W Kobrow
■ Military aircraft finance & leasing. Freight & passenger air carrier (CNDA). Composite airframe design. [98]

80-06470 Contract Professionals, Inc. (CPI)
4141 West Walton Boulevard, Waterford, MI 48329
☎ (+1 248) 673 38 00 ✆ (800) 228 48 03
[E] cpi@cpijobs.com [W] http://www.cpijobs.com
⊖ (+1 248) 673 59 92
♣ $ 23,000,000 (1996) ♟ 350 (1997) ⌂ 16,000 sq ft
President: Steven York
Sales and Marketing Director: Gene Kowalski
Purchasing Manager: Mark McGinnis
■ Supplies engineers, designers, technical writers, logistics engineers, proposal writers, programmers, systems analysts for project duration. Areas include tactical vehicles, aerospace, electronics, integrated logistics support and technical publications. [98]

80-06730 Criminal Intelligence Administration
38-11 Ditmars Boulevard, Suite 116, Astoria, NY 11105
☎ (+1 917) 491 33 24 ✆ (800) 792 71 81
[E] 102532.1073@compuserve.com
[W] http://www.iypn.com/intelligence
⊖ (+1 718) 626 41 15
♣ $ 1,780,000 (1996) ♟ 65
President: Tim O'Brien
Marketing Director: John Anders
■ International counterintelligence/counterterrorism firm. All employees are either active or retired members of the US intelligence community. [98]

80-06810 CTA Inc.
6116 Executive Boulevard, Suite 800,
Rockville, MD 20852
☎ (+1 301) 816 12 00 [E] info@cta.com
[W] http://www.cta.com ⊖ (+1 301) 816 14 16
♣ $ 180,000,000 (1996) ♟ 1,250 (1997)
President and Chief Executive Officer: C E (Tom) Velez
President, Space and Telecommunications Company: Ricardo de Bastos
President, Information Technology Services Company: Raymond V McMillan
Vice President of Corporate Communications: David H Johnson
■ Technical and commercial representation of US and overseas firms to defence agencies and prime contractors. Program management and assistance. Systems integration. Technical liaison. [98]

80-06893 Leo A. Daly
8600 Indian Hills Drive, Omaha, NE 68114
☎ (+1 402) 391 81 11 ⊖ (+1 402) 391 85 64
♣ $ 70,000,000 (1997) ♟ 715 (1997)
Chairman and President: Leo A Daly
Senior Vice President and Director Aviation Services: Joseph D Vaccaro
Senior Vice President: Edward F Cambridge
Vice President: Charles D Dalluge
Marketing Manager: Michael J Riordan
■ International planning, architecture, engineering & interior design. [98]

80-06898 Dan-Am Shipping and Chartering Inc.
118 Cardinal Drive Ext. 104, Wilmington, NC 28405
☎ (+1 910) 392 43 81 ⊠ DANAM ⊠ 197385
⊖ (+1 910) 392 43 85
♣ US$ 20,000,000 (1996) ♟ 15 (1996) ⌂ 3,000 sq ft
President: Michael S Steenberg
Sales and Marketing Director: Joseph M Travers
Operations and Technical Director: S Michaels
Operations: Ms Tina Christianson
■ Independent vessel operators specialising in international transport of all dangerous and hazardous goods, as well as projects worldwide. Service includes exclusive full and part charter. → 85
Other Offices:
UK, Monaco, Cyprus [98]

80-06900 Danzas Corporation
(† Danzas Inc.)
Group HQ
3650 131st Avenue SE, PO Box 53370,
Seattle, WA 98006
☎ (+1 425) 649 93 39 ✆ (800) 426 59 62
⊠ DZUSBLVHQ [W] http://www.danzas.com
⊠ 4979609 (itt) ⊖ (+1 425) 649 49 40
♣ $ 5,000,000,000 (1996) ♟ 15,000 (1996)
President and Chief Executive Officer: Jack Edwards
Vice President Human Resources: Linda Findlay
Vice President Logistics: Rolf Noetzli
Vice President Domestic: Chuck Nelson
Corporate Communications: Lesley Rollo
Sales and Marketing: Arnie Goldstein
■ Int'l freight forwarder & customs house broker specializing in FMS activities, via Ocean & Air. Offices in most major USA ports. Fully licenced by all appropriate United States government agencies. [98]

80-07100 Data Search Associates
17220 Newhope St, Suite 107, PO Box 8361,
Fountain Valley, CA 92708
☎ (+1 714) 540 9386 ⊖ (+1 714) 540 9396
President: Ted G. Nicholas
Vice President: Ms Rita C. Rossi
■ Cost analysis consulting on missiles & aircraft; publishers of missiles & aircraft data books. [97]

80-07200 John M. Davis & Associates
PO Box 8795, Wichita, KS 67208-0795
☎ (+1 316) 685 79 29
5 (1997)
Managing Director: John M Davis
Marketing Director: Jorge Ellis
Operations Director: Jesus Martinez
■ Consultancy in availability of aviation equipment, assistance in purchasing and organisation review. 98

80-07400 DCS Corp.
1330 Braddock Place, Alexandria, VA 22314
☎ (+1 703) 683 8430 ⊖ (+1 703) 684 7229
♣ $ 27 000 000 (1995) 320 51 000 sq ft
President: James T. Wood
Vice President: Jackie Junkin
Vice President: David Rubin
VP Business Development: Walter J. Bank
■ High technology R&D; systems analysis; system integration, systems engineering in fields of electro-optics, avionics, airborne & shipboard weapon systems, air armament, electronic warfare, automatic test equipment, laboratory & field T&E, modelling & simulation; machine vision systems for quality control; vehicle location systems.
● Route 4, PO Box 10G, Lexington Park, MD 20653
☎ (+1 301) 862 2390 ⊖ (+1 301) 863 0324
Dept Mgr: William Nelson
● PO Box 785, Ridgecrest, CA 93556
☎ (+1 619) 375 7859 ⊖ (+1 619) 375 0281
Manager: Larry C. Smith
● 500 Northwest Plaza, Suite 1050, St. Louis, MO 63074
☎ (+1 314) 739 7755 ⊖ (+1 314) 739 8262
Manager: Gerald T. Hipp 97

80-07450 DDC-I Inc
(a subsidiary of DDC-I A/S, Denmark)
400 North 5th Street, Suite 1150, Phoenix, AZ 85004
☎ (+1 602) 275 7172 ✆ (800) 221 8643
⊖ (+1 602) 252 6054
♧ US$ 341,973 ♣ US$ 2,760,065 20
President: Ole Oest
Sales & Marketing Dir: Ole Oest
Operations Director: Sean Van Doren
■ Develops and markets high performance Ada cross compiler systems and tools to common microprocessor targets for the development of real-time safety critical embedded systems. DDC-I Ada run-time systems are FAA certifiable, and are used in many military and commercial aerospace products, including the Boeing 777. 98

80-07700 Defense Logistics Supply Corp.
1735 Jefferson Davis Highway, Suite 103,
Arlington, VA 22202
☎ (+1 703) 413 42 00 ⊖ (+1 703) 413 06 25
Vice President: A Kundanmal
Accounts Department: D Kundanmal
Manager: H Chacon 98

80-08020 Delex Systems Inc. (DELEX)
1953 Gallows Road, Suite 700, Vienna, VA 22182-3991
☎ (+1 703) 734 83 00, extension 204
W http://www.delex.com ⊖ (+1 703) 893 53 38
♣ US$ 14,200,000 (1996) 175 (1997)
32,765 (1997)
President and Chief Executive Officer: H L Piper
Vice President: D R Bensing
Purchasing Manager: Gerald M Lanager
Executive Assistant: Ms Patti Tarallo
■ Test and evaluation, software development and quality assurance, technical publications development for ASW, ASUW, AAW, tactical training, ILS, Research and Development maintenance training, flight training, pilot training, AQP services. 98

80-08300 Designers & Planners Inc.
(a subs. of British Maritime Technology Ltd)
2120 Washington Boulevard, Suite 200,
Arlington, VA 22204
☎ (+1 703) 920 7070 ⊖ (+1 703) 920 7177
80
President: John Gallagher
■ Maritime technical services covering naval architecture & marine structures, maritime safety, health, environmental protection, software development & testing. 97

80-08480 DJH Engineering Center Inc. (DJHEC)
540 Arapeen Drive #209, Salt Lake City, UT 84108
☎ (+1 801) 583 39 34 E djhec@mail.aros.net
⊖ (+1 801) 583 67 21
♣ $ 1,500,000 (1996) 17 (1997) 6,150 m2
President: Dusan J Hvolka
Operations Director: Reid Oliver
Sales and Marketing Director: Bill Bruno
Accountant: Donald R Campbell
■ Advanced mobility engineering; drive trains, transmission, axles, and wheel end assemblies. Aerospace/aviation engineering services for design and evaluation for the following: components and sub assemblies, stress analysis, ground support equipment, assembly tooling, detail tooling. 98

80-08500 Document Engineering Co.
15210 Stagg St, Van Nuys, CA 91405-1002
☎ (+1 818) 782 1010 ✆ (800) 645 7732
⊖ (+1 818) 782 2374
13
CEO/President: Eric D. Molinari
■ Provide re-prints of government specifications & standards as well as industry standards. Provide update services, QPL sets, AN/AND/MS sets with updates. Sales agent in US for Standards Infodisk (CD-ROM Bibliographic standards DB). 97

80-08600 John W. Drake Transportation Consultant
3909 Somerset Place, West Lafayette, IN 47906-8855
☎ (+1 317) 463 5882 TWX 650 2329 491
⊖ (+1 317) 463 4610
12 (1996)
Proprietor: Dr John W. Drake
Research Asst: Ms Bonnie Gray
■ Market research regarding civil & military aircraft, engines & related products. Aircraft selection studies for airliner & corporate fleet operators. Safety evaluations. 97

80-08670 DUAL Incorporated
(† Dual & Associates Inc.)
30 Skyline Drive, Lake Mary, FL 32746
☎ (+1 407) 333 8880 ⊖ (+1 407) 333 8850
♣ $ 30 000 000 300 75 000 sq ft
President & CEO: Fred J. Dual (Jr)
VP Business Development: Gary R. Tyler
Sr VP Research & Development: Hank Okaski
Vice President: Mike Palamar
Vice President: Doug Story
Business Operations Dir: Tom Wilke
General Mgr: Ed Harris
■ Technical support services: Advanced software engineering; systems integration; hardware/software simulator engineering; training/training systems support. → 80
● 4799 Cuba Millington Road, Millington, TN 38053
☎ (+1 901) 873 0076 ⊖ (+1 901) 872 6738
● 1225 Jefferson Davis Hwy. Suite 1300,
Arlington, VA 22202
☎ (+1 703) 416 0999 ⊖ (+1 703) 416 2774
● 15040 Avenida Venusta, Unit 187,
San Diego, CA 92128
☎ (+1 619) 673 9505 ⊖ (+1 619) 673 6738
● 2101 Wilson Blvd, #600, Arlington, VA 22201
☎ (+1 703) 527 3500 ⊖ (+1 703) 527 0829
● 745 C Great Mills Road, Lexington Park, MD 20653
☎ (+1 301) 870 3960 ⊖ (+1 301) 862 5043
● 1331 Gemini Street, Suite 270, Houston, TX 77058
☎ (+1 713) 486 1984 ⊖ (+1 713) 480 2445
● 318 Pendleton Way, Oakland, CA 94621
☎ (+1 510) 430 9001-2 ⊖ (+1 510) 430 9307
● 1792 Greensward Quay, Virginia Beach, VA 23454
☎ (+1 804) 496 2863
● 2203-2 East 151st Street, Carmel, IN 46227
☎ (+1 317) 306 3471 ⊖ (+1 317) 306 4258 97

80-08730 Dyna East Corp.
3620 Horizon Drive, King of Prussia, PA 19406
☎ (+1 610) 270 9900 ⊖ (+1 610) 270 9744
25 16 000 sq ft
President: Dr Pei C. Chou
Vice President: George C.H. Chou
VP Engineering: Robert D. Ciccarelli
Business Development Dir: Thomas P. Giaquinto (Jr)
■ Engineering design, fabrication, evaluation testing & production of advanced chemical energy warheads including Explosively Formed Penetrator (EFP) & tandem warhead systems. Development of state-of-the-art armour protection systems. Development of ballistic computer codes. 97

80-09450 EG&G Washington Analytical Services Center
1396 Piccard Drive, Rockville, MD 20850
☎ (+1 301) 840 30 00 ⊖ (+1 301) 840 07 85
General Mgr: Randy Wotring
■ Engineering, scientific & technical support services for defence, energy, space & ocean programmes; customized instrumentation & field services to US & foreign governments. 98

80-09750 Electronic Development Associates Inc. (EDA)
1 Westcliff Drive, Dix Hills, NY 11746-5618
☎ (+1 516) 673 38 81 E eda@eda-design.com
⊖ (+1 516) 673 59 79
President: Leonard Zuckerman
■ Electronics R&D, prototypes & manufacture drawings, switching power supplies, electro-optics, smart instruments. 98

80-09900 emf-emi Control
(† Interference Control Technologies)
(a subs. of Don White Consultants Inc.)
6193 Finchingfield Road, Gainesville, VA 22065
☎ (+1 540) 347 0030 E emfemi@summit.net
W http://www.emf-emi.com 899165 dwcigaiv
⊖ (+1 540) 347 5813
President: Donald R.J. White
Marketing Mgr: Ms Peggy Settle
■ EMC & EMI consultants. 97

80-09950 Emprise Corp.
830 Franklin Court, Marietta, GA 30067-8939
☎ (+1 770) 425 1420 ⊖ (+1 770) 425 1425
♣ $ 3 500 000 20
President: Ronald A. DuBose
Managing Dir: Donald Yelton
Sales/Marketing Dir: Michael Kemether
■ Turnkey supplier of test facilities & equipment. Specialized experience in design, engineering, software development, fabrication, assembly & installation of instrumentation, data acquisition & control systems, test stand equipment & test support systems. 97

80-10150 Engineering & Professional Services Inc. (EPS)
78 Apple St, Tinton Falls, NJ 07724
☎ (+1 908) 747 8277 ⊖ (+1 908) 530 4726
210
President: Francesco A. Musorrafiti
Sr VP Admin & Finance: Ms Antoinette M. Musorrafiti
Sr VP Operations: Richard P. Bordone
VP Engineering: David O. Linthicum
VP Field Engineering: Charles F. White
VP European Operations: John Kessel
Contracts Dir: Joseph M. Buccieri
■ Technical professional services & project support ranging from mission/requirements analysis to the concept formulation, design, development & implementation of major systems for the government, military, business & industry.
● 3633 Wheeler Road, Suite 280, Augusta, GA 30909
☎ (+1 706) 868 1380 ⊖ (+1 706) 868 1379
VP Field Engineering: Charles F. White
■ Field services, involving systems engineering & analysis, operations, maintenance & logistic services, including antennas, airborne radar, secure-combat net radio, fiber optics, intrusion alarm & telephone systems. Install & maintain computer facilities. Provides third party maintenance & field training. 97

80-10350 EPI Technologies Inc.
(† Environmental Processing Inc.)
2901 Summit Avenue, Suite 400, Plano, TX 75074
☎ (+1 972) 398 55 00 ⊖ (+1 972) 398 55 01
100 35,000 sq ft
Chairman, President and Chief Executive Officer: John L Pierce
Executive Vice President, Chief Financial Officer and Secretary: Lawrence F Kern
Vice President of Manufacturing: Phil Gier

Vice President of Quality Assurance: Bob Zagotta
Vice President of KGD Division: Ken Smith
Director of Sales and Marketing: Michael S Fisher
☎(+1 972) 398 55 08 E mfisher@cyberramp.net
Purchasing Manager: Tim Cook
■ Electronic component testing, conditioning, DPA and failure analysis. Known Good Die (KGD); bare die burn-in and test services. 98

80-11000 F&F Associates Inc.
PO Box 9335, Arlington, VA 22219
☎(+1 703) 271 00 61 ⊖ (+1 703) 271 09 63
President: Donald R Fowler E dfowler@bmpcoe.org
■ Consultant firm providing research analysis, policy formulation and planning services in manufacturing, industrial base, technology transfer and modernization. 98

80-11350 Ferro-Ceramic Grinding Inc.
247 Water Street, Wakefield, MA 01880
☎(+1 617) 245 18 33 ✆ (800) 282 18 33
E ferroceram@aol.com ⊖ (+1 617) 246 14 29
♣ $ 3,000,000 (1996) 👥 19
President: Carl Strasnick
Vice President: Phillip Strasnick
Operations Director: Frank Ciarlante
Director Advertising: Thomas Register
Sales and Marketing Director: Costa Sideridis
■ Insulators for Patriot, radar. 98

80-11600 Flight International Inc. Group Headquarters
(a subsidiary of The Flight International Group Inc.)
Newport News/Williamsburg Airport,
Newport News, VA 23602
☎(+1 757) 886 55 00 ✆ (800) 358 46 85 (FLT INTL)
E flight-mktg@fltintl.com W http://www.fltintl.com
⊖ (+1 757) 874 74 81
♣ $ 12,000,000 (1996) 👥 15 (1997) 🏠 100,000 sq ft
Chairman and President: David E Sandlin
Vice President and Chief Operating Officer: David R Sharp
Chief Financial Officer: Wayne M Richmon
Marketing Administrator: Ms Dru Chittum
■ Commercial contract air services for US Government agencies, foreign Governments and civilian clientele.
Provides under government contracts specialized military readiness training such as electronic warfare simulation, radar surveillance exercises and target towing. 98

80-11630 Flight Management International Corporation
14669 Keeneland Circle, North Potomac, MD 20878
☎(+1 301) 548 90 56 E fmicorp@erols.com
⊖ (+1 301) 840 73 16
♣ $ 240,000 (1996) 👥 4 🏠 1,500 sq ft
President: Steven Ponicarchuk
■ Consulting services to the aviation industry including airports, aviation businesses and government organisations. Problem diagnosis, opportunity specifications, analysis and forecast of decision alternatives, task performance, strategic recommendations, follow-up evaluations, long-range planning and marketing strategies. 98

80-11650 FlightSafety International Inc.
Marine Air Terminal, La Guardia Airport,
Flushing, NY 11371
☎(+1 718) 565 4100 ⊠ 66586 ⊖ (+1 718) 565 4134
♣ $ 271 743 000 (1994) 👥 1 000
President & Chairman: Albert L. Ueltschi
Executive VP & Dir: Bruce N. Whitman
VP Finance & Treasurer: Kenneth Motschwiller
VP Government Affairs: Elmer G. Gleske
VP Marketing: James S. Waugh
Government Contracts & Training Dir: Bill Nugent
Communications Dir: Roger Ritchie
Asst Dir Regulatory Compliance Standards: John Sauers
Maintenance Training Dir: Bonnie Hendrix
National Sales Mgr: Scott Fera
Business Development Mgr: Mike Croitoru
Subsidiaries:
FlightSafety Canada Ltd → Canada 80
FlightSafety Services Corp. → 80
Affiliates:
Centre d'Instruction Falcon AMD-BA, France
Simulation Systems Div.
2700 North Hemlock Circle, Broken Arrow, OK 74012
☎(+1 918) 251 0500 ⊠ 492384 ⊖ (+1 918) 251 5597
👥 250
Chairman & CEO: A.L. Ueltschi
General Mgr: Dennis Gulasy
■ Manufacture training equipment electronics; aircraft simulators, vehicle simulators, cockpit procedures trainers, information displays, automated testing devices, educational training aids, flight & systems trainers.
Instructional Systems Div.
8900 Trinity Blvd, Hurst, TX 76053
☎(+1 817) 595 5450 ⊖ (+1 817) 595 5479
General Mgr: Mike Lee
■ Industrial training programs & audio visual training aids.
FlightSafety Academy
2805 Airport Blvd, PO Box 2708, Vero Beach, FL 32961
☎(+1 407) 567 5178 ✆ (800) 800 1411
⊖ (+1 407) 567 5228
Manager: Peter Banfield
Bethany Learning Center
Wiley Post Airport, 7310 NW 50th St, PO Box 1640,
Bethany, OK 73008
☎(+1 405) 495 6400 ⊖ (+1 405) 495 6404
Manager: Ted Finck
Daleville Learning Center
600 Industrial Blvd, Daleville, AL 36322
☎(+1 205) 598 4485 ⊖ (+1 205) 598 4488
Manager: Britt Hoskins
Houston Learning Center
Wm P.Hobby Airport, 7525 Fauna St, Houston, TX 77061
☎(+1 713) 644 1521 ⊖ (+1 713) 644 2118
Manager: Al Johnson
Alliance Flight Training Academy
2250 Alliance Blvd, Fort Worth, TX 76177
☎(+1 817) 491 9699 ✆ (800) 791 1414
⊖ (+1 817) 491 4002
Manager: John Nashe
Atlanta Learning Center
1804 Hyannis Court, College Park, GA 30337
☎(+1 404) 991 6064 ✆ (800) 889 7616
⊖ (+1 404) 991 5959
Manager: Ken Hesse
Cincinati Airline Learning Center
1600 Dolwick Drive, Erlanger, KY 41018
☎(+1 606) 283 2345 ⊖ (+1 606) 283 2362
Director: Don Gall
Dothan Learning Center
600 FlightSafety Drive, Dothan, AL 36303
☎(+1 334) 983 5652 ⊖ (+1 334) 983 1393
Director: Britt Hoskins
Fort Worth Learning Center
9601 Trinity Blvd, Fort Worth, TX 76053
☎(+1 817) 282 2557 ⊖ (+1 817) 282 8543
Director: John Nasche
Greater Philadelphia
New Castle County Airport, 155 N. Dupont Hwy,
New Castle, DE 19720
☎(+1 302) 328 7548 ⊖ (+1 302) 322 6664
Director: Barry Massey
LaGuardia Airline Learning Center
Marine Air Terminal, LaGuardia Airport,
Flushing, NY 11371
☎(+1 718) 565 4100 ⊖ (+1 718) 565 4174
Director: Terry Hibler
Lakeland Flight Academy
2949 Medulla Road, Lakeland, FL 33811
☎(+1 813) 646 5037 ⊖ (+1 813) 644 6211
Director: Weldon Burnett
Long Beach Learning Center
Long Beach Municipal Airport,
4330 Donald Douglas Drive, Long Beach, CA 90808
☎(+1 310) 420 7733 ✆ (800) 487 7670
⊖ (+1 310) 429 1226
Director: Jim Downey
Miami Airline Learning Center
4800 N.W. 36th Street, Miami, FL 33122
☎(+1 305) 871 8625 ⊖ (+1 305) 871 8659
Director: Berton Beach
St Louis Learning Center
Lambert-St Louis Int'l Airport, 6185 Aviation Drive,
St Louis, MO 63134-0888
☎(+1 314) 731 2040 ⊖ (+1 314) 731 3077
Salt Lake City Airline
Learning Center, 201 North 2200 West,
Salt Lake City, UT 84116
☎(+1 801) 355 3901 ⊖ (+1 801) 355 3801
Director: J.C. Quesadacéspedes
San Antonio Airline
San Antonio International Airport, 9027 Airport Blvd,
San Antonio, TX 78216
☎(+1 210) 826 6358 ✆ (800) 889 7917
⊖ (+1 210) 826 4008
Director: Donnie Nunn
Savannah Learning Center
Travis Field, PO Box 2307, Savannah, GA 31402
☎(+1 912) 964 6421 ✆ (800) 625 9369
⊖ (+1 912) 964 6430
Director: Al Krvsy
Seattle Airline Learning Center
1505 South 192nd Street, Seattle, WA 98148
☎(+1 206) 243 9081 ⊖ (+1 206) 243 0357
Director: Harvey McCarter
Teterboro Learning Center
Teterboro Airport, 100 Moonachie Avenue,
Moonachie, NJ 07074
☎(+1 201) 939 1810 ✆ (800) 827 8058
⊖ (+1 201) 939 7341
Director: Tom Mahoney
Toledo Learning Center
Toledo Express Airport, 11600 West Airport Service Rd,
Swanton, OH 43558
☎(+1 419) 865 0551 ✆ (800) 497 4023
⊖ (+1 419) 865 0754
Director: Jack Ellis
Tucson Learning Center
Tucson International airport, 1071E. Aero Park Blvd,
Tucson, AZ 85706
☎(+1 602) 889 9538 ✆ (800) 203 5627
⊖ (+1 602) 889 9619
Director: Randy Lewis
West Palm Beach Learning Center
3887 Southern Blvd, West Palm Beach, FL 33406
☎(+1 407) 686 7677 ✆ (800) 769 6763
⊖ (+1 407) 689 7719
Director: Pat Knott
Wichita (Beech) Learning Center
9720 E. Central Avenue, Wichita, KS 67206
☎(+1 316) 685 4949 ✆ (800) 488 3747
⊖ (+1 316) 685 2476
Director: Jim Boots
Wichita (Beech) Maintenance
9525 East Central Avenue, Wichita, KS 67206
☎(+1 316) 685 5510 ✆ (800) 808 0976
⊖ (+1 316) 685 2448
Director: Jim Sparks
Wichita (Cessna) Learning Center
1951 Airport Road, Wichita, KS 67209
☎(+1 316) 943 2140 ✆ (800) 227 5656
⊖ (+1 316) 943 1017
Director: Charlie Geer
Wichita (Cessna) Maintenance
Learning Center, 1962 Midfield Road, Wichita, KS 67209
☎(+1 316) 945 0123 ✆ (800) 491 9796
⊖ (+1 316) 945 0161
Director: John Hansen
Wichita (Citation) Learning Center
1851 Airport Rd, Wichita, KS 67209
☎(+1 316) 943 3214 ✆ (800) 488 3214
⊖ (+1 316) 943 7651
Director: Sid Tucker
Wichita (Learjet) Learning Center
2 Learjet Way, Wichita, KS 67209
☎(+1 316) 943 3394 ✆ (800) 491 9807
⊖ (+1 316) 943 0314
Director: Randy Lewis 97

80-11720 Fluor Daniel Inc.
(a subs. of Fluor Corp.)
3333 Michelson Dr., Irvine, CA 92730
☎(+1 202) 955 9316 ⊖ (+1 202) 833 1630
✧ $ 173 026 000 (1995) ♣ $ 9 301 384 000 (1995)
👥 41 678
Managing Dir: Ron G. Peterson
Vice President: Dick Behrman ☎(+1 703) 247 7722
Operations/Technical Dir: R.J. Parker
Sales/Marketing Dir: Richard C. Forrester (PhD)
■ Provides worldwide engineering, construction, procurement, project management, operations & maintenance services to defence, aerospace, & government agencies. Services include project management; feasibility & engineering studies; planning, engineering studies; planning engineering, design; reliability; availability & maintainability (RAM) analysis; environmental studies & design; construction & construction management; procurement; telecommunications design, studies, & project management; operations & maintenance.
Washington Office
800 Connecticut Avenue N.W., Ste 600,
Washington, DC 20006 97

Contact Reference

80-11760 Forecast International/DMS
(† Forecast Associates)
22 Commerce Road, Newtown, CT 06470
☎ (+1 203) 426 08 00 ✆ (800) 451 49 75
E fidms@ibm.net W http://www.forecast1.com
Telex 467615
Fax (+1 203) 426 02 23; 426 19 64; 426 42 62
♣ $ 12,000,000 (1996) 62 (1997) 21,000 sq ft
President: Edward M Nebinger
Sales and Marketing Director: Sheward Hagerty
Editorial/Research Director: Michael Khatana
■ Specialist publisher of research studies & forecasts in aerospace/defence, automotive, transportation & power systems. 98

80-12150 Garber International Associates Inc. (GIA)
1800 Diagonal Rd, Suite 510,
Alexandria, VA 22314-2840
☎ (+1 703) 299 1236 Fax (+1 703) 299 1239
Chairman & CEO: Dr V. Garber
Public Affairs Officer: Ms M. Jurach
■ Defence consulting services across a full range of armament & C3I co-operative programmes including US, European & Pacific Rim efforts. 97

80-12550 General Offshore Corporation
112 South Alfred Street, Alexandria, VA 22314
☎ (+1 703) 684 77 76 Fax (+1 703) 684 89 22
♣ $ 30,000,000 300 (worldwide)
Managing Dir: C. Lundberg
Sales and Marketing Director: J Schweigert
■ Marine engineering, submarine cable installation & burial, beach civil works, underwater range maintenance & installation, vessel operation & maintence. 98

80-12850 Global Engineering Documents
(a division of Information Handling Services)
15 Inverness Way East, Englewood, CO 80150
☎ (+1 303) 792 21 81 ✆ (800) 624 3974
Fax (+1 303) 792 21 92
55
Vice President of Operations: Tom Littman
Worldwide Sales Director: Ms Kate Sackett
■ Worldwide, military, industrial and government specifications, standards, codes, regulations and publications. 98

80-12950 Global Security Complex
5750 Herring Road, PO Box 580, Arvin, CA 93203
☎ (+1 805) 845 70 11 ✆ (800) 417 6542
E global@lightspeed.net Fax (+1 805) 845 79 45
♣ US$ 1,000,000 (1993) 10 20,000 m2
President and Managing Director: Ed R Hoffman
President and Managing Director: Ms Kathleen Hoffman
Operations Director: Frank Bay
Marketing: Ms Kathleen McGinnis Hoffman
■ Global Security Complex: 640 acre training facility for law enforcement, military and other public safety agencies. Facilities include 2,000 sq ft live fire shooting house and 6-story "war tower" building as well as two FAA-approved runways and both rifle and pistol ranges.
Manufacture, master distributor and representative of T.E.A. Communications gear, Hoffman Road Spike Systems, Tasco optical goods, Laser Products flashlights and tactical lights, Pelican Products waterproof cases and lights, Casto batons and other related products. These products are sold to the US Military, US Government agencies as well as law enforcement/publicity safety agencies. Also sell products to international agencies worldwide. → 75 → 85 → 90 98

80-13000 GlomalMart
(a subsidiary of Kennedy Marketing Communications Corp.)
PO Box 667, Long Beach Airport, CA 90801
☎ (+1 562) 498 43 20 Fax (+1 562) 597 22 75
President: Dr T W Kennedy
■ Aviation online services. 98

80-13120 GRC International Inc.
1900 Gallows Road, Vienna, VA 22182
☎ (+1 703) 506 50 00 W http://www.grci.com
Fax (+1 703) 506 92 41
♣ $ 125,000,000 (1996) 1,272 (1997)
President and Chief Executive Officer: James Roth
Executive Vice President and Chief Operating Officer: Gary Denman
Vice President, Treasurer and Chief Financial Officer: Ronald Alexander
Vice President and General Counsel: Thomas E McCabe
Vice President and Director of Contracts and Purchasing: Ms Mary Jo Cippel
Vice President and Director of Administration Services: Ashley Smith
Director of Corporate Communications: Wayne Jackson 98

80-13155 Green River Insurance Company
(a subsidiary of Kawasaki Heavy Industries Ltd, Japan)
84 Pine Street - 600 Financial Plaza, PO Box 5010,
Burlington, VT 05402-5010
☎ (+1 802) 860 04 44 Fax (+1 802) 860 04 40 98

80-13170 Greenwich Air Services Inc.
PO Box 522187, Miami, FL 33152-2187
4590 North West 36th Street, Miami 33122
☎ (+1 305) 526 70 00 ✆ (800) 736 08 00
Fax (+1 305) 526 70 05
US$ 273,595,000 (1996)
♣ US$ 397,913,000 (1996)
3,218 (1996) (including subsidiaries)
Chairman and Chief Executive Officer: Eugene P Conese (Jr)
President and Chief Operating Officer: Eugene P Conese (Jr)
Senior Vice President of Operations: Gerry Waltman
Vice President of Sales and Marketing: William C Madison
Vice President of Purchasing: Scott Campbell
■ Repair, maintenance, modification services for aircraft, jet engines and systems/components. Repair and overhaul of industrial and marine gas turbine engines.

Greenwich Air Services - Texas, LP
(a subsidiary of Greenwich Air Services, Inc.)
9311 Reeves Street, Dallas, TX 75235
☎ (+1 214) 956 50 00 ✆ (800) 956 50 02
W http://www.businesswire.com/cnn/gasia. htm
Fax (+1 214) 956 57 23
Vice President annd General Manager: James W Phillips
Director of Corporate Communications: Gus Whitcomb
■ Repair, overhaul and testing of aircraft engines and associated fuel, air and oil components and accessories. 98

80-13310 G E Capital Spacenet Services, Inc
(a subs. of GTE Corp.)
1700 Old Meadow Road, McLean, VA 22102
☎ (+1 703) 848 1000 Fax (+1 703) 848 0004
President: Gino Picasso
Vice President of Marketing: Andrew Wong
Operations Director: Joe Chisholm
Sales Director: Don Zebe
■ Satellite communications. Telecommunications services and systems to industry, news organisations, educational institutions and government agencies. 98

80-13350 GTT Industries Inc.
5655 Lindero Canyon Rd, Ste 421,
Westlake Village, CA 91362-4016
☎ (+1 818) 707 3162 Fax (+1 818) 707 0079
40
President & CEO: Kamill R. Hilberth
VP Contracts: Barry I. Harwood
Marketing Dir: John K. Scully
■ Systems & software engineering company specialising in automatic test & support system technology. 97

80-13600 Hale Group International
(† Nathan Hale Group)
4094 Majestic Lane, Fairfax, VA 22033
☎ (+1 703) 691 31 87 E halegrp@tidalwave.net
Fax (+1 703) 691 31 87
5 (1997)
President: Dennis Dorn
■ Explosives, propellant and ordnance plant design. 98

80-13910 Heliponents Inc.
1808 West Third Street, Tempe, AZ 85281
☎ (+1 602) 731 95 97 Fax (+1 602) 966 85 08
♣ $ 2,000,000 5
President: Ted Johnson
Vice President: Martin Doss
Sales: Sean Sinclair
■ Bell Approved Customer Service Facility for component overhauls, limited airframe repair and starter generator overhauls. 98

80-13930 Hellier Technical Training & Consulting
(a div. of Hellier)
3710 Artesia Ave, Fullerton, CA 92633
☎ (+1 714) 562 9227 Fax (+1 714) 562 9228
♣ $ 350 000 (1994) 3
Chairman: Charles J. Hellier
President: George V. Crowe
General Mgr: Richard A. Harrison
■ Non destructive testing and quality control training courses. 97

80-13950 H&H Shipping Co. Inc.
120 Sylvan Ave, PO Box 1796,
Englewood Cliffs, NJ 07632
☎ (+1 201) 585 2900 HANDHSHIP Telex 130043 flortex
Fax (+1 201) 585 0501
14 2 000 sq ft
President: H. Heilpern
Sales Marketing Dir: L. Heilpern 97

80-13960 Hi-Rel Laboratories Inc.
6116 North Freeway, Spokane, WA 99207
☎ (+1 509) 325 58 00 Fax (+1 509) 325 95 08
♣ $ 2,900,000 19
President: John R. Davaney
Sales & Marketing Dir: Trevor Devaney
General Mgr: John Level
Purchasing Mgr: Barbara Devaney
■ Destructive physical analysis & failure analysis. 98

80-14340 Holmatro Industrial & Rescue Equipment Inc
(a subsidiary of NV Holmatro, Netherlands)
1110 Benfield Boulevard, Millersville, MD 21108
☎ (+1 410) 987 66 83 Fax (+1 410) 987 16 38
♣ $ 5,000,000 40
President: Giff Swayne
Sales Department: Bill Chalgren
■ Sales and service of hydraulic rescue tools and pneumatic rescue systems in the North and South America and The Caribbean basin. → 90 98

80-14450 Horizons Technology Inc. (HTI)
700 Technology Park Drive, Billerica, MA 01821-4196
☎ (+1 508) 663 66 00 W http://www.horizons.com/
Fax (+1 508) 663 83 57
♣ $ 50,000,000 (1996) 300
Chief Executive Officer: Dr James Palmer
President: Earl Pontius
Business Development Manager: Joe Bisognano
Marketing Coordinator: Ms Beth McGreery
■ Software development, systems integration, the entire range of acquisition disciplines and day to day technical and program management support. Information technology including systems analysis and design, network services, database development, programming, Internet consulting and multimedia implementation.

Boston
700 Technology Park Drive, Billerica, MA 01821-4196
☎ (+1 508) 663 66 00 Fax (+1 508) 663 83 57
Florida
1800 Penn Street, Suite 4B, Melbourne, FL 32901-2625
☎ (+1 407) 728 04 81 Fax (+1 407) 728 04 83
Virginia
1725 Duke Street, Suite 620, Alexandria, VA 22314
☎ (+1 703) 684 58 50 Fax (+1 703) 684 33 80 98

80-14480 Horrigan Analytics
1460 North Sandburg Terrace, Suite 1001,
Chicago, IL 60610
☎ (+1 312) 642 82 72
6
Owner: Timothy J Horrigan
Director of Operations and Communications: Dr Cynthia L Bathurst
■ Mathematical research & analysis directed to understanding the relationship between the behaviour of weapon systems in use in combat & their hardware characteristics. Areas include ground combat, undersea warfare, naval mine warfare & air defence. 98

80-14500 William C. Houze & Co. Headquarters
48-249 Vista De Nopal, La Quinta, CA 92253-9979
☎ (+1 760) 564 64 00 ⊖ (+1 760) 564 84 29
Partner: William C Houze
Partner: Geoffry C Houze
■ Executive search consulting services internationally to companies in aerospace, defence electronics, computers, general aviation, telecommunications, automotives, academe and other related high-tech industries in accordance with the standards of AESC.
UK Affiliate:
Dennis & Gemmill International Ltd → UK 80 [98]

80-14900 C. Raymond Hunt Associates Inc.
69 Long Wharf, Boston, MA 02110
☎ (+1 617) 742 56 69 [E] crhunt@tiac.net
⊖ (+1 617) 742 63 54
👥 10
President: John H. Deknatel
Naval Architect: Winn Willard
■ Naval architects, specialising in the design & engineering of high-speed aluminium, steel or glass fibre reinforced plastic patrol boats. Creator of the Deep-V hull design.
Subsidiary:
H/RI (Hunt/Railton International) [98]

80-15080 IMI Trading USA Inc.
485 Madison Ave, 14th Floor, New York, NY 10022
☎ (+1 212) 836 9528 ⊖ (+1 212) 644 6717
President: Yehuda Porath
Purchasing Mgr: Dan Halevy
Office Mgr: Simcha Aharon
■ Procurement Services. [97]

80-15530 Information Handling Services Inc.
(a division of IHS Group Inc.)
15 Inverness Way East, Englewood, CO 80112
☎ (+1 303) 790 06 00 ext. 009
Ø (800) 716 34 47 ext. 009 (800) 716 64 47 (fax)
[E] info@ihs.com [W] http://www.ihs.com
♣ $ 320,000,000 (1996) 👥 2,000 (worldwide, 1997)
🏢 314,000 m2
President and Chief Operating Officer: Pat Romich
President, Chief Operating Officer and Chief Financial Officer: L Christopher Meyer
President of International Sales Operations: Robert C Wing
Chief Executive Officer: Michael J Timbers
Senior Director Of Marketing: Pat Wimberley
Director of Marketing: Ms Carol Berryman
Marketing, Communication and Event Specialist: Ms Donna Sybert
■ Supplier of value added information services. High indexed information services containing product data, specifications, standards, regulations, codes and other documents used by the technical, engineering and regulatory markets. [98]

80-15870 Intel Corp.
5000 West Chandler Boulevard, Chandler, AZ 85226
☎ (+1 602) 554 80 80 [T] TWX 810 8539 219
⊖ (+1 602) 961 26 71
President: A S Grove
Public Affairs Manager: Ms Shelly Esque
Intel Magnetics Operation
3065 Bowers Avenue, Santa Clara CA 95051 [98]

80-15950 Interactive Development Environments Inc. (IDE)
595 Market St, 10th floor, San Francisco, CA 94105
☎ (+1 415) 543 0900 Ø (800) 888 4331
[T] 256697 ide ur ⊖ (+1 415) 543 0145
♣ $ 25 000 000 (1994) 👥 150 (1995)
CEO: Chris Kenber
Operations/Technology Dir: Peter Pircher
Contact: Ms Laura Fontana
■ Develops & markets Software through a family of integrated software development environments that supports the entire development process. IDE combines Software through Pictures products & its services with other development tools. IDE also provides technical support, training & consulting services. [97]

80-16050 Interdyne Corp.
(† International Trade Operations Inc.)
3775 Vest Mill Road, PO Box 24846, Winston-Salem, NC 27114
☎ (+1 910) 659 95 33 ⊖ (+1 910) 659 07 03
♣ US$ 4,500,000 (1993) 👥 23 🏢 5,300 sq ft
President: J L Hester
Vice President: R M Hester
Secretary and Treasurer: S L Hester
■ Engineering systems integration, worldwide suppliers of turnkey facilities, support and material. Active in Middle East, Russia, Poland and Egypt with US Army Corps of Engineers.
Aircraft and helicopter refueling systems, carts and pump systems. [98]

80-16100 Inter-National Research Institute Inc. (INRI)
12200 Sunrise Valley Dr, Ste 300, Reston, VA 22091
☎ (+1 703) 715 9605 ⊖ (+1 703) 715 9607
♣ $ 34 000 000 👥 335
President & CEO: Dr Frank P. Engel
■ Computer software company specialising in command & control systems for the military, developing the architectural framework of command, control, communications & intelligence systems. [97]

80-16800 IPECO Inc
(a subsidiary of Ipeco Aerospace Ltd)
3882 Del Amo Boulevard Suite 604, Torrance, CA 90503
☎ (+1 310) 371 44 77 ⊖ (+1 310) 371 56 60
👥 9
President: Chris Johnson
Vice President and General Manager: Terry Glover
■ Marketing and product support for pilot and aircrew seats. [98]

80-17080
Jane's Information Group US Operation (JIG)
(a member of Jane's Information Group, UK)

Jane's INFORMATION GROUP

1340 Braddock Place, Suite 300,
Alexandria, VA 22314-1651
☎ (+1 703) 683 37 00
[W] http://www.janes.com/janes.html
⊖ (+1 703) 836 00 29
General Manager: Joe Bria
Senior Vice President, Product Sales: Joe McHale
Vice President, Business Development: Ms Sally Velthaus [E] velthaus@janes.com
Director, Advertising, Sales and Marketing: Ms Kim Hanson
Marketing Director: Ms Deborah Chiao
Controller: Ms Anju Chopra
West Coast Office
17310 Red Hill Avenue, Suite 370, Irvine, CA 92714
☎ (+1 714) 724 08 68 ⊖ (+1 714) 724 15 76
Manager, Product Sales Western Region: Robert Petty [98]

80-17300 JWK International Corp.
7617 Little River Turnpike, Annandale, VA 22003
☎ (+1 703) 750 0500 ⊖ (+1 703) 750 1986
Chairman & CEO: Jay W. Khim
President & COO: Harrison Young (II)
■ Specializes in both defense and non-defense related services in the areas of training and education, engineering services, logistics engineering, systems integration, software engineering and facilites management. [97]

80-17500 Kaman Electromagnetics Corp.
(a subs. of Kaman Corp.)
2 Fox Road, Hudson, MA 01749
☎ (+1 508) 562 2933 ⊖ (+1 508) 562 6830
President: Dr G.L. Roderick
Marketing Mgr: P.D. Tolman
■ Development of advance technology electromagnetic devices, launchers, high power density alternators, generators & motor & power electronics. Vehicle & ship electric drive systems. [97]

80-17600 Kaman Sciences Corp.
(a subsidiary of Kaman Corporation)
1500 Garden of the Gods Road, PO Box 7463, Colorado Springs, CO 80933-7463
☎ (+1 719) 599 15 00 [W] http://www.ksc.com
⊖ (+1 719) 599 19 42
♣ $ 125,000,000 (1996) 👥 1,250 (1996) 🏢 50,000 m2
President: Dr J Lamar Allen
Vice President of Plans and Marketing: Robert L Harter
Vice President, Operations Director: Dr Charles Eklund
Purchasing Manager: Barry Houtz
■ Serves government and industry customers with analysis, testing, system engineering and technology integration. Includes C3I analysis and systems engineering, software overhaul and maintenance, simulation and modeling, effects analysis of nuclear directed and kinetic energy weapons. [98]

80-17630 Kapos Associates Inc.
1101 Wilson Blvd, Ste 1900, Arlington, VA 22209-2248
☎ (+1 703) 528 4575 ⊖ (+1 703) 276 1264
👥 25
President: Ervin Kapos
Vice President: Ms Cordellia Scruggs
■ Operations research & systems analysis, including mission analysis, requirements studies, effectiveness assessments, test & evaluation, program analysis & support - particularly in maritime application areas. [97]

80-17655 Kawasaki Motors Finance Corporation
(a subsidiary of Kawasaki Heavy Industries Ltd, Japan)
9950 Jeronimo Road, Irvine, CA 92718-2016
☎ (+1 714) 770 04 00 ⊖ (+1 714) 460 56 28 [98]

80-17660 Kay & Associates Inc.
3820 Ventura, Arlington Heights, IL 60004
☎ (+1 847) 255 84 44 Ø (800) 323 41 81
[E] rainbd@kayinc.com [T] 9106870764
⊖ (+1 847) 255 68 18
👥 1,000 (1997)
President: Gregory Kay
Vice President Operations: Bradley J Kay
Vice President Finance: Ms Dianna K Chinn
Vice President Business Development: Lowell M McMann
Vice President Staff Services: George F Wrobell
■ Organizational & intermediate aircraft maintenance incl. airframes, engines, aircraft weapons systems & ground support; metrology services; instruction & training. Engineering consultants. [98]

80-17680 Kennedy Marketing Communications Corp.
World Trade Centre Towers, PO Box 667, Long Beach, CA 90801
☎ (+1 310) 599 0900 ⊖ (+1 310) 597 2275
President: Dr Ted W. Kennedy
■ Computer hardware & software, development for aerospace & military applications. International trade & strategic metals marketing.
Divisions:
Technology Transfer
Kennedy Films
GlobalMart [97]

80-17780 Keystone Helicopter Corp.
1420 Phoenixville Pike, West Chester, PA 19380
☎ (+1 610) 644 44 30 Ø (800) 633 15 55
[E] sales@keystone-helicopter.com
[W] http://www.keystone-helicopter.com
⊖ (+1 610) 644 76 81
♣ $ 25,000,000 (1995) 👥 195 🏢 35,600 sq ft
President: Peter Wright (Jr)
Vice President Airframe Maintenance: Dave Ford
Vice President Sales and Marketing: Rick Hinkle
Vice President Engine Services: John Loney
Customer Service Manager: Tim Wright

Contact Reference

■ Authorised by major airframe & engine manufactures. Full completion house: FAA certified interiors; contract operator for turnkey EMS flight management.
Division:
Keystone Helicopter Services
Keystone Flight Services
Keystone Engine Services
KeysTech [98]

80-17900 Korea Express USA Inc.
(a subs. of The Korea Express Co. Ltd)
901 Castle Rd, Secaucus, NJ 07094
☎(+1 201) 863 7505; (+1 212) 432 1890
⊖ (+1 201) 863 5036
♧ $ 250 000 ♣ $ 9 556 000 ⁂ 70 ⌂ 105 000 sq ft
President: Sang Kyu Kim
Vice President: John J. Fay
FMS Manager: D S Leigh
Director: Eock Soo Oh
■ United Airfreight Services located in New York, Carson, California, & San Francisco is the Airfreight Division of Korea Express USA Inc.
● 709 E. Walnut St, Carson, CA 90745
☎(+1 310) 532 3333 ⊠ 854731 ⊖ (+1 310) 532 9350
Managing Director: J K Lee
General Manager: H C Lee
FMS Manager: David Chang
● 129 Sylvester Road, So.San Francisco, CA 94080
☎(+1 415) 583 3440 ⊠ 5101004103 unafsfo
⊖ (+1 415) 583 3440
General Mgr: J.S. Kim [97]

80-18100 Labinal Aero Defense Systems Inc
(a subsidiary of Labinal, France)
2709 Forum Drive, Grand Prairie, TX 75051-7027
☎(+1 214) 606 98 83 ⊖ (+1 214) 606 88 74
⁂ 60
Marketing Manager: Chris Christensen
■ Provides North American marketing, customer service and product support for the LABINAL group of companies (Acro Systems Division) The Grand Prairie (Dallas), Texas facility is an FAA approved repair station for products manufactured by:
LABINAL-DSA: Wiring harnesses, tyre pressure/brake temperature monitoring systems, actuators, starter-generators and lighting systems.
TECHNOFAN: Airborne blowers, ECS cooling fans, brake cooling fans, air valves and integrated environmental control systems.
SOFRANCE: Fluid filtration products.
® France: Labinal, Sofrance, Technofan [98]

80-18330 LME
(† Logistics Management Engineering Inc.)
201 Defense Highway, PO Box 6637,
Annapolis, MD 21401
☎(+1 410) 573 11 25 Ⓔ lmeinc@aol.com
⊖ (+1 301) 261 82 58
♣ US$ 6,400,000 (1997) ⁂ 80 (1997)
President and Chief Operating Officer: Stuart F Hall
Sales and Marketing Director: James R Maitland
■ Provide logistics, engineering, technical support services in all phases of the acquisition process: research, development, testing, production and in-service support. Engineering and technical services, include program planning, studies and analysis, design assurance, test and evaluation, data documentation and field engineering. Largest product line: Aviation Life Support Equipment (ALSE). Also provide complete test and evaluation services including design and fabrication of test vehicles. [98]

80-18390 Lockheed Martin Aircraft Center (LMAC)
(a subsidiary of Lockheed Martin Corp.)
244 Terminal Road, Greenville, SC 29605
☎(+1 864) 299 76 38 ⊖ (+1 864) 299 76 29
⁂ 1,200 (1996) ⌂ 300,000 sq m
President: G E Albers ☎(+1 803) 299 71 20
Vice President and General Manager: Rick Ribich
☎(+1 803) 299 72 01
Director of New Business Development: Kennon D Hines
■ Aircraft maintenance and modification services for military, government and commercial customers. [98]

80-18500 Lockheed Martin Space Mission Systems and Services (LESC)
(† Lockheed Engineering and Sciences Co.)
(a subsidiary of Lockheed Martin Corp.)
2625 Bay Area Boulevard, Houston, TX 77058-8999
☎(+1 281) 283 4400 Ⓦ http://www.lmsg.lmco.com
⊖ (+1 281) 283 46 60
⁂ 5,600 (1997)
President and Corporate Vice President: Clint Denny
Executive Vice President: J Honeycutt
Public Relations Director: Ms Wendy Morrison Owen
Operations Director: Mike Cyrus
Sales and Marketing Director: Ken Asbury
■ Engineering and scientific support services; management of government technical facilities; spacecraft avionics; test range instrumentation; computer sciences support and management. [98]

80-18530 Lockheed Martin Aircraft and Logistics Centre
(a subsidiary of Lockheed Martin Corporation)
107 Frederick Street, Greenville, SC 29607-6594
☎(+1 864) 422 63 03 ⊖ (+1 817) 548 23 19
President: Daniel Patterson
Communications Manager: David Jewell
■ Technical, administration, training and management assistance and services on a worldwide basis. Maintenance, repair, modification and logistic support of air and land, electronic, mechanical, electromechanical, avionic, hydraulic systems, equipment and devices. [98]

80-18540 Lockheed Martin Services Group
(a subsidiary of Lockheed Martin Corp.)
2339 Route 70 West, Cherry Hill, NJ 08358-0001
☎(+1 609) 486 50 04 ⊖ (+1 609) 486 58 05
⁂ 11,000
President: Michael F Camardo
Vice President of Finance: Jeffrey A Cook
Vice President of Human Resources: Kenneth J Disken
Vice President and General Counsel: Neal J Murray
Vice President of System Support Services: David J Posek
Vice President of Business Development: Kenneth S Morton
Vice President of Plant Combat System Services: Stanley J Pryzby
Vice President of Information Support Services: Ms Linda R Renfro
Vice President of Computing and Communication Services: Thomas B Shafer
Vice president of Engineering and Science Services: James C Adamson
Public Affairs Manager: Ms Wendy Morrison
■ Engineering and technical services in the following areas: scientific and engineering; design and systems engineering and integration; information management, software design and development, computer and supercomputing centre operations; operation and maintenance services including installation and modification of communications, command control and instrumentation facilities; logistics support, ILS planning, life-cycle support, documentation and training; assembly, test and checkout services, built-to-print and other production support for electronic test, instrumentation and weapon systems and subsystems. [98]

80-18612 Lockridge Mapstrap Co.
12525 Overbrook Dr, Reno, NV 89511
☎(+1 702) 853 6001 ⊖ (+1 702) 853 6009
Owner: James E. Lockridge
■ R&D consulting; cockpit design, manufacture special lightweight flexible kneeboard devices for fighter aircrews. → 75 [97]

80-18618 Logica Inc
(a subsidiary of Logica plc, UK)
32 Hartwell Avenue, Lexington, MA 02173
☎(+1 617) 476 80 00 ⊖ (+1 617) 476 80 10
New York Office
5 Penn Plaza - 21st Floor, New York, NY 10001
☎(+1 212) 629 34 56 ⊖ (+1 212) 631 02 69
Fort Lauderdale Office
800 Corporate Drive - Suite 600,
Fort Lauderdale, FL 33334
☎(+1 305) 938 82 88 ⊖ (+1 305) 772 84 75
San Francisco Office
353 Sacramento Street - Suite 2300,
San Francisco, CA 94111-3650
☎(+1 415) 288 52 00 ⊖ (+1 415) 288 52 99
Los Angeles Office
1801 Avenue of the Stars, Suite 1415,
Los Angeles, CA 90067-5808
☎(+1 310) 551 06 60 ⊖ (+1 310) 785 05 05
Williamsburg Office
229 Mclaws Circle, Williamsburg, VA 23185
☎(+1 1804) 229 44 33 ⊖ (+1 1804) 229 88 92
Synercom
2500 City West Boulevard, Suite 1100,
Houston, Texas 77042
☎(+1 713) 954 70 00 ⊖ (+1 713) 785 08 80 [98]

80-18625 Logicon Geodynamics Inc.
(a subsidiary of Logicon, Inc., USA)
6053 West Century Boulevard, PO Box 881239,
Los Angeles, CA 90009-1239
☎(+1 310) 384 11 28 ⊖ (+1 310) 645 00 70
♣ US$ 46,000,000 (1997) ⁂ 370 (1997) ⌂ 90,000 sq ft
President and Chief Executive Officer: Robert L Paulson
☎(+1 310) 348 11 28
Vice President of Advanced Programs: Paul J Henrikson
Sales and Marketing Director: John Clarry
■ Provides information engineering services and products for government, commercial and international information systems. Specializes in custom end-to-end systems requirements definition, engineering, software development, comercial-off-the- shelf (COTS) hardware and software integration, test, training and maintenance for its customer base. [98]

80-18800 Logistics Services International Inc.
6200 Lake Gray Boulevard, Jacksonville, FL 32244-5826
☎(+1 904) 771 21 00 ✆ (800) 759 09 53
⊠ TWX 810 9349 131 ⊖ (+1 904) 771 19 13
♣ $ 21,000,000 ⁂ 285 ⌂ 48,600 sq ft
President: M J McKenny
Chief Executive Officer: J A McKinney
Chief Financial Officer: W S Rosander
Senior Vice President of Operations: C H Frey
Director of Administration: R J Boatright
Ⓔ rboatright@lsijax.com
■ Provides operational and maintenance training, including computer based interactive courseware and logistic support on numerous out-of-production and current US weapon systems worldwide. [98]

80-18900 Lucas Aerospace Customer Support Americas
30 Van Nostrand Ave, Englewood, NJ 07631
☎(+1 201) 567 6400 ⊠ 135374 lucasaero egw
⊖ (+1 201) 894 1965
VP & General Mgr: Glenn Brown
Customer Service Mgr: Ms Janet Milone
Human Resources Mgr: Peter Rees
Finance Mgr: Ms Angela Elmo
Business Mgr: John Heath
Commercial Mgr: Glenn Pedrick
Sr Field Service Technical Rep.: Acmar Singh
Sr Field Service Technical Rep: Oscar Ramsingh
■ FAA-approved facilities for overhaul & repair, complete service support. FAA Repair Station No. 1X1R234K.
● c/o American A'lines,3800 N.Mingo Rd,
PO Box 582809, Mail Drop 44, Tulsa, OK 74158-2809
☎(+1 918) 836 1106 ⊖ (+1 918) 832 8664
Sr Field Service Technical Rep.: Pete Polczynski
■ Technical support. [97]

80-19200 Management Analysis Inc. (MAI)
8150 Leesburg Pike, Suite 1100, Vienna, VA 22182
☎(+1 703) 506 05 05 ⊖ (+1 703) 506 14 36
♧ US$ 844,259 ♣ US$ 3,865,038 (1997) ⁂ 49 (1997)
⌂ 13,400 sq ft
Chairman and Chief Executive Officer: T Arthur Smith
President: Arthur L Smith
Advertising Manager: Ms Margaret Reynolds
■ Cost effective studies on electronic, armoured and helicopter systems. Extensive cost data bases and cost models on military systems. Productivity measurement and work standard development studies for military and manufacturing clients. Economic analysis of major defence logistics systems. Force cost modelling. Activity based costing. Pvitatisation studies. [98]

80-19680 Management Consulting & Research Inc. (MCR)
5165 Farm Credit Drive, Ste 300, McLean, VA 22102

☎(+1 703) 506 4600 ⊜ (+1 703) 506 4621
👥 100
President & CEO: Dr Gerald R. McNichols
VP Business Operations: Wm. Michael Bryan
VP Programme Operations: William Palmer
VP Programme Operations: Robert C. Owens
VP Programme Operations: R. Qivera
VP Programme Operations: Neil Albert
Corporate Business Development Mgr: Donald Meyer
Purchasing Mgr: Sandie Meek
Assistant to CEO & President: Ms Pat Wells
■ Consulting services in areas of management science, weapon system schedule analysis, acquisition & logistics research, risk & uncertainty analysis, cost & economic analysis, cost growth studies, financial/ budget analysis, design-to-cost/life cycle cost investigations, cost trade-off analyses & cost, operational effectiveness analyses. Develops cost models, cost databases, software/hardware cost estimates, automated resource management systems & database management systems. [97]

80-19780 M International Export (MINTEX)
1301 Dolley Madison Blvd (Rte 123), McLean, VA 22101
☎(+1 703) 448 44 00 [E] mintex@internetmci.com
Telex 64112 mintex ⊜ (+1 703) 448 44 08
👥 29
President: Richard D McLonn
Vice President of Corporate Development and Technology: Jan P Rapalski
Director of Finance and General Manager: Mehdi Protzuk
Sales and Marketing Manager: Alan Vordell

McTurbine Inc.
1902 Holly Road, Corpus Christi, TX 78417
☎(+1 512) 851 12 90 [E] mcturbine@intcomm.net
⊜ (+1 512) 851 93 22
General Manager: Jerry Wade [98]

80-20050 Maritime Contractors Inc.
201 Harris Ave, Bellingham, WA 98225-7018
☎(+1 360) 647 0080
👥 165
President: Romy V. Mendoza
CEO: Dale White
Projects & PR Dir: George Kolakowski
■ Drydocking, repair & modification to all classes of vessels. [97]

80-20300 Masco Service Corp.
1200 North Kimball Avenue, Southlake, TX 76092
☎(+1 817) 481 82 05 ⊜ (+1 817) 488 89 18
👥 40 (1997) 🏠 3,000 (1997)
President: Ronald B Massey
■ Repair, overhaul, exchange and sale of helicopter engine and flight instruments; specialising in all turbine-powered Bell aircraft, Sikorsky S-76, Eurocopter BO105 and BK117.
● 109 Pine Park Drive, Lafayette, LA 70508
☎(+1 318) 237 32 77 ⊜ (+1 318) 233 35 18
Branch Manager: Jerry P Martinez [98]

80-20350 MATHSOFT
101 Main Street, Cambridge, MA 02142
☎(+1 617) 577 10 17 [W] http://www.mathsoft.com
⊜ (+1 617) 577 88 29
👥 135
Chief Executive Officer: Charles Digales
Vice President Sales and Marketing: Christian Randles
Marketing Communications Manager: Ms Ellen Koup [98]

80-20500 McCallie Associates Inc. (MAI)
1501 J.F. Kennedy Drive, Bellevue, NE 68005
☎(+1 402) 291 22 03 ⊜ (+1 402) 291 82 21
♣ $ 3,500,000 (1996) 👥 60 (1996)
President: Bob McCallie
Vice President Federal Services: Ms Jennifer Maassen [E] jmaassen@mccallie.com
■ System engineering & integration; software & hardware engineering; database; graphics, systems & application programming; network management; quality assurance. [98]

80-20800 John J. McMullen Associates Inc.
(a subsidiary of Talley Industries)
Two World Trade Center, Suite 1510,
New York, NY 10048
☎(+1 212) 466 2200 ⊜ (+1 212) 466 2282
👥 900 (USA)
President & CEO: P. Thomas Diamant
VP Marketing: D. McMullen
■ Naval architecture & marine engineering for all types of naval & commercial ships. [98]

80-20870 Mercury Aviation
Reno Jet Center
(a subsidiary of Mercury Air Group Inc.)
655 South Rock Boulevard, Reno, NV 89502
☎(+1 702) 858 73 00 ✆ (800) 223 8433
✉ http://www.mercuryaircentres.com
⊜ (+1 702) 858 73 34
♣ US$ 6,500,000 (1996) 👥 205 🏠 30,000 m2
Managing Director: Dan McDyre
Sales and Marketing Director: Shelly Doyle
Public Relations Manager: Wesley Daniels
Purchasing Manager: Dan McDyre
■ DoD military fuel contract. [98]

80-20950 Metal Improvement Company Inc.
(a subs. of Curtiss-Wright Corporation)
10 Forest Avenue, Paramus, NJ 07652
☎(+1 201) 843 78 00 ⊜ (+1 201) 843 34 60
👥 1,100
President: Gerald Nachman
Sr VP Marketing: James Daly
■ Shot peening and peen forming. [98]

80-21080 Micro Data Base Systems Inc.
1305 Cumberland Avenue, PO Box 2438,
West Lafayette, IN 47906
☎(+1 765) 463 72 00 ✆ (800) 445 63 27
[E] info@mdbs.com [W] http://www.mdbs.com
⊜ (+1 765) 463 12 34
👥 60 (1997)
President: Kevin Castleberry
Sales and Marketing Director: Ms Beth Decker-Colnitis
Advertising Manager: Ms Ele Scott
Purchasing Manager: Dave Cicardo
■ Titanium - High performance network model database management system.
KnowledgeMan - Integrated development environment with 4GL and RDBMS.
GURU - Integrated expert system development environment with 4GL and RDBMS. Object/1 - Object-oriented development environment for GUI applications. [98]

80-21320 MJI Associates Inc.
PO Box 40, Centerport, NY 11721
☎(+1 516) 261 21 41 [E] hq@mjiteam.com
[W] http://www.mjiteam.com ⊜ (+1 516) 261 21 49
👥 25
President: Michael Ianelli
Executive Vice President: Klive S Caiola
Sales & Marketing Dir: Anthony Pacilio
■ International management, business development and training consulting group which assists companies in planning, developing and managing major programme & proposal efforts within defence & aerospace. [98]

80-21360 ST Mobile Aerospace Engineering Inc.
(† Mobile Aerospace Engineering Inc.)
(a subs. of Singapore Technology Ltd)
2100 Ninth St, Brookley Complex, Mobile, AL 36615
☎(+1 334) 438 8888 ⊜ (+1 334) 438 8892
👥 800 🏠 31 000 sq m
President & CEO: Bob L.H. Tan
Sales & Marketing Dir: Stephen M. Lim
Operations/Technical Dir: Ronnie L.H. Koh
Purchasing Mgr: Barry Karr
■ Maintenance & modification of commercial narrow- and wide-bodied aircraft. Design engineering & in-house DERs. [97]

80-21380 MuSE Technologies Inc.
1601 Randolph SE, Suite 210, Albuquerque, NM 87106
☎(+1 505) 843 6873 ⊜ (+1 505) 766 9123
VP Marketing & Sales, Govt. Systems Dir: Arlan Andrews [E] aandrews@musetech.com
■ Synthetic environment software used in synthetic environments, simulation, virtual reality & human-computer interfaces. [97]

80-21400 National Airmotive Corp.
(a subsidiary of National Airmotive Corporation)
2508 Palm Drive, Long Beach, CA 90806
☎(+1 310) 981 27 23 ✆ (800) 468 48 36
⊜ (+1 310) 595 17 35
♣ $ 12,000,000 👥 35 🏠 30,000 sq m
Vice President of Marketing and Sales: Joe Ghanthos
Director of Operations: Guy M Davies
■ Turbine engine service and maintenance. [98]

80-21650 National Systems Management Corp. (NSM)
4600-H Pinecrest Office Park Drive, Alexandria VA 22312
☎(+1 703) 941 90 21 [E] busdevlo@nsm.com
⊜ (+1 703) 256 18 11
♧ $ 2,000,000 (1996) ♣ $ 7,500,000 (1996)
👥 150 (1997)
Managing Director: Ms Sandra H Raftery
Director of Business Deveolpment: Bobby G Christy
Operations Director: Chris Lamb
Purchasing Manager: Ms Cindy Fritter
■ Provides systems engineering, integrated logistics, production management, acquisition documentation and management information support to the military services. NSM's capabilities cover the entire life span of weapon systems from concept exploration through engineering development, production, deployment. In the Foreign Military Sales (FMS) arena. NSM provides case management support from inception to closeout for aircraft, weaponry and ancillary equipment. [98]

80-21730 New Helvetia Trade Group
2215 J Street, Sacramento, CA 95816
☎(+1 916) 446 27 37 ⊜ (+1 916) 446 07 22
President: R Stilwell
Managing Director: E L Faust [98]

80-21750 Newport Aeronautical Sales (NAS)
1524 Monrovia Avenue, Newport Beach, CA 92663-2807
☎(+1 714) 574 41 00 [E] newportaero@primenet.com
[W] http://www.newportaero.com ⊜ (+1 714) 454 82 88
President: George M Posey
Office Manager: Ms Betty Foster
■ Suppliers of military and commercial technical manuals (operation, maintenance, illustrated parts catalogues) and other government publications. [98]

80-21780 Nichols Research Corp. (NRC)
4040 South Memorial Parkway, PO Box 400002,
Huntsville, AL 35815-1502
☎(+1 205) 883 11 40 ✆ (800) 477 11 70
[E] info@nichols.com [W] http://www.nichols.com
⊜ (+1 205) 880 03 67
♧ $ 113,000,000 (1996) ♣ $ 242,000,000 (1996)
👥 2,000 (1997) 🏠 43,200 m2
Chairman and Chief Executive Officer: Chris H Horgen
President and Chief Operating Officer: Michael J Mruz
Chief Financial Officer: Allen E Dillard
Sales and Marketing Director: James M Coward
Director of Business Development: Frank M "Pete" Gray
Purchasing Manager: Ms Laura Glynn
■ Primary service areas, information systems development and delivery of turnkey information systems including software development. Computer system integration-architecture and delivery of integrated computer systems. Information systems support engineering and analytical support, modification, or development of information systems.
Missile and tactical systems engineering and technologies engineering, analysis and testing in support of tactical, TMD, NMD, air defence, command and control, foreign weapon systems and subsystems.
Sensor, surveillance and avionics engineering and analysis. Design of sensors and avionics for weapons, space surveillance and target acquisition systems and sunsystems including distributed interactive simulations (DIS) and virtual prototyping. [98]

80-21800 Norsam Technologies
61 Rover Boulevard, Los Alamos, NM 87544
☎(+1 505) 672 09 20 ✆ (505) 672 09 21
[E] jane@norsam.com [W] http://www.norsam.com
⊜ (+1 505) 672 09 22
👥 15 🏠 4,500 sq ft
President: John Bishop
Chief Operating Officer: Homi Fatemi
Vice President of Operations: Alex Jourabchi
Sales Director: Ms Elizabeth Rohrbach

Contact Reference

Senior Software Engineer: Mark Eberle
Co-Inventor: Roger Stutz
Staff Scientist: Dr Andrew Bartlett
Public Relations: Steve Suckman
■ Micro-machining services for security applications. Ion beam machines used for IC device modifications and mini-scribing identification tags and serial numbers on equipment. [98]

80-21840 Northrop Grumman Corporation (ESSD)
(† Westinghouse Electronic Systems International Co.)
Electronic Sensors and Systems Division
PO Box 451, A275, Baltimore, MD 21203
☎ (+1 410) 765 27 00 [W] http://www.essd.com
[W] http://www.northropgrum.com ⊖ (+1 410) 765 58 77
👥 7,000+
President, Electronic Sensors and Systems Integration: C Lloyd Carpenter ☎ (+1 410) 765 49 52
■ Aerostat radars, airborne fire control radars, air traffic control radars, worldwide air defence systems, command and control systems, electronic countermeasures, electronic support centres, integrated logistics support, shipboard radar systems and tactical surveillance radars. [98]

80-22000 Nottingham International
(† Sherwood International Export Corp.)
325 North Beverly Drive, Beverly Hills, CA 90210
☎ (+1 310) 274 67 00 ⊖ (+1 310) 274 15 18
♧ US$ 18,000,000 👥 30 🏭 35,000 sq ft
President: Michael D Kokin
■ Provides design, development, retrofit, spare parts, support, training, licencing, conversion, modernisation and dieselisation to assist governments throughout the world with their ordnance requirements. [98]

80-22372 O'Neil & Associates Inc.
425 N. Findlay St, Dayton, OH 45404-2203
☎ (+1 513) 461 1852 ⊖ (+1 513) 228 0135
♧ $ 4 000 000 (1995) ♣ $ 8 500 000 (1995)
👥 140 (1995) 🏭 60 000 sq ft
President: John E. Staten
Marketing Dir: Thomas R. Milligan (CPL) ☎ (+1 513) 461 1602 ext. 3900
Operations/Technical Dir: Robert J. Heilman
Purchasing Mgr: John Schomberg
■ Technical manuals, training materials, graphic arts, provisioning data, & logistics support analysis (LSA/LSAR) services to DoD contractors & directly to military agencies. Electronic publishing & CD-ROM data conversion. [97]

80-22600 Orman Associates Inc.
(† General Technology Systems Inc.)
Suite 350, 15200 Shady Grove Road, Rockville, MD 20850
☎ (+1 301) 670 28 17 [E] or2withdog@aol.com
⊖ (+1 301) 840 39 01
Chief Executive Officer: Dr Stanley Orman ☎ (+1 301) 670 28 17
Vice President: Maj Gen Eugene Fox (retd) ☎ (+1 703) 734 18 11
Associate: Lieut Gen R D Hammond (retd) ☎ (+1 703) 425 66 43
Associate: Col C F Stewart (retd) ☎ (+1 703) 323 86 43
■ Consultancy, market surveys and strategic studies, strategic and theatre defence, national and international alliances and joint ventures in high technology industries. [98]

80-23060 Pacific Propeller Inc. (PPI)
5802 South 228th Street, PO Box 1187, Kent, WA 98035-1187
☎ (+1 253) 872 77 67 ✆ (800) 722 77 67 ☒ 320368
⊖ (+1 253) 872 72 21
👥 60 🏭 60,000 sq ft
President: Jim Matheson
Quality Assurance Manager: Mike Thornton
■ Overhaul/repair & parts support for all aircraft propeller systems; specialize in larger propeller such as used on C-130, L-188, CV580 & commuters such as DHC-7/8, ATR 42/72 and EMB 120. [98]

80-23080 Pacific Ship Repair & Fabrication
1625 Rigel Street, PO Box 13428, San Diego, CA 92113-3808
☎ (+1 619) 232 32 00 [E] pacship@pacship.com
⊖ (+1 619) 232 20 70
♣ $ 38,000,000 (1996) 👥 450 🏭 215,000 m2
President and Chief Executive Officer: David L Bain
Chief Financial Officer: David Moore
Vice President: Gary Thomas
Operations Director: Mike South
Sales and Marketing Director: Hugh Webster
Purchasing Manager: Charlie Jenkins
■ Ship repair. [98]

80-23700 Parsons Power Group Inc.
(† Parsons Main Inc.)
(a subsidiary of the Parsons Corp.)
Prudential Center, Boston, MA 02199
☎ (+1 617) 262 32 00 ☒ 4430035, 940598
👥 500
Operations Director: Ms Deborah Gustafson
Sales and Marketing Director: Ronald Cochran
■ Study, architectural, design, engineering, procurement, construction & project management services for hydroelectric, thermal & other types of power generation; power transmission & distribution systems; water resource projects; environmental & hazardous materials projects. [98]

80-23750 Parsons Engineering Science
(a subsidiary of Parsons Corporation)
100 West Walnut Street, Pasadena, CA 91124
☎ (+1 818) 440 60 00 ✉ ENGINSCI ARIA
[W] http://www.parsons.com ⊖ (+1 818) 440 61 95
👥 1,905
President: David Bernstein
Sales and Marketing Director: James E Thrash
Southern Operations Manager: David Burstein
Northern Operations Manager: W Gary Christopher
Western Operations manager: Allan L Udin
Corporate Relations: Ms Patricia Gallemore
■ Environmental engineering services including water and wastewater treatment, hazardous and industrial waste management, marine waste disposal, solid waste management and air pollution control. [98]

80-23780 Pasha Publications Inc.
1616 North Fort Myer Dr, Suite 1000, Arlington, VA 22209-3107
1616 North Fort Meyer Drive, Suite 1000
☎ (+1 703) 528 12 44 ✆ (800) 424 29 08
[W] http://www.pasha.com ☒ 248852 pasur
⊖ (+1 703) 528 12 53
Publisher: Tod Sedgwick
Publisher Gas Division: Ms Joan Henderson
Publisher Defence Division: Harry Baiden
Publisher Religion Division: Ms Kathy Thorne
Operations Director: Ms Sheila Manes
Marketing Defence Division: Ms Clair Hill
Marketing Environmental Division: Ms Mary Anvari
Marketing Gas Division: Ms Delores Baisden
Marketing Energy Division: Ms Julie Johnson
Marketing: Shawn Simmons
■ Publisher of reports, newsletters and magazines for the defence, environmental, gas and energy sectors. [98]

80-24250 Peterson Builders Inc. (PBI)
41 North Third Avenue, Sturgeon Bay, WI 54235
☎ (+1 414) 743 55 74 [E] pinc@mail.wiscnet.net
[W] http://www.pbinc.com ⊖ (+1 414) 743 60 89
👥 150
Chairman of the Board: E L Peterson
President: James M Stawicki
Executive Secretary: Ms Carol Russell
■ Ship repair & logistic support. [98]

80-24300 Piper Pacific International
2020 14th Street North, Suite 701, Arlington, VA 22201-2515
☎ (+1 703) 524 75 56 ✉ piperpacif@aol.com
⊖ (+1 703) 524 75 59
👥 5 (1997)
President: Dr W Stephen Piper
Vice President: Ms Wara Madzimmugo
Vice President: Ms Naomi Wiegler
■ Programme consultants on international high-technology and defence related technology co-operation and transfer with primary focus on the Pacific Basin. [98]

80-24500 PLS Ballistic Service
200 Finney Avenue, Churchville, MD 21028
☎ (+1 410) 734 7890 ⊖ (+1 410) 734 4166
♣ $ 50 000 👥 3
Consultant Exterior Ballistics: C.H.L. Lebegern
■ Ballistics consultancy service covering all aspects from gun to target.Wound ballistics. NATO standards to safe and suitable for service status. Qualification testing to NATO 4224.

Aberdeen Plant
325 Rogers Street, Aberdeen, MD 21001
☎ (+1 410) 273 7507
Consultant: J. Sperrazza [98]

80-24680 Praxair Surface Technologies Inc.
1500 Polco St, PO Box 24184, Indianapolis, IN 46224
☎ (+1 317) 240 2500 ✆ (800) 822 72 84 ☒ 27413
⊖ (+1 317) 240 2380
Business Mgr: John E. Land
Marketing Dir: Bruce Ryman
Advertising Mgr: Martin Holdsworth
■ Metal & ceramic coatings that provide abrasion resistance, wear resistance, thermal barrier & sulfidation resistance. [97]

80-24700 PRB Associates Inc.
43865 Airport View Drive, Hollywood, MD 20636-9760
☎ (+1 301) 373 23 60 [W] http://www.prb.net
⊖ (+1 301) 373 34 21
♣ US$ 27,200,000 (1996) 👥 260 (1996) 🏭 75,500 sq ft
President and Chief Executive Officer: Lawrence M Schadegg
General Manager: R A Bos
Vice President of Business Development: Matthew G Duranske
Administration Assistant: Ms Christine M Magee
■ Software and systems engineering for electronic warfare, anti-submarine warfare and C3I applications, focusing on tactical mission support and battle management.

Crystal City Office
1745 Jefferson Davis Highway, Suite 609, Arlington, VA 22202-3402
☎ (+1 703) 413 44 55 ⊖ (+1 703) 413 44 60

Camarillo Office
80 North Wood Road, Suite 302, Camarillo, CA 93010
☎ (+1 805) 987 88 20; 987 68 32
⊖ (+1 805) 987 31 36

Oak Harbour Office
3165 North Goldie Road, Oak Harbor, WA 98277-2732
☎ (+1 206) 679 15 41 ⊖ (+1 206) 675 67 65

Niceville Office
912 South Palm Boulevard, Suite B, Niceville, Fl 32578
☎ (+1 904) 651 41 11 ⊖ (+1 904) 651 57 66

Devoe & Matthews, LC
348 Miracle Strip Mile, SW, Suite 23, Fort Walton Beach, FL 32548-5263
☎ (+1 904) 244 32 21 ⊖ (+1 904) 244 50 21

Simwright, Inc.
135 East John Sims Parkway, Niceville, FL 32578
☎ (+1 904) 678 80 40 ⊖ (+1 904) 729 88 07 [98]

80-24800 PRC Inc.
1500 PRC Drive, McLean, VA 22102
☎ (+1 703) 556 11 11 ⊖ (+1 703) 556 10 69
♣ $ 883,000,000 (1994) 👥 7,000 (1995)
Vice President and Division President: Leonard M Pomata
Senior Vice President of Sales and Marketing: Austin Yerks
Director of Marketing Communications: Steve Lagana
■ Information technology systems and services for the US military and civil governments. [98]

80-24875 Prism Color Group
(† Dayton Tech Art Co.)
(a subsidiary of Dayton Tech Art Co.)
1329 Stanley Avenue, Dayton, OH 45404
☎ (+1 937) 461 47 20 [E] prismcolor@earthlink.com
⊖ (+1 937) 228 34 75
♧ $ 320,000 (1996) ♣ $ 700,000 (1996) 👥 8 (1997)
🏭 700 m2
President: Alex J. Chestnut
■ Prepress service for the printing industry: 4-colour process negatives, computer enhanced photographs. [98]

80-24900 Proaction
5940 Grey Rock Rd, Agoura Hills, CA 91301
☎ (+1 818) 706 2200 E proaction@aol.com
⊜ (+1 818) 706 2271
♣ $ 750 000 ♟ 4 ⌂ 175 sq m
President: G. Miller
■ Management consulting & education firm. Specialise in A+D manufacturing business systems. 97

80-24930 Proprietary Software Systems
429 Santa Monica Boulevard, Ste 430,
Santa Monica, CA 90401-3401
429 Santa Monica Boulevard, Suite 430
☎ (+1 310) 394 52 33 ⊜ (+1 310) 393 31 22
♧ US$ 500,000 ♣ US$ 400,000 (1994) ♟ 7
President and Chief Executive Officer: Richard Gilinsky
✉ rglinsky@pro-software.com
Engineering Director: Dale Rankin ☎ (+1 206) 241 69 22
■ Defence related embedded computer system software translators. (Ada, C Jovial, Java). 98

80-24955 Public Safety Group Inc.
(† Harvey J. McGeorge & Associates Inc.)
12608-B Lake Ridge Drive, PO Box 4494,
Woodbridge, VA 22192
☎ (+1 703) 491 52 36 ⊜ (+1 703) 491 85 05
♟ 6 ⌂ 2,400 sq ft
President: Harvey J McGeorge II
Computing Systems Manager: Ms Barbara Dyer
■ Chemical and biological warfare effects analyses, study services and software development. 98

80-25150 QUANTICS incorporated
801 Springdale Drive, Exton, PA 19341-2866
☎ (+1 610) 993 99 93 E users@quantics.com
⊜ (+1 610) 524 74 29
♟ 16 (1996)
President: Dr J J Bolmarcich
Operations Manager: J M Fullman
■ QUANTICS is an amalgamation of scientists & military performing operations research & specializing in analytic models of warfare for munition stockpile planning. 98

80-25420 R&B Aircraft Supply Inc.
6848 Farmdale Avenue, PO Box 3065,
North Hollywood 91609-0065
☎ (+1 818) 764 39 10 E rbaircraft@aol.com
⊜ (+1 818) 765 24 36
♣ $ 7,500,000 (1996) ♟ 20 (1997)
President: John J Rogers
Secretary, Treasurer and Vice President of Sales: Ronald D Vickrey
Sales and Marketing Director: Max Meredith
Purchasing Manager: Ms Nina Sicuro
■ Supplier of aircraft spare parts, airframe, engine, accessories, avionics and instruments for aircraft. 98

80-25450 Ramapo Helicopters Inc
5 Airport Road, Morristown, NJ 07960
☎ (+1 201) 984 2370 ⊜ (+1 201) 984 7431
♧ $ 3 500 000 ♣ $ 1 500 000 ♟ 10 ⌂ 15 000 sq m
President: Richard Eimert
■ Minor & major maintenance, repair & overhaul. Spare parts service. Repair station for Bell, AS 350 & AS 355, Sikorsky S-76 & Agusta 109 helicopters. 98

80-25600 Rational Software Corporation
2800 San Tomas Expressway,
Santa Clara, CA 95051-0951
☎ (+1 408) 496 3600 ✆ (800) RAT 1212
⊜ (+1 408) 496 3636
♣ $ 73 000 000 (1995) ♟ 520
President: Paul D. Levy
VP Sales: John R. Lovitt
Dir Public Relations: Kara Myers
■ Comprehensive development environment that supports modern software engineering practices & the Ada programming language. Also provides technical engineering services that aid in the transition to modern software technology. 97

80-25650 Raytheon Aerospace
(† Beech Aerospace Services Inc.)
(a subs. of Raytheon Aircraft)
555 Industrial Drive South, Madison, MS 39110-9073
☎ (+1 601) 856 2274 ⊠ 504132 basi milt jks
⊜ (+1 601) 856 8006
♟ 2 600 (worldwide)
President: Daniel A. Grafton
■ Aircraft material and maintenance support services worldwide. 97

80-25880 Repairtech International Inc.
16134 Saticoy Street, Van Nuys, CA 91406
☎ (+1 818) 989 26 81 E rti4u@aol.com
⊜ (+1 818) 989 43 58
♟ 12
President: Stanley H Bennett
Operations Director: T H Deitrick
■ Repair of military and commercial aircraft parts. Machining, grinding, honing, lapping, plasma spray, welding and plating. FAA licenced. 98

80-25900 Research Associates of Syracuse (RAS)
6780 Northern Boulevard, Suite 100,
East Syracuse, NY 13057
☎ (+1 315) 463 22 66 W http://www.ras.com/
⊜ (+1 315) 455 80 37
♧ $ 220,000 (1997) ♣ $ 1,600,000 (1997)
♟ 20 (1997)
President: Robert Shields
Vice President and Chief Scientist: Richard Wiley
E dickwiley@ras.com
■ Research ELINT, EW, sensors. Software development, EW modelling. Seminars in ELINT/EW/RADAR. 98

80-25920 Resource Consultants Inc.
(a subs. of Gilbert Associates Inc.)
1960 Gallows Rd, Vienna, VA 22182-3824
☎ (+1 703) 893 6120 ⊜ (+1 703) 893 0917
♣ $ 70 000 000 (1995) ♟ 1 000
President: Ronald S. Newlan
VP Sales & Marketing: George Troendle
VP Admin & Corp. Strategy: Edward K. Walker (Jr)
Dir Public Relations: Robert Duarte
■ Response & support to the US Government & defence industries in the areas of studies & analysis; logistic, engineering & communications support, combat systems simulation, ADP systems design & installation. Provide advice on career transition of military & government civilian personnel to private corporate life. 97

80-25950 Retlif Testing Laboratories
795 Marconi Avenue, Ronkonkoma, NY 11779-7231
☎ (+1 516) 737 15 00 ⊜ (+1 516) 737 14 97
♟ 30
President: Walter A Poggi
Managing Director: Richard Reitz
Vice President Sales and Marketing: Thomas Volpe
Operations Director: William Hayes
Purchasing Manager: Ms Marilyn Poggi
■ Full service EMC & environmental simulation testing laboratory providing testing services to all currently used military & commercial EMC & ESS test specifications. 98

80-26200 RKI/Cleanroom Services
PO Box 16372, Rochester NY 14616
☎ (+1 716) 621 69 46 ✆ (800) 447 47 54
E clnrmervs@aol.com
W http://www.cleanroomservices.com
⊜ (+1 716) 621 27 78
⌂ 10,000 m2
President: Ralph Kraft
Managing Director: Wallace Baker
Operations Director: David Bugdan
Sales and Marketing Director: Charles Williams
Purchasing Manager: Douglas Shelton
■ Provision of turnkey design/build cleanroom systems, fan powered hepe filter modules, softwall cleanrooms, all snoillary supplies and equipment for cleanrooms and laminar flow workstations. Consulting, training seminars and materials, quality assurance programs for cleanroom construction, airflow management and visualization studies, testing and certification and fabric testing. 98

80-26280 Richard T. Lyons Associates Inc.
31 South Abingdon Street, Arlington, VA 22204
☎ (+1 703) 685 14 20 E rlyons@capaccess.org
⊜ (+1 703) 685 02 26
♟ 3 ⌂ 700 sq ft
President: Richard T Lyons
■ International business consultant in export/import specialising in foreign military sales to the Middle East and Brazil. 98

80-26400 Rohr Aero Services Inc.
(a subsidiary of Rohr Inc.)
16150 South Greeno Road, Fairhope, AL 36532
☎ (+1 334) 928 32 20 ⊜ (+1 334) 928 28 49
♧ US$ 8,000,000 (1997) ♣ US$ 20,000,000 (1997)
♟ 114 (1997) ⌂ 122,000 sq ft
General Manager: Harold G Burkholder
Operations Director: Rick H Ates
Sales and Marketing Director: Robert A Procter
Purchasing Manager: George E Delhomme
■ Repair, overhaul and modification of nacelles, thrust reversers for commercial and business aircraft. 98

80-26450 M. Rosenblatt & Son Inc.
Executive Office
350 Broadway, New York, NY 10013-9966
☎ (+1 212) 431 6900 ⊜ (+1 212) 334 0837
♟ 650
Chairman: Lester Rosenblatt
Vice Chairman: Perry W. Nelson
President: Rosenblatt Bruce
■ Naval architectural & marine engineering services covering feasibility studies & conceptual, contract & detail design for the construction, conversion, overhaul & repair of all classes of naval combatants & auxiliaries & high performance craft. Also, integrated logistics support (ILS) & technical manuals.

Washington, DC Office
2341 Jefferson Davis Highway, Suite 500,
Arlington, VA 22202
☎ (+1 703) 415 7800 ⊜ (+1 703) 415 7828
VP & Mgr: Amos Baki

Bremerton, WA
245 Fourth St, Ste 301, Bremerton, WA 98310-1891
☎ (+1 206) 479 9920 ⊜ (+1 206) 373 0632
Manager: Thomas Long

Charleston, SC
PO Box 71902, N. Charleston, SC 29415-1902
☎ (+1 803) 744 1686 ⊜ (+1 803) 747 8099
Asst Vice President & Mgr: Stephen T. Dunlap

Newport News, VA
813 Diligence Drive, Ste 100, Newport News, VA 23606
☎ (+1 804) 873 0611 ⊜ (+1 804) 873 0622
Asst Vice President & Mgr: Feridun K. Serim

San Diego, CA
1007 Fifth Ave, Ste 300, San Diego, CA 92101
☎ (+1 619) 238 1300 ⊜ (+1 619) 238 0070
VP & Mgr: David M. Krepchin

San Fransisco, CA
620 Folsom St, Ste 301, San Francisco, CA 94107
☎ (+1 415) 777 0500 ⊜ (+1 415) 777 4015
VP & Mgr: Stephen Halpern

Virginia Beach, VA
5700 Thurston Ave, Ste 204, Virginia Beach, VA 23455
☎ (+1 804) 460 4449 ⊜ (+1 804) 464 2801
Asst Vice President & Mgr: David Pellett 97

80-26500 Sabreliner Corp.
Pierre Laclede Center, Suite 1500,
7733 Forsyth Boulevard, Clayton, MO 63105-1821
☎ (+1 314) 863 68 80 ✆ (800) 325 46 63
E sabrecorp@worldnet.att.net ⊜ (+1 314) 863 68 87
♣ US$ 200,000,000 (1997) ♟ 2,000 (1997)
⌂ 2,000,000 sq ft
President: F Holmes Lamoreux
Executive Vice President and Chief Operations Officer: Jerry L Leath
Senior Vice President and Chief Financial Officer: Rodney E Olson
Vice President of Government Marketing: Gene L Harbula
Vice President of Engineering: Bob D Hanks
Business Development Director: D J Meier
International Sales Manager: Davis E Damba
■ Repair and overhaul of military aircraft engines, components and accessories. 98

80-26680 Safety Consulting Engineers Inc. (SCE)
2131 Hammond Drive, Schaumburg, IL 60173
☎ (+1 847) 925 81 00 ⊜ (+1 847) 925 81 20
President: C James Dahn
Business Manager: Ms Sophia B McAskill
■ Consulting, testing, R&D company providing fire and explosion hazards evaluations involving chemical, propellant, pyrotechnic and explosives manufacturing. 98

Contact Reference

80-26700 **Safway Steel Products Inc.**
N14 W 23883 Stone Ridge Drive, Suite 400,
Waukesha, WI 53188
☎ (+1 414) 523 65 00 ⊜ (+1 414) 523 65 95
♣ $ 100,000,000 (1997) 👥 1,000 (1997)
President: Marc Wilson
Operations Director: Jerry Johns
Marketing Manager: Deanne Algaier
Purchasing Manager: Gary Schneiden
■ Sectional steel scaffolds, systems scaffolding, tube & clamp scaffolds, manual & powered swing stages. Shoring equipment, including post shores & sectional, heavy duty adjustable shoring. Recreational outdoor bleachers. → 90 [98]

80-26850 **SAR-ASSIST**
One Island Lane, Greenwich, CT 06830
☎ (+1 203) 869 13 22 ⊜ (+1 203) 869 13 22
Principal Dir: Ms Gloria W. Heath
■ Consultants in disaster/distress response, safety, rescue, survival in space, air, sea & land operations. [98]

80-26870 **Sas R&D Services Inc.**
(† Sas Research & Development Services Inc.)
2714 South West 183rd Avenue,
Miramar, FL 33029-5178
☎ (+1 954) 432 23 45 ⊜ (+1 954) 432 41 41
♧ $ 100,000 (1996) ♣ $ 5,000,000 (1996)
👥 17 (1996) 🏭 7,500 m2
Chief Executive Officer: Ted Sas [E] tedsas@sasrad.com
Office Manager: Ms Sophia Sas
■ Supply of anti-narcotic, IED/EOD, riot control and mine clearance equipment. Procurement services. [98]

80-26890 **Schneider Sohn & Associates Inc. (SSAI)**
181 Post Road West, PO Box 5237, Westport, CT 06881
☎ (+1 203) 454 21 47 ⊜ (+1 203) 227 95 90
♣ $ 1,000,000 (1995) 👥 4
Chairman and Chief Executive Officer: William Schneider (Jr) ☎ (+1 703) 524 55 22
President and Chief Operating Officer: Stephen Sohn
■ International trade & financial advisory organisation. Strategic marketing, investment, technology transfer, trade financing (including offsets & countertrade) leasing, financing through government issued note purchasing & other innovative approaches supporting the global sales effort of medium & large advanced technology companies. Merchant/investment banking services. Strategic alliances. Offices in Washington, D.C., Westport, Connecticut, USA.

1925 North Lynn Street
Washington, DC Office
Arlington, VA 22209
☎ (+1 703) 524 55 22 ⊜ (+1 703) 524 23 42 [98]

80-26900 **Schneider Specialized Carriers Inc.**
(a Schneider National Co.)
PO Box 2300, Green Bay, WI 54306
☎ (+1 414) 592 6888 ✆ (800) 482 2737 ▣ 290788
⊜ (+1 414) 592 6822
President: John Lanigan
VP & General Mgr: Larry Chaplin
VP Import/Exports: Herb Hilken
VP Pricing & Business Eng.: Tom Anderson
Sales Support Dir: Jim Apfel
Customer Service Dir: John Gutschow
■ Motor carrier services between points in US, Canada & Mexico. [97]

80-27250 **SEAL Laboratories**
(a subsidiary of Kleinet Industries Inc.)
250 North Nash Street, El Segundo, CA 90245-4529
☎ (+1 310) 322 20 11 ✆ (800) SEALLAB
[W] http://www.home.earthlink.net/~seallabs
⊜ (+1 310) 322 22 43
👥 31 (1997) 🏭 2,400 sq m
President: Roland Marti
Vice President and General Manager: Dr Arun Kumar
■ Destructive Physical Analysis (DPA) of electronic components; failure analysis; SEM, EDX, AES, ESCA, SIMS, FTIR and RGA services; consultation in metallurgy, materials and electronics. [98]

80-27300 **Seamodal Transport Corp.**
221 Main Street, Suite 765, San Francisco, CA 94105
☎ (+1 415) 543 74 20 [E] efvy@mcimail.com
▣ TWX 910 372 8406 ⊜ (+1 415) 543 21 37
♣ $ 25,000,000 👥 50
President: N E Vacakis
Executive Vice President: G L Fischer
Marketing Manager: F K Villanueva
■ International freight forwarder and customs broker involved in transportation logistics of military equipment and spares for friendly countries purchasing under foreign military sales program or through direct commercial buys.

● 24 Vendue Range, Suite 107, Charleston, SC 29401
☎ (+1 803) 577 37 42 ⊜ (+1 803) 723 11 13
District Manager: D L Seithel

● 3158 Des Plaines Avenue, River Road, Suite 24, Des Plaines, IL 60018
3158 Des Plaines Avenue, (River Road), Suite 24
☎ (+1 847) 299 61 70 ⊜ (+1 847) 299 61 75
District Manager: R Archia

● 249 East Ocean Boulevard, Suite 506, Long Beach, CA 90802
☎ (+1 310) 432 40 88 ⊜ (+1 310) 432 33 08
District Manager: J Urbano

● Plume Center West, Suite 520, 100 West Plume Street, Norfolk, VA 23510
☎ (+1 757) 625 19 28 ⊜ (+1 757) 625 01 46
District Manager: C N Byrd

● 221 Main Street, Suite 765, San Francisco, CA 94105
☎ (+1 415) 543 74 20 ⊜ (+1 415) 543 21 37
Regional Manager: A V Dolojan
District Manager: C M Roque

● 3544 Arden Road, Hayward, CA 94545
☎ (+1 510) 785 57 00 ⊜ (+1 510) 785 75 06
Manager of FMS Operations: O G Dolojan [98]

80-27750 **SEMCOR Inc.**
815 East Gate Drive, Mount Laurel, NJ 08054
☎ (+1 609) 234 67 00 [W] http://www.semcor.com
⊜ (+1 609) 234 76 46
♣ $ 81,000,000 (1996) 👥 1,000 + (1997)
Director Business Development: Donald R Furmanski
Director Contracting: Thomas J Karr
Operations Director: Thomas J Brennan
Vice President Sales and Marketing: John C Phillips
■ Provides engineering & program management support services to DoD, NATO & industry. Professional staff specializes in applying systems engineering & management expertise to research, development, test, evaluation, acquisition & life-cycle support of weapons systems. [98]

80-28000 **George G. Sharp Inc.**
100 Church Street, New York, NY 10007
☎ (+1 212) 732 28 00 ⊜ (+1 212) 732 28 09
♣ $ 32,000,000 (1995) 👥 350 🏭 10,000 sq m
Chairman of the Board/Technical Dir: I. Hilary Rolih
President: Allen Chin
Asst to the President: Robert H. Kiefer
■ Marine transportation, naval architect & marine engineering firm. Support of the US navy, coast guard & maritime administration. Management consulting services to firms & municipalities in areas of marine business development & port operations.

● 635 Slaters Lane, Suite 200, Alexandria, VA 22314
☎ (+1 703) 548 44 00 ⊜ (+1 703) 548 45 92

● 209 South Park Drive, Virginia Beach, VA 23462
☎ (+1 757) 499 41 25 ⊜ (+1 757) 671 93 01

● 2 Echelon Plaza, Suite 260, New Jersey, NY 08043
☎ (+1 609) 772 08 88 ⊜ (+1 609) 772 24 29 [98]

80-28160 **SimuFlite Training International**
Dallas/Fort Worth Int'l Airport, PO Box 619119,
Dallas, TX 75261
☎ (+1 972) 456 80 00 [E] info@simuflite.com
⊜ (+1 972) 456 83 13
👥 200
President: Charles C Carson II
Vice President of Finance and Strategic Planning: A B Liles
Vice President of Sales and Marketing: Jeffery G Roberts
Vice President of Training Services: Eugene Haggerty
Managing Director of Courseware and Communications: Mrs Allison Blankenship
Managing Director of Technica Services: Thomas Ferranti
■ Provider of advanced aviation training services for professional pilots and maintenance personnel operating turbine-powered aircraft in corporate, government and military service. [98]

80-28200 **SJA International Inc.**
1001 Connecticut Ave, NW, Ste 401,
Washington, DC 20036
☎ (+1 202) 223 4001 ⊜ (+1 202) 457 6566
👥 5
President: Samuel J. Agger
Vice President: Ms Ellen Walp
Vice President: Ms April Callahan
■ Assists US & European companies in transportation & communications systems, law enforcement identification, military systems, advanced industrial equipment & manufacturing & construction techniques. Specialises in transfer of European technology to USA. [97]

80-28250 **Skyways Inc.**
PO Box 19121, Irvine, CA 92713-9121
☎ (+1 714) 476 84 44 [E] intlferry@aol.com
⊜ (+1 714) 775 84 04
♣ $ 3,700,000 (1994) 👥 26
President: John J Barrett
■ Worldwide flight delivery of military and civilian jet aircraft for US and foreign government agencies, specialising in trans-oceanic delivery of current manufactured tactical aircraft. [98]

80-29050 **Space Electronics Inc.**
4031 Sorrento Valley Boulevard, San Diego, CA 92121
☎ (+1 619) 452 41 66; 452 41 67.
[W] http://www.spaceelectronics.com
⊜ (+1 619) 452 54 99
♣ $ 10,600,000 (1996) 👥 75 (1997) 🏭 9,000 sq m
President: David J Strobel
Managing Director: Robert Czajkowski
Sales and Marketing Director: David Czajkowski
Engineering Manager: Ms Janet Patterson
Operations Manager: Paul Blevins
Advertising Manager: Bruce Lawrence
Quality Assurance Manager: Ms Janice Wakefield
Purchasing Manager: Chris Blanch
■ Provides engineering services for space and defence contractors in the microelectronics/parts arena, RAD-PAK spare IC's, rad. hard IC's, parts screening, DPA, failure analysis, hybrids, radiation testing and parts procurement programmes. [98]

80-29200 **SPC International Corp.**
(a subsidiary of System Planning Corporation)
1000 Wilson Boulevard Department J,
Arlington, VA 22209-2211
☎ (+1 703) 351 85 64 [W] http://www.sysplan.com
▣ TWX 710 9550 678 ⊜ (+1 703) 351 85 67
Chief Executive Officer: Dr Dov S Zakheim
Vice President Operations: Richard N Smull
Director of Project Development: William Malamud
Assistant Director of Business Development and Export Administration: Paul Segretto
■ Assists in developing defence and related high-tech business initiatives with US and international partners. Establishes co-operative or teaming arrangements and provides strategic market planning. Additional services include political and economic assessments and a complete range of cost analyses for high technology programs. [98]

80-29250 **Spectral Intelligence Corp. (SI)**
Station Square 3, Suite 206, Paoli, PA 19301
☎ (+1 610) 889 1349 ✆ (800) 889 8909
⊜ (+1 610) 889 0600
President: Dr Kenneth N. Davis (PhD)
Sales Dir: Morris Anderson
Operations Dir: April Frazier
■ Software guide to chemical warfare. Laboratory information & management systems. Artificial intelligence neural networks. [97]

80-29350 **Spectrum Associates Inc.**
7800 Lowmoor Court, Virginia 22153
☎ (+1 703) 451 74 39 ⊜ (+1 703) 451 74 39
👥 5 (1997) 🏭 1 250 sq ft
President: Kenneth S Brower

■ Feasibility studies and development of conceptual designs of new warships for international clients and comparative engineering analyses of US and international warships. Defence consultants specialising in net assesments and the interaction of technology on the future battlefield. Possess extensive database on international warship design data. Ship design synthesis computer programs for patrol boats, DPVs, replenishment ships, RO-ROs, fast attack craft, corvettes, frigates, destroyers, cruisers, aircraft carriers and submarines. to develop conceptual designs, conduct trade studies and reverse engineering design studies. 98

80-29400 Spire Corp.
One Patriots Park, Bedford, MA 01730-2396
☎ (+1 617) 275 60 00 E email@spirecorp.com
W http://www.spirecorp.com ⊖ (+1 617) 275 74 70
♣ $ 17,300,000 (1996) 👥 140 (1997)
President and Chief Executive Officer: Roger G Little
Marketing Services: Ms Anne P Preston
■ Products and services for optoelectronics, photovoltaics and biomaterials markets. Compound semiconductor epitaxial structures and optoelectronic components, photovoltaics manufacturing equipment, processing services for biomedical devices. 98

80-29500 Sprint
(† Sprint/Carolina Telephone)
14111 Capital Boulevard, Wake Forest, NC 27587
☎ (+1 919) 554 74 49 W http://www.sprint.com
⊖ (+1 919) 554 74 74
President: William E McDonald
Director of Corporate Communications: Mrs Janis S Dempster
Manager of Staff Operations: Ms Ricky Day
■ Telephone services. 98

80-29550 Squadron International Ltd
(a div. of IWI)
18 Eklund Blvd, Nesconset, NY 11767
☎ (+1 516) 293 9494 ⊖ (+1 516) 293 3630
👥 16
President: Marc D. Lewis
Dir Marketing: Robert G. Wilk
Sr Analyst: Dean Pappas
■ Publish market intelligence reports on a consultant basis, on the military capabilities of third world nations (inventories, military manufacturing base, potential capabilities) & the international electronics retrofit market for military aircraft & warships. 97

80-29600 SRI International World Headquarters
(† Stanford Research Institute)
333 Ravenswood Avenue, Menlo Park, CA 94025-3493
☎ (+1 415) 326 62 00 ✉ SRI INTL MPK
W http://www.sri.com ☒ 334486; TWX 910 3732 046
⊖ (+1 415) 326 55 12
♣ $ 326,000,000 (1996) 👥 1,500 (1997)
President and Chief Executive Officer: Dr William P Sommers
Executive Vice President: Paul Jorgensen
Executive Vice President: John McHenry
Executive Vice President: Harold Kruth
Corporate Marketing and Communications: Ms Alice Galloway
■ Basic and applied research and consulting in engineering, technology, physical and life sciences, management and social systems and policy studies.
Subsidiary organization:
David Sarnoff Research Center → 60
SRI Consulting →
International Regional Headquarters:
SRI International → UK 80
Other offices worldwide:
Milan, Zurich, Tokyo, South Korea.
Washington DC Office
1611 North Kent Street, Arlington, VA 22209
☎ (+1 703) 524 20 53 ⊖ (+1 703) 524 85 69 98

80-29650 SST/Burlington Microelectronics
(† Solid State Testing Inc.)
56 Middlesex Turnpike, Burlington, MA 01803-4921
☎ (+1 617) 272 0972 E sstmicro@ix.netcom.com
W http://www.sstmicro.com ⊖ (+1 617) 273 4896
👥 150
President: Fred Momayez
Vice President Sales and Marketing: James Roberts
Operations Director: John H Bishop
Purchasing Manager: Ms Diane Flood
■ Independent testing laboratory for screening & qualification of electronic components such as integrated circuits, diodes & transistors, resistors & capacitors to military specification. Assembly of custom circuits, quick-turn builds. 98

80-29700 Stanford Technology Corp.
57 Poplar Street, PO Box 2100 D, Glenbrook, CT 06906
☎ (+1 203) 348 40 80 ⊖ (+1 203) 327 52 25
♣ $ 500,000 👥 5 🏭 700 sq m
President: Charles C Cullari
■ Qualification and acceptance testing on aerospace and military products. 98

80-30100 Strat*X International
(a subsidiary of Strat*X, France)
222 Third Street, Cambridge, MA 02142
☎ (+1 617) 494 82 82 W http://www.stratx.com
⊖ (+1 617) 494 14 21
👥 8
Managing Director: Scott A Imperatore
Marketing Manager: Ms Debra Lavender
■ Specialises in the development and implementation of tailored management development programme in business and marketing strategy. The aim is to improve clients' performance by improving their managers' abilities to formulate and implement successful market led strategies. 98

80-30400 STV Group Inc.
205 West Welsh Drive, Douglassville, PA 19518
☎ (+1 610) 385 82 00 E stvinc@netaxs.com
W http://www.stvinc.com ⊖ (+1 610) 385 85 00
♣ $ 94,073,000 (1996) 👥 1,000 (1997)
Chairman and Chief Executive Officer: M Haratunian
President: D M Servedio
Purchasing Manager: Walter Cleaver
Manager of Corporate Communications: Ms Debra B Trace
■ Aerospace: security systems; program management; engineering development, reliability, availability, maintainability, documentation; procurement; integrated logistics support; facility design; communications; environmental; infrastructure.
STV Incorporated
225 Park Avenue South, New York 10003
☎ (+1 212) 777 44 00 ⊖ (+1 212) 529 52 37
President: D M Servedio
STV Incorporated
205 Welsh Drive, Douglassville, PA 19518
☎ (+1 610) 385 82 00 ⊖ (+1 610) 385 85 01
Senior Vice President: Whitney A Sanders (II)
STV Environmental Inc.
205 Welsh Drive, Douglassville, PA 19518
☎ (+1 610) 385 82 00 ⊖ (+1 610) 385 85 01
Senior Vice President: Roger Zyma
STV Architects
205 Welsh Drive, Douglassville, PA 19518
☎ (+1 610) 385 82 00 ⊖ (+1 610) 385 85 01
Senior Vice President: Robert W Darlington
STV Construction Services
205 Welsh Drive, Douglassville, PA 19518
☎ (+1 610) 385 82 00 ⊖ (+1 610) 385 85 01
Senior Vice President: James J Fetterolf
STV International
225 Park Avenue South, New York, NY 10003
☎ (+1 212) 777 44 00 ⊖ (+1 212) 529 52 37
Senior Vice Prtesident: Michael R Santoro
STV/Silver & Ziskind
233 Park Avenue South, New york, NY 10003
☎ (+1 212) 477 19 00 ⊖ (+1 212) 473 27 80
President: David M Ziskind 98

80-30600 S.V. Marketing
6189 Wellington Commons Drive, Alexandria, VA 22310
Dir Region Int'l/5: Prof. Sully Valetash 97

80-30650 Sverdrup Corp.
13723 Riverport Drive, Maryland Heights, MO 63043
☎ (+1 314) 436 76 00 ⊖ (+1 314) 770 51 05
♣ $ 626,000,000 (1995) 👥 5,000 (group)
Chairman and Chief Executive Officer: R E Beumer
President: J C Uselton
Senior Vice President and Chief Financial Officer: R J Messey
Senior Vice President Secretary and General Counsel: A S Morrisson
Senior Vice President and President of Sverdrup Technology Inc.: R W Butler
Senior VVice President and President of Sverdrup Facilities Inc.: K G Kreikemeier
Senior Vice President and President of Sverdrup Civil Inc: Dr H G Schwartz (Jr)
Vice President of Corporate Communications: J L Bryan
■ Complete planning, design, construction and programme management services for: defence/aerospace applications (including military bases, systems and facilities); specialised launch, test and manufacturing facilities; advanced technology and research facilities and operation/maintenance of major test facilities.
Sverdrup Corp. Subsidiaries:
Sverdrup Technology Inc. → 80
Sverdrup Civil Inc.
Sverdrup Facilities Inc.
Sverdrup Environmental Inc.
Sverdrup Investments Inc. 98

80-30680 Sverdrup Technology Inc.
(a subsidiary of Sverdrup Corp.)
600 William Northern Boulevard, PO Box 884, Tullahoma, TN 37388
☎ (+1 615) 455 64 00 W http://www.sverdrup.com
⊖ (+1 615) 393 62 12
👥 3,180 (1997)
Chairman: J C Uselton ☎ (+1 615) 455 64 00
President: R W Butler ☎ (+1 615) 455 64 00
Senior Vice President: R F Starr (Jr)
Senior Vice President and General Manager of AEDC Group: Dr W F Kimzey ☎ (+1 615) 454 30 00
Senior Vice President and Manager of Technology Group: A L Baer
Senior Vice President and General Manager of Naval Warfare Center, Weapons Group: Dr D E Bowyer ☎ (+1 619) 446 15 49
Vice President and General Manager of Marshall SFC Group: Dr N W Haars ☎ (+1 205) 971 95 50
Vice President and General Manager of TEAS Group/Elgin AFB: Dr B D Arnold (Jr) ☎ (+1 904) 678 20 01
General Manager of Ames Division: R A Marmol ☎ (+1 415) 604 59 27
General Manager of Langley Research Centre Division: Dr P E Sensmeier ☎ (+1 804) 827 17 86
General Manager of Operations Support Services Division: L W Hickok ☎ (+1 805) 275 61 00
Divisional Manager of Advanced Systems Group: M S Williams ☎ (+1 513) 429 50 56
General Manager of Systems Support Group: F M Gay ☎ (+1 904) 863 77 00
Marketing Specialist: Ms Diana B Williams
■ Provides professional engineering service to support development, integration, test and acquisition of weapons systems. Development, operations and maintenance of government test facilities with the research and development of defence systems. 98

80-30750 Synectics Corp.
10400 Eaton Place, Ste 200, Fairfax, VA 22030-2208
☎ (+1 703) 385 0190 ⊖ (+1 703) 385 1987
♤ $ 5 000 000 (1995) ♣ $ 24 000 000 (1995) 👥 140
🏭 20 000 sq m
President: Dr James W. Altman
Sr Exec. Assistant: Wilfredo Serrano
■ Consultancy, systems analysis & threat assessment services. Data processing, software development & technical support services. 97

80-31720 Teal Group Corp.
3900 University Drive, Suite 220, Fairfax, VA 22030
☎ (+1 703) 385 1992 ⊖ (+1 703) 691 9591
President: William C. Storey (Jr)
Operations Dir: Tim H. Storey
Sr Analyst: Frank Dahm
Analyst: Richard Aboulafia
Analyst: Marco Caceres
Analyst: Renée Gentry
Analyst: Steve Zaloga
Analyst: David Rochwell
■ International defence & aerospace consulting services in the areas of competitive intelligence, market analysis, programme & defence budget forecasting. Proprietary as well as subscription services available.
Sales Offices:
Eastern Region, Tel (+1 703) 573 5374
Western Region, Tel (+1 310) 434 7159
Central Region, Tel (+1 513) 247 9350 97

80-31800 TECHMedia Corp.
1804 La Cima, San Clemente, CA 92672
☎(+1 714) 498 39 00 🄴 propconslt@aol.com
⊖ (+1 714) 498 39 83
Managing Partner Proposal Services: Thomas Kaplin
Managing Partner: John McCosker
Managing Partner Administration: John Bradley
■ Proposal consulting services [98]

80-31850 Techmatics Inc.
12450 Fair Lakes Circle, Suite 800, Fairfax, VA 22039
☎(+1 703) 802 83 00 🅆 http://www.thematics.com
⊖ (+1 703) 802 04 12
♣ $ 58,000,000 (1997) 🙵 750 (1997)
Executive Vice President: Walter S Szczypinski
☎(+1 703) 802 23 55
Vice President of Finance: Michael Maraghy
☎(+1 703) 802 23 41
■ Engineering services and information technology including systems engineering and integration, program co-ordination, imaging applications, database development and optical disc technology. [98]

80-31930 Technology Service Corp.
2950 31st Street, Santa Monica, CA 90405
☎(+1 310) 450 97 55 🅆 http://www.tcs.com
⊖ (+1 310) 452 31 75
♣ US$ 24,000,000 (1997) 🙵 200 (1997)
President: Robert Graziano
Vice President: Dr Russell Lefevre 🄴 rlefevre@tsc.com
■ Defense related services: radar, sonar, electro-optical and electronic warfare.
Washington Division
962 Wayne Avenue, Suite 800, Silver Spring, MD 20910
☎(+1 301) 565 29 70 [98]

80-32700 Tetra Corp.
3701 Hawkins Street North East,
Albuquerque, NM 87109-4512
☎(+1 505) 345 86 23 🄴 tetra@rt66.com
⊖ (+1 505) 345 73 18
♣ $ 1,500,000 (1996) 🙵 15 (1996)
Chief Executive Officer: William M Moeny
Manager of Contracts and Administration: D Olings
■ Research and development in pulsed power science and engineering and pulsed electric lasers. [98]

80-32800 Thompson Aerospace
(† Thompson Aircraft Tire Corp.)
7775 NW 12th Street, Miami, FL 33126
☎(+1 305) 592 3530 ☒ TWX 156250532
⊖ (+1 305) 592 0485
♣ $ 35 000 000 (1994) 🙵 300 (1994) 🏭 5 400 sq m
President: Anthony Battaglia
VP Technical Svcs: H.E. Davis
VP Operations: Mike Mackovich
VP Finance: S. O'Meally
Regional Sales Mgr: R. Seguera
Sales & Marketing Dir: R. Riera
Regional Sales Mgr: Bill Wendler
Contract Admin Dir: Jack Kercheval
Customer Service Mgr: Ms Mayra Consuegra
■ Aircraft tyre retreading & holographic non-destructive inspection, new aircraft tyre distributor, aircraft wheel & brake service & overhaul w/automated Eddy Current inspection.
® Japan: Bridgestone Tires [97]

80-33080 TMSI Incorporated
(a subs. of TSB International)
Archway 60 Office Centre,
9321 Midlothian Turnpike, Ste C, Richmond, VA 23235
☎(+1 804) 272 6900
Managing Dir: Ray Donnelly ☎(+1 804) 272 2912
Contact: C. Begley
Management Services
PO Box 1957, Cary, NC 23235
☎(+1 919) 233 5335 [97]

80-33100 Top Line Co. (TLC)
6443 Gerard Court, Falls Church, VA 22043
☎(+1 703) 538 45 88
President: Theodore H Smith
■ Produces printed and electronic information service tracking the US DoD congressional process and identifying aerospace market trends; consulting on aerospace market planning. [98]

80-33152 Tracor Applied Sciences Inc.
(a subsidiary of Tracor Inc.)
1601 Research Boulevard, Rockville, MD 20850
☎(+1 301) 838 60 00 ⊖ (+1 301) 838 61 01
President: K Bruce Hamilton
■ Provides complex technical expertise primarily in the areas of communication, systems integration, test and evaluation, engineering analysis and maintenance planning for navy vessels, including advanced combatants, the AEGIS class cruisers and destroyers. Other products and services include acquisition support to navy headquarters, anti-submarine warfare research and development, environmental and industrial hygiene services to government and industry, as well as capabilities in the areas of underwater acoustics, signal-processing, air traffic control and identification friend-or-foe.
Analysis & Applied Research Division
6500 Tracor Lane, Austin, TX 78725-2050
☎(+1 512) 929 20 53 ⊖ (+1 512) 929 40 89
Division Vice President: Homer E Wieland
Electronics Systems Division
402 Cottonwood Parkway, California, MD 20619
☎(+1 301) 862 93 00 ⊖ (+1 301) 862 93 30
Division Vice President: Robert Weiland
Energy & Environmental Services Division
1601 Research Boulevard, Rockville, MD 20850
☎(+1 301) 838 62 04 ⊖ (+1 301) 838 62 87
Division Vice President: K Bruce Hamilton (acting)
Vitro Services Corporation
557 Mary Esther Cut-Off,
Fort Walton Beach, FL 32548-4090
☎(+1 904) 244 77 11 ⊖ (+1 904) 244 77 60
Division Vice President: Richard Manley [98]

80-33154 Tracor Information Systems Company
1601 Research Boulevard, Rockville, MD 20850-3173
☎(+1 301) 838 68 00 ⊖ (+1 301) 838 68 50
President: Barry G Campbell
→ 75
Cordant, Inc.
11400 Commerce Park Drive, Reston, VA 20190
☎(+1 703) 758 70 00 ⊖ (+1 703) 758 70 47
President: Kenneth L Johnson
■ Provides information technology solutions primarily to the federal marketplace and focusses on the design, development and integration of a variety of information systems for applications including medical, public access, records management/workflow and Computer-Aided Design/Geographic Information Systems.
Quality Systems, Inc. (QSI)
8201 Greensboro Drive, Suite 1200, McLean, VA 22102
☎(+1 703) 847 58 20 ⊖ (+1 703) 847 58 87
President: Bill Shernit
■ Design information systems for predominantly classified customers. Provide systems engine engineering and integration, database design, applications development, open source integration, commercial off-the-shelf customisation network design and training.
Software Center of Excellence (SCOE)
1601 Research Boulevard, Rockville, MD 20850-3173
☎(+1 301) 838 68 00 ⊖ (+1 301) 838 68 05
Vice President: Franklin E Herman
■ Delivers more than one hundred specialised software engineering courses to train customers and employees in new technology. [98]

80-33550 Turbo Power Inc.
14820 N.W. 60th Avenue, Miami Lakes, FL 33014
☎(+1 305) 820 3225 ✆ 800 344 3387
⊖ (+1 305) 820 0404
♣ $ 22 800 000 🙵 88
President: Andrew J. Brose
Sales & Marketing Dir: Francisco Guillen
■ Overhaul & repair of Allison 501/T56 gas turbine engines. Compressor overhaul, turbine replacement, gearbox overhaul. Engine sales & exchange. Complete accessory overhaul. Pratt & Whitney PT6A, PT6T & repair including accessories. [97]

80-33600 United Beechcraft Inc.
(a subs. of Raytheon Aircraft)
PO Box 2966, Wichita, KS 67201
☎(+1 310) 676 5201
President: John Willis
■ A network of 12 Fixed Base Operations (FBOs) throughout the USA that sells and services aircraft from airport locations. [97]

80-34100 Vanguard Research Inc. (VRI)
10306 Eaton Place, Suite 450, Fairfax, VA 22030-2201
☎(+1 703) 934 63 00 ⊖ (+1 703) 273 93 98
♣ US$ 12,400,000 (1997) 🙵 150 (1997)
President and Chief Executive Officer: Mel Chaskin
Senior Vice President: John F Kantak
Vice President of Business Development: Nick Judge
✉ njudge@vriffx.com
Vice President: John H Summer
■ Research on C3I and SDI
Colorado Springs Division
5050 Edison drive, Suite 102,
Colorado Springs, CO 80915
☎(+1 719) 596 11 74 ⊖ (+1 719) 596 01 83
🙵 16
President and General Manager: Lee Morgan
■ Provides support to the Strategic Defence Initiative organization's National Test Bed Joint Program Office (NTBJPO) and provides technical and analytical support to the United States Space Command (AFSPACECOM).
Omaha Division
119 West Mission Avenue, Suite D, Bellevue, NE 68005
☎(+1 402) 291 34 75 ⊖ (+1 402) 291 36 95
🙵 27
General Manager: George Wheeler
Marketing Director: Bob Hohler
■ Supports Military Command Control (C2), information management, strategic integrated operations planning, weapons planning and automated aircrew mission planning.
Washington Division
10306 Eaton Place, Suite 450, Fairfax, VA 22030-2201
☎(+1 703) 934 63 00 ⊖ (+1 703) 273 93 98
🙵 32
Marketing Director: Danny Holtzman
General Manager: Gordon Smith•
■ Products and services include systems engineering and analysis, technical support and services, modelling and simulation development, management information system and software development, MANPRINT analysis and specialised product research and development. [98]

80-34290 Video Ordnance Inc.
1 West 37th Street, New York, NY 10018
☎(+1 212) 575 05 50
🄴 74127.2020@compuserve.com
⊖ (+1 212) 730 00 10
President: Ms Marlene Cardin
Director: Paul Cardin
■ Production of video & TV programmes for the aerospace & defence industry. [98]

80-34350 Vinnell Corp.
(a subsidiary of BDM Holdings Inc.)
12150 East Monument Drive, Suite 800,
Fairfax, VA 22033
☎(+1 703) 385 45 44 ⊖ (+1 703) 385 37 26
♣ $ 100,000,000
President and Chief Executive Officer: Roy V Woodle
Assistant Vice President and Corporate Secretary: Kevin M O'Melia
■ Technical services and military training, logistics support, maintenance and support of advanced management systems.
Transportation Center
Logistics & Transportation Services Inc.
DeSoto Ind. Park, 1117 Wilson Dr., Baltimore, MD 21223
Desoto Industrial Park, 1117 Wilson Drive
☎(+1 301) 644 69 00 ☒ 87950 vnltrnsctr bal
■ Transportation & freight forwarding services. [98]

80-34400 Vitro Corp.
(a subsidiary of Tracor Inc.)
1601 Research Boulevard, Rockville, MD 20850-3173
☎(+1 301) 738 40 00 ⊖ (+1 301) 738 43 23
🏭 1,300 sq ft
President and Chief Executive Officer: Barry G Campbell
Senior Vice President of Business Development: John R Gantt
Senior Vice President of Administration: Arthur A Rossi
Senior Vice President of Finance: Gary P Burris
Vice President, Systems Application Branch: William W Coe
Director of Federal Aviation Administration Business: Thomas P Messier
Director of Marketing Communications: Ms Paula Sandin
☎(+1 301) 738 46 53

■ Provides a full spectrum of engineering and technical services, including systems engineering, software engineering and technical and management support for the US Department of Defense, FAA, NASA, other federal agencies and the intelligence community. Provide full life-cycle support including system design, integration and test; software engineering, tool development and test; logistics systems analysis and applications. [98]

80-34441 Vosper Inc
PO Box 2096, Oak Harbour, WA 98277-6096
☎ (+1 206) 675 2089 ⊠ 361273 ⊖ (+1 206) 675 5937
♣ £ 200,000 ♟ 5
Vice President: Paul Gillon (Jr)
■ Small warship design/engrg consultancy providing design on technical & support services to defence authorities & shipyards engaged in the operation & construction of light frigates, corvettes patrol craft & high performance vessels. [98]

80-34460 VRC Corp.
68 Long Court, Ste 2A, Thousand Oaks, CA 91360
☎ (+1 805) 496 3549 ⊖ (+1 805) 496 7191
♣ $ 7 000 000 (1995) ♟ 14
President: Donald Vreuls
Technical Dir: Carlisle (Dick) King
Purchasing Mgr: Ms Shirley R. Conway
■ Professional services in test & evaluation, simulation and modelling, software development, user-system interface design, training & human performance assessment. [97]

80-34500 Walashek Industrial & Marine Inc.
3236 16th Ave, Seattle, WA 98134-1024
☎ (+1 206) 624 2880 ✆ (800) 886 7706
⊖ (+1 206) 624 3324
♧ $ 853 469 ♣ $ 12 682 581 ♟ 150 ⌂ 55 000 sq ft
President: Gail Walashek
Chief Exec. Officer: Michael Walashek
Comptroller: Dorothy Michaelson
General Mgr: Michael Walashek
Operations Mgr: James Doud
Ship Repair Mgr: Steven Rosebrook
Chief Estimator: Kenneth Messick
QA Mgr: Joseph Ripollone
■ Ship/dockside repair, pipe fitting, electronics, boiler/condenser/heat exchanger repair & overhaul.

San Diego Office
595 "C" Street, Chula Vista, CA 91910
☎ (+1 619) 498 1711 ⊖ (+1 619) 498 0710
Manager: Ron Gray
Office Mgr: Laurie Daughn
QA Mgr: Charles Perry
Field Foreman: Paul Hicks
Shop Foreman: Fred Wallen

Portland Office
5555 N. Channel Drive, Building 10, Portland, OR 97217
☎ (+1 503) 285 9926 ⊖ (+1 503) 285 9934
Manager: Dennis Railing

Mobile Office
5785 Plantation Road, Theodore, AL 36582
☎ (+1 334) 653 6900 ⊖ (+1 334) 653 8700
Manager: Bradford White
Office Mgr: Marty Manley
Production Superintendent: Jerry McCants [97]

80-34650 Welt International Corp.
1413 K Street NW, Ste 1400, Washington, DC 20005
☎ (+1 202) 371 0555 ⊠ 281409 taoa ur
⊖ (+1 202) 408 9369
President: Leo G.B. Welt
■ Specialising in offset, counter-trade & counter-purchase fulfilment. Consultants/advisors in these fields. [97]

80-35300 World Trade Transport Corp. (WTT)
Washington/Dulles Int'l Airport, PO Box 17042, Washington, DC 20041
☎ (+1 703) 450 6800
⊠ TWX 710 833 0685 wuttian hndn
⊖ (+1 703) 450 4002
♟ 49 ⌂ 48 000 sq m
President: Henri R. Youngblood

WTT Customs House Brokerage
Washington/Dulles Int'l Airport,
Cargo Bldg 2, PO Box 17042, Washington, DC 20041
☎ (+1 703) 471 4487 ⊠ TWX 710 833 0685
⊖ (+1 703) 450 4002
Marketing Mgr: Jack P. Moore

■ Warehousing, shipping (air & ocean), military spec. packaging, customs house clearance, exhibit logistics. [97]

80-35400 Wornall Electronics Inc.
8901 Washington, Kansas City, MO 64114-3533
☎ (+1 816) 333 62 99 [E] wornall@wornall.com
⊖ (+1 816) 333 27 96
♟ 20 ⌂ 20,000 sq ft
President: Jack L Meltzer
Operations and Technical Director: Glenn Kessler
Sales and Marketing Director: Ed Fried
■ Specialize in design, development and installation of large-scale, integrated security and control systems. [98]

80-36650 Zernow Technical Services Inc.
425 West Bonita, Suite 208, San Dimas, CA 91773
☎ (+1 909) 592 40 02 ⊖ (+1 909) 592 40 02
♟ 2 (1997)
President: Dr Louis Zernow ☎ (+1 909) 592 40 02
■ Design, analysis and experimental evaluation lethality and vulnerability, high strain rate behaviour of materials. [98]

85 Stockists, Distributors, Spare Parts Suppliers

85-00005 Aerial Specialists Inc.
1760 State Street, PO Box 085114, Racine, WI 53408
☎ (+1 414) 835 16 16; 632 01 23
⊖ (+1 414) 632 24 58
♣ $ 1,000,000+ ♟ 7
Managing Dir: Craig A. Ihde
■ Maintenance - high level areas. Stock aerial work platforms. Truck mounted booms & vertical lifts (scissor). [98]

85-00010 Aero Products Associates Inc.
1808 West Third Street, Tempe, AZ 85281
☎ (+1 602) 829 67 89 ⊖ (+1 602) 966 85 08
♣ $ 5,000,000 (1996) ♟ 6 ⌂ 10,500 sq ft
President: Ted Johnson
Sales: Sean Sinclair
Administrator: Ms Amy Johnson-Skalla
■ Distributor for Bell Helicopter airframe and component spares and aftermarket accessories. [98]

85-00030 Aero Quality Sales Co. Inc.
(a division of Western Sky Industries Inc)
47 Harbor View Avenue, Stamford, CT 06902
☎ (+1 203) 351 85 00 ✆ (800) 288 19 54
[W] http://www.aeroquality.com ⊖ (+1 203) 351 85 55
♣ $ 20,000,000 (1996) ♟ 27 (1997) ⌂ 15,000 m2
Vice President: John W Downey
[E] jdowney@poboxes.com
Operations Director: Mark Vilas
Sales and Marketing Director: George Drake
Purchasing Manager: Eric Masna

■ Nickel cadmium aircraft batteries. Ground support equipment including AC and DC ground power units, towbars, jacks and hydraulic units. [98]

85-00040 Aerospace Bearing Support Inc. (ABS)
(† Ruart Bearing Company)
(a subsidiary of Banner Aerospace, Inc.)
11953 Challenger Court, Moorpark, CA 93021-7119
☎ (+1 805) 531 00 01 ✆ (800) 641 8800
[W] http://www.banner.com ⊖ (+1 805) 531 00 05
♣ $ 20,000,000+ ♟ 32 (1998) ⌂ 12,000 sq ft
President and Chief Executive Officer: William C Connell
Executive Vice President and Chief Operating Officer: Ralph Foster
Vice President of Finance: Ms Susan Weikel
Vice President: John Knapp
Advertising Manager: Warren Wilson
Purchasing Manager: Ms Tammy Macaluso
■ Aircraft airframe control bearings needle and track roller bearings, spherical and rod end bearings, miniature and instrument bearings, ball and roller bearings, thrust bearings and journal/sleeve self lubricating bushings. [98]

85-00110 Air Technology Inc.
2884 S. Horseshoe Dr., Naples, FL 34104
☎ (+1 941) 643 0011 ⊖ (+1 941) 643 0940
♟ 12 ⌂ 17 000 sq ft
President: Mike Turner
Engineering Dir: R. Turner
UK Representative: Tony Soames ☎ (+44 749) 673 485
■ Worldwide stocking distributor & maintenance overhaul center for AlliedSignal T53, T55, Alf502, Allison T56, T63, GE J85, T700, CT7 PW Pt6 series engines, C130 aircraft & UH-1H helicopters. [97]

85-00124 American Link Belt Co.
242 Centre St, Trenton, NJ 08611
☎ (+1 609) 392 8014 ⊖ (+1 609) 392 3006
President: Louis Reaback
■ Supply aircraft cargo air delivery systems; clothing and field items, specialized requirements for military items. [97]

85-00125 American West Corp.
PO Box 3124, Manhattan Beach, CA 90266
☎ (+1 310) 318 27 33 ✆ (800) 367 44 96
⊖ (+1 310) 374 73 71
General Manager: William Hedderich
[E] billhedderich@msn.com
Sales and Marketing Director: Jon Angelet
Purchasing Manager: Ms Kristal Fortride
■ Distribute solid state and rotary frequency converters, avionics/laboratory power supplies. Variable frequency power sources and UPS, mil. airborne DC and AC power supplies and converters.
® USA: Advanced Control Systems, Harmonics Ltd, Holec Inc., InterSet West, MTE, Mag-Tek, SafeTech Inc., Wire Benders [98]

85-00180 ASC Industries
(a division of ASC International Inc.)
1227 Corporate Drive West, PO Box 5068, Arlington, TX 76006
☎ (+1 817) 640 13 00 [E] ascind@fastlane.net ⊠ 758276
⊖ (+1 817) 649 26 85
♟ 50 ⌂ 50,000 sq ft
President: Ollin R Taylor
Vice President and General Manager: Bill Foster
Operations Director: Craig Booth
Export Sales Manager: Greg Vinson
Purchasing Manager: Ms Lynn Cook
■ Large stocking distributor of aircraft fasteners (Hardware).
® USA: Adel Fasteners, Anillo, Camloc Fasteners, Fairchild Fasteners, Greer-Smyrna, Kaynar, Parco, Republic, Rexnord, SPS Technologies, TridairFasteners, Unbrako, Valley Todeco [98]

85-00181 Associated Industries
(a division of Arnold A. Semler Inc.)
11347 Vanowen Street, North Hollywood, CA 91605
☎ (+1 818) 760 10 00 [E] talk@associated-ind.com
[W] http://www.associated-ind.com ⊖ (+1 818) 760 21 42
♣ $ 12,000,000 (1996) ♟ 55 (1997) ⌂ 30,000 sq ft
President: Arnold A Semler
International Marketing Director: Ravi Achar

Technical Director: Robert Lee
Purchasing Manager: Sherin Faiez
■ Manufacturer and distribütor of US Government defence products including tactical military communications for portable, ground and vehicle systems, radio link, wire communications, antennas, intercom sets, CVC helmets, installation kits, accessories, ancillary items, specialised test equipment and spare parts. Telecommunications, secure voice encryption, analogue-digital, encrypted cellular and land line telephones, digital central office switches and secure fax. US Government and certified MIL-I-45208A sub-contractor for US and international defence industries. After sales service, system integration, licencing, offset and co-production programmes. 98

85-00182 Atlantic Aviation Aircraft Support Group Inc.
(† Atlantic Aviation Supply Co. Inc.)
(a subsidiary of Atlantic Aviation Corp.)
PO Box 15000, Wilmington, DE 19850
☎(+1 302) 322 72 84 ⊖ (+1 302) 323 01 95
👥 500
Chief Executive Officer: Ben E Waide (III)
Senior Vice President: John A Moyn
Marketing Director: John Elwood
Public Relations Manager: Franklin S Eyster (II)
■ Airframe support for wide variety of aircraft and helicopters. Distribution of electronics, engines, avionics, airframes, electrical parts and accessories. → 80
● 1-A Parkway Business Center, 200-C Lisa Drive, New Castle, DE 19720
☎(+1 302) 322 43 88; 322 37 88
⊖ (+1 302) 324 84 34 98

85-00185 Avia-Sonic Corp.
7921-105 Queenair Dr, Montgomery Airpark, Gaithersburg, MD 20879
☎(+1 301) 977 4900 ⊖ (+1 301) 990 0954
♧ $ 250 000 ♣ $ 750 000 👥 4 🏭 3 000 sq ft
President: Jorge G. Silva
■ Military & commercial airframe & engine parts & accessories. 97

85-00188 Aviall Distribution Services
(a division of Aviall Services Inc)
2075 Diplomat Drive, Dallas, TX 75234
☎(+1 214) 406 20 00 ⊖ (+1 214) 406 20 71
Managing Director: Eric Anderson
Operations and Technical Director: Dan Curtin
Marketing Director: James T. Quinn
Aviall Aerospace
2527 Willowbrook Road, Dallas, TX 75220-4401
☎(+1 214) 956 3400 ⊖ (+1 214) 956 3566
■ Specializes in distributing aerospace fasteners to major aircraft manufacturers & subcontractors. 97

85-00190 Aviation Supply
7935 Tujunga Avenue, North Hollywood, CA 91605
☎(+1 818) 765 59 79 ✉ aviach ⊠ 194415
⊖ (+1 818) 764 19 19
♣ US$ 8,250,000 👥 4 🏭 4,000 m2
Chief of Advertising: C Forrand
■ Commercial and military aircraft parts sales. Consignment inventories/inventories bought. Consulting. Purchasing agents. Specializing in repairables, avionics, instruments, air data. 98

85-00195 Aviojet Corp.
76 Brookside Drive, Upper Saddle River, NJ 07458
☎(+1 201) 825 31 11 ✉ AVIOJET
⊖ (+1 201) 825 69 50
🏭 10,000 sq ft
President & General Mgr: Antonio Ignoffo
Sales Mgr: Anthony Ignoffo (Jr)
Treasurer: Ms Melissa Laera
Purchasing Mgr: Joseph E. Smith
Office Mgr: Ms Rita Ignoffo
Controller: James Matteotti
Customer Service: Paul N. Gatto
■ Supplier of aircraft & engine spares, military & commercial. Manufacture representative. 98

85-00210 Bethpage Associates
PO Box 1799, Costa Mesa, CA 92628-1799
☎(+1 714) 979 1162 ⊖ (+1 714) 979 1165
♣ $ 698 240 (1995) 👥 4 (1996) 🏭 2 500 sq ft
President: Beth Page
VP & Sales Mgr: Jerry Page
Vice President: Gerrad Page
■ Stocking, distributor of electron tubes, semiconductor devices & related avionics military components. 97

85-00273 Caliber Sales Engineering Inc.
3320 North West 53rd Street, Suite 204, Fort Lauderdale, FL 33309
☎(+1 954) 486 23 32 ⊖ (+1 954) 486 13 22
♣ $ 2,500,000 (1997) 👥 5 (1997) 🏭 1,750 m2
President: David R Mazoff
Advertising Manager: Ryan V Heath
Purchasing Manager: Marcela Rojas
■ Sales of aviation, marine and defence equipment including aircraft helicopter spares, ground support equipment and safety/survival equipment. Component maintenance support on ground based radar systems. 98

85-00274 California Eastern Laboratories
4590 Patrick Henry Drive, Santa Clara, CA 95056-0964
☎(+1 408) 988 35 00 W http://www.cel.com
⊖ (+1 408) 988 02 79
♣ US$ 62,000,000 (1997) 👥 180 (1997) 🏭 4,500 m2
President: Jerry A Arden
Vice President of Sales and Marketing: Leonard H Lea
Advertising Director: Ms Jenifer Jacobs
Sales Director: J Barry
Marketing Director: E Schmidt
■ RF and microwave semiconductors, low noise and power GaAs fet's, low noise and power silicon bipolar transistors, GaAs and SI MMIC'S, high reliability screened products. Photo couplers, interupters, IR IR LEDs photo detectors. 98

85-00276 CASA Aircraft USA Inc.
(a subs. of Construcciones Aeronáuticas SA (CASA))
3810 Concorde Parkway, Ste 1000, Chantilly, VA 22021
☎(+1 703) 802 1000 ⊠ 322763 casa inc
⊖ (+1 703) 802 1025
Chairman & CEO: Juan Alonso
President: Douglas A. Bacon
VP Product Support: Robert Wood
■ Sole distributors of C-212 and CN-235 aircraft in USA, Canada and the Caribbean. → 80
Product Support Facilities
Avion Business Park, 3810 Concorde Parkway, Ste 1000, Chantilly, VA 22021
☎(+1 703) 802 1017 ⊠ 322763 casa prod suppt
⊖ (+1 703) 802 1351
Vice President: Robert Wood 97

85-00277 Certified Aircraft Parts Inc
2870 Stirling Road, Hollywood, FL 33020-1199
☎(+1 954) 925 09 66 E cap@icanect.net
⊖ (+1 954) 925 09 88
♣ US$ 30,000,000 (1997) 👥 59 (1997) 🏭 2,755 m2
Vice President and General Manager: R C Moore
☎(+1 954) 925 09 66 ext. 132
Vice President Government Programmes: Bill Irvine
☎(+1 954) 925 09 66 ext. 158
Vice President, Commercial: Jim Bowden
☎(+1 954) 925 09 66 ext. 163
Vice President, Sales: Rod D'Elia
☎(+1 954) 925 09 66 ext. 131
International Marketing Manager: Bill Walls
☎(+1 954) 925 09 66 ext. 117
■ Stockist of L-100/C-130 aircraft parts, repair and overhaul of components. 98

85-00278 Chicago Army Navy Surplus Co.
605 W. Odgen Avenue, Downers Grove, IL 60515-2939
☎(+1 630) 969 17 86; 17 87 E sitco@mcs.com
W http://www.chiarmy.com ⊖ (+1 630) 969 17 88
♣ $ 500,000 (1997) 👥 4 🏭 2,300 sq m
President/Owner: Robert Scott
■ Supply of new & used military clothing & equipment. Will undertake sourcing as required. 98

85-00280 Colonial Trading Corp.
111 Franklin St, Alexandria, VA 22314-3840
☎(+1 703) 549 4042 ⊖ (+1 703) 548 2870
♣ $ 1 900 000 👥 7
President: Stanley J. Bak
■ Exporter of military ordnance spare parts including various weapon & torpedo services. 97

85-00282 Columbus Equipment Co.
50 Kingston Avenue, Columbus, OH 43207-2436
☎(+1 614) 443 65 41 E cec@ascinet.com
W http://www.columbus-equipment.com
⊖ (+1 614) 443 82 57
♧ $ 14,000,000 (1997) ♣ $ 94,000,000 (1997)
👥 190 (1997) 🏭 55,000 sq m
Managing Director: Robert Olejniczak
Operations Director: Michael Sarrey
Sales and Marketing Director: Tom Stivison
Government Contracts Manager: Eric Tubbs
■ Heavy construction equipment, parts & services. Sales & leasing. Cranes, Bulldozers, loaders, backhoes, rollers, pavers. 98

85-00285 Commonwealth Trading Corporation
8232 Kings Arm Drive, PO Box 6280, Alexandria, VA 22306-0280
☎(+1 703) 799 0836 ⊖ (+1 703) 799 0837
President: Harvison W. Hunt
VP Sales: Glenn E. Daniels
■ US DoD certified contractor & supplier of military equipment & spares for army, navy & air force applications. 97

85-00300 C-Way Industries Inc.
(a subsidiary of Aerotech International Inc.)
400 South LaSalle, Spencer, WI 54479
☎(+1 715) 659 44 61 ⊖ (+1 715) 659 45 57
👥 3
Managing Director: Paul R Sigurdson
Sales and Marketing Director: Ms Penny Schmutzler
■ Supplier of aircraft ground support equipment. 98

85-00350 Hunter Industries Inc.
4 Stickney Terrace, PO Box 1200, Hampton, NH 03842
☎(+1 603) 926 88 81 ✆ (800) 468 80 00
⊖ (+1 603) 926 78 55
President: Douglass Hunter E doug@ddaircraft.com
Managing Dir: Jan Staalenburg
Sales Dir: Stephanie Bolduc
■ Distributor of stainless fasteners to the general aviation aircraft industry (AN-MS MIL-SPEC). Originator of the S/S Kits (stainless exterior trim screw & washer kits. 98

85-00370 Da-Green Electronics
37 Main St, South River, NJ 08882-0267
☎(+1 908) 254 2735 ✆ (800) 343 2677
⊖ (+1 908) 254 9172
♧ $ $ 2 000 000 (1995) ♣ $ 4 000 000 (1995)
👥 30 (1995) 🏭 3 700 sq m
President: Barry E. Greenberg
Sales & Marketing Dir: Ken Clark
Purchasing Mgr: Marc Gable
■ Stockist & distributor of connectors & tools for military & civil industries. 97

85-00440 Derco Aerospace Inc.
8000 W. Tower Avenue, PO Box 25549, Milwaukee, WI 53225
☎(+1 414) 355 30 66
W http://www.dercoaerospace.com ⊠ 260342
⊖ (+1 414) 355 61 29
♣ $ 81,000,000 (1996) 👥 170 (1997) 🏭 95,000 sq m
President: Eric Dermond
Vice President: Max Dermond
General Mgr: Mark Hoehnen
Dir Sales & Customer Service: Lisa Champagne Hade
Dir Materials & O.E.M. Support Services: Paul McBride
Marketing Communications Supervisor: Ms Michele Miller
■ Specialist in the distribution & maintenance of Original Equipment Manufacturer (OEM) spares & components for the C130, P-3, F-4, A-4, F-16, F-5 and F-111. 98

85-00450 D.J.Associates Inc
8411 SO.Zero Street, Fort Smith, AR 72903-7097
☎(+1 501) 452 39 87 E dja@pia.com
⊖ (+1 501) 452 77 52
♧ US$ 500,000 👥 10 (1997) 🏭 4,000 sq ft
President and Chief Executive Officer: Ms I J Beck
Vice President: M P Beck
Secretary and Treasurer: Ms I J Beck
■ AN/MS/NAS and Military spec. hardware for parachutes, cargo restraint and electronics.
Wholly Owned Subs::
Transcon Manufacturing and Amrad Division 98

85-00520 Encore Computer Corp.
6901 W.Sunrise Blvd, Fort Lauderdale, FL 33313
☎(+1 945) 587 2900 ✆ (800) 933 6267
⊖ (+1 945) 797 5793
650 250 000 sq ft
Chairman & CEO: Kenneth G. Fisher
President: Rowland H. Thomas (Jr)
VP Marketing: Michael Parker
VP US Storage Sales: Timothy Logan
VP Int'l Real Time Applications Sales: Robert DiNanno
VP Development: George S. Teixeira
VP Development: Ziya Aral
VP Human Resources & Corp. Relations: Charles S. Anderson
Marketing Communications Dir: Ms Lynne Connors
■ Leading supplier of open, scalable computer & storage systems for data center and real time applications. 97

85-00530 Evair Associates Inc.
(† Evans Industries Inc.)
Wolters Industrial Park, bldg 209,
Mineral Wells, TX 76067
☎(+1 817) 328 1373 ⊖ (+1 817) 325 9234
$ 1 500 000 6 32 000 sq ft
President: Richard Neil Evans
Vice President: Lorin Evans
Secretary: Ms Mary L. Capps
General Mgr: John Wesley Evans
■ Manufacture & supply aircraft spares & components for McDonnell-Douglas F-4 & General Dynamics F-16 aircraft. Sub-speciality aircraft armament systems 97

85-00600 Flame Enterprises Inc.
20945 Osborne Street, Canoga Park, CA 91304
☎(+1 818) 700 29 05 ✆ (800) 854 22 55
⊖ (+1 818) 700 91 68
US$ 6,000,000 (1995) ♣ US$ 11,800,000 (1995)
28 (1995) 20,000 m2
President and Chief Executive Officer: Michael V Epstein
Operations and Technical Director: Neil M Rostholder
Sales Manager: Bob Baker
Purchasing Manager: Andrew Franke
■ Stocking distributor of relays, switches and circuit breakers. 98

85-00710 Gemini International Corp.
11 Hacienda Dr., Arcadia, CA 91006-2347
☎(+1 818) 355 1016 ⊖ (+1 818) 355 7357
3
President: Fred Jahnke
■ Military specification, rugged printers, plotters, rugged data acquisition systems. 97

85-00800 Hardware Specialty Co. Inc. (HS)
48-75 36th Street, Long Island City, NY 11101
☎(+1 718) 361 93 93 E hardspeci@juno.com
125468 ⊖ (+1 718) 482 88 16
♣ $ 48,000,000 (1995) 262 135,000 sq ft
Vice President: Arthur Z. Cohen
Managing Dir: Edward Kaufman
Purchasing Mgr: Gail Steets
■ Authorised stockist for US military/ aerospace fasteners & electronic hardware. Over 250 000 different items in stock. Approved by major international aerospace & military electronic industry companies. 98

85-00822 Carl G. Hokanson & Associates Inc.
Hokanair
16021 Woodvale Rd, Encino, CA 91436
☎(+1 818) 907 0432 ⊖ (+1 808) 788 9920
5
President: Carl G. Hokanson
■ Management consultants, aviation equipment & leasing, parts & equipment sourcing for overseas customers. Manufacturers representantive: wheelchocks, ground equipment, aircraft parts, location of difficult item/parts for military aircraft & systems. 97

85-00835 Imrex Company Inc.
307 East Shore Road, Great Neck, NY 11023
☎(+1 516) 466 52 10 EXPORTECH
⊖ (+1 516) 773 44 74
Chairman: Gidon Armon
President: Jack Armon
Vice president: Harry Armon
Vice President: Alan Weiss
■ Worldwide distributors of military & commercial components & spares for aircraft, missiles, ordnance including engines, airframe, raw material, avionics, hardware electronics, instruments. Overhaul services. → 80 98

85-00920 International Arms Sales Corporation
(a div. of Dion & Lee Inc.)
319 Ellis Street, Suite # A, Augusta, GA 30901
☎(+1 706) 722 7229 ⊖ (+1 706) 722 7229
3 3 600 sq m
Managing Dir: Michael D. Nelson
Sales Dir: D. Lee Nelson
■ Military hardware, uniforms, military food supplies & tactical police weaponary & hardware. 97

85-00950 International Helicopter Parts
1905 Aviation Boulevard, Lincoln, CA 95648-9557
☎(+1 916) 645 89 19 ✆ (800) 447 46 85
E ihp@ihpww.com W http://www.ihpww.com
⊖ (+1 916) 645 79 91
♣ $ 5,500,000 (1996) 15 (1996) 24,000 sq ft
Vice President and Managing Director: J F Koshinsky
Sales and Marketing Director: Barry Quintarelli
Operations and Technical Director: Harry Ferguson
Purchasing Manager: Charlie Jones
■ Sikorsky and Bell helicopter parts specialists and stockist. FAA Repair Station QRSR885K. 98

85-00990 Interstate Wire Co. Inc.
10355 Sanden Drive, Dallas, TX 75238
☎(+1 214) 553 1311 ✆ (800) 527 0010
TWX 910 960 5084 ⊖ (+1 214) 348 7106
60 000 sq m
President: Jeff Fleming
VP Sales & Marketing: Chris B. Brandon
■ Distribute Teflon & Tefzel hi-temperature wire & cable, MIL-SPEC & commercial tubing & associated services. 97

85-01000 Intro Corporation
12252 South West 131 Avenue, Miami, FL 33186
☎(+1 305) 251 89 95 ✆ (800) 722 22 33
⊖ (+1 305) 251 02 53
President: Doug Chagnon
■ Distribute electrical connectors, circular, rack and panel, D-subminiatures and audio connectors. Boeing and commercial applications. 98

85-01010 Jensen Tools Inc.
(a division of Stanley)
7815 S 46th Street, Phoenix, AZ 85044-5399
☎(+1 602) 968 62 31 ✆ (800) 426 11 94; 366 96 62
W http://www.jensentools.com ⊖ (+1 602) 438 16 90
150 (1997)
Vice President and General Manager: Peter Sturman
Vice President of Marketing and Sales: Gary Treiber
Marketing Services Manager: Gary Lapointe
Supervisor of International Sales: Ms Mary S Sliger
■ Tools and tool kits for servicing electronic products in the defence industry. 98

85-01050 Jet Engineering Inc.
PO Box 36100, Pensacola, FL 32516-6100
1621 Dog Track Road, Pensacola, FL 32506
☎(+1 904) 453 31 46 702413 ⊖ (+1 904) 453 85 16
♣ $ 4,000,000 14 22,500 sq ft
President: Robert C Walls
■ Spare parts for Lockheed, C-130/L100; Allison T-56/501, Pratt & Whitney PT6/PT6T engines. 98

85-01080 JMS Electronics Inc.
13125 Saticoy Street, North Hollywood, CA 91605-3403
☎(+1 818) 764 33 03 E jmselex@ix.netcom.com
⊖ (+1 818) 764 48 75
12
Chief Executive Officer and President: Martin E Klopert
Vice President of Sales and Marketing: Mrs Dru Woodward
■ Distributor of electronic components, specializing in Diodes, transistors and integrated circuits. 98

85-01120 Jonas Aircraft & Arms Co. Inc.
225 Broadway, New York, NY 10007
☎(+1 212) 619 0330 ⊖ (+1 212) 619 2743
President: G.S. Steinemann
Director: Robert S. Jonas
■ Export distributors for aircraft & engines, parts & accessories. Sales worldwide. Lycoming distributors, Western Hemisphere distributors. Pilatus BN-2 Islander & Trislander.
® Switzerland: Pilatus Aircraft
UK: Automotive Products Co, Pilatus Britten Norman, Rose Bearing
Miami Office
Chapel Trail Commerce Center,
911 NW 209th Avenue, Ste 113,
Pembroke Pines, FL 33029
☎(+1 954) 430 0450 ⊖ (+1 954) 430 5624 97

85-01150 John Jovino Co. Inc.
5 Centre Market Place, New York, NY 10013
☎(+1 212) 925 48 81 620384 jovco
⊖ (+1 212) 966 49 86
28 14,500 sq ft
Managing Director: Charles Hu
President: Anthony Imperato
Purchasing Manager: Terry Rush
■ Distr. of firearms, police & military equipt → 80
Associated company:
AMAC Inc → 75 98

85-01240 MAC Aerospace Corporation
14301-I Sullyfield Circle, Chantilly, VA 22021-1630
☎(+1 703) 502 8300 ⊖ (+1 703) 502 8303
♣ $ 9 000 000 20 20 000 sq ft
President: J. Rodríguez
Marketing Dir: S. White (Jr)
General Mgr: L. Colareta
Marketing Co-ordinator: A. Mitchum
■ International supplier of space parts for military aircraft & weapon system. Approved OEM vendor. Authorised OEM distributor. 97

85-01250 Manufacturers Export Organization
9707 Mason Ave, Chatsworth, CA 91311
☎(+1 818) 998 2527 683987 elmfg uw
⊖ (+1 818) 882 8483
$ 500 000 (1995) ♣ $ 5 000 000 9 10 000 sq ft
President: Sam Ghoraishy
Sales Dir: Fred Ball
Operations Dir: Ms Selina Waters
Purchasing Mgr: Ms Maryam Tah
■ Supplier of aircraft spares, commercial & military: Boeing, Douglas, Lockheed. Avionics, electronics, engine & airframe parts. Accessory & standard parts. Manufacturing capability for sheet metal & precision machined parts. → 75 97

85-01300 Mardave Co.
1845 E. Hwy 199, Azle, TX 76082
☎(+1 817) 677 2918 ⊖ (+1 817) 677 5658
4 10 000 sq ft
President: M. Stepich
Vice President: Bob Radecki
■ Stockist of F111, F4, B52, KC-135 airframe hardware, electrical, instrument, accessories, electronics. 97

85-01325 Messier-Dowty Customer Support Center - Americas
(† Dowty Aerospace Customer Support Ctr)
(a member of the TI Group)
PO Box 49, Sterling, VA 20167
☎(+1 703) 450 8200; (+1 202) 928 1269 AOG
⊖ (+1 703) 430 4932
Commercial Mgr: Richard Dogaer
Technical Mgr: Jim Aaron
■ Provides spare parts, AOG service and technical support for Messier-Dowty & Dowty Aerospace in the Americas. 98

85-01343 Militech International
PO Box 410, Ashburn, VA 22011
☎(+1 703) 729 57 00 ⊖ (+1 703) 729 67 00
♣ $ 4,900,000 (1996) 6
President: James A Oberdick E oberdick@erols.com
Sales and Marketing Director: Ms Linda Birchfield
Operations Director: Scott Oberdick
Purchasing Manager: Ms Lisa Nelson
■ Supplier of military small arms ammunition, tactical personal armour, anti-riot equipment, individual clothing and equipage for military and special forces units worldwide. 98

Contact Reference

85-01345 **Montecito Trading Co.**
710 North Second Street, St Louis, MO 63102
☎ (+1 314) 421 1712 ⊠ 415649 ⊖ (+1 314) 421 1714
Chairman: Dale E. Cloutier
Export Dir: Stephen C. Mitchell
■ Military aviation parts, equipment, chemicals & tools. [97]

85-01370 **National Audio-Visual Supply**
Route 121E, PO Box 116, Grafton, VT 05146
☎ (+1 802) 869 61 16 ✆ (800) 222 0109
[W] http://www.natav.com ⊖ (+1 802) 869 61 19
👥 26
President: Scott E. Heller
VP Sales & Marketing: Robert Singleton
■ Complete line of audio-visual & video equipment, accessories & supplies for large & small presentation, briefing & training applications. [97]

85-01375 **NavSymm Positioning Systems**
85 West Tasman Drive, San José, CA 95134
✆ (888) 367 79 66 ⊖ (+1 408) 428 79 98
👥 5
Marketing and Sales Director: Joe Blanda
■ Military GPS Receivers, 6 and 12 channel versions. XR5 kernel-OEM GPS boardsets. UHF datalink. RTK systems. [98]

85-01396 **Nortek Repair Center**
(† NorTek Aviation Inc.)
7550 North West 26th Street, Miami, FL 33122
☎ (+1 305) 593 60 38 ⊠ 807334 nortekmia
⊖ (+1 305) 593 94 83
👥 120 🏢 150 m2
President: J R Montalvo (III)
■ FAA and JAA certified repair station specializing in instrument and hydraulic accessories. Sale of rotable and expendable components. [98]

85-01397 **Omicron Steel Products Co Inc.**
(The Omicron Group)
PO Box 150326, Kew Gardens, NY 11415-0326
☎ (+1 718) 805 3400 ⊖ (+1 718) 805 3401
♧ $ 250 000 000 (1995) ♣ $ 1 000 000 (1995) 👥 6
🏢 15 000 sq m
President: Jeffrey S. Blaugrund
Operations/Technical Dir: Nancy Mint
Sales Marketing Dir: Scott Reynolds
Purchasing Mgr: John Burrell
■ Suppliers of spares for general aviation & air transport, military hardware & electronics, commercial test & measuring equipment, industrial electronics spares. Military safety & survival equipment, industrial equipment, conveyor systems, industrial casters, industrial racks. [97]

85-01398 **Pac-Rim Aviation**
1227 North Division Street, Spokane, WA 99202-1729
☎ (+1 509) 326 15 20 [E] migdriver@juno.com
[W] http://www.data -by-design.com/pac-rim1.htm
⊖ (+1 509) 326 14 90
♣ US$ 10,000,000 (1996) 👥 3
Vice President Aviation Group: Mike Marks
■ International trade and commerce supplying everything from transportation, construction, industrial, manufacturing, and food processing to aviation. [98]

85-01399 **Palomar Products Inc**
(a subs. of Palomar Technologies Inc.)
2051 Palomar Airport Road, Carlsbad, CA 92009
☎ (+1 619) 931 3222 ⊖ (+1 619) 931 5198
♣ $ 100 000 000 (1995) 👥 300
Managing Dir: Gary E. Gist
CEO: Parke Hess
Marketing Dir: Rick Moore
Sales Dir: Val Policky
■ Image displays: monochrome/colour sunlight readable (CRT/LCD); biocular/monocular displays audio distribution systems/ communication distribution systems, digital switching systems. Automatic test & wire bonding equipment.

East Coast Field Office
329 Marba Road, Severna Park, MD 21146
☎ (+1 410) 647 0434 ⊖ (+1 410) 647 6157
President AML Wireless Systems: Cliff Gorby
President HID Corporation: Don Nelson [97]

85-01420 **Pars International Corporation**
PO Box 37184, Louisville, KY 40233
☎ (+1 502) 363 00 00 [E] pars@ntr.net
[W] http://www.ntr.ntr/~pars ⊠ 856841 pars intl ud
⊖ (+1 502) 367 89 90
Director: Saeid Shafizadeh
■ Importer/exporter of high explosives, destructive devices, firearms and ammunition. Owner/operator of US Customs high security bonded warehouse. Specialising in law enforcement and commercial markets. [98]

85-01470 **A.E. Petsche Co.Inc.**
2112 West Division, Arlington, TX 76012
☎ (+1 817) 461 94 73 [W] http://www.aepetsche.com
⊠ 758290; 910-8904905 ⊖ (+1 817) 277 28 87
♣ $ 50,000,000 (1997) 👥 80 (1997) 🏢 23,200 m2
Managing Director: Glen Davidson
Marketing Director: Lee Carmody
Operations Director: Fred Savage
■ Distribution of aircraft and aerospace, high performance wire and cable. Military and OEM specified products available from stock. Harness management products, including shrinkable tubing, termination devices, cables ties, spiral wrap and lacing cord. Multiple stocking locations in North America.
® USA: Raychem Corporation, Specialty Cable, Teledyne Thermatics, Tensolite Division of Carlisle, Times Fiber [98]

85-01522 **Rotadata Telecommunications Inc**
1211 Semoran Boulevard, Suite 101,
Casselberry, FL 32707
☎ (+1 407) 679 12 74 ⊖ (+1 407) 679 84 55
® UK: Rotadata Ltd [98]

85-01800 **SEA Wire & Cable Inc.**
110 Celtic Circle, PO Box 647, Madison, AL 35758
☎ (+1 205) 772 96 16 ✆ (800) 633 7210
[E] seawc@traveller.com ⊖ (+1 205) 772 74 02
👥 82 (1997) 🏢 4,000 sq m
President: Bob Wills
Sales and Marketing Director: Jerry Sandlin
Operations Director: Marty Clark
Purchasing Manager: Ms Judy Germany
■ Supplier of Mil-Spec wire, cable, tubing and ties. Wire cutting and stripping. Wire twisting, cut tubing, cut and marked tubing and wire printing.
● 2525 Ridgmar Boulevard, Suite 208,
Forth Worth, TX 76116
☎ (+1 817) 377 39 04 ⊖ (+1 817) 377 39 24 [98]

85-01920 **Schweizer Aircraft Corp.**
PO Box 147, Elmira, NY 14902
☎ (+1 607) 739 38 21 [E] schwzer@aol.com
[W] http://www.schweizer-aircraft.com
⊠ 932459 schweizer bigf ⊖ (+1 607) 796 24 88
👥 400
President: W. Stuart Schweizer
Exec. Vice President: Leslie E. Schweizer
Exec. Vice President: Paul H. Schweizer
Sales Manager: David Savage
Programme Mgr: Rocky Peters
Marketing Mgr: Ms Barbara J. Tweedt
■ Manufacture aircraft, sailplanes, helicopters, aircraft parts & assemblies. [98]

85-01990 **Smith Industries Defense Systems International**
(† SLI International)
(a division of Smiths Industries, UK)
One Circle West, PO Box 10377,
Stamford, CT 06904-2377
☎ (+1 203) 351 83 00 ⊖ (+1 203) 351 83 05
President: J.B. Hines
VP Marketing: N. Cugliari
■ Int'l marketing of military displays & systems.
® UK: Smiths Industries A&D Syst.
USA: Smiths Industries A&D Syst. [98]

85-02050 **Spot Image Corp.**
(a subsidiary of Spot Image SA, France)
1897 Preston White Drive, Reston, VA 20191-4368
☎ (+1 703) 715 31 00 ✆ (800) ASK SPOT
[W] http://www.spot.com ⊖ (+1 703) 648 18 13
👥 50 (1997)
President: Ted Nanz
Vice President of Sales: Dan Carayiannis
Vice President of Finance: Neal Carney
Vice President of Technical Operations: Ron Hodge
Marketing and Communications Manager: Clark Nelson
New Products Manager: Mark Whalley
Marketing Specialist: Ms Colleen Hanley
■ Spot satellite imagery and derived products including models and land classifications. [98]

85-02095 **Tam Electronics Inc.**
10 Conover Place, Little Silver, NJ 07739
☎ (+1 908) 530 07 68 ✆ (800) 845 98 60
[E] tam-electronics@worldnet.att.net
⊖ (+1 908) 530 08 65
♧ $ 350,000 (1997) ♣ $ 1,800,000 (1997) 👥 5 (1997)
🏢 1,500 m2
President: Christian F Decker
President: W Gary Fendelander
■ Spare parts. Audio telecommunications, cables, wires, connectors, handsets, headsets. All parts by NSN or P/N. → 80 [98]

85-02350 **Trace International Group Ltd TRACE Worldwide Corp.**
1720 South Catalina Avenue, Redondo Beach, CA 90277
☎ (+1 310) 316 23 00 [E] tigl@traceww.com
⊖ (+1 310) 316 23 25
👥 45
President and Chief Executive Officer: John Douglass
Executive Vice President: Robert Mabli
Vice President of Operations: Rori Wells
■ Global representation network, export distribution and project management. Headquarters in Los Angeles, regional centers in London, Moscow and Tokyo. Nineteen marketing and engineering offices worldwide. Aircraft products and systems include onboard aircraft power distribution, commercial and defence electronics, electromechanical systems, onboard lighting, rescue and evacuation de-icing, airline seats and cabin equipment. [98]

85-02450 **United Alloys Inc. Aircraft Division (UAI)**
3398 Leonis Boulevard, Vernon, CA 90058
☎ (+1 213) 264 51 01 ✆ (800) 624 80 73
⊖ (+1 213) 262 11 72
Vice President: Arthur R Sacks
General Manager: F I Mackie
■ Metals distributor specialising in Titanium, Nickel and Cobalt base alloys, refractory metals. [98]

85-02600 **US Airmotive Inc.**
(a subsidiary of US Airmotive Holdings, Inc.)
PO Box 522514, Miami Springs, FL 33152
5439 Norht West 36th Street, Miami Springs, FL 33166
☎ (+1 305) 885 49 91 ✉ USAIR
[E] usair1@ix.netcom.com
[W] http://www.montacargas.com ⊠ 6811141 usair
⊖ (+1 305) 887 24 05
♧ $ 1,600,000 (1996) ♣ $ 10,000,000 (1996)
👥 65 (1996) 🏢 51,000 sq ft
Chairman: Edward A Kruse
President: John Kruse
Vice President and General Manager: Frank G Cale
Executive Vice President: Frank J Bortunk
Assistant Vice President: David Smouse
Director of Advertising and Public Relations: Ms Maria Pérez
Comptroller: John E Kruszewski
Product Manager: Bernie Gil
■ Supplier of parts and accessories for airframe, engines and ground equipment. Brokers and consultants. Manufacture aircraft ground power cables. → 75

Orlando FL Branch
1815 Acme Street, Orlando, FL 32805-3758
☎ (+1 407) 843 25 65 ⊖ (+1 407) 648 58 45
Vice President: David Hensley [98]

85-02610 **US Airmotive Forklifts, Inc**
7855 North West 77th Avenue, Miami, FL 33166
☎ (+1 305) 885 43 13 ⊖ (+1 305) 883 67 07
Branch Manager: Ronald Paredo
■ Distributor for Linde-Baker and TCM forklift Trucks. [98]

85-03500 **Walker & Associates Inc.**
(a subsidiary of the Walker Group Inc.)
7129 Old Highway 52, PO Box 1029,
Welcome, NC 27374-1029

☎ (+1 910) 731 63 91 ✆ (800) 472 17 46
👥 175 (1995)
Chief Operating Officer: Tim Lawing
Executive Vice President, General Manager: Farley Davis
Special Projects Manager: Lee Thorne
■ Telecommunications equipment. 98

85-03900 Wesco Manufacturing Inc.

(a subsidiary of Wesco Aircraft)
Plaza Building, 10 Powerhouse Road,
Roslyn Heights, NY 11577-0127
☎ (+1 516) 621 60 60 ⊖ (+1 516) 621 60 62
♣ $ 120,000,000 👥 300 🏭 160,000 sq ft
Chairman: Randy Snyder
President: Jay Israel
Vice President: Lee Israel
■ Major components and spare parts for US manufactured military equipment. 98

85-03950 Westair International USA Inc.

(a subsidiary of Aero Tec Data)
PO Box 771, Monument, CO 80132
☎ (+1 719) 481 22 86
E westair.aerotecdata@worldnet.att.net
⊖ (+1 719) 481 22 03
👥 7 🏭 7,500 sq ft
President: Jim Cullen
Operations and Technical Director: Kevin Cullen
Sales and Marketing Director: Grace Wrabley
Purchasing Manager: Ms Grace Cullen
■ Cargo and tactical aircraft sales. Spares support. Aircraft ferrying worldwide. Technical manuals, flight maintenance, IPB, airframe, engine, civilian/military aircraft and helicopters. 98

90 Sales Agents, Sales & Liaison Offices

Companies represented:

3M, USA . . . 75-63590
3M, USA . . . 75-73860
Aardvark Group, UK . . . 75-31560
Acoustics Products Corp., USA . . . 90-00920
ACR Electronics, USA . . . 90-20730
ACRO Industries, USA . . . 90-23760
Adel Fasteners, USA . . . 85-00180
Advance Hivolt, UK . . . 90-00200
Advanced Control Systems, USA . . . 85-00125
Advanced Switch Technology, USA . . . 90-05020
Aerospace Control Products, USA . . . 90-20730
Aerospace Engineering Support Corp., USA . . . 90-15530
Aerospatiale, France . . . 90-00260
Aerotech Inc., Canada . . . 90-00275
Air N Sea Pty Ltd, Australia . . . 75-53030
Airborne Div., USA . . . 90-15530
Airspace Products Ltd, UK . . . 75-53030
Alcan Aluminium, Canada . . . 75-73860
Alcan Aluminium, USA . . . 75-63590
Alcan Aluminium, USA . . . 75-73860
Alfred Karcher GmbH & Co, Germany . . . 75-05940
Alfred Karcher GmbH & Co, Germany . . . 90-11100
Allen Aircraft Products, USA . . . 90-20730
Allison Engine Company, USA . . . 90-00240
Alternative Fuel Systems Rexworks Int'l, USA . . . 90-16725
American Aviation, Israel . . . 75-53030
American Minerals, USA . . . 75-63590
American Minerals, USA . . . 75-73860
Amesco, USA . . . 90-03740
AMS Ground Support, France . . . 80-00845
Anillo, USA . . . 85-00180
Ansaldo, Italy . . . 75-83570
ArconConsultants, Ireland . . . 90-03740
Armtech Industries, USA . . . 90-14600
Army Material Command, Denmark . . . 90-03740
Ashot Ashkelon Ind., Israel . . . 90-01350
Assemblies Inc., USA . . . 90-20100
Automotive Products Co, UK . . . 85-01120
Avesta Nyby, Sweden . . . 75-63590
Avesta Nyby, Sweden . . . 75-73860
Aviation Services Int'l SA, Spain . . . 90-03680
Aviel, USA . . . 90-05020
Avimo Electro-Optics Pte Ltd, Singapore . . . 90-01835
Avon Inflatables Ltd, UK . . . 90-08990
Avon Inflatables Ltd UK, USA . . . 90-01880
Baikowski Int'l, Pechiney, France . . . 75-73860
Baikowski International, Pechiney, France . . . 75-63590
BaseWest, USA . . . 90-20730
Baxter Electro und Motorentchnik, Germany . . . 90-03740
BEI Sensors and Systems, USA . . . 90-14600
Beretta SpA, Italy . . . 75-13605
BF Goodrich, USA . . . 90-14600
Bharat Electronics, India . . . 90-02380
BP Solar Limited, UK . . . 90-02600
Breeze Eastern, USA . . . 90-20730
Bridgestone Tires, Japan . . . 80-32800
British Steel Engineering Steels, UK . . . 90-02750
Brüel & Kjaer, Denmark . . . 90-02800
C-Tech Ltd, Canada . . . 90-03000
Cabot, Canada . . . 75-63590
Cabot, Canada . . . 75-73860
California Instruments Inc., USA . . . 90-20220
Camloc Fasteners, USA . . . 85-00180
Canadair Defence Systems Division, Canada . . . 90-03080
Carlton Technologies, USA . . . 90-14600
Cartwright Electronics Inc., USA . . . 90-20220
Cerac, Canada . . . 75-63590
Cerac, Canada . . . 75-73860
Cerberus-Guinard SA, France . . . 90-03280
Cessna Aircraft, USA . . . 90-14600
Cessna Aircraft Raytheon, USA . . . 90-14600
Chelton Electrostatics Ltd, UK . . . 75-18120
Chelton/Rayan, France . . . 75-18120
Cleveland Trencher Co., USA . . . 90-16725
Cleveland Wheel & Brake, USA . . . 90-15530
Co-Operative Industries, USA . . . 90-20730
Conax Florida Corp., USA . . . 90-20730
Control Products Corp., USA . . . 90-20730
Cook Airtronic, USA . . . 90-20730
Crane Lear Romec, USA . . . 90-14600
Crompton Instruments Ltd, UK . . . 90-03750
CSA Ltd, UK . . . 90-03760
Custom Electronics Inc, USA . . . 90-03800
Custom Electronics Inc, USA . . . 90-08055
Custom Electronics Inc, USA . . . 90-09050
Custom Electronics Inc, USA . . . 90-13130
Custom Electronics Inc, USA . . . 90-14200
Custom Electronics Inc, USA . . . 90-16300
Custom Electronics Inc, USA . . . 90-18800
Custom Electronics Inc, USA . . . 90-20500
Custom Power Systems, USA . . . 90-05020
Customs Electronics Inc, USA . . . 90-16722
Daewoo Corp, Korea (ROK) . . . 90-03900
Daicel Chemical Industries, Japan . . . 75-28325
Daicel Chemical Industries Ltd, Japan . . . 90-04010
David Sarnoff, USA . . . 90-05020
Dayton-Granger, USA . . . 90-20730
De Gussa, Metallgesellschaft, Germany . . . 75-63590
De Gussa, Metallgesellschaft, Germany . . . 75-73860
DEC Technologies, USA . . . 90-20730
Die-Matic Products Inc., USA . . . 90-20730
Dornier GmbH, Germany . . . 90-04660
Dow Chemical, Canada . . . 75-63590
Dow Chemical, Canada . . . 75-73860
Dupont, Canada . . . 75-63590
Dupont, Canada . . . 75-73860
Dynapower/Stratopower (Vickers), USA . . . 90-20730
E.H.Industries Ltd, UK . . . 90-05300
EEV Ltd, UK . . . 75-29500
ELI/ELBO, Italy . . . 90-19130
Elkem, Canada . . . 75-63590
Elkem, Canada . . . 75-73860
Embraer, Brazil . . . 90-05520
EMI Technologies Inc., USA . . . 90-20220
Engelhard, Canada . . . 75-63590
Engelhard, Canada . . . 75-73860
Engine Systems Division, BFGoodrich, USA . . . 90-20730
Engineered Magnetics, USA . . . 90-00920
ESDU International plc, UK . . . 90-05820
Eurocopter, Germany . . . 75-05790
Eurocopter, France . . . 75-05790
Fairchild Fasteners, USA . . . 85-00180
Fairey Arlon Ltd, UK . . . 75-33330
Fairey Hydraulics Ltd, UK . . . 90-06040
Filter Microwave, USA . . . 90-05020
FL Jennings, USA . . . 90-13440
Frank & Schulte, Germany . . . 75-63590
Frank & Schulte, Germany . . . 75-73860
Fuji Industries Co. Ltd, Japan . . . 90-06830
G&H Technology, USA . . . 90-20730
GAF, Canada . . . 75-63590
GAF, Canada . . . 75-73860
GEC Marconi Aerospace, USA . . . 90-20730
GEC-Marconi Avionics Ltd, UK . . . 75-37413
Gentex Int'l Inc., USA . . . 90-20730
Getac Corp., Taiwan . . . 90-07290
GKN Westland plc, UK . . . 90-39670
Gloves, USA . . . 90-20730
Greer-Smyrna, USA . . . 85-00180
Gull Electronics., USA . . . 90-15530
Hadland Photonics Ltd, UK . . . 90-08030
Handy & Harmon, Canada . . . 75-63590
Handy & Harmon, Canada . . . 75-73860
Harmonics Ltd, USA . . . 85-00125
Hart Metals, USA . . . 75-63590
Hart Metals, USA . . . 75-73860
Hayes Targets, USA . . . 90-20730
Heckler & Koch, Germany . . . 75-42860
Hoegannes, USA . . . 75-63590
Hoegannes, USA . . . 75-73860
Holec Inc., USA . . . 85-00125
Honeywell Inc., USA . . . 90-15530
Hughes-Treitler, USA . . . 90-20730
Hybrid-Tek Inc., USA . . . 90-20100
ILC Dover, USA . . . 90-20730
IMC Magnetics, USA . . . 90-20730
INCO, Canada . . . 75-63590
INCO, Canada . . . 75-73860
Inertia Switch, Inc., USA . . . 90-00920
Int'l Target Systems Corp, USA . . . 90-20220
InterSet West, USA . . . 85-00125
Irvin Aerospace Canada Limited, Canada . . . 90-09690
Israel Aircraft Industries, Israel . . . 90-09760
Ivex Corp, Canada . . . 90-14600
J.P. Sauer & Sohn GmbH, Germany . . . 75-80700
JDA Enterprises Corporation, Japan . . . 90-03740
JDR Cable Systems BV, Netherlands . . . 90-10100
Johnson Matthey, UK . . . 75-63590
Johnson Matthey, UK . . . 75-73860
Johnson Patching Plants, USA . . . 90-16725
Kaynar, USA . . . 85-00180
Krauss-Maffei AG, Germany . . . 90-11700
La Roche, USA . . . 75-63590
La Roche, USA . . . 75-73860
Labinal, France . . . 80-18100
Lapes Equipment, USA . . . 90-20730
Lear Astronics, USA . . . 90-14600
Lee Labs, USA . . . 90-05020
Lentern (Aircraft)Ltd, UK . . . 90-12350
Litton ILSD, USA . . . 90-20730
Lloyd Instruments Ltd., UK . . . 90-03400
LMS International, Belgium . . . 90-12600
Logistics Pty Ltd, Australia . . . 80-00250
Lohmann & Stolterfoth, Germany . . . 90-19150
Lucas Aerospace, USA . . . 90-14600
Lucas Western Inc., USA . . . 90-15530
Lucky Gold Star Components, USA . . . 90-05020
Luwa Ltd, Switzerland . . . 90-13020
Mag-Tek, USA . . . 85-00125
Manitou BF, France . . . 90-11110
Marathon Battery Co., USA . . . 90-20730
Marconi Instruments Ltd, UK . . . 90-13075
Marotta Scientific Controls, USA . . . 90-00920
MCLI, USA . . . 90-05020
Metric Systems, USA . . . 90-05020
Micro Metalsmiths, UK . . . 90-11900
Micro Metalsmiths, UK . . . 90-13300
Micro Metalsmiths Ltd, UK . . . 90-06870
Micro Metalsmiths Ltd, UK . . . 90-20222
Microsystems Inc., USA . . . 90-20220
Microwave Communications Laboratories, USA . . . 90-20100
Microwave Devices Inc., USA . . . 90-05020
Mitsubishi Steel, Japan . . . 75-63590
Mitsubishi Steel, Japan . . . 75-73860
Moog Inc, USA . . . 90-14600
Morgen Manufacturing Co., USA . . . 90-16725
MTE, USA . . . 85-00125
Mustang Survival, USA . . . 90-20730
NATO Supply Centere, Luxembourg . . . 90-03740
Network Science, USA . . . 90-05020
Network Sciences, USA . . . 90-20100
Nichols, USA . . . 90-15530
NKI BV, Netherlands . . . 90-14530
Noricum Trading, Cyprus . . . 90-03740
Northern Aero Industries, USA . . . 90-00920
Northern Airborne Technology Ltd, Canada . . . 75-18120
Onboard Systems, USA . . . 90-20730

United States of America

Optische Werke G.Rodenstock, Germany . . . 75-75980
Oscilliquartz, USA . . . 90-05020
Oxley Developments Co.Ltd, UK . . . 75-67870
Ozone Industries, USA . . . 90-20730
Parco, USA . . . 85-00180
Parker Hannifin, USA . . . 90-15530
Paxman Diesels Ltd, UK . . . 90-05840
Penny & Giles Data Systems Ltd, UK . . . 90-16220
Pfizer Inc., USA . . . 75-63590
Pfizer Inc., USA . . . 75-73860
Pilatus Aircraft, Switzerland . . . 85-01120
Pilatus Britten Norman, UK . . . 85-01120
Pilkington Optronics Ltd, UK . . . 75-70100
Pilkington P.E. Ltd, UK . . . 75-70100
Pilkington Thorn Optronics, UK . . . 75-70100
PNI International Co. Ltd, Thailand . . . 90-03740
Portescap, Switzerland . . . 90-16690
Prodan, Germany . . . 90-19130
Quality Industries Inc., USA . . . 90-16725
Racal Avionics Limited, UK . . . 90-17000
Racal Electronics Plc, UK . . . 75-72380
Racal Survey, UK . . . 90-18022
Rada Electronic Industries, Israel . . . 90-18030
Radiall, France . . . 90-18035
Raychem Corporation, USA . . . 85-01470
Reliance Gear, UK . . . 90-19130
Republic, USA . . . 85-00180
Rexnord, USA . . . 85-00180
Reynolds Metals Co., USA . . . 75-63590
Reynolds Metals Co., USA . . . 75-73860
Rose Bearing, UK . . . 85-01120
Rotadata Ltd, UK . . . 85-01522
RTG Inc., USA . . . 90-20100
SafeTech Inc., USA . . . 85-00125
SAFT, France . . . 75-77550
Saunders-Roe, UK . . . 90-19090
Schottel-Werft, Germany . . . 90-19150
Scott Aviation Div., USA . . . 90-20730
Senstar Corp., Canada . . . 90-19250
Serck Heat Transfer Ltd, UK . . . 90-19270
Sextant Avionique SA, France . . . 90-19310
Shaw Aero Development, USA . . . 90-00920
ShinMaya Industries Ltd., Japan . . . 90-19320
Sierra Research, USA . . . 90-14600
SIG, Switzerland . . . 75-80700
Singapore Elec & Eng Ltd, Singapore . . . 90-03740
SKB Trading NV, Belgium . . . 75-53030
Smiths Industries A&D Syst., UK . . . 85-01990
Smiths Industries A&D Syst., USA . . . 85-01990
Sofrance, France . . . 80-18100
Southwest Aerospace Corp., USA . . . 90-20220
Southwest Automotive Singapore, Singapore . . . 75-53030
Southwest Products Co., USA . . . 90-00920
Spar Aerospace Ltd, Canada . . . 90-14600
Specialty Cable, USA . . . 85-01470
SPS Technologies, USA . . . 85-00180
SSI, USA . . . 90-05020
SSS Gears Ltd, UK . . . 75-83570
Staiger Mohilo, Germany . . . 90-19130
Stainless Steel Products, USA . . . 90-20730
Sterer Vickers, USA . . . 90-14600
Steyr-Daimler-Puch, Austria . . . 90-07860
Stop-Choc GmbH, Germany . . . 75-80457
Stop-Choc Ltd, UK . . . 75-80457
Stratoflex, USA . . . 90-00920
Strite Industries Ltd, Canada . . . 90-23760
Submersible Systems, USA . . . 90-20730
Sumitomo Precision Prods, Japan . . . 90-19852
Symetrics, USA . . . 90-00920
Symetrics, USA . . . 90-20730
Systems Research Labs, USA . . . 90-20730
TAM, USA . . . 75-63590
TAM, USA . . . 75-73860
Technetics Corporation, USA . . . 90-00920
Technofan, France . . . 80-18100
Techspace Aero, Belgium . . . 90-29250
Tedeco, USA . . . 90-20730
Tel America, Japan . . . 75-53030
Telair, USA . . . 90-14600
Teledyne CAE, USA . . . 90-14600
Teledyne Thermatics, USA . . . 85-01470
Teleflex Syneravia, France . . . 90-23475
Tensolite Division of Carlisle, USA . . . 85-01470
TERMA Elektronik AS, Denmark . . . 90-20270
Testek, USA . . . 90-14600
The Marconi Company Ltd, UK . . . 90-06890
Thermoid, USA . . . 90-20730
Thiem Industries, USA . . . 90-15530
Thomson-Saginaw, USA . . . 90-20730
Tiernay Metals, USA . . . 90-01180
Times Fiber, USA . . . 85-01470
TIP-O-LINER, Germany . . . 90-19130
Tow-Targets Inc., USA . . . 90-20730
Trans-Tech Inc., USA . . . 75-63590
Trans-Tech Inc., USA . . . 75-73860
TridairFasteners, USA . . . 85-00180
Trimos-Sylvac Task Force, Switzerland . . . 90-06700
Turbomeca, France . . . 75-91300
Turner Gold & France, USA . . . 90-05020
Ubem Ltd, Turkey . . . 90-03740
Unbrako, USA . . . 85-00180
United Aircraft, USA . . . 90-15530
United Emmissions Catalyst, USA . . . 90-00920
Valley Todeco, USA . . . 85-00180
Vibrachoc SA, Spain . . . 75-80457
Vibrachoc SA, France . . . 75-80457
Vickers SMD, USA . . . 90-14600
Wagner Electric, USA . . . 90-00920
Wall Colmonoy, USA . . . 75-63590
Wall Colmonoy, USA . . . 75-73860
Washington Mills Abrasive, USA . . . 75-63590
Washington Mills Abrasive, USA . . . 75-73860
Waterman Hydraulics, USA . . . 90-00920
Whittaker Controls Inc., USA . . . 90-15530
Whittam Precision, UK . . . 90-19130
Wire Benders, USA . . . 85-00125
Woerner Engineering Inc., USA . . . 90-00920
XL Microwave, USA . . . 90-05020
Zinc Corp. of America, USA . . . 75-73860
Zinc Corporation of America, USA . . . 75-63590

Sales agents/representatives listed elsewhere:

Aero International Associates . . . 80-00800
Aircraft Instrument & Radio Co. Inc. . . . 80-01600
Allied Defense Industries Inc. . . . 80-02000
Avirex Ltd . . . 75-11250
Commercial Marketing Assistance . . . 80-05855
EFW Inc. . . . 75-29520
Foxtronics Inc. . . . 75-35752
Global Security Complex . . . 80-12950
Holmatro Inc . . . 80-14340
Instrumentation Marketing Corp. . . . 90-09030
Interpoint A Crane Company . . . 75-46870
Kongsberg Simrad Inc . . . 75-53160
NanoMaterials, Inc. . . . 75-63590
Polytronic Inc . . . 75-70870
Reade Advanced Materials . . . 75-73860
Safway Steel Products Inc. . . . 80-26700
Second Chance Body Armor Inc. . . . 75-79000
The Robar Companies Inc. . . . 75-75630

90-00040 ABB Lummus Global Inc
1515 Broad Street, Bloomfield, NJ 07003-3096
☎(+1 201) 893 1515 ⊖ (+1 201) 893 2000
♣ $ 1,257,000,000 👥 2,883 [98]

90-00050 ABB Vetco Gray Inc
10777 North West Freeway, PO Box 2291,
Houston, TX 77252
☎(+1 713) 681 4685 ⊖ (+1 713) 683 2471
♣ $ 485,000,000 (1996) 👥 2,196 [98]

90-00150 ACR Systems Inc.
20961 Hollberry Ct., Ashburn, VA 22011
☎(+1 703) 724 0030 ⊖ (+1 703) 729 6956
👥 3
CEO/President: Omer S. Esener
VP Sales & Marketing: Richard C. Peterson
VP Public Relations: Ms Janine Sacco
■ Helicopter (Bell, Sikorsky & Hughes) & aircraft (Northrop, F-5 & NF-5) equipment, parts & accessories (body & engine), radar (Marconi) components & parts. [97]

90-00200 Advance Hivolt Sales Office
11 Sixth Road, Woburn, MA 01801-1744
■ High voltage DC power supplies.
® UK: Advance Hivolt [98]

90-00240 Aeromaritime Mediterranean Corporation
(† Aeromaritime Inc.)
4115 Pleasant Valley Road, Suite 720,
Chantilly, VA 22021
☎(+1 703) 631 3111 ⊖ (+1 703) 631 3144
⊕ $ 1 700 000 (1995) ♣ $ 5 500 000 (1995) 👥 40
⌂ 1 500 sq m
President: E.S. (Gene) Courson
Treasurer: Larry Marcus
■ Export sales of weapon systems, including airborne & ground-based systems. Authorised maintenance center for Allison Engine Company, model 250 series gas turbine engines for the Middle East & Africa.
® USA: Allison Engine Company [97]

90-00260 Aerospatiale Inc.
(a subsidiary of Aerospatiale, France)
1101 15th Street NW - Suite 800,
Washington, DC 20005
☎(+1 202) 293 06 50 ⊖ (+1 202) 429 06 38
President: Gregory Bradford
® France: Aerospatiale [98]

90-00275 Aerotech Inc.
231 - 23rd Street, PO Box 329, Pembina, ND 58271
☎(+1 701) 825 63 63 ⊠ 755272
Operations Mgr: Paul Sigurdsen
■ Herman Nelson heater specialist.
® Canada: Aerotech Inc.
● Highway 5 West, Cavalier, ND 58220
☎(+1 701) 265 8577 [98]

90-00430 Agusta Aerospace Corp.
(a subsidiary of Agusta, Italy)
3050 Red Lion Road, Philadelphia, PA 19114
☎(+1 215) 281 14 00 ⊖ (+1 215) 281 04 47
👥 75
President and Chief Executive Officer: Dr Robert J Budica
Vice President Business Development: Flaminio Cianci
Market Analysis/Strategic Planning Director: Louis P Bartolotta
Vice President Finance and Administration: Vincent Genovese
Procurement Director: Vincenzo Corradi
Marketing and Sales Director: Robert A Brant
Regional Marketing Manager (West): Peter Menza
Regional Marketing Manager (Midwest): John W May
Regional Marketing Manager (North East): Robert Caldwell
Regional Marketing Manager (South East): William Kirk [98]

90-00870 All-System Aerospace International Inc.
29 Central Avenue, Hauppauge, NY 11788
☎(+1 516) 582 92 00 [E] sales@allsystem.com
[W] http://www.allsystem.com ⊖ (+1 516) 582 92 06
President: Karl G Zacek
Operations Director: Ms Vera B Zacek
Sales and Marketing Director: Christian Gerard
Purchasing Manager: Ms Kathy Bonifazio
■ Export sales new & used helicopters, spares support for Bell, Boeing Vertol & Sikorsky helicopters & for F-4, F-104, C 130, F-111 & P3 aircraft. E-3A. [98]

90-00920 Allied International Corp.
7 Hill Street, Bedford Hills, NY 10507
☎(+1 914) 241 69 00 ⊠ ALIEDINTER New York
[E] 103226.3243@compuserve.com
[W] http://www.alliedinter.com ⊠ 824022
⊖ (+1 914) 241 69 85
♣ $ 5,000,000 (1997) 👥 17 (1997) ⌂ 10,000 sq ft
President: Alexander E Ulmann
Senior Vice President: Adolfo Silva-Sadder
■ Consultants, representatives, systems integrators & distributors for aerospace & defence industries. → 80
® USA: Acoustics Products Corp., Engineered Magnetics, Inertia Switch, Inc., Marotta Scientific Controls, Northern Aero Industries, Shaw Aero Development, Southwest Products Co., Stratoflex, Symetrics, Technetics Corporation, United Emmissions Catalyst, Wagner Electric, Waterman Hydraulics, Woerner Engineering Inc.
International Offices:
France, Germany, Indonesia, Italy, Japan, Spain, Sweden, Thailand, Taiwan, UK [98]

90-01090 Ambient Technologies Inc.
(† I.D.E.)
Western Hemisphere Office
2999 Northeast 191 Str, Suite 407,
North Miami Beach, FL 33180
☎ (+1 305) 937 0610 ⊖ (+1 305) 937 2137
President: P. Elovic [97]

90-01115 American Aerospace & Defense Industries Inc. (AADI)
550 Mamaroneck Ave, Harrison, NY 10528
☎ (+1 914) 698 9385
✉ AEROEXPORT HARRISONNEWYORKSTATE
⊖ (+1 914) 698 3979
President: Hans G. Hollander
■ International marketing support & exhibition services for the aerospace & defence industry. [97]

90-01180 AMESCO
American Military Equipment & Spares Co.
(a subsidiary of Logistic Services International)
1271 Stoner Avenue, Suite 201, Los Angeles, CA 90025
☎ (+1 310) 473 73 73 ⊠ 4949272 mil ex
⊖ (+1 310) 996 10 80
♣ $ 800,000 (1994) ⁂ 6 (1995) ⌂ 1,400 sq m
President: Nejat Munisoglu
Contract Manager: E M Mitchell
■ Worldwide sales and supply of spares for F-16, F-5, F-4, A-4, S-2, P-3, C 130, Cessna Aircraft and Bell Helicopter. Specialty areas: avionics, engine and airframe spares, brake components, hydraulic and precision machine parts. Test sets (engine/avionics), communications equipment (land-based).
® USA: Tiernay Metals [98]

90-01300 The Armstrong Monitoring Corp.
810 Proctor Avenue, Ogdensburg, New York 13669
☎ (+1 800) 465 6777 [97]

90-01350 Ashot USA
(a subs. of Ashot Ashkelon Ind., Israel)
485 Madison Ave. 14th Fl., New York, NY 10022
☎ (+1 212) 223 4050 ⊖ (+1 212) 223 4156
® Israel: Ashot Ashkelon Ind. [97]

90-01450 Wamco
11555-A Coley River Circle, Fountain Valley, CA 92708
☎ (+1 714) 545 55 60 Ⓔ wamco@worldnet.att.net
⊖ (+1 714) 545 60 93
♣ US$ 27,000,000 ⁂ 27
President: Michael Matthews
Executive Vice President: Ms Sheila Sherman
■ Night vision goggle compatible cockpit lighting. Subminiature lamps. Direct view filament display. [98]

90-01600 Assembly & Manufacturing Systems Inc.
(a subsidiary of SGI International)
2222 Shasta Way, Simi Valley, CA 93065
☎ (+1 805) 583 89 61 ⊖ (+1 805) 583 04 42
♣ $ 6,800,000 (1997) ⁂ 50 (1997) ⌂ 1,850 m2
President and Chief Executive Officer: Dominick Savo
Vice President of Operations: Amir Khiabani
Sales Manager: Carl Timmerman
Purchasing Manager: Davé Burrows [98]

90-01650 Astro Aerospace Corp.
(a subsidiary of Spar Aerospace Ltd, Canada)
6384 Via Real, Carpinteria, CA 93013-2993
☎ (+1 805) 684 66 41 ⊖ (+1 805) 684 33 72
President: J T Conlan [98]

90-01700 Atlantic Aviation Corp.
New Castle County Airport, PO Box 15000,
Wilmington, DE 19850-5000
☎ (+1 302) 322 70 00 ✆ (800) 441 9390
⊖ (+1 302) 322 72 27
⁂ 500
Chairman and Chief Executive Officer: Ben E Waide
Senior Vice President, Treasurer: Frank C Zirnkilton (Jr)
Senior Vice President of Aircraft Services: John A Moyn
Vice President of Sales: John C Elwood
Subsidiary:
Atlantic Aviation Aircraft Support Group, Inc.

Head Office
PO Box 15000, Wilmington, DE 19850
☎ (+1 302) 322 70 00 ⊖ (+1 302) 322 72 83
Aircraft Services Division
PO Box 15000, Wilmington, DE 19850-5000
☎ (+1 302) 322 70 00 ⊖ (+1 302) 322 72 83
Vice President and General Manager: Mark Willey
☎ (+1 302) 322 75 26
Aircraft Maintenance Manager: John Hankammer
☎ (+1 302) 322 72 90
Avionics Manager: Sumitte Forseka
☎ (+1 302) 322 75 69
Sales Manager: Gary Fender ☎ (+1 302) 322 72 07
● WP Hobby Airport, 7930 Airport Blvd,
Houston, TX 77061-4199
☎ (+1 713) 644 64 31
Manager: Robert M Clark ☎ (+1 713) 644 64 31 [98]

90-01835 Avimo US Inc
6911 South Sixty-Sixth East Avenue, Suite 207,
Tulsa, OK 74133
☎ (+1 918) 299 4781 ⊠ 286176 avsus ur
⊖ (+1 918) 299 0453
Marketing & Business Development: Robert H. Hogrefe
® Singapore: Avimo Electro-Optics Pte Ltd [97]

90-01880 Avon Marine Inc.
(† Avon Seagull Marine)
11215 Young River Avenue, Fountain Valley, Ca 92708
☎ (+1 714) 444 92 44 ✆ (800) 854 75 95
Ⓦ http://www.avonmarine.com ⊖ (+1 714) 444 92 69
♣ $ 10,000,000 (1995) ⁂ 17 (1997) ⌂ 15,000 sq m
Managing Director: Alan Morgan
Sales and Marketing Director: Lesley Hamilton
Operations and Technical Manager: L W Curtis
Purchasing Manager: Chet Radd
® USA: Avon Inflatables Ltd UK [98]

90-02380 Bharat Electronics Ltd
(a division of Bharat Electronics Ltd, India)
53-55 Hilton Avenue, Garden City, NY 11530
☎ (+1 516) 248 40 20 ✉ BELUSA GARDENCITY NY
Ⓔ belnyro@ix.netcom.com
⊖ (+1 516) 741 58 94; 877 79 07
⁂ 5 (1997)
Chief Regional Manager: T R Prasad
® India: Bharat Electronics [98]

90-02600 BP Solar Inc
2300 North Watney Way, Fairfield, CA 94533
☎ (+1 707) 428 7800 ⊖ (+1 707) 428 7878
Managing Dir: Richard Sowter
Operations/Technical Dir: Mark Hammonds
Sales/Marketing Dir: Sherwin McDonald
Purchasing Mgr: Ms Jan Foltz
® UK: BP Solar Limited [98]

90-02750 British Steel Engineering Steels Sales Inc
(a subsidiary of British Steel Engineering Steels, UK)
475 North Martingale Road, Suite 400, Schaumburg, IL
☎ (+1 847) 619 90 90 ⊖ (+1 847) 619 90 95
® UK: British Steel Engineering Steels [98]

90-02800 Brüel & Kjaer Instruments Inc.
B&K Instruments
(a div. of Spectris Technologies Inc.)
2364 Park Central Blvd, Decatur, MA 30035
☎ (+1 404) 981 3998 ✆ (800) 332 2040
⊖ (+1 404) 808 7818
■ Instrumentation for sound, vibration, illumination, thermal environment, toxic gas, signal analysis.
® Denmark: Brüel & Kjaer [97]

90-03000 C-Tech Inc.
PO Box 949, Rooseveltown, NY 13683
☎ (+1 315) 764 9088 ⊖ (+1 315) 764 9090
® Canada: C-Tech Ltd [97]

90-03080 Canadair Defence Systems Division
(a member of Bombardier Aerospace Corp.)
1111 Jefferson Davis Highway, Suite 510,
Arlington, VA 22202-4306
☎ (+1 703) 413 02 10 ⊖ (+1 703) 413 05 92
International Marketing Director: S Michael Hughes
■ The marketing office has a focus on unmanned aerial vehicles, landing period designators, aircraft maintenance and repair and overhaul for the US and international markets. It also provides support to the other activities of the Defense Systems Division.
® Canada: Canadair Defence Systems Division [98]

90-03200 CATIC (USA) Inc.
9645 Telstar Ave, El Monte, CA 91731
☎ (+1 818) 575 6566 ⊖ (+1 818) 575 6599 [97]

90-03280 Cerberus Pyrotronics
(† Pyrotronics)
(partner with Cerberus Guinard SA, France)
118 Teakwood, Lee's Summit, MO 64064
☎ (+1 816) 373 3935 ⊖ (+1 816) 373 7851
Key National Accounts Mgr: Dale Hasty
® France: Cerberus-Guinard SA [97]

90-03360 Charlotte Aircraft Corp.
Delta Air Base, PO Box 25555,
Charlotte, NC 28229-5555
☎ (+1 704) 537 02 12 ⊖ (+1 704) 537 79 10
⁂ 35 ⌂ 200,000 sq ft
President: H J Caldwell (Jr)
Senior Vice President and General Manager: M L McCurry
Sales Director: Ms Marion Hicks
Sales Director: Harry Blume
Administrator: Ms Ruth S Cheshire
■ Sale of transport-type aircraft parts and components. Engine components. [98]

90-03400 John Chatillon & Sons Inc.
7609 Business Park Drive, Greenboro, NC 27409
☎ (+1 910) 668 0841 ✆ (800) 241 5867
⊖ (+1 910) 668 9163
Sales Manager: Dick Palke
® UK: Lloyd Instruments Ltd. [98]

90-03450 Cleco Systems Paxton & Vierling Steel
(a subsidiary of Cleco Limited, UK)
PO Box 1085, Omaha, NE
☎ (+1 712) 347 51 00 ⊖ (+1 712) 347 61 66 [98]

90-03580 Columbia Electronics International Inc.
3720 Groves Place, PO Box 960, Somis, CA 93066-0960
☎ (+1 805) 386 2312 ✉ COLUMBIA ⊠ 286980
⊖ (+1 805) 386 2314
President: Paul P. Keys
■ Supplier of US Government surplus new & reconditioned equipment & spares. [97]

90-03640 Commerce Overseas Corp.
500 Executive Blvd, Elmsford, NY 10523
☎ (+1 914) 592 2100 ✉ OVERCOMER
✉ OVERCOMER Elmsford ⊠ 131654; 131656
⊖ (+1 914) 592 2109
⁂ 25 ⌂ 7 000 sq ft
President: Christopher J. Garville
General Mgr: Michael R. Mallozzi
Traffic Mgr: Rudy Lobo
Military Components Corp.
(a div. of Commerce Overseas Corp.)
500 Executive Blvd, Elmsford, NY 10523
☎ (+1 914) 592 2100 ✉ OVERCOMER Elmsford
⊠ 131654 ⊖ (+1 914) 592 2109
President: Christopher J. Garville
■ Supply & support the maintenance of US Army tanks, sighting & fire control equipment, vehicles, communications, small arms & artillery equipment. Also optical & electrosystems, vehicle & aircraft engines. [97]

90-03665 Computing Applications Software Technology Inc. (CAST)
(a subs. of Pacer Systems Inc.)
900 Technology Park Drive, Billerica, MA 01821
☎ (+1 508) 667 8720 ⊖ (+1 508) 667 8873
President: John Clark
Marketing Mgr: Richard Gibson [97]

Contact Reference

90-03680 Cooper Aviation International Inc.
2149 E Pratt, Elk Grove Village, IL 60007
☎(+1 312) 364 2600 ⊠ 270617
👥 45
President: Marty Cooper
VP Marketing & Sales: Ken Cooper
■ General aviation stockist & distributor.
® Spain: Aviation Services Int'l SA
[97]

90-03690 Cosworth Racing Inc.
(† Cosworth Engineering Inc.)
(a subsidiary of Cosworth Racing, UK)
3031 Fujita Street, Torrance, CA 90505
☎(+1 310) 534 13 90 ⊖ (+1 310) 534 26 31
♣ $ 28 608 (1995) 👥 54 (1997) 🏭 23,363 sq ft
Executive Vice President: Ian E Bisco
Purchasing Manager: Ken Jacobs
→ 80
[98]

90-03740 Crescent International
200 North Strong Avenue, PO Box 623,
Lindenhurst, NY 11757-0623
☎(+1 516) 957 58 97
⊖ (+1 516) 957 59 08; 244 76 31
♣ $ 1,600,000 (1996) 👥 8 (1997) 🏭 15,000 sq ft
President: David C Howe [E] dchoweAAworldnet.att.net
■ Spare parts for armoured vehicles.
® Cyprus: Noricum Trading
Denmark: Army Material Command
Germany: Baxter Electro und Motorentchnik
Ireland: ArconConsultants
Japan: JDA Enterprises Corporation
Luxembourg: NATO Supply Centere
Singapore: Singapore Elec & Eng Ltd
Thailand: PNI International Co. Ltd
Turkey: Ubem Ltd
USA: Amesco
Autodyne Manufacturing Co. Inc., USA:
Cardwell Condenser Corp., USA
Sonetronics Inc., USA
Relays Inc., USA
Universal Relay Company, USA
Neuzeitliche Electro und Motorentechnick GmbH, Germany
Baxter Electronics, Germany
TE.LE.GI Sarl, Italy
S.E.I.F. Sarl, Italy
Ubem Ltd, Turkey
[98]

90-03750 Crompton Instruments Inc.
1640 Airport Road Boulevard, Suite 109,
Kennesaw, GA 30144
☎(+1 404) 425 8903 ⊖ (+1 404) 423 7194
♣ £ 6 000 000 👥 21
General Mgr: R. Annis-Brown
® UK: Crompton Instruments Ltd
[98]

90-03760 C&S Antennas Inc.
(a subsidiary of Comsat RSI-CSA Antennas)
45250 Severn Way, Sterling, VA 20166-8913
☎(+1 703) 430 70 73 ⊖ (+1 703) 430 70 75
👥 3
President: Ken A Hayward
Vice President of Sales and Marketing: William R Whittington
Operations and Technical Director: Brian Collins
Office Administrator: Ms Jean Hardison
■ Tactical communications antennas and telescopic masts; broadband HF antennas and HF active loop systems; tactical UHF satellite antennas and radio relay UHF antennas, HF direction finding systems.
® UK: CSA Ltd
[98]

90-03800 Custom Electronics Inc
Boca Raton, FL 33498
☎(+1 407) 487 7594 ⊖ (+1 407) 487 7594
® USA: Custom Electronics Inc
[97]

90-03900 Daewoo International (America) Corp
New Jersey (Headquarters)
85 Challenger Road, Ridgefield Park,
New Jersey, NJ 07660-2114
☎(+1 201) 229 45 00 ⊠ 126092 daewoo
⊖ (+1 201) 440 22 44
® Korea (ROK): Daewoo Corp

New York Office
104 W 40th Street - 10th Floor, New York, NY 10018
☎(+1 212) 391 15 70 ⊖ (+1 212) 391 18 82
Chicago Office
3206 North Kennicott Avenue, Arlington, IL 60004
☎(+1 847) 590 98 00 ⊠ 6871670 dwa chg
⊖ (+1 847) 590 98 14
Miami Office
5200 Blue Lagoon Drive Suite 770, Miami, FL 33126
☎(+1 305) 267 00 91 ⊠ 6811212 dwa mia
⊖ (+1 305) 267 00 16
Dallas Office
2120 Hutton Drive Suite 800, Carrollton, TX 75006
☎(+1 214) 481 16 16 ⊖ (+1 214) 481 16 20
Los Angeles Office
1055 West Victoria Street, Compton, CA 90220
☎(+1 310) 884 58 00 ⊠ 6831818 dwa la
⊖ (+1 310) 669 20 20
Seattle Office
Plaza Center Suite 1105, 10900 North East 8th Street,
Bellevue, WA 98004
☎(+1 206) 454 14 25 ⊠ 6838093 dwa sea
⊖ (+1 206) 453 70 04
San Francisco Office
1820 Gateway Drive - Suite 230, San Mateo, CA 94404
☎(+1 415) 341 08 88 ⊖ (+1 415) 341 43 77
[98]

90-04005 Dage Corp.
1011 High Ridge Road, Stamford, CT 06905
☎(+1 203) 461 90 00 [E] emailus@dage.com
⊖ (+1 203) 461 97 95
👥 20
President: Peter Sterling
Executive Vice President: John Everett
■ Full service export management. Supplies custom designed, catalogue & spare parts components & subsystems for the commercial & military markets: communications of all types, cellular, line-of-sight, mobile earth stations, satellites. Radars; ground based, fixed & mobile, ship-borne, missile & space. Electronic warfare system components. TV, radio transmitter & receiver parts. Linear & ring accelerator components. Industrial heating products.
Represent:
Arc Technologies
Atlantic Microwave
Bliley Electric
Custom Microwave
DB Products
EMF Systems
EMR Corporation
Enon Microwave
Herley MDI
Lark Engineering
Mac Technology
MBF Microwave
Mega Industries
Microwave Development Co. (MDC)
Microwave Development Labs. (MDL)
Microwave Device Technology
Microwave Resources Inc. (MRI)
Microwave Solutions
Omniyig Inc.
Reactel
Renaissance Electronics
Resin Systems
Res-Net Microwave
Telgaas Inc.
TRX Division, Barry TelGaAs Inc. Industries
UB Corporation
[98]

90-04010 Daicel (USA) Inc.
One Parker Plaza, 400 Kelby Street, Fort Lee, NJ 07024
☎(+1 201) 461 4466 ⊖ (+1 201) 461 2776
® Japan: Daicel Chemical Industries Ltd
[97]

90-04020 Dallas Aerospace Inc.
(† Avionics Components & Equipment Inc.)
(a subs. of Banner Aerospace Inc.)
1875 N. I-35E, Carrollton, TX 75006-3761
☎(+1 214) 245 9633 ⊘ (800) 527 3722
⊖ (+1 214) 245 9634
♣ $ 49 000 000 (1996) 👥 58 🏭 152 000 sq ft
President: Herman Schulz ☎(+1 214) 245 9633
Vice President: John Entenza ☎(+1 214) 245 9633
Controller: Ted Knust
CFO: Terry Goodnight
[97]

90-04100 Dana Corporation
Dana World Trade Division
PO BOX 405, Toledo, OH 43697
1609 Henthorne Drive, Maumee, OH 43537
☎(+1 419) 867 2100 ⊖ (+1 419) 867 2118
President: Carlos Montejo
[97]

90-04180 Dassault Falcon Jet Corp
(a subsidiary of Dassault Aviation Group, France)
Teterboro Airport, PO Box 2000,
South Hackensack, NJ 07606
☎(+1 201) 440 67 00 ⊖ (+1 201) 967 44 75
👥 823 (1994)
Chairman: Serge Dassault
President and Chief Executive Officer: Jean-François Georges
President: Jean Rosanvallon
Senior Vice President, Customer Support: Peter L Ginochio
Sr VP Finance & Administration: J.M. Young
Sr VP & General Counsel: Matthew A. Boyle
VP Internatinal Marketing: Ronald A. Zielinski
VP Marketing, Eastern US: W.R. Desiderio
VP Marketing, Western US: N.J. MacDonald
Sales Director, Latin America: Jean Michel Jacob
Sales Director, South East Asia, Australia, New Zealand: Michael Fies
Communications Dir: John H. House
Sales Director, Northern Asia: Jeff Habib
Sales Mgr, Military Government: James T. Hurley
Little Rock Division
Adams Field, Little Rock, AR 72206
☎(+1 501) 372 52 54 ⊠ 536246 ⊖ (+1 501) 372 58 50
Vice President and General Manager: William C Bracas
[98]

90-04660 Dornier of North America Inc
Washington Office
1350 I Street N W - Suite 800, Washington, DC 20005
☎(+1 202) 408 56 00 ⊖ (+1 202) 408 48 93
® Germany: Dornier GmbH
[98]

90-04980 Dytec Inc.
800 East Northwest Highway, Suite 614,
Palatine, IL 60067
☎(+1 847) 776 89 90 [E] dytecil@aol.com
⊖ (+1 847) 776 89 95
Sales Representative: Martin Linke
Indiana Office
7212 North Shadeland Avenue, Suite 220,
Indianapolis, IN 46250
☎(+1 317) 578 04 74 ⊖ (+1 317) 578 04 76
President: Dan Connors
[98]

90-05020 EBA Microwave
33122 Mesa Vista, Dana Point, CA 92629
☎(+1 714) 240 81 81
[W] http://www.home.earthlink.net/~bdreyer/
⊖ (+1 714) 240 81 81 then # 3
♣ US$ 2,000,000 (1996)
President: Elliot Bell [E] ebell99090@aol.com
® USA: Advanced Switch Technology, Aviel, Custom Power Systems, David Sarnoff, Filter Microwave, Lee Labs, Lucky Gold Star Components, MCLI, Metric Systems, Microwave Devices Inc., Network Science, Oscilliquartz, SSI, Turner Gold & France, XL Microwave
[98]

90-05150 EFI Corporation
(a subsidiary of The Racal Corporation)
5357 Randall Place, Fremont, CA 94538
☎(+1 510) 353 99 00 ⊖ (+1 510) 353 99 39
Director of Operations: Robert Berrisford
■ Manufacturer of composite pressure vessel cylinders.
[98]

90-05300 EH Industries Inc.
(a subsidiary of E H Industries Ltd, UK)
1735 Jeff Davis Highway, Suite 805, Arlington, VA 22202
☎(+1 703) 412 80 00 ⊖ (+1 703) 412 80 06
Chairman: G Orsi
President: Thomas Battaglia
Marketing Co-ordinator: Julie Meehan
■ EH-101 market development and sales in USA and Canada.
® UK: E.H.Industries Ltd
[98]

90-05450 Elitex International Inc.
7 Stonycrest Road, White Plains, NY 10604
☎ (+1 914) 949 45 88 ⊖ (+1 914) 949 45 88
President: Svato Kudej
■ Lacing tapes for tying harnesses/cables together in electronic, aircraft and defence industries; parachute cords. 98

90-05520 Embraer Aircraft Corp. (EAC)
(a subs. of Embraer SA, Brazil)
276 SW 34th Street, Fort Lauderdale, FL 33315-3688
☎ (+1 954) 359 3700 ⊠ 201262 (Spare parts)
⊖ (+1 954) 359 3701
♣ $ 36 000 000 (1993) 👥 78
President: Samuel Hill
VP Sales: Mark P. Hale
® Brazil: Embraer 97

90-05810 Erickson Aerospace
16638 NE 48th Court, Redmond, WA 98052-5433
☎ (+1 206) 881 1814 ⊠ 329473 ⊖ (+1 206) 881 3082
President: Randy Erickson 97

90-05820 ESDU International
PO Box 1633, Manassas, VA 22110
⊖ (+1 703) 330 1642
® UK: ESDU International plc 98

90-05840 European Gas Turbines Inc. Paxman Diesels Div.
(† Ruston Gas Turbines Inc.)
15950 Park Row, Houston, TX 77084
☎ (+1 713) 578 7828 ⊖ (+1 713) 492 8277
Division Mgr: Steve Morrow
® UK: Paxman Diesels Ltd 97

90-05885 Export Consultants Corp. (ECC)
250 Lackland Drive Suite 6, PO Box 308, Middlesex, NJ 08846
☎ (+1 732) 469 07 00 ⊖ (+1 732) 469 91 05
Vice President: M.A. Althausen
VP International Sales: L.M. Brinson
■ Military fixed wing & helicopter spares, accessories, avionics, instruments, tools, engine spares, airborne & non-airborne UHF/VHF communications, walkie-talkies, antennas & related equipment. Also ground support equipment. → 85 98

90-06040 Fairey Hydraulics Inc.
6400 West Snowville Road, Cleveland, OH 44141-3248
☎ (+1 216) 838 15 00 ⊖ (+1 216) 838 15 80
President: Paul Maloney
■ US product support operation for Fairey Hydraulics Ltd (UK), including FAA approved repair & overhaul services.
® UK: Fairey Hydraulics Ltd 98

90-06700 Fred V. Fowler Co Inc.
66 Rowe Street, Newton, MA 02166
☎ (+1 617) 332 70 04 ✆ (800) 788 23 53
W http://www.fvfowler.com ⊖ (+1 617) 332 41 37
♧ $ 1,500,000 (1997) ♣ $ 20,000,000 (1996)
👥 50 (1997) 🏭 26,000 sq ft
President: Fred V Fowler (Jr)
Operations Director: Uldis Ozolins
Technical Director: Monticello Abrams
Sales and Marketing Director: David Francis
Purchasing Manager: Ms Kim Kablin
® Switzerland: Trimos-Sylvac Task Force 98

90-06830 Fuji Industries Corp Los Angeles Office
2811 Wilshire Boulevard, Santa Monica, CA 90403
☎ (+1 213) 829 47 74 ⊠ 652476 ⊖ (+1 213) 828 11 72
Executive Vice President: Masakazu Kuramoto
® Japan: Fuji Industries Co. Ltd 98

90-06870 Fusion Inc.
14248 Santa Fe Trail, Shawnee Mission, KS 66215
☎ (+1 913) 888 4200 ⊖ (+1 913) 888 1762
♣ $ 7 000 000 👥 5
President: Robert B. Clipsham
■ Passive microwave component & sub-system sales together with non-ferrous investment casting, die casting, injection molded & powdered metal engineering shapes.
® UK: Micro Metalsmiths Ltd 97

90-06887 GEC ALSTHOM Sales Network Inc.
(† The English Electric Corp.)
(† GEC ALSTHOM International Inc.)
(Jointly owned by The General Electric Company plc and Alcatel Alsthom)
4 Skyline Drive, Hawthorne, NY 10532-2160
☎ (+1 914) 347 5155 ⊖ (+1 914) 347 5432
President: Paul J. Jancek
Communications Mgr: Jean Cevallos
■ World leader in energy & rail transport. 97

90-06890 GEC-Marconi
1111 Jefferson Davis Hwy, Ste 800, Crystal Gateway 3, Arlington, VA 22202
☎ (+1 703) 416 6582 ⊖ (+1 703) 416 0135
President: Bernard Rosser
Exec. Secretary to President US Ops: Ms Kathleen Fiamengo
■ Project management, communications, command & control, microelectronics, ATE, test instruments, guided weapons, guidance systems, EW, satellite communications, space systems, radar, secure radio systems, spacecraft, torpedoes, mines, sonar, underwater surveillance, avionics & telecommunication systems.
® UK: The Marconi Company Ltd 97

90-07290 Getac Inc.
460 Hindry Drive, Unit G, Inglewood, CA 90301
☎ (+1 310) 410 14 24 ⊖ (+1 310) 410 14 92
Vice President: Joe Chernoff
® Taiwan: Getac Corp. 98

90-07860 GSI Inc.
(† Gun South Inc.)
108 Morrow Ave, PO Box 129, Trussville, AL 35173
☎ (+1 205) 655 8299 ⊠ 782113 gunsouth bhm
⊖ (+1 205) 655 7078
👥 11 (1996) 🏭 700 sq m
President: Donald F. Wood
Dir Military Programs: Einar Hoff
■ US agent for Steyr-Daimler-Puch AG for small arms sales & services.
® Austria: Steyr-Daimler-Puch 97

90-08030 Hadland Photonics Inc.
(a subsidiary of Photonics Holdongs Ltd, UK)
20480 Pacifica Drive, Suite D, Cupertino, CA 95014
☎ (+1 408) 996 09 01 ✆ (800) 248 46 86
W http://www.hadlandphotonics.com
⊖ (+1 408) 996 08 89
♣ $ 2,000,000 (1996) 👥 50 (worldwide, 1997)
🏭 2,200 m2 (worldwide)
President: Frank Kosel
Managing Director: Heddwyn Davies
Operations Director: David Taylor
Sales and Marketing Director: Andrew Hilton
Purchasing Manager: Ms Jenny Grainger
® UK: Hadland Photonics Ltd 98

90-08035 Hägglunds Vehicle AB
(a subsidiary of Hägglunds Vehicle AB, Sweden)
Hagglunds Liaison Office
35418 Mound Road, Sterling Heights, MI 48310
☎ (+1 810) 264 82 80 ✆ (800) 856 96 39
E hagglunds.vehicle@worldnet.att.net
⊖ (+1 810) 268 93 20
👥 4
Vice President Business: Urban Lundberg
Operations Director: Ms Sherry Ireland
Purchasing Manager: Ms Tracy Hoffmeister
■ Products represented include the Bv206 All Terrain Carrier, Combat Vehicle 90 (CV90), recovery units & devices. 98

90-08055 Arthur Harris & Assoc. Inc
875 Walnut Street, Suite 300, Cary, NC 27511
☎ (+1 919) 460 8660 ⊖ (+1 919) 460 1129
® USA: Custom Electronics Inc 97

90-08098 HCH Supply Co., Inc.
101 Glynn Way, Houston, TX 77056
☎ (+1 713) 439 14 60 ⊠ 792235 hch hou
⊖ (+1 713) 439 14 61
Contact: Ms Jeanie Colias
Representing:
GEC - Marconi
Skoda Export
Omnipol
Paulhun Electric Mfg Co.
Hughes Redifusion Simulation
Kelvin Hughes
Williams Fairey Ltd
SBFI
Control Techniques (CT)
Keystone
Tuma Turbo Mach
Wonderware Corporation 98

90-08990 Imtra Corp.
30 Samuel Barnett Blvd, N. Bed. Ind., New Bedford, MA 02745-1205
30 Samuel Barnett Boulevard, North Bed. Ind.
⊠ 294138
President: William H Farnham
® UK: Avon Inflatables Ltd 98

90-09020 Industri Pesawat Terbang Nusantara North America Office (IPTN)
IPTN Building - Tukwila, 1035 Andover Park West Suite B, Seattle, WA 98188-7681
☎ (+1 206) 575 65 07 ⊖ (+1 206) 575 03 18 98

90-09030 Instrumentation Marketing Corp.
(a subsidiary of Photo-Sonic Inc.)
820 South Mariposa Street, Burbank, CA 91506
☎ (+1 818) 842 21 41 E info@photosonics.com
⊖ (+1 818) 842 26 10
♧ US$ 1,000,000 (1997) ♣ US$ 4,000,000 (1997)
👥 8 (1997) 🏭 3,000 sq m
President: Philip Kiel
Sales and Marketing Director: Andrew Bridges
■ High speed film & video systems distributor. High speed cameras for instrumentation motion analysis. → 90 98

90-09050 Integrated Technology Inc.
182 Main Street, Salem, NH 03079
☎ (+1 603) 898 6333 ⊖ (+1 603) 898 6895
® USA: Custom Electronics Inc 97

90-09550 International Procurement Services Inc.
Suite 909, 7200 Wisconsin Ave, Bethesda, MD 20814-4811
☎ (+1 301) 652 0067 ⊖ (+1 301) 951 8808
♣ $ 2 000 000 (1996) 👥 5 (1996)
CEO/President: Pedro Lynch
Public Relations Dir: Cesar Lynch
Operations Dir: Claudio Ortiz
■ Modems, sonar systems, transducers, electronic components, fixed wing & helicopter components/systems, mock-ups & training aids, computers, audio-intelligence devices, naval systems/spare parts, hydrographic, navigation & oceanographic equipment, airport/heliport lighting devices, radar, ESM, ECM & ECCM systems, nuclear radiation monitoring systems, marine aids for navigation, survival equipment. 97

90-09690 Irvin Aerospace Canada Limited
(† Irvin Industries Canada Ltd)
US Office
PO Box 1163, Buffalo, NY 14205-1163
☎ (+1 716) 852 33 62 ⊖ (800) 665 07 02
President and Chief Executive Officer: C L Corbin
Manager Sales and Marketing: Ms Elizabeth Matchett
Purchasing: Edward J Johnson
Marketing Assistant: Ms Pam Dobrensky
® Canada: Irvin Aerospace Canada Limited 98

90-09760 Israel Aircraft Industries International Inc.
50 West 23rd Street, New York, NY 10010

☎(+1 212) 620 44 00 ⊠ 125180; 640381
⊖ (+1 212) 242 17 73; 620 17 99
President: D Onn
® Israel: Israel Aircraft Industries

Washington DC Office
1700 North Moore Street, Suite 1210,
Arlington, VA 22209
☎(+1 703) 875 37 77; 875 37 22
⊖ (+1 703) 875 37 40
Vice President Marketing: I Sharon
Vice President Public Affairs: M G Klemow [98]

90-09900 Jacman Aircraft Inc.
8950 Fullbright Avenue, Chatsworth, CA 91311
☎(+1 818) 407 0555 ⊖ (+1 818) 407 0060
♣ $ 6 500 000 (1995) 👥 21 🏭 28 000 sq m
General Mgr: Michael Lonker
■ Distribute Boeing, MDD, Lockheed & MIL-SPEC aircraft parts, specifications in survival equipment (ELT, life rafts, vests). [97]

90-10100 JDR Cable Systems Inc
(† De Regt Special Cable Inc)
12600 Exchange Drive - Suite 200, Stafford, TX 77477
☎(+1 713) 240 66 00 ⊖ (+1 713) 240 66 01
Contact: Colin Zak
® Netherlands: JDR Cable Systems BV [98]

90-10130 JGW International Ltd
10640 Main Street, Suite 200, Fairfax, VA 22030
☎(+1 703) 352 34 00 🄴 jgwintl@dgs.dgsys.com
⊖ (+1 703) 385 64 70
♣ $ 600,000 (1997) 👥 13 (1997) 🏭 1,500 m2
Chairman: John G Wilson
President: Andrew M Wilson
Vice President: John M O'Neil
→ 80 [98]

90-11100 Alfred Karcher Inc
1600 Cottontail Lane, PO Box 6910, Somerset, NJ 08875
☎(+1 908) 356 11 99 ⊖ (+1 908) 356 41 51
Vice President Sales and Marketing: T Dyszkiewicz
® Germany: Alfred Karcher GmbH & Co [98]

90-11106 KBK Inc
330 Madison Avenue - 26th Floor,
New York, NY 10017-5091
☎(+1 212) 687 85 64 ⊖ (+1 212) 972 92 12

Los Angeles Representative Office
801 South Figueroa Street Suite 1060,
Los Angeles, CA 90017
☎(+1 213) 891 00 11 ⊖ (+1 213) 892 16 16

Atlanta Representative Office
100 Galleria Parkway Suite 1850, Atlanta, GA 30339
☎(+1 770) 984 07 07 ⊖ (+1 770) 984 10 04

Cedar Rapids Representative Office
122 Park Place North East Suite F,
Cedar Rapids, IA 52402-2019
☎(+1 319) 393 23 06 ⊖ (+1 319) 393 23 21

Michigan Representative Office
4950 West Dickman Road Suite B-6, Battle Creek 49015
☎(+1 616) 964 05 70 ⊖ (+1 616) 964 13 10 [98]

90-11110 KD Manitou Inc
(a subsidiary of Manitou BF, France)
PO Box 154009, Waco, TX 76715-4009
☎(+1 817) 799 02 32 ✆ (800) 433 33 04
⊖ (+1 817) 799 43 33
👥 100
Executive Vice President: Serge Bosché
■ Rough terrain forklifts.
® France: Manitou BF [98]

90-11700 Krauss-Maffei Corp
(a division of Krauss-Maffei AG, Germany)
7095 Industrial Road, PO Box 6270,
Florence, KY 41042-6270
☎(+1 606) 283 02 00 ⊠ 988334 ⊖ (+1 606) 283 02 91
Chairman: Burkhard Wollschläger
® Germany: Krauss-Maffei AG [98]

90-11900 Kylar Inc
(† MHR Sales Inc.)
4902 Orchard Green Place, Fort Wayne, IN 46804-6591
☎(+1 219) 432 7716 ⊖ (+1 219) 432 1734
® UK: Micro Metalsmiths [97]

90-12350 Lentern International Inc.
(a subsidiary of Lentern (Aircraft) Ltd)
590 SW 34th St, Ft Lauderdale International Airport,
Fort Lauderdale, FL 33315
☎(+1 305) 359 16 94; 947 90 74
⊖ (+1 305) 359 16 98
President: R. Richards
® UK: Lentern (Aircraft)Ltd [98]

90-12600 LMS North America
5455 Corporate Drive - Suite 303, Troy, MI 48098
☎(+1 810) 952 56 64 ⊖ (+1 810) 952 16 10
👥 20
Sales Manager: R Boyd
→ 80
® Belgium: LMS International [98]

90-13000 Logistic Services International
1271 Stoner Avenue, Suite 201, Los Angeles, CA 90025
☎(+1 310) 473 73 73 ⊖ (+1 310) 996 10 80
Contract Manager: E M Mitchell [98]

90-13020 Luwa Bahnson Inc.
3901 West Point Blvd, Winston-Salem, NC 27103
☎(+1 910) 760 3111 ⊖ (+1 910) 768 0283
👥 625
Sales & Marketing Mgr: Thomas F. Riley
■ Shelter equipment including blast valves, overpressure valves, blast doors, NBC filters & ventilating units.
® Switzerland: Luwa Ltd [97]

90-13030 MGP Instruments, Inc.
(a subsidiary of MPG Finance)
5000 Highlands Parkway, Suite 150, Smyrna, GA 30082
☎(+1 770) 432 27 44 ✉ http://www.mpgi.com
⊖ (+1 770) 432 91 79
♣ US$ 8,000,000 (1997) 👥 35 (1997)
Vice President: Loïc Gate
Sales and Marketing Director: Bob Manley
General Manager: K Rosanski [98]

90-13075 Marconi Instruments Inc
2301 Horizon Drive, Fort Worth, TX 76177
☎(+1 817) 224 92 00 ⊖ (+1 817) 224 92 01
President: Jim Koehn
■ Manufacture of communications test and measurement equipment, specialising in the field of RF signal generation modulation meters, microwave analysers, power meters, counters and sources, spectrum analysers, digital test equipment (PCM), TV test instrumentation.
® UK: Marconi Instruments Ltd [98]

90-13130 Masco Electronics Inc.
4622 West Slauson Avenue, Los Angeles, CA 90043
® USA: Custom Electronics Inc [97]

90-13300 Metal-Tech Inc.
(† Wayland Engineering Sales)
PO Box Drawer 1786, Clemmons, NC 27012
☎(+1 901) 940 2580 ⊖ (+1 901) 940 5970
Contact: Andy Hansen
® UK: Micro Metalsmiths [97]

90-13440 Milano Brothers International Corp.
378 SW 12th Ave, Deerfield Beach, FL 33442-3106
☎(+1 305) 420 5000 ⊖ (+1 305) 420 5005
♣ $ 16 000 000 👥 40 🏭 11 000 sq ft
President: Robert Milano
Vice President: Piero Guidugli
■ Supply of electronic components & related products to foreign governments, defense contractors & sub-contractors. → 80 [97]

90-13550 MLRS International Corporation (MIC)
M/S WT-49, PO Box 650003, Dallas, TX 75265-0003
President: Ed Ulery
■ Marketing and Sales forces operate in conjunction with: Lockheed Martin Vought Systems (LMVS), USA
Diehl GmbH and Co, Germany
concentrating on the MLRS System and the Phase II Rocket respectively. [98]

90-13600 Monitor Air Supply
(a partner of Honos Ltd, Portugal)
17401 Nichols Street, Unit N,
Huntington Beach, CA 92647
☎(+1 714) 841 56 70 ⊖ (+1 714) 841 50 93
👥 1
Owner: Scott Brown [98]

90-14160 Navy Arms Co.
(a subs. of Service Group Ltd)
689 Bergen Blvd, Ridgefield, NJ 07657
☎(+1 201) 945 2500 ⊠ 134619 ser arm ride
⊖ (+1 201) 945 6859
♣ $ 9 000 000 (1995) 👥 50 🏭 35 000 sq ft
President: Val Forgett (Jr)
Sales/Marketing Dir: William O'Melia
Operations/Technical Dir: Paul W. Reed
■ Supplier of small arms & small arms ammo & involved in countertrade accepting used or obsolete small arms or small arms ammo in trade for new material. [97]

90-14200 Nelson Electronics Inc.
3992 Columbia Pike, Ellicott City, MD 21043
☎(+1 410) 465 1946 ⊖ (+1 410) 465 1378
® USA: Custom Electronics Inc [97]

90-14470 Nissho Iwai Aerospace (America) Corp.
Rockefeller Center, 1211 Ave of the Americas,
New York, NY 10036
☎(+1 212) 704 6915 ✉ NISSHOIWAI NEWYORKNY
⊠ 232641; 233567; 12329 ⊖ (+1 212) 704 6543
President: K. Sotsuka
Manager: Masayuki Haga
■ Export trading of aircraft subsystems, parts & defence equipment.

Los Angeles Office
Arco Center, 1055 West Seventh Street, Ste 3200,
Los Angeles, CA 90017
☎(+1 213) 688 0678 ⊖ (+1 213) 688 0603
Representative: N. Ozeki

St Louis Office
1854 Craig Road, St Louis, MO 63146
☎(+1 314) 576 9911 ⊖ (+1 314) 576 9960
Representative: K. Sawada, T. Kita, S. Araoka [97]

90-14530 NKI Inc
Overlook III, 2859 Paces Ferry Road, Suite 800, Georgia
☎(+1 404) 434 22 88 ⊖ (+1 404) 436 10 02
President: G Aerts
Executive Vice President: P F M M Langeveld
® Netherlands: NKI BV [98]

90-14600 Nozaki America Inc
1325 Avenue of the Americas, 26th Floor,
New York, NY 10019
☎(+1 212) 424 15 00 ⊖ (+1 212) 757 80 54
President: Mr Shimizan
® Canada: Ivex Corp, Spar Aerospace Ltd
USA: Armtech Industries, BEI Sensors and Systems, BF Goodrich, Carlton Technologies, Cessna Aircraft, Cessna Aircraft Raytheon, Crane Lear Romec, Lear Astronics, Lucas Aerospace, Moog Inc, Sierra Research, Sterer Vickers, Telair, Teledyne CAE, Testek, Vickers SMD [98]

90-15530 Okura Aerospace USA, Inc
(a subsidiary of Okura Aerospace Co. Ltd, Japan)
550 N. Continental Blvd, Ste 280, El Segundo, CA 90245
☎(+1 310) 535 9000 ⊠ 181506 ⊖ (+1 310) 535 9100
Controller: Steven Caraco
® USA: Aerospace Engineering Support Corp., Airborne Div., Cleveland Wheel & Brake, Gull Electronics., Honeywell Inc., Lucas Western Inc., Nichols, Parker Hannifin, Thiem Industries, United Aircraft, Whittaker Controls Inc. [97]

90-16220 Penny & Giles Data Systems Inc.
(a subsidiary of Penny & Giles Data Systems Ltd, UK)
12401 Middlebrook Road, Suite 240,
Germantown, MD 20874-1525
☎ (+1 301) 428 06 00 ✆ (800) 243 25 72
[W] http://www.penny-giles.co.uk ⊜ (+1 301) 353 01 29
Operations Manager: Richard A Gloor
® UK: Penny & Giles Data Systems Ltd [98]

90-16300 P&M Electronics
1261 Lincoln Avenue Suite 210, San Josè, CA 95125
☎ (+1 408) 287 0402 ⊜ (+1 408) 287 2110
® USA: Custom Electronics Inc [97]

90-16690 Portescap US Inc.
(a subs. of Portescap, Switzerland)
36 Central Ave, Hauppauge, NY 11788
☎ (+1 516) 234 3900 ⊜ (+1 516) 234 3986
👥 15
Managing Dir: Peter Thornton
® Switzerland: Portescap [97]

90-16722 Pro Active Sales
52 Warren Way, Burlington Township, NJ 08016
☎ (+1 609) 386 9014 ⊜ (+1 609) 386 7037
® USA: Customs Electronics Inc [97]

90-16725 Projects International Development Corp. (PIDC)
2840 Northup Way A104, Bellevue, WA 98004
☎ (+1 206) 883 8212 ⊠ 181162 ⊜ (+1 206) 883 7846
⚘ $ 500 000 ♣ $ 1 500 000 👥 6 🏭 2 000 sq ft
President: M. Shabrawi
Managing Dir: Salam Mostafa
Sales & Marketing Dir: A. K. EL Shabrawy
Purchasing Mgr: J. Hicks
■ Military vehicles, military support systems, defence marketing and consultancy.
® USA: Alternative Fuel Systems Rexworks Int'l, Cleveland Trencher Co., Johnson Patching Plants, Morgen Manufacturing Co., Quality Industries Inc. [97]

90-16727 Prosig USA Inc
81 Bassett Highway, Suite 7, PO Box 969,
Dover, NJ 07802-0969
☎ (+1 201) 366 3999 ⊜ (+1 201) 366 4949
Director: George Chang
■ Software and hardware engineering company specializing in the integration of computer systems to real-time data acquisition and signal processing applications. → 75 → 80
● 8320 Ravine Drive, Westland, MI 48185
☎ (+1 313) 513 82 69
Sales Manager: Adam Callens [98]

90-17000 Racal Avionics Inc
(a subsidiary of The Racal Corp, UK)
8851 Monard Drive, Silver Spring, MD 20910-1816
☎ (+1 301) 495 66 95 [E] ravinc@clark.net
⊜ (+1 301) 585 75 78
♣ $ 1,609,652,000 (1996)
Vice President and General Manager: Steven L Kemp
Manager, Technical Support: Martin Richardson
■ Supplier of airborne avionic systems including flight and navigation systems for military, and general aviation. Satellite communication systems.
® UK: Racal Avionics Limited [98]

90-18016 Racal Recorders Inc.
(a subsidiary of The Racal Corporation)
15375 Barranca Parkway, Irvine, CA 92618
☎ (+1 714) 727 34 44 ✆ (800) 847 12 26
[E] recorders@usa.racal.com ⊜ (+1 714) 727 17 74
Vice President and General Manager: John R Cummings
Sales Director: Dr David Nutton
Purchasing Manager: Ms Carol Kalahar
■ Reel-to-reel and VHS analog recorders reproducers; digital formatters; voice logging reel-to-reel and VHS recorders reproducers: instant recall recorders; rapid access recorders. [98]

90-18022 Racal NCS Inc.
(† Racal Survey USA Inc.)
(a subsidiary of The Racal Corporation)
3624 Westchase Drive, Houston, TX 77042
☎ (+1 713) 784 44 82 ⊠ 775194 decsurvey hou
⊜ (+1 713) 784 81 62
Vice President: Anthony Harrison
■ Supplier of specialised marine and survey systems and services.
® UK: Racal Survey [98]

90-18030 Rada Electronic Industries Inc.
(a subsidiary of Rada Electronic Industries Ltd)
80 Express Street, Plainview, NY 11803
☎ (+1 516) 938 78 78 ⊜ (+1 516) 938 41 41
♣ $ 34,000,000 (1994)
President: H Bodinger
Manager: Gabriel Sered
® Israel: Rada Electronic Industries [98]

90-18035 Radiall Inc
(a subsidiary of Radiall, France)
150 Long Beach Boulevard, Stratford, CT 06497
☎ (+1 203) 386 10 30 ⊠ 311796 ⊜ (+1 203) 375 38 08
Managing Director: Etienne Lamairesse
® France: Radiall [98]

90-18040 Rafael USA Inc.
(† Rafael Armament Development Authority)
4455 Connecticut Ave. N.W., Suite B400,
Washington, DC 20008
☎ (+1 202) 895 5300 ⊜ (+1 202) 895 5298
President & CEO: Dr Eytan Youdelevity
[E] yudi@ix.netcom.com
Represent:
Rafael Armament Development Authority [97]

90-18400 Rheintech Inc.
6832 Old Dominion Drive, Suite 206, McLean, VA 22101
☎ (+1 703) 448 04 40 [E] vfwinc@erols.com
⊜ (+1 703) 356 63 94
President: Volney F Warner
Vice President of Marketing: Orest Kaczmarskyj
Office Manager: Ms Emiley Seymour [98]

90-18630 Rolls-Royce North America Inc.
(a subsidiary of Rolls Royce Plc)
11911 Freedom Drive, Reston, VA 20190
☎ (+1 703) 834 17 00 ⊜ (+1 703) 709 60 87
President and Chief Executive Officer: John Sandford
Senior Vice President of Commercial Marketing: Jack Phillips
Vice President, Commercial: D J Whetton
Vice President of Corporate Communications: R N Baugniet
Vice President of Human Resources: Ms Becky Blackman
Vice President of Personnel, Administration and General Counsel: T P Dale
● c/o American Airlines, Hangar 3,
Washington National Airport, Washington, DC 20001
☎ (+1 703) 419 21 49 ⊜ (+1 703) 419 21 50
Senior Service Representative: Allan D Broom
Arlington Virginia Office
Suite 270, 1530 Wilson Boulevard, Arlington, VA 22209
☎ (+1 703) 351 09 00 ⊜ (+1 703) 351 76 21
Senior Vice President of Military Aero Programs: H L Garrett
USMC Programme Director: Michael Gladwin
Navy Projects Director: Michael B Ferguson
Army Projects Director: George Sibert
General Manager of Military Programs: William McGuth
● 301-116th Avenue S E, Suite 211, Bellevue, WA 98004
301-116th Avenue South East, Suite 211
☎ (+1 206) 451 80 08 ⊜ (+1 206) 451 83 18
Vice President of Seattle Operations: Chris Hornblower
Atlanta Engineering
2675 Paces Ferry Road, Suite 150, Atlanta, GA 30339
☎ (+1 404) 436 7900 ⊜ (+1 404) 436 8729
Vice President of International Combustion Division/RRIPI: Steve Deabler
● Langley Road, Mcas Cherry Point, NC 28533
☎ (+1 919) 447 15 95 ⊜ (+1 919) 447 63 13
Senior Service Representative: Peter A Moore
Long Beach, CA
c/o Douglas Aircraft Co. MC 36-42, 3855,
Lakewood Boulevard, Long Beach, CA 90846
☎ (+1 310) 593 71 83; 593 68 30 ⊠ 674357
⊜ (+1 310) 593 53 52
US Adour Operations Director: Ray Barnes
Patuxent River, MD
PO Box 37, Patuxent, MD 20670
☎ (+1 301) 863 50 11; 863 50 12
⊠ TWX 510 6012 701
Senior Service Representative: C Francis
St Louis, MO
16414 Prestwick Place Lane, Wildwood, MO 63011
☎ (+1 314) 458 94 43 ⊜ (+1 314) 458 95 20
Vice President: Ken Bushell
Yuma, AZ
PO Box 12515, Yuma, AZ 85365
☎ (+1 602) 341 33 51; 341 15 81
⊜ (+1 602) 344 04 90
Senior Service Representative: James D Corcoran [98]

90-18800 Rozner Sales
400 West NW Highway, Arlington Heights, IL 60004
☎ (+1 708) 577 0067 ⊜ (+1 708) 398 7620
® USA: Custom Electronics Inc [97]

90-19090 SRB Technologies Inc. (SRBT)
2580 Landmark Drive, PO Box 25267,
Winston-Salem, NC 27114-5267
☎ (+1 910) 659 26 10 ✆ (800) 552 00 98
[E] srbt@ix.netcom.com ⊜ (+1 910) 768 77 20
President: Brian Pullen
Military Sales Executive: Ms Martha D Cavanaugh
■ Betalight military markers and illumination devices.
® UK: Saunders-Roe [98]

Contact Reference

90-19130 Schlenker Enterprises Ltd
5143 Electric Avenue, Hillside, IL 60162
☎ (+1 708) 449 57 00 [E] jschlenk@aol.com
⊜ (+1 708) 449 57 03
♣ $ 3,000,000 (1990)
President: M Schlenker
Advertising Manager: J Slawecka
Sales Co-ordinator: D Kasnick
■ Scales, torque measuring systems, clutches, brakes & clutch-brakes, linear motion bearings & ball screws AC&DC servo systems. Precision gears, gear boxes & racks for precision actuation &/or measuring systems. Applications include azimuth & elevation position feedback, robot joints where low backlash & high accuracy are demanded. → 85
® Germany: Prodan, Staiger Mohilo, TIP-O-LINER
Italy: ELI/ELBO
UK: Reliance Gear, Whittam Precision [98]

90-19150 Schottel North America
1505 Corbin Ave, Hammond, LA 70403
☎ (+1 504) 542 7174 ⊜ (+1 504) 542 5347
Managing Dir: Ken Robbins
® Germany: Lohmann & Stolterfoth, Schottel-Werft [97]

90-19250 Senstar-Stellar Inc.
Building Five, 101 Billerica Park, North Billerica, MA 01862
☎ (+1 508) 670 06 00 ⊜ (+1 508) 670 98 69
Vice President US Operations: Bill Evenson
® Canada: Senstar Corp. [98]

90-19270 Serck Heat Transfer
North American Office
426 East North St, Suite 29, Waukesha, WI 53188
☎ (+1 414) 547 2150 ⊜ (+1 414) 547 2281
® UK: Serck Heat Transfer Ltd [97]

90-19310 SEXTANT Avionique Inc Beacon Centre
(a subsidiary of SEXTANT Avionique SA, France)
1924 NW 84th Avenue, Miami, FL 33126
☎ (+1 305) 597 63 00 ⊜ (+1 305) 597 63 63
👥 60
General Manager: Willy Moses
■ Sales office for Sextant Avionique products including flight/engine instruments, attitude indicators, automatic pilots for helicopters, actuators and air data benches. Customer and product support services. → 80
® France: Sextant Avionique SA [98]

90-19320 ShinMaywa (America) Ltd
Chicago Office
1603 Barclay Blvd, Buffalo Grove, IL 60089
☎(+1 708) 808 1281 ⊜ (+1 708) 808 1286
® Japan: ShinMaya Industries Ltd.
Los Angeles Branch
6 Centrepoint Drive, Suite 200 La Palma,
Los Angeles, CA 90623
☎(+1 714) 670 7550 ⊜ (+1 714) 670 8065 [97]

90-19330 Shiposh Engineering Ltd
9001 51st Avenue, Elmhust, Queens, New York 11373
☎(+1 718) 592 8031
Contact: Micha Shiposh [98]

90-19420 Singapore Aerospace International
(a subsidiary of Singapore Technologies Aerospace)
217 East Alameda Avenue, Suite 204,
Burbank, CA 91502-2621
☎(+1 818) 846 63 33 [E] staerosp@aol.com
⊜ (+1 818) 846 38 12
Regional Manager for North America: Federiko Yap [98]

90-19580 Snap-on Tools International Ltd
8100 SW 81 Drive - Suite 290, Miami 33143-6684
☎(+1 305) 271 09 44 ✉ SNAPONMIA ⊠ 441760
Division Manager: Arthur Rasmussen [98]

90-19700 Spar Aerospace U.S. Limited
(a subsidiary of Spar Aerospace Ltd, Canada)
One Rodney Square, PO Box 551, Wilmington, DE 19899
☎(+1 302) 658 65 41 ⊜ (+1 302) 658 65 48 [98]

90-19852 Sumitomo Precision Products Co.Ltd
345 Park Avenue, New York, NY 10154
☎(+1 212) 826 36 34 ⊠ 224328 ⊜ (+1 212) 826 43 90
Manager: Alex Matsuyuki
® Japan: Sumitomo Precision Prods [98]

90-20100 Tauber-Dreyer Corporation
17074 Dearborn St, Northridge, CA 91325-2635
☎(+1 818) 993 0048 ⊜ (+1 818) 993 5758
♣ $ 1 000 000 (1995) 👥 2 🏠 400 sq m
Contact: Bob Dreyer
® USA: Assemblies Inc., Hybrid-Tek Inc., Microwave Communications Laboratories, Network Sciences, RTG Inc. [97]

90-20200 TechMan International Corp. Inc.
TechMan Centre, Route 20,
Charlton City, MA 01508-0727
☎(+1 508) 248 3211 ⊜ (+1 508) 248 3113
♧ $ 1 000 000 ♣ $ 15 000 000 👥 43 (worldwide)
🏠 12 000 sq ft
Chairman & CEO: Dr M. Mahmud Awan
Marketing Dir: James Maher
Advertising Mgr: Farooq Ali
■ Distribution & marketing of defense equipment. Agent for US & European defense contractors. [97]

90-20220 Technical Development & Marketing Inc. (TDM)
4883 Ronson Court, Suite I, San Diego, CA 92111
☎(+1 619) 292 1331 (1996) ⊜ (+1 619) 292 0645
♧ $ 500 000 (1996) ♣ $ 3 000 000 (1995) 👥 5 (1996)
🏠 1 500 sq ft
President: Bernard D. Paul
Marketing Dir: Richard D. Grafton
Purchasing Mgr: R.G. Madrigal
■ Manufacturer representative/distributor for export of aerial tow target systems (including two-way reeling machines, scoring systems - miss distance indicator) and procurement specialist for aerospace/electronic (ground and airborne) equipment.
® USA: California Instruments Inc., Cartwright Electronics Inc., EMI Technologies Inc., Int'l Target Systems Corp, Microsystems Inc., Southwest Aerospace Corp. [97]

90-20222 Technical Marketing Services
3313 Industrial 25th Street, Fort Pierce, FL 34946
☎(+1 561) 466 31 13 [E] techmark@gate.net
⊜ (+1 561) 466 31 16
President: Mike Hogan
Administration Officer: Ms Kathryn Chandler
® UK: Micro Metalsmiths Ltd [98]

90-20255 Telo Inc.
8341 Redmac Street, Port Richey, FL 34668
☎(+1 813) 848 18 10 [E] telo@microd.com
⊜ (+1 813) 848 19 91
🏠 2,000 sq ft
President and General Manager: Ms Lynn G Voelkle
Secretary: Ms Janey Rochek
■ Hardware and components for the defence and aerospace industry. [98]

90-20270 TERMA Electronics Inc.
PO Box 3035, McLean, VA 22103
☎(+1 703) 759 0046 ⊜ (+1 703) 759 0256
👥 1
■ High-quality electronic products. International space & aerospace programmes & advanced land-based & seaborne communications & radar systems.
® Denmark: TERMA Elektronik AS [97]

90-20300 Thomson-CSF Inc.
(a subsidiary of Thomson-CSF, France)
99 Canal Center Plaza 450, Alexandria, VA 22314-1588
☎(+1 703) 838 96 85 [W] http://www.thompson-csf.com
⊜ (+1 703) 838 16 88
👥 6 (1997)
Office Manager: Ms Susan Hept ☎(+1 703) 838 56 47 [98]

90-20500 Tomco Reps
PO Box 55454, Houston, TX 77255
☎(+1 713) 468 6716 ⊜ (+1 713) 458 6716
® USA: Custom Electronics Inc [97]

90-20730 Transaero Inc.
80 Crossways Park Drive, Woodbury, NY 11797
☎(+1 516) 921 74 00 [W] http://www.transaeroinc.com
⊠ 967734 ⊜ (+1 516) 921 74 07
♣ $ 60,000,000 (1997) 👥 80 (1997) 🏠 32,000 sq ft
President: Shell Youngwall
Vice President: Charles Kane
Vice President Business Development: Perry K Youngwall
Operations Director: Richard Kane
Airline Sales Director: Anthony Cartelli
US Government/Military Sales Director: A Fiero
Life Support Director: E McDonald
■ Airframe & engine spares, life support & survival equipment, oxygen systems, batteries, electro-optics, target systems, helicopter spares, communications systems, helmets, hydraulic systems, thermocouples & heat exchangers. → 85
® USA: ACR Electronics, Aerospace Control Products, Allen Aircraft Products, BaseWest, Breeze Eastern, Co-Operative Industries, Conax Florida Corp., Control Products Corp., Cook Airtronic, DEC Technologies, Dayton-Granger, Die-Matic Products Inc., Dynapower/Stratopower (Vickers), Engine Systems Division, BFGoodrich, G&H Technology, GEC Marconi Aerospace, Gentex Int'l Inc., Gloves, Hayes Targets, Hughes-Treitler, ILC Dover, IMC Magnetics, Lapes Equipment, Litton ILSD, Marathon Battery Co., Mustang Survival, Onboard Systems, Ozone Industries, Scott Aviation Div., Stainless Steel Products, Submersible Systems, Symetrics, Systems Research Labs, Tedeco, Thermoid, Thomson-Saginaw, Tow-Targets Inc.
International sales offices:
Australia, Canada, Dubai, Ecuador, France, Germany, Israel, Malaysia, Sweden, Thailand, UK.
Domestic sales offices:
Florida, Washington, San Francisco.
Transaero Washington
PO Box 4081, Silver Springs, MD 20914
☎(+1 301) 384 37 38 ⊜ (+1 301) 384 27 25
Manager: Gerald Citrano
Transaero California
1286 39th Avenue, San Francisco, CA 94121
☎(+1 415) 242 99 00 ⊜ (+1 415) 242 99 01
Manager: Rich McCarty [98]

90-20860 UBE Industries (America) Inc
666 Fifth Avenue, New York, NY 10103
☎(+1 212) 765 58 65 ⊜ (+1 212) 765 52 63 [98]

90-22040 Vaisala Inc
(a subs. of Vaisala Oy, Finland)
100 Commerce Way, Woburn, MA 01801
☎(+1 617) 933 4500 ⊜ (+1 617) 933 8029
President: Steven H. Chansky
Marketing Mgr: Ms Dorothy Marshall
Artais Division
7450 Industrial Parkway, Plain City, OH 43065-9005
☎(+1 614) 873 68 80 ⊜ (+1 614) 873 68 90
Director: Charles Shanklin
■ Airport observation systems. [98]

90-23050 Waltech Group Inc.
96 Vreeland Avenue, South Hackensack, NJ 07606
☎(+1 201) 641 73 00 ⊜ (+1 201) 641 72 29
♣ $ 3,000,000 👥 10 🏠 2,000 sq m
President: Kobus Reyneke [E] kobus@tribeca.105.com
Dir Sales/Marketing: Michael Fay
■ Procurement of spare parts and equipment for US and worldwide distribution. [98]

90-23100 Weco Inc. Aerospace Systems
Corporate Headquarters
1020 Airport Road, Lincoln, CA 95648
☎(+1 916) 645 89 61 [E] wecoinc@inreach.com
⊜ (+1 916) 645 15 47
👥 35
President: William Weygandt
Vice President: Wayne Burk
Technical Director: Mike Maupin
Burbank Division
4206 West Burbank Boulevard, PO Box 6547,
Burbank, CA 91505
☎(+1 818) 845 87 11 ⊜ (+1 818) 845 29 72
Manager: Tony Zito
■ Aircraft electrical accessories. instruments and ground support equipment. Sales; service (overhaul); exchange military and commercial aircraft. [98]

90-23475 Wintermark Corp.
PO Box 366, Woodbury, NY 11797
☎(+1 516) 692 24 02 ⊜ (+1 516) 692 24 11
👥 1 (1997)
President: Grant W Schleich
■ Provides marketing capability in the US for various foreign companies.
® France: Teleflex Syneravia [98]

90-23500 World Aviation Associates
(a member of The NORDAM Group)
PO Box 3365, Tulsa, OK 74101
☎(+1 918) 243 67 32 [E] waatul@aol.com ⊠ 158105
⊜ (+1 918) 247 67 33
Managing Director: Andrew Allendorf
Sales and Marketing Director: F Rodney Miller
Manager of Financial and Information Systems: Ms Karen S Doyle [98]

90-23760 World Sales Development (WSD)
2607 South Woodland Blvd, Ste 255, Deland, FL 32720
2607 South Woodland Boulevard, Suite 255
☎(+1 904) 228 27 27 ⊜ (+1 904) 228 00 11
♣ $ 2,200,000 (1996) 👥 4 (1996)
President: John F Doherty
® Canada: Strite Industries Ltd
USA: ACRO Industries [98]

90-29250 Techspace Aero Sales and Marketing Office
(† FN Moteurs)
(a subsidiary of Techspace Aero, Belgium)
11107 Wurzbach Road, Suite 104,
San Antonio, TX 78230
☎(+1 210) 690 88 48; 690 88 49
⊜ (+1 210) 690 13 46
Managing Director: V van Poppel
Sales and Marketing: Theo J van Poppel
® Belgium: Techspace Aero [98]

90-39670 GKN Westland Inc.
(a subsidiary of GKN Westland Helicopters Ltd)
1735 Jefferson Davis Hwy, Suite 805,
Arlington, VA 22202
☎(+1 703) 412 80 00 ▣ 901152 ⊖ (+1 703) 412 80 06
President and Chief Executive Officer: Thomas Battaglia
Marketing Coordinator: Ms Julie M Meechan
■ Sales, Marketing and customer support for GKN Westland plc in North America.
® UK: GKN Westland plc 98

Uruguay

Head of State:
Julio María Sanguinetti

10 Head of State

10-10000 Office of the President
Casa de Gobierno, Edificio Libertad,
Avenida Luis Alberto de Herrera 3350, 1600 Montevideo
☎(+598 2) 47 21 10
President of the Republic: Dr Julio María Sanguinetti
Vice President: Dr Hugo Batalla
Secretary to the President: Dr Elías Bluth
Minister of External Relations: Ing Alvaro Ramos
☎(+598 2) 92 21 32; 92 13 49 (fax)
Military Secretary: Brig Gen José P Malaquin
Director of Public Relations: Daniel Fletis 98

10-15000 Office of the Vice President
Palacio Legislativo, Montevideo
☎(+598 2) 23 45 10 ⊖ (+598 2) 29 96 32
Vice President: Dr Hugo Batalla 98

12 Ministries

12-10000 Ministry of National Defence
Ministerio Defensa Nacional
Edificio Gral Artigas, Avenida 8 de Octubre 2628,
11600 Montevideo
☎(+598 2) 80 97 07; 80 71 09 ▣ 23317 dinaco uy
Minister: Raúl Iturría 98

12-20000 Ministry of Foreign Affairs
Avenida 18 de Julio 1205, 11100 Montevideo
☎(+598 2) 92 10 08 ▣ 22074; 6682 dipcia uy
⊖ (+598 2) 92 13 27
Minister: Alvaro Ramos 98

12-40000 Ministry of Interior
Mercedes 993, 11100 Montevideo
Minister: Didier Opertti 98

12-50000 Ministry of Economy & Finance
Colonia 1089, Piso 3, 11100 Montevideo
☎(+598 2) 92 10 17 ▣ 6269 minecon uy
⊖ (+598 2) 92 12 77
Minister: Luis Mosca 98

15 Parliament

15-10000 Senate
Camara de Senadores
Palacio Legislativo, Montevideo 98

15-20000 House of Representatives
Palacio Legislativo, Montevideo 98

31 Embassies & Attachés

31-00590 Embassy of France
853 Avenida Uruguay, PO Box 290, 11400 Montevideo
☎(+582 2) 92 00 77 ▣ 26986 ambafra uy
⊖ (+582 2) 92 37 11
Ambassador: H.E. Mr Pierre Charasse
Deputy Chief of Mission: M.Patrick Flot
Commercial Counsellor's Office
Calle Lord Ponsomby 2506, 11600 Montevideo
☎(+598 2) 78 03 82; 78 03 88 ▣ 26656 comatta uy
⊖ (+598 2) 79 41 43
Commercial Counsellor: M.Robert Villemin 97

31-00990 Embassy of the Republic of Korea
Antonio Costa 3555, 11300 Montevideo
☎(+598 2) 62 04 35; 62 05 66; 62 77 76
⊖ (+598 2) 68 34 36
Ambassador: H E Mr Young Sik Kim
First Secretary: Hyung Pil Kim
Third Secretary and Vice Consul: Seung Lae Hwang 98

31-01270 Royal Netherlands Embassy
Leyenda Patria 2880, P2, Apt 202, PO Box 1519,
Montevideo
☎(+598 2) 71 29 56/57/58 ✉ HOLLANDIA Montevideo
▣ 22273 ned mon uy ⊖ (+598 2) 71 33 01
Ambassador: H.E. Mr C.J. Vreedenburgh
First Secretary: H.J. Simons 97

31-01740 Embassy of Switzerland
Calle Federico H.Abadie 2936/40, P11, PO Box 12261,
11300 Montevideo
☎(+598 2) 70 43 15; 70 46 10
E swissmtv@multi.com.uy ⊖ (+598 2) 71 50 31
Ambassador: H E Mr Urs Stemmler
Head of Chancery: Jean-Louis Giddey 98

31-01890 Embassy of the United Kingdom
Calle Marco Bruto 1073, PO Box 16024,
11300 Montevideo
☎(+598 2) 62 36 30/50 ⊖ (+598 2) 62 78 15
Ambassador: H.E. Mr R.A.M. Hendrie
Deputy Head of Mission: R.W. James
Defence Attaché: Gp Capt. Andrew Ades (RAF)
Second Secretary (Commercial): D. Carter 97

31-01900 Embassy of the United States of America
Calle Lauro Muller 1776, Montevideo
☎(+598 2) 23 60 61; 48 77 77 ⊖ (+598 2) 48 86 11
Ambassador: H E Mr Thomas J Dodd
Deputy Chief of Mission: Ms Nancy Mason
Economic and Commercial Counsellor: Johnathan Farrar
Public Affairs Officer: Peter M Brennen
Chief, Office of Defense Co-operation: Col Kenneth 98

75 Manufacturing Companies

75-05000 Kolon International Corp. Montevideo Branch
Cerro Largo 1163, Piso 11, AP 1109, Montevideo
⊖ (+598 2) 40 98 57
■ Manufacture textiles for military use in South Korea. Authorised exporter for the Korean Government. 98

Uzbekistan

Head of State:
Islom Karimov

10 Head of State

10-10000 Office of the President
Tashkent
☎(+7 3712) 39 81 31
President: Islom Karimov 98

11 Prime Minister

11-10000 Office of the Chairman of the Cabinet of Ministers
PL V I Lenina, Tashkent 8
Chairman Supreme Assembly: Erkin Khalilov
Prime Minister: Otkir Sultonov
First Deputy Prime Minister: Ismoil Jorabekov
Deputy Prime Minister: Viktor Chzhen
Deputy Prime Minister: Dilbar Ghulomova
Deputy Prime Minister: Qayim Haqqulov
Deputy Prime Minister: Alisher Azizkhojayev
Deputy Prime Minister: Rustam Yunusov
Deputy Prime Minister: Murat Sharifkhojayev
Deputy Prime Minister: Mirabror Usmonov 98

12 Ministries

12-10000 Ministry of Defence
Kh Abdoullaev Avenue 100, 700000 Tashkent
☎(+7 3712) 69 35 67
Minister: Maj Gen Rustam U Akhmedov 98

12-20000 Ministry of Foreign Affairs
5 Mustakillik Maidoni, 700029 Tashkent
☎(+7 3712) 33 64 75 ▣ 116343 mid su
⊖ (+7 3712) 39 43 48
Minister: Abdulaziz Komilov 98

12-30000 Ministry for Foreign Economic Relations
Lunacharskoe Shosse 75, 700 077 Tashkent
☎(+7 3712) 68 92 56 ▣ 116122 atlas
⊖ (+7 3712) 34 60 01; 68 74 44
Minister: Elyor Ganiyev 98

12-70000 Ministry of Communications
A Tolstoi 1, 700 000 Tashkent
☎(+7 3712) 33 66 66 ▣ 116149 zuuk
Minister: Abduvahid Jarabayev 98

Uzbekistan

29 Customs & Excise

29-00500 State Tax Committee
Khalkiar Dustligi Avenue 35, 700135 Tashkent [98]

31 Embassies & Attachés

31-00920 Embassy of Japan
Street Sadyk & Bull, Azimov 52/1, 700047 Tashkent
☎ (+7 3712) 33 51 42 ⊖ (+7 3712) 89 15 14 [97]

31-01890 Embassy of the United Kingdom
Ulitsa Gogolya, 700000 Tashkent
☎ (+7 3712) 33 88 47; 40 62 88; 40 64 51
⊖ (+7 3712) 40 65 49; 40 65 75
Ambassador: H E Miss Barbara Hay
Deputy Chief of Mission: K R Allan
Third Secretary: P R Knott
British Council Director: M S Melbourne
Defence Attaché: Air Cdre M L Feenan [98]

90 Sales Agents, Sales & Liaison Offices

90-05000 Daewoo Corp Tashkent Office
Furkat Street - 1st Block, Building N°1 - 8th Floor, 700027 Tashkent
☎ (+7 3272) 33 33 77 ▣ 251331 dwala su
⊖ (+7 3272) 33 59 83 [98]

Vanuatu

Head of State:
Jean Marie Leye

10 Head of State

10-10000 Office of the President
President's Office, Private Mail Bag, Port Vila
President: Jean Marie Leye [98]

11 Prime Minister

11-10000 Office of the Prime Minister
PO Box 124, Port Vila
▣ 1040 vangov nh
Prime Minister: Serge Vohor
Deputy Prime Minister: Barak Sope [98]

12 Ministries

12-20000 Department of Foreign Affairs
PO Box 124, Port Vila
☎ (+678) 224 13 ▣ 1040 vangov nh
Minister of Foreign Affairs: Vidal Soksok [98]

12-40000 Ministry of Home Affairs
Port Vila
Minister: Robert Karie [98]

26 Police (Civil)

26-10000 Vanuatu Police Headquarters
PMB 014, Port Vila
☎ (+678) 222 22 ▣ 07711127 vpfnoc ⊖ (+678) 228 00
Commissioner: Luc Siba
Foreign Relations Dir: Jean Sese
Minister: Maxim Carlo Korman
Police Minister: Charlie Nako
Deputy Minister: Donald Kalpokas
Public Informatin Officer: Pakoa Samuel [97]

31 Embassies & Attachés

31-00590 Embassy of France
Rue Higginson, PO Box 60, Port Vila
☎ (+678) 226 27; 223 53; 228 16 ⊖ (+678) 226 95
Ambassador: H E Mr Jean-Claude Moreau
Military/Defense Attaché: Capt Le Pivain
Commercial Officer: Claude Foulon
Public Information Officer: Philippe Latapie [98]

31-01890 United Kingdom High Commission
KPMG House - rue Pasteur, PO Box 567, Port Vila
☎ (+678) 231 00 ⊖ (+678) 236 51
High Commissioner: H E Mr Malcolm Hilson
Deputy High Commissioner: Ms C E Everson
Aid Management Attaché: A Harper [98]

Venezuela

Head of State:
Rafael Caldera Rodríguez

10 Head of State

10-10000 Office of the President
Palacio de Miraflores, Avenida Urdaneta, 1010 Caracas
☎ (+58 2) 81 08 11; 81 08 11 30 ▣ 21162
⊖ (+58 2) 810 36 44
President of the Republic: Rafael Caldera Rodríguez
Minister Secretariat of the Presidency: Asdrubal Aguiar [98]

12 Ministries

12-10000 Ministry of Defence
Fuerte Tiuna, Conejo Blanco, El Valle, 1090 Caracas
☎ (+58 2) 662 12 53; 693 14 04 ▣ 22598
Minister of Defence: VAdm Tito Manglio Rincon Bravo [98]

12-20000 Ministry of Foreign Affairs
1010 Caracas
Edificio MRE, PB, Avenue Urdaneta Esquina de Carmelitas
☎ (+58 2) 81 85 21; 81 85 22; 81 85 23; 81 85 24; 81 85 25; 81 85 26; 81 85 27; 81 85 28; 81 85 29 ▣ 22721; 28085
⊖ (+58 2) 83 36 33
Minister of Foreign Affairs: Miguel Angel Burelli Rivas [98]

12-40000 Ministry of Interior Affairs
Esquina de Carmelitas, Avenida Urdaneta, 1010 Caracas
Minister of Interior: Guillermo Adueza [98]

15 Parliament

15-10000 Senate
Senado
Congreso de la Republica, 1010 Caracas [98]

15-20000 Chamber of Deputies
Camara de Diputados
Congreso de la Republica, 1010 Caracas [98]

31 Embassies & Attachés

31-00100 Embassy of Austria
Av La Estancia, Ed. Torre las Mercedes, PO Box 61381, Caracas 106
Edificio Torre Las Mercedes, Piso 4, Officina 408, Chuao, Apdo. 61.381, 1060-A Caracas
☎ (+58 2) 91 38 63; 91 39 79; 92 29 56
E e.austria@cantv.net
▣ 23 435 aust vc; 23 435 oebot vc ⊖ (+58 2) 92 95 08
Ambassador: H E Dr Johannes Skriwan
Second Secretary and Vice Consul: Peter Janisch
Commercial Counsellor: Peter Rattinger
Attaché: Bita Rasoulian
Deputy Attaché: Ms Silvia Zarazik [98]

31-00150 Embassy of Belgium
Caracas
Quinta "La Azulita" Avda. 11 Entre Transversales 6 & 7 Altamira
☎ (+58 2) 262 04 21; 262 14 45; 261 93 97 ▣ 21223
⊖ (+58 2) 261 03 09; 261 13 33
Ambassador: H E Mr Ronald De Langhe
Counsellor, Deputy Chief of Mission, Public Affairs Officer: Dr Frank van de Caen
Commercial Counsellor: Piet Morisse [98]

31-00310 Embassy of Canada
PO Box 62302, 1060-A Caracas
Edificio Torre Europe, 7th Floor, Avenida Francisco de Miranda, Campo Alegre, Caracas
☎ (+58 2) 951 61 66; 951 61 67; 951 61 68
✉ DOMCAN Caracas E crcas@crcas01.x400.gc.ca
▣ 23377 domca vc ⊖ (+58 2) 951 49 50
Ambassador: H E Mr Yves Gagnon
Counsellor: Douglas McCue
Commercial Counsellor: Georges Lemieux
Attaché: G Compagne [98]

31-00510 Embassy of the Arab Republic of Egypt
Urb. Las Mercedes, Calle Guiacaipuro, Quinta Maribel, Sec. San Roman, Caracas
☎ (+58 2) 993 15 55; 92 62 59 ▣ 21135 boust vc
⊖ (+58 2) 933 15 55
Ambassador: H E Mr Mohamed Fadel Weheda
Counsellor: Moustafa M Ahmad Moustafa
Diplomatic Attaché: Ahmed Sabry
Administrative Attaché: Atef El Araby Soliman [98]

31-00590 Embassy of France
Calle Madrid con Trinidad, Las Mercedes, PO Box 60385, 1060 Caracas
☎ (+58 2) 993 66 66 ⊖ (+58 2) 993 34 83
Ambassador: H.E. Mr Henri Vignal
Deputy Chief of Mission: Didier Ortolland
☎ (+58 2) 993 94 80 ☎ (+58 2) 993 55 20 (fax)
Armed Forces Attaché: Lt Col.(Army) Etienne Banal
☎ (+58 2) 91 57 08
Press Attaché: Jean-Luc Gregoire
Office of the Commercial Counsellor
PO Box 61616, 1060 Caracas
☎ (+58 2) 285 34 55; 285 35 35 ✉ COMATTA Caracas
⊖ (+58 2) 285 23 42
Commercial Counsellor: M. Herve Le Roy [97]

31-00660 Embassy of the Federal Republic of Germany
Avda San Juan Bosco, Altamira, PO Box 2078, Caracas
☎ (+58 2) 261 01 81; 12 05; 22 29 ▣ 23262 aaccs vc
⊖ (+58 2) 261 06 41

Ambassador: H.E. Mr Eike E. Bracklo
Commercial Counsellor: Dr Hans Pliseyka [97]

31-00840 Embassy of the Republic of Indonesia
Avenue El Paseo, Quinta la Trinidad, Prados del Este, 1080 Caracas
☎ (+58 2) 976 27 25 ✉ 27 862 indos vc
⊖ (+58 2) 976 05 50
Ambassador: H E Dr Amin Rianom
Minister, Councellor, Political Affairs: Achmad Wirahadikusumah
First Secretary of Aministration and Consular Affairs: Lefman Sembiring
Second Secretary of Economic and Commercial Affairs: Dr Amir Radjab Harahap
Second Secretary of Economic and Cultural Affairs: Dr Robertas Irawan
Second Secretary of Administration: Wisnu Hutomo
Third Secretary of Communications: Mrs Jaswadiyah Supardjono [98]

31-00890 Embassy of Italy
Calle Soroca Edf.Atrium Ph, El Rosal, Caracas
☎ (+58 2) 952 73 11 ✉ 29986 ⊖ (+58 2) 952 49 60
Ambassador: H.E. Carlo Civiletti
Minister Counsellor: Dr Luigi Mercolini
Military, Naval & Air Attaché: Capt. (N) Michele Tari
First Secretary: Dr Paolo De Nicolo
First Secretary: Dr Bruno Antonio Pasquino
Cultural Attaché: Dr Angelo Rodolfo Manenti [97]

31-01270 Royal Netherlands Embassy
PO Box 62 286, 1060 A Caracas
☎ (+58 2) 263 36 22 [E] nlgovcar@internet.ve
✉ 23 177 holanda ⊖ (+58 2) 263 04 62
Ambassador: H E Mr C M M H R van Hanswijck de Jongo
Deputy Chief of Mission: J A M Geison
Commercial Sectretary and Public Information Officer: T A Reintjes [98]

31-01820 Embassy of the Republic of Trinidad and Tobago
PO Box Apartado del Este 61322, 1060A Caracas
Quinta Serrana, 4a Avenida Entre 7a y 8a Transversales, Altamira
☎ (+58 2) 261 37 48; 261 57 96; 261 47 72
⊖ (+58 2) 261 98 01
Ambassador: H E Mr Philip Sealy [98]

31-01890 Embassy of the United Kingdom
Embajada Britanics, Apartado 1246, 1010-A Caracas
Edificio Torre Las Mercedes (Piso 3), Avenida La Estancia, 1060 Caracas
☎ (+58 2) 993 41 11; 993 42 24 ✉ 23468 proca vc
⊖ (+58 2) 993 99 89
Ambassador: H E Mr R D Wilkinson
Deputy Chief of Mission: P Hare
Defence Attaché: Capt K Ridland (RN)
First Secretary, Commercial: A Goodworth
Third Secretary, Commercial: C Gracey [98]

90 Sales Agents, Sales & Liaison Offices

Companies represented:

Balzers AG, Liechtenstein	90-01960
CAE Electronics International, Canada	90-01300
CEIS T.M., France	90-01300
Hughey & Phillips Inc, USA	90-01300
ION Tract Instruments, UK	90-01300
LBA Systems Ltd, UK	90-01300
Powermark Inc., USA	90-01300
Racal Recorders Ltd, UK	90-01300
Sicamex, France	90-01300
Tana Wire Manufacturing Inc, USA	90-01300
Waverley Electronics, UK→ Netherlands	80-18300

90-00500 Daewoo Corp Caracas Office
Edificio Ceturia Piso 6 Oficinas 6a, Avenida Venezuela El Rosal, Caracas
☎ (+58 2) 953 60 06 ⊖ (+58 2) 953 15 34 [98]

90-01300 Michel Frochot Representaciones Industriales
Edificio Santana, Oficina 34, Aptdo 2648, 1010A Caracas
☎ (+58 2) 545 62 20; 545 62 91 ✉ CAROBA ✉ 21205
⊖ (+58 2) 545 64 12
General Manager: Michel Frochot
■ Interavia Representatives in Venezuela
® Canada: CAE Electronics International
France: CEIS T.M., Sicamex
UK: ION Tract Instruments, LBA Systems Ltd, Racal Recorders Ltd
USA: Hughey & Phillips Inc, Powermark Inc., Tana Wire Manufacturing Inc [98]

90-01960 Secotec SA
PO Box 3452, 1010A Caracas
☎ (+58 2) 573 02 70 [E] secotec@cantv.net ✉ 21652
⊖ (+58 2) 573 19 32
Managing Director: Nicolas Loskootoff
Director: Robert Weisshaar
Sales and Marketing Director: Ms Valery Dazun
® Liechtenstein: Balzers AG [98]

Vietnam

Head of State:
Nong Duc Manh

12 Ministries

12-10000 Ministry of National Defence
Hanoi
Minister: Gen Doan Khue [98]

12-20000 Ministry of Foreign Affairs
1 Ton That Dam, Hanoi
☎ (+84) 592 01 ✉ 4238 bnz vt
Minister: Nyugen Manh Cam [98]

12-40000 Ministry of the Interior
Rue Ngo Quyen, Hanoi
Minister: Maj Gen Le Minh Hong [98]

15 Parliament

15-10000 National Assembly
Quoc Hoi
Hanoi
Chairman of National Assembly: Nong Duc Manh [98]

31 Embassies & Attachés

31-00840 Embassy of the Republic of Indonesia
50 Pho Ngo Quyen, Hanoi
☎ (+84) 825 33 53; 30 84 ✉ 411434 indoha vt
⊖ (+84) 825 92 74
Ambassador: H.E. Mr Djafar Assegaff
Counsellor Political: Mr Rahardjo
Defence Attaché: Col Lumban Sianipar
First Secretary Economic: Mr Subijaksono
First Secretary Information: Bunyan Saptomo
First Secretary Administration: Purwohadi Atmodjo
Second Secretary Consular: Adhy Soesanto
Third Secretary Administration: Sjachril Abusamah [97]

31-00890 Embassy of Italy
9 Le Phong Hieu, PO Box 26, Hanoi 9
☎ (+84) 825 62 56; 62 46 ✉ 411416 itam vt
⊖ (+84) 826 76 02
Ambassador: H.E. Mr Mario Vittorio Zamboni di Salerano
First Secretary Economic & Commercial Affairs: Carlo Formosa
Consular Attaché: Andreina Fontana
Admin Attaché: Riccardo de Martino
Commercial Attaché: Livio Angeloni
Attaché: Pierino Marsella
Attaché: Giorgio Marconi
Attaché: Ms Annamaria Bonicïolli [97]

31-01110 Embassy of Malaysia
A-3 Van Phuc, Hanoi
☎ (+84) 845 33 71; 823 35 20 ✉ 411311 mala vn
⊖ (+84) 823 21 66
Ambassador: H.E. Mr Cheah Sam Kip
Deputy Chief of Mission: Siddiq Firdause Mohd Ali
Third Secretary: Hares Lisot [97]

31-01210 Embassy of Mongolia
Messica Jan Fu 39, Hanoi
☎ (+84) 521 51; 530 09 [98]

31-01350 Embassy of the Sultanate of Oman
Mahatma Gandi Cad. 63, VN-06700 Gaziosmanpasa, Ankara
☎ (+90 312) 447 06 30-31 ✉ 46507 ombe tr
⊖ (+90 312) 447 06 36
Ambassador: H E Mr Bashir Saleem Faraj Al-Faraj
Counsellor: Talal Al-Said
First Secretary: Saleh Suleiman Ahmed Al-Harthy [98]

31-01420 Embassy of the Philippines
27-B Tran Hung Dao St, Hanoi
☎ (+84 4) 825 78 73; 825 79 48; 826 18 26
✉ AMBAPHIL HANOI ⊖ (+84 4) 826 57 60
Ambassador: H.E. Ms Rosalinda V. Tirona
Third Secretary & Vice Consul & Economic Officer: Alex V. Lamadrid
Attaché: Mrs Norma R. Duran
Attaché: Ruben F. de Jesus
Attaché: Solomon E. Balmeo [97]

31-01480 Embassy of Romania
5 Le Hong Phong, Hanoi
☎ (+84) 520 14 ⊖ (+84) 309 22
Ambassador: H.E. Mr Valeriu Arteni [97]

31-01600 Royal Embassy of Saudi Arabia
Turan Emeksiz Sok. No. 6 G.O.P., VN-06700 Ankara
☎ (+90 312) 468 55 40-46 ⊖ (+90 312) 427 48 86
Ambassador: H E Mr Naji S Mufti
First Secretary: Khalid O Abd Rabuh
Military Attaché: Col Khalid Al Suney
Fiirst Secretary: Saud Saad Al Thobaiti
Second Secretary: Jamal Nasreddin [98]

31-01780 Royal Thai Embassy
63-65 Hoang Dieu St, Hanoi
☎ (+84 4) 823 50 92/94 ⊖ (+84 4) 823 50 88
Ambassador: H.E. Mr Chalermpon Ake-uru
Minister: Satchaphand Atthakor
Minister Counsellor: Choncharoen Devahastin Na Ayudhaya
Second Secretary Information: Mongkol Visitstump
Office of Commercial Affairs
108 Le Duan St, Hanoi
☎ (+84 4) 851 07 81/82 ⊖ (+84 4) 851 07 22
Commercial Counsellor: Sintorn Guntawong
Office of Defence/Army Attaché
40B Truong Chinh St, Hanoi
☎ (+84 4) 869 16 65 ⊖ (+84 4) 869 16 65
Defence & Army Attaché: Col Patiphon Kulapichitr [97]

31-01890 Embassy of the United Kingdom
Central Building, 31 Hai Ba Trung, Hanoi
☎ (+84 4) 825 25 10; 826 75 60; 826 75 61; 826 75 62; 826 75 63; 826 75 64; 825 75 65
✉ PRODROME HANOI
⊖ (+84 4) 826 57 62 (chancery/commercial); 825 23 49 (consular/management)

Contact Reference

Ambassador: H.E. Mr P.K. Williams (CMG)
First Secretary Commercial & Deputy Head of Mission: J.R. McCulloch
Public Information Officer: Giles Lever

Commercial Office
100 Tue Tinh, Hanoi
☎ (+84 4) 822 94 55/6 ⊖ (+84 4) 822 94 57
Second Secretary Commercial: M. Garrett 98

31-01900 Embassy of the USA
7 Lang Ha Road, Ba Ding District, PSC 461, Box 400, Hanoi
☎ (+84 4) 843 15 00 ⊖ (+84 4) 835 04 84; 843 15 10
Ambassador: H E Peter Peterson 98

31-02010 Embassy of the Federal Republic of Yugoslavia
47 Tran Phu Street, PO Box 29, Hanoi
☎ (+84) 548 64; 546 66
Chargé d'Affaires a.i.: Milos Bogicevic 97

75 Manufacturing Companies

75-01000 DAEHA Company Ltd
(a subsidiary of Daewoo Corporation, South Korea)
Level 2, Daeha Business Center-Office Tower, VN-Cau Giay Ba Binh, Hanoi
Ngoc Khanh/Lieu Giai Street
☎ (+84 4) 834 25 16; 834 25 17; 834 58 37
⊖ (+84 4) 834 57 44 98

75-01010 Daewoo Corp Hanoi Office
9th Floor Office Tower, Daeha Business Center, Hanoi
Ngoc Khanh/Lieu Giai Street
☎ (+84 4) 831 52 00; 831 52 01; 831 52 02
⊖ (+84 4) 831 54 79 98

75-01015 Daewoo Corp Ho Chi Minh City
Unit 6C1 Hannam Building, 65 Nyugen Du Street, VN-District 1 Ho Chi Minh City
☎ (+84 8) 822 91 60; 822 91 61 ⊖ (+84 8) 822 91 62 98

75-05000 Transfield Pty Ltd
1 Trinh Hoai Duc Street, Dong Da District, Hanoi
☎ (+84 4) 823 00 29; 823 77 15 ⊖ (+84 4) 823 24 06 98

90 Sales Agents, Sales & Liaison Offices

90-00500 BP Solar Vietnam
13 Ho Xuan Huong, Hanoi
☎ (+84 4) 826 53 25 ⊖ (+84 4) 826 61 56 98

90-05000 Daewoo Corp Vietnam Representative Office
8F Office Tower Daeha Business Ctr, 360 Kim Ma Street, Ba Dinh District, Hanoi
☎ (+84 4) 831 52 20 ⊠ 411296 daewoo vt
⊖ (+84 4) 831 78 89

Ho Chi Minh Branch
74 Hai ba Trung Street, District 1, Hochiminh City
☎ (+84 8) 822 43 13 ⊠ 811277 daewoo vt
⊖ (+84 8) 822 43 23 98

British Virgin Islands

Head of State:
H.M. Queen Elizabeth II

10 Head of State

10-10000 Office of the Governor
PO Box 702, Road Town, Tortola
☎ (+1 809) 494 23 45; 494 35 20
⊖ (+1 809) 494 55 82
Governor: D P R MacKilligin
Deputy Governor: E Georges
Assistant to the Governor: Carl Garn
Personal Assistant to the Governor: Ms Susan McCarthy 98

Yemen

Head of State:
Lieut Gen Ali Abdullah Salih

10 Head of State

10-10000 Office of the President
Zubairy Street, Sana'a
☎ (+967 2) 713 92; 741 91-95 ⊠ 2422 riasah ye
President: Lieut Gen Ali Abdallah Salih
Vice President: Maj Gen Abd al-Rab Mansur la Hadi 98

11 Prime Minister

11-10000 Office of the Prime Minister
Street of 26th September, Sana'a
☎ (+967 2) 730 92; 730 95
Prime Minister: Dr Faraj Said bin Ghanim
Deputy Prime Minister: Dr Abd al-Karim Ali al Iryani
Minister of State for Cabinet Affairs: Ahmad Ali al- Bishari 98

12 Ministries

12-10000 Ministry of Defense
Sana'a
Defence Minister: Brig Gen Muhammad Dayfallah Muhammad 98

12-20000 Ministry of Foreign Affairs
Alolofi Square, Sana'a
☎ (+967 2) 735 34; 734 30 ⊠ 2216 alkhareji ye
Minister of Foreign Affairs: Dr Abd al-Karim Ali al Iryani 98

12-40000 Ministry of the Interior
Almatar Road, PO Box 1110, Sana'a
☎ (+967 2) 726 77, 771 61 ⊠ 2274 intposa ye
Minister: Brig Gen Husayn Muhammad al Arab 98

31 Embassies & Attachés

31-00890 Embassy of Italy
Via Safiah Janubia, PO Box 1151, Sana'a
☎ (+967 1) 269 16; 26 56 16 ⊖ (+967 1) 26 61 37 98

31-01890 Embassy of the United Kingdom
129 Haddah Road, PO Box 1287, Sana'a
☎ (+967 1) 26 40 81; 26 40 82; 26 40 83
⊖ (+967) 126 30 59
Ambassador: H E Mr Victor J Henderson
Deputy Chief of Mission: D A Pearce
Air Attaché: Wg Cdr Chris Troke (Resident in Riyadh)
Defence and Military Attaché: Brig W E Strong (Resident in Riyadh)
Naval Attaché: Cdr I A Vosper (RN-Resident in Riyadh)
British Council Director: B M McSharry
Second Secretary Commercial, Economical and Political: J Downer
Third Secretary, Management: C Sturgeon
Third Secretary, Immigration: M Hardy 98

Fed. Republic of Yugoslavia

Head of State:
Zoran Lilic

10 Head of State

10-10000 Office of the President
Nemanjina 11, 11000 Belgrade
☎ (+381 11) 33 42 81 ⊠ 11448 siv yu
President: Zoran Lilic 98

11 Prime Minister

11-10000 Office of the Prime Minister
Nemanjina 11, 11000 Belgrade
Prime Minister: Radoje Kontic
Deputy Prime Minister: Uros Klikovac
Deputy Prime Minister: Danko Djunic
Deputy Prime Minister: Vladan Kutlesic 98

12 Ministries

12-10000 Ministry of National Defence
Nemanjina 11, 11000 Belgrade
☎ (+381 11) 68 52 40 ⊖ (+381 11) 68 21 67; 65 61 22
Minister: Col Gen Pavle Bulatovic 98

12-20000 Ministry of Foreign Affairs
Kneza Milosa 24-26, 11000 Belgrade
☎ (+38 11) 68 25 55; 68 26 66 ⊠ 11143; 11173
⊖ (+38 11) 68 35 64; 68 26 68
Minister: Milan Milutinovic ☎ (+38 11) 68 45 82
Deputy Minister: Dobroslav Vejzovic ☎ (+38 11) 68 14 52
Chief of Staff: Miroslav Buljancevic 98

12-40000 Ministry of Internal Affairs
Kneza Milosa 103, 11000 Belgrade
☎ (+38 11) 68 42 66 ⊖ (+38 11) 64 18 67
Minister: Zoran Sokolovic 98

12-50000 Federal Ministry for Transportation and Communications
Bulevar AVNOJ-a 104, 11070 Novi Beograd
☎(+381 11) 60 45 76 ⊠ 12062 dfca
⊖ (+381 11) 222 39 46
Minister: Svetolik Kostadinovic
Deputy Minister: Budimir Saranovic
Secretary: Zoran Karaicic [98]

31 Embassies & Attachés

31-00090 Australian Embassy
13 Cika Ljubina, 11000 Belgrade
☎(+381 11) 62 46 55; 63 22 61
✉AUSTEMBA Belgrade ⊖ (+381 11) 62 40 29
Chargé d'Affaires: Noel D. Campbell
First Secretary: Donald Foley
Second Secretary: Richard Fogarty [97]

31-00580 Embassy of Finland
Bircaninova 29, 11000 Belgrade
☎(+381 11) 64 63 22, 64 61 66 ⊠ 11707 finem yu
Ambassador: H E Mr Ilpo Mannninen
☎(+381 11) 68 38 80
Second Secretary: Mrs Pirkko Kaikkonen
Attaché, Administrative Affairs: Mauri Lindberg [98]

31-00660 Embassy of the Federal Republic of Germany
Ulica Kneza Milosa 74-76, PO Box 304, 11001 Belgrade
☎(+381 11) 64 57 55
[E] 106112.3427@compuserve.com ⊠ 11107
⊖ (+381 11) 65 69 89
Ambassador: H E Mr Wilfried Gruber
Chargé d'Affaires ai and Commercial Counsellor: Dr Rolf Krause
Defence Attaché: Lt Col Nikolaus von Wittenburg
Defence Attaché: Lieut Col Nikolaus von Wittenburg
Third Secretary: Michael Ahrens
Public Information Officer: Dirk Lölke [98]

31-00670 Embassy of Ghana
Ognjena Price 50, Belgrade
☎(+381 11) 444 24 45 ⊠ 11 720 ganemb yu
⊖ (+381 11) 43 63 14
Ambassador: H E Mr Samuel Ansah-Antwi
Third Secretary: Joshua Nyarko [98]

31-00690 Embassy of Greece
Francuska 33, Belgrade
☎(+381 11) 62 62 34 ⊠ 11361 ⊖ (+381 11) 62 10 22 [98]

31-00920 Embassy of Japan
Genex Apartments, Vladimira Popovica 6, 11070 Belgrade
☎(+381 11) 222 14 34; 222 22 64; 222 37 30; 222 38 98; 222 44 38 ⊖ (+381 11) 14 56 90 [97]

31-01730 Embassy of Sweeden
Ulica Pariska 7, PO Box 725, 11001 Belgrade
☎(+38 11) 62 64 22 [E] swedeemb@eunet.yu
⊠ 11595 svensk yu ⊖ (+38 11) 62 64 92
Ambassador: H E Mr Mats Staffansson
Third Secretary: Annet Karlson Fouda [98]

31-01740 Embassy of Switzerland
Bircaninova 27, PO Box 817, 11001 Belgrade
☎(+38 11) 64 68 43; 64 69 74
✉AMBASUISSE Belgrade ⊠ 11383 yu ambas [98]

31-01840 Embassy of the Republic of Turkey
Proleterskih brigada 1, 11000 Belgrade
☎(+38 11) 323 554 31; 323 54 32 [E] turem@eunet-yu
⊖ (+38 11) 323 54 33
Ambaassador: Alev Kilic
First Secretary: Hakan Olcay
Military, Naval and Air Attaché: Col Etem Baskurt
Administrative Attaché: Zeki Öztürk
Administrative Attaché: Mrs Mehves Olcay
Attaché: Mehmet Sahin
Attaché: Yuksel Aras
Attaché: Muammer Yigit
Third Secretary and Vice Consul: Mustafa Kemal Basa
Commercial Counsellor's Office
Strahinjica Bana 47/IV, Belgrade
☎(+38 11) 63 03 03 ⊖ (+38 11) 63 03 03 [98]

31-01890 Embassy of the United Kingdom
Generala Zdanova 46, 11000 Belgrade
☎(+38 11) 64 50 34; 64 50 55 [E] ukembbg@eunet.yu
⊠ 11468 yu probel ⊖ (+38 11) 65 96 51
Chargé d'Affaires ai: I A Roberts
Deputy Head of Mission: A C Crombie
Defence and Military Attaché: Col J H Crosland
Defence Attaché: Staff Sgt I B Duncan
Commercial Officer: D Slinn
Public Information Officer: J Braithwaite [98]

40 Government Defense Sales/Procurement Organizations

40-00600 Yugoimport - SDPR
(† Federal Directorate of Supply & Procurement (SDPR))
Bulevar umetnosti 2, PO Box 89, 11070 Novi Beograd
☎(+381 11) 22 22 743; 22 23 697; 22 21 284
✉JUSDPR [E] jugoimport@sezampro.yu
⊠ 71000; 72566 jusdpr
⊖ (+381 11) 32 48 748; 32 48 791; 32 48 738
♧ US$ 4,101,189 (1996) ♣ US$ 6,864,885 (1996)
ꝏ 300 ⌂ 8,980 m2
Director General: Gen Jocan Cekovic
Marketing Director: Jelica Micanovic
☎(+381 11) 13 02 63
Commercial Director: Miomir Kalezic
☎(+381 11) 13 65 26
Operations Director: Slobodan Sario
■ Transfer of advanced technologies
Project management; export and import of defence equipment, consumer goods and services; research and development, manufacture, joint venture. [98]

65 Industry, Trade & Armed Forces Associations

65-01200 Institute of Physics
(† Institut Za Fiziku)
Maksima Gorkog 118, Pregrevica, Beograd-Zemun
☎(+38 11) 10 71 07; 10 80 46 ⊠ 11002
⊖ (+38 11) 10 81 98
Managing Director: Drugan Popvic
Librarian: Olivera Obracevic [98]

75 Manufacturing Companies

75-00475 Kompanija Sloboda D D (SDD)
(† Sloboda DP)
Ulitsa ratka Mitrovica b b, 32000 Cacak
☎(+381 32) 621 41 ⊠ 13822 ⊖ (+38 32) 263 90
♣ US$ 50,000,000 (1990) ꝏ 2,300 (1997)
Managing Director: Radomir Ljujic ☎(+381 32) 633 30
Marketing Director: Tomislav Kovacevic
■ Aircraft, anti-aircraft and naval guns, ammunition, (20-57 mm), artillery and tank guns ammunition (76-155 mm), anti-tank and fragmentation rifle grenades, anti-tank rockets and other pyrotechnic material, machined parts and pyrotechnic items for missiles. Hand shock grenade, shock round (38mm), illuminating cartridge (38mm), rubber bullet round (38mm), signal cartridge (38mm). [98]

75-00610 Prva Petoletka Industrie Hidraulike i Pneumatike (PPIHP)
37240 Trstenik
☎(+381 37) 71 12 22; 71 42 22
✉PETOLETKA Namenska
⊠ 17429; 17519; 17529 ppt yu ⊖ (+381 37) 71 37 90
ꝏ 14,000 ⌂ 350,000 sq m
Managing Director: Slobodan Jakovlsvig
Head of Directorate for Sales: Miroslav Novakovic
■ Development and manufacture of hydraulics, special purpose hydraulics, cylinders, servo steering gears, brake technology industrial pneumatics, lubrication technology, refrigerating devices and systems, vacuum techniques, cranes and manipulators, tools, machine accessories, connecting elements, devices and components, swivel bearings, sealing elements for applications in general industry, transport vehicles. [97]

75-00690 Shipyard Beograd
Savski Nasip 7, PO Box 45, 11071 Belgrade
☎(+38 11) 15 50 20 ⊠ 11386 ⊖ (+38 11) 16 59 34
Managing Dir: Milanko Krsmanovic ☎(+38 11) 15 52 87
Technical Dir: Dusan Dinic
Commercial Mgr: Vasiljevic Neboisà ☎(+38 11) 15 45 64
Purchasing Mgr: Bjegovic Velimir
■ Ship construction & repair. [97]

75-01900 Teleoptik Ziroskopi
Cara Dusna br 139-141, Beograd-Zemun
☎(+381 11) 61 45 22 ⊠ 12346
⊖ (+381 11) 61 54 39; 61 51 32
♣ UK£ 1,000,000 (1996) ꝏ 803 (1997) ⌂ 14,000 sq m
Managing Director: Radovan Andjelkovic
Operations Director: Ms Natasa Jovanovic
Marketing Director: Srboiljub Miladinovic
Purchasing Manager: Pop-Tobic Jovan [98]

80 Service Companies & Organizations

80-02500 Vazduhoplovno Remonti Zavod Moma Stanojlovic
Nesvrstanih Zemalja 160, Beograd
☎(+38 11) 848 16 01; 61 74 55 ⊠ 11675
⊖ (+38 11) 61 94 83; 84 86 236
ꝏ 1,300 (1997) ⌂ 56,625 m2
General Manager: Vladet Milovanovic
Technical Director: Miroslav Markovic
Sales Director: Miodrag Celebic
Purchasing Manager: Radenko Hedeljkovic
Public Relations and Marketing Manager: Branko Milasinovic ☎(+381 11) 61 90 27
■ Overhaul:
Helicopters, transport aircraft, fighter aircraft, light combat aircraft, light piston aircraft, turboprop and turboshaft engines, piston engines, OTO and diesel engines, aircraft armament, A/C servicing and maintenance devices, aircraft instruments, radio navigation instruments and aids, electro equipment and avionics, photo and reconnaissance equipment, flaying equipment, oxygen equipment, communication equipment, missile technique maintenance equipment.
Parts reconditioning:
Manufacture: Spare parts for aircraft and vehicles, sintered brake and friction elements, TJ engine thermal insulation blankets, A/C and MTD heat exchangers, composite parts, trainers and training equipment, training and other aids.
Design engineering, development, engineering: Modifications and updating MTD, overhauling plants, workshops, shops, technological lines.
Calibration of measuring devices: Calibration of instruments, servicing of instruments. [98]

Zambia

Head of State:
Frederick Chiluba

10 Head of State

10-10000 Office of the President
State House, PO Box 30208, Lusaka
☎ (+260 1) 21 18 33 ☒ 41460
President of the Republic: Frederick Chiluba
Vice President: Brig Gen Godfrey Miyanda [98]

12 Ministries

12-10000 Ministry of Defense
PO Box 31931, Lusaka
☎ (+260 1) 25 23 66 ☒ 40476
Minister of Defense: Benjamin Mwila [98]

12-20000 Ministry of Foreign Affairs
PO Box 50069, Lusaka
☎ (+260 1) 22 90 07; 22 90 20 ☒ 41290 mfalus za
⊖ (+260 1) 25 02 40; 25 28 67
Minister: Lawrence Shimba [98]

12-30000 Ministry of Commerce Industry and Foreign Trade
PO Box 31 968, Lusaka
☎ (+260 1) 21 37 67
Minister: Alfeyo Hambayi [98]

12-40000 Ministry of Home Affairs
PO Box 31862, Lusaka
☎ (+260 1) 21 35 05
Minister: Chitalu Sampa [98]

31 Embassies & Attachés

31-00310 Canadian High Commission
5199 United Nations Ave, PO Box 31313, Lusaka
☎ (+260 1) 25 08 33 ⊖ (+260 1) 25 41 76
High Commissioner: H.E. Ms Mary Mosser
Deputy High Commissioner: Sean Sunderland
Defence Attaché: Lt Col Chris Sprouk
Commercial Officer: Shilpa Patel [97]

31-00590 Embassy of France
74 Independence Avenue, PO Box 30 062, Lusaka
☎ (+260 1) 25 13 22 ⊖ (+260 1) 25 44 74
Ambassador: H E Mr Zhilippe Perrier de la Bâthie [98]

31-00830 High Commission of India
PO Box 32111, Lusaka
1 Pandit Nehru Rd., Lusaka
☎ (+260 1) 25 31 52/59/60; 25 30 66
☒ 41420 hicomind za ⊖ (+260 1) 25 41 18
High Commissioner: H.E. Mr Rinzing Wangdi
First Secretary: O.P. Bajaj
Second Secretary/Defence Adviser: S. Bhattacharjee [97]

31-00890 Embassy of Italy
Diplomatic Triangle, Embassy Park, Plot 5211, Lusaka
☎ (+260 1) 25 07 55 ☒ 43380 ⊖ (+260 1) 25 49 29
Ambassador: H E Dr Margherita Costa
Consular Attaché: Dr Nunzio Maria Caldarelli
Commercial Attaché: Ms Vita Maria Valluzzi [98]

31-01730 Swedish Embassy
Haile Selassie Avenue, Longacres, PO Box 30 788, Lusaka
☎ (+260 1) 25 17 11 Ⓔ swedemb@zamnet.zm
☒ 41820 svensk za ⊖ (+260 1) 25 40 49
Ambassador: H E Ms Kristina Svensson
Counsellor: Christer Agren
Senior Administration Officer: Karl Norheden [98]

31-01890 United Kingdom High Commission
10 Independence Avenue, PO Box 50050,
15101 Ridgeway, Lusaka
☎ (+260 1) 25 11 33 ☒ 41150 ukrep za
⊖ (+260 1) 25 37 98
High Commissioner: H E Mr P M Nixon
Deputy High Commissioner: B S Jones
First Secretary Commercial/Consular: R Clark
First Secretary Development and Economic: M T Murray
Management Officer: P R Seaby [98]

31-01900 Embassy of the United States of America
PO Box 31617, Lusaka
Corner Independence and United Nations Avenues
☎ (+260 1) 22 29 93; 22 29 94 Ⓔ usiszam@zamnet.zm
☒ 41970 amemb za ⊖ (+260 1) 22 65 23
Ambassador: H E Mr Arlene Render
Deputy Chief of Mission: Stevenson McIlvaine
Commercial Counsellor: Andrew Critton
Procurement Officer: Sherman Williamson
Public Affairs Officer: Ms Ruby Apsler
Political and Military Attaché: Edward Stafford [98]

31-02010 Embassy of the Federal Republic of Yugoslavia
Plot 5216 Diplomatic Triangle, PO Box 31 180, Lusaka
☎ (+260 1) 25 02 47 ⊖ (+260 1) 25 38 43
Chargé d'Affaires: Stanimir Jovanovic
☎ (+260 1) 26 32 56 [97]

Zimbabwe

Head of State:
Robert Gabriel Mugabe

10 Head of State

10-10000 Office of the Executive President
PO Box 368, Harare
☎ (+263 4) 266 66 ☒ 4476
Executive President: Robert Gabriel Mugabe
Vice-President: Simon Vengai Muzenda
Minister of State in the President's Office: Cephas Msipa [98]

12 Ministries

12-10000 Ministry of Defense
Private Bag 7713, Causeway, Harare
Samora Machel Avenue, Munhumutapa Building
☎ (+263 4) 70 01 55 ✉ DEFENCE HARARE ☒ 24824
⊖ (+263 4) 79 67 62
Minister: Moven Mahachi [98]

12-20000 Ministry of Foreign Affairs
Samora Machel Avenue, PO Box 4240, Harare
☎ (+263 4) 79 08 71
Minister: Stanislaus Mudenge [98]

12-40000 Ministry of Home Affairs
Livingstone Avenue, PO Box 7703, Causeway, Haare
☎ (+263 4) 70 36 41; 38 94; 70 32 90
⊖ (+263 4) 72 67 16
Minister: Dumiso Dabengwa
Private Secretary to the Minister: Ms B Dube
Permanent Secretary of State: T E Chigudu
☎ (+263 4) 72 08 51
Commissioner of Police: A M Chihuri ☎ (+263 4) 70 09 18
Deputy Commisioner Administration: P Mhike
☎ (+263 4) 72 66 27
Deputy Commisioner: E Z Ngwenya ☎ (+263 4) 72 55 04
Deputy Commisioner Personnel: D C Blake
☎ (+263 4) 72 35 44
Deputy Commisioner Crime: G T Matanga
☎ (+263 4) 72 35 44
Executive Director National Museum and Monuments: D Munjeri ☎ (+263 4) 75 17 97
Deputy Secretary: F M Maponsa ☎ (+263 4) 70 58 80
Under Secretary: F Dembetembe ☎ (+263 4) 79 27 06
Chief Immigration Officer: E H Mugwadi
☎ (+263 4) 79 01 93
Registrar General: T T Mudede ☎ (+263 4) 72 55 40 [98]

25 Defense Forces

25-20000 Air Force Headquarters
Private Bag 7721, Causeway, Harare
☎ (+263 4) 73 00 61 ☒ 24318 ⊖ (+263 4) 70 83 18
Commander Zimbabwe Air Force: AM Perence Shiri
Chief of Staff (Support Services): AVM Henry Muchena
Director General Operations: AVM Fred Vinyu
Chief of Staff Operations: AVM Ian M Harvey
Director General Support Services: Air Cdre Mike T Moyo [98]

31 Embassies & Attachés

31-00090 Australian High Commission
PO Box 4541, Harare
Corner Samora Machel Avenue and J Nyerere Way
☎ (+263 4) 75 77 74 ⊖ (+263 4) 75 77 70
High Commissioner: H E Mr P F Peters
Attaché, Vice Consul: L Van Leeuwen [98]

31-00460 Royal Danish Embassy
1st Floor, UDC-Centre, 59 Union Ave, PO Box 4711, Harare
1st Floor, UDC-Centre, 59 Union Avenue
☎ (+263 4) 75 81 85 ✉ AMBADANE HARARE
☒ 24677 ambdk zw ⊖ (+263 4) 75 81 89
Ambassador: H E Mr Erik Fiil
Attaché: A M Christensen [98]

31-00660 Embassy of the Federal Republic of Germany
14 Samora Machel Avenue, PO Box 2168, Harare
☎ (+263 4) 73 19 55; 73 19 58 ☒ 24609 dipger zw
⊖ (+263 4) 79 06 80
Ambassador: H E Dr Norwin Graf Leutrum
Deputy Chief of Mission and Commercial Counsellor: Dr Christoph Müller [98]

31-00690 Embassy of Greece
8 Deary Avenue, PO Box 4809, Belgravia, Harare
☎ (+263 4) 79 32 08 ☒ 24790 ⊖ (+263 4) 72 27 74 [98]

31-00890 Embassy of Italy
7 Bartholomew Close, PO Box 1062, Greendale, Harare
☎ (+263 4) 49 72 00; 49 73 73; 49 81 90; 49 72 79
Ⓔ amitalia@samara.co.zw ☒ 24380 italdpl zw
⊖ (+263 4) 49 81 99
Ambassador: H E Dr Lucaa Brofferio
First Secretary: Dr Paol Munzi
Military Attaché: Col Franceso Criscoli (Resident in Maputo) ☎ (+263 4) 49 05 03; 6-414 MILIT MO (telex) [98]

31-01270 **Royal Netherlands Embassy**
2 Arden Rd, PO Box HG 601, Harare
☎ (+263 4) 77 67 01/4 Telex 24357 nedamb zw
Fax (+263 4) 77 67 00 [97]

31-01440 **Embassy of Portugal**
10 Samora Machel Avenue, PO Box 406, Harare
☎ (+263 4) 72 34 02 Telex 4714 emport zw
Fax (+263 4) 72 22 91
Ambassador: H.E. Dr José Ernst Henzler Vieira Branco
Counsellor: Dr Manuel Lobo Antunes
Commercial Attaché: Jorge de Jesus Cavaco Encarnacao
Admin Officer: Miss Joaquina Maria Guerreiro [97]

31-01680 **Embassy of Spain**
16 Phillips Avenue, Belgravia, PO Box 3300, Harare
☎ (+263 4) 73 86 81; 73 86 82; 73 86 83
Fax (+263 4) 79 54 40
Ambassador: H E Mr Tomas Gragera
First Secretary: Fernando Ruigomez [98]

31-01740 **Embassy of Switzerland**
9 Lanark Road, Belgravia, PO Box 3440, Harare
☎ (+263 4) 70 39 97/98 Fax (+263 4) 79 49 25
Ambassador: H E Dr C Krieg Polejack [98]

31-01890 **United Kingdom High Commission**
PO Box 4490, Harare
Corner House, Samora Machel Avenue, Takawira Street, Harare
☎ (+263 4) 77 29 90; 77 74 00 Cable UKREPHRE
Telex 24607 ukrep zw Fax (+263 4) 77 46 17
High Commissioner: H E Mr Martin Williams (CMG)
Deputy High Commissioner: T I Hay-Campbell
Defence Adviser: Col A J Reed Screen
First Secretary, Management: J Anderson
First Secretary, Consular: G M Johnson
First Secretary, Commercial: A R Ashcroft
Second Secretary, Aid: Miss D J Pearey
Second Secretary, Political: D Quarrey
Second Secretary, Commercial: A S Harrison
Third Secretary, Information: W S Robertson
Third Secretary, Consular: A H Bowyer
Third Secretary, Management: A I McGlone [98]

90 Sales Agents, Sales & Liaison Offices

Companies represented:

Communications Systems, Zimbabwe . 90-01950
The General Electric Co. plc, UK 90-01950

90-00400 **CATIC**
Representative Office
8 Uplands Close, Highlands, Harare
☎ (+263 4) 466 07 Fax (+263 4) 466 07 [97]

90-00500 **Daewoo Corp Harare Office**
Fidelity Life Tower 12th Floor, Ra Leigh Street, Harare
☎ (+263 4) 78 10 09 Telex 24747 zw Fax (+263 4) 78 10 09 [98]

90-01950 **GEC Zimbabwe (Pvt) Ltd**
PO Box 845, Harare
☎ (+263 4) 79 36 41 Telex 22177 gec zw
Fax (+263 4) 73 79 79
Managing Dir: C. Myers
■ Supply, installation & service of PABX systems. Marketing of electronic systems.
® UK: The General Electric Co. plc
Zimbabwe: Communications Systems

Contact Reference

Notes

TRUST

Product / Service Index

This section lists over 3,500 product and service designations at present covered in the coding system. It provides a quick means of finding the suppliers of any particular product or service. Having found the product or service you are looking for, simply note the code number printed on the right of the column, turn to the section «Product–Supplier Link» and look up this code number in the numerical sequence. There you will find a list of names of manufacturers or suppliers, the country in which they are located, and the Contact Reference page/key number under which full contact information will be found.

Product/Service Designation	Code
A	
AA gun systems, ground-based	515 41 10
AA gun systems, ground-based, self-propelled	515 41 15
AA gun systems, naval	515 31 10
AA guns, automatic, naval	512 50 32 10
AA guns, automatic, self-propelled	512 50 42 15
AA guns, automatic, vehicle-mounted	512 50 42 15
AA guns, ground-based, automatic	512 50 42 10
AA missile systems, ground-based	515 41 20
AA missile systems, ground-based, low / medium altitude	515 41 24
AA missile systems, ground-based, low altitude	515 41 22
AA missile systems, ground-based, medium / high altitude	515 41 26
AA missile systems, low / medium altitude, naval	515 31 24
AA missile systems, low-altitude, naval	515 31 22
AA missile systems, naval	515 31 20
AA weapon systems, ground-based	515 41
Ablution facilities, warship	231 30 10
Ablution units, field	622 10
ABS systems, aircraft brake	221 40 60 90 12
Accelerometers	421 90 10
Accessories, airborne display equipment	455 20 80
Accessories, airborne EW equipment	461 20 80
Accessories, aircraft airframe	221 40 80
Accessories, aircraft bomb	523 20 80
Accessories, aircraft engine	321 80
Accessories, ammunition	521 80
Accessories, communications equipment	446 80
Accessories, communications equipment, airborne	446 20 80
Accessories, communications equipment, airborne radio	446 20 80 30
Accessories, communications equipment, ground-based	446 40 80
Accessories, communications equipment, ground-based radio	446 40 80 30
Accessories, communications equipment, naval	446 30 80
Accessories, communications equipment, naval radio	446 30 80 30
Accessories, diesel-electric propulsion systems, submarine	339 10 80
Accessories, diesel engine, non-armoured vehicle	341 50 10 80
Accessories, diesel engine, submarine	339 11 80
Accessories, diesel engine, tracked armoured vehicle	341 10 10 80
Accessories, diesel engine, wheeled armoured vehicle	341 30 10 80
Accessories, display equipment, ground-based	455 40 80
Accessories, electric propulsion systems, tracked armoured vehicle	341 10 30 80
Accessories, EW equipment, airborne	461 20 80
Accessories, EW equipment, ground-based	461 40 80
Accessories, gas turbine engine, non-armoured vehicle	341 50 25 80
Accessories, gas turbine engine, tracked armoured vehicle	341 10 25 80
Accessories, ground-based EW equipment	461 40 80
Accessories, gun	512 50 80
Accessories, helicopter	222 80
Accessories, helicopter engine	322 80
Accessories, multifuel engine, non-armoured vehicle	341 50 15 80
Accessories, multifuel engine, tracked armoured vehicle	341 10 15 80
Accessories, multifuel engine, wheeled armoured vehicle	341 30 15 80
Accessories, naval EW equipment	461 30 80
Accessories, naval mine	513 38
Accessories, non-armoured vehicle	241 50 80
Accessories, non-armoured vehicle propulsion system-related	341 50 80
Accessories, petrol engine, non-armoured vehicle	341 50 20 80
Accessories, petrol engine, tracked armoured vehicle	341 10 20 80
Accessories, petrol engine, wheeled armoured vehicle	341 30 20 80
Accessories, radar	411 18
Accessories, radio communications equipment, airborne	446 20 80 30
Accessories, radio communications equipment, ground-based	446 40 80 30
Accessories, radio communications equipment, naval	446 30 80 30
Accessories, recording equipment	456 80
Accessories, RPV / Drone	223 80
Accessories, RPV / Drone engine	323 80
Accessories, shipborne display equipment	455 30 80
Accessories, sighting equipment	517 40 60 80
Accessories, sonar	411 28
Accessories, surface ship propulsion system-related	331 80
Accessories, tracked armoured vehicle	241 10 80
Accessories, tracked armoured vehicle propulsion system-related	341 10 80
Accessories, wheeled armoured vehicle	241 30 80
Accessories, wheeled armoured vehicle propulsion system-related	341 30 80
Accessory drive shafts	221 40 80 05 50
Accessory drives, aircraft engine	321 80 05
Accessory drives, airframe-mounted	221 40 80 05
Accessory drives, helicopter engine	322 80 05
Accident data recorders, airborne	456 20 30
Accident data recorders, aircraft	221 10 70 35
Accommodation, field	621
Accommodation units, portable	611 23 10
Accoutrements, medals, insignia	741
Acoustic beacons, underwater	423 30 70 10 10
Acoustic communications equipment, underwater	446 30 70 60
Acoustic countermeasures launchers	461 30 70 50
Acoustic countermeasures systems, naval shipborne	461 30 11
Acoustic countermeasures systems, submarine launched	461 30 11 60
Acoustic data processing systems, airborne	450 20 60
Acoustic mines, sea	513 30 05
Acoustic navigation equipment, shipborne	421 30 05 70
Acoustic position fixing equipment, naval	423 30 70 10
Acoustic position fixing systems, naval	423 30 10
Acoustic surveillance equipment, ground-based	432 40 70 10
Acoustic surveillance equipment, naval	432 30 70 10
Acoustic surveillance systems, ground-based	432 40 10
Acoustic surveillance systems, naval	432 30 10
Acoustic target location & tracking systems & equipment	411 30
Acoustic target location & tracking systems & equipment, underwater	411 33
Acoustic test ranges	617 15 10
Action information & communications systems, shipborne	454 30 10
Action information & communications systems, shipborne (see code 454.30.10)	231 20 05
Active armour plates, tracked armoured vehicle	241 10 80 11
Actuation systems, missile fin	511 29 20
Actuators, aircraft	221 40 70 10
Actuators, special	527 60 30
Add-on armour, tracked armoured vehicle	241 10 80 10
Add-on armour, wheeled armoured vehicle	241 30 80 10
Add-on armour kits, non-armoured vehicle	241 50 80 10
Additives, ammunition	851 10
Advisory services	910
Aerial refuelling equipment, aircraft	221 20 40 70 10
Aerial refuelling systems	221 20 40
Aerial refuelling systems, boom type	221 20 40 10
Aerial refuelling systems, probe & drogue type	221 20 40 50
Aerial target systems	616 12 67
Aeromedical systems & equipment	615 20 10
Aerosols, anti-riot	491 65 10
AEW radars	411 12 20
AFVs, tracked	141 10 05
Aiming & target designation systems, aircraft	221 20 50
Aiming circles, ground-based	423 40 70 10 10
Aiming circles, gyro, ground-based	423 40 70 35 10
Aiming lights	517 40 60 80 10
Aiming systems, aircraft weapon	221 20 50 90
Air bases, airports, turnkey	610 20
Air charging sets	622 50 20

Product-Service Index

Product / Service Index

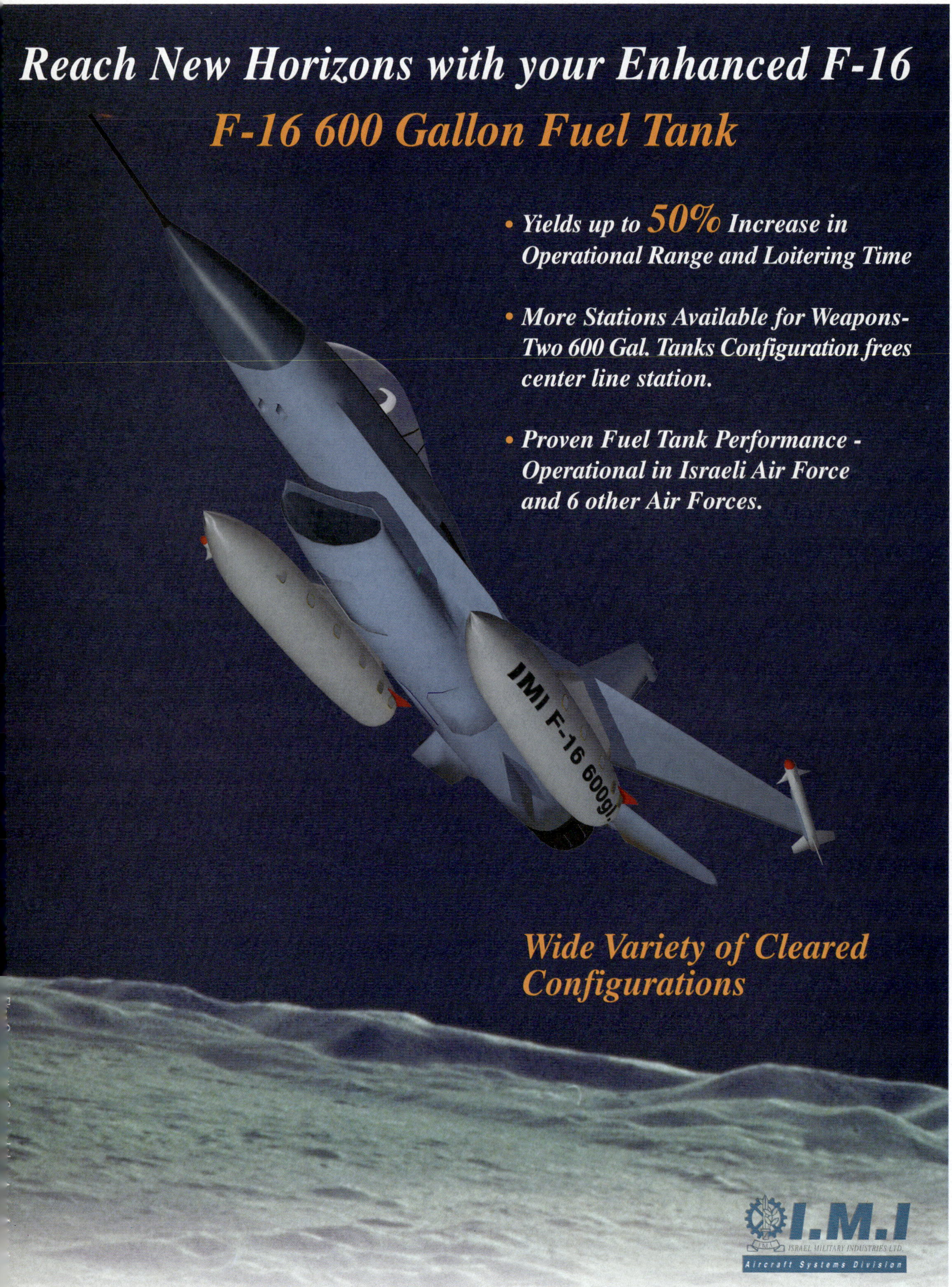
Reach New Horizons with your Enhanced F-16
F-16 600 Gallon Fuel Tank
• Yields up to 50% Increase in Operational Range and Loitering Time
• More Stations Available for Weapons- Two 600 Gal. Tanks Configuration frees center line station.
• Proven Fuel Tank Performance - Operational in Israeli Air Force and 6 other Air Forces.
IMI F-16 600gl.
Wide Variety of Cleared Configurations
I.M.I
ISRAEL MILITARY INDUSTRIES LTD.
Aircraft Systems Division

Aircraft elevators / lifts, aircraft carrier	231 20 10 05
Aircraft engine accessories	321 80
Aircraft engine accessory drives	321 80 05
Aircraft engine component maintenance, repair & overhaul services	945 25 90
Aircraft engine components	321 90
Aircraft engine control systems & equipment	221 10 20
Aircraft engine health monitoring equipment	617 50 10
Aircraft engine maintenance, repair, overhaul services	945 25
Aircraft engines, piston	321 30
Aircraft engines, rocket / ramjet	321 40
Aircraft engines, turbojet / turbofan	321 10
Aircraft engines, turboshaft / turboprop	321 20
Aircraft fire warning & protection systems	221 40 40
Aircraft flight control equipment	221 10 70
Aircraft flight control systems	221 10
Aircraft flight management systems	221 10 30
Aircraft flight recorders	221 10 70 35
Aircraft flight training services	980 22
Aircraft floats	221 40 70 40
Aircraft fly-by-wire systems	221 10 12
Aircraft fuel system equipment	221 40 45 70
Aircraft fuel systems	221 40 45
Aircraft fuel tanks	221 40 45 74
Aircraft ground handling & loading equipment, forward airfield, mobile	625 20 75
Aircraft ground handling & support equipment	612 20 75
Aircraft gun ammunition	521 20 30
Aircraft guns, 20mm & over	512 50 20
Aircraft/helicopter rocket pods	512 82 10
Aircraft hydraulic equipment	221 40 50 70
Aircraft hydraulic systems	221 40 50
Aircraft instrument repair, overhaul & maintenance services	945 20 15
Aircraft instruments	221 10 70 30
Aircraft landing gear	221 40 60
Aircraft landing lights	221 40 65 70 20
Aircraft lifting bags, inflatable	612 20 80 10
Aircraft lighting equipment	221 40 65 70
Aircraft lighting systems	221 40 65
Aircraft loading vehicles, airfield	141 70 40 05 12
Aircraft machine guns	512 35
Aircraft maintenance, repair, overhaul services	945 20
Aircraft maintenance facilities & equipment, field mobile / transportable	625 20 70
Aircraft maintenance systems & equipment	612 20 70
Aircraft maintenance trainers	616 32
Aircraft mission-related equipment (airframe integrated)	221 20 70
Aircraft mission-related systems, (airframe integrated)	221 20
Aircraft navigation lights	221 40 65 70 30
Aircraft pneumatic systems	221 40 68
Aircraft propellers	321 50
Aircraft propulsion	321
Aircraft radomes	221 40 90 60
Aircraft recovery lifting bags, inflatable	612 20 80 10
Aircraft refuelling equipment	221 20 40 70
Aircraft refuelling systems	221 20 40
Aircraft refuelling systems, boom type	221 20 40 10
Aircraft refuelling systems, probe & drogue type	221 20 40 50
Aircraft rockets	521 20 60
Aircraft searchlights	221 20 70 50
Aircraft seats	221 30 70 60
Aircraft systems & equipment	221
Aircraft towing / positioning tractors	141 70 30 05
Aircraft transparencies	221 40 90 70
Aircraft tyres	221 40 60 90 70
Aircraft wheels	221 40 60 90 90
Aircraft windows	221 40 90 90
Aircrew helmets	712 10 10
Aircrew seats	221 30 70 61
Aircrew seats, helicopter	222 30 70 61
Airfield approach & landing aids	612 20 45
Airfield approach & landing aids, forward airfield, mobile / transportable	625 20 45
Airfield approach lights	612 20 50 91
Airfield attack missiles, air to surface	511 22 05
Airfield attack submunitions	524 20 10
Airfield construction services	941 10
Airfield control towers, mobile / transportable	625 20 20 75
Airfield emergency systems & equipment	612 20 80
Airfield firefighting equipment	612 20 80 30
Airfield firefighting equipment, stationary	612 20 80 30 40
Airfield firefighting vehicles	612 20 80 39
Airfield floodlights	612 20 50 93
Airfield lighting equipment	612 20 50 90
Airfield lighting equipment, forward airfield, mobile / transportable	625 20 50
Airfield lighting systems & equipment	612 20 50
Airfield movement control equipment	612 20 25 90
Airfield movement control systems & equipment	612 20 25
Airfield precision approach path indicators (PAPI)	612 20 45 10
Airfield radars, ground-based, fixed	411 14 10 05
Airfield rapid runway repair equipment	625 20 10 60
Airfield runway / taxiway lights	612 20 50 92
Airfield runway de-icing equipment	612 20 10 20
Airfield runway fog clearance systems & equipment	612 20 10 10
Airfield runway grooving machines	612 20 10 35
Airfield runway lighting equipment, forward airfield, mobile / transportable	625 20 50 10
Airfield runway lighting systems	612 20 50 10
Airfield runway-related systems & equipment	612 20 10
Airfield runway-related systems & equipment, mobile / transportable	625 20 10
Airfield runway rubber deposit removal machines	612 20 10 32
Airfield runway snow clearance equipment	612 20 10 15
Airfield runway snow ploughs	612 20 10 15 50
Airfield runway sweepers	612 20 10 30
Airfield surveillance radars, ground-based, fixed	411 14 10 05 30
Airfield vehicles	141 70 40 05
Airfield vehicles, air conditioning	141 70 40 05 10
Airfield vehicles, aircraft loading	141 70 40 05 12
Airfield vehicles, crash / rescue	141 70 40 05 20
Airfield vehicles, firefighting	141 70 40 05 40
Airfield vehicles, food service	141 70 40 05 45
Airfield vehicles, refuelling	141 70 40 05 50
Airframe accessories, aircraft	221 40 80
Airframe components, aircraft	221 40 90
Airframe maintenance, repair, overhaul services, aircraft	945 20 10
Airscrews, surface effect craft	333 50
Airship / Dirigible, Balloon equipment	224 70
Airship / Dirigible engines, piston	324 10
Airship / Dirigible propulsion	324
Airship / Dirigible systems	224
Airships / Dirigibles	124 10
Airships / Dirigibles, Balloons	124
Airstairs vehicles, airfield	141 70 40 05 15
Alignment equipment, optical, airborne radar	617 70 60 20
Alignment equipment, optical, shipborne radar	617 70 60 30
All-weather landing systems, aircraft	221 10 15
Altimeters	221 10 70 30 10
Altimeters, barometric	221 10 70 30 10 20
Altimeters, radar	221 10 70 30 10 50
Altimeters, radio	221 10 70 30 10 60
Altitude chambers, hypobaric	615 20 14
Aluminium alloy plates	822 22
Aluminium alloy sections	822 12
Aluminium armour plates	822 22 10
Aluminium bronze wire, bar, rods	822 12 10
Ambulance kits, aircraft	221 30 70 70
Ambulance kits, helicopter	222 30 70 70
Ambulances, armoured half-tracked	141 20 35
Ambulances, armoured tracked	141 10 35
Ambulances, armoured wheeled	141 30 35
Ambulances, non-armoured wheeled	141 70 21 35
Ammunition	521
Ammunition, AFV	521 40 30
Ammunition, aircraft	521 20
Ammunition, aircraft cannon	521 20 30
Ammunition, aircraft gun	521 20 30
Ammunition, aircraft gun, 20mm	521 20 30 20
Ammunition, aircraft gun, 30mm	521 20 30 30
Ammunition, artillery, 105mm	521 40 10 10
Ammunition, artillery, 130mm	521 40 10 12
Ammunition, artillery, 155mm	521 40 10 15
Ammunition, artillery, 203mm	521 40 10 20
Ammunition, artillery, 90mm	521 40 10 09
Ammunition, artillery, ground-based	521 40 10
Ammunition, automatic gun, 20mm, ground-based	521 40 15 20
Ammunition, automatic gun, 20mm, naval	521 30 15 20
Ammunition, automatic gun, 25mm, ground-based	521 40 15 25
Ammunition, automatic gun, 25mm, naval	521 30 15 25
Ammunition, automatic gun, 30mm, ground-based	521 40 15 30
Ammunition, automatic gun, 30mm, naval	521 30 15 30
Ammunition, automatic gun, 35mm, ground-based	521 40 15 35
Ammunition, automatic gun, 35mm, naval	521 30 15 35
Ammunition, automatic gun, ground-based	521 40 15
Ammunition, automatic gun, naval, 20-35mm	521 30 15
Ammunition, Explosives, Propellants, Pyrotechnics	520
Ammunition, fighting vehicle	521 40 30
Ammunition, fighting vehicle, 105-130mm	521 40 38

Product-Service Index

Product / Service Index

Product / Service Index

B

Product / Service Index

D

Product-Service Index

E

Product-Service Index

Product / Service Index

F

G

Product / Service Index

H

I

Product / Service Index

Product-Service Index

M

N

Product-Service Index

P

Product-Service Index

Product / Service Index

Product-Service Index

T

Product-Service Index

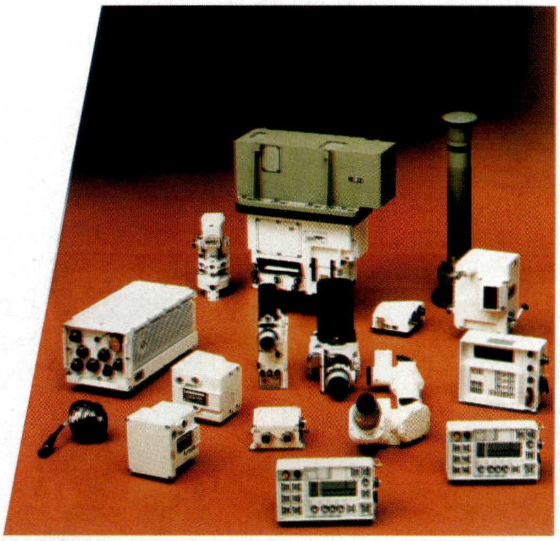

Product-Service Index

Product / Service Index

W

X

Z

Notes

Notes

Platforms / Primary systems

Spaceborne Platforms

111 Spacecraft

Suppliers	Country	Page	Key No.
Aerospatiale	France	131	75-00600
Fuji Heavy Industries Ltd	Japan	251	75-01100
Khrunichev State Research &	Russia	334	55-10400
Lockheed Martin Corp.	USA	591	75-55635
Lockheed Martin Missiles & Space	USA	592	75-55640
Matra Marconi Space UK, Ltd	UK	425	74-06900
Mitsubishi Heavy Industries Ltd	Japan	253	75-05800
Northrop Grumman Corporation	USA	599	75-65450
Programmed Composites Inc.	USA	606	75-71500
Rockwell	USA	610	75-75930
Sener Ingenieria y Sistemas SA	Spain	367	75-05200
SONACA SA	Belgium	47	75-06100
The Boeing Company	USA	560	75-14250

111 12 Spacecraft, manned, earth-orbital

Suppliers	Country	Page	Key No.
Aerospatiale	France	131	75-00600
Khrunichev State Research &	Russia	334	55-10400
Lockheed Martin Corp.	USA	591	75-55635
Matra Marconi Space UK, Ltd	UK	425	74-06900
Mitsubishi Heavy Industries Ltd	Japan	253	75-05800
Rockwell	USA	610	75-75930
Sener Ingenieria y Sistemas SA	Spain	367	75-05200
The Boeing Company	USA	560	75-14250

112 Satellites

Suppliers	Country	Page	Key No.
Aerospatiale	France	131	75-00600
Computer Resources International A/S.	Denmark	104	80-00100
Cubic Defense Systems Inc.	USA	567	75-22900
Dornier GmbH	Germany	167	75-06700
Israel Aircraft Industries Ltd	Israel	219	75-10200
Korolev Rocket Space Corp	Russia	334	55-10450
Labtech Ltd	UK	464	75-34280
Lockheed Martin Missiles & Space	USA	592	75-55640
Matra Défense-Espace	France	139	75-26000
Matra Marconi Space UK, Ltd	UK	425	74-06900
Motorola Ltd	UK	469	75-41050
Northrop Grumman Corporation	USA	599	75-65450
NPO Prikladnoi Mechaniki	Russia	338	75-14600
Rockwell	USA	610	75-75930
Rutherford Appleton Laboratory	UK	420	60-01850
Space and Defence Division	France	131	75-00600
Space Systems Finland Ltd	Finland	118	75-02100
Tadiran Spectralink	Israel	222	75-21550
The Boeing Company	USA	560	75-14250

112 21 Satellites, communication

Suppliers	Country	Page	Key No.
Aerospatiale	France	131	75-00600
China National Precision Machinery	China PR	88	55-00265
Computer Resources International A/S.	Denmark	104	80-00100
Cranfield University	UK	419	60-00350
Cubic Defense Systems Inc.	USA	567	75-22900
Harris Corporation	USA	581	75-42050
Kaman Sciences Corp.	USA	637	80-17600
Korolev Rocket Space Corp	Russia	334	55-10450
Labtech Ltd	UK	464	75-34280
Matra Défense-Espace	France	139	75-26000
Matra Marconi Space UK, Ltd	UK	425	74-06900
Motorola Ltd	UK	469	75-41050
NPO Prikladnoi Mechaniki	Russia	338	75-14600
Rutherford Appleton Laboratory	UK	420	60-01850
SED Systems Inc	Canada	79	80-01937
Siemens Atea NV	Belgium	47	75-05950
Space and Defence Division	France	131	75-00600
Spar Aerospace Ltd	Canada	76	75-13000

112 22 Satellites, meteorological

Suppliers	Country	Page	Key No.
Aerospatiale	France	131	75-00600
Cranfield University	UK	419	60-00350
Explorocean Technology Ltd	UK	502	80-09870
ITT Defense Ltd	UK	505	80-13570
Matra Défense-Espace	France	139	75-26000
Matra Marconi Space UK, Ltd	UK	425	74-06900
Rutherford Appleton Laboratory	UK	420	60-01850

112 23 Satellites, navigation

Suppliers	Country	Page	Key No.
Cranfield University	UK	419	60-00350
GEC-Marconi Avionics Ltd	UK	454	75-25320
Kearfott Guidance & Navigation Corp.	USA	588	75-51760
Matra Défense-Espace	France	139	75-26000
NPO Prikladnoi Mechaniki	Russia	338	75-14600
Rockwell	USA	610	75-75930
SED Systems Inc	Canada	79	80-01937
Sener Ingenieria y Sistemas SA	Spain	367	75-05200
Sperry Marine Inc.	USA	617	75-83250

112 24 Satellites, reconnaissance

Suppliers	Country	Page	Key No.
Cranfield University	UK	419	60-00350
GEC-Marconi Research Centre	UK	419	60-00700
Litton Industries Inc.	USA	590	75-55400
Matra Défense-Espace	France	139	75-26000
SED Systems Inc	Canada	79	80-01937

113 Space launchers & boosters

Suppliers	Country	Page	Key No.
Aerospatiale	France	131	75-00600
Cranfield University	UK	419	60-00350
FiatAvio SpA	Italy	234	75-09112
Khrunichev State Research &	Russia	334	55-10400
Lockheed Martin Corp.	USA	591	75-55635
Mitsubishi Heavy Industries Ltd	Japan	253	75-05800
Nissan Motor Co. Ltd	Japan	254	75-06790
NPO Molniya plc	Russia	337	75-13380
Rockwell	USA	610	75-75930
Royal Ordnance	UK	481	75-50620
Sener Ingenieria y Sistemas SA	Spain	367	75-05200
Space and Defence Division	France	131	75-00600
The Boeing Company	USA	560	75-14250

113 10 Rockets, space launcher

Suppliers	Country	Page	Key No.
Aerospatiale	France	131	75-00600
FiatAvio SpA	Italy	234	75-09112
Khrunichev State Research &	Russia	334	55-10400
Korolev Rocket Space Corp	Russia	334	55-10450
Lockheed Martin Corp.	USA	591	75-55635
McDonnell Douglas Aerospace	USA	594	75-59480
Nissan Motor Co. Ltd	Japan	254	75-06790
Royal Ordnance	UK	481	75-50620
Sener Ingenieria y Sistemas SA	Spain	367	75-05200
Space and Defence Division	France	131	75-00600

113 12 Rockets, space launcher, large payload

Suppliers	Country	Page	Key No.
Aerospatiale	France	131	75-00600
FiatAvio SpA	Italy	234	75-09112
Khrunichev State Research &	Russia	334	55-10400
Lockheed Martin Corp.	USA	591	75-55635
Nissan Motor Co. Ltd	Japan	254	75-06790

Sener Ingenieria y Sistemas SA	Spain	*367*	75-05200
The Boeing Company	USA	*560*	75-14250

113 14 Rockets, space launcher, medium payload

Aerospatiale	France	*131*	75-00600
FiatAvio SpA	Italy	*234*	75-09112
Khrunichev State Research &	Russia	*334*	55-10400
Lockheed Martin Corp.	USA	*591*	75-55635
Mitsubishi Heavy Industries Ltd	Japan	*253*	75-05800
Nissan Motor Co. Ltd	Japan	*254*	75-06790
Royal Ordnance	UK	*481*	75-50620
Sener Ingenieria y Sistemas SA	Spain	*367*	75-05200

113 18 20 Boosters, space launcher, strap-on

Aerospatiale	France	*131*	75-00600
FiatAvio SpA	Italy	*234*	75-09112
Lockheed Martin Corp.	USA	*591*	75-55635
Nissan Motor Co. Ltd	Japan	*254*	75-06790
NPO Energomash	Russia	*338*	75-14300
NPO Molniya plc	Russia	*337*	75-13380
Royal Ordnance	UK	*481*	75-50620

114 Rockets, research / sounding

Aerojet	USA	*547*	75-01500
Arab Organization for Ind.	Egypt	*109*	55-01700
Avibras Aeroespacial SA	Brazil	*55*	75-00968
Bristol Aerospace Ltd	Canada	*69*	75-01500
Cranfield University	UK	*419*	60-00350
Dornier GmbH	Germany	*167*	75-06700
FiatAvio SpA	Italy	*234*	75-09112
Nissan Motor Co. Ltd	Japan	*254*	75-06790
Royal Ordnance	UK	*481*	75-50620
Universal Propulsion Co. Inc.	USA	*625*	75-92700

114 05 Rockets, research

Arab Organization for Ind.	Egypt	*109*	55-01700
Bristol Aerospace Ltd	Canada	*69*	75-01500
Cranfield University	UK	*419*	60-00350
FiatAvio SpA	Italy	*234*	75-09112
Matra Marconi Space UK, Ltd	UK	*425*	74-06900
Nissan Motor Co. Ltd	Japan	*254*	75-06790
Royal Ordnance	UK	*481*	75-50620
Rutherford Appleton Laboratory	UK	*420*	60-01850
SED Systems Inc	Canada	*79*	80-01937
Universal Propulsion Co. Inc.	USA	*625*	75-92700

114 10 Rockets, meteorological sounding

Avibras Aeroespacial SA	Brazil	*55*	75-00968
Nissan Motor Co. Ltd	Japan	*254*	75-06790
Royal Ordnance	UK	*481*	75-50620
SED Systems Inc	Canada	*79*	80-01937
Universal Propulsion Co. Inc.	USA	*625*	75-92700

Airborne Platforms

121 Aircraft

A S Yakovlev Design Bureau	Russia	*339*	75-78000
Advanced Aerodynamics	USA	*546*	75-00790
Aereon Corp.	USA	*630*	80-00550
Aermacchi SpA	Italy	*230*	75-00500
Aero Boero SA	Argentina	*4*	75-00200
Aero Vodochody Ltd	Czech Republic	*98*	55-00200
Aeromot Ind.Mecanico-Metalurgica SA	Brazil	*55*	75-00635
Airbus Industrie	France	*129*	74-00100
Aircraft Division	France	*131*	75-00600
Alenia Aerospazio	**Italy**	***231***	**75-01160**
Alenia SpA	Italy	*232*	75-01170
Antonov	Ukraine	*405*	55-00100
Arab Organization for Ind.	Egypt	*109*	55-01700
Avia Ltd Scientific Industrial Union	Russia	*336*	75-01150
Bell-Boeing	USA	*544*	74-00300
British Aerospace Military Aircraft	UK	*436*	75-08886
Burkhart Grob Luft und Raumfahrt	Germany	*163*	74-04300
Cessna Aircraft Company	USA	*563*	75-17750
China National Aero-Tech I/E Corp.	China PR	*88*	55-00200
Construcciones Aeronáuticas SA	Spain	*364*	55-00120
Dassault Aviation Group	France	*135*	75-12975
Dornier GmbH	Germany	*167*	75-06700
Dornier Luftfahrt GmbH	Germany	*168*	75-08860
Ekran Belarusskoye PO	Belarus	*38*	75-00400
Embraer	Brazil	*54*	55-05000
Empresa Nacional de Aeronáutica	Chile	*84*	55-00550
Eurofighter Jagdflugzeug GmbH	Germany	*163*	74-05850
FFA Flugzeugwerke Altenrhein AG	Switzerland	*386*	75-04350
Finavitec Oy	Finland	*119*	80-00580
Fuji Heavy Industries Ltd	Japan	*251*	75-01100
Guizhou Aviation Industry Corp.	China PR	*88*	55-00800
Gulfstream Aerospace Corporation	USA	*580*	75-40950
Harbin Dongan Engine Company	China PR	*88*	55-00850
HB Flugtechnik GmbH	Austria	*33*	75-01225
Ilyushin Design Bureau	Russia	*333*	55-09000
Ingenieurbüro Flugwesen &	Switzerland	*388*	80-00800
INIMEX	**Belgium**	***46***	**75-04650**
Innotech Aviation Ltd.	Canada	*78*	80-00960
Kawasaki Heavy Industries Ltd	Japan	*252*	75-04350
Lambert Aircraft Engineering	Belgium	*46*	75-04750
Learjet Inc.	USA	*590*	75-54430
Lockheed Martin Aeronautical Systems	USA	*592*	75-55637
Lockheed Martin Aircraft Argentina	Argentina	*4*	80-08000
Mitsubishi Heavy Industries Ltd	Japan	*253*	75-05800
Moscow Aircraft Production	Russia	*334*	55-13050
Nanchang Aircraft Company	China PR	*88*	55-00900
Northrop Grumman Corporation	USA	*599*	75-65450
Pacific Aerospace Corp. Ltd	New Zealand	*302*	75-01600
Pakistan Aeronautical Complex	Pakistan	*315*	55-01600
Panavia Aircraft GmbH	Germany	*164*	74-10000
Partenavia	Italy	*238*	75-18700
Pilatus Aircraft Ltd	Switzerland	*387*	75-10100
Pilatus Britten-Norman Ltd	UK	*474*	75-45400
PT Industri Pesawat Terbang	Indonesia	*209*	55-00300
Raytheon Aircraft Company	USA	*607*	75-73200
Reims-Aviation SA	France	*141*	75-32100
Rinaldo Piaggio SpA	Italy	*238*	75-18790
SABCA	Belgium	*46*	75-05590
Sadler Aircraft Corp.	USA	*612*	75-77345
Scaled Composites Inc.	USA	*541*	60-19160
Shaanxi Aircraft Manufacturing Co.	China PR	*88*	55-01900
Shenyang Aircraft Corp.	China PR	*88*	55-01950
ShinMaywa Industries Ltd	Japan	*254*	75-07800
Skyfox Aviation Ltd.	Australia	*17*	75-08300
Slingsby Aviation Ltd.	UK	*484*	75-54410
SOCATA	France	*145*	75-36350
SONACA SA	Belgium	*47*	75-06100
Sukhoi Design Bureau	Russia	*334*	55-18500
Swiss Aircraft & Systems Co	Switzerland	*384*	55-00100
Taganrog JSC	Russia	*336*	60-04500
The Boeing Company	USA	*559*	75-14250
The New Piper Aircraft Inc.	USA	*604*	75-70250
Tupolev JSC	Russia	*334*	55-19500
TUSAS Aerospace Industries Inc.	Turkey	*403*	75-02120
Wolfsberg Aircraft Corporation NV	Belgium	*48*	75-40000
Xian Aircraft Company	China PR	*88*	55-02400

121 03 Aircraft, anti-tank

Dornier Luftfahrt GmbH	Germany	*168*	75-08860

121 06 Aircraft, ASW / maritime patrol

Aero International (Regional)	France	*149*	80-00180
Aircraft Division	France	*131*	75-00600
Alenia Aerospazio	**Italy**	***231***	**75-01160**
Alenia SpA	Italy	*232*	75-01170
Bombardier Inc	Canada	*69*	75-01900
Dassault Aviation Group	France	*135*	75-12975
Dornier Luftfahrt GmbH	Germany	*168*	75-08860
Ilyushin Design Bureau	Russia	*333*	55-09000
Kawasaki Heavy Industries Ltd	Japan	*252*	75-04350
Learjet Inc.	USA	*590*	75-54430
Lockheed Martin Aeronautical Systems	USA	*592*	75-55637
Northrop Grumman Corporation	USA	*599*	75-65450
Pilatus Britten-Norman Ltd	UK	*474*	75-45400
Shaanxi Aircraft Manufacturing Co.	China PR	*88*	55-01900
Taganrog JSC	Russia	*336*	60-04500
The Boeing Company	USA	*559*	75-14250

121 06 10 Aircraft, ASW / maritime patrol, carrier-based

Aereon Corp.	USA	630	80-00550
Lockheed Martin Aeronautical Systems	USA	592	75-55637
Northrop Grumman Corporation.	USA	599	75-65450

121 06 20 Aircraft, ASW / maritime patrol, land-based

Aereon Corp.	USA	630	80-00550
Bombardier Inc	Canada	69	75-01900
Construcciones Aeronáuticas SA	Spain	364	55-00120
Dassault Aviation Group	France	135	75-12975
Dornier Luftfahrt GmbH	Germany	168	75-08860
Embraer	Brazil	54	55-05000
Gulfstream Aerospace Corporation	USA	580	75-40950
Israel Aircraft Industries Ltd	Israel	219	75-10200
Kawasaki Heavy Industries Ltd	Japan	252	75-04350
Learjet Inc.	USA	590	75-54430
Lockheed Martin Aeronautical Systems	USA	592	75-55637
Pilatus Britten-Norman Ltd	UK	474	75-45400
Taganrog JSC	Russia	336	60-04500
The Boeing Company	USA	559	75-14250

121 09 Aircraft, bomber

China National Aero-Tech I/E Corp.	China PR	88	55-00200
Dassault Aviation Group	France	135	75-12975
Dornier Luftfahrt GmbH	Germany	168	75-08860
Northrop Grumman Corporation.	USA	599	75-65450
The Boeing Company	USA	559	75-14250
Tupolev JSC.	Russia	334	55-19500
Xian Aircraft Company	China PR	88	55-02400

121 09 10 Aircraft, bomber, strategic

Dassault Aviation Group	France	135	75-12975
The Boeing Company	USA	559	75-14250
Tupolev JSC.	Russia	334	55-19500

121 09 20 Aircraft, bomber, tactical

Alenia Aerospazio	**Italy**	**231**	**75-01160**
Alenia SpA	Italy	232	75-01170
Dassault Aviation Group	France	135	75-12975
Dornier Luftfahrt GmbH	Germany	168	75-08860
Israel Aircraft Industries Ltd	Israel	219	75-10200
Northrop Grumman Corporation.	USA	599	75-65450
Pakistan Aeronautical Complex	Pakistan	315	55-01600
Sukhoi Design Bureau	Russia	334	55-18500

121 12 Aircraft, close air support (see code 121.28)

Alenia Aerospazio	**Italy**	**231**	**75-01160**
Alenia SpA	Italy	232	75-01170
British Aerospace Military Aircraft	UK	436	75-08886
Dassault Aviation Group	France	135	75-12975
Dornier Luftfahrt GmbH	Germany	168	75-08860
Embraer	Brazil	54	55-05000
INIMEX	**Belgium**	**46**	**75-04650**
Israel Aircraft Industries Ltd	Israel	219	75-10200
Mitsubishi Heavy Industries Ltd	Japan	253	75-05800
Northrop Grumman Corporation.	USA	599	75-65450
Pakistan Aeronautical Complex	Pakistan	315	55-01600
Scaled Composites Inc.	USA	541	60-19160
Sukhoi Design Bureau	Russia	334	55-18500
Swiss Aircraft & Systems Co	Switzerland	384	55-00100
The New Piper Aircraft Inc.	USA	604	75-70250

121 15 Aircraft, command & control

Aereon Corp.	USA	630	80-00550
British Aerospace Military Aircraft	UK	436	75-08886
Cranfield Aerospace Ltd	UK	500	80-07200
Cranfield University	UK	419	60-00350
Dassault Aviation Group	France	135	75-12975
EH Industries Ltd	UK	425	74-05350
Farsound Engineering Ltd	UK	450	75-22420
Gulfstream Aerospace Corporation	USA	580	75-40950
Israel Aircraft Industries Ltd	Israel	219	75-10200
Labtech Ltd	UK	464	75-34280
Northrop Grumman Corporation.	USA	599	75-65450
Slingsby Aviation Ltd.	UK	484	75-54410
The Boeing Company	USA	559	75-14250

121 16 Aircraft, communications intelligence (COMINT)

Dassault Aviation Group	France	135	75-12975
Dornier Luftfahrt GmbH	Germany	168	75-08860
Israel Aircraft Industries Ltd	Israel	219	75-10200
Northrop Grumman Corporation.	USA	599	75-65450
Pilatus Britten-Norman Ltd	UK	474	75-45400

121 18 Aircraft, counter-insurgency

Construcciones Aeronáuticas SA	Spain	364	55-00120
Embraer	Brazil	54	55-05000
Israel Aircraft Industries Ltd	Israel	219	75-10200
Learjet Inc.	USA	590	75-54430
Lockheed Martin Aircraft Argentina.	Argentina	4	80-08000
Northrop Grumman Corporation.	USA	599	75-65450

121 21 Aircraft, early warning & control

Aereon Corp.	USA	630	80-00550
Alenia SpA	Italy	232	75-01170
British Aerospace Military Aircraft	UK	436	75-08886
Cranfield Aerospace Ltd	UK	500	80-07200
Dornier Luftfahrt GmbH	Germany	168	75-08860
EH Industries Ltd	UK	425	74-05350
Ilyushin Design Bureau	Russia	333	55-09000
Northrop Grumman Corporation.	USA	599	75-65450
Pilatus Britten-Norman Ltd	UK	474	75-45400
The Boeing Company	USA	559	75-14250

121 24 Aircraft, electronic warfare

Aircraft Division	France	131	75-00600
Bombardier Inc	Canada	69	75-01900
British Aerospace Military Aircraft	UK	436	75-08886
Construcciones Aeronáuticas SA	Spain	364	55-00120
Cranfield Aerospace Ltd	UK	500	80-07200
Cranfield University	UK	419	60-00350
Dassault Aviation Group	France	135	75-12975
Dornier Luftfahrt GmbH	Germany	168	75-08860
EH Industries Ltd	UK	425	74-05350
FR Aviation Limited	UK	503	80-10850
Gulfstream Aerospace Corporation	USA	580	75-40950
Innotech Aviation Ltd.	Canada	78	80-00960
Israel Aircraft Industries Ltd	Israel	219	75-10200
Kaman Sciences Corp.	USA	637	80-17600
Labtech Ltd	UK	464	75-34280
Learjet Inc.	USA	590	75-54430
Northrop Grumman Corporation.	USA	599	75-65450
Panavia Aircraft GmbH	Germany	164	74-10000
Pilatus Britten-Norman Ltd	UK	474	75-45400

121 27 Aircraft, fighter, air defense

Alenia Aerospazio	**Italy**	**231**	**75-01160**
Alenia SpA	Italy	232	75-01170
British Aerospace Military Aircraft	UK	436	75-08886
Dassault Aviation Group	France	135	75-12975
Dornier Luftfahrt GmbH	Germany	168	75-08860
Eurofighter Jagdflugzeug GmbH.	Germany	163	74-05850
HAL Nasik Div.	India	204	55-01000
Hindustan Aeronautics Ltd.	India	204	55-01000
INIMEX	**Belgium**	**46**	**75-04650**
Israel Aircraft Industries Ltd	Israel	219	75-10200
Lockheed Martin Aeronautical Systems	USA	592	75-55637
Lockheed Martin Tactical Aircraft	USA	592	75-55650
McDonnell Douglas Aerospace	USA	594	75-59480
Mitsubishi Heavy Industries Ltd	Japan	253	75-05800
Moscow Aircraft Production	Russia	334	55-13050
Northrop Grumman Corporation.	USA	599	75-65450
Panavia Aircraft GmbH	Germany	164	74-10000
Saab AB Gripen	Sweden	378	75-09200
SABCA	Belgium	46	75-05590
Shenyang Aircraft Corp.	China PR	88	55-01950

Sukhoi Design Bureau	Russia	*334*	55-18500
Swiss Aircraft & Systems Co	Switzerland	*384*	55-00100

121 28 Aircraft, fighter / ground attack (see code 121.12)

Aermacchi SpA	Italy	*230*	75-00500
Arab Organization for Ind.	Egypt	*109*	55-01700
British Aerospace Military Aircraft	UK	*436*	75-08886
China National Aero-Tech I/E Corp.	China PR	*88*	55-00200
Dassault Aviation Group	France	*135*	75-12975
Dornier Luftfahrt GmbH	Germany	*168*	75-08860
Embraer	Brazil	*54*	55-05000
Eurofighter Jagdflugzeug GmbH.	Germany	*163*	74-05850
INIMEX	**Belgium**	***46***	**75-04650**
Israel Aircraft Industries Ltd	Israel	*219*	75-10200
Lockheed Martin Aircraft Argentina.	Argentina	*4*	80-08000
McDonnell Douglas Aerospace	USA	*594*	75-59480
Mitsubishi Heavy Industries Ltd	Japan	*253*	75-05800
Moscow Aircraft Production.	Russia	*334*	55-13050
Nanchang Aircraft Company	China PR	*88*	55-00900
Northrop Grumman Corporation.	USA	*599*	75-65450
Pakistan Aeronautical Complex	Pakistan	*315*	55-01600
Panavia Aircraft GmbH	Germany	*164*	74-10000
Samsung Aerospace Industries Ltd.	Korea (ROK)	*266*	75-04630
Scaled Composites Inc.	USA	*541*	60-19160
Sukhoi Design Bureau	Russia	*334*	55-18500
Swiss Aircraft & Systems Co	Switzerland	*384*	55-00100

121 30 Aircraft, forward air control / observation

Dornier Luftfahrt GmbH	Germany	*168*	75-08860
Northrop Grumman Corporation.	USA	*599*	75-65450
Partenavia	Italy	*238*	75-18700
Pilatus Aircraft Ltd	Switzerland	*387*	75-10100
Scaled Composites Inc.	USA	*541*	60-19160

121 32 Aircraft, multi-role

Alenia Aerospazio	**Italy**	***231***	**75-01160**
Alenia SpA	Italy	*232*	75-01170
British Aerospace Military Aircraft	UK	*436*	75-08886
Dassault Aviation Group	France	*135*	75-12975
Eurofighter Jagdflugzeug GmbH.	Germany	*163*	74-05850
Learjet Inc.	USA	*590*	75-54430
McDonnell Douglas Aerospace	USA	*594*	75-59480
Moscow Aircraft Production.	Russia	*334*	55-13050
Pakistan Aeronautical Complex	Pakistan	*315*	55-01600
Panavia Aircraft GmbH	Germany	*164*	74-10000
Saab AB Gripen	Sweden	*378*	75-09200
Shenyang Aircraft Corp.	China PR	*88*	55-01950
Xian Aircraft Company	China PR	*88*	55-02400

121 33 Aircraft, naval strike / fighter

Dassault Aviation Group	France	*135*	75-12975
McDonnell Douglas Aerospace	USA	*594*	75-59480
Moscow Aircraft Production.	Russia	*334*	55-13050
Northrop Grumman Corporation.	USA	*599*	75-65450
Panavia Aircraft GmbH	Germany	*164*	74-10000

121 33 70 Aircraft, naval strike / fighter, V/STOL

A S Yakovlev Design Bureau	Russia	*339*	75-78000
British Aerospace Military Aircraft	UK	*436*	75-08886
McDonnell Douglas Aerospace	USA	*594*	75-59480

121 36 Aircraft, offshore patrol / coastal surveillance

Aereon Corp.	USA	*630*	80-00550
Bombardier Inc	Canada	*69*	75-01900
Burkhart Grob Luft und Raumfahrt.	Germany	*163*	74-04300
Construcciones Aeronáuticas SA	Spain	*364*	55-00120
Dassault Aviation Group	France	*135*	75-12975
Dornier Luftfahrt GmbH	Germany	*168*	75-08860
Gulfstream Aerospace Corporation	USA	*580*	75-40950
Israel Aircraft Industries Ltd	Israel	*219*	75-10200
Learjet Inc.	USA	*590*	75-54430
Lockheed Martin Aeronautical Systems	USA	*592*	75-55637
Partenavia	Italy	*238*	75-18700
Pilatus Britten-Norman Ltd	UK	*474*	75-45400
Taganrog JSC	Russia	*336*	60-04500

121 39 Aircraft, reconnaissance

Burkhart Grob Luft und Raumfahrt.	Germany	*163*	74-04300
Dassault Aviation Group	France	*135*	75-12975
Dornier Luftfahrt GmbH	Germany	*168*	75-08860
Eagle Aircraft	Australia	*15*	75-05000
Gulfstream Aerospace Corporation	USA	*580*	75-40950
Learjet Inc.	USA	*590*	75-54430
LET a.s.	Czech Republic	*99*	75-00550
Lockheed Martin Aeronautical Systems	USA	*592*	75-55637
Lockheed Martin Skunk Works	USA	*591*	75-55636
Northrop Grumman Corporation.	USA	*599*	75-65450
Panavia Aircraft GmbH	Germany	*164*	74-10000
Partenavia	Italy	*238*	75-18700
Pilatus Britten-Norman Ltd	UK	*474*	75-45400
Raytheon E-Systems	USA	*608*	75-73302
Saab AB Gripen	Sweden	*378*	75-09200
Sadler Aircraft Corp.	USA	*612*	75-77345
SOCATA	France	*145*	75-36350
Swiss Aircraft & Systems Co	Switzerland	*384*	55-00100

121 39 20 Aircraft, reconnaissance, high altitude

Dornier Luftfahrt GmbH	Germany	*168*	75-08860
Lockheed Martin Aeronautical Systems	USA	*592*	75-55637

121 42 Aircraft, search & rescue

Aero International (Regional).	France	*149*	80-00180
Aircraft Division	France	*131*	75-00600
Bombardier Inc	Canada	*69*	75-01900
Burkhart Grob Luft und Raumfahrt.	Germany	*163*	74-04300
Construcciones Aeronáuticas SA	Spain	*364*	55-00120
Dassault Aviation Group	France	*135*	75-12975
Dornier Luftfahrt GmbH	Germany	*168*	75-08860
Glatzer Industries Corp.	USA	*579*	75-39760
Gulfstream Aerospace Corporation	USA	*580*	75-40950
Learjet Inc.	USA	*590*	75-54430
Northrop Grumman Corporation.	USA	*599*	75-65450
Partenavia	Italy	*238*	75-18700
Pilatus Aircraft Ltd	Switzerland	*387*	75-10100
Pilatus Britten-Norman Ltd	UK	*474*	75-45400
ShinMaywa Industries Ltd	Japan	*254*	75-07800
SOCATA	France	*145*	75-36350
Taganrog JSC	Russia	*336*	60-04500

121 43 Aircraft, special operations

Aereon Corp.	USA	*630*	80-00550
Burkhart Grob Luft und Raumfahrt.	Germany	*163*	74-04300
Dornier Luftfahrt GmbH	Germany	*168*	75-08860
Innotech Aviation Ltd.	Canada	*78*	80-00960
Israel Aircraft Industries Ltd	Israel	*219*	75-10200
Learjet Inc.	USA	*590*	75-54430
Lockheed Martin Aeronautical Systems	USA	*592*	75-55637
Lockheed Martin Skunk Works	USA	*591*	75-55636
Raytheon Aircraft Company	USA	*607*	75-73200
Scaled Composites Inc.	USA	*541*	60-19160
Swiss Aircraft & Systems Co	Switzerland	*384*	55-00100

121 45 Aircraft, tactical strike / attack

Aermacchi SpA	Italy	*230*	75-00500
AMX International Ltd	Brazil	*55*	74-00100
British Aerospace Military Aircraft	UK	*436*	75-08886
Dassault Aviation Group	France	*135*	75-12975
HAL Aircraft Div.	India	*204*	55-01000
Israel Aircraft Industries Ltd	Israel	*219*	75-10200
McDonnell Douglas Aerospace	USA	*594*	75-59480
Moscow Aircraft Production.	Russia	*334*	55-13050
Northrop Grumman Corporation.	USA	*599*	75-65450
Panavia Aircraft GmbH	Germany	*164*	74-10000
Saab AB Gripen	Sweden	*378*	75-09200

121 45 70 Aircraft, tactical strike / attack, V/STOL

British Aerospace Military Aircraft	UK	*436*	75-08886
McDonnell Douglas Aerospace	USA	*594*	75-59480

121 48 Aircraft, tanker, air refuelling, strategic

Dornier Luftfahrt GmbH	Germany	*168*	75-08860
The Boeing Company	USA	*559*	75-14250

121 51 Aircraft, target towing

Dassault Aviation Group	France	*135*	75-12975
Dornier Luftfahrt GmbH	Germany	*168*	75-08860
Hayes Targets	USA	*603*	75-69135
Learjet Inc.	USA	*590*	75-54430
Pilatus Aircraft Ltd	Switzerland	*387*	75-10100
Swiss Aircraft & Systems Co	Switzerland	*384*	55-00100

121 53 Aircraft, trainer

Aermacchi SpA	Italy	*230*	75-00500
Aero Boero SA	Argentina	*4*	75-00200
Aero Vodochody Ltd	Czech Republic	*98*	55-00200
Aerostar SA	Romania	*331*	80-00100
Aircraft Division	France	*131*	75-00600
Arab Organization for Ind.	Egypt	*109*	55-01700
British Aerospace Military Aircraft	UK	*436*	75-08886
Burkhart Grob Luft und Raumfahrt	Germany	*163*	74-04300
China National Aero-Tech I/E Corp.	China PR	*88*	55-00200
Chincul	Argentina	*4*	75-03451
Dassault Aviation Group	France	*135*	75-12975
Dornier Luftfahrt GmbH	Germany	*168*	75-08860
Eagle Aircraft	Australia	*15*	75-05000
Embraer	Brazil	*54*	55-05000
Empresa Nacional de Aeronáutica	Chile	*84*	55-00550
FFA Flugzeugwerke Altenrhein AG	Switzerland	*386*	75-04350
Fuji Heavy Industries Ltd	Japan	*251*	75-01100
Guizhou Aviation Industry Corp.	China PR	*88*	55-00800
HB Flugtechnik GmbH	Austria	*33*	75-01225
Hindustan Aeronautics Ltd	India	*204*	55-01000
Israel Aircraft Industries Ltd	Israel	*219*	75-10200
Kader Factory for Developed Ind.	Egypt	*109*	55-01700
Learjet Inc.	USA	*590*	75-54430
Moscow Aircraft Production	Russia	*334*	55-13050
Northrop Grumman Corporation	USA	*599*	75-65450
Pacific Aerospace Corp. Ltd	New Zealand	*302*	75-01600
Pakistan Aeronautical Complex	Pakistan	*315*	55-01600
Partenavia	Italy	*238*	75-18700
Pezetel	Poland	*325*	80-01000
Pilatus Aircraft Ltd	Switzerland	*387*	75-10100
SIAI Marchetti Sarl	Italy	*238*	75-21950
Skyfox Aviation Ltd.	Australia	*17*	75-08300
Sukhoi Design Bureau	Russia	*334*	55-18500
The New Piper Aircraft Inc.	USA	*604*	75-70250

121 53 10 Aircraft, trainer, advanced

Aermacchi SpA	Italy	*230*	75-00500
Aero Vodochody Ltd	Czech Republic	*98*	55-00200
Arab Organization for Ind.	Egypt	*109*	55-01700
British Aerospace Military Aircraft	UK	*436*	75-08886
Dassault Aviation Group	France	*135*	75-12975
Dornier Luftfahrt GmbH	Germany	*168*	75-08860
Embraer	Brazil	*54*	55-05000
Empresa Nacional de Aeronáutica	Chile	*84*	55-00550
Fuji Heavy Industries Ltd	Japan	*251*	75-01100
Learjet Inc.	USA	*590*	75-54430
Lockheed Martin Aircraft Argentina	Argentina	*4*	80-08000
Mitsubishi Heavy Industries Ltd	Japan	*253*	75-05800
Moscow Aircraft Production	Russia	*334*	55-13050
Pezetel	Poland	*325*	80-01000
Samsung Aerospace Industries Ltd	Korea (ROK)	*266*	75-04630
Scaled Composites Inc.	USA	*541*	60-19160
Swiss Aircraft & Systems Co	Switzerland	*384*	55-00100
The New Piper Aircraft Inc.	USA	*604*	75-70250

121 53 50 Aircraft, trainer, primary

Aero Boero SA	Argentina	*4*	75-00200
Arab Organization for Ind.	Egypt	*109*	55-01700
Eagle Aircraft	Australia	*15*	75-05000
Empresa Nacional de Aeronáutica	Chile	*84*	55-00550
FFA Flugzeugwerke Altenrhein AG	Switzerland	*386*	75-04350
Fuji Heavy Industries Ltd	Japan	*251*	75-01100
Pakistan Aeronautical Complex	Pakistan	*315*	55-01600
Pilatus Aircraft Ltd	Switzerland	*387*	75-10100
Slingsby Aviation Ltd.	UK	*484*	75-54410
The New Piper Aircraft Inc.	USA	*604*	75-70250

121 54 Aircraft, trainer / light strike

Aero Vodochody Ltd	Czech Republic	*98*	55-00200
Arab Organization for Ind.	Egypt	*109*	55-01700
British Aerospace Military Aircraft	UK	*436*	75-08886
Cessna Aircraft Company	USA	*563*	75-17750
Chincul	Argentina	*4*	75-03451
Construcciones Aeronáuticas SA	Spain	*364*	55-00120
Dassault Aviation Group	France	*135*	75-12975
Embraer	Brazil	*54*	55-05000
Empresa Nacional de Aeronáutica	Chile	*84*	55-00550
Fuji Heavy Industries Ltd	Japan	*251*	75-01100
Gulfstream Aerospace Corporation	USA	*580*	75-40950
HAL Aircraft Div.	India	*204*	55-01000
INIMEX	**Belgium**	***46***	**75-04650**
Pacific Aerospace Corp. Ltd	New Zealand	*302*	75-01600
Raytheon Aircraft Company	USA	*607*	75-73200
SOCATA	France	*145*	75-36350
The New Piper Aircraft Inc.	USA	*604*	75-70250

121 57 Aircraft, transport

Airbus Industrie	France	*129*	74-00100
Alenia Aerospazio	**Italy**	***231***	**75-01160**
Alenia SpA	Italy	*232*	75-01170
Antonov	Ukraine	*405*	55-00100
Bombardier Inc	Canada	*69*	75-01900
Construcciones Aeronáuticas SA	Spain	*364*	55-00120
Dornier Luftfahrt GmbH	Germany	*168*	75-08860
Embraer	Brazil	*54*	55-05000
Gulfstream Aerospace Corporation	USA	*580*	75-40950
Ilyushin Design Bureau	Russia	*333*	55-09000
Israel Aircraft Industries Ltd	Israel	*219*	75-10200
Learjet Inc.	USA	*590*	75-54430
Lockheed Martin Aeronautical Systems	USA	*592*	75-55637
McDonnell Douglas Aerospace	USA	*594*	75-59480
Shaanxi Aircraft Manufacturing Co.	China PR	*88*	55-01900
The Boeing Company	USA	*559*	75-14250
Xian Aircraft Company	China PR	*88*	55-02400

121 57 10 Aircraft, transport, communications / liaison

Aero Boero SA	Argentina	*4*	75-00200
Aero International (Regional)	France	*149*	80-00180
Bombardier Inc	Canada	*69*	75-01900
Cessna Aircraft Company	USA	*563*	75-17750
Construcciones Aeronáuticas SA	Spain	*364*	55-00120
Dassault Aviation Group	France	*135*	75-12975
Embraer	Brazil	*54*	55-05000
Gulfstream Aerospace Corporation	USA	*580*	75-40950
Israel Aircraft Industries Ltd	Israel	*219*	75-10200
Pilatus Aircraft Ltd	Switzerland	*387*	75-10100
Pilatus Britten-Norman Ltd	UK	*474*	75-45400
Raytheon Aircraft Company	USA	*607*	75-73200
Sino Swearingen Aircraft Company	USA	*615*	75-80880
SOCATA	France	*145*	75-36350

121 57 20 Aircraft, transport, long-range strategic

Antonov	Ukraine	*405*	55-00100
Lockheed Martin Aeronautical Systems	USA	*592*	75-55637
The Boeing Company	USA	*559*	75-14250

121 57 30 Aircraft, transport, medical evacuation

Aero Innovations Inc.	USA	*546*	75-01108
Bombardier Inc	Canada	*69*	75-01900
Construcciones Aeronáuticas SA	Spain	*364*	55-00120
Dassault Aviation Group	France	*135*	75-12975
Dornier Luftfahrt GmbH	Germany	*168*	75-08860

Gulfstream Aerospace Corporation	USA	*580*	75-40950
Innotech Aviation Ltd.	Canada	*78*	80-00960
Israel Aircraft Industries Ltd	Israel	*219*	75-10200
Learjet Inc.	USA	*590*	75-54430
LET a.s.	Czech Republic	*99*	75-00550
Pilatus Aircraft Ltd	Switzerland	*387*	75-10100
Pilatus Britten-Norman Ltd	UK	*474*	75-45400

121 57 40 Aircraft, transport, naval COD

Aereon Corp.	USA	*630*	80-00550
Alenia SpA	Italy	*232*	75-01170

121 57 50 Aircraft, transport, tactical

Aereon Corp.	USA	*630*	80-00550
Aircraft Division	France	*131*	75-00600
Alenia Aerospazio	**Italy**	***231***	**75-01160**
Alenia SpA	Italy	*232*	75-01170
Hindustan Aeronautics Ltd.	India	*204*	55-01000
Lockheed Martin Aeronautical Systems	USA	*592*	75-55637
McDonnell Douglas Aerospace	USA	*594*	75-59480
Scaled Composites Inc.	USA	*541*	60-19160

121 57 51 Aircraft, transport, tactical, light

Construcciones Aeronáuticas SA	Spain	*364*	55-00120
Israel Aircraft Industries Ltd	Israel	*219*	75-10200
Pilatus Britten-Norman Ltd	UK	*474*	75-45400
PT Industri Pesawat Terbang	Indonesia	*209*	55-00300

121 57 52 Aircraft, transport, tactical, ultra-light

Research Institute for	Ukraine	*406*	60-00990

121 57 60 Aircraft, transport, troop

Alenia Aerospazio	**Italy**	***231***	**75-01160**
Alenia SpA	Italy	*232*	75-01170
Construcciones Aeronáuticas SA	Spain	*364*	55-00120
Dornier Luftfahrt GmbH	Germany	*168*	75-08860
Embraer	Brazil	*54*	55-05000
LET a.s.	Czech Republic	*99*	75-00550
The Boeing Company	USA	*559*	75-14250

121 57 70 Aircraft, transport, utility

Aereon Corp.	USA	*630*	80-00550
Aero International (Regional)	France	*149*	80-00180
Bombardier Inc	Canada	*69*	75-01900
Construcciones Aeronáuticas SA	Spain	*364*	55-00120
Dornier Luftfahrt GmbH	Germany	*168*	75-08860
Harbin Dongan Engine Company	China PR	*88*	55-00850
Israel Aircraft Industries Ltd	Israel	*219*	75-10200
Learjet Inc.	USA	*590*	75-54430
Lockheed Martin Aeronautical Systems	USA	*592*	75-55637
Partenavia	Italy	*238*	75-18700
Pilatus Aircraft Ltd	Switzerland	*387*	75-10100
Reims-Aviation SA	France	*141*	75-32100
SIAI Marchetti Sarl	Italy	*238*	75-21950

121 60 Aircraft, V/STOL (see code 121.33.70 / 121.45.70)

Bell-Boeing	USA	*544*	74-00300
British Aerospace Military Aircraft	UK	*436*	75-08886
McDonnell Douglas Aerospace	USA	*594*	75-59480
Moller International	USA	*597*	75-62410

121 90 Aircraft, other types

Aeromot Ind.Mecanico-Metalurgica SA	Brazil	*55*	75-00635
Avia Ltd Scientific Industrial Union	Russia	*336*	75-01150
Consortium Aviaspectrans JSC	Russia	*336*	74-01200
Douglas Aircraft Company	USA	*569*	75-27200
Industria Aeronautica Neiva S.A.	Brazil	*56*	75-09460
Moscow Aircraft Production	Russia	*334*	55-13050
Scaled Composites Inc.	USA	*541*	60-19160
ShinMaywa Industries Ltd	Japan	*254*	75-07800
Sukhoi Design Bureau	Russia	*334*	55-18500

122 Helicopters

Agusta	Italy	*231*	75-01000
American Eurocopter	USA	*552*	75-05790
Bell Helicopter Textron	USA	*558*	75-13100
Boeing Defense & Space Group	USA	*559*	75-14250
Enstrom Helicopter Corp.	USA	*574*	75-31750
Eurocopter Canada Ltd	Canada	*71*	75-05050
Eurocopter Deutschland GmbH	Germany	*163*	74-05800
Eurocopter SA	France	*130*	74-00500
Fuji Heavy Industries Ltd	Japan	*251*	75-01100
GKN Westland Helicopters Ltd	UK	*455*	75-25650
HAL Helicopter Div.	India	*204*	55-01000
Harbin Dongan Engine Company	China PR	*88*	55-00850
Helibras	Brazil	*56*	75-08250
Helwan Factory for Developed	Egypt	*110*	74-06000
INIMEX	**Belgium**	***46***	**75-04650**
Israel Aircraft Industries Ltd	Israel	*219*	75-10200
Kaman Aerospace Corp.	USA	*587*	75-51500
Kamov JSC	Russia	*337*	75-10200
Kawasaki Heavy Industries Ltd	Japan	*252*	75-04350
Kazan Helicopters JSC	Russia	*337*	75-11000
MIL Moscow Helicopter Plant	Russia	*337*	75-13300
Mitsubishi Heavy Industries Ltd	Japan	*253*	75-05800
NH Industries Sarl.	France	*130*	74-01200
PT Industri Pesawat Terbang	Indonesia	*209*	55-00300
Robinson Helicopter Co.	USA	*610*	75-75650
Sikorsky Aircraft	USA	*615*	75-80900
SONACA SA	Belgium	*47*	75-06100
Swiss Aircraft & Systems Co	Switzerland	*384*	55-00100
The Boeing Company	USA	*559*	75-14250
TUSAS Aerospace Industries Inc.	Turkey	*403*	75-02120
Veetol Helicopters Pty Ltd	Australia	*23*	75-22450

122 05 Helicopters, anti-tank

Agusta	Italy	*231*	75-01000
Boeing Defense & Space Group	USA	*559*	75-14250
Eurocopter Deutschland GmbH	Germany	*163*	74-05800
Eurocopter SA	France	*130*	74-00500
Fuji Heavy Industries Ltd	Japan	*251*	75-01100
INIMEX	**Belgium**	***46***	**75-04650**
Sikorsky Aircraft	USA	*615*	75-80900
The Boeing Company	USA	*559*	75-14250

122 10 Helicopters, anti-submarine

Agusta	Italy	*231*	75-01000
Boeing Defense & Space Group	USA	*559*	75-14250
EH Industries Ltd	UK	*425*	74-05350
GKN Westland Helicopters Ltd	UK	*455*	75-25650
Kaman Aerospace Corp.	USA	*587*	75-51500
Kamov JSC	Russia	*337*	75-10200
Mitsubishi Heavy Industries Ltd	Japan	*253*	75-05800
Sikorsky Aircraft	USA	*615*	75-80900

122 15 Helicopters, armed attack / gunship

Agusta	Italy	*231*	75-01000
Bell Helicopter Textron	USA	*558*	75-13100
Eurocopter SA	France	*130*	74-00500
GKN Westland Helicopters Ltd	UK	*455*	75-25650
Helibras	Brazil	*56*	75-08250
INIMEX	**Belgium**	***46***	**75-04650**
Kazan Helicopters JSC	Russia	*337*	75-11000
MIL Moscow Helicopter Plant	Russia	*337*	75-13300
Sikorsky Aircraft	USA	*615*	75-80900
The Boeing Company	USA	*559*	75-14250
WSK-PZL Swidnik SA	Poland	*325*	75-10000

122 20 Helicopters, assault / combat, tactical

Agusta	Italy	*231*	75-01000
EH Industries Ltd	UK	*425*	74-05350
Fuji Heavy Industries Ltd	Japan	*251*	75-01100
GKN Westland Helicopters Ltd	UK	*455*	75-25650

INIMEX	**Belgium**	*46*	**75-04650**
Kamov JSC	Russia	*337*	75-10200
Kazan Helicopters JSC	Russia	*337*	75-11000
MIL Moscow Helicopter Plant	Russia	*337*	75-13300
Sikorsky Aircraft	USA	*615*	75-80900
WSK-PZL Swidnik SA	Poland	*325*	75-10000

122 25 Helicopters, command post

Agusta	Italy	*231*	75-01000
EH Industries Ltd	UK	*425*	74-05350
Kazan Helicopters JSC	Russia	*337*	75-11000
Sikorsky Aircraft	USA	*615*	75-80900

122 30 Helicopters, general purpose / multi-role

Agusta	Italy	*231*	75-01000
American Eurocopter	USA	*552*	75-05790
Arab Organization for Ind.	Egypt	*109*	55-01700
Bell Helicopter Textron	USA	*558*	75-13100
Boeing Defense & Space Group	USA	*559*	75-14250
EH Industries Ltd	UK	*425*	74-05350
Eurocopter Canada Ltd	Canada	*71*	75-05050
Eurocopter Deutschland GmbH	Germany	*163*	74-05800
Fuji Heavy Industries Ltd	Japan	*251*	75-01100
GKN Westland Helicopters Ltd	UK	*455*	75-25650
HAL Helicopter Div.	India	*204*	55-01000
Harbin Dongan Engine Company	China PR	*88*	55-00850
Helwan Factory for Developed	Egypt	*110*	74-06000
INIMEX	**Belgium**	*46*	**75-04650**
Kaman Aerospace Corp.	USA	*587*	75-51500
Kamov JSC	Russia	*337*	75-10200
Kawasaki Heavy Industries Ltd	Japan	*252*	75-04350
Kazan Helicopters JSC	Russia	*337*	75-11000
MIL Moscow Helicopter Plant	Russia	*337*	75-13300
NH Industries Sarl	France	*130*	74-01200
PT Industri Pesawat Terbang	Indonesia	*209*	55-00300
Robinson Helicopter Co.	USA	*610*	75-75650
Sikorsky Aircraft	USA	*615*	75-80900
Swiss Aircraft & Systems Co	Switzerland	*384*	55-00100
The Boeing Company	USA	*559*	75-14250
WSK-PZL Swidnik SA	Poland	*325*	75-10000

122 40 Helicopters, mine warfare

Boeing Defense & Space Group	USA	*559*	75-14250
EH Industries Ltd	UK	*425*	74-05350
Fuji Heavy Industries Ltd	Japan	*251*	75-01100
Sikorsky Aircraft	USA	*615*	75-80900
WSK-PZL Swidnik SA	Poland	*325*	75-10000

122 45 Helicopters, observation / scout

Agusta	Italy	*231*	75-01000
Arab Organization for Ind.	Egypt	*109*	55-01700
Bell Helicopter Textron	USA	*558*	75-13100
Boeing Defense & Space Group	USA	*559*	75-14250
Enstrom Helicopter Corp.	USA	*574*	75-31750
Eurocopter Canada Ltd	Canada	*71*	75-05050
Fuji Heavy Industries Ltd	Japan	*251*	75-01100
GKN Westland Helicopters Ltd	UK	*455*	75-25650
Sikorsky Aircraft	USA	*615*	75-80900
The Boeing Company	USA	*559*	75-14250

122 50 Helicopters, search & rescue

Agusta	Italy	*231*	75-01000
American Eurocopter	USA	*552*	75-05790
Bell Helicopter Textron	USA	*558*	75-13100
Boeing Defense & Space Group	USA	*559*	75-14250
EH Industries Ltd	UK	*425*	74-05350
Eurocopter Canada Ltd	Canada	*71*	75-05050
GKN Westland Helicopters Ltd	UK	*455*	75-25650
Glatzer Industries Corp.	USA	*579*	75-39760
INIMEX	**Belgium**	*46*	**75-04650**
Kaman Aerospace Corp.	USA	*587*	75-51500
Kazan Helicopters JSC	Russia	*337*	75-11000
Mitsubishi Heavy Industries Ltd	Japan	*253*	75-05800
Swiss Aircraft & Systems Co	Switzerland	*384*	55-00100
WSK-PZL Swidnik SA	Poland	*325*	75-10000

122 55 Helicopters, transport

Agusta	Italy	*231*	75-01000
Bell Helicopter Textron	USA	*558*	75-13100
Boeing Defense & Space Group	USA	*559*	75-14250
EH Industries Ltd	UK	*425*	74-05350
Eurocopter Canada Ltd	Canada	*71*	75-05050
Eurocopter Deutschland GmbH	Germany	*163*	74-05800
Eurocopter SA	France	*130*	74-00500
Fuji Heavy Industries Ltd	Japan	*251*	75-01100
HAL Helicopter Div.	India	*204*	55-01000
Kawasaki Heavy Industries Ltd	Japan	*252*	75-04350
Kazan Helicopters JSC	Russia	*337*	75-11000
MIL Moscow Helicopter Plant	Russia	*337*	75-13300
NH Industries Sarl	France	*130*	74-01200
Sikorsky Aircraft	USA	*615*	75-80900
Swiss Aircraft & Systems Co	Switzerland	*384*	55-00100
The Boeing Company	USA	*559*	75-14250
WSK-PZL Swidnik SA	Poland	*325*	75-10000

122 55 10 Helicopters, transport, heavy duty / heavy lift

Agusta	Italy	*231*	75-01000
Boeing Defense & Space Group	USA	*559*	75-14250
EH Industries Ltd	UK	*425*	74-05350
GKN Westland Helicopters Ltd	UK	*455*	75-25650
Sikorsky Aircraft	USA	*615*	75-80900
The Boeing Company	USA	*559*	75-14250

122 55 15 Helicopters, heavy lift, crane type

Boeing Defense & Space Group	USA	*559*	75-14250
MIL Moscow Helicopter Plant	Russia	*337*	75-13300

122 55 20 Helicopters, transport, medium

Agusta	Italy	*231*	75-01000
Bell Helicopter Textron	USA	*558*	75-13100
Boeing Defense & Space Group	USA	*559*	75-14250
EH Industries Ltd	UK	*425*	74-05350
Eurocopter SA	France	*130*	74-00500
Fuji Heavy Industries Ltd	Japan	*251*	75-01100
Kazan Helicopters JSC	Russia	*337*	75-11000
NH Industries Sarl	France	*130*	74-01200
Sikorsky Aircraft	USA	*615*	75-80900

122 55 30 Helicopters, transport, light / liaison

Agusta	Italy	*231*	75-01000
Bell Helicopter Textron	USA	*558*	75-13100
Enstrom Helicopter Corp.	USA	*574*	75-31750
Fuji Heavy Industries Ltd	Japan	*251*	75-01100
Sikorsky Aircraft	USA	*615*	75-80900
Veetol Helicopters Pty Ltd	Australia	*23*	75-22450

122 90 Helicopters, other types

Enstrom Helicopter Corp.	USA	*574*	75-31750
HAL Helicopter Div.	India	*204*	55-01000
Sikorsky Aircraft	USA	*615*	75-80900
The Boeing Company	USA	*559*	75-14250
Wallis Autogyros	UK	*492*	75-64150

123 Remotely Piloted Vehicles (RPVs) & Drones

Advanced Aerodynamics	USA	*546*	75-00790
ATS Aerospace Inc.	Canada	*68*	75-00752
Bombardier Inc	Canada	*69*	75-01900
Dornier GmbH	Germany	*167*	75-06700
Dornier Luftfahrt GmbH	Germany	*168*	75-08860
Enics Research Center	Russia	*335*	60-01200
Flight Refuelling Limited	UK	*451*	75-23850
Fuji Heavy Industries Ltd	Japan	*251*	75-01100
General Atomics	USA	*578*	75-37447
Israel Aircraft Industries Ltd	Israel	*219*	75-10200
Israel Military Industries Ltd	Israel	*220*	75-10300
Japan Aircraft Manufacturing Co. Ltd.	Japan	*252*	75-03200
Lockheed Martin Corp.	USA	*591*	75-55635
Oerlikon-Contraves AG.	Switzerland	*387*	75-09250

Pacer Systems Inc.	USA	*602*	75-68000
Pilkington Optronics Ltd	UK	*426*	74-10000
R&D Aeronautical Engineering Co.	USA	*607*	75-72430
Raytheon Aircraft Company	USA	*607*	75-73200
Sener Ingenieria y Sistemas SA	Spain	*367*	75-05200
Silver Arrow	Israel	*222*	75-20400
Swiss Aircraft & Systems Co	Switzerland	*384*	55-00100
Teledyne Ryan Aeronautical	USA	*621*	75-87560
The Boeing Company	USA	*559*	75-14250
Wallis Autogyros	UK	*492*	75-64150

123 05 RPV systems

Advanced Technologies &	S. Africa	*357*	75-00500
Alden Sales Co.	USA	*548*	75-04410
Bombardier Inc	Canada	*69*	75-01900
Chapy Corporation	USA	*563*	75-18065
Dornier Luftfahrt GmbH	Germany	*168*	75-08860
EL-OP Electro Optics Industries Ltd	Israel	*219*	75-06300
Fuji Heavy Industries Ltd	Japan	*251*	75-01100
GEC-Marconi Avionics Ltd	UK	*454*	75-25320
Indal Technologies Inc	Canada	*72*	75-07740
Israel Military Industries Ltd	Israel	*220*	75-10300
Kaman Sciences Corp.	USA	*637*	80-17600
Lear Astronics Corp.	USA	*589*	75-54390
Matra Défense-Espace	France	*139*	75-26000
Northrop Grumman Corporation	USA	*599*	75-65450
Oerlikon-Contraves AG.	Switzerland	*387*	75-09250
Pacer Systems Inc.	USA	*602*	75-68000
Pilkington Optronics Ltd	UK	*426*	74-10000
Prototypa-ZM, s.r.o.	Czech Republic	*98*	60-01000
R&D Aeronautical Engineering Co.	USA	*607*	75-72430
Sener Ingenieria y Sistemas SA	Spain	*367*	75-05200
Silver Arrow	Israel	*222*	75-20400
Target Technology Ltd	UK	*487*	75-58660
TUSAS Aerospace Industries Inc.	Turkey	*403*	75-02120
Vanguard Research Inc.	USA	*644*	80-34100
Xi CAD North America Inc.	USA	*628*	75-97760

123 10 RPVs

Advanced Aerodynamics	USA	*546*	75-00790
Advanced Technologies &	S. Africa	*357*	75-00500
AeroSpace Technologies of Australia	Australia	*11*	75-00340
Bombardier Inc	Canada	*69*	75-01900
Chapy Corporation	USA	*563*	75-18065
Dornier Luftfahrt GmbH	Germany	*168*	75-08860
Fuji Heavy Industries Ltd	Japan	*251*	75-01100
GEC-Marconi Avionics Ltd	UK	*454*	75-25320
Israel Military Industries Ltd	Israel	*220*	75-10300
Kaman Sciences Corp.	USA	*637*	80-17600
Kentron	S. Africa	*358*	75-09000
Lear Astronics Corp.	USA	*589*	75-54390
Moller International	USA	*597*	75-62410
Northrop Grumman Corporation	USA	*599*	75-65450
Oerlikon-Contraves AG.	Switzerland	*387*	75-09250
Prototypa-ZM, s.r.o.	Czech Republic	*98*	60-01000
R&D Aeronautical Engineering Co.	USA	*607*	75-72430
Raytheon E-Systems	USA	*608*	75-73302
Scaled Composites Inc.	USA	*541*	60-19160
Silver Arrow	Israel	*222*	75-20400
Slingsby Aviation Ltd.	UK	*484*	75-54410
St Bernard Composites Ltd	UK	*486*	75-56222
Swiss Aircraft & Systems Co	Switzerland	*384*	55-00100
Target Technology Ltd	UK	*487*	75-58660
Teledyne Ryan Aeronautical	USA	*621*	75-87560
TUSAS Aerospace Industries Inc.	Turkey	*403*	75-02120
Vanguard Research Inc.	USA	*644*	80-34100
Wallis Autogyros	UK	*492*	75-64150
Xi CAD North America Inc.	USA	*628*	75-97760

123 10 10 RPVs, battlefield observation

Advanced Technologies &	S. Africa	*357*	75-00500
Bombardier Inc	Canada	*69*	75-01900
Dornier Luftfahrt GmbH	Germany	*168*	75-08860
EL-OP Electro Optics Industries Ltd	Israel	*219*	75-06300
GEC-Marconi Avionics Ltd	UK	*454*	75-25320
Lear Astronics Corp.	USA	*589*	75-54390
Northrop Grumman Corporation	USA	*599*	75-65450
Oerlikon-Contraves AG.	Switzerland	*387*	75-09250
Vanguard Research Inc.	USA	*644*	80-34100

123 10 11 RPVs, battlefield observation, fixed wing

ATS Aerospace Inc.	Canada	*68*	75-00752
Dornier Luftfahrt GmbH	Germany	*168*	75-08860
Flight Refuelling Limited	UK	*451*	75-23850
GEC-Marconi Avionics Ltd	UK	*454*	75-25320
Lear Astronics Corp.	USA	*589*	75-54390
Northrop Grumman Corporation	USA	*599*	75-65450
Oerlikon-Contraves AG.	Switzerland	*387*	75-09250
Raytheon E-Systems	USA	*608*	75-73302
SAGEM SA	**France**	***143***	**75-33950**
Swiss Aircraft & Systems Co	Switzerland	*384*	55-00100
Target Technology Ltd	UK	*487*	75-58660

123 10 12 RPVs, battlefield observation, rotary wing

Bombardier Inc	Canada	*69*	75-01900
Dornier Luftfahrt GmbH	Germany	*168*	75-08860
Flight Refuelling Limited	UK	*451*	75-23850
Target Technology Ltd	UK	*487*	75-58660

123 10 20 RPVs, tactical reconnaissance

Bombardier Inc	Canada	*69*	75-01900
Dornier Luftfahrt GmbH	Germany	*168*	75-08860
GEC-Marconi Avionics Ltd	UK	*454*	75-25320
Lear Astronics Corp.	USA	*589*	75-54390
Northrop Grumman Corporation	USA	*599*	75-65450
Oerlikon-Contraves AG.	Switzerland	*387*	75-09250
Pacer Systems Inc.	USA	*602*	75-68000
R&D Aeronautical Engineering Co.	USA	*607*	75-72430
Raytheon E-Systems	USA	*608*	75-73302
Vanguard Research Inc.	USA	*644*	80-34100

123 10 30 RPVs, tactical strike

AeroSpace Technologies of Australia	Australia	*11*	75-00340
Lear Astronics Corp.	USA	*589*	75-54390
Raytheon Aircraft Company	USA	*607*	75-73200
Vanguard Research Inc.	USA	*644*	80-34100

123 10 40 RPVs, target (including unmanned target aircraft)

AeroSpace Technologies of Australia	Australia	*11*	75-00340
Arab Organization for Ind.	Egypt	*109*	55-01700
Cranfield Aerospace Ltd	UK	*500*	80-07200
Cranfield University	UK	*419*	60-00350
Designability Ltd	UK	*501*	80-08450
Dornier Luftfahrt GmbH	Germany	*168*	75-08860
Flight Refuelling Limited	UK	*451*	75-23850
FR Aviation Limited	UK	*503*	80-10850
Hayes Targets	USA	*603*	75-69135
Japan Aircraft Manufacturing Co. Ltd.	Japan	*252*	75-03200
JDR Cable Systems Ltd	UK	*462*	75-32480
Kader Factory for Developed Ind.	Egypt	*109*	55-01700
Kaman Sciences Corp.	USA	*637*	80-17600
Lear Astronics Corp.	USA	*589*	75-54390
MASS Consultants Limited	UK	*506*	80-15650
Northrop Grumman Corporation	USA	*599*	75-65450
Pearpoint Ltd	UK	*473*	75-44270
St Bernard Composites Ltd	UK	*486*	75-56222
Target Technology Ltd	UK	*487*	75-58660
Trimble Navigation Europe Ltd	UK	*489*	75-61140

123 10 50 Mini-RPVs

Alden Sales Co.	USA	*548*	75-04410
Bombardier Inc	Canada	*69*	75-01900
Chapy Corporation	USA	*563*	75-18065
Dornier Luftfahrt GmbH	Germany	*168*	75-08860
Flight Refuelling Limited	UK	*451*	75-23850
Israel Aircraft Industries Ltd	Israel	*219*	75-10200
Israel Military Industries Ltd	Israel	*220*	75-10300
Kaman Sciences Corp.	USA	*637*	80-17600
Lear Astronics Corp.	USA	*589*	75-54390
Northrop Grumman Corporation	USA	*599*	75-65450
Oerlikon-Contraves AG.	Switzerland	*387*	75-09250
R&D Aeronautical Engineering Co.	USA	*607*	75-72430

Swiss Aircraft & Systems Co	Switzerland	*384*	55-00100
Xi CAD North America Inc.	USA	*628*	75-97760

123 10 90 RPVs, other types

Flight Refuelling Limited	UK	*451*	75-23850
Israel Military Industries Ltd	Israel	*220*	75-10300
Kaman Sciences Corp.	USA	*637*	80-17600
Lear Astronics Corp.	USA	*589*	75-54390
Vanguard Research Inc.	USA	*644*	80-34100

123 20 Drones

Advanced Aerodynamics	USA	*546*	75-00790
Bombardier Inc	Canada	*69*	75-01900
Dornier Luftfahrt GmbH	Germany	*168*	75-08860
Fuji Heavy Industries Ltd	Japan	*251*	75-01100
Lear Astronics Corp.	USA	*589*	75-54390
Meteor CAE SpA	Italy	*237*	75-15400
Oerlikon-Contraves AG.	Switzerland	*387*	75-09250
Pakistan Aeronautical Complex	Pakistan	*315*	55-01600
R&D Aeronautical Engineering Co.	USA	*607*	75-72430
Raytheon Aircraft Company	USA	*607*	75-73200
Sadler Aircraft Corp.	USA	*612*	75-77345
SAGEM SA	**France**	***143***	**75-33950**
Scaled Composites Inc.	USA	*541*	60-19160
Teledyne Ryan Aeronautical	USA	*621*	75-87560
The Boeing Company	USA	*559*	75-14250
TUSAS Aerospace Industries Inc.	Turkey	*403*	75-02120

123 20 10 Drones, reconnaissance

Bombardier Inc	Canada	*69*	75-01900
Dornier GmbH	Germany	*167*	75-06700
Dornier Luftfahrt GmbH	Germany	*168*	75-08860
Lear Astronics Corp.	USA	*589*	75-54390
Meteor CAE SpA	Italy	*237*	75-15400
Oerlikon-Contraves AG.	Switzerland	*387*	75-09250
Sadler Aircraft Corp.	USA	*612*	75-77345

123 20 10 50 Drones, reconnaissance, propeller-driven

Dornier Luftfahrt GmbH	Germany	*168*	75-08860
Lear Astronics Corp.	USA	*589*	75-54390
Oerlikon-Contraves AG.	Switzerland	*387*	75-09250
Swiss Aircraft & Systems Co	Switzerland	*384*	55-00100
The Boeing Company	USA	*559*	75-14250

123 20 10 60 Drones, reconnaissance, rocket-powered

Lear Astronics Corp.	USA	*589*	75-54390

123 20 11 Drones, reconnaissance / observation, tethered

Dornier Luftfahrt GmbH	Germany	*168*	75-08860
Lear Astronics Corp.	USA	*589*	75-54390
Oerlikon-Contraves AG.	Switzerland	*387*	75-09250

123 20 20 Drones, target (see code 616.12.77.30)

Aeromot Ind.Mecanico-Metalurgica SA	Brazil	*55*	75-00635
AeroSpace Technologies of Australia	Australia	*11*	75-00340
Balmoral Group Ltd	UK	*433*	75-05420
B.I. Engineering	UK	*434*	75-06630
Cranfield Aerospace Ltd	UK	*500*	80-07200
Cranfield University	UK	*419*	60-00350
Designability Ltd	UK	*501*	80-08450
Dornier Luftfahrt GmbH	Germany	*168*	75-08860
Flight Refuelling Limited	UK	*451*	75-23850
FR Aviation Limited	UK	*503*	80-10850
Fuji Heavy Industries Ltd	Japan	*251*	75-01100
Hayes Targets	USA	*603*	75-69135
Hovermarine International Ltd	UK	*459*	75-30200
Lear Astronics Corp.	USA	*589*	75-54390
Meteor CAE SpA	Italy	*237*	75-15400
OTB	S. Africa	*357*	60-00500
Pakistan Aeronautical Complex	Pakistan	*315*	55-01600
R&D Aeronautical Engineering Co.	USA	*607*	75-72430
Raytheon Aircraft Company	USA	*607*	75-73200
Société Europeenne de Propulsion	France	*144*	75-34850
St Bernard Composites Ltd	UK	*486*	75-56222
Target Technology Ltd	UK	*487*	75-58660
Teledyne Ryan Aeronautical	USA	*621*	75-87560
TUSAS Aerospace Industries Inc.	Turkey	*403*	75-02120

123 20 21 Drones, target, subsonic

Arab Organization for Ind.	Egypt	*109*	55-01700
ATS Aerospace Inc.	Canada	*68*	75-00752
Cranfield Aerospace Ltd	UK	*500*	80-07200
CSE Aviation Ltd	UK	*500*	80-07600
Designability Ltd	UK	*501*	80-08450
Dornier Luftfahrt GmbH	Germany	*168*	75-08860
Flight Refuelling Limited	UK	*451*	75-23850
FR Aviation Limited	UK	*503*	80-10850
Fuji Heavy Industries Ltd	Japan	*251*	75-01100
Hayes Targets	USA	*603*	75-69135
Hovermarine International Ltd	UK	*459*	75-30200
JDR Cable Systems Ltd	UK	*462*	75-32480
Lear Astronics Corp.	USA	*589*	75-54390
Pakistan Aeronautical Complex	Pakistan	*315*	55-01600
Raytheon Aircraft Company	USA	*607*	75-73200
Swiss Aircraft & Systems Co	Switzerland	*384*	55-00100
Target Technology Ltd	UK	*487*	75-58660

123 20 22 Drones, target, supersonic

ATS Aerospace Inc.	Canada	*68*	75-00752
Hayes Targets	USA	*603*	75-69135
Raytheon Aircraft Company	USA	*607*	75-73200
Target Technology Ltd	UK	*487*	75-58660

123 20 50 Mini-drones

Alden Sales Co.	USA	*548*	75-04410
Lear Astronics Corp.	USA	*589*	75-54390
Pakistan Aeronautical Complex	Pakistan	*315*	55-01600
The Boeing Company	USA	*559*	75-14250
Yamaha Motor Co Ltd	Japan	*255*	75-25000

123 20 90 Drones, other types

Bombardier Inc	Canada	*69*	75-01900
Meteor CAE SpA	Italy	*237*	75-15400

123 50 Unmanned air vehicles (UAVs)

Aereon Corp.	USA	*630*	80-00550
CONTROP Precision Technologies Ltd	Israel	*218*	75-04170
Daedalus Research Inc.	USA	*567*	75-23360
Electronics & Aviation Systems Ltd.	Israel	*219*	75-05800
Enics Research Center	Russia	*335*	60-01200
Kaman Sciences Corp.	USA	*637*	80-17600
Kennedy Aerospace Corp.	USA	*540*	60-11111
Orion Aviation	USA	*602*	75-67500
Pacer Systems Inc.	USA	*602*	75-68000
Pathfinder Systems Inc.	USA	*603*	75-68995
Scaled Composites Inc.	USA	*541*	60-19160
Silver Arrow	Israel	*222*	75-20400
STN ATLAS Elektronik GmbH	Germany	*165*	75-01150
Target Technology Ltd	UK	*487*	75-58660
Technical Products Group Inc.	USA	*620*	75-86764
Vanguard Research Inc.	USA	*644*	80-34100
Xi CAD North America Inc.	USA	*628*	75-97760

124 Airships / Dirigibles, Balloons

Advanced Hybrid Aircraft Ltd	Australia	*11*	75-00200
American Blimp Corporation	USA	*552*	75-05540
Cameron Balloons Ltd	UK	*438*	75-11400
China National Aero-Tech I/E Corp.	China PR	*88*	55-00200
Gesellschaft zur Enticklung und	Germany	*169*	75-10700
Lear Astronics Corp.	USA	*589*	75-54390
Lockheed Martin Services Group	USA	*638*	80-18540
Number 605 Research Institute	China PR	*89*	75-01450
Pan Atlantic Aerospace Corporation	Canada	*74*	75-10770

Research Institute for	Ukraine	406	60-00990
UPship	USA	625	75-92915
US Lighter Than Air Corporation	USA	625	75-93000
Zeppelin Luftschifftechnik GmbH	Germany	177	75-34780
21st Century Airships Inc	Canada	67	75-00010

124 10 Airships / Dirigibles

Aereon Corp.	USA	630	80-00550
American Blimp Corporation	USA	552	75-05540
Cameron Balloons Ltd	UK	438	75-11400
Gesellschaft zur Enticklung und	Germany	169	75-10700
Number 605 Research Institute	China PR	89	75-01450
Pan Atlantic Aerospace Corporation	Canada	74	75-10770
UPship	USA	625	75-92915
US Lighter Than Air Corporation	USA	625	75-93000
Zeppelin Luftschifftechnik GmbH	Germany	177	75-34780
21st Century Airships Inc	Canada	67	75-00010

124 20 Balloons

Aérazur	France	131	75-00500
Airborne Industries	UK	428	75-01550
Bachmann SA	France	132	75-04900
Cameron Balloons Ltd	UK	438	75-11400
Gesellschaft zur Enticklung und	Germany	169	75-10700
Number 605 Research Institute	China PR	89	75-01450
Research Institute for	Ukraine	406	60-00990

124 21 Balloons, tethered

Airborne Industries	UK	428	75-01550
Bachmann SA	France	132	75-04900
BCB International Ltd	UK	433	75-06075
Israel Product Research Co. Ltd	Israel	219	75-10000
Lockheed Martin Services Group	USA	638	80-18540
Number 605 Research Institute	China PR	89	75-01450

124 21 20 Balloons, parachutist training

Airborne Industries	UK	428	75-01550
Bachmann SA	France	132	75-04900

124 21 30 Balloons, observation

Bachmann SA	France	132	75-04900

124 22 10 Balloons, meteorological

Nansen SA	Brazil	57	75-14000
SED Systems Inc	Canada	79	80-01937
Vaisala Oy	Finland	119	75-02800

Sea-based Platforms

131 Warships, surface combattant craft

ASC Shipbuilding Pty Limited	Australia	12	75-01530
ASMAR	Chile	84	55-00150
Australian Submarine Corporation	Australia	12	75-01910
Avondale Industries Inc.	USA	556	75-11400
Bath Iron Works Corp.	USA	557	75-12200
Bender Shipbuilding & Repair Co.	USA	558	75-13300
Blohm + Voss GmbH	Germany	165	75-02800
Bollinger Shipyard Inc.	USA	560	75-14600
Brazan SA	Spain	364	55-00140
C. Raymond Hunt Associates Inc.	USA	637	80-14900
Cenzin Co. Ltd.	Poland	324	75-00200
Cougar Limited	UK	443	75-15250
Fincantieri	Italy	234	75-09250
Finnyards Ltd	Finland	118	75-00600
Garden Reach Shipbuilders	India	203	55-00300
George G. Sharp Inc.	USA	642	80-28000
Hitachi Zosen Co. Ltd.	Japan	251	75-02390
Hong Leong-Lürssen Shipyard	Malaysia	277	75-00055
INIMEX	**Belgium**	**46**	**75-04650**
Ishikawajima-Harima H.I. Co. Ltd	Japan	251	75-02800
Israel Shipyards Ltd.	Israel	220	75-10600
Kvaerner Mandal A/S	Norway	309	75-01100
Kvaerner Masa-Yards Inc	**Finland**	**118**	**75-01370**
Litton Industries Inc.	USA	590	75-55400
Marinette Marine Corp.	USA	594	75-58280
Mazagon Dock Ltd	India	204	55-01200
McTay Marine Limited	UK	467	75-38230
Mitsubishi Heavy Industries Ltd	Japan	253	75-05800
Newport News Shipbuilding	USA	599	75-64300
Penang Shipbuilding Corporation	Malaysia	277	75-00800
Royal Schelde	Netherlands	294	75-06150
Saint John Shipbuilding Ltd	Canada	75	75-12300
Sener Ingenieria y Sistemas SA	Spain	367	75-05200
Sperry Marine Inc.	USA	617	75-83250
Swiftships Inc.	USA	619	75-85350
Tacoma Boatbuilding Co.	USA	619	75-86000
Thyssen Rheinstahl Technik GmbH	Germany	175	75-31320
TTL Dynamics	UK	490	75-61510
Vickers Shipbuilding & Eng Ltd	UK	491	75-63100
Vosper Thornycroft	UK	492	75-63800

131 10 Aircraft carriers

Brazan SA	Spain	364	55-00140
Fincantieri	Italy	234	75-09250
Newport News Shipbuilding	USA	599	75-64300
Vickers Shipbuilding & Eng Ltd	UK	491	75-63100

131 20 Cruisers

Fincantieri	Italy	234	75-09250
George G. Sharp Inc.	USA	642	80-28000
Litton Industries Inc.	USA	590	75-55400
Newport News Shipbuilding	USA	599	75-64300
Vickers Shipbuilding & Eng Ltd	UK	491	75-63100

131 30 Destroyers, escorts

Avondale Industries Inc.	USA	556	75-11400
Bath Iron Works Corp.	USA	557	75-12200
Blohm + Voss GmbH	Germany	165	75-02800
Brazan SA	Spain	364	55-00140
Daewoo Heavy Industries Ltd	Korea (ROK)	265	75-00600
Davie Industries Inc.	Canada	70	75-04040
Fincantieri	Italy	234	75-09250
George G. Sharp Inc.	USA	642	80-28000
Hitachi Zosen Co. Ltd.	Japan	251	75-02390
Hyundai Heavy Industries Co. Ltd	Korea (ROK)	265	75-01600
INIMEX	**Belgium**	**46**	**75-04650**
Ishikawajima-Harima H.I. Co. Ltd	Japan	251	75-02800
Litton Industries Inc.	USA	590	75-55400
Mazagon Dock Ltd	India	204	55-01200
Mitsubishi Heavy Industries Ltd	Japan	253	75-05800
Vickers Shipbuilding & Eng Ltd	UK	491	75-63100
Yarrow Shipbuilders Ltd	UK	495	75-67500

131 40 Frigates, avisos, corvettes

ASC Shipbuilding Pty Limited	Australia	12	75-01530
Avondale Industries Inc.	USA	556	75-11400
Bath Iron Works Corp.	USA	557	75-12200
Blohm + Voss GmbH	Germany	165	75-02800
Brazan SA	Spain	364	55-00140
Chantiers de l'Atlantique	France	133	75-08850
Daewoo Heavy Industries Ltd	Korea (ROK)	265	75-00600
Davie Industries Inc.	Canada	70	75-04040
Fincantieri	Italy	234	75-09250
Garden Reach Shipbuilders	India	203	55-00300
George G. Sharp Inc.	USA	642	80-28000
Howaldtswerke Deutsche Werft AG	Germany	170	75-13400
Hyundai	Korea (ROK)	265	75-01540
Hyundai Heavy Industries Co. Ltd	Korea (ROK)	265	75-01600
INIMEX	**Belgium**	**46**	**75-04650**
Ishikawajima-Harima H.I. Co. Ltd	Japan	251	75-02800
Karlskronavarvet AB	Sweden	377	75-07400
Litton Industries Inc.	USA	590	75-55400
Mazagon Dock Ltd	India	204	55-01200
PT PAL Indonesia (PERSERO)	Indonesia	210	75-00500
Royal Schelde	Netherlands	294	75-06150
Saint John Shipbuilding Ltd	Canada	75	75-12300
SIMA PERU	Peru	319	55-01950

Stocznia Polnocna SA	Poland	*325*	75-02800
Tacoma Boatbuilding Co.	USA	*619*	75-86000
Vickers Shipbuilding & Eng Ltd	UK	*491*	75-63100
Yarrow Shipbuilders Ltd	UK	*495*	75-67500

131 50 Mine warfare ships

Abeking & Rasmussen	Germany	*164*	75-00060
Avondale Industries Inc.	USA	*556*	75-11400
George G. Sharp Inc.	USA	*642*	80-28000
Hitachi Zosen Co. Ltd.	Japan	*251*	75-02390
INIMEX	**Belgium**	***46***	**75-04650**
Karlskronavarvet AB	Sweden	*377*	75-07400
Kvaerner Mandal A/S	Norway	*309*	75-01100
Kvaerner Masa-Yards Inc	**Finland**	***118***	**75-01370**
STN ATLAS Elektronik GmbH	Germany	*165*	75-01150
Tacoma Boatbuilding Co.	USA	*619*	75-86000
Transfield Shipbuilding WA	Australia	*22*	75-20812
Van der Giessen-de Noord	Netherlands	*296*	75-07850
Vickers Shipbuilding & Eng Ltd	UK	*491*	75-63100
Yarrow Shipbuilders Ltd	UK	*495*	75-67500

131 51 Minehunters

ASC Shipbuilding Pty Limited	Australia	*12*	75-01530
Avondale Industries Inc.	USA	*556*	75-11400
Azimut SpA	Italy	*232*	75-02360
George G. Sharp Inc.	USA	*642*	80-28000
Halmatic Ltd	UK	*456*	75-27700
Hitachi Zosen Co. Ltd.	Japan	*251*	75-02390
Hytech BV	Netherlands	*293*	75-03530
INIMEX	**Belgium**	***46***	**75-04650**
Intermarine SpA	Italy	*235*	75-11900
Kvaerner Mandal A/S	Norway	*309*	75-01100
NKI	Netherlands	*294*	75-05100
PT PAL Indonesia (PERSERO)	Indonesia	*210*	75-00500
Tacoma Boatbuilding Co.	USA	*619*	75-86000
Transfield Shipbuilding WA	Australia	*22*	75-20812
Vickers Shipbuilding & Eng Ltd	UK	*491*	75-63100

131 52 Minelayers

Azimut SpA	Italy	*232*	75-02360
George G. Sharp Inc.	USA	*642*	80-28000
Hitachi Zosen Co. Ltd.	Japan	*251*	75-02390
Hytech BV	Netherlands	*293*	75-03530
Hyundai Heavy Industries Co. Ltd	Korea (ROK)	*265*	75-01600
INIMEX	**Belgium**	***46***	**75-04650**
Intermarine SpA	Italy	*235*	75-11900
Transfield Shipbuilding WA	Australia	*22*	75-20812
Vickers Shipbuilding & Eng Ltd	UK	*491*	75-63100

131 53 Minesweepers

Azimut SpA	Italy	*232*	75-02360
Davie Industries Inc.	Canada	*70*	75-04040
George G. Sharp Inc.	USA	*642*	80-28000
Hitachi Zosen Co. Ltd.	Japan	*251*	75-02390
Hytech BV	Netherlands	*293*	75-03530
INIMEX	**Belgium**	***46***	**75-04650**
Intermarine SpA	Italy	*235*	75-11900
Kvaerner Mandal A/S	Norway	*309*	75-01100
Tacoma Boatbuilding Co.	USA	*619*	75-86000
Transfield Shipbuilding WA	Australia	*22*	75-20812
Vickers Shipbuilding & Eng Ltd	UK	*491*	75-63100

131 60 Patrol ships, armed

ACH	France	*130*	75-00150
ASC Shipbuilding Pty Limited	Australia	*12*	75-01530
Bender Shipbuilding & Repair Co.	USA	*558*	75-13300
Blohm + Voss GmbH	Germany	*165*	75-02800
Brazan SA	Spain	*364*	55-00140
Cenzin Co. Ltd.	Poland	*324*	75-00200
Chantiers de l'Atlantique	France	*133*	75-08850
Daewoo Heavy Industries Ltd	Korea (ROK)	*265*	75-00600
Danyard A/S	Denmark	*102*	75-04300
Emsworth Shipyard	UK	*448*	75-20750
FBM Marine Limited	UK	*450*	75-22485
Fincantieri	Italy	*234*	75-09250
Garden Reach Shipbuilders	India	*203*	55-00300
George G. Sharp Inc.	USA	*642*	80-28000
Goa Shipyard Ltd	India	*203*	55-00400
Hanjin Heavy Industries Co. Ltd	Korea (ROK)	*265*	75-01505
Hong Leong-Lürssen Shipyard	Malaysia	*277*	75-00055
Hyundai Heavy Industries Co. Ltd	Korea (ROK)	*265*	75-01600
INIMEX	**Belgium**	***46***	**75-04650**
Intermarine SpA	Italy	*235*	75-11900
Ishikawajima-Harima H.I. Co. Ltd	Japan	*251*	75-02800
Israel Aircraft Industries Ltd	Israel	*219*	75-10200
Israel Shipyards Ltd.	Israel	*220*	75-10600
Karlskronavarvet AB	Sweden	*377*	75-07400
Kvaerner Mandal A/S	Norway	*309*	75-01100
Kvaerner Masa-Yards Inc	**Finland**	***118***	**75-01370**
Magnum Marine Corp.	USA	*593*	75-57400
Mazagon Dock Ltd	India	*204*	55-01200
NQEA Australia Pty Ltd	Australia	*19*	75-14890
Penang Shipbuilding Corporation	Malaysia	*277*	75-00800
Rodman Polyships SA	Spain	*367*	75-05000
RTK Marine	UK	*481*	75-50650
Souter Shipyard Ltd	UK	*485*	75-55212
Stocznia Polnocna SA	Poland	*325*	75-02800
Tacoma Boatbuilding Co.	USA	*619*	75-86000
Transfield Shipbuilding WA	Australia	*22*	75-20812
Vickers Shipbuilding & Eng Ltd	UK	*491*	75-63100
Yarrow Shipbuilders Ltd	UK	*495*	75-67500

131 70 Fast patrol boats, armed

Abeking & Rasmussen	Germany	*164*	75-00060
ACH	France	*130*	75-00150
ASC Shipbuilding Pty Limited	Australia	*12*	75-01530
Azimut SpA	Italy	*232*	75-02360
Bender Shipbuilding & Repair Co.	USA	*558*	75-13300
Blohm + Voss GmbH	Germany	*165*	75-02800
Bollinger Shipyard Inc.	USA	*560*	75-14600
Brazan SA	Spain	*364*	55-00140
C. Raymond Hunt Associates Inc.	USA	*637*	80-14900
Cougar Limited	UK	*443*	75-15250
Cygnus Marine Ltd	UK	*444*	75-16500
Daewoo Heavy Industries Ltd	Korea (ROK)	*265*	75-00600
Damen Shipyards	Netherlands	*292*	75-01500
Danyard A/S	Denmark	*102*	75-04300
Emsworth Shipyard	UK	*448*	75-20750
FBM Marine Limited	UK	*450*	75-22485
Fincantieri	Italy	*234*	75-09250
Garden Reach Shipbuilders	India	*203*	55-00300
George G. Sharp Inc.	USA	*642*	80-28000
Goa Shipyard Ltd	India	*203*	55-00400
Halmatic Ltd	UK	*456*	75-27700
Halter Marine Group	USA	*580*	75-41700
Hanjin Heavy Industries Co. Ltd	Korea (ROK)	*265*	75-01505
Hong Leong-Lürssen Shipyard	Malaysia	*277*	75-00055
Hyundai	Korea (ROK)	*265*	75-01540
Hyundai Heavy Industries Co. Ltd	Korea (ROK)	*265*	75-01600
INIMEX	**Belgium**	***46***	**75-04650**
Intermarine SpA	Italy	*235*	75-11900
Ishikawajima-Harima H.I. Co. Ltd	Japan	*251*	75-02800
Israel Aircraft Industries Ltd	Israel	*219*	75-10200
Israel Shipyards Ltd.	Israel	*220*	75-10600
Karlskronavarvet AB	Sweden	*377*	75-07400
Kvaerner Mandal A/S	Norway	*309*	75-01100
Magnum Marine Corp.	USA	*593*	75-57400
NQEA Australia Pty Ltd	Australia	*19*	75-14890
Penang Shipbuilding Corporation	Malaysia	*277*	75-00800
PT PAL Indonesia (PERSERO)	Indonesia	*210*	75-00500
Rodman Polyships SA	Spain	*367*	75-05000
Royal Schelde	Netherlands	*294*	75-06150
RTK Marine	UK	*481*	75-50650
Saint John Shipbuilding Ltd	Canada	*75*	75-12300
SIMA PERU	Peru	*319*	55-01950
Souter Shipyard Ltd	UK	*485*	75-55212
Stocznia Polnocna SA	Poland	*325*	75-02800
Svendborg Vaerft A/S	Denmark	*104*	75-19800
Tacoma Boatbuilding Co.	USA	*619*	75-86000
Transfield Shipbuilding WA	Australia	*22*	75-20812
Tremlett Boat Sales Ltd	UK	*489*	75-61075
TTL Dynamics	UK	*490*	75-61510
Vickers Shipbuilding & Eng Ltd	UK	*491*	75-63100

131 70 50 Coastal patrol / surveillance craft

ACH	France	*130*	75-00150

ASC Shipbuilding Pty Limited	Australia	*12*	75-01530
Blohm + Voss GmbH	Germany	*165*	75-02800
C. Raymond Hunt Associates Inc.	USA	*637*	80-14900
Carbrasmar Ind. Com Ltda	Brazil	*55*	75-03030
Colvic Craft plc	UK	*441*	75-14200
Cougar Limited	UK	*443*	75-15250
Damen Shipyards	Netherlands	*292*	75-01500
Emsworth Shipyard	UK	*448*	75-20750
FBM Marine Limited	UK	*450*	75-22485
Garden Reach Shipbuilders	India	*203*	55-00300
George G. Sharp Inc.	USA	*642*	80-28000
Goa Shipyard Ltd	India	*203*	55-00400
Halter Marine Group	USA	*580*	75-41700
Hong Leong-Lürssen Shipyard	Malaysia	*277*	75-00055
INIMEX	**Belgium**	***46***	**75-04650**
Israel Shipyards Ltd.	Israel	*220*	75-10600
Marr Technical Services Ltd	UK	*506*	80-15620
Penang Shipbuilding Corporation	Malaysia	*277*	75-00800
Rodman Polyships SA	Spain	*367*	75-05000
Stocznia Polnocna SA	Poland	*325*	75-02800
Storebro Bruks AB	Sweden	*379*	75-10600
Svendborg Vaerft A/S	Denmark	*104*	75-19800
Tacoma Boatbuilding Co.	USA	*619*	75-86000
TTL Dynamics	UK	*490*	75-61510
Vickers Shipbuilding & Eng Ltd	UK	*491*	75-63100

131 80 Riverine warfare craft

Azimut SpA	Italy	*232*	75-02360
Bender Shipbuilding & Repair Co.	USA	*558*	75-13300
Boston Whaler Inc.	USA	*560*	75-14760
C. Raymond Hunt Associates Inc.	USA	*637*	80-14900
Cougar Limited	UK	*443*	75-15250
Daewoo Heavy Industries Ltd	Korea (ROK)	*265*	75-00600
Emsworth Shipyard	UK	*448*	75-20750
FBM Marine Limited	UK	*450*	75-22485
Halter Marine Group	USA	*580*	75-41700
Hong Leong-Lürssen Shipyard	Malaysia	*277*	75-00055
INIMEX	**Belgium**	***46***	**75-04650**
Israel Aircraft Industries Ltd	Israel	*219*	75-10200
Magnum Marine Corp.	USA	*593*	75-57400
NQEA Australia Pty Ltd	Australia	*19*	75-14890
Penang Shipbuilding Corporation	Malaysia	*277*	75-00800
RTK Marine	UK	*481*	75-50650
Transfield Shipbuilding WA	Australia	*22*	75-20812
Tremlett Boat Sales Ltd	UK	*489*	75-61075
Yarrow Shipbuilders Ltd	UK	*495*	75-67500

131 90 Warships, other types

Blohm + Voss GmbH	Germany	*165*	75-02800
Cenzin Co. Ltd.	Poland	*324*	75-00200
INIMEX	**Belgium**	***46***	**75-04650**
Kvaerner Masa-Yards Inc	**Finland**	***118***	**75-01370**
Litton Industries Inc.	USA	*590*	75-55400
Penang Shipbuilding Corporation	Malaysia	*277*	75-00800
Vickers Shipbuilding & Eng Ltd	UK	*491*	75-63100

132 Amphibious warfare ships / craft

ASC Shipbuilding Pty Limited	Australia	*12*	75-01530
C. Raymond Hunt Associates Inc.	USA	*637*	80-14900
George G. Sharp Inc.	USA	*642*	80-28000
Howaldtswerke Deutsche Werft AG	Germany	*170*	75-13400
Hyundai Heavy Industries Co. Ltd	Korea (ROK)	*265*	75-01600
INIMEX	**Belgium**	***46***	**75-04650**
Ishikawajima-Harima H.I. Co. Ltd	Japan	*251*	75-02800
Litton Industries Inc.	USA	*590*	75-55400
Ocean Dynamics	UK	*471*	75-42900
Sperry Marine Inc.	USA	*617*	75-83250
Tacoma Boatbuilding Co.	USA	*619*	75-86000
Vickers Shipbuilding & Eng Ltd	UK	*491*	75-63100

132 10 Amphibious command ships

George G. Sharp Inc.	USA	*642*	80-28000
Hong Leong-Lürssen Shipyard	Malaysia	*277*	75-00055
Ishikawajima-Harima H.I. Co. Ltd	Japan	*251*	75-02800
Litton Industries Inc.	USA	*590*	75-55400
Vickers Shipbuilding & Eng Ltd	UK	*491*	75-63100
Vosper Thornycroft	UK	*492*	75-63800

132 20 Amphibious assault ships

ASC Shipbuilding Pty Limited	Australia	*12*	75-01530
George G. Sharp Inc.	USA	*642*	80-28000
Ishikawajima-Harima H.I. Co. Ltd	Japan	*251*	75-02800
Israel Shipyards Ltd.	Israel	*220*	75-10600
Litton Industries Inc.	USA	*590*	75-55400
Vickers Shipbuilding & Eng Ltd	UK	*491*	75-63100

132 30 Amphibious transport ships

ASC Shipbuilding Pty Limited	Australia	*12*	75-01530
Avondale Industries Inc.	USA	*556*	75-11400
Daewoo Heavy Industries Ltd	Korea (ROK)	*265*	75-00600
George G. Sharp Inc.	USA	*642*	80-28000
Ishikawajima-Harima H.I. Co. Ltd	Japan	*251*	75-02800
NQEA Australia Pty Ltd	Australia	*19*	75-14890
Vickers Shipbuilding & Eng Ltd	UK	*491*	75-63100

132 40 Landing support ships

Avondale Industries Inc.	USA	*556*	75-11400
Daewoo Heavy Industries Ltd	Korea (ROK)	*265*	75-00600
Emsworth Shipyard	UK	*448*	75-20750
George G. Sharp Inc.	USA	*642*	80-28000
Halter Marine Group	USA	*580*	75-41700
Hyundai	Korea (ROK)	*265*	75-01540
Hyundai Heavy Industries Co. Ltd	Korea (ROK)	*265*	75-01600
INIMEX	**Belgium**	***46***	**75-04650**
Ishikawajima-Harima H.I. Co. Ltd	Japan	*251*	75-02800
National Steel & Shipbuilding Co.	USA	*598*	75-63900
Penang Shipbuilding Corporation	Malaysia	*277*	75-00800
Souter Shipyard Ltd	UK	*485*	75-55212
Tacoma Boatbuilding Co.	USA	*619*	75-86000
Vickers Shipbuilding & Eng Ltd	UK	*491*	75-63100
Yarrow Shipbuilders Ltd	UK	*495*	75-67500

132 50 Landing ships & craft

Daewoo Heavy Industries Ltd	Korea (ROK)	*265*	75-00600
Emsworth Shipyard	UK	*448*	75-20750
FBM Marine Limited	UK	*450*	75-22485
George G. Sharp Inc.	USA	*642*	80-28000
Hong Leong-Lürssen Shipyard	Malaysia	*277*	75-00055
Howaldtswerke Deutsche Werft AG	Germany	*170*	75-13400
Hyundai	Korea (ROK)	*265*	75-01540
Hyundai Heavy Industries Co. Ltd	Korea (ROK)	*265*	75-01600
INIMEX	**Belgium**	***46***	**75-04650**
Ishikawajima-Harima H.I. Co. Ltd	Japan	*251*	75-02800
NQEA Australia Pty Ltd	Australia	*19*	75-14890
Penang Shipbuilding Corporation	Malaysia	*277*	75-00800
SIMA PERU	Peru	*319*	55-01950
Tacoma Boatbuilding Co.	USA	*619*	75-86000
Transfield Shipbuilding WA	Australia	*22*	75-20812
Vickers Shipbuilding & Eng Ltd	UK	*491*	75-63100
Yarrow Shipbuilders Ltd	UK	*495*	75-67500

132 52 Landing craft

ASC Shipbuilding Pty Limited	Australia	*12*	75-01530
Avondale Industries Inc.	USA	*556*	75-11400
Azimut SpA	Italy	*232*	75-02360
Cygnus Marine Ltd	UK	*444*	75-16500
Daewoo Heavy Industries Ltd	Korea (ROK)	*265*	75-00600
Emsworth Shipyard	UK	*448*	75-20750
FBM Marine Limited	UK	*450*	75-22485
George G. Sharp Inc.	USA	*642*	80-28000
Halmatic Ltd	UK	*456*	75-27700
Halter Marine Group	USA	*580*	75-41700
Hong Leong-Lürssen Shipyard	Malaysia	*277*	75-00055
INIMEX	**Belgium**	***46***	**75-04650**
Israel Shipyards Ltd.	Israel	*220*	75-10600
McTay Marine Limited	UK	*467*	75-38230
NQEA Australia Pty Ltd	Australia	*19*	75-14890
Penang Shipbuilding Corporation	Malaysia	*277*	75-00800
Rodman Polyships SA	Spain	*367*	75-05000
RTK Marine	UK	*481*	75-50650
Souter Shipyard Ltd	UK	*485*	75-55212
Stocznia Polnocna SA	Poland	*325*	75-02800
Tacoma Boatbuilding Co.	USA	*619*	75-86000

Textron Marine & Land Systems	USA	*622*	75-88660
Transfield Shipbuilding WA	Australia	*22*	75-20812

132 55 Landing ships

Avondale Industries Inc.	USA	*556*	75-11400
Chantiers de l'Atlantique	France	*133*	75-08850
Daewoo Heavy Industries Ltd	Korea (ROK)	*265*	75-00600
FBM Marine Limited	UK	*450*	75-22485
George G. Sharp Inc.	USA	*642*	80-28000
Hong Leong-Lürssen Shipyard	Malaysia	*277*	75-00055
INIMEX	**Belgium**	***46***	**75-04650**
NQEA Australia Pty Ltd	Australia	*19*	75-14890
Penang Shipbuilding Corporation	Malaysia	*277*	75-00800
Rodman Polyships SA	Spain	*367*	75-05000
Tacoma Boatbuilding Co.	USA	*619*	75-86000
Transfield Shipbuilding WA	Australia	*22*	75-20812
Vickers Shipbuilding & Eng Ltd	UK	*491*	75-63100

132 55 50 Landing ship tanks (LSTs)

FBM Marine Limited	UK	*450*	75-22485
Garden Reach Shipbuilders	India	*203*	55-00300
George G. Sharp Inc.	USA	*642*	80-28000
Hyundai Heavy Industries Co. Ltd	Korea (ROK)	*265*	75-01600
INIMEX	**Belgium**	***46***	**75-04650**
McTay Marine Limited	UK	*467*	75-38230
Penang Shipbuilding Corporation	Malaysia	*277*	75-00800
Vickers Shipbuilding & Eng Ltd	UK	*491*	75-63100

132 60 Assault boats, raiding craft

ASC Shipbuilding Pty Limited	Australia	*12*	75-01530
Avon Inflatables Ltd	UK	*432*	75-05100
Azimut SpA	Italy	*232*	75-02360
C. Raymond Hunt Associates Inc.	USA	*637*	80-14900
Cougar Limited	UK	*443*	75-15250
Damen Shipyards	Netherlands	*292*	75-01500
Emsworth Shipyard	UK	*448*	75-20750
FBM Marine Limited	UK	*450*	75-22485
Halter Marine Group	USA	*580*	75-41700
Hong Leong-Lürssen Shipyard	Malaysia	*277*	75-00055
INIMEX	**Belgium**	***46***	**75-04650**
Israel Shipyards Ltd.	Israel	*220*	75-10600
Ocean Dynamics	UK	*471*	75-42900
Rodman Polyships SA	Spain	*367*	75-05000
RTK Marine	UK	*481*	75-50650
Sillinger	France	*145*	75-35660
Souter Shipyard Ltd	UK	*485*	75-55212
Transfield Shipbuilding WA	Australia	*22*	75-20812
Tremlett Boat Sales Ltd	UK	*489*	75-61075
TTL Dynamics	UK	*490*	75-61510

132 61 Assault boats

Avon Inflatables Ltd	UK	*432*	75-05100
Azimut SpA	Italy	*232*	75-02360
C. Raymond Hunt Associates Inc.	USA	*637*	80-14900
Cougar Limited	UK	*443*	75-15250
Damen Shipyards	Netherlands	*292*	75-01500
Emsworth Shipyard	UK	*448*	75-20750
Halmatic Ltd	UK	*456*	75-27700
Hong Leong-Lürssen Shipyard	Malaysia	*277*	75-00055
INIMEX	**Belgium**	***46***	**75-04650**
ML Lifeguard Equipment Ltd	UK	*469*	75-40230
Rodman Polyships SA	Spain	*367*	75-05000
Sillinger	France	*145*	75-35660
Souter Shipyard Ltd	UK	*485*	75-55212
Transfield Shipbuilding WA	Australia	*22*	75-20812

132 65 Raiding craft

Avon Inflatables Ltd	UK	*432*	75-05100
C. Raymond Hunt Associates Inc.	USA	*637*	80-14900
Cougar Limited	UK	*443*	75-15250
Emsworth Shipyard	UK	*448*	75-20750
Halmatic Ltd	UK	*456*	75-27700
Hong Leong-Lürssen Shipyard	Malaysia	*277*	75-00055
INIMEX	**Belgium**	***46***	**75-04650**
Rodman Polyships SA	Spain	*367*	75-05000
Sillinger	France	*145*	75-35660
TTL Dynamics	UK	*490*	75-61510
Zodiac International	France	*149*	75-47310

132 70 Inflatable craft

Avon Inflatables Ltd	UK	*432*	75-05100
Bachmann SA	France	*132*	75-04900
C. Raymond Hunt Associates Inc.	USA	*637*	80-14900
E.P. Barrus Ltd	UK	*433*	75-05900
Halmatic Ltd	UK	*456*	75-27700
INIMEX	**Belgium**	***46***	**75-04650**
ML Lifeguard Equipment Ltd	UK	*469*	75-40230
Nautika Ltda	Brazil	*57*	75-14150
Novamarine 2 SpA	Italy	*237*	75-16630
Ocean Dynamics	UK	*471*	75-42900
Olympic Hellas Papapanagiotoy	Greece	*191*	75-02170
Sillinger	France	*145*	75-35660
Zodiac International	France	*149*	75-47310

133 Surface effect ships & craft

Air Vehicles Ltd	UK	*429*	75-02150
Blohm + Voss GmbH	Germany	*165*	75-02800
Tacoma Boatbuilding Co.	USA	*619*	75-86000

133 10 Surface effect ships

ACH	France	*130*	75-00150
Blohm + Voss GmbH	Germany	*165*	75-02800
Halter Marine Group	USA	*580*	75-41700
Hong Leong-Lürssen Shipyard	Malaysia	*277*	75-00055
NKI	Netherlands	*294*	75-05100
Textron Marine & Land Systems	USA	*622*	75-88660

133 20 Surface effect craft

Hong Leong-Lürssen Shipyard	Malaysia	*277*	75-00055
Hovermarine International Ltd	UK	*459*	75-30200
Tacoma Boatbuilding Co.	USA	*619*	75-86000
Textron Marine & Land Systems	USA	*622*	75-88660

133 20 10 Hovercraft, patrol

Griffon Hovercraft Ltd	UK	*456*	75-27025
NQEA Australia Pty Ltd	Australia	*19*	75-14890
Slingsby Aviation Ltd.	UK	*484*	75-54410
Textron Marine & Land Systems	USA	*622*	75-88660

133 20 20 Hovercraft, amphibious attack

Griffon Hovercraft Ltd	UK	*456*	75-27025
NQEA Australia Pty Ltd	Australia	*19*	75-14890
Textron Marine & Land Systems	USA	*622*	75-88660

133 20 30 Hovercraft, amphibious landing / transport

Air Vehicles Ltd	UK	*429*	75-02150
Griffon Hovercraft Ltd	UK	*456*	75-27025
NQEA Australia Pty Ltd	Australia	*19*	75-14890
Slingsby Aviation Ltd.	UK	*484*	75-54410
Textron Marine & Land Systems	USA	*622*	75-88660

133 20 40 Air cushion craft, small multi-role

Griffon Hovercraft Ltd	UK	*456*	75-27025
Sea Tiger Marine Pty Ltd	Australia	*21*	75-19350
Textron Marine & Land Systems	USA	*622*	75-88660

134 Hydrofoils

Fincantieri	Italy	*234*	75-09250
Hitachi Zosen Co. Ltd.	Japan	*251*	75-02390
Hyundai Heavy Industries Co. Ltd	Korea (ROK)	*265*	75-01600
Israel Shipyards Ltd.	Israel	*220*	75-10600

Product–Supplier Link

134 10 Hydrofoils, armed patrol

Fincantieri	Italy	234	75-09250
Israel Shipyards Ltd.	Israel	220	75-10600

134 60 Hydrofoils, other types

EDO Corporation	USA	571	75-29250
Hitachi Zosen Co. Ltd.	Japan	251	75-02390
PT PAL Indonesia (PERSERO)	Indonesia	210	75-00500

135 Mission support ships, auxiliaries

ASC Shipbuilding Pty Limited	Australia	12	75-01530
Australian Submarine Corporation	Australia	12	75-01910
Avondale Industries Inc.	USA	556	75-11400
Bender Shipbuilding & Repair Co.	USA	558	75-13300
Blohm + Voss GmbH	Germany	165	75-02800
Deggendorfer Werft & Eisenbau GmbH.	Germany	167	75-05800
Fincantieri	Italy	234	75-09250
George G. Sharp Inc.	USA	642	80-28000
Hitachi Zosen Co. Ltd.	Japan	251	75-02390
Hong Leong-Lürssen Shipyard	Malaysia	277	75-00055
Hyundai Heavy Industries Co. Ltd	Korea (ROK)	265	75-01600
Israel Shipyards Ltd.	Israel	220	75-10600
Kvaerner Mandal A/S	Norway	309	75-01100
Kvaerner Masa-Yards Inc	**Finland**	**118**	**75-01370**
Marinette Marine Corp.	USA	594	75-58280
McTay Marine Limited	UK	467	75-38230
Nichols Brothers Boat Builders Inc.	USA	599	75-64380
Royal Schelde	Netherlands	294	75-06150
Santierul Naval Constanta SA	Romania	330	75-00750
Sperry Marine Inc.	USA	617	75-83250
Stocznia Szczecinska SA.	Poland	325	75-02850
Swiftships Inc.	USA	619	75-85350
Tacoma Boatbuilding Co.	USA	619	75-86000
Vickers Shipbuilding & Eng Ltd	UK	491	75-63100

135 10 Mission support ships

Bender Shipbuilding & Repair Co.	USA	558	75-13300
Datamat Ingegneria dei Sistemi SpA	Italy	233	75-06500
Fincantieri	Italy	234	75-09250
George G. Sharp Inc.	USA	642	80-28000
Hyundai	Korea (ROK)	265	75-01540
INIMEX	**Belgium**	**46**	**75-04650**
Ishikawajima-Harima H.I. Co. Ltd	Japan	251	75-02800
Kvaerner Masa-Yards Inc	**Finland**	**118**	**75-01370**
McTay Marine Limited	UK	467	75-38230
Stocznia Szczecinska SA.	Poland	325	75-02850
Svendborg Vaerft A/S	Denmark	104	75-19800
Vickers Shipbuilding & Eng Ltd	UK	491	75-63100
Yarrow Shipbuilders Ltd	UK	495	75-67500

135 10 30 Depot ships, submarine

George G. Sharp Inc.	USA	642	80-28000
INIMEX	**Belgium**	**46**	**75-04650**
Vickers Shipbuilding & Eng Ltd	UK	491	75-63100

135 20 Fleet auxiliaries

ACH	France	130	75-00150
Avondale Industries Inc.	USA	556	75-11400
Fincantieri	Italy	234	75-09250
George G. Sharp Inc.	USA	642	80-28000
Hyundai	Korea (ROK)	265	75-01540
Hyundai Heavy Industries Co. Ltd	Korea (ROK)	265	75-01600
Ishikawajima-Harima H.I. Co. Ltd	Japan	251	75-02800
National Steel & Shipbuilding Co.	USA	598	75-63900
Royal Schelde	Netherlands	294	75-06150
SIMA PERU	Peru	319	55-01950
Vickers Shipbuilding & Eng Ltd	UK	491	75-63100

135 20 20 Cable ships

ACH	France	130	75-00150
George G. Sharp Inc.	USA	642	80-28000
INIMEX	**Belgium**	**46**	**75-04650**
Kvaerner Masa-Yards Inc	**Finland**	**118**	**75-01370**
Svendborg Vaerft A/S	Denmark	104	75-19800

135 20 35 Icebreakers

ASC Shipbuilding Pty Limited	Australia	12	75-01530
Avondale Industries Inc.	USA	556	75-11400
George G. Sharp Inc.	USA	642	80-28000
INIMEX	**Belgium**	**46**	**75-04650**
McTay Marine Limited	UK	467	75-38230
Svendborg Vaerft A/S	Denmark	104	75-19800
Tacoma Boatbuilding Co.	USA	619	75-86000
Vancouver Shipyards Co. Ltd	Canada	77	75-14435

135 20 40 Mooring, salvage & boom vessels

Emsworth Shipyard	UK	448	75-20750
FBM Marine Limited	UK	450	75-22485
George G. Sharp Inc.	USA	642	80-28000
Halter Marine Group	USA	580	75-41700
Hong Leong-Lürssen Shipyard	Malaysia	277	75-00055
Hyundai Heavy Industries Co. Ltd	Korea (ROK)	265	75-01600
INIMEX	**Belgium**	**46**	**75-04650**
McTay Marine Limited	UK	467	75-38230
Sillinger	France	145	75-35660
Svendborg Vaerft A/S	Denmark	104	75-19800

135 20 40 60 Rescue ships, submarine

Fincantieri	Italy	234	75-09250
George G. Sharp Inc.	USA	642	80-28000
Hyundai Heavy Industries Co. Ltd	Korea (ROK)	265	75-01600
INIMEX	**Belgium**	**46**	**75-04650**
Transfield Shipbuilding WA	Australia	22	75-20812
Vickers Shipbuilding & Eng Ltd	UK	491	75-63100

135 20 50 Research & survey ships

ACH	France	130	75-00150
Avondale Industries Inc.	USA	556	75-11400
Cygnus Marine Ltd	UK	444	75-16500
Fincantieri	Italy	234	75-09250
George G. Sharp Inc.	USA	642	80-28000
Goa Shipyard Ltd	India	203	55-00400
Halmatic Ltd	UK	456	75-27700
Hitachi Zosen Co. Ltd.	Japan	251	75-02390
Hong Leong-Lürssen Shipyard	Malaysia	277	75-00055
Hyundai Heavy Industries Co. Ltd	Korea (ROK)	265	75-01600
INIMEX	**Belgium**	**46**	**75-04650**
Marr Technical Services Ltd	UK	506	80-15620
McTay Marine Limited	UK	467	75-38230
NQEA Australia Pty Ltd	Australia	19	75-14890
Penang Shipbuilding Corporation	Malaysia	277	75-00800
Sillinger	France	145	75-35660
Stocznia Polnocna SA	Poland	325	75-02800
Stocznia Szczecinska SA.	Poland	325	75-02850
Svendborg Vaerft A/S	Denmark	104	75-19800
Tacoma Boatbuilding Co.	USA	619	75-86000
Transfield Shipbuilding WA	Australia	22	75-20812
Vickers Shipbuilding & Eng Ltd	UK	491	75-63100

135 20 60 Stores support ships

ASC Shipbuilding Pty Limited	Australia	12	75-01530
Chantiers de l'Atlantique	France	133	75-08850
Davie Industries Inc.	Canada	70	75-04040
George G. Sharp Inc.	USA	642	80-28000
National Steel & Shipbuilding Co.	USA	598	75-63900
Svendborg Vaerft A/S	Denmark	104	75-19800

135 20 70 Ships, fleet replenishment

ACH	France	130	75-00150
Avondale Industries Inc.	USA	556	75-11400
Fincantieri	Italy	234	75-09250
George G. Sharp Inc.	USA	642	80-28000
Hitachi Zosen Co. Ltd.	Japan	251	75-02390
Hyundai Heavy Industries Co. Ltd	Korea (ROK)	265	75-01600
INIMEX	**Belgium**	**46**	**75-04650**

McTay Marine Limited	UK	467	75-38230
National Steel & Shipbuilding Co.	USA	598	75-63900
Svendborg Vaerft A/S	Denmark	104	75-19800
Vickers Shipbuilding & Eng Ltd	UK	491	75-63100

135 20 70 90 Underway replenishment ships

ACH	France	130	75-00150
Avondale Industries Inc.	USA	556	75-11400
Daewoo Heavy Industries Ltd	Korea (ROK)	265	75-00600
Fincantieri	Italy	234	75-09250
George G. Sharp Inc.	USA	642	80-28000
Hitachi Zosen Co. Ltd.	Japan	251	75-02390
Hyundai Heavy Industries Co. Ltd	Korea (ROK)	265	75-01600
INIMEX	**Belgium**	***46***	**75-04650**
National Steel & Shipbuilding Co.	USA	598	75-63900
Svendborg Vaerft A/S	Denmark	104	75-19800

135 20 74 Tankers, fleet replenishment

ACH	France	130	75-00150
Avondale Industries Inc.	USA	556	75-11400
Daewoo Heavy Industries Ltd	Korea (ROK)	265	75-00600
Davie Industries Inc.	Canada	70	75-04040
Fincantieri	Italy	234	75-09250
Garden Reach Shipbuilders	India	203	55-00300
George G. Sharp Inc.	USA	642	80-28000
Hyundai Heavy Industries Co. Ltd	Korea (ROK)	265	75-01600
INIMEX	**Belgium**	***46***	**75-04650**
National Steel & Shipbuilding Co.	USA	598	75-63900
Sillinger	France	145	75-35660
Svendborg Vaerft A/S	Denmark	104	75-19800

135 20 75 Tankers, freighting

ACH	France	130	75-00150
Astilleros Espanoles SA	Spain	365	75-01400
Avondale Industries Inc.	USA	556	75-11400
Deggendorfer Werft & Eisenbau GmbH.	Germany	167	75-05800
Fincantieri	Italy	234	75-09250
George G. Sharp Inc.	USA	642	80-28000
INIMEX	**Belgium**	***46***	**75-04650**
McTay Marine Limited	UK	467	75-38230
Santierul Naval Constanta SA	Romania	330	75-00750
Stocznia Szczecinska SA.	Poland	325	75-02850
Svendborg Vaerft A/S	Denmark	104	75-19800

135 20 79 Tenders, multi-purpose

ACH	France	130	75-00150
George G. Sharp Inc.	USA	642	80-28000

135 20 80 Tenders, submarine

George G. Sharp Inc.	USA	642	80-28000
Vickers Shipbuilding & Eng Ltd	UK	491	75-63100

135 20 84 Torpedo recovery vessels

INIMEX	**Belgium**	***46***	**75-04650**
McTay Marine Limited	UK	467	75-38230
Tacoma Boatbuilding Co.	USA	619	75-86000
Transfield Shipbuilding WA	Australia	22	75-20812

135 20 85 Training ships

Blohm + Voss GmbH	Germany	165	75-02800
Daewoo Heavy Industries Ltd	Korea (ROK)	265	75-00600
Emsworth Shipyard	UK	448	75-20750
George G. Sharp Inc.	USA	642	80-28000
Goa Shipyard Ltd	India	203	55-00400
Hong Leong-Lürssen Shipyard	Malaysia	277	75-00055
INIMEX	**Belgium**	***46***	**75-04650**
Kvaerner Masa-Yards Inc	**Finland**	***118***	**75-01370**
Marr Technical Services Ltd	UK	506	80-15620
McTay Marine Limited	UK	467	75-38230
Penang Shipbuilding Corporation	Malaysia	277	75-00800
Stocznia Szczecinska SA.	Poland	325	75-02850
Svendborg Vaerft A/S	Denmark	104	75-19800
Tacoma Boatbuilding Co.	USA	619	75-86000
Vickers Shipbuilding & Eng Ltd	UK	491	75-63100

135 20 87 Trials ships

George G. Sharp Inc.	USA	642	80-28000
Hyundai Heavy Industries Co. Ltd	Korea (ROK)	265	75-01600
INIMEX	**Belgium**	***46***	**75-04650**
Marr Technical Services Ltd	UK	506	80-15620
McTay Marine Limited	UK	467	75-38230
Svendborg Vaerft A/S	Denmark	104	75-19800
Vickers Shipbuilding & Eng Ltd	UK	491	75-63100

135 20 90 Tugs

ASC Shipbuilding Pty Limited	Australia	12	75-01530
Australian Submarine Corporation	Australia	12	75-01910
Avondale Industries Inc.	USA	556	75-11400
Cygnus Marine Ltd	UK	444	75-16500
Daewoo Heavy Industries Ltd	Korea (ROK)	265	75-00600
Damen Shipyards	Netherlands	292	75-01500
Deggendorfer Werft & Eisenbau GmbH.	Germany	167	75-05800
Garden Reach Shipbuilders	India	203	55-00300
George G. Sharp Inc.	USA	642	80-28000
Goa Shipyard Ltd	India	203	55-00400
Halter Marine Group	USA	580	75-41700
Hong Leong-Lürssen Shipyard	Malaysia	277	75-00055
INIMEX	**Belgium**	***46***	**75-04650**
Israel Shipyards Ltd.	Israel	220	75-10600
McTay Marine Limited	UK	467	75-38230
Nichols Brothers Boat Builders Inc.	USA	599	75-64380
Penang Shipbuilding Corporation	Malaysia	277	75-00800
Stocznia Polnocna SA	Poland	325	75-02800
Stocznia Szczecinska SA.	Poland	325	75-02850
Svendborg Vaerft A/S	Denmark	104	75-19800
Tacoma Boatbuilding Co.	USA	619	75-86000
Transfield Shipbuilding WA	Australia	22	75-20812

135 20 92 Tugs, harbour

ASC Shipbuilding Pty Limited	Australia	12	75-01530
Avondale Industries Inc.	USA	556	75-11400
Cygnus Marine Ltd	UK	444	75-16500
Daewoo Heavy Industries Ltd	Korea (ROK)	265	75-00600
Garden Reach Shipbuilders	India	203	55-00300
George G. Sharp Inc.	USA	642	80-28000
Goa Shipyard Ltd	India	203	55-00400
Halter Marine Group	USA	580	75-41700
Hong Leong-Lürssen Shipyard	Malaysia	277	75-00055
INIMEX	**Belgium**	***46***	**75-04650**
Israel Shipyards Ltd.	Israel	220	75-10600
McTay Marine Limited	UK	467	75-38230
Nichols Brothers Boat Builders Inc.	USA	599	75-64380
Penang Shipbuilding Corporation	Malaysia	277	75-00800
Stocznia Szczecinska SA.	Poland	325	75-02850
Svendborg Vaerft A/S	Denmark	104	75-19800
Tacoma Boatbuilding Co.	USA	619	75-86000
Transfield Shipbuilding WA	Australia	22	75-20812

135 20 94 Tugs, ocean-going

ASC Shipbuilding Pty Limited	Australia	12	75-01530
Avondale Industries Inc.	USA	556	75-11400
Daewoo Heavy Industries Ltd	Korea (ROK)	265	75-00600
Garden Reach Shipbuilders	India	203	55-00300
George G. Sharp Inc.	USA	642	80-28000
Halter Marine Group	USA	580	75-41700
Hong Leong-Lürssen Shipyard	Malaysia	277	75-00055
Hyundai Heavy Industries Co. Ltd	Korea (ROK)	265	75-01600
INIMEX	**Belgium**	***46***	**75-04650**
Israel Shipyards Ltd.	Israel	220	75-10600
McTay Marine Limited	UK	467	75-38230
Penang Shipbuilding Corporation	Malaysia	277	75-00800
Stocznia Szczecinska SA.	Poland	325	75-02850
Svendborg Vaerft A/S	Denmark	104	75-19800
Tacoma Boatbuilding Co.	USA	619	75-86000
Transfield Shipbuilding WA	Australia	22	75-20812

135 50 Intelligence collection ships

George G. Sharp Inc.	USA	*642*	80-28000
Svendborg Vaerft A/S	Denmark	*104*	75-19800

136 Coastguard, Customs & Police vessels / craft

Avondale Industries Inc.	USA	*556*	75-11400
Blohm + Voss GmbH	Germany	*165*	75-02800
Bollinger Shipyard Inc.	USA	*560*	75-14600
Boston Whaler Inc.	USA	*560*	75-14760
C. Raymond Hunt Associates Inc.	USA	*637*	80-14900
Colvic Craft plc	UK	*441*	75-14200
Cougar Limited	UK	*443*	75-15250
Emsworth Shipyard	UK	*448*	75-20750
FBM Marine Limited	UK	*450*	75-22485
Fincantieri	Italy	*234*	75-09250
George G. Sharp Inc.	USA	*642*	80-28000
Hitachi Zosen Co. Ltd.	Japan	*251*	75-02390
Hong Leong-Lürssen Shipyard	Malaysia	*277*	75-00055
Hyundai	Korea (ROK)	*265*	75-01540
Hyundai Heavy Industries Co. Ltd	Korea (ROK)	*265*	75-01600
INIMEX	**Belgium**	***46***	**75-04650**
Ishikawajima-Harima H.I. Co. Ltd	Japan	*251*	75-02800
Israel Aircraft Industries Ltd	Israel	*219*	75-10200
Israel Shipyards Ltd.	Israel	*220*	75-10600
Kvaerner Mandal A/S	Norway	*309*	75-01100
Kvaerner Masa-Yards Inc	**Finland**	***118***	**75-01370**
Marinette Marine Corp.	USA	*594*	75-58280
Nichols Brothers Boat Builders Inc.	USA	*599*	75-64380
Ocean Dynamics	UK	*471*	75-42900
Penang Shipbuilding Corporation	Malaysia	*277*	75-00800
Rodman Polyships SA	Spain	*367*	75-05000
SeaArk Marine Inc.	USA	*613*	75-78900
Sperry Marine Inc.	USA	*617*	75-83250
Tacoma Boatbuilding Co.	USA	*619*	75-86000
Transfield Shipbuilding WA	Australia	*22*	75-20812
TTL Dynamics	UK	*490*	75-61510
Vosper Thornycroft	UK	*492*	75-63800

136 10 Coastguard vessels

ACH	France	*130*	75-00150
Avondale Industries Inc.	USA	*556*	75-11400
Azimut SpA	Italy	*232*	75-02360
Blohm + Voss GmbH	Germany	*165*	75-02800
Colvic Craft plc	UK	*441*	75-14200
Daewoo Heavy Industries Ltd	Korea (ROK)	*265*	75-00600
Danyard A/S	Denmark	*102*	75-04300
FBM Marine Limited	UK	*450*	75-22485
Fincantieri	Italy	*234*	75-09250
Finnyards Ltd	Finland	*118*	75-00600
Garden Reach Shipbuilders	India	*203*	55-00300
George G. Sharp Inc.	USA	*642*	80-28000
Halter Marine Group	USA	*580*	75-41700
Hanjin Heavy Industries Co. Ltd	Korea (ROK)	*265*	75-01505
Hitachi Zosen Co. Ltd.	Japan	*251*	75-02390
Hong Leong-Lürssen Shipyard	Malaysia	*277*	75-00055
Hyundai Heavy Industries Co. Ltd	Korea (ROK)	*265*	75-01600
INIMEX	**Belgium**	***46***	**75-04650**
Ishikawajima-Harima H.I. Co. Ltd	Japan	*251*	75-02800
Karlskronavarvet AB	Sweden	*377*	75-07400
Kvaerner Mandal A/S	Norway	*309*	75-01100
Kvaerner Masa-Yards Inc	**Finland**	***118***	**75-01370**
Magnum Marine Corp.	USA	*593*	75-57400
Marr Technical Services Ltd	UK	*506*	80-15620
Mazagon Dock Ltd	India	*204*	55-01200
McTay Marine Limited	UK	*467*	75-38230
NQEA Australia Pty Ltd	Australia	*19*	75-14890
Penang Shipbuilding Corporation	Malaysia	*277*	75-00800
Rodman Polyships SA	Spain	*367*	75-05000
Sea Tiger Marine Pty Ltd	Australia	*21*	75-19350
Sillinger	France	*145*	75-35660
Souter Shipyard Ltd	UK	*485*	75-55212
Storebro Bruks AB	Sweden	*379*	75-10600
Svendborg Vaerft A/S	Denmark	*104*	75-19800
Tacoma Boatbuilding Co.	USA	*619*	75-86000
Textron Marine & Land Systems	USA	*622*	75-88660
Transfield Shipbuilding WA	Australia	*22*	75-20812
Tremlett Boat Sales Ltd	UK	*489*	75-61075
TTL Dynamics	UK	*490*	75-61510
Yarrow Shipbuilders Ltd	UK	*495*	75-67500

136 10 20 Cutters, coastguard

Colvic Craft plc	UK	*441*	75-14200
Cougar Limited	UK	*443*	75-15250
Daewoo Heavy Industries Ltd	Korea (ROK)	*265*	75-00600
Danyard A/S	Denmark	*102*	75-04300
Emsworth Shipyard	UK	*448*	75-20750
FBM Marine Limited	UK	*450*	75-22485
George G. Sharp Inc.	USA	*642*	80-28000
Halmatic Ltd	UK	*456*	75-27700
INIMEX	**Belgium**	***46***	**75-04650**
McTay Marine Limited	UK	*467*	75-38230
NQEA Australia Pty Ltd	Australia	*19*	75-14890
Rodman Polyships SA	Spain	*367*	75-05000
Svendborg Vaerft A/S	Denmark	*104*	75-19800
Tacoma Boatbuilding Co.	USA	*619*	75-86000
Transfield Shipbuilding WA	Australia	*22*	75-20812

136 10 22 Cutters, coastguard, high endurance

Avondale Industries Inc.	USA	*556*	75-11400
Cougar Limited	UK	*443*	75-15250
Daewoo Heavy Industries Ltd	Korea (ROK)	*265*	75-00600
Danyard A/S	Denmark	*102*	75-04300
FBM Marine Limited	UK	*450*	75-22485
George G. Sharp Inc.	USA	*642*	80-28000
INIMEX	**Belgium**	***46***	**75-04650**
McTay Marine Limited	UK	*467*	75-38230
NQEA Australia Pty Ltd	Australia	*19*	75-14890
Penang Shipbuilding Corporation	Malaysia	*277*	75-00800
Rodman Polyships SA	Spain	*367*	75-05000
Sillinger	France	*145*	75-35660
Svendborg Vaerft A/S	Denmark	*104*	75-19800
Tacoma Boatbuilding Co.	USA	*619*	75-86000
Transfield Shipbuilding WA	Australia	*22*	75-20812

136 20 Customs / Police patrol craft

ASC Shipbuilding Pty Limited	Australia	*12*	75-01530
Avon Inflatables Ltd	UK	*432*	75-05100
Azimut SpA	Italy	*232*	75-02360
Blohm + Voss GmbH	Germany	*165*	75-02800
Bollinger Shipyard Inc.	USA	*560*	75-14600
Boston Whaler Inc.	USA	*560*	75-14760
C. Raymond Hunt Associates Inc.	USA	*637*	80-14900
Cougar Limited	UK	*443*	75-15250
Cygnus Marine Ltd	UK	*444*	75-16500
Daewoo Heavy Industries Ltd	Korea (ROK)	*265*	75-00600
Danyard A/S	Denmark	*102*	75-04300
Emsworth Shipyard	UK	*448*	75-20750
FBM Marine Limited	UK	*450*	75-22485
Garden Reach Shipbuilders	India	*203*	55-00300
George G. Sharp Inc.	USA	*642*	80-28000
Halmatic Ltd	UK	*456*	75-27700
Halter Marine Group	USA	*580*	75-41700
Hanjin Heavy Industries Co. Ltd	Korea (ROK)	*265*	75-01505
Hitachi Zosen Co. Ltd.	Japan	*251*	75-02390
Hong Leong-Lürssen Shipyard	Malaysia	*277*	75-00055
Humber Fabrications (Hull) Ltd.	UK	*459*	75-30580
INIMEX	**Belgium**	***46***	**75-04650**
Ishikawajima-Harima H.I. Co. Ltd	Japan	*251*	75-02800
Japan Aircraft Manufacturing Co. Ltd.	Japan	*252*	75-03200
Karlskronavarvet AB	Sweden	*377*	75-07400
Kvaerner Mandal A/S	Norway	*309*	75-01100
Magnum Marine Corp.	USA	*593*	75-57400
McTay Marine Limited	UK	*467*	75-38230
Napco International Inc.	USA	*598*	75-63600
NQEA Australia Pty Ltd	Australia	*19*	75-14890
Ocean Dynamics	UK	*471*	75-42900
Penang Shipbuilding Corporation	Malaysia	*277*	75-00800
Rodman Polyships SA	Spain	*367*	75-05000
RTK Marine	UK	*481*	75-50650
SeaArk Marine Inc.	USA	*613*	75-78900
Sillinger	France	*145*	75-35660
Souter Shipyard Ltd	UK	*485*	75-55212
Storebro Bruks AB	Sweden	*379*	75-10600
Transfield Shipbuilding WA	Australia	*22*	75-20812
Tremlett Boat Sales Ltd	UK	*489*	75-61075
TTL Dynamics	UK	*490*	75-61510

136 90 Coastguard, Customs & Police vessels / craft, other types

Avon Inflatables Ltd	UK	*432*	75-05100
Avondale Industries Inc.	USA	*556*	75-11400
Azimut SpA	Italy	*232*	75-02360
Blohm + Voss GmbH	Germany	*165*	75-02800
C. Raymond Hunt Associates Inc.	USA	*637*	80-14900
Cougar Limited	UK	*443*	75-15250
Cygnus Marine Ltd	UK	*444*	75-16500
Daewoo Heavy Industries Ltd	Korea (ROK)	*265*	75-00600
Danyard A/S	Denmark	*102*	75-04300
Emsworth Shipyard	UK	*448*	75-20750
FBM Marine Limited	UK	*450*	75-22485
George G. Sharp Inc.	USA	*642*	80-28000
Halmatic Ltd	UK	*456*	75-27700
Hong Leong-Lürssen Shipyard	Malaysia	*277*	75-00055
Humber Fabrications (Hull) Ltd	UK	*459*	75-30580
INIMEX	**Belgium**	***46***	**75-04650**
Intermarine SpA	Italy	*235*	75-11900
Israel Aircraft Industries Ltd	Israel	*219*	75-10200
Israel Shipyards Ltd	Israel	*220*	75-10600
Japan Aircraft Manufacturing Co. Ltd	Japan	*252*	75-03200
Kvaerner Mandal A/S	Norway	*309*	75-01100
Kvaerner Masa-Yards Inc	**Finland**	***118***	**75-01370**
Magnum Marine Corp.	USA	*593*	75-57400
McTay Marine Limited	UK	*467*	75-38230
Napco International Inc.	USA	*598*	75-63600
NQEA Australia Pty Ltd	Australia	*19*	75-14890
Penang Shipbuilding Corporation	Malaysia	*277*	75-00800
Rodman Polyships SA	Spain	*367*	75-05000
Sea Tiger Marine Pty Ltd	Australia	*21*	75-19350
Sillinger	France	*145*	75-35660
Souter Shipyard Ltd	UK	*485*	75-55212
Transfield Shipbuilding WA	Australia	*22*	75-20812
Tremlett Boat Sales Ltd	UK	*489*	75-61075
TTL Dynamics	UK	*490*	75-61510

137 Launches & other small craft

Aqua Bell Ltd	UK	*431*	75-03600
C. Raymond Hunt Associates Inc.	USA	*637*	80-14900
Cougar Limited	UK	*443*	75-15250
Cygnus Marine Ltd	UK	*444*	75-16500
Halmatic Ltd	UK	*456*	75-27700
Hitachi Zosen Co. Ltd.	Japan	*251*	75-02390
Hong Leong-Lürssen Shipyard	Malaysia	*277*	75-00055
Hyundai	Korea (ROK)	*265*	75-01540
INIMEX	**Belgium**	***46***	**75-04650**
Ishikawajima-Harima H.I. Co. Ltd	Japan	*251*	75-02800
Israel Aircraft Industries Ltd	Israel	*219*	75-10200
Israel Shipyards Ltd	Israel	*220*	75-10600
Malakoff & Wm. Moore Ltd	UK	*466*	75-37065
Marinette Marine Corp.	USA	*594*	75-58280
Nichols Brothers Boat Builders Inc.	USA	*599*	75-64380
Penang Shipbuilding Corporation	Malaysia	*277*	75-00800
SeaArk Marine Inc.	USA	*613*	75-78900
Stocznia Polnocna SA	Poland	*325*	75-02800
Tacoma Boatbuilding Co.	USA	*619*	75-86000
Tremlett Boat Sales Ltd	UK	*489*	75-61075
Vosper Thornycroft	UK	*492*	75-63800

137 40 Pilot boats

Aqua Bell Ltd	UK	*431*	75-03600
ASC Shipbuilding Pty Limited	Australia	*12*	75-01530
C. Raymond Hunt Associates Inc.	USA	*637*	80-14900
Emsworth Shipyard	UK	*448*	75-20750
FBM Marine Limited	UK	*450*	75-22485
Garden Reach Shipbuilders	India	*203*	55-00300
Halmatic Ltd	UK	*456*	75-27700
Halter Marine Group	USA	*580*	75-41700
Hitachi Zosen Co. Ltd.	Japan	*251*	75-02390
Hong Leong-Lürssen Shipyard	Malaysia	*277*	75-00055
INIMEX	**Belgium**	***46***	**75-04650**
Israel Shipyards Ltd	Israel	*220*	75-10600
McTay Marine Limited	UK	*467*	75-38230
Nichols Brothers Boat Builders Inc.	USA	*599*	75-64380
Penang Shipbuilding Corporation	Malaysia	*277*	75-00800
Rodman Polyships SA	Spain	*367*	75-05000

137 50 Workboats

Avon Inflatables Ltd	UK	*432*	75-05100
Azimut SpA	Italy	*232*	75-02360
Beaufort Air-Sea Equipment Ltd	UK	*433*	75-06100
Boston Whaler Inc.	USA	*560*	75-14760
C. Raymond Hunt Associates Inc.	USA	*637*	80-14900
Cougar Limited	UK	*443*	75-15250
Cygnus Marine Ltd	UK	*444*	75-16500
Daewoo Heavy Industries Ltd	Korea (ROK)	*265*	75-00600
Damen Shipyards	Netherlands	*292*	75-01500
Emsworth Shipyard	UK	*448*	75-20750
FBM Marine Limited	UK	*450*	75-22485
Garden Reach Shipbuilders	India	*203*	55-00300
Halmatic Ltd	UK	*456*	75-27700
Halter Marine Group	USA	*580*	75-41700
Hitachi Zosen Co. Ltd.	Japan	*251*	75-02390
Hong Leong-Lürssen Shipyard	Malaysia	*277*	75-00055
Humber Fabrications (Hull) Ltd	UK	*459*	75-30580
INIMEX	**Belgium**	***46***	**75-04650**
Malakoff & Wm. Moore Ltd	UK	*466*	75-37065
McTay Marine Limited	UK	*467*	75-38230
Nichols Brothers Boat Builders Inc.	USA	*599*	75-64380
Penang Shipbuilding Corporation	Malaysia	*277*	75-00800
Rodman Polyships SA	Spain	*367*	75-05000
RTK Marine	UK	*481*	75-50650
Sea Tiger Marine Pty Ltd	Australia	*21*	75-19350
Sillinger	France	*145*	75-35660
Souter Shipyard Ltd	UK	*485*	75-55212
Textron Marine & Land Systems	USA	*622*	75-88660
Transfield Shipbuilding WA	Australia	*22*	75-20812

139 Submarines, submersibles

Australian Submarine Corporation	Australia	*12*	75-01910
COSMOS SpA	Italy	*233*	75-05700
Fincantieri	Italy	*234*	75-09250
Howaldtswerke Deutsche Werft AG	Germany	*170*	75-13400
Int'l Submarine Engrg.	Canada	*73*	75-07780
Kockums Industries AB	Sweden	*377*	75-07500
Mitsubishi Heavy Industries Ltd	Japan	*253*	75-05800
Newport News Shipbuilding	USA	*599*	75-64300
RDM Submarines BV	Netherlands	*294*	75-05770
Sener Ingenieria y Sistemas SA	Spain	*367*	75-05200
Sperry Marine Inc.	USA	*617*	75-83250
STN ATLAS Elektronik GmbH	Germany	*165*	75-01150
Vickers Shipbuilding & Eng Ltd	UK	*491*	75-63100

139 10 Submarines

Australian Submarine Corporation	Australia	*12*	75-01910
Brazan SA	Spain	*364*	55-00140
Comex Pro.	France	*134*	75-10200
Daewoo Heavy Industries Ltd	Korea (ROK)	*265*	75-00600
Fincantieri	Italy	*234*	75-09250
Howaldtswerke Deutsche Werft AG	Germany	*170*	75-13400
Kockums Industries AB	Sweden	*377*	75-07500
Mazagon Dock Ltd	India	*204*	55-01200
Mitsubishi Heavy Industries Ltd	Japan	*253*	75-05800
Newport News Shipbuilding	USA	*599*	75-64300
PT PAL Indonesia (PERSERO)	Indonesia	*210*	75-00500
RDM Submarines BV	Netherlands	*294*	75-05770
Vickers Shipbuilding & Eng Ltd	UK	*491*	75-63100

139 10 10 Submarines, attack

Australian Submarine Corporation	Australia	*12*	75-01910
Comex Pro.	France	*134*	75-10200
Fincantieri	Italy	*234*	75-09250
Howaldtswerke Deutsche Werft AG	Germany	*170*	75-13400
INIMEX	**Belgium**	***46***	**75-04650**
Kockums Industries AB	Sweden	*377*	75-07500
Newport News Shipbuilding	USA	*599*	75-64300
RDM Submarines BV	Netherlands	*294*	75-05770
Vickers Shipbuilding & Eng Ltd	UK	*491*	75-63100

139 10 20 Submarines, ballistic missile

Newport News Shipbuilding	USA	*599*	75-64300
Vickers Shipbuilding & Eng Ltd	UK	*491*	75-63100

139 10 30 Submarines, light

Comex Pro.	France	*134*	75-10200
Howaldtswerke Deutsche Werft AG	Germany	*170*	75-13400
Int'l Submarine Engrg.	Canada	*73*	75-07780
Vickers Shipbuilding & Eng Ltd	UK	*491*	75-63100

139 10 40 Mini-submarines

Comex Pro.	France	*134*	75-10200
ECA SA	France	*136*	75-14050
Goa Shipyard Ltd	India	*203*	55-00400
Int'l Submarine Engrg.	Canada	*73*	75-07780
Ocean Dynamics	UK	*471*	75-42900
STN ATLAS Elektronik GmbH	Germany	*165*	75-01150
Vickers Shipbuilding & Eng Ltd	UK	*491*	75-63100

139 10 50 Submarines, rescue

Comex Pro.	France	*134*	75-10200
Hytech BV	Netherlands	*293*	75-03530
Kockums Industries AB	Sweden	*377*	75-07500
Slingsby Engineering Ltd	UK	*484*	75-54420
STN ATLAS Elektronik GmbH	Germany	*165*	75-01150
Vickers Shipbuilding & Eng Ltd	UK	*491*	75-63100

139 10 60 Submarines, transport

STN ATLAS Elektronik GmbH	Germany	*165*	75-01150
Vickers Shipbuilding & Eng Ltd	UK	*491*	75-63100

139 10 70 Submarines, wet (diver transport vehicles)

Comex Pro.	France	*134*	75-10200
COSMOS SpA.	Italy	*233*	75-05700
ECA SA	France	*136*	75-14050
Hytech BV	Netherlands	*293*	75-03530
Int'l Submarine Engrg.	Canada	*73*	75-07780
Slingsby Engineering Ltd	UK	*484*	75-54420
STN ATLAS Elektronik GmbH	Germany	*165*	75-01150

139 20 Submersibles

Bofors Underwater Systems AB	Sweden	*376*	75-03052
Deep Ocean Engineering Inc.	USA	*568*	75-24820
Int'l Submarine Engrg.	Canada	*73*	75-07780
Perry Technologies	USA	*603*	75-69360
Slingsby Engineering Ltd	UK	*484*	75-54420
STN ATLAS Elektronik GmbH	Germany	*165*	75-01150
Vickers Shipbuilding & Eng Ltd	UK	*491*	75-63100

139 20 50 Submersible vehicles, manned

Comex Pro.	France	*134*	75-10200
Deep Ocean Engineering Inc.	USA	*568*	75-24820
Slingsby Engineering Ltd	UK	*484*	75-54420
STN ATLAS Elektronik GmbH	Germany	*165*	75-01150
Vickers Shipbuilding & Eng Ltd	UK	*491*	75-63100

139 20 60 Submersible vehicles, remote-operated

Benthos Inc.	USA	*558*	75-13540
Bofors Underwater Systems AB	Sweden	*376*	75-03052
Comex Pro.	France	*134*	75-10200
Deep Ocean Engineering Inc.	USA	*568*	75-24820
ECA SA	France	*136*	75-14050
GKN Westland-Sitec GmbH	Germany	*169*	75-10900
Hydrovision Ltd	UK	*460*	75-30795
MacArtney A/S	Denmark	*103*	75-12800
Perry Technologies	USA	*603*	75-69360
Riva Calzoni SpA	Italy	*238*	75-20500
Slingsby Engineering Ltd	UK	*484*	75-54420
Soil Machine Dynamics Ltd	UK	*485*	75-54870
STN ATLAS Elektronik GmbH	Germany	*165*	75-01150
Thomson Marconi Sonar Limited	UK	*426*	74-20000
Vickers Shipbuilding & Eng Ltd	UK	*491*	75-63100

139 30 Seabed contact vehicles, remote-controlled

Benthos Inc.	USA	*558*	75-13540
Bofors Underwater Systems AB	Sweden	*376*	75-03052
Comex Pro.	France	*134*	75-10200
ECA SA	France	*136*	75-14050
GKN Westland-Sitec GmbH	Germany	*169*	75-10900
Int'l Submarine Engrg.	Canada	*73*	75-07780
Slingsby Engineering Ltd	UK	*484*	75-54420
Soil Machine Dynamics Ltd	UK	*485*	75-54870

Land-based Platforms

141 10 Vehicles, armoured, tracked

A Goninan & Co. Ltd	Australia	*16*	75-07760
Alvis Industries Ltd	UK	*430*	75-02850
Astra	Italy	*232*	75-01800
Cenzin Co. Ltd.	Poland	*324*	75-00200
CIC International Ltd	USA	*563*	75-18570
Eyal Engineering & Industrial Co.	Israel	*219*	75-07200
Giat Industries	France	*126*	55-00100
GKN Defence Ltd	**UK**	***454***	**75-25600**
INIMEX	**Belgium**	***46***	**75-04650**
Kaman Sciences Corp.	USA	*637*	80-17600
Kioleides SA	Greece	*190*	75-01982
Krauss-Maffei Wehrtechnik GmbH	Germany	*171*	75-16345
MaK Systemgesellschaft mbH.	Germany	*172*	75-19500
Mitsubishi Heavy Industries Ltd	Japan	*253*	75-05800
Napco International Inc.	USA	*598*	75-63600
Patria Vehicles Oy	Finland	*118*	75-01710
Reumech-OMC	S. Africa	*359*	75-18265
Royal Ordnance	UK	*481*	75-50620
Santa Barbara SA	Spain	*364*	55-01000
Sleeman Engineering	UK	*484*	75-54400
SP aerospace and vehicle systems bv	**Netherlands**	***295***	**75-06860**
Steyr-Daimler-Puch Spezialfahrzeug	Austria	*34*	75-02810
Swiss Federal Armament Works	Switzerland	*384*	55-00250
Thyssen Henschel	Germany	*175*	75-31000
Vickers Defence Systems	UK	*491*	75-63040
Vickers Shipbuilding & Eng Ltd	UK	*491*	75-63100

141 10 05 Tanks (armoured fighting vehicles)

Alvis Industries Ltd	UK	*430*	75-02850
Cenzin Co. Ltd.	Poland	*324*	75-00200
Giat Industries	France	*126*	55-00100
INIMEX	**Belgium**	***46***	**75-04650**
Kaman Sciences Corp.	USA	*637*	80-17600
Krauss-Maffei Wehrtechnik GmbH	Germany	*171*	75-16345
Mitsubishi Heavy Industries Ltd	Japan	*253*	75-05800
Napco International Inc.	USA	*598*	75-63600
NORINCO	China PR	*88*	55-00280
OTO Breda	Italy	*237*	75-18100
Royal Ordnance	UK	*481*	75-50620
Santa Barbara SA	Spain	*364*	55-01000
Swiss Federal Armament Works	Switzerland	*384*	55-00250
Thyssen Henschel	Germany	*175*	75-31000
United Defense L.P.	USA	*624*	75-92050
Vickers Defence Systems	UK	*491*	75-63040
ZTS TEES Martin s.p.	Slovakia	*353*	75-01500

141 10 05 10 Tanks, heavy

Giat Industries	France	*126*	55-00100
INIMEX	**Belgium**	***46***	**75-04650**
Krauss-Maffei Wehrtechnik GmbH	Germany	*171*	75-16345
Napco International Inc.	USA	*598*	75-63600
Royal Ordnance	UK	*481*	75-50620
Swiss Federal Armament Works	Switzerland	*384*	55-00250
Vickers Defence Systems	UK	*491*	75-63040
ZTS TEES Martin s.p.	Slovakia	*353*	75-01500

141 10 05 20 Tanks, light

Alvis Industries Ltd	UK	*430*	75-02850
CIC International Ltd	USA	*563*	75-18570
GKN Defence Ltd	**UK**	***454***	**75-25600**
INIMEX	**Belgium**	***46***	**75-04650**

Krauss-Maffei Wehrtechnik GmbH	Germany	*171*	75-16345
Napco International Inc.	USA	*598*	75-63600
Royal Ordnance	UK	*481*	75-50620
Thyssen Henschel	Germany	*175*	75-31000
United Defense L.P.	USA	*624*	75-92050
Vickers Defence Systems	UK	*491*	75-63040

141 10 05 30 Tanks, medium

Bernardini SA	Brazil	*55*	75-02120
Giat Industries	France	*126*	55-00100
INIMEX	**Belgium**	***46***	**75-04650**
Krauss-Maffei Wehrtechnik GmbH	Germany	*171*	75-16345
Mitsubishi Heavy Industries Ltd	Japan	*253*	75-05800
Napco International Inc.	USA	*598*	75-63600
Reumech-OMC	S. Africa	*359*	75-18265
Swiss Federal Armament Works	Switzerland	*384*	55-00250
Thyssen Henschel	Germany	*175*	75-31000
United Defense L.P.	USA	*624*	75-92050
Vickers Defence Systems	UK	*491*	75-63040

141 10 10 Anti-tank vehicles, armoured

Alvis Industries Ltd	UK	*430*	75-02850
General Motors of Canada Ltd	Canada	*71*	75-05900
Giat Industries	France	*126*	55-00100
GKN Defence Ltd	**UK**	***454***	**75-25600**
INIMEX	**Belgium**	***46***	**75-04650**
Krauss-Maffei Wehrtechnik GmbH	Germany	*171*	75-16345
Napco International Inc.	USA	*598*	75-63600
Royal Ordnance	UK	*481*	75-50620
SP aerospace and vehicle systems bv	**Netherlands**	***295***	**75-06860**
Vickers Shipbuilding & Eng Ltd	UK	*491*	75-63100

141 10 15 Anti-aircraft vehicles, armoured tracked

Giat Industries	France	*126*	55-00100
GKN Defence Ltd	**UK**	***454***	**75-25600**
INIMEX	**Belgium**	***46***	**75-04650**
Napco International Inc.	USA	*598*	75-63600
Oerlikon-Contraves AG.	Switzerland	*387*	75-09250
OTO Breda	Italy	*237*	75-18100
Vickers Shipbuilding & Eng Ltd	UK	*491*	75-63100

141 10 20 Infantry fighting vehicles, armoured tracked

A Goninan & Co. Ltd	Australia	*16*	75-07760
Alvis Industries Ltd	UK	*430*	75-02850
Giat Industries	France	*126*	55-00100
GKN Defence Ltd	**UK**	***454***	**75-25600**
Hägglunds Moelv AS	Norway	*309*	75-00880
Hägglunds Vehicle AB	Sweden	*377*	75-05800
INIMEX	**Belgium**	***46***	**75-04650**
Krauss-Maffei Wehrtechnik GmbH	Germany	*171*	75-16345
Napco International Inc.	USA	*598*	75-63600
NORINCO	China PR	*88*	55-00280
OTO Breda	Italy	*237*	75-18100
SP aerospace and vehicle systems bv	**Netherlands**	***295***	**75-06860**
Thyssen Henschel	Germany	*175*	75-31000
United Defense L.P.	USA	*624*	75-92050
Vickers Shipbuilding & Eng Ltd	UK	*491*	75-63100

141 10 25 Personnel carriers, armoured tracked

A Goninan & Co. Ltd	Australia	*16*	75-07760
Alvis Industries Ltd	UK	*430*	75-02850
Giat Industries	France	*126*	55-00100
GKN Defence Ltd	**UK**	***454***	**75-25600**
Hägglunds Vehicle AB	Sweden	*377*	75-05800
Hellenic Vehicle Industry SA	Greece	*189*	75-01780
INIMEX	**Belgium**	***46***	**75-04650**
Krauss-Maffei Wehrtechnik GmbH	Germany	*171*	75-16345
Napco International Inc.	USA	*598*	75-63600
OTO Breda	Italy	*237*	75-18100
Royal Ordnance	UK	*481*	75-50620
Santa Barbara SA	Spain	*364*	55-01000
Sleeman Engineering	UK	*484*	75-54400
SP aerospace and vehicle systems bv	**Netherlands**	***295***	**75-06860**
Suspension & Parts Industries Ltd	Israel	*222*	75-21320
United Defense L.P.	USA	*624*	75-92050
Vickers Shipbuilding & Eng Ltd	UK	*491*	75-63100

141 10 30 Mortar carriers, armoured tracked

Alvis Industries Ltd	UK	*430*	75-02850
GKN Defence Ltd	**UK**	***454***	**75-25600**
Hägglunds Moelv AS	Norway	*309*	75-00880
Hägglunds Vehicle AB	Sweden	*377*	75-05800
INIMEX	**Belgium**	***46***	**75-04650**
Krauss-Maffei Wehrtechnik GmbH	Germany	*171*	75-16345
Lohr Industrie	France	*139*	75-25200
Napco International Inc.	USA	*598*	75-63600
Royal Ordnance	UK	*481*	75-50620
SP aerospace and vehicle systems bv	**Netherlands**	***295***	**75-06860**
United Defense L.P.	USA	*624*	75-92050
Vickers Shipbuilding & Eng Ltd	UK	*491*	75-63100

141 10 35 Ambulances, armoured tracked

Alvis Industries Ltd	UK	*430*	75-02850
Giat Industries	France	*126*	55-00100
GKN Defence Ltd	**UK**	***454***	**75-25600**
Hägglunds Vehicle AB	Sweden	*377*	75-05800
INIMEX	**Belgium**	***46***	**75-04650**
Krauss-Maffei Wehrtechnik GmbH	Germany	*171*	75-16345
Napco International Inc.	USA	*598*	75-63600
NORINCO	China PR	*88*	55-00280
Santa Barbara SA	Spain	*364*	55-01000
SP aerospace and vehicle systems bv	**Netherlands**	***295***	**75-06860**
Thyssen Henschel	Germany	*175*	75-31000
United Defense L.P.	USA	*624*	75-92050
Vickers Shipbuilding & Eng Ltd	UK	*491*	75-63100

141 10 40 Command post / radio vehicles, armoured tracked

Alvis Industries Ltd	UK	*430*	75-02850
Giat Industries	France	*126*	55-00100
GKN Defence Ltd	**UK**	***454***	**75-25600**
Hägglunds Vehicle AB	Sweden	*377*	75-05800
Harris Corporation	USA	*581*	75-42050
INIMEX	**Belgium**	***46***	**75-04650**
Krauss-Maffei Wehrtechnik GmbH	Germany	*171*	75-16345
Napco International Inc.	USA	*598*	75-63600
SP aerospace and vehicle systems bv	**Netherlands**	***295***	**75-06860**
United Defense L.P.	USA	*624*	75-92050
Vickers Shipbuilding & Eng Ltd	UK	*491*	75-63100

141 10 45 Recovery vehicles, armoured tracked

Alvis Industries Ltd	UK	*430*	75-02850
Eyal Engineering & Industrial Co.	Israel	*219*	75-07200
Giat Industries	France	*126*	55-00100
GKN Defence Ltd	**UK**	***454***	**75-25600**
Hägglunds Vehicle AB	Swéden	*377*	75-05800
INIMEX	**Belgium**	***46***	**75-04650**
Jung Jungenthal GmbH	Germany	*170*	75-15130
Krauss-Maffei Wehrtechnik GmbH	Germany	*171*	75-16345
Mitsubishi Heavy Industries Ltd	Japan	*253*	75-05800
Napco International Inc.	USA	*598*	75-63600
NORINCO	China PR	*88*	55-00280
Sleeman Engineering	UK	*484*	75-54400
SP aerospace and vehicle systems bv	**Netherlands**	***295***	**75-06860**
United Defense L.P.	USA	*624*	75-92050
Vickers Defence Systems	UK	*491*	75-63040
VOP 025 Novy Jicín, s.p.	Czech Republic	*98*	55-05000
ZTS TEES Martin s.p.	Slovakia	*353*	75-01500

141 10 50 Workshop vehicles, armoured tracked

GKN Defence Ltd	**UK**	***454***	**75-25600**
INIMEX	**Belgium**	***46***	**75-04650**

Napco International Inc.	USA	598	75-63600
Sleeman Engineering	UK	484	75-54400
United Defense L.P.	USA	624	75-92050
Vickers Defence Systems	UK	491	75-63040
Vickers Shipbuilding & Eng Ltd	UK	491	75-63100

141 10 55 Bridging vehicles, armoured tracked (see code 627.10.05)

CNIM	France	133	75-09990
Eyal Engineering & Industrial Co.	Israel	219	75-07200
INIMEX	**Belgium**	**46**	**75-04650**
MaK Systemgesellschaft mbH	Germany	172	75-19500
Napco International Inc.	USA	598	75-63600
NORINCO	China PR	88	55-00280
Royal Ordnance	UK	481	75-50620
United Defense L.P.	USA	624	75-92050
Vickers Defence Systems	UK	491	75-63040
ZTS TEES Martin s.p.	Slovakia	353	75-01500

141 10 60 Combat engineer vehicles, armoured tracked

Eyal Engineering & Industrial Co.	Israel	219	75-07200
Giat Industries	France	126	55-00100
INIMEX	**Belgium**	**46**	**75-04650**
Krauss-Maffei Wehrtechnik GmbH	Germany	171	75-16345
Napco International Inc.	USA	598	75-63600
Royal Ordnance	UK	481	75-50620
SP aerospace and vehicle systems bv	**Netherlands**	**295**	**75-06860**
United Defense L.P.	USA	624	75-92050
Vickers Defence Systems	UK	491	75-63040

141 10 65 Logistics / cargo carrier vehicles, armoured tracked

Alvis Industries Ltd	UK	430	75-02850
GKN Defence Ltd	**UK**	**454**	**75-25600**
Hägglunds Vehicle AB	Sweden	377	75-05800
INIMEX	**Belgium**	**46**	**75-04650**
Napco International Inc.	USA	598	75-63600
NORINCO	China PR	88	55-00280
Sleeman Engineering	UK	484	75-54400
SP aerospace and vehicle systems bv	**Netherlands**	**295**	**75-06860**
United Defense L.P.	USA	624	75-92050
Vickers Shipbuilding & Eng Ltd	UK	491	75-63100

141 10 70 Prime movers, armoured tracked

Napco International Inc.	USA	598	75-63600
SP aerospace and vehicle systems bv	**Netherlands**	**295**	**75-06860**

141 10 75 Very light vehicles, armoured tracked

Hägglunds Vehicle AB	Sweden	377	75-05800
Lohr Industrie	France	139	75-25200
Napco International Inc.	USA	598	75-63600
Patria Vehicles Oy	Finland	118	75-01710
Sleeman Engineering	UK	484	75-54400
Thyssen Henschel	Germany	175	75-31000

141 20 Vehicles, armoured, half-tracked

Eyal Engineering & Industrial Co.	Israel	219	75-07200
Napco International Inc.	USA	598	75-63600
Precimetal Fonderie de Précision SA	Belgium	46	75-05500

141 20 25 Personnel carriers, armoured half-tracked

Napco International Inc.	USA	598	75-63600
Nottingham International	USA	640	80-22000
Sleeman Engineering	UK	484	75-54400

141 20 28 Reconnaissance vehicles, armoured half-tracked

Napco International Inc.	USA	598	75-63600
Nottingham International	USA	640	80-22000

141 20 30 Weapon carriers, armoured half-tracked

Napco International Inc.	USA	598	75-63600
Nottingham International	USA	640	80-22000

141 20 35 Ambulances, armoured half-tracked

Napco International Inc.	USA	598	75-63600
Nottingham International	USA	640	80-22000

141 20 50 Other vehicles, armoured half-tracked

Eyal Engineering & Industrial Co.	Israel	219	75-07200
Napco International Inc.	USA	598	75-63600
Nottingham International	USA	640	80-22000

141 30 Vehicles, armoured, wheeled

AM General Corporation	USA	552	75-05110
Arab Organization for Ind.	Egypt	109	55-01700
Avibras Aeroespacial SA	Brazil	55	75-00968
Courtaulds Aerospace	UK	443	75-15330
Emergency One Inc.	USA	573	75-31100
Engineering Equipment Co.	USA	574	75-31600
Giat Industries	France	126	55-00100
GKN Defence Ltd	**UK**	**454**	**75-25600**
INIMEX	**Belgium**	**46**	**75-04650**
IVECO Fiat SpA	Italy	235	75-13230
Kader Factory for Developed Ind.	Egypt	109	55-01700
MaK Systemgesellschaft mbH	Germany	172	75-19500
MOWAG	Switzerland	387	75-08900
Napco International Inc.	USA	598	75-63600
Panhard & Levassor SA	France	140	75-28750
Patria Vehicles Oy	Finland	118	75-01710
Penman Engineering Ltd	**UK**	**473**	**75-44400**
Protection Development Int'l Corp.	USA	606	75-71670
Q-Systems International Inc.	USA	606	75-71960
Reumech-OMC	S. Africa	359	75-18265
Royal Ordnance	UK	481	75-50620
SP aerospace and vehicle systems bv	**Netherlands**	**295**	**75-06860**
Steyr-Daimler-Puch Spezialfahrzeug	Austria	34	75-02810
TECTRAN	Brazil	58	75-20205
Textron Marine & Land Systems	USA	622	75-88660
Thyssen Henschel	Germany	175	75-31000
Vickers Shipbuilding & Eng Ltd	UK	491	75-63100

141 30 02 Vehicles, armoured, wheeled, amphibious

Alvis Industries Ltd	UK	430	75-02850
AV Technology LLC	USA	556	75-11405
Carrozzeria Boneschi Sarl	Italy	233	75-05000
General Motors of Canada Ltd	Canada	71	75-05900
Giat Industries	France	126	55-00100
GKN Defence Ltd	**UK**	**454**	**75-25600**
INIMEX	**Belgium**	**46**	**75-04650**
Napco International Inc.	USA	598	75-63600
Patria Vehicles Oy	Finland	118	75-01710
Thyssen Henschel	Germany	175	75-31000

141 30 10 Anti-tank vehicles, armoured wheeled

AV Technology LLC	USA	556	75-11405
General Motors of Canada Ltd	Canada	71	75-05900
Giat Industries	France	126	55-00100
GKN Defence Ltd	**UK**	**454**	**75-25600**
INIMEX	**Belgium**	**46**	**75-04650**
Israel Aircraft Industries Ltd	Israel	219	75-10200
Iveco Pegaso SA	Spain	364	55-00250
Napco International Inc.	USA	598	75-63600
Panhard & Levassor SA	France	140	75-28750
Renault Véhicules Industriels SA	France	142	75-32300

Reumech-OMC	S. Africa	*359*	75-18265
Royal Ordnance	UK	*481*	75-50620

141 30 15 Anti-aircraft vehicles, armoured wheeled

AV Technology LLC	USA	*556*	75-11405
General Motors of Canada Ltd	Canada	*71*	75-05900
GKN Defence Ltd	**UK**	***454***	**75-25600**
INIMEX	**Belgium**	***46***	**75-04650**
Krauss-Maffei Wehrtechnik GmbH	Germany	*171*	75-16345
Napco International Inc.	USA	*598*	75-63600
Nottingham International	USA	*640*	80-22000
Oerlikon-Contraves AG	Switzerland	*387*	75-09250
OTO Breda	Italy	*237*	75-18100
Panhard & Levassor SA	France	*140*	75-28750
Santa Barbara SA	Spain	*364*	55-01000

141 30 18 Fire support vehicles, armoured wheeled

AV Technology LLC	USA	*556*	75-11405
General Motors of Canada Ltd	Canada	*71*	75-05900
Giat Industries	France	*126*	55-00100
GKN Defence Ltd	**UK**	***454***	**75-25600**
INIMEX	**Belgium**	***46***	**75-04650**
Iveco Pegaso SA	Spain	*364*	55-00250
Napco International Inc.	USA	*598*	75-63600
Panhard & Levassor SA	France	*140*	75-28750
Penman Engineering Ltd	**UK**	***473***	**75-44400**
Protection Development Int'l Corp.	USA	*606*	75-71670
Reumech-OMC	S. Africa	*359*	75-18265
Thyssen Henschel	Germany	*175*	75-31000

141 30 20 Infantry fighting vehicles, armoured wheeled

AV Technology LLC	USA	*556*	75-11405
General Motors of Canada Ltd	Canada	*71*	75-05900
Giat Industries	France	*126*	55-00100
GKN Defence Ltd	**UK**	***454***	**75-25600**
INIMEX	**Belgium**	***46***	**75-04650**
Israel Aircraft Industries Ltd	Israel	*219*	75-10200
Iveco Pegaso SA	Spain	*364*	55-00250
Napco International Inc.	USA	*598*	75-63600
NORINCO	China PR	*88*	55-00280
Nottingham International	USA	*640*	80-22000
Panhard & Levassor SA	France	*140*	75-28750
Protection Development Int'l Corp.	USA	*606*	75-71670
Reumech-OMC	S. Africa	*359*	75-18265
Thyssen Henschel	Germany	*175*	75-31000
Vickers Shipbuilding & Eng Ltd	UK	*491*	75-63100

141 30 25 Personnel carriers, armoured wheeled

Arab Organization for Ind.	Egypt	*109*	55-01700
Asia Motors Co Ltd	Korea (ROK)	*265*	75-00100
AV Technology LLC	USA	*556*	75-11405
Avibras Aeroespacial SA	Brazil	*55*	75-00968
Carrozzeria Boneschi Sarl	Italy	*233*	75-05000
General Motors of Canada Ltd	Canada	*71*	75-05900
Giat Industries	France	*126*	55-00100
GKN Defence Ltd	**UK**	***454***	**75-25600**
Glover Webb Ltd	UK	*455*	75-25700
INIMEX	**Belgium**	***46***	**75-04650**
Israel Aircraft Industries Ltd	Israel	*219*	75-10200
Iveco Pegaso SA	Spain	*364*	55-00250
Kader Factory for Developed Ind.	Egypt	*109*	55-01700
MOWAG	Switzerland	*387*	75-08900
Napco International Inc.	USA	*598*	75-63600
OTO Breda	Italy	*237*	75-18100
Panhard & Levassor SA	France	*140*	75-28750
Patria Vehicles Oy	Finland	*118*	75-01710
Penman Engineering Ltd	**UK**	***473***	**75-44400**
Protection Development Int'l Corp.	USA	*606*	75-71670
Reumech-OMC	S. Africa	*359*	75-18265
Santa Barbara SA	Spain	*364*	55-01000
Sleeman Engineering	UK	*484*	75-54400
Suspension & Parts Industries Ltd	Israel	*222*	75-21320
Vickers Defence Systems	UK	*491*	75-63040
Vickers Shipbuilding & Eng Ltd	UK	*491*	75-63100

141 30 28 Reconnaissance / scout cars, armoured wheeled

Alvis Industries Ltd	UK	*430*	75-02850
Arab Organization for Ind.	Egypt	*109*	55-01700
AV Technology LLC	USA	*556*	75-11405
General Motors of Canada Ltd	Canada	*71*	75-05900
Giat Industries	France	*126*	55-00100
GKN Defence Ltd	**UK**	***454***	**75-25600**
Glover Webb Ltd	UK	*455*	75-25700
INIMEX	**Belgium**	***46***	**75-04650**
Israel Aircraft Industries Ltd	Israel	*219*	75-10200
Iveco Pegaso SA	Spain	*364*	55-00250
Lohr Industrie	France	*139*	75-25200
MOWAG	Switzerland	*387*	75-08900
Napco International Inc.	USA	*598*	75-63600
Panhard & Levassor SA	France	*140*	75-28750
Protection Development Int'l Corp.	USA	*606*	75-71670
Reumech-OMC	S. Africa	*359*	75-18265
Royal Ordnance	UK	*481*	75-50620
Santa Barbara SA	Spain	*364*	55-01000
Vickers Shipbuilding & Eng Ltd	UK	*491*	75-63100

141 30 30 Mortar carriers, armoured wheeled

AV Technology LLC	USA	*556*	75-11405
General Motors of Canada Ltd	Canada	*71*	75-05900
GKN Defence Ltd	**UK**	***454***	**75-25600**
INIMEX	**Belgium**	***46***	**75-04650**
Iveco Pegaso SA	Spain	*364*	55-00250
Lohr Industrie	France	*139*	75-25200
Napco International Inc.	USA	*598*	75-63600
Panhard & Levassor SA	France	*140*	75-28750
Reumech-OMC	S. Africa	*359*	75-18265
Santa Barbara SA	Spain	*364*	55-01000
Vickers Shipbuilding & Eng Ltd	UK	*491*	75-63100

141 30 35 Ambulances, armoured wheeled

AV Technology LLC	USA	*556*	75-11405
Carrozzeria Boneschi Sarl	Italy	*233*	75-05000
Empl Fahrzeugwerk GesmbH	Austria	*32*	75-00920
General Motors of Canada Ltd	Canada	*71*	75-05900
Giat Industries	France	*126*	55-00100
GKN Defence Ltd	**UK**	***454***	**75-25600**
Glover Webb Ltd	UK	*455*	75-25700
Israel Aircraft Industries Ltd	Israel	*219*	75-10200
Iveco Pegaso SA	Spain	*364*	55-00250
Napco International Inc.	USA	*598*	75-63600
Panhard & Levassor SA	France	*140*	75-28750
Patria Vehicles Oy	Finland	*118*	75-01710
Penman Engineering Ltd	**UK**	***473***	**75-44400**
Reumech-OMC	S. Africa	*359*	75-18265
Thyssen Henschel	Germany	*175*	75-31000
Vickers Shipbuilding & Eng Ltd	UK	*491*	75-63100
Wheeled Coach Industries	USA	*627*	75-96350

141 30 38 Battlefield observation / surveillance vehicles, armoured wheeled

Automotive Industries Ltd	Israel	*217*	75-02500
AV Technology LLC	USA	*556*	75-11405
CIC International Ltd	USA	*563*	75-18570
General Motors of Canada Ltd	Canada	*71*	75-05900
Giat Industries	France	*126*	55-00100
GKN Defence Ltd	**UK**	***454***	**75-25600**
INIMEX	**Belgium**	***46***	**75-04650**
Israel Aircraft Industries Ltd	Israel	*219*	75-10200
Napco International Inc.	USA	*598*	75-63600
Panhard & Levassor SA	France	*140*	75-28750
Protection Development Int'l Corp.	USA	*606*	75-71670
Vickers Shipbuilding & Eng Ltd	UK	*491*	75-63100

141 30 40 Command post / radio vehicles, armoured wheeled

Arab Organization for Ind.	Egypt	*109*	55-01700
AV Technology LLC	USA	*556*	75-11405
Avibras Aeroespacial SA	Brazil	*55*	75-00968
Carrozzeria Boneschi Sarl	Italy	*233*	75-05000
CIC International Ltd	USA	*563*	75-18570

General Motors of Canada Ltd	Canada	*71*	75-05900
Giat Industries	France	*126*	55-00100
GKN Defence Ltd	**UK**	***454***	**75-25600**
Glover Webb Ltd	UK	*455*	75-25700
Harris Corporation	USA	*581*	75-42050
INIMEX	**Belgium**	***46***	**75-04650**
Israel Aircraft Industries Ltd	Israel	*219*	75-10200
Iveco Pegaso SA	Spain	*364*	55-00250
Lohr Industrie	France	*139*	75-25200
MOWAG	Switzerland	*387*	75-08900
Napco International Inc.	USA	*598*	75-63600
Panhard & Levassor SA	France	*140*	75-28750
Patria Vehicles Oy	Finland	*118*	75-01710
Penman Engineering Ltd	**UK**	***473***	**75-44400**
Protection Development Int'l Corp.	USA	*606*	75-71670
Reumech-OMC	S. Africa	*359*	75-18265
Vickers Shipbuilding & Eng Ltd	UK	*491*	75-63100

141 30 45 Recovery vehicles, armoured wheeled

AV Technology LLC	USA	*556*	75-11405
General Motors of Canada Ltd	Canada	*71*	75-05900
GKN Defence Ltd	**UK**	***454***	**75-25600**
Iveco Pegaso SA	Spain	*364*	55-00250
Napco International Inc.	USA	*598*	75-63600
Vickers Shipbuilding & Eng Ltd	UK	*491*	75-63100

141 30 50 Workshop vehicles, wheeled armoured

AV Technology LLC	USA	*556*	75-11405
General Motors of Canada Ltd	Canada	*71*	75-05900
Glover Webb Ltd	UK	*455*	75-25700
Napco International Inc.	USA	*598*	75-63600
Panhard & Levassor SA	France	*140*	75-28750
Vickers Shipbuilding & Eng Ltd	UK	*491*	75-63100

141 30 55 Bridging vehicles, armoured wheeled

CNIM	France	*133*	75-09990
Lohr Industrie	France	*139*	75-25200
Napco International Inc.	USA	*598*	75-63600

141 30 59 Vehicles, armoured, wheeled, utility

AV Technology LLC	USA	*556*	75-11405
Avibras Aeroespacial SA	Brazil	*55*	75-00968
General Motors of Canada Ltd	Canada	*71*	75-05900
GKN Defence Ltd	**UK**	***454***	**75-25600**
Israel Aircraft Industries Ltd	Israel	*219*	75-10200
MOWAG	Switzerland	*387*	75-08900
Napco International Inc.	USA	*598*	75-63600
Penman Engineering Ltd	**UK**	***473***	**75-44400**
Protection Development Int'l Corp.	USA	*606*	75-71670
Reumech-OMC	S. Africa	*359*	75-18265
Royal Ordnance	UK	*481*	75-50620
Vehicle Systems Development Corp.	USA	*625*	75-93570

141 30 65 Logistics / cargo carriers, armoured wheeled

Asia Motors Co Ltd	Korea (ROK)	*265*	75-00100
AV Technology LLC	USA	*556*	75-11405
General Motors of Canada Ltd	Canada	*71*	75-05900
Israel Aircraft Industries Ltd	Israel	*219*	75-10200
Iveco Pegaso SA	Spain	*364*	55-00250
Napco International Inc.	USA	*598*	75-63600
NORINCO	China PR	*88*	55-00280
Panhard & Levassor SA	France	*140*	75-28750

141 30 70 Gendarmerie vehicles, armoured wheeled

Arab Organization for Ind.	Egypt	*109*	55-01700
AV Technology LLC	USA	*556*	75-11405
Avibras Aeroespacial SA	Brazil	*55*	75-00968
Carrozzeria Boneschi Sarl	Italy	*233*	75-05000
Emergency One Inc.	USA	*573*	75-31100
Glover Webb Ltd	UK	*455*	75-25700
Israel Aircraft Industries Ltd	Israel	*219*	75-10200
Iveco Pegaso SA	Spain	*364*	55-00250
Lohr Industrie	France	*139*	75-25200
Napco International Inc.	USA	*598*	75-63600
Panhard & Levassor SA	France	*140*	75-28750
Penman Engineering Ltd	**UK**	***473***	**75-44400**
Reumech-OMC	S. Africa	*359*	75-18265

141 30 72 Internal security vehicles, armoured wheeled

AV Technology LLC	USA	*556*	75-11405
Emergency One Inc.	USA	*573*	75-31100
Giat Industries	France	*126*	55-00100
GKN Defence Ltd	**UK**	***454***	**75-25600**
Glover Webb Ltd	UK	*455*	75-25700
Israel Aircraft Industries Ltd	Israel	*219*	75-10200
Iveco Pegaso SA	Spain	*364*	55-00250
Lohr Industrie	France	*139*	75-25200
Napco International Inc.	USA	*598*	75-63600
Panhard & Levassor SA	France	*140*	75-28750
Protection Development Int'l Corp.	USA	*606*	75-71670
Reumech-OMC	S. Africa	*359*	75-18265
Royal Ordnance	UK	*481*	75-50620
Sleeman Engineering	UK	*484*	75-54400
Vehicle Systems Development Corp.	USA	*625*	75-93570

141 30 75 Armoured vehicles, wheeled, special purpose, non-military

AV Technology LLC	USA	*556*	75-11405
Carrozzeria Boneschi Sarl	Italy	*233*	75-05000
Engineering Equipment Co.	USA	*574*	75-31600
Giat Industries	France	*126*	55-00100
Glover Webb Ltd	UK	*455*	75-25700
Israel Aircraft Industries Ltd	Israel	*219*	75-10200
Napco International Inc.	USA	*598*	75-63600
O'Gara-Hess & Eisenhardt Armoring	USA	*601*	75-66400
Penman Engineering Ltd	**UK**	***473***	**75-44400**
Protection Development Int'l Corp.	USA	*606*	75-71670
Repetti & Montiglio SpA	Italy	*238*	75-20120
Vehicle Systems Development Corp.	USA	*625*	75-93570

141 30 75 50 Armoured limousines

Carrozzeria Boneschi Sarl	Italy	*233*	75-05000
Glover Webb Ltd	UK	*455*	75-25700
Mondiale Security Ltd	UK	*469*	75-40548
Napco International Inc.	USA	*598*	75-63600
O'Gara-Hess & Eisenhardt Armoring	USA	*601*	75-66400
Protection Development Int'l Corp.	USA	*606*	75-71670
Silent Partner Body Armor Inc.	USA	*615*	75-80930
Société Labbe SA	France	*146*	75-40500
Vehicle Systems Development Corp.	USA	*625*	75-93570

141 30 75 70 Armoured security / bullion vans, wheeled

Bedwas Bodyworks	UK	*434*	75-06140
Carrozzeria Boneschi Sarl	Italy	*233*	75-05000
Glover Webb Ltd	UK	*455*	75-25700
GM-Holden's Automotive	Australia	*16*	75-07520
Lightweight Body Armour	UK	*464*	75-35650
Napco International Inc.	USA	*598*	75-63600
O'Gara-Hess & Eisenhardt Armoring	USA	*601*	75-66400
Penman Engineering Ltd	**UK**	***473***	**75-44400**
Protection Development Int'l Corp.	USA	*606*	75-71670
Silent Partner Body Armor Inc.	USA	*615*	75-80930
Sleeman Engineering	UK	*484*	75-54400
Société Labbe SA	France	*146*	75-40500
Vehicle Systems Development Corp.	USA	*625*	75-93570

141 30 79 Vehicles, wheeled armoured, other types

Bedwas Bodyworks	UK	*434*	75-06140
CCS International Ltd	USA	*562*	75-17250
Emergency One Inc.	USA	*573*	75-31100
General Motors of Canada Ltd	Canada	*71*	75-05900
Giat Industries	France	*126*	55-00100
GKN Defence Ltd	**UK**	***454***	**75-25600**
Glover Webb Ltd	UK	*455*	75-25700
Lightweight Body Armour	UK	*464*	75-35650
Napco International Inc.	USA	*598*	75-63600
OTO Breda	Italy	*237*	75-18100
Panhard & Levassor SA	France	*140*	75-28750

Penman Engineering Ltd	**UK**	***473***	**75-44400**
Protection Development Int'l Corp.	USA	*606*	75-71670

141 50 Vehicles, non-armoured, tracked

Bombardier Inc.	Canada	*69*	75-01400
Foremost Industries Inc.	Canada	*71*	75-05620
Giat Industries	France	*126*	55-00100
Hägglunds Moelv AS	Norway	*309*	75-00880
Hägglunds Vehicle AB	Sweden	*377*	75-05800
M.A. Bigland (Preparations) Ltd	UK	*434*	75-06700
Napco International Inc.	USA	*598*	75-63600
O'Gara-Hess & Eisenhardt Armoring	USA	*601*	75-66400

141 50 10 Vehicles, over-snow, tracked

Bombardier Inc.	Canada	*69*	75-01400
Foremost Industries Inc.	Canada	*71*	75-05620
Hägglunds Moelv AS	Norway	*309*	75-00880
Hägglunds Vehicle AB	Sweden	*377*	75-05800
Napco International Inc.	USA	*598*	75-63600
Ontario Drive & Gear Ltd.	Canada	*74*	75-10400
Patria Vehicles Oy	Finland	*118*	75-01710

141 60 Vehicles, non-armoured, half-tracked

Napco International Inc.	USA	*598*	75-63600

141 70 Vehicles, non-armoured, wheeled

Alvis Unipower Ltd	UK	*430*	75-02863
AM General Corporation	USA	*552*	75-05110
Arab American Vehicles Co.	Egypt	*109*	74-00100
Automobiles Peugeot	France	*140*	75-29600
Automotive Industries Ltd	Israel	*217*	75-02500
Auverland/Sovamag	France	*132*	75-04520
Aveling Barford (Machines) plc	UK	*432*	75-04550
Bartlett Lifting Devices Inc.	USA	*557*	75-12060
Bedford Trucks	UK	*434*	75-06135
BMY-Wheeled Vehicles	USA	*560*	75-14855
Bombardier Inc.	Canada	*69*	75-01400
Bucher-Guyer Ltd, Vehicles	Switzerland	*385*	75-01580
Carmichael International Ltd	UK	*438*	75-11700
Daimler-Benz AG	**Germany**	***166***	**75-05600**
Eagle Industrial Truck Mfg. Inc.	USA	*570*	75-28430
Emergency One Inc.	USA	*573*	75-31100
Foremost Industries Inc.	Canada	*71*	75-05620
FWD International Inc.	USA	*577*	75-36680
Giat Industries	France	*126*	55-00100
Gontrailer SL	Spain	*367*	75-05660
Hellenic Vehicle Industry SA	Greece	*189*	75-01780
Heller Truck Body Corp.	USA	*581*	75-43000
Howard Porter Pty Ltd	Australia	*17*	75-08720
IVECO Fiat SpA	Italy	*235*	75-13230
Iveco Pegaso SA	Spain	*364*	55-00250
MAN Nutzfahrzeuge AG	Germany	*172*	75-19640
Mercedes-Benz (Australia) Pty Ltd	Australia	*19*	75-13355
Napco International Inc.	USA	*598*	75-63600
Oshkosh Truck Corp.	USA	*602*	75-67550
Panhard & Levassor SA	France	*140*	75-28750
Reynolds Boughton Ltd	UK	*479*	75-49340
Robert Wright & Son (Coachworks) Ltd	UK	*494*	75-67150
ShinMaywa Industries Ltd	Japan	*254*	75-07800
SP aerospace and vehicle systems bv	**Netherlands**	***295***	**75-06860**
SPL Ordnance Ltd.	Israel	*225*	80-01935
Static Engineering Pty Ltd	Australia	*21*	75-19780
Stewart & Stevenson Services Inc.	USA	*618*	75-84180
Stinar Corp.	USA	*618*	75-84220
Terberg Benschop BV	Netherlands	*296*	75-07600
Volvo Truck Corporation	Sweden	*379*	75-11800

141 70 10 Logistics vehicles, non-armoured wheeled

Alvis Unipower Ltd	UK	*430*	75-02863
AM General Corporation	USA	*552*	75-05110
Arab American Vehicles Co.	Egypt	*109*	74-00100
Asia Motors Co Ltd	Korea (ROK)	*265*	75-00100
Automotive Industries Ltd	Israel	*217*	75-02500
Auverland/Sovamag	France	*132*	75-04520
Bedford Trucks	UK	*434*	75-06135
BMY-Wheeled Vehicles	USA	*560*	75-14855
Bombardier Inc.	Canada	*69*	75-01400
Bucher-Guyer Ltd, Vehicles	Switzerland	*385*	75-01580
DAF Trucks NV	Netherlands	*292*	75-01400
Daimler-Benz AG	**Germany**	***166***	**75-05600**
Faun GmbH	Germany	*168*	75-08900
Foden Trucks	UK	*451*	75-24100
Foremost Industries Inc.	Canada	*71*	75-05620
FWD International Inc.	USA	*577*	75-36680
GM-Holden's Automotive	Australia	*16*	75-07520
Hellenic Vehicle Industry SA	Greece	*189*	75-01780
Heller Truck Body Corp.	USA	*581*	75-43000
Howard Porter Pty Ltd	Australia	*17*	75-08720
Hyundai Motor Co. Ltd	Korea (ROK)	*266*	75-01602
Iveco Pegaso SA	Spain	*364*	55-00250
JC Bamford Excavators Ltd	UK	*462*	75-32297
Leyland Trucks Ltd	UK	*464*	75-35400
M.A. Bigland (Preparations) Ltd	UK	*434*	75-06700
MAN Nutzfahrzeuge AG	Germany	*172*	75-19640
Napco International Inc.	USA	*598*	75-63600
Ontario Drive & Gear Ltd.	Canada	*74*	75-10400
Oshkosh Truck Corp.	USA	*602*	75-67550
Osterreichische Automobilfabrik	Austria	*33*	75-01820
Patria Vehicles Oy	Finland	*118*	75-01710
Renault Véhicules Industriels SA	France	*142*	75-32300
Stewart & Stevenson Services Inc.	USA	*618*	75-84180
Steyr-Daimler-Puch Fahrzeugtechnik	Austria	*34*	75-02805
Systems & Electronics Inc.	USA	*619*	75-85650
Volvo Truck and Bus Ltd	UK	*492*	75-63610
Volvo Truck Corporation	Sweden	*379*	75-11800

141 70 10 10 Logistics vehicles, non-armoured wheeled, heavy

Bedford Trucks	UK	*434*	75-06135
Bombardier Inc.	Canada	*69*	75-01400
Foden Trucks	UK	*451*	75-24100
Foremost Industries Inc.	Canada	*71*	75-05620
FWD International Inc.	USA	*577*	75-36680
GM-Holden's Automotive	Australia	*16*	75-07520
Hellenic Vehicle Industry SA	Greece	*189*	75-01780
Howard Porter Pty Ltd	Australia	*17*	75-08720
Iveco Pegaso SA	Spain	*364*	55-00250
Leyland Trucks Ltd	UK	*464*	75-35400
MAN Nutzfahrzeuge AG	Germany	*172*	75-19640
Napco International Inc.	USA	*598*	75-63600
Oshkosh Truck Corp.	USA	*602*	75-67550
Patria Vehicles Oy	Finland	*118*	75-01710
Systems & Electronics Inc.	USA	*619*	75-85650
Terberg Benschop BV	Netherlands	*296*	75-07600
Volvo Truck and Bus Ltd	UK	*492*	75-63610

141 70 10 20 Logistics vehicles, non-armoured wheeled, medium

Automotive Industries Ltd	Israel	*217*	75-02500
Bedford Trucks	UK	*434*	75-06135
BMY-Wheeled Vehicles	USA	*560*	75-14855
Bombardier Inc.	Canada	*69*	75-01400
Daimler-Benz AG	**Germany**	***166***	**75-05600**
FWD International Inc.	USA	*577*	75-36680
GM-Holden's Automotive	Australia	*16*	75-07520
Hellenic Vehicle Industry SA	Greece	*189*	75-01780
Iveco Pegaso SA	Spain	*364*	55-00250
JC Bamford Excavators Ltd	UK	*462*	75-32297
Leyland Trucks Ltd	UK	*464*	75-35400
MAN Nutzfahrzeuge AG	Germany	*172*	75-19640
Napco International Inc.	USA	*598*	75-63600
Patria Vehicles Oy	Finland	*118*	75-01710
Reynolds Boughton Ltd	UK	*479*	75-49340
Systems & Electronics Inc.	USA	*619*	75-85650

141 70 10 30 Logistics vehicles, non-armoured wheeled, light

AM General Corporation	USA	*552*	75-05110
Auverland/Sovamag	France	*132*	75-04520
Bedford Trucks	UK	*434*	75-06135
Bernardini SA	Brazil	*55*	75-02120
Bombardier Inc.	Canada	*69*	75-01400

Bucher-Guyer Ltd, Vehicles	Switzerland	*385*	75-01580
Daimler-Benz AG	**Germany**	***166***	**75-05600**
GM-Holden's Automotive	Australia	*16*	75-07520
Hellenic Vehicle Industry SA	Greece	*189*	75-01780
Iveco Pegaso SA	Spain	*364*	55-00250
Leyland Trucks Ltd	UK	*464*	75-35400
MAN Nutzfahrzeuge AG	Germany	*172*	75-19640
Mercedes-Benz Türk A.S.	Turkey	*403*	75-01400
Napco International Inc.	USA	*598*	75-63600
Ontario Drive & Gear Ltd.	Canada	*74*	75-10400
Stewart & Stevenson Services Inc.	USA	*618*	75-84180
Supacat Ltd	UK	*487*	75-57820
Systems & Electronics Inc.	USA	*619*	75-85650

141 70 10 30 70 Trucks, pick-up

Arab American Vehicles Co.	Egypt	*109*	74-00100
BMC Sanayi ve Ticaret AS	Turkey	*402*	75-00150
Capral Aluminium Ltd.	Australia	*14*	75-03170
Daimler-Benz AG	**Germany**	***166***	**75-05600**
GM-Holden's Automotive	Australia	*16*	75-07520
Hellenic Vehicle Industry SA	Greece	*189*	75-01780
Napco International Inc.	USA	*598*	75-63600
ShinMaywa Industries Ltd	Japan	*254*	75-07800
Stephan Wood Products Inc.	USA	*628*	75-97280
Stinar Corp.	USA	*618*	75-84220

141 70 10 40 Logistics vehicles, non-armoured wheeled, off-road

AM General Corporation	USA	*552*	75-05110
Arab Organization for Ind.	Egypt	*109*	55-01700
Auverland/Sovamag	France	*132*	75-04520
Aveling Barford (Machines) plc	UK	*432*	75-04550
Bedford Trucks	UK	*434*	75-06135
BMY-Wheeled Vehicles	USA	*560*	75-14855
Bombardier Inc.	Canada	*69*	75-01400
Bucher-Guyer Ltd, Vehicles	Switzerland	*385*	75-01580
Crane Fruehauf Ltd	UK	*443*	75-15750
Daimler-Benz AG	**Germany**	***166***	**75-05600**
Foden Trucks	UK	*451*	75-24100
Foremost Industries Inc.	Canada	*71*	75-05620
FWD International Inc.	USA	*577*	75-36680
GM-Holden's Automotive	Australia	*16*	75-07520
Hellenic Vehicle Industry SA	Greece	*189*	75-01780
Iveco Pegaso SA	Spain	*364*	55-00250
MAN Nutzfahrzeuge AG	Germany	*172*	75-19640
Mercedes-Benz (Australia) Pty Ltd	Australia	*19*	75-13355
Ontario Drive & Gear Ltd.	Canada	*74*	75-10400
Oshkosh Truck Corp.	USA	*602*	75-67550
Patria Vehicles Oy	Finland	*118*	75-01710
Renault Véhicules Industriels SA	France	*142*	75-32300
Supacat Ltd	UK	*487*	75-57820
Systems & Electronics Inc.	USA	*619*	75-85650
Terberg Benschop BV	Netherlands	*296*	75-07600

141 70 10 41 Logistics vehicles, non-armoured wheeled, off-road, amphibious

Daimler-Benz AG	**Germany**	***166***	**75-05600**
Glover Webb Ltd	UK	*455*	75-25700
Ontario Drive & Gear Ltd.	Canada	*74*	75-10400
Supacat Ltd	UK	*487*	75-57820

141 70 20 Auxiliary & liaison vehicles, non-armoured, wheeled

Arab American Vehicles Co.	Egypt	*109*	74-00100
Asia Motors Co Ltd	Korea (ROK)	*265*	75-00100
Auto Crane Co.	USA	*556*	75-10360
Automobiles Peugeot	France	*140*	75-29600
Automotive Industries Ltd	Israel	*217*	75-02500
Auverland/Sovamag	France	*132*	75-04520
Bombardier Inc.	Canada	*69*	75-01400
Bucher-Guyer Ltd, Vehicles	Switzerland	*385*	75-01580
DAF Trucks NV	Netherlands	*292*	75-01400
GM-Holden's Automotive	Australia	*16*	75-07520
Hyundai Motor Co. Ltd	Korea (ROK)	*266*	75-01602
Lohr Industrie	France	*139*	75-25200
MAN Nutzfahrzeuge AG	Germany	*172*	75-19640
Mercedes-Benz Türk A.S.	Turkey	*403*	75-01400
Ontario Drive & Gear Ltd.	Canada	*74*	75-10400
Steyr-Daimler-Puch Fahrzeugtechnik	Austria	*34*	75-02805
Terberg Benschop BV	Netherlands	*296*	75-07600
Volvo Truck Corporation	Sweden	*379*	75-11800

141 70 21 Vehicles, auxiliary

Auto Crane Co.	USA	*556*	75-10360
Auverland/Sovamag	France	*132*	75-04520
Bombardier Inc.	Canada	*69*	75-01400
Bucher-Guyer Ltd, Vehicles	Switzerland	*385*	75-01580
Capral Aluminium Ltd.	Australia	*14*	75-03170
GM-Holden's Automotive	Australia	*16*	75-07520
Hadaikan Automotive Assemblies	Israel	*219*	75-08205
MAN Nutzfahrzeuge AG	Germany	*172*	75-19640
Ontario Drive & Gear Ltd.	Canada	*74*	75-10400
Renault Véhicules Industriels SA	France	*142*	75-32300
Reynolds Boughton Ltd	UK	*479*	75-49340

141 70 21 35 Ambulances, non-armoured wheeled

AM General Corporation	USA	*552*	75-05110
Arab Organization for Ind.	Egypt	*109*	55-01700
Auverland/Sovamag	France	*132*	75-04520
Bucher-Guyer Ltd, Vehicles	Switzerland	*385*	75-01580
Carrozzeria Boneschi Sarl	Italy	*233*	75-05000
Daimler-Benz AG	**Germany**	***166***	**75-05600**
Glover Webb Ltd	UK	*455*	75-25700
GM-Holden's Automotive	Australia	*16*	75-07520
Iveco Pegaso SA	Spain	*364*	55-00250
MMB International Ltd	UK	*507*	80-16895
O'Gara-Hess & Eisenhardt Armoring	USA	*601*	75-66400
Ontario Drive & Gear Ltd.	Canada	*74*	75-10400
Pilcher-Greene Ltd	UK	*474*	75-45450
Singapore Technologies Automotive	Singapore	*348*	75-04280
Wheeled Coach Industries	USA	*627*	75-96350

141 70 25 Vehicles, liaison

Auto Crane Co.	USA	*556*	75-10360
Auverland/Sovamag	France	*132*	75-04520
BMC Sanayi ve Ticaret AS	Turkey	*402*	75-00150
Bombardier Inc.	Canada	*69*	75-01400
Bucher-Guyer Ltd, Vehicles	Switzerland	*385*	75-01580
GM-Holden's Automotive	Australia	*16*	75-07520
Panhard & Levassor SA	France	*140*	75-28750
Robert Wright & Son (Coachworks) Ltd	UK	*494*	75-67150
Stewart & Stevenson Services Inc.	USA	*618*	75-84180

141 70 25 42 Buses, coaches

Carrozzeria Boneschi Sarl	Italy	*233*	75-05000
Dennis Specialist Vehicles	UK	*445*	75-17570
EvoBus GmbH.	Germany	*168*	75-08720
Glover Webb Ltd	UK	*455*	75-25700
GM-Holden's Automotive	Australia	*16*	75-07520
Hadaikan Automotive Assemblies	Israel	*219*	75-08205
Hellenic Vehicle Industry SA	Greece	*189*	75-01780
Iveco Pegaso SA	Spain	*364*	55-00250
MAN Nutzfahrzeuge AG	Germany	*172*	75-19640
Mercedes-Benz (Australia) Pty Ltd	Australia	*19*	75-13355
Pilcher-Greene Ltd	UK	*474*	75-45450
Robert Wright & Son (Coachworks) Ltd	UK	*494*	75-67150
Scania AB	Sweden	*378*	75-10200

141 70 26 Vehicles, liaison, off-road

Auverland/Sovamag	France	*132*	75-04520
Bombardier Inc.	Canada	*69*	75-01400
Bucher-Guyer Ltd, Vehicles	Switzerland	*385*	75-01580
GM-Holden's Automotive	Australia	*16*	75-07520
Mercedes-Benz (Australia) Pty Ltd	Australia	*19*	75-13355
Ontario Drive & Gear Ltd.	Canada	*74*	75-10400
Stewart & Stevenson Services Inc.	USA	*618*	75-84180

141 70 26 40 Vehicles, light liaison

Automobiles Peugeot	France	*140*	75-29600

Bombardier Inc.	Canada	*69*	75-01400
GM-Holden's Automotive	Australia	*16*	75-07520
Hadaikan Automotive Assemblies	Israel	*219*	75-08205
Stewart & Stevenson Services Inc.	USA	*618*	75-84180

141 70 26 45 40 Vehicles, light liaison, off-road

Arab American Vehicles Co.	Egypt	*109*	74-00100
Bombardier Inc.	Canada	*69*	75-01400
Bucher-Guyer Ltd, Vehicles	Switzerland	*385*	75-01580
GM-Holden's Automotive	Australia	*16*	75-07520
Lohr Industrie	France	*139*	75-25200
Ontario Drive & Gear Ltd.	Canada	*74*	75-10400
Stewart & Stevenson Services Inc.	USA	*618*	75-84180

141 70 28 Light combat / scout vehicles, non-armoured wheeled

AM General Corporation	USA	*552*	75-05110
Glover Webb Ltd	UK	*455*	75-25700
Lohr Industrie	France	*139*	75-25200
Ontario Drive & Gear Ltd.	Canada	*74*	75-10400
Panhard & Levassor SA	France	*140*	75-28750
Stewart & Stevenson Services Inc.	USA	*618*	75-84180

141 70 30 Tractors / prime movers, non-armoured wheeled

Auto Crane Co.	USA	*556*	75-10360
DAF Trucks NV	Netherlands	*292*	75-01400
Daimler-Benz AG	**Germany**	***166***	**75-05600**
Electricars Ltd	UK	*448*	75-19870
Faun GmbH	Germany	*168*	75-08900
Foden Trucks	UK	*451*	75-24100
FWD International Inc.	USA	*577*	75-36680
GM-Holden's Automotive	Australia	*16*	75-07520
M.A. Bigland (Preparations) Ltd	UK	*434*	75-06700
MAN Nutzfahrzeuge AG	Germany	*172*	75-19640
Nicolas Industrie SA	France	*140*	75-27800
Oshkosh Truck Corp.	USA	*602*	75-67550
Penman Engineering Ltd	**UK**	***473***	**75-44400**
Terberg Benschop BV	Netherlands	*296*	75-07600
Volvo Truck Corporation	Sweden	*379*	75-11800

141 70 30 02 Tractors, agricultural type, military usage

Daimler-Benz AG	**Germany**	***166***	**75-05600**

141 70 30 05 Tractors, aircraft towing / positioning (see code 625.20.75.80)

E C Hallam Engineering (Leicester)	UK	*456*	75-27430
Eagle Industrial Truck Mfg. Inc.	USA	*570*	75-28430
Echtman Engineering Corp. Ltd	Israel	*218*	75-05050
Electricars Ltd	UK	*448*	75-19870
FWD International Inc.	USA	*577*	75-36680
Hamech Ltd	UK	*456*	75-27730
JC Bamford Excavators Ltd	UK	*462*	75-32297
Lektro Inc.	USA	*590*	75-54630
MOL Cy NV	Belgium	*46*	75-04998
Perry Engineering	Australia	*20*	75-16460
Schopf Maschinenbau GmbH	Germany	*174*	75-26600
Stewart & Stevenson Services Inc.	USA	*618*	75-84180

141 70 30 10 Tractors, artillery

DAF Trucks NV	Netherlands	*292*	75-01400
Iveco Pegaso SA	Spain	*364*	55-00250
Osterreichische Automobilfabrik	Austria	*33*	75-01820
Renault Véhicules Industriels SA	France	*142*	75-32300
Volvo Truck Corporation	Sweden	*379*	75-11800

141 70 30 20 Tractors, semi-trailer

DAF Trucks NV	Netherlands	*292*	75-01400
Foden Trucks	UK	*451*	75-24100
FWD International Inc.	USA	*577*	75-36680
GM-Holden's Automotive	Australia	*16*	75-07520
Iveco Pegaso SA	Spain	*364*	55-00250
Lohr Industrie	France	*139*	75-25200
MAN Nutzfahrzeuge AG	Germany	*172*	75-19640
Oshkosh Truck Corp.	USA	*602*	75-67550
Patria Vehicles Oy	Finland	*118*	75-01710
Terberg Benschop BV	Netherlands	*296*	75-07600

141 70 30 22 Tractors, semi-trailer, heavy duty

Bartlett Lifting Devices Inc.	USA	*557*	75-12060
DAF Trucks NV	Netherlands	*292*	75-01400
Faun GmbH	Germany	*168*	75-08900
Foden Trucks	UK	*451*	75-24100
FWD International Inc.	USA	*577*	75-36680
Iveco Pegaso SA	Spain	*364*	55-00250
Lohr Industrie	France	*139*	75-25200
MAN Nutzfahrzeuge AG	Germany	*172*	75-19640
Nicolas Industrie SA	France	*140*	75-27800
Oshkosh Truck Corp.	USA	*602*	75-67550
Patria Vehicles Oy	Finland	*118*	75-01710
Renault Véhicules Industriels SA	France	*142*	75-32300
Terberg Benschop BV	Netherlands	*296*	75-07600
Volvo Truck Corporation	Sweden	*379*	75-11800

141 70 30 24 Tractors, semi-trailer, medium duty

Bartlett Lifting Devices Inc.	USA	*557*	75-12060
DAF Trucks NV	Netherlands	*292*	75-01400
Faun GmbH	Germany	*168*	75-08900
FWD International Inc.	USA	*577*	75-36680
GM-Holden's Automotive	Australia	*16*	75-07520
Iveco Pegaso SA	Spain	*364*	55-00250
MAN Nutzfahrzeuge AG	Germany	*172*	75-19640
Patria Vehicles Oy	Finland	*118*	75-01710
Volvo Truck Corporation	Sweden	*379*	75-11800

141 70 30 26 Tractors, semi-trailer, light duty

GM-Holden's Automotive	Australia	*16*	75-07520
Volvo Truck Corporation	Sweden	*379*	75-11800

141 70 40 Vehicles, special purpose, non-armoured wheeled

AAI Corp.	USA	*545*	75-00020
AM General Corporation	USA	*552*	75-05110
Auto Crane Co.	USA	*556*	75-10360
Automobiles Peugeot	France	*140*	75-29600
Bombardier Inc.	Canada	*69*	75-01400
Bucher-Guyer Ltd, Vehicles	Switzerland	*385*	75-01580
Capral Aluminium Ltd.	Australia	*14*	75-03170
Carrozzeria Boneschi Sarl	Italy	*233*	75-05000
CSC (Specialised Vehicles) Ltd	UK	*444*	75-16220
Daewoo Heavy Industries Ltd	Korea (ROK)	*265*	75-00600
DAF Trucks NV	Netherlands	*292*	75-01400
Foremost Industries Inc.	Canada	*71*	75-05620
FWD International Inc.	USA	*577*	75-36680
Glover Webb Ltd	UK	*455*	75-25700
Hyundai Motor Co. Ltd	Korea (ROK)	*266*	75-01602
Lohr Industrie	France	*139*	75-25200
MAN Nutzfahrzeuge AG	Germany	*172*	75-19640
Marchetti Autogru SpA	Italy	*236*	75-14400
Nicolas Industrie SA	France	*140*	75-27800
O'Gara-Hess & Eisenhardt Armoring	USA	*601*	75-66400
Ontario Drive & Gear Ltd.	Canada	*74*	75-10400
Oshkosh Truck Corp.	USA	*602*	75-67550
Osterreichische Automobilfabrik	Austria	*33*	75-01820
Renault Véhicules Industriels SA	France	*142*	75-32300
Saval NV	Belgium	*47*	75-05650
Snorkel	USA	*616*	75-82170
SPL Ordnance Ltd.	Israel	*225*	80-01935
Static Engineering Pty Ltd	Australia	*21*	75-19780
Stinar Corp.	USA	*618*	75-84220
Supacat Ltd	UK	*487*	75-57820
Volvo Truck Corporation	Sweden	*379*	75-11800

141 70 40 02 Vehicles, mine clearance, expendable

STN ATLAS Elektronik GmbH	Germany	*165*	75-01150

141 70 40 05 Vehicles, airfield

Bedwas Bodyworks	UK	*434*	75-06140
Hadaikan Automotive Assemblies	Israel	*219*	75-08205
Oshkosh Truck Corp.	USA	*602*	75-67550
Ros Roca SA	Spain	*367*	75-05030
Static Engineering Pty Ltd	Australia	*21*	75-19780
Stewart & Stevenson Services Inc.	USA	*618*	75-84180
Systems & Electronics Inc.	USA	*619*	75-85650

141 70 40 05 10 Vehicles, airfield, air conditioning

Aircraft Maintenance Support	UK	*428*	75-01580
A/S Dantherm	**Denmark**	***102***	**75-04100**

141 70 40 05 12 Vehicles, airfield, aircraft loading

AAI Corp.	USA	*545*	75-00020
ADE-HML Limited	UK	*427*	75-00710
CSC (Specialised Vehicles) Ltd	UK	*444*	75-16220
GM-Holden's Automotive	Australia	*16*	75-07520
Ros Roca SA	Spain	*367*	75-05030
Smith's Electric Vehicles Ltd.	UK	*485*	75-54650
Static Engineering Pty Ltd	Australia	*21*	75-19780
Stewart & Stevenson Services Inc.	USA	*618*	75-84180
Systems & Electronics Inc.	USA	*619*	75-85650
Terberg Benschop BV	Netherlands	*296*	75-07600

141 70 40 05 15 Vehicles, airfield, airstairs

GM-Holden's Automotive	Australia	*16*	75-07520
Smith's Electric Vehicles Ltd.	UK	*485*	75-54650
Static Engineering Pty Ltd	Australia	*21*	75-19780
Stinar Corp.	USA	*618*	75-84220

141 70 40 05 20 Vehicles, airfield, crash / rescue

Alvis Unipower Ltd	UK	*430*	75-02863
Austral Specialised Vehicles	Australia	*12*	75-01710
Constructeurs Associés de Matériels	France	*134*	75-10280
Emergency One Inc.	USA	*573*	75-31100
FWD International Inc.	USA	*577*	75-36680
Glover Webb Ltd	UK	*455*	75-25700
Mills-Tui Ltd	New Zealand	*302*	75-01350
Ontario Drive & Gear Ltd.	Canada	*74*	75-10400
Oshkosh Truck Corp.	USA	*602*	75-67550
Perry Engineering	Australia	*20*	75-16460
Société Industrielle pour le	France	*146*	75-40470

141 70 40 05 40 Vehicles, airfield, firefighting (see code 612.20.80.39)

Bucher-Guyer Ltd, Vehicles	Switzerland	*385*	75-01580
Carmichael International Ltd	UK	*438*	75-11700
Dennis Specialist Vehicles	UK	*445*	75-17570
Emergency One Inc.	USA	*573*	75-31100
FWD International Inc.	USA	*577*	75-36680
Gruppo Nuova MA.NA.RO.	Italy	*235*	75-10730
Hellenic Vehicle Industry SA	Greece	*189*	75-01780
Mills-Tui Ltd	New Zealand	*302*	75-01350
Nordic Systems Inc.	Canada	*74*	75-09950
Ontario Drive & Gear Ltd.	Canada	*74*	75-10400
Oshkosh Truck Corp.	USA	*602*	75-67550
Patria Vehicles Oy	Finland	*118*	75-01710
Perry Engineering	Australia	*20*	75-16460
Pilcher-Greene Ltd	UK	*474*	75-45450
Reynolds Boughton Ltd	UK	*479*	75-49340
Saval NV	Belgium	*47*	75-05650
Société Industrielle pour le	France	*146*	75-40470

141 70 40 05 45 Vehicles, airfield, food service

CSC (Specialised Vehicles) Ltd	UK	*444*	75-16220
EFE Endüstri ve Ticaret A.S.	Turkey	*402*	75-00530
GM-Holden's Automotive	Australia	*16*	75-07520
Ros Roca SA	Spain	*367*	75-05030
Static Engineering Pty Ltd	Australia	*21*	75-19780

141 70 40 05 50 Vehicles, airfield, refuelling

ADE-HML Limited	UK	*427*	75-00710
Crane Fruehauf Ltd	UK	*443*	75-15750
Foden Trucks	UK	*451*	75-24100
Fuelquip Services Ltd.	New Zealand	*302*	75-00700
FWD International Inc.	USA	*577*	75-36680
GM-Holden's Automotive	Australia	*16*	75-07520
Gruppo Nuova MA.NA.RO.	Italy	*235*	75-10730
Iveco Pegaso SA	Spain	*364*	55-00250
Oshkosh Truck Corp.	USA	*602*	75-67550
Ros Roca SA	Spain	*367*	75-05030
Simon Access (UK) Ltd.	UK	*484*	75-53908
Thompson Carmichael Ltd	UK	*488*	75-59525

141 70 40 05 60 Vehicles, airfield, engine starter / generating

ADE-HML Limited	UK	*427*	75-00710
Arvico Corporation	USA	*555*	75-09240
FWD International Inc.	USA	*577*	75-36680
Stewart & Stevenson Services Inc.	USA	*618*	75-84180

141 70 40 10 Vehicles, crane

Atlas Hydraulic Loaders Ltd	UK	*432*	75-04300
Auto Crane Co.	USA	*556*	75-10360
Belotti SpA	Italy	*232*	75-02640
Farid Industrie SA	Italy	*234*	75-08630
Grove Europe Ltd	**UK**	***456***	**75-27030**
Iveco Pegaso SA	Spain	*364*	55-00250
Marchetti Autogru SpA	Italy	*236*	75-14400
MOL Cy NV	Belgium	*46*	75-04998
ShinMaywa Industries Ltd	Japan	*254*	75-07800
Terberg Benschop BV	Netherlands	*296*	75-07600

141 70 40 40 Vehicles, firefighting (excluding airfield - see code 612.40.80.39)

Carmichael International Ltd	UK	*438*	75-11700
Dennis Specialist Vehicles	UK	*445*	75-17570
Engineering Equipment Co.	USA	*574*	75-31600
Faun GmbH	Germany	*168*	75-08900
FWD International Inc.	USA	*577*	75-36680
Giat Industries	France	*126*	55-00100
GM-Holden's Automotive	Australia	*16*	75-07520
Hadaikan Automotive Assemblies	Israel	*219*	75-08205
Hellenic Vehicle Industry SA	Greece	*189*	75-01780
Iveco Pegaso SA	Spain	*364*	55-00250
Mills-Tui Ltd	New Zealand	*302*	75-01350
Nordic Systems Inc.	Canada	*74*	75-09950
Northrop Grumman Corporation	USA	*599*	75-65450
Patria Vehicles Oy	Finland	*118*	75-01710
Pilcher-Greene Ltd	UK	*474*	75-45450
Renault Véhicules Industriels SA	France	*142*	75-32300
Robert Aebi AG	Switzerland	*385*	75-00070
SPL Ordnance Ltd	Israel	*225*	80-01935

141 70 40 60 Vehicles, semi-trailer

Capral Aluminium Ltd.	Australia	*14*	75-03170
Crane Fruehauf Ltd	UK	*443*	75-15750
Daimler-Benz AG	**Germany**	***166***	**75-05600**
EFE Endüstri ve Ticaret A.S.	Turkey	*402*	75-00530
Gontrailer SL	Spain	*367*	75-05660
Hellenic Vehicle Industry SA	Greece	*189*	75-01780
MAN Nutzfahrzeuge AG	Germany	*172*	75-19640
Patria Vehicles Oy	Finland	*118*	75-01710
Recrusul SA	Brazil	*57*	75-18340
Static Engineering Pty Ltd	Australia	*21*	75-19780
Systems & Electronics Inc.	USA	*619*	75-85650
Thompson Carmichael Ltd	UK	*488*	75-59525
Urdan Industries (USA) Inc.	USA	*625*	75-92930

141 70 40 62 Vehicles, semi-trailer, heavy equipment transport

Crane Fruehauf Ltd	UK	*443*	75-15750
Daimler-Benz AG	**Germany**	***166***	**75-05600**
EFE Endüstri ve Ticaret A.S.	Turkey	*402*	75-00530

Gontrailer SL	Spain	*367*	75-05660
Hellenic Vehicle Industry SA	Greece	*189*	75-01780
Kioleides SA	Greece	*190*	75-01982
MAN Nutzfahrzeuge AG	Germany	*172*	75-19640
Nicolas Industrie SA	France	*140*	75-27800
Oshkosh Truck Corp.	USA	*602*	75-67550
Patria Vehicles Oy	Finland	*118*	75-01710
Static Engineering Pty Ltd	Australia	*21*	75-19780
Systems & Electronics Inc.	USA	*619*	75-85650
Urdan Industries (USA) Inc.	USA	*625*	75-92930

141 70 40 62 50 Tank transporters

Alvis Unipower Ltd	UK	*430*	75-02863
Auto Crane Co.	USA	*556*	75-10360
Crane Fruehauf Ltd	UK	*443*	75-15750
DAF Trucks NV	Netherlands	*292*	75-01400
Daimler-Benz AG	**Germany**	***166***	**75-05600**
EFE Endüstri ve Ticaret A.S.	Turkey	*402*	75-00530
Faun GmbH	Germany	*168*	75-08900
Foden Trucks	UK	*451*	75-24100
FWD International Inc.	USA	*577*	75-36680
Gontrailer SL	Spain	*367*	75-05660
Hellenic Vehicle Industry SA	Greece	*189*	75-01780
Iveco Pegaso SA	Spain	*364*	55-00250
Kioleides SA	Greece	*190*	75-01982
Lohr Industrie	France	*139*	75-25200
MOL Cy NV	Belgium	*46*	75-04998
Nicolas Industrie SA	France	*140*	75-27800
Oshkosh Truck Corp.	USA	*602*	75-67550
Patria Vehicles Oy	Finland	*118*	75-01710
Renault Véhicules Industriels SA	France	*142*	75-32300
Systems & Electronics Inc.	USA	*619*	75-85650
Terberg Benschop BV	Netherlands	*296*	75-07600

141 70 40 64 Vehicles, semi-trailer, medium duty

Capral Aluminium Ltd.	Australia	*14*	75-03170
Crane Fruehauf Ltd	UK	*443*	75-15750
EFE Endüstri ve Ticaret A.S.	Turkey	*402*	75-00530
Electricars Ltd	UK	*448*	75-19870
Gontrailer SL	Spain	*367*	75-05660
Kioleides SA	Greece	*190*	75-01982
Patria Vehicles Oy	Finland	*118*	75-01710
Systems & Electronics Inc.	USA	*619*	75-85650
Urdan Industries (USA) Inc.	USA	*625*	75-92930

141 70 40 66 Vehicles, semi-trailer, light duty

Capral Aluminium Ltd.	Australia	*14*	75-03170
Crane Fruehauf Ltd	UK	*443*	75-15750
EFE Endüstri ve Ticaret A.S.	Turkey	*402*	75-00530
Electricars Ltd	UK	*448*	75-19870
Gontrailer SL	Spain	*367*	75-05660
Systems & Electronics Inc.	USA	*619*	75-85650
Urdan Industries (USA) Inc.	USA	*625*	75-92930

141 70 40 70 Vehicles, recovery, non-armoured, wheeled

Alvis Unipower Ltd	UK	*430*	75-02863
Bucher-Guyer Ltd, Vehicles	Switzerland	*385*	75-01580
DAF Trucks NV	Netherlands	*292*	75-01400
EKA Limited	UK	*448*	75-19810
Faun GmbH	Germany	*168*	75-08900
FWD International Inc.	USA	*577*	75-36680
Hägglunds Moelv AS	Norway	*309*	75-00880
Iveco Pegaso SA	Spain	*364*	55-00250
Lohr Industrie	France	*139*	75-25200
M.A. Bigland (Preparations) Ltd	UK	*434*	75-06700
Oshkosh Truck Corp.	USA	*602*	75-67550
Renault Véhicules Industriels SA	France	*142*	75-32300
SP aerospace and vehicle systems bv	**Netherlands**	***295***	**75-06860**
Terberg Benschop BV	Netherlands	*296*	75-07600
Urdan Industries (USA) Inc.	USA	*625*	75-92930

141 70 40 70 70 Recovery vehicle equipment

Daimler-Benz AG	**Germany**	***166***	**75-05600**
EKA Limited	UK	*448*	75-19810
Farid Industrie SA	Italy	*234*	75-08630
Foden Trucks	UK	*451*	75-24100
Jung Jungenthal GmbH	Germany	*170*	75-15130
Lohr Industrie	France	*139*	75-25200
M.A. Bigland (Preparations) Ltd	UK	*434*	75-06700
MFC Survival Ltd	UK	*467*	75-38900
Renown Engineering Ltd	UK	*478*	75-49227
Reynolds Boughton Ltd	UK	*479*	75-49340
SP aerospace and vehicle systems bv	**Netherlands**	***295***	**75-06860**

141 70 40 70 77 Recovery units, vehicle-mounted

EKA Limited	UK	*448*	75-19810
Farid Industrie SA	Italy	*234*	75-08630
Lohr Industrie	France	*139*	75-25200
Oshkosh Truck Corp.	USA	*602*	75-67550
Reynolds Boughton Ltd	UK	*479*	75-49340
SP aerospace and vehicle systems bv	**Netherlands**	***295***	**75-06860**

141 70 40 80 Vehicles, tanker

Alvis Unipower Ltd	UK	*430*	75-02863
Bombardier Inc.	Canada	*69*	75-01400
Capral Aluminium Ltd.	Australia	*14*	75-03170
DAF Trucks NV	Netherlands	*292*	75-01400
Daimler-Benz AG	**Germany**	***166***	**75-05600**
EFE Endüstri ve Ticaret A.S.	Turkey	*402*	75-00530
FWD International Inc.	USA	*577*	75-36680
GM-Holden's Automotive	Australia	*16*	75-07520
Oshkosh Truck Corp.	USA	*602*	75-67550
Reynolds Boughton Ltd	UK	*479*	75-49340
Thompson Carmichael Ltd	UK	*488*	75-59525

141 70 40 80 10 Vehicles, army bulk refuelling

FWD International Inc.	USA	*577*	75-36680
Thompson Carmichael Ltd	UK	*488*	75-59525

141 70 55 Bridging vehicles, non-armoured wheeled

Alvis Unipower Ltd	UK	*430*	75-02863
CNIM	France	*133*	75-09990
Daimler-Benz AG	**Germany**	***166***	**75-05600**
Foden Trucks	UK	*451*	75-24100
Lohr Industrie	France	*139*	75-25200
MAN Gutehoffnungshütte AG	Germany	*172*	75-19620
SP aerospace and vehicle systems bv	**Netherlands**	***295***	**75-06860**
Systems & Electronics Inc.	USA	*619*	75-85650

141 70 75 Very light vehicles, non-armoured, wheeled

BMC Sanayi ve Ticaret AS	Turkey	*402*	75-00150
Bombardier Inc.	Canada	*69*	75-01400
Capral Aluminium Ltd.	Australia	*14*	75-03170
Carrozzeria Boneschi Sarl	Italy	*233*	75-05000
Daimler-Benz AG	**Germany**	***166***	**75-05600**
Giat Industries	France	*126*	55-00100
GM-Holden's Automotive	Australia	*16*	75-07520
Hadaikan Automotive Assemblies	Israel	*219*	75-08205
Israel Military Industries Ltd	Israel	*220*	75-10300
Lohr Industrie	France	*139*	75-25200
Ontario Drive & Gear Ltd	Canada	*74*	75-10400
Still GmbH	Germany	*175*	75-29000
Systems & Electronics Inc.	USA	*619*	75-85650

143 Trailers, semi-trailers

Aarding BV	Netherlands	*291*	75-00020
Arab Organization for Ind.	Egypt	*109*	55-01700
Aveling Barford (Machines) plc	UK	*432*	75-04550
De Filippi Andrea Sas	Italy	*233*	75-06550
EFE Endüstri ve Ticaret A.S.	Turkey	*402*	75-00530
Genco Mobile Systems BV	Netherlands	*293*	75-02340
Goldhofer Fahrzeugwerk GmbH & Co.	Germany	*169*	75-11000
Gontrailer SL	Spain	*367*	75-05660
Hägglunds Moelv AS	Norway	*309*	75-00880

Heller Truck Body Corp.	USA	*581*	75-43000
Kalyn-Siebert Inc.	USA	*587*	75-51450
Kioleides SA.	Greece	*190*	75-01982
Kutaisi Automobile Plant Production	Georgia	*156*	55-05000
Lohr Industrie	France	*139*	75-25200
Nicolas Industrie SA	France	*140*	75-27800
Norca Machinery Company	USA	*599*	75-64800
NU Van Technology Inc.	USA	*600*	75-65807
Oldbury Components Ltd	UK	*472*	75-42985
Oshkosh Truck Corp.	USA	*602*	75-67550
Reynolds Boughton Ltd	UK	*479*	75-49340
RM Trailers Ltd	UK	*479*	75-49860
Static Engineering Pty Ltd	Australia	*21*	75-19780
Systems & Electronics Inc.	USA	*619*	75-85650
Talbert Manufacturing Inc.	USA	*619*	75-86080
Terberg Benschop BV	Netherlands	*296*	75-07600
Western Trailers	UK	*493*	75-65370

143 10 Trailers, drawbar

Arab Organization for Ind.	Egypt	*109*	55-01700
Automotive Industries Ltd	Israel	*217*	75-02500
Crane Fruehauf Ltd	UK	*443*	75-15750
De Filippi Andrea Sas.	Italy	*233*	75-06550
EFE Endüstri ve Ticaret A.S.	Turkey	*402*	75-00530
Genco Mobile Systems BV.	Netherlands	*293*	75-02340
GKN Defence Ltd	**UK**	***454***	**75-25600**
Goldhofer Fahrzeugwerk GmbH & Co.	Germany	*169*	75-11000
Gontrailer SL	Spain	*367*	75-05660
Hägglunds Moelv AS	Norway	*309*	75-00880
Kioleides SA.	Greece	*190*	75-01982
Lohr Industrie	France	*139*	75-25200
M.A. Bigland (Preparations) Ltd	UK	*434*	75-06700
Norca Machinery Company	USA	*599*	75-64800
Oldbury Components Ltd	UK	*472*	75-42985
Oshkosh Truck Corp.	USA	*602*	75-67550
Renown Engineering Ltd	UK	*478*	75-49227
Reynolds Boughton Ltd	UK	*479*	75-49340
RM Trailers Ltd	UK	*479*	75-49860
Rubery Owen Group Services Ltd	UK	*510*	80-29350
Silver Eagle Manufacturing Co.	USA	*615*	75-81020
Static Engineering Pty Ltd	Australia	*21*	75-19780
Terberg Benschop BV	Netherlands	*296*	75-07600
Urdan Industries Ltd	Israel	*223*	75-23210
Vammas Ltd	**Finland**	***117***	**55-02220**

143 20 Trailers, powered axle

Crane Fruehauf Ltd	UK	*443*	75-15750
Genco Mobile Systems BV.	Netherlands	*293*	75-02340
Goldhofer Fahrzeugwerk GmbH & Co.	Germany	*169*	75-11000
Gontrailer SL	Spain	*367*	75-05660
Lohr Industrie	France	*139*	75-25200
Nicolas Industrie SA	France	*140*	75-27800
Reynolds Boughton Ltd	UK	*479*	75-49340
Rubery Owen Group Services Ltd	UK	*510*	80-29350
Static Engineering Pty Ltd	Australia	*21*	75-19780

143 30 Semi-trailers

Capral Aluminium Ltd.	Australia	*14*	75-03170
Crane Fruehauf Ltd	UK	*443*	75-15750
De Filippi Andrea Sas.	Italy	*233*	75-06550
EFE Endüstri ve Ticaret A.S.	Turkey	*402*	75-00530
Goldhofer Fahrzeugwerk GmbH & Co.	Germany	*169*	75-11000
Gontrailer SL	Spain	*367*	75-05660
Kioleides SA.	Greece	*190*	75-01982
Lohr Industrie	France	*139*	75-25200
Nicolas Industrie SA	France	*140*	75-27800
NU Van Technology Inc.	USA	*600*	75-65807
Oshkosh Truck Corp.	USA	*602*	75-67550
RM Trailers Ltd	UK	*479*	75-49860
Rubery Owen Group Services Ltd	UK	*510*	80-29350
Sanvas SA Indústria Metal-Mecânica	Brazil	*57*	75-19140
Static Engineering Pty Ltd	Australia	*21*	75-19780
Thompson Carmichael Ltd	UK	*488*	75-59525
Urdan Industries Ltd	Israel	*223*	75-23210

143 30 10 Semi-trailers, light duty

Capral Aluminium Ltd.	Australia	*14*	75-03170
Crane Fruehauf Ltd	UK	*443*	75-15750
De Filippi Andrea Sas.	Italy	*233*	75-06550
EFE Endüstri ve Ticaret A.S.	Turkey	*402*	75-00530
Genco Mobile Systems BV.	Netherlands	*293*	75-02340
Goldhofer Fahrzeugwerk GmbH & Co.	Germany	*169*	75-11000
Gontrailer SL	Spain	*367*	75-05660
Indústrias de Defesa	Portugal	*327*	55-00950
Kioleides SA.	Greece	*190*	75-01982
Lohr Industrie	France	*139*	75-25200
RM Trailers Ltd	UK	*479*	75-49860
Rubery Owen Group Services Ltd	UK	*510*	80-29350
Static Engineering Pty Ltd	Australia	*21*	75-19780

143 30 20 Semi-trailers, medium duty

Capral Aluminium Ltd.	Australia	*14*	75-03170
Crane Fruehauf Ltd	UK	*443*	75-15750
De Filippi Andrea Sas.	Italy	*233*	75-06550
EFE Endüstri ve Ticaret A.S.	Turkey	*402*	75-00530
Genco Mobile Systems BV.	Netherlands	*293*	75-02340
Goldhofer Fahrzeugwerk GmbH & Co.	Germany	*169*	75-11000
Gontrailer SL	Spain	*367*	75-05660
Indústrias de Defesa	Portugal	*327*	55-00950
Kioleides SA.	Greece	*190*	75-01982
Lohr Industrie	France	*139*	75-25200
Nicolas Industrie SA	France	*140*	75-27800
Oshkosh Truck Corp.	USA	*602*	75-67550
Rubery Owen Group Services Ltd	UK	*510*	80-29350
Static Engineering Pty Ltd	Australia	*21*	75-19780

143 30 30 Semi-trailers, heavy duty

Capral Aluminium Ltd.	Australia	*14*	75-03170
Crane Fruehauf Ltd	UK	*443*	75-15750
De Filippi Andrea Sas.	Italy	*233*	75-06550
EFE Endüstri ve Ticaret A.S.	Turkey	*402*	75-00530
Genco Mobile Systems BV.	Netherlands	*293*	75-02340
Goldhofer Fahrzeugwerk GmbH & Co.	Germany	*169*	75-11000
Gontrailer SL	Spain	*367*	75-05660
Kalyn-Siebert Inc.	USA	*587*	75-51450
Kioleides SA.	Greece	*190*	75-01982
Lohr Industrie	France	*139*	75-25200
Nicolas Industrie SA	France	*140*	75-27800
Norca Machinery Company	USA	*599*	75-64800
Oshkosh Truck Corp.	USA	*602*	75-67550
Rubery Owen Group Services Ltd	UK	*510*	80-29350
Static Engineering Pty Ltd	Australia	*21*	75-19780
Systems & Electronics Inc.	USA	*619*	75-85650
Talbert Manufacturing Inc.	USA	*619*	75-86080

143 30 31 Semi-trailers, heavy duty, ampliroll type

Genco Mobile Systems BV.	Netherlands	*293*	75-02340
Gontrailer SL	Spain	*367*	75-05660
Kioleides SA.	Greece	*190*	75-01982
Lohr Industrie	France	*139*	75-25200
Norca Machinery Company	USA	*599*	75-64800
Oshkosh Truck Corp.	USA	*602*	75-67550
Reynolds Boughton Ltd	UK	*479*	75-49340
Static Engineering Pty Ltd	Australia	*21*	75-19780

143 30 33 Semi-trailers, heavy duty, flat-bed

Capral Aluminium Ltd.	Australia	*14*	75-03170
Crane Fruehauf Ltd	UK	*443*	75-15750
De Filippi Andrea Sas.	Italy	*233*	75-06550
EFE Endüstri ve Ticaret A.S.	Turkey	*402*	75-00530
Genco Mobile Systems BV.	Netherlands	*293*	75-02340
Goldhofer Fahrzeugwerk GmbH & Co.	Germany	*169*	75-11000
Gontrailer SL	Spain	*367*	75-05660
Kalyn-Siebert Inc.	USA	*587*	75-51450
Kioleides SA.	Greece	*190*	75-01982
Lohr Industrie	France	*139*	75-25200
Norca Machinery Company	USA	*599*	75-64800
Oshkosh Truck Corp.	USA	*602*	75-67550
SP aerospace and vehicle systems bv	**Netherlands**	***295***	**75-06860**
Static Engineering Pty Ltd	Australia	*21*	75-19780
Systems & Electronics Inc.	USA	*619*	75-85650

143 30 35 Semi-trailers, heavy duty, liquid transport

Aarding BV	Netherlands	*291*	75-00020
Capral Aluminium Ltd.	Australia	*14*	75-03170
Crane Fruehauf Ltd	UK	*443*	75-15750
EFE Endüstri ve Ticaret A.S.	Turkey	*402*	75-00530
Genco Mobile Systems BV.	Netherlands	*293*	75-02340
Gontrailer SL	Spain	*367*	75-05660
Kioleides SA.	Greece	*190*	75-01982
Lohr Industrie	France	*139*	75-25200
SP aerospace and vehicle systems bv	**Netherlands**	***295***	**75-06860**
Static Engineering Pty Ltd	Australia	*21*	75-19780
Thompson Carmichael Ltd.	UK	*488*	75-59525
Western Trailers	UK	*493*	75-65370

143 30 38 Semi-trailers, heavy duty, tank transporter

Crane Fruehauf Ltd	UK	*443*	75-15750
De Filippi Andrea Sas.	Italy	*233*	75-06550
Gontrailer SL	Spain	*367*	75-05660
Kalyn-Siebert Inc.	USA	*587*	75-51450
Kioleides SA.	Greece	*190*	75-01982
Lohr Industrie	France	*139*	75-25200
Nicolas Industrie SA	France	*140*	75-27800
Norca Machinery Company	USA	*599*	75-64800
NU Van Technology Inc.	USA	*600*	75-65807
Oshkosh Truck Corp.	USA	*602*	75-67550
SP aerospace and vehicle systems bv	**Netherlands**	***295***	**75-06860**
Static Engineering Pty Ltd	Australia	*21*	75-19780
Systems & Electronics Inc..	USA	*619*	75-85650

145 Railway locomotives & rolling stock

A Goninan & Co. Ltd	Australia	*16*	75-07760
CIC International Ltd	USA	*563*	75-18570
Companhia Comércio e Construções	Brazil	*56*	75-03230
Kioleides SA.	Greece	*190*	75-01982
Walkers Ltd	Australia	*23*	75-23090

145 10 Railway locomotives

A Goninan & Co. Ltd	Australia	*16*	75-07760
Kioleides SA.	Greece	*190*	75-01982
SIMA PERU	Peru	*319*	55-01950
Walkers Ltd	Australia	*23*	75-23090

145 11 Locomotives, diesel

A Goninan & Co. Ltd	Australia	*16*	75-07760
Walkers Ltd	Australia	*23*	75-23090

145 11 01 Locomotives, diesel-electric

A Goninan & Co. Ltd	Australia	*16*	75-07760
Companhia Comércio e Construções	Brazil	*56*	75-03230
Hyundai	Korea (ROK)	*265*	75-01540
Kioleides SA.	Greece	*190*	75-01982
Walkers Ltd	Australia	*23*	75-23090

145 11 02 Locomotives, diesel-hydraulic

A Goninan & Co. Ltd	Australia	*16*	75-07760

145 12 Locomotives, electric

A Goninan & Co. Ltd	Australia	*16*	75-07760
Hyundai	Korea (ROK)	*265*	75-01540
Walkers Ltd	Australia	*23*	75-23090

145 20 Railway multiple units

A Goninan & Co. Ltd	Australia	*16*	75-07760
Companhia Comércio e Construções	Brazil	*56*	75-03230

145 50 Railway rolling stock

A Goninan & Co. Ltd	Australia	*16*	75-07760
AAI Corp.	USA	*545*	75-00020
CIC International Ltd	USA	*563*	75-18570
Hanjin Heavy Industries Co. Ltd	Korea (ROK)	*265*	75-01505
Kioleides SA.	Greece	*190*	75-01982
SIMA PERU	Peru	*319*	55-01950
Walkers Ltd	Australia	*23*	75-23090

145 51 Coaches, baggage cars, railway

A Goninan & Co. Ltd	Australia	*16*	75-07760
Companhia Comércio e Construções	Brazil	*56*	75-03230
Hyundai	Korea (ROK)	*265*	75-01540
Walkers Ltd	Australia	*23*	75-23090

145 52 Freight wagons, freight cars

A Goninan & Co. Ltd	Australia	*16*	75-07760
CIC International Ltd	USA	*563*	75-18570
Companhia Comércio e Construções	Brazil	*56*	75-03230
Hyundai	Korea (ROK)	*265*	75-01540
Walkers Ltd	Australia	*23*	75-23090

145 53 Track maintenance machines, railway

Kango Ltd	UK	*462*	75-33210

145 53 60 Track ballasting machines, railway

Kango Ltd	UK	*462*	75-33210

145 53 70 Tracklaying machines, railway

Companhia Comércio e Construções	Brazil	*56*	75-03230

145 54 Breakdown cranes, railway

CIC International Ltd	USA	*563*	75-18570
Grove Europe Ltd	**UK**	***456***	**75-27030**

Platform Systems & Equipment (excluding Propulsion Systems)

Space Platform Systems & Equipment

211 Spacecraft systems & equipment

AEA Technology plc	UK	*496*	80-00480
Aerojet.	USA	*547*	75-01500
Aerospatiale.	France	*131*	75-00600
Aircraft Tanks Ltd	UK	*429*	75-01730
Amptek Inc.	USA	*553*	75-06880
Anite Systems Ltd	UK	*497*	80-01240
Avica Equipment	UK	*432*	75-04870
Cambridge Consultants Ltd	UK	*499*	80-05300
CJB Developments Ltd.	UK	*440*	75-13430
Com Dev Europe	UK	*441*	75-14300
Cubic Defense Systems Inc.	USA	*567*	75-22900
Dowty Aerospace Wolverhampton.	UK	*446*	75-18589
Fokker Space & Systems	Netherlands	*293*	75-02325
GEC-Marconi Research Centre.	UK	*419*	60-00700
Hartron Corporation	Ukraine.	*406*	75-03000
Howden Aircontrol Ltd	UK	*459*	75-30240
Howden Fluid Systems.	USA	*582*	75-44400
Intertechnique SA.	France	*137*	75-23000
ITT Aerospace/Communications Div.	USA	*586*	75-49455
Kearfott Guidance & Navigation Corp.	USA	*588*	75-51760
Korolev Rocket Space Corp	Russia	*334*	55-10450
Litton Industries Inc.	USA	*591*	75-55400
Lockheed Martin Corp.	USA	*591*	75-55635

Lockheed Martin Space Mission	USA	*638*	80-18500
Matra Défense-Espace	France	*139*	75-26000
Matra Marconi Space UK, Ltd	UK	*425*	74-06900
McDonnell Douglas Aerospace	USA	*594*	75-59480
Microtecnica-Torino SpA	Italy	*237*	75-15500
Mitsubishi Heavy Industries Ltd	Japan	*253*	75-05800
Mitsubishi Precision Co. Ltd	Japan	*253*	75-06000
Moog Controls Ltd	UK	*469*	75-40670
Moog Italiana Srl	Italy	*237*	75-16230
MPB Technologies Inc	Canada	*74*	75-09855
Oerlikon-Contraves AG	Switzerland	*387*	75-09250
Parker Bertea Aerospace	USA	*602*	75-68900
Pressure Systems Inc.	USA	*605*	75-71438
Programmed Composites Inc.	USA	*606*	75-71500
Riva Calzoni SpA	Italy	*238*	75-20500
Saab Ericsson Space AB	Sweden	*378*	75-09150
SAGEM SA	**France**	***143***	**75-33950**
Sanders	USA	*612*	75-77850
Schaeffer Magnetics Inc.	USA	*612*	75-78356
SED Systems Inc	Canada	*79*	80-01937
Sener Ingenieria y Sistemas SA	Spain	*367*	75-05200
SEXTANT Avionique SA	France	*144*	75-35000
Sheldahl Inc.	USA	*614*	75-80150
Sira Ltd.	UK	*484*	75-54250
Smiths Aerospace & Defense	USA	*615*	75-82090
Société Anonyme d'Études	France	*145*	75-37100
Spar Aerospace Ltd	Canada	*76*	75-13000
TERMA Elektronik AS	Denmark	*104*	75-20300
The Boeing Company	USA	*560*	75-14250
THERMOCOAX	France	*147*	75-44930
Tracor Inc	USA	*623*	75-90000
Warner Electric	USA	*627*	75-95105

212 Satellite systems & equipment

AB Electronics (Rogerstone) Ltd	UK	*426*	75-00080
Aerojet	USA	*547*	75-01500
Aerospatiale	France	*131*	75-00600
Aircraft Tanks Ltd	UK	*429*	75-01730
Alcatel Bell	Belgium	*44*	75-00055
Alcatel Bell-SDT	Belgium	*44*	75-00060
Amptek Inc.	USA	*553*	75-06880
Anite Systems Ltd	UK	*497*	80-01240
Avica Equipment	UK	*432*	75-04870
California Microwave Inc.	USA	*561*	75-16350
CJB Developments Ltd	UK	*440*	75-13430
Com Dev Europe	UK	*441*	75-14300
Cranfield University	UK	*419*	60-00350
Cubic Defense Systems Inc.	USA	*567*	75-22900
Custom Electronics Inc.	USA	*567*	75-23085
Daimler-Benz Aerospace France SAS	**France**	***134***	**75-12950**
Dowty Aerospace Wolverhampton	UK	*446*	75-18589
Fabbrica Italiana Apparecchiature	Italy	*234*	75-08550
Fokker Space & Systems	Netherlands	*293*	75-02325
GEC-Marconi Avionics Ltd	UK	*454*	75-25320
GEC-Marconi Research Centre	UK	*419*	60-00700
Harris Corporation	USA	*581*	75-42050
Hartron Corporation	Ukraine	*406*	75-03000
Hughes Microelectronics Europe Ltd	UK	*459*	75-30480
Intertechnique SA	France	*137*	75-23000
Ishikawajima-Harima H.I. Co. Ltd	Japan	*251*	75-02800
Japan Aviation Electronics Ind. Ltd	Japan	*252*	75-03300
JDR Cable Systems Ltd	UK	*462*	75-32480
Korolev Rocket Space Corp	Russia	*334*	55-10450
Latécoère	France	*138*	75-24200
Lockheed Martin Corp.	USA	*591*	75-55635
Lockheed Martin Space Mission	USA	*638*	80-18500
Matra Défense-Espace	France	*139*	75-26000
Matra Marconi Space UK, Ltd	UK	*425*	74-06900
Microtecnica-Torino SpA	Italy	*237*	75-15500
Moog Controls Ltd	UK	*469*	75-40670
Moog Italiana Srl	Italy	*237*	75-16230
MPB Technologies Inc	Canada	*74*	75-09855
NPO Prikladnoi Mechaniki	Russia	*338*	75-14600
Oerlikon-Contraves AG	Switzerland	*387*	75-09250
Parker Bertea Aerospace	USA	*602*	75-68900
Penny & Giles Data Systems Ltd	UK	*473*	75-44560
Pilkington Space Technology	UK	*426*	74-10000
Pilkington Thorn Optronics	UK	*426*	74-10000
Programmed Composites Inc.	USA	*606*	75-71500
Racal Antennas Ltd	UK	*476*	75-47656
Racal Radar Defence Systems Ltd.	UK	*477*	75-47688
Rockwell	USA	*610*	75-75930
Rutherford Appleton Laboratory	UK	*420*	60-01850
Saab Ericsson Space AB	Sweden	*378*	75-09150
SAGEM SA	**France**	***143***	**75-33950**
Sanders	USA	*612*	75-77850
Science Systems (Industrial) Ltd	UK	*510*	80-31600
Sea Tel Inc.	USA	*613*	75-78930
SED Systems Inc	Canada	*79*	80-01937
Sener Ingenieria y Sistemas SA	Spain	*367*	75-05200
SEXTANT Avionique SA	France	*144*	75-35000
SFIM	**France**	***144***	**75-35080**
Sheldahl Inc.	USA	*614*	75-80150
Sira Ltd.	UK	*484*	75-54250
Société Anonyme d'Études	France	*145*	75-37100
Space Systems Finland Ltd	Finland	*118*	75-02100
Spar Aerospace Ltd	Canada	*76*	75-13000
Sperry Marine Inc.	USA	*617*	75-83250
Teldix GmbH	Germany	*175*	75-30100
TERMA Elektronik AS	Denmark	*104*	75-20300
THERMOCOAX	France	*147*	75-44930
Tracor Inc	USA	*623*	75-90000

213 Space launcher systems & equipment

Aerospatiale	France	*131*	75-00600
Alcatel Bell	Belgium	*44*	75-00055
Alcatel Bell-SDT	Belgium	*44*	75-00060
Avica Equipment	UK	*432*	75-04870
Bradford Cylinders Ltd	UK	*435*	75-07850
EASAMS Ltd	UK	*501*	80-09000
Fokker Space & Systems	Netherlands	*293*	75-02325
Forward Industries Ltd	UK	*452*	75-24200
GEC-Marconi Avionics Ltd	UK	*454*	75-25320
Haley & Weller Ltd	UK	*456*	75-27400
Hartron Corporation	Ukraine	*406*	75-03000
Insumat Ltd	UK	*461*	75-31660
Intertechnique SA	France	*137*	75-23000
Japan Aviation Electronics Ind. Ltd	Japan	*252*	75-03300
Korolev Rocket Space Corp	Russia	*334*	55-10450
Lockheed Martin Corp.	USA	*591*	75-55635
Matra Défense-Espace	France	*139*	75-26000
Matra Marconi Space UK, Ltd	UK	*425*	74-06900
McDonnell Douglas Aerospace	USA	*594*	75-59480
Microtecnica-Torino SpA	Italy	*237*	75-15500
Mitsubishi Heavy Industries Ltd	Japan	*253*	75-05800
Moog Controls Ltd	UK	*469*	75-40670
Nissan Motor Co. Ltd	Japan	*254*	75-06790
Oerlikon-Contraves AG	Switzerland	*387*	75-09250
Parker Bertea Aerospace	USA	*602*	75-68900
Programmed Composites Inc.	USA	*606*	75-71500
Saab Ericsson Space AB	Sweden	*378*	75-09150
SABCA	Belgium	*46*	75-05590
SAGEM SA	**France**	***143***	**75-33950**
Sener Ingenieria y Sistemas SA	Spain	*367*	75-05200
SEXTANT Avionique SA	France	*144*	75-35000
SFIM	**France**	***144***	**75-35080**
Société Anonyme d'Études	France	*145*	75-37100
Société d'Applications des Machines	France	*143*	75-34300
TERMA Elektronik AS	Denmark	*104*	75-20300
The Boeing Company	USA	*560*	75-14250
THERMOCOAX	France	*147*	75-44930
Votaw Precision Technologies Inc.	USA	*626*	75-94580

214 Rocket, research / sounding, systems & equipment

Aerojet	USA	*547*	75-01500
Avibras Aeroespacial SA	Brazil	*55*	75-00968
Avica Equipment	UK	*432*	75-04870
Bressel SA	Spain	*366*	75-01670
Bristol Aerospace Ltd	Canada	*69*	75-01500
Cranfield University	UK	*419*	60-00350
Fokker Space & Systems	Netherlands	*293*	75-02325
Intertechnique SA	France	*137*	75-23000
Ishikawajima-Harima H.I. Co. Ltd	Japan	*251*	75-02800
Japan Aviation Electronics Ind. Ltd	Japan	*252*	75-03300
Mitsubishi Heavy Industries Ltd	Japan	*253*	75-05800
Moog Controls Ltd	UK	*469*	75-40670
Moog Italiana Srl	Italy	*237*	75-16230
Nissan Motor Co. Ltd	Japan	*254*	75-06790
Penny & Giles Data Systems Ltd	UK	*473*	75-44560
Royal Ordnance	UK	*481*	75-50620
Rutherford Appleton Laboratory	UK	*420*	60-01850
Saab Ericsson Space AB	Sweden	*378*	75-09150
SAGEM SA	**France**	***143***	**75-33950**

SED Systems Inc	Canada	79	80-01937
Sener Ingenieria y Sistemas SA	Spain	367	75-05200
SFIM	**France**	**144**	**75-35080**
Sodankyla Geophysical Observatory	Finland	117	60-03500
THERMOCOAX	France	147	75-44930
Tracor Inc	USA	623	75-90000
Votaw Precision Technologies Inc.	USA	626	75-94580

Airborne Platform Systems & Equipment

221 Aircraft systems & equipment

AAR Corp	USA	629	80-00035
AeroSpace Technologies of Australia	Australia	11	75-00340
AlliedSignal Aerospace Canada	Canada	68	75-00470
Ametek Inc.	USA	553	75-06510
Anite Systems Ltd	UK	497	80-01240
APPH Ltd	UK	431	75-03415
Astronautics Corp. of America	USA	555	75-10000
AVIAC Technologies	France	132	75-04600
Avica Equipment	UK	432	75-04870
Barber-Colman Co.	USA	557	75-11960
Baxter Woodhouse & Taylor Ltd	UK	433	75-06030
Bestobell Aviation	UK	434	75-06600
BFGoodrich Avionics Systems Inc	USA	559	75-13771
Boeing Australia Limited	Australia	12	75-02500
Bowmar Instrument Corp.	USA	560	75-14860
Bridport-Quin Pty Ltd	Australia	13	75-02746
British Aerospace Australia Ltd	Australia	13	75-02750
British Aerospace (Systems &	UK	436	75-08890
Bruce Industries Inc.	USA	561	75-15360
BTR Aerospace Australia	Australia	13	75-02815
Bülow Air Target Systems AB	Sweden	376	75-03600
Canadian Marconi Co.	Canada	69	75-02600
Chelton Inc.	USA	563	75-18120
CIC International Ltd	USA	563	75-18570
Computing Devices Company Ltd	UK	442	75-14700
Concorde Battery Corp.	USA	566	75-20845
Cranfield Aerospace Ltd	UK	500	80-07200
Cranfield University	UK	419	60-00350
CSE Aviation Ltd	UK	500	80-07600
Cubic Defense Systems Inc.	USA	567	75-22900
Daimler-Benz Aerospace France SAS	**France**	**134**	**75-12950**
Darchem Engineering Ltd.	UK	444	75-16750
Dornier Luftfahrt GmbH	Germany	168	75-08860
Dowty Aerospace Wolverhampton	UK	446	75-18589
DRS Technologies, Inc.	USA	570	75-27400
Dukane Corp.	USA	570	75-27650
Dunlop Aviation Division	UK	447	75-19000
EL-OP Electro Optics Industries Ltd	Israel	219	75-06300
Eldec Corp.	USA	572	75-29660
Electrodynamics Inc.	USA	572	75-29690
Electronic Space Systems Corp.	USA	572	75-30100
Envirovac Inc.	USA	574	75-31920
Ericsson Microwave Systems AB	**Sweden**	**377**	**75-04890**
Esterline Technologies	USA	574	75-32360
Fabbrica Italiana Apparecchiature	Italy	234	75-08550
Fairey Hydraulics	UK	450	75-22255
Farsound Engineering Ltd	UK	450	75-22420
Flight International Inc.	USA	635	80-11600
FPT Industries Ltd	UK	452	75-24400
GEC-Marconi Avionics Ltd	UK	454	75-25320
GKN Westland Technologies Ltd	UK	455	75-25690
Howden Aircontrol Ltd	UK	459	75-30240
Howden Fluid Systems	USA	582	75-44400
Howden Wade Ltd	UK	459	75-30260
IMP Aerospace Limited	Canada	72	75-07730
Intertechnique (Aerospace) Ltd	UK	461	75-31948
Israel Aircraft Industries Ltd	Israel	220	75-10200
Japan Aviation Electronics Ind. Ltd	Japan	252	75-03300
Kearfott Guidance & Navigation Corp.	USA	588	75-51760
Kidde-Graviner Ltd	UK	463	75-33880
Kollsman	USA	589	75-53150
Labinal Aero Defense Systems Inc	USA	638	80-18100
Lear Astronics Corp.	USA	589	75-54390
LITEF GmbH	**Germany**	**171**	**75-18600**
Litton Industries Inc.	USA	590	75-55400
Logic SpA	Italy	236	75-13800
Lucas Aerospace FCS Paris	France	139	75-25390
Lucas Aerospace Power Transmission	USA	592	75-56265
Lucas Western Inc.	USA	592	75-56300
Magnaghi Milano SpA	Italy	236	75-13980
Messier-Dowty Electronics	Canada	74	75-09650
Microtecnica-Torino SpA	Italy	237	75-15500
Microturbo SA	France	140	75-27000
Mitsubishi Heavy Industries Ltd	Japan	253	75-05800
Mitsubishi Precision Co. Ltd	Japan	253	75-06000
Moog Controls Ltd	UK	469	75-40670
Moog Italiana Srl	Italy	237	75-16230
Nanoquest Defence Products Ltd	UK	470	75-41890
Nord-Micro Elektronik Feinmechanik	Germany	173	75-22700
Normalair-Garrett Ltd	UK	425	74-07000
Northrop Grumman Corporation	USA	599	75-65450
Novatronics Inc	Canada	74	75-10300
Pakistan Aeronautical Complex	Pakistan	315	55-01600
Parker Bertea Aerospace	USA	602	75-68900
Penny & Giles Drives Technology Ltd	UK	474	75-44563
Pilkington Aerospace Ltd	UK	474	75-45500
Pilkington Optronics Ltd	UK	426	74-10000
R. Alkan (UK) Ltd	UK	429	75-02680
Racal Avionics Ltd	UK	477	75-47660
Racal Radar Defence Systems Ltd	UK	477	75-47688
Raytheon E-Systems	USA	608	75-73302
Revue THOMMEN AG	Switzerland	387	75-11100
Rogerson Aircraft Corp.	USA	611	75-76030
Saab Dynamics AB	Sweden	378	75-09100
SABCA	Belgium	46	75-05590
Safe Flight Instrument Corp.	USA	612	75-77400
SAGEM SA	**France**	**143**	**75-33950**
Sanders	USA	612	75-77850
Schlumberger Industries	France	143	75-34432
SEXTANT Avionique SA	France	144	75-35000
SFIM	**France**	**144**	**75-35080**
Sheldahl Inc.	USA	614	75-80150
Sierra Technologies Inc.	USA	614	75-80640
Smiths Aerospace & Defense	USA	615	75-82090
Smiths Industries Aerospace	UK	485	75-54700
SP aerospace and vehicle systems bv	**Netherlands**	**295**	**75-06860**
Survival Products Inc.	USA	618	75-85200
Swiss Aircraft & Systems Co	Switzerland	384	55-00100
Technofan SA	France	147	75-44300
Textron Systems Division	USA	621	75-88655
THERMOCOAX	France	147	75-44930
Tracor Inc	USA	623	75-90000
TransInstruments	UK	489	75-60960
Tumo Ltd	Finland	119	75-02650
Universal Propulsion Co. Inc.	USA	625	75-92700
Vickers Aerospace Components	UK	491	75-63020
Vickers Systems Division TRINOVA Ltd	UK	491	75-63150
Warner Electric	USA	627	75-95105
Warren-Knight Instrument Co.	USA	627	75-95150

221 10 Aircraft flight control systems

Alden Sales Co.	USA	548	75-04410
Anite Systems Ltd	UK	497	80-01240
Astronautics Corp. of America	USA	555	75-10000
Avionics Specialities	USA	556	75-11200
BFGoodrich Avionics Systems Inc	USA	559	75-13771
Bodenseewerk Gerätetechnik GmbH	Germany	165	75-03000
Boeing Australia Limited	Australia	12	75-02500
Bowmar Instrument Corp.	USA	560	75-14860
C-RAN Corp.	USA	566	75-22400
Canadian Marconi Co.	Canada	69	75-02600
CIC International Ltd	USA	563	75-18570
Cranfield Aerospace Ltd	UK	500	80-07200
Cranfield University	UK	419	60-00350
Curtiss-Wright Flight Systems Inc.	USA	567	75-23055
Dassault Aviation Group	France	135	75-12975
Design Bureau of Industrial	Russia	337	75-03200
Dornier Luftfahrt GmbH	Germany	168	75-08860
Dowty Aerospace Wolverhampton	UK	446	75-18589
EL-OP Electro Optics Industries Ltd	Israel	219	75-06300
Engineered Systems Co.	USA	574	75-31590
Esterline Technologies	USA	574	75-32360
Fairey Hydraulics	UK	450	75-22255
Farsound Engineering Ltd	UK	450	75-22420
GEC-Marconi Avionics Ltd	UK	454	75-25320
Interface Products Inc.	USA	585	75-47300
Israel Aircraft Industries Ltd	Israel	220	75-10200
Japan Aviation Electronics Ind. Ltd	Japan	252	75-03300
Kearfott Guidance & Navigation Corp.	USA	588	75-51760
Kidde-Fenwal	USA	588	75-52510

Litton Industries Inc.	USA	*590*	75-55400
Lucas Aerospace FCS Paris	France	*139*	75-25390
Lucas Aerospace Power Transmission	USA	*592*	75-56265
Menasco Aerospace Ltd	Canada	*74*	75-09600
Messier-Dowty Electronics	Canada	*74*	75-09650
Microtecnica-Torino SpA	Italy	*237*	75-15500
Mitsubishi Heavy Industries Ltd	Japan	*253*	75-05800
Moog Controls Ltd	UK	*469*	75-40670
Moog Inc.	USA	*597*	75-62550
Moog Italiana Srl	Italy	*237*	75-16230
Nord-Micro Elektronik Feinmechanik	Germany	*173*	75-22700
Novatronics Inc	Canada	*74*	75-10300
Parker Bertea Aerospace	USA	*602*	75-68900
Raytheon E-Systems	USA	*608*	75-73302
Revue THOMMEN AG	Switzerland	*387*	75-11100
Rockwell	USA	*610*	75-75930
Rockwell International Co.	Italy	*233*	75-05200
SEXTANT Avionique SA	France	*144*	75-35000
SFIM	**France**	***144***	**75-35080**
Smiths Aerospace & Defense	USA	*615*	75-82090
Smiths Industries Aerospace	UK	*485*	75-54700
Société d'Applications des Machines	France	*143*	75-34300
TERMA Elektronik AS	Denmark	*104*	75-20300
Tokyo Aircraft Instrument Co Ltd	Japan	*254*	75-08600
Tracor Inc	USA	*623*	75-90000

221 10 03 De-icing systems, aircraft, pneumatic

Bestobell Aviation	UK	*434*	75-06600
Boeing Australia Limited	Australia	*12*	75-02500
DNE, Inc.	USA	*569*	75-27000
IMI Marston Ltd	UK	*460*	75-31155
Tempo Instruments Inc	USA	*621*	75-87900

221 10 10 Automatic flight control systems, aircraft

AeroSpace Technologies of Australia	Australia	*11*	75-00340
Anite Systems Ltd	UK	*497*	80-01240
Astronautics Corp. of America	USA	*555*	75-10000
BFGoodrich Avionics Systems Inc	USA	*559*	75-13771
Bodenseewerk Gerätetechnik GmbH	Germany	*165*	75-03000
Boeing Australia Limited	Australia	*12*	75-02500
Design Bureau of Industrial	Russia	*337*	75-03200
EL-OP Electro Optics Industries Ltd	Israel	*219*	75-06300
GEC-Marconi Avionics Inc.	USA	*577*	75-37413
GEC-Marconi Avionics Ltd	UK	*454*	75-25320
Israel Aircraft Industries Ltd	Israel	*219*	75-10200
Japan Aviation Electronics Ind. Ltd	Japan	*252*	75-03300
Menasco Aerospace Ltd	Canada	*74*	75-09600
Moog Controls Ltd	UK	*469*	75-40670
Moog Inc.	USA	*597*	75-62550
Moog Italiana Srl	Italy	*237*	75-16230
Nord-Micro Elektronik Feinmechanik	Germany	*173*	75-22700
Novatronics Inc	Canada	*74*	75-10300
Parker Bertea Aerospace	USA	*602*	75-68900
Recon/Optical Inc.	USA	*608*	75-73930
Rockwell	USA	*610*	75-75930
Rockwell International Co.	Italy	*233*	75-05200
SEXTANT Avionique SA	France	*144*	75-35000
SFIM	**France**	***144***	**75-35080**
TERMA Elektronik AS	Denmark	*104*	75-20300
Tokyo Aircraft Instrument Co Ltd	Japan	*254*	75-08600
Tracor Inc	USA	*623*	75-90000

221 10 11 Automatic pilots, aircraft

Astronautics Corp. of America	USA	*555*	75-10000
BFGoodrich Avionics Systems Inc	USA	*559*	75-13771
Bodenseewerk Gerätetechnik GmbH	Germany	*165*	75-03000
Boeing Australia Limited	Australia	*12*	75-02500
GEC-Marconi Avionics Inc.	USA	*577*	75-37413
GEC-Marconi Avionics Ltd	UK	*454*	75-25320
Japan Aviation Electronics Ind. Ltd	Japan	*252*	75-03300
Meteor CAE SpA	Italy	*237*	75-15400
Moog Italiana Srl	Italy	*237*	75-16230
Parker Bertea Aerospace	USA	*602*	75-68900
Rockwell	USA	*610*	75-75930
Rockwell International Co.	Italy	*233*	75-05200
SEXTANT Avionique SA	France	*144*	75-35000
SFIM	**France**	***144***	**75-35080**
Tokyo Aircraft Instrument Co Ltd	Japan	*254*	75-08600

221 10 12 Aircraft fly-by-wire systems

Bodenseewerk Gerätetechnik GmbH	Germany	*165*	75-03000
Cranfield Aerospace Ltd	UK	*500*	80-07200
Cranfield University	UK	*419*	60-00350
Dassault Aviation Group	France	*135*	75-12975
Dornier Luftfahrt GmbH	Germany	*168*	75-08860
Dowty Aerospace Wolverhampton	UK	*446*	75-18589
Fairey Hydraulics	UK	*450*	75-22255
GEC-Marconi Avionics Inc.	USA	*577*	75-37413
GEC-Marconi Avionics Ltd	UK	*454*	75-25320
Japan Aviation Electronics Ind. Ltd	Japan	*252*	75-03300
Lear Astronics Corp.	USA	*589*	75-54390
Lucas Aerospace FCS Paris	France	*139*	75-25390
Lucas Aerospace Power Transmission	USA	*592*	75-56265
Lucas Western Inc.	USA	*592*	75-56300
Menasco Aerospace Ltd	Canada	*74*	75-09600
Messier-Dowty Electronics	Canada	*74*	75-09650
Moog Controls Ltd	UK	*469*	75-40670
Moog Inc.	USA	*597*	75-62550
Moog Japan Ltd.	Japan	*253*	75-06100
Novatronics Inc	Canada	*74*	75-10300
Parker Bertea Aerospace	USA	*602*	75-68900
SEXTANT Avionique SA	France	*144*	75-35000
Vickers Inc.	USA	*626*	75-93820
Vickers Systems	Germany	*176*	75-32830
Vickers Systems SA	France	*148*	75-46930
Vickers Systems Trinova SpA	Italy	*239*	75-27150

221 10 15 All-weather landing systems, aircraft

Alden Sales Co.	USA	*548*	75-04410
Beaufort Air-Sea Equipment Ltd	UK	*433*	75-06100
Boeing Australia Limited	Australia	*12*	75-02500
CIC International Ltd	USA	*563*	75-18570
Dornier Luftfahrt GmbH	Germany	*168*	75-08860
Fabbrica Italiana Apparecchiature	Italy	*234*	75-08550
GEC-Marconi Avionics Inc.	USA	*577*	75-37413
GEC-Marconi Avionics Ltd	UK	*454*	75-25320
Rockwell	USA	*610*	75-75930
Rockwell International Co.	Italy	*233*	75-05200
SEXTANT Avionique SA	France	*144*	75-35000
Warren-Knight Instrument Co.	USA	*627*	75-95150

221 10 17 Ground proximity warning systems, aircraft

GEC-Marconi Avionics Inc.	USA	*577*	75-37413
Israel Aircraft Industries Ltd	Israel	*219*	75-10200
Japan Aviation Electronics Ind. Ltd	Japan	*252*	75-03300
Kaiser Electronics	USA	*587*	75-51350
Rockwell	USA	*610*	75-75930
TERMA Elektronik AS	Denmark	*104*	75-20300
Tokyo Aircraft Instrument Co Ltd	Japan	*254*	75-08600

221 10 20 Aircraft engine control systems & equipment

Astronautics Corp. of America	USA	*555*	75-10000
Bodenseewerk Gerätetechnik GmbH	Germany	*165*	75-03000
Boeing Australia Limited	Australia	*12*	75-02500
Bowmar Instrument Corp.	USA	*560*	75-14860
Canadian Marconi Co.	Canada	*69*	75-02600
Dornier Luftfahrt GmbH	Germany	*168*	75-08860
Dowty Controls	UK	*446*	75-18592
ELECMA	France	*136*	75-14400
Farsound Engineering Ltd	UK	*450*	75-22420
GEC-Marconi Avionics Inc.	USA	*577*	75-37413
GEC-Marconi Avionics Ltd	UK	*454*	75-25320
Gull Electronic Systems Div.	USA	*580*	75-41130
Litton Industries Inc.	USA	*590*	75-55400
Lucas Aerospace Power Transmission	USA	*592*	75-56265
Meggitt Avionics (UK)	UK	*467*	75-38340
Microtecnica-Torino SpA	Italy	*237*	75-15500
Moog Controls Ltd	UK	*469*	75-40670
Moog Inc.	USA	*597*	75-62550
Parker Bertea Aerospace	USA	*602*	75-68900
Rockwell	USA	*610*	75-75930
RSL Electronics Ltd	Israel	*222*	75-18900
Safe Flight Instrument Corp.	USA	*612*	75-77400
Smiths Aerospace & Defense	USA	*615*	75-82090
Sundstrand Aerospace	USA	*618*	75-85105
THERMOCOAX	France	*147*	75-44930
Tuthill Corporation	USA	*624*	75-91320

221 10 30 Aircraft flight management systems

Astronautics Corp. of America	USA	555	75-10000
Boeing Australia Limited	Australia	12	75-02500
Canadian Marconi Co.	Canada	69	75-02600
CIC International Ltd	USA	563	75-18570
Computer Resources International A/S.	Denmark	104	80-00100
Dornier Luftfahrt GmbH	Germany	168	75-08860
Farsound Engineering Ltd	UK	450	75-22420
GEC-Marconi Avionics Inc.	USA	577	75-37413
GEC-Marconi Avionics Ltd	UK	454	75-25320
Gull Electronic Systems Div.	USA	580	75-41130
Northrop Grumman Corporation.	USA	599	75-65450
Pointer Inc.	USA	604	75-70810
Racal Avionics Ltd	UK	477	75-47660
Racal Radar Defence Systems Ltd.	UK	477	75-47688
Rockwell.	USA	610	75-75930
Rockwell International Co.	Italy.	233	75-05200
SEXTANT Avionique SA.	France	144	75-35000
Smiths Aerospace & Defense	USA	615	75-82090
Teledyne Controls	USA	620	75-87518
Tokyo Aircraft Instrument Co Ltd	Japan	254	75-08600

221 10 70 Aircraft flight control equipment

Alden Sales Co.	USA	548	75-04410
Anite Systems Ltd	UK	497	80-01240
Astronautics Corp. of America	USA	555	75-10000
AVIAC Technologies	France	132	75-04600
Avionics Specialities	USA	556	75-11200
BFGoodrich Avionics Systems Inc	USA	559	75-13771
Bodenseewerk Gerätetechnik GmbH	Germany	165	75-03000
CIC International Ltd	USA	563	75-18570
Curtiss-Wright Flight Systems Inc.	USA	567	75-23055
Dornier Luftfahrt GmbH	Germany	168	75-08860
Dowty Aerospace Wolverhampton	UK	446	75-18589
EL-OP Electro Optics Industries Ltd	Israel	219	75-06300
Eldec Corp.	USA	572	75-29660
Esterline Technologies	USA	574	75-32360
Fairey Hydraulics	UK	450	75-22255
GEC-Marconi Avionics Ltd	UK	454	75-25320
Japan Aviation Electronics Ind. Ltd	Japan	252	75-03300
Kaiser Electroprecision	USA	587	75-51350
Kavlico Corp.	USA	588	75-51640
Kearfott Guidance & Navigation Corp.	USA	588	75-51760
Kollsman.	USA	589	75-53150
Liebherr-Aerospace Lindenberg GmbH	Germany	171	75-18300
Logica UK Limited.	UK	506	80-14350
Magnaghi Milano SpA	Italy.	236	75-13980
Menasco Aerospace Ltd	Canada.	74	75-09600
Microtecnica-Torino SpA	Italy.	237	75-15500
Moog Controls Ltd	UK	469	75-40670
Moog Inc.	USA	597	75-62550
Moog Japan Ltd.	Japan	253	75-06100
Page Aerospace Ltd	UK	472	75-43505
Parker Bertea Aerospace.	USA	602	75-68900
Racal Radar Defence Systems Ltd.	UK	477	75-47688
Raytheon Aircraft Company	USA	607	75-73200
Raytheon E-Systems	USA	608	75-73302
Rockwell.	USA	610	75-75930
Rockwell International Co.	Italy.	233	75-05200
Rogerson Aircraft Corp.	USA	611	75-76030
SABCA	Belgium	46	75-05590
SAGEM SA	**France**	**143**	**75-33950**
SEXTANT Avionique SA.	France	144	75-35000
SFIM	**France**	**144**	**75-35080**
TERMA Elektronik AS	Denmark	104	75-20300
THERMOCOAX	France	147	75-44930
Tokyo Aircraft Instrument Co Ltd	Japan	254	75-08600
Universal Avionics Systems Corp	USA	625	75-92650

221 10 70 20 Aircraft control columns

GEC-Marconi Avionics Ltd	UK	454	75-25320
Imco Industries Ltd	Israel	219	75-09400
Interface Products Inc.	USA	585	75-47300
Page Aerospace Ltd	UK	472	75-43505

221 10 70 30 Flight instruments / avionics, aircraft

Advanced Technologies &.	S. Africa	357	75-00500
AIM Enterprises Inc.	USA	630	80-01350
Alden Sales Co.	USA	548	75-04410
AlliedSignal Aerospace Canada	Canada	68	75-00470
Ametek Inc.	USA	553	75-06510
Astronautics Corp. of America	USA	555	75-10000
Aviapribor Corp	Russia	336	75-01200
Avionics Specialities	USA	556	75-11200
Bavaria Avionik Technologie GmbH.	Germany	165	75-02040
Becker Flugfunkwerk GmbH	Germany	165	75-02200
BFGoodrich Aerospace	USA	558	75-13770
BFGoodrich Avionics Systems Inc	USA	559	75-13771
Bodenseewerk Gerätetechnik GmbH	Germany	165	75-03000
Boeing Australia Limited	Australia	12	75-02500
Bospec Neue Technologien	Germany	166	75-03420
Canadian Marconi Co.	Canada	69	75-02600
Century Electronics	USA	562	75-17385
CIC International Ltd	USA	563	75-18570
CSE Aviation Ltd	UK	500	80-07600
Diehl GmbH & Co	Germany	167	75-06500
Ekran Belaruskoye PO	Belarus	38	75-00400
EL-OP Electro Optics Industries Ltd	Israel	219	75-06300
Esterline Technologies	USA	574	75-32360
Flightline Aircraft Instruments	USA	575	75-34840
GEC-Marconi Avionics Ltd	UK	454	75-25320
Grimes Aerospace Company	USA	580	75-40400
Gull Electronic Systems Div.	USA	580	75-41130
INDRA	Spain	366	75-03800
Interface Products Inc.	USA	585	75-47300
Japan Aviation Electronics Ind. Ltd	Japan	252	75-03300
JP Instruments Inc.	USA	587	75-51200
Kaiser Electronics	USA	587	75-51350
Kazan Design Bureau	Russia	333	55-10350
Kearfott Guidance & Navigation Corp.	USA	588	75-51760
Kollsman.	USA	589	75-53150
Litton Industries Inc.	USA	590	75-55400
Logica UK Limited.	UK	506	80-14350
Meggitt Avionics Inc.	USA	595	75-59842
Meggitt Avionics (UK)	UK	467	75-38340
Merex Inc.	USA	595	75-60250
Microtecnica-Torino SpA	Italy.	237	75-15500
Norwegian Defence Technology A/S.	Norway	310	75-02050
Pacer Systems Inc.	USA	602	75-68000
Page Aerospace Ltd	UK	472	75-43505
Penny & Giles Aerospace Ltd	UK	473	75-44550
Phaostron Instrument & Electronic Co	USA	603	75-69500
Racal Avionics Ltd	UK	477	75-47660
Racal Radar Defence Systems Ltd.	UK	477	75-47688
Revue THOMMEN AG	Switzerland	387	75-11100
Rockwell.	USA	610	75-75930
Rockwell International Co.	Italy.	233	75-05200
Rogerson Aircraft Corp.	USA	611	75-76030
Rogerson Kratos	USA	611	75-76035
Rosemount Aerospace Inc.	USA	611	75-76650
Safe Flight Instrument Corp.	USA	612	75-77400
SAGEM SA	**France**	**143**	**75-33950**
SEXTANT Avionique SA.	France	144	75-35000
SFIM	**France**	**144**	**75-35080**
Smiths Aerospace & Defense	USA	615	75-82090
Teldix GmbH	Germany	175	75-30100
Tempo Instruments Inc	USA	621	75-87900
THERMOCOAX	France	147	75-44930
Thompson Aero Products	USA	622	75-88890
Tokyo Aircraft Instrument Co Ltd	Japan	254	75-08600
Tracor Inc	USA	623	75-90000

221 10 70 30 10 Altimeters

Alden Sales Co.	USA	548	75-04410
Arab Organization for Ind.	Egypt	109	55-01700
Aviapribor Corp	Russia	336	75-01200
Bavaria Avionik Technologie GmbH.	Germany	165	75-02040
Boeing Australia Limited	Australia	12	75-02500
CIC International Ltd	USA	563	75-18570
Explorocean Technology Ltd	UK	502	80-09870
GEC-Marconi Avionics Ltd	UK	454	75-25320
Kollsman.	USA	589	75-53150
Litton Industries Inc.	USA	590	75-55400
Meggitt Avionics (UK)	UK	467	75-38340
Merex Inc.	USA	595	75-60250
Rogerson Aircraft Corp.	USA	611	75-76030
Smiths Aerospace & Defense	USA	615	75-82090
Tokyo Aircraft Instrument Co Ltd	Japan	254	75-08600

221 10 70 30 10 20 Altimeters, barometric

Alden Sales Co.	USA	548	75-04410
Bavaria Avionik Technologie GmbH.	Germany	165	75-02040
CIC International Ltd	USA	563	75-18570
Explorocean Technology Ltd	UK	502	80-09870
Flightline Aircraft Instruments	USA	575	75-34840
Litton Industries Inc.	USA	590	75-55400
Meggitt Avionics (UK)	UK	467	75-38340
Merex Inc.	USA	595	75-60250
Tokyo Aircraft Instrument Co Ltd	Japan	254	75-08600

221 10 70 30 10 50 Altimeters, radar

Alden Sales Co.	USA	548	75-04410
Boeing Australia Limited	Australia	12	75-02500
CIC International Ltd	USA	563	75-18570
GEC-Marconi Avionics Inc.	USA	577	75-37413
GEC-Marconi Avionics Ltd	UK	454	75-25320
Japan Aviation Electronics Ind. Ltd	Japan	252	75-03300
Merex Inc.	USA	595	75-60250

221 10 70 30 10 60 Altimeters, radio

Arab Organization for Ind.	Egypt	109	55-01700
Boeing Australia Limited	Australia	12	75-02500
Ericsson Defence Systems Pty Ltd	Australia	16	75-05820
GEC-Marconi Avionics Inc.	USA	577	75-37413
Rockwell International Co.	Italy	233	75-05200

221 10 70 30 15 Attitude reference systems

Alden Sales Co.	USA	548	75-04410
Astronautics C.A. Ltd	Israel	217	75-02400
Astronautics Corp. of America	USA	555	75-10000
BFGoodrich Avionics Systems Inc	USA	559	75-13771
Boeing Australia Limited	Australia	12	75-02500
GEC-Marconi Avionics Inc.	USA	577	75-37413
GEC-Marconi Avionics Ltd	UK	454	75-25320
Humphrey Inc.	USA	583	75-45050
Israel Aircraft Industries Ltd	Israel	220	75-10200
Japan Aviation Electronics Ind. Ltd	Japan	252	75-03300
Kearfott Guidance & Navigation Corp.	USA	588	75-51760
LITEF GmbH	**Germany**	**171**	**75-18600**
Litton Industries Inc.	USA	590	75-55400
Rockwell International Co.	Italy	233	75-05200
SAGEM SA	**France**	**143**	**75-33950**
SFIM	**France**	**144**	**75-35080**
Smiths Aerospace & Defense	USA	615	75-82090
Swiss Aircraft & Systems Co	Switzerland	384	55-00100

221 10 70 30 50 Receivers, landing system, aircraft

Alden Sales Co.	USA	548	75-04410
Boeing Australia Limited	Australia	12	75-02500
Canadian Marconi Co.	Canada	69	75-02600
GEC-Marconi Avionics Inc.	USA	577	75-37413
GEC-Marconi Avionics Ltd	UK	454	75-25320
Rockwell International Co.	Italy	233	75-05200

221 10 70 30 54 Receivers, ILS, aircraft

Becker Flugfunkwerk GmbH	Germany	165	75-02200
Boeing Australia Limited	Australia	12	75-02500
GEC-Marconi Avionics Inc.	USA	577	75-37413
GEC-Marconi Avionics Ltd	UK	454	75-25320
Raytheon Cossor Electronics Ltd	UK	478	75-48367
Rockwell	USA	610	75-75930
Rockwell International Co.	Italy	233	75-05200

221 10 70 30 55 Receivers, MLS, aircraft

Boeing Australia Limited	Australia	12	75-02500
Canadian Marconi Co.	Canada	69	75-02600
Fabbrica Italiana Apparecchiature	Italy	234	75-08550
GEC-Marconi Avionics Inc.	USA	577	75-37413
GEC-Marconi Avionics Ltd	UK	454	75-25320
Marconi SpA DTM	Italy	236	75-14550
Rockwell	USA	610	75-75930
Smiths Aerospace & Defense	USA	615	75-82090

221 10 70 30 60 Thunderstorm mapping equipment, aircraft, passive

BFGoodrich Aerospace	USA	559	75-13770
Boeing Australia Limited	Australia	12	75-02500
GEC-Marconi Avionics Inc.	USA	577	75-37413

221 10 70 35 Aircraft flight recorders

Anite Systems Ltd	UK	497	80-01240
Boeing Australia Limited	Australia	12	75-02500
CIC International Ltd	USA	563	75-18570
DRS Precision Echo, Inc.	USA	570	75-27360
GEC-Marconi Avionics Inc.	USA	577	75-37413
GEC-Marconi Avionics Ltd	UK	454	75-25320
Japan Aviation Electronics Ind. Ltd	Japan	252	75-03300
Penny & Giles Aerospace Ltd	UK	473	75-44550
Rockwell	USA	610	75-75930
Rogerson Aircraft Corp.	USA	611	75-76030
SAGEM SA	**France**	**143**	**75-33950**
Schlumberger Industries	France	143	75-34432
SFIM	**France**	**144**	**75-35080**
Smiths Aerospace & Defense	USA	615	75-82090
The Flight Data Company Ltd	UK	502	80-10600
Tokyo Aircraft Instrument Co Ltd	Japan	254	75-08600
TransInstruments	UK	489	75-60960
Universal Avionics Systems Corp	USA	625	75-92650

221 10 70 40 Head-up displays, aircraft

Astronautics Corp. of America	USA	555	75-10000
Avimo Ltd	UK	432	75-04950
BFGoodrich Avionics Systems Inc	USA	559	75-13771
Bospec Neue Technologien	Germany	166	75-03420
China National Aero-Tech I/E Corp.	China PR	88	55-00200
CIC International Ltd	USA	563	75-18570
Daimler-Benz Aerospace France SAS	**France**	**134**	**75-12950**
Delft Sensor Systems	**Belgium**	**46**	**75-02500**
Delft Sensor Systems	Netherlands	292	75-01590
DRS Technologies, Inc.	USA	570	75-27400
EL-OP Electro Optics Industries Ltd	Israel	219	75-06300
Fabbrica Italiana Apparecchiature	Italy	234	75-08550
GEC-Marconi Avionics Inc.	USA	577	75-37413
GEC-Marconi Avionics Ltd	UK	454	75-25320
Hughes Elcan Optical Technologies	Canada	72	75-07650
Kaiser Electronics	USA	587	75-51350
Pilkington Thorn Optronics	UK	426	74-10000
Planar Advance Inc.	UK	475	75-45735
Rogerson Aircraft Corp.	USA	611	75-76030
SEXTANT Avionique SA.	France	144	75-35000
SRL Operations	USA	617	75-83550
Teldix GmbH	Germany	175	75-30100

221 10 70 50 Map displays, aircraft

Astronautics Corp. of America	USA	555	75-10000
Eventide Inc.	USA	574	75-32655
GEC-Marconi Avionics Inc.	USA	577	75-37413
GEC-Marconi Avionics Ltd	UK	454	75-25320
Logicon Geodynamics Inc.	USA	638	80-18625
Rogerson Aircraft Corp.	USA	611	75-76030
SEXTANT Avionique SA.	France	144	75-35000
Smiths Aerospace & Defense	USA	615	75-82090
Teldix GmbH	Germany	175	75-30100

221 10 70 60 Parachutes, aircraft braking

Autoflug GmbH & Co	Germany	165	75-01400
Engineered Systems Co.	USA	574	75-31590
Irvin Aerospace Limited	UK	462	75-32050
Irvin Industries Inc.	USA	586	75-48960
Pioneer Aerospace Corp	USA	604	75-70200
Sächsische Spezialkonfektion GmbH	Germany	174	75-25880
Sala International Pty Ltd	Australia	21	75-18980
Syndex Aerospace Inc.	USA	619	75-85540

221 20 Aircraft mission-related systems, (airframe integrated)

Aerea SpA	Italy	*230*	75-00280
Aero Union Corp.	USA	*546*	75-01118
Aircraft Materials Ltd	UK	*428*	75-01650
Anite Systems Ltd	UK	*497*	80-01240
Astronautics Corp. of America	USA	*555*	75-10000
Boeing Australia Limited	Australia	*12*	75-02500
Breeze Eastern Corp.	USA	*560*	75-15110
Bridport-Quin Pty Ltd	Australia	*13*	75-02746
Bülow Air Target Systems AB.	Sweden	*376*	75-03600
Canadian Marconi Co.	Canada	*69*	75-02600
CIC International Ltd	USA	*563*	75-18570
Computing Devices Company Ltd	UK	*442*	75-14700
Cranfield Aerospace Ltd	UK	*500*	80-07200
Cranfield University	UK	*419*	60-00350
CSE Aviation Ltd	UK	*500*	80-07600
Cubic Defense Systems Inc.	USA	*567*	75-22900
DRS Technologies, Inc.	USA	*570*	75-27400
Elbit Systems Ltd	Israel	*218*	75-05300
Ericsson Microwave Systems AB	**Sweden**	***377***	**75-04890**
Flight International Inc.	USA	*635*	80-11600
GEC-Marconi Avionics Inc.	USA	*577*	75-37413
GEC-Marconi Avionics Ltd	UK	*454*	75-25320
Hayes Targets	USA	*603*	75-69135
Israel Aircraft Industries Ltd	Israel	*220*	75-10200
Litton Industries Inc.	USA	*590*	75-55400
Lockheed Martin Missiles & Space	USA	*592*	75-55640
Lucas Aerospace Power Transmission	USA	*592*	75-56265
Lucas Western Inc.	USA	*592*	75-56300
Nanoquest Defence Products Ltd	UK	*470*	75-41890
Northrop Grumman Corporation	USA	*599*	75-65450
Pemco World Air Services	USA	*603*	75-69135
Racal Avionics Ltd	UK	*477*	75-47660
Racal Radar Defence Systems Ltd.	UK	*477*	75-47688
Raytheon E-Systems	USA	*608*	75-73302
SAGEM SA	**France**	***143***	**75-33950**
Sanders	USA	*612*	75-77850
Sargent Fletcher Inc.	USA	*612*	75-78150
SEXTANT Avionique SA.	France	*144*	75-35000
SFIM	**France**	***144***	**75-35080**
Smiths Industries Aerospace	UK	*485*	75-54700
Static Engineering Pty Ltd	Australia	*21*	75-19780
TransInstruments	UK	*489*	75-60960
Western Design Howden	USA	*627*	75-95800

221 20 05 Mission avionics, aircraft

Advanced Technologies &	S. Africa	*357*	75-00500
Aerostar SA	Romania	*331*	80-00100
Anite Systems Ltd	UK	*497*	80-01240
Astronautics Corp. of America	USA	*555*	75-10000
Boeing Australia Limited	Australia	*12*	75-02500
Canadian Marconi Co.	Canada	*69*	75-02600
Dornier Luftfahrt GmbH	Germany	*168*	75-08860
DRS Technologies, Inc.	USA	*570*	75-27400
Ericsson Microwave Systems AB	**Sweden**	***377***	**75-04890**
Fabbrica Italiana Apparecchiature	Italy	*234*	75-08550
Flight International Inc.	USA	*635*	80-11600
GEC-Marconi Avionics Inc.	USA	*577*	75-37413
GEC-Marconi Avionics Ltd	UK	*454*	75-25320
Israel Aircraft Industries Ltd	Israel	*220*	75-10200
Kaiser Electronics	USA	*587*	75-51350
LITEF GmbH	**Germany**	***171***	**75-18600**
Litton Industries Inc.	USA	*590*	75-55400
Lockheed Martin Tactical Aircraft	USA	*592*	75-55650
Racal Radar Defence Systems Ltd.	UK	*477*	75-47688
Rockwell	USA	*610*	75-75930
Rockwell International Co.	Italy	*233*	75-05200
SAGEM SA	**France**	***143***	**75-33950**
SEXTANT Avionique SA.	France	*144*	75-35000
SFIM	**France**	***144***	**75-35080**
Sierra Technologies Inc.	USA	*614*	75-80640
Smiths Aerospace & Defense	USA	*615*	75-82090
Teldix GmbH	Germany	*175*	75-30100
TransInstruments	UK	*489*	75-60960
Universal Avionics Systems Corp	USA	*625*	75-92650

221 20 05 10 Mission avionics, AEW aircraft

Anite Systems Ltd	UK	*497*	80-01240
Astronautics Corp. of America	USA	*555*	75-10000
Boeing Australia Limited	Australia	*12*	75-02500
CIC International Ltd	USA	*563*	75-18570
Dornier Luftfahrt GmbH	Germany	*168*	75-08860
DRS Technologies, Inc.	USA	*570*	75-27400
Ericsson Microwave Systems AB	**Sweden**	***377***	**75-04890**
Fabbrica Italiana Apparecchiature	Italy	*234*	75-08550
GEC-Marconi Avionics Inc.	USA	*577*	75-37413
GEC-Marconi Avionics Ltd	UK	*454*	75-25320
LITEF GmbH	**Germany**	***171***	**75-18600**
Litton Industries Inc.	USA	*590*	75-55400
Racal Radar Defence Systems Ltd.	UK	*477*	75-47688
Rockwell	USA	*610*	75-75930
SAGEM SA	**France**	***143***	**75-33950**
Smiths Aerospace & Defense	USA	*615*	75-82090
TransInstruments	UK	*489*	75-60960

221 20 05 12 Mission avionics, ASW aircraft

Aircraft Appliances & Equipment Ltd	Canada	*68*	75-00430
Anite Systems Ltd	UK	*497*	80-01240
Astronautics Corp. of America	USA	*555*	75-10000
Boeing Australia Limited	Australia	*12*	75-02500
Canadian Marconi Co.	Canada	*69*	75-02600
Dornier Luftfahrt GmbH	Germany	*168*	75-08860
DRS Technologies, Inc.	USA	*570*	75-27400
EDO Corporation	USA	*571*	75-29250
Fabbrica Italiana Apparecchiature	Italy	*234*	75-08550
GEC-Marconi Avionics Inc.	USA	*577*	75-37413
GEC-Marconi Avionics Ltd	UK	*454*	75-25320
Litton Industries Inc.	USA	*590*	75-55400
NEC Corporation	Japan	*253*	75-06350
Pacer Systems Inc.	USA	*602*	75-68000
PRB Associates Inc.	USA	*640*	80-24700
Racal Radar Defence Systems Ltd.	UK	*477*	75-47688
Rockwell	USA	*610*	75-75930
Rockwell International Co.	Italy	*233*	75-05200
SAGEM SA	**France**	***143***	**75-33950**
SEXTANT Avionique SA.	France	*144*	75-35000
Smiths Aerospace & Defense	USA	*615*	75-82090
TransInstruments	UK	*489*	75-60960

221 20 10 Cargo air delivery systems, aircraft

AAI Corp.	USA	*545*	75-00020
Aerial Machine & Tool Corp.	USA	*546*	75-01105
Ancra CHS BV.	Netherlands	*291*	75-00510
Bridport Aviation Products	UK	*435*	75-08400
Irvin Aerospace Limited	UK	*462*	75-32050
Para-Flite Inc.	USA	*602*	75-68800
Pioneer Aerospace Corp	USA	*604*	75-70200
Static Engineering Pty Ltd	Australia	*21*	75-19780
Van Riemsdijk Rotterdam	Netherlands	*296*	75-07890

221 20 10 70 Cargo air delivery equipment, aircraft

Aerial Machine & Tool Corp.	USA	*546*	75-01105
Aircraft Materials Ltd	UK	*428*	75-01650
Ancra CHS BV.	Netherlands	*291*	75-00510
Bridport Aviation Products	UK	*435*	75-08400
Capewell Aerial Delivery Systems	USA	*576*	75-36300
FXC Corp.	USA	*577*	75-36700
Irvin Aerospace Limited	UK	*462*	75-32050
MIL-BASE Industries Inc.	USA	*596*	75-61530
Pioneer Aerospace Corp	USA	*604*	75-70200
Static Engineering Pty Ltd	Australia	*21*	75-19780
Van Riemsdijk Rotterdam	Netherlands	*296*	75-07890

221 20 12 Cargo handling & stowage systems, aircraft

AAR Corp	USA	*629*	80-00035
Aerial Machine & Tool Corp.	USA	*546*	75-01105
Aircraft Materials Ltd	UK	*428*	75-01650
Ancra CHS BV.	Netherlands	*291*	75-00510
Bavaria Avionik Technologie GmbH.	Germany	*165*	75-02040
Breeze Eastern Corp.	USA	*560*	75-15110
Bridport Aviation Products	UK	*435*	75-08400
Irvin Aerospace Limited	UK	*462*	75-32050
Lucas Western Inc.	USA	*592*	75-56300
MIL-BASE Industries Inc.	USA	*596*	75-61530
Package Control (UK) Ltd.	UK	*472*	75-43450
Sala International Pty Ltd.	Australia	*21*	75-18980

SpanSet Inter AG	Switzerland	*387*	75-13550
Static Engineering Pty Ltd	Australia	*21*	75-19780
Van Riemsdijk Rotterdam	Netherlands	*296*	75-07890

221 20 12 70 Cargo handling & stowage equipment, aircraft

Aerial Machine & Tool Corp.	USA	*546*	75-01105
AIM Aviation (Jecco) Ltd.	UK	*428*	75-01530
Aircraft Materials Ltd	UK	*428*	75-01650
Ancra CHS BV.	Netherlands	*291*	75-00510
Bavaria Avionik Technologie GmbH.	Germany	*165*	75-02040
Breeze Eastern Corp.	USA	*560*	75-15110
Bridport Aviation Products	UK	*435*	75-08400
Irvin Aerospace Limited	UK	*462*	75-32050
Lucas Western Inc.	USA	*592*	75-56300
MIL-BASE Industries Inc.	USA	*596*	75-61530
Package Control (UK) Ltd.	UK	*472*	75-43450
ShinMaywa Industries Ltd	Japan	*254*	75-07800
SpanSet Inter AG	Switzerland	*387*	75-13550
Static Engineering Pty Ltd	Australia	*21*	75-19780
Van Riemsdijk Rotterdam	Netherlands	*296*	75-07890
Vickers Aerospace Components	UK	*491*	75-63020

221 20 12 70 50 Load restraint equipment, aircraft cargo

Aerial Machine & Tool Corp.	USA	*546*	75-01105
AIM Aviation (Jecco) Ltd.	UK	*428*	75-01530
Aircraft Materials Ltd	UK	*428*	75-01650
Ancra CHS BV.	Netherlands	*291*	75-00510
Bridport Aviation Products	UK	*435*	75-08400
Bridport-Quin Pty Ltd	Australia	*13*	75-02746
Cargo Aids	UK	*438*	75-11620
Irvin Aerospace Limited	UK	*462*	75-32050
Irvin Aerospace SpA	Italy	*235*	75-12200
Lite Flite Aps	Denmark	*103*	75-12000
Lucas Western Inc.	USA	*592*	75-56300
Package Control (UK) Ltd.	UK	*472*	75-43450
R. Alkan (UK) Ltd	UK	*429*	75-02680
Sala International Pty Ltd.	Australia	*21*	75-18980
SpanSet Inter AG	Switzerland	*387*	75-13550
Van Riemsdijk Rotterdam	Netherlands	*296*	75-07890
WB Chain Co Ltd	UK	*439*	75-12440

221 20 30 Fuel pods, aircraft, external

Aero Union Corp.	USA	*546*	75-01118
Aircraft Tanks Ltd	UK	*429*	75-01730
Arab Organization for Ind.	Egypt	*109*	55-01700
BP Chemicals Ltd	UK	*435*	75-07790
Dassault Aviation Group	France	*135*	75-12975
Dornier Luftfahrt GmbH	Germany	*168*	75-08860
EDO Corporation	USA	*571*	75-29250
Flight Refuelling Limited	UK	*451*	75-23850
Israel Military Industries Ltd	Israel	*220*	75-10300
Mandra Plant	Greece	*190*	75-02000
Pakistan Aeronautical Complex	Pakistan	*315*	55-01600
Sargent Fletcher Inc.	USA	*612*	75-78150
Sumitomo Electric Industries Ltd	Japan	*254*	75-07950
Swiss Aircraft & Systems Co	Switzerland	*384*	55-00100
Technical Products Group Inc.	USA	*620*	75-86764

221 20 40 Refuelling systems, aircraft

CIC International Ltd	USA	*563*	75-18570
Cranfield University	UK	*419*	60-00350
Dornier Luftfahrt GmbH	Germany	*168*	75-08860
Gull Electronic Systems Div.	USA	*580*	75-41130
Intertechnique SA.	France	*137*	75-23000
Logistic Projects	S. Africa	*358*	75-09800
MOL Cy NV	Belgium	*46*	75-04998
Sargent Fletcher Inc.	USA	*612*	75-78150

221 20 40 10 Refuelling systems, boom type, aircraft

Dornier Luftfahrt GmbH	Germany	*168*	75-08860
Gull Electronic Systems Div.	USA	*580*	75-41130
Intertechnique SA.	France	*137*	75-23000

221 20 40 50 Refuelling systems, probe & drogue type, aircraft

Flight Refuelling Limited	UK	*451*	75-23850
Giat Industries	France	*126*	55-00100
Gull Electronic Systems Div.	USA	*580*	75-41130
Intertechnique SA.	France	*137*	75-23000
Northrop Grumman Corporation	USA	*599*	75-65450

221 20 40 70 Refuelling equipment, aircraft

Aeroquip Corp.	USA	*547*	75-02350
Aircraft Appliances & Equipment Ltd	Canada	*68*	75-00430
Crane Fruehauf Ltd	UK	*443*	75-15750
Dabico Inc.	USA	*567*	75-23145
DNE, Inc.	USA	*569*	75-27000
Dornier Luftfahrt GmbH	Germany	*168*	75-08860
Flight Refuelling Limited	UK	*451*	75-23850
GEC-Marconi Avionics Ltd	UK	*454*	75-25320
Giat Industries	France	*126*	55-00100
Gruppo Nuova MA.NA.RO.	Italy	*235*	75-10730
Gull Electronic Systems Div.	USA	*580*	75-41130
Intertechnique SA.	France	*137*	75-23000
Zwicky Engineering Ltd	UK	*495*	75-67900

221 20 40 70 10 Air refuelling equipment, aircraft

DNE, Inc.	USA	*569*	75-27000
Dornier Luftfahrt GmbH	Germany	*168*	75-08860
GEC-Marconi Avionics Ltd	UK	*454*	75-25320

221 20 40 75 Refuelling pods, aircraft

Aero Union Corp.	USA	*546*	75-01118
EDO Corporation	USA	*571*	75-29250
Flight Refuelling Limited	UK	*451*	75-23850
Intertechnique SA.	France	*137*	75-23000

221 20 50 Target designation & aiming systems, aircraft

Boeing Australia Limited	Australia	*12*	75-02500
CI Systems Ltd	Israel	*218*	75-03750
Ericsson Microwave Systems AB	**Sweden**	***377***	**75-04890**
Fabbrica Italiana Apparecchiature	Italy	*234*	75-08550
GEC-Marconi Avionics Inc.	USA	*577*	75-37413
GEC-Marconi Avionics Ltd	UK	*454*	75-25320
Israel Aircraft Industries Ltd	Israel	*220*	75-10200
Kaiser Electronics	USA	*587*	75-51350
Rockwell	USA	*610*	75-75930
SAGEM SA	**France**	***143***	**75-33950**
Smiths Aerospace & Defense	USA	*615*	75-82090
Thomson-CSF Optronique	France	*148*	75-45137

221 20 50 90 Weapon aiming systems, aircraft

Aircraft Tanks Ltd	UK	*429*	75-01730
Astronautics Corp. of America	USA	*555*	75-10000
Avimo Ltd	UK	*432*	75-04950
Boeing Australia Limited	Australia	*12*	75-02500
Canadian Marconi Co.	Canada	*69*	75-02600
CI Systems Ltd	Israel	*218*	75-03750
Coherent Optics (Europe) Ltd	UK	*441*	75-14090
Cranfield University	UK	*419*	60-00350
Ericsson Microwave Systems AB	**Sweden**	***377***	**75-04890**
Fabbrica Italiana Apparecchiature	Italy	*234*	75-08550
Frazier Aviation Inc.	USA	*576*	75-35970
GEC-Marconi Avionics Inc.	USA	*577*	75-37413
GEC-Marconi Avionics Ltd	UK	*454*	75-25320
GEC-Marconi Research Centre	UK	*419*	60-00700
Israel Aircraft Industries Ltd	Israel	*220*	75-10200
Kaiser Electronics	USA	*587*	75-51350
Labtech Ltd	UK	*464*	75-34280
Nanoquest Defence Products Ltd	UK	*470*	75-41890
Pilkington Thorn Optronics	UK	*426*	74-10000
Polhemus Inc.	USA	*604*	75-70855
Rockwell	USA	*610*	75-75930
SAGEM SA	**France**	***143***	**75-33950**
Sener Ingenieria y Sistemas SA	Spain	*367*	75-05200
Smiths Aerospace & Defense	USA	*615*	75-82090
Warren-Knight Instrument Co.	USA	*627*	75-95150

221 20 55 Target towing systems, aircraft

Aerial Machine & Tool Corp.	USA	*546*	75-01105
Air Target Sweden AB	Sweden	*375*	75-00850
Bülow Air Target Systems AB.	Sweden	*376*	75-03600
CIC International Ltd	USA	*563*	75-18570
Cranfield Aerospace Ltd	UK	*500*	80-07200
Dornier Luftfahrt GmbH	Germany	*168*	75-08860
Flight International Inc.	USA	*635*	80-11600
Hayes Targets	USA	*603*	75-69135
Hill Industries Inc.	USA	*582*	75-43465
Miller Industries Towing Equipment	USA	*596*	75-61655
Precision Standard Inc.	USA	*605*	75-71365
R. Alkan (UK) Ltd	UK	*429*	75-02680
SECAPEM	France	*143*	75-34750
Southwest Aerospace	USA	*616*	75-82470
Tracor Inc	USA	*623*	75-90000

221 20 55 70 Target towing equipment, aircraft

Aerial Machine & Tool Corp.	USA	*546*	75-01105
Air Target Sweden AB	Sweden	*375*	75-00850
Bülow Air Target Systems AB.	Sweden	*376*	75-03600
Cranfield Aerospace Ltd	UK	*500*	80-07200
CSE Aviation Ltd	UK	*500*	80-07600
Dornier Luftfahrt GmbH	Germany	*168*	75-08860
Flight International Inc.	USA	*635*	80-11600
Hayes Targets	USA	*603*	75-69135
Hill Industries Inc.	USA	*582*	75-43465
MIL-BASE Industries Inc.	USA	*596*	75-61530
R. Alkan (UK) Ltd	UK	*429*	75-02680

221 20 55 75 Target towing winches, aircraft

Air Target Sweden AB	Sweden	*375*	75-00850
Breeze Eastern Corp.	USA	*560*	75-15110
Bülow Air Target Systems AB.	Sweden	*376*	75-03600
Cranfield Aerospace Ltd	UK	*500*	80-07200
CSE Aviation Ltd	UK	*500*	80-07600
Dornier Luftfahrt GmbH	Germany	*168*	75-08860
Flight International Inc.	USA	*635*	80-11600
Hayes Targets	USA	*603*	75-69135
Hill Industries Inc.	USA	*582*	75-43465
Lucas Western Inc.	USA	*592*	75-56300
MIL-BASE Industries Inc.	USA	*596*	75-61530
Superwinch Ltd	UK	*487*	75-57950

221 20 58 Stores management systems, aircraft

Air-Log Ltd	UK	*429*	75-01800
Boeing Australia Limited	Australia	*12*	75-02500
Dornier Luftfahrt GmbH	Germany	*168*	75-08860
Elbit Systems Ltd	Israel	*218*	75-05300
GEC-Marconi Avionics Inc.	USA	*577*	75-37413
GEC-Marconi Avionics Ltd	UK	*454*	75-25320
Logic SpA	Italy	*236*	75-13800
Lucas Western Inc.	USA	*592*	75-56300
Page Aerospace Ltd	UK	*472*	75-43505
R. Alkan (UK) Ltd	UK	*429*	75-02680
SAGEM SA	**France**	***143***	**75-33950**
Smiths Aerospace & Defense	USA	*615*	75-82090
Sparton Technology Inc.	USA	*616*	75-82950

221 20 60 Weapon carrier & release systems, aircraft

AAI Corp.	USA	*545*	75-00020
Aerea Luftfahrtgeräte GmbH.	Germany	*164*	75-00260
Aerea SpA.	Italy	*230*	75-00280
EDO Corporation	USA	*571*	75-29250
Elbit Systems Ltd	Israel	*218*	75-05300
GKN Westland Technologies Ltd	UK	*455*	75-25690
Hunting Engineering Ltd	UK	*459*	75-30700
INIMEX	**Belgium**	***46***	**75-04650**
Israel Military Industries Ltd	Israel	*220*	75-10300
Lucas Western Inc.	USA	*592*	75-56300
Normalair-Garrett Ltd	UK	*425*	74-07000
R. Alkan (UK) Ltd	UK	*429*	75-02680
Royal Ordnance	UK	*481*	75-50620

221 20 60 70 Weapon carrier & release equipment, aircraft

AAI Corp.	USA	*545*	75-00020
Aerea Luftfahrtgeräte GmbH.	Germany	*164*	75-00260
Aerea SpA.	Italy	*230*	75-00280
Aerial Machine & Tool Corp.	USA	*546*	75-01105
Computing Devices Company Ltd	UK	*442*	75-14700
GKN Westland Technologies Ltd	UK	*455*	75-25690
Hill Industries Inc.	USA	*582*	75-43465
Hunting Engineering Ltd	UK	*459*	75-30700
INIMEX	**Belgium**	***46***	**75-04650**
Israel Military Industries Ltd	Israel	*220*	75-10300
Lucas Western Inc.	USA	*592*	75-56300
MIL-BASE Industries Inc.	USA	*596*	75-61530
Minebea Co. Ltd	Japan	*253*	75-05600
Normalair-Garrett Ltd	UK	*425*	74-07000
R. Alkan (UK) Ltd	UK	*429*	75-02680
Royal Ordnance	UK	*481*	75-50620

221 20 60 73 Weapon ejectors, aircraft

AAI Corp.	USA	*545*	75-00020
Aerea SpA.	Italy	*230*	75-00280
EDO Corporation	USA	*571*	75-29250
Eurodef Aerospace & Defence	Greece	*189*	75-01295
GKN Westland Technologies Ltd	UK	*455*	75-25690
Hill Industries Inc.	USA	*582*	75-43465
INIMEX	**Belgium**	***46***	**75-04650**
Israel Military Industries Ltd	Israel	*220*	75-10300
Lucas Aerospace Power Transmission	USA	*592*	75-56265
Lucas Western Inc.	USA	*592*	75-56300
Normalair-Garrett Ltd	UK	*425*	74-07000
R. Alkan (UK) Ltd	UK	*429*	75-02680

221 20 65 Weapon loading systems, aircraft

AAI Corp.	USA	*545*	75-00020
Air-Log Ltd	UK	*429*	75-01800
Israel Military Industries Ltd	Israel	*220*	75-10300
Mandra Plant	Greece	*190*	75-02000
MIL-BASE Industries Inc.	USA	*596*	75-61530
Riva Calzoni SpA	Italy	*238*	75-20500

221 20 65 70 Weapon loading equipment, aircraft

AAI Corp.	USA	*545*	75-00020
Aarding BV	Netherlands	*291*	75-00020
Air-Log Ltd	UK	*429*	75-01800
MIL-BASE Industries Inc.	USA	*596*	75-61530
R. Alkan (UK) Ltd	UK	*429*	75-02680
Riva Calzoni SpA	Italy	*238*	75-20500
Western Design Howden	USA	*627*	75-95800

221 20 65 79 Winches, weapon loading, aircraft

Ancra CHS BV.	Netherlands	*291*	75-00510
Breeze Eastern Corp.	USA	*560*	75-15110
Israel Military Industries Ltd	Israel	*220*	75-10300
Lucas Aerospace FCS Paris	France	*139*	75-25390
Lucas Western Inc.	USA	*592*	75-56300
R. Alkan (UK) Ltd	UK	*429*	75-02680

221 20 70 Aircraft mission-related equipment (airframe integrated)

Cranfield Aerospace Ltd	UK	*500*	80-07200
Elettronica Aster SpA	Italy	*233*	75-07000
Logica UK Limited	UK	*506*	80-14350
Lucas Aerospace Power Transmission	USA	*592*	75-56265
Luminator Aircraft Products	USA	*593*	75-56450
R. Alkan (UK) Ltd	UK	*429*	75-02680
Royal Ordnance	UK	*481*	75-50620
TransInstruments	UK	*489*	75-60960
Tumo Ltd	Finland	*119*	75-02650

221 20 70 05 Aircraft armour

AEA Technology plc	UK	*496*	80-00480

Permali Gloucester Ltd	UK	*474*	75-44850
Royal Ordnance	UK	*481*	75-50620

221 20 70 50 Aircraft searchlights

Luminator Aircraft Products	USA	*593*	75-56450

221 20 70 92 Weapon pylons, aircraft

AAI Corp.	USA	*545*	75-00020
Air-Log Ltd	UK	*429*	75-01800
EDO Corporation	USA	*571*	75-29250
Eurodef Aerospace & Defence	Greece	*189*	75-01295
Hill Industries Inc.	USA	*582*	75-43465
Israel Aircraft Industries Ltd	Israel	*219*	75-10200
Israel Military Industries Ltd	Israel	*220*	75-10300
Lucas Western Inc.	USA	*592*	75-56300
Mandra Plant	Greece	*190*	75-02000
R. Alkan (UK) Ltd	UK	*429*	75-02680
Sargent Fletcher Inc.	USA	*612*	75-78150
Swiss Aircraft & Systems Co	Switzerland	*384*	55-00100

221 30 Aircraft crew / passenger related systems

Aerospace Syst. & Comp.	USA	*547*	75-02501
Airscrew Howden Ltd	UK	*429*	75-02050
AlliedSignal Aerospace Canada	Canada	*68*	75-00470
Beaufort Air-Sea Equipment Ltd	UK	*433*	75-06100
Envirovac Inc.	USA	*574*	75-31920
Frazier Aviation Inc.	USA	*576*	75-35970
GEC-Marconi Avionics Ltd	UK	*454*	75-25320
H. Koch & Sons Co.	USA	*589*	75-53030
Howden Aircontrol Ltd	UK	*459*	75-30240
Intertechnique SA.	France	*137*	75-23000
Liebherr-Aerospace Toulouse SA	France	*138*	75-24950
Litton Industries Inc.	USA	*590*	75-55400
Nord-Micro Elektronik Feinmechanik	Germany	*173*	75-22700
Raytheon E-Systems	USA	*608*	75-73302
SFIM	**France**	***144***	**75-35080**
TAT Industries Ltd	Israel	*223*	75-22000

221 30 05 Environmental control systems, aircraft

Aerospace Syst. & Comp.	USA	*547*	75-02501
Airscrew Howden Ltd	UK	*429*	75-02050
AlliedSignal Aerospace Canada	Canada	*68*	75-00470
Barber-Colman Co.	USA	*557*	75-11960
Baxter Woodhouse & Taylor Ltd	UK	*433*	75-06030
Bestobell Aviation	UK	*434*	75-06600
Boeing Australia Limited	Australia	*12*	75-02500
Dornier Luftfahrt GmbH	Germany	*168*	75-08860
Envirovac Inc.	USA	*574*	75-31920
GKN Westland Technologies Ltd	UK	*455*	75-25690
Graseby Dynamics Ltd	UK	*455*	75-26590
Howden Aircontrol Ltd	UK	*459*	75-30240
Howden Sirocco	France	*137*	75-21700
Howden Wade Ltd	UK	*459*	75-30260
Intertechnique SA.	France	*137*	75-23000
Liebherr-Aerospace Toulouse SA	France	*138*	75-24950
Microtecnica-Torino SpA	Italy	*237*	75-15500
Munters Cargocaire	USA	*597*	75-63445
Nord-Micro Elektronik Feinmechanik	Germany	*173*	75-22700
Normalair-Garrett Australia	Australia	*19*	75-14870
Normalair-Garrett Ltd	UK	*425*	74-07000
Parker Bertea Aerospace	USA	*602*	75-68900
Spar Aviation Services	Canada	*76*	75-13020
Sundstrand Aerospace	USA	*618*	75-85105

221 30 05 10 Air conditioning systems, aircraft

Aerospace Syst. & Comp.	USA	*547*	75-02501
A/S Dantherm	**Denmark**	***102***	**75-04100**
Bestobell Aviation	UK	*434*	75-06600
Frazier Aviation Inc.	USA	*576*	75-35970
GKN Westland Technologies Ltd	UK	*455*	75-25690
Howden Aircontrol Ltd	UK	*459*	75-30240
Howden Fluid Systems	USA	*582*	75-44400
Howden Wade Ltd	UK	*459*	75-30260
Intertechnique SA.	France	*137*	75-23000
Liebherr-Aerospace Lindenberg GmbH	Germany	*171*	75-18300
Liebherr-Aerospace Toulouse SA	France	*138*	75-24950
Microtecnica-Torino SpA	Italy	*237*	75-15500
Normalair-Garrett Australia	Australia	*19*	75-14870
Normalair-Garrett Ltd	UK	*425*	74-07000
Parker Bertea Aerospace	USA	*602*	75-68900
Spar Aviation Services	Canada	*76*	75-13020
The Hymatic Engineering Company Ltd	UK	*460*	75-30900
Trilectron Industries Inc.	USA	*623*	75-90920

221 30 05 40 Oxygen systems, aircraft

Aerospace Syst. & Comp.	USA	*547*	75-02501
Dornier Luftfahrt GmbH	Germany	*168*	75-08860
EROS GIE	France	*137*	75-19500
GKN Westland Technologies Ltd	UK	*455*	75-25690
Intertechnique SA.	France	*137*	75-23000
Kidde-Graviner Ltd	UK	*463*	75-33880
Litton Industries Inc.	USA	*590*	75-55400
Meggitt Oxygen Systems	UK	*467*	75-38360
Normalair-Garrett Australia	Australia	*19*	75-14870
Normalair-Garrett Ltd	UK	*425*	74-07000
Parker Bertea Aerospace	USA	*603*	75-68900
Puritan-Bennett Aero Systems Co.	USA	*606*	75-71890
Scott Aviation	USA	*613*	75-78850
SFIM	**France**	***144***	**75-35080**
Tokyo Aircraft Instrument Co Ltd	Japan	*254*	75-08600

221 30 05 60 Pressurization systems, aircraft

Bestobell Aviation	UK	*434*	75-06600
Dornier Luftfahrt GmbH	Germany	*168*	75-08860
GKN Westland Technologies Ltd	UK	*455*	75-25690
Howden Aircontrol Ltd	UK	*459*	75-30240
Howden Wade Ltd	UK	*459*	75-30260
Intertechnique SA.	France	*137*	75-23000
Kollsman	USA	*589*	75-53150
Liebherr-Aerospace Toulouse SA	France	*138*	75-24950
Litton Industries Inc.	USA	*590*	75-55400
Microtecnica-Torino SpA	Italy	*237*	75-15500
Nord-Micro Elektronik Feinmechanik	Germany	*173*	75-22700
Normalair-Garrett Australia	Australia	*19*	75-14870
Normalair-Garrett Ltd	UK	*425*	74-07000
RSL Electronics Ltd	Israel	*222*	75-18900
Spar Aviation Services	Canada	*76*	75-13020

221 30 40 Intercommunication systems, aircraft

Becker Flugfunkwerk GmbH	Germany	*165*	75-02200
Boeing Australia Limited	Australia	*12*	75-02500
Daimler-Benz Aerospace France SAS	**France**	***134***	**75-12950**
David Clark Co. Inc.	USA	*564*	75-18850
Davies Industrial Communications Ltd	UK	*444*	75-17010
Dowty Controls	UK	*446*	75-18592
GEC-Marconi Avionics Ltd	UK	*454*	75-25320
Northern Airborne Technology Ltd	Canada	*74*	75-10145
West Electronics	USA	*627*	75-95750

221 30 70 Aircraft crew / passenger related equipment

Aerial Machine & Tool Corp.	USA	*546*	75-01105
Aircraft Materials Ltd	UK	*428*	75-01650
EROS GIE	France	*137*	75-19500
Frazier Aviation Inc.	USA	*576*	75-35970
H. Koch & Sons Co.	USA	*589*	75-53030
Teledyne Ryan Aeronautical McCormick	USA	*621*	75-87555
Universal Propulsion Co. Inc.	USA	*625*	75-92700

221 30 70 50 Safety harness, aircrew

Aerial Machine & Tool Corp.	USA	*546*	75-01105
AIM Aviation (Jecco) Ltd.	UK	*428*	75-01530
AIM Enterprises Inc.	USA	*630*	80-01350
Aircraft Materials Ltd	UK	*428*	75-01650
Autoflug GmbH & Co	Germany	*165*	75-01400
Capewell Aerial Delivery Systems	USA	*576*	75-36300
H. Koch & Sons Co.	USA	*589*	75-53030
Irvin Aerospace Canada Limited	Canada	*73*	75-08300
Irvin Aerospace Limited	UK	*462*	75-32050

MIL-BASE Industries Inc.	USA	*596*	75-61530
Morse Controls Ltd	UK	*469*	75-40945
Sala International Pty Ltd	Australia	*21*	75-18980
Universal Propulsion Co. Inc.	USA	*625*	75-92700

221 30 70 60 Seats, aircraft

Aeromot Ind.Mecanico-Metalurgica SA	Brazil	*55*	75-00635
AIM Aviation (Jecco) Ltd.	UK	*428*	75-01530
Aircraft Materials Ltd	UK	*428*	75-01650
ARA Inc.	USA	*554*	75-08370
IPECO Inc	USA	*637*	80-16800
Israel Aircraft Industries Ltd	Israel	*219*	75-10200
Meggitt Composites	UK	*467*	75-38345
Oro Manufacturing Co.	USA	*602*	75-67525
Simula Government Products, Inc.	USA	*615*	75-81250

221 30 70 61 Seats, aircrew

Aircraft Tanks Ltd	UK	*429*	75-01730
ARA Inc.	USA	*554*	75-08370
Frazier Aviation Inc.	USA	*576*	75-35970
Ipeco Europe	UK	*461*	75-32040
IPECO Inc	USA	*637*	80-16800
Israel Aircraft Industries Ltd	Israel	*219*	75-10200
Morse Controls Ltd	UK	*469*	75-40945
Oro Manufacturing Co.	USA	*602*	75-67525

221 30 70 63 Seats, ejection, aircraft

Aérazur	France	*131*	75-00500
AIM Enterprises Inc.	USA	*630*	80-01350
Autoflug GmbH & Co	Germany	*165*	75-01400
Daicel Chemical Industries Ltd	Japan	*250*	75-00950
Haley & Weller Ltd	UK	*456*	75-27400
Israel Aircraft Industries Ltd	Israel	*219*	75-10200
Litton Industries Inc.	USA	*590*	75-55400
Royal Ordnance	UK	*481*	75-50620
SICAMB SpA	Italy	*238*	75-22020
Société d'Exploitation	France	*130*	74-01800
Swiss Aircraft & Systems Co	Switzerland	*384*	55-00100
Syndex Aerospace Inc.	USA	*619*	75-85540
Talley Defense Systems Inc.	USA	*619*	75-86120
Universal Propulsion Co. Inc.	USA	*625*	75-92700

221 30 70 65 Extraction systems, low-performance aircraft

Aerial Machine & Tool Corp.	USA	*546*	75-01105
Pioneer Aerospace Corp	USA	*604*	75-70200
Royal Ordnance	UK	*481*	75-50620
Universal Propulsion Co. Inc.	USA	*625*	75-92700

221 30 70 70 Ambulance kits, aircraft

Ingenieurbüro Flugwesen &	Switzerland	*388*	80-00800

221 40 Aircraft airframe-related systems

Actronics Inc.	USA	*545*	75-00560
Aerospace Syst. & Comp.	USA	*547*	75-02501
AlliedSignal Aerospace Canada	Canada	*68*	75-00470
APPH Ltd	UK	*431*	75-03415
Avica Equipment	UK	*432*	75-04870
British Filters Ltd	UK	*436*	75-09030
Canadian Marconi Co.	Canada	*69*	75-02600
Cranfield Aerospace Ltd	UK	*500*	80-07200
Dialight Corp.	USA	*569*	75-26440
Dynex/Rivett Inc.	USA	*570*	75-28350
Eldec Corp.	USA	*572*	75-29660
Esterline Technologies	USA	*574*	75-32360
Flight International Inc.	USA	*635*	80-11600
Flight Products Inc.	USA	*575*	75-34700
Frazier Aviation Inc.	USA	*576*	75-35970
Hoffman Engineering Corp.	USA	*582*	75-43980
Liebherr-Aerospace Toulouse SA	France	*138*	75-24950
Lucas Aerospace Power Transmission	USA	*592*	75-56265
Luminator Aircraft Products	USA	*593*	75-56450
Microturbo Inc	USA	*596*	75-61170
Nanoquest Defence Products Ltd	UK	*470*	75-41890
Purolator Products Co.	USA	*606*	75-71900
SFIM	**France**	***144***	**75-35080**
Sierracin/Sylmar Corporation	USA	*614*	75-80650
SP aerospace and vehicle systems bv	**Netherlands**	***295***	**75-06860**
Survival Products Inc.	USA	*618*	75-85200
THERMOCOAX	France	*147*	75-44930

221 40 05 Auxiliary power units, aircraft

Air-Log Ltd	UK	*429*	75-01800
Aircraft Parts Corp.	USA	*548*	75-03440
AlliedSignal Aerospace GmbH	Germany	*164*	75-00610
Boeing Australia Limited	Australia	*12*	75-02500
Merex Inc.	USA	*595*	75-60250
Microturbo Inc	USA	*596*	75-61170
Microturbo Ltd	UK	*468*	75-39160
Newton Derby Ltd	UK	*470*	75-42520
Normalair-Garrett Australia	Australia	*19*	75-14870
Sirples Sarl	Italy	*239*	75-22580
Spar Aviation Services	Canada	*76*	75-13020
Sundstrand Aerospace	USA	*618*	75-85105
Sundstrand Power Systems	USA	*618*	75-85120
Teledyne Continental Motors	USA	*620*	75-87506

221 40 10 De-icing systems, aircraft

Bestobell Aviation	UK	*434*	75-06600
Boeing Australia Limited	Australia	*12*	75-02500
Cox & Company Inc.	USA	*566*	75-22300
DNE, Inc.	USA	*569*	75-27000
Dunlop Aviation Division	UK	*447*	75-19000
Eldec Corp.	USA	*572*	75-29660
Engineered Fabrics Corp.	USA	*573*	75-31565
Mack Electric Devices Inc.	USA	*593*	75-56850
Magnaghi Milano SpA	Italy	*236*	75-13980
Ontic Engineering & Mfg Inc.	USA	*601*	75-66875
Paulstra SNC Hutchinson	France	*140*	75-29100
Pilkington Aerospace Ltd	UK	*474*	75-45500
Rosemount Aerospace Inc.	USA	*611*	75-76650
Spar Aviation Services	Canada	*76*	75-13020
Tempo Instruments Inc	USA	*621*	75-87900
Zwicky Engineering Ltd	UK	*495*	75-67900

221 40 10 05 De-icing systems, aircraft, electro-thermal

Dunlop Aviation Division	UK	*447*	75-19000

221 40 30 Electrical systems, aircraft

AIM Enterprises Inc.	USA	*630*	80-01350
Aircraft Parts Corp.	USA	*548*	75-03440
AlliedSignal Aerospace Canada	Canada	*68*	75-00470
Canadian Marconi Co.	Canada	*69*	75-02600
Century Electronics	USA	*562*	75-17385
Daimler-Benz Aerospace France SAS	**France**	***134***	**75-12950**
ECE GmbH	Germany	*167*	75-07200
Eldec Corp.	USA	*572*	75-29660
GEC-Marconi Aerospace Ltd	UK	*453*	75-25300
GSI Corporation	USA	*580*	75-40715
Icore International Ltd	UK	*460*	75-31080
IMP Aerospace Limited	Canada	*72*	75-07730
Labinal	France	*138*	75-23900
Labinal Aero Defense Systems Inc	USA	*638*	80-18100
Lucas Aerospace FCS Paris	France	*139*	75-25390
Lucas Aerospace Power Systems	USA	*592*	75-56260
Newton Derby Ltd	UK	*470*	75-42520
Sirples Sarl	Italy	*239*	75-22580
Spar Aviation Services	Canada	*76*	75-13020
Sundstrand Aerospace	USA	*618*	75-85105
Walter Dittel GmbH	Germany	*167*	75-06510
White Products BV	Netherlands	*296*	75-09100

221 40 40 Fire warning & protection systems, aircraft

Cerberus-Guinard SA	France	*133*	75-08800
Deugra Ges. für Brandschutzsysteme	Germany	*167*	75-05900
ECE GmbH	Germany	*167*	75-07200

Farsound Engineering Ltd	UK	450	75-22420
GEC-Marconi Avionics Ltd	UK	454	75-25320
Gull Electronic Systems Div.	USA	580	75-41130
Howden Wade Ltd	UK	459	75-30260
Hughes Electronics Corp.	USA	583	75-44865
Kidde-Fenwal	USA	588	75-52510
Kidde-Graviner Ltd	UK	463	75-33880
Lucas Aerospace FCS Paris	France	139	75-25390
Merex Inc.	USA	595	75-60250
Spectronix Ltd	Israel	222	75-21100
THERMOCOAX	France	147	75-44930
Walter Kidde Aerospace	USA	588	75-52500

221 40 45 Fuel systems, aircraft

Actronics Inc.	USA	545	75-00560
AIM Enterprises Inc.	USA	630	80-01350
Cranfield Aerospace Ltd	UK	500	80-07200
Cranfield University	UK	419	60-00350
Flight Refuelling Limited	UK	451	75-23850
GEC-Marconi Aerospace Ltd	UK	453	75-25300
Gull Electronic Systems Div.	USA	580	75-41130
Intertechnique (Aerospace) Ltd	UK	461	75-31948
Intertechnique SA.	France	137	75-23000
Meggitt Avionics Inc.	USA	595	75-59842
Moog Inc.	USA	597	75-62550
Nord-Micro Elektronik Feinmechanik	Germany	173	75-22700
Novatronics Inc	Canada	74	75-10300
Ontic Engineering & Mfg Inc.	USA	601	75-66875
Smiths Aerospace & Defense	USA	615	75-82090
Sundstrand Aerospace	USA	618	75-85105
Teledyne Controls	USA	620	75-87518
Veeder-Root Co.	USA	625	75-93520
Vickers Inc.	USA	626	75-93820
Vickers Systems	Germany	176	75-32830
Vickers Systems SA	France	148	75-46930
Vickers Systems Trinova SpA.	Italy	239	75-27150
Western Filter Corp.	USA	627	75-95950

221 40 45 70 Fuel system equipment, aircraft

Actronics Inc.	USA	545	75-00560
Aeroquip Corp.	USA	547	75-02350
Aerospace Syst. & Comp.	USA	547	75-02501
AIM Enterprises Inc.	USA	630	80-01350
Allied Screw Products Inc.	USA	549	75-04690
Argo-Tech Corp.	USA	555	75-08630
British Filters Ltd	UK	436	75-09030
Cranfield Aerospace Ltd	UK	500	80-07200
Dowty Seals Ltd	UK	446	75-18620
Eldec Corp.	USA	572	75-29660
Farsound Engineering Ltd	UK	450	75-22420
GEC-Marconi Aerospace Ltd	UK	453	75-25300
GEC-Marconi Avionics Ltd	UK	454	75-25320
GKN Westland Technologies Ltd	UK	455	75-25690
GKN Westland-Sitec GmbH	Germany	169	75-10900
Gull Electronic Systems Div.	USA	580	75-41130
Hammonds Technical Services Inc.	USA	580	75-41800
Howden Fluid Systems	USA	582	75-44400
Intertechnique SA.	France	137	75-23000
Microtecnica-Torino SpA	Italy	237	75-15500
Moog Controls Ltd	UK	469	75-40670
Novatronics Inc	Canada	74	75-10300
Parker Bertea Aerospace	USA	602	75-68900
Purolator Products Co.	USA	606	75-71900
Secondo Mona SpA	Italy	237	75-15900
Sundstrand Aerospace	USA	618	75-85105
TAT Industries Ltd	Israel	223	75-22000
Vickers Aerospace Components	UK	491	75-63020
Vickers Systems Division TRINOVA Ltd	UK	491	75-63150

221 40 45 74 Fuel tanks, aircraft

Aerea SpA.	Italy	230	75-00280
Aircraft Tanks Ltd	UK	429	75-01730
BP Chemicals Ltd	UK	435	75-07790
Cranfield Aerospace Ltd	UK	500	80-07200
Cranfield University	UK	419	60-00350
EDO Corporation	USA	571	75-29250
Engineered Fabrics Corp.	USA	573	75-31565
Eurodef Aerospace & Defence	Greece	189	75-01295
FPT Industries Ltd	UK	452	75-24400
Intertechnique SA.	France	137	75-23000
Israel Aircraft Industries Ltd	Israel	219	75-10200
Israel Military Industries Ltd	Israel	220	75-10300
Magnaghi Napoli SpA	Italy	236	75-14000
MOL Cy NV	Belgium	46	75-04998
Sumitomo Electric Industries Ltd	Japan	254	75-07950
Technical Products Group Inc.	USA	620	75-86764

221 40 45 74 10 Fuel tanks, aircraft, flexible

Aero Tec Laboratories Inc.	USA	546	75-01117
Continental AG Werk Uniroyal.	Germany	176	75-32090
Engineered Fabrics Corp.	USA	573	75-31565
FPT Inc.	USA	576	75-35760
FPT Industries Ltd	UK	452	75-24400
Sèkur SpA	Italy	238	75-21650
Sumitomo Electric Industries Ltd	Japan	254	75-07950
Technical Products Group Inc.	USA	620	75-86764

221 40 50 Hydraulic systems, aircraft

Abex GmbH Aerohydraul	Germany	164	75-00105
APPH Ltd	UK	431	75-03415
Dowty Aerospace Yakima	USA	569	75-27280
Fairey Hydraulics	UK	450	75-22255
Frazier Aviation Inc.	USA	576	75-35970
GKN Westland Technologies Ltd	UK	455	75-25690
Howden Fluid Systems	USA	582	75-44400
Israel Aircraft Industries Ltd	Israel	219	75-10200
Lucas Aerospace	Canada	73	75-09140
Lucas Aerospace FCS Paris	France	139	75-25390
Lucas Aerospace Power Transmission	USA	592	75-56265
Magnaghi Milano SpA	Italy	236	75-13980
Mannesmann Rexroth Ltd	UK	466	75-37100
Mannesmann Rexroth SA	France	139	75-25500
M.C. Aerospace Corp.	USA	594	75-59370
Merex Inc.	USA	595	75-60250
Messier-Bugatti	France	140	75-26400
Moog Inc.	USA	597	75-62550
Ontic Engineering & Mfg Inc.	USA	601	75-66875
Parker Bertea Aerospace	USA	602	75-68900
Prva Petoletka Industrie	Yugoslavia	661	75-00610
Riva Calzoni SpA	Italy	238	75-20500
SABCA	Belgium	46	75-05590
Spar Aviation Services	Canada	76	75-13020
Sundstrand Aerospace	USA	618	75-85105
Vickers Inc.	USA	626	75-93820
Vickers Systems	Germany	176	75-32830
Vickers Systems SA	France	148	75-46930
Vickers Systems Trinova SpA.	Italy	239	75-27150

221 40 50 50 Hydraulic power supply units, aircraft

Abex GmbH Aerohydraul	Germany	164	75-00105
Dayton T. Brown Inc.	USA	632	80-04920
Frazier Aviation Inc.	USA	576	75-35970
Lucas Aerospace Power Transmission	USA	592	75-56265
Magnaghi Milano SpA	Italy	236	75-13980
M.C. Aerospace Corp.	USA	594	75-59370
Spar Aviation Services	Canada	76	75-13020

221 40 50 70 Hydraulic equipment, aircraft

Abex GmbH Aerohydraul	Germany	164	75-00105
Aeroquip Aerospace Marine Group	UK	428	75-01101
Aeroquip Corp.	USA	547	75-02350
AIM Enterprises Inc.	USA	630	80-01350
APPH Ltd	UK	431	75-03415
British Filters Ltd	UK	436	75-09030
BTR Aerospace Australia	Australia	13	75-02815
Busak + Shamban Inc.	USA	561	75-16006
Crane Co.	USA	566	75-22452
Donaldson Co. Inc.	USA	569	75-27050
Dowty Aerospace Wolverhampton	UK	446	75-18589
Dowty Seals Ltd	UK	446	75-18620
Dunlop Aviation Division	UK	447	75-19000
Dynex/Rivett Inc.	USA	570	75-28350
Elettronica Aster SpA	Italy	233	75-07000
Fairey Hydraulics	UK	450	75-22255
Fairey Microfiltrex	UK	450	75-22300

Frazier Aviation Inc.	USA	*576*	75-35970
GKN Westland Technologies Ltd	UK	*455*	75-25690
Gull Electronic Systems Div.	USA	*580*	75-41130
Howden Fluid Systems	USA	*582*	75-44400
Invar Manufacturing Ltd	Canada	*73*	75-08220
Israel Aircraft Industries Ltd	Israel	*219*	75-10200
Liebherr-Aerospace Lindenberg GmbH	Germany	*171*	75-18300
Lucas Aerospace FCS Paris	France	*139*	75-25390
Lucas Aerospace Power Transmission	USA	*592*	75-56265
Magnaghi Milano SpA	Italy	*236*	75-13980
Magnaghi Napoli SpA	Italy	*236*	75-14000
Mannesmann Rexroth Ltd	UK	*466*	75-37100
Mannesmann Rexroth SA	France	*139*	75-25500
M.C. Aerospace Corp.	USA	*594*	75-59370
Menasco Aerospace Ltd	Canada	*74*	75-09600
Merex Inc.	USA	*595*	75-60250
Messier-Bugatti	France	*140*	75-26400
Messier-Dowty Inc	Canada	*74*	75-09655
Microtecnica-Torino SpA	Italy	*237*	75-15500
Moog Controls Ltd	UK	*469*	75-40670
Moog Japan Ltd.	Japan	*253*	75-06100
Normalair-Garrett Australia	Australia	*19*	75-14870
Normalair-Garrett Ltd.	UK	*425*	74-07000
Parker Bertea Aerospace	USA	*602*	75-68900
Prva Petoletka Industrie	Yugoslavia	*661*	75-00610
Riva Calzoni SpA	Italy	*238*	75-20500
Secondo Mona SpA	Italy	*237*	75-15900
SP aerospace and vehicle systems bv	**Netherlands**	***295***	**75-06860**
Spar Aviation Services	Canada	*76*	75-13020
Sundstrand Aerospace	USA	*618*	75-85105
Vickers Inc.	USA	*626*	75-93820
Vickers Systems	Germany	*176*	75-32830
Vickers Systems Division TRINOVA Ltd	UK	*491*	75-63150
Vickers Systems SA	France	*148*	75-46930
Vickers Systems Trinova SpA	Italy	*239*	75-27150
Zwicky Engineering Ltd	UK	*495*	75-67900

221 40 50 70 50 Hydraulic system filters, aircraft

AIM Enterprises Inc.	USA	*630*	80-01350
APPH (Bolton) Ltd.	UK	*431*	75-03415
British Filters Ltd	UK	*436*	75-09030
Fairey Hydraulics	UK	*450*	75-22255
Labinal	France	*138*	75-23900
Magnaghi Milano SpA	Italy	*236*	75-13980
Mannesmann Rexroth SA	France	*139*	75-25500
Pall Aerospace	UK	*472*	75-43517
Prva Petoletka Industrie	Yugoslavia	*661*	75-00610
Spar Aviation Services	Canada	*76*	75-13020
Vickers Systems Division TRINOVA Ltd	UK	*491*	75-63150

221 40 60 Landing gear, aircraft

Aerostar SA	Romania	*331*	80-00100
AIM Enterprises Inc.	USA	*630*	80-01350
APPH Ltd	UK	*431*	75-03415
BFGoodrich Aerospace	USA	*558*	75-13770
Elettronica Aster SpA	Italy	*233*	75-07000
Fairey Hydraulics	UK	*450*	75-22255
Frazier Aviation Inc.	USA	*576*	75-35970
Heroux Inc	Canada	*72*	75-07400
Israel Aircraft Industries Ltd	Israel	*219*	75-10200
Liebherr-Aerospace Lindenberg GmbH	Germany	*171*	75-18300
Magnaghi Napoli SpA	Italy	*236*	75-14000
Menasco Aerospace Ltd	Canada	*74*	75-09600
Merex Inc.	USA	*595*	75-60250
Messier-Dowty Inc	Canada	*74*	75-09655
Messier-Dowty Limited	UK	*467*	75-38700
Messier-Dowty SAS	France	*140*	75-26500
Ontic Engineering & Mfg Inc.	USA	*601*	75-66875
Prva Petoletka Industrie	Yugoslavia	*661*	75-00610
Revue THOMMEN AG	Switzerland	*387*	75-11100
SP aerospace and vehicle systems bv	**Netherlands**	***295***	**75-06860**

221 40 60 90 Landing gear components, aircraft

APPH Ltd	UK	*431*	75-03415
Bowmar Instrument Corp.	USA	*560*	75-14860
BTR Aerospace Australia	Australia	*13*	75-02815
Busak + Shamban Inc.	USA	*561*	75-16006
Constructions Mécaniques des Vosges	France	*134*	75-12300
Dowty Seals Ltd	UK	*446*	75-18620
Eldec Corp.	USA	*572*	75-29660
Fairey Hydraulics	UK	*450*	75-22255
Frazier Aviation Inc.	USA	*576*	75-35970
Goodyear Great Britain Ltd.	UK	*455*	75-26135
Invar Manufacturing Ltd	Canada	*73*	75-08220
Lucas Western Inc.	USA	*592*	75-56300
Magnaghi Milano SpA	Italy	*236*	75-13980
Magnaghi Napoli SpA	Italy	*236*	75-14000
M.C. Aerospace Corp.	USA	*594*	75-59370
Menasco Aerospace Ltd	Canada	*74*	75-09600
Merex Inc.	USA	*595*	75-60250
Messier-Dowty Inc	Canada	*74*	75-09655
Messier-Dowty Limited	UK	*467*	75-38700
Messier-Dowty SAS	France	*140*	75-26500
Moog Controls Ltd	UK	*469*	75-40670
Moog Inc.	USA	*597*	75-62550
Parker Bertea Aerospace	USA	*602*	75-68900
Prva Petoletka Industrie	Yugoslavia	*661*	75-00610
Revue THOMMEN AG	Switzerland	*387*	75-11100
Riva Calzoni SpA	Italy	*238*	75-20500
SP aerospace and vehicle systems bv	**Netherlands**	***295***	**75-06860**
Vehicle Systems Development Corp.	USA	*625*	75-93570
Voss Industries Inc.	USA	*626*	75-94550
Warner Electric	USA	*627*	75-95105

221 40 60 90 10 Brakes, aircraft

AIM Enterprises Inc.	USA	*630*	80-01350
Airscrew Howden Ltd	UK	*429*	75-02050
Allied Screw Products Inc.	USA	*549*	75-04690
BFGoodrich Aerospace	USA	*558*	75-13770
Brassinter SA Indústria e Comércio	Brazil	*55*	75-02290
BTR Aerospace Australia	Australia	*13*	75-02815
Constructions Mécaniques des Vosges	France	*134*	75-12300
Dunlop Aviation Division	UK	*447*	75-19000
Elettronica Aster SpA	Italy	*233*	75-07000
Eurodef Aerospace & Defence	Greece	*189*	75-01295
Israel Aircraft Industries Ltd	Israel	*219*	75-10200
Magnaghi Milano SpA	Italy	*236*	75-13980
Merex Inc.	USA	*595*	75-60250
Messier-Bugatti	France	*140*	75-26400
Parker Bertea Aerospace	USA	*602*	75-68900
Plan Aero Ltd	Greece	*191*	75-02230
Prva Petoletka Industrie	Yugoslavia	*661*	75-00610
Sielman SA	Greece	*191*	75-02360
Thompson Aerospace	USA	*644*	80-32800

221 40 60 90 12 ABS systems, aircraft brake

Crane Co.	USA	*566*	75-22452
Israel Aircraft Industries Ltd	Israel	*219*	75-10200
Messier-Bugatti	France	*140*	75-26400
Moog Inc.	USA	*597*	75-62550
Ontic Engineering & Mfg Inc.	USA	*601*	75-66875
Thompson Aerospace	USA	*644*	80-32800

221 40 60 90 70 Tyres, aircraft

Bridgestone Corp	Japan	*250*	75-00700
Dunlop & Olympic Tyres	Australia	*15*	75-04930
Dunlop Aircraft Tyres Limited	UK	*447*	75-19005
Global Airtech International	USA	*579*	75-39769
Goodyear Aviation Products	France	*137*	75-19800
Goodyear Great Britain Ltd.	UK	*455*	75-26135
Goodyear Tire & Rubber Co.	USA	*579*	75-40050
Michelin	France	*140*	75-26600
Novatração Artefatos	Brazil	*57*	75-14780
Scott Aviation	USA	*613*	75-78850
Sime Aircraft Tyre Co. Sdn Bhd	Malaysia	*278*	75-01800
Thompson Aerospace	USA	*644*	80-32800

221 40 60 90 90 Wheels, aircraft

Allied Screw Products Inc.	USA	*549*	75-04690
BFGoodrich Aerospace	USA	*558*	75-13770
BTR Aerospace Australia	Australia	*13*	75-02815

Dunlop Aviation Division	UK	*447*	75-19000
Elettronica Aster SpA	Italy	*233*	75-07000
Global Airtech International	USA	*579*	75-39769
Magnaghi Milano SpA	Italy	*236*	75-13980
Messier-Bugatti	France	*140*	75-26400
Parker Bertea Aerospace	USA	*602*	75-68900
Prva Petoletka Industrie	Yugoslavia	*661*	75-00610
Scott Aviation	USA	*613*	75-78850

221 40 65 Lighting systems, aircraft

ADB NV/SA	Belgium	*45*	75-00070
BFGoodrich Aerospace	USA	*558*	75-13770
Bruce Industries Inc.	USA	*561*	75-15360
DNE, Inc.	USA	*569*	75-27000
ECE GmbH	Germany	*167*	75-07200
Grimes Aerospace Company	USA	*580*	75-40400
H. Koch & Sons Co.	USA	*589*	75-53030
Hoffman Engineering Corp.	USA	*582*	75-43980
Luminescent Systems	Belgium	*48*	80-01400
Luminescent Systems Inc.	USA	*593*	75-56480
Page Aerospace Ltd	UK	*472*	75-43505
Soderberg Manufacturing Co. Inc.	USA	*616*	75-82200
Spar Aviation Services	Canada	*76*	75-13020
Survival Products Inc.	USA	*618*	75-85200
Tempo Instruments Inc	USA	*621*	75-87900
White Products BV	Netherlands	*296*	75-09100

221 40 65 70 Lighting equipment, aircraft

Bruce Industries Inc.	USA	*561*	75-15360
Control Products Corp.	USA	*566*	75-21928
ECE GmbH	Germany	*167*	75-07200
Eldec Corp.	USA	*572*	75-29660
Hella Aerospace GmbH	Germany	*169*	75-11975
Hoffman Engineering Corp.	USA	*582*	75-43980
Korry Electronics Co.	USA	*589*	75-53250
Luminescent Systems Inc.	USA	*593*	75-56480
Page Aerospace Ltd	UK	*472*	75-43505
Soderberg Manufacturing Co. Inc.	USA	*616*	75-82200
Spar Aviation Services	Canada	*76*	75-13020
Spectrolab Inc.	USA	*617*	75-83100
Survival Products Inc.	USA	*618*	75-85200
White Products BV	Netherlands	*296*	75-09100

221 40 65 70 10 Beacons, rotating, aircraft

Bruce Industries Inc.	USA	*561*	75-15360
Control Products Corp.	USA	*566*	75-21928
Hella Aerospace GmbH	Germany	*169*	75-11975
Soderberg Manufacturing Co. Inc.	USA	*616*	75-82200
Spar Aviation Services	Canada	*76*	75-13020

221 40 65 70 20 Landing lights, aircraft

Bruce Industries Inc.	USA	*561*	75-15360
Control Products Corp.	USA	*566*	75-21928
Soderberg Manufacturing Co. Inc.	USA	*616*	75-82200
Spar Aviation Services	Canada	*76*	75-13020
Strumech Versatower Ltd	UK	*486*	75-57530

221 40 65 70 30 Navigation lights, aircraft

Bruce Industries Inc.	USA	*561*	75-15360
Control Products Corp.	USA	*566*	75-21928
Hella Aerospace GmbH	Germany	*169*	75-11975
Luminator Aircraft Products	USA	*593*	75-56450
Soderberg Manufacturing Co. Inc.	USA	*616*	75-82200
Spar Aviation Services	Canada	*76*	75-13020
Teleflex-Syneravia SA	France	*147*	75-44470

221 40 65 70 40 Strobe lights, aircraft

Control Products Corp.	USA	*566*	75-21928
Hella Aerospace GmbH	Germany	*169*	75-11975
Soderberg Manufacturing Co. Inc.	USA	*616*	75-82200
Spar Aviation Services	Canada	*76*	75-13020

221 40 65 70 50 Panels, illuminated, aircraft

Boeing Australia Limited	Australia	*12*	75-02500
Bospec Neue Technologien	Germany	*166*	75-03420
Control Products Corp.	USA	*566*	75-21928
Dialight Corp.	USA	*569*	75-26440
ECE GmbH	Germany	*167*	75-07200
Electrodynamics Inc.	USA	*572*	75-29690
Esterline Technologies	USA	*574*	75-32360
Interface Products Inc.	USA	*585*	75-47300
Korry Electronics Co.	USA	*589*	75-53250
Northern Airborne Technology Ltd	Canada	*74*	75-10145
Saunders-Roe Ltd.	UK	*482*	75-51850
Soderberg Manufacturing Co. Inc.	USA	*616*	75-82200
TEC Electrical Components Ltd	UK	*487*	75-58840
Transicoil Inc.	USA	*623*	75-90750

221 40 67 Monitoring systems, airframe / engine

Aircraft Controls Group	USA	*628*	75-97290
Astronautics C.A. Ltd	Israel	*217*	75-02400
Bavaria Avionik Technologie GmbH.	Germany	*165*	75-02040
BEI Sensors & Systems Co.	USA	*558*	75-12740
Boeing Australia Limited	Australia	*12*	75-02500
Eldec Corp.	USA	*572*	75-29660
ELECMA	France	*136*	75-14400
Esterline Technologies	USA	*574*	75-32360
GEC-Marconi Avionics Ltd	UK	*454*	75-25320
Meggitt Avionics (UK)	UK	*467*	75-38340
RSL Electronics Ltd	Israel	*222*	75-18900
SFIM	**France**	***144***	**75-35080**
Smiths Aerospace & Defense	USA	*615*	75-82090
Tempo Instruments Inc	USA	*621*	75-87900
Transicoil Inc.	USA	*623*	75-90750
TransInstruments	UK	*489*	75-60960
UCC International Limited	UK	*490*	75-61700
Vibro-Meter SA	Switzerland	*388*	75-14710
Vickers Systems Division TRINOVA Ltd	UK	*491*	75-63150
Woodward Governor Company	USA	*628*	75-97290

221 40 68 Pneumatic systems, aircraft

Aerospace Syst. & Comp.	USA	*547*	75-02501
Avica Equipment	UK	*432*	75-04870
Bestobell Aviation	UK	*434*	75-06600
CompAir Luchard SA	France	*134*	75-10250
GEC-Marconi Aerospace Ltd	UK	*453*	75-25300
GKN Westland Technologies Ltd	UK	*455*	75-25690
Insumat Ltd	UK	*461*	75-31660
Intertechnique SA	France	*137*	75-23000
M.C. Aerospace Corp.	USA	*594*	75-59370
Merex Inc.	USA	*595*	75-60250
Ontic Engineering & Mfg Inc.	USA	*601*	75-66875
Parker Bertea Aerospace	USA	*602*	75-68900
Prva Petoletka Industrie	Yugoslavia	*661*	75-00610
Spar Aviation Services	Canada	*76*	75-13020
Voss Industries Inc.	USA	*626*	75-94550

221 40 70 Aircraft airframe-related equipment

AIM Enterprises Inc.	USA	*630*	80-01350
Allied Screw Products Inc.	USA	*549*	75-04690
AlliedSignal Aerospace Canada	Canada	*68*	75-00470
APPH Ltd	UK	*431*	75-03415
AVIAC Technologies	France	*132*	75-04600
Cessna Aircraft Company	USA	*563*	75-17750
Eldec Corp.	USA	*572*	75-29660
Flight International Inc.	USA	*635*	80-11600
Frazier Aviation Inc.	USA	*576*	75-35970
Lucas Aerospace Power Transmission	USA	*592*	75-56265
Merex Inc.	USA	*595*	75-60250
Messier-Dowty Limited	UK	*467*	75-38700
Riva Calzoni SpA	Italy	*238*	75-20500
SP aerospace and vehicle systems bv	**Netherlands**	***295***	**75-06860**
Survival Products Inc.	USA	*618*	75-85200
Tumo Ltd	Finland	*119*	75-02650

221 40 70 10 Actuators, aircraft

AIM Enterprises Inc.	USA	*630*	80-01350
Allied Screw Products Inc.	USA	*549*	75-04690
APPH Ltd	UK	*431*	75-03415
Astronautics Corp. of America	USA	*555*	75-10000
AVIAC Technologies	France	*132*	75-04600
Barber-Colman Co.	USA	*557*	75-11960
BFGoodrich Avionics Systems Inc	USA	*559*	75-13771
Curtiss-Wright Flight Systems Inc.	USA	*567*	75-23055
Dowty Aerospace Wolverhampton	UK	*446*	75-18589
ECE GmbH	Germany	*167*	75-07200
Elettronica Aster SpA	Italy	*233*	75-07000
Fairey Hydraulics	UK	*450*	75-22255
Fokker Special Products BV	Netherlands	*293*	75-02330
Frazier Aviation Inc.	USA	*576*	75-35970
GEC-Marconi Aerospace Inc.	USA	*577*	75-37412
GEC-Marconi Avionics Ltd	UK	*454*	75-25320
Hunting Engineering Ltd	UK	*459*	75-30700
INIMEX	**Belgium**	***46***	**75-04650**
Israel Military Industries Ltd	Israel	*220*	75-10300
Kearfott Guidance & Navigation Corp.	USA	*588*	75-51760
Labinal	France	*138*	75-23900
Labinal Aero Defense Systems Inc	USA	*638*	80-18100
Lucas Aerospace	Canada	*73*	75-09140
Lucas Aerospace FCS Paris	France	*139*	75-25390
Lucas Aerospace Power Systems	USA	*592*	75-56260
Lucas Aerospace Power Transmission	USA	*592*	75-56265
Lucas Western Inc.	USA	*592*	75-56300
Meggitt Avionics (UK)	UK	*467*	75-38340
Messier-Dowty Inc	Canada	*74*	75-09655
Moog Controls Ltd	UK	*469*	75-40670
Moog Inc.	USA	*597*	75-62550
Muirhead Vactric Components Ltd	UK	*470*	75-41404
Nord-Micro Elektronik Feinmechanik	Germany	*173*	75-22700
Ontic Engineering & Mfg Inc.	USA	*601*	75-66875
Parker Bertea Aerospace	USA	*602*	75-68900
Prva Petoletka Industrie	Yugoslavia	*661*	75-00610
Ratier-Figeac SA	France	*141*	75-31800
Riva Calzoni SpA	Italy	*238*	75-20500
SEXTANT Avionique SA	France	*144*	75-35000
Société Française de Moteurs à	France	*145*	75-35100
SP aerospace and vehicle systems bv	**Netherlands**	***295***	**75-06860**
Spar Aviation Services	Canada	*76*	75-13020
Teldix GmbH	Germany	*175*	75-30100
Transicoil Inc.	USA	*623*	75-90750
Vickers Aerospace Components	UK	*491*	75-63020

221 40 70 20 Crash position indicators, aircraft

AlliedSignal Aerospace Canada	Canada	*68*	75-00470
Dukane Corp.	USA	*570*	75-27650
Oceano Instruments (UK) Ltd	UK	*508*	80-19050
Pointer Avionics	Canada	*75*	75-11300
Smiths Aerospace & Defense	USA	*615*	75-82090

221 40 70 30 Batteries, aircraft (see code 633.20.10)

BFGoodrich Avionics Systems Inc	USA	*559*	75-13771
Concorde Battery Corp.	USA	*566*	75-20845
Eldec Corp.	USA	*572*	75-29660
Hawker Energy Products Ltd	UK	*457*	75-28490
Saft America Inc.	USA	*612*	75-77550
Saft Nife AB	Sweden	*378*	75-09850
Saft Nife Ltd	UK	*481*	75-51470
VHB Industrial Batteries Ltd	Canada	*77*	75-14540
Walter Dittel GmbH	Germany	*167*	75-06510

221 40 70 35 Fatigue monitoring equipment, airframe

Hadland Photonics Ltd	UK	*456*	75-27200
Teldix GmbH	Germany	*175*	75-30100

221 40 70 40 Floats, seaplane

Aircraft Tanks Ltd	UK	*429*	75-01730
Eurovinil Industries SpA	Italy	*234*	75-08400
SEI Industries Ltd	Canada	*76*	75-12750
Survival Products Inc.	USA	*618*	75-85200

221 40 80 Airframe accessories, aircraft

AIM Enterprises Inc.	USA	*630*	80-01350
Associated Aircraft Mfg & Sales Inc.	USA	*555*	75-09500
Canadian Marconi Co.	Canada	*69*	75-02600
Frazier Aviation Inc.	USA	*576*	75-35970
Hindustan Aeronautics Ltd.	India	*204*	55-01000
H.R. Smith (Microwaves) Ltd	UK	*484*	75-54590
Lucas Aerospace FCS Paris	France	*139*	75-25390
Lucas Aerospace Power Transmission	USA	*592*	75-56265
Olympic Tool Hellas SA	Greece	*191*	75-02180
Tumo Ltd	Finland	*119*	75-02650
Vantage Associates Inc.	USA	*625*	75-93250

221 40 80 05 Accessory drives, airframe-mounted

Lucas Western Inc.	USA	*592*	75-56300
M.C. Aerospace Corp.	USA	*594*	75-59370
Spar Aviation Services	Canada	*76*	75-13020
Vickers Inc.	USA	*626*	75-93830
Vickers Systems	Germany	*176*	75-32830
Vickers Systems SA	France	*148*	75-46930
Vickers Systems Trinova SpA	Italy	*239*	75-27150

221 40 80 05 50 Accessory drive shafts

Firth Rixson plc	UK	*451*	75-23600
Horstman Defence Systems Ltd	UK	*459*	75-29920
Lucas Aerospace Power Transmission	USA	*592*	75-56265
Technical Products Group Inc.	USA	*620*	75-86764

221 40 80 10 Antennas, aircraft (see code 446.20.80.30.10)

Canadian Marconi Co.	Canada	*69*	75-02600
Chelton Inc.	USA	*563*	75-18120
Foxtronics Inc.	USA	*576*	75-35752
GEC Marconi Hazeltine Corp	USA	*578*	75-37415
General Electronics Communications	Greece	*189*	75-01300
Herley-Vega Systems	USA	*581*	75-43090
H.R. Smith (Microwaves) Ltd	UK	*484*	75-54590
Litton Industries Inc.	USA	*590*	75-55400
MA/COM Antenna & Cable Div.	USA	*593*	75-56953
Specmat Technologies Inc.	USA	*616*	75-83055
UB Corp.	USA	*624*	75-91380
Vantage Associates Inc.	USA	*625*	75-93250
Walter Dittel GmbH	Germany	*167*	75-06510

221 40 80 15 Arrester hooks, aircraft

Merex Inc.	USA	*595*	75-60250
WB Chain Co Ltd	UK	*439*	75-12440

221 40 90 Airframe components, aircraft

Aeromax Industries Inc.	USA	*547*	75-01920
AIM Enterprises Inc.	USA	*630*	80-01350
Associated Aircraft Mfg & Sales Inc.	USA	*555*	75-09500
Aviquei Aerospace	Brazil	*55*	75-00975
BFGoodrich Aerospace	USA	*558*	75-13770
Busak + Shamban Inc.	USA	*561*	75-16006
Cessna Aircraft Company	USA	*563*	75-17750
Chicopee Manufacturing Limited	Canada	*70*	75-03200
China National Aero-Tech I/E Corp.	China PR	*88*	55-00200
Cleveland-Guest (Engineering) Ltd	UK	*440*	75-13810
Darchem Engineering Ltd.	UK	*444*	75-16750
FFA Flugzeugwerke Altenrhein AG	Switzerland	*386*	75-04350
Fleet Industries Ltd	Canada	*71*	75-05500
Frazier Aviation Inc.	USA	*576*	75-35970
GEC-Marconi Aerospace Ltd	UK	*453*	75-25300
GKN Westland Aerospace Ltd.	UK	*454*	75-25625
Glover Machine Company	USA	*579*	75-39820
Hitchiner Manufacturing Co. Inc.	USA	*582*	75-43550
Howden Fluid Systems	USA	*582*	75-44400
IMP Aerospace Limited	Canada	*72*	75-07730
Industria Aeronautica Neiva S.A.	Brazil	*56*	75-09460
Insumat Ltd	UK	*461*	75-31660
Invar Manufacturing Ltd	Canada	*73*	75-08220
Jehier SA	France	*137*	75-23200
Manitoba Aerospace Initiative	Canada	*73*	75-09215

Merex Inc.	USA	*595*	75-60250
Normalair-Garrett Australia	Australia	*19*	75-14870
Olympic Tool Hellas SA	Greece	*191*	75-02180
Per Udsen Co. Aircraft Industry A/S	Denmark	*104*	75-21300
Polycast Technology Corp.	USA	*604*	75-70858
Raytheon E-Systems	USA	*608*	75-73302
Reil Industrial Enterprises Ltd	Canada	*75*	75-12000
Riva Calzoni SpA	Italy	*238*	75-20500
SIAI Marchetti Sarl	Italy	*238*	75-21950
SONACA SA	Belgium	*47*	75-06100
T E D Engineering Australia Limited	Australia	*22*	75-20300
Teledyne Advanced Materials	USA	*621*	75-87547
The NORDAM Group	USA	*599*	75-64880
Thomas Instrument & Machine Co. Inc.	USA	*622*	75-88860
Transicoil Inc.	USA	*623*	75-90750
Tumo Ltd	Finland	*119*	75-02650
TUSAS Aerospace Industries Inc.	Turkey	*403*	75-02120
Vantage Associates Inc.	USA	*625*	75-93250

221 40 90 60 Radomes, aircraft

AeroSpace Technologies of Australia	Australia	*11*	75-00340
AIM Enterprises Inc.	USA	*630*	80-01350
Chelton (Electrostatics) Limited	UK	*439*	75-12700
Complas Limited	UK	*442*	75-14543
Daimler-Benz Aerospace France SAS	**France**	***134***	**75-12950**
Dassault Aviation Group	France	*135*	75-12975
Electronic Space Systems Corp.	USA	*572*	75-30100
ERA Technology Ltd	UK	*502*	80-09600
ESSCO Collins Ltd	Ireland	*214*	75-00100
Frazier Aviation Inc.	USA	*576*	75-35970
General Electronics Communications	Greece	*189*	75-01300
Ratier-Figeac SA	France	*141*	75-31800
Raytheon E-Systems	USA	*608*	75-73302
Sea Tel Inc.	USA	*613*	75-78930
St Bernard Composites Ltd	UK	*486*	75-56222
Technical Products Group Inc.	USA	*620*	75-86764
UB Corp.	USA	*624*	75-91380
Vantage Associates Inc.	USA	*625*	75-93250

221 40 90 70 Transparencies, aircraft

AIM Enterprises Inc.	USA	*630*	80-01350
ICL Technical Plastics Ltd	UK	*460*	75-31060
Pilkington Aerospace Ltd	UK	*474*	75-45500
Polycast Technology Corp.	USA	*604*	75-70858

221 40 90 70 20 Canopies, aircraft

AeroSpace Technologies of Australia	Australia	*11*	75-00340
AIM Enterprises Inc.	USA	*630*	80-01350
Dassault Aviation Group	France	*135*	75-12975
Pilkington Aerospace Ltd	UK	*474*	75-45500
Verres Industriels SA	Switzerland	*388*	75-14700

221 40 90 90 Windows, aircraft

AIM Enterprises Inc.	USA	*630*	80-01350
Hughes Danbury Optical Sys.	USA	*582*	75-44857
ICL Technical Plastics Ltd	UK	*460*	75-31060
Litton Industries Inc.	USA	*590*	75-55400
Pilkington Aerospace Ltd	UK	*474*	75-45500

222 Helicopter systems & equipment

AIM Enterprises Inc.	USA	*630*	80-01350
Ametek Inc.	USA	*553*	75-06510
Anite Systems Ltd	UK	*497*	80-01240
APPH Ltd	UK	*431*	75-03415
Avibras Aeroespacial SA	Brazil	*55*	75-00968
Avimo Ltd	UK	*432*	75-04950
Baxter Woodhouse & Taylor Ltd	UK	*433*	75-06030
Boeing Australia Limited	Australia	*12*	75-02500
Bristow Helicopters Ltd	UK	*499*	80-03950
British Aerospace (Systems &	UK	*436*	75-08890
Canadian Marconi Co.	Canada	*69*	75-02600
CIC International Ltd	USA	*563*	75-18570
Computing Devices Canada	Canada	*70*	75-03600
Concorde Battery Corp.	USA	*566*	75-20845
Connecticut Machine Products Inc.	USA	*566*	75-21050
Darchem Engineering Ltd.	UK	*444*	75-16750
Dowty Aerospace Wolverhampton	UK	*446*	75-18589
DRS Technologies, Inc.	USA	*570*	75-27400
ECE GmbH	Germany	*167*	75-07200
EL-OP Electro Optics Industries Ltd	Israel	*219*	75-06300
Fabbrica Italiana Apparecchiature	Italy	*234*	75-08550
Fairey Hydraulics	UK	*450*	75-22255
FPT Industries Ltd	UK	*452*	75-24400
Fuji Heavy Industries Ltd	Japan	*251*	75-01100
GEC-Marconi Avionics Ltd	UK	*454*	75-25320
Howden Aircontrol Ltd	UK	*459*	75-30240
IMP Aerospace Limited	Canada	*72*	75-07730
Indal Technologies Inc	Canada	*72*	75-07740
Israel Aircraft Industries Ltd	Israel	*220*	75-10200
Japan Aviation Electronics Ind. Ltd	Japan	*252*	75-03300
Kearfott Guidance & Navigation Corp.	USA	*588*	75-51760
Labinal Aero Defense Systems Inc	USA	*638*	80-18100
LITEF GmbH	**Germany**	***171***	**75-18600**
Litton Industries Inc.	USA	*590*	75-55400
Logic SpA	Italy	*236*	75-13800
Logica UK Limited	UK	*506*	80-14350
Lucas Aerospace Power Transmission	USA	*592*	75-56265
Lucas Western Inc.	USA	*592*	75-56300
Magnaghi Milano SpA	Italy	*236*	75-13980
Messier-Dowty Electronics	Canada	*74*	75-09650
Mitsubishi Heavy Industries Ltd	Japan	*253*	75-05800
R. Alkan (UK) Ltd	UK	*429*	75-02680
Revue THOMMEN AG	Switzerland	*387*	75-11100
Saab Dynamics AB	Sweden	*378*	75-09100
SEXTANT Avionique SA	France	*144*	75-35000
SFIM	**France**	***144***	**75-35080**
SP aerospace and vehicle systems bv	**Netherlands**	***295***	**75-06860**
Survival Products Inc.	USA	*618*	75-85200
Technofan SA	France	*147*	75-44300
Tracor Inc	USA	*623*	75-90000
TransInstruments	UK	*489*	75-60960

222 10 Helicopter flight control systems

Alden Sales Co.	USA	*548*	75-04410
Anite Systems Ltd	UK	*497*	80-01240
Astronautics Corp. of America	USA	*555*	75-10000
Bodenseewerk Gerätetechnik GmbH	Germany	*165*	75-03000
Bristow Helicopters Ltd	UK	*499*	80-03950
Canadian Marconi Co.	Canada	*69*	75-02600
Design Bureau of Industrial	Russia	*337*	75-03200
Dowty Aerospace Wolverhampton	UK	*446*	75-18589
Fairey Hydraulics	UK	*450*	75-22255
GEC-Marconi Avionics Ltd	UK	*454*	75-25320
Japan Aviation Electronics Ind. Ltd	Japan	*252*	75-03300
Litton Industries Inc.	USA	*590*	75-55400
Lucas Aerospace FCS Paris	France	*139*	75-25390
Lucas Aerospace Power Transmission	USA	*592*	75-56265
Menasco Aerospace Ltd	Canada	*74*	75-09600
Mitsubishi Heavy Industries Ltd	Japan	*253*	75-05800
Moog Controls Ltd	UK	*469*	75-40670
Moog Inc.	USA	*597*	75-62550
Nord-Micro Elektronik Feinmechanik	Germany	*173*	75-22700
Parker Bertea Aerospace	USA	*602*	75-68900
Raytheon E-Systems	USA	*608*	75-73302
Revue THOMMEN AG	Switzerland	*387*	75-11100
SEXTANT Avionique SA	France	*144*	75-35000
SFIM	**France**	***144***	**75-35080**
Smiths Industries Aerospace	UK	*485*	75-54700
Société d'Applications des Machines	France	*143*	75-34300

222 10 10 Flight stabilisation systems, helicopter

Alden Sales Co.	USA	*548*	75-04410
Anite Systems Ltd	UK	*497*	80-01240
Astronautics Corp. of America	USA	*555*	75-10000
Bristow Helicopters Ltd	UK	*499*	80-03950
Fairey Hydraulics	UK	*450*	75-22255
GEC-Marconi Avionics Inc.	USA	*577*	75-37413
GEC-Marconi Avionics Ltd	UK	*454*	75-25320
Japan Aviation Electronics Ind. Ltd	Japan	*252*	75-03300
Lucas Aerospace FCS Paris	France	*139*	75-25390
Microtecnica-Torino SpA	Italy	*237*	75-15500
Moog Inc.	USA	*597*	75-62550
Nord-Micro Elektronik Feinmechanik	Germany	*173*	75-22700
Parker Bertea Aerospace	USA	*602*	75-68900

SEXTANT Avionique SA.	France	144	75-35000
SFIM	**France**	**144**	**75-35080**

222 10 20 Engine control systems, helicopter

Anite Systems Ltd	UK	497	80-01240
Bodenseewerk Gerätetechnik GmbH	Germany	165	75-03000
Connecticut Machine Products Inc.	USA	566	75-21050
ELECMA	France	136	75-14400
Froude Engineering Inc.	USA	576	75-36340
GEC-Marconi Avionics Inc.	USA	577	75-37413
GEC-Marconi Avionics Ltd	UK	454	75-25320
Lucas Aerospace FCS Paris	France	139	75-25390
Microtecnica-Torino SpA	Italy	237	75-15500
Moog Inc.	USA	597	75-62550
Moog Italiana Srl	Italy	237	75-16230
RSL Electronics Ltd	Israel	222	75-18900
Tuthill Corporation	USA	624	75-91320

222 10 30 De-icing systems, helicopter engine inlet

Boeing Australia Limited	Australia	12	75-02500
Cox & Company Inc.	USA	566	75-22300
DNE, Inc.	USA	569	75-27000
Dunlop Aviation Division	UK	447	75-19000
Engineered Fabrics Corp.	USA	573	75-31565
Paulstra SNC Hutchinson	France	140	75-29100
Sumitomo Electric Industries Ltd	Japan	254	75-07950
Technical Products Group Inc.	USA	620	75-86764
Tempo Instruments Inc	USA	621	75-87900

222 10 32 Anti-icing systems, helicopter engine inlet

Boeing Australia Limited	Australia	12	75-02500
DNE, Inc.	USA	569	75-27000
Engineered Fabrics Corp.	USA	573	75-31565
Logic SpA	Italy	236	75-13800
Parker Bertea Aerospace.	USA	602	75-68900
Paulstra SNC Hutchinson	France	140	75-29100
Sumitomo Electric Industries Ltd	Japan	254	75-07950
Tempo Instruments Inc	USA	621	75-87900

222 10 35 De-icing systems, helicopter rotor blade

Boeing Australia Limited	Australia	12	75-02500
Daimler-Benz Aerospace France SAS	**France**	**134**	**75-12950**
DNE, Inc.	USA	569	75-27000
Dunlop Aviation Division	UK	447	75-19000
ECE GmbH	Germany	167	75-07200
Engineered Fabrics Corp.	USA	573	75-31565
Paulstra SNC Hutchinson	France	140	75-29100

222 10 38 Ice detectors, helicopter

Boeing Australia Limited	Australia	12	75-02500
DNE, Inc.	USA	569	75-27000
Penny & Giles Drives Technology Ltd	UK	474	75-44563
Tempo Instruments Inc	USA	621	75-87900
Vibro-Meter SA	Switzerland	388	75-14710

222 10 60 Helicopter rotor brakes

Elettronica Aster SpA	Italy	233	75-07000
Merex Inc.	USA	595	75-60250
Parker Bertea Aerospace.	USA	602	75-68900

222 10 70 Helicopter flight control equipment

Anite Systems Ltd	UK	497	80-01240
Astronautics Corp. of America	USA	555	75-10000
AVIAC Technologies	France	132	75-04600
Barber-Colman Co.	USA	557	75-11960
BFGoodrich Avionics Systems Inc	USA	559	75-13771
Bodenseewerk Gerätetechnik GmbH	Germany	165	75-03000
Bristow Helicopters Ltd	UK	499	80-03950
Dowty Aerospace Wolverhampton	UK	446	75-18589
GEC-Marconi Avionics Ltd	UK	454	75-25320
Humphrey Inc.	USA	583	75-45050
Japan Aviation Electronics Ind. Ltd	Japan	252	75-03300
Kearfott Guidance & Navigation Corp.	USA	588	75-51760
Logica UK Limited.	UK	506	80-14350
Lucas Aerospace FCS Paris	France	139	75-25390
Magnaghi Milano SpA	Italy	236	75-13980
Menasco Aerospace Ltd	Canada	74	75-09600
Moog Controls Ltd	UK	469	75-40670
Moog Inc.	USA	597	75-62550
Moog Italiana Srl	Italy	237	75-16230
Moog Japan Ltd.	Japan	253	75-06100
Nord-Micro Elektronik Feinmechanik	Germany	173	75-22700
Pacer Systems Inc.	USA	602	75-68000
Parker Bertea Aerospace.	USA	602	75-68900
Raytheon E-Systems	USA	608	75-73302
SEXTANT Avionique SA.	France	144	75-35000
SFIM	**France**	**144**	**75-35080**
Smiths Aerospace & Defense	USA	615	75-82090

222 10 70 10 Helicopter air data measuring systems

Alden Sales Co.	USA	548	75-04410
Anite Systems Ltd	UK	497	80-01240
Arcom Control Systems Ltd	UK	431	75-03700
Astronautics Corp. of America	USA	555	75-10000
Boeing Australia Limited	Australia	12	75-02500
GEC-Marconi Avionics Inc.	USA	577	75-37413
GEC-Marconi Avionics Ltd	UK	454	75-25320
Pacer Systems Inc.	USA	602	75-68000
Transicoil Inc.	USA	623	75-90750
TransInstruments	UK	489	75-60960

222 10 70 30 Flight instruments, helicopter

Alden Sales Co.	USA	548	75-04410
Ametek Inc.	USA	553	75-06510
Astronautics C.A. Ltd	Israel	217	75-02400
Astronautics Corp. of America	USA	555	75-10000
BFGoodrich Avionics Systems Inc	USA	559	75-13771
Bodenseewerk Gerätetechnik GmbH	Germany	165	75-03000
Boeing Australia Limited	Australia	12	75-02500
Bospec Neue Technologien	Germany	166	75-03420
Canadian Marconi Co.	Canada	69	75-02600
GEC-Marconi Avionics Inc.	USA	577	75-37413
Humphrey Inc.	USA	583	75-45050
Japan Aviation Electronics Ind. Ltd	Japan	252	75-03300
Litton Industries Inc.	USA	590	75-55400
Logica UK Limited.	UK	506	80-14350
Meggitt Avionics (UK)	UK	467	75-38340
Merex Inc.	USA	595	75-60250
Microtecnica-Torino SpA	Italy	237	75-15500
Pacer Systems Inc.	USA	602	75-68000
Racal Avionics Ltd	UK	477	75-47660
Rockwell International Co.	Italy	233	75-05200
SEXTANT Avionique SA.	France	144	75-35000
SFIM	**France**	**144**	**75-35080**
Smiths Aerospace & Defense	USA	615	75-82090
Transicoil Inc.	USA	623	75-90750

222 10 70 30 60 Thunderstorm mapping equipment, helicopter, passive

BFGoodrich Aerospace	USA	559	75-13770
GEC-Marconi Avionics Inc.	USA	577	75-37413

222 10 70 50 Heading & attitude reference systems, helicopter

Ametek Inc.	USA	553	75-06510
Astronautics C.A. Ltd	Israel	217	75-02400
Astronautics Corp. of America	USA	555	75-10000
Boeing Australia Limited	Australia	12	75-02500
GEC-Marconi Avionics Inc.	USA	577	75-37413
GEC-Marconi Avionics Ltd	UK	454	75-25320
Israel Aircraft Industries Ltd	Israel	220	75-10200
Japan Aviation Electronics Ind. Ltd	Japan	252	75-03300
Kearfott Guidance & Navigation Corp.	USA	588	75-51760
LITEF GmbH	**Germany**	**171**	**75-18600**
Litton Industries Inc.	USA	590	75-55400
Merex Inc.	USA	595	75-60250
SAGEM SA	**France**	**143**	**75-33950**

SFIM	**France**	***144***	**75-35080**
Smiths Aerospace & Defense	USA	*615*	75-82090

222 10 70 70 Trackers, helicopter rotor blade

Merex Inc.	USA	*595*	75-60250
Recon/Optical Inc.	USA	*608*	75-73930

222 20 Helicopter mission-related systems

BAeSEMA Ltd.	UK	*498*	80-03070
Breeze Eastern Corp.	USA	*560*	75-15110
Bristow Helicopters Ltd	UK	*499*	80-03950
Canadian Marconi Co.	Canada	*69*	75-02600
Computing Devices Canada	Canada	*70*	75-03600
Computing Devices Company Ltd	UK	*442*	75-14700
Cubic Defense Systems Inc.	USA	*567*	75-22900
Daimler-Benz Aerospace France SAS	**France**	***134***	**75-12950**
DRS Technologies, Inc.	USA	*570*	75-27400
Fabbrica Italiana Apparecchiature	Italy	*234*	75-08550
Fibertek	USA	*575*	75-34330
Fraser-Volpe Corp.	USA	*576*	75-35950
GEC-Marconi Avionics Ltd	UK	*454*	75-25320
Logica UK Limited	UK	*506*	80-14350
Lucas Western Inc.	USA	*592*	75-56300
Northern Airborne Technology Ltd	Canada	*74*	75-10145
Pilkington Optronics Ltd	UK	*426*	74-10000
Racal Avionics Ltd	UK	*477*	75-47660
Royal Ordnance	UK	*481*	75-50620
Saab Dynamics AB	Sweden	*378*	75-09100
SAGEM SA	**France**	***143***	**75-33950**
Sanders	USA	*612*	75-77850
SEXTANT Avionique SA	France	*144*	75-35000
SFIM	**France**	***144***	**75-35080**
Smiths Aerospace & Defense	USA	*615*	75-82090

222 20 30 Fuel tanks, external, helicopter

AIM Enterprises Inc.	USA	*630*	80-01350
EDO Corporation	USA	*571*	75-29250
Fibertek	USA	*575*	75-34330
Israel Military Industries Ltd	Israel	*220*	75-10300
SEI Industries Ltd	Canada	*76*	75-12750
Sumitomo Electric Industries Ltd	Japan	*254*	75-07950

222 20 50 Target designation & aiming systems, helicopter

Boeing Australia Limited	Australia	*12*	75-02500
CI Systems Ltd	Israel	*218*	75-03750
Daimler-Benz Aerospace France SAS	**France**	***134***	**75-12950**
Fabbrica Italiana Apparecchiature	Italy	*234*	75-08550
Fraser-Volpe Corp.	USA	*576*	75-35950
GEC-Marconi Avionics Inc.	USA	*577*	75-37413
GEC-Marconi Avionics Ltd	UK	*454*	75-25320
Hayes Targets	USA	*603*	75-69135
Israel Aircraft Industries Ltd	Israel	*220*	75-10200
Litton Industries Inc.	USA	*590*	75-55400
Northrop Grumman Corporation	USA	*599*	75-65450
Rockwell	USA	*610*	75-75930
Saab Dynamics AB	Sweden	*378*	75-09100
SAGEM SA	**France**	***143***	**75-33950**
SFIM	**France**	***144***	**75-35080**
Smiths Aerospace & Defense	USA	*615*	75-82090

222 20 50 10 Sights, roof-mounted, helicopter

Alan Gordon Enterprises Inc.	USA	*579*	75-40060
Daimler-Benz Aerospace France SAS	**France**	***134***	**75-12950**
Fabbrica Italiana Apparecchiature	Italy	*234*	75-08550
GEC-Marconi Avionics Inc.	USA	*577*	75-37413
GEC-Marconi Avionics Ltd	UK	*454*	75-25320
Israel Aircraft Industries Ltd	Israel	*220*	75-10200
Kaiser Electronics	USA	*587*	75-51350
Saab Dynamics AB	Sweden	*378*	75-09100
SEXTANT Avionique SA	France	*144*	75-35000
SFIM	**France**	***144***	**75-35080**

222 20 50 17 Sights, roof-mounted, helicopter, thermal imaging

Bospec Neue Technologien	Germany	*166*	75-03420
CI Systems Ltd	Israel	*218*	75-03750
CIC International Ltd	USA	*563*	75-18570
Fabbrica Italiana Apparecchiature	Italy	*234*	75-08550
GEC-Marconi Avionics Inc.	USA	*577*	75-37413
GEC-Marconi Avionics Ltd	UK	*454*	75-25320
Israel Aircraft Industries Ltd	Israel	*220*	75-10200
Kaiser Electronics	USA	*587*	75-51350
Northrop Grumman Corporation	USA	*599*	75-65450
Saab Dynamics AB	Sweden	*378*	75-09100
SFIM	**France**	***144***	**75-35080**

222 20 60 Weapon delivery / release systems, helicopter

AAI Corp.	USA	*545*	75-00020
Aerea Luftfahrtgeräte GmbH	Germany	*164*	75-00260
Aerea SpA	Italy	*230*	75-00280
Alenia Difesa	**Italy**	***231***	**75-01165**
Astronautics Corp. of America	USA	*555*	75-10000
Canadian Marconi Co.	Canada	*69*	75-02600
Computing Devices Company Ltd	UK	*442*	75-14700
Daimler-Benz Aerospace France SAS	**France**	***134***	**75-12950**
EDO Corporation	USA	*571*	75-29250
GEC-Marconi Avionics Inc.	USA	*577*	75-37413
GEC-Marconi Avionics Ltd	UK	*454*	75-25320
GKN Westland Technologies Ltd	UK	*455*	75-25690
Hunting Engineering Ltd	UK	*459*	75-30700
Kaiser Electronics	USA	*587*	75-51350
Logic SpA	Italy	*236*	75-13800
Lucas Western Inc.	USA	*592*	75-56300
Matra Défense-Espace	France	*139*	75-26000
Normalair-Garrett Ltd	UK	*425*	74-07000
R. Alkan (UK) Ltd	UK	*429*	75-02680
Royal Ordnance	UK	*481*	75-50620
Smiths Aerospace & Defense	USA	*615*	75-82090

222 20 60 70 Stores carrier & release equipment, helicopter

AAI Corp.	USA	*545*	75-00020
Aerea Luftfahrtgeräte GmbH	Germany	*164*	75-00260
Aerea SpA	Italy	*230*	75-00280
EDO Corporation	USA	*571*	75-29250
GEC-Marconi Avionics Inc.	USA	*577*	75-37413
GKN Westland Technologies Ltd	UK	*455*	75-25690
Hunting Engineering Ltd	UK	*459*	75-30700
Lucas Western Inc.	USA	*592*	75-56300
Normalair-Garrett Ltd	UK	*425*	74-07000
R. Alkan (UK) Ltd	UK	*429*	75-02680
Royal Ordnance	UK	*481*	75-50620
Talley Defense Systems Inc.	USA	*619*	75-86120

222 20 60 73 Ejectors, helicopter weapon

GKN Westland Technologies Ltd	UK	*455*	75-25690
Hill Industries Inc.	USA	*582*	75-43465
Lucas Western Inc.	USA	*592*	75-56300
Normalair-Garrett Ltd	UK	*425*	74-07000
R. Alkan (UK) Ltd	UK	*429*	75-02680
Royal Ordnance	UK	*481*	75-50620
Talley Defense Systems Inc.	USA	*619*	75-86120

222 20 70 Helicopter mission-related equipment

AIM Enterprises Inc.	USA	*630*	80-01350
Bridport Aviation Products	UK	*435*	75-08400
Bristow Helicopters Ltd	UK	*499*	80-03950
Canadian Marconi Co.	Canada	*69*	75-02600
Computing Devices Canada	Canada	*70*	75-03600
DRS Technologies, Inc.	USA	*570*	75-27400
Fabbrica Italiana Apparecchiature	Italy	*234*	75-08550
Lucas Aerospace FCS Paris	France	*139*	75-25390
Northern Airborne Technology Ltd	Canada	*74*	75-10145
Racal Avionics Ltd	UK	*477*	75-47660
SEXTANT Avionique SA	France	*144*	75-35000
Thomson Marconi Sonar Limited	UK	*426*	74-20000
TransInstruments	UK	*489*	75-60960

222 20 70 20 Cargo air delivery equipment, helicopter

Aerial Machine & Tool Corp.	USA	*546*	75-01105
Ancra CHS BV.	Netherlands	*291*	75-00510
Breeze Eastern Corp.	USA	*560*	75-15110
Bridport Aviation Products.	UK	*435*	75-08400
Cargo Aids	UK	*438*	75-11620
Intamar Logistics Inc.	USA	*585*	75-47050
Irvin Aerospace Canada Limited	Canada	*73*	75-08300
Irvin Aerospace Limited	UK	*462*	75-32050
Lucas Western Inc.	USA	*592*	75-56300
MIL-BASE Industries Inc.	USA	*596*	75-61530
SEI Industries Ltd	Canada	*76*	75-12750

222 20 70 22 Cargo slings, helicopter

Aerial Machine & Tool Corp.	USA	*546*	75-01105
AIM Aviation (Jecco) Ltd.	UK	*428*	75-01530
Ancra CHS BV.	Netherlands	*291*	75-00510
Bridport Aviation Products.	UK	*435*	75-08400
Cargo Aids	UK	*438*	75-11620
H. Koch & Sons Co.	USA	*589*	75-53030
Intamar Logistics Inc.	USA	*585*	75-47050
Irvin Aerospace Canada Limited	Canada	*73*	75-08300
Irvin Aerospace Limited	UK	*462*	75-32050
Irvin Industries Inc.	USA	*586*	75-48960
MIL-BASE Industries Inc.	USA	*596*	75-61530
Parsons Chain Co. Ltd	UK	*473*	75-44035
SpanSet Inter AG	Switzerland	*387*	75-13550
WB Chain Co Ltd	UK	*439*	75-12440

222 20 70 50 Searchlights, helicopter

Bristow Helicopters Ltd	UK	*499*	80-03950
Spectrolab Inc.	USA	*617*	75-83100

222 20 70 92 Weapon pylons, helicopter

Aerea Luftfahrtgeräte GmbH.	Germany	*164*	75-00260
Aerea SpA.	Italy	*230*	75-00280
Israel Military Industries Ltd	Israel	*220*	75-10300
Lucas Western Inc.	USA	*592*	75-56300
R. Alkan (UK) Ltd	UK	*429*	75-02680

222 20 70 95 Winches, helicopter

Air Target Sweden AB	Sweden	*375*	75-00850
A1Marketing	UK	*426*	75-00025
Bristow Helicopters Ltd	UK	*499*	80-03950
Hill Industries Inc.	USA	*582*	75-43465
Lucas Western Inc.	USA	*592*	75-56300

222 20 70 95 50 Rescue winches, helicopter

Breeze Eastern Corp.	USA	*560*	75-15110
Lucas Aerospace FCS Paris	France	*139*	75-25390
Lucas Aerospace Singapore Pte Ltd	Singapore	*349*	80-01200
Lucas Western Inc.	USA	*592*	75-56300

222 20 70 95 60 Winches, helicopter dunking sonar

A1Marketing	UK	*426*	75-00025
Breeze Eastern Corp.	USA	*560*	75-15110
Bristow Helicopters Ltd	UK	*499*	80-03950
Indal Technologies Inc	Canada	*72*	75-07740
Thomson Marconi Sonar Limited	UK	*426*	74-20000
Thomson Sintra	France	*148*	75-45135

222 30 Helicopter crew / passenger-related systems

Aerospace Syst. & Comp.	USA	*547*	75-02501
AIM Aviation (Jecco) Ltd.	UK	*428*	75-01530
AIM Enterprises Inc.	USA	*630*	80-01350
Airscrew Howden Ltd	UK	*429*	75-02050
AlliedSignal Aerospace Canada	Canada	*68*	75-00470
Avica Equipment	UK	*432*	75-04870
Beaufort Air-Sea Equipment Ltd.	UK	*433*	75-06100
GEC-Marconi Avionics Ltd	UK	*454*	75-25320
H. Koch & Sons Co.	USA	*589*	75-53030
Howden Aircontrol Ltd.	UK	*459*	75-30240
Liebherr-Aerospace Toulouse SA.	France	*138*	75-24950
Northern Airborne Technology Ltd	Canada	*74*	75-10145
SFIM	**France**	***144***	**75-35080**
Skyline Industries Inc.	USA	*615*	75-81550

222 30 05 Environmental control systems, helicopter

Aerospace Syst. & Comp.	USA	*547*	75-02501
Airscrew Howden Ltd	UK	*429*	75-02050
AlliedSignal Aerospace Canada	Canada	*68*	75-00470
Avica Equipment	UK	*432*	75-04870
Barber-Colman Co.	USA	*557*	75-11960
Baxter Woodhouse & Taylor Ltd	UK	*433*	75-06030
GKN Westland Technologies Ltd	UK	*455*	75-25690
Howden Aircontrol Ltd.	UK	*459*	75-30240
Intertechnique SA.	France	*137*	75-23000
Liebherr-Aerospace Toulouse SA.	France	*138*	75-24950
Microtecnica-Torino SpA	Italy	*237*	75-15500
Munters Cargocaire.	USA	*597*	75-63445
Normalair-Garrett Ltd.	UK	*425*	74-07000
Parker Bertea Aerospace.	USA	*602*	75-68900
Spar Aviation Services	Canada	*76*	75-13020
Sundstrand Aerospace.	USA	*618*	75-85105

222 30 05 10 Air conditioning systems, helicopter

Aerospace Syst. & Comp.	USA	*547*	75-02501
AlliedSignal Aerospace Canada	Canada	*68*	75-00470
Avica Equipment	UK	*432*	75-04870
GKN Westland Technologies Ltd	UK	*455*	75-25690
Howden Aircontrol Ltd.	UK	*459*	75-30240
Howden Fluid Systems.	USA	*582*	75-44400
Intertechnique SA.	France	*137*	75-23000
Liebherr-Aerospace Toulouse SA.	France	*138*	75-24950
Microtecnica-Torino SpA	Italy	*237*	75-15500
Normalair-Garrett Ltd.	UK	*425*	74-07000
Parker Bertea Aerospace.	USA	*602*	75-68900
Spar Aviation Services	Canada	*76*	75-13020

222 30 05 40 Oxygen systems, helicopter

Aerospace Syst. & Comp.	USA	*547*	75-02501
EROS GIE	France	*137*	75-19500
GKN Westland Technologies Ltd	UK	*455*	75-25690
Intertechnique SA.	France	*137*	75-23000
Kidde-Graviner Ltd	UK	*463*	75-33880
Litton Industries Inc.	USA	*590*	75-55400
Meggitt Oxygen Systems	UK	*467*	75-38360
Normalair-Garrett Ltd.	UK	*425*	74-07000
Parker Bertea Aerospace.	USA	*603*	75-68900
Puritan-Bennett Aero Systems Co.	USA	*606*	75-71890
Scott Aviation.	USA	*613*	75-78850
Scott Health and Safety	USA	*613*	75-78850
SFIM	**France**	***144***	**75-35080**
Tokyo Aircraft Instrument Co Ltd.	Japan	*254*	75-08600

222 30 40 Helicopter internal communication systems

Davies Industrial Communications Ltd	UK	*444*	75-17010
GEC-Marconi Avionics Ltd.	UK	*454*	75-25320
Northern Airborne Technology Ltd	Canada	*74*	75-10145
Tempo Instruments Inc	USA	*621*	75-87900

222 30 70 Helicopter crew / passenger related equipment

EROS GIE	France	*137*	75-19500
H. Koch & Sons Co.	USA	*589*	75-53030

222 30 70 60 Seats, helicopter

Aeromot Ind.Mecanico-Metalurgica SA	Brazil	*55*	75-00635
ARA Inc.	USA	*554*	75-08370
Cyclone Aviation Products Ltd	Israel	*218*	75-04250
FPT Industries Ltd	UK	*452*	75-24400
Skyline Industries Inc.	USA	*615*	75-81550
St Bernard Composites Ltd	UK	*486*	75-56222

Sumitomo Electric Industries Ltd	Japan	*254*	75-07950
Vermont Composites, Inc.	USA	*625*	75-93450

222 30 70 61 Seats, aircrew, helicopter

ARA Inc.	USA	*554*	75-08370
Cyclone Aviation Products Ltd	Israel	*218*	75-04250
Skyline Industries Inc.	USA	*615*	75-81550
Société d'Exploitation	France	*130*	74-01800
St Bernard Composites Ltd	UK	*486*	75-56222
Vermont Composites, Inc.	USA	*625*	75-93450

222 30 70 61 10 Seats, armoured, helicopter crew

Autoflug GmbH & Co	Germany	*165*	75-01400
Plasan Sasa	Israel	*221*	75-17510
Skyline Industries Inc.	USA	*615*	75-81550
St Bernard Composites Ltd	UK	*486*	75-56222

222 30 70 65 Safety harnesses, helicopter

Aerial Machine & Tool Corp.	USA	*546*	75-01105
AIM Aviation (Jecco) Ltd.	UK	*428*	75-01530
Aircraft Materials Ltd	UK	*428*	75-01650
Ancra CHS BV.	Netherlands	*291*	75-00510
H. Koch & Sons Co.	USA	*589*	75-53030
Irvin Aerospace Canada Limited	Canada	*73*	75-08300
Irvin Aerospace Limited	UK	*462*	75-32050
MIL-BASE Industries Inc.	USA	*596*	75-61530
Skyline Industries Inc.	USA	*615*	75-81550
White Products BV	Netherlands	*296*	75-09100

222 30 70 70 Ambulance kits, helicopter

Ingenieurbüro Flugwesen &	Switzerland	*388*	80-00800

222 40 Helicopter airframe-related systems

Aerospace Syst. & Comp.	USA	*547*	75-02501
AIM Enterprises Inc.	USA	*630*	80-01350
APPH Ltd	UK	*431*	75-03415
Canadian Marconi Co.	Canada	*69*	75-02600
Daimler-Benz Aerospace France SAS	**France**	***134***	**75-12950**
Dowty Aerospace Wolverhampton	UK	*446*	75-18589
Eldec Corp.	USA	*572*	75-29660
H. Koch & Sons Co.	USA	*589*	75-53030
Merex Inc.	USA	*595*	75-60250
Skyline Industries Inc.	USA	*615*	75-81550
SP aerospace and vehicle systems bv	**Netherlands**	***295***	**75-06860**
Survival Products Inc.	USA	*618*	75-85200

222 40 30 Electrical systems, helicopter

Aircraft Parts Corp.	USA	*548*	75-03440
AlliedSignal Aerospace Canada	Canada	*68*	75-00470
Century Electronics	USA	*562*	75-17385
Daimler-Benz Aerospace France SAS	**France**	***134***	**75-12950**
ECE GmbH	Germany	*167*	75-07200
Eldec Corp.	USA	*572*	75-29660
GSI Corporation	USA	*580*	75-40715
Icore International Ltd	UK	*460*	75-31080
Labinal	France	*138*	75-23900
Lucas Aerospace FCS Paris	France	*139*	75-25390
Merex Inc.	USA	*595*	75-60250
Newton Derby Ltd	UK	*470*	75-42520
Spar Aviation Services	Canada	*76*	75-13020
Teledyne Continental Motors	USA	*620*	75-87506
Walter Dittel GmbH	Germany	*167*	75-06510
White Products BV	Netherlands	*296*	75-09100

222 40 45 Fuel systems, helicopter

Aeroquip Corp.	USA	*547*	75-02350
Aerospace Syst. & Comp.	USA	*547*	75-02501
AIM Enterprises Inc.	USA	*630*	80-01350
British Filters Ltd	UK	*436*	75-09030
Connecticut Machine Products Inc.	USA	*566*	75-21050
Continental AG Werk Uniroyal	Germany	*176*	75-32090
Flight Refuelling Limited	UK	*451*	75-23850
Howden Fluid Systems	USA	*582*	75-44400
Microtecnica-Torino SpA	Italy	*237*	75-15500
Ontic Engineering & Mfg Inc.	USA	*601*	75-66875
Secondo Mona SpA	Italy	*237*	75-15900
Vickers Inc.	USA	*626*	75-93820
Vickers Systems	Germany	*176*	75-32830
Vickers Systems Division TRINOVA Ltd	UK	*491*	75-63150
Vickers Systems SA	France	*148*	75-46930
Vickers Systems Trinova SpA.	Italy	*239*	75-27150

222 40 50 Hydraulic systems, helicopter

Abex GmbH Aerohydraul	Germany	*164*	75-00105
Aeroquip Corp.	USA	*547*	75-02350
AIM Enterprises Inc.	USA	*630*	80-01350
APPH Ltd	UK	*431*	75-03415
British Filters Ltd	UK	*436*	75-09030
Dowty Aerospace Wolverhampton	UK	*446*	75-18589
Fairey Hydraulics	UK	*450*	75-22255
GKN Westland Technologies Ltd	UK	*455*	75-25690
Howden Fluid Systems	USA	*582*	75-44400
Lucas Aerospace FCS Paris	France	*139*	75-25390
Lucas Aerospace Power Transmission	USA	*592*	75-56265
Magnaghi Milano SpA	Italy	*236*	75-13980
Mannesmann Rexroth Ltd	UK	*466*	75-37100
Mannesmann Rexroth SA	France	*139*	75-25500
M.C. Aerospace Corp.	USA	*594*	75-59370
Menasco Aerospace Ltd	Canada	*74*	75-09600
Messier-Bugatti	France	*140*	75-26400
Moog Controls Ltd	UK	*469*	75-40670
Ontic Engineering & Mfg Inc.	USA	*601*	75-66875
Parker Bertea Aerospace	USA	*602*	75-68900
Riva Calzoni SpA	Italy	*238*	75-20500
Secondo Mona SpA	Italy	*237*	75-15900
SP aerospace and vehicle systems bv	**Netherlands**	***295***	**75-06860**
Spar Aviation Services	Canada	*76*	75-13020
Vickers Systems Division TRINOVA Ltd	UK	*491*	75-63150

222 40 60 Landing gear, helicopter

AIM Enterprises Inc.	USA	*630*	80-01350
APPH Ltd	UK	*431*	75-03415
Fairey Hydraulics	UK	*450*	75-22255
Heroux Inc	Canada	*72*	75-07400
Magnaghi Milano SpA	Italy	*236*	75-13980
Menasco Aerospace Ltd	Canada	*74*	75-09600
Merex Inc.	USA	*595*	75-60250
Messier-Dowty Electronics	Canada	*74*	75-09650
Messier-Dowty Inc	Canada	*74*	75-09655
Messier-Dowty Limited	UK	*467*	75-38700
Messier-Dowty SAS	France	*140*	75-26500
National Forge Co.	USA	*598*	75-63850
Parker Bertea Aerospace	USA	*602*	75-68900
Riva Calzoni SpA	Italy	*238*	75-20500
Skyline Industries Inc.	USA	*615*	75-81550
SP aerospace and vehicle systems bv	**Netherlands**	***295***	**75-06860**

222 40 65 Lighting systems, helicopter

Bruce Industries Inc.	USA	*561*	75-15360
ECE GmbH	Germany	*167*	75-07200
H. Koch & Sons Co.	USA	*589*	75-53030
Hoffman Engineering Corp.	USA	*582*	75-43980
Luminescent Systems Inc.	USA	*593*	75-56480
Page Aerospace Ltd	UK	*472*	75-43505
Soderberg Manufacturing Co. Inc.	USA	*616*	75-82200
Spar Aviation Services	Canada	*76*	75-13020
Survival Products Inc.	USA	*618*	75-85200
Tempo Instruments Inc	USA	*621*	75-87900
White Products BV	Netherlands	*296*	75-09100

222 40 65 70 Lighting equipment, helicopter

Bruce Industries Inc.	USA	561	75-15360
DNE, Inc.	USA	569	75-27000
ECE GmbH	Germany	167	75-07200
Eldec Corp.	USA	572	75-29660
H. Koch & Sons Co.	USA	589	75-53030
Hoffman Engineering Corp.	USA	582	75-43980
Luminescent Systems Inc.	USA	593	75-56480
Soderberg Manufacturing Co. Inc.	USA	616	75-82200
Spar Aviation Services	Canada	76	75-13020
Survival Products Inc.	USA	618	75-85200
Tempo Instruments Inc	USA	621	75-87900
White Products BV	Netherlands	296	75-09100

222 40 65 75 Lights, helicopter

Bruce Industries Inc.	USA	561	75-15360
Luminescent Systems Inc.	USA	593	75-56480
Oxley Developments Co. Ltd	UK	472	75-43400
Soderberg Manufacturing Co. Inc.	USA	616	75-82200
Spar Aviation Services	Canada	76	75-13020
Survival Products Inc.	USA	618	75-85200

222 40 70 Helicopter equipment, airframe-related

AIM Enterprises Inc.	USA	630	80-01350
Avica Equipment	UK	432	75-04870
Bristow Helicopters Ltd	UK	499	80-03950
Canadian Marconi Co.	Canada	69	75-02600
Connecticut Machine Products Inc.	USA	566	75-21050
Cyclone Aviation Products Ltd	Israel	218	75-04250
Darchem Engineering Ltd.	UK	444	75-16750
Eldec Corp.	USA	572	75-29660
Survival Products Inc.	USA	618	75-85200

222 40 70 30 Batteries, helicopter

Concorde Battery Corp.	USA	566	75-20845
Eldec Corp.	USA	572	75-29660
Global Airtech International	USA	579	75-39769
Hawker Energy Products Ltd	UK	457	75-28490
Saft America Inc.	USA	612	75-77550

222 40 70 40 Flotation gear, helicopter

Aérazur	France	131	75-00500
Aerospace and Electronics Division	USA	606	75-71472
Air Cruisers Co	USA	548	75-03600
Bristow Helicopters Ltd	UK	499	80-03950
Eurodef Aerospace & Defence	Greece	189	75-01295
FPT Inc.	USA	576	75-35760
FPT Industries Ltd	UK	452	75-24400
Irvin Aerospace Canada Limited	Canada	73	75-08300
Kidde-Graviner Ltd	UK	463	75-33880
Merex Inc.	USA	595	75-60250
Moog Controls Ltd	UK	469	75-40670
Sèkur SpA	Italy	238	75-21650
SMR Technologies Inc.	USA	615	75-82100
Survival Products Inc.	USA	618	75-85200
Technical Products Group Inc.	USA	620	75-86764
Zodiac International	France	149	75-47310

222 40 70 60 Floats, helicopter

Aérazur	France	131	75-00500
Air Cruisers Co	USA	548	75-03600
Eurovinil Industries SpA	Italy	234	75-08400
FPT Industries Ltd	UK	452	75-24400
Merex Inc.	USA	595	75-60250

222 40 90 60 Radomes, helicopter

Chelton (Electrostatics) Limited	UK	439	75-12700
Complas Limited	UK	442	75-14543
Norton Performance Plastics Corp.	USA	600	75-65520
Raytheon E-Systems	USA	608	75-73302
Sea Tel Inc.	USA	613	75-78930
St Bernard Composites Ltd	UK	486	75-56222
Vermont Composites, Inc.	USA	625	75-93450

222 40 90 70 20 Canopies, helicopter

AIM Enterprises Inc.	USA	630	80-01350
Technical Products Group Inc.	USA	620	75-86764

222 40 90 90 Windows, helicopter

AIM Enterprises Inc.	USA	630	80-01350
Hughes Danbury Optical Sys.	USA	582	75-44857
M&N Aerospace	USA	597	75-62240

222 80 Helicopter accessories

AIM Enterprises Inc.	USA	630	80-01350
Associated Aircraft Mfg & Sales Inc.	USA	555	75-09500
Avica Equipment	UK	432	75-04870
Breeze Eastern Corp.	USA	560	75-15110
Canadian Marconi Co.	Canada	69	75-02600
Connecticut Machine Products Inc.	USA	566	75-21050
Horstman Defence Systems Ltd.	UK	459	75-29920
Lucas Aerospace Power Transmission	USA	592	75-56265
M&N Aerospace	USA	597	75-62240
Merex Inc.	USA	595	75-60250
Penny & Giles Aerospace Ltd	UK	473	75-44550
Riva Calzoni SpA	Italy	238	75-20500
Skyline Industries Inc.	USA	615	75-81550
Spectrolab Inc.	USA	617	75-83100
Survival Products Inc.	USA	618	75-85200

222 90 Helicopter components

AIM Enterprises Inc.	USA	630	80-01350
Associated Aircraft Mfg & Sales Inc.	USA	555	75-09500
Avica Equipment	UK	432	75-04870
Barber-Colman Co.	USA	557	75-11960
Breeze Eastern Corp.	USA	560	75-15110
Busak + Shamban Inc.	USA	561	75-16006
Canadian Marconi Co.	Canada	69	75-02600
Connecticut Machine Products Inc.	USA	566	75-21050
Darchem Engineering Ltd.	UK	444	75-16750
Giat Industries	France	126	55-00100
Hardcote Technologies	USA	580	75-41900
Howden Fluid Systems	USA	582	75-44400
IMP Aerospace Limited	Canada	72	75-07730
Jehier SA	France	137	75-23200
Litton Industries Inc.	USA	590	75-55400
Lord SA	France	150	80-01350
M&N Aerospace	USA	597	75-62240
Manitoba Aerospace Initiative	Canada	73	75-09215
Merex Inc.	USA	595	75-60250
Muirhead Vactric Components Ltd	UK	470	75-41404
Penny & Giles Aerospace Ltd	UK	473	75-44550
Reil Industrial Enterprises Ltd	Canada	75	75-12000
Riva Calzoni SpA	Italy	238	75-20500
Skyline Industries Inc.	USA	615	75-81550
SP aerospace and vehicle systems bv	**Netherlands**	**295**	**75-06860**
Technical Products Group Inc.	USA	620	75-86764
Vermont Composites, Inc.	USA	625	75-93450
Voss Industries Inc.	USA	626	75-94550
ZF Luftfahrttechnik GmbH	Germany	177	75-36000

223 RPV, Drone systems & equipment

AAI Corp.	USA	545	75-00020
Alden Sales Co.	USA	548	75-04410
Astronautics Corp. of America	USA	555	75-10000
Aydin Corp.	USA	556	75-11450
Bombardier Inc	Canada	69	75-01900
CAC Systèmes	France	133	75-07680
Canadian Marconi Co.	Canada	69	75-02600
China National Aero-Tech I/E Corp.	China PR	88	55-00200
Cranfield Aerospace Ltd	UK	500	80-07200
Fabbrica Italiana Apparecchiature	Italy	234	75-08550
Fuji Heavy Industries Ltd	Japan	251	75-01100
GEC-Marconi Avionics Ltd	UK	454	75-25320
GEC-Marconi Hazeltine Corp.	USA	577	75-37414

Israel Aircraft Industries Ltd	Israel	*220*	75-10200
Israel Military Industries Ltd	Israel	*220*	75-10300
Japan Aviation Electronics Ind. Ltd	Japan	*252*	75-03300
Kearfott Guidance & Navigation Corp.	USA	*588*	75-51760
Lear Astronics Corp.	USA	*589*	75-54390
Lockheed Martin Corp.	USA	*591*	75-55635
Matra Défense-Espace	France	*139*	75-26000
Meteor CAE SpA	Italy.	*237*	75-15400
Moog Inc.	USA	*597*	75-62550
Moog Italiana Srl	Italy.	*237*	75-16230
Northrop Grumman Corporation.	USA	*599*	75-65450
Oerlikon-Contraves AG.	Switzerland	*387*	75-09250
Raytheon E-Systems	USA	*608*	75-73302
Revue THOMMEN AG	Switzerland	*387*	75-11100
SFIM	**France**	***144***	**75-35080**
Swiss Aircraft & Systems Co	Switzerland	*384*	55-00100
Tracor Inc	USA	*623*	75-90000
Universal Propulsion Co. Inc.	USA	*625*	75-92700
Vanguard Research Inc.	USA	*644*	80-34100
Xi CAD North America Inc.	USA	*628*	75-97760

223 10 RPV / Drone flight control systems

AAI Corp.	USA	*545*	75-00020
AeroSpace Technologies of Australia	Australia	*11*	75-00340
Alden Sales Co.	USA	*548*	75-04410
Astronautics Corp. of America	USA	*555*	75-10000
Aydin Corp.	USA	*556*	75-11450
Bombardier Inc	Canada	*69*	75-01900
CAC Systèmes	France	*133*	75-07680
Canadian Marconi Co.	Canada	*69*	75-02600
GEC-Marconi Avionics Inc.	USA	*577*	75-37413
GEC-Marconi Avionics Ltd	UK	*454*	75-25320
GEC-Marconi Hazeltine Corp.	USA	*577*	75-37414
Hayes Targets.	USA	*603*	75-69135
Herley-Vega Systems	USA	*581*	75-43090
Japan Aviation Electronics Ind. Ltd	Japan	*252*	75-03300
Kearfott Guidance & Navigation Corp.	USA	*588*	75-51760
Lear Astronics Corp.	USA	*589*	75-54390
Moog Controls Ltd	UK	*469*	75-40670
Moog Inc.	USA	*597*	75-62550
Moog Italiana Srl	Italy.	*237*	75-16230
Northrop Grumman Corporation.	USA	*599*	75-65450
Oerlikon-Contraves AG.	Switzerland	*387*	75-09250
Revue THOMMEN AG	Switzerland	*387*	75-11100
Riva Calzoni SpA	Italy.	*238*	75-20500
SAGEM SA	**France**	***143***	**75-33950**
SFIM	**France**	***144***	**75-35080**
Smiths Aerospace & Defense	USA	*615*	75-82090
Swiss Aircraft & Systems Co	Switzerland	*384*	55-00100
Target Technology Ltd	UK	*487*	75-58660
Tracor Inc	USA	*623*	75-90000
Vanguard Research Inc.	USA	*644*	80-34100
Vega-Cantley Instrument Company	UK	*490*	75-62740

223 10 70 RPV / Drone flight control equipment

AAI Corp.	USA	*545*	75-00020
AeroSpace Technologies of Australia	Australia	*11*	75-00340
Astronautics Corp. of America	USA	*555*	75-10000
AVIAC Technologies	France	*132*	75-04600
Aydin Corp.	USA	*556*	75-11450
CAC Systèmes	France	*133*	75-07680
GEC-Marconi Avionics Inc.	USA	*577*	75-37413
GEC-Marconi Avionics Ltd	UK	*454*	75-25320
Hayes Targets.	USA	*603*	75-69135
Herley-Vega Systems	USA	*581*	75-43090
Humphrey Inc.	USA	*583*	75-45050
Israel Aircraft Industries Ltd	Israel	*220*	75-10200
Israel Military Industries Ltd	Israel	*220*	75-10300
Japan Aviation Electronics Ind. Ltd	Japan	*252*	75-03300
Lear Astronics Corp.	USA	*589*	75-54390
Moog Controls Ltd	UK	*469*	75-40670
Moog Inc.	USA	*597*	75-62550
Northrop Grumman Corporation.	USA	*599*	75-65450
Oerlikon-Contraves AG.	Switzerland	*387*	75-09250
Riva Calzoni SpA	Italy.	*238*	75-20500
SAGEM SA	**France**	***143***	**75-33950**
SFIM	**France**	***144***	**75-35080**
Smiths Aerospace & Defense	USA	*615*	75-82090
Swiss Aircraft & Systems Co	Switzerland	*384*	55-00100
Target Technology Ltd	UK	*487*	75-58660
Vega-Cantley Instrument Company	UK	*490*	75-62740

223 10 70 50 Recovery parachutes, RPV / Drone

AAI Corp.	USA	*545*	75-00020
FXC Corp.	USA	*577*	75-36700
Irvin Aerospace Canada Limited	Canada	*73*	75-08300
Irvin Aerospace Limited	UK	*462*	75-32050
Irvin Aerospace SpA	Italy.	*235*	75-12200
Irvin Industries Inc.	USA	*586*	75-48960
Northrop Grumman Corporation.	USA	*599*	75-65450
Pioneer Aerospace Corp	USA	*604*	75-70200
Research Institute for	Ukraine.	*406*	60-00990
Sächsische Spezialkonfektion GmbH	Germany	*174*	75-25880
Strong Enterprises	USA	*618*	75-84500

223 20 RPV / Drone mission systems

AAI Corp.	USA	*545*	75-00020
Bombardier Inc	Canada	*69*	75-01900
CAC Systèmes	France	*133*	75-07680
Canadian Marconi Co.	Canada	*69*	75-02600
Cartwright Electronics Inc.	USA	*562*	75-16910
Computing Devices Company Ltd	UK	*442*	75-14700
Cubic Defense Systems Inc.	USA	*567*	75-22900
GEC-Marconi Avionics Inc.	USA	*577*	75-37413
GEC-Marconi Avionics Ltd	UK	*454*	75-25320
Herley-Vega Systems	USA	*581*	75-43090
Israel Aircraft Industries Ltd	Israel	*220*	75-10200
Lear Astronics Corp.	USA	*589*	75-54390
Northrop Grumman Corporation.	USA	*599*	75-65450
Oerlikon-Contraves AG.	Switzerland	*387*	75-09250
Racal Communications Inc.	USA	*606*	75-72380
Raytheon E-Systems	USA	*608*	75-73302
SAGEM SA	**France**	***143***	**75-33950**
Sanders	USA	*612*	75-77850
SFIM	**France**	***144***	**75-35080**
Vega-Cantley Instrument Company	UK	*490*	75-62740
Xi CAD North America Inc.	USA	*628*	75-97760

223 20 60 RPV / Drone reconnaissance systems

AAI Corp.	USA	*545*	75-00020
Bombardier Inc	Canada	*69*	75-01900
CAC Systèmes	France	*133*	75-07680
GEC-Marconi Avionics Inc.	USA	*577*	75-37413
GEC-Marconi Avionics Ltd	UK	*454*	75-25320
Israel Aircraft Industries Ltd	Israel	*220*	75-10200
Lear Astronics Corp.	USA	*589*	75-54390
Litton Industries Inc.	USA	*591*	75-55400
Northrop Grumman Corporation.	USA	*599*	75-65450
Oerlikon-Contraves AG.	Switzerland	*387*	75-09250
Racal Communications Inc.	USA	*606*	75-72380
Raytheon E-Systems	USA	*608*	75-73302
Recon/Optical Inc.	USA	*608*	75-73930
Vega-Cantley Instrument Company	UK	*490*	75-62740
Zeiss-Eltro Optronic GmbH	Germany	*177*	75-34700

223 20 70 RPV / Drone mission equipment

AAI Corp.	USA	*545*	75-00020
Air Target Sweden AB	Sweden	*375*	75-00850
Bombardier Inc	Canada	*69*	75-01900
Cartwright Electronics Inc.	USA	*562*	75-16910
GEC-Marconi Avionics Inc.	USA	*577*	75-37413
GEC-Marconi Avionics Ltd	UK	*454*	75-25320
Herley-Vega Systems	USA	*581*	75-43090
Israel Aircraft Industries Ltd	Israel	*220*	75-10200
Lear Astronics Corp.	USA	*589*	75-54390
Meteor CAE SpA	Italy.	*237*	75-15400
Northrop Grumman Corporation.	USA	*599*	75-65450
Oerlikon-Contraves AG.	Switzerland	*387*	75-09250
Racal Communications Inc.	USA	*606*	75-72380
Raytheon E-Systems	USA	*608*	75-73302
SAGEM SA	**France**	***143***	**75-33950**
Sparton of Canada Ltd	Canada	*77*	75-13100
Swiss Aircraft & Systems Co	Switzerland	*384*	55-00100
Vega-Cantley Instrument Company	UK	*490*	75-62740
Xi CAD North America Inc.	USA	*628*	75-97760

223 20 70 60 RPV / Drone reconnaissance equipment

AAI Corp.	USA	*545*	75-00020

Canadian Marconi Co.	Canada	*69*	75-02600
GEC-Marconi Avionics Inc.	USA	*577*	75-37413
GEC-Marconi Avionics Ltd	UK	*454*	75-25320
Israel Aircraft Industries Ltd	Israel	*220*	75-10200
Lear Astronics Corp.	USA	*589*	75-54390
Litton Industries Inc.	USA	*591*	75-55400
MA/COM Antenna & Cable Div.	USA	*593*	75-56953
Racal Communications Inc.	USA	*606*	75-72380
Raytheon E-Systems	USA	*608*	75-73302
SAGEM SA	**France**	***143***	**75-33950**
Vega-Cantley Instrument Company	UK	*490*	75-62740
Zeiss-Eltro Optronic GmbH	Germany	*177*	75-34700

223 40 RPV / Drone airframe-related systems

AAI Corp.	USA	*545*	75-00020
Bombardier Inc	Canada	*69*	75-01900
Lear Astronics Corp.	USA	*589*	75-54390
Riva Calzoni SpA	Italy	*238*	75-20500
SAGEM SA	**France**	***143***	**75-33950**
Scaled Composites Inc.	USA	*541*	60-19160
Target Technology Ltd	UK	*487*	75-58660
Technical Products Group Inc.	USA	*620*	75-86764
Vanguard Research Inc.	USA	*644*	80-34100
Xi CAD North America Inc.	USA	*628*	75-97760

223 40 60 Landing gear, RPV / Drone

AAI Corp.	USA	*545*	75-00020
Menasco Aerospace Ltd	Canada	*74*	75-09600
Riva Calzoni SpA	Italy	*238*	75-20500
Scaled Composites Inc.	USA	*541*	60-19160
Swiss Aircraft & Systems Co	Switzerland	*384*	55-00100

223 40 70 RPV / Drone airframe-related equipment

AAI Corp.	USA	*545*	75-00020
Cyclone Aviation Products Ltd	Israel	*218*	75-04250
Fabbrica Italiana Apparecchiature	Italy	*234*	75-08550
Israel Aircraft Industries Ltd	Israel	*219*	75-10200
Riva Calzoni SpA	Italy	*238*	75-20500
SAGEM SA	**France**	***143***	**75-33950**
Target Technology Ltd	UK	*487*	75-58660

223 40 70 40 Flotation gear, RPV / Drone

AAI Corp.	USA	*545*	75-00020
Aerospace and Electronics Division	USA	*606*	75-71472
Air Cruisers Co	USA	*548*	75-03600
Irvin Aerospace Canada Limited	Canada	*73*	75-08300
Irvin Aerospace SpA	Italy	*235*	75-12200
Lear Astronics Corp.	USA	*589*	75-54390
Northrop Grumman Corporation	USA	*599*	75-65450

223 80 RPV / Drone accessories

AAI Corp.	USA	*545*	75-00020
Aerospace Syst. & Comp.	USA	*547*	75-02501
Bombardier Inc	Canada	*69*	75-01900
Lear Astronics Corp.	USA	*589*	75-54390
Northrop Grumman Corporation	USA	*599*	75-65450
Riva Calzoni SpA	Italy	*238*	75-20500
Southwest Aerospace	USA	*616*	75-82470
Swiss Aircraft & Systems Co	Switzerland	*384*	55-00100
Target Technology Ltd	UK	*487*	75-58660

223 80 40 RPV / Drone launchers

AAI Corp.	USA	*545*	75-00020
Bombardier Inc	Canada	*69*	75-01900
Hayes Targets	USA	*603*	75-69135
Israel Aircraft Industries Ltd	Israel	*219*	75-10200
Lohr Industrie	France	*139*	75-25200
Northrop Grumman Corporation	USA	*599*	75-65450
Riva Calzoni SpA	Italy	*238*	75-20500
Royal Ordnance	UK	*481*	75-50620
Swiss Aircraft & Systems Co	Switzerland	*384*	55-00100
Target Technology Ltd	UK	*487*	75-58660
Universal Propulsion Co. Inc.	USA	*625*	75-92700

223 90 RPV / Drone components

AAI Corp.	USA	*545*	75-00020
Aerospace Syst. & Comp.	USA	*547*	75-02501
Bombardier Inc	Canada	*69*	75-01900
Busak + Shamban Inc.	USA	*561*	75-16006
Continental AG Werk Uniroyal	Germany	*176*	75-32090
Cranfield Aerospace Ltd	UK	*500*	80-07200
GEC-Marconi Aerospace Ltd	UK	*453*	75-25300
Howmet Corporation	USA	*582*	75-44550
Lear Astronics Corp.	USA	*589*	75-54390
MA/COM Antenna & Cable Div.	USA	*593*	75-56953
Muirhead Vactric Components Ltd	UK	*470*	75-41404
Northrop Grumman Corporation	USA	*599*	75-65450
Riva Calzoni SpA	Italy	*238*	75-20500
Royal Ordnance	UK	*481*	75-50620
SAGEM SA	**France**	***143***	**75-33950**
Scaled Composites Inc.	USA	*541*	60-19160
Swiss Aircraft & Systems Co	Switzerland	*384*	55-00100
Target Technology Ltd	UK	*487*	75-58660
Vanguard Research Inc.	USA	*644*	80-34100

224 Airship / Dirigible systems

GEC-Marconi Avionics Ltd	UK	*454*	75-25320

224 70 Airship / Dirigible, Balloon equipment

Clarke Chapman Marine	UK	*440*	75-13530
GEC-Marconi Avionics Ltd	UK	*454*	75-25320
Tracor Inc	USA	*623*	75-90000

Seaborne Platform Systems & Equipment

231 Warship, surface combattant craft systems & equipment

Alenia-Elsag	Italy	*230*	74-00100
ATS Aerospace Inc.	Canada	*68*	75-00752
Australian Defence Industries Ltd	Australia	*9*	55-01000
Avica Equipment	UK	*432*	75-04870
Blohm + Voss GmbH	Germany	*165*	75-02800
Boeing Australia Limited	Australia	*12*	75-02500
Bowmar Instrument Corp.	USA	*560*	75-14860
British Aerospace (Systems &	UK	*436*	75-08890
Cincinnati Electronics Corp.	USA	*564*	75-18640
Clarke Chapman Marine	UK	*440*	75-13530
Computing Devices Canada	Canada	*70*	75-03600
Concorde Battery Corp.	USA	*566*	75-20845
Cubic Defense Systems Inc.	USA	*567*	75-22900
Dowty Aerospace Wolverhampton	UK	*446*	75-18589
DRS Technologies, Inc.	USA	*570*	75-27400
ECE GmbH	Germany	*167*	75-07200
Envirovac Inc.	USA	*574*	75-31920
Esterline Technologies	USA	*574*	75-32360
Fenner Fluid Power	UK	*450*	75-22550
GEC-Marconi Avionics Ltd	UK	*454*	75-25320
GEC-Marconi Communications Ltd	UK	*454*	75-25335
George G. Sharp Inc.	USA	*642*	80-28000
Graseby Dynamics Ltd	UK	*455*	75-26590
Hitachi Zosen Co. Ltd.	Japan	*251*	75-02390
Howden Aircontrol Ltd	UK	*459*	75-30240
Howden Wade Ltd	UK	*459*	75-30260
Hunting Engineering Ltd	UK	*459*	75-30700
Hyundai Electrical Engineering	Korea (ROK)	*265*	75-01550
Indal Technologies Inc	Canada	*72*	75-07740
Ishikawajima-Harima H.I. Co. Ltd	Japan	*251*	75-02800
ITT Avionics	USA	*586*	75-49455
J-Tec Associates Inc.	USA	*587*	75-51250
Kearfott Guidance & Navigation Corp.	USA	*588*	75-51760
Labinal Aero Defense Systems Inc	USA	*638*	80-18100
Litton Industries Inc.	USA	*590*	75-55400
Logica UK Limited	UK	*506*	80-14350
Lucas Aerospace Power Transmission	USA	*592*	75-56265
Meggitt Avionics (UK)	UK	*467*	75-38340
Motorola Ltd	UK	*469*	75-41050
Officine Panerai SpA	Italy	*237*	75-17500
Pacer Systems Inc.	USA	*602*	75-68000
RAMFAN Corporation	USA	*607*	75-72700

Saab Dynamics AB	Sweden	*378*	75-09100
Saft Singapore Pte Ltd	Singapore	*348*	75-03400
SELIN Sistemi Sarl	Italy	*238*	75-21750
S.G. Brown	UK	*437*	75-10330
SITEP SpA	Italy	*239*	75-22595
Société d'Applications des Machines	France	*143*	75-34300
Strachan & Henshaw Ltd	UK	*512*	80-47400
Technofan SA	France	*147*	75-44300
Teldix GmbH	Germany	*175*	75-30100
Thomson Marconi Sonar Limited	UK	*426*	74-20000
Tirfor Ltd	UK	*488*	75-60150
Tracor Inc	USA	*623*	75-90000
United Defense L.P.	USA	*624*	75-92050
Van Rietschoten & Houwens	Netherlands	*296*	75-07900
Vosper Thornycroft	UK	*492*	75-63800
Warren-Knight Instrument Co.	USA	*627*	75-95150

231 10 Warship control systems

Alenia-Elsag	Italy	*230*	74-00100
Australian Defence Industries Ltd	Australia	*9*	55-01000
BAeSEMA Ltd.	UK	*498*	80-03070
Blohm + Voss GmbH	Germany	*165*	75-02800
Bowmar Instrument Corp.	USA	*560*	75-14860
George G. Sharp Inc.	USA	*642*	80-28000
Hyde Products Inc.	USA	*583*	75-45280
Ishikawajima-Harima H.I. Co. Ltd	Japan	*251*	75-02800
Kearfott Guidance & Navigation Corp.	USA	*588*	75-51760
Litton Industries Inc.	USA	*590*	75-55400
SAGEM SA	**France**	***143***	**75-33950**
S.G. Brown	UK	*437*	75-10330
Société d'Applications des Machines	France	*143*	75-34300
Svendborg Vaerft A/S	Denmark	*104*	75-19800
TERMA Elektronik AS	Denmark	*104*	75-20300
United Defense L.P.	USA	*624*	75-92050
Van Rietschoten & Houwens	Netherlands	*296*	75-07900
Vickers Shipbuilding & Eng Ltd	UK	*491*	75-63100
Vosper Thornycroft	UK	*492*	75-63800
Whippendell Electrical Ltd	UK	*493*	75-65800

231 10 02 Automation systems, warship

George G. Sharp Inc.	USA	*642*	80-28000
Telektron Ltd	UK	*488*	75-59200
Vickers Shipbuilding & Eng Ltd	UK	*491*	75-63100

231 10 05 Bridge control systems, warship

George G. Sharp Inc.	USA	*642*	80-28000
Kearfott Guidance & Navigation Corp.	USA	*588*	75-51760
Kobelt Manufacturing Co. Ltd	Canada	*73*	75-08770
Litton Industries Inc.	USA	*590*	75-55400
Radamec Control Systems Ltd	UK	*477*	75-47755
Vickers Shipbuilding & Eng Ltd	UK	*491*	75-63100

231 10 10 Steering control systems & equipment, warship

ACH	France	*130*	75-00150
GEC-Marconi Avionics Ltd	UK	*454*	75-25320
George G. Sharp Inc.	USA	*642*	80-28000
Hitachi Zosen Co. Ltd.	Japan	*251*	75-02390
Hyde Products Inc.	USA	*583*	75-45280
Kearfott Guidance & Navigation Corp.	USA	*588*	75-51760
Kobelt Manufacturing Co. Ltd	Canada	*73*	75-08770
Menasco Aerospace Ltd	Canada	*74*	75-09600
Saab Dynamics AB	Sweden	*378*	75-09100
S.G. Brown	UK	*437*	75-10330
SITEP SpA	Italy	*239*	75-22595
Vickers Inc.	USA	*626*	75-93820
Vickers Shipbuilding & Eng Ltd	UK	*491*	75-63100
Vickers Systems	Germany	*176*	75-32830
Vickers Systems SA	France	*148*	75-46930
Vickers Systems Trinova SpA	Italy	*239*	75-27150
Vosper Thornycroft	UK	*492*	75-63800

231 10 11 Automatic pilot systems, warship

George G. Sharp Inc.	USA	*642*	80-28000
Hitachi Zosen Co. Ltd.	Japan	*251*	75-02390
S.G. Brown	UK	*437*	75-10330
Vickers Shipbuilding & Eng Ltd	UK	*491*	75-63100

231 10 12 Stabilization systems, warship

ACH	France	*130*	75-00150
George G. Sharp Inc.	USA	*642*	80-28000
Jered Brown Brothers Inc.	USA	*586*	75-50380
Muirhead Vactric Components Ltd	UK	*470*	75-41404
S.G. Brown	UK	*437*	75-10330
Sperry Marine SpA	Italy	*239*	75-24400
Vickers Inc.	USA	*626*	75-93820
Vickers Shipbuilding & Eng Ltd	UK	*491*	75-63100
Vickers Systems	Germany	*176*	75-32830
Vickers Systems Trinova SpA	Italy	*239*	75-27150
Vosper Thornycroft	UK	*492*	75-63800

231 10 20 Machinery control systems, warship

George G. Sharp Inc.	USA	*642*	80-28000
Litton Industries Inc.	USA	*590*	75-55400
Officine Panerai SpA	Italy	*237*	75-17500
Telektron Ltd	UK	*488*	75-59200
Vickers Shipbuilding & Eng Ltd	UK	*491*	75-63100
Vosper Thornycroft	UK	*492*	75-63800

231 10 70 Warship control equipment

Aeronautical & General Instr. Ltd	UK	*428*	75-01050
Bowmar Instrument Corp.	USA	*560*	75-14860
Computing Devices Canada	Canada	*70*	75-03600
Electric Tachometer Corp.	USA	*572*	75-29680
George G. Sharp Inc.	USA	*642*	80-28000
Horstman Defence Systems Ltd.	UK	*459*	75-29920
Howaldtswerke Deutsche Werft AG	Germany	*170*	75-13400
Hunting Engineering Ltd	UK	*459*	75-30700
Logica UK Limited.	UK	*506*	80-14350
Svendborg Vaerft A/S	Denmark	*104*	75-19800

231 10 70 10 Binnacles, naval

George G. Sharp Inc.	USA	*642*	80-28000
Litton Industries Inc.	USA	*590*	75-55400
W.F. Stanley & Co. Ltd	UK	*486*	75-56350

231 10 70 20 Warship control consoles

Astronautics C.A. Ltd	Israel	*217*	75-02400
Bowmar Instrument Corp.	USA	*560*	75-14860
George G. Sharp Inc.	USA	*642*	80-28000
S.G. Brown	UK	*437*	75-10330

231 10 70 22 Damage control consoles, warship

George G. Sharp Inc.	USA	*642*	80-28000
Litton Industries Inc.	USA	*590*	75-55400

231 10 70 25 Machinery control & monitoring consoles, warship

George G. Sharp Inc.	USA	*642*	80-28000
Hunting Engineering Ltd	UK	*459*	75-30700
Litton Industries Inc.	USA	*590*	75-55400
Telektron Ltd	UK	*488*	75-59200
Thomson Marconi Sonar Limited	UK	*426*	74-20000
Vosper Thornycroft	UK	*492*	75-63800

231 10 70 30 Instruments & measuring equipment, warship control

Aeronautical & General Instr. Ltd	UK	*428*	75-01050
ATS Aerospace Inc.	Canada	*68*	75-00752
Bowmar Instrument Corp.	USA	*560*	75-14860
Brookes & Gatehouse Ltd	UK	*437*	75-10100
Crompton Instruments Ltd.	UK	*443*	75-16070
Electric Tachometer Corp.	USA	*572*	75-29680
George G. Sharp Inc.	USA	*642*	80-28000

Horstman Defence Systems Ltd	UK	459	75-29920
Interface Products Inc.	USA	585	75-47300
Kearfott Guidance & Navigation Corp.	USA	588	75-51760
Litton Industries Inc.	USA	590	75-55400
Marine Data Ltd	UK	466	75-37288
Meggitt Avionics (UK)	UK	467	75-38340
Newport News Shipbuilding	USA	599	75-64300
S.G. Brown	UK	437	75-10330
SITEP SpA	Italy	239	75-22595
Telektron Ltd	UK	488	75-59200
TransInstruments	UK	489	75-60960
Warren-Knight Instrument Co.	USA	627	75-95150
W.F. Stanley & Co. Ltd	UK	486	75-56350

231 10 70 30 50 Logs, ship

Aeronautical & General Instr. Ltd	UK	428	75-01050
SAGEM SA	**France**	**143**	**75-33950**
Sperry Marine SpA	Italy	239	75-24400

231 10 70 30 53 Logs, ship, electromagnetic

Aeronautical & General Instr. Ltd	UK	428	75-01050
SAGEM SA	**France**	**143**	**75-33950**

231 10 70 30 54 Logs, ship, impeller type

Brookes & Gatehouse Ltd	UK	437	75-10100

231 20 Warship mission-related systems

Anite Systems Ltd	UK	497	80-01240
Australian Defence Industries Ltd	Australia	9	55-01000
BAeSEMA Ltd	UK	498	80-03070
Breeze Eastern Corp.	USA	560	75-15110
Cubic Defense Systems Inc.	USA	567	75-22900
Graseby Dynamics Ltd	UK	455	75-26590
Hunting Engineering Ltd	UK	459	75-30700
Ishikawajima-Harima H.I. Co. Ltd	Japan	251	75-02800
ITT Avionics	USA	586	75-49455
Litton Industries Inc.	USA	590	75-55400
Lucas Western Inc.	USA	592	75-56300
Pilkington Optronics Ltd	UK	426	74-10000
Riva Calzoni SpA	Italy	238	75-20500
SAGEM SA	**France**	**143**	**75-33950**
Sanders	USA	612	75-77850
Strachan & Henshaw Ltd	UK	512	80-47400
Vickers Shipbuilding & Eng Ltd	UK	491	75-63100

231 20 05 Action information & communications systems, shipborne (see code 454.30.10)

Anite Systems Ltd	UK	497	80-01240
Boeing Australia Limited	Australia	12	75-02500
Computing Devices Canada	Canada	70	75-03600
GEC-Marconi Communications Ltd	UK	454	75-25335
Hunting Engineering Ltd	UK	459	75-30700
Litton Industries Inc.	USA	590	75-55400
Nera Telecommunications	UK	470	75-42318
Raytheon E-Systems	USA	608	75-73302
Vickers Shipbuilding & Eng Ltd	UK	491	75-63100

231 20 10 Aircraft deck handling & support systems, warship

AAI Corp.	USA	545	75-00020
Blohm + Voss GmbH	Germany	165	75-02800
Fairey Hydraulics	UK	450	75-22255
George G. Sharp Inc.	USA	642	80-28000
Grove Europe Ltd	**UK**	**456**	**75-27030**
Indal Technologies Inc	Canada	72	75-07740
Jered Brown Brothers Inc.	USA	586	75-50380
MacTaggart Scott & Co. Ltd	UK	466	75-36900
MIL-BASE Industries Inc.	USA	596	75-61530
Officine Panerai SpA	Italy	237	75-17500
Strachan & Henshaw Ltd	UK	512	80-47400
Vehicle Systems Development Corp.	USA	625	75-93570
Vickers Inc.	USA	626	75-93830
Vickers Systems	Germany	176	75-32830
Vickers Systems Trinova SpA	Italy	239	75-27150

231 20 10 05 Aircraft elevators / lifts, aircraft carrier

AAI Corp.	USA	545	75-00020
George G. Sharp Inc.	USA	642	80-28000
Jered Brown Brothers Inc.	USA	586	75-50380
MacTaggart Scott & Co. Ltd	UK	466	75-36900
Newport News Shipbuilding	USA	599	75-64300
Strachan & Henshaw Ltd	UK	512	80-47400

231 20 10 10 Aircraft arrester gear, aircraft carrier

MacTaggart Scott & Co. Ltd	UK	466	75-36900
MIL-BASE Industries Inc.	USA	596	75-61530
National Forge Co.	USA	598	75-63850
Newport News Shipbuilding	USA	599	75-64300

231 20 10 20 Catapult systems, aircraft carrier

MIL-BASE Industries Inc.	USA	596	75-61530
National Forge Co.	USA	598	75-63850

231 20 10 50 Helicopter haul-down systems, warship

Breeze Eastern Corp.	USA	560	75-15110
Indal Technologies Inc	Canada	72	75-07740
MIL-BASE Industries Inc.	USA	596	75-61530

231 20 10 52 Helicopter hangars, warship

Blohm + Voss GmbH	Germany	165	75-02800
Garden Reach Shipbuilders	India	203	55-00300
Indal Technologies Inc	Canada	72	75-07740
Israel Aircraft Industries Ltd	Israel	219	75-10200
MAFO Howden BV	Netherlands	294	75-04350

231 20 10 55 Helicopter deck restraint systems, warship

Cargo Aids	UK	438	75-11620
Fairey Hydraulics	UK	450	75-22255
Indal Technologies Inc	Canada	72	75-07740
MacTaggart Scott & Co. Ltd	UK	466	75-36900
MIL-BASE Industries Inc.	USA	596	75-61530
Riva Calzoni SpA	Italy	238	75-20500
SpanSet Inter AG	Switzerland	387	75-13550

231 20 10 57 Helicopter landing platforms, warship

AIM Enterprises Inc.	USA	630	80-01350
Blohm + Voss GmbH	Germany	165	75-02800
Indal Technologies Inc	Canada	72	75-07740
Strachan & Henshaw Ltd	UK	512	80-47400
Vickers Shipbuilding & Eng Ltd	UK	491	75-63100

231 20 10 60 Helicopter ship-to-air refuelling systems, warship

Flight Refuelling Limited	UK	451	75-23850

231 20 20 Air traffic control systems, aircraft carrier

ATS Aerospace Inc.	Canada	68	75-00752
Denro Inc.	USA	568	75-25651
GEC-Marconi Avionics Inc.	USA	577	75-37413
Park Air Electronics Ltd	UK	472	75-43750
Raytheon Cossor Electronics Ltd	UK	478	75-48367

231 20 20 05 Visual approach landing aids, shipborne

Aeronautical & General Instr. Ltd	UK	428	75-01050
Officine Panerai SpA	Italy	237	75-17500

231 20 20 05 10 Visual approach landing aids, aircraft, aircraft carrier

Aeronautical & General Instr. Ltd	UK	428	75-01050
Officine Panerai SpA	Italy	237	75-17500
Soderberg Manufacturing Co. Inc.	USA	616	75-82200
Vega Industries Ltd	New Zealand	303	75-02250

231 20 20 05 30 Visual approach landing aids, helicopter, shipborne

Aeronautical & General Instr. Ltd	UK	428	75-01050
Cray Marine	UK	443	75-15915
Indal Technologies Inc	Canada	72	75-07740
Officine Panerai SpA	Italy	237	75-17500
Vega Industries Ltd	New Zealand	303	75-02250

231 20 20 10 Automatic landing systems, shipborne

GEC-Marconi Avionics Inc.	USA	577	75-37413

231 20 20 10 10 Carrier approach landing aids, automatic

GEC-Marconi Avionics Inc.	USA	577	75-37413

231 20 30 Depth charge launch systems, warship

AAI Corp.	USA	545	75-00020

231 20 40 Torpedo launch systems, warship

AAI Corp.	USA	545	75-00020
David Brown Defence Equipment Ltd	UK	437	75-10150
Thomson Sintra	France	148	75-45135
United Defense L.P.	USA	624	75-92050
Whitehead SpA	Italy	240	75-27600

231 20 40 70 Torpedo launch equipment, warship

AAI Corp.	USA	545	75-00020
David Brown Defence Equipment Ltd	UK	437	75-10150
Strachan & Henshaw Ltd	UK	512	80-47400
Thomson Sintra	France	148	75-45135
United Defense L.P.	USA	624	75-92050
Whitehead SpA	Italy	240	75-27600

231 20 40 70 10 Torpedo launch cradles, warship

Halmatic Ltd	UK	456	75-27700
United Defense L.P.	USA	624	75-92050
Whitehead SpA	Italy	240	75-27600

231 20 40 70 30 Torpedo tubes, warship

Australian Defence Industries Ltd	Australia	9	55-01000
National Forge Co.	USA	598	75-63850
Strachan & Henshaw Ltd	UK	512	80-47400
Thomson Sintra	France	148	75-45135
United Defense L.P.	USA	624	75-92050
Whitehead SpA	Italy	240	75-27600

231 20 50 Washdown systems, anti-fallout, warship

Blohm + Voss GmbH	Germany	165	75-02800

231 20 55 Weapon handling systems, warship

AAI Corp.	USA	545	75-00020
GEC ALSTHOM Engineering Systems Ltd	UK	453	75-25225
Hitachi Zosen Co. Ltd.	Japan	251	75-02390
Jered Brown Brothers Inc.	USA	586	75-50380
MacTaggart Scott & Co. Ltd	UK	466	75-36900
Riva Calzoni SpA	Italy	238	75-20500
SpanSet Inter AG	Switzerland	387	75-13550

231 20 55 70 Weapon handling equipment, warship

Aerial Machine & Tool Corp.	USA	546	75-01105
British Filters Ltd	UK	436	75-09030
Indal Technologies Inc	Canada	72	75-07740
Jered Brown Brothers Inc.	USA	586	75-50380
MIL-BASE Industries Inc.	USA	596	75-61530
SpanSet Inter AG	Switzerland	387	75-13550

231 20 58 Wind sensing systems, warship

Brookes & Gatehouse Ltd	UK	437	75-10100
Meggitt Avionics (UK)	UK	467	75-38340
Oceano Instruments (UK) Ltd	UK	508	80-19050
Officine Panerai SpA	Italy	237	75-17500
SITEP SpA.	Italy	239	75-22595
Vaisala Oy	Finland	119	75-02800
Wilh. Lambrecht GmbH	Germany	171	75-17350

231 20 58 70 Wind sensing equipment, warship

Aerial Machine & Tool Corp.	USA	546	75-01105
Brookes & Gatehouse Ltd	UK	437	75-10100
Meggitt Avionics (UK)	UK	467	75-38340
Oceano Instruments (UK) Ltd	UK	508	80-19050
Officine Panerai SpA	Italy	237	75-17500
SITEP SpA.	Italy	239	75-22595
Vaisala Oy	Finland	119	75-02800
Wilh. Lambrecht GmbH	Germany	171	75-17350

231 20 70 Warship mission-related equipment

Australian Defence Industries Ltd	Australia	9	55-01000
Hunting Engineering Ltd	UK	459	75-30700

231 20 70 10 Armour systems, ship

DTC International	USA	570	75-27620
George G. Sharp Inc.	USA	642	80-28000
Permali Gloucester Ltd	UK	474	75-44850

231 20 70 50 Searchlights / spotlights, naval

Ibak Helmut Hunger GmbH & Co KG.	Germany	170	75-13580
SAGEM SA	**France**	**143**	**75-33950**
Vega Industries Ltd	New Zealand	303	75-02250

231 20 70 55 Signalling equipment, warship

ATS Aerospace Inc.	Canada	68	75-00752
Étienne Lacroix	France	138	75-24050
Ibak Helmut Hunger GmbH & Co KG.	Germany	170	75-13580
Officine Panerai SpA	Italy	237	75-17500
Vega Industries Ltd	New Zealand	303	75-02250

231 30 Warship crew-related facilities & equipment

Airscrew Howden Ltd	UK	429	75-02050
A.P. Metalcraft Ltd	UK	430	75-03410
Blohm + Voss GmbH	Germany	165	75-02800
Envirovac Inc.	USA	574	75-31920
Exstar International Corp.	USA	575	75-33000
Howden Aircontrol Ltd	UK	459	75-30240
Hunting Engineering Ltd	UK	459	75-30700
IDE Technologies Ltd	Israel	219	75-09170
Ishikawajima-Harima H.I. Co. Ltd	Japan	251	75-02800
Litton Industries Inc.	USA	590	75-55400
Maine Group Ltd	UK	466	75-37050
Mechanical Equipment Co. Inc.	USA	595	75-59815
RAMFAN Corporation	USA	607	75-72700
SITEP SpA.	Italy	239	75-22595
Svendborg Vaerft A/S	Denmark	104	75-19800

231 30 05 Environmental control systems, warship

Airscrew Howden Ltd	UK	429	75-02050
Barber-Colman Co.	USA	557	75-11960

CJB Developments Ltd	UK	*440*	75-13430
Graseby Dynamics Ltd	UK	*455*	75-26590
Howden Aircontrol Ltd	UK	*459*	75-30240
Munters Cargocaire	USA	*597*	75-63445
Stork Bronswerk BV	Netherlands	*295*	75-07200
Svendborg Vaerft A/S	Denmark	*104*	75-19800
The Armstrong Monitoring Corp.	Canada	*68*	75-00740
Vickers Shipbuilding & Eng Ltd	UK	*491*	75-63100

231 30 05 10 Air conditioning systems, warship

Airflow Co.	USA	*548*	75-03650
Blohm + Voss GmbH	Germany	*165*	75-02800
Howden Aircontrol Ltd	UK	*459*	75-30240
Howden Compressors Ltd	UK	*459*	75-30250
Kinetics Ltd	Israel	*220*	75-11320
New Philadelphia Fan Company	USA	*587*	75-51130
RAMFAN Corporation	USA	*607*	75-72700
Stork Bronswerk BV	Netherlands	*295*	75-07200
Svendborg Vaerft A/S	Denmark	*104*	75-19800
van Buuren-van Swaay BV	Netherlands	*296*	75-07830
Vickers Shipbuilding & Eng Ltd	UK	*491*	75-63100
VSG Verfahrenstechnik	Germany	*176*	75-33030

231 30 20 Fresh water production plant, warship

Aircraft Appliances & Equipment Ltd	Canada	*68*	75-00430
Exstar International Corp.	USA	*575*	75-33000
SCAM SpA	Italy	*238*	75-21530
Svendborg Vaerft A/S	Denmark	*104*	75-19800

231 30 30 Furniture, mess & accommodation, warship

ACH	France	*130*	75-00150
A.P. Metalcraft Ltd	UK	*430*	75-03410
Blohm + Voss GmbH	Germany	*165*	75-02800
Maine Group Ltd	UK	*466*	75-37050
Svendborg Vaerft A/S	Denmark	*104*	75-19800

231 30 40 Warship internal communication systems

Hunting Engineering Ltd	UK	*459*	75-30700
LARIMART SpA	Italy	*236*	75-13500
Litton Industries Inc.	USA	*590*	75-55400
Svendborg Vaerft A/S	Denmark	*104*	75-19800

231 30 42 Warship TV distribution systems

Hunting Engineering Ltd	UK	*459*	75-30700
Litton Industries Inc.	USA	*590*	75-55400
Moonraker Australia Pty Ltd	Australia	*19*	75-13770
Naval Electronics Inc.	USA	*598*	75-64025
Svendborg Vaerft A/S	Denmark	*104*	75-19800

231 30 43 Laundry facilities, warship

Svendborg Vaerft A/S	Denmark	*104*	75-19800

231 30 45 Medical facilities, warship

Svendborg Vaerft A/S	Denmark	*104*	75-19800

231 30 50 Messing facilities, warship

George G. Sharp Inc.	USA	*642*	80-28000
Svendborg Vaerft A/S	Denmark	*104*	75-19800

231 30 60 Waste treatment systems, warship

AEA Technology plc	UK	*496*	80-00480
Envirovac Inc.	USA	*574*	75-31920
Garden Reach Shipbuilders	India	*203*	55-00300
George G. Sharp Inc.	USA	*642*	80-28000
Howden Wade Ltd	UK	*459*	75-30260
Svendborg Vaerft A/S	Denmark	*104*	75-19800

231 30 63 Waste disposal systems, warship

Arteco Srl	Italy	*232*	75-01770
Envirovac Inc.	USA	*574*	75-31920
George G. Sharp Inc.	USA	*642*	80-28000
Milldale Limited	UK	*468*	75-39500
Strachan & Henshaw Ltd	UK	*512*	80-47400
Svendborg Vaerft A/S	Denmark	*104*	75-19800

231 30 66 Waste shredders, warship

George G. Sharp Inc.	USA	*642*	80-28000
Milldale Limited	UK	*468*	75-39500
Svendborg Vaerft A/S	Denmark	*104*	75-19800

231 40 Warship systems, other

APPH (Bolton) Ltd.	UK	*431*	75-03415
Australian Defence Industries Ltd	Australia	*9*	55-01000
Avica Equipment	UK	*432*	75-04870
British Filters Ltd	UK	*436*	75-09030
Corrintec Ltd	UK	*442*	75-15120
DRS Technologies, Inc.	USA	*570*	75-27400
Eldec Corp.	USA	*572*	75-29660
Elesco Oy	Finland	*118*	75-00300
Gates Hydraulics Ltd	UK	*453*	75-25210
Hunting Engineering Ltd	UK	*459*	75-30700
Ishikawajima-Harima H.I. Co. Ltd	Japan	*251*	75-02800
Ralbo Inc.	USA	*607*	75-72690
SAGEM SA	**France**	***143***	**75-33950**
SELIN Sistemi Sarl	Italy	*238*	75-21750
Tecalemit Systems Ltd.	UK	*487*	75-58860

231 40 10 Audio systems, warship command & control

Boeing Australia Limited	Australia	*12*	75-02500
Cincinnati Electronics Corp.	USA	*564*	75-18640
GEC-Marconi Communications Ltd	UK	*454*	75-25335
George G. Sharp Inc.	USA	*642*	80-28000
Hunting Engineering Ltd	UK	*459*	75-30700
Litton Industries Inc.	USA	*590*	75-55400
Stanilite Electronics Pty Ltd	Australia	*21*	75-19690

231 40 10 70 Audio equipment, warship command & control

Boeing Australia Limited	Australia	*12*	75-02500
Cincinnati Electronics Corp.	USA	*564*	75-18640
Davies Industrial Communications Ltd	UK	*444*	75-17010
George G. Sharp Inc.	USA	*642*	80-28000
Hunting Engineering Ltd	UK	*459*	75-30700
Litton Industries Inc.	USA	*590*	75-55400

231 40 20 Cathodic protection systems, warship

Cathelco Limited	UK	*439*	75-12100
Corrintec Ltd	UK	*442*	75-15120
George G. Sharp Inc.	USA	*642*	80-28000
Svendborg Vaerft A/S	Denmark	*104*	75-19800

231 40 22 Cranes, material handling, warship

AAI Corp.	USA	*545*	75-00020
ACH	France	*130*	75-00150
Clarke Chapman Marine	UK	*440*	75-13530
George G. Sharp Inc.	USA	*642*	80-28000
Jered Brown Brothers Inc.	USA	*586*	75-50380
Schat-Harding BV	Netherlands	*295*	75-06280
Svendborg Vaerft A/S	Denmark	*104*	75-19800
Tirfor Ltd	UK	*488*	75-60150

231 40 25 Data bus systems, warship

Aeronautical & General Instr. Ltd	UK	*428*	75-01050
Blohm + Voss GmbH	Germany	*165*	75-02800
Boeing Australia Limited	Australia	*12*	75-02500
GEC-Marconi Communications Ltd	UK	*454*	75-25335
George G. Sharp Inc.	USA	*642*	80-28000

Litton Industries Inc.	USA	590	75-55400
Logica UK Limited	UK	506	80-14350
Meggitt Avionics (UK)	UK	467	75-38340
SAGEM SA	**France**	**143**	**75-33950**
Trompeter Electronics Inc.	USA	623	75-91070
United Defense L.P.	USA	624	75-92050
Vickers Shipbuilding & Eng Ltd	UK	491	75-63100

231 40 27 Degaussing systems, warship

Davis (W.R.) Engineering Ltd	Canada	70	75-04050
Elesco Oy	Finland	118	75-00300
GEC-Marconi Communications Ltd	UK	454	75-25335
George G. Sharp Inc.	USA	642	80-28000
Thomson Sintra	France	148	75-45135

231 40 30 Electrical systems, warship

Century Electronics	USA	562	75-17385
Combimac BV	Netherlands	292	75-01200
ECE GmbH	Germany	167	75-07200
Eldec Corp.	USA	572	75-29660
Electropar Ltd	New Zealand	302	75-00500
EUROCONTROL SpA	Italy	233	75-08200
Field & Grant Ltd	UK	451	75-23000
George G. Sharp Inc.	USA	642	80-28000
GSI Corporation	USA	580	75-40715
Holec Machines & Apparaten BV	Netherlands	293	75-03110
Icore International Ltd	UK	460	75-31080
Mawdsley's Ltd	UK	466	75-37900
Meggitt Avionics (UK)	UK	467	75-38340
Newton Derby Ltd	UK	470	75-42520
TERMA Elektronik AS	Denmark	104	75-20300
United Defense L.P.	USA	624	75-92050
White Products BV	Netherlands	296	75-09100

231 40 30 50 Lighting systems, warship

CRL Technologies	Canada	70	75-03760
ECE GmbH	Germany	167	75-07200
Electropar Ltd	New Zealand	302	75-00500
George G. Sharp Inc.	USA	642	80-28000
Luminescent Systems Inc.	USA	593	75-56480
Westair Ltd	UK	493	75-65410

231 40 30 60 Electrical generating plant, warship

Astrosyn International Technology	UK	432	75-04235
EUROCONTROL SpA	Italy	233	75-08200
Field & Grant Ltd	UK	451	75-23000
GEC ALSTHOM Medium Turbo Machines	UK	453	75-25243
George G. Sharp Inc.	USA	642	80-28000
Holec Machines & Apparaten BV	Netherlands	293	75-03110
MAN B&W Diesel AG	Germany	172	75-19610
Marine Electric Systems Inc.	USA	593	75-58000
Mawdsley's Ltd	UK	466	75-37900
Newton Derby Ltd	UK	470	75-42520
SELIN Sistemi Sarl	Italy	238	75-21750
Solar Turbines	USA	616	75-82300
Stewart & Stevenson Services Inc.	USA	618	75-84180

231 40 40 Fire protection systems, warship

Deugra Ges. für Brandschutzsysteme	Germany	167	75-05900
George G. Sharp Inc.	USA	642	80-28000
Hughes Electronics Corp.	USA	583	75-44865
Kidde-Fenwal	USA	588	75-52510
Nordic Systems Inc	Canada	74	75-09950
Saval BV	Netherlands	295	75-06200
Spectronix Ltd	Israel	222	75-21100
Svendborg Vaerft A/S	Denmark	104	75-19800
Victaulic Co. of America	USA	626	75-93840
Walter Kidde Aerospace	USA	588	75-52500

231 40 45 Fuel systems, warship

Aircraft Appliances & Equipment Ltd	Canada	68	75-00430
British Filters Ltd	UK	436	75-09030
George G. Sharp Inc.	USA	642	80-28000
Imo Industries Inc.	USA	584	75-46150
Svendborg Vaerft A/S	Denmark	104	75-19800

231 40 50 Hydraulic systems, warship

ACH	France	130	75-00150
APPH (Bolton) Ltd.	UK	431	75-03415
British Filters Ltd	UK	436	75-09030
Caley Ocean Systems	UK	438	75-11260
Dowty Aerospace Wolverhampton	UK	446	75-18589
Forward Industries Ltd	UK	452	75-24200
Garden Reach Shipbuilders	India	203	55-00300
GEC ALSTHOM Engineering Systems Ltd	UK	453	75-25225
George G. Sharp Inc.	USA	642	80-28000
Jered Brown Brothers Inc.	USA	586	75-50380
Kinetics Ltd	Israel	220	75-11320
MacTaggart Scott & Co. Ltd	UK	466	75-36900
Mannesmann Rexroth SA	France	139	75-25500
Messier-Dowty Inc	Canada	74	75-09655
Poclain Hydraulics SA	France	141	75-30402
Prva Petoletka Industrie	Yugoslavia	661	75-00610
Strachan & Henshaw Ltd	UK	512	80-47400
Svendborg Vaerft A/S	Denmark	104	75-19800
United Defense L.P.	USA	624	75-92050
Vickers Inc.	USA	626	75-93820
Vickers Systems	Germany	176	75-32830
Vickers Systems Trinova SpA	Italy	239	75-27150
Vosper Thornycroft	UK	492	75-63800

231 40 50 70 Hydraulic equipment, warship

Abex GmbH Aerohydraul	Germany	164	75-00105
Aeroquip Corp.	USA	547	75-02350
APPH (Bolton) Ltd.	UK	431	75-03415
Bradford Cylinders Ltd	UK	435	75-07850
Dowty Aerospace Wolverhampton	UK	446	75-18589
Fenner Fluid Power	UK	450	75-22550
Forward Industries Ltd	UK	452	75-24200
Gates Hydraulics Ltd	UK	453	75-25210
George G. Sharp Inc.	USA	642	80-28000
GKN Westland Technologies Ltd	UK	455	75-25690
Hitachi Zosen Co. Ltd.	Japan	251	75-02390
Imo Industries Inc.	USA	584	75-46150
Jered Brown Brothers Inc.	USA	586	75-50380
Kinetics Ltd	Israel	220	75-11320
MacTaggart Scott & Co. Ltd	UK	466	75-36900
Mannesmann Rexroth SA	France	139	75-25500
Moog Controls Ltd	UK	469	75-40670
Moog Japan Ltd.	Japan	253	75-06100
Normalair-Garrett Ltd	UK	425	74-07000
Prva Petoletka Industrie	Yugoslavia	661	75-00610
Ralbo Inc.	USA	607	75-72690
Svendborg Vaerft A/S	Denmark	104	75-19800
Tirfor Ltd	UK	488	75-60150
Vickers Systems Division TRINOVA Ltd	UK	491	75-63150
Vosper Thornycroft	UK	492	75-63800

231 40 50 70 50 Hydraulic system filters, warship

Aircraft Appliances & Equipment Ltd	Canada	68	75-00430
British Filters Ltd	UK	436	75-09030
Fairey Arlon Ltd	UK	450	75-22200
George G. Sharp Inc.	USA	642	80-28000
Mannesmann Rexroth SA	France	139	75-25500
Prva Petoletka Industrie	Yugoslavia	661	75-00610
Svendborg Vaerft A/S	Denmark	104	75-19800

231 40 55 Refrigeration plant, warship

George G. Sharp Inc.	USA	642	80-28000
J&E Hall Limited	UK	456	75-27420
Stork Bronswerk BV	Netherlands	295	75-07200
Svendborg Vaerft A/S	Denmark	104	75-19800
VSG Verfahrenstechnik	Germany	176	75-33030

231 40 60 Steam-raising plant, warship

George G. Sharp Inc.	USA	*642*	80-28000
Svendborg Vaerft A/S	Denmark	*104*	75-19800

231 40 70 Warship equipment, other

A.P. Metalcraft Ltd	UK	*430*	75-03410
Australian Defence Industries Ltd	Australia	*9*	55-01000
Howden Compressors Ltd	UK	*459*	75-30250
Howden Sirocco	France	*137*	75-21700
Hyde Products Inc.	USA	*583*	75-45280
IDE Technologies Ltd	Israel	*219*	75-09170
Jehier SA	France	*137*	75-23200
Litton Industries Inc.	USA	*590*	75-55400
Lucas Aerospace Power Transmission	USA	*592*	75-56265
Svendborg Vaerft A/S	Denmark	*104*	75-19800
U E F Chesterfield Cylinders	UK	*490*	75-61780
Van Rietschoten & Houwens	Netherlands	*296*	75-07900
Wegmann & Co GmbH	Germany	*177*	75-33700

231 40 70 10 Anchors, ship

A. Noble & Son Ltd	Australia	*19*	75-14690
Balmoral Group Ltd	UK	*433*	75-05420
George G. Sharp Inc.	USA	*642*	80-28000
Svendborg Vaerft A/S	Denmark	*104*	75-19800

231 40 70 10 30 Anchor chains, ships

Ancra CHS BV	Netherlands	*291*	75-00510
Balmoral Group Ltd	UK	*433*	75-05420
George G. Sharp Inc.	USA	*642*	80-28000
PWB Anchor Ltd	Australia	*20*	75-17000
Svendborg Vaerft A/S	Denmark	*104*	75-19800

231 40 70 20 Compressors, marine

Compair (Australasia) Ltd	Australia	*14*	75-03870
CompAir BroomWade Ltd	UK	*442*	75-14520
George G. Sharp Inc.	USA	*642*	80-28000
Spar Aviation Services	Canada	*76*	75-13020
Svendborg Vaerft A/S	Denmark	*104*	75-19800

231 40 70 30 Batteries, warship

Concorde Battery Corp.	USA	*566*	75-20845
Eldec Corp.	USA	*572*	75-29660
George G. Sharp Inc.	USA	*642*	80-28000
Hawker Energy Products Ltd	UK	*457*	75-28490
Saft Singapore Pte Ltd	Singapore	*348*	75-03400

231 40 70 33 Coalescers, marine gas turbine, naval ship

George G. Sharp Inc.	USA	*642*	80-28000
Svendborg Vaerft A/S	Denmark	*104*	75-19800

231 40 70 35 Compressors, air / gas, surface warship

Compair (Australasia) Ltd	Australia	*14*	75-03870
CompAir Reavell Ltd	UK	*442*	75-14530
George G. Sharp Inc.	USA	*642*	80-28000
Hamworthy Marine	UK	*457*	75-27850
Hatlapa Uetersener Maschinenfabrik	Germany	*169*	75-11735
Howden Compressors Ltd	UK	*459*	75-30250
Svendborg Vaerft A/S	Denmark	*104*	75-19800
Weir Engineering Pty Ltd	Australia	*23*	75-23300

231 40 70 37 Cranes, marine

AAI Corp.	USA	*545*	75-00020
Atlas Hydraulic Loaders Ltd	UK	*432*	75-04300
George G. Sharp Inc.	USA	*642*	80-28000
Jered Brown Brothers Inc.	USA	*586*	75-50380
Loadlift Equipment Ltd	New Zealand	*302*	75-01250
Optima Hydraulics	S. Africa	*359*	75-15500
Schat-Harding BV	Netherlands	*295*	75-06280
Sevastopol JSC	Ukraine	*407*	75-06500
Stothert & Pitt	UK	*512*	80-47200
Svendborg Vaerft A/S	Denmark	*104*	75-19800

231 40 70 38 Couplings, flexible, marine

Aeroquip Corp.	USA	*547*	75-02350
Avica Equipment	UK	*432*	75-04870
George G. Sharp Inc.	USA	*642*	80-28000
Lucas Aerospace Power Transmission	USA	*592*	75-56265
MacTaggart Scott & Co. Ltd	UK	*466*	75-36900
Svendborg Vaerft A/S	Denmark	*104*	75-19800
Vickers Shipbuilding & Eng Ltd	UK	*491*	75-63100

231 40 70 40 Incineration equipment, warship

George G. Sharp Inc.	USA	*642*	80-28000
Hamworthy Marine	UK	*457*	75-27850
Svendborg Vaerft A/S	Denmark	*104*	75-19800

231 40 70 45 Machinery, deck, naval ship

AAI Corp.	USA	*545*	75-00020
Champion Friction Co.	USA	*563*	75-17975
Garden Reach Shipbuilders	India	*203*	55-00300
George G. Sharp Inc.	USA	*642*	80-28000
Hitachi Zosen Co. Ltd.	Japan	*251*	75-02390
Hyde Products Inc.	USA	*583*	75-45280
Jered Brown Brothers Inc.	USA	*586*	75-50380
MacGregor (Aus) Pty Ltd	Australia	*18*	75-13000
Messier-Dowty Inc	Canada	*74*	75-09655
Svendborg Vaerft A/S	Denmark	*104*	75-19800

231 40 70 50 Pumps, marine

Combimac BV	Netherlands	*292*	75-01200
Fenner Fluid Power	UK	*450*	75-22550
Garbarino Pompe SpA	Italy	*234*	75-09530
George G. Sharp Inc.	USA	*642*	80-28000
Hamworthy Marine	UK	*457*	75-27850
HMD Seal/Less Pumps Ltd	UK	*458*	75-29470
Imo Industries Inc.	USA	*584*	75-46150
Ingersoll-Dresser Pumps (UK) Ltd	UK	*461*	75-31450
Lucas Aerospace Power Transmission	USA	*592*	75-56265
Plenty Ltd	UK	*475*	75-46050
Prosser/Enpo	USA	*574*	75-31670
Svendborg Vaerft A/S	Denmark	*104*	75-19800
United Defense L.P.	USA	*624*	75-92050
Urdan Industries (USA) Inc.	USA	*625*	75-92930
Weir Engineering Pty Ltd	Australia	*23*	75-23300
Weir Pumps Ltd	UK	*493*	75-65100

231 40 70 63 Separators, oily water, shipborne

Aircraft Appliances & Equipment Ltd	Canada	*68*	75-00430
George G. Sharp Inc.	USA	*642*	80-28000
Hamworthy Marine	UK	*457*	75-27850
Harco Manufacturing Co. Inc.	USA	*580*	75-41870
Perry Engineering	Australia	*20*	75-16460
Svendborg Vaerft A/S	Denmark	*104*	75-19800

231 40 70 66 Stores, naval

A.P. Metalcraft Ltd	UK	*430*	75-03410
George G. Sharp Inc.	USA	*642*	80-28000
Svendborg Vaerft A/S	Denmark	*104*	75-19800

231 40 70 90 Winches, marine

ACH	France	*130*	75-00150
A1Marketing	UK	*426*	75-00025
Champion Friction Co.	USA	*563*	75-17975
Dowty Aerospace Wolverhampton	UK	*446*	75-18589
dp Manufacturing Inc.	USA	*569*	75-27320
Garden Reach Shipbuilders	India	*203*	55-00300
GEC ALSTHOM Engineering Systems Ltd	UK	*453*	75-25225
George G. Sharp Inc.	USA	*642*	80-28000

Jered Brown Brothers Inc.	USA	586	75-50380
MacArtney A/S	Denmark	103	75-12800
MacTaggart Scott & Co. Ltd	UK	466	75-36900
Mannesmann Rexroth SA	France	139	75-25500
Messier-Dowty Inc	Canada	74	75-09655
Poclain Hydraulics SA	France	141	75-30402
PWB Anchor Ltd	Australia	20	75-17000
Schat-Harding BV	Netherlands	295	75-06280
Strachan & Henshaw Ltd	UK	512	80-47400
Svendborg Vaerft A/S	Denmark	104	75-19800
Vickers Inc.	USA	626	75-93830
Vickers Systems	Germany	176	75-32830
Vickers Systems SA	France	148	75-46930
Vickers Systems Trinova SpA	Italy	239	75-27150

231 40 70 92 Window wipers, marine

George G. Sharp Inc.	USA	642	80-28000
Hepworth Marine International	UK	458	75-29045
Hutchwilco Limited	New Zealand	302	75-00860
Svendborg Vaerft A/S	Denmark	104	75-19800
Wynn Marine Ltd	UK	495	75-67300

231 50 Mine warfare ship systems & equipment

Boeing Australia Limited	Australia	12	75-02500
Crestbury Ltd	UK	443	75-16030
Dowty Aerospace Wolverhampton	UK	446	75-18589
EUROCONTROL SpA	Italy	233	75-08200
Fabbrica Italiana Apparecchiature	Italy	234	75-08550
George G. Sharp Inc.	USA	642	80-28000
Meggitt Avionics (UK)	UK	467	75-38340
Riva Calzoni SpA	Italy	238	75-20500
SAGEM SA	**France**	**143**	**75-33950**
SELIN Sistemi Sarl	Italy	238	75-21750
S.G. Brown	UK	437	75-10330
Thomson Sintra	France	148	75-45135

231 50 10 Mine warfare ship control systems

Boeing Australia Limited	Australia	12	75-02500
Fabbrica Italiana Apparecchiature	Italy	234	75-08550
George G. Sharp Inc.	USA	642	80-28000
Kobelt Manufacturing Co. Ltd	Canada	73	75-08770
MacDonald Dettwiler Limited	Australia	25	80-12990
Riva Calzoni SpA	Italy	238	75-20500
Thomson Sintra	France	148	75-45135

231 50 10 70 Mine warfare ship control equipment

Boeing Australia Limited	Australia	12	75-02500
EDO Corporation	USA	571	75-29250
Fabbrica Italiana Apparecchiature	Italy	234	75-08550
George G. Sharp Inc.	USA	642	80-28000
Kobelt Manufacturing Co. Ltd	Canada	73	75-08770
Riva Calzoni SpA	Italy	238	75-20500
SAGEM SA	**France**	**143**	**75-33950**
SELIN Sistemi Sarl	Italy	238	75-21750

231 50 10 73 Mine warfare ship steering control equipment

George G. Sharp Inc.	USA	642	80-28000
Jered Brown Brothers Inc.	USA	586	75-50380
Kobelt Manufacturing Co. Ltd	Canada	73	75-08770

231 50 10 75 Mine warfare ship manoeuvring units

George G. Sharp Inc.	USA	642	80-28000
Riva Calzoni SpA	Italy	238	75-20500

231 50 20 Mine warfare ship mission-related systems

Benthos Inc.	USA	558	75-13540
Datamat Ingegneria dei Sistemi SpA	Italy	233	75-06500
Dowty Aerospace Wolverhampton	UK	446	75-18589
DRS Technologies, Inc.	USA	570	75-27400
EDO Corporation	USA	571	75-29250
Exstar International Corp.	USA	575	75-33000
Fabbrica Italiana Apparecchiature	Italy	234	75-08550
MacDonald Dettwiler Limited	Australia	25	80-12990
Marr Technical Services Ltd	UK	506	80-15620
Pacer Systems Inc.	USA	602	75-68000
Thomson Sintra	France	148	75-45135

231 50 20 60 Mine warfare ship sensors (see code 411.23)

Thomson Sintra	France	148	75-45135

231 50 20 70 Mine warfare ship mission-related equipment

Crestbury Ltd	UK	443	75-16030
Dowty Aerospace Wolverhampton	UK	446	75-18589
E.P. Barrus Ltd	UK	433	75-05900
Exstar International Corp.	USA	575	75-33000
Fabbrica Italiana Apparecchiature	Italy	234	75-08550
Hitachi Zosen Co. Ltd.	Japan	251	75-02390
Jered Brown Brothers Inc.	USA	586	75-50380
Lucas Western Inc.	USA	592	75-56300
MIL-BASE Industries Inc.	USA	596	75-61530
Riva Calzoni SpA	Italy	238	75-20500
Thomson Sintra	France	148	75-45135

231 50 20 70 20 Cranes, anti-magnetic, mine warfare ship

George G. Sharp Inc.	USA	642	80-28000
Jered Brown Brothers Inc.	USA	586	75-50380
Lucas Western Inc.	USA	592	75-56300
MacTaggart Scott & Co. Ltd	UK	466	75-36900
Morgan Marine	USA	597	75-62625
Riva Calzoni SpA	Italy	238	75-20500
Schat-Harding BV	Netherlands	295	75-06280

231 70 FPB / light combattant craft systems & equipment

Alenia-Elsag	Italy	230	74-00100
British Aerospace (Systems &	UK	436	75-08890
Ishikawajima-Harima H.I. Co. Ltd	Japan	251	75-02800
ITT Avionics	USA	586	75-49455
Meggitt Avionics (UK)	UK	467	75-38340
SAGEM SA	**France**	**143**	**75-33950**
Sanders	USA	612	75-77850
S.G. Brown	UK	437	75-10330
Van Rietschoten & Houwens	Netherlands	296	75-07900

231 70 10 FPB control systems

Davies Industrial Communications Ltd	UK	444	75-17010
Kobelt Manufacturing Co. Ltd	Canada	73	75-08770

231 70 20 47 Torpedo launch equipment, FPB

AAI Corp.	USA	545	75-00020

231 70 20 47 30 Torpedo tubes, FPB

Technical Products Group Inc.	USA	620	75-86764

231 70 40 FPB systems & equipment, general

S.G. Brown	UK	437	75-10330
Wegmann & Co GmbH	Germany	177	75-33700

232 Amphibious Warfare ship / landing craft systems & equipment

AAI Corp.	USA	545	75-00020
APPH (Bolton) Ltd.	UK	431	75-03415
Cubic Defense Systems Inc.	USA	567	75-22900
Horstman Defence Systems Ltd.	UK	459	75-29920

Lucas Aerospace Power Transmission	USA	592	75-56265
MacTaggart Scott & Co. Ltd	UK	466	75-36900
Pacer Systems Inc.	USA	602	75-68000
Riva Calzoni SpA	Italy	238	75-20500

233 Surface effect craft systems & equipment

Avon Technical Products	UK	432	75-05150
Cubic Defense Systems Inc.	USA	567	75-22900
Envirovac Inc.	USA	574	75-31920
FPT Inc.	USA	576	75-35760
George G. Sharp Inc.	USA	642	80-28000
Horstman Defence Systems Ltd.	UK	459	75-29920
Lucas Aerospace Power Transmission	USA	592	75-56265
Meggitt Avionics (UK)	UK	467	75-38340
United Defense L.P.	USA	624	75-92050
Van Rietschoten & Houwens	Netherlands	296	75-07900

233 70 60 Skirts, hovercraft

Avon Technical Products	UK	432	75-05150
George G. Sharp Inc.	USA	642	80-28000
SMR Technologies Inc.	USA	615	75-82100

234 Hydrofoil systems & equipment

Alenia-Elsag	Italy	230	74-00100
APPH (Bolton) Ltd.	UK	431	75-03415
Cubic Defense Systems Inc.	USA	567	75-22900
Horstman Defence Systems Ltd.	UK	459	75-29920
Howden Fluid Systems	USA	582	75-44400
Magnaghi Milano SpA	Italy	236	75-13980
Meggitt Avionics (UK)	UK	467	75-38340
Riva Calzoni SpA	Italy	238	75-20500

235 Mission support ship / Auxiliary ship systems & equipment

Alenia-Elsag	Italy	230	74-00100
British Aerospace (Systems &	UK	436	75-08890
Envirovac Inc.	USA	574	75-31920
George G. Sharp Inc.	USA	642	80-28000
IDE Technologies Ltd	Israel	219	75-09170
Ishikawajima-Harima H.I. Co. Ltd	Japan	251	75-02800
ITT Avionics	USA	586	75-49455
Meggitt Avionics (UK)	UK	467	75-38340
Van Rietschoten & Houwens	Netherlands	296	75-07900

235 10 Mission support / Auxiliary ship control systems

George G. Sharp Inc.	USA	642	80-28000

235 20 Mission support / Auxiliary ship mission systems

Cubic Defense Systems Inc.	USA	567	75-22900
George G. Sharp Inc.	USA	642	80-28000
Ishikawajima-Harima H.I. Co. Ltd	Japan	251	75-02800
ITT Avionics	USA	586	75-49455

235 21 Fuel transfer systems, ship-to-ship

George G. Sharp Inc.	USA	642	80-28000
MacTaggart Scott & Co. Ltd	UK	466	75-36900

235 22 Logistics transfer systems, ship-to-ship

ACH	France	130	75-00150
George G. Sharp Inc.	USA	642	80-28000
Lucas Western Inc.	USA	592	75-56300
MacTaggart Scott & Co. Ltd	UK	466	75-36900
SpanSet Inter AG	Switzerland	387	75-13550

235 23 Logistics transfer systems, ship-to-shore

George G. Sharp Inc.	USA	642	80-28000
SpanSet Inter AG	Switzerland	387	75-13550

235 25 Crew transfer systems, ship-to-ship

George G. Sharp Inc.	USA	642	80-28000

235 40 Mission support ship / Auxiliary ship systems, other

George G. Sharp Inc.	USA	642	80-28000

236 Coastguard / Customs / Police vessel & craft systems & equipment

Alenia-Elsag	Italy	230	74-00100
Avon Inflatables Ltd	UK	432	75-05100
British Aerospace (Systems &	UK	436	75-08890
Canadian Marconi Co.	Canada	69	75-02600
Cougar Limited	UK	443	75-15250
Cubic Defense Systems Inc.	USA	567	75-22900
Envirovac Inc.	USA	574	75-31920
E.P. Barrus Ltd	UK	433	75-05900
Exstar International Corp.	USA	575	75-33000
Horstman Defence Systems Ltd.	UK	459	75-29920
Ishikawajima-Harima H.I. Co. Ltd	Japan	251	75-02800
Israel Aircraft Industries Ltd	Israel	219	75-10200
Kobelt Manufacturing Co. Ltd	Canada	73	75-08770
Northern Airborne Technology Ltd	Canada	74	75-10145
Raytheon E-Systems	USA	608	75-73302
Spectrolab Inc.	USA	617	75-83100
Stringer Welding & Machine Inc.	USA	618	75-84435
Textron Marine & Land Systems	USA	622	75-88660
Van Rietschoten & Houwens	Netherlands	296	75-07900

237 Launch / small craft systems & equipment

Alenia-Elsag	Italy	230	74-00100
Australian Defence Industries Ltd	Australia	9	55-01000
British Aerospace (Systems &	UK	436	75-08890
Cougar Limited	UK	443	75-15250
Cubic Defense Systems Inc.	USA	567	75-22900
Envirovac Inc.	USA	574	75-31920
E.P. Barrus Ltd	UK	433	75-05900
Ishikawajima-Harima H.I. Co. Ltd	Japan	251	75-02800
Israel Aircraft Industries Ltd	Israel	219	75-10200
Kobelt Manufacturing Co. Ltd	Canada	73	75-08770
M.A. Bigland (Preparations) Ltd	UK	434	75-06700
Meggitt Avionics (UK)	UK	467	75-38340
Van Rietschoten & Houwens	Netherlands	296	75-07900

239 Submarine / submersible systems & equipment

AAI Corp.	USA	545	75-00020
Alenia-Elsag	Italy	230	74-00100
Anite Systems Ltd	UK	497	80-01240
Avica Equipment	UK	432	75-04870
British Aerospace (Systems &	UK	436	75-08890
COSMOS SpA	Italy	233	75-05700
Cubic Defense Systems Inc.	USA	567	75-22900
ECE GmbH	Germany	167	75-07200
Envirovac Inc.	USA	574	75-31920
EUROCONTROL SpA	Italy	233	75-08200
Fenner Fluid Power	UK	450	75-22550
Haux-Life-Support GmbH	Germany	169	75-11970
Howaldtswerke Deutsche Werft AG	Germany	170	75-13400
Howden Aircontrol Ltd	UK	459	75-30240
Hytech BV	Netherlands	293	75-03530
Kongsberg Simrad Inc	USA	589	75-53160
Lucas Aerospace FCS Paris	France	139	75-25390
Lucas Aerospace Power Transmission	USA	592	75-56265
Meggitt Avionics (UK)	UK	467	75-38340
Officine Panerai SpA	Italy	237	75-17500
Raytheon Anschütz GmbH	Germany	173	75-24650
RDM Submarines BV	Netherlands	294	75-05770
Riva Calzoni SpA	Italy	238	75-20500
Saab Dynamics AB	Sweden	378	75-09100
Saft Singapore Pte Ltd	Singapore	348	75-03400
SAGEM SA	**France**	**143**	**75-33950**
Sanders	USA	612	75-77850
SELIN Sistemi Sarl	Italy	238	75-21750
S.G. Brown	UK	437	75-10330
Sperry Marine Inc.	USA	617	75-83250
Stanley Plastics Ltd	UK	486	75-56355

Strachan & Henshaw Ltd	UK	*512*	80-47400
Teldix GmbH	Germany	*175*	75-30100
Thomson Sintra	France	*148*	75-45135
Van Rietschoten & Houwens	Netherlands	*296*	75-07900
Vickers Shipbuilding & Eng Ltd	UK	*491*	75-63100
Zeiss-Eltro Optronic GmbH	Germany	*177*	75-34700

239 10 Submarine control systems

Alenia-Elsag	Italy	*230*	74-00100
Anite Systems Ltd	UK	*497*	80-01240
Australian Defence Industries Ltd	Australia	*9*	55-01000
Boeing Australia Limited	Australia	*12*	75-02500
Howaldtswerke Deutsche Werft AG	Germany	*170*	75-13400
Int'l Submarine Engrg	Canada	*73*	75-07780
Officine Panerai SpA	Italy	*237*	75-17500
Raytheon Anschütz GmbH	Germany	*173*	75-24650
Saab Dynamics AB	Sweden	*378*	75-09100
SAGEM SA	**France**	***143***	**75-33950**
S.G. Brown	UK	*437*	75-10330
Société d'Applications des Machines	France	*143*	75-34300
Van Rietschoten & Houwens	Netherlands	*296*	75-07900
Vickers Shipbuilding & Eng Ltd	UK	*491*	75-63100
Whippendell Electrical Ltd	UK	*493*	75-65800

239 10 10 Depth & course control systems, submarine

Int'l Submarine Engrg	Canada	*73*	75-07780
Raytheon Anschütz GmbH	Germany	*173*	75-24650
Saab Dynamics AB	Sweden	*378*	75-09100
SAGEM SA	**France**	***143***	**75-33950**
The Dewey Electronics Corp	USA	*569*	75-26100

239 10 11 Automatic pilot systems, submarine

NEC Corporation	Japan	*253*	75-06350
Saab Dynamics AB	Sweden	*378*	75-09100
SAGEM SA	**France**	***143***	**75-33950**
S.G. Brown	UK	*437*	75-10330
Slingsby Engineering Ltd	UK	*484*	75-54420

239 10 20 Machinery control systems, submarine

Curtiss-Wright Flow Control	USA	*619*	75-86366
Officine Panerai SpA	Italy	*237*	75-17500
Raytheon Anschütz GmbH	Germany	*173*	75-24650

239 10 70 Submarine control equipment

Astronautics C.A. Ltd	Israel	*217*	75-02400
Australian Defence Industries Ltd	Australia	*9*	55-01000
Catsaros Automation SA	Greece	*187*	75-00700
Lucas Aerospace FCS Paris	France	*139*	75-25390
Officine Panerai SpA	Italy	*237*	75-17500
Saab Dynamics AB	Sweden	*378*	75-09100
SAGEM SA	**France**	***143***	**75-33950**

239 10 70 30 Instruments, submarine control

Boeing Australia Limited	Australia	*12*	75-02500
Crompton Instruments Ltd	UK	*443*	75-16070
Meggitt Avionics (UK)	UK	*467*	75-38340
Saab Dynamics AB	Sweden	*378*	75-09100
SAGEM SA	**France**	***143***	**75-33950**
Tempo Instruments Inc	USA	*621*	75-87900
The Dewey Electronics Corp	USA	*569*	75-26100
Transfield Shipbuilding WA	Australia	*22*	75-20812

239 20 Submarine mission systems

Australian Defence Industries Ltd	Australia	*9*	55-01000
COSMOS SpA	Italy	*233*	75-05700
Pilkington Optronics Ltd	UK	*426*	74-10000
RDM Submarines BV	Netherlands	*294*	75-05770
Riva Calzoni SpA	Italy	*238*	75-20500
SAGEM SA	**France**	***143***	**75-33950**
Sanders	USA	*612*	75-77850
Strachan & Henshaw Ltd	UK	*512*	80-47400
Thomson Sintra	France	*148*	75-45135
Vickers Shipbuilding & Eng Ltd	UK	*491*	75-63100

239 20 15 Mine expulsion systems, submarine

Strachan & Henshaw Ltd	UK	*512*	80-47400
Vickers Shipbuilding & Eng Ltd	UK	*491*	75-63100

239 20 20 Missile launch systems, submarine

Strachan & Henshaw Ltd	UK	*512*	80-47400
Vickers Shipbuilding & Eng Ltd	UK	*491*	75-63100

239 20 23 Missile tubes, submarine

National Forge Co.	USA	*598*	75-63850
RDM Submarines BV	Netherlands	*294*	75-05770
Strachan & Henshaw Ltd	UK	*512*	80-47400
Technical Products Group Inc	USA	*620*	75-86764
Vickers Shipbuilding & Eng Ltd	UK	*491*	75-63100

239 20 30 Periscopes, submarine

Eloptro	S. Africa	*358*	75-05680
Kongsberg Simrad Inc	USA	*589*	75-53160
National Forge Co.	USA	*598*	75-63850
Pilkington Optronics Ltd	UK	*426*	74-10000
SAGEM SA	**France**	***143***	**75-33950**
SFIM ODS	France	*144*	75-35085
Zeiss-Eltro Optronic GmbH	Germany	*177*	75-34700

239 20 35 Snorkel systems, submarine

COSMOS SpA	Italy	*233*	75-05700
MacTaggart Scott & Co. Ltd	UK	*466*	75-36900
Riva Calzoni SpA	Italy	*238*	75-20500
Transfield Shipbuilding WA	Australia	*22*	75-20812

239 20 40 Torpedo launch systems & equipment, submarine

COSMOS SpA	Italy	*233*	75-05700
Messier-Dowty Limited	UK	*467*	75-38700
MSI-Defence Systems Ltd	UK	*469*	75-41280
Strachan & Henshaw Ltd	UK	*512*	80-47400
Tempo Instruments Inc	USA	*621*	75-87900
Thomson Sintra	France	*148*	75-45135
Vickers Shipbuilding & Eng Ltd	UK	*491*	75-63100
Whitehead SpA	Italy	*240*	75-27600

239 20 41 Torpedo launch cradles, submarine

Strachan & Henshaw Ltd	UK	*512*	80-47400
Transfield Shipbuilding WA	Australia	*22*	75-20812
Vickers Shipbuilding & Eng Ltd	UK	*491*	75-63100
Whitehead SpA	Italy	*240*	75-27600

239 20 43 Torpedo tubes, submarine

Australian Defence Industries Ltd	Australia	*9*	55-01000
COSMOS SpA	Italy	*233*	75-05700
RDM Submarines BV	Netherlands	*294*	75-05770
Strachan & Henshaw Ltd	UK	*512*	80-47400
Technical Products Group Inc	USA	*620*	75-86764
Transfield Shipbuilding WA	Australia	*22*	75-20812
Whitehead SpA	Italy	*240*	75-27600

239 20 70 Submarine mission equipment

Australian Defence Industries Ltd	Australia	*9*	55-01000
Hytech BV	Netherlands	*293*	75-03530
Transfield Shipbuilding WA	Australia	*22*	75-20812
Vickers Shipbuilding & Eng Ltd	UK	*491*	75-63100

239 30 Submarine crew-related facilities & equipment

Beaufort Air-Sea Equipment Ltd	UK	*433*	75-06100
British Filters Ltd	UK	*436*	75-09030
Daimler-Benz Aerospace France SAS	**France**	***134***	**75-12950**
Maine Group Ltd	UK	*466*	75-37050
Vickers Shipbuilding & Eng Ltd	UK	*491*	75-63100

239 30 05 Environmental control systems, submarine

Airscrew Howden Ltd	UK	*429*	75-02050
Bradford Cylinders Ltd	UK	*435*	75-07850
British Filters Ltd	UK	*436*	75-09030
CJB Developments Ltd	UK	*440*	75-13430
Envirovac Inc.	USA	*574*	75-31920
Graseby Dynamics Ltd	UK	*455*	75-26590
Howden Aircontrol Ltd	UK	*459*	75-30240
Hytech BV	Netherlands	*293*	75-03530
The Armstrong Monitoring Corp.	Canada	*68*	75-00740

239 30 40 Submarine internal communication systems

Boeing Australia Limited	Australia	*12*	75-02500
Daimler-Benz Aerospace France SAS	**France**	***134***	**75-12950**
Davies Industrial Communications Ltd	UK	*444*	75-17010
Northern Airborne Technology Ltd	Canada	*74*	75-10145
Vickers Shipbuilding & Eng Ltd	UK	*491*	75-63100

239 40 Submarine systems, other

Eldec Corp.	USA	*572*	75-29660
Horstman Defence Systems Ltd.	UK	*459*	75-29920
Int'l Submarine Engrg.	Canada	*73*	75-07780
MacTaggart Scott & Co. Ltd	UK	*466*	75-36900
Nautronix Ltd	UK	*470*	75-41900
SELIN Sistemi Sarl	Italy	*238*	75-21750
Van Rietschoten & Houwens	Netherlands	*296*	75-07900

239 40 20 Chilled water systems, submarine

Avica Equipment	UK	*432*	75-04870
Transfield Shipbuilding WA	Australia	*22*	75-20812
van Buuren-van Swaay BV	Netherlands	*296*	75-07830
VSG Verfahrenstechnik	Germany	*176*	75-33030
Worcester Controls (UK) Ltd	UK	*494*	75-67000

239 40 30 Electrical systems, submarine

Century Electronics	USA	*562*	75-17385
Combimac BV	Netherlands	*292*	75-01200
ECE GmbH	Germany	*167*	75-07200
Eldec Corp.	USA	*572*	75-29660
EUROCONTROL SpA	Italy	*233*	75-08200
GSI Corporation	USA	*580*	75-40715
Holec Machines & Apparaten BV	Netherlands	*293*	75-03110
Horstman Defence Systems Ltd.	UK	*459*	75-29920
Icore International Ltd	UK	*460*	75-31080
Latécoère	France	*138*	75-24200
Mawdsley's Ltd	UK	*466*	75-37900
Newton Derby Ltd	UK	*470*	75-42520
Saab Dynamics AB	Sweden	*378*	75-09100
Transfield Shipbuilding WA	Australia	*22*	75-20812
Tronic Electronic Services Ltd	UK	*489*	75-61330
White Products BV	Netherlands	*296*	75-09100

239 40 30 50 Lighting systems, submarine

ECE GmbH	Germany	*167*	75-07200
Luminescent Systems Inc.	USA	*593*	75-56480
Transfield Shipbuilding WA	Australia	*22*	75-20812

239 40 30 60 Electrical generating plant, submarine

EUROCONTROL SpA	Italy	*233*	75-08200
Holec Machines & Apparaten BV	Netherlands	*293*	75-03110
Mawdsley's Ltd	UK	*466*	75-37900
SELIN Sistemi Sarl	Italy	*238*	75-21750
Transfield Shipbuilding WA	Australia	*22*	75-20812

239 40 40 Fire protection systems, submarine

Nordic Systems Inc	Canada	*74*	75-09950
Spectronix Ltd	Israel	*222*	75-21100
Worcester Controls (UK) Ltd	UK	*494*	75-67000

239 40 45 Fuel systems, submarine

Aeroquip Corp.	USA	*547*	75-02350
Aircraft Appliances & Equipment Ltd	Canada	*68*	75-00430
British Filters Ltd	UK	*436*	75-09030
Transfield Shipbuilding WA	Australia	*22*	75-20812
Worcester Controls (UK) Ltd	UK	*494*	75-67000

239 40 70 Submarine equipment, general

AAI Corp.	USA	*545*	75-00020
Australian Defence Industries Ltd	Australia	*9*	55-01000
Eldec Corp.	USA	*572*	75-29660
Exstar International Corp.	USA	*575*	75-33000
GNB Australia Limited	Australia	*16*	75-07700
Jered Brown Brothers Inc.	USA	*586*	75-50380
Riva Calzoni SpA	Italy	*238*	75-20500
U E F Chesterfield Cylinders	UK	*490*	75-61780
Vickers Shipbuilding & Eng Ltd	UK	*491*	75-63100

239 40 70 10 Batteries, submarine (see code 633.30.10.60)

Chloride Industrial Batteries Ltd	UK	*457*	75-28490
Eldec Corp.	USA	*572*	75-29660
GNB Australia Limited	Australia	*16*	75-07700
Hawker Energy Products Ltd	UK	*457*	75-28490
Saft Nife AB	Sweden	*378*	75-09850
Saft Singapore Pte Ltd	Singapore	*348*	75-03400
Transfield Shipbuilding WA	Australia	*22*	75-20812
VHB Industrial Batteries Ltd	Canada	*77*	75-14540

239 40 70 20 Compressors, submarine

Compair (Australasia) Ltd	Australia	*14*	75-03870
Transfield Shipbuilding WA	Australia	*22*	75-20812

239 40 70 38 Couplings, submarine

Avica Equipment	UK	*432*	75-04870
Lucas Aerospace Power Transmission	USA	*592*	75-56265

239 40 70 45 Submarine masts

MacTaggart Scott & Co. Ltd	UK	*466*	75-36900
MSI-Defence Systems Ltd	UK	*469*	75-41280
Pilkington Optronics Ltd	UK	*426*	74-10000
Riva Calzoni SpA	Italy	*238*	75-20500
Sperry Marine Inc.	USA	*617*	75-83250
Technical Products Group Inc.	USA	*620*	75-86764
Transfield Shipbuilding WA	Australia	*22*	75-20812

239 40 70 50 Pumps, submarine

Combimac BV	Netherlands	*292*	75-01200
Fenner Fluid Power	UK	*450*	75-22550
Ingersoll-Dresser Pumps (UK) Ltd	UK	*461*	75-31450
Lucas Aerospace FCS Paris	France	*139*	75-25390
Lucas Aerospace Power Transmission	USA	*592*	75-56265
MacTaggart Scott & Co. Ltd	UK	*466*	75-36900
Mawdsley's Ltd	UK	*466*	75-37900
Transfield Shipbuilding WA	Australia	*22*	75-20812
Urdan Industries (USA) Inc.	USA	*625*	75-92930
Weir Engineering Pty Ltd	Australia	*23*	75-23300

Product–Supplier Link

Land-based Platform Systems & equipment

241 Vehicle systems & equipment

AAI Corp.	USA	545	75-00020
Air-Log Ltd	UK	429	75-01800
Allied Research Corp.	USA	549	75-04670
APPH (Bolton) Ltd.	UK	431	75-03415
Ashot Ashkelon Industries Ltd	Israel	217	75-02100
Australian Defence Industries Ltd	Australia	9	55-01000
Avon Technical Products	UK	432	75-05150
Bernardini SA	Brazil	55	75-02120
Bowmar Instrument Corp.	USA	560	75-14860
British Aerospace (Systems &	UK	436	75-08890
CIC International Ltd	USA	563	75-18570
Cincinnati Electronics Corp.	USA	564	75-18640
Commercial Hydraulics Kontak Ltd	UK	441	75-14415
Concorde Battery Corp.	USA	566	75-20845
Cubic Defense Systems Inc.	USA	567	75-22900
Dynex/Rivett Inc.	USA	570	75-28350
Dytecna Ltd	UK	501	80-08900
ECE GmbH	Germany	167	75-07200
EL-OP Electro Optics Industries Ltd	Israel	219	75-06300
Ericsson Microwave Systems AB	**Sweden**	**377**	**75-04890**
Fabbrica Italiana Apparecchiature	Italy	234	75-08550
Fenner Fluid Power	UK	450	75-22550
FPT Industries Ltd	UK	452	75-24400
Fraser-Volpe Corp.	USA	576	75-35950
GEC-Marconi Avionics Ltd	UK	454	75-25320
Helio Mirror Co. Ltd	UK	457	75-28900
Horstman Defence Systems Ltd	UK	459	75-29920
Howard Porter Pty Ltd	Australia	17	75-08720
Howden Aircontrol Ltd	UK	459	75-30240
Instro Precision Ltd	UK	461	75-31630
Israel Aircraft Industries Ltd	Israel	220	75-10200
J-Tec Associates Inc.	USA	587	75-51250
KUKA Wehrtechnik GmbH	Germany	171	75-17200
Labinal Aero Defense Systems Inc	USA	638	80-18100
Lucas Aerospace Power Transmission	USA	592	75-56265
Lucas Western Inc.	USA	592	75-56300
M. Paz Logistics Ltd	Israel	224	80-01600
Mambrini Equipamentos Rodoviarios	Brazil	57	75-13260
Meggitt Avionics (UK)	UK	467	75-38340
Meggitt Composites	UK	467	75-38345
Messier-Dowty Electronics	Canada	74	75-09650
Microturbo SA	France	140	75-27000
Mitsubishi Precision Co. Ltd	Japan	253	75-06000
Nanoquest Defence Products Ltd	UK	470	75-41890
Napco International Inc.	USA	598	75-63600
Novatraçáo Artefatos	Brazil	57	75-14780
Palbam Metal Works	Israel	221	75-17000
Patria Vehicles Oy	Finland	118	75-01710
Pilkington Aerospace Ltd	UK	474	75-45500
Raytheon E-Systems	USA	608	75-73302
Reunert Technology Systems Ltd	S. Africa	359	75-18300
Rockwell	USA	610	75-75930
SELIN Sistemi Sarl	Italy	238	75-21750
SEXTANT Avionique SA	France	144	75-35000
SFIM	**France**	**144**	**75-35080**
Silent Partner Body Armor Inc.	USA	615	75-80930
SP aerospace and vehicle systems bv	**Netherlands**	**295**	**75-06860**
Stephan Wood Products Inc.	USA	628	75-97280
Steyr Antriebstechnik	Austria	34	75-02800
Technofan SA	France	147	75-44300
Tecmotiv Corporation	Canada	77	75-13510
Teldix GmbH	Germany	175	75-30100
Transfield Pty Ltd	Australia	22	75-20810
Vickers Shipbuilding & Eng Ltd	UK	491	75-63100
Warren-Knight Instrument Co.	USA	627	75-95150

241 10 Tracked armoured vehicle systems & equipment

AAI Corp.	USA	545	75-00020
Air-Log Ltd	UK	429	75-01800
Alenia Difesa	**Italy**	**231**	**75-01165**
Allied Research Corp.	USA	549	75-04670
AlliedSignal Aerospace Canada	Canada	68	75-00470
APPH (Bolton) Ltd.	UK	431	75-03415
Avon Technical Products	UK	432	75-05150
Bernardini SA	Brazil	55	75-02120
British Aerospace (Systems &	UK	436	75-08890
Clouth Gummiwerke AG	Germany	166	75-04900
Commercial Hydraulics Kontak Ltd	UK	441	75-14415
DRS Technologies, Inc.	USA	570	75-27400
Ericsson Microwave Systems AB	**Sweden**	**377**	**75-04890**
Fraser-Volpe Corp.	USA	576	75-35950
Giat Industries	France	126	55-00100
Helio Mirror Co. Ltd	UK	457	75-28900
Howden Aircontrol Ltd	UK	459	75-30240
Hughes Danbury Optical Sys.	USA	582	75-44857
Hunting Engineering Ltd	UK	459	75-30700
Israel Aircraft Industries Ltd	Israel	220	75-10200
J-Tec Associates Inc.	USA	587	75-51250
Kearfott Guidance & Navigation Corp.	USA	588	75-51760
Kinetics Ltd	Israel	220	75-11320
Krauss-Maffei Wehrtechnik GmbH	Germany	171	75-16345
KUKA Wehrtechnik GmbH	Germany	171	75-17200
Lucas Aerospace Power Transmission	USA	592	75-56265
Lucas Western Inc.	USA	592	75-56300
M. Paz Logistics Ltd	Israel	224	80-01600
MacOptic SA	Greece	190	75-01999
Meggitt Avionics (UK)	UK	467	75-38340
Meggitt Composites	UK	467	75-38345
Microturbo SA	France	140	75-27000
MOH-9	S. Africa	358	75-11500
MOWAG	Switzerland	387	75-08900
Nanoquest Defence Products Ltd	UK	470	75-41890
Napco International Inc.	USA	598	75-63600
Norca Machinery Company	USA	599	75-64800
Ralbo Inc.	USA	607	75-72690
Reunert Technology Systems Ltd	S. Africa	359	75-18300
SAGEM SA	**France**	**143**	**75-33950**
Sèkur SpA	Italy	238	75-21650
SELIN Sistemi Sarl	Italy	238	75-21750
SEXTANT Avionique SA	France	144	75-35000
Smiths Industries Aerospace	UK	485	75-54700
SP aerospace and vehicle systems bv	**Netherlands**	**295**	**75-06860**
Stephan Wood Products Inc.	USA	628	75-97280
Tecmotiv Corporation	Canada	77	75-13510
Transfield Pty Ltd	Australia	22	75-20810
United Defense L.P.	USA	624	75-92050
Urdan Industries Ltd	Israel	223	75-23210
Vickers Shipbuilding & Eng Ltd	UK	491	75-63100
Wegmann & Co GmbH	Germany	177	75-33700

241 10 10 Tracked armoured vehicle control systems

AAI Corp.	USA	545	75-00020
APPH (Bolton) Ltd.	UK	431	75-03415
BAeSEMA Ltd.	UK	498	80-03070
Fairey Hydraulics	UK	450	75-22255
Hunting Engineering Ltd	UK	459	75-30700
Israel Aircraft Industries Ltd	Israel	219	75-10200
Kidde-Fenwal	USA	588	75-52510
Kinetics Ltd	Israel	220	75-11320
Merex Inc.	USA	595	75-60250
Moog Controls Ltd	UK	469	75-40670
Napco International Inc.	USA	598	75-63600
Reunert Technology Systems Ltd	S. Africa	359	75-18300
Société d'Applications des Machines	France	143	75-34300
United Defense L.P.	USA	624	75-92050
Vickers Defence Systems	UK	491	75-63040

241 10 10 05 Braking systems, tracked armoured vehicle

AAI Corp.	USA	545	75-00020
APPH (Bolton) Ltd.	UK	431	75-03415
David Brown Defence Equipment Ltd	UK	437	75-10150
Kinetics Ltd	Israel	220	75-11320
Kobelt Manufacturing Co. Ltd	Canada	73	75-08770
Napco International Inc.	USA	598	75-63600
Sielman SA	Greece	191	75-02360
Tuthill Corporation	USA	624	75-91320

241 10 10 10 Steering systems, tracked armoured vehicle

AAI Corp.	USA	545	75-00020
APPH (Bolton) Ltd.	UK	431	75-03415
David Brown Defence Equipment Ltd	UK	437	75-10150
Israel Aircraft Industries Ltd	Israel	219	75-10200
Menasco Aerospace Ltd	Canada	74	75-09600

Merex Inc.	USA	*595*	75-60250
Moog Inc.	USA	*597*	75-62550
Napco International Inc.	USA	*598*	75-63600
Norca Machinery Company	USA	*599*	75-64800
Poclain Hydraulics SA	France	*141*	75-30402
Self-Changing Gears Ltd	UK	*482*	75-52700
Sielman SA	Greece	*191*	75-02360
Tuthill Corporation	USA	*624*	75-91320
ZF Great Britain Ltd	UK	*495*	75-67720

241 10 10 10 40 Steering systems, hydrostatic, tracked armoured vehicle

AAI Corp.	USA	*545*	75-00020
David Brown Defence Equipment Ltd	UK	*437*	75-10150
Moog Inc.	USA	*597*	75-62550
Napco International Inc.	USA	*598*	75-63600
Norca Machinery Company	USA	*599*	75-64800
Poclain Hydraulics SA	France	*141*	75-30402
Self-Changing Gears Ltd	UK	*482*	75-52700
Sielman SA	Greece	*191*	75-02360

241 10 20 Tracked armoured vehicle mission systems

Bofors AB	Sweden	*375*	74-00020
Cubic Defense Systems Inc.	USA	*567*	75-22900
Fraser-Volpe Corp.	USA	*576*	75-35950
Israel Military Industries Ltd	Israel	*220*	75-10300
Kearfott Guidance & Navigation Corp.	USA	*588*	75-51760
Kinetics Ltd	Israel	*220*	75-11320
Lucas Western Inc.	USA	*592*	75-56300
M. Paz Logistics Ltd	Israel	*224*	80-01600
Moog Inc.	USA	*597*	75-62550
Napco International Inc.	USA	*598*	75-63600
Pilkington Optronics Ltd	UK	*426*	74-10000
Raytheon E-Systems	USA	*608*	75-73302
ROTZLER GmbH & Co	Germany	*174*	75-25800
SAGEM SA	**France**	***143***	**75-33950**
Sanders	USA	*612*	75-77850
SFIM	**France**	***144***	**75-35080**
Société d'Applications des Machines	France	*143*	75-34300
United Defense L.P.	USA	*624*	75-92050
Vickers Shipbuilding & Eng Ltd	UK	*491*	75-63100
Western Design Howden	USA	*627*	75-95800

241 10 22 Weapon-related systems, tracked armoured vehicle

AAI Corp.	USA	*545*	75-00020
Air-Log Ltd	UK	*429*	75-01800
Alenia Difesa	**Italy**	***231***	**75-01165**
Allied Research Corp.	USA	*549*	75-04670
Delft Sensor Systems	**Belgium**	***46***	**75-02500**
Delft Sensor Systems	Netherlands	*292*	75-01590
Fairey Hydraulics	UK	*450*	75-22255
Fraser-Volpe Corp.	USA	*576*	75-35950
GEC-Marconi Avionics Inc.	USA	*577*	75-37413
Hunting Engineering Ltd	UK	*459*	75-30700
Kinetics Ltd	Israel	*220*	75-11320
KUKA Wehrtechnik GmbH	Germany	*171*	75-17200
Lucas Western Inc.	USA	*592*	75-56300
M. Paz Logistics Ltd	Israel	*224*	80-01600
Moog Inc.	USA	*597*	75-62550
Napco International Inc.	USA	*598*	75-63600
Nottingham International	USA	*640*	80-22000
Reunert Technology Systems Ltd	S. Africa	*359*	75-18300
Transfield Pty Ltd	Australia	*22*	75-20810
United Defense L.P.	USA	*624*	75-92050

241 10 22 10 Automatic gun loading systems, tracked armoured vehicle

AAI Corp.	USA	*545*	75-00020
Air-Log Ltd	UK	*429*	75-01800
David Brown Defence Equipment Ltd	UK	*437*	75-10150
Fairey Hydraulics	UK	*450*	75-22255
Giat Industries	France	*126*	55-00100
Howden Aircontrol Ltd	UK	*459*	75-30240
KUKA Wehrtechnik GmbH	Germany	*171*	75-17200
M. Paz Logistics Ltd	Israel	*224*	80-01600

Moog Inc.	USA	*597*	75-62550
Napco International Inc.	USA	*598*	75-63600
United Defense L.P.	USA	*624*	75-92050
Vickers Shipbuilding & Eng Ltd	UK	*491*	75-63100
Wegmann & Co GmbH	Germany	*177*	75-33700
Western Design Howden	USA	*627*	75-95800

241 10 22 30 Fire control systems, tracked armoured vehicle (see code 415.40.20.80)

AAI Corp.	USA	*545*	75-00020
Astronautics Corp. of America	USA	*555*	75-10000
Avimo Ltd	UK	*432*	75-04950
Bharat Electronics Ltd	India	*203*	55-00200
Computing Devices Canada	Canada	*70*	75-03600
Delft Sensor Systems	**Belgium**	***46***	**75-02500**
Delft Sensor Systems	Netherlands	*292*	75-01590
EL-OP Electro Optics Industries Ltd	Israel	*219*	75-06300
Eloptro	S. Africa	*358*	75-05680
EUROCONTROL SpA	Italy	*233*	75-08200
Fabbrica Italiana Apparecchiature	Italy	*234*	75-08550
Fraser-Volpe Corp.	USA	*576*	75-35950
GEC-Marconi Avionics Inc.	USA	*577*	75-37413
Kearfott Guidance & Navigation Corp.	USA	*588*	75-51760
Kinetics Ltd	Israel	*220*	75-11320
Kollsman	USA	*589*	75-53150
Litton Industries Inc.	USA	*590*	75-55400
M. Paz Logistics Ltd	Israel	*224*	80-01600
Merex Inc.	USA	*595*	75-60250
Napco International Inc.	USA	*598*	75-63600
Norca Machinery Company	USA	*599*	75-64800
Pilkington Optronics Ltd	UK	*426*	74-10000
Przemyslowy Instytut Telekomunikacji	**Poland**	***325***	**75-02650**
SAGEM SA	**France**	***143***	**75-33950**
SFIM	**France**	***144***	**75-35080**
Vickers Shipbuilding & Eng Ltd	UK	*491*	75-63100
Warren-Knight Instrument Co.	USA	*627*	75-95150

241 10 22 70 Mission-related equipment, tracked armoured vehicle

AAI Corp.	USA	*545*	75-00020
Analog Modules Inc.	USA	*553*	75-07140
A.P. Metalcraft Ltd	UK	*430*	75-03410
Bavaria Avionik Technologie GmbH.	Germany	*165*	75-02040
Ericsson Microwave Systems AB	**Sweden**	***377***	**75-04890**
GEC-Marconi Avionics Ltd	UK	*454*	75-25320
Horstman Defence Systems Ltd.	UK	*459*	75-29920
Hunting Engineering Ltd	UK	*459*	75-30700
KUKA Wehrtechnik GmbH	Germany	*171*	75-17200
M. Paz Logistics Ltd	Israel	*224*	80-01600
Napco International Inc.	USA	*598*	75-63600
Pilkington Optronics Ltd	UK	*426*	74-10000
Precision Sensors	USA	*605*	75-71360
SFIM	**France**	***144***	**75-35080**
SP aerospace and vehicle systems bv	**Netherlands**	***295***	**75-06860**
Vickers Defence Systems	UK	*491*	75-63040
Vickers Shipbuilding & Eng Ltd	UK	*491*	75-63100
Wegmann USA Inc.	USA	*627*	75-95543

241 10 22 70 10 Ammunition racks & containers, tracked armoured vehicle

A.P. Metalcraft Ltd	UK	*430*	75-03410
E-A-R Specialty Composites	USA	*571*	75-28500
Fairey Hydraulics	UK	*450*	75-22255
Israel Military Industries Ltd	Israel	*220*	75-10300
M. Paz Logistics Ltd	Israel	*224*	80-01600
Napco International Inc.	USA	*598*	75-63600
Norca Machinery Company	USA	*599*	75-64800
Ordex Inc.	USA	*601*	75-67350
Permali Gloucester Ltd	UK	*474*	75-44850
Strachan & Henshaw Ltd	UK	*512*	80-47400
United Defense L.P.	USA	*624*	75-92050
Vickers Shipbuilding & Eng Ltd	UK	*491*	75-63100
Wegmann & Co GmbH	Germany	*177*	75-33700
Wegmann USA Inc.	USA	*627*	75-95543

Product–Supplier Link

241 10 22 70 20 Cupolas, tracked armoured vehicle

AAI Corp.	USA	*545*	75-00020
Alenia Difesa	**Italy**	***231***	**75-01165**
Giat Industries	France	*126*	55-00100
Hebron & Medlock Engineering Ltd	UK	*457*	75-28860
Helio Mirror Co. Ltd	UK	*457*	75-28900
KUKA Wehrtechnik GmbH	Germany	*171*	75-17200
M. Paz Logistics Ltd	Israel	*224*	80-01600
Napco International Inc.	USA	*598*	75-63600
Norca Machinery Company	USA	*599*	75-64800
Nottingham International	USA	*640*	80-22000
SP aerospace and vehicle systems bv	**Netherlands**	***295***	**75-06860**
United Defense L.P.	USA	*624*	75-92050
Urdan Industries Ltd	Israel	*223*	75-23210
Urdan Industries (USA) Inc.	USA	*625*	75-92930
Vickers Shipbuilding & Eng Ltd	UK	*491*	75-63100

241 10 22 70 22 Cranes, hydraulic, weapon / ammunition handling, tracked armoured vehicle

AAI Corp.	USA	*545*	75-00020
Auto Crane Co.	USA	*556*	75-10360
Grove Europe Ltd	**UK**	***456***	**75-27030**
KUKA Wehrtechnik GmbH	Germany	*171*	75-17200
M. Paz Logistics Ltd	Israel	*224*	80-01600
Napco International Inc.	USA	*598*	75-63600
United Defense L.P.	USA	*624*	75-92050
Vickers Inc.	USA	*626*	75-93830
Vickers Systems	Germany	*176*	75-32830
Vickers Systems Trinova SpA	Italy	*239*	75-27150

241 10 22 70 30 Firing ports, tracked armoured vehicle

Helio Mirror Co. Ltd	UK	*457*	75-28900
INIMEX	**Belgium**	***46***	**75-04650**
M. Paz Logistics Ltd	Israel	*224*	80-01600
Napco International Inc.	USA	*598*	75-63600
United Defense L.P.	USA	*624*	75-92050
Wegmann & Co GmbH	Germany	*177*	75-33700

241 10 22 70 40 Gearboxes, gun control, tracked armoured vehicle

AIM Enterprises Inc.	USA	*630*	80-01350
Ashot Ashkelon Industries Ltd	Israel	*217*	75-02100
David Brown Defence Equipment Ltd	UK	*437*	75-10150
EUROCONTROL SpA	Italy	*233*	75-08200
Giat Industries	France	*126*	55-00100
Hebron & Medlock Engineering Ltd	UK	*457*	75-28860
Horstman Defence Systems Ltd	UK	*459*	75-29920
Israel Military Industries Ltd	Israel	*220*	75-10300
Kinetics Ltd	Israel	*220*	75-11320
Merex Inc.	USA	*595*	75-60250
MSI-Defence Systems Ltd	UK	*469*	75-41280
Napco International Inc.	USA	*598*	75-63600
Norca Machinery Company	USA	*599*	75-64800
Self-Changing Gears Ltd	UK	*482*	75-52700
SELIN Sistemi Sarl	Italy	*238*	75-21750

241 10 22 70 55 Ring mounts, tracked armoured vehicle

Hägglunds Moelv AS	Norway	*309*	75-00880
KUKA Wehrtechnik GmbH	Germany	*171*	75-17200
Napco International Inc.	USA	*598*	75-63600
Norca Machinery Company	USA	*599*	75-64800
Nottingham International	USA	*640*	80-22000
Stringer Welding & Machine Inc.	USA	*618*	75-84435
Theodor Rapp GmbH Co	Germany	*173*	75-24600

241 10 22 70 70 Turrets, tracked armoured vehicle

AAI Corp.	USA	*545*	75-00020
Alenia Difesa	**Italy**	***231***	**75-01165**
AV Technology LLC	USA	*556*	75-11405
Delco Systems Operations	USA	*568*	75-25360
Helio Mirror Co. Ltd	UK	*457*	75-28900
Israel Military Industries Ltd	Israel	*220*	75-10300
KUKA Wehrtechnik GmbH	Germany	*171*	75-17200
M. Paz Logistics Ltd	Israel	*224*	80-01600
MaK Systemgesellschaft mbH	Germany	*172*	75-19500
Napco International Inc.	USA	*598*	75-63600
Norca Machinery Company	USA	*599*	75-64800
Nottingham International	USA	*640*	80-22000
Rheinmetall Waffe & Munition GmbH	Germany	*173*	75-24950
SP aerospace and vehicle systems bv	**Netherlands**	***295***	**75-06860**
Textron Marine & Land Systems	USA	*622*	75-88660
Transfield Pty Ltd	Australia	*22*	75-20810
United Defense L.P.	USA	*624*	75-92050
Urdan Industries Ltd	Israel	*223*	75-23210
Urdan Industries (USA) Inc.	USA	*625*	75-92930
Vickers Shipbuilding & Eng Ltd	UK	*491*	75-63100
Wegmann & Co GmbH	Germany	*177*	75-33700

241 10 22 70 70 30 Gun turrets, tracked armoured vehicle

Alenia Difesa	**Italy**	***231***	**75-01165**
Giat Industries	France	*126*	55-00100
Hägglunds Vehicle AB	Sweden	*377*	75-05800
Israel Military Industries Ltd	Israel	*220*	75-10300
KUKA Wehrtechnik GmbH	Germany	*171*	75-17200
M. Paz Logistics Ltd	Israel	*224*	80-01600
Nottingham International	USA	*640*	80-22000
Rheinmetall Waffe & Munition GmbH	Germany	*173*	75-24950
SP aerospace and vehicle systems bv	**Netherlands**	***295***	**75-06860**
Textron Marine & Land Systems	USA	*622*	75-88660
Transfield Pty Ltd	Australia	*22*	75-20810
Vickers Shipbuilding & Eng Ltd	UK	*491*	75-63100

241 10 22 70 70 30 10 Gun turrets, anti-aircraft, tracked armoured vehicle

Alenia Difesa	**Italy**	***231***	**75-01165**
Giat Industries	France	*126*	55-00100
Israel Military Industries Ltd	Israel	*220*	75-10300
KUKA Wehrtechnik GmbH	Germany	*171*	75-17200
M. Paz Logistics Ltd	Israel	*224*	80-01600
Nottingham International	USA	*640*	80-22000
Textron Marine & Land Systems	USA	*622*	75-88660

241 10 25 Rangefinder systems, tracked armoured vehicle

Arab International Optronics Co.	Egypt	*109*	74-00500
Delft Sensor Systems	**Belgium**	***46***	**75-02500**
Delft Sensor Systems	Netherlands	*292*	75-01590
DRS Technologies, Inc.	USA	*570*	75-27400
EL-OP Electro Optics Industries Ltd	Israel	*219*	75-06300
GEC-Marconi Avionics Inc.	USA	*577*	75-37413
GEC-Marconi Avionics Ltd	UK	*454*	75-25320
GEC-Marconi Materials Technology Ltd	UK	*419*	60-00550
Litton Industries Inc.	USA	*590*	75-55400
M. Paz Logistics Ltd	Israel	*224*	80-01600
Merex Inc.	USA	*595*	75-60250
Nanoquest Defence Products Ltd	UK	*470*	75-41890
Napco International Inc.	USA	*598*	75-63600

241 10 25 40 Rangefinder systems, tracked armoured vehicle, laser

Analog Modules Inc.	USA	*553*	75-07140
Arab International Optronics Co.	Egypt	*109*	74-00500
Avimo Ltd	UK	*432*	75-04950
Bharat Electronics Ltd	India	*203*	55-00200
Bospec Neue Technologien	Germany	*166*	75-03420
Delft Sensor Systems	**Belgium**	***46***	**75-02500**
Delft Sensor Systems	Netherlands	*292*	75-01590
EL-OP Electro Optics Industries Ltd	Israel	*219*	75-06300
Eloptro	S. Africa	*358*	75-05680
Ericsson Microwave Systems AB	**Sweden**	***377***	**75-04890**
Fabbrica Italiana Apparecchiature	Italy	*234*	75-08550
GEC-Marconi Avionics Inc.	USA	*577*	75-37413
GEC-Marconi Avionics Ltd	UK	*454*	75-25320
GEC-Marconi Materials Technology Ltd	UK	*419*	60-00550
Instro Precision Ltd	UK	*461*	75-31630
Kollsman	USA	*589*	75-53150
Litton Industries Inc.	USA	*590*	75-55400

M. Paz Logistics Ltd	Israel	*224*	80-01600
Nanoquest Defence Products Ltd	UK	*470*	75-41890
Napco International Inc.	USA	*598*	75-63600

241 10 25 50 Rangefinder systems, tracked armoured vehicle, optical

Bospec Neue Technologien	Germany	*166*	75-03420
Computer Optics Inc.	USA	*565*	75-20680
Delft Sensor Systems	**Belgium**	***46***	**75-02500**
Delft Sensor Systems	Netherlands	*292*	75-01590
DRS Technologies, Inc.	USA	*570*	75-27400
EL-OP Electro Optics Industries Ltd	Israel	*219*	75-06300
GEC-Marconi Avionics Inc..	USA	*577*	75-37413
GEC-Marconi Avionics Ltd	UK	*454*	75-25320
Kollsman	USA	*589*	75-53150
Litton Industries Inc.	USA	*590*	75-55400
M. Paz Logistics Ltd	Israel	*224*	80-01600
Nanoquest Defence Products Ltd	UK	*470*	75-41890
Napco International Inc.	USA	*598*	75-63600

241 10 27 50 Searchlights, tracked armoured vehicle

M. Paz Logistics Ltd	Israel	*224*	80-01600
Napco International Inc.	USA	*598*	75-63600

241 10 27 60 Sensors, tracked armoured vehicle

GEC-Marconi Avionics Inc..	USA	*577*	75-37413
Kollsman	USA	*589*	75-53150
Litton Industries Inc.	USA	*590*	75-55400
Napco International Inc.	USA	*598*	75-63600
Pilkington Optronics Ltd	UK	*426*	74-10000
Precision Sensors	USA	*605*	75-71360
Zeiss-Eltro Optronic GmbH	Germany	*177*	75-34700

241 10 27 60 40 Sensors, laser, tracked armoured vehicle

Avimo Ltd	UK	*432*	75-04950
Elettronica (UK) Ltd	UK	*448*	75-20475
Ericsson Microwave Systems AB	**Sweden**	***377***	**75-04890**
Fabbrica Italiana Apparecchiature	Italy	*234*	75-08550
GEC-Marconi Avionics Inc..	USA	*577*	75-37413
GEC-Marconi Avionics Ltd	UK	*454*	75-25320
Hughes Danbury Optical Sys.	USA	*582*	75-44857
Kollsman	USA	*589*	75-53150
Litton Industries Inc.	USA	*590*	75-55400
Napco International Inc.	USA	*598*	75-63600
Pilkington Optronics Ltd	UK	*426*	74-10000
Zeiss-Eltro Optronic GmbH	Germany	*177*	75-34700

241 10 27 60 90 Sensors, wind, tracked armoured vehicle

GEC-Marconi Avionics Inc..	USA	*577*	75-37413
J-Tec Associates Inc.	USA	*587*	75-51250
Kollsman	USA	*589*	75-53150
Meggitt Avionics (UK)	UK	*467*	75-38340
Napco International Inc.	USA	*598*	75-63600
Zeiss-Eltro Optronic GmbH	Germany	*177*	75-34700

241 10 27 65 Sights, tracked armoured vehicle

Arab International Optronics Co.	Egypt	*109*	74-00500
Avimo Ltd	UK	*432*	75-04950
Delco Systems Operations	USA	*568*	75-25360
EUROCONTROL SpA	Italy	*233*	75-08200
Fraser-Volpe Corp.	USA	*576*	75-35950
GEC-Marconi Avionics Inc..	USA	*577*	75-37413
Helio Mirror Co. Ltd	UK	*457*	75-28900
M. Paz Logistics Ltd	Israel	*224*	80-01600
Nanoquest Defence Products Ltd	UK	*470*	75-41890
Napco International Inc.	USA	*598*	75-63600
Pilkington Optronics Ltd	UK	*426*	74-10000
Recon/Optical Inc.	USA	*608*	75-73930
SAGEM SA	**France**	***143***	**75-33950**
SFIM	**France**	***144***	**75-35080**
SFIM ODS	France	*144*	75-35085
W.F. Stanley & Co. Ltd	UK	*486*	75-56350
Zeiss-Eltro Optronic GmbH	Germany	*177*	75-34700

241 10 27 65 50 Sights, periscopic, tracked armoured vehicle

Arab International Optronics Co.	Egypt	*109*	74-00500
Avimo Ltd	UK	*432*	75-04950
Delft Sensor Systems	**Belgium**	***46***	**75-02500**
Delft Sensor Systems	Netherlands	*292*	75-01590
Fraser-Volpe Corp.	USA	*576*	75-35950
GEC-Marconi Avionics Inc..	USA	*577*	75-37413
Helio Mirror Co. Ltd	UK	*457*	75-28900
M. Paz Logistics Ltd	Israel	*224*	80-01600
Microtecnica-Torino SpA	Italy	*237*	75-15500
Nanoquest Defence Products Ltd	UK	*470*	75-41890
Napco International Inc.	USA	*598*	75-63600
Pilkington Optronics Ltd	UK	*426*	74-10000
SAGEM SA	**France**	***143***	**75-33950**
SFIM	**France**	***144***	**75-35080**
SFIM ODS	France	*144*	75-35085
W.F. Stanley & Co. Ltd	UK	*486*	75-56350
Zeiss-Eltro Optronic GmbH	Germany	*177*	75-34700

241 10 27 65 60 Sights, telescopic, tracked armoured vehicle

Arab International Optronics Co.	Egypt	*109*	74-00500
Avimo Ltd	UK	*432*	75-04950
Fraser-Volpe Corp.	USA	*576*	75-35950
GEC-Marconi Avionics Inc..	USA	*577*	75-37413
M. Paz Logistics Ltd	Israel	*224*	80-01600
Nanoquest Defence Products Ltd	UK	*470*	75-41890
Napco International Inc.	USA	*598*	75-63600
Pilkington Optronics Ltd	UK	*426*	74-10000
SFIM ODS	France	*144*	75-35085
Urals Optical & Mechanical Plant	Russia	*339*	75-53000
Warren-Knight Instrument Co.	USA	*627*	75-95150
W.F. Stanley & Co. Ltd	UK	*486*	75-56350
Zeiss-Eltro Optronic GmbH	Germany	*177*	75-34700

241 10 27 80 Smoke pots, tracked armoured vehicle

Engineered Air Systems Inc..	USA	*573*	75-31560
Étienne Lacroix	France	*138*	75-24050
Haley & Weller Ltd	UK	*456*	75-27400
Helio Mirror Co. Ltd	UK	*457*	75-28900
Napco International Inc.	USA	*598*	75-63600
Pains-Wessex Ltd	UK	*472*	75-43510
Wegmann & Co GmbH	Germany	*177*	75-33700

241 10 27 90 Vision blocks, tracked armoured vehicle

DRS Technologies, Inc.	USA	*570*	75-27400
Eloptro	S. Africa	*358*	75-05680
Helio Mirror Co. Ltd	UK	*457*	75-28900
M. Paz Logistics Ltd	Israel	*224*	80-01600
MacOptic SA	Greece	*190*	75-01999
Nanoquest Defence Products Ltd	UK	*470*	75-41890
Napco International Inc.	USA	*598*	75-63600
Norca Machinery Company	USA	*599*	75-64800
SAGEM SA	**France**	***143***	**75-33950**
SFIM	**France**	***144***	**75-35080**
Sielman SA	Greece	*191*	75-02360

241 10 28 Turret stabilisation systems

EUROCONTROL SpA	Italy	*233*	75-08200
Israel Aircraft Industries Ltd	Israel	*220*	75-10200
Kinetics Ltd	Israel	*220*	75-11320
M. Paz Logistics Ltd	Israel	*224*	80-01600
Messier-Dowty Inc	Canada	*74*	75-09655
Moog Japan Ltd.	Japan	*253*	75-06100
Napco International Inc.	USA	*598*	75-63600
Norca Machinery Company	USA	*599*	75-64800
SELIN Sistemi Sarl	Italy	*238*	75-21750

241 10 28 30 Gun stabilisation systems

EUROCONTROL SpA	Italy	*233*	75-08200
GEC-Marconi Avionics Ltd	UK	*454*	75-25320
GEC-Marconi Hazeltine Corp.	USA	*577*	75-37414
Giat Industries	France	*126*	55-00100

Kinetics Ltd	Israel	*220*	75-11320
M. Paz Logistics Ltd	Israel	*224*	80-01600
Messier-Dowty Inc	Canada	*74*	75-09655
Moog Inc.	USA	*597*	75-62550
Moog Japan Ltd.	Japan	*253*	75-06100
Napco International Inc.	USA	*598*	75-63600
Nottingham International	USA	*640*	80-22000
Pilkington Optronics Ltd	UK	*426*	74-10000
SAGEM SA	**France**	***143***	**75-33950**
SELIN Sistemi Sarl	Italy	*238*	75-21750

241 10 30 Tracked armoured vehicle crew-related systems

Aerospace Syst. & Comp.	USA	*547*	75-02501
AlliedSignal Aerospace Canada	Canada	*68*	75-00470
British Filters Ltd	UK	*436*	75-09030
Giat Industries	France	*126*	55-00100
Helio Mirror Co. Ltd	UK	*457*	75-28900
Hunting Engineering Ltd	UK	*459*	75-30700
Kinetics Ltd	Israel	*220*	75-11320
Napco International Inc.	USA	*598*	75-63600
Stone India Ltd	India	*205*	75-01950
United Defense L.P.	USA	*624*	75-92050
Vickers Shipbuilding & Eng Ltd	UK	*491*	75-63100

241 10 30 30 Fume extraction units, crew compartment, tracked armoured vehicle

AAI Corp.	USA	*545*	75-00020
Aerospace Syst. & Comp.	USA	*547*	75-02501
Airscrew Howden Ltd	UK	*429*	75-02050
British Filters Ltd	UK	*436*	75-09030
EUROCONTROL SpA	Italy	*233*	75-08200
Howden Aircontrol Ltd	UK	*459*	75-30240
Napco International Inc.	USA	*598*	75-63600

241 10 30 35 Heaters, crew compartment, tracked armoured vehicle

AAI Corp.	USA	*545*	75-00020
Aerospace Syst. & Comp.	USA	*547*	75-02501
Aircraft Maintenance Support	UK	*428*	75-01580
Airscrew Howden Ltd	UK	*429*	75-02050
A/S Dantherm	**Denmark**	***102***	**75-04100**
Electrothermal Engineering Ltd	UK	*448*	75-20400
Gallay Ltd	UK	*453*	75-25100
Global Thermoelectric Inc.	Canada	*71*	75-06200
Howden Aircontrol Ltd	UK	*459*	75-30240
Hunter Manufacturing	USA	*583*	75-45150
Kinetics Ltd	Israel	*220*	75-11320
Microturbo Ltd	UK	*468*	75-39160
Napco International Inc.	USA	*598*	75-63600
Stone India Ltd	India	*205*	75-01950
Teledyne Brown Engineering	USA	*620*	75-87508
THERMOCOAX	France	*147*	75-44930

241 10 30 40 Internal communication systems, tracked armoured vehicle

AAI Corp.	USA	*545*	75-00020
Cincinnati Electronics Corp.	USA	*564*	75-18640
Davies Industrial Communications Ltd	UK	*444*	75-17010
Giat Industries	France	*126*	55-00100
Hunting Engineering Ltd	UK	*459*	75-30700
Napco International Inc.	USA	*598*	75-63600

241 10 30 50 NBC filtration systems, tracked armoured vehicle

AAF International BV	Netherlands	*291*	75-00005
AAI Corp.	USA	*545*	75-00020
Airscrew Howden Ltd	UK	*429*	75-02050
AlliedSignal Aerospace Canada	Canada	*68*	75-00470
British Filters Ltd	UK	*436*	75-09030
Engineered Air Systems Inc.	USA	*573*	75-31560
EROS GIE	France	*137*	75-19500
Fairey Microfiltrex	UK	*450*	75-22300
Giat Industries	France	*126*	55-00100
GKN Westland Technologies Ltd	UK	*455*	75-25690
Graseby Dynamics Ltd	UK	*455*	75-26590
Howden Aircontrol Ltd	UK	*459*	75-30240
Hunting Engineering Ltd	UK	*459*	75-30700
Kinetics Ltd	Israel	*220*	75-11320
Napco International Inc.	USA	*598*	75-63600
Normalair-Garrett Ltd	UK	*425*	74-07000
Pall Land & Marine	UK	*472*	75-43520
Parker Bertea Aerospace	USA	*603*	75-68900
Sèkur SpA	Italy	*238*	75-21650
Stella-Meta	UK	*486*	75-56780
Vickers Shipbuilding & Eng Ltd	UK	*491*	75-63100

241 10 30 55 Air conditioning systems, tracked armoured vehicle

AAI Corp.	USA	*545*	75-00020
Aerospace Syst. & Comp.	USA	*547*	75-02501
Airflow Co.	USA	*548*	75-03650
Airscrew Howden Ltd	UK	*429*	75-02050
Gallay Ltd	UK	*453*	75-25100
GKN Westland Technologies Ltd	UK	*455*	75-25690
Howden Aircontrol Ltd	UK	*459*	75-30240
Howden Fluid Systems	USA	*582*	75-44400
Hunting Engineering Ltd	UK	*459*	75-30700
Kinetics Ltd	Israel	*220*	75-11320
Microtecnica-Torino SpA	Italy	*237*	75-15500
Microturbo Ltd	UK	*468*	75-39160
Napco International Inc.	USA	*598*	75-63600
Normalair-Garrett Ltd	UK	*425*	74-07000
Parker Bertea Aerospace	USA	*603*	75-68900
Stone India Ltd	India	*205*	75-01950
Vickers Shipbuilding & Eng Ltd	UK	*491*	75-63100

241 10 30 70 Tracked armoured vehicle crew-related equipment

A.P. Metalcraft Ltd	UK	*430*	75-03410
Kinetics Ltd	Israel	*220*	75-11320
M. Paz Logistics Ltd	Israel	*224*	80-01600
Napco International Inc.	USA	*598*	75-63600
Stephan Wood Products Inc.	USA	*628*	75-97280

241 10 30 70 60 Crew seats, tracked armoured vehicle

AAI Corp.	USA	*545*	75-00020
Hebron & Medlock Engineering Ltd	UK	*457*	75-28860
M. Paz Logistics Ltd	Israel	*224*	80-01600
Napco International Inc.	USA	*598*	75-63600
Plastic Fabricating Co. Inc.	USA	*604*	75-70335
Stephan Wood Products Inc.	USA	*628*	75-97280
Vickers Shipbuilding & Eng Ltd	UK	*491*	75-63100

241 10 40 Tracked armoured vehicle systems, other

AAI Corp.	USA	*545*	75-00020
AlliedSignal Aerospace Canada	Canada	*68*	75-00470
APPH (Bolton) Ltd.	UK	*431*	75-03415
Dytecna Ltd	UK	*501*	80-08900
Eldec Corp.	USA	*572*	75-29660
Gates Hydraulics Ltd	UK	*453*	75-25210
Giat Industries	France	*126*	55-00100
Horstman Defence Systems Ltd	UK	*459*	75-29920
Howden Sirocco	France	*137*	75-21700
Hunting Engineering Ltd	UK	*459*	75-30700
Israel Military Industries Ltd	Israel	*220*	75-10300
Kinetics Ltd	Israel	*220*	75-11320
Lucas Aerospace Power Transmission	USA	*592*	75-56265
M. Paz Logistics Ltd	Israel	*224*	80-01600
Midcon Cables Co.	USA	*596*	75-61300
Napco International Inc.	USA	*598*	75-63600
SELIN Sistemi Sarl	Italy	*238*	75-21750
Stone India Ltd	India	*205*	75-01950
Vickers Shipbuilding & Eng Ltd	UK	*491*	75-63100

241 10 40 30 Electrical systems, tracked armoured vehicle

AAI Corp.	USA	*545*	75-00020
AIM Enterprises Inc.	USA	*630*	80-01350

AlliedSignal Aerospace Canada	Canada	68	75-00470
Bradley Electronics Ltd	UK	435	75-07860
Commercial Hydraulics Kontak Ltd	UK	441	75-14415
ECE GmbH	Germany	167	75-07200
Eldec Corp.	USA	572	75-29660
Electrothermal Engineering Ltd	UK	448	75-20400
EUROCONTROL SpA	Italy	233	75-08200
GSI Corporation	USA	580	75-40715
Hebron & Medlock Engineering Ltd	UK	457	75-28860
Hunting Engineering Ltd	UK	459	75-30700
Icore International Ltd	UK	460	75-31080
Imco Industries Ltd	Israel	219	75-09400
Labinal	France	138	75-23900
M. Paz Logistics Ltd	Israel	224	80-01600
Microturbo Ltd	UK	468	75-39160
Napco International Inc.	USA	598	75-63600
Newton Derby Ltd	UK	470	75-42520
Norca Machinery Company	USA	599	75-64800
Reunert Technology Systems Ltd	S. Africa	359	75-18300
SELIN Sistemi Sarl	Italy	238	75-21750
Sielman SA	Greece	191	75-02360
Stewart & Stevenson Services Inc.	USA	618	75-84180
Stone India Ltd	India	205	75-01950
Teledyne Continental Motors	USA	620	75-87506
Tempo Instruments Inc	USA	621	75-87900
Vickers Shipbuilding & Eng Ltd	UK	491	75-63100
Walter Dittel GmbH	Germany	167	75-06510

241 10 40 30 40 Lighting systems, tracked armoured vehicle

ECE GmbH	Germany	167	75-07200
Luminescent Systems Inc.	USA	593	75-56480
M. Paz Logistics Ltd	Israel	224	80-01600
Napco International Inc.	USA	598	75-63600
Norca Machinery Company	USA	599	75-64800
Premier Hazard Systems Inc	USA	605	75-71395
Premier Hazard Systems Pty Ltd	Australia	20	75-16865

241 10 40 30 40 40 Lighting, interior, tracked armoured vehicle

Bruce Industries Inc.	USA	561	75-15360
Luminescent Systems Inc.	USA	593	75-56480
M. Paz Logistics Ltd	Israel	224	80-01600
Napco International Inc.	USA	598	75-63600
Premier Hazard Systems Inc	USA	605	75-71395
Premier Hazard Systems Pty Ltd	Australia	20	75-16865
Saunders-Roe Ltd	UK	482	75-51850
Soderberg Manufacturing Co. Inc.	USA	616	75-82200
Westair Ltd	UK	493	75-65410

241 10 40 30 50 Slip rings, tracked armoured vehicle

AAI Corp.	USA	545	75-00020
EUROCONTROL SpA	Italy	233	75-08200
IDM Electronics Ltd	UK	460	75-31100
Litton Industries Inc.	USA	590	75-55400
Mécanex SA	Switzerland	386	75-08300
Napco International Inc.	USA	598	75-63600
Norca Machinery Company	USA	599	75-64800
Reunert Technology Systems Ltd	S. Africa	359	75-18300
Vorkauf SA	Spain	367	75-07000
Wegmann & Co GmbH	Germany	177	75-33700

241 10 40 30 50 70 Slip rings, tracked armoured vehicle turret

AAI Corp.	USA	545	75-00020
EUROCONTROL SpA	Italy	233	75-08200
IDM Electronics Ltd	UK	460	75-31100
Litton Industries Inc.	USA	590	75-55400
Napco International Inc.	USA	598	75-63600
Norca Machinery Company	USA	599	75-64800
Vorkauf SA	Spain	367	75-07000
Wegmann & Co GmbH	Germany	177	75-33700

241 10 40 30 60 Wiring harnesses, electrical system, tracked armoured vehicle

AAI Corp.	USA	545	75-00020
Electrothermal Engineering Ltd	UK	448	75-20400
Hebron & Medlock Engineering Ltd	UK	457	75-28860
Hunting Engineering Ltd	UK	459	75-30700
Icore International Inc.	USA	584	75-45890
Icore International Ltd	UK	460	75-31080
Ionix High Performance Wiring	UK	461	75-32007
Kembrey Wiring Systems Ltd	UK	463	75-33525
Midcon Cables Co.	USA	596	75-61300
Napco International Inc.	USA	598	75-63600
Norca Machinery Company	USA	599	75-64800
Reunert Technology Systems Ltd	S. Africa	359	75-18300
Sielman SA	Greece	191	75-02360
Theodor Rapp GmbH Co	Germany	173	75-24600
Wegmann & Co GmbH	Germany	177	75-33700
Westwire Harnessing Limited	UK	493	75-65700
White Products BV	Netherlands	296	75-09100

241 10 40 40 Fire extinguishing systems, tracked armoured vehicle

Deugra Ges. für Brandschutzsysteme	Germany	167	75-05900
Hughes Electronics Corp.	USA	583	75-44865
Kidde-Graviner Ltd	UK	463	75-33880
Lehavot Industries	Israel	220	75-12200
M. Paz Logistics Ltd	Israel	224	80-01600
Napco International Inc.	USA	598	75-63600
Spectrex Inc.	USA	617	75-83095
Spectronix Ltd	Israel	222	75-21100
Trelleborg Viking A/S	Norway	311	75-02950
Walter Kidde Aerospace	USA	588	75-52500

241 10 40 45 Fuel systems, tracked armoured vehicle

Aerospace Syst. & Comp.	USA	547	75-02501
British Filters Ltd	UK	436	75-09030
Continental AG Werk Uniroyal	Germany	176	75-32090
FPT Industries Ltd	UK	452	75-24400
GEC-Marconi Avionics Ltd	UK	454	75-25320
Kinetics Ltd	Israel	220	75-11320
Napco International Inc.	USA	598	75-63600
Palbam Metal Works	Israel	221	75-17000
Secondo Mona SpA	Italy	237	75-15900
Stone India Ltd	India	205	75-01950

241 10 40 50 Hydraulic systems, tracked armoured vehicle

AAI Corp.	USA	545	75-00020
AIM Enterprises Inc.	USA	630	80-01350
APPH (Bolton) Ltd.	UK	431	75-03415
Commercial Intertech Corp.	USA	564	75-20050
Fairey Hydraulics	UK	450	75-22255
Forward Industries Ltd	UK	452	75-24200
FPT Industries Ltd	UK	452	75-24400
GEREP GmbH	Germany	168	75-10300
GKN Westland Technologies Ltd	UK	455	75-25690
Israel Aircraft Industries Ltd	Israel	219	75-10200
Kinetics Ltd	Israel	220	75-11320
Mannesmann Rexroth Ltd	UK	466	75-37100
Mannesmann Rexroth SA	France	139	75-25500
M.C. Aerospace Corp.	USA	594	75-59370
Menasco Aerospace Ltd	Canada	74	75-09600
Messier-Dowty Inc	Canada	74	75-09655
Messier-Dowty Limited	UK	467	75-38700
Moog Inc.	USA	597	75-62550
Napco International Inc.	USA	598	75-63600
Poclain Hydraulics SA	France	141	75-30402
Prva Petoletka Industrie	Yugoslavia	661	75-00610
Stewart & Stevenson Services Inc.	USA	618	75-84180
Vickers Inc.	USA	626	75-93820
Vickers Systems	Germany	176	75-32830
Vickers Systems Trinova SpA	Italy	239	75-27150

241 10 40 50 50 Hydraulic motors, turret slewing, tracked armoured vehicle

AAI Corp.	USA	545	75-00020

Abex GmbH Aerohydraul	Germany	*164*	75-00105
Commercial Intertech Corp.	USA	*564*	75-20050
Kinetics Ltd	Israel	*220*	75-11320
Mannesmann Rexroth	Germany	*172*	75-20005
Mannesmann Rexroth Ltd	UK	*466*	75-37100
Mannesmann Rexroth SA	France	*139*	75-25500
M.C. Aerospace Corp.	USA	*594*	75-59370
Messier-Dowty Limited	UK	*467*	75-38700
Napco International Inc.	USA	*598*	75-63600
Norca Machinery Company	USA	*599*	75-64800
Poclain Hydraulics SA	France	*141*	75-30402
Prva Petoletka Industrie	Yugoslavia	*661*	75-00610
Theodor Rapp GmbH Co	Germany	*173*	75-24600
Vickers Inc.	USA	*626*	75-93820
Vickers Systems	Germany	*176*	75-32830
Vickers Systems Division TRINOVA Ltd	UK	*491*	75-63150
Vickers Systems Trinova SpA.	Italy	*239*	75-27150

241 10 40 50 70 Hydraulic equipment, tracked armoured vehicle

Abex GmbH Aerohydraul	Germany	*164*	75-00105
Aeroquip Corp.	USA	*547*	75-02350
APPH (Bolton) Ltd.	UK	*431*	75-03415
Bradford Cylinders Ltd	UK	*435*	75-07850
Busak + Shamban Inc.	USA	*561*	75-16006
Commercial Intertech Corp.	USA	*564*	75-20050
Donaldson Co. Inc.	USA	*569*	75-27050
Dynex/Rivett Inc.	USA	*570*	75-28350
Fairey Hydraulics	UK	*450*	75-22255
Fairey Microfiltrex	UK	*450*	75-22300
Forward Industries Ltd	UK	*452*	75-24200
Gates Hydraulics Ltd	UK	*453*	75-25210
Giat Industries	France	*126*	55-00100
GKN Westland Technologies Ltd	UK	*455*	75-25690
Horstman Defence Systems Ltd.	UK	*459*	75-29920
Kinetics Ltd	Israel	*220*	75-11320
Mannesmann Rexroth Ltd	UK	*466*	75-37100
Mannesmann Rexroth SA	France	*139*	75-25500
M.C. Aerospace Corp.	USA	*594*	75-59370
Menasco Aerospace Ltd	Canada	*74*	75-09600
Messier-Dowty Limited	UK	*467*	75-38700
Moog Japan Ltd.	Japan	*253*	75-06100
Napco International Inc.	USA	*598*	75-63600
Norca Machinery Company	USA	*599*	75-64800
Parker Bertea Aerospace	USA	*602*	75-68900
Poclain Hydraulics SA	France	*141*	75-30402
Prva Petoletka Industrie	Yugoslavia	*661*	75-00610
Ralbo Inc.	USA	*607*	75-72690
SP aerospace and vehicle systems bv	**Netherlands**	***295***	**75-06860**
Theodor Rapp GmbH Co	Germany	*173*	75-24600
Vickers Inc.	USA	*626*	75-93820
Vickers Systems	Germany	*176*	75-32830
Vickers Systems Division TRINOVA Ltd	UK	*491*	75-63150
Vickers Systems Trinova SpA.	Italy	*239*	75-27150

241 10 40 50 70 50 Hydraulic pumps, tracked armoured vehicle

Abex GmbH Aerohydraul	Germany	*164*	75-00105
APPH (Bolton) Ltd.	UK	*431*	75-03415
Commercial Intertech Corp.	USA	*564*	75-20050
Dynex/Rivett Inc.	USA	*570*	75-28350
Fenner Fluid Power	UK	*450*	75-22550
Forward Industries Ltd	UK	*452*	75-24200
Giat Industries	France	*126*	55-00100
Howden Fluid Systems	USA	*582*	75-44400
Kinetics Ltd	Israel	*220*	75-11320
Lucas Aerospace FCS Paris	France	*139*	75-25390
Mannesmann Rexroth	Germany	*172*	75-20005
Mannesmann Rexroth Ltd	UK	*466*	75-37100
Mannesmann Rexroth SA	France	*139*	75-25500
M.C. Aerospace Corp.	USA	*594*	75-59370
Napco International Inc.	USA	*598*	75-63600
Norca Machinery Company	USA	*599*	75-64800
Poclain Hydraulics SA	France	*141*	75-30402
Prva Petoletka Industrie	Yugoslavia	*661*	75-00610
Urdan Industries (USA) Inc.	USA	*625*	75-92930
Vickers Inc.	USA	*626*	75-93830
Vickers Systems	Germany	*176*	75-32830
Vickers Systems Division TRINOVA Ltd	UK	*491*	75-63150
Vickers Systems Trinova SpA.	Italy	*239*	75-27150

241 10 40 60 Suspension systems, tracked armoured vehicle

AAI Corp.	USA	*545*	75-00020
Air-Log Ltd	UK	*429*	75-01800
Ashot Ashkelon Industries Ltd	Israel	*217*	75-02100
GEREP GmbH	Germany	*168*	75-10300
Horstman Defence Systems Ltd.	UK	*459*	75-29920
Kinetics Ltd	Israel	*220*	75-11320
Loc Performance Products Inc.	USA	*592*	75-55710
Lucas Aerospace FCS Paris	France	*139*	75-25390
Lucas Aerospace Power Transmission	USA	*592*	75-56265
M. Paz Logistics Ltd	Israel	*224*	80-01600
Messier-Dowty Limited	UK	*467*	75-38700
Napco International Inc.	USA	*598*	75-63600
Norca Machinery Company	USA	*599*	75-64800
Novatração Artefatos	Brazil	*57*	75-14780
Suspension & Parts Industries Ltd	Israel	*222*	75-21320
Urdan Industries Ltd	Israel	*223*	75-23210
Urdan Industries (USA) Inc.	USA	*625*	75-92930
Van Halteren Metaal BV	Netherlands	*296*	75-07870
Vickers Defence Systems	UK	*491*	75-63040

241 10 40 70 Tracked armoured vehicle equipment, other

Aero Tec Laboratories Inc.	USA	*546*	75-01117
Dytecna Ltd	UK	*501*	80-08900
Helio Mirror Co. Ltd.	UK	*457*	75-28900
IDE Technologies Ltd.	Israel	*219*	75-09170
Israel Military Industries Ltd	Israel	*220*	75-10300
Karipidis Bros. SA	Greece	*190*	75-01980
Kinetics Ltd	Israel	*220*	75-11320
KUKA Wehrtechnik GmbH	Germany	*171*	75-17200
M. Paz Logistics Ltd	Israel	*224*	80-01600
Napco International Inc.	USA	*598*	75-63600
Norca Machinery Company	USA	*599*	75-64800
Permali Gloucester Ltd	UK	*474*	75-44850
Tecalemit Systems Ltd	UK	*487*	75-58860
United Defense L.P.	USA	*624*	75-92050
Urdan Industries (USA) Inc.	USA	*625*	75-92930

241 10 40 70 30 Electrical equipment, tracked armoured vehicle

AAI Corp.	USA	*545*	75-00020
AlliedSignal Aerospace Canada	Canada	*68*	75-00470
Bradley Electronics Ltd	UK	*435*	75-07860
Commercial Hydraulics Kontak Ltd	UK	*441*	75-14415
ECE GmbH	Germany	*167*	75-07200
Electrothermal Engineering Ltd	UK	*448*	75-20400
EUROCONTROL SpA	Italy	*233*	75-08200
GEC-Marconi Avionics Ltd	UK	*454*	75-25320
Hebron & Medlock Engineering Ltd	UK	*457*	75-28860
Helio Mirror Co. Ltd.	UK	*457*	75-28900
M. Paz Logistics Ltd	Israel	*224*	80-01600
Napco International Inc.	USA	*598*	75-63600
Reunert Technology Systems Ltd.	S. Africa	*359*	75-18300
SELIN Sistemi Sarl	Italy	*238*	75-21750
Vickers Aerospace Components	UK	*491*	75-63020

241 10 40 70 30 30 Batteries, tracked armoured vehicle

Concorde Battery Corp.	USA	*566*	75-20845
Hawker Energy Products Ltd	UK	*457*	75-28490
Napco International Inc.	USA	*598*	75-63600
Saft	France	*142*	75-33920
Saft America Inc	Mexico	*282*	75-18000
Saft America Inc.	USA	*612*	75-77550
Saft Korea Co Ltd	Korea (ROK)	*266*	75-04300
Saft Nife AB	Sweden	*378*	75-09850
Saft Nife Argentina SA	Argentina	*4*	75-09000
Saft Nife AS	Norway	*310*	75-02400
Saft Nife BV	Netherlands	*294*	75-06160
Saft Nife Corp	Canada	*75*	75-12290
Saft Nife GmbH	Austria	*34*	75-02142
Saft Nife GmbH	Germany	*174*	75-25901

Saft Nife Iberica SL	Spain	367	75-05100
Saft Nife Ltd	UK	481	75-51470
Saft Nife M.E. Ltd	Cyprus	96	75-05000
Saft Nife Oy	Finland	118	75-01960
Saft Nife Power Systems Australia	Australia	20	75-18950
Saft Singapore Pte Ltd	Singapore	348	75-03400
Vulcan Batteries Ltd	Israel	223	75-23290
Yuasa Automotive Batteries (Europe)	UK	495	75-67560

241 10 40 70 50 Instruments, tracked armoured vehicle

Benchmark Physical Laboratories Ltd	UK	434	75-06195
Futronix Inc.	USA	577	75-36650
Humphrey Inc.	USA	583	75-45050
Napco International Inc.	USA	598	75-63600

241 10 80 Tracked armoured vehicle accessories

AAI Corp.	USA	545	75-00020
Åkers Krutbruk Protection AB	Sweden	375	75-01100
Giat Industries	France	126	55-00100
Horstman Defence Systems Ltd	UK	459	75-29920
Israel Military Industries Ltd	Israel	220	75-10300
Kinetics Ltd	Israel	220	75-11320
M. Paz Logistics Ltd	Israel	224	80-01600
Napco International Inc.	USA	598	75-63600
Ordex Inc.	USA	601	75-67350
Silent Partner Body Armor Inc.	USA	615	75-80930
Stephan Wood Products Inc.	USA	628	75-97280
Urdan Industries (USA) Inc.	USA	625	75-92930

241 10 80 10 Armour plates, add-on, tracked armoured vehicle

AAI Corp.	USA	545	75-00020
AEA Technology plc	UK	496	80-00480
Åkers Krutbruk Protection AB	Sweden	375	75-01100
Clouth Gummiwerke AG	Germany	166	75-04900
DTC International	USA	570	75-27620
M. Paz Logistics Ltd	Israel	224	80-01600
Meggitt Composites	UK	467	75-38345
Napco International Inc.	USA	598	75-63600
National Forge Co.	USA	598	75-63850
Palbam Metal Works	Israel	221	75-17000
Permali Gloucester Ltd	UK	474	75-44850
Plasan Sasa	Israel	221	75-17510
Q-Systems International Inc.	USA	606	75-71960
Silent Partner Body Armor Inc.	USA	615	75-80930
Urdan Industries Ltd	Israel	223	75-23210
Urdan Industries (USA) Inc.	USA	625	75-92930
Vibrachoc SA	France	148	75-46900
Vickers Defence Systems	UK	491	75-63040

241 10 80 11 Armour plates, add-on, active, tracked armoured vehicle

AAI Corp.	USA	545	75-00020
M. Paz Logistics Ltd	Israel	224	80-01600
Napco International Inc.	USA	598	75-63600
Norca Machinery Company	USA	599	75-64800
Permali Gloucester Ltd	UK	474	75-44850
Plasan Sasa	Israel	221	75-17510

241 10 80 60 Skirt plates, tracked armoured vehicle

AAI Corp.	USA	545	75-00020
M. Paz Logistics Ltd	Israel	224	80-01600
Napco International Inc.	USA	598	75-63600
Permali Gloucester Ltd	UK	474	75-44850
Renown Engineering Ltd	UK	478	75-49227
Urdan Industries Ltd	Israel	223	75-23210

241 10 80 62 Skirt plates, metal, tracked armoured vehicle

AAI Corp.	USA	545	75-00020
M. Paz Logistics Ltd	Israel	224	80-01600
Renown Engineering Ltd	UK	478	75-49227
Urdan Industries (USA) Inc.	USA	625	75-92930

241 10 80 64 Skirt plates, rubber, tracked armoured vehicle

AAI Corp.	USA	545	75-00020
Avon Spencer Moulton.	France	132	75-04750
Avon Technical Products	UK	432	75-05150
Hutchinson s.n.c.	France	137	75-21900
M. Paz Logistics Ltd	Israel	224	80-01600
Novatraçáo Artefatos	Brazil	57	75-14780

241 10 80 65 Swimming accessories, tracked armoured vehicle

AAI Corp.	USA	545	75-00020
Avon Technical Products	UK	432	75-05150
Crestbury Ltd	UK	443	75-16030
Novatraçáo Artefatos	Brazil	57	75-14780
Stephan Wood Products Inc.	USA	628	75-97280

241 10 80 90 Wading accessories, tracked armoured vehicle

Avon Technical Products	UK	432	75-05150

241 10 80 95 Winches, tracked armoured vehicle

EKA Limited	UK	448	75-19810
Horstman Defence Systems Ltd	UK	459	75-29920
Mannesmann Rexroth Ltd	UK	466	75-37100
Mannesmann Rexroth SA	France	139	75-25500
Norca Machinery Company	USA	599	75-64800
Plumett Ltd	UK	475	75-46340
Poclain Hydraulics SA	France	141	75-30402
Urdan Industries (USA) Inc.	USA	625	75-92930

241 10 90 Tracked armoured vehicle components

APPH (Bolton) Ltd.	UK	431	75-03415
Armored Vehicle's Spares Ltd.	Israel	217	75-01700
BTR Silvertown Ltd	UK	437	75-10830
Diehl GmbH & Co	Germany	167	75-06500
GEREP GmbH	Germany	168	75-10300
Giat Industries	France	126	55-00100
Hebron & Medlock Engineering Ltd	UK	457	75-28860
Hitchiner Manufacturing Co. Inc.	USA	582	75-43550
Honeywell Control Systems Ltd	UK	458	75-29730
Horstman Defence Systems Ltd	UK	459	75-29920
Israel Military Industries Ltd	Israel	220	75-10300
Kanakas Ltd	Greece	190	75-01970
Kinetics Ltd	Israel	220	75-11320
Lyon & Brandfield Ltd.	UK	465	75-36720
M. Paz Logistics Ltd	Israel	224	80-01600
MaK Systemgesellschaft mbH.	Germany	172	75-19500
Matren Ltd.	Greece	190	75-02010
Napco International Inc.	USA	598	75-63600
Norca Machinery Company	USA	599	75-64800
Specialty Bar Products	USA	616	75-83005
UFC Co.	USA	624	75-91382
United Defense L.P.	USA	624	75-92050
Urdan Industries (USA) Inc.	USA	625	75-92930
William Cook George Blair Ltd	UK	442	75-14920

241 10 90 10 05 Components, braking system, tracked armoured vehicle

AIM Enterprises Inc.	USA	630	80-01350
APPH (Bolton) Ltd.	UK	431	75-03415
Busak + Shamban Inc.	USA	561	75-16006
Champion Friction Co.	USA	563	75-17975
David Brown Defence Equipment Ltd	UK	437	75-10150
Kinetics Ltd	Israel	220	75-11320
Lucas Aerospace FCS Paris	France	139	75-25390
Mannesmann Rexroth Ltd	UK	466	75-37100
Napco International Inc.	USA	598	75-63600
Norca Machinery Company	USA	599	75-64800
Sielman SA	Greece	191	75-02360
Stringer Welding & Machine Inc.	USA	618	75-84435
UFC Co.	USA	624	75-91382
United Defense L.P.	USA	624	75-92050

Product-Supplier Link

Product–Supplier Link

241 10 90 10 10 Components, steering system, tracked armoured vehicle

AIM Enterprises Inc.	USA	*630*	80-01350
APPH (Bolton) Ltd.	UK	*431*	75-03415
Champion Friction Co.	USA	*563*	75-17975
Commercial Intertech Corp.	USA	*564*	75-20050
Dowty Seals Ltd	UK	*446*	75-18620
Giat Industries	France	*126*	55-00100
Kinetics Ltd	Israel	*220*	75-11320
Norca Machinery Company	USA	*599*	75-64800
Sielman SA	Greece	*191*	75-02360
Stringer Welding & Machine Inc.	USA	*618*	75-84435
United Defense L.P.	USA	*624*	75-92050

241 10 90 70 Components, running gear, tracked armoured vehicle

Air-Log Ltd	UK	*429*	75-01800
Ashot Ashkelon Industries Ltd	Israel	*217*	75-02100
Avon Spencer Moulton	France	*132*	75-04750
Avon Technical Products	UK	*432*	75-05150
BTR Silvertown Ltd	UK	*437*	75-10830
Busak + Shamban Inc.	USA	*561*	75-16006
Champion Friction Co.	USA	*563*	75-17975
Commercial Intertech Corp.	USA	*564*	75-20050
Dowty Seals Ltd	UK	*446*	75-18620
Horstman Defence Systems Ltd.	UK	*459*	75-29920
Kinetics Ltd	Israel	*220*	75-11320
Lucas Aerospace FCS Paris	France	*139*	75-25390
Napco International Inc.	USA	*598*	75-63600
Norca Machinery Company	USA	*599*	75-64800
Renown Engineering Ltd	UK	*478*	75-49227
Stringer Welding & Machine Inc.	USA	*618*	75-84435
United Defense L.P.	USA	*624*	75-92050
William Cook George Blair Ltd	UK	*442*	75-14920

241 10 90 76 Road wheels, tracked armoured vehicle

Air-Log Ltd	UK	*429*	75-01800
Avon Spencer Moulton	France	*132*	75-04750
Avon Technical Products	UK	*432*	75-05150
BTR Silvertown Ltd	UK	*437*	75-10830
Clouth Gummiwerke AG	Germany	*166*	75-04900
GKN Defence Ltd	**UK**	***454***	**75-25600**
Loc Performance Products Inc.	USA	*592*	75-55710
M. Paz Logistics Ltd	Israel	*224*	80-01600
Napco International Inc.	USA	*598*	75-63600
Norca Machinery Company	USA	*599*	75-64800
Novatraçáo Artefatos	Brazil	*57*	75-14780
Renown Engineering Ltd	UK	*478*	75-49227
Suspension & Parts Industries Ltd	Israel	*222*	75-21320
United Defense L.P.	USA	*624*	75-92050
Urdan Industries Ltd	Israel	*223*	75-23210
Urdan Industries (USA) Inc.	USA	*625*	75-92930
Van Halteren Metaal BV	Netherlands	*296*	75-07870
William Cook George Blair Ltd	UK	*442*	75-14920

241 10 90 77 Tracks, armoured vehicle

AAI Corp.	USA	*545*	75-00020
Air-Log Ltd	UK	*429*	75-01800
Australian Defence Industries Ltd	Australia	*9*	55-01000
Avon Technical Products	UK	*432*	75-05150
Diehl GmbH & Co	Germany	*167*	75-06500
Diehl Remscheid GmbH & Co	Germany	*167*	75-06505
M. Paz Logistics Ltd	Israel	*224*	80-01600
Merex Inc.	USA	*595*	75-60250
Napco International Inc.	USA	*598*	75-63600
Norca Machinery Company	USA	*599*	75-64800
Novatraçáo Artefatos	Brazil	*57*	75-14780
Soucy International Inc.	Canada	*76*	75-12980
Specialty Bar Products	USA	*616*	75-83005
Trelleborg Viking A/S	Norway	*311*	75-02950
United Defense L.P.	USA	*624*	75-92050
Urdan Industries Ltd	Israel	*223*	75-23210
Urdan Industries (USA) Inc.	USA	*625*	75-92930
Vorkauf SA	Spain	*367*	75-07000
William Cook George Blair Ltd	UK	*442*	75-14920

241 10 90 77 40 Track links, armoured vehicle

AAI Corp.	USA	*545*	75-00020
Air-Log Ltd	UK	*429*	75-01800
Avon Spencer Moulton	France	*132*	75-04750
CIC International Ltd	USA	*563*	75-18570
Diehl Remscheid GmbH & Co	Germany	*167*	75-06505
M. Paz Logistics Ltd	Israel	*224*	80-01600
Merex Inc.	USA	*595*	75-60250
Napco International Inc.	USA	*598*	75-63600
Norca Machinery Company	USA	*599*	75-64800
Novatraçáo Artefatos	Brazil	*57*	75-14780
Papadopoulos & Co Ltd	Greece	*191*	75-02195
Specialty Bar Products	USA	*616*	75-83005
Suspension & Parts Industries Ltd	Israel	*222*	75-21320
United Defense L.P.	USA	*624*	75-92050
Urdan Industries Ltd	Israel	*223*	75-23210
Urdan Industries (USA) Inc.	USA	*625*	75-92930
William Cook George Blair Ltd	UK	*442*	75-14920

241 10 90 77 50 Track pads, armoured vehicle

AAI Corp.	USA	*545*	75-00020
AIM Enterprises Inc.	USA	*630*	80-01350
Armored Vehicle's Spares Ltd.	Israel	*217*	75-01700
Avon Spencer Moulton	France	*132*	75-04750
Avon Technical Products	UK	*432*	75-05150
BTR Silvertown Ltd	UK	*437*	75-10830
Clouth Gummiwerke AG	Germany	*166*	75-04900
Karipidis Bros. SA	Greece	*190*	75-01980
K.Z.B. Technical Rubber Products Ltd	Israel	*220*	75-12190
M. Paz Logistics Ltd	Israel	*224*	80-01600
Merex Inc.	USA	*595*	75-60250
Napco International Inc.	USA	*598*	75-63600
Norca Machinery Company	USA	*599*	75-64800
Novatraçáo Artefatos	Brazil	*57*	75-14780
Sielman SA	Greece	*191*	75-02360
Specialty Bar Products	USA	*616*	75-83005
Suspension & Parts Industries Ltd	Israel	*222*	75-21320
Synapco Industries Ltd	New Zealand	*303*	75-01950
United Defense L.P.	USA	*624*	75-92050
Urdan Industries Ltd	Israel	*223*	75-23210
Urdan Industries (USA) Inc.	USA	*625*	75-92930
William Cook George Blair Ltd	UK	*442*	75-14920

241 10 90 77 55 Track pins, armoured vehicle

AAI Corp.	USA	*545*	75-00020
Avon Spencer Moulton	France	*132*	75-04750
Avon Technical Products	UK	*432*	75-05150
Diehl Remscheid GmbH & Co	Germany	*167*	75-06505
M. Paz Logistics Ltd	Israel	*224*	80-01600
Merex Inc.	USA	*595*	75-60250
Napco International Inc.	USA	*598*	75-63600
Norca Machinery Company	USA	*599*	75-64800
Novatraçáo Artefatos	Brazil	*57*	75-14780
Specialty Bar Products	USA	*616*	75-83005
Stringer Welding & Machine Inc.	USA	*618*	75-84435
Suspension & Parts Industries Ltd	Israel	*222*	75-21320
United Defense L.P.	USA	*624*	75-92050
Urdan Industries Ltd	Israel	*223*	75-23210
Urdan Industries (USA) Inc.	USA	*625*	75-92930
William Cook George Blair Ltd	UK	*442*	75-14920

241 10 90 77 60 Track rollers, armoured vehicle

AAI Corp.	USA	*545*	75-00020
Avon Spencer Moulton	France	*132*	75-04750
BTR Silvertown Ltd	UK	*437*	75-10830
Diehl Remscheid GmbH & Co	Germany	*167*	75-06505
M. Paz Logistics Ltd	Israel	*224*	80-01600
Merex Inc.	USA	*595*	75-60250
Napco International Inc.	USA	*598*	75-63600
Norca Machinery Company	USA	*599*	75-64800
Specialty Bar Products	USA	*616*	75-83005
Suspension & Parts Industries Ltd	Israel	*222*	75-21320
United Defense L.P.	USA	*624*	75-92050
Urdan Industries Ltd	Israel	*223*	75-23210
Urdan Industries (USA) Inc.	USA	*625*	75-92930
William Cook George Blair Ltd	UK	*442*	75-14920

241 10 90 77 70 Drive sprockets, armoured vehicle

AAI Corp.	USA	*545*	75-00020
Air-Log Ltd	UK	*429*	75-01800
BTR Silvertown Ltd	UK	*437*	75-10830
Loc Performance Products Inc.	USA	*592*	75-55710
M. Paz Logistics Ltd	Israel	*224*	80-01600
Napco International Inc.	USA	*598*	75-63600
Norca Machinery Company	USA	*599*	75-64800
United Defense L.P.	USA	*624*	75-92050
Urdan Industries Ltd	Israel	*223*	75-23210
Urdan Industries (USA) Inc.	USA	*625*	75-92930
Vickers Defence Systems	UK	*491*	75-63040
William Cook George Blair Ltd	UK	*442*	75-14920

241 30 Wheeled armoured vehicle systems & equipment

Alenia Difesa	**Italy**	***231***	**75-01165**
Allied Research Corp.	USA	*549*	75-04670
APPH (Bolton) Ltd.	UK	*431*	75-03415
Baxter Woodhouse & Taylor Ltd	UK	*433*	75-06030
Bernardini SA	Brazil	*55*	75-02120
Commercial Hydraulics Kontak Ltd	UK	*441*	75-14415
Cubic Defense Systems Inc.	USA	*567*	75-22900
Fraser-Volpe Corp.	USA	*576*	75-35950
Helio Mirror Co. Ltd	UK	*457*	75-28900
Howden Aircontrol Ltd	UK	*459*	75-30240
Hughes Danbury Optical Sys.	USA	*582*	75-44857
Israel Aircraft Industries Ltd	Israel	*220*	75-10200
Israel Military Industries Ltd	Israel	*220*	75-10300
J-Tec Associates Inc.	USA	*587*	75-51250
Labinal Aero Defense Systems Inc	USA	*638*	80-18100
Lucas Aerospace Power Transmission	USA	*592*	75-56265
M. Paz Logistics Ltd	Israel	*224*	80-01600
MacOptic SA	Greece	*190*	75-01999
Meggitt Composites	UK	*467*	75-38345
Microturbo SA	France	*140*	75-27000
MOWAG	Switzerland	*387*	75-08900
Napco International Inc.	USA	*598*	75-63600
Northrop Grumman Corporation	USA	*599*	75-65450
Patria Vehicles Oy	Finland	*118*	75-01710
SAGEM SA	**France**	***143***	**75-33950**
SFIM	**France**	***144***	**75-35080**
Silent Partner Body Armor Inc.	USA	*615*	75-80930
SP aerospace and vehicle systems bv	**Netherlands**	***295***	**75-06860**
Technofan SA	France	*147*	75-44300
Tecmotiv Corporation	Canada	*77*	75-13510

241 30 10 Wheeled armoured vehicle control systems

APPH (Bolton) Ltd.	UK	*431*	75-03415
Fairey Hydraulics	UK	*450*	75-22255
Hunting Engineering Ltd	UK	*459*	75-30700
Kinetics Ltd	Israel	*220*	75-11320
M. Paz Logistics Ltd	Israel	*224*	80-01600
Napco International Inc.	USA	*598*	75-63600
Norca Machinery Company	USA	*599*	75-64800
Protection Development Int'l Corp.	USA	*606*	75-71670
Smiths Industries Aerospace	UK	*485*	75-54700
Société d'Applications des Machines.	France	*143*	75-34300
SP aerospace and vehicle systems bv	**Netherlands**	***295***	**75-06860**

241 30 10 05 Braking systems & equipment, wheeled armoured vehicle

AIM Enterprises Inc.	USA	*630*	80-01350
APPH (Bolton) Ltd.	UK	*431*	75-03415
Champion Friction Co.	USA	*563*	75-17975
Kinetics Ltd	Israel	*220*	75-11320
Mannesmann Rexroth Ltd	UK	*466*	75-37100
Napco International Inc.	USA	*598*	75-63600
Prva Petoletka Industrie	Yugoslavia	*661*	75-00610
Sielman SA	Greece	*191*	75-02360
Tuthill Corporation	USA	*624*	75-91320
Wabco Automotive UK Ltd	UK	*492*	75-64000

241 30 10 10 Steering systems & equipment, wheeled armoured vehicle

APPH (Bolton) Ltd.	UK	*431*	75-03415
Champion Friction Co.	USA	*563*	75-17975
Giat Industries	France	*126*	55-00100
Napco International Inc.	USA	*598*	75-63600
Prva Petoletka Industrie	Yugoslavia	*661*	75-00610
Rolls-Royce Motor Cars Ltd	UK	*480*	75-50270
SP aerospace and vehicle systems bv	**Netherlands**	***295***	**75-06860**
Tuthill Corporation	USA	*624*	75-91320
ZF Friedrichshafen AG	Germany	*177*	75-35000
ZF Great Britain Ltd	UK	*495*	75-67720

241 30 20 Wheeled armoured vehicle mission systems

AAI Corp.	USA	*545*	75-00020
Bofors AB	Sweden	*375*	74-00020
Cubic Defense Systems Inc.	USA	*567*	75-22900
Fraser-Volpe Corp.	USA	*576*	75-35950
General Motors of Canada Ltd	Canada	*71*	75-05900
Horstman Defence Systems Ltd	UK	*459*	75-29920
KUKA Wehrtechnik GmbH	Germany	*171*	75-17200
Lucas Western Inc.	USA	*592*	75-56300
M. Paz Logistics Ltd	Israel	*224*	80-01600
Napco International Inc.	USA	*598*	75-63600
Norca Machinery Company	USA	*599*	75-64800
ROTZLER GmbH & Co	Germany	*174*	75-25800
SAGEM SA	**France**	***143***	**75-33950**
Sanders	USA	*612*	75-77850
SFIM	**France**	***144***	**75-35080**
Société d'Applications des Machines.	France	*143*	75-34300

241 30 22 Weapon-related systems, wheeled armoured vehicle

AAI Corp.	USA	*545*	75-00020
Alenia Difesa	**Italy**	***231***	**75-01165**
AV Technology LLC	USA	*556*	75-11405
Fairey Hydraulics	UK	*450*	75-22255
Fraser-Volpe Corp.	USA	*576*	75-35950
General Motors of Canada Ltd	Canada	*71*	75-05900
Horstman Defence Systems Ltd	UK	*459*	75-29920
Kinetics Ltd	Israel	*220*	75-11320
KUKA Wehrtechnik GmbH	Germany	*171*	75-17200
M. Paz Logistics Ltd	Israel	*224*	80-01600
Napco International Inc.	USA	*598*	75-63600
Northrop Grumman Corporation	USA	*599*	75-65450
Protection Development Int'l Corp.	USA	*606*	75-71670
United Defense L.P.	USA	*624*	75-92050

241 30 22 10 Automatic gun loading systems, wheeled armoured vehicle

AAI Corp.	USA	*545*	75-00020
Fairey Hydraulics	UK	*450*	75-22255
Giat Industries	France	*126*	55-00100
Israel Aircraft Industries Ltd	Israel	*219*	75-10200
KUKA Wehrtechnik GmbH	Germany	*171*	75-17200
Lucas Aerospace FCS Paris	France	*139*	75-25390
M. Paz Logistics Ltd	Israel	*224*	80-01600
United Defense L.P.	USA	*624*	75-92050
Wegmann & Co GmbH	Germany	*177*	75-33700

241 30 22 30 Fire control systems, wheeled armoured vehicle (see code 415.40.20)

AAI Corp.	USA	*545*	75-00020
AV Technology LLC	USA	*556*	75-11405
Avimo Ltd	UK	*432*	75-04950
Computing Devices Canada	Canada	*70*	75-03600
Delft Sensor Systems	**Belgium**	***46***	**75-02500**
Delft Sensor Systems	Netherlands	*292*	75-01590
Fabbrica Italiana Apparecchiature	Italy	*234*	75-08550
Fraser-Volpe Corp.	USA	*576*	75-35950
GEC-Marconi Avionics Inc.	USA	*577*	75-37413
Kearfott Guidance & Navigation Corp.	USA	*588*	75-51760
Kollsman	USA	*589*	75-53150
M. Paz Logistics Ltd	Israel	*224*	80-01600
Moog Controls Ltd	UK	*469*	75-40670

Napco International Inc.	USA	598	75-63600
Protection Development Int'l Corp.	USA	606	75-71670
Przemyslowy Instytut Telekomunikacji	**Poland**	**325**	**75-02650**
SFIM	**France**	**144**	**75-35080**
Warren-Knight Instrument Co.	USA	627	75-95150

241 30 22 70 Mission-related equipment, wheeled armoured vehicle

AAI Corp.	USA	545	75-00020
Analog Modules Inc.	USA	553	75-07140
A.P. Metalcraft Ltd	UK	430	75-03410
AV Technology LLC	USA	556	75-11405
Hunting Engineering Ltd	UK	459	75-30700
Israel Military Industries Ltd	Israel	220	75-10300
KUKA Wehrtechnik GmbH	Germany	171	75-17200
M. Paz Logistics Ltd	Israel	224	80-01600
Napco International Inc.	USA	598	75-63600
Norca Machinery Company	USA	599	75-64800
Northrop Grumman Corporation	USA	599	75-65450
Pilkington Optronics Ltd	UK	426	74-10000
Precision Sensors	USA	605	75-71360
SAGEM SA	**France**	**143**	**75-33950**
SFIM	**France**	**144**	**75-35080**
SP aerospace and vehicle systems bv	**Netherlands**	**295**	**75-06860**
Wegmann & Co GmbH	Germany	177	75-33700
Wegmann USA Inc.	USA	627	75-95543

241 30 22 70 10 Ammunition racks & containers, wheeled armoured vehicle

AAI Corp.	USA	545	75-00020
A.P. Metalcraft Ltd	UK	430	75-03410
AV Technology LLC	USA	556	75-11405
Israel Military Industries Ltd	Israel	220	75-10300
M. Paz Logistics Ltd	Israel	224	80-01600
Napco International Inc.	USA	598	75-63600
Norca Machinery Company	USA	599	75-64800
Ordex Inc.	USA	601	75-67350
Stringer Welding & Machine Inc.	USA	618	75-84435
United Defense L.P.	USA	624	75-92050
Wegmann & Co GmbH	Germany	177	75-33700

241 30 22 70 20 Cupolas, wheeled armoured vehicle

AAI Corp.	USA	545	75-00020
Alenia Difesa	**Italy**	**231**	**75-01165**
AV Technology LLC	USA	556	75-11405
Giat Industries	France	126	55-00100
Hebron & Medlock Engineering Ltd	UK	457	75-28860
Helio Mirror Co. Ltd	UK	457	75-28900
INIMEX	**Belgium**	**46**	**75-04650**
KUKA Wehrtechnik GmbH	Germany	171	75-17200
M. Paz Logistics Ltd	Israel	224	80-01600
Napco International Inc.	USA	598	75-63600
Norca Machinery Company	USA	599	75-64800
SP aerospace and vehicle systems bv	**Netherlands**	**295**	**75-06860**
United Defense L.P.	USA	624	75-92050
Urdan Industries Ltd	Israel	223	75-23210
Urdan Industries (USA) Inc.	USA	625	75-92930

241 30 22 70 22 Cranes, hydraulic, weapon / ammunition handling, wheeled armoured vehicle

Auto Crane Co.	USA	556	75-10360
Grove Europe Ltd	**UK**	**456**	**75-27030**
Kinetics Ltd	Israel	220	75-11320
M. Paz Logistics Ltd	Israel	224	80-01600
Napco International Inc.	USA	598	75-63600
Norca Machinery Company	USA	599	75-64800
United Defense L.P.	USA	624	75-92050
Vickers Inc.	USA	626	75-93820
Vickers Systems	Germany	176	75-32830
Vickers Systems Trinova SpA	Italy	239	75-27150

241 30 22 70 30 Firing ports, wheeled armoured vehicle

Étienne Lacroix	France	138	75-24050
Helio Mirror Co. Ltd	UK	457	75-28900
M. Paz Logistics Ltd	Israel	224	80-01600
Napco International Inc.	USA	598	75-63600
Wegmann & Co GmbH	Germany	177	75-33700

241 30 22 70 55 Ring mounts, wheeled armoured vehicle

Hägglunds Vehicle AB	Sweden	377	75-05800
Helio Mirror Co. Ltd	UK	457	75-28900
KUKA Wehrtechnik GmbH	Germany	171	75-17200
Napco International Inc.	USA	598	75-63600
Norca Machinery Company	USA	599	75-64800
Stringer Welding & Machine Inc.	USA	618	75-84435
Theodor Rapp GmbH Co	Germany	173	75-24600

241 30 22 70 70 Turrets, wheeled armoured vehicle

AAI Corp.	USA	545	75-00020
Alenia Difesa	**Italy**	**231**	**75-01165**
AV Technology LLC	USA	556	75-11405
Delco Systems Operations	USA	568	75-25360
Euromissile GIE	France	130	74-00535
General Motors of Canada Ltd	Canada	71	75-05900
Hägglunds Vehicle AB	Sweden	377	75-05800
Hebron & Medlock Engineering Ltd	UK	457	75-28860
Helio Mirror Co. Ltd	UK	457	75-28900
INIMEX	**Belgium**	**46**	**75-04650**
KUKA Wehrtechnik GmbH	Germany	171	75-17200
M. Paz Logistics Ltd	Israel	224	80-01600
MaK Systemgesellschaft mbH	Germany	172	75-19500
Napco International Inc.	USA	598	75-63600
Norca Machinery Company	USA	599	75-64800
SP aerospace and vehicle systems bv	**Netherlands**	**295**	**75-06860**
Textron Marine & Land Systems	USA	622	75-88660
United Defense L.P.	USA	624	75-92050
Urdan Industries Ltd	Israel	223	75-23210
Wegmann & Co GmbH	Germany	177	75-33700

241 30 25 Rangefinder systems, wheeled armoured vehicle

Delft Sensor Systems	**Belgium**	**46**	**75-02500**
Delft Sensor Systems	Netherlands	292	75-01590
GEC-Marconi Avionics Inc.	USA	577	75-37413
GEC-Marconi Avionics Ltd	UK	454	75-25320
M. Paz Logistics Ltd	Israel	224	80-01600
Nanoquest Defence Products Ltd	UK	470	75-41890
Napco International Inc.	USA	598	75-63600
Norca Machinery Company	USA	599	75-64800
OM=	USA	591	75-55400
Pilkington Optronics Ltd	UK	426	74-10000
Zeiss-Eltro Optronic GmbH	Germany	177	75-34700

241 30 25 40 Rangefinder systems, wheeled armoured vehicle, laser

Analog Modules Inc.	USA	553	75-07140
Avimo Ltd	UK	432	75-04950
Bodenseewerk Gerätetechnik GmbH	Germany	165	75-03000
Delft Sensor Systems	**Belgium**	**46**	**75-02500**
Delft Sensor Systems	Netherlands	292	75-01590
Eloptro	S. Africa	358	75-05680
Fabbrica Italiana Apparecchiature	Italy	234	75-08550
GEC-Marconi Avionics Inc.	USA	577	75-37413
GEC-Marconi Avionics Ltd	UK	454	75-25320
Instro Precision Ltd	UK	461	75-31630
Kollsman	USA	589	75-53150
M. Paz Logistics Ltd	Israel	224	80-01600
Nanoquest Defence Products Ltd	UK	470	75-41890
Napco International Inc.	USA	598	75-63600
Norca Machinery Company	USA	599	75-64800
Pilkington Optronics Ltd	UK	426	74-10000
Zeiss-Eltro Optronic GmbH	Germany	177	75-34700

241 30 27 50 Searchlights, wheeled armoured vehicle

Havis-Shields Equipment Corp.	USA	581	75-42700
M. Paz Logistics Ltd	Israel	224	80-01600
Napco International Inc.	USA	598	75-63600
Norca Machinery Company	USA	599	75-64800

241 30 27 60 Sensors, wheeled armoured vehicle

GEC-Marconi Avionics Inc.	USA	577	75-37413
J-Tec Associates Inc.	USA	587	75-51250
Kollsman	USA	589	75-53150
Napco International Inc.	USA	598	75-63600
Norca Machinery Company	USA	599	75-64800
Northrop Grumman Corporation	USA	599	75-65450
Pilkington Optronics Ltd	UK	426	74-10000
Precision Sensors	USA	605	75-71360
Zeiss-Eltro Optronic GmbH	Germany	177	75-34700

241 30 27 60 40 Sensors, laser, wheeled armoured vehicle

Avimo Ltd	UK	432	75-04950
Elettronica (UK) Ltd	UK	448	75-20475
Fabbrica Italiana Apparecchiature	Italy	234	75-08550
GEC-Marconi Avionics Inc.	USA	577	75-37413
GEC-Marconi Avionics Ltd	UK	454	75-25320
Hughes Danbury Optical Sys.	USA	582	75-44857
Kollsman	USA	589	75-53150
Napco International Inc.	USA	598	75-63600
Norca Machinery Company	USA	599	75-64800
Pilkington Optronics Ltd	UK	426	74-10000
Racal Communications Inc.	USA	606	75-72380
SRI International	USA	643	80-29600
Zeiss-Eltro Optronic GmbH	Germany	177	75-34700

241 30 27 60 90 Sensors, wind, wheeled armoured vehicle

Brookes & Gatehouse Ltd	UK	437	75-10100
GEC-Marconi Avionics Inc.	USA	577	75-37413
J-Tec Associates Inc.	USA	587	75-51250
Kollsman	USA	589	75-53150
Napco International Inc.	USA	598	75-63600

241 30 27 65 Sights, wheeled armoured vehicle

Arab International Optronics Co.	Egypt	109	74-00500
Avimo Ltd	UK	432	75-04950
Delco Systems Operations	USA	568	75-25360
EUROCONTROL SpA	Italy	233	75-08200
Fraser-Volpe Corp.	USA	576	75-35950
GEC-Marconi Avionics Inc.	USA	577	75-37413
Helio Mirror Co. Ltd	UK	457	75-28900
Kollsman	USA	589	75-53150
M. Paz Logistics Ltd	Israel	224	80-01600
Nanoquest Defence Products Ltd	UK	470	75-41890
Napco International Inc.	USA	598	75-63600
Norca Machinery Company	USA	599	75-64800
SAGEM SA	**France**	**143**	**75-33950**
SFIM	**France**	**144**	**75-35080**
SFIM ODS	France	144	75-35085
W.F. Stanley & Co. Ltd	UK	486	75-56350
Zeiss-Eltro Optronic GmbH	Germany	177	75-34700

241 30 27 65 50 Sights, periscopic, wheeled armoured vehicle

Arab International Optronics Co.	Egypt	109	74-00500
Avimo Ltd	UK	432	75-04950
CIC International Ltd	USA	563	75-18570
Delft Sensor Systems	**Belgium**	**46**	**75-02500**
Delft Sensor Systems	Netherlands	292	75-01590
Fraser-Volpe Corp.	USA	576	75-35950
GEC-Marconi Avionics Inc.	USA	577	75-37413
Helio Mirror Co. Ltd	UK	457	75-28900
M. Paz Logistics Ltd	Israel	224	80-01600
Microtecnica-Torino SpA	Italy	237	75-15500
Nanoquest Defence Products Ltd	UK	470	75-41890
Napco International Inc.	USA	598	75-63600
Norca Machinery Company	USA	599	75-64800
Pilkington Optronics Ltd	UK	426	74-10000
SAGEM SA	**France**	**143**	**75-33950**
SFIM	**France**	**144**	**75-35080**
SFIM ODS	France	144	75-35085
Warren-Knight Instrument Co.	USA	627	75-95150
W.F. Stanley & Co. Ltd	UK	486	75-56350
Zeiss-Eltro Optronic GmbH	Germany	177	75-34700

241 30 27 65 60 Sights, telescopic, wheeled armoured vehicle

Arab International Optronics Co.	Egypt	109	74-00500
Avimo Ltd	UK	432	75-04950
Fraser-Volpe Corp.	USA	576	75-35950
GEC-Marconi Avionics Inc.	USA	577	75-37413
M. Paz Logistics Ltd	Israel	224	80-01600
Nanoquest Defence Products Ltd	UK	470	75-41890
Napco International Inc.	USA	598	75-63600
Norca Machinery Company	USA	599	75-64800
Pilkington Optronics Ltd	UK	426	74-10000
SFIM ODS	France	144	75-35085
W.F. Stanley & Co. Ltd	UK	486	75-56350
Zeiss-Eltro Optronic GmbH	Germany	177	75-34700

241 30 27 80 Smoke pots, wheeled armoured vehicle

Engineered Air Systems Inc.	USA	573	75-31560
Étienne Lacroix	France	138	75-24050
Haley & Weller Ltd	UK	456	75-27400
Helio Mirror Co. Ltd	UK	457	75-28900
Napco International Inc.	USA	598	75-63600
Norca Machinery Company	USA	599	75-64800
Wegmann & Co GmbH	Germany	177	75-33700

241 30 27 90 Vision blocks, wheeled armoured vehicle

Eloptro	S. Africa	358	75-05680
Helio Mirror Co. Ltd	UK	457	75-28900
M. Paz Logistics Ltd	Israel	224	80-01600
MacOptic SA	Greece	190	75-01999
Nanoquest Defence Products Ltd	UK	470	75-41890
Napco International Inc.	USA	598	75-63600
Norca Machinery Company	USA	599	75-64800
SAGEM SA	**France**	**143**	**75-33950**
SFIM	**France**	**144**	**75-35080**
Sielman SA	Greece	191	75-02360

241 30 30 Wheeled armoured vehicle crew-related systems

Aerospace Syst. & Comp.	USA	547	75-02501
Airscrew Howden Ltd	UK	429	75-02050
AlliedSignal Aerospace Canada	Canada	68	75-00470
ECE GmbH	Germany	167	75-07200
Giat Industries	France	126	55-00100
Helio Mirror Co. Ltd	UK	457	75-28900
Kinetics Ltd	Israel	220	75-11320
M. Paz Logistics Ltd	Israel	224	80-01600
Napco International Inc.	USA	598	75-63600
Norca Machinery Company	USA	599	75-64800
Stephan Wood Products Inc.	USA	628	75-97280

241 30 30 30 Fume extraction units, crew compartment, wheeled armoured vehicle

Aerospace Syst. & Comp.	USA	547	75-02501
Airscrew Howden Ltd	UK	429	75-02050
British Filters Ltd	UK	436	75-09030
EUROCONTROL SpA	Italy	233	75-08200
Howden Aircontrol Ltd	UK	459	75-30240
Napco International Inc.	USA	598	75-63600

241 30 30 35 Heaters, crew compartment, wheeled armoured vehicle

Aerospace Syst. & Comp.	USA	547	75-02501
Airscrew Howden Ltd	UK	429	75-02050
A/S Dantherm	**Denmark**	**102**	**75-04100**
Electrothermal Engineering Ltd	UK	448	75-20400
Gallay Ltd	UK	453	75-25100
Global Thermoelectric Inc.	Canada	71	75-06200
Howden Aircontrol Ltd	UK	459	75-30240
Hunter Manufacturing	USA	583	75-45150
Kinetics Ltd	Israel	220	75-11320
Microturbo Ltd	UK	468	75-39160
Napco International Inc.	USA	598	75-63600
Norca Machinery Company	USA	599	75-64800

Parker Bertea Aerospace	USA	*603*	75-68900
Teledyne Brown Engineering	USA	*620*	75-87508

241 30 30 40 Internal communication systems, wheeled armoured vehicle

Airscrew Howden Ltd	UK	*429*	75-02050
Cincinnati Electronics Corp.	USA	*564*	75-18640
David Clark Co. Inc.	USA	*564*	75-18850
Davies Industrial Communications Ltd	UK	*444*	75-17010
Hunting Engineering Ltd	UK	*459*	75-30700
LARIMART SpA	Italy	*236*	75-13500
Napco International Inc.	USA	*598*	75-63600
White Products BV	Netherlands	*296*	75-09100

241 30 30 50 NBC filtration systems, wheeled armoured vehicle

AAF International BV	Netherlands	*291*	75-00005
Airscrew Howden Ltd	UK	*429*	75-02050
AlliedSignal Aerospace Canada	Canada	*68*	75-00470
British Filters Ltd	UK	*436*	75-09030
Combimac BV	Netherlands	*292*	75-01200
Engineered Air Systems Inc.	USA	*573*	75-31560
Fairey Microfiltrex	UK	*450*	75-22300
Giat Industries	France	*126*	55-00100
GKN Westland Technologies Ltd	UK	*455*	75-25690
Graseby Dynamics Ltd	UK	*455*	75-26590
Howden Aircontrol Ltd	UK	*459*	75-30240
Hunting Engineering Ltd	UK	*459*	75-30700
Kinetics Ltd	Israel	*220*	75-11320
Lucas Aerospace FCS Paris	France	*139*	75-25390
Napco International Inc.	USA	*598*	75-63600
Norca Machinery Company	USA	*599*	75-64800
Normalair-Garrett Ltd	UK	*425*	74-07000
Sèkur SpA	Italy	*238*	75-21650
Stork Bronswerk BV	Netherlands	*295*	75-07200

241 30 30 55 Air conditioning systems, wheeled armoured vehicle

Aerospace Syst. & Comp.	USA	*547*	75-02501
Airflow Co.	USA	*548*	75-03650
Airscrew Howden Ltd	UK	*429*	75-02050
AlliedSignal Aerospace Canada	Canada	*68*	75-00470
Gallay Ltd	UK	*453*	75-25100
GKN Westland Technologies Ltd	UK	*455*	75-25690
Howden Aircontrol Ltd	UK	*459*	75-30240
Howden Fluid Systems	USA	*582*	75-44400
Hunting Engineering Ltd	UK	*459*	75-30700
Kinetics Ltd	Israel	*220*	75-11320
Microtecnica-Torino SpA	Italy	*237*	75-15500
Microturbo Ltd	UK	*468*	75-39160
Napco International Inc.	USA	*598*	75-63600
Normalair-Garrett Ltd	UK	*425*	74-07000
Parker Bertea Aerospace	USA	*603*	75-68900
Stork Bronswerk BV	Netherlands	*295*	75-07200

241 30 30 70 Wheeled armoured vehicle crew-related equipment

A.P. Metalcraft Ltd	UK	*430*	75-03410
Hebron & Medlock Engineering Ltd	UK	*457*	75-28860
Kinetics Ltd	Israel	*220*	75-11320
M. Paz Logistics Ltd	Israel	*224*	80-01600
Napco International Inc.	USA	*598*	75-63600
Norca Machinery Company	USA	*599*	75-64800

241 30 30 70 60 Seats, crew, wheeled armoured vehicle

General Motors of Canada Ltd	Canada	*71*	75-05900
Hebron & Medlock Engineering Ltd	UK	*457*	75-28860
M. Paz Logistics Ltd	Israel	*224*	80-01600
Napco International Inc.	USA	*598*	75-63600
Norca Machinery Company	USA	*599*	75-64800
Plastic Fabricating Co. Inc.	USA	*604*	75-70335
Stephan Wood Products Inc.	USA	*628*	75-97280

241 30 40 Wheeled armoured vehicle systems, other

AlliedSignal Aerospace Canada	Canada	*68*	75-00470
APPH (Bolton) Ltd.	UK	*431*	75-03415
ECE GmbH	Germany	*167*	75-07200
Eldec Corp.	USA	*572*	75-29660
Gates Hydraulics Ltd	UK	*453*	75-25210
Horstman Defence Systems Ltd.	UK	*459*	75-29920
Howden Sirocco	France	*137*	75-21700
Lucas Aerospace Power Transmission	USA	*592*	75-56265
M. Paz Logistics Ltd	Israel	*224*	80-01600
Midcon Cables Co.	USA	*596*	75-61300
Napco International Inc.	USA	*598*	75-63600
Norca Machinery Company	USA	*599*	75-64800
Patria Vehicles Oy	Finland	*118*	75-01710
SELIN Sistemi Sarl	Italy	*238*	75-21750

241 30 40 30 Electrical systems & equipment, wheeled armoured vehicle

AIM Enterprises Inc.	USA	*630*	80-01350
AlliedSignal Aerospace Canada	Canada	*68*	75-00470
Bradley Electronics Ltd	UK	*435*	75-07860
Commercial Hydraulics Kontak Ltd	UK	*441*	75-14415
ECE GmbH	Germany	*167*	75-07200
Eldec Corp.	USA	*572*	75-29660
Electrothermal Engineering Ltd	UK	*448*	75-20400
EUROCONTROL SpA	Italy	*233*	75-08200
GSI Corporation	USA	*580*	75-40715
Hebron & Medlock Engineering Ltd	UK	*457*	75-28860
Hunting Engineering Ltd	UK	*459*	75-30700
M. Paz Logistics Ltd	Israel	*224*	80-01600
Napco International Inc.	USA	*598*	75-63600
Newton Derby Ltd	UK	*470*	75-42520
Norca Machinery Company	USA	*599*	75-64800
SELIN Sistemi Sarl	Italy	*238*	75-21750
Sielman SA	Greece	*191*	75-02360
Stewart & Stevenson Services Inc.	USA	*618*	75-84180
Teledyne Continental Motors	USA	*620*	75-87506
Vickers Aerospace Components	UK	*491*	75-63020
Walter Dittel GmbH	Germany	*167*	75-06510
White Products BV	Netherlands	*296*	75-09100

241 30 40 30 40 Lighting systems, wheeled armoured vehicle

Bruce Industries Inc.	USA	*561*	75-15360
CRL Technologies	Canada	*70*	75-03760
ECE GmbH	Germany	*167*	75-07200
Hunting Engineering Ltd	UK	*459*	75-30700
Luminescent Systems Inc.	USA	*593*	75-56480
M. Paz Logistics Ltd	Israel	*224*	80-01600
Napco International Inc.	USA	*598*	75-63600
Norca Machinery Company	USA	*599*	75-64800
Westair Ltd	UK	*493*	75-65410

241 30 40 30 40 40 Lighting, interior, wheeled armoured vehicle

Bruce Industries Inc.	USA	*561*	75-15360
CRL Technologies	Canada	*70*	75-03760
Hunting Engineering Ltd	UK	*459*	75-30700
Luminescent Systems Inc.	USA	*593*	75-56480
M. Paz Logistics Ltd	Israel	*224*	80-01600
Napco International Inc.	USA	*598*	75-63600
Norca Machinery Company	USA	*599*	75-64800
Saunders-Roe Ltd.	UK	*482*	75-51850
Soderberg Manufacturing Co. Inc.	USA	*616*	75-82200
Westair Ltd	UK	*493*	75-65410

241 30 40 30 50 Slip rings, wheeled armoured vehicle

AAI Corp.	USA	*545*	75-00020
EUROCONTROL SpA	Italy	*233*	75-08200
IDM Electronics Ltd	UK	*460*	75-31100
Litton Industries Inc.	USA	*590*	75-55400
Mécanex SA	Switzerland	*386*	75-08300
Napco International Inc.	USA	*598*	75-63600
Norca Machinery Company	USA	*599*	75-64800
Vorkauf SA	Spain	*367*	75-07000
Wegmann & Co GmbH	Germany	*177*	75-33700

241 30 40 30 50 70 Slip rings, wheeled armoured vehicle turret

AAI Corp.	USA	*545*	75-00020
EUROCONTROL SpA	Italy	*233*	75-08200
IDM Electronics Ltd	UK	*460*	75-31100
Litton Industries Inc.	USA	*590*	75-55400
Napco International Inc.	USA	*598*	75-63600
Norca Machinery Company	USA	*599*	75-64800
Vorkauf SA	Spain	*367*	75-07000
Wegmann & Co GmbH	Germany	*177*	75-33700

241 30 40 30 60 Wiring harnesses, electrical system, wheeled armoured vehicle

AV Technology LLC	USA	*556*	75-11405
Electrothermal Engineering Ltd	UK	*448*	75-20400
Hebron & Medlock Engineering Ltd	UK	*457*	75-28860
Hunting Engineering Ltd	UK	*459*	75-30700
Icore International Inc.	USA	*584*	75-45890
Icore International Ltd	UK	*460*	75-31080
Ionix High Performance Wiring	UK	*461*	75-32007
Israel Aircraft Industries Ltd	Israel	*219*	75-10200
Kembrey Wiring Systems Ltd	UK	*463*	75-33525
Labinal	France	*138*	75-23900
M. Paz Logistics Ltd	Israel	*224*	80-01600
Midcon Cables Co.	USA	*596*	75-61300
Napco International Inc.	USA	*598*	75-63600
Norca Machinery Company	USA	*599*	75-64800
Sielman SA	Greece	*191*	75-02360
Theodor Rapp GmbH Co	Germany	*173*	75-24600
Wegmann & Co GmbH	Germany	*177*	75-33700
Westwire Harnessing Limited	UK	*493*	75-65700
White Products BV	Netherlands	*296*	75-09100

241 30 40 40 Fire extinguishing systems, wheeled armoured vehicle

Deugra Ges. für Brandschutzsysteme	Germany	*167*	75-05900
Hughes Electronics Corp.	USA	*583*	75-44865
Kidde-Graviner Ltd	UK	*463*	75-33880
M. Paz Logistics Ltd	Israel	*224*	80-01600
Napco International Inc.	USA	*598*	75-63600
Norca Machinery Company	USA	*599*	75-64800
Spectrex Inc.	USA	*617*	75-83095
Spectronix Ltd	Israel	*222*	75-21100
Walter Kidde Aerospace	USA	*588*	75-52500

241 30 40 45 Fuel systems, wheeled armoured vehicle

Aerospace Syst. & Comp.	USA	*547*	75-02501
Continental AG Werk Uniroyal	Germany	*176*	75-32090
FPT Industries Ltd	UK	*452*	75-24400
GUD Manufacturing Co. Pty Ltd	Australia	*17*	75-07885
Kinetics Ltd	Israel	*220*	75-11320
M. Paz Logistics Ltd	Israel	*224*	80-01600
Napco International Inc.	USA	*598*	75-63600
Norca Machinery Company	USA	*599*	75-64800
Palbam Metal Works	Israel	*221*	75-17000
Secondo Mona SpA	Italy	*237*	75-15900

241 30 40 50 Hydraulic systems, wheeled armoured vehicle

AIM Enterprises Inc.	USA	*630*	80-01350
APPH (Bolton) Ltd.	UK	*431*	75-03415
AV Technology LLC	USA	*556*	75-11405
British Filters Ltd	UK	*436*	75-09030
Commercial Intertech Corp.	USA	*564*	75-20050
Fairey Hydraulics	UK	*450*	75-22255
Forward Industries Ltd	UK	*452*	75-24200
GEREP GmbH	Germany	*168*	75-10300
Kinetics Ltd	Israel	*220*	75-11320
M. Paz Logistics Ltd	Israel	*224*	80-01600
Mannesmann Rexroth Ltd	UK	*466*	75-37100
Mannesmann Rexroth SA	France	*139*	75-25500
M.C. Aerospace Corp.	USA	*594*	75-59370
Menasco Aerospace Ltd	Canada	*74*	75-09600
Messier-Dowty Limited	UK	*467*	75-38700
Napco International Inc.	USA	*598*	75-63600
Norca Machinery Company	USA	*599*	75-64800
Poclain Hydraulics SA	France	*141*	75-30402
Prva Petoletka Industrie	Yugoslavia	*661*	75-00610
Stewart & Stevenson Services Inc.	USA	*618*	75-84180
Vickers Inc.	USA	*626*	75-93820
Vickers Systems	Germany	*176*	75-32830
Vickers Systems Division TRINOVA Ltd	UK	*491*	75-63150
Vickers Systems Trinova SpA	Italy	*239*	75-27150

241 30 40 50 50 Hydraulic motors, turret slewing, wheeled armoured vehicle

Abex GmbH Aerohydraul	Germany	*164*	75-00105
AIM Enterprises Inc.	USA	*630*	80-01350
Commercial Intertech Corp.	USA	*564*	75-20050
Kinetics Ltd	Israel	*220*	75-11320
Lucas Aerospace FCS Paris	France	*139*	75-25390
Mannesmann Rexroth Ltd	UK	*466*	75-37100
Mannesmann Rexroth SA	France	*139*	75-25500
M.C. Aerospace Corp.	USA	*594*	75-59370
Messier-Dowty Limited	UK	*467*	75-38700
Napco International Inc.	USA	*598*	75-63600
Norca Machinery Company	USA	*599*	75-64800
Poclain Hydraulics SA	France	*141*	75-30402
Prva Petoletka Industrie	Yugoslavia	*661*	75-00610
Vickers Inc.	USA	*626*	75-93820
Vickers Systems	Germany	*176*	75-32830
Vickers Systems Division TRINOVA Ltd	UK	*491*	75-63150
Vickers Systems Trinova SpA	Italy	*239*	75-27150

241 30 40 50 70 Hydraulic equipment, wheeled armoured vehicle

Abex GmbH Aerohydraul	Germany	*164*	75-00105
Aeroquip Corp.	USA	*547*	75-02350
AIM Enterprises Inc.	USA	*630*	80-01350
APPH (Bolton) Ltd.	UK	*431*	75-03415
Bradford Cylinders Ltd	UK	*435*	75-07850
Commercial Intertech Corp.	USA	*564*	75-20050
Donaldson Co. Inc.	USA	*569*	75-27050
Dynex/Rivett Inc.	USA	*570*	75-28350
Fairey Hydraulics	UK	*450*	75-22255
Fairey Microfiltrex	UK	*450*	75-22300
Forward Industries Ltd	UK	*452*	75-24200
Gates Hydraulics Ltd	UK	*453*	75-25210
GEREP GmbH	Germany	*168*	75-10300
Giat Industries	France	*126*	55-00100
Karipidis Bros. SA	Greece	*190*	75-01980
Kinetics Ltd	Israel	*220*	75-11320
M. Paz Logistics Ltd	Israel	*224*	80-01600
Mannesmann Rexroth Ltd	UK	*466*	75-37100
Mannesmann Rexroth SA	France	*139*	75-25500
M.C. Aerospace Corp.	USA	*594*	75-59370
Menasco Aerospace Ltd	Canada	*74*	75-09600
Messier-Dowty Limited	UK	*467*	75-38700
Moog Japan Ltd.	Japan	*253*	75-06100
Napco International Inc.	USA	*598*	75-63600
Norca Machinery Company	USA	*599*	75-64800
Parker Bertea Aerospace	USA	*602*	75-68900
Poclain Hydraulics SA	France	*141*	75-30402
Prva Petoletka Industrie	Yugoslavia	*661*	75-00610
SP aerospace and vehicle systems bv	**Netherlands**	***295***	**75-06860**
Theodor Rapp GmbH Co	Germany	*173*	75-24600
Vickers Inc.	USA	*626*	75-93820
Vickers Systems	Germany	*176*	75-32830
Vickers Systems Division TRINOVA Ltd	UK	*491*	75-63150
Vickers Systems Trinova SpA	Italy	*239*	75-27150

241 30 40 50 70 50 Hydraulic pumps, wheeled armoured vehicle

Abex GmbH Aerohydraul	Germany	*164*	75-00105
AIM Enterprises Inc.	USA	*630*	80-01350
APPH (Bolton) Ltd.	UK	*431*	75-03415
Commercial Intertech Corp.	USA	*564*	75-20050
Dynex/Rivett Inc.	USA	*570*	75-28350
Fenner Fluid Power	UK	*450*	75-22550
Forward Industries Ltd	UK	*452*	75-24200

Giat Industries	France	126	55-00100
Kinetics Ltd	Israel	220	75-11320
Mannesmann Rexroth Ltd	UK	466	75-37100
Mannesmann Rexroth SA	France	139	75-25500
M.C. Aerospace Corp.	USA	594	75-59370
Napco International Inc.	USA	598	75-63600
Norca Machinery Company	USA	599	75-64800
Poclain Hydraulics SA	France	141	75-30402
Prva Petoletka Industrie	Yugoslavia	661	75-00610
Urdan Industries (USA) Inc.	USA	625	75-92930
Vickers Inc.	USA	626	75-93820
Vickers Systems	Germany	176	75-32830
Vickers Systems Division TRINOVA Ltd	UK	491	75-63150
Vickers Systems Trinova SpA	Italy	239	75-27150

241 30 40 60 Suspension systems, wheeled armoured vehicle

General Motors of Canada Ltd	Canada	71	75-05900
GEREP GmbH	Germany	168	75-10300
Horstman Defence Systems Ltd	UK	459	75-29920
Kinetics Ltd	Israel	220	75-11320
Lucas Aerospace Power Transmission	USA	592	75-56265
M. Paz Logistics Ltd	Israel	224	80-01600
Napco International Inc.	USA	598	75-63600
Norca Machinery Company	USA	599	75-64800
Patria Vehicles Oy	Finland	118	75-01710
Urdan Industries (USA) Inc.	USA	625	75-92930

241 30 40 70 Wheeled armoured vehicle equipment, other

AM General Corporation	USA	552	75-05110
Hebron & Medlock Engineering Ltd	UK	457	75-28860
Helio Mirror Co. Ltd	UK	457	75-28900
Kinetics Ltd	Israel	220	75-11320
KUKA Wehrtechnik GmbH	Germany	171	75-17200
M. Paz Logistics Ltd	Israel	224	80-01600
Napco International Inc.	USA	598	75-63600
Norca Machinery Company	USA	599	75-64800
Permali Gloucester Ltd	UK	474	75-44850
SELIN Sistemi Sarl	Italy	238	75-21750

241 30 40 70 30 Electrical equipment, wheeled armoured vehicle

Benchmark Physical Laboratories Ltd	UK	434	75-06195
Bradley Electronics Ltd	UK	435	75-07860
Century Electronics	USA	562	75-17385
Commercial Hydraulics Kontak Ltd	UK	441	75-14415
ECE GmbH	Germany	167	75-07200
EUROCONTROL SpA	Italy	233	75-08200
Helio Mirror Co. Ltd	UK	457	75-28900
M. Paz Logistics Ltd	Israel	224	80-01600
Microturbo Ltd	UK	468	75-39160
Moog Controls Ltd	UK	469	75-40670
Napco International Inc.	USA	598	75-63600
Newton Derby Ltd	UK	470	75-42520
Norca Machinery Company	USA	599	75-64800
SELIN Sistemi Sarl	Italy	238	75-21750

241 30 40 70 30 30 Batteries, wheeled armoured vehicle

Concorde Battery Corp.	USA	566	75-20845
Hawker Energy Products Ltd	UK	457	75-28490
Napco International Inc.	USA	598	75-63600
Saft	France	142	75-33920
Saft America Inc	Mexico	282	75-18000
Saft America Inc.	USA	612	75-77550
Saft Korea Co Ltd	Korea (ROK)	266	75-04300
Saft Nife AB	Sweden	378	75-09850
Saft Nife Argentina SA	Argentina	4	75-09000
Saft Nife AS	Norway	310	75-02400
Saft Nife BV	Netherlands	294	75-06160
Saft Nife Corp	Canada	75	75-12290
Saft Nife GmbH	Austria	34	75-02142
Saft Nife GmbH	Germany	174	75-25901
Saft Nife Iberica SL	Spain	367	75-05100
Saft Nife Ltd	UK	481	75-51470
Saft Nife M.E. Ltd	Cyprus	96	75-05000
Saft Nife Oy	Finland	118	75-01960
Saft Nife Power Systems Australia	Australia	20	75-18950
Saft Singapore Pte Ltd	Singapore	348	75-03400
Vulcan Batteries Ltd	Israel	223	75-23290
Yuasa Automotive Batteries (Europe)	UK	495	75-67560

241 30 40 70 50 Instruments, wheeled armoured vehicle

Benchmark Physical Laboratories Ltd	UK	434	75-06195
Crompton Instruments Ltd	UK	443	75-16070
ECE GmbH	Germany	167	75-07200
Humphrey Inc.	USA	583	75-45050
M. Paz Logistics Ltd	Israel	224	80-01600
Napco International Inc.	USA	598	75-63600
Norca Machinery Company	USA	599	75-64800

241 30 80 Wheeled armoured vehicle accessories

AAI Corp.	USA	545	75-00020
Åkers Krutbruk Protection AB	Sweden	375	75-01100
APPH (Bolton) Ltd.	UK	431	75-03415
Hebron & Medlock Engineering Ltd	UK	457	75-28860
Kinetics Ltd	Israel	220	75-11320
M. Paz Logistics Ltd	Israel	224	80-01600
Napco International Inc.	USA	598	75-63600
Norca Machinery Company	USA	599	75-64800
Patria Vehicles Oy	Finland	118	75-01710
Stringer Welding & Machine Inc.	USA	618	75-84435

241 30 80 10 Armour plates, add-on, wheeled armoured vehicle

AAI Corp.	USA	545	75-00020
AEA Technology plc	UK	496	80-00480
Åkers Krutbruk Protection AB	Sweden	375	75-01100
Barrday Inc.	Canada	68	75-01080
Blohm + Voss GmbH	Germany	165	75-02800
M. Paz Logistics Ltd	Israel	224	80-01600
Meggitt Composites	UK	467	75-38345
Napco International Inc.	USA	598	75-63600
Palbam Metal Works	Israel	221	75-17000
Permali Gloucester Ltd	UK	474	75-44850
Silent Partner Body Armor Inc.	USA	615	75-80930
Vibrachoc SA	France	148	75-46900

241 30 80 20 Towing equipment, wheeled armoured vehicle

Bradley Doublelock Ltd	UK	435	75-07855

241 30 80 65 Swimming accessories, wheeled armoured vehicle

Napco International Inc.	USA	598	75-63600
Norca Machinery Company	USA	599	75-64800
Novatraçáo Artefatos	Brazil	57	75-14780
Patria Vehicles Oy	Finland	118	75-01710
Stephan Wood Products Inc.	USA	628	75-97280

241 30 80 90 Wading accessories, wheeled armoured vehicle

AAI Corp.	USA	545	75-00020
Napco International Inc.	USA	598	75-63600

241 30 80 95 Winches, wheeled armoured vehicle

Horstman Defence Systems Ltd	UK	459	75-29920
Mannesmann Rexroth Ltd	UK	466	75-37100
Mannesmann Rexroth SA	France	139	75-25500
Napco International Inc.	USA	598	75-63600
Norca Machinery Company	USA	599	75-64800
Patria Vehicles Oy	Finland	118	75-01710
Reynolds Boughton (Devon) Ltd	UK	479	75-49340

241 30 90 Wheeled armoured vehicle components

APPH (Bolton) Ltd.	UK	*431*	75-03415
Diehl GmbH & Co	Germany	*167*	75-06500
General Kinetics Engineering Corp.	Canada	*71*	75-05890
GEREP GmbH	Germany	*168*	75-10300
Giat Industries	France	*126*	55-00100
Hebron & Medlock Engineering Ltd	UK	*457*	75-28860
Kinetics Ltd	Israel	*220*	75-11320
M. Paz Logistics Ltd	Israel	*224*	80-01600
MaK Systemgesellschaft mbH	Germany	*172*	75-19500
Napco International Inc.	USA	*598*	75-63600
Norca Machinery Company	USA	*599*	75-64800
Patria Vehicles Oy	Finland	*118*	75-01710

241 30 90 10 05 Components, braking system, wheeled armoured vehicle

APPH (Bolton) Ltd.	UK	*431*	75-03415
Busak + Shamban Inc.	USA	*561*	75-16006
Giat Industries	France	*126*	55-00100
Kinetics Ltd	Israel	*220*	75-11320
Mannesmann Rexroth Ltd	UK	*466*	75-37100
Napco International Inc.	USA	*598*	75-63600
Norca Machinery Company	USA	*599*	75-64800
Patria Vehicles Oy	Finland	*118*	75-01710
Prva Petoletka Industrie	Yugoslavia	*661*	75-00610

241 30 90 70 Tyres, wheeled armoured vehicle

Goodyear Tire & Rubber Co.	USA	*579*	75-40050
Michelin	France	*140*	75-26600
Napco International Inc.	USA	*598*	75-63600
Patria Vehicles Oy	Finland	*118*	75-01710

241 30 90 75 Tyre safety devices, wheeled armoured vehicles

Compounding Ingredients Ltd	UK	*442*	75-14545
Napco International Inc.	USA	*598*	75-63600
Novatraçáo Artefatos	Brazil	*57*	75-14780

241 30 90 77 Tyres, runflat, wheeled armoured vehicle

Hutchinson s.n.c.	France	*137*	75-21900
Napco International Inc.	USA	*598*	75-63600
Novatraçáo Artefatos	Brazil	*57*	75-14780
Silent Partner Body Armor Inc.	USA	*615*	75-80930

241 30 90 90 Windows, bullet-proof, wheeled armoured vehicle

ICL Technical Plastics Ltd	UK	*460*	75-31060
M. Paz Logistics Ltd	Israel	*224*	80-01600
Napco International Inc.	USA	*598*	75-63600
Pilkington Aerospace Ltd	UK	*474*	75-45500
Protection Development Int'l Corp.	USA	*606*	75-71670
Silent Partner Body Armor Inc.	USA	*615*	75-80930
Verres Industriels SA	Switzerland	*388*	75-14700

241 50 Non-armoured vehicle systems & equipment

APPH (Bolton) Ltd.	UK	*431*	75-03415
Bernardini SA	Brazil	*55*	75-02120
Bombardier Inc.	Canada	*69*	75-01400
Cubic Defense Systems Inc.	USA	*567*	75-22900
Giat Industries	France	*126*	55-00100
Howden Aircontrol Ltd	UK	*459*	75-30240
IDE Technologies Ltd	Israel	*219*	75-09170
Israel Military Industries Ltd	Israel	*220*	75-10300
Labinal Aero Defense Systems Inc	USA	*638*	80-18100
Messier-Dowty Electronics	Canada	*74*	75-09650
Northrop Grumman Corporation	USA	*599*	75-65450
Patria Vehicles Oy	Finland	*118*	75-01710
SELIN Sistemi Sarl	Italy	*238*	75-21750
SEXTANT Avionique SA	France	*144*	75-35000
Silent Partner Body Armor Inc.	USA	*615*	75-80930
Technofan SA	France	*147*	75-44300
Tecmotiv Corporation	Canada	*77*	75-13510

241 50 10 Non-armoured vehicle control systems

APPH (Bolton) Ltd.	UK	*431*	75-03415
Hunting Engineering Ltd	UK	*459*	75-30700
Société d'Applications des Machines	France	*143*	75-34300

241 50 10 05 Braking systems & equipment, non-armoured vehicle

APPH (Bolton) Ltd.	UK	*431*	75-03415
Champion Friction Co.	USA	*563*	75-17975
Kinetics Ltd	Israel	*220*	75-11320
Mannesmann Rexroth Ltd	UK	*466*	75-37100
Messier-Dowty Electronics	Canada	*74*	75-09650
PBR Automotive Pty Ltd	Australia	*20*	75-16245
Prva Petoletka Industrie	Yugoslavia	*661*	75-00610
Sielman SA	Greece	*191*	75-02360
Wabco Automotive UK Ltd	UK	*492*	75-64000

241 50 10 10 Steering systems & equipment, non-armoured vehicle

APPH (Bolton) Ltd.	UK	*431*	75-03415
Giat Industries	France	*126*	55-00100
Menasco Aerospace Ltd	Canada	*74*	75-09600
Prva Petoletka Industrie	Yugoslavia	*661*	75-00610
ZF Friedrichshafen AG	Germany	*177*	75-35000

241 50 20 Non-armoured vehicle mission systems

Cubic Defense Systems Inc.	USA	*567*	75-22900
Israel Military Industries Ltd	Israel	*220*	75-10300
Sanders	USA	*612*	75-77850

241 50 20 70 Non-armoured vehicle mission equipment

Israel Military Industries Ltd	Israel	*220*	75-10300
Patria Vehicles Oy	Finland	*118*	75-01710
SP aerospace and vehicle systems bv	**Netherlands**	***295***	**75-06860**
Stephan Wood Products Inc.	USA	*628*	75-97280

241 50 20 70 10 Bodywork, vehicle, special purpose

Ancra CHS BV	Netherlands	*291*	75-00510
CSC (Specialised Vehicles) Ltd	UK	*444*	75-16220
Howard Porter Pty Ltd	Australia	*17*	75-08720
Mambrini Equipamentos Rodoviarios	Brazil	*57*	75-13260
Patria Vehicles Oy	Finland	*118*	75-01710
Permali Gloucester Ltd	UK	*474*	75-44850
Pilcher-Greene Ltd	UK	*474*	75-45450
SP aerospace and vehicle systems bv	**Netherlands**	***295***	**75-06860**

241 50 20 70 30 Cranes, non-armoured vehicle mounted

Atlas Hydraulic Loaders Ltd	UK	*432*	75-04300
Auto Crane Co.	USA	*556*	75-10360
EKA Limited	UK	*448*	75-19810
Grove Europe Ltd	**UK**	***456***	**75-27030**

241 50 20 70 30 50 Cranes, loading, non-armoured vehicle mounted

Atlas Hydraulic Loaders Ltd	UK	*432*	75-04300
Auto Crane Co.	USA	*556*	75-10360
EKA Limited	UK	*448*	75-19810
Grove Europe Ltd	**UK**	***456***	**75-27030**
Partek Cargotec Ltd	UK	*473*	75-44055

241 50 20 70 30 90 Cranes, wrecking, non-armoured vehicle mounted

EKA Limited	UK	*448*	75-19810
Grove Europe Ltd	**UK**	***456***	**75-27030**

Partek Cargotec Ltd	UK	*473*	75-44055
SP aerospace and vehicle systems bv	**Netherlands**	***295***	**75-06860**

241 50 30 Non-armoured vehicle crew-related systems

Israel Military Industries Ltd	Israel	*220*	75-10300
Stephan Wood Products Inc.	USA	*628*	75-97280
Teledyne Brown Engineering	USA	*620*	75-87508

241 50 30 55 Air conditioning systems, non-armoured vehicle

Aircraft Maintenance Support	UK	*428*	75-01580
Airflow Co.	USA	*548*	75-03650
Airscrew Howden Ltd	UK	*429*	75-02050
A/S Dantherm	**Denmark**	***102***	**75-04100**
ECE GmbH	Germany	*167*	75-07200
Gallay Ltd	UK	*453*	75-25100
Howden Aircontrol Ltd	UK	*459*	75-30240
Howden Fluid Systems	USA	*582*	75-44400
Kinetics Ltd	Israel	*220*	75-11320
Lucas Aerospace FCS Paris	France	*139*	75-25390
Microtecnica-Torino SpA	Italy	*237*	75-15500
Microturbo Ltd	UK	*468*	75-39160

241 50 40 Non-armoured vehicle systems, other

AlliedSignal Aerospace Canada	Canada	*68*	75-00470
APPH (Bolton) Ltd.	UK	*431*	75-03415
Bartlett Lifting Devices Inc.	USA	*557*	75-12060
Bradley Electronics Ltd	UK	*435*	75-07860
Eldec Corp.	USA	*572*	75-29660
Gates Hydraulics Ltd	UK	*453*	75-25210
Hebron & Medlock Engineering Ltd	UK	*457*	75-28860
Israel Military Industries Ltd	Israel	*220*	75-10300
Patria Vehicles Oy	Finland	*118*	75-01710
SELIN Sistemi Sarl	Italy	*238*	75-21750
Teledyne Continental Motors	USA	*620*	75-87506

241 50 40 30 Electrical systems, non-armoured vehicle

AlliedSignal Aerospace Canada	Canada	*68*	75-00470
ECE GmbH	Germany	*167*	75-07200
Eldec Corp.	USA	*572*	75-29660
GSI Corporation	USA	*580*	75-40715
Hebron & Medlock Engineering Ltd	UK	*457*	75-28860
Hunting Engineering Ltd	UK	*459*	75-30700
Icore International Ltd	UK	*460*	75-31080
Newton Derby Ltd	UK	*470*	75-42520
SELIN Sistemi Sarl	Italy	*238*	75-21750
Sielman SA	Greece	*191*	75-02360

241 50 40 30 40 Lighting systems, non-armoured vehicle

Bruce Industries Inc.	USA	*561*	75-15360
ECE GmbH	Germany	*167*	75-07200
Hunting Engineering Ltd	UK	*459*	75-30700
Luminescent Systems Inc.	USA	*593*	75-56480
Westair Ltd	UK	*493*	75-65410

241 50 40 30 40 40 Lighting, interior, non-armoured vehicle

Bruce Industries Inc.	USA	*561*	75-15360
Hunting Engineering Ltd	UK	*459*	75-30700
Luminescent Systems Inc.	USA	*593*	75-56480
Soderberg Manufacturing Co. Inc.	USA	*616*	75-82200
Westair Ltd	UK	*493*	75-65410

241 50 40 45 Fuel systems, non-armoured vehicle

Continental AG Werk Uniroyal	Germany	*176*	75-32090

241 50 40 50 70 Hydraulic equipment, non-armoured vehicle

Aeroquip Corp.	USA	*547*	75-02350
APPH (Bolton) Ltd.	UK	*431*	75-03415
Bartlett Lifting Devices Inc.	USA	*557*	75-12060
Bradford Cylinders Ltd	UK	*435*	75-07850
Dynex/Rivett Inc.	USA	*570*	75-28350
Forward Industries Ltd	UK	*452*	75-24200
Gates Hydraulics Ltd	UK	*453*	75-25210
Giat Industries	France	*126*	55-00100
Karipidis Bros. SA	Greece	*190*	75-01980
Kinetics Ltd	Israel	*220*	75-11320
Mannesmann Rexroth Ltd	UK	*466*	75-37100
Mannesmann Rexroth SA	France	*139*	75-25500
M.C. Aerospace Corp.	USA	*594*	75-59370
Menasco Aerospace Ltd	Canada	*74*	75-09600
Messier-Dowty Limited	UK	*467*	75-38700
Moog Japan Ltd.	Japan	*253*	75-06100
Parker Bertea Aerospace	USA	*602*	75-68900
Partek Cargotec Ltd	UK	*473*	75-44055
Poclain Hydraulics SA	France	*141*	75-30402
Prva Petoletka Industrie	Yugoslavia	*661*	75-00610
SP aerospace and vehicle systems bv	**Netherlands**	***295***	**75-06860**
Vickers Inc.	USA	*626*	75-93820
Vickers Systems	Germany	*176*	75-32830
Vickers Systems Division TRINOVA Ltd	UK	*491*	75-63150
Vickers Systems Trinova SpA	Italy	*239*	75-27150

241 50 40 50 70 50 Hydraulic pumps, non-armoured vehicle

APPH (Bolton) Ltd.	UK	*431*	75-03415
Fenner Fluid Power	UK	*450*	75-22550
Giat Industries	France	*126*	55-00100
Howden Fluid Systems	USA	*582*	75-44400
Kinetics Ltd	Israel	*220*	75-11320
Mannesmann Rexroth Ltd	UK	*466*	75-37100
Mannesmann Rexroth SA	France	*139*	75-25500
M.C. Aerospace Corp.	USA	*594*	75-59370
Poclain Hydraulics SA	France	*141*	75-30402
Prva Petoletka Industrie	Yugoslavia	*661*	75-00610
Vickers Inc.	USA	*626*	75-93830
Vickers Systems	Germany	*176*	75-32830
Vickers Systems Division TRINOVA Ltd	UK	*491*	75-63150
Vickers Systems Trinova SpA	Italy	*239*	75-27150

241 50 40 70 Non-armoured vehicle equipment,other

Bartlett Lifting Devices Inc.	USA	*557*	75-12060
Giat Industries	France	*126*	55-00100
Helio Mirror Co. Ltd	UK	*457*	75-28900
IDE Technologies Ltd	Israel	*219*	75-09170
Patria Vehicles Oy	Finland	*118*	75-01710

241 50 40 70 30 Electrical equipment, non-armoured vehicle

Australian Defence Industries Ltd	Australia	*9*	55-01000
Century Electronics	USA	*562*	75-17385
Eldec Corp.	USA	*572*	75-29660
Electrothermal Engineering Ltd	UK	*448*	75-20400
SELIN Sistemi Sarl	Italy	*238*	75-21750
Vickers Aerospace Components	UK	*491*	75-63020

241 50 40 70 30 30 Batteries, non-armoured vehicle (see code 633.40.10.80)

Concorde Battery Corp.	USA	*566*	75-20845
Eldec Corp.	USA	*572*	75-29660
Exide Batteries Ltd	UK	*449*	75-22050
Hawker Energy Products Ltd	UK	*457*	75-28490
Saft	France	*142*	75-33920
Saft America Inc	Mexico	*282*	75-18000
Saft America Inc.	USA	*612*	75-77550
Saft Korea Co Ltd	Korea (ROK)	*266*	75-04300
Saft Nife AB	Sweden	*378*	75-09850
Saft Nife Argentina SA	Argentina	*4*	75-09000

Saft Nife AS	Norway	310	75-02400
Saft Nife BV	Netherlands	294	75-06160
Saft Nife Corp	Canada	75	75-12290
Saft Nife GmbH	Austria	34	75-02142
Saft Nife GmbH	Germany	174	75-25901
Saft Nife Iberica SL	Spain	367	75-05100
Saft Nife Ltd	UK	481	75-51470
Saft Nife M.E. Ltd	Cyprus	96	75-05000
Saft Nife Oy	Finland	118	75-01960
Saft Nife Power Systems Australia	Australia	20	75-18950
Saft Singapore Pte Ltd	Singapore	348	75-03400
Vulcan Batteries Ltd	Israel	223	75-23290
Yuasa Automotive Batteries (Europe)	UK	495	75-67560

241 50 80 Accessories, non-armoured vehicle

AGM Container Controls Inc.	USA	547	75-02870
Åkers Krutbruk Protection AB	Sweden	375	75-01100
Barrday Inc.	Canada	68	75-01080
Fibertek	USA	575	75-34330
Hebron & Medlock Engineering Ltd	UK	457	75-28860
M. Paz Logistics Ltd	Israel	224	80-01600
Patria Vehicles Oy	Finland	118	75-01710
Silent Partner Body Armor Inc.	USA	615	75-80930
Stephan Wood Products Inc.	USA	628	75-97280

241 50 80 10 Armour kits, add-on, non-armoured vehicle

AEA Technology plc	UK	496	80-00480
Åkers Krutbruk Protection AB	Sweden	375	75-01100
Barrday Inc.	Canada	68	75-01080
Cochrane Steel Products (PTY) Ltd	S. Africa	357	75-03000
Fibertek	USA	575	75-34330
Glover Webb Ltd	UK	455	75-25700
M. Paz Logistics Ltd	Israel	224	80-01600
Permali Gloucester Ltd	UK	474	75-44850
Silent Partner Body Armor Inc.	USA	615	75-80930
TBA Textiles	UK	487	75-58740
Urdan Industries (USA) Inc.	USA	625	75-92930

241 50 80 20 Towing equipment, non-armoured vehicle

B. Dixon-Bate Ltd	UK	445	75-18250
Bradley Doublelock Ltd	UK	435	75-07855

241 50 80 30 Covers, canvas, vehicle body

Barrday Inc.	Canada	68	75-01080
Military Logistics MFC	USA	596	75-61550
Stephan Wood Products Inc.	USA	628	75-97280

241 50 80 95 Winches, non-armoured vehicle

dp Manufacturing Inc.	USA	569	75-27320
EKA Limited	UK	448	75-19810
FEVA - Maquinas Ferdinand Vaders SA	Brazil	56	75-06240
Horstman Defence Systems Ltd.	UK	459	75-29920
Mannesmann Rexroth Ltd	UK	466	75-37100
Mannesmann Rexroth SA	France	139	75-25500
Patria Vehicles Oy	Finland	118	75-01710
Poclain Hydraulics SA	France	141	75-30402
Ramsey Winch Co.	USA	607	75-72830
Reynolds Boughton Ltd	UK	479	75-49340
Superwinch Ltd	UK	487	75-57950

241 50 90 Components, non-armoured vehicle

Ancra CHS BV	Netherlands	291	75-00510
APPH (Bolton) Ltd.	UK	431	75-03415
Busak + Shamban Inc.	USA	561	75-16006
Fluorocarbon Company Limited	UK	451	75-24020
Grantex Hellenic Brakes Industry SA	Greece	189	75-01400
Hebron & Medlock Engineering Ltd	UK	457	75-28860
Hella Manufacturing Co. Pty Ltd	Australia	17	75-08370
Interkat SA	Greece	190	75-01960
Oldbury Components Ltd	UK	472	75-42985
Patria Vehicles Oy	Finland	118	75-01710
Permali Gloucester Ltd	UK	474	75-44850
Singapore Technologies Automotive	Singapore	348	75-04280
Stringer Welding & Machine Inc.	USA	618	75-84435

241 50 90 10 05 Components, braking system, non-armoured vehicle

APPH (Bolton) Ltd.	UK	431	75-03415
Grantex Hellenic Brakes Industry SA	Greece	189	75-01400
Kinetics Ltd	Israel	220	75-11320
Patria Vehicles Oy	Finland	118	75-01710
Prva Petoletka Industrie	Yugoslavia	661	75-00610
Stringer Welding & Machine Inc.	USA	618	75-84435

241 50 90 70 Tyres, non-armoured vehicle

Goodyear Tire & Rubber Co.	USA	579	75-40050
Michelin	France	140	75-26600

241 50 90 77 Tyres, runflat, non-armoured vehicle

Glover Webb Ltd	UK	455	75-25700
Novatraçáo Artefatos	Brazil	57	75-14780
Silent Partner Body Armor Inc.	USA	615	75-80930

241 50 90 90 Windows, non-armoured vehicle

Glover Webb Ltd	UK	455	75-25700
ICL Technical Plastics Ltd	UK	460	75-31060

Platform Propulsion

Space Platform / Launcher Propulsion

311 Spacecraft & satellite propulsion

AEA Technology plc	UK	496	80-00480
Aerojet	USA	547	75-01500
Aerospace and Electronics Division	USA	606	75-71472
Arde Inc.	USA	554	75-08500
Atlantic Research Corp.	USA	555	75-10150
China National Precision Machinery	China PR	88	55-00265
Cranfield University	UK	419	60-00350
Matra Marconi Space UK, Ltd	UK	425	74-06900
Moog Controls Ltd	UK	469	75-40670
Nissan Motor Co. Ltd.	Japan	254	75-06790
Pratt & Whitney	USA	605	75-71150
Royal Ordnance	UK	481	75-50620
Universal Propulsion Co. Inc.	USA	625	75-92700

311 10 Rocket motors, spacecraft

Aerospace and Electronics Division	USA	606	75-71472
Cranfield University	UK	419	60-00350
Nissan Motor Co. Ltd.	Japan	254	75-06790
Pratt & Whitney	USA	605	75-71150
R&D Institute of Mechanical	Russia	335	60-04000
Rohr Inc.	USA	611	75-76150
Royal Ordnance	UK	481	75-50620

311 20 Vernier motors, satellite

Aerospace and Electronics Division	USA	606	75-71472
Cranfield University	UK	419	60-00350
Matra Marconi Space UK, Ltd	UK	425	74-06900
Royal Ordnance	UK	481	75-50620
Universal Propulsion Co. Inc.	USA	625	75-92700

312 Space vehicle launcher / booster propulsion

Aerojet	USA	547	75-01500
Atlantic Research Corp.	USA	555	75-10150
Cranfield University	UK	419	60-00350

Fuji Heavy Industries Ltd	Japan	*251*	75-01100
Moog Controls Ltd	UK	*469*	75-40670
Moog Inc.	USA	*597*	75-62550
Nissan Motor Co. Ltd.	Japan	*254*	75-06790
NPO Molniya plc	Russia	*337*	75-13380
Royal Ordnance	UK	*481*	75-50620
SNPE	France	*145*	75-36315
Société Europeenne de Propulsion	France	*144*	75-34850

312 10 Rocket motors, space vehicle launcher / booster

Aerospace and Electronics Division	USA	*606*	75-71472
Cranfield University	UK	*419*	60-00350
McDonnell Douglas Aerospace	USA	*594*	75-59480
Nissan Motor Co. Ltd.	Japan	*254*	75-06790
NPO Energomash	Russia	*338*	75-14300
NPO Molniya plc	Russia	*337*	75-13380
Pratt & Whitney	USA	*605*	75-71150
Royal Ordnance	UK	*481*	75-50620

312 90 Components, rocket motor, space launcher / booster

Aerospace Forgings Ltd	UK	*428*	75-01160
Aerospace Metal Composites Limited	UK	*428*	75-01180
Arde Inc.	USA	*554*	75-08500
Astech Manufacturing Inc.	USA	*555*	75-09700
Avica Equipment	UK	*432*	75-04870
Avionic Components Ltd.	UK	*432*	75-04970
Busak + Shamban Inc.	USA	*561*	75-16006
Cranfield University	UK	*419*	60-00350
Darchem Engineering Ltd.	UK	*444*	75-16750
Doncasters plc	UK	*446*	75-18410
Dowty Aerospace Wolverhampton	UK	*446*	75-18589
EG&G Sealol Ltd	UK	*448*	75-19675
Farsound Engineering Ltd	UK	*450*	75-22420
Fiat Avio SpA	Italy	*234*	75-09110
Firth Rixson plc	UK	*451*	75-23600
Fuji Heavy Industries Ltd	Japan	*251*	75-01100
Holscot Industrial Linings Ltd	UK	*458*	75-29655
Hughes Microelectronics Europe Ltd	UK	*459*	75-30480
Insumat Ltd	UK	*461*	75-31660
Iver Precision Engineers Ltd	UK	*462*	75-32230
Jehier SA	France	*137*	75-23200
KEC Limited	UK	*463*	75-33700
Moog Controls Ltd	UK	*469*	75-40670
Moog Inc.	USA	*597*	75-62550
Moog Japan Ltd.	Japan	*253*	75-06100
Muirhead Vactric Components Ltd	UK	*470*	75-41404
National Forge Co.	USA	*598*	75-63850
Nissan Motor Co. Ltd.	Japan	*254*	75-06790
NPO Energomash	Russia	*338*	75-14300
NPO Molniya plc	Russia	*337*	75-13380
NV Bekaert SA	Belgium	*45*	75-01300
Parker Bertea Aerospace	USA	*602*	75-68900
Price & Orphin Limited	UK	*509*	80-24250
R&D Institute of Mechanical	Russia	*335*	60-04000
Royal Ordnance	UK	*481*	75-50620
Scaled Composites Inc.	USA	*541*	60-19160
Technical Products Group Inc.	USA	*620*	75-86764
Volvo Aero Corporation	Sweden	*379*	75-11810
Voss Industries Inc.	USA	*626*	75-94550
Votaw Precision Technologies Inc.	USA	*626*	75-94580
Wyman-Gordon Ltd	UK	*494*	75-67280

314 Rocket, research / sounding, propulsion

Aerojet	USA	*547*	75-01500
Atlantic Research Corp.	USA	*555*	75-10150
Bristol Aerospace Ltd	Canada	*69*	75-01500
Cranfield University	UK	*419*	60-00350
Fuji Heavy Industries Ltd	Japan	*251*	75-01100
Ishikawajima-Harima H.I. Co. Ltd	Japan	*251*	75-02800
Israel Military Industries Ltd	Israel	*220*	75-10300
Israel Military Industries Rocket Systems Division	**Israel**	***220***	**75-10310**
Matra Marconi Space UK, Ltd	UK	*425*	74-06900
Nissan Motor Co. Ltd.	Japan	*254*	75-06790
Royal Ordnance	UK	*481*	75-50620
SNPE	France	*145*	75-36315
Société Europeenne de Propulsion	France	*144*	75-34850
Universal Propulsion Co. Inc.	USA	*625*	75-92700

314 10 Rocket motors, research / sounding rocket

Aerospace and Electronics Division	USA	*606*	75-71472
Cranfield University	UK	*419*	60-00350
Israel Military Industries Ltd	Israel	*220*	75-10300
Israel Military Industries Rocket Systems Division	**Israel**	***220***	**75-10310**
Nissan Motor Co. Ltd.	Japan	*254*	75-06790
Rohr Inc.	USA	*611*	75-76150
Royal Ordnance	UK	*481*	75-50620
Universal Propulsion Co. Inc.	USA	*625*	75-92700

314 90 Components, rocket motor, research / sounding rocket

Aerospace and Electronics Division	USA	*606*	75-71472
Aerospace Forgings Ltd	UK	*428*	75-01160
Aerospace Metal Composites Limited	UK	*428*	75-01180
Arde Inc.	USA	*554*	75-08500
Avica Equipment	UK	*432*	75-04870
Avionic Components Ltd.	UK	*432*	75-04970
Busak + Shamban Inc.	USA	*561*	75-16006
Darchem Engineering Ltd.	UK	*444*	75-16750
EG&G Sealol Ltd	UK	*448*	75-19675
Firth Rixson plc	UK	*451*	75-23600
Israel Military Industries Ltd	Israel	*220*	75-10300
Israel Military Industries Rocket Systems Division	**Israel**	***220***	**75-10310**
Iver Precision Engineers Ltd	UK	*462*	75-32230
Jehier SA	France	*137*	75-23200
J.S. Chinn & Co. Ltd	UK	*439*	75-13045
KEC Limited	UK	*463*	75-33700
Mibert Precision Ltd	UK	*468*	75-38920
Moog Controls Ltd	UK	*469*	75-40670
Moog Japan Ltd.	Japan	*253*	75-06100
Muirhead Vactric Components Ltd	UK	*470*	75-41404
National Forge Co.	USA	*598*	75-63850
Nissan Motor Co. Ltd.	Japan	*254*	75-06790
Oxley Developments Co. Ltd	UK	*472*	75-43400
Parker Bertea Aerospace	USA	*602*	75-68900
Price & Orphin Limited	UK	*509*	80-24250
Royal Ordnance	UK	*481*	75-50620
Scaled Composites Inc.	USA	*541*	60-19160
Technical Products Group Inc.	USA	*620*	75-86764
Tempo Instruments Inc	USA	*621*	75-87900
Universal Propulsion Co. Inc.	USA	*625*	75-92700
Wyman-Gordon Ltd	UK	*494*	75-67280

Airborne Platform Propulsion

321 Aircraft propulsion

AAR Defense Systems	USA	*629*	80-00040
Aero Systems Engineering Inc.	USA	*546*	75-01116
Aeronca Inc.	USA	*547*	75-02200
Allied Screw Products Inc.	USA	*549*	75-04690
Anite Systems Ltd	UK	*497*	80-01240
Associated Aerospace Activities Inc.	USA	*555*	75-09450
Bet Shemesh Engines Ltd	Israel	*218*	75-03200
Bowmar Instrument Corp.	USA	*560*	75-14860
CFM International Inc.	USA	*544*	74-00400
China National Aero-Tech I/E Corp.	China PR	*88*	55-00200
Darchem Engineering Ltd.	UK	*444*	75-16750
Dynamic Metal Products Co.	USA	*570*	75-28240
Esterline Technologies	USA	*574*	75-32360
Eurojet Turbo GmbH	Germany	*164*	74-05900
Fiat Avio SpA	Italy	*234*	75-09110
Ishikawajima-Harima H.I. Co. Ltd	Japan	*251*	75-02800
Kawasaki Heavy Industries Ltd	Japan	*252*	75-04350
Microturbo Inc	USA	*596*	75-61170
Microturbo Ltd	UK	*468*	75-39160
Microturbo SA	France	*140*	75-27000
Mitsubishi Heavy Industries Ltd	Japan	*253*	75-05800
Molding Solutions	USA	*597*	75-62400
Moog Inc.	USA	*597*	75-62550
MTU München GmbH	Germany	*172*	75-21801
Palmer Manufacturing Co Inc.	USA	*602*	75-68620
Pratt & Whitney	USA	*605*	75-71150
Pratt & Whitney Canada Inc	Canada	*75*	75-11400
Purolator Products Co.	USA	*606*	75-71900
Rolls-Royce (Canada) Ltd.	Canada	*75*	75-12120

Rolls-Royce Mil. Aero Engines Ltd	UK	*480*	75-50265
Royal Ordnance	UK	*481*	75-50620
SNECMA	France	*145*	75-36305
TAT Industries Ltd	Israel	*223*	75-22000
Techspace Aero SA	Belgium	*47*	75-06200
Telair International	USA	*620*	75-87400
Turbo-Union Ltd	Germany	*164*	74-20000
Turbomeca	France	*148*	75-46000
Universal Propulsion Co. Inc.	USA	*625*	75-92700
Wall Colomonoy Corp.	USA	*626*	75-94900

321 10 Turbojet / turbofan engines, aircraft

AAR Corp	USA	*629*	80-00035
Aero Systems Engineering Inc.	USA	*546*	75-01116
AOI Engine Factory	Egypt	*109*	55-01700
Associated Aerospace Activities Inc.	USA	*555*	75-09450
Bet Shemesh Engines Ltd	Israel	*218*	75-03200
CFM International Inc.	USA	*544*	74-00400
Eurojet Turbo GmbH	Germany	*164*	74-05900
Fiat Avio SpA	Italy	*234*	75-09110
Groupement Turboméca SNECMA	France	*130*	74-00800
Guizhou Aviation Industry Corp.	China PR	*88*	55-00800
HAL Engine Div.	India	*204*	55-01000
Hindustan Aeronautics Ltd.	India	*204*	55-01000
Ishikawajima-Harima H.I. Co. Ltd	Japan	*251*	75-02800
JSC Aviadvigatel	Russia	*339*	80-00180
Kawasaki Heavy Industries Ltd	Japan	*252*	75-04350
Lyulka-Saturn Inc	Russia	*337*	75-12000
Merex Inc.	USA	*595*	75-60250
Microturbo Inc	USA	*596*	75-61170
Microturbo Ltd	UK	*468*	75-39160
Microturbo SA	France	*140*	75-27000
Mitsubishi Heavy Industries Ltd	Japan	*253*	75-05800
MTU München GmbH	Germany	*172*	75-21801
Pratt & Whitney Canada Inc	Canada	*75*	75-11400
Rinaldo Piaggio SpA	Italy	*238*	75-18790
Rolls-Royce (Canada) Ltd.	Canada	*75*	75-12120
Rolls-Royce Mil. Aero Engines Ltd	UK	*480*	75-50265
Rolls-Royce plc	UK	*479*	75-50250
Rolls-Royce Turbomeca	UK	*426*	74-18000
Samsung Aerospace Industries Ltd.	Korea (ROK)	*266*	75-04630
Sener Ingenieria y Sistemas SA	Spain	*367*	75-05200
SNECMA	France	*145*	75-36305
Sverdrup Technology Inc.	USA	*643*	80-30680
Techspace Aero SA	Belgium	*47*	75-06200
Teledyne Ryan Aeronautical	USA	*621*	75-87560
Turbo-Union Ltd	Germany	*164*	74-20000
Turbo-Union Ltd	UK	*426*	74-20700
Turbomeca	France	*148*	75-46000
Volvo Aero Corporation	Sweden	*379*	75-11810
Williams International	USA	*628*	75-96950

321 20 Turboprop / turboshaft engines, aircraft

Aero Systems Engineering Inc.	USA	*546*	75-01116
AOI Engine Factory	Egypt	*109*	55-01700
Eurojet Turbo GmbH	Germany	*164*	74-05900
Fiat Avio SpA	Italy	*234*	75-09110
Firth Rixson plc	UK	*451*	75-23600
HAL Engine Div.	India	*204*	55-01000
Ishikawajima-Harima H.I. Co. Ltd	Japan	*251*	75-02800
JSC Aviadvigatel	Russia	*339*	80-00180
Lyulka-Saturn Inc	Russia	*337*	75-12000
Merex Inc.	USA	*595*	75-60250
Mitsubishi Heavy Industries Ltd	Japan	*253*	75-05800
MTU München GmbH	Germany	*172*	75-21801
Pratt & Whitney	USA	*605*	75-71150
Pratt & Whitney Canada Inc	Canada	*75*	75-11400
Rolls-Royce (Canada) Ltd.	Canada	*75*	75-12120
Rolls-Royce plc	UK	*479*	75-50250
Samsung Aerospace Industries Ltd.	Korea (ROK)	*266*	75-04630
SNECMA	France	*145*	75-36305
Sverdrup Technology Inc.	USA	*643*	80-30680
Techspace Aero SA	Belgium	*47*	75-06200

321 30 Piston engines, aircraft

Aero Systems Engineering Inc.	USA	*546*	75-01116
Aerostar SA	Romania	*331*	80-00100
Bombardier-Rotax GmbH	Austria	*32*	75-00660
Merex Inc.	USA	*595*	75-60250
Rolls-Royce (Canada) Ltd.	Canada	*75*	75-12120
Teledyne Continental Motors	USA	*620*	75-87516

321 40 Rocket / ramjet engines, aircraft

Aero Systems Engineering Inc.	USA	*546*	75-01116
Ishikawajima-Harima H.I. Co. Ltd	Japan	*251*	75-02800
Merex Inc.	USA	*595*	75-60250
Rolls-Royce Mil. Aero Engines Ltd	UK	*480*	75-50265
Royal Ordnance	UK	*481*	75-50620

321 40 50 Rocket motors, JATO, aircraft

Universal Propulsion Co. Inc.	USA	*625*	75-92700

321 50 Propellers, aircraft

Arrowprop Co. Inc.	USA	*555*	75-09150
Daedalus Research Inc.	USA	*567*	75-23360
Doncasters plc	UK	*446*	75-18410
Dowty Aerospace Propellers	UK	*446*	75-18587
Merex Inc.	USA	*595*	75-60250
Messier-Dowty Limited	UK	*467*	75-38700
Ratier-Figeac SA	France	*141*	75-31800

321 80 Accessories, aircraft engine

AAR Corp	USA	*629*	80-00035
AAR Defense Systems	USA	*629*	80-00040
Aircraft Controls Group	USA	*628*	75-97290
Associated Aerospace Activities Inc.	USA	*555*	75-09450
British Filters Ltd	UK	*436*	75-09030
Esterline Technologies	USA	*574*	75-32360
GEC-Marconi Avionics Ltd	UK	*454*	75-25320
Gull Electronic Systems Div.	USA	*580*	75-41130
Howden Fan Company Pty Ltd	Australia	*17*	75-08740
Insumat Ltd	UK	*461*	75-31660
Lucas Aerospace Power Transmission	USA	*592*	75-56265
Merex Inc.	USA	*595*	75-60250
Microturbo SA	France	*140*	75-27000
Moog Italiana Srl	Italy	*237*	75-16230
Nord-Micro Elektronik Feinmechanik	Germany	*173*	75-22700
Ontic Engineering & Mfg Inc.	USA	*601*	75-66875
Parker Bertea Aerospace	USA	*603*	75-68900
Serck Aviation Div.	UK	*482*	75-52860
Telair International	USA	*620*	75-87400
Tuthill Corporation	USA	*624*	75-91320
Walter Kidde Aerospace	USA	*588*	75-52500
Woodward Governor Company	USA	*628*	75-97290

321 80 05 Accessory drives, aircraft engine

Fiat Avio SpA	Italy	*234*	75-09110
Lucas Aerospace Power Transmission	USA	*592*	75-56265
Lucas Western Inc.	USA	*592*	75-56300
Merex Inc.	USA	*595*	75-60250
Messier-Dowty Limited	UK	*467*	75-38700

321 80 10 Anti-vibration mountings, aircraft

BTR Silvertown Ltd	UK	*437*	75-10830
Esterline Technologies	USA	*574*	75-32360
James Walker & Co. Ltd	UK	*492*	75-64120
Loggers Rubbertechniek BV	Netherlands	*293*	75-04300
Lord SA	France	*150*	80-01350
Moog Inc.	USA	*597*	75-62550
Stop-Choc Ltd.	UK	*486*	75-57345

321 90 Components, aircraft engine

Aero Systems Engineering Inc.	USA	*546*	75-01116
Aeromax Industries Inc.	USA	*547*	75-01920
Aeronca Inc.	USA	*547*	75-02200
Aerospace Syst. & Comp.	USA	*547*	75-02501
A.O. Henton Engineering Co Ltd.	UK	*458*	75-29040
Associated Aerospace Activities Inc.	USA	*555*	75-09450
Astech Manufacturing Inc.	USA	*555*	75-09700
Barnes Aerospace Group	USA	*557*	75-12000

Bet Shemesh Engines Ltd	Israel	*218*	75-03200
Bowmar Instrument Corp.	USA	*560*	75-14860
BP Chemicals Ltd	UK	*435*	75-07790
Busak + Shamban Inc.	USA	*561*	75-16006
Chromalloy France	France	*133*	75-09650
Cleveland-Guest (Engineering) Ltd	UK	*440*	75-13810
Darchem Engineering Ltd.	UK	*444*	75-16750
Defontaine	France	*135*	75-13160
Dynamic Metal Products Co.	USA	*570*	75-28240
EG&G Sealol Ltd	UK	*448*	75-19675
Eldim BV	Netherlands	*297*	80-05500
Electro-Methods Inc.	USA	*572*	75-29800
Firth Rixson plc	UK	*451*	75-23600
Fluorocarbon Company Limited	UK	*451*	75-24020
Hammonds Technical Services Inc.	USA	*580*	75-41800
Heroux Inc	Canada	*72*	75-07400
Howmet Corporation	USA	*582*	75-44550
Industria de Turbo Propulsores SA	Spain	*366*	75-03870
Interturbine Germany GmbH	Germany	*170*	75-14630
Jehier SA	France	*137*	75-23200
J.S. Chinn & Co. Ltd	UK	*439*	75-13045
LORI Heat Transfer Operations	USA	*592*	75-56090
Lucas Aerospace FCS Paris	France	*139*	75-25390
Manitoba Aerospace Initiative	Canada	*73*	75-09215
Masbe Corporation Ltd	Taiwan	*392*	75-14000
McMellon Bros Inc.	USA	*595*	75-59710
Merex Inc.	USA	*595*	75-60250
Mibert Precision Ltd	UK	*468*	75-38920
Microturbo Inc	USA	*596*	75-61170
Molding Solutions	USA	*597*	75-62400
Motores Rolls-Royce Ltda	Brazil	*58*	80-20000
Norsk Jetmotor A/S	Norway	*310*	75-01950
Ontic Engineering & Mfg Inc.	USA	*601*	75-66875
Palmer Manufacturing Co Inc.	USA	*602*	75-68620
Parker Bertea Aerospace	USA	*603*	75-68900
Price & Orphin Limited	UK	*509*	80-24250
Purolator Products Co.	USA	*606*	75-71900
Rohr Europe	France	*142*	75-33000
Rohr Inc.	USA	*611*	75-76150
Samsung Aerospace Industries Ltd	Korea (ROK)	*266*	75-04630
Sequa Corp.	USA	*613*	75-79490
Standard Steel	USA	*617*	75-83700
Teadit Juntas SA	Brazil	*55*	75-00910
Technical Products Group Inc.	USA	*620*	75-86764
Teledyne Ryan Aeronautical	USA	*621*	75-87560
The NORDAM Group	USA	*599*	75-64880
Tuthill Corporation	USA	*624*	75-91320
Volvo Aero Corporation	Sweden	*379*	75-11810
Voss Industries Inc.	USA	*626*	75-94550
Walbar Canada	Canada	*77*	75-15200
Wall Colomonoy Corp.	USA	*626*	75-94900
Westport Development & Manufacturing	USA	*627*	75-96280

321 90 30 Engine fuel control & ignition systems, aircraft

AAR Corp	USA	*629*	80-00035
Aero Systems Engineering Inc.	USA	*546*	75-01116
Aerospace Syst. & Comp.	USA	*547*	75-02501
AIM Enterprises Inc.	USA	*630*	80-01350
Aircraft Controls Group	USA	*628*	75-97290
Allied Screw Products Inc.	USA	*549*	75-04690
Anite Systems Ltd	UK	*497*	80-01240
Bowmar Instrument Corp.	USA	*560*	75-14860
GEC-Marconi Aerospace Ltd	UK	*453*	75-25300
GEC-Marconi Avionics Inc.	USA	*577*	75-37413
GEC-Marconi Avionics Ltd	UK	*454*	75-25320
Hammonds Technical Services Inc.	USA	*580*	75-41800
Kearsley Airways Ltd.	UK	*463*	75-33370
Lucas Aerospace	Canada	*73*	75-09140
Merex Inc.	USA	*595*	75-60250
Microtecnica-Torino SpA	Italy	*237*	75-15500
Moog Inc.	USA	*597*	75-62550
Moog Italiana Srl	Italy	*237*	75-16230
Moog Japan Ltd.	Japan	*253*	75-06100
Nord-Micro Elektronik Feinmechanik	Germany	*173*	75-22700
Ontic Engineering & Mfg Inc.	USA	*601*	75-66875
Palmer Manufacturing Co Inc.	USA	*602*	75-68620
Parker Bertea Aerospace	USA	*602*	75-68900
Parker Hannifin Corporation	USA	*603*	75-68910
Sener Ingenieria y Sistemas SA	Spain	*367*	75-05200
Woodward Governor Company	USA	*628*	75-97290

322 Helicopter propulsion

AAR Defense Systems	USA	*629*	80-00040
Aero Systems Engineering Inc.	USA	*546*	75-01116
Anite Systems Ltd	UK	*497*	80-01240
Arab British Engine Co.	Egypt	*109*	74-00300
Bowmar Instrument Corp.	USA	*560*	75-14860
Darchem Engineering Ltd.	UK	*444*	75-16750
Fuji Heavy Industries Ltd	Japan	*251*	75-01100
GKN Westland Aerospace Transmissions	UK	*455*	75-25630
Hardcote Technologies	USA	*580*	75-41900
Ishikawajima-Harima H.I. Co. Ltd	Japan	*251*	75-02800
Mitsubishi Heavy Industries Ltd	Japan	*253*	75-05800
Moog Inc.	USA	*597*	75-62550
MTU München GmbH	Germany	*172*	75-21801
Pratt & Whitney	USA	*605*	75-71150
Pratt & Whitney Canada Inc	Canada	*75*	75-11400
SP aerospace and vehicle systems bv	**Netherlands**	***295***	**75-06860**

322 10 Turboshaft engines, helicopter

Aero Systems Engineering Inc.	USA	*546*	75-01116
AOI Engine Factory	Egypt	*109*	55-01700
Arab British Engine Co.	Egypt	*109*	74-00300
Fiat Avio SpA	Italy	*234*	75-09110
HAL Engine Div.	India	*204*	55-01000
Ishikawajima-Harima H.I. Co. Ltd	Japan	*251*	75-02800
Merex Inc.	USA	*595*	75-60250
Mitsubishi Heavy Industries Ltd	Japan	*253*	75-05800
MTU München GmbH	Germany	*172*	75-21801
Pratt & Whitney Canada Inc	Canada	*75*	75-11400
Rinaldo Piaggio SpA	Italy	*238*	75-18790
Rolls-Royce (Canada) Ltd.	Canada	*75*	75-12120
Rolls-Royce plc	UK	*479*	75-50250
Rolls-Royce Turbomeca	UK	*426*	74-18000
Samsung Aerospace Industries Ltd	Korea (ROK)	*266*	75-04630
Sverdrup Technology Inc.	USA	*643*	80-30680
Techspace Aero SA	Belgium	*47*	75-06200
Turbomeca	France	*148*	75-46000

322 20 Piston engines, helicopter

Aero Systems Engineering Inc.	USA	*546*	75-01116
Merex Inc.	USA	*595*	75-60250

322 50 Rotor blades, helicopter

Fuji Heavy Industries Ltd	Japan	*251*	75-01100
Hardcote Technologies	USA	*580*	75-41900

322 70 Transmissions, helicopter

Aero Systems Engineering Inc.	USA	*546*	75-01116
Fiat Avio SpA	Italy	*234*	75-09110
Fuji Heavy Industries Ltd	Japan	*251*	75-01100
GKN Westland Aerospace Transmissions	UK	*455*	75-25630
Merex Inc.	USA	*595*	75-60250
SP aerospace and vehicle systems bv	**Netherlands**	***295***	**75-06860**
Sverdrup Technology Inc.	USA	*643*	80-30680
ZF Friedrichshafen AG	Germany	*177*	75-35000

322 70 40 Gearboxes, helicopter transmission

Aero Systems Engineering Inc.	USA	*546*	75-01116
Fiat Avio SpA	Italy	*234*	75-09110
Fuji Heavy Industries Ltd	Japan	*251*	75-01100
GKN Westland Aerospace Transmissions	UK	*455*	75-25630
Merex Inc.	USA	*595*	75-60250
SP aerospace and vehicle systems bv	**Netherlands**	***295***	**75-06860**
Sverdrup Technology Inc.	USA	*643*	80-30680
The Purdy Corp.	USA	*606*	75-71850

322 80 Accessories, helicopter engine

Aero Systems Engineering Inc.	USA	*546*	75-01116
Aircraft Parts Corp.	USA	*548*	75-03440
Esterline Technologies	USA	*574*	75-32360
GKN Westland Aerospace Transmissions	UK	*455*	75-25630
Howden Fan Company Pty Ltd	Australia	*17*	75-08740
Insumat Ltd	UK	*461*	75-31660
Lucas Aerospace Power Transmission	USA	*592*	75-56265
Merex Inc.	USA	*595*	75-60250
Ontic Engineering & Mfg Inc.	USA	*601*	75-66875
Parker Bertea Aerospace	USA	*603*	75-68900
Serck Aviation Div.	UK	*482*	75-52860
Walter Kidde Aerospace	USA	*588*	75-52500
Woodward Governor Company	USA	*628*	75-97290

322 80 05 Accessory drives, helicopter engine

Aero Systems Engineering Inc.	USA	*546*	75-01116
Fiat Avio SpA	Italy	*234*	75-09110
Lucas Aerospace Power Transmission	USA	*592*	75-56265
Lucas Western Inc.	USA	*592*	75-56300
Merex Inc.	USA	*595*	75-60250

322 80 10 Anti-vibration mountings, helicopter

AIM Enterprises Inc.	USA	*630*	80-01350
BTR Silvertown Ltd	UK	*437*	75-10830
Esterline Technologies	USA	*574*	75-32360
Fuji Heavy Industries Ltd	Japan	*251*	75-01100
Loggers Rubbertechniek BV	Netherlands	*293*	75-04300
Lord SA	France	*150*	80-01350
Merex Inc.	USA	*595*	75-60250
Moog Inc.	USA	*597*	75-62550
Moog Japan Ltd.	Japan	*253*	75-06100
Saab Dynamics AB	Sweden	*378*	75-09100
Stop-Choc Ltd.	UK	*486*	75-57345

322 90 Components, helicopter engine

AAR Defense Systems	USA	*629*	80-00040
Aero Systems Engineering Inc.	USA	*546*	75-01116
Aerospace Syst. & Comp.	USA	*547*	75-02501
AIM Enterprises Inc.	USA	*630*	80-01350
A.O. Henton Engineering Co Ltd.	UK	*458*	75-29040
British Filters Ltd	UK	*436*	75-09030
Busak + Shamban Inc.	USA	*561*	75-16006
Chromalloy France	France	*133*	75-09650
Darchem Engineering Ltd.	UK	*444*	75-16750
EG&G Sealol Ltd	UK	*448*	75-19675
Fluorocarbon Company Limited	UK	*451*	75-24020
Fuji Heavy Industries Ltd	Japan	*251*	75-01100
GKN Westland Aerospace Transmissions	UK	*455*	75-25630
Howmet Corporation	USA	*582*	75-44550
Jehier SA	France	*137*	75-23200
LORI Heat Transfer Operations	USA	*592*	75-56090
Merex Inc.	USA	*595*	75-60250
Mibert Precision Ltd	UK	*468*	75-38920
Moog Inc.	USA	*597*	75-62550
National Forge Co.	USA	*598*	75-63850
Ontic Engineering & Mfg Inc.	USA	*601*	75-66875
Parker Bertea Aerospace	USA	*603*	75-68900
Price & Orphin Limited	UK	*509*	80-24250
Samsung Aerospace Industries Ltd	Korea (ROK)	*266*	75-04630
Technical Products Group Inc.	USA	*620*	75-86764
The Purdy Corp.	USA	*606*	75-71850
Tuthill Corporation	USA	*624*	75-91320
Voss Industries Inc.	USA	*626*	75-94550
Walbar Canada	Canada	*77*	75-15200

322 90 30 Engine fuel control & ignition systems, helicopter

Aero Systems Engineering Inc.	USA	*546*	75-01116
Aerospace Syst. & Comp.	USA	*547*	75-02501
AIM Enterprises Inc.	USA	*630*	80-01350
Anite Systems Ltd	UK	*497*	80-01240
Bowmar Instrument Corp.	USA	*560*	75-14860
GEC-Marconi Avionics Inc.	USA	*577*	75-37413
Lucas Aerospace	Canada	*73*	75-09140
Merex Inc.	USA	*595*	75-60250
Microtecnica-Torino SpA	Italy	*237*	75-15500
Moog Inc.	USA	*597*	75-62550
Moog Italiana Srl	Italy	*237*	75-16230
Ontic Engineering & Mfg Inc.	USA	*601*	75-66875
Parker Bertea Aerospace	USA	*602*	75-68900
Woodward Governor Company	USA	*628*	75-97290

323 RPV, Drone propulsion

Microturbo Ltd	UK	*468*	75-39160
Microturbo SA	France	*140*	75-27000
Mitsubishi Heavy Industries Ltd	Japan	*253*	75-05800
Royal Ordnance	UK	*481*	75-50620
Sundstrand Power Systems	USA	*618*	75-85120
Target Technology Ltd	UK	*487*	75-58660
Universal Propulsion Co. Inc.	USA	*625*	75-92700

323 10 Engines, turbojet, RPV / Drone

GKN Westland Technologies Ltd	UK	*455*	75-25690
Hindustan Aeronautics Ltd.	India	*203*	55-01000
Ishikawajima-Harima H.I. Co. Ltd	Japan	*251*	75-02800
Microturbo Ltd	UK	*468*	75-39160
Microturbo SA	France	*140*	75-27000
Mitsubishi Heavy Industries Ltd	Japan	*253*	75-05800
Rolls-Royce (Canada) Ltd.	Canada	*75*	75-12120
Rolls-Royce plc	UK	*479*	75-50250
Sundstrand Power Systems	USA	*618*	75-85120
Teledyne Ryan Aeronautical	USA	*621*	75-87560
Williams International	USA	*628*	75-96950

323 20 Engines, turboprop / turboshaft, RPV / Drone

Rolls-Royce (Canada) Ltd.	Canada	*75*	75-12120
Williams International	USA	*628*	75-96950

323 30 Engines, piston, RPV / Drone

Cranfield University	UK	*419*	60-00350
GKN Westland Technologies Ltd	UK	*455*	75-25690
Target Technology Ltd	UK	*487*	75-58660
Yamaha Motor Co Ltd	Japan	*255*	75-25000

323 35 Engines, pulse jet, RPV / Drone

Thermo-Jet Standard Inc.	USA	*622*	75-88700

323 40 Engines, rocket, RPV / Drone

Royal Ordnance	UK	*481*	75-50620
Universal Propulsion Co. Inc.	USA	*625*	75-92700

323 80 Accessories, RPV / Drone engine

Aerospace Syst. & Comp.	USA	*547*	75-02501
British Filters Ltd	UK	*436*	75-09030
Parker Bertea Aerospace	USA	*603*	75-68900
Target Technology Ltd	UK	*487*	75-58660
Woodward Governor Company	USA	*628*	75-97290

323 90 Components, RPV / Drone engine

Aerospace Syst. & Comp.	USA	*547*	75-02501
EG&G Sealol Ltd	UK	*448*	75-19675
Howmet Corporation	USA	*582*	75-44550
Parker Bertea Aerospace	USA	*603*	75-68900
Target Technology Ltd	UK	*487*	75-58660
Teledyne Ryan Aeronautical	USA	*621*	75-87560
Tuthill Corporation	USA	*624*	75-91320
Woodward Governor Company	USA	*628*	75-97290

324 Airship / Dirigible propulsion

CRM SpA Motori Marini	Italy	*233*	75-06300

Product–Supplier Link

324 10 Engines, piston, airship / dirigible

CRM SpA Motori Marini	Italy	*233*	75-06300

Sea-based Platform Propulsion

331 Surface ship propulsion

Australian Defence Industries Ltd	Australia	*9*	55-01000
Brunton's Propellers Ltd	UK	*437*	75-10400
CRM SpA Motori Marini	Italy	*233*	75-06300
Darchem Engineering Ltd.	UK	*444*	75-16750
Detroit Diesel Corp.	USA	*568*	75-25830
Deütz Motor GmbH	Germany	*170*	75-15900
Deutz Motor Industrie Motoren GmbH	Germany	*167*	75-06350
Fiat Avio SpA	Italy	*234*	75-09110
Fincantieri	Italy	*234*	75-09250
GEC ALSTHOM Large Machines Ltd	UK	*453*	75-25240
GEC ALSTHOM Medium Turbo Machines	UK	*453*	75-25243
GEC ALSTHOM Ruston Diesels Ltd	UK	*453*	75-25255
George G. Sharp Inc.	USA	*642*	80-28000
Hitachi Zosen Co. Ltd.	Japan	*251*	75-02390
Holec Machines & Apparaten BV	Netherlands	*293*	75-03110
Hyundai Engine & Machinery Co. Ltd	Korea (ROK)	*265*	75-01570
Ishikawajima-Harima H.I. Co. Ltd	Japan	*251*	75-02800
Kvaerner Energy AS	Norway	*309*	75-01093
Kvaerner Masa-Yards Inc	**Finland**	***118***	**75-01370**
Lucas Aerospace Power Transmission	USA	*592*	75-56265
Mermaid Marine Engines Ltd.	UK	*467*	75-38550
Michigan Wheel Corp.	USA	*596*	75-60930
MTU Friedrichschafen GmbH	Germany	*172*	75-21802
Newport News Shipbuilding	USA	*599*	75-64300
Pratt & Whitney Canada Inc	Canada	*75*	75-11400
Royal Schelde	Netherlands	*294*	75-06150
Schottel-Werft	Germany	*174*	75-26700
SSS Clutch Co. Inc.	USA	*617*	75-83570
Stork-Wärtsilä Diesel BV	Netherlands	*295*	75-07300
Sulzer Hydro Wyss GmbH	Germany	*175*	75-29450
Svendborg Vaerft A/S	Denmark	*104*	75-19800
Teadit Juntas SA	Brazil	*55*	75-00910
TTL Dynamics	UK	*490*	75-61510
VM MOTORI SpA	Italy	*239*	75-27400
Vokes Ltd	UK	*492*	75-63450

331 10 Engines, marine diesel, surface ship

Caterpillar of Australia Ltd	Australia	*14*	75-03240
CRM SpA Motori Marini	Italy	*233*	75-06300
Cummins Engine Co. Inc.	USA	*567*	75-22980
Dalian Marine Diesel Works	China PR	*88*	75-00350
Detroit Diesel Corp.	USA	*568*	75-25830
Detroit Diesel-Allison Australia Pty	Australia	*15*	75-04450
Deütz Motor GmbH	Germany	*170*	75-15900
Fincantieri	Italy	*234*	75-09250
FP Diesel LP	USA	*576*	75-35753
Garden Reach Shipbuilders	India	*203*	55-00300
GEC ALSTHOM Diesels Ltd	UK	*453*	75-25220
GEC ALSTHOM Paxman Diesels Ltd	UK	*453*	75-25245
GEC ALSTHOM Ruston Diesels Ltd	UK	*453*	75-25255
George G. Sharp Inc.	USA	*642*	80-28000
Hitachi Zosen Co. Ltd.	Japan	*251*	75-02390
Hyundai	Korea (ROK)	*265*	75-01540
Ishikawajima-Harima H.I. Co. Ltd	Japan	*251*	75-02800
MAN B&W Diesel AG	Germany	*172*	75-19610
Manises Diesel Engine Co.	Spain	*366*	75-04500
Merex Inc.	USA	*595*	75-60250
Mermaid Marine Engines Ltd.	UK	*467*	75-38550
Mirrlees Blackstone Limited	UK	*468*	75-40050
MTU Friedrichschafen GmbH	Germany	*172*	75-21802
NORINCO	China PR	*88*	55-00280
Perkins Engines (Shrews.)	UK	*474*	75-44750
Rolls-Royce Motor Cars Ltd	UK	*480*	75-50270
SEMT Pielstick	France	*144*	75-34843
Stewart & Stevenson Services Inc.	USA	*618*	75-84180
Stork-Wärtsilä Diesel BV.	Netherlands	*295*	75-07300
Varity Perkins	UK	*474*	75-44750
Wartsila Diesel France SA	France	*149*	75-47070

331 10 90 Components, marine diesel engine, surface ship

A.O. Henton Engineering Co Ltd.	UK	*458*	75-29040
Australian Defence Industries Ltd	Australia	*9*	55-01000
BTR Silvertown Ltd	UK	*437*	75-10830
Busak + Shamban Inc.	USA	*561*	75-16006
Darchem Engineering Ltd.	UK	*444*	75-16750
Fincantieri	Italy	*234*	75-09250
Fluorocarbon Company Limited	UK	*451*	75-24020
FP Diesel LP	USA	*576*	75-35753
GEC ALSTHOM Ruston Diesels Ltd	UK	*453*	75-25255
George G. Sharp Inc.	USA	*642*	80-28000
Industrial Acoustics Company Ltd	UK	*460*	75-31300
James Walker & Co. Ltd	UK	*492*	75-64120
Merex Inc.	USA	*595*	75-60250
National Forge Co.	USA	*598*	75-63850
Perkins Engines (Shrews.)	UK	*474*	75-44750
Premaberg Manufacturing Ltd	UK	*475*	75-46600
SEMT Pielstick	France	*144*	75-34843
Stewart & Stevenson Services Inc.	USA	*618*	75-84180
Stork-Wärtsilä Diesel BV.	Netherlands	*295*	75-07300
Stringer Welding & Machine Inc.	USA	*618*	75-84435
Teadit Juntas SA	Brazil	*55*	75-00910
Tuthill Corporation	USA	*624*	75-91320

331 20 Engines, marine gas turbine, surface ship

Fiat Avio SpA	Italy	*234*	75-09110
GEC ALSTHOM Medium Turbo Machines	UK	*453*	75-25243
George G. Sharp Inc.	USA	*642*	80-28000
Ishikawajima-Harima H.I. Co. Ltd	Japan	*251*	75-02800
Merex Inc.	USA	*595*	75-60250
MTU do Brasil Ltda	Brazil	*57*	75-13900
MTU Friedrichschafen GmbH	Germany	*172*	75-21802
Pratt & Whitney Canada Inc	Canada	*75*	75-11400
Rolls-Royce (Canada) Ltd.	Canada	*75*	75-12120
Rolls-Royce plc	UK	*479*	75-50250
Solar Turbines	USA	*616*	75-82300
Stewart & Stevenson Services Inc.	USA	*618*	75-84180
TTL Dynamics	UK	*490*	75-61510

331 20 90 Components, marine gas turbine, surface ship

Barnes Aerospace Group	USA	*557*	75-12000
BTR Silvertown Ltd	UK	*437*	75-10830
Busak + Shamban Inc.	USA	*561*	75-16006
Chromalloy France	France	*133*	75-09650
Darchem Engineering Ltd.	UK	*444*	75-16750
Fiat Avio SpA	Italy	*234*	75-09110
GEC ALSTHOM Medium Turbo Machines	UK	*453*	75-25243
George G. Sharp Inc.	USA	*642*	80-28000
Industrial Acoustics Company Ltd	UK	*460*	75-31300
James Walker & Co. Ltd	UK	*492*	75-64120
J.S. Chinn & Co. Ltd	UK	*439*	75-13045
Merex Inc.	USA	*595*	75-60250
National Forge Co.	USA	*598*	75-63850
Premaberg Manufacturing Ltd	UK	*475*	75-46600
SSS Clutch Co. Inc.	USA	*617*	75-83570
Tuthill Corporation	USA	*624*	75-91320
Woodward Governor Company	USA	*628*	75-97290

331 30 Nuclear propulsion systems, surface ship

AEA Technology plc	UK	*496*	80-00480
Framatome	France	*136*	75-18405
George G. Sharp Inc.	USA	*642*	80-28000
International Combustion Ltd	UK	*461*	75-31800
Ishikawajima-Harima H.I. Co. Ltd	Japan	*251*	75-02800
Newport News Shipbuilding	USA	*599*	75-64300

331 40 Steam turbines, surface ship propulsion

Fincantieri	Italy	*234*	75-09250
Framatome	France	*136*	75-18405
GEC ALSTHOM Medium Turbo Machines	UK	*453*	75-25243
George G. Sharp Inc.	USA	*642*	80-28000
Ishikawajima-Harima H.I. Co. Ltd	Japan	*251*	75-02800

SSS Clutch Co. Inc.	USA	617	75-83570
Weir Pumps Ltd.	UK	493	75-65100

331 40 90 Components, steam turbine, surface ship

A.O. Henton Engineering Co Ltd.	UK	458	75-29040
BTR Silvertown Ltd	UK	437	75-10830
Busak + Shamban Inc.	USA	561	75-16006
Fincantieri	Italy	234	75-09250
Fluorocarbon Company Limited	UK	451	75-24020
Framatome	France	136	75-18405
GEC ALSTHOM Medium Turbo Machines	UK	453	75-25243
George G. Sharp Inc.	USA	642	80-28000
International Combustion Ltd	UK	461	75-31800
James Walker & Co. Ltd	UK	492	75-64120
National Forge Co.	USA	598	75-63850
SSS Clutch Co. Inc.	USA	617	75-83570
Teadit Juntas SA	Brazil	55	75-00910

331 50 Propellers, surface ship

acbLIPS SA	France	129	74-00050
Australian Defence Industries Ltd	Australia	9	55-01000
Brunton's Propellers Ltd	UK	437	75-10400
Dowty Aerospace Propellers	UK	446	75-18587
F. Bamford & Co. Ltd	UK	433	75-05450
Fincantieri	Italy	234	75-09250
George G. Sharp Inc.	USA	642	80-28000
Ishikawajima-Harima H.I. Co. Ltd	Japan	251	75-02800
Kamewa Group	Sweden	377	75-07300
Lips BV	Netherlands	293	75-04200
Lips Propellers	USA	590	75-55250
Mermaid Marine Engines Ltd.	UK	467	75-38550
Michigan Wheel Corp.	USA	596	75-60930
Newport News Shipbuilding	USA	599	75-64300
Schottel-Werft	Germany	174	75-26700
Stone Manganese Marine Ltd	UK	486	75-57335
Sulzer Hydro Wyss GmbH	Germany	175	75-29450

331 60 Thruster units, surface ship

acbLIPS SA	France	129	74-00050
Fincantieri	Italy	234	75-09250
George G. Sharp Inc.	USA	642	80-28000
Lips BV	Netherlands	293	75-04200
Mannesmann Rexroth Ltd	UK	466	75-37100
Schottel France Sarl	France	143	75-34450
Schottel-Werft	Germany	174	75-26700
Vosper Thornycroft	UK	492	75-63800

331 70 Propulsion transmission systems, surface ship

acbLIPS SA	France	129	74-00050
Australian Defence Industries Ltd	Australia	9	55-01000
Brunton's Propellers Ltd	UK	437	75-10400
Fincantieri	Italy	234	75-09250
George G. Sharp Inc.	USA	642	80-28000
Ishikawajima-Harima H.I. Co. Ltd	Japan	251	75-02800
Lucas Aerospace Power Transmission	USA	592	75-56265
Mannesmann Rexroth Ltd	UK	466	75-37100
Newport News Shipbuilding	USA	599	75-64300
RENK AG	Germany	173	75-24700
Schottel France Sarl	France	143	75-34450
Schottel-Werft	Germany	174	75-26700
SSS Clutch Co. Inc.	USA	617	75-83570
SSS Gears Ltd.	UK	486	75-56220
Stork-Wärtsilä Diesel BV	Netherlands	295	75-07300
The Cincinnati Gear Co.	USA	564	75-18675
Twin Disc Inc.	USA	624	75-91350
ZF Friedrichshafen AG	Germany	177	75-35000
ZF Great Britain Ltd	UK	495	75-67720

331 70 40 Propulsion gearing, surface ship, main

Fincantieri	Italy	234	75-09250
GEC ALSTHOM Gears Ltd	UK	453	75-25230
George G. Sharp Inc.	USA	642	80-28000
Ishikawajima-Harima H.I. Co. Ltd	Japan	251	75-02800
Mannesmann Rexroth Ltd	UK	466	75-37100
Mannesmann Rexroth SA	France	139	75-25500
Newport News Shipbuilding	USA	599	75-64300
Royal Schelde	Netherlands	294	75-06150
Schottel France Sarl	France	143	75-34450
Schottel-Werft	Germany	174	75-26700
SSS Clutch Co. Inc.	USA	617	75-83570
SSS Gears Ltd.	UK	486	75-56220
The Cincinnati Gear Co.	USA	564	75-18675
Twin Disc Inc.	USA	624	75-91350
Vickers Shipbuilding & Eng Ltd	UK	491	75-63100
ZF Friedrichshafen AG	Germany	177	75-35000
ZF Great Britain Ltd	UK	495	75-67720

331 70 50 Transmission shaft assemblies, surface ship

Australian Defence Industries Ltd	Australia	9	55-01000
Brunton's Propellers Ltd	UK	437	75-10400
F. Bamford & Co. Ltd	UK	433	75-05450
Fincantieri	Italy	234	75-09250
Flexibox Ltd	UK	451	75-23800
George G. Sharp Inc.	USA	642	80-28000
Ishikawajima-Harima H.I. Co. Ltd	Japan	251	75-02800
Lucas Aerospace Power Transmission	USA	592	75-56265
National Forge Co.	USA	598	75-63850
Newport News Shipbuilding	USA	599	75-64300
Schottel France Sarl	France	143	75-34450
SSS Clutch Co. Inc.	USA	617	75-83570
Sulzer Hydro Wyss GmbH	Germany	175	75-29450
Technical Products Group Inc.	USA	620	75-86764
Vickers Shipbuilding & Eng Ltd	UK	491	75-63100

331 70 90 Propulsion transmission components, surface ship

Brunton's Propellers Ltd	UK	437	75-10400
BTR Silvertown Ltd	UK	437	75-10830
Fincantieri	Italy	234	75-09250
Flexibox Ltd	UK	451	75-23800
GEC ALSTHOM Gears Ltd	UK	453	75-25230
George G. Sharp Inc.	USA	642	80-28000
International Combustion Ltd	UK	461	75-31800
Ishikawajima-Harima H.I. Co. Ltd	Japan	251	75-02800
Loggers Rubbertechniek BV	Netherlands	293	75-04300
Lucas Aerospace Power Transmission	USA	592	75-56265
Mannesmann Rexroth Ltd	UK	466	75-37100
Michell Bearings	UK	468	75-38935
National Forge Co.	USA	598	75-63850
Newport News Shipbuilding	USA	599	75-64300
Schottel France Sarl	France	143	75-34450
SSS Clutch Co. Inc.	USA	617	75-83570
SSS Gears Ltd.	UK	486	75-56220
The Cincinnati Gear Co.	USA	564	75-18675
Twin Disc Inc.	USA	624	75-91350
Vickers Shipbuilding & Eng Ltd	UK	491	75-63100

331 80 Accessories, surface ship propulsion system-related

Brunton's Propellers Ltd	UK	437	75-10400
BTR Silvertown Ltd	UK	437	75-10830
Environmental Elements Corp.	USA	574	75-31850
George G. Sharp Inc.	USA	642	80-28000
Industrial Acoustics Company Ltd	UK	460	75-31300
Insumat Ltd	UK	461	75-31660
Jehier SA	France	137	75-23200
John Crane Inc.	USA	566	75-22500
Kobelt Manufacturing Co. Ltd	Canada	73	75-08770
Newport News Shipbuilding	USA	599	75-64300
Vokes Ltd	UK	492	75-63450
Walter Kidde Aerospace	USA	588	75-52500

331 80 10 Anti-vibration mountings, ship

Avon Spencer Moulton	France	132	75-04750
BTR Silvertown Ltd	UK	437	75-10830
Dunlop Ltd	UK	447	75-18900
George G. Sharp Inc.	USA	642	80-28000
Intermarine SpA	Italy	235	75-11900
James Walker & Co. Ltd	UK	492	75-64120
Loggers Rubbertechniek BV	Netherlands	293	75-04300
Newport News Shipbuilding	USA	599	75-64300

Stop-Choc Ltd.	UK	486	75-57345
Stork-Wärtsilä Diesel BV	Netherlands	295	75-07300

333 Surface effect craft propulsion

Aerojet	USA	547	75-01500
Air Vehicles Ltd	UK	429	75-02150
Lucas Aerospace Power Transmission	USA	592	75-56265
Mermaid Marine Engines Ltd.	UK	467	75-38550
MTU Friedrichshafen GmbH	Germany	172	75-21802
SSS Clutch Co. Inc.	USA	617	75-83570
Stewart & Stevenson Services Inc.	USA	618	75-84180
Stork-Wärtsilä Diesel BV	Netherlands	295	75-07300

333 10 Marine engines, surface effect craft propulsion

Cummins Engine Co. Inc.	USA	567	75-22980
E.P. Barrus Ltd	UK	433	75-05900
GEC ALSTHOM Paxman Diesels Ltd	UK	453	75-25245
Mermaid Marine Engines Ltd.	UK	467	75-38550
MTU Friedrichshafen GmbH	Germany	172	75-21802
SEMT Pielstick	France	144	75-34843
Stewart & Stevenson Services Inc.	USA	618	75-84180
Stork-Wärtsilä Diesel BV	Netherlands	295	75-07300
TTL Dynamics	UK	490	75-61510

333 12 Engines, marine lift fan, surface effect craft

Air Vehicles Ltd	UK	429	75-02150
MTU Friedrichshafen GmbH	Germany	172	75-21802
Stewart & Stevenson Services Inc.	USA	618	75-84180

333 15 Engines, marine turboshaft, surface effect craft

Pratt & Whitney Canada Inc	Canada	75	75-11400
Solar Turbines	USA	616	75-82300
Stewart & Stevenson Services Inc.	USA	618	75-84180
TTL Dynamics	UK	490	75-61510

333 50 Propellers (airscrews), surface effect craft

Arrowprop Co. Inc.	USA	555	75-09150
Dowty Aerospace Propellers	UK	446	75-18587
Ratier-Figeac SA	France	141	75-31800

333 55 Lift fans, surface effect craft

Air Vehicles Ltd	UK	429	75-02150
Howden Sirocco	France	137	75-21700
Textron Marine & Land Systems	USA	622	75-88660

333 70 Propulsion transmission systems, surface effect ship / craft

acbLIPS SA	France	129	74-00050
Detroit Diesel Corp.	USA	568	75-25830
Lucas Aerospace Power Transmission	USA	592	75-56265
MTU do Brasil Ltda	Brazil	57	75-13900
The Cincinnati Gear Co.	USA	564	75-18675
ZF Friedrichshafen AG	Germany	177	75-35000
ZF Great Britain Ltd	UK	495	75-67720

333 70 90 Propulsion transmission components, surface effect ship / craft

BTR Silvertown Ltd	UK	437	75-10830
Flexibox Ltd	UK	451	75-23800
GEC ALSTHOM Gears Ltd	UK	453	75-25230
Loggers Rubbertechniek BV	Netherlands	293	75-04300
Lucas Aerospace Power Transmission	USA	592	75-56265
National Forge Co.	USA	598	75-63850
Parker Bertea Aerospace	USA	603	75-68900
SSS Clutch Co. Inc.	USA	617	75-83570
Textron Marine & Land Systems	USA	622	75-88660
The Cincinnati Gear Co.	USA	564	75-18675

334 Hydrofoil propulsion

Aerojet	USA	547	75-01500
Lucas Aerospace Power Transmission	USA	592	75-56265
Mermaid Marine Engines Ltd.	UK	467	75-38550
MTU Friedrichshafen GmbH	Germany	172	75-21802
Stewart & Stevenson Services Inc.	USA	618	75-84180

334 10 Engines, marine diesel, hydrofoil propulsion

Cummins Engine Co. Inc.	USA	567	75-22980
Detroit Diesel Corp.	USA	568	75-25830
FP Diesel LP	USA	576	75-35753
GEC ALSTHOM Paxman Diesels Ltd	UK	453	75-25245
Mermaid Marine Engines Ltd.	UK	467	75-38550
MTU Friedrichshafen GmbH	Germany	172	75-21802
Stewart & Stevenson Services Inc.	USA	618	75-84180
Stork-Wärtsilä Diesel BV	Netherlands	295	75-07300
Volvo Penta UK Ltd	UK	492	75-63600

334 20 Engines, marine gas turbine, hydrofoil propulsion

MTU Friedrichshafen GmbH	Germany	172	75-21802
Rolls-Royce (Canada) Ltd.	Canada	75	75-12120
Rolls-Royce plc	UK	479	75-50250
Stewart & Stevenson Services Inc.	USA	618	75-84180

334 30 Waterjet propulsion units, hydrofoil

acbLIPS SA	France	129	74-00050
Kamewa Group	Sweden	377	75-07300
North American Marine Jet	USA	599	75-65210
Schottel France Sarl	France	143	75-34450

334 30 90 Components, waterjet propulsion units, hydrofoil

A.O. Henton Engineering Co Ltd.	UK	458	75-29040
BTR Silvertown Ltd	UK	437	75-10830
Busak + Shamban Inc.	USA	561	75-16006
Fluorocarbon Company Limited	UK	451	75-24020
National Forge Co.	USA	598	75-63850

334 50 Propellers, hydrofoil

acbLIPS SA	France	129	74-00050
Dowty Aerospace Propellers	UK	446	75-18587
Lips BV	Netherlands	293	75-04200

334 70 Propulsion transmission systems, hydrofoil

acbLIPS SA	France	129	74-00050
Lucas Aerospace Power Transmission	USA	592	75-56265
The Cincinnati Gear Co.	USA	564	75-18675
Volvo Penta UK Ltd	UK	492	75-63600
ZF Friedrichshafen AG	Germany	177	75-35000
ZF Great Britain Ltd	UK	495	75-67720

334 70 90 Propulsion transmission components, hydrofoil

BTR Silvertown Ltd	UK	437	75-10830
Flexibox Ltd	UK	451	75-23800
GEC ALSTHOM Gears Ltd	UK	453	75-25230
Lucas Aerospace Power Transmission	USA	592	75-56265
National Forge Co.	USA	598	75-63850
SSS Clutch Co. Inc.	USA	617	75-83570
The Cincinnati Gear Co.	USA	564	75-18675
Volvo Penta UK Ltd	UK	492	75-63600

337 Launch / small craft propulsion

AB Volvo Penta	Sweden	379	75-11850
E.P. Barrus Ltd	UK	433	75-05900
Lister-Petter Ltd.	UK	465	75-36070
Michigan Wheel Corp.	USA	596	75-60930
MTU Friedrichshafen GmbH	Germany	172	75-21802
MWM Motores Diesel Ltda	Brazil	57	75-13980

Schottel-Werft	Germany	*174*	75-26700
Stewart & Stevenson Services Inc.	USA	*618*	75-84180
Volvo Penta UK Ltd	UK	*492*	75-63600

337 10 Engines, launch / small craft, on-board

AB Volvo Penta	Sweden	*379*	75-11850
Cummins Engine Co. Inc.	USA	*567*	75-22980
Detroit Diesel Corp.	USA	*568*	75-25830
E.P. Barrus Ltd	UK	*433*	75-05900
Lister-Petter Ltd	UK	*465*	75-36070
Volvo Penta UK Ltd	UK	*492*	75-63600

337 11 Diesel engines, launch / small craft

AB Volvo Penta	Sweden	*379*	75-11850
Caterpillar Inc.	USA	*562*	75-17100
Detroit Diesel Corp.	USA	*568*	75-25830
Deütz Motor GmbH	Germany	*170*	75-15900
E.P. Barrus Ltd	UK	*433*	75-05900
FP Diesel LP	USA	*576*	75-35753
GEC ALSTHOM Paxman Diesels Ltd	UK	*453*	75-25245
Lister-Petter Ltd	UK	*465*	75-36070
Mermaid Marine Engines Ltd	UK	*467*	75-38550
MTU Friedrichschafen GmbH	Germany	*172*	75-21802
MWM Motores Diesel Ltda	Brazil	*57*	75-13980
Perkins Engines (Shrews.)	UK	*474*	75-44750
Stewart & Stevenson Services Inc.	USA	*618*	75-84180
Volvo Penta UK Ltd	UK	*492*	75-63600

337 12 Petrol engines, launch / small craft

Volvo Penta UK Ltd	UK	*492*	75-63600

337 12 50 Outboard motors, launch / small craft

E.P. Barrus Ltd	UK	*433*	75-05900

337 20 Launch engines, petrol, outboard

E.P. Barrus Ltd	UK	*433*	75-05900

337 50 Propellers, launch / small craft

acbLIPS SA	France	*129*	74-00050
Brunton's Propellers Ltd	UK	*437*	75-10400
E.P. Barrus Ltd	UK	*433*	75-05900
F. Bamford & Co. Ltd	UK	*433*	75-05450
Lips BV	Netherlands	*293*	75-04200
Michigan Wheel Corp.	USA	*596*	75-60930
Schottel-Werft	Germany	*174*	75-26700
Volvo Penta UK Ltd	UK	*492*	75-63600

339 Submarine, submersible propulsion

COSMOS SpA	Italy	*233*	75-05700
GEC ALSTHOM Medium Turbo Machines	UK	*453*	75-25243
Holec Machines & Apparaten BV	Netherlands	*293*	75-03110
Int'l Submarine Engrg.	Canada	*73*	75-07780
Lucas Aerospace Power Transmission	USA	*592*	75-56265
Michigan Wheel Corp.	USA	*596*	75-60930
MTU Friedrichshafen GmbH	Germany	*172*	75-21802
RDM Submarines BV	Netherlands	*294*	75-05770
Riva Calzoni SpA	Italy	*238*	75-20500
SELIN Sistemi Sarl	Italy	*238*	75-21750
Vickers Shipbuilding & Eng Ltd	UK	*491*	75-63100
Vokes Ltd	UK	*492*	75-63450

339 10 Submarine propulsion systems, diesel-electric

COSMOS SpA	Italy	*233*	75-05700
RDM Submarines BV	Netherlands	*294*	75-05770
SELIN Sistemi Sarl	Italy	*238*	75-21750
Stork-Wärtsilä Diesel BV	Netherlands	*295*	75-07300
Vickers Shipbuilding & Eng Ltd	UK	*491*	75-63100

339 10 80 Accessories, diesel-electric propulsion systems, submarine

Aeroquip Corp.	USA	*547*	75-02350
SELIN Sistemi Sarl	Italy	*238*	75-21750
Stork-Wärtsilä Diesel BV	Netherlands	*295*	75-07300

339 11 Engines, diesel, submarine

Detroit Diesel Corp.	USA	*568*	75-25830
Fincantieri	Italy	*234*	75-09250
GEC ALSTHOM Paxman Diesels Ltd	UK	*453*	75-25245
Hedemora Diesel AB	Sweden	*377*	75-06030
MTU Friedrichschafen GmbH	Germany	*172*	75-21802
Newport News Shipbuilding	USA	*599*	75-64300
SEMT Pielstick	France	*144*	75-34843
Stork-Wärtsilä Diesel BV	Netherlands	*295*	75-07300

339 11 80 Accessories, diesel engine, submarine

Aeroquip Corp.	USA	*547*	75-02350
Fincantieri	Italy	*234*	75-09250
Insumat Ltd	UK	*461*	75-31660
Newport News Shipbuilding	USA	*599*	75-64300
Riva Calzoni SpA	Italy	*238*	75-20500
Vokes Ltd	UK	*492*	75-63450

339 12 Submarine propulsion motors, electric

GEC ALSTHOM Large Machines Ltd	UK	*453*	75-25240
Holec Machines & Apparaten BV	Netherlands	*293*	75-03110
Newport News Shipbuilding	USA	*599*	75-64300

339 20 Nuclear propulsion systems, submarine

AEA Technology plc	UK	*496*	80-00480
Framatome	France	*136*	75-18405
Newport News Shipbuilding	USA	*599*	75-64300
Rolls-Royce plc	UK	*479*	75-50250
Vickers Shipbuilding & Eng Ltd	UK	*491*	75-63100

339 21 Steam turbines, nuclear submarine propulsion

GEC ALSTHOM Medium Turbo Machines	UK	*453*	75-25243
Newport News Shipbuilding	USA	*599*	75-64300
Rolls-Royce & Associates Ltd	UK	*510*	80-29100

339 50 Propellers, submarine / submersible

acbLIPS SA	France	*129*	74-00050
Brunton's Propellers Ltd	UK	*437*	75-10400
F. Bamford & Co. Ltd	UK	*433*	75-05450
Fincantieri	Italy	*234*	75-09250
Kamewa Group	Sweden	*377*	75-07300
Lips BV	Netherlands	*293*	75-04200
Michigan Wheel Corp.	USA	*596*	75-60930
Newport News Shipbuilding	USA	*599*	75-64300
Stone Manganese Marine Ltd	UK	*486*	75-57335

339 51 Propellers, submarine

acbLIPS SA	France	*129*	74-00050
Kamewa Group	Sweden	*377*	75-07300
Lips BV	Netherlands	*293*	75-04200
Newport News Shipbuilding	USA	*599*	75-64300
Stone Manganese Marine Ltd	UK	*486*	75-57335

339 52 Propellers, submersible

acbLIPS SA	France	*129*	74-00050
Brunton's Propellers Ltd	UK	*437*	75-10400
Lips BV	Netherlands	*293*	75-04200
Newport News Shipbuilding	USA	*599*	75-64300
Stone Manganese Marine Ltd	UK	*486*	75-57335

339 70 Propulsion transmission systems, submarine

acbLIPS SA	France	129	74-00050
Lucas Aerospace Power Transmission	USA	592	75-56265
MIL Systems	Canada	79	80-01360
Newport News Shipbuilding	USA	599	75-64300
RDM Submarines BV	Netherlands	294	75-05770
Vickers Shipbuilding & Eng Ltd	UK	491	75-63100

339 70 30 Propulsion drives, submersible

EUROCONTROL SpA	Italy	233	75-08200
Newport News Shipbuilding	USA	599	75-64300

339 70 40 Gearing, main propulsion, submarine

GEC ALSTHOM Gears Ltd	UK	453	75-25230
Mannesmann Rexroth Ltd	UK	466	75-37100
Newport News Shipbuilding	USA	599	75-64300
Vickers Shipbuilding & Eng Ltd	UK	491	75-63100

339 70 50 Transmission shaft assemblies, submarine

Flexibox Ltd	UK	451	75-23800
Lucas Aerospace Power Transmission	USA	592	75-56265
Newport News Shipbuilding	USA	599	75-64300
Technical Products Group Inc.	USA	620	75-86764
Vickers Shipbuilding & Eng Ltd	UK	491	75-63100

339 70 90 Propulsion transmission components, submarine

BTR Silvertown Ltd	UK	437	75-10830
EG&G Sealol Ltd	UK	448	75-19675
Flexibox Ltd	UK	451	75-23800
GEC ALSTHOM Gears Ltd	UK	453	75-25230
Lucas Aerospace Power Transmission	USA	592	75-56265
Mannesmann Rexroth Ltd	UK	466	75-37100
Michell Bearings	UK	468	75-38935
National Forge Co.	USA	598	75-63850
Newport News Shipbuilding	USA	599	75-64300
Vickers Shipbuilding & Eng Ltd	UK	491	75-63100

Land-based Platform Propulsion

341 Armoured vehicle propulsion

David Brown Defence Equipment Ltd	UK	437	75-10150
FP Diesel LP	USA	576	75-35753
Hercules Engines Company	USA	581	75-43080
Howden Aircontrol Ltd	UK	459	75-30240
Howden Wade Ltd	UK	459	75-30260
IVECO Fiat SpA	Italy	235	75-13230
MTU Friedrichshafen GmbH	Germany	172	75-21802
MWM Motores Diesel Ltda	Brazil	57	75-13980
Schottel-Werft	Germany	174	75-26700
SELIN Sistemi Sarl	Italy	238	75-21750
Vokes Ltd	UK	492	75-63450

341 10 Tracked armoured vehicle propulsion

AAI Corp.	USA	545	75-00020
Avica Equipment	UK	432	75-04870
British Filters Ltd	UK	436	75-09030
Cummins Engine Co. Inc.	USA	567	75-22980
DAF Trucks NV	Netherlands	292	75-01400
David Brown Defence Equipment Ltd	UK	437	75-10150
Detroit Diesel Corp.	USA	568	75-25830
FP Diesel LP	USA	576	75-35753
Giat Industries	France	126	55-00100
Hercules Engines Company	USA	581	75-43080
Howden Aircontrol Ltd	UK	459	75-30240
IVECO Fiat SpA	Italy	235	75-13230
LaCombe Ind. e Com. de Turbos Ltda	Brazil	56	75-12050
SELIN Sistemi Sarl	Italy	238	75-21750
Teadit Juntas SA	Brazil	55	75-00910
Varity Perkins	UK	474	75-44750

341 10 10 Diesel engines, tracked armoured vehicle

AAI Corp.	USA	545	75-00020
Cummins Engine Co. Inc.	USA	567	75-22980
DAF Trucks NV	Netherlands	292	75-01400
Detroit Diesel Corp.	USA	568	75-25830
FP Diesel LP	USA	576	75-35753
Hammonds Technical Services Inc.	USA	580	75-41800
Hercules Engines Company	USA	581	75-43080
Merex Inc.	USA	595	75-60250
MTU Friedrichshafen GmbH	Germany	172	75-21802
NORINCO	China PR	88	55-00280
Perkins Engines (Shrews.)	UK	474	75-44750
Stewart & Stevenson Services Inc.	USA	618	75-84180
Varity Perkins	UK	474	75-44750
Wartsila France SA	France	149	75-47090

341 10 10 80 Accessories, diesel engine, tracked armoured vehicle

AAI Corp.	USA	545	75-00020
FP Diesel LP	USA	576	75-35753
Hammonds Technical Services Inc.	USA	580	75-41800
Hercules Engines Company	USA	581	75-43080
Howden Wade Ltd	UK	459	75-30260
Kinetics Ltd	Israel	220	75-11320
Merex Inc.	USA	595	75-60250
Stewart & Stevenson Services Inc.	USA	618	75-84180
Turbomeca	France	148	75-46000
Walter Kidde Aerospace	USA	588	75-52500

341 10 10 90 Components, diesel engine, tracked armoured vehicle

AAI Corp.	USA	545	75-00020
A.O. Henton Engineering Co Ltd.	UK	458	75-29040
Busak + Shamban Inc.	USA	561	75-16006
Fluorocarbon Company Limited	UK	451	75-24020
FP Diesel LP	USA	576	75-35753
Hercules Engines Company	USA	581	75-43080
LaCombe Ind. e Com. de Turbos Ltda	Brazil	56	75-12050
Merex Inc.	USA	595	75-60250
National Forge Co.	USA	598	75-63850
Ordex Inc.	USA	601	75-67350
Perkins Engines (Shrews.)	UK	474	75-44750
Stewart & Stevenson Services Inc.	USA	618	75-84180
Teadit Juntas SA	Brazil	55	75-00910
Voss Industries Inc.	USA	626	75-94550
Warsaw Mechanical Works "Wuzetem"	Poland	324	55-20000

341 10 10 90 20 Cooling equipment, diesel engine, tracked armoured vehicle

AAI Corp.	USA	545	75-00020
FP Diesel LP	USA	576	75-35753
Gallay Ltd	UK	453	75-25100
General Thermodynamics Inc.	USA	578	75-39010
Hercules Engines Company	USA	581	75-43080
Howden Aircontrol Ltd	UK	459	75-30240
Merex Inc.	USA	595	75-60250
Microtecnica-Torino SpA	Italy	237	75-15500
Serck Heat Transfer	UK	483	75-52880

341 10 10 90 40 Air filters, diesel engine, tracked armoured vehicle

British Filters Ltd	UK	436	75-09030
Coopers Filters Ltd	UK	442	75-14950
Donaldson Co. Inc.	USA	569	75-27050
Farr Europe	UK	450	75-22410
FP Diesel LP	USA	576	75-35753
Gallay Ltd	UK	453	75-25100
GUD Manufacturing Co. Pty Ltd	Australia	17	75-07885
Hercules Engines Company	USA	581	75-43080
Kinetics Ltd	Israel	220	75-11320

341 10 10 90 45 Fuel / oil filters, diesel engine, tracked armoured vehicle

British Filters Ltd	UK	*436*	75-09030
Coopers Filters Ltd	UK	*442*	75-14950
Donaldson Co. Inc.	USA	*569*	75-27050
FP Diesel LP	USA	*576*	75-35753
Hercules Engines Company	USA	*581*	75-43080
Kinetics Ltd	Israel	*220*	75-11320
Labinal	France	*138*	75-23900
Vokes Ltd	UK	*492*	75-63450

341 10 15 Multifuel engines, tracked armoured vehicle

Detroit Diesel Corp.	USA	*568*	75-25830
Hercules Engines Company	USA	*581*	75-43080
MTU Friedrichschafen GmbH	Germany	*172*	75-21802

341 10 15 80 Accessories, multifuel engine, tracked armoured vehicle

Aerospace Syst. & Comp.	USA	*547*	75-02501
Hercules Engines Company	USA	*581*	75-43080
Insumat Ltd	UK	*461*	75-31660
Ordex Inc.	USA	*601*	75-67350
Walter Kidde Aerospace	USA	*588*	75-52500

341 10 15 90 Components, multifuel engine, tracked armoured vehicle

Aeroquip Corp.	USA	*547*	75-02350
Aerospace Syst. & Comp.	USA	*547*	75-02501
A.O. Henton Engineering Co Ltd.	UK	*458*	75-29040
Busak + Shamban Inc.	USA	*561*	75-16006
Fluorocarbon Company Limited	UK	*451*	75-24020
Hercules Engines Company	USA	*581*	75-43080
National Forge Co.	USA	*598*	75-63850
Ordex Inc.	USA	*601*	75-67350
Voss Industries Inc.	USA	*626*	75-94550
Warsaw Mechanical Works "Wuzetem"	Poland	*324*	55-20000

341 10 15 90 20 Cooling equipment, multifuel engine, tracked armoured vehicle

Aerospace Syst. & Comp.	USA	*547*	75-02501
Gallay Ltd	UK	*453*	75-25100
Hercules Engines Company	USA	*581*	75-43080
Howden Aircontrol Ltd.	UK	*459*	75-30240
International Radiator Services Ltd	UK	*461*	75-31840
Merex Inc.	USA	*595*	75-60250
Microtecnica-Torino SpA	Italy.	*237*	75-15500
Parker Bertea Aerospace.	USA	*603*	75-68900
Serck Heat Transfer	UK	*483*	75-52880

341 10 15 90 40 Air filters, multifuel engine, tracked armoured vehicle

British Filters Ltd	UK	*436*	75-09030
Coopers Filters Ltd	UK	*442*	75-14950
Donaldson Co. Inc.	USA	*569*	75-27050
Farr Europe	UK	*450*	75-22410
Gallay Ltd	UK	*453*	75-25100
GUD Manufacturing Co. Pty Ltd	Australia	*17*	75-07885
Hercules Engines Company	USA	*581*	75-43080
Kinetics Ltd	Israel	*220*	75-11320

341 10 15 90 45 Fuel / oil filters, multifuel engine, tracked armoured vehicle

British Filters Ltd	UK	*436*	75-09030
Coopers Filters Ltd	UK	*442*	75-14950
Hercules Engines Company	USA	*581*	75-43080
Kinetics Ltd	Israel	*220*	75-11320
Labinal	France	*138*	75-23900
Ordex Inc.	USA	*601*	75-67350

341 10 20 80 Accessories, petrol engine, tracked armoured vehicle

Walter Kidde Aerospace	USA	*588*	75-52500

341 10 20 90 Components, petrol engine, tracked armoured vehicle

Aeroquip Corp.	USA	*547*	75-02350
BTR Silvertown Ltd	UK	*437*	75-10830
Busak + Shamban Inc.	USA	*561*	75-16006
Fluorocarbon Company Limited	UK	*451*	75-24020
Voss Industries Inc.	USA	*626*	75-94550
Warsaw Mechanical Works "Wuzetem"	Poland	*324*	55-20000

341 10 20 90 20 Cooling equipment, petrol engine, tracked armoured vehicle

Gallay Ltd	UK	*453*	75-25100
Howden Aircontrol Ltd.	UK	*459*	75-30240
Microtecnica-Torino SpA	Italy.	*237*	75-15500
Parker Bertea Aerospace.	USA	*603*	75-68900

341 10 20 90 40 Air filters, petrol engine, tracked armoured vehicle

British Filters Ltd	UK	*436*	75-09030
Coopers Filters Ltd	UK	*442*	75-14950
Donaldson Co. Inc.	USA	*569*	75-27050
Farr Europe	UK	*450*	75-22410
Gallay Ltd	UK	*453*	75-25100
GUD Manufacturing Co. Pty Ltd	Australia	*17*	75-07885
Kinetics Ltd	Israel	*220*	75-11320

341 10 20 90 45 Fuel / oil filters, petrol engine, tracked armoured vehicle

British Filters Ltd	UK	*436*	75-09030
Coopers Filters Ltd	UK	*442*	75-14950
Donaldson Co. Inc.	USA	*569*	75-27050
Kinetics Ltd	Israel	*220*	75-11320
Labinal	France	*138*	75-23900

341 10 25 Gas turbine engines, tracked armoured vehicle

Aero Systems Engineering Inc.	USA	*546*	75-01116
Hardcote Technologies	USA	*580*	75-41900
Merex Inc.	USA	*595*	75-60250
Rolls-Royce (Canada) Ltd.	Canada	*75*	75-12120
Williams International	USA	*628*	75-96950

341 10 25 80 Accessories, gas turbine engine, tracked armoured vehicle

Aero Systems Engineering Inc.	USA	*546*	75-01116
Aerospace Syst. & Comp.	USA	*547*	75-02501
Insumat Ltd	UK	*461*	75-31660
Merex Inc.	USA	*595*	75-60250
Microturbo Ltd	UK	*468*	75-39160
Parker Bertea Aerospace.	USA	*603*	75-68900
Walter Kidde Aerospace	USA	*588*	75-52500

341 10 25 90 Components, gas turbine engine, tracked armoured vehicle

Aero Systems Engineering Inc.	USA	*546*	75-01116
Aeroquip Corp.	USA	*547*	75-02350
Aerospace Syst. & Comp.	USA	*547*	75-02501
A.O. Henton Engineering Co Ltd.	UK	*458*	75-29040
BTR Silvertown Ltd	UK	*437*	75-10830
Busak + Shamban Inc.	USA	*561*	75-16006
Fluorocarbon Company Limited	UK	*451*	75-24020
Hardcote Technologies	USA	*580*	75-41900
Howmet Corporation	USA	*582*	75-44550
J.S. Chinn & Co. Ltd	UK	*439*	75-13045
Mibert Precision Ltd	UK	*468*	75-38920

Microturbo Ltd	UK	*468*	75-39160
National Forge Co.	USA	*598*	75-63850
Parker Bertea Aerospace	USA	*603*	75-68900
Price & Orphin Limited	UK	*509*	80-24250
Voss Industries Inc.	USA	*626*	75-94550

341 10 30 Electric propulsion systems, tracked armoured vehicle

Aero Systems Engineering Inc.	USA	*546*	75-01116
Kaman Electromagnetics Corp.	USA	*637*	80-17500
SELIN Sistemi Sarl	Italy	*238*	75-21750

341 10 30 80 Accessories, electric propulsion systems, tracked armoured vehicle

SELIN Sistemi Sarl	Italy	*238*	75-21750

341 10 30 90 Components, electric propulsion systems, tracked armoured vehicle

A.O. Henton Engineering Co Ltd.	UK	*458*	75-29040
David Brown Defence Equipment Ltd	UK	*437*	75-10150
SELIN Sistemi Sarl	Italy	*238*	75-21750

341 10 35 Water propulsion systems, tracked armoured vehicle

Detroit Diesel Corp.	USA	*568*	75-25830
Schottel-Werft	Germany	*174*	75-26700

341 10 35 10 Water propulsion systems, propeller-driven, tracked armoured vehicle

Detroit Diesel Corp.	USA	*568*	75-25830

341 10 35 10 90 Components, water propulsion systems, propeller, tracked armoured vehicle

BTR Silvertown Ltd	UK	*437*	75-10830
Busak + Shamban Inc.	USA	*561*	75-16006
Fluorocarbon Company Limited	UK	*451*	75-24020

341 10 35 20 Water propulsion systems, waterjet-type, tracked armoured vehicle

David Brown Defence Equipment Ltd	UK	*437*	75-10150
Detroit Diesel Corp.	USA	*568*	75-25830

341 10 35 20 90 Components, waterjet propulsion system, tracked armoured vehicle

A.O. Henton Engineering Co Ltd.	UK	*458*	75-29040
David Brown Defence Equipment Ltd	UK	*437*	75-10150

341 10 70 Transmission systems, tracked armoured vehicle

Alvis Industries Ltd	UK	*430*	75-02850
Ashot Ashkelon Industries Ltd	Israel	*217*	75-02100
Champion Friction Co.	USA	*563*	75-17975
DAF Trucks NV	Netherlands	*292*	75-01400
David Brown Defence Equipment Ltd	UK	*437*	75-10150
Lucas Aerospace Power Transmission	USA	*592*	75-56265
Mannesmann Rexroth SA	France	*139*	75-25500
Self-Changing Gears Ltd	UK	*482*	75-52700
ZF Friedrichshafen AG	Germany	*177*	75-35000
ZF Great Britain Ltd	UK	*495*	75-67720

341 10 79 Transmission system components, tracked armoured vehicle

Alvis Industries Ltd	UK	*430*	75-02850
Ashot Ashkelon Industries Ltd	Israel	*217*	75-02100
David Brown Defence Equipment Ltd	UK	*437*	75-10150
David Brown Gear Industries Ltd	Australia	*13*	75-02800
Giat Industries	France	*126*	55-00100
Kinetics Ltd	Israel	*220*	75-11320
Lucas Aerospace Power Transmission	USA	*592*	75-56265
Mannesmann Rexroth SA	France	*139*	75-25500
RENK AG	Germany	*173*	75-24700
Self-Changing Gears Ltd	UK	*482*	75-52700

341 10 79 40 Gearboxes, tracked armoured vehicle

Alvis Transmission Ltd	UK	*430*	75-02862
Ashot Ashkelon Industries Ltd	Israel	*217*	75-02100
David Brown Defence Equipment Ltd	UK	*437*	75-10150
Giat Industries	France	*126*	55-00100
RENK AG	Germany	*173*	75-24700
Self-Changing Gears Ltd	UK	*482*	75-52700
ZF Friedrichshafen AG	Germany	*177*	75-35000
ZF Great Britain Ltd	UK	*495*	75-67720

341 10 80 Accessories, tracked armoured vehicle propulsion system-related

David Brown Defence Equipment Ltd	UK	*437*	75-10150
Harco Manufacturing Co. Inc.	USA	*580*	75-41870
Kinetics Ltd	Israel	*220*	75-11320
Self-Changing Gears Ltd	UK	*482*	75-52700

341 10 80 10 Anti-vibration mountings, tracked armoured vehicle

Avon Spencer Moulton	France	*132*	75-04750
BTR Silvertown Ltd	UK	*437*	75-10830
James Walker & Co. Ltd	UK	*492*	75-64120
Karipidis Bros. SA	Greece	*190*	75-01980
Loggers Rubbertechniek BV	Netherlands	*293*	75-04300
Stop-Choc Ltd.	UK	*486*	75-57345
Vibrachoc SA	France	*148*	75-46900

341 30 Wheeled armoured vehicle propulsion

DAF Trucks NV	Netherlands	*292*	75-01400
Detroit Diesel Corp.	USA	*568*	75-25830
Deutz Motor Industrie Motoren GmbH	Germany	*167*	75-06350
FP Diesel LP	USA	*576*	75-35753
Hercules Engines Company	USA	*581*	75-43080
Howden Aircontrol Ltd	UK	*459*	75-30240
IVECO Fiat SpA	Italy	*235*	75-13230
LaCombe Ind. e Com. de Turbos Ltda	Brazil	*56*	75-12050
MAN Nutzfahrzeuge AG	Germany	*172*	75-19640
Rolls-Royce Motor Cars Ltd	UK	*480*	75-50270
SELIN Sistemi Sarl	Italy	*238*	75-21750
Teadit Juntas SA	Brazil	*55*	75-00910
Varity Perkins	UK	*474*	75-44750

341 30 10 Diesel engines, wheeled armoured vehicle

Caterpillar Inc.	USA	*562*	75-17100
Cummins Engine Co. Inc.	USA	*567*	75-22980
DAF Trucks NV	Netherlands	*292*	75-01400
Detroit Diesel Corp.	USA	*568*	75-25830
FP Diesel LP	USA	*576*	75-35753
GM-Holden's Automotive	Australia	*16*	75-07520
Hammonds Technical Services Inc.	USA	*580*	75-41800
Hercules Engines Company	USA	*581*	75-43080
MAN Nutzfahrzeuge AG	Germany	*172*	75-19640
Merex Inc.	USA	*595*	75-60250
NORINCO	China PR	*88*	55-00280
Perkins Engines (Shrews.)	UK	*474*	75-44750
Rolls-Royce Motor Cars Ltd	UK	*480*	75-50270
Stewart & Stevenson Services Inc.	USA	*618*	75-84180
Varity Perkins	UK	*474*	75-44750

341 30 10 80 Accessories, diesel engine, wheeled armoured vehicle

Donaldson Co. Inc.	USA	*569*	75-27050
FP Diesel LP	USA	*576*	75-35753
Hammonds Technical Services Inc.	USA	*580*	75-41800

Hercules Engines Company	USA	*581*	75-43080
Hispano Suiza	France	*137*	75-21300
Howden Wade Ltd	UK	*459*	75-30260
Kinetics Ltd	Israel	*220*	75-11320
Merex Inc.	USA	*595*	75-60250
Stewart & Stevenson Services Inc.	USA	*618*	75-84180
Walter Kidde Aerospace	USA	*588*	75-52500

341 30 10 90 Components, diesel engine, wheeled armoured vehicle

Aeroquip Corp.	USA	*547*	75-02350
A.O. Henton Engineering Co Ltd.	UK	*458*	75-29040
Busak + Shamban Inc.	USA	*561*	75-16006
Fluorocarbon Company Limited	UK	*451*	75-24020
FP Diesel LP	USA	*576*	75-35753
Hercules Engines Company	USA	*581*	75-43080
Hispano Suiza	France	*137*	75-21300
Kinetics Ltd	Israel	*220*	75-11320
LaCombe Ind. e Com. de Turbos Ltda	Brazil	*56*	75-12050
Merex Inc.	USA	*595*	75-60250
Ordex Inc.	USA	*601*	75-67350
Perkins Engines (Shrews.)	UK	*474*	75-44750
Stewart & Stevenson Services Inc.	USA	*618*	75-84180
Teadit Juntas SA	Brazil	*55*	75-00910
Warsaw Mechanical Works "Wuzetem"	Poland	*324*	55-20000

341 30 10 90 20 Cooling equipment, diesel engine, wheeled armoured vehicle

Air International Transit	Australia	*11*	75-00450
FP Diesel LP	USA	*576*	75-35753
Gallay Ltd	UK	*453*	75-25100
General Thermodynamics Inc.	USA	*578*	75-39010
Hercules Engines Company	USA	*581*	75-43080
Howden Aircontrol Ltd	UK	*459*	75-30240
Microtecnica-Torino SpA	Italy	*237*	75-15500
Serck Heat Transfer	UK	*483*	75-52880

341 30 15 Multifuel engines, wheeled armoured vehicle

Detroit Diesel Corp.	USA	*568*	75-25830
Hercules Engines Company	USA	*581*	75-43080
Stewart & Stevenson Services Inc.	USA	*618*	75-84180
Vickers Defence Systems	UK	*491*	75-63040

341 30 15 80 Accessories, multifuel engine, wheeled armoured vehicle

Aerospace Syst. & Comp.	USA	*547*	75-02501
Autocar Electrical Equipment Ltd	UK	*432*	75-04440
Donaldson Co. Inc.	USA	*569*	75-27050
Hercules Engines Company	USA	*581*	75-43080
Insumat Ltd	UK	*461*	75-31660
Parker Bertea Aerospace	USA	*603*	75-68900
Walter Kidde Aerospace	USA	*588*	75-52500

341 30 15 90 Components, multifuel engine, wheeled armoured vehicle

Aeroquip Corp.	USA	*547*	75-02350
Aerospace Syst. & Comp.	USA	*547*	75-02501
A.O. Henton Engineering Co Ltd.	UK	*458*	75-29040
BTR Silvertown Ltd	UK	*437*	75-10830
Fluorocarbon Company Limited	UK	*451*	75-24020
Hercules Engines Company	USA	*581*	75-43080
Parker Bertea Aerospace	USA	*603*	75-68900
Voss Industries Inc.	USA	*626*	75-94550
Warsaw Mechanical Works "Wuzetem"	Poland	*324*	55-20000

341 30 15 90 20 Cooling equipment, multifuel engine, wheeled armoured vehicle

Aerospace Syst. & Comp.	USA	*547*	75-02501
Air International Transit	Australia	*11*	75-00450
Gallay Ltd	UK	*453*	75-25100

Hercules Engines Company	USA	*581*	75-43080
Howden Aircontrol Ltd	UK	*459*	75-30240
Microtecnica-Torino SpA	Italy	*237*	75-15500
Parker Bertea Aerospace	USA	*603*	75-68900

341 30 20 Petrol engines, wheeled armoured vehicle

Elgin Industries	USA	*573*	75-30800
Rolls-Royce Motor Cars Ltd	UK	*480*	75-50270
Vickers Defence Systems	UK	*491*	75-63040

341 30 20 80 Accessories, petrol engine, wheeled armoured vehicle

Autocar Electrical Equipment Ltd	UK	*432*	75-04440
Donaldson Co. Inc.	USA	*569*	75-27050
Walter Kidde Aerospace	USA	*588*	75-52500

341 30 20 90 Components, petrol engine, wheeled armoured vehicle

Aeroquip Corp.	USA	*547*	75-02350
A.O. Henton Engineering Co Ltd.	UK	*458*	75-29040
BTR Silvertown Ltd	UK	*437*	75-10830
Busak + Shamban Inc.	USA	*561*	75-16006
Elgin Industries	USA	*573*	75-30800
Fluorocarbon Company Limited	UK	*451*	75-24020
Voss Industries Inc.	USA	*626*	75-94550
Warsaw Mechanical Works "Wuzetem"	Poland	*324*	55-20000

341 30 20 90 20 Cooling equipment, petrol engine, wheeled armoured vehicle

Air International Transit	Australia	*11*	75-00450
Gallay Ltd	UK	*453*	75-25100
Howden Aircontrol Ltd	UK	*459*	75-30240
Microtecnica-Torino SpA	Italy	*237*	75-15500
Parker Bertea Aerospace	USA	*603*	75-68900

341 30 30 Electric propulsion systems, wheeled armoured vehicle

Holec Machines & Apparaten BV	Netherlands	*293*	75-03110
Kaman Electromagnetics Corp.	USA	*637*	80-17500
SELIN Sistemi Sarl	Italy	*238*	75-21750

341 30 30 90 Components, electric propulsion system, wheeled armoured vehicle

A.O. Henton Engineering Co Ltd.	UK	*458*	75-29040
BTR Silvertown Ltd	UK	*437*	75-10830
Holec Machines & Apparaten BV	Netherlands	*293*	75-03110
SELIN Sistemi Sarl	Italy	*238*	75-21750

341 30 35 Water propulsion & steering systems, wheeled armoured vehicle

Schottel-Werft	Germany	*174*	75-26700

341 30 35 10 Water propulsion & steering systems, propeller, wheeled armoured vehicle

Schottel France Sarl	France	*143*	75-34450
Schottel-Werft	Germany	*174*	75-26700

341 30 35 20 Water propulsion systems, waterjet-type, wheeled armoured vehicle

Schottel France Sarl	France	*143*	75-34450
Schottel-Werft	Germany	*174*	75-26700

341 30 70 Transmission systems, wheeled armoured vehicle

David Brown Defence Equipment Ltd	UK	*437*	75-10150
Mannesmann Rexroth SA	France	*139*	75-25500
RENK AG	Germany	*173*	75-24700
Self-Changing Gears Ltd	UK	*482*	75-52700
ZF Friedrichshafen AG	Germany	*177*	75-35000
ZF Great Britain Ltd	UK	*495*	75-67720

341 30 79 Transmission components, wheeled armoured vehicle

David Brown Defence Equipment Ltd	UK	*437*	75-10150
David Brown Gear Industries Ltd	Australia	*13*	75-02800
Donaldson Co. Inc.	USA	*569*	75-27050
Flexibox Ltd	UK	*451*	75-23800
Giat Industries	France	*126*	55-00100
Mannesmann Rexroth SA	France	*139*	75-25500
RENK AG	Germany	*173*	75-24700
Self-Changing Gears Ltd	UK	*482*	75-52700

341 30 79 30 Differentials, wheeled armoured vehicle

General Motors of Canada Ltd	Canada	*71*	75-05900
Self-Changing Gears Ltd	UK	*482*	75-52700
ZF Friedrichshafen AG	Germany	*177*	75-35000
ZF Great Britain Ltd	UK	*495*	75-67720

341 30 79 40 Gearboxes, wheeled armoured vehicle

Alvis Transmission Ltd	UK	*430*	75-02862
General Motors of Canada Ltd	Canada	*71*	75-05900
Giat Industries	France	*126*	55-00100
Lucas Western Inc.	USA	*592*	75-56300
RENK AG	Germany	*173*	75-24700
Self-Changing Gears Ltd	UK	*482*	75-52700
ZF Friedrichshafen AG	Germany	*177*	75-35000
ZF Great Britain Ltd	UK	*495*	75-67720

341 30 80 Accessories, wheeled armoured vehicle propulsion system-related

Donaldson Co. Inc.	USA	*569*	75-27050
Harco Manufacturing Co. Inc.	USA	*580*	75-41870
Hebron & Medlock Engineering Ltd	UK	*457*	75-28860
Kinetics Ltd	Israel	*220*	75-11320
Riva Calzoni SpA	Italy	*238*	75-20500

341 30 80 10 Anti-vibration mountings, wheeled armoured vehicle

Avon Spencer Moulton	France	*132*	75-04750
BTR Silvertown Ltd	UK	*437*	75-10830
James Walker & Co. Ltd	UK	*492*	75-64120
Karipidis Bros. SA	Greece	*190*	75-01980
Loggers Rubbertechniek BV	Netherlands	*293*	75-04300
Stop-Choc Ltd	UK	*486*	75-57345
Vibrachoc SA	France	*148*	75-46900

341 50 Non-armoured vehicle propulsion

AB Volvo Penta	Sweden	*379*	75-11850
Cummins Engine Co. Inc.	USA	*567*	75-22980
DAF Trucks NV	Netherlands	*292*	75-01400
Detroit Diesel Corp.	USA	*568*	75-25830
Deutz Motor Industrie Motoren GmbH	Germany	*167*	75-06350
FP Diesel LP	USA	*576*	75-35753
Hercules Engines Company	USA	*581*	75-43080
Howden Aircontrol Ltd	UK	*459*	75-30240
IVECO Fiat SpA	Italy	*235*	75-13230
Kutaisi Automobile Plant Production	Georgia	*156*	55-05000
LaCombe Ind. e Com. de Turbos Ltda	Brazil	*56*	75-12050
Lister-Petter Ltd	UK	*465*	75-36070
MWM Motores Diesel Ltda	Brazil	*57*	75-13980
Pratt & Whitney Canada Inc	Canada	*75*	75-11400
Purolator Products Co.	USA	*606*	75-71900
SELIN Sistemi Sarl	Italy	*238*	75-21750
Teadit Juntas SA	Brazil	*55*	75-00910

341 50 10 Diesel engines, non-armoured vehicle

Caterpillar Inc.	USA	*562*	75-17100
Caterpillar of Australia Ltd	Australia	*14*	75-03240
Cummins Engine Co. Inc.	USA	*567*	75-22980
DAF Trucks NV	Netherlands	*292*	75-01400
Detroit Diesel Corp.	USA	*568*	75-25830
Deutz Corp.	USA	*569*	75-25950
Deütz Motor GmbH	Germany	*170*	75-15900
FP Diesel LP	USA	*576*	75-35753
GM-Holden's Automotive	Australia	*16*	75-07520
Hammonds Technical Services Inc.	USA	*580*	75-41800
Hercules Engines Company	USA	*581*	75-43080
IVECO Fiat SpA	Italy	*235*	75-13230
Kutaisi Automobile Plant Production	Georgia	*156*	55-05000
Lister-Petter Ltd	UK	*465*	75-36070
MWM Motores Diesel Ltda	Brazil	*57*	75-13980
NORINCO	China PR	*88*	55-00280
Perkins Engines (Shrews.)	UK	*474*	75-44750
Stewart & Stevenson Services Inc.	USA	*618*	75-84180

341 50 10 80 Accessories, diesel engine, non-armoured vehicle

Donaldson Co. Inc.	USA	*569*	75-27050
FP Diesel LP	USA	*576*	75-35753
Hammonds Technical Services Inc.	USA	*580*	75-41800
Hercules Engines Company	USA	*581*	75-43080
Howden Wade Ltd	UK	*459*	75-30260

341 50 10 90 Components, diesel engine, non-armoured vehicle

Aeroquip Corp.	USA	*547*	75-02350
A.O. Henton Engineering Co Ltd.	UK	*458*	75-29040
Busak + Shamban Inc.	USA	*561*	75-16006
FP Diesel LP	USA	*576*	75-35753
Gallay Ltd	UK	*453*	75-25100
Hercules Engines Company	USA	*581*	75-43080
LaCombe Ind. e Com. de Turbos Ltda	Brazil	*56*	75-12050
National Forge Co.	USA	*598*	75-63850
Olympic Filter SA	Greece	*191*	75-02165
Ordex Inc.	USA	*601*	75-67350
Perkins Engines (Shrews.)	UK	*474*	75-44750
Purolator Products Co.	USA	*606*	75-71900
Stewart & Stevenson Services Inc.	USA	*618*	75-84180
Teadit Juntas SA	Brazil	*55*	75-00910
Voss Industries Inc.	USA	*626*	75-94550

341 50 10 90 20 Cooling equipment, diesel engine, non-armoured vehicle

Air International Transit	Australia	*11*	75-00450
FP Diesel LP	USA	*576*	75-35753
Gallay Ltd	UK	*453*	75-25100
General Thermodynamics Inc.	USA	*578*	75-39010
Hercules Engines Company	USA	*581*	75-43080
Howden Aircontrol Ltd	UK	*459*	75-30240
International Radiator Services Ltd	UK	*461*	75-31840
Microtecnica-Torino SpA	Italy	*237*	75-15500
Serck Heat Transfer	UK	*483*	75-52880

341 50 15 Multifuel engines, non-armoured vehicle

Detroit Diesel Corp.	USA	*568*	75-25830
Hercules Engines Company	USA	*581*	75-43080
Vickers Defence Systems	UK	*491*	75-63040

341 50 15 80 Accessories, multifuel engine, non-armoured vehicle

Aerospace Syst. & Comp.	USA	*547*	75-02501
Autocar Electrical Equipment Ltd	UK	*432*	75-04440
Donaldson Co. Inc.	USA	*569*	75-27050
Hercules Engines Company	USA	*581*	75-43080

341 50 15 90 Components, multifuel engine, non-armoured vehicle

Aeroquip Corp.	USA	547	75-02350
Aerospace Syst. & Comp.	USA	547	75-02501
Busak + Shamban Inc.	USA	561	75-16006
Fluorocarbon Company Limited	UK	451	75-24020
Hercules Engines Company	USA	581	75-43080
Ordex Inc.	USA	601	75-67350
Purolator Products Co.	USA	606	75-71900
Voss Industries Inc.	USA	626	75-94550

341 50 15 90 20 Cooling equipment, multifuel engine, non-armoured vehicle

Aerospace Syst. & Comp.	USA	547	75-02501
Air International Transit	Australia	11	75-00450
Gallay Ltd	UK	453	75-25100
Hercules Engines Company	USA	581	75-43080
Howden Aircontrol Ltd	UK	459	75-30240
International Radiator Services Ltd	UK	461	75-31840
Microtecnica-Torino SpA	Italy	237	75-15500

341 50 20 Petrol engines, non-armoured vehicle

Elgin Industries	USA	573	75-30800
Hercules Engines Company	USA	581	75-43080
Orbital Engine Company (Australia)	Australia	25	80-15830
Vickers Defence Systems	UK	491	75-63040

341 50 20 80 Accessories, petrol engine, non-armoured vehicle

Autocar Electrical Equipment Ltd	UK	432	75-04440
Donaldson Co. Inc.	USA	569	75-27050
Hercules Engines Company	USA	581	75-43080
Ordex Inc.	USA	601	75-67350

341 50 20 90 Components, petrol engine, non-armoured vehicle

Aeroquip Corp.	USA	547	75-02350
Busak + Shamban Inc.	USA	561	75-16006
Elgin Industries	USA	573	75-30800
Fluorocarbon Company Limited	UK	451	75-24020
Hercules Engines Company	USA	581	75-43080
Orbital Engine Company (Australia)	Australia	25	80-15830
Ordex Inc.	USA	601	75-67350
Purolator Products Co.	USA	606	75-71900
Voss Industries Inc.	USA	626	75-94550
Warsaw Mechanical Works "Wuzetem"	Poland	324	55-20000

341 50 20 90 20 Cooling equipment, petrol engine, non-armoured vehicle

Air International Transit	Australia	11	75-00450
Gallay Ltd	UK	453	75-25100
Hercules Engines Company	USA	581	75-43080
Howden Aircontrol Ltd	UK	459	75-30240
Microtecnica-Torino SpA	Italy	237	75-15500

341 50 25 Gas turbine engines, non-armoured vehicle

Hardcote Technologies	USA	580	75-41900
Pratt & Whitney Canada Inc	Canada	75	75-11400

341 50 25 80 Accessories, gas turbine engine, non-armoured vehicle

Aerospace Syst. & Comp.	USA	547	75-02501
Donaldson Co. Inc.	USA	569	75-27050

341 50 25 90 Components, gas turbine engine, non-armoured vehicle

Aeroquip Corp.	USA	547	75-02350
Aerospace Syst. & Comp.	USA	547	75-02501
BTR Silvertown Ltd	UK	437	75-10830
Busak + Shamban Inc.	USA	561	75-16006
Fluorocarbon Company Limited	UK	451	75-24020
Global Airtech International	USA	579	75-39769
Hardcote Technologies	USA	580	75-41900
Howmet Corporation	USA	582	75-44550
Mibert Precision Ltd	UK	468	75-38920
Price & Orphin Limited	UK	509	80-24250
Voss Industries Inc.	USA	626	75-94550

341 50 30 Electric propulsion systems, non-armoured vehicle

Holec Machines & Apparaten BV	Netherlands	293	75-03110
Kaman Electromagnetics Corp.	USA	637	80-17500
SELIN Sistemi Sarl	Italy	238	75-21750

341 50 35 Water propulsion units, non-armoured vehicle

Schottel-Werft	Germany	174	75-26700

341 50 35 10 Water propulsion systems, propeller-driven, non-armoured vehicle

Schottel France Sarl	France	143	75-34450
Schottel-Werft	Germany	174	75-26700

341 50 35 10 90 Components, water propulsion systems, propeller, non-armoured vehicle

A.O. Henton Engineering Co Ltd.	UK	458	75-29040
BTR Silvertown Ltd	UK	437	75-10830
Busak + Shamban Inc.	USA	561	75-16006
Schottel France Sarl	France	143	75-34450

341 50 35 20 Waterjet propulsion units, non-armoured vehicle

David Brown Defence Equipment Ltd	UK	437	75-10150
Schottel France Sarl	France	143	75-34450
Schottel-Werft	Germany	174	75-26700

341 50 70 Transmission systems, non-armoured vehicle

David Brown Defence Equipment Ltd	UK	437	75-10150
Mannesmann Rexroth SA	France	139	75-25500
Self-Changing Gears Ltd	UK	482	75-52700
ZF Friedrichshafen AG	Germany	177	75-35000
ZF Great Britain Ltd	UK	495	75-67720

341 50 79 Transmission components, non-armoured vehicle

David Brown Defence Equipment Ltd	UK	437	75-10150
David Brown Gear Industries Ltd	Australia	13	75-02800
Flexibox Ltd	UK	451	75-23800
Mannesmann Rexroth SA	France	139	75-25500
Self-Changing Gears Ltd	UK	482	75-52700

341 50 79 30 Differentials, non-armoured vehicle

David Brown Defence Equipment Ltd	UK	437	75-10150
Self-Changing Gears Ltd	UK	482	75-52700
ZF Friedrichshafen AG	Germany	177	75-35000
ZF Great Britain Ltd	UK	495	75-67720

341 50 79 40 Gearboxes, non-armoured vehicle

Alvis Transmission Ltd	UK	430	75-02862
David Brown Defence Equipment Ltd	UK	437	75-10150
Giat Industries	France	126	55-00100

Lucas Western Inc.	USA	*592*	75-56300
Self-Changing Gears Ltd	UK	*482*	75-52700
ZF Friedrichshafen AG	Germany	*177*	75-35000
ZF Great Britain Ltd	UK	*495*	75-67720

341 50 80 Accessories, non-armoured vehicle propulsion system-related

Bartlett Lifting Devices Inc.	USA	*557*	75-12060
Donaldson Co. Inc.	USA	*569*	75-27050
Harco Manufacturing Co. Inc.	USA	*580*	75-41870

341 50 80 10 Anti-vibration mountings, non-armoured vehicle

Avon Spencer Moulton	France	*132*	75-04750
BTR Silvertown Ltd	UK	*437*	75-10830
James Walker & Co. Ltd	UK	*492*	75-64120
Karipidis Bros. SA	Greece	*190*	75-01980
Loggers Rubbertechniek BV	Netherlands	*293*	75-04300
Stop-Choc Ltd.	UK	*486*	75-57345
Vibrachoc SA	France	*148*	75-46900

345 10 Diesel engines, railway locomotive

Cummins Engine Co. Inc.	USA	*567*	75-22980
Mirrlees Blackstone Limited	UK	*468*	75-40050
Perkins Engines (Shrews.)	UK	*474*	75-44750

345 20 Diesel-electric propulsion systems, railway locomotive

David Brown Defence Equipment Ltd	UK	*437*	75-10150
Mirrlees Blackstone Limited	UK	*468*	75-40050

345 40 Electric propulsion systems, railway locomotive

Holec Machines & Apparaten BV	Netherlands	*293*	75-03110

Weapon Propulsion

351 Missile propulsion systems

Aero Systems Engineering Inc.	USA	*546*	75-01116
Aerojet	USA	*547*	75-01500
AIM Inc.	USA	*547*	75-03020
Atlantic Research Corp.	USA	*555*	75-10150
Bayern-Chemie	Germany	*165*	75-02100
Boulder Scientific Co.	USA	*560*	75-14780
Chloride Industrial Batteries Ltd	UK	*457*	75-28490
Cranfield University	UK	*419*	60-00350
FMI Composites Ltd	UK	*451*	75-24040
Hawker Energy Products Ltd	UK	*457*	75-28490
Hughes Microelectronics Europe Ltd	UK	*459*	75-30480
Ishikawajima-Harima H.I. Co. Ltd	Japan	*251*	75-02800
Israel Military Industries Ltd	Israel	*220*	75-10300
Israel Military Industries Rocket Systems Division	**Israel**	***220***	**75-10310**
Matra BAe Dynamics	UK	*425*	74-06850
Microturbo Ltd	UK	*468*	75-39160
Microturbo SA	France	*140*	75-27000
ML Wallop Defence Systems Ltd	UK	*469*	75-40240
Nissan Motor Co. Ltd	Japan	*254*	75-06790
Rafael	Israel	*216*	55-01800
Raufoss Technology AS	Norway	*310*	75-02200
Rolls-Royce Mil. Aero Engines Ltd	UK	*480*	75-50265
Rolls-Royce Turbomeca	UK	*426*	74-18000
Royal Ordnance	UK	*481*	75-50620
SIMA PERU	Peru	*319*	55-01950
SNPE	France	*145*	75-36315
Somchem	S. Africa	*359*	75-19200
Universal Propulsion Co. Inc.	USA	*625*	75-92700

351 10 Rocket motors, missile

Aerojet	USA	*547*	75-01500
AIM Inc.	USA	*547*	75-03020
Arab Organization for Ind.	Egypt	*109*	55-01700
Australian Defence Industries Ltd	Australia	*9*	55-01000
Bristol Aerospace Ltd	Canada	*69*	75-01500
Cranfield University	UK	*419*	60-00350
GEL Engineering Ltd	UK	*454*	75-25500
Ishikawajima-Harima H.I. Co. Ltd	Japan	*251*	75-02800
Israel Military Industries Ltd	Israel	*220*	75-10300
Israel Military Industries Rocket Systems Division	**Israel**	***220***	**75-10310**
ML Wallop Defence Systems Ltd	UK	*469*	75-40240
Nissan Motor Co. Ltd	Japan	*254*	75-06790
Rafael	Israel	*216*	55-01800
Raufoss Technology AS	Norway	*310*	75-02200
Royal Ordnance	UK	*481*	75-50620
Somchem	S. Africa	*359*	75-19200
Sverdrup Technology Inc.	USA	*643*	80-30680

351 11 Rocket motors, liquid fuel, missile

Aerojet	USA	*547*	75-01500
Cranfield University	UK	*419*	60-00350
Ishikawajima-Harima H.I. Co. Ltd	Japan	*251*	75-02800
Royal Ordnance	UK	*481*	75-50620
Sequa Corp.	USA	*613*	75-79490
Sverdrup Technology Inc.	USA	*643*	80-30680
Volvo Aero Corporation	Sweden	*379*	75-11810

351 12 Rocket motors, solid fuel, missile

Aerojet	USA	*547*	75-01500
Arab Organization for Ind.	Egypt	*109*	55-01700
Australian Defence Industries Ltd	Australia	*9*	55-01000
Bayern-Chemie	Germany	*165*	75-02100
Boulder Scientific Co.	USA	*560*	75-14780
Bristol Aerospace Ltd	Canada	*69*	75-01500
Cranfield University	UK	*419*	60-00350
Daicel Chemical Industries Ltd	Japan	*250*	75-00950
INIMEX	**Belgium**	***46***	**75-04650**
Ishikawajima-Harima H.I. Co. Ltd	Japan	*251*	75-02800
Israel Military Industries Ltd	Israel	*220*	75-10300
Israel Military Industries Rocket Systems Division	**Israel**	***220***	**75-10310**
ML Aviation Ltd	UK	*469*	75-40200
ML Wallop Defence Systems Ltd	UK	*469*	75-40240
Nissan Motor Co. Ltd	Japan	*254*	75-06790
Rafael	Israel	*216*	55-01800
Raufoss Technology AS	Norway	*310*	75-02200
Royal Ordnance	UK	*481*	75-50620
Samsung Aerospace Industries Ltd	Korea (ROK)	*266*	75-04630
Sequa Corp.	USA	*613*	75-79490
SNPE	France	*145*	75-36315
Somchem	S. Africa	*359*	75-19200
Sverdrup Technology Inc.	USA	*643*	80-30680
Talley Defense Systems Inc.	USA	*619*	75-86120
Universal Propulsion Co. Inc.	USA	*625*	75-92700

351 19 Components, rocket motor, missile

Aeroquip Corp.	USA	*547*	75-02350
Aerospace Forgings Ltd	UK	*428*	75-01160
Aerospace Metal Composites Limited	UK	*428*	75-01180
Aircraft Tanks Ltd	UK	*429*	75-01730
A.O. Henton Engineering Co Ltd.	UK	*458*	75-29040
Arab Organization for Ind.	Egypt	*109*	55-01700
Avionic Components Ltd.	UK	*432*	75-04970
Busak + Shamban Inc.	USA	*561*	75-16006
Cranfield University	UK	*419*	60-00350
Darchem Engineering Ltd.	UK	*444*	75-16750
Doncasters plc	UK	*446*	75-18410
EG&G Sealol Ltd	UK	*448*	75-19675
Emco Inc.	USA	*573*	75-30985
Fairey Hydraulics	UK	*450*	75-22255
Farsound Engineering Ltd	UK	*450*	75-22420
Firth Rixson plc	UK	*451*	75-23600
FMI Composites Ltd	UK	*451*	75-24040
Hollow Extrusions Ltd	UK	*458*	75-29652
Holscot Industrial Linings Ltd	UK	*458*	75-29655
Howmet Corporation	USA	*582*	75-44550
Howmet Limited	UK	*459*	75-30280
Insumat Ltd	UK	*461*	75-31660
Israel Military Industries Ltd	Israel	*220*	75-10300

Israel Military Industries Rocket Systems Division	**Israel**	*220*	**75-10310**
Iver Precision Engineers Ltd	UK	*462*	75-32230
J.S. Chinn & Co. Ltd	UK	*439*	75-13045
KEC Limited	UK	*463*	75-33700
Moog Controls Ltd	UK	*469*	75-40670
Muirhead Vactric Components Ltd	UK	*470*	75-41404
National Forge Co.	USA	*598*	75-63850
Nissan Motor Co. Ltd.	Japan	*254*	75-06790
Oxley Developments Co. Ltd	UK	*472*	75-43400
Parker Bertea Aerospace.	USA	*602*	75-68900
Price & Orphin Limited	UK	*509*	80-24250
Pyle-National	UK	*476*	75-47290
Rafael	Israel	*216*	55-01800
Rohr Inc.	USA	*611*	75-76150
Royal Ordnance	UK	*481*	75-50620
Royal Ordnance Speciality Metals	UK	*481*	75-50630
Somchem	S. Africa	*359*	75-19200
Technical Products Group Inc.	USA	*620*	75-86764
The Northern Rubber Company Ltd	UK	*471*	75-42740
Universal Propulsion Co. Inc.	USA	*625*	75-92700
Voss Industries Inc.	USA	*626*	75-94550
Wyman-Gordon Ltd	UK	*494*	75-67280
Yewlands Engineering Co. Ltd.	UK	*495*	75-67520

351 20 Ramjet engines, missile

Bayern-Chemie	Germany	*165*	75-02100
Cranfield University	UK	*419*	60-00350
GEL Engineering Ltd	UK	*454*	75-25500
Ishikawajima-Harima H.I. Co. Ltd	Japan	*251*	75-02800
Rafael	Israel	*216*	55-01800
Rolls-Royce Mil. Aero Engines Ltd	UK	*480*	75-50265
Rolls-Royce plc	UK	*479*	75-50250
Rolls-Royce Turbomeca	UK	*426*	74-18000
Royal Ordnance	UK	*481*	75-50620
Somchem	S. Africa	*359*	75-19200
Volvo Aero Corporation	Sweden	*379*	75-11810

351 20 90 Components, ramjet engine, missile

Aerospace Forgings Ltd	UK	*428*	75-01160
Aerospace Metal Composites Limited	UK	*428*	75-01180
Aircraft Equipment International Ltd	UK	*428*	75-01560
Aircraft Tanks Ltd	UK	*429*	75-01730
A.O. Henton Engineering Co Ltd.	UK	*458*	75-29040
Avionic Components Ltd.	UK	*432*	75-04970
BTR Silvertown Ltd	UK	*437*	75-10830
Busak + Shamban Inc.	USA	*561*	75-16006
Cranfield University	UK	*419*	60-00350
Darchem Engineering Ltd.	UK	*444*	75-16750
Firth Rixson plc	UK	*451*	75-23600
Fluorocarbon Company Limited	UK	*451*	75-24020
Hughes Microelectronics Europe Ltd	UK	*459*	75-30480
Insumat Ltd	UK	*461*	75-31660
Iver Precision Engineers Ltd	UK	*462*	75-32230
J.S. Chinn & Co. Ltd	UK	*439*	75-13045
KEC Limited	UK	*463*	75-33700
Mibert Precision Ltd	UK	*468*	75-38920
Moog Controls Ltd	UK	*469*	75-40670
Muirhead Vactric Components Ltd	UK	*470*	75-41404
Parker Bertea Aerospace.	USA	*603*	75-68900
Price & Orphin Limited	UK	*509*	80-24250
Rolls-Royce Mil. Aero Engines Ltd	UK	*480*	75-50265
Rolls-Royce Turbomeca	UK	*426*	74-18000
Royal Ordnance	UK	*481*	75-50620
Royal Ordnance Speciality Metals	UK	*481*	75-50630
Somchem	S. Africa	*359*	75-19200
Technical Products Group Inc.	USA	*620*	75-86764
Volvo Aero Corporation	Sweden	*379*	75-11810
Yewlands Engineering Co. Ltd.	UK	*495*	75-67520

351 30 Turbojet engines, missile

Cranfield University	UK	*419*	60-00350
Ishikawajima-Harima H.I. Co. Ltd	Japan	*251*	75-02800
Microturbo Ltd	UK	*468*	75-39160
Microturbo SA	France	*140*	75-27000
Rolls-Royce (Canada) Ltd.	Canada	*75*	75-12120
Rolls-Royce Mil. Aero Engines Ltd	UK	*480*	75-50265
Rolls-Royce Turbomeca	UK	*426*	74-18000
Sundstrand Power Systems	USA	*618*	75-85120
Teledyne Ryan Aeronautical	USA	*621*	75-87560
Turbomeca	France	*148*	75-46000

351 30 90 Components, turbojet engine, missile

Aerospace Forgings Ltd	UK	*428*	75-01160
Aerospace Metal Composites Limited	UK	*428*	75-01180
Aircraft Equipment International Ltd	UK	*428*	75-01560
Aircraft Tanks Ltd	UK	*429*	75-01730
A.O. Henton Engineering Co Ltd.	UK	*458*	75-29040
Avionic Components Ltd.	UK	*432*	75-04970
Busak + Shamban Inc.	USA	*561*	75-16006
Cleveland-Guest (Engineering) Ltd	UK	*440*	75-13810
Cranfield University	UK	*419*	60-00350
Darchem Engineering Ltd.	UK	*444*	75-16750
Deloro Stellite Ltd.	UK	*445*	75-17460
Doncasters Monk Bridge	UK	*446*	75-18450
Doncasters plc	UK	*446*	75-18410
EG&G Sealol Ltd	UK	*448*	75-19675
Firth Rixson plc	UK	*451*	75-23600
Fluorocarbon Company Limited	UK	*451*	75-24020
Howmet Corporation	USA	*582*	75-44550
Howmet Limited	UK	*459*	75-30280
Insumat Ltd	UK	*461*	75-31660
Iver Precision Engineers Ltd	UK	*462*	75-32230
J.S. Chinn & Co. Ltd	UK	*439*	75-13045
KEC Limited	UK	*463*	75-33700
Mibert Precision Ltd	UK	*468*	75-38920
Microturbo Ltd	UK	*468*	75-39160
Moog Controls Ltd	UK	*469*	75-40670
Muirhead Vactric Components Ltd	UK	*470*	75-41404
National Forge Co.	USA	*598*	75-63850
Parker Bertea Aerospace.	USA	*603*	75-68900
Price & Orphin Limited	UK	*509*	80-24250
Pyle-National	UK	*476*	75-47290
Rohr Inc.	USA	*611*	75-76150
Rolls-Royce Mil. Aero Engines Ltd	UK	*480*	75-50265
Rolls-Royce Turbomeca	UK	*426*	74-18000
Royal Ordnance Speciality Metals	UK	*481*	75-50630
Serck Aviation Div.	UK	*482*	75-52860
Technical Products Group Inc.	USA	*620*	75-86764
The Barden Corporation (UK) Ltd	UK	*433*	75-05630
Yewlands Engineering Co. Ltd.	UK	*495*	75-67520

352 Torpedo propulsion systems

Microturbo Ltd	UK	*468*	75-39160
SELIN Sistemi Sarl	Italy	*238*	75-21750
STN ATLAS Elektronik GmbH	Germany	*165*	75-01150

352 20 Electric propulsion systems, torpedo

Newton Derby Ltd	UK	*470*	75-42520
SELIN Sistemi Sarl	Italy	*238*	75-21750

352 40 Steam propulsion systems, torpedo

Framatome	France	*136*	75-18405
Microturbo Ltd	UK	*468*	75-39160

Mission-related Systems & Equipment

Target Location, Tracking, Identification & Ranging

411 Target location & tracking

Advanced System Architectures Ltd	UK	*496*	80-00420
Alenia-Elsag	Italy	*230*	74-00100
Andrew Ltd	UK	*430*	75-03156
Aydin Corp.	USA	*556*	75-11450
Bharat Electronics Ltd	India	*203*	55-00200
Boeing Australia Limited	Australia	*12*	75-02500
British Aerospace (Systems &	UK	*436*	75-08890

Product-Supplier Link

Cambridge Consultants Ltd	UK	*499*	80-05300
Cambridge Neurodynamics Ltd	UK	*499*	80-05320
Central Research Laboratories Ltd	UK	*499*	80-05895
CIC International Ltd	USA	*563*	75-18570
Cincinnati Electronics Corp.	USA	*564*	75-18640
Coherent Optics (Europe) Ltd	UK	*441*	75-14090
Communications & Power Industries	USA	*565*	75-20393
Compagnie Générale de Géophysique	France	*134*	75-11335
Computing Devices Canada	Canada	*70*	75-03600
Continental Electronics Corp.	USA	*566*	75-21800
Cranfield University	UK	*419*	60-00350
Dassault Electronique SA (Group)	**France**	***135***	**75-12980**
Data Cell Ltd	UK	*500*	80-07720
Delft Sensor Systems	**Belgium**	***46***	**75-02500**
Delft Sensor Systems	Netherlands	*292*	75-01590
DRS Technologies, Inc.	USA	*570*	75-27400
EG&G Inc.	USA	*571*	75-29555
EL-OP Electro Optics Industries Ltd	Israel	*219*	75-06300
ELECMA	France	*136*	75-14400
Elettronica (UK) Ltd	UK	*448*	75-20475
Ericsson Microwave Systems AB	**Sweden**	***377***	**75-04890**
EUROCONTROL SpA	Italy	*233*	75-08200
Fabbrica Italiana Apparecchiature	Italy	*234*	75-08550
Flight International Inc.	USA	*635*	80-11600
GEC-Marconi Avionics Ltd	UK	*454*	75-25320
GEC-Marconi Research Centre	UK	*419*	60-00700
GEM Elettronica Srl	Italy	*235*	75-10100
Hughes Elcan Optical Technologies	Canada	*72*	75-07650
Hydroscience Technologies Inc.	USA	*584*	75-45550
INIMEX	**Belgium**	***46***	**75-04650**
Interstate Electronics	USA	*586*	75-48700
ITT Avionics	USA	*586*	75-49455
ITT Gilfillan	USA	*586*	75-49455
Kaman Sciences Corp.	USA	*637*	80-17600
Leica AG	Switzerland	*386*	75-07310
Litton Industries Inc.	USA	*590*	75-55400
Lockheed Martin Corp.	USA	*591*	75-55635
Lockheed Martin Missiles & Space	USA	*592*	75-55640
MacDonald Dettwiler and Associates	Canada	*78*	80-01310
Matra BAe Dynamics	UK	*425*	74-06850
Meteor CAE SpA	Italy	*237*	75-15400
Morphyspribor	Russia	*335*	60-02730
MSI-Defence Systems Ltd	UK	*469*	75-41280
NORINCO	China PR	*88*	55-00280
Northrop Grumman Corporation	USA	*640*	80-21840
Octec Ltd	UK	*471*	75-42950
Oerlikon-Contraves SpA	Italy	*237*	75-16800
Pearpoint Ltd	UK	*473*	75-44270
Pilkington Thorn Optronics	UK	*426*	74-10000
R. Alkan (UK) Ltd	UK	*429*	75-02680
Racal Communications Inc.	USA	*606*	75-72380
Racal Radar Defence Systems Ltd	UK	*477*	75-47688
Raytheon Canada Ltd	Canada	*75*	75-11700
Raytheon E-Systems	USA	*608*	75-73302
Reson Offshore Ltd	UK	*478*	75-49260
Roke Manor Research Ltd	UK	*420*	60-01800
SABCA	Belgium	*46*	75-05590
SAGEM SA	**France**	***143***	**75-33950**
Sanders	USA	*612*	75-77850
Sener Ingenieria y Sistemas SA	Spain	*367*	75-05200
SFIM	**France**	***144***	**75-35080**
Shorts Missile Systems Ltd	**UK**	***483***	**75-53550**
Siemens Plessey Systems	UK	*483*	75-53770
Siemens Switzerland Ltd	Switzerland	*387*	75-11980
Sierra Technologies Inc.	USA	*614*	75-80640
Sperry Marine Inc.	USA	*617*	75-83250
SRI International	USA	*643*	80-29600
Tadiran Ltd	Israel	*222*	75-21600
Thomson Marconi Sonar Limited	UK	*426*	74-20000
Thomson Sintra	France	*148*	75-45135
Thomson-CSF	France	*148*	75-45100
THORN Lighting AB	Sweden	*379*	75-11280
Trimble Navigation Europe Ltd	UK	*489*	75-61140
TSS (UK) Ltd	UK	*489*	75-61395
Winkelmann (UK) Ltd	UK	*494*	75-66600
Zeta	USA	*629*	75-98190
Znamya Production Group	Ukraine	*407*	75-10000

411 10 Radar target location & tracking systems & equipment

Advanced System Architectures Ltd	UK	*496*	80-00420
AIL Systems Inc.	USA	*547*	75-02950
Alenia-Elsag	Italy	*230*	74-00100
Andrew Sensor Products	USA	*554*	75-07452
APT Radar Systems Ltd	UK	*431*	75-03560
ASMAR	Chile	*84*	55-00150
Bharat Electronics Ltd	India	*203*	55-00200
Boeing Australia Limited	Australia	*12*	75-02500
Cambridge Neurodynamics Ltd	UK	*499*	80-05320
Canadian Marconi Co.	Canada	*69*	75-02600
CEA Technologies Pty Ltd	Australia	*24*	80-03100
Central Research Laboratories Ltd	UK	*499*	80-05895
Cenzin Co. Ltd	Poland	*324*	75-00200
CIC International Ltd	USA	*563*	75-18570
Continental Electronics Corp.	USA	*566*	75-21800
Contraves GmbH	Germany	*166*	75-05200
Cranfield Aerospace Ltd	UK	*500*	80-07200
Cranfield University	UK	*419*	60-00350
Custom Electronics Inc.	USA	*567*	75-23085
Daimler-Benz Aerospace France SAS	**France**	***134***	**75-12950**
Dassault Electronique SA (Group)	**France**	***135***	**75-12980**
ELECMA	France	*136*	75-14400
Electronic Space Systems Corp.	USA	*572*	75-30100
ELTA Electronics Industries Ltd	Israel	*220*	75-10200
Ericsson Microwave Systems AB	**Sweden**	***377***	**75-04890**
Ericsson Radar A/S	Norway	*309*	75-00860
ESSCO Collins Ltd	Ireland	*214*	75-00100
Fabbrica Italiana Apparecchiature	Italy	*234*	75-08550
Force Computers GmbH	Germany	*168*	75-09270
GEC Marconi Hazeltine Corp	USA	*578*	75-37415
GEC-Marconi Avionics Inc.	USA	*577*	75-37413
GEC-Marconi Avionics Ltd	UK	*454*	75-25320
GEC-Marconi Materials Technology Ltd	UK	*419*	60-00600
GEC-Marconi Research Centre	UK	*419*	60-00700
GEM Elettronica Srl	Italy	*235*	75-10100
Herley-Vega Systems	USA	*581*	75-43090
Hollandse Signaalapparaten BV	Netherlands	*293*	75-03300
Howden Aircontrol Ltd	UK	*459*	75-30240
Howden Fluid Systems	USA	*582*	75-44400
H.R. Smith (Microwaves) Ltd	UK	*484*	75-54590
IDS Ingegneria dei Sistemi SpA	Italy	*240*	80-00920
INDRA	Spain	*366*	75-03800
INIMEX	**Belgium**	***46***	**75-04650**
Israel Aircraft Industries Ltd	Israel	*219*	75-10200
Italtel Spa	Italy	*235*	75-13205
ITT Aerospace/Communications Div.	USA	*586*	75-49455
Kaman Sciences Corp.	USA	*637*	80-17600
Labtech Ltd	UK	*464*	75-34280
Litton Canada Ltd	Canada	*73*	75-09090
Litton Industries Inc.	USA	*591*	75-55400
Lockheed Martin Corp.	USA	*591*	75-55635
MacDonald Dettwiler and Associates	Canada	*78*	80-01310
MPB Technologies Inc	Canada	*74*	75-09855
Nera Telecommunications	UK	*470*	75-42318
Northrop Grumman Corporation	USA	*599*	75-65450
Oerlikon-Contraves AG	Switzerland	*387*	75-09250
Oerlikon-Contraves SpA	Italy	*237*	75-16800
Philips Traffic & Engineering	Australia	*20*	75-16550
Racal Communications Inc.	USA	*606*	75-72380
Racal Radar Defence Systems Ltd	UK	*477*	75-47688
Radio Research Instrument Co. Inc.	USA	*607*	75-72600
Raytheon E-Systems	USA	*608*	75-73302
Reunert Technology Systems Ltd	S. Africa	*359*	75-18300
Roke Manor Research Ltd	UK	*420*	60-01800
Sanders	USA	*612*	75-77850
Schrack Aerospace GmbH	Austria	*34*	75-02195
Siemens Aktiengesellschaft	Germany	*175*	75-27700
Siemens Plessey Systems	UK	*483*	75-53770
Siemens Switzerland Ltd	Switzerland	*387*	75-11980
Spar Aerospace Ltd	Canada	*76*	75-13000
Sperry Marine Inc.	USA	*617*	75-83250
SRI International	USA	*643*	80-29600
Technology Service Corp.	USA	*644*	80-31930
TERMA Elektronik AS	Denmark	*104*	75-20300
Thomson-CSF	France	*147*	75-45100
Znamya Production Group	Ukraine	*407*	75-10000

411 12 Radars, airborne, target location & tracking

Andrew Sensor Products	USA	*554*	75-07452
CIC International Ltd	USA	*563*	75-18570
Cranfield Aerospace Ltd	UK	*500*	80-07200
Cranfield University	UK	*419*	60-00350
Daimler-Benz Aerospace France SAS	**France**	***134***	**75-12950**
ELECMA	France	*136*	75-14400

ELTA Electronics Industries Ltd	Israel	*220*	75-10200
Ericsson Microwave Systems AB	**Sweden**	***377***	**75-04890**
Fabbrica Italiana Apparecchiature	Italy	*234*	75-08550
GEC-Marconi Avionics Ltd	UK	*454*	75-25320
GEC-Marconi Research Centre	UK	*419*	60-00700
INIMEX	**Belgium**	***46***	**75-04650**
Labtech Ltd	UK	*464*	75-34280
MacDonald Dettwiler and Associates	Canada	*78*	80-01310
Nera Telecommunications	UK	*470*	75-42318
Norden Systems Inc.	USA	*599*	75-64900
Northrop Grumman Corporation	USA	*640*	80-21840
Philips Traffic & Engineering	Australia	*20*	75-16550
Racal Communications Inc.	USA	*606*	75-72380
Racal Radar Defence Systems Ltd	UK	*477*	75-47688
Radio Research Instrument Co. Inc.	USA	*607*	75-72600
Roke Manor Research Ltd	UK	*420*	60-01800
Spar Aerospace Ltd	Canada	*76*	75-13000
SRI International	USA	*643*	80-29600
Technology Service Corp.	USA	*644*	80-31930
TERMA Elektronik AS	Denmark	*104*	75-20300
Znamya Production Group	Ukraine	*407*	75-10000

411 12 20 Radars, airborne, early warning

CIC International Ltd	USA	*563*	75-18570
Daimler-Benz Aerospace France SAS	**France**	***134***	**75-12950**
Ericsson Microwave Systems AB	**Sweden**	***377***	**75-04890**
Fabbrica Italiana Apparecchiature	Italy	*234*	75-08550
GEC Marconi Hazeltine Corp	USA	*578*	75-37415
GEC-Marconi Avionics Ltd	UK	*454*	75-25320
Labtech Ltd	UK	*464*	75-34280
Northrop Grumman Corporation	USA	*640*	80-21840
Radio Research Instrument Co. Inc.	USA	*607*	75-72600
Roke Manor Research Ltd	UK	*420*	60-01800
Spar Aerospace Ltd	Canada	*76*	75-13000
Technology Service Corp.	USA	*644*	80-31930

411 12 30 Radars, airborne, fire control & target designation

Andrew Sensor Products	USA	*554*	75-07452
Canadian Marconi Co.	Canada	*69*	75-02600
CIC International Ltd	USA	*563*	75-18570
Daimler-Benz Aerospace France SAS	**France**	***134***	**75-12950**
ELTA Electronics Industries Ltd	Israel	*220*	75-10200
Ericsson Microwave Systems AB	**Sweden**	***377***	**75-04890**
Fabbrica Italiana Apparecchiature	Italy	*234*	75-08550
GEC-Marconi Avionics Ltd	UK	*454*	75-25320
Israel Aircraft Industries Ltd	Israel	*219*	75-10200
Labtech Ltd	UK	*464*	75-34280
Northrop Grumman Corporation	USA	*599*	75-65450
Racal Communications Inc.	USA	*606*	75-72380
Radio Research Instrument Co. Inc.	USA	*607*	75-72600
Technology Service Corp.	USA	*644*	80-31930
Znamya Production Group	Ukraine	*407*	75-10000

411 12 50 Radars, airborne, maritime search

Array Systems Computing Inc.	Canada	*77*	80-00060
Canadian Marconi Co.	Canada	*69*	75-02600
CIC International Ltd	USA	*563*	75-18570
Daimler-Benz Aerospace France SAS	**France**	***134***	**75-12950**
Ericsson Microwave Systems AB	**Sweden**	***377***	**75-04890**
Fabbrica Italiana Apparecchiature	Italy	*234*	75-08550
Flight International Inc.	USA	*635*	80-11600
GEC-Marconi Avionics Ltd	UK	*454*	75-25320
GEC-Marconi Research Centre	UK	*419*	60-00700
Labtech Ltd	UK	*464*	75-34280
Litton Industries Inc.	USA	*590*	75-55400
Litton Systems Canada Ltd	Canada	*73*	75-09100
MacDonald Dettwiler Limited	Australia	*25*	80-12990
Northrop Grumman Corporation	USA	*640*	80-21840
Przemyslowy Instytut Telekomunikacji	**Poland**	***325***	**75-02650**
Racal Communications Inc.	USA	*606*	75-72380
Radio Research Instrument Co. Inc.	USA	*607*	75-72600
Roke Manor Research Ltd	UK	*420*	60-01800
Technology Service Corp.	USA	*644*	80-31930
TERMA Elektronik AS	Denmark	*104*	75-20300

411 12 60 Radars, airborne surveillance & target acquisition

Andrew Sensor Products	USA	*554*	75-07452
CIC International Ltd	USA	*563*	75-18570
Cranfield University	UK	*419*	60-00350
ELTA Electronics Industries Ltd	Israel	*220*	75-10200
Ericsson Defence Systems Pty Ltd	Australia	*16*	75-05820
Ericsson Microwave Systems AB	**Sweden**	***377***	**75-04890**
Fabbrica Italiana Apparecchiature	Italy	*234*	75-08550
GEC Marconi Hazeltine Corp	USA	*578*	75-37415
GEC-Marconi Avionics Ltd	UK	*454*	75-25320
GEC-Marconi Research Centre	UK	*419*	60-00700
Labtech Ltd	UK	*464*	75-34280
Litton Systems Canada Ltd	Canada	*73*	75-09100
MacDonald Dettwiler and Associates	Canada	*78*	80-01310
Metric Systems Corp.	USA	*596*	75-60800
Northrop Grumman Corporation	USA	*599*	75-65450
Racal Communications Inc.	USA	*606*	75-72380
Racal Radar Defence Systems Ltd	UK	*477*	75-47688
Roke Manor Research Ltd	UK	*420*	60-01800
Technology Service Corp.	USA	*644*	80-31930

411 12 70 Radars, airborne, tactical

AIL Systems Inc.	USA	*547*	75-02950
Andrew Sensor Products	USA	*554*	75-07452
Canadian Marconi Co.	Canada	*69*	75-02600
CIC International Ltd	USA	*563*	75-18570
Daimler-Benz Aerospace France SAS	**France**	***134***	**75-12950**
ELTA Electronics Industries Ltd	Israel	*220*	75-10200
Ericsson Microwave Systems AB	**Sweden**	***377***	**75-04890**
Fabbrica Italiana Apparecchiature	Italy	*234*	75-08550
GEC-Marconi Avionics Ltd	UK	*454*	75-25320
Labtech Ltd	UK	*464*	75-34280
MacDonald Dettwiler and Associates	Canada	*78*	80-01310
Northrop Grumman Corporation	USA	*640*	80-21840
Racal Communications Inc.	USA	*606*	75-72380
Radio Research Instrument Co. Inc.	USA	*607*	75-72600
Technology Service Corp.	USA	*644*	80-31930
Znamya Production Group	Ukraine	*407*	75-10000

411 12 72 Radars, airborne, tactical, air-to-air

CIC International Ltd	USA	*563*	75-18570
ELTA Electronics Industries Ltd	Israel	*220*	75-10200
Ericsson Microwave Systems AB	**Sweden**	***377***	**75-04890**
Fabbrica Italiana Apparecchiature	Italy	*234*	75-08550
GEC Marconi Hazeltine Corp	USA	*578*	75-37415
GEC-Marconi Avionics Ltd	UK	*454*	75-25320
Israel Aircraft Industries Ltd	Israel	*219*	75-10200
Labtech Ltd	UK	*464*	75-34280
Northrop Grumman Corporation	USA	*640*	80-21840
Radio Research Instrument Co. Inc.	USA	*607*	75-72600
Technology Service Corp.	USA	*644*	80-31930

Product-Supplier Link

Product–Supplier Link

411 12 73 Radars, airborne, side-looking

Andrew Sensor Products	USA	*554*	75-07452
CIC International Ltd	USA	*563*	75-18570
Cranfield University	UK	*419*	60-00350
ELTA Electronics Industries Ltd	Israel	*220*	75-10200
Ericsson Defence Systems Pty Ltd	Australia	*16*	75-05820
Ericsson Microwave Systems AB	**Sweden**	***377***	**75-04890**
Fabbrica Italiana Apparecchiature	Italy	*234*	75-08550
GEC-Marconi Avionics Ltd	UK	*454*	75-25320
GEC-Marconi Research Centre	UK	*419*	60-00700
Labtech Ltd	UK	*464*	75-34280
MacDonald Dettwiler and Associates	Canada	*78*	80-01310
Radio Research Instrument Co. Inc.	USA	*607*	75-72600
Technology Service Corp.	USA	*644*	80-31930

411 12 74 Radars, airborne, tactical, air-to-surface

AIL Systems Inc.	USA	*547*	75-02950
CIC International Ltd	USA	*563*	75-18570
Ericsson Microwave Systems AB	**Sweden**	***377***	**75-04890**
Fabbrica Italiana Apparecchiature	Italy	*234*	75-08550
GEC Marconi Hazeltine Corp	USA	*578*	75-37415
GEC-Marconi Avionics Ltd	UK	*454*	75-25320
MacDonald Dettwiler and Associates	Canada	*78*	80-01310
Northrop Grumman Corporation	USA	*640*	80-21840
Radio Research Instrument Co. Inc.	USA	*607*	75-72600
Technology Service Corp.	USA	*644*	80-31930

411 12 74 30 Radars, airborne, tactical, air-to-surface, helicopter

AIL Systems Inc.	USA	*547*	75-02950
CIC International Ltd	USA	*563*	75-18570
Fabbrica Italiana Apparecchiature	Italy	*234*	75-08550
GEC-Marconi Avionics Ltd	UK	*454*	75-25320
MacDonald Dettwiler and Associates	Canada	*78*	80-01310
Northrop Grumman Corporation	USA	*640*	80-21840
Racal Communications Inc.	USA	*606*	75-72380
Technology Service Corp.	USA	*644*	80-31930

411 12 76 Radars, airborne, tactical, multi-mode

AIL Systems Inc.	USA	*547*	75-02950
CIC International Ltd	USA	*563*	75-18570
ELTA Electronics Industries Ltd	Israel	*220*	75-10200
Ericsson Microwave Systems AB	**Sweden**	***377***	**75-04890**
Fabbrica Italiana Apparecchiature	Italy	*234*	75-08550
GEC-Marconi Avionics Ltd	UK	*454*	75-25320
MacDonald Dettwiler and Associates	Canada	*78*	80-01310
Northrop Grumman Corporation	USA	*640*	80-21840
Philips Traffic & Engineering	Australia	*20*	75-16550
Radio Research Instrument Co. Inc.	USA	*607*	75-72600
Technology Service Corp.	USA	*644*	80-31930

411 12 80 Radars, airborne, terrain-following

CIC International Ltd	USA	*563*	75-18570
Dassault Electronique SA (Group)	**France**	***135***	**75-12980**
Fabbrica Italiana Apparecchiature	Italy	*234*	75-08550
Northrop Grumman Corporation	USA	*640*	80-21840
Philips Traffic & Engineering	Australia	*20*	75-16550
Radio Research Instrument Co. Inc.	USA	*607*	75-72600
Technology Service Corp.	USA	*644*	80-31930

411 12 90 Radars, airborne, weather

Andrew Sensor Products	USA	*554*	75-07452
BFGoodrich Avionics Systems Inc	USA	*559*	75-13771
CIC International Ltd	USA	*563*	75-18570
Ericsson Defence Systems Pty Ltd	Australia	*16*	75-05820
Fabbrica Italiana Apparecchiature	Italy	*234*	75-08550
Radio Research Instrument Co. Inc.	USA	*607*	75-72600
Rockwell International Co.	Italy	*233*	75-05200
Technology Service Corp.	USA	*644*	80-31930

411 13 Radars, naval, target location & tracking

Alenia-Elsag	Italy	*230*	74-00100
Andrew Sensor Products	USA	*554*	75-07452
Bharat Electronics Ltd	India	*203*	55-00200
Canadian Marconi Co.	Canada	*69*	75-02600
CelsiusTech Electronics AB	Sweden	*376*	75-03720
CIC International Ltd	USA	*563*	75-18570
Contraves GmbH	Germany	*166*	75-05200
Cranfield University	UK	*419*	60-00350
Elettronica (UK) Ltd	UK	*448*	75-20475
ELTA Electronics Industries Ltd	Israel	*220*	75-10200
Ericsson Microwave Systems AB	**Sweden**	***377***	**75-04890**
GEC Marconi Hazeltine Corp	USA	*578*	75-37415
Hollandse Signaalapparaten BV	Netherlands	*293*	75-03300
ITT Gilfillan	USA	*586*	75-49455
Kelvin Hughes Ltd	UK	*463*	75-33453
Labtech Ltd	UK	*464*	75-34280
Norden Systems Inc.	USA	*599*	75-64900
Northrop Grumman Corporation	USA	*640*	80-21840
Oerlikon-Contraves AG	Switzerland	*387*	75-09250
Oerlikon-Contraves SpA	Italy	*237*	75-16800
Radio Research Instrument Co. Inc.	USA	*607*	75-72600
Roke Manor Research Ltd	UK	*420*	60-01800
Sanders	USA	*612*	75-77850
Science Systems (Industrial) Ltd	UK	*510*	80-31600
Siemens Plessey Systems	UK	*483*	75-53770
Siemens Switzerland Ltd	Switzerland	*387*	75-11980
Sperry Marine Inc.	USA	*617*	75-83250
Technology Service Corp.	USA	*644*	80-31930
Znamya Production Group	Ukraine	*407*	75-10000

411 13 30 Radars, naval, fire control & target designation

Andrew Sensor Products	USA	*554*	75-07452
Bharat Electronics Ltd	India	*203*	55-00200
CelsiusTech Electronics AB	Sweden	*376*	75-03720
CIC International Ltd	USA	*563*	75-18570
ELTA Electronics Industries Ltd	Israel	*220*	75-10200
Ericsson Microwave Systems AB	**Sweden**	***377***	**75-04890**
GEC-Marconi Avionics Ltd	UK	*454*	75-25320
Hollandse Signaalapparaten BV	Netherlands	*293*	75-03300
Israel Aircraft Industries Ltd	Israel	*219*	75-10200
ITT Aerospace/Communications Div.	USA	*586*	75-49455
Labtech Ltd	UK	*464*	75-34280
Northrop Grumman Corporation	USA	*640*	80-21840
Oerlikon-Contraves SpA	Italy	*237*	75-16800
Philips Traffic & Engineering	Australia	*20*	75-16550
Radio Research Instrument Co. Inc.	USA	*607*	75-72600
Roke Manor Research Ltd	UK	*420*	60-01800
Sanders	USA	*612*	75-77850
Siemens Plessey Systems	UK	*483*	75-53770
Siemens Switzerland Ltd	Switzerland	*387*	75-11980
Sperry Marine Inc.	USA	*617*	75-83250
Technology Service Corp.	USA	*644*	80-31930
TERMA Elektronik AS	Denmark	*104*	75-20300
Thomson-CSF	France	*148*	75-45100

411 13 40 Radars, naval, general purpose

Andrew Sensor Products	USA	*554*	75-07452
CelsiusTech Electronics AB	Sweden	*376*	75-03720
CIC International Ltd	USA	*563*	75-18570
ELTA Electronics Industries Ltd	Israel	*220*	75-10200
Hollandse Signaalapparaten BV	Netherlands	*293*	75-03300
Israel Aircraft Industries Ltd	Israel	*219*	75-10200
ITT Aerospace/Communications Div.	USA	*586*	75-49455
Kelvin Hughes Ltd	UK	*463*	75-33453
Northrop Grumman Corporation	USA	*640*	80-21840
Racal Communications Inc.	USA	*606*	75-72380
Radio Research Instrument Co. Inc.	USA	*607*	75-72600
Siemens Plessey Systems	UK	*483*	75-53770
Spar Aerospace Ltd	Canada	*76*	75-13000
Sperry Marine Inc.	USA	*617*	75-83250
Sperry Marine SpA	Italy	*239*	75-24400
Technology Service Corp.	USA	*644*	80-31930
TERMA Elektronik AS	Denmark	*104*	75-20300
Znamya Production Group	Ukraine	*407*	75-10000

411 13 45 Radars, naval, missile detection

Andrew Sensor Products	USA	*554*	75-07452
CelsiusTech Electronics AB	Sweden	*376*	75-03720
CIC International Ltd	USA	*563*	75-18570
ELTA Electronics Industries Ltd	Israel	*220*	75-10200

Ericsson Microwave Systems AB	**Sweden**	377	**75-04890**
GEC-Marconi Research Centre	UK	419	60-00700
ITT Gilfillan	USA	586	75-49455
Labtech Ltd	UK	464	75-34280
Oerlikon-Contraves SpA	Italy	237	75-16800
Racal Radar Defence Systems Ltd	UK	477	75-47688
Roke Manor Research Ltd	UK	420	60-01800
Siemens Plessey Systems	UK	483	75-53770
Siemens Switzerland Ltd	Switzerland	387	75-11980
Sperry Marine Inc.	USA	617	75-83250
Technology Service Corp.	USA	644	80-31930
Thomson-CSF	France	148	75-45100

411 13 50 Radars, naval, tactical

Bharat Electronics Ltd	India	203	55-00200
CelsiusTech Electronics AB	Sweden	376	75-03720
CIC International Ltd	USA	563	75-18570
ELTA Electronics Industries Ltd	Israel	220	75-10200
Israel Aircraft Industries Ltd	Israel	219	75-10200
ITT Gilfillan	USA	586	75-49455
Kelvin Hughes Ltd	UK	463	75-33453
Northrop Grumman Corporation	USA	640	80-21840
Oerlikon-Contraves SpA	Italy	237	75-16800
Racal Communications Inc.	USA	606	75-72380
Radio Research Instrument Co. Inc.	USA	607	75-72600
Siemens Plessey Systems	UK	483	75-53770
Sperry Marine Inc.	USA	617	75-83250
Technology Service Corp.	USA	644	80-31930
Thomson-CSF	France	148	75-45100

411 13 70 Radars, naval, surveillance

Andrew Sensor Products	USA	554	75-07452
Bharat Electronics Ltd	India	203	55-00200
Canadian Marconi Co.	Canada	69	75-02600
CelsiusTech Electronics AB	Sweden	376	75-03720
CIC International Ltd	USA	563	75-18570
ELTA Electronics Industries Ltd	Israel	220	75-10200
Ericsson Microwave Systems AB	**Sweden**	377	**75-04890**
Hollandse Signaalapparaten BV	Netherlands	293	75-03300
Israel Aircraft Industries Ltd	Israel	219	75-10200
ITT Aerospace/Communications Div.	USA	586	75-49455
ITT Gilfillan	USA	586	75-49455
Kelvin Hughes Ltd	UK	463	75-33453
Northrop Grumman Corporation	USA	640	80-21840
Oerlikon-Contraves SpA	Italy	237	75-16800
Racal Communications Inc.	USA	606	75-72380
Racal Radar Defence Systems Ltd	UK	477	75-47688
Radio Research Instrument Co. Inc.	USA	607	75-72600
Siemens Plessey Systems	UK	483	75-53770
Sperry Marine Inc.	USA	617	75-83250
SRI International	USA	643	80-29600
STN ATLAS Elektronik GmbH	Germany	165	75-01150
Technology Service Corp.	USA	644	80-31930
TERMA Elektronik AS	Denmark	104	75-20300
Thomson-CSF	France	148	75-45100

411 13 71 Radars, naval, air surveillance

Andrew Canada Inc.	Canada	68	75-00600
Andrew Sensor Products	USA	554	75-07452
Bharat Electronics Ltd	India	203	55-00200
CelsiusTech Electronics AB	Sweden	376	75-03720
CIC International Ltd	USA	563	75-18570
ELTA Electronics Industries Ltd	Israel	220	75-10200
Ericsson Microwave Systems AB	**Sweden**	377	**75-04890**
GEC Marconi Hazeltine Corp	USA	578	75-37415
Hollandse Signaalapparaten BV	Netherlands	293	75-03300
Israel Aircraft Industries Ltd	Israel	219	75-10200
ITT Aerospace/Communications Div.	USA	586	75-49455
ITT Defense Ltd	UK	505	80-13570
ITT Gilfillan	USA	586	75-49455
NEC Corporation	Japan	253	75-06350
Northrop Grumman Corporation	USA	640	80-21840
Oerlikon-Contraves SpA	Italy	237	75-16800
Radio Research Instrument Co. Inc.	USA	607	75-72600
Siemens Plessey Systems	UK	483	75-53770
SRI International	USA	643	80-29600
STN ATLAS Elektronik GmbH	Germany	165	75-01150
Technology Service Corp.	USA	644	80-31930
TERMA Elektronik AS	Denmark	104	75-20300
Thomson-CSF	France	148	75-45100

411 13 72 Radars, naval, combined air & surface surveillance

Alenia-Elsag	Italy	230	74-00100
Andrew Sensor Products	USA	554	75-07452
Canadian Marconi Co.	Canada	69	75-02600
CelsiusTech Electronics AB	Sweden	376	75-03720
CIC International Ltd	USA	563	75-18570
ELTA Electronics Industries Ltd	Israel	220	75-10200
Ericsson Microwave Systems AB	**Sweden**	377	**75-04890**
Hollandse Signaalapparaten BV	Netherlands	293	75-03300
Israel Aircraft Industries Ltd	Israel	219	75-10200
Northrop Grumman Corporation	USA	640	80-21840
Radio Research Instrument Co. Inc.	USA	607	75-72600
Siemens Plessey Systems	UK	483	75-53770
SRI International	USA	643	80-29600
STN ATLAS Elektronik GmbH	Germany	165	75-01150
Technology Service Corp.	USA	644	80-31930
TERMA Elektronik AS	Denmark	104	75-20300
Thomson-CSF	France	148	75-45100

411 13 77 Radars, naval, surface surveillance

Alenia-Elsag	Italy	230	74-00100
Bharat Electronics Ltd	India	203	55-00200
Canadian Marconi Co.	Canada	69	75-02600
CelsiusTech Electronics AB	Sweden	376	75-03720
CIC International Ltd	USA	563	75-18570
ELTA Electronics Industries Ltd	Israel	220	75-10200
Ericsson Microwave Systems AB	**Sweden**	377	**75-04890**
Hollandse Signaalapparaten BV	Netherlands	293	75-03300
Kelvin Hughes Ltd	UK	463	75-33453
Northrop Grumman Corporation	USA	640	80-21840
Racal Communications Inc.	USA	606	75-72380
Radio Research Instrument Co. Inc.	USA	607	75-72600
Siemens Plessey Systems	UK	483	75-53770
STN ATLAS Elektronik GmbH	Germany	165	75-01150
Technology Service Corp.	USA	644	80-31930
TERMA Elektronik AS	Denmark	104	75-20300
Thomson-CSF	France	148	75-45100

411 13 90 Radars, submarine

Canadian Marconi Co.	Canada	69	75-02600
CelsiusTech Electronics AB	Sweden	376	75-03720
Kelvin Hughes Ltd	UK	463	75-33453
Technology Service Corp.	USA	644	80-31930

411 14 Radars, ground-based, target location & tracking

AIL Systems Inc.	USA	547	75-02950
Alenia-Elsag	Italy	230	74-00100
Andrew Canada Inc.	Canada	68	75-00600
Andrew Sensor Products	USA	554	75-07452
APT Radar Systems Ltd	UK	431	75-03560
Bharat Electronics Ltd	India	203	55-00200
Carter Manufacturing Corp.	USA	562	75-16800
CIC International Ltd	USA	563	75-18570
Continental Electronics Corp.	USA	566	75-21800
Contraves GmbH	Germany	166	75-05200
Cranfield University	UK	419	60-00350
Daimler-Benz Aerospace France SAS	**France**	134	**75-12950**
Elettronica (UK) Ltd	UK	448	75-20475
ELTA Electronics Industries Ltd	Israel	220	75-10200
Ericsson Microwave Systems AB	**Sweden**	377	**75-04890**
ESSCO Collins Ltd	Ireland	214	75-00100
GEC-Marconi Research Centre	UK	419	60-00700
Hollandse Signaalapparaten BV	Netherlands	293	75-03300
INIMEX	**Belgium**	46	**75-04650**
Israel Aircraft Industries Ltd	Israel	219	75-10200
ITT Gilfillan	USA	586	75-49455
KEL Aerospace Pty Ltd	Australia	18	75-11440
Kelvin Hughes Ltd	UK	463	75-33453
Labtech Ltd	UK	464	75-34280
MPB Technologies Inc	Canada	74	75-09855
Northrop Grumman Corporation	USA	640	80-21840
Racal Communications Inc.	USA	606	75-72380

Product-Supplier Link

Racal Radar Defence Systems Ltd.	UK	*477*	75-47688
Radio Research Instrument Co. Inc.	USA	*607*	75-72600
Roke Manor Research Ltd	UK	*420*	60-01800
Sanders	USA	*612*	75-77850
Siemens Aktiengesellschaft	Germany	*175*	75-27700
Siemens Plessey Systems	UK	*483*	75-53770
Siemens Switzerland Ltd	Switzerland	*387*	75-11980
Sierra Technologies Inc.	USA	*614*	75-80640
Sperry Marine Inc.	USA	*617*	75-83250
Technology Service Corp.	USA	*644*	80-31930
Thomson-CSF	France	*148*	75-45100
Znamya Production Group	Ukraine	*407*	75-10000

411 14 10 Radars, ground-based, fixed

Andrew Canada Inc.	Canada	*68*	75-00600
Andrew Sensor Products	USA	*554*	75-07452
APT Radar Systems Ltd	UK	*431*	75-03560
Bharat Electronics Ltd	India	*203*	55-00200
CIC International Ltd	USA	*563*	75-18570
Daimler-Benz Aerospace France SAS	**France**	***134***	**75-12950**
Electronic Space Systems Corp.	USA	*572*	75-30100
ELTA Electronics Industries Ltd	Israel	*220*	75-10200
Ericsson Microwave Systems AB	**Sweden**	***377***	**75-04890**
ESSCO Collins Ltd	Ireland	*214*	75-00100
Hollandse Signaalapparaten BV	Netherlands	*293*	75-03300
Israel Aircraft Industries Ltd	Israel	*219*	75-10200
ITT Aerospace/Communications Div.	USA	*586*	75-49455
ITT Gilfillan	USA	*586*	75-49455
Labtech Ltd	UK	*464*	75-34280
NEC Corporation	Japan	*253*	75-06350
Northrop Grumman Corporation	USA	*640*	80-21840
Radio Research Instrument Co. Inc.	USA	*607*	75-72600
Roke Manor Research Ltd	UK	*420*	60-01800
Rutherford Appleton Laboratory	UK	*420*	60-01850
Sanders	USA	*612*	75-77850
Siemens Plessey Systems	UK	*483*	75-53770
Technology Service Corp.	USA	*644*	80-31930
TERMA Elektronik AS	Denmark	*104*	75-20300
Thomson-CSF	France	*148*	75-45100
TIW Systems Inc.	USA	*623*	75-89755

411 14 10 05 Radars, ground-based, fixed, airfield

Andrew Canada Inc.	Canada	*68*	75-00600
Andrew Sensor Products	USA	*554*	75-07452
APT Radar Systems Ltd	UK	*431*	75-03560
CIC International Ltd	USA	*563*	75-18570
Daimler-Benz Aerospace France SAS	**France**	***134***	**75-12950**
ELTA Electronics Industries Ltd	Israel	*220*	75-10200
Israel Aircraft Industries Ltd	Israel	*219*	75-10200
ITT Defense Ltd	UK	*505*	80-13570
ITT Gilfillan	USA	*586*	75-49455
NEC Corporation	Japan	*253*	75-06350
Northrop Grumman Corporation	USA	*640*	80-21840
Radio Research Instrument Co. Inc.	USA	*607*	75-72600
Technology Service Corp.	USA	*644*	80-31930
TERMA Elektronik AS	Denmark	*104*	75-20300
Thomson-CSF	France	*148*	75-45100

411 14 10 05 10 Radars, ground-based, fixed, air traffic control

Andrew Sensor Products	USA	*554*	75-07452
APT Radar Systems Ltd	UK	*431*	75-03560
CIC International Ltd	USA	*563*	75-18570
Daimler-Benz Aerospace France SAS	**France**	***134***	**75-12950**
ELTA Electronics Industries Ltd	Israel	*220*	75-10200
Fabbrica Italiana Apparecchiature	Italy	*234*	75-08550
IDS Ingegneria dei Sistemi SpA	Italy	*240*	80-00920
Israel Aircraft Industries Ltd	Israel	*219*	75-10200
ITT Aerospace/Communications Div.	USA	*586*	75-49455
ITT Gilfillan	USA	*586*	75-49455
NEC Corporation	Japan	*253*	75-06350
Northrop Grumman Corporation	USA	*640*	80-21840
Racal Radar Defence Systems Ltd	UK	*477*	75-47688
Radio Research Instrument Co. Inc.	USA	*607*	75-72600
Raytheon Canada Ltd	Canada	*75*	75-11700
Technology Service Corp.	USA	*644*	80-31930
TERMA Elektronik AS	Denmark	*104*	75-20300
Thomson-CSF	France	*148*	75-45100

411 14 10 05 20 Radars, ground-based, fixed, airfield precision approach

Andrew Canada Inc.	Canada	*68*	75-00600
APT Radar Systems Ltd	UK	*431*	75-03560
CIC International Ltd	USA	*563*	75-18570
Daimler-Benz Aerospace France SAS	**France**	***134***	**75-12950**
ELTA Electronics Industries Ltd	Israel	*220*	75-10200
Fabbrica Italiana Apparecchiature	Italy	*234*	75-08550
ITT Aerospace/Communications Div.	USA	*586*	75-49455
ITT Defense Ltd	UK	*505*	80-13570
ITT Gilfillan	USA	*586*	75-49455
NEC Corporation	Japan	*253*	75-06350
Northrop Grumman Corporation	USA	*640*	80-21840
Radio Research Instrument Co. Inc.	USA	*607*	75-72600
Raytheon Canada Ltd	Canada	*75*	75-11700
Technology Service Corp.	USA	*644*	80-31930
TERMA Elektronik AS	Denmark	*104*	75-20300
Thomson-CSF	France	*148*	75-45100

411 14 10 05 30 Radars, ground-based, fixed, airfield surveillance

Andrew Sensor Products	USA	*554*	75-07452
APT Radar Systems Ltd	UK	*431*	75-03560
CIC International Ltd	USA	*563*	75-18570
Daimler-Benz Aerospace France SAS	**France**	***134***	**75-12950**
ELTA Electronics Industries Ltd	Israel	*220*	75-10200
Fabbrica Italiana Apparecchiature	Italy	*234*	75-08550
Israel Aircraft Industries Ltd	Israel	*219*	75-10200
ITT Aerospace/Communications Div.	USA	*586*	75-49455
ITT Gilfillan	USA	*586*	75-49455
Kelvin Hughes Ltd	UK	*463*	75-33453
Labtech Ltd	UK	*464*	75-34280
NEC Corporation	Japan	*253*	75-06350
Northrop Grumman Corporation	USA	*640*	80-21840
Racal Communications Inc.	USA	*606*	75-72380
Radar/Digital Systems	USA	*607*	75-72460
Radio Research Instrument Co. Inc.	USA	*607*	75-72600
Roke Manor Research Ltd	UK	*420*	60-01800
Technology Service Corp.	USA	*644*	80-31930
TERMA Elektronik AS	Denmark	*104*	75-20300
Thomson-CSF	France	*148*	75-45100

411 14 10 05 40 Radars, ground-based, fixed, ATC airspace surveillance

CIC International Ltd	USA	*563*	75-18570
ELTA Electronics Industries Ltd	Israel	*220*	75-10200
IDS Ingegneria dei Sistemi SpA	Italy	*240*	80-00920
Israel Aircraft Industries Ltd	Israel	*219*	75-10200
ITT Aerospace/Communications Div.	USA	*586*	75-49455
ITT Gilfillan	USA	*586*	75-49455
Northrop Grumman Corporation	USA	*640*	80-21840
Radar/Digital Systems	USA	*607*	75-72460
Radio Research Instrument Co. Inc.	USA	*607*	75-72600
Raytheon Canada Ltd	Canada	*75*	75-11700
Raytheon Cossor Electronics Ltd	UK	*478*	75-48367
Technology Service Corp.	USA	*644*	80-31930
Thomson-CSF	France	*148*	75-45100

411 14 10 10 Radars, ground-based, fixed, air defense

Andrew Sensor Products	USA	*554*	75-07452
Bharat Electronics Ltd	India	*203*	55-00200
CIC International Ltd	USA	*563*	75-18570
Daimler-Benz Aerospace France SAS	**France**	***134***	**75-12950**
ELTA Electronics Industries Ltd	Israel	*220*	75-10200
Ericsson Microwave Systems AB	**Sweden**	***377***	**75-04890**
Fabbrica Italiana Apparecchiature	Italy	*234*	75-08550
GEC Marconi Hazeltine Corp	USA	*578*	75-37415
Israel Aircraft Industries Ltd	Israel	*219*	75-10200
ITT Aerospace/Communications Div.	USA	*586*	75-49455
ITT Gilfillan	USA	*586*	75-49455

Labtech Ltd	UK	*464*	75-34280
NEC Corporation	Japan	*253*	75-06350
Northrop Grumman Corporation	USA	*640*	80-21840
Racal Communications Inc.	USA	*606*	75-72380
Radio Research Instrument Co. Inc.	USA	*607*	75-72600
Roke Manor Research Ltd	UK	*420*	60-01800
Sanders	USA	*612*	75-77850
Siemens Plessey Systems	UK	*483*	75-53770
Technology Service Corp.	USA	*644*	80-31930
Thomson-CSF	France	*148*	75-45100
Znamya Production Group	Ukraine	*407*	75-10000

411 14 10 12 Radars, ground-based, fixed, air defense, medium range

Andrew Sensor Products	USA	*554*	75-07452
Bharat Electronics Ltd	India	*203*	55-00200
CIC International Ltd	USA	*563*	75-18570
Cranfield University	UK	*419*	60-00350
Daimler-Benz Aerospace France SAS	**France**	***134***	**75-12950**
ELTA Electronics Industries Ltd	Israel	*220*	75-10200
Ericsson Microwave Systems AB	**Sweden**	***377***	**75-04890**
Israel Aircraft Industries Ltd	Israel	*219*	75-10200
ITT Aerospace/Communications Div.	USA	*586*	75-49455
ITT Gilfillan	USA	*586*	75-49455
Labtech Ltd	UK	*464*	75-34280
Northrop Grumman Corporation	USA	*640*	80-21840
Racal Communications Inc.	USA	*606*	75-72380
Radio Research Instrument Co. Inc.	USA	*607*	75-72600
Roke Manor Research Ltd	UK	*420*	60-01800
Siemens Plessey Systems	UK	*483*	75-53770
Technology Service Corp.	USA	*644*	80-31930
Thomson-CSF	France	*148*	75-45100

411 14 10 14 Radars, ground-based, fixed, air defense, long range

Andrew Sensor Products	USA	*554*	75-07452
Bharat Electronics Ltd	India	*203*	55-00200
CIC International Ltd	USA	*563*	75-18570
Cranfield University	UK	*419*	60-00350
Daimler-Benz Aerospace France SAS	**France**	***134***	**75-12950**
ITT Aerospace/Communications Div.	USA	*586*	75-49455
ITT Gilfillan	USA	*586*	75-49455
Labtech Ltd	UK	*464*	75-34280
NEC Corporation	Japan	*253*	75-06350
Northrop Grumman Corporation	USA	*640*	80-21840
Radio Research Instrument Co. Inc.	USA	*607*	75-72600
Roke Manor Research Ltd	UK	*420*	60-01800
Siemens Plessey Systems	UK	*483*	75-53770
SRI International	USA	*643*	80-29600
Technology Service Corp.	USA	*644*	80-31930
Thomson-CSF	France	*148*	75-45100

411 14 10 16 Radars, ground-based, fixed, air defense, over-the-horizon

Andrew Ltd	UK	*430*	75-03156
Andrew Sensor Products	USA	*554*	75-07452
CIC International Ltd	USA	*563*	75-18570
KEL Aerospace Pty Ltd	Australia	*18*	75-11440
Labtech Ltd	UK	*464*	75-34280
Northrop Grumman Corporation	USA	*640*	80-21840
Radio Research Instrument Co. Inc.	USA	*607*	75-72600
SRI International	USA	*643*	80-29600
Technology Service Corp.	USA	*644*	80-31930

411 14 10 18 Radars, ground-based, fixed, space surveillance

Andrew Sensor Products	USA	*554*	75-07452
CIC International Ltd	USA	*563*	75-18570
Labtech Ltd	UK	*464*	75-34280
Matra Marconi Space UK, Ltd	UK	*425*	74-06900
Northrop Grumman Corporation	USA	*640*	80-21840
Radio Research Instrument Co. Inc.	USA	*607*	75-72600
Technology Service Corp.	USA	*644*	80-31930

411 14 10 30 Radars, ground-based, fixed, coastal defense

Alenia-Elsag	Italy	*230*	74-00100
Andrew Sensor Products	USA	*554*	75-07452
CIC International Ltd	USA	*563*	75-18570
Denbridge Digital	USA	*609*	75-75500
ELTA Electronics Industries Ltd	Israel	*220*	75-10200
Ericsson Microwave Systems AB	**Sweden**	***377***	**75-04890**
Fabbrica Italiana Apparecchiature	Italy	*234*	75-08550
Hollandse Signaalapparaten BV	Netherlands	*293*	75-03300
Israel Aircraft Industries Ltd	Israel	*219*	75-10200
ITT Aerospace/Communications Div.	USA	*586*	75-49455
ITT Gilfillan	USA	*586*	75-49455
Northrop Grumman Corporation	USA	*640*	80-21840
Radio Research Instrument Co. Inc.	USA	*607*	75-72600
Siemens Plessey Systems	UK	*483*	75-53770
Technology Service Corp.	USA	*644*	80-31930
TERMA Elektronik AS	Denmark	*104*	75-20300
Thomson-CSF	France	*148*	75-45100

411 14 10 50 Radars, ground-based, meteorological

Advanced Technical Corp.	USA	*546*	75-01000
Andrew Canada Inc.	Canada	*68*	75-00600
APT Radar Systems Ltd	UK	*431*	75-03560
Bharat Electronics Ltd	India	*203*	55-00200
CIC International Ltd	USA	*563*	75-18570
Fabbrica Italiana Apparecchiature	Italy	*234*	75-08550
KEL Aerospace Pty Ltd	Australia	*18*	75-11440
Radio Research Instrument Co. Inc.	USA	*607*	75-72600
Remtech SA	France	*141*	75-32230
Technology Service Corp.	USA	*644*	80-31930
TERMA Elektronik AS	Denmark	*104*	75-20300
Thomson-CSF	France	*148*	75-45100
TIW Systems Inc.	USA	*623*	75-89755

411 14 30 Radars, ground-based, mobile / transportable

Andrew Sensor Products	USA	*554*	75-07452
APT Radar Systems Ltd	UK	*431*	75-03560
Bharat Electronics Ltd	India	*203*	55-00200
CelsiusTech Electronics AB	Sweden	*376*	75-03720
CIC International Ltd	USA	*563*	75-18570
ELTA Electronics Industries Ltd	Israel	*220*	75-10200
Ericsson Microwave Systems AB	**Sweden**	***377***	**75-04890**
Ericsson Radar A/S	Norway	*309*	75-00860
GEC-Marconi Research Centre	UK	*419*	60-00700
Israel Aircraft Industries Ltd	Israel	*219*	75-10200
ITT Aerospace/Communications Div.	USA	*586*	75-49455
ITT Gilfillan	USA	*586*	75-49455
K&L Microwave Inc.	USA	*589*	75-52880
Labtech Ltd	UK	*464*	75-34280
MPB Technologies Inc	Canada	*74*	75-09855
NEC Corporation	Japan	*253*	75-06350
Northrop Grumman Corporation	USA	*640*	80-21840
Radio Research Instrument Co. Inc.	USA	*607*	75-72600
Roke Manor Research Ltd	UK	*420*	60-01800
Siemens Aktiengesellschaft	Germany	*175*	75-27700
Siemens Plessey Systems	UK	*483*	75-53770
Siemens Switzerland Ltd	Switzerland	*387*	75-11980
Technology Service Corp.	USA	*644*	80-31930
Thomson-CSF	France	*148*	75-45100

411 14 30 05 10 Radars, mobile, air traffic control

Andrew Sensor Products	USA	*554*	75-07452
CIC International Ltd	USA	*563*	75-18570
ELTA Electronics Industries Ltd	Israel	*220*	75-10200
IDS Ingegneria dei Sistemi SpA	Italy	*240*	80-00920
Israel Aircraft Industries Ltd	Israel	*219*	75-10200
ITT Aerospace/Communications Div.	USA	*586*	75-49455
ITT Gilfillan	USA	*586*	75-49455
NEC Corporation	Japan	*253*	75-06350
Northrop Grumman Corporation	USA	*640*	80-21840
Radio Research Instrument Co. Inc.	USA	*607*	75-72600
Raytheon Canada Ltd	Canada	*75*	75-11700
Technology Service Corp.	USA	*644*	80-31930
Thomson-CSF	France	*148*	75-45100

Product-Supplier Link

411 14 30 05 20 Radars, mobile, airfield precision approach

APT Radar Systems Ltd	UK	*431*	75-03560
CIC International Ltd	USA	*563*	75-18570
ITT Aerospace/Communications Div.	USA	*586*	75-49455
ITT Gilfillan	USA	*586*	75-49455
NEC Corporation	Japan	*253*	75-06350
Northrop Grumman Corporation	USA	*640*	80-21840
Radio Research Instrument Co. Inc.	USA	*607*	75-72600
Raytheon Canada Ltd	Canada	*75*	75-11700
Raytheon Cossor Electronics Ltd	UK	*478*	75-48367
Technology Service Corp.	USA	*644*	80-31930
Thomson-CSF	France	*148*	75-45100

411 14 30 10 Radars, mobile, artillery locating

Bharat Electronics Ltd	India	*203*	55-00200
CIC International Ltd	USA	*563*	75-18570
Ericsson Defence Systems Pty Ltd	Australia	*16*	75-05820
Ericsson Microwave Systems AB	**Sweden**	***377***	**75-04890**
Ericsson Radar A/S	Norway	*309*	75-00860
ITT Gilfillan	USA	*586*	75-49455
Northrop Grumman Corporation	USA	*640*	80-21840
Radio Research Instrument Co. Inc.	USA	*607*	75-72600
Siemens Aktiengesellschaft	Germany	*175*	75-27700
Technology Service Corp.	USA	*644*	80-31930

411 14 30 20 Radars, ground-based, mobile, fire control

Andrew Sensor Products	USA	*554*	75-07452
Bharat Electronics Ltd	India	*203*	55-00200
CelsiusTech Electronics AB	Sweden	*376*	75-03720
CIC International Ltd	USA	*563*	75-18570
Ericsson Microwave Systems AB	**Sweden**	***377***	**75-04890**
Hollandse Signaalapparaten BV	Netherlands	*293*	75-03300
Labtech Ltd	UK	*464*	75-34280
Northrop Grumman Corporation	USA	*640*	80-21840
Oerlikon-Contraves SpA	Italy	*237*	75-16800
Radio Research Instrument Co. Inc.	USA	*607*	75-72600
Roke Manor Research Ltd	UK	*420*	60-01800
Siemens Aktiengesellschaft	Germany	*175*	75-27700
Siemens Switzerland Ltd	Switzerland	*387*	75-11980
Technology Service Corp.	USA	*644*	80-31930
Thomson-CSF	France	*148*	75-45100

411 14 30 30 Radars, ground-based, mobile, coastal defense

Andrew Sensor Products	USA	*554*	75-07452
CelsiusTech Electronics AB	Sweden	*376*	75-03720
CIC International Ltd	USA	*563*	75-18570
ELTA Electronics Industries Ltd	Israel	*220*	75-10200
Ericsson Microwave Systems AB	**Sweden**	***377***	**75-04890**
ITT Gilfillan	USA	*586*	75-49455
Northrop Grumman Corporation	USA	*640*	80-21840
Siemens Plessey Systems	UK	*483*	75-53770
Technology Service Corp.	USA	*644*	80-31930
Thomson-CSF	France	*148*	75-45100

411 14 30 40 Radars, mobile, hostile fire indicating

CIC International Ltd	USA	*563*	75-18570
Ericsson Radar A/S	Norway	*309*	75-00860
Northrop Grumman Corporation	USA	*640*	80-21840
Radio Research Instrument Co. Inc.	USA	*607*	75-72600
Technology Service Corp.	USA	*644*	80-31930

411 14 30 45 Radars, mobile, meteorological

Advanced Technical Corp.	USA	*546*	75-01000
CIC International Ltd	USA	*563*	75-18570
Ericsson Defence Systems Pty Ltd	Australia	*16*	75-05820
Radio Research Instrument Co. Inc.	USA	*607*	75-72600
Technology Service Corp.	USA	*644*	80-31930
Vaisala Oy	Finland	*119*	75-02800

411 14 30 50 Radars, mobile, mortar locating

Bharat Electronics Ltd	India	*203*	55-00200
CIC International Ltd	USA	*563*	75-18570
Ericsson Radar A/S	Norway	*309*	75-00860
Fabbrica Italiana Apparecchiature	Italy	*234*	75-08550
Radio Research Instrument Co. Inc.	USA	*607*	75-72600
Technology Service Corp.	USA	*644*	80-31930

411 14 30 70 Radars, mobile, surveillance

Andrew Sensor Products	USA	*554*	75-07452
CIC International Ltd	USA	*563*	75-18570
CONTROP Precision Technologies Ltd	Israel	*218*	75-04170
Cranfield University	UK	*419*	60-00350
Daimler-Benz Aerospace France SAS	**France**	***134***	**75-12950**
Ericsson Defence Systems Pty Ltd	Australia	*16*	75-05820
GEC-Marconi Research Centre	UK	*419*	60-00700
Hollandse Signaalapparaten BV	Netherlands	*293*	75-03300
ITT Aerospace/Communications Div.	USA	*586*	75-49455
ITT Gilfillan	USA	*586*	75-49455
Labtech Ltd	UK	*464*	75-34280
MPB Technologies Inc	Canada	*74*	75-09855
NEC Corporation	Japan	*253*	75-06350
Northrop Grumman Corporation	USA	*640*	80-21840
Oerlikon-Contraves SpA	Italy	*237*	75-16800
Racal Communications Inc.	USA	*606*	75-72380
Racal Radar Defence Systems Ltd.	UK	*477*	75-47688
Radio Research Instrument Co. Inc.	USA	*607*	75-72600
Roke Manor Research Ltd	UK	*420*	60-01800
Siemens Plessey Systems	UK	*483*	75-53770
Technology Service Corp.	USA	*644*	80-31930

411 14 30 72 Radars, mobile, air surveillance

Andrew Sensor Products	USA	*554*	75-07452
Bharat Electronics Ltd	India	*203*	55-00200
CelsiusTech Electronics AB	Sweden	*376*	75-03720
CIC International Ltd	USA	*563*	75-18570
Daimler-Benz Aerospace France SAS	**France**	***134***	**75-12950**
ELTA Electronics Industries Ltd	Israel	*220*	75-10200
Ericsson Microwave Systems AB	**Sweden**	***377***	**75-04890**
GEC-Marconi Research Centre	UK	*419*	60-00700
Hollandse Signaalapparaten BV	Netherlands	*293*	75-03300
Israel Aircraft Industries Ltd	Israel	*219*	75-10200
ITT Aerospace/Communications Div.	USA	*586*	75-49455
ITT Gilfillan	USA	*586*	75-49455
Labtech Ltd	UK	*464*	75-34280
MPB Technologies Inc	Canada	*74*	75-09855
NEC Corporation	Japan	*253*	75-06350
Northrop Grumman Corporation	USA	*640*	80-21840
Racal Communications Inc.	USA	*606*	75-72380
Radio Research Instrument Co. Inc.	USA	*607*	75-72600
Roke Manor Research Ltd	UK	*420*	60-01800
Sanders	USA	*612*	75-77850
Siemens Aktiengesellschaft	Germany	*175*	75-27700
Siemens Plessey Systems	UK	*483*	75-53770
Technology Service Corp.	USA	*644*	80-31930
Thomson-CSF	France	*148*	75-45100

411 14 30 72 10 Radars, mobile, air surveillance, low-level

Andrew Sensor Products	USA	*554*	75-07452
Belgian Advanced Technology	Belgium	*45*	75-01320
Bharat Electronics Ltd	India	*203*	55-00200
CIC International Ltd	USA	*563*	75-18570
ELTA Electronics Industries Ltd	Israel	*220*	75-10200
Ericsson Microwave Systems AB	**Sweden**	***377***	**75-04890**
GEC-Marconi Research Centre	UK	*419*	60-00700
Hollandse Signaalapparaten BV	Netherlands	*293*	75-03300
Israel Aircraft Industries Ltd	Israel	*219*	75-10200
ITT Aerospace/Communications Div.	USA	*586*	75-49455
ITT Gilfillan	USA	*586*	75-49455
Labtech Ltd	UK	*464*	75-34280
Lear Astronics Corp.	USA	*589*	75-54390
Litton Systems Canada Ltd	Canada	*73*	75-09100
Northrop Grumman Corporation	USA	*640*	80-21840
Oerlikon-Contraves SpA	Italy	*237*	75-16800
Racal Communications Inc.	USA	*606*	75-72380

Racal Radar Defence Systems Ltd.	UK	*477*	75-47688
Radio Research Instrument Co. Inc.	USA	*607*	75-72600
Roke Manor Research Ltd	UK	*420*	60-01800
Sanders	USA	*612*	75-77850
Siemens Aktiengesellschaft	Germany	*175*	75-27700
Siemens Plessey Systems	UK	*483*	75-53770
Siemens Switzerland Ltd.	Switzerland	*387*	75-11980
Technology Service Corp.	USA	*644*	80-31930
Thomson-CSF.	France	*147*	75-45100

411 14 30 72 20 Radars, mobile, air surveillance, medium-high altitude

Andrew Sensor Products	USA	*554*	75-07452
Bharat Electronics Ltd	India	*203*	55-00200
CIC International Ltd	USA	*563*	75-18570
Ericsson Microwave Systems AB	**Sweden**	***377***	**75-04890**
GEC-Marconi Research Centre.	UK	*419*	60-00700
ITT Aerospace/Communications Div.	USA	*586*	75-49455
ITT Gilfillan	USA	*586*	75-49455
Labtech Ltd	UK	*464*	75-34280
Northrop Grumman Corporation.	USA	*640*	80-21840
Radio Research Instrument Co. Inc.	USA	*607*	75-72600
Siemens Plessey Systems	UK	*483*	75-53770
Technology Service Corp.	USA	*644*	80-31930

411 14 30 74 Radars, mobile, battlefield surveillance

Belgian Advanced Technology	Belgium	*45*	75-01320
Bharat Electronics Ltd	India	*203*	55-00200
CIC International Ltd	USA	*563*	75-18570
ELTA Electronics Industries Ltd	Israel	*220*	75-10200
GEC Marconi Hazeltine Corp	USA	*578*	75-37415
Hollandse Signaalapparaten BV	Netherlands	*293*	75-03300
INIMEX	**Belgium**	***46***	**75-04650**
Israel Aircraft Industries Ltd	Israel	*219*	75-10200
NEC Corporation	Japan	*253*	75-06350
Northrop Grumman Corporation.	USA	*640*	80-21840
Racal Communications Inc.	USA	*606*	75-72380
Racal Radar Defence Systems Ltd.	UK	*477*	75-47688
Radio Research Instrument Co. Inc.	USA	*607*	75-72600
Technology Service Corp.	USA	*644*	80-31930
Thomson-CSF.	France	*148*	75-45100

411 14 40 Radars, ground-based, air-portable

Andrew Sensor Products	USA	*554*	75-07452
CIC International Ltd	USA	*563*	75-18570
ELTA Electronics Industries Ltd	Israel	*220*	75-10200
Israel Aircraft Industries Ltd	Israel	*219*	75-10200
Northrop Grumman Corporation.	USA	*640*	80-21840
Racal Communications Inc.	USA	*606*	75-72380
Radio Research Instrument Co. Inc.	USA	*607*	75-72600
Sanders	USA	*612*	75-77850
Siemens Plessey Systems	UK	*483*	75-53770
Technology Service Corp.	USA	*644*	80-31930
Thomson-CSF.	France	*148*	75-45100

411 14 50 Radars, ground-based, man-portable

Andrew Sensor Products	USA	*554*	75-07452
ASMAR	Chile	*84*	55-00150
CIC International Ltd	USA	*563*	75-18570
Israel Aircraft Industries Ltd	Israel	*219*	75-10200
Racal Communications Inc.	USA	*606*	75-72380
Racal Radar Defence Systems Ltd.	UK	*477*	75-47688
Radio Research Instrument Co. Inc.	USA	*607*	75-72600
Technology Service Corp.	USA	*644*	80-31930
Vega-Cantley Instrument Company	UK	*490*	75-62740

411 14 50 55 Radars, moving target indication, man-portable

ASMAR	Chile	*84*	55-00150
CIC International Ltd	USA	*563*	75-18570
Israel Aircraft Industries Ltd	Israel	*219*	75-10200
Radio Research Instrument Co. Inc.	USA	*607*	75-72600
Technology Service Corp.	USA	*644*	80-31930

411 14 80 Radars, ground-based, transportable

Andrew Sensor Products	USA	*554*	75-07452
APT Radar Systems Ltd	UK	*431*	75-03560
Bharat Electronics Ltd	India	*203*	55-00200
CIC International Ltd	USA	*563*	75-18570
ELTA Electronics Industries Ltd	Israel	*220*	75-10200
Israel Aircraft Industries Ltd	Israel	*219*	75-10200
ITT Aerospace/Communications Div.	USA	*586*	75-49455
ITT Gilfillan	USA	*586*	75-49455
Lear Astronics Corp.	USA	*589*	75-54390
Northrop Grumman Corporation.	USA	*640*	80-21840
Radio Research Instrument Co. Inc.	USA	*607*	75-72600
Rafael	Israel	*216*	55-01800
Sanders	USA	*612*	75-77850
Siemens Plessey Systems	UK	*483*	75-53770
Technology Service Corp.	USA	*644*	80-31930

411 14 80 20 Radars, ground-based, transportable, containerised

Alenia SpA	Italy	*232*	75-01170
APT Radar Systems Ltd	UK	*431*	75-03560
Bharat Electronics Ltd	India	*203*	55-00200
CIC International Ltd	USA	*563*	75-18570
ELTA Electronics Industries Ltd	Israel	*220*	75-10200
Northrop Grumman Corporation.	USA	*640*	80-21840
Radio Research Instrument Co. Inc.	USA	*607*	75-72600
Siemens Aktiengesellschaft	Germany	*175*	75-27700
Siemens Plessey Systems	UK	*483*	75-53770
Technology Service Corp.	USA	*644*	80-31930

411 18 Radar accessories

Advanced Technical Corp.	USA	*546*	75-01000
Aerospace Syst. & Comp.	USA	*547*	75-02501
Airscrew Howden Ltd	UK	*429*	75-02050
Andrew Sensor Products	USA	*554*	75-07452
APT Radar Systems Ltd	UK	*431*	75-03560
Bharat Electronics Ltd	India	*203*	55-00200
British Sarozal Ltd.	UK	*436*	75-09570
Chemring Countermeasures	UK	*439*	75-12800
CIC International Ltd	USA	*563*	75-18570
ELTA Electronics Industries Ltd	Israel	*220*	75-10200
Emerson and Cuming Microwave	USA	*573*	75-31150
ESSCO Collins Ltd	Ireland	*214*	75-00100
Fabbrica Italiana Apparecchiature	Italy	*234*	75-08550
Flight Refuelling Limited	UK	*451*	75-23850
Francis & Lewis International Ltd	UK	*452*	75-24500
Howden Fluid Systems	USA	*582*	75-44400
IDS Ingegneria dei Sistemi SpA	Italy	*240*	80-00920
Instro Precision Ltd	UK	*461*	75-31630
KEL Aerospace Pty Ltd.	Australia	*18*	75-11440
Kelvin Hughes Ltd.	UK	*463*	75-33453
Labtech Ltd	UK	*464*	75-34280
MA/COM Antenna & Cable Div.	USA	*593*	75-56953
Millitech Corp.	USA	*596*	75-61665
NEC Corporation	Japan	*253*	75-06350
Nera Telecommunications	UK	*470*	75-42318
Northrop Grumman Corporation.	USA	*640*	80-21840
Optima Enclosures Ltd	UK	*472*	75-43135
Philips Traffic & Engineering	Australia	*20*	75-16550
Roke Manor Research Ltd	UK	*420*	60-01800
Rollet H. & Co. Ltd	UK	*479*	75-50208
Serck Aviation Div.	UK	*482*	75-52860
Shimadzu Corporation	Japan	*254*	75-07600
Siemens Plessey Systems	UK	*483*	75-53770
Siemens Switzerland Ltd	Switzerland	*387*	75-11980
Sperry Marine Inc.	USA	*617*	75-83250
Technical Products Group Inc.	USA	*620*	75-86764
Technology Service Corp.	USA	*644*	80-31930
Thomson-CSF.	France	*148*	75-45100
THORN Microwave Devices Ltd	UK	*488*	75-59760
Vacuum Reflex Ltd	UK	*490*	75-62600
Wessex Electronics Ltd	UK	*493*	75-65333
Znamya Production Group	Ukraine	*407*	75-10000

411 19 Radar components

Advanced Technical Corp.	USA	*546*	75-01000
Aerospace Metal Composites Limited	UK	*428*	75-01180
Albacom Ltd	UK	*429*	75-02372

Alenia SpA	Italy	232	75-01170
Andrew Canada Inc.	Canada	68	75-00600
Andrew Sensor Products	USA	554	75-07452
A.O. Henton Engineering Co Ltd.	UK	458	75-29040
APT Radar Systems Ltd	UK	431	75-03560
Atlantic Microwave Corporation	USA	555	75-10145
Avionic Components Ltd.	UK	432	75-04970
Bharat Electronics Ltd	India	203	55-00200
British Sarozal Ltd.	UK	436	75-09570
Carter Manufacturing Corp.	USA	562	75-16800
Chelton (Electrostatics) Limited	UK	439	75-12700
CIC International Ltd	USA	563	75-18570
Com Dev Europe	UK	441	75-14300
Communications & Power Industries	Canada	70	75-03550
Covergie SA	France	134	75-12400
EG&G Sealol Ltd	UK	448	75-19675
ELTA Electronics Industries Ltd	Israel	220	75-10200
ERA Technology Ltd	UK	502	80-09600
Fabbrica Italiana Apparecchiature	Italy	234	75-08550
Fluorocarbon Company Limited	UK	451	75-24020
Force Computers GmbH	Germany	168	75-09270
Francis & Lewis International Ltd	UK	452	75-24500
Frequency Electronics Inc.	USA	576	75-36190
GEC-Marconi Avionics Ltd	UK	454	75-25320
GEC-Marconi Materials Technology Ltd	UK	419	60-00600
GEC-Marconi Research Centre	UK	419	60-00700
Holscot Industrial Linings Ltd	UK	458	75-29655
Hughes Microelectronics Europe Ltd	UK	459	75-30480
IDS Ingegneria dei Sistemi SpA	Italy	240	80-00920
Iver Precision Engineers Ltd	UK	462	75-32230
Jennings Technology Corp.	USA	586	75-50350
KEL Aerospace Pty Ltd	Australia	18	75-11440
Kelvin Hughes Ltd.	UK	463	75-33453
MA/COM Antenna & Cable Div.	USA	593	75-56953
Micro Metalsmiths Ltd	UK	468	75-39010
Millitech Corp.	USA	596	75-61665
NEC Corporation	Japan	253	75-06350
Nera Telecommunications	UK	470	75-42318
Northrop Grumman Corporation	USA	640	80-21840
Oxley Developments Co. Ltd	UK	472	75-43400
Premier Microwave Corp.	USA	605	75-71400
Pyle-National	UK	476	75-47290
Quasar Microwave Technology Ltd	UK	476	75-47450
Racal-MESL Ltd	UK	477	75-47682
Roke Manor Research Ltd	UK	420	60-01800
Rollet H. & Co. Ltd	UK	479	75-50208
Rotech Components Ltd	UK	481	75-50575
Siemens Plessey Systems	UK	483	75-53770
Siemens Switzerland Ltd	Switzerland	387	75-11980
Sperry Marine Inc.	USA	617	75-83250
Technology Service Corp.	USA	644	80-31930
Teledyne Electronic Technologies	USA	620	75-87522
Thomson-CSF	France	148	75-45100
THORN Microwave Devices Ltd	UK	488	75-59760
Wessex Electronics Ltd	UK	493	75-65333
Zeta	USA	629	75-98190
Znamya Production Group	Ukraine	407	75-10000

411 19 10 Radar antennas

Advanced Technical Corp.	USA	546	75-01000
AEA Technology plc	UK	496	80-00480
Alenia SpA	Italy	232	75-01170
Andrew Canada Inc.	Canada	68	75-00600
Andrew Ltd	UK	430	75-03156
Andrew Sensor Products	USA	554	75-07452
APT Radar Systems Ltd	UK	431	75-03560
Atlantic Microwave Corporation	USA	555	75-10145
Bharat Electronics Ltd	India	203	55-00200
Chelton (Electrostatics) Limited	UK	439	75-12700
CIC International Ltd	USA	563	75-18570
Cranfield University	UK	419	60-00350
Elettronica (UK) Ltd	UK	448	75-20475
ELTA Electronics Industries Ltd	Israel	220	75-10200
ERA Technology Ltd	UK	502	80-09600
ESSCO Collins Ltd	Ireland	214	75-00100
Fleet Industries Ltd	Canada	71	75-05500
GEC-Marconi Avionics Ltd	UK	454	75-25320
GEC-Marconi Research Centre	UK	419	60-00700
Hollandse Signaalapparaten BV	Netherlands	293	75-03300
H.R. Smith (Microwaves) Ltd	UK	484	75-54590
IDS Ingegneria dei Sistemi SpA	Italy	240	80-00920
ITT Gilfillan	USA	586	75-49455
John Curran Ltd	UK	444	75-16430
KEL Aerospace Pty Ltd	Australia	18	75-11440
Kelvin Hughes Ltd	UK	463	75-33453
Labtech Ltd	UK	464	75-34280
MA/COM Antenna & Cable Div.	USA	593	75-56953
Micro Metalsmiths Ltd	UK	468	75-39010
Millitech Corp.	USA	596	75-61665
NEC Corporation	Japan	253	75-06350
Northrop Grumman Corporation	USA	640	80-21840
Premier Microwave Corp.	USA	605	75-71400
Racal Radar Defence Systems Ltd	UK	477	75-47688
Reunert Technology Systems Ltd	S. Africa	359	75-18300
Roke Manor Research Ltd	UK	420	60-01800
Rollet H. & Co. Ltd	UK	479	75-50208
Siemens Aktiengesellschaft	Germany	175	75-27700
Siemens Plessey Systems	UK	483	75-53770
Sperry Marine Inc.	USA	617	75-83250
Technology Service Corp.	USA	644	80-31930
Thomson-CSF	France	148	75-45100
Znamya Production Group	Ukraine	407	75-10000

411 20 Sonar target location & tracking systems & equipment

Alenia-Elsag	Italy	230	74-00100
AlliedSignal Ltd	UK	430	75-02747
Bharat Electronics Ltd	India	203	55-00200
Boeing Australia Limited	Australia	12	75-02500
Bowmar Instrument Corp.	USA	560	75-14860
Breeze Eastern Corp.	USA	560	75-15110
C-Tech Ltd.	Canada	70	75-03800
Daimler-Benz Aerospace France SAS	**France**	**134**	**75-12950**
DRS Technologies, Inc.	USA	570	75-27400
EG&G Inc.	USA	571	75-29555
Fabbrica Italiana Apparecchiature	Italy	234	75-08550
Finnyards Electronics Ltd	Finland	118	75-00600
Force Computers GmbH	Germany	168	75-09270
Fugro-UDI Ltd	UK	452	75-24885
GEC-Marconi Avionics Ltd	UK	454	75-25320
Graseby Dynamics Ltd	UK	455	75-26590
Horstman Defence Systems Ltd	UK	459	75-29920
Imagenex Technology Corp.	Canada	72	75-07728
INDRA	Spain	366	75-03800
Lockheed Martin Corp.	USA	591	75-55635
Massa Products Corp.	USA	594	75-58970
Northrop Grumman Corporation	USA	640	80-21840
Ocean Data Equipment Corp.	USA	601	75-65860
Oki Electric Industry Co. Ltd	Japan	254	75-07200
R. Alkan (UK) Ltd	UK	429	75-02680
Radio Research Instrument Co. Inc.	USA	607	75-72600
Reson A/S	Denmark	103	75-18300
Reson Offshore Ltd	UK	478	75-49260
Sanders	USA	612	75-77850
Sparton of Canada Ltd	Canada	77	75-13100
Sperry Marine Inc.	USA	617	75-83250
STN ATLAS Elektronik GmbH	Germany	165	75-01150
Technology Service Corp.	USA	644	80-31930
Thomson Marconi Sonar Limited	UK	426	74-20000
Thomson Sintra	France	148	75-45135
Triton Technology Inc.	USA	623	75-91050

411 22 Sonars & sonobuoys, air-launched

AlliedSignal Ltd	UK	430	75-02747
Daimler-Benz Aerospace France SAS	**France**	**134**	**75-12950**
EDO Corporation	USA	571	75-29250
Graseby Dynamics Ltd	UK	455	75-26590
Hermes Electronics Ltd	Canada	72	75-07300
NEC Corporation	Japan	253	75-06350
Sparton Electronics	USA	616	75-82900
Sparton of Canada Ltd	Canada	77	75-13100
Thomson Marconi Sonar Limited	UK	426	74-20000
Thomson Sintra	France	148	75-45135

411 22 10 Sonars, helicopter, dunking

AlliedSignal Ltd	UK	430	75-02747
Computing Devices Canada	Canada	70	75-03600
Fabbrica Italiana Apparecchiature	Italy	234	75-08550
Massa Products Corp.	USA	594	75-58970
NEC Corporation	Japan	253	75-06350

Sanders	USA	612	75-77850
Simrad Mesotech Ltd.	Canada	76	75-12795
Thomson Marconi Sonar Limited	UK	426	74-20000
Thomson Sintra	France	148	75-45135

411 22 20 Sonobuoys

Boeing Australia Limited	Australia	12	75-02500
Bofors SA Marine AB	Sweden	376	75-03040
EDO Corporation	USA	571	75-29250
GEC Marconi Hazeltine Corp	USA	578	75-37415
Giat Industries	France	126	55-00100
Graseby Dynamics Ltd	UK	455	75-26590
Hermes Electronics Ltd.	Canada	72	75-07300
NEC Corporation	Japan	253	75-06350
Oki Electric Industry Co. Ltd	Japan	254	75-07200
Sparton Electronics	USA	616	75-82900
Thomson Sintra	France	148	75-45135
Whitehead SpA	Italy	240	75-27600

411 23 Sonar target location & tracking systems, shipborne

C-Tech Ltd.	Canada	70	75-03800
DRS Technologies, Inc.	USA	570	75-27400
EDO Corporation	USA	571	75-29250
EG&G Inc.	USA	571	75-29555
Finnyards Electronics Ltd	Finland	118	75-00600
Fugro-UDI Ltd	UK	452	75-24885
Graseby Dynamics Ltd	UK	455	75-26590
Imagenex Technology Corp.	Canada	72	75-07728
Logica UK Limited	UK	506	80-14350
MacArtney A/S	Denmark	103	75-12800
MacDonald Dettwiler and Associates	Canada	78	80-01310
Marine Electronics Ltd	UK	466	75-37290
Morphyspribor	Russia	335	60-02730
NEC Corporation	Japan	253	75-06350
Safare-Crouzet SA	France	142	75-33900
Sanders	USA	612	75-77850
Simrad Norge A/S	Norway	310	75-02630
STN ATLAS Elektronik GmbH	Germany	165	75-01150
Technology Service Corp.	USA	644	80-31930
Thomson Marconi Sonar Limited	UK	426	74-20000
Thomson Sintra	France	148	75-45135
Triton Technology Inc.	USA	623	75-91050

411 23 10 Sonars, area defense

Fugro-UDI Ltd	UK	452	75-24885
Graseby Dynamics Ltd	UK	455	75-26590
NEC Corporation	Japan	253	75-06350
Simrad Mesotech Ltd.	Canada	76	75-12795
Simrad Norge A/S	Norway	310	75-02630
STN ATLAS Elektronik GmbH	Germany	165	75-01150
Thomson Sintra	France	148	75-45135

411 23 20 Sonars, ship hull-mounted

Alenia-Elsag	Italy	230	74-00100
Computing Devices Canada	Canada	70	75-03600
EDO Corporation	USA	571	75-29250
Explorocean Technology Ltd	UK	502	80-09870
Graseby Dynamics Ltd	UK	455	75-26590
Litton Industries Inc.	USA	591	75-55400
Marine Electronics Ltd	UK	466	75-37290
Massa Products Corp.	USA	594	75-58970
Morphyspribor	Russia	335	60-02730
MSI-Defence Systems Ltd	UK	469	75-41280
NEC Corporation	Japan	253	75-06350
Reson Offshore Ltd	UK	478	75-49260
Safare-Crouzet SA	France	142	75-33900
Sanders	USA	612	75-77850
Simrad Norge A/S	Norway	310	75-02630
Sperry Marine Inc.	USA	617	75-83250
STN ATLAS Elektronik GmbH	Germany	165	75-01150
Technology Service Corp.	USA	644	80-31930
Thomson Marconi Sonar Limited	UK	426	74-20000
Thomson Sintra	France	148	75-45135

411 23 22 Sonars, ship hull-mounted, passive

Bharat Electronics Ltd	India	203	55-00200
Explorocean Technology Ltd	UK	502	80-09870
Fugro-UDI Ltd	UK	452	75-24885
Safare-Crouzet SA	France	142	75-33900
Simrad Norge A/S	Norway	310	75-02630
STN ATLAS Elektronik GmbH	Germany	165	75-01150
Technology Service Corp.	USA	644	80-31930
Thomson Marconi Sonar Limited	UK	426	74-20000

411 23 25 Sonars, hull-mounted, retractable

EDO Corporation	USA	571	75-29250
Fugro-UDI Ltd	UK	452	75-24885
Graseby Dynamics Ltd	UK	455	75-26590
Indal Technologies Inc	Canada	72	75-07740
Marine Electronics Ltd	UK	466	75-37290
Massa Products Corp.	USA	594	75-58970
Simrad Norge A/S	Norway	310	75-02630
Technology Service Corp.	USA	644	80-31930
Thomson Sintra	France	148	75-45135

411 23 26 Sonars, hull-mounted, small ship

Fugro-UDI Ltd	UK	452	75-24885
Graseby Dynamics Ltd	UK	455	75-26590
Marine Electronics Ltd	UK	466	75-37290
Massa Products Corp.	USA	594	75-58970
MSI-Defence Systems Ltd	UK	469	75-41280
NEC Corporation	Japan	253	75-06350
Simrad Norge A/S	Norway	310	75-02630
STN ATLAS Elektronik GmbH	Germany	165	75-01150
Technology Service Corp.	USA	644	80-31930
Thomson Marconi Sonar Limited	UK	426	74-20000
Thomson Sintra	France	148	75-45135

411 23 27 Sonars, hull-mounted, variable depth

Bharat Electronics Ltd	India	203	55-00200
Fugro-UDI Ltd	UK	452	75-24885
International Transducer Corp.	USA	585	75-48650
Marine Electronics Ltd	UK	466	75-37290
Simrad Norge A/S	Norway	310	75-02630
Sparton of Canada Ltd	Canada	77	75-13100
STN ATLAS Elektronik GmbH	Germany	165	75-01150
Technology Service Corp.	USA	644	80-31930
Thomson Marconi Sonar Limited	UK	426	74-20000
Thomson Sintra	France	148	75-45135

411 23 30 Sonars, mine-hunting

Alenia-Elsag	Italy	230	74-00100
EDO Corporation	USA	571	75-29250
EG&G Inc.	USA	571	75-29555
Fabbrica Italiana Apparecchiature	Italy	234	75-08550
Fugro-UDI Ltd	UK	452	75-24885
Klein Associates Inc.	USA	589	75-52900
MacArtney A/S	Denmark	103	75-12800
MacDonald Dettwiler and Associates	Canada	78	80-01310
Marine Electronics Ltd	UK	466	75-37290
Massa Products Corp.	USA	594	75-58970
MSI-Defence Systems Ltd	UK	469	75-41280
NEC Corporation	Japan	253	75-06350
Reson Offshore Ltd	UK	478	75-49260
Simrad Mesotech Ltd.	Canada	76	75-12795
Simrad Norge A/S	Norway	310	75-02630
STN ATLAS Elektronik GmbH	Germany	165	75-01150
Thomson Marconi Sonar Limited	UK	426	74-20000
Thomson Sintra	France	148	75-45135

411 23 70 Sonar target location & tracking systems, submarine

Boeing Australia Limited	Australia	12	75-02500
DRS Technologies, Inc.	USA	570	75-27400
GEC Marconi Hazeltine Corp	USA	578	75-37415
International Transducer Corp.	USA	585	75-48650
Logica UK Limited	UK	506	80-14350
MacArtney A/S	Denmark	103	75-12800

NEC Corporation	Japan	*253*	75-06350
Simrad Norge A/S	Norway	*310*	75-02630
STN ATLAS Elektronik GmbH	Germany	*165*	75-01150
Technology Service Corp.	USA	*644*	80-31930
Thomson Marconi Sonar Limited	UK	*426*	74-20000
Thomson Sintra	France	*148*	75-45135

411 23 80 Sonars, towed array

AlliedSignal Ltd	UK	*430*	75-02747
Computing Devices Canada	Canada	*70*	75-03600
EDO Corporation	USA	*571*	75-29250
Engineered Fabrics Corp.	USA	*573*	75-31565
Hermes Electronics Ltd.	Canada	*72*	75-07300
Hydroscience Technologies Inc.	USA	*584*	75-45550
International Transducer Corp.	USA	*585*	75-48650
Klein Associates Inc.	USA	*589*	75-52900
Litton Industries Inc.	USA	*590*	75-55400
Marine Acoustics Ltd.	UK	*466*	75-37286
Massa Products Corp.	USA	*594*	75-58970
NEC Corporation	Japan	*253*	75-06350
Sanders	USA	*612*	75-77850
Simrad Norge A/S	Norway	*310*	75-02630
STN ATLAS Elektronik GmbH	Germany	*165*	75-01150
Technology Service Corp.	USA	*644*	80-31930
Thomson Marconi Sonar Limited	UK	*426*	74-20000
Thomson Sintra	France	*148*	75-45135

411 23 90 Sonars, variable depth

EDO Corporation	USA	*571*	75-29250
Fabbrica Italiana Apparecchiature	Italy	*234*	75-08550
International Transducer Corp.	USA	*585*	75-48650
MacArtney A/S	Denmark	*103*	75-12800
Massa Products Corp.	USA	*594*	75-58970
NEC Corporation	Japan	*253*	75-06350
Sanders	USA	*612*	75-77850
Simrad Norge A/S	Norway	*310*	75-02630
STN ATLAS Elektronik GmbH	Germany	*165*	75-01150
Technology Service Corp.	USA	*644*	80-31930
Thomson Marconi Sonar Limited	UK	*426*	74-20000
Thomson Sintra	France	*148*	75-45135

411 28 Sonar accessories

David Clark Co. Inc.	USA	*564*	75-18850
EDO Corporation	USA	*571*	75-29250
Fabbrica Italiana Apparecchiature	Italy	*234*	75-08550
Fugro-UDI Ltd	UK	*452*	75-24885
Graseby Dynamics Ltd	UK	*455*	75-26590
GSI Corporation	USA	*580*	75-40715
Hydroscience Technologies Inc.	USA	*584*	75-45550
Klein Associates Inc.	USA	*589*	75-52900
Lucas Western Inc.	USA	*592*	75-56300
Permali Gloucester Ltd	UK	*474*	75-44850
Sensitised Coatings Ltd	UK	*482*	75-52800
Simrad Mesotech Ltd.	Canada	*76*	75-12795
Simrad Norge A/S	Norway	*310*	75-02630
Sparton Electronics	USA	*616*	75-82900
Sparton of Canada Ltd	Canada	*77*	75-13100
STN ATLAS Elektronik GmbH	Germany	*165*	75-01150
Thomson Marconi Sonar Limited	UK	*426*	74-20000
Thomson Sintra	France	*148*	75-45135

411 28 10 Array processors, sonar

Computing Devices Canada	Canada	*70*	75-03600
DRS Technologies, Inc.	USA	*570*	75-27400
Fugro-UDI Ltd	UK	*452*	75-24885
MacDonald Dettwiler and Associates	Canada	*78*	80-01310
Sanders	USA	*612*	75-77850
Simrad Norge A/S	Norway	*310*	75-02630
STN ATLAS Elektronik GmbH	Germany	*165*	75-01150
Technology Service Corp.	USA	*644*	80-31930
Thomson Marconi Sonar Limited	UK	*426*	74-20000

411 28 30 Sonar domes

Indal Technologies Inc	Canada	*72*	75-07740
Simrad Norge A/S	Norway	*310*	75-02630
STN ATLAS Elektronik GmbH	Germany	*165*	75-01150
Thomson Marconi Sonar Limited	UK	*426*	74-20000
Thomson Sintra	France	*148*	75-45135

411 28 40 Sonobuoy launchers

R. Alkan (UK) Ltd	UK	*429*	75-02680
Sparton of Canada Ltd	Canada	*77*	75-13100

411 28 50 Sonar towed array handling systems

ACH	France	*130*	75-00150
Boeing Australia Limited	Australia	*12*	75-02500
Breeze Eastern Corp.	USA	*560*	75-15110
Dowty Aerospace Wolverhampton	UK	*446*	75-18589
EDO Corporation	USA	*571*	75-29250
Indal Technologies Inc	Canada	*72*	75-07740
MacArtney A/S	Denmark	*103*	75-12800
MacTaggart Scott & Co. Ltd	UK	*466*	75-36900
STN ATLAS Elektronik GmbH	Germany	*165*	75-01150
Strachan & Henshaw Ltd	UK	*512*	80-47400
Thomson Marconi Sonar Limited	UK	*426*	74-20000

411 28 60 Sonar tow cable assemblies

Cogent	UK	*500*	80-06250
GSI Corporation	USA	*580*	75-40715
Horstman Defence Systems Ltd.	UK	*459*	75-29920
Hydrocable Systems	UK	*460*	75-30790
Indal Technologies Inc	Canada	*72*	75-07740
Klein Associates Inc.	USA	*589*	75-52900
MacArtney A/S	Denmark	*103*	75-12800
Simrad Norge A/S	Norway	*310*	75-02630
Slingsby Engineering Ltd	UK	*484*	75-54420
STN ATLAS Elektronik GmbH	Germany	*165*	75-01150
Thomson Marconi Sonar Limited	UK	*426*	74-20000

411 29 Sonar components

C-Tech Ltd.	Canada	*70*	75-03800
EG&G Reticon	USA	*571*	75-29562
Explorocean Technology Ltd	UK	*502*	80-09870
Fluorocarbon Company Limited	UK	*451*	75-24020
Force Computers GmbH	Germany	*168*	75-09270
Frequency Electronics Inc.	USA	*576*	75-36190
Fugro-UDI Ltd	UK	*452*	75-24885
GSI Corporation	USA	*580*	75-40715
Klein Associates Inc.	USA	*589*	75-52900
Massa Products Corp.	USA	*594*	75-58970
NEC Corporation	Japan	*253*	75-06350
Permali Gloucester Ltd	UK	*474*	75-44850
Racal-MESL Ltd	UK	*477*	75-47682
Simrad Mesotech Ltd.	Canada	*76*	75-12795
Simrad Norge A/S	Norway	*310*	75-02630
STN ATLAS Elektronik GmbH	Germany	*165*	75-01150
Teledyne Electronic Technologies	USA	*620*	75-87522
Thomson Marconi Sonar Limited	UK	*426*	74-20000
Thomson Sintra	France	*148*	75-45135

411 29 60 Sonar transducers

EDO Corporation	USA	*571*	75-29250
Explorocean Technology Ltd	UK	*502*	80-09870
Fabbrica Italiana Apparecchiature	Italy	*234*	75-08550
Fugro-UDI Ltd	UK	*452*	75-24885
GEC Marconi Hazeltine Corp	USA	*578*	75-37415
Klein Associates Inc.	USA	*589*	75-52900
Marine Acoustics Ltd.	UK	*466*	75-37286
NEC Corporation	Japan	*253*	75-06350
Sanders	USA	*612*	75-77850
Simrad Norge A/S	Norway	*310*	75-02630
Sonar Research & Development Ltd	UK	*485*	75-55170
Sparton Electronics	USA	*616*	75-82900
Sparton of Canada Ltd	Canada	*77*	75-13100
STN ATLAS Elektronik GmbH	Germany	*165*	75-01150
Thomson Marconi Sonar Limited	UK	*426*	74-20000

411 30 Acoustic target location & tracking systems & equipment

Compagnie Générale de Géophysique	France	*134*	75-11335
DRS Technologies, Inc.	USA	*570*	75-27400
EG&G Inc.	USA	*571*	75-29555
Elesco Oy	Finland	*118*	75-00300
GEC Marconi Hazeltine Corp	USA	*578*	75-37415
GEC-Marconi Avionics Ltd	UK	*454*	75-25320
Howden Fluid Systems	USA	*582*	75-44400
Hydroscience Technologies Inc.	USA	*584*	75-45550
Israel Aircraft Industries Ltd	Israel	*219*	75-10200
Lockheed Martin Corp.	USA	*591*	75-55635
Nautronix Ltd	UK	*470*	75-41900
Northrop Grumman Corporation	USA	*640*	80-21840
Pacer Systems Inc.	USA	*602*	75-68000
Raytheon E-Systems	USA	*608*	75-73302
Reson A/S	Denmark	*103*	75-18300
Safare-Crouzet SA	France	*142*	75-33900
Sanders	USA	*612*	75-77850
STN ATLAS Elektronik GmbH	Germany	*165*	75-01150
Technology Service Corp.	USA	*644*	80-31930
THERMOCOAX	France	*147*	75-44930
Thomson Marconi Sonar Limited	UK	*426*	74-20000
Thomson Sintra	France	*148*	75-45135
THORN Lighting AB	Sweden	*379*	75-11280

411 30 70 Sound ranging equipment, artillery location

Boeing Australia Limited	Australia	*12*	75-02500
Davies Industrial Communications Ltd	UK	*444*	75-17010
Oceano Instruments (UK) Ltd	UK	*508*	80-19050
Siemens Plessey Systems	UK	*483*	75-53770
STN ATLAS Elektronik GmbH	Germany	*165*	75-01150
Thomson Sintra	France	*148*	75-45135

411 33 Underwater target location & tracking systems & equipment, acoustic

A1Marketing	UK	*426*	75-00025
Benthos Inc.	USA	*558*	75-13540
Boeing Australia Limited	Australia	*12*	75-02500
Bofors SA Marine AB	Sweden	*376*	75-03040
Cogent	UK	*500*	80-06250
Computing Devices Canada	Canada	*70*	75-03600
Dukane Corp.	USA	*570*	75-27650
EDO Corporation	USA	*571*	75-29250
EG&G Inc.	USA	*571*	75-29555
Elesco Oy	Finland	*118*	75-00300
Fugro-UDI Ltd	UK	*452*	75-24885
MacArtney A/S	Denmark	*103*	75-12800
Marine Electronics Ltd	UK	*466*	75-37290
Nautronix Ltd	UK	*470*	75-41900
NEC Corporation	Japan	*253*	75-06350
Oceano Instruments (UK) Ltd	UK	*508*	80-19050
Officine Panerai SpA	Italy	*237*	75-17500
Pacer Systems Inc.	USA	*602*	75-68000
Safare-Crouzet SA	France	*142*	75-33900
Sanders	USA	*612*	75-77850
Simrad Mesotech Ltd.	Canada	*76*	75-12795
Simrad Norge A/S	Norway	*310*	75-02630
STN ATLAS Elektronik GmbH	Germany	*165*	75-01150
Technology Service Corp.	USA	*644*	80-31930
Thomson Marconi Sonar Limited	UK	*426*	74-20000
Thomson Sintra	France	*148*	75-45135

411 33 30 Hydrophone arrays, underwater

Benthos Inc.	USA	*558*	75-13540
Boeing Australia Limited	Australia	*12*	75-02500
C-Tech Ltd.	Canada	*70*	75-03800
Dukane Corp.	USA	*570*	75-27650
EDO Corporation	USA	*571*	75-29250
Elesco Oy	Finland	*118*	75-00300
Explorocean Technology Ltd	UK	*502*	80-09870
GEC Marconi Hazeltine Corp	USA	*578*	75-37415
International Transducer Corp.	USA	*585*	75-48650
Litton Industries Inc.	USA	*590*	75-55400
Marine Electronics Ltd	UK	*466*	75-37290
Massa Products Corp.	USA	*594*	75-58970
NEC Corporation	Japan	*253*	75-06350
Sanders	USA	*612*	75-77850
Simrad Norge A/S	Norway	*310*	75-02630
Sparton Electronics	USA	*616*	75-82900
STN ATLAS Elektronik GmbH	Germany	*165*	75-01150
THERMOCOAX	France	*147*	75-44930
Thomson Marconi Sonar Limited	UK	*426*	74-20000
Thomson Sintra	France	*148*	75-45135
Underwater & Environmental	UK	*490*	75-61900

411 33 50 Sonar intercept receivers

Boeing Australia Limited	Australia	*12*	75-02500
Safare-Crouzet SA	France	*142*	75-33900
STN ATLAS Elektronik GmbH	Germany	*165*	75-01150

411 40 Electro-optical target location & tracking systems & equipment

AAI Corp.	USA	*545*	75-00020
Advanced System Architectures Ltd	UK	*496*	80-00420
Andrew Ltd	UK	*430*	75-03156
Andrew Sensor Products	USA	*554*	75-07452
Angénieux	France	*132*	75-02700
Arcata Associates Inc.	USA	*554*	75-08400
Avimo Ltd	UK	*432*	75-04950
Boeing Australia Limited	Australia	*12*	75-02500
Bospec Neue Technologien	Germany	*166*	75-03420
British Aerospace (Systems &	UK	*436*	75-08890
Central Design Bureau	Ukraine	*406*	75-00800
Central Research Laboratories Ltd	UK	*499*	80-05895
Coherent Optics (Europe) Ltd	UK	*441*	75-14090
Computer Optics Inc.	USA	*565*	75-20680
Contraves GmbH	Germany	*166*	75-05200
Cranfield Aerospace Ltd	UK	*500*	80-07200
Cranfield University	UK	*419*	60-00350
Data Cell Ltd.	UK	*500*	80-07720
Delft Sensor Systems	**Belgium**	***46***	**75-02500**
Delft Sensor Systems	Netherlands	*292*	75-01590
Diamond Electronics Inc.	USA	*569*	75-26450
EL-OP Electro Optics Industries Ltd	Israel	*219*	75-06300
EUROCONTROL SpA	Italy	*233*	75-08200
Fabbrica Italiana Apparecchiature	Italy	*234*	75-08550
GEC-Marconi Avionics Ltd	UK	*454*	75-25320
GEC-Marconi Hazeltine Corp.	USA	*577*	75-37414
GEC-Marconi Research Centre	UK	*419*	60-00700
GEM Elettronica Srl	Italy	*235*	75-10100
Hadland Photonics Ltd	UK	*456*	75-27200
Hollandse Signaalapparaten BV	Netherlands	*293*	75-03300
Howden Aircontrol Ltd	UK	*459*	75-30240
Hughes Elcan Optical Technologies	Canada	*72*	75-07650
Instro Precision Ltd	UK	*461*	75-31630
Iscan Inc.	USA	*586*	75-49100
Israel Aircraft Industries Ltd	Israel	*219*	75-10200
Kaman Sciences Corp.	USA	*637*	80-17600
Kearfott Guidance & Navigation Corp.	USA	*588*	75-51760
Labtech Ltd	UK	*464*	75-34280
Litton Industries Inc.	USA	*590*	75-55400
Lockheed Martin Corp.	USA	*591*	75-55635
Matra Défense-Espace	France	*139*	75-26000
Matra Marconi Space UK, Ltd	UK	*425*	74-06900
McLennan Marine Ltd.	UK	*467*	75-38080
MSI-Defence Systems Ltd	UK	*469*	75-41280
Nanoquest Defence Products Ltd	UK	*470*	75-41890
Navtech Systems Limited	UK	*507*	80-18000
Night Vision Equipment Co. Inc.	USA	*599*	75-64390
Northrop Grumman Corporation	USA	*599*	75-65450
Octec Ltd	UK	*471*	75-42950
Oerlikon-Contraves AG.	Switzerland	*387*	75-09250
Omitec Electro-Optics Ltd	UK	*472*	75-43050
Pilkington Optronics Ltd	UK	*426*	74-10000
Pilkington Thorn Optronics	UK	*426*	74-10000
Radamec Defence Systems Ltd	UK	*477*	75-47750
Raytheon E-Systems	USA	*608*	75-73302
Roke Manor Research Ltd	UK	*420*	60-01800
Saab Dynamics AB	Sweden	*378*	75-09100
SAGEM SA	**France**	***143***	**75-33950**
Sanders	USA	*612*	75-77850
Sener Ingenieria y Sistemas SA	Spain	*367*	75-05200
SFIM	**France**	***144***	**75-35080**
Siemens Aktiengesellschaft	Germany	*175*	75-27700
Sira Ltd.	UK	*484*	75-54250
SRI International	USA	*643*	80-29600
STN ATLAS Elektronik GmbH	Germany	*165*	75-01150
Technology Service Corp.	USA	*644*	80-31930

Tracor Inc	USA	623	75-90000
Vision Research Co.	UK	491	75-63360

411 40 20 Electro-optical target location & tracking systems, airborne

AAI Corp.	USA	545	75-00020
Andrew Ltd	UK	430	75-03156
Andrew Sensor Products	USA	554	75-07452
British Aerospace (Systems &	UK	436	75-08890
Central Design Bureau	Ukraine	406	75-00800
Central Research Laboratories Ltd	UK	499	80-05895
Coherent Optics (Europe) Ltd	UK	441	75-14090
Computer Optics Inc.	USA	565	75-20680
Contraves Inc	USA	566	75-21850
Cranfield Aerospace Ltd	UK	500	80-07200
EL-OP Electro Optics Industries Ltd	Israel	219	75-06300
Fabbrica Italiana Apparecchiature	Italy	234	75-08550
GEC-Marconi Avionics Ltd	UK	454	75-25320
GEC-Marconi Research Centre	UK	419	60-00700
Hughes Elcan Optical Technologies	Canada	72	75-07650
Israel Aircraft Industries Ltd	Israel	220	75-10200
ITT Aerospace/Communications Div.	USA	586	75-49455
Kaiser Electronics	USA	587	75-51350
Kollsman	USA	589	75-53150
Litton Industries Inc.	USA	590	75-55400
Microtecnica-Torino SpA	Italy	237	75-15500
Nanoquest Defence Products Ltd	UK	470	75-41890
Navtech Systems Limited	UK	507	80-18000
Night Vision Equipment Co. Inc.	USA	599	75-64390
Northrop Grumman Corporation	USA	599	75-65450
Octec Ltd	UK	471	75-42950
OM=	USA	591	75-55400
Omitec Electro-Optics Ltd	UK	472	75-43050
Pilkington Optronics Ltd	UK	426	74-10000
Pilkington Thorn Optronics	UK	426	74-10000
Roke Manor Research Ltd	UK	420	60-01800
Rutherford Appleton Laboratory	UK	420	60-01850
Saab Dynamics AB	Sweden	378	75-09100
Sener Ingenieria y Sistemas SA	Spain	367	75-05200
SFIM	**France**	**144**	**75-35080**
SFIM ODS	France	144	75-35085
Siemens Aktiengesellschaft	Germany	175	75-27700
Sira Ltd.	UK	484	75-54250
Technology Service Corp.	USA	644	80-31930

411 40 20 70 Electro-optical targeting / navigation pods, airborne

AAI Corp.	USA	545	75-00020
Andrew Sensor Products	USA	554	75-07452
British Aerospace (Systems &	UK	436	75-08890
CI Systems Ltd	Israel	218	75-03750
Coherent Optics (Europe) Ltd	UK	441	75-14090
Computer Optics Inc.	USA	565	75-20680
Cranfield Aerospace Ltd	UK	500	80-07200
Delft Sensor Systems	**Belgium**	**46**	**75-02500**
Delft Sensor Systems	Netherlands	292	75-01590
EL-OP Electro Optics Industries Ltd	Israel	219	75-06300
Elettronica (UK) Ltd	UK	448	75-20475
Fabbrica Italiana Apparecchiature	Italy	234	75-08550
GEC-Marconi Avionics Ltd	UK	454	75-25320
GEC-Marconi Research Centre	UK	419	60-00700
Hughes Danbury Optical Sys.	USA	582	75-44857
Israel Aircraft Industries Ltd	Israel	220	75-10200
Kollsman	USA	589	75-53150
Litton Industries Inc.	USA	590	75-55400
Omitec Electro-Optics Ltd	UK	472	75-43050
Pilkington Thorn Optronics	UK	426	74-10000
Rafael	Israel	216	55-01800
SAGEM SA	**France**	**143**	**75-33950**
Sener Ingenieria y Sistemas SA	Spain	367	75-05200
Technical Products Group Inc.	USA	620	75-86764
Technology Service Corp.	USA	644	80-31930

411 40 28 Television target tracking systems, airborne

AAI Corp.	USA	545	75-00020
Andrew Sensor Products	USA	554	75-07452
Arcata Associates Inc.	USA	554	75-08400
Coherent Optics (Europe) Ltd	UK	441	75-14090
Computer Optics Inc.	USA	565	75-20680
Contemporary Systems Design Ltd	UK	442	75-14860
Contraves Inc	USA	566	75-21850
Data Cell Ltd	UK	500	80-07720
EL-OP Electro Optics Industries Ltd	Israel	219	75-06300
Fabbrica Italiana Apparecchiature	Italy	234	75-08550
GEC-Marconi Avionics Ltd	UK	454	75-25320
GEC-Marconi Research Centre	UK	419	60-00700
Iscan Inc.	USA	586	75-49100
Israel Aircraft Industries Ltd	Israel	220	75-10200
Kaiser Electronics	USA	587	75-51350
Navtech Systems Limited	UK	507	80-18000
Northrop Grumman Corporation	USA	599	75-65450
Octec Ltd	UK	471	75-42950
Pearpoint Ltd	UK	473	75-44270
Pilkington Thorn Optronics	UK	426	74-10000
Regis SpA	Italy	238	75-20110
Roke Manor Research Ltd	UK	420	60-01800
SFIM	**France**	**144**	**75-35080**
Technology Service Corp.	USA	644	80-31930

411 40 30 Electro-optical target location & tracking systems, shipborne

AAI Corp.	USA	545	75-00020
Alenia Difesa	**Italy**	**231**	**75-01165**
Andrew Canada Inc.	Canada	68	75-00600
Andrew Ltd	UK	430	75-03156
Andrew Sensor Products	USA	554	75-07452
British Aerospace (Systems &	UK	436	75-08890
Central Research Laboratories Ltd	UK	499	80-05895
CI Systems Ltd	Israel	218	75-03750
Coherent Optics (Europe) Ltd	UK	441	75-14090
Computer Optics Inc.	USA	565	75-20680
Contraves Inc	USA	566	75-21850
EL-OP Electro Optics Industries Ltd	Israel	219	75-06300
EUROCONTROL SpA	Italy	233	75-08200
Fabbrica Italiana Apparecchiature	Italy	234	75-08550
GEC-Marconi Avionics Ltd	UK	454	75-25320
GEC-Marconi Research Centre	UK	419	60-00700
Hollandse Signaalapparaten BV	Netherlands	293	75-03300
Israel Aircraft Industries Ltd	Israel	220	75-10200
Kollsman	USA	589	75-53150
Kongsberg Simrad Ltd	UK	463	75-34100
Litton Industries Inc.	USA	590	75-55400
McLennan Marine Ltd.	UK	467	75-38080
Microtecnica-Torino SpA	Italy	237	75-15500
MSI-Defence Systems Ltd	UK	469	75-41280
Nanoquest Defence Products Ltd	UK	470	75-41890
Night Vision Equipment Co. Inc.	USA	599	75-64390
Octec Ltd	UK	471	75-42950
OM=	USA	591	75-55400
Omitec Electro-Optics Ltd	UK	472	75-43050
Pearpoint Ltd	UK	473	75-44270
Pilkington Optronics Ltd	UK	426	74-10000
Pilkington Thorn Optronics	UK	426	74-10000
Radamec Defence Systems Ltd	UK	477	75-47750
Roke Manor Research Ltd	UK	420	60-01800
Saab Dynamics AB	Sweden	378	75-09100
SAGEM SA	**France**	**143**	**75-33950**
SFIM	**France**	**144**	**75-35080**
SFIM ODS	France	144	75-35085
Sira Ltd.	UK	484	75-54250
Spar Aerospace Ltd	Canada	76	75-13000
STN ATLAS Elektronik GmbH	Germany	165	75-01150
Technical Products Group Inc.	USA	620	75-86764
Technology Service Corp.	USA	644	80-31930
Thomson-CSF	France	148	75-45100

411 40 38 Television target tracking systems, shipborne

Andrew Ltd	UK	430	75-03156
Andrew Sensor Products	USA	554	75-07452
British Aerospace (Systems &	UK	436	75-08890
Coherent Optics (Europe) Ltd	UK	441	75-14090
Computer Optics Inc.	USA	565	75-20680
Contemporary Systems Design Ltd	UK	442	75-14860
Contraves Inc	USA	566	75-21850
CSE Aviation Ltd	UK	500	80-07600
Data Cell Ltd	UK	500	80-07720
EL-OP Electro Optics Industries Ltd	Israel	219	75-06300
EUROCONTROL SpA	Italy	233	75-08200
Fabbrica Italiana Apparecchiature	Italy	234	75-08550
GEC-Marconi Avionics Ltd	UK	454	75-25320

GEC-Marconi Research Centre	UK	*419*	60-00700
Hollandse Signaalapparaten BV	Netherlands	*293*	75-03300
Iscan Inc.	USA	*586*	75-49100
Israel Aircraft Industries Ltd	Israel	*220*	75-10200
Kongsberg Simrad Ltd	UK	*463*	75-34100
McLennan Marine Ltd.	UK	*467*	75-38080
Octec Ltd	UK	*471*	75-42950
Pearpoint Ltd	UK	*473*	75-44270
Pilkington Thorn Optronics	UK	*426*	74-10000
Radamec Defence Systems Ltd	UK	*477*	75-47750
Roke Manor Research Ltd	UK	*420*	60-01800
Saab Dynamics AB	Sweden	*378*	75-09100
SAGEM SA	**France**	***143***	**75-33950**
SFIM	**France**	***144***	**75-35080**
STN ATLAS Elektronik GmbH	Germany	*165*	75-01150
Technical Products Group Inc.	USA	*620*	75-86764
Technology Service Corp.	USA	*644*	80-31930

411 40 40 Electro-optical target location & tracking systems, ground-based

AAI Corp.	USA	*545*	75-00020
Andrew Canada Inc.	Canada	*68*	75-00600
Andrew Ltd	UK	*430*	75-03156
Andrew Sensor Products	USA	*554*	75-07452
Avimo Ltd	UK	*432*	75-04950
British Aerospace (Systems &	UK	*436*	75-08890
Central Research Laboratories Ltd	UK	*499*	80-05895
CI Systems Ltd	Israel	*218*	75-03750
Coherent Optics (Europe) Ltd	UK	*441*	75-14090
Computer Optics Inc.	USA	*565*	75-20680
Contraves Inc	USA	*566*	75-21850
Cranfield University	UK	*419*	60-00350
Delco Systems Operations	USA	*568*	75-25360
Delft Sensor Systems	**Belgium**	***46***	**75-02500**
Delft Sensor Systems	Netherlands	*292*	75-01590
EL-OP Electro Optics Industries Ltd	Israel	*219*	75-06300
EUROCONTROL SpA	Italy	*233*	75-08200
Fabbrica Italiana Apparecchiature	Italy	*234*	75-08550
GEC-Marconi Avionics Ltd	UK	*454*	75-25320
GEC-Marconi Research Centre	UK	*419*	60-00700
Hollandse Signaalapparaten BV	Netherlands	*293*	75-03300
Instro Precision Ltd	UK	*461*	75-31630
Israel Aircraft Industries Ltd	Israel	*220*	75-10200
JAI A/S	Denmark	*103*	75-09400
Kearfott Guidance & Navigation Corp.	USA	*588*	75-51760
Kollsman	USA	*589*	75-53150
Kongsberg Simrad Ltd	UK	*463*	75-34100
Litton Industries Inc.	USA	*590*	75-55400
Logica UK Limited	UK	*506*	80-14350
McLennan Marine Ltd.	UK	*467*	75-38080
Microtecnica-Torino SpA	Italy	*237*	75-15500
Nanoquest Defence Products Ltd	UK	*470*	75-41890
Night Vision Equipment Co. Inc.	USA	*599*	75-64390
Northrop Grumman Corporation	USA	*599*	75-65450
Octec Ltd	UK	*471*	75-42950
OM=	USA	*591*	75-55400
Omi Corp.	USA	*601*	75-66550
Omitec Electro-Optics Ltd	UK	*472*	75-43050
Pearpoint Ltd	UK	*473*	75-44270
Pilkington Optronics Ltd	UK	*426*	74-10000
Pilkington Thorn Optronics	UK	*426*	74-10000
Radamec Defence Systems Ltd	UK	*477*	75-47750
Roke Manor Research Ltd	UK	*420*	60-01800
Rutherford Appleton Laboratory	UK	*420*	60-01850
SAGEM SA	**France**	***143***	**75-33950**
SFIM	**France**	***144***	**75-35080**
SFIM ODS	France	*144*	75-35085
Sira Ltd.	UK	*484*	75-54250
STN ATLAS Elektronik GmbH	Germany	*165*	75-01150
Technical Products Group Inc.	USA	*620*	75-86764
Technology Service Corp.	USA	*644*	80-31930
Thomson-CSF	France	*148*	75-45100
Warren-Knight Instrument Co.	USA	*627*	75-95150
Zeiss-Eltro Optronic GmbH	Germany	*177*	75-34700

411 40 48 Television target tracking systems, ground-based

Andrew Sensor Products	USA	*554*	75-07452
British Aerospace (Systems &	UK	*436*	75-08890
Coherent Optics (Europe) Ltd	UK	*441*	75-14090
Computer Optics Inc.	USA	*565*	75-20680
Contemporary Systems Design Ltd	UK	*442*	75-14860
Contraves Inc	USA	*566*	75-21850
Cranfield University	UK	*419*	60-00350
EL-OP Electro Optics Industries Ltd	Israel	*219*	75-06300
EUROCONTROL SpA	Italy	*233*	75-08200
Fabbrica Italiana Apparecchiature	Italy	*234*	75-08550
GEC-Marconi Avionics Ltd	UK	*454*	75-25320
GEC-Marconi Research Centre	UK	*419*	60-00700
Hollandse Signaalapparaten BV	Netherlands	*293*	75-03300
Insight Vision Systems Ltd	UK	*461*	75-31530
Instro Precision Ltd	UK	*461*	75-31630
Iscan Inc.	USA	*586*	75-49100
Israel Aircraft Industries Ltd	Israel	*220*	75-10200
JAI A/S	Denmark	*103*	75-09400
Kongsberg Simrad Ltd	UK	*463*	75-34100
Northrop Grumman Corporation	USA	*599*	75-65450
Octec Ltd	UK	*471*	75-42950
Pearpoint Ltd	UK	*473*	75-44270
Pilkington Thorn Optronics	UK	*426*	74-10000
Radamec Defence Systems Ltd	UK	*477*	75-47750
Regis SpA	Italy	*238*	75-20110
Roke Manor Research Ltd	UK	*420*	60-01800
Saab Dynamics AB	Sweden	*378*	75-09100
SAGEM SA	**France**	***143***	**75-33950**
SFIM	**France**	***144***	**75-35080**
STN ATLAS Elektronik GmbH	Germany	*165*	75-01150
Technology Service Corp.	USA	*644*	80-31930
Vision Research Co.	UK	*491*	75-63360

411 50 Optical target tracking systems & equipment

AAI Corp.	USA	*545*	75-00020
Andrew Sensor Products	USA	*554*	75-07452
Angénieux	France	*132*	75-02700
Boeing Australia Limited	Australia	*12*	75-02500
British Aerospace (Systems &	UK	*436*	75-08890
Central Design Bureau	Ukraine	*406*	75-00800
Coherent Optics (Europe) Ltd	UK	*441*	75-14090
Computer Optics Inc.	USA	*565*	75-20680
Cranfield University	UK	*419*	60-00350
Data Cell Ltd	UK	*500*	80-07720
Delft Sensor Systems	**Belgium**	***46***	**75-02500**
Delft Sensor Systems	Netherlands	*292*	75-01590
EG&G Inc.	USA	*571*	75-29555
EL-OP Electro Optics Industries Ltd	Israel	*219*	75-06300
Fabbrica Italiana Apparecchiature	Italy	*234*	75-08550
GEC-Marconi Avionics Inc.	USA	*577*	75-37413
GEC-Marconi Research Centre	UK	*419*	60-00700
Hadland Photonics Ltd	UK	*456*	75-27200
Heinsoldt Systemtechnick GmbH	Germany	*171*	75-17800
Helio Mirror Co. Ltd	UK	*457*	75-28900
Hollandse Signaalapparaten BV	Netherlands	*293*	75-03300
Instro Precision Ltd	UK	*461*	75-31630
Israel Aircraft Industries Ltd	Israel	*219*	75-10200
ITT Aerospace/Communications Div.	USA	*586*	75-49455
Kongsberg Simrad Ltd	UK	*463*	75-34100
Litton Industries Inc.	USA	*590*	75-55400
Lockheed Martin Corp.	USA	*591*	75-55635
MS Instruments plc	UK	*469*	75-41265
MSI-Defence Systems Ltd	UK	*469*	75-41280
Nanoquest Defence Products Ltd	UK	*470*	75-41890
Octec Ltd	UK	*471*	75-42950
Oerlikon-Contraves AG.	Switzerland	*387*	75-09250
Radamec Defence Systems Ltd	UK	*477*	75-47750
Raytheon E-Systems	USA	*608*	75-73302
Sanders	USA	*612*	75-77850
SFIM	**France**	***144***	**75-35080**
Sira Ltd.	UK	*484*	75-54250
Urals Optical & Mechanical Plant	Russia	*339*	75-53000

411 50 30 Optical target tracking systems, shipborne

AAI Corp.	USA	*545*	75-00020
Alan Gordon Enterprises Inc.	USA	*579*	75-40060
Alenia Difesa	**Italy**	***231***	**75-01165**
Andrew Canada Inc.	Canada	*68*	75-00600
Andrew Sensor Products	USA	*554*	75-07452
British Aerospace (Systems &	UK	*436*	75-08890
Coherent Optics (Europe) Ltd	UK	*441*	75-14090
Computer Optics Inc.	USA	*565*	75-20680
Contraves Inc	USA	*566*	75-21850
CSE Aviation Ltd	UK	*500*	80-07600

EL-OP Electro Optics Industries Ltd	Israel	*219*	75-06300
Fabbrica Italiana Apparecchiature	Italy	*234*	75-08550
GEC-Marconi Research Centre	UK	*419*	60-00700
Hollandse Signaalapparaten BV	Netherlands	*293*	75-03300
Israel Aircraft Industries Ltd	Israel	*220*	75-10200
Kollsman	USA	*589*	75-53150
Kongsberg Simrad Ltd	UK	*463*	75-34100
MSI-Defence Systems Ltd	UK	*469*	75-41280
Nanoquest Defence Products Ltd	UK	*470*	75-41890
Octec Ltd	UK	*471*	75-42950
Pilkington Optronics Ltd	UK	*426*	74-10000
Pilkington Thorn Optronics	UK	*426*	74-10000
Radamec Defence Systems Ltd	UK	*477*	75-47750
SAGEM SA	**France**	***143***	**75-33950**
SFIM	**France**	***144***	**75-35080**
Sira Ltd.	UK	*484*	75-54250
Urals Optical & Mechanical Plant	Russia	*339*	75-53000

411 50 40 Optical target tracking systems, ground-based

AAI Corp.	USA	*545*	75-00020
Alan Gordon Enterprises Inc.	USA	*579*	75-40060
Andrew Canada Inc.	Canada	*68*	75-00600
Andrew Sensor Products	USA	*554*	75-07452
Arab International Optronics Co.	Egypt	*109*	74-00500
Armalon Ltd	UK	*431*	75-03729
British Aerospace (Systems &	UK	*436*	75-08890
Coherent Optics (Europe) Ltd	UK	*441*	75-14090
Computer Optics Inc.	USA	*565*	75-20680
Contraves Inc	USA	*566*	75-21850
Cranfield University	UK	*419*	60-00350
EL-OP Electro Optics Industries Ltd	Israel	*219*	75-06300
Fabbrica Italiana Apparecchiature	Italy	*234*	75-08550
GEC-Marconi Research Centre	UK	*419*	60-00700
Helio Mirror Co. Ltd.	UK	*457*	75-28900
Hollandse Signaalapparaten BV	Netherlands	*293*	75-03300
Instro Precision Ltd	UK	*461*	75-31630
ITT Aerospace/Communications Div.	USA	*586*	75-49455
Kongsberg Simrad Ltd	UK	*463*	75-34100
Litton Industries Inc.	USA	*590*	75-55400
Nanoquest Defence Products Ltd	UK	*470*	75-41890
Northrop Grumman Corporation	USA	*599*	75-65450
Octec Ltd	UK	*471*	75-42950
Photo-Sonics Inc.	USA	*604*	75-69800
Pilkington Optronics Ltd	UK	*426*	74-10000
Pilkington Thorn Optronics	UK	*426*	74-10000
SAGEM SA	**France**	***143***	**75-33950**
SFIM	**France**	***144***	**75-35080**
Sira Ltd.	UK	*484*	75-54250
Warren-Knight Instrument Co.	USA	*627*	75-95150
Zeiss-Eltro Optronic GmbH	Germany	*177*	75-34700

411 60 Magnetic anomaly target location & tracking systems

Advanced Technical Corp.	USA	*546*	75-01000
Associated Aircraft Mfg & Sales Inc.	USA	*555*	75-09500
CAE Electronics Ltd	Canada	*69*	75-01770
EG&G Inc.	USA	*571*	75-29555
Scintrex Ltd	Canada	*76*	75-12550
SEXTANT Avionique SA	France	*144*	75-35000
TSS (UK) Ltd	UK	*489*	75-61395

411 60 70 50 Magnetometers

Advanced Technical Corp.	USA	*546*	75-01000
Associated Aircraft Mfg & Sales Inc.	USA	*555*	75-09500
Bartington Instruments Ltd	UK	*433*	75-05940
EG&G Inc.	USA	*571*	75-29555
Foerster UK Ltd	UK	*451*	75-24150
Humphrey Inc.	USA	*583*	75-45050
KVH Industries Inc.	USA	*589*	75-53670
Scintrex Ltd	Canada	*76*	75-12550
Sodankyla Geophysical Observatory	Finland	*117*	60-03500
Watson Industries Inc.	USA	*627*	75-95210

411 62 Magnetic anomaly target location & tracking systems, airborne

Advanced Technical Corp.	USA	*546*	75-01000
Associated Aircraft Mfg & Sales Inc.	USA	*555*	75-09500
CAE Electronics Ltd	Canada	*69*	75-01770

411 70 Infra-red target tracking systems

AAI Corp.	USA	*545*	75-00020
Alenia-Elsag	Italy	*230*	74-00100
Andrew Sensor Products	USA	*554*	75-07452
Angénieux	France	*132*	75-02700
Boeing Australia Limited	Australia	*12*	75-02500
British Aerospace (Systems &	UK	*436*	75-08890
Central Research Laboratories Ltd	UK	*499*	80-05895
Cincinnati Electronics Corp.	USA	*564*	75-18640
Coherent Optics (Europe) Ltd	UK	*441*	75-14090
Computer Optics Inc.	USA	*565*	75-20680
Contraves GmbH	Germany	*166*	75-05200
Contraves Inc	USA	*566*	75-21850
Delft Sensor Systems	**Belgium**	***46***	**75-02500**
Delft Sensor Systems	Netherlands	*292*	75-01590
EG&G Inc.	USA	*571*	75-29555
EL-OP Electro Optics Industries Ltd	Israel	*219*	75-06300
Elettronica (UK) Ltd	UK	*448*	75-20475
Fabbrica Italiana Apparecchiature	Italy	*234*	75-08550
GEC-Marconi Avionics Inc.	USA	*577*	75-37413
GEC-Marconi Avionics Ltd	UK	*454*	75-25320
GEC-Marconi Infra-Red Limited	UK	*454*	75-25338
GEC-Marconi Research Centre	UK	*419*	60-00700
Hadland Photonics Ltd	UK	*456*	75-27200
Hollandse Signaalapparaten BV	Netherlands	*293*	75-03300
Hughes Elcan Optical Technologies	Canada	*72*	75-07650
IFI Srl	Italy	*235*	75-11220
Instro Precision Ltd	UK	*461*	75-31630
Israel Aircraft Industries Ltd	Israel	*220*	75-10200
Kollsman	USA	*589*	75-53150
Litton Industries Inc.	USA	*590*	75-55400
Lockheed Martin Corp.	USA	*591*	75-55635
McLennan Marine Ltd.	UK	*467*	75-38080
Northrop Grumman Corporation	USA	*599*	75-65450
Octec Ltd	UK	*471*	75-42950
Pearpoint Ltd	UK	*473*	75-44270
Pilkington Thorn Optronics	UK	*426*	74-10000
Racal Communications Inc.	USA	*606*	75-72380
Rafael	Israel	*216*	55-01800
Raytheon E-Systems	USA	*608*	75-73302
Roke Manor Research Ltd	UK	*420*	60-01800
Sabre Computers Int'l Ltd	UK	*481*	75-51150
SAGEM SA	**France**	***143***	**75-33950**
Sanders	USA	*612*	75-77850
SFIM	**France**	***144***	**75-35080**
Sira Ltd.	UK	*484*	75-54250
SNPE	France	*145*	75-36315
STN ATLAS Elektronik GmbH	Germany	*165*	75-01150
Technical Products Group Inc.	USA	*620*	75-86764
Technology Service Corp.	USA	*644*	80-31930

411 70 90 60 Sensors, target tracking & surveillance, infra-red

AAI Corp.	USA	*545*	75-00020
Andrew Sensor Products	USA	*554*	75-07452
Bodenseewerk Gerätetechnik GmbH	Germany	*165*	75-03000
Central Research Laboratories Ltd	UK	*499*	80-05895
CI Systems Ltd	Israel	*218*	75-03750
Cincinnati Electronics Corp.	USA	*564*	75-18640
Computer Optics Inc.	USA	*565*	75-20680
Contraves Inc	USA	*566*	75-21850
Cranfield Aerospace Ltd	UK	*500*	80-07200
Cranfield University	UK	*419*	60-00350
Delft Sensor Systems	**Belgium**	***46***	**75-02500**
Delft Sensor Systems	Netherlands	*292*	75-01590
EG&G Inc.	USA	*571*	75-29555
EG&G Reticon	USA	*571*	75-29562
EL-OP Electro Optics Industries Ltd	Israel	*219*	75-06300
Elettronica (UK) Ltd	UK	*448*	75-20475
Fabbrica Italiana Apparecchiature	Italy	*234*	75-08550
GEC-Marconi Avionics Ltd	UK	*454*	75-25320
GEC-Marconi Infra-Red Limited	UK	*454*	75-25338
GEC-Marconi Research Centre	UK	*419*	60-00700
Hollandse Signaalapparaten BV	Netherlands	*293*	75-03300
Instro Precision Ltd	UK	*461*	75-31630
Israel Aircraft Industries Ltd	Israel	*220*	75-10200
Kollsman	USA	*589*	75-53150
McLennan Marine Ltd.	UK	*467*	75-38080

Octec Ltd	UK	471	75-42950
Pearpoint Ltd	UK	473	75-44270
Pilkington Optronics Ltd	UK	426	74-10000
Pilkington Thorn Optronics	UK	426	74-10000
Rafael	Israel	216	55-01800
Regis SpA	Italy	238	75-20110
Roke Manor Research Ltd	UK	420	60-01800
Saab Dynamics AB	Sweden	378	75-09100
SAGEM SA	**France**	**143**	**75-33950**
Sira Ltd.	UK	484	75-54250
STN ATLAS Elektronik GmbH	Germany	165	75-01150
Technology Service Corp.	USA	644	80-31930
W. Vinten Ltd	UK	491	75-63353
Zeiss-Eltro Optronic GmbH	Germany	177	75-34700

411 80 Thermal target tracking systems

AAI Corp.	USA	545	75-00020
Boeing Australia Limited	Australia	12	75-02500
Central Research Laboratories Ltd	UK	499	80-05895
Coherent Optics (Europe) Ltd	UK	441	75-14090
Contraves GmbH	Germany	166	75-05200
Delft Sensor Systems	**Belgium**	**46**	**75-02500**
Delft Sensor Systems	Netherlands	292	75-01590
EL-OP Electro Optics Industries Ltd	Israel	219	75-06300
Fabbrica Italiana Apparecchiature	Italy	234	75-08550
GEC-Marconi Avionics Inc..	USA	577	75-37413
GEC-Marconi Avionics Ltd	UK	454	75-25320
GEC-Marconi Infra-Red Limited	UK	454	75-25338
GEC-Marconi Research Centre	UK	419	60-00700
Kollsman	USA	589	75-53150
Litton Industries Inc.	USA	590	75-55400
Lockheed Martin Corp.	USA	591	75-55635
McLennan Marine Ltd.	UK	467	75-38080
Octec Ltd	UK	471	75-42950
Pilkington Optronics Ltd	UK	426	74-10000
Pilkington Thorn Optronics	UK	426	74-10000
Roke Manor Research Ltd	UK	420	60-01800
SABCA	Belgium	46	75-05590
SAGEM SA	**France**	**143**	**75-33950**
SFIM	**France**	**144**	**75-35080**
Sira Ltd.	UK	484	75-54250
Technical Products Group Inc..	USA	620	75-86764
Thomson Marconi Sonar Limited	UK	426	74-20000

411 80 90 60 Sensors, target tracking & surveillance, thermal imaging

AAI Corp.	USA	545	75-00020
Andrew Sensor Products	USA	554	75-07452
Central Research Laboratories Ltd	UK	499	80-05895
CI Systems Ltd	Israel	218	75-03750
Coherent Optics (Europe) Ltd	UK	441	75-14090
Delft Sensor Systems	**Belgium**	**46**	**75-02500**
Delft Sensor Systems	Netherlands	292	75-01590
EL-OP Electro Optics Industries Ltd	Israel	219	75-06300
Elettronica (UK) Ltd	UK	448	75-20475
Elmwood Sensors Inc.	USA	573	75-30850
Fabbrica Italiana Apparecchiature	Italy	234	75-08550
GEC-Marconi Avionics Ltd	UK	454	75-25320
GEC-Marconi Infra-Red Limited	UK	454	75-25338
GEC-Marconi Materials Technology Ltd	UK	419	60-00600
GEC-Marconi Research Centre	UK	419	60-00700
Instro Precision Ltd	UK	461	75-31630
Kollsman	USA	589	75-53150
Kongsberg Simrad Ltd	UK	463	75-34100
McLennan Marine Ltd.	UK	467	75-38080
Pilkington Optronics Ltd	UK	426	74-10000
Pilkington Thorn Optronics	UK	426	74-10000
Regis SpA	Italy	238	75-20110
Roke Manor Research Ltd	UK	420	60-01800
SABCA	Belgium	46	75-05590
SAGEM SA	**France**	**143**	**75-33950**
Sira Ltd.	UK	484	75-54250
Thomson Marconi Sonar Limited	UK	426	74-20000
Zeiss-Eltro Optronic GmbH	Germany	177	75-34700

411 90 Combined sensor target tracking systems

Advanced System Architectures Ltd	UK	496	80-00420
Andrew Ltd	UK	430	75-03156
Boeing Australia Limited	Australia	12	75-02500
British Aerospace (Systems &	UK	436	75-08890
Cambridge Consultants Ltd	UK	499	80-05300
Cincinnati Electronics Corp.	USA	564	75-18640
Contraves GmbH	Germany	166	75-05200
Cranfield Aerospace Ltd	UK	500	80-07200
Delft Sensor Systems	**Belgium**	**46**	**75-02500**
Delft Sensor Systems	Netherlands	292	75-01590
EDS Defence Ltd	UK	501	80-09200
EL-OP Electro Optics Industries Ltd	Israel	219	75-06300
Elettronica (UK) Ltd	UK	448	75-20475
Fabbrica Italiana Apparecchiature	Italy	234	75-08550
GEC-Marconi Avionics Inc..	USA	577	75-37413
GEC-Marconi Avionics Ltd	UK	454	75-25320
GEC-Marconi Research Centre	UK	419	60-00700
Hollandse Signaalapparaten BV	Netherlands	293	75-03300
Instro Precision Ltd	UK	461	75-31630
Israel Aircraft Industries Ltd	Israel	220	75-10200
Lockheed Martin Corp.	USA	591	75-55635
McLennan Marine Ltd.	UK	467	75-38080
Northrop Grumman Corporation	USA	599	75-65450
Octec Ltd	UK	471	75-42950
Pilkington Optronics Ltd	UK	426	74-10000
Pilkington Thorn Optronics	UK	426	74-10000
Roke Manor Research Ltd	UK	420	60-01800
Rutherford Appleton Laboratory	UK	420	60-01850
SAGEM SA	**France**	**143**	**75-33950**
SFIM	**France**	**144**	**75-35080**
Technology Service Corp.	USA	644	80-31930

411 92 Combined sensor target tracking systems, airborne

Advanced System Architectures Ltd	UK	496	80-00420
Andrew Canada Inc.	Canada	68	75-00600
Andrew Ltd	UK	430	75-03156
Andrew Sensor Products	USA	554	75-07452
Cincinnati Electronics Corp.	USA	564	75-18640
Cranfield Aerospace Ltd	UK	500	80-07200
Cranfield University	UK	419	60-00350
EL-OP Electro Optics Industries Ltd	Israel	219	75-06300
Elettronica (UK) Ltd	UK	448	75-20475
Fabbrica Italiana Apparecchiature	Italy	234	75-08550
GEC-Marconi Avionics Ltd	UK	454	75-25320
Israel Aircraft Industries Ltd	Israel	220	75-10200
Kaiser Electronics	USA	587	75-51350
Litton Industries Inc.	USA	591	75-55400
Northrop Grumman Corporation	USA	599	75-65450
Octec Ltd	UK	471	75-42950
Pilkington Optronics Ltd	UK	426	74-10000
Pilkington Thorn Optronics	UK	426	74-10000
Radamec Defence Systems Ltd	UK	477	75-47750
Rutherford Appleton Laboratory	UK	420	60-01850
SAGEM SA	**France**	**143**	**75-33950**
SFIM	**France**	**144**	**75-35080**
Technology Service Corp.	USA	644	80-31930

411 93 Combined sensor target tracking systems, shipborne

AAI Corp.	USA	545	75-00020
Advanced System Architectures Ltd	UK	496	80-00420
Andrew Canada Inc.	Canada	68	75-00600
Andrew Ltd	UK	430	75-03156
Andrew Sensor Products	USA	554	75-07452
British Aerospace (Systems &	UK	436	75-08890
Contraves GmbH	Germany	166	75-05200
EL-OP Electro Optics Industries Ltd	Israel	219	75-06300
Elettronica (UK) Ltd	UK	448	75-20475
Fabbrica Italiana Apparecchiature	Italy	234	75-08550
GEC-Marconi Avionics Ltd	UK	454	75-25320
Hollandse Signaalapparaten BV	Netherlands	293	75-03300
Israel Aircraft Industries Ltd	Israel	220	75-10200
McLennan Marine Ltd.	UK	467	75-38080
MK Services Ltd.	UK	468	75-40170
MSI-Defence Systems Ltd	UK	469	75-41280
Octec Ltd	UK	471	75-42950
Pilkington Optronics Ltd	UK	426	74-10000
Pilkington Thorn Optronics	UK	426	74-10000
Radamec Defence Systems Ltd	UK	477	75-47750
Regis SpA	Italy	238	75-20110
Roke Manor Research Ltd	UK	420	60-01800
Saab Dynamics AB	Sweden	378	75-09100

SAGEM SA	**France**	***143***	**75-33950**
SFIM	**France**	***144***	**75-35080**
Technology Service Corp.	USA	*644*	80-31930
Thomson Marconi Sonar Limited	UK	*426*	74-20000
Thomson-CSF	France	*148*	75-45100

411 94 Combined sensor target tracking systems, ground-based

AAI Corp.	USA	*545*	75-00020
Advanced System Architectures Ltd	UK	*496*	80-00420
Andrew Canada Inc.	Canada	*68*	75-00600
Andrew Ltd	UK	*430*	75-03156
Andrew Sensor Products	USA	*554*	75-07452
British Aerospace (Systems &	UK	*436*	75-08890
Contraves GmbH	Germany	*166*	75-05200
Cranfield University	UK	*419*	60-00350
EL-OP Electro Optics Industries Ltd	Israel	*219*	75-06300
Fabbrica Italiana Apparecchiature	Italy	*234*	75-08550
GEC-Marconi Avionics Ltd	UK	*454*	75-25320
Hollandse Signaalapparaten BV	Netherlands	*293*	75-03300
Instro Precision Ltd	UK	*461*	75-31630
Israel Aircraft Industries Ltd	Israel	*220*	75-10200
JAI A/S	Denmark	*103*	75-09400
McLennan Marine Ltd	UK	*467*	75-38080
Octec Ltd	UK	*471*	75-42950
Pilkington Optronics Ltd	UK	*426*	74-10000
Pilkington Thorn Optronics	UK	*426*	74-10000
Radamec Defence Systems Ltd	UK	*477*	75-47750
Regis SpA	Italy	*238*	75-20110
Roke Manor Research Ltd	UK	*420*	60-01800
SAGEM SA	**France**	***143***	**75-33950**
SFIM	**France**	***144***	**75-35080**
Technology Service Corp.	USA	*644*	80-31930
Thomson-CSF	France	*148*	75-45100
Vega-Cantley Instrument Company	UK	*490*	75-62740
Vision Research Co.	UK	*491*	75-63360

412 Target identification

Aeromaritime Systembau GmbH	Germany	*164*	75-00290
Bharat Electronics Ltd	India	*203*	55-00200
Boeing Australia Limited	Australia	*12*	75-02500
CEA Technologies Pty Ltd	Australia	*24*	80-03100
Fabbrica Italiana Apparecchiature	Italy	*234*	75-08550
GEC Marconi Hazeltine Corp	USA	*578*	75-37415
ITT Avionics	USA	*586*	75-49455
Litton Industries Inc.	USA	*590*	75-55400
Radio Research Instrument Co. Inc.	USA	*607*	75-72600
Raytheon Cossor Electronics Ltd	UK	*478*	75-48367
SABCA	Belgium	*46*	75-05590
Sanders	USA	*612*	75-77850
Siemens Switzerland Ltd	Switzerland	*387*	75-11980
STAREC	France	*147*	75-42775

412 10 IFF systems & equipment

Advanced Technical Corp.	USA	*546*	75-01000
Aeromaritime Systembau GmbH	Germany	*164*	75-00290
AIL Systems Inc.	USA	*547*	75-02950
Alenia-Elsag	Italy	*230*	74-00100
Associated Aircraft Mfg & Sales Inc.	USA	*555*	75-09500
Bharat Electronics Ltd	India	*203*	55-00200
Boeing Australia Limited	Australia	*12*	75-02500
Fabbrica Italiana Apparecchiature	Italy	*234*	75-08550
GEC Marconi Hazeltine Corp	USA	*578*	75-37415
GEC-Marconi Avionics Inc.	USA	*577*	75-37413
Italtel Spa	Italy	*235*	75-13205
ITT Avionics	USA	*586*	75-49455
Litton Industries Inc.	USA	*590*	75-55400
Radio Research Instrument Co. Inc.	USA	*607*	75-72600
Rafael	Israel	*216*	55-01800
Raytheon Cossor Electronics Ltd	UK	*478*	75-48367
Sanders	USA	*612*	75-77850
Siemens Aktiengesellschaft	Germany	*175*	75-27700
Siemens Switzerland Ltd	Switzerland	*387*	75-11980
Thomson-CSF	France	*148*	75-45100

412 10 20 IFF systems & equipment, airborne

Advanced Technical Corp.	USA	*546*	75-01000
Associated Aircraft Mfg & Sales Inc.	USA	*555*	75-09500
Boeing Australia Limited	Australia	*12*	75-02500
GEC Marconi Hazeltine Corp	USA	*578*	75-37415
Herley-Vega Systems	USA	*581*	75-43090
Hughes Danbury Optical Sys.	USA	*582*	75-44857
Litton Industries Inc.	USA	*590*	75-55400
Radio Research Instrument Co. Inc.	USA	*607*	75-72600
Raytheon Cossor Electronics Ltd	UK	*478*	75-48367
SABCA	Belgium	*46*	75-05590
Schrack Aerospace GmbH	Austria	*34*	75-02195
Siemens Aktiengesellschaft	Germany	*175*	75-27700

412 10 20 60 IFF radar receivers, airborne

Advanced Technical Corp.	USA	*546*	75-01000
Associated Aircraft Mfg & Sales Inc.	USA	*555*	75-09500
Boeing Australia Limited	Australia	*12*	75-02500
Racal-MESL Ltd	UK	*477*	75-47682
Radio Research Instrument Co. Inc.	USA	*607*	75-72600
Rafael	Israel	*216*	55-01800
Raytheon Cossor Electronics Ltd	UK	*478*	75-48367
STAREC	France	*147*	75-42775
Toyo Communication Equipment Co	Japan	*255*	75-10000
Vega-Cantley Instrument Company	UK	*490*	75-62740

412 10 20 65 IFF radar interrogators, airborne

Advanced Technical Corp.	USA	*546*	75-01000
AIL Systems Inc.	USA	*547*	75-02950
Alenia-Elsag	Italy	*230*	74-00100
Associated Aircraft Mfg & Sales Inc.	USA	*555*	75-09500
Boeing Australia Limited	Australia	*12*	75-02500
GEC Marconi Hazeltine Corp	USA	*578*	75-37415
Italtel Spa	Italy	*235*	75-13205
Litton Industries Inc.	USA	*590*	75-55400
Radio Research Instrument Co. Inc.	USA	*607*	75-72600
Raytheon Cossor Electronics Ltd	UK	*478*	75-48367
Siemens Aktiengesellschaft	Germany	*175*	75-27700

412 10 20 70 IFF radar transponders, airborne

Advanced Technical Corp.	USA	*546*	75-01000
AIL Systems Inc.	USA	*547*	75-02950
Alenia-Elsag	Italy	*230*	74-00100
Associated Aircraft Mfg & Sales Inc.	USA	*555*	75-09500
Boeing Australia Limited	Australia	*12*	75-02500
Dassault Electronique SA (Group)	**France**	***135***	**75-12980**
Fabbrica Italiana Apparecchiature	Italy	*234*	75-08550
GEC Marconi Hazeltine Corp	USA	*578*	75-37415
Italtel Spa	Italy	*235*	75-13205
Litton Industries Inc.	USA	*590*	75-55400
Radio Research Instrument Co. Inc.	USA	*607*	75-72600
Raytheon Cossor Electronics Ltd	UK	*478*	75-48367
Siemens Aktiengesellschaft	Germany	*175*	75-27700
Toyo Communication Equipment Co	Japan	*255*	75-10000
Vega-Cantley Instrument Company	UK	*490*	75-62740

412 10 30 IFF systems & equipment, shipborne

Advanced Technical Corp.	USA	*546*	75-01000
Aeromaritime Systembau GmbH	Germany	*164*	75-00290
Boeing Australia Limited	Australia	*12*	75-02500
GEC Marconi Hazeltine Corp	USA	*578*	75-37415
Litton Industries Inc.	USA	*590*	75-55400
Radio Research Instrument Co. Inc.	USA	*607*	75-72600
Raytheon Cossor Electronics Ltd	UK	*478*	75-48367

412 10 30 60 IFF radar receivers, shipborne

Advanced Technical Corp.	USA	*546*	75-01000
Boeing Australia Limited	Australia	*12*	75-02500
CelsiusTech Electronics AB	Sweden	*376*	75-03720
Racal-MESL Ltd	UK	*477*	75-47682
Radio Research Instrument Co. Inc.	USA	*607*	75-72600
Raytheon Cossor Electronics Ltd	UK	*478*	75-48367

412 10 30 65 IFF radar interrogators, shipborne

Boeing Australia Limited	Australia	*12*	75-02500

GEC Marconi Hazeltine Corp	USA	*578*	75-37415
Radio Research Instrument Co. Inc.	USA	*607*	75-72600
Raytheon Cossor Electronics Ltd	UK	*478*	75-48367
Siemens Aktiengesellschaft	Germany	*175*	75-27700
Toyo Communication Equipment Co	Japan	*255*	75-10000

412 10 30 70 IFF radar transponders, shipborne

Boeing Australia Limited	Australia	*12*	75-02500
GEC Marconi Hazeltine Corp	USA	*578*	75-37415
Radio Research Instrument Co. Inc.	USA	*607*	75-72600
Raytheon Cossor Electronics Ltd	UK	*478*	75-48367
Siemens Aktiengesellschaft	Germany	*175*	75-27700
Toyo Communication Equipment Co	Japan	*255*	75-10000

412 10 40 IFF systems & equipment, ground-based

Advanced Technical Corp.	USA	*546*	75-01000
Bharat Electronics Ltd	India	*203*	55-00200
Boeing Australia Limited	Australia	*12*	75-02500
GEC Marconi Hazeltine Corp	USA	*578*	75-37415
Hughes Danbury Optical Sys.	USA	*582*	75-44857
Radio Research Instrument Co. Inc.	USA	*607*	75-72600
Rafael	Israel	*216*	55-01800
Raytheon Cossor Electronics Ltd	UK	*478*	75-48367
Schrack Aerospace GmbH	Austria	*34*	75-02195
Siemens Aktiengesellschaft	Germany	*175*	75-27700
Siemens Switzerland Ltd	Switzerland	*387*	75-11980
Thomson-CSF	France	*148*	75-45100

412 10 40 60 IFF radar receivers, ground-based

Advanced Technical Corp.	USA	*546*	75-01000
Bharat Electronics Ltd	India	*203*	55-00200
Boeing Australia Limited	Australia	*12*	75-02500
NEC Corporation	Japan	*253*	75-06350
Racal-MESL Ltd	UK	*477*	75-47682
Radio Research Instrument Co. Inc.	USA	*607*	75-72600
Raytheon Cossor Electronics Ltd	UK	*478*	75-48367

412 10 40 65 IFF radar interrogators, ground-based

Advanced Technical Corp.	USA	*546*	75-01000
Bharat Electronics Ltd	India	*203*	55-00200
Boeing Australia Limited	Australia	*12*	75-02500
GEC Marconi Hazeltine Corp	USA	*578*	75-37415
NEC Corporation	Japan	*253*	75-06350
Radio Research Instrument Co. Inc.	USA	*607*	75-72600
Raytheon Cossor Electronics Ltd	UK	*478*	75-48367
Siemens Aktiengesellschaft	Germany	*175*	75-27700
Siemens Switzerland Ltd	Switzerland	*387*	75-11980
Thomson-CSF	France	*148*	75-45100

412 10 40 70 IFF radar transponders, ground-based

Advanced Technical Corp.	USA	*546*	75-01000
Boeing Australia Limited	Australia	*12*	75-02500
GEC Marconi Hazeltine Corp	USA	*578*	75-37415
Radio Research Instrument Co. Inc.	USA	*607*	75-72600
Raytheon Cossor Electronics Ltd	UK	*478*	75-48367
Siemens Aktiengesellschaft	Germany	*175*	75-27700

412 20 Radiometry systems & equipment

Alcatel Bell	Belgium	*44*	75-00055
Com Dev Europe	UK	*441*	75-14300
ERA Technology Ltd	UK	*502*	80-09600
GEC-Marconi Research Centre	UK	*419*	60-00700
Instro Precision Ltd	UK	*461*	75-31630
Millitech Corp.	USA	*596*	75-61665
MPB Technologies Inc	Canada	*74*	75-09855
Pilkington Thorn Optronics	UK	*426*	74-10000
Roke Manor Research Ltd	UK	*420*	60-01800
Sanders	USA	*612*	75-77850
Sira Ltd.	UK	*484*	75-54250
Steatite Insulations Ltd	UK	*486*	75-56450
Technology Service Corp.	USA	*644*	80-31930
Znamya Production Group	Ukraine	*407*	75-10000

412 20 20 Radiometry systems & equipment, airborne

CI Systems Ltd	Israel	*218*	75-03750
Cranfield Aerospace Ltd	UK	*500*	80-07200
GEC-Marconi Research Centre	UK	*419*	60-00700
Intertechnique (Aerospace) Ltd	UK	*461*	75-31948
Millitech Corp.	USA	*596*	75-61665
MPB Technologies Inc	Canada	*74*	75-09855
Pilkington Thorn Optronics	UK	*426*	74-10000
Roke Manor Research Ltd	UK	*420*	60-01800
Rutherford Appleton Laboratory	UK	*420*	60-01850
Technology Service Corp.	USA	*644*	80-31930
Znamya Production Group	Ukraine	*407*	75-10000

412 20 30 Radiometry systems & equipment, shipborne

Advanced Technical Corp.	USA	*546*	75-01000
CI Systems Ltd	Israel	*218*	75-03750
Intertechnique (Aerospace) Ltd	UK	*461*	75-31948
Millitech Corp.	USA	*596*	75-61665
Pilkington Thorn Optronics	UK	*426*	74-10000
Roke Manor Research Ltd	UK	*420*	60-01800
Technology Service Corp.	USA	*644*	80-31930
Znamya Production Group	Ukraine	*407*	75-10000

412 20 40 Radiometry systems & equipment, ground-based

Alcatel Bell	Belgium	*44*	75-00055
CI Systems Ltd	Israel	*218*	75-03750
GEC-Marconi Research Centre	UK	*419*	60-00700
Instro Precision Ltd	UK	*461*	75-31630
Intertechnique (Aerospace) Ltd	UK	*461*	75-31948
Millitech Corp.	USA	*596*	75-61665
MPB Technologies Inc	Canada	*74*	75-09855
Pilkington Thorn Optronics	UK	*426*	74-10000
Roke Manor Research Ltd	UK	*420*	60-01800
Sira Ltd.	UK	*484*	75-54250
Steatite Insulations Ltd	UK	*486*	75-56450
Technology Service Corp.	USA	*644*	80-31930
Znamya Production Group	Ukraine	*407*	75-10000

413 Target rangefinding / heightfinding

Alenia-Elsag	Italy	*230*	74-00100
Avimo Ltd	UK	*432*	75-04950
Bharat Electronics Ltd	India	*203*	55-00200
Central Research Laboratories Ltd	UK	*499*	80-05895
CIC International Ltd	USA	*563*	75-18570
Contraves GmbH	Germany	*166*	75-05200
Cubic Defense Systems Inc.	USA	*567*	75-22900
Delft Sensor Systems	**Belgium**	***46***	**75-02500**
Delft Sensor Systems	Netherlands	*292*	75-01590
EL-OP Electro Optics Industries Ltd	Israel	*219*	75-06300
Fabbrica Italiana Apparecchiature	Italy	*234*	75-08550
GEC-Marconi Avionics Inc.	USA	*577*	75-37413
GEC-Marconi Avionics Ltd	UK	*454*	75-25320
Hughes Elcan Optical Technologies	Canada	*72*	75-07650
KDI Precision Products Inc.	USA	*588*	75-51705
Kearfott Guidance & Navigation Corp.	USA	*588*	75-51760
Labtech Ltd	UK	*464*	75-34280
Litton Industries Inc.	USA	*590*	75-55400
M. Paz Logistics Ltd	Israel	*224*	80-01600
Microelectronics, guidance and	Turkey	*402*	55-00050
Oerlikon-Contraves AG.	Switzerland	*387*	75-09250
Roke Manor Research Ltd	UK	*420*	60-01800
Sanders	USA	*612*	75-77850
Technology Service Corp.	USA	*644*	80-31930
Thomson Sintra	France	*148*	75-45135

413 10 Radar rangefinding / heightfinding systems & equipment

Alenia-Elsag	Italy	*230*	74-00100
Bharat Electronics Ltd	India	*203*	55-00200
Central Research Laboratories Ltd	UK	*499*	80-05895
CIC International Ltd	USA	*563*	75-18570
Cubic Defense Systems Inc.	USA	*567*	75-22900
GEM Elettronica Srl	Italy	*235*	75-10100
ITT Aerospace/Communications Div.	USA	*586*	75-49455
KDI Precision Products Inc.	USA	*588*	75-51705

Labtech Ltd	UK	464	75-34280
M. Paz Logistics Ltd	Israel	224	80-01600
Oerlikon-Contraves AG.	Switzerland	387	75-09250
Power Technology Inc.	USA	605	75-71090
Radio Research Instrument Co. Inc.	USA	607	75-72600
Roke Manor Research Ltd	UK	420	60-01800
Sanders	USA	612	75-77850
Technology Service Corp.	USA	644	80-31930
TERMA Elektronik AS	Denmark	104	75-20300
Thomson-CSF.	France	148	75-45100
Znamya Production Group	Ukraine	407	75-10000

413 12 Radars, heightfinding

CIC International Ltd	USA	563	75-18570
Fabbrica Italiana Apparecchiature	Italy	234	75-08550
GEC-Marconi Research Centre	UK	419	60-00700
KDI Precision Products Inc.	USA	588	75-51705
Labtech Ltd	UK	464	75-34280
Racal-MESL Ltd	UK	477	75-47682
Radio Research Instrument Co. Inc.	USA	607	75-72600
Roke Manor Research Ltd	UK	420	60-01800
Rutherford Appleton Laboratory	UK	420	60-01850
Shorts Missile Systems Ltd	**UK**	**483**	**75-53550**
Technology Service Corp.	USA	644	80-31930
Thomson-CSF.	France	148	75-45100

413 12 30 Radars, heightfinding, shipborne

CIC International Ltd	USA	563	75-18570
Contraves GmbH	Germany	166	75-05200
Labtech Ltd	UK	464	75-34280
Radio Research Instrument Co. Inc.	USA	607	75-72600
Roke Manor Research Ltd	UK	420	60-01800
Technology Service Corp.	USA	644	80-31930

413 12 40 Radars, heightfinding, ground-based

Bharat Electronics Ltd	India	203	55-00200
CIC International Ltd	USA	563	75-18570
Contraves GmbH	Germany	166	75-05200
Labtech Ltd	UK	464	75-34280
Radio Research Instrument Co. Inc.	USA	607	75-72600
Roke Manor Research Ltd	UK	420	60-01800
Technology Service Corp.	USA	644	80-31930
Thomson-CSF.	France	148	75-45100

413 14 Radars, rangefinding

CIC International Ltd	USA	563	75-18570
KDI Precision Products Inc.	USA	588	75-51705
Labtech Ltd	UK	464	75-34280
Oerlikon-Contraves AG.	Switzerland	387	75-09250
Radio Research Instrument Co. Inc.	USA	607	75-72600
Roke Manor Research Ltd	UK	420	60-01800
Technology Service Corp.	USA	644	80-31930
Thomson-CSF.	France	148	75-45100

413 14 20 Radars, rangefinding, airborne

CIC International Ltd	USA	563	75-18570
Cranfield Aerospace Ltd	UK	500	80-07200
Fabbrica Italiana Apparecchiature	Italy	234	75-08550
Radio Research Instrument Co. Inc.	USA	607	75-72600
Technology Service Corp.	USA	644	80-31930

413 14 30 Radars, rangefinding, shipborne

CIC International Ltd	USA	563	75-18570
Contraves GmbH	Germany	166	75-05200
Radio Research Instrument Co. Inc.	USA	607	75-72600
Roke Manor Research Ltd	UK	420	60-01800
Technology Service Corp.	USA	644	80-31930

413 14 40 Radars, rangefinding, ground-based

CIC International Ltd	USA	563	75-18570
Contraves GmbH	Germany	166	75-05200
Radio Research Instrument Co. Inc.	USA	607	75-72600
Roke Manor Research Ltd	UK	420	60-01800
Technology Service Corp.	USA	644	80-31930
Thomson-CSF.	France	148	75-45100

413 20 Rangefinding systems, sonar

Alenia-Elsag	Italy	230	74-00100
Sanders	USA	612	75-77850
Thomson Sintra	France	148	75-45135

413 40 Electro-optical rangefinding systems & equipment

Arab International Optronics Co.	Egypt	109	74-00500
Avimo Ltd	UK	432	75-04950
Bospec Neue Technologien	Germany	166	75-03420
British Aerospace Australia Ltd	Australia	13	75-02750
CIC International Ltd	USA	563	75-18570
Coherent Optics (Europe) Ltd	UK	441	75-14090
Computer Optics Inc.	USA	565	75-20680
Contraves GmbH	Germany	166	75-05200
Contraves Inc	USA	566	75-21850
Cranfield Aerospace Ltd	UK	500	80-07200
Davin Optical Ltd	UK	444	75-17100
Delft Sensor Systems	**Belgium**	**46**	**75-02500**
Delft Sensor Systems	Netherlands	292	75-01590
EL-OP Electro Optics Industries Ltd	Israel	219	75-06300
Eloptro	S. Africa	358	75-05680
Ericsson Defence Systems Pty Ltd	Australia	16	75-05820
GEC-Marconi Avionics Ltd	UK	454	75-25320
GEC-Marconi Materials Technology Ltd	UK	419	60-00550
GEC-Marconi Research Centre	UK	419	60-00700
Hughes Elcan Optical Technologies	Canada	72	75-07650
Instro Precision Ltd	UK	461	75-31630
KVH Industries Inc.	USA	589	75-53670
Leica AG.	Switzerland	386	75-07310
Litton Industries Inc.	USA	590	75-55400
M. Paz Logistics Ltd	Israel	224	80-01600
Microtecnica-Torino SpA	Italy	237	75-15500
Nanoquest Defence Products Ltd	UK	470	75-41890
Northrop Grumman Corporation	USA	599	75-65450
Octec Ltd	UK	471	75-42950
Oerlikon-Contraves AG.	Switzerland	387	75-09250
Omitec Electro-Optics Ltd	UK	472	75-43050
Pilkington Optronics Ltd	UK	426	74-10000
Pilkington Thorn Optronics	UK	426	74-10000
Power Technology Inc.	USA	605	75-71090
Roke Manor Research Ltd	UK	420	60-01800
Technical Products Group Inc.	USA	620	75-86764
Zeiss-Eltro Optronic GmbH	Germany	177	75-34700

413 40 40 Rangefinders, infra-red

CIC International Ltd	USA	563	75-18570
Davin Optical Ltd	UK	444	75-17100
Delft Sensor Systems	**Belgium**	**46**	**75-02500**
Delft Sensor Systems	Netherlands	292	75-01590
Eloptro	S. Africa	358	75-05680
GEC-Marconi Materials Technology Ltd	UK	419	60-00550
Leica AG.	Switzerland	386	75-07310
M. Paz Logistics Ltd	Israel	224	80-01600
Merex Inc.	USA	595	75-60250
Night Vision Equipment Co. Inc.	USA	599	75-64390
OM=	USA	591	75-55400
Pilkington Optronics Ltd	UK	426	74-10000
Pilkington Thorn Optronics	UK	426	74-10000
Roke Manor Research Ltd	UK	420	60-01800
Technical Products Group Inc.	USA	620	75-86764
Zeiss-Eltro Optronic GmbH	Germany	177	75-34700

413 40 43 Rangefinders, infra-red, shipborne

CIC International Ltd	USA	563	75-18570
Contraves GmbH	Germany	166	75-05200
Davin Optical Ltd	UK	444	75-17100
GEC-Marconi Avionics Ltd	UK	454	75-25320
Kollsman	USA	589	75-53150
Night Vision Equipment Co. Inc.	USA	599	75-64390
Pilkington Optronics Ltd	UK	426	74-10000
Pilkington Thorn Optronics	UK	426	74-10000

Roke Manor Research Ltd	UK	420	60-01800
Technical Products Group Inc.	USA	620	75-86764

413 40 44 Rangefinders, infra-red, ground-based

CIC International Ltd	USA	563	75-18570
Contraves GmbH	Germany	166	75-05200
Davin Optical Ltd	UK	444	75-17100
Delft Sensor Systems	**Belgium**	**46**	**75-02500**
Delft Sensor Systems	Netherlands	292	75-01590
Eloptro	S. Africa	358	75-05680
GEC-Marconi Avionics Ltd	UK	454	75-25320
GEC-Marconi Materials Technology Ltd	UK	419	60-00550
Instro Precision Ltd	UK	461	75-31630
Kollsman	USA	589	75-53150
M. Paz Logistics Ltd	Israel	224	80-01600
Nanoquest Defence Products Ltd	UK	470	75-41890
Night Vision Equipment Co. Inc.	USA	599	75-64390
Pilkington Optronics Ltd	UK	426	74-10000
Pilkington Thorn Optronics	UK	426	74-10000
Roke Manor Research Ltd	UK	420	60-01800
Technical Products Group Inc.	USA	620	75-86764
Zeiss-Eltro Optronic GmbH	Germany	177	75-34700

413 40 50 Rangefinders, laser

Alenia-Elsag	Italy	230	74-00100
Angénieux	France	132	75-02700
Avimo Ltd	UK	432	75-04950
Bharat Electronics Ltd	India	203	55-00200
CIC International Ltd	USA	563	75-18570
Compagnie Industrielle des Lasers	France	134	75-10240
Davin Optical Ltd	UK	444	75-17100
Delft Sensor Systems	**Belgium**	**46**	**75-02500**
Delft Sensor Systems	Netherlands	292	75-01590
Econ Optics Mechanics SA	Greece	188	75-01070
EL-OP Electro Optics Industries Ltd	Israel	219	75-06300
Eloptro	S. Africa	358	75-05680
Fabbrica Italiana Apparecchiature	Italy	234	75-08550
GEC-Marconi Avionics Ltd	UK	454	75-25320
GEC-Marconi Materials Technology Ltd	UK	419	60-00550
GEC-Marconi Research Centre	UK	419	60-00700
Hall & Watts (Defence Optics) Group	**UK**	**456**	**75-27650**
Hughes Elcan Optical Technologies	Canada	72	75-07650
Iskra Agency Ltd	UK	462	75-32060
Leica AG	Switzerland	386	75-07310
Lem Norma GmbH	Austria	33	75-01615
Litton Industries Inc.	USA	590	75-55400
M. Paz Logistics Ltd	Israel	224	80-01600
Merex Inc.	USA	595	75-60250
Microelectronics, guidance and	Turkey	402	55-00050
Nanoquest Defence Products Ltd	UK	470	75-41890
Pilkington Optronics Ltd	UK	426	74-10000
Pilkington Thorn Optronics	UK	426	74-10000
Power Technology Inc.	USA	605	75-71090
Roke Manor Research Ltd	UK	420	60-01800
SAGEM SA	**France**	**143**	**75-33950**
Sanders	USA	612	75-77850
Simrad Optronics ASA	Norway	310	75-02620
Transvaro A.S.	Turkey	403	75-02080
Zeiss-Eltro Optronic GmbH	Germany	177	75-34700

413 40 52 Rangefinders, laser, airborne

China National Aero-Tech I/E Corp.	China PR	88	55-00200
CIC International Ltd	USA	563	75-18570
Compagnie Industrielle des Lasers	France	134	75-10240
Cranfield Aerospace Ltd	UK	500	80-07200
Davin Optical Ltd	UK	444	75-17100
EL-OP Electro Optics Industries Ltd	Israel	219	75-06300
Eloptro	S. Africa	358	75-05680
Fabbrica Italiana Apparecchiature	Italy	234	75-08550
GEC-Marconi Avionics Ltd	UK	454	75-25320
GEC-Marconi Materials Technology Ltd	UK	419	60-00550
GEC-Marconi Research Centre	UK	419	60-00700
Litton Industries Inc.	USA	590	75-55400
M. Paz Logistics Ltd	Israel	224	80-01600
Pilkington Thorn Optronics	UK	426	74-10000
Power Technology Inc.	USA	605	75-71090
Roke Manor Research Ltd	UK	420	60-01800

413 40 53 Rangefinders, laser, shipborne

CIC International Ltd	USA	563	75-18570
Compagnie Industrielle des Lasers	France	134	75-10240
Contraves GmbH	Germany	166	75-05200
Davin Optical Ltd	UK	444	75-17100
EL-OP Electro Optics Industries Ltd	Israel	219	75-06300
Eloptro	S. Africa	358	75-05680
Fabbrica Italiana Apparecchiature	Italy	234	75-08550
GEC-Marconi Research Centre	UK	419	60-00700
Litton Industries Inc.	USA	590	75-55400
M. Paz Logistics Ltd	Israel	224	80-01600
Pilkington Optronics Ltd	UK	426	74-10000
Pilkington Thorn Optronics	UK	426	74-10000
Power Technology Inc.	USA	605	75-71090
Roke Manor Research Ltd	UK	420	60-01800
Simrad Optronics ASA	Norway	310	75-02620
Zeiss-Eltro Optronic GmbH	Germany	177	75-34700

413 40 54 Rangefinders, laser, ground-based

Armalon Ltd	UK	431	75-03729
Avimo Electro-Optics Pte Ltd	Singapore	348	75-00800
Avimo Ltd	UK	432	75-04950
Bharat Electronics Ltd	India	203	55-00200
CIC International Ltd	USA	563	75-18570
Compagnie Industrielle des Lasers	France	134	75-10240
Contraves GmbH	Germany	166	75-05200
Davin Optical Ltd	UK	444	75-17100
Delft Sensor Systems	**Belgium**	**46**	**75-02500**
Delft Sensor Systems	Netherlands	292	75-01590
EL-OP Electro Optics Industries Ltd	Israel	219	75-06300
Eloptro	S. Africa	358	75-05680
Fabbrica Italiana Apparecchiature	Italy	234	75-08550
GEC-Marconi Avionics Ltd	UK	454	75-25320
GEC-Marconi Materials Technology Ltd	UK	419	60-00550
GEC-Marconi Research Centre	UK	419	60-00700
Leica AG	Switzerland	386	75-07310
Lem Norma GmbH	Austria	33	75-01615
Litton Industries Inc.	USA	590	75-55400
M. Paz Logistics Ltd	Israel	224	80-01600
Merex Inc.	USA	595	75-60250
Nanoquest Defence Products Ltd	UK	470	75-41890
NORINCO	China PR	88	55-00280
Pilkington Thorn Optronics	UK	426	74-10000
Power Technology Inc.	USA	605	75-71090
Roke Manor Research Ltd	UK	420	60-01800
Siemens Aktiengesellschaft	Germany	175	75-27700
Simrad Optronics ASA	Norway	310	75-02620
Transvaro A.S.	Turkey	403	75-02080
Zeiss-Eltro Optronic GmbH	Germany	177	75-34700

413 50 Optical rangefinding systems & equipment

AAI Corp.	USA	545	75-00020
Angénieux	France	132	75-02700
Bospec Neue Technologien	Germany	166	75-03420
Central Research Laboratories Ltd	UK	499	80-05895
CIC International Ltd	USA	563	75-18570
Coherent Optics (Europe) Ltd	UK	441	75-14090
Computer Optics Inc.	USA	565	75-20680
Contraves Inc	USA	566	75-21850
Davin Optical Ltd	UK	444	75-17100
Delft Sensor Systems	**Belgium**	**46**	**75-02500**
Delft Sensor Systems	Netherlands	292	75-01590
Eloptro	S. Africa	358	75-05680
GEC-Marconi Research Centre	UK	419	60-00700
Hall & Watts (Defence Optics) Group	**UK**	**456**	**75-27650**
Hensoldt AG	Germany	169	75-12450
Hughes Elcan Optical Technologies	Canada	72	75-07650
Instro Precision Ltd	UK	461	75-31630
Israel Aircraft Industries Ltd	Israel	220	75-10200
Kearfott Guidance & Navigation Corp.	USA	588	75-51760
KVH Industries Inc.	USA	589	75-53670
M. Paz Logistics Ltd	Israel	224	80-01600
Microtecnica-Torino SpA	Italy	237	75-15500
Nanoquest Defence Products Ltd	UK	470	75-41890
Omitec Electro-Optics Ltd	UK	472	75-43050
Penny & Giles Data Systems Ltd	UK	473	75-44560
Pilkington Optronics Ltd	UK	426	74-10000
Pilkington Thorn Optronics	UK	426	74-10000
Recon/Optical Inc.	USA	608	75-73930

Sanders	USA	*612*	75-77850
Urals Optical & Mechanical Plant	Russia	*339*	75-53000
Zeiss-Eltro Optronic GmbH	Germany	*177*	75-34700

413 50 30 Optical rangefinding systems, shipborne

Coherent Optics (Europe) Ltd	UK	*441*	75-14090
Davin Optical Ltd	UK	*444*	75-17100
Eloptro	S. Africa	*358*	75-05680
GEC-Marconi Research Centre	UK	*419*	60-00700
Hughes Elcan Optical Technologies	Canada	*72*	75-07650
Israel Aircraft Industries Ltd	Israel	*220*	75-10200
M. Paz Logistics Ltd	Israel	*224*	80-01600
Omitec Electro-Optics Ltd	UK	*472*	75-43050
Pilkington Optronics Ltd	UK	*426*	74-10000
Pilkington Thorn Optronics	UK	*426*	74-10000
Urals Optical & Mechanical Plant	Russia	*339*	75-53000

413 50 40 Optical rangefinding systems, ground-based

CIC International Ltd	USA	*563*	75-18570
Coherent Optics (Europe) Ltd	UK	*441*	75-14090
Davin Optical Ltd	UK	*444*	75-17100
Delft Sensor Systems	**Belgium**	***46***	**75-02500**
Delft Sensor Systems	Netherlands	*292*	75-01590
GEC-Marconi Research Centre	UK	*419*	60-00700
Hughes Elcan Optical Technologies	Canada	*72*	75-07650
Israel Aircraft Industries Ltd	Israel	*220*	75-10200
M. Paz Logistics Ltd	Israel	*224*	80-01600
Omitec Electro-Optics Ltd	UK	*472*	75-43050
Pilkington Optronics Ltd	UK	*426*	74-10000
Pilkington Thorn Optronics	UK	*426*	74-10000
Zeiss-Eltro Optronic GmbH	Germany	*177*	75-34700

413 50 70 Optical rangefinding equipment, ground-based

Coherent Optics (Europe) Ltd	UK	*441*	75-14090
Davin Optical Ltd	UK	*444*	75-17100
Delft Sensor Systems	**Belgium**	***46***	**75-02500**
Delft Sensor Systems	Netherlands	*292*	75-01590
EL-OP Electro Optics Industries Ltd	Israel	*219*	75-06300
GEC-Marconi Research Centre	UK	*419*	60-00700
Hughes Elcan Optical Technologies	Canada	*72*	75-07650
Instro Precision Ltd	UK	*461*	75-31630
M. Paz Logistics Ltd	Israel	*224*	80-01600
Nanoquest Defence Products Ltd	UK	*470*	75-41890
Omitec Electro-Optics Ltd	UK	*472*	75-43050
Pilkington Optronics Ltd	UK	*426*	74-10000
Pilkington Thorn Optronics	UK	*426*	74-10000
Zeiss-Eltro Optronic GmbH	Germany	*177*	75-34700

414 Weapon guidance systems & equipment

Aircraft Tanks Ltd	UK	*429*	75-01730
Alenia-Elsag	Italy	*230*	74-00100
Angénieux	France	*132*	75-02700
Boeing Australia Limited	Australia	*12*	75-02500
Bofors AB	Sweden	*375*	74-00020
British Aerospace (Systems &	UK	*436*	75-08890
CIC International Ltd	USA	*563*	75-18570
Contraves GmbH	Germany	*166*	75-05200
Cranfield University	UK	*419*	60-00350
Daimler-Benz Aerospace France SAS	**France**	***134***	**75-12950**
Diehl GmbH & Co	Germany	*167*	75-06500
DRS Technologies, Inc.	USA	*570*	75-27400
EPC Group Ltd.	UK	*449*	75-21150
Fabbrica Italiana Apparecchiature	Italy	*234*	75-08550
GEC-Marconi Avionics Inc..	USA	*577*	75-37413
GEC-Marconi Avionics Ltd	UK	*454*	75-25320
GEC-Marconi Materials Technology Ltd	UK	*419*	60-00550
GEC-Marconi Research Centre	UK	*419*	60-00700
Iscan Inc.	USA	*586*	75-49100
Israel Aircraft Industries Ltd	Israel	*219*	75-10200
Iver Precision Engineers Ltd	UK	*462*	75-32230
Kaman Sciences Corp.	USA	*637*	80-17600
Kearfott Guidance & Navigation Corp.	USA	*588*	75-51760
L'Garde Inc.	USA	*590*	75-55060
Litton Industries Inc.	USA	*590*	75-55400
Lockheed Martin Corp.	USA	*591*	75-55635
M. Paz Logistics Ltd	Israel	*224*	80-01600
MASS Consultants Limited	UK	*506*	80-15650
Motorola Ltd	UK	*469*	75-41050
Northrop Grumman Corporation	USA	*640*	80-21840
Oerlikon-Contraves AG	Switzerland	*387*	75-09250
Pearpoint Ltd	UK	*473*	75-44270
Pilkington Optronics Ltd	UK	*426*	74-10000
Pilkington Thorn Optronics	UK	*426*	74-10000
Racal Radar Defence Systems Ltd.	UK	*477*	75-47688
Ring Sights Holdings Co. Ltd.	UK	*479*	75-49700
Roke Manor Research Ltd	UK	*420*	60-01800
SAGEM SA	**France**	***143***	**75-33950**
Sanders	USA	*612*	75-77850
Sener Ingenieria y Sistemas SA	Spain	*367*	75-05200
SFIM	**France**	***144***	**75-35080**
Shorts Missile Systems Ltd	**UK**	***483***	**75-53550**
Signal Computing Ltd.	UK	*511*	80-36050
Smiths Aerospace & Defense	USA	*615*	75-82090
STN ATLAS Elektronik GmbH	Germany	*165*	75-01150
Technology Service Corp.	USA	*644*	80-31930
Thomson Sintra	France	*148*	75-45135
United Defense L.P.	USA	*624*	75-92050

414 10 Radar weapon guidance systems

Alenia SpA	Italy	*232*	75-01170
Alenia-Elsag	Italy	*230*	74-00100
CelsiusTech AB	Sweden	*376*	75-03710
CIC International Ltd	USA	*563*	75-18570
Contraves GmbH	Germany	*166*	75-05200
Cranfield University	UK	*419*	60-00350
Daimler-Benz Aerospace France SAS	**France**	***134***	**75-12950**
Dassault Electronique SA (Group)	**France**	***135***	**75-12980**
Motorola Ltd	UK	*469*	75-41050
Philips Traffic & Engineering	Australia	*20*	75-16550
Polhemus Inc.	USA	*604*	75-70855
Racal Radar Defence Systems Ltd.	UK	*477*	75-47688
Radio Research Instrument Co. Inc.	USA	*607*	75-72600
Roke Manor Research Ltd	UK	*420*	60-01800
Sanders	USA	*612*	75-77850
Shorts Missile Systems Ltd	**UK**	***483***	**75-53550**
Siemens Aktiengesellschaft	Germany	*175*	75-27700
Siemens Plessey Systems	UK	*483*	75-53770
Technology Service Corp.	USA	*644*	80-31930
Thomson-CSF	France	*148*	75-45100

414 20 Sonar weapon guidance systems

Bofors SA Marine AB.	Sweden	*376*	75-03040
Logica UK Limited.	UK	*506*	80-14350
Radio Research Instrument Co. Inc.	USA	*607*	75-72600
STN ATLAS Elektronik GmbH	Germany	*165*	75-01150
Technology Service Corp.	USA	*644*	80-31930
Thomson Sintra	France	*148*	75-45135

414 20 10 Sonar weapon guidance systems, active

Bofors SA Marine AB.	Sweden	*376*	75-03040
Logica UK Limited.	UK	*506*	80-14350
STN ATLAS Elektronik GmbH	Germany	*165*	75-01150
Technology Service Corp.	USA	*644*	80-31930
Thomson Sintra	France	*148*	75-45135

414 20 20 Sonar weapon guidance systems, passive

Litton Industries Inc.	USA	*590*	75-55400
Logica UK Limited.	UK	*506*	80-14350
STN ATLAS Elektronik GmbH	Germany	*165*	75-01150
Technology Service Corp.	USA	*644*	80-31930
Thomson Sintra	France	*148*	75-45135

414 20 30 Sonar weapon guidance systems, active / passive

Bofors SA Marine AB.	Sweden	*376*	75-03040
Logica UK Limited.	UK	*506*	80-14350
STN ATLAS Elektronik GmbH	Germany	*165*	75-01150
Technology Service Corp.	USA	*644*	80-31930
Thomson Sintra	France	*148*	75-45135

414 30 Radio command weapon guidance systems

Cranfield University	UK	*419*	60-00350
Daimler-Benz Aerospace France SAS	**France**	***134***	**75-12950**
Kollsman	USA	*589*	75-53150
Racal Antennas Ltd	UK	*476*	75-47656
Racal Radar Defence Systems Ltd	UK	*477*	75-47688
Shorts Missile Systems Ltd	**UK**	***483***	**75-53550**
Technology Service Corp.	USA	*644*	80-31930

414 40 Electro-optical weapon guidance systems

Bodenseewerk Gerätetechnik GmbH	Germany	*165*	75-03000
CIC International Ltd	USA	*563*	75-18570
Coherent Optics (Europe) Ltd	UK	*441*	75-14090
Contraves GmbH	Germany	*166*	75-05200
Cranfield University	UK	*419*	60-00350
Davin Optical Ltd	UK	*444*	75-17100
Delco Systems Operations	USA	*568*	75-25360
DRS Technologies, Inc.	USA	*570*	75-27400
EUROCONTROL SpA	Italy	*233*	75-08200
GEC-Marconi Avionics Ltd	UK	*454*	75-25320
GEC-Marconi Materials Technology Ltd	UK	*419*	60-00550
GEC-Marconi Research Centre	UK	*419*	60-00700
JAI A/S	Denmark	*103*	75-09400
Kollsman	USA	*589*	75-53150
Litton Industries Inc.	USA	*590*	75-55400
McLennan Marine Ltd.	UK	*467*	75-38080
MSI-Defence Systems Ltd	UK	*469*	75-41280
Night Vision Equipment Co. Inc.	USA	*599*	75-64390
Northrop Grumman Corporation	USA	*599*	75-65450
Octec Ltd	UK	*471*	75-42950
Pilkington Optronics Ltd	UK	*426*	74-10000
Pilkington Thorn Optronics	UK	*426*	74-10000
Radamec Defence Systems Ltd	UK	*477*	75-47750
Roke Manor Research Ltd	UK	*420*	60-01800
SAGEM SA	**France**	***143***	**75-33950**
SFIM ODS	France	*144*	75-35085
Shorts Missile Systems Ltd	**UK**	***483***	**75-53550**
Siemens Aktiengesellschaft	Germany	*175*	75-27700
Technology Service Corp.	USA	*644*	80-31930
Zeiss-Eltro Optronic GmbH	Germany	*177*	75-34700

414 40 80 Sights, helmet-mounted, aircraft pilot

Bospec Neue Technologien	Germany	*166*	75-03420
Central Design Bureau	Ukraine	*406*	75-00800
CIC International Ltd	USA	*563*	75-18570
Delft Sensor Systems	**Belgium**	***46***	**75-02500**
Delft Sensor Systems	Netherlands	*292*	75-01590
DRS Technologies, Inc.	USA	*570*	75-27400
GEC-Marconi Avionics Ltd	UK	*454*	75-25320
Hughes Elcan Optical Technologies	Canada	*72*	75-07650
Kaiser Electronics	USA	*587*	75-51350
Kollsman	USA	*589*	75-53150
M. Paz Logistics Ltd	Israel	*224*	80-01600
Pilkington Optronics Ltd	UK	*426*	74-10000
Polhemus Inc.	USA	*604*	75-70855
SFIM ODS	France	*144*	75-35085

414 44 Infra-red weapon guidance systems

AAI Corp.	USA	*545*	75-00020
Angénieux	France	*132*	75-02700
Central Research Laboratories Ltd	UK	*499*	80-05895
CI Systems Ltd	Israel	*218*	75-03750
CIC International Ltd	USA	*563*	75-18570
Coherent Optics (Europe) Ltd	UK	*441*	75-14090
Contraves GmbH	Germany	*166*	75-05200
Davin Optical Ltd	UK	*444*	75-17100
Delft Sensor Systems	**Belgium**	***46***	**75-02500**
Delft Sensor Systems	Netherlands	*292*	75-01590
Fabbrica Italiana Apparecchiature	Italy	*234*	75-08550
GEC-Marconi Avionics Ltd	UK	*454*	75-25320
GEC-Marconi Materials Technology Ltd	UK	*419*	60-00550
GEC-Marconi Research Centre	UK	*419*	60-00700
Hughes Elcan Optical Technologies	Canada	*72*	75-07650
INIMEX	**Belgium**	***46***	**75-04650**
Kollsman	USA	*589*	75-53150
Litton Industries Inc.	USA	*590*	75-55400
Nanoquest Defence Products Ltd	UK	*470*	75-41890
Octec Ltd	UK	*471*	75-42950
Omitec Electro-Optics Ltd	UK	*472*	75-43050
Pilkington Optronics Ltd	UK	*426*	74-10000
Pilkington Thorn Optronics	UK	*426*	74-10000
Polhemus Inc.	USA	*604*	75-70855
Roke Manor Research Ltd	UK	*420*	60-01800
SAGEM SA	**France**	***143***	**75-33950**
Sener Ingenieria y Sistemas SA	Spain	*367*	75-05200
SFIM ODS	France	*144*	75-35085
SNPE	France	*145*	75-36315
Technology Service Corp.	USA	*644*	80-31930

414 44 30 Thermal cameras, weapon guidance

Central Research Laboratories Ltd	UK	*499*	80-05895
CIC International Ltd	USA	*563*	75-18570
Coherent Optics (Europe) Ltd	UK	*441*	75-14090
Contraves GmbH	Germany	*166*	75-05200
Davin Optical Ltd	UK	*444*	75-17100
Delft Sensor Systems	**Belgium**	***46***	**75-02500**
Delft Sensor Systems	Netherlands	*292*	75-01590
Ericsson Defence Systems Pty Ltd	Australia	*16*	75-05820
Fabbrica Italiana Apparecchiature	Italy	*234*	75-08550
GEC-Marconi Avionics Ltd	UK	*454*	75-25320
GEC-Marconi Research Centre	UK	*419*	60-00700
Insight Vision Systems Ltd	UK	*461*	75-31530
Kollsman	USA	*589*	75-53150
Nanoquest Defence Products Ltd	UK	*470*	75-41890
Pilkington Optronics Ltd	UK	*426*	74-10000
Pilkington Thorn Optronics	UK	*426*	74-10000
Regis SpA	Italy	*238*	75-20110
SAGEM SA	**France**	***143***	**75-33950**
Simrad Optronics ASA	Norway	*310*	75-02620

414 44 80 Infra-red target markers

Davin Optical Ltd	UK	*444*	75-17100
Delft Sensor Systems	**Belgium**	***46***	**75-02500**
Delft Sensor Systems	Netherlands	*292*	75-01590
GEC-Marconi Research Centre	UK	*419*	60-00700
ILEE AG	Switzerland	*386*	75-06300
L'Garde Inc.	USA	*590*	75-55060
Litton Industries Inc.	USA	*590*	75-55400
M. Paz Logistics Ltd	Israel	*224*	80-01600
Night Vision Equipment Co. Inc.	USA	*599*	75-64390
Pilkington Thorn Optronics	UK	*426*	74-10000
Technical Products Group Inc.	USA	*620*	75-86764

414 45 Laser weapon guidance systems

Bofors AB	Sweden	*375*	74-00020
CIC International Ltd	USA	*563*	75-18570
Coherent Optics (Europe) Ltd	UK	*441*	75-14090
Compagnie Industrielle des Lasers	France	*134*	75-10240
Cranfield University	UK	*419*	60-00350
Davin Optical Ltd	UK	*444*	75-17100
Delco Systems Operations	USA	*568*	75-25360
Delft Sensor Systems	**Belgium**	***46***	**75-02500**
Delft Sensor Systems	Netherlands	*292*	75-01590
Fabbrica Italiana Apparecchiature	Italy	*234*	75-08550
GEC-Marconi Avionics Ltd	UK	*454*	75-25320
GEC-Marconi Materials Technology Ltd	UK	*419*	60-00550
GEC-Marconi Research Centre	UK	*419*	60-00700
ILEE AG	Switzerland	*386*	75-06300
Kollsman	USA	*589*	75-53150
Litton Industries Inc.	USA	*590*	75-55400
M. Paz Logistics Ltd	Israel	*224*	80-01600
Northrop Grumman Corporation	USA	*599*	75-65450
Oerlikon-Contraves AG	Switzerland	*387*	75-09250
Pilkington Optronics Ltd	UK	*426*	74-10000
Pilkington Thorn Optronics	UK	*426*	74-10000
Polhemus Inc.	USA	*604*	75-70855
Sener Ingenieria y Sistemas SA	Spain	*367*	75-05200
Shorts Missile Systems Ltd	**UK**	***483***	**75-53550**
Zeiss-Eltro Optronic GmbH	Germany	*177*	75-34700

414 45 80 Lasers, target illuminating

CIC International Ltd	USA	*563*	75-18570
Davin Optical Ltd	UK	*444*	75-17100
Delft Sensor Systems	**Belgium**	***46***	**75-02500**

Delft Sensor Systems	Netherlands	*292*	75-01590
Eloptro	S. Africa	*358*	75-05680
Fabbrica Italiana Apparecchiature	Italy	*234*	75-08550
GEC-Marconi Avionics Ltd	UK	*454*	75-25320
GEC-Marconi Research Centre	UK	*419*	60-00700
Israel Aircraft Industries Ltd	Israel	*220*	75-10200
Kollsman	USA	*589*	75-53150
Litton Industries Inc.	USA	*590*	75-55400
M. Paz Logistics Ltd	Israel	*224*	80-01600
Merex Inc.	USA	*595*	75-60250
Night Vision Equipment Co. Inc.	USA	*599*	75-64390
Pilkington Thorn Optronics	UK	*426*	74-10000

414 45 80 50 Laser pistols, target illuminating

Delft Sensor Systems	**Belgium**	***46***	**75-02500**
Delft Sensor Systems	Netherlands	*292*	75-01590
ILEE AG	Switzerland	*386*	75-06300
Litton Industries Inc.	USA	*590*	75-55400
M. Paz Logistics Ltd	Israel	*224*	80-01600

414 48 Television weapon guidance systems

AAI Corp.	USA	*545*	75-00020
Alden Sales Co.	USA	*548*	75-04410
Angénieux	France	*132*	75-02700
British Aerospace (Systems &	UK	*436*	75-08890
Coherent Optics (Europe) Ltd	UK	*441*	75-14090
Contraves GmbH	Germany	*166*	75-05200
Cranfield University	UK	*419*	60-00350
Diamond Electronics Inc.	USA	*569*	75-26450
GEC-Marconi Avionics Ltd	UK	*454*	75-25320
GEC-Marconi Research Centre	UK	*419*	60-00700
Hadland Photonics Ltd	UK	*456*	75-27200
Insight Vision Systems Ltd	UK	*461*	75-31530
Instro Precision Ltd	UK	*461*	75-31630
Iscan Inc.	USA	*586*	75-49100
Israel Aircraft Industries Ltd	Israel	*220*	75-10200
M. Paz Logistics Ltd	Israel	*224*	80-01600
Octec Ltd	UK	*471*	75-42950
Pearpoint Ltd	UK	*473*	75-44270
Pilkington Thorn Optronics	UK	*426*	74-10000
Polhemus Inc.	USA	*604*	75-70855
Radamec Defence Systems Ltd	UK	*477*	75-47750
Regis SpA	Italy	*238*	75-20110
Roke Manor Research Ltd	UK	*420*	60-01800
Shorts Missile Systems Ltd	**UK**	***483***	**75-53550**

414 50 Optical weapon guidance systems

Angénieux	France	*132*	75-02700
Bospec Neue Technologien	Germany	*166*	75-03420
British Aerospace (Systems &	UK	*436*	75-08890
Central Design Bureau	Ukraine	*406*	75-00800
Coherent Optics (Europe) Ltd	UK	*441*	75-14090
Contraves GmbH	Germany	*166*	75-05200
Cranfield University	UK	*419*	60-00350
Davin Optical Ltd	UK	*444*	75-17100
DRS Technologies, Inc.	USA	*570*	75-27400
Fabbrica Italiana Apparecchiature	Italy	*234*	75-08550
GEC-Marconi Research Centre	UK	*419*	60-00700
Hughes Elcan Optical Technologies	Canada	*72*	75-07650
INIMEX	**Belgium**	***46***	**75-04650**
Israel Aircraft Industries Ltd	Israel	*219*	75-10200
Litton Industries Inc.	USA	*590*	75-55400
M. Paz Logistics Ltd	Israel	*224*	80-01600
Nanoquest Defence Products Ltd	UK	*470*	75-41890
Omitec Electro-Optics Ltd	UK	*472*	75-43050
Pilkington Optronics Ltd	UK	*426*	74-10000
Pilkington Thorn Optronics	UK	*426*	74-10000
Polhemus Inc.	USA	*604*	75-70855
Ring Sights Holdings Co. Ltd.	UK	*479*	75-49700
Sener Ingenieria y Sistemas SA	Spain	*367*	75-05200
SFIM ODS	France	*144*	75-35085
Shorts Missile Systems Ltd	**UK**	***483***	**75-53550**
Zeiss-Eltro Optronic GmbH	Germany	*177*	75-34700

414 50 30 Fibre optic weapon guidance systems

AAI Corp.	USA	*545*	75-00020
Cogent	UK	*500*	80-06250
Cranfield University	UK	*419*	60-00350
EDO Corporation	USA	*571*	75-29250
GEC-Marconi Research Centre	UK	*419*	60-00700
Israel Product Research Co. Ltd	Israel	*219*	75-10000
Litton Industries Inc.	USA	*590*	75-55400
Pilkington Thorn Optronics	UK	*426*	74-10000

414 60 Inertial weapon guidance systems

Astronautics Corp. of America	USA	*555*	75-10000
British Aerospace (Systems &	UK	*436*	75-08890
Coherent Optics (Europe) Ltd	UK	*441*	75-14090
Cranfield University	UK	*419*	60-00350
GEC-Marconi Avionics Ltd	UK	*454*	75-25320
Kearfott Guidance & Navigation Corp.	USA	*588*	75-51760
Litton Industries Inc.	USA	*590*	75-55400
Northrop Grumman Corporation	USA	*599*	75-65450
SAGEM SA	**France**	***143***	**75-33950**
Sener Ingenieria y Sistemas SA	Spain	*367*	75-05200
SFIM	**France**	***144***	**75-35080**
Smiths Aerospace & Defense	USA	*615*	75-82090

414 70 Thermal weapon guidance systems

CI Systems Ltd	Israel	*218*	75-03750
Coherent Optics (Europe) Ltd	UK	*441*	75-14090
Contraves GmbH	Germany	*166*	75-05200
Delft Sensor Systems	**Belgium**	***46***	**75-02500**
Delft Sensor Systems	Netherlands	*292*	75-01590
Fabbrica Italiana Apparecchiature	Italy	*234*	75-08550
GEC-Marconi Avionics Ltd	UK	*454*	75-25320
GEC-Marconi Research Centre	UK	*419*	60-00700
Insight Vision Systems Ltd	UK	*461*	75-31530
Instro Precision Ltd	UK	*461*	75-31630
Israel Aircraft Industries Ltd	Israel	*220*	75-10200
JAI A/S	Denmark	*103*	75-09400
Nanoquest Defence Products Ltd	UK	*470*	75-41890
Octec Ltd	UK	*471*	75-42950
Pilkington Optronics Ltd	UK	*426*	74-10000
Pilkington Thorn Optronics	UK	*426*	74-10000
SFIM ODS	France	*144*	75-35085

414 90 Wire guidance systems, weapon

Cogent	UK	*500*	80-06250

415 Weapon fire control systems

AAI Corp.	USA	*545*	75-00020
Alenia-Elsag	Italy	*230*	74-00100
Bharat Electronics Ltd	India	*203*	55-00200
Bressel SA	Spain	*366*	75-01670
British Aerospace (Systems &	UK	*436*	75-08890
Canadian Marconi Co.	Canada	*69*	75-02600
CIC International Ltd	USA	*563*	75-18570
Computing Devices Canada	Canada	*70*	75-03600
Computing Devices Company Ltd	UK	*442*	75-14700
Contraves GmbH	Germany	*166*	75-05200
Delft Sensor Systems	**Belgium**	***46***	**75-02500**
Delft Sensor Systems	Netherlands	*292*	75-01590
EL-OP Electro Optics Industries Ltd	Israel	*219*	75-06300
Elbit Systems Ltd	Israel	*218*	75-05300
EUROCONTROL SpA	Italy	*233*	75-08200
Fraser-Volpe Corp.	USA	*576*	75-35950
GEC-Marconi Avionics Inc.	USA	*577*	75-37413
Hollandse Signaalapparaten BV	Netherlands	*293*	75-03300
Hughes Elcan Optical Technologies	Canada	*72*	75-07650
INDRA	Spain	*366*	75-03800
Instro Precision Ltd	UK	*461*	75-31630
Iscan Inc.	USA	*586*	75-49100
Kollsman	USA	*589*	75-53150
Lockheed Martin Corp.	USA	*591*	75-55635
M. Paz Logistics Ltd	Israel	*224*	80-01600
Matra Défense-Espace	France	*139*	75-26000
NORINCO	China PR	*88*	55-00280
Oerlikon-Contraves AG.	Switzerland	*387*	75-09250
Oerlikon-Contraves SpA	Italy	*237*	75-16800
Pilkington Optronics Ltd	UK	*426*	74-10000
Recon/Optical Inc.	USA	*608*	75-73930
Reunert Technology Systems Ltd	S. Africa	*359*	75-18300
SABCA	Belgium	*46*	75-05590

SFIM	**France**	**144**	**75-35080**
Siemens Aktiengesellschaft	Germany	175	75-27700
Smiths Aerospace & Defense	USA	615	75-82090
STN ATLAS Elektronik GmbH	Germany	165	75-01150
Technology Service Corp.	USA	644	80-31930
Thomson Sintra	France	148	75-45135
United Defense L.P.	USA	624	75-92050
Vickers Shipbuilding & Eng Ltd	UK	491	75-63100
Warren-Knight Instrument Co.	USA	627	75-95150
Zeiss-Eltro Optronic GmbH	Germany	177	75-34700

415 20 Weapon fire control systems, airborne

Air-Log Ltd	UK	429	75-01800
Astronautics Corp. of America	USA	555	75-10000
Coherent Optics (Europe) Ltd	UK	441	75-14090
Computing Devices Company Ltd	UK	442	75-14700
Cranfield University	UK	419	60-00350
EL-OP Electro Optics Industries Ltd	Israel	219	75-06300
Fabbrica Italiana Apparecchiature	Italy	234	75-08550
Fraser-Volpe Corp.	USA	576	75-35950
GEC-Marconi Avionics Inc.	USA	577	75-37413
GEC-Marconi Avionics Ltd	UK	454	75-25320
Iscan Inc.	USA	586	75-49100
Pilkington Thorn Optronics	UK	426	74-10000
SFIM	**France**	**144**	**75-35080**
Smiths Aerospace & Defense	USA	615	75-82090
Sparton Technology Inc.	USA	616	75-82950
Technology Service Corp.	USA	644	80-31930
Thomson Sintra	France	148	75-45135

415 20 10 Combined weapon fire control systems, aircraft

Computing Devices Company Ltd	UK	442	75-14700
Elbit Systems Ltd	Israel	218	75-05300
Fabbrica Italiana Apparecchiature	Italy	234	75-08550
GEC-Marconi Avionics Ltd	UK	454	75-25320
Kaiser Electronics	USA	587	75-51350
Matra Défense-Espace	France	139	75-26000
Smiths Aerospace & Defense	USA	615	75-82090
Sparton Technology Inc.	USA	616	75-82950
Technology Service Corp.	USA	644	80-31930

415 20 20 Gun fire control systems, aircraft

Computing Devices Company Ltd	UK	442	75-14700
Kaiser Electronics	USA	587	75-51350
Smiths Aerospace & Defense	USA	615	75-82090

415 20 30 Missile fire control systems, aircraft

Alenia SpA	Italy	232	75-01170
Canadian Marconi Co.	Canada	69	75-02600
Coherent Optics (Europe) Ltd	UK	441	75-14090
Computing Devices Company Ltd	UK	442	75-14700
Cranfield University	UK	419	60-00350
Kaiser Electronics	USA	587	75-51350
Matra Défense-Espace	France	139	75-26000
Pilkington Thorn Optronics	UK	426	74-10000
Smiths Aerospace & Defense	USA	615	75-82090
Technology Service Corp.	USA	644	80-31930

415 20 40 Fire control systems, aircraft-launched torpedo

Smiths Aerospace & Defense	USA	615	75-82090
Thomson Sintra	France	148	75-45135
Whitehead SpA	Italy	240	75-27600

415 20 70 Fire control equipment, airborne weapon

Air-Log Ltd	UK	429	75-01800
Fabbrica Italiana Apparecchiature	Italy	234	75-08550
Fraser-Volpe Corp.	USA	576	75-35950
Iscan Inc.	USA	586	75-49100
Kaiser Electronics	USA	587	75-51350
Smiths Aerospace & Defense	USA	615	75-82090
Sparton Technology Inc.	USA	616	75-82950

415 20 70 10 Fire control equipment, aircraft gun

AAI Corp.	USA	545	75-00020
EUROCONTROL SpA	Italy	233	75-08200
Iscan Inc.	USA	586	75-49100
Kaiser Electronics	USA	587	75-51350
Smiths Aerospace & Defense	USA	615	75-82090

415 20 70 20 Fire control equipment, missile, airborne

AAI Corp.	USA	545	75-00020
Kaiser Electronics	USA	587	75-51350
Pilkington Thorn Optronics	UK	426	74-10000
Smiths Aerospace & Defense	USA	615	75-82090

415 30 Weapon fire control systems, naval shipborne

AAI Corp.	USA	545	75-00020
Alenia-Elsag	Italy	230	74-00100
Cincinnati Electronics Corp.	USA	564	75-18640
Coherent Optics (Europe) Ltd	UK	441	75-14090
Contraves GmbH	Germany	166	75-05200
Elbit Systems Ltd	Israel	218	75-05300
EUROCONTROL SpA	Italy	233	75-08200
GEC-Marconi Avionics Inc.	USA	577	75-37413
GEC-Marconi Avionics Ltd	UK	454	75-25320
Hollandse Signaalapparaten BV	Netherlands	293	75-03300
Matra Défense-Espace	France	139	75-26000
MSI-Defence Systems Ltd	UK	469	75-41280
Norwegian Defence Technology A/S	Norway	310	75-02050
Oerlikon-Contraves AG	Switzerland	387	75-09250
Oerlikon-Contraves SpA	Italy	237	75-16800
Philips Traffic & Engineering	Australia	20	75-16550
Pilkington Thorn Optronics	UK	426	74-10000
Radamec Defence Systems Ltd	UK	477	75-47750
SFIM	**France**	**144**	**75-35080**
STN ATLAS Elektronik GmbH	Germany	165	75-01150
Technology Service Corp.	USA	644	80-31930
Thomson Sintra	France	148	75-45135

415 30 05 ASW fire control systems, naval shipborne

Hollandse Signaalapparaten BV	Netherlands	293	75-03300
STN ATLAS Elektronik GmbH	Germany	165	75-01150
Technology Service Corp.	USA	644	80-31930
Thomson Sintra	France	148	75-45135

415 30 10 Fire control systems, combined weapon, naval

Alenia-Elsag	Italy	230	74-00100
CelsiusTech AB	Sweden	376	75-03710
GEC-Marconi Avionics Ltd	UK	454	75-25320
Hollandse Signaalapparaten BV	Netherlands	293	75-03300
MSI-Defence Systems Ltd	UK	469	75-41280
Oerlikon-Contraves SpA	Italy	237	75-16800
Radamec Defence Systems Ltd	UK	477	75-47750
STN ATLAS Elektronik GmbH	Germany	165	75-01150
Technology Service Corp.	USA	644	80-31930
Thomson-CSF	France	148	75-45100

415 30 12 Combined weapon fire control systems, naval ship

AAI Corp.	USA	545	75-00020
Alenia-Elsag	Italy	230	74-00100
CelsiusTech Electronics AB	Sweden	376	75-03720
Hollandse Signaalapparaten BV	Netherlands	293	75-03300
Israel Aircraft Industries Ltd	Israel	219	75-10200
MSI-Defence Systems Ltd	UK	469	75-41280
Radamec Defence Systems Ltd	UK	477	75-47750
STN ATLAS Elektronik GmbH	Germany	165	75-01150
Technology Service Corp.	USA	644	80-31930
Thomson-CSF	France	148	75-45100

415 30 12 50 Fire directors, optronic, combined weapon, naval

Cincinnati Electronics Corp.	USA	564	75-18640
EL-OP Electro Optics Industries Ltd	Israel	219	75-06300

MSI-Defence Systems Ltd	UK	469	75-41280
Radamec Defence Systems Ltd	UK	477	75-47750
SFIM	**France**	**144**	**75-35080**
Thomson-CSF	France	148	75-45100

415 30 14 Combined weapon fire control systems, submarine

CelsiusTech AB	Sweden	376	75-03710
STN ATLAS Elektronik GmbH	Germany	165	75-01150
Thomson Sintra	France	148	75-45135

415 30 20 Gun fire control systems, naval

AAI Corp.	USA	545	75-00020
Alenia-Elsag	Italy	230	74-00100
Bofors AB	Sweden	375	74-00020
CelsiusTech AB	Sweden	376	75-03710
EUROCONTROL SpA	Italy	233	75-08200
GEC-Marconi Avionics Ltd	UK	454	75-25320
Hollandse Signaalapparaten BV	Netherlands	293	75-03300
Israel Aircraft Industries Ltd	Israel	219	75-10200
Kollsman	USA	589	75-53150
MSI-Defence Systems Ltd	UK	469	75-41280
Oerlikon-Contraves SpA	Italy	237	75-16800
Radamec Defence Systems Ltd	UK	477	75-47750
SAGEM SA	**France**	**143**	**75-33950**
Sanders	USA	612	75-77850
Technology Service Corp.	USA	644	80-31930
Thomson-CSF	France	148	75-45100
United Defense L.P.	USA	624	75-92050

415 30 20 50 Fire directors, optronic, naval gun

Cincinnati Electronics Corp.	USA	564	75-18640
GEC-Marconi Avionics Ltd	UK	454	75-25320
MSI-Defence Systems Ltd	UK	469	75-41280
Thomson-CSF	France	148	75-45100

415 30 30 Missile fire control systems, naval

Alenia SpA	Italy	232	75-01170
Alenia-Elsag	Italy	230	74-00100
British Aerospace (Systems &	UK	436	75-08890
CelsiusTech AB	Sweden	376	75-03710
GEC-Marconi Avionics Ltd	UK	454	75-25320
Hollandse Signaalapparaten BV	Netherlands	293	75-03300
Israel Aircraft Industries Ltd	Israel	219	75-10200
Matra Défense-Espace	France	139	75-26000
MSI-Defence Systems Ltd	UK	469	75-41280
Oerlikon-Contraves SpA	Italy	237	75-16800
Pilkington Thorn Optronics	UK	426	74-10000
Technology Service Corp.	USA	644	80-31930
United Defense L.P.	USA	624	75-92050

415 30 40 Torpedo fire control systems, naval ship

Alenia-Elsag	Italy	230	74-00100
CelsiusTech AB	Sweden	376	75-03710
Hollandse Signaalapparaten BV	Netherlands	293	75-03300
Norwegian Defence Technology A/S	Norway	310	75-02050
STN ATLAS Elektronik GmbH	Germany	165	75-01150
Thomson Sintra	France	148	75-45135

415 30 42 Fire control systems, torpedo, naval ship-launched

Alenia-Elsag	Italy	230	74-00100
CelsiusTech AB	Sweden	376	75-03710
Hollandse Signaalapparaten BV	Netherlands	293	75-03300
STN ATLAS Elektronik GmbH	Germany	165	75-01150
Thomson Sintra	France	148	75-45135
Whitehead SpA	Italy	240	75-27600

415 30 44 Torpedo fire control systems, submarine

CelsiusTech AB	Sweden	376	75-03710
Norwegian Defence Technology A/S	Norway	310	75-02050
STN ATLAS Elektronik GmbH	Germany	165	75-01150
Thomson Sintra	France	148	75-45135
Whitehead SpA	Italy	240	75-27600

415 30 70 Fire control equipment, naval weapon

AAI Corp.	USA	545	75-00020
Hollandse Signaalapparaten BV	Netherlands	293	75-03300
Kinetics Ltd	Israel	220	75-11320
MSI-Defence Systems Ltd	UK	469	75-41280
Nottingham International	USA	640	80-22000
Sanders	USA	612	75-77850
SELIN Sistemi Sarl	Italy	238	75-21750
Technology Service Corp.	USA	644	80-31930

415 30 70 10 Fire control equipment, naval gun

AAI Corp.	USA	545	75-00020
Alenia-Elsag	Italy	230	74-00100
Hollandse Signaalapparaten BV	Netherlands	293	75-03300
MSI-Defence Systems Ltd	UK	469	75-41280
Oerlikon-Contraves SpA	Italy	237	75-16800
Radamec Defence Systems Ltd	UK	477	75-47750
Sanders	USA	612	75-77850
SELIN Sistemi Sarl	Italy	238	75-21750
Technology Service Corp.	USA	644	80-31930

415 30 70 20 Fire control equipment, naval missile

AAI Corp.	USA	545	75-00020
Alenia-Elsag	Italy	230	74-00100
Hollandse Signaalapparaten BV	Netherlands	293	75-03300
MSI-Defence Systems Ltd	UK	469	75-41280
Oerlikon-Contraves SpA	Italy	237	75-16800
SELIN Sistemi Sarl	Italy	238	75-21750
Technology Service Corp.	USA	644	80-31930

415 30 70 30 Fire control equipment, naval torpedo

Alenia-Elsag	Italy	230	74-00100
Hollandse Signaalapparaten BV	Netherlands	293	75-03300
STN ATLAS Elektronik GmbH	Germany	165	75-01150
Thomson Sintra	France	148	75-45135
Whitehead SpA	Italy	240	75-27600

415 40 Weapon fire control systems, ground-based

AAI Corp.	USA	545	75-00020
Avibras Aeroespacial SA	Brazil	55	75-00968
Bressel SA.	Spain	366	75-01670
CIC International Ltd	USA	563	75-18570
Contraves GmbH	Germany	166	75-05200
Delco Systems Operations	USA	568	75-25360
Delft Sensor Systems	**Belgium**	**46**	**75-02500**
Delft Sensor Systems	Netherlands	292	75-01590
EUROCONTROL SpA	Italy	233	75-08200
Fraser-Volpe Corp.	USA	576	75-35950
GEC-Marconi Avionics Inc.	USA	577	75-37413
Hollandse Signaalapparaten BV	Netherlands	293	75-03300
Hughes Elcan Optical Technologies	Canada	72	75-07650
Instro Precision Ltd	UK	461	75-31630
Kearfott Guidance & Navigation Corp.	USA	588	75-51760
M. Paz Logistics Ltd	Israel	224	80-01600
Matra Défense-Espace	France	139	75-26000
Oerlikon-Contraves AG.	Switzerland	387	75-09250
Oerlikon-Contraves SpA	Italy	237	75-16800
Pilkington Optronics Ltd	UK	426	74-10000
Racal Radar Defence Systems Ltd.	UK	477	75-47688
Recon/Optical Inc.	USA	608	75-73930
Reunert Technology Systems Ltd	S. Africa	359	75-18300
SABCA	Belgium	46	75-05590
SFIM	**France**	**144**	**75-35080**
Société d'Applications des Machines	France	143	75-34300
STN ATLAS Elektronik GmbH	Germany	165	75-01150
Technology Service Corp.	USA	644	80-31930
Thomson-CSF	France	148	75-45100
Vickers Shipbuilding & Eng Ltd	UK	491	75-63100
Zeiss-Eltro Optronic GmbH	Germany	177	75-34700

415 40 10 Combined weapon fire control systems, ground-based

AAI Corp.	USA	*545*	75-00020
Arab International Optronics Co.	Egypt	*109*	74-00500
Delft Sensor Systems	**Belgium**	***46***	**75-02500**
Delft Sensor Systems	Netherlands	*292*	75-01590
Giat Industries	France	*126*	55-00100
M. Paz Logistics Ltd	Israel	*224*	80-01600
Oerlikon-Contraves SpA	Italy	*237*	75-16800
Pilkington Optronics Ltd	UK	*426*	74-10000
Radamec Defence Systems Ltd	UK	*477*	75-47750
Reunert Technology Systems Ltd	S. Africa	*359*	75-18300
Thomson-CSF	France	*148*	75-45100
Vickers Shipbuilding & Eng Ltd	UK	*491*	75-63100

415 40 20 Gun fire control systems, ground-based

AAI Corp.	USA	*545*	75-00020
Arab International Optronics Co.	Egypt	*109*	74-00500
Avibras Aeroespacial SA	Brazil	*55*	75-00968
Bofors AB	Sweden	*375*	74-00020
CIC International Ltd	USA	*563*	75-18570
Computing Devices Canada	Canada	*70*	75-03600
Delft Sensor Systems	**Belgium**	***46***	**75-02500**
Delft Sensor Systems	Netherlands	*292*	75-01590
EL-OP Electro Optics Industries Ltd	Israel	*219*	75-06300
EUROCONTROL SpA	Italy	*233*	75-08200
Helio Mirror Co. Ltd	UK	*457*	75-28900
Hollandse Signaalapparaten BV	Netherlands	*293*	75-03300
Hughes Elcan Optical Technologies	Canada	*72*	75-07650
Instro Precision Ltd	UK	*461*	75-31630
Israel Aircraft Industries Ltd	Israel	*219*	75-10200
Litton Industries Inc.	USA	*590*	75-55400
M. Paz Logistics Ltd	Israel	*224*	80-01600
NORINCO	China PR	*88*	55-00280
Oerlikon-Contraves SpA	Italy	*237*	75-16800
Racal Radar Defence Systems Ltd	UK	*477*	75-47688
Radamec Defence Systems Ltd	UK	*477*	75-47750
Recon/Optical Inc.	USA	*608*	75-73930
Reunert Technology Systems Ltd	S. Africa	*359*	75-18300
SABCA	Belgium	*46*	75-05590
SAGEM SA	**France**	***143***	**75-33950**
Société d'Applications des Machines	France	*143*	75-34300
Thomson-CSF	France	*148*	75-45100

415 40 20 10 Gun fire control systems, anti-aircraft

AAI Corp.	USA	*545*	75-00020
Avibras Aeroespacial SA	Brazil	*55*	75-00968
Bofors AB	Sweden	*375*	74-00020
Dassault Electronique SA (Group)	**France**	***135***	**75-12980**
EL-OP Electro Optics Industries Ltd	Israel	*219*	75-06300
EUROCONTROL SpA	Italy	*233*	75-08200
Fraser-Volpe Corp.	USA	*576*	75-35950
Hall & Watts (Defence Optics) Group	**UK**	***456***	**75-27650**
Helio Mirror Co. Ltd	UK	*457*	75-28900
Hollandse Signaalapparaten BV	Netherlands	*293*	75-03300
Israel Aircraft Industries Ltd	Israel	*219*	75-10200
Litton Industries Inc.	USA	*590*	75-55400
Oerlikon-Contraves SpA	Italy	*237*	75-16800
Radamec Defence Systems Ltd	UK	*477*	75-47750
Recon/Optical Inc.	USA	*608*	75-73930
Saab Dynamics AB	Sweden	*378*	75-09100
SABCA	Belgium	*46*	75-05590
Siemens Aktiengesellschaft	Germany	*175*	75-27700
Société d'Applications des Machines	France	*143*	75-34300
Thomson-CSF	France	*148*	75-45100

415 40 20 20 Artillery fire control systems

AAI Corp.	USA	*545*	75-00020
Arab International Optronics Co.	Egypt	*109*	74-00500
Astronautics Corp. of America	USA	*555*	75-10000
Avibras Aeroespacial SA	Brazil	*55*	75-00968
Bofors AB	Sweden	*375*	74-00020
CIC International Ltd	USA	*563*	75-18570
Computing Devices Canada	Canada	*70*	75-03600
Delft Sensor Systems	**Belgium**	***46***	**75-02500**
Delft Sensor Systems	Netherlands	*292*	75-01590
Econ Optics Mechanics SA	Greece	*188*	75-01070
EUROCONTROL SpA	Italy	*233*	75-08200
Giat Industries	France	*126*	55-00100
Hall & Watts (Defence Optics) Group	**UK**	***456***	**75-27650**
Helio Mirror Co. Ltd	UK	*457*	75-28900
Hughes Elcan Optical Technologies	Canada	*72*	75-07650
Instro Precision Ltd	UK	*461*	75-31630
Israel Aircraft Industries Ltd	Israel	*219*	75-10200
Kearfott Guidance & Navigation Corp.	USA	*588*	75-51760
Litton Industries Inc.	USA	*590*	75-55400
M. Paz Logistics Ltd	Israel	*224*	80-01600
Norwegian Defence Technology A/S	Norway	*310*	75-02050
Oerlikon-Contraves AG.	Switzerland	*387*	75-09250
Oerlikon-Contraves SpA	Italy	*237*	75-16800
Omi Corp.	USA	*601*	75-66550
Reunert Technology Systems Ltd	S. Africa	*359*	75-18300
Société d'Applications des Machines	France	*143*	75-34300
Technology Service Corp.	USA	*644*	80-31930
Thomson-CSF	France	*148*	75-45100
Vickers Shipbuilding & Eng Ltd	UK	*491*	75-63100
Warren-Knight Instrument Co.	USA	*627*	75-95150

415 40 20 50 Mortar fire control systems

AAI Corp.	USA	*545*	75-00020
CIC International Ltd	USA	*563*	75-18570
Hall & Watts (Defence Optics) Group	**UK**	***456***	**75-27650**
Hughes Elcan Optical Technologies	Canada	*72*	75-07650
Instro Precision Ltd	UK	*461*	75-31630
Israel Aircraft Industries Ltd	Israel	*219*	75-10200
M. Paz Logistics Ltd	Israel	*224*	80-01600
Technology Service Corp.	USA	*644*	80-31930

415 40 20 80 Tank fire control systems (see code 241.10.22.30)

AAI Corp.	USA	*545*	75-00020
Arab International Optronics Co.	Egypt	*109*	74-00500
Astronautics C.A. Ltd	Israel	*217*	75-02400
Astronautics Corp. of America	USA	*555*	75-10000
Avimo Ltd	UK	*432*	75-04950
Bharat Electronics Ltd	India	*203*	55-00200
CIC International Ltd	USA	*563*	75-18570
Computing Devices Canada	Canada	*70*	75-03600
Delft Sensor Systems	**Belgium**	***46***	**75-02500**
Delft Sensor Systems	Netherlands	*292*	75-01590
EL-OP Electro Optics Industries Ltd	Israel	*219*	75-06300
Elbit Systems Ltd	Israel	*218*	75-05300
Eloptro	S. Africa	*358*	75-05680
EUROCONTROL SpA	Italy	*233*	75-08200
Hughes Elcan Optical Technologies	Canada	*72*	75-07650
Israel Aircraft Industries Ltd	Israel	*219*	75-10200
Kinetics Ltd	Israel	*220*	75-11320
Kollsman	USA	*589*	75-53150
Litton Industries Inc.	USA	*590*	75-55400
M. Paz Logistics Ltd	Israel	*224*	80-01600
Omi Corp.	USA	*601*	75-66550
Pilkington Optronics Ltd	UK	*426*	74-10000
Przemyslowy Instytut Telekomunikacji	**Poland**	***325***	**75-02650**
Reunert Technology Systems Ltd	S. Africa	*359*	75-18300
SABCA	Belgium	*46*	75-05590
SAGEM SA	**France**	***143***	**75-33950**
SFIM	**France**	***144***	**75-35080**
Société d'Applications des Machines	France	*143*	75-34300
STN ATLAS Elektronik GmbH	Germany	*165*	75-01150
United Defense L.P.	USA	*624*	75-92050
Zeiss-Eltro Optronic GmbH	Germany	*177*	75-34700

415 40 30 Missile fire control systems, ground-based

AAI Corp.	USA	*545*	75-00020
Alenia SpA	Italy	*232*	75-01170
Avibras Aeroespacial SA	Brazil	*55*	75-00968
Bofors AB	Sweden	*375*	74-00020
British Aerospace (Systems &	UK	*436*	75-08890
Cranfield University	UK	*419*	60-00350
Fraser-Volpe Corp.	USA	*576*	75-35950
Hollandse Signaalapparaten BV	Netherlands	*293*	75-03300
Litton Industries Inc.	USA	*590*	75-55400
M. Paz Logistics Ltd	Israel	*224*	80-01600
Matra Défense-Espace	France	*139*	75-26000
Omi Corp.	USA	*601*	75-66550

Pilkington Thorn Optronics	UK	*426*	74-10000
Radio Research Instrument Co. Inc.	USA	*607*	75-72600
Shorts Missile Systems Ltd	**UK**	***483***	**75-53550**
Société d'Applications des Machines	France	*143*	75-34300
Technology Service Corp.	USA	*644*	80-31930
Thomson-CSF	France	*148*	75-45100
United Defense L.P.	USA	*624*	75-92050

415 40 30 10 Missile fire control systems, ground-based anti-aircraft

Alenia SpA	Italy	*232*	75-01170
Avibras Aeroespacial SA	Brazil	*55*	75-00968
Bofors AB	Sweden	*375*	74-00020
British Aerospace (Systems &	UK	*436*	75-08890
Cranfield University	UK	*419*	60-00350
Hollandse Signaalapparaten BV	Netherlands	*293*	75-03300
M. Paz Logistics Ltd	Israel	*224*	80-01600
Matra Défense-Espace	France	*139*	75-26000
Norwegian Defence Technology A/S	Norway	*310*	75-02050
Oerlikon-Contraves SpA	Italy	*237*	75-16800
Pilkington Thorn Optronics	UK	*426*	74-10000
Radio Research Instrument Co. Inc.	USA	*607*	75-72600
Recon/Optical Inc.	USA	*608*	75-73930
Shorts Missile Systems Ltd	**UK**	***483***	**75-53550**
Société d'Applications des Machines	France	*143*	75-34300
Technology Service Corp.	USA	*644*	80-31930
Thomson-CSF	France	*148*	75-45100
United Defense L.P.	USA	*624*	75-92050

415 40 30 20 Missile fire control systems, ground-based, anti-ICBM

Pilkington Thorn Optronics	UK	*426*	74-10000
Technology Service Corp.	USA	*644*	80-31930
United Defense L.P.	USA	*624*	75-92050

415 40 30 30 Missile fire control systems, ground-based, coastal defense

Hollandse Signaalapparaten BV	Netherlands	*293*	75-03300
Israel Aircraft Industries Ltd	Israel	*219*	75-10200
Matra Défense-Espace	France	*139*	75-26000
Pilkington Thorn Optronics	UK	*426*	74-10000
Shorts Missile Systems Ltd	**UK**	***483***	**75-53550**
Technology Service Corp.	USA	*644*	80-31930
United Defense L.P.	USA	*624*	75-92050

415 40 30 40 Missile fire control systems, anti-tank, ground-based

Bofors AB	Sweden	*375*	74-00020
Fraser-Volpe Corp.	USA	*576*	75-35950
Instro Precision Ltd	UK	*461*	75-31630
M. Paz Logistics Ltd	Israel	*224*	80-01600
Omi Corp.	USA	*601*	75-66550
Recon/Optical Inc.	USA	*608*	75-73930
Technology Service Corp.	USA	*644*	80-31930
White Products BV	Netherlands	*296*	75-09100
Zeiss-Eltro Optronic GmbH	Germany	*177*	75-34700

415 40 30 50 Missile fire control systems, tactical, ground-based

Alenia SpA	Italy	*232*	75-01170
Cranfield University	UK	*419*	60-00350
Pilkington Thorn Optronics	UK	*426*	74-10000
Société d'Applications des Machines	France	*143*	75-34300
Technology Service Corp.	USA	*644*	80-31930

415 40 30 60 Fire control systems, strategic missile, ground-based

Octec Ltd	UK	*471*	75-42950
Pilkington Optronics Ltd	UK	*426*	74-10000
Pilkington Thorn Optronics	UK	*426*	74-10000
Technology Service Corp.	USA	*644*	80-31930

415 40 70 Fire control equipment, ground-based weapon

AAI Corp.	USA	*545*	75-00020
British Aerospace (Systems &	UK	*436*	75-08890
Chubb Alarms Ltd	UK	*440*	75-13150
Computing Devices Canada	Canada	*70*	75-03600
Davies Industrial Communications Ltd	UK	*444*	75-17010
DSD Ltd	UK	*501*	80-08700
EUROCONTROL SpA	Italy	*233*	75-08200
Fraser-Volpe Corp.	USA	*576*	75-35950
Instro Precision Ltd	UK	*461*	75-31630
J-Tec Associates Inc.	USA	*587*	75-51250
Kinetics Ltd	Israel	*220*	75-11320
M. Paz Logistics Ltd	Israel	*224*	80-01600
Octec Ltd	UK	*471*	75-42950
Pilkington Optronics Ltd	UK	*426*	74-10000
Pilkington Thorn Optronics	UK	*426*	74-10000
SELIN Sistemi Sarl	Italy	*238*	75-21750
Shorts Missile Systems Ltd	**UK**	***483***	**75-53550**
Société d'Applications des Machines	France	*143*	75-34300
Technology Service Corp.	USA	*644*	80-31930

415 40 70 10 Fire control equipment, artillery, ground-based

AAI Corp.	USA	*545*	75-00020
Bressel SA	Spain	*366*	75-01670
Computing Devices Canada	Canada	*70*	75-03600
Delft Sensor Systems	**Belgium**	***46***	**75-02500**
Delft Sensor Systems	Netherlands	*292*	75-01590
EUROCONTROL SpA	Italy	*233*	75-08200
Giat Industries	France	*126*	55-00100
Instro Precision Ltd	UK	*461*	75-31630
M. Paz Logistics Ltd	Israel	*224*	80-01600
Pilkington Optronics Ltd	UK	*426*	74-10000
SAGEM SA	**France**	***143***	**75-33950**
SELIN Sistemi Sarl	Italy	*238*	75-21750
Société d'Applications des Machines	France	*143*	75-34300
Technology Service Corp.	USA	*644*	80-31930

415 40 70 20 Fire control equipment, missile, ground-based

AAI Corp.	USA	*545*	75-00020
Fraser-Volpe Corp.	USA	*576*	75-35950
Instro Precision Ltd	UK	*461*	75-31630
Litton Industries Inc.	USA	*590*	75-55400
M. Paz Logistics Ltd	Israel	*224*	80-01600
Pilkington Optronics Ltd	UK	*426*	74-10000
Pilkington Thorn Optronics	UK	*426*	74-10000
SELIN Sistemi Sarl	Italy	*238*	75-21750
Shorts Missile Systems Ltd	**UK**	***483***	**75-53550**
Société d'Applications des Machines	France	*143*	75-34300
Technology Service Corp.	USA	*644*	80-31930

Navigation, Direction Finding, Position Fixing

421 Navigation systems & equipment

AAR Corp	USA	*629*	80-00035
Advanced Technical Corp.	USA	*546*	75-01000
Andrew Canada Inc.	Canada	*68*	75-00600
Anite Systems Ltd	UK	*497*	80-01240
ARC Industries Inc.	USA	*554*	75-08380
Astronautics Corp. of America	USA	*555*	75-10000
Ausrire Scientific & Research	Russia	*335*	60-00090
Balmoral Group Ltd	UK	*433*	75-05420
BFGoodrich Aerospace	USA	*558*	75-13770
Bharat Electronics Ltd	India	*203*	55-00200
Bodenseewerk Gerätetechnik GmbH	Germany	*165*	75-03000
Boeing Australia Limited	Australia	*12*	75-02500
British Aerospace (Systems &	UK	*436*	75-08890
Canadian Marconi Co.	Canada	*69*	75-02600
Computing Devices Company Ltd	UK	*442*	75-14700
Consortium Aviaspectrans JSC	Russia	*336*	74-01200
Cubic Defense Systems Inc.	USA	*567*	75-22900
Elan GmbH.	Germany	*168*	75-07550
Elbit Systems Ltd	Israel	*218*	75-05300
Explorocean Technology Ltd	UK	*502*	80-09870

Fabbrica Italiana Apparecchiature	Italy	*234*	75-08550
GEC-Marconi Avionics Ltd	UK	*454*	75-25320
Hindustan Aeronautics Ltd	India	*204*	55-01000
Hispano Radio Maritima SA	Spain	*367*	80-00840
Instro Precision Ltd	UK	*461*	75-31630
Israel Aircraft Industries Ltd	Israel	*220*	75-10200
ITT Aerospace/Communications Div.	USA	*586*	75-49455
ITT Avionics	USA	*586*	75-49455
Japan Aviation Electronics Ind. Ltd	Japan	*252*	75-03300
Jeppesen & Co GmbH	Germany	*170*	75-15050
Kearfott Guidance & Navigation Corp.	USA	*588*	75-51760
LITEF GmbH	**Germany**	***171***	**75-18600**
Litton Industries Inc.	USA	*590*	75-55400
Litton Systems Canada Ltd	Canada	*73*	75-09100
Lockheed Martin Corp.	USA	*591*	75-55635
M. Paz Logistics Ltd	Israel	*224*	80-01600
Magnavox Electronic Systems Company	USA	*593*	75-57120
Mils Elektronik GmbH & Co KG	**Austria**	***33***	**75-01700**
Motorola Ltd	UK	*469*	75-41050
Nautronix Ltd	Australia	*19*	75-14300
Navigation Data Systems Inc.	USA	*598*	75-64050
Navigation Science Corp.	USA	*598*	75-64070
Navstar Systems Ltd	UK	*470*	75-41955
Northrop Grumman Corporation	USA	*599*	75-65450
NPO Geofizika	Russia	*338*	75-14315
Pelorus Navigation Systems Inc.	Canada	*75*	75-11035
Polaris International Ltd	UK	*508*	80-23700
Racal Avionics Ltd	UK	*477*	75-47660
Racal Survey Ltd-Products	UK	*509*	80-26320
Raytheon Aircraft Company	USA	*607*	75-73200
Raytheon E-Systems	USA	*608*	75-73302
Revue THOMMEN AG	Switzerland	*387*	75-11100
Rockwell	USA	*610*	75-75930
SAGEM SA	**France**	***143***	**75-33950**
Seatex A/S	Norway	*310*	75-02410
SEXTANT Avionique SA	France	*144*	75-35000
SFIM	**France**	***144***	**75-35080**
S.G. Brown	UK	*437*	75-10330
Smiths Aerospace & Defense	USA	*615*	75-82090
Sperry Marine Inc.	USA	*617*	75-83250
STAREC	France	*147*	75-42775
STN ATLAS Elektronik GmbH	Germany	*165*	75-01150
Stocker & Yale Inc.	USA	*618*	75-84250
Technology Service Corp.	USA	*644*	80-31930
Teldix GmbH	Germany	*175*	75-30100
Teledyne Brown Engineering	USA	*620*	75-87526
Toyo Communication Equipment Co	Japan	*255*	75-10000
Trimble Navigation Europe Ltd	UK	*489*	75-61140
Universal Avionics Systems Corp	USA	*625*	75-92650
Vickers Shipbuilding & Eng Ltd	UK	*491*	75-63100
Warren-Knight Instrument Co.	USA	*627*	75-95150
Znamya Production Group	Ukraine	*407*	75-10000

421 10 Navigation systems, spaceborne

Astronautics Corp. of America	USA	*555*	75-10000
Cubic Defense Systems Inc.	USA	*567*	75-22900
GEC-Marconi Avionics Ltd	UK	*454*	75-25320
Kearfott Guidance & Navigation Corp.	USA	*588*	75-51760
Labtech Ltd	UK	*464*	75-34280
Litton Industries Inc.	USA	*590*	75-55400
Navigation Data Systems Inc.	USA	*598*	75-64050
Northrop Grumman Corporation	USA	*599*	75-65450
NPO Geofizika	Russia	*338*	75-14315
Rockwell	USA	*610*	75-75930
SAGEM SA	**France**	***143***	**75-33950**
Signal Computing Ltd	UK	*511*	80-36050
Smiths Aerospace & Defense	USA	*615*	75-82090

421 10 10 Astro-navigation systems, spaceborne

Astronautics Corp. of America	USA	*555*	75-10000
Kearfott Guidance & Navigation Corp.	USA	*588*	75-51760
Litton Industries Inc.	USA	*590*	75-55400
Northrop Grumman Corporation	USA	*599*	75-65450
NPO Geofizika	Russia	*338*	75-14315
Rutherford Appleton Laboratory	UK	*420*	60-01850

421 10 70 Navigation equipment, spaceborne

Astronautics Corp. of America	USA	*555*	75-10000
Cranfield University	UK	*419*	60-00350
GEC-Marconi Avionics Ltd	UK	*454*	75-25320
Israel Aircraft Industries Ltd	Israel	*220*	75-10200
Kearfott Guidance & Navigation Corp.	USA	*588*	75-51760
Labtech Ltd	UK	*464*	75-34280
Matra Marconi Space UK, Ltd	UK	*425*	74-06900
Motorola Ltd	UK	*469*	75-41050
NPO Geofizika	Russia	*338*	75-14315
SAGEM SA	**France**	***143***	**75-33950**
Smiths Aerospace & Defense	USA	*615*	75-82090
Société Anonyme d'Études	France	*145*	75-37100

421 20 Navigation systems, airborne

AAR Corp	USA	*629*	80-00035
Advanced Technical Corp.	USA	*546*	75-01000
Alden Sales Co.	USA	*548*	75-04410
Anite Systems Ltd	UK	*497*	80-01240
Astronautics Corp. of America	USA	*555*	75-10000
Aviapribor Corp.	Russia	*336*	75-01200
Becker Flugfunkwerk GmbH	Germany	*165*	75-02200
BFGoodrich Aerospace	USA	*558*	75-13770
Boeing Australia Limited	Australia	*12*	75-02500
British Aerospace (Systems &	UK	*436*	75-08890
Canadian Marconi Co.	Canada	*69*	75-02600
Computing Devices Company Ltd	UK	*442*	75-14700
Cubic Defense Systems Inc.	USA	*567*	75-22900
Dassault Electronique SA (Group)	**France**	***135***	**75-12980**
Diehl GmbH & Co	Germany	*167*	75-06500
Elan GmbH	Germany	*168*	75-07550
Eventide Inc.	USA	*574*	75-32655
GEC-Marconi Avionics Inc.	USA	*577*	75-37413
GEC-Marconi Avionics Ltd	UK	*454*	75-25320
Hindustan Aeronautics Ltd	India	*204*	55-01000
H.R. Smith (Microwaves) Ltd	UK	*484*	75-54590
Israel Aircraft Industries Ltd	Israel	*220*	75-10200
ITT Avionics	USA	*586*	75-49455
Japan Aviation Electronics Ind. Ltd	Japan	*252*	75-03300
Kearfott Guidance & Navigation Corp.	USA	*588*	75-51760
Kentron	S. Africa	*358*	75-09000
LITEF GmbH	**Germany**	***171***	**75-18600**
Litton Industries Inc.	USA	*590*	75-55400
Litton Systems Canada Ltd	Canada	*73*	75-09100
Megapulse Inc.	USA	*595*	75-59830
Microelectronics, guidance and	Turkey	*402*	55-00050
Motorola Ltd	UK	*469*	75-41050
Northrop Grumman Corporation	USA	*599*	75-65450
Novatronics Inc	Canada	*74*	75-10300
Pelorus Navigation Systems Inc.	Canada	*75*	75-11035
Racal Avionics Ltd	UK	*477*	75-47660
Racal Radar Defence Systems Ltd	UK	*477*	75-47688
Raytheon E-Systems	USA	*608*	75-73302
Revue THOMMEN AG	Switzerland	*387*	75-11100
Rockwell	USA	*610*	75-75930
Rockwell International Co.	Italy	*233*	75-05200
SAGEM SA	**France**	***143***	**75-33950**
SEXTANT Avionique SA	France	*144*	75-35000
SFIM	**France**	***144***	**75-35080**
Sierra Technologies Inc.	USA	*614*	75-80640
Smiths Aerospace & Defense	USA	*615*	75-82090
Smiths Industries Aerospace	UK	*485*	75-54700
Teledyne Brown Engineering	USA	*620*	75-87526
Trimble Navigation Europe Ltd	UK	*489*	75-61140
Universal Avionics Systems Corp	USA	*625*	75-92650

421 20 10 Astro-navigation systems, airborne

Astronautics Corp. of America	USA	*555*	75-10000
Kearfott Guidance & Navigation Corp.	USA	*588*	75-51760
Litton Industries Inc.	USA	*590*	75-55400

421 20 10 70 Astro-navigation equipment, airborne

Kearfott Guidance & Navigation Corp.	USA	*588*	75-51760
Litton Industries Inc.	USA	*590*	75-55400

421 20 20 Inertial navigation systems, airborne

Aeroelectrónica Industria	Brazil	*55*	75-00585
Alden Sales Co.	USA	*548*	75-04410
Astronautics Corp. of America	USA	*555*	75-10000

Aviapribor Corp.	Russia	*336*	75-01200
British Aerospace (Systems &	UK	*436*	75-08890
Cranfield Aerospace Ltd	UK	*500*	80-07200
GEC-Marconi Avionics Ltd	UK	*454*	75-25320
Israel Aircraft Industries Ltd	Israel	*220*	75-10200
Japan Aviation Electronics Ind. Ltd	Japan	*252*	75-03300
Jeppesen UK Ltd	UK	*505*	80-13850
Kearfott Guidance & Navigation Corp.	USA	*588*	75-51760
LITEF GmbH	**Germany**	***171***	**75-18600**
Litton Industries Inc.	USA	*590*	75-55400
Litton Italia SpA	Italy	*236*	75-13700
Litton Systems Canada Ltd	Canada	*73*	75-09100
Microelectronics, guidance and	Turkey	*402*	55-00050
Northrop Grumman Corporation	USA	*599*	75-65450
Norwegian Defence Technology A/S	Norway	*310*	75-02050
SAGEM SA	**France**	***143***	**75-33950**
SEXTANT Avionique SA	France	*144*	75-35000
SFIM	**France**	***144***	**75-35080**
Smiths Aerospace & Defense	USA	*615*	75-82090

421 20 20 70 Inertial navigation equipment, airborne

Alden Sales Co.	USA	*548*	75-04410
Astronautics Corp. of America	USA	*555*	75-10000
British Aerospace (Systems &	UK	*436*	75-08890
Cranfield Aerospace Ltd	UK	*500*	80-07200
GEC-Marconi Avionics Ltd	UK	*454*	75-25320
Israel Aircraft Industries Ltd	Israel	*220*	75-10200
Japan Aviation Electronics Ind. Ltd	Japan	*252*	75-03300
Kearfott Guidance & Navigation Corp.	USA	*588*	75-51760
LITEF GmbH	**Germany**	***171***	**75-18600**
Litton Industries Inc.	USA	*590*	75-55400
Litton Systems Canada Ltd	Canada	*73*	75-09100
SAGEM SA	**France**	***143***	**75-33950**
SEXTANT Avionique SA	France	*144*	75-35000
SFIM	**France**	***144***	**75-35080**
Smiths Aerospace & Defense	USA	*615*	75-82090

421 20 20 90 Inertial navigation system components, airborne

Aerospace Metal Composites Limited	UK	*428*	75-01180
Airscrew Howden Ltd	UK	*429*	75-02050
Astronautics Corp. of America	USA	*555*	75-10000
Avionic Components Ltd.	UK	*432*	75-04970
Bodenseewerk Gerätetechnik GmbH	Germany	*165*	75-03000
British Aerospace (Systems &	UK	*436*	75-08890
Cranfield Aerospace Ltd	UK	*500*	80-07200
Frequency Electronics Inc.	USA	*576*	75-36190
GEC-Marconi Avionics Ltd	UK	*454*	75-25320
GEC-Marconi Research Centre	UK	*419*	60-00700
Israel Aircraft Industries Ltd	Israel	*219*	75-10200
Iver Precision Engineers Ltd	UK	*462*	75-32230
Japan Aviation Electronics Ind. Ltd	Japan	*252*	75-03300
Jeppesen UK Ltd	UK	*505*	80-13850
Kearfott Guidance & Navigation Corp.	USA	*588*	75-51760
KEC Limited	UK	*463*	75-33700
LITEF GmbH	**Germany**	***171***	**75-18600**
Litton Industries Inc.	USA	*590*	75-55400
Litton Italia SpA	Italy	*236*	75-13700
Northrop Grumman Corporation	USA	*599*	75-65450
Penny & Giles Drives Technology Ltd	UK	*474*	75-44563
Rotech Components Ltd	UK	*481*	75-50575
SAGEM SA	**France**	***143***	**75-33950**
SEXTANT Avionique SA	France	*144*	75-35000
SFIM	**France**	***144***	**75-35080**
Smiths Aerospace & Defense	USA	*615*	75-82090
The Barden Corporation (UK) Ltd	UK	*433*	75-05630

421 20 23 Heading & vertical reference systems, airborne

Alden Sales Co.	USA	*548*	75-04410
Astronautics Corp. of America	USA	*555*	75-10000
BFGoodrich Avionics Systems Inc	USA	*559*	75-13771
Bodenseewerk Gerätetechnik GmbH	Germany	*165*	75-03000
China National Aero-Tech I/E Corp.	China PR	*88*	55-00200
GEC-Marconi Avionics Ltd	UK	*454*	75-25320
Israel Aircraft Industries Ltd	Israel	*220*	75-10200
Japan Aviation Electronics Ind. Ltd	Japan	*252*	75-03300
Kearfott Guidance & Navigation Corp.	USA	*588*	75-51760
LITEF GmbH	**Germany**	***171***	**75-18600**
Litton Industries Inc.	USA	*590*	75-55400
Northrop Grumman Corporation	USA	*599*	75-65450
Rockwell International Co.	Italy	*233*	75-05200
SAGEM SA	**France**	***143***	**75-33950**
SFIM	**France**	***144***	**75-35080**
Smiths Aerospace & Defense	USA	*615*	75-82090
Tokyo Aircraft Instrument Co Ltd	Japan	*254*	75-08600
Trimble Navigation Europe Ltd	UK	*489*	75-61140

421 20 24 Heading repeater equipment, aircraft

Israel Aircraft Industries Ltd	Israel	*220*	75-10200
Kearfott Guidance & Navigation Corp.	USA	*588*	75-51760
Litton Industries Inc.	USA	*590*	75-55400
SEXTANT Avionique SA	France	*144*	75-35000
SFIM	**France**	***144***	**75-35080**
Tokyo Aircraft Instrument Co Ltd	Japan	*254*	75-08600

421 20 25 Night navigation & targeting systems, airborne, infra-red

GEC-Marconi Avionics Ltd	UK	*454*	75-25320
Interface Products Inc.	USA	*585*	75-47300
Kaiser Electronics	USA	*587*	75-51350
Litton Industries Inc.	USA	*590*	75-55400
SAGEM SA	**France**	***143***	**75-33950**
Zeiss-Eltro Optronic GmbH	Germany	*177*	75-34700

421 20 30 Radar navigation systems, airborne

GEC-Marconi Research Centre	UK	*419*	60-00700
Radio Research Instrument Co. Inc.	USA	*607*	75-72600
Znamya Production Group	Ukraine	*407*	75-10000

421 20 30 70 Navigation radars, airborne

Andrew Canada Inc.	Canada	*68*	75-00600
Daimler-Benz Aerospace France SAS	**France**	***134***	**75-12950**
Israel Aircraft Industries Ltd	Israel	*219*	75-10200
Znamya Production Group	Ukraine	*407*	75-10000

421 20 40 Radio navigation systems, airborne

Alden Sales Co.	USA	*548*	75-04410
Becker Flugfunkwerk GmbH	Germany	*165*	75-02200
Cubic Defense Systems Inc.	USA	*567*	75-22900
GEC-Marconi Avionics Ltd	UK	*454*	75-25320
Litton Industries Inc.	USA	*591*	75-55400
Marconi SpA DTM	Italy	*236*	75-14550
Rockwell International Co.	Italy	*233*	75-05200
SEXTANT Avionique SA	France	*144*	75-35000
Sierra Technologies Inc.	USA	*614*	75-80640

421 20 40 10 Decca navigation systems, airborne

Racal Avionics Ltd	UK	*477*	75-47660
Racal Communications Inc.	USA	*606*	75-72380
Racal Radar Defence Systems Ltd.	UK	*477*	75-47688

421 20 40 11 Decca-Doppler navigation systems, airborne

Racal Avionics Ltd	UK	*477*	75-47660
Racal Communications Inc.	USA	*606*	75-72380
Racal Radar Defence Systems Ltd.	UK	*477*	75-47688

421 20 40 15 Distance measuring equipment (DME)

Aerocom/Squitter	USA	*546*	75-01150
Alcatel Air Navigation Systems GmbH	Germany	*164*	75-00535
Marconi SpA DTM	Italy	*236*	75-14550
Raytheon E-Systems	USA	*608*	75-73302
Rockwell International Co.	Italy	*233*	75-05200
Thorn DNT Airfield Lighting Pty Ltd	Australia	*22*	75-20540

421 20 40 20 Doppler navigation systems, airborne

Aviapribor Corp	Russia	*336*	75-01200
Canadian Marconi Co.	Canada	*69*	75-02600
GEC-Marconi Avionics Ltd	UK	*454*	75-25320
Marconi SpA DTM	Italy	*236*	75-14550
Racal Avionics Ltd	UK	*477*	75-47660
Racal Radar Defence Systems Ltd	UK	*477*	75-47688

421 20 40 25 LORAN systems, airborne

Navigation Data Systems Inc.	USA	*598*	75-64050
Racal Avionics Ltd	UK	*477*	75-47660
Racal Radar Defence Systems Ltd	UK	*477*	75-47688
Universal Avionics Systems Corp	USA	*625*	75-92650
Vega-Cantley Instrument Company	UK	*490*	75-62740

421 20 40 30 Omega / VLF navigation systems, airborne

Canadian Marconi Co.	Canada	*69*	75-02600
Litton Industries Inc.	USA	*590*	75-55400
Racal Avionics Ltd	UK	*477*	75-47660
SEXTANT Avionique SA	France	*144*	75-35000
Universal Avionics Systems Corp	USA	*625*	75-92650

421 20 40 40 TACAN systems, airborne

Advanced Technical Corp.	USA	*546*	75-01000
BFGoodrich Aerospace	USA	*559*	75-13770
GEC-Marconi Avionics Ltd	UK	*454*	75-25320
ITT Aerospace/Communications Div.	USA	*586*	75-49455
NEC Corporation	Japan	*253*	75-06350
Racal Avionics Ltd	UK	*477*	75-47660
Rockwell	USA	*610*	75-75930
Rockwell International Co.	Italy	*233*	75-05200
Sierra Technologies Inc.	USA	*614*	75-80640

421 20 40 70 Navigation equipment, airborne radio

Advanced Technical Corp.	USA	*546*	75-01000
Becker Flugfunkwerk GmbH	Germany	*165*	75-02200
BFGoodrich Aerospace	USA	*559*	75-13770
Cubic Defense Systems Inc.	USA	*567*	75-22900
GEC-Marconi Avionics Ltd	UK	*454*	75-25320
Marconi SpA DTM	Italy	*236*	75-14550
Racal Avionics Ltd	UK	*477*	75-47660
Rockwell International Co.	Italy	*233*	75-05200
Rockwell-Collins France SA	France	*142*	75-32800
Universal Avionics Systems Corp	USA	*625*	75-92650

421 20 40 80 Radio navigation equipment accessories, airborne

Advanced Technical Corp.	USA	*546*	75-01000
Becker Flugfunkwerk GmbH	Germany	*165*	75-02200
H.R. Smith (Microwaves) Ltd	UK	*484*	75-54590
Teledyne Brown Engineering	USA	*620*	75-87526

421 20 40 90 Radio navigation equipment components, airborne

Advanced Technical Corp.	USA	*546*	75-01000
Frequency Electronics Inc.	USA	*576*	75-36190
STAREC	France	*147*	75-42775
Teledyne Brown Engineering	USA	*620*	75-87526

421 20 50 Satellite navigation systems, airborne

Astronautics Corp. of America	USA	*555*	75-10000
Canadian Marconi Co.	Canada	*69*	75-02600
Chelton (Electrostatics) Limited	UK	*439*	75-12700
Cranfield Aerospace Ltd	UK	*500*	80-07200
Fabbrica Italiana Apparecchiature	Italy	*234*	75-08550
Israel Aircraft Industries Ltd	Israel	*220*	75-10200
Motorola Ltd	UK	*469*	75-41050
Navstar Systems Ltd	UK	*470*	75-41955
Racal Avionics Ltd	UK	*477*	75-47660
Racal Radar Defence Systems Ltd	UK	*477*	75-47688
Racal-MESL Ltd	UK	*477*	75-47682
Raytheon Cossor Electronics Ltd	UK	*478*	75-48367
Rockwell International Co.	Italy	*233*	75-05200
SAGEM SA	**France**	***143***	**75-33950**
SEXTANT Avionique SA	France	*144*	75-35000
Steatite Insulations Ltd	UK	*486*	75-56450
Trimble Navigation Europe Ltd	UK	*489*	75-61140
Universal Avionics Systems Corp	USA	*625*	75-92650

421 20 60 Terrain comparison navigation systems, airborne

Fabbrica Italiana Apparecchiature	Italy	*234*	75-08550
GEC-Marconi Avionics Ltd	UK	*454*	75-25320
SAGEM SA	**France**	***143***	**75-33950**

421 20 70 Navigation equipment, airborne, other

Advanced Technical Corp.	USA	*546*	75-01000
Astronautics Corp. of America	USA	*555*	75-10000
Boeing Australia Limited	Australia	*12*	75-02500
Brunton Co.	USA	*561*	75-15460
Canadian Marconi Co.	Canada	*69*	75-02600
GEC-Marconi Avionics Inc.	USA	*577*	75-37413
GEC-Marconi Avionics Ltd	UK	*454*	75-25320
Interface Products Inc.	USA	*585*	75-47300
Israel Aircraft Industries Ltd	Israel	*219*	75-10200
Japan Aviation Electronics Ind. Ltd	Japan	*252*	75-03300
Kearfott Guidance & Navigation Corp.	USA	*588*	75-51760
Litton Italia SpA	Italy	*236*	75-13700
MTC Industries	Israel	*221*	75-15450
SAGEM SA	**France**	***143***	**75-33950**
SFIM	**France**	***144***	**75-35080**
Smiths Aerospace & Defense	USA	*615*	75-82090
Universal Avionics Systems Corp	USA	*625*	75-92650

421 20 70 10 Compasses, airborne

ARC Industries Inc.	USA	*554*	75-08380
Brunton Co.	USA	*561*	75-15460
Kearfott Guidance & Navigation Corp.	USA	*588*	75-51760
KVH Industries Inc.	USA	*589*	75-53670
Rockwell International Co.	Italy	*233*	75-05200
SEXTANT Avionique SA	France	*144*	75-35000
SFIM	**France**	***144***	**75-35080**
SIRS Navigation Ltd.	UK	*484*	75-54300
Tokyo Aircraft Instrument Co Ltd	Japan	*254*	75-08600

421 20 70 11 Compasses, gyro, airborne

Alden Sales Co.	USA	*548*	75-04410
Israel Aircraft Industries Ltd	Israel	*220*	75-10200
Kearfott Guidance & Navigation Corp.	USA	*588*	75-51760
Rockwell International Co.	Italy	*233*	75-05200
SFIM	**France**	***144***	**75-35080**
Tokyo Aircraft Instrument Co Ltd	Japan	*254*	75-08600

421 20 70 12 Compasses, magnetic, airborne

Aeronautical & General Instr. Ltd	UK	*428*	75-01050
Alden Sales Co.	USA	*548*	75-04410
ARC Industries Inc.	USA	*554*	75-08380
Brunton Co.	USA	*561*	75-15460
KVH Industries Inc.	USA	*589*	75-53670
Rockwell International Co.	Italy	*233*	75-05200
SEXTANT Avionique SA	France	*144*	75-35000
SFIM	**France**	***144***	**75-35080**
SIRS Navigation Ltd.	UK	*484*	75-54300
Tokyo Aircraft Instrument Co Ltd	Japan	*254*	75-08600

421 20 70 12 60 Compasses, magnetic, standby, airborne

Aeronautical & General Instr. Ltd	UK	*428*	75-01050
Israel Aircraft Industries Ltd	Israel	*220*	75-10200
KVH Industries Inc.	USA	*589*	75-53670
SIRS Navigation Ltd.	UK	*484*	75-54300

421 20 70 13 Compasses, radio, airborne

Alden Sales Co.	USA	*548*	75-04410
GEC-Marconi Avionics Ltd	UK	*454*	75-25320
Rockwell International Co.	Italy	*233*	75-05200

421 20 70 20 Gyroscopes, airborne

AAR Corp	USA	*629*	80-00035
Astronautics Corp. of America	USA	*555*	75-10000
Autoflug GmbH & Co	Germany	*165*	75-01400
Bodenseewerk Gerätetechnik GmbH	Germany	*165*	75-03000
Central Design Bureau	Ukraine	*406*	75-00800
Elettronica Aster SpA	Italy	*233*	75-07000
Humphrey Inc.	USA	*583*	75-45050
Israel Aircraft Industries Ltd	Israel	*219*	75-10200
Japan Aviation Electronics Ind. Ltd	Japan	*252*	75-03300
Kearfott Guidance & Navigation Corp.	USA	*588*	75-51760
Meggitt Avionics (UK)	UK	*467*	75-38340
Microtecnica-Torino SpA	Italy	*237*	75-15500
MTC Industries	Israel	*221*	75-15450
Rockwell International Co.	Italy	*233*	75-05200
SAGEM SA	**France**	***143***	**75-33950**
SEXTANT Avionique SA	France	*144*	75-35000
SFIM	**France**	***144***	**75-35080**
SFIM Industries Deutschland GmbH	Germany	*175*	75-27495
Smiths Aerospace & Defense	USA	*615*	75-82090
Trimble Navigation Europe Ltd	UK	*489*	75-61140

421 30 Navigation systems, shipborne

Advanced Technical Corp.	USA	*546*	75-01000
Anite Systems Ltd	UK	*497*	80-01240
Bharat Electronics Ltd	India	*203*	55-00200
Boeing Australia Limited	Australia	*12*	75-02500
British Aerospace (Systems &	UK	*436*	75-08890
Cubic Defense Systems Inc.	USA	*567*	75-22900
Explorocean Technology Ltd	UK	*502*	80-09870
GEC-Marconi Avionics Inc.	USA	*577*	75-37413
GEC-Marconi Avionics Ltd	UK	*454*	75-25320
Israel Aircraft Industries Ltd	Israel	*219*	75-10200
ITT Aerospace/Communications Div.	USA	*586*	75-49455
ITT Avionics	USA	*586*	75-49455
Kearfott Guidance & Navigation Corp.	USA	*588*	75-51760
Kelvin Hughes Ltd	UK	*463*	75-33453
LITEF GmbH	**Germany**	***171***	**75-18600**
Litton Industries Inc.	USA	*590*	75-55400
Litton Systems Canada Ltd	Canada	*73*	75-09100
M. Paz Logistics Ltd	Israel	*224*	80-01600
Magnavox Electronic Systems Company	USA	*593*	75-57120
Marine Electronics Ltd	UK	*466*	75-37290
Megapulse Inc.	USA	*595*	75-59830
MSI-Defence Systems Ltd	UK	*469*	75-41280
Nautronix Ltd	Australia	*19*	75-14300
Navigation Data Systems Inc.	USA	*598*	75-64050
Racal Survey Ltd-Products	UK	*509*	80-26320
Raytheon Anschütz GmbH	Germany	*173*	75-24650
Rockwell	USA	*610*	75-75930
SAGEM SA	**France**	***143***	**75-33950**
Scorpio Navigation Services	UK	*510*	80-31900
S.G. Brown	UK	*437*	75-10330
STN ATLAS Elektronik GmbH	Germany	*165*	75-01150
Teldix GmbH	Germany	*175*	75-30100
Vickers Shipbuilding & Eng Ltd	UK	*491*	75-63100

421 30 05 70 Acoustic navigation equipment, shipborne

A1Marketing	UK	*426*	75-00025
Explorocean Technology Ltd	UK	*502*	80-09870
Fugro-UDI Ltd	UK	*452*	75-24885
Marine Electronics Ltd	UK	*466*	75-37290
Nautronix Ltd	Australia	*19*	75-14300
Oceano Instruments (UK) Ltd	UK	*508*	80-19050
Simrad Norge A/S	Norway	*310*	75-02630
STN ATLAS Elektronik GmbH	Germany	*165*	75-01150

421 30 05 71 Echo-sounders, shipborne

Brookes & Gatehouse Ltd	UK	*437*	75-10100
Finnyards Electronics Ltd	Finland	*118*	75-00600
Kelvin Hughes Ltd	UK	*463*	75-33453
Litton Industries Inc.	USA	*590*	75-55400
Marine Electronics Ltd	UK	*466*	75-37290
Massa Products Corp.	USA	*594*	75-58970
Navitronic Systems AS	Denmark	*103*	75-14090
Ocean Data Equipment Corp.	USA	*601*	75-65860
Simrad Norge A/S	Norway	*310*	75-02630
STN ATLAS Elektronik GmbH	Germany	*165*	75-01150

421 30 10 70 Astro-navigation equipment, shipborne

Litton Industries Inc.	USA	*590*	75-55400
SAGEM SA	**France**	***143***	**75-33950**

421 30 20 Inertial navigation systems, shipborne

Bradley Electronics Ltd	UK	*435*	75-07860
British Aerospace (Systems &	UK	*436*	75-08890
GEC-Marconi Avionics Ltd	UK	*454*	75-25320
Israel Aircraft Industries Ltd	Israel	*219*	75-10200
Kearfott Guidance & Navigation Corp.	USA	*588*	75-51760
LITEF GmbH	**Germany**	***171***	**75-18600**
Litton Industries Inc.	USA	*590*	75-55400
Litton Systems Canada Ltd	Canada	*73*	75-09100
Raytheon Anschütz GmbH	Germany	*173*	75-24650
SAGEM SA	**France**	***143***	**75-33950**
S.G. Brown	UK	*437*	75-10330
Sperry Marine Inc.	USA	*617*	75-83250
Sperry Marine SpA	Italy	*239*	75-24400

421 30 20 90 Inertial navigation system components, shipborne

Astronautics Corp. of America	USA	*555*	75-10000
Frequency Electronics Inc.	USA	*576*	75-36190
GEC-Marconi Avionics Ltd	UK	*454*	75-25320
Israel Aircraft Industries Ltd	Israel	*219*	75-10200
Kearfott Guidance & Navigation Corp.	USA	*588*	75-51760
LITEF GmbH	**Germany**	***171***	**75-18600**
Litton Industries Inc.	USA	*590*	75-55400
SAGEM SA	**France**	***143***	**75-33950**
S.G. Brown	UK	*437*	75-10330
Sperry Marine Inc.	USA	*617*	75-83250
Sperry Marine SpA	Italy	*239*	75-24400

421 30 23 Heading & vertical reference systems, shipborne

Alden Sales Co.	USA	*548*	75-04410
Astronautics Corp. of America	USA	*555*	75-10000
GEC-Marconi Avionics Ltd	UK	*454*	75-25320
Kearfott Guidance & Navigation Corp.	USA	*588*	75-51760
KVH Industries Inc.	USA	*589*	75-53670
LITEF GmbH	**Germany**	***171***	**75-18600**
Litton Industries Inc.	USA	*590*	75-55400
Measurement Devices Ltd	UK	*467*	75-38275
Raytheon Anschütz GmbH	Germany	*173*	75-24650
SAGEM SA	**France**	***143***	**75-33950**
S.G. Brown	UK	*437*	75-10330
Sperry Marine Inc.	USA	*617*	75-83250
Tokyo Aircraft Instrument Co Ltd	Japan	*254*	75-08600
Trimble Navigation Europe Ltd	UK	*489*	75-61140

421 30 24 Heading repeater equipment, shipborne

Brookes & Gatehouse Ltd	UK	*437*	75-10100
Kearfott Guidance & Navigation Corp.	USA	*588*	75-51760
KVH Industries Inc.	USA	*589*	75-53670
LITEF GmbH	**Germany**	***171***	**75-18600**
Measurement Devices Ltd	UK	*467*	75-38275
Meggitt Avionics (UK)	UK	*467*	75-38340
Raytheon Anschütz GmbH	Germany	*173*	75-24650
SAGEM SA	**France**	***143***	**75-33950**
S.G. Brown	UK	*437*	75-10330
Sperry Marine Inc.	USA	*617*	75-83250
Tokyo Aircraft Instrument Co Ltd	Japan	*254*	75-08600

421 30 30 70 Navigation radars, shipborne

Bharat Electronics Ltd	India	203	55-00200
Canadian Marconi Co.	Canada	69	75-02600
CelsiusTech Electronics AB	Sweden	376	75-03720
Israel Aircraft Industries Ltd	Israel	219	75-10200
ITT Aerospace/Communications Div.	USA	586	75-49455
Kelvin Hughes Ltd.	UK	463	75-33453
Radio Research Instrument Co. Inc.	USA	607	75-72600
Sperry Marine Inc.	USA	617	75-83250
STN ATLAS Elektronik GmbH	Germany	165	75-01150
TERMA Elektronik AS	Denmark	104	75-20300
Znamya Production Group	Ukraine	407	75-10000

421 30 40 Radio navigation systems, shipborne

Cubic Defense Systems Inc.	USA	567	75-22900
Fugro-UDI Ltd	UK	452	75-24885
Racal Survey Ltd-Products	UK	509	80-26320
STN ATLAS Elektronik GmbH	Germany	165	75-01150
Teledyne Brown Engineering	USA	620	75-87526
Znamya Production Group	Ukraine	407	75-10000

421 30 40 70 Radio navigation equipment, shipborne

Cubic Defense Systems Inc.	USA	567	75-22900
Fugro-UDI Ltd	UK	452	75-24885
Racal Survey Ltd-Products	UK	509	80-26320
STAREC	France	147	75-42775
STN ATLAS Elektronik GmbH	Germany	165	75-01150
Teledyne Brown Engineering	USA	620	75-87526
Znamya Production Group	Ukraine	407	75-10000

421 30 50 Satellite navigation systems, shipborne

Canadian Marconi Co.	Canada	69	75-02600
Fugro-UDI Ltd	UK	452	75-24885
Israel Aircraft Industries Ltd	Israel	220	75-10200
ITT Aerospace/Communications Div.	USA	586	75-49455
Litton Industries Inc.	USA	590	75-55400
M. Paz Logistics Ltd	Israel	224	80-01600
Magnavox Electronic Systems Company	USA	593	75-57120
Navigation Data Systems Inc.	USA	598	75-64050
Navstar Systems Ltd	UK	470	75-41955
Raytheon Cossor Electronics Ltd	UK	478	75-48367
SAGEM SA	**France**	**143**	**75-33950**
Steatite Insulations Ltd	UK	486	75-56450
Trimble Navigation Europe Ltd	UK	489	75-61140

421 30 50 70 Satellite navigation equipment, shipborne

Fabbrica Italiana Apparecchiature	Italy	234	75-08550
Fugro-UDI Ltd	UK	452	75-24885
ITT Aerospace/Communications Div.	USA	586	75-49455
Litton Industries Inc.	USA	590	75-55400
M. Paz Logistics Ltd	Israel	224	80-01600
Magnavox Electronic Systems Company	USA	593	75-57120
Navigation Data Systems Inc.	USA	598	75-64050
Navstar Systems Ltd	UK	470	75-41955
Nera Telecommunications	UK	470	75-42318
SAGEM SA	**France**	**143**	**75-33950**
Trimble Navigation Europe Ltd	UK	489	75-61140

421 30 70 Navigation equipment, shipborne, other

Boeing Australia Limited	Australia	12	75-02500
Brunton Co.	USA	561	75-15460
GEC-Marconi Avionics Inc.	USA	577	75-37413
GEC-Marconi Avionics Ltd	UK	454	75-25320
ITT Aerospace/Communications Div.	USA	586	75-49455
Kearfott Guidance & Navigation Corp.	USA	588	75-51760
M. Paz Logistics Ltd	Israel	224	80-01600
Raytheon Anschütz GmbH	Germany	173	75-24650
SAGEM SA	**France**	**143**	**75-33950**
S.G. Brown	UK	437	75-10330
Simrad Norge A/S	Norway	310	75-02630

421 30 70 10 Compasses, shipborne

Brookes & Gatehouse Ltd	UK	437	75-10100
Brunton Co.	USA	561	75-15460
GEC-Marconi Avionics Ltd	UK	454	75-25320
Kearfott Guidance & Navigation Corp.	USA	588	75-51760
KVH Industries Inc.	USA	589	75-53670
Litton Industries Inc.	USA	590	75-55400
M. Paz Logistics Ltd	Israel	224	80-01600
Marine Data Ltd	UK	466	75-37288
Microtecnica-Torino SpA	Italy	237	75-15500
Raytheon Anschütz GmbH	Germany	173	75-24650
SAGEM SA	**France**	**143**	**75-33950**
S.G. Brown	UK	437	75-10330
SIRS Navigation Ltd.	UK	484	75-54300
W.F. Stanley & Co. Ltd	UK	486	75-56350

421 30 70 11 Compasses, gyro, shipborne

Fugro-UDI Ltd	UK	452	75-24885
Israel Aircraft Industries Ltd	Israel	220	75-10200
Kearfott Guidance & Navigation Corp.	USA	588	75-51760
LITEF GmbH	**Germany**	**171**	**75-18600**
Litton Industries Inc.	USA	590	75-55400
Marine Data Ltd	UK	466	75-37288
Measurement Devices Ltd	UK	467	75-38275
Microtecnica-Torino SpA	Italy	237	75-15500
Raytheon Anschütz GmbH	Germany	173	75-24650
SAGEM SA	**France**	**143**	**75-33950**
S.G. Brown	UK	437	75-10330
Sperry Marine SpA	Italy	239	75-24400

421 30 70 12 Compasses, magnetic, shipborne

Aeronautical & General Instr. Ltd	UK	428	75-01050
Brookes & Gatehouse Ltd	UK	437	75-10100
Brunton Co.	USA	561	75-15460
KVH Industries Inc.	USA	589	75-53670
Litton Industries Inc.	USA	590	75-55400
M. Paz Logistics Ltd	Israel	224	80-01600
Marine Data Ltd	UK	466	75-37288
SIRS Navigation Ltd.	UK	484	75-54300
W.F. Stanley & Co. Ltd	UK	486	75-56350

421 30 70 20 Gyroscopes, shipborne

Astronautics Corp. of America	USA	555	75-10000
Kearfott Guidance & Navigation Corp.	USA	588	75-51760
Litton Industries Inc.	USA	590	75-55400
Microtecnica-Torino SpA	Italy	237	75-15500
Raytheon Anschütz GmbH	Germany	173	75-24650
SAGEM SA	**France**	**143**	**75-33950**
S.G. Brown	UK	437	75-10330

421 39 Navigation systems, submarine

Boeing Australia Limited	Australia	12	75-02500
GEC-Marconi Avionics Inc.	USA	577	75-37413
ITT Aerospace/Communications Div.	USA	586	75-49455
Julie Research Laboratories	USA	587	75-51265
LITEF GmbH	**Germany**	**171**	**75-18600**
M. Paz Logistics Ltd	Israel	224	80-01600
Magnavox Electronic Systems Company	USA	593	75-57120
MSI-Defence Systems Ltd	UK	469	75-41280
Officine Panerai SpA	Italy	237	75-17500
SAGEM SA	**France**	**143**	**75-33950**
S.G. Brown	UK	437	75-10330
Teldix GmbH	Germany	175	75-30100
Teledyne Brown Engineering	USA	620	75-87526
Vickers Shipbuilding & Eng Ltd	UK	491	75-63100

421 39 70 Navigation equipment, submarine

Astronautics Corp. of America	USA	555	75-10000
Julie Research Laboratories	USA	587	75-51265
Kelvin Hughes Ltd.	UK	463	75-33453
LITEF GmbH	**Germany**	**171**	**75-18600**
Litton Industries Inc.	USA	590	75-55400
M. Paz Logistics Ltd	Israel	224	80-01600

Magnavox Electronic Systems Company	USA	*593*	75-57120
MSI-Defence Systems Ltd	UK	*469*	75-41280
SAGEM SA	**France**	***143***	**75-33950**
S.G. Brown	UK	*437*	75-10330
Teledyne Brown Engineering	USA	*620*	75-87526
Vickers Shipbuilding & Eng Ltd	UK	*491*	75-63100

421 40 Navigation systems, ground-based

Aerocom/Squitter	USA	*546*	75-01150
Bodenseewerk Gerätetechnik GmbH	Germany	*165*	75-03000
Boeing Australia Limited	Australia	*12*	75-02500
Fernau Avionics Ltd	UK	*450*	75-22600
GEC-Marconi Avionics Inc.	USA	*577*	75-37413
GEC-Marconi Avionics Ltd	UK	*454*	75-25320
Israel Aircraft Industries Ltd	Israel	*219*	75-10200
ITT Avionics	USA	*586*	75-49455
Japan Aviation Electronics Ind. Ltd	Japan	*252*	75-03300
Jeppesen UK Ltd	UK	*505*	80-13850
Kearfott Guidance & Navigation Corp.	USA	*588*	75-51760
LITEF GmbH	**Germany**	***171***	**75-18600**
Litton Industries Inc.	USA	*590*	75-55400
M. Paz Logistics Ltd	Israel	*224*	80-01600
Magnavox Electronic Systems Company	USA	*593*	75-57120
Megapulse Inc.	USA	*595*	75-59830
Navigation Science Corp.	USA	*598*	75-64070
Northrop Grumman Corporation	USA	*599*	75-65450
Polaris International Ltd	UK	*508*	80-23700
Pyser-SGI Ltd	UK	*476*	75-47300
Racal Avionics Ltd	UK	*477*	75-47660
Raytheon Aircraft Company	USA	*607*	75-73200
Raytheon E-Systems	USA	*608*	75-73302
Rockwell	USA	*610*	75-75930
Rohde & Schwarz GmbH & Co KG	Germany	*174*	75-25400
SAGEM SA	**France**	***143***	**75-33950**
SFIM	**France**	***144***	**75-35080**

421 40 10 Navigation systems, man-portable

Israel Aircraft Industries Ltd	Israel	*220*	75-10200
Litton Industries Inc.	USA	*590*	75-55400
M. Paz Logistics Ltd	Israel	*224*	80-01600
Pyser-SGI Ltd	UK	*476*	75-47300
Racal Radar Defence Systems Ltd	UK	*477*	75-47688
Raytheon E-Systems	USA	*608*	75-73302
Rockwell	USA	*610*	75-75930
Stocker & Yale Inc.	USA	*618*	75-84250
Trimble Navigation Europe Ltd	UK	*489*	75-61140

421 40 10 70 Navigation equipment, man-portable

Fugro-UDI Ltd	UK	*452*	75-24885
Israel Aircraft Industries Ltd	Israel	*220*	75-10200
M. Paz Logistics Ltd	Israel	*224*	80-01600
Racal Radar Defence Systems Ltd	UK	*477*	75-47688
Rockwell	USA	*610*	75-75930
Rohde & Schwarz GmbH & Co KG	Germany	*174*	75-25400
Stocker & Yale Inc.	USA	*618*	75-84250
Trimble Navigation Europe Ltd	UK	*489*	75-61140
W.F. Stanley & Co. Ltd	UK	*486*	75-56350

421 40 20 Navigation systems, vehicle

Astronautics Corp. of America	USA	*555*	75-10000
Belgian Advanced Technology	Belgium	*45*	75-01320
Bodenseewerk Gerätetechnik GmbH	Germany	*165*	75-03000
GEC-Marconi Avionics Ltd	UK	*454*	75-25320
Israel Aircraft Industries Ltd	Israel	*219*	75-10200
Japan Aviation Electronics Ind. Ltd	Japan	*252*	75-03300
Kearfott Guidance & Navigation Corp.	USA	*588*	75-51760
KVH Industries Inc.	USA	*589*	75-53670
LITEF GmbH	**Germany**	***171***	**75-18600**
Litton Industries Inc.	USA	*590*	75-55400
Litton Italia SpA	Italy	*236*	75-13700
M. Paz Logistics Ltd	Israel	*224*	80-01600
Magnavox Electronic Systems Company	USA	*593*	75-57120
Northrop Grumman Corporation	USA	*599*	75-65450
Racal Radar Defence Systems Ltd	UK	*477*	75-47688
Raytheon E-Systems	USA	*608*	75-73302
Rockwell	USA	*610*	75-75930
Rohde & Schwarz GmbH & Co KG	Germany	*174*	75-25400
SAGEM SA	**France**	***143***	**75-33950**
SEXTANT Avionique SA	France	*144*	75-35000
SFIM	**France**	***144***	**75-35080**
Teldix GmbH	Germany	*175*	75-30100
Trimble Navigation Europe Ltd	UK	*489*	75-61140
Xi CAD North America Inc.	USA	*628*	75-97760

421 40 20 40 Navigation systems, vehicle, radio

Magnavox Electronic Systems Company	USA	*593*	75-57120
Rockwell	USA	*610*	75-75930

421 40 20 50 Navigation systems, vehicle, satellite

Astronautics Corp. of America	USA	*555*	75-10000
Fugro-UDI Ltd	UK	*452*	75-24885
M. Paz Logistics Ltd	Israel	*224*	80-01600
Magnavox Electronic Systems Company	USA	*593*	75-57120
SAGEM SA	**France**	***143***	**75-33950**
Trimble Navigation Europe Ltd	UK	*489*	75-61140

421 40 20 70 Vehicle navigation equipment

Astronautics Corp. of America	USA	*555*	75-10000
Bodenseewerk Gerätetechnik GmbH	Germany	*165*	75-03000
Delco Systems Operations	USA	*568*	75-25360
GEC-Marconi Avionics Ltd	UK	*454*	75-25320
Israel Aircraft Industries Ltd	Israel	*219*	75-10200
Japan Aviation Electronics Ind. Ltd	Japan	*252*	75-03300
Kearfott Guidance & Navigation Corp.	USA	*588*	75-51760
KVH Industries Inc.	USA	*589*	75-53670

LITEF GmbH	**Germany**	**171**	**75-18600**
M. Paz Logistics Ltd	Israel	224	80-01600
Magnavox Electronic Systems Company	USA	593	75-57120
Northrop Grumman Corporation	USA	599	75-65450
Rockwell	USA	610	75-75930
Rohde & Schwarz GmbH & Co KG	Germany	174	75-25400
SAGEM SA	**France**	**143**	**75-33950**
SFIM	**France**	**144**	**75-35080**
Xi CAD North America Inc.	USA	628	75-97760

421 40 70 Navigation equipment, ground-based, other

Advanced Technical Corp.	USA	546	75-01000
Andrew Canada Inc.	Canada	68	75-00600
GEC-Marconi Avionics Inc.	USA	577	75-37413
Israel Aircraft Industries Ltd	Israel	220	75-10200
Japan Aviation Electronics Ind. Ltd	Japan	252	75-03300
Kearfott Guidance & Navigation Corp.	USA	588	75-51760
M. Paz Logistics Ltd	Israel	224	80-01600
MTC Industries	Israel	221	75-15450
Pelorus Navigation Systems Inc.	Canada	75	75-11035
Raytheon Aircraft Company	USA	607	75-73200
Raytheon E-Systems	USA	608	75-73302
SAGEM SA	**France**	**143**	**75-33950**
Teldix GmbH	Germany	175	75-30100
Wilcox Electric Inc.	USA	627	75-96730
Xi CAD North America Inc.	USA	628	75-97760

421 40 70 20 Gyroscopes, ground-based

Bodenseewerk Gerätetechnik GmbH	Germany	165	75-03000
Central Design Bureau	Ukraine	406	75-00800
Humphrey Inc.	USA	583	75-45050
Instro Precision Ltd	UK	461	75-31630
Israel Aircraft Industries Ltd	Israel	219	75-10200
Japan Aviation Electronics Ind. Ltd	Japan	252	75-03300
Litton Industries Inc.	USA	590	75-55400
MTC Industries	Israel	221	75-15450
Northrop Grumman Corporation	USA	599	75-65450
SAGEM SA	**France**	**143**	**75-33950**
Teldix GmbH	Germany	175	75-30100

421 50 Navigation aids, ground-based

Andrew Canada Inc.	Canada	68	75-00600
Balmoral Group Ltd	UK	433	75-05420
Canadian Marconi Co.	Canada	69	75-02600
CORIS	France	134	75-12450
Instro Precision Ltd	UK	461	75-31630
Israel Aircraft Industries Ltd	Israel	220	75-10200
M. Paz Logistics Ltd	Israel	224	80-01600
Navstar Systems Ltd	UK	470	75-41955
NEC Corporation	Japan	253	75-06350
Pelorus Navigation Systems Inc.	Canada	75	75-11035
Pyser-SGI Ltd	UK	476	75-47300
Thomson-CSF	France	148	75-45100
Thorn DNT Airfield Lighting Pty Ltd	Australia	22	75-20540
Wilcox Electric Inc.	USA	627	75-96730
Xi CAD North America Inc.	USA	628	75-97760

421 50 70 TACAN navaid systems, ground-based

Advanced Technical Corp.	USA	546	75-01000
Alcatel Air Navigation Systems GmbH	Germany	164	75-00535
Associated Aircraft Mfg & Sales Inc.	USA	555	75-09500
CORIS	France	134	75-12450
Fernau Avionics Ltd	UK	450	75-22600
NEC Corporation	Japan	253	75-06350
Racal Radar Defence Systems Ltd.	UK	477	75-47688

421 50 80 VOR navaid systems, ground-based

Advanced Technical Corp.	USA	546	75-01000
Aerocom/Squitter	USA	546	75-01150
Alcatel Air Navigation Systems GmbH	Germany	164	75-00535
Andrew Canada Inc.	Canada	68	75-00600
Canadian Marconi Co.	Canada	69	75-02600
CORIS	France	134	75-12450
Navigation Science Corp.	USA	598	75-64070
NEC Corporation	Japan	253	75-06350
Northrop Grumman Corporation	USA	599	75-65450
Racal Avionics Ltd	UK	477	75-47660
Racal Radar Defence Systems Ltd.	UK	477	75-47688
Thomson-CSF	France	148	75-45100
Thorn DNT Airfield Lighting Pty Ltd	Australia	22	75-20540
Wilcox Electric Inc.	USA	627	75-96730

421 50 85 VORTAC navaid systems, ground-based

Advanced Technical Corp.	USA	546	75-01000
CORIS	France	134	75-12450
NEC Corporation	Japan	253	75-06350
Raytheon E-Systems	USA	608	75-73302

421 90 Components, navigation systems & equipment

British Sarozal Ltd.	UK	436	75-09570
Explorocean Technology Ltd	UK	502	80-09870
General Electronics Communications	Greece	189	75-01300
Humphrey Inc.	USA	583	75-45050
Jennings Technology Corp.	USA	586	75-50350
Kearfott Guidance & Navigation Corp.	USA	588	75-51760
KVH Industries Inc.	USA	589	75-53670
LITEF GmbH	**Germany**	**171**	**75-18600**
Litton Industries Inc.	USA	590	75-55400
M. Paz Logistics Ltd	Israel	224	80-01600
Nordic VLSI A/S	Norway	310	75-01640
Racal-MESL Ltd	UK	477	75-47682
SAGEM SA	**France**	**143**	**75-33950**
SFIM	**France**	**144**	**75-35080**
Teldix GmbH	Germany	175	75-30100
Warren-Knight Instrument Co.	USA	627	75-95150
Wilcox Electric Inc.	USA	627	75-96730

421 90 10 Accelerometers

Elettronica Aster SpA	Italy	233	75-07000
Humphrey Inc.	USA	583	75-45050
Israel Aircraft Industries Ltd	Israel	220	75-10200
Japan Aviation Electronics Ind. Ltd	Japan	252	75-03300
Kearfott Guidance & Navigation Corp.	USA	588	75-51760
LITEF GmbH	**Germany**	**171**	**75-18600**
SAGEM SA	**France**	**143**	**75-33950**
SFIM	**France**	**144**	**75-35080**
Sperry Marine SpA	Italy	239	75-24400

422 Direction finding systems & equipment

AAR Corp	USA	629	80-00035
Advanced Technical Corp.	USA	546	75-01000
Andrew Canada Inc.	Canada	68	75-00600
Boeing Australia Limited	Australia	12	75-02500
CEA Technologies Pty Ltd	Australia	24	80-03100
Cubic Communications Inc.	USA	567	75-22850
Cubic Defense Systems Inc.	USA	567	75-22900
Daimler-Benz Aerospace France SAS	**France**	**134**	**75-12950**
Études Electroniques Informatiques	France	150	80-00540
Fernau Avionics Ltd	UK	450	75-22600
GEC-Marconi Avionics Inc.	USA	577	75-37413
GEC-Marconi Avionics Ltd	UK	454	75-25320
GEM Elettronica Srl	Italy	235	75-10100
Giat Industries	France	126	55-00100
Instro Precision Ltd	UK	461	75-31630
Japan Aviation Electronics Ind. Ltd	Japan	252	75-03300
Kearfott Guidance & Navigation Corp.	USA	588	75-51760
KVH Industries Inc.	USA	589	75-53670
Leica AG	Switzerland	386	75-07310
Lockheed Martin Corp.	USA	591	75-55635
M. Paz Logistics Ltd	Israel	224	80-01600
Northern Airborne Technology Ltd	Canada	74	75-10145
Raytheon E-Systems	USA	608	75-73302
Rockwell	USA	610	75-75930
Rockwell International Co.	Italy	233	75-05200
Rohde & Schwarz GmbH & Co KG	Germany	174	75-25400
SAGEM SA	**France**	**143**	**75-33950**
Sanders	USA	612	75-77850
SRI International	USA	643	80-29600
STAREC	France	147	75-42775
Tadiran Ltd	Israel	222	75-21600

Product-Supplier Link

Watkins-Johnson Co.	USA	*627*	75-95200
W.F. Stanley & Co. Ltd.	UK	*486*	75-56350
Zeta	USA	*629*	75-98190

422 20 Direction finding & homing systems, airborne

Advanced Technical Corp.	USA	*546*	75-01000
Chelton (Electrostatics) Limited	UK	*439*	75-12700
Emergency Beacon Corp.	USA	*573*	75-31050
Études Electroniques Informatiques.	France	*150*	80-00540
Glatzer Industries Corp.	USA	*579*	75-39760
Israel Aircraft Industries Ltd	Israel	*220*	75-10200
Northern Airborne Technology Ltd	Canada	*74*	75-10145
Raytheon E-Systems	USA	*608*	75-73302
Rockwell International Co.	Italy	*233*	75-05200
SAGEM SA	**France**	***143***	**75-33950**
Sanders	USA	*612*	75-77850
SRI International	USA	*643*	80-29600
Tadiran Ltd	Israel	*222*	75-21600
Watkins-Johnson Co.	USA	*627*	75-95200
Zeta	USA	*629*	75-98190

422 20 20 10 North-seeking gyro systems, airborne

Alden Sales Co.	USA	*548*	75-04410
Humphrey Inc.	USA	*583*	75-45050
Israel Aircraft Industries Ltd	Israel	*220*	75-10200
Kearfott Guidance & Navigation Corp.	USA	*588*	75-51760
LITEF GmbH	**Germany**	***171***	**75-18600**
SAGEM SA	**France**	***143***	**75-33950**

422 20 70 Direction finding & homing equipment, airborne

Advanced Technical Corp.	USA	*546*	75-01000
Associated Aircraft Mfg & Sales Inc.	USA	*555*	75-09500
Aydin Corp.	USA	*556*	75-11450
Becker Flugfunkwerk GmbH	Germany	*165*	75-02200
Emergency Beacon Corp.	USA	*573*	75-31050
GEC-Marconi Avionics Ltd	UK	*454*	75-25320
Glatzer Industries Corp.	USA	*579*	75-39760
M. Paz Logistics Ltd	Israel	*224*	80-01600
Marconi SpA DTM	Italy	*236*	75-14550
Rockwell International Co.	Italy	*233*	75-05200
SRI International	USA	*643*	80-29600
STAREC	France	*147*	75-42775
Tadiran Ltd	Israel	*222*	75-21600
Tokyo Aircraft Instrument Co Ltd	Japan	*254*	75-08600
Watkins-Johnson Co.	USA	*627*	75-95200

422 20 70 60 Sonobuoy homing & direction finding receivers

Associated Aircraft Mfg & Sales Inc.	USA	*555*	75-09500

422 30 Direction finding systems, naval

Amper Programas SA	Spain	*365*	75-00480
Becker Flugfunkwerk GmbH	Germany	*165*	75-02200
Cubic Communications Inc.	USA	*567*	75-22850
Cubic Defense Systems Inc.	USA	*567*	75-22900
Daimler-Benz Aerospace France SAS	**France**	***134***	**75-12950**
Emergency Beacon Corp.	USA	*573*	75-31050
Études Electroniques Informatiques.	France	*150*	80-00540
Fernau Avionics Ltd	UK	*450*	75-22600
GEM Elettronica Srl	Italy	*235*	75-10100
Glatzer Industries Corp.	USA	*579*	75-39760
KVH Industries Inc.	USA	*589*	75-53670
M. Paz Logistics Ltd	Israel	*224*	80-01600
MA/COM Antenna & Cable Div.	USA	*593*	75-56953
Radamec Control Systems Ltd	UK	*477*	75-47755
Rohde & Schwarz Canada Inc	Canada	*79*	80-01825
Rohde & Schwarz GmbH & Co KG.	Germany	*174*	75-25400
Sanders	USA	*612*	75-77850
SRI International	USA	*643*	80-29600
State College Operations.	USA	*617*	75-83990
Tadiran Ltd	Israel	*222*	75-21600
Watkins-Johnson Co.	USA	*627*	75-95200
Zeta	USA	*629*	75-98190

422 30 20 10 North-seeking gyro systems, shipborne

Humphrey Inc.	USA	*583*	75-45050
Israel Aircraft Industries Ltd	Israel	*220*	75-10200
LITEF GmbH	**Germany**	***171***	**75-18600**
SAGEM SA	**France**	***143***	**75-33950**

422 30 70 Direction finding equipment, naval

Becker Flugfunkwerk GmbH	Germany	*165*	75-02200
Daimler-Benz Aerospace France SAS	**France**	***134***	**75-12950**
Israel Aircraft Industries Ltd	Israel	*220*	75-10200
KVH Industries Inc.	USA	*589*	75-53670
M. Paz Logistics Ltd	Israel	*224*	80-01600
Moonraker Australia Pty Ltd	Australia	*19*	75-13770
Rohde & Schwarz GmbH & Co KG.	Germany	*174*	75-25400
Sanders	USA	*612*	75-77850
Servo Corp. of America	USA	*613*	75-79650
Tadiran Ltd	Israel	*222*	75-21600
Watkins-Johnson Co.	USA	*627*	75-95200
Zeta	USA	*629*	75-98190

422 40 Direction finding systems, ground-based

Andrew Canada Inc.	Canada	*68*	75-00600
CORIS	France	*134*	75-12450
Cubic Communications Inc.	USA	*567*	75-22850
Emergency Beacon Corp.	USA	*573*	75-31050
Études Electroniques Informatiques.	France	*150*	80-00540
Fernau Avionics Ltd	UK	*450*	75-22600
GEM Elettronica Srl	Italy	*235*	75-10100
Giat Industries	France	*126*	55-00100
Glatzer Industries Corp.	USA	*579*	75-39760
Kearfott Guidance & Navigation Corp.	USA	*588*	75-51760
KEL Aerospace Pty Ltd	Australia	*18*	75-11440
Litton Industries Inc.	USA	*590*	75-55400
MA/COM Antenna & Cable Div.	USA	*593*	75-56953
MTC Industries	Israel	*221*	75-15450
Rohde & Schwarz Canada Inc	Canada	*79*	80-01825
Rohde & Schwarz France Telecom	France	*142*	75-32900
Rohde & Schwarz GmbH & Co KG.	Germany	*174*	75-25400
SAGEM SA	**France**	***143***	**75-33950**
Sanders	USA	*612*	75-77850
SRI International	USA	*643*	80-29600
Tadiran Ltd	Israel	*222*	75-21600
Teldix GmbH	Germany	*175*	75-30100
Warren-Knight Instrument Co.	USA	*627*	75-95150
Watkins-Johnson Co.	USA	*627*	75-95200
Zeta	USA	*629*	75-98190

422 40 20 10 North-seeking gyro systems, ground-based

Bodenseewerk Gerätetechnik GmbH	Germany	*165*	75-03000
Humphrey Inc.	USA	*583*	75-45050
Instro Precision Ltd	UK	*461*	75-31630
Japan Aviation Electronics Ind. Ltd	Japan	*252*	75-03300
Kearfott Guidance & Navigation Corp.	USA	*588*	75-51760
LITEF GmbH	**Germany**	***171***	**75-18600**
MTC Industries	Israel	*221*	75-15450
SAGEM SA	**France**	***143***	**75-33950**
Teldix GmbH	Germany	*175*	75-30100

422 40 20 30 Inertial reference systems, ground-based

Japan Aviation Electronics Ind. Ltd	Japan	*252*	75-03300
Kearfott Guidance & Navigation Corp.	USA	*588*	75-51760
LITEF GmbH	**Germany**	***171***	**75-18600**
Litton Industries Inc.	USA	*590*	75-55400
Northrop Grumman Corporation.	USA	*599*	75-65450
SAGEM SA	**France**	***143***	**75-33950**
SFIM Industries Deutschland GmbH.	Germany	*175*	75-27495

422 40 40 Radio direction finding systems, ground-based

Amper Programas SA	Spain	*365*	75-00480
Andrew Canada Inc.	Canada	*68*	75-00600
CORIS	France	*134*	75-12450
Daimler-Benz Aerospace France SAS	**France**	***134***	**75-12950**

Fernau Avionics Ltd	UK	450	75-22600
KEL Aerospace Pty Ltd	Australia	18	75-11440
Racal Communications Inc.	USA	606	75-72380
Rohde & Schwarz Canada Inc	Canada	79	80-01825
Rohde & Schwarz GmbH & Co KG	Germany	174	75-25400
Sanders	USA	612	75-77850
Servo Corp. of America	USA	613	75-79650
South Midlands Communications Ltd	UK	485	75-55213
SRI International	USA	643	80-29600
Tadiran Ltd	Israel	222	75-21600
Watkins-Johnson Co.	USA	627	75-95200

422 40 40 70 Radio direction finding equipment, ground-based

Amper Programas SA	Spain	365	75-00480
Andrew Canada Inc.	Canada	68	75-00600
Canadian Marconi Co.	Canada	69	75-02600
Daimler-Benz Aerospace France SAS	**France**	**134**	**75-12950**
Delfin Systems	USA	568	75-25390
Fernau Avionics Ltd	UK	450	75-22600
KEL Aerospace Pty Ltd	Australia	18	75-11440
Moonraker Australia Pty Ltd	Australia	19	75-13770
Racal Communications Inc.	USA	606	75-72380
Rohde & Schwarz Canada Inc	Canada	79	80-01825
Rohde & Schwarz GmbH & Co KG	Germany	174	75-25400
Sanders	USA	612	75-77850
Servo Corp. of America	USA	613	75-79650
South Midlands Communications Ltd	UK	485	75-55213
STAREC	France	147	75-42775
Tadiran Ltd	Israel	222	75-21600
Watkins-Johnson Co.	USA	627	75-95200

422 40 70 Direction finding equipment, ground-based

Andrew Canada Inc.	Canada	68	75-00600
Brookes & Gatehouse Ltd	UK	437	75-10100
Brunton Co.	USA	561	75-15460
Cubic Communications Inc.	USA	567	75-22850
Emergency Beacon Corp.	USA	573	75-31050
Fernau Avionics Ltd	UK	450	75-22600
Giat Industries	France	126	55-00100
Kearfott Guidance & Navigation Corp.	USA	588	75-51760
KVH Industries Inc.	USA	589	75-53670
Leica AG	Switzerland	386	75-07310
M. Paz Logistics Ltd	Israel	224	80-01600
Racal Communications Inc.	USA	606	75-72380
Rohde & Schwarz Canada Inc	Canada	79	80-01825
SAGEM SA	**France**	**143**	**75-33950**
Servo Corp. of America	USA	613	75-79650
South Midlands Communications Ltd	UK	485	75-55213
W.F. Stanley & Co. Ltd	UK	486	75-56350
Zeta	USA	629	75-98190

422 40 70 10 Compasses, ground-based

Brookes & Gatehouse Ltd	UK	437	75-10100
Brunton Co.	USA	561	75-15460
Hall & Watts (Defence Optics) Group	**UK**	**456**	**75-27650**
Instro Precision Ltd	UK	461	75-31630
Israel Aircraft Industries Ltd	Israel	220	75-10200
Kearfott Guidance & Navigation Corp.	USA	588	75-51760
KVH Industries Inc.	USA	589	75-53670
Litton Industries Inc.	USA	590	75-55400
M. Paz Logistics Ltd	Israel	224	80-01600
Pyser-SGI Ltd	UK	476	75-47300
SAGEM SA	**France**	**143**	**75-33950**
W.F. Stanley & Co. Ltd	UK	486	75-56350

422 40 70 11 Compasses, gyro, ground-based

Israel Aircraft Industries Ltd	Israel	220	75-10200
Kearfott Guidance & Navigation Corp.	USA	588	75-51760
Leica AG	Switzerland	386	75-07310
LITEF GmbH	**Germany**	**171**	**75-18600**
Litton Industries Inc.	USA	590	75-55400
SAGEM SA	**France**	**143**	**75-33950**
Teldix GmbH	Germany	175	75-30100

422 40 70 12 Compasses, magnetic, ground-based

Aeronautical & General Instr. Ltd	UK	428	75-01050
ARC Industries Inc.	USA	554	75-08380
BCB International Ltd	UK	433	75-06075
Brunton Co.	USA	561	75-15460
Hall & Watts (Defence Optics) Group	**UK**	**456**	**75-27650**
Instro Precision Ltd	UK	461	75-31630
Israel Aircraft Industries Ltd	Israel	220	75-10200
KVH Industries Inc.	USA	589	75-53670
Leica AG	Switzerland	386	75-07310
Litton Industries Inc.	USA	590	75-55400
M. Paz Logistics Ltd	Israel	224	80-01600
Pyser-SGI Ltd	UK	476	75-47300
RECTA SA	Switzerland	387	75-10900
Sidem (UK) Ltd	UK	511	80-35850
Stocker & Yale Inc.	USA	618	75-84250
W.F. Stanley & Co. Ltd	UK	486	75-56350

423 Position fixing systems & equipment

AAR Corp	USA	629	80-00035
Azimuth Ltd	Israel	217	75-02610
Bavaria Avionik Technologie GmbH	Germany	165	75-02040
Boeing Australia Limited	Australia	12	75-02500
Cubic Defense Systems Inc.	USA	567	75-22900
Del Norte Technology Inc.	USA	568	75-25365
Garmin International	USA	577	75-36920
GEC-Marconi Avionics Inc.	USA	577	75-37413
GeoAcoustics Ltd.	UK	454	75-25405
Instro Precision Ltd	UK	461	75-31630
Israel Aircraft Industries Ltd	Israel	220	75-10200
ITT Avionics	USA	586	75-49455
Leica AG	Switzerland	386	75-07310
Litton Industries Inc.	USA	590	75-55400
Lockheed Martin Corp.	USA	591	75-55635
M. Paz Logistics Ltd	Israel	224	80-01600
Magnavox Electronic Systems Company	USA	593	75-57120
Motorola Ltd	UK	469	75-41050
Navigation Data Systems Inc.	USA	598	75-64050
Racal Avionics Ltd	UK	477	75-47660
Racal Survey Ltd-Products	UK	509	80-26320
SAGEM SA	**France**	**143**	**75-33950**
Sanders	USA	612	75-77850
Securicor Datatrak Ltd	UK	482	75-52390
STAREC	France	147	75-42775
Universal Avionics Systems Corp.	USA	625	75-92650
Vickers Shipbuilding & Eng Ltd	UK	491	75-63100
Warren-Knight Instrument Co.	USA	627	75-95150

423 20 Airborne position fixing systems

Andrew Sensor Products	USA	554	75-07452
Azimuth Ltd	Israel	217	75-02610
Bavaria Avionik Technologie GmbH	Germany	165	75-02040
Cubic Defense Systems Inc.	USA	567	75-22900
Israel Aircraft Industries Ltd	Israel	220	75-10200
ITT Avionics	USA	586	75-49455
Litton Industries Inc.	USA	590	75-55400
Northrop Grumman Corporation	USA	599	75-65450
Racal Avionics Ltd	UK	477	75-47660
Sanders	USA	612	75-77850
Smiths Aerospace & Defense	USA	615	75-82090
Teldix GmbH	Germany	175	75-30100
Universal Avionics Systems Corp.	USA	625	75-92650

423 20 30 Global positioning systems, airborne

Azimuth Ltd	Israel	217	75-02610
Canadian Marconi Co.	Canada	69	75-02600
Cranfield Aerospace Ltd	UK	500	80-07200
Cranfield University	UK	419	60-00350
Del Norte Technology Inc.	USA	568	75-25365
Flight Refuelling Limited	UK	451	75-23850
FR Aviation Limited	UK	503	80-10850
Garmin International	USA	577	75-36920
GEC-Marconi Research Centre	UK	419	60-00700
Interstate Electronics	USA	586	75-48700
Israel Aircraft Industries Ltd	Israel	220	75-10200
Jeppesen UK Ltd	UK	505	80-13850

Product-Supplier Link

Labtech Ltd	UK	*464*	75-34280
Litton Industries Inc.	USA	*590*	75-55400
Motorola Ltd	UK	*469*	75-41050
Navstar Systems Ltd	UK	*470*	75-41955
Navtech Systems Limited	UK	*507*	80-18000
Raytheon Cossor Electronics Ltd	UK	*478*	75-48367
Rockwell International Co.	Italy	*233*	75-05200
SAGEM SA	**France**	***143***	**75-33950**
SRI International	USA	*643*	80-29600
Steatite Insulations Ltd	UK	*486*	75-56450
Trimble Navigation Europe Ltd	UK	*489*	75-61140
Universal Avionics Systems Corp	USA	*625*	75-92650

423 20 70 Airborne position fixing equipment

Andrew Sensor Products	USA	*554*	75-07452
Azimuth Ltd	Israel	*217*	75-02610
Bavaria Avionik Technologie GmbH.	Germany	*165*	75-02040
Cubic Defense Systems Inc.	USA	*567*	75-22900
Litton Industries Inc.	USA	*590*	75-55400
Navstar Systems Ltd	UK	*470*	75-41955
Northrop Grumman Corporation	USA	*599*	75-65450
Racal Avionics Ltd	UK	*477*	75-47660
Smiths Aerospace & Defense	USA	*615*	75-82090
Universal Avionics Systems Corp	USA	*625*	75-92650

423 20 70 30 GPS receivers, airborne

Alden Sales Co.	USA	*548*	75-04410
Azimuth Ltd	Israel	*217*	75-02610
Canadian Marconi Co.	Canada	*69*	75-02600
Fabbrica Italiana Apparecchiature	Italy	*234*	75-08550
Garmin (Europe) Ltd	UK	*453*	75-25150
Garmin International	USA	*577*	75-36920
GEM Elettronica Srl	Italy	*235*	75-10100
Interstate Electronics	USA	*586*	75-48700
Litton Industries Inc.	USA	*590*	75-55400
M. Paz Logistics Ltd	Israel	*224*	80-01600
Motorola Ltd	UK	*469*	75-41050
Navstar Systems Ltd	UK	*470*	75-41955
Rockwell International Co.	Italy	*233*	75-05200
Rokar International Ltd	Israel	*222*	75-18805
SAGEM SA	**France**	***143***	**75-33950**
Universal Avionics Systems Corp	USA	*625*	75-92650

423 20 70 30 50 GPS & Loran-C receivers, airborne

Azimuth Ltd	Israel	*217*	75-02610
Litton Industries Inc.	USA	*590*	75-55400
Navigation Data Systems Inc.	USA	*598*	75-64050
Universal Avionics Systems Corp	USA	*625*	75-92650

423 30 Naval position fixing systems

BAeSEMA Ltd.	UK	*498*	80-03070
Cubic Defense Systems Inc.	USA	*567*	75-22900
GeoAcoustics Ltd.	UK	*454*	75-25405
Litton Industries Inc.	USA	*590*	75-55400
M. Paz Logistics Ltd	Israel	*224*	80-01600
Magnavox Electronic Systems Company	USA	*593*	75-57120
Megapulse Inc.	USA	*595*	75-59830
Motorola Ltd	UK	*469*	75-41050
Nautronix Ltd	Australia	*19*	75-14300
SAGEM SA	**France**	***143***	**75-33950**
Sanders	USA	*612*	75-77850
Servo Corp. of America	USA	*613*	75-79650

423 30 10 Naval position fixing systems, acoustic

Dukane Corp.	USA	*570*	75-27650
Fugro-UDI Ltd	UK	*452*	75-24885
GeoAcoustics Ltd.	UK	*454*	75-25405
Oceano Instruments (UK) Ltd	UK	*508*	80-19050

423 30 30 Global positioning systems, naval

Azimuth Ltd	Israel	*217*	75-02610
Cranfield University	UK	*419*	60-00350
Del Norte Technology Inc.	USA	*568*	75-25365
Fugro-UDI Ltd	UK	*452*	75-24885
Israel Aircraft Industries Ltd	Israel	*220*	75-10200
Kelvin Hughes Ltd.	UK	*463*	75-33453
Labtech Ltd	UK	*464*	75-34280
M. Paz Logistics Ltd	Israel	*224*	80-01600
Marine Data Ltd	UK	*466*	75-37288
Motorola Ltd	UK	*469*	75-41050
Navigation Data Systems Inc.	USA	*598*	75-64050
Navstar Systems Ltd	UK	*470*	75-41955
Raytheon Cossor Electronics Ltd	UK	*478*	75-48367
SAGEM SA	**France**	***143***	**75-33950**
Scorpio Navigation Services	UK	*510*	80-31900
Steatite Insulations Ltd	UK	*486*	75-56450
Trimble Navigation Europe Ltd	UK	*489*	75-61140

423 30 50 Naval position fixing systems, radio

Fugro-UDI Ltd	UK	*452*	75-24885

423 30 70 Naval position fixing equipment

Cubic Defense Systems Inc.	USA	*567*	75-22900
Fugro-UDI Ltd	UK	*452*	75-24885
Magnavox Electronic Systems Company	USA	*593*	75-57120
SAGEM SA	**France**	***143***	**75-33950**
Simrad Norge A/S	Norway	*310*	75-02630

423 30 70 10 Naval position fixing equipment, acoustic

Fugro-UDI Ltd	UK	*452*	75-24885
GeoAcoustics Ltd.	UK	*454*	75-25405

423 30 70 10 10 Acoustic beacons, underwater

A1Marketing	UK	*426*	75-00025
Benthos Inc.	USA	*558*	75-13540
Dukane Corp.	USA	*570*	75-27650
GeoAcoustics Ltd.	UK	*454*	75-25405
Marine Electronics Ltd	UK	*466*	75-37290
Sparton Electronics	USA	*616*	75-82900

423 30 70 30 GPS receivers, naval

Bharat Electronics Ltd	India	*203*	55-00200
Fabbrica Italiana Apparecchiature	Italy	*234*	75-08550
GEM Elettronica Srl	Italy	*235*	75-10100
M. Paz Logistics Ltd	Israel	*224*	80-01600
Navstar Systems Ltd	UK	*470*	75-41955
SAGEM SA	**France**	***143***	**75-33950**
Sperry Marine SpA	Italy	*239*	75-24400
STAREC	France	*147*	75-42775

423 40 Ground-based position fixing systems

Cubic Defense Systems Inc.	USA	*567*	75-22900
Kearfott Guidance & Navigation Corp.	USA	*588*	75-51760
Leica AG	Switzerland	*386*	75-07310
Litton Industries Inc.	USA	*590*	75-55400
M. Paz Logistics Ltd	Israel	*224*	80-01600
Measurement Devices Ltd	UK	*467*	75-38275
Megapulse Inc.	USA	*595*	75-59830
Motorola Ltd	UK	*469*	75-41050
Northrop Grumman Corporation	USA	*599*	75-65450
SAGEM SA	**France**	***143***	**75-33950**
Sanders	USA	*612*	75-77850

423 40 30 Global positioning systems, ground-based

Azimuth Ltd	Israel	*217*	75-02610
BCB International Ltd	UK	*433*	75-06075
Cranfield Aerospace Ltd	UK	*500*	80-07200
Cranfield University	UK	*419*	60-00350
Del Norte Technology Inc.	USA	*568*	75-25365
EASAMS Ltd	UK	*501*	80-09000
Francis & Lewis International Ltd	UK	*452*	75-24500
Fugro-UDI Ltd	UK	*452*	75-24885

Interstate Electronics	USA	*586*	75-48700
Jeppesen UK Ltd	UK	*505*	80-13850
Labtech Ltd	UK	*464*	75-34280
M. Paz Logistics Ltd	Israel	*224*	80-01600
Motorola Ltd	UK	*469*	75-41050
Raytheon Cossor Electronics Ltd	UK	*478*	75-48367
SAGEM SA	**France**	***143***	**75-33950**
Scorpio Navigation Services	UK	*510*	80-31900
Securicor Datatrak Ltd	UK	*482*	75-52390
Steatite Insulations Ltd	UK	*486*	75-56450
Trimble Navigation Europe Ltd	UK	*489*	75-61140
TRL Technology Ltd	UK	*514*	80-58700

423 40 35 Gyroscope systems, automatic, ground-based

Central Design Bureau	Ukraine	*406*	75-00800
Japan Aviation Electronics Ind. Ltd	Japan	*252*	75-03300
Kearfott Guidance & Navigation Corp.	USA	*588*	75-51760
Leica AG	Switzerland	*386*	75-07310
Northrop Grumman Corporation	USA	*599*	75-65450
Saab Dynamics AB	Sweden	*378*	75-09100
SAGEM SA	**France**	***143***	**75-33950**
Teldix GmbH	Germany	*175*	75-30100

423 40 50 Ground-based position fixing systems, radio

Azimuth Ltd	Israel	*217*	75-02610
CORIS	France	*134*	75-12450
Fernau Avionics Ltd	UK	*450*	75-22600
Motorola Ltd	UK	*469*	75-41050
SRI International	USA	*643*	80-29600

423 40 60 Survey stations, combined, ground-based

CORIS	France	*134*	75-12450
Instro Precision Ltd	UK	*461*	75-31630
Kearfott Guidance & Navigation Corp.	USA	*588*	75-51760
Leica AG	Switzerland	*386*	75-07310
Warren-Knight Instrument Co.	USA	*627*	75-95150

423 40 70 Ground-based position fixing equipment

Leica AG	Switzerland	*386*	75-07310
M. Paz Logistics Ltd	Israel	*224*	80-01600
Magnavox Electronic Systems Company	USA	*593*	75-57120
Measurement Devices Ltd	UK	*467*	75-38275
Motorola Ltd	UK	*469*	75-41050
Teldix GmbH	Germany	*175*	75-30100

423 40 70 10 Angle measuring equipment, ground-based

Arab International Optronics Co.	Egypt	*109*	74-00500
Fugro-UDI Ltd	UK	*452*	75-24885
GAP Instrument Corp.	USA	*577*	75-36900
Hall & Watts (Defence Optics) Group	**UK**	***456***	**75-27650**
Humphrey Inc.	USA	*583*	75-45050
Instro Precision Ltd	UK	*461*	75-31630
Leica AG	Switzerland	*386*	75-07310
M. Paz Logistics Ltd	Israel	*224*	80-01600
Measurement Devices Ltd	UK	*467*	75-38275
Warren-Knight Instrument Co.	USA	*627*	75-95150

423 40 70 10 10 Aiming circles, ground-based

Arab International Optronics Co.	Egypt	*109*	74-00500
Hall & Watts (Defence Optics) Group	**UK**	***456***	**75-27650**
Instro Precision Ltd	UK	*461*	75-31630
Leica AG	Switzerland	*386*	75-07310
Theodor Rapp GmbH Co	Germany	*173*	75-24600
Vickers Shipbuilding & Eng Ltd	UK	*491*	75-63100
Warren-Knight Instrument Co.	USA	*627*	75-95150

423 40 70 10 30 Goniometers, ground-based

Arab International Optronics Co.	Egypt	*109*	74-00500
Hall & Watts (Defence Optics) Group	**UK**	***456***	**75-27650**
Instro Precision Ltd	UK	*461*	75-31630
Leica AG	Switzerland	*386*	75-07310
SAGEM SA	**France**	***143***	**75-33950**
SFIM ODS	France	*144*	75-35085

423 40 70 10 60 Theodolites, ground-based

Andrew Sensor Products	USA	*554*	75-07452
Fugro-UDI Ltd	UK	*452*	75-24885
Hall & Watts (Defence Optics) Group	**UK**	***456***	**75-27650**
Instro Precision Ltd	UK	*461*	75-31630
Leica AG	Switzerland	*386*	75-07310
Measurement Devices Ltd	UK	*467*	75-38275
Warren-Knight Instrument Co.	USA	*627*	75-95150

423 40 70 30 GPS receivers, ground-based

Allen Osborne Assoc., Inc.	USA	*548*	75-04610
Azimuth Ltd	Israel	*217*	75-02610
Cranfield Aerospace Ltd	UK	*500*	80-07200
Cranfield University	UK	*419*	60-00350
Fabbrica Italiana Apparecchiature	Italy	*234*	75-08550
Fugro-UDI Ltd	UK	*452*	75-24885
Garmin (Europe) Ltd	UK	*453*	75-25150
Garmin International	USA	*577*	75-36920
GEC-Marconi Research Centre	UK	*419*	60-00700
GEM Elettronica Srl	Italy	*235*	75-10100
Hall & Watts (Defence Optics) Group	**UK**	***456***	**75-27650**
Interstate Electronics	USA	*586*	75-48700
Labtech Ltd	UK	*464*	75-34280
M. Paz Logistics Ltd	Israel	*224*	80-01600
Marine Data Ltd	UK	*466*	75-37288
Matra Marconi Space UK, Ltd	UK	*425*	74-06900
Motorola Ltd	UK	*469*	75-41050
Navstar Systems Ltd	UK	*470*	75-41955
Navtech Systems Limited	UK	*507*	80-18000
Racal-MESL Ltd	UK	*477*	75-47682
Raytheon Cossor Electronics Ltd	UK	*478*	75-48367
SAGEM SA	**France**	***143***	**75-33950**
Scorpio Navigation Services	UK	*510*	80-31900
Signal Computing Ltd.	UK	*511*	80-36050
STAREC	France	*147*	75-42775
Steatite Insulations Ltd	UK	*486*	75-56450
Trimble Navigation Europe Ltd	UK	*489*	75-61140

423 40 70 35 Gyroscopic position-fixing equipment, ground-based

Alden Sales Co.	USA	*548*	75-04410
Bodenseewerk Gerätetechnik GmbH	Germany	*165*	75-03000
Central Design Bureau	Ukraine	*406*	75-00800
Instro Precision Ltd	UK	*461*	75-31630
Israel Aircraft Industries Ltd	Israel	*220*	75-10200
Kearfott Guidance & Navigation Corp.	USA	*588*	75-51760
Leica AG	Switzerland	*386*	75-07310
Measurement Devices Ltd	UK	*467*	75-38275
SAGEM SA	**France**	***143***	**75-33950**
SFIM Industries Deutschland GmbH	Germany	*175*	75-27495
Teldix GmbH	Germany	*175*	75-30100

423 40 70 35 10 Gyro aiming circles, ground-based

Hall & Watts (Defence Optics) Group	**UK**	***456***	**75-27650**
Japan Aviation Electronics Ind. Ltd	Japan	*252*	75-03300
Leica AG	Switzerland	*386*	75-07310
SIRS Navigation Ltd.	UK	*484*	75-54300
Vickers Shipbuilding & Eng Ltd	UK	*491*	75-63100

Reconnaissance & Surveillance

431 Reconnaissance systems & equipment

Beloma	Belarus	*38*	75-00200

California Microwave Inc.	USA	*561*	75-16350
CEA Technologies Pty Ltd	Australia	*24*	80-03100
Computing Devices Company Ltd	UK	*442*	75-14700
CONTROP Precision Technologies Ltd	Israel	*218*	75-04170
Dassault Aviation Group	France	*135*	75-12975
Fabbrica Italiana Apparecchiature	Italy	*234*	75-08550
GEC-Marconi Avionics Inc.	USA	*577*	75-37413
GEC-Marconi Avionics Ltd	UK	*454*	75-25320
Hughes Danbury Optical Sys.	USA	*582*	75-44857
II-VI Inc.	USA	*624*	75-91355
Innotech Aviation Ltd.	Canada	*78*	80-00960
Israel Aircraft Industries Ltd	Israel	*220*	75-10200
Litton Industries Inc.	USA	*590*	75-55400
Lockheed Martin Corp.	USA	*591*	75-55635
Lockheed Martin Missiles & Space	USA	*592*	75-55640
MacDonald Dettwiler and Associates	Canada	*78*	80-01310
MA/COM Antenna & Cable Div.	USA	*593*	75-56953
Motorola Ltd	UK	*469*	75-41050
MPB Technologies Inc	Canada	*74*	75-09855
Northrop Grumman Corporation	USA	*599*	75-65450
Pearpoint Ltd	UK	*473*	75-44270
Pilkington Optronics Ltd	UK	*426*	74-10000
Prostab International Ltd	UK	*476*	75-46870
Pyser-SGI Ltd	UK	*476*	75-47300
Racal Communications Inc.	USA	*606*	75-72380
Raytheon E-Systems	USA	*608*	75-73302
Recon/Optical Inc.	USA	*608*	75-73930
Rohde & Schwarz France Telecom	France	*142*	75-32900
Sanders	USA	*612*	75-77850
SFIM	**France**	***144***	**75-35080**
Sira Ltd.	UK	*484*	75-54250
Tadiran Ltd	Israel	*222*	75-21600
Technology Service Corp.	USA	*644*	80-31930
Thomson-CSF Optronique	France	*148*	75-45137
W. Vinten Ltd	UK	*491*	75-63353

431 10 Reconnaissance systems, satellite-based

Beloma	Belarus	*38*	75-00200
Coherent Optics (Europe) Ltd	UK	*441*	75-14090
Cranfield University	UK	*419*	60-00350
GEC-Marconi Research Centre	UK	*419*	60-00700
Matra Marconi Space UK, Ltd	UK	*425*	74-06900
Sanders	USA	*612*	75-77850
Sira Ltd.	UK	*484*	75-54250
Technology Service Corp.	USA	*644*	80-31930

431 10 50 Optical reconnaissance systems, spaceborne

Aerospatiale	France	*131*	75-00600
Beloma	Belarus	*38*	75-00200
Coherent Optics (Europe) Ltd	UK	*441*	75-14090
Cranfield University	UK	*419*	60-00350
Eloptro	S. Africa	*358*	75-05680
GEC-Marconi Research Centre	UK	*419*	60-00700
Matra Marconi Space UK, Ltd	UK	*425*	74-06900
Pilkington Thorn Optronics	UK	*426*	74-10000
SFIM ODS	France	*144*	75-35085
Sira Ltd.	UK	*484*	75-54250
Technology Service Corp.	USA	*644*	80-31930

431 17 Reconnaissance equipment, satellite

Coherent Optics (Europe) Ltd	UK	*441*	75-14090
Com Dev Europe	UK	*441*	75-14300
Cranfield University	UK	*419*	60-00350
MacDonald Dettwiler and Associates	Canada	*78*	80-01310
Matra Marconi Space UK, Ltd	UK	*425*	74-06900
Nera Telecommunications	UK	*470*	75-42318
Pilkington Thorn Optronics	UK	*426*	74-10000
Sanders	USA	*612*	75-77850
Sira Ltd.	UK	*484*	75-54250
Société Anonyme d'Études	France	*145*	75-37100
Technology Service Corp.	USA	*644*	80-31930

431 20 Reconnaissance systems, airborne

AAI Corp.	USA	*545*	75-00020
Aeroelectrónica Industria	Brazil	*55*	75-00585
Alan Gordon Enterprises Inc.	USA	*579*	75-40060
Computing Devices Company Ltd	UK	*442*	75-14700
Contraves Inc	USA	*566*	75-21850
CONTROP Precision Technologies Ltd	Israel	*218*	75-04170
Dassault Aviation Group	France	*135*	75-12975
GEC-Marconi Avionics Ltd	UK	*454*	75-25320
Innotech Aviation Ltd.	Canada	*78*	80-00960
Israel Aircraft Industries Ltd	Israel	*220*	75-10200
Litton Industries Inc.	USA	*590*	75-55400
MacDonald Dettwiler and Associates	Canada	*78*	80-01310
Northrop Grumman Corporation	USA	*599*	75-65450
Racal Communications Inc.	USA	*606*	75-72380
Recon/Optical Inc.	USA	*608*	75-73930
Sanders	USA	*612*	75-77850
SFIM	**France**	***144***	**75-35080**
Tadiran Ltd	Israel	*222*	75-21600
Technology Service Corp.	USA	*644*	80-31930
Thomson-CSF Optronique	France	*148*	75-45137
W. Vinten Ltd	UK	*491*	75-63353
Zeiss-Eltro Optronic GmbH	Germany	*177*	75-34700

431 20 30 Reconnaissance systems, airborne, infra-red

AAI Corp.	USA	*545*	75-00020
Computing Devices Company Ltd	UK	*442*	75-14700
Contraves Inc	USA	*566*	75-21850
GEC-Marconi Avionics Ltd	UK	*454*	75-25320
II-VI Inc.	USA	*624*	75-91355
Innotech Aviation Ltd.	Canada	*78*	80-00960
Israel Aircraft Industries Ltd	Israel	*220*	75-10200
Litton Industries Inc.	USA	*590*	75-55400
Northrop Grumman Corporation	USA	*599*	75-65450
Pilkington Optronics Ltd	UK	*426*	74-10000
Pilkington Thorn Optronics	UK	*426*	74-10000
Recon/Optical Inc.	USA	*608*	75-73930
SFIM	**France**	***144***	**75-35080**
Technology Service Corp.	USA	*644*	80-31930
Thomson-CSF Optronique	France	*148*	75-45137
W. Vinten Ltd	UK	*491*	75-63353

431 20 30 10 Reconnaissance systems, airborne, infra-red, forward-looking

AAI Corp.	USA	*545*	75-00020
Computing Devices Company Ltd	UK	*442*	75-14700
Eloptro	S. Africa	*358*	75-05680
GEC-Marconi Avionics Ltd	UK	*454*	75-25320
II-VI Inc.	USA	*624*	75-91355
Israel Aircraft Industries Ltd	Israel	*220*	75-10200
Litton Industries Inc.	USA	*590*	75-55400
Pilkington Optronics Ltd	UK	*426*	74-10000
Pilkington Thorn Optronics	UK	*426*	74-10000
Sira Ltd.	UK	*484*	75-54250
Technology Service Corp.	USA	*644*	80-31930
Thomson-CSF Optronique	France	*148*	75-45137
Zeiss-Eltro Optronic GmbH	Germany	*177*	75-34700

431 20 40 Reconnaissance systems, airborne, laser

Computing Devices Company Ltd	UK	*442*	75-14700
Hughes Danbury Optical Sys.	USA	*582*	75-44857
Innotech Aviation Ltd.	Canada	*78*	80-00960
Israel Aircraft Industries Ltd	Israel	*220*	75-10200
Litton Industries Inc.	USA	*590*	75-55400
Pilkington Optronics Ltd	UK	*426*	74-10000
Sira Ltd.	UK	*484*	75-54250
Thomson-CSF Optronique	France	*148*	75-45137
W. Vinten Ltd	UK	*491*	75-63353

431 20 50 Reconnaissance systems, airborne, photographic

Alan Gordon Enterprises Inc.	USA	*579*	75-40060
Charles A. Hulcher Co. Inc.	USA	*583*	75-45000
Computing Devices Company Ltd	UK	*442*	75-14700
Dassault Aviation Group	France	*135*	75-12975
Innotech Aviation Ltd.	Canada	*78*	80-00960
Israel Aircraft Industries Ltd	Israel	*220*	75-10200
Meggitt Avionics (UK)	UK	*467*	75-38340
Recon/Optical Inc.	USA	*608*	75-73930
Thomson-CSF Optronique	France	*148*	75-45137
W. Vinten Ltd	UK	*491*	75-63353

431 25 Reconnaissance pods, airborne

Dassault Aviation Group	France	135	75-12975
Delft Sensor Systems	**Belgium**	**46**	**75-02500**
Delft Sensor Systems	Netherlands	292	75-01590
GEC-Marconi Avionics Ltd	UK	454	75-25320
Israel Aircraft Industries Ltd	Israel	220	75-10200
Litton Industries Inc.	USA	590	75-55400
Technology Service Corp.	USA	644	80-31930
Thomson-CSF Optronique	France	148	75-45137
W. Vinten Ltd	UK	491	75-63353

431 27 Reconnaissance equipment, airborne

Bavaria Keytronic Technologie GmbH	Germany	165	75-02050
Computing Devices Company Ltd	UK	442	75-14700
Fraser-Volpe Corp.	USA	576	75-35950
GEC-Marconi Avionics Ltd	UK	454	75-25320
Howden Fluid Systems	USA	582	75-44400
Hughes Elcan Optical Technologies	Canada	72	75-07650
Israel Aircraft Industries Ltd	Israel	220	75-10200
Litton Industries Inc.	USA	590	75-55400
MacDonald Dettwiler and Associates	Canada	78	80-01310
MA/COM Antenna & Cable Div.	USA	593	75-56953
Pearpoint Ltd	UK	473	75-44270
Raytheon E-Systems	USA	608	75-73302
SAGEM SA	**France**	**143**	**75-33950**
Sanders	USA	612	75-77850
Tadiran Ltd	Israel	222	75-21600
Technology Service Corp.	USA	644	80-31930
Thomson-CSF Optronique	France	148	75-45137
W. Vinten Ltd	UK	491	75-63353
Zeiss-Eltro Optronic GmbH	Germany	177	75-34700

431 27 50 Reconnaissance equipment, airborne, photographic

Alan Gordon Enterprises Inc.	USA	579	75-40060
Hughes Elcan Optical Technologies	Canada	72	75-07650
Innotech Aviation Ltd	Canada	78	80-00960
Tadiran Ltd	Israel	222	75-21600
Thomson-CSF Optronique	France	148	75-45137
W. Vinten Ltd	UK	491	75-63353

431 27 50 10 Reconnaissance cameras, airborne, film

Alan Gordon Enterprises Inc.	USA	579	75-40060
Bavaria Keytronic Technologie GmbH	Germany	165	75-02050
Carl Zeiss Canada Ltd	Canada	70	75-02900
GEC-Marconi Avionics Ltd	UK	454	75-25320
Photo-Sonics Inc.	USA	604	75-69800
Recon/Optical Inc.	USA	608	75-73930
Thomson-CSF Optronique	France	148	75-45137
W. Vinten Ltd	UK	491	75-63353

431 27 50 10 94 Lenses, reconnaissance camera, airborne

Alan Gordon Enterprises Inc.	USA	579	75-40060
Fraser-Volpe Corp.	USA	576	75-35950
Pearpoint Ltd	UK	473	75-44270
Photo-Sonics Inc.	USA	604	75-69800
Recon/Optical Inc.	USA	608	75-73930
Regis SpA	Italy	238	75-20110
SFIM ODS	France	144	75-35085
Thomson-CSF Optronique	France	148	75-45137

431 27 50 15 Reconnaissance cameras, airborne, television

Computing Devices Company Ltd	UK	442	75-14700
Fabbrica Italiana Apparecchiature	Italy	234	75-08550
Pearpoint Ltd	UK	473	75-44270
Regis SpA	Italy	238	75-20110
Tadiran Ltd	Israel	222	75-21600
W. Vinten Ltd	UK	491	75-63353

431 27 50 20 Reconnaissance camera controls, airborne

Alan Gordon Enterprises Inc.	USA	579	75-40060
Fabbrica Italiana Apparecchiature	Italy	234	75-08550
Pearpoint Ltd	UK	473	75-44270
Recon/Optical Inc.	USA	608	75-73930

431 27 50 25 Reconnaissance camera mounts, airborne

Alan Gordon Enterprises Inc.	USA	579	75-40060
Fabbrica Italiana Apparecchiature	Italy	234	75-08550
SAGEM SA	**France**	**143**	**75-33950**

431 27 50 30 Electronic flash night illumination systems

Recon/Optical Inc.	USA	608	75-73930

431 27 50 60 Cameras, radar recording, airborne

Alan Gordon Enterprises Inc.	USA	579	75-40060

431 27 50 80 Viewfinders, reconnaissance / navigation, airborne

Alan Gordon Enterprises Inc.	USA	579	75-40060
Litton Industries Inc.	USA	590	75-55400
Recon/Optical Inc.	USA	608	75-73930

431 27 90 Video recorders, airborne

GEC-Marconi Avionics Ltd	UK	454	75-25320
Prostab International Ltd	UK	476	75-46870
Thomson-CSF Optronique	France	148	75-45137

431 28 Film, airborne reconnaissance camera

Agfa-Gevaert NV	Belgium	45	75-00100
Alan Gordon Enterprises Inc.	USA	579	75-40060
CONTROP Precision Technologies Ltd	Israel	218	75-04170
Thomson-CSF Optronique	France	148	75-45137

431 30 Reconnaissance systems, naval

Alan Gordon Enterprises Inc.	USA	579	75-40060
Computing Devices Company Ltd	UK	442	75-14700
Israel Aircraft Industries Ltd	Israel	220	75-10200
Litton Industries Inc.	USA	590	75-55400
Pilkington Optronics Ltd	UK	426	74-10000
Racal Communications Inc.	USA	606	75-72380
Raytheon E-Systems	USA	608	75-73302
Recon/Optical Inc.	USA	608	75-73930
Saab Dynamics AB	Sweden	378	75-09100
Sanders	USA	612	75-77850
Technology Service Corp.	USA	644	80-31930

431 37 Reconnaissance equipment, naval

Alan Gordon Enterprises Inc.	USA	579	75-40060
Int'l Submarine Engrg.	Canada	73	75-07780
Litton Industries Inc.	USA	590	75-55400
MA/COM Antenna & Cable Div.	USA	593	75-56953
Pilkington Optronics Ltd	UK	426	74-10000
Racal Communications Inc.	USA	606	75-72380
Recon/Optical Inc.	USA	608	75-73930
Sanders	USA	612	75-77850
Technology Service Corp.	USA	644	80-31930

431 40 Reconnaissance systems, ground-based

Alan Gordon Enterprises Inc.	USA	579	75-40060
Computing Devices Company Ltd	UK	442	75-14700
Israel Aircraft Industries Ltd	Israel	220	75-10200
Litton Industries Inc.	USA	590	75-55400
Racal Communications Inc.	USA	606	75-72380
Raytheon E-Systems	USA	608	75-73302
Rohde & Schwarz France Telecom	France	142	75-32900
Sanders	USA	612	75-77850
Technology Service Corp.	USA	644	80-31930

431 40 60 Reconnaissance systems, ground-based, tethered

Technology Service Corp.	USA	*644*	80-31930

431 47 Reconnaissance equipment, ground-based

Alan Gordon Enterprises Inc.	USA	*579*	75-40060
Contraves Inc	USA	*566*	75-21850
Litton Industries Inc.	USA	*590*	75-55400
Logica UK Limited.	UK	*506*	80-14350
MA/COM Antenna & Cable Div.	USA	*593*	75-56953
Racal Communications Inc.	USA	*606*	75-72380
Recon/Optical Inc.	USA	*608*	75-73930
Rohde & Schwarz France Telecom	France	*142*	75-32900
Sanders	USA	*612*	75-77850
Sparton Technology Inc.	USA	*616*	75-82950
Technology Service Corp.	USA	*644*	80-31930

432 Surveillance systems & equipment (excluding Night Vision - see code 433)

Advanced Vision Systems	UK	*427*	75-00900
Benthos Inc.	USA	*558*	75-13540
Bospec Neue Technologien	Germany	*166*	75-03420
Broadcast Microwave Services Inc.	USA	*560*	75-15250
CCS International Ltd	USA	*562*	75-17250
CEA Technologies Pty Ltd	Australia	*24*	80-03100
Cincinnati Electronics Corp.	USA	*564*	75-18640
Cohu Inc.	USA	*564*	75-19400
Computer Optics Inc.	USA	*565*	75-20680
CPAD Technologies Inc.	Canada	*78*	80-00386
Cubic Defense Systems Inc.	USA	*567*	75-22900
DM Aerospace Ltd	UK	*444*	75-17300
DRS Technologies, Inc.	USA	*570*	75-27400
Fraser-Volpe Corp.	USA	*576*	75-35950
GEC Marconi Hazeltine Corp	USA	*578*	75-37415
GEC-Marconi Avionics Inc.	USA	*577*	75-37413
GEC-Marconi Avionics Ltd	UK	*454*	75-25320
Instro Precision Ltd	UK	*461*	75-31630
JAI A/S	Denmark	*103*	75-09400
Kongsberg Simrad Ltd	UK	*463*	75-34100
Litton Industries Inc.	USA	*590*	75-55400
Lockheed Martin Corp.	USA	*591*	75-55635
M. Paz Logistics Ltd	Israel	*224*	80-01600
Motorola Ltd	UK	*469*	75-41050
Nanoquest Defence Products Ltd	UK	*470*	75-41890
NORINCO	China PR	*88*	55-00280
NOVO Corporation	Russia	*338*	75-14200
Pearpoint Ltd	UK	*473*	75-44270
Pilkington Optronics Ltd	UK	*426*	74-10000
PK Electronic International	USA	*604*	75-70300
Pyser-SGI Ltd	UK	*476*	75-47300
Raytheon E-Systems	USA	*608*	75-73302
Recon/Optical Inc.	USA	*608*	75-73930
RGB Spectrum	USA	*609*	75-75000
Sanders	USA	*612*	75-77850
SFIM	**France**	***144***	**75-35080**
Shield International Ltd	UK	*511*	80-35500
Shorrock Integrated Systems	UK	*483*	75-53450
Sira Ltd.	UK	*484*	75-54250
Spar Aerospace Ltd	Canada	*76*	75-13000
Special Electronic Security	Israel	*222*	75-21050
Spectronic Denmark	Denmark	*103*	75-19650
SRI International	USA	*643*	80-29600
Warren-Knight Instrument Co.	USA	*627*	75-95150
Watkins-Johnson Co.	USA	*627*	75-95200

432 10 Surveillance systems, spaceborne

Aerojet	USA	*547*	75-01500
Coherent Optics (Europe) Ltd	UK	*441*	75-14090
Cranfield University	UK	*419*	60-00350
GEC-Marconi Research Centre	UK	*419*	60-00700
Matra Marconi Space UK, Ltd	UK	*425*	74-06900
Navtech Systems Limited	UK	*507*	80-18000
Sanders	USA	*612*	75-77850
Sira Ltd.	UK	*484*	75-54250
Technology Service Corp.	USA	*644*	80-31930

432 10 70 Spaceborne surveillance equipment

Coherent Optics (Europe) Ltd	UK	*441*	75-14090
Cranfield University	UK	*419*	60-00350
Navtech Systems Limited	UK	*507*	80-18000
Pilkington Thorn Optronics	UK	*426*	74-10000
Rutherford Appleton Laboratory	UK	*420*	60-01850
Sira Ltd.	UK	*484*	75-54250
Spar Aerospace Ltd	Canada	*76*	75-13000

432 20 Surveillance systems, airborne

AAI Corp.	USA	*545*	75-00020
Andrew Sensor Products	USA	*554*	75-07452
Broadcast Microwave Services Inc.	USA	*560*	75-15250
Computing Devices Company Ltd	UK	*442*	75-14700
Contraves Inc	USA	*566*	75-21850
Cubic Defense Systems Inc.	USA	*567*	75-22900
DRS Technologies, Inc.	USA	*570*	75-27400
Elettronica (UK) Ltd	UK	*448*	75-20475
ELTA Electronics Industries Ltd	Israel	*220*	75-10200
GEC Marconi Hazeltine Corp	USA	*578*	75-37415
GEC-Marconi Avionics Ltd	UK	*454*	75-25320
Hawkeye Surveillance Systems Ltd	UK	*503*	80-12240
Hughes Danbury Optical Sys.	USA	*582*	75-44857
Inframetrics Inc.	USA	*585*	75-46750
Insight Vision Systems Ltd	UK	*461*	75-31530
Israel Aircraft Industries Ltd	Israel	*219*	75-10200
Kaiser Electronics	USA	*587*	75-51350
Litton Industries Inc.	USA	*590*	75-55400
MacDonald Dettwiler and Associates	Canada	*78*	80-01310
MA/COM Antenna & Cable Div.	USA	*593*	75-56953
Pulnix America Inc.	USA	*606*	75-71815
Rand Instrument Corp.	USA	*607*	75-72900
Raytheon E-Systems	USA	*608*	75-73302
Recon/Optical Inc.	USA	*608*	75-73930
RGB Spectrum	USA	*609*	75-75000
Sanders	USA	*612*	75-77850
SFIM	**France**	***144***	**75-35080**
Siemens Aktiengesellschaft	Germany	*175*	75-27700
Spar Aerospace Ltd	Canada	*76*	75-13000
Sparton of Canada Ltd	Canada	*77*	75-13100
SRI International	USA	*643*	80-29600
Technology Service Corp.	USA	*644*	80-31930
Tracor Inc	USA	*623*	75-90000
Universal Avionics Systems Corp.	USA	*625*	75-92650
Watkins-Johnson Co.	USA	*627*	75-95200
Zeiss-Eltro Optronic GmbH	Germany	*177*	75-34700

432 20 30 Surveillance systems, airborne, radio frequency

Cincinnati Electronics Corp.	USA	*564*	75-18640
Cubic Defense Systems Inc.	USA	*567*	75-22900
Elettronica (UK) Ltd	UK	*448*	75-20475
ELTA Electronics Industries Ltd	Israel	*220*	75-10200
GEC-Marconi Avionics Ltd	UK	*454*	75-25320
Icom (UK) Ltd	UK	*460*	75-31070
MacDonald Dettwiler and Associates	Canada	*78*	80-01310
MPB Technologies Inc	Canada	*74*	75-09855
Technology Service Corp.	USA	*644*	80-31930
Vega-Cantley Instrument Company	UK	*490*	75-62740
Watkins-Johnson Co.	USA	*627*	75-95200

432 20 35 Surveillance systems, airborne, electro-optical

AAI Corp.	USA	*545*	75-00020
Andrew Sensor Products	USA	*554*	75-07452
Computer Optics Inc.	USA	*565*	75-20680
Contraves Inc	USA	*566*	75-21850
Delft Sensor Systems	**Belgium**	***46***	**75-02500**
Delft Sensor Systems	Netherlands	*292*	75-01590
Elettronica (UK) Ltd	UK	*448*	75-20475
GEC-Marconi Avionics Ltd	UK	*454*	75-25320
Hawkeye Surveillance Systems Ltd	UK	*503*	80-12240
Hughes Danbury Optical Sys.	USA	*582*	75-44857
Israel Aircraft Industries Ltd	Israel	*220*	75-10200
JAI A/S	Denmark	*103*	75-09400
Kaiser Electronics	USA	*587*	75-51350
Litton Industries Inc.	USA	*590*	75-55400
Recon/Optical Inc.	USA	*608*	75-73930
SFIM	**France**	***144***	**75-35080**

Sira Ltd.	UK	*484*	75-54250
Spar Aerospace Ltd	Canada	*76*	75-13000
SRI International	USA	*643*	80-29600
Technology Service Corp.	USA	*644*	80-31930
Thomson-CSF Optronique	France	*148*	75-45137
W. Vinten Ltd	UK	*491*	75-63353
Zeiss-Eltro Optronic GmbH	Germany	*177*	75-34700

432 20 35 70 Surveillance systems, airborne, television

AAI Corp.	USA	*545*	75-00020
Computer Optics Inc.	USA	*565*	75-20680
Computing Devices Company Ltd	UK	*442*	75-14700
Diamond Electronics Inc.	USA	*569*	75-26450
GEC-Marconi Avionics Ltd	UK	*454*	75-25320
Graflex Inc	USA	*579*	75-40310
Insight Vision Systems Ltd	UK	*461*	75-31530
Israel Aircraft Industries Ltd	Israel	*220*	75-10200
Navtech Systems Limited	UK	*507*	80-18000
Pearpoint Ltd	UK	*473*	75-44270
Regis SpA	Italy	*238*	75-20110
RGB Spectrum	USA	*609*	75-75000
Vega-Cantley Instrument Company	UK	*490*	75-62740

432 20 70 Surveillance equipment, airborne

AAI Corp.	USA	*545*	75-00020
Andrew Sensor Products	USA	*554*	75-07452
Cubic Defense Systems Inc.	USA	*567*	75-22900
ELTA Electronics Industries Ltd	Israel	*220*	75-10200
GEC Marconi Hazeltine Corp	USA	*578*	75-37415
GEC-Marconi Avionics Ltd	UK	*454*	75-25320
Inframetrics Inc.	USA	*585*	75-46750
Israel Aircraft Industries Ltd	Israel	*220*	75-10200
JAI A/S	Denmark	*103*	75-09400
Litton Industries Inc.	USA	*590*	75-55400
MA/COM Antenna & Cable Div.	USA	*593*	75-56953
Pulnix America Inc.	USA	*606*	75-71815
Recon/Optical Inc.	USA	*608*	75-73930
RGB Spectrum	USA	*609*	75-75000
Rohde & Schwarz France Telecom	France	*142*	75-32900
SAGEM SA	**France**	***143***	**75-33950**
Sanders	USA	*612*	75-77850
Technology Service Corp.	USA	*644*	80-31930
Watkins-Johnson Co.	USA	*627*	75-95200

432 20 70 30 Surveillance equipment, radio frequency, airborne

Cincinnati Electronics Corp.	USA	*564*	75-18640
Cubic Defense Systems Inc.	USA	*567*	75-22900
Elettronica (UK) Ltd	UK	*448*	75-20475
ELTA Electronics Industries Ltd	Israel	*220*	75-10200
MacDonald Dettwiler and Associates	Canada	*78*	80-01310
Racal Communications Inc.	USA	*606*	75-72380
Vega-Cantley Instrument Company	UK	*490*	75-62740
Vision Research Co.	UK	*491*	75-63360
Watkins-Johnson Co.	USA	*627*	75-95200

432 20 70 35 70 Television cameras, airborne

AAI Corp.	USA	*545*	75-00020
Cohu Inc.	USA	*564*	75-19400
EEV Ltd	UK	*448*	75-19590
Fabbrica Italiana Apparecchiature	Italy	*234*	75-08550
GEC-Marconi Avionics Ltd	UK	*454*	75-25320
Graflex Inc	USA	*579*	75-40310
Insight Vision Systems Ltd	UK	*461*	75-31530
Israel Aircraft Industries Ltd	Israel	*220*	75-10200
Mekel Engineering Inc.	USA	*595*	75-59850
Pearpoint Ltd	UK	*473*	75-44270
Pulnix America Inc.	USA	*606*	75-71815
Regis SpA	Italy	*238*	75-20110

432 30 Surveillance systems, naval

Alenia-Elsag	Italy	*230*	74-00100
British Aerospace (Systems &	UK	*436*	75-08890
C-Tech Ltd.	Canada	*70*	75-03800
Cubic Defense Systems Inc.	USA	*567*	75-22900
DRS Technologies, Inc.	USA	*570*	75-27400
EL-OP Electro Optics Industries Ltd	Israel	*219*	75-06300
Elesco Oy	Finland	*118*	75-00300
Elettronica (UK) Ltd	UK	*448*	75-20475
Fraser-Volpe Corp.	USA	*576*	75-35950
GEC Marconi Hazeltine Corp	USA	*578*	75-37415
GEC-Marconi Avionics Ltd	UK	*454*	75-25320
Int'l Submarine Engrg.	Canada	*73*	75-07780
Israel Aircraft Industries Ltd	Israel	*219*	75-10200
JAI A/S	Denmark	*103*	75-09400
KEL Aerospace Pty Ltd	Australia	*18*	75-11440
Litton Industries Inc.	USA	*590*	75-55400
M. Paz Logistics Ltd	Israel	*224*	80-01600
MA/COM Antenna & Cable Div.	USA	*593*	75-56953
Pulnix America Inc.	USA	*606*	75-71815
Racal Communications Inc.	USA	*606*	75-72380
Radamec Defence Systems Ltd	UK	*477*	75-47750
Rand Instrument Corp.	USA	*607*	75-72900
Raytheon E-Systems	USA	*608*	75-73302
Recon/Optical Inc.	USA	*608*	75-73930
RGB Spectrum	USA	*609*	75-75000
Sanders	USA	*612*	75-77850
SFIM	**France**	***144***	**75-35080**
Spar Aerospace Ltd	Canada	*76*	75-13000
SRI International	USA	*643*	80-29600
STN ATLAS Elektronik GmbH	Germany	*165*	75-01150
Technology Service Corp.	USA	*644*	80-31930
Watkins-Johnson Co.	USA	*627*	75-95200

432 30 10 Surveillance systems, acoustic, naval

C-Tech Ltd.	Canada	*70*	75-03800
DRS Technologies, Inc.	USA	*570*	75-27400
Elesco Oy	Finland	*118*	75-00300
Fugro-UDI Ltd	UK	*452*	75-24885
NEC Corporation	Japan	*253*	75-06350
Technology Service Corp.	USA	*644*	80-31930

432 30 10 30 Hydrophone arrays, naval

C-Tech Ltd.	Canada	*70*	75-03800
Elesco Oy	Finland	*118*	75-00300
International Transducer Corp.	USA	*585*	75-48650
Litton Industries Inc.	USA	*590*	75-55400
NEC Corporation	Japan	*253*	75-06350
Sparton Electronics	USA	*616*	75-82900
Underwater & Environmental	UK	*490*	75-61900

432 30 30 Surveillance systems, naval, radio frequency

Cubic Defense Systems Inc.	USA	*567*	75-22900
Elettronica (UK) Ltd	UK	*448*	75-20475
ELTA Electronics Industries Ltd	Israel	*220*	75-10200
Icom (UK) Ltd	UK	*460*	75-31070
KEL Aerospace Pty Ltd	Australia	*18*	75-11440
Litton Industries Inc.	USA	*590*	75-55400
Racal Communications Inc.	USA	*606*	75-72380
Rohde & Schwarz GmbH & Co KG	Germany	*174*	75-25400
Sanders	USA	*612*	75-77850
Technology Service Corp.	USA	*644*	80-31930
Vega-Cantley Instrument Company	UK	*490*	75-62740
Watkins-Johnson Co.	USA	*627*	75-95200

432 30 35 Surveillance systems, naval, electro-optical

Alenia-Elsag	Italy	*230*	74-00100
Andrew Sensor Products	USA	*554*	75-07452
Computer Optics Inc.	USA	*565*	75-20680
Delft Sensor Systems	**Belgium**	***46***	**75-02500**
Delft Sensor Systems	Netherlands	*292*	75-01590
EUROCONTROL SpA	Italy	*233*	75-08200
GEC-Marconi Avionics Ltd	UK	*454*	75-25320
Israel Aircraft Industries Ltd	Israel	*220*	75-10200
M. Paz Logistics Ltd	Israel	*224*	80-01600
Radamec Defence Systems Ltd	UK	*477*	75-47750
Recon/Optical Inc.	USA	*608*	75-73930
SAGEM SA	**France**	***143***	**75-33950**
SFIM	**France**	***144***	**75-35080**
Sira Ltd.	UK	*484*	75-54250
Spar Aerospace Ltd	Canada	*76*	75-13000
SRI International	USA	*643*	80-29600

Technology Service Corp.	USA	*644*	80-31930
Thomson-CSF Optronique	France	*148*	75-45137
Zeiss-Eltro Optronic GmbH	Germany	*177*	75-34700

432 30 35 70 Surveillance systems, naval, television

Alenia-Elsag	Italy	*230*	74-00100
Computer Optics Inc.	USA	*565*	75-20680
Diamond Electronics Inc.	USA	*569*	75-26450
DRS Technologies, Inc.	USA	*570*	75-27400
GEC-Marconi Avionics Ltd	UK	*454*	75-25320
Israel Aircraft Industries Ltd	Israel	*220*	75-10200
JAI A/S	Denmark	*103*	75-09400
Kongsberg Simrad Ltd	UK	*463*	75-34100
Navtech Systems Limited	UK	*507*	80-18000
Radamec Defence Systems Ltd	UK	*477*	75-47750
Regis SpA	Italy	*238*	75-20110
RGB Spectrum	USA	*609*	75-75000
SITEP SpA	Italy	*239*	75-22595
Television Installation Services	UK	*488*	75-59250
Vega-Cantley Instrument Company	UK	*490*	75-62740

432 30 65 Surveillance systems, multi-sensor, ship-based

Elettronica (UK) Ltd	UK	*448*	75-20475
GEC-Marconi Avionics Ltd	UK	*454*	75-25320
RGB Spectrum	USA	*609*	75-75000

432 30 70 Surveillance equipment, naval

British Aerospace (Systems &	UK	*436*	75-08890
Cubic Defense Systems Inc.	USA	*567*	75-22900
ELTA Electronics Industries Ltd	Israel	*220*	75-10200
GEC-Marconi Avionics Ltd	UK	*454*	75-25320
Israel Aircraft Industries Ltd	Israel	*220*	75-10200
Litton Industries Inc.	USA	*590*	75-55400
M. Paz Logistics Ltd	Israel	*224*	80-01600
MA/COM Antenna & Cable Div.	USA	*593*	75-56953
Pulnix America Inc.	USA	*606*	75-71815
Radamec Defence Systems Ltd	UK	*477*	75-47750
Recon/Optical Inc.	USA	*608*	75-73930
RGB Spectrum	USA	*609*	75-75000
Rohde & Schwarz France Telecom	France	*142*	75-32900
Safare-Crouzet SA	France	*142*	75-33900
SAGEM SA	**France**	***143***	**75-33950**
Sanders	USA	*612*	75-77850
Sparton of Canada Ltd	Canada	*77*	75-13100
STN ATLAS Elektronik GmbH	Germany	*165*	75-01150
Tadiran Ltd	Israel	*222*	75-21600
Watkins-Johnson Co.	USA	*627*	75-95200

432 30 70 10 Surveillance equipment, acoustic, naval

C-Tech Ltd.	Canada	*70*	75-03800
DRS Technologies, Inc.	USA	*570*	75-27400
GEC Marconi Hazeltine Corp	USA	*578*	75-37415
NEC Corporation	Japan	*253*	75-06350
Sparton Electronics	USA	*616*	75-82900
Technology Service Corp.	USA	*644*	80-31930

432 30 70 30 Surveillance equipment, electronic, radio frequency

Cubic Defense Systems Inc.	USA	*567*	75-22900
Datong Electronics Ltd	UK	*444*	75-16885
ELTA Electronics Industries Ltd	Israel	*220*	75-10200
Hagen	UK	*456*	75-27220
Litton Industries Inc.	USA	*590*	75-55400
MA/COM Antenna & Cable Div.	USA	*593*	75-56953
Sanders	USA	*612*	75-77850
Vega-Cantley Instrument Company	UK	*490*	75-62740
Vision Research Co.	UK	*491*	75-63360
Watkins-Johnson Co.	USA	*627*	75-95200

432 30 70 35 Surveillance equipment, electro-optical, naval

Andrew Sensor Products	USA	*554*	75-07452
Delft Sensor Systems	**Belgium**	***46***	**75-02500**
Delft Sensor Systems	Netherlands	*292*	75-01590
Eloptro	S. Africa	*358*	75-05680
GEC-Marconi Avionics Ltd	UK	*454*	75-25320
M. Paz Logistics Ltd	Israel	*224*	80-01600
Pulnix America Inc.	USA	*606*	75-71815
Radamec Defence Systems Ltd	UK	*477*	75-47750
Recon/Optical Inc.	USA	*608*	75-73930
SAGEM SA	**France**	***143***	**75-33950**
SFIM	**France**	***144***	**75-35080**
Spar Aerospace Ltd	Canada	*76*	75-13000
Technology Service Corp.	USA	*644*	80-31930
Zeiss-Eltro Optronic GmbH	Germany	*177*	75-34700

432 30 70 35 70 Television cameras, naval

Cohu Inc.	USA	*564*	75-19400
Diamond Electronics Inc.	USA	*569*	75-26450
Fabbrica Italiana Apparecchiature	Italy	*234*	75-08550
GEC-Marconi Avionics Ltd	UK	*454*	75-25320
Insight Vision Systems Ltd	UK	*461*	75-31530
Kongsberg Simrad Ltd	UK	*463*	75-34100
Pulnix America Inc.	USA	*606*	75-71815
Radamec Defence Systems Ltd	UK	*477*	75-47750
Regis SpA	Italy	*238*	75-20110
SITEP SpA	Italy	*239*	75-22595

432 30 70 35 72 Television cameras, naval, underwater

Cohu Inc.	USA	*564*	75-19400
DeepSea Power & Light Inc.	USA	*568*	75-24840
Diamond Electronics Inc.	USA	*569*	75-26450
Fugro-UDI Ltd	UK	*452*	75-24885
Ibak Helmut Hunger GmbH & Co KG.	Germany	*170*	75-13580
Insight Vision Systems Ltd	UK	*461*	75-31530
Kongsberg Simrad Inc	USA	*589*	75-53160
Kongsberg Simrad Ltd	UK	*463*	75-34100
Pulnix America Inc.	USA	*606*	75-71815
Regis SpA	Italy	*238*	75-20110

432 30 70 50 Surveillance equipment, optical, naval

Litton Industries Inc.	USA	*590*	75-55400
Radamec Defence Systems Ltd	UK	*477*	75-47750
SFIM	**France**	***144***	**75-35080**

432 30 70 52 Binoculars, naval

Avimo Ltd	UK	*432*	75-04950
Brunton Co.	USA	*561*	75-15460
Fraser-Volpe Corp.	USA	*576*	75-35950
M. Paz Logistics Ltd	Israel	*224*	80-01600
Pyser-SGI Ltd	UK	*476*	75-47300
Warren-Knight Instrument Co.	USA	*627*	75-95150
W.F. Stanley & Co. Ltd.	UK	*486*	75-56350

432 30 70 57 Telescopes, naval

Fraser-Volpe Corp.	USA	*576*	75-35950
M. Paz Logistics Ltd	Israel	*224*	80-01600
W.F. Stanley & Co. Ltd.	UK	*486*	75-56350

432 40 Surveillance systems, ground-based

Amper Programas SA	Spain	*365*	75-00480
Chubb Alarms Ltd	UK	*440*	75-13150
Computing Devices Company Ltd.	UK	*442*	75-14700
Contraves Inc	USA	*566*	75-21850
Cubic Defense Systems Inc.	USA	*567*	75-22900
Delft Sensor Systems	**Belgium**	***46***	**75-02500**
Delft Sensor Systems	Netherlands	*292*	75-01590
DM Aerospace Ltd	UK	*444*	75-17300
EL-OP Electro Optics Industries Ltd	Israel	*219*	75-06300
Elettronica (UK) Ltd	UK	*448*	75-20475
Fraser-Volpe Corp.	USA	*576*	75-35950
GEC Marconi Hazeltine Corp	USA	*578*	75-37415
GEC-Marconi Avionics Ltd	UK	*454*	75-25320
Insight Vision Systems Ltd	UK	*461*	75-31530
Israel Aircraft Industries Ltd	Israel	*219*	75-10200

KEL Aerospace Pty Ltd	Australia	18	75-11440
Litton Industries Inc.	USA	590	75-55400
M. Paz Logistics Ltd	Israel	224	80-01600
MA/COM Antenna & Cable Div.	USA	593	75-56953
NOVO Corporation	Russia	338	75-14200
P W Allen & Co Ltd	UK	429	75-02740
Pilkington Optronics Ltd	UK	426	74-10000
Pulnix America Inc.	USA	606	75-71815
Racal-Comsec Ltd	UK	477	75-47665
Radamec Defence Systems Ltd	UK	477	75-47750
Rand Instrument Corp.	USA	607	75-72900
Raytheon E-Systems	USA	608	75-73302
RGB Spectrum	USA	609	75-75000
Rohde & Schwarz GmbH & Co KG	Germany	174	75-25400
Sanders	USA	612	75-77850
Seaboard Electronics Co.	USA	613	75-78910
Siemens Aktiengesellschaft	Germany	175	75-27700
Spar Aerospace Ltd	Canada	76	75-13000
SRI International	USA	643	80-29600
Traffipax-Vertrieb GmbH	Germany	176	75-31450
Watkins-Johnson Co.	USA	627	75-95200
Winkelmann (UK) Ltd	UK	494	75-66600
Wood & Douglas	UK	494	75-66790
Zettler GmbH Kommunikation und	Germany	177	75-34930

432 40 10 Surveillance systems, acoustic, ground-based

Audiotel International Ltd	UK	432	75-04400
Davies Industrial Communications Ltd	UK	444	75-17010

432 40 30 Surveillance systems, ground-based, radio frequency

Amper Programas SA	Spain	365	75-00480
Audiotel International Ltd	UK	432	75-04400
Cincinnati Electronics Corp.	USA	564	75-18640
Cubic Defense Systems Inc.	USA	567	75-22900
Elettronica (UK) Ltd	UK	448	75-20475
ELTA Electronics Industries Ltd	Israel	220	75-10200
GEC Marconi Hazeltine Corp	USA	578	75-37415
Hagen	UK	456	75-27220
Icom (UK) Ltd	UK	460	75-31070
KEL Aerospace Pty Ltd	Australia	18	75-11440
MPB Technologies Inc	Canada	74	75-09855
NEC Corporation	Japan	253	75-06350
Racal Communications Inc.	USA	606	75-72380
Rohde & Schwarz GmbH & Co KG	Germany	174	75-25400
Sanders	USA	612	75-77850
Seaboard Electronics Co.	USA	613	75-78910
Siemens Aktiengesellschaft	Germany	175	75-27700
Technology Service Corp.	USA	644	80-31930
Traffipax-Vertrieb GmbH	Germany	176	75-31450
Vega-Cantley Instrument Company	UK	490	75-62740
Watkins-Johnson Co.	USA	627	75-95200
Winkelmann (UK) Ltd	UK	494	75-66600

432 40 35 Surveillance systems, electro-optical, ground-based

Andrew Sensor Products	USA	554	75-07452
Arab International Optronics Co.	Egypt	109	74-00500
Computer Optics Inc.	USA	565	75-20680
Contraves Inc	USA	566	75-21850
Delft Sensor Systems	**Belgium**	**46**	**75-02500**
Delft Sensor Systems	Netherlands	292	75-01590
EL-OP Electro Optics Industries Ltd	Israel	219	75-06300
Eloptro	S. Africa	358	75-05680
EUROCONTROL SpA	Italy	233	75-08200
GEC-Marconi Avionics Ltd	UK	454	75-25320
Hawkeye Surveillance Systems Ltd	UK	503	80-12240
Israel Aircraft Industries Ltd	Israel	220	75-10200
Kollsman	USA	589	75-53150
M. Paz Logistics Ltd	Israel	224	80-01600
Pearpoint Ltd	UK	473	75-44270
Pilkington Optronics Ltd	UK	426	74-10000
Pilkington Thorn Optronics	UK	426	74-10000
Radamec Defence Systems Ltd	UK	477	75-47750
Sira Ltd.	UK	484	75-54250
Spar Aerospace Ltd	Canada	76	75-13000
Technology Service Corp.	USA	644	80-31930
Visual Methods Inc.	USA	626	75-94320
Zeiss-Eltro Optronic GmbH	Germany	177	75-34700

432 40 35 70 Surveillance systems, television, ground-based

CCS International Ltd	USA	562	75-17250
Chubb Alarms Ltd	UK	440	75-13150
Diamond Electronics Inc.	USA	569	75-26450
EL-OP Electro Optics Industries Ltd	Israel	219	75-06300
GEC-Marconi Avionics Ltd	UK	454	75-25320
Graflex Inc	USA	579	75-40310
Guardall Alarms New Zealand Ltd	New Zealand	302	75-00780
Israel Aircraft Industries Ltd	Israel	220	75-10200
Kollsman	USA	589	75-53150
Kongsberg Simrad Ltd	UK	463	75-34100
Navtech Systems Limited	UK	507	80-18000
Pearpoint Ltd	UK	473	75-44270
Pulnix America Inc.	USA	606	75-71815
Radamec Defence Systems Ltd	UK	477	75-47750
Regis SpA	Italy	238	75-20110
Television Installation Services	UK	488	75-59250
The Dindima Group Pty Ltd	Australia	15	75-04580
Vega-Cantley Instrument Company	UK	490	75-62740
Vision Products Pty Ltd	Australia	23	75-25000
Visual Methods Inc.	USA	626	75-94320

432 40 40 Surveillance systems, ground-based, infra-red

Avimo Ltd	UK	432	75-04950
Chubb Alarms Ltd	UK	440	75-13150
Contraves Inc	USA	566	75-21850
Delft Sensor Systems	**Belgium**	**46**	**75-02500**
Delft Sensor Systems	Netherlands	292	75-01590
EL-OP Electro Optics Industries Ltd	Israel	219	75-06300
Elettronica (UK) Ltd	UK	448	75-20475
Eloptro	S. Africa	358	75-05680
GEC-Marconi Avionics Ltd	UK	454	75-25320
Kollsman	USA	589	75-53150
Pearpoint Ltd	UK	473	75-44270
Radamec Defence Systems Ltd	UK	477	75-47750
Regis SpA	Italy	238	75-20110
Spar Aerospace Ltd	Canada	76	75-13000
Sparton Technology Inc.	USA	616	75-82950
The Dindima Group Pty Ltd	Australia	15	75-04580
Visual Methods Inc.	USA	626	75-94320
Zeiss-Eltro Optronic GmbH	Germany	177	75-34700

432 40 50 Surveillance systems, optical, ground-based

Andrew Sensor Products	USA	554	75-07452
Arab International Optronics Co.	Egypt	109	74-00500
Computer Optics Inc.	USA	565	75-20680
Contraves Inc	USA	566	75-21850
Delft Sensor Systems	**Belgium**	**46**	**75-02500**
Delft Sensor Systems	Netherlands	292	75-01590
Israel Aircraft Industries Ltd	Israel	220	75-10200
Kollsman	USA	589	75-53150
Litton Industries Inc.	USA	590	75-55400
M. Paz Logistics Ltd	Israel	224	80-01600
Pearpoint Ltd	UK	473	75-44270
Sira Ltd.	UK	484	75-54250
Traffipax-Vertrieb GmbH	Germany	176	75-31450
Warren-Knight Instrument Co.	USA	627	75-95150
Zeiss-Eltro Optronic GmbH	Germany	177	75-34700

432 40 55 Surveillance systems, seismic, ground-based

Chubb Alarms Ltd	UK	440	75-13150
Contraves Inc	USA	566	75-21850
Geoquip Ltd	UK	454	75-25415
Kollsman	USA	589	75-53150
Racal-Comsec Ltd	UK	477	75-47665
Sparton Technology Inc.	USA	616	75-82950

432 40 60 Surveillance systems, thermal imaging, ground-based

Delft Sensor Systems	**Belgium**	**46**	**75-02500**
Delft Sensor Systems	Netherlands	292	75-01590
EL-OP Electro Optics Industries Ltd	Israel	219	75-06300
Eloptro	S. Africa	358	75-05680
GEC-Marconi Avionics Ltd	UK	454	75-25320
Israel Aircraft Industries Ltd	Israel	220	75-10200

Kongsberg Simrad Ltd	UK	*463*	75-34100
Opgal Optronic Industry Ltd	Israel	*221*	75-16200
Pearpoint Ltd	UK	*473*	75-44270
Pilkington Optronics Ltd	UK	*426*	74-10000
Pilkington Thorn Optronics	UK	*426*	74-10000
Regis SpA	Italy	*238*	75-20110
Sira Ltd	UK	*484*	75-54250
The Dindima Group Pty Ltd	Australia	*15*	75-04580
Visual Methods Inc.	USA	*626*	75-94320
Zeiss-Eltro Optronic GmbH	Germany	*177*	75-34700

432 40 65 Surveillance systems, multi-sensor, ground-based

EL-OP Electro Optics Industries Ltd	Israel	*219*	75-06300
Elettronica (UK) Ltd	UK	*448*	75-20475
GEC-Marconi Avionics Ltd	UK	*454*	75-25320
Instro Precision Ltd	UK	*461*	75-31630
Sparton Technology Inc.	USA	*616*	75-82950
Technology Service Corp.	USA	*644*	80-31930
Visual Methods Inc.	USA	*626*	75-94320

432 40 70 Surveillance equipment, ground-based

Angénieux	France	*132*	75-02700
Azienda Torinese Elettronica	Italy	*232*	75-02300
Canadian Marconi Co.	Canada	*69*	75-02600
Chubb Alarms Ltd	UK	*440*	75-13150
Computing Devices Company Ltd	UK	*442*	75-14700
Contemporary Systems Design Ltd	UK	*442*	75-14860
Contraves Inc	USA	*566*	75-21850
Cubic Defense Systems Inc.	USA	*567*	75-22900
Delft Sensor Systems	**Belgium**	***46***	**75-02500**
Delft Sensor Systems	Netherlands	*292*	75-01590
DM Aerospace Ltd	UK	*444*	75-17300
GEC Marconi Hazeltine Corp	USA	*578*	75-37415
GEC-Marconi Avionics Ltd	UK	*454*	75-25320
Israel Aircraft Industries Ltd	Israel	*220*	75-10200
Litton Industries Inc.	USA	*590*	75-55400
M. Paz Logistics Ltd	Israel	*224*	80-01600
MA/COM Antenna & Cable Div.	USA	*593*	75-56953
Nanoquest Defence Products Ltd	UK	*470*	75-41890
P W Allen & Co Ltd	UK	*429*	75-02740
Pearpoint Ltd	UK	*473*	75-44270
Pulnix America Inc.	USA	*606*	75-71815
Pyser-SGI Ltd	UK	*476*	75-47300
Racal Communications Inc.	USA	*606*	75-72380
Radamec Defence Systems Ltd	UK	*477*	75-47750
Rohde & Schwarz France Telecom	France	*142*	75-32900
SAGEM SA	**France**	***143***	**75-33950**
Sanders	USA	*612*	75-77850
SFIM	**France**	***144***	**75-35080**
Surveillance Technology Group	UK	*487*	75-57955
Watkins-Johnson Co.	USA	*627*	75-95200
Winkelmann (UK) Ltd	UK	*494*	75-66600
Zettler GmbH Kommunikation und	Germany	*177*	75-34930

432 40 70 10 Surveillance equipment, acoustic, ground-based

CCS International Ltd	USA	*562*	75-17250
Chubb Alarms Ltd	UK	*440*	75-13150
Davies Industrial Communications Ltd	UK	*444*	75-17010
Lorraine Electronics	UK	*465*	75-36500
Racal Recorders Ltd	UK	*477*	75-47689

432 40 70 30 Surveillance equipment, electronic, ground-based

Advanced Vision Systems	UK	*427*	75-00900
Andrew Ltd	UK	*430*	75-03156
Audiotel International Ltd	UK	*432*	75-04400
Chubb Alarms Ltd	UK	*440*	75-13150
Coherent Optics (Europe) Ltd	UK	*441*	75-14090
Cranfield University	UK	*419*	60-00350
Cubic Defense Systems Inc.	USA	*567*	75-22900
Data Cell Ltd	UK	*500*	80-07720
ERA Technology Ltd	UK	*502*	80-09600
EW Simulation Technology Ltd	UK	*449*	75-22150
Francis & Lewis International Ltd	UK	*452*	75-24500
Fugro-UDI Ltd	UK	*452*	75-24885
GEC-Marconi Research Centre	UK	*419*	60-00700
Geoquip Ltd	UK	*454*	75-25415
Instro Precision Ltd	UK	*461*	75-31630
Kongsberg Simrad Ltd	UK	*463*	75-34100
MA/COM Antenna & Cable Div.	USA	*593*	75-56953
Oerlikon-Contraves SpA	Italy	*237*	75-16800
Penny & Giles Data Systems Ltd	UK	*473*	75-44560
PK Electronic International Ltd	UK	*475*	75-45720
Racal Radio Ltd	UK	*477*	75-47698
Racal Recorders Ltd	UK	*477*	75-47689
Racal-Comsec Ltd	UK	*477*	75-47665
Sanders	USA	*612*	75-77850
Siemens Plessey Systems	UK	*483*	75-53770
Sonic Communications (Int.) Ltd.	UK	*485*	75-55202
Sparton Technology Inc.	USA	*616*	75-82950
Spectronic Denmark	Denmark	*103*	75-19650
Tadiran Ltd	Israel	*222*	75-21600
TRL Technology Ltd	UK	*514*	80-58700
Vega-Cantley Instrument Company	UK	*490*	75-62740
Watkins-Johnson Co.	USA	*627*	75-95200
Winkelmann (UK) Ltd	UK	*494*	75-66600
Wood & Douglas	UK	*494*	75-66790

432 40 70 35 Surveillance equipment, electro-optical, ground-based

ACAL Auriema Ltd	UK	*427*	75-00290
Advanced Vision Systems	UK	*427*	75-00900
Andrew Sensor Products	USA	*554*	75-07452
Arab International Optronics Co.	Egypt	*109*	74-00500
British Aerospace (Systems &	UK	*436*	75-08890
Coherent Optics (Europe) Ltd	UK	*441*	75-14090
Computer Optics Inc.	USA	*565*	75-20680
Contemporary Systems Design Ltd	UK	*442*	75-14860
Contraves Inc	USA	*566*	75-21850
Cranfield University	UK	*419*	60-00350
Custom Cameras Ltd	UK	*444*	75-16444
Data Cell Ltd	UK	*500*	80-07720
Davin Optical Ltd	UK	*444*	75-17100
Delft Sensor Systems	**Belgium**	***46***	**75-02500**
Delft Sensor Systems	Netherlands	*292*	75-01590
DRS Technologies, Inc.	USA	*570*	75-27400
Eloptro	S. Africa	*358*	75-05680
Francis & Lewis International Ltd	UK	*452*	75-24500
Fraser-Volpe Corp.	USA	*576*	75-35950
Fugro-UDI Ltd	UK	*452*	75-24885
GEC-Marconi Avionics Ltd	UK	*454*	75-25320
GEC-Marconi Research Centre	UK	*419*	60-00700
Instro Precision Ltd	UK	*461*	75-31630
Israel Aircraft Industries Ltd	Israel	*220*	75-10200
JAI A/S	Denmark	*103*	75-09400
Kongsberg Simrad Ltd	UK	*463*	75-34100
Litton Industries Inc.	USA	*590*	75-55400
Lyon & Brandfield Ltd.	UK	*465*	75-36720
M. Paz Logistics Ltd	Israel	*224*	80-01600
Nanoquest Defence Products Ltd	UK	*470*	75-41890
Octec Ltd	UK	*471*	75-42950
Oerlikon-Contraves SpA	Italy	*237*	75-16800
Pilkington Thorn Optronics	UK	*426*	74-10000
PK Electronic International Ltd	UK	*475*	75-45720
Radamec Defence Systems Ltd	UK	*477*	75-47750
SAGEM SA	**France**	***143***	**75-33950**
Sira Ltd	UK	*484*	75-54250
Spar Aerospace Ltd	Canada	*76*	75-13000
TRL Technology Ltd	UK	*514*	80-58700
Visual Methods Inc.	USA	*626*	75-94320
Zeiss-Eltro Optronic GmbH	Germany	*177*	75-34700

432 40 70 35 70 Television cameras, ground-based

Chubb Alarms Ltd	UK	*440*	75-13150
Cohu Inc.	USA	*564*	75-19400
EEV Ltd	UK	*448*	75-19590
Fabbrica Italiana Apparecchiature	Italy	*234*	75-08550
Fugro-UDI Ltd	UK	*452*	75-24885
GEC-Marconi Avionics Ltd	UK	*454*	75-25320
Israel Aircraft Industries Ltd	Israel	*220*	75-10200
JAI A/S	Denmark	*103*	75-09400
Kongsberg Simrad Ltd	UK	*463*	75-34100
Pulnix America Inc.	USA	*606*	75-71815
Radamec Defence Systems Ltd	UK	*477*	75-47750
Regis SpA	Italy	*238*	75-20110

Remsdaq Ltd	UK	478	75-49210
Security Research Ltd	UK	482	75-52450
Visual Methods Inc.	USA	626	75-94320

432 40 70 40 Surveillance equipment, infra-red, ground-based

British Aerospace (Systems &	UK	436	75-08890
Central Research Laboratories Ltd	UK	499	80-05895
Chubb Alarms Ltd	UK	440	75-13150
Civil Defence Supply	UK	440	75-13350
Coherent Optics (Europe) Ltd	UK	441	75-14090
Computing Devices Company Ltd	UK	442	75-14700
Contraves Inc	USA	566	75-21850
Cranfield University	UK	419	60-00350
Davin Optical Ltd	UK	444	75-17100
Delft Sensor Systems	**Belgium**	**46**	**75-02500**
Delft Sensor Systems	Netherlands	292	75-01590
DRS Technologies, Inc.	USA	570	75-27400
Elettronica (UK) Ltd	UK	448	75-20475
FJW Optical Systems Inc.	USA	575	75-34480
Francis & Lewis International Ltd	UK	452	75-24500
GEC-Marconi Avionics Ltd	UK	454	75-25320
GEC-Marconi Research Centre	UK	419	60-00700
Inframetrics Inc.	USA	585	75-46750
Insight Vision Systems Ltd	UK	461	75-31530
Instro Precision Ltd	UK	461	75-31630
Israel Aircraft Industries Ltd	Israel	220	75-10200
JAI A/S	Denmark	103	75-09400
Kollsman	USA	589	75-53150
Nanoquest Defence Products Ltd	UK	470	75-41890
Octec Ltd	UK	471	75-42950
Pilkington Optronics Ltd	UK	426	74-10000
Pilkington Thorn Optronics	UK	426	74-10000
Racal Communications Inc.	USA	606	75-72380
Racal-Comsec Ltd	UK	477	75-47665
SAGEM SA	**France**	**143**	**75-33950**
Sira Ltd.	UK	484	75-54250
Spar Aerospace Ltd	Canada	76	75-13000
Sparton Technology Inc.	USA	616	75-82950
The Dindima Group Pty Ltd	Australia	15	75-04580
Thomson-CSF Optronique	France	148	75-45137
TIRAX AG	Switzerland	387	75-14410
TRL Technology Ltd	UK	514	80-58700
Visual Methods Inc.	USA	626	75-94320

432 40 70 40 20 Binoculars, thermal infra-red imaging, ground-based

Delft Sensor Systems	**Belgium**	**46**	**75-02500**
Delft Sensor Systems	Netherlands	292	75-01590
GEC-Marconi Avionics Ltd	UK	454	75-25320
Sira Ltd.	UK	484	75-54250

432 40 70 40 30 Cameras, thermal infra-red imaging, ground-based

Delft Sensor Systems	**Belgium**	**46**	**75-02500**
Delft Sensor Systems	Netherlands	292	75-01590
EEV Canada Ltd	Canada	71	75-04700
Eloptro	S. Africa	358	75-05680
GEC-Marconi Avionics Ltd	UK	454	75-25320
Insight Vision Systems Ltd	UK	461	75-31530
JAI A/S	Denmark	103	75-09400
Litton Industries Inc.	USA	590	75-55400
Pilkington Optronics Ltd	UK	426	74-10000
Regis SpA	Italy	238	75-20110
SAGEM SA	**France**	**143**	**75-33950**
Signaal USFA	Netherlands	295	75-06500
Sira Ltd.	UK	484	75-54250
Thomson-CSF Optronique	France	148	75-45137
Visual Methods Inc.	USA	626	75-94320
Zeiss-Eltro Optronic GmbH	Germany	177	75-34700

432 40 70 40 70 Surveillance equipment, infra-red, vehicle-mounted

Avimo Ltd	UK	432	75-04950
Civil Defence Supply	UK	440	75-13350
Delft Sensor Systems	**Belgium**	**46**	**75-02500**
Delft Sensor Systems	Netherlands	292	75-01590
EL-OP Electro Optics Industries Ltd	Israel	219	75-06300
Eloptro	S. Africa	358	75-05680
GEC-Marconi Avionics Ltd	UK	454	75-25320
Insight Vision Systems Ltd	UK	461	75-31530
Israel Aircraft Industries Ltd	Israel	220	75-10200
JAI A/S	Denmark	103	75-09400
Pilkington Optronics Ltd	UK	426	74-10000
Racal Communications Inc.	USA	606	75-72380
SAGEM SA	**France**	**143**	**75-33950**
SFIM	**France**	**144**	**75-35080**
Sira Ltd.	UK	484	75-54250
Visual Methods Inc.	USA	626	75-94320
Zeiss-Eltro Optronic GmbH	Germany	177	75-34700

432 40 70 50 Surveillance equipment, optical, ground-based

Advanced Vision Systems	UK	427	75-00900
Andrew Sensor Products	USA	554	75-07452
Computer Optics Inc.	USA	565	75-20680
Contraves Inc	USA	566	75-21850
Delft Sensor Systems	**Belgium**	**46**	**75-02500**
Delft Sensor Systems	Netherlands	292	75-01590
Israel Aircraft Industries Ltd	Israel	220	75-10200
Litton Industries Inc.	USA	590	75-55400
M. Paz Logistics Ltd	Israel	224	80-01600
Nanoquest Defence Products Ltd	UK	470	75-41890
Olympus Industrial	UK	472	75-43020
Pearpoint Ltd	UK	473	75-44270
Pilkington Optronics Ltd	UK	426	74-10000
Security Research Ltd	UK	482	75-52450
Sira Ltd.	UK	484	75-54250
W.F. Stanley & Co. Ltd.	UK	486	75-56350
Zeiss-Eltro Optronic GmbH	Germany	177	75-34700

432 40 70 52 Binoculars, army

BCB International Ltd	UK	433	75-06075
Brunton Co.	USA	561	75-15460
Deben Group Industries Ltd	UK	444	75-17270
Heinsoldt Systemtechnick GmbH	Germany	171	75-17800
M. Paz Logistics Ltd	Israel	224	80-01600
Pyser-SGI Ltd	UK	476	75-47300
Sidem (UK) Ltd	UK	511	80-35850
W.F. Stanley & Co. Ltd.	UK	486	75-56350

432 40 70 54 Observation systems, long range optical, ground-based

Armalon Ltd	UK	431	75-03729
Contraves Inc	USA	566	75-21850
EL-OP Electro Optics Industries Ltd	Israel	219	75-06300
Israel Aircraft Industries Ltd	Israel	220	75-10200
JAI A/S	Denmark	103	75-09400
Litton Industries Inc.	USA	590	75-55400
M. Paz Logistics Ltd	Israel	224	80-01600
Nanoquest Defence Products Ltd	UK	470	75-41890
Pyser-SGI Ltd	UK	476	75-47300
Visual Methods Inc.	USA	626	75-94320
Zeiss-Eltro Optronic GmbH	Germany	177	75-34700

432 40 70 55 Periscopes, trench

M. Paz Logistics Ltd	Israel	224	80-01600
Nanoquest Defence Products Ltd	UK	470	75-41890
W.F. Stanley & Co. Ltd.	UK	486	75-56350

432 40 70 60 Surveillance equipment, seismic, ground-based

Chubb Alarms Ltd	UK	440	75-13150
Geoquip Ltd	UK	454	75-25415
Japan Aviation Electronics Ind. Ltd	Japan	252	75-03300
Night Vision Equipment Co. Inc.	USA	599	75-64390
Racal-Comsec Ltd	UK	477	75-47665

433 Night vision systems & equipment

AlliedSignal Aerospace Canada	Canada	*68*	75-00470
Beloma	Belarus	*38*	75-00200
Benthos Inc.	USA	*558*	75-13540
Boeing Australia Limited	Australia	*12*	75-02500
Bospec Neue Technologien	Germany	*166*	75-03420
CCS International Ltd	USA	*562*	75-17250
CIC International Ltd	USA	*563*	75-18570
Coherent Optics (Europe) Ltd	UK	*441*	75-14090
Daimler-Benz Aerospace France SAS	**France**	***134***	**75-12950**
Delft Sensor Systems	**Belgium**	***46***	**75-02500**
Delft Sensor Systems	Netherlands	*292*	75-01590
DRS Technologies, Inc.	USA	*570*	75-27400
ECE GmbH	Germany	*167*	75-07200
EEV Canada Ltd	Canada	*71*	75-04700
EEV Inc.	USA	*571*	75-29500
EEV Ltd	UK	*448*	75-19590
EG&G Inc.	USA	*571*	75-29555
EL-OP Electro Optics Industries Ltd	Israel	*219*	75-06300
Electro Optical Industries Inc.	USA	*573*	75-30450
Electrodynamics Inc.	USA	*572*	75-29690
Electrophysics Corp.	USA	*573*	75-30500
Eloptro	S. Africa	*358*	75-05680
Fraser-Volpe Corp.	USA	*576*	75-35950
GEC-Marconi Avionics Inc.	USA	*577*	75-37413
GEC-Marconi Avionics Ltd	UK	*454*	75-25320
GEC-Marconi Infra-Red Limited	UK	*454*	75-25338
IFI Srl	Italy	*235*	75-11220
Inframetrics Inc.	USA	*585*	75-46750
INIMEX	**Belgium**	***46***	**75-04650**
Instro Precision Ltd	UK	*461*	75-31630
IRT Infrarot-Technik Eiselt	Germany	*170*	75-14640
Israel Aircraft Industries Ltd	Israel	*220*	75-10200
ITT Night Vision	USA	*586*	75-49455
Leica AG	Switzerland	*386*	75-07310
Litton Industries Inc.	USA	*590*	75-55400
Lockheed Martin Corp.	USA	*591*	75-55635
M. Paz Logistics Ltd	Israel	*224*	80-01600
Night Vision Equipment Co. Inc.	USA	*599*	75-64390
NOVO Corporation	Russia	*338*	75-14200
Optechs Korea Inc	Korea (ROK)	*266*	75-03750
Optical Systems Tech., Inc.	USA	*601*	75-66965
Pearpoint Ltd	UK	*473*	75-44270
Pilkington Optronics Ltd	UK	*426*	74-10000
Pilkington Thorn Optronics	UK	*426*	74-10000
Polytec-PI SA	France	*141*	75-30450
Pyser-SGI Ltd	UK	*476*	75-47300
Rafael	Israel	*216*	55-01800
Rand Instrument Corp.	USA	*607*	75-72900
Saab Dynamics AB	Sweden	*378*	75-09100
SAGEM SA	**France**	***143***	**75-33950**
SFIM	**France**	***144***	**75-35080**
SFIM ODS	France	*144*	75-35085
Shield International Ltd	UK	*511*	80-35500
Siemens Switzerland Ltd	Switzerland	*387*	75-11980
Signaal USFA	Netherlands	*295*	75-06500
Sira Ltd.	UK	*484*	75-54250
Spar Aerospace Ltd	Canada	*76*	75-13000
SRL Operations	USA	*617*	75-83550
Technical Products Group Inc.	USA	*620*	75-86764
Technology Service Corp.	USA	*644*	80-31930
Thomson-CSF Optronique	France	*148*	75-45137

433 20 Night vision systems, airborne

AAI Corp.	USA	*545*	75-00020
Andrew Sensor Products	USA	*554*	75-07452
CIC International Ltd	USA	*563*	75-18570
CRL Technologies	Canada	*70*	75-03760
Daimler-Benz Aerospace France SAS	**France**	***134***	**75-12950**
Delft Sensor Systems	**Belgium**	***46***	**75-02500**
Delft Sensor Systems	Netherlands	*292*	75-01590
ECE GmbH	Germany	*167*	75-07200
EG&G Inc.	USA	*571*	75-29555
EL-OP Electro Optics Industries Ltd	Israel	*219*	75-06300
Electrophysics Corp.	USA	*573*	75-30500
GEC-Marconi Avionics Inc.	USA	*577*	75-37413
GEC-Marconi Avionics Ltd	UK	*454*	75-25320
Insight Vision Systems Ltd	UK	*461*	75-31530
Israel Aircraft Industries Ltd	Israel	*220*	75-10200
ITT Night Vision	USA	*586*	75-49455
JAI A/S	Denmark	*103*	75-09400
Kaiser Electronics	USA	*587*	75-51350
Litton Industries Inc.	USA	*590*	75-55400
M. Paz Logistics Ltd	Israel	*224*	80-01600
Night Vision Equipment Co. Inc.	USA	*599*	75-64390
Northrop Grumman Corporation	USA	*599*	75-65450
Oxley Developments Co. Ltd	UK	*472*	75-43400
Pilkington Optronics Ltd	UK	*426*	74-10000
Pulnix America Inc.	USA	*606*	75-71815
Rand Instrument Corp.	USA	*607*	75-72900
Saab Dynamics AB	Sweden	*378*	75-09100
SFIM	**France**	***144***	**75-35080**
SFIM ODS	France	*144*	75-35085
Spar Aerospace Ltd	Canada	*76*	75-13000
Spectrolab Inc.	USA	*617*	75-83100
SRL Operations	USA	*617*	75-83550
Thomson-CSF Optronique	France	*148*	75-45137
Zeiss-Eltro Optronic GmbH	Germany	*177*	75-34700

433 20 60 Night vision systems, thermal imaging, airborne

AAI Corp.	USA	*545*	75-00020
EL-OP Electro Optics Industries Ltd	Israel	*219*	75-06300
Fabbrica Italiana Apparecchiature	Italy	*234*	75-08550
GEC-Marconi Avionics Inc.	USA	*577*	75-37413
GEC-Marconi Avionics Ltd	UK	*454*	75-25320
GEC-Marconi Infra-Red Limited	UK	*454*	75-25338
Inframetrics Inc.	USA	*585*	75-46750
Insight Vision Systems Ltd	UK	*461*	75-31530
Iskra Agency Ltd	UK	*462*	75-32060
Israel Aircraft Industries Ltd	Israel	*220*	75-10200
JAI A/S	Denmark	*103*	75-09400
Kaiser Electronics	USA	*587*	75-51350
Litton Industries Inc.	USA	*590*	75-55400
M. Paz Logistics Ltd	Israel	*224*	80-01600
Pilkington Optronics Ltd	UK	*426*	74-10000
Pilkington Thorn Optronics	UK	*426*	74-10000
Regis SpA	Italy	*238*	75-20110
Saab Dynamics AB	Sweden	*378*	75-09100
SFIM	**France**	***144***	**75-35080**
SFIM ODS	France	*144*	75-35085
Sira Ltd.	UK	*484*	75-54250
Thomson-CSF Optronique	France	*148*	75-45137

433 20 70 Night vision equipment, airborne

Canadian Marconi Co.	Canada	*69*	75-02600
CIC International Ltd	USA	*563*	75-18570
Delft Sensor Systems	**Belgium**	***46***	**75-02500**
Delft Sensor Systems	Netherlands	*292*	75-01590
ECE GmbH	Germany	*167*	75-07200
Electrophysics Corp.	USA	*573*	75-30500
GEC-Marconi Avionics Inc.	USA	*577*	75-37413
GEC-Marconi Avionics Ltd	UK	*454*	75-25320
Insight Vision Systems Ltd	UK	*461*	75-31530
Interface Products Inc.	USA	*585*	75-47300
Israel Aircraft Industries Ltd	Israel	*220*	75-10200
ITT Defense Ltd.	UK	*505*	80-13570
JAI A/S	Denmark	*103*	75-09400
Kaiser Electronics	USA	*587*	75-51350
Litton Industries Inc.	USA	*590*	75-55400
M. Paz Logistics Ltd	Israel	*224*	80-01600
Night Vision Equipment Co. Inc.	USA	*599*	75-64390
Northrop Grumman Corporation	USA	*599*	75-65450
Oxley Developments Co. Ltd	UK	*472*	75-43400
Oxley Inc.	USA	*602*	75-67870
Pilkington Optronics Ltd	UK	*426*	74-10000
Pulnix America Inc.	USA	*606*	75-71815
Saab Dynamics AB	Sweden	*378*	75-09100
SAGEM SA	**France**	***143***	**75-33950**
Saunders-Roe Ltd.	UK	*482*	75-51850
SFIM	**France**	***144***	**75-35080**
SFIM ODS	France	*144*	75-35085
Siemens Aktiengesellschaft	Germany	*175*	75-27700
Signaal USFA	Netherlands	*295*	75-06500
Sira Ltd.	UK	*484*	75-54250
Soderberg Manufacturing Co. Inc.	USA	*616*	75-82200
Spar Aerospace Ltd	Canada	*76*	75-13000
Spectrolab Inc.	USA	*617*	75-83100
SRL Operations	USA	*617*	75-83550
Thomson-CSF Optronique	France	*148*	75-45137

433 20 70 30 Goggles, night vision, aircrew

Bharat Electronics Ltd	India	203	55-00200
CIC International Ltd	USA	563	75-18570
Delft Sensor Systems	**Belgium**	**46**	**75-02500**
Delft Sensor Systems	Netherlands	292	75-01590
EL-OP Electro Optics Industries Ltd	Israel	219	75-06300
GEC-Marconi Avionics Inc.	USA	577	75-37413
GEC-Marconi Avionics Ltd	UK	454	75-25320
Hadland Photonics Ltd	UK	456	75-27200
Hughes Microelectronics Europe Ltd	UK	459	75-30480
ITT Defense Ltd	UK	505	80-13570
ITT Night Vision	USA	586	75-49455
Kaiser Electronics	USA	587	75-51350
Litton Industries Inc.	USA	590	75-55400
M. Paz Logistics Ltd	Israel	224	80-01600
Night Vision Equipment Co. Inc.	USA	599	75-64390
SFIM ODS	France	144	75-35085
Signaal USFA	Netherlands	295	75-06500
Sira Ltd.	UK	484	75-54250
Thomson-CSF Optronique	France	148	75-45137

433 30 Night vision systems, naval

Delft Sensor Systems	**Belgium**	**46**	**75-02500**
Delft Sensor Systems	Netherlands	292	75-01590
ECE GmbH	Germany	167	75-07200
Electrophysics Corp.	USA	573	75-30500
GEC-Marconi Avionics Inc.	USA	577	75-37413
GEC-Marconi Avionics Ltd	UK	454	75-25320
Indal Technologies Inc	Canada	72	75-07740
Israel Aircraft Industries Ltd	Israel	220	75-10200
JAI A/S	Denmark	103	75-09400
Litton Industries Inc.	USA	590	75-55400
M. Paz Logistics Ltd	Israel	224	80-01600
McLennan Marine Ltd	UK	467	75-38080
Night Vision Equipment Co. Inc.	USA	599	75-64390
Northrop Grumman Corporation	USA	599	75-65450
Omi Corp.	USA	601	75-66550
Pearpoint Ltd	UK	473	75-44270
Pilkington Optronics Ltd	UK	426	74-10000
Pulnix America Inc.	USA	606	75-71815
Radamec Defence Systems Ltd	UK	477	75-47750
Rand Instrument Corp.	USA	607	75-72900
Regis SpA	Italy	238	75-20110
SAGEM SA	**France**	**143**	**75-33950**
Saunders-Roe Ltd	UK	482	75-51850
SFIM ODS	France	144	75-35085
Signaal USFA	Netherlands	295	75-06500
Spectrolab Inc.	USA	617	75-83100
Technical Products Group Inc.	USA	620	75-86764
Technology Service Corp.	USA	644	80-31930
Thomson-CSF Optronique	France	148	75-45137

433 30 35 Night vision systems, electro-optical, low-light, naval

Delft Sensor Systems	**Belgium**	**46**	**75-02500**
Delft Sensor Systems	Netherlands	292	75-01590
ECE GmbH	Germany	167	75-07200
Electrophysics Corp.	USA	573	75-30500
EUROCONTROL SpA	Italy	233	75-08200
Fabbrica Italiana Apparecchiature	Italy	234	75-08550
GEC-Marconi Avionics Ltd	UK	454	75-25320
Insight Vision Systems Ltd	UK	461	75-31530
IRT Infrarot-Technik Eiselt	Germany	170	75-14640
Israel Aircraft Industries Ltd	Israel	220	75-10200
JAI A/S	Denmark	103	75-09400
Litton Industries Inc.	USA	590	75-55400
M. Paz Logistics Ltd	Israel	224	80-01600
Night Vision Equipment Co. Inc.	USA	599	75-64390
Norca Machinery Company	USA	599	75-64800
Pearpoint Ltd	UK	473	75-44270
Pilkington Optronics Ltd	UK	426	74-10000
Radamec Defence Systems Ltd	UK	477	75-47750
SFIM ODS	France	144	75-35085
Signaal USFA	Netherlands	295	75-06500
Sira Ltd.	UK	484	75-54250
Technical Products Group Inc.	USA	620	75-86764
Technology Service Corp.	USA	644	80-31930

433 30 35 40 Image intensifier night vision systems, naval

Bharat Electronics Ltd	India	203	55-00200
Delft Sensor Systems	**Belgium**	**46**	**75-02500**
Delft Sensor Systems	Netherlands	292	75-01590
Diamond Electronics Inc.	USA	569	75-26450
Electrophysics Corp.	USA	573	75-30500
Insight Vision Systems Ltd	UK	461	75-31530
JAI A/S	Denmark	103	75-09400
Litton Industries Inc.	USA	590	75-55400
Lyon & Brandfield Ltd.	UK	465	75-36720
M. Paz Logistics Ltd	Israel	224	80-01600
Night Vision Equipment Co. Inc.	USA	599	75-64390
Omi Corp.	USA	601	75-66550
Pilkington Optronics Ltd	UK	426	74-10000
Pilkington Thorn Optronics	UK	426	74-10000
Pulnix America Inc.	USA	606	75-71815
Rand Instrument Corp.	USA	607	75-72900

433 30 35 50 Low-light television (LLTV) systems, naval

Diamond Electronics Inc.	USA	569	75-26450
Electrophysics Corp.	USA	573	75-30500
Fabbrica Italiana Apparecchiature	Italy	234	75-08550
GEC-Marconi Avionics Ltd	UK	454	75-25320
Insight Vision Systems Ltd	UK	461	75-31530
JAI A/S	Denmark	103	75-09400
M. Paz Logistics Ltd	Israel	224	80-01600
Polytec-PI SA	France	141	75-30450
Rand Instrument Corp.	USA	607	75-72900
Regis SpA	Italy	238	75-20110

433 30 35 60 Night vision systems, thermal imaging, shipborne

Delft Sensor Systems	**Belgium**	**46**	**75-02500**
Delft Sensor Systems	Netherlands	292	75-01590
EL-OP Electro Optics Industries Ltd	Israel	219	75-06300
Electrophysics Corp.	USA	573	75-30500
Fabbrica Italiana Apparecchiature	Italy	234	75-08550
GEC-Marconi Avionics Ltd	UK	454	75-25320
GEC-Marconi Infra-Red Limited	UK	454	75-25338
Insight Vision Systems Ltd	UK	461	75-31530
Israel Aircraft Industries Ltd	Israel	220	75-10200
M. Paz Logistics Ltd	Israel	224	80-01600
Omi Corp.	USA	601	75-66550
Pilkington Optronics Ltd	UK	426	74-10000
Regis SpA	Italy	238	75-20110
SAGEM SA	**France**	**143**	**75-33950**
SFIM ODS	France	144	75-35085
Signaal USFA	Netherlands	295	75-06500
Simrad Optronics ASA	Norway	310	75-02620
Sira Ltd.	UK	484	75-54250
Technical Products Group Inc.	USA	620	75-86764

433 30 70 Night vision equipment, naval

Civil Defence Supply	UK	440	75-13350
Delft Sensor Systems	**Belgium**	**46**	**75-02500**
Delft Sensor Systems	Netherlands	292	75-01590
Electrophysics Corp.	USA	573	75-30500
GEC-Marconi Avionics Ltd	UK	454	75-25320
Indal Technologies Inc	Canada	72	75-07740
Insight Vision Systems Ltd	UK	461	75-31530
Interface Products Inc.	USA	585	75-47300
Israel Aircraft Industries Ltd	Israel	220	75-10200
Litton Industries Inc.	USA	590	75-55400
M. Paz Logistics Ltd	Israel	224	80-01600
Night Vision Equipment Co. Inc.	USA	599	75-64390
Pilkington Optronics Ltd	UK	426	74-10000
Rand Instrument Corp.	USA	607	75-72900
Regis SpA	Italy	238	75-20110
SAGEM SA	**France**	**143**	**75-33950**
SFIM ODS	France	144	75-35085
Signaal USFA	Netherlands	295	75-06500
Soderberg Manufacturing Co. Inc.	USA	616	75-82200
Spectrolab Inc.	USA	617	75-83100
Technical Products Group Inc.	USA	620	75-86764

433 30 70 35 Night vision equipment, electro-optical, low-light, naval

Computer Optics Inc.	USA	*565*	75-20680
Delft Sensor Systems	**Belgium**	***46***	**75-02500**
Delft Sensor Systems	Netherlands	*292*	75-01590
DRS Technologies, Inc.	USA	*570*	75-27400
Electrophysics Corp.	USA	*573*	75-30500
Fabbrica Italiana Apparecchiature	Italy	*234*	75-08550
GEC-Marconi Avionics Ltd	UK	*454*	75-25320
Insight Vision Systems Ltd	UK	*461*	75-31530
IRT Infrarot-Technik Eiselt	Germany	*170*	75-14640
Israel Aircraft Industries Ltd	Israel	*220*	75-10200
Litton Industries Inc.	USA	*590*	75-55400
M. Paz Logistics Ltd	Israel	*224*	80-01600
Night Vision Equipment Co. Inc.	USA	*599*	75-64390
Pearpoint Ltd	UK	*473*	75-44270
Pilkington Optronics Ltd	UK	*426*	74-10000
Polytec-PI SA	France	*141*	75-30450
Radamec Defence Systems Ltd	UK	*477*	75-47750
Rand Instrument Corp.	USA	*607*	75-72900
SFIM ODS	France	*144*	75-35085
Signaal USFA	Netherlands	*295*	75-06500
Sira Ltd.	UK	*484*	75-54250

433 30 70 35 20 Binoculars, night vision, naval

Bharat Electronics Ltd	India	*203*	55-00200
Delft Sensor Systems	**Belgium**	***46***	**75-02500**
Delft Sensor Systems	Netherlands	*292*	75-01590
Litton Industries Inc.	USA	*590*	75-55400
M. Paz Logistics Ltd	Israel	*224*	80-01600
Night Vision Equipment Co. Inc.	USA	*599*	75-64390
Rand Instrument Corp.	USA	*607*	75-72900
SFIM ODS	France	*144*	75-35085
Signaal USFA	Netherlands	*295*	75-06500
Simrad Optronics ASA	Norway	*310*	75-02620
Sira Ltd.	UK	*484*	75-54250
Transvaro A.S.	Turkey	*403*	75-02080

433 30 70 35 30 Goggles, low-light vision, underwater

Litton Industries Inc.	USA	*590*	75-55400
M. Paz Logistics Ltd	Israel	*224*	80-01600
Pilkington Optronics Ltd	UK	*426*	74-10000
Rand Instrument Corp.	USA	*607*	75-72900

433 30 70 35 40 Image intensifier night vision equipment, naval

Delft Sensor Systems	**Belgium**	***46***	**75-02500**
Delft Sensor Systems	Netherlands	*292*	75-01590
Electrophysics Corp.	USA	*573*	75-30500
GEC-Marconi Avionics Ltd	UK	*454*	75-25320
Insight Vision Systems Ltd	UK	*461*	75-31530
JAI A/S	Denmark	*103*	75-09400
Litton Industries Inc.	USA	*590*	75-55400
M. Paz Logistics Ltd	Israel	*224*	80-01600
Night Vision Equipment Co. Inc.	USA	*599*	75-64390
Omi Corp.	USA	*601*	75-66550
Pilkington Optronics Ltd	UK	*426*	74-10000
Polytec-PI SA	France	*141*	75-30450
Pulnix America Inc.	USA	*606*	75-71815
Rand Instrument Corp.	USA	*607*	75-72900

433 30 70 35 50 Television cameras, low-light, shipborne

Cohu Inc.	USA	*564*	75-19400
Diamond Electronics Inc.	USA	*569*	75-26450
EEV Ltd	UK	*448*	75-19590
Electrophysics Corp.	USA	*573*	75-30500
Fabbrica Italiana Apparecchiature	Italy	*234*	75-08550
GEC-Marconi Avionics Ltd	UK	*454*	75-25320
Insight Vision Systems Ltd	UK	*461*	75-31530
JAI A/S	Denmark	*103*	75-09400
M. Paz Logistics Ltd	Israel	*224*	80-01600
Pearpoint Ltd	UK	*473*	75-44270
Polytec-PI SA	France	*141*	75-30450
Pulnix America Inc.	USA	*606*	75-71815
Radamec Defence Systems Ltd	UK	*477*	75-47750
Rand Instrument Corp.	USA	*607*	75-72900
Regis SpA	Italy	*238*	75-20110

433 40 Night vision systems, ground-based

AlliedSignal Aerospace Canada	Canada	*68*	75-00470
Avimo Ltd	UK	*432*	75-04950
Bharat Electronics Ltd	India	*203*	55-00200
CCS International Ltd	USA	*562*	75-17250
CIC International Ltd	USA	*563*	75-18570
Contraves Inc	USA	*566*	75-21850
Daimler-Benz Aerospace France SAS	**France**	***134***	**75-12950**
Davin Optical Ltd	UK	*444*	75-17100
Delft Sensor Systems	**Belgium**	***46***	**75-02500**
Delft Sensor Systems	Netherlands	*292*	75-01590
DRS Technologies, Inc.	USA	*570*	75-27400
ECE GmbH	Germany	*167*	75-07200
EEV Canada Ltd	Canada	*71*	75-04700
EEV Inc.	USA	*571*	75-29500
EEV Ltd	UK	*448*	75-19590
EL-OP Electro Optics Industries Ltd	Israel	*219*	75-06300
Electrophysics Corp.	USA	*573*	75-30500
Eloptro	S. Africa	*358*	75-05680
Fraser-Volpe Corp.	USA	*576*	75-35950
GEC-Marconi Avionics Inc.	USA	*577*	75-37413
GEC-Marconi Avionics Ltd	UK	*454*	75-25320
Hensoldt AG	Germany	*169*	75-12450
Hoffman Engineering Corp.	USA	*582*	75-43980
Hughes Elcan Optical Technologies	Canada	*72*	75-07650
Insight Vision Systems Ltd	UK	*461*	75-31530
Israel Aircraft Industries Ltd	Israel	*220*	75-10200
ITT Night Vision	USA	*586*	75-49455
JAI A/S	Denmark	*103*	75-09400
Litton Industries Inc.	USA	*590*	75-55400
M. Paz Logistics Ltd	Israel	*224*	80-01600
McLennan Marine Ltd.	UK	*467*	75-38080
NEC Corporation	Japan	*253*	75-06350
Night Vision Equipment Co. Inc.	USA	*599*	75-64390
NORINCO	China PR	*88*	55-00280
Optechs Korea Inc	Korea (ROK)	*266*	75-03750
Pearpoint Ltd	UK	*473*	75-44270
Pilkington Optronics Ltd	UK	*426*	74-10000
Polytec-PI SA	France	*141*	75-30450
Przemyslowy Instytut Telekomunikacji	**Poland**	***325***	**75-02650**
Pulnix America Inc.	USA	*606*	75-71815
Pyser-SGI Ltd	UK	*476*	75-47300
Rand Instrument Corp.	USA	*607*	75-72900
Regis SpA	Italy	*238*	75-20110
SABCA	Belgium	*46*	75-05590
SAGEM SA	**France**	***143***	**75-33950**
SFIM	**France**	***144***	**75-35080**
Siemens Switzerland Ltd	Switzerland	*387*	75-11980
Signaal USFA	Netherlands	*295*	75-06500
Spar Aerospace Ltd	Canada	*76*	75-13000
Spectrolab Inc.	USA	*617*	75-83100
Technical Products Group Inc.	USA	*620*	75-86764
Thomson-CSF Optronique	France	*148*	75-45137
Zeiss-Eltro Optronic GmbH	Germany	*177*	75-34700

433 40 35 Night vision systems, electro-optical, low-light, ground-based

Advanced Vision Systems	UK	*427*	75-00900
AlliedSignal Aerospace Canada	Canada	*68*	75-00470
Andrew Sensor Products	USA	*554*	75-07452
Arab International Optronics Co.	Egypt	*109*	74-00500
Avimo Ltd	UK	*432*	75-04950
Bharat Electronics Ltd	India	*203*	55-00200
CIC International Ltd	USA	*563*	75-18570
Computer Optics Inc.	USA	*565*	75-20680
Contraves Inc	USA	*566*	75-21850
Davin Optical Ltd	UK	*444*	75-17100
Delft Sensor Systems	**Belgium**	***46***	**75-02500**
Delft Sensor Systems	Netherlands	*292*	75-01590
DRS Technologies, Inc.	USA	*570*	75-27400
ECE GmbH	Germany	*167*	75-07200
EL-OP Electro Optics Industries Ltd	Israel	*219*	75-06300
Electrophysics Corp.	USA	*573*	75-30500
Eloptro	S. Africa	*358*	75-05680
Hensoldt AG	Germany	*169*	75-12450
IRT Infrarot-Technik Eiselt	Germany	*170*	75-14640
Israel Aircraft Industries Ltd	Israel	*220*	75-10200
JAI A/S	Denmark	*103*	75-09400

Litton Industries Inc.	USA	*590*	75-55400
M. Paz Logistics Ltd	Israel	*224*	80-01600
Night Vision Equipment Co. Inc.	USA	*599*	75-64390
Pearpoint Ltd	UK	*473*	75-44270
Pilkington Optronics Ltd	UK	*426*	74-10000
Pilkington Thorn Optronics	UK	*426*	74-10000
Polytec-PI SA	France	*141*	75-30450
Radamec Defence Systems Ltd	UK	*477*	75-47750
Rand Instrument Corp.	USA	*607*	75-72900
SFIM ODS	France	*144*	75-35085
Sira Ltd.	UK	*484*	75-54250
Technical Products Group Inc.	USA	*620*	75-86764
Visual Methods Inc.	USA	*626*	75-94320

433 40 35 40 Image intensifier night vision systems, ground-based

Advanced Vision Systems	UK	*427*	75-00900
AlliedSignal Aerospace Canada	Canada	*68*	75-00470
Arab International Optronics Co.	Egypt	*109*	74-00500
Avimo Ltd	UK	*432*	75-04950
Bharat Electronics Ltd	India	*203*	55-00200
CIC International Ltd	USA	*563*	75-18570
Davin Optical Ltd	UK	*444*	75-17100
Delft Sensor Systems	**Belgium**	***46***	**75-02500**
Delft Sensor Systems	Netherlands	*292*	75-01590
Diamond Electronics Inc.	USA	*569*	75-26450
EEV Canada Ltd	Canada	*71*	75-04700
EL-OP Electro Optics Industries Ltd	Israel	*219*	75-06300
Electrophysics Corp.	USA	*573*	75-30500
Eloptro	S. Africa	*358*	75-05680
GEC-Marconi Avionics Ltd	UK	*454*	75-25320
Hall & Watts (Defence Optics) Group	**UK**	***456***	**75-27650**
Insight Vision Systems Ltd	UK	*461*	75-31530
JAI A/S	Denmark	*103*	75-09400
Leica AG	Switzerland	*386*	75-07310
Litton Industries Inc.	USA	*590*	75-55400
Lyon & Brandfield Ltd.	UK	*465*	75-36720
M. Paz Logistics Ltd	Israel	*224*	80-01600
NEC Corporation	Japan	*253*	75-06350
Night Vision Equipment Co. Inc.	USA	*599*	75-64390
Optical Systems Tech., Inc.	USA	*601*	75-66965
Pearpoint Ltd	UK	*473*	75-44270
Pilkington Optronics Ltd	UK	*426*	74-10000
Pilkington Thorn Optronics	UK	*426*	74-10000
Polytec-PI SA	France	*141*	75-30450
Pulnix America Inc.	USA	*606*	75-71815
Rand Instrument Corp.	USA	*607*	75-72900
SFIM ODS	France	*144*	75-35085
Simrad Optronics ASA	Norway	*310*	75-02620
Sira Ltd.	UK	*484*	75-54250
Visual Methods Inc.	USA	*626*	75-94320

433 40 35 50 Low light television systems, ground-based

Advanced Vision Systems	UK	*427*	75-00900
Delft Sensor Systems	**Belgium**	***46***	**75-02500**
Delft Sensor Systems	Netherlands	*292*	75-01590
Diamond Electronics Inc.	USA	*569*	75-26450
Electrophysics Corp.	USA	*573*	75-30500
Fabbrica Italiana Apparecchiature	Italy	*234*	75-08550
Hensoldt AG	Germany	*169*	75-12450
Insight Vision Systems Ltd	UK	*461*	75-31530
JAI A/S	Denmark	*103*	75-09400
Litton Industries Inc.	USA	*590*	75-55400
M. Paz Logistics Ltd	Israel	*224*	80-01600
Pearpoint Ltd	UK	*473*	75-44270
Polytec-PI SA	France	*141*	75-30450
Pulnix America Inc.	USA	*606*	75-71815
Radamec Defence Systems Ltd	UK	*477*	75-47750
Rand Instrument Corp.	USA	*607*	75-72900
Regis SpA	Italy	*238*	75-20110
The Dindima Group Pty Ltd	Australia	*15*	75-04580
Visual Methods Inc.	USA	*626*	75-94320

433 40 35 60 Night vision systems, thermal imaging, ground-based

CIC International Ltd	USA	*563*	75-18570
Computing Devices Company Ltd	UK	*442*	75-14700
Contraves Inc	USA	*566*	75-21850
Delft Sensor Systems	**Belgium**	***46***	**75-02500**
Delft Sensor Systems	Netherlands	*292*	75-01590
DRS Technologies, Inc.	USA	*570*	75-27400
EL-OP Electro Optics Industries Ltd	Israel	*219*	75-06300
Electrophysics Corp.	USA	*573*	75-30500
GEC-Marconi Avionics Ltd	UK	*454*	75-25320
GEC-Marconi Infra-Red Limited	UK	*454*	75-25338
Inframetrics Inc.	USA	*585*	75-46750
Insight Vision Systems Ltd	UK	*461*	75-31530
IRT Infrarot-Technik Eiselt	Germany	*170*	75-14640
Israel Aircraft Industries Ltd	Israel	*220*	75-10200
JAI A/S	Denmark	*103*	75-09400
Leica AG	Switzerland	*386*	75-07310
Litton Industries Inc.	USA	*590*	75-55400
M. Paz Logistics Ltd	Israel	*224*	80-01600
Pearpoint Ltd	UK	*473*	75-44270
Pilkington Optronics Ltd	UK	*426*	74-10000
Pilkington Thorn Optronics	UK	*426*	74-10000
Regis SpA	Italy	*238*	75-20110
SAGEM SA	**France**	***143***	**75-33950**
SFIM ODS	France	*144*	75-35085
Siemens Switzerland Ltd	Switzerland	*387*	75-11980
Signaal USFA	Netherlands	*295*	75-06500
Sira Ltd.	UK	*484*	75-54250
Visual Methods Inc.	USA	*626*	75-94320

433 40 70 Night vision equipment, ground-based

AlliedSignal Aerospace Canada	Canada	*68*	75-00470
Bharat Electronics Ltd	India	*203*	55-00200
CIC International Ltd	USA	*563*	75-18570
Civil Defence Supply	UK	*440*	75-13350
Computing Devices Company Ltd	UK	*442*	75-14700
Contraves Inc	USA	*566*	75-21850
Davin Optical Ltd	UK	*444*	75-17100
Delft Sensor Systems	**Belgium**	***46***	**75-02500**
Delft Sensor Systems	Netherlands	*292*	75-01590
DRS Technologies, Inc.	USA	*570*	75-27400
EEV Canada Ltd	Canada	*71*	75-04700
EEV Inc.	USA	*571*	75-29500
EEV Ltd	UK	*448*	75-19590
Electrophysics Corp.	USA	*573*	75-30500
Eloptro	S. Africa	*358*	75-05680
GEC-Marconi Avionics Ltd	UK	*454*	75-25320
Hoffman Engineering Corp.	USA	*582*	75-43980
Insight Vision Systems Ltd	UK	*461*	75-31530
Israel Aircraft Industries Ltd	Israel	*220*	75-10200
ITT Defense Ltd	UK	*505*	80-13570
ITT Night Vision	USA	*586*	75-49455
JAI A/S	Denmark	*103*	75-09400
Leica AG	Switzerland	*386*	75-07310
Litton Industries Inc.	USA	*590*	75-55400
M. Paz Logistics Ltd	Israel	*224*	80-01600
NEC Corporation	Japan	*253*	75-06350
Night Vision Equipment Co. Inc.	USA	*599*	75-64390
Optechs Korea Inc	Korea (ROK)	*266*	75-03750
Pearpoint Ltd	UK	*473*	75-44270
Pilkington Optronics Ltd	UK	*426*	74-10000
Rand Instrument Corp.	USA	*607*	75-72900
SAGEM SA	**France**	***143***	**75-33950**
Saunders-Roe Ltd	UK	*482*	75-51850
SFIM ODS	France	*144*	75-35085
Siemens Switzerland Ltd	Switzerland	*387*	75-11980
Simrad Optronics ASA	Norway	*310*	75-02620
Soderberg Manufacturing Co. Inc.	USA	*616*	75-82200
Spectrolab Inc.	USA	*617*	75-83100
Technical Products Group Inc.	USA	*620*	75-86764
Visual Methods Inc.	USA	*626*	75-94320

433 40 70 35 Night vision equipment, electro-optical, ground-based

AlliedSignal Aerospace Canada	Canada	*68*	75-00470
Arab International Optronics Co.	Egypt	*109*	74-00500
Armalon Ltd	UK	*431*	75-03729
CIC International Ltd	USA	*563*	75-18570
Davin Optical Ltd	UK	*444*	75-17100
Delft Sensor Systems	**Belgium**	***46***	**75-02500**
Delft Sensor Systems	Netherlands	*292*	75-01590
DRS Technologies, Inc.	USA	*570*	75-27400
Electrophysics Corp.	USA	*573*	75-30500
Eloptro	S. Africa	*358*	75-05680
GEC-Marconi Avionics Ltd	UK	*454*	75-25320

Product-Supplier Link

Insight Vision Systems Ltd	UK	461	75-31530
Instro Precision Ltd	UK	461	75-31630
IRT Infrarot-Technik Eiselt	Germany	170	75-14640
JAI A/S	Denmark	103	75-09400
Leica AG	Switzerland	386	75-07310
Litton Industries Inc.	USA	590	75-55400
M. Paz Logistics Ltd	Israel	224	80-01600
Night Vision Equipment Co. Inc.	USA	599	75-64390
Pilkington Optronics Ltd	UK	426	74-10000
Pilkington Thorn Optronics	UK	426	74-10000
Radamec Defence Systems Ltd	UK	477	75-47750
Rand Instrument Corp.	USA	607	75-72900
SAGEM SA	**France**	**143**	**75-33950**
SFIM ODS	France	144	75-35085
Siemens Switzerland Ltd	Switzerland	387	75-11980
Simrad Optronics ASA	Norway	310	75-02620
Sira Ltd.	UK	484	75-54250
Technical Products Group Inc.	USA	620	75-86764
Visual Methods Inc.	USA	626	75-94320

433 40 70 35 10 Monoculars, night vision, ground-based

Arab International Optronics Co.	Egypt	109	74-00500
CIC International Ltd	USA	563	75-18570
Davin Optical Ltd	UK	444	75-17100
EEV Canada Ltd	Canada	71	75-04700
EEV Inc.	USA	571	75-29500
EEV Ltd	UK	448	75-19590
Electrophysics Corp.	USA	573	75-30500
Fraser-Volpe Corp.	USA	576	75-35950
GEC-Marconi Avionics Ltd	UK	454	75-25320
ITT Defense Ltd	UK	505	80-13570
Leica AG	Switzerland	386	75-07310
Litton Industries Inc.	USA	590	75-55400
M. Paz Logistics Ltd	Israel	224	80-01600
NEC Corporation	Japan	253	75-06350
Night Vision Equipment Co. Inc.	USA	599	75-64390
Pyser-SGI Ltd	UK	476	75-47300
Rand Instrument Corp.	USA	607	75-72900
SFIM ODS	France	144	75-35085
Simrad Optronics ASA	Norway	310	75-02620

433 40 70 35 20 Binoculars, night vision, ground-based

Bharat Electronics Ltd	India	203	55-00200
Brunton Co.	USA	561	75-15460
CIC International Ltd	USA	563	75-18570
Delft Sensor Systems	**Belgium**	**46**	**75-02500**
Delft Sensor Systems	Netherlands	292	75-01590
Eloptro	S. Africa	358	75-05680
GEC-Marconi Avionics Ltd	UK	454	75-25320
ITT Defense Ltd	UK	505	80-13570
Leica AG	Switzerland	386	75-07310
Litton Industries Inc.	USA	590	75-55400
M. Paz Logistics Ltd	Israel	224	80-01600
NEC Corporation	Japan	253	75-06350
Night Vision Equipment Co. Inc.	USA	599	75-64390
Optical Systems Tech., Inc.	USA	601	75-66965
Rand Instrument Corp.	USA	607	75-72900
SFIM ODS	France	144	75-35085
Simrad Optronics ASA	Norway	310	75-02620
Thomson-CSF Optronique	France	148	75-45137
Transvaro A.S.	Turkey	403	75-02080

433 40 70 35 30 Goggles, night vision, ground-based

AlliedSignal Aerospace Canada	Canada	68	75-00470
Arab International Optronics Co.	Egypt	109	74-00500
Avimo Electro-Optics Pte Ltd	Singapore	348	75-00800
Avimo Ltd	UK	432	75-04950
Bharat Electronics Ltd	India	203	55-00200
CIC International Ltd	USA	563	75-18570
Delft Sensor Systems	**Belgium**	**46**	**75-02500**
Delft Sensor Systems	Netherlands	292	75-01590
Econ Optics Mechanics SA	Greece	188	75-01070
EL-OP Electro Optics Industries Ltd	Israel	219	75-06300
Eloptro	S. Africa	358	75-05680
GEC-Marconi Avionics Ltd	UK	454	75-25320
Hoffman Engineering Corp.	USA	582	75-43980
IRT Infrarot-Technik Eiselt	Germany	170	75-14640
ITT Defense Ltd	UK	505	80-13570
ITT Night Vision	USA	586	75-49455
Lanarealm Ltd	UK	464	75-34430
Leica AG	Switzerland	386	75-07310
Litton Industries Inc.	USA	590	75-55400
M. Paz Logistics Ltd	Israel	224	80-01600
NEC Corporation	Japan	253	75-06350
Night Vision Equipment Co. Inc.	USA	599	75-64390
Optical Systems Tech., Inc.	USA	601	75-66965
Pilkington Thorn Optronics	UK	426	74-10000
SFIM ODS	France	144	75-35085
Simrad Optronics ASA	Norway	310	75-02620
Sira Ltd.	UK	484	75-54250
Thomson-CSF Optronique	France	148	75-45137

433 40 70 35 50 Periscopes, night vision, ground-based

AlliedSignal Aerospace Canada	Canada	68	75-00470
Arab International Optronics Co.	Egypt	109	74-00500
Avimo Ltd	UK	432	75-04950
CIC International Ltd	USA	563	75-18570
Delft Sensor Systems	**Belgium**	**46**	**75-02500**
Delft Sensor Systems	Netherlands	292	75-01590
Fraser-Volpe Corp.	USA	576	75-35950
GEC-Marconi Avionics Ltd	UK	454	75-25320
IRT Infrarot-Technik Eiselt	Germany	170	75-14640
Leica AG	Switzerland	386	75-07310
M. Paz Logistics Ltd	Israel	224	80-01600
Norca Machinery Company	USA	599	75-64800
P W Allen & Co Ltd	UK	429	75-02740
SFIM ODS	France	144	75-35085

433 40 70 35 50 30 Periscopes, day / night driving, armoured vehicle

AlliedSignal Aerospace Canada	Canada	68	75-00470
Arab International Optronics Co.	Egypt	109	74-00500
Avimo Ltd	UK	432	75-04950
CIC International Ltd	USA	563	75-18570
Delft Sensor Systems	**Belgium**	**46**	**75-02500**
Delft Sensor Systems	Netherlands	292	75-01590
Econ Optics Mechanics SA	Greece	188	75-01070
EL-OP Electro Optics Industries Ltd	Israel	219	75-06300
Eloptro	S. Africa	358	75-05680
Fraser-Volpe Corp.	USA	576	75-35950
GEC-Marconi Avionics Ltd	UK	454	75-25320
Hall & Watts (Defence Optics) Group	**UK**	**456**	**75-27650**
M. Paz Logistics Ltd	Israel	224	80-01600
Norca Machinery Company	USA	599	75-64800
Omi Corp.	USA	601	75-66550
Optechs Korea Inc	Korea (ROK)	266	75-03750
Pilkington Optronics Ltd	UK	426	74-10000
SFIM ODS	France	144	75-35085
Signaal USFA	Netherlands	295	75-06500

433 40 70 35 60 Television cameras, low-light, ground-based

Cohu Inc.	USA	564	75-19400
Delft Sensor Systems	**Belgium**	**46**	**75-02500**
Delft Sensor Systems	Netherlands	292	75-01590
Diamond Electronics Inc.	USA	569	75-26450
EEV Canada Ltd	Canada	71	75-04700
EEV Inc.	USA	571	75-29500
EEV Ltd	UK	448	75-19590
Electrophysics Corp.	USA	573	75-30500
Fabbrica Italiana Apparecchiature	Italy	234	75-08550
Insight Vision Systems Ltd	UK	461	75-31530
Israel Aircraft Industries Ltd	Israel	220	75-10200
JAI A/S	Denmark	103	75-09400
M. Paz Logistics Ltd	Israel	224	80-01600
Optical Systems Tech., Inc.	USA	601	75-66965
Pearpoint Ltd	UK	473	75-44270
Polytec-PI SA	France	141	75-30450
Radamec Defence Systems Ltd	UK	477	75-47750
Regis SpA	Italy	238	75-20110

433 40 70 35 70 Sights, night vision, ground-based

AAI Corp.	USA	545	75-00020
AlliedSignal Aerospace Canada	Canada	68	75-00470
Arab International Optronics Co.	Egypt	109	74-00500

Avimo Electro-Optics Pte Ltd	Singapore	348	75-00800
CIC International Ltd	USA	563	75-18570
Davin Optical Ltd	UK	444	75-17100
Delft Sensor Systems	**Belgium**	**46**	**75-02500**
Delft Sensor Systems	Netherlands	292	75-01590
EL-OP Electro Optics Industries Ltd	Israel	219	75-06300
GEC-Marconi Avionics Ltd	UK	454	75-25320
Insight Vision Systems Ltd	UK	461	75-31530
IRT Infrarot-Technik Eiselt	Germany	170	75-14640
Israel Aircraft Industries Ltd	Israel	220	75-10200
JAI A/S	Denmark	103	75-09400
Laser & Atomic Research & Devt	Croatia	94	60-00450
Litton Industries Inc.	USA	590	75-55400
M. Paz Logistics Ltd	Israel	224	80-01600
Night Vision Equipment Co. Inc.	USA	599	75-64390
Pilkington Optronics Ltd	UK	426	74-10000
Pilkington Thorn Optronics	UK	426	74-10000
SAGEM SA	**France**	**143**	**75-33950**
Saunders-Roe Ltd	UK	482	75-51850
SFIM ODS	France	144	75-35085

433 40 70 35 70 10 Sights, night vision, armoured vehicle

AAI Corp.	USA	545	75-00020
Arab International Optronics Co.	Egypt	109	74-00500
Avimo Electro-Optics Pte Ltd	Singapore	348	75-00800
Bharat Electronics Ltd	India	203	55-00200
CIC International Ltd	USA	563	75-18570
Delft Sensor Systems	**Belgium**	**46**	**75-02500**
Delft Sensor Systems	Netherlands	292	75-01590
EL-OP Electro Optics Industries Ltd	Israel	219	75-06300
Eloptro	S. Africa	358	75-05680
GEC-Marconi Avionics Ltd	UK	454	75-25320
IRT Infrarot-Technik Eiselt	Germany	170	75-14640
Israel Aircraft Industries Ltd	Israel	220	75-10200
ITT Night Vision	USA	586	75-49455
JAI A/S	Denmark	103	75-09400
Litton Industries Inc.	USA	590	75-55400
M. Paz Logistics Ltd	Israel	224	80-01600
Omi Corp.	USA	601	75-66550
Pilkington Optronics Ltd	UK	426	74-10000
Pilkington Thorn Optronics	UK	426	74-10000
SAGEM SA	**France**	**143**	**75-33950**
SFIM	**France**	**144**	**75-35080**
SFIM ODS	France	144	75-35085
Signaal USFA	Netherlands	295	75-06500
Spar Aerospace Ltd	Canada	76	75-13000
Zeiss-Eltro Optronic GmbH	Germany	177	75-34700

433 40 70 35 70 30 Sights, night vision, hand-held

AlliedSignal Aerospace Canada	Canada	68	75-00470
Arab International Optronics Co.	Egypt	109	74-00500
Bharat Electronics Ltd	India	203	55-00200
CIC International Ltd	USA	563	75-18570
Davin Optical Ltd	UK	444	75-17100
Delft Sensor Systems	**Belgium**	**46**	**75-02500**
Delft Sensor Systems	Netherlands	292	75-01590
EEV Ltd	UK	448	75-19590
EL-OP Electro Optics Industries Ltd	Israel	219	75-06300
Eloptro	S. Africa	358	75-05680
Fraser-Volpe Corp.	USA	576	75-35950
GEC-Marconi Avionics Ltd	UK	454	75-25320
Insight Vision Systems Ltd	UK	461	75-31530
IRT Infrarot-Technik Eiselt	Germany	170	75-14640
Israel Aircraft Industries Ltd	Israel	220	75-10200
ITT Defense Ltd	UK	505	80-13570
ITT Night Vision	USA	586	75-49455
Litton Industries Inc.	USA	590	75-55400
M. Paz Logistics Ltd	Israel	224	80-01600
NEC Corporation	Japan	253	75-06350
Night Vision Equipment Co. Inc.	USA	599	75-64390
Pilkington Thorn Optronics	UK	426	74-10000
SFIM ODS	France	144	75-35085

433 40 70 35 70 60 Sights, night vision, small arms

AAI Corp.	USA	545	75-00020
Accuracy International Ltd	UK	427	75-00360
Arab International Optronics Co.	Egypt	109	74-00500
Avimo Electro-Optics Pte Ltd	Singapore	348	75-00800
CIC International Ltd	USA	563	75-18570
Davin Optical Ltd	UK	444	75-17100
Delft Sensor Systems	**Belgium**	**46**	**75-02500**
Delft Sensor Systems	Netherlands	292	75-01590
EEV Canada Ltd	Canada	71	75-04700
EEV Inc.	USA	571	75-29500
EEV Ltd	UK	448	75-19590
EL-OP Electro Optics Industries Ltd	Israel	219	75-06300
Eloptro	S. Africa	358	75-05680
GEC-Marconi Avionics Ltd	UK	454	75-25320
Hensoldt AG	Germany	169	75-12450
Inframetrics Inc.	USA	585	75-46750
IRT Infrarot-Technik Eiselt	Germany	170	75-14640
ITT Night Vision	USA	586	75-49455
Litton Industries Inc.	USA	590	75-55400
M. Paz Logistics Ltd	Israel	224	80-01600
Night Vision Equipment Co. Inc.	USA	599	75-64390
Pilkington Thorn Optronics	UK	426	74-10000
Saunders-Roe Ltd	UK	482	75-51850
SFIM ODS	France	144	75-35085
Signaal USFA	Netherlands	295	75-06500

433 40 70 35 70 70 Sights, night, thermal imaging, ground-based

AlliedSignal Aerospace Canada	Canada	68	75-00470
Avimo Ltd	UK	432	75-04950
CIC International Ltd	USA	563	75-18570
Delft Sensor Systems	**Belgium**	**46**	**75-02500**
Delft Sensor Systems	Netherlands	292	75-01590
EL-OP Electro Optics Industries Ltd	Israel	219	75-06300
Eloptro	S. Africa	358	75-05680
GEC-Marconi Avionics Ltd	UK	454	75-25320
GEC-Marconi Infra-Red Limited	UK	454	75-25338
Insight Vision Systems Ltd	UK	461	75-31530
JAI A/S	Denmark	103	75-09400
M. Paz Logistics Ltd	Israel	224	80-01600
Pilkington Optronics Ltd	UK	426	74-10000
Pilkington Thorn Optronics	UK	426	74-10000
Regis SpA	Italy	238	75-20110
SAGEM SA	**France**	**143**	**75-33950**
Sira Ltd.	UK	484	75-54250
Zeiss-Eltro Optronic GmbH	Germany	177	75-34700

433 40 70 35 70 78 Sights, day / night, thermal imaging, armoured vehicle

AlliedSignal Aerospace Canada	Canada	68	75-00470
Avimo Ltd	UK	432	75-04950
CIC International Ltd	USA	563	75-18570
Delft Sensor Systems	**Belgium**	**46**	**75-02500**
Delft Sensor Systems	Netherlands	292	75-01590
EL-OP Electro Optics Industries Ltd	Israel	219	75-06300
Eloptro	S. Africa	358	75-05680
GEC-Marconi Avionics Ltd	UK	454	75-25320
Inframetrics Inc.	USA	585	75-46750
Insight Vision Systems Ltd	UK	461	75-31530
M. Paz Logistics Ltd	Israel	224	80-01600
Omi Corp.	USA	601	75-66550
Pilkington Optronics Ltd	UK	426	74-10000
Pilkington Thorn Optronics	UK	426	74-10000
Regis SpA	Italy	238	75-20110
SAGEM SA	**France**	**143**	**75-33950**
SFIM	**France**	**144**	**75-35080**
SFIM ODS	France	144	75-35085
Sira Ltd.	UK	484	75-54250
Zeiss-Eltro Optronic GmbH	Germany	177	75-34700

433 90 Night vision equipment components

Bertan High Voltage	USA	558	75-13720
CIC International Ltd	USA	563	75-18570
Computing Devices Company Ltd	UK	442	75-14700
Delft Sensor Systems	**Belgium**	**46**	**75-02500**
Delft Sensor Systems	Netherlands	292	75-01590
DRS Technologies, Inc.	USA	570	75-27400
ECE GmbH	Germany	167	75-07200
EEV Ltd	UK	448	75-19590
EG&G Reticon	USA	571	75-29562
EL-OP Electro Optics Industries Ltd	Israel	219	75-06300
Electrodynamics Inc.	USA	572	75-29690
Electrophysics Corp.	USA	573	75-30500
Eloptro	S. Africa	358	75-05680

Product-Supplier Link

GEC-Marconi Infra-Red Limited	UK	*454*	75-25338
Hale Hamilton (Valves) Ltd	UK	*456*	75-27350
Hughes Danbury Optical Sys.	USA	*582*	75-44857
Indal Technologies Inc	Canada	*72*	75-07740
Instro Precision Ltd	UK	*461*	75-31630
Interface Products Inc.	USA	*585*	75-47300
IRT Infrarot-Technik Eiselt	Germany	*170*	75-14640
Litton Industries Inc.	USA	*590*	75-55400
M. Paz Logistics Ltd	Israel	*224*	80-01600
Night Vision Equipment Co. Inc.	USA	*599*	75-64390
Norotos Inc.	USA	*599*	75-65000
Optechs Korea Inc	Korea (ROK)	*266*	75-03750
Oxley Developments Co. Ltd	UK	*472*	75-43400
Oxley Inc.	USA	*602*	75-67870
Page Aerospace Ltd	UK	*472*	75-43505
Pilkington Optronics Ltd	UK	*426*	74-10000
Pilkington Thorn Optronics	UK	*426*	74-10000
Pulnix America Inc.	USA	*606*	75-71815
Saunders-Roe Ltd.	UK	*482*	75-51850
SFIM ODS	France	*144*	75-35085
Visual Methods Inc.	USA	*626*	75-94320

435 Photo / image processing systems & equipment

Alan Gordon Enterprises Inc.	USA	*579*	75-40060
Aydin Corp.	USA	*556*	75-11450
Benthos Inc.	USA	*558*	75-13540
Carl Zeiss Canada Ltd	Canada	*70*	75-02900
Computing Devices Company Ltd.	UK	*442*	75-14700
Litton Industries Inc.	USA	*590*	75-55400
Lockheed Martin Missiles & Space	USA	*592*	75-55640
M. Paz Logistics Ltd	Israel	*224*	80-01600
Mercury Computer Systems Inc.	USA	*595*	75-60220
Photon Beard Ltd	UK	*474*	75-45260
Raytheon E-Systems	USA	*608*	75-73302
W. Vinten Ltd	UK	*491*	75-63353

435 10 Photo / image processing systems

Agfa-Gevaert Ltd	UK	*428*	75-01475
Alan Gordon Enterprises Inc.	USA	*579*	75-40060
British Aerospace (Systems &	UK	*436*	75-08890
Cambridge Consultants Ltd	UK	*499*	80-05300
Cambridge Neurodynamics Ltd	UK	*499*	80-05320
Cartographic Engineering Ltd	UK	*439*	75-11900
Central Research Laboratories Ltd	UK	*499*	80-05895
Computing Devices Company Ltd.	UK	*442*	75-14700
Cranfield University	UK	*419*	60-00350
Custom Cameras Ltd	UK	*444*	75-16444
Data Cell Ltd.	UK	*500*	80-07720
Davin Optical Ltd	UK	*444*	75-17100
GEC-Marconi Research Centre.	UK	*419*	60-00700
Hunting Engineering Ltd	UK	*459*	75-30700
Logica UK Limited.	UK	*506*	80-14350
M. Paz Logistics Ltd	Israel	*224*	80-01600
Matra Marconi Space UK, Ltd	UK	*425*	74-06900
Metrum Information Storage Ltd	UK	*467*	75-38890
Octec Ltd	UK	*471*	75-42950
PhotoTelesis	USA	*604*	75-69810
SAIC Ltd.	UK	*510*	80-30220
Scientific Computers Ltd	UK	*510*	80-31550
Strategic Imaging Systems	UK	*486*	75-57515

435 10 10 Photographic laboratories

Alan Gordon Enterprises Inc.	USA	*579*	75-40060

435 10 11 Photographic laboratories, airportable

Alan Gordon Enterprises Inc.	USA	*579*	75-40060

435 10 70 Photo / image processing equipment

Agfa-Gevaert NV	Belgium	*45*	75-00100
Alan Gordon Enterprises Inc.	USA	*579*	75-40060
Bertan High Voltage	USA	*558*	75-13720
Computing Devices Company Ltd.	UK	*442*	75-14700
Litton Industries Inc.	USA	*590*	75-55400
Logica UK Limited.	UK	*506*	80-14350
M. Paz Logistics Ltd	Israel	*224*	80-01600
Mekel Engineering Inc.	USA	*595*	75-59850
PhotoTelesis	USA	*604*	75-69810

435 10 73 Electronic image processing equipment

Agfa-Gevaert Ltd	UK	*428*	75-01475
Audiotel International Ltd	UK	*432*	75-04400
British Aerospace (Systems &	UK	*436*	75-08890
Cambridge Neurodynamics Ltd	UK	*499*	80-05320
Central Research Laboratories Ltd	UK	*499*	80-05895
Computing Devices Company Ltd.	UK	*442*	75-14700
Cranfield University	UK	*419*	60-00350
Custom Cameras Ltd	UK	*444*	75-16444
Data Cell Ltd.	UK	*500*	80-07720
Davin Optical Ltd	UK	*444*	75-17100
GEC-Marconi Research Centre.	UK	*419*	60-00700
Laser-Scan	UK	*506*	80-14210
Logica UK Limited.	UK	*506*	80-14350
Mercury Computer Systems Inc.	USA	*595*	75-60220
Metrum Information Storage Ltd	UK	*467*	75-38890
MS Instruments plc	UK	*469*	75-41265
PhotoTelesis	USA	*604*	75-69810
Primary Image Ltd	UK	*476*	75-46790
Rutherford Appleton Laboratory	UK	*420*	60-01850
Strategic Imaging Systems	UK	*486*	75-57515
The Dindima Group Pty Ltd	Australia	*15*	75-04580

435 10 74 Image enhancement equipment

Cambridge Neurodynamics Ltd	UK	*499*	80-05320
Central Research Laboratories Ltd	UK	*499*	80-05895
Colorado Video Inc.	USA	*564*	75-19730
Computing Devices Company Ltd.	UK	*442*	75-14700
Cranfield University	UK	*419*	60-00350
Custom Cameras Ltd	UK	*444*	75-16444
Data Cell Ltd.	UK	*500*	80-07720
GEC-Marconi Research Centre.	UK	*419*	60-00700
Insight Vision Systems Ltd	UK	*461*	75-31530
Litton Industries Inc.	USA	*590*	75-55400
Logica UK Limited.	UK	*506*	80-14350
Marine Data Ltd	UK	*466*	75-37288
MK Services Ltd.	UK	*468*	75-40170
MS Instruments plc	UK	*469*	75-41265
NORINCO	China PR	*88*	55-00280
PhotoTelesis	USA	*604*	75-69810
Rutherford Appleton Laboratory	UK	*420*	60-01850
Security Research Ltd	UK	*482*	75-52450
The Dindima Group Pty Ltd	Australia	*15*	75-04580

435 10 76 Photographic processing equipment

Agfa-Gevaert NV	Belgium	*45*	75-00100
Alan Gordon Enterprises Inc.	USA	*579*	75-40060

436 Photo / image interpretation systems & equipment

Agfa-Gevaert Ltd	UK	*428*	75-01475
Aydin Corp.	USA	*556*	75-11450
Cambridge Neurodynamics Ltd	UK	*499*	80-05320
Carl Zeiss Canada Ltd	Canada	*70*	75-02900
Cartographic Engineering Ltd	UK	*439*	75-11900
Central Research Laboratories Ltd	UK	*499*	80-05895
Computing Devices Company Ltd.	UK	*442*	75-14700
Cranfield University	UK	*419*	60-00350
Data Cell Ltd.	UK	*500*	80-07720
Davin Optical Ltd	UK	*444*	75-17100
EG&G Inc.	USA	*571*	75-29555
GEC-Marconi Research Centre.	UK	*419*	60-00700
Litton Industries Inc.	USA	*590*	75-55400
MacDonald Dettwiler and Associates.	Canada	*78*	80-01310
Matra Marconi Space UK, Ltd	UK	*425*	74-06900
Rutherford Appleton Laboratory	UK	*420*	60-01850
SAIC Ltd.	UK	*510*	80-30220
Scientific Computers Ltd	UK	*510*	80-31550
Texas Optoelectronics Inc.	USA	*621*	75-88400
The Richards Corp.	USA	*609*	75-75350
Visual Instrumentation Corp.	USA	*626*	75-94300

436 10 Photo / image interpretation systems

Agfa-Gevaert Ltd	UK	*428*	75-01475
Alan Gordon Enterprises Inc.	USA	*579*	75-40060
Cambridge Neurodynamics Ltd	UK	*499*	80-05320
Cartographic Engineering Ltd	UK	*439*	75-11900
Central Research Laboratories Ltd	UK	*499*	80-05895
Cincinnati Electronics Corp.	USA	*564*	75-18640
Computing Devices Company Ltd	UK	*442*	75-14700
Cranfield University	UK	*419*	60-00350
Data Cell Ltd	UK	*500*	80-07720
EG&G Inc.	USA	*571*	75-29555
GEC-Marconi Research Centre	UK	*419*	60-00700
Hunting Engineering Ltd	UK	*459*	75-30700
Laser-Scan	UK	*506*	80-14210
Litton Industries Inc.	USA	*590*	75-55400
Logica UK Limited	UK	*506*	80-14350
Matra Marconi Space UK, Ltd	UK	*425*	74-06900
Photo-Sonics Inc.	USA	*604*	75-69800
PhotoTelesis	USA	*604*	75-69810
Primary Image Ltd	UK	*476*	75-46790
RGB Spectrum	USA	*609*	75-75000
The Richards Corp.	USA	*609*	75-75350
Visual Instrumentation Corp.	USA	*626*	75-94300

436 70 Photo / image interpretation equipment

Agfa-Gevaert Ltd	UK	*428*	75-01475
Alan Gordon Enterprises Inc.	USA	*579*	75-40060
Cartographic Engineering Ltd	UK	*439*	75-11900
Computing Devices Company Ltd	UK	*442*	75-14700
Cranfield University	UK	*419*	60-00350
Data Cell Ltd	UK	*500*	80-07720
GEC-Marconi Research Centre	UK	*419*	60-00700
Litton Industries Inc.	USA	*590*	75-55400
Logica UK Limited	UK	*506*	80-14350
Photo-Sonics Inc.	USA	*604*	75-69800
Photon Beard Ltd	UK	*474*	75-45260
PhotoTelesis	USA	*604*	75-69810
RGB Spectrum	USA	*609*	75-75000
The Richards Corp.	USA	*609*	75-75350
Visual Instrumentation Corp.	USA	*626*	75-94300

Command, Control, Communications & Information

441 Command, control, communications & information systems

Alcatel Bell	Belgium	*44*	75-00055
Alcatel España SA	Spain	*365*	75-00470
Alenia-Elsag	Italy	*230*	74-00100
American Management Systems	USA	*630*	80-02350
Amper Programas SA	Spain	*365*	75-00480
Anite Systems Ltd	UK	*497*	80-01240
Aspect Computing Pty Ltd	Australia	*24*	80-01150
Astronautics Corp. of America	USA	*555*	75-10000
Audiotel International Ltd	UK	*432*	75-04400
Aydin Corp.	USA	*556*	75-11450
Barrett Communications Pty Ltd	Australia	*12*	75-02050
Boeing Australia Limited	Australia	*12*	75-02500
California Microwave Inc.	USA	*561*	75-16350
Calspan SRL Technology	USA	*540*	60-03000
Cambridge Consultants Ltd	UK	*499*	80-05300
Canadian Marconi Co.	Canada	*69*	75-02600
CEA Technologies Pty Ltd	Australia	*24*	80-03100
CelsiusTech AB	Sweden	*376*	75-03710
Central Research Laboratories Ltd	UK	*499*	80-05895
CIC International Ltd	USA	*563*	75-18570
Cincinnati Electronics Corp.	USA	*564*	75-18640
Computing Devices Company Ltd	UK	*442*	75-14700
Cranfield University	UK	*419*	60-00350
Cubic Defense Systems Inc.	USA	*567*	75-22900
Cyberchron Corp.	USA	*567*	75-23120
Daimler-Benz Aerospace France SAS	**France**	***134***	**75-12950**
Data Sciences Ltd	UK	*500*	80-07740
Datamat Ingegneria dei Sistemi SpA	Italy	*233*	75-06500
Delft Sensor Systems	**Belgium**	***46***	**75-02500**
Delft Sensor Systems	Netherlands	*292*	75-01590
Denro Inc.	USA	*568*	75-25651
Dopra Systems Integration Ltd	UK	*446*	75-18400
ECE GmbH	Germany	*167*	75-07200
EDO Corporation	USA	*571*	75-29250
EDS Defence Ltd	UK	*501*	80-09200
ELTA Electronics Industries Ltd	Israel	*220*	75-10200
Fluor Daniel Inc.	USA	*635*	80-11720
GEC Marconi Hazeltine Corp	USA	*578*	75-37415
GEC-Marconi Communications Ltd	UK	*454*	75-25335
GEC-Marconi Hazeltine Corp.	USA	*577*	75-37414
GEC-Marconi Research Centre	UK	*419*	60-00700
Giat Industries	France	*126*	55-00100
Global Communications Co.	USA	*579*	75-39775
GMV SA	Spain	*367*	80-00800
Hardigg Cases	USA	*580*	75-41930
Harris Corporation	USA	*581*	75-42050
Hollandse Signaalapparaten BV	Netherlands	*293*	75-03300
ICL DESC	UK	*501*	80-08400
Israel Aircraft Industries Ltd	Israel	*219*	75-10200
Israel Military Industries Ltd	Israel	*220*	75-10300
ITT Aerospace/Communications Div.	USA	*586*	75-49455
ITT Avionics	USA	*586*	75-49455
ITT Gilfillan	USA	*586*	75-49455
JPS Communications Inc.	USA	*587*	75-51220
Kaman Sciences Corp.	USA	*637*	80-17600
Litton Industries Inc.	USA	*590*	75-55400
Litton Italia SpA	Italy	*236*	75-13700
Lockheed Martin Corp.	USA	*591*	75-55635
Lockheed Martin Missiles & Space	USA	*592*	75-55640
Logica SA/NV	Belgium	*48*	80-01000
Logica UK Limited	UK	*506*	80-14350
MacDonald Dettwiler and Associates	Canada	*78*	80-01310
MA/COM Antenna & Cable Div.	USA	*593*	75-56953
Messier-Dowty Electronics	Canada	*74*	75-09650
Mitel Telecom Ltd	UK	*468*	75-40150
National Aerospace Laboratory	Netherlands	*290*	60-01400
Pacer Systems Inc.	USA	*602*	75-68000
Philips Traffic & Engineering	Australia	*20*	75-16550
PRB Associates Inc.	USA	*640*	80-24700
PRC Inc.	USA	*640*	80-24800
Racal Antennas Ltd	UK	*476*	75-47656
Racal Avionics Ltd	UK	*477*	75-47660
Racal Communications Inc.	USA	*606*	75-72380
Racal Radar Defence Systems Ltd	UK	*477*	75-47688
Radstone Technology plc	UK	*477*	75-47810
Raytheon E-Systems	USA	*608*	75-73302
Redifon MEL Limited	UK	*478*	75-48530
Remsdaq Ltd	UK	*478*	75-49210
Reunert Technology Systems Ltd	S. Africa	*359*	75-18300
RGB Spectrum	USA	*609*	75-75000
Rockwell	USA	*610*	75-75930
Sanders	USA	*612*	75-77850
Security Research Ltd	UK	*482*	75-52450
Shorts Missile Systems Ltd	**UK**	***483***	**75-53550**
Siemens Atea NV	Belgium	*47*	75-05950
Siemens Plessey Electronic Systems	Australia	*21*	75-19475
Siemens Plessey Systems	UK	*483*	75-53770
Sierra Technologies Inc.	USA	*614*	75-80640
Signaal Communications	Netherlands	*295*	75-06350
SRI International	USA	*643*	80-29600
Synectics Corp.	USA	*643*	80-30750
Tadiran Ltd	Israel	*222*	75-21600
Technology Service Corp.	USA	*644*	80-31930
TERMA Elektronik AS	Denmark	*104*	75-20300
The Boeing Company	USA	*559*	75-14250
Thomson-CSF	France	*148*	75-45100
Van Rietschoten & Houwens	Netherlands	*296*	75-07900
Vickers Shipbuilding & Eng Ltd	UK	*491*	75-63100
Zeta	USA	*629*	75-98190

441 20 Command, control, communications & information systems, airborne

Astronautics Corp. of America	USA	*555*	75-10000
Aydin Corp.	USA	*556*	75-11450
Chelton (Electrostatics) Limited	UK	*439*	75-12700
Cincinnati Electronics Corp.	USA	*564*	75-18640
Cranfield Aerospace Ltd	UK	*500*	80-07200
Cranfield University	UK	*419*	60-00350
Cubic Defense Systems Inc.	USA	*567*	75-22900
Cyberchron Corp.	USA	*567*	75-23120
EASAMS Ltd	UK	*501*	80-09000
ECE GmbH	Germany	*167*	75-07200
EDO Corporation	USA	*571*	75-29250
EDS Defence Ltd	UK	*501*	80-09200

Elbit Systems Ltd	Israel	218	75-05300
ELTA Electronics Industries Ltd	Israel	220	75-10200
GEC Marconi Hazeltine Corp	USA	578	75-37415
GEC-Marconi Hazeltine Corp.	USA	577	75-37414
Global Communications Co.	USA	579	75-39775
Icom (UK) Ltd	UK	460	75-31070
Israel Aircraft Industries Ltd	Israel	219	75-10200
ITT Aerospace/Communications Div.	USA	586	75-49455
ITT Avionics	USA	586	75-49455
ITT Defense Ltd.	UK	505	80-13570
Litton Industries Inc.	USA	590	75-55400
Lockheed Martin Corp.	USA	591	75-55635
Lynwood	UK	465	75-36700
MA/COM Antenna & Cable Div.	USA	593	75-56953
MASS Consultants Limited	UK	506	80-15650
Messier-Dowty Electronics	Canada	74	75-09650
National Aerospace Laboratory	Netherlands	290	60-01400
Northern Airborne Technology Ltd	Canada	74	75-10145
PhotoTelesis	USA	604	75-69810
Racal Acoustics Ltd	UK	476	75-47653
Racal Avionics Ltd	UK	477	75-47660
Racal Radar Defence Systems Ltd.	UK	477	75-47688
Raytheon E-Systems	USA	608	75-73302
Reunert Technology Systems Ltd.	S. Africa	359	75-18300
RGB Spectrum	USA	609	75-75000
Rockwell	USA	610	75-75930
Rockwell International Co.	Italy	233	75-05200
Rutherford Appleton Laboratory	UK	420	60-01850
SAIC Ltd.	UK	510	80-30220
Sanders	USA	612	75-77850
Siemens Atea NV	Belgium	47	75-05950
Signal Computing Ltd.	UK	511	80-36050
SRI International	USA	643	80-29600
Tadiran Ltd	Israel	222	75-21600
Technology Service Corp.	USA	644	80-31930
The Boeing Company	USA	559	75-14250

441 30 Command, control, communications & information systems, naval

Alenia-Elsag	Italy	230	74-00100
Anite Systems Ltd	UK	497	80-01240
Aspect Computing Pty Ltd.	Australia	24	80-01150
Astronautics Corp. of America	USA	555	75-10000
Aydin Corp.	USA	556	75-11450
CelsiusTech AB	Sweden	376	75-03710
Cincinnati Electronics Corp.	USA	564	75-18640
Computing Devices Company Ltd.	UK	442	75-14700
Concurrent Computer Corporation Ltd	UK	442	75-14810
Cranfield University	UK	419	60-00350
Cubic Defense Systems Inc.	USA	567	75-22900
Cyberchron Corp.	USA	567	75-23120
Denro Inc.	USA	568	75-25651
EASAMS Ltd	UK	501	80-09000
ECE GmbH	Germany	167	75-07200
EDO Corporation	USA	571	75-29250
EDS Defence Ltd	UK	501	80-09200
Elbit Systems Ltd	Israel	218	75-05300
GEC Marconi Hazeltine Corp	USA	578	75-37415
GEC-Marconi Communications Ltd	UK	454	75-25335
General Offshore Corporation	USA	636	80-12550
General Offshore (UK) Ltd	UK	503	80-11670
Global Communications Co.	USA	579	75-39775
Harris Corporation	USA	581	75-42050
Hollandse Signaalapparaten BV	Netherlands	293	75-03300
Hunting Engineering Ltd	UK	459	75-30700
ICL DESC	UK	501	80-08400
ICL Enterprises	UK	504	80-13100
Icom (UK) Ltd	UK	460	75-31070
Israel Aircraft Industries Ltd	Israel	219	75-10200
ITT Aerospace/Communications Div.	USA	586	75-49455
ITT Avionics	USA	586	75-49455
JPS Communications Inc.	USA	587	75-51220
Kelvin Hughes Ltd.	UK	463	75-33453
Litton Industries Inc.	USA	590	75-55400
Litton Systems Canada Ltd	Canada	73	75-09100
Logica UK Limited.	UK	506	80-14350
Lynwood	UK	465	75-36700
MacDonald Dettwiler and Associates.	Canada	78	80-01310
MA/COM Antenna & Cable Div.	USA	593	75-56953
MASS Consultants Limited	UK	506	80-15650
Nautronix Ltd	UK	470	75-41900
Northern Airborne Technology Ltd	Canada	74	75-10145
PhotoTelesis	USA	604	75-69810
PRB Associates Inc.	USA	640	80-24700
Racal Radar Defence Systems Ltd.	UK	477	75-47688
Raytheon E-Systems	USA	608	75-73302
Redifon MEL Limited	UK	478	75-48530
RGB Spectrum	USA	609	75-75000
Rockwell	USA	610	75-75930
Saab Dynamics AB	Sweden	378	75-09100
SAIC Ltd.	UK	510	80-30220
Sanders	USA	612	75-77850
Science Systems (Industrial) Ltd	UK	510	80-31600
SED Systems Inc	Canada	79	80-01937
Siemens Atea NV	Belgium	47	75-05950
Siemens Plessey Systems	UK	483	75-53770
Signal Computing Ltd.	UK	511	80-36050
SRI International	USA	643	80-29600
Tadiran Ltd	Israel	222	75-21600
Technology Service Corp.	USA	644	80-31930
TERMA Elektronik AS	Denmark	104	75-20300
Thomson Marconi Sonar Limited	UK	426	74-20000
Van Rietschoten & Houwens	Netherlands	296	75-07900
Vickers Shipbuilding & Eng Ltd	UK	491	75-63100

441 40 Command, control, communications & information systems, ground

Alcatel Bell	Belgium	44	75-00055
Alifabs Design & Construction Ltd	UK	496	80-01000
Anite Systems Ltd	UK	497	80-01240
Aspect Computing Pty Ltd.	Australia	24	80-01150
Astronautics Corp. of America	USA	555	75-10000
Aydin Corp.	USA	556	75-11450
CAE	UK	438	75-11125
California Microwave Inc.	USA	561	75-16350
Calspan SRL Technology	USA	540	60-03000
Canadian Marconi Co.	Canada	69	75-02600
CelsiusTech AB	Sweden	376	75-03710
Central Research Laboratories Ltd	UK	499	80-05895
CIC International Ltd	USA	563	75-18570
Cincinnati Electronics Corp.	USA	564	75-18640
Clark Masts Teksam Ltd	UK	440	75-13550
Computing Devices Company Ltd.	UK	442	75-14700
Concurrent Computer Corporation Ltd	UK	442	75-14810
CORIS	France	134	75-12450
Cranfield University	UK	419	60-00350
Cubic Defense Systems Inc.	USA	567	75-22900
Cyberchron Corp.	USA	567	75-23120
Denro Inc.	USA	568	75-25651
Dopra Systems Integration Ltd	UK	446	75-18400
DSD Ltd	UK	501	80-08700
Dytecna Ltd	UK	501	80-08900
EASAMS Ltd	UK	501	80-09000
ECE GmbH	Germany	167	75-07200
EDO Corporation	USA	571	75-29250
EDS Defence Ltd	UK	501	80-09200
Elbit Systems Ltd	Israel	218	75-05300
GEC Marconi Hazeltine Corp	USA	578	75-37415
GEC-Marconi Communications Ltd	UK	454	75-25335
GEC-Marconi Research Centre.	UK	419	60-00700
Global Communications Co.	USA	579	75-39775
GMV SA	Spain	367	80-00800
Harris Corporation	USA	581	75-42050
Hollandse Signaalapparaten BV	Netherlands	293	75-03300
Hunting Engineering Ltd	UK	459	75-30700
ICL DESC	UK	501	80-08400
ICL Enterprises	UK	504	80-13100
Icom (UK) Ltd	UK	460	75-31070
Israel Aircraft Industries Ltd	Israel	219	75-10200
Israel Military Industries Ltd	Israel	220	75-10300
ITT Aerospace/Communications Div.	USA	586	75-49455
ITT Defense Ltd.	UK	505	80-13570
JPS Communications Inc.	USA	587	75-51220
Laser-Scan	UK	506	80-14210
Litton Industries Inc.	USA	590	75-55400
Logica UK Limited.	UK	506	80-14350
Lynwood	UK	465	75-36700
MacDonald Dettwiler and Associates.	Canada	78	80-01310
MA/COM Antenna & Cable Div.	USA	593	75-56953
MASS Consultants Limited	UK	506	80-15650
National Aerospace Laboratory	Netherlands	290	60-01400
NEC Corporation	Japan	253	75-06350
Panorama Antennas Ltd	UK	472	75-43660
PhotoTelesis	USA	604	75-69810
PRB Associates Inc.	USA	640	80-24700
Racal Antennas Ltd.	UK	476	75-47656

Racal Communications Inc.	USA	606	75-72380
Racal Radar Defence Systems Ltd.	UK	477	75-47688
Racal Radio Ltd	UK	477	75-47698
Racal-Airtech Ltd	UK	476	75-47655
Raytheon E-Systems	USA	608	75-73302
Redifon MEL Limited	UK	478	75-48530
Remsdaq Ltd	UK	478	75-49210
Reunert Technology Systems Ltd	S. Africa	359	75-18300
RGB Spectrum	USA	609	75-75000
Rockwell	USA	610	75-75930
Rutherford Appleton Laboratory	UK	420	60-01850
SAIC Ltd	UK	510	80-30220
Sanders	USA	612	75-77850
Science Systems (Industrial) Ltd	UK	510	80-31600
Securicor Datatrak Ltd	UK	482	75-52390
Security Research Ltd	UK	482	75-52450
SED Systems Inc	Canada	79	80-01937
Shorts Missile Systems Ltd	**UK**	**483**	**75-53550**
Siemens Aktiengesellschaft	Germany	175	75-27700
Siemens Atea NV	Belgium	47	75-05950
Siemens Plessey Systems	UK	483	75-53770
Signal Computing Ltd	UK	511	80-36050
SR Comms Management Ltd	UK	512	80-46000
SRI International	USA	643	80-29600
Sverdrup Corp.	USA	643	80-30650
Sverdrup Technology Inc.	USA	643	80-30680
Tadiran Ltd	Israel	222	75-21600
Technology Service Corp.	USA	644	80-31930
Telegenix Inc.	USA	621	75-87610
TERMA Elektronik AS	Denmark	104	75-20300
The Boeing Company	USA	559	75-14250
Thomson-CSF	France	148	75-45100
TRL Technology Ltd	UK	514	80-58700
Van Rietschoten & Houwens	Netherlands	296	75-07900

442 Command & Control systems

Alenia SpA	Italy	232	75-01170
Alenia-Elsag	Italy	230	74-00100
Anite Systems Ltd	UK	497	80-01240
Arcata Associates Inc.	USA	554	75-08400
Aspect Computing Pty Ltd	Australia	24	80-01150
Astronautics Corp. of America	USA	555	75-10000
Atlantis Aerospace Corp.	Canada	68	75-00750
Aydin Corp.	USA	556	75-11450
Boeing Australia Limited	Australia	12	75-02500
British Aerospace Australia Ltd	Australia	13	75-02750
Calspan SRL Technology	USA	540	60-03000
Canadian Marconi Co.	Canada	69	75-02600
CIC International Ltd	USA	563	75-18570
Cincinnati Electronics Corp.	USA	564	75-18640
Cubic Defense Systems Inc.	USA	567	75-22900
Cyberchron Corp.	USA	567	75-23120
Data Sciences Ltd.	UK	500	80-07740
Datamat Ingegneria dei Sistemi SpA	Italy	233	75-06500
Denro Inc.	USA	568	75-25651
Elbit Systems Ltd	Israel	218	75-05300
ELTA Electronics Industries Ltd	Israel	220	75-10200
Fluor Daniel Inc.	USA	635	80-11720
GEC Marconi Hazeltine Corp	USA	578	75-37415
Global Communications Co.	USA	579	75-39775
Herley-Vega Systems	USA	581	75-43090
Hunting Engineering Ltd	UK	459	75-30700
INDRA	Spain	366	75-03800
Ingenieria de Sistemas	Spain	367	80-00900
Israel Aircraft Industries Ltd	Israel	219	75-10200
ITT Aerospace/Communications Div.	USA	586	75-49455
Kaman Sciences Corp.	USA	637	80-17600
Litton Industries Inc.	USA	590	75-55400
Lockheed Martin Corp.	USA	591	75-55635
Logica SA/NV	Belgium	48	80-01000
Lynwood	UK	465	75-36700
MacDonald Dettwiler and Associates	Canada	78	80-01310
Marconi SpA DTM	Italy	236	75-14550
Motorola Ltd	UK	469	75-41050
National Aerospace Laboratory	Netherlands	290	60-01400
Navigation Data Systems Inc.	USA	598	75-64050
Norden Systems Inc.	USA	599	75-64900
NORINCO	China PR	88	55-00280
Northrop Grumman Corporation	USA	640	80-21840
PRB Associates Inc.	USA	640	80-24700
PRC Inc.	USA	640	80-24800
Racal Communications Inc.	USA	606	75-72380
Racal Radar Defence Systems Ltd.	UK	477	75-47688
Raytheon E-Systems	USA	608	75-73302
Redifon MEL Limited	UK	478	75-48530
Reunert Technology Systems Ltd	S. Africa	359	75-18300
RGB Spectrum	USA	609	75-75000
Rockwell	USA	610	75-75930
Sanders	USA	612	75-77850
Siemens Aktiengesellschaft	Germany	175	75-27700
Siemens Atea NV	Belgium	47	75-05950
Siemens Plessey Electronic Systems	Australia	21	75-19475
Siemens Plessey Systems	UK	483	75-53770
SRI International	USA	643	80-29600
State College Operations	USA	617	75-83990
Tadiran Ltd	Israel	222	75-21600
Technology Service Corp.	USA	644	80-31930
The Boeing Company	USA	559	75-14250
Van Rietschoten & Houwens	Netherlands	296	75-07900
VDO-Luftfahrtgeräte Werk GmbH.	Germany	176	75-32600

442 20 Command & control systems, airborne

Aeroelectrónica Industria	Brazil	55	75-00585
Arcata Associates Inc.	USA	554	75-08400
Aspect Computing Pty Ltd	Australia	24	80-01150
Astronautics Corp. of America	USA	555	75-10000
Aydin Corp.	USA	556	75-11450
British Aerospace Australia Ltd	Australia	13	75-02750
Central Research Laboratories Ltd	UK	499	80-05895
CIC International Ltd	USA	563	75-18570
Cincinnati Electronics Corp.	USA	564	75-18640
Cubic Defense Systems Inc.	USA	567	75-22900
Cyberchron Corp.	USA	567	75-23120
EASAMS Ltd	UK	501	80-09000
ELTA Electronics Industries Ltd	Israel	220	75-10200
GEC Marconi Hazeltine Corp	USA	578	75-37415
GEC-Marconi Research Centre	UK	419	60-00700
Global Communications Co.	USA	579	75-39775
Israel Aircraft Industries Ltd	Israel	219	75-10200
ITT Aerospace/Communications Div.	USA	586	75-49455
ITT Avionics	USA	586	75-49455
Litton Industries Inc.	USA	590	75-55400
Lynwood	UK	465	75-36700
Motorola Ltd	UK	469	75-41050
National Aerospace Laboratory	Netherlands	290	60-01400
Navigation Data Systems Inc.	USA	598	75-64050
Northern Airborne Technology Ltd	Canada	74	75-10145
Northrop Grumman Corporation	USA	640	80-21840
PRB Associates Inc.	USA	640	80-24700
Racal Radar Defence Systems Ltd.	UK	477	75-47688
Raytheon E-Systems	USA	608	75-73302
RGB Spectrum	USA	609	75-75000
Rockwell	USA	610	75-75930
Rutherford Appleton Laboratory	UK	420	60-01850
SAIC Ltd	UK	510	80-30220
Sanders	USA	612	75-77850
Siemens Atea NV	Belgium	47	75-05950
SRI International	USA	643	80-29600
Tadiran Ltd	Israel	222	75-21600
Technology Service Corp.	USA	644	80-31930
The Boeing Company	USA	559	75-14250
The Titan Corp	USA	622	75-89620
VDO-Luftfahrtgeräte Werk GmbH.	Germany	176	75-32600

442 20 70 Command & control equipment, airborne

Anite Systems Ltd	UK	497	80-01240
Arcata Associates Inc.	USA	554	75-08400
Astronautics Corp. of America	USA	555	75-10000
Aydin Corp.	USA	556	75-11450
British Aerospace Australia Ltd	Australia	13	75-02750
Cincinnati Electronics Corp.	USA	564	75-18640
Cyberchron Corp.	USA	567	75-23120
ELTA Electronics Industries Ltd	Israel	220	75-10200
GEC Marconi Hazeltine Corp	USA	578	75-37415
Howden Fluid Systems	USA	582	75-44400
ITT Avionics	USA	586	75-49455
Litton Industries Inc.	USA	590	75-55400
Logica UK Limited	UK	506	80-14350
Northrop Grumman Corporation	USA	640	80-21840
PhotoTelesis	USA	604	75-69810
Racal Communications Inc.	USA	606	75-72380
Racal Radar Defence Systems Ltd.	UK	477	75-47688
Raytheon E-Systems	USA	608	75-73302
RGB Spectrum	USA	609	75-75000

Rockwell	USA	*610*	75-75930
Sanders	USA	*612*	75-77850
Siemens Atea NV	Belgium	*47*	75-05950
Siemens Plessey Systems	UK	*483*	75-53770
Tadiran Ltd	Israel	*222*	75-21600
Technology Service Corp.	USA	*644*	80-31930

442 30 Command & control systems, naval

Alenia-Elsag	Italy	*230*	74-00100
Aspect Computing Pty Ltd	Australia	*24*	80-01150
Astronautics Corp. of America	USA	*555*	75-10000
Aydin Corp.	USA	*556*	75-11450
BAeSEMA Ltd.	UK	*498*	80-03070
CAE	USA	*438*	75-11125
Calspan SRL Technology	USA	*540*	60-03000
Cambridge Consultants Ltd	UK	*499*	80-05300
CelsiusTech AB	Sweden	*376*	75-03710
Central Research Laboratories Ltd	UK	*499*	80-05895
Cincinnati Electronics Corp.	USA	*564*	75-18640
Concurrent Computer Corporation Ltd	UK	*442*	75-14810
Cray Marine	UK	*443*	75-15915
Cubic Defense Systems Inc.	USA	*567*	75-22900
Cyberchron Corp.	USA	*567*	75-23120
Datamat Ingegneria dei Sistemi SpA	Italy	*233*	75-06500
Davies Industrial Communications Ltd	UK	*444*	75-17010
EASAMS Ltd	UK	*501*	80-09000
GEC Marconi Hazeltine Corp	USA	*578*	75-37415
General Offshore Corporation	USA	*636*	80-12550
General Offshore (UK) Ltd	UK	*503*	80-11670
Global Communications Co.	USA	*579*	75-39775
Hunting Engineering Ltd	UK	*459*	75-30700
Israel Aircraft Industries Ltd	Israel	*219*	75-10200
ITT Aerospace/Communications Div.	USA	*586*	75-49455
ITT Avionics	USA	*586*	75-49455
ITT Gilfillan	USA	*586*	75-49455
JDR Cable Systems Ltd	UK	*462*	75-32480
Kelvin Hughes Ltd.	UK	*463*	75-33453
Litton Industries Inc.	USA	*590*	75-55400
Logica UK Limited.	UK	*506*	80-14350
Lynwood	UK	*465*	75-36700
MacDonald Dettwiler and Associates.	Canada	*78*	80-01310
Navigation Data Systems Inc.	USA	*598*	75-64050
Northrop Grumman Corporation.	USA	*640*	80-21840
Park Air Electronics Ltd	UK	*472*	75-43750
PRB Associates Inc.	USA	*640*	80-24700
Racal Radar Defence Systems Ltd.	UK	*477*	75-47688
Raytheon E-Systems	USA	*608*	75-73302
RGB Spectrum	USA	*609*	75-75000
Rockwell	USA	*610*	75-75930
Saab Dynamics AB	Sweden	*378*	75-09100
SAIC Ltd.	UK	*510*	80-30220
Sanders	USA	*612*	75-77850
SED Systems Inc	Canada	*79*	80-01937
Siemens Atea NV	Belgium	*47*	75-05950
Siemens Plessey Systems	UK	*483*	75-53770
SRI International	USA	*643*	80-29600
Tadiran Ltd	Israel	*222*	75-21600
Technology Service Corp.	USA	*644*	80-31930
TERMA Elektronik AS	Denmark	*104*	75-20300
The Titan Corp	USA	*622*	75-89620
Van Rietschoten & Houwens	Netherlands	*296*	75-07900
Whippendell Electrical Ltd	UK	*493*	75-65800
WS Atkins plc.	UK	*497*	80-01550

442 30 70 Command & control equipment, naval

Anite Systems Ltd	UK	*497*	80-01240
Astronautics Corp. of America	USA	*555*	75-10000
Aydin Corp.	USA	*556*	75-11450
CelsiusTech AB	Sweden	*376*	75-03710
Cincinnati Electronics Corp.	USA	*564*	75-18640
Cyberchron Corp.	USA	*567*	75-23120
GEC Marconi Hazeltine Corp	USA	*578*	75-37415
Int'l Submarine Engrg.	Canada	*73*	75-07780
ITT Avionics	USA	*586*	75-49455
Litton Industries Inc.	USA	*590*	75-55400
Logica UK Limited.	UK	*506*	80-14350
Northrop Grumman Corporation.	USA	*640*	80-21840
PhotoTelesis	USA	*604*	75-69810
Racal Communications Inc.	USA	*606*	75-72380
Raytheon E-Systems	USA	*608*	75-73302
RGB Spectrum	USA	*609*	75-75000
Rockwell	USA	*610*	75-75930
Sanders	USA	*612*	75-77850
SED Systems Inc	Canada	*79*	80-01937
Siemens Atea NV	Belgium	*47*	75-05950
Siemens Plessey Systems	UK	*483*	75-53770
Tadiran Ltd	Israel	*222*	75-21600
Technology Service Corp.	USA	*644*	80-31930

442 40 Command & control systems, ground-based

Alenia SpA	Italy	*232*	75-01170
Anite Systems Ltd	UK	*497*	80-01240
Aspect Computing Pty Ltd	Australia	*24*	80-01150
Astronautics Corp. of America	USA	*555*	75-10000
Aydin Corp.	USA	*556*	75-11450
CelsiusTech AB	Sweden	*376*	75-03710
Central Research Laboratories Ltd	UK	*499*	80-05895
CIC International Ltd	USA	*563*	75-18570
Cincinnati Electronics Corp.	USA	*564*	75-18640
Concurrent Computer Corporation Ltd	UK	*442*	75-14810
CORIS	France	*134*	75-12450
Cubic Defense Systems Inc.	USA	*567*	75-22900
Cyberchron Corp.	USA	*567*	75-23120
Dytecna Ltd	UK	*501*	80-08900
EASAMS Ltd	UK	*501*	80-09000
Giat Industries	France	*126*	55-00100
Global Communications Co.	USA	*579*	75-39775
Harris Corporation	USA	*581*	75-42050
Herley-Vega Systems	USA	*581*	75-43090
Hunting Engineering Ltd	UK	*459*	75-30700
Israel Aircraft Industries Ltd	Israel	*219*	75-10200
ITT Aerospace/Communications Div.	USA	*586*	75-49455
ITT Gilfillan	USA	*586*	75-49455
Laser-Scan	UK	*506*	80-14210
Litton Industries Inc.	USA	*590*	75-55400
Logica UK Limited.	UK	*506*	80-14350
Lynwood	UK	*465*	75-36700
MacDonald Dettwiler and Associates.	Canada	*78*	80-01310
Marconi SpA DTM	Italy	*236*	75-14550
Motorola Ltd	UK	*469*	75-41050
Navigation Data Systems Inc.	USA	*598*	75-64050
NEC Corporation	Japan	*253*	75-06350
Northrop Grumman Corporation.	USA	*640*	80-21840
Oerlikon-Contraves SpA	Italy	*237*	75-16800
Park Air Electronics Ltd	UK	*472*	75-43750
Racal Antennas Ltd	UK	*476*	75-47656
Racal Radar Defence Systems Ltd.	UK	*477*	75-47688
Racal Radio Ltd	UK	*477*	75-47698
Raytheon E-Systems	USA	*608*	75-73302
Reunert Technology Systems Ltd	S. Africa	*359*	75-18300
RGB Spectrum	USA	*609*	75-75000
Rockwell	USA	*610*	75-75930
Rutherford Appleton Laboratory	UK	*420*	60-01850
SAIC Ltd	UK	*510*	80-30220
Sanders	USA	*612*	75-77850
SED Systems Inc	Canada	*79*	80-01937
Serco Aviation Services	UK	*511*	80-34150
Shorts Missile Systems Ltd	**UK**	***483***	**75-53550**
Siemens Aktiengesellschaft	Germany	*175*	75-27700
Siemens Atea NV	Belgium	*47*	75-05950
Siemens Plessey Systems	UK	*483*	75-53770
Sierra Technologies Inc.	USA	*614*	75-80640
SRI International	USA	*643*	80-29600
Sverdrup Corp.	USA	*643*	80-30650
Tadiran Ltd	Israel	*222*	75-21600
Technology Service Corp.	USA	*644*	80-31930
TERMA Elektronik AS	Denmark	*104*	75-20300
The Boeing Company	USA	*559*	75-14250
The Titan Corp	USA	*622*	75-89620
Thomson-CSF	France	*148*	75-45100
TRL Technology Ltd	UK	*514*	80-58700
Van Rietschoten & Houwens	Netherlands	*296*	75-07900
WS Atkins plc.	UK	*497*	80-01550

442 40 50 Battlefield information & control systems

CelsiusTech AB	Sweden	*376*	75-03710
CIC International Ltd	USA	*563*	75-18570
Cyberchron Corp.	USA	*567*	75-23120
GEC Marconi Hazeltine Corp	USA	*578*	75-37415
Hunting Engineering Ltd	UK	*459*	75-30700
Laser-Scan	UK	*506*	80-14210

Marconi SpA DTM	Italy	*236*	75-14550
Motorola Ltd	UK	*469*	75-41050
NEC Corporation	Japan	*253*	75-06350
PhotoTelesis	USA	*604*	75-69810
RGB Spectrum	USA	*609*	75-75000
Sanders	USA	*612*	75-77850
SFIM	**France**	***144***	**75-35080**
Siemens Atea NV	Belgium	*47*	75-05950
Technology Service Corp.	USA	*644*	80-31930

442 40 70 Command & control equipment, ground-based

Alenia SpA	Italy	*232*	75-01170
Anite Systems Ltd	UK	*497*	80-01240
Astronautics Corp. of America	USA	*555*	75-10000
Aydin Corp.	USA	*556*	75-11450
Calspan SRL Technology	USA	*540*	60-03000
CelsiusTech AB	Sweden	*376*	75-03710
CIC International Ltd	USA	*563*	75-18570
Cincinnati Electronics Corp.	USA	*564*	75-18640
CORIS	France	*134*	75-12450
Cyberchron Corp.	USA	*567*	75-23120
Davies Industrial Communications Ltd	UK	*444*	75-17010
Elesco Oy	Finland	*118*	75-00300
GEC Marconi Hazeltine Corp	USA	*578*	75-37415
ICL DESC	UK	*501*	80-08400
Laser-Scan	UK	*506*	80-14210
Litton Industries Inc.	USA	*590*	75-55400
Logica UK Limited	UK	*506*	80-14350
Marconi SpA DTM	Italy	*236*	75-14550
NEC Corporation	Japan	*253*	75-06350
Northrop Grumman Corporation	USA	*640*	80-21840
PhotoTelesis	USA	*604*	75-69810
Racal Communications Inc.	USA	*606*	75-72380
Raytheon E-Systems	USA	*608*	75-73302
Reunert Technology Systems Ltd	S. Africa	*359*	75-18300
RGB Spectrum	USA	*609*	75-75000
Rockwell	USA	*610*	75-75930
Sanders	USA	*612*	75-77850
SED Systems Inc	Canada	*79*	80-01937
SFIM	**France**	***144***	**75-35080**
Siemens Atea NV	Belgium	*47*	75-05950
Siemens Plessey Systems	UK	*483*	75-53770
Tadiran Ltd	Israel	*222*	75-21600
Technology Service Corp.	USA	*644*	80-31930
Telegenix Inc.	USA	*621*	75-87610

446 Communications systems & equipment

A Goninan & Co. Ltd	Australia	*16*	75-07760
Adams-Russell BV	Ireland	*214*	75-00020
Advanced Technical Corp.	USA	*546*	75-01000
Aerea SpA	Italy	*230*	75-00280
Aeromaritime Systembau GmbH	Germany	*164*	75-00290
Alcatel Bell	Belgium	*44*	75-00055
Alcatel Bell-SDT	Belgium	*44*	75-00060
Alcatel España SA	Spain	*365*	75-00470
Alcatel Telecom Norway A/S	Norway	*308*	75-00090
Anco SA	Greece	*187*	75-00200
Andrew Canada Inc.	Canada	*68*	75-00600
Applied Signal Technology	USA	*554*	75-08210
Arab Organization for Ind.	Egypt	*109*	55-01700
Arinc	USA	*631*	80-03130
Astron Corp.	USA	*555*	75-09980
AudioPack Sound Systems	USA	*556*	75-10240
Aydin Corp.	USA	*556*	75-11450
Barrett Communications Pty Ltd.	Australia	*12*	75-02050
Boeing Australia Limited	Australia	*12*	75-02500
British Sarozal Ltd.	UK	*436*	75-09570
Calian Technology Ltd	Canada	*69*	75-01800
California Microwave Inc.	USA	*561*	75-16350
Canadian Marconi Co.	Canada	*69*	75-02600
Catsaros Automation SA	Greece	*187*	75-00700
CIC International Ltd	USA	*563*	75-18570
Cincinnati Electronics Corp.	USA	*564*	75-18640
Clark Masts Teksam Ltd	UK	*440*	75-13550
Codan Pty Ltd	Australia	*14*	75-03620
Communications-Applied Tech.	USA	*564*	75-20290
Comstream Corporation	USA	*565*	75-20780
Cubic Defense Systems Inc.	USA	*567*	75-22900
Cylink Corp.	USA	*567*	75-23134
Cypher Research Laboratories	Australia	*10*	60-00300
Daimler-Benz Aerospace France SAS	**France**	***134***	**75-12950**
Datron World Communications Inc.	USA	*568*	75-24020
David Clark Co. Inc.	USA	*564*	75-18850
Davies Industrial Communications Ltd	UK	*444*	75-17010
Deutsche System-Technik GmbH	Germany	*167*	75-06320
DNE, Inc.	USA	*569*	75-27000
Eddystone Radio Ltd	UK	*447*	75-19370
EDS Defence Ltd	UK	*501*	80-09200
EG&G Power Systems	USA	*571*	75-29558
Eidsvoll Electronics AS	Norway	*309*	75-00850
Elan GmbH.	Germany	*168*	75-07550
Electro-Metrics Inc.	USA	*572*	75-29810
ELTA Electronics Industries Ltd	Israel	*220*	75-10200
Ericsson Defence Systems Pty Ltd	Australia	*16*	75-05820
FEI Microwave Inc.	USA	*575*	75-34080
Fluor Daniel Inc.	USA	*635*	80-11720
Fockas Pantelis Ltd	Greece	*189*	75-01293
Force Computers (UK) Ltd	UK	*451*	75-24155
Fujitsu Ltd	Japan	*251*	75-01400
Gandalf Technologies Inc.	Canada	*71*	75-05650
GEC Marconi Hazeltine Corp	USA	*578*	75-37415
GEC-Marconi Avionics Ltd	UK	*454*	75-25320
GEC-Marconi Communications Ltd	UK	*454*	75-25335
Global Communications Co.	USA	*579*	75-39775
Harris Corporation	USA	*581*	75-42050
Hispano Radio Maritima SA	Spain	*367*	80-00840
Hunting Engineering Ltd	UK	*459*	75-30700
Infocom Systems A/S	Denmark	*103*	75-09200
Intelect Inc.	USA	*585*	75-47160
International Data Sciences Inc.	USA	*585*	75-47865
Italtel Spa	Italy	*235*	75-13205
ITT Aerospace/Communications Div.	USA	*586*	75-49455
ITT Avionics	USA	*586*	75-49455
Joyce Telectronics Corp.	USA	*587*	75-51150
Kaman Sciences Corp.	USA	*637*	80-17600
Litton Industries Inc.	USA	*590*	75-55400
Lockheed Martin Corp.	USA	*591*	75-55635
Logica SA/NV.	Belgium	*48*	80-01000
Logica UK Limited.	UK	*506*	80-14350
MA/COM Antenna & Cable Div.	USA	*593*	75-56953
Magnavox Electronic Systems Company	USA	*593*	75-57120
Matrix Systems Ltd.	Canada	*73*	75-09507
MCL Inc.	USA	*595*	75-59700
MILCOM International Ltd	Greece	*191*	75-02060
Motorola Ltd	UK	*469*	75-41050
Multitone Electronics plc	UK	*470*	75-41700
Northern Airborne Technology Ltd	Canada	*74*	75-10145
Northrop Grumman Corporation.	USA	*640*	80-21840
NOVO Corporation	Russia	*338*	75-14200
Oki Electric Industry Co. Ltd	Japan	*254*	75-07200
Omnitécnica Sarl	Portugal	*327*	75-15000
Ortel Corp.	USA	*602*	75-67530
Polyot Research and Production Co.	Russia	*338*	75-14700
PRC Inc.	USA	*640*	80-24800
PT Centronix	Indonesia	*210*	75-00200
Racal Avionics Ltd	UK	*477*	75-47660
Raytheon Canada Ltd.	Canada	*75*	75-11700
Raytheon Cossor Electronics Ltd	UK	*478*	75-48367
Raytheon E-Systems	USA	*608*	75-73302
Redifon MEL Limited	UK	*478*	75-48530
Remmele Engineering Inc.	USA	*609*	75-74400
Reunert Technology Systems Ltd.	S. Africa	*359*	75-18300
RF International Inc.	USA	*609*	75-74970
Rockwell.	USA	*610*	75-75930
Rohde & Schwarz France Telecom	France	*142*	75-32900
Safare-Crouzet SA	France	*142*	75-33900
Sanders	USA	*612*	75-77850
Scitec (EMA) Limited	UK	*482*	75-52160
Siemens Atea NV	Belgium	*47*	75-05950
Siemens plc	UK	*483*	75-53750
Siemens Plessey Electronic Systems	Australia	*21*	75-19475
Siemens Plessey Systems	UK	*483*	75-53770
Siemens Switzerland Ltd	Switzerland	*387*	75-11980
Signal Technology Corp.	USA	*618*	75-84240
Signature Industries Ltd	UK	*484*	75-53785
Smiths Aerospace & Defense	USA	*615*	75-82090
Sonic Communications (Int.) Ltd.	UK	*485*	75-55202
South Midlands Communications Ltd	UK	*485*	75-55213
Spar Aerospace Ltd	Canada	*76*	75-13000
STAREC	France	*147*	75-42775
Technical Materiel Corp.	USA	*619*	75-86760
Technitrol Inc.	USA	*620*	75-86800
Teledyne Controls	USA	*620*	75-87518

The Boeing Company	USA	*559*	75-14250
Thomson-CSF	France	*147*	75-45100
Vega-Cantley Instrument Company	UK	*490*	75-62740
ViaSat Inc.	USA	*626*	75-93790
Watkins-Johnson Co.	USA	*627*	75-95200
Zeta	USA	*629*	75-98190
Znamya Production Group	Ukraine	*407*	75-10000

446 10 Communications systems, spaceborne

Alcatel Bell	Belgium	*44*	75-00055
Aydin Corp.	USA	*556*	75-11450
CIC International Ltd	USA	*563*	75-18570
Daimler-Benz Aerospace France SAS	**France**	***134***	**75-12950**
Global Communications Co.	USA	*579*	75-39775
Matra Marconi Space UK, Ltd	UK	*425*	74-06900
Motorola Ltd	UK	*469*	75-41050
Pilkington Thorn Optronics	UK	*426*	74-10000
Rockwell	USA	*610*	75-75930
Rutherford Appleton Laboratory	UK	*420*	60-01850
Sanders	USA	*612*	75-77850
Siemens Atea NV	Belgium	*47*	75-05950
Smiths Aerospace & Defense	USA	*615*	75-82090
Spar Aerospace Ltd	Canada	*76*	75-13000

446 10 30 Communications systems, spaceborne, radio

Alcatel Bell	Belgium	*44*	75-00055
CIC International Ltd	USA	*563*	75-18570
GEC-Marconi Research Centre	UK	*419*	60-00700
Global Communications Co.	USA	*579*	75-39775
Matra Marconi Space UK, Ltd	UK	*425*	74-06900
Motorola Ltd	UK	*469*	75-41050
Smiths Aerospace & Defense	USA	*615*	75-82090

446 10 70 Communications equipment, spaceborne

Alcatel Bell	Belgium	*44*	75-00055
CIC International Ltd	USA	*563*	75-18570
Cubic Defense Systems Inc.	USA	*567*	75-22900
David Clark Co. Inc.	USA	*564*	75-18850
Frequency Electronics Inc.	USA	*576*	75-36190
Global Communications Co.	USA	*579*	75-39775
ITT Aerospace/Communications Div.	USA	*586*	75-49455
Labtech Ltd	UK	*464*	75-34280
MA/COM Antenna & Cable Div.	USA	*593*	75-56953
Motorola Ltd	UK	*469*	75-41050
Pilkington Thorn Optronics	UK	*426*	74-10000
Rockwell	USA	*610*	75-75930
Rutherford Appleton Laboratory	UK	*420*	60-01850
Sanders	USA	*612*	75-77850
Smiths Aerospace & Defense	USA	*615*	75-82090
Spar Aerospace Ltd	Canada	*76*	75-13000
Watkins-Johnson Co.	USA	*627*	75-95200

446 10 70 30 Communications equipment, spaceborne, radio

Alcatel Bell	Belgium	*44*	75-00055
CIC International Ltd	USA	*563*	75-18570
Smiths Aerospace & Defense	USA	*615*	75-82090
STAREC	France	*147*	75-42775
Watkins-Johnson Co.	USA	*627*	75-95200

446 20 Communications systems, airborne

Advanced Technical Corp.	USA	*546*	75-01000
Aeroelectrónica Industria	Brazil	*55*	75-00585
Applied Electro Mechanics Inc.	USA	*554*	75-08130
Associated Aircraft Mfg & Sales Inc.	USA	*555*	75-09500
Aydin Corp.	USA	*556*	75-11450
BAMS Export Office	Belgium	*44*	74-00500
Becker Flugfunkwerk GmbH	Germany	*165*	75-02200
Chelton (Electrostatics) Limited	UK	*439*	75-12700
CIC International Ltd	USA	*563*	75-18570
Codan Pty Ltd	Australia	*14*	75-03620
Cranfield Aerospace Ltd	UK	*500*	80-07200
Cranfield University	UK	*419*	60-00350
Cubic Defense Systems Inc.	USA	*567*	75-22900
David Clark Co. Inc.	USA	*564*	75-18850
Davies Industrial Communications Ltd	UK	*444*	75-17010
Elan GmbH.	Germany	*168*	75-07550
ELTA Electronics Industries Ltd	Israel	*220*	75-10200
GEC Marconi Hazeltine Corp	USA	*578*	75-37415
GEC-Marconi Avionics Ltd	UK	*454*	75-25320
GEC-Marconi Research Centre	UK	*419*	60-00700
Global Communications Co.	USA	*579*	75-39775
Interface Products Inc.	USA	*585*	75-47300
Italtel Spa	Italy	*235*	75-13205
ITT Aerospace/Communications Div.	USA	*586*	75-49455
ITT Avionics	USA	*586*	75-49455
Labtech Ltd	UK	*464*	75-34280
Litton Industries Inc.	USA	*590*	75-55400
MCL Inc.	USA	*595*	75-59700
Metraplex Corp.	USA	*595*	75-60700
Motorola Ltd	UK	*469*	75-41050
Northern Airborne Technology Ltd	Canada	*74*	75-10145
Penny & Giles Drives Technology Ltd	UK	*474*	75-44563
PhotoTelesis	USA	*604*	75-69810
Racal Acoustics Ltd	UK	*476*	75-47653
Racal Radar Defence Systems Ltd.	UK	*477*	75-47688
Raytheon E-Systems	USA	*608*	75-73302
Rockwell	USA	*610*	75-75930
Rockwell International Co.	Italy	*233*	75-05200
Rohde & Schwarz France Telecom	France	*142*	75-32900
SAGEM SA	**France**	***143***	**75-33950**
Sanders	USA	*612*	75-77850
Siemens Atea NV	Belgium	*47*	75-05950
Smiths Aerospace & Defense	USA	*615*	75-82090
Tadiran Ltd	Israel	*222*	75-21600
Technology Service Corp.	USA	*644*	80-31930
The Boeing Company	USA	*559*	75-14250
Walter Dittel GmbH	Germany	*167*	75-06510

446 20 30 Communications systems, airborne, radio

Amper Programas SA	Spain	*365*	75-00480
Arab Organization for Ind.	Egypt	*109*	55-01700
Becker Flugfunkwerk GmbH	Germany	*165*	75-02200
CelsiusTech Electronics AB	Sweden	*376*	75-03720
CIC International Ltd	USA	*563*	75-18570
Cincinnati Electronics Corp.	USA	*564*	75-18640
Cubic Defense Systems Inc.	USA	*567*	75-22900
ELTA Electronics Industries Ltd	Israel	*220*	75-10200
Frederick Electronics	USA	*576*	75-36000
GEC-Marconi Avionics Ltd	UK	*454*	75-25320
Global Communications Co.	USA	*579*	75-39775
ITT Aerospace/Communications Div.	USA	*586*	75-49455
Litton Industries Inc.	USA	*590*	75-55400
Northern Airborne Technology Ltd	Canada	*74*	75-10145
Reunert Technology Systems Ltd	S. Africa	*359*	75-18300
Rockwell International Co.	Italy	*233*	75-05200
Smiths Aerospace & Defense	USA	*615*	75-82090
Tadiran Ltd	Israel	*222*	75-21600
Technology Service Corp.	USA	*644*	80-31930
The Boeing Company	USA	*559*	75-14250
Walter Dittel GmbH	Germany	*167*	75-06510

446 20 30 30 Data transmission systems, airborne

Aydin Corp.	USA	*556*	75-11450
Calian Technology Ltd	Canada	*69*	75-01800
CIC International Ltd	USA	*563*	75-18570
Cubic Defense Systems Inc.	USA	*567*	75-22900
ELTA Electronics Industries Ltd	Israel	*220*	75-10200
ERA Technology Ltd	UK	*502*	80-09600
GEC Marconi Hazeltine Corp	USA	*578*	75-37415
GEC-Marconi Avionics Ltd	UK	*454*	75-25320
Global Communications Co.	USA	*579*	75-39775
ITT Aerospace/Communications Div.	USA	*586*	75-49455
Litton Industries Inc.	USA	*590*	75-55400
Navigation Data Systems Inc.	USA	*598*	75-64050
Northern Airborne Technology Ltd	Canada	*74*	75-10145
Polyot Research and Production Co.	Russia	*338*	75-14700
Racal Avionics Ltd	UK	*477*	75-47660
Rockwell International Co.	Italy	*233*	75-05200
Tadiran Ltd	Israel	*222*	75-21600
Technology Service Corp.	USA	*644*	80-31930
The Boeing Company	USA	*559*	75-14250
TransInstruments	UK	*489*	75-60960
Vega-Cantley Instrument Company	UK	*490*	75-62740

446 20 30 50 Satellite communications systems, airborne

British Aerospace Australia Ltd	Australia	*13*	75-02750
Canadian Marconi Co.	Canada	*69*	75-02600
Litton Industries Inc.	USA	*590*	75-55400
Motorola Ltd	UK	*469*	75-41050
Nera Telecommunications	UK	*470*	75-42318
Rockwell International Co.	Italy	*233*	75-05200
SED Systems Inc	Canada	*79*	80-01937
Technology Service Corp.	USA	*644*	80-31930
Toyo Communication Equipment Co	Japan	*255*	75-10000
ViaSat Inc.	USA	*626*	75-93790

446 20 30 60 Telemetry systems, airborne

Aydin Corp.	USA	*556*	75-11450
CIC International Ltd	USA	*563*	75-18570
Cincinnati Electronics Corp.	USA	*564*	75-18640
ELECMA	France	*136*	75-14400
ELTA Electronics Industries Ltd	Israel	*220*	75-10200
Global Communications Co.	USA	*579*	75-39775
Intertechnique SA.	France	*137*	75-23000
Litton Industries Inc.	USA	*590*	75-55400
Measurement Devices Ltd	UK	*467*	75-38275
Metraplex Corp.	USA	*595*	75-60700
Motorola Ltd	UK	*469*	75-41050
Polyot Research and Production Co.	Russia	*338*	75-14700
SED Systems Inc	Canada	*79*	80-01937
Sierra Technologies Inc.	USA	*614*	75-80640
Technology Service Corp.	USA	*644*	80-31930
TransInstruments	UK	*489*	75-60960
Vega-Cantley Instrument Company	UK	*490*	75-62740

446 20 40 Communications systems, airborne, telecopier / facsimile

CIC International Ltd	USA	*563*	75-18570
DNE, Inc.	USA	*569*	75-27000
Litton Industries Inc.	USA	*590*	75-55400
SAGEM SA	**France**	***143***	**75-33950**
Siemens Atea NV	Belgium	*47*	75-05950

446 20 60 Communications systems, airborne, teleprinter / telex

CIC International Ltd	USA	*563*	75-18570
DNE, Inc.	USA	*569*	75-27000
Frederick Electronics	USA	*576*	75-36000
HAL Communications Corp.	USA	*580*	75-41450
SAGEM SA	**France**	***143***	**75-33950**
The Boeing Company	USA	*559*	75-14250

446 20 70 Communications equipment, airborne

Advanced Technical Corp.	USA	*546*	75-01000
Amper Programas SA	Spain	*365*	75-00480
Arab Organization for Ind.	Egypt	*109*	55-01700
Associated Aircraft Mfg & Sales Inc.	USA	*555*	75-09500
Becker Flugfunkwerk GmbH	Germany	*165*	75-02200
CIC International Ltd	USA	*563*	75-18570
Cubic Defense Systems Inc.	USA	*567*	75-22900
Daimler-Benz Aerospace France SAS	**France**	***134***	**75-12950**
David Clark Co. Inc.	USA	*564*	75-18850
DNE, Inc.	USA	*569*	75-27000
Elan GmbH.	Germany	*168*	75-07550
Electronic Space Systems Corp.	USA	*572*	75-30100
Elfon Ltd	Greece	*188*	75-01150
ELTA Electronics Industries Ltd	Israel	*220*	75-10200
GEC-Marconi Avionics Ltd	UK	*454*	75-25320
Howden Fluid Systems	USA	*582*	75-44400
H.R. Smith (Microwaves) Ltd	UK	*484*	75-54590
Interface Products Inc.	USA	*585*	75-47300
Israel Aircraft Industries Ltd	Israel	*219*	75-10200
ITT Aerospace/Communications Div.	USA	*586*	75-49455
Litton Industries Inc.	USA	*590*	75-55400
MCL Inc.	USA	*595*	75-59700
Motorola Ltd	UK	*469*	75-41050
Northern Airborne Technology Ltd	Canada	*74*	75-10145
PhotoTelesis	USA	*604*	75-69810
Polyot Research and Production Co.	Russia	*338*	75-14700
Racal Radio Ltd	UK	*477*	75-47698
R.E.M.F. Radiocoms-Avionics	France	*141*	75-32200
RJR Industries	USA	*579*	75-39720
Roanwell Corp.	USA	*609*	75-75600
Rockwell	USA	*610*	75-75930
Rockwell International Co.	Italy	*233*	75-05200
Sanders	USA	*612*	75-77850
Smiths Aerospace & Defense	USA	*615*	75-82090
Technology Service Corp.	USA	*644*	80-31930

446 20 70 30 Communications equipment, radio, airborne

Advanced Technical Corp.	USA	*546*	75-01000
Amper Programas SA	Spain	*365*	75-00480
Arab Organization for Ind.	Egypt	*109*	55-01700
Associated Aircraft Mfg & Sales Inc.	USA	*555*	75-09500
Becker Flugfunkwerk GmbH	Germany	*165*	75-02200
CIC International Ltd	USA	*563*	75-18570
Cincinnati Electronics Corp.	USA	*564*	75-18640
Cubic Defense Systems Inc.	USA	*567*	75-22900
ELTA Electronics Industries Ltd	Israel	*220*	75-10200
Frederick Electronics	USA	*576*	75-36000
GEC-Marconi Avionics Ltd	UK	*454*	75-25320
Global Communications Co.	USA	*579*	75-39775
Icom (UK) Ltd	UK	*460*	75-31070
ITT Aerospace/Communications Div.	USA	*586*	75-49455
Litton Industries Inc.	USA	*590*	75-55400
Marconi SpA DTM	Italy	*236*	75-14550
Motorola Ltd	UK	*469*	75-41050
NEC Corporation	Japan	*253*	75-06350
Northern Airborne Technology Ltd	Canada	*74*	75-10145
Racal Radio Ltd	UK	*477*	75-47698
Reunert Technology Systems Ltd	S. Africa	*359*	75-18300
Rockwell International Co.	Italy	*233*	75-05200
Smiths Aerospace & Defense	USA	*615*	75-82090
Technology Service Corp.	USA	*644*	80-31930
Walter Dittel GmbH.	Germany	*167*	75-06510

446 20 70 30 20 Communications sets, airborne radio

Advanced Technical Corp.	USA	*546*	75-01000
Arab Organization for Ind.	Egypt	*109*	55-01700
Associated Aircraft Mfg & Sales Inc.	USA	*555*	75-09500
Becker Flugfunkwerk GmbH	Germany	*165*	75-02200
CIC International Ltd	USA	*563*	75-18570
Cincinnati Electronics Corp.	USA	*564*	75-18640
GEC-Marconi Avionics Ltd	UK	*454*	75-25320
Global Communications Co.	USA	*579*	75-39775
ITT Aerospace/Communications Div.	USA	*586*	75-49455
Litton Industries Inc.	USA	*590*	75-55400
Marconi SpA DTM	Italy	*236*	75-14550
Motorola Ltd	UK	*469*	75-41050
Northern Airborne Technology Ltd	Canada	*74*	75-10145
Reunert Technology Systems Ltd	S. Africa	*359*	75-18300
Rockwell International Co.	Italy	*233*	75-05200
Walter Dittel GmbH.	Germany	*167*	75-06510

446 20 70 30 20 50 Intercom sets, airborne radio

Arab Organization for Ind.	Egypt	*109*	55-01700
Associated Aircraft Mfg & Sales Inc.	USA	*555*	75-09500
Becker Flugfunkwerk GmbH	Germany	*165*	75-02200
Communications-Applied Tech.	USA	*564*	75-20290
Davies Industrial Communications Ltd	UK	*444*	75-17010
GEC-Marconi Avionics Ltd	UK	*454*	75-25320
Northern Airborne Technology Ltd	Canada	*74*	75-10145

446 20 70 30 30 Data transmission equipment, airborne

Aydin Corp.	USA	*556*	75-11450
CIC International Ltd	USA	*563*	75-18570
Cubic Defense Systems Inc.	USA	*567*	75-22900
GEC-Marconi Avionics Ltd	UK	*454*	75-25320
Global Communications Co.	USA	*579*	75-39775
ITT Aerospace/Communications Div.	USA	*586*	75-49455
Litton Industries Inc.	USA	*590*	75-55400
Marconi SpA DTM	Italy	*236*	75-14550
Navigation Data Systems Inc.	USA	*598*	75-64050
Rockwell International Co.	Italy	*233*	75-05200
Rockwell-Collins France SA	France	*142*	75-32800
SAGEM SA	**France**	***143***	**75-33950**

Tadiran Ltd	Israel	222	75-21600
TransInstruments	UK	489	75-60960

446 20 70 30 40 Satellite communications subsystems, airborne

FEI Microwave Inc.	USA	575	75-34080
Nera Telecommunications	UK	470	75-42318
Rockwell International Co.	Italy	233	75-05200

446 20 70 30 50 Satellite communications equipment, airborne

Canadian Marconi Co.	Canada	69	75-02600
Motorola Ltd	UK	469	75-41050
Nera Telecommunications	UK	470	75-42318
Rockwell International Co.	Italy	233	75-05200
SP Radio A/S	Denmark	103	75-19600

446 20 70 30 50 50 Satellite communications terminals, airborne

Motorola Ltd	UK	469	75-41050
Nera Telecommunications	UK	470	75-42318
Rockwell International Co.	Italy	233	75-05200

446 20 70 30 50 60 Satellite communications transceivers, airborne

Rockwell International Co.	Italy	233	75-05200

446 20 70 30 60 Telemetry sets, airborne

Aydin Corp.	USA	556	75-11450
CIC International Ltd	USA	563	75-18570
ELECMA	France	136	75-14400
Global Communications Co.	USA	579	75-39775
Herley-Vega Systems	USA	581	75-43090
Intertechnique SA	France	137	75-23000
Sierra Technologies Inc.	USA	614	75-80640
STAREC	France	147	75-42775
TransInstruments	UK	489	75-60960
Vega-Cantley Instrument Company	UK	490	75-62740
Wood & Douglas	UK	494	75-66790

446 20 70 46 Communication equipment, airborne, teleprinter

HAL Communications Corp.	USA	580	75-41450

446 20 70 46 50 Teleprinters, airborne

SAGEM SA	France	143	75-33950

446 20 70 60 Speech recognition equipment, airborne

CIC International Ltd	USA	563	75-18570
Logica UK Limited	UK	506	80-14350

446 20 70 65 Voice conditioning units, airborne

Davies Industrial Communications Ltd	UK	444	75-17010
SAIT Systems SA	Belgium	47	75-05600

446 20 80 Communications equipment accessories, airborne

Advanced Technical Corp.	USA	546	75-01000
Associated Aircraft Mfg & Sales Inc.	USA	555	75-09500
CIC International Ltd	USA	563	75-18570
Davies Industrial Communications Ltd	UK	444	75-17010
DRS Precision Echo, Inc.	USA	570	75-27360
Litton Industries Inc.	USA	590	75-55400
Naval Electronics Inc.	USA	598	75-64025
Northern Airborne Technology Ltd	Canada	74	75-10145
Trompeter Electronics Inc.	USA	623	75-91070
Watkins-Johnson Co.	USA	627	75-95200

446 20 80 30 Communications equipment accessories, airborne radio

Advanced Technical Corp.	USA	546	75-01000
Associated Aircraft Mfg & Sales Inc.	USA	555	75-09500
CIC International Ltd	USA	563	75-18570
David Clark Co. Inc.	USA	564	75-18850
Davies Industrial Communications Ltd	UK	444	75-17010
DRS Precision Echo, Inc.	USA	570	75-27360
Frederick Electronics	USA	576	75-36000
Global Communications Co.	USA	579	75-39775
Howden Fluid Systems	USA	582	75-44400
Icom (UK) Ltd	UK	460	75-31070
Motorola Ltd	UK	469	75-41050
Northern Airborne Technology Ltd	Canada	74	75-10145
Racal Acoustics Ltd	UK	476	75-47653
Reunert Technology Systems Ltd	S. Africa	359	75-18300
Trompeter Electronics Inc.	USA	623	75-91070
Walter Dittel GmbH	Germany	167	75-06510
Watkins-Johnson Co.	USA	627	75-95200

446 20 80 30 10 Antennas, radio, airborne

Advanced Technical Corp.	USA	546	75-01000
Chelton Inc.	USA	563	75-18120
CIC International Ltd	USA	563	75-18570
Cincinnati Electronics Corp.	USA	564	75-18640
Datron/Transco Inc.	USA	568	75-24000
ELTA Electronics Industries Ltd	Israel	220	75-10200
ERA Technology Ltd	UK	502	80-09600
Foxtronics Inc.	USA	576	75-35752
General Electronics Communications	Greece	189	75-01300
Global Communications Co.	USA	579	75-39775
H.R. Smith Group	UK	484	75-54590
H.R. Smith (Microwaves) Ltd	UK	484	75-54590
Litton Industries Inc.	USA	590	75-55400
MA/COM Antenna & Cable Div.	USA	593	75-56953
Polyot Research and Production Co.	Russia	338	75-14700
R.A. Miller Industries Inc.	USA	596	75-61650
Racal Antennas Ltd	UK	476	75-47656
RJR Industries	USA	579	75-39720
Rockwell International Co.	Italy	233	75-05200
Sensor Systems Inc.	USA	613	75-79400
STAREC	France	147	75-42775
UB Corp.	USA	624	75-91380
Vega-Cantley Instrument Company	UK	490	75-62740
Walter Dittel GmbH	Germany	167	75-06510
Watkins-Johnson Co.	USA	627	75-95200

446 20 80 30 30 Headsets, airborne radio

Acousticom Corp.	USA	545	75-00380
Bose BV	Netherlands	291	75-00790
CIC International Ltd	USA	563	75-18570
Clement Clarke International	UK	440	75-13771
Davies Industrial Communications Ltd	UK	444	75-17010
Hagen	UK	456	75-27220
Icom (UK) Ltd	UK	460	75-31070

446 20 80 30 30 10 Headsets, airborne radio, active noise reduction

Bose BV	Netherlands	291	75-00790
Bose Corporation	USA	560	75-14700

446 20 80 30 50 Microphones, airborne communications

CIC International Ltd	USA	563	75-18570
Clement Clarke International	UK	440	75-13771
Davies Industrial Communications Ltd	UK	444	75-17010
Hagen	UK	456	75-27220

446 30 Communications systems, naval

Aeromaritime Systembau GmbH	Germany	164	75-00290

Amper Programas SA	Spain	*365*	75-00480
Applied Electro Mechanics Inc.	USA	*554*	75-08130
Barrett Communications Pty Ltd.	Australia	*12*	75-02050
Bharat Electronics Ltd	India	*203*	55-00200
Boeing Australia Limited	Australia	*12*	75-02500
CelsiusTech Electronics AB	Sweden	*376*	75-03720
CIC International Ltd	USA	*563*	75-18570
Codalex Ltd	Canada	*70*	75-03400
Codan Pty Ltd	Australia	*14*	75-03620
Cubic Defense Systems Inc.	USA	*567*	75-22900
Datamat Ingegneria dei Sistemi SpA	Italy	*233*	75-06500
David Clark Co. Inc.	USA	*564*	75-18850
Davies Industrial Communications Ltd	UK	*444*	75-17010
Eddystone Radio Ltd	UK	*447*	75-19370
ELTA Electronics Industries Ltd	Israel	*220*	75-10200
Fluor Daniel Inc.	USA	*635*	80-11720
GEC-Marconi Communications Ltd	UK	*454*	75-25335
Global Communications Co.	USA	*579*	75-39775
Graphico Electronics Group	Canada	*71*	75-06460
Graseby Dynamics Ltd	UK	*455*	75-26590
Harris Corporation	USA	*581*	75-42050
Howden Fluid Systems	USA	*582*	75-44400
Hunting Engineering Ltd	UK	*459*	75-30700
Israel Aircraft Industries Ltd	Israel	*219*	75-10200
Italtel Spa	Italy	*235*	75-13205
ITT Aerospace/Communications Div.	USA	*586*	75-49455
ITT Avionics	USA	*586*	75-49455
JPS Communications Inc.	USA	*587*	75-51220
KEL Aerospace Pty Ltd	Australia	*18*	75-11440
Litton Industries Inc.	USA	*590*	75-55400
Magnavox Electronic Systems Company	USA	*593*	75-57120
MCL Inc.	USA	*595*	75-59700
Northern Airborne Technology Ltd	Canada	*74*	75-10145
Ocean Technology Systems	USA	*601*	75-66020
Park Air Electronics Ltd	UK	*472*	75-43750
PhotoTelesis	USA	*604*	75-69810
PRC Inc.	USA	*640*	80-24800
Racal Communications Inc.	USA	*606*	75-72380
Raytheon E-Systems	USA	*608*	75-73302
Redifon MEL Limited	UK	*478*	75-48530
Roanwell Corp.	USA	*609*	75-75600
Rockwell	USA	*610*	75-75930
Rockwell International Co.	Italy	*233*	75-05200
Safare-Crouzet SA	France	*142*	75-33900
SAIT Systems SA	Belgium	*47*	75-05600
Sanders	USA	*612*	75-77850
Siemens Aktiengesellschaft	Germany	*175*	75-27700
Siemens Atea NV	Belgium	*47*	75-05950
Siemens Plessey Systems	UK	*483*	75-53770
Signaal Communications	Netherlands	*295*	75-06350
Sperry Marine Inc.	USA	*617*	75-83250
Tadiran Ltd	Israel	*222*	75-21600
Technology Service Corp.	USA	*644*	80-31930
Tracor Inc	USA	*623*	75-90000
ViaSat Inc.	USA	*626*	75-93790

446 30 10 Communications systems, shipborne, optical

CIC International Ltd	USA	*563*	75-18570
Global Communications Co.	USA	*579*	75-39775
Hunting Engineering Ltd	UK	*459*	75-30700
iMPath Networks Inc.	Canada	*77*	75-46151
Litton Industries Inc.	USA	*590*	75-55400
Redifon MEL Limited	UK	*478*	75-48530
RGB Spectrum	USA	*609*	75-75000
Siemens Atea NV	Belgium	*47*	75-05950
Technology Service Corp.	USA	*644*	80-31930

446 30 15 Communications systems, shipborne, infra-red

CIC International Ltd	USA	*563*	75-18570
Global Communications Co.	USA	*579*	75-39775
Litton Industries Inc.	USA	*590*	75-55400
Technology Service Corp.	USA	*644*	80-31930

446 30 20 Communications systems, shipborne, laser

CIC International Ltd	USA	*563*	75-18570
Global Communications Co.	USA	*579*	75-39775
Litton Industries Inc.	USA	*590*	75-55400
Technology Service Corp.	USA	*644*	80-31930

446 30 30 Communications systems, shipborne, radio

Aeromaritime Systembau GmbH	Germany	*164*	75-00290
Amper Programas SA	Spain	*365*	75-00480
Bharat Electronics Ltd	India	*203*	55-00200
Chapman & Smith Ltd	UK	*439*	75-12500
CIC International Ltd	USA	*563*	75-18570
Cincinnati Electronics Corp.	USA	*564*	75-18640
Cubic Defense Systems Inc.	USA	*567*	75-22900
Datron World Communications Inc.	USA	*568*	75-24020
ELTA Electronics Industries Ltd	Israel	*220*	75-10200
Frederick Electronics	USA	*576*	75-36000
GEC-Marconi Communications Ltd	UK	*454*	75-25335
Global Communications Co.	USA	*579*	75-39775
Graphico Electronics Group	Canada	*71*	75-06460
Harris Corporation	USA	*581*	75-42050
Icom (UK) Ltd	UK	*460*	75-31070
ITT Aerospace/Communications Div.	USA	*586*	75-49455
Jotron (UK) Ltd	UK	*462*	75-32895
JPS Communications Inc.	USA	*587*	75-51220
KEL Aerospace Pty Ltd	Australia	*18*	75-11440
Litton Industries Inc.	USA	*590*	75-55400
Park Air Electronics Ltd	UK	*472*	75-43750
Racal Communications Inc.	USA	*606*	75-72380
Redifon MEL Limited	UK	*478*	75-48530
SAIT Systems SA	Belgium	*47*	75-05600
SED Systems Inc	Canada	*79*	80-01937
Siemens Aktiengesellschaft	Germany	*175*	75-27700
Signaal Communications	Netherlands	*295*	75-06350
SP Radio A/S	Denmark	*103*	75-19600
Tadiran Ltd	Israel	*222*	75-21600
Technology Service Corp.	USA	*644*	80-31930
Toyo Communication Equipment Co	Japan	*255*	75-10000

446 30 30 30 Data transmission systems, shipborne

Aeromaritime Systembau GmbH	Germany	*164*	75-00290
Aydin Corp.	USA	*556*	75-11450
CIC International Ltd	USA	*563*	75-18570
Cogent	UK	*500*	80-06250
Computer Resources International A/S.	Denmark	*104*	80-00100
Cubic Defense Systems Inc.	USA	*567*	75-22900
DNE, Inc.	USA	*569*	75-27000
ERA Technology Ltd	UK	*502*	80-09600
Frederick Electronics	USA	*576*	75-36000
Global Communications Co.	USA	*579*	75-39775
Graphico Electronics Group	Canada	*71*	75-06460
Harris Corporation	USA	*581*	75-42050
Hunting Engineering Ltd	UK	*459*	75-30700
LARIMART SpA	Italy	*236*	75-13500
Litton FiberCom	USA	*590*	75-55370
Litton Industries Inc.	USA	*590*	75-55400
Meggitt Avionics (UK)	UK	*467*	75-38340
Redifon MEL Limited	UK	*478*	75-48530
SAGEM SA	**France**	***143***	**75-33950**
SAIT Systems SA	Belgium	*47*	75-05600
SED Systems Inc	Canada	*79*	80-01937
Siemens Aktiengesellschaft	Germany	*175*	75-27700
Siemens Atea NV	Belgium	*47*	75-05950
Siemens Plessey Systems	UK	*483*	75-53770
Signaal Communications	Netherlands	*295*	75-06350
Tadiran Ltd	Israel	*222*	75-21600
Technology Service Corp.	USA	*644*	80-31930

446 30 30 50 Satellite communications systems, shipborne

CIC International Ltd	USA	*563*	75-18570
GEC-Marconi Communications Ltd	UK	*454*	75-25335
Harris Corporation	USA	*581*	75-42050
NEC Corporation	Japan	*253*	75-06350
Nera Telecommunications	UK	*470*	75-42318
Redifon MEL Limited	UK	*478*	75-48530
Sea Tel Inc.	USA	*613*	75-78930
Technology Service Corp.	USA	*644*	80-31930
ViaSat Inc.	USA	*626*	75-93790

446 30 30 60 Telemetry systems, shipborne

Aydin Corp.	USA	*556*	75-11450
CIC International Ltd	USA	*563*	75-18570
Global Communications Co.	USA	*579*	75-39775

Litton Industries Inc.	USA	*590*	75-55400
Navigation Data Systems Inc.	USA	*598*	75-64050
STAREC	France	*147*	75-42775
Technology Service Corp.	USA	*644*	80-31930
TIW Systems Inc.	USA	*623*	75-89755
TransInstruments	UK	*489*	75-60960
Vega-Cantley Instrument Company	UK	*490*	75-62740

446 30 40 Communications systems, shipborne, telephone

Aeromaritime Systembau GmbH	Germany	*164*	75-00290
CIC International Ltd	USA	*563*	75-18570
Cogent	UK	*500*	80-06250
Frederick Electronics	USA	*576*	75-36000
Global Communications Co.	USA	*579*	75-39775
Graphico Electronics Group	Canada	*71*	75-06460
Graseby Dynamics Ltd	UK	*455*	75-26590
Harris Corporation	USA	*581*	75-42050
Litton Industries Inc.	USA	*590*	75-55400
Redifon MEL Limited	UK	*478*	75-48530
SAGEM SA	**France**	***143***	**75-33950**
Siemens Atea NV	Belgium	*47*	75-05950
Stanilite Electronics Pty Ltd	Australia	*21*	75-19690
Tadiran Ltd	Israel	*222*	75-21600
Technology Service Corp.	USA	*644*	80-31930

446 30 40 40 Communications systems, shipborne, telephone, intercom

Aeromaritime Systembau GmbH	Germany	*164*	75-00290
CIC International Ltd	USA	*563*	75-18570
Davies Industrial Communications Ltd	UK	*444*	75-17010
Frederick Electronics	USA	*576*	75-36000
Global Communications Co.	USA	*579*	75-39775
Graphico Electronics Group	Canada	*71*	75-06460
LARIMART SpA	Italy	*236*	75-13500
Litton Industries Inc.	USA	*590*	75-55400
Multitone Electronics plc	UK	*470*	75-41700
Northern Airborne Technology Ltd	Canada	*74*	75-10145
Redifon MEL Limited	UK	*478*	75-48530
Siemens Atea NV	Belgium	*47*	75-05950
Stanilite Electronics Pty Ltd	Australia	*21*	75-19690
Tadiran Ltd	Israel	*222*	75-21600
Technology Service Corp.	USA	*644*	80-31930
Telegenix Inc.	USA	*621*	75-87610

446 30 40 60 Communications systems, shipborne, radiotelephone

Amper Programas SA	Spain	*365*	75-00480
CIC International Ltd	USA	*563*	75-18570
Cubic Defense Systems Inc.	USA	*567*	75-22900
Electronic Marketing Co. SpA	Italy	*233*	75-06860
GEC-Marconi Communications Ltd	UK	*454*	75-25335
Global Communications Co.	USA	*579*	75-39775
Graphico Electronics Group	Canada	*71*	75-06460
Harris Corporation	USA	*581*	75-42050
Icom (UK) Ltd	UK	*460*	75-31070
JPS Communications Inc.	USA	*587*	75-51220
LARIMART SpA	Italy	*236*	75-13500
Litton Industries Inc.	USA	*590*	75-55400
Redifon MEL Limited	UK	*478*	75-48530
Siemens Aktiengesellschaft	Germany	*175*	75-27700
SP Radio A/S	Denmark	*103*	75-19600
Sunair Electronics Inc.	USA	*618*	75-84950
Tadiran Ltd	Israel	*222*	75-21600
Technology Service Corp.	USA	*644*	80-31930
Telegenix Inc.	USA	*621*	75-87610

446 30 40 90 Communications systems, underwater telephone

CIC International Ltd	USA	*563*	75-18570
Cogent	UK	*500*	80-06250
Nautronix Ltd	UK	*470*	75-41900
NEC Corporation	Japan	*253*	75-06350
Redifon MEL Limited	UK	*478*	75-48530
Slingsby Engineering Ltd	UK	*484*	75-54420
Technology Service Corp.	USA	*644*	80-31930

446 30 42 Communications systems, shipborne, telegraphic

CIC International Ltd	USA	*563*	75-18570
Cubic Defense Systems Inc.	USA	*567*	75-22900
Frederick Electronics	USA	*576*	75-36000
Global Communications Co.	USA	*579*	75-39775
HAL Communications Corp.	USA	*580*	75-41450
Harris Corporation	USA	*581*	75-42050
Redifon MEL Limited	UK	*478*	75-48530
SAGEM SA	**France**	***143***	**75-33950**
Siemens Aktiengesellschaft	Germany	*175*	75-27700
SP Radio A/S	Denmark	*103*	75-19600
Sunair Electronics Inc.	USA	*618*	75-84950
Tadiran Ltd	Israel	*222*	75-21600
Technology Service Corp.	USA	*644*	80-31930

446 30 42 90 Communications systems, underwater telegraph

GEC-Marconi Communications Ltd	UK	*454*	75-25335
NEC Corporation	Japan	*253*	75-06350
Redifon MEL Limited	UK	*478*	75-48530
Technology Service Corp.	USA	*644*	80-31930

446 30 46 Communications systems, shipborne, teleprinter / telex

Aeromaritime Systembau GmbH	Germany	*164*	75-00290
Datron World Communications Inc.	USA	*568*	75-24020
Frederick Electronics	USA	*576*	75-36000
Global Communications Co.	USA	*579*	75-39775
Harris Corporation	USA	*581*	75-42050
Redifon MEL Limited	UK	*478*	75-48530
SAGEM SA	**France**	***143***	**75-33950**
Siemens Aktiengesellschaft	Germany	*175*	75-27700
SP Radio A/S	Denmark	*103*	75-19600
Technology Service Corp.	USA	*644*	80-31930

446 30 48 Communications systems, shipborne, telecopier / facsimile

Aeromaritime Systembau GmbH	Germany	*164*	75-00290
CIC International Ltd	USA	*563*	75-18570
DNE, Inc.	USA	*569*	75-27000
Global Communications Co.	USA	*579*	75-39775
Graphico Electronics Group	Canada	*71*	75-06460
Harris Corporation	USA	*581*	75-42050
PhotoTelesis	USA	*604*	75-69810
Redifon MEL Limited	UK	*478*	75-48530
SAGEM SA	**France**	***143***	**75-33950**
Technology Service Corp.	USA	*644*	80-31930

446 30 60 Communications systems, naval, underwater

Aeromaritime Systembau GmbH	Germany	*164*	75-00290
CIC International Ltd	USA	*563*	75-18570
Global Communications Co.	USA	*579*	75-39775
Graseby Dynamics Ltd	UK	*455*	75-26590
Nautronix Ltd	UK	*470*	75-41900
Redifon MEL Limited	UK	*478*	75-48530
Slingsby Engineering Ltd	UK	*484*	75-54420
Technology Service Corp.	USA	*644*	80-31930

446 30 63 Communications systems, diver, underwater

Davies Industrial Communications Ltd	UK	*444*	75-17010
Global Communications Co.	USA	*579*	75-39775
Graseby Dynamics Ltd	UK	*455*	75-26590
Interspiro Ltd	UK	*461*	75-31945
MacArtney A/S	Denmark	*103*	75-12800
Nautronix Ltd	UK	*470*	75-41900
Ocean Technology Systems	USA	*601*	75-66020
Slingsby Engineering Ltd	UK	*484*	75-54420
Technology Service Corp.	USA	*644*	80-31930

446 30 66 Communications systems, submarine

Aeromaritime Systembau GmbH	Germany	*164*	75-00290

CIC International Ltd	USA	*563*	75-18570
Cogent	UK	*500*	80-06250
Daimler-Benz Aerospace France SAS	**France**	***134***	**75-12950**
Davies Industrial Communications Ltd	UK	*444*	75-17010
Fugro-UDI Ltd	UK	*452*	75-24885
GEC-Marconi Communications Ltd	UK	*454*	75-25335
Global Communications Co.	USA	*579*	75-39775
Graseby Dynamics Ltd	UK	*455*	75-26590
Nautronix Ltd	UK	*470*	75-41900
Northrop Grumman Corporation	USA	*599*	75-65450
Redifon MEL Limited	UK	*478*	75-48530
Slingsby Engineering Ltd	UK	*484*	75-54420
Technology Service Corp.	USA	*644*	80-31930

446 30 70 Communications equipment, shipborne

Amper Programas SA	Spain	*365*	75-00480
CIC International Ltd	USA	*563*	75-18570
Codalex Ltd	Canada	*70*	75-03400
Communications & Power Industries	USA	*565*	75-20393
Cubic Defense Systems Inc.	USA	*567*	75-22900
Daimler-Benz Aerospace France SAS	**France**	***134***	**75-12950**
David Clark Co. Inc.	USA	*564*	75-18850
Davies Industrial Communications Ltd	UK	*444*	75-17010
Elfon Ltd	Greece	*188*	75-01150
ELTA Electronics Industries Ltd	Israel	*220*	75-10200
GEC-Marconi Communications Ltd	UK	*454*	75-25335
Global Communications Co.	USA	*579*	75-39775
Graphico Electronics Group	Canada	*71*	75-06460
Graseby Dynamics Ltd	UK	*455*	75-26590
Harris Corporation	USA	*581*	75-42050
Howden Fluid Systems	USA	*582*	75-44400
Hunting Engineering Ltd	UK	*459*	75-30700
Israel Aircraft Industries Ltd	Israel	*219*	75-10200
JPS Communications Inc.	USA	*587*	75-51220
KEL Aerospace Pty Ltd	Australia	*18*	75-11440
Litton Industries Inc.	USA	*590*	75-55400
Magnavox Electronic Systems Company	USA	*593*	75-57120
MCL Inc.	USA	*595*	75-59700
Park Air Electronics Ltd	UK	*472*	75-43750
PhotoTelesis	USA	*604*	75-69810
Racal Radio Ltd	UK	*477*	75-47698
Redifon MEL Limited	UK	*478*	75-48530
RGB Spectrum	USA	*609*	75-75000
RJR Industries	USA	*579*	75-39720
Rockwell	USA	*610*	75-75930
Sanders	USA	*612*	75-77850
Siemens Aktiengesellschaft	Germany	*175*	75-27700
Siemens Atea NV	Belgium	*47*	75-05950
Siemens Plessey Systems	UK	*483*	75-53770
Signaal Communications	Netherlands	*295*	75-06350
SP Radio A/S	Denmark	*103*	75-19600
Spar Aerospace Ltd	Canada	*76*	75-13000
Sunair Electronics Inc.	USA	*618*	75-84950

446 30 70 10 Communications equipment, shipborne, optical

CIC International Ltd	USA	*563*	75-18570
Cogent	UK	*500*	80-06250
Global Communications Co.	USA	*579*	75-39775
Hunting Engineering Ltd	UK	*459*	75-30700
iMPath Networks Inc.	Canada	*77*	75-46151
Litton Industries Inc.	USA	*590*	75-55400
Redifon MEL Limited	UK	*478*	75-48530
RGB Spectrum	USA	*609*	75-75000
Siemens Atea NV	Belgium	*47*	75-05950

446 30 70 15 Communications equipment, shipborne, infra-red

CIC International Ltd	USA	*563*	75-18570
Global Communications Co.	USA	*579*	75-39775
Litton Industries Inc.	USA	*590*	75-55400
RGB Spectrum	USA	*609*	75-75000
Siemens Atea NV	Belgium	*47*	75-05950

446 30 70 20 Communications equipment, shipborne, laser

CIC International Ltd	USA	*563*	75-18570
Global Communications Co.	USA	*579*	75-39775
Litton Industries Inc.	USA	*590*	75-55400

446 30 70 30 Communications equipment, shipborne, radio

Amper Programas SA	Spain	*365*	75-00480
CIC International Ltd	USA	*563*	75-18570
Cincinnati Electronics Corp.	USA	*564*	75-18640
Cubic Defense Systems Inc.	USA	*567*	75-22900
Datron World Communications Inc.	USA	*568*	75-24020
ELTA Electronics Industries Ltd	Israel	*220*	75-10200
Frederick Electronics	USA	*576*	75-36000
GEC-Marconi Communications Ltd	UK	*454*	75-25335
Global Communications Co.	USA	*579*	75-39775
Graphico Electronics Group	Canada	*71*	75-06460
Harris Corporation	USA	*581*	75-42050
JPS Communications Inc.	USA	*587*	75-51220
KEL Aerospace Pty Ltd	Australia	*18*	75-11440
Litton Industries Inc.	USA	*590*	75-55400
Marconi SpA DTM	Italy	*236*	75-14550
Multitone Electronics plc	UK	*470*	75-41700
Park Air Electronics Ltd	UK	*472*	75-43750
Racal Communications Inc.	USA	*606*	75-72380
Racal Radio Ltd	UK	*477*	75-47698
Redifon MEL Limited	UK	*478*	75-48530
Siemens Aktiengesellschaft	Germany	*175*	75-27700
Signaal Communications	Netherlands	*295*	75-06350
SP Radio A/S	Denmark	*103*	75-19600
Sunair Electronics Inc.	USA	*618*	75-84950
Tadiran Ltd	Israel	*222*	75-21600
Telegenix Inc.	USA	*621*	75-87610

446 30 70 30 20 Communications sets, shipborne, radio

Amper Programas SA	Spain	*365*	75-00480
Chapman & Smith Ltd	UK	*439*	75-12500
CIC International Ltd	USA	*563*	75-18570
Cincinnati Electronics Corp.	USA	*564*	75-18640
Cubic Defense Systems Inc.	USA	*567*	75-22900
Datron World Communications Inc.	USA	*568*	75-24020
Electronic Marketing Co. SpA	Italy	*233*	75-06860
ELTA Electronics Industries Ltd	Israel	*220*	75-10200
GEC-Marconi Communications Ltd	UK	*454*	75-25335
Global Communications Co.	USA	*579*	75-39775
Graphico Electronics Group	Canada	*71*	75-06460
Harris Corporation	USA	*581*	75-42050
Icom (UK) Ltd	UK	*460*	75-31070
Northern Airborne Technology Ltd	Canada	*74*	75-10145
Park Air Electronics Ltd	UK	*472*	75-43750
Radio Research Instrument Co. Inc.	USA	*607*	75-72600
Redifon MEL Limited	UK	*478*	75-48530
Siemens Aktiengesellschaft	Germany	*175*	75-27700
Signaal Communications	Netherlands	*295*	75-06350
Tadiran Ltd	Israel	*222*	75-21600

446 30 70 30 20 35 Radio sets, naval, hand-held

Amper Programas SA	Spain	*365*	75-00480
CIC International Ltd	USA	*563*	75-18570
Cincinnati Electronics Corp.	USA	*564*	75-18640
Racal Radio Ltd	UK	*477*	75-47698

446 30 70 30 20 50 Intercom sets, radio, shipborne

Chapman & Smith Ltd	UK	*439*	75-12500
CIC International Ltd	USA	*563*	75-18570
Communications-Applied Tech.	USA	*564*	75-20290
Davies Industrial Communications Ltd	UK	*444*	75-17010
Hunting Engineering Ltd	UK	*459*	75-30700
Icom (UK) Ltd	UK	*460*	75-31070
Northern Airborne Technology Ltd	Canada	*74*	75-10145
Redifon MEL Limited	UK	*478*	75-48530
Stanilite Electronics Pty Ltd	Australia	*21*	75-19690

446 30 70 30 30 Data transmission equipment, shipborne

Aydin Corp.	USA	*556*	75-11450
CIC International Ltd	USA	*563*	75-18570
Codalex Ltd	Canada	*70*	75-03400
Cogent	UK	*500*	80-06250
Cubic Defense Systems Inc.	USA	*567*	75-22900
Frederick Electronics	USA	*576*	75-36000
Global Communications Co.	USA	*579*	75-39775
Graphico Electronics Group	Canada	*71*	75-06460
Harris Corporation	USA	*581*	75-42050
Hunting Engineering Ltd	UK	*459*	75-30700
LARIMART SpA	Italy	*236*	75-13500
Navigation Data Systems Inc.	USA	*598*	75-64050
Redifon MEL Limited	UK	*478*	75-48530
SAGEM SA	**France**	***143***	**75-33950**
Sperry Marine Inc.	USA	*617*	75-83250
Tadiran Ltd	Israel	*222*	75-21600

446 30 70 30 40 Satellite communications subsystems, shipborne

Nera Telecommunications	UK	*470*	75-42318

446 30 70 30 50 Satellite communications equipment, shipborne

GEC-Marconi Communications Ltd	UK	*454*	75-25335
Harris Corporation	USA	*581*	75-42050
Magnavox Electronic Systems Company	USA	*593*	75-57120
NEC Corporation	Japan	*253*	75-06350
Nera Telecommunications	UK	*470*	75-42318
SAIT Systems SA	Belgium	*47*	75-05600
SP Radio A/S	Denmark	*103*	75-19600
Sperry Marine Inc.	USA	*617*	75-83250
TIW Systems Inc.	USA	*623*	75-89755

446 30 70 30 50 50 Satellite communications terminals, shipborne

CIC International Ltd	USA	*563*	75-18570
GEC-Marconi Communications Ltd	UK	*454*	75-25335
Harris Corporation	USA	*581*	75-42050
Magnavox Electronic Systems Company	USA	*593*	75-57120
SAIT Systems SA	Belgium	*47*	75-05600
Sperry Marine Inc.	USA	*617*	75-83250

446 30 70 30 50 60 Satellite communications transceivers, shipborne

CIC International Ltd	USA	*563*	75-18570
Harris Corporation	USA	*581*	75-42050
SAIT Systems SA	Belgium	*47*	75-05600
Sperry Marine Inc.	USA	*617*	75-83250

446 30 70 30 60 Telemetry sets, shipborne

Aydin Corp.	USA	*556*	75-11450
CIC International Ltd	USA	*563*	75-18570
Cincinnati Electronics Corp.	USA	*564*	75-18640
Global Communications Co.	USA	*579*	75-39775
STAREC	France	*147*	75-42775
TransInstruments	UK	*489*	75-60960
Vega-Cantley Instrument Company	UK	*490*	75-62740
Wood & Douglas	UK	*494*	75-66790

446 30 70 30 70 Troposcatter terminals, shipborne

CIC International Ltd	USA	*563*	75-18570
Codalex Ltd	Canada	*70*	75-03400
GEC-Marconi Communications Ltd	UK	*454*	75-25335
Global Communications Co.	USA	*579*	75-39775

446 30 70 40 Communications equipment, shipborne, telephone

CIC International Ltd	USA	*563*	75-18570
Cogent	UK	*500*	80-06250
Davies Industrial Communications Ltd	UK	*444*	75-17010
Frederick Electronics	USA	*576*	75-36000
Global Communications Co.	USA	*579*	75-39775
Graphico Electronics Group	Canada	*71*	75-06460
JPS Communications Inc.	USA	*587*	75-51220
LARIMART SpA	Italy	*236*	75-13500
Redifon MEL Limited	UK	*478*	75-48530
Siemens Atea NV	Belgium	*47*	75-05950
Spar Aerospace Ltd	Canada	*76*	75-13000
Tadiran Ltd	Israel	*222*	75-21600
Telegenix Inc.	USA	*621*	75-87610

446 30 70 40 50 Telephone sets, ship-based

CIC International Ltd	USA	*563*	75-18570
Cogent	UK	*500*	80-06250
Redifon MEL Limited	UK	*478*	75-48530
Siemens Atea NV	Belgium	*47*	75-05950

446 30 70 42 Communications equipment, shipborne, telegraphic

CIC International Ltd	USA	*563*	75-18570
Codalex Ltd	Canada	*70*	75-03400
Cubic Defense Systems Inc.	USA	*567*	75-22900
Frederick Electronics	USA	*576*	75-36000
Global Communications Co.	USA	*579*	75-39775
HAL Communications Corp.	USA	*580*	75-41450
Harris Corporation	USA	*581*	75-42050
Redifon MEL Limited	UK	*478*	75-48530
Siemens Aktiengesellschaft	Germany	*175*	75-27700
SP Radio A/S	Denmark	*103*	75-19600
Sunair Electronics Inc.	USA	*618*	75-84950
Tadiran Ltd	Israel	*222*	75-21600

446 30 70 46 Communications equipment, shipborne, teleprinter

CIC International Ltd	USA	*563*	75-18570
DNE, Inc.	USA	*569*	75-27000
Frederick Electronics	USA	*576*	75-36000
GEC-Marconi Communications Ltd	UK	*454*	75-25335
Global Communications Co.	USA	*579*	75-39775
HAL Communications Corp.	USA	*580*	75-41450
Harris Corporation	USA	*581*	75-42050
SAGEM SA	**France**	***143***	**75-33950**
Siemens Aktiengesellschaft	Germany	*175*	75-27700
SP Radio A/S	Denmark	*103*	75-19600

446 30 70 46 50 Teleprinters, shipborne

Raytheon Cossor Electronics Ltd	UK	*478*	75-48367
SAGEM SA	**France**	***143***	**75-33950**

446 30 70 48 Communications equipment, shipborne, telecopier / facsimile

CIC International Ltd	USA	*563*	75-18570
Global Communications Co.	USA	*579*	75-39775
Graphico Electronics Group	Canada	*71*	75-06460
Harris Corporation	USA	*581*	75-42050
SAGEM SA	**France**	***143***	**75-33950**

446 30 70 60 Communications equipment, naval, underwater

CIC International Ltd	USA	*563*	75-18570
Cogent	UK	*500*	80-06250
Global Communications Co.	USA	*579*	75-39775
Graseby Dynamics Ltd	UK	*455*	75-26590
Nautronix Ltd	UK	*470*	75-41900
Redifon MEL Limited	UK	*478*	75-48530

446 30 70 66 Communications equipment, submarine

CIC International Ltd	USA	*563*	75-18570
Cogent	UK	*500*	80-06250
Delta Electronics Inc.	USA	*568*	75-25550
Hunting Engineering Ltd	UK	*459*	75-30700
Nautronix Ltd	UK	*470*	75-41900
Redifon MEL Limited	UK	*478*	75-48530
Signaal Communications	Netherlands	*295*	75-06350

446 30 70 80 Communications equipment, voice amplification, naval

CIC International Ltd	USA	*563*	75-18570
Davies Industrial Communications Ltd	UK	*444*	75-17010
Redifon MEL Limited	UK	*478*	75-48530

446 30 80 Communications equipment accessories, naval

British Sarozal Ltd	UK	*436*	75-09570
CIC International Ltd	USA	*563*	75-18570
Cubic Defense Systems Inc.	USA	*567*	75-22900
Davies Industrial Communications Ltd	UK	*444*	75-17010
Elbit Systems Ltd	Israel	*218*	75-05300
Electronic Marketing Co. SpA	Italy	*233*	75-06860
Frederick Electronics	USA	*576*	75-36000
Harris Corporation	USA	*581*	75-42050
Israel Aircraft Industries Ltd	Israel	*219*	75-10200
Naval Electronics Inc.	USA	*598*	75-64025
Racal Acoustics Ltd	UK	*476*	75-47653
Redifon MEL Limited	UK	*478*	75-48530
Signaal Communications	Netherlands	*295*	75-06350
Tadiran Ltd	Israel	*222*	75-21600
Trompeter Electronics Inc.	USA	*623*	75-91070

446 30 80 30 Communications equipment accessories, naval radio

British Sarozal Ltd	UK	*436*	75-09570
CIC International Ltd	USA	*563*	75-18570
Cubic Defense Systems Inc.	USA	*567*	75-22900
Davies Industrial Communications Ltd	UK	*444*	75-17010
Delta Electronics Inc.	USA	*568*	75-25550
Frederick Electronics	USA	*576*	75-36000
GEC-Marconi Communications Ltd	UK	*454*	75-25335
Global Communications Co.	USA	*579*	75-39775
Harris Corporation	USA	*581*	75-42050
Icom (UK) Ltd	UK	*460*	75-31070
LARIMART SpA	Italy	*236*	75-13500
Moonraker Australia Pty Ltd	Australia	*19*	75-13770
Naval Electronics Inc.	USA	*598*	75-64025
Park Air Electronics Ltd	UK	*472*	75-43750
Racal Acoustics Ltd	UK	*476*	75-47653
Redifon MEL Limited	UK	*478*	75-48530
Signaal Communications	Netherlands	*295*	75-06350
Tadiran Ltd	Israel	*222*	75-21600
Trompeter Electronics Inc.	USA	*623*	75-91070

446 30 80 30 10 Antennas, radio communications, shipborne

Allen Telecom Inc.	USA	*548*	75-04630
Andrew Canada Inc.	Canada	*68*	75-00600
CEA Technologies Pty Ltd	Australia	*24*	80-03100
CIC International Ltd	USA	*563*	75-18570
Comrod A/S	Norway	*309*	75-00550
Cubic Defense Systems Inc.	USA	*567*	75-22900
Electronic Marketing Co. SpA	Italy	*233*	75-06860
ELTA Electronics Industries Ltd	Israel	*220*	75-10200
GEC-Marconi Communications Ltd	UK	*454*	75-25335
General Electronics Communications	Greece	*189*	75-01300
Global Communications Co.	USA	*579*	75-39775
Harris Corporation	USA	*581*	75-42050
Herley-Vega Systems	USA	*581*	75-43090
H.R. Smith Group	UK	*484*	75-54590
Icom (UK) Ltd	UK	*460*	75-31070
Jaybeam Ltd	UK	*462*	75-32400
Jotron (UK) Ltd	UK	*462*	75-32895
KEL Aerospace Pty Ltd	Australia	*18*	75-11440
Moonraker Australia Pty Ltd	Australia	*19*	75-13770
Naval Electronics Inc.	USA	*598*	75-64025
R.A. Miller Industries Inc.	USA	*596*	75-61650
Racal Antennas Ltd	UK	*476*	75-47656
Racal Communications Inc.	USA	*606*	75-72380
Renair Antenna	UK	*478*	75-49225
RJR Industries	USA	*579*	75-39720
Shakespeare Products Group	USA	*614*	75-79800
South Midlands Communications Ltd	UK	*485*	75-55213
STAREC	France	*147*	75-42775
Vantage Associates Inc.	USA	*625*	75-93250
Watkins-Johnson Co.	USA	*627*	75-95200

446 30 80 30 10 30 Antennas, dish, shipborne communications

Andrew Canada Inc.	Canada	*68*	75-00600
CIC International Ltd	USA	*563*	75-18570
ERA Technology Ltd	UK	*502*	80-09600
General Electronics Communications	Greece	*189*	75-01300
H.R. Smith Group	UK	*484*	75-54590
Jaybeam Ltd	UK	*462*	75-32400
South Midlands Communications Ltd	UK	*485*	75-55213
STAREC	France	*147*	75-42775
Vermont Composites, Inc.	USA	*625*	75-93450

446 30 80 30 10 60 Antennas, rod, shipborne communications

Comrod A/S	Norway	*309*	75-00550
ERA Technology Ltd	UK	*502*	80-09600
GEC-Marconi Communications Ltd	UK	*454*	75-25335
General Electronics Communications	Greece	*189*	75-01300
H.R. Smith Group	UK	*484*	75-54590
Icom (UK) Ltd	UK	*460*	75-31070
Jaybeam Ltd	UK	*462*	75-32400
KEL Aerospace Pty Ltd	Australia	*18*	75-11440
MA/COM Antenna & Cable Div.	USA	*593*	75-56953
Moonraker Australia Pty Ltd	Australia	*19*	75-13770
Naval Electronics Inc.	USA	*598*	75-64025
Racal Antennas Ltd	UK	*476*	75-47656
South Midlands Communications Ltd	UK	*485*	75-55213
STAREC	France	*147*	75-42775

446 30 80 30 10 65 Antennas, radio communications, submarine

CEA Technologies Pty Ltd	Australia	*24*	80-03100
ERA Technology Ltd	UK	*502*	80-09600
H.R. Smith Group	UK	*484*	75-54590
Watkins-Johnson Co.	USA	*627*	75-95200

446 30 80 30 10 70 Antennas, TV, marine

CEA Technologies Pty Ltd	Australia	*24*	80-03100
ERA Technology Ltd	UK	*502*	80-09600
H.R. Smith Group	UK	*484*	75-54590
Moonraker Australia Pty Ltd	Australia	*19*	75-13770
Naval Electronics Inc.	USA	*598*	75-64025
Racal Antennas Ltd	UK	*476*	75-47656
South Midlands Communications Ltd	UK	*485*	75-55213

446 30 80 30 30 Headsets, naval communications

Davies Industrial Communications Ltd	UK	*444*	75-17010

446 30 80 30 50 Microphones, naval communications

Graphico Electronics Group	Canada	*71*	75-06460

446 40 Communications systems, ground-based

A Goninan & Co. Ltd	Australia	*16*	75-07760
Alcatel España SA	Spain	*365*	75-00470
Alcatel Telecom Norway A/S	Norway	*308*	75-00090
Amper Programas SA	Spain	*365*	75-00480
AN/COM Industries Inc.	USA	*553*	75-07300
Andrew Canada Inc.	Canada	*68*	75-00600
Aydin Corp.	USA	*556*	75-11450

Barrett Communications Pty Ltd.	Australia	12	75-02050
Bharat Electronics Ltd	India	203	55-00200
Boeing Australia Limited	Australia	12	75-02500
Canadian Marconi Co.	Canada	69	75-02600
CelsiusTech Electronics AB	Sweden	376	75-03720
CIC International Ltd	USA	563	75-18570
Codan Pty Ltd	Australia	14	75-03620
Cogent	UK	500	80-06250
Comstream Corporation	USA	565	75-20780
Conrac Elektron GmbH	Germany	166	75-05100
Cubic Defense Systems Inc.	USA	567	75-22900
Cylink Corp.	USA	567	75-23134
Cypher Research Laboratories	Australia	10	60-00300
Daimler-Benz Aerospace France SAS	**France**	**134**	**75-12950**
David Clark Co. Inc.	USA	564	75-18850
Davies Industrial Communications Ltd	UK	444	75-17010
Econ Optics Mechanics SA	Greece	188	75-01070
Eddystone Radio Ltd	UK	447	75-19370
EDS Defence Ltd	UK	501	80-09200
Elbit Systems Ltd	Israel	218	75-05300
ELTA Electronics Industries Ltd	Israel	220	75-10200
Ericsson Defence Systems Pty Ltd	Australia	16	75-05820
Firan Corp.	Canada	71	75-05400
Fluor Daniel Inc.	USA	635	80-11720
GEC Marconi Hazeltine Corp	USA	578	75-37415
GEC-Marconi Communications Ltd	UK	454	75-25335
Global Communications Co.	USA	579	75-39775
Harris Corporation	USA	581	75-42050
Hunting Engineering Ltd	UK	459	75-30700
Instro Precision Ltd	UK	461	75-31630
Intelect Inc.	USA	585	75-47160
Israel Aircraft Industries Ltd	Israel	219	75-10200
Italtel Spa	Italy	235	75-13205
ITT Aerospace/Communications Div.	USA	586	75-49455
ITT Avionics	USA	586	75-49455
Jotron (UK) Ltd	UK	462	75-32895
JPS Communications Inc.	USA	587	75-51220
KEL Aerospace Pty Ltd	Australia	18	75-11440
Litton Industries Inc.	USA	590	75-55400
Logica UK Limited	UK	506	80-14350
MA/COM Antenna & Cable Div.	USA	593	75-56953
MCL Inc.	USA	595	75-59700
Med-Eng Systems Inc.	Canada	74	75-09550
Meteor Communications Corp.	USA	595	75-60600
Microelectronics, guidance and	Turkey	402	55-00050
Motorola Ltd	UK	469	75-41050
MPB Technologies Inc	Canada	74	75-09855
Multitone Electronics plc	UK	470	75-41700
Napco International Inc.	USA	598	75-63600
NFT Ericsson Communications ANS	Norway	308	74-01000
Philips Traffic & Engineering	Australia	20	75-16550
Racal Acoustics Ltd	UK	476	75-47653
Racal Communications Inc.	USA	606	75-72380
Racal Radio Ltd	UK	477	75-47698
Racal-Comsec Ltd	UK	477	75-47665
Racal-Datacom Ltd	UK	477	75-47670
Raytheon E-Systems	USA	608	75-73302
Redifon MEL Limited	UK	478	75-48530
R.E.M.F. Radiocoms-Avionics	France	141	75-32200
Reunert Technology Systems Ltd	S. Africa	359	75-18300
Roanwell Corp.	USA	609	75-75600
Rockwell	USA	610	75-75930
Rockwell International Co.	Italy	233	75-05200
SAIT Systems SA	Belgium	47	75-05600
Sanders	USA	612	75-77850
Siemens Aktiengesellschaft	Germany	175	75-27700
Siemens Atea NV	Belgium	47	75-05950
Siemens Plessey Electronic Systems	Australia	21	75-19475
Siemens Plessey Systems	UK	483	75-53770
Siemens Switzerland Ltd	Switzerland	387	75-11980
Signaal Communications	Netherlands	295	75-06350
Signature Industries Ltd	UK	484	75-53785
Sperry Marine Inc.	USA	617	75-83250
Tadiran Com. Ltd.	Israel	222	75-21500
Tadiran Ltd	Israel	222	75-21600
Telegenix Inc.	USA	621	75-87610
The Boeing Company	USA	559	75-14250
The Titan Corp	USA	622	75-89620
VHB Industrial Batteries Ltd	Canada	77	75-14540
ViaSat Inc.	USA	626	75-93790
Videotek Inc.	USA	626	75-94090
Wicker-Bürki AG	Switzerland	388	75-15300

446 40 10 Communications systems, ground-based, optical

ADC Telecommunications Inc.	USA	545	75-00730
CIC International Ltd	USA	563	75-18570
Cogent	UK	500	80-06250
Cranfield University	UK	419	60-00350
Francis & Lewis International Ltd	UK	452	75-24500
GEC-Marconi Research Centre	UK	419	60-00700
Global Communications Co.	USA	579	75-39775
Hunting Engineering Ltd	UK	459	75-30700
Instrument Services Ltd	UK	504	80-13328
Litton Industries Inc.	USA	590	75-55400
MPB Technologies Inc	Canada	74	75-09855
Pilkington Thorn Optronics	UK	426	74-10000
Redifon MEL Limited	UK	478	75-48530
RGB Spectrum	USA	609	75-75000
SAS Communications Ltd	UK	482	75-51800
Siemens Atea NV	Belgium	47	75-05950
Siemens Switzerland Ltd	Switzerland	387	75-11980
Thomson Marconi Sonar Limited	UK	426	74-20000

446 40 15 Communications systems, ground-based, infra-red

CIC International Ltd	USA	563	75-18570
Francis & Lewis International Ltd	UK	452	75-24500
GEC-Marconi Research Centre	UK	419	60-00700
Global Communications Co.	USA	579	75-39775
Litton Industries Inc.	USA	590	75-55400
MPB Technologies Inc	Canada	74	75-09855
Pilkington Thorn Optronics	UK	426	74-10000
RGB Spectrum	USA	609	75-75000
Siemens Atea NV	Belgium	47	75-05950

446 40 20 Communications systems, ground-based, laser

CIC International Ltd	USA	563	75-18570
Francis & Lewis International Ltd	UK	452	75-24500
GEC-Marconi Research Centre	UK	419	60-00700
Global Communications Co.	USA	579	75-39775
Litton Industries Inc.	USA	590	75-55400
MPB Technologies Inc	Canada	74	75-09855

446 40 30 Communications systems, ground-based, radio

Amper Programas SA	Spain	365	75-00480
AN/COM Industries Inc.	USA	553	75-07300
Andrew Canada Inc.	Canada	68	75-00600
Andrew Ltd	UK	430	75-03156
Aydin Corp.	USA	556	75-11450
Becker Flugfunkwerk GmbH	Germany	165	75-02200
Bharat Electronics Ltd	India	203	55-00200
CAE	UK	438	75-11125
Canadian Marconi Co.	Canada	69	75-02600
CIC International Ltd	USA	563	75-18570
Cincinnati Electronics Corp.	USA	564	75-18640
Cogent	UK	500	80-06250
Cranfield Aerospace Ltd	UK	500	80-07200
Cranfield University	UK	419	60-00350
Cubic Defense Systems Inc.	USA	567	75-22900
Cylink Corp.	USA	567	75-23134
Cypher Research Laboratories	Australia	10	60-00300
Datron World Communications Inc.	USA	568	75-24020
DSD Ltd	UK	501	80-08700
Dytecna Ltd	UK	501	80-08900
EASAMS Ltd	UK	501	80-09000
EDS Defence Ltd	UK	501	80-09200
ELTA Electronics Industries Ltd	Israel	220	75-10200
Ericsson Defence Systems Pty Ltd	Australia	16	75-05820
Fernau Avionics Ltd	UK	450	75-22600
Francis & Lewis International Ltd	UK	452	75-24500
Frederick Electronics	USA	576	75-36000
GEC Marconi Hazeltine Corp	USA	578	75-37415
GEC-Marconi Communications Ltd	UK	454	75-25335
GEC-Marconi Research Centre	UK	419	60-00700
Global Communications Co.	USA	579	75-39775
Graphico Electronics Group	Canada	71	75-06460
Harris Corporation	USA	581	75-42050

Hunting Engineering Ltd	UK	*459*	75-30700
Icom (UK) Ltd	UK	*460*	75-31070
Infocom Systems A/S	Denmark	*103*	75-09200
Interspiro Ltd	UK	*461*	75-31945
ITT Aerospace/Communications Div.	USA	*586*	75-49455
ITT Defense Ltd	UK	*505*	80-13570
Jaybeam Ltd	UK	*462*	75-32400
Jotron (UK) Ltd	UK	*462*	75-32895
JPS Communications Inc.	USA	*587*	75-51220
KEL Aerospace Pty Ltd	Australia	*18*	75-11440
LARIMART SpA	Italy	*236*	75-13500
Litton Industries Inc.	USA	*590*	75-55400
Metraplex Corp.	USA	*595*	75-60700
Microelectronics, guidance and	Turkey	*402*	55-00050
Motorola Ltd	UK	*469*	75-41050
Multitone Electronics plc	UK	*470*	75-41700
Napco International Inc.	USA	*598*	75-63600
Panorama Antennas Ltd	UK	*472*	75-43660
Park Air Electronics Ltd	UK	*472*	75-43750
Philips Traffic & Engineering	Australia	*20*	75-16550
Racal Acoustics Ltd	UK	*476*	75-47653
Racal Antennas Ltd	UK	*476*	75-47656
Racal Radio Ltd	UK	*477*	75-47698
Raytheon Cossor Electronics Ltd	UK	*478*	75-48367
Redifon MEL Limited	UK	*478*	75-48530
Reunert Technology Systems Ltd	S. Africa	*359*	75-18300
SAGEM SA	**France**	***143***	**75-33950**
SAIT Systems SA	Belgium	*47*	75-05600
Security Research Ltd	UK	*482*	75-52450
SED Systems Inc	Canada	*79*	80-01937
Siemens Aktiengesellschaft	Germany	*175*	75-27700
Siemens Plessey Systems	UK	*483*	75-53770
Siemens Switzerland Ltd	Switzerland	*387*	75-11980
Signaal Communications	Netherlands	*295*	75-06350
Signature Industries Ltd	UK	*484*	75-53785
Sunair Electronics Inc.	USA	*618*	75-84950
Tadiran Com. Ltd.	Israel	*222*	75-21500
Tadiran Ltd	Israel	*222*	75-21600
TERMA Elektronik AS	Denmark	*104*	75-20300
The Boeing Company	USA	*559*	75-14250
Toyo Communication Equipment Co	Japan	*255*	75-10000
TRL Technology Ltd	UK	*514*	80-58700
Vega-Cantley Instrument Company	UK	*490*	75-62740
Walter Dittel GmbH	Germany	*167*	75-06510
Wood & Douglas	UK	*494*	75-66790
WS Atkins plc	UK	*497*	80-01550

446 40 30 20 Communications systems, ground-based, tactical radio

Amper Programas SA	Spain	*365*	75-00480
AN/COM Industries Inc.	USA	*553*	75-07300
Andrew Canada Inc.	Canada	*68*	75-00600
Andrew Ltd	UK	*430*	75-03156
Ascom Systec Ltd	Switzerland	*385*	75-00540
Becker Flugfunkwerk GmbH	Germany	*165*	75-02200
Bharat Electronics Ltd	India	*203*	55-00200
Bradley Electronics Ltd	UK	*435*	75-07860
CAE	UK	*438*	75-11125
Cincinnati Electronics Corp.	USA	*564*	75-18640
Codalex Ltd	Canada	*70*	75-03400
Cogent	UK	*500*	80-06250
Computing Devices Canada	Canada	*70*	75-03600
Comtech Systems Inc.	USA	*565*	75-20810
Cranfield University	UK	*419*	60-00350
Datron World Communications Inc.	USA	*568*	75-24020
DSD Ltd	UK	*501*	80-08700
Dytecna Ltd	UK	*501*	80-08900
EASAMS Ltd	UK	*501*	80-09000
EDS Defence Ltd	UK	*501*	80-09200
ELTA Electronics Industries Ltd	Israel	*220*	75-10200
Francis & Lewis International Ltd	UK	*452*	75-24500
GEC-Marconi Communications Ltd	UK	*454*	75-25335
GEC-Marconi Research Centre	UK	*419*	60-00700
General Electronics Communications	Greece	*189*	75-01300
Graphico Electronics Group	Canada	*71*	75-06460
Harris Corporation	USA	*581*	75-42050
Hunting Engineering Ltd	UK	*459*	75-30700
Infocom Systems A/S	Denmark	*103*	75-09200
ITT Defense Ltd	UK	*505*	80-13570
Jotron (UK) Ltd	UK	*462*	75-32895
Kachina Communications Inc.	USA	*587*	75-51280
KEL Aerospace Pty Ltd	Australia	*18*	75-11440
Litton Industries Inc.	USA	*590*	75-55400
Marconi SpA DTM	Italy	*236*	75-14550
Microelectronics, guidance and	Turkey	*402*	55-00050
NFT Ericsson Communications ANS	Norway	*308*	74-01000
Northern Airborne Technology Ltd	Canada	*74*	75-10145
Panorama Antennas Ltd	UK	*472*	75-43660
Park Air Electronics Ltd	UK	*472*	75-43750
PhotoTelesis	USA	*604*	75-69810
Racal Acoustics Ltd	UK	*476*	75-47653
Racal Antennas Ltd	UK	*476*	75-47656
Racal Communications Inc.	USA	*606*	75-72380
Racal Radio Ltd	UK	*477*	75-47698
Reunert Technology Systems Ltd	S. Africa	*359*	75-18300
SAIT Systems SA	Belgium	*47*	75-05600
Siemens Aktiengesellschaft	Germany	*175*	75-27700
Siemens Plessey Systems	UK	*483*	75-53770
TRL Technology Ltd	UK	*514*	80-58700

446 40 30 20 20 Combat radio systems, ground-based

Amper Programas SA	Spain	*365*	75-00480
AN/COM Industries Inc.	USA	*553*	75-07300
Andrew Canada Inc.	Canada	*68*	75-00600
Ascom Systec Ltd	Switzerland	*385*	75-00540
BAMS Export Office	Belgium	*44*	74-00500
Bharat Electronics Ltd	India	*203*	55-00200
CIC International Ltd	USA	*563*	75-18570
Cincinnati Electronics Corp.	USA	*564*	75-18640
Cogent	UK	*500*	80-06250
Cypher Research Laboratories	Australia	*10*	60-00300
Datron World Communications Inc.	USA	*568*	75-24020
EDS Defence Ltd	UK	*501*	80-09200
GEC-Marconi Communications Ltd	UK	*454*	75-25335
Harris Corporation	USA	*581*	75-42050
Hunting Engineering Ltd	UK	*459*	75-30700
Infocom Systems A/S	Denmark	*103*	75-09200
ITT Defense Ltd	UK	*505*	80-13570
Litton Industries Inc.	USA	*590*	75-55400
Marconi SpA DTM	Italy	*236*	75-14550
Microelectronics, guidance and	Turkey	*402*	55-00050
Motorola Ltd	UK	*469*	75-41050
NFT Ericsson Communications ANS	Norway	*308*	74-01000
Racal Communications Inc.	USA	*606*	75-72380
Racal Radio Ltd	UK	*477*	75-47698
SAIT Systems SA	Belgium	*47*	75-05600
Siemens Aktiengesellschaft	Germany	*175*	75-27700
Siemens Plessey Electronic Systems	Australia	*21*	75-19475
Siemens Plessey Systems	UK	*483*	75-53770
Signaal Communications	Netherlands	*295*	75-06350

446 40 30 20 55 Microwave radio systems, ground-based

Andrew Canada Inc.	Canada	*68*	75-00600
Aydin Corp.	USA	*556*	75-11450
Bharat Electronics Ltd	India	*203*	55-00200
CIC International Ltd	USA	*563*	75-18570
Cogent	UK	*500*	80-06250
Comtech Systems Inc.	USA	*565*	75-20810
Cylink Corp.	USA	*567*	75-23134
FEI Microwave Inc.	USA	*575*	75-34080
GEC-Marconi Communications Ltd	UK	*454*	75-25335
Harris Corporation	USA	*581*	75-42050
Litton Industries Inc.	USA	*590*	75-55400
MA/COM Antenna & Cable Div.	USA	*593*	75-56953
Marconi SpA DTM	Italy	*236*	75-14550
SED Systems Inc	Canada	*79*	80-01937
Siemens Aktiengesellschaft	Germany	*175*	75-27700
Vega-Cantley Instrument Company	UK	*490*	75-62740
Vision Research Co.	UK	*491*	75-63360
Wood & Douglas	UK	*494*	75-66790
Znamya Production Group	Ukraine	*407*	75-10000

446 40 30 30 Data transmission systems, ground-based

Alcatel España SA	Spain	*365*	75-00470
Alcatel Telecom Norway A/S	Norway	*308*	75-00090
Andrew Canada Inc.	Canada	*68*	75-00600
Aydin Corp.	USA	*556*	75-11450
Bharat Electronics Ltd	India	*203*	55-00200
CAE	UK	*438*	75-11125
California Microwave Inc.	USA	*561*	75-16350
Canadian Marconi Co.	Canada	*69*	75-02600
CIC International Ltd	USA	*563*	75-18570

Coastcom	USA	564	75-19195
Cogent	UK	500	80-06250
Computer Resources International A/S	Denmark	104	80-00100
Cubic Defense Systems Inc.	USA	567	75-22900
Cylink Corp.	USA	567	75-23134
Cypher Research Laboratories	Australia	10	60-00300
Frederick Electronics	USA	576	75-36000
GEC-Marconi Communications Ltd	UK	454	75-25335
Global Communications Co.	USA	579	75-39775
Graphico Electronics Group	Canada	71	75-06460
Harris Corporation	USA	581	75-42050
Husky Computers Limited	UK	460	75-30770
Intertechnique SA	France	137	75-23000
ITT Aerospace/Communications Div.	USA	586	75-49455
LARIMART SpA	Italy	236	75-13500
Litton Industries Inc.	USA	590	75-55400
Logica UK Limited	UK	506	80-14350
Marconi SpA DTM	Italy	236	75-14550
Meteor Communications Corp.	USA	595	75-60600
Microelectronics, guidance and	Turkey	402	55-00050
Navigation Data Systems Inc.	USA	598	75-64050
Racal-Datacom Ltd	UK	477	75-47670
Raytheon E-Systems	USA	608	75-73302
Reunert Technology Systems Ltd	S. Africa	359	75-18300
SED Systems Inc	Canada	79	80-01937
Siemens Aktiengesellschaft	Germany	175	75-27700
Siemens Atea NV	Belgium	47	75-05950
Siemens Plessey Systems	UK	483	75-53770
Siemens Switzerland Ltd	Switzerland	387	75-11980
Tadiran Ltd	Israel	222	75-21600
Telegenix Inc.	USA	621	75-87610
Telxon Ltd	UK	488	75-59305
TransInstruments	UK	489	75-60960
Vega-Cantley Instrument Company	UK	490	75-62740

446 40 30 50 Satellite communications systems, ground-based

A Goninan & Co. Ltd	Australia	16	75-07760
Andrew Canada Inc.	Canada	68	75-00600
Bharat Electronics Ltd	India	203	55-00200
Calian Technology Ltd	Canada	69	75-01800
California Microwave Inc.	USA	561	75-16350
CIC International Ltd	USA	563	75-18570
Cincinnati Electronics Corp.	USA	564	75-18640
Computer Resources International A/S	Denmark	104	80-00100
Comstream Corporation	USA	565	75-20780
GEC-Marconi Communications Ltd	UK	454	75-25335
Harris Corporation	USA	581	75-42050
Interstate Electronics	USA	586	75-48700
Litton Industries Inc.	USA	590	75-55400
Magnavox Electronic Systems Company	USA	593	75-57120
Motorola Ltd	UK	469	75-41050
Multipoint Communications Ltd	UK	470	75-41690
Nera Telecommunications	UK	470	75-42318
Raytheon Canada Ltd	Canada	75	75-11700
SAIT Systems SA	Belgium	47	75-05600
SED Systems Inc	Canada	79	80-01937
Siemens Atea NV	Belgium	47	75-05950
South Midlands Communications Ltd	UK	485	75-55213
SP Radio A/S	Denmark	103	75-19600
TIW Systems Inc.	USA	623	75-89755
Vega-Cantley Instrument Company	UK	490	75-62740
ViaSat Inc.	USA	626	75-93790

446 40 30 60 Telemetry systems, ground-based

Ateliers de Constructions du Centre	France	132	75-03700
Aydin Corp.	USA	556	75-11450
California Microwave Inc.	USA	561	75-16350
CIC International Ltd	USA	563	75-18570
Cylink Corp.	USA	567	75-23134
Eidsvoll Electronics AS	Norway	309	75-00850
Global Communications Co.	USA	579	75-39775
Intertechnique SA	France	137	75-23000
Litton Industries Inc.	USA	590	75-55400
Logica UK Limited	UK	506	80-14350
Measurement Devices Ltd	UK	467	75-38275
Meteor CAE SpA	Italy	237	75-15400
Metraplex Corp.	USA	595	75-60700
NEC Corporation	Japan	253	75-06350
SED Systems Inc	Canada	79	80-01937
TIW Systems Inc.	USA	623	75-89755
TransInstruments	UK	489	75-60960
Vega-Cantley Instrument Company	UK	490	75-62740
Wood & Douglas	UK	494	75-66790

446 40 30 80 Voice communications control systems, ground-based

Amper Programas SA	Spain	365	75-00480
Andrew Canada Inc.	Canada	68	75-00600
California Microwave Inc.	USA	561	75-16350
CIC International Ltd	USA	563	75-18570
Cincinnati Electronics Corp.	USA	564	75-18640
Computer Resources International A/S	Denmark	104	80-00100
Cylink Corp.	USA	567	75-23134
Cypher Research Laboratories	Australia	10	60-00300
Davies Industrial Communications Ltd	UK	444	75-17010
EDS Defence Ltd	UK	501	80-09200
Graphico Electronics Group	Canada	71	75-06460
Jotron (UK) Ltd	UK	462	75-32895
JPS Communications Inc.	USA	587	75-51220
Litton Industries Inc.	USA	590	75-55400
Motorola Ltd	UK	469	75-41050
SED Systems Inc	Canada	79	80-01937
Siemens Atea NV	Belgium	47	75-05950
Siemens Switzerland Ltd	Switzerland	387	75-11980
SR Comms Management Ltd	UK	512	80-46000
Tadiran Com. Ltd.	Israel	222	75-21500

446 40 40 Communications systems, ground-based, telephone

Alcatel Telecom Norway A/S	Norway	308	75-00090
Amper Programas SA	Spain	365	75-00480
AN/COM Industries Inc.	USA	553	75-07300
Ascom Systec Ltd	Switzerland	385	75-00540
CAE	UK	438	75-11125
CIC International Ltd	USA	563	75-18570
Codalex Ltd	Canada	70	75-03400
Cogent	UK	500	80-06250
Comstream Corporation	USA	565	75-20780
Cylink Corp.	USA	567	75-23134
Ericsson Defence Systems Pty Ltd	Australia	16	75-05820
Frederick Electronics	USA	576	75-36000
Global Communications Co.	USA	579	75-39775
Graphico Electronics Group	Canada	71	75-06460
Harris Corporation	USA	581	75-42050
ITT Aerospace/Communications Div.	USA	586	75-49455
LARIMART SpA	Italy	236	75-13500
Marconi SpA DTM	Italy	236	75-14550
Microelectronics, guidance and	Turkey	402	55-00050
Philips Traffic & Engineering	Australia	20	75-16550
Racal Communications Inc.	USA	606	75-72380
Redifon MEL Limited	UK	478	75-48530
Reunert Technology Systems Ltd	S. Africa	359	75-18300
SAGEM SA	**France**	**143**	**75-33950**
Siemens Aktiengesellschaft	Germany	175	75-27700
Siemens Atea NV	Belgium	47	75-05950
Siemens Plessey Systems	UK	483	75-53770
Siemens Switzerland Ltd	Switzerland	387	75-11980
Spar Aerospace Ltd	Canada	76	75-13000
SR Comms Management Ltd	UK	512	80-46000
Tadiran Ltd	Israel	222	75-21600
Telegenix Inc.	USA	621	75-87610
Telxon Ltd	UK	488	75-59305
Thomson-CSF	France	147	75-45100

446 40 40 30 Communications centrals, ground-based, telephone

Ascom Systec Ltd	Switzerland	385	75-00540
CAE	UK	438	75-11125
CIC International Ltd	USA	563	75-18570
Cogent	UK	500	80-06250
Denro Inc.	USA	568	75-25651
Graphico Electronics Group	Canada	71	75-06460
Med-Eng Systems Inc.	Canada	74	75-09550
Redifon MEL Limited	UK	478	75-48530
SAGEM SA	**France**	**143**	**75-33950**
Siemens Atea NV	Belgium	47	75-05950
SR Comms Management Ltd	UK	512	80-46000

446 40 42 Communications systems, ground-based, telegraphic

Codalex Ltd	Canada	*70*	75-03400
Computer Resources International A/S.	Denmark	*104*	80-00100
Cubic Defense Systems Inc.	USA	*567*	75-22900
Frederick Electronics	USA	*576*	75-36000
Global Communications Co.	USA	*579*	75-39775
Racal-Datacom Ltd	UK	*477*	75-47670
Redifon MEL Limited	UK	*478*	75-48530
Reunert Technology Systems Ltd	S. Africa	*359*	75-18300
Siemens Aktiengesellschaft	Germany	*175*	75-27700
Siemens Plessey Systems	UK	*483*	75-53770
Siemens Switzerland Ltd	Switzerland	*387*	75-11980
Signaal Communications	Netherlands	*295*	75-06350
Tadiran Ltd	Israel	*222*	75-21600

446 40 46 Communications systems, ground-based, teleprinter / telex

Arinc	USA	*631*	80-03130
Bharat Electronics Ltd	India	*203*	55-00200
CAE	UK	*438*	75-11125
CIC International Ltd	USA	*563*	75-18570
Computer Resources International A/S.	Denmark	*104*	80-00100
Datron World Communications Inc.	USA	*568*	75-24020
Frederick Electronics	USA	*576*	75-36000
Global Communications Co.	USA	*579*	75-39775
HAL Communications Corp.	USA	*580*	75-41450
Harris Corporation	USA	*581*	75-42050
Inovex Technology Ltd	UK	*504*	80-13310
Raytheon Cossor Electronics Ltd	UK	*478*	75-48367
Redifon MEL Limited	UK	*478*	75-48530
SAGEM SA	**France**	***143***	**75-33950**
Siemens Aktiengesellschaft	Germany	*175*	75-27700
Siemens Plessey Systems	UK	*483*	75-53770
Sunair Electronics Inc.	USA	*618*	75-84950
Vaisala Oy	Finland	*119*	75-02800

446 40 46 50 Message centres, automated

Alcatel Telecom Norway A/S	Norway	*308*	75-00090
Computer Resources International A/S.	Denmark	*104*	80-00100
ICL DESC	UK	*501*	80-08400
SR Comms Management Ltd	UK	*512*	80-46000

446 40 48 Communications systems, ground-based, telecopier / facsimile

Computer Resources International A/S.	Denmark	*104*	80-00100
Global Communications Co.	USA	*579*	75-39775
Graphico Electronics Group	Canada	*71*	75-06460
Litton Industries Inc.	USA	*590*	75-55400
PhotoTelesis	USA	*604*	75-69810
Redifon MEL Limited	UK	*478*	75-48530
SAGEM SA	**France**	***143***	**75-33950**
Siemens Aktiengesellschaft	Germany	*175*	75-27700
Siemens Plessey Systems	UK	*483*	75-53770

446 40 50 Communications systems, ground-based, video

Colorado Video Inc.	USA	*564*	75-19730
Computer Resources International A/S.	Denmark	*104*	80-00100
Comstream Corporation	USA	*565*	75-20780
Harris Corporation	USA	*581*	75-42050
Hunting Engineering Ltd	UK	*459*	75-30700
Videotek Inc.	USA	*626*	75-94090
Winkelmann (UK) Ltd	UK	*494*	75-66600

446 40 60 Public address systems, ground-based

Applied Electro Mechanics Inc.	USA	*554*	75-08130
Chrysanthacopoulos SA	Greece	*188*	75-00910
Davies Industrial Communications Ltd	UK	*444*	75-17010
Redifon MEL Limited	UK	*478*	75-48530

446 40 70 Communications equipment, ground-based

A Goninan & Co. Ltd	Australia	*16*	75-07760
ADC Telecommunications Inc.	USA	*545*	75-00730
Aerocom/Squitter	USA	*546*	75-01150
Alcatel Bell	Belgium	*44*	75-00055
Alcatel Telecom Norway A/S	Norway	*308*	75-00090
AlliedSignal Aerospace Canada	Canada	*68*	75-00470
Amper Programas SA	Spain	*365*	75-00480
Analog Modules Inc.	USA	*553*	75-07140
AN/COM Industries Inc.	USA	*553*	75-07300
Andrew Canada Inc.	Canada	*68*	75-00600
Azienda Torinese Elettronica	Italy	*232*	75-02300
Broadcast Electronics Inc.	USA	*560*	75-15240
Calian Technology Ltd	Canada	*69*	75-01800
California Microwave Inc.	USA	*561*	75-16350
Canadian Marconi Co.	Canada	*69*	75-02600
CIC International Ltd	USA	*563*	75-18570
Codalex Ltd	Canada	*70*	75-03400
Communications & Power Industries	USA	*565*	75-20393
Comstream Corporation	USA	*565*	75-20780
Conrac Elektron GmbH	Germany	*166*	75-05100
Cubic Defense Systems Inc.	USA	*567*	75-22900
Daimler-Benz Aerospace France SAS	**France**	***134***	**75-12950**
Dare Electronics	USA	*567*	75-23580
David Clark Co. Inc.	USA	*564*	75-18850
Davies Industrial Communications Ltd	UK	*444*	75-17010
Delta Electronics Inc.	USA	*568*	75-25550
DNE, Inc.	USA	*569*	75-27000
Electronic Space Systems Corp.	USA	*572*	75-30100
Elfon Ltd	Greece	*188*	75-01150
ELTA Electronics Industries Ltd	Israel	*220*	75-10200
Global Communications Co.	USA	*579*	75-39775
GoldStar Electric Co. Ltd	Korea (ROK)	*265*	75-01400
Graphico Electronics Group	Canada	*71*	75-06460
Harris Corporation	USA	*581*	75-42050
Howden Fluid Systems	USA	*582*	75-44400
Hunting Engineering Ltd	UK	*459*	75-30700
ITT Aerospace/Communications Div.	USA	*586*	75-49455
Jay-El Products Inc.	USA	*586*	75-50100
Jotron Electronics A/S	Norway	*309*	75-01030
Jotron (UK) Ltd	UK	*462*	75-32895
JPS Communications Inc.	USA	*587*	75-51220
KEL Aerospace Pty Ltd	Australia	*18*	75-11440
Litton Industries Inc.	USA	*590*	75-55400
MA/COM Antenna & Cable Div.	USA	*593*	75-56953
MCL Inc.	USA	*595*	75-59700
Megapulse Inc.	USA	*595*	75-59830
Microelectronics, guidance and	Turkey	*402*	55-00050
Ortel Corp.	USA	*602*	75-67530
Philips Traffic & Engineering	Australia	*20*	75-16550
Racal Communications Inc.	USA	*606*	75-72380
Racal-Datacom Ltd	UK	*477*	75-47670
Raytheon Cossor Electronics Ltd	UK	*478*	75-48367
Redifon MEL Limited	UK	*478*	75-48530
RGB Spectrum	USA	*609*	75-75000
RJR Industries	USA	*579*	75-39720
SAIT Systems SA	Belgium	*47*	75-05600
Sanders	USA	*612*	75-77850
Sea Tel Inc.	USA	*613*	75-78930
Siemens Aktiengesellschaft	Germany	*175*	75-27700
Siemens Atea NV	Belgium	*47*	75-05950
Siemens Plessey Systems	UK	*483*	75-53770
Siemens Switzerland Ltd	Switzerland	*387*	75-11980
Signaal Communications	Netherlands	*295*	75-06350
Sunair Electronics Inc.	USA	*618*	75-84950
Telegenix Inc.	USA	*621*	75-87610
Walter Dittel GmbH	Germany	*167*	75-06510

446 40 70 10 Communications equipment, ground-based, optical

ADC Telecommunications Inc.	USA	*545*	75-00730
CIC International Ltd	USA	*563*	75-18570
ELTA Electronics Industries Ltd	Israel	*220*	75-10200
Global Communications Co.	USA	*579*	75-39775
Hunting Engineering Ltd	UK	*459*	75-30700
MPB Technologies Inc	Canada	*74*	75-09855
Siemens Atea NV	Belgium	*47*	75-05950

446 40 70 10 30 Communications equipment, ground-based, fibre-optic

ADC Telecommunications Inc.	USA	545	75-00730
Alcatel Bell	Belgium	44	75-00055
Analog Modules Inc.	USA	553	75-07140
Canadian Marconi Co.	Canada	69	75-02600
CIC International Ltd	USA	563	75-18570
Cogent	UK	500	80-06250
Fotec Inc.	USA	576	75-35680
Hunting Engineering Ltd	UK	459	75-30700
iMPath Networks Inc.	Canada	77	75-46151
MPB Technologies Inc	Canada	74	75-09855
Ortel Corp.	USA	602	75-67530
Redifon MEL Limited	UK	478	75-48530
Siemens Atea NV	Belgium	47	75-05950
Siemens Switzerland Ltd	Switzerland	387	75-11980
Tadiran Ltd	Israel	222	75-21600

446 40 70 15 Communications equipment, ground-based, infra-red

CIC International Ltd	USA	563	75-18570
Global Communications Co.	USA	579	75-39775
Hughes Electronics Corp.	USA	583	75-44865
MPB Technologies Inc	Canada	74	75-09855

446 40 70 20 Communications equipment, ground-based, laser

CIC International Ltd	USA	563	75-18570
GEC-Marconi Research Centre	UK	419	60-00700
Global Communications Co.	USA	579	75-39775
MPB Technologies Inc	Canada	74	75-09855
Pilkington Thorn Optronics	UK	426	74-10000

446 40 70 30 Communications equipment, radio, ground-based

Aerocom/Squitter	USA	546	75-01150
Alcatel Bell	Belgium	44	75-00055
Amper Programas SA	Spain	365	75-00480
Andrew Canada Inc.	Canada	68	75-00600
Aydin Corp.	USA	556	75-11450
Becker Flugfunkwerk GmbH	Germany	165	75-02200
Bharat Electronics Ltd	India	203	55-00200
Canadian Marconi Co.	Canada	69	75-02600
CIC International Ltd	USA	563	75-18570
Cincinnati Electronics Corp.	USA	564	75-18640
Communications & Power Industries	Canada	70	75-03550
Cubic Defense Systems Inc.	USA	567	75-22900
Datron World Communications Inc.	USA	568	75-24020
Electronic Marketing Co. SpA	Italy	233	75-06860
ELTA Electronics Industries Ltd	Israel	220	75-10200
Fernau Avionics Ltd	UK	450	75-22600
Frederick Electronics	USA	576	75-36000
GEC-Marconi Communications Ltd	UK	454	75-25335
Global Communications Co.	USA	579	75-39775
Graphico Electronics Group	Canada	71	75-06460
Hagen	UK	456	75-27220
Harris Corporation	USA	581	75-42050
ITT Aerospace/Communications Div.	USA	586	75-49455
JPS Communications Inc.	USA	587	75-51220
LARIMART SpA	Italy	236	75-13500
MA/COM Antenna & Cable Div.	USA	593	75-56953
Microelectronics, guidance and	Turkey	402	55-00050
Multitone Electronics plc	UK	470	75-41700
Napco International Inc.	USA	598	75-63600
NEC Corporation	Japan	253	75-06350
Philips Traffic & Engineering	Australia	20	75-16550
Racal Communications Inc.	USA	606	75-72380
Racal Radio Ltd	UK	477	75-47698
Redifon MEL Limited	UK	478	75-48530
Reunert Technology Systems Ltd	S. Africa	359	75-18300
SAIT Systems SA	Belgium	47	75-05600
Seaboard Electronics Co.	USA	613	75-78910
Siemens Aktiengesellschaft	Germany	175	75-27700
Siemens Plessey Systems	UK	483	75-53770
Signaal Communications	Netherlands	295	75-06350
Signature Industries Ltd	UK	484	75-53785
South Midlands Communications Ltd	UK	485	75-55213
Sunair Electronics Inc.	USA	618	75-84950
Tadiran Ltd	Israel	222	75-21600
Telegenix Inc.	USA	621	75-87610
Walter Dittel GmbH	Germany	167	75-06510

446 40 70 30 20 Communications sets, radio, ground-based

Aerocom/Squitter	USA	546	75-01150
Alcatel Bell	Belgium	44	75-00055
Alcatel Bell-SDT	Belgium	44	75-00060
Amper Programas SA	Spain	365	75-00480
AN/COM Industries Inc.	USA	553	75-07300
Ascom Systec Ltd	Switzerland	385	75-00540
Becker Flugfunkwerk GmbH	Germany	165	75-02200
Bharat Electronics Ltd	India	203	55-00200
Canadian Marconi Co.	Canada	69	75-02600
CIC International Ltd	USA	563	75-18570
Cincinnati Electronics Corp.	USA	564	75-18640
Cubic Defense Systems Inc.	USA	567	75-22900
Cylink Corp.	USA	567	75-23134
Datron World Communications Inc.	USA	568	75-24020
Electronic Marketing Co. SpA	Italy	233	75-06860
ELTA Electronics Industries Ltd	Israel	220	75-10200
Fernau Avionics Ltd	UK	450	75-22600
Frederick Electronics	USA	576	75-36000
GEC-Marconi Communications Ltd	UK	454	75-25335
Global Communications Co.	USA	579	75-39775
Graphico Electronics Group	Canada	71	75-06460
Harris Corporation	USA	581	75-42050
Icom (UK) Ltd	UK	460	75-31070
ITT Aerospace/Communications Div.	USA	586	75-49455
Jotron (UK) Ltd	UK	462	75-32895
LARIMART SpA	Italy	236	75-13500
Microelectronics, guidance and	Turkey	402	55-00050
Napco International Inc.	USA	598	75-63600
NEC Corporation	Japan	253	75-06350
Racal Communications Inc.	USA	606	75-72380
Racal Radio Ltd	UK	477	75-47698
Redifon MEL Limited	UK	478	75-48530
Reunert Technology Systems Ltd	S. Africa	359	75-18300
SAIT Systems SA	Belgium	47	75-05600
Security Research Ltd	UK	482	75-52450
Siemens Aktiengesellschaft	Germany	175	75-27700
Signaal Communications	Netherlands	295	75-06350
Signature Industries Ltd	UK	484	75-53785
South Midlands Communications Ltd	UK	485	75-55213
Sunair Electronics Inc.	USA	618	75-84950
Tadiran Com. Ltd.	Israel	222	75-21500
Tadiran Ltd	Israel	222	75-21600
Walter Dittel GmbH	Germany	167	75-06510

446 40 70 30 20 20 Combat radios, ground-based

Amper Programas SA	Spain	365	75-00480
AN/COM Industries Inc.	USA	553	75-07300
BAMS Export Office	Belgium	44	74-00500
Bharat Electronics Ltd	India	203	55-00200
Cincinnati Electronics Corp.	USA	564	75-18640
Datron World Communications Inc.	USA	568	75-24020
ELTA Electronics Industries Ltd	Israel	220	75-10200
GEC-Marconi Communications Ltd	UK	454	75-25335
Harris Corporation	USA	581	75-42050
Marconi SpA DTM	Italy	236	75-14550
Microelectronics, guidance and	Turkey	402	55-00050
Motorola Ltd	UK	469	75-41050
Napco International Inc.	USA	598	75-63600
Racal Communications Inc.	USA	606	75-72380
Racal Radio Ltd	UK	477	75-47698
SAIT Systems SA	Belgium	47	75-05600
Siemens Plessey Systems	UK	483	75-53770
Watkins-Johnson Co.	USA	627	75-95200

446 40 70 30 20 35 Radio sets, ground-based, hand-held

Alcatel Bell	Belgium	44	75-00055
Amper Programas SA	Spain	365	75-00480
AN/COM Industries Inc.	USA	553	75-07300
Ascom Systec Ltd	Switzerland	385	75-00540
Audiotel International Ltd	UK	432	75-04400
Bharat Electronics Ltd	India	203	55-00200
Chapman & Smith Ltd	UK	439	75-12500

CIC International Ltd	USA	563	75-18570
Cincinnati Electronics Corp.	USA	564	75-18640
Electronic Marketing Co. SpA	Italy	233	75-06860
Icom (UK) Ltd	UK	460	75-31070
Jotron (UK) Ltd	UK	462	75-32895
LARIMART SpA	Italy	236	75-13500
Microelectronics, guidance and	Turkey	402	55-00050
Motorola Ltd	UK	469	75-41050
Napco International Inc.	USA	598	75-63600
Philips Traffic & Engineering	Australia	20	75-16550
Racal Communications Inc.	USA	606	75-72380
Racal Radio Ltd	UK	477	75-47698
Reunert Technology Systems Ltd	S. Africa	359	75-18300
SAIT Systems SA	Belgium	47	75-05600
Security Research Ltd	UK	482	75-52450
Signature Industries Ltd	UK	484	75-53785
South Midlands Communications Ltd	UK	485	75-55213
Tadiran Ltd	Israel	222	75-21600

446 40 70 30 20 35 90 Walkie-talkie sets, radio, ground-based

Amper Programas SA	Spain	365	75-00480
AN/COM Industries Inc.	USA	553	75-07300
Audiotel International Ltd	UK	432	75-04400
CIC International Ltd	USA	563	75-18570
Cincinnati Electronics Corp.	USA	564	75-18640
Icom (UK) Ltd	UK	460	75-31070
Jotron (UK) Ltd	UK	462	75-32895
LARIMART SpA	Italy	236	75-13500
Napco International Inc.	USA	598	75-63600
Racal Communications Inc.	USA	606	75-72380
SAIT Systems SA	Belgium	47	75-05600
South Midlands Communications Ltd	UK	485	75-55213

446 40 70 30 20 50 Manpack radios, ground-based

Alcatel Bell	Belgium	44	75-00055
Amper Programas SA	Spain	365	75-00480
AN/COM Industries Inc.	USA	553	75-07300
BAMS Export Office	Belgium	44	74-00500
Becker Flugfunkwerk GmbH	Germany	165	75-02200
Bharat Electronics Ltd	India	203	55-00200
Canadian Marconi Co.	Canada	69	75-02600
Chapman & Smith Ltd	UK	439	75-12500
CIC International Ltd	USA	563	75-18570
Cincinnati Electronics Corp.	USA	564	75-18640
Datron World Communications Inc.	USA	568	75-24020
GEC-Marconi Communications Ltd	UK	454	75-25335
Harris Corporation	USA	581	75-42050
Jotron (UK) Ltd	UK	462	75-32895
Kachina Communications Inc.	USA	587	75-51280
Microelectronics, guidance and	Turkey	402	55-00050
Motorola Ltd	UK	469	75-41050
Napco International Inc.	USA	598	75-63600
Racal Communications Inc.	USA	606	75-72380
Racal Radio Ltd	UK	477	75-47698
Reunert Technology Systems Ltd	S. Africa	359	75-18300
SAIT Systems SA	Belgium	47	75-05600
Siemens Aktiengesellschaft	Germany	175	75-27700
Siemens Plessey Systems	UK	483	75-53770
South Midlands Communications Ltd	UK	485	75-55213
Tadiran Ltd	Israel	222	75-21600
Watkins-Johnson Co.	USA	627	75-95200

446 40 70 30 20 55 Microwave radio sets, ground-based

Bharat Electronics Ltd	India	203	55-00200
CIC International Ltd	USA	563	75-18570
GEC-Marconi Communications Ltd	UK	454	75-25335
Harris Corporation	USA	581	75-42050
Marconi SpA DTM	Italy	236	75-14550
Napco International Inc.	USA	598	75-63600
Siemens Aktiengesellschaft	Germany	175	75-27700
Vision Research Co.	UK	491	75-63360
Watkins-Johnson Co.	USA	627	75-95200

446 40 70 30 20 60 Radio test sets, ground-based

AN/COM Industries Inc.	USA	553	75-07300
CIC International Ltd	USA	563	75-18570
ELTA Electronics Industries Ltd	Israel	220	75-10200
Hewlett-Packard Company	USA	581	75-43200
Marconi Instruments Ltd	UK	466	75-37270
Microelectronics, guidance and	Turkey	402	55-00050
Motorola Ltd	UK	469	75-41050
Napco International Inc.	USA	598	75-63600
Watkins-Johnson Co.	USA	627	75-95200

446 40 70 30 20 65 Vehicle radio sets

Alcatel Bell	Belgium	44	75-00055
Amper Programas SA	Spain	365	75-00480
AN/COM Industries Inc.	USA	553	75-07300
BAMS Export Office	Belgium	44	74-00500
Becker Flugfunkwerk GmbH	Germany	165	75-02200
Bharat Electronics Ltd	India	203	55-00200
CIC International Ltd	USA	563	75-18570
Cincinnati Electronics Corp.	USA	564	75-18640
Datron World Communications Inc.	USA	568	75-24020
David Clark Co. Inc.	USA	564	75-18850
Electronic Marketing Co. SpA	Italy	233	75-06860
ELTA Electronics Industries Ltd	Israel	220	75-10200
Harris Corporation	USA	581	75-42050
Icom (UK) Ltd	UK	460	75-31070
Jotron (UK) Ltd	UK	462	75-32895
Kachina Communications Inc.	USA	587	75-51280
Marconi SpA DTM	Italy	236	75-14550
Microelectronics, guidance and	Turkey	402	55-00050
Motorola Ltd	UK	469	75-41050
Napco International Inc.	USA	598	75-63600
Racal Communications Inc.	USA	606	75-72380
Racal Radio Ltd	UK	477	75-47698
Reunert Technology Systems Ltd	S. Africa	359	75-18300
Siemens Aktiengesellschaft	Germany	175	75-27700
Siemens Plessey Systems	UK	483	75-53770
Signaal Communications	Netherlands	295	75-06350
Signature Industries Ltd	UK	484	75-53785
Tadiran Ltd	Israel	222	75-21600

446 40 70 30 20 66 Intercom sets, radio, ground-based vehicle

Amper Programas SA	Spain	365	75-00480
AN/COM Industries Inc.	USA	553	75-07300
CIC International Ltd	USA	563	75-18570
Davies Industrial Communications Ltd	UK	444	75-17010
Denro Inc.	USA	568	75-25651
Hagen	UK	456	75-27220
Harris Corporation	USA	581	75-42050
Hunting Engineering Ltd	UK	459	75-30700
Icom (UK) Ltd	UK	460	75-31070
Napco International Inc.	USA	598	75-63600
Northern Airborne Technology Ltd	Canada	74	75-10145
Racal Acoustics Ltd	UK	476	75-47653
Signaal Communications	Netherlands	295	75-06350

446 40 70 30 20 70 Radio transmitters, high-powered

Amper Programas SA	Spain	365	75-00480
Broadcast Electronics Inc.	USA	560	75-15240
CIC International Ltd	USA	563	75-18570
Cincinnati Electronics Corp.	USA	564	75-18640
Dare Electronics	USA	567	75-23580
Datron World Communications Inc.	USA	568	75-24020
Electronic Marketing Co. SpA	Italy	233	75-06860
GEC-Marconi Communications Ltd	UK	454	75-25335
Icom (UK) Ltd	UK	460	75-31070
Jotron (UK) Ltd	UK	462	75-32895
KEL Aerospace Pty Ltd	Australia	18	75-11440
Napco International Inc.	USA	598	75-63600
Racal Communications Inc.	USA	606	75-72380
Redifon MEL Limited	UK	478	75-48530
SAGEM SA	**France**	**143**	**75-33950**
SAIT Systems SA	Belgium	47	75-05600
Signaal Communications	Netherlands	295	75-06350
Signature Industries Ltd	UK	484	75-53785
Tadiran Ltd	Israel	222	75-21600
Technitrol Inc.	USA	620	75-86800

Product–Supplier Link

446 40 70 30 30 Data transmission sets, ground-based

Alcatel Bell	Belgium	*44*	75-00055
Amper Programas SA	Spain	*365*	75-00480
Canadian Marconi Co.	Canada	*69*	75-02600
CIC International Ltd	USA	*563*	75-18570
Cincinnati Electronics Corp.	USA	*564*	75-18640
Cubic Defense Systems Inc.	USA	*567*	75-22900
Cylink Corp.	USA	*567*	75-23134
Denro Inc.	USA	*568*	75-25651
Frederick Electronics	USA	*576*	75-36000
Global Communications Co.	USA	*579*	75-39775
Graphico Electronics Group	Canada	*71*	75-06460
ICL DESC	UK	*501*	80-08400
Icom (UK) Ltd	UK	*460*	75-31070
Intertechnique SA.	France	*137*	75-23000
ITT Aerospace/Communications Div.	USA	*586*	75-49455
LARIMART SpA	Italy	*236*	75-13500
Logica UK Limited.	UK	*506*	80-14350
Microelectronics, guidance and	Turkey	*402*	55-00050
Napco International Inc.	USA	*598*	75-63600
Racal Communications Inc.	USA	*606*	75-72380
Raytheon Cossor Electronics Ltd	UK	*478*	75-48367
Siemens Aktiengesellschaft	Germany	*175*	75-27700
Siemens Plessey Systems	UK	*483*	75-53770
Signaal Communications	Netherlands	*295*	75-06350
Stanilite Electronics Pty Ltd	Australia	*21*	75-19690
Sunair Electronics Inc.	USA	*618*	75-84950
Tadiran Ltd	Israel	*222*	75-21600
Telxon Ltd	UK	*488*	75-59305
Vega-Cantley Instrument Company	UK	*490*	75-62740

446 40 70 30 30 30 Data entry devices, ground-based

CIC International Ltd	USA	*563*	75-18570
Husky Computers Limited	UK	*460*	75-30770
ICL DESC	UK	*501*	80-08400
Jay-El Products Inc.	USA	*586*	75-50100

446 40 70 30 40 Satellite communication subsystems, ground-based

Andrew Canada Inc.	Canada	*68*	75-00600
Andrew Ltd	UK	*430*	75-03156
Calian Technology Ltd	Canada	*69*	75-01800
Cranfield University	UK	*419*	60-00350
Denis Ferranti Meters Ltd	UK	*450*	75-22850
ERA Technology Ltd	UK	*502*	80-09600
FEI Microwave Inc.	USA	*575*	75-34080
Francis & Lewis International Ltd	UK	*452*	75-24500
GEC-Marconi Communications Ltd	UK	*454*	75-25335
Motorola Ltd	UK	*469*	75-41050
Multipoint Communications Ltd	UK	*470*	75-41690
Nera Telecommunications	UK	*470*	75-42318
Park Air Electronics Ltd	UK	*472*	75-43750
Racal Antennas Ltd	UK	*476*	75-47656
Racal Radio Ltd	UK	*477*	75-47698
Raytheon Canada Ltd	Canada	*75*	75-11700
Rollet H. & Co. Ltd	UK	*479*	75-50208
Rutherford Appleton Laboratory	UK	*420*	60-01850
Siemens Plessey Systems	UK	*483*	75-53770
Signal Processors Ltd	UK	*483*	75-53780
Sonic Communications (Int.) Ltd.	UK	*485*	75-55202
STAREC	France	*147*	75-42775
TDSi Micro Applications Ltd	UK	*513*	80-51000
TIW Systems Inc..	USA	*623*	75-89755
TRL Technology Ltd	UK	*514*	80-58700

446 40 70 30 50 Satellite communications equipment, ground-based

Andrew Canada Inc.	Canada	*68*	75-00600
Andrew Ltd	UK	*430*	75-03156
Bharat Electronics Ltd	India	*203*	55-00200
California Microwave Inc.	USA	*561*	75-16350
Celwave R.F. A/S.	Denmark	*102*	75-03230
CIC International Ltd	USA	*563*	75-18570
Communications & Power Industries	USA	*565*	75-20393
Comstream Corporation	USA	*565*	75-20780
Francis & Lewis International Ltd	UK	*452*	75-24500
GEC-Marconi Communications Ltd	UK	*454*	75-25335
GEC-Marconi Research Centre	UK	*419*	60-00700
Harris Corporation	USA	*581*	75-42050
Logica UK Limited.	UK	*506*	80-14350
Magnavox Electronic Systems Company	USA	*593*	75-57120
Matra Marconi Space UK, Ltd	UK	*425*	74-06900
Moonraker Australia Pty Ltd	Australia	*19*	75-13770
Napco International Inc.	USA	*598*	75-63600
Nera Telecommunications	UK	*470*	75-42318
Racal Antennas Ltd	UK	*476*	75-47656
Racal Radio Ltd	UK	*477*	75-47698
Raytheon Canada Ltd	Canada	*75*	75-11700
Rutherford Appleton Laboratory	UK	*420*	60-01850
Sea Tel Inc.	USA	*613*	75-78930
Siemens Plessey Systems	UK	*483*	75-53770
South Midlands Communications Ltd.	UK	*485*	75-55213
STAREC	France	*147*	75-42775
TIW Systems Inc..	USA	*623*	75-89755
Vega-Cantley Instrument Company	UK	*490*	75-62740

446 40 70 30 50 50 Satellite communications terminals, ground-based

Alcatel Bell	Belgium	*44*	75-00055
Alden Electronics Incorporated	USA	*548*	75-04400
Andrew Canada Inc.	Canada	*68*	75-00600
Andrew Ltd	UK	*430*	75-03156
Bharat Electronics Ltd	India	*203*	55-00200
California Microwave Inc.	USA	*561*	75-16350
CIC International Ltd	USA	*563*	75-18570
Francis & Lewis International Ltd	UK	*452*	75-24500
GEC-Marconi Communications Ltd	UK	*454*	75-25335
Global Communications Co.	USA	*579*	75-39775
Harris Corporation	USA	*581*	75-42050
Logica UK Limited.	UK	*506*	80-14350
Magnavox Electronic Systems Company	USA	*593*	75-57120
Matra Marconi Space UK, Ltd	UK	*425*	74-06900
Nera Telecommunications	UK	*470*	75-42318
Racal Antennas Ltd	UK	*476*	75-47656
Racal Radio Ltd	UK	*477*	75-47698
Raytheon Canada Ltd	Canada	*75*	75-11700
Redifon MEL Limited	UK	*478*	75-48530
Rutherford Appleton Laboratory	UK	*420*	60-01850
Siemens Plessey Systems	UK	*483*	75-53770
South Midlands Communications Ltd.	UK	*485*	75-55213
Titan Linkabit	USA	*623*	75-89650
TIW Systems Inc..	USA	*623*	75-89755
TRL Technology Ltd	UK	*514*	80-58700
Vega-Cantley Instrument Company	UK	*490*	75-62740

446 40 70 30 50 50 50 Satellite communications terminals, ground-based, mobile

Andrew Canada Inc.	Canada	*68*	75-00600
Andrew Ltd	UK	*430*	75-03156
Bharat Electronics Ltd	India	*203*	55-00200
California Microwave Inc.	USA	*561*	75-16350
Cranfield University	UK	*419*	60-00350
GEC-Marconi Communications Ltd	UK	*454*	75-25335
Multipoint Communications Ltd	UK	*470*	75-41690
Nera Telecommunications	UK	*470*	75-42318
Racal Antennas Ltd	UK	*476*	75-47656
Racal Radio Ltd	UK	*477*	75-47698
Siemens Plessey Systems	UK	*483*	75-53770
South Midlands Communications Ltd.	UK	*485*	75-55213
TRL Technology Ltd	UK	*514*	80-58700

446 40 70 30 50 60 Satellite communications transceivers, ground-based

California Microwave Inc.	USA	*561*	75-16350
CIC International Ltd	USA	*563*	75-18570
GEC-Marconi Communications Ltd	UK	*454*	75-25335
Harris Corporation	USA	*581*	75-42050
Matra Marconi Space UK, Ltd	UK	*425*	74-06900
Motorola Ltd	UK	*469*	75-41050
Multipoint Communications Ltd	UK	*470*	75-41690
Nera Telecommunications	UK	*470*	75-42318
TRL Technology Ltd	UK	*514*	80-58700
Vega-Cantley Instrument Company	UK	*490*	75-62740

446 40 70 30 60 Telemetry sets, ground-based

Alcatel Bell	Belgium	*44*	75-00055
Aydin Corp.	USA	*556*	75-11450
CAE	UK	*438*	75-11125
Cambridge Consultants Ltd	UK	*499*	80-05300
CIC International Ltd	USA	*563*	75-18570
Cincinnati Electronics Corp.	USA	*564*	75-18640
Contemporary Systems Design Ltd	UK	*442*	75-14860
GEC-Marconi Research Centre	UK	*419*	60-00700
Global Communications Co.	USA	*579*	75-39775
Instro Precision Ltd	UK	*461*	75-31630
Intertechnique SA	France	*137*	75-23000
Jotron (UK) Ltd	UK	*462*	75-32895
Logica UK Limited	UK	*506*	80-14350
MASS Consultants Limited	UK	*506*	80-15650
Measurement Devices Ltd	UK	*467*	75-38275
NEC Corporation	Japan	*253*	75-06350
Remsdaq Ltd	UK	*478*	75-49210
Rutherford Appleton Laboratory	UK	*420*	60-01850
Security Research Ltd	UK	*482*	75-52450
Shorts Missile Systems Ltd	**UK**	***483***	**75-53550**
Steatite Insulations Ltd	UK	*486*	75-56450
Vega-Cantley Instrument Company	UK	*490*	75-62740

446 40 70 30 70 Troposcatter terminals, ground-based

Bharat Electronics Ltd	India	*203*	55-00200
California Microwave Inc.	USA	*561*	75-16350
CIC International Ltd	USA	*563*	75-18570
Comtech Systems Inc.	USA	*565*	75-20810
GEC-Marconi Communications Ltd	UK	*454*	75-25335
GEC-Marconi Research Centre	UK	*419*	60-00700
Global Communications Co.	USA	*579*	75-39775
Harris Corporation	USA	*581*	75-42050
Marconi SpA DTM	Italy	*236*	75-14550

446 40 70 30 72 Troposcatter terminals, ground-based, mobile

Bharat Electronics Ltd	India	*203*	55-00200
California Microwave Inc.	USA	*561*	75-16350
CIC International Ltd	USA	*563*	75-18570
GEC-Marconi Communications Ltd	UK	*454*	75-25335
GEC-Marconi Research Centre	UK	*419*	60-00700
Global Communications Co.	USA	*579*	75-39775
Harris Corporation	USA	*581*	75-42050
Marconi SpA DTM	Italy	*236*	75-14550

446 40 70 40 Communications equipment, ground-based, telephone

Alcatel Bell	Belgium	*44*	75-00055
Alcatel Telecom Norway A/S	Norway	*308*	75-00090
Ascom Systec Ltd	Switzerland	*385*	75-00540
CAE	UK	*438*	75-11125
CIC International Ltd	USA	*563*	75-18570
Coastcom	USA	*564*	75-19195
CORIS	France	*134*	75-12450
Davies Industrial Communications Ltd	UK	*444*	75-17010
Ericsson Defence Systems Pty Ltd	Australia	*16*	75-05820
Frederick Electronics	USA	*576*	75-36000
Global Communications Co.	USA	*579*	75-39775
Graphico Electronics Group	Canada	*71*	75-06460
JPS Communications Inc.	USA	*587*	75-51220
LARIMART SpA	Italy	*236*	75-13500
Microelectronics, guidance and	Turkey	*402*	55-00050
Racal Acoustics Ltd	UK	*476*	75-47653
Redifon MEL Limited	UK	*478*	75-48530
Reunert Technology Systems Ltd	S. Africa	*359*	75-18300
SAGEM SA	**France**	***143***	**75-33950**
Siemens Aktiengesellschaft	Germany	*175*	75-27700
Siemens Atea NV	Belgium	*47*	75-05950
Siemens Plessey Systems	UK	*483*	75-53770
Siemens Switzerland Ltd	Switzerland	*387*	75-11980
Spar Aerospace Ltd	Canada	*76*	75-13000
Stanilite Electronics Pty Ltd	Australia	*21*	75-19690
Tadiran Ltd	Israel	*222*	75-21600

446 40 70 40 50 Telephone sets, ground-based

Ascom Systec Ltd	Switzerland	*385*	75-00540
CIC International Ltd	USA	*563*	75-18570
Racal Acoustics Ltd	UK	*476*	75-47653
Redifon MEL Limited	UK	*478*	75-48530
SAGEM SA	**France**	***143***	**75-33950**
Siemens Atea NV	Belgium	*47*	75-05950

446 40 70 40 50 30 Telephone sets, field

AN/COM Industries Inc.	USA	*553*	75-07300
Ascom Systec Ltd	Switzerland	*385*	75-00540
CIC International Ltd	USA	*563*	75-18570
Davies Industrial Communications Ltd	UK	*444*	75-17010
Microelectronics, guidance and	Turkey	*402*	55-00050
Racal Acoustics Ltd	UK	*476*	75-47653
Siemens Atea NV	Belgium	*47*	75-05950
Siemens Switzerland Ltd	Switzerland	*387*	75-11980
SISTEL	Portugal	*327*	75-18300

446 40 70 40 60 Telephone switchboards, ground-based

Bharat Electronics Ltd	India	*203*	55-00200
Graphico Electronics Group	Canada	*71*	75-06460
Mitel Telecom Ltd.	UK	*468*	75-40150
Racal Acoustics Ltd	UK	*476*	75-47653
Redifon MEL Limited	UK	*478*	75-48530
Siemens Aktiengesellschaft	Germany	*175*	75-27700
Siemens Atea NV	Belgium	*47*	75-05950

446 40 70 40 60 60 Tactical switchboards, ground-based

AN/COM Industries Inc.	USA	*553*	75-07300
Bharat Electronics Ltd	India	*203*	55-00200
CIC International Ltd	USA	*563*	75-18570
Denro Inc.	USA	*568*	75-25651
Graphico Electronics Group	Canada	*71*	75-06460
Infocom Systems A/S	Denmark	*103*	75-09200
Marconi SpA DTM	Italy	*236*	75-14550
Siemens Aktiengesellschaft	Germany	*175*	75-27700
Siemens Atea NV	Belgium	*47*	75-05950

446 40 70 40 70 Telephone terminals, ground-based

CIC International Ltd	USA	*563*	75-18570
Redifon MEL Limited	UK	*478*	75-48530
SAGEM SA	**France**	***143***	**75-33950**
Siemens Atea NV	Belgium	*47*	75-05950
Siemens Switzerland Ltd	Switzerland	*387*	75-11980

446 40 70 40 70 50 Telephone terminals, mobile, ground-based

CIC International Ltd	USA	*563*	75-18570
Codalex Ltd	Canada	*70*	75-03400
SAGEM SA	**France**	***143***	**75-33950**
Siemens Atea NV	Belgium	*47*	75-05950
Siemens Switzerland Ltd	Switzerland	*387*	75-11980

446 40 70 40 70 60 Telephone terminals, tactical

CIC International Ltd	USA	*563*	75-18570
Marconi SpA DTM	Italy	*236*	75-14550
Siemens Atea NV	Belgium	*47*	75-05950
Siemens Switzerland Ltd	Switzerland	*387*	75-11980

446 40 70 42 Communications equipment, ground-based, telegraphic

Alcatel Bell	Belgium	*44*	75-00055
CAE	UK	*438*	75-11125
CIC International Ltd	USA	*563*	75-18570
Cubic Defense Systems Inc.	USA	*567*	75-22900
Frederick Electronics	USA	*576*	75-36000
Global Communications Co.	USA	*579*	75-39775
Graphico Electronics Group	Canada	*71*	75-06460
HAL Communications Corp.	USA	*580*	75-41450

Marconi SpA DTM	Italy	*236*	75-14550
Redifon MEL Limited	UK	*478*	75-48530
Siemens Aktiengesellschaft	Germany	*175*	75-27700
Siemens Plessey Systems	UK	*483*	75-53770

446 40 70 46 Communications equipment, ground-based, teleprinter

Alcatel Bell	Belgium	*44*	75-00055
CAE	UK	*438*	75-11125
CIC International Ltd	USA	*563*	75-18570
Frederick Electronics	USA	*576*	75-36000
Global Communications Co.	USA	*579*	75-39775
HAL Communications Corp.	USA	*580*	75-41450
Harris Corporation	USA	*581*	75-42050
ICL DESC	UK	*501*	80-08400
Raytheon Cossor Electronics Ltd	UK	*478*	75-48367
SAGEM SA	**France**	***143***	**75-33950**
Siemens Aktiengesellschaft	Germany	*175*	75-27700
Siemens Plessey Systems	UK	*483*	75-53770
Vaisala Oy	Finland	*119*	75-02800

446 40 70 46 40 Message preparation terminals, ground-based

Bharat Electronics Ltd	India	*203*	55-00200
CIC International Ltd	USA	*563*	75-18570
HAL Communications Corp.	USA	*580*	75-41450

446 40 70 46 50 Teleprinters, ground-based

CIC International Ltd	USA	*563*	75-18570
Harris Corporation	USA	*581*	75-42050

446 40 70 48 Communications equipment, ground-based, telecopier / facsimile

Alcatel Bell	Belgium	*44*	75-00055
Alcatel Bell-SDT	Belgium	*44*	75-00060
Alden Electronics Incorporated	USA	*548*	75-04400
Canadian Marconi Co.	Canada	*69*	75-02600
CIC International Ltd	USA	*563*	75-18570
Global Communications Co.	USA	*579*	75-39775
Graphico Electronics Group	Canada	*71*	75-06460
Osterreichisches Forschungszentrum	Austria	*32*	60-05000
SAGEM SA	**France**	***143***	**75-33950**
Siemens Plessey Systems	UK	*483*	75-53770

446 40 70 80 Communications equipment, voice amplification, ground-based

AN/COM Industries Inc.	USA	*553*	75-07300
AudioPack Sound Systems	USA	*556*	75-10240
CAE	UK	*438*	75-11125
CIC International Ltd	USA	*563*	75-18570
Davies Industrial Communications Ltd	UK	*444*	75-17010
Racal Communications Inc.	USA	*606*	75-72380

446 40 80 Communications equipment accessories, ground-based

ADC Telecommunications Inc.	USA	*545*	75-00730
Aerea SpA	Italy	*230*	75-00280
Alcatel Bell	Belgium	*44*	75-00055
AN/COM Industries Inc.	USA	*553*	75-07300
Bren-Tronics Inc.	USA	*560*	75-15170
CAE	UK	*438*	75-11125
CIC International Ltd	USA	*563*	75-18570
Davies Industrial Communications Ltd	UK	*444*	75-17010
Elbit Systems Ltd	Israel	*218*	75-05300
Electronic Marketing Co. SpA	Italy	*233*	75-06860
G.E. Mortley, Sprague & Co. Ltd	UK	*469*	75-41000
GEC-Marconi Communications Ltd	UK	*454*	75-25335
General Electronics Communications	Greece	*189*	75-01300
Hagen	UK	*456*	75-27220
Harris Corporation	USA	*581*	75-42050
Hunting Engineering Ltd	UK	*459*	75-30700
Litton FiberCom	USA	*590*	75-55370
Microelectronics, guidance and	Turkey	*402*	55-00050
Naval Electronics Inc.	USA	*598*	75-64025
Racal Acoustics Ltd	UK	*476*	75-47653
Racal Communications Inc.	USA	*606*	75-72380
Redifon MEL Limited	UK	*478*	75-48530
Siemens Plessey Systems	UK	*483*	75-53770
Tramec SA	France	*148*	75-45500
Trompeter Electronics Inc.	USA	*623*	75-91070

446 40 80 30 Communications equipment accessories, radio, ground-based

Advanced Semiconductor Inc.	USA	*546*	75-00950
Alcatel Bell	Belgium	*44*	75-00055
Amper Programas SA	Spain	*365*	75-00480
AN/COM Industries Inc.	USA	*553*	75-07300
Bharat Electronics Ltd	India	*203*	55-00200
CAE	UK	*438*	75-11125
CIC International Ltd	USA	*563*	75-18570
Conrac Elektron GmbH	Germany	*166*	75-05100
David Clark Co. Inc.	USA	*564*	75-18850
Davies Industrial Communications Ltd	UK	*444*	75-17010
Electronic Marketing Co. SpA	Italy	*233*	75-06860
Elekluft GmbH	Germany	*178*	80-00537
Frederick Electronics	USA	*576*	75-36000
G.E. Mortley, Sprague & Co. Ltd	UK	*469*	75-41000
General Electronics Communications	Greece	*189*	75-01300
Global Communications Co.	USA	*579*	75-39775
Hagen	UK	*456*	75-27220
Harris Corporation	USA	*581*	75-42050
Naval Electronics Inc.	USA	*598*	75-64025
Racal Acoustics Ltd	UK	*476*	75-47653
Racal Communications Inc.	USA	*606*	75-72380
Radio Research Instrument Co. Inc.	USA	*607*	75-72600
Redifon MEL Limited	UK	*478*	75-48530
Reunert Technology Systems Ltd	S. Africa	*359*	75-18300
Siemens Aktiengesellschaft	Germany	*175*	75-27700
South Midlands Communications Ltd	UK	*485*	75-55213
Sunair Electronics Inc.	USA	*618*	75-84950
Tadiran Ltd	Israel	*222*	75-21600
TERMA Elektronik AS	Denmark	*104*	75-20300
Tramec SA	France	*148*	75-45500
Trompeter Electronics Inc.	USA	*623*	75-91070

446 40 80 30 05 Antenna masts, radio, ground-based

AB Wibe	Sweden	*379*	75-13000
Allen Osborne Assoc., Inc.	USA	*548*	75-04610
AN/COM Industries Inc.	USA	*553*	75-07300
Andrew Canada Inc.	Canada	*68*	75-00600
Bharat Electronics Ltd	India	*203*	55-00200
CIC International Ltd	USA	*563*	75-18570
Clark Masts Teksam Ltd	UK	*440*	75-13550
Comrod A/S	Norway	*309*	75-00550
CORIS	France	*134*	75-12450
Electronic Marketing Co. SpA	Italy	*233*	75-06860
Francis & Lewis International Ltd	UK	*452*	75-24500
Garofoli Costruzioni Meccaniche SpA	Italy	*235*	75-09550
GEC-Marconi Communications Ltd	UK	*454*	75-25335
General Electronics Communications	Greece	*189*	75-01300
Genk Technical Products NV	Belgium	*46*	75-04000
Global Communications Co.	USA	*579*	75-39775
Harris Corporation	USA	*581*	75-42050
Instro Precision Ltd	UK	*461*	75-31630
Jaybeam Ltd	UK	*462*	75-32400
LARIMART SpA	Italy	*236*	75-13500
MFD International Ltd	UK	*506*	80-16450
Moonraker Australia Pty Ltd	Australia	*19*	75-13770
Racal Antennas Ltd	UK	*476*	75-47656
Racal Communications Inc.	USA	*606*	75-72380
South Midlands Communications Ltd	UK	*485*	75-55213
St Bernard Composites Ltd	UK	*486*	75-56222
Strumech Versatower Ltd	UK	*486*	75-57530
Technical Products Group Inc.	USA	*620*	75-86764
TERMA Elektronik AS	Denmark	*104*	75-20300
The Will-Burt Co.	USA	*628*	75-96830
Tramec SA	France	*148*	75-45500
Victrix Ltd	Canada	*77*	75-15000
Watkins-Johnson Co.	USA	*627*	75-95200
Wicker-Bürki AG	Switzerland	*388*	75-15300
Z. Rabiner Ltd	Israel	*221*	75-17840

446 40 80 30 10 Antennas, radio, ground-based

Allen Telecom Inc.	USA	*548*	75-04630
AN/COM Industries Inc.	USA	*553*	75-07300
Andrew Canada Inc.	Canada	*68*	75-00600
Astron Corp.	USA	*555*	75-09980
Avibras Aeroespacial SA	Brazil	*55*	75-00968
Bharat Electronics Ltd	India	*203*	55-00200
CIC International Ltd	USA	*563*	75-18570
Comrod A/S	Norway	*309*	75-00550
CORIS	France	*134*	75-12450
Electro-Metrics Inc.	USA	*572*	75-29810
Electronic Marketing Co. SpA	Italy	*233*	75-06860
ELTA Electronics Industries Ltd	Israel	*220*	75-10200
ERA Technology Ltd	UK	*502*	80-09600
Garofoli Costruzioni Meccaniche SpA	Italy	*235*	75-09550
GEC-Marconi Communications Ltd	UK	*454*	75-25335
General Electronics Communications	Greece	*189*	75-01300
Global Communications Co.	USA	*579*	75-39775
Harris Corporation	USA	*581*	75-42050
Hermes Electronics Ltd.	Canada	*72*	75-07300
H.R. Smith Group	UK	*484*	75-54590
Jaybeam Ltd	UK	*462*	75-32400
KEL Aerospace Pty Ltd.	Australia	*18*	75-11440
MA/COM Antenna & Cable Div.	USA	*593*	75-56953
MFD International Ltd.	UK	*506*	80-16450
Moonraker Australia Pty Ltd	Australia	*19*	75-13770
Panorama Antennas Ltd	UK	*472*	75-43660
R.A. Miller Industries Inc.	USA	*596*	75-61650
Racal Antennas Ltd	UK	*476*	75-47656
Racal Communications Inc.	USA	*606*	75-72380
Renair Antenna	UK	*478*	75-49225
RJR Industries	USA	*579*	75-39720
Shakespeare Products Group	USA	*614*	75-79800
Siemens Aktiengesellschaft	Germany	*175*	75-27700
South Midlands Communications Ltd	UK	*485*	75-55213
St Bernard Composites Ltd	UK	*486*	75-56222
STAREC	France	*147*	75-42775
Tadiran Ltd	Israel	*222*	75-21600
Technitrol Inc.	USA	*620*	75-86800
TERMA Elektronik AS	Denmark	*104*	75-20300
Walter Dittel GmbH	Germany	*167*	75-06510
Watkins-Johnson Co.	USA	*627*	75-95200
Wicker-Bürki AG	Switzerland	*388*	75-15300

446 40 80 30 10 30 Antennas, dish, ground-based

A Goninan & Co. Ltd	Australia	*16*	75-07760
Aerea SpA.	Italy	*230*	75-00280
Andrew Canada Inc.	Canada	*68*	75-00600
Bharat Electronics Ltd	India	*203*	55-00200
Celwave R.F. A/S	Denmark	*102*	75-03230
CIC International Ltd	USA	*563*	75-18570
Comtech Systems Inc.	USA	*565*	75-20810
CORIS	France	*134*	75-12450
ERA Technology Ltd	UK	*502*	80-09600
ESSCO Collins Ltd	Ireland	*214*	75-00100
Gabriel Electronics Inc.	USA	*577*	75-36715
GEC-Marconi Communications Ltd	UK	*454*	75-25335
General Electronics Communications	Greece	*189*	75-01300
Harris Corporation	USA	*581*	75-42050
Jaybeam Ltd	UK	*462*	75-32400
John Curran Ltd	UK	*444*	75-16430
MFD International Ltd.	UK	*506*	80-16450
Racal Antennas Ltd	UK	*476*	75-47656
Siemens Aktiengesellschaft	Germany	*175*	75-27700
STAREC	France	*147*	75-42775
Textron Systems Division	USA	*621*	75-88655
TIW Systems Inc.	USA	*623*	75-89755
Vermont Composites, Inc.	USA	*625*	75-93450

446 40 80 30 12 Antennas, radio, vehicle-mounted

Allen Telecom Inc.	USA	*548*	75-04630
AlliedSignal Aerospace Canada	Canada	*68*	75-00470
AN/COM Industries Inc.	USA	*553*	75-07300
Bharat Electronics Ltd	India	*203*	55-00200
CIC International Ltd	USA	*563*	75-18570
Comrod A/S	Norway	*309*	75-00550
Electronic Marketing Co. SpA	Italy	*233*	75-06860
ELTA Electronics Industries Ltd	Israel	*220*	75-10200
ERA Technology Ltd	UK	*502*	80-09600
GEC-Marconi Communications Ltd	UK	*454*	75-25335
General Electronics Communications	Greece	*189*	75-01300
Genk Technical Products NV	Belgium	*46*	75-04000
Global Communications Co.	USA	*579*	75-39775
Harris Corporation	USA	*581*	75-42050
H.R. Smith Group	UK	*484*	75-54590
Icom (UK) Ltd	UK	*460*	75-31070
Jaybeam Ltd	UK	*462*	75-32400
LARIMART SpA	Italy	*236*	75-13500
MA/COM Antenna & Cable Div.	USA	*593*	75-56953
MFD International Ltd.	UK	*506*	80-16450
Moonraker Australia Pty Ltd	Australia	*19*	75-13770
Panorama Antennas Ltd	UK	*472*	75-43660
Racal Antennas Ltd	UK	*476*	75-47656
RJR Industries	USA	*579*	75-39720
Rohde & Schwarz Canada Inc	Canada	*79*	80-01825
Shakespeare Products Group	USA	*614*	75-79800
South Midlands Communications Ltd	UK	*485*	75-55213
St Bernard Composites Ltd	UK	*486*	75-56222
STAREC	France	*147*	75-42775
Tadiran Ltd	Israel	*222*	75-21600
TERMA Elektronik AS	Denmark	*104*	75-20300
The Will-Burt Co.	USA	*628*	75-96830
Walter Dittel GmbH	Germany	*167*	75-06510

446 40 80 30 16 Tuners, radio antenna, ground-based

AN/COM Industries Inc.	USA	*553*	75-07300
CIC International Ltd	USA	*563*	75-18570
Delta Electronics Inc.	USA	*568*	75-25550
Electronic Marketing Co. SpA	Italy	*233*	75-06860
ERA Technology Ltd	UK	*502*	80-09600
General Electronics Communications	Greece	*189*	75-01300
Global Communications Co.	USA	*579*	75-39775
Icom (UK) Ltd	UK	*460*	75-31070
Moonraker Australia Pty Ltd	Australia	*19*	75-13770
Racal Antennas Ltd	UK	*476*	75-47656
Redifon MEL Limited	UK	*478*	75-48530
South Midlands Communications Ltd	UK	*485*	75-55213
STAREC	France	*147*	75-42775
TERMA Elektronik AS	Denmark	*104*	75-20300
UB Corp.	USA	*624*	75-91380
Watkins-Johnson Co.	USA	*627*	75-95200

446 40 80 30 30 Headsets, ground-based communications

Acousticom Corp.	USA	*545*	75-00380
Akustische und Kino-geräte GmbH	Austria	*32*	75-00075
Bose BV	Netherlands	*291*	75-00790
Davies Industrial Communications Ltd	UK	*444*	75-17010
Helmet Integrated Systems Ltd	UK	*458*	75-29000
Icom (UK) Ltd	UK	*460*	75-31070
South Midlands Communications Ltd	UK	*485*	75-55213

446 40 80 30 30 10 Headsets, ground based communications, active noise reduction

Bose BV	Netherlands	*291*	75-00790
Bose Corporation	USA	*560*	75-14700

446 40 80 30 50 Microphones, ground-based communications

Akustische und Kino-geräte GmbH	Austria	*32*	75-00075
Davies Industrial Communications Ltd	UK	*444*	75-17010
Hagen	UK	*456*	75-27220
Helmet Integrated Systems Ltd	UK	*458*	75-29000
Icom (UK) Ltd	UK	*460*	75-31070
South Midlands Communications Ltd	UK	*485*	75-55213

446 50 50 Intercom systems (tri-service)

Codalex Ltd	Canada	*70*	75-03400
Communications-Applied Tech.	USA	*564*	75-20290
Davies Industrial Communications Ltd	UK	*444*	75-17010
Hagen	UK	*456*	75-27220
LARIMART SpA	Italy	*236*	75-13500

Northern Airborne Technology Ltd	Canada	*74*	75-10145
Redifon MEL Limited	UK	*478*	75-48530

446 80 Communications equipment accessories

Advanced Semiconductor Inc.	USA	*546*	75-00950
Akustische und Kino-geräte GmbH	Austria	*32*	75-00075
Andrew Canada Inc.	Canada	*68*	75-00600
Applied Signal Technology	USA	*554*	75-08210
AudioPack Sound Systems	USA	*556*	75-10240
Bren-Tronics Inc.	USA	*560*	75-15170
British Sarozal Ltd.	UK	*436*	75-09570
CAE	UK	*438*	75-11125
California Microwave Inc.	USA	*561*	75-16350
David Clark Co. Inc.	USA	*564*	75-18850
Electro Impulse Laboratory Inc.	USA	*572*	75-29700
Electronic Marketing Co. SpA	Italy	*233*	75-06860
Frederick Electronics	USA	*576*	75-36000
GEC-Marconi Communications Ltd	UK	*454*	75-25335
Giat Industries	France	*126*	55-00100
Harris Corporation	USA	*581*	75-42050
IBV Multipower Ltd	UK	*470*	75-41695
Intelect Inc.	USA	*585*	75-47160
Jotron (UK) Ltd	UK	*462*	75-32895
KEL Aerospace Pty Ltd	Australia	*18*	75-11440
Litton FiberCom	USA	*590*	75-55370
NEC Corporation	Japan	*253*	75-06350
Racal Communications Inc.	USA	*606*	75-72380
Redifon MEL Limited	UK	*478*	75-48530
Sensitised Coatings Ltd	UK	*482*	75-52800
Siemens Atea NV	Belgium	*47*	75-05950
Siemens Plessey Systems	UK	*483*	75-53770
Signaal Communications	Netherlands	*295*	75-06350
SR Comms Management Ltd	UK	*512*	80-46000
Tadiran Ltd	Israel	*222*	75-21600
Trompeter Electronics Inc.	USA	*623*	75-91070

446 80 50 Modems

Applied Signal Technology	USA	*554*	75-08210
Aydin Corp.	USA	*556*	75-11450
CAE	UK	*438*	75-11125
California Microwave Inc.	USA	*561*	75-16350
GEC-Marconi Communications Ltd	UK	*454*	75-25335
HAL Communications Corp.	USA	*580*	75-41450
Instrument Services Ltd	UK	*504*	80-13328
Litton FiberCom	USA	*590*	75-55370
NEC Corporation	Japan	*253*	75-06350
Redifon MEL Limited	UK	*478*	75-48530
Rockwell International Co.	Italy	*233*	75-05200
Siemens Atea NV	Belgium	*47*	75-05950
Sunair Electronics Inc.	USA	*618*	75-84950
Telegenix Inc.	USA	*621*	75-87610
The Titan Corp	USA	*622*	75-89620
Titan Linkabit	USA	*623*	75-89650

446 90 Communications equipment components

ADC Telecommunications Inc.	USA	*545*	75-00730
Advanced Semiconductor Inc.	USA	*546*	75-00950
Alcatel Bell-SDT	Belgium	*44*	75-00060
Andrew Canada Inc.	Canada	*68*	75-00600
Applied Signal Technology	USA	*554*	75-08210
Bharat Electronics Ltd	India	*203*	55-00200
Bren-Tronics Inc.	USA	*560*	75-15170
British Aerospace (Systems &	UK	*436*	75-08890
British Sarozal Ltd.	UK	*436*	75-09570
CAE	UK	*438*	75-11125
Canadian Marconi Co.	Canada	*69*	75-02600
Chelton (Electrostatics) Limited	UK	*439*	75-12700
Cincinnati Electronics Corp.	USA	*564*	75-18640
Com Dev Europe	UK	*441*	75-14300
Cubic Defense Systems Inc.	USA	*567*	75-22900
Davies Industrial Communications Ltd	UK	*444*	75-17010
Delta Electronics Inc.	USA	*568*	75-25550
EG&G Reticon	USA	*571*	75-29562
Elcom Systems Inc.	USA	*572*	75-29650
Electronic Marketing Co. SpA	Italy	*233*	75-06860
GEC-Marconi Communications Ltd	UK	*454*	75-25335
Giat Industries	France	*126*	55-00100
Global Communications Co.	USA	*579*	75-39775
Intelect Inc.	USA	*585*	75-47160
ITT Avionics	USA	*586*	75-49455
Jennings Technology Corp.	USA	*586*	75-50350
KEL Aerospace Pty Ltd	Australia	*18*	75-11440
LARIMART SpA	Italy	*236*	75-13500
Litton Industries Inc.	USA	*590*	75-55400
Metric Systems Corp.	USA	*596*	75-60800
Micro Metalsmiths Ltd	UK	*468*	75-39010
Millitech Corp.	USA	*596*	75-61665
NEC Corporation	Japan	*253*	75-06350
Netcom Ltd	Greece	*191*	75-02150
Nordic VLSI A/S	Norway	*310*	75-01640
Ortel Corp.	USA	*602*	75-67530
Pointer Inc.	USA	*604*	75-70810
Presentey Engineering	Canada	*75*	75-11500
Quasar Microwave Technology Ltd	UK	*476*	75-47450
Racal-MESL Ltd	UK	*477*	75-47682
Raytheon E-Systems	USA	*608*	75-73302
Redifon MEL Limited	UK	*478*	75-48530
Rockwell	USA	*610*	75-75930
SAGEM SA	**France**	***143***	**75-33950**
Sanders	USA	*612*	75-77850
Siemens plc	UK	*483*	75-53750
Signaal Communications	Netherlands	*295*	75-06350
SR Comms Management Ltd	UK	*512*	80-46000
Tadiran Ltd	Israel	*222*	75-21600
Teledyne Electronic Technologies	USA	*620*	75-87522
The Titan Corp	USA	*622*	75-89620
Trompeter Electronics Inc.	USA	*623*	75-91070
Videotek Inc.	USA	*626*	75-94090
Walter Dittel GmbH	Germany	*167*	75-06510

447 Communications security systems & equipment

Alcatel Bell	Belgium	*44*	75-00055
Alcatel Bell-SDT	Belgium	*44*	75-00060
Alcatel Telecom Norway A/S	Norway	*308*	75-00090
Amper Programas SA	Spain	*365*	75-00480
Barrett Communications Pty Ltd.	Australia	*12*	75-02050
Boeing Australia Limited	Australia	*12*	75-02500
CCS International Ltd	USA	*562*	75-17250
CIC International Ltd	USA	*563*	75-18570
Cincinnati Electronics Corp.	USA	*564*	75-18640
Cogent	UK	*500*	80-06250
Crypto AG	Switzerland	*385*	75-02400
Cylink Corp.	USA	*567*	75-23134
Denian Electronic Ltd	UK	*445*	75-17530
Eidsvoll Electronics AS	Norway	*309*	75-00850
Elektron SA	France	*136*	75-14700
Fluor Daniel Inc.	USA	*635*	80-11720
Global Communications Co.	USA	*579*	75-39775
Honeywell Control Systems Ltd	UK	*458*	75-29730
Ilex Systems Inc.	USA	*584*	75-46080
Ingenieria de Sistemas	Spain	*367*	80-00900
Israel Aircraft Industries Ltd	Israel	*219*	75-10200
Israel Product Research Co. Ltd	Israel	*219*	75-10000
Italtel Spa	Italy	*235*	75-13205
ITT Aerospace/Communications Div.	USA	*586*	75-49455
James Hardie Building Services Ltd	Australia	*17*	75-10178
Litton Industries Inc.	USA	*590*	75-55400
Lockheed Martin Corp.	USA	*591*	75-55635
Logica SA/NV	Belgium	*48*	80-01000
Mils Elektronik GmbH & Co KG	**Austria**	***33***	**75-01700**
NOVO Corporation	Russia	*338*	75-14200
PK Electronic International Ltd	UK	*475*	75-45720
Racal-Airtech Ltd	UK	*476*	75-47655
Racal-Comsec Ltd	UK	*477*	75-47665
Raytheon Cossor Electronics Ltd	UK	*478*	75-48367
Raytheon E-Systems	USA	*608*	75-73302
Rockwell International Co.	Italy	*233*	75-05200
SAGEM SA	**France**	***143***	**75-33950**
Siemens Aktiengesellschaft	Germany	*175*	75-27700
Siemens Atea NV	Belgium	*47*	75-05950
Smiths Aerospace & Defense	USA	*615*	75-82090
Technical Comms Corp.	USA	*619*	75-86750
TST Tele Security Timmann	Germany	*176*	75-31700
Vitro Corp.	USA	*644*	80-34400
Wood & Douglas	UK	*494*	75-66790
Zettler GmbH Kommunikation und	Germany	*177*	75-34930

447 10 Communications security systems, spaceborne

Matra Marconi Space UK, Ltd	UK	*425*	74-06900

447 10 10 Secure communication links, space to ground

Cranfield University	UK	*419*	60-00350
Denis Ferranti Meters Ltd	UK	*450*	75-22850
Matra Marconi Space UK, Ltd	UK	*425*	74-06900

447 20 Communications security systems, airborne

CIC International Ltd	USA	*563*	75-18570
Cogent	UK	*500*	80-06250
Crypto AG	Switzerland	*385*	75-02400
Cylink Corp.	USA	*567*	75-23134
Daimler-Benz Aerospace France SAS	**France**	***134***	**75-12950**
ELTA Electronics Industries Ltd	Israel	*220*	75-10200
Global Communications Co.	USA	*579*	75-39775
Israel Aircraft Industries Ltd	Israel	*219*	75-10200
ITT Aerospace/Communications Div.	USA	*586*	75-49455
Motorola Ltd	UK	*469*	75-41050
PhotoTelesis	USA	*604*	75-69810
Rockwell International Co.	Italy	*233*	75-05200
Smiths Aerospace & Defense	USA	*615*	75-82090
Technical Comms Corp.	USA	*619*	75-86750

447 20 70 Communications security equipment, airborne

CIC International Ltd	USA	*563*	75-18570
Crypto AG	Switzerland	*385*	75-02400
Cylink Corp.	USA	*567*	75-23134
ELTA Electronics Industries Ltd	Israel	*220*	75-10200
Global Communications Co.	USA	*579*	75-39775
Israel Aircraft Industries Ltd	Israel	*219*	75-10200
ITT Aerospace/Communications Div.	USA	*586*	75-49455
Motorola Ltd	UK	*469*	75-41050
PhotoTelesis	USA	*604*	75-69810
SAGEM SA	**France**	***143***	**75-33950**
Smiths Aerospace & Defense	USA	*615*	75-82090
Technical Comms Corp.	USA	*619*	75-86750
TST Tele Security Timmann	Germany	*176*	75-31700

447 20 70 30 Datex communications security equipment, airborne

CIC International Ltd	USA	*563*	75-18570
Cylink Corp.	USA	*567*	75-23134
Daimler-Benz Aerospace France SAS	**France**	***134***	**75-12950**
Global Communications Co.	USA	*579*	75-39775
Motorola Ltd	UK	*469*	75-41050
SAGEM SA	**France**	***143***	**75-33950**
Technical Comms Corp.	USA	*619*	75-86750
TST Tele Security Timmann	Germany	*176*	75-31700

447 20 70 50 Telex communications security equipment, airborne

CIC International Ltd	USA	*563*	75-18570
Daimler-Benz Aerospace France SAS	**France**	***134***	**75-12950**
Frederick Electronics	USA	*576*	75-36000
HAL Communications Corp.	USA	*580*	75-41450
Motorola Ltd	UK	*469*	75-41050
Technical Comms Corp.	USA	*619*	75-86750
TST Tele Security Timmann	Germany	*176*	75-31700

447 20 70 80 Voice communications security equipment, airborne

CIC International Ltd	USA	*563*	75-18570
Crypto AG	Switzerland	*385*	75-02400
Cylink Corp.	USA	*567*	75-23134
Cypher Research Laboratories	Australia	*10*	60-00300
Daimler-Benz Aerospace France SAS	**France**	***134***	**75-12950**
Eidsvoll Electronics AS	Norway	*309*	75-00850
ELTA Electronics Industries Ltd	Israel	*220*	75-10200
Global Communications Co.	USA	*579*	75-39775
ITT Aerospace/Communications Div.	USA	*586*	75-49455
Motorola Ltd	UK	*469*	75-41050
Racal Communications Inc.	USA	*606*	75-72380
SAGEM SA	**France**	***143***	**75-33950**
Smiths Aerospace & Defense	USA	*615*	75-82090
Technical Comms Corp.	USA	*619*	75-86750
TST Tele Security Timmann	Germany	*176*	75-31700

447 20 70 85 Voice communications security equipment, airborne radio

CIC International Ltd	USA	*563*	75-18570
Cincinnati Electronics Corp.	USA	*564*	75-18640
Cylink Corp.	USA	*567*	75-23134
ELTA Electronics Industries Ltd	Israel	*220*	75-10200
Icom (UK) Ltd	UK	*460*	75-31070
Motorola Ltd	UK	*469*	75-41050
NEC Corporation	Japan	*253*	75-06350
Racal Communications Inc.	USA	*606*	75-72380
Smiths Aerospace & Defense	USA	*615*	75-82090
Technical Comms Corp.	USA	*619*	75-86750
TST Tele Security Timmann	Germany	*176*	75-31700

447 30 Communications security systems, naval

Amper Programas SA	Spain	*365*	75-00480
CIC International Ltd	USA	*563*	75-18570
Cogent	UK	*500*	80-06250
Crypto AG	Switzerland	*385*	75-02400
Cylink Corp.	USA	*567*	75-23134
Daimler-Benz Aerospace France SAS	**France**	***134***	**75-12950**
ELTA Electronics Industries Ltd	Israel	*220*	75-10200
Frederick Electronics	USA	*576*	75-36000
Global Communications Co.	USA	*579*	75-39775
Graphico Electronics Group	Canada	*71*	75-06460
Israel Aircraft Industries Ltd	Israel	*219*	75-10200
ITT Aerospace/Communications Div.	USA	*586*	75-49455
Mils Elektronik GmbH & Co KG	**Austria**	***33***	**75-01700**
Omnisec AG	Switzerland	*387*	75-09330
PhotoTelesis	USA	*604*	75-69810
Racal-Comsec Ltd	UK	*477*	75-47665
Raytheon Cossor Electronics Ltd	UK	*478*	75-48367
Siemens Atea NV	Belgium	*47*	75-05950
Technical Comms Corp.	USA	*619*	75-86750

447 30 70 Communications security equipment, naval

Amper Programas SA	Spain	*365*	75-00480
CIC International Ltd	USA	*563*	75-18570
Crypto AG	Switzerland	*385*	75-02400
Cylink Corp.	USA	*567*	75-23134
DNE, Inc.	USA	*569*	75-27000
ELTA Electronics Industries Ltd	Israel	*220*	75-10200
Frederick Electronics	USA	*576*	75-36000
Global Communications Co.	USA	*579*	75-39775
Graphico Electronics Group	Canada	*71*	75-06460
ITT Aerospace/Communications Div.	USA	*586*	75-49455
Mils Elektronik GmbH & Co KG	**Austria**	***33***	**75-01700**
Omnisec AG	Switzerland	*387*	75-09330
Racal-Comsec Ltd	UK	*477*	75-47665
SAGEM SA	**France**	***143***	**75-33950**
Technical Comms Corp.	USA	*619*	75-86750
TST Tele Security Timmann	Germany	*176*	75-31700

447 30 70 30 Datex communications security equipment, naval

Amper Programas SA	Spain	*365*	75-00480
CIC International Ltd	USA	*563*	75-18570
Cylink Corp.	USA	*567*	75-23134
Global Communications Co.	USA	*579*	75-39775
Mils Elektronik GmbH & Co KG	**Austria**	***33***	**75-01700**
NEC Corporation	Japan	*253*	75-06350
Omnisec AG	Switzerland	*387*	75-09330
SAGEM SA	**France**	***143***	**75-33950**
Siemens Atea NV	Belgium	*47*	75-05950
Technical Comms Corp.	USA	*619*	75-86750
TST Tele Security Timmann	Germany	*176*	75-31700

Product–Supplier Link

447 30 70 50 Telex communications security equipment, naval

CIC International Ltd	USA	*563*	75-18570
Crypto AG	Switzerland	*385*	75-02400
Daimler-Benz Aerospace France SAS	**France**	***134***	**75-12950**
Frederick Electronics	USA	*576*	75-36000
Global Communications Co.	USA	*579*	75-39775
HAL Communications Corp.	USA	*580*	75-41450
Mils Elektronik GmbH & Co KG	**Austria**	***33***	**75-01700**
Omnisec AG.	Switzerland	*387*	75-09330
Racal-Comsec Ltd	UK	*477*	75-47665
SP Radio A/S	Denmark	*103*	75-19600
Technical Comms Corp.	USA	*619*	75-86750
TST Tele Security Timmann	Germany	*176*	75-31700

447 30 70 80 Voice communications security equipment, naval

Amper Programas SA	Spain	*365*	75-00480
CIC International Ltd	USA	*563*	75-18570
Crypto AG	Switzerland	*385*	75-02400
Cylink Corp.	USA	*567*	75-23134
Daimler-Benz Aerospace France SAS	**France**	***134***	**75-12950**
ELTA Electronics Industries Ltd	Israel	*220*	75-10200
Global Communications Co.	USA	*579*	75-39775
Graphico Electronics Group	Canada	*71*	75-06460
ITT Aerospace/Communications Div.	USA	*586*	75-49455
Mils Elektronik GmbH & Co KG	**Austria**	***33***	**75-01700**
Omnisec AG.	Switzerland	*387*	75-09330
Racal Communications Inc.	USA	*606*	75-72380
Racal-Comsec Ltd	UK	*477*	75-47665
SAGEM SA	**France**	***143***	**75-33950**
Siemens Atea NV	Belgium	*47*	75-05950
SP Radio A/S	Denmark	*103*	75-19600
Tadiran Ltd	Israel	*222*	75-21600
Technical Comms Corp.	USA	*619*	75-86750
TST Tele Security Timmann	Germany	*176*	75-31700

447 30 70 85 Voice communications security equipment, naval radio

Amper Programas SA	Spain	*365*	75-00480
CIC International Ltd	USA	*563*	75-18570
Continental Electronics Corp.	USA	*566*	75-21800
ELTA Electronics Industries Ltd	Israel	*220*	75-10200
Graphico Electronics Group	Canada	*71*	75-06460
Icom (UK) Ltd	UK	*460*	75-31070
Mils Elektronik GmbH & Co KG	**Austria**	***33***	**75-01700**
Omnisec AG.	Switzerland	*387*	75-09330
Racal Communications Inc.	USA	*606*	75-72380
Racal-Comsec Ltd	UK	*477*	75-47665
Siemens Atea NV	Belgium	*47*	75-05950
SP Radio A/S	Denmark	*103*	75-19600
Tadiran Ltd	Israel	*222*	75-21600
Technical Comms Corp.	USA	*619*	75-86750
TST Tele Security Timmann	Germany	*176*	75-31700

447 40 Communications security systems, ground-based

ADC Telecommunications Inc.	USA	*545*	75-00730
Alcatel Bell	Belgium	*44*	75-00055
Alcatel Bell-SDT	Belgium	*44*	75-00060
Alcatel Telecom Norway A/S	Norway	*308*	75-00090
Amper Programas SA	Spain	*365*	75-00480
Canadian Marconi Co.	Canada	*69*	75-02600
CIC International Ltd	USA	*563*	75-18570
Cogent	UK	*500*	80-06250
Computer Resources International A/S.	Denmark	*104*	80-00100
CORIS	France	*134*	75-12450
Crypto AG	Switzerland	*385*	75-02400
Cylink Corp.	USA	*567*	75-23134
Daimler-Benz Aerospace France SAS	**France**	***134***	**75-12950**
ELTA Electronics Industries Ltd	Israel	*220*	75-10200
Ericsson Defence Systems Pty Ltd	Australia	*16*	75-05820
Global Communications Co.	USA	*579*	75-39775
Graphico Electronics Group	Canada	*71*	75-06460
Hagen	UK	*456*	75-27220
Harris Corporation	USA	*581*	75-42050
Israel Aircraft Industries Ltd	Israel	*219*	75-10200
Israel Product Research Co. Ltd	Israel	*219*	75-10000
ITT Aerospace/Communications Div.	USA	*586*	75-49455
ITT Defense Ltd	UK	*505*	80-13570
Microelectronics, guidance and	Turkey	*402*	55-00050
Mils Elektronik GmbH & Co KG	**Austria**	***33***	**75-01700**
Motorola Ltd	UK	*469*	75-41050
NOVO Corporation	Russia	*338*	75-14200
Omnisec AG.	Switzerland	*387*	75-09330
PhotoTelesis	USA	*604*	75-69810
PK Electronic International Ltd	UK	*475*	75-45720
Racal Communications Inc.	USA	*606*	75-72380
Racal Radio Ltd	UK	*477*	75-47698
Racal-Comsec Ltd	UK	*477*	75-47665
Racal-Datacom Ltd	UK	*477*	75-47670
SAIT Systems SA	Belgium	*47*	75-05600
Siemens Atea NV	Belgium	*47*	75-05950
Technical Comms Corp.	USA	*619*	75-86750
The Titan Corp	USA	*622*	75-89620
TST Tele Security Timmann	Germany	*176*	75-31700

447 40 10 Secure communications links, ground to space

Cranfield University	UK	*419*	60-00350
Denis Ferranti Meters Ltd	UK	*450*	75-22850
Francis & Lewis International Ltd	UK	*452*	75-24500
Siemens Plessey Systems	UK	*483*	75-53770

447 40 70 Communications security equipment, ground-based

ADC Telecommunications Inc.	USA	*545*	75-00730
Amper Programas SA	Spain	*365*	75-00480
Bharat Electronics Ltd	India	*203*	55-00200
Chubb Alarms Ltd	UK	*440*	75-13150
CIC International Ltd	USA	*563*	75-18570
Cogent	UK	*500*	80-06250
Continental Electronics Corp.	USA	*566*	75-21800
Crypto AG	Switzerland	*385*	75-02400
Cylink Corp.	USA	*567*	75-23134
Denian Electronic Ltd	UK	*445*	75-17530
Elektron SA	France	*136*	75-14700
Elesco Oy	Finland	*118*	75-00300
ELTA Electronics Industries Ltd	Israel	*220*	75-10200
Global Communications Co.	USA	*579*	75-39775
Graphico Electronics Group	Canada	*71*	75-06460
Harris Corporation	USA	*581*	75-42050
ICL DESC	UK	*501*	80-08400
Israel Product Research Co. Ltd	Israel	*219*	75-10000
ITT Aerospace/Communications Div.	USA	*586*	75-49455
Microelectronics, guidance and	Turkey	*402*	55-00050
Mils Elektronik GmbH & Co KG	**Austria**	***33***	**75-01700**
Motorola Ltd	UK	*469*	75-41050
Omnisec AG.	Switzerland	*387*	75-09330
Racal Communications Inc.	USA	*606*	75-72380
Racal-Comsec Ltd	UK	*477*	75-47665
SAGEM SA	**France**	***143***	**75-33950**
Siemens Aktiengesellschaft	Germany	*175*	75-27700
Siemens Atea NV	Belgium	*47*	75-05950
SP Radio A/S	Denmark	*103*	75-19600
Tadiran Ltd	Israel	*222*	75-21600
Technical Comms Corp.	USA	*619*	75-86750
The Titan Corp	USA	*622*	75-89620
TST Tele Security Timmann	Germany	*176*	75-31700
Winkelmann (UK) Ltd.	UK	*494*	75-66600

447 40 70 30 Datex communications security equipment, ground-based

Alcatel Telecom Norway A/S	Norway	*308*	75-00090
Amper Programas SA	Spain	*365*	75-00480
CIC International Ltd	USA	*563*	75-18570
Cylink Corp.	USA	*567*	75-23134
Daimler-Benz Aerospace France SAS	**France**	***134***	**75-12950**
Global Communications Co.	USA	*579*	75-39775
Honeywell Control Systems Ltd.	UK	*458*	75-29730
Microelectronics, guidance and	Turkey	*402*	55-00050
Mils Elektronik GmbH & Co KG	**Austria**	***33***	**75-01700**
Omnisec AG.	Switzerland	*387*	75-09330
Racal Communications Inc.	USA	*606*	75-72380
Racal-Comsec Ltd	UK	*477*	75-47665

Racal-Datacom Ltd	UK	*477*	75-47670
Reunert Technology Systems Ltd	S. Africa	*359*	75-18300
SAGEM SA	**France**	***143***	**75-33950**
Siemens Aktiengesellschaft	Germany	*175*	75-27700
Siemens Atea NV	Belgium	*47*	75-05950
Technical Comms Corp.	USA	*619*	75-86750
TST Tele Security Timmann	Germany	*176*	75-31700

447 40 70 50 Telex communications security equipment, ground-based

Alcatel Bell-SDT	Belgium	*44*	75-00060
Bharat Electronics Ltd	India	*203*	55-00200
CIC International Ltd	USA	*563*	75-18570
Crypto AG	Switzerland	*385*	75-02400
Daimler-Benz Aerospace France SAS	**France**	***134***	**75-12950**
Frederick Electronics	USA	*576*	75-36000
Global Communications Co.	USA	*579*	75-39775
HAL Communications Corp.	USA	*580*	75-41450
Mils Elektronik GmbH & Co KG	**Austria**	***33***	**75-01700**
Omnisec AG	Switzerland	*387*	75-09330
Racal-Comsec Ltd	UK	*477*	75-47665
Racal-Datacom Ltd	UK	*477*	75-47670
SAGEM SA	**France**	***143***	**75-33950**
SP Radio A/S	Denmark	*103*	75-19600
Technical Comms Corp.	USA	*619*	75-86750
TST Tele Security Timmann	Germany	*176*	75-31700

447 40 70 60 Telefax communications security equipment, ground-based

Bharat Electronics Ltd	India	*203*	55-00200
Crypto AG	Switzerland	*385*	75-02400
Cylink Corp.	USA	*567*	75-23134
Elektron SA	France	*136*	75-14700
Graphico Electronics Group	Canada	*71*	75-06460
Ilex Systems Inc.	USA	*584*	75-46080
Motorola Ltd	UK	*469*	75-41050
Omnisec AG	Switzerland	*387*	75-09330
PhotoTelesis	USA	*604*	75-69810
PK Electronic International Ltd	UK	*475*	75-45720
SAGEM SA	**France**	***143***	**75-33950**
Technical Comms Corp.	USA	*619*	75-86750
TST Tele Security Timmann	Germany	*176*	75-31700
Winkelmann (UK) Ltd	UK	*494*	75-66600

447 40 70 80 Voice communications security equipment, ground-based

Alcatel Bell-SDT	Belgium	*44*	75-00060
Amper Programas SA	Spain	*365*	75-00480
Ascom Systec Ltd	Switzerland	*385*	75-00540
Audiotel International Ltd	UK	*432*	75-04400
Bharat Electronics Ltd	India	*203*	55-00200
CIC International Ltd	USA	*563*	75-18570
Cogent	UK	*500*	80-06250
Continental Electronics Corp.	USA	*566*	75-21800
Crypto AG	Switzerland	*385*	75-02400
Cylink Corp.	USA	*567*	75-23134
Cypher Research Laboratories	Australia	*10*	60-00300
Daimler-Benz Aerospace France SAS	**France**	***134***	**75-12950**
Elesco Oy	Finland	*118*	75-00300
ELTA Electronics Industries Ltd	Israel	*220*	75-10200
Frederick Electronics	USA	*576*	75-36000
Global Communications Co.	USA	*579*	75-39775
Graphico Electronics Group	Canada	*71*	75-06460
Hagen	UK	*456*	75-27220
ITT Aerospace/Communications Div.	USA	*586*	75-49455
Jotron (UK) Ltd	UK	*462*	75-32895
Microelectronics, guidance and	Turkey	*402*	55-00050
Mils Elektronik GmbH & Co KG	**Austria**	***33***	**75-01700**
Motorola Ltd	UK	*469*	75-41050
NEC Corporation	Japan	*253*	75-06350
NOVO Corporation	Russia	*338*	75-14200
Omnisec AG	Switzerland	*387*	75-09330
Racal Communications Inc.	USA	*606*	75-72380
Racal-Comsec Ltd	UK	*477*	75-47665
Reunert Technology Systems Ltd	S. Africa	*359*	75-18300
SAGEM SA	**France**	***143***	**75-33950**
Siemens Aktiengesellschaft	Germany	*175*	75-27700
Siemens Atea NV	Belgium	*47*	75-05950
SP Radio A/S	Denmark	*103*	75-19600
Technical Comms Corp.	USA	*619*	75-86750
TST Tele Security Timmann	Germany	*176*	75-31700

447 40 70 85 Voice communications security equipment, ground-based radio

Amper Programas SA	Spain	*365*	75-00480
Ascom Systec Ltd	Switzerland	*385*	75-00540
Audiotel International Ltd	UK	*432*	75-04400
Bharat Electronics Ltd	India	*203*	55-00200
CIC International Ltd	USA	*563*	75-18570
Cincinnati Electronics Corp.	USA	*564*	75-18640
Continental Electronics Corp.	USA	*566*	75-21800
CORIS	France	*134*	75-12450
Crypto AG	Switzerland	*385*	75-02400
Cylink Corp.	USA	*567*	75-23134
Denian Electronic Ltd	UK	*445*	75-17530
ELTA Electronics Industries Ltd	Israel	*220*	75-10200
Graphico Electronics Group	Canada	*71*	75-06460
Hagen	UK	*456*	75-27220
Harris Corporation	USA	*581*	75-42050
Icom (UK) Ltd	UK	*460*	75-31070
Jotron (UK) Ltd	UK	*462*	75-32895
Microelectronics, guidance and	Turkey	*402*	55-00050
Mils Elektronik GmbH & Co KG	**Austria**	***33***	**75-01700**
Motorola Ltd	UK	*469*	75-41050
Omnisec AG	Switzerland	*387*	75-09330
Racal Communications Inc.	USA	*606*	75-72380
Racal-Comsec Ltd	UK	*477*	75-47665
Reunert Technology Systems Ltd	S. Africa	*359*	75-18300
SAIT Systems SA	Belgium	*47*	75-05600
Siemens Aktiengesellschaft	Germany	*175*	75-27700
SP Radio A/S	Denmark	*103*	75-19600
Tadiran Ltd	Israel	*222*	75-21600
Technical Comms Corp.	USA	*619*	75-86750
TST Tele Security Timmann	Germany	*176*	75-31700
Winkelmann (UK) Ltd	UK	*494*	75-66600

447 40 70 86 Voice communications security equipment, telephone, ground-based

Amper Programas SA	Spain	*365*	75-00480
Audiotel International Ltd	UK	*432*	75-04400
CIC International Ltd	USA	*563*	75-18570
Cogent	UK	*500*	80-06250
CORIS	France	*134*	75-12450
Crypto AG	Switzerland	*385*	75-02400
Cylink Corp.	USA	*567*	75-23134
Denian Electronic Ltd	UK	*445*	75-17530
Elektron SA	France	*136*	75-14700
ELTA Electronics Industries Ltd	Israel	*220*	75-10200
Graphico Electronics Group	Canada	*71*	75-06460
Microelectronics, guidance and	Turkey	*402*	55-00050
Mils Elektronik GmbH & Co KG	**Austria**	***33***	**75-01700**
Motorola Ltd	UK	*469*	75-41050
NEC Corporation	Japan	*253*	75-06350
Omnisec AG	Switzerland	*387*	75-09330
Racal-Comsec Ltd	UK	*477*	75-47665
SAGEM SA	**France**	***143***	**75-33950**
Siemens Aktiengesellschaft	Germany	*175*	75-27700
Siemens Atea NV	Belgium	*47*	75-05950
SP Radio A/S	Denmark	*103*	75-19600
Tadiran Ltd	Israel	*222*	75-21600
Technical Comms Corp.	USA	*619*	75-86750
TST Tele Security Timmann	Germany	*176*	75-31700
Winkelmann (UK) Ltd	UK	*494*	75-66600

448 Electronic security systems & equipment

Audiotel International Ltd	UK	*432*	75-04400
Catsaros Automation SA	Greece	*187*	75-00700
CCS International Ltd	USA	*562*	75-17250
CIC International Ltd	USA	*563*	75-18570
Cogent	UK	*500*	80-06250
Crypto AG	Switzerland	*385*	75-02400
Dynasound Inc.	USA	*570*	75-28265
EG&G Inc.	USA	*571*	75-29555
Geoquip Ltd	UK	*454*	75-25415
Global Communications Co.	USA	*579*	75-39775

Hagen	UK	*456*	75-27220
ITP Ltd	Israel	*220*	75-10800
James Hardie Building Services Ltd	Australia	*17*	75-10178
Kidde-Graviner Ltd	UK	*463*	75-33880
M. Paz Logistics Ltd	Israel	*224*	80-01600
Microelectronics, guidance and	Turkey	*402*	55-00050
MPB Technologies Inc	Canada	*74*	75-09855
Raytheon E-Systems	USA	*608*	75-73302
RGB Spectrum	USA	*609*	75-75000
Sas R&D Services Inc.	USA	*642*	80-26870
Spectronic Denmark	Denmark	*103*	75-19650
Strategic Technologies Pte Ltd	**Singapore**	***348***	**75-04400**
Television Installation Services	UK	*488*	75-59250
Vitro Corp.	USA	*644*	80-34400
Winkelmann (UK) Ltd	UK	*494*	75-66600
Zettler GmbH Kommunikation und	Germany	*177*	75-34930

448 40 Electronic security systems

AB Electronics (Rogerstone) Ltd	UK	*426*	75-00080
Agfa-Gevaert Ltd	UK	*428*	75-01475
Audiotel International Ltd	UK	*432*	75-04400
CEM Systems Ltd	UK	*439*	75-12270
Central Research Laboratories Ltd	UK	*499*	80-05895
Chubb Alarms Ltd	UK	*440*	75-13150
CIC International Ltd	USA	*563*	75-18570
Cincinnati Electronics Corp.	USA	*564*	75-18640
Cogent	UK	*500*	80-06250
Cotag International	UK	*443*	75-15235
Cranfield University	UK	*419*	60-00350
Data Cell Ltd	UK	*500*	80-07720
Dytecna Ltd	UK	*501*	80-08900
EDS Defence Ltd	UK	*501*	80-09200
EG&G Inc.	USA	*571*	75-29555
Encrypta Electronics Ltd	UK	*448*	75-20770
FHP International plc	UK	*502*	80-10300
Geoquip Ltd	UK	*454*	75-25415
Global Communications Co.	USA	*579*	75-39775
James Hardie Building Services Ltd	Australia	*17*	75-10178
Karl International Ltd	UK	*463*	75-33240
Litton Industries Inc.	USA	*590*	75-55400
Lynwood	UK	*465*	75-36700
M. Paz Logistics Ltd	Israel	*224*	80-01600
Mackenzie Tribbeck Associates	UK	*506*	80-15150
Microelectronics, guidance and	Turkey	*402*	55-00050
MPB Technologies Inc	Canada	*74*	75-09855
Octec Ltd	UK	*471*	75-42950
Oxley Developments Co. Ltd	UK	*472*	75-43400
Parity Solutions Ltd	UK	*508*	80-20300
Pilkington Thorn Optronics	UK	*426*	74-10000
PK Electronic International Ltd	UK	*475*	75-45720
Radio Research Instrument Co. Inc.	USA	*607*	75-72600
Remsdaq Ltd	UK	*478*	75-49210
RGB Spectrum	USA	*609*	75-75000
SAIC Ltd	UK	*510*	80-30220
Seaboard Electronics Co.	USA	*613*	75-78910
Security Research Ltd	UK	*482*	75-52450
Shorrock Integrated Systems	UK	*483*	75-53450
Siemens Plessey Systems	UK	*483*	75-53770
Sparton Technology Inc.	USA	*616*	75-82950
SRI International	USA	*643*	80-29600
Strategic Imaging Systems	UK	*486*	75-57515
Sverdrup Corp.	USA	*643*	80-30650
TIRAX AG	Switzerland	*387*	75-14410
Vitro Corp.	USA	*644*	80-34400
Zettler GmbH Kommunikation und	Germany	*177*	75-34930

448 40 70 Electronic security equipment

Audiotel International Ltd	UK	*432*	75-04400
Chubb Alarms Ltd	UK	*440*	75-13150
CIC International Ltd	USA	*563*	75-18570
Cincinnati Electronics Corp.	USA	*564*	75-18640
Cogent	UK	*500*	80-06250
EG&G Inc.	USA	*571*	75-29555
Encrypta Electronics Ltd	UK	*448*	75-20770
Geoquip Ltd	UK	*454*	75-25415
Global Communications Co.	USA	*579*	75-39775
Hagen	UK	*456*	75-27220
James Hardie Building Services Ltd	Australia	*17*	75-10178
Litton Industries Inc.	USA	*590*	75-55400
M. Paz Logistics Ltd	Israel	*224*	80-01600
Microelectronics, guidance and	Turkey	*402*	55-00050
PK Electronic International	Germany	*173*	75-23600
PK Electronic International	USA	*604*	75-70300
PK Electronic International Ltd	UK	*475*	75-45720
Pulnix America Inc.	USA	*606*	75-71815
Remsdaq Ltd	UK	*478*	75-49210
RGB Spectrum	USA	*609*	75-75000
Seaboard Electronics Co.	USA	*613*	75-78910
Security Research Ltd	UK	*482*	75-52450
Sparton Technology Inc.	USA	*616*	75-82950
TIRAX AG	Switzerland	*387*	75-14410
Vitro Corp.	USA	*644*	80-34400
Winkelmann (UK) Ltd	UK	*494*	75-66600

448 40 70 30 Bug detectors, electronic

Audiotel International Ltd	UK	*432*	75-04400
CCS International Ltd	USA	*562*	75-17250
CIC International Ltd	USA	*563*	75-18570
Datong Electronics Ltd	UK	*444*	75-16885
Global Communications Co.	USA	*579*	75-39775
Security Research Ltd	UK	*482*	75-52450
Special Electronic Security	Israel	*222*	75-21050
Spectronic Denmark	Denmark	*103*	75-19650
Winkelmann (UK) Ltd	UK	*494*	75-66600

448 40 70 90 Wiretap detectors, telephone

Audiotel International Ltd	UK	*432*	75-04400
CIC International Ltd	USA	*563*	75-18570
Global Communications Co.	USA	*579*	75-39775
Security Research Ltd	UK	*482*	75-52450
Special Electronic Security	Israel	*222*	75-21050
Winkelmann (UK) Ltd	UK	*494*	75-66600

Data Processing, Display & Recording

450 Data processing systems

Alenia SpA	Italy	*232*	75-01170
American Management Systems	USA	*630*	80-02350
Anite Networks Ltd	Australia	*11*	75-01293
Anite Systems Ltd	UK	*497*	80-01240
Arcata Associates Inc.	USA	*554*	75-08400
Array Systems Computing Inc.	Canada	*77*	80-00060
Aspect Computing Pty Ltd	Australia	*24*	80-01150
Aydin Corp.	USA	*556*	75-11450
Calspan SRL Technology	USA	*540*	60-03000
CEA Technologies Pty Ltd	Australia	*24*	80-03100
Central Research Laboratories Ltd	UK	*499*	80-05895
Concurrent Computer Corporation Ltd	UK	*442*	75-14810
Covergie SA	France	*134*	75-12400
Cranfield Aerospace Ltd	UK	*500*	80-07200
Cranfield University	UK	*419*	60-00350
Data Cell Ltd	UK	*500*	80-07720
Data Sciences Ltd	UK	*500*	80-07740
Datacom Inc.	USA	*567*	75-23645
Datamat Ingegneria dei Sistemi SpA	Italy	*233*	75-06500
EASAMS Ltd	UK	*501*	80-09000
EDS Defence Ltd	UK	*501*	80-09200
Eldec Corp.	USA	*572*	75-29660
ELICS	France	*136*	75-14800
ELTA Electronics Industries Ltd	Israel	*220*	75-10200
Ericsson Microwave Systems AB	**Sweden**	***377***	**75-04890**
EUROCONTROL SpA	Italy	*233*	75-08200
GEC-Marconi Avionics Inc.	USA	*577*	75-37413
GEC-Marconi Communications Ltd	UK	*454*	75-25335
GET Engineering Corp.	USA	*578*	75-39540
GKN Westland Technologies Ltd	UK	*455*	75-25690
GMV SA	Spain	*367*	80-00800
Graphico Electronics Group	Canada	*71*	75-06460
Hewlett-Packard (Canada) Ltd	Canada	*72*	75-07520
Hollandse Signaalapparaten BV	Netherlands	*293*	75-03300
ICL DESC	UK	*501*	80-08400
ICL Enterprises	UK	*504*	80-13100
Intermec UK Ltd	UK	*504*	80-13345
Israel Military Industries Ltd	Israel	*220*	75-10300
Jeppesen & Co GmbH	Germany	*170*	75-15050
Kaman Sciences Corp.	USA	*637*	80-17600
Kinetic Systems Corp.	USA	*588*	75-52660
LITEF GmbH	**Germany**	***171***	**75-18600**

Litton Industries Inc.	USA	*590*	75-55400
Lockheed Martin Corp.	USA	*591*	75-55635
Logica SA/NV.	Belgium	*48*	80-01000
Logica UK Limited.	UK	*506*	80-14350
Logicon Geodynamics Inc.	USA	*638*	80-18625
MacDonald Dettwiler	Malaysia	*278*	80-00270
MacDonald Dettwiler and Associates.	Canada	*78*	80-01310
Mackenzie Tribbeck Associates.	UK	*506*	80-15150
MASS Consultants Limited	UK	*506*	80-15650
Meggitt Avionics (UK)	UK	*467*	75-38340
Mitsubishi Precision Co. Ltd	Japan	*253*	75-06000
Modcomp	USA	*597*	75-62295
Motorola Ltd	UK	*469*	75-41050
Nord-Micro Elektronik Feinmechanik	Germany	*173*	75-22700
Nordic VLSI A/S	Norway	*310*	75-01640
Normalair-Garrett Ltd.	UK	*425*	74-07000
Pacer Systems Inc.	USA	*602*	75-68000
Parity Solutions Ltd.	UK	*508*	80-20300
PRC Inc.	USA	*640*	80-24800
Prosig Computer Consultants Ltd	UK	*476*	75-46865
R-R Industrial Controls Ltd	UK	*510*	80-29340
Radstone Technology Corp	USA	*607*	75-72610
Radstone Technology plc	UK	*477*	75-47810
Raytheon E-Systems	USA	*608*	75-73302
Rutherford Appleton Laboratory	UK	*420*	60-01850
SAIC Ltd.	UK	*510*	80-30220
Science Systems (Industrial) Ltd	UK	*510*	80-31600
Shade (Computer Services) Ltd	UK	*511*	80-34570
Siemens A/S	Norway	*310*	75-02450
Siemens Plessey Systems	UK	*483*	75-53770
Signal Computing Ltd.	UK	*511*	80-36050
Signal Technology Corp.	USA	*618*	75-84240
SISDEF Ltda	Chile	*85*	80-01940
STN ATLAS Elektronik GmbH.	Germany	*165*	75-01150
Tadiran Spectralink	Israel	*222*	75-21550
Technology Service Corp.	USA	*644*	80-31930
Telxon Ltd	UK	*488*	75-59305
TERMA Elektronik AS	Denmark	*104*	75-20300
The Flight Data Company Ltd	UK	*502*	80-10600
The Titan Corp	USA	*622*	75-89620
Thomson Marconi Sonar Limited	UK	*426*	74-20000
Thomson-CSF.	France	*148*	75-45100
Triad Group plc	UK	*514*	80-58600
Van Rietschoten & Houwens	Netherlands	*296*	75-07900
Vickers Shipbuilding & Eng Ltd	UK	*491*	75-63100

450 20 Data processing systems, airborne

Anite Systems Ltd	UK	*497*	80-01240
Arcata Associates Inc.	USA	*554*	75-08400
Aydin Corp.	USA	*556*	75-11450
Calspan SRL Technology.	USA	*540*	60-03000
Computing Devices Company Ltd	UK	*442*	75-14700
Computing Devices International	USA	*565*	75-20730
Eldec Corp.	USA	*572*	75-29660
ELTA Electronics Industries Ltd	Israel	*220*	75-10200
Ericsson Microwave Systems AB	**Sweden**	***377***	**75-04890**
GEC-Marconi Avionics Ltd	UK	*454*	75-25320
LITEF GmbH	**Germany**	***171***	**75-18600**
Litton Industries Inc.	USA	*590*	75-55400
Logica UK Limited.	UK	*506*	80-14350
Logicon Geodynamics Inc.	USA	*638*	80-18625
MacDonald Dettwiler and Associates.	Canada	*78*	80-01310
MacDonald Dettwiler Limited	Australia	*25*	80-12990
Nord-Micro Elektronik Feinmechanik	Germany	*173*	75-22700
Northrop Grumman Corporation.	USA	*640*	80-21840
Radstone Technology Corp	USA	*607*	75-72610
Rockwell-Collins France SA	France	*142*	75-32800
SISDEF Ltda	Chile	*85*	80-01940
Tadiran Spectralink	Israel	*222*	75-21550
Technology Service Corp.	USA	*644*	80-31930
Teldix GmbH	Germany	*175*	75-30100
The Titan Corp	USA	*622*	75-89620
Thomson Sintra.	France	*148*	75-45135

450 20 50 Data processing systems, airborne radar

Arcata Associates Inc.	USA	*554*	75-08400
Computing Devices International	USA	*565*	75-20730
ELTA Electronics Industries Ltd	Israel	*220*	75-10200
Ericsson Microwave Systems AB	**Sweden**	***377***	**75-04890**
MacDonald Dettwiler and Associates.	Canada	*78*	80-01310
MacDonald Dettwiler Limited	Australia	*25*	80-12990
Technology Service Corp.	USA	*644*	80-31930

450 20 60 Sonobuoy data processing systems, airborne

Computing Devices Canada	Canada	*70*	75-03600
Computing Devices International	USA	*565*	75-20730
ELTA Electronics Industries Ltd	Israel	*220*	75-10200
GEC-Marconi Avionics Ltd	UK	*454*	75-25320
Logica UK Limited.	UK	*506*	80-14350
NEC Corporation	Japan	*253*	75-06350
Pacer Systems Inc.	USA	*602*	75-68000
Technology Service Corp.	USA	*644*	80-31930
Thomson Sintra.	France	*148*	75-45135

450 30 Data processing systems, naval

Anite Systems Ltd	UK	*497*	80-01240
Aspect Computing Pty Ltd.	Australia	*24*	80-01150
Computing Devices International	USA	*565*	75-20730
Eldec Corp.	USA	*572*	75-29660
ELICS.	France	*136*	75-14800
GET Engineering Corp.	USA	*578*	75-39540
Graphico Electronics Group	Canada	*71*	75-06460
Hollandse Signaalapparaten BV	Netherlands	*293*	75-03300
LITEF GmbH	**Germany**	***171***	**75-18600**
Litton Industries Inc.	USA	*590*	75-55400
Logica UK Limited.	UK	*506*	80-14350
MacDonald Dettwiler and Associates.	Canada	*78*	80-01310
MacDonald Dettwiler Limited	Australia	*25*	80-12990
Meggitt Avionics (UK)	UK	*467*	75-38340
Northrop Grumman Corporation.	USA	*640*	80-21840
Prosig Computer Consultants Ltd	UK	*476*	75-46865
Radstone Technology Corp	USA	*607*	75-72610
Siemens Plessey Systems	UK	*483*	75-53770
SISDEF Ltda	Chile	*85*	80-01940
STN ATLAS Elektronik GmbH.	Germany	*165*	75-01150
Technology Service Corp.	USA	*644*	80-31930
Teldix GmbH	Germany	*175*	75-30100
Telxon Ltd	UK	*488*	75-59305
The Recorder Co.	USA	*579*	75-40360
The Titan Corp	USA	*622*	75-89620
Thomson Marconi Sonar Limited	UK	*426*	74-20000
Vickers Shipbuilding & Eng Ltd	UK	*491*	75-63100

450 30 50 Data processing systems, naval radar

Calspan SRL Technology.	USA	*540*	60-03000
Hollandse Signaalapparaten BV	Netherlands	*293*	75-03300
Siemens Plessey Systems	UK	*483*	75-53770
Sigma Signals Corporation.	USA	*615*	75-80790
Technology Service Corp.	USA	*644*	80-31930

450 30 60 Sonar data processing systems, naval

Computing Devices International	USA	*565*	75-20730
EDO Corporation	USA	*571*	75-29250
ELICS.	France	*136*	75-14800
Logica UK Limited.	UK	*506*	80-14350
MacDonald Dettwiler and Associates.	Canada	*78*	80-01310
Pacer Systems Inc.	USA	*602*	75-68000
STN ATLAS Elektronik GmbH.	Germany	*165*	75-01150
Technology Service Corp.	USA	*644*	80-31930
Thomson Marconi Sonar Limited	UK	*426*	74-20000
Thomson Sintra.	France	*148*	75-45135

450 40 Data processing systems, ground-based

Alenia SpA	Italy	*232*	75-01170
Anite Systems Ltd	UK	*497*	80-01240
Aspect Computing Pty Ltd.	Australia	*24*	80-01150
Avibras Aeroespacial SA	Brazil	*55*	75-00968
Aydin Corp.	USA	*556*	75-11450
Calspan SRL Technology.	USA	*540*	60-03000
Central Research Laboratories Ltd	UK	*499*	80-05895
Computer Resources International A/S.	Denmark	*104*	80-00100
Computing Devices International	USA	*565*	75-20730
Concurrent Computer Corporation Ltd	UK	*442*	75-14810
Cranfield Aerospace Ltd	UK	*500*	80-07200
Cranfield University.	UK	*419*	60-00350

Data Cell Ltd	UK	*500*	80-07720
DSD Ltd	UK	*501*	80-08700
EASAMS Ltd	UK	*501*	80-09000
GKN Westland Technologies Ltd	UK	*455*	75-25690
GMV SA	Spain	*367*	80-00800
Graphico Electronics Group	Canada	*71*	75-06460
Hollandse Signaalapparaten BV	Netherlands	*293*	75-03300
ICL DESC	UK	*501*	80-08400
Intermec UK Ltd	UK	*504*	80-13345
Laser-Scan	UK	*506*	80-14210
LITEF GmbH	**Germany**	***171***	**75-18600**
Logica UK Limited	UK	*506*	80-14350
MacDonald Dettwiler and Associates	Canada	*78*	80-01310
MacDonald Dettwiler Limited	Australia	*25*	80-12990
Mackenzie Tribbeck Associates	UK	*506*	80-15150
MASS Consultants Limited	UK	*506*	80-15650
Microelectronics, guidance and	Turkey	*402*	55-00050
Modcomp	USA	*597*	75-62295
Northrop Grumman Corporation	USA	*640*	80-21840
Penny & Giles Data Systems Ltd	UK	*473*	75-44560
R-R Industrial Controls Ltd	UK	*510*	80-29340
Radstone Technology Corp	USA	*607*	75-72610
Radstone Technology plc	UK	*477*	75-47810
Rutherford Appleton Laboratory	UK	*420*	60-01850
SAIC Ltd	UK	*510*	80-30220
Securicor Datatrak Ltd	UK	*482*	75-52390
Siemens Aktiengesellschaft	Germany	*175*	75-27700
Siemens Plessey Systems	UK	*483*	75-53770
SISDEF Ltda	Chile	*85*	80-01940
Tadiran Ltd	Israel	*222*	75-21600
Tadiran Spectralink	Israel	*222*	75-21550
Technology Service Corp.	USA	*644*	80-31930
Telxon Ltd	UK	*488*	75-59305
The Flight Data Company Ltd	UK	*502*	80-10600
The Recorder Co.	USA	*579*	75-40360
The Titan Corp	USA	*622*	75-89620
Thomson-CSF	France	*148*	75-45100
Triad Group plc	UK	*514*	80-58600
Xi CAD North America Inc.	USA	*628*	75-97760

450 40 10 Artillery data processing systems, ground-based

Avibras Aeroespacial SA	Brazil	*55*	75-00968
Israel Aircraft Industries Ltd	Israel	*219*	75-10200
LITEF GmbH	**Germany**	***171***	**75-18600**
Litton Industries Inc.	USA	*590*	75-55400
Logica UK Limited	UK	*506*	80-14350
MacDonald Dettwiler and Associates	Canada	*78*	80-01310
Microelectronics, guidance and	Turkey	*402*	55-00050
Siemens Plessey Systems	UK	*483*	75-53770
Tadiran Ltd	Israel	*222*	75-21600
Technology Service Corp.	USA	*644*	80-31930
TERMA Elektronik AS	Denmark	*104*	75-20300

450 40 70 Satellite data processing systems, ground-based

Computing Devices International	USA	*565*	75-20730
Concurrent Computer Corporation Ltd	UK	*442*	75-14810
Cranfield University	UK	*419*	60-00350
GMV SA	Spain	*367*	80-00800
Laser-Scan	UK	*506*	80-14210
Litton Industries Inc.	USA	*590*	75-55400
Logica UK Limited	UK	*506*	80-14350
MacDonald Dettwiler and Associates	Canada	*78*	80-01310
MacDonald Dettwiler Limited	Australia	*25*	80-12990
Mackenzie Tribbeck Associates	UK	*506*	80-15150
Matra Marconi Space UK, Ltd	UK	*425*	74-06900
Navstar Systems Ltd	UK	*470*	75-41955
Rutherford Appleton Laboratory	UK	*420*	60-01850
SAIC Ltd	UK	*510*	80-30220
Science Systems (Industrial) Ltd	UK	*510*	80-31600
SED Systems Inc	Canada	*79*	80-01937
Siemens Plessey Systems	UK	*483*	75-53770
Technology Service Corp.	USA	*644*	80-31930
The Dindima Group Pty Ltd	Australia	*15*	75-04580
Triad Group plc	UK	*514*	80-58600

451 Computers

Advanced Technology Ltd	Israel	*217*	75-00400
Alenia-Elsag	Italy	*230*	74-00100
Anite Systems Ltd	UK	*497*	80-01240
Aspect Computing Pty Ltd	Australia	*24*	80-01150
Astronautics Corp. of America	USA	*555*	75-10000
Bering Technology Inc.	USA	*558*	75-13700
Bodenseewerk Gerätetechnik GmbH	Germany	*165*	75-03000
Computing Devices Canada	Canada	*70*	75-03600
Computing Devices Company Ltd	UK	*442*	75-14700
Computing Devices International	USA	*565*	75-20730
Daimler-Benz Aerospace France SAS	**France**	***134***	**75-12950**
Dassault Electronique SA (Group)	**France**	***135***	**75-12980**
Datametrics Corporation	USA	*567*	75-23700
Del Mar Avionics	USA	*568*	75-25362
DY-4 Systems Inc.	Canada	*71*	75-04650
EDS Defence Ltd	UK	*501*	80-09200
Elbit Systems Ltd	Israel	*218*	75-05300
Ericsson Defence Systems Pty Ltd	Australia	*16*	75-05820
Ericsson Microwave Systems AB	**Sweden**	***377***	**75-04890**
European Data Systems	UK	*449*	75-21583
Force Computers GmbH	Germany	*168*	75-09270
Force Computers (UK) Ltd	UK	*451*	75-24155
Gandalf Technologies Inc.	Canada	*71*	75-05650
GEC-Marconi Avionics Inc.	USA	*577*	75-37413
GEC-Marconi Avionics Ltd	UK	*454*	75-25320
GEM Elettronica Srl	Italy	*235*	75-10100
Getac Corp	Taiwan	*392*	75-08000
Hollandse Signaalapparaten BV	Netherlands	*293*	75-03300
Howaldtswerke Deutsche Werft AG	Germany	*170*	75-13400
Husky Computers Limited	UK	*460*	75-30770
Instrument Services Ltd	UK	*504*	80-13328
Israel Aircraft Industries Ltd	Israel	*220*	75-10200
Japan Aviation Electronics Ind. Ltd	Japan	*252*	75-03300
Kearfott Guidance & Navigation Corp.	USA	*588*	75-51760
Kennedy Marketing Communications	USA	*637*	80-17680
LITEF GmbH	**Germany**	***171***	**75-18600**
Litton Industries Inc.	USA	*590*	75-55400
Lockheed Martin Corp.	USA	*591*	75-55635
Logica SA/NV	Belgium	*48*	80-01000
Mercury Computer Systems Inc.	USA	*595*	75-60220
Miltope Corp.	USA	*596*	75-61810
NAI Technologies Inc.	USA	*598*	75-63580
Northrop Grumman Corporation	USA	*640*	80-21840
PRC Inc.	USA	*640*	80-24800
Programming Research Ltd	UK	*509*	80-24700
Radstone Technology Corp	USA	*607*	75-72610
Radstone Technology plc	UK	*477*	75-47810
Rafael	Israel	*216*	55-01800
Rational Software Corporation	USA	*641*	80-25600
Raytheon Cossor Electronics Ltd	UK	*478*	75-48367
RGB Spectrum	USA	*609*	75-75000
SAGEM SA	**France**	***143***	**75-33950**
SAIC Ltd	UK	*510*	80-30220
Sanders	USA	*612*	75-77850
SEXTANT Avionique SA	France	*144*	75-35000
Siemens A/S	Norway	*310*	75-02450
SISDEF Ltda	Chile	*85*	80-01940
STN ATLAS Elektronik GmbH	Germany	*165*	75-01150
Synectics Corp.	USA	*643*	80-30750
The Titan Corp	USA	*622*	75-89620
Titan Electronics	USA	*623*	75-89630
VDO-Luftfahrtgeräte Werk GmbH	Germany	*176*	75-32600
WPI Termiflex Corp.	USA	*628*	75-97600
Xi CAD North America Inc.	USA	*628*	75-97760
Zitel Corporation	USA	*629*	75-98260

451 10 Computers, spaceborne

Astronautics Corp. of America	USA	*555*	75-10000
Computing Devices International	USA	*565*	75-20730
Cranfield University	UK	*419*	60-00350
Israel Aircraft Industries Ltd	Israel	*220*	75-10200
Kaiser Electronics	USA	*587*	75-51350
Kearfott Guidance & Navigation Corp.	USA	*588*	75-51760
Litton Industries Inc.	USA	*590*	75-55400
Matra Marconi Space UK, Ltd	UK	*425*	74-06900
Microtecnica-Torino SpA	Italy	*237*	75-15500
Rutherford Appleton Laboratory	UK	*420*	60-01850
SAGEM SA	**France**	***143***	**75-33950**
SEXTANT Avionique SA	France	*144*	75-35000
Space Systems Finland Ltd	Finland	*118*	75-02100
The Titan Corp	USA	*622*	75-89620

451 20 Computers, airborne

Astronautics C.A. Ltd	Israel	*217*	75-02400
Astronautics Corp. of America	USA	*555*	75-10000
Bodenseewerk Gerätetechnik GmbH	Germany	*165*	75-03000
Computing Devices Company Ltd	UK	*442*	75-14700
Computing Devices International	USA	*565*	75-20730
Cyberchron Corp.	USA	*567*	75-23120
Daimler-Benz Aerospace France SAS	**France**	***134***	**75-12950**
Elbit Systems Ltd	Israel	*218*	75-05300
Eldec Corp.	USA	*572*	75-29660
ELTA Electronics Industries Ltd	Israel	*220*	75-10200
Ericsson Microwave Systems AB	**Sweden**	***377***	**75-04890**
GEC-Marconi Avionics Ltd	UK	*454*	75-25320
GEC-Marconi Hazeltine Corp.	USA	*577*	75-37414
Intertechnique SA	France	*137*	75-23000
Israel Aircraft Industries Ltd	Israel	*219*	75-10200
Kaiser Electronics	USA	*587*	75-51350
Kearfott Guidance & Navigation Corp.	USA	*588*	75-51760
LITEF GmbH	**Germany**	***171***	**75-18600**
Litton Industries Inc.	USA	*590*	75-55400
Microtecnica-Torino SpA	Italy	*237*	75-15500
Miltope Corp.	USA	*596*	75-61810
Nord-Micro Elektronik Feinmechanik	Germany	*173*	75-22700
Racal Avionics Ltd	UK	*477*	75-47660
Radstone Technology Corp	USA	*607*	75-72610
Radstone Technology plc	UK	*477*	75-47810
SEXTANT Avionique SA	France	*144*	75-35000
Smiths Aerospace & Defense	USA	*615*	75-82090
Teldix GmbH	Germany	*175*	75-30100
The Titan Corp	USA	*622*	75-89620
VDO-Luftfahrtgeräte Werk GmbH	Germany	*176*	75-32600

451 20 10 Computers, air data, airborne

Astronautics C.A. Ltd	Israel	*217*	75-02400
Astronautics Corp. of America	USA	*555*	75-10000
Computing Devices Company Ltd	UK	*442*	75-14700
GEC-Marconi Avionics Ltd	UK	*454*	75-25320
Intertechnique SA	France	*137*	75-23000
Kearfott Guidance & Navigation Corp.	USA	*588*	75-51760
Kollsman	USA	*589*	75-53150
LITEF GmbH	**Germany**	***171***	**75-18600**
Litton Industries Inc.	USA	*590*	75-55400
Microtecnica-Torino SpA	Italy	*237*	75-15500
Miltope Corp.	USA	*596*	75-61810
Nord-Micro Elektronik Feinmechanik	Germany	*173*	75-22700
Pacer Systems Inc.	USA	*602*	75-68000
Racal Avionics Ltd	UK	*477*	75-47660
Radstone Technology plc	UK	*477*	75-47810
Rockwell International Co.	Italy	*233*	75-05200
Rosemount Aerospace Inc.	USA	*611*	75-76650
SEXTANT Avionique SA	France	*144*	75-35000
Smiths Aerospace & Defense	USA	*615*	75-82090
Teldix GmbH	Germany	*175*	75-30100
The Titan Corp	USA	*622*	75-89620
Tokyo Aircraft Instrument Co Ltd	Japan	*254*	75-08600

451 20 20 Computers, airborne flight control

Astronautics Corp. of America	USA	*555*	75-10000
Bodenseewerk Gerätetechnik GmbH	Germany	*165*	75-03000
Eldec Corp.	USA	*572*	75-29660
GEC-Marconi Avionics Ltd	UK	*454*	75-25320
Intertechnique SA	France	*137*	75-23000
Israel Aircraft Industries Ltd	Israel	*219*	75-10200
Kearfott Guidance & Navigation Corp.	USA	*588*	75-51760
Litton Industries Inc.	USA	*590*	75-55400
Miltope Corp.	USA	*596*	75-61810
Nord-Micro Elektronik Feinmechanik	Germany	*173*	75-22700
Racal Avionics Ltd	UK	*477*	75-47660
Radstone Technology plc	UK	*477*	75-47810
Rockwell International Co.	Italy	*233*	75-05200
Smiths Aerospace & Defense	USA	*615*	75-82090
Teldix GmbH	Germany	*175*	75-30100
The Titan Corp	USA	*622*	75-89620

451 20 30 Computers, airborne navigation

Astronautics Corp. of America	USA	*555*	75-10000
Bodenseewerk Gerätetechnik GmbH	Germany	*165*	75-03000
Computing Devices International	USA	*565*	75-20730
Elbit Systems Ltd	Israel	*218*	75-05300
GEC-Marconi Avionics Ltd	UK	*454*	75-25320
Intertechnique SA	France	*137*	75-23000
Israel Aircraft Industries Ltd	Israel	*219*	75-10200
Japan Aviation Electronics Ind. Ltd	Japan	*252*	75-03300
Kearfott Guidance & Navigation Corp.	USA	*588*	75-51760
LITEF GmbH	**Germany**	***171***	**75-18600**
Litton Industries Inc.	USA	*590*	75-55400
Marconi SpA DTM	Italy	*236*	75-14550
Miltope Corp.	USA	*596*	75-61810
NEC Corporation	Japan	*253*	75-06350
Racal Avionics Ltd	UK	*477*	75-47660
Radstone Technology plc	UK	*477*	75-47810
Rockwell International Co.	Italy	*233*	75-05200
SAGEM SA	**France**	***143***	**75-33950**
SEXTANT Avionique SA	France	*144*	75-35000
Smiths Aerospace & Defense	USA	*615*	75-82090

451 20 40 Computers, airborne weapon aiming / fire control

Astronautics Corp. of America	USA	*555*	75-10000
Bodenseewerk Gerätetechnik GmbH	Germany	*165*	75-03000
Computing Devices Company Ltd	UK	*442*	75-14700
Computing Devices International	USA	*565*	75-20730
Elbit Systems Ltd	Israel	*218*	75-05300
Ericsson Microwave Systems AB	**Sweden**	***377***	**75-04890**
GEC-Marconi Avionics Ltd	UK	*454*	75-25320
Intertechnique SA	France	*137*	75-23000
Israel Aircraft Industries Ltd	Israel	*219*	75-10200
Kaiser Electronics	USA	*587*	75-51350
Kearfott Guidance & Navigation Corp.	USA	*588*	75-51760
LITEF GmbH	**Germany**	***171***	**75-18600**
Litton Industries Inc.	USA	*590*	75-55400
Miltope Corp.	USA	*596*	75-61810
NEC Corporation	Japan	*253*	75-06350
Radstone Technology plc	UK	*477*	75-47810
SAGEM SA	**France**	***143***	**75-33950**
Smiths Aerospace & Defense	USA	*615*	75-82090
Teldix GmbH	Germany	*175*	75-30100
The Titan Corp	USA	*622*	75-89620

451 30 Computers, naval

Cyberchron Corp.	USA	*567*	75-23120
Daimler-Benz Aerospace France SAS	**France**	***134***	**75-12950**
Force Computers GmbH	Germany	*168*	75-09270
GET Engineering Corp.	USA	*578*	75-39540
Hollandse Signaalapparaten BV	Netherlands	*293*	75-03300
Husky Computers Limited	UK	*460*	75-30770
Israel Aircraft Industries Ltd	Israel	*219*	75-10200
LITEF GmbH	**Germany**	***171***	**75-18600**
Litton Industries Inc.	USA	*590*	75-55400
Miltope Corp.	USA	*596*	75-61810
Radstone Technology Corp	USA	*607*	75-72610
Radstone Technology plc	UK	*477*	75-47810
Siemens Plessey Systems	UK	*483*	75-53770
SISDEF Ltda	Chile	*85*	80-01940
STN ATLAS Elektronik GmbH	Germany	*165*	75-01150
TERMA Elektronik AS	Denmark	*104*	75-20300
The Recorder Co.	USA	*579*	75-40360
The Titan Corp	USA	*622*	75-89620

451 30 30 Computers, navigation, shipborne

Force Computers GmbH	Germany	*168*	75-09270
Israel Aircraft Industries Ltd	Israel	*220*	75-10200
Litton Industries Inc.	USA	*590*	75-55400
NEC Corporation	Japan	*253*	75-06350
Radstone Technology plc	UK	*477*	75-47810
Teldix GmbH	Germany	*175*	75-30100
TERMA Elektronik AS	Denmark	*104*	75-20300

451 30 40 Computers, naval fire control

Computing Devices International	USA	*565*	75-20730
Force Computers GmbH	Germany	*168*	75-09270
Hollandse Signaalapparaten BV	Netherlands	*293*	75-03300
Israel Aircraft Industries Ltd	Israel	*219*	75-10200

Litton Industries Inc.	USA	*590*	75-55400
Radstone Technology plc	UK	*477*	75-47810
SISDEF Ltda	Chile	*85*	80-01940
STN ATLAS Elektronik GmbH	Germany	*165*	75-01150

451 40 Computers, ground-based

Alenia SpA	Italy	*232*	75-01170
Applied Dynamics International	USA	*554*	75-08100
Cambridge Parallel Processing	USA	*562*	75-16480
Computing Devices International	USA	*565*	75-20730
Cyberchron Corp.	USA	*567*	75-23120
Daimler-Benz Aerospace France SAS	**France**	***134***	**75-12950**
Force Computers GmbH	Germany	*168*	75-09270
Hollandse Signaalapparaten BV	Netherlands	*293*	75-03300
Husky Computers Limited	UK	*460*	75-30770
ICL DESC	UK	*501*	80-08400
ICL Enterprises	UK	*504*	80-13100
Jeppesen UK Ltd	UK	*505*	80-13850
Kearfott Guidance & Navigation Corp.	USA	*588*	75-51760
LITEF GmbH	**Germany**	***171***	**75-18600**
Litton Industries Inc.	USA	*590*	75-55400
Logica UK Limited	UK	*506*	80-14350
Mercury Computer Systems Inc.	USA	*595*	75-60220
Miltope Corp.	USA	*596*	75-61810
Modcomp	USA	*597*	75-62295
Radstone Technology Corp	USA	*607*	75-72610
Radstone Technology plc	UK	*477*	75-47810
SAIC Ltd	UK	*510*	80-30220
Scientific Computers Ltd	UK	*510*	80-31550
Siemens Plessey Systems	UK	*483*	75-53770
Stanilite Electronics Pty Ltd	Australia	*21*	75-19690
The Titan Corp	USA	*622*	75-89620
Xi CAD North America Inc.	USA	*628*	75-97760

451 40 20 Computers, ballistic

Computing Devices Canada	Canada	*70*	75-03600
Concurrent Computer Corporation Ltd	UK	*442*	75-14810
Cranfield University	UK	*419*	60-00350
EUROCONTROL SpA	Italy	*233*	75-08200
Hollandse Signaalapparaten BV	Netherlands	*293*	75-03300
Litton Industries Inc.	USA	*590*	75-55400
MS Instruments plc	UK	*469*	75-41265
Octec Ltd	UK	*471*	75-42950
Radstone Technology plc	UK	*477*	75-47810

451 40 20 30 Computers, ballistic, hand-held

Computing Devices Canada	Canada	*70*	75-03600
Litton Industries Inc.	USA	*590*	75-55400

451 40 25 Computers, general purpose, ground-based

Alenia SpA	Italy	*232*	75-01170
Amextra GmbH	Germany	*177*	80-00200
Computing Devices International	USA	*565*	75-20730
Cray Research Inc.	USA	*566*	75-22570
ELTA Electronics Industries Ltd	Israel	*220*	75-10200
Hollandse Signaalapparaten BV	Netherlands	*293*	75-03300
Husky Computers Limited	UK	*460*	75-30770
Israel Aircraft Industries Ltd	Israel	*219*	75-10200
Litton Industries Inc.	USA	*590*	75-55400
Marconi SpA DTM	Italy	*236*	75-14550
Modcomp	USA	*597*	75-62295
Radstone Technology plc	UK	*477*	75-47810
SAGEM SA	**France**	***143***	**75-33950**
SAIC Ltd	UK	*510*	80-30220
Siemens Plessey Systems	UK	*483*	75-53770
SISDEF Ltda	Chile	*85*	80-01940
Stanilite Electronics Pty Ltd	Australia	*21*	75-19690
The Recorder Co.	USA	*579*	75-40360
The Titan Corp	USA	*622*	75-89620

451 40 30 Computers, navigation, ground-based

Force Computers GmbH	Germany	*168*	75-09270
GEC-Marconi Avionics Ltd	UK	*454*	75-25320
Israel Aircraft Industries Ltd	Israel	*219*	75-10200
Japan Aviation Electronics Ind. Ltd	Japan	*252*	75-03300
Kearfott Guidance & Navigation Corp.	USA	*588*	75-51760
Litton Industries Inc.	USA	*590*	75-55400
Radstone Technology plc	UK	*477*	75-47810
SAIC Ltd	UK	*510*	80-30220
Siemens Plessey Systems	UK	*483*	75-53770
Xi CAD North America Inc.	USA	*628*	75-97760

451 40 40 Computers, fire control, ground-based

Astronautics Corp. of America	USA	*555*	75-10000
Avibras Aeroespacial SA	Brazil	*55*	75-00968
Computing Devices Canada	Canada	*70*	75-03600
Elbit Systems Ltd	Israel	*218*	75-05300
EUROCONTROL SpA	Italy	*233*	75-08200
Force Computers GmbH	Germany	*168*	75-09270
Hollandse Signaalapparaten BV	Netherlands	*293*	75-03300
Israel Aircraft Industries Ltd	Israel	*219*	75-10200
Kearfott Guidance & Navigation Corp.	USA	*588*	75-51760
Litton Industries Inc.	USA	*590*	75-55400
Radstone Technology plc	UK	*477*	75-47810
SAIC Ltd	UK	*510*	80-30220
STN ATLAS Elektronik GmbH	Germany	*165*	75-01150
The Titan Corp	USA	*622*	75-89620

451 40 40 10 Computers, artillery fire control, ground-based

Astronautics Corp. of America	USA	*555*	75-10000
Avibras Aeroespacial SA	Brazil	*55*	75-00968
Computing Devices Canada	Canada	*70*	75-03600
Elbit Systems Ltd	Israel	*218*	75-05300
Force Computers GmbH	Germany	*168*	75-09270
Israel Aircraft Industries Ltd	Israel	*219*	75-10200
Kearfott Guidance & Navigation Corp.	USA	*588*	75-51760
Litton Industries Inc.	USA	*590*	75-55400
NORINCO	China PR	*88*	55-00280
Radstone Technology plc	UK	*477*	75-47810
SAIC Ltd	UK	*510*	80-30220

451 40 40 50 Computers, missile fire control, ground-based

Force Computers GmbH	Germany	*168*	75-09270
Hollandse Signaalapparaten BV	Netherlands	*293*	75-03300
Litton Industries Inc.	USA	*590*	75-55400
Radstone Technology plc	UK	*477*	75-47810

451 40 40 60 Computers, tank fire control, ground-based

Astronautics C.A. Ltd	Israel	*217*	75-02400
Astronautics Corp. of America	USA	*555*	75-10000
Computing Devices Canada	Canada	*70*	75-03600
Delft Sensor Systems	**Belgium**	***46***	**75-02500**
Delft Sensor Systems	Netherlands	*292*	75-01590
Hollandse Signaalapparaten BV	Netherlands	*293*	75-03300
Radstone Technology plc	UK	*477*	75-47810
SAIC Ltd	UK	*510*	80-30220
STN ATLAS Elektronik GmbH	Germany	*165*	75-01150

451 40 40 80 Computers, vehicle navigation, ground-based

Astronautics Corp. of America	USA	*555*	75-10000
Bodenseewerk Gerätetechnik GmbH	Germany	*165*	75-03000
Computing Devices Canada	Canada	*70*	75-03600
Force Computers GmbH	Germany	*168*	75-09270
GEC-Marconi Avionics Ltd	UK	*454*	75-25320
Israel Aircraft Industries Ltd	Israel	*220*	75-10200
Japan Aviation Electronics Ind. Ltd	Japan	*252*	75-03300
Litton Industries Inc.	USA	*590*	75-55400
Radstone Technology plc	UK	*477*	75-47810
SAGEM SA	**France**	***143***	**75-33950**
SAIC Ltd	UK	*510*	80-30220
Teldix GmbH	Germany	*175*	75-30100
Xi CAD North America Inc.	USA	*628*	75-97760

451 45 Minicomputers

Bering Technology Inc.	USA	*558*	75-13700
Hewlett-Packard (Canada) Ltd	Canada	*72*	75-07520
ICL DESC	UK	*501*	80-08400
ICL Enterprises	UK	*504*	80-13100

451 50 Supercomputers

Cambridge Consultants Ltd	UK	*499*	80-05300
Central Research Laboratories Ltd	UK	*499*	80-05895
Concurrent Computer Corporation Ltd	UK	*442*	75-14810
Cranfield University	UK	*419*	60-00350
Cray Research Inc.	USA	*566*	75-22570
Hewlett-Packard (Canada) Ltd	Canada	*72*	75-07520
NEL	UK	*419*	60-01400
Rutherford Appleton Laboratory	UK	*420*	60-01850

451 50 50 Superminicomputers

Cambridge Parallel Processing	USA	*562*	75-16480
Cray Research Inc.	USA	*566*	75-22570
Hewlett-Packard (Canada) Ltd	Canada	*72*	75-07520

451 70 Computer software

Advanced Technology Ltd	Israel	*217*	75-00400
AEA Technology plc	UK	*496*	80-00480
Alenia SpA	Italy	*232*	75-01170
Amextra GmbH	Germany	*177*	80-00200
Anite Systems Ltd	UK	*497*	80-01240
Aonix Europe Ltd	UK	*497*	80-01251
Array Systems Computing Inc.	Canada	*77*	80-00060
Aspect Computing Pty Ltd	Australia	*24*	80-01150
Avibras Aeroespacial SA	Brazil	*55*	75-00968
Aydin Corp.	USA	*556*	75-11450
BAeSEMA Ltd.	UK	*498*	80-03070
BRUHN NewTech A/S	Denmark	*104*	80-00050
CHAM (Concentration, Heat	UK	*499*	80-05900
Computer Sciences Corp.	USA	*565*	75-20720
Computing Devices Company Ltd	UK	*442*	75-14700
Computing Devices International	USA	*565*	75-20730
Cray Research Inc.	USA	*566*	75-22570
Dassault Electronique SA (Group)	**France**	***135***	**75-12980**
Datamat Ingegneria dei Sistemi SpA	Italy	*233*	75-06500
DDC-I A/S	Denmark	*104*	80-00400
DDC-I Inc	USA	*634*	80-07450
DY-4 Systems Inc.	Canada	*71*	75-04650
EASAMS Ltd	UK	*501*	80-09000
EDS Defence Ltd	UK	*501*	80-09200
Eiva A/S	Denmark	*104*	80-00500
ESDU International plc	UK	*502*	80-09700
EUROCONTROL SpA	Italy	*233*	75-08200
Explosivos Alaveses SA	Spain	*366*	75-03050
E2S NV	Belgium	*48*	80-00600
Franz Inc.	USA	*576*	75-35850
Gandalf Technologies Inc.	Canada	*71*	75-05650
GEM Elettronica Srl	Italy	*235*	75-10100
GMV SA	Spain	*367*	80-00800
GTT Industries Inc.	USA	*636*	80-13350
High Integrity Systems Ltd	UK	*458*	75-29250
Hollandse Signaalapparaten BV	Netherlands	*293*	75-03300
Howaldtswerke Deutsche Werft AG	Germany	*170*	75-13400
Hyundai	Korea (ROK)	*265*	75-01540
ICL DESC	UK	*501*	80-08400
ICL Enterprises	UK	*504*	80-13100
INRI UK Ltd	UK	*504*	80-13320
Interactive Dev. Environments	USA	*637*	80-15950
Interactive Development	UK	*504*	80-13330
Israel Aircraft Industries Ltd	Israel	*219*	75-10200
Japan Aviation Electronics Ind. Ltd	Japan	*252*	75-03300
J.S. Chinn Project Engineering Ltd	UK	*440*	75-13060
Kaiser Electronics	USA	*587*	75-51350
Kockums Computer Systems (UK) Ltd	UK	*505*	80-14180
Laser-Scan	UK	*506*	80-14210
LITEF GmbH	**Germany**	***171***	**75-18600**
Litton Industries Inc.	USA	*590*	75-55400
Logica UK Limited	UK	*506*	80-14350
Logicon Geodynamics Inc.	USA	*638*	80-18625
MacDonald Dettwiler and Associates	Canada	*78*	80-01310
MacNeal-Schwendler	Germany	*178*	80-01280
Mantix Systems Ltd	UK	*506*	80-15180
MATHSOFT	USA	*639*	80-20350
Micro Data Base Systems Inc.	USA	*639*	80-21080
Modcomp	USA	*597*	75-62295
MuSE Technologies Inc.	USA	*639*	80-21380
NEC Corporation	Japan	*253*	75-06350
Nordic VLSI A/S	Norway	*310*	75-01640
Optim Electronics Corp.	USA	*601*	75-66980
Persoft Inc.	USA	*603*	75-69380
Phase Two Industries	USA	*603*	75-69560
Praxis Critical Systems Ltd	UK	*509*	80-23930
PRC Inc.	USA	*640*	80-24800
Primary Image Ltd	UK	*476*	75-46790
Programming Research Ltd	UK	*509*	80-24700
Proprietary Software Systems	USA	*641*	80-24930
Prosig Computer Consultants Ltd	UK	*476*	75-46865
Public Safety Group Inc.	USA	*641*	80-24955
Racal-Airtech Ltd	UK	*476*	75-47655
Radstone Technology Corp	USA	*607*	75-72610
Rational Software Corporation	USA	*641*	80-25600
Raytheon E-Systems	USA	*608*	75-73302
Rohde & Schwarz Canada Inc	Canada	*79*	80-01825
SAC Technographic	UK	*510*	80-30100
Sageware Inc.	USA	*612*	75-77670
Sanders	USA	*612*	75-77850
Shade (Computer Services) Ltd	UK	*511*	80-34570
Siemens Aktiengesellschaft	Germany	*175*	75-27700
Siemens Plessey Systems	UK	*483*	75-53770
Sigma Signals Corporation	USA	*615*	75-80790
SISDEF Ltda	Chile	*85*	80-01940
Sonak Systems & Software SA	Greece	*193*	80-01970
Space Systems Finland Ltd	Finland	*118*	75-02100
STN ATLAS Elektronik GmbH	Germany	*165*	75-01150
Synectics Corp.	USA	*643*	80-30750
Systematic Software Engineering Ltd.	UK	*513*	80-49480
TA Group Ltd	UK	*513*	80-50110
Telxon Ltd	UK	*488*	75-59305
The Flight Data Company Ltd	UK	*502*	80-10600
The Titan Corp	USA	*622*	75-89620
Thomson Software Products	USA	*552*	75-04950
Van Rietschoten & Houwens	Netherlands	*296*	75-07900
Xi CAD North America Inc.	USA	*628*	75-97760
XL Computing Corp.	USA	*628*	75-98000
Zuken-Redac Ltd	UK	*495*	75-67800

451 80 Computer peripherals

Amextra GmbH	Germany	*177*	80-00200
Astro-Med Inc.	USA	*555*	75-09955
Astronautics Corp. of America	USA	*555*	75-10000
Computing Devices International	USA	*565*	75-20730
Cyberchron Corp.	USA	*567*	75-23120
Datametrics Corporation	USA	*567*	75-23700
Dicoll Electronics Limited	UK	*445*	75-18040
ELTA Electronics Industries Ltd	Israel	*220*	75-10200
Emco Inc.	USA	*573*	75-30985
European Data Systems	UK	*449*	75-21583
Husky Computers Limited	UK	*460*	75-30770
ICL DESC	UK	*501*	80-08400
ICL Enterprises	UK	*504*	80-13100
LARIMART SpA	Italy	*236*	75-13500
Laser-Scan	UK	*506*	80-14210
Measurement Systems Inc.	USA	*595*	75-59800
Miltope Corp.	USA	*596*	75-61810
Modcomp	USA	*597*	75-62295
NAI Technologies Inc.	USA	*598*	75-63580
Penny & Giles Position Sensors Ltd	UK	*474*	75-44570
Perex Ltd	UK	*474*	75-44680
Planar America Inc.	USA	*604*	75-70320
Prosig Computer Consultants Ltd	UK	*476*	75-46865
Radstone Technology Corp	USA	*607*	75-72610
RGB Spectrum	USA	*609*	75-75000
Ruggedtronics	USA	*611*	75-77015
SAIC Ltd	UK	*510*	80-30220
Siemens Plessey Systems	UK	*483*	75-53770
STN ATLAS Elektronik GmbH	Germany	*165*	75-01150
The Graphics Co. of America	USA	*579*	75-40360
Velleman Switch NV	Belgium	*47*	75-07000
Xi CAD North America Inc.	USA	*628*	75-97760
XL Computing Corp.	USA	*628*	75-98000

451 90 Computer components

Amextra GmbH	Germany	*177*	80-00200

Analysis Management Systems	S. Africa	*357*	75-00770
Bering Technology Inc.	USA	*558*	75-13700
Bospec Neue Technologien	Germany	*166*	75-03420
DY-4 Systems Inc.	Canada	*71*	75-04650
ELTA Electronics Industries Ltd	Israel	*220*	75-10200
Emco Inc.	USA	*573*	75-30985
European Data Systems	UK	*449*	75-21583
Force Computers GmbH	Germany	*168*	75-09270
ICL Enterprises	UK	*504*	80-13100
M-Systems Flash Disk Pioneers	Israel	*220*	75-12600
Mannesmann Tally Corp.	USA	*593*	75-57650
Modcomp	USA	*597*	75-62295
Mountain Optech Inc.	USA	*597*	75-63000
Nordic VLSI A/S	Norway	*310*	75-01640
North Hills Signal Processing	USA	*599*	75-65300
Prosig Computer Consultants Ltd	UK	*476*	75-46865
Radstone Technology Corp	USA	*607*	75-72610
Schlumberger Industries	France	*143*	75-34432
STN ATLAS Elektronik GmbH	Germany	*165*	75-01150
Teledyne Electronic Technologies	USA	*620*	75-87522
The Graphics Co. of America	USA	*579*	75-40360
The Recorder Co.	USA	*579*	75-40360
Zitel Corporation	USA	*629*	75-98260

453 Data handling & transmission systems, combined

Anite Networks Ltd	Australia	*11*	75-01293
Aydin Corp.	USA	*556*	75-11450
Cogent	UK	*500*	80-06250
Communications & Power Industries	USA	*565*	75-20393
Computing Devices Company Ltd	UK	*442*	75-14700
Cubic Defense Systems Inc.	USA	*567*	75-22900
Daimler-Benz Aerospace France SAS	**France**	***134***	**75-12950**
Eldec Corp.	USA	*572*	75-29660
GEM Elettronica Srl	Italy	*235*	75-10100
Hadland Photonics Ltd	UK	*456*	75-27200
Hollandse Signaalapparaten BV	Netherlands	*293*	75-03300
Kinetic Systems Corp.	USA	*588*	75-52660
Litton Industries Inc.	USA	*590*	75-55400
Lockheed Martin Corp.	USA	*591*	75-55635
Logica SA/NV	Belgium	*48*	80-01000
Logica UK Limited	UK	*506*	80-14350
Matrix Systems Ltd	Canada	*73*	75-09507
Meggitt Avionics (UK)	UK	*467*	75-38340
Nordic VLSI A/S	Norway	*310*	75-01640
Northrop Grumman Corporation	USA	*640*	80-21840
Optical Cable Corp.	USA	*601*	75-66952
PRC Inc.	USA	*640*	80-24800
Racal Communications Inc.	USA	*606*	75-72380
Raytheon E-Systems	USA	*608*	75-73302
RGB Spectrum	USA	*609*	75-75000
Sanders	USA	*612*	75-77850
Siemens Plessey Systems	UK	*483*	75-53770
Smiths Aerospace & Defense	USA	*615*	75-82090
Smiths Industries Aerospace	UK	*485*	75-54700
STN ATLAS Elektronik GmbH	Germany	*165*	75-01150
Telxon Ltd	UK	*488*	75-59305
Vickers Shipbuilding & Eng Ltd	UK	*491*	75-63100

454 Data handling & display systems, combined

Alenia SpA	Italy	*232*	75-01170
Anite Systems Ltd	UK	*497*	80-01240
Aspect Computing Pty Ltd	Australia	*24*	80-01150
Aydin Corp.	USA	*556*	75-11450
Calspan SRL Technology	USA	*540*	60-03000
Computing Devices Canada	Canada	*70*	75-03600
Computing Devices Company Ltd	UK	*442*	75-14700
Conrac Elektron GmbH	Germany	*166*	75-05100
Cubic Defense Systems Inc.	USA	*567*	75-22900
DRS Technologies, Inc.	USA	*570*	75-27400
Eldec Corp.	USA	*572*	75-29660
GEC-Marconi Avionics Ltd	UK	*454*	75-25320
GEM Elettronica Srl	Italy	*235*	75-10100
GET Engineering Corp.	USA	*578*	75-39540
Hadland Photonics Ltd	UK	*456*	75-27200
Hollandse Signaalapparaten BV	Netherlands	*293*	75-03300
InfoLink Technologies Inc.	USA	*585*	75-46703
Interstate Electronics	USA	*586*	75-48700
Israel Aircraft Industries Ltd	Israel	*219*	75-10200
ITT Aerospace/Communications Div.	USA	*586*	75-49455
Litton Industries Inc.	USA	*590*	75-55400
Lockheed Martin Corp.	USA	*591*	75-55635
Lockheed Martin Missiles & Space	USA	*592*	75-55640
Logic SpA	Italy	*236*	75-13800
Logica SA/NV	Belgium	*48*	80-01000
MacDonald Dettwiler and Associates	Canada	*78*	80-01310
Meggitt Avionics (UK)	UK	*467*	75-38340
Northrop Grumman Corporation	USA	*640*	80-21840
PRC Inc.	USA	*640*	80-24800
Prosig Computer Consultants Ltd	UK	*476*	75-46865
Racal Communications Inc.	USA	*606*	75-72380
Raytheon E-Systems	USA	*608*	75-73302
Recon/Optical Inc.	USA	*608*	75-73930
RGB Spectrum	USA	*609*	75-75000
Sanders	USA	*612*	75-77850
Siemens plc	UK	*483*	75-53750
Siemens Plessey Electronic Systems	Australia	*21*	75-19475
Siemens Plessey Systems	UK	*483*	75-53770
SISDEF Ltda	Chile	*85*	80-01940
STN ATLAS Elektronik GmbH	Germany	*165*	75-01150
Synectics Corp.	USA	*643*	80-30750
TERMA Elektronik AS	Denmark	*104*	75-20300
Vickers Shipbuilding & Eng Ltd	UK	*491*	75-63100

454 20 Data handling & display systems, combined, airborne

Anite Systems Ltd	UK	*497*	80-01240
Aydin Corp.	USA	*556*	75-11450
Computing Devices Company Ltd	UK	*442*	75-14700
DRS Technologies, Inc.	USA	*570*	75-27400
Eldec Corp.	USA	*572*	75-29660
ELTA Electronics Industries Ltd	Israel	*220*	75-10200
GEC-Marconi Avionics Ltd	UK	*454*	75-25320
Intertechnique SA	France	*137*	75-23000
Israel Aircraft Industries Ltd	Israel	*219*	75-10200
Kaiser Electronics	USA	*587*	75-51350
Litton Industries Inc.	USA	*590*	75-55400
Logic SpA	Italy	*236*	75-13800
MacDonald Dettwiler and Associates	Canada	*78*	80-01310
Recon/Optical Inc.	USA	*608*	75-73930
RGB Spectrum	USA	*609*	75-75000
Sanders	USA	*612*	75-77850
Teldix GmbH	Germany	*175*	75-30100

454 30 Data handling & display systems, combined, shipborne

Aeronautical & General Instr. Ltd	UK	*428*	75-01050
Anite Systems Ltd	UK	*497*	80-01240
Aspect Computing Pty Ltd	Australia	*24*	80-01150
Calspan SRL Technology	USA	*540*	60-03000
Computing Devices Canada	Canada	*70*	75-03600
DRS Technologies, Inc.	USA	*570*	75-27400
GET Engineering Corp.	USA	*578*	75-39540
Hollandse Signaalapparaten BV	Netherlands	*293*	75-03300
Israel Aircraft Industries Ltd	Israel	*219*	75-10200
Litton Industries Inc.	USA	*590*	75-55400
MacDonald Dettwiler and Associates	Canada	*78*	80-01310
Meggitt Avionics (UK)	UK	*467*	75-38340
RGB Spectrum	USA	*609*	75-75000
Sanders	USA	*612*	75-77850
Siemens Plessey Systems	UK	*483*	75-53770
SISDEF Ltda	Chile	*85*	80-01940
STN ATLAS Elektronik GmbH	Germany	*165*	75-01150
Teldix GmbH	Germany	*175*	75-30100
Thomson-CSF	France	*148*	75-45100

454 30 10 Action information & communications systems, shipborne

Computing Devices Canada	Canada	*70*	75-03600
GEC-Marconi Communications Ltd	UK	*454*	75-25335
Hollandse Signaalapparaten BV	Netherlands	*293*	75-03300
Hunting Engineering Ltd	UK	*459*	75-30700
Litton FiberCom	USA	*590*	75-55370
Litton Industries Inc.	USA	*590*	75-55400
MacDonald Dettwiler and Associates	Canada	*78*	80-01310
Nera Telecommunications	UK	*470*	75-42318
RGB Spectrum	USA	*609*	75-75000
SED Systems Inc	Canada	*79*	80-01937
Simrad Norge A/S	Norway	*310*	75-02630

454 40 Data handling & display systems, combined, ground-based

Aerobel Defence Technology Ltd	UK	*496*	80-00520
Ampex Data Systems Corporation	USA	*553*	75-06750
Anite Systems Ltd	UK	*497*	80-01240
Aydin Corp.	USA	*556*	75-11450
Belgian Advanced Technology	Belgium	*45*	75-01320
Bharat Electronics Ltd	India	*203*	55-00200
CAE	UK	*438*	75-11125
Calspan SRL Technology	USA	*540*	60-03000
Cambridge Consultants Ltd	UK	*499*	80-05300
Central Research Laboratories Ltd	UK	*499*	80-05895
Computer Resources International A/S	Denmark	*104*	80-00100
Computing Devices Canada	Canada	*70*	75-03600
Concurrent Computer Corporation Ltd	UK	*442*	75-14810
Cranfield Aerospace Ltd	UK	*500*	80-07200
Cranfield University	UK	*419*	60-00350
Data Cell Ltd	UK	*500*	80-07720
DRS Technologies, Inc.	USA	*570*	75-27400
Dynamic Graphics Ltd	UK	*501*	80-08850
EASAMS Ltd	UK	*501*	80-09000
Hollandse Signaalapparaten BV	Netherlands	*293*	75-03300
ICL DESC	UK	*501*	80-08400
InfoLink Technologies Inc.	USA	*585*	75-46703
Intertechnique SA	France	*137*	75-23000
Israel Aircraft Industries Ltd	Israel	*219*	75-10200
ITT Aerospace/Communications Div.	USA	*586*	75-49455
Laser-Scan	UK	*506*	80-14210
Litton Industries Inc.	USA	*590*	75-55400
Logica UK Limited	UK	*506*	80-14350
Logicon Geodynamics Inc.	USA	*638*	80-18625
MacDonald Dettwiler and Associates	Canada	*78*	80-01310
Mackenzie Tribbeck Associates	UK	*506*	80-15150
MASS Consultants Limited	UK	*506*	80-15650
Matra Marconi Space UK, Ltd	UK	*425*	74-06900
Normalair-Garrett Ltd	UK	*425*	74-07000
Octec Ltd	UK	*471*	75-42950
Oxley Developments Co. Ltd	UK	*472*	75-43400
Penny & Giles Data Systems Ltd	UK	*473*	75-44560
Racal Communications Inc.	USA	*606*	75-72380
Racal Recorders Ltd	UK	*477*	75-47689
Racal-MESL Ltd	UK	*477*	75-47682
Radamec Control Systems Ltd	UK	*477*	75-47755
Radstone Technology plc	UK	*477*	75-47810
Recon/Optical Inc.	USA	*608*	75-73930
RGB Spectrum	USA	*609*	75-75000
Rutherford Appleton Laboratory	UK	*420*	60-01850
SAIC Ltd	UK	*510*	80-30220
Sanders	USA	*612*	75-77850
Science Systems (Industrial) Ltd	UK	*510*	80-31600
Siemens Aktiengesellschaft	Germany	*175*	75-27700
Siemens Plessey Systems	UK	*483*	75-53770
SISDEF Ltda	Chile	*85*	80-01940
STN ATLAS Elektronik GmbH	Germany	*165*	75-01150
Strategic Imaging Systems	UK	*486*	75-57515
Tadiran Ltd	Israel	*222*	75-21600
TDSi Micro Applications Ltd	UK	*513*	80-51000
TERMA Elektronik AS	Denmark	*104*	75-20300
The Flight Data Company Ltd	UK	*502*	80-10600
Thomson-CSF	France	*148*	75-45100
Triad Group plc	UK	*514*	80-58600
Vega-Cantley Instrument Company	UK	*490*	75-62740
Xi CAD North America Inc.	USA	*628*	75-97760

455 Display systems & equipment

Alenia SpA	Italy	*232*	75-01170
Allard Industries Inc.	USA	*548*	75-04580
Amper Programas SA	Spain	*365*	75-00480
Anite Systems Ltd	UK	*497*	80-01240
Astronautics Corp. of America	USA	*555*	75-10000
Aydin Corp.	USA	*556*	75-11450
Barco Chromatics	USA	*557*	75-11920
Boeing Australia Limited	Australia	*12*	75-02500
Bospec Neue Technologien	Germany	*166*	75-03420
Brimar Inc.	USA	*560*	75-15210
Calspan SRL Technology	USA	*540*	60-03000
Catsaros Automation SA	Greece	*187*	75-00700
CEA Technologies Pty Ltd	Australia	*24*	80-03100
Computing Devices Canada	Canada	*70*	75-03600
Computing Devices Company Ltd	UK	*442*	75-14700
Conrac Elektron GmbH	Germany	*166*	75-05100
Conrac Systems Inc	USA	*566*	75-21520
Cubic Defense Systems Inc.	USA	*567*	75-22900
Cyberchron Corp.	USA	*567*	75-23120
DRS Technologies, Inc.	USA	*570*	75-27400
EDO Corporation	USA	*571*	75-29250
Elbit Systems Ltd	Israel	*218*	75-05300
Eldec Corp.	USA	*572*	75-29660
Electronic Systems Group	USA	*573*	75-30220
Ericsson Microwave Systems AB	**Sweden**	***377***	**75-04890**
Fabbrica Italiana Apparecchiature	Italy	*234*	75-08550
GEC Marconi Hazeltine Corp	USA	*578*	75-37415
GEC-Marconi Avionics Inc.	USA	*577*	75-37413
GEC-Marconi Avionics Ltd	UK	*454*	75-25320
GEM Elettronica Srl	Italy	*235*	75-10100
Hadland Photonics Ltd	UK	*456*	75-27200
Interstate Electronics	USA	*586*	75-48700
Japan Aviation Electronics Ind. Ltd	Japan	*252*	75-03300
Kaiser Optical Systems Inc.	USA	*587*	75-51355
Kearfott Guidance & Navigation Corp.	USA	*588*	75-51760
Litton Industries Inc.	USA	*590*	75-55400
Litton Systems Canada Ltd	Canada	*73*	75-09100
Logic SpA	Italy	*236*	75-13800
Lynwood	UK	*465*	75-36700
Meggitt Avionics (UK)	UK	*467*	75-38340
Metheus Corporation	UK	*467*	75-38800
Miltope Corp.	USA	*596*	75-61810
Mitsubishi Precision Co. Ltd	Japan	*253*	75-06000
Motorola Ltd	UK	*469*	75-41050
MSI-Defence Systems Ltd	UK	*469*	75-41280
Northrop Grumman Corporation	USA	*640*	80-21840
Oracle Telecomputing Inc.	Canada	*74*	75-10430
Philips Traffic & Engineering	Australia	*20*	75-16550
Planar Advance Inc.	UK	*475*	75-45735
Planar America Inc.	USA	*604*	75-70320
Planar International Ltd	Finland	*118*	75-01720
Polhemus Inc.	USA	*604*	75-70855
Prostab International Ltd	UK	*476*	75-46870
Racal Communications Inc.	USA	*606*	75-72380
Raytheon Canada Ltd	Canada	*75*	75-11700
Raytheon E-Systems	USA	*608*	75-73302
RGB Spectrum	USA	*609*	75-75000
Sanders	USA	*612*	75-77850
Scientific Computers Ltd	UK	*510*	80-31550
SEOS Displays	USA	*613*	75-79450
SEOS Displays Ltd	UK	*482*	75-52830
Siemens plc	UK	*483*	75-53750
Siemens Plessey Systems	UK	*483*	75-53770
Synectics Corp.	USA	*643*	80-30750
Tadiran Ltd	Israel	*222*	75-21600
Technology Applications & Services	USA	*620*	75-86890
Teledyne Controls	USA	*620*	75-87518
Thomson Sintra	France	*148*	75-45135
Thomson-CSF	France	*148*	75-45100
VDO-Luftfahrtgeräte Werk GmbH.	Germany	*176*	75-32600
Vickers Shipbuilding & Eng Ltd	UK	*491*	75-63100

455 20 Display systems, airborne

Alden Electronics Incorporated	USA	*548*	75-04400
Anite Systems Ltd	UK	*497*	80-01240
Astronautics C.A. Ltd	Israel	*217*	75-02400
Astronautics Corp. of America	USA	*555*	75-10000
Aydin Corp.	USA	*556*	75-11450
Barco NV	Belgium	*45*	75-01120
Canadian Marconi Co.	Canada	*69*	75-02600
Computing Devices Company Ltd	UK	*442*	75-14700
Conrac Systems Inc	USA	*566*	75-21520
Cubic Defense Systems Inc.	USA	*567*	75-22900
DRS Technologies, Inc.	USA	*570*	75-27400
Elbit Systems Ltd	Israel	*218*	75-05300
Eldec Corp.	USA	*572*	75-29660
Ericsson Microwave Systems AB	**Sweden**	***377***	**75-04890**
Fabbrica Italiana Apparecchiature	Italy	*234*	75-08550
GEC Marconi Hazeltine Corp	USA	*578*	75-37415
GEC-Marconi Avionics Ltd	UK	*454*	75-25320
GEM Elettronica Srl	Italy	*235*	75-10100
Japan Aviation Electronics Ind. Ltd	Japan	*252*	75-03300
Kaiser Electronics	USA	*587*	75-51350
Kaiser Optical Systems Inc.	USA	*587*	75-51355
Kearfott Guidance & Navigation Corp.	USA	*588*	75-51760
Litton Industries Inc.	USA	*590*	75-55400
Litton Systems Canada Ltd	Canada	*73*	75-09100
Logic SpA	Italy	*236*	75-13800
Miltope Corp.	USA	*596*	75-61810
Regis SpA	Italy	*238*	75-20110

Rockwell International Co.	Italy	233	75-05200
Sanders	USA	612	75-77850
SEXTANT Avionique SA.	France	144	75-35000
Siemens Aktiengesellschaft	Germany	175	75-27700
Siemens plc	UK	483	75-53750
Smiths Aerospace & Defense	USA	615	75-82090
Smiths Industries Aerospace	UK	485	75-54700
Teldix GmbH	Germany	175	75-30100
Teledyne Controls	USA	620	75-87518
Teledyne Electronic Technologies	USA	620	75-87522
VDO-Luftfahrtgeräte Werk GmbH.	Germany	176	75-32600

455 20 70 Display equipment, airborne

Amper Programas SA	Spain	365	75-00480
Astronautics C.A. Ltd	Israel	217	75-02400
Astronautics Corp. of America	USA	555	75-10000
Canadian Marconi Co.	Canada	69	75-02600
Computing Devices Canada	Canada	70	75-03600
Conrac Systems Inc	USA	566	75-21520
CRL Technologies	Canada	70	75-03760
Elbit Systems Ltd	Israel	218	75-05300
Fabbrica Italiana Apparecchiature	Italy	234	75-08550
GEC-Marconi Avionics Ltd	UK	454	75-25320
Interface Products Inc.	USA	585	75-47300
Israel Aircraft Industries Ltd	Israel	220	75-10200
Japan Aviation Electronics Ind. Ltd	Japan	252	75-03300
Kaiser Electronics	USA	587	75-51350
Kearfott Guidance & Navigation Corp.	USA	588	75-51760
Litton Industries Inc.	USA	590	75-55400
Litton Systems Canada Ltd	Canada	73	75-09100
Miltope Corp.	USA	596	75-61810
Planar Advance Inc.	UK	475	75-45735
Regis SpA	Italy	238	75-20110
Sanders	USA	612	75-77850
SEXTANT Avionique SA.	France	144	75-35000
Teledyne Electronic Technologies	USA	620	75-87522
Thomson Sintra	France	148	75-45135
Tokyo Aircraft Instrument Co Ltd	Japan	254	75-08600

455 20 70 20 Display consoles, airborne

Amper Programas SA	Spain	365	75-00480
Astronautics Corp. of America	USA	555	75-10000
Conrac Systems Inc	USA	566	75-21520
Fabbrica Italiana Apparecchiature	Italy	234	75-08550
GEC Marconi Hazeltine Corp	USA	578	75-37415
GEC-Marconi Avionics Ltd	UK	454	75-25320
Israel Aircraft Industries Ltd	Israel	220	75-10200
Kaiser Electronics	USA	587	75-51350
Litton Industries Inc.	USA	590	75-55400
Lynwood	UK	465	75-36700
Miltope Corp.	USA	596	75-61810
Sanders	USA	612	75-77850

455 20 80 Display equipment accessories, airborne

Fabbrica Italiana Apparecchiature	Italy	234	75-08550
GEC-Marconi Avionics Ltd	UK	454	75-25320
Giat Industries	France	126	55-00100
Kaiser Electronics	USA	587	75-51350
Litton Industries Inc.	USA	590	75-55400
Perex Ltd	UK	474	75-44680
Regis SpA	Italy	238	75-20110
Saunders-Roe Ltd	UK	482	75-51850
Teledyne Electronic Technologies	USA	620	75-87522

455 30 Display systems, naval

Aeronautical & General Instr. Ltd	UK	428	75-01050
Anite Systems Ltd	UK	497	80-01240
Astronautics C.A. Ltd	Israel	217	75-02400
Astronautics Corp. of America	USA	555	75-10000
Barco NV	Belgium	45	75-01120
Calspan SRL Technology	USA	540	60-03000
Computing Devices Canada	Canada	70	75-03600
Conrac Systems Inc	USA	566	75-21520
DRS Technologies, Inc.	USA	570	75-27400
EDO Corporation	USA	571	75-29250
Eldec Corp.	USA	572	75-29660
Fabbrica Italiana Apparecchiature	Italy	234	75-08550
Hollandse Signaalapparaten BV	Netherlands	293	75-03300
Israel Aircraft Industries Ltd	Israel	219	75-10200
Kearfott Guidance & Navigation Corp.	USA	588	75-51760
Meggitt Avionics (UK)	UK	467	75-38340
Motorola Ltd	UK	469	75-41050
Regis SpA	Italy	238	75-20110
RGB Spectrum	USA	609	75-75000
Sanders	USA	612	75-77850
Siemens Plessey Systems	UK	483	75-53770
STN ATLAS Elektronik GmbH	Germany	165	75-01150
Tadiran Ltd	Israel	222	75-21600
Teledyne Electronic Technologies	USA	620	75-87522
TERMA Elektronik AS	Denmark	104	75-20300
Thomson Marconi Sonar Limited	UK	426	74-20000

455 30 70 Display equipment, naval

Aeronautical & General Instr. Ltd	UK	428	75-01050
Astronautics C.A. Ltd	Israel	217	75-02400
Astronautics Corp. of America	USA	555	75-10000
Barco NV	Belgium	45	75-01120
Computing Devices Canada	Canada	70	75-03600
Conrac Systems Inc	USA	566	75-21520
EDO Corporation	USA	571	75-29250
Eldec Corp.	USA	572	75-29660
Fabbrica Italiana Apparecchiature	Italy	234	75-08550
Hollandse Signaalapparaten BV	Netherlands	293	75-03300
Interface Products Inc.	USA	585	75-47300
Israel Aircraft Industries Ltd	Israel	219	75-10200
Kearfott Guidance & Navigation Corp.	USA	588	75-51760
Litton Industries Inc.	USA	590	75-55400
Meggitt Avionics (UK)	UK	467	75-38340
NEC Corporation	Japan	253	75-06350
RGB Spectrum	USA	609	75-75000
Sanders	USA	612	75-77850
Siemens Plessey Systems	UK	483	75-53770
STN ATLAS Elektronik GmbH	Germany	165	75-01150
Tadiran Ltd	Israel	222	75-21600
Teledyne Electronic Technologies	USA	620	75-87522
TERMA Elektronik AS	Denmark	104	75-20300
Thomson Marconi Sonar Limited	UK	426	74-20000

455 30 70 20 Display consoles, naval

Astronautics C.A. Ltd	Israel	217	75-02400
Astronautics Corp. of America	USA	555	75-10000
Bharat Electronics Ltd	India	203	55-00200
Computing Devices Canada	Canada	70	75-03600
Conrac Systems Inc	USA	566	75-21520
Devtek Corporation	Canada	71	75-04300
EDO Corporation	USA	571	75-29250
Fabbrica Italiana Apparecchiature	Italy	234	75-08550
GEC-Marconi Avionics Ltd	UK	454	75-25320
Hollandse Signaalapparaten BV	Netherlands	293	75-03300
Kearfott Guidance & Navigation Corp.	USA	588	75-51760
Litton Industries Inc.	USA	590	75-55400
Lynwood	UK	465	75-36700
NEC Corporation	Japan	253	75-06350
Sanders	USA	612	75-77850
STN ATLAS Elektronik GmbH	Germany	165	75-01150
Thomson Marconi Sonar Limited	UK	426	74-20000

455 30 70 50 Plotting tables, naval

MSI-Defence Systems Ltd	UK	469	75-41280
The Recorder Co.	USA	579	75-40360
Xi CAD North America Inc.	USA	628	75-97760

455 30 80 Display equipment accessories, shipborne

Aeronautical & General Instr. Ltd	UK	428	75-01050
Computing Devices Canada	Canada	70	75-03600
Meggitt Avionics (UK)	UK	467	75-38340
Perex Ltd	UK	474	75-44680
Regis SpA	Italy	238	75-20110

455 40 Display systems, ground-based

Alenia SpA	Italy	232	75-01170
APT Radar Systems Ltd	UK	431	75-03560

Astronautics C.A. Ltd	Israel	*217*	75-02400
Astronautics Corp. of America	USA	*555*	75-10000
Barco NV	Belgium	*45*	75-01120
Calspan SRL Technology	USA	*540*	60-03000
Computer Resources International A/S.	Denmark	*104*	80-00100
Computing Devices Canada	Canada	*70*	75-03600
Conrac Systems Inc	USA	*566*	75-21520
CORIS	France	*134*	75-12450
DRS Technologies, Inc.	USA	*570*	75-27400
Eldec Corp.	USA	*572*	75-29660
Hollandse Signaalapparaten BV	Netherlands	*293*	75-03300
ICL DESC	UK	*501*	80-08400
INDRA	Spain	*366*	75-03800
Israel Aircraft Industries Ltd	Israel	*220*	75-10200
Litton Industries Inc.	USA	*590*	75-55400
Motorola Ltd	UK	*469*	75-41050
Plexsys Interface Products, Inc.	USA	*604*	75-70400
Raytheon Canada Ltd	Canada	*75*	75-11700
Raytheon Cossor Electronics Ltd	UK	*478*	75-48367
Regis SpA	Italy	*238*	75-20110
RGB Spectrum	USA	*609*	75-75000
Sanders	USA	*612*	75-77850
SEOS Displays	USA	*613*	75-79450
SEOS Displays Ltd	UK	*482*	75-52830
Siemens Aktiengesellschaft	Germany	*175*	75-27700
Siemens Plessey Systems	UK	*483*	75-53770
Tadiran Ltd	Israel	*222*	75-21600
TERMA Elektronik AS	Denmark	*104*	75-20300
Thomson-CSF	France	*148*	75-45100
Vickers Shipbuilding & Eng Ltd	UK	*491*	75-63100
WPI Termiflex Corp.	USA	*628*	75-97600
Xi CAD North America Inc.	USA	*628*	75-97760

455 40 70 Display equipment, ground-based

Alenia SpA	Italy	*232*	75-01170
Astronautics C.A. Ltd	Israel	*217*	75-02400
Astronautics Corp. of America	USA	*555*	75-10000
Barco NV	Belgium	*45*	75-01120
Computing Devices Canada	Canada	*70*	75-03600
Conrac Elektron GmbH	Germany	*166*	75-05100
Conrac Systems Inc	USA	*566*	75-21520
Eldec Corp.	USA	*572*	75-29660
Hollandse Signaalapparaten BV	Netherlands	*293*	75-03300
Interface Products Inc.	USA	*585*	75-47300
Israel Aircraft Industries Ltd	Israel	*219*	75-10200
Laser-Scan	UK	*506*	80-14210
Litton Industries Inc.	USA	*590*	75-55400
NEC Corporation	Japan	*253*	75-06350
Racal Communications Inc.	USA	*606*	75-72380
Regis SpA	Italy	*238*	75-20110
RGB Spectrum	USA	*609*	75-75000
Sanders	USA	*612*	75-77850
SEOS Displays Ltd	UK	*482*	75-52830
Siemens Plessey Systems	UK	*483*	75-53770
Tadiran Ltd	Israel	*222*	75-21600
Teldix GmbH	Germany	*175*	75-30100
TERMA Elektronik AS	Denmark	*104*	75-20300
Thomson-CSF	France	*148*	75-45100
Vickers Shipbuilding & Eng Ltd	UK	*491*	75-63100
WPI Termiflex Corp.	USA	*628*	75-97600

455 40 70 20 Display consoles, ground-based

Alenia SpA	Italy	*232*	75-01170
Astronautics C.A. Ltd	Israel	*217*	75-02400
Bharat Electronics Ltd	India	*203*	55-00200
Computing Devices Canada	Canada	*70*	75-03600
Conrac Systems Inc	USA	*566*	75-21520
CORIS	France	*134*	75-12450
Hollandse Signaalapparaten BV	Netherlands	*293*	75-03300
Litton Industries Inc.	USA	*590*	75-55400
Lynwood	UK	*465*	75-36700
NEC Corporation	Japan	*253*	75-06350
Plexsys Interface Products, Inc.	USA	*604*	75-70400
Racal Communications Inc.	USA	*606*	75-72380
Sanders	USA	*612*	75-77850
Siemens Aktiengesellschaft	Germany	*175*	75-27700
Thomson-CSF	France	*148*	75-45100
Vickers Shipbuilding & Eng Ltd	UK	*491*	75-63100
WPI Termiflex Corp.	USA	*628*	75-97600

455 40 80 Display equipment accessories, ground-based

Amextra GmbH	Germany	*177*	80-00200
Bertan High Voltage	USA	*558*	75-13720
Litton Industries Inc.	USA	*590*	75-55400
Metheus Corporation	UK	*467*	75-38800
Perex Ltd	UK	*474*	75-44680
Racal Communications Inc.	USA	*606*	75-72380
Radar/Digital Systems	USA	*607*	75-72460
Regis SpA	Italy	*238*	75-20110
SEOS Displays Ltd	UK	*482*	75-52830
Teledyne Electronic Technologies	USA	*620*	75-87522

455 40 80 50 Printer plotters, colour

Amextra GmbH	Germany	*177*	80-00200
Oki Systems (UK) Ltd.	UK	*471*	75-42967
SAGEM SA	**France**	***143***	**75-33950**
The Recorder Co.	USA	*579*	75-40360

456 Recording systems & equipment

Alden Electronics Incorporated	USA	*548*	75-04400
Ampex Data Systems Corporation	USA	*553*	75-06750
Angus Electronics Co.	USA	*554*	75-07630
Anite Systems Ltd	UK	*497*	80-01240
ATS Aerospace Inc.	Canada	*68*	75-00752
Avalon Electronics Ltd	UK	*432*	75-04485
Bristow Helicopters Ltd	UK	*499*	80-03950
Catsaros Automation SA	Greece	*187*	75-00700
CCS International Ltd	USA	*562*	75-17250
CEA Technologies Pty Ltd	Australia	*24*	80-03100
Computing Devices Company Ltd	UK	*442*	75-14700
Dassault Electronique SA (Group)	**France**	***135***	**75-12980**
DRS Technologies, Inc.	USA	*570*	75-27400
GEC-Marconi Avionics Inc.	USA	*577*	75-37413
GEC-Marconi Avionics Ltd	UK	*454*	75-25320
GEM Elettronica Srl	Italy	*235*	75-10100
Graff Electronic Machines Ltd	UK	*455*	75-26450
Hadland Photonics Ltd	UK	*456*	75-27200
Japan Aviation Electronics Ind. Ltd	Japan	*252*	75-03300
Kaman Sciences Corp.	USA	*637*	80-17600
Kinetic Systems Corp.	USA	*588*	75-52660
Litton Industries Inc.	USA	*590*	75-55400
Neal Recording Systems Ltd	UK	*470*	75-42270
Penny & Giles Data Systems Ltd	UK	*473*	75-44560
Precision Filters Inc.	USA	*605*	75-71330
Prosig Computer Consultants Ltd	UK	*476*	75-46865
Prostab International Ltd	UK	*476*	75-46870
Racal Recorders Ltd	UK	*477*	75-47689
Raytheon E-Systems	USA	*608*	75-73302
Reproduction Technologies Inc.	USA	*609*	75-74550
RGB Spectrum	USA	*609*	75-75000
Sanders	USA	*612*	75-77850
Schlumberger Industries	France	*143*	75-34432
SFIM	**France**	***144***	**75-35080**
SISDEF Ltda	Chile	*85*	80-01940
Smiths Aerospace & Defense	USA	*615*	75-82090
The Recorder Co.	USA	*579*	75-40360
TransInstruments	UK	*489*	75-60960
W. Vinten Ltd	UK	*491*	75-63353

456 10 Recording systems, spaceborne

AIT Recorders Ltd	UK	*429*	75-02200
Avalon Electronics Ltd	UK	*432*	75-04485
Cranfield University	UK	*419*	60-00350
L'Garde Inc.	USA	*590*	75-55060
Matra Marconi Space UK, Ltd	UK	*425*	74-06900
Miltope Corp.	USA	*596*	75-61810
Racal Recorders Ltd	UK	*477*	75-47689
Steatite Insulations Ltd	UK	*486*	75-56450
TransInstruments	UK	*489*	75-60960

456 10 70 Recording equipment, spaceborne

AIT Recorders Ltd	UK	*429*	75-02200
Avalon Electronics Ltd	UK	*432*	75-04485
Cranfield University	UK	*419*	60-00350
L'Garde Inc.	USA	*590*	75-55060
Matra Marconi Space UK, Ltd	UK	*425*	74-06900

Metrum Information Storage Ltd	UK	*467*	75-38890
Normalair-Garrett Ltd	UK	*425*	74-07000
Octec Ltd	UK	*471*	75-42950
Odetics Inc.	USA	*601*	75-66050
Photo-Sonics Inc.	USA	*604*	75-69800
Racal Recorders Ltd	UK	*477*	75-47689
SAGEM SA	**France**	***143***	**75-33950**
Steatite Insulations Ltd	UK	*486*	75-56450
The Recorder Co.	USA	*579*	75-40360
TransInstruments	UK	*489*	75-60960

456 20 Recording systems, airborne

Bristow Helicopters Ltd	UK	*499*	80-03950
Canadian Marconi Co.	Canada	*69*	75-02600
Computing Devices Company Ltd	UK	*442*	75-14700
DRS Precision Echo, Inc.	USA	*570*	75-27360
DRS Technologies, Inc.	USA	*570*	75-27400
GEC-Marconi Avionics Ltd	UK	*454*	75-25320
GEM Elettronica Srl	Italy	*235*	75-10100
GKN Westland Technologies Ltd	UK	*455*	75-25690
Hadland Photonics Ltd	UK	*456*	75-27200
Japan Aviation Electronics Ind. Ltd	Japan	*252*	75-03300
Litton Industries Inc.	USA	*590*	75-55400
Miltope Corp.	USA	*596*	75-61810
Normalair-Garrett Ltd	UK	*425*	74-07000
Regis SpA	Italy	*238*	75-20110
RGB Spectrum	USA	*609*	75-75000
Sanders	USA	*612*	75-77850
Schlumberger Industries	France	*143*	75-34432
SFIM	**France**	***144***	**75-35080**
Smiths Aerospace & Defense	USA	*615*	75-82090
Tokyo Aircraft Instrument Co Ltd	Japan	*254*	75-08600
TransInstruments	UK	*489*	75-60960
W. Vinten Ltd	UK	*491*	75-63353

456 20 30 Flight recorders, aircraft (see code 221.10.70.35)

Anite Systems Ltd	UK	*497*	80-01240
Bristow Helicopters Ltd	UK	*499*	80-03950
DRS Precision Echo, Inc.	USA	*570*	75-27360
GEC-Marconi Avionics Ltd	UK	*454*	75-25320
GKN Westland Technologies Ltd	UK	*455*	75-25690
Japan Aviation Electronics Ind. Ltd	Japan	*252*	75-03300
L'Garde Inc.	USA	*590*	75-55060
Penny & Giles Aerospace Ltd	UK	*473*	75-44550
Sanders	USA	*612*	75-77850
Schlumberger Industries	France	*143*	75-34432
SFIM	**France**	***144***	**75-35080**
Smiths Aerospace & Defense	USA	*615*	75-82090
The Flight Data Company Ltd	UK	*502*	80-10600
TransInstruments	UK	*489*	75-60960
Universal Avionics Systems Corp	USA	*625*	75-92650
W. Vinten Ltd	UK	*491*	75-63353

456 20 60 Tactical data recorders, airborne

DRS Technologies, Inc.	USA	*570*	75-27400
L'Garde Inc.	USA	*590*	75-55060
SAGEM SA	**France**	***143***	**75-33950**
Sanders	USA	*612*	75-77850
SFIM	**France**	***144***	**75-35080**

456 20 70 Recording equipment, airborne

Avalon Electronics Ltd	UK	*432*	75-04485
Bavaria Keytronic Technologie GmbH	Germany	*165*	75-02050
Datametrics Corporation	USA	*567*	75-23700
DRS Precision Echo, Inc.	USA	*570*	75-27360
DRS Technologies, Inc.	USA	*570*	75-27400
GEC-Marconi Avionics Ltd	UK	*454*	75-25320
GKN Westland Technologies Ltd	UK	*455*	75-25690
L'Garde Inc.	USA	*590*	75-55060
Normalair-Garrett Ltd	UK	*425*	74-07000
Penny & Giles Aerospace Ltd	UK	*473*	75-44550
Penny & Giles Data Systems Ltd	UK	*473*	75-44560
Photo-Sonics Inc.	USA	*604*	75-69800
Prosig Computer Consultants Ltd	UK	*476*	75-46865
Racal Recorders Ltd	UK	*477*	75-47689
Regis SpA	Italy	*238*	75-20110

SAGEM SA	**France**	***143***	**75-33950**
Sanders	USA	*612*	75-77850
Schlumberger Industries	France	*143*	75-34432
SFIM	**France**	***144***	**75-35080**
Smiths Aerospace & Defense	USA	*615*	75-82090
Tokyo Aircraft Instrument Co Ltd	Japan	*254*	75-08600
TransInstruments	UK	*489*	75-60960
W. Vinten Ltd	UK	*491*	75-63353

456 30 Recording systems, naval

AIT Recorders Ltd	UK	*429*	75-02200
Ampex Data Systems Corporation	USA	*553*	75-06750
ATS Aerospace Inc.	Canada	*68*	75-00752
DRS Precision Echo, Inc.	USA	*570*	75-27360
DRS Technologies, Inc.	USA	*570*	75-27400
GEC-Marconi Avionics Ltd	UK	*454*	75-25320
Hollandse Signaalapparaten BV	Netherlands	*293*	75-03300
L'Garde Inc.	USA	*590*	75-55060
Meggitt Avionics (UK)	UK	*467*	75-38340
Miltope Corp.	USA	*596*	75-61810
Racal Recorders Ltd	UK	*477*	75-47689
Regis SpA	Italy	*238*	75-20110
RGB Spectrum	USA	*609*	75-75000
SISDEF Ltda	Chile	*85*	80-01940
STN ATLAS Elektronik GmbH	Germany	*165*	75-01150
The Recorder Co.	USA	*579*	75-40360
Thomson Marconi Sonar Limited	UK	*426*	74-20000
TransInstruments	UK	*489*	75-60960

456 30 70 Recording equipment, naval

AIT Recorders Ltd	UK	*429*	75-02200
Ampex Data Systems Corporation	USA	*553*	75-06750
ATS Aerospace Inc.	Canada	*68*	75-00752
Datametrics Corporation	USA	*567*	75-23700
DRS Precision Echo, Inc.	USA	*570*	75-27360
DRS Technologies, Inc.	USA	*570*	75-27400
GEC-Marconi Avionics Ltd	UK	*454*	75-25320
Hollandse Signaalapparaten BV	Netherlands	*293*	75-03300
Japan Aviation Electronics Ind. Ltd	Japan	*252*	75-03300
L'Garde Inc.	USA	*590*	75-55060
Meggitt Avionics (UK)	UK	*467*	75-38340
Penny & Giles Aerospace Ltd	UK	*473*	75-44550
Penny & Giles Data Systems Ltd	UK	*473*	75-44560
Photo-Sonics Inc.	USA	*604*	75-69800
Prosig Computer Consultants Ltd	UK	*476*	75-46865
Racal Recorders Ltd	UK	*477*	75-47689
Regis SpA	Italy	*238*	75-20110
RGB Spectrum	USA	*609*	75-75000
SAGEM SA	**France**	***143***	**75-33950**
STN ATLAS Elektronik GmbH	Germany	*165*	75-01150
TransInstruments	UK	*489*	75-60960

456 40 Recording systems, ground-based

AIT Recorders Ltd	UK	*429*	75-02200
Ampex Data Systems Corporation	USA	*553*	75-06750
ATS Aerospace Inc.	Canada	*68*	75-00752
Audiotel International Ltd	UK	*432*	75-04400
Automatic Systems Laboratories Ltd	UK	*432*	75-04470
CORIS	France	*134*	75-12450
Del Mar Avionics	USA	*568*	75-25362
DRS Precision Echo, Inc.	USA	*570*	75-27360
DRS Technologies, Inc.	USA	*570*	75-27400
L'Garde Inc.	USA	*590*	75-55060
Miltope Corp.	USA	*596*	75-61810
Neal Recording Systems Ltd	UK	*470*	75-42270
Penny & Giles Data Systems Ltd	UK	*473*	75-44560
Prostab International Ltd	UK	*476*	75-46870
Racal Recorders Ltd	UK	*477*	75-47689
Regis SpA	Italy	*238*	75-20110
RGB Spectrum	USA	*609*	75-75000
SISDEF Ltda	Chile	*85*	80-01940
STN ATLAS Elektronik GmbH	Germany	*165*	75-01150
TERMA Elektronik AS	Denmark	*104*	75-20300
TransInstruments	UK	*489*	75-60960
Xi CAD North America Inc.	USA	*628*	75-97760

456 40 70 Recording equipment, ground-based

AIT Recorders Ltd	UK	429	75-02200
Ampex Data Systems Corporation	USA	553	75-06750
ATS Aerospace Inc.	Canada	68	75-00752
Audiotel International Ltd	UK	432	75-04400
Del Mar Avionics	USA	568	75-25362
DRS Precision Echo, Inc.	USA	570	75-27360
DRS Technologies, Inc.	USA	570	75-27400
L'Garde Inc.	USA	590	75-55060
Neal Recording Systems Ltd	UK	470	75-42270
Penny & Giles Aerospace Ltd	UK	473	75-44550
Penny & Giles Data Systems Ltd	UK	473	75-44560
Prosig Computer Consultants Ltd	UK	476	75-46865
Prostab International Ltd	UK	476	75-46870
Racal Recorders Ltd	UK	477	75-47689
Regis SpA	Italy	238	75-20110
RGB Spectrum	USA	609	75-75000
SAGEM SA	**France**	**143**	**75-33950**
STN ATLAS Elektronik GmbH	Germany	165	75-01150
TERMA Elektronik AS	Denmark	104	75-20300
The Recorder Co.	USA	579	75-40360
TransInstruments	UK	489	75-60960
Xi CAD North America Inc.	USA	628	75-97760

456 80 Recording equipment accessories

Ampex Data Systems Corporation	USA	553	75-06750
Audiotel International Ltd	UK	432	75-04400
DRS Precision Echo, Inc.	USA	570	75-27360
Graff Electronic Machines Ltd	UK	455	75-26450
Neal Recording Systems Ltd	UK	470	75-42270
Penny & Giles Data Systems Ltd	UK	473	75-44560
Sensitised Coatings Ltd	UK	482	75-52800
Smiths Aerospace & Defense	USA	615	75-82090
The Graphics Co. of America	USA	579	75-40360
The Recorder Co.	USA	579	75-40360
TransInstruments	UK	489	75-60960

456 90 Recording equipment components

Ampex Data Systems Corporation	USA	553	75-06750
Del Mar Avionics	USA	568	75-25362
DRS Precision Echo, Inc.	USA	570	75-27360
Neal Recording Systems Ltd	UK	470	75-42270
Penny & Giles Data Systems Ltd	UK	473	75-44560
Teledyne Electronic Technologies	USA	620	75-87522
The Graphics Co. of America	USA	579	75-40360
TransInstruments	UK	489	75-60960

Electronic Warfare & Countermeasures

461 Electronic warfare systems & equipment

AIL Systems Inc.	USA	547	75-02950
Amherst Systems Inc.	USA	553	75-06600
Amper Programas SA	Spain	365	75-00480
Andrew Canada Inc.	Canada	68	75-00600
Anite Systems Ltd	UK	497	80-01240
Applied Microwave Energy Inc.	USA	554	75-08165
ATS Aerospace Inc.	Canada	68	75-00752
Boeing Australia Limited	Australia	12	75-02500
British Aerospace (Systems &	UK	436	75-08890
CAE Elektronik GmbH	Germany	166	75-04500
Calspan SRL Technology	USA	540	60-03000
CEA Technologies Pty Ltd	Australia	24	80-03100
Chelton (Electrostatics) Limited	UK	439	75-12700
Chrysanthacopoulos SA	Greece	188	75-00910
Cincinnati Electronics Corp.	USA	564	75-18640
Circuitwise Electronics Ltd	UK	440	75-13300
Comptek Federal Systems Inc.	USA	565	75-20650
Condor Systems Inc.	USA	566	75-20950
Cubic Defense Systems Inc.	USA	567	75-22900
Custom Electronics Inc.	USA	567	75-23085
Cyberchron Corp.	USA	567	75-23120
Daimler-Benz Aerospace France SAS	**France**	**134**	**75-12950**
Data Sciences Ltd	UK	500	80-07740
Datacom Inc.	USA	567	75-23645
Desarrollo de Tecnologias y Sistemas	Chile	84	55-00550
EDS Defence Ltd	UK	501	80-09200
EG&G Inc.	USA	571	75-29555
Elbit Systems Ltd	Israel	218	75-05300
Elekluft GmbH	Germany	178	80-00537
Elettronica GmbH	Germany	168	75-07565
Elettronica SpA	Italy	233	75-07050
Elettronica (UK) Ltd	UK	448	75-20475
Elisra Electronic Systems Ltd	Israel	219	75-06150
ELTA Electronics Industries Ltd	Israel	220	75-10200
Ericsson Microwave Systems AB	**Sweden**	**377**	**75-04890**
Étienne Lacroix	France	138	75-24050
Flight International Inc.	USA	635	80-11600
GEC-Marconi Communications Ltd	UK	454	75-25335
Grinaker System Technologies	S. Africa	360	80-06500
Howden Fluid Systems	USA	582	75-44400
H.R. Smith (Microwaves) Ltd	UK	484	75-54590
Hughes Danbury Optical Sys.	USA	582	75-44857
INDRA	Spain	366	75-03800
INIMEX	**Belgium**	**46**	**75-04650**
Israel Aircraft Industries Ltd	Israel	219	75-10200
ITT Aerospace/Communications Div.	USA	586	75-49455
ITT Avionics	USA	586	75-49455
Kaman Sciences Corp.	USA	637	80-17600
Litton Industries Inc.	USA	590	75-55400
Lockheed Martin Canada Inc.	Canada	73	75-09110
Lockheed Martin Corp.	USA	591	75-55635
Logica UK Limited	UK	506	80-14350
MA/COM Antenna & Cable Div.	USA	593	75-56953
MCL Inc.	USA	595	75-59700
Mikes Microwave Electronic Systems	Turkey	403	75-01450
Motorola Ltd	UK	469	75-41050
Noise Com Inc.	USA	599	75-64760
Northrop Grumman Corporation	USA	599	75-65450
P W Allen & Co Ltd	UK	429	75-02740
PRB Associates Inc.	USA	640	80-24700
R. Alkan (UK) Ltd	UK	429	75-02680
Racal Communications Inc.	USA	606	75-72380
Racal Radar Defence Systems Ltd	UK	477	75-47688
Rafael	Israel	216	55-01800
Raytheon E-Systems	USA	608	75-73302
Research Associates of Syracuse	USA	641	80-25900
Rodale Electronics Inc.	USA	611	75-75970
Rohde & Schwarz France Telecom	France	142	75-32900
Rohde & Schwarz GmbH & Co KG	Germany	174	75-25400
Sanders	USA	612	75-77850
Siemens Plessey Electronic Systems	Australia	21	75-19475
Siemens Plessey Systems	UK	483	75-53770
Sierra Technologies Inc.	USA	614	75-80640
Sperry Marine Inc.	USA	617	75-83250
SRI International	USA	643	80-29600
STAREC	France	147	75-42775
State College Operations	USA	617	75-83990
Tadiran Ltd	Israel	222	75-21600
Technology Service Corp.	USA	644	80-31930
Teledyne Electronic Technologies	USA	621	75-87534
Teledyne Ryan Aeronautical	USA	621	75-87560
Textron Systems Division	USA	621	75-88655
Thomson-CSF	France	147	75-45100
Tracor Inc	USA	623	75-90000
Vickers Shipbuilding & Eng Ltd	UK	491	75-63100
Watkins-Johnson Co.	USA	627	75-95200
Zeta	USA	629	75-98190

461 20 Electronic warfare systems, airborne

AIL Systems Inc.	USA	547	75-02950
Anite Systems Ltd	UK	497	80-01240
Applied Microwave Energy Inc.	USA	554	75-08165
Cartwright Electronics Inc.	USA	562	75-16910
Cincinnati Electronics Corp.	USA	564	75-18640
Condor Systems Inc.	USA	566	75-20950
Daimler-Benz Aerospace France SAS	**France**	**134**	**75-12950**
Elettronica GmbH	Germany	168	75-07565
Elettronica SpA	Italy	233	75-07050
Elettronica (UK) Ltd	UK	448	75-20475
Elisra Electronic Systems Ltd	Israel	219	75-06150
ELTA Electronics Industries Ltd	Israel	220	75-10200
Étienne Lacroix	France	138	75-24050
Israel Aircraft Industries Ltd	Israel	219	75-10200
ITT Aerospace/Communications Div.	USA	586	75-49455
ITT Defense Ltd	UK	505	80-13570
Kollsman	USA	589	75-53150
Litton Industries Inc.	USA	590	75-55400
MCL Inc.	USA	595	75-59700

Product-Supplier Link

Mikes Microwave Electronic Systems	Turkey	*403*	75-01450
ML Aviation Ltd	UK	*469*	75-40200
Northrop Grumman Corporation	USA	*599*	75-65450
Racal Radar Defence Systems Ltd	UK	*477*	75-47688
Rafael	Israel	*216*	55-01800
Raytheon E-Systems	USA	*608*	75-73302
Rodale Electronics Inc	USA	*611*	75-75970
Sanders	USA	*612*	75-77850
Siemens Aktiengesellschaft	Germany	*175*	75-27700
SRI International	USA	*643*	80-29600
SRL Operations	USA	*617*	75-83550
State College Operations	USA	*617*	75-83990
Tadiran Ltd	Israel	*222*	75-21600
Technical Products Group Inc	USA	*620*	75-86764
Technology Service Corp	USA	*644*	80-31930
Tracor Inc	USA	*623*	75-90000

461 20 05 Electronic warfare suites, airborne

CelsiusTech Electronics AB	Sweden	*376*	75-03720
Condor Systems Inc	USA	*566*	75-20950
Daimler-Benz Aerospace France SAS	**France**	***134***	**75-12950**
Elettronica SpA	Italy	*233*	75-07050
Elettronica (UK) Ltd	UK	*448*	75-20475
Elisra Electronic Systems Ltd	Israel	*219*	75-06150
ELTA Electronics Industries Ltd	Israel	*220*	75-10200
Flight International Inc	USA	*635*	80-11600
Israel Aircraft Industries Ltd	Israel	*219*	75-10200
ITT Aerospace/Communications Div.	USA	*586*	75-49455
Litton Industries Inc	USA	*590*	75-55400
Mikes Microwave Electronic Systems	Turkey	*403*	75-01450
Northrop Grumman Corporation	USA	*599*	75-65450
Racal Radar Defence Systems Ltd	UK	*477*	75-47688
Rafael	Israel	*216*	55-01800
Rodale Electronics Inc	USA	*611*	75-75970
Sanders	USA	*612*	75-77850
Technology Service Corp	USA	*644*	80-31930

461 20 10 Electronic warfare systems, airborne, active

Anite Systems Ltd	UK	*497*	80-01240
Cartwright Electronics Inc	USA	*562*	75-16910
Daimler-Benz Aerospace France SAS	**France**	***134***	**75-12950**
Elettronica SpA	Italy	*233*	75-07050
Elettronica (UK) Ltd	UK	*448*	75-20475
Elisra Electronic Systems Ltd	Israel	*219*	75-06150
ELTA Electronics Industries Ltd	Israel	*220*	75-10200
Ericsson Microwave Systems AB	**Sweden**	***377***	**75-04890**
Israel Aircraft Industries Ltd	Israel	*219*	75-10200
ITT Aerospace/Communications Div.	USA	*586*	75-49455
ITT Defense Ltd	UK	*505*	80-13570
L'Garde Inc	USA	*590*	75-55060
Litton Industries Inc	USA	*590*	75-55400
Mikes Microwave Electronic Systems	Turkey	*403*	75-01450
Motorola Ltd	UK	*469*	75-41050
Northrop Grumman Corporation	USA	*599*	75-65450
Racal Radar Defence Systems Ltd	UK	*477*	75-47688
Rodale Electronics Inc	USA	*611*	75-75970
Sanders	USA	*612*	75-77850
Siemens Aktiengesellschaft	Germany	*175*	75-27700
Technical Products Group Inc	USA	*620*	75-86764
Technology Service Corp	USA	*644*	80-31930
Tracor Inc	USA	*623*	75-90000

461 20 13 Electronic countermeasures (ECM) systems, airborne

Calspan SRL Technology	USA	*540*	60-03000
Cincinnati Electronics Corp	USA	*564*	75-18640
Circuitwise Electronics Ltd	UK	*440*	75-13300
Com Dev Europe	UK	*441*	75-14300
EG&G Inc	USA	*571*	75-29555
Elettronica GmbH	Germany	*168*	75-07565
Elettronica SpA	Italy	*233*	75-07050
Elettronica (UK) Ltd	UK	*448*	75-20475
Elisra Electronic Systems Ltd	Israel	*219*	75-06150
ELTA Electronics Industries Ltd	Israel	*220*	75-10200
Ericsson Microwave Systems AB	**Sweden**	***377***	**75-04890**
Étienne Lacroix	France	*138*	75-24050
Flight International Inc	USA	*635*	80-11600
ITT Defense Ltd	UK	*505*	80-13570
Kollsman	USA	*589*	75-53150
Litton Industries Inc	USA	*590*	75-55400
MA/COM Antenna & Cable Div.	USA	*593*	75-56953
Mikes Microwave Electronic Systems	Turkey	*403*	75-01450
NEC Corporation	Japan	*253*	75-06350
Northrop Grumman Corporation	USA	*599*	75-65450
Racal Radar Defence Systems Ltd	UK	*477*	75-47688
Rodale Electronics Inc	USA	*611*	75-75970
Rokar International Ltd	Israel	*222*	75-18805
Sanders	USA	*612*	75-77850
Siemens Aktiengesellschaft	Germany	*175*	75-27700
Tadiran Ltd	Israel	*222*	75-21600
Technical Products Group Inc	USA	*620*	75-86764
W. Vinten Ltd	UK	*491*	75-63353
Znamya Production Group	Ukraine	*407*	75-10000

461 20 14 Infra-red countermeasures systems, airborne, active

ATS Aerospace Inc	Canada	*68*	75-00752
British Aerospace (Systems &	UK	*436*	75-08890
Cincinnati Electronics Corp	USA	*564*	75-18640
Cranfield University	UK	*419*	60-00350
Davin Optical Ltd	UK	*444*	75-17100
Étienne Lacroix	France	*138*	75-24050
Fothergill Engineered Fabrics Ltd	UK	*452*	75-24380
GEC-Marconi Research Centre	UK	*419*	60-00700
ITT Aerospace/Communications Div.	USA	*586*	75-49455
Kollsman	USA	*589*	75-53150
L'Garde Inc	USA	*590*	75-55060
Litton Industries Inc	USA	*590*	75-55400
ML Aviation Ltd	UK	*469*	75-40200
Northrop Grumman Corporation	USA	*599*	75-65450
Pilkington Thorn Optronics	UK	*426*	74-10000
Sanders	USA	*612*	75-77850
Tracor Inc	USA	*623*	75-90000

461 20 15 Electronic warfare systems, airborne, active, pod-mounted

Calspan SRL Technology	USA	*540*	60-03000
Com Dev Europe	UK	*441*	75-14300
Daimler-Benz Aerospace France SAS	**France**	***134***	**75-12950**
Elettronica SpA	Italy	*233*	75-07050
Elettronica (UK) Ltd	UK	*448*	75-20475
ELTA Electronics Industries Ltd	Israel	*220*	75-10200
Ericsson Microwave Systems AB	**Sweden**	***377***	**75-04890**
Israel Aircraft Industries Ltd	Israel	*219*	75-10200
ITT Aerospace/Communications Div.	USA	*586*	75-49455
Litton Industries Inc	USA	*590*	75-55400
Northrop Grumman Corporation	USA	*599*	75-65450
Racal Radar Defence Systems Ltd	UK	*477*	75-47688
Rodale Electronics Inc	USA	*611*	75-75970
Sanders	USA	*612*	75-77850
Tracor Inc	USA	*623*	75-90000

461 20 15 40 Infra-red countermeasures pods, airborne active

Cincinnati Electronics Corp	USA	*564*	75-18640
INIMEX	**Belgium**	***46***	**75-04650**
Litton Industries Inc	USA	*590*	75-55400
Northrop Grumman Corporation	USA	*599*	75-65450
Sanders	USA	*612*	75-77850
Tracor Inc	USA	*623*	75-90000

461 20 17 Electronic warfare equipment, airborne, active

Amherst Systems Inc	USA	*553*	75-06600
Anite Systems Ltd	UK	*497*	80-01240
Cincinnati Electronics Corp	USA	*564*	75-18640
Communications & Power Industries	USA	*565*	75-20393
Daimler-Benz Aerospace France SAS	**France**	***134***	**75-12950**
Dassault Electronique SA (Group)	**France**	***135***	**75-12980**
EG&G Inc	USA	*571*	75-29555
Elettronica GmbH	Germany	*168*	75-07565
Elettronica SpA	Italy	*233*	75-07050

Elettronica (UK) Ltd	UK	*448*	75-20475
Elisra Electronic Systems Ltd	Israel	*219*	75-06150
ELTA Electronics Industries Ltd	Israel	*220*	75-10200
Flight International Inc.	USA	*635*	80-11600
Howden Fluid Systems	USA	*582*	75-44400
Israel Aircraft Industries Ltd	Israel	*219*	75-10200
ITT Aerospace/Communications Div.	USA	*586*	75-49455
L'Garde Inc.	USA	*590*	75-55060
Litton Industries Inc.	USA	*590*	75-55400
Mikes Microwave Electronic Systems	Turkey	*403*	75-01450
Northrop Grumman Corporation	USA	*599*	75-65450
Racal Radar Defence Systems Ltd	UK	*477*	75-47688
Rodale Electronics Inc.	USA	*611*	75-75970
Sanders	USA	*612*	75-77850
Siemens Aktiengesellschaft	Germany	*175*	75-27700
Tracor Inc	USA	*623*	75-90000

461 20 17 60 Transmitting / Jamming / Deception equipment, airborne

British Aerospace Australia Ltd	Australia	*13*	75-02750
Cincinnati Electronics Corp.	USA	*564*	75-18640
Circuitwise Electronics Ltd	UK	*440*	75-13300
Daimler-Benz Aerospace France SAS	**France**	***134***	**75-12950**
Elettronica GmbH	Germany	*168*	75-07565
Elettronica SpA	Italy	*233*	75-07050
Elettronica (UK) Ltd	UK	*448*	75-20475
ELTA Electronics Industries Ltd	Israel	*220*	75-10200
Grinaker System Technologies	S. Africa	*360*	80-06500
Israel Aircraft Industries Ltd	Israel	*219*	75-10200
ITT Aerospace/Communications Div.	USA	*586*	75-49455
ITT Defense Ltd	UK	*505*	80-13570
L'Garde Inc.	USA	*590*	75-55060
MacDonald Dettwiler and Associates	Canada	*78*	80-01310
MA/COM Antenna & Cable Div.	USA	*593*	75-56953
Mikes Microwave Electronic Systems	Turkey	*403*	75-01450
Motorola Ltd	UK	*469*	75-41050
Northrop Grumman Corporation	USA	*599*	75-65450
Racal Radar Defence Systems Ltd	UK	*477*	75-47688
Rodale Electronics Inc.	USA	*611*	75-75970
Sanders	USA	*612*	75-77850
Siemens Aktiengesellschaft	Germany	*175*	75-27700
Sierra Technologies Inc.	USA	*614*	75-80640
SRI International	USA	*643*	80-29600
Tracor Inc	USA	*623*	75-90000

461 20 17 60 40 Jammers, airborne ECM

Cincinnati Electronics Corp.	USA	*564*	75-18640
Circuitwise Electronics Ltd	UK	*440*	75-13300
Daimler-Benz Aerospace France SAS	**France**	***134***	**75-12950**
Elettronica SpA	Italy	*233*	75-07050
Elettronica (UK) Ltd	UK	*448*	75-20475
Elisra Electronic Systems Ltd	Israel	*219*	75-06150
ELTA Electronics Industries Ltd	Israel	*220*	75-10200
Flight International Inc.	USA	*635*	80-11600
Israel Aircraft Industries Ltd	Israel	*219*	75-10200
ITT Aerospace/Communications Div.	USA	*586*	75-49455
ITT Defense Ltd	UK	*505*	80-13570
Noise Com Inc.	USA	*599*	75-64760
Northrop Grumman Corporation	USA	*599*	75-65450
Racal Communications Inc.	USA	*606*	75-72380
Racal Radar Defence Systems Ltd	UK	*477*	75-47688
Rodale Electronics Inc.	USA	*611*	75-75970
Sanders	USA	*612*	75-77850
SRI International	USA	*643*	80-29600
Tracor Inc	USA	*623*	75-90000

461 20 17 60 50 Optical countermeasures, airborne

Davin Optical Ltd	UK	*444*	75-17100
EG&G Inc.	USA	*571*	75-29555
GEC-Marconi Research Centre	UK	*419*	60-00700
L'Garde Inc.	USA	*590*	75-55060
Omitec Electro-Optics Ltd	UK	*472*	75-43050
Pilkington Thorn Optronics	UK	*426*	74-10000
Sanders	USA	*612*	75-77850

461 20 17 60 60 Transmitters, airborne ECM

Cincinnati Electronics Corp.	USA	*564*	75-18640
Daimler-Benz Aerospace France SAS	**France**	***134***	**75-12950**
Elettronica GmbH	Germany	*168*	75-07565
Elettronica SpA	Italy	*233*	75-07050
Elettronica (UK) Ltd	UK	*448*	75-20475
ELTA Electronics Industries Ltd	Israel	*220*	75-10200
Israel Aircraft Industries Ltd	Israel	*219*	75-10200
Northrop Grumman Corporation	USA	*599*	75-65450
Racal Radar Defence Systems Ltd	UK	*477*	75-47688
Rodale Electronics Inc.	USA	*611*	75-75970
Sanders	USA	*612*	75-77850
Tracor Inc	USA	*623*	75-90000

461 20 20 Electronic warfare systems, airborne, passive

Anite Systems Ltd	UK	*497*	80-01240
Com Dev Europe	UK	*441*	75-14300
Condor Systems Inc.	USA	*566*	75-20950
Contraves Inc	USA	*566*	75-21850
Daimler-Benz Aerospace France SAS	**France**	***134***	**75-12950**
Elbit Systems Ltd	Israel	*218*	75-05300
Elettronica GmbH	Germany	*168*	75-07565
Elettronica SpA	Italy	*233*	75-07050
Elettronica (UK) Ltd	UK	*448*	75-20475
Elisra Electronic Systems Ltd	Israel	*219*	75-06150
ELTA Electronics Industries Ltd	Israel	*220*	75-10200
EW Simulation Technology Ltd	UK	*449*	75-22150
Grinaker System Technologies	S. Africa	*360*	80-06500
Israel Aircraft Industries Ltd	Israel	*219*	75-10200
ITT Aerospace/Communications Div.	USA	*586*	75-49455
Litton Industries Inc.	USA	*590*	75-55400
Northrop Grumman Corporation	USA	*640*	80-21840
Pacer Systems Inc.	USA	*602*	75-68000
Racal Communications Inc.	USA	*606*	75-72380
Racal Radar Defence Systems Ltd	UK	*477*	75-47688
Sanders	USA	*612*	75-77850
SRI International	USA	*643*	80-29600
Tadiran Ltd	Israel	*222*	75-21600
Technology Service Corp.	USA	*644*	80-31930
Tracor Inc	USA	*623*	75-90000
Zeta	USA	*629*	75-98190

461 20 20 30 Electronic intelligence systems, airborne, passive

AB Electronics (Rogerstone) Ltd	UK	*426*	75-00080
Andrew Ltd	UK	*430*	75-03156
CelsiusTech Electronics AB	Sweden	*376*	75-03720
Central Research Laboratories Ltd	UK	*499*	80-05895
Condor Systems Inc.	USA	*566*	75-20950
Contraves Inc	USA	*566*	75-21850
Daimler-Benz Aerospace France SAS	**France**	***134***	**75-12950**
Elbit Systems Ltd	Israel	*218*	75-05300
Elettronica SpA	Italy	*233*	75-07050
Elisra Electronic Systems Ltd	Israel	*219*	75-06150
ELTA Electronics Industries Ltd	Israel	*220*	75-10200
GEC-Marconi Research Centre	UK	*419*	60-00700
Litton Industries Inc.	USA	*590*	75-55400
MA/COM Antenna & Cable Div.	USA	*593*	75-56953
MASS Consultants Limited	UK	*506*	80-15650
Racal Communications Inc.	USA	*606*	75-72380
Racal Radar Defence Systems Ltd	UK	*477*	75-47688
Sanders	USA	*612*	75-77850
State College Operations	USA	*617*	75-83990
Tadiran Ltd	Israel	*222*	75-21600
Technology Service Corp.	USA	*644*	80-31930
THORN Microwave Devices Ltd	UK	*488*	75-59760
Tracor Inc	USA	*623*	75-90000
Watkins-Johnson Co.	USA	*627*	75-95200

461 20 20 32 Communications intelligence systems, airborne, passive

Andrew Ltd	UK	*430*	75-03156
Central Research Laboratories Ltd	UK	*499*	80-05895
Condor Systems Inc.	USA	*566*	75-20950
Cranfield University	UK	*419*	60-00350

Daimler-Benz Aerospace France SAS	**France**	*134*	**75-12950**
Elettronica SpA	Italy	*233*	75-07050
ELTA Electronics Industries Ltd	Israel	*220*	75-10200
GEC-Marconi Research Centre	UK	*419*	60-00700
Israel Aircraft Industries Ltd	Israel	*219*	75-10200
Litton Industries Inc.	USA	*590*	75-55400
MA/COM Antenna & Cable Div.	USA	*593*	75-56953
NEC Corporation	Japan	*253*	75-06350
Sanders	USA	*612*	75-77850
State College Operations	USA	*617*	75-83990
Tadiran Ltd	Israel	*222*	75-21600
Technology Service Corp.	USA	*644*	80-31930
Watkins-Johnson Co.	USA	*627*	75-95200
Zeta	USA	*629*	75-98190

461 20 22 Electronic support measures (ESM) systems, airborne

British Aerospace Australia Ltd	Australia	*13*	75-02750
Elettronica GmbH	Germany	*168*	75-07565
Elettronica SpA	Italy	*233*	75-07050
Elettronica (UK) Ltd	UK	*448*	75-20475
Elisra Electronic Systems Ltd	Israel	*219*	75-06150
ELTA Electronics Industries Ltd	Israel	*220*	75-10200
EW Simulation Technology Ltd	UK	*449*	75-22150
ITT Defense Ltd.	UK	*505*	80-13570
Litton Industries Inc.	USA	*590*	75-55400
Lockheed Martin Canada Inc.	Canada	*73*	75-09110
Pacer Systems Inc.	USA	*602*	75-68000
Racal Communications Inc.	USA	*606*	75-72380
Racal Radar Defence Systems Ltd.	UK	*477*	75-47688
Sanders	USA	*612*	75-77850
State College Operations	USA	*617*	75-83990
Tadiran Ltd	Israel	*222*	75-21600
Technology Service Corp.	USA	*644*	80-31930
Watkins-Johnson Co.	USA	*627*	75-95200

461 20 24 Infra-red search & warning systems, airborne, passive

Cincinnati Electronics Corp.	USA	*564*	75-18640
Litton Industries Inc.	USA	*590*	75-55400
Marconi SpA DTM	Italy	*236*	75-14550
Moked Engineering (1969) Ltd	Israel	*221*	75-15200
Northrop Grumman Corporation	USA	*599*	75-65450
Technology Service Corp.	USA	*644*	80-31930
Tracor Inc	USA	*623*	75-90000

461 20 27 Electronic warfare equipment, airborne, passive

Anite Systems Ltd	UK	*497*	80-01240
Com Dev Europe	UK	*441*	75-14300
Condor Systems Inc.	USA	*566*	75-20950
Daimler-Benz Aerospace France SAS	**France**	*134*	**75-12950**
Elettronica SpA	Italy	*233*	75-07050
Elettronica (UK) Ltd	UK	*448*	75-20475
Elisra Electronic Systems Ltd	Israel	*219*	75-06150
ELTA Electronics Industries Ltd	Israel	*220*	75-10200
Flight International Inc.	USA	*635*	80-11600
GEC-Marconi Avionics Ltd	UK	*454*	75-25320
Howden Fluid Systems	USA	*582*	75-44400
Israel Aircraft Industries Ltd	Israel	*219*	75-10200
ITT Aerospace/Communications Div.	USA	*586*	75-49455
Litton Industries Inc.	USA	*590*	75-55400
MA/COM Antenna & Cable Div.	USA	*593*	75-56953
Northrop Grumman Corporation	USA	*640*	80-21840
Racal Communications Inc.	USA	*606*	75-72380
Racal Radar Defence Systems Ltd.	UK	*477*	75-47688
Sanders	USA	*612*	75-77850
Tadiran Ltd	Israel	*222*	75-21600
Tracor Inc	USA	*623*	75-90000
Watkins-Johnson Co.	USA	*627*	75-95200

461 20 27 30 Direction Finding / Homing equipment, airborne EW

Cincinnati Electronics Corp.	USA	*564*	75-18640
Condor Systems Inc.	USA	*566*	75-20950
Elettronica SpA	Italy	*233*	75-07050
Elettronica (UK) Ltd	UK	*448*	75-20475
Elisra Electronic Systems Ltd	Israel	*219*	75-06150
ELTA Electronics Industries Ltd	Israel	*220*	75-10200
Études Electroniques Informatiques	France	*150*	80-00540
EW Simulation Technology Ltd	UK	*449*	75-22150
GEC-Marconi Avionics Ltd	UK	*454*	75-25320
Israel Aircraft Industries Ltd	Israel	*219*	75-10200
Litton Industries Inc.	USA	*590*	75-55400
MA/COM Antenna & Cable Div.	USA	*593*	75-56953
Racal Communications Inc.	USA	*606*	75-72380
Racal Radar Defence Systems Ltd.	UK	*477*	75-47688
Tadiran Ltd	Israel	*222*	75-21600
Watkins-Johnson Co.	USA	*627*	75-95200
Zeta	USA	*629*	75-98190

461 20 27 50 Receiving / Detecting / Warning equipment, airborne EW

Andrew Canada Inc.	Canada	*68*	75-00600
Cincinnati Electronics Corp.	USA	*564*	75-18640
Com Dev Europe	UK	*441*	75-14300
Condor Systems Inc.	USA	*566*	75-20950
Elettronica SpA	Italy	*233*	75-07050
Elisra Electronic Systems Ltd	Israel	*219*	75-06150
ELTA Electronics Industries Ltd	Israel	*220*	75-10200
EW Simulation Technology Ltd	UK	*449*	75-22150
GEC-Marconi Avionics Ltd	UK	*454*	75-25320
Hughes Danbury Optical Sys.	USA	*582*	75-44857
Israel Aircraft Industries Ltd	Israel	*219*	75-10200
Litton Industries Inc.	USA	*590*	75-55400
MA/COM Antenna & Cable Div.	USA	*593*	75-56953
Northrop Grumman Corporation	USA	*599*	75-65450
Racal Communications Inc.	USA	*606*	75-72380
Racal Radar Defence Systems Ltd.	UK	*477*	75-47688
State College Operations	USA	*617*	75-83990
Tadiran Ltd	Israel	*222*	75-21600
Tracor Inc	USA	*623*	75-90000
Watkins-Johnson Co.	USA	*627*	75-95200

461 20 27 55 Receiving equipment, airborne EW

Andrew Canada Inc.	Canada	*68*	75-00600
Com Dev Europe	UK	*441*	75-14300
Condor Systems Inc.	USA	*566*	75-20950
Elettronica SpA	Italy	*233*	75-07050
Elettronica (UK) Ltd	UK	*448*	75-20475
Elisra Electronic Systems Ltd	Israel	*219*	75-06150
ELTA Electronics Industries Ltd	Israel	*220*	75-10200
EW Simulation Technology Ltd	UK	*449*	75-22150
Icom (UK) Ltd	UK	*460*	75-31070
Israel Aircraft Industries Ltd	Israel	*219*	75-10200
Litton Industries Inc.	USA	*590*	75-55400
MA/COM Antenna & Cable Div.	USA	*593*	75-56953
Racal Communications Inc.	USA	*606*	75-72380
Racal Radar Defence Systems Ltd.	UK	*477*	75-47688
Sanders	USA	*612*	75-77850
State College Operations	USA	*617*	75-83990
Tadiran Ltd	Israel	*222*	75-21600
Tracor Inc	USA	*623*	75-90000
Watkins-Johnson Co.	USA	*627*	75-95200

461 20 27 55 30 Receivers, electronic intelligence, airborne

Andrew Canada Inc.	Canada	*68*	75-00600
Condor Systems Inc.	USA	*566*	75-20950
Elettronica SpA	Italy	*233*	75-07050
Elettronica (UK) Ltd	UK	*448*	75-20475
Elisra Electronic Systems Ltd	Israel	*219*	75-06150
ELTA Electronics Industries Ltd	Israel	*220*	75-10200
Israel Aircraft Industries Ltd	Israel	*219*	75-10200
Litton Industries Inc.	USA	*590*	75-55400
MA/COM Antenna & Cable Div.	USA	*593*	75-56953
Racal Communications Inc.	USA	*606*	75-72380
Racal Radar Defence Systems Ltd.	UK	*477*	75-47688
Sanders	USA	*612*	75-77850
Tadiran Ltd	Israel	*222*	75-21600
Technology Service Corp.	USA	*644*	80-31930
Watkins-Johnson Co.	USA	*627*	75-95200

461 20 27 55 32 Receivers, communications intelligence, airborne

Andrew Canada Inc.	Canada	*68*	75-00600
Condor Systems Inc.	USA	*566*	75-20950
Daimler-Benz Aerospace France SAS	**France**	***134***	**75-12950**
Elettronica SpA	Italy	*233*	75-07050
Elettronica (UK) Ltd	UK	*448*	75-20475
Elisra Electronic Systems Ltd	Israel	*219*	75-06150
ELTA Electronics Industries Ltd	Israel	*220*	75-10200
Israel Aircraft Industries Ltd	Israel	*219*	75-10200
Litton Industries Inc.	USA	*590*	75-55400
MA/COM Antenna & Cable Div.	USA	*593*	75-56953
NEC Corporation	Japan	*253*	75-06350
Racal Communications Inc.	USA	*606*	75-72380
Sanders	USA	*612*	75-77850
Tadiran Ltd	Israel	*222*	75-21600
Technology Service Corp.	USA	*644*	80-31930
Watkins-Johnson Co.	USA	*627*	75-95200

461 20 27 55 40 Receivers, infra-red warning, airborne

Cincinnati Electronics Corp.	USA	*564*	75-18640
Elettronica (UK) Ltd	UK	*448*	75-20475
ITT Aerospace/Communications Div.	USA	*586*	75-49455
Kollsman	USA	*589*	75-53150
Litton Industries Inc.	USA	*590*	75-55400
Northrop Grumman Corporation	USA	*599*	75-65450

461 20 27 55 45 Receivers, laser warning, airborne

GEC-Marconi Avionics Ltd	UK	*454*	75-25320
Hughes Danbury Optical Sys.	USA	*582*	75-44857
Kollsman	USA	*589*	75-53150
Litton Industries Inc.	USA	*590*	75-55400

461 20 27 55 50 Receivers, missile alert & launch warning, EW

Cincinnati Electronics Corp.	USA	*564*	75-18640
Elettronica SpA	Italy	*233*	75-07050
Elettronica (UK) Ltd	UK	*448*	75-20475
Elisra Electronic Systems Ltd	Israel	*219*	75-06150
ELTA Electronics Industries Ltd	Israel	*220*	75-10200
Litton Industries Inc.	USA	*590*	75-55400
Technology Service Corp.	USA	*644*	80-31930
Tracor Inc	USA	*623*	75-90000

461 20 27 55 60 Radar homing & warning receivers

Elettronica SpA	Italy	*233*	75-07050
Elettronica (UK) Ltd	UK	*448*	75-20475
ELTA Electronics Industries Ltd	Israel	*220*	75-10200
ITT Aerospace/Communications Div.	USA	*586*	75-49455
Litton Industries Inc.	USA	*590*	75-55400
Tracor Inc	USA	*623*	75-90000

461 20 27 55 61 Receivers, radar warning, airborne EW

Com Dev Europe	UK	*441*	75-14300
Daimler-Benz Aerospace France SAS	**France**	***134***	**75-12950**
Elettronica SpA	Italy	*233*	75-07050
Elettronica (UK) Ltd	UK	*448*	75-20475
ELTA Electronics Industries Ltd	Israel	*220*	75-10200
EW Simulation Technology Ltd	UK	*449*	75-22150
Litton Industries Inc.	USA	*590*	75-55400
Metric Systems Corp.	USA	*596*	75-60800
Northrop Grumman Corporation	USA	*599*	75-65450
Tadiran Ltd	Israel	*222*	75-21600
Tracor Inc	USA	*623*	75-90000

461 20 70 Electronic warfare equipment, airborne, other types

AIL Systems Inc.	USA	*547*	75-02950
Amherst Systems Inc.	USA	*553*	75-06600
Chelton (Electrostatics) Limited	UK	*439*	75-12700
Chemring Countermeasures	UK	*439*	75-12800
Flight International Inc.	USA	*635*	80-11600
Israel Military Industries Ltd	Israel	*220*	75-10300
ITT Avionics	USA	*586*	75-49455
MCL Inc.	USA	*595*	75-59700
R. Alkan (UK) Ltd	UK	*429*	75-02680
Sanders	USA	*612*	75-77850
Tracor Inc	USA	*623*	75-90000

461 20 70 20 Chaff dispensing equipment, airborne

CelsiusTech Electronics AB	Sweden	*376*	75-03720
Chemring Countermeasures	UK	*439*	75-12800
Étienne Lacroix	France	*138*	75-24050
Flight International Inc.	USA	*635*	80-11600
FR Aviation Limited	UK	*503*	80-10850
GEC ALSTHOM Engineering Systems Ltd	UK	*453*	75-25225
Israel Military Industries Ltd	Israel	*220*	75-10300
ML Aviation Ltd	UK	*469*	75-40200
Northrop Grumman Corporation	USA	*599*	75-65450
R. Alkan (UK) Ltd	UK	*429*	75-02680
Southwest Aerospace	USA	*616*	75-82470
Tracor Inc	USA	*623*	75-90000
W. Vinten Ltd	UK	*491*	75-63353

461 20 70 30 Flare dispensing equipment, airborne

CelsiusTech Electronics AB	Sweden	*376*	75-03720
Étienne Lacroix	France	*138*	75-24050
Eurodef Aerospace & Defence	Greece	*189*	75-01295
Israel Military Industries Ltd	Israel	*220*	75-10300
R. Alkan (UK) Ltd	UK	*429*	75-02680
Tracor Inc	USA	*623*	75-90000
W. Vinten Ltd	UK	*491*	75-63353

461 20 80 Electronic warfare equipment accessories, airborne

British Sarozal Ltd.	UK	*436*	75-09570
Elettronica SpA	Italy	*233*	75-07050
Elisra Electronic Systems Ltd	Israel	*219*	75-06150
Emerson and Cuming Microwave	USA	*573*	75-31150
Hewlett-Packard (Canada) Ltd	Canada	*72*	75-07520
Howden Fluid Systems	USA	*582*	75-44400
H.R. Smith (Microwaves) Ltd	UK	*484*	75-54590
MA/COM Antenna & Cable Div.	USA	*593*	75-56953
Sanders	USA	*612*	75-77850
Technology Service Corp.	USA	*644*	80-31930
Tracor Inc	USA	*623*	75-90000
Trompeter Electronics Inc.	USA	*623*	75-91070

461 20 80 20 Expendables, air-launched EW

CelsiusTech Electronics AB	Sweden	*376*	75-03720
Chemring Countermeasures	UK	*439*	75-12800
Étienne Lacroix	France	*138*	75-24050
Fernau Avionics Ltd	UK	*450*	75-22600
FR Aviation Limited	UK	*503*	80-10850
GEC-Marconi Research Centre	UK	*419*	60-00700
INIMEX	**Belgium**	***46***	**75-04650**
Irvin Aerospace Limited	UK	*462*	75-32050
Israel Military Industries Ltd	Israel	*220*	75-10300
Israel Military Industries Rocket Systems Division	**Israel**	***220***	**75-10310**
ML Aviation Ltd	UK	*469*	75-40200
Pains-Wessex Ltd	UK	*472*	75-43510
Sanders	USA	*612*	75-77850
Technology Service Corp.	USA	*644*	80-31930
THORN Microwave Devices Ltd	UK	*488*	75-59760
Tracor Inc	USA	*623*	75-90000

461 20 80 22 Chaff, air-launched

CelsiusTech Electronics AB	Sweden	*376*	75-03720
Chemring Countermeasures	UK	*439*	75-12800
Compagnie Industriel des Reflecteurs	France	*134*	75-11450
Étienne Lacroix	France	*138*	75-24050
INIMEX	**Belgium**	***46***	**75-04650**
Israel Military Industries Ltd	Israel	*220*	75-10300

Product-Supplier Link

Israel Military Industries Rocket Systems Division	**Israel**	***220***	**75-10310**
NV Bekaert SA	Belgium	*45*	75-01300
Pains-Wessex Ltd	UK	*472*	75-43510
Tracor Inc	USA	*623*	75-90000

461 20 80 24 Flares, air-launched decoy

CelsiusTech Electronics AB	Sweden	*376*	75-03720
Étienne Lacroix	France	*138*	75-24050
INIMEX	**Belgium**	***46***	**75-04650**
Israel Military Industries Ltd	Israel	*220*	75-10300
Israel Military Industries Rocket Systems Division	**Israel**	***220***	**75-10310**
NV Bekaert SA	Belgium	*45*	75-01300
Pains-Wessex Ltd	UK	*472*	75-43510
Tracor Inc	USA	*623*	75-90000

461 30 Electronic warfare systems, naval shipborne

Alenia Difesa	**Italy**	***231***	**75-01165**
Alenia-Elsag	Italy	*230*	74-00100
Amper Programas SA	Spain	*365*	75-00480
Andrew Canada Inc.	Canada	*68*	75-00600
Applied Microwave Energy Inc.	USA	*554*	75-08165
Condor Systems Inc.	USA	*566*	75-20950
Dassault Electronique SA (Group)	**France**	***135***	**75-12980**
Elettronica SpA	Italy	*233*	75-07050
Elettronica (UK) Ltd	UK	*448*	75-20475
Elisra Electronic Systems Ltd	Israel	*219*	75-06150
ELTA Electronics Industries Ltd	Israel	*220*	75-10200
Étienne Lacroix	France	*138*	75-24050
ITT Aerospace/Communications Div.	USA	*586*	75-49455
ITT Avionics	USA	*586*	75-49455
Litton Industries Inc.	USA	*590*	75-55400
MCL Inc.	USA	*595*	75-59700
ML Aviation Ltd	UK	*469*	75-40200
ML Wallop Defence Systems Ltd	UK	*469*	75-40240
Northrop Grumman Corporation	USA	*640*	80-21840
Philips Traffic & Engineering	Australia	*20*	75-16550
Rafael	Israel	*216*	55-01800
Raytheon E-Systems	USA	*608*	75-73302
Rohde & Schwarz GmbH & Co KG	Germany	*174*	75-25400
Sanders	USA	*612*	75-77850
Siemens Aktiengesellschaft	Germany	*175*	75-27700
Siemens Plessey Systems	UK	*483*	75-53770
SRI International	USA	*643*	80-29600
Tadiran Ltd	Israel	*222*	75-21600
Technology Service Corp.	USA	*644*	80-31930
Tracor Inc	USA	*623*	75-90000

461 30 05 Electronic warfare suites, shipborne

CelsiusTech Electronics AB	Sweden	*376*	75-03720
Condor Systems Inc.	USA	*566*	75-20950
Elettronica SpA	Italy	*233*	75-07050
Elettronica (UK) Ltd	UK	*448*	75-20475
Elisra Electronic Systems Ltd	Israel	*219*	75-06150
ELTA Electronics Industries Ltd	Israel	*220*	75-10200
Israel Aircraft Industries Ltd	Israel	*219*	75-10200
PRB Associates Inc.	USA	*640*	80-24700
Racal Radar Defence Systems Ltd.	UK	*477*	75-47688
Rafael	Israel	*216*	55-01800
Technology Service Corp.	USA	*644*	80-31930
Tracor Inc	USA	*623*	75-90000

461 30 10 Electronic warfare systems, naval, active

Daimler-Benz Aerospace France SAS	**France**	***134***	**75-12950**
EDO Corporation	USA	*571*	75-29250
Elettronica SpA	Italy	*233*	75-07050
Elettronica (UK) Ltd	UK	*448*	75-20475
Elisra Electronic Systems Ltd	Israel	*219*	75-06150
ELTA Electronics Industries Ltd	Israel	*220*	75-10200
Racal Communications Inc.	USA	*606*	75-72380
Racal Radar Defence Systems Ltd.	UK	*477*	75-47688
Sanders	USA	*612*	75-77850
SRI International	USA	*643*	80-29600
Technology Service Corp.	USA	*644*	80-31930
Thomson Sintra	France	*148*	75-45135

461 30 11 Acoustic countermeasures systems, naval shipborne

AlliedSignal Ltd	UK	*430*	75-02747
EDO Corporation	USA	*571*	75-29250
NEC Corporation	Japan	*253*	75-06350
Signature Management	UK	*484*	75-53790
Technology Service Corp.	USA	*644*	80-31930
Thomson Sintra	France	*148*	75-45135

461 30 11 60 Acoustic countermeasures systems, submarine launched

Associated Aircraft Mfg & Sales Inc.	USA	*555*	75-09500
EDO Corporation	USA	*571*	75-29250
Rafael	Israel	*216*	55-01800
Signature Management	UK	*484*	75-53790
Technology Service Corp.	USA	*644*	80-31930
Thomson Sintra	France	*148*	75-45135

461 30 13 Electronic countermeasures (ECM) systems, naval shipborne

Amper Programas SA	Spain	*365*	75-00480
Bharat Electronics Ltd	India	*203*	55-00200
CelsiusTech Electronics AB	Sweden	*376*	75-03720
Daimler-Benz Aerospace France SAS	**France**	***134***	**75-12950**
Elettronica SpA	Italy	*233*	75-07050
Elettronica (UK) Ltd	UK	*448*	75-20475
Elisra Electronic Systems Ltd	Israel	*219*	75-06150
ELTA Electronics Industries Ltd	Israel	*220*	75-10200
Israel Aircraft Industries Ltd	Israel	*219*	75-10200
Litton Industries Inc.	USA	*591*	75-55400
Racal Radar Defence Systems Ltd.	UK	*477*	75-47688
Saab Dynamics AB	Sweden	*378*	75-09100
Sanders	USA	*612*	75-77850
Siemens Plessey Systems	UK	*483*	75-53770
Technology Service Corp.	USA	*644*	80-31930
Znamya Production Group	Ukraine	*407*	75-10000

461 30 17 Electronic warfare equipment, naval, active

Daimler-Benz Aerospace France SAS	**France**	***134***	**75-12950**
Dassault Electronique SA (Group)	**France**	***135***	**75-12980**
Elettronica SpA	Italy	*233*	75-07050
Elettronica (UK) Ltd	UK	*448*	75-20475
Elisra Electronic Systems Ltd	Israel	*219*	75-06150
ELTA Electronics Industries Ltd	Israel	*220*	75-10200
Litton Industries Inc.	USA	*590*	75-55400
Racal Radar Defence Systems Ltd.	UK	*477*	75-47688
Sanders	USA	*612*	75-77850
Siemens Aktiengesellschaft	Germany	*175*	75-27700
Siemens Plessey Systems	UK	*483*	75-53770
Technology Service Corp.	USA	*644*	80-31930

461 30 17 60 Transmitting / Jamming / Deception equipment, naval

Amper Programas SA	Spain	*365*	75-00480
Cincinnati Electronics Corp.	USA	*564*	75-18640
Circuitwise Electronics Ltd.	UK	*440*	75-13300
Continental Electronics Corp.	USA	*566*	75-21800
Elettronica SpA	Italy	*233*	75-07050
Elettronica (UK) Ltd	UK	*448*	75-20475
Elisra Electronic Systems Ltd	Israel	*219*	75-06150
ELTA Electronics Industries Ltd	Israel	*220*	75-10200
Grinaker System Technologies	S. Africa	*360*	80-06500
Israel Aircraft Industries Ltd	Israel	*219*	75-10200
MA/COM Antenna & Cable Div.	USA	*593*	75-56953
Racal Communications Inc.	USA	*606*	75-72380
Racal Radar Defence Systems Ltd.	UK	*477*	75-47688
Sanders	USA	*612*	75-77850
SRI International	USA	*643*	80-29600
Technology Service Corp.	USA	*644*	80-31930

461 30 17 60 40 Jammers, shipborne ECM

Amper Programas SA	Spain	*365*	75-00480

Bharat Electronics Ltd	India	*203*	55-00200
Cincinnati Electronics Corp.	USA	*564*	75-18640
Circuitwise Electronics Ltd	UK	*440*	75-13300
Daimler-Benz Aerospace France SAS	**France**	***134***	**75-12950**
Elettronica SpA	Italy	*233*	75-07050
Elettronica (UK) Ltd	UK	*448*	75-20475
Elisra Electronic Systems Ltd	Israel	*219*	75-06150
ELTA Electronics Industries Ltd	Israel	*220*	75-10200
Israel Aircraft Industries Ltd	Israel	*219*	75-10200
Noise Com Inc.	USA	*599*	75-64760
Racal Radar Defence Systems Ltd.	UK	*477*	75-47688
Sanders	USA	*612*	75-77850
Siemens Aktiengesellschaft	Germany	*175*	75-27700
SRI International	USA	*643*	80-29600
Technology Service Corp.	USA	*644*	80-31930

461 30 17 60 60 Transmitters, naval ECM

Cincinnati Electronics Corp.	USA	*564*	75-18640
Daimler-Benz Aerospace France SAS	**France**	***134***	**75-12950**
Elettronica SpA	Italy	*233*	75-07050
Elettronica (UK) Ltd	UK	*448*	75-20475
Elisra Electronic Systems Ltd	Israel	*219*	75-06150
ELTA Electronics Industries Ltd	Israel	*220*	75-10200
Israel Aircraft Industries Ltd	Israel	*219*	75-10200
Racal Radar Defence Systems Ltd.	UK	*477*	75-47688
Sanders	USA	*612*	75-77850
Technology Service Corp.	USA	*644*	80-31930

461 30 20 Electronic warfare systems, naval, passive

Amper Programas SA	Spain	*365*	75-00480
Andrew Canada Inc.	Canada	*68*	75-00600
Calspan SRL Technology	USA	*540*	60-03000
CEA Technologies Pty Ltd	Australia	*24*	80-03100
Condor Systems Inc.	USA	*566*	75-20950
Cubic Defense Systems Inc.	USA	*567*	75-22900
Daimler-Benz Aerospace France SAS	**France**	***134***	**75-12950**
EDO Corporation	USA	*571*	75-29250
Elettronica SpA	Italy	*233*	75-07050
Elettronica (UK) Ltd	UK	*448*	75-20475
Elisra Electronic Systems Ltd	Israel	*219*	75-06150
ELTA Electronics Industries Ltd	Israel	*220*	75-10200
Grinaker System Technologies	S. Africa	*360*	80-06500
Israel Military Industries Ltd	Israel	*220*	75-10300
Litton Industries Inc.	USA	*591*	75-55400
Racal Radar Defence Systems Ltd.	UK	*477*	75-47688
Sanders	USA	*612*	75-77850
Siemens Plessey Systems	UK	*483*	75-53770
Tadiran Ltd	Israel	*222*	75-21600
Technology Service Corp.	USA	*644*	80-31930
Tracor Inc	USA	*623*	75-90000

461 30 20 30 Electronic intelligence systems, naval, passive

Amper Programas SA	Spain	*365*	75-00480
Andrew Canada Inc.	Canada	*68*	75-00600
Condor Systems Inc.	USA	*566*	75-20950
Cubic Defense Systems Inc.	USA	*567*	75-22900
Daimler-Benz Aerospace France SAS	**France**	***134***	**75-12950**
Elettronica SpA	Italy	*233*	75-07050
Elettronica (UK) Ltd	UK	*448*	75-20475
Elisra Electronic Systems Ltd	Israel	*219*	75-06150
ELTA Electronics Industries Ltd	Israel	*220*	75-10200
MA/COM Antenna & Cable Div.	USA	*593*	75-56953
Rohde & Schwarz GmbH & Co KG.	Germany	*174*	75-25400
Sanders	USA	*612*	75-77850
Siemens Plessey Systems	UK	*483*	75-53770
State College Operations	USA	*617*	75-83990
Tadiran Ltd	Israel	*222*	75-21600
Technology Service Corp.	USA	*644*	80-31930
Watkins-Johnson Co.	USA	*627*	75-95200

461 30 20 32 Communications intelligence systems, shipborne, passive

Amper Programas SA	Spain	*365*	75-00480
Andrew Canada Inc.	Canada	*68*	75-00600
Condor Systems Inc.	USA	*566*	75-20950
Daimler-Benz Aerospace France SAS	**France**	***134***	**75-12950**
Elettronica SpA	Italy	*233*	75-07050
Elettronica (UK) Ltd	UK	*448*	75-20475
ELTA Electronics Industries Ltd	Israel	*220*	75-10200
Israel Aircraft Industries Ltd	Israel	*219*	75-10200
MA/COM Antenna & Cable Div.	USA	*593*	75-56953
Rohde & Schwarz GmbH & Co KG.	Germany	*174*	75-25400
Sanders	USA	*612*	75-77850
State College Operations	USA	*617*	75-83990
Tadiran Ltd	Israel	*222*	75-21600
Technology Service Corp.	USA	*644*	80-31930
Watkins-Johnson Co.	USA	*627*	75-95200

461 30 22 Electronic support measures (ESM) systems, naval, shipborne

Amper Programas SA	Spain	*365*	75-00480
Andrew Canada Inc.	Canada	*68*	75-00600
Elettronica SpA	Italy	*233*	75-07050
Elettronica (UK) Ltd	UK	*448*	75-20475
Elisra Electronic Systems Ltd	Israel	*219*	75-06150
ELTA Electronics Industries Ltd	Israel	*220*	75-10200
EW Simulation Technology Ltd	UK	*449*	75-22150
Israel Aircraft Industries Ltd	Israel	*219*	75-10200
Litton Industries Inc.	USA	*591*	75-55400
Lockheed Martin Canada Inc.	Canada	*73*	75-09110
NEC Corporation	Japan	*253*	75-06350
Racal Radar Defence Systems Ltd.	UK	*477*	75-47688
Rohde & Schwarz GmbH & Co KG.	Germany	*174*	75-25400
Sanders	USA	*612*	75-77850
Siemens Aktiengesellschaft	Germany	*175*	75-27700
Siemens Plessey Systems	UK	*483*	75-53770
State College Operations	USA	*617*	75-83990
Tadiran Ltd	Israel	*222*	75-21600
Technology Service Corp.	USA	*644*	80-31930
Watkins-Johnson Co.	USA	*627*	75-95200

461 30 24 Infra-red search & warning systems, shipborne, passive

SAGEM SA	**France**	***143***	**75-33950**
Technology Service Corp.	USA	*644*	80-31930

461 30 27 Electronic warfare equipment, naval, passive

Andrew Canada Inc.	Canada	*68*	75-00600
CEA Technologies Pty Ltd	Australia	*24*	80-03100
Condor Systems Inc.	USA	*566*	75-20950
Daimler-Benz Aerospace France SAS	**France**	***134***	**75-12950**
Elettronica SpA	Italy	*233*	75-07050
Elettronica (UK) Ltd	UK	*448*	75-20475
Elisra Electronic Systems Ltd	Israel	*219*	75-06150
ELTA Electronics Industries Ltd	Israel	*220*	75-10200
Howden Fluid Systems	USA	*582*	75-44400
KEL Aerospace Pty Ltd	Australia	*18*	75-11440
Laser & Atomic Research & Devt	Croatia	*94*	60-00450
Litton Industries Inc.	USA	*590*	75-55400
Racal Communications Inc.	USA	*606*	75-72380
Racal Radar Defence Systems Ltd.	UK	*477*	75-47688
Rohde & Schwarz GmbH & Co KG.	Germany	*174*	75-25400
Sanders	USA	*612*	75-77850
Siemens Plessey Systems	UK	*483*	75-53770
Sperry Marine Inc.	USA	*617*	75-83250
Technology Service Corp.	USA	*644*	80-31930
Watkins-Johnson Co.	USA	*627*	75-95200

461 30 27 30 Direction Finding / Homing equipment, shipborne EW

Andrew Canada Inc.	Canada	*68*	75-00600
Condor Systems Inc.	USA	*566*	75-20950
Elettronica SpA	Italy	*233*	75-07050
Elisra Electronic Systems Ltd	Israel	*219*	75-06150

Product-Supplier Link

Études Electroniques Informatiques	France	*150*	80-00540
EW Simulation Technology Ltd	UK	*449*	75-22150
Israel Aircraft Industries Ltd	Israel	*219*	75-10200
Litton Industries Inc.	USA	*591*	75-55400
MA/COM Antenna & Cable Div.	USA	*593*	75-56953
Racal Communications Inc.	USA	*606*	75-72380
Rohde & Schwarz GmbH & Co KG	Germany	*174*	75-25400
Sanders	USA	*612*	75-77850
Sperry Marine Inc.	USA	*617*	75-83250
Tadiran Ltd	Israel	*222*	75-21600
Watkins-Johnson Co.	USA	*627*	75-95200

461 30 27 50 Receiving / Detecting / Warning equipment, shipborne EW

Andrew Canada Inc.	Canada	*68*	75-00600
Condor Systems Inc.	USA	*566*	75-20950
Elettronica SpA	Italy	*233*	75-07050
Elisra Electronic Systems Ltd	Israel	*219*	75-06150
ELTA Electronics Industries Ltd	Israel	*220*	75-10200
EW Simulation Technology Ltd	UK	*449*	75-22150
GEC-Marconi Avionics Ltd	UK	*454*	75-25320
Israel Aircraft Industries Ltd	Israel	*219*	75-10200
Litton Industries Inc.	USA	*591*	75-55400
MA/COM Antenna & Cable Div.	USA	*593*	75-56953
NEC Corporation	Japan	*253*	75-06350
Siemens Aktiengesellschaft	Germany	*175*	75-27700
Sperry Marine Inc.	USA	*617*	75-83250
Watkins-Johnson Co.	USA	*627*	75-95200

461 30 27 55 Receiving equipment, shipborne EW

Andrew Canada Inc.	Canada	*68*	75-00600
Condor Systems Inc.	USA	*566*	75-20950
Daimler-Benz Aerospace France SAS	**France**	***134***	**75-12950**
Elettronica SpA	Italy	*233*	75-07050
Elettronica (UK) Ltd	UK	*448*	75-20475
Elisra Electronic Systems Ltd	Israel	*219*	75-06150
ELTA Electronics Industries Ltd	Israel	*220*	75-10200
EW Simulation Technology Ltd	UK	*449*	75-22150
Israel Aircraft Industries Ltd	Israel	*219*	75-10200
KEL Aerospace Pty Ltd	Australia	*18*	75-11440
Litton Industries Inc.	USA	*591*	75-55400
MA/COM Antenna & Cable Div.	USA	*593*	75-56953
NEC Corporation	Japan	*253*	75-06350
Racal Communications Inc.	USA	*606*	75-72380
Racal Radar Defence Systems Ltd.	UK	*477*	75-47688
Rohde & Schwarz GmbH & Co KG	Germany	*174*	75-25400
Sanders	USA	*612*	75-77850
Sperry Marine Inc.	USA	*617*	75-83250
Watkins-Johnson Co.	USA	*627*	75-95200

461 30 27 55 32 Receivers, communications intelligence, shipborne

Amper Programas SA	Spain	*365*	75-00480
Andrew Canada Inc.	Canada	*68*	75-00600
Condor Systems Inc.	USA	*566*	75-20950
Daimler-Benz Aerospace France SAS	**France**	***134***	**75-12950**
Elettronica SpA	Italy	*233*	75-07050
Elettronica (UK) Ltd	UK	*448*	75-20475
ELTA Electronics Industries Ltd	Israel	*220*	75-10200
KEL Aerospace Pty Ltd	Australia	*18*	75-11440
MA/COM Antenna & Cable Div.	USA	*593*	75-56953
Racal Communications Inc.	USA	*606*	75-72380
Rohde & Schwarz GmbH & Co KG	Germany	*174*	75-25400
Sanders	USA	*612*	75-77850
Sperry Marine Inc.	USA	*617*	75-83250
Tadiran Ltd	Israel	*222*	75-21600
Watkins-Johnson Co.	USA	*627*	75-95200

461 30 27 55 40 Receivers, infra-red warning, naval shipborne

Elettronica (UK) Ltd	UK	*448*	75-20475
Laser & Atomic Research & Devt	Croatia	*94*	60-00450

461 30 70 Electronic warfare equipment, naval, other types

Alenia Difesa	**Italy**	***231***	**75-01165**
Daimler-Benz Aerospace France SAS	**France**	***134***	**75-12950**
Elisra Electronic Systems Ltd	Israel	*219*	75-06150
Institut Dr Förster GmbH & Co KG	Germany	*170*	75-14500
ITT Avionics	USA	*586*	75-49455
MCL Inc.	USA	*595*	75-59700
Rafael	Israel	*216*	55-01800
Raytheon E-Systems	USA	*608*	75-73302
Sperry Marine Inc.	USA	*617*	75-83250
Sperry Marine SpA	Italy	*239*	75-24400
Tracor Inc	USA	*623*	75-90000
Vickers Shipbuilding & Eng Ltd	UK	*491*	75-63100

461 30 70 20 Chaff dispensing systems, naval shipborne

Alenia Difesa	**Italy**	***231***	**75-01165**
CelsiusTech Electronics AB	Sweden	*376*	75-03720
Daimler-Benz Aerospace France SAS	**France**	***134***	**75-12950**
Étienne Lacroix	France	*138*	75-24050
Rafael	Israel	*216*	55-01800
Tracor Inc	USA	*623*	75-90000
Vickers Shipbuilding & Eng Ltd	UK	*491*	75-63100

461 30 70 30 Flare dispensing equipment, shipborne

Alenia Difesa	**Italy**	***231***	**75-01165**
CelsiusTech Electronics AB	Sweden	*376*	75-03720
Étienne Lacroix	France	*138*	75-24050
Tracor Inc	USA	*623*	75-90000

461 30 70 50 Launchers, acoustic countermeasures

Strachan & Henshaw Ltd	UK	*512*	80-47400

461 30 80 Electronic warfare equipment accessories, naval

Elisra Electronic Systems Ltd	Israel	*219*	75-06150
General Electronics Communications	Greece	*189*	75-01300
Hewlett-Packard (Canada) Ltd	Canada	*72*	75-07520
H.R. Smith (Microwaves) Ltd	UK	*484*	75-54590
Institut Dr Förster GmbH & Co KG	Germany	*170*	75-14500
Israel Military Industries Ltd	Israel	*220*	75-10300
MA/COM Antenna & Cable Div.	USA	*593*	75-56953
Radamec Defence Systems Ltd	UK	*477*	75-47750
Raytheon E-Systems	USA	*608*	75-73302
Sanders	USA	*612*	75-77850
Tracor Inc	USA	*623*	75-90000
Trompeter Electronics Inc.	USA	*623*	75-91070

461 30 80 20 Expendables, naval EW

Israel Military Industries Ltd	Israel	*220*	75-10300
Sanders	USA	*612*	75-77850
Tracor Inc	USA	*623*	75-90000

461 30 80 22 Chaff, naval

Compagnie Industriel des Reflecteurs	France	*134*	75-11450
Étienne Lacroix	France	*138*	75-24050
Israel Military Industries Ltd	Israel	*220*	75-10300
Israel Military Industries Rocket Systems Division	**Israel**	***220***	**75-10310**
Pains-Wessex Ltd	UK	*472*	75-43510
Tracor Inc	USA	*623*	75-90000

461 30 80 24 Flares, ship-launched decoy

Étienne Lacroix	France	*138*	75-24050
Israel Military Industries Ltd	Israel	*220*	75-10300
Israel Military Industries Rocket Systems Division	**Israel**	***220***	**75-10310**

Pains-Wessex Ltd	UK	*472*	75-43510
Tracor Inc	USA	*623*	75-90000

461 39 Electronic support measures (ESM) systems, submarine

Daimler-Benz Aerospace France SAS	**France**	***134***	**75-12950**
Elettronica SpA	Italy	*233*	75-07050
Elettronica (UK) Ltd	UK	*448*	75-20475
Elisra Electronic Systems Ltd	Israel	*219*	75-06150
Litton Industries Inc.	USA	*591*	75-55400
Racal Radar Defence Systems Ltd.	UK	*477*	75-47688
Sperry Marine Inc.	USA	*617*	75-83250
Sperry Marine SpA	Italy	*239*	75-24400
Watkins-Johnson Co.	USA	*627*	75-95200

461 40 Electronic warfare systems, ground-based

AIL Systems Inc.	USA	*547*	75-02950
Amper Programas SA	Spain	*365*	75-00480
Andrew Canada Inc.	Canada	*68*	75-00600
Condor Systems Inc.	USA	*566*	75-20950
Elettronica SpA	Italy	*233*	75-07050
Elettronica (UK) Ltd	UK	*448*	75-20475
Elisra Electronic Systems Ltd	Israel	*219*	75-06150
ELTA Electronics Industries Ltd	Israel	*220*	75-10200
Israel Aircraft Industries Ltd	Israel	*219*	75-10200
ITT Avionics	USA	*586*	75-49455
Litton Industries Inc.	USA	*590*	75-55400
Logica UK Limited	UK	*506*	80-14350
MA/COM Antenna & Cable Div.	USA	*593*	75-56953
MCL Inc.	USA	*595*	75-59700
Northrop Grumman Corporation	USA	*640*	80-21840
P W Allen & Co Ltd	UK	*429*	75-02740
Racal Radar Defence Systems Ltd.	UK	*477*	75-47688
Racal Radio Ltd	UK	*477*	75-47698
Radio Research Instrument Co. Inc.	USA	*607*	75-72600
Raytheon E-Systems	USA	*608*	75-73302
Rohde & Schwarz GmbH & Co KG.	Germany	*174*	75-25400
Sanders	USA	*612*	75-77850
Siemens Plessey Systems	UK	*483*	75-53770
Sierra Technologies Inc.	USA	*614*	75-80640
Sperry Marine Inc.	USA	*617*	75-83250
SRI International	USA	*643*	80-29600
Tadiran Ltd	Israel	*222*	75-21600
Teledyne Electronic Technologies	USA	*621*	75-87534
Tracor Inc	USA	*623*	75-90000
Zeta	USA	*629*	75-98190

461 40 10 Electronic warfare systems, ground-based, active

Amper Programas SA	Spain	*365*	75-00480
Daimler-Benz Aerospace France SAS	**France**	***134***	**75-12950**
Elettronica SpA	Italy	*233*	75-07050
Elisra Electronic Systems Ltd	Israel	*219*	75-06150
ELTA Electronics Industries Ltd	Israel	*220*	75-10200
Israel Aircraft Industries Ltd	Israel	*219*	75-10200
MA/COM Antenna & Cable Div.	USA	*593*	75-56953
Northrop Grumman Corporation	USA	*640*	80-21840
Racal Radar Defence Systems Ltd.	UK	*477*	75-47688
Rohde & Schwarz GmbH & Co KG.	Germany	*174*	75-25400
Sanders	USA	*612*	75-77850
Siemens Plessey Systems	UK	*483*	75-53770
Sperry Marine Inc.	USA	*617*	75-83250
Tadiran Ltd	Israel	*222*	75-21600

461 40 13 Electronic countermeasures systems, ground-based

Cincinnati Electronics Corp.	USA	*564*	75-18640
Daimler-Benz Aerospace France SAS	**France**	***134***	**75-12950**
Elettronica SpA	Italy	*233*	75-07050
Elettronica (UK) Ltd	UK	*448*	75-20475
Elisra Electronic Systems Ltd	Israel	*219*	75-06150
ELTA Electronics Industries Ltd	Israel	*220*	75-10200
Israel Aircraft Industries Ltd	Israel	*219*	75-10200
Racal Radar Defence Systems Ltd.	UK	*477*	75-47688
Racal Radio Ltd	UK	*477*	75-47698
Rohde & Schwarz GmbH & Co KG.	Germany	*174*	75-25400
Sanders	USA	*612*	75-77850
Siemens Plessey Systems	UK	*483*	75-53770
Sperry Marine Inc.	USA	*617*	75-83250
Tadiran Ltd	Israel	*222*	75-21600
Znamya Production Group	Ukraine	*407*	75-10000

461 40 17 Electronic warfare equipment, ground-based, active

Amper Programas SA	Spain	*365*	75-00480
Cincinnati Electronics Corp.	USA	*564*	75-18640
Daimler-Benz Aerospace France SAS	**France**	***134***	**75-12950**
Elettronica SpA	Italy	*233*	75-07050
Elettronica (UK) Ltd	UK	*448*	75-20475
Elisra Electronic Systems Ltd	Israel	*219*	75-06150
ELTA Electronics Industries Ltd	Israel	*220*	75-10200
Howden Fluid Systems	USA	*582*	75-44400
Israel Aircraft Industries Ltd	Israel	*219*	75-10200
KEL Aerospace Pty Ltd	Australia	*18*	75-11440
Litton Industries Inc.	USA	*590*	75-55400
Northrop Grumman Corporation	USA	*640*	80-21840
Racal Radar Defence Systems Ltd.	UK	*477*	75-47688
Radio Research Instrument Co. Inc.	USA	*607*	75-72600
Rohde & Schwarz GmbH & Co KG.	Germany	*174*	75-25400
Sanders	USA	*612*	75-77850
Siemens Plessey Systems	UK	*483*	75-53770
Tadiran Ltd	Israel	*222*	75-21600

461 40 17 60 Transmitting / Jamming / Deception equipment, ground-based

Bharat Electronics Ltd	India	*203*	55-00200
Chrysanthacopoulos SA	Greece	*188*	75-00910
Cincinnati Electronics Corp.	USA	*564*	75-18640
Circuitwise Electronics Ltd	UK	*440*	75-13300
Continental Electronics Corp.	USA	*566*	75-21800
Daimler-Benz Aerospace France SAS	**France**	***134***	**75-12950**
Elettronica SpA	Italy	*233*	75-07050
Elettronica (UK) Ltd	UK	*448*	75-20475
ELTA Electronics Industries Ltd	Israel	*220*	75-10200
Fernau Avionics Ltd	UK	*450*	75-22600
Grinaker System Technologies	S. Africa	*360*	80-06500
Israel Aircraft Industries Ltd	Israel	*219*	75-10200
P W Allen & Co Ltd	UK	*429*	75-02740
Sanders	USA	*612*	75-77850
Siemens Plessey Systems	UK	*483*	75-53770
STAREC	France	*147*	75-42775
Tadiran Ltd	Israel	*222*	75-21600

461 40 17 60 40 Jammers, ground-based ECM

Amper Programas SA	Spain	*365*	75-00480
Chrysanthacopoulos SA	Greece	*188*	75-00910
Cincinnati Electronics Corp.	USA	*564*	75-18640
Circuitwise Electronics Ltd	UK	*440*	75-13300
Daimler-Benz Aerospace France SAS	**France**	***134***	**75-12950**
Elettronica SpA	Italy	*233*	75-07050
Elettronica (UK) Ltd	UK	*448*	75-20475
ELTA Electronics Industries Ltd	Israel	*220*	75-10200
Fernau Avionics Ltd	UK	*450*	75-22600
Israel Aircraft Industries Ltd	Israel	*219*	75-10200
Noise Com Inc.	USA	*599*	75-64760
Racal Radar Defence Systems Ltd.	UK	*477*	75-47688
Racal Radio Ltd	UK	*477*	75-47698
Rohde & Schwarz GmbH & Co KG.	Germany	*174*	75-25400
Sanders	USA	*612*	75-77850
Siemens Plessey Systems	UK	*483*	75-53770
Tadiran Ltd	Israel	*222*	75-21600
Thomson-CSF	France	*147*	75-45100

461 40 17 60 60 Transmitters, ground-based ECM

APT Radar Systems Ltd	UK	*431*	75-03560
Cincinnati Electronics Corp.	USA	*564*	75-18640

Daimler-Benz Aerospace France SAS	**France**	***134***	**75-12950**
Elettronica SpA	Italy	*233*	75-07050
Elettronica (UK) Ltd	UK	*448*	75-20475
Elisra Electronic Systems Ltd	Israel	*219*	75-06150
ELTA Electronics Industries Ltd	Israel	*220*	75-10200
Israel Aircraft Industries Ltd	Israel	*219*	75-10200
KEL Aerospace Pty Ltd	Australia	*18*	75-11440
Radio Research Instrument Co. Inc.	USA	*607*	75-72600
Sanders	USA	*612*	75-77850
Siemens Plessey Systems	UK	*483*	75-53770
Tadiran Ltd	Israel	*222*	75-21600
Technology Service Corp.	USA	*644*	80-31930

461 40 20 Electronic warfare systems, ground-based, passive

Amper Programas SA	Spain	*365*	75-00480
Andrew Canada Inc.	Canada	*68*	75-00600
Cincinnati Electronics Corp.	USA	*564*	75-18640
Com Dev Europe	UK	*441*	75-14300
Comptek Federal Systems Inc.	USA	*565*	75-20650
Condor Systems Inc.	USA	*566*	75-20950
Elettronica SpA	Italy	*233*	75-07050
Elettronica (UK) Ltd	UK	*448*	75-20475
Elisra Electronic Systems Ltd	Israel	*219*	75-06150
ELTA Electronics Industries Ltd	Israel	*220*	75-10200
Grinaker System Technologies	S. Africa	*360*	80-06500
Israel Aircraft Industries Ltd	Israel	*219*	75-10200
Litton Industries Inc.	USA	*591*	75-55400
Racal Radar Defence Systems Ltd.	UK	*477*	75-47688
Racal Radio Ltd	UK	*477*	75-47698
Rohde & Schwarz GmbH & Co KG.	Germany	*174*	75-25400
Sanders	USA	*612*	75-77850
Siemens Aktiengesellschaft	Germany	*175*	75-27700
Siemens Plessey Systems	UK	*483*	75-53770
Tadiran Ltd	Israel	*222*	75-21600
Watkins-Johnson Co.	USA	*627*	75-95200

461 40 20 30 Electronic intelligence systems, ground-based, passive

Amper Programas SA	Spain	*365*	75-00480
Andrew Canada Inc.	Canada	*68*	75-00600
Comptek Federal Systems Inc.	USA	*565*	75-20650
Condor Systems Inc.	USA	*566*	75-20950
Contraves Inc	USA	*566*	75-21850
Elettronica SpA	Italy	*233*	75-07050
Elisra Electronic Systems Ltd	Israel	*219*	75-06150
ELTA Electronics Industries Ltd	Israel	*220*	75-10200
Litton Industries Inc.	USA	*591*	75-55400
Racal Communications Inc.	USA	*606*	75-72380
Racal Radar Defence Systems Ltd.	UK	*477*	75-47688
Rohde & Schwarz GmbH & Co KG.	Germany	*174*	75-25400
Sanders	USA	*612*	75-77850
Siemens Plessey Systems	UK	*483*	75-53770
State College Operations	USA	*617*	75-83990
Tadiran Ltd	Israel	*222*	75-21600
Technology Service Corp.	USA	*644*	80-31930
Tracor Inc	USA	*623*	75-90000
Watkins-Johnson Co.	USA	*627*	75-95200

461 40 20 32 Communications intelligence systems, ground-based, passive

Amper Programas SA	Spain	*365*	75-00480
Andrew Canada Inc.	Canada	*68*	75-00600
Comptek Federal Systems Inc.	USA	*565*	75-20650
Condor Systems Inc.	USA	*566*	75-20950
Davies Industrial Communications Ltd	UK	*444*	75-17010
Elettronica SpA	Italy	*233*	75-07050
Elisra Electronic Systems Ltd	Israel	*219*	75-06150
ELTA Electronics Industries Ltd	Israel	*220*	75-10200
Israel Aircraft Industries Ltd	Israel	*219*	75-10200
Litton Industries Inc.	USA	*591*	75-55400
MA/COM Antenna & Cable Div.	USA	*593*	75-56953
NEC Corporation	Japan	*253*	75-06350
Racal Communications Inc.	USA	*606*	75-72380
Racal Radio Ltd	UK	*477*	75-47698
Rohde & Schwarz GmbH & Co KG.	Germany	*174*	75-25400
Sanders	USA	*612*	75-77850
Tadiran Ltd	Israel	*222*	75-21600
Technology Service Corp.	USA	*644*	80-31930
Tracor Inc	USA	*623*	75-90000
Watkins-Johnson Co.	USA	*627*	75-95200

461 40 20 36 Radar identification / location systems, ground-based

Andrew Canada Inc.	Canada	*68*	75-00600
APT Radar Systems Ltd	UK	*431*	75-03560
British Sarozal Ltd.	UK	*436*	75-09570
Condor Systems Inc.	USA	*566*	75-20950
Elettronica SpA	Italy	*233*	75-07050
Elettronica (UK) Ltd	UK	*448*	75-20475
Elisra Electronic Systems Ltd	Israel	*219*	75-06150
ELTA Electronics Industries Ltd	Israel	*220*	75-10200
IDS Ingegneria dei Sistemi SpA	Italy	*240*	80-00920
Racal Radar Defence Systems Ltd.	UK	*477*	75-47688
Sanders	USA	*612*	75-77850
Siemens Plessey Systems	UK	*483*	75-53770
Technology Service Corp.	USA	*644*	80-31930
Watkins-Johnson Co.	USA	*627*	75-95200
Znamya Production Group	Ukraine	*407*	75-10000

461 40 22 Electronic support measures (ESM) systems, ground-based

Amper Programas SA	Spain	*365*	75-00480
Andrew Canada Inc.	Canada	*68*	75-00600
Chrysanthacopoulos SA	Greece	*188*	75-00910
Elettronica SpA	Italy	*233*	75-07050
Elettronica (UK) Ltd	UK	*448*	75-20475
Elisra Electronic Systems Ltd	Israel	*219*	75-06150
ELTA Electronics Industries Ltd	Israel	*220*	75-10200
EW Simulation Technology Ltd	UK	*449*	75-22150
Racal Communications Inc.	USA	*606*	75-72380
Racal Radar Defence Systems Ltd.	UK	*477*	75-47688
Racal Radio Ltd	UK	*477*	75-47698
Rohde & Schwarz GmbH & Co KG.	Germany	*174*	75-25400
Sanders	USA	*612*	75-77850
Siemens Plessey Systems	UK	*483*	75-53770
State College Operations	USA	*617*	75-83990
Tadiran Ltd	Israel	*222*	75-21600
Technology Service Corp.	USA	*644*	80-31930
Watkins-Johnson Co.	USA	*627*	75-95200

461 40 24 Infra-red search & warning systems, ground-based, passive

Delft Sensor Systems	**Belgium**	***46***	**75-02500**
Delft Sensor Systems	Netherlands	*292*	75-01590
ITT Defense Ltd.	UK	*505*	80-13570
Moked Engineering (1969) Ltd	Israel	*221*	75-15200
Pulnix America Inc.	USA	*606*	75-71815
SAGEM SA	**France**	***143***	**75-33950**
Technology Service Corp.	USA	*644*	80-31930

461 40 27 Electronic warfare equipment, ground-based, passive

Amherst Systems Inc.	USA	*553*	75-06600
Andrew Canada Inc.	Canada	*68*	75-00600
CEA Technologies Pty Ltd	Australia	*24*	80-03100
Com Dev Europe	UK	*441*	75-14300
Comptek Federal Systems Inc.	USA	*565*	75-20650
Condor Systems Inc.	USA	*566*	75-20950
Elettronica SpA	Italy	*233*	75-07050
Elettronica (UK) Ltd	UK	*448*	75-20475
Elisra Electronic Systems Ltd	Israel	*219*	75-06150
ELTA Electronics Industries Ltd	Israel	*220*	75-10200
GEC-Marconi Avionics Ltd	UK	*454*	75-25320
KEL Aerospace Pty Ltd	Australia	*18*	75-11440
Laser & Atomic Research & Devt	Croatia	*94*	60-00450
Litton Industries Inc.	USA	*590*	75-55400
MA/COM Antenna & Cable Div.	USA	*593*	75-56953
Racal Radar Defence Systems Ltd.	UK	*477*	75-47688
Rohde & Schwarz GmbH & Co KG.	Germany	*174*	75-25400
Sanders	USA	*612*	75-77850
Siemens Plessey Systems	UK	*483*	75-53770
Watkins-Johnson Co.	USA	*627*	75-95200

461 40 27 30 Direction Finding / Homing equipment, ground-based, passive EW

Andrew Canada Inc.	Canada	68	75-00600
Condor Systems Inc.	USA	566	75-20950
Elettronica SpA	Italy	233	75-07050
Elettronica (UK) Ltd	UK	448	75-20475
Elisra Electronic Systems Ltd	Israel	219	75-06150
ELTA Electronics Industries Ltd	Israel	220	75-10200
Études Electroniques Informatiques	France	150	80-00540
EW Simulation Technology Ltd	UK	449	75-22150
Israel Aircraft Industries Ltd	Israel	219	75-10200
MA/COM Antenna & Cable Div.	USA	593	75-56953
Racal Communications Inc.	USA	606	75-72380
Racal Radio Ltd	UK	477	75-47698
Rohde & Schwarz Canada Inc	Canada	79	80-01825
Rohde & Schwarz GmbH & Co KG	Germany	174	75-25400
Sanders	USA	612	75-77850
Siemens Plessey Systems	UK	483	75-53770
STAREC	France	147	75-42775
Tadiran Ltd	Israel	222	75-21600
Watkins-Johnson Co.	USA	627	75-95200

461 40 27 50 Receiving / Detecting / Warning equipment, ground-based EW

Amherst Systems Inc.	USA	553	75-06600
Andrew Canada Inc.	Canada	68	75-00600
Condor Systems Inc.	USA	566	75-20950
Elettronica SpA	Italy	233	75-07050
Elettronica (UK) Ltd	UK	448	75-20475
ELTA Electronics Industries Ltd	Israel	220	75-10200
GEC-Marconi Avionics Ltd	UK	454	75-25320
Israel Aircraft Industries Ltd	Israel	219	75-10200
MA/COM Antenna & Cable Div.	USA	593	75-56953
Racal Communications Inc.	USA	606	75-72380
Racal Radar Defence Systems Ltd	UK	477	75-47688
Siemens Plessey Systems	UK	483	75-53770
Tadiran Ltd	Israel	222	75-21600
Watkins-Johnson Co.	USA	627	75-95200

461 40 27 55 Receiving equipment, ground-based, passive EW

Andrew Canada Inc.	Canada	68	75-00600
Condor Systems Inc.	USA	566	75-20950
Elettronica SpA	Italy	233	75-07050
Elettronica (UK) Ltd	UK	448	75-20475
Elisra Electronic Systems Ltd	Israel	219	75-06150
ELTA Electronics Industries Ltd	Israel	220	75-10200
GEC-Marconi Communications Ltd	UK	454	75-25335
Israel Aircraft Industries Ltd	Israel	219	75-10200
KEL Aerospace Pty Ltd	Australia	18	75-11440
MA/COM Antenna & Cable Div.	USA	593	75-56953
NEC Corporation	Japan	253	75-06350
Racal Communications Inc.	USA	606	75-72380
Racal Radar Defence Systems Ltd	UK	477	75-47688
Rohde & Schwarz Canada Inc	Canada	79	80-01825
Rohde & Schwarz GmbH & Co KG	Germany	174	75-25400
Sanders	USA	612	75-77850
Siemens Plessey Systems	UK	483	75-53770
State College Operations	USA	617	75-83990
Tadiran Ltd	Israel	222	75-21600
THORN Microwave Devices Ltd	UK	488	75-59760
Watkins-Johnson Co.	USA	627	75-95200

461 40 27 55 32 Receivers, communications intelligence, ground-based

Amper Programas SA	Spain	365	75-00480
Andrew Canada Inc.	Canada	68	75-00600
Condor Systems Inc.	USA	566	75-20950
Elettronica SpA	Italy	233	75-07050
ELTA Electronics Industries Ltd	Israel	220	75-10200
GEC-Marconi Communications Ltd	UK	454	75-25335
Israel Aircraft Industries Ltd	Israel	219	75-10200
KEL Aerospace Pty Ltd	Australia	18	75-11440
MA/COM Antenna & Cable Div.	USA	593	75-56953
NEC Corporation	Japan	253	75-06350
Racal Communications Inc.	USA	606	75-72380
Rohde & Schwarz Canada Inc	Canada	79	80-01825
Rohde & Schwarz GmbH & Co KG	Germany	174	75-25400
Sanders	USA	612	75-77850
State College Operations	USA	617	75-83990
Tadiran Ltd	Israel	222	75-21600
Watkins-Johnson Co.	USA	627	75-95200

461 40 27 55 40 Receivers, infra-red warning, ground-based

Laser & Atomic Research & Devt	Croatia	94	60-00450

461 40 70 Electronic warfare equipment, ground-based, other types

British Aerospace (Systems &	UK	436	75-08890
Daimler-Benz Aerospace France SAS	**France**	**134**	**75-12950**
ITT Avionics	USA	586	75-49455
MA/COM Antenna & Cable Div.	USA	593	75-56953
MCL Inc.	USA	595	75-59700
Raytheon E-Systems	USA	608	75-73302
Sanders	USA	612	75-77850
Tracor Inc	USA	623	75-90000

461 40 70 20 Chaff dispensing equipment, ground-based

Daimler-Benz Aerospace France SAS	**France**	**134**	**75-12950**

461 40 80 Electronic warfare equipment accessories, ground-based

Bren-Tronics Inc.	USA	560	75-15170
General Electronics Communications	Greece	189	75-01300
Hewlett-Packard (Canada) Ltd	Canada	72	75-07520
Litton Industries Inc.	USA	590	75-55400
MA/COM Antenna & Cable Div.	USA	593	75-56953
Raytheon E-Systems	USA	608	75-73302
Rohde & Schwarz France Telecom	France	142	75-32900
Sanders	USA	612	75-77850
Tracor Inc	USA	623	75-90000
Trompeter Electronics Inc.	USA	623	75-91070

461 40 80 20 Expendables, ground-launched EW

Sanders	USA	612	75-77850

461 40 80 22 Chaff, ground-launched

Pains-Wessex Ltd	UK	472	75-43510

461 70 Electronic warfare equipment, other

AIL Systems Inc.	USA	547	75-02950
CAE Elektronic GmbH	Germany	166	75-04500
Cartwright Electronics Inc.	USA	562	75-16910
Communications & Power Industries	USA	565	75-20393
Dassault Electronique SA (Group)	**France**	**135**	**75-12980**
Datron/Transco Inc.	USA	568	75-24000
ELTA Electronics Industries Ltd	Israel	220	75-10200
ITT Avionics	USA	586	75-49455
KEL Aerospace Pty Ltd	Australia	18	75-11440
Litton Industries Inc.	USA	590	75-55400
MA/COM Antenna & Cable Div.	USA	593	75-56953
MCL Inc.	USA	595	75-59700
Rohde & Schwarz France Telecom	France	142	75-32900
Rohde & Schwarz GmbH & Co KG	Germany	174	75-25400
Sanders	USA	612	75-77850
Tracor Inc	USA	623	75-90000
Watkins-Johnson Co.	USA	627	75-95200

461 90 Electronic warfare equipment components

British Sarozal Ltd.	UK	436	75-09570
Century Electronics	USA	562	75-17385
Communications & Power Industries	USA	565	75-20393
Datron/Transco Inc.	USA	568	75-24000

EEV Inc.	USA	571	75-29500
EEV Ltd	UK	448	75-19590
EG&G Reticon	USA	571	75-29562
Elekluft GmbH	Germany	178	80-00537
Elettronica SpA	Italy	233	75-07050
Elettronica (UK) Ltd	UK	448	75-20475
Elisra Electronic Systems Ltd	Israel	219	75-06150
ELTA Electronics Industries Ltd	Israel	220	75-10200
ERA Technology Ltd	UK	502	80-09600
Jennings Technology Corp.	USA	586	75-50350
Litton Industries Inc.	USA	590	75-55400
MA/COM Antenna & Cable Div.	USA	593	75-56953
Micro Metalsmiths Ltd	UK	468	75-39010
Nordic VLSI A/S	Norway	310	75-01640
Parker Bertea Aerospace	USA	603	75-68900
Premier Microwave Corp.	USA	605	75-71400
Racal-MESL Ltd	UK	477	75-47682
Raytheon E-Systems	USA	608	75-73302
Rohde & Schwarz France Telecom	France	142	75-32900
Sanders	USA	612	75-77850
Teledyne Electronic Technologies	USA	620	75-87522
THORN Microwave Devices Ltd	UK	488	75-59760
Tracor Inc	USA	623	75-90000
Trompeter Electronics Inc.	USA	623	75-91070
Watkins-Johnson Co.	USA	627	75-95200
Zeta	USA	629	75-98190

Mine Warfare & Countermeasures

471 Minelaying systems

AAI Corp.	USA	545	75-00020
Bofors SA Marine AB.	Sweden	376	75-03040
Consorzio SMIN.	Italy	230	74-03700
Fuji Heavy Industries Ltd	Japan	251	75-01100
NORINCO	China PR	88	55-00280
Polaris International Ltd	UK	508	80-23700
R. Alkan (UK) Ltd	UK	429	75-02680
Valsella Meccanotecnica SpA	Italy	239	75-26800

471 20 Minelaying systems, airborne

Explosivos Alaveses SA	Spain	366	75-03050
Fuji Heavy Industries Ltd	Japan	251	75-01100
R. Alkan (UK) Ltd	UK	429	75-02680

471 22 Minelaying systems, airborne, land scatter

AAI Corp.	USA	545	75-00020
Explosivos Alaveses SA	Spain	366	75-03050
R. Alkan (UK) Ltd	UK	429	75-02680

471 30 Minelaying systems, naval shipborne

Bofors SA Marine AB.	Sweden	376	75-03040
Consorzio SMIN.	Italy	230	74-03700
Strachan & Henshaw Ltd	UK	512	80-47400

471 37 Minelaying equipment, naval shipborne

Bofors SA Marine AB.	Sweden	376	75-03040
Consorzio SMIN.	Italy	230	74-03700
Kobelt Manufacturing Co. Ltd	Canada	73	75-08770
Strachan & Henshaw Ltd	UK	512	80-47400

471 40 Minelaying systems, ground-based

AAI Corp.	USA	545	75-00020
Dynamit Nobel GmbH	Germany	167	75-07100
Fuji Heavy Industries Ltd	Japan	251	75-01100
Valsella Meccanotecnica SpA	Italy	239	75-26800

471 42 Minelaying systems, anti-personnel, ground-based

AAI Corp.	USA	545	75-00020
Valsella Meccanotecnica SpA	Italy	239	75-26800

471 44 Minelaying systems, anti-tank, ground-based

AAI Corp.	USA	545	75-00020
Valsella Meccanotecnica SpA	Italy	239	75-26800

471 47 Minelaying equipment, ground-based

Saunders-Roe Ltd.	UK	482	75-51850
Valsella Meccanotecnica SpA	Italy	239	75-26800

471 47 50 Minelaying ploughs

Explosivos Alaveses SA	Spain	366	75-03050
Pearson Engineering Ltd	UK	473	75-44330

471 47 60 Minelaying equipment, vehicle-mounted

AAI Corp.	USA	545	75-00020
Valsella Meccanotecnica SpA	Italy	239	75-26800

472 Minehunting & detection systems

Aardvark Clear Mine Ltd	UK	426	75-00034
Alenia-Elsag	Italy	230	74-00100
AlliedSignal Ltd	UK	430	75-02747
BAeSEMA Ltd.	UK	498	80-03070
Benthos Inc.	USA	558	75-13540
Consorzio SMIN.	Italy	230	74-03700
Cubic Defense Systems Inc.	USA	567	75-22900
DRS Technologies, Inc.	USA	570	75-27400
Fabbrica Italiana Apparecchiature	Italy	234	75-08550
Institut Dr Förster GmbH & Co KG.	Germany	170	75-14500
Int'l Submarine Engrg.	Canada	73	75-07780
MacDonald Dettwiler and Associates.	Canada	78	80-01310
Mechem	S. Africa	358	75-11000
Microelectronics, guidance and	Turkey	402	55-00050
Northrop Grumman Corporation	USA	640	80-21840
P W Allen & Co Ltd	UK	429	75-02740
Recon/Optical Inc.	USA	608	75-73930
SAIC Ltd	UK	510	80-30220
SPL Ordnance Ltd.	Israel	225	80-01935
STN ATLAS Elektronik GmbH	Germany	165	75-01150
Thomson Sintra	France	148	75-45135
Vallon GmbH	**Germany**	**176**	**75-32150**

472 20 Minehunting & detection systems, airborne

DRS Technologies, Inc.	USA	570	75-27400
EDO Corporation	USA	571	75-29250
MPB Technologies Inc	Canada	74	75-09855

472 27 Minehunting & detection equipment, airborne

DRS Technologies, Inc.	USA	570	75-27400

472 30 Minehunting & detection systems, naval

Bofors SA Marine AB.	Sweden	376	75-03040
C-Tech Ltd.	Canada	70	75-03800
Consorzio SMIN.	Italy	230	74-03700
Cubic Defense Systems Inc.	USA	567	75-22900
Dowty Aerospace Wolverhampton	UK	446	75-18589
DRS Technologies, Inc.	USA	570	75-27400
EDO Corporation	USA	571	75-29250
Eiva A/S	Denmark	104	80-00500
ENDECO/YSI Inc.	USA	573	75-31540
Exstar International Corp.	USA	575	75-33000
Fabbrica Italiana Apparecchiature	Italy	234	75-08550
Fugro-UDI Ltd	UK	452	75-24885
Klein Associates Inc.	USA	589	75-52900
MacDonald Dettwiler and Associates.	Canada	78	80-01310
Marine Electronics Ltd	UK	466	75-37290
NEC Corporation	Japan	253	75-06350
STN ATLAS Elektronik GmbH	Germany	165	75-01150
Thomson Marconi Sonar Limited	UK	426	74-20000
Thomson Sintra	France	148	75-45135
Vallon GmbH	**Germany**	**176**	**75-32150**

472 37 Minehunting & detection equipment, naval

Alenia-Elsag	Italy	230	74-00100
AlliedSignal Ltd	UK	430	75-02747
Bofors SA Marine AB	Sweden	376	75-03040
Dowty Aerospace Wolverhampton	UK	446	75-18589
DRS Technologies, Inc.	USA	570	75-27400
ENDECO/YSI Inc.	USA	573	75-31540
Fabbrica Italiana Apparecchiature	Italy	234	75-08550
Gayrobot Undersea Technology	Switzerland	386	75-04700
Klein Associates Inc.	USA	589	75-52900
MacArtney A/S	Denmark	103	75-12800
MacTaggart Scott & Co. Ltd	UK	466	75-36900
Marine Electronics Ltd	UK	466	75-37290
NEC Corporation	Japan	253	75-06350
Oceano Instruments (UK) Ltd	UK	508	80-19050
Slingsby Engineering Ltd	UK	484	75-54420
STN ATLAS Elektronik GmbH	Germany	165	75-01150
Thomson Marconi Sonar Limited	UK	426	74-20000
Thomson Sintra	France	148	75-45135
Vallon GmbH	**Germany**	**176**	**75-32150**

472 39 Minehunting & detection equipment components, naval

Bofors SA Marine AB	Sweden	376	75-03040
Dowty Aerospace Wolverhampton	UK	446	75-18589
Exstar International Corp.	USA	575	75-33000
Fabbrica Italiana Apparecchiature	Italy	234	75-08550
Fugro-UDI Ltd	UK	452	75-24885
Ibak Helmut Hunger GmbH & Co KG	Germany	170	75-13580
Klein Associates Inc.	USA	589	75-52900
MacTaggart Scott & Co. Ltd	UK	466	75-36900
Marine Electronics Ltd	UK	466	75-37290
NEC Corporation	Japan	253	75-06350
Regis SpA	Italy	238	75-20110
Sparton Electronics	USA	616	75-82900
Thomson Marconi Sonar Limited	UK	426	74-20000
Thomson Sintra	France	148	75-45135

472 40 Minehunting & detection systems, ground-based

Förster Imaden	Brazil	56	75-06650
Institut Dr Förster GmbH & Co KG	Germany	170	75-14500
Microelectronics, guidance and	Turkey	402	55-00050
Pearson Engineering Ltd	UK	473	75-44330
Schiebel Elektronische Geraete	Austria	34	75-02180
SPL Ordnance Ltd	Israel	225	80-01935
Thomson Sintra	France	148	75-45135
Vallon GmbH	**Germany**	**176**	**75-32150**
White's Electronics (UK) Ltd	UK	494	75-65950

472 47 Minehunting & detection equipment, ground-based

GoldStar Electric Co. Ltd	Korea (ROK)	265	75-01400
Institut Dr Förster GmbH & Co KG	Germany	170	75-14500
Microelectronics, guidance and	Turkey	402	55-00050
Pearson Engineering Ltd	UK	473	75-44330
Schiebel Elektronische Geraete	Austria	34	75-02180
SPL Ordnance Ltd	Israel	225	80-01935
Theodor Rapp GmbH Co	Germany	173	75-24600
Thomson Sintra	France	148	75-45135
Vallon GmbH	**Germany**	**176**	**75-32150**

472 47 30 Mine detectors, ground-based

Aardvark Clear Mine Ltd	UK	426	75-00034
ERA Technology Ltd	UK	502	80-09600
Foerster UK Ltd	UK	451	75-24150
Microelectronics, guidance and	Turkey	402	55-00050
P W Allen & Co Ltd	UK	429	75-02740
Protovale (Oxford) Ltd	UK	476	75-47000
REMOTEC Inc.	USA	609	75-74430
Schiebel Elektronische Geraete	Austria	34	75-02180
SPL Ordnance Ltd	Israel	225	80-01935
Vallon GmbH	**Germany**	**176**	**75-32150**

472 47 30 50 Mine detectors, metallic

Cubic Defense Systems Inc.	USA	567	75-22900
Forster-France Sarl	France	136	75-18300
Guartel Ltd	UK	456	75-27130
Microelectronics, guidance and	Turkey	402	55-00050
Protovale (Oxford) Ltd	UK	476	75-47000
Schiebel Elektronische Geraete	Austria	34	75-02180
SPL Ordnance Ltd	Israel	225	80-01935
Vallon GmbH	**Germany**	**176**	**75-32150**

472 47 30 60 Mine detectors, non-metallic

Cubic Defense Systems Inc.	USA	567	75-22900
Microelectronics, guidance and	Turkey	402	55-00050
Schiebel Elektronische Geraete	Austria	34	75-02180
SIMA PERU	Peru	319	55-01950
SPL Ordnance Ltd	Israel	225	80-01935

472 47 30 70 Mine detectors, metallic / non-metallic

Microelectronics, guidance and	Turkey	402	55-00050
Schiebel Elektronische Geraete	Austria	34	75-02180
SPL Ordnance Ltd	Israel	225	80-01935
Theodor Rapp GmbH Co	Germany	173	75-24600
Vallon GmbH	**Germany**	**176**	**75-32150**
White's Electronics (UK) Ltd	UK	494	75-65950

472 47 30 80 Mine detectors, road, vehicle-mounted

Reunert Technology Systems Ltd	S. Africa	359	75-18300
SPL Ordnance Ltd	Israel	225	80-01935
Vallon GmbH	**Germany**	**176**	**75-32150**

473 Mine clearance systems

Aardvark Clear Mine Ltd	UK	426	75-00034
Alenia-Elsag	Italy	230	74-00100
AlliedSignal Ltd	UK	430	75-02747
Breeze Eastern Corp.	USA	560	75-15110
Draka Kabel BV	Netherlands	292	75-01650
ECA SA	France	136	75-14050
Elesco Oy	Finland	118	75-00300
Ensign Bickford Co.	USA	574	75-31700
GEC ALSTHOM Engineering Systems Ltd	UK	453	75-25225
Hitachi Zosen Co. Ltd	Japan	251	75-02390
Int'l Submarine Engrg.	Canada	73	75-07780
Israel Military Industries Ltd	Israel	220	75-10300
Mechem	S. Africa	358	75-11000
Northrop Grumman Corporation	USA	640	80-21840
Pains-Wessex Ltd	UK	472	75-43510
Pearson Engineering Ltd	UK	473	75-44330
Precision Machining Inc.	USA	605	75-71350
Riva Calzoni SpA	Italy	238	75-20500
Rotherham Precision Engineers Ltd	UK	481	75-50590
Simrad Norge A/S	Norway	310	75-02630
SPL Ordnance Ltd	Israel	225	80-01935
STN ATLAS Elektronik GmbH	Germany	165	75-01150
Whitehead SpA	Italy	240	75-27600

473 20 Mine clearance systems, helicopter-towed

Breeze Eastern Corp.	USA	*560*	75-15110
EDO Corporation	USA	*571*	75-29250
MIL-BASE Industries Inc.	USA	*596*	75-61530

473 27 Mine clearance equipment, helicopter-towed

Breeze Eastern Corp.	USA	*560*	75-15110
EDO Corporation	USA	*571*	75-29250
MIL-BASE Industries Inc.	USA	*596*	75-61530
STN ATLAS Elektronik GmbH	Germany	*165*	75-01150

473 30 Mine clearance systems, naval shipborne

Bofors SA Marine AB.	Sweden	*376*	75-03040
Elesco Oy	Finland	*118*	75-00300
Hitachi Zosen Co. Ltd.	Japan	*251*	75-02390
Oceano Instruments (UK) Ltd	UK	*508*	80-19050
Permali Gloucester Ltd	UK	*474*	75-44850
Riva Calzoni SpA	Italy	*238*	75-20500
STN ATLAS Elektronik GmbH	Germany	*165*	75-01150
Thomson Sintra	France	*148*	75-45135

473 30 60 Mine clearance systems, naval, towed

Bofors SA Marine AB.	Sweden	*376*	75-03040
Cogent	UK	*500*	80-06250
ECA SA	France	*136*	75-14050
Elesco Oy	Finland	*118*	75-00300
Hitachi Zosen Co. Ltd.	Japan	*251*	75-02390
MIL-BASE Industries Inc.	USA	*596*	75-61530
Thomson Sintra	France	*148*	75-45135

473 30 64 Mine clearance systems, naval, towed, influence sweep

Bofors SA Marine AB.	Sweden	*376*	75-03040
Elesco Oy	Finland	*118*	75-00300
Thomson Sintra	France	*148*	75-45135

473 30 66 Mine clearance systems, naval, towed, wire sweep

Bofors SA Marine AB.	Sweden	*376*	75-03040
ECA SA	France	*136*	75-14050
Hitachi Zosen Co. Ltd.	Japan	*251*	75-02390
MIL-BASE Industries Inc.	USA	*596*	75-61530
Rotherham Precision Engineers Ltd	UK	*481*	75-50590
Thomson Sintra	France	*148*	75-45135

473 30 80 Mine clearance vehicles, underwater remote-controlled

Bofors Underwater Systems AB	Sweden	*376*	75-03052
Deep Ocean Engineering Inc.	USA	*568*	75-24820
ECA SA	France	*136*	75-14050
Gayrobot Undersea Technology	Switzerland	*386*	75-04700
MacArtney A/S	Denmark	*103*	75-12800
Permali Gloucester Ltd	UK	*474*	75-44850
Riva Calzoni SpA	Italy	*238*	75-20500
STN ATLAS Elektronik GmbH	Germany	*165*	75-01150
Thomson Marconi Sonar Limited	UK	*426*	74-20000
Thomson Sintra	France	*148*	75-45135

473 37 Mine clearance equipment, naval shipborne

Bofors SA Marine AB.	Sweden	*376*	75-03040
Crestbury Ltd	UK	*443*	75-16030
Interspiro AB	Sweden	*377*	75-06950
Interspiro Ltd	UK	*461*	75-31945
Med-Eng Systems Inc.	Canada	*74*	75-09550
Oceano Instruments (UK) Ltd	UK	*508*	80-19050
Rotherham Precision Engineers Ltd	UK	*481*	75-50590
STN ATLAS Elektronik GmbH	Germany	*165*	75-01150
Thomson Marconi Sonar Limited	UK	*426*	74-20000
Thomson Sintra	France	*148*	75-45135

473 37 80 Mine clearance equipment accessories, naval shipborne

Bofors SA Marine AB.	Sweden	*376*	75-03040
Crestbury Ltd	UK	*443*	75-16030
Jered Brown Brothers Inc.	USA	*586*	75-50380
Oceano Instruments (UK) Ltd	UK	*508*	80-19050
Thomson Sintra	France	*148*	75-45135

473 37 80 90 Winches, mine clearance equipment, naval shipborne

Bofors SA Marine AB.	Sweden	*376*	75-03040
Jered Brown Brothers Inc.	USA	*586*	75-50380
Lucas Western Inc.	USA	*592*	75-56300
MacArtney A/S	Denmark	*103*	75-12800
MacTaggart Scott & Co. Ltd	UK	*466*	75-36900
Strachan & Henshaw Ltd	UK	*512*	80-47400

473 37 90 Mine clearance equipment components, naval shipborne

Bofors SA Marine AB.	Sweden	*376*	75-03040
Jered Brown Brothers Inc.	USA	*586*	75-50380
Thomson Sintra	France	*148*	75-45135

473 37 90 20 Cables, mine clearance equipment, naval

Bofors SA Marine AB.	Sweden	*376*	75-03040
Draka Kabel BV	Netherlands	*292*	75-01650
JDR Cable Systems BV	Netherlands	*293*	75-03600
MacArtney A/S	Denmark	*103*	75-12800

473 40 Mine clearance systems, ground-based

Aardvark Clear Mine Ltd	UK	*426*	75-00034
Comet GmbH	Germany	*166*	75-05000
Diehl GmbH & Co	Germany	*167*	75-06500
Ensign Bickford Co.	USA	*574*	75-31700
Glover Webb Ltd	UK	*455*	75-25700
Israel Aircraft Industries Ltd	Israel	*219*	75-10200
Israel Military Industries Ltd	Israel	*220*	75-10300
MaK Systemgesellschaft mbH.	Germany	*172*	75-19500
Mining Resource Engineering Ltd	Canada	*74*	75-09770
Pains-Wessex Ltd	UK	*472*	75-43510
Pearson Engineering Ltd	UK	*473*	75-44330
Somchem	S. Africa	*359*	75-19200
SPL Ordnance Ltd.	Israel	*225*	80-01935
Thomson Sintra	France	*148*	75-45135

473 40 13 Mine clearance systems, ground-based, explosive hose type

Ensign Bickford Co.	USA	*574*	75-31700
Pains-Wessex Ltd	UK	*472*	75-43510
Somchem	S. Africa	*359*	75-19200
SPL Ordnance Ltd.	Israel	*225*	80-01935

473 40 50 Mine clearance systems, ground-based, remote controlled vehicle type

Aardvark Clear Mine Ltd	UK	*426*	75-00034
Glover Webb Ltd	UK	*455*	75-25700
SPL Ordnance Ltd.	Israel	*225*	80-01935

473 47 Mine clearance equipment, ground-based

AAI Corp.	USA	*545*	75-00020
Aardvark Clear Mine Ltd	UK	*426*	75-00034
Emco Inc.	USA	*573*	75-30985
Glover Webb Ltd	UK	*455*	75-25700
Israel Aircraft Industries Ltd	Israel	*219*	75-10200
Israel Military Industries Ltd	Israel	*220*	75-10300
Laser & Atomic Research & Devt	Croatia	*94*	60-00450
Mining Resource Engineering Ltd	Canada	*74*	75-09770
Nissan Motor Co. Ltd.	Japan	*254*	75-06790
Saunders-Roe Ltd.	UK	*482*	75-51850
Somchem	S. Africa	*359*	75-19200

SPL Ordnance Ltd.	Israel	225	80-01935
Urdan Industries (USA) Inc.	USA	625	75-92930

473 47 10 Mine clearance equipment, ground-based, armoured vehicle-mounted

Aardvark Clear Mine Ltd	UK	426	75-00034
Engineered Air Systems Inc.	USA	573	75-31560
Glover Webb Ltd	UK	455	75-25700
Israel Military Industries Ltd	Israel	220	75-10300
Pearson Engineering Ltd	UK	473	75-44330
Somchem	S. Africa	359	75-19200
SPL Ordnance Ltd.	Israel	225	80-01935
Urdan Industries Ltd	Israel	223	75-23210
Urdan Industries (USA) Inc.	USA	625	75-92930

473 47 10 50 Mine clearance equipment, flail type, armoured vehicle-mounted

Aardvark Clear Mine Ltd	UK	426	75-00034
Glover Webb Ltd	UK	455	75-25700
Hägglunds Moelv AS	Norway	309	75-00880
SPL Ordnance Ltd.	Israel	225	80-01935
Urdan Industries (USA) Inc.	USA	625	75-92930

473 47 30 Mine clearance equipment, ground-based, explosive tube type, rigid

CMS Inc.	USA	564	75-19170
Israel Military Industries Ltd	Israel	220	75-10300
Nissan Motor Co. Ltd.	Japan	254	75-06790
SPL Ordnance Ltd.	Israel	225	80-01935

473 47 50 Mine disposal equipment, ground-based

Haley & Weller Ltd	UK	456	75-27400
Pearson Engineering Ltd	UK	473	75-44330
SPL Ordnance Ltd.	Israel	225	80-01935

474 Minefield marking systems

Hunting Engineering Ltd	UK	459	75-30700
Israel Military Industries Ltd	Israel	220	75-10300
Pearson Engineering Ltd	UK	473	75-44330
SPL Ordnance Ltd.	Israel	225	80-01935

474 27 Minefield marking equipment, air-launched

Pearson Engineering Ltd	UK	473	75-44330

474 27 10 Minefield marker buoys, air-launched

MIL-BASE Industries Inc.	USA	596	75-61530

474 37 Minefield marking equipment, naval

Dukane Corp.	USA	570	75-27650
MIL-BASE Industries Inc.	USA	596	75-61530

474 37 10 Minefield marker buoys, naval

MIL-BASE Industries Inc.	USA	596	75-61530

474 40 Minefield marking systems, ground-based

Hunting Engineering Ltd	UK	459	75-30700
Israel Military Industries Ltd	Israel	220	75-10300
Pearson Engineering Ltd	UK	473	75-44330
Saunders-Roe Ltd.	UK	482	75-51850
SPL Ordnance Ltd.	Israel	225	80-01935
Theodor Rapp GmbH Co	Germany	173	75-24600

474 47 Minefield marking equipment, ground-based

Hunting Engineering Ltd	UK	459	75-30700
Israel Military Industries Ltd	Israel	220	75-10300
Pearson Engineering Ltd	UK	473	75-44330
Saunders-Roe Ltd.	UK	482	75-51850
SPL Ordnance Ltd.	Israel	225	80-01935
Theodor Rapp GmbH Co	Germany	173	75-24600

474 47 40 Minefield marking equipment, ground-based, illuminated

Israel Military Industries Ltd	Israel	220	75-10300
Pearson Engineering Ltd	UK	473	75-44330
Saunders-Roe Ltd.	UK	482	75-51850
SPL Ordnance Ltd.	Israel	225	80-01935

NBC Countermeasures Equipment

481 NBC Detection & monitoring systems

AEA Technology plc	UK	496	80-00480
Åkers Krutbruk Protection AB	Sweden	375	75-01100
Anachemia Canada Inc.	Canada	68	75-00500
Anite Systems Ltd	UK	497	80-01240
Calspan SRL Technology	USA	540	60-03000
Environics Oy	Finland	118	75-00500
Environmental Technologies Group Inc.	USA	574	75-31860
Ericsson Defence Systems Pty Ltd	Australia	16	75-05820
General Motors of Canada Ltd.	Canada	71	75-05900
Giat Industries	France	126	55-00100
Jasmin Simtec Ltd	UK	462	75-32370
NORINCO	China PR	88	55-00280
Nuclear Associates Division	USA	600	75-65730
Nuclear Research Corp.	USA	600	75-65770
Odenwald Werke Rittersbach	Germany	173	75-22850
O.I. Analytical	USA	600	75-65810
Richmond Packaging (UK) Ltd	UK	479	75-49630
Scintrex Ltd	Canada	76	75-12550
Sensidyne Inc.	USA	613	75-79320
Siemens Environmental Systems Ltd	UK	483	75-53700
SKC Inc.	USA	615	75-81480
SRI International	USA	643	80-29600
Technical Products Group Inc.	USA	620	75-86764

481 20 NBC Detection & monitoring systems, airborne

Anachemia Canada Inc.	Canada	68	75-00500
Siemens Environmental Systems Ltd	UK	483	75-53700
SRI International	USA	643	80-29600
Technical Products Group Inc.	USA	620	75-86764

481 27 NBC Detection & monitoring equipment, airborne

Anachemia Canada Inc.	Canada	68	75-00500
Graseby Dynamics Ltd	UK	455	75-26590
Siemens Environmental Systems Ltd	UK	483	75-53700
SRI International	USA	643	80-29600
Technical Products Group Inc.	USA	620	75-86764

481 30 NBC Detection & monitoring systems, naval

Anachemia Canada Inc.	Canada	68	75-00500
Environics Oy	Finland	118	75-00500
Siemens Environmental Systems Ltd	UK	483	75-53700
Technical Products Group Inc.	USA	620	75-86764

481 37 NBC Detection & monitoring equipment, naval

Anachemia Canada Inc.	Canada	68	75-00500
Environics Oy	Finland	118	75-00500
Graseby Dynamics Ltd	UK	455	75-26590
Siemens Environmental Systems Ltd	UK	483	75-53700
Technical Products Group Inc.	USA	620	75-86764

Product–Supplier Link

481 40 NBC Detection & monitoring systems, ground-based

AEA Technology plc	UK	*496*	80-00480
Åkers Krutbruk Protection AB	Sweden	*375*	75-01100
Anachemia Canada Inc.	Canada	*68*	75-00500
British Nuclear Fuels plc	**UK**	***499***	**80-04175**
Calspan SRL Technology	USA	*540*	60-03000
Environics Oy	Finland	*118*	75-00500
Giat Industries	France	*126*	55-00100
Laser & Atomic Research & Devt	Croatia	*94*	60-00450
Metis Construction & Trade Co. Inc.	Turkey	*403*	80-01330
Odenwald Werke Rittersbach	Germany	*173*	75-22850
O.I. Analytical	USA	*600*	75-65810
Scintrex Ltd	Canada	*76*	75-12550
Siemens Environmental Systems Ltd	UK	*483*	75-53700
SRI International	USA	*643*	80-29600
Technical Products Group Inc.	USA	*620*	75-86764

481 40 50 Nuclear radiation measuring & monitoring systems

AEA Technology plc	UK	*496*	80-00480
British Nuclear Fuels plc	**UK**	***499***	**80-04175**
Fluor Daniel Inc.	USA	*635*	80-11720
Machine Technology Ltd	UK	*465*	75-36870
Metis Construction & Trade Co. Inc.	Turkey	*403*	80-01330
Scintrex Ltd	Canada	*76*	75-12550
Siemens Environmental Systems Ltd	UK	*483*	75-53700
Technical Products Group Inc.	USA	*620*	75-86764

481 47 NBC Detection & monitoring equipment, ground-based

AEA Technology plc	UK	*496*	80-00480
Åkers Krutbruk Protection AB	Sweden	*375*	75-01100
Anachemia Canada Inc.	Canada	*68*	75-00500
Anite Systems Ltd	UK	*497*	80-01240
British Nuclear Fuels plc	**UK**	***499***	**80-04175**
Calspan SRL Technology	USA	*540*	60-03000
Environics Oy	Finland	*118*	75-00500
Environmental Technologies Group Inc	USA	*574*	75-31860
Graseby Dynamics Ltd	UK	*455*	75-26590
Machine Technology Ltd	UK	*465*	75-36870
Scintrex Ltd	Canada	*76*	75-12550
Siemens Environmental Systems Ltd	UK	*483*	75-53700
SRI International	USA	*643*	80-29600
Technical Products Group Inc.	USA	*620*	75-86764

481 47 10 Air sampling system

AEA Technology plc	UK	*496*	80-00480
Anachemia Canada Inc.	Canada	*68*	75-00500
British Nuclear Fuels plc	**UK**	***499***	**80-04175**
Ludlum Measurements Inc.	USA	*592*	75-56360
O.I. Analytical	USA	*600*	75-65810
Sensidyne Inc.	USA	*613*	75-79320
Staplex Air Sampler Division	USA	*617*	75-83952
Technical Products Group Inc.	USA	*620*	75-86764

481 47 50 Nuclear radiation measuring & monitoring equipment

AEA Technology plc	UK	*496*	80-00480
AIL Systems Inc.	USA	*547*	75-02950
Bertan High Voltage	USA	*558*	75-13720
British Nuclear Fuels plc	**UK**	***499***	**80-04175**
Dosimeter	USA	*569*	75-27120
GEC-Marconi Materials Technology Ltd	UK	*419*	60-00550
Ludlum Measurements Inc.	USA	*592*	75-56360
Machine Technology Ltd	UK	*465*	75-36870
National Nuclear Corporation	USA	*598*	75-63870
Nuclear Associates Division	USA	*600*	75-65730
Nuclear Research Corp.	USA	*600*	75-65770
N.Wood Counter Laboratory Inc.	USA	*628*	75-97278
Osterreichisches Forschungszentrum	Austria	*32*	60-05000
Quantrad Sensor	USA	*606*	75-72250
Rados Technology Oy	Finland	*118*	75-01780
Scintrex Ltd	Canada	*76*	75-12550
SE International Inc.	USA	*613*	75-79225
Siemens Environmental Systems Ltd	UK	*483*	75-53700
Technical Products Group Inc.	USA	*620*	75-86764

481 47 50 30 Dose rate meters, radiation monitoring

AEA Technology plc	UK	*496*	80-00480
British Nuclear Fuels plc	**UK**	***499***	**80-04175**
Dosimeter	USA	*569*	75-27120
GEC-Marconi Materials Technology Ltd	UK	*419*	60-00550
Ludlum Measurements Inc.	USA	*592*	75-56360
Machine Technology Ltd	UK	*465*	75-36870
National Nuclear Corporation	USA	*598*	75-63870
Nuclear Associates Division	USA	*600*	75-65730
Odenwald Werke Rittersbach	Germany	*173*	75-22850
Osterreichisches Forschungszentrum	Austria	*32*	60-05000
Rados Technology Oy	Finland	*118*	75-01780
Scintrex Ltd	Canada	*76*	75-12550
Siemens Environmental Systems Ltd	UK	*483*	75-53700
Technical Products Group Inc.	USA	*620*	75-86764

482 NBC Protection & prevention systems

AAI Corp.	USA	*545*	75-00020
AEA Technology plc	UK	*496*	80-00480
Allied Research Corp.	USA	*549*	75-04670
Anachemia Canada Inc.	Canada	*68*	75-00500
Bachmann SA	France	*132*	75-04900
Calspan SRL Technology	USA	*540*	60-03000
CORIS	France	*134*	75-12450
EROS GIE	France	*137*	75-19500
GEL Engineering Ltd	UK	*454*	75-25500
Giat Industries	France	*126*	55-00100
Howden Aircontrol Ltd	UK	*459*	75-30240
Hunter Manufacturing	USA	*583*	75-45150
Intertechnique SA	France	*137*	75-23000
Kinetics Ltd	Israel	*220*	75-11320
L'hotellier Montrichard	France	*138*	75-24900
MDH Limited	UK	*467*	75-38270
NII Stali	Russia	*338*	75-14100
Normalair-Garrett Ltd	UK	*425*	74-07000
Odenwald Werke Rittersbach	Germany	*173*	75-22850
Shalon Chemical Industries Ltd	Israel	*222*	75-20000
Silent Partner Body Armor Inc.	USA	*615*	75-80930
Technical Products Group Inc.	USA	*620*	75-86764
Vokes Ltd	UK	*492*	75-63450

482 20 NBC Protection & prevention systems, airborne

AAF International BV	Netherlands	*291*	75-00005
American Cleaning Manufacturing Co.	USA	*553*	75-05940
Anachemia Canada Inc.	Canada	*68*	75-00500
Calspan SRL Technology	USA	*540*	60-03000
Carleton Life Support Technologies	Canada	*70*	75-02950
EROS GIE	France	*137*	75-19500
Giat Industries	France	*126*	55-00100
Howden Aircontrol Ltd	UK	*459*	75-30240
Hunter Manufacturing	USA	*583*	75-45150
Kinetics Ltd	Israel	*220*	75-11320
Normalair-Garrett Ltd	UK	*425*	74-07000
Parker Bertea Aerospace	USA	*603*	75-68900
Racal Health & Safety Group	UK	*477*	75-47675
Sutcliffe Speakman Carbons Ltd	UK	*487*	75-58000
Technical Products Group Inc.	USA	*620*	75-86764

482 30 NBC Protection & prevention systems, naval

AAF International BV	Netherlands	*291*	75-00005
American Cleaning Manufacturing Co.	USA	*553*	75-05940
Anachemia Canada Inc.	Canada	*68*	75-00500
Calspan SRL Technology	USA	*540*	60-03000
Howden Aircontrol Ltd	UK	*459*	75-30240
Interspiro Ltd	UK	*461*	75-31945
Intertechnique SA	France	*137*	75-23000
Kinetics Ltd	Israel	*220*	75-11320
MDH Limited	UK	*467*	75-38270
Normalair-Garrett Ltd	UK	*425*	74-07000
Parker Bertea Aerospace	USA	*603*	75-68900
Racal Health & Safety Group	UK	*477*	75-47675
Stork Bronswerk BV	Netherlands	*295*	75-07200
Sutcliffe Speakman Carbons Ltd	UK	*487*	75-58000
Technical Products Group Inc.	USA	*620*	75-86764

482 37 NBC Protection & prevention equipment, naval

AAF International BV	Netherlands	*291*	75-00005
American Cleaning Manufacturing Co.	USA	*553*	75-05940
Anachemia Canada Inc.	Canada	*68*	75-00500
Calspan SRL Technology	USA	*540*	60-03000
Howden Aircontrol Ltd	UK	*459*	75-30240
Interspiro Ltd	UK	*461*	75-31945
Intertechnique SA	France	*137*	75-23000
Kinetics Ltd	Israel	*220*	75-11320
MDH Limited	UK	*467*	75-38270
Parker Bertea Aerospace	USA	*603*	75-68900
Racal Health & Safety Group	UK	*477*	75-47675
Stork Bronswerk BV	Netherlands	*295*	75-07200
Sutcliffe Speakman Carbons Ltd	UK	*487*	75-58000
Technical Products Group Inc.	USA	*620*	75-86764
van Buuren-van Swaay BV	Netherlands	*296*	75-07830

482 40 NBC Protection & prevention systems, ground-based

American Cleaning Manufacturing Co.	USA	*553*	75-05940
Anachemia Canada Inc.	Canada	*68*	75-00500
Andair AG	Switzerland	*385*	75-00500
Calspan SRL Technology	USA	*540*	60-03000
CORIS	France	*134*	75-12450
Étienne Lacroix	France	*138*	75-24050
Howden Aircontrol Ltd	UK	*459*	75-30240
Hunter Manufacturing	USA	*583*	75-45150
Interspiro Ltd	UK	*461*	75-31945
Intertechnique SA	France	*137*	75-23000
Irvin Aerospace Limited	UK	*462*	75-32050
J&D Wilkie Ltd	UK	*494*	75-66240
Kinetics Ltd	Israel	*220*	75-11320
MDH Limited	UK	*467*	75-38270
Normalair-Garrett Ltd	UK	*425*	74-07000
Odenwald Werke Rittersbach	Germany	*173*	75-22850
Parker Bertea Aerospace	USA	*603*	75-68900
Racal Health & Safety Group	UK	*477*	75-47675
SRI International	USA	*643*	80-29600
Sutcliffe Speakman Carbons Ltd	UK	*487*	75-58000
Technical Products Group Inc.	USA	*620*	75-86764

482 40 60 NBC shelters

AEA Technology plc	UK	*496*	80-00480
American Cleaning Manufacturing Co.	USA	*553*	75-05940
Andrew Mitchell & Co. Ltd	UK	*468*	75-40100
Calspan SRL Technology	USA	*540*	60-03000
CORIS	France	*134*	75-12450
Engineered Air Systems Inc.	USA	*573*	75-31560
Giat Industries	France	*126*	55-00100
Hunter Manufacturing	USA	*583*	75-45150
J&D Wilkie Ltd	UK	*494*	75-66240
Kinetics Ltd	Israel	*220*	75-11320
L'hotellier Montrichard	France	*138*	75-24900
Odenwald Werke Rittersbach	Germany	*173*	75-22850
Société des Tuyaux Bonna	France	*146*	75-41520
Sutcliffe Speakman Carbons Ltd	UK	*487*	75-58000
Technical Products Group Inc.	USA	*620*	75-86764
Weatherhaven	Canada	*77*	75-15220
White Cross Rubber Products Ltd.	UK	*494*	75-65857

482 40 60 70 NBC shelter equipment

Air International Transit	Australia	*11*	75-00450
Airflow Co.	USA	*548*	75-03650
American Cleaning Manufacturing Co.	USA	*553*	75-05940
Andair AG	Switzerland	*385*	75-00500
Calspan SRL Technology	USA	*540*	60-03000
CJB Developments Ltd	UK	*440*	75-13430
Daloc Sheltec AB	Sweden	*376*	75-04000
Engineered Air Systems Inc.	USA	*573*	75-31560
Giat Industries	France	*126*	55-00100
GKN Westland Technologies Ltd	UK	*455*	75-25690
Graseby Dynamics Ltd	UK	*455*	75-26590
Howden Aircontrol Ltd	UK	*459*	75-30240
Howden Fan Company Pty Ltd	Australia	*17*	75-08740
J&D Wilkie Ltd	UK	*494*	75-66240
Kinetics Ltd	Israel	*220*	75-11320
MDH Limited	UK	*467*	75-38270
Normalair-Garrett Ltd	UK	*425*	74-07000
Odenwald Werke Rittersbach	Germany	*173*	75-22850
Parker Bertea Aerospace	USA	*603*	75-68900
Stork Bronswerk BV	Netherlands	*295*	75-07200
Sutcliffe Speakman Carbons Ltd	UK	*487*	75-58000

482 47 NBC Protection & prevention equipment, ground-based

Air International Transit	Australia	*11*	75-00450
Anachemia Canada Inc.	Canada	*68*	75-00500
Andair AG	Switzerland	*385*	75-00500
Armalon Ltd	UK	*431*	75-03729
Bachmann SA	France	*132*	75-04900
Calspan SRL Technology	USA	*540*	60-03000
GEL Engineering Ltd	UK	*454*	75-25500
General Motors of Canada Ltd	Canada	*71*	75-05900
Giat Industries	France	*126*	55-00100
GKN Westland Technologies Ltd	UK	*455*	75-25690
Howden Aircontrol Ltd	UK	*459*	75-30240
Hunter Manufacturing	USA	*583*	75-45150
Interspiro Ltd	UK	*461*	75-31945
Intertechnique SA	France	*137*	75-23000
J&D Wilkie Ltd	UK	*494*	75-66240
Kinetics Ltd	Israel	*220*	75-11320
MDH Limited	UK	*467*	75-38270
Normalair-Garrett Ltd	UK	*425*	74-07000
Odenwald Werke Rittersbach	Germany	*173*	75-22850
Parker Bertea Aerospace	USA	*603*	75-68900
Paul Boyé	France	*133*	75-07020
Shalon Chemical Industries Ltd	Israel	*222*	75-20000
Siebe Gorman & Co. Ltd	UK	*483*	75-53600
Silent Partner Body Armor Inc.	USA	*615*	75-80930
SRI International	USA	*643*	80-29600
Sutcliffe Speakman Carbons Ltd	UK	*487*	75-58000
Technical Products Group Inc.	USA	*620*	75-86764
Theodor Rapp GmbH Co	Germany	*173*	75-24600
Vorkauf SA	Spain	*367*	75-07000

482 47 60 Filters, NBC shelter

Air International Transit	Australia	*11*	75-00450
Airflow Co.	USA	*548*	75-03650
Andair AG	Switzerland	*385*	75-00500
CORIS	France	*134*	75-12450
Daloc Sheltec AB	Sweden	*376*	75-04000
Donaldson Co. Inc.	USA	*569*	75-27050
Engineered Air Systems Inc.	USA	*573*	75-31560
Flodins Filter AB	Sweden	*377*	75-05300
Giat Industries	France	*126*	55-00100
GKN Westland Technologies Ltd	UK	*455*	75-25690
Howden Aircontrol Ltd	UK	*459*	75-30240
Hunter Manufacturing	USA	*583*	75-45150
MDH Limited	UK	*467*	75-38270
Normalair-Garrett Ltd	UK	*425*	74-07000
Odenwald Werke Rittersbach	Germany	*173*	75-22850
Pall Land & Marine	UK	*472*	75-43520
Racal Filter Technologies Ltd	Canada	*75*	75-11600
Shalon Chemical Industries Ltd	Israel	*222*	75-20000
Stork Bronswerk BV	Netherlands	*295*	75-07200
Sutcliffe Speakman Carbons Ltd	UK	*487*	75-58000
Vokes Ltd	UK	*492*	75-63450

483 NBC Decontamination systems

AAI Corp.	USA	*545*	75-00020
Alfred Karcher GmbH & Co	Germany	*170*	75-15650
Bachmann SA	France	*132*	75-04900
Calspan SRL Technology	USA	*540*	60-03000
GEL Engineering Ltd	UK	*454*	75-25500
Giat Industries	France	*126*	55-00100
Karcher France	France	*138*	75-23500
Karl H. Høie & Co. A/S	Norway	*309*	75-00970
Odenwald Werke Rittersbach	Germany	*173*	75-22850
Richmond Packaging (UK) Ltd	UK	*479*	75-49630
SRI International	USA	*643*	80-29600
Technical Products Group Inc.	USA	*620*	75-86764

483 27 NBC Decontamination equipment, aircraft

American Cleaning Manufacturing Co.	USA	*553*	75-05940
Engineered Air Systems Inc.	USA	*573*	75-31560

Karcher France	France	*138*	75-23500
Technical Products Group Inc.	USA	*620*	75-86764

483 30 NBC Decontamination systems, naval

American Cleaning Manufacturing Co.	USA	*553*	75-05940
Karcher France	France	*138*	75-23500
Technical Products Group Inc.	USA	*620*	75-86764

483 37 NBC Decontamination equipment, naval

American Cleaning Manufacturing Co.	USA	*553*	75-05940
Giat Industries	France	*126*	55-00100
Technical Products Group Inc.	USA	*620*	75-86764

483 40 NBC Decontamination systems, ground-based

AAI Corp.	USA	*545*	75-00020
American Cleaning Manufacturing Co.	USA	*553*	75-05940
Engineered Air Systems Inc.	USA	*573*	75-31560
Giat Industries	France	*126*	55-00100
Karcher France	France	*138*	75-23500
Metis Construction & Trade Co. Inc.	Turkey	*403*	80-01330
Odenwald Werke Rittersbach	Germany	*173*	75-22850
Technical Products Group Inc.	USA	*620*	75-86764

483 47 NBC Decontamination equipment, ground-based

AAI Corp.	USA	*545*	75-00020
American Cleaning Manufacturing Co.	USA	*553*	75-05940
Bachmann SA	France	*132*	75-04900
BCB International Ltd	UK	*433*	75-06075
Denis Ferranti Meters Ltd	UK	*450*	75-22850
Engineered Air Systems Inc.	USA	*573*	75-31560
GEL Engineering Ltd	UK	*454*	75-25500
Giat Industries	France	*126*	55-00100
Karcher France	France	*138*	75-23500
Karl H. Høie & Co. A/S	Norway	*309*	75-00970
Medesign GmbH	Germany	*172*	75-20370
NIS Ltd	UK	*471*	75-42555
Odenwald Werke Rittersbach	Germany	*173*	75-22850
Richmond Packaging (UK) Ltd	UK	*479*	75-49630
SRI International	USA	*643*	80-29600
Technical Products Group Inc.	USA	*620*	75-86764

485 Radioactive waste handling & disposal systems

AEA Technology plc	UK	*496*	80-00480
British Nuclear Fuels plc	**UK**	***499***	**80-04175**
Fluor Daniel Inc.	USA	*635*	80-11720
Giat Industries	France	*126*	55-00100
NIS Ltd	UK	*471*	75-42555

485 30 Radioactive waste disposal systems

AEA Technology plc	UK	*496*	80-00480
British Nuclear Fuels plc	**UK**	***499***	**80-04175**
Fluor Daniel Inc.	USA	*635*	80-11720
NIS Ltd	UK	*471*	75-42555

485 40 70 Radioactive waste handling equipment

AEA Technology plc	UK	*496*	80-00480
British Nuclear Fuels plc	**UK**	***499***	**80-04175**
Fluor Daniel Inc.	USA	*635*	80-11720
NIS Ltd	UK	*471*	75-42555
PD Technical Mouldings plc	UK	*473*	75-44200

489 NBC countermeasures systems & equipment, other types

AEA Technology plc	UK	*496*	80-00480
AlliedSignal Aerospace Canada	Canada	*68*	75-00470
American Cleaning Manufacturing Co.	USA	*553*	75-05940
Busak + Shamban Inc.	USA	*561*	75-16006
Charcoal Cloth (International) Ltd	UK	*439*	75-12550
Kinetics Ltd	Israel	*220*	75-11320
Parker Bertea Aerospace	USA	*603*	75-68900
STI	USA	*618*	75-85230
Theodor Rapp GmbH Co	Germany	*173*	75-24600

Security, Law Enforcement and Internal Security Equipment

491 Anti-riot & crowd control weapons & equipment

Aardvark Tactical Inc.	USA	*545*	75-00122
Accuracy International Ltd	UK	*427*	75-00360
Allied Research Corp.	USA	*549*	75-04670
Armour of America Inc.	USA	*555*	75-08850
Barrday Inc.	Canada	*68*	75-01080
Brock's Explosives Ltd	UK	*437*	75-09980
Charcoal Cloth (International) Ltd	UK	*439*	75-12550
Civil Defence Supply	UK	*440*	75-13350
Grenadier Engineering Inc.	USA	*580*	75-40390
Haley & Weller Ltd	UK	*456*	75-27400
Hiatt & Company Ltd	UK	*458*	75-29180
Israel Military Industries Ltd	Israel	*220*	75-10300
Israel Product Research Co. Ltd	Israel	*219*	75-10000
Law Enforcement International Ltd	UK	*464*	75-34870
Le Protecteur	France	*138*	75-24440
Lightweight Body Armour	UK	*464*	75-35650
Londesborough Security Equipment plc	UK	*465*	75-36300
Meggitt Composites	UK	*467*	75-38345
P.A.D. International	UK	*472*	75-43470
Pains-Wessex Ltd	UK	*472*	75-43510
Pietro Beretta SpA	Italy	*232*	75-02800
R/M Equipment Inc.	USA	*609*	75-75470
Royal Ordnance	UK	*481*	75-50620
SAE Alsetex	France	*142*	75-33810
Silent Partner Body Armor Inc.	USA	*615*	75-80930
Sleeman Engineering	UK	*484*	75-54400
SNPE	France	*145*	75-36315
South Central Ordnance Co.	USA	*616*	75-82400
SPL Ordnance Ltd.	Israel	*225*	80-01935
Swartklip Products	S. Africa	*359*	75-19500
Webley & Scott Ltd	UK	*493*	75-64950
ZARC International Inc.	USA	*628*	75-98050

491 05 Ammunition, riot control

AAI Corp.	USA	*545*	75-00020
Aardvark Tactical Inc.	USA	*545*	75-00122
Accuracy International Ltd	UK	*427*	75-00360
Allied Research Corp.	USA	*549*	75-04670
Armalon Ltd	UK	*431*	75-03729
Brock's Explosives Ltd	UK	*437*	75-09980
Combined Systems Inc.	USA	*564*	75-19893
Haley & Weller Ltd	UK	*456*	75-27400
Israel Military Industries Ltd	Israel	*220*	75-10300
Israel Product Research Co. Ltd	Israel	*219*	75-10000
Manroy Engineering Ltd	**UK**	***466***	**75-37165**
ML Lifeguard Equipment Ltd	UK	*469*	75-40230
Nottingham International	USA	*640*	80-22000
Pains-Wessex Ltd	UK	*472*	75-43510
Pietro Beretta SpA	Italy	*232*	75-02800
Royal Ordnance	UK	*481*	75-50620
Simad SpA	Italy	*230*	74-23350
South Central Ordnance Co.	USA	*616*	75-82400
SPL Ordnance Ltd.	Israel	*225*	80-01935
Swartklip Products	S. Africa	*359*	75-19500

491 05 10 Ampoules, CN / CS

AAI Corp.	USA	*545*	75-00020
Combined Systems Inc.	USA	*564*	75-19893
Haley & Weller Ltd	UK	*456*	75-27400
Israel Product Research Co. Ltd	Israel	*219*	75-10000
SPL Ordnance Ltd.	Israel	*225*	80-01935

491 05 50 Grenades, riot control (see code 522.40.10)

AAI Corp.	USA	*545*	75-00020
Aardvark Tactical Inc.	USA	*545*	75-00122
Allied Research Corp.	USA	*549*	75-04670
Brock's Explosives Ltd	UK	*437*	75-09980

Grenadier Engineering Inc.	USA	*580*	75-40390
Haley & Weller Ltd	UK	*456*	75-27400
Israel Military Industries Ltd	Israel	*220*	75-10300
Israel Product Research Co. Ltd	Israel	*219*	75-10000
Manroy Engineering Ltd	**UK**	***466***	**75-37165**
Nottingham International	USA	*640*	80-22000
Pains-Wessex Ltd	UK	*472*	75-43510
Royal Ordnance	UK	*481*	75-50620
SAE Alsetex	France	*142*	75-33810
Simad SpA	Italy	*230*	74-23350
South Central Ordnance Co.	USA	*616*	75-82400
SPL Ordnance Ltd.	Israel	*225*	80-01935
Swartklip Products	S. Africa	*359*	75-19500

491 10 Batons, electronic

Le Protecteur	France	*138*	75-24440
NII Stali	Russia	*338*	75-14100
South Central Ordnance Co.	USA	*616*	75-82400

491 60 Riot shields (see code 719.60)

Armour of America Inc.	USA	*555*	75-08850
Barrday Inc.	Canada	*68*	75-01080
Civil Defence Supply	UK	*440*	75-13350
Defense Technology Corporation of	USA	*568*	75-24980
Forjas Taurus SA	Brazil	*56*	75-06640
Hiatt & Company Ltd	UK	*458*	75-29180
H.P.P. (UK) Ltd	UK	*459*	75-30360
Israel Product Research Co. Ltd	Israel	*219*	75-10000
Lightweight Body Armour	UK	*464*	75-35650
ML Lifeguard Equipment Ltd	UK	*469*	75-40230
Pains-Wessex Ltd	UK	*472*	75-43510
RBR Armour Ltd.	UK	*478*	75-48370
Royal Ordnance	UK	*481*	75-50620
Sas R&D Services Inc.	USA	*642*	80-26870
Silent Partner Body Armor Inc.	USA	*615*	75-80930
Sleeman Engineering	UK	*484*	75-54400
South Central Ordnance Co.	USA	*616*	75-82400
SPL Ordnance Ltd.	Israel	*225*	80-01935

491 65 Sprays / dispensers, anti-riot

Aardvark Tactical Inc.	USA	*545*	75-00122
Israel Product Research Co. Ltd	Israel	*219*	75-10000
Le Protecteur	France	*138*	75-24440
Pains-Wessex Ltd	UK	*472*	75-43510
SAE Alsetex	France	*142*	75-33810
SNPE	France	*145*	75-36315
South Central Ordnance Co.	USA	*616*	75-82400
SPL Ordnance Ltd.	Israel	*225*	80-01935
ZARC International Inc.	USA	*628*	75-98050

491 65 10 Aerosols, anti-riot

Aardvark Tactical Inc.	USA	*545*	75-00122
Combined Systems Inc.	USA	*564*	75-19893
Israel Product Research Co. Ltd	Israel	*219*	75-10000
Le Protecteur	France	*138*	75-24440
Pains-Wessex Ltd	UK	*472*	75-43510
SNPE	France	*145*	75-36315
South Central Ordnance Co.	USA	*616*	75-82400
SPL Ordnance Ltd.	Israel	*225*	80-01935
Swartklip Products	S. Africa	*359*	75-19500
ZARC International Inc.	USA	*628*	75-98050

491 90 Weapons, anti-riot

AAI Corp.	USA	*545*	75-00020
Aardvark Tactical Inc.	USA	*545*	75-00122
Accuracy International Ltd	UK	*427*	75-00360
Armalon Ltd	UK	*431*	75-03729
Helston Gunsmiths	UK	*458*	75-29010
Israel Military Industries Ltd	Israel	*220*	75-10300
Londesborough Security Equipment plc	UK	*465*	75-36300
Manroy Engineering Ltd	**UK**	***466***	**75-37165**
ML Lifeguard Equipment Ltd	UK	*469*	75-40230
Nottingham International	USA	*640*	80-22000
Pains-Wessex Ltd	UK	*472*	75-43510
Reunert Technology Systems Ltd	S. Africa	*359*	75-18300
Royal Ordnance	UK	*481*	75-50620
South Central Ordnance Co.	USA	*616*	75-82400
SPL Ordnance Ltd.	Israel	*225*	80-01935
ZARC International Inc.	USA	*628*	75-98050

491 90 20 Cannons, water

Israel Product Research Co. Ltd	Israel	*219*	75-10000
SPL Ordnance Ltd.	Israel	*225*	80-01935

491 90 35 Grenade launchers, anti-riot

AAI Corp.	USA	*545*	75-00020
Aardvark Tactical Inc.	USA	*545*	75-00122
Grenadier Engineering Inc.	USA	*580*	75-40390
Indústrias de Defesa	Portugal	*327*	55-00950
Jonathan Arthur Ciener Inc.	USA	*564*	75-18590
ML Lifeguard Equipment Ltd	UK	*469*	75-40230
Nottingham International	USA	*640*	80-22000
Pains-Wessex Ltd	UK	*472*	75-43510
R/M Equipment Inc.	USA	*609*	75-75470
Royal Ordnance	UK	*481*	75-50620
SAE Alsetex	France	*142*	75-33810
South Central Ordnance Co.	USA	*616*	75-82400
SPL Ordnance Ltd.	Israel	*225*	80-01935
Swartklip Products	S. Africa	*359*	75-19500

491 90 40 Guns, anti-riot

Aardvark Tactical Inc.	USA	*545*	75-00122
Accuracy International Ltd	UK	*427*	75-00360
Armalon Ltd	UK	*431*	75-03729
FN Herstal SA	Belgium	*46*	75-03600
FN Manufacturing Inc	USA	*576*	75-35320
Helston Gunsmiths	UK	*458*	75-29010
Manroy Engineering Ltd	**UK**	***466***	**75-37165**
ML Lifeguard Equipment Ltd	UK	*469*	75-40230
Nottingham International	USA	*640*	80-22000
Pains-Wessex Ltd	UK	*472*	75-43510
Reunert Technology Systems Ltd.	S. Africa	*359*	75-18300
R/M Equipment Inc.	USA	*609*	75-75470
Royal Ordnance	UK	*481*	75-50620
SNPE	France	*145*	75-36315
South Central Ordnance Co.	USA	*616*	75-82400
SPL Ordnance Ltd.	Israel	*225*	80-01935
Webley & Scott Ltd.	UK	*493*	75-64950

491 90 45 Shotguns, anti-riot

AAI Corp.	USA	*545*	75-00020
Aardvark Tactical Inc.	USA	*545*	75-00122
Armalon Ltd	UK	*431*	75-03729
Helston Gunsmiths	UK	*458*	75-29010
Jonathan Arthur Ciener Inc.	USA	*564*	75-18590
J.P. Sauer & Sohn GmbH.	Germany	*174*	75-26050
Law Enforcement International Ltd	UK	*464*	75-34870
Manroy Engineering Ltd	**UK**	***466***	**75-37165**
Nottingham International	USA	*640*	80-22000
Pietro Beretta SpA	Italy	*232*	75-02800
Republic Arms (Pty) Ltd	**S. Africa**	***359***	**75-18210**
South Central Ordnance Co.	USA	*616*	75-82400
SPL Ordnance Ltd.	Israel	*225*	80-01935

491 90 45 50 Stun guns, anti-riot

Aardvark Tactical Inc.	USA	*545*	75-00122
Israel Military Industries Ltd	Israel	*220*	75-10300
South Central Ordnance Co.	USA	*616*	75-82400
SPL Ordnance Ltd.	Israel	*225*	80-01935

491 90 50 Pistols, anti-riot

Aardvark Tactical Inc.	USA	*545*	75-00122
Accuracy International Ltd	UK	*427*	75-00360
Grenadier Engineering Inc.	USA	*580*	75-40390
Helston Gunsmiths	UK	*458*	75-29010
Israel Military Industries Ltd	Israel	*220*	75-10300
Manroy Engineering Ltd	**UK**	***466***	**75-37165**
Pains-Wessex Ltd	UK	*472*	75-43510

R/M Equipment Inc.	USA	*609*	75-75470
SNPE	France	*145*	75-36315
South Central Ordnance Co.	USA	*616*	75-82400
SPL Ordnance Ltd.	Israel	*225*	80-01935
Webley & Scott Ltd.	UK	*493*	75-64950

492 Explosives / weapons search equipment

AEA Technology plc	UK	*496*	80-00480
Catsaros Automation SA	Greece	*187*	75-00700
CCS International Ltd	USA	*562*	75-17250
EG&G Inc.	USA	*571*	75-29555
Elektron SA	France	*136*	75-14700
Haley & Weller Ltd	UK	*456*	75-27400
ION Track Instruments	UK	*461*	75-32000
Israel Military Industries Ltd	Israel	*220*	75-10300
P W Allen & Co Ltd	UK	*429*	75-02740
P.A.D. International	UK	*472*	75-43470
PK Electronic International Ltd	UK	*475*	75-45720
S&D Group	UK	*510*	80-32000
SPL Ordnance Ltd	Israel	*225*	80-01935
Vallon GmbH	**Germany**	***176***	**75-32150**

492 30 Explosives / weapons search equipment, fixed

CCS International Ltd	USA	*562*	75-17250
EG&G Inc.	USA	*571*	75-29555
Infinetics Inc.	USA	*584*	75-46700
Israel Military Industries Ltd	Israel	*220*	75-10300
P W Allen & Co Ltd	UK	*429*	75-02740
SPL Ordnance Ltd.	Israel	*225*	80-01935
Vallon GmbH	**Germany**	***176***	**75-32150**

492 36 Explosives search equipment, fixed, sniffer type

Israel Military Industries Ltd	Israel	*220*	75-10300
SPL Ordnance Ltd.	Israel	*225*	80-01935

492 39 Explosives / weapons search equipment, fixed, X-ray

Audiotel International Ltd	UK	*432*	75-04400
CCS International Ltd	USA	*562*	75-17250
Corrigan Technologies Ltd	Canada	*78*	80-00380
EG&G Astrophysics Research Ltd.	UK	*448*	75-19670
EG&G Inc.	USA	*571*	75-29555
Israel Military Industries Ltd	Israel	*220*	75-10300
J.S. Chinn Project Engineering Ltd	UK	*440*	75-13060
P W Allen & Co Ltd	UK	*429*	75-02740
Rapiscan Security Products Ltd	UK	*477*	75-48305
Security Defense Systems Corp.	USA	*613*	75-79080
SPL Ordnance Ltd.	Israel	*225*	80-01935
Todd Research Ltd	UK	*488*	75-60360

492 50 Explosives / weapons search equipment, portable

AEA Technology plc	UK	*496*	80-00480
Audiotel International Ltd	UK	*432*	75-04400
Corrigan Technologies Ltd	Canada	*78*	80-00380
Henotex (1981) Inc	Canada	*72*	75-07100
Israel Military Industries Ltd	Israel	*220*	75-10300
P W Allen & Co Ltd	UK	*429*	75-02740
Rapiscan Security Products Ltd	UK	*477*	75-48305
Security Defense Systems Corp.	USA	*613*	75-79080
SPL Ordnance Ltd.	Israel	*225*	80-01935
Vallon GmbH	**Germany**	***176***	**75-32150**

492 56 Explosives search equipment, portable, sniffer type

Audiotel International Ltd	UK	*432*	75-04400
Graseby Dynamics Ltd	UK	*455*	75-26590
ION Track Instruments	UK	*461*	75-32000
Israel Military Industries Ltd	Israel	*220*	75-10300
PK Electronic International Ltd	UK	*475*	75-45720
Rapiscan Security Products Ltd	UK	*477*	75-48305
SPL Ordnance Ltd	Israel	*225*	80-01935

493 Explosives disposal equipment

AAI Corp.	USA	*545*	75-00020
Alvis Logistics Ltd	UK	*430*	75-02860
Armour of America Inc.	USA	*555*	75-08850
CPAD Technologies Inc.	Canada	*78*	80-00386
Davies Industrial Communications Ltd	UK	*444*	75-17010
Domain Defence Systems Pty Ltd.	Australia	*25*	80-04000
DTC International	USA	*570*	75-27620
Elektron SA	France	*136*	75-14700
Fluor Daniel Inc.	USA	*635*	80-11720
Haley & Weller Ltd	UK	*456*	75-27400
Institut Dr Förster GmbH & Co KG.	Germany	*170*	75-14500
ION Track Instruments	UK	*461*	75-32000
Med-Eng Systems Inc.	Canada	*74*	75-09550
Mining Resource Engineering Ltd	Canada	*74*	75-09770
NABCO Inc.	USA	*598*	75-63550
P W Allen & Co Ltd	UK	*429*	75-02740
P.A.D. International	UK	*472*	75-43470
Pains-Wessex Ltd	UK	*472*	75-43510
PK Electronic International Ltd	UK	*475*	75-45720
Proparms Ltd	Canada	*75*	75-11540
Royal Ordnance	UK	*481*	75-50620
S&D Group	UK	*510*	80-32000
Sas R&D Services Inc.	USA	*642*	80-26870
South Central Ordnance Co.	USA	*616*	75-82400
SPL Ordnance Ltd.	Israel	*225*	80-01935

493 20 Containers, explosives disposal

NABCO Inc.	USA	*598*	75-63550
Silent Partner Body Armor Inc.	USA	*615*	75-80930
South Central Ordnance Co.	USA	*616*	75-82400
SPL Ordnance Ltd.	Israel	*225*	80-01935

493 80 Bomb disposal robot vehicles, remote-controlled

AAI Corp.	USA	*545*	75-00020
Alvis Industries Ltd	UK	*430*	75-02850
Alvis Logistics Ltd	UK	*430*	75-02860
Cybernétix	France	*134*	75-12910
Framatome	France	*136*	75-18405
Giat Industries	France	*126*	55-00100
Glover Webb Ltd	UK	*455*	75-25700
Israel Military Industries Ltd	Israel	*220*	75-10300
Kentree Ltd	Ireland	*215*	75-00250
Med-Eng Systems Inc.	Canada	*74*	75-09550
REMOTEC Inc.	USA	*609*	75-74430
Royal Ordnance	UK	*481*	75-50620
SPL Ordnance Ltd.	Israel	*225*	80-01935

495 Police & Gendarmerie equipment

Applied Holographics plc	UK	*431*	75-03520
Avon Technical Products	UK	*432*	75-05150
Bernardini SA	Brazil	*55*	75-02120
Checkmate Industries Ltd	Hong Kong SAR	*198*	75-00200
Criminal Research Products Inc.	USA	*566*	75-22590
Davies Industrial Communications Ltd	UK	*444*	75-17010
DM Aerospace Ltd	UK	*444*	75-17300
FN Herstal SA	Belgium	*46*	75-03600
GEC-Marconi Communications Ltd	UK	*454*	75-25335
Giat Industries	France	*126*	55-00100
Hiatt & Company Ltd	UK	*458*	75-29180
Hobson & Sons (London) Ltd	UK	*458*	75-29550
H.P.P. (UK) Ltd	UK	*459*	75-30360
Instro Precision Ltd	UK	*461*	75-31630
International Procurement Services	UK	*461*	75-31820
Israel Military Industries Ltd	Israel	*220*	75-10300
Le Protecteur	France	*138*	75-24440
Lightweight Body Armour	UK	*464*	75-35650
Londesborough Security Equipment plc	UK	*465*	75-36300
M. Paz Logistics Ltd	Israel	*224*	80-01600
Meggitt Composites	UK	*467*	75-38345
Neal Recording Systems Ltd.	UK	*470*	75-42270
P W Allen & Co Ltd	UK	*429*	75-02740
P.A.D. International	UK	*472*	75-43470
Philips Traffic & Engineering	Australia	*20*	75-16550
PK Electronic International Ltd	UK	*475*	75-45720
Q-Systems International Inc.	USA	*606*	75-71960
Rabintex Industries Ltd	Israel	*221*	75-17850

RGB Spectrum	USA	*609*	75-75000
SAIT Systems SA	Belgium	*47*	75-05600
SFIM	**France**	***144***	**75-35080**
Shield International Ltd	UK	*511*	80-35500
Sigma Security Devices Ltd	UK	*483*	75-53778
Silent Partner Body Armor Inc.	USA	*615*	75-80930
Smith & Wesson	USA	*615*	75-82060
South Central Ordnance Co.	USA	*616*	75-82400
SPL Ordnance Ltd.	Israel	*225*	80-01935
ZARC International Inc.	USA	*628*	75-98050

495 20 Communications systems, Police & Gendarmerie

Audiotel International Ltd	UK	*432*	75-04400
CSE Aviation Ltd	UK	*500*	80-07600
David Clark Co. Inc.	USA	*564*	75-18850
Davies Industrial Communications Ltd	UK	*444*	75-17010
GEC-Marconi Communications Ltd	UK	*454*	75-25335
Hagen	UK	*456*	75-27220
Harris Corporation	USA	*581*	75-42050
Israel Military Industries Ltd	Israel	*220*	75-10300
Logica SA/NV.	Belgium	*48*	80-01000
Logica UK Limited.	UK	*506*	80-14350
Microelectronics, guidance and	Turkey	*402*	55-00050
MPB Technologies Inc	Canada	*74*	75-09855
Northern Airborne Technology Ltd	Canada	*74*	75-10145
Philips Traffic & Engineering	Australia	*20*	75-16550
PRC Inc.	USA	*640*	80-24800
SAIT Systems SA	Belgium	*47*	75-05600
Security Research Ltd	UK	*482*	75-52450

495 20 70 Communications equipment, Police & Gendarmerie

Audiotel International Ltd	UK	*432*	75-04400
CSE Aviation Ltd	UK	*500*	80-07600
David Clark Co. Inc.	USA	*564*	75-18850
Davies Industrial Communications Ltd	UK	*444*	75-17010
Hagen	UK	*456*	75-27220
Harris Corporation	USA	*581*	75-42050
Logica SA/NV.	Belgium	*48*	80-01000
Microelectronics, guidance and	Turkey	*402*	55-00050
Moonraker Australia Pty Ltd	Australia	*19*	75-13770
Neal Recording Systems Ltd	UK	*470*	75-42270
Northern Airborne Technology Ltd	Canada	*74*	75-10145
Racal Acoustics Ltd	UK	*476*	75-47653
SAIT Systems SA	Belgium	*47*	75-05600
Security Research Ltd	UK	*482*	75-52450

495 20 70 30 Radios, portable, Police & Gendarmerie

Audiotel International Ltd	UK	*432*	75-04400
Icom (UK) Ltd	UK	*460*	75-31070
SAIT Systems SA	Belgium	*47*	75-05600
Security Research Ltd	UK	*482*	75-52450

495 20 70 30 20 Radios, vehicle-mounted, Police & Gendarmerie

Davies Industrial Communications Ltd	UK	*444*	75-17010
Icom (UK) Ltd	UK	*460*	75-31070
Northern Airborne Technology Ltd	Canada	*74*	75-10145
SAIT Systems SA	Belgium	*47*	75-05600

495 30 Fingerprinting equipment

Criminal Research Products Inc.	USA	*566*	75-22590
PK Electronic International Ltd	UK	*475*	75-45720

495 35 Fingerprint recognition equipment

Criminal Research Products Inc.	USA	*566*	75-22590
Logica UK Limited.	UK	*506*	80-14350
Remsdaq Ltd	UK	*478*	75-49210

495 40 Forced-entry equipment, building, Police & Gendarmerie

Sigma Security Devices Ltd	UK	*483*	75-53778

495 80 Police & Gendarmerie vehicles

Carrozzeria Boneschi Sarl	Italy	*233*	75-05000
Giat Industries	France	*126*	55-00100
Glover Webb Ltd	UK	*455*	75-25700
GM-Holden's Automotive	Australia	*16*	75-07520
South Central Ordnance Co.	USA	*616*	75-82400
Vehicle Systems Development Corp.	USA	*625*	75-93570

495 80 10 Police & Gendarmerie buses / coaches

Carrozzeria Boneschi Sarl	Italy	*233*	75-05000
Glover Webb Ltd	UK	*455*	75-25700
GM-Holden's Automotive	Australia	*16*	75-07520
South Central Ordnance Co.	USA	*616*	75-82400

495 80 40 Police patrol cars (see also code 141.70.80.10)

Carrozzeria Boneschi Sarl	Italy	*233*	75-05000

495 80 50 Police & Gendarmerie vehicles, armoured anti-riot

AV Technology LLC	USA	*556*	75-11405
Bernardini SA	Brazil	*55*	75-02120
Carrozzeria Boneschi Sarl	Italy	*233*	75-05000
Giat Industries	France	*126*	55-00100
Glover Webb Ltd	UK	*455*	75-25700
Gruppo Nuova MA.NA.RO.	Italy	*235*	75-10730
MOL Cy NV	Belgium	*46*	75-04998
South Central Ordnance Co.	USA	*616*	75-82400
SPL Ordnance Ltd.	Israel	*225*	80-01935

495 80 55 Police & Gendarmerie vehicles, bomb disposal

Glover Webb Ltd	UK	*455*	75-25700
Kentree Ltd	Ireland	*215*	75-00250
South Central Ordnance Co.	USA	*616*	75-82400
SPL Ordnance Ltd.	Israel	*225*	80-01935

495 80 70 Police & Gendarmerie vehicle equipment

Carrozzeria Boneschi Sarl	Italy	*233*	75-05000
Civil Defence Supply	UK	*440*	75-13350
Glover Webb Ltd	UK	*455*	75-25700
SFIM	**France**	***144***	**75-35080**
South Central Ordnance Co.	USA	*616*	75-82400
SPL Ordnance Ltd.	Israel	*225*	80-01935

495 90 Weapons, Police / Gendarmerie

FN Herstal SA	Belgium	*46*	75-03600
Heckler & Koch Inc	USA	*581*	75-42860
Helston Gunsmiths	UK	*458*	75-29010
Israel Military Industries Ltd	Israel	*220*	75-10300
Jonathan Arthur Ciener Inc.	USA	*564*	75-18590
Pains-Wessex Ltd	UK	*472*	75-43510
Smith & Wesson	USA	*615*	75-82060
South Central Ordnance Co.	USA	*616*	75-82400
ZARC International Inc.	USA	*628*	75-98050

495 90 50 Pistols, Police / Gendarmerie

Accuracy International Ltd	UK	*427*	75-00360
Armalon Ltd	UK	*431*	75-03729
FN Herstal SA	Belgium	*46*	75-03600
Heckler & Koch Inc	USA	*581*	75-42860
INIMEX	**Belgium**	***46***	**75-04650**
Israel Military Industries Ltd	Israel	*220*	75-10300
Nottingham International	USA	*640*	80-22000

Smith & Wesson	USA	*615*	75-82060
South Central Ordnance Co.	USA	*616*	75-82400

495 90 60 Rifles, Police / Gendarmerie

Accuracy International Ltd	UK	*427*	75-00360
Armalon Ltd	UK	*431*	75-03729
FN Herstal SA	Belgium	*46*	75-03600
Heckler & Koch Inc	USA	*581*	75-42860
Indústrias de Defesa	Portugal	*327*	55-00950
INIMEX	**Belgium**	***46***	**75-04650**
Israel Military Industries Ltd	Israel	*220*	75-10300
Nottingham International	USA	*640*	80-22000
South Central Ordnance Co.	USA	*616*	75-82400
Steyr Mannlicher AG & Co KG	Austria	*34*	75-02820

495 90 70 Shotguns, Police / Gendarmerie

Accuracy International Ltd	UK	*427*	75-00360
Armalon Ltd	UK	*431*	75-03729
Companhia Brasileira de Cartuchos	Brazil	*55*	75-03150
FN Herstal SA	Belgium	*46*	75-03600
Heckler & Koch Inc	USA	*581*	75-42860
Helston Gunsmiths	UK	*458*	75-29010
Indústrias de Defesa	Portugal	*327*	55-00950
Israel Military Industries Ltd	Israel	*220*	75-10300
J.P. Sauer & Sohn GmbH	Germany	*174*	75-26050
Nottingham International	USA	*640*	80-22000
Republic Arms (Pty) Ltd	**S. Africa**	***359***	**75-18210**
Reunert Technology Systems Ltd	S. Africa	*359*	75-18300
South Central Ordnance Co.	USA	*616*	75-82400

495 99 Other Police & Gendarmerie equipment

Armour of America Inc.	USA	*555*	75-08850
Beta Company	USA	*558*	75-13750
Chubb Safe Equipment Co.	UK	*440*	75-13205
Civil Defence Supply	UK	*440*	75-13350
GEC-Marconi Avionics Ltd	UK	*454*	75-25320
Hiatt & Company Ltd	UK	*458*	75-29180
ML Lifeguard Equipment Ltd	UK	*469*	75-40230
NOVO Corporation	Russia	*338*	75-14200
Price Western Leather Co. Ltd.	UK	*476*	75-46755
SFIM	**France**	***144***	**75-35080**
SIMA PERU	Peru	*319*	55-01950
Vision Research Co.	UK	*491*	75-63360

495 99 60 Surveillance & identification systems

Computing Devices Canada	Canada	*70*	75-03600
Davin Optical Ltd	UK	*444*	75-17100
Delft Sensor Systems	**Belgium**	***46***	**75-02500**
Delft Sensor Systems	Netherlands	*292*	75-01590
DM Aerospace Ltd	UK	*444*	75-17300
Elettronica (UK) Ltd	UK	*448*	75-20475
Fraser-Volpe Corp.	USA	*576*	75-35950
GEC-Marconi Avionics Ltd	UK	*454*	75-25320
Graflex Inc	USA	*579*	75-40310
Icom (UK) Ltd	UK	*460*	75-31070
Insight Vision Systems Ltd	UK	*461*	75-31530
Instro Precision Ltd	UK	*461*	75-31630
M. Paz Logistics Ltd	Israel	*224*	80-01600
P W Allen & Co Ltd	UK	*429*	75-02740
PK Electronic International Ltd	UK	*475*	75-45720
RGB Spectrum	USA	*609*	75-75000
Sparton Technology Inc.	USA	*616*	75-82950
Vanguard Research Inc.	USA	*644*	80-34100

495 99 60 40 Surveillance & identification systems, infra-red

Delft Sensor Systems	**Belgium**	***46***	**75-02500**
Delft Sensor Systems	Netherlands	*292*	75-01590
Elettronica (UK) Ltd	UK	*448*	75-20475
GEC-Marconi Avionics Ltd	UK	*454*	75-25320
Insight Vision Systems Ltd	UK	*461*	75-31530
Litton Industries Inc.	USA	*590*	75-55400
M. Paz Logistics Ltd	Israel	*224*	80-01600
MPB Technologies Inc	Canada	*74*	75-09855
RGB Spectrum	USA	*609*	75-75000
SAGEM SA	**France**	***143***	**75-33950**
Sparton Technology Inc.	USA	*616*	75-82950

495 99 60 40 20 Surveillance & identification systems, infra-red, airborne

GEC-Marconi Avionics Ltd	UK	*454*	75-25320
Insight Vision Systems Ltd	UK	*461*	75-31530
M. Paz Logistics Ltd	Israel	*224*	80-01600
RGB Spectrum	USA	*609*	75-75000
Thomson-CSF Optronique	France	*148*	75-45137

495 99 60 40 30 Surveillance & identification systems, infra-red, naval

Insight Vision Systems Ltd	UK	*461*	75-31530
M. Paz Logistics Ltd	Israel	*224*	80-01600
RGB Spectrum	USA	*609*	75-75000
SAGEM SA	**France**	***143***	**75-33950**

495 99 60 40 40 Surveillance & identification systems, infra-red, ground-based

Custom Cameras Ltd	UK	*444*	75-16444
Delft Sensor Systems	**Belgium**	***46***	**75-02500**
Delft Sensor Systems	Netherlands	*292*	75-01590
GEC-Marconi Avionics Ltd	UK	*454*	75-25320
Insight Vision Systems Ltd	UK	*461*	75-31530
M. Paz Logistics Ltd	Israel	*224*	80-01600
RGB Spectrum	USA	*609*	75-75000
SAGEM SA	**France**	***143***	**75-33950**
Sparton Technology Inc.	USA	*616*	75-82950

495 99 60 60 Surveillance & identification systems, TV

Custom Cameras Ltd	UK	*444*	75-16444
Davin Optical Ltd	UK	*444*	75-17100
Diamond Electronics Inc.	USA	*569*	75-26450
DM Aerospace Ltd	UK	*444*	75-17300
Graflex Inc	USA	*579*	75-40310
Insight Vision Systems Ltd	UK	*461*	75-31530
Instro Precision Ltd	UK	*461*	75-31630
M. Paz Logistics Ltd	Israel	*224*	80-01600
P W Allen & Co Ltd	UK	*429*	75-02740
PK Electronic International Ltd	UK	*475*	75-45720
RGB Spectrum	USA	*609*	75-75000
Vanguard Research Inc.	USA	*644*	80-34100

495 99 60 60 20 Surveillance & identification systems, TV, airborne

Diamond Electronics Inc.	USA	*569*	75-26450
GEC-Marconi Avionics Ltd	UK	*454*	75-25320
Graflex Inc	USA	*579*	75-40310
Insight Vision Systems Ltd	UK	*461*	75-31530
M. Paz Logistics Ltd	Israel	*224*	80-01600

511 24 41 Missiles, surface to air, ground-based, air defense

Alenia SpA	Italy	*232*	75-01170
Australian Defence Industries Ltd	Australia	*9*	55-01000
Bofors AB	Sweden	*375*	74-00020
Chungshan Inst. of Science & Tech.	Taiwan	*392*	60-01000
Cranfield University	UK	*419*	60-00350
GEL Engineering Ltd	UK	*454*	75-25500
INIMEX	**Belgium**	***46***	**75-04650**
Lockheed Martin Missiles & Space	USA	*592*	75-55640
Matra BAe Dynamics	UK	*425*	74-06850
Matra Défense-Espace	France	*139*	75-26000
Oerlikon-Contraves AG.	Switzerland	*387*	75-09250
Rafael	Israel	*216*	55-01800
Royal Ordnance	UK	*481*	75-50620
The Boeing Company	USA	*560*	75-14250

511 24 41 10 Missiles, surface to air, ground-based, air defense, low-level

Alenia SpA	Italy	*232*	75-01170
Arab Organization for Ind.	Egypt	*109*	55-01700
Avibras Aeroespacial SA	Brazil	*55*	75-00968
Bofors AB	Sweden	*375*	74-00020
Cranfield University	UK	*419*	60-00350
Euromissile GIE	France	*130*	74-00535
GEL Engineering Ltd	UK	*454*	75-25500
INIMEX	**Belgium**	***46***	**75-04650**
Matra BAe Dynamics	UK	*425*	74-06850
Matra Défense-Espace	France	*139*	75-26000
Oerlikon-Contraves AG.	Switzerland	*387*	75-09250
Rafael	Israel	*216*	55-01800
Royal Ordnance	UK	*481*	75-50620
Shorts Missile Systems Ltd	**UK**	***483***	**75-53550**

511 24 41 12 Missiles, surface to air, ground-based, air defense, low-level, man-portable

Arab Organization for Ind.	Egypt	*109*	55-01700
Bofors AB	Sweden	*375*	74-00020
INIMEX	**Belgium**	***46***	**75-04650**
Matra BAe Dynamics	UK	*425*	74-06850
Matra Défense-Espace	France	*139*	75-26000
Microelectronics, guidance and	Turkey	*402*	55-00050
Royal Ordnance	UK	*481*	75-50620
Sakr Factory for Developed Ind.	Egypt	*109*	55-01700
Shorts Missile Systems Ltd	**UK**	***483***	**75-53550**

511 24 41 14 Missiles, surface to air, ground-based, air defense, low-level, mobile

Alenia SpA	Italy	*232*	75-01170
Bofors AB	Sweden	*375*	74-00020
INIMEX	**Belgium**	***46***	**75-04650**
Matra BAe Dynamics	UK	*425*	74-06850
Matra Défense-Espace	France	*139*	75-26000
Oerlikon-Contraves AG.	Switzerland	*387*	75-09250
Royal Ordnance	UK	*481*	75-50620
Shorts Missile Systems Ltd	**UK**	***483***	**75-53550**

511 24 41 14 10 Missiles, surface to air, ground-based, air defense, low-level, trailer-mounted

Alenia SpA	Italy	*232*	75-01170
Matra BAe Dynamics	UK	*425*	74-06850
Matra Défense-Espace	France	*139*	75-26000
Oerlikon-Contraves AG.	Switzerland	*387*	75-09250
Royal Ordnance	UK	*481*	75-50620

511 24 41 14 20 Missiles, surface to air, ground-based, air defense, low-level, vehicle-mounted

Cranfield University	UK	*419*	60-00350
Eurosam GIE.	France	*130*	74-00600
GEL Engineering Ltd	UK	*454*	75-25500
INIMEX	**Belgium**	***46***	**75-04650**
Matra BAe Dynamics	UK	*425*	74-06850
Matra Défense-Espace	France	*139*	75-26000
Oerlikon-Contraves AG.	Switzerland	*387*	75-09250
Rafael	Israel	*216*	55-01800
Shorts Missile Systems Ltd	**UK**	***483***	**75-53550**
Swiss Aircraft & Systems Co	Switzerland	*384*	55-00100

511 24 41 20 Missiles, surface to air, ground-based, air defense, low-medium altitude

Alenia SpA	Italy	*232*	75-01170
Cranfield University	UK	*419*	60-00350
GEL Engineering Ltd	UK	*454*	75-25500
INIMEX	**Belgium**	***46***	**75-04650**
Matra BAe Dynamics	UK	*425*	74-06850
Matra Défense-Espace	France	*139*	75-26000
Oerlikon-Contraves AG.	Switzerland	*387*	75-09250

511 24 41 24 Missiles, surface to air, ground-based, air defense, low-medium altitude, mobile

Cranfield University	UK	*419*	60-00350
Eurosam GIE.	France	*130*	74-00600
GEL Engineering Ltd	UK	*454*	75-25500
INIMEX	**Belgium**	***46***	**75-04650**
Matra BAe Dynamics	UK	*425*	74-06850
Matra Défense-Espace	France	*139*	75-26000
Mitsubishi Heavy Industries Ltd	Japan	*253*	75-05800
Oerlikon-Contraves AG.	Switzerland	*387*	75-09250

511 24 41 30 Missiles, surface to air, ground-based, air defense, high altitude

Cranfield University	UK	*419*	60-00350
Eurosam GIE.	France	*130*	74-00600
GEL Engineering Ltd	UK	*454*	75-25500
Hughes Missile Systems Company	USA	*583*	75-44865

511 24 41 32 Missiles, surface to air, ground-based, air defense, high altitude, anti-missile

Cranfield University	UK	*419*	60-00350
Eurosam GIE.	France	*130*	74-00600
GEL Engineering Ltd	UK	*454*	75-25500

511 25 Missiles & missile systems, surface to surface

Bofors AB	Sweden	*375*	74-00020
Cenzin Co. Ltd.	Poland	*324*	75-00200
CIC International Ltd	USA	*563*	75-18570
Daimler-Benz Aerospace France SAS	**France**	***134***	**75-12950**
INIMEX	**Belgium**	***46***	**75-04650**
Israel Military Industries Ltd	Israel	*220*	75-10300
Matra BAe Dynamics	UK	*425*	74-06850
Matra Défense-Espace	France	*139*	75-26000
The Boeing Company	USA	*560*	75-14250
United Defense L.P.	USA	*624*	75-92050

511 25 30 Missiles, surface to surface, ship-based

Chungshan Inst. of Science & Tech.	Taiwan	*392*	60-01000
GEL Engineering Ltd	UK	*454*	75-25500
Matra BAe Dynamics	UK	*425*	74-06850
Matra Défense-Espace	France	*139*	75-26000
Saab Dynamics AB	Sweden	*378*	75-09100

511 25 31 Missiles, surface to surface, ship-based, anti-ship

INIMEX	**Belgium**	***46***	**75-04650**
Israel Aircraft Industries Ltd	Israel	*219*	75-10200
Matra BAe Dynamics	UK	*425*	74-06850
Matra Défense-Espace	France	*139*	75-26000
Mitsubishi Heavy Industries Ltd	Japan	*253*	75-05800
Norwegian Defence Technology A/S.	Norway	*310*	75-02050
Saab Dynamics AB	Sweden	*378*	75-09100

511 25 35 Missiles, surface to surface, ship-based, land target

GEL Engineering Ltd	UK	*454*	75-25500

511 25 35 10 Missiles, surface to surface, ship-based, cruise

GEL Engineering Ltd	UK	*454*	75-25500

511 25 38 Missiles, surface to surface, ship-based, strategic

GEL Engineering Ltd	UK	*454*	75-25500

511 25 38 20 Missiles, surface to surface, ship-based, strategic, medium range

GEL Engineering Ltd	UK	*454*	75-25500

511 25 40 Missiles & missile systems, surface to surface, ground-based

Avibras Aeroespacial SA	Brazil	*55*	75-00968
CIC International Ltd	USA	*563*	75-18570
Fairey Hydraulics	UK	*450*	75-22255
GEL Engineering Ltd	UK	*454*	75-25500
Israel Military Industries Ltd	Israel	*220*	75-10300
Matra BAe Dynamics	UK	*425*	74-06850
The Boeing Company	USA	*560*	75-14250
United Defense L.P.	USA	*624*	75-92050

511 25 41 Missiles, surface to surface, ground-based, anti-tank

Arab Organization for Ind.	Egypt	*109*	55-01700
Avibras Aeroespacial SA	Brazil	*55*	75-00968
Bofors AB	Sweden	*375*	74-00020
Cenzin Co. Ltd.	Poland	*324*	75-00200
CIC International Ltd	USA	*563*	75-18570
Euromissile Dynamics Group	France	*130*	74-00530
Euromissile GIE	France	*130*	74-00535
Hughes Missile Systems Company	USA	*583*	75-44865
INIMEX	**Belgium**	***46***	**75-04650**
Israel Military Industries Ltd	Israel	*220*	75-10300
Luchaire Défense SA	France	*139*	75-25400
Matra BAe Dynamics	UK	*425*	74-06850
Royal Ordnance	UK	*481*	75-50620
Société Europeenne de Propulsion	France	*144*	75-34850
Swiss Aircraft & Systems Co	Switzerland	*384*	55-00100

511 25 41 20 Missiles, surface to surface, ground-based, anti-tank, man-portable

Arab Organization for Ind.	Egypt	*109*	55-01700
CIC International Ltd	USA	*563*	75-18570
Euromissile Dynamics Group	France	*130*	74-00530
Euromissile GIE	France	*130*	74-00535
Hunting Engineering Ltd	UK	*459*	75-30700
Israel Military Industries Ltd	Israel	*220*	75-10300
Matra BAe Dynamics	UK	*425*	74-06850
Royal Ordnance	UK	*481*	75-50620
Swiss Aircraft & Systems Co	Switzerland	*384*	55-00100

511 25 41 40 Missiles, surface to surface, ground-based, anti-tank, vehicle-mounted

Arab Organization for Ind.	Egypt	*109*	55-01700
Avibras Aeroespacial SA	Brazil	*55*	75-00968
CIC International Ltd	USA	*563*	75-18570
Euromissile GIE	France	*130*	74-00535
INIMEX	**Belgium**	***46***	**75-04650**
Israel Military Industries Ltd	Israel	*220*	75-10300
Matra BAe Dynamics	UK	*425*	74-06850
NORINCO	China PR	*88*	55-00280
Royal Ordnance	UK	*481*	75-50620
Swiss Aircraft & Systems Co	Switzerland	*384*	55-00100

511 25 42 Missiles, surface to surface, artillery tactical / battlefield support

Arab Organization for Ind.	Egypt	*109*	55-01700
Avibras Aeroespacial SA	Brazil	*55*	75-00968
China National Precision Machinery	China PR	*88*	55-00265
CIC International Ltd	USA	*563*	75-18570
INIMEX	**Belgium**	***46***	**75-04650**
Israel Military Industries Ltd	Israel	*220*	75-10300

511 25 43 Missiles, multi-purpose assault, surface to surface

Arab Organization for Ind.	Egypt	*109*	55-01700
CIC International Ltd	USA	*563*	75-18570
INIMEX	**Belgium**	***46***	**75-04650**
Israel Military Industries Ltd	Israel	*220*	75-10300

511 25 44 Missiles, surface to surface, coastal defense

Avibras Aeroespacial SA	Brazil	*55*	75-00968
CIC International Ltd	USA	*563*	75-18570
INIMEX	**Belgium**	***46***	**75-04650**
Matra BAe Dynamics	UK	*425*	74-06850
Matra Défense-Espace	France	*139*	75-26000
Saab Dynamics AB	Sweden	*378*	75-09100

511 25 45 Missiles, surface to surface, cruise, ground-launched

CIC International Ltd	USA	*563*	75-18570
Cranfield University	UK	*419*	60-00350
GEL Engineering Ltd	UK	*454*	75-25500

511 25 48 Missiles, surface to surface, strategic, ground-based

CIC International Ltd	USA	*563*	75-18570
GEL Engineering Ltd	UK	*454*	75-25500
The Boeing Company	USA	*560*	75-14250

511 25 48 10 Missiles, surface to surface, strategic, intercontinental ballistic

CIC International Ltd	USA	*563*	75-18570
GEL Engineering Ltd	UK	*454*	75-25500
Space and Defence Division	France	*131*	75-00600

511 25 48 20 Missiles, surface to surface, strategic, medium range, land-based

Aerospatiale	France	*131*	75-00600
CIC International Ltd	USA	*563*	75-18570
GEL Engineering Ltd	UK	*454*	75-25500
Labtech Ltd	UK	*464*	75-34280

511 25 50 Missiles, surface to surface, tactical ballistic

Aerospatiale	France	*131*	75-00600
GEL Engineering Ltd	UK	*454*	75-25500
Hunting Engineering Ltd	UK	*459*	75-30700
Space and Defence Division	France	*131*	75-00600

511 26 Missiles, surface to subsurface

AeroSpace Technologies of Australia	Australia	*11*	75-00340

511 28 Missiles, subsurface to surface

Aerospatiale	France	*131*	75-00600
GEL Engineering Ltd	UK	*454*	75-25500

511 28 38 Missiles, subsurface to surface, strategic

Aerospatiale	France	*131*	75-00600

GEL Engineering Ltd	UK	454	75-25500
Hunting Engineering Ltd	UK	459	75-30700

511 29 Missile components

Accurate Energetic Systems, LLC	USA	545	75-00330
Aeronca Inc.	USA	547	75-02200
Alenia SpA	Italy	232	75-01170
AlliedSignal Aerospace Canada	Canada	68	75-00470
A.O. Henton Engineering Co Ltd	UK	458	75-29040
Astech Manufacturing Inc.	USA	555	75-09700
Avibras Aeroespacial SA	Brazil	55	75-00968
Bodenseewerk Gerätetechnik GmbH	Germany	165	75-03000
Busak + Shamban Inc.	USA	561	75-16006
Carter Manufacturing Corp.	USA	562	75-16800
CIC International Ltd	USA	563	75-18570
Daimler-Benz Aerospace France SAS	**France**	**134**	**75-12950**
Diehl GmbH & Co	Germany	167	75-06500
Dowty Aerospace Wolverhampton	UK	446	75-18589
Eurometaal NV	Netherlands	292	75-02200
Fairey Hydraulics	UK	450	75-22255
Floturn Inc.	USA	576	75-35060
Fluorocarbon Company Limited	UK	451	75-24020
Fokker Space & Systems	Netherlands	293	75-02325
Fokker Special Products BV	Netherlands	293	75-02330
GEC-Marconi Avionics Ltd	UK	454	75-25320
General Pneumatics	USA	578	75-38950
Global Defense Products Inc.	USA	579	75-39770
Hollow Extrusions Ltd	UK	458	75-29652
Israel Military Industries Ltd	Israel	220	75-10300
Japan Aviation Electronics Ind. Ltd	Japan	252	75-03300
Kern Engrng & Mfg Corp.	USA	588	75-52330
L'Garde Inc.	USA	590	75-55060
Litton Industries Inc.	USA	590	75-55400
Lockheed Martin Missiles & Space	USA	592	75-55640
Lucas Aerospace FCS Paris	France	139	75-25390
MA/COM Antenna & Cable Div.	USA	593	75-56953
MAN Technologie AG	Germany	172	75-19650
Matra BAe Dynamics	UK	425	74-06850
Matra Défense-Espace	France	139	75-26000
Microelectronics, guidance and	Turkey	402	55-00050
Moog Inc.	USA	597	75-62550
Moog Italiana Srl	Italy	237	75-16230
Moog Japan Ltd.	Japan	253	75-06100
Motorola Ltd	UK	469	75-41050
National Forge Co.	USA	598	75-63850
Nera Telecommunications	UK	470	75-42318
Normalair-Garrett Ltd	UK	425	74-07000
Norton Performance Plastics Corp.	USA	600	75-65520
Paulstra SNC Hutchinson	France	140	75-29100
Riva Calzoni SpA	Italy	238	75-20500
Royal Ordnance	UK	481	75-50620
Ruggieri Technologie	France	142	75-33500
SFIM	**France**	**144**	**75-35080**
Société d'Études, Réalisations et	France	144	75-34885
Textron Systems Division	USA	621	75-88655
Thomas Instrument & Machine Co. Inc.	USA	622	75-88860
Universal Propulsion Co. Inc.	USA	625	75-92700
Voss Industries Inc.	USA	626	75-94550
Votaw Precision Technologies Inc.	USA	626	75-94580
West Electronics	USA	627	75-95750

511 29 20 Missile fin actuation systems

Israel Military Industries Ltd	Israel	220	75-10300
Litton Industries Inc.	USA	590	75-55400
Moog Inc.	USA	597	75-62550
Normalair-Garrett Ltd	UK	425	74-07000
Universal Propulsion Co. Inc.	USA	625	75-92700

511 29 30 Missile guidance systems

Alenia SpA	Italy	232	75-01170
Avibras Aeroespacial SA	Brazil	55	75-00968
Bodenseewerk Gerätetechnik GmbH	Germany	165	75-03000
CIC International Ltd	USA	563	75-18570
Cranfield University	UK	419	60-00350
Cray Marine	UK	443	75-15915
Daimler-Benz Aerospace France SAS	**France**	**134**	**75-12950**
EASAMS Ltd	UK	501	80-09000
GEL Engineering Ltd	UK	454	75-25500
Hughes Microelectronics Europe Ltd	UK	459	75-30480
Israel Military Industries Ltd	Israel	220	75-10300
L'Garde Inc.	USA	590	75-55060
Matra Défense-Espace	France	139	75-26000
Microelectronics, guidance and	Turkey	402	55-00050
Nera Telecommunications	UK	470	75-42318
SAGEM SA	**France**	**143**	**75-33950**
Sakr Factory for Developed Ind.	Egypt	109	55-01700

511 29 33 Missile guidance systems, fibre optic

CIC International Ltd	USA	563	75-18570
Cranfield University	UK	419	60-00350
GEC-Marconi Research Centre	UK	419	60-00700
GEL Engineering Ltd	UK	454	75-25500
Israel Military Industries Ltd	Israel	220	75-10300
Litton Industries Inc.	USA	590	75-55400

511 29 34 Missile guidance systems, inertial

Bodenseewerk Gerätetechnik GmbH	Germany	165	75-03000
British Aerospace (Systems &	UK	436	75-08890
Cranfield University	UK	419	60-00350
GEC-Marconi Research Centre	UK	419	60-00700
GEL Engineering Ltd	UK	454	75-25500
Israel Military Industries Ltd	Israel	220	75-10300
Japan Aviation Electronics Ind. Ltd	Japan	252	75-03300
Litton Industries Inc.	USA	590	75-55400
Matra BAe Dynamics	UK	425	74-06850
SAGEM SA	**France**	**143**	**75-33950**
SFIM	**France**	**144**	**75-35080**

511 29 35 Missile guidance systems, infra-red

Avibras Aeroespacial SA	Brazil	55	75-00968
Bodenseewerk Gerätetechnik GmbH	Germany	165	75-03000
Cranfield University	UK	419	60-00350
GEC-Marconi Avionics Ltd	UK	454	75-25320
GEC-Marconi Research Centre	UK	419	60-00700
GEL Engineering Ltd	UK	454	75-25500
Israel Military Industries Ltd	Israel	220	75-10300
Matra BAe Dynamics	UK	425	74-06850
Matra Défense-Espace	France	139	75-26000
Microelectronics, guidance and	Turkey	402	55-00050
Octec Ltd	UK	471	75-42950
SAGEM SA	**France**	**143**	**75-33950**

511 29 36 Missile guidance systems, laser

Bodenseewerk Gerätetechnik GmbH	Germany	165	75-03000
CIC International Ltd	USA	563	75-18570
Cranfield University	UK	419	60-00350
GEC-Marconi Avionics Ltd	UK	454	75-25320
GEC-Marconi Research Centre	UK	419	60-00700
GEL Engineering Ltd	UK	454	75-25500
Israel Military Industries Ltd	Israel	220	75-10300
Litton Industries Inc.	USA	590	75-55400
Matra BAe Dynamics	UK	425	74-06850
Shorts Missile Systems Ltd	**UK**	**483**	**75-53550**

511 29 37 Missile guidance systems, radio command

Avibras Aeroespacial SA	Brazil	55	75-00968
CIC International Ltd	USA	563	75-18570
Cincinnati Electronics Corp.	USA	564	75-18640
Daimler-Benz Aerospace France SAS	**France**	**134**	**75-12950**
Israel Military Industries Ltd	Israel	220	75-10300
Motorola Ltd	UK	469	75-41050

511 29 38 Missile guidance systems, terrain comparison (TERCOM)

British Aerospace (Systems &	UK	436	75-08890
Cranfield University	UK	419	60-00350
GEC-Marconi Research Centre	UK	419	60-00700
GEL Engineering Ltd	UK	454	75-25500
Israel Military Industries Ltd	Israel	220	75-10300

Labtech Ltd	UK	464	75-34280
Matra BAe Dynamics	UK	425	74-06850
SAGEM SA	**France**	**143**	**75-33950**

511 29 39 Missile guidance systems, wire

Avibras Aeroespacial SA	Brazil	55	75-00968
CIC International Ltd	USA	563	75-18570
Matra BAe Dynamics	UK	425	74-06850
NORINCO	China PR	88	55-00280

511 29 40 Missile homing heads

Alenia SpA	Italy	232	75-01170
Analysis Management Systems	S. Africa	357	75-00770
Avibras Aeroespacial SA	Brazil	55	75-00968
Bodenseewerk Gerätetechnik GmbH	Germany	165	75-03000
Central Design Bureau	Ukraine	406	75-00800
Cranfield University	UK	419	60-00350
GEC-Marconi Avionics Ltd	UK	454	75-25320
GEL Engineering Ltd	UK	454	75-25500
Israel Military Industries Ltd	Israel	220	75-10300
Matra BAe Dynamics	UK	425	74-06850
Moscow Research Institute Agat	Russia	335	60-02780

511 29 40 90 Missile homing head components

Albacom Ltd	UK	429	75-02372
Analysis Management Systems	S. Africa	357	75-00770
Bodenseewerk Gerätetechnik GmbH	Germany	165	75-03000
Central Design Bureau	Ukraine	406	75-00800
ERA Technology Ltd	UK	502	80-09600
GEC-Marconi Avionics Ltd	UK	454	75-25320
Israel Military Industries Ltd	Israel	220	75-10300
Litton Industries Inc.	USA	590	75-55400
Moscow Research Institute Agat	Russia	335	60-02780
Norton Performance Plastics Corp.	USA	600	75-65520

511 29 90 Missile warheads

Avibras Aeroespacial SA	Brazil	55	75-00968
CIC International Ltd	USA	563	75-18570
Daicel Chemical Industries Ltd	Japan	250	75-00950
Ensign Bickford Co.	USA	574	75-31700
GEL Engineering Ltd	UK	454	75-25500
Hunting Engineering Ltd	UK	459	75-30700
Israel Military Industries Ltd	Israel	220	75-10300
Matra Défense-Espace	France	139	75-26000
Raufoss Technology AS	Norway	310	75-02200
Royal Ordnance	UK	481	75-50620
Shorts Missile Systems Ltd	**UK**	**483**	**75-53550**
Société d'Études, Réalisations et	France	144	75-34885
Wyman-Gordon Ltd	UK	494	75-67280

511 30 Munitions, terminally-guided

Aerojet	USA	547	75-01500
Analysis Management Systems	S. Africa	357	75-00770
Bodenseewerk Gerätetechnik GmbH	Germany	165	75-03000
CIC International Ltd	USA	563	75-18570
Diehl GmbH & Co	Germany	167	75-06500
Eurometaal NV	Netherlands	292	75-02200
Hunting Engineering Ltd	UK	459	75-30700
Israel Military Industries Ltd	Israel	220	75-10300
Matra BAe Dynamics	UK	425	74-06850
Santa Barbara SA	Spain	364	55-01000
Sener Ingenieria y Sistemas SA	Spain	367	75-05200

511 30 10 Artillery shells, terminally-guided

Bodenseewerk Gerätetechnik GmbH	Germany	165	75-03000
CIC International Ltd	USA	563	75-18570
Israel Military Industries Ltd	Israel	220	75-10300
Sener Ingenieria y Sistemas SA	Spain	367	75-05200

511 31 Sub-munitions, terminally guided

Aerojet	USA	547	75-01500

Bodenseewerk Gerätetechnik GmbH	Germany	165	75-03000
Hunting Engineering Ltd	UK	459	75-30700

511 40 Torpedoes

AAI Corp.	USA	545	75-00020
AlliedSignal Ltd	UK	430	75-02747
Bofors Underwater Systems AB	Sweden	376	75-03050
INIMEX	**Belgium**	**46**	**75-04650**
Northrop Grumman Corporation	USA	599	75-65450
STN ATLAS Elektronik GmbH	Germany	165	75-01150
The Dewey Electronics Corp.	USA	569	75-26100
Whitehead SpA	Italy	240	75-27600

511 41 Torpedoes, air-launched

Bofors Underwater Systems AB	Sweden	376	75-03050
Whitehead SpA	Italy	240	75-27600

511 41 10 Torpedoes, air-launched, lightweight

Bofors Underwater Systems AB	Sweden	376	75-03050
Whitehead SpA	Italy	240	75-27600

511 42 Torpedoes, ship / submarine-launched

Bofors Underwater Systems AB	Sweden	376	75-03050
Whitehead SpA	Italy	240	75-27600

511 45 Torpedoes, practice

Bofors Underwater Systems AB	Sweden	376	75-03050
The Dewey Electronics Corp.	USA	569	75-26100
Whitehead SpA	Italy	240	75-27600

511 49 Torpedo components

Air-Log Ltd	UK	429	75-01800
A.O. Henton Engineering Co Ltd.	UK	458	75-29040
Busak + Shamban Inc.	USA	561	75-16006
Eurometaal NV	Netherlands	292	75-02200
Fairey Hydraulics	UK	450	75-22255
GEC-Marconi Avionics Ltd	UK	454	75-25320
Japan Aviation Electronics Ind. Ltd	Japan	252	75-03300
Moog Controls Ltd	UK	469	75-40670
Moog Italiana Srl	Italy	237	75-16230
Moog Japan Ltd.	Japan	253	75-06100
Parker Bertea Aerospace	USA	603	75-68900
SFIM	**France**	**144**	**75-35080**
The Dewey Electronics Corp.	USA	569	75-26100
Voss Industries Inc.	USA	626	75-94550
Whitehead SpA	Italy	240	75-27600

511 49 30 Torpedo guidance systems

GEC-Marconi Avionics Ltd	UK	454	75-25320
Japan Aviation Electronics Ind. Ltd	Japan	252	75-03300
SAGEM SA	**France**	**143**	**75-33950**
SFIM	**France**	**144**	**75-35080**
Whitehead SpA	Italy	240	75-27600

511 49 40 Torpedo homing heads

Alenia-Elsag	Italy	230	74-00100
NEC Corporation	Japan	253	75-06350

511 49 90 Torpedo warheads

AAI Corp.	USA	545	75-00020
Eurometaal NV	Netherlands	292	75-02200
The Dewey Electronics Corp.	USA	569	75-26100

512 Barrelled weapons

AAI Corp.	USA	545	75-00020

Allied Research Corp.	USA	*549*	75-04670
Amadeo Rossi SA	Brazil	*57*	75-18780
ARES Inc.	USA	*555*	75-08600
Arms Corporation of the Philippines	Philippines	*320*	75-00150
ASAI AG.	Switzerland	*385*	75-00950
Australian Defence Industries Ltd	Australia	*9*	55-01000
Bernardini SA	Brazil	*55*	75-02120
Bressel SA.	Spain	*366*	75-01670
Bushmaster Firearms Inc.	USA	*561*	75-16007
Century International Arms Inc.	USA	*633*	80-05400
Cenzin Co. Ltd.	Poland	*324*	75-00200
Ceska Zbrojovka	Czech Republic	*98*	75-00300
Chamberlain Manufacturing Corp.	USA	*563*	75-17950
CIC International Ltd	USA	*563*	75-18570
Companhia Brasileira de Cartuchos	Brazil	*55*	75-03150
Daudsons Armoury	Pakistan	*315*	75-00500
D.J.Litt (Firearms) Ltd	UK	*465*	75-36080
Explosivos Alaveses SA	Spain	*366*	75-03050
Far East Industrial Supply Co Inc.	Philippines	*320*	75-00800
FN Herstal SA	Belgium	*46*	75-03600
FN Manufacturing Inc	USA	*576*	75-35320
Forjas Taurus SA	Brazil	*56*	75-06640
Giat Industries	France	*126*	55-00100
Gibbs Rifle Co.	USA	*579*	75-39570
Heckler & Koch Inc	USA	*581*	75-42860
Hellenic Arms Industry	Greece	*187*	55-00800
Howa Machinery Ltd	Japan	*251*	75-02550
INIMEX	**Belgium**	***46***	**75-04650**
Israel Military Industries Ltd	Israel	*220*	75-10300
Law Enforcement International Ltd	UK	*464*	75-34870
Llama Gabilondo	Spain	*366*	75-04200
Manroy Engineering Ltd	**UK**	***466***	**75-37165**
Minebea Co. Ltd	Japan	*253*	75-05600
Noricum	Portugal	*327*	75-14000
NORINCO	China PR	*88*	55-00280
Oerlikon-Contraves AG.	Switzerland	*387*	75-09250
Ordnance Development and Engineering	Singapore	*348*	75-02800
Peregrine Industries Inc.	USA	*603*	75-69280
Pietro Beretta SpA	Italy	*232*	75-02800
RAMO Manufacturing Inc.	USA	*607*	75-72810
Reunert Technology Systems Ltd	S. Africa	*359*	75-18300
Rheinmetall Industrie	Germany	*173*	75-24900
Royal Ordnance	UK	*481*	75-50620
Sabre Computers Int'l Ltd	UK	*481*	75-51150
Sako Ltd	Finland	*118*	75-01980
Santa Barbara SA	Spain	*364*	55-01000
SIMA PERU	Peru	*319*	55-01950
South Central Ordnance Co.	USA	*616*	75-82400
Steyr Mannlicher AG & Co KG	Austria	*34*	75-02820
T&T Technology Trading Ltd	Switzerland	*388*	75-14600
Tamares Ltd.	UK	*487*	75-58500
Vickers Shipbuilding & Eng Ltd	UK	*491*	75-63100

512 10 Pistols, automatics

Accuracy International Ltd	UK	*427*	75-00360
Amadeo Rossi SA	Brazil	*57*	75-18780
Armalon Ltd	UK	*431*	75-03729
ASAI AG.	Switzerland	*385*	75-00950
ASTRA Sport SA	Spain	*365*	75-01500
Benelli Armi SpA	Italy	*232*	75-02700
Beretta USA Corp.	USA	*558*	75-13605
Century International Arms Inc.	USA	*633*	80-05400
Ceska Zbrojovka	Czech Republic	*98*	75-00300
Daudsons Armoury	Pakistan	*315*	75-00500
Emco Inc.	USA	*573*	75-30985
Fabbrica d'Armi Fratt.Tanfoglio SRL	Italy	*239*	75-25020
FN Herstal SA	Belgium	*46*	75-03600
FN Manufacturing Inc	USA	*576*	75-35320
Forjas Taurus SA	Brazil	*56*	75-06640
Giat Industries	France	*126*	55-00100
Gibbs Rifle Co.	USA	*579*	75-39570
Hämmerli AG	Switzerland	*386*	75-05850
Heckler & Koch GmbH	Germany	*169*	75-11972
Heckler & Koch Inc	USA	*581*	75-42860
Hellenic Arms Industry	Greece	*187*	55-00800
Helston Gunsmiths	UK	*458*	75-29010
INIMEX	**Belgium**	***46***	**75-04650**
Israel Military Industries Ltd	Israel	*220*	75-10300
J.P. Sauer & Sohn GmbH.	Germany	*174*	75-26050
Korth GmbH	Germany	*171*	75-16200
Llama Gabilondo	Spain	*366*	75-04200
Makina ve Kimya Endüstrisi	Turkey	*402*	55-01310
Manroy Engineering Ltd	**UK**	***466***	**75-37165**
New Helvetia Trade Group	USA	*639*	80-21730
Nottingham International	USA	*640*	80-22000
Parker-Hale Ltd	UK	*473*	75-43850
Peregrine Industries Inc.	USA	*603*	75-69280
Pietro Beretta SpA	Italy	*232*	75-02800
Precimetal Fonderie de Précision SA	Belgium	*46*	75-05500
SIGARMS Inc.	USA	*614*	75-80700
Smith & Wesson	USA	*615*	75-82060
South Central Ordnance Co.	USA	*616*	75-82400
Steyr Mannlicher AG & Co KG	Austria	*34*	75-02820
Taurus International Mfg Inc.	USA	*619*	75-86385
The Robar Companies Inc.	USA	*609*	75-75630
Wildey Inc.	USA	*628*	75-96800

512 10 60 Pistols, automatics, silenced

Accuracy International Ltd	UK	*427*	75-00360
Armalon Ltd	UK	*431*	75-03729
ASAI AG.	Switzerland	*385*	75-00950
INIMEX	**Belgium**	***46***	**75-04650**
Israel Military Industries Ltd	Israel	*220*	75-10300
Jonathan Arthur Ciener Inc.	USA	*564*	75-18590
Law Enforcement International Ltd	UK	*464*	75-34870
South Central Ordnance Co.	USA	*616*	75-82400

512 20 Rifles, carbines

Accuracy International Ltd	UK	*427*	75-00360
Amadeo Rossi SA	Brazil	*57*	75-18780
ARES Inc.	USA	*555*	75-08600
Armalon Ltd	UK	*431*	75-03729
Beretta USA Corp.	USA	*558*	75-13605
Browning SA	Belgium	*45*	75-01950
Bushmaster Firearms Inc.	USA	*561*	75-16007
Century International Arms Inc.	USA	*633*	80-05400
Chartered Firearms Industries Pte	Singapore	*348*	75-01150
Companhia Brasileira de Cartuchos	Brazil	*55*	75-03150
Daudsons Armoury	Pakistan	*315*	75-00500
D.J.Litt (Firearms) Ltd	UK	*465*	75-36080
FN Herstal SA	Belgium	*46*	75-03600
FN Manufacturing Inc	USA	*576*	75-35320
Giat Industries	France	*126*	55-00100
Gibbs Rifle Co.	USA	*579*	75-39570
Heckler & Koch GmbH	Germany	*169*	75-11972
Heckler & Koch Inc	USA	*581*	75-42860
Hellenic Arms Industry	Greece	*187*	55-00800
Helston Gunsmiths	UK	*458*	75-29010
Indústrias de Defesa	Portugal	*327*	55-00950
INIMEX	**Belgium**	***46***	**75-04650**
Israel Military Industries Ltd	Israel	*220*	75-10300
Law Enforcement International Ltd	UK	*464*	75-34870
Manroy Engineering Ltd	**UK**	***466***	**75-37165**
New Helvetia Trade Group	USA	*639*	80-21730
Nottingham International	USA	*640*	80-22000
Ordnance Development and Engineering	Singapore	*348*	75-02800
Pietro Beretta SpA	Italy	*232*	75-02800
Precimetal Fonderie de Précision SA	Belgium	*46*	75-05500
Santa Barbara SA	Spain	*364*	55-01000
SIGARMS Inc.	USA	*614*	75-80700
South Central Ordnance Co.	USA	*616*	75-82400
Steyr Mannlicher AG & Co KG	Austria	*34*	75-02820
The Robar Companies Inc.	USA	*609*	75-75630
Voere Kufsteiner Gerätebau	Austria	*34*	75-03330

512 20 60 Rifles, silenced

Accuracy International Ltd	UK	*427*	75-00360
Armalon Ltd	UK	*431*	75-03729
Heckler & Koch Inc	USA	*581*	75-42860
INIMEX	**Belgium**	***46***	**75-04650**
Jonathan Arthur Ciener Inc.	USA	*564*	75-18590
Law Enforcement International Ltd	UK	*464*	75-34870
Manroy Engineering Ltd	**UK**	***466***	**75-37165**
Sako Ltd	Finland	*118*	75-01980
South Central Ordnance Co.	USA	*616*	75-82400
Steyr Mannlicher AG & Co KG	Austria	*34*	75-02820

Product–Supplier Link

512 21 Rifles, automatic

AAI Corp.	USA	*545*	75-00020
Accuracy International Ltd	UK	*427*	75-00360
Australian Defence Industries Ltd	Australia	*9*	55-01000
Bushmaster Firearms Inc.	USA	*561*	75-16007
Century International Arms Inc.	USA	*633*	80-05400
Chartered Firearms Industries Pte	Singapore	*348*	75-01150
Emco Inc.	USA	*573*	75-30985
Eurometaal NV	Netherlands	*292*	75-02200
FN Herstal SA	Belgium	*46*	75-03600
Giat Industries	France	*126*	55-00100
Grenadier Engineering Inc.	USA	*580*	75-40390
Heckler & Koch Inc	USA	*581*	75-42860
Helston Gunsmiths	UK	*458*	75-29010
INIMEX	**Belgium**	***46***	**75-04650**
Israel Military Industries Ltd	Israel	*220*	75-10300
Makina ve Kimya Endüstrisi	Turkey	*402*	55-01310
Manroy Engineering Ltd	**UK**	***466***	**75-37165**
NORINCO	China PR	*88*	55-00280
Nottingham International	USA	*640*	80-22000
Ordnance Development and Engineering	Singapore	*348*	75-02800
Peregrine Industries Inc.	USA	*603*	75-69280
Pietro Beretta SpA	Italy	*232*	75-02800
Robert W. Hart & Son Inc.	USA	*581*	75-42450
Royal Ordnance	UK	*481*	75-50620
South Central Ordnance Co.	USA	*616*	75-82400
Steyr Mannlicher AG & Co KG	Austria	*34*	75-02820
US Army Equipment Guns and Ammo	Belgium	*47*	75-06345

512 22 Rifles, bolt action

Accuracy International Ltd	UK	*427*	75-00360
Armalon Ltd	UK	*431*	75-03729
Century International Arms Inc.	USA	*633*	80-05400
Companhia Brasileira de Cartuchos	Brazil	*55*	75-03150
FN Herstal SA	Belgium	*46*	75-03600
Gibbs Rifle Co.	USA	*579*	75-39570
Helston Gunsmiths	UK	*458*	75-29010
INIMEX	**Belgium**	***46***	**75-04650**
J.P. Sauer & Sohn GmbH	Germany	*174*	75-26050
Manroy Engineering Ltd	**UK**	***466***	**75-37165**
Nottingham International	USA	*640*	80-22000
Pietro Beretta SpA	Italy	*232*	75-02800
Sako Ltd	Finland	*118*	75-01980
South Central Ordnance Co.	USA	*616*	75-82400
Steyr Mannlicher AG & Co KG	Austria	*34*	75-02820
US Army Equipment Guns and Ammo	Belgium	*47*	75-06345

512 26 Rifles, sniper

Accuracy International Ltd	UK	*427*	75-00360
Amadeo Rossi SA	Brazil	*57*	75-18780
Armalon Ltd	UK	*431*	75-03729
Century International Arms Inc.	USA	*633*	80-05400
FN Herstal SA	Belgium	*46*	75-03600
Gibbs Rifle Co.	USA	*579*	75-39570
Heckler & Koch GmbH	Germany	*169*	75-11972
Heckler & Koch Inc	USA	*581*	75-42860
INIMEX	**Belgium**	***46***	**75-04650**
Israel Military Industries Ltd	Israel	*220*	75-10300
J.P. Sauer & Sohn GmbH	Germany	*174*	75-26050
Law Enforcement International Ltd	UK	*464*	75-34870
Manroy Engineering Ltd	**UK**	***466***	**75-37165**
Military Logistics MFC	USA	*596*	75-61550
NORINCO	China PR	*88*	55-00280
Nottingham International	USA	*640*	80-22000
Peregrine Industries Inc.	USA	*603*	75-69280
Pietro Beretta SpA	Italy	*232*	75-02800
Robert W. Hart & Son Inc.	USA	*581*	75-42450
Royal Ordnance	UK	*481*	75-50620
Sako Ltd	Finland	*118*	75-01980
South Central Ordnance Co.	USA	*616*	75-82400
Steyr Mannlicher AG & Co KG	Austria	*34*	75-02820
The Robar Companies Inc.	USA	*609*	75-75630
US Army Equipment Guns and Ammo	Belgium	*47*	75-06345
Voere Kufsteiner Gerätebau	Austria	*34*	75-03330

512 27 Shotguns, military

Accuracy International Ltd	UK	*427*	75-00360
Amadeo Rossi SA	Brazil	*57*	75-18780
Armalon Ltd	UK	*431*	75-03729
Century International Arms Inc.	USA	*633*	80-05400
Daudsons Armoury	Pakistan	*315*	75-00500
FN Herstal SA	Belgium	*46*	75-03600
Heckler & Koch Inc	USA	*581*	75-42860
Helston Gunsmiths	UK	*458*	75-29010
INIMEX	**Belgium**	***46***	**75-04650**
Jonathan Arthur Ciener Inc.	USA	*564*	75-18590
J.P. Sauer & Sohn GmbH	Germany	*174*	75-26050
Law Enforcement International Ltd	UK	*464*	75-34870
Manroy Engineering Ltd	**UK**	***466***	**75-37165**
Nottingham International	USA	*640*	80-22000
Pietro Beretta SpA	Italy	*232*	75-02800
Reunert Technology Systems Ltd	S. Africa	*359*	75-18300
South Central Ordnance Co.	USA	*616*	75-82400

512 30 Machine guns

AAI Corp.	USA	*545*	75-00020
Aircraft Equipment International Ltd	UK	*428*	75-01560
ARES Inc.	USA	*555*	75-08600
Australian Defence Industries Ltd	Australia	*9*	55-01000
Century International Arms Inc.	USA	*633*	80-05400
Ceska Zbrojovka	Czech Republic	*98*	75-00300
Chamberlain Manufacturing Corp.	USA	*563*	75-17950
Chartered Firearms Industries Pte	Singapore	*348*	75-01150
CIC International Ltd	USA	*563*	75-18570
FN Herstal SA	Belgium	*46*	75-03600
FN Manufacturing Inc	USA	*576*	75-35320
Heckler & Koch GmbH	Germany	*169*	75-11972
Heckler & Koch Inc	USA	*581*	75-42860
Hellenic Arms Industry	Greece	*187*	55-00800
Helston Gunsmiths	UK	*458*	75-29010
Indústrias de Defesa	Portugal	*327*	55-00950
INIMEX	**Belgium**	***46***	**75-04650**
Israel Military Industries Ltd	Israel	*220*	75-10300
Makina ve Kimya Endüstrisi	Turkey	*402*	55-01310
Manroy Engineering Ltd	**UK**	***466***	**75-37165**
Military Systems Group Inc.	USA	*596*	75-61570
Napco International Inc.	USA	*598*	75-63600
New Helvetia Trade Group	USA	*639*	80-21730
NORINCO	China PR	*88*	55-00280
Nottingham International	USA	*640*	80-22000
Ordnance Development and Engineering	Singapore	*348*	75-02800
Pietro Beretta SpA	Italy	*232*	75-02800
Precimetal Fonderie de Précision SA	Belgium	*46*	75-05500
RAMO Manufacturing Inc.	USA	*607*	75-72810
Rheinmetall Industrie	Germany	*173*	75-24900
Royal Ordnance	UK	*481*	75-50620
Saco Defense Inc.	USA	*612*	75-77340
Santa Barbara SA	Spain	*364*	55-01000
South Central Ordnance Co.	USA	*616*	75-82400
Steyr Mannlicher AG & Co KG	Austria	*34*	75-02820

512 31 Machine guns, light

AAI Corp.	USA	*545*	75-00020
Aircraft Equipment International Ltd	UK	*428*	75-01560
Australian Defence Industries Ltd	Australia	*9*	55-01000
Ceska Zbrojovka	Czech Republic	*98*	75-00300
Chamberlain Manufacturing Corp.	USA	*563*	75-17950
Chartered Firearms Industries Pte	Singapore	*348*	75-01150
CIC International Ltd	USA	*563*	75-18570
FN Herstal SA	Belgium	*46*	75-03600
Giat Industries	France	*126*	55-00100
Heckler & Koch Inc	USA	*581*	75-42860
Helston Gunsmiths	UK	*458*	75-29010
INIMEX	**Belgium**	***46***	**75-04650**
Israel Military Industries Ltd	Israel	*220*	75-10300
Manroy Engineering Ltd	**UK**	***466***	**75-37165**
Napco International Inc.	USA	*598*	75-63600
NORINCO	China PR	*88*	55-00280
Nottingham International	USA	*640*	80-22000
Ordnance Development and Engineering	Singapore	*348*	75-02800
Pietro Beretta SpA	Italy	*232*	75-02800
Precimetal Fonderie de Précision SA	Belgium	*46*	75-05500
RAMO Manufacturing Inc.	USA	*607*	75-72810
Royal Ordnance	UK	*481*	75-50620
Saco Defense Inc.	USA	*612*	75-77340
Santa Barbara SA	Spain	*364*	55-01000

South Central Ordnance Co.	USA	616	75-82400
Steyr Mannlicher AG & Co KG	Austria	34	75-02820

512 32 Machine guns, medium

AAI Corp.	USA	545	75-00020
Chamberlain Manufacturing Corp.	USA	563	75-17950
FN Herstal SA	Belgium	46	75-03600
Grenadier Engineering Inc.	USA	580	75-40390
Helston Gunsmiths	UK	458	75-29010
INIMEX	**Belgium**	**46**	**75-04650**
Israel Military Industries Ltd	Israel	220	75-10300
Manroy Engineering Ltd	**UK**	**466**	**75-37165**
Military Systems Group Inc.	USA	596	75-61570
Napco International Inc.	USA	598	75-63600
NORINCO	China PR	88	55-00280
Nottingham International	USA	640	80-22000
Pietro Beretta SpA	Italy	232	75-02800
Precimetal Fonderie de Précision SA	Belgium	46	75-05500
RAMO Manufacturing Inc.	USA	607	75-72810
Royal Ordnance	UK	481	75-50620
Saco Defense Inc.	USA	612	75-77340
South Central Ordnance Co.	USA	616	75-82400

512 33 Machine guns, heavy

AAI Corp.	USA	545	75-00020
Aircraft Equipment International Ltd	UK	428	75-01560
Chamberlain Manufacturing Corp.	USA	563	75-17950
Chartered Firearms Industries Pte	Singapore	348	75-01150
FN Herstal SA	Belgium	46	75-03600
Helston Gunsmiths	UK	458	75-29010
INIMEX	**Belgium**	**46**	**75-04650**
Israel Military Industries Ltd	Israel	220	75-10300
Manroy Engineering Ltd	**UK**	**466**	**75-37165**
Military Systems Group Inc.	USA	596	75-61570
Napco International Inc.	USA	598	75-63600
NORINCO	China PR	88	55-00280
Nottingham International	USA	640	80-22000
Ordnance Development and Engineering	Singapore	348	75-02800
Precimetal Fonderie de Précision SA	Belgium	46	75-05500
RAMO Manufacturing Inc.	USA	607	75-72810
Royal Ordnance	UK	481	75-50620
Saco Defense Inc.	USA	612	75-77340
South Central Ordnance Co.	USA	616	75-82400

512 35 Machine guns, aircraft

AIM Enterprises Inc.	USA	630	80-01350
Aircraft Equipment International Ltd	UK	428	75-01560
FN Herstal SA	Belgium	46	75-03600
INIMEX	**Belgium**	**46**	**75-04650**
Israel Military Industries Ltd	Israel	220	75-10300
Manroy Engineering Ltd	**UK**	**466**	**75-37165**
Military Systems Group Inc.	USA	596	75-61570
Napco International Inc.	USA	598	75-63600
Royal Ordnance	UK	481	75-50620
Saco Defense Inc.	USA	612	75-77340

512 38 Machine guns, vehicle-mounted

Aircraft Equipment International Ltd	UK	428	75-01560
Chamberlain Manufacturing Corp.	USA	563	75-17950
Chartered Firearms Industries Pte	Singapore	348	75-01150
FN Herstal SA	Belgium	46	75-03600
Heckler & Koch Inc	USA	581	75-42860
INIMEX	**Belgium**	**46**	**75-04650**
Israel Military Industries Ltd	Israel	220	75-10300
JSC Tulamashzavod	Russia	334	55-19450
Manroy Engineering Ltd	**UK**	**466**	**75-37165**
Military Systems Group Inc.	USA	596	75-61570
Napco International Inc.	USA	598	75-63600
Nottingham International	USA	640	80-22000
Ordnance Development and Engineering	Singapore	348	75-02800
Pietro Beretta SpA	Italy	232	75-02800
Royal Ordnance	UK	481	75-50620
Saco Defense Inc.	USA	612	75-77340
South Central Ordnance Co.	USA	616	75-82400

512 40 Submachine guns

Beretta USA Corp.	USA	558	75-13605
Ceska Zbrojovka	Czech Republic	98	75-00300
Chamberlain Manufacturing Corp.	USA	563	75-17950
FN Herstal SA	Belgium	46	75-03600
F.R. Ordnance International Ltd	UK	452	75-24480
Heckler & Koch GmbH	Germany	169	75-11972
Heckler & Koch Inc	USA	581	75-42860
Hellenic Arms Industry	Greece	187	55-00800
Helston Gunsmiths	UK	458	75-29010
Indústria de Material Bélico	Brazil	56	75-09190
INIMEX	**Belgium**	**46**	**75-04650**
Israel Military Industries Ltd	Israel	220	75-10300
Makina ve Kimya Endüstrisi	Turkey	402	55-01310
Manroy Engineering Ltd	**UK**	**466**	**75-37165**
New Helvetia Trade Group	USA	639	80-21730
NORINCO	China PR	88	55-00280
Nottingham International	USA	640	80-22000
Pietro Beretta SpA	Italy	232	75-02800
Royal Ordnance	UK	481	75-50620
South Central Ordnance Co.	USA	616	75-82400
Steyr Mannlicher AG & Co KG	Austria	34	75-02820

512 50 Guns, 20mm & over

AIM Enterprises Inc.	USA	630	80-01350
Aircraft Equipment International Ltd	UK	428	75-01560
Allied Research Corp.	USA	549	75-04670
ARES Inc.	USA	555	75-08600
Bernardini SA	Brazil	55	75-02120
Bofors AB	Sweden	375	74-00020
Century International Arms Inc.	USA	633	80-05400
Chamberlain Manufacturing Corp.	USA	563	75-17950
CIC International Ltd	USA	563	75-18570
Cockerill Mechanical Industries	Belgium	46	75-02410
Giat Industries	France	126	55-00100
Hellenic Arms Industry	Greece	187	55-00800
INIMEX	**Belgium**	**46**	**75-04650**
Israel Aircraft Industries Ltd	Israel	219	75-10200
NORINCO	China PR	88	55-00280
Oerlikon-Contraves AG	Switzerland	387	75-09250
Precimetal Fonderie de Précision SA	Belgium	46	75-05500
Rheinmetall Industrie	Germany	173	75-24900

512 50 20 Guns, aircraft, 20mm & over

Aircraft Equipment International Ltd	UK	428	75-01560
Bernardini SA	Brazil	55	75-02120
Chamberlain Manufacturing Corp.	USA	563	75-17950
Giat Industries	France	126	55-00100
INIMEX	**Belgium**	**46**	**75-04650**
Israel Aircraft Industries Ltd	Israel	219	75-10200
Israel Military Industries Ltd	Israel	220	75-10300
JSC Tulamashzavod	Russia	334	55-19450
Mauser-Werke	Germany	172	75-20300
Nottingham International	USA	640	80-22000

512 50 22 Guns, automatic, aircraft / helicopter

Aircraft Equipment International Ltd	UK	428	75-01560
Bernardini SA	Brazil	55	75-02120
Chamberlain Manufacturing Corp.	USA	563	75-17950
Daewoo Heavy Industries Ltd	Korea (ROK)	265	75-00600
Giat Industries	France	126	55-00100
INIMEX	**Belgium**	**46**	**75-04650**
Israel Aircraft Industries Ltd	Israel	219	75-10200
Israel Military Industries Ltd	Israel	220	75-10300
JSC Tulamashzavod	Russia	334	55-19450
Mauser-Werke	Germany	172	75-20300
Military Systems Group Inc.	USA	596	75-61570
Precimetal Fonderie de Précision SA	Belgium	46	75-05500
Saco Defense Inc.	USA	612	75-77340

512 50 22 10 Guns, automatic, aircraft / helicopter, podded

Chamberlain Manufacturing Corp.	USA	563	75-17950
Giat Industries	France	126	55-00100
Israel Aircraft Industries Ltd	Israel	219	75-10200
Israel Military Industries Ltd	Israel	220	75-10300

512 50 30 Guns, naval, 20mm & over

Aircraft Equipment International Ltd	UK	428	75-01560
Alenia Difesa	**Italy**	**231**	**75-01165**
Australian Defence Industries Ltd	Australia	9	55-01000
Bofors AB	Sweden	375	74-00020
Chamberlain Manufacturing Corp.	USA	563	75-17950
Giat Industries	France	126	55-00100
INIMEX	**Belgium**	**46**	**75-04650**
Israel Aircraft Industries Ltd	Israel	219	75-10200
Israel Military Industries Ltd	Israel	220	75-10300
JSC Tulamashzavod	Russia	334	55-19450
MSI-Defence Systems Ltd	UK	469	75-41280
Nottingham International	USA	640	80-22000
United Defense L.P.	USA	624	75-92050
Vickers Shipbuilding & Eng Ltd	UK	491	75-63100

512 50 32 Guns, naval, automatic

Alenia Difesa	**Italy**	**231**	**75-01165**
JSC Tulamashzavod	Russia	334	55-19450
Military Systems Group Inc.	USA	596	75-61570
OTO Breda	Italy	237	75-18100
United Defense L.P.	USA	624	75-92050

512 50 32 10 Guns, naval, automatic, anti-aircraft

Alenia Difesa	**Italy**	**231**	**75-01165**
JSC Tulamashzavod	Russia	334	55-19450
United Defense L.P.	USA	624	75-92050

512 50 32 20 Guns, naval, automatic, anti-missile

Alenia Difesa	**Italy**	**231**	**75-01165**
Bofors AB	Sweden	375	74-00020
INIMEX	**Belgium**	**46**	**75-04650**
Israel Aircraft Industries Ltd	Israel	219	75-10200
JSC Tulamashzavod	Russia	334	55-19450
United Defense L.P.	USA	624	75-92050

512 50 32 50 Guns, naval, automatic, multi-role

Alenia Difesa	**Italy**	**231**	**75-01165**
INIMEX	**Belgium**	**46**	**75-04650**
JSC Tulamashzavod	Russia	334	55-19450
Military Systems Group Inc.	USA	596	75-61570
MSI-Defence Systems Ltd	UK	469	75-41280
United Defense L.P.	USA	624	75-92050

512 50 37 Guns, naval, surface target

Alenia Difesa	**Italy**	**231**	**75-01165**
JSC Tulamashzavod	Russia	334	55-19450
OTO Breda	Italy	237	75-18100
Rheinmetall Industrie	Germany	173	75-24900
United Defense L.P.	USA	624	75-92050

512 50 37 10 Guns, naval, surface target, 40-100mm

Alenia Difesa	**Italy**	**231**	**75-01165**
Australian Defence Industries Ltd	Australia	9	55-01000
Israel Military Industries Ltd	Israel	220	75-10300
National Forge Co.	USA	598	75-63850
Saco Defense Inc.	USA	612	75-77340

512 50 37 20 Guns, naval, surface target, 105-195mm

Australian Defence Industries Ltd	Australia	9	55-01000
National Forge Co.	USA	598	75-63850
OTO Breda	Italy	237	75-18100
United Defense L.P.	USA	624	75-92050
Vickers Shipbuilding & Eng Ltd	UK	491	75-63100

512 50 37 30 Guns, naval, surface target, 200mm & over

National Forge Co.	USA	598	75-63850
United Defense L.P.	USA	624	75-92050

512 50 40 Guns, ground-based, 20mm & over

AAI Corp.	USA	545	75-00020
Aircraft Equipment International Ltd	UK	428	75-01560
Alenia Difesa	**Italy**	**231**	**75-01165**
Allied Research Corp.	USA	549	75-04670
Bernardini SA	Brazil	55	75-02120
Chamberlain Manufacturing Corp.	USA	563	75-17950
Cockerill Mechanical Industries	Belgium	46	75-02410
Giat Industries	France	126	55-00100
Hellenic Arms Industry	Greece	187	55-00800
INIMEX	**Belgium**	**46**	**75-04650**
Israel Aircraft Industries Ltd	Israel	219	75-10200
Israel Military Industries Ltd	Israel	220	75-10300
JSC Tulamashzavod	Russia	334	55-19450
Noricum	Portugal	327	75-14000
Royal Ordnance	UK	481	75-50620
Saco Defense Inc.	USA	612	75-77340
T&T Technology Trading Ltd	Switzerland	388	75-14600
Vickers Shipbuilding & Eng Ltd	UK	491	75-63100

512 50 41 Guns, ground-based, anti-tank

Allied Research Corp.	USA	549	75-04670
Bofors AB	Sweden	375	74-00020
Giat Industries	France	126	55-00100
INIMEX	**Belgium**	**46**	**75-04650**
Israel Aircraft Industries Ltd	Israel	219	75-10200
MECAR SA	Belgium	46	75-04930
NORINCO	China PR	88	55-00280
Royal Ordnance	UK	481	75-50620
Santa Barbara SA	Spain	364	55-01000

512 50 41 10 Guns, anti-tank, vehicle-mounted

Allied Research Corp.	USA	549	75-04670
Cockerill Mechanical Industries	Belgium	46	75-02410
Giat Industries	France	126	55-00100
Israel Military Industries Ltd	Israel	220	75-10300
MECAR SA	Belgium	46	75-04930
Royal Ordnance	UK	481	75-50620

512 50 42 Guns, ground-based, automatic, 20mm & over

Alenia Difesa	**Italy**	**231**	**75-01165**
Australian Defence Industries Ltd	Australia	9	55-01000
Bofors AB	Sweden	375	74-00020
Cockerill Mechanical Industries	Belgium	46	75-02410
Giat Industries	France	126	55-00100
Hellenic Arms Industry	Greece	187	55-00800
INIMEX	**Belgium**	**46**	**75-04650**
Israel Aircraft Industries Ltd	Israel	219	75-10200
NORINCO	China PR	88	55-00280
Saco Defense Inc.	USA	612	75-77340

512 50 42 10 Guns, ground-based, automatic, anti-aircraft

Alenia Difesa	**Italy**	**231**	**75-01165**
Australian Defence Industries Ltd	Australia	9	55-01000
Bofors AB	Sweden	375	74-00020
CIC International Ltd	USA	563	75-18570
Cockerill Mechanical Industries	Belgium	46	75-02410
Giat Industries	France	126	55-00100
Hellenic Arms Industry	Greece	187	55-00800
Israel Aircraft Industries Ltd	Israel	219	75-10200
Rheinmetall Industrie	Germany	173	75-24900

512 50 42 15 Guns, vehicle-mounted, automatic, anti-aircraft

Alenia Difesa	**Italy**	**231**	**75-01165**
CIC International Ltd	USA	563	75-18570
Cockerill Mechanical Industries	Belgium	46	75-02410
Israel Aircraft Industries Ltd	Israel	219	75-10200
JSC Tulamashzavod	Russia	334	55-19450
KUKA Wehrtechnik GmbH	Germany	171	75-17200
Military Systems Group Inc.	USA	596	75-61570

512 50 42 20 Guns, ground-based, automatic, anti-tank

CIC International Ltd	USA	563	75-18570
Cockerill Mechanical Industries	Belgium	46	75-02410
Rheinmetall Industrie	Germany	173	75-24900

512 50 42 25 Guns, vehicle-mounted, automatic, anti-tank

CIC International Ltd	USA	563	75-18570
Cockerill Mechanical Industries	Belgium	46	75-02410
Giat Industries	France	126	55-00100
United Defense L.P.	USA	624	75-92050

512 50 42 30 Guns, ground-based, automatic, surface target

Alenia Difesa	**Italy**	**231**	**75-01165**
CIC International Ltd	USA	563	75-18570
Hellenic Arms Industry	Greece	187	55-00800
Israel Aircraft Industries Ltd	Israel	219	75-10200
Saco Defense Inc.	USA	612	75-77340
United Defense L.P.	USA	624	75-92050

512 50 42 35 Guns, vehicle-mounted, automatic, surface target

Alenia Difesa	**Italy**	**231**	**75-01165**
Chamberlain Manufacturing Corp.	USA	563	75-17950
CIC International Ltd	USA	563	75-18570
Cockerill Mechanical Industries	Belgium	46	75-02410
Giat Industries	France	126	55-00100
Israel Aircraft Industries Ltd	Israel	219	75-10200
Military Systems Group Inc.	USA	596	75-61570
Saco Defense Inc.	USA	612	75-77340
United Defense L.P.	USA	624	75-92050

512 50 42 40 Guns, ground-based, automatic, multi-role

Alenia Difesa	**Italy**	**231**	**75-01165**
CIC International Ltd	USA	563	75-18570
Hellenic Arms Industry	Greece	187	55-00800
Israel Aircraft Industries Ltd	Israel	219	75-10200
Saco Defense Inc.	USA	612	75-77340
United Defense L.P.	USA	624	75-92050

512 50 42 45 Guns, vehicle-mounted, automatic, multi-role

Alenia Difesa	**Italy**	**231**	**75-01165**
Allied Research Corp.	USA	549	75-04670
ARES Inc.	USA	555	75-08600
Bofors AB	Sweden	375	74-00020
Chamberlain Manufacturing Corp.	USA	563	75-17950
CIC International Ltd	USA	563	75-18570
Cockerill Mechanical Industries	Belgium	46	75-02410
Giat Industries	France	126	55-00100
Hellenic Arms Industry	Greece	187	55-00800
INIMEX	**Belgium**	**46**	**75-04650**
Israel Aircraft Industries Ltd	Israel	219	75-10200
JSC Tulamashzavod	Russia	334	55-19450
Mauser-Werke	Germany	172	75-20300
Military Systems Group Inc.	USA	596	75-61570
Saco Defense Inc.	USA	612	75-77340
Santa Barbara SA	Spain	364	55-01000
United Defense L.P.	USA	624	75-92050

512 50 44 Guns, coastal defense

Bofors AB	Sweden	375	74-00020
Vickers Shipbuilding & Eng Ltd	UK	491	75-63100

512 50 45 Guns, field

Alenia Difesa	**Italy**	**231**	**75-01165**
Allied Research Corp.	USA	549	75-04670
Australian Defence Industries Ltd	Australia	9	55-01000
Bofors AB	Sweden	375	74-00020
CIC International Ltd	USA	563	75-18570
Giat Industries	France	126	55-00100
INIMEX	**Belgium**	**46**	**75-04650**
OTO Breda	Italy	237	75-18100
Tamares Ltd	UK	487	75-58500
United Defense L.P.	USA	624	75-92050
Vammas Ltd	**Finland**	**117**	**55-02220**
Vickers Shipbuilding & Eng Ltd	UK	491	75-63100

512 50 45 10 Guns, field, light

Alenia Difesa	**Italy**	**231**	**75-01165**
Australian Defence Industries Ltd	Australia	9	55-01000
CIC International Ltd	USA	563	75-18570
Giat Industries	France	126	55-00100
Nottingham International	USA	640	80-22000
United Defense L.P.	USA	624	75-92050

512 50 45 20 Guns, field, medium

Alenia Difesa	**Italy**	**231**	**75-01165**
Allied Research Corp.	USA	549	75-04670
Australian Defence Industries Ltd	Australia	9	55-01000
CIC International Ltd	USA	563	75-18570
Giat Industries	France	126	55-00100
INIMEX	**Belgium**	**46**	**75-04650**
Nottingham International	USA	640	80-22000
Tamares Ltd	UK	487	75-58500
United Defense L.P.	USA	624	75-92050
Vickers Shipbuilding & Eng Ltd	UK	491	75-63100

512 50 45 60 Guns, field, self-propelled

Giat Industries	France	126	55-00100
INIMEX	**Belgium**	**46**	**75-04650**
Nottingham International	USA	640	80-22000
Royal Ordnance	UK	481	75-50620
Vickers Shipbuilding & Eng Ltd	UK	491	75-63100

512 50 46 Howitzers

Australian Defence Industries Ltd	Australia	9	55-01000
Bernardini SA	Brazil	55	75-02120
INIMEX	**Belgium**	**46**	**75-04650**
Noricum	Portugal	327	75-14000
NORINCO	China PR	88	55-00280
Nottingham International	USA	640	80-22000
Ordnance Development and Engineering	Singapore	348	75-02800
Swiss Federal Armament Works	Switzerland	384	55-00250
T&T Technology Trading Ltd	Switzerland	388	75-14600
Tamares Ltd	UK	487	75-58500
United Defense L.P.	USA	624	75-92050

Vammas Ltd	**Finland**	***117***	**55-02220**
Vickers Shipbuilding & Eng Ltd	UK	*491*	75-63100

512 50 46 30 Howitzers, field

Australian Defence Industries Ltd	Australia	*9*	55-01000
Giat Industries	France	*126*	55-00100
INIMEX	**Belgium**	***46***	**75-04650**
Noricum	Portugal	*327*	75-14000
Nottingham International	USA	*640*	80-22000
Ordnance Development and Engineering	Singapore	*348*	75-02800
Rheinmetall Industrie	Germany	*173*	75-24900
T&T Technology Trading Ltd	Switzerland	*388*	75-14600
Tamares Ltd	UK	*487*	75-58500
United Defense L.P.	USA	*624*	75-92050
Vammas Ltd	**Finland**	***117***	**55-02220**
Vickers Shipbuilding & Eng Ltd	UK	*491*	75-63100

512 50 46 37 Howitzers, field, towed

Australian Defence Industries Ltd	Australia	*9*	55-01000
Bernardini SA	Brazil	*55*	75-02120
Bofors AB	Sweden	*375*	74-00020
Giat Industries	France	*126*	55-00100
Noricum	Portugal	*327*	75-14000
Nottingham International	USA	*640*	80-22000
Ordnance Development and Engineering	Singapore	*348*	75-02800
OTO Breda	Italy	*237*	75-18100
Rheinmetall Industrie	Germany	*173*	75-24900
Royal Ordnance	UK	*481*	75-50620
T&T Technology Trading Ltd	Switzerland	*388*	75-14600
Tamares Ltd	UK	*487*	75-58500
United Defense L.P.	USA	*624*	75-92050
Vickers Shipbuilding & Eng Ltd	UK	*491*	75-63100

512 50 46 50 Howitzers, pack

INIMEX	**Belgium**	***46***	**75-04650**
Nottingham International	USA	*640*	80-22000
OTO Breda	Italy	*237*	75-18100
T&T Technology Trading Ltd	Switzerland	*388*	75-14600
United Defense L.P.	USA	*624*	75-92050
Vickers Shipbuilding & Eng Ltd	UK	*491*	75-63100

512 50 46 70 Howitzers, self-propelled, tracked

Allied Research Corp.	USA	*549*	75-04670
Bofors AB	Sweden	*375*	74-00020
Giat Industries	France	*126*	55-00100
INIMEX	**Belgium**	***46***	**75-04650**
Nottingham International	USA	*640*	80-22000
OTO Breda	Italy	*237*	75-18100
Rheinmetall Industrie	Germany	*173*	75-24900
United Defense L.P.	USA	*624*	75-92050
Vickers Shipbuilding & Eng Ltd	UK	*491*	75-63100

512 50 47 Guns, tank / armoured vehicle

Allied Research Corp.	USA	*549*	75-04670
ARES Inc.	USA	*555*	75-08600
Bernardini SA	Brazil	*55*	75-02120
Bofors AB	Sweden	*375*	74-00020
Cockerill Mechanical Industries	Belgium	*46*	75-02410
Giat Industries	France	*126*	55-00100
INIMEX	**Belgium**	***46***	**75-04650**
Israel Aircraft Industries Ltd	Israel	*219*	75-10200
Rheinmetall Industrie	Germany	*173*	75-24900
Santa Barbara SA	Spain	*364*	55-01000
Swiss Federal Armament Works	Switzerland	*384*	55-00250
United Defense L.P.	USA	*624*	75-92050

512 50 80 Gun accessories

Beta Company	USA	*558*	75-13750
Century International Arms Inc.	USA	*633*	80-05400
Chartered Firearms Industries Pte	Singapore	*348*	75-01150
Helston Gunsmiths	UK	*458*	75-29010
Howden Fan Company Pty Ltd	Australia	*17*	75-08740
Jonathan Arthur Ciener Inc.	USA	*564*	75-18590
MWG Co Inc.	USA	*597*	75-63490
RAMO Manufacturing Inc.	USA	*607*	75-72810
Royal Ordnance	UK	*481*	75-50620
United Defense L.P.	USA	*624*	75-92050

512 50 80 50 Plastic magazines, automatic rifle

Beta Company	USA	*558*	75-13750
Chartered Firearms Industries Pte	Singapore	*348*	75-01150
Nottingham International	USA	*640*	80-22000

512 50 80 60 Silencers, weapon

Accuracy International Ltd	UK	*427*	75-00360
ASAI AG.	Switzerland	*385*	75-00950
Brügger & Thomet Manufacturing Co	Switzerland	*385*	75-01450
Jonathan Arthur Ciener Inc.	USA	*564*	75-18590
Law Enforcement International Ltd	UK	*464*	75-34870
Parker-Hale Ltd	UK	*473*	75-43850

512 50 90 Gun components

Accuracy International Ltd	UK	*427*	75-00360
Alenia Difesa	**Italy**	***231***	**75-01165**
Amadeo Rossi SA	Brazil	*57*	75-18780
A.O. Henton Engineering Co Ltd.	UK	*458*	75-29040
Australian Defence Industries Ltd	Australia	*9*	55-01000
Beldam Crossley Ltd	UK	*434*	75-06170
Century International Arms Inc.	USA	*633*	80-05400
Cenzin Co. Ltd.	Poland	*324*	75-00200
Chamberlain Manufacturing Corp.	USA	*563*	75-17950
Chartered Firearms Industries Pte	Singapore	*348*	75-01150
Cockerill Mechanical Industries	Belgium	*46*	75-02410
Eurometaal NV	Netherlands	*292*	75-02200
Fluorocarbon Company Limited	UK	*451*	75-24020
Giat Industries	France	*126*	55-00100
Helston Gunsmiths	UK	*458*	75-29010
Hitchiner Manufacturing Co. Inc.	USA	*582*	75-43550
Indústrias de Defesa	Portugal	*327*	55-00950
Israel Military Industries Ltd	Israel	*220*	75-10300
National Forge Co.	USA	*598*	75-63850
RAMO Manufacturing Inc.	USA	*607*	75-72810
Royal Ordnance	UK	*481*	75-50620
Saco Defense Inc.	USA	*612*	75-77340
Saunders-Roe Ltd.	UK	*482*	75-51850
United Defense L.P.	USA	*624*	75-92050

512 50 90 10 Gun barrels

Accuracy International Ltd	UK	*427*	75-00360
Alenia Difesa	**Italy**	***231***	**75-01165**
Amadeo Rossi SA	Brazil	*57*	75-18780
Australian Defence Industries Ltd	Australia	*9*	55-01000
Canons Delcour SA	Belgium	*46*	75-02200
Chamberlain Manufacturing Corp.	USA	*563*	75-17950
Chartered Firearms Industries Pte	Singapore	*348*	75-01150
Cockerill Mechanical Industries	Belgium	*46*	75-02410
Hellenic Arms Industry	Greece	*187*	55-00800
Indústrias de Defesa	Portugal	*327*	55-00950
Israel Military Industries Ltd	Israel	*220*	75-10300
National Forge Co.	USA	*598*	75-63850
Parker-Hale Ltd	UK	*473*	75-43850
RAMO Manufacturing Inc.	USA	*607*	75-72810
Robert W. Hart & Son Inc.	USA	*581*	75-42450
Royal Ordnance	UK	*481*	75-50620
Saco Defense Inc.	USA	*612*	75-77340
Vammas Ltd	**Finland**	***117***	**55-02220**

512 50 90 16 Gun barrels, tank

Australian Defence Industries Ltd	Australia	*9*	55-01000
Cockerill Mechanical Industries	Belgium	*46*	75-02410
INIMEX	**Belgium**	***46***	**75-04650**
Israel Military Industries Ltd	Israel	*220*	75-10300
Makina ve Kimya Endüstrisi	Turkey	*402*	55-01310
National Forge Co.	USA	*598*	75-63850
Nottingham International	USA	*640*	80-22000
Royal Ordnance	UK	*481*	75-50620

512 60 Grenade launchers / dischargers

ARAM	S. Africa	*357*	75-00795
Chamberlain Manufacturing Corp.	USA	*563*	75-17950
Chartered Firearms Industries Pte	Singapore	*348*	75-01150
Giat Industries	France	*126*	55-00100
Grenadier Engineering Inc.	USA	*580*	75-40390
Helio Mirror Co. Ltd	UK	*457*	75-28900
INIMEX	**Belgium**	***46***	**75-04650**
R/M Equipment Inc.	USA	*609*	75-75470
Saco Defense Inc.	USA	*612*	75-77340
SAE Alsetex	France	*142*	75-33810
Wegmann & Co GmbH	Germany	*177*	75-33700
Wegmann USA Inc.	USA	*627*	75-95543

512 61 Grenade launchers, ground base-plate mounted

Grenadier Engineering Inc.	USA	*580*	75-40390
Wegmann & Co GmbH	Germany	*177*	75-33700

512 62 Grenade launchers, hand-held

ARAM	S. Africa	*357*	75-00795
Chartered Firearms Industries Pte	Singapore	*348*	75-01150
Giat Industries	France	*126*	55-00100
Grenadier Engineering Inc.	USA	*580*	75-40390
Indústrias de Defesa	Portugal	*327*	55-00950
Jonathan Arthur Ciener Inc.	USA	*564*	75-18590
R/M Equipment Inc.	USA	*609*	75-75470

512 63 Grenade launchers, rifle

AAI Corp.	USA	*545*	75-00020
Allied Research Corp.	USA	*549*	75-04670
Australian Defence Industries Ltd	Australia	*9*	55-01000
Giat Industries	France	*126*	55-00100
Grenadier Engineering Inc.	USA	*580*	75-40390
Heckler & Koch GmbH	Germany	*169*	75-11972
INIMEX	**Belgium**	***46***	**75-04650**
Jonathan Arthur Ciener Inc.	USA	*564*	75-18590
Luchaire Défense SA	France	*139*	75-25400
MECAR SA	Belgium	*46*	75-04930
R/M Equipment Inc.	USA	*609*	75-75470

512 64 Grenade launchers / dischargers, vehicle-mounted

AAI Corp.	USA	*545*	75-00020
Chamberlain Manufacturing Corp.	USA	*563*	75-17950
Giat Industries	France	*126*	55-00100
Helio Mirror Co. Ltd	UK	*457*	75-28900
Saco Defense Inc.	USA	*612*	75-77340
Wegmann & Co GmbH	Germany	*177*	75-33700
Wegmann USA Inc.	USA	*627*	75-95543

512 70 Mortars

AAI Corp.	USA	*545*	75-00020
Alenia Difesa	**Italy**	***231***	**75-01165**
Allied Research Corp.	USA	*549*	75-04670
Daudsons Armoury	Pakistan	*315*	75-00500
Dynamit Nobel Graz GmbH	Austria	*32*	75-00800
INIMEX	**Belgium**	***46***	**75-04650**
Makina ve Kimya Endüstrisi	Turkey	*402*	55-01310
MECAR SA	Belgium	*46*	75-04930
NORINCO	China PR	*88*	55-00280
Nottingham International	USA	*640*	80-22000
Tamares Ltd	UK	*487*	75-58500
United Defense L.P.	USA	*624*	75-92050
Wejra A/S	Denmark	*104*	75-23300

512 73 Mortars, naval

Bofors AB	Sweden	*375*	74-00020

512 74 Mortars, ground-based

Alenia Difesa	**Italy**	***231***	**75-01165**
Allied Research Corp.	USA	*549*	75-04670
CIC International Ltd	USA	*563*	75-18570
Daudsons Armoury	Pakistan	*315*	75-00500
Royal Ordnance	UK	*481*	75-50620
Saunders-Roe Ltd	UK	*482*	75-51850
Tamares Ltd	UK	*487*	75-58500
Wejra A/S	Denmark	*104*	75-23300

512 74 10 Mortars, infantry

Allied Research Corp.	USA	*549*	75-04670
CIC International Ltd	USA	*563*	75-18570
Daudsons Armoury	Pakistan	*315*	75-00500
Indústrias de Defesa	Portugal	*327*	55-00950
INIMEX	**Belgium**	***46***	**75-04650**
Nottingham International	USA	*640*	80-22000
Saunders-Roe Ltd	UK	*482*	75-51850
Tamares Ltd	UK	*487*	75-58500
United Defense L.P.	USA	*624*	75-92050
Wejra A/S	Denmark	*104*	75-23300

512 74 11 Mortars, infantry, light

Alenia Difesa	**Italy**	***231***	**75-01165**
CIC International Ltd	USA	*563*	75-18570
Daudsons Armoury	Pakistan	*315*	75-00500
Indústrias de Defesa	Portugal	*327*	55-00950
Nottingham International	USA	*640*	80-22000
Saunders-Roe Ltd	UK	*482*	75-51850
Tamares Ltd	UK	*487*	75-58500
Vammas Ltd	**Finland**	***117***	**55-02220**
Wejra A/S	Denmark	*104*	75-23300

512 74 12 Mortars, infantry, medium

CIC International Ltd	USA	*563*	75-18570
Daewoo Heavy Industries Ltd	Korea (ROK)	*265*	75-00600
Daudsons Armoury	Pakistan	*315*	75-00500
Dynamit Nobel Graz GmbH	Austria	*32*	75-00800
Indústrias de Defesa	Portugal	*327*	55-00950
MECAR SA	Belgium	*46*	75-04930
Nottingham International	USA	*640*	80-22000
Saunders-Roe Ltd	UK	*482*	75-51850
Tamares Ltd	UK	*487*	75-58500
Vammas Ltd	**Finland**	***117***	**55-02220**
Wejra A/S	Denmark	*104*	75-23300

512 74 13 Mortars, infantry, heavy

CIC International Ltd	USA	*563*	75-18570
Daudsons Armoury	Pakistan	*315*	75-00500
Dynamit Nobel Graz GmbH	Austria	*32*	75-00800
National Forge Co.	USA	*598*	75-63850
Nottingham International	USA	*640*	80-22000
Saunders-Roe Ltd	UK	*482*	75-51850
Tamares Ltd	UK	*487*	75-58500
Vammas Ltd	**Finland**	***117***	**55-02220**
Wejra A/S	Denmark	*104*	75-23300

512 74 30 Mortars, vehicle turret-mounted

Bofors AB	Sweden	*375*	74-00020
CIC International Ltd	USA	*563*	75-18570
Wejra A/S	Denmark	*104*	75-23300

512 74 45 Mortars, vehicle mounted

Royal Ordnance	UK	*481*	75-50620

512 80 Rocket systems

AAI Corp.	USA	*545*	75-00020
Avibras Aeroespacial SA	Brazil	*55*	75-00968
Bofors AB	Sweden	*375*	74-00020
Bressel SA	Spain	*366*	75-01670
Bristol Aerospace Ltd	Canada	*69*	75-01500
CIC International Ltd	USA	*563*	75-18570
Emco Inc.	USA	*573*	75-30985

Product-Supplier Link

Eurodef Aerospace & Defence	Greece	*189*	75-01295
FN Manufacturing Inc	USA	*576*	75-35320
Forges de Zeebrugge SA	Belgium	*46*	75-03750
G. Shiposh, Engineering Works Ltd	Israel	*222*	75-20250
INIMEX	**Belgium**	***46***	**75-04650**
Israel Military Industries Ltd	Israel	*220*	75-10300
Japan Aircraft Manufacturing Co. Ltd	Japan	*252*	75-03200
Lucas Western Inc.	USA	*592*	75-56300
MECAR SA	Belgium	*46*	75-04930
NORINCO	China PR	*88*	55-00280
Precision Standard Inc.	USA	*605*	75-71365
Royal Ordnance	UK	*481*	75-50620
Sakr Factory for Developed Ind.	Egypt	*109*	55-01700
Textron Systems Division	USA	*621*	75-88655

512 82 Rocket systems, airborne

Avibras Aeroespacial SA	Brazil	*55*	75-00968
Bristol Aerospace Ltd	Canada	*69*	75-01500
Eurodef Aerospace & Defence	Greece	*189*	75-01295
Forges de Zeebrugge SA	Belgium	*46*	75-03750
Israel Military Industries Ltd	Israel	*220*	75-10300

512 82 10 Rocket pods, aircraft / helicopter (see code 515.20)

Aerea Luftfahrtgeräte GmbH.	Germany	*164*	75-00260
Aerea SpA.	Italy	*230*	75-00280
Avibras Aeroespacial SA	Brazil	*55*	75-00968
Bofors AB	Sweden	*375*	74-00020
Eurodef Aerospace & Defence	Greece	*189*	75-01295
Forges de Zeebrugge SA	Belgium	*46*	75-03750
Giat Industries	France	*126*	55-00100
Israel Military Industries Ltd	Israel	*220*	75-10300
Lucas Western Inc.	USA	*592*	75-56300
Precimetal Fonderie de Précision SA	Belgium	*46*	75-05500

512 83 Rocket systems, naval

Alenia Difesa	**Italy**	***231***	**75-01165**
Alenia-Elsag	Italy	*230*	74-00100
Bofors AB	Sweden	*375*	74-00020
Somchem	S. Africa	*359*	75-19200
United Defense L.P.	USA	*624*	75-92050

512 84 Rocket systems, ground-based

Avibras Aeroespacial SA	Brazil	*55*	75-00968
GEL Engineering Ltd	UK	*454*	75-25500
Giat Industries	France	*126*	55-00100
Haley & Weller Ltd	UK	*456*	75-27400
Israel Military Industries Ltd	Israel	*220*	75-10300
ML Aviation Ltd	UK	*469*	75-40200
Royal Ordnance	UK	*481*	75-50620
Sakr Factory for Developed Ind.	Egypt	*109*	55-01700

512 84 10 Rocket systems, artillery

Bressel SA.	Spain	*366*	75-01670
CIC International Ltd	USA	*563*	75-18570
Dynamit Nobel GmbH	Germany	*167*	75-07100
Forges de Zeebrugge SA	Belgium	*46*	75-03750
Giat Industries	France	*126*	55-00100
INIMEX	**Belgium**	***46***	**75-04650**
Israel Military Industries Ltd	Israel	*220*	75-10300
Israel Military Industries Rocket Systems Division	**Israel**	***220***	**75-10310**
Sakr Factory for Developed Ind.	Egypt	*109*	55-01700
Société Europeenne de Propulsion	France	*144*	75-34850
Somchem	S. Africa	*359*	75-19200
State Research & Association	Russia	*334*	55-18325
Wegmann & Co GmbH	Germany	*177*	75-33700

512 84 11 Rocket systems, artillery, towed

Avibras Aeroespacial SA	Brazil	*55*	75-00968
CIC International Ltd	USA	*563*	75-18570
Israel Military Industries Ltd	Israel	*220*	75-10300
Israel Military Industries Rocket Systems Division	**Israel**	***220***	**75-10310**
Somchem	S. Africa	*359*	75-19200

512 84 12 Rocket systems, artillery, vehicle-mounted

Avibras Aeroespacial SA	Brazil	*55*	75-00968
CIC International Ltd	USA	*563*	75-18570
Forges de Zeebrugge SA	Belgium	*46*	75-03750
Israel Military Industries Ltd	Israel	*220*	75-10300
Israel Military Industries Rocket Systems Division	**Israel**	***220***	**75-10310**
Japan Aircraft Manufacturing Co. Ltd	Japan	*252*	75-03200
Somchem	S. Africa	*359*	75-19200
TECTRAN	Brazil	*58*	75-20205
Wegmann & Co GmbH	Germany	*177*	75-33700

512 84 20 Rocket launchers, man-portable

Alenia Difesa	**Italy**	***231***	**75-01165**
Dynamit Nobel GmbH	Germany	*167*	75-07100
Hunting Engineering Ltd	UK	*459*	75-30700
INIMEX	**Belgium**	***46***	**75-04650**
Instalaza SA	Spain	*366*	75-03900
Israel Military Industries Ltd	Israel	*220*	75-10300
Israel Military Industries Rocket Systems Division	**Israel**	***220***	**75-10310**
Lucas Western Inc.	USA	*592*	75-56300
MECAR SA	Belgium	*46*	75-04930
NORINCO	China PR	*88*	55-00280
Sakr Factory for Developed Ind.	Egypt	*109*	55-01700
Somchem	S. Africa	*359*	75-19200
Talley Defense Systems Inc.	USA	*619*	75-86120
Technical Products Group Inc.	USA	*620*	75-86764

513 Mines

AAI Corp.	USA	*545*	75-00020
Cenzin Co. Ltd.	Poland	*324*	75-00200
CIC International Ltd	USA	*563*	75-18570
Dynamit Nobel Graz GmbH.	Austria	*32*	75-00800
Explosivos Alaveses SA	Spain	*366*	75-03050
Graseby Dynamics Ltd	UK	*455*	75-26590
IBQ Industrias Químicas Ltda	Brazil	*55*	75-02700
INIMEX	**Belgium**	***46***	**75-04650**
MIL-BASE Industries Inc.	USA	*596*	75-61530
Pakistan Ordnance Factories	Pakistan	*314*	55-00100
Royal Ordnance	UK	*481*	75-50620
SAE Alsetex	France	*142*	75-33810
SIMA PERU	Peru	*319*	55-01950
SPL Ordnance Ltd.	Israel	*225*	80-01935
STN ATLAS Elektronik GmbH	Germany	*165*	75-01150
Textron Systems Division	USA	*621*	75-88655
The Dewey Electronics Corp.	USA	*569*	75-26100
Valsella Meccanotecnica SpA	Italy	*239*	75-26800
Vickers Shipbuilding & Eng Ltd	UK	*491*	75-63100

513 20 Mines, air-launched

CMS Inc.	USA	*564*	75-19170
Dynamit Nobel GmbH	Germany	*167*	75-07100
Graseby Dynamics Ltd	UK	*455*	75-26590
Royal Ordnance	UK	*481*	75-50620
Valsella Meccanotecnica SpA	Italy	*239*	75-26800

513 21 Mines, air-launched, anti-personnel

CMS Inc.	USA	*564*	75-19170
Graseby Dynamics Ltd	UK	*455*	75-26590
Valsella Meccanotecnica SpA	Italy	*239*	75-26800

513 22 Mines, air-launched, anti-tank

CMS Inc.	USA	*564*	75-19170
Explosivos Alaveses SA	Spain	*366*	75-03050
Graseby Dynamics Ltd	UK	*455*	75-26590
Valsella Meccanotecnica SpA	Italy	*239*	75-26800

513 30 Mines, naval

Bofors SA Marine AB	Sweden	376	75-03040
COSMOS SpA	Italy	233	75-05700
Explosivos Alaveses SA	Spain	366	75-03050
Hitachi Zosen Co. Ltd.	Japan	251	75-02390
Royal Ordnance	UK	481	75-50620
SIMA PERU	Peru	319	55-01950
STN ATLAS Elektronik GmbH	Germany	165	75-01150
Thomson Sintra	France	148	75-45135
Whitehead SpA	Italy	240	75-27600

513 30 05 Mines, sea, acoustic

AlliedSignal Ltd	UK	430	75-02747
Hitachi Zosen Co. Ltd.	Japan	251	75-02390

513 30 10 Mines, sea, bottom / seabed

Bofors AB	Sweden	375	74-00020
Bofors SA Marine AB	Sweden	376	75-03040
COSMOS SpA	Italy	233	75-05700
Eurometaal NV	Netherlands	292	75-02200
The Dewey Electronics Corp.	USA	569	75-26100
Thomson Sintra	France	148	75-45135
Whitehead SpA	Italy	240	75-27600

513 30 20 Mines, sea, self-propelled

Bofors SA Marine AB	Sweden	376	75-03040

513 30 22 Mines, sea, remotely-controlled

Bofors SA Marine AB	Sweden	376	75-03040

513 30 30 Mines, sea, tethered

Bofors AB	Sweden	375	74-00020
Bofors SA Marine AB	Sweden	376	75-03040

513 32 Mines, sea, limpet

COSMOS SpA	Italy	233	75-05700

513 35 Mines, naval, practice

Bofors AB	Sweden	375	74-00020
Bofors SA Marine AB	Sweden	376	75-03040
COSMOS SpA	Italy	233	75-05700
The Dewey Electronics Corp.	USA	569	75-26100
Thomson Sintra	France	148	75-45135
Whitehead SpA	Italy	240	75-27600

513 37 Mines, river

Bofors SA Marine AB	Sweden	376	75-03040

513 38 Mine accessories, naval

Bofors SA Marine AB	Sweden	376	75-03040
Sparton Electronics	USA	616	75-82900
The Dewey Electronics Corp.	USA	569	75-26100

513 39 Mine components, naval

Bodenseewerk Gerätetechnik GmbH	Germany	165	75-03000
Bofors SA Marine AB	Sweden	376	75-03040
MIL-BASE Industries Inc.	USA	596	75-61530
Sparton Electronics	USA	616	75-82900
The Dewey Electronics Corp.	USA	569	75-26100

513 40 Mines, land launched / laid

Aerojet	USA	547	75-01500
CIC International Ltd	USA	563	75-18570
Dynamit Nobel GmbH	Germany	167	75-07100
Giat Industries	France	126	55-00100
Pakistan Ordnance Factories	Pakistan	314	55-00100
SPL Ordnance Ltd.	Israel	225	80-01935
Textron Systems Division	USA	621	75-88655
Valsella Meccanotecnica SpA	Italy	239	75-26800

513 41 Mines, land, anti-personnel

CIC International Ltd	USA	563	75-18570
Graseby Dynamics Ltd	UK	455	75-26590
IBQ Industrias Químicas Ltda	Brazil	55	75-02700
INIMEX	**Belgium**	**46**	**75-04650**
Makina ve Kimya Endüstrisi	Turkey	402	55-01310
Pakistan Ordnance Factories	Pakistan	314	55-00100
SIMA PERU	Peru	319	55-01950
SPL Ordnance Ltd.	Israel	225	80-01935
Valsella Meccanotecnica SpA	Italy	239	75-26800

513 41 10 Mines, land, anti-personnel, Claymore type

CIC International Ltd	USA	563	75-18570
INIMEX	**Belgium**	**46**	**75-04650**
Pakistan Ordnance Factories	Pakistan	314	55-00100
SPL Ordnance Ltd.	Israel	225	80-01935
Valsella Meccanotecnica SpA	Italy	239	75-26800

513 41 20 Mines, land, anti-personnel, scatter

CIC International Ltd	USA	563	75-18570
Emco Inc.	USA	573	75-30985
Explosivos Alaveses SA	Spain	366	75-03050
Graseby Dynamics Ltd	UK	455	75-26590
SPL Ordnance Ltd.	Israel	225	80-01935
Valsella Meccanotecnica SpA	Italy	239	75-26800

513 42 Mines, land, anti-tank

CIC International Ltd	USA	563	75-18570
Dynamit Nobel GmbH	Germany	167	75-07100
Dynamit Nobel Graz GmbH.	Austria	32	75-00800
Eurometaal NV	Netherlands	292	75-02200
Explosivos Alaveses SA	Spain	366	75-03050
Giat Industries	France	126	55-00100
Graseby Dynamics Ltd	UK	455	75-26590
IBQ Industrias Químicas Ltda	Brazil	55	75-02700
INIMEX	**Belgium**	**46**	**75-04650**
INTERTECHNIK Techn. Produktionen	Austria	33	75-01430
Makina ve Kimya Endüstrisi	Turkey	402	55-01310
SAE Alsetex	France	142	75-33810
SIMA PERU	Peru	319	55-01950
SPL Ordnance Ltd.	Israel	225	80-01935
Valsella Meccanotecnica SpA	Italy	239	75-26800

513 42 10 Mines, land, anti-tank, buried

CIC International Ltd	USA	563	75-18570
Explosivos Alaveses SA	Spain	366	75-03050
IBQ Industrias Químicas Ltda	Brazil	55	75-02700
SAE Alsetex	France	142	75-33810
SPL Ordnance Ltd.	Israel	225	80-01935
Valsella Meccanotecnica SpA	Italy	239	75-26800

513 42 20 Mines, land, anti-tank, surface

CIC International Ltd	USA	563	75-18570
Dynamit Nobel GmbH	Germany	167	75-07100
Explosivos Alaveses SA	Spain	366	75-03050
Graseby Dynamics Ltd	UK	455	75-26590
SAE Alsetex	France	142	75-33810
SPL Ordnance Ltd.	Israel	225	80-01935
Textron Systems Division	USA	621	75-88655
Valsella Meccanotecnica SpA	Italy	239	75-26800

513 42 21 Mines, land, anti-tank, surface, off-route

CMS Inc.	USA	564	75-19170

Giat Industries	France	*126*	55-00100
Graseby Dynamics Ltd	UK	*455*	75-26590
Manurhin Défense	France	*139*	75-25630
SPL Ordnance Ltd	Israel	*225*	80-01935

513 43 Mines, land, anti-vehicle, surface

CIC International Ltd	USA	*563*	75-18570
Dynamit Nobel Graz GmbH	Austria	*32*	75-00800
SAE Alsetex	France	*142*	75-33810
SPL Ordnance Ltd	Israel	*225*	80-01935
Textron Systems Division	USA	*621*	75-88655
Valsella Meccanotecnica SpA	Italy	*239*	75-26800

513 44 Mines, land, general purpose

CIC International Ltd	USA	*563*	75-18570
Giat Industries	France	*126*	55-00100
SAE Alsetex	France	*142*	75-33810
SPL Ordnance Ltd	Israel	*225*	80-01935
Valsella Meccanotecnica SpA	Italy	*239*	75-26800

513 45 Mines, land, noise-generating

Ruggieri Technologie	France	*142*	75-33500
SPL Ordnance Ltd	Israel	*225*	80-01935

513 46 Mines, land, illuminating

CIC International Ltd	USA	*563*	75-18570
Dynamit Nobel Graz GmbH	Austria	*32*	75-00800
Giat Industries	France	*126*	55-00100
Pakistan Ordnance Factories	Pakistan	*314*	55-00100
Ruggieri Technologie	France	*142*	75-33500
Saunders-Roe Ltd	UK	*482*	75-51850
Simad SpA	Italy	*230*	74-23350
SPL Ordnance Ltd	Israel	*225*	80-01935
Valsella Meccanotecnica SpA	Italy	*239*	75-26800

513 47 Mines, land, bounding

CIC International Ltd	USA	*563*	75-18570
Explosivos Alaveses SA	Spain	*366*	75-03050
Pakistan Ordnance Factories	Pakistan	*314*	55-00100
SPL Ordnance Ltd	Israel	*225*	80-01935
Valsella Meccanotecnica SpA	Italy	*239*	75-26800

513 48 Mines, land, non-metallic

IBQ Industrias Químicas Ltda	Brazil	*55*	75-02700
SPL Ordnance Ltd	Israel	*225*	80-01935
Valsella Meccanotecnica SpA	Italy	*239*	75-26800

513 49 60 Sensors, ground-launched anti-personnel mine

CIC International Ltd	USA	*563*	75-18570
SPL Ordnance Ltd	Israel	*225*	80-01935

513 50 Mine systems

AAI Corp	USA	*545*	75-00020
Cenzin Co. Ltd	Poland	*324*	75-00200
CIC International Ltd	USA	*563*	75-18570
Explosivos Alaveses SA	Spain	*366*	75-03050
Hunting Engineering Ltd	UK	*459*	75-30700
R. Alkan (UK) Ltd	UK	*429*	75-02680
Rheinmetall Waffe & Munition GmbH	Germany	*173*	75-24950
The Dewey Electronics Corp	USA	*569*	75-26100
Valsella Meccanotecnica SpA	Italy	*239*	75-26800
Vickers Shipbuilding & Eng Ltd	UK	*491*	75-63100

513 50 20 Mine systems, air-launched

CMS Inc	USA	*564*	75-19170
Explosivos Alaveses SA	Spain	*366*	75-03050
Hunting Engineering Ltd	UK	*459*	75-30700
Valsella Meccanotecnica SpA	Italy	*239*	75-26800

513 50 22 Mine systems, helicopter-launched

Explosivos Alaveses SA	Spain	*366*	75-03050
R. Alkan (UK) Ltd	UK	*429*	75-02680
Valsella Meccanotecnica SpA	Italy	*239*	75-26800

513 50 30 Mine systems, naval

Bofors SA Marine AB	Sweden	*376*	75-03040
Hunting Engineering Ltd	UK	*459*	75-30700
The Dewey Electronics Corp	USA	*569*	75-26100

513 50 40 Mine systems, ground-based

AAI Corp	USA	*545*	75-00020
Hunting Engineering Ltd	UK	*459*	75-30700
Rheinmetall Waffe & Munition GmbH	Germany	*173*	75-24950
Valsella Meccanotecnica SpA	Italy	*239*	75-26800
Vickers Shipbuilding & Eng Ltd	UK	*491*	75-63100

513 50 42 Mine systems, vehicle launched / laid

CIC International Ltd	USA	*563*	75-18570
Valsella Meccanotecnica SpA	Italy	*239*	75-26800
Vickers Shipbuilding & Eng Ltd	UK	*491*	75-63100

513 50 42 10 Mine systems, vehicle launched / laid, bar mine type

CIC International Ltd	USA	*563*	75-18570
Valsella Meccanotecnica SpA	Italy	*239*	75-26800
Vickers Shipbuilding & Eng Ltd	UK	*491*	75-63100

513 50 42 60 Mine systems, vehicle launched / laid, scatter type

AAI Corp	USA	*545*	75-00020
CIC International Ltd	USA	*563*	75-18570
Hunting Engineering Ltd	UK	*459*	75-30700
Valsella Meccanotecnica SpA	Italy	*239*	75-26800
Vickers Shipbuilding & Eng Ltd	UK	*491*	75-63100

515 Weapon systems

AAI Corp	USA	*545*	75-00020
Aerojet	USA	*547*	75-01500
Alenia Difesa	**Italy**	***231***	**75-01165**
British Aerospace (Systems &	UK	*436*	75-08890
Calspan SRL Technology	USA	*540*	60-03000
Dynamit Nobel GmbH	Germany	*167*	75-07100
FN Herstal SA	Belgium	*46*	75-03600
Forges de Zeebrugge SA	Belgium	*46*	75-03750
Fuji Heavy Industries Ltd	Japan	*251*	75-01100
Hellenic Arms Industry	Greece	*187*	55-00800
Hunting Engineering Ltd	UK	*459*	75-30700
Israel Aircraft Industries Ltd	Israel	*219*	75-10200
Israel Military Industries Ltd	Israel	*220*	75-10300
JSC Tulamashzavod	Russia	*334*	55-19450
Lockheed Martin Corp	USA	*591*	75-55635
Lucas Western Inc	USA	*592*	75-56300
Matra BAe Dynamics	UK	*425*	74-06850
Oerlikon-Contraves AG	Switzerland	*387*	75-09250
R. Alkan (UK) Ltd	UK	*429*	75-02680
Royal Ordnance	UK	*481*	75-50620
State Research Institute	Russia	*336*	60-04150
The Boeing Company	USA	*560*	75-14250
Thomson Sintra	France	*148*	75-45135
Vickers Shipbuilding & Eng Ltd	UK	*491*	75-63100

515 20 Weapon systems, aircraft

Aerojet	USA	*547*	75-01500
AIM Enterprises Inc	USA	*630*	80-01350
Eurodef Aerospace & Defence	Greece	*189*	75-01295

FN Herstal SA	Belgium	*46*	75-03600
FN Manufacturing Inc	USA	*576*	75-35320
Forges de Zeebrugge SA	Belgium	*46*	75-03750
Hunting Engineering Ltd	UK	*459*	75-30700
Oerlikon-Contraves AG.	Switzerland	*387*	75-09250
Royal Ordnance	UK	*481*	75-50620

515 20 50 Weapon pods, aircraft

Aerea Luftfahrtgeräte GmbH.	Germany	*164*	75-00260
Aerea SpA.	Italy.	*230*	75-00280
Dassault Aviation Group	France	*135*	75-12975
FN Herstal SA	Belgium	*46*	75-03600
Forges de Zeebrugge SA	Belgium	*46*	75-03750
Giat Industries	France	*126*	55-00100
R. Alkan (UK) Ltd	UK	*429*	75-02680
Royal Ordnance	UK	*481*	75-50620

515 20 55 Machine gun pods, aircraft

Aerea Luftfahrtgeräte GmbH.	Germany	*164*	75-00260
Aerea SpA.	Italy.	*230*	75-00280
FN Herstal SA	Belgium	*46*	75-03600
Giat Industries	France	*126*	55-00100
Royal Ordnance	UK	*481*	75-50620
Stringer Welding & Machine Inc.	USA	*618*	75-84435

515 20 57 Rocket pods, aircraft (see code 512.82.10)

Aerea SpA.	Italy.	*230*	75-00280
Dassault Aviation Group	France	*135*	75-12975
FN Herstal SA	Belgium	*46*	75-03600
Forges de Zeebrugge SA	Belgium	*46*	75-03750
Fuji Heavy Industries Ltd	Japan	*251*	75-01100
Royal Ordnance	UK	*481*	75-50620

515 24 Weapon systems, helicopter

Aerea SpA.	Italy.	*230*	75-00280
AIM Enterprises Inc.	USA	*630*	80-01350
Avibras Aeroespacial SA	Brazil	*55*	75-00968
CMS Inc.	USA	*564*	75-19170
FN Herstal SA	Belgium	*46*	75-03600
Forges de Zeebrugge SA	Belgium	*46*	75-03750
Fuji Heavy Industries Ltd	Japan	*251*	75-01100
Military Systems Group Inc.	USA	*596*	75-61570
Nottingham International.	USA	*640*	80-22000
Precimetal Fonderie de Précision SA	Belgium	*46*	75-05500

515 24 50 Weapon pods, helicopter

Aerea SpA.	Italy.	*230*	75-00280
FN Herstal SA	Belgium	*46*	75-03600
Forges de Zeebrugge SA	Belgium	*46*	75-03750
Fuji Heavy Industries Ltd	Japan	*251*	75-01100
Giat Industries	France	*126*	55-00100
Nottingham International.	USA	*640*	80-22000
R. Alkan (UK) Ltd	UK	*429*	75-02680

515 24 55 Machine gun pods, helicopter

Aerea SpA.	Italy.	*230*	75-00280
AIM Enterprises Inc.	USA	*630*	80-01350
Alenia Difesa	**Italy**	***231***	**75-01165**
FN Herstal SA	Belgium	*46*	75-03600
Giat Industries	France	*126*	55-00100
Nottingham International.	USA	*640*	80-22000
Stringer Welding & Machine Inc.	USA	*618*	75-84435

515 24 57 Rocket pods, helicopter (see code 512.82.10)

Aerea SpA.	Italy.	*230*	75-00280
FN Herstal SA	Belgium	*46*	75-03600
Forges de Zeebrugge SA	Belgium	*46*	75-03750
Fuji Heavy Industries Ltd	Japan	*251*	75-01100

515 30 Weapon systems, naval

AAI Corp.	USA	*545*	75-00020
AeroSpace Technologies of Australia	Australia	*11*	75-00340
Alenia Difesa	**Italy**	***231***	**75-01165**
Alenia-Elsag	Italy.	*230*	74-00100
FN Herstal SA	Belgium	*46*	75-03600
FN Manufacturing Inc	USA	*576*	75-35320
Hunting Engineering Ltd	UK	*459*	75-30700
JSC Tulamashzavod	Russia	*334*	55-19450
Matra BAe Dynamics	UK	*425*	74-06850
MSI-Defence Systems Ltd	UK	*469*	75-41280
Oerlikon-Contraves AG.	Switzerland	*387*	75-09250
The Boeing Company.	USA	*560*	75-14250
Thomson Sintra	France	*148*	75-45135
Vickers Shipbuilding & Eng Ltd	UK	*491*	75-63100

515 31 Weapon systems, naval, anti-aircraft

Alenia-Elsag	Italy.	*230*	74-00100
FN Herstal SA	Belgium	*46*	75-03600
Israel Aircraft Industries Ltd	Israel	*219*	75-10200
JSC Tulamashzavod	Russia	*334*	55-19450
Matra BAe Dynamics	UK	*425*	74-06850
Military Systems Group Inc.	USA	*596*	75-61570
Oerlikon-Contraves AG.	Switzerland	*387*	75-09250
United Defense L.P..	USA	*624*	75-92050

515 31 10 Weapon systems, naval, anti-aircraft gun

Alenia Difesa	**Italy**	***231***	**75-01165**
Brazan SA	Spain	*364*	55-00140
FN Herstal SA	Belgium	*46*	75-03600
JSC Tulamashzavod	Russia	*334*	55-19450
MSI-Defence Systems Ltd	UK	*469*	75-41280
Oerlikon-Contraves AG.	Switzerland	*387*	75-09250
United Defense L.P..	USA	*624*	75-92050

515 31 20 Weapon systems, naval, anti-aircraft missile

Israel Aircraft Industries Ltd	Israel	*219*	75-10200
JSC Tulamashzavod	Russia	*334*	55-19450
Matra BAe Dynamics	UK	*425*	74-06850
United Defense L.P..	USA	*624*	75-92050

515 31 22 Weapon systems, naval, anti-aircraft missile, low-altitude

Israel Aircraft Industries Ltd	Israel	*219*	75-10200
JSC Tulamashzavod	Russia	*334*	55-19450
Matra BAe Dynamics	UK	*425*	74-06850
United Defense L.P..	USA	*624*	75-92050

515 31 24 Weapon systems, naval, anti-aircraft missile, low / medium altitude

Israel Aircraft Industries Ltd	Israel	*219*	75-10200
JSC Tulamashzavod	Russia	*334*	55-19450
Matra BAe Dynamics	UK	*425*	74-06850
United Defense L.P..	USA	*624*	75-92050

515 31 30 Weapon systems, naval, anti-aircraft, containerised

Israel Aircraft Industries Ltd	Israel	*219*	75-10200
Matra BAe Dynamics	UK	*425*	74-06850
United Defense L.P..	USA	*624*	75-92050

515 33 Weapon systems, naval, anti-missile

Israel Aircraft Industries Ltd	Israel	*219*	75-10200
JSC Tulamashzavod	Russia	*334*	55-19450
Oerlikon-Contraves AG.	Switzerland	*387*	75-09250
Oerlikon-Contraves SpA	Italy.	*237*	75-16800
United Defense L.P..	USA	*624*	75-92050

515 33 10 Weapon systems, naval, anti-missile, gun-based

Alenia Difesa	**Italy**	*231*	**75-01165**
Israel Aircraft Industries Ltd	Israel	*219*	75-10200
JSC Tulamashzavod	Russia	*334*	55-19450
Matra BAe Dynamics	UK	*425*	74-06850
MSI-Defence Systems Ltd	UK	*469*	75-41280
Oerlikon-Contraves AG.	Switzerland	*387*	75-09250
United Defense L.P..	USA	*624*	75-92050

515 33 20 Weapon systems, naval, anti-missile, missile-based

Israel Aircraft Industries Ltd	Israel	*219*	75-10200
JSC Tulamashzavod	Russia	*334*	55-19450
Matra BAe Dynamics	UK	*425*	74-06850
Matra Défense-Espace	France	*139*	75-26000
United Defense L.P..	USA	*624*	75-92050

515 35 Weapon systems, naval, anti-submarine

Northrop Grumman Corporation.	USA	*599*	75-65450
State Research & Association	Russia	*334*	55-18325
The Boeing Company.	USA	*560*	75-14250
Thomson Sintra.	France	*148*	75-45135
United Defense L.P..	USA	*624*	75-92050

515 37 Weapon systems, naval, surface target

Alenia Difesa	**Italy**	*231*	**75-01165**
Matra BAe Dynamics	UK	*425*	74-06850
MSI-Defence Systems Ltd	UK	*469*	75-41280
United Defense L.P..	USA	*624*	75-92050

515 37 30 Weapon systems, naval, surface target, containerised

Matra BAe Dynamics	UK	*425*	74-06850
Thomson Sintra.	France	*148*	75-45135
United Defense L.P..	USA	*624*	75-92050

515 38 Weapon systems, naval, containerised

Matra BAe Dynamics	UK	*425*	74-06850
Oerlikon-Contraves AG.	Switzerland	*387*	75-09250
United Defense L.P..	USA	*624*	75-92050

515 40 Weapon systems, ground-based

AAI Corp.	USA	*545*	75-00020
Aerojet.	USA	*547*	75-01500
Fábricas y Maestranzas del Ejercito	Chile	*84*	55-00695
FN Herstal SA	Belgium	*46*	75-03600
Forges de Zeebrugge SA	Belgium	*46*	75-03750
Helio Mirror Co. Ltd.	UK	*457*	75-28900
Hunting Engineering Ltd	UK	*459*	75-30700
Israel Aircraft Industries Ltd	Israel	*219*	75-10200
Israel Military Industries Ltd	Israel	*220*	75-10300
Manurhin Défense	France	*139*	75-25630
Matra BAe Dynamics	UK	*425*	74-06850
Oerlikon-Contraves AG.	Switzerland	*387*	75-09250
The Boeing Company.	USA	*560*	75-14250

515 41 Weapon systems, ground-based, anti-aircraft

Alenia Difesa	**Italy**	*231*	**75-01165**
FN Herstal SA	Belgium	*46*	75-03600
Forges de Zeebrugge SA	Belgium	*46*	75-03750
Helio Mirror Co. Ltd.	UK	*457*	75-28900
Hellenic Arms Industry	Greece	*187*	55-00800
Mandra Plant	Greece	*190*	75-02000
Matra BAe Dynamics	UK	*425*	74-06850
Oerlikon-Contraves AG.	Switzerland	*387*	75-09250
The Boeing Company.	USA	*560*	75-14250
United Defense L.P..	USA	*624*	75-92050

515 41 10 Weapon systems, ground-based, anti-aircraft gun

Alenia Difesa	**Italy**	*231*	**75-01165**
FN Herstal SA	Belgium	*46*	75-03600
Giat Industries.	France	*126*	55-00100
Helio Mirror Co. Ltd.	UK	*457*	75-28900
Hellenic Arms Industry	Greece	*187*	55-00800
Israel Aircraft Industries Ltd	Israel	*219*	75-10200

515 41 15 Weapon systems, ground-based, anti-aircraft gun, self-propelled

Israel Aircraft Industries Ltd	Israel	*219*	75-10200

515 41 20 Weapon systems, ground-based, anti-aircraft missile

Matra BAe Dynamics	UK	*425*	74-06850
The Boeing Company.	USA	*560*	75-14250
United Defense L.P..	USA	*624*	75-92050

515 41 22 Weapon systems, ground-based, anti-aircraft missile, low-altitude

Matra BAe Dynamics	UK	*425*	74-06850
The Boeing Company.	USA	*560*	75-14250
United Defense L.P..	USA	*624*	75-92050

515 41 24 Weapon systems, ground-based, anti-aircraft missile, low / medium altitude

Avibras Aeroespacial SA	Brazil	*55*	75-00968
Matra BAe Dynamics	UK	*425*	74-06850
United Defense L.P..	USA	*624*	75-92050

515 41 26 Weapon systems, ground-based, anti-aircraft missile, medium / high altitude

United Defense L.P..	USA	*624*	75-92050

515 41 28 Weapon systems, ground-based, anti-aircraft missile, self-propelled

Matra BAe Dynamics	UK	*425*	74-06850
United Defense L.P..	USA	*624*	75-92050

515 42 Weapon systems, ground-based, anti-tank

Alenia Difesa	**Italy**	*231*	**75-01165**
Allied Research Corp.	USA	*549*	75-04670
Dynamit Nobel GmbH	Germany	*167*	75-07100
Forges de Zeebrugge SA	Belgium	*46*	75-03750
Helio Mirror Co. Ltd.	UK	*457*	75-28900
Israel Military Industries Ltd	Israel	*220*	75-10300
Manurhin Défense	France	*139*	75-25630
Oerlikon-Contraves AG.	Switzerland	*387*	75-09250
Somchem	S. Africa	*359*	75-19200
United Defense L.P..	USA	*624*	75-92050

515 42 10 Weapon systems, ground-based, anti-tank gun

Allied Research Corp.	USA	*549*	75-04670
Israel Military Industries Ltd	Israel	*220*	75-10300

515 42 15 Weapon systems, ground-based, anti-tank gun, self-propelled

Allied Research Corp.	USA	*549*	75-04670
United Defense L.P..	USA	*624*	75-92050

515 42 20 Weapon systems, ground-based, anti-tank missile

CMS Inc.	USA	*564*	75-19170
Helio Mirror Co. Ltd.	UK	*457*	75-28900
Israel Military Industries Ltd	Israel	*220*	75-10300
Matra BAe Dynamics	UK	*425*	74-06850

515 42 25 Weapon systems, ground-based, anti-tank missile, self-propelled

Israel Military Industries Ltd	Israel	*220*	75-10300
Matra BAe Dynamics	UK	*425*	74-06850

515 48 Weapon systems, ground-based, containerised

Matra BAe Dynamics	UK	*425*	74-06850
Oerlikon-Contraves AG.	Switzerland	*387*	75-09250

517 Weapon-related systems & equipment

AAI Corp.	USA	*545*	75-00020
Aerea SpA.	Italy.	*230*	75-00280
AIM Enterprises Inc.	USA	*630*	80-01350
Alenia SpA	Italy.	*232*	75-01170
Boeing Australia Limited	Australia	*12*	75-02500
Bofors AB	Sweden	*375*	74-00020
Break-Free Inc.	USA	*560*	75-14990
BTR Aerospace Australia	Australia	*13*	75-02815
Calspan SRL Technology	USA	*540*	60-03000
CMI - Ballarat	Australia	*14*	75-03600
Computing Devices Company Ltd	UK	*442*	75-14700
EG&G Inc.	USA	*571*	75-29555
Elbit Systems Ltd	Israel	*218*	75-05300
FN Herstal SA	Belgium	*46*	75-03600
Fuji Heavy Industries Ltd	Japan	*251*	75-01100
Glaser Safety Slug Inc.	USA	*579*	75-39750
Heckler & Koch Inc	USA	*581*	75-42860
Instro Precision Ltd	UK	*461*	75-31630
Israel Aircraft Industries Ltd	Israel	*219*	75-10200
Israel Military Industries Ltd	Israel	*220*	75-10300
Lockheed Martin Corp.	USA	*591*	75-55635
Lucas Western Inc.	USA	*592*	75-56300
M. Paz Logistics Ltd	Israel	*224*	80-01600
Matra BAe Dynamics	UK	*425*	74-06850
Oerlikon-Contraves AG.	Switzerland	*387*	75-09250
R. Alkan (UK) Ltd	UK	*429*	75-02680
Riva Calzoni SpA	Italy.	*238*	75-20500
Royal Ordnance	UK	*481*	75-50620
SELIN Sistemi Sarl	Italy.	*238*	75-21750
SFIM	**France**	***144***	**75-35080**
Sparton Technology Inc.	USA	*616*	75-82950
Stringer Welding & Machine Inc.	USA	*618*	75-84435
Summit Machine Builders Corp.	USA	*618*	75-84900
Sverdrup Corp.	USA	*643*	80-30650
Tamares Ltd.	UK	*487*	75-58500
United Defense L.P.	USA	*624*	75-92050
Vickers Shipbuilding & Eng Ltd	UK	*491*	75-63100

517 20 Weapon-related systems & equipment, airborne

AIM Enterprises Inc.	USA	*630*	80-01350
Air-Log Ltd	UK	*429*	75-01800
Alenia SpA	Italy.	*232*	75-01170
Cargo Aids	UK	*438*	75-11620
Computing Devices Company Ltd	UK	*442*	75-14700
Elbit Systems Ltd	Israel	*218*	75-05300
Electro-Optical Systems Group	USA	*572*	75-29812
Fibertek	USA	*575*	75-34330
FN Herstal SA	Belgium	*46*	75-03600
Israel Aircraft Industries Ltd	Israel	*219*	75-10200
Lucas Western Inc.	USA	*592*	75-56300
Matra BAe Dynamics	UK	*425*	74-06850
Royal Ordnance	UK	*481*	75-50620
SFIM	**France**	***144***	**75-35080**
Sparton Technology Inc.	USA	*616*	75-82950
Sverdrup Corp.	USA	*643*	80-30650
Sverdrup Technology Inc.	USA	*643*	80-30680

517 20 40 Weapon handling & loading systems, airborne

Air-Log Ltd	UK	*429*	75-01800
Avibras Aeroespacial SA	Brazil	*55*	75-00968
FN Herstal SA	Belgium	*46*	75-03600
Howden Sirocco	France	*137*	75-21700
Israel Aircraft Industries Ltd	Israel	*219*	75-10200
Royal Ordnance	UK	*481*	75-50620
Sparton Technology Inc.	USA	*616*	75-82950

517 20 50 Weapon launching systems

Aerea SpA.	Italy.	*230*	75-00280
Avibras Aeroespacial SA	Brazil	*55*	75-00968
GKN Westland Technologies Ltd	UK	*455*	75-25690
Israel Military Industries Ltd	Israel	*220*	75-10300
Lucas Western Inc.	USA	*592*	75-56300
R. Alkan (UK) Ltd	UK	*429*	75-02680
Recon/Optical Inc.	USA	*608*	75-73930
Royal Ordnance	UK	*481*	75-50620
Somchem	S. Africa	*359*	75-19200
Sparton Technology Inc.	USA	*616*	75-82950
Sverdrup Corp.	USA	*643*	80-30650
Sverdrup Technology Inc.	USA	*643*	80-30680

517 20 50 30 Dispensers, weapon, airborne

Aerea Luftfahrtgeräte GmbH.	Germany	*164*	75-00260
Aerea SpA.	Italy.	*230*	75-00280
Aerospace and Electronics Division.	USA	*606*	75-71472
CMS Inc.	USA	*564*	75-19170
GKN Westland Technologies Ltd	UK	*455*	75-25690
Hunting Engineering Ltd	UK	*459*	75-30700
Israel Military Industries Ltd	Israel	*220*	75-10300
Normalair-Garrett Ltd.	UK	*425*	74-07000
R. Alkan (UK) Ltd	UK	*429*	75-02680
Rheinmetall Industrie	Germany	*173*	75-24900
Royal Ordnance	UK	*481*	75-50620
Technical Products Group Inc.	USA	*620*	75-86764

517 20 50 32 Dispensers, airborne countermeasures

Aerea Luftfahrtgeräte GmbH.	Germany	*164*	75-00260
Aerospace and Electronics Division.	USA	*606*	75-71472
GKN Westland Technologies Ltd	UK	*455*	75-25690
Israel Military Industries Ltd	Israel	*220*	75-10300
ML Wallop Defence Systems Ltd	UK	*469*	75-40240
R. Alkan (UK) Ltd	UK	*429*	75-02680
Southwest Aerospace	USA	*616*	75-82470
W. Vinten Ltd.	UK	*491*	75-63353

517 20 50 35 Submunitions dispensers, airborne

Aerospace and Electronics Division.	USA	*606*	75-71472
CMS Inc.	USA	*564*	75-19170
Étienne Lacroix	France	*138*	75-24050
Hunting Engineering Ltd	UK	*459*	75-30700
R. Alkan (UK) Ltd	UK	*429*	75-02680
Royal Ordnance	UK	*481*	75-50620
Technical Products Group Inc.	USA	*620*	75-86764

517 20 50 50 Missile launchers, airborne

Aerea SpA.	Italy.	*230*	75-00280
Fibertek	USA	*575*	75-34330
Israel Military Industries Ltd	Israel	*220*	75-10300
R. Alkan (UK) Ltd	UK	*429*	75-02680

517 20 60 Weapon sights, airborne

FN Herstal SA	Belgium	*46*	75-03600
Matra BAe Dynamics	UK	*425*	74-06850
Military Systems Group Inc.	USA	*596*	75-61570
Recon/Optical Inc.	USA	*608*	75-73930
Ring Sights Holdings Co. Ltd.	UK	*479*	75-49700
SFIM	**France**	***144***	**75-35080**

517 20 61 Weapon sights, aircraft

FN Herstal SA	Belgium	*46*	75-03600
Recon/Optical Inc.	USA	*608*	75-73930
Ring Sights Holdings Co. Ltd.	UK	*479*	75-49700

517 20 61 10 Bomb sights, aircraft

517 20 61 30 Gun sights, aircraft

Ring Sights Holdings Co. Ltd.	UK	*479*	75-49700

517 20 63 Weapon sights, helicopter

FN Herstal SA	Belgium	*46*	75-03600
Matra BAe Dynamics	UK	*425*	74-06850
Recon/Optical Inc.	USA	*608*	75-73930
Ring Sights Holdings Co. Ltd.	UK	*479*	75-49700
SFIM	**France**	***144***	**75-35080**

517 20 63 30 Weapon sights, helicopter, gyro-stabilized

FN Herstal SA	Belgium	*46*	75-03600
Matra BAe Dynamics	UK	*425*	74-06850
Recon/Optical Inc.	USA	*608*	75-73930
SFIM	**France**	***144***	**75-35080**

517 20 70 Weapon-related equipment, airborne

FN Herstal SA	Belgium	*46*	75-03600
Royal Ordnance	UK	*481*	75-50620

517 30 Weapon-related systems & equipment, shipborne

Air-Log Ltd	UK	*429*	75-01800
Alenia Difesa	**Italy**	***231***	**75-01165**
Australian Defence Industries Ltd	Australia	*9*	55-01000
Cargo Aids	UK	*438*	75-11620
EG&G Inc.	USA	*571*	75-29555
Fibertek	USA	*575*	75-34330
FN Herstal SA	Belgium	*46*	75-03600
Israel Aircraft Industries Ltd	Israel	*219*	75-10200
Oerlikon-Contraves AG.	Switzerland	*387*	75-09250
Riva Calzoni SpA	Italy	*238*	75-20500
Royal Ordnance	UK	*481*	75-50620
SELIN Sistemi Sarl	Italy	*238*	75-21750
Strachan & Henshaw Ltd	UK	*512*	80-47400
Stringer Welding & Machine Inc.	USA	*618*	75-84435
The Dewey Electronics Corp.	USA	*569*	75-26100
United Defense L.P.	USA	*624*	75-92050
Vickers Shipbuilding & Eng Ltd	UK	*491*	75-63100

517 30 30 Weapon embarcation systems, shipborne

Forward Industries Ltd	UK	*452*	75-24200
Strachan & Henshaw Ltd	UK	*512*	80-47400

517 30 30 50 Weapon embarcation equipment, submarine

Forward Industries Ltd	UK	*452*	75-24200
Strachan & Henshaw Ltd	UK	*512*	80-47400

517 30 40 Weapon handling & reloading systems, shipborne

Air-Log Ltd	UK	*429*	75-01800
Forward Industries Ltd	UK	*452*	75-24200
Howden Fan Company Pty Ltd	Australia	*17*	75-08740
Howden Sirocco	France	*137*	75-21700
Israel Aircraft Industries Ltd	Israel	*219*	75-10200
MIL-BASE Industries Inc.	USA	*596*	75-61530
Riva Calzoni SpA	Italy	*238*	75-20500
Strachan & Henshaw Ltd	UK	*512*	80-47400
The Dewey Electronics Corp.	USA	*569*	75-26100
United Defense L.P.	USA	*624*	75-92050

517 30 50 Weapon mounts, naval

Australian Defence Industries Ltd	Australia	*9*	55-01000
FN Herstal SA	Belgium	*46*	75-03600
Israel Aircraft Industries Ltd	Israel	*219*	75-10200
Military Systems Group Inc.	USA	*596*	75-61570
MSI-Defence Systems Ltd	UK	*469*	75-41280
United Defense L.P.	USA	*624*	75-92050

517 30 50 30 Gun mounts, naval

Alenia Difesa	**Italy**	***231***	**75-01165**
Australian Defence Industries Ltd	Australia	*9*	55-01000
KUKA Wehrtechnik GmbH	Germany	*171*	75-17200
Military Systems Group Inc.	USA	*596*	75-61570
MSI-Defence Systems Ltd	UK	*469*	75-41280
Sidem (UK) Ltd	UK	*511*	80-35850
Stop-Choc Ltd.	UK	*486*	75-57345
Stringer Welding & Machine Inc.	USA	*618*	75-84435
Textron Marine & Land Systems	USA	*622*	75-88660
United Defense L.P.	USA	*624*	75-92050
Vickers Shipbuilding & Eng Ltd	UK	*491*	75-63100

517 30 50 30 80 Gun mount equipment, naval

Australian Defence Industries Ltd	Australia	*9*	55-01000
Military Systems Group Inc.	USA	*596*	75-61570
MSI-Defence Systems Ltd	UK	*469*	75-41280
Sidem (UK) Ltd	UK	*511*	80-35850
Stringer Welding & Machine Inc.	USA	*618*	75-84435
United Defense L.P.	USA	*624*	75-92050
Vickers Shipbuilding & Eng Ltd	UK	*491*	75-63100

517 30 50 30 87 Slip rings, gun mount, naval

EUROCONTROL SpA	Italy	*233*	75-08200
IDM Electronics Ltd	UK	*460*	75-31100
Litton Industries Inc.	USA	*590*	75-55400
United Defense L.P.	USA	*624*	75-92050
Vickers Shipbuilding & Eng Ltd	UK	*491*	75-63100

517 30 50 36 Gun turrets, naval

Alenia Difesa	**Italy**	***231***	**75-01165**
Australian Defence Industries Ltd	Australia	*9*	55-01000
Urdan Industries (USA) Inc.	USA	*625*	75-92930
Vickers Shipbuilding & Eng Ltd	UK	*491*	75-63100

517 30 50 36 80 Gun turret equipment, naval

EUROCONTROL SpA	Italy	*233*	75-08200
SELIN Sistemi Sarl	Italy	*238*	75-21750
Vickers Shipbuilding & Eng Ltd	UK	*491*	75-63100

517 30 50 36 87 Slip rings, gun turret, naval

EUROCONTROL SpA	Italy	*233*	75-08200
IDM Electronics Ltd	UK	*460*	75-31100
Litton Industries Inc.	USA	*590*	75-55400
Vickers Shipbuilding & Eng Ltd	UK	*491*	75-63100

517 30 50 50 Missile mounts, naval

Australian Defence Industries Ltd	Australia	*9*	55-01000
Fibertek	USA	*575*	75-34330
Stop-Choc Ltd.	UK	*486*	75-57345
United Defense L.P.	USA	*624*	75-92050
Vickers Shipbuilding & Eng Ltd	UK	*491*	75-63100

517 30 50 60 Missile launching systems

Domain Defence Systems Pty Ltd	Australia	*25*	80-04000
Forward Industries Ltd	UK	*452*	75-24200

GEC ALSTHOM Engineering Systems Ltd	UK	*453*	75-25225
GEL Engineering Ltd	UK	*454*	75-25500
Remmele Engineering Inc.	USA	*609*	75-74400
Reynolds Boughton Ltd	UK	*479*	75-49340
Shorts Missile Systems Ltd	**UK**	***483***	**75-53550**
Vickers Shipbuilding & Eng Ltd	UK	*491*	75-63100

517 30 50 68 Missile launching systems, vertical

GEL Engineering Ltd	UK	*454*	75-25500
Reynolds Boughton Ltd	UK	*479*	75-49340

517 30 50 70 Launch containers, missile

Air-Log Ltd	UK	*429*	75-01800
Israel Military Industries Ltd	Israel	*220*	75-10300
Technical Products Group Inc.	USA	*620*	75-86764
Vickers Shipbuilding & Eng Ltd	UK	*491*	75-63100

517 30 60 Weapon sights, naval

EG&G Inc.	USA	*571*	75-29555
EUROCONTROL SpA	Italy	*233*	75-08200
Hensoldt AG	Germany	*169*	75-12450
Military Systems Group Inc.	USA	*596*	75-61570
Recon/Optical Inc.	USA	*608*	75-73930
Ring Sights Holdings Co. Ltd.	UK	*479*	75-49700
SAGEM SA	**France**	***143***	**75-33950**
SFIM	**France**	***144***	**75-35080**
Stringer Welding & Machine Inc.	USA	*618*	75-84435
W.F. Stanley & Co. Ltd	UK	*486*	75-56350

517 30 70 Weapon-related equipment, naval

Nanoquest Defence Products Ltd	UK	*470*	75-41890
Stringer Welding & Machine Inc.	USA	*618*	75-84435
The Dewey Electronics Corp.	USA	*569*	75-26100
United Defense L.P.	USA	*624*	75-92050

517 40 Weapon-related systems & equipment, ground-based

AAI Corp.	USA	*545*	75-00020
Air-Log Ltd	UK	*429*	75-01800
Allied Research Corp.	USA	*549*	75-04670
Atomic Energy Corp	S. Africa	*357*	75-01000
Brownells Inc.	USA	*560*	75-15300
Cargo Aids	UK	*438*	75-11620
EG&G Inc.	USA	*571*	75-29555
Fuji Heavy Industries Ltd	Japan	*251*	75-01100
Israel Military Industries Ltd	Israel	*220*	75-10300
M. Paz Logistics Ltd	Israel	*224*	80-01600
Nanoquest Defence Products Ltd	UK	*470*	75-41890
Oerlikon-Contraves AG.	Switzerland	*387*	75-09250
Ring Sights Ltd	UK	*509*	80-28200
SELIN Sistemi Sarl	Italy	*238*	75-21750
SFIM	**France**	***144***	**75-35080**
Sverdrup Corp.	USA	*643*	80-30650
Sverdrup Technology Inc.	USA	*643*	80-30680
Tamares Ltd.	UK	*487*	75-58500
Theodor Rapp GmbH Co	Germany	*173*	75-24600
United Defense L.P.	USA	*624*	75-92050
Wegmann USA Inc.	USA	*627*	75-95543

517 40 40 Weapon handling & reloading systems, ground-based

Air-Log Ltd	UK	*429*	75-01800
EG&G Inc.	USA	*571*	75-29555
Howden Sirocco	France	*137*	75-21700
KUKA Wehrtechnik GmbH	Germany	*171*	75-17200
MIL-BASE Industries Inc.	USA	*596*	75-61530
Sverdrup Corp.	USA	*643*	80-30650
Sverdrup Technology Inc.	USA	*643*	80-30680
Tamares Ltd.	UK	*487*	75-58500
The Dewey Electronics Corp.	USA	*569*	75-26100
United Defense L.P.	USA	*624*	75-92050

517 40 40 30 Ammunition handling & reloading systems, gun

Air-Log Ltd	UK	*429*	75-01800
Boss Group Limited	UK	*435*	75-07750
Tamares Ltd.	UK	*487*	75-58500
United Defense L.P.	USA	*624*	75-92050

517 40 40 37 Ammunition handling & reloading systems, tank gun

Air-Log Ltd	UK	*429*	75-01800
Fairey Hydraulics	UK	*450*	75-22255
Howden Sirocco	France	*137*	75-21700
Israel Aircraft Industries Ltd	Israel	*219*	75-10200
United Defense L.P.	USA	*624*	75-92050

517 40 40 50 Missile handling & reloading systems, ground-based

Air-Log Ltd	UK	*429*	75-01800
Alenia SpA	Italy	*232*	75-01170
Boss Group Limited	UK	*435*	75-07750
Sverdrup Corp.	USA	*643*	80-30650
Sverdrup Technology Inc.	USA	*643*	80-30680
Systems & Electronics Inc.	USA	*619*	75-85650
United Defense L.P.	USA	*624*	75-92050

517 40 50 Weapon mounts, ground-based

AAI Corp.	USA	*545*	75-00020
Engineered Air Systems Inc.	USA	*573*	75-31560
Glover Webb Ltd	UK	*455*	75-25700
Instro Precision Ltd	UK	*461*	75-31630
Lucas Western Inc.	USA	*592*	75-56300
Manroy Engineering Ltd	**UK**	***466***	**75-37165**
Renown Engineering Ltd	UK	*478*	75-49227
Sidem (UK) Ltd	UK	*511*	80-35850
Stop-Choc Ltd.	UK	*486*	75-57345
Stringer Welding & Machine Inc.	USA	*618*	75-84435
Theodor Rapp GmbH Co	Germany	*173*	75-24600
United Defense L.P.	USA	*624*	75-92050

517 40 50 30 Gun mounts, ground-based

Alenia Difesa	**Italy**	***231***	**75-01165**
Giat Industries	France	*126*	55-00100
Glover Webb Ltd	UK	*455*	75-25700
Instro Precision Ltd	UK	*461*	75-31630
Manroy Engineering Ltd	**UK**	***466***	**75-37165**
Robert W. Hart & Son Inc.	USA	*581*	75-42450
SELIN Sistemi Sarl	Italy	*238*	75-21750
Sidem (UK) Ltd	UK	*511*	80-35850
Stop-Choc Ltd.	UK	*486*	75-57345
Stringer Welding & Machine Inc.	USA	*618*	75-84435
Theodor Rapp GmbH Co	Germany	*173*	75-24600

517 40 50 30 60 Gun mounts, ground-based, small arms

Giat Industries	France	*126*	55-00100
Glover Webb Ltd	UK	*455*	75-25700
Heckler & Koch Inc	USA	*581*	75-42860
Helston Gunsmiths	UK	*458*	75-29010
Manroy Engineering Ltd	**UK**	***466***	**75-37165**
Sidem (UK) Ltd	UK	*511*	80-35850
Stop-Choc Ltd.	UK	*486*	75-57345
Stringer Welding & Machine Inc.	USA	*618*	75-84435

517 40 50 30 70 Tripods, machine gun

Glover Webb Ltd	UK	*455*	75-25700
Helston Gunsmiths	UK	*458*	75-29010
Manroy Engineering Ltd	**UK**	***466***	**75-37165**
Sidem (UK) Ltd	UK	*511*	80-35850
Stringer Welding & Machine Inc.	USA	*618*	75-84435
Theodor Rapp GmbH Co	Germany	*173*	75-24600

Product–Supplier Link

517 40 50 50 Missile launchers, ground-based

Alenia SpA	Italy	232	75-01170
Fibertek	USA	575	75-34330
Fuji Heavy Industries Ltd	Japan	251	75-01100
Instro Precision Ltd	UK	461	75-31630
SELIN Sistemi Sarl	Italy	238	75-21750
Systems & Electronics Inc.	USA	619	75-85650
Technical Products Group Inc.	USA	620	75-86764

517 40 60 Weapon sights, ground-based

Avimo Electro-Optics Pte Ltd	Singapore	348	75-00800
Computing Devices Canada	Canada	70	75-03600
Deben Group Industries Ltd	UK	444	75-17270
Delft Sensor Systems	**Belgium**	**46**	**75-02500**
Delft Sensor Systems	Netherlands	292	75-01590
D.J.Litt (Firearms) Ltd	UK	465	75-36080
EPC Group Ltd.	UK	449	75-21150
FN Herstal SA	Belgium	46	75-03600
Hall & Watts (Defence Optics) Group	**UK**	**456**	**75-27650**
Hensoldt AG	Germany	169	75-12450
Lem Norma GmbH	Austria	33	75-01615
Lyon & Brandfield Ltd.	UK	465	75-36720
M. Paz Logistics Ltd	Israel	224	80-01600
Manroy Engineering Ltd	**UK**	**466**	**75-37165**
Military Systems Group Inc.	USA	596	75-61570
Nanoquest Defence Products Ltd	UK	470	75-41890
Optechs Korea Inc	Korea (ROK)	266	75-03750
Pilkington Thorn Optronics	UK	426	74-10000
Recon/Optical Inc.	USA	608	75-73930
Ring Sights Holdings Co. Ltd.	UK	479	75-49700
SAGEM SA	**France**	**143**	**75-33950**
Saunders-Roe Ltd.	UK	482	75-51850
Tamares Ltd.	UK	487	75-58500
Theodor Rapp GmbH Co	Germany	173	75-24600
Zeiss-Eltro Optronic GmbH	Germany	177	75-34700

517 40 60 10 Sights, armoured vehicle weapon

Delft Sensor Systems	**Belgium**	**46**	**75-02500**
Delft Sensor Systems	Netherlands	292	75-01590
Eloptro	S. Africa	358	75-05680
Hall & Watts (Defence Optics) Group	**UK**	**456**	**75-27650**
Heinsoldt Systemtechnick GmbH	Germany	171	75-17800
Hensoldt AG	Germany	169	75-12450
M. Paz Logistics Ltd	Israel	224	80-01600
Military Systems Group Inc.	USA	596	75-61570
Nanoquest Defence Products Ltd	UK	470	75-41890
Ring Sights Holdings Co. Ltd.	UK	479	75-49700
SAGEM SA	**France**	**143**	**75-33950**
SFIM	**France**	**144**	**75-35080**
W.F. Stanley & Co. Ltd	UK	486	75-56350

517 40 60 15 Artillery sighting systems, optical

Delft Sensor Systems	**Belgium**	**46**	**75-02500**
Delft Sensor Systems	Netherlands	292	75-01590
Eloptro	S. Africa	358	75-05680
Hall & Watts (Defence Optics) Group	**UK**	**456**	**75-27650**
Heinsoldt Systemtechnick GmbH	Germany	171	75-17800
Hensoldt AG	Germany	169	75-12450
Instro Precision Ltd	UK	461	75-31630
M. Paz Logistics Ltd	Israel	224	80-01600
Ring Sights Holdings Co. Ltd.	UK	479	75-49700
SAGEM SA	**France**	**143**	**75-33950**
Tamares Ltd.	UK	487	75-58500
W.F. Stanley & Co. Ltd	UK	486	75-56350
Zeiss-Eltro Optronic GmbH	Germany	177	75-34700

517 40 60 40 Mortar sighting systems, optical

Econ Optics Mechanics SA	Greece	188	75-01070
Hall & Watts (Defence Optics) Group	**UK**	**456**	**75-27650**
Hensoldt AG	Germany	169	75-12450
Instro Precision Ltd	UK	461	75-31630
M. Paz Logistics Ltd	Israel	224	80-01600
Optechs Korea Inc	Korea (ROK)	266	75-03750
Ring Sights Holdings Co. Ltd.	UK	479	75-49700
SAGEM SA	**France**	**143**	**75-33950**
Saunders-Roe Ltd.	UK	482	75-51850
Tamares Ltd.	UK	487	75-58500
W.F. Stanley & Co. Ltd	UK	486	75-56350

517 40 60 70 Sights, rifle

Accuracy International Ltd	UK	427	75-00360
Armalon Ltd	UK	431	75-03729
Bharat Electronics Ltd	India	203	55-00200
Deben Group Industries Ltd	UK	444	75-17270
Delft Sensor Systems	**Belgium**	**46**	**75-02500**
Delft Sensor Systems	Netherlands	292	75-01590
Eloptro	S. Africa	358	75-05680
Hall & Watts (Defence Optics) Group	**UK**	**456**	**75-27650**
Heckler & Koch Inc	USA	581	75-42860
Hensoldt AG	Germany	169	75-12450
Insight Vision Systems Ltd	UK	461	75-31530
Lyon & Brandfield Ltd.	UK	465	75-36720
M. Paz Logistics Ltd	Israel	224	80-01600
Manroy Engineering Ltd	**UK**	**466**	**75-37165**
Parker-Hale Ltd	UK	473	75-43850
Ring Sights Holdings Co. Ltd.	UK	479	75-49700
Ring Sights Ltd	UK	509	80-28200
Robert W. Hart & Son Inc.	USA	581	75-42450
SAGEM SA	**France**	**143**	**75-33950**
Saunders-Roe Ltd.	UK	482	75-51850
W.F. Stanley & Co. Ltd	UK	486	75-56350

517 40 60 75 Sights, laser, gun

Accuracy International Ltd	UK	427	75-00360
ARSOC Sprl	Belgium	45	75-00520
Bharat Electronics Ltd	India	203	55-00200
Delft Sensor Systems	**Belgium**	**46**	**75-02500**
Delft Sensor Systems	Netherlands	292	75-01590
Law Enforcement International Ltd	UK	464	75-34870
M. Paz Logistics Ltd	Israel	224	80-01600
Oerlikon-Contraves AG.	Switzerland	387	75-09250
Pilkington Optronics Ltd	UK	426	74-10000
Power Technology Inc.	USA	605	75-71090
Rapid Acquisition Targeting Systems.	USA	607	75-72920
Transvaro A.S.	Turkey	403	75-02080

517 40 60 80 Accessories, sighting equipment

Accuracy International Ltd	UK	427	75-00360
Aimpoint AB	Sweden	375	75-00800
Armalon Ltd	UK	431	75-03729
ARSOC Sprl	Belgium	45	75-00520
Hall & Watts (Defence Optics) Group	**UK**	**456**	**75-27650**
Hesco Inc.	USA	581	75-43150
Instro Precision Ltd	UK	461	75-31630
Litton Industries Inc.	USA	590	75-55400
M. Paz Logistics Ltd	Israel	224	80-01600
MWG Co Inc.	USA	597	75-63490
Optechs Korea Inc	Korea (ROK)	266	75-03750
Ring Sights Holdings Co. Ltd.	UK	479	75-49700
Ring Sights Ltd	UK	509	80-28200
SFIM	**France**	**144**	**75-35080**
Zeiss-Eltro Optronic GmbH	Germany	177	75-34700

517 40 60 80 10 Aiming lights

Aimpoint AB	Sweden	375	75-00800
Armalon Ltd	UK	431	75-03729
ARSOC Sprl	Belgium	45	75-00520
M. Paz Logistics Ltd	Israel	224	80-01600

517 40 60 80 30 Graticule illumination units

Armalon Ltd	UK	431	75-03729
Hall & Watts (Defence Optics) Group	**UK**	**456**	**75-27650**
Hensoldt AG	Germany	169	75-12450
Hesco Inc.	USA	581	75-43150

M. Paz Logistics Ltd	Israel	*224*	80-01600
Ring Sights Holdings Co. Ltd.	UK	*479*	75-49700
Saunders-Roe Ltd.	UK	*482*	75-51850

517 40 70 Weapon-related equipment, ground-based

Royal Ordnance	UK	*481*	75-50620
Summit Machine Builders Corp.	USA	*618*	75-84900
Tamares Ltd.	UK	*487*	75-58500
Theodor Rapp GmbH Co	Germany	*173*	75-24600
W.H.J. Fagg & Son Ltd.	UK	*450*	75-22190

517 40 70 15 Artillery / mortar support instruments

Hall & Watts (Defence Optics) Group	**UK**	***456***	**75-27650**
Instro Precision Ltd	UK	*461*	75-31630
Tamares Ltd.	UK	*487*	75-58500
United Defense L.P.	USA	*624*	75-92050
W.F. Stanley & Co. Ltd	UK	*486*	75-56350

517 40 70 90 Zeroing devices, small weapon, ground-based

Armalon Ltd	UK	*431*	75-03729
Hall & Watts (Defence Optics) Group	**UK**	***456***	**75-27650**
Hensoldt AG	Germany	*169*	75-12450
Instro Precision Ltd	UK	*461*	75-31630
Ring Sights Holdings Co. Ltd.	UK	*479*	75-49700
Tamares Ltd.	UK	*487*	75-58500

517 40 90 Weapon-related components, ground-based

ARAM	S. Africa	*357*	75-00795
Bushmaster Firearms Inc.	USA	*561*	75-16007
CMI - Ballarat	Australia	*14*	75-03600
EG&G Inc.	USA	*571*	75-29555
Fluorocarbon Company Limited	UK	*451*	75-24020
Giat Industries	France	*126*	55-00100
JFS Inc.	USA	*586*	75-50920
Lucas Western Inc.	USA	*592*	75-56300
National Forge Co.	USA	*598*	75-63850
Papadopoulos & Co Ltd	Greece	*191*	75-02195
Republic Arms (Pty) Ltd	**S. Africa**	***359***	**75-18210**
Royal Ordnance	UK	*481*	75-50620
Summit Machine Builders Corp.	USA	*618*	75-84900
Textron Systems Division	USA	*621*	75-88655
Theodor Rapp GmbH Co	Germany	*173*	75-24600

517 50 Containerised electronic units, weapon-related

Air-Log Ltd	UK	*429*	75-01800
EG&G Inc.	USA	*571*	75-29555
Sparton Technology Inc.	USA	*616*	75-82950
Summit Machine Builders Corp.	USA	*618*	75-84900

518 Modernization kits, weapon

Giat Industries	France	*126*	55-00100
Israel Aircraft Industries Ltd	Israel	*219*	75-10200
Israel Military Industries Ltd	Israel	*220*	75-10300
Krauss-Maffei Wehrtechnik GmbH	Germany	*171*	75-16345
M. Paz Logistics Ltd	Israel	*224*	80-01600
Nottingham International	USA	*640*	80-22000
Oerlikon-Contraves AG	Switzerland	*387*	75-09250
Rafael	Israel	*216*	55-01800
Royal Ordnance	UK	*481*	75-50620
South Central Ordnance Co.	USA	*616*	75-82400
Vickers Shipbuilding & Eng Ltd	UK	*491*	75-63100
Wegmann & Co GmbH	Germany	*177*	75-33700

518 40 Modernization kits, armoured fighting vehicle

Allied Research Corp.	USA	*549*	75-04670
Avibras Aeroespacial SA	Brazil	*55*	75-00968
Econ Optics Mechanics SA	Greece	*188*	75-01070
Giat Industries	France	*126*	55-00100
Israel Aircraft Industries Ltd	Israel	*219*	75-10200
Israel Military Industries Ltd	Israel	*220*	75-10300
Kinetics Ltd	Israel	*220*	75-11320
Krauss-Maffei Wehrtechnik GmbH	Germany	*171*	75-16345
M. Paz Logistics Ltd	Israel	*224*	80-01600
Nimda Co. Ltd.	Israel	*224*	80-01400
Norca Machinery Company	USA	*599*	75-64800
Nottingham International	USA	*640*	80-22000
Rafael	Israel	*216*	55-01800
SAGEM SA	**France**	***143***	**75-33950**
Singapore Technologies Automotive	Singapore	*348*	75-04280
South Central Ordnance Co.	USA	*616*	75-82400
Tecmotiv Corporation	Canada	*77*	75-13510
Vickers Shipbuilding & Eng Ltd	UK	*491*	75-63100
Wegmann & Co GmbH	Germany	*177*	75-33700

Ammunition, Explosives, Propellants, Pyrotechnics

521 Ammunition

AAI Corp.	USA	*545*	75-00020
Accuracy International Ltd	UK	*427*	75-00360
Aerojet	USA	*547*	75-01500
Allied Ordnance of Singapore Pte Ltd	Singapore	*347*	75-00500
Allied Research Corp.	USA	*549*	75-04670
Arcus Co.	Bulgaria	*61*	55-00060
Armalon Ltd	UK	*431*	75-03729
Arms Corporation of the Philippines	Philippines	*320*	75-00150
Artefatos Eléctricos e Mecânicos	Brazil	*55*	74-00905
Australian Defence Industries Ltd	Australia	*9*	55-01000
Bofors AB	Sweden	*375*	74-00020
Bofors Carl Gustaf AB	Sweden	*376*	75-02950
Bristol Aerospace Ltd	Canada	*69*	75-01500
Browning SA	Belgium	*45*	75-01950
Calspan SRL Technology	USA	*540*	60-03000
Century International Arms Inc.	USA	*633*	80-05400
Chamberlain Manufacturing Corp.	USA	*563*	75-17950
CIC International Ltd	USA	*563*	75-18570
Cime Bocuze	France	*133*	75-09660
Companhia Brasileira de Cartuchos	Brazil	*55*	75-03150
Conjay Arms Co. Ltd	UK	*442*	75-14855
Daudsons Armoury	Pakistan	*315*	75-00500
Diehl GmbH & Co	Germany	*167*	75-06500
Dynamit Nobel GmbH	Germany	*167*	75-07100
Dynamit Nobel Graz GmbH	Austria	*32*	75-00800
Eley Ltd	UK	*448*	75-20550
Esterline Technologies	USA	*574*	75-32360
Eurometaal NV	Netherlands	*292*	75-02200
Explosivos Alaveses SA	Spain	*366*	75-03050
Fabrica Española de Cartucheria SL	Spain	*366*	75-03300
Fabrica Militar "Frey Luis Beltran"	Argentina	*4*	55-06160
FN Manufacturing Inc.	USA	*576*	75-35320
Forges de Zeebrugge SA	Belgium	*46*	75-03750
Giat Industries	France	*126*	55-00100
Glaser Safety Slug Inc.	USA	*579*	75-39750
Greek Powder & Cartridge Co. SA	Greece	*189*	75-01500
Hellenic Arms Industry	Greece	*187*	55-00800
Helston Gunsmiths	UK	*458*	75-29010

Hirtenberger AG	Austria	33	75-01250
Hornady Manufacturing Co.	USA	582	75-44300
IBQ Industrias Químicas Ltda	Brazil	55	75-02700
Indústrias de Defesa	Portugal	327	55-00950
INIMEX	**Belgium**	**46**	**75-04650**
Instalaza SA	Spain	366	75-03900
Interstate Technical Products	UK	461	75-31946
INTERTECHNIK Techn. Produktionen	Austria	33	75-01430
Israel Military Industries Ltd	Israel	220	75-10300
JSC Tulamashzavod	Russia	334	55-19450
Kompanija Sloboda D D	Yugoslavia	661	75-00475
Laboratoire Arcane	France	138	75-24010
Lapua Oy	Finland	118	75-01430
Lockheed Martin Corp.	USA	591	75-55635
Luchaire Défense SA	France	139	75-25400
Makina ve Kimya Endüstrisi	Turkey	402	55-01310
Manroy Engineering Ltd	**UK**	**466**	**75-37165**
Manurhin Défense	France	139	75-25630
Matra Défense-Espace	France	139	75-26000
MECAR SA	Belgium	46	75-04930
Merex Inc.	USA	595	75-60250
NORINCO	China PR	88	55-00280
Nottingham International	USA	640	80-22000
Nuclear Metals Inc.	USA	600	75-65760
NWM de Kruithoorn BV	Netherlands	294	75-05200
Oerlikon-Contraves AG.	Switzerland	387	75-09250
Olin Australia Ltd	Australia	19	75-15642
Ordnance & Tactical Systems Division	USA	606	75-71472
Pains-Wessex Ltd	UK	472	75-43510
Pakistan Ordnance Factories	Pakistan	314	55-00100
Parker-Hale Ltd	UK	473	75-43850
Primex Technologies, Inc	USA	606	75-71472
Pyrospace	France	141	75-31300
Raufoss Technology AS	Norway	310	75-02200
Rheinmetall Industrie	Germany	173	75-24900
Robert W. Hart & Son Inc.	USA	581	75-42450
Royal Ordnance	UK	481	75-50620
SAE Alsetex	France	142	75-33810
Sako Ltd	Finland	118	75-01980
Sandeep Metalcraft Pvt Ltd	India	205	75-01900
Santa Barbara SA	Spain	364	55-01000
Sellier & Bellot a.s.	Czech Republic	99	75-00900
Sener Ingenieria y Sistemas SA	Spain	367	75-05200
Simunition	Canada	67	74-19100
SNC Industrial Tech.	Canada	76	75-12960
SNPE	France	145	75-36315
Société Nouvelle PRB	Belgium	47	75-06050
Somchem	S. Africa	359	75-19200
South Central Ordnance Co.	USA	616	75-82400
Southern Ammunition Co. Inc.	USA	616	75-82440
Strategic Technologies Pte Ltd	**Singapore**	**348**	**75-04400**
Summit Machine Builders Corp.	USA	618	75-84900
Swartklip Products	S. Africa	359	75-19500
Tamares Ltd	UK	487	75-58500
The Robar Companies Inc.	USA	609	75-75630
Tlocznia Metali "Pressta"	Poland	324	55-10000
Vammas Ltd	**Finland**	**117**	**55-02220**
Vickers Shipbuilding & Eng Ltd	UK	491	75-63100
Vihtavuori Oy	Finland	119	75-02850
Wejra A/S	Denmark	104	75-23300
ZVS Dubnica Jas	Slovakia	353	75-02000

521 20 Ammunition, aircraft

Australian Defence Industries Ltd	Australia	9	55-01000
Cime Bocuze	France	133	75-09660
Explosivos Alaveses SA	Spain	366	75-03050
Fabrica Española de Cartucheria SL	Spain	366	75-03300
Forges de Zeebrugge SA	Belgium	46	75-03750
Greek Powder & Cartridge Co. SA	Greece	189	75-01500
Hellenic Arms Industry	Greece	187	55-00800
INIMEX	**Belgium**	**46**	**75-04650**
Israel Military Industries Ltd	Israel	220	75-10300
Manroy Engineering Ltd	**UK**	**466**	**75-37165**
Manurhin Défense	France	139	75-25630
Nottingham International	USA	640	80-22000
NWM de Kruithoorn BV	Netherlands	294	75-05200
Oerlikon-Contraves AG.	Switzerland	387	75-09250
Ordnance & Tactical Systems Division	USA	606	75-71472
Orlando Technology Inc.	USA	540	60-16000
Pakistan Ordnance Factories	Pakistan	314	55-00100
Primex Technologies, Inc	USA	606	75-71472
Royal Ordnance	UK	481	75-50620

521 20 30 Ammunition, aircraft gun

Australian Defence Industries Ltd	Australia	9	55-01000
Cime Bocuze	France	133	75-09660
Companhia Brasileira de Cartuchos	Brazil	55	75-03150
Fabrica Española de Cartucheria SL	Spain	366	75-03300
Hellenic Arms Industry	Greece	187	55-00800
INIMEX	**Belgium**	**46**	**75-04650**
Israel Military Industries Ltd	Israel	220	75-10300
Manurhin Défense	France	139	75-25630
Nottingham International	USA	640	80-22000
NWM de Kruithoorn BV	Netherlands	294	75-05200
Ordnance & Tactical Systems Division	USA	606	75-71472
Pakistan Ordnance Factories	Pakistan	314	55-00100
Primex Technologies, Inc	USA	606	75-71472
Royal Ordnance	UK	481	75-50620

521 20 30 20 Ammunition, aircraft gun, 20mm

Australian Defence Industries Ltd	Australia	9	55-01000
Century International Arms Inc.	USA	633	80-05400
Cime Bocuze	France	133	75-09660
Companhia Brasileira de Cartuchos	Brazil	55	75-03150
Eurometaal NV	Netherlands	292	75-02200
Explosivos Alaveses SA	Spain	366	75-03050
Fabrica Española de Cartucheria SL	Spain	366	75-03300
Greek Powder & Cartridge Co. SA	Greece	189	75-01500
Indústrias de Defesa	Portugal	327	55-00950
INIMEX	**Belgium**	**46**	**75-04650**
Israel Military Industries Ltd	Israel	220	75-10300
Lapua Oy	Finland	118	75-01430
Manurhin Défense	France	139	75-25630
Merex Inc.	USA	595	75-60250
Nottingham International	USA	640	80-22000
Nuclear Metals Inc.	USA	600	75-65760
NWM de Kruithoorn BV	Netherlands	294	75-05200
Ordnance & Tactical Systems Division	USA	606	75-71472
Pakistan Ordnance Factories	Pakistan	314	55-00100
Primex Technologies, Inc	USA	606	75-71472
Royal Ordnance	UK	481	75-50620
SNC Industrial Tech.	Canada	76	75-12960

521 20 30 30 Ammunition, aircraft gun, 30mm

AAI Corp.	USA	545	75-00020
Cime Bocuze	France	133	75-09660
Companhia Brasileira de Cartuchos	Brazil	55	75-03150
Eurometaal NV	Netherlands	292	75-02200
Fabrica Española de Cartucheria SL	Spain	366	75-03300
Hellenic Arms Industry	Greece	187	55-00800
INIMEX	**Belgium**	**46**	**75-04650**
Israel Military Industries Ltd	Israel	220	75-10300
Manurhin Défense	France	139	75-25630
Merex Inc.	USA	595	75-60250
Nuclear Metals Inc.	USA	600	75-65760
Ordnance & Tactical Systems Division	USA	606	75-71472
Pakistan Ordnance Factories	Pakistan	314	55-00100
Primex Technologies, Inc	USA	606	75-71472
Royal Ordnance	UK	481	75-50620

521 20 60 Rockets, aircraft

AAI Corp.	USA	545	75-00020
Artefatos Eléctricos e Mecânicos	Brazil	55	74-00905
Avibras Aeroespacial SA	Brazil	55	75-00968
Bristol Aerospace Ltd	Canada	69	75-01500
Cime Bocuze	France	133	75-09660
CMS Inc.	USA	564	75-19170
Eurometaal NV	Netherlands	292	75-02200
Fabrica Española de Cartucheria SL	Spain	366	75-03300
Forges de Zeebrugge SA	Belgium	46	75-03750
INIMEX	**Belgium**	**46**	**75-04650**
Israel Military Industries Ltd	Israel	220	75-10300
Makina ve Kimya Endüstrisi	Turkey	402	55-01310
Matra Défense-Espace	France	139	75-26000
Royal Ordnance	UK	481	75-50620
Somchem	S. Africa	359	75-19200

521 30 Ammunition, naval

AAI Corp.	USA	545	75-00020

Australian Defence Industries Ltd	Australia	9	55-01000
Cime Bocuze	France	133	75-09660
Companhia Brasileira de Cartuchos	Brazil	55	75-03150
Eurometaal NV	Netherlands	292	75-02200
Greek Powder & Cartridge Co. SA	Greece	189	75-01500
Indústrias de Defesa	Portugal	327	55-00950
INIMEX	**Belgium**	***46***	**75-04650**
Manurhin Défense	France	139	75-25630
NWM de Kruithoorn BV	Netherlands	294	75-05200
Oerlikon-Contraves AG.	Switzerland	387	75-09250

521 30 15 Ammunition, automatic gun, naval, 20-35mm

Australian Defence Industries Ltd	Australia	9	55-01000
Cime Bocuze	France	133	75-09660
Eurometaal NV	Netherlands	292	75-02200
Hellenic Arms Industry	Greece	187	55-00800
INIMEX	**Belgium**	***46***	**75-04650**
Manurhin Défense	France	139	75-25630
NWM de Kruithoorn BV	Netherlands	294	75-05200
Ordnance & Tactical Systems Division	USA	606	75-71472
Primex Technologies, Inc	USA	606	75-71472

521 30 15 20 Ammunition, automatic gun, 20mm, naval

Australian Defence Industries Ltd	Australia	9	55-01000
Cime Bocuze	France	133	75-09660
Companhia Brasileira de Cartuchos	Brazil	55	75-03150
INIMEX	**Belgium**	***46***	**75-04650**
Lapua Oy	Finland	118	75-01430
Manurhin Défense	France	139	75-25630
Nuclear Metals Inc.	USA	600	75-65760
NWM de Kruithoorn BV	Netherlands	294	75-05200
Ordnance & Tactical Systems Division	USA	606	75-71472
Primex Technologies, Inc	USA	606	75-71472

521 30 15 25 Ammunition, automatic gun, 25mm, naval

Cime Bocuze	France	133	75-09660
INIMEX	**Belgium**	***46***	**75-04650**
Manurhin Défense	France	139	75-25630
Nuclear Metals Inc.	USA	600	75-65760
NWM de Kruithoorn BV	Netherlands	294	75-05200
Ordnance & Tactical Systems Division	USA	606	75-71472
Primex Technologies, Inc	USA	606	75-71472

521 30 15 30 Ammunition, automatic gun, 30mm, naval

Cime Bocuze	France	133	75-09660
Eurometaal NV	Netherlands	292	75-02200
Hellenic Arms Industry	Greece	187	55-00800
INIMEX	**Belgium**	***46***	**75-04650**
Manurhin Défense	France	139	75-25630
NWM de Kruithoorn BV	Netherlands	294	75-05200
Ordnance & Tactical Systems Division	USA	606	75-71472
Primex Technologies, Inc	USA	606	75-71472

521 30 15 35 Ammunition, automatic gun, 35mm, naval

Cime Bocuze	France	133	75-09660
INIMEX	**Belgium**	***46***	**75-04650**
Manurhin Défense	France	139	75-25630
Ordnance & Tactical Systems Division	USA	606	75-71472
Primex Technologies, Inc	USA	606	75-71472

521 30 30 Ammunition, naval gun, non-automatic

Cime Bocuze	France	133	75-09660
Manurhin Défense	France	139	75-25630
Ordnance & Tactical Systems Division	USA	606	75-71472
Primex Technologies, Inc	USA	606	75-71472
Rheinmetall Industrie	Germany	173	75-24900
Textron Systems Division	USA	621	75-88655

521 30 36 Ammunition, naval gun, 40-100mm

Cime Bocuze	France	133	75-09660
INIMEX	**Belgium**	***46***	**75-04650**
Luchaire Défense SA	France	139	75-25400
Manurhin Défense	France	139	75-25630
Nottingham International	USA	640	80-22000
NWM de Kruithoorn BV	Netherlands	294	75-05200

521 30 36 40 Ammunition, naval, 40mm

Allied Ordnance of Singapore Pte Ltd.	Singapore	347	75-00500
Australian Defence Industries Ltd	Australia	9	55-01000
Cime Bocuze	France	133	75-09660
Eurometaal NV	Netherlands	292	75-02200
Greek Powder & Cartridge Co. SA	Greece	189	75-01500
INIMEX	**Belgium**	***46***	**75-04650**
Manurhin Défense	France	139	75-25630
Nottingham International	USA	640	80-22000
NWM de Kruithoorn BV	Netherlands	294	75-05200
Ordnance & Tactical Systems Division	USA	606	75-71472
Primex Technologies, Inc	USA	606	75-71472

521 30 36 76 Ammunition, naval, 76mm

Cime Bocuze	France	133	75-09660
Eurometaal NV	Netherlands	292	75-02200
Explosivos Alaveses SA	Spain	366	75-03050
Greek Powder & Cartridge Co. SA	Greece	189	75-01500
INIMEX	**Belgium**	***46***	**75-04650**
Israel Military Industries Ltd	Israel	220	75-10300
SNC Industrial Tech.	Canada	76	75-12960

521 30 38 Ammunition, naval, 105-130mm

AAI Corp.	USA	545	75-00020
Australian Defence Industries Ltd	Australia	9	55-01000
Cime Bocuze	France	133	75-09660
G P Embelton & Co. Pty Ltd	Australia	16	75-05740
Indústrias de Defesa	Portugal	327	55-00950
INIMEX	**Belgium**	***46***	**75-04650**
Makina ve Kimya Endüstrisi	Turkey	402	55-01310
Vammas Ltd	**Finland**	***117***	**55-02220**

521 30 38 27 Ammunition, naval, 127mm

Cime Bocuze	France	133	75-09660
Explosivos Alaveses SA	Spain	366	75-03050
INIMEX	**Belgium**	46	75-04650

521 30 38 45 Ammunition, naval, 4.5in

AAI Corp.	USA	545	75-00020
Cime Bocuze	France	133	75-09660
Eurometaal NV	Netherlands	292	75-02200
INIMEX	**Belgium**	***46***	**75-04650**
Vickers Shipbuilding & Eng Ltd	UK	491	75-63100

521 30 39 Ammunition, naval, over 130mm

AAI Corp.	USA	545	75-00020
Cime Bocuze	France	133	75-09660

521 30 60 Rockets, naval

AAI Corp.	USA	545	75-00020
Cime Bocuze	France	133	75-09660
CMS Inc.	USA	564	75-19170
INIMEX	**Belgium**	***46***	**75-04650**
Makina ve Kimya Endüstrisi	Turkey	402	55-01310

521 40 Ammunition, ground-based weapon

AAI Corp.	USA	545	75-00020
Allied Research Corp.	USA	549	75-04670
Arcus Co.	Bulgaria	61	55-00060
Australian Defence Industries Ltd	Australia	9	55-01000
Chamberlain Manufacturing Corp.	USA	563	75-17950
CIC International Ltd	USA	563	75-18570
Cime Bocuze	France	133	75-09660
Companhia Brasileira de Cartuchos	Brazil	55	75-03150

Daudsons Armoury	Pakistan	*315*	75-00500
Esterline Technologies	USA	*574*	75-32360
Eurometaal NV	Netherlands	*292*	75-02200
Explosivos Alaveses SA	Spain	*366*	75-03050
Hirtenberger AG	Austria	*33*	75-01250
Indústrias de Defesa	Portugal	*327*	55-00950
INIMEX	**Belgium**	***46***	**75-04650**
Israel Military Industries Ltd	Israel	*220*	75-10300
JSC Tulamashzavod	Russia	*334*	55-19450
Lockheed Martin Corp.	USA	*591*	75-55635
Nottingham International	USA	*640*	80-22000
NWM de Kruithoorn BV	Netherlands	*294*	75-05200
Oerlikon-Contraves AG	Switzerland	*387*	75-09250
Raufoss Technology AS	Norway	*310*	75-02200
Sener Ingenieria y Sistemas SA	Spain	*367*	75-05200
Société Nouvelle PRB	Belgium	*47*	75-06050
Tamares Ltd	UK	*487*	75-58500
ZVS Dubnica Jas	Slovakia	*353*	75-02000

521 40 10 Ammunition, artillery, ground-based

AAI Corp.	USA	*545*	75-00020
Allied Research Corp.	USA	*549*	75-04670
Arcus Co.	Bulgaria	*61*	55-00060
Australian Defence Industries Ltd	Australia	*9*	55-01000
Chamberlain Manufacturing Corp.	USA	*563*	75-17950
CIC International Ltd	USA	*563*	75-18570
Cime Bocuze	France	*133*	75-09660
Eurometaal NV	Netherlands	*292*	75-02200
Explosivos Alaveses SA	Spain	*366*	75-03050
Giat Industries	France	*126*	55-00100
Greek Powder & Cartridge Co. SA	Greece	*189*	75-01500
Hellenic Arms Industry	Greece	*187*	55-00800
Hirtenberger AG	Austria	*33*	75-01250
INIMEX	**Belgium**	***46***	**75-04650**
Israel Military Industries Ltd	Israel	*220*	75-10300
Manroy Engineering Ltd	**UK**	***466***	**75-37165**
Merex Inc.	USA	*595*	75-60250
Nottingham International	USA	*640*	80-22000
Orlando Technology Inc.	USA	*540*	60-16000
Pakistan Ordnance Factories	Pakistan	*314*	55-00100
Raufoss Technology AS	Norway	*310*	75-02200
Rheinmetall Industrie	Germany	*173*	75-24900
SAE Alsetex	France	*142*	75-33810
Tamares Ltd	UK	*487*	75-58500
Tlocznia Metali "Pressta"	Poland	*324*	55-10000
Vammas Ltd	**Finland**	***117***	**55-02220**
ZVS Dubnica Jas	Slovakia	*353*	75-02000

521 40 10 09 Ammunition, artillery, 90mm

AAI Corp.	USA	*545*	75-00020
Allied Research Corp.	USA	*549*	75-04670
Chamberlain Manufacturing Corp.	USA	*563*	75-17950
CIC International Ltd	USA	*563*	75-18570
Cime Bocuze	France	*133*	75-09660
Greek Powder & Cartridge Co. SA	Greece	*189*	75-01500
INIMEX	**Belgium**	***46***	**75-04650**
Israel Military Industries Ltd	Israel	*220*	75-10300
Kompanija Sloboda D D	Yugoslavia	*661*	75-00475
Nottingham International	USA	*640*	80-22000
Orlando Technology Inc.	USA	*540*	60-16000

521 40 10 10 Ammunition, artillery, 105mm

AAI Corp.	USA	*545*	75-00020
Allied Research Corp.	USA	*549*	75-04670
Australian Defence Industries Ltd	Australia	*9*	55-01000
Chamberlain Manufacturing Corp.	USA	*563*	75-17950
CIC International Ltd	USA	*563*	75-18570
Cime Bocuze	France	*133*	75-09660
Explosivos Alaveses SA	Spain	*366*	75-03050
Greek Powder & Cartridge Co. SA	Greece	*189*	75-01500
Hellenic Arms Industry	Greece	*187*	55-00800
Hirtenberger AG	Austria	*33*	75-01250
Indústrias de Defesa	Portugal	*327*	55-00950
INIMEX	**Belgium**	***46***	**75-04650**
Israel Military Industries Ltd	Israel	*220*	75-10300
Kompanija Sloboda D D	Yugoslavia	*661*	75-00475
Makina ve Kimya Endüstrisi	Turkey	*402*	55-01310
Nottingham International	USA	*640*	80-22000
Orlando Technology Inc.	USA	*540*	60-16000
Pakistan Ordnance Factories	Pakistan	*314*	55-00100
SNC Industrial Tech.	Canada	*76*	75-12960
Vammas Ltd	**Finland**	***117***	**55-02220**

521 40 10 12 Ammunition, artillery, 130mm

INIMEX	**Belgium**	***46***	**75-04650**
SAE Alsetex	France	*142*	75-33810

521 40 10 15 Ammunition, artillery, 155mm

AAI Corp.	USA	*545*	75-00020
Allied Research Corp.	USA	*549*	75-04670
Chamberlain Manufacturing Corp.	USA	*563*	75-17950
CIC International Ltd	USA	*563*	75-18570
Cime Bocuze	France	*133*	75-09660
Esterline Technologies	USA	*574*	75-32360
Explosivos Alaveses SA	Spain	*366*	75-03050
Greek Powder & Cartridge Co. SA	Greece	*189*	75-01500
Hellenic Arms Industry	Greece	*187*	55-00800
Hirtenberger AG	Austria	*33*	75-01250
Indústrias de Defesa	Portugal	*327*	55-00950
Israel Military Industries Ltd	Israel	*220*	75-10300
Kompanija Sloboda D D	Yugoslavia	*661*	75-00475
Nottingham International	USA	*640*	80-22000
Ordnance & Tactical Systems Division	USA	*606*	75-71472
Orlando Technology Inc.	USA	*540*	60-16000
Pakistan Ordnance Factories	Pakistan	*314*	55-00100
Primex Technologies, Inc	USA	*606*	75-71472
SAE Alsetex	France	*142*	75-33810
SNC Industrial Tech.	Canada	*76*	75-12960
Somchem	S. Africa	*359*	75-19200
Tamares Ltd	UK	*487*	75-58500
Vammas Ltd	**Finland**	***117***	**55-02220**
Vickers Shipbuilding & Eng Ltd	UK	*491*	75-63100

521 40 10 20 Ammunition, artillery, 203mm

Allied Research Corp.	USA	*549*	75-04670
Chamberlain Manufacturing Corp.	USA	*563*	75-17950
CIC International Ltd	USA	*563*	75-18570
Explosivos Alaveses SA	Spain	*366*	75-03050
Hellenic Arms Industry	Greece	*187*	55-00800
INIMEX	**Belgium**	***46***	**75-04650**
Israel Military Industries Ltd	Israel	*220*	75-10300

521 40 15 Ammunition, automatic gun, ground-based

AAI Corp.	USA	*545*	75-00020
CIC International Ltd	USA	*563*	75-18570
Cime Bocuze	France	*133*	75-09660
INIMEX	**Belgium**	***46***	**75-04650**
NORINCO	China PR	*88*	55-00280

521 40 15 20 Ammunition, automatic gun, 20mm, ground-based

CIC International Ltd	USA	*563*	75-18570
Cime Bocuze	France	*133*	75-09660
Explosivos Alaveses SA	Spain	*366*	75-03050
Greek Powder & Cartridge Co. SA	Greece	*189*	75-01500
INIMEX	**Belgium**	***46***	**75-04650**
Manurhin Défense	France	*139*	75-25630
Nottingham International	USA	*640*	80-22000
Nuclear Metals Inc.	USA	*600*	75-65760
NWM de Kruithoorn BV	Netherlands	*294*	75-05200
Ordnance & Tactical Systems Division	USA	*606*	75-71472
Primex Technologies, Inc	USA	*606*	75-71472
SNC Industrial Tech.	Canada	*76*	75-12960

521 40 15 25 Ammunition, automatic gun, 25mm, ground-based

CIC International Ltd	USA	*563*	75-18570
Cime Bocuze	France	*133*	75-09660
INIMEX	**Belgium**	***46***	**75-04650**
Israel Military Industries Ltd	Israel	*220*	75-10300
Manurhin Défense	France	*139*	75-25630
Nottingham International	USA	*640*	80-22000

Nuclear Metals Inc.	USA	*600*	75-65760
NWM de Kruithoorn BV	Netherlands	*294*	75-05200
Ordnance & Tactical Systems Division	USA	*606*	75-71472
Primex Technologies, Inc	USA	*606*	75-71472

521 40 15 30 Ammunition, automatic gun, 30mm, ground-based

CIC International Ltd	USA	*563*	75-18570
Cime Bocuze	France	*133*	75-09660
Hellenic Arms Industry	Greece	*187*	55-00800
INIMEX	**Belgium**	***46***	**75-04650**
Israel Military Industries Ltd	Israel	*220*	75-10300
Manurhin Défense	France	*139*	75-25630
Nottingham International	USA	*640*	80-22000
Nuclear Metals Inc.	USA	*600*	75-65760
Ordnance & Tactical Systems Division	USA	*606*	75-71472
Primex Technologies, Inc	USA	*606*	75-71472
Royal Ordnance	UK	*481*	75-50620

521 40 15 35 Ammunition, automatic gun, 35mm, ground-based

CIC International Ltd	USA	*563*	75-18570
Cime Bocuze	France	*133*	75-09660
INIMEX	**Belgium**	***46***	**75-04650**
Manurhin Défense	France	*139*	75-25630
Nottingham International	USA	*640*	80-22000
NWM de Kruithoorn BV	Netherlands	*294*	75-05200
Ordnance & Tactical Systems Division	USA	*606*	75-71472
Primex Technologies, Inc	USA	*606*	75-71472
SNC Industrial Tech.	Canada	*76*	75-12960

521 40 30 Ammunition, fighting vehicle

AAI Corp.	USA	*545*	75-00020
Allied Research Corp.	USA	*549*	75-04670
CIC International Ltd	USA	*563*	75-18570
Cime Bocuze	France	*133*	75-09660
Giat Industries	France	*126*	55-00100
Hirtenberger AG	Austria	*33*	75-01250
INIMEX	**Belgium**	***46***	**75-04650**
Israel Military Industries Ltd	Israel	*220*	75-10300
Makina ve Kimya Endüstrisi	Turkey	*402*	55-01310
Manurhin Défense	France	*139*	75-25630
NWM de Kruithoorn BV	Netherlands	*294*	75-05200
Royal Ordnance	UK	*481*	75-50620
Société Nouvelle PRB	Belgium	*47*	75-06050
Tlocznia Metali "Pressta"	Poland	*324*	55-10000

521 40 32 Ammunition, fighting vehicle, under 20mm

Australian Defence Industries Ltd	Australia	*9*	55-01000
CIC International Ltd	USA	*563*	75-18570
Cime Bocuze	France	*133*	75-09660
Companhia Brasileira de Cartuchos	Brazil	*55*	75-03150
Hirtenberger AG	Austria	*33*	75-01250
INIMEX	**Belgium**	***46***	**75-04650**
Manurhin Défense	France	*139*	75-25630
Ordnance & Tactical Systems Division	USA	*606*	75-71472
Primex Technologies, Inc	USA	*606*	75-71472
SNC Industrial Tech.	Canada	*76*	75-12960

521 40 34 Ammunition, fighting vehicle, under 40mm

CIC International Ltd	USA	*563*	75-18570
Cime Bocuze	France	*133*	75-09660
Manurhin Défense	France	*139*	75-25630
NWM de Kruithoorn BV	Netherlands	*294*	75-05200
Ordnance & Tactical Systems Division	USA	*606*	75-71472
Primex Technologies, Inc	USA	*606*	75-71472

521 40 34 20 Ammunition, fighting vehicle, 20mm

CIC International Ltd	USA	*563*	75-18570
Cime Bocuze	France	*133*	75-09660
Lapua Oy	Finland	*118*	75-01430
Manurhin Défense	France	*139*	75-25630
NWM de Kruithoorn BV	Netherlands	*294*	75-05200
Ordnance & Tactical Systems Division	USA	*606*	75-71472
Primex Technologies, Inc	USA	*606*	75-71472

521 40 34 25 Ammunition, fighting vehicle, 25mm

CIC International Ltd	USA	*563*	75-18570
Cime Bocuze	France	*133*	75-09660
Israel Military Industries Ltd	Israel	*220*	75-10300
Manurhin Défense	France	*139*	75-25630
NWM de Kruithoorn BV	Netherlands	*294*	75-05200
Ordnance & Tactical Systems Division	USA	*606*	75-71472
Primex Technologies, Inc	USA	*606*	75-71472

521 40 34 30 Ammunition, fighting vehicle, 30mm

CIC International Ltd	USA	*563*	75-18570
Cime Bocuze	France	*133*	75-09660
Israel Military Industries Ltd	Israel	*220*	75-10300
Manurhin Défense	France	*139*	75-25630
Ordnance & Tactical Systems Division	USA	*606*	75-71472
Primex Technologies, Inc	USA	*606*	75-71472
Royal Ordnance	UK	*481*	75-50620

521 40 34 35 Ammunition, fighting vehicle, 35mm

CIC International Ltd	USA	*563*	75-18570
Cime Bocuze	France	*133*	75-09660
Manurhin Défense	France	*139*	75-25630
NWM de Kruithoorn BV	Netherlands	*294*	75-05200
Ordnance & Tactical Systems Division	USA	*606*	75-71472
Primex Technologies, Inc	USA	*606*	75-71472

521 40 36 Ammunition, fighting vehicle, 40-100mm

Allied Research Corp.	USA	*549*	75-04670
CIC International Ltd	USA	*563*	75-18570
Cime Bocuze	France	*133*	75-09660
Israel Military Industries Ltd	Israel	*220*	75-10300
JSC Tulamashzavod	Russia	*334*	55-19450
Manurhin Défense	France	*139*	75-25630
MECAR SA	Belgium	*46*	75-04930
NWM de Kruithoorn BV	Netherlands	*294*	75-05200
Pakistan Ordnance Factories	Pakistan	*314*	55-00100
Rheinmetall Industrie	Germany	*173*	75-24900

521 40 36 76 Ammunition, fighting vehicle, 76mm

Allied Research Corp.	USA	*549*	75-04670
CIC International Ltd	USA	*563*	75-18570
Cime Bocuze	France	*133*	75-09660
Nottingham International	USA	*640*	80-22000
Royal Ordnance	UK	*481*	75-50620
SNC Industrial Tech.	Canada	*76*	75-12960

521 40 36 90 Ammunition, fighting vehicle, 90mm

Allied Research Corp.	USA	*549*	75-04670
CIC International Ltd	USA	*563*	75-18570
Cime Bocuze	France	*133*	75-09660
Giat Industries	France	*126*	55-00100
INIMEX	**Belgium**	***46***	**75-04650**
Israel Military Industries Ltd	Israel	*220*	75-10300
MECAR SA	Belgium	*46*	75-04930
Nottingham International	USA	*640*	80-22000

521 40 38 Ammunition, fighting vehicle, 105-130mm

Allied Research Corp.	USA	*549*	75-04670
CIC International Ltd	USA	*563*	75-18570
Cime Bocuze	France	*133*	75-09660
INIMEX	**Belgium**	***46***	**75-04650**
Israel Military Industries Ltd	Israel	*220*	75-10300
MECAR SA	Belgium	*46*	75-04930
Nottingham International	USA	*640*	80-22000
Nuclear Metals Inc.	USA	*600*	75-65760
Rheinmetall Industrie	Germany	*173*	75-24900
SNC Industrial Tech.	Canada	*76*	75-12960

521 40 38 10 Ammunition, fighting vehicle, 105mm

Allied Research Corp.	USA	*549*	75-04670
CIC International Ltd	USA	*563*	75-18570
Cime Bocuze	France	*133*	75-09660
Giat Industries	France	*126*	55-00100
Israel Military Industries Ltd	Israel	*220*	75-10300
Nottingham International	USA	*640*	80-22000
Ordnance & Tactical Systems Division	USA	*606*	75-71472
Pakistan Ordnance Factories	Pakistan	*314*	55-00100
Primex Technologies, Inc	USA	*606*	75-71472
Royal Ordnance	UK	*481*	75-50620
SNC Industrial Tech.	Canada	*76*	75-12960

521 40 38 12 Ammunition, fighting vehicle, 120mm

Allied Research Corp.	USA	*549*	75-04670
Cime Bocuze	France	*133*	75-09660
Esterline Technologies	USA	*574*	75-32360
Giat Industries	France	*126*	55-00100
Nottingham International	USA	*640*	80-22000
Nuclear Metals Inc.	USA	*600*	75-65760
Ordnance & Tactical Systems Division	USA	*606*	75-71472
Primex Technologies, Inc	USA	*606*	75-71472
Royal Ordnance	UK	*481*	75-50620

521 40 39 Ammunition, fighting vehicle, over 130mm

CIC International Ltd	USA	*563*	75-18570
Cime Bocuze	France	*133*	75-09660
INIMEX	**Belgium**	***46***	**75-04650**
Israel Military Industries Ltd	Israel	*220*	75-10300

521 40 40 Ammunition, mortar, ground-based

Allied Research Corp.	USA	*549*	75-04670
Australian Defence Industries Ltd	Australia	*9*	55-01000
CIC International Ltd	USA	*563*	75-18570
Hellenic Arms Industry	Greece	*187*	55-00800
INIMEX	**Belgium**	***46***	**75-04650**
Israel Military Industries Ltd	Israel	*220*	75-10300
MECAR SA	Belgium	*46*	75-04930
NORINCO	China PR	*88*	55-00280
Pakistan Ordnance Factories	Pakistan	*314*	55-00100
Raufoss Technology AS	Norway	*310*	75-02200
Swartklip Products	S. Africa	*359*	75-19500
Tamares Ltd	UK	*487*	75-58500
Vammas Ltd	**Finland**	***117***	**55-02220**
Wejra A/S	Denmark	*104*	75-23300

521 40 40 12 Ammunition, mortar, ground-based, 120mm

Allied Research Corp.	USA	*549*	75-04670
Analysis Management Systems	S. Africa	*357*	75-00770
CIC International Ltd	USA	*563*	75-18570
Explosivos Alaveses SA	Spain	*366*	75-03050
Hirtenberger AG	Austria	*33*	75-01250
Indústrias de Defesa	Portugal	*327*	55-00950
Instalaza SA	Spain	*366*	75-03900
Israel Military Industries Ltd	Israel	*220*	75-10300
Makina ve Kimya Endüstrisi	Turkey	*402*	55-01310
Pakistan Ordnance Factories	Pakistan	*314*	55-00100
SAE Alsetex	France	*142*	75-33810
Tamares Ltd	UK	*487*	75-58500
Vammas Ltd	**Finland**	***117***	**55-02220**
Wejra A/S	Denmark	*104*	75-23300

521 40 40 16 Ammunition, mortar, ground-based, 160mm

Tamares Ltd	UK	*487*	75-58500
Wejra A/S	Denmark	*104*	75-23300

521 40 40 52 Ammunition, mortar, ground-based, 52mm

Israel Military Industries Ltd	Israel	*220*	75-10300

521 40 40 60 Ammunition, mortar, ground-based, 60mm

Chamberlain Manufacturing Corp.	USA	*563*	75-17950
CIC International Ltd	USA	*563*	75-18570
Cime Bocuze	France	*133*	75-09660
Esterline Technologies	USA	*574*	75-32360
Explosivos Alaveses SA	Spain	*366*	75-03050
Hirtenberger AG	Austria	*33*	75-01250
Indústrias de Defesa	Portugal	*327*	55-00950
INIMEX	**Belgium**	***46***	**75-04650**
Israel Military Industries Ltd	Israel	*220*	75-10300
Nottingham International	USA	*640*	80-22000
Pakistan Ordnance Factories	Pakistan	*314*	55-00100
SAE Alsetex	France	*142*	75-33810
SNC Industrial Tech.	Canada	*76*	75-12960
Swartklip Products	S. Africa	*359*	75-19500
Tamares Ltd	UK	*487*	75-58500
Wejra A/S	Denmark	*104*	75-23300

521 40 40 81 Ammunition, mortar, ground-based, 81mm

Allied Research Corp.	USA	*549*	75-04670
Australian Defence Industries Ltd	Australia	*9*	55-01000
Chamberlain Manufacturing Corp.	USA	*563*	75-17950
CIC International Ltd	USA	*563*	75-18570
Cime Bocuze	France	*133*	75-09660
Esterline Technologies	USA	*574*	75-32360
Explosivos Alaveses SA	Spain	*366*	75-03050
Greek Powder & Cartridge Co. SA	Greece	*189*	75-01500
Hirtenberger AG	Austria	*33*	75-01250
Indústrias de Defesa	Portugal	*327*	55-00950
INIMEX	**Belgium**	***46***	**75-04650**
Israel Military Industries Ltd	Israel	*220*	75-10300
Makina ve Kimya Endüstrisi	Turkey	*402*	55-01310
MECAR SA	Belgium	*46*	75-04930
Nottingham International	USA	*640*	80-22000
Pains-Wessex Ltd	UK	*472*	75-43510
Pakistan Ordnance Factories	Pakistan	*314*	55-00100
SAE Alsetex	France	*142*	75-33810
SNC Industrial Tech.	Canada	*76*	75-12960
Tamares Ltd	UK	*487*	75-58500
Vammas Ltd	**Finland**	***117***	**55-02220**
Wejra A/S	Denmark	*104*	75-23300

521 40 45 Ammunition, practice, ground-based

Allied Research Corp.	USA	*549*	75-04670
Australian Defence Industries Ltd	Australia	*9*	55-01000
CIC International Ltd	USA	*563*	75-18570
Giat Industries	France	*126*	55-00100
Hirtenberger AG	Austria	*33*	75-01250
INIMEX	**Belgium**	***46***	**75-04650**
Instalaza SA	Spain	*366*	75-03900
Israel Military Industries Ltd	Israel	*220*	75-10300
Lapua Oy	Finland	*118*	75-01430
Manurhin Défense	France	*139*	75-25630
Nottingham International	USA	*640*	80-22000
NWM de Kruithoorn BV	Netherlands	*294*	75-05200
OTEC Support Services Ltd	UK	*508*	80-19600
Royal Ordnance	UK	*481*	75-50620
SNC Industrial Tech.	Canada	*76*	75-12960
Swartklip Products	S. Africa	*359*	75-19500
Tamares Ltd	UK	*487*	75-58500
Wejra A/S	Denmark	*104*	75-23300

521 40 45 10 Ammunition, practice, small arms

Accuracy International Ltd	UK	*427*	75-00360
Armalon Ltd	UK	*431*	75-03729
CIC International Ltd	USA	*563*	75-18570
Eley Ltd	UK	*448*	75-20550
Giat Industries	France	*126*	55-00100
Greek Powder & Cartridge Co. SA	Greece	*189*	75-01500
Hirtenberger AG	Austria	*33*	75-01250
Indústrias de Defesa	Portugal	*327*	55-00950
INIMEX	**Belgium**	***46***	**75-04650**
Israel Military Industries Ltd	Israel	*220*	75-10300
Lapua Oy	Finland	*118*	75-01430
Manroy Engineering Ltd	**UK**	***466***	**75-37165**
Nottingham International	USA	*640*	80-22000
Royal Ordnance	UK	*481*	75-50620
Simunition	Canada	*67*	74-19100

SNC Industrial Tech.	Canada	*76*	75-12960
Swartklip Products	S. Africa	*359*	75-19500

521 40 45 50 Ammunition, practice, large calibre

Allied Research Corp.	USA	*549*	75-04670
Australian Defence Industries Ltd	Australia	*9*	55-01000
Chamberlain Manufacturing Corp.	USA	*563*	75-17950
CIC International Ltd	USA	*563*	75-18570
Greek Powder & Cartridge Co. SA	Greece	*189*	75-01500
Hirtenberger AG	Austria	*33*	75-01250
IBQ Industrias Químicas Ltda	Brazil	*55*	75-02700
INIMEX	**Belgium**	***46***	**75-04650**
Israel Military Industries Ltd	Israel	*220*	75-10300
Nottingham International	USA	*640*	80-22000
Ordnance & Tactical Systems Division	USA	*606*	75-71472
Primex Technologies, Inc	USA	*606*	75-71472
Royal Ordnance	UK	*481*	75-50620
Simad SpA	Italy.	*230*	74-23350
SNC Industrial Tech.	Canada	*76*	75-12960
Vammas Ltd	**Finland**	***117***	**55-02220**

521 40 45 50 50 Projectiles, radar augmented, training

Analysis Management Systems	S. Africa	*357*	75-00770

521 40 50 Ammunition, riot control (see code 491.05)

AAI Corp.	USA	*545*	75-00020
Aardvark Tactical Inc.	USA	*545*	75-00122
Allied Research Corp.	USA	*549*	75-04670
Armalon Ltd	UK	*431*	75-03729
Browning SA	Belgium	*45*	75-01950
CIC International Ltd	USA	*563*	75-18570
Combined Systems Inc.	USA	*564*	75-19893
Eley Ltd	UK	*448*	75-20550
Haley & Weller Ltd	UK	*456*	75-27400
Israel Military Industries Ltd	Israel	*220*	75-10300
Pains-Wessex Ltd	UK	*472*	75-43510
Swartklip Products	S. Africa	*359*	75-19500

521 40 60 Rockets, artillery

AAI Corp.	USA	*545*	75-00020
Arcus Co.	Bulgaria	*61*	55-00060
Avibras Aeroespacial SA	Brazil	*55*	75-00968
Chamberlain Manufacturing Corp.	USA	*563*	75-17950
CIC International Ltd	USA	*563*	75-18570
Explosivos Alaveses SA	Spain	*366*	75-03050
Forges de Zeebrugge SA	Belgium	*46*	75-03750
INIMEX	**Belgium**	***46***	**75-04650**
Israel Military Industries Ltd	Israel	*220*	75-10300
Makina ve Kimya Endüstrisi	Turkey	*402*	55-01310
Nissan Motor Co. Ltd	Japan	*254*	75-06790
Somchem	S. Africa	*359*	75-19200
State Research & Association	Russia	*334*	55-18325

521 40 65 Rockets, anti-tank

Aerospace and Electronics Division	USA	*606*	75-71472
Allied Research Corp.	USA	*549*	75-04670
CIC International Ltd	USA	*563*	75-18570
Forges de Zeebrugge SA	Belgium	*46*	75-03750
Hellenic Arms Industry	Greece	*187*	55-00800
Instalaza SA	Spain	*366*	75-03900
Israel Military Industries Ltd	Israel	*220*	75-10300
Kompanija Sloboda D D	Yugoslavia	*661*	75-00475
Luchaire Défense SA	France	*139*	75-25400
Manurhin Défense	France	*139*	75-25630
Pakistan Ordnance Factories	Pakistan	*314*	55-00100
Raufoss Technology AS	Norway	*310*	75-02200
Royal Ordnance	UK	*481*	75-50620
Somchem	S. Africa	*359*	75-19200

521 50 Ammunition, small arms

Accuracy International Ltd	UK	*427*	75-00360
Arcus Co.	Bulgaria	*61*	55-00060
Australian Defence Industries Ltd	Australia	*9*	55-01000
Bofors AB	Sweden	*375*	74-00020
Bofors Carl Gustaf AB	Sweden	*376*	75-02950
Century International Arms Inc.	USA	*633*	80-05400
CIC International Ltd	USA	*563*	75-18570
Companhia Brasileira de Cartuchos	Brazil	*55*	75-03150
Conjay Arms Co. Ltd	UK	*442*	75-14855
Daudsons Armoury	Pakistan	*315*	75-00500
Dynamit Nobel Graz GmbH.	Austria	*32*	75-00800
Giat Industries	France	*126*	55-00100
Glaser Safety Slug Inc.	USA	*579*	75-39750
Greek Powder & Cartridge Co. SA	Greece	*189*	75-01500
Helston Gunsmiths	UK	*458*	75-29010
Hirtenberger AG	Austria	*33*	75-01250
Hornady Manufacturing Co.	USA	*582*	75-44300
Indústrias de Defesa	Portugal	*327*	55-00950
INIMEX	**Belgium**	***46***	**75-04650**
Interstate Technical Products	UK	*461*	75-31946
Israel Military Industries Ltd	Israel	*220*	75-10300
KJM Fabritek Inc.	USA	*589*	75-52877
Laboratoire Arcane	France	*138*	75-24010
Lapua Oy.	Finland	*118*	75-01430
Merex Inc.	USA	*595*	75-60250
New Generation Ammunition	S. Africa	*359*	75-15200
New Helvetia Trade Group	USA	*639*	80-21730
Olin Australia Ltd	Australia	*19*	75-15642
Parker-Hale Ltd	UK	*473*	75-43850
Royal Ordnance	UK	*481*	75-50620
Sako Ltd	Finland	*118*	75-01980
Simunition	Canada	*67*	74-19100
SNC Industrial Tech.	Canada	*76*	75-12960
SNPE	France	*145*	75-36315
South Central Ordnance Co.	USA	*616*	75-82400
Southern Ammunition Co. Inc.	USA	*616*	75-82440
Swartklip Products	S. Africa	*359*	75-19500

521 50 10 Ammunition, pistol / revolver

Accuracy International Ltd	UK	*427*	75-00360
Arcus Co.	Bulgaria	*61*	55-00060
Armalon Ltd	UK	*431*	75-03729
Australian Defence Industries Ltd	Australia	*9*	55-01000
Century International Arms Inc.	USA	*633*	80-05400
CIC International Ltd	USA	*563*	75-18570
Companhia Brasileira de Cartuchos	Brazil	*55*	75-03150
Conjay Arms Co. Ltd	UK	*442*	75-14855
Daudsons Armoury	Pakistan	*315*	75-00500
Giat Industries	France	*126*	55-00100
Glaser Safety Slug Inc.	USA	*579*	75-39750
Helston Gunsmiths	UK	*458*	75-29010
Hirtenberger AG	Austria	*33*	75-01250
Hornady Manufacturing Co.	USA	*582*	75-44300
Indústrias de Defesa	Portugal	*327*	55-00950
Interstate Technical Products	UK	*461*	75-31946
Israel Military Industries Ltd	Israel	*220*	75-10300
Laboratoire Arcane	France	*138*	75-24010
Lapua Oy.	Finland	*118*	75-01430
Makina ve Kimya Endüstrisi	Turkey	*402*	55-01310
Merex Inc.	USA	*595*	75-60250
Nottingham International	USA	*640*	80-22000
Robert W. Hart & Son Inc.	USA	*581*	75-42450
Royal Ordnance	UK	*481*	75-50620
Sako Ltd	Finland	*118*	75-01980
Sellier & Bellot a.s.	Czech Republic	*99*	75-00900
SNC Industrial Tech.	Canada	*76*	75-12960
South Central Ordnance Co.	USA	*616*	75-82400
Southern Ammunition Co. Inc.	USA	*616*	75-82440
Swartklip Products	S. Africa	*359*	75-19500

521 50 20 Ammunition, rifle

Accuracy International Ltd	UK	*427*	75-00360
Armalon Ltd	UK	*431*	75-03729
Australian Defence Industries Ltd	Australia	*9*	55-01000
Century International Arms Inc.	USA	*633*	80-05400
CIC International Ltd	USA	*563*	75-18570
Companhia Brasileira de Cartuchos	Brazil	*55*	75-03150
Conjay Arms Co. Ltd	UK	*442*	75-14855
Giat Industries	France	*126*	55-00100
Glaser Safety Slug Inc.	USA	*579*	75-39750
Helston Gunsmiths	UK	*458*	75-29010
Hirtenberger AG	Austria	*33*	75-01250
Indústrias de Defesa	Portugal	*327*	55-00950
INIMEX	**Belgium**	***46***	**75-04650**

Interstate Technical Products	UK	*461*	75-31946
Israel Military Industries Ltd	Israel	*220*	75-10300
KJM Fabritek Inc.	USA	*589*	75-52877
Lapua Oy	Finland	*118*	75-01430
Merex Inc.	USA	*595*	75-60250
Nottingham International	USA	*640*	80-22000
Olin Australia Ltd	Australia	*19*	75-15642
Robert W. Hart & Son Inc.	USA	*581*	75-42450
Royal Ordnance	UK	*481*	75-50620
Sako Ltd	Finland	*118*	75-01980
SNC Industrial Tech.	Canada	*76*	75-12960
South Central Ordnance Co.	USA	*616*	75-82400
Southern Ammunition Co. Inc.	USA	*616*	75-82440

521 50 30 Cartridges, rifle grenade launching

Aerospace and Electronics Division	USA	*606*	75-71472
Allied Research Corp.	USA	*549*	75-04670
Dynamit Nobel Graz GmbH.	Austria	*32*	75-00800
Giat Industries	France	*126*	55-00100
Indústrias de Defesa	Portugal	*327*	55-00950
Israel Military Industries Ltd	Israel	*220*	75-10300
Merex Inc.	USA	*595*	75-60250
Nottingham International	USA	*640*	80-22000
South Central Ordnance Co.	USA	*616*	75-82400
Swartklip Products	S. Africa	*359*	75-19500

521 80 Ammunition accessories

AAI Corp.	USA	*545*	75-00020
Accuracy International Ltd	UK	*427*	75-00360
Century International Arms Inc.	USA	*633*	80-05400
CIC International Ltd	USA	*563*	75-18570
Indústrias de Defesa	Portugal	*327*	55-00950
Israel Military Industries Ltd	Israel	*220*	75-10300
Nottingham International	USA	*640*	80-22000
Royal Ordnance	UK	*481*	75-50620
Standard Armament Inc.	USA	*617*	75-83650

521 80 10 Belt links, automatic weapon

CIC International Ltd	USA	*563*	75-18570
Diehl GmbH & Co	Germany	*167*	75-06500
Greek Powder & Cartridge Co. SA	Greece	*189*	75-01500
Indústrias de Defesa	Portugal	*327*	55-00950
INIMEX	**Belgium**	***46***	**75-04650**
Israel Military Industries Ltd	Israel	*220*	75-10300
Merex Inc.	USA	*595*	75-60250
Nottingham International	USA	*640*	80-22000
Royal Ordnance	UK	*481*	75-50620
SNC Industrial Tech.	Canada	*76*	75-12960

521 80 20 Charger clips, automatic weapon

Beta Company	USA	*558*	75-13750
CIC International Ltd	USA	*563*	75-18570
Giat Industries	France	*126*	55-00100
Greek Powder & Cartridge Co. SA	Greece	*189*	75-01500
Royal Ordnance	UK	*481*	75-50620
SNC Industrial Tech.	Canada	*76*	75-12960

521 90 Ammunition components

AAI Corp.	USA	*545*	75-00020
Accuracy International Ltd	UK	*427*	75-00360
AEA Technology plc	UK	*496*	80-00480
Amron L.L.C.	USA	*553*	75-06883
Aquarius Fabrique International Pvt	India	*204*	75-00160
Armtec Defense Products Co.	USA	*555*	75-08950
Australian Defence Industries Ltd	Australia	*9*	55-01000
Bofors AB	Sweden	*375*	74-00020
Calspan SRL Technology	USA	*540*	60-03000
Chamberlain Manufacturing Corp.	USA	*563*	75-17950
Chartered Ammunition Industries	Singapore	*348*	75-01090
CIC International Ltd	USA	*563*	75-18570
Combined Systems Inc.	USA	*564*	75-19893
Companhia Brasileira de Cartuchos	Brazil	*55*	75-03150
Dyna East Corp.	USA	*634*	80-08730
Dynamit Nobel GmbH	Germany	*167*	75-07100
Eley Ltd	UK	*448*	75-20550
Esterline Technologies	USA	*574*	75-32360
Eurometaal NV	Netherlands	*292*	75-02200
Floturn Inc.	USA	*576*	75-35060
Giat Industries	France	*126*	55-00100
Hirtenberger AG	Austria	*33*	75-01250
Israel Military Industries Ltd	Israel	*220*	75-10300
Lapua Oy	Finland	*118*	75-01430
Luchaire Défense SA	France	*139*	75-25400
MECAR SA	Belgium	*46*	75-04930
Nuclear Metals Inc.	USA	*600*	75-65760
NWM de Kruithoorn BV	Netherlands	*294*	75-05200
Oerlikon-Contraves AG.	Switzerland	*387*	75-09250
Ordnance & Tactical Systems Division	USA	*606*	75-71472
Primex Technologies, Inc	USA	*606*	75-71472
Raufoss Technology AS	Norway	*310*	75-02200
Republic Arms (Pty) Ltd	**S. Africa**	***359***	**75-18210**
Royal Ordnance	UK	*481*	75-50620
Santa Barbara SA	Spain	*364*	55-01000
Simunition	Canada	*67*	74-19100
SNPE	France	*145*	75-36315
Somchem	S. Africa	*359*	75-19200
Tamares Ltd	UK	*487*	75-58500
Vihtavuori Oy	Finland	*119*	75-02850

521 90 10 Ammunition cores

AEA Technology plc	UK	*496*	80-00480
CIC International Ltd	USA	*563*	75-18570
Cime Bocuze	France	*133*	75-09660
Hirtenberger AG	Austria	*33*	75-01250
Israel Military Industries Ltd	Israel	*220*	75-10300
Nuclear Metals Inc.	USA	*600*	75-65760
NWM de Kruithoorn BV	Netherlands	*294*	75-05200
Royal Ordnance	UK	*481*	75-50620
Royal Ordnance Speciality Metals	UK	*481*	75-50630
Teledyne Advanced Materials	USA	*620*	75-87523

521 90 20 Penetrators, kinetic energy

AEA Technology plc	UK	*496*	80-00480
Ashot Ashkelon Industries Ltd	Israel	*217*	75-02100
CIC International Ltd	USA	*563*	75-18570
Cime Bocuze	France	*133*	75-09660
Hirtenberger AG	Austria	*33*	75-01250
Israel Military Industries Ltd	Israel	*220*	75-10300
NWM de Kruithoorn BV	Netherlands	*294*	75-05200
Royal Ordnance	UK	*481*	75-50620
Royal Ordnance Speciality Metals	UK	*481*	75-50630
Teledyne Advanced Materials	USA	*620*	75-87523

521 90 25 Penetrators, chemical energy

Dyna East Corp.	USA	*634*	80-08730

521 90 30 Cartridge cases, ammunition

Accuracy International Ltd	UK	*427*	75-00360
Amron L.L.C.	USA	*553*	75-06883
Armtec Defense Products Co.	USA	*555*	75-08950
Chartered Ammunition Industries	Singapore	*348*	75-01090
C.I.A.M.-E.A. Sarl	Italy	*233*	75-05150
CIC International Ltd	USA	*563*	75-18570
Combined Systems Inc.	USA	*564*	75-19893
Companhia Brasileira de Cartuchos	Brazil	*55*	75-03150
Dynamit Nobel GmbH	Germany	*167*	75-07100
Esterline Technologies	USA	*574*	75-32360
Eurometaal NV	Netherlands	*292*	75-02200
Fabrica Militar "Frey Luis Beltran"	Argentina	*4*	55-06160
Giat Industries	France	*126*	55-00100
Greek Powder & Cartridge Co. SA	Greece	*189*	75-01500
Hirtenberger AG	Austria	*33*	75-01250
Indústrias de Defesa	Portugal	*327*	55-00950
Israel Military Industries Ltd	Israel	*220*	75-10300
KJM Fabritek Inc.	USA	*589*	75-52877
Lapua Oy	Finland	*118*	75-01430
Manurhin Défense	France	*139*	75-25630
Merex Inc.	USA	*595*	75-60250
NWM de Kruithoorn BV	Netherlands	*294*	75-05200
Ordnance & Tactical Systems Division	USA	*606*	75-71472
Primex Technologies, Inc	USA	*606*	75-71472
Raufoss Technology AS	Norway	*310*	75-02200

Royal Ordnance	UK	*481*	75-50620
Simunition	Canada	*67*	74-19100
SNC Industrial Tech.	Canada	*76*	75-12960
Vammas Ltd	**Finland**	***117***	**55-02220**

521 90 30 30 Cartridge cases, combustible

Allied Research Corp.	USA	*549*	75-04670
Armtec Defense Products Co.	USA	*555*	75-08950
CIC International Ltd	USA	*563*	75-18570
Combined Systems Inc.	USA	*564*	75-19893
Esterline Technologies	USA	*574*	75-32360
Fabrica Militar "Frey Luis Beltran"	Argentina	*4*	55-06160
Giat Industries	France	*126*	55-00100
Israel Military Industries Ltd	Israel	*220*	75-10300
Somchem	S. Africa	*359*	75-19200

521 90 60 Sabots, discarding

AAI Corp.	USA	*545*	75-00020
Allied Research Corp.	USA	*549*	75-04670
CIC International Ltd	USA	*563*	75-18570
Cime Bocuze	France	*133*	75-09660
Israel Military Industries Ltd	Israel	*220*	75-10300
NWM de Kruithoorn BV	Netherlands	*294*	75-05200
Royal Ordnance	UK	*481*	75-50620
SNC Industrial Tech.	Canada	*76*	75-12960
Technical Products Group Inc.	USA	*620*	75-86764

521 95 Ammunition assembly & filling

Allied Research Corp.	USA	*549*	75-04670
Armalon Ltd	UK	*431*	75-03729
CMS Inc.	USA	*564*	75-19170
Eurometaal NV	Netherlands	*292*	75-02200
Giat Industries	France	*126*	55-00100
Haley & Weller Ltd	UK	*456*	75-27400
Indústrias de Defesa	Portugal	*327*	55-00950
Interstate Technical Products	UK	*461*	75-31946
Israel Military Industries Ltd	Israel	*220*	75-10300
Lapua Oy	Finland	*118*	75-01430
NWM de Kruithoorn BV	Netherlands	*294*	75-05200
Oerlikon-Contraves AG.	Switzerland	*387*	75-09250
Pains-Wessex Ltd	UK	*472*	75-43510
Royal Ordnance	UK	*481*	75-50620
SNC Industrial Tech.	Canada	*76*	75-12960
Summit Machine Builders Corp.	USA	*618*	75-84900
Tamares Ltd.	UK	*487*	75-58500

522 Grenades

AAI Corp.	USA	*545*	75-00020
Allied Research Corp.	USA	*549*	75-04670
Amron L.L.C.	USA	*553*	75-06883
Australian Defence Industries Ltd	Australia	*9*	55-01000
Brock's Explosives Ltd	UK	*437*	75-09980
Cenzin Co. Ltd.	Poland	*324*	75-00200
CIC International Ltd	USA	*563*	75-18570
Cime Bocuze	France	*133*	75-09660
Dynamit Nobel Graz GmbH.	Austria	*32*	75-00800
Elviemek SA.	Greece	*188*	75-01200
Étienne Lacroix	France	*138*	75-24050
Explosivos Alaveses SA	Spain	*366*	75-03050
FN Manufacturing Inc	USA	*576*	75-35320
Giat Industries	France	*126*	55-00100
Grenadier Engineering Inc.	USA	*580*	75-40390
Haley & Weller Ltd	UK	*456*	75-27400
IBQ Industrias Químicas Ltda	Brazil	*55*	75-02700
INIMEX	**Belgium**	***46***	**75-04650**
Instalaza SA	Spain	*366*	75-03900
INTERTECHNIK Techn. Produktionen	Austria	*33*	75-01430
Israel Military Industries Ltd	Israel	*220*	75-10300
Israel Product Research Co. Ltd	Israel	*219*	75-10000
Kompanija Sloboda D D	Yugoslavia	*661*	75-00475
Lockheed Martin Corp.	USA	*591*	75-55635
Martin Electronics Inc.	USA	*594*	75-58540
MECAR SA	Belgium	*46*	75-04930
ML Wallop Defence Systems Ltd	UK	*469*	75-40240
NORINCO	China PR	*88*	55-00280
Nottingham International	USA	*640*	80-22000
Pains-Wessex Ltd	UK	*472*	75-43510
Pakistan Ordnance Factories	Pakistan	*314*	55-00100
Royal Ordnance	UK	*481*	75-50620
Ruggieri Technologie	France	*142*	75-33500
SAE Alsetex	France	*142*	75-33810
SIMA PERU	Peru	*319*	55-01950
South Central Ordnance Co.	USA	*616*	75-82400
SPL Ordnance Ltd.	Israel	*225*	80-01935
Swartklip Products	S. Africa	*359*	75-19500

522 40 10 Grenades, anti-riot (internal security)

Allied Research Corp.	USA	*549*	75-04670
Brock's Explosives Ltd	UK	*437*	75-09980
CIC International Ltd	USA	*563*	75-18570
Grenadier Engineering Inc.	USA	*580*	75-40390
Haley & Weller Ltd	UK	*456*	75-27400
Israel Military Industries Ltd	Israel	*220*	75-10300
Nottingham International	USA	*640*	80-22000
Pains-Wessex Ltd	UK	*472*	75-43510
Royal Ordnance	UK	*481*	75-50620
Simad SpA	Italy	*230*	74-23350
South Central Ordnance Co.	USA	*616*	75-82400
SPL Ordnance Ltd.	Israel	*225*	80-01935
Swartklip Products	S. Africa	*359*	75-19500

522 40 10 30 Grenades, anti-riot, hand-launched

Aardvark Tactical Inc.	USA	*545*	75-00122
Brock's Explosives Ltd	UK	*437*	75-09980
CIC International Ltd	USA	*563*	75-18570
Haley & Weller Ltd	UK	*456*	75-27400
Israel Military Industries Ltd	Israel	*220*	75-10300
Israel Product Research Co. Ltd	Israel	*219*	75-10000
Nottingham International	USA	*640*	80-22000
Pains-Wessex Ltd	UK	*472*	75-43510
Royal Ordnance	UK	*481*	75-50620
Simad SpA	Italy	*230*	74-23350
South Central Ordnance Co.	USA	*616*	75-82400
SPL Ordnance Ltd.	Israel	*225*	80-01935
Swartklip Products	S. Africa	*359*	75-19500

522 40 10 32 Grenades, anti-riot, CS irritant smoke, hand-launched

AAI Corp.	USA	*545*	75-00020
Aardvark Tactical Inc.	USA	*545*	75-00122
Brock's Explosives Ltd	UK	*437*	75-09980
CIC International Ltd	USA	*563*	75-18570
Combined Systems Inc.	USA	*564*	75-19893
Defense Technology Corporation of	USA	*568*	75-24980
Haley & Weller Ltd	UK	*456*	75-27400
Israel Product Research Co. Ltd	Israel	*219*	75-10000
Nottingham International	USA	*640*	80-22000
Pains-Wessex Ltd	UK	*472*	75-43510
Royal Ordnance	UK	*481*	75-50620
SAE Alsetex	France	*142*	75-33810
Simad SpA	Italy	*230*	74-23350
South Central Ordnance Co.	USA	*616*	75-82400
SPL Ordnance Ltd.	Israel	*225*	80-01935

522 40 10 33 Grenades, anti-riot, explosive dye, hand-launched

Aardvark Tactical Inc.	USA	*545*	75-00122
Haley & Weller Ltd	UK	*456*	75-27400
Israel Product Research Co. Ltd	Israel	*219*	75-10000
South Central Ordnance Co.	USA	*616*	75-82400

522 40 10 34 Grenades, anti-riot, tear gas CN, hand-launched

AAI Corp.	USA	*545*	75-00020
Aardvark Tactical Inc.	USA	*545*	75-00122
CIC International Ltd	USA	*563*	75-18570
Combined Systems Inc.	USA	*564*	75-19893
Defense Technology Corporation of	USA	*568*	75-24980
Israel Product Research Co. Ltd	Israel	*219*	75-10000
Nottingham International	USA	*640*	80-22000
SAE Alsetex	France	*142*	75-33810
Simad SpA	Italy	*230*	74-23350

South Central Ordnance Co.	USA	616	75-82400
SPL Ordnance Ltd.	Israel	225	80-01935

522 40 10 35 Distraction grenades, hand-launched

Brock's Explosives Ltd	UK	437	75-09980
Haley & Weller Ltd	UK	456	75-27400
Martin Electronics Inc.	USA	594	75-58540
Nottingham International	USA	640	80-22000
Pains-Wessex Ltd	UK	472	75-43510
South Central Ordnance Co.	USA	616	75-82400
SPL Ordnance Ltd.	Israel	225	80-01935

522 40 10 36 Stun grenades, hand-launched

Brock's Explosives Ltd	UK	437	75-09980
CIC International Ltd	USA	563	75-18570
Haley & Weller Ltd	UK	456	75-27400
Israel Product Research Co. Ltd	Israel	219	75-10000
Nottingham International	USA	640	80-22000
Pains-Wessex Ltd	UK	472	75-43510
Royal Ordnance	UK	481	75-50620
SAE Alsetex	France	142	75-33810
South Central Ordnance Co.	USA	616	75-82400
SPL Ordnance Ltd.	Israel	225	80-01935
Swartklip Products	S. Africa	359	75-19500

522 40 10 50 Grenades, anti-riot, weapon-launched

Aardvark Tactical Inc.	USA	545	75-00122
Allied Research Corp.	USA	549	75-04670
CIC International Ltd	USA	563	75-18570
Grenadier Engineering Inc.	USA	580	75-40390
Haley & Weller Ltd	UK	456	75-27400
Israel Military Industries Ltd	Israel	220	75-10300
Israel Product Research Co. Ltd	Israel	219	75-10000
Nottingham International	USA	640	80-22000
Pains-Wessex Ltd	UK	472	75-43510
Simad SpA	Italy	230	74-23350
South Central Ordnance Co.	USA	616	75-82400
SPL Ordnance Ltd.	Israel	225	80-01935
Swartklip Products	S. Africa	359	75-19500

522 40 10 52 Grenades, anti-riot, CS irritant smoke, weapon-launched

AAI Corp.	USA	545	75-00020
Aardvark Tactical Inc.	USA	545	75-00122
Allied Research Corp.	USA	549	75-04670
CIC International Ltd	USA	563	75-18570
Defense Technology Corporation of	USA	568	75-24980
Grenadier Engineering Inc.	USA	580	75-40390
Haley & Weller Ltd	UK	456	75-27400
Israel Product Research Co. Ltd	Israel	219	75-10000
Nottingham International	USA	640	80-22000
Pains-Wessex Ltd	UK	472	75-43510
Royal Ordnance	UK	481	75-50620
SAE Alsetex	France	142	75-33810
Simad SpA	Italy	230	74-23350
South Central Ordnance Co.	USA	616	75-82400
SPL Ordnance Ltd.	Israel	225	80-01935
Swartklip Products	S. Africa	359	75-19500

522 40 10 54 Grenades, anti-riot, tear gas CN, weapon-launched

AAI Corp.	USA	545	75-00020
Aardvark Tactical Inc.	USA	545	75-00122
Allied Research Corp.	USA	549	75-04670
CIC International Ltd	USA	563	75-18570
Defense Technology Corporation of	USA	568	75-24980
Grenadier Engineering Inc.	USA	580	75-40390
Israel Product Research Co. Ltd	Israel	219	75-10000
SAE Alsetex	France	142	75-33810
Simad SpA	Italy	230	74-23350
South Central Ordnance Co.	USA	616	75-82400
SPL Ordnance Ltd.	Israel	225	80-01935

522 40 30 Grenades, combat

Allied Research Corp.	USA	549	75-04670
Australian Defence Industries Ltd	Australia	9	55-01000
CIC International Ltd	USA	563	75-18570
Dynamit Nobel Graz GmbH	Austria	32	75-00800
Giat Industries	France	126	55-00100
Haley & Weller Ltd	UK	456	75-27400
Israel Military Industries Ltd	Israel	220	75-10300
MECAR SA	Belgium	46	75-04930
Nottingham International	USA	640	80-22000
Pains-Wessex Ltd	UK	472	75-43510
Pakistan Ordnance Factories	Pakistan	314	55-00100
Royal Ordnance	UK	481	75-50620
Ruggieri Technologie	France	142	75-33500
SAE Alsetex	France	142	75-33810
SNPE	France	145	75-36315
SPL Ordnance Ltd.	Israel	225	80-01935
Swartklip Products	S. Africa	359	75-19500

522 40 30 20 Grenades, defensive

CIC International Ltd	USA	563	75-18570
Giat Industries	France	126	55-00100
Haley & Weller Ltd	UK	456	75-27400
Israel Military Industries Ltd	Israel	220	75-10300
Nottingham International	USA	640	80-22000
SPL Ordnance Ltd.	Israel	225	80-01935

522 40 30 30 Grenades, combat, hand-launched

Australian Defence Industries Ltd	Australia	9	55-01000
Giat Industries	France	126	55-00100
Haley & Weller Ltd	UK	456	75-27400
Israel Military Industries Ltd	Israel	220	75-10300
Nottingham International	USA	640	80-22000
Pakistan Ordnance Factories	Pakistan	314	55-00100
Royal Ordnance	UK	481	75-50620
SPL Ordnance Ltd.	Israel	225	80-01935
Swartklip Products	S. Africa	359	75-19500

522 40 30 31 Grenades, combat, hand-launched, anti-personnel

Australian Defence Industries Ltd	Australia	9	55-01000
Giat Industries	France	126	55-00100
Haley & Weller Ltd	UK	456	75-27400
Israel Military Industries Ltd	Israel	220	75-10300
Nottingham International	USA	640	80-22000
SAE Alsetex	France	142	75-33810
SPL Ordnance Ltd.	Israel	225	80-01935
Swartklip Products	S. Africa	359	75-19500

522 40 30 32 Grenades, combat, hand-launched, blast effect

Haley & Weller Ltd	UK	456	75-27400
Israel Military Industries Ltd	Israel	220	75-10300
SAE Alsetex	France	142	75-33810
SPL Ordnance Ltd.	Israel	225	80-01935

522 40 30 34 Grenades, combat, hand-launched, blinding effect (flash)

Haley & Weller Ltd	UK	456	75-27400
Israel Military Industries Ltd	Israel	220	75-10300
SAE Alsetex	France	142	75-33810
SPL Ordnance Ltd.	Israel	225	80-01935

522 40 30 35 Grenades, combat, hand-launched, distraction

Brock's Explosives Ltd	UK	437	75-09980
Haley & Weller Ltd	UK	456	75-27400
Israel Military Industries Ltd	Israel	220	75-10300
SPL Ordnance Ltd.	Israel	225	80-01935

522 40 30 38 Grenades, combat, hand-launched, incendiary

Australian Defence Industries Ltd	Australia	9	55-01000
Explosivos Alaveses SA	Spain	366	75-03050
Haley & Weller Ltd	UK	456	75-27400
Israel Military Industries Ltd	Israel	220	75-10300
Pains-Wessex Ltd	UK	472	75-43510
Pakistan Ordnance Factories	Pakistan	314	55-00100
SPL Ordnance Ltd	Israel	225	80-01935
Swartklip Products	S. Africa	359	75-19500

522 40 30 50 Grenades, combat, weapon-launched

Allied Research Corp.	USA	549	75-04670
CIC International Ltd	USA	563	75-18570
Giat Industries	France	126	55-00100
Grenadier Engineering Inc.	USA	580	75-40390
Haley & Weller Ltd	UK	456	75-27400
Israel Military Industries Ltd	Israel	220	75-10300
MECAR SA	Belgium	46	75-04930
Swartklip Products	S. Africa	359	75-19500

522 40 30 52 Grenades, combat, rifle-launched

Allied Research Corp.	USA	549	75-04670
CIC International Ltd	USA	563	75-18570
Giat Industries	France	126	55-00100
Instalaza SA	Spain	366	75-03900
Israel Military Industries Ltd	Israel	220	75-10300
MECAR SA	Belgium	46	75-04930
Pakistan Ordnance Factories	Pakistan	314	55-00100
Swartklip Products	S. Africa	359	75-19500

522 40 30 52 10 Grenades, combat, rifle-launched, anti-tank

Allied Research Corp.	USA	549	75-04670
CIC International Ltd	USA	563	75-18570
Giat Industries	France	126	55-00100
Israel Military Industries Ltd	Israel	220	75-10300
MECAR SA	Belgium	46	75-04930

522 40 30 52 20 Grenades, combat, rifle-launched, anti-personnel / anti-vehicle (AP/AV)

Allied Research Corp.	USA	549	75-04670
CIC International Ltd	USA	563	75-18570
Giat Industries	France	126	55-00100
Israel Military Industries Ltd	Israel	220	75-10300
SAE Alsetex	France	142	75-33810
Swartklip Products	S. Africa	359	75-19500

522 40 30 52 30 Grenades, combat, rifle-launched, smoke

Allied Research Corp.	USA	549	75-04670
CIC International Ltd	USA	563	75-18570
Dynamit Nobel Graz GmbH.	Austria	32	75-00800
Giat Industries	France	126	55-00100
Haley & Weller Ltd	UK	456	75-27400
Israel Military Industries Ltd	Israel	220	75-10300
Pains-Wessex Ltd	UK	472	75-43510
Swartklip Products	S. Africa	359	75-19500

522 40 30 52 50 Grenades, combat, rifle-launched, illuminating

Allied Research Corp.	USA	549	75-04670
Dynamit Nobel Graz GmbH.	Austria	32	75-00800
Giat Industries	France	126	55-00100
Israel Military Industries Ltd	Israel	220	75-10300

522 40 30 54 Grenades, combat, hand or rifle-launched

Allied Research Corp.	USA	549	75-04670
Eurometaal NV	Netherlands	292	75-02200
Giat Industries	France	126	55-00100
Israel Military Industries Ltd	Israel	220	75-10300
Swartklip Products	S. Africa	359	75-19500

522 40 30 70 Grenades, combat, vehicle-launched

CIC International Ltd	USA	563	75-18570
Giat Industries	France	126	55-00100
Haley & Weller Ltd	UK	456	75-27400
Israel Military Industries Ltd	Israel	220	75-10300

522 40 30 71 Grenades, combat, vehicle-launched, anti-personnel

CIC International Ltd	USA	563	75-18570
Giat Industries	France	126	55-00100
Haley & Weller Ltd	UK	456	75-27400
Israel Military Industries Ltd	Israel	220	75-10300
SIMA PERU	Peru	319	55-01950

522 40 30 73 Grenades, illuminating, vehicle-launched

Giat Industries	France	126	55-00100
Haley & Weller Ltd	UK	456	75-27400
Israel Military Industries Ltd	Israel	220	75-10300

522 40 30 76 Grenades, smoke, vehicle-launched

Haley & Weller Ltd	UK	456	75-27400
IBQ Industrias Químicas Ltda	Brazil	55	75-02700
Israel Military Industries Ltd	Israel	220	75-10300
Pains-Wessex Ltd	UK	472	75-43510
Royal Ordnance	UK	481	75-50620
Swartklip Products	S. Africa	359	75-19500

522 40 70 Grenades, practice

Allied Research Corp.	USA	549	75-04670
Australian Defence Industries Ltd	Australia	9	55-01000
Brock's Explosives Ltd	UK	437	75-09980
Eurometaal NV	Netherlands	292	75-02200
Giat Industries	France	126	55-00100
Haley & Weller Ltd	UK	456	75-27400
Israel Military Industries Ltd	Israel	220	75-10300
MECAR SA	Belgium	46	75-04930
Pains-Wessex Ltd	UK	472	75-43510
Royal Ordnance	UK	481	75-50620
Ruggieri Technologie	France	142	75-33500
SAE Alsetex	France	142	75-33810
Simad SpA	Italy	230	74-23350
Swartklip Products	S. Africa	359	75-19500
Zen Technologies Ltd	India	205	75-03000

522 40 70 30 Grenades, practice, hand-launched

Australian Defence Industries Ltd	Australia	9	55-01000
Brock's Explosives Ltd	UK	437	75-09980
Étienne Lacroix	France	138	75-24050
Giat Industries	France	126	55-00100
Haley & Weller Ltd	UK	456	75-27400
Israel Military Industries Ltd	Israel	220	75-10300
Pains-Wessex Ltd	UK	472	75-43510
Royal Ordnance	UK	481	75-50620
SAE Alsetex	France	142	75-33810
Simad SpA	Italy	230	74-23350
Swartklip Products	S. Africa	359	75-19500

522 40 70 50 Grenades, practice, rifle-launched

Allied Research Corp.	USA	549	75-04670
Giat Industries	France	126	55-00100
Pains-Wessex Ltd	UK	472	75-43510
Royal Ordnance	UK	481	75-50620
Simad SpA	Italy	230	74-23350
Swartklip Products	S. Africa	359	75-19500

523 Bombs, depth charges

AAI Corp.	USA	*545*	75-00020
Aerea SpA.	Italy	*230*	75-00280
Arab Organization for Ind.	Egypt	*109*	55-01700
CIC International Ltd	USA	*563*	75-18570
Daudsons Armoury	Pakistan	*315*	75-00500
ECA SA	France	*136*	75-14050
Eurometaal NV	Netherlands	*292*	75-02200
Israel Military Industries Ltd	Israel	*220*	75-10300
Lockheed Martin Missiles & Space	USA	*592*	75-55640
Royal Ordnance	UK	*481*	75-50620
Sargent Fletcher Inc.	USA	*612*	75-78150

523 20 Bombs, aircraft

Aerea SpA.	Italy	*230*	75-00280
Analysis Management Systems	S. Africa	*357*	75-00770
Arab Organization for Ind.	Egypt	*109*	55-01700
Australian Defence Industries Ltd	Australia	*9*	55-01000
Avibras Aeroespacial SA	Brazil	*55*	75-00968
Bofors SA Marine AB.	Sweden	*376*	75-03040
Daudsons Armoury	Pakistan	*315*	75-00500
Explosivos Alaveses SA	Spain	*366*	75-03050
Fabrica Española de Cartucheria SL	Spain	*366*	75-03300
Hunting Engineering Ltd	UK	*459*	75-30700
INIMEX	**Belgium**	***46***	**75-04650**
Israel Military Industries Ltd	Israel	*220*	75-10300
Kader Factory for Developed Ind.	Egypt	*109*	55-01700
Matra Défense-Espace	France	*139*	75-26000
NORINCO	China PR	*88*	55-00280
Orlando Technology Inc.	USA	*540*	60-16000
Pakistan Ordnance Factories	Pakistan	*314*	55-00100
Reunert Technology Systems Ltd	S. Africa	*359*	75-18300
Reutech Defence Industries	S. Africa	*359*	75-18400
Royal Ordnance	UK	*481*	75-50620
Sargent Fletcher Inc.	USA	*612*	75-78150

523 20 05 Bombs, aircraft, anti-armour

Hunting Engineering Ltd	UK	*459*	75-30700
Israel Military Industries Ltd	Israel	*220*	75-10300
Royal Ordnance	UK	*481*	75-50620
Textron Systems Division	USA	*621*	75-88655

523 20 10 Bombs, aircraft, cluster

Artefatos Eléctricos e Mecânicos	Brazil	*55*	74-00905
Avibras Aeroespacial SA	Brazil	*55*	75-00968
CIC International Ltd	USA	*563*	75-18570
Eurodef Aerospace & Defence	Greece	*189*	75-01295
Explosivos Alaveses SA	Spain	*366*	75-03050
Hunting Engineering Ltd	UK	*459*	75-30700
INIMEX	**Belgium**	***46***	**75-04650**
International Technology SA	Spain	*366*	75-03950
Israel Military Industries Ltd	Israel	*220*	75-10300
Orlando Technology Inc.	USA	*540*	60-16000
Rafael	Israel	*216*	55-01800
Royal Ordnance	UK	*481*	75-50620

523 20 20 Bombs, aircraft, fragmentation

Avibras Aeroespacial SA	Brazil	*55*	75-00968
INIMEX	**Belgium**	***46***	**75-04650**
International Technology SA	Spain	*366*	75-03950
Israel Military Industries Ltd	Israel	*220*	75-10300
Reutech Defence Industries	S. Africa	*359*	75-18400
Royal Ordnance	UK	*481*	75-50620

523 20 30 Bombs, aircraft, general purpose

Arab Organization for Ind.	Egypt	*109*	55-01700
Australian Defence Industries Ltd	Australia	*9*	55-01000
Avibras Aeroespacial SA	Brazil	*55*	75-00968
Daudsons Armoury	Pakistan	*315*	75-00500
Explosivos Alaveses SA	Spain	*366*	75-03050
INIMEX	**Belgium**	***46***	**75-04650**
International Technology SA	Spain	*366*	75-03950
Israel Military Industries Ltd	Israel	*220*	75-10300
Pakistan Ordnance Factories	Pakistan	*314*	55-00100
Royal Ordnance	UK	*481*	75-50620

523 20 35 Bombs, aircraft, guided (see code 511.10)

Analysis Management Systems	S. Africa	*357*	75-00770
Israel Military Industries Ltd	Israel	*220*	75-10300
Royal Ordnance	UK	*481*	75-50620

523 20 40 Bombs, aircraft, incendiary

Eurodef Aerospace & Defence	Greece	*189*	75-01295
Explosivos Alaveses SA	Spain	*366*	75-03050
INIMEX	**Belgium**	***46***	**75-04650**
International Technology SA	Spain	*366*	75-03950
Israel Military Industries Ltd	Israel	*220*	75-10300
Reutech Defence Industries	S. Africa	*359*	75-18400
Royal Ordnance	UK	*481*	75-50620
Sargent Fletcher Inc.	USA	*612*	75-78150

523 20 45 Bombs, aircraft, practice

Aerea SpA.	Italy	*230*	75-00280
Arab Organization for Ind.	Egypt	*109*	55-01700
Artefatos Eléctricos e Mecânicos	Brazil	*55*	74-00905
Australian Defence Industries Ltd	Australia	*9*	55-01000
Daudsons Armoury	Pakistan	*315*	75-00500
Étienne Lacroix	France	*138*	75-24050
Explosivos Alaveses SA	Spain	*366*	75-03050
Fabrica Española de Cartucheria SL	Spain	*366*	75-03300
Hunting Engineering Ltd	UK	*459*	75-30700
INIMEX	**Belgium**	***46***	**75-04650**
International Technology SA	Spain	*366*	75-03950
Israel Military Industries Ltd	Israel	*220*	75-10300
Kader Factory for Developed Ind.	Egypt	*109*	55-01700
National Forge Co.	USA	*598*	75-63850
Royal Ordnance	UK	*481*	75-50620
Swiss Aircraft & Systems Co	Switzerland	*384*	55-00100
Victrix Ltd	Canada	*77*	75-15000

523 20 50 Bombs, aircraft, retarded

Arab Organization for Ind.	Egypt	*109*	55-01700
Artefatos Eléctricos e Mecânicos	Brazil	*55*	74-00905
Explosivos Alaveses SA	Spain	*366*	75-03050
Hunting Engineering Ltd	UK	*459*	75-30700
International Technology SA	Spain	*366*	75-03950
Israel Military Industries Ltd	Israel	*220*	75-10300
Reutech Defence Industries	S. Africa	*359*	75-18400
Royal Ordnance	UK	*481*	75-50620
Swiss Aircraft & Systems Co	Switzerland	*384*	55-00100

523 20 55 Bombs, aircraft, runway penetration

Avibras Aeroespacial SA	Brazil	*55*	75-00968
Explosivos Alaveses SA	Spain	*366*	75-03050
Hunting Engineering Ltd	UK	*459*	75-30700
Israel Military Industries Ltd	Israel	*220*	75-10300
Kader Factory for Developed Ind.	Egypt	*109*	55-01700
Matra Défense-Espace	France	*139*	75-26000
National Forge Co.	USA	*598*	75-63850
Royal Ordnance	UK	*481*	75-50620
Swiss Aircraft & Systems Co	Switzerland	*384*	55-00100

523 20 65 Bombs, aircraft, smoke

Explosivos Alaveses SA	Spain	*366*	75-03050
INIMEX	**Belgium**	***46***	**75-04650**
Israel Military Industries Ltd	Israel	*220*	75-10300
Royal Ordnance	UK	*481*	75-50620

523 20 80 Bomb accessories, aircraft

Explosivos Alaveses SA	Spain	*366*	75-03050
INIMEX	**Belgium**	***46***	**75-04650**
International Technology SA	Spain	*366*	75-03950
Israel Military Industries Ltd	Israel	*220*	75-10300
Royal Ordnance	UK	*481*	75-50620

523 20 83 Guidance kits, bomb

Israel Military Industries Ltd	Israel	220	75-10300

523 20 86 Retarder kits, bomb

Hunting Engineering Ltd	UK	459	75-30700
International Technology SA	Spain	366	75-03950
Irvin Aerospace Limited	UK	462	75-32050
Irvin Industries Inc.	USA	586	75-48960

523 30 Bombs, naval depth

Bofors SA Marine AB.	Sweden	376	75-03040
ECA SA	France	136	75-14050
Eurometaal NV	Netherlands	292	75-02200
NORINCO	China PR	88	55-00280
Pains-Wessex Ltd	UK	472	75-43510

523 30 10 Bombs, naval depth, mortar type

Bofors AB	Sweden	375	74-00020
Bofors SA Marine AB.	Sweden	376	75-03040

523 30 20 Depth bombs, naval, drum type

Bofors SA Marine AB.	Sweden	376	75-03040

523 90 Bomb components

Australian Defence Industries Ltd	Australia	9	55-01000
Bofors SA Marine AB.	Sweden	376	75-03040
Busak + Shamban Inc.	USA	561	75-16006
Daudsons Armoury	Pakistan	315	75-00500
EG&G Reticon	USA	571	75-29562
Giat Industries	France	126	55-00100
Global Defense Products Inc.	USA	579	75-39770
Haley & Weller Ltd	UK	456	75-27400
Hollow Extrusions Ltd	UK	458	75-29652
National Forge Co.	USA	598	75-63850
Royal Ordnance	UK	481	75-50620
SNPE	France	145	75-36315

524 Submunitions, non-guided

AAI Corp.	USA	545	75-00020
Amron L.L.C.	USA	553	75-06883
Dynamit Nobel GmbH	Germany	167	75-07100
IBQ Industrias Químicas Ltda	Brazil	55	75-02700
Instalaza SA	Spain	366	75-03900
Israel Military Industries Ltd	Israel	220	75-10300
Royal Ordnance	UK	481	75-50620
Santa Barbara SA	Spain	364	55-01000
State Research & Association	Russia	334	55-18325
Textron Systems Division	USA	621	75-88655

524 20 Submunitions, air-launched

Dynamit Nobel GmbH	Germany	167	75-07100
Hunting Engineering Ltd	UK	459	75-30700
Israel Military Industries Ltd	Israel	220	75-10300
Royal Ordnance	UK	481	75-50620
State Research & Association	Russia	334	55-18325

524 20 10 Submunitions, air-launched, airfield attack

Forges de Zeebrugge SA	Belgium	46	75-03750
Hunting Engineering Ltd	UK	459	75-30700
Israel Military Industries Ltd	Israel	220	75-10300
Royal Ordnance	UK	481	75-50620

524 20 90 Components, submunitions, air-launched

Giat Industries	France	126	55-00100
Hunting Engineering Ltd	UK	459	75-30700
Israel Military Industries Ltd	Israel	220	75-10300
Moog Controls Ltd	UK	469	75-40670
Orlando Technology Inc.	USA	540	60-16000
Royal Ordnance	UK	481	75-50620

524 20 98 Warheads, submunitions, air-launched

CMS Inc.	USA	564	75-19170
Forges de Zeebrugge SA	Belgium	46	75-03750
Hunting Engineering Ltd	UK	459	75-30700
Israel Military Industries Ltd	Israel	220	75-10300
Luchaire Défense SA	France	139	75-25400
Orlando Technology Inc.	USA	540	60-16000
Royal Ordnance	UK	481	75-50620
Somchem	S. Africa	359	75-19200

524 20 98 70 Warheads, submunitions, air-launched, training

CMS Inc.	USA	564	75-19170
Eurodef Aerospace & Defence	Greece	189	75-01295
Israel Military Industries Ltd	Israel	220	75-10300
Orlando Technology Inc.	USA	540	60-16000
Royal Ordnance	UK	481	75-50620

524 30 Submunitions, naval

Israel Military Industries Ltd	Israel	220	75-10300

524 40 Submunitions, ground-based

Emco Inc.	USA	573	75-30985
IBQ Industrias Químicas Ltda	Brazil	55	75-02700
Israel Military Industries Ltd	Israel	220	75-10300

525 Detonators, fuzes, arming & safety devices

AAI Corp.	USA	545	75-00020
Arab Organization for Ind.	Egypt	109	55-01700
Arcus Co.	Bulgaria	61	55-00060
Australian Defence Industries Ltd	Australia	9	55-01000
Bowmar Instrument Corp.	USA	560	75-14860
Bressel SA.	Spain	366	75-01670
Bulova Technologies L.L.C.	USA	561	75-15600
CIC International Ltd	USA	563	75-18570
Combined Systems Inc.	USA	564	75-19893
Custom Electronics Inc.	USA	567	75-23085
Daudsons Armoury	Pakistan	315	75-00500
Davey Bickford S.N.C.	France	135	75-13000
Dynamit Nobel GmbH	Germany	167	75-07100
Dynamit Nobel Graz GmbH.	Austria	32	75-00800
EG&G Inc.	USA	571	75-29555
Eley Ltd	UK	448	75-20550
Explosivos Alaveses SA	Spain	366	75-03050
FUCHS Electronics	S. Africa	358	75-06000
Haley & Weller Ltd	UK	456	75-27400
INIMEX	**Belgium**	**46**	**75-04650**
Israel Military Industries Ltd	Israel	220	75-10300
Israel Military Industries Rocket Systems Division	**Israel**	**220**	**75-10310**
KDI Precision Products Inc.	USA	588	75-51705
Kidde-Graviner Ltd	UK	463	75-33880
Lucas Western Inc.	USA	592	75-56300
Luchaire Défense SA	France	139	75-25400
Microelectronics, guidance and	Turkey	402	55-00050
Miltec SA	Switzerland	387	75-08498
Motorola Ltd	UK	469	75-41050
OEA Aerospace Inc.	USA	601	75-66100
Oerlikon-Contraves AG.	Switzerland	387	75-09250
Pakistan Ordnance Factories.	Pakistan	314	55-00100
Reshef Technologies Ltd	Israel	222	75-18600
Rheinmetall Waffe & Munition GmbH	Germany	173	75-24950
Rotherham Precision Engineers Ltd	UK	481	75-50590
Royal Ordnance	UK	481	75-50620
SAE Alsetex.	France	142	75-33810
Santa Barbara SA	Spain	364	55-01000
Sellier & Bellot a.s.	Czech Republic	99	75-00900
Signaal USFA	Netherlands	295	75-06500
Smiths Aerospace & Defense	USA	615	75-82090
Somchem	S. Africa	359	75-19200
SPL Ordnance Ltd.	Israel	225	80-01935

Tamares Ltd	UK	*487*	75-58500
Tracor Inc	USA	*623*	75-90000
Valsella Meccanotecnica SpA	Italy	*239*	75-26800
Vammas Ltd	**Finland**	***117***	**55-02220**
Vihtavuori Oy	Finland	*119*	75-02850

525 10 Detonators

Australian Defence Industries Ltd	Australia	*9*	55-01000
CIC International Ltd	USA	*563*	75-18570
Combined Systems Inc.	USA	*564*	75-19893
Davey Bickford S.N.C.	France	*135*	75-13000
EEV Canada Ltd	Canada	*71*	75-04700
EG&G Inc.	USA	*571*	75-29555
EMS-PATVAG AG	Switzerland	*386*	75-03650
Ensign Bickford Co.	USA	*574*	75-31700
Explosivos Alaveses SA	Spain	*366*	75-03050
Greek Powder & Cartridge Co. SA	Greece	*189*	75-01500
INIMEX	**Belgium**	***46***	**75-04650**
Israel Military Industries Ltd	Israel	*220*	75-10300
Israel Military Industries Rocket Systems Division	**Israel**	***220***	**75-10310**
Merex Inc.	USA	*595*	75-60250
Oerlikon-Contraves AG.	Switzerland	*387*	75-09250
Royal Ordnance	UK	*481*	75-50620
SPL Ordnance Ltd.	Israel	*225*	80-01935
Walter Kidde Aerospace	USA	*588*	75-52500

525 10 20 Detonating cord

Brock's Explosives Ltd	UK	*437*	75-09980
Davey Bickford S.N.C.	France	*135*	75-13000
EG&G Inc.	USA	*571*	75-29555
Ensign Bickford Co.	USA	*574*	75-31700
Explosivos Alaveses SA	Spain	*366*	75-03050
Haley & Weller Ltd	UK	*456*	75-27400
Israel Military Industries Ltd	Israel	*220*	75-10300
Israel Military Industries Rocket Systems Division	**Israel**	***220***	**75-10310**
Pains-Wessex Ltd	UK	*472*	75-43510
Royal Ordnance	UK	*481*	75-50620
Société Suisse des Explosifs	Switzerland	*387*	75-13500
SPL Ordnance Ltd.	Israel	*225*	80-01935

525 20 Fuzing systems

AAI Corp.	USA	*545*	75-00020
Arcus Co.	Bulgaria	*61*	55-00060
Australian Defence Industries Ltd	Australia	*9*	55-01000
Avibras Aeroespacial SA	Brazil	*55*	75-00968
BDL Systems Ltd	UK	*433*	75-06080
Bofors SA Marine AB.	Sweden	*376*	75-03040
Bowmar Instrument Corp.	USA	*560*	75-14860
Bressel SA.	Spain	*366*	75-01670
Bulova Technologies L.L.C.	USA	*561*	75-15600
Combined Systems Inc.	USA	*564*	75-19893
Davey Bickford S.N.C.	France	*135*	75-13000
ECA SA	France	*136*	75-14050
EMS-PATVAG AG	Switzerland	*386*	75-03650
Ensign Bickford Co.	USA	*574*	75-31700
Étienne Lacroix	France	*138*	75-24050
Explosivos Alaveses SA	Spain	*366*	75-03050
FIC Corp.	USA	*575*	75-34340
Forges de Zeebrugge SA	Belgium	*46*	75-03750
FUCHS Electronics	S. Africa	*358*	75-06000
Giat Industries	France	*126*	55-00100
Graseby Dynamics Ltd	UK	*455*	75-26590
Indústrias de Defesa	Portugal	*327*	55-00950
Israel Military Industries Ltd	Israel	*220*	75-10300
Junghans Feinwerktechnik.	Germany	*170*	75-15120
KDI Precision Products Inc.	USA	*588*	75-51705
Machine Technology Ltd	UK	*465*	75-36870
Martin Electronics Inc.	USA	*594*	75-58540
Merex Inc.	USA	*595*	75-60250
Microelectronics, guidance and	Turkey	*402*	55-00050
Miltec SA	Switzerland	*387*	75-08498
Motorola Ltd	UK	*469*	75-41050
Oerlikon-Contraves AG.	Switzerland	*387*	75-09250
Reunert Technology Systems Ltd	S. Africa	*359*	75-18300
Reutech Defence Industries	S. Africa	*359*	75-18400
Royal Ordnance	UK	*481*	75-50620
Sellier & Bellot a.s.	Czech Republic	*99*	75-00900
Smiths Aerospace & Defense	USA	*615*	75-82090
Somchem	S. Africa	*359*	75-19200
SPL Ordnance Ltd.	Israel	*225*	80-01935
Tamares Ltd.	UK	*487*	75-58500
Textron Systems Division	USA	*621*	75-88655

525 20 10 Fuzing systems, ammunition

AAI Corp.	USA	*545*	75-00020
Australian Defence Industries Ltd	Australia	*9*	55-01000
Bowmar Instrument Corp.	USA	*560*	75-14860
Bressel SA.	Spain	*366*	75-01670
Bulova Technologies L.L.C.	USA	*561*	75-15600
CIC International Ltd	USA	*563*	75-18570
Combined Systems Inc.	USA	*564*	75-19893
Davey Bickford S.N.C.	France	*135*	75-13000
Emco Inc.	USA	*573*	75-30985
Explosivos Alaveses SA	Spain	*366*	75-03050
Giat Industries	France	*126*	55-00100
Graseby Dynamics Ltd	UK	*455*	75-26590
Haley & Weller Ltd	UK	*456*	75-27400
Israel Military Industries Ltd	Israel	*220*	75-10300
Junghans Feinwerktechnik.	Germany	*170*	75-15120
Miltec SA	Switzerland	*387*	75-08498
Motorola Ltd	UK	*469*	75-41050
Nottingham International	USA	*640*	80-22000
Reunert Technology Systems Ltd	S. Africa	*359*	75-18300
Royal Ordnance	UK	*481*	75-50620
SNC Industrial Tech.	Canada	*76*	75-12960
SPL Ordnance Ltd.	Israel	*225*	80-01935
Tamares Ltd.	UK	*487*	75-58500

525 20 20 Fuzing systems, bomb

AAI Corp.	USA	*545*	75-00020
Bowmar Instrument Corp.	USA	*560*	75-14860
Bulova Technologies L.L.C.	USA	*561*	75-15600
CIC International Ltd	USA	*563*	75-18570
Daudsons Armoury	Pakistan	*315*	75-00500
Davey Bickford S.N.C.	France	*135*	75-13000
Étienne Lacroix	France	*138*	75-24050
Explosivos Alaveses SA	Spain	*366*	75-03050
FUCHS Electronics	S. Africa	*358*	75-06000
Graseby Dynamics Ltd	UK	*455*	75-26590
Haley & Weller Ltd	UK	*456*	75-27400
Israel Military Industries Ltd	Israel	*220*	75-10300
Miltec SA	Switzerland	*387*	75-08498
Motorola Ltd	UK	*469*	75-41050
Nottingham International	USA	*640*	80-22000
Reunert Technology Systems Ltd	S. Africa	*359*	75-18300
Reutech Defence Industries	S. Africa	*359*	75-18400
Royal Ordnance	UK	*481*	75-50620
SPL Ordnance Ltd.	Israel	*225*	80-01935

525 20 30 Fuzing systems, grenade

AAI Corp.	USA	*545*	75-00020
Australian Defence Industries Ltd	Australia	*9*	55-01000
Bowmar Instrument Corp.	USA	*560*	75-14860
Bulova Technologies L.L.C.	USA	*561*	75-15600
Dynamit Nobel Graz GmbH.	Austria	*32*	75-00800
Giat Industries	France	*126*	55-00100
Haley & Weller Ltd	UK	*456*	75-27400
Israel Military Industries Ltd	Israel	*220*	75-10300
KDI Precision Products Inc.	USA	*588*	75-51705
Martin Electronics Inc.	USA	*594*	75-58540
Nottingham International	USA	*640*	80-22000
Royal Ordnance	UK	*481*	75-50620
SAE Alsetex.	France	*142*	75-33810
SPL Ordnance Ltd.	Israel	*225*	80-01935

525 20 40 Fuzing systems, mine

AAI Corp.	USA	*545*	75-00020
Bofors SA Marine AB.	Sweden	*376*	75-03040
Bowmar Instrument Corp.	USA	*560*	75-14860
Bulova Technologies L.L.C.	USA	*561*	75-15600
Davey Bickford S.N.C.	France	*135*	75-13000
Dynamit Nobel Graz GmbH.	Austria	*32*	75-00800
Giat Industries	France	*126*	55-00100
Graseby Dynamics Ltd	UK	*455*	75-26590

Israel Military Industries Ltd	Israel	*220*	75-10300
Leafield Engineering Ltd	UK	*464*	75-34910
Manurhin Défense	France	*139*	75-25630
Miltec SA	Switzerland	*387*	75-08498
Royal Ordnance	UK	*481*	75-50620
SAE Alsetex	France	*142*	75-33810
SPL Ordnance Ltd	Israel	*225*	80-01935
Valsella Meccanotecnica SpA	Italy	*239*	75-26800

525 20 50 Fuzing systems, missile

AAI Corp.	USA	*545*	75-00020
Bowmar Instrument Corp.	USA	*560*	75-14860
Bulova Technologies L.L.C.	USA	*561*	75-15600
Cranfield University	UK	*419*	60-00350
Davey Bickford S.N.C.	France	*135*	75-13000
EMS-PATVAG AG	Switzerland	*386*	75-03650
Ensign Bickford Co.	USA	*574*	75-31700
F.R. Ordnance International Ltd	UK	*452*	75-24480
FUCHS Electronics	S. Africa	*358*	75-06000
Graseby Dynamics Ltd	UK	*455*	75-26590
Israel Military Industries Ltd	Israel	*220*	75-10300
Junghans Feinwerktechnik	Germany	*170*	75-15120
Labtech Ltd	UK	*464*	75-34280
Leafield Engineering Ltd	UK	*464*	75-34910
Manurhin Défense	France	*139*	75-25630
Miltec SA	Switzerland	*387*	75-08498
ML Aviation Ltd	UK	*469*	75-40200
Motorola Ltd	UK	*469*	75-41050
Royal Ordnance	UK	*481*	75-50620
Shorts Missile Systems Ltd	**UK**	***483***	**75-53550**

525 20 60 Fuzing systems, rocket

AAI Corp.	USA	*545*	75-00020
Bowmar Instrument Corp.	USA	*560*	75-14860
Bressel SA	Spain	*366*	75-01670
Bulova Technologies L.L.C.	USA	*561*	75-15600
Davey Bickford S.N.C.	France	*135*	75-13000
EMS-PATVAG AG	Switzerland	*386*	75-03650
Étienne Lacroix	France	*138*	75-24050
Forges de Zeebrugge SA	Belgium	*46*	75-03750
FUCHS Electronics	S. Africa	*358*	75-06000
Giat Industries	France	*126*	55-00100
Graseby Dynamics Ltd	UK	*455*	75-26590
Israel Military Industries Ltd	Israel	*220*	75-10300
Junghans Feinwerktechnik	Germany	*170*	75-15120
Miltec SA	Switzerland	*387*	75-08498
Royal Ordnance	UK	*481*	75-50620
Smiths Aerospace & Defense	USA	*615*	75-82090
Somchem	S. Africa	*359*	75-19200

525 20 70 Fuzes

AAI Corp.	USA	*545*	75-00020
Alsons Industries (PVT) Ltd	Pakistan	*315*	75-00100
Arab Organization for Ind.	Egypt	*109*	55-01700
Arcus Co.	Bulgaria	*61*	55-00060
Australian Defence Industries Ltd	Australia	*9*	55-01000
Bowmar Instrument Corp.	USA	*560*	75-14860
Bressel SA	Spain	*366*	75-01670
Bulova Technologies L.L.C.	USA	*561*	75-15600
CET Technologies	Singapore	*348*	75-01080
Combined Systems Inc.	USA	*564*	75-19893
Cranfield University	UK	*419*	60-00350
Davey Bickford S.N.C.	France	*135*	75-13000
Diehl GmbH & Co	Germany	*167*	75-06500
Dixi-Fresard Microtechniques SA	France	*135*	75-13680
Dynamit Nobel Graz GmbH	Austria	*32*	75-00800
EMS-PATVAG AG	Switzerland	*386*	75-03650
F.R. Ordnance International Ltd	UK	*452*	75-24480
FUCHS Electronics	S. Africa	*358*	75-06000
GEC-Marconi Research Centre	UK	*419*	60-00700
Giat Industries	France	*126*	55-00100
Graseby Dynamics Ltd	UK	*455*	75-26590
Greek Powder & Cartridge Co. SA	Greece	*189*	75-01500
Indústrias de Defesa	Portugal	*327*	55-00950
INIMEX	**Belgium**	***46***	**75-04650**
International Technology SA	Spain	*366*	75-03950
Israel Military Industries Ltd	Israel	*220*	75-10300
Junghans Feinwerktechnik	Germany	*170*	75-15120
Labtech Ltd	UK	*464*	75-34280
Makina ve Kimya Endüstrisi	Turkey	*402*	55-01310
Micron Instruments Private Limited	India	*205*	75-01200
Miltec SA	Switzerland	*387*	75-08498
Motorola Ltd	UK	*469*	75-41050
Noptel Oy	Finland	*118*	75-01650
Nottingham International	USA	*640*	80-22000
Orlando Technology Inc.	USA	*540*	60-16000
Pakistan Ordnance Factories	Pakistan	*314*	55-00100
Republic Arms (Pty) Ltd	**S. Africa**	***359***	**75-18210**
Reshef Technologies Ltd	Israel	*222*	75-18600
Reunert Technology Systems Ltd	S. Africa	*359*	75-18300
Reutech Defence Industries	S. Africa	*359*	75-18400
Rheinmetall Waffe & Munition GmbH	Germany	*173*	75-24950
Royal Ordnance	UK	*481*	75-50620
Sandeep Metalcraft Pvt Ltd	India	*205*	75-01900
SPL Ordnance Ltd	Israel	*225*	80-01935
Tamares Ltd	UK	*487*	75-58500
Vammas Ltd	**Finland**	***117***	**55-02220**

525 20 70 20 Fuzes, contact

AAI Corp.	USA	*545*	75-00020
Australian Defence Industries Ltd	Australia	*9*	55-01000
Bowmar Instrument Corp.	USA	*560*	75-14860
Bressel SA	Spain	*366*	75-01670
Bulova Technologies L.L.C.	USA	*561*	75-15600
Cranfield University	UK	*419*	60-00350
Davey Bickford S.N.C.	France	*135*	75-13000
Indústrias de Defesa	Portugal	*327*	55-00950
Israel Military Industries Ltd	Israel	*220*	75-10300
Junghans Feinwerktechnik	Germany	*170*	75-15120
Labtech Ltd	UK	*464*	75-34280
Reshef Technologies Ltd	Israel	*222*	75-18600
Reunert Technology Systems Ltd	S. Africa	*359*	75-18300
Reutech Defence Industries	S. Africa	*359*	75-18400
Royal Ordnance	UK	*481*	75-50620
Shorts Missile Systems Ltd	**UK**	***483***	**75-53550**
SPL Ordnance Ltd	Israel	*225*	80-01935
Tamares Ltd	UK	*487*	75-58500
Vammas Ltd	**Finland**	***117***	**55-02220**

525 20 70 30 Fuzes, cord type

Davey Bickford S.N.C.	France	*135*	75-13000
Haley & Weller Ltd	UK	*456*	75-27400
Israel Military Industries Ltd	Israel	*220*	75-10300
Reshef Technologies Ltd	Israel	*222*	75-18600
Royal Ordnance	UK	*481*	75-50620
SPL Ordnance Ltd	Israel	*225*	80-01935

525 20 70 40 Fuzes, optical

AAI Corp.	USA	*545*	75-00020
Cranfield University	UK	*419*	60-00350
Davey Bickford S.N.C.	France	*135*	75-13000
GEC-Marconi Research Centre	UK	*419*	60-00700
Machine Technology Ltd	UK	*465*	75-36870
Reshef Technologies Ltd	Israel	*222*	75-18600
Royal Ordnance	UK	*481*	75-50620
SPL Ordnance Ltd	Israel	*225*	80-01935

525 20 70 50 Fuzes, proximity

AAI Corp.	USA	*545*	75-00020
Arab Organization for Ind.	Egypt	*109*	55-01700
Australian Defence Industries Ltd	Australia	*9*	55-01000
Bressel SA	Spain	*366*	75-01670
Bulova Technologies L.L.C.	USA	*561*	75-15600
CET Technologies	Singapore	*348*	75-01080
CIC International Ltd	USA	*563*	75-18570
Cranfield University	UK	*419*	60-00350
Davey Bickford S.N.C.	France	*135*	75-13000
Explosivos Alaveses SA	Spain	*366*	75-03050
FUCHS Electronics	S. Africa	*358*	75-06000
Giat Industries	France	*126*	55-00100
Israel Military Industries Ltd	Israel	*220*	75-10300
Junghans Feinwerktechnik	Germany	*170*	75-15120
KDI Precision Products Inc.	USA	*588*	75-51705
Labtech Ltd	UK	*464*	75-34280
Machine Technology Ltd	UK	*465*	75-36870
Microelectronics, guidance and	Turkey	*402*	55-00050

Motorola Ltd	UK	*469*	75-41050
Nottingham International	USA	*640*	80-22000
Reshef Technologies Ltd	Israel	*222*	75-18600
Reunert Technology Systems Ltd	S. Africa	*359*	75-18300
Rheinmetall Waffe & Munition GmbH	Germany	*173*	75-24950
Royal Ordnance	UK	*481*	75-50620
Shorts Missile Systems Ltd	**UK**	***483***	**75-53550**
Signaal USFA	Netherlands	*295*	75-06500
SPL Ordnance Ltd	Israel	*225*	80-01935
Tamares Ltd	UK	*487*	75-58500
Vammas Ltd	**Finland**	***117***	**55-02220**

525 20 70 60 Fuzes, time

AAI Corp.	USA	*545*	75-00020
Bowmar Instrument Corp.	USA	*560*	75-14860
Bressel SA.	Spain	*366*	75-01670
Bulova Technologies L.L.C.	USA	*561*	75-15600
CET Technologies	Singapore	*348*	75-01080
CIC International Ltd	USA	*563*	75-18570
Cranfield University	UK	*419*	60-00350
Davey Bickford S.N.C.	France	*135*	75-13000
Forges de Zeebrugge SA	Belgium	*46*	75-03750
Greek Powder & Cartridge Co. SA	Greece	*189*	75-01500
Haley & Weller Ltd	UK	*456*	75-27400
Indústrias de Defesa	Portugal	*327*	55-00950
International Technology SA	Spain	*366*	75-03950
Israel Military Industries Ltd	Israel	*220*	75-10300
Junghans Feinwerktechnik	Germany	*170*	75-15120
Miltec SA	Switzerland	*387*	75-08498
Nottingham International	USA	*640*	80-22000
Reshef Technologies Ltd	Israel	*222*	75-18600
Reunert Technology Systems Ltd	S. Africa	*359*	75-18300
Rotherham Precision Engineers Ltd	UK	*481*	75-50590
Royal Ordnance	UK	*481*	75-50620
SPL Ordnance Ltd	Israel	*225*	80-01935
Tamares Ltd	UK	*487*	75-58500

525 30 Arming & safety devices

AAI Corp.	USA	*545*	75-00020
Air-Log Ltd	UK	*429*	75-01800
Arcus Co.	Bulgaria	*61*	55-00060
Australian Defence Industries Ltd	Australia	*9*	55-01000
Bofors SA Marine AB.	Sweden	*376*	75-03040
Bowmar Instrument Corp.	USA	*560*	75-14860
Bressel SA.	Spain	*366*	75-01670
Bulova Technologies L.L.C.	USA	*561*	75-15600
CIC International Ltd	USA	*563*	75-18570
Davey Bickford S.N.C.	France	*135*	75-13000
EG&G Inc.	USA	*571*	75-29555
Emco Inc.	USA	*573*	75-30985
EMS-PATVAG AG	Switzerland	*386*	75-03650
Graseby Dynamics Ltd	UK	*455*	75-26590
Israel Military Industries Ltd	Israel	*220*	75-10300
Junghans Feinwerktechnik	Germany	*170*	75-15120
KDI Precision Products Inc.	USA	*588*	75-51705
Kidde-Graviner Ltd	UK	*463*	75-33880
Miltec SA	Switzerland	*387*	75-08498
Oerlikon-Contraves AG.	Switzerland	*387*	75-09250
Pyromeca	France	*141*	75-31200
Royal Ordnance	UK	*481*	75-50620
Sparton Technology Inc.	USA	*616*	75-82950
SPL Ordnance Ltd	Israel	*225*	80-01935
Syndex Aerospace Inc.	USA	*619*	75-85540
Textron Systems Division	USA	*621*	75-88655

525 40 Firing devices

AAI Corp.	USA	*545*	75-00020
Aircraft Tanks Ltd	UK	*429*	75-01730
Armtec Defense Products Co.	USA	*555*	75-08950
Bowmar Instrument Corp.	USA	*560*	75-14860
Bressel SA.	Spain	*366*	75-01670
Bulova Technologies L.L.C.	USA	*561*	75-15600
Cranfield University	UK	*419*	60-00350
EG&G Inc.	USA	*571*	75-29555
Explosivos Alaveses SA	Spain	*366*	75-03050
Haley & Weller Ltd	UK	*456*	75-27400
INIMEX	**Belgium**	***46***	**75-04650**
Israel Military Industries Ltd	Israel	*220*	75-10300
Leafield Engineering Ltd	UK	*464*	75-34910

Oceano Instruments (UK) Ltd	UK	*508*	80-19050
Oerlikon-Contraves AG.	Switzerland	*387*	75-09250
Pains-Wessex Ltd	UK	*472*	75-43510
Rotherham Precision Engineers Ltd	UK	*481*	75-50590
Royal Ordnance	UK	*481*	75-50620
SPL Ordnance Ltd	Israel	*225*	80-01935
Walter Kidde Aerospace	USA	*588*	75-52500

525 90 Fuze components

AAI Corp.	USA	*545*	75-00020
Bowmar Instrument Corp.	USA	*560*	75-14860
Bressel SA.	Spain	*366*	75-01670
Bulova Technologies L.L.C.	USA	*561*	75-15600
Busak + Shamban Inc.	USA	*561*	75-16006
Cold Formed Products Ltd	UK	*441*	75-14100
Combined Systems Inc.	USA	*564*	75-19893
Cranfield University	UK	*419*	60-00350
Emco Inc.	USA	*573*	75-30985
EMS-PATVAG AG	Switzerland	*386*	75-03650
Explosivos Alaveses SA	Spain	*366*	75-03050
FIC Corp.	USA	*575*	75-34340
GEC-Marconi Research Centre	UK	*419*	60-00700
Giat Industries	France	*126*	55-00100
Graseby Dynamics Ltd	UK	*455*	75-26590
Haley & Weller Ltd	UK	*456*	75-27400
Israel Military Industries Ltd	Israel	*220*	75-10300
Junghans Feinwerktechnik	Germany	*170*	75-15120
KDI Precision Products Inc.	USA	*588*	75-51705
Leafield Engineering Ltd	UK	*464*	75-34910
Nottingham International	USA	*640*	80-22000
Oerlikon-Contraves AG.	Switzerland	*387*	75-09250
Reshef Technologies Ltd	Israel	*222*	75-18600
Rotherham Precision Engineers Ltd	UK	*481*	75-50590
Royal Ordnance	UK	*481*	75-50620
SPL Ordnance Ltd	Israel	*225*	80-01935
THORN Microwave Devices Ltd	UK	*488*	75-59760

526 Explosives, powders, propellants

Accurate Energetic Systems, LLC	USA	*545*	75-00330
Armalon Ltd	UK	*431*	75-03729
Australian Defence Industries Ltd	Australia	*9*	55-01000
Bofors Explosives AB	Sweden	*376*	75-03000
Brock's Explosives Ltd	UK	*437*	75-09980
Calspan SRL Technology	USA	*540*	60-03000
Cenzin Co. Ltd	Poland	*324*	75-00200
Chartered Ammunition Industries	Singapore	*348*	75-01090
CIC International Ltd	USA	*563*	75-18570
Combat Simulation Systems (Aust)	Australia	*14*	75-03750
Companhia Brasileira de Cartuchos	Brazil	*55*	75-03150
Davey Bickford S.N.C.	France	*135*	75-13000
Dynamit Nobel GmbH	Germany	*167*	75-07100
Explosivos Alaveses SA	Spain	*366*	75-03050
Haley & Weller Ltd	UK	*456*	75-27400
IBQ Industrias Químicas Ltda	Brazil	*55*	75-02700
INIMEX	**Belgium**	***46***	**75-04650**
Israel Military Industries Ltd	Israel	*220*	75-10300
NORINCO	China PR	*88*	55-00280
Nottingham International	USA	*640*	80-22000
Oerlikon-Contraves AG.	Switzerland	*387*	75-09250
Pakistan Ordnance Factories	Pakistan	*314*	55-00100
Pyrospace	France	*141*	75-31300
Rheinmetall Waffe & Munition GmbH	Germany	*173*	75-24950
Royal Ordnance	UK	*481*	75-50620
SAE Alsetex	France	*142*	75-33810
Santa Barbara SA	Spain	*364*	55-01000
SIMA PERU	Peru	*319*	55-01950
SNPE	France	*145*	75-36315
Somchem	S. Africa	*359*	75-19200
SPL Ordnance Ltd	Israel	*225*	80-01935
Synthesia a.s.	Czech Republic	*99*	75-01000
Tracor Inc	USA	*623*	75-90000
Valsella Meccanotecnica SpA	Italy	*239*	75-26800
Vihtavuori Oy	Finland	*119*	75-02850
Wah Nobel (Pvt) Ltd	Pakistan	*315*	75-02300

526 10 Explosives

Accurate Energetic Systems, LLC	USA	*545*	75-00330
Australian Defence Industries Ltd	Australia	*9*	55-01000
Avibras Aeroespacial SA	Brazil	*55*	75-00968

Bofors AB	Sweden	375	74-00020
Bofors Explosives AB	Sweden	376	75-03000
Brock's Explosives Ltd	UK	437	75-09980
Broco Inc	USA	560	75-15260
Cenzin Co. Ltd.	Poland	324	75-00200
Clearway Explosives Services Ltd.	UK	500	80-06150
Combat Simulation Systems (Aust)	Australia	14	75-03750
Cranfield University	UK	419	60-00350
Davey Bickford S.N.C.	France	135	75-13000
Dynamit Nobel GmbH	Germany	167	75-07100
Dynamit Nobel Graz GmbH.	Austria	32	75-00800
Dyno Industrier ASA	Norway	309	75-00600
Elviemek SA.	Greece	188	75-01200
EMS-Dottikon AG.	Switzerland	386	75-03610
Explosivos Alaveses SA	Spain	366	75-03050
Greek Powder & Cartridge Co. SA	Greece	189	75-01500
Haley & Weller Ltd	UK	456	75-27400
Hellenic Arms Industry	Greece	187	55-00800
Israel Military Industries Ltd	Israel	220	75-10300
Leafield Engineering Ltd	UK	464	75-34910
Makina ve Kimya Endüstrisi	Turkey	402	55-01310
Merex Inc.	USA	595	75-60250
NanoMaterials, Inc.	USA	598	75-63590
NORINCO	China PR	88	55-00280
Nottingham International	USA	640	80-22000
Oerlikon-Contraves AG.	Switzerland	387	75-09250
Ordnance & Tactical Systems Division	USA	606	75-71472
Pains-Wessex Ltd	UK	472	75-43510
Pakistan Ordnance Factories.	Pakistan	314	55-00100
Primex Technologies, Inc	USA	606	75-71472
Reade Advanced Materials	USA	608	75-73860
Royal Ordnance	UK	481	75-50620
Santa Barbara SA	Spain	364	55-01000
SNPE	France	145	75-36315
Société Suisse des Explosifs	Switzerland	387	75-13500
Somchem	S. Africa	359	75-19200
SPL Ordnance Ltd.	Israel	225	80-01935
Synthesia a.s.	Czech Republic	99	75-01000
Vihtavuori Oy	Finland	119	75-02850
Wah Nobel (Pvt) Ltd	Pakistan	315	75-02300

526 10 50 Explosives, plastic

Bofors Explosives AB	Sweden	376	75-03000
Brock's Explosives Ltd	UK	437	75-09980
Broco Inc	USA	560	75-15260
Clearway Explosives Services Ltd.	UK	500	80-06150
Cranfield University	UK	419	60-00350
Dyno Industrier ASA	Norway	309	75-00600
Ensign Bickford Co.	USA	574	75-31700
Explosivos Alaveses SA	Spain	366	75-03050
Haley & Weller Ltd	UK	456	75-27400
IBQ Industrias Químicas Ltda	Brazil	55	75-02700
Israel Military Industries Ltd	Israel	220	75-10300
Nottingham International	USA	640	80-22000
Pains-Wessex Ltd	UK	472	75-43510
Royal Ordnance	UK	481	75-50620
Somchem	S. Africa	359	75-19200
SPL Ordnance Ltd.	Israel	225	80-01935

526 10 50 30 Charges, cutting

Brock's Explosives Ltd	UK	437	75-09980
Broco Inc	USA	560	75-15260
Explosivos Alaveses SA	Spain	366	75-03050
Haley & Weller Ltd	UK	456	75-27400
IBQ Industrias Químicas Ltda	Brazil	55	75-02700
Israel Military Industries Ltd	Israel	220	75-10300
Mining Resource Engineering Ltd	Canada	74	75-09770
Nottingham International	USA	640	80-22000
Pains-Wessex Ltd	UK	472	75-43510
Pyromeca	France	141	75-31200
Royal Ordnance	UK	481	75-50620
SAE Alsetex.	France	142	75-33810
SNC Industrial Tech.	Canada	76	75-12960
SPL Ordnance Ltd.	Israel	225	80-01935

526 10 50 32 Charges, demolition

Accurate Energetic Systems, LLC.	USA	545	75-00330
Australian Defence Industries Ltd	Australia	9	55-01000
Brock's Explosives Ltd	UK	437	75-09980
Broco Inc	USA	560	75-15260
CMS Inc.	USA	564	75-19170
Dynamit Nobel Graz GmbH.	Austria	32	75-00800
Dyno Industrier ASA	Norway	309	75-00600
Ensign Bickford Co.	USA	574	75-31700
Explosivos Alaveses SA	Spain	366	75-03050
Haley & Weller Ltd	UK	456	75-27400
IBQ Industrias Químicas Ltda	Brazil	55	75-02700
Israel Military Industries Ltd	Israel	220	75-10300
Mining Resource Engineering Ltd	Canada	74	75-09770
Nottingham International	USA	640	80-22000
Pains-Wessex Ltd	UK	472	75-43510
Royal Ordnance	UK	481	75-50620
SAE Alsetex.	France	142	75-33810
SNC Industrial Tech.	Canada	76	75-12960
SPL Ordnance Ltd.	Israel	225	80-01935
Valsella Meccanotecnica SpA	Italy	239	75-26800
Wah Nobel (Pvt) Ltd	Pakistan	315	75-02300

526 10 50 35 Charges, mine disposal

Brock's Explosives Ltd	UK	437	75-09980
CMS Inc.	USA	564	75-19170
Explosivos Alaveses SA	Spain	366	75-03050
Haley & Weller Ltd	UK	456	75-27400
Israel Military Industries Ltd	Israel	220	75-10300
Mining Resource Engineering Ltd	Canada	74	75-09770
Royal Ordnance	UK	481	75-50620
SAE Alsetex.	France	142	75-33810
SPL Ordnance Ltd.	Israel	225	80-01935
Valsella Meccanotecnica SpA	Italy	239	75-26800
Whitehead SpA	Italy	240	75-27600

526 10 50 37 Charges, slab

Australian Defence Industries Ltd	Australia	9	55-01000
Brock's Explosives Ltd	UK	437	75-09980
Israel Military Industries Ltd	Israel	220	75-10300
Pains-Wessex Ltd	UK	472	75-43510
Royal Ordnance	UK	481	75-50620
SAE Alsetex.	France	142	75-33810
SPL Ordnance Ltd.	Israel	225	80-01935

526 10 60 Explosives, sheet

Brock's Explosives Ltd	UK	437	75-09980
Broco Inc	USA	560	75-15260
Israel Military Industries Ltd	Israel	220	75-10300
Mining Resource Engineering Ltd	Canada	74	75-09770
Pains-Wessex Ltd	UK	472	75-43510
Royal Ordnance	UK	481	75-50620
SPL Ordnance Ltd.	Israel	225	80-01935

526 30 Explosive devices

Australian Defence Industries Ltd	Australia	9	55-01000
Avibras Aeroespacial SA	Brazil	55	75-00968
Brock's Explosives Ltd	UK	437	75-09980
Broco Inc	USA	560	75-15260
Chartered Ammunition Industries	Singapore	348	75-01090
CMS Inc.	USA	564	75-19170
Davey Bickford S.N.C.	France	135	75-13000
Haley & Weller Ltd	UK	456	75-27400
IBQ Industrias Químicas Ltda	Brazil	55	75-02700
Luchaire Défense SA	France	139	75-25400
Mining Resource Engineering Ltd	Canada	74	75-09770
OEA Aerospace Inc.	USA	601	75-66100
Pains-Wessex Ltd	UK	472	75-43510
Pakistan Ordnance Factories.	Pakistan	314	55-00100
Pyromeca	France	141	75-31200
Royal Ordnance	UK	481	75-50620
Santa Barbara SA	Spain	364	55-01000
SNPE	France	145	75-36315
Somchem	S. Africa	359	75-19200
SPL Ordnance Ltd.	Israel	225	80-01935
Syndex Aerospace Inc.	USA	619	75-85540

526 40 Powders

Armalon Ltd	UK	431	75-03729

Bofors Explosives AB	Sweden	*376*	75-03000
Calspan SRL Technology	USA	*540*	60-03000
Davey Bickford S.N.C.	France	*135*	75-13000
Dyno Industrier ASA	Norway	*309*	75-00600
Greek Powder & Cartridge Co. SA	Greece	*189*	75-01500
Israel Military Industries Ltd	Israel	*220*	75-10300
Merex Inc.	USA	*595*	75-60250
Muiden Chemie International BV	Netherlands	*294*	75-04600
NanoMaterials, Inc.	USA	*598*	75-63590
Nottingham International	USA	*640*	80-22000
Reade Advanced Materials	USA	*608*	75-73860
Royal Ordnance	UK	*481*	75-50620
Somchem	S. Africa	*359*	75-19200
Synthesia a.s.	Czech Republic	*99*	75-01000

526 40 50 Powders, igniters / ignition charges

Bayern-Chemie	Germany	*165*	75-02100
Calspan SRL Technology	USA	*540*	60-03000
Cranfield University	UK	*419*	60-00350
Davey Bickford S.N.C.	France	*135*	75-13000
Haley & Weller Ltd	UK	*456*	75-27400
Interstate Technical Products	UK	*461*	75-31946
Israel Military Industries Ltd	Israel	*220*	75-10300
Leafield Engineering Ltd	UK	*464*	75-34910
Magnesium Elektron	UK	*466*	75-37000
ML Aviation Ltd	UK	*469*	75-40200
Muiden Chemie International BV	Netherlands	*294*	75-04600
NanoMaterials, Inc.	USA	*598*	75-63590
Nottingham International	USA	*640*	80-22000
O.T.P. Ballistics	UK	*508*	80-19700
Pains-Wessex Ltd	UK	*472*	75-43510
Reade Advanced Materials	USA	*608*	75-73860
Royal Ordnance	UK	*481*	75-50620
Somchem	S. Africa	*359*	75-19200
The ALPOCO Group	UK	*430*	75-02762

526 50 Propellants, propellant charges

Accurate Energetic Systems, LLC	USA	*545*	75-00330
Allied Research Corp.	USA	*549*	75-04670
Armalon Ltd	UK	*431*	75-03729
Armtec Defense Products Co.	USA	*555*	75-08950
Australian Defence Industries Ltd	Australia	*9*	55-01000
Avibras Aeroespacial SA	Brazil	*55*	75-00968
Bayern-Chemie	Germany	*165*	75-02100
Bofors AB	Sweden	*375*	74-00020
Bofors Explosives AB	Sweden	*376*	75-03000
Bristol Aerospace Ltd	Canada	*69*	75-01500
Broco Inc	USA	*560*	75-15260
Calspan SRL Technology	USA	*540*	60-03000
CIC International Ltd	USA	*563*	75-18570
Daicel Chemical Industries Ltd	Japan	*250*	75-00950
Davey Bickford S.N.C.	France	*135*	75-13000
Dynamit Nobel GmbH	Germany	*167*	75-07100
Dyno Industrier ASA	Norway	*309*	75-00600
Eurometaal NV	Netherlands	*292*	75-02200
FiatAvio SpA	Italy	*234*	75-09112
Greek Powder & Cartridge Co. SA	Greece	*189*	75-01500
Israel Military Industries Ltd	Israel	*220*	75-10300
Luchaire Défense SA	France	*139*	75-25400
Makina ve Kimya Endüstrisi	Turkey	*402*	55-01310
Muiden Chemie International BV	Netherlands	*294*	75-04600
Nottingham International	USA	*640*	80-22000
Ordnance & Tactical Systems Division	USA	*606*	75-71472
Primex Physics International	USA	*605*	75-71460
Primex Technologies, Inc	USA	*606*	75-71472
Rheinmetall Waffe & Munition GmbH	Germany	*173*	75-24950
Royal Ordnance	UK	*481*	75-50620
Santa Barbara SA	Spain	*364*	55-01000
SIMA PERU	Peru	*319*	55-01950
SNPE	France	*145*	75-36315
Somchem	S. Africa	*359*	75-19200
Synthesia a.s.	Czech Republic	*99*	75-01000
Uniroyal Chemical Company Inc.	USA	*624*	75-91800
Vihtavuori Oy	Finland	*119*	75-02850

526 51 Propellants

Accurate Energetic Systems, LLC	USA	*545*	75-00330
Armalon Ltd	UK	*431*	75-03729
Australian Defence Industries Ltd	Australia	*9*	55-01000
Avibras Aeroespacial SA	Brazil	*55*	75-00968
Bofors Explosives AB	Sweden	*376*	75-03000
Broco Inc	USA	*560*	75-15260
CIC International Ltd	USA	*563*	75-18570
Daicel Chemical Industries Ltd	Japan	*250*	75-00950
Davey Bickford S.N.C.	France	*135*	75-13000
Dyno Industrier ASA	Norway	*309*	75-00600
INIMEX	**Belgium**	***46***	**75-04650**
Israel Military Industries Ltd	Israel	*220*	75-10300
Muiden Chemie International BV	Netherlands	*294*	75-04600
NanoMaterials, Inc.	USA	*598*	75-63590
Nissan Motor Co. Ltd	Japan	*254*	75-06790
NORINCO	China PR	*88*	55-00280
Nottingham International	USA	*640*	80-22000
Ordnance & Tactical Systems Division	USA	*606*	75-71472
Pakistan Ordnance Factories	Pakistan	*314*	55-00100
Primex Physics International	USA	*605*	75-71460
Primex Technologies, Inc	USA	*606*	75-71472
Reade Advanced Materials	USA	*608*	75-73860
Royal Ordnance	UK	*481*	75-50620
Somchem	S. Africa	*359*	75-19200
Synthesia a.s.	Czech Republic	*99*	75-01000

526 51 10 Propellants, liquid

Davey Bickford S.N.C.	France	*135*	75-13000
Israel Military Industries Ltd	Israel	*220*	75-10300
Primex Physics International	USA	*605*	75-71460
Royal Ordnance	UK	*481*	75-50620

526 51 20 Propellants, powdered

Accurate Energetic Systems, LLC	USA	*545*	75-00330
Armalon Ltd	UK	*431*	75-03729
Bofors Explosives AB	Sweden	*376*	75-03000
CIC International Ltd	USA	*563*	75-18570
Davey Bickford S.N.C.	France	*135*	75-13000
Dyno Industrier ASA	Norway	*309*	75-00600
Israel Military Industries Ltd	Israel	*220*	75-10300
Muiden Chemie International BV	Netherlands	*294*	75-04600
NanoMaterials, Inc.	USA	*598*	75-63590
Ordnance & Tactical Systems Division	USA	*606*	75-71472
Primex Physics International	USA	*605*	75-71460
Primex Technologies, Inc	USA	*606*	75-71472
Reade Advanced Materials	USA	*608*	75-73860
Royal Ordnance	UK	*481*	75-50620
Somchem	S. Africa	*359*	75-19200

526 51 30 Propellants, solid

Atlantic Research Corp.	USA	*555*	75-10150
Australian Defence Industries Ltd	Australia	*9*	55-01000
Avibras Aeroespacial SA	Brazil	*55*	75-00968
Bayern-Chemie	Germany	*165*	75-02100
Bofors Explosives AB	Sweden	*376*	75-03000
CIC International Ltd	USA	*563*	75-18570
Davey Bickford S.N.C.	France	*135*	75-13000
Dyno Industrier ASA	Norway	*309*	75-00600
Israel Military Industries Ltd	Israel	*220*	75-10300
Israel Military Industries Rocket Systems Division	**Israel**	***220***	**75-10310**
Muiden Chemie International BV	Netherlands	*294*	75-04600
Nissan Motor Co. Ltd	Japan	*254*	75-06790
Ordnance & Tactical Systems Division	USA	*606*	75-71472
Primex Physics International	USA	*605*	75-71460
Primex Technologies, Inc	USA	*606*	75-71472
Royal Ordnance	UK	*481*	75-50620
SIMA PERU	Peru	*319*	55-01950
Somchem	S. Africa	*359*	75-19200
Synthesia a.s.	Czech Republic	*99*	75-01000
Talley Defense Systems Inc.	USA	*619*	75-86120

526 55 Propellant charges, bagged

Allied Research Corp.	USA	*549*	75-04670
Armtec Defense Products Co.	USA	*555*	75-08950
Australian Defence Industries Ltd	Australia	*9*	55-01000
CIC International Ltd	USA	*563*	75-18570
Davey Bickford S.N.C.	France	*135*	75-13000
Dyno Industrier ASA	Norway	*309*	75-00600
Greek Powder & Cartridge Co. SA	Greece	*189*	75-01500

Israel Military Industries Ltd	Israel	*220*	75-10300
Muiden Chemie International BV	Netherlands	*294*	75-04600
Royal Ordnance	UK	*481*	75-50620
Somchem	S. Africa	*359*	75-19200

526 55 10 Propellant charges, bagged, tank ammunition

Allied Research Corp.	USA	*549*	75-04670
Armtec Defense Products Co.	USA	*555*	75-08950
CIC International Ltd	USA	*563*	75-18570
Davey Bickford S.N.C.	France	*135*	75-13000
Greek Powder & Cartridge Co. SA	Greece	*189*	75-01500
Israel Military Industries Ltd	Israel	*220*	75-10300
Muiden Chemie International BV	Netherlands	*294*	75-04600
Royal Ordnance	UK	*481*	75-50620

526 55 20 Propellant charges, foamed

Israel Military Industries Ltd	Israel	*220*	75-10300

527 Pyrotechnics & related equipment

AAI Corp.	USA	*545*	75-00020
Accurate Energetic Systems, LLC	USA	*545*	75-00330
Australian Defence Industries Ltd	Australia	*9*	55-01000
Bayern-Chemie	Germany	*165*	75-02100
BDL Systems Ltd	UK	*433*	75-06080
Brock's Explosives Ltd	UK	*437*	75-09980
Capco Inc.	USA	*562*	75-16650
Chartered Ammunition Industries	Singapore	*348*	75-01090
Combat Simulation Systems (Aust)	Australia	*14*	75-03750
Combined Systems Inc.	USA	*564*	75-19893
Davey Bickford S.N.C.	France	*135*	75-13000
D.J.Litt (Firearms) Ltd	UK	*465*	75-36080
Eley Ltd	UK	*448*	75-20550
Elviemek SA	Greece	*188*	75-01200
Étienne Lacroix	France	*138*	75-24050
Eurometaal NV	Netherlands	*292*	75-02200
FiatAvio SpA	Italy	*234*	75-09112
Haley & Weller Ltd	UK	*456*	75-27400
Heckler & Koch Inc	USA	*581*	75-42860
Hutchwilco Limited	New Zealand	*302*	75-00860
IBQ Industrias Químicas Ltda	Brazil	*55*	75-02700
Israel Military Industries Ltd	Israel	*220*	75-10300
Israel Military Industries Rocket Systems Division	**Israel**	***220***	**75-10310**
Kompanija Sloboda D D	Yugoslavia	*661*	75-00475
Luchaire Défense SA	France	*139*	75-25400
Manurhin Défense	France	*139*	75-25630
ML Wallop Defence Systems Ltd	UK	*469*	75-40240
Nico-Pyrotechnik	Germany	*173*	75-22500
Nissan Motor Co. Ltd	Japan	*254*	75-06790
Nottingham International	USA	*640*	80-22000
Pains-Wessex Ltd	UK	*472*	75-43510
Pyrospace	France	*141*	75-31300
Raufoss Technology AS	Norway	*310*	75-02200
Royal Ordnance	UK	*481*	75-50620
Ruggieri Technologie	France	*142*	75-33500
Société d'Études, Réalisations et	France	*144*	75-34885
SPL Ordnance Ltd	Israel	*225*	80-01935
Summit Machine Builders Corp.	USA	*618*	75-84900
Swartklip Products	S. Africa	*359*	75-19500
Synthesia a.s.	Czech Republic	*99*	75-01000
Tracor Inc	USA	*623*	75-90000
Valsella Meccanotecnica SpA	Italy	*239*	75-26800

527 05 Pyrotechnics, decoy (countermeasures)

Cranfield University	UK	*419*	60-00350
Davey Bickford S.N.C.	France	*135*	75-13000
Étienne Lacroix	France	*138*	75-24050
Haley & Weller Ltd	UK	*456*	75-27400
Hunting Engineering Ltd	UK	*459*	75-30700
Israel Military Industries Ltd	Israel	*220*	75-10300
ML Aviation Ltd	UK	*469*	75-40200
ML Wallop Defence Systems Ltd	UK	*469*	75-40240
Nottingham International	USA	*640*	80-22000
Pains-Wessex Ltd	UK	*472*	75-43510
Royal Ordnance	UK	*481*	75-50620
Swartklip Products	S. Africa	*359*	75-19500
Tracor Inc	USA	*623*	75-90000

527 05 20 Pyrotechnics, decoy (countermeasures), air-launched

Cranfield University	UK	*419*	60-00350
Davey Bickford S.N.C.	France	*135*	75-13000
Étienne Lacroix	France	*138*	75-24050
Israel Military Industries Ltd	Israel	*220*	75-10300
Israel Military Industries Rocket Systems Division	**Israel**	***220***	**75-10310**
Leafield Engineering Ltd	UK	*464*	75-34910
ML Aviation Ltd	UK	*469*	75-40200
ML Wallop Defence Systems Ltd	UK	*469*	75-40240
Nottingham International	USA	*640*	80-22000
Pains-Wessex Ltd	UK	*472*	75-43510
Royal Ordnance	UK	*481*	75-50620
Swartklip Products	S. Africa	*359*	75-19500
Tracor Inc	USA	*623*	75-90000

527 05 20 10 Pyrotechnics, decoy, infra-red, air-launched

Cranfield University	UK	*419*	60-00350
Davey Bickford S.N.C.	France	*135*	75-13000
Étienne Lacroix	France	*138*	75-24050
INIMEX	**Belgium**	***46***	**75-04650**
Israel Military Industries Ltd	Israel	*220*	75-10300
Israel Military Industries Rocket Systems Division	**Israel**	***220***	**75-10310**
ML Aviation Ltd	UK	*469*	75-40200
ML Wallop Defence Systems Ltd	UK	*469*	75-40240
Pains-Wessex Ltd	UK	*472*	75-43510
Royal Ordnance	UK	*481*	75-50620
Swartklip Products	S. Africa	*359*	75-19500
Tracor Inc	USA	*623*	75-90000

527 05 30 Pyrotechnics, decoy (countermeasures), ship-launched

Davey Bickford S.N.C.	France	*135*	75-13000
Étienne Lacroix	France	*138*	75-24050
Israel Military Industries Ltd	Israel	*220*	75-10300
Pains-Wessex Ltd	UK	*472*	75-43510
Tracor Inc	USA	*623*	75-90000

527 05 30 10 Pyrotechnics, decoy, infra-red, ship-launched

Davey Bickford S.N.C.	France	*135*	75-13000
Étienne Lacroix	France	*138*	75-24050
INIMEX	**Belgium**	***46***	**75-04650**
Israel Military Industries Ltd	Israel	*220*	75-10300
Pains-Wessex Ltd	UK	*472*	75-43510
Tracor Inc	USA	*623*	75-90000

527 05 40 Pyrotechnics, decoy (countermeasures), ground-launched

Davey Bickford S.N.C.	France	*135*	75-13000
Étienne Lacroix	France	*138*	75-24050
Haley & Weller Ltd	UK	*456*	75-27400
Israel Military Industries Ltd	Israel	*220*	75-10300
Pains-Wessex Ltd	UK	*472*	75-43510

527 10 Pyrotechnics, illuminating

Accurate Energetic Systems, LLC	USA	*545*	75-00330
Allied Research Corp.	USA	*549*	75-04670
Australian Defence Industries Ltd	Australia	*9*	55-01000
Bofors AB	Sweden	*375*	74-00020
Combined Systems Inc.	USA	*564*	75-19893
Comet GmbH	Germany	*166*	75-05000
Davey Bickford S.N.C.	France	*135*	75-13000
Diehl GmbH & Co	Germany	*167*	75-06500
Étienne Lacroix	France	*138*	75-24050
Haley & Weller Ltd	UK	*456*	75-27400
IBQ Industrias Químicas Ltda	Brazil	*55*	75-02700
Israel Military Industries Ltd	Israel	*220*	75-10300
Makina ve Kimya Endüstrisi	Turkey	*402*	55-01310
Merex Inc.	USA	*595*	75-60250
ML Wallop Defence Systems Ltd	UK	*469*	75-40240
Nico-Pyrotechnik	Germany	*173*	75-22500

Pains-Wessex Ltd	UK	*472*	75-43510
Pakistan Ordnance Factories	Pakistan	*314*	55-00100
Royal Ordnance	UK	*481*	75-50620
Ruggieri Technologie	France	*142*	75-33500
Simad SpA	Italy	*230*	74-23350
Swartklip Products	S. Africa	*359*	75-19500
Tracor Inc	USA	*623*	75-90000

527 10 20 Pyrotechnics, illuminating, air-launched

Davey Bickford S.N.C.	France	*135*	75-13000
Étienne Lacroix	France	*138*	75-24050
Israel Military Industries Ltd	Israel	*220*	75-10300
Merex Inc.	USA	*595*	75-60250
Pains-Wessex Ltd	UK	*472*	75-43510
Rafael	Israel	*216*	55-01800
Royal Ordnance	UK	*481*	75-50620
Valsella Meccanotecnica SpA	Italy	*239*	75-26800

527 10 30 Pyrotechnics, illuminating, ship-launched

Davey Bickford S.N.C.	France	*135*	75-13000
Étienne Lacroix	France	*138*	75-24050
Israel Military Industries Ltd	Israel	*220*	75-10300
Merex Inc.	USA	*595*	75-60250
Pains-Wessex Ltd	UK	*472*	75-43510
Sala International Pty Ltd.	Australia	*21*	75-18980

527 10 40 Pyrotechnics, illuminating, ground-launched

Allied Research Corp.	USA	*549*	75-04670
Davey Bickford S.N.C.	France	*135*	75-13000
Étienne Lacroix	France	*138*	75-24050
Haley & Weller Ltd	UK	*456*	75-27400
IBQ Industrias Químicas Ltda	Brazil	*55*	75-02700
Israel Military Industries Ltd	Israel	*220*	75-10300
Merex Inc.	USA	*595*	75-60250
Nottingham International	USA	*640*	80-22000
Pains-Wessex Ltd	UK	*472*	75-43510
Royal Ordnance	UK	*481*	75-50620
Ruggieri Technologie	France	*142*	75-33500
Swartklip Products	S. Africa	*359*	75-19500

527 10 42 Battlefield illuminating systems, ground-launched

Allied Research Corp.	USA	*549*	75-04670
Combined Systems Inc.	USA	*564*	75-19893
Haley & Weller Ltd	UK	*456*	75-27400
IBQ Industrias Químicas Ltda	Brazil	*55*	75-02700
Israel Military Industries Ltd	Israel	*220*	75-10300
Merex Inc.	USA	*595*	75-60250
Pains-Wessex Ltd	UK	*472*	75-43510
Sala International Pty Ltd.	Australia	*21*	75-18980
Swartklip Products	S. Africa	*359*	75-19500

527 15 Pyrotechnics, marking

Brock's Explosives Ltd	UK	*437*	75-09980
Comet GmbH	Germany	*166*	75-05000
Davey Bickford S.N.C.	France	*135*	75-13000
Étienne Lacroix	France	*138*	75-24050
Haley & Weller Ltd	UK	*456*	75-27400
Israel Military Industries Ltd	Israel	*220*	75-10300
MECAR SA	Belgium	*46*	75-04930
Merex Inc.	USA	*595*	75-60250
Pains-Wessex Ltd	UK	*472*	75-43510
Ruggieri Technologie	France	*142*	75-33500
SAE Alsetex.	France	*142*	75-33810
Simad SpA	Italy	*230*	74-23350
Swartklip Products	S. Africa	*359*	75-19500
Victrix Ltd	Canada	*77*	75-15000

527 15 20 Pyrotechnics, marking, air-launched

Brock's Explosives Ltd	UK	*437*	75-09980
CMS Inc.	USA	*564*	75-19170
Davey Bickford S.N.C.	France	*135*	75-13000
Étienne Lacroix	France	*138*	75-24050
Israel Military Industries Ltd	Israel	*220*	75-10300
Pains-Wessex Ltd	UK	*472*	75-43510
Royal Ordnance	UK	*481*	75-50620
Ruggieri Technologie	France	*142*	75-33500

527 15 30 Pyrotechnics, marking, ship-launched

Brock's Explosives Ltd	UK	*437*	75-09980
CMS Inc.	USA	*564*	75-19170
Davey Bickford S.N.C.	France	*135*	75-13000
Israel Military Industries Ltd	Israel	*220*	75-10300
Pains-Wessex Ltd	UK	*472*	75-43510
Ruggieri Technologie	France	*142*	75-33500

527 15 40 Pyrotechnics, marking, ground-launched

Brock's Explosives Ltd	UK	*437*	75-09980
CMS Inc.	USA	*564*	75-19170
Davey Bickford S.N.C.	France	*135*	75-13000
Étienne Lacroix	France	*138*	75-24050
Haley & Weller Ltd	UK	*456*	75-27400
Israel Military Industries Ltd	Israel	*220*	75-10300
Pains-Wessex Ltd	UK	*472*	75-43510
Ruggieri Technologie	France	*142*	75-33500
SAE Alsetex.	France	*142*	75-33810
Swartklip Products	S. Africa	*359*	75-19500

527 20 Pyrotechnics, screening

AAI Corp.	USA	*545*	75-00020
Bofors AB	Sweden	*375*	74-00020
Brock's Explosives Ltd	UK	*437*	75-09980
Davey Bickford S.N.C.	France	*135*	75-13000
Étienne Lacroix	France	*138*	75-24050
Haley & Weller Ltd	UK	*456*	75-27400
IBQ Industrias Químicas Ltda	Brazil	*55*	75-02700
Israel Military Industries Ltd	Israel	*220*	75-10300
Luchaire Défense SA	France	*139*	75-25400
MECAR SA	Belgium	*46*	75-04930
Merex Inc.	USA	*595*	75-60250
ML Wallop Defence Systems Ltd	UK	*469*	75-40240
Nico-Pyrotechnik	Germany	*173*	75-22500
Pains-Wessex Ltd	UK	*472*	75-43510
Royal Ordnance	UK	*481*	75-50620
Ruggieri Technologie	France	*142*	75-33500
SAE Alsetex.	France	*142*	75-33810
Simad SpA	Italy	*230*	74-23350
Swartklip Products	S. Africa	*359*	75-19500
Tracor Inc	USA	*623*	75-90000

527 20 20 Pyrotechnics, screening, air-launched

Davey Bickford S.N.C.	France	*135*	75-13000
Étienne Lacroix	France	*138*	75-24050
Israel Military Industries Ltd	Israel	*220*	75-10300
Merex Inc.	USA	*595*	75-60250
Rafael	Israel	*216*	55-01800
Royal Ordnance	UK	*481*	75-50620

527 20 30 Pyrotechnics, screening, ship-launched

Davey Bickford S.N.C.	France	*135*	75-13000
Étienne Lacroix	France	*138*	75-24050
Israel Military Industries Ltd	Israel	*220*	75-10300

527 20 40 Pyrotechnics, screening, ground-launched

Brock's Explosives Ltd	UK	*437*	75-09980
Davey Bickford S.N.C.	France	*135*	75-13000
Étienne Lacroix	France	*138*	75-24050
Haley & Weller Ltd	UK	*456*	75-27400
IBQ Industrias Químicas Ltda	Brazil	*55*	75-02700
Israel Military Industries Ltd	Israel	*220*	75-10300
Pains-Wessex Ltd	UK	*472*	75-43510
Pakistan Ordnance Factories.	Pakistan	*314*	55-00100
Ruggieri Technologie	France	*142*	75-33500
SAE Alsetex.	France	*142*	75-33810
Swartklip Products	S. Africa	*359*	75-19500

527 25 Pyrotechnics, signalling

Brock's Explosives Ltd	UK	437	75-09980
Comet GmbH	Germany	166	75-05000
Davey Bickford S.N.C.	France	135	75-13000
Diehl GmbH & Co	Germany	167	75-06500
Eley Ltd	UK	448	75-20550
Étienne Lacroix	France	138	75-24050
Haley & Weller Ltd	UK	456	75-27400
Heckler & Koch Inc	USA	581	75-42860
Hutchwilco Limited	New Zealand	302	75-00860
IBQ Industrias Químicas Ltda	Brazil	55	75-02700
Israel Military Industries Ltd	Israel	220	75-10300
MECAR SA	Belgium	46	75-04930
Merex Inc.	USA	595	75-60250
ML Aviation Ltd	UK	469	75-40200
ML Wallop Defence Systems Ltd	UK	469	75-40240
Nico-Pyrotechnik	Germany	173	75-22500
Pains-Wessex Ltd	UK	472	75-43510
Royal Ordnance	UK	481	75-50620
Ruggieri Technologie	France	142	75-33500
Simad SpA	Italy	230	74-23350
Tracor Inc	USA	623	75-90000

527 25 20 Pyrotechnics, signalling, air-launched

Brock's Explosives Ltd	UK	437	75-09980
Davey Bickford S.N.C.	France	135	75-13000
Étienne Lacroix	France	138	75-24050
Israel Military Industries Ltd	Israel	220	75-10300
Pains-Wessex Ltd	UK	472	75-43510
Royal Ordnance	UK	481	75-50620
Ruggieri Technologie	France	142	75-33500

527 25 30 Pyrotechnics, signalling, ship-launched

Brock's Explosives Ltd	UK	437	75-09980
Davey Bickford S.N.C.	France	135	75-13000
Étienne Lacroix	France	138	75-24050
IBQ Industrias Químicas Ltda	Brazil	55	75-02700
Israel Military Industries Ltd	Israel	220	75-10300
ML Aviation Ltd	UK	469	75-40200
Pains-Wessex Ltd	UK	472	75-43510
Ruggieri Technologie	France	142	75-33500

527 25 39 Pyrotechnics, signalling, sub-launched distress

Brock's Explosives Ltd	UK	437	75-09980
Israel Military Industries Ltd	Israel	220	75-10300
ML Wallop Defence Systems Ltd	UK	469	75-40240
Swartklip Products	S. Africa	359	75-19500

527 25 40 Pyrotechnics, signalling, ground-launched

Brock's Explosives Ltd	UK	437	75-09980
Davey Bickford S.N.C.	France	135	75-13000
Étienne Lacroix	France	138	75-24050
Haley & Weller Ltd	UK	456	75-27400
Heckler & Koch Inc	USA	581	75-42860
IBQ Industrias Químicas Ltda	Brazil	55	75-02700
Israel Military Industries Ltd	Israel	220	75-10300
Pains-Wessex Ltd	UK	472	75-43510
Royal Ordnance	UK	481	75-50620
Ruggieri Technologie	France	142	75-33500
Swartklip Products	S. Africa	359	75-19500

527 30 Pyrotechnics, training & simulation

BDL Systems Ltd	UK	433	75-06080
Brock's Explosives Ltd	UK	437	75-09980
Combat Simulation Systems (Aust)	Australia	14	75-03750
Combined Systems Inc.	USA	564	75-19893
Comet GmbH	Germany	166	75-05000
Davey Bickford S.N.C.	France	135	75-13000
Eurodef Aerospace & Defence	Greece	189	75-01295
Haley & Weller Ltd	UK	456	75-27400
Hunting Engineering Ltd	UK	459	75-30700
Israel Military Industries Ltd	Israel	220	75-10300
ML Wallop Defence Systems Ltd	UK	469	75-40240
Nico-Pyrotechnik	Germany	173	75-22500
Pains-Wessex Ltd	UK	472	75-43510
Royal Ordnance	UK	481	75-50620
Ruggieri Technologie	France	142	75-33500
SAE Alsetex	France	142	75-33810
Simad SpA	Italy	230	74-23350
Tracor Inc	USA	623	75-90000

527 30 20 Pyrotechnics, training & simulation, air-launched

Brock's Explosives Ltd	UK	437	75-09980
CMS Inc.	USA	564	75-19170
Davey Bickford S.N.C.	France	135	75-13000
Eurodef Aerospace & Defence	Greece	189	75-01295
Israel Military Industries Ltd	Israel	220	75-10300
Pains-Wessex Ltd	UK	472	75-43510

527 30 30 Pyrotechnics, training & simulation, ship-launched

Brock's Explosives Ltd	UK	437	75-09980
CMS Inc.	USA	564	75-19170
Davey Bickford S.N.C.	France	135	75-13000
Israel Military Industries Ltd	Israel	220	75-10300
ML Aviation Ltd	UK	469	75-40200
Pains-Wessex Ltd	UK	472	75-43510

527 30 40 Pyrotechnics, training & simulation, ground-launched

BDL Systems Ltd	UK	433	75-06080
Brock's Explosives Ltd	UK	437	75-09980
CMS Inc.	USA	564	75-19170
Davey Bickford S.N.C.	France	135	75-13000
Haley & Weller Ltd	UK	456	75-27400
Israel Military Industries Ltd	Israel	220	75-10300
Pains-Wessex Ltd	UK	472	75-43510
Ruggieri Technologie	France	142	75-33500
SAE Alsetex	France	142	75-33810

527 40 Pyrotechnics, crowd control

Brock's Explosives Ltd	UK	437	75-09980
Davey Bickford S.N.C.	France	135	75-13000
Étienne Lacroix	France	138	75-24050
Haley & Weller Ltd	UK	456	75-27400
Israel Military Industries Ltd	Israel	220	75-10300
ML Wallop Defence Systems Ltd	UK	469	75-40240
Nico-Pyrotechnik	Germany	173	75-22500
Pains-Wessex Ltd	UK	472	75-43510
Ruggieri Technologie	France	142	75-33500
Simad SpA	Italy	230	74-23350
SPL Ordnance Ltd	Israel	225	80-01935

527 60 Pyromechanical devices

Australian Defence Industries Ltd	Australia	9	55-01000
Bofors AB	Sweden	375	74-00020
Comet GmbH	Germany	166	75-05000
Dassault Aviation Group	France	135	75-12975
Davey Bickford S.N.C.	France	135	75-13000
EG&G Inc.	USA	571	75-29555
Étienne Lacroix	France	138	75-24050
Haley & Weller Ltd	UK	456	75-27400
Hunting Engineering Ltd	UK	459	75-30700
INIMEX	**Belgium**	**46**	**75-04650**
Israel Military Industries Ltd	Israel	220	75-10300
Merex Inc.	USA	595	75-60250
ML Wallop Defence Systems Ltd	UK	469	75-40240
Nico-Pyrotechnik	Germany	173	75-22500
Pyromeca	France	141	75-31200
Royal Ordnance	UK	481	75-50620
Ruggieri Technologie	France	142	75-33500
SNPE	France	145	75-36315
SPL Ordnance Ltd	Israel	225	80-01935
Syndex Aerospace Inc.	USA	619	75-85540
Tracor Inc	USA	623	75-90000
Universal Propulsion Co. Inc.	USA	625	75-92700

Product–Supplier Link

527 60 10 Cable cutters

Broco Inc	USA	560	75-15260
Dassault Aviation Group	France	135	75-12975
Davey Bickford S.N.C.	France	135	75-13000
ECA SA	France	136	75-14050
EG&G Inc.	USA	571	75-29555
Haley & Weller Ltd	UK	456	75-27400
Hellermann Electric	UK	458	75-28970
Israel Military Industries Ltd	Israel	220	75-10300
Israel Military Industries Rocket Systems Division	**Israel**	**220**	**75-10310**
Leafield Engineering Ltd	UK	464	75-34910
Merex Inc.	USA	595	75-60250
Pains-Wessex Ltd	UK	472	75-43510
Pyromeca	France	141	75-31200
SPL Ordnance Ltd.	Israel	225	80-01935
Syndex Aerospace Inc..	USA	619	75-85540
Universal Propulsion Co. Inc.	USA	625	75-92700

527 60 20 Explosive links

Broco Inc	USA	560	75-15260
Dassault Aviation Group	France	135	75-12975
Davey Bickford S.N.C.	France	135	75-13000
EG&G Inc.	USA	571	75-29555
Haley & Weller Ltd	UK	456	75-27400
Israel Military Industries Ltd	Israel	220	75-10300
Israel Military Industries Rocket Systems Division	**Israel**	**220**	**75-10310**
Leafield Engineering Ltd	UK	464	75-34910
Merex Inc.	USA	595	75-60250
Pyromeca	France	141	75-31200
SPL Ordnance Ltd.	Israel	225	80-01935
Syndex Aerospace Inc..	USA	619	75-85540
Universal Propulsion Co. Inc.	USA	625	75-92700

527 60 30 Actuators, special

Barber-Colman Co.	USA	557	75-11960
Combined Systems Inc.	USA	564	75-19893
Dassault Aviation Group	France	135	75-12975
Davey Bickford S.N.C.	France	135	75-13000
EG&G Inc.	USA	571	75-29555
EMS-PATVAG AG	Switzerland	386	75-03650
GEC-Marconi Aerospace Inc.	USA	577	75-37412
Haley & Weller Ltd	UK	456	75-27400
INIMEX	**Belgium**	**46**	**75-04650**
Israel Military Industries Ltd	Israel	220	75-10300
Israel Military Industries Rocket Systems Division	**Israel**	**220**	**75-10310**
Leafield Engineering Ltd	UK	464	75-34910
Lucas Aerospace Power Systems	USA	592	75-56260
Merex Inc.	USA	595	75-60250
Minebea Co. Ltd	Japan	253	75-05600
ML Wallop Defence Systems Ltd	UK	469	75-40240
Networks Electronic Corp.	USA	598	75-64260
Oceano Instruments (UK) Ltd	UK	508	80-19050
Pyromeca	France	141	75-31200
Riva Calzoni SpA	Italy	238	75-20500
Royal Ordnance	UK	481	75-50620
Ruggieri Technologie	France	142	75-33500
SPL Ordnance Ltd.	Israel	225	80-01935
Syndex Aerospace Inc..	USA	619	75-85540
Telair International	USA	620	75-87400
Teleflex-Syneravia SA	France	147	75-44470
Umbra Cuscinetti SpA	Italy	239	75-26630
Universal Propulsion Co. Inc.	USA	625	75-92700

527 70 Pyrotechnics-related equipment

Bayern-Chemie	Germany	165	75-02100
BDL Systems Ltd	UK	433	75-06080
Brock's Explosives Ltd	UK	437	75-09980
Comet GmbH	Germany	166	75-05000
Davey Bickford S.N.C.	France	135	75-13000
Haley & Weller Ltd	UK	456	75-27400
Israel Military Industries Ltd	Israel	220	75-10300
Israel Military Industries Rocket Systems Division	**Israel**	**220**	**75-10310**
Kidde-Graviner Ltd	UK	463	75-33880
Manurhin Défense	France	139	75-25630
ML Wallop Defence Systems Ltd	UK	469	75-40240
Nico-Pyrotechnik	Germany	173	75-22500
Pains-Wessex Ltd	UK	472	75-43510
Pyromeca	France	141	75-31200
Royal Ordnance	UK	481	75-50620
Société d'Études, Réalisations et	France	144	75-34885
SPL Ordnance Ltd.	Israel	225	80-01935
Summit Machine Builders Corp.	USA	618	75-84900
Universal Propulsion Co. Inc.	USA	625	75-92700

527 70 05 Pyrotechnic initiators

Bayern-Chemie	Germany	165	75-02100
BDL Systems Ltd	UK	433	75-06080
Brock's Explosives Ltd	UK	437	75-09980
Dassault Aviation Group	France	135	75-12975
Giat Industries	France	126	55-00100
Haley & Weller Ltd	UK	456	75-27400
Israel Military Industries Ltd	Israel	220	75-10300
Israel Military Industries Rocket Systems Division	**Israel**	**220**	**75-10310**
Manurhin Défense	France	139	75-25630
Merex Inc.	USA	595	75-60250
Oceano Instruments (UK) Ltd	UK	508	80-19050
Pains-Wessex Ltd	UK	472	75-43510
Pyromeca	France	141	75-31200
Royal Ordnance	UK	481	75-50620
Société d'Études, Réalisations et	France	144	75-34885
SPL Ordnance Ltd.	Israel	225	80-01935
Universal Propulsion Co. Inc.	USA	625	75-92700

527 70 10 Pyrotechnic launchers

Bayern-Chemie	Germany	165	75-02100
Davey Bickford S.N.C.	France	135	75-13000
Haley & Weller Ltd	UK	456	75-27400
Israel Military Industries Ltd	Israel	220	75-10300
Israel Military Industries Rocket Systems Division	**Israel**	**220**	**75-10310**
Leafield Engineering Ltd	UK	464	75-34910
Merex Inc.	USA	595	75-60250
Pains-Wessex Ltd	UK	472	75-43510
Pyromeca	France	141	75-31200
Royal Ordnance	UK	481	75-50620
Strachan & Henshaw Ltd.	UK	512	80-47400
Universal Propulsion Co. Inc.	USA	625	75-92700

527 70 10 50 Pistols, signalling

Haley & Weller Ltd	UK	456	75-27400
Heckler & Koch Inc	USA	581	75-42860
Helston Gunsmiths	UK	458	75-29010
Israel Military Industries Ltd	Israel	220	75-10300
Pains-Wessex Ltd	UK	472	75-43510

527 70 20 Pyrotechnic line-throwers

Allied Research Corp.	USA	549	75-04670
Davey Bickford S.N.C.	France	135	75-13000
Étienne Lacroix	France	138	75-24050
Haley & Weller Ltd	UK	456	75-27400
Israel Military Industries Ltd	Israel	220	75-10300
Pains-Wessex Ltd	UK	472	75-43510

Mission Support Facilities & Equipment

Fixed Base Facilities & Equipment

610 20 Air bases, airports, turnkey

Airport Engineering Hugo Marom Ltd.	Israel	223	80-00100
Campenon Bernard SGE	France	133	75-07780
CORIS	France	134	75-12450
Costain Middle East	UAE	407	75-00300
Fluor Daniel Inc.	USA	635	80-11720

John Laing International Ltd	UK	*505*	80-13870
Metis Construction & Trade Co. Inc.	Turkey	*403*	80-01330
MFD International Ltd.	UK	*506*	80-16450
Sverdrup Corp.	USA	*643*	80-30650
Taylor Woodrow Construction Ltd	UK	*513*	80-50300
Taylor Woodrow Management	UK	*513*	80-50400
Thorn DNT Airfield Lighting Pty Ltd	Australia	*22*	75-20540

610 30 Naval bases, turnkey

Campenon Bernard SGE	France	*133*	75-07780
Costain Civil Engineering Ltd.	UK	*443*	75-15213
Costain Middle East	UAE	*407*	75-00300
Fluor Daniel Inc.	USA	*635*	80-11720
Graflex Inc	USA	*579*	75-40310
John Laing International Ltd	UK	*505*	80-13870
Metis Construction & Trade Co. Inc.	Turkey	*403*	80-01330
MFD International Ltd.	UK	*506*	80-16450
Sverdrup Corp.	USA	*643*	80-30650
Taylor Woodrow Construction Ltd	UK	*513*	80-50300
Taylor Woodrow Management	UK	*513*	80-50400
Weatherhaven	Canada	*77*	75-15220

610 40 Army bases, turnkey

Campenon Bernard SGE	France	*133*	75-07780
Costain Civil Engineering Ltd.	UK	*443*	75-15213
Costain Middle East	UAE	*407*	75-00300
Fluor Daniel Inc.	USA	*635*	80-11720
John Laing International Ltd	UK	*505*	80-13870
Metis Construction & Trade Co. Inc.	Turkey	*403*	80-01330
MFD International Ltd.	UK	*506*	80-16450
PRC Inc.	USA	*640*	80-24800
Sverdrup Corp.	USA	*643*	80-30650
Taylor Woodrow Construction Ltd	UK	*513*	80-50300
Taylor Woodrow Management	UK	*513*	80-50400
Weatherhaven	Canada	*77*	75-15220

611 Buildings & structures, permanent / semi-permanent & temporary

Airport Engineering Hugo Marom Ltd.	Israel	*223*	80-00100
Armalon Ltd.	UK	*431*	75-03729
Bachmann SA	France	*132*	75-04900
Balmoral Group Ltd	UK	*433*	75-05420
Caledonian Mining Co. Ltd	UK	*438*	75-11250
Campenon Bernard SGE	France	*133*	75-07780
Costain Middle East	UAE	*407*	75-00300
Electronic Space Systems Corp.	USA	*572*	75-30100
Fluor Daniel Inc.	USA	*635*	80-11720
GEL Engineering Ltd	UK	*454*	75-25500
Gontrailer SL	Spain	*367*	75-05660
Hyundai	Korea (ROK)	*265*	75-01540
Indal Technologies Inc	Canada	*72*	75-07740
John Laing International Ltd	UK	*505*	80-13870
Karl International Ltd	UK	*463*	75-33240
Metis Construction & Trade Co. Inc.	Turkey	*403*	80-01330
MFD International Ltd.	UK	*506*	80-16450
Morgan Marine Ltd	UK	*469*	75-40865
NNC Ltd	UK	*507*	80-18500
Rollalong.	UK	*479*	75-50180
Rubb Buildings Ltd	UK	*481*	75-50750
Siemens Plessey Systems	UK	*483*	75-53770
Sprung Instant Structures Ltd	Canada	*77*	75-13200
Sverdrup Corp.	USA	*643*	80-30650
Taylor Woodrow Construction Ltd	UK	*513*	80-50300
Weatherhaven	Canada	*77*	75-15220

611 10 Buildings & structures, permanent / semi-permanent

Armalon Ltd.	UK	*431*	75-03729
Balmoral Group Ltd	UK	*433*	75-05420
Caledonian Mining Co. Ltd	UK	*438*	75-11250
COES (Derby) Ltd	UK	*441*	75-14070
Electronic Space Systems Corp.	USA	*572*	75-30100
Fluor Daniel Inc.	USA	*635*	80-11720
Gontrailer SL	Spain	*367*	75-05660
John Laing International Ltd	UK	*505*	80-13870
Metis Construction & Trade Co. Inc.	Turkey	*403*	80-01330
MFD International Ltd.	UK	*506*	80-16450
Morgan Marine Ltd	UK	*469*	75-40865
NNC Ltd	UK	*507*	80-18500
Rollalong.	UK	*479*	75-50180
Shenango Steel Buildings Inc.	USA	*614*	75-80205
Sprung Instant Structures Ltd	Canada	*77*	75-13200
Sverdrup Corp.	USA	*643*	80-30650
Taylor Woodrow Construction Ltd	UK	*513*	80-50300
Taylor Woodrow Management	UK	*513*	80-50400
Weatherhaven	Canada	*77*	75-15220

611 11 Buildings & structures, conventional

Armalon Ltd.	UK	*431*	75-03729
Fluor Daniel Inc.	USA	*635*	80-11720
Gontrailer SL	Spain	*367*	75-05660
John Laing International Ltd	UK	*505*	80-13870
Metis Construction & Trade Co. Inc.	Turkey	*403*	80-01330
MFD International Ltd.	UK	*506*	80-16450
Sverdrup Corp.	USA	*643*	80-30650
Taylor Woodrow Construction Ltd	UK	*513*	80-50300
Taylor Woodrow Management	UK	*513*	80-50400

611 12 Buildings & structures, prefabricated concrete

Caledonian Mining Co. Ltd	UK	*438*	75-11250
Fluor Daniel Inc.	USA	*635*	80-11720
Gontrailer SL	Spain	*367*	75-05660
Hyundai	Korea (ROK)	*265*	75-01540
John Laing International Ltd	UK	*505*	80-13870
Metis Construction & Trade Co. Inc.	Turkey	*403*	80-01330
MFD International Ltd.	UK	*506*	80-16450
Sverdrup Corp.	USA	*643*	80-30650
Taylor Woodrow Management	UK	*513*	80-50400

611 12 10 Revetments, protective modular

Airport Engineering Hugo Marom Ltd.	Israel	*223*	80-00100
Fluor Daniel Inc.	USA	*635*	80-11720
John Laing International Ltd	UK	*505*	80-13870
Metis Construction & Trade Co. Inc.	Turkey	*403*	80-01330
MFD International Ltd.	UK	*506*	80-16450
Spantech Pty Ltd	Australia	*21*	75-19573
Sverdrup Corp.	USA	*643*	80-30650
Taylor Woodrow Management	UK	*513*	80-50400

611 12 20 Emplacements, hardened

Caledonian Mining Co. Ltd	UK	*438*	75-11250
CORIS	France	*134*	75-12450
Fluor Daniel Inc.	USA	*635*	80-11720
John Laing International Ltd	UK	*505*	80-13870
Metis Construction & Trade Co. Inc.	Turkey	*403*	80-01330
MFD International Ltd.	UK	*506*	80-16450
Sverdrup Corp.	USA	*643*	80-30650
Taylor Woodrow Management	UK	*513*	80-50400

611 13 Buildings & structures, steel-framed

Armalon Ltd.	UK	*431*	75-03729
C. Rokas SA	Greece	*191*	75-02310
Commercial Intertech Corp.	USA	*564*	75-20050
Electronic Space Systems Corp.	USA	*572*	75-30100
John Laing International Ltd	UK	*505*	80-13870
Metis Construction & Trade Co. Inc.	Turkey	*403*	80-01330
MFD International Ltd.	UK	*506*	80-16450
Rollalong.	UK	*479*	75-50180
Shenango Steel Buildings Inc.	USA	*614*	75-80205
Sprung Instant Structures Ltd	Canada	*77*	75-13200
Sverdrup Corp.	USA	*643*	80-30650
Taylor Woodrow Management	UK	*513*	80-50400
Weatherhaven	Canada	*77*	75-15220

611 14 Buildings & structures, wooden

Armalon Ltd.	UK	*431*	75-03729
Hyundai	Korea (ROK)	*265*	75-01540
John Laing International Ltd	UK	*505*	80-13870
MFD International Ltd.	UK	*506*	80-16450

Rollalong	UK	479	75-50180
Sverdrup Corp.	USA	643	80-30650

611 20 Buildings & structures, temporary

Aimsco Inc.	USA	547	75-03050
COES (Derby) Ltd	UK	441	75-14070
Gontrailer SL	Spain	367	75-05660
John Laing International Ltd	UK	505	80-13870
Metis Construction & Trade Co. Inc.	Turkey	403	80-01330
MFD International Ltd.	UK	506	80-16450
Rollalong	UK	479	75-50180
Sprung Instant Structures Ltd	Canada	77	75-13200
Sverdrup Corp.	USA	643	80-30650
Weatherhaven	Canada	77	75-15220

611 21 Buildings & structures, lightweight framed (see code 621.10)

COES (Derby) Ltd	UK	441	75-14070
Electronic Space Systems Corp.	USA	572	75-30100
John Laing International Ltd	UK	505	80-13870
Metis Construction & Trade Co. Inc.	Turkey	403	80-01330
MFD International Ltd.	UK	506	80-16450
Morgan Marine Ltd	UK	469	75-40865
Sprung Instant Structures Ltd	Canada	77	75-13200
Sverdrup Corp.	USA	643	80-30650
Taylor Woodrow Management	UK	513	80-50400
Technical Products Group Inc.	USA	620	75-86764
Weatherhaven	Canada	77	75-15220

611 22 Buildings & structures, inflatable

Autoflug GmbH & Co.	Germany	165	75-01400
Electronic Space Systems Corp.	USA	572	75-30100
Eurovinil Industries SpA	Italy	234	75-08400
Weatherhaven	Canada	77	75-15220

611 23 Buildings & structures, transportable

Balmoral Group Ltd	UK	433	75-05420
COES (Derby) Ltd	UK	441	75-14070
Electronic Space Systems Corp.	USA	572	75-30100
Engineered Air Systems Inc.	USA	573	75-31560
Fluor Daniel Inc.	USA	635	80-11720
Gontrailer SL	Spain	367	75-05660
John Laing International Ltd	UK	505	80-13870
MFD International Ltd.	UK	506	80-16450
Morgan Marine Ltd	UK	469	75-40865
Rollalong	UK	479	75-50180
Sprung Instant Structures Ltd	Canada	77	75-13200
Technical Products Group Inc.	USA	620	75-86764
Weatherhaven	Canada	77	75-15220

611 23 10 Portable accommodation units

Balmoral Group Ltd	UK	433	75-05420
Engineered Air Systems Inc.	USA	573	75-31560
Fluor Daniel Inc.	USA	635	80-11720
H.C. Slingsby plc	UK	484	75-54430
John Laing International Ltd	UK	505	80-13870
MFD International Ltd.	UK	506	80-16450
Morgan Marine Ltd	UK	469	75-40865
Rollalong	UK	479	75-50180
Sprung Instant Structures Ltd	Canada	77	75-13200
Technical Products Group Inc.	USA	620	75-86764
Weatherhaven	Canada	77	75-15220

611 90 Components, building

Air International Transit	Australia	11	75-00450
Armalon Ltd	UK	431	75-03729
Fluor Daniel Inc.	USA	635	80-11720
G Kull AG, Lunor	Switzerland	386	75-07200
National Bullet Proof Inc.	USA	598	75-63800
Shenango Steel Buildings Inc.	USA	614	75-80205

611 90 10 Safety signs, building

Focal Signs Ltd	UK	451	75-24095

611 90 30 Doors, aircraft hangar

Booth Industries Limited	UK	435	75-07670
C. Rokas SA	Greece	191	75-02310
Envirodoor Markus Ltd	UK	449	75-21090
Fluor Daniel Inc.	USA	635	80-11720
Indal Technologies Inc	Canada	72	75-07740

612 Operations support facilities & equipment, fixed base

ADE-HML Limited	UK	427	75-00710
Air-Log Ltd	UK	429	75-01800
Algram Group Ltd	UK	429	75-02625
ATS Aerospace Inc.	Canada	68	75-00752
Australian Defence Industries Ltd	Australia	9	55-01000
Bachmann SA	France	132	75-04900
Calspan SRL Technology	USA	540	60-03000
Campenon Bernard SGE	France	133	75-07780
Cincinnati Electronics Corp.	USA	564	75-18640
Clark Masts Teksam Ltd	UK	440	75-13550
Columbus Jack Corp.	USA	564	75-19850
Conrac Elektron GmbH	Germany	166	75-05100
CORIS	France	134	75-12450
Fenner Fluid Power	UK	450	75-22550
Fluor Daniel Inc.	USA	635	80-11720
Fuji Heavy Industries Ltd	Japan	251	75-01100
GAAM Fire & Rescue	Australia	16	75-07000
GEL Engineering Ltd	UK	454	75-25500
Gilbert Gilkes & Gordon Ltd	UK	454	75-25430
Haux-Life-Support GmbH	Germany	169	75-11970
Howden Aircontrol Ltd	UK	459	75-30240
Hunting Aviation Ltd	UK	504	80-12785
Hydroscience Technologies Inc.	USA	584	75-45550
Indal Technologies Inc	Canada	72	75-07740
James Hardie Building Services Ltd	Australia	17	75-10178
Litton Canada Ltd	Canada	73	75-09090
Litton Industries Inc.	USA	590	75-55400
Marshall of Cambridge Aerospace Ltd	UK	466	75-37530
Metalite Aviation Lighting	UK	467	75-38750
Northrop Grumman Corporation	USA	599	75-65450
Park Air Electronics Ltd	UK	472	75-43750
Pilatus Aircraft Ltd	Switzerland	387	75-10100
Pronal SA	France	141	75-31100
Raytheon E-Systems	USA	608	75-73302
Reynolds Boughton Ltd	UK	479	75-49340
Rockwell	USA	610	75-75930
Saval NV	Belgium	47	75-05650
Seaward International Inc.	USA	613	75-78940
ShinMaywa Industries Ltd	Japan	254	75-07800
Siemens Plessey Systems	UK	483	75-53770
Snorkel	USA	616	75-82170
Static Engineering Pty Ltd	Australia	21	75-19780
Sverdrup Corp.	USA	643	80-30650
Sverdrup Technology Inc.	USA	643	80-30680
The Boeing Company	USA	559	75-14250
THORN Lighting Ltd.	UK	488	75-59750
United Defense L.P.	USA	624	75-92050

612 10 Operations support facilities & equipment, space base

British Sarozal Ltd.	UK	436	75-09570
Calspan SRL Technology	USA	540	60-03000
Fluor Daniel Inc.	USA	635	80-11720
GMV SA	Spain	367	80-00800
Lockheed Martin Corp.	USA	591	75-55635
Logica UK Limited	UK	506	80-14350
Matra Marconi Space UK, Ltd	UK	425	74-06900
Siemens Plessey Systems	UK	483	75-53770
Sverdrup Corp.	USA	643	80-30650
Sverdrup Technology Inc.	USA	643	80-30680

612 20 Operations support facilities & equipment, air base / airport

ADE-HML Limited	UK	427	75-00710

Advanced Ground Systems Engineering	USA	546	75-00905
Air-Log Ltd	UK	429	75-01800
Airport Engineering Hugo Marom Ltd	Israel	223	80-00100
Ansul Incorporated	USA	554	75-07810
Arvico Corporation	USA	555	75-09240
Australian Defence Industries Ltd	Australia	9	55-01000
British Filters Ltd	UK	436	75-09030
British Sarozal Ltd	UK	436	75-09570
Carmichael International Ltd	UK	438	75-11700
Columbus Jack Corp.	USA	564	75-19850
Emergency One Inc.	USA	573	75-31100
Fluor Daniel Inc.	USA	635	80-11720
General Electrodynamics Corp.	USA	578	75-38500
Howden Aircontrol Ltd	UK	459	75-30240
Litton Industries Inc.	USA	591	75-55400
Logica UK Limited	UK	506	80-14350
Luminator Aircraft Products	USA	593	75-56450
Northrop Grumman Corporation	USA	599	75-65450
Park Air Electronics Ltd	UK	472	75-43750
Raytheon E-Systems	USA	608	75-73302
Saval NV	Belgium	47	75-05650
Sverdrup Corp.	USA	643	80-30650
Sverdrup Technology Inc.	USA	643	80-30680
Taylor Woodrow Management	UK	513	80-50400

612 20 10 Airfield runway-related systems & equipment

ABB Leitungsbau GmbH	Germany	164	75-00055
Aeronautical & General Instr. Ltd	UK	428	75-01050
CORIS	France	134	75-12450
ERA a.s.	Czech Republic	98	75-00400
Presentey Engineering	Canada	75	75-11500
Racal Avionics Ltd	UK	477	75-47660
Schmidt UK Limited	UK	482	75-52070
Sverdrup Corp.	USA	643	80-30650
Tapley Instrumentation	UK	487	75-58650
Thorn DNT Airfield Lighting Pty Ltd	Australia	22	75-20540

612 20 10 05 Aircraft arrester barriers, air base

Aérazur	France	131	75-00500
Marshall of Cambridge Aerospace Ltd	UK	466	75-37530
SpanSet Inter AG	Switzerland	387	75-13550

612 20 10 15 Airfield runway snow clearance equipment

Aveling Barford (Machines) plc	UK	432	75-04550
Bunce (Ashbury) Ltd	UK	438	75-10870

612 20 10 15 50 Airfield runway snow ploughs

Aveling Barford (Machines) plc	UK	432	75-04550
Bunce (Ashbury) Ltd	UK	438	75-10870
EFE Endüstri ve Ticaret A.S.	Turkey	402	75-00530

612 20 10 20 Airfield runway de-icing equipment

Bunce (Ashbury) Ltd	UK	438	75-10870

612 20 10 30 Airfield runway sweepers

Bunce (Ashbury) Ltd	UK	438	75-10870
Schmidt UK Limited	UK	482	75-52070

612 20 10 32 Airfield runway rubber deposit removal machines

Elf Atochem	Australia	16	75-05630
Von Arx Ltd	Switzerland	388	75-14750

612 20 10 35 Airfield runway grooving machines

Von Arx Ltd	Switzerland	388	75-14750

612 20 10 70 Runway visual range (RVR) systems, airfield

Aeronautical & General Instr. Ltd	UK	428	75-01050
Jenoptik Impulsphysik GmbH	Germany	170	75-14800
Thorn DNT Airfield Lighting Pty Ltd	Australia	22	75-20540

612 20 20 Air traffic control systems & equipment, air base

Aeronautical & General Instr. Ltd	UK	428	75-01050
Ausrire Scientific & Research	Russia	335	60-00090
CORIS	France	134	75-12450
Datamat Ingegneria dei Sistemi SpA	Italy	233	75-06500
Denro Inc.	USA	568	75-25651
EDS Defence Ltd	UK	501	80-09200
ERA a.s.	Czech Republic	98	75-00400
Ericsson Defence Systems Pty Ltd	Australia	16	75-05820
Fernau Avionics Ltd	UK	450	75-22600
Firan Corp.	Canada	71	75-05400
INDRA	Spain	366	75-03800
Matrix Systems Ltd	Canada	73	75-09507
Park Air Electronics Ltd	UK	472	75-43750
Presentey Engineering	Canada	75	75-11500
Racal Avionics Ltd	UK	477	75-47660
Redifon MEL Limited	UK	478	75-48530
Telegenix Inc.	USA	621	75-87610

612 20 20 70 Air traffic control towers, fixed base

Alifabs Design & Construction Ltd	UK	496	80-01000
CORIS	France	134	75-12450
Datamat Ingegneria dei Sistemi SpA	Italy	233	75-06500
Denro Inc.	USA	568	75-25651
Fluor Daniel Inc.	USA	635	80-11720
Park Air Electronics Ltd	UK	472	75-43750
Sverdrup Corp.	USA	643	80-30650

612 20 20 75 Air traffic control towers, mobile (see code 625.20.20.75)

Alifabs Design & Construction Ltd	UK	496	80-01000
ATS Aerospace Inc.	Canada	68	75-00752
CSC (Specialised Vehicles) Ltd	UK	444	75-16220
Denro Inc.	USA	568	75-25651
Park Air Electronics Ltd	UK	472	75-43750

612 20 20 90 Air traffic control equipment, air base

Ausrire Scientific & Research	Russia	335	60-00090
Becker Flugfunkwerk GmbH	Germany	165	75-02200
CORIS	France	134	75-12450
ERA a.s.	Czech Republic	98	75-00400
Fernau Avionics Ltd	UK	450	75-22600
Jotron (UK) Ltd	UK	462	75-32895
Laser-Scan	UK	506	80-14210
Racal Avionics Ltd	UK	477	75-47660
Racal Recorders Ltd	UK	477	75-47689
Telegenix Inc.	USA	621	75-87610

612 20 20 90 20 Air traffic control consoles

Alifabs Design & Construction Ltd	UK	496	80-01000
ATS Aerospace Inc.	Canada	68	75-00752
Becker Flugfunkwerk GmbH	Germany	165	75-02200
CORIS	France	134	75-12450
Graphico Electronics Group	Canada	71	75-06460
MacDonald Dettwiler and Associates	Canada	78	80-01310
Presentey Engineering	Canada	75	75-11500
Telegenix Inc.	USA	621	75-87610

612 20 25 Airfield movement control systems & equipment

ERA a.s.	Czech Republic	98	75-00400
Presentey Engineering	Canada	75	75-11500

Product-Supplier Link

612 20 25 90 Airfield movement control equipment

ERA a.s.	Czech Republic	98	75-00400

612 20 30 Meteorological systems & equipment, air base

Advanced Technical Corp.	USA	546	75-01000
Alden Electronics Inc.	USA	548	75-04350
Ausrire Scientific & Research	Russia	335	60-00090
Becker Flugfunkwerk GmbH	Germany	165	75-02200
Belfort Instrument Co.	USA	558	75-12800
Brookes & Gatehouse Ltd	UK	437	75-10100
Catsaros Automation SA	Greece	187	75-00700
CORIS	France	134	75-12450
Datamat Ingegneria dei Sistemi SpA	Italy	233	75-06500
Feedback Instruments	UK	450	75-22520
Fluor Daniel Inc.	USA	635	80-11720
Hall & Watts (Defence Optics) Group	**UK**	**456**	**75-27650**
Hobeco Sociedade Consignatária Ltda	Brazil	56	75-08600
Jenoptik Impulsphysik GmbH	Germany	170	75-14800
Jotron (UK) Ltd	UK	462	75-32895
Meggitt Avionics (UK)	UK	467	75-38340
Remtech SA	France	141	75-32230
Thorn DNT Airfield Lighting Pty Ltd	Australia	22	75-20540
Vaisala Artais Division	USA	625	75-93087
Vaisala Oy	Finland	119	75-02800

612 20 30 90 Meteorological equipment, air base

Advanced Technical Corp.	USA	546	75-01000
Alden Electronics Inc.	USA	548	75-04350
Brookes & Gatehouse Ltd	UK	437	75-10100
Catsaros Automation SA	Greece	187	75-00700
Datamat Ingegneria dei Sistemi SpA	Italy	233	75-06500
Hall & Watts (Defence Optics) Group	**UK**	**456**	**75-27650**
Jenoptik Impulsphysik GmbH	Germany	170	75-14800
Jotron (UK) Ltd	UK	462	75-32895
Meggitt Avionics (UK)	UK	467	75-38340
Thorn DNT Airfield Lighting Pty Ltd	Australia	22	75-20540
Vaisala Oy	Finland	119	75-02800

612 20 40 Fuel, oil & lubricants storage systems, airfield

British Filters Ltd	UK	436	75-09030
Fluor Daniel Inc.	USA	635	80-11720
GNY Equipment L.L.C.	USA	579	75-39950
Metis Construction & Trade Co. Inc.	Turkey	403	80-01330
SEI Industries Ltd	Canada	76	75-12750
Sverdrup Corp.	USA	643	80-30650
Taylor Woodrow Construction Ltd	UK	513	80-50300

612 20 40 70 Fuel, oil & lubricants storage equipment, air base

British Filters Ltd	UK	436	75-09030
Fluor Daniel Inc.	USA	635	80-11720
GNY Equipment L.L.C.	USA	579	75-39950
SEI Industries Ltd	Canada	76	75-12750
Titan Aviation	France	148	75-45200

612 20 40 70 50 Fuel, oil & lubricant pumps, air base

Fenner Fluid Power	UK	450	75-22550
Gilbert Gilkes & Gordon Ltd	UK	454	75-25430
GNY Equipment L.L.C.	USA	579	75-39950
Hamworthy Marine	UK	457	75-27850
Ingersoll-Dresser Pumps (UK) Ltd	UK	461	75-31450
PD Technical Mouldings plc	UK	473	75-44200
Plenty Ltd	UK	475	75-46050
SIHI Australia Pty Ltd	Australia	21	75-19505
Weir Engineering Pty Ltd	Australia	23	75-23300

612 20 40 70 60 Fuel, oil & lubricants storage tanks, air base

GNY Equipment L.L.C.	USA	579	75-39950
Metis Construction & Trade Co. Inc.	Turkey	403	80-01330
PD Technical Mouldings plc	UK	473	75-44200
Pronal SA	France	141	75-31100
SEI Industries Ltd	Canada	76	75-12750
SIHI Australia Pty Ltd	Australia	21	75-19505
White Cross Rubber Products Ltd.	UK	494	75-65857

612 20 43 Fuel, oil & lubricants delivery systems, air base

Aljac Engineering Ltd	UK	429	75-02650
British Filters Ltd	UK	436	75-09030
Dabico Inc.	USA	567	75-23145
GNY Equipment L.L.C.	USA	579	75-39950
Pronal SA	France	141	75-31100
SIHI Australia Pty Ltd	Australia	21	75-19505
Taylor Woodrow Construction Ltd	UK	513	80-50300
Titan Aviation	France	148	75-45200
Victaulic Co. of America	USA	626	75-93840

612 20 43 40 Hydrant refuelling systems, air base

Dabico Inc.	USA	567	75-23145
GEC-Marconi Avionics Ltd	UK	454	75-25320
MITM Consulting Engineers	Israel	224	80-01350
SIHI Australia Pty Ltd	Australia	21	75-19505
Sverdrup Corp.	USA	643	80-30650
Titan Aviation	France	148	75-45200
Victaulic Co. of America	USA	626	75-93840

612 20 43 70 Fuel, oil & lubricants delivery equipment, air base

ADE-HML Limited	UK	427	75-00710
Aeroquip Corp.	USA	547	75-02350
British Filters Ltd	UK	436	75-09030
Dabico Inc.	USA	567	75-23145
GEC-Marconi Avionics Ltd	UK	454	75-25320
Gilbert Gilkes & Gordon Ltd	UK	454	75-25430
GNY Equipment L.L.C.	USA	579	75-39950
Premier Fuel Systems Limited	UK	475	75-46655
SIHI Australia Pty Ltd	Australia	21	75-19505
Titan Aviation	France	148	75-45200

612 20 45 Airfield approach & landing aids

ABB Leitungsbau GmbH	Germany	164	75-00055
Aeronautical & General Instr. Ltd	UK	428	75-01050
Alcatel Air Navigation Systems GmbH	Germany	164	75-00535
AMA Pvte Ltd	India	204	75-00130
Canadian Marconi Co.	Canada	69	75-02600
Compagnie Industrielle des Lasers	France	134	75-10240
CORIS	France	134	75-12450
ERNI Licht-Technik AG	Switzerland	386	75-03900
Fernau Avionics Ltd	UK	450	75-22600
GEC-Marconi Hazeltine Corp.	USA	577	75-37414
GMV SA	Spain	367	80-00800
Metalite Aviation Lighting	UK	467	75-38750
NEC Corporation	Japan	253	75-06350
Officine Panerai SpA	Italy	237	75-17500
Presentey Engineering	Canada	75	75-11500
Racal Avionics Ltd	UK	477	75-47660
Thorn DNT Airfield Lighting Pty Ltd	Australia	22	75-20540
Wilcox Electric Inc.	USA	627	75-96730

612 20 45 10 Airfield precision approach path indicators (PAPI)

ABB Leitungsbau GmbH	Germany	164	75-00055
AMA Pvte Ltd	India	204	75-00130
ERNI Licht-Technik AG	Switzerland	386	75-03900
Metalite Aviation Lighting	UK	467	75-38750
Research Engineers	UK	478	75-49250
Thorn DNT Airfield Lighting Pty Ltd	Australia	22	75-20540

612 20 45 20 Ground controlled approach systems & equipment, airfield

Aeronautical & General Instr. Ltd	UK	428	75-01050

Canadian Marconi Co.	Canada	*69*	75-02600
NEC Corporation	Japan	*253*	75-06350
Thorn DNT Airfield Lighting Pty Ltd	Australia	*22*	75-20540

612 20 45 30 Marker beacons, radio

Aeroelectrónica Industria	Brazil	*55*	75-00585
Elan GmbH.	Germany	*168*	75-07550
Racal Avionics Ltd	UK	*477*	75-47660
Thorn DNT Airfield Lighting Pty Ltd	Australia	*22*	75-20540
Wilcox Electric Inc.	USA	*627*	75-96730

612 20 45 40 Instrument landing systems (ILS), airfield

Alcatel Air Navigation Systems GmbH	Germany	*164*	75-00535
Indal Technologies Inc	Canada	*72*	75-07740
NEC Corporation	Japan	*253*	75-06350
Racal Avionics Ltd	UK	*477*	75-47660
Thorn DNT Airfield Lighting Pty Ltd	Australia	*22*	75-20540
Wilcox Electric Inc.	USA	*627*	75-96730

612 20 45 45 Microwave landing systems (MLS), airfield

Alcatel Air Navigation Systems GmbH	Germany	*164*	75-00535
Andrew Canada Inc.	Canada	*68*	75-00600
GEC-Marconi Avionics Ltd	UK	*454*	75-25320
GEC-Marconi Hazeltine Corp.	USA	*577*	75-37414
NEC Corporation	Japan	*253*	75-06350
Textron Systems Division	USA	*621*	75-88655

612 20 50 Airfield lighting systems & equipment

ABB Leitungsbau GmbH	Germany	*164*	75-00055
AMA Pvte Ltd	India	*204*	75-00130
Ameriel Inc.	USA	*553*	75-06500
CORIS	France	*134*	75-12450
ERNI Licht-Technik AG	Switzerland	*386*	75-03900
Metalite Aviation Lighting	UK	*467*	75-38750
Metis Construction & Trade Co. Inc.	Turkey	*403*	80-01330
Officine Panerai SpA	Italy	*237*	75-17500
Strumech Versatower Ltd	UK	*486*	75-57530
Sverdrup Corp.	USA	*643*	80-30650
Thorn DNT Airfield Lighting Pty Ltd	Australia	*22*	75-20540
THORN Lighting Ltd.	UK	*488*	75-59750

612 20 50 10 Airfield runway lighting systems

ABB Leitungsbau GmbH	Germany	*164*	75-00055
AMA Pvte Ltd	India	*204*	75-00130
ERNI Licht-Technik AG	Switzerland	*386*	75-03900
Metalite Aviation Lighting	UK	*467*	75-38750
Metis Construction & Trade Co. Inc.	Turkey	*403*	80-01330
Microelectronics Ltd	Israel	*221*	75-14250
Officine Panerai SpA	Italy	*237*	75-17500
Sverdrup Corp.	USA	*643*	80-30650
Thorn DNT Airfield Lighting Pty Ltd	Australia	*22*	75-20540
THORN Lighting Ltd.	UK	*488*	75-59750

612 20 50 90 Airfield lighting equipment

ABB Leitungsbau GmbH	Germany	*164*	75-00055
AMA Pvte Ltd	India	*204*	75-00130
ERNI Licht-Technik AG	Switzerland	*386*	75-03900
Metalite Aviation Lighting	UK	*467*	75-38750
OCEM SpA	Italy	*237*	75-16750
Officine Panerai SpA	Italy	*237*	75-17500
Smith & Prince Ltd	UK	*485*	75-54630
Strumech Versatower Ltd	UK	*486*	75-57530
Thorn DNT Airfield Lighting Pty Ltd	Australia	*22*	75-20540
THORN Lighting Ltd.	UK	*488*	75-59750

612 20 50 91 Airfield approach lights

ABB Leitungsbau GmbH	Germany	*164*	75-00055
AMA Pvte Ltd	India	*204*	75-00130
ERNI Licht-Technik AG	Switzerland	*386*	75-03900
Metalite Aviation Lighting	UK	*467*	75-38750
Officine Panerai SpA	Italy	*237*	75-17500
Thorn DNT Airfield Lighting Pty Ltd	Australia	*22*	75-20540

612 20 50 92 Airfield runway / taxiway lights

ABB Leitungsbau GmbH	Germany	*164*	75-00055
AMA Pvte Ltd	India	*204*	75-00130
ERNI Licht-Technik AG	Switzerland	*386*	75-03900
Metalite Aviation Lighting	UK	*467*	75-38750
Officine Panerai SpA	Italy	*237*	75-17500
Thorn DNT Airfield Lighting Pty Ltd	Australia	*22*	75-20540

612 20 50 93 Airfield floodlights

ABB Leitungsbau GmbH	Germany	*164*	75-00055
Engineered Air Systems Inc.	USA	*573*	75-31560
Luminator Aircraft Products	USA	*593*	75-56450
Officine Panerai SpA	Italy	*237*	75-17500
Smith & Prince Ltd	UK	*485*	75-54630
Strumech Versatower Ltd	UK	*486*	75-57530
Thorn DNT Airfield Lighting Pty Ltd	Australia	*22*	75-20540

612 20 70 Aircraft maintenance systems & equipment

Abbey Equipment Pty Ltd	Australia	*11*	75-00100
Access International Ltd	UK	*427*	75-00300
ADE-HML Limited	UK	*427*	75-00710
Aero Systems Engineering Inc.	USA	*546*	75-01116
Atlas Copco Airpower NV	Belgium	*45*	75-00900
Commercial Hydraulics Keelavite Ltd	UK	*441*	75-14410
Dabico Inc.	USA	*567*	75-23145
Dayton T. Brown Inc.	USA	*632*	80-04920
ELECMA	France	*136*	75-14400
Elf Atochem	Australia	*16*	75-05630
Fastening Systems International Inc.	USA	*575*	75-33870
GEC-Marconi Avionics Ltd	UK	*454*	75-25320
Hodge Clemco Ltd	UK	*458*	75-29590
Interdyne Corp.	USA	*637*	80-16050
Nanoquest Defence Products Ltd	UK	*470*	75-41890
Nea-Lindberg A/S	Denmark	*103*	75-14300
Newton Derby Ltd	UK	*470*	75-42520
Normalair-Garrett Australia	Australia	*19*	75-14870
Smith & Prince Ltd	UK	*485*	75-54630
Société d'Études et de Fabrications	France	*143*	75-34810
Trompeter Electronics Inc.	USA	*623*	75-91070
Vacu-Blast Ltd	UK	*490*	75-62550
Vapormatt Ltd	UK	*490*	75-62630
Zwick Energy Research Organization	USA	*629*	75-98400

612 20 70 20 Battery chargers, aircraft battery

Air-Log Ltd	UK	*429*	75-01800
Automation NV	Belgium	*45*	75-00950
Engineered Magnetics Inc.	USA	*573*	75-31568
Meggitt Avionics (UK)	UK	*467*	75-38340
Newton Derby Ltd	UK	*470*	75-42520
Trilectron Industries Inc.	USA	*623*	75-90920

612 20 70 60 Rechargeable battery maintenance systems, aircraft

Dabico Inc.	USA	*567*	75-23145

612 20 70 70 Battery maintenance equipment, aircraft

Air-Log Ltd	UK	*429*	75-01800
Dabico Inc.	USA	*567*	75-23145
Meggitt Avionics (UK)	UK	*467*	75-38340
Newton Derby Ltd	UK	*470*	75-42520

612 20 70 90 Work platforms, elevating

Abbey Equipment Pty Ltd	Australia	*11*	75-00100
Access International Ltd	UK	*427*	75-00300
Air-Log Ltd	UK	*429*	75-01800
Aircraft Maintenance Support	UK	*428*	75-01580
EFE Endüstri ve Ticaret A.S.	Turkey	*402*	75-00530
Grove Europe Ltd	**UK**	***456***	**75-27030**

H.C. Slingsby plc	UK	484	75-54430
Howard Porter Pty Ltd	Australia	17	75-08720
Robert Mitchell Inc.	Canada	74	75-09800
Static Engineering Pty Ltd	Australia	21	75-19780

612 20 70 90 50 Work platforms, elevating, mobile

Abbey Equipment Pty Ltd	Australia	11	75-00100
Aerotech International Inc	Canada	68	75-00300
Air-Log Ltd	UK	429	75-01800
Aircraft Maintenance Support	UK	428	75-01580
EFE Endüstri ve Ticaret A.S.	Turkey	402	75-00530
Grove Europe Ltd	**UK**	**456**	**75-27030**
H.C. Slingsby plc	UK	484	75-54430
Howard Porter Pty Ltd	Australia	17	75-08720
Robert Mitchell Inc.	Canada	74	75-09800
Snorkel	USA	616	75-82170
Static Engineering Pty Ltd	Australia	21	75-19780

612 20 75 Aircraft ground handling & support equipment

Aarding BV	Netherlands	291	75-00020
Access International Ltd	UK	427	75-00300
Advanced Ground Systems Engineering	USA	546	75-00905
Aero Systems Engineering Inc.	USA	546	75-01116
Aerotech International Inc	Canada	68	75-00300
Air Logistics Corp.	USA	548	75-03700
Air-Log Ltd	UK	429	75-01800
Aircraft Equipment Overhauls & Sales	Australia	23	80-00650
Aircraft Maintenance Support	UK	428	75-01580
Algram Group Ltd.	UK	429	75-02625
Appareillages et Matériels de	France	132	75-03050
Australian Defence Industries Ltd	Australia	9	55-01000
BEI Sensors & Systems Co.	USA	558	75-12740
Columbus Jack Corp.	USA	564	75-19850
Commercial Hydraulics Keelavite Ltd	UK	441	75-14410
Dabico Inc.	USA	567	75-23145
Didsbury Engineering Co. Ltd	UK	445	75-18050
Eagle Industrial Truck Mfg. Inc.	USA	570	75-28430
Electrothermal Engineering Ltd	UK	448	75-20400
EPS Logistics Technology Ltd	UK	502	80-09550
Flight Refuelling Limited	UK	451	75-23850
Fuji Heavy Industries Ltd	Japan	251	75-01100
GEC-Marconi Avionics Ltd	UK	454	75-25320
General Electrodynamics Corp.	USA	578	75-38500
General Electronics Communications	Greece	189	75-01300
GNY Equipment L.L.C.	USA	579	75-39950
Hamech Ltd	UK	456	75-27730
Hunting Aviation Ltd	UK	504	80-12785
Indal Technologies Inc	Canada	72	75-07740
Logistic Projects	S. Africa	358	75-09800
Lukassen Flughafentechnic GmbH	Germany	171	75-18950
Meggitt Avionics (UK)	UK	467	75-38340
Omori Seikoki Co. Ltd	Japan	254	75-07310
Pacific Aerospace Corp. Ltd	New Zealand	302	75-01600
Parker Bertea Aerospace	USA	602	75-68900
Regent Mfg Inc.	USA	609	75-74200
SEI Industries Ltd	Canada	76	75-12750
Simon Access (UK) Ltd.	UK	484	75-53908
Spar Aviation Services	Canada	76	75-13020
Spec Tool Co.	USA	616	75-83050
Static Engineering Pty Ltd	Australia	21	75-19780
Strachan & Henshaw Ltd	UK	512	80-47400
Teledyne Fluid Systems	USA	621	75-87570
Thorn DNT Airfield Lighting Pty Ltd	Australia	22	75-20540
Zwick Energy Research Organization	USA	629	75-98400

612 20 75 30 Lifting equipment, aircraft cargo

Aerial Machine & Tool Corp.	USA	546	75-01105
Air-Log Ltd	UK	429	75-01800
Haacon Hebetechnik GmbH	Germany	169	75-11400
Hamech Ltd	UK	456	75-27730
Kinetics Ltd	Israel	220	75-11320
Sellick Equipment Ltd.	Canada	76	75-12755
SpanSet Inter AG	Switzerland	387	75-13550
SpanSet Ltd	UK	485	75-55230
Static Engineering Pty Ltd	Australia	21	75-19780
Stewart & Stevenson Services Inc.	USA	618	75-84180

612 20 75 40 Loading docks, aircraft

Howard Porter Pty Ltd	Australia	17	75-08720
Static Engineering Pty Ltd	Australia	21	75-19780
Sverdrup Corp.	USA	643	80-30650

612 20 75 50 Loaders, aircraft cargo

Air-Log Ltd	UK	429	75-01800
Aircraft Maintenance Support	UK	428	75-01580
Equipos Industriales de Manutención	Spain	366	75-02850
Haacon Hebetechnik GmbH	Germany	169	75-11400
Robert Mitchell Inc.	Canada	74	75-09800
Static Engineering Pty Ltd	Australia	21	75-19780

612 20 75 50 10 Loaders, aircraft cargo, self-propelled

Air-Log Ltd	UK	429	75-01800
Aircraft Maintenance Support	UK	428	75-01580
Perry Engineering	Australia	20	75-16460
Static Engineering Pty Ltd	Australia	21	75-19780

612 20 75 60 Scissor lifts, aircraft loading

Air-Log Ltd	UK	429	75-01800
Aircraft Maintenance Support	UK	428	75-01580
EFE Endüstri ve Ticaret A.S.	Turkey	402	75-00530
Grove Europe Ltd	**UK**	**456**	**75-27030**
Howard Porter Pty Ltd	Australia	17	75-08720
Regent Mfg Inc.	USA	609	75-74200
Static Engineering Pty Ltd	Australia	21	75-19780
Strachan & Henshaw Ltd	UK	512	80-47400

612 20 75 70 Servicing trolleys, aircraft ground handling

ADE-HML Limited	UK	427	75-00710
Air-Log Ltd	UK	429	75-01800
Aircraft Maintenance Support	UK	428	75-01580
Forward Industries Ltd	UK	452	75-24200
Howden Aircontrol Ltd	UK	459	75-30240
Lista Ltd	Switzerland	386	75-07350
Regent Mfg Inc.	USA	609	75-74200
Spar Aviation Services	Canada	76	75-13020
Static Engineering Pty Ltd	Australia	21	75-19780
Zwick Energy Research Organization	USA	629	75-98400

612 20 75 72 Servicing vehicles, aircraft ground handling

ADE-HML Limited	UK	427	75-00710
GNY Equipment L.L.C.	USA	579	75-39950
Regent Mfg Inc.	USA	609	75-74200
Static Engineering Pty Ltd	Australia	21	75-19780
Thorn DNT Airfield Lighting Pty Ltd	Australia	22	75-20540

612 20 75 74 Starting systems, jet engine, centralised

Arvico Corporation	USA	555	75-09240
Atlas Copco Airpower NV	Belgium	45	75-00900
Dabico Inc.	USA	567	75-23145
Fuji Heavy Industries Ltd	Japan	251	75-01100
MITM Consulting Engineers	Israel	224	80-01350
Parker Bertea Aerospace	USA	602	75-68900

612 20 75 75 Starter trucks, aircraft engine

ADE-HML Limited	UK	427	75-00710
Arvico Corporation	USA	555	75-09240
Atlas Copco Airpower NV	Belgium	45	75-00900
Houchin Aerospace Ltd	UK	459	75-30100
MaK Systemgesellschaft mbH.	Germany	172	75-19500
Stewart & Stevenson Services Inc.	USA	618	75-84180
Trilectron Industries Inc.	USA	623	75-90920

612 20 75 76 Starter trolleys, aircraft engine

ADE-HML Limited	UK	427	75-00710
Air-Log Ltd	UK	429	75-01800

Arvico Corporation	USA	555	75-09240
Atlas Copco Airpower NV	Belgium	45	75-00900
Foxtronics Inc.	USA	576	75-35752
Houchin Aerospace Ltd	UK	459	75-30100
MaK Systemgesellschaft mbH	Germany	172	75-19500
Stewart & Stevenson Services Inc.	USA	618	75-84180

612 20 75 80 Tractors, aircraft towing (see code 141.70)

Eagle Industrial Truck Mfg. Inc.	USA	570	75-28430
Echtman Engineering Corp. Ltd	Israel	218	75-05050
Hamech Ltd	UK	456	75-27730
Miller Industries Towing Equipment	USA	596	75-61655
Perry Engineering	Australia	20	75-16460
Static Engineering Pty Ltd	Australia	21	75-19780
Stewart & Stevenson Services Inc.	USA	618	75-84180
Thorn DNT Airfield Lighting Pty Ltd	Australia	22	75-20540

612 20 80 Airfield emergency systems & equipment

ATS Aerospace Inc.	Canada	68	75-00752
Chubb Fire Products & Systems Ltd	UK	440	75-13200
Clark Masts Teksam Ltd	UK	440	75-13550
Sabo France	France	142	75-33650
Saval NV	Belgium	47	75-05650
SEI Industries Ltd	Canada	76	75-12750
Thorn DNT Airfield Lighting Pty Ltd	Australia	22	75-20540

612 20 80 10 Lifting bags, inflatable, aircraft recovery

Eurovinil Industries SpA	Italy	234	75-08400
Manfred Vetter GmbH & Co	Germany	176	75-32800
MFC Survival Ltd	UK	467	75-38900
Pronal SA	France	141	75-31100
Regent Mfg Inc.	USA	609	75-74200

612 20 80 30 Firefighting equipment, airfield

Ansul Incorporated	USA	554	75-07810
Carmichael International Ltd	UK	438	75-11700
Chubb Fire Products & Systems Ltd	UK	440	75-13200
GAAM Fire & Rescue	Australia	16	75-07000
GD Protective Clothing and Equipment	Ireland	214	75-00120
GNY Equipment L.L.C.	USA	579	75-39950
Godiva Ltd	UK	455	75-25900
Haynes & Cann Ltd	UK	457	75-28750
Interspiro Ltd	UK	461	75-31945
Lehavot Industries	Israel	220	75-12200
MFC Survival Ltd	UK	467	75-38900
Nordic Systems Inc.	Canada	74	75-09950
PD Technical Mouldings plc	UK	473	75-44200
Phillips & Smith Ltd	New Zealand	303	75-01650
Protection Development Int'l Corp.	USA	606	75-71670
Rosenbauer International AG	Austria	34	75-02132
Sabo France	France	142	75-33650
Saval BV	Netherlands	295	75-06200
Saval NV	Belgium	47	75-05650
SEI Industries Ltd	Canada	76	75-12750
Silvani Antincendi SpA	Italy	239	75-22250
Snorkel	USA	616	75-82170
Victaulic Co. of America	USA	626	75-93840

612 20 80 30 30 Fire extinguishers, mobile, airfield

Chubb Australia Ltd.	Australia	14	75-03400
Chubb Fire Products & Systems Ltd	UK	440	75-13200
Gloria Werke H.-Schulte-Frankenfeld	Germany	169	75-10950
Lehavot Industries	Israel	220	75-12200
Nordic Systems Inc.	Canada	74	75-09950
Rosenbauer International AG	Austria	34	75-02132
Saval BV	Netherlands	295	75-06200

612 20 80 30 35 Fire extinguishers, portable, airfield

Ansul Incorporated	USA	554	75-07810
Chubb Australia Ltd.	Australia	14	75-03400
Chubb Fire Products & Systems Ltd	UK	440	75-13200
Gloria Werke H.-Schulte-Frankenfeld	Germany	169	75-10950
Lehavot Industries	Israel	220	75-12200
Metalcraft Inc.	USA	595	75-60385
Rosenbauer International AG	Austria	34	75-02132
Saval BV	Netherlands	295	75-06200
SIHI Australia Pty Ltd	Australia	21	75-19505

612 20 80 30 40 Firefighting equipment, stationary, airfield

Dabico Inc.	USA	567	75-23145
GAAM Fire & Rescue	Australia	16	75-07000
Giat Industries	France	126	55-00100
Godiva Ltd	UK	455	75-25900
Lehavot Industries	Israel	220	75-12200
Nordic Systems Inc	Canada	74	75-09950
PD Technical Mouldings plc	UK	473	75-44200
Rosenbauer International AG	Austria	34	75-02132
Sabo France	France	142	75-33650
SEI Industries Ltd	Canada	76	75-12750
SIHI Australia Pty Ltd	Australia	21	75-19505

612 20 80 30 50 Fire pumps, airfield

GAAM Fire & Rescue	Australia	16	75-07000
Gilbert Gilkes & Gordon Ltd	UK	454	75-25430
Godiva Ltd	UK	455	75-25900
Hamworthy Marine	UK	457	75-27850
Ingersoll-Dresser Pumps (UK) Ltd	UK	461	75-31450
Rosenbauer International AG	Austria	34	75-02132
SIHI Australia Pty Ltd	Australia	21	75-19505
Weir Engineering Pty Ltd	Australia	23	75-23300

612 20 80 39 Firefighting vehicles, airfield (see code 141.70.40.05.40)

Austral Specialised Vehicles	Australia	12	75-01710
Carmichael International Ltd	UK	438	75-11700
Dennis Specialist Vehicles	UK	445	75-17570
Emergency One Inc.	USA	573	75-31100
Giat Industries	France	126	55-00100
Nordic Systems Inc	Canada	74	75-09950
Patria Vehicles Oy	Finland	118	75-01710
PD Technical Mouldings plc	UK	473	75-44200
Pilcher-Greene Ltd	UK	474	75-45450
Reynolds Boughton Ltd	UK	479	75-49340
Rosenbauer International AG	Austria	34	75-02132
Simon Access (UK) Ltd.	UK	484	75-53908

612 30 Operations support facilities & equipment, naval base

Ansul Incorporated	USA	554	75-07810
British Filters Ltd	UK	436	75-09030
Calspan SRL Technology	USA	540	60-03000
Hydroscience Technologies Inc.	USA	584	75-45550
Indal Technologies Inc	Canada	72	75-07740
Litton Industries Inc.	USA	591	75-55400
Logica UK Limited	UK	506	80-14350
Raytheon E-Systems	USA	608	75-73302
Strachan & Henshaw Ltd	UK	512	80-47400
Sverdrup Corp.	USA	643	80-30650
Sverdrup Technology Inc.	USA	643	80-30680
Taylor Woodrow Management	UK	513	80-50400

612 30 20 Ship berthing & mooring systems & equipment

A. Noble & Son Ltd	Australia	19	75-14690
Balmoral Group Ltd	UK	433	75-05420
British Filters Ltd	UK	436	75-09030
Fluor Daniel Inc.	USA	635	80-11720
Hippo Marine Products Ltd	UK	458	75-29430
MacTaggart Scott & Co. Ltd	UK	466	75-36900
Seaward International Inc.	USA	613	75-78940
Strachan & Henshaw Ltd	UK	512	80-47400
Sverdrup Corp.	USA	643	80-30650

612 30 25 Compression chambers, naval static

Divex Ltd.	UK	445	75-18220
Haux-Life-Support GmbH	Germany	169	75-11970

Sverdrup Corp.	USA	643	80-30650
Sverdrup Technology Inc.	USA	643	80-30680

612 30 30 Degaussing facilities & equipment

Institut Dr Förster GmbH & Co KG	Germany	170	75-14500

612 30 30 70 Degaussing range equipment

Fluor Daniel Inc.	USA	635	80-11720
Permali Gloucester Ltd	UK	474	75-44850

612 30 40 Fuel, oil & lubricants storage systems, naval base

British Filters Ltd	UK	436	75-09030
Fluor Daniel Inc.	USA	635	80-11720
GNY Equipment L.L.C.	USA	579	75-39950
Metis Construction & Trade Co. Inc.	Turkey	403	80-01330
SEI Industries Ltd	Canada	76	75-12750
Sverdrup Corp.	USA	643	80-30650
Taylor Woodrow Construction Ltd	UK	513	80-50300

612 30 40 70 Fuel, oil & lubricants storage equipment, naval base

GNY Equipment L.L.C.	USA	579	75-39950
SEI Industries Ltd	Canada	76	75-12750

612 30 40 70 50 Fuel, oil & lubricant pumps, naval base

Fenner Fluid Power	UK	450	75-22550
GNY Equipment L.L.C.	USA	579	75-39950
Hamworthy Marine	UK	457	75-27850
H.C. Slingsby plc	UK	484	75-54430
Imo Industries Inc.	USA	584	75-46150
Ingersoll-Dresser Pumps (UK) Ltd	UK	461	75-31450
PD Technical Mouldings plc	UK	473	75-44200
Plenty Ltd	UK	475	75-46050
SIHI Australia Pty Ltd	Australia	21	75-19505
Weir Engineering Pty Ltd	Australia	23	75-23300

612 30 40 70 60 Fuel, oil & lubricants storage tanks, naval base

Fluor Daniel Inc.	USA	635	80-11720
GNY Equipment L.L.C.	USA	579	75-39950
Metis Construction & Trade Co. Inc.	Turkey	403	80-01330
PD Technical Mouldings plc	UK	473	75-44200
Pronal SA	France	141	75-31100
SEI Industries Ltd	Canada	76	75-12750
White Cross Rubber Products Ltd	UK	494	75-65857

612 30 43 Fuel, oil & lubricants delivery systems , naval base

Aljac Engineering Ltd	UK	429	75-02650
British Filters Ltd	UK	436	75-09030
GNY Equipment L.L.C.	USA	579	75-39950
PD Technical Mouldings plc	UK	473	75-44200
Pronal SA	France	141	75-31100
SIHI Australia Pty Ltd	Australia	21	75-19505
Taylor Woodrow Construction Ltd	UK	513	80-50300
Victaulic Co. of America	USA	626	75-93840

612 30 43 70 Fuel, oil & lubricants delivery equipment, naval base

Aeroquip Corp.	USA	547	75-02350
Dabico Inc.	USA	567	75-23145
Dunlop Oil & Marine	UK	447	75-19050
GEC-Marconi Avionics Ltd	UK	454	75-25320
GNY Equipment L.L.C.	USA	579	75-39950
PD Technical Mouldings plc	UK	473	75-44200
SIHI Australia Pty Ltd	Australia	21	75-19505

612 30 45 Hydrographic data acquisition systems

Hydroscience Technologies Inc.	USA	584	75-45550
MacDonald Dettwiler and Associates	Canada	78	80-01310
Navitronic Systems AS	Denmark	103	75-14090
Van Rietschoten & Houwens	Netherlands	296	75-07900
Wilcoxon Research	USA	627	75-96750

612 30 50 Lighting systems & equipment, naval base

AMA Pvte Ltd	India	204	75-00130
Havis-Shields Equipment Corp.	USA	581	75-42700
Sverdrup Corp.	USA	643	80-30650

612 30 70 Maintenance & repair facilities & equipment, naval

Astilleros Espanoles SA	Spain	365	75-01400
Australian Defence Industries Ltd	Australia	9	55-01000
Dabico Inc.	USA	567	75-23145
Froude Engineering Inc.	USA	576	75-36340
GEC-Marconi Avionics Ltd	UK	454	75-25320
Hellenic Shipyards Co.	Greece	187	55-00850
Hodge Clemco Ltd	UK	458	75-29590
Hydroscience Technologies Inc.	USA	584	75-45550
Morska Stocznia Remontowa SA	Poland	324	75-02600
Santierul Naval Constanta SA	Romania	330	75-00750
Stocznia Szczecinska SA.	Poland	325	75-02850
Sverdrup Corp.	USA	643	80-30650
Sverdrup Technology Inc.	USA	643	80-30680
Tulcea Shipyard	Romania	331	75-01200
Vapormatt Ltd	UK	490	75-62630
Viktor Lenac Ship Repair	Croatia	95	75-08000

612 30 75 30 Lifting equipment, naval base

Didsbury Engineering Co. Ltd	UK	445	75-18050
Hamech Ltd	UK	456	75-27730
Parsons Chain Co. Ltd	UK	473	75-44035
SpanSet Inter AG	Switzerland	387	75-13550
Stothert & Pitt	UK	512	80-47200
Strachan & Henshaw Ltd	UK	512	80-47400

612 30 75 32 Cranes, dockside

Grove Europe Ltd	**UK**	**456**	**75-27030**
KCI Carruthers Ltd	UK	463	75-33300
Stothert & Pitt	UK	512	80-47200

612 30 80 Emergency & safety systems & equipment, naval base

Chubb Fire Products & Systems Ltd	UK	440	75-13200
Elts Unlimited Inc.	USA	573	75-30900
Emergency Beacon Corp.	USA	573	75-31050
Glatzer Industries Corp.	USA	579	75-39760
Interspiro Ltd	UK	461	75-31945
Manfred Vetter GmbH & Co	Germany	176	75-32800
Metis Construction & Trade Co. Inc.	Turkey	403	80-01330
Mustang Survival (USA) Corp	USA	597	75-63480
Sabo France	France	142	75-33650
Saval NV	Belgium	47	75-05650
The Armstrong Monitoring Corp.	Canada	68	75-00740
Zellweger Analytics Ltd	UK	495	75-67600

612 30 80 30 Firefighting equipment, naval base

Ansul Incorporated	USA	554	75-07810
Chubb Fire Products & Systems Ltd	UK	440	75-13200
Dennis Specialist Vehicles	UK	445	75-17570
Godiva Ltd	UK	455	75-25900
Interspiro Ltd	UK	461	75-31945
Lehavot Industries	Israel	220	75-12200
Nordic Systems Inc	Canada	74	75-09950
PD Technical Mouldings plc	UK	473	75-44200
Phillips & Smith Ltd	New Zealand	303	75-01650
Rosenbauer International AG	Austria	34	75-02132
Sabo France	France	142	75-33650
Saval NV	Belgium	47	75-05650

SEI Industries Ltd	Canada	76	75-12750
Silvani Antincendi SpA	Italy	239	75-22250
Snorkel	USA	616	75-82170
Victaulic Co. of America	USA	626	75-93840

612 30 80 30 30 Fire extinguishers, mobile, naval base

Chubb Fire Products & Systems Ltd	UK	440	75-13200
Gloria Werke H.-Schulte-Frankenfeld	Germany	169	75-10950
Lehavot Industries	Israel	220	75-12200
Nordic Systems Inc	Canada	74	75-09950
Rosenbauer International AG	Austria	34	75-02132
SIHI Australia Pty Ltd	Australia	21	75-19505

612 30 80 30 35 Fire extinguishers, portable, naval base

Chubb Fire Products & Systems Ltd	UK	440	75-13200
Gloria Werke H.-Schulte-Frankenfeld	Germany	169	75-10950
Lehavot Industries	Israel	220	75-12200
Metalcraft Inc.	USA	595	75-60385
Resil Industria e Comercio Ltda	Brazil	57	75-18400
Rosenbauer International AG	Austria	34	75-02132

612 30 80 30 40 Firefighting equipment, stationary, naval base

Dabico Inc.	USA	567	75-23145
GAAM Fire & Rescue	Australia	16	75-07000
Godiva Ltd	UK	455	75-25900
Lehavot Industries	Israel	220	75-12200
Nordic Systems Inc	Canada	74	75-09950
PD Technical Mouldings plc	UK	473	75-44200
Rosenbauer International AG	Austria	34	75-02132
Sabo France	France	142	75-33650
SEI Industries Ltd	Canada	76	75-12750
SIHI Australia Pty Ltd	Australia	21	75-19505

612 30 80 30 50 Fire pumps, naval base

GAAM Fire & Rescue	Australia	16	75-07000
Gilbert Gilkes & Gordon Ltd	UK	454	75-25430
Godiva Ltd	UK	455	75-25900
Hamworthy Marine	UK	457	75-27850
Ingersoll-Dresser Pumps (UK) Ltd	UK	461	75-31450
Rosenbauer International AG	Austria	34	75-02132
SIHI Australia Pty Ltd	Australia	21	75-19505
Urdan Industries (USA) Inc.	USA	625	75-92930
Weir Engineering Pty Ltd	Australia	23	75-23300

612 30 80 39 Firefighting vehicles, naval base (see code 141.70.40.40)

Dennis Specialist Vehicles	UK	445	75-17570
Nordic Systems Inc	Canada	74	75-09950
PD Technical Mouldings plc	UK	473	75-44200
Pilcher-Greene Ltd	UK	474	75-45450
Rosenbauer International AG	Austria	34	75-02132

612 40 Operations support facilities & equipment, army base

Ansul Incorporated	USA	554	75-07810
Calspan SRL Technology	USA	540	60-03000
Environmental Technologies Group Inc	USA	574	75-31860
Gates Hydraulics Ltd	UK	453	75-25210
L'hotellier Montrichard	France	138	75-24900
Litton Industries Inc.	USA	591	75-55400
Logica UK Limited	UK	506	80-14350
Nanoquest Defence Products Ltd	UK	470	75-41890
Raytheon E-Systems	USA	608	75-73302
Sverdrup Corp.	USA	643	80-30650
Taylor Woodrow Management	UK	513	80-50400

612 40 30 Meteorological systems & equipment, ground-based

Alden Electronics Incorporated	USA	548	75-04400
Analysis Management Systems	S. Africa	357	75-00770
Ausrire Scientific & Research	Russia	335	60-00090
Becker Flugfunkwerk GmbH	Germany	165	75-02200
Belfort Instrument Co.	USA	558	75-12800
CAE Elektronic GmbH	Germany	166	75-04500
Datamat Ingegneria dei Sistemi SpA	Italy	233	75-06500
Feedback Instruments	UK	450	75-22520
Fluor Daniel Inc.	USA	635	80-11720
Gill Instruments Ltd	UK	454	75-25450
Hall & Watts (Defence Optics) Group	**UK**	**456**	**75-27650**
Hobeco Sociedade Consignatária Ltda	Brazil	56	75-08600
Jenoptik Impulsphysik GmbH	Germany	170	75-14800
MacDonald Dettwiler and Associates	Canada	78	80-01310
Meggitt Avionics (UK)	UK	467	75-38340
Osterreichisches Forschungszentrum	Austria	32	60-05000
Remtech SA	France	141	75-32230
Reunert Technology Systems Ltd	S. Africa	359	75-18300
R.M. Young Co.	USA	628	75-98020
SITEP SpA	Italy	239	75-22595
Teledyne Brown Engineering	USA	620	75-87508
Vaisala Oy	Finland	119	75-02800

612 40 40 Fuel, oil & lubricants storage systems, army base

British Filters Ltd	UK	436	75-09030
Fluor Daniel Inc.	USA	635	80-11720
GNY Equipment L.L.C.	USA	579	75-39950
Metis Construction & Trade Co. Inc.	Turkey	403	80-01330
PD Technical Mouldings plc	UK	473	75-44200
SEI Industries Ltd	Canada	76	75-12750
Sverdrup Corp.	USA	643	80-30650
Taylor Woodrow Construction Ltd	UK	513	80-50300

612 40 40 70 Fuel, oil & lubricants storage equipment, army base

GNY Equipment L.L.C.	USA	579	75-39950
PD Technical Mouldings plc	UK	473	75-44200
SEI Industries Ltd	Canada	76	75-12750

612 40 40 70 50 Fuel, oil & lubricant pumps, army base

GNY Equipment L.L.C.	USA	579	75-39950
Hamworthy Marine	UK	457	75-27850
H.C. Slingsby plc	UK	484	75-54430
Imo Industries Inc.	USA	584	75-46150
Ingersoll-Dresser Pumps (UK) Ltd	UK	461	75-31450
PD Technical Mouldings plc	UK	473	75-44200
Plenty Ltd	UK	475	75-46050
Weir Engineering Pty Ltd	Australia	23	75-23300

612 40 40 70 60 Fuel, oil & lubricants storage tanks, army base

British Filters Ltd	UK	436	75-09030
Fluor Daniel Inc.	USA	635	80-11720
GNY Equipment L.L.C.	USA	579	75-39950
Metis Construction & Trade Co. Inc.	Turkey	403	80-01330
PD Technical Mouldings plc	UK	473	75-44200
Pronal SA	France	141	75-31100
SEI Industries Ltd	Canada	76	75-12750
Sverdrup Corp.	USA	643	80-30650
White Cross Rubber Products Ltd	UK	494	75-65857

612 40 43 Fuel, oil & lubricants delivery systems , army base

Dabico Inc.	USA	567	75-23145
Emco Wheaton Australia Pty Ltd	Australia	16	75-05760
Fluor Daniel Inc.	USA	635	80-11720
GNY Equipment L.L.C.	USA	579	75-39950
PD Technical Mouldings plc	UK	473	75-44200
Pronal SA	France	141	75-31100
Sverdrup Corp.	USA	643	80-30650
Taylor Woodrow Construction Ltd	UK	513	80-50300
Victaulic Co. of America	USA	626	75-93840

612 40 43 70 Fuel, oil & lubricants delivery equipment, army base

Dabico Inc.	USA	*567*	75-23145
GEC-Marconi Avionics Ltd	UK	*454*	75-25320
GNY Equipment L.L.C.	USA	*579*	75-39950
PD Technical Mouldings plc	UK	*473*	75-44200

612 40 50 Lighting systems & equipment, army base

Bruce Industries Inc.	USA	*561*	75-15360
Fluor Daniel Inc.	USA	*635*	80-11720
General Electronics Communications	Greece	*189*	75-01300
Havis-Shields Equipment Corp.	USA	*581*	75-42700
Smith & Prince Ltd	UK	*485*	75-54630
Strumech Versatower Ltd	UK	*486*	75-57530
Sverdrup Corp.	USA	*643*	80-30650

612 40 70 Maintenance facilities & equipment, vehicle, army base

CSC (Specialised Vehicles) Ltd	UK	*444*	75-16220
Froude Engineering Inc.	USA	*576*	75-36340
Gates Hydraulics Ltd	UK	*453*	75-25210
GEC-Marconi Avionics Ltd	UK	*454*	75-25320
Hodge Clemco Ltd	UK	*458*	75-29590
Lista Ltd	Switzerland	*386*	75-07350
Nanoquest Defence Products Ltd	UK	*470*	75-41890
Sverdrup Corp.	USA	*643*	80-30650
Vapormatt Ltd	UK	*490*	75-62630
Weatherhaven	Canada	*77*	75-15220
Wellman Garage Equipment	UK	*493*	75-65130

612 40 70 20 Battery chargers, vehicle battery

Automation NV	Belgium	*45*	75-00950
General Electronics Communications	Greece	*189*	75-01300
H.C. Slingsby plc	UK	*484*	75-54430
Meggitt Avionics (UK)	UK	*467*	75-38340
Newton Derby Ltd	UK	*470*	75-42520

612 40 70 60 Rechargeable battery maintenance systems, vehicle

Dabico Inc.	USA	*567*	75-23145
LARIMART SpA	Italy	*236*	75-13500
Newton Derby Ltd	UK	*470*	75-42520

612 40 70 70 Battery maintenance equipment, vehicle

LARIMART SpA	Italy	*236*	75-13500
Newton Derby Ltd	UK	*470*	75-42520

612 40 75 30 Lifting equipment, army base

Kinetics Ltd	Israel	*220*	75-11320
L'hotellier Montrichard	France	*138*	75-24900
Mamor Works	USA	*593*	75-57600
Manfred Vetter GmbH & Co	Germany	*176*	75-32800
Parsons Chain Co. Ltd	UK	*473*	75-44035
Sellick Equipment Ltd.	Canada	*76*	75-12755
SpanSet Inter AG	Switzerland	*387*	75-13550
SpanSet Ltd	UK	*485*	75-55230
Stothert & Pitt	UK	*512*	80-47200

612 40 75 32 Container lifting equipment, army base

Grove Europe Ltd	**UK**	***456***	**75-27030**
SpanSet Inter AG	Switzerland	*387*	75-13550
SpanSet Ltd	UK	*485*	75-55230
Stothert & Pitt	UK	*512*	80-47200

612 40 80 Emergency & safety systems & equipment, army base

Chubb Fire Products & Systems Ltd	UK	*440*	75-13200
Elts Unlimited Inc.	USA	*573*	75-30900
Emergency Beacon Corp.	USA	*573*	75-31050
Interspiro Ltd	UK	*461*	75-31945
L'hotellier Montrichard	France	*138*	75-24900
Manfred Vetter GmbH & Co	Germany	*176*	75-32800
Medesign GmbH	Germany	*172*	75-20370
Metis Construction & Trade Co. Inc.	Turkey	*403*	80-01330
Mustang Survival (USA) Corp	USA	*597*	75-63480
Saval NV	Belgium	*47*	75-05650
Sverdrup Corp.	USA	*643*	80-30650
The Armstrong Monitoring Corp.	Canada	*68*	75-00740
Zellweger Analytics Ltd	UK	*495*	75-67600

612 40 80 30 Firefighting equipment, army base

Ansul Incorporated	USA	*554*	75-07810
Chubb Fire Products & Systems Ltd	UK	*440*	75-13200
Dennis Specialist Vehicles	UK	*445*	75-17570
GAAM Fire & Rescue	Australia	*16*	75-07000
GNY Equipment L.L.C.	USA	*579*	75-39950
Godiva Ltd	UK	*455*	75-25900
Interspiro Ltd	UK	*461*	75-31945
James Hardie Engineering	New Zealand	*303*	80-00900
Lehavot Industries	Israel	*220*	75-12200
Nordic Systems Inc	Canada	*74*	75-09950
PD Technical Mouldings plc	UK	*473*	75-44200
Phillips & Smith Ltd	New Zealand	*303*	75-01650
Rosenbauer International AG	Austria	*34*	75-02132
Sabo France	France	*142*	75-33650
Saval NV	Belgium	*47*	75-05650
SEI Industries Ltd	Canada	*76*	75-12750
Silvani Antincendi SpA	Italy	*239*	75-22250
Snorkel	USA	*616*	75-82170

612 40 80 30 30 Fire extinguishers, mobile, army base

Chubb Fire Products & Systems Ltd	UK	*440*	75-13200
Lehavot Industries	Israel	*220*	75-12200
Nordic Systems Inc	Canada	*74*	75-09950
Phillips & Smith Ltd	New Zealand	*303*	75-01650
Rosenbauer International AG	Austria	*34*	75-02132

612 40 80 30 35 Fire extinguishers, portable, army base

Ansul Incorporated	USA	*554*	75-07810
Chubb Australia Ltd.	Australia	*14*	75-03400
Chubb Fire Products & Systems Ltd	UK	*440*	75-13200
James Hardie Engineering	New Zealand	*303*	80-00900
Lehavot Industries	Israel	*220*	75-12200
Metalcraft Inc.	USA	*595*	75-60385
Phillips & Smith Ltd	New Zealand	*303*	75-01650
Resil Industria e Comercio Ltda	Brazil	*57*	75-18400
Rosenbauer International AG	Austria	*34*	75-02132

612 40 80 30 40 Firefighting equipment, stationary, army base

Dabico Inc.	USA	*567*	75-23145
GAAM Fire & Rescue	Australia	*16*	75-07000
Godiva Ltd	UK	*455*	75-25900
Lehavot Industries	Israel	*220*	75-12200
Nordic Systems Inc	Canada	*74*	75-09950
PD Technical Mouldings plc	UK	*473*	75-44200
Rosenbauer International AG	Austria	*34*	75-02132
Sabo France	France	*142*	75-33650
SEI Industries Ltd	Canada	*76*	75-12750
SIHI Australia Pty Ltd	Australia	*21*	75-19505

612 40 80 30 50 Fire pumps, army base

GAAM Fire & Rescue	Australia	*16*	75-07000
Gilbert Gilkes & Gordon Ltd	UK	*454*	75-25430
Godiva Ltd	UK	*455*	75-25900
Hamworthy Marine	UK	*457*	75-27850
Ingersoll-Dresser Pumps (UK) Ltd	UK	*461*	75-31450
Rosenbauer International AG	Austria	*34*	75-02132
SIHI Australia Pty Ltd	Australia	*21*	75-19505
Urdan Industries (USA) Inc.	USA	*625*	75-92930
Weir Engineering Pty Ltd	Australia	*23*	75-23300

612 40 80 39 Firefighting vehicles, army base (see code 141.70.40.40)

Dennis Specialist Vehicles	UK	*445*	75-17570
Nordic Systems Inc.	Canada	*74*	75-09950
PD Technical Mouldings plc	UK	*473*	75-44200
Protection Development Int'l Corp.	USA	*606*	75-71670
Reynolds Boughton Ltd	UK	*479*	75-49340
Rosenbauer International AG	Austria	*34*	75-02132
SPL Ordnance Ltd.	Israel	*225*	80-01935

613 Logistics systems & equipment

Advanced Technology Ltd.	Israel	*217*	75-00400
Australian Defence Industries Ltd	Australia	*9*	55-01000
Avery Berkel	UK	*432*	75-04560
Avon Technical Products	UK	*432*	75-05150
Bartlett Lifting Devices Inc.	USA	*557*	75-12060
Boss Group Limited	UK	*435*	75-07750
Case (United Kingdom) Ltd.	UK	*439*	75-12050
Cleco Limited	UK	*440*	75-13750
Cubic Defense Systems Inc.	USA	*567*	75-22900
Danish Camp Supply	Denmark	*102*	75-04050
Ebs-Ray Pumps Pty Ltd.	Australia	*15*	75-05260
EPS Logistics Technology Ltd	UK	*502*	80-09550
GEC-Marconi Avionics Ltd	UK	*454*	75-25320
GEL Engineering Ltd	UK	*454*	75-25500
Gerlinger Carrier Co.	USA	*578*	75-39530
Giat Industries	France	*126*	55-00100
Gilbert Gilkes & Gordon Ltd	UK	*454*	75-25430
Haacon Hebetechnik GmbH	Germany	*169*	75-11400
Hyster Europe Ltd.	UK	*460*	75-30950
Ingenieria de Sistemas	Spain	*367*	80-00900
ITT Avionics	USA	*586*	75-49455
L'hotellier Montrichard	France	*138*	75-24900
Litton Industries Inc.	USA	*590*	75-55400
Logan Fenamec (UK) Ltd	UK	*465*	75-36270
Logtek (Logistic Technologies)	S. Africa	*360*	80-09950
M. Paz Logistics Ltd	Israel	*224*	80-01600
Megkon	S. Africa	*360*	80-10000
Northrop Grumman Corporation	USA	*640*	80-21840
Polymathic Engineering Co. Ltd	UK	*475*	75-46350
Raytheon E-Systems	USA	*608*	75-73302
Reynolds Boughton Ltd	UK	*479*	75-49340
Siemens Plessey Systems	UK	*483*	75-53770
Strachan & Henshaw Ltd	UK	*512*	80-47400
Sverdrup Corp.	USA	*643*	80-30650
United Defense L.P.	USA	*624*	75-92050
Western Design Howden	USA	*627*	75-95800

613 10 Munition storage & handling systems & equipment

ACH	France	*130*	75-00150
Air-Log Ltd	UK	*429*	75-01800
Australian Defence Industries Ltd	Australia	*9*	55-01000
Aveling Barford (Machines) plc	UK	*432*	75-04550
Avery Berkel	UK	*432*	75-04560
Case (United Kingdom) Ltd.	UK	*439*	75-12050
C.I.A.M.-E.A. Sarl	Italy	*233*	75-05150
L'hotellier Montrichard	France	*138*	75-24900
Matren Ltd.	Greece	*190*	75-02010
Metis Construction & Trade Co. Inc.	Turkey	*403*	80-01330
Reynolds Boughton (Devon) Ltd.	UK	*479*	75-49340
Reynolds Boughton Ltd	UK	*479*	75-49340
Spar Aviation Services	Canada	*76*	75-13020
Sparton Technology Inc.	USA	*616*	75-82950
St Bernard Composites Ltd	UK	*486*	75-56222
Still GmbH	Germany	*175*	75-29000
Strachan & Henshaw Ltd	UK	*512*	80-47400
Sverdrup Corp.	USA	*643*	80-30650
Sverdrup Technology Inc.	USA	*643*	80-30680
Western Design Howden	USA	*627*	75-95800

613 10 22 Cranes, hydraulic, munition storage & handling

Air-Log Ltd	UK	*429*	75-01800
Auto Crane Co.	USA	*556*	75-10360
Dassault Aviation Group	France	*135*	75-12975
Grove Europe Ltd	**UK**	***456***	**75-27030**
Kinetics Ltd	Israel	*220*	75-11320
Morgan Marine	USA	*597*	75-62625
Regent Mfg Inc.	USA	*609*	75-74200
Reynolds Boughton Ltd	UK	*479*	75-49340
Strachan & Henshaw Ltd	UK	*512*	80-47400

613 10 70 Trolleys, weapon handling / loading / transporting

Air-Log Ltd	UK	*429*	75-01800
Aircraft Maintenance Support	UK	*428*	75-01580
Didsbury Engineering Co. Ltd	UK	*445*	75-18050
EPS Logistics Technology Ltd	UK	*502*	80-09550
MIL-BASE Industries Inc.	USA	*596*	75-61530
Regent Mfg Inc.	USA	*609*	75-74200
Strachan & Henshaw Ltd	UK	*512*	80-47400

613 20 Cargo storage & handling systems & equipment

Aerial Machine & Tool Corp.	USA	*546*	75-01105
Air-Log Ltd	UK	*429*	75-01800
Ancra CHS BV.	Netherlands	*291*	75-00510
Avery Berkel	UK	*432*	75-04560
Bartlett Lifting Devices Inc.	USA	*557*	75-12060
Boss Group Limited	UK	*435*	75-07750
Cargo Aids	UK	*438*	75-11620
Davis Aircraft Products Co. Inc.	USA	*568*	75-24290
Gerlinger Carrier Co.	USA	*578*	75-39530
Howard Porter Pty Ltd	Australia	*17*	75-08720
Robert Mitchell Inc.	Canada	*74*	75-09800
SpanSet Ltd	UK	*485*	75-55230
Stothert & Pitt	UK	*512*	80-47200
Sverdrup Corp.	USA	*643*	80-30650
Systems & Electronics Inc.	USA	*619*	75-85650

613 20 20 Cargo conveying systems & equipment

Aerial Machine & Tool Corp.	USA	*546*	75-01105
Aero Union Corp.	USA	*546*	75-01118
Air-Log Ltd	UK	*429*	75-01800
Ancra CHS BV.	Netherlands	*291*	75-00510
Danish Camp Supply	Denmark	*102*	75-04050
Logan Fenamec (UK) Ltd	UK	*465*	75-36270
Package Control (UK) Ltd.	UK	*472*	75-43450
Stothert & Pitt	UK	*512*	80-47200
Sverdrup Corp.	USA	*643*	80-30650

613 20 50 Cargo lifting & stacking systems & equipment

Access International Ltd	UK	*427*	75-00300
Aerial Machine & Tool Corp.	USA	*546*	75-01105
Auto Crane Co.	USA	*556*	75-10360
Case (United Kingdom) Ltd.	UK	*439*	75-12050
Equipos Industriales de Manutención	Spain	*366*	75-02850
Gerlinger Carrier Co.	USA	*578*	75-39530
Haacon Hebetechnik GmbH	Germany	*169*	75-11400
SpanSet Ltd	UK	*485*	75-55230
Still GmbH	Germany	*175*	75-29000
Stothert & Pitt	UK	*512*	80-47200
Sverdrup Corp.	USA	*643*	80-30650
WB Chain Co Ltd	UK	*439*	75-12440

613 20 50 05 Container transfer systems

Bartlett Lifting Devices Inc.	USA	*557*	75-12060
Belotti SpA	Italy	*232*	75-02640
Bromma Inc.	USA	*560*	75-15270
Danish Camp Supply	Denmark	*102*	75-04050
EKA Limited	UK	*448*	75-19810
Gerlinger Carrier Co.	USA	*578*	75-39530
Haacon Hebetechnik GmbH	Germany	*169*	75-11400
Klaus Transportsysteme GmbH & Co KG	Germany	*165*	75-00930
L'hotellier Montrichard	France	*138*	75-24900
Reynolds Boughton (Devon) Ltd.	UK	*479*	75-49340
Stothert & Pitt	UK	*512*	80-47200
Sverdrup Corp.	USA	*643*	80-30650

613 20 50 10 Fork lift trucks (see code 627.60.40)

Aveling Barford (Machines) plc	UK	*432*	75-04550

Boss Group Limited	UK	435	75-07750
Hamech Ltd	UK	456	75-27730
Hyster Europe Ltd.	UK	460	75-30950
JC Bamford Excavators Ltd	UK	462	75-32297
Loadlift Equipment Ltd	New Zealand	302	75-01250
Sellick Equipment Ltd.	Canada	76	75-12755
Still GmbH	Germany	175	75-29000

613 20 50 20 Hand lift trucks

Boss Group Limited	UK	435	75-07750
Polymathic Engineering Co. Ltd	UK	475	75-46350
Still GmbH	Germany	175	75-29000

613 20 50 50 Loaders, cargo, mobile

Air-Log Ltd	UK	429	75-01800
Ancra CHS BV.	Netherlands	291	75-00510
Equipos Industriales de Manutención	Spain	366	75-02850
Komatsu Hanomag Aktiengesellschaft	Germany	171	75-16185

613 30 Logistics support systems, electronic

Air-Log Ltd	UK	429	75-01800
Ancra CHS BV.	Netherlands	291	75-00510
British Sarozal Ltd.	UK	436	75-09570
GEC-Marconi Avionics Ltd	UK	454	75-25320
Giat Industries	France	126	55-00100
Ingenieria de Sistemas	Spain	367	80-00900
L'hotellier Montrichard	France	138	75-24900
Siemens Plessey Systems	UK	483	75-53770
Sverdrup Corp.	USA	643	80-30650

613 40 Petroleum logistics systems & equipment (see code 612)

British Filters Ltd	UK	436	75-09030
PD Technical Mouldings plc	UK	473	75-44200
Pronal SA	France	141	75-31100
Victaulic Co. of America	USA	626	75-93840

613 40 70 Petroleum logistics equipment

PD Technical Mouldings plc	UK	473	75-44200

613 40 70 10 Flexible barges, petroleum products transport

Pronal SA	France	141	75-31100
SEI Industries Ltd	Canada	76	75-12750

613 40 70 50 Pumps, ground-based

Clyde Sturtevant Ltd	UK	441	75-13868
Ebs-Ray Pumps Pty Ltd.	Australia	15	75-05260
Engineered Air Systems Inc.	USA	573	75-31560
Gilbert Gilkes & Gordon Ltd	UK	454	75-25430
Hamworthy Marine	UK	457	75-27850
Imo Industries Inc.	USA	584	75-46150
Imo Pump AB	Sweden	377	75-06500
Ingersoll-Dresser Pumps (UK) Ltd	UK	461	75-31450
PD Technical Mouldings plc	UK	473	75-44200
Plenty Ltd	UK	475	75-46050
Spar Aviation Services	Canada	76	75-13020
Urdan Industries (USA) Inc.	USA	625	75-92930
Zwicky Engineering Ltd	UK	495	75-67900

613 70 Logistics equipment, airborne

Bridport Aviation Products	UK	435	75-08400
Fabbrica Italiana Apparecchiature	Italy	234	75-08550
Irvin Aerospace Limited	UK	462	75-32050
SpanSet Inter AG	Switzerland	387	75-13550
Sverdrup Technology Inc.	USA	643	80-30680

613 70 10 Logistics equipment, air drop

Bridport Aviation Products	UK	435	75-08400
Irvin Aerospace Canada Limited	Canada	73	75-08300
Irvin Aerospace Limited	UK	462	75-32050
Irvin Aerospace SpA	Italy	235	75-12200
Package Control (UK) Ltd.	UK	472	75-43450

613 70 10 10 Aircraft cargo stowage & release equipment, air supply drop

Aerial Machine & Tool Corp.	USA	546	75-01105
Ancra CHS BV.	Netherlands	291	75-00510
Bridport Aviation Products	UK	435	75-08400
Equipos Industriales de Manutención	Spain	366	75-02850
Intamar Logistics Inc.	USA	585	75-47050
Irvin Aerospace Canada Limited	Canada	73	75-08300
Irvin Aerospace Limited	UK	462	75-32050
Irvin Aerospace SpA	Italy	235	75-12200

613 70 10 50 Pallets / pads, air drop

Irvin Aerospace Canada Limited	Canada	73	75-08300
Irvin Aerospace Limited	UK	462	75-32050
Irvin Aerospace SpA	Italy	235	75-12200
Technical Products Group Inc.	USA	620	75-86764

613 70 10 55 Parachutes, air supply drop

C.I.A.M.-E.A. Sarl	Italy	233	75-05150
Irvin Aerospace Canada Limited	Canada	73	75-08300
Irvin Aerospace Limited	UK	462	75-32050
Irvin Aerospace SpA	Italy	235	75-12200
Irvin Industries Inc.	USA	586	75-48960
North American Aerodynamics Inc.	USA	599	75-65200
Research Institute for	Ukraine	406	60-00990
Sächsische Spezialkonfektion GmbH	Germany	174	75-25880

613 70 40 Logistics equipment, ground-based

AB Carl Munters	Sweden	377	75-08230
Air-Log Ltd	UK	429	75-01800
Aircraft Appliances & Equipment Ltd	Canada	68	75-00430
Ancra CHS BV.	Netherlands	291	75-00510
EPS Logistics Technology Ltd	UK	502	80-09550
GEC-Marconi Avionics Ltd	UK	454	75-25320
Giat Industries	France	126	55-00100
M. Paz Logistics Ltd	Israel	224	80-01600
SpanSet Inter AG	Switzerland	387	75-13550

613 70 40 30 Flexible liquid containers

Avon Technical Products	UK	432	75-05150
Pronal SA	France	141	75-31100
SEI Industries Ltd	Canada	76	75-12750

614 Base security systems & equipment

ABB Leitungsbau GmbH	Germany	164	75-00055
Alpha Zaicon Tech. Inc.	Canada	68	75-00475
Bolloré Protection	France	133	75-06970
Booth Industries Limited	UK	435	75-07670
Caledonian Mining Co. Ltd	UK	438	75-11250
CCS International Ltd	USA	562	75-17250
CEM Systems Ltd	UK	439	75-12270
Chubb Alarms Ltd	UK	440	75-13150
Chubb Safe Equipment Co.	UK	440	75-13205
Chubb Security plc	UK	440	75-13207
Cochrane Steel Products (PTY) Ltd	S. Africa	357	75-03000
Cotag International	UK	443	75-15235
Dornier GmbH	Germany	167	75-06700
EF International SA	Switzerland	385	75-03260
EG&G Inc.	USA	571	75-29555
Elkosta VKS Dr Dörnemann GmbH & Co	Germany	168	75-07570
Fairmile Fencing Ltd	UK	450	75-22370
Garrett Metal Detectors	USA	577	75-36950
GEL Engineering Ltd	UK	454	75-25500
Grinaker System Technologies	S. Africa	360	80-06500

Guardall Ltd	UK	*456*	75-27100
Ilco Unican Inc.	Canada	*72*	75-07725
Insulgard Corp.	USA	*585*	75-47020
Israel Military Industries Ltd	Israel	*220*	75-10300
Israel Product Research Co. Ltd	Israel	*219*	75-10000
James Hardie Building Services Ltd	Australia	*17*	75-10178
James Hardie Engineering	New Zealand	*303*	80-00900
J.S. Chinn Project Engineering Ltd	UK	*440*	75-13060
Karl International Ltd	UK	*463*	75-33240
La-Vey Security International Ltd	UK	*506*	80-14215
Litton Industries Inc.	USA	*590*	75-55400
Londesborough Security Equipment plc	UK	*465*	75-36300
M. Paz Logistics Ltd	Israel	*224*	80-01600
Morgan Marine Ltd	UK	*469*	75-40865
P W Allen & Co Ltd	UK	*429*	75-02740
Perimeter Products Inc.	USA	*603*	75-69290
PK Electronic International Ltd	UK	*475*	75-45720
Racal Communications Inc.	USA	*606*	75-72380
Racal-Comsec Ltd	UK	*477*	75-47665
Raytheon E-Systems	USA	*608*	75-73302
Seaboard Electronics Co.	USA	*613*	75-78910
Security Defense Systems Corp.	USA	*613*	75-79080
Senstar-Stellar Corp	Canada	*76*	75-12760
Shorrock Integrated Systems	UK	*483*	75-53450
Siemens Aktiengesellschaft	Germany	*175*	75-27700
SISDEF Ltda	Chile	*85*	80-01940
Sleeman Engineering	UK	*484*	75-54400
Sparton Technology Inc.	USA	*616*	75-82950
SPL Ordnance Ltd	Israel	*225*	80-01935
Sverdrup Corp.	USA	*643*	80-30650
THORN Security Ltd	UK	*488*	75-59770
Tomsed Corp.	USA	*623*	75-89778
Vidisco Ltd	UK	*491*	75-63205
Vision Products Pty Ltd	Australia	*23*	75-25000
Visual Methods Inc.	USA	*626*	75-94320
Vitro Corp.	USA	*644*	80-34400
W.H. Barker & Sons Engineers Ltd	UK	*493*	75-65750
Witters & Co	USA	*628*	75-97150
Zettler GmbH Kommunikation und	Germany	*177*	75-34930

614 10 Perimeter security systems

ABB Leitungsbau GmbH	Germany	*164*	75-00055
BDL Systems Ltd	UK	*433*	75-06080
Bolloré Protection	France	*133*	75-06970
Caledonian Mining Co. Ltd	UK	*438*	75-11250
Canadian Marconi Co.	Canada	*69*	75-02600
Chicago Bullet Proof Systems	USA	*563*	75-18200
Chubb Alarms Ltd	UK	*440*	75-13150
Chubb Security plc	UK	*440*	75-13207
Computing Devices Company Ltd	UK	*442*	75-14700
Dabico Inc.	USA	*567*	75-23145
Diamond Electronics Inc.	USA	*569*	75-26450
EG&G Inc.	USA	*571*	75-29555
Elkosta VKS Dr Dörnemann GmbH & Co	Germany	*168*	75-07570
Fairmile Fencing Ltd	UK	*450*	75-22370
General Electronics Communications	Greece	*189*	75-01300
General Motors of Canada Ltd	Canada	*71*	75-05900
Geoquip Ltd	UK	*454*	75-25415
Grinaker System Technologies	S. Africa	*360*	80-06500
Guardall Ltd	UK	*456*	75-27100
Israel Product Research Co. Ltd	Israel	*219*	75-10000
Kollsman	USA	*589*	75-53150
Litton Industries Inc.	USA	*590*	75-55400
M. Paz Logistics Ltd	Israel	*224*	80-01600
Magal Security Systems Ltd	Israel	*221*	75-12650
Metis Construction & Trade Co. Inc.	Turkey	*403*	80-01330
Perimeter Products Inc.	USA	*603*	75-69290
PK Electronic International Ltd	UK	*475*	75-45720
Racal Communications Inc.	USA	*606*	75-72380
Remsdaq Ltd	UK	*478*	75-49210
Seaboard Electronics Co.	USA	*613*	75-78910
Senstar-Stellar Corp	Canada	*76*	75-12760
Shorrock Integrated Systems	UK	*483*	75-53450
SISDEF Ltda	Chile	*85*	80-01940
Sparton Technology Inc.	USA	*616*	75-82950
SPL Ordnance Ltd	Israel	*225*	80-01935
Strumech Versatower Ltd	UK	*486*	75-57530
THORN Security Ltd	UK	*488*	75-59770
TIRAX AG	Switzerland	*387*	75-14410
Vision Products Pty Ltd	Australia	*23*	75-25000
Vitro Corp.	USA	*644*	80-34400

614 10 30 Perimeter fences

Alpha Zaicon Tech. Inc.	Canada	*68*	75-00475
Caledonian Mining Co. Ltd	UK	*438*	75-11250
Elkosta VKS Dr Dörnemann GmbH & Co	Germany	*168*	75-07570
Eusung Industrial Co. Ltd	Korea (ROK)	*265*	75-01200
Fairmile Fencing Ltd	UK	*450*	75-22370
General Electronics Communications	Greece	*189*	75-01300
Grinaker System Technologies	S. Africa	*360*	80-06500
Guardall Ltd	UK	*456*	75-27100
Israel Product Research Co. Ltd	Israel	*219*	75-10000
Metis Construction & Trade Co. Inc.	Turkey	*403*	80-01330
SPL Ordnance Ltd	Israel	*225*	80-01935

614 10 30 50 Perimeter fences, high security

Bolloré Protection	France	*133*	75-06970
Elkosta VKS Dr Dörnemann GmbH & Co	Germany	*168*	75-07570
Fairmile Fencing Ltd	UK	*450*	75-22370
Karl International Ltd	UK	*463*	75-33240
Metis Construction & Trade Co. Inc.	Turkey	*403*	80-01330
SPL Ordnance Ltd	Israel	*225*	80-01935

614 10 50 Perimeter patrol vehicles, autonomous robot

Framatome	France	*136*	75-18405
General Motors of Canada Ltd	Canada	*71*	75-05900

614 10 60 Sensor / detector systems, perimeter security

ABB Leitungsbau GmbH	Germany	*164*	75-00055
Chubb Alarms Ltd	UK	*440*	75-13150
Delft Sensor Systems	**Belgium**	***46***	**75-02500**
Delft Sensor Systems	Netherlands	*292*	75-01590
EG&G Inc.	USA	*571*	75-29555
Eltec Instruments Inc.	USA	*573*	75-30880
Foxtronics Inc.	USA	*576*	75-35752
Geoquip Ltd	UK	*454*	75-25415
Giat Industries	France	*126*	55-00100
Guardall Ltd	UK	*456*	75-27100
Israel Product Research Co. Ltd	Israel	*219*	75-10000
Karl International Ltd	UK	*463*	75-33240
Kollsman	USA	*589*	75-53150
Litton Industries Inc.	USA	*590*	75-55400
M. Paz Logistics Ltd	Israel	*224*	80-01600
Magal Security Systems Ltd	Israel	*221*	75-12650
Metis Construction & Trade Co. Inc.	Turkey	*403*	80-01330
Perimeter Products Inc.	USA	*603*	75-69290
Pulnix America Inc.	USA	*606*	75-71815
Racal Communications Inc.	USA	*606*	75-72380
Racal-Comsec Ltd	UK	*477*	75-47665
Remsdaq Ltd	UK	*478*	75-49210
Senstar-Stellar Corp	Canada	*76*	75-12760
Shorrock Integrated Systems	UK	*483*	75-53450
SISDEF Ltda	Chile	*85*	80-01940
Sparton Technology Inc.	USA	*616*	75-82950
SPL Ordnance Ltd	Israel	*225*	80-01935
Sverdrup Corp.	USA	*643*	80-30650
THORN Security Ltd	UK	*488*	75-59770
Trans Security Systems & Technology	Israel	*223*	75-23055
Vision Products Pty Ltd	Australia	*23*	75-25000
Vision Research Co.	UK	*491*	75-63360
Vision Systems Ltd	Australia	*23*	75-22890
Vitro Corp.	USA	*644*	80-34400

614 20 Personnel-related security systems

Applied Holographics plc	UK	*431*	75-03520
Bolloré Protection	France	*133*	75-06970
CEM Systems Ltd	UK	*439*	75-12270
Chubb Alarms Ltd	UK	*440*	75-13150
Civil Defence Supply	UK	*440*	75-13350
Cotag International	UK	*443*	75-15235
Diamond Electronics Inc.	USA	*569*	75-26450
EG&G Astrophysics Research Ltd	UK	*448*	75-19670
EG&G Inc.	USA	*571*	75-29555
Elkosta VKS Dr Dörnemann GmbH & Co	Germany	*168*	75-07570
Ilco Unican Inc.	Canada	*72*	75-07725

Litton Industries Inc.	USA	590	75-55400
M. Paz Logistics Ltd	Israel	224	80-01600
Metis Construction & Trade Co. Inc.	Turkey	403	80-01330
National Bullet Proof Inc.	USA	598	75-63800
Remsdaq Ltd	UK	478	75-49210
Seaboard Electronics Co.	USA	613	75-78910
Security Defense Systems Corp.	USA	613	75-79080
SPL Ordnance Ltd.	Israel	225	80-01935
Sverdrup Corp.	USA	643	80-30650
Synergistics Inc.	USA	619	75-85550
Tomsed Corp.	USA	623	75-89778
Visual Methods Inc.	USA	626	75-94320
Vitro Corp.	USA	644	80-34400

614 20 10 Personnel identification systems

Chubb Alarms Ltd	UK	440	75-13150
Cotag International	UK	443	75-15235
Diamond Electronics Inc.	USA	569	75-26450
Litton Industries Inc.	USA	590	75-55400
Metis Construction & Trade Co. Inc.	Turkey	403	80-01330
Remsdaq Ltd	UK	478	75-49210
Synergistics Inc.	USA	619	75-85550
Visual Methods Inc.	USA	626	75-94320

614 20 20 Personnel X-ray check equipment

CCS International Ltd	USA	562	75-17250
EG&G Astrophysics Research Ltd.	UK	448	75-19670
EG&G Inc.	USA	571	75-29555
SPL Ordnance Ltd.	Israel	225	80-01935
Todd Research Ltd	UK	488	75-60360

614 20 30 Metal detection equipment, fixed

Corrigan Technologies Ltd	Canada	78	80-00380
EG&G Inc.	USA	571	75-29555
Garrett Metal Detectors	USA	577	75-36950
Henotex (1981) Inc	Canada	72	75-07100
Infinetics Inc.	USA	584	75-46700
M. Paz Logistics Ltd	Israel	224	80-01600
P W Allen & Co Ltd	UK	429	75-02740
PK Electronic International Ltd	UK	475	75-45720
Security Defense Systems Corp.	USA	613	75-79080
Sparton Technology Inc.	USA	616	75-82950
SPL Ordnance Ltd.	Israel	225	80-01935

614 20 40 Metal detection equipment, portable

Adams Electronics	UK	427	75-00580
Corrigan Technologies Ltd	Canada	78	80-00380
CPAD Technologies Inc.	Canada	78	80-00386
Garrett Metal Detectors	USA	577	75-36950
Graseby Dynamics Ltd	UK	455	75-26590
Henotex (1981) Inc	Canada	72	75-07100
Infinetics Inc.	USA	584	75-46700
M. Paz Logistics Ltd	Israel	224	80-01600
P W Allen & Co Ltd	UK	429	75-02740
PK Electronic International Ltd	UK	475	75-45720
Sas R&D Services Inc.	USA	642	80-26870
Sparton Technology Inc.	USA	616	75-82950
SPL Ordnance Ltd.	Israel	225	80-01935

614 30 Closed circuit TV systems

BDL Systems Ltd	UK	433	75-06080
Chubb Alarms Ltd	UK	440	75-13150
Diamond Electronics Inc.	USA	569	75-26450
Frazier Aviation Inc.	USA	576	75-35970
JAI A/S	Denmark	103	75-09400
James Hardie Engineering	New Zealand	303	80-00900
La-Vey Security International Ltd	UK	506	80-14215
Litton Industries Inc.	USA	590	75-55400
Metis Construction & Trade Co. Inc.	Turkey	403	80-01330
Naval Electronics Inc.	USA	598	75-64025
Photo-Scan Systems Ltd.	UK	508	80-21700
PK Electronic International Ltd	UK	475	75-45720
Puritan-Bennett Aero Systems Co.	USA	606	75-71890
Remsdaq Ltd	UK	478	75-49210
Security Research Ltd	UK	482	75-52450
Shorrock Integrated Systems	UK	483	75-53450
SITEP SpA.	Italy	239	75-22595
Strumech Versatower Ltd	UK	486	75-57530
THORN Security Ltd	UK	488	75-59770
Vision Products Pty Ltd	Australia	23	75-25000
Vision Research Co.	UK	491	75-63360
Vision Systems Ltd	Australia	23	75-22890

614 40 Bunkers, security

Karl International Ltd	UK	463	75-33240
Metis Construction & Trade Co. Inc.	Turkey	403	80-01330
Spantech Pty Ltd	Australia	21	75-19573

614 50 Enclosures, shielded EMI / NEMP / Tempest

AlliedSignal Aerospace Canada	Canada	68	75-00470
Compac	USA	565	75-20400
Industrial Acoustics Company Ltd	UK	460	75-31300
Metis Construction & Trade Co. Inc.	Turkey	403	80-01330
Morgan Marine Ltd	UK	469	75-40865
PK Electronic International Ltd	UK	475	75-45720
Presentey Engineering	Canada	75	75-11500
Sverdrup Corp.	USA	643	80-30650
Technical Products Group Inc.	USA	620	75-86764
The Graphics Co. of America	USA	579	75-40360

614 50 25 Doors, security

Booth Industries Limited	UK	435	75-07670
Chubb Safe Equipment Co.	UK	440	75-13205
Frazier Aviation Inc.	USA	576	75-35970
Insulgard Corp.	USA	585	75-47020
Metis Construction & Trade Co. Inc.	Turkey	403	80-01330
National Bullet Proof Inc.	USA	598	75-63800
Remsdaq Ltd	UK	478	75-49210
Witters & Co	USA	628	75-97150

614 50 30 Doors, EMP / blast resistant

Booth Industries Limited	UK	435	75-07670
Daloc Sheltec AB	Sweden	376	75-04000
Envirodoor Markus Ltd	UK	449	75-21090
G Kull AG, Lunor	Switzerland	386	75-07200
L&F Industries	USA	589	75-53690
Metis Construction & Trade Co. Inc.	Turkey	403	80-01330
Morgan Marine Ltd	UK	469	75-40865
Overly Manufacturing Co.	USA	602	75-67750
Presentey Engineering	Canada	75	75-11500
Sleeman Engineering	UK	484	75-54400

614 50 40 Filters, NEMP

EEV Canada Ltd	Canada	71	75-04700
G Kull AG, Lunor	Switzerland	386	75-07200
Metis Construction & Trade Co. Inc.	Turkey	403	80-01330
Presentey Engineering	Canada	75	75-11500

614 60 Protection systems, harbour

BTR Silvertown Ltd	UK	437	75-10830
Geoquip Ltd	UK	454	75-25415
Israel Product Research Co. Ltd	Israel	219	75-10000
SpanSet Inter AG	Switzerland	387	75-13550
Sparton Technology Inc.	USA	616	75-82950
Sverdrup Corp.	USA	643	80-30650

615 Personnel-related facilities & equipment, fixed base

Avery Berkel	UK	432	75-04560
Envirovac Inc.	USA	574	75-31920
GEL Engineering Ltd	UK	454	75-25500
Indústrias de Defesa	Portugal	327	55-00950
InfoLink Technologies Inc.	USA	585	75-46703
Maine Group Ltd	UK	466	75-37050
Pre-Mac (Kent) Ltd	UK	476	75-46660
Sverdrup Corp.	USA	643	80-30650
UO Equipment Co.	USA	625	75-92870

615 05 Barracks equipment, fixed base

Maine Group Ltd	UK	466	75-37050

615 05 20 Barrack beds

Maine Group Ltd	UK	466	75-37050

615 10 Catering facilities & equipment, fixed base

A E Atherton & Sons Pty Ltd	Australia	12	75-01570
Avery Berkel	UK	432	75-04560
Exstar International Corp.	USA	575	75-33000
GEL Engineering Ltd	UK	454	75-25500
Lennys Commercial Kitchens Pty Ltd	Australia	18	75-12370
Pland	UK	475	75-45740
Plastico Ltd	UK	475	75-45950
Progress-Werk Oberkirch AG	Germany	173	75-24320
Stott Benham	UK	486	75-57460

615 10 40 Kitchens, fully equipped, fixed base

A E Atherton & Sons Pty Ltd	Australia	12	75-01570
Carrozzeria Boneschi Sarl	Italy	233	75-05000
Indústrias de Defesa	Portugal	327	55-00950
Lennys Commercial Kitchens Pty Ltd	Australia	18	75-12370
Palbam Metal Works	Israel	221	75-17000
Stott Benham	UK	486	75-57460

615 10 70 Kitchen equipment, fixed base

A E Atherton & Sons Pty Ltd	Australia	12	75-01570
Avery Berkel	UK	432	75-04560
Lennys Commercial Kitchens Pty Ltd	Australia	18	75-12370
Michsaf	Israel	221	75-14200
Palbam Metal Works	Israel	221	75-17000
Stott Benham	UK	486	75-57460

615 15 Drinking water treatment plants, fixed base

Everpure Inc.	USA	575	75-32680
Exstar International Corp.	USA	575	75-33000
Pre-Mac (Kent) Ltd	UK	476	75-46660

615 20 Medical facilities & equipment, fixed base

Adam Medical Products Ltd	Israel	217	75-00250
Avery Berkel	UK	432	75-04560
GEL Engineering Ltd	UK	454	75-25500
Malaysian Oxygen Berhad	Malaysia	277	75-00060
Medesign GmbH	Germany	172	75-20370
M.G. Electric (Colchester) Ltd	UK	468	75-38912
Seward Medical Limited	UK	483	75-53050
Sonomed Inc.	USA	616	75-82360
Sverdrup Corp.	USA	643	80-30650
Todd Research Ltd	UK	488	75-60360
Urals Optical & Mechanical Plant	Russia	339	75-53000
Vitalograph Ltd	UK	492	75-63395
W&H Dentalwerk Bürmoos GesmbH	Austria	32	75-00750

615 20 10 Aeromedical systems & equipment

Aero Innovations Inc.	USA	546	75-01108
Cyclone Aviation Products Ltd	Israel	218	75-04250
Medesign GmbH	Germany	172	75-20370

615 21 Hospitals, base

International Hospitals Group	UK	505	80-13410
Medesign GmbH	Germany	172	75-20370
Sverdrup Corp.	USA	643	80-30650
UO Equipment Co.	USA	625	75-92870

615 21 70 Hospital equipment, base

A E Atherton & Sons Pty Ltd	Australia	12	75-01570
Arjo Ltd.	UK	431	75-03715
Avery Berkel	UK	432	75-04560
Medesign GmbH	Germany	172	75-20370
Seward Medical Limited	UK	483	75-53050
Sonomed Inc.	USA	616	75-82360

615 21 70 10 Hospital beds

Arjo Ltd.	UK	431	75-03715
Hoskins Medical Equipment Ltd	UK	459	75-30000

615 21 70 50 Operating theatre equipment, base hospital

Carrozzeria Boneschi Sarl	Italy	233	75-05000
Eschmann Bros. & Walsh Ltd	UK	449	75-21500
Seward Medical Limited	UK	483	75-53050
Sonomed Inc.	USA	616	75-82360
Vickers Medical Ltd.	UK	514	80-65500

615 21 70 90 Ward furniture, hospital

Hoskins Medical Equipment Ltd	UK	459	75-30000
Medesign GmbH	Germany	172	75-20370
Seward Medical Limited	UK	483	75-53050

615 21 80 Medical supplies

BCB International Ltd	UK	433	75-06075
Malaysian Oxygen Berhad	Malaysia	277	75-00060
Medesign GmbH	Germany	172	75-20370
Seton Healthcare Group plc	UK	483	75-53000
Seward Medical Limited	UK	483	75-53050
Solvay Duphar BV	Netherlands	295	75-06850
UO Equipment Co.	USA	625	75-92870
Urals Optical & Mechanical Plant	Russia	339	75-53000

615 30 Furniture, barracks

A.P. Metalcraft Ltd	UK	430	75-03410
Hoskins Medical Equipment Ltd	UK	459	75-30000
Maine Group Ltd	UK	466	75-37050

615 50 Office equipment & supplies

Chubb Safe Equipment Co.	UK	440	75-13205
Eurochart	UK	449	75-21580
Gichner Systems Group	USA	579	75-39610
H.C. Slingsby plc	UK	484	75-54430
Lista International Corp.	USA	590	75-55350
Maine Group Ltd	UK	466	75-37050
McQueen Ltd	UK	467	75-38200
Oki Systems (UK) Ltd	UK	471	75-42967
Samas Roneo Ltd	UK	481	75-51650

615 50 30 Colour printers, large format

Oki Systems (UK) Ltd	UK	471	75-42967

615 50 50 Document handling machines

InfoLink Technologies Inc.	USA	585	75-46703

615 50 55 Document shredding machines

Data Disposal & Destruction	UK	444	75-16820
H.C. Slingsby plc	UK	484	75-54430
International Procurement Services	UK	461	75-31820
L&F Industries	USA	589	75-53690
Security Engineered Machinery	USA	613	75-79100

615 50 60 Stationery, forms

The Graphics Co. of America	USA	579	75-40360

615 70 Sewage treatment plants, fixed base

Envirovac Inc.	USA	*574*	75-31920
Exceltech International Corp.	USA	*575*	75-32725
Hamworthy Marine	UK	*457*	75-27850
Interdyne Corp.	USA	*637*	80-16050
Sverdrup Corp.	USA	*643*	80-30650
Taylor Woodrow Construction Ltd	UK	*513*	80-50300

615 70 50 Sewage treatment plants, packaged

Exceltech International Corp.	USA	*575*	75-32725
Hamworthy Marine	UK	*457*	75-27850
Interdyne Corp.	USA	*637*	80-16050

616 Training facilities & equipment

Advanced Technology Ltd	Israel	*217*	75-00400
Aerobel Defence Technology Ltd	UK	*496*	80-00520
Amherst Systems Inc.	USA	*553*	75-06600
Anite Systems Ltd	UK	*497*	80-01240
ATS Aerospace Inc.	Canada	*68*	75-00752
Aviation Simulation Technology Inc.	USA	*556*	75-11000
British Aerospace (Systems &	UK	*436*	75-08890
Bülow Air Target Systems AB	Sweden	*376*	75-03600
CAE Elektronic GmbH	Germany	*166*	75-04500
Calspan SRL Technology	USA	*540*	60-03000
Caswell International Corp.	USA	*562*	75-16990
Cenzin Co. Ltd.	Poland	*324*	75-00200
Cincinnati Electronics Corp.	USA	*564*	75-18640
Computing Devices Company Ltd	UK	*442*	75-14700
Conrac Elektron GmbH	Germany	*166*	75-05100
Cubic Defense Systems Inc.	USA	*567*	75-22900
Data Sciences Ltd.	UK	*500*	80-07740
Datamat Ingegneria dei Sistemi SpA	Italy	*233*	75-06500
Del Mar Avionics	USA	*568*	75-25362
Deutsche System-Technik GmbH	Germany	*167*	75-06320
Doron Precision Systems Inc.	USA	*569*	75-27110
DRS Technologies, Inc.	USA	*570*	75-27400
Dynamit Nobel GmbH	Germany	*167*	75-07100
EDS Defence Ltd	UK	*501*	80-09200
Elbit Systems Ltd	Israel	*218*	75-05300
Elettronica GmbH	Germany	*168*	75-07565
Ericsson Microwave Systems AB	**Sweden**	***377***	**75-04890**
Evans & Sutherland Computer Corp.	USA	*574*	75-32650
EW Simulation Technology Ltd	UK	*449*	75-22150
FN Herstal SA	Belgium	*46*	75-03600
FN Manufacturing Inc	USA	*576*	75-35320
Fokker Space & Systems	Netherlands	*293*	75-02325
Frasca International Inc.	USA	*576*	75-35900
Gajon Institute of Technology SpA	Italy	*229*	60-00080
GEC-Marconi Avionics Ltd	UK	*454*	75-25320
GMV SA	Spain	*367*	80-00800
Gunpell & Co. Ltd	UK	*456*	75-27170
Hadland Photonics Ltd	UK	*456*	75-27200
Haux-Life-Support GmbH	Germany	*169*	75-11970
Herley-Vega Systems	USA	*581*	75-43090
Hughes Training, Inc	USA	*583*	75-44900
Hydroscience Technologies Inc.	USA	*584*	75-45550
INDRA	Spain	*366*	75-03800
IVEX Corp.	USA	*586*	75-49950
Krauss-Maffei Wehrtechnik GmbH	Germany	*171*	75-16345
Laser & Atomic Research & Devt	Croatia	*94*	60-00450
Litton Industries Inc.	USA	*590*	75-55400
MacDonald Dettwiler and Associates	Canada	*78*	80-01310
Microelectronics, guidance and	Turkey	*402*	55-00050
Mitsubishi Precision Co. Ltd	Japan	*253*	75-06000
MS Instruments plc	UK	*469*	75-41265
Nanoquest Defence Products Ltd	UK	*470*	75-41890
Oerlikon-Contraves AG.	Switzerland	*387*	75-09250
Pacer Systems Inc.	USA	*602*	75-68000
Pains-Wessex Ltd	UK	*472*	75-43510
Pennant International Limited	UK	*473*	75-44450
Perceptronics Inc..	USA	*603*	75-69270
Polyformes Ltd	UK	*475*	75-46348
Polytronic AG	Switzerland	*387*	75-10250
Polytronic Inc	USA	*604*	75-70870
PRB Associates Inc.	USA	*640*	80-24700
Primary Image Ltd	UK	*476*	75-46790
Raytheon E-Systems	USA	*608*	75-73302
Reflectone Inc	USA	*608*	75-74150
Reflectone UK Ltd.	UK	*478*	75-48380
RGB Spectrum	USA	*609*	75-75000
Rockwell	USA	*610*	75-75930
Ruggieri Technologie	France	*142*	75-33500
Sargent Fletcher Inc.	USA	*612*	75-78150
SEOS Displays Ltd	UK	*482*	75-52830
Shield International Ltd	UK	*511*	80-35500
Ship Analytics Inc.	USA	*614*	75-80455
Shorts Missile Systems Ltd	**UK**	***483***	**75-53550**
SISDEF Ltda	Chile	*85*	80-01940
Sogitec	France	*146*	75-42050
Somchem	S. Africa	*359*	75-19200
SP aerospace and vehicle systems bv	**Netherlands**	***295***	**75-06860**
Spartanics	USA	*616*	75-82850
SPL Ordnance Ltd.	Israel	*225*	80-01935
STN ATLAS Elektronik GmbH	Germany	*165*	75-01150
Strategic Technologies Pte Ltd	**Singapore**	***348***	**75-04400**
Sverdrup Corp.	USA	*643*	80-30650
Sverdrup Technology Inc.	USA	*643*	80-30680
Tamares Ltd	UK	*487*	75-58500
Tector Visual Systems	UK	*488*	75-59020
Teledyne Brown Engineering	USA	*620*	75-87510
Telegenix Inc.	USA	*621*	75-87610
The Boeing Company	USA	*559*	75-14250
Thomson Sintra	France	*148*	75-45135
Thomson Training & Simulation Ltd.	UK	*488*	75-59567
TIRAX AG	Switzerland	*387*	75-14410
Trenton Technical Services Ltd	UK	*489*	75-61110
UK Training Ltd	UK	*514*	80-60000
United Defense L.P.	USA	*624*	75-92050
Van Rietschoten & Houwens	Netherlands	*296*	75-07900
Vickers Shipbuilding & Eng Ltd	UK	*491*	75-63100
Wegmann & Co GmbH	Germany	*177*	75-33700
Zen Technologies Ltd	India	*205*	75-03000

616 10 Mission & weapon training systems & simulators

Anite Systems Ltd	UK	*497*	80-01240
ATS Aerospace Inc.	Canada	*68*	75-00752
Aviation Simulation Technology Inc.	USA	*556*	75-11000
BVR Technologies Ltd	Israel	*218*	75-03350
CAE Electronics Ltd	Canada	*69*	75-01770
CAE Elektronic GmbH	Germany	*166*	75-04500
CAE Invertron Ltd.	UK	*438*	75-11145
Calspan SRL Technology	USA	*540*	60-03000
Cenzin Co. Ltd.	Poland	*324*	75-00200
Cubic Defense Systems Inc.	USA	*567*	75-22900
Dassault Aviation Group	France	*135*	75-12975
DRS Technologies, Inc.	USA	*570*	75-27400
Dynamit Nobel GmbH	Germany	*167*	75-07100
Elbit Systems Ltd	Israel	*218*	75-05300
Evans & Sutherland Computer Corp.	USA	*574*	75-32650
EW Simulation Technology Ltd	UK	*449*	75-22150
FATS, Inc.	USA	*575*	75-33877
FN Manufacturing Inc	USA	*576*	75-35320
Force Computers GmbH	Germany	*168*	75-09270
Frasca International Inc.	USA	*576*	75-35900
GDI Simulation	France	*137*	75-19700
GEC-Marconi Avionics Ltd	UK	*454*	75-25320
GEC-Marconi Hazeltine Corp.	USA	*577*	75-37414
Hayes Targets	USA	*603*	75-69135
Hughes Electronics Corp.	USA	*583*	75-44865
INDRA	Spain	*366*	75-03800
IVEX Corp.	USA	*586*	75-49950
Krauss-Maffei Wehrtechnik GmbH	Germany	*171*	75-16345
Litton Industries Inc.	USA	*590*	75-55400
Logica UK Limited	UK	*506*	80-14350
Microturbo SA	France	*140*	75-27000
Oerlikon-Contraves AG.	Switzerland	*387*	75-09250
Ogle Design Ltd	UK	*508*	80-19400
Pacer Systems Inc.	USA	*602*	75-68000
Pennant International Limited	UK	*473*	75-44450
Perceptronics Inc..	USA	*603*	75-69270
Polytronic AG	Switzerland	*387*	75-10250
Polytronic Inc	USA	*604*	75-70870
Primary Image Ltd	UK	*476*	75-46790
Reflectone Inc	USA	*608*	75-74150
Reflectone UK Ltd.	UK	*478*	75-48380
RGB Spectrum	USA	*609*	75-75000
Saab Training Systems AB	Sweden	*378*	75-09250
SEOS Displays Ltd	UK	*482*	75-52830
Shield International Ltd	UK	*511*	80-35500
Ship Analytics Inc.	USA	*614*	75-80455
Shorts Missile Systems Ltd	**UK**	***483***	**75-53550**
SISDEF Ltda	Chile	*85*	80-01940

Sogitec	France	*146*	75-42050
Spitz Inc.	USA	*617*	75-83350
STN ATLAS Elektronik GmbH	Germany	*165*	75-01150
Strategic Technologies Pte Ltd	**Singapore**	***348***	**75-04400**
Sverdrup Corp.	USA	*643*	80-30650
Sverdrup Technology Inc.	USA	*643*	80-30680
Tadiran Ltd	Israel	*222*	75-21600
Tamares Ltd	UK	*487*	75-58500
Tector Visual Systems	UK	*488*	75-59020
The Boeing Company	USA	*559*	75-14250
Thomson Sintra	France	*148*	75-45135
Thomson Training & Simulation Ltd.	UK	*488*	75-59567
Thomson-CSF	France	*147*	75-45100
Trenton Technical Services Ltd	UK	*489*	75-61110
Vickers Shipbuilding & Eng Ltd	UK	*491*	75-63100
Wegmann & Co GmbH	Germany	*177*	75-33700

616 12 Mission & weapon training systems & simulators, aircraft-related

Anite Systems Ltd	UK	*497*	80-01240
CAE Electronics Ltd	Canada	*69*	75-01770
CAE Elektronic GmbH	Germany	*166*	75-04500
Calspan SRL Technology	USA	*540*	60-03000
Cubic Defense Systems Inc.	USA	*567*	75-22900
Dassault Aviation Group	France	*135*	75-12975
Datamat Ingegneria dei Sistemi SpA	Italy	*233*	75-06500
Evans & Sutherland Computer Corp.	USA	*574*	75-32650
Force Computers GmbH	Germany	*168*	75-09270
Frasca International Inc.	USA	*576*	75-35900
Hayes Targets	USA	*603*	75-69135
Hughes Electronics Corp.	USA	*583*	75-44865
Industrieanlagen-Betriebs GmbH	Germany	*178*	80-00945
IVEX Corp.	USA	*586*	75-49950
Kennedy Aerospace Corp.	USA	*540*	60-11111
Litton Industries Inc.	USA	*591*	75-55400
Logica UK Limited	UK	*506*	80-14350
Metric Systems Corp.	USA	*596*	75-60800
Ogle Design Ltd	UK	*508*	80-19400
Pacer Systems Inc.	USA	*602*	75-68000
Pennant International Limited	UK	*473*	75-44450
Perceptronics Inc.	USA	*603*	75-69270
Primary Image Ltd	UK	*476*	75-46790
Reflectone Inc	USA	*608*	75-74150
Reflectone UK Ltd.	UK	*478*	75-48380
RGB Spectrum	USA	*609*	75-75000
SBS Technologies Inc.	USA	*612*	75-78330
SECAPEM	France	*143*	75-34750
SEOS Displays Ltd	UK	*482*	75-52830
Sogitec	France	*146*	75-42050
Spitz Inc.	USA	*617*	75-83350
Strategic Technologies Pte Ltd	**Singapore**	***348***	**75-04400**
Sverdrup Corp.	USA	*643*	80-30650
Tector Visual Systems	UK	*488*	75-59020
The Boeing Company	USA	*559*	75-14250
Thomson Training & Simulation Ltd.	UK	*488*	75-59567
Thomson-CSF	France	*147*	75-45100

616 12 10 Air combat training systems

BVR Technologies Ltd	Israel	*218*	75-03350
CAE Elektronic GmbH	Germany	*166*	75-04500
Cartwright Electronics Inc.	USA	*562*	75-16910
Cincinnati Electronics Corp.	USA	*564*	75-18640
Cubic Defense Systems Inc.	USA	*567*	75-22900
EDS Defence Ltd	UK	*501*	80-09200
Evans & Sutherland Computer Corp.	USA	*574*	75-32650
Frasca International Inc.	USA	*576*	75-35900
GEC-Marconi Avionics Ltd	UK	*454*	75-25320
Hayes Targets	USA	*603*	75-69135
IVEX Corp.	USA	*586*	75-49950
Metric Systems Corp.	USA	*596*	75-60800
Perceptronics Inc.	USA	*603*	75-69270
Reflectone Inc	USA	*608*	75-74150
Reflectone UK Ltd.	UK	*478*	75-48380
RGB Spectrum	USA	*609*	75-75000
Sogitec	France	*146*	75-42050
Tector Visual Systems	UK	*488*	75-59020

616 12 20 Ranges, air-to-ground weapon training

CAE Elektronic GmbH	Germany	*166*	75-04500
Cartwright Electronics Inc.	USA	*562*	75-16910
Evans & Sutherland Computer Corp.	USA	*574*	75-32650
MS Instruments plc	UK	*469*	75-41265
Sabre Computers Int'l Ltd	UK	*481*	75-51150
Sogitec	France	*146*	75-42050
Strategic Technologies Pte Ltd	**Singapore**	***348***	**75-04400**
Sverdrup Technology Inc.	USA	*643*	80-30680

616 12 60 Simulators, aircraft mission-related

BVR Technologies Ltd	Israel	*218*	75-03350
CAE Elektronic GmbH	Germany	*166*	75-04500
Calspan SRL Technology	USA	*540*	60-03000
Cincinnati Electronics Corp.	USA	*564*	75-18640
Dassault Aviation Group	France	*135*	75-12975
Datamat Ingegneria dei Sistemi SpA	Italy	*233*	75-06500
Emro Engineering Co. Inc.	USA	*573*	75-31450
Evans & Sutherland Computer Corp.	USA	*574*	75-32650
Force Computers GmbH	Germany	*168*	75-09270
Frasca International Inc.	USA	*576*	75-35900
GEC-Marconi Avionics Ltd	UK	*454*	75-25320
Haux-Life-Support GmbH	Germany	*169*	75-11970
IVEX Corp.	USA	*586*	75-49950
Litton Industries Inc.	USA	*591*	75-55400
Logica UK Limited	UK	*506*	80-14350
Logicon Geodynamics Inc.	USA	*638*	80-18625
National Aerospace Laboratory	Netherlands	*290*	60-01400
Ogle Design Ltd	UK	*508*	80-19400
Pacer Systems Inc.	USA	*602*	75-68000
Pennant International Limited	UK	*473*	75-44450
Perceptronics Inc.	USA	*603*	75-69270
Primary Image Ltd	UK	*476*	75-46790
Reflectone Inc	USA	*608*	75-74150
Reflectone UK Ltd.	UK	*478*	75-48380
RGB Spectrum	USA	*609*	75-75000
SEOS Displays Ltd	UK	*482*	75-52830
Ship Analytics Inc.	USA	*614*	75-80455
Sogitec	France	*146*	75-42050
Strategic Technologies Pte Ltd	**Singapore**	***348***	**75-04400**
Sverdrup Technology Inc.	USA	*643*	80-30680
Tector Visual Systems	UK	*488*	75-59020
The Boeing Company	USA	*559*	75-14250
Thomson Training & Simulation Ltd.	UK	*488*	75-59567
Thomson-CSF	France	*147*	75-45100

616 12 60 10 Simulators, air combat training

BVR Technologies Ltd	Israel	*218*	75-03350
CAE Elektronic GmbH	Germany	*166*	75-04500
Calspan SRL Technology	USA	*540*	60-03000
Cincinnati Electronics Corp.	USA	*564*	75-18640
EASAMS Ltd	UK	*501*	80-09000
Emro Engineering Co. Inc.	USA	*573*	75-31450
Evans & Sutherland Computer Corp.	USA	*574*	75-32650
Force Computers GmbH	Germany	*168*	75-09270
Frasca International Inc.	USA	*576*	75-35900
IVEX Corp.	USA	*586*	75-49950
Kennedy Aerospace Corp.	USA	*540*	60-11111
Perceptronics Inc.	USA	*603*	75-69270

Reflectone Inc	USA	608	75-74150
Reflectone UK Ltd.	UK	478	75-48380
SEOS Displays Ltd	UK	482	75-52830
Serco Aviation Services	UK	511	80-34150
Sogitec	France	146	75-42050
Spitz Inc..	USA	617	75-83350
Tector Visual Systems	UK	488	75-59020

616 12 60 25 Simulators, EW training, aircraft-related

CAE Elektronic GmbH	Germany	166	75-04500
Calian Technology Ltd	Canada	69	75-01800
Calspan SRL Technology	USA	540	60-03000
Cincinnati Electronics Corp.	USA	564	75-18640
ELTA Electronics Industries Ltd	Israel	220	75-10200
Evans & Sutherland Computer Corp.	USA	574	75-32650
Force Computers GmbH	Germany	168	75-09270
Frasca International Inc.	USA	576	75-35900
Litton Industries Inc.	USA	590	75-55400
MPD Technologies Inc..	USA	597	75-63125
Pacer Systems Inc.	USA	602	75-68000
Phase Two Industries	USA	603	75-69560
Primary Image Ltd	UK	476	75-46790
Reflectone Inc	USA	608	75-74150
Reflectone UK Ltd.	UK	478	75-48380
Ship Analytics Inc.	USA	614	75-80455
Sogitec	France	146	75-42050
Southwest Aerospace	USA	616	75-82470
Thomson Training & Simulation Ltd.	UK	488	75-59567

616 12 60 40 Simulators, helicopter combat

CAE Electronics Ltd	Canada	69	75-01770
Evans & Sutherland Computer Corp.	USA	574	75-32650
Force Computers GmbH	Germany	168	75-09270
Frasca International Inc.	USA	576	75-35900
IVEX Corp.	USA	586	75-49950
Reflectone UK Ltd.	UK	478	75-48380
SBS Technologies Inc.	USA	612	75-78330
SEOS Displays Ltd	UK	482	75-52830
Spitz Inc..	USA	617	75-83350
Tector Visual Systems	UK	488	75-59020

616 12 60 90 Simulator components, aircraft related

AMI Instruments Inc.	USA	553	75-06620
Applied Science Laboratories	USA	554	75-08200
CAE Elektronic GmbH	Germany	166	75-04500
Carco Electronics	USA	562	75-16700
Cranfield Aerospace Ltd	UK	500	80-07200
Emro Engineering Co. Inc.	USA	573	75-31450
Evans & Sutherland Computer Corp.	USA	574	75-32650
Frasca International Inc.	USA	576	75-35900
Ogle Design Ltd	UK	508	80-19400
Pandect Precision Components Ltd	UK	472	75-43650
Pennant International Limited	UK	473	75-44450
Primary Image Ltd	UK	476	75-46790
Reflectone UK Ltd.	UK	478	75-48380
RGB Spectrum	USA	609	75-75000
Riva Calzoni SpA	Italy	238	75-20500
SEOS Displays	USA	613	75-79450
SEOS Displays Ltd	UK	482	75-52830
Tector Visual Systems	UK	488	75-59020

616 12 60 90 80 Visual systems, aircraft simulator

AMI Instruments Inc.	USA	553	75-06620
CAE Elektronic GmbH	Germany	166	75-04500
Emro Engineering Co. Inc.	USA	573	75-31450
Evans & Sutherland Computer Corp.	USA	574	75-32650
Force Computers GmbH	Germany	168	75-09270
Frasca International Inc.	USA	576	75-35900
Hughes Electronics Corp.	USA	583	75-44865
IVEX Corp.	USA	586	75-49950
Kaiser Electronics	USA	587	75-51350
Mitsubishi Precision Co. Ltd	Japan	253	75-06000
Primary Image Ltd	UK	476	75-46790
Reflectone UK Ltd.	UK	478	75-48380
RGB Spectrum	USA	609	75-75000
SBS Technologies Inc.	USA	612	75-78330
SEOS Displays Ltd	UK	482	75-52830
Ship Analytics Inc.	USA	614	75-80455
Sira Ltd.	UK	484	75-54250
STN ATLAS Elektronik GmbH	Germany	165	75-01150
Tector Visual Systems	UK	488	75-59020

616 12 67 Target systems, airborne

ATS Aerospace Inc.	Canada	68	75-00752
Evans & Sutherland Computer Corp.	USA	574	75-32650
Hayes Targets	USA	603	75-69135
Southwest Aerospace	USA	616	75-82470
Target Technology Ltd	UK	487	75-58660

616 12 70 Mission & weapon training equipment, aircraft-related

Anite Systems Ltd	UK	497	80-01240
ATS Aerospace Inc.	Canada	68	75-00752
Bülow Air Target Systems AB.	Sweden	376	75-03600
BVR Technologies Ltd	Israel	218	75-03350
Calspan SRL Technology	USA	540	60-03000
Elettronica GmbH	Germany	168	75-07565
Evans & Sutherland Computer Corp.	USA	574	75-32650
GEC-Marconi Avionics Ltd	UK	454	75-25320
Hayes Targets	USA	603	75-69135
Ogle Design Ltd	UK	508	80-19400
Pains-Wessex Ltd	UK	472	75-43510
Perceptronics Inc..	USA	603	75-69270
Primary Image Ltd	UK	476	75-46790
Reflectone Inc	USA	608	75-74150
Reflectone UK Ltd.	UK	478	75-48380
Sargent Fletcher Inc.	USA	612	75-78150
SECAPEM	France	143	75-34750
Swiss Aircraft & Systems Co	Switzerland	384	55-00100
Tector Visual Systems	UK	488	75-59020
The Boeing Company.	USA	559	75-14250
Thomson Training & Simulation Ltd.	UK	488	75-59567

616 12 75 Training missiles, aircraft

ATS Aerospace Inc.	Canada	68	75-00752
Eurodef Aerospace & Defence	Greece	189	75-01295
Evans & Sutherland Computer Corp.	USA	574	75-32650
Kennedy Aerospace Corp.	USA	540	60-11111
Ogle Design Ltd	UK	508	80-19400
Sargent Fletcher Inc.	USA	612	75-78150

616 12 77 Targets, airborne

Air Target Sweden AB	Sweden	375	75-00850
ATS Aerospace Inc.	Canada	68	75-00752
Bülow Air Target Systems AB.	Sweden	376	75-03600
Evans & Sutherland Computer Corp.	USA	574	75-32650
Hayes Targets	USA	603	75-69135
Raytheon Aircraft Company	USA	607	75-73200
Swiss Aircraft & Systems Co	Switzerland	384	55-00100
Target Technology Ltd	UK	487	75-58660

616 12 77 30 Target drones, airborne (see codes 123.20.20-22)

Aeromot Ind.Mecanico-Metalurgica SA	Brazil	55	75-00635
Bülow Air Target Systems AB.	Sweden	376	75-03600
Evans & Sutherland Computer Corp.	USA	574	75-32650
Hayes Targets	USA	603	75-69135
Meteor CAE SpA	Italy	237	75-15400
Raytheon Aircraft Company	USA	607	75-73200
Swiss Aircraft & Systems Co	Switzerland	384	55-00100
Target Technology Ltd	UK	487	75-58660
TUSAS Aerospace Industries Inc.	Turkey	403	75-02120

616 12 77 60 Targets, towed, airborne

Air Target Services Pty Ltd	Australia	24	80-00840
Air Target Sweden AB	Sweden	375	75-00850
Bülow Air Target Systems AB.	Sweden	376	75-03600
Evans & Sutherland Computer Corp.	USA	574	75-32650
Flight Refuelling Limited	UK	451	75-23850

Hayes Targets	USA	603	75-69135
Ness Manufacturing Inc.	USA	598	75-64240
Pains-Wessex Ltd	UK	472	75-43510
Swiss Aircraft & Systems Co	Switzerland	384	55-00100

616 12 77 63 Targets, towed, airborne, drogue type

Air Target Sweden AB	Sweden	375	75-00850
Autoflug GmbH & Co	Germany	165	75-01400
Bülow Air Target Systems AB	Sweden	376	75-03600
Hayes Targets	USA	603	75-69135
Pioneer Aerospace Corp	USA	604	75-70200
Sabre Computers Int'l Ltd	UK	481	75-51150

616 12 77 65 Targets, towed, airborne, panel type

Bülow Air Target Systems AB	Sweden	376	75-03600
Pioneer Aerospace Corp	USA	604	75-70200
SECAPEM	France	143	75-34750
Southwest Aerospace	USA	616	75-82470

616 12 77 70 Target equipment, airborne

Air Target Services Pty Ltd	Australia	24	80-00840
ATS Aerospace Inc.	Canada	68	75-00752
Bülow Air Target Systems AB	Sweden	376	75-03600
Hayes Targets	USA	603	75-69135
Southwest Aerospace	USA	616	75-82470

616 12 77 75 Parachutes, air target recovery

Bülow Air Target Systems AB	Sweden	376	75-03600
Irvin Aerospace Limited	UK	462	75-32050
Irvin Aerospace SpA	Italy	235	75-12200
Irvin Industries Inc.	USA	586	75-48960
Pioneer Aerospace Corp	USA	604	75-70200
Sächsische Spezialkonfektion GmbH	Germany	174	75-25880
Sala International Pty Ltd	Australia	21	75-18980

616 12 77 76 Scoring systems, air target

Air Target Services Pty Ltd	Australia	24	80-00840
Air Target Sweden AB	Sweden	375	75-00850
Bülow Air Target Systems AB	Sweden	376	75-03600
Cartwright Electronics Inc.	USA	562	75-16910
ELECMA	France	136	75-14400
Emro Engineering Co. Inc.	USA	573	75-31450
Hayes Targets	USA	603	75-69135
Sabre Computers Int'l Ltd	UK	481	75-51150
SECAPEM	France	143	75-34750
SEXTANT Avionique SA	France	144	75-35000
Swiss Aircraft & Systems Co	Switzerland	384	55-00100

616 12 77 76 50 Scoring sensors, air target

Bülow Air Target Systems AB	Sweden	376	75-03600
Cartwright Electronics Inc.	USA	562	75-16910
Hayes Targets	USA	603	75-69135
Sabre Computers Int'l Ltd	UK	481	75-51150
SEXTANT Avionique SA	France	144	75-35000
Société Anonyme d'Études	France	145	75-37100

616 12 77 76 55 Scoring sensors, radar doppler, air target

Cartwright Electronics Inc.	USA	562	75-16910
ELECMA	France	136	75-14400
Hayes Targets	USA	603	75-69135

616 12 80 Radar jammer operator training equipment

Amherst Systems Inc.	USA	553	75-06600
Calian Technology Ltd	Canada	69	75-01800
Elettronica GmbH	Germany	168	75-07565
Ericsson Microwave Systems AB	**Sweden**	**377**	**75-04890**
Evans & Sutherland Computer Corp.	USA	574	75-32650
Hughes Training, Inc	USA	583	75-44900
MPD Technologies Inc.	USA	597	75-63125
Ogle Design Ltd	UK	508	80-19400
Reflectone UK Ltd.	UK	478	75-48380

616 12 85 Radar threat generators, tactical

Amherst Systems Inc.	USA	553	75-06600
ATS Aerospace Inc.	Canada	68	75-00752
CAE Invertron Ltd.	UK	438	75-11145
Evans & Sutherland Computer Corp.	USA	574	75-32650
EW Simulation Technology Ltd	UK	449	75-22150
Labtech Ltd	UK	464	75-34280
Litton Industries Inc.	USA	591	75-55400
Metric Systems Corp.	USA	596	75-60800
MPD Technologies Inc.	USA	597	75-63125
Reflectone UK Ltd.	UK	478	75-48380

616 13 Mission & weapon training systems & simulators, naval

Anite Systems Ltd	UK	497	80-01240
ATS Aerospace Inc.	Canada	68	75-00752
Calspan SRL Technology	USA	540	60-03000
Cubic Defense Systems Inc.	USA	567	75-22900
Datamat Ingegneria dei Sistemi SpA	Italy	233	75-06500
DRS Technologies, Inc.	USA	570	75-27400
Evans & Sutherland Computer Corp.	USA	574	75-32650
EW Simulation Technology Ltd	UK	449	75-22150
Force Computers GmbH	Germany	168	75-09270
Haux-Life-Support GmbH	Germany	169	75-11970
Hughes Electronics Corp.	USA	583	75-44865
Litton Industries Inc.	USA	591	75-55400
Ogle Design Ltd	UK	508	80-19400
Pacer Systems Inc.	USA	602	75-68000
Pennant International Limited	UK	473	75-44450
Primary Image Ltd	UK	476	75-46790
RGB Spectrum	USA	609	75-75000
SAC Technographic	UK	510	80-30100
Ship Analytics Inc.	USA	614	75-80455
SISDEF Ltda	Chile	85	80-01940
Sogitec	France	146	75-42050
STN ATLAS Elektronik GmbH	Germany	165	75-01150
Sverdrup Corp.	USA	643	80-30650
Thomson Marconi Sonar Limited	UK	426	74-20000
Thomson Sintra	France	148	75-45135
Thomson-CSF	France	147	75-45100
ViaSat Inc.	USA	626	75-93790
Vickers Shipbuilding & Eng Ltd	UK	491	75-63100

616 13 60 Mission & weapon training simulators, naval

Amherst Systems Inc.	USA	553	75-06600
ATS Aerospace Inc.	Canada	68	75-00752
Bharat Electronics Ltd	India	203	55-00200
Calspan SRL Technology	USA	540	60-03000
Cubic Defense Systems Inc.	USA	567	75-22900
Datamat Ingegneria dei Sistemi SpA	Italy	233	75-06500
DRS Technologies, Inc.	USA	570	75-27400
ECA SA	France	136	75-14050
Evans & Sutherland Computer Corp.	USA	574	75-32650
Force Computers GmbH	Germany	168	75-09270
Kongsberg Simulation & Training	Norway	309	75-01060
Litton Industries Inc.	USA	591	75-55400
Logica UK Limited	UK	506	80-14350
Norcontrol A/S	Norway	310	75-01630
Ogle Design Ltd	UK	508	80-19400
Pacer Systems Inc.	USA	602	75-68000
Primary Image Ltd	UK	476	75-46790
Reflectone UK Ltd.	UK	478	75-48380
Ship Analytics Inc.	USA	614	75-80455
SISDEF Ltda	Chile	85	80-01940
STN ATLAS Elektronik GmbH	Germany	165	75-01150
Thomson-CSF	France	147	75-45100

616 13 60 10 Simulators, ASW training, naval

Datamat Ingegneria dei Sistemi SpA	Italy	233	75-06500
DRS Technologies, Inc.	USA	570	75-27400
Evans & Sutherland Computer Corp.	USA	574	75-32650
Force Computers GmbH	Germany	168	75-09270
GEC-Marconi Avionics Ltd	UK	454	75-25320
NEC Corporation	Japan	253	75-06350

Ogle Design Ltd	UK	508	80-19400
Pacer Systems Inc.	USA	602	75-68000
Primary Image Ltd	UK	476	75-46790
Reflectone UK Ltd.	UK	478	75-48380
Ship Analytics Inc.	USA	614	75-80455
Tadiran Ltd	Israel	222	75-21600
Thomson Marconi Sonar Limited	UK	426	74-20000
Thomson Sintra	France	148	75-45135

616 13 60 20 Simulators, C3I training, naval

Amherst Systems Inc.	USA	553	75-06600
Calspan SRL Technology	USA	540	60-03000
Evans & Sutherland Computer Corp.	USA	574	75-32650
Force Computers GmbH	Germany	168	75-09270
HAL Communications Corp.	USA	580	75-41450
Ogle Design Ltd	UK	508	80-19400
Pathfinder Systems Inc.	USA	603	75-68995
Primary Image Ltd	UK	476	75-46790
Reflectone UK Ltd.	UK	478	75-48380
RGB Spectrum	USA	609	75-75000
Ship Analytics Inc.	USA	614	75-80455
ViaSat Inc.	USA	626	75-93790

616 13 60 23 Simulators, naval diving

Haux-Life-Support GmbH	Germany	169	75-11970
Ogle Design Ltd	UK	508	80-19400

616 13 60 25 Simulators, electronic warfare training, naval

Amherst Systems Inc.	USA	553	75-06600
Calspan SRL Technology	USA	540	60-03000
Datamat Ingegneria dei Sistemi SpA	Italy	233	75-06500
EW Simulation Technology Ltd	UK	449	75-22150
Force Computers GmbH	Germany	168	75-09270
Litton Industries Inc.	USA	590	75-55400
MPD Technologies Inc.	USA	597	75-63125
NEC Corporation	Japan	253	75-06350
Ogle Design Ltd	UK	508	80-19400
Primary Image Ltd	UK	476	75-46790
Reflectone UK Ltd.	UK	478	75-48380
Ship Analytics Inc.	USA	614	75-80455

616 13 64 Simulators, mission & weapon training, naval shipborne

Amherst Systems Inc.	USA	553	75-06600
Calspan SRL Technology	USA	540	60-03000
Evans & Sutherland Computer Corp.	USA	574	75-32650
Force Computers GmbH	Germany	168	75-09270
Litton Industries Inc.	USA	590	75-55400
MPD Technologies Inc.	USA	597	75-63125
Ogle Design Ltd	UK	508	80-19400
Primary Image Ltd	UK	476	75-46790
Reflectone UK Ltd.	UK	478	75-48380
Ship Analytics Inc.	USA	614	75-80455

616 13 66 Simulators, mission & weapon training, naval shore-based

Amherst Systems Inc.	USA	553	75-06600
Bharat Electronics Ltd	India	203	55-00200
Calspan SRL Technology	USA	540	60-03000
Evans & Sutherland Computer Corp.	USA	574	75-32650
Litton Industries Inc.	USA	590	75-55400
MPD Technologies Inc.	USA	597	75-63125
Ogle Design Ltd	UK	508	80-19400
Primary Image Ltd	UK	476	75-46790
Reflectone UK Ltd.	UK	478	75-48380
Ship Analytics Inc.	USA	614	75-80455
Vickers Shipbuilding & Eng Ltd	UK	491	75-63100

616 13 66 50 Simulators, radar training, naval shore-based

Amherst Systems Inc.	USA	553	75-06600
Atlantis Aerospace Corp.	Canada	68	75-00750
Bharat Electronics Ltd	India	203	55-00200
Calspan SRL Technology	USA	540	60-03000
ECC Simulation Ltd	UK	447	75-19280
Evans & Sutherland Computer Corp.	USA	574	75-32650
MPD Technologies Inc.	USA	597	75-63125
Norcontrol A/S	Norway	310	75-01630
Ogle Design Ltd	UK	508	80-19400
Primary Image Ltd	UK	476	75-46790
Reflectone UK Ltd.	UK	478	75-48380
Ship Analytics Inc.	USA	614	75-80455
Sogitec	France	146	75-42050
STN ATLAS Elektronik GmbH	Germany	165	75-01150
Vickers Shipbuilding & Eng Ltd	UK	491	75-63100

616 13 66 60 Simulators, submarine control, shore-based

Atlantis Aerospace Corp.	Canada	68	75-00750
Evans & Sutherland Computer Corp.	USA	574	75-32650
NEC Corporation	Japan	253	75-06350
Ogle Design Ltd	UK	508	80-19400
Pennant International Limited	UK	473	75-44450
Primary Image Ltd	UK	476	75-46790
Ship Analytics Inc.	USA	614	75-80455
STN ATLAS Elektronik GmbH	Germany	165	75-01150
Vickers Shipbuilding & Eng Ltd	UK	491	75-63100

616 13 66 65 Simulators, sonar training, naval shore-based

Datamat Ingegneria dei Sistemi SpA	Italy	233	75-06500
Electronic Systems Group	USA	573	75-30220
Force Computers GmbH	Germany	168	75-09270
Kongsberg Simulation & Training	Norway	309	75-01060
NEC Corporation	Japan	253	75-06350
Ogle Design Ltd	UK	508	80-19400
Pacer Systems Inc.	USA	602	75-68000
Primary Image Ltd	UK	476	75-46790
Reflectone UK Ltd.	UK	478	75-48380
Ship Analytics Inc.	USA	614	75-80455
STN ATLAS Elektronik GmbH	Germany	165	75-01150
Thomson Marconi Sonar Limited	UK	426	74-20000
Thomson Sintra	France	148	75-45135
Vickers Shipbuilding & Eng Ltd	UK	491	75-63100

616 13 70 Mission & weapon training equipment, naval

ATS Aerospace Inc.	Canada	68	75-00752
Bülow Air Target Systems AB.	Sweden	376	75-03600
Calspan SRL Technology	USA	540	60-03000
DRS Technologies, Inc.	USA	570	75-27400
Evans & Sutherland Computer Corp.	USA	574	75-32650
Hughes Training, Inc	USA	583	75-44900
Ogle Design Ltd	UK	508	80-19400
Primary Image Ltd	UK	476	75-46790
Sogitec	France	146	75-42050
STN ATLAS Elektronik GmbH	Germany	165	75-01150
Thomson Sintra	France	148	75-45135
Thomson-CSF	France	147	75-45100

616 13 77 Targets, naval

ATS Aerospace Inc.	Canada	68	75-00752
Bülow Air Target Systems AB.	Sweden	376	75-03600
ECA SA	France	136	75-14050
Flight Refuelling Limited	UK	451	75-23850
Hayes Targets	USA	603	75-69135
Int'l Submarine Engrg.	Canada	73	75-07780
Sogitec	France	146	75-42050
STN ATLAS Elektronik GmbH	Germany	165	75-01150

616 13 77 20 Target craft, naval, remotely-controlled

Flight Refuelling Limited	UK	451	75-23850
Hayes Targets	USA	603	75-69135

616 13 77 80 Targets, underwater, remotely-controlled, mobile

ECA SA	France	*136*	75-14050
Thomson Sintra	France	*148*	75-45135

616 13 77 80 10 ASW training targets, mobile, expendable

Thomson Sintra	France	*148*	75-45135

616 14 Mission & weapon training systems & simulators, army-related

Aerobel Defence Technology Ltd	UK	*496*	80-00520
Anite Systems Ltd	UK	*497*	80-01240
ATS Aerospace Inc.	Canada	*68*	75-00752
BDL Systems Ltd	UK	*433*	75-06080
CAE Elektronic GmbH	Germany	*166*	75-04500
CAE Invertron Ltd.	UK	*438*	75-11145
Calspan SRL Technology	USA	*540*	60-03000
Dynamit Nobel GmbH	Germany	*167*	75-07100
Elbit Systems Ltd	Israel	*218*	75-05300
Evans & Sutherland Computer Corp.	USA	*574*	75-32650
Fabbrica Italiana Apparecchiature	Italy	*234*	75-08550
FN Herstal SA	Belgium	*46*	75-03600
FN Manufacturing Inc	USA	*576*	75-35320
Force Computers GmbH	Germany	*168*	75-09270
GDI Simulation	France	*137*	75-19700
Indústrias de Defesa	Portugal	*327*	55-00950
Infocom Systems A/S	Denmark	*103*	75-09200
Krauss-Maffei Wehrtechnik GmbH	Germany	*171*	75-16345
Laser & Atomic Research & Devt	Croatia	*94*	60-00450
Logica UK Limited	UK	*506*	80-14350
MS Instruments plc	UK	*469*	75-41265
Nanoquest Defence Products Ltd	UK	*470*	75-41890
Oscmar International Ltd	New Zealand	*302*	75-01550
Pains-Wessex Ltd	UK	*472*	75-43510
Pathfinder Systems Inc.	USA	*603*	75-68995
Primary Image Ltd	UK	*476*	75-46790
Reflectone UK Ltd.	UK	*478*	75-48380
RGB Spectrum	USA	*609*	75-75000
SABCA	Belgium	*46*	75-05590
Ship Analytics Inc.	USA	*614*	75-80455
SISDEF Ltda	Chile	*85*	80-01940
Sogitec	France	*146*	75-42050
SPL Ordnance Ltd.	Israel	*225*	80-01935
STN ATLAS Elektronik GmbH	Germany	*165*	75-01150
Strategic Technologies Pte Ltd	**Singapore**	***348***	**75-04400**
Tamares Ltd.	UK	*487*	75-58500
Tector Visual Systems	UK	*488*	75-59020
Thermodyne International Ltd	USA	*622*	75-88690
Thomson-CSF	France	*147*	75-45100
TIRAX AG	Switzerland	*387*	75-14410
Victrix Ltd	Canada	*77*	75-15000
Wegmann USA Inc.	USA	*627*	75-95543
Zen Technologies Ltd	India	*205*	75-03000

616 14 20 Training systems, air defense combat

Cartwright Electronics Inc.	USA	*562*	75-16910
Cincinnati Electronics Corp.	USA	*564*	75-18640
Elettronica GmbH	Germany	*168*	75-07565
Evans & Sutherland Computer Corp.	USA	*574*	75-32650
Infocom Systems A/S	Denmark	*103*	75-09200
MPD Technologies Inc.	USA	*597*	75-63125
Reflectone UK Ltd.	UK	*478*	75-48380
Tector Visual Systems	UK	*488*	75-59020

616 14 30 Training systems, air defense gunnery

Air Target Sweden AB	Sweden	*375*	75-00850
ATS Aerospace Inc.	Canada	*68*	75-00752
Bülow Air Target Systems AB.	Sweden	*376*	75-03600
Cartwright Electronics Inc.	USA	*562*	75-16910
Evans & Sutherland Computer Corp.	USA	*574*	75-32650
GDI Simulation	France	*137*	75-19700
Infocom Systems A/S	Denmark	*103*	75-09200
Krauss-Maffei Wehrtechnik GmbH	Germany	*171*	75-16345
Laser & Atomic Research & Devt	Croatia	*94*	60-00450
NORINCO	China PR	*88*	55-00280
Ogle Design Ltd	UK	*508*	80-19400
Pains-Wessex Ltd	UK	*472*	75-43510
Primary Image Ltd	UK	*476*	75-46790
Reflectone UK Ltd.	UK	*478*	75-48380
Tadiran Ltd	Israel	*222*	75-21600
Tector Visual Systems	UK	*488*	75-59020

616 14 40 Mission & weapon training systems, infantry

CAE Invertron Ltd.	UK	*438*	75-11145
Elbit Systems Ltd	Israel	*218*	75-05300
Evans & Sutherland Computer Corp.	USA	*574*	75-32650
FN Herstal SA	Belgium	*46*	75-03600
GDI Simulation	France	*137*	75-19700
Laser & Atomic Research & Devt	Croatia	*94*	60-00450
MS Instruments plc	UK	*469*	75-41265
Noptel Oy	Finland	*118*	75-01650
Nottingham International	USA	*640*	80-22000
Primary Image Ltd	UK	*476*	75-46790
Reflectone UK Ltd.	UK	*478*	75-48380
SISDEF Ltda	Chile	*85*	80-01940
Somchem	S. Africa	*359*	75-19200
Spartanics	USA	*616*	75-82850
SPL Ordnance Ltd.	Israel	*225*	80-01935
Tamares Ltd.	UK	*487*	75-58500
TIRAX AG	Switzerland	*387*	75-14410

616 14 46 Ranges, target, infantry training

Armalon Ltd	UK	*431*	75-03729
Caswell International Corp.	USA	*562*	75-16990
FN Herstal SA	Belgium	*46*	75-03600
Hämmerli AG	Switzerland	*386*	75-05850
McQueen Ltd	UK	*467*	75-38200
MS Instruments plc	UK	*469*	75-41265
Polytronic AG	Switzerland	*387*	75-10250
Polytronic Inc	USA	*604*	75-70870
Sabre Computers Int'l Ltd	UK	*481*	75-51150
Strategic Technologies Pte Ltd	**Singapore**	***348***	**75-04400**

616 14 46 70 Range equipment, infantry training

Armalon Ltd	UK	*431*	75-03729
Caswell International Corp.	USA	*562*	75-16990
Drello Ing P. Drewell GmbH & Co KG	Germany	*167*	75-06900
FN Herstal SA	Belgium	*46*	75-03600
Herley-Vega Systems	USA	*581*	75-43090
Laser & Atomic Research & Devt	Croatia	*94*	60-00450
McQueen Ltd	UK	*467*	75-38200
MS Instruments plc	UK	*469*	75-41265
Noptel Oy	Finland	*118*	75-01650
Sabre Computers Int'l Ltd	UK	*481*	75-51150
Strategic Technologies Pte Ltd	**Singapore**	***348***	**75-04400**

616 14 47 Range training systems, ground forces

Caswell International Corp.	USA	*562*	75-16990
FN Herstal SA	Belgium	*46*	75-03600
MS Instruments plc	UK	*469*	75-41265
Noptel Oy	Finland	*118*	75-01650
Polytronic AG	Switzerland	*387*	75-10250
Polytronic Inc	USA	*604*	75-70870
Sabre Computers Int'l Ltd	UK	*481*	75-51150
SPL Ordnance Ltd.	Israel	*225*	80-01935
Strategic Technologies Pte Ltd	**Singapore**	***348***	**75-04400**
TIRAX AG	Switzerland	*387*	75-14410

616 14 48 Small arms training systems

Armalon Ltd	UK	*431*	75-03729
Caswell International Corp.	USA	*562*	75-16990
FATS, Inc.	USA	*575*	75-33877
FN Herstal SA	Belgium	*46*	75-03600
Jonathan Arthur Ciener Inc.	USA	*564*	75-18590
Laser & Atomic Research & Devt	Croatia	*94*	60-00450
MS Instruments plc	UK	*469*	75-41265
Noptel Oy	Finland	*118*	75-01650
Oscmar International Ltd	New Zealand	*302*	75-01550
Sabre Computers Int'l Ltd	UK	*481*	75-51150
SNC Industrial Tech.	Canada	*76*	75-12960

Spartanics	USA	616	75-82850
TIRAX AG	Switzerland	387	75-14410

616 14 50 Mission & weapon training systems, tank crew

Aerobel Defence Technology Ltd	UK	496	80-00520
CAE Invertron Ltd	UK	438	75-11145
Evans & Sutherland Computer Corp.	USA	574	75-32650
Fabbrica Italiana Apparecchiature	Italy	234	75-08550
GDI Simulation	France	137	75-19700
Krauss-Maffei Wehrtechnik GmbH	Germany	171	75-16345
Laser & Atomic Research & Devt	Croatia	94	60-00450
Nanoquest Defence Products Ltd	UK	470	75-41890
Nottingham International	USA	640	80-22000
Ogle Design Ltd	UK	508	80-19400
Perceptronics Inc.	USA	603	75-69270
Primary Image Ltd	UK	476	75-46790
Reflectone UK Ltd.	UK	478	75-48380
SABCA	Belgium	46	75-05590
SEOS Displays Ltd	UK	482	75-52830
Tadiran Ltd	Israel	222	75-21600
Tector Visual Systems	UK	488	75-59020
Wegmann & Co GmbH	Germany	177	75-33700

616 14 50 50 Tank gunnery training aids

Aerobel Defence Technology Ltd	UK	496	80-00520
Giat Industries	France	126	55-00100
Hebron & Medlock Engineering Ltd	UK	457	75-28860
Pains-Wessex Ltd	UK	472	75-43510
Primary Image Ltd	UK	476	75-46790
Reflectone UK Ltd.	UK	478	75-48380
SEOS Displays Ltd	UK	482	75-52830
Tector Visual Systems	UK	488	75-59020
Victrix Ltd	Canada	77	75-15000

616 14 60 Simulators, army-related

Aerobel Defence Technology Ltd	UK	496	80-00520
BRUHN NewTech A/S	Denmark	104	80-00050
CAE Invertron Ltd.	UK	438	75-11145
Contraves Inc	USA	566	75-21850
ECC Simulation Ltd	UK	447	75-19280
Elbit Systems Ltd	Israel	218	75-05300
Evans & Sutherland Computer Corp.	USA	574	75-32650
Force Computers GmbH	Germany	168	75-09270
Hunting Engineering Ltd	UK	459	75-30700
Infocom Systems A/S	Denmark	103	75-09200
Laser & Atomic Research & Devt	Croatia	94	60-00450
Logica UK Limited.	UK	506	80-14350
Ogle Design Ltd	UK	508	80-19400
Oscmar International Ltd	New Zealand	302	75-01550
Pacer Systems Inc.	USA	602	75-68000
Pains-Wessex Ltd	UK	472	75-43510
Pathfinder Systems Inc.	USA	603	75-68995
Perceptronics Inc.	USA	603	75-69270
Primary Image Ltd	UK	476	75-46790
Reflectone UK Ltd.	UK	478	75-48380
RGB Spectrum	USA	609	75-75000
SEOS Displays Ltd	UK	482	75-52830
Ship Analytics Inc.	USA	614	75-80455
Tector Visual Systems	UK	488	75-59020
Thomson-CSF.	France	147	75-45100
TIRAX AG	Switzerland	387	75-14410
Zen Technologies Ltd	India	205	75-03000

616 14 60 10 Simulators, battlefield

BDL Systems Ltd	UK	433	75-06080
Evans & Sutherland Computer Corp.	USA	574	75-32650
General Motors of Canada Ltd	Canada	71	75-05900
Giat Industries	France	126	55-00100
Infocom Systems A/S	Denmark	103	75-09200
Logica UK Limited.	UK	506	80-14350
Logicon Geodynamics Inc.	USA	638	80-18625
Ogle Design Ltd	UK	508	80-19400
Oscmar International Ltd	New Zealand	302	75-01550
Pains-Wessex Ltd	UK	472	75-43510
Pathfinder Systems Inc.	USA	603	75-68995
Perceptronics Inc.	USA	603	75-69270
Primary Image Ltd	UK	476	75-46790
Reflectone UK Ltd.	UK	478	75-48380
Ship Analytics Inc.	USA	614	75-80455
TIRAX AG	Switzerland	387	75-14410

616 14 60 20 Simulators, C3I training, ground-based

Comptek Federal Systems Inc.	USA	565	75-20650
Evans & Sutherland Computer Corp.	USA	574	75-32650
HAL Communications Corp.	USA	580	75-41450
Hunting Engineering Ltd	UK	459	75-30700
Logica UK Limited.	UK	506	80-14350
Ogle Design Ltd	UK	508	80-19400
Pathfinder Systems Inc.	USA	603	75-68995
Perceptronics Inc.	USA	603	75-69270
Primary Image Ltd	UK	476	75-46790
Reflectone UK Ltd.	UK	478	75-48380
RGB Spectrum	USA	609	75-75000

616 14 60 25 Simulators, EW training, ground-based

Comlab Telecommunications Inc.	Canada	70	75-03540
Comptek Federal Systems Inc.	USA	565	75-20650
EW Simulation Technology Ltd	UK	449	75-22150
Logica UK Limited.	UK	506	80-14350
Ogle Design Ltd	UK	508	80-19400
Phase Two Industries	USA	603	75-69560
Primary Image Ltd	UK	476	75-46790
Reflectone UK Ltd.	UK	478	75-48380
Reunert Technology Systems Ltd.	S. Africa	359	75-18300

616 14 60 51 Simulators, tank combat

BVR Technologies Ltd	Israel	218	75-03350
CAE Invertron Ltd.	UK	438	75-11145
Contraves Inc	USA	566	75-21850
Elbit Systems Ltd	Israel	218	75-05300
Evans & Sutherland Computer Corp.	USA	574	75-32650
Fabbrica Italiana Apparecchiature	Italy	234	75-08550
Force Computers GmbH	Germany	168	75-09270
GDI Simulation	France	137	75-19700
Giat Industries	France	126	55-00100
Hughes Training, Inc	USA	583	75-44900
Krauss-Maffei Wehrtechnik GmbH	Germany	171	75-16345
Nanoquest Defence Products Ltd	UK	470	75-41890
Ogle Design Ltd	UK	508	80-19400
Pains-Wessex Ltd	UK	472	75-43510
Pathfinder Systems Inc.	USA	603	75-68995
Perceptronics Inc.	USA	603	75-69270
Primary Image Ltd	UK	476	75-46790
Reflectone UK Ltd.	UK	478	75-48380
Ruggieri Technologie	France	142	75-33500
SEOS Displays Ltd	UK	482	75-52830
Ship Analytics Inc.	USA	614	75-80455
STN ATLAS Elektronik GmbH	Germany	165	75-01150
Tadiran Ltd	Israel	222	75-21600
Thomson-CSF.	France	147	75-45100

616 14 60 52 Simulators, tank driving

Evans & Sutherland Computer Corp.	USA	574	75-32650
Force Computers GmbH	Germany	168	75-09270
Hughes Training, Inc	USA	583	75-44900
Krauss-Maffei Wehrtechnik GmbH	Germany	171	75-16345
Ogle Design Ltd	UK	508	80-19400
Primary Image Ltd	UK	476	75-46790
Reflectone UK Ltd.	UK	478	75-48380
SEOS Displays Ltd	UK	482	75-52830
Ship Analytics Inc.	USA	614	75-80455
STN ATLAS Elektronik GmbH	Germany	165	75-01150
Thomson-CSF.	France	147	75-45100

616 14 60 80 Simulators, weapon training

BVR Technologies Ltd	Israel	218	75-03350
CAE Invertron Ltd.	UK	438	75-11145
ECC International Corp	USA	571	75-28980
EUROCONTROL SpA	Italy	233	75-08200
Evans & Sutherland Computer Corp.	USA	574	75-32650
Hebron & Medlock Engineering Ltd	UK	457	75-28860

Hughes Training, Inc	USA	*583*	75-44900
Krauss-Maffei Wehrtechnik GmbH	Germany	*171*	75-16345
Laser & Atomic Research & Devt	Croatia	*94*	60-00450
Ogle Design Ltd	UK	*508*	80-19400
Pains-Wessex Ltd	UK	*472*	75-43510
Primary Image Ltd	UK	*476*	75-46790
Reflectone UK Ltd.	UK	*478*	75-48380
SEOS Displays Ltd	UK	*482*	75-52830
Spartanics	USA	*616*	75-82850
Tector Visual Systems	UK	*488*	75-59020
TIRAX AG	Switzerland	*387*	75-14410
Trenton Technical Services Ltd	UK	*489*	75-61110
Zen Technologies Ltd	India	*205*	75-03000

616 14 60 80 50 Simulators, small arms training

Indústrias de Defesa	Portugal	*327*	55-00950
Laser & Atomic Research & Devt	Croatia	*94*	60-00450
Noptel Oy	Finland	*118*	75-01650
Ogle Design Ltd	UK	*508*	80-19400
Oscmar International Ltd	New Zealand	*302*	75-01550
Primary Image Ltd	UK	*476*	75-46790
Spartanics	USA	*616*	75-82850
TIRAX AG	Switzerland	*387*	75-14410
Trenton Technical Services Ltd	UK	*489*	75-61110
Zen Technologies Ltd	India	*205*	75-03000

616 14 60 85 Simulators, weapon effects

Aardvark Clear Mine Ltd	UK	*426*	75-00034
Caswell International Corp.	USA	*562*	75-16990
Combined Systems Inc.	USA	*564*	75-19893
Oscmar International Ltd	New Zealand	*302*	75-01550
Theodor Rapp GmbH Co	Germany	*173*	75-24600

616 14 65 Target systems, ground-based

Advanced Training Systems Inc.	USA	*546*	75-01030
Armalon Ltd	UK	*431*	75-03729
Caswell International Corp.	USA	*562*	75-16990
Contraves Inc	USA	*566*	75-21850
Delft Sensor Systems	**Belgium**	***46***	**75-02500**
Delft Sensor Systems	Netherlands	*292*	75-01590
McQueen Ltd	UK	*467*	75-38200
MS Instruments plc	UK	*469*	75-41265
Sabre Computers Int'l Ltd	UK	*481*	75-51150
Strategic Technologies Pte Ltd	**Singapore**	***348***	**75-04400**
Wegmann USA Inc.	USA	*627*	75-95543

616 14 70 Mission & weapon training equipment, army-related

Calspan SRL Technology	USA	*540*	60-03000
Capewell Aerial Delivery Systems	USA	*576*	75-36300
Evans & Sutherland Computer Corp.	USA	*574*	75-32650
FN Herstal SA	Belgium	*46*	75-03600
Gunpell & Co. Ltd	UK	*456*	75-27170
Israel Military Industries Ltd	Israel	*220*	75-10300
Nanoquest Defence Products Ltd	UK	*470*	75-41890
Nichols Research Corp.	USA	*639*	80-21780
Ogle Design Ltd	UK	*508*	80-19400
Pains-Wessex Ltd	UK	*472*	75-43510
Pathfinder Systems Inc.	USA	*603*	75-68995
Perceptronics Inc.	USA	*603*	75-69270
Primary Image Ltd	UK	*476*	75-46790
Reflectone UK Ltd.	UK	*478*	75-48380
Ruggieri Technologie	France	*142*	75-33500
SAE Alsetex	France	*142*	75-33810
Securesearch Inc.	Canada	*76*	75-12600
SNC Industrial Tech.	Canada	*76*	75-12960
Somchem	S. Africa	*359*	75-19200
Swiss Aircraft & Systems Co	Switzerland	*384*	55-00100
Tector Visual Systems	UK	*488*	75-59020

616 14 70 50 Parachutes, paratroop training

Intamar Logistics Inc.	USA	*585*	75-47050
Irvin Aerospace Canada Limited	Canada	*73*	75-08300
Irvin Aerospace Limited	UK	*462*	75-32050
North American Aerodynamics Inc.	USA	*599*	75-65200
Pioneer Aerospace Corp	USA	*604*	75-70200
Sächsische Spezialkonfektion GmbH	Germany	*174*	75-25880
Sala International Pty Ltd	Australia	*21*	75-18980
Strong Enterprises	USA	*618*	75-84500

616 14 70 55 Parachute landing fall trainers

Irvin Aerospace Canada Limited	Canada	*73*	75-08300
Ogle Design Ltd	UK	*508*	80-19400

616 14 70 60 Simulator equipment, gunnery training, armoured vehicle crew

Bharat Electronics Ltd	India	*203*	55-00200
CAE Invertron Ltd.	UK	*438*	75-11145
Evans & Sutherland Computer Corp.	USA	*574*	75-32650
Force Computers GmbH	Germany	*168*	75-09270
Hebron & Medlock Engineering Ltd	UK	*457*	75-28860
Krauss-Maffei Wehrtechnik GmbH	Germany	*171*	75-16345
Nanoquest Defence Products Ltd	UK	*470*	75-41890
Perceptronics Inc.	USA	*603*	75-69270
Primary Image Ltd	UK	*476*	75-46790
Rafael	Israel	*216*	55-01800
Reflectone UK Ltd.	UK	*478*	75-48380
Ruggieri Technologie	France	*142*	75-33500
Tector Visual Systems	UK	*488*	75-59020

616 14 70 65 Weapon simulators, gunnery training, anti-tank missile

CAE Invertron Ltd.	UK	*438*	75-11145
Calspan SRL Technology	USA	*540*	60-03000
EUROCONTROL SpA	Italy	*233*	75-08200
Israel Military Industries Ltd	Israel	*220*	75-10300
IVEX Corp.	USA	*586*	75-49950
Krauss-Maffei Wehrtechnik GmbH	Germany	*171*	75-16345
Ogle Design Ltd	UK	*508*	80-19400
Pains-Wessex Ltd	UK	*472*	75-43510
Perceptronics Inc.	USA	*603*	75-69270
Primary Image Ltd	UK	*476*	75-46790
Reflectone UK Ltd.	UK	*478*	75-48380
Swiss Aircraft & Systems Co	Switzerland	*384*	55-00100

616 14 70 68 Weapon effects simulators, artillery

Combined Systems Inc.	USA	*564*	75-19893
Ogle Design Ltd	UK	*508*	80-19400

616 14 70 70 Targets, training, ground-based

Advanced Training Systems Inc.	USA	*546*	75-01030
ARSOC Sprl	Belgium	*45*	75-00520
Caswell International Corp.	USA	*562*	75-16990
FN Herstal SA	Belgium	*46*	75-03600
Hämmerli AG	Switzerland	*386*	75-05850
Kaman Sciences Corp.	USA	*637*	80-17600
McQueen Ltd	UK	*467*	75-38200
MS Instruments plc	UK	*469*	75-41265
PD Technical Mouldings plc	UK	*473*	75-44200
Polyformes Ltd	UK	*475*	75-46348
Shield International Ltd	UK	*511*	80-35500
Thermodyne International Ltd	USA	*622*	75-88690
Victrix Ltd	Canada	*77*	75-15000

616 14 70 74 Targets, inflatable, visual acquisition training

Airborne Industries	UK	*428*	75-01550
Eurovinil Industries SpA	Italy	*234*	75-08400

616 14 70 77 Targets, weapon training, ground-based

Advanced Training Systems Inc.	USA	*546*	75-01030
ARSOC Sprl	Belgium	*45*	75-00520
Caswell International Corp.	USA	*562*	75-16990
FN Herstal SA	Belgium	*46*	75-03600
Hämmerli AG	Switzerland	*386*	75-05850
McQueen Ltd	UK	*467*	75-38200

MS Instruments plc	UK	469	75-41265
PD Technical Mouldings plc	UK	473	75-44200
Polyformes Ltd	UK	475	75-46348
R.X. Products Ltd	UK	481	75-51020
Sabre Computers Int'l Ltd	UK	481	75-51150
Thermodyne International Ltd	USA	622	75-88690
Victrix Ltd	Canada	77	75-15000

616 14 70 77 50 Targets, weapon training, paper

Accuracy International Ltd	UK	427	75-00360
Advanced Training Systems Inc.	USA	546	75-01030
Armalon Ltd	UK	431	75-03729
FN Herstal SA	Belgium	46	75-03600
McQueen Ltd	UK	467	75-38200
Sabre Computers Int'l Ltd	UK	481	75-51150

616 14 70 77 60 Targets, weapon training, remote controlled, ground-based

Advanced Training Systems Inc.	USA	546	75-01030
Caswell International Corp.	USA	562	75-16990
FN Herstal SA	Belgium	46	75-03600
Kaman Sciences Corp.	USA	637	80-17600
McQueen Ltd	UK	467	75-38200
Sabre Computers Int'l Ltd	UK	481	75-51150
Shield International Ltd	UK	511	80-35500
Thermodyne International Ltd	USA	622	75-88690
Victrix Ltd	Canada	77	75-15000

616 14 70 77 68 Targets, weapon training, moving projected

Caswell International Corp.	USA	562	75-16990
Evans & Sutherland Computer Corp.	USA	574	75-32650
GEC Marconi Hazeltine Corp	USA	578	75-37415
Primary Image Ltd	UK	476	75-46790

616 14 70 77 70 Targets, weapon training, static projected

Caswell International Corp.	USA	562	75-16990
Primary Image Ltd	UK	476	75-46790

616 14 70 77 72 Targets, weapon training, cine projected

Caswell International Corp.	USA	562	75-16990

616 14 70 77 76 Targets, weapon training, slide projected

Caswell International Corp.	USA	562	75-16990

616 14 70 77 78 Targets, weapon training, video projected

CAE Invertron Ltd	UK	438	75-11145
Caswell International Corp.	USA	562	75-16990
Evans & Sutherland Computer Corp.	USA	574	75-32650
Primary Image Ltd	UK	476	75-46790
SBS Technologies Inc.	USA	612	75-78330
Sogitec	France	146	75-42050
Tadiran Ltd	Israel	222	75-21600

616 14 70 77 80 Target accessories, ground-based

Advanced Training Systems Inc.	USA	546	75-01030
Armalon Ltd	UK	431	75-03729
Berleburger Schaumstoffwerk GmbH	Germany	165	75-02470
Caswell International Corp.	USA	562	75-16990
FN Herstal SA	Belgium	46	75-03600
McQueen Ltd	UK	467	75-38200
Nottingham International	USA	640	80-22000
PD Technical Mouldings plc	UK	473	75-44200

616 14 70 77 89 Target towing winches, ground-based

Superwinch Ltd	UK	487	75-57950

616 14 77 76 Scoring systems, ground-based target

Advanced Training Systems Inc.	USA	546	75-01030
Cartwright Electronics Inc.	USA	562	75-16910
Caswell International Corp.	USA	562	75-16990
Del Mar Avionics	USA	568	75-25362
Drello Ing P. Drewell GmbH & Co KG	Germany	167	75-06900
MS Instruments plc	UK	469	75-41265
Primary Image Ltd	UK	476	75-46790
Sabre Computers Int'l Ltd	UK	481	75-51150
Strategic Technologies Pte Ltd	**Singapore**	**348**	**75-04400**

616 15 Mission & weapon training systems, police / law enforcement agency

Advanced Training Systems Inc.	USA	546	75-01030
Caswell International Corp.	USA	562	75-16990
Evans & Sutherland Computer Corp.	USA	574	75-32650
Gunpell & Co. Ltd	UK	456	75-27170
Hughes Training, Inc	USA	583	75-44900
Israel Military Industries Ltd	Israel	220	75-10300
Ogle Design Ltd	UK	508	80-19400
Polyformes Ltd	UK	475	75-46348
Primary Image Ltd	UK	476	75-46790
SBS Technologies Inc.	USA	612	75-78330
Shield International Ltd	UK	511	80-35500
SPL Ordnance Ltd	Israel	225	80-01935
Thought Technology Ltd	Canada	77	75-13770
TIRAX AG	Switzerland	387	75-14410
Trenton Technical Services Ltd	UK	489	75-61110

616 15 46 Ranges, shooting, police weapon training

Advanced Training Systems Inc.	USA	546	75-01030
Armalon Ltd	UK	431	75-03729
Berleburger Schaumstoffwerk GmbH	Germany	165	75-02470
Caswell International Corp.	USA	562	75-16990
Gunpell & Co. Ltd	UK	456	75-27170
Hämmerli AG	Switzerland	386	75-05850
Israel Military Industries Ltd	Israel	220	75-10300
MS Instruments plc	UK	469	75-41265
Polytronic AG	Switzerland	387	75-10250
Polytronic Inc	USA	604	75-70870
Shield International Ltd	UK	511	80-35500
SISDEF Ltda	Chile	85	80-01940
Thought Technology Ltd	Canada	77	75-13770
TIRAX AG	Switzerland	387	75-14410

616 15 46 70 Range equipment, police weapon training

Advanced Training Systems Inc.	USA	546	75-01030
Armalon Ltd	UK	431	75-03729
Caswell International Corp.	USA	562	75-16990
Evans & Sutherland Computer Corp.	USA	574	75-32650
Israel Military Industries Ltd	Israel	220	75-10300
MS Instruments plc	UK	469	75-41265
Noptel Oy	Finland	118	75-01650
Polyformes Ltd	UK	475	75-46348
Shield International Ltd	UK	511	80-35500
Strategic Technologies Pte Ltd	**Singapore**	**348**	**75-04400**
Thought Technology Ltd	Canada	77	75-13770
TIRAX AG	Switzerland	387	75-14410

616 15 70 Mission & weapon training equipment, police

Advanced Training Systems Inc.	USA	546	75-01030
Caswell International Corp.	USA	562	75-16990
Evans & Sutherland Computer Corp.	USA	574	75-32650
Hughes Training, Inc	USA	583	75-44900
Ogle Design Ltd	UK	508	80-19400
Pacer Systems Inc.	USA	602	75-68000
Primary Image Ltd	UK	476	75-46790
SBS Technologies Inc.	USA	612	75-78330
Spartanics	USA	616	75-82850
SPL Ordnance Ltd	Israel	225	80-01935

TIRAX AG	Switzerland	*387*	75-14410
Trenton Technical Services Ltd	UK	*489*	75-61110

616 20 Platform system operation & procedures trainers & simulators

Applied Data Sciences Inc.	USA	*554*	75-08050
Atlantis Aerospace Corp.	Canada	*68*	75-00750
CAE Electronics Ltd	Canada	*69*	75-01770
Calspan SRL Technology	USA	*540*	60-03000
Cubic Defense Systems Inc.	USA	*567*	75-22900
Evans & Sutherland Computer Corp.	USA	*574*	75-32650
General Motors of Canada Ltd	Canada	*71*	75-05900
Hughes Electronics Corp.	USA	*583*	75-44865
Ishikawajima-Harima H.I. Co. Ltd	Japan	*251*	75-02800
Ogle Design Ltd	UK	*508*	80-19400
Pennant International Limited	UK	*473*	75-44450
Primary Image Ltd	UK	*476*	75-46790
Reflectone Inc	USA	*608*	75-74150
Reflectone UK Ltd.	UK	*478*	75-48380
SAC Technographic	UK	*510*	80-30100
Sogitec	France	*146*	75-42050
SP aerospace and vehicle systems bv	**Netherlands**	***295***	**75-06860**
Strachan & Henshaw Ltd	UK	*512*	80-47400
Sverdrup Corp.	USA	*643*	80-30650
Thomson Training & Simulation Ltd.	UK	*488*	75-59567
United Defense L.P..	USA	*624*	75-92050
Van Rietschoten & Houwens	Netherlands	*296*	75-07900

616 22 Platform system operation & procedures trainers / simulators, aircraft

Atlantis Aerospace Corp.	Canada	*68*	75-00750
CAE Electronics Ltd	Canada	*69*	75-01770
Calspan SRL Technology	USA	*540*	60-03000
Computing Devices Company Ltd	UK	*442*	75-14700
Cubic Defense Systems Inc.	USA	*567*	75-22900
Emro Engineering Co. Inc.	USA	*573*	75-31450
Evans & Sutherland Computer Corp.	USA	*574*	75-32650
Hughes Electronics Corp.	USA	*583*	75-44865
Hughes Training, Inc	USA	*583*	75-44900
NEC Corporation	Japan	*253*	75-06350
Northrop Grumman Corporation.	USA	*599*	75-65450
Ogle Design Ltd	UK	*508*	80-19400
Pennant International Limited	UK	*473*	75-44450
Primary Image Ltd	UK	*476*	75-46790
Reflectone UK Ltd.	UK	*478*	75-48380
SBS Technologies Inc.	USA	*612*	75-78330
Sogitec	France	*146*	75-42050
Sonda Aviation Enterprises Limited	UK	*512*	80-40000
Thomson Training & Simulation Ltd.	UK	*488*	75-59567

616 22 10 Cockpit procedures trainers

Atlantis Aerospace Corp.	Canada	*68*	75-00750
Emro Engineering Co. Inc.	USA	*573*	75-31450
Evans & Sutherland Computer Corp.	USA	*574*	75-32650
Frasca International Inc.	USA	*576*	75-35900
Ogle Design Ltd	UK	*508*	80-19400
Pennant International Limited	UK	*473*	75-44450
Primary Image Ltd	UK	*476*	75-46790
Reflectone UK Ltd.	UK	*478*	75-48380
Sogitec	France	*146*	75-42050
Thomson Training & Simulation Ltd.	UK	*488*	75-59567

616 22 30 Flight simulators, aircraft

Applied Data Sciences Inc.	USA	*554*	75-08050
Atlantis Aerospace Corp.	Canada	*68*	75-00750
CAE Electronics Ltd	Canada	*69*	75-01770
Calspan SRL Technology	USA	*540*	60-03000
Emro Engineering Co. Inc.	USA	*573*	75-31450
Evans & Sutherland Computer Corp.	USA	*574*	75-32650
Force Computers GmbH	Germany	*168*	75-09270
Frasca International Inc.	USA	*576*	75-35900
Hughes Electronics Corp.	USA	*583*	75-44865
Ogle Design Ltd	UK	*508*	80-19400
Pezetel	Poland	*325*	80-01000
Primary Image Ltd	UK	*476*	75-46790
Reflectone UK Ltd.	UK	*478*	75-48380
SBS Technologies Inc.	USA	*612*	75-78330
SEOS Displays Ltd	UK	*482*	75-52830
Sogitec	France	*146*	75-42050
Sonda Aviation Enterprises Limited	UK	*512*	80-40000
Thomson Training & Simulation Ltd.	UK	*488*	75-59567

616 22 30 50 Flight simulators, helicopter

Atlantis Aerospace Corp.	Canada	*68*	75-00750
CAE Electronics Ltd	Canada	*69*	75-01770
Evans & Sutherland Computer Corp.	USA	*574*	75-32650
Force Computers GmbH	Germany	*168*	75-09270
Frasca International Inc.	USA	*576*	75-35900
Logica UK Limited	UK	*506*	80-14350
Ogle Design Ltd	UK	*508*	80-19400
Reflectone UK Ltd.	UK	*478*	75-48380
SBS Technologies Inc.	USA	*612*	75-78330
SEOS Displays Ltd	UK	*482*	75-52830
Sogitec	France	*146*	75-42050
Thomson Training & Simulation Ltd.	UK	*488*	75-59567

616 23 Platform system operation & procedures trainers / simulators, naval

Atlantis Aerospace Corp.	Canada	*68*	75-00750
CAE Electronics Ltd	Canada	*69*	75-01770
Cubic Defense Systems Inc.	USA	*567*	75-22900
ECC Simulation Ltd	UK	*447*	75-19280
Evans & Sutherland Computer Corp.	USA	*574*	75-32650
Frasca International Inc.	USA	*576*	75-35900
Norcontrol A/S	Norway	*310*	75-01630
Ogle Design Ltd	UK	*508*	80-19400
Primary Image Ltd	UK	*476*	75-46790
Reflectone UK Ltd.	UK	*478*	75-48380
SAC Technographic	UK	*510*	80-30100
SISDEF Ltda	Chile	*85*	80-01940
Strachan & Henshaw Ltd	UK	*512*	80-47400
Thomson Training & Simulation Ltd.	UK	*488*	75-59567

616 23 30 Diver training systems

Hughes Training, Inc	USA	*583*	75-44900
Pains-Wessex Ltd	UK	*472*	75-43510

616 23 60 Ship handling simulators

Bharat Electronics Ltd	India	*203*	55-00200
CAE Electronics Ltd	Canada	*69*	75-01770
Calspan SRL Technology	USA	*540*	60-03000
Evans & Sutherland Computer Corp.	USA	*574*	75-32650
Force Computers GmbH	Germany	*168*	75-09270
Hughes Training, Inc	USA	*583*	75-44900
Norcontrol A/S	Norway	*310*	75-01630
Ogle Design Ltd	UK	*508*	80-19400
Reflectone UK Ltd.	UK	*478*	75-48380
SAC Technographic	UK	*510*	80-30100
SEOS Displays Ltd	UK	*482*	75-52830
Ship Analytics Inc.	USA	*614*	75-80455
STN ATLAS Elektronik GmbH	Germany	*165*	75-01150
Thomson Training & Simulation Ltd.	UK	*488*	75-59567

616 23 65 Submarine escape trainers

Ogle Design Ltd	UK	*508*	80-19400

616 24 Platform system operation & procedures trainers / simulators, army

Atlantis Aerospace Corp.	Canada	*68*	75-00750
CAE Electronics Ltd	Canada	*69*	75-01770
Calspan SRL Technology	USA	*540*	60-03000
Evans & Sutherland Computer Corp.	USA	*574*	75-32650
Hughes Training, Inc	USA	*583*	75-44900
Ogle Design Ltd	UK	*508*	80-19400
Primary Image Ltd	UK	*476*	75-46790
Reflectone UK Ltd.	UK	*478*	75-48380
SP aerospace and vehicle systems bv	**Netherlands**	***295***	**75-06860**

616 30 Platform & system maintenance trainers

Atlantis Aerospace Corp.	Canada	*68*	75-00750
CAE Electronics Ltd	Canada	*69*	75-01770
CAE Elektronik GmbH	Germany	*166*	75-04500
Calspan SRL Technology	USA	*540*	60-03000
Cubic Defense Systems Inc.	USA	*567*	75-22900
ECC International Corp	USA	*571*	75-28980
ECC Simulation Ltd	UK	*447*	75-19280
Hughes Training, Inc	USA	*583*	75-44900
Ishikawajima-Harima H.I. Co. Ltd	Japan	*251*	75-02800
Japan Aircraft Manufacturing Co. Ltd.	Japan	*252*	75-03200
Ogle Design Ltd	UK	*508*	80-19400
Pennant International Limited	UK	*473*	75-44450
Primary Image Ltd	UK	*476*	75-46790
Reflectone Inc.	USA	*608*	75-74150
Reflectone UK Ltd.	UK	*478*	75-48380
SBS Technologies Inc.	USA	*612*	75-78330
Sogitec	France	*146*	75-42050
SP aerospace and vehicle systems bv	**Netherlands**	***295***	**75-06860**
STN ATLAS Elektronik GmbH	Germany	*165*	75-01150
Strachan & Henshaw Ltd	UK	*512*	80-47400
The Boeing Company	USA	*559*	75-14250
Thomson Training & Simulation Ltd.	UK	*488*	75-59567
United Defense L.P.	USA	*624*	75-92050
Van Rietschoten & Houwens	Netherlands	*296*	75-07900

616 32 Platform & system maintenance trainers, aircraft-related

Atlantis Aerospace Corp.	Canada	*68*	75-00750
CAE Electronics Ltd	Canada	*69*	75-01770
CAE Elektronik GmbH	Germany	*166*	75-04500
Calspan SRL Technology	USA	*540*	60-03000
Cubic Defense Systems Inc.	USA	*567*	75-22900
ECC Simulation Ltd	UK	*447*	75-19280
Hughes Training, Inc	USA	*583*	75-44900
Ogle Design Ltd	UK	*508*	80-19400
Pennant International Limited	UK	*473*	75-44450
Primary Image Ltd	UK	*476*	75-46790
Reflectone UK Ltd.	UK	*478*	75-48380
SBS Technologies Inc.	USA	*612*	75-78330
Thomson Training & Simulation Ltd.	UK	*488*	75-59567

616 33 Platform & system maintenance trainers, naval

Atlantis Aerospace Corp.	Canada	*68*	75-00750
Calspan SRL Technology	USA	*540*	60-03000
Cubic Defense Systems Inc.	USA	*567*	75-22900
ECC Simulation Ltd	UK	*447*	75-19280
Hughes Training, Inc	USA	*583*	75-44900
Ogle Design Ltd	UK	*508*	80-19400
Pennant International Limited	UK	*473*	75-44450
Primary Image Ltd	UK	*476*	75-46790
Reflectone UK Ltd.	UK	*478*	75-48380
SAC Technographic	UK	*510*	80-30100
Strachan & Henshaw Ltd	UK	*512*	80-47400

616 34 Platform & system maintenance trainers, army-related

Atlantis Aerospace Corp.	Canada	*68*	75-00750
CAE Elektronik GmbH	Germany	*166*	75-04500
Calspan SRL Technology	USA	*540*	60-03000
Cubic Defense Systems Inc.	USA	*567*	75-22900
ECC Simulation Ltd	UK	*447*	75-19280
Hughes Training, Inc	USA	*583*	75-44900
Krauss-Maffei Wehrtechnik GmbH	Germany	*171*	75-16345
Ogle Design Ltd	UK	*508*	80-19400
Pennant International Limited	UK	*473*	75-44450
Primary Image Ltd	UK	*476*	75-46790
Reflectone UK Ltd.	UK	*478*	75-48380
SP aerospace and vehicle systems bv	**Netherlands**	***295***	**75-06860**

616 34 60 Maintenance simulators, army-related

Atlantis Aerospace Corp.	Canada	*68*	75-00750
ECC Simulation Ltd	UK	*447*	75-19280
Hughes Training, Inc	USA	*583*	75-44900
Krauss-Maffei Wehrtechnik GmbH	Germany	*171*	75-16345
Pennant International Limited	UK	*473*	75-44450
Primary Image Ltd	UK	*476*	75-46790
Reflectone UK Ltd.	UK	*478*	75-48380
RGB Spectrum	USA	*609*	75-75000
Sogitec	France	*146*	75-42050
SP aerospace and vehicle systems bv	**Netherlands**	***295***	**75-06860**

616 34 60 90 Maintenance simulators, weapon system

Atlantis Aerospace Corp.	Canada	*68*	75-00750
CAE Elektronik GmbH	Germany	*166*	75-04500
ECC Simulation Ltd	UK	*447*	75-19280
Hughes Training, Inc	USA	*583*	75-44900
Krauss-Maffei Wehrtechnik GmbH	Germany	*171*	75-16345
Ogle Design Ltd	UK	*508*	80-19400
Pennant International Limited	UK	*473*	75-44450
Primary Image Ltd	UK	*476*	75-46790
Reflectone UK Ltd.	UK	*478*	75-48380
RGB Spectrum	USA	*609*	75-75000
SEOS Displays Ltd	UK	*482*	75-52830
Sogitec	France	*146*	75-42050

616 40 Physical training equipment

Powersport International Ltd	UK	*475*	75-46500
RGB Spectrum	USA	*609*	75-75000
Thought Technology Ltd	Canada	*77*	75-13770

616 50 Technical training equipment

CAE Electronics Ltd	Canada	*69*	75-01770
CAE Elektronik GmbH	Germany	*166*	75-04500
Calspan SRL Technology	USA	*540*	60-03000
Conrac Elektron GmbH	Germany	*166*	75-05100
Elettronica GmbH	Germany	*168*	75-07565
Evans & Sutherland Computer Corp.	USA	*574*	75-32650
Gajon Institute of Technology SpA	Italy	*229*	60-00080
Hughes Electronics Corp.	USA	*583*	75-44865
Hughes Training, Inc	USA	*583*	75-44900
Interspiro Ltd	UK	*461*	75-31945
Ishikawajima-Harima H.I. Co. Ltd	Japan	*251*	75-02800
Microelectronics, guidance and	Turkey	*402*	55-00050
Northrop Grumman Corporation	USA	*599*	75-65450
Ogle Design Ltd	UK	*508*	80-19400
Plint & Partners Ltd	UK	*475*	75-46330
RGB Spectrum	USA	*609*	75-75000
SISDEF Ltda	Chile	*85*	80-01940
United Defense L.P.	USA	*624*	75-92050

616 55 Educational training equipment

Ascot Systems	UK	*431*	75-04050
CAE Electronics Ltd	Canada	*69*	75-01770
CAE Elektronik GmbH	Germany	*166*	75-04500
Calspan SRL Technology	USA	*540*	60-03000
Chrysanthacopoulos SA	Greece	*188*	75-00910
Harkness Hall Ltd	UK	*457*	75-28050
Hughes Electronics Corp.	USA	*583*	75-44865
Hughes Training, Inc	USA	*583*	75-44900
Microelectronics, guidance and	Turkey	*402*	55-00050
Ogle Design Ltd	UK	*508*	80-19400
Pennant International Limited	UK	*473*	75-44450
Perceptronics Inc.	USA	*603*	75-69270
Plint & Partners Ltd	UK	*475*	75-46330
Telegenix Inc.	USA	*621*	75-87610
Wilson & Garden Ltd	UK	*494*	75-66450

616 60 Equipment operator trainers

ATS Aerospace Inc.	Canada	*68*	75-00752
Belgian Advanced Technology	Belgium	*45*	75-01320
CAE	UK	*438*	75-11125
CAE Electronics Ltd	Canada	*69*	75-01770
CAE Elektronik GmbH	Germany	*166*	75-04500
Calspan SRL Technology	USA	*540*	60-03000
Elettronica GmbH	Germany	*168*	75-07565
Evans & Sutherland Computer Corp.	USA	*574*	75-32650
Pacer Systems Inc.	USA	*602*	75-68000

Primary Image Ltd	UK	476	75-46790
Reflectone UK Ltd.	UK	478	75-48380
Reunert Technology Systems Ltd	S. Africa	359	75-18300
Ship Analytics Inc.	USA	614	75-80455
Siemens Environmental Systems Ltd	UK	483	75-53700
SISDEF Ltda	Chile	85	80-01940

616 60 10 ATC simulators

ATS Aerospace Inc.	Canada	68	75-00752
Denro Inc.	USA	568	75-25651
EDS Defence Ltd	UK	501	80-09200
Evans & Sutherland Computer Corp.	USA	574	75-32650
Hughes Training, Inc	USA	583	75-44900
Oracle Telecomputing Inc.	Canada	74	75-10430
Plexsys Interface Products, Inc.	USA	604	75-70400
Primary Image Ltd	UK	476	75-46790
Reflectone UK Ltd.	UK	478	75-48380
SEOS Displays Ltd	UK	482	75-52830

616 60 20 Communications training equipment

ATS Aerospace Inc.	Canada	68	75-00752
CAE	UK	438	75-11125
Davies Industrial Communications Ltd	UK	444	75-17010
Hughes Training, Inc	USA	583	75-44900
Primary Image Ltd	UK	476	75-46790
Reflectone UK Ltd.	UK	478	75-48380
Reunert Technology Systems Ltd	S. Africa	359	75-18300
Tadiran Ltd	Israel	222	75-21600
Telegenix Inc.	USA	621	75-87610

616 60 50 Machine tool simulators

CAE Elektronic GmbH	Germany	166	75-04500
Primary Image Ltd	UK	476	75-46790

616 70 Training equipment, tri-service, general

Calspan SRL Technology	USA	540	60-03000
Evans & Sutherland Computer Corp.	USA	574	75-32650
Hughes Training, Inc	USA	583	75-44900
Perceptronics Inc.	USA	603	75-69270
Primary Image Ltd	UK	476	75-46790
Spitz Inc.	USA	617	75-83350
Thought Technology Ltd	Canada	77	75-13770

616 80 Training equipment accessories

Evans & Sutherland Computer Corp.	USA	574	75-32650
FFA Flugzeugwerke Altenrhein AG	Switzerland	386	75-04350
Harkness Hall Ltd	UK	457	75-28050
Jonathan Arthur Ciener Inc.	USA	564	75-18590
Nobo (UK) Limited	UK	471	75-42645
Primary Image Ltd	UK	476	75-46790

616 90 Training equipment, other

Atlantis Aerospace Corp.	Canada	68	75-00750
CAE	UK	438	75-11125
CAE Electronics Ltd	Canada	69	75-01770
CAE Elektronic GmbH	Germany	166	75-04500
Calspan SRL Technology	USA	540	60-03000
Cubic Defense Systems Inc.	USA	567	75-22900
Dynamit Nobel GmbH	Germany	167	75-07100
Evans & Sutherland Computer Corp.	USA	574	75-32650
Hebron & Medlock Engineering Ltd	UK	457	75-28860
Japan Aircraft Manufacturing Co. Ltd.	Japan	252	75-03200
John Lawson Display & Technical	UK	464	75-34875
Jonathan Arthur Ciener Inc.	USA	564	75-18590
Krauss-Maffei Wehrtechnik GmbH	Germany	171	75-16345
Litton Industries Inc.	USA	590	75-55400
Pilatus Aircraft Ltd	Switzerland	387	75-10100
Powersport International Ltd	UK	475	75-46500
Primary Image Ltd	UK	476	75-46790
Reflectone Inc	USA	608	75-74150
RGB Spectrum	USA	609	75-75000
R.X. Products Ltd	UK	481	75-51020
SAE Alsetex	France	142	75-33810
SAGEM SA	France	143	75-33950
Sparton Technology Inc.	USA	616	75-82950
Spitz Inc.	USA	617	75-83350
Symtron Systems Inc.	USA	619	75-85530
Tamares Ltd.	UK	487	75-58500
The Boeing Company	USA	559	75-14250
Thomson-CSF	France	147	75-45100
Thought Technology Ltd	Canada	77	75-13770
Van Rietschoten & Houwens	Netherlands	296	75-07900

616 90 20 Training systems, computer generated imagery, visual

Aerobel Defence Technology Ltd	UK	496	80-00520
Atlantis Aerospace Corp.	Canada	68	75-00750
CAE Elektronic GmbH	Germany	166	75-04500
Evans & Sutherland Computer Corp.	USA	574	75-32650
Force Computers GmbH	Germany	168	75-09270
Hughes Electronics Corp.	USA	583	75-44865
IVEX Corp.	USA	586	75-49950
Primary Image Ltd	UK	476	75-46790
Reflectone UK Ltd.	UK	478	75-48380
RGB Spectrum	USA	609	75-75000
SAC Technographic	UK	510	80-30100
SBS Technologies Inc.	USA	612	75-78330
Ship Analytics Inc.	USA	614	75-80455
Spitz Inc.	USA	617	75-83350

616 90 50 Training systems, interactive videodisc

Hughes Electronics Corp.	USA	583	75-44865
Krauss-Maffei Wehrtechnik GmbH	Germany	171	75-16345
Perceptronics Inc.	USA	603	75-69270
Reflectone UK Ltd.	UK	478	75-48380
RGB Spectrum	USA	609	75-75000
SAC Technographic	UK	510	80-30100
SBS Technologies Inc.	USA	612	75-78330
Ship Analytics Inc.	USA	614	75-80455
Sogitec	France	146	75-42050
Tadiran Ltd	Israel	222	75-21600
Thomson Training & Simulation Ltd.	UK	488	75-59567

616 90 70 Training/simulation systems, virtual-reality

Aerobel Defence Technology Ltd	UK	496	80-00520
Sonda Aviation Enterprises Limited	UK	512	80-40000

617 Test, inspection, measurement, calibration & alignment

AACO Inc.	USA	545	75-00010
ADE-HML Limited	UK	427	75-00710
Advanced Electromagnetics	USA	546	75-00895
Advanced Technical Corp.	USA	546	75-01000
Aermacchi SpA	Italy	230	75-00500
Aircraft Appliances & Equipment Ltd	Canada	68	75-00430
Allied Research Corp.	USA	549	75-04670
Andrew Canada Inc.	Canada	68	75-00600
Angus Electronics Co.	USA	554	75-07630
Appareillages et Matériels de	France	132	75-03050
ATS Aerospace Inc.	Canada	68	75-00752
Autoflug GmbH & Co	Germany	165	75-01400
B&W Engineering Corp.	USA	561	75-16009
Bauer Aerospace	USA	557	75-12250
Bowers Group	UK	435	75-07770
Bristow Helicopters Ltd	UK	499	80-03950
Budenberg Gauges (Australia) Pty Ltd	Australia	14	75-02860
Canadian Marconi Co.	Canada	69	75-02600
Carco Electronics	USA	562	75-16700
Central Design Bureau	Ukraine	406	75-00800
Central Engineering Company	USA	562	75-17290
Cincinnati Electronics Corp.	USA	564	75-18640
Circle-C Transformers	Australia	14	75-03450
Computing Devices Company Ltd	UK	442	75-14700
Cranfield University	UK	419	60-00350
Datacom Inc.	USA	567	75-23645
Dowty Controls	UK	446	75-18592
EG&G Inc.	USA	571	75-29555
EIP Microwave Inc.	USA	572	75-29590
Elan GmbH.	Germany	168	75-07550
Elbit Systems Ltd	Israel	218	75-05300

Product-Supplier Link

Electro Optical Industries Inc.	USA	573	75-30450
Emprise Corp.	USA	634	80-09950
Everest Imaging Ltd	USA	574	75-32657
Fabbrica Italiana Apparecchiature	Italy	234	75-08550
Fokker Space & Systems	Netherlands	293	75-02325
Froude Consine	UK	452	75-24800
GEC-Marconi Avionics Ltd	UK	454	75-25320
GEC-Marconi Materials Technology Ltd	UK	419	60-00550
General Electrodynamics Corp.	USA	578	75-38500
Guildline Instruments Inc.	USA	580	75-40900
Haux-Life-Support GmbH	Germany	169	75-11970
Hocking NDT Ltd	UK	458	75-29580
Hoffman Engineering Corp.	USA	582	75-43980
Howden Wade Ltd	UK	459	75-30260
Indústrias de Defesa	Portugal	327	55-00950
International Machine & Tool Corp.	USA	585	75-46860
Israel Military Industries Ltd	Israel	220	75-10300
ISVR Consultancy Services	UK	505	80-13520
J & E Hofmann Engineering Pty Ltd	Australia	17	75-08630
Julie Research Laboratories	USA	587	75-51265
Kaiser Electroprecision	USA	587	75-51350
Kaman Electromagnetics Corp.	USA	637	80-17500
Kearfott Guidance & Navigation Corp.	USA	588	75-51760
Lenox Instrument Co.	USA	590	75-54670
Leonard Farnell & Co. Ltd	UK	450	75-22400
Litton Industries Inc.	USA	590	75-55400
Lucas Aerospace Power Transmission	USA	592	75-56265
Lucas Aftermarket Operations	UK	465	75-36655
MA/COM Antenna & Cable Div.	USA	593	75-56953
Marconi Instruments Ltd	UK	466	75-37270
Marine Electric Systems Inc.	USA	593	75-58000
McNab Inc.	USA	595	75-59720
MS Instruments plc	UK	469	75-41265
MTU München GmbH	Germany	172	75-21801
Nanoquest Defence Products Ltd	UK	470	75-41890
Northrop Grumman Corporation	USA	599	75-65450
NPO Geofizika	Russia	338	75-14315
Oerlikon-Contraves AG.	Switzerland	387	75-09250
Philips Calibration Services	Australia	25	80-16490
Philips Traffic & Engineering	Australia	20	75-16550
Polytec-PI SA	France	141	75-30450
Prosig Computer Consultants Ltd	UK	476	75-46865
QES Ltd	UK	509	80-25000
Quantum Inspection & Testing Ltd	Canada	79	80-01800
Retlif Testing Laboratories	USA	641	80-25950
Rolls-Royce MatEval	UK	480	75-50260
Sanders	USA	612	75-77850
SELIN Sistemi Sarl	Italy	238	75-21750
SFIM	**France**	**144**	**75-35080**
Siemens plc	UK	483	75-53750
SISDEF Ltda	Chile	85	80-01940
SST/Burlington Microelectronics	USA	643	80-29650
Stanford Research Systems Inc.	USA	617	75-83750
STN ATLAS Elektronik GmbH	Germany	165	75-01150
Stork RMO BV.	Netherlands	295	75-07220
Summit Machine Builders Corp.	USA	618	75-84900
Sverdrup Corp.	USA	643	80-30650
Sverdrup Technology Inc.	USA	643	80-30680
Tamares Ltd	UK	487	75-58500
Technitrol Inc.	USA	620	75-86800
Tel-Instrument Electronics Corp.	USA	621	75-87860
Teledyne Brown Engineering	USA	620	75-87510
Teledyne Controls	USA	620	75-87518
Teradyne Limited	UK	488	75-59318
Tracor Inc	USA	623	75-90000
Turbo Lufttechnik GmbH	Germany	176	75-31900
Vazduhoplovno Remonti Zavod	Yugoslavia	661	80-02500
Vernon Gauging Systems Ltd	UK	491	75-62750
Vickers Shipbuilding & Eng Ltd	UK	491	75-63100
Vickers Systems Division TRINOVA Ltd	UK	491	75-63150
Vipac Engineers & Scientists Ltd	Australia	26	80-22580
Warren-Knight Instrument Co.	USA	627	75-95150
Wegmann & Co GmbH	Germany	177	75-33700
Weinschel Corporation	USA	627	75-95544
Zeiss-Eltro Optronic GmbH	Germany	177	75-34700

617 10 Test installations, large fixed

A. Boët SA	France	133	75-06945
ADE-HML Limited	UK	427	75-00710
Advanced Research & Applications	USA	546	75-00940
AEA Technology plc	UK	496	80-00480
Aero Systems Engineering Inc.	USA	546	75-01116
ATS Aerospace Inc.	Canada	68	75-00752
Calspan SRL Technology	USA	540	60-03000
Commercial Hydraulics Keelavite Ltd	UK	441	75-14410
Cranfield Aerospace Ltd	UK	500	80-07200
Cranfield University	UK	419	60-00350
David Brown Defence Equipment Ltd	UK	437	75-10150
Dayton T. Brown Inc.	USA	632	80-04920
Froude Consine	UK	452	75-24800
GEC-Marconi Materials Technology Ltd	UK	454	75-25340
GEL Engineering Ltd	UK	454	75-25500
Industrial Acoustics Company Ltd	UK	460	75-31300
Industrieanlagen-Betriebs GmbH	Germany	178	80-00945
ISVR Consultancy Services	UK	505	80-13520
Magnus Engineering & Maintenance Ltd	Israel	221	75-12750
Northrop Grumman Corporation	USA	599	75-65450
Philips Calibration Services	Australia	25	80-16490
Quantum Inspection & Testing Ltd	Canada	79	80-01800
Reliant Design Development Co Ltd	UK	478	75-49170
Retlif Testing Laboratories	USA	641	80-25950
Rolls-Royce MatEval	UK	480	75-50260
Sanders	USA	612	75-77850
Sverdrup Corp.	USA	643	80-30650
Swiss Aircraft & Systems Co	Switzerland	384	55-00100
Tel-Instrument Electronics Corp.	USA	621	75-87860
Turbo Lufttechnik GmbH	Germany	176	75-31900
Vipac Engineers & Scientists Ltd	Australia	26	80-22580

617 10 20 Test cells, jet engine

A. Boët SA	France	133	75-06945
Aarding BV	Netherlands	291	75-00020
Calspan SRL Technology	USA	540	60-03000
Cranfield University	UK	419	60-00350
Environmental Elements Corp.	USA	574	75-31850
Froude Engineering Inc.	USA	576	75-36340
Industrial Acoustics Company Ltd	UK	460	75-31300
Ishikawajima Noise Control Co. Ltd	Japan	252	75-02820
John Curran Ltd	UK	444	75-16430
Magnus Engineering & Maintenance Ltd	Israel	221	75-12750
MTU München GmbH	Germany	172	75-21801
Sverdrup Corp.	USA	643	80-30650
Sverdrup Technology Inc.	USA	643	80-30680
Swiss Aircraft & Systems Co	Switzerland	384	55-00100

617 11 Anechoic chambers

Advanced Electromagnetics	USA	546	75-00895
Andrew Canada Inc.	Canada	68	75-00600
ATS Aerospace Inc.	Canada	68	75-00752
Calspan SRL Technology	USA	540	60-03000
Contraves Inc	USA	566	75-21850
Emerson and Cuming Microwave	Belgium	46	75-02900
Froude Consine	UK	452	75-24800
GEC-Marconi Avionics Ltd	UK	454	75-25320
GEC-Marconi Materials Technology Ltd	UK	454	75-25340
Industrial Acoustics Company Ltd	UK	460	75-31300
ISVR Consultancy Services	UK	505	80-13520
Litton Industries Inc.	USA	590	75-55400
National Aerospace Laboratory	Netherlands	290	60-01400
RFI Industries Pty Ltd	Australia	20	75-18500
Sound Attenuators Ltd	UK	485	75-55208
Sverdrup Corp.	USA	643	80-30650
Sverdrup Technology Inc.	USA	643	80-30680
Vipac Engineers & Scientists Ltd	Australia	26	80-22580

617 12 Wind tunnels

Aero Systems Engineering Inc.	USA	546	75-01116
ATS Aerospace Inc.	Canada	68	75-00752
Calspan SRL Technology	USA	540	60-03000
Cranfield Aerospace Ltd	UK	500	80-07200
Cranfield University	UK	419	60-00350
National Aerospace Laboratory	Netherlands	290	60-01400
Northrop Grumman Corporation	USA	599	75-65450
Sverdrup Corp.	USA	643	80-30650
Sverdrup Technology Inc.	USA	643	80-30680
Swiss Aircraft & Systems Co	Switzerland	384	55-00100
Turbo Lufttechnik GmbH	Germany	176	75-31900
Vipac Engineers & Scientists Ltd	Australia	26	80-22580

617 13 Environmental test installations

Calspan SRL Technology	USA	*540*	60-03000
Cincinnati Electronics Corp.	USA	*564*	75-18640
CMS Inc.	USA	*564*	75-19170
ERA Technology Ltd	UK	*502*	80-09600
Froude Engineering Inc.	USA	*576*	75-36340
GEC-Marconi Avionics Ltd	UK	*454*	75-25320
Litton Industries Inc.	USA	*591*	75-55400
Montford Instruments Ltd	UK	*469*	75-40600
NPO Geofizika	Russia	*338*	75-14315
Sverdrup Corp.	USA	*643*	80-30650
The Armstrong Monitoring Corp.	Canada	*68*	75-00740
Turbo Lufttechnik GmbH	Germany	*176*	75-31900
Vipac Engineers & Scientists Ltd	Australia	*26*	80-22580

617 14 Test basins

Aero Systems Engineering Inc.	USA	*546*	75-01116
Davis (W.R.) Engineering Ltd	Canada	*70*	75-04050
SSPA Maritime Consulting AB	Sweden	*380*	80-01960
Sverdrup Corp.	USA	*643*	80-30650

617 15 Test ranges

AEA Technology plc	UK	*496*	80-00480
Allied Research Corp.	USA	*549*	75-04670
Andrew Canada Inc.	Canada	*68*	75-00600
Armalon Ltd	UK	*431*	75-03729
ATS Aerospace Inc.	Canada	*68*	75-00752
British Aerospace Military Aircraft	UK	*436*	75-08886
Calspan SRL Technology	USA	*540*	60-03000
Cranfield University	UK	*419*	60-00350
EG&G Inc.	USA	*571*	75-29555
FR Aviation Limited	UK	*503*	80-10850
Litton Industries Inc.	USA	*590*	75-55400
MASS Consultants Limited	UK	*506*	80-15650
Octec Ltd	UK	*471*	75-42950
OTB	S. Africa	*357*	60-00500
Ridsdale Ranges	UK	*509*	80-28100
Sabre Computers Int'l Ltd	UK	*481*	75-51150
SAIC Ltd	UK	*510*	80-30220
Sverdrup Corp.	USA	*643*	80-30650
Sverdrup Technology Inc.	USA	*643*	80-30680
Thomson Marconi Sonar Limited	UK	*426*	74-20000
Turbo Lufttechnik GmbH	Germany	*176*	75-31900
Vipac Engineers & Scientists Ltd	Australia	*26*	80-22580
WS Atkins plc.	UK	*497*	80-01550

617 15 10 Test ranges, acoustic

ISVR Consultancy Services	UK	*505*	80-13520
Sverdrup Corp.	USA	*643*	80-30650
Sverdrup Technology Inc.	USA	*643*	80-30680
Thomson Marconi Sonar Limited	UK	*426*	74-20000

617 15 15 Test ranges, ammunition

Allied Research Corp.	USA	*549*	75-04670
Armalon Ltd	UK	*431*	75-03729
ATS Aerospace Inc.	Canada	*68*	75-00752
Calspan SRL Technology	USA	*540*	60-03000
CMS Inc.	USA	*564*	75-19170
Drello Ing P. Drewell GmbH & Co KG	Germany	*167*	75-06900

617 15 40 Test ranges, missile

FR Aviation Limited	UK	*503*	80-10850
MASS Consultants Limited	UK	*506*	80-15650
OTB	S. Africa	*357*	60-00500
Sverdrup Corp.	USA	*643*	80-30650
Sverdrup Technology Inc.	USA	*643*	80-30680
Swiss Aircraft & Systems Co	Switzerland	*384*	55-00100
WS Atkins plc.	UK	*497*	80-01550

617 15 60 Test ranges, underwater

ATS Aerospace Inc.	Canada	*68*	75-00752
ISVR Consultancy Services	UK	*505*	80-13520
Thomson Marconi Sonar Limited	UK	*426*	74-20000

617 15 70 Test range equipment

ATS Aerospace Inc.	Canada	*68*	75-00752
Drello Ing P. Drewell GmbH & Co KG	Germany	*167*	75-06900
Interstate Electronics	USA	*586*	75-48700
ISVR Consultancy Services	UK	*505*	80-13520
Laser-Scan	UK	*506*	80-14210
MS Instruments plc	UK	*469*	75-41265
Oceano Instruments (UK) Ltd	UK	*508*	80-19050
Sabre Computers Int'l Ltd	UK	*481*	75-51150
SFIM	**France**	***144***	**75-35080**
Thomson Marconi Sonar Limited	UK	*426*	74-20000

617 15 70 50 Range instrumentation

Applied Measurement Aust Pty Ltd	Australia	*12*	75-01420
Armalon Ltd	UK	*431*	75-03729
ATS Aerospace Inc.	Canada	*68*	75-00752
Calspan SRL Technology	USA	*540*	60-03000
Drello Ing P. Drewell GmbH & Co KG	Germany	*167*	75-06900
Interstate Electronics	USA	*586*	75-48700
MS Instruments plc	UK	*469*	75-41265
Sabre Computers Int'l Ltd	UK	*481*	75-51150
SFIM	**France**	***144***	**75-35080**
Sverdrup Corp.	USA	*643*	80-30650
Thomson Marconi Sonar Limited	UK	*426*	74-20000

617 20 Test systems

AAI/ACL Technologies Inc.	USA	*545*	75-00360
Advanced Research & Applications	USA	*546*	75-00940
Advanced Technical Corp.	USA	*546*	75-01000
AIM Enterprises Inc.	USA	*630*	80-01350
Anritsu Wiltron	USA	*554*	75-07800
Automatic Systems Laboratories Ltd	UK	*432*	75-04470
Bauer Aerospace	USA	*557*	75-12250
Calspan SRL Technology	USA	*540*	60-03000
Carco Electronics	USA	*562*	75-16700
Cincinnati Electronics Corp.	USA	*564*	75-18640
Commercial Hydraulics Keelavite Ltd	UK	*441*	75-14410
Computing Devices Company Ltd	UK	*442*	75-14700
Contraves Inc	USA	*566*	75-21850
David Brown Defence Equipment Ltd	UK	*437*	75-10150
EG&G Inc.	USA	*571*	75-29555
Elan GmbH.	Germany	*168*	75-07550
Environmental Elements Corp.	USA	*574*	75-31850
ERA Technology Ltd	UK	*502*	80-09600
Fabbrica Italiana Apparecchiature	Italy	*234*	75-08550
FFV Aerotech AB	Sweden	*379*	80-00600
Froude Consine	UK	*452*	75-24800
GEC-Marconi Avionics Ltd	UK	*454*	75-25320
Hewlett-Packard (Canada) Ltd	Canada	*72*	75-07520
Hewlett-Packard Company	USA	*581*	75-43200
Hocking NDT Ltd	UK	*458*	75-29580
Hoffman Engineering Corp.	USA	*582*	75-43980
Industrieanlagen-Betriebs GmbH	Germany	*178*	80-00945
International Machine & Tool Corp.	USA	*585*	75-46860
Intertechnique SA.	France	*137*	75-23000
ISVR Consultancy Services	UK	*505*	80-13520
John Curran Ltd	UK	*444*	75-16430
Kelsey Instruments Ltd	UK	*463*	75-33455
Krauss-Maffei Wehrtechnik GmbH	Germany	*171*	75-16345
Lloyd Instruments Ltd	UK	*465*	75-36150
Lucas Aerospace Power Transmission	USA	*592*	75-56265
Marconi Instruments Ltd	UK	*466*	75-37270
Microelectronics, guidance and	Turkey	*402*	55-00050
Racal Instruments Inc.	USA	*606*	75-72402
Reliant Design Development Co Ltd	UK	*478*	75-49170
Rolls-Royce MatEval	UK	*480*	75-50260
Senso-Metrics Inc.	USA	*613*	75-79340
SEXTANT Avionique SA.	France	*144*	75-35000
Stanford Research Systems Inc.	USA	*617*	75-83750
Sverdrup Corp.	USA	*643*	80-30650
Tel-Instrument Electronics Corp.	USA	*621*	75-87860
Thomson Marconi Sonar Limited	UK	*426*	74-20000
Trevin Products Pty Ltd	Australia	*22*	75-20840
Vipac Engineers & Scientists Ltd	Australia	*26*	80-22580
Wayne Kerr Electronics	UK	*492*	75-64860
Weinschel Corporation	USA	*627*	75-95544

Product–Supplier Link

617 20 20 Test systems, fixed

AAI/ACL Technologies Inc.	USA	*545*	75-00360
Advanced Research & Applications	USA	*546*	75-00940
Air-Log Ltd	UK	*429*	75-01800
Anritsu Wiltron	USA	*554*	75-07800
Commercial Hydraulics Keelavite Ltd	UK	*441*	75-14410
Digital Technology Inc.	USA	*569*	75-26650
Electro Optical Industries Inc.	USA	*573*	75-30450
ERA Technology Ltd	UK	*502*	80-09600
Fabbrica Italiana Apparecchiature	Italy	*234*	75-08550
Froude Consine	UK	*452*	75-24800
Froude Engineering Inc.	USA	*576*	75-36340
GEC-Marconi Avionics Ltd	UK	*454*	75-25320
Hewlett-Packard (Canada) Ltd	Canada	*72*	75-07520
Indústrias de Defesa	Portugal	*327*	55-00950
Intertechnique SA.	France	*137*	75-23000
Krauss-Maffei Wehrtechnik GmbH	Germany	*171*	75-16345
Lloyd Instruments Ltd	UK	*465*	75-36150
Lucas Aerospace Power Transmission	USA	*592*	75-56265
Marconi Instruments Ltd	UK	*466*	75-37270
Microelectronics, guidance and	Turkey	*402*	55-00050
Reliant Design Development Co Ltd.	UK	*478*	75-49170
Rolls-Royce MatEval	UK	*480*	75-50260
Sira Ltd.	UK	*484*	75-54250
Sverdrup Corp.	USA	*643*	80-30650
Technitrol Inc.	USA	*620*	75-86800
Tel-Instrument Electronics Corp.	USA	*621*	75-87860
Thomson Marconi Sonar Limited	UK	*426*	74-20000
Vipac Engineers & Scientists Ltd	Australia	*26*	80-22580
Wegmann & Co GmbH	Germany	*177*	75-33700

617 20 20 30 Test systems, EMP

AEA Technology plc	UK	*496*	80-00480
Aero Systems Engineering Inc.	USA	*546*	75-01116
Cincinnati Electronics Corp.	USA	*564*	75-18640
EG&G Inc.	USA	*571*	75-29555
Froude Consine	UK	*452*	75-24800
GEC-Marconi Avionics Ltd	UK	*454*	75-25320
National Aerospace Laboratory	Netherlands	*290*	60-01400
Reliant Design Development Co Ltd.	UK	*478*	75-49170
RFI Industries Pty Ltd	Australia	*20*	75-18500
Sverdrup Corp.	USA	*643*	80-30650

617 20 40 Test systems, mobile

AAI/ACL Technologies Inc.	USA	*545*	75-00360
AEA Technology plc	UK	*496*	80-00480
Air-Log Ltd	UK	*429*	75-01800
Anritsu Wiltron	USA	*554*	75-07800
Calspan SRL Technology.	USA	*540*	60-03000
Commercial Hydraulics Keelavite Ltd	UK	*441*	75-14410
Drello Ing P. Drewell GmbH & Co KG	Germany	*167*	75-06900
ERA Technology Ltd	UK	*502*	80-09600
Fabbrica Italiana Apparecchiature	Italy	*234*	75-08550
Froude Consine	UK	*452*	75-24800
GEC-Marconi Avionics Ltd	UK	*454*	75-25320
Hewlett-Packard (Canada) Ltd	Canada	*72*	75-07520
Intertechnique SA.	France	*137*	75-23000
Marconi Instruments Ltd	UK	*466*	75-37270
Reliant Design Development Co Ltd.	UK	*478*	75-49170
Rolls-Royce MatEval	UK	*480*	75-50260
SEXTANT Avionique SA.	France	*144*	75-35000
Sira Ltd.	UK	*484*	75-54250
Technitrol Inc.	USA	*620*	75-86800
Tel-Instrument Electronics Corp.	USA	*621*	75-87860
Thomson Marconi Sonar Limited	UK	*426*	74-20000
Vipac Engineers & Scientists Ltd	Australia	*26*	80-22580
Wegmann & Co GmbH	Germany	*177*	75-33700

617 30 Test & check-out equipment

AAI/ACL Technologies Inc.	USA	*545*	75-00360
Acutronic (Switzerland) AG	Switzerland	*385*	75-00050
ADE-HML Limited.	UK	*427*	75-00710
Advance Hivolt	UK	*427*	75-00903
Advanced Technical Corp.	USA	*546*	75-01000
Air-Log Ltd	UK	*429*	75-01800
Aircraft Appliances & Equipment Ltd	Canada	*68*	75-00430
Anritsu Wiltron	USA	*554*	75-07800
Applied Measurement Aust Pty Ltd.	Australia	*12*	75-01420
Atlantis Aerospace Corp.	Canada	*68*	75-00750
Ballard Technology Inc.	USA	*557*	75-11800
Barfield Inc.	USA	*557*	75-11930
Bauer Aerospace	USA	*557*	75-12250
Bradley Electronics Ltd	UK	*435*	75-07860
Calspan SRL Technology.	USA	*540*	60-03000
Central Engineering Company	USA	*562*	75-17290
Codalex Ltd	Canada.	*70*	75-03400
Commercial Hydraulics Keelavite Ltd	UK	*441*	75-14410
Computing Devices Company Ltd.	UK	*442*	75-14700
Contraves Inc	USA	*566*	75-21850
CTF Systems Inc.	Canada	*70*	75-03810
Dowty Controls	UK	*446*	75-18592
Edgcumbe Instruments Ltd	UK	*447*	75-19410
Edison Aerospace & Defense	USA	*571*	75-29100
ELECMA	France	*136*	75-14400
Emprise Corp.	USA	*634*	80-09950
Fabbrica Italiana Apparecchiature	Italy.	*234*	75-08550
Frederick Duffield Pty Ltd	Australia	*16*	75-06840
Froude Consine	UK	*452*	75-24800
Froude Engineering Inc.	USA	*576*	75-36340
GAP Instrument Corp.	USA	*577*	75-36900
GEC-Marconi Avionics Ltd	UK	*454*	75-25320
Graseby Dynamics Ltd	UK	*455*	75-26590
Hewlett-Packard Company	USA	*581*	75-43200
Hydroscience Technologies Inc.	USA	*584*	75-45550
Ideal Aerosmith Inc.	USA	*584*	75-45900
Imco Industries Ltd	Israel	*219*	75-09400
International Machine & Tool Corp.	USA	*585*	75-46860
Intertechnique SA.	France	*137*	75-23000
Israel Aircraft Industries Ltd	Israel	*220*	75-10200
Japan Aviation Electronics Ind. Ltd	Japan	*252*	75-03300
Kearfott Guidance & Navigation Corp.	USA	*588*	75-51760
Komatsu Hanomag Aktiengesellschaft	Germany	*171*	75-16185
Krauss-Maffei Wehrtechnik GmbH	Germany	*171*	75-16345
Lockheed Martin Corp.	USA	*591*	75-55635
Lucas Aerospace Power Transmission	USA	*592*	75-56265
Marconi Instruments Ltd	UK	*466*	75-37270
Merex Inc.	USA	*595*	75-60250
Nansen SA	Brazil	*57*	75-14000
North Hills Signal Processing	USA	*599*	75-65300
Northrop Grumman Corporation.	USA	*599*	75-65450
Penny & Giles Drives Technology Ltd.	UK	*474*	75-44563
Plint & Partners Ltd	UK	*475*	75-46330
Power Technology Inc.	USA	*605*	75-71090
Quantum Inspection & Testing Ltd	Canada	*79*	80-01800
Racal Instruments Ltd	UK	*477*	75-47676
Reliant Design Development Co Ltd.	UK	*478*	75-49170
SAJ-Froude Test Plant Pvt Ltd	India	*204*	74-01000
SELIN Sistemi Sarl	Italy.	*238*	75-21750
Senso-Metrics Inc.	USA	*613*	75-79340
Sparton Technology Inc.	USA	*616*	75-82950
Stanford Research Systems Inc.	USA	*617*	75-83750
Sverdrup Corp.	USA	*643*	80-30650
Sverdrup Technology Inc.	USA	*643*	80-30680
Technitrol Inc.	USA	*620*	75-86800
Tel-Instrument Electronics Corp.	USA	*621*	75-87860
Teledyne Fluid Systems	USA	*621*	75-87570
Teradyne Limited	UK	*488*	75-59318
The Boeing Company.	USA	*559*	75-14250
Toyo Communication Equipment Co	Japan	*255*	75-10000
Trevin Products Pty Ltd	Australia	*22*	75-20840
Triplett Corporation	USA	*623*	75-90970
Vipac Engineers & Scientists Ltd	Australia	*26*	80-22580
Watkins-Johnson Co.	USA	*627*	75-95200
Wavetek GmbH	Germany	*176*	75-33620
Wayne Kerr Electronics	UK	*492*	75-64860
Wegmann & Co GmbH	Germany	*177*	75-33700

617 30 10 Test equipment, automatic

AAI/ACL Technologies Inc.	USA	*545*	75-00360
AAR Allen Group International	Netherlands	*296*	80-00050
Acutronic (Switzerland) AG	Switzerland	*385*	75-00050
Advanced Research & Applications	USA	*546*	75-00940
Air-Log Ltd	UK	*429*	75-01800
Anritsu Wiltron	USA	*554*	75-07800
Atlantis Aerospace Corp.	Canada	*68*	75-00750
B&W Engineering Corp.	USA	*561*	75-16009
Bavaria Keytronic Technologie GmbH	Germany	*165*	75-02050
Bertan High Voltage	USA	*558*	75-13720
Calspan SRL Technology.	USA	*540*	60-03000
Commercial Hydraulics Keelavite Ltd	UK	*441*	75-14410
Contraves Inc	USA	*566*	75-21850

Dayton T. Brown Inc.	USA	*632*	80-04920
ELECMA	France	*136*	75-14400
ELTA Electronics Industries Ltd	Israel	*220*	75-10200
Fabbrica Italiana Apparecchiature	Italy	*234*	75-08550
Froude Consine	UK	*452*	75-24800
Froude Engineering Inc.	USA	*576*	75-36340
GAP Instrument Corp.	USA	*577*	75-36900
GEC-Marconi Avionics Ltd	UK	*454*	75-25320
Giat Industries	France	*126*	55-00100
Graseby Dynamics Ltd	UK	*455*	75-26590
Hewlett-Packard (Canada) Ltd	Canada	*72*	75-07520
Hewlett-Packard Company	USA	*581*	75-43200
Ideal Aerosmith Inc.	USA	*584*	75-45900
INDRA	Spain	*366*	75-03800
Industrial Acoustics Company Ltd	UK	*460*	75-31300
Israel Aircraft Industries Ltd	Israel	*220*	75-10200
Japan Aviation Electronics Ind. Ltd	Japan	*252*	75-03300
Kearfott Guidance & Navigation Corp.	USA	*588*	75-51760
Krauss-Maffei Wehrtechnik GmbH	Germany	*171*	75-16345
Litton Industries Inc.	USA	*590*	75-55400
Logic SpA	Italy	*236*	75-13800
Logica UK Limited	UK	*506*	80-14350
Lucas Aerospace Power Transmission	USA	*592*	75-56265
Marconi Instruments Ltd	UK	*466*	75-37270
Merex Inc.	USA	*595*	75-60250
MPD Technologies Inc.	USA	*597*	75-63125
Pentastar Electronics Inc.	USA	*603*	75-69260
QES Ltd	UK	*509*	80-25000
Racal Instruments Ltd	UK	*477*	75-47676
Reliant Design Development Co Ltd	UK	*478*	75-49170
Rockwell-Collins France SA	France	*142*	75-32800
Rolls-Royce MatEval	UK	*480*	75-50260
Sanders	USA	*612*	75-77850
SEXTANT Avionique SA	France	*144*	75-35000
Sparton Technology Inc.	USA	*616*	75-82950
Teledyne Fluid Systems	USA	*621*	75-87570
The Graphics Co. of America	USA	*579*	75-40360
Watkins-Johnson Co.	USA	*627*	75-95200
Wayne Kerr Electronics	UK	*492*	75-64860

617 30 20 Test & check-out equipment, fixed

AAI/ACL Technologies Inc.	USA	*545*	75-00360
ADE-HML Limited	UK	*427*	75-00710
Advanced Research & Applications	USA	*546*	75-00940
Air-Log Ltd	UK	*429*	75-01800
Anritsu Wiltron	USA	*554*	75-07800
Applied Measurement Aust Pty Ltd	Australia	*12*	75-01420
Baganoff Engineering Inc.	USA	*557*	75-11660
Commercial Hydraulics Keelavite Ltd	UK	*441*	75-14410
ERA Technology Ltd	UK	*502*	80-09600
Froude Consine	UK	*452*	75-24800
GEC-Marconi Avionics Ltd	UK	*454*	75-25320
Hewlett-Packard Company	USA	*581*	75-43200
Intertechnique SA	France	*137*	75-23000
Israel Aircraft Industries Ltd	Israel	*220*	75-10200
Krauss-Maffei Wehrtechnik GmbH	Germany	*171*	75-16345
Litton Industries Inc.	USA	*591*	75-55400
Lucas Aerospace Power Transmission	USA	*592*	75-56265
Marconi Instruments Ltd	UK	*466*	75-37270
Merex Inc.	USA	*595*	75-60250
MPD Technologies Inc.	USA	*597*	75-63125
QES Ltd	UK	*509*	80-25000
Reliant Design Development Co Ltd	UK	*478*	75-49170
Sira Ltd.	UK	*484*	75-54250
Strachan & Henshaw Ltd	UK	*512*	80-47400
Sverdrup Corp.	USA	*643*	80-30650
Technitrol Inc.	USA	*620*	75-86800
Tel-Instrument Electronics Corp.	USA	*621*	75-87860
Teledyne Fluid Systems	USA	*621*	75-87570
Vipac Engineers & Scientists Ltd	Australia	*26*	80-22580
Wayne Kerr Electronics	UK	*492*	75-64860

617 30 20 60 Test stands, fixed

AAI/ACL Technologies Inc.	USA	*545*	75-00360
AAR Allen Group International	Netherlands	*296*	80-00050
Aircraft Appliances & Equipment Ltd	Canada	*68*	75-00430
Appareillages et Matériels de	France	*132*	75-03050
Commercial Hydraulics Keelavite Ltd	UK	*441*	75-14410
David Brown Defence Equipment Ltd	UK	*437*	75-10150
Dayton T. Brown Inc.	USA	*632*	80-04920
Emprise Corp.	USA	*634*	80-09950
Froude Consine	UK	*452*	75-24800
Froude Engineering Inc.	USA	*576*	75-36340
Hewlett-Packard Company	USA	*581*	75-43200
Ideal Aerosmith Inc.	USA	*584*	75-45900
Industrial Acoustics Company Ltd	UK	*460*	75-31300
Israel Aircraft Industries Ltd	Israel	*220*	75-10200
Krauss-Maffei Wehrtechnik GmbH	Germany	*171*	75-16345
L&F Industries	USA	*589*	75-53690
Litton Industries Inc.	USA	*591*	75-55400
Lucas Aerospace Power Transmission	USA	*592*	75-56265
Lucas Aftermarket Operations	UK	*465*	75-36655
Prva Petoletka Industrie	Yugoslavia	*661*	75-00610
Reliant Design Development Co Ltd	UK	*478*	75-49170
SELIN Sistemi Sarl	Italy	*238*	75-21750
Sverdrup Corp.	USA	*643*	80-30650
Tecmotiv Corporation	Canada	*77*	75-13510
Teledyne Fluid Systems	USA	*621*	75-87570
Vickers Systems Division TRINOVA Ltd	UK	*491*	75-63150
ZF Luftfahrttechnik GmbH	Germany	*177*	75-36000

617 30 40 Test & check-out equipment, mobile

AAI/ACL Technologies Inc.	USA	*545*	75-00360
Acutronic (Switzerland) AG	Switzerland	*385*	75-00050
ADE-HML Limited	UK	*427*	75-00710
Aermacchi SpA	Italy	*230*	75-00500
Air-Log Ltd	UK	*429*	75-01800
Baganoff Engineering Inc.	USA	*557*	75-11660
Bradley Electronics Ltd	UK	*435*	75-07860
Commercial Hydraulics Keelavite Ltd	UK	*441*	75-14410
Computing Devices Company Ltd	UK	*442*	75-14700
Edgcumbe Instruments Ltd	UK	*447*	75-19410
ELECMA	France	*136*	75-14400
Froude Engineering Inc.	USA	*576*	75-36340
GAP Instrument Corp.	USA	*577*	75-36900
Intertechnique SA	France	*137*	75-23000
Krauss-Maffei Wehrtechnik GmbH	Germany	*171*	75-16345
Lucas Aerospace Power Transmission	USA	*592*	75-56265
Mannesmann Rexroth Ltd	UK	*466*	75-37100
Marconi Instruments Ltd	UK	*466*	75-37270
MPD Technologies Inc.	USA	*597*	75-63125
Northrop Grumman Corporation	USA	*599*	75-65450
Omori Seikoki Co. Ltd	Japan	*254*	75-07310
Pentastar Electronics Inc.	USA	*603*	75-69260
Quantum Inspection & Testing Ltd	Canada	*79*	80-01800
Reliant Design Development Co Ltd	UK	*478*	75-49170
Sanders	USA	*612*	75-77850
Siemens plc	UK	*483*	75-53750
Sira Ltd.	UK	*484*	75-54250
Strachan & Henshaw Ltd	UK	*512*	80-47400
Tel-Instrument Electronics Corp.	USA	*621*	75-87860
Teledyne Fluid Systems	USA	*621*	75-87570
Vickers Systems Division TRINOVA Ltd	UK	*491*	75-63150
Vipac Engineers & Scientists Ltd	Australia	*26*	80-22580
Wavetek GmbH	Germany	*176*	75-33620
Wayne Kerr Electronics	UK	*492*	75-64860

617 30 40 30 Hydraulic test equipment, mobile

AAI/ACL Technologies Inc.	USA	*545*	75-00360
ADE-HML Limited	UK	*427*	75-00710
Aermacchi SpA	Italy	*230*	75-00500
Aeroquip Corp.	USA	*547*	75-02350
Applied Measurement Aust Pty Ltd	Australia	*12*	75-01420
Baganoff Engineering Inc.	USA	*557*	75-11660
Commercial Hydraulics Keelavite Ltd	UK	*441*	75-14410
Dayton T. Brown Inc.	USA	*632*	80-04920
Forward Industries Ltd	UK	*452*	75-24200
Froude Consine	UK	*452*	75-24800
Howden Fan Company Pty Ltd	Australia	*17*	75-08740
Howden Wade Ltd	UK	*459*	75-30260
Kinetics Ltd	Israel	*220*	75-11320
Logistic Projects	S. Africa	*358*	75-09800
Lucas Aerospace Power Transmission	USA	*592*	75-56265
Mannesmann Rexroth SA	France	*139*	75-25500
Messier-Bugatti	France	*140*	75-26400
Omori Seikoki Co. Ltd	Japan	*254*	75-07310
Prva Petoletka Industrie	Yugoslavia	*661*	75-00610
Quantum Inspection & Testing Ltd	Canada	*79*	80-01800
Regent Mfg Inc.	USA	*609*	75-74200
Reliant Design Development Co Ltd	UK	*478*	75-49170

Riva Calzoni SpA	Italy	238	75-20500
Spar Aviation Services	Canada	76	75-13020
Teledyne Fluid Systems	USA	621	75-87570
Theodor Rapp GmbH Co	Germany	173	75-24600
Vickers Systems Division TRINOVA Ltd	UK	491	75-63150

617 30 40 60 Test stands, mobile field

AAI/ACL Technologies Inc.	USA	545	75-00360
AAR Allen Group International	Netherlands	296	80-00050
Aero Systems Engineering Inc.	USA	546	75-01116
Aircraft Appliances & Equipment Ltd	Canada	68	75-00430
Commercial Hydraulics Keelavite Ltd	UK	441	75-14410
David Brown Defence Equipment Ltd	UK	437	75-10150
Dayton T. Brown Inc.	USA	632	80-04920
Froude Consine	UK	452	75-24800
Froude Engineering Inc.	USA	576	75-36340
Hewlett-Packard Company	USA	581	75-43200
Hydroscience Technologies Inc.	USA	584	75-45550
Industrial Acoustics Company Ltd	UK	460	75-31300
Lucas Aftermarket Operations	UK	465	75-36655
Messier-Bugatti	France	140	75-26400
Omori Seikoki Co. Ltd	Japan	254	75-07310
Reliant Design Development Co Ltd	UK	478	75-49170
SELIN Sistemi Sarl	Italy	238	75-21750
Teledyne Fluid Systems	USA	621	75-87570

617 30 50 Test & check-out equipment, portable

AAI/ACL Technologies Inc.	USA	545	75-00360
Aero Systems Engineering Inc.	USA	546	75-01116
Air-Log Ltd	UK	429	75-01800
Aircraft Appliances & Equipment Ltd	Canada	68	75-00430
American Induction HTG Corp.	USA	553	75-05920
Anritsu Wiltron	USA	554	75-07800
ATS Aerospace Inc.	Canada	68	75-00752
Baganoff Engineering Inc.	USA	557	75-11660
Bradley Electronics Ltd	UK	435	75-07860
Codalex Ltd	Canada	70	75-03400
Commercial Hydraulics Keelavite Ltd	UK	441	75-14410
Digital Technology Inc.	USA	569	75-26650
Edgcumbe Instruments Ltd	UK	447	75-19410
ELECMA	France	136	75-14400
Electro Optical Industries Inc.	USA	573	75-30450
ERA Technology Ltd	UK	502	80-09600
Explorocean Technology Ltd	UK	502	80-09870
GAP Instrument Corp.	USA	577	75-36900
GEC-Marconi Avionics Ltd	UK	454	75-25320
Graseby Dynamics Ltd	UK	455	75-26590
Hewlett-Packard Company	USA	581	75-43200
Intertechnique SA	France	137	75-23000
Japan Aircraft Manufacturing Co. Ltd	Japan	252	75-03200
Krauss-Maffei Wehrtechnik GmbH	Germany	171	75-16345
Lucas Aftermarket Operations	UK	465	75-36655
Marconi Instruments Ltd	UK	466	75-37270
MPD Technologies Inc.	USA	597	75-63125
Northrop Grumman Corporation	USA	599	75-65450
Omori Seikoki Co. Ltd	Japan	254	75-07310
Quantum Inspection & Testing Ltd	Canada	79	80-01800
Reliant Design Development Co Ltd	UK	478	75-49170
Revue THOMMEN AG	Switzerland	387	75-11100
Sanders	USA	612	75-77850
Siemens plc	UK	483	75-53750
Techtest Ltd	UK	484	75-54590
Tel-Instrument Electronics Corp.	USA	621	75-87860
Teledyne Fluid Systems	USA	621	75-87570
THORN Microwave Devices Ltd	UK	488	75-59760
Vipac Engineers & Scientists Ltd	Australia	26	80-22580
Watkins-Johnson Co.	USA	627	75-95200
Wavetek GmbH	Germany	176	75-33620
Wayne Kerr Electronics	UK	492	75-64860

617 30 56 Test sets, portable

AAI/ACL Technologies Inc.	USA	545	75-00360
Advance Hivolt	UK	427	75-00903
Advanced Technical Corp.	USA	546	75-01000
AEA Technology plc	UK	496	80-00480
Aero Systems Engineering Inc.	USA	546	75-01116
Air-Log Ltd	UK	429	75-01800
Anritsu Wiltron	USA	554	75-07800
Associated Aircraft Mfg & Sales Inc.	USA	555	75-09500
AVO International	USA	556	75-11300
Canadian Marconi Co.	Canada	69	75-02600
CI Systems Ltd	Israel	218	75-03750
Codalex Ltd	Canada	70	75-03400
Commercial Hydraulics Keelavite Ltd	UK	441	75-14410
ELECMA	France	136	75-14400
Explorocean Technology Ltd	UK	502	80-09870
FT Technologies Ltd	UK	452	75-24880
GEC-Marconi Avionics Ltd	UK	454	75-25320
Hewlett-Packard Company	USA	581	75-43200
Hocking NDT Ltd	UK	458	75-29580
Howden Wade Ltd	UK	459	75-30260
Krauss-Maffei Wehrtechnik GmbH	Germany	171	75-16345
Magnus Engineering & Maintenance Ltd	Israel	221	75-12750
Omori Seikoki Co. Ltd	Japan	254	75-07310
QES Ltd	UK	509	80-25000
Reliant Design Development Co Ltd	UK	478	75-49170
SEXTANT Avionique SA	France	144	75-35000
Technical Products Group Inc.	USA	620	75-86764
Watkins-Johnson Co.	USA	627	75-95200
Wayne Kerr Electronics	UK	492	75-64860

617 40 Test instrumentation

AAI/ACL Technologies Inc.	USA	545	75-00360
AAR Allen Group International	Netherlands	296	80-00050
Acutronic Ltd	UK	427	75-00520
Aero Systems Engineering Inc.	USA	546	75-01116
American Aerospace Controls Inc.	USA	552	75-05450
Anritsu Wiltron	USA	554	75-07800
Applied Measurement Aust Pty Ltd	Australia	12	75-01420
Australian Calibrating Services	Australia	12	75-01785
Automatic Systems Laboratories Ltd	UK	432	75-04470
AVO International	USA	556	75-11300
Bauer Aerospace	USA	557	75-12250
Bertan High Voltage	USA	558	75-13720
Cadex Electronics Inc.	Canada	69	75-01750
Calspan SRL Technology	USA	540	60-03000
Circle-C Transformers	Australia	14	75-03450
Commercial Hydraulics Keelavite Ltd	UK	441	75-14410
Contraves Inc	USA	566	75-21850
EG&G Inc.	USA	571	75-29555
Electro Optical Industries Inc.	USA	573	75-30450
Emprise Corp.	USA	634	80-09950
Froude Consine	UK	452	75-24800
FT Technologies Ltd	UK	452	75-24880
GEC-Marconi Avionics Ltd	UK	454	75-25320
Hewlett-Packard (Canada) Ltd	Canada	72	75-07520
Hewlett-Packard Company	USA	581	75-43200
Industrial Acoustics Company Ltd	UK	460	75-31300
Interstate Electronics	USA	586	75-48700
ISVR Consultancy Services	UK	505	80-13520
Lloyd Instruments Ltd	UK	465	75-36150
Marconi Instruments Ltd	UK	466	75-37270
Meggitt Avionics (UK)	UK	467	75-38340
Novamarine Instruments Pty Ltd	Australia	19	75-14885
Phase Two Industries	USA	603	75-69560
Philips Calibration Services	Australia	25	80-16490
Power Technology Inc.	USA	605	75-71090
Prosig Computer Consultants Ltd	UK	476	75-46865
QES Ltd	UK	509	80-25000
Quantum Inspection & Testing Ltd	Canada	79	80-01800
Racal Instruments Inc.	USA	606	75-72402
RFI Industries Pty Ltd	Australia	20	75-18500
Rolls-Royce MatEval	UK	480	75-50260
Schaffner EMC Ltd	UK	482	75-51970
Senso-Metrics Inc.	USA	613	75-79340
SFIM	**France**	**144**	**75-35080**
SISDEF Ltda	Chile	85	80-01940
Sverdrup Corp.	USA	643	80-30650
Sverdrup Technology Inc.	USA	643	80-30680
Tektronix Canada Inc.	Canada	77	75-13540
Tel-Instrument Electronics Corp.	USA	621	75-87860
The Graphics Co. of America	USA	579	75-40360
Vipac Engineers & Scientists Ltd	Australia	26	80-22580
Watkins-Johnson Co.	USA	627	75-95200
Wavetek Corporation	USA	627	75-95270
Wayne Kerr Electronics	UK	492	75-64860
Weinschel Corporation	USA	627	75-95544

617 40 39 Test instrumentation, underwater

Aero Systems Engineering Inc.	USA	*546*	75-01116
EG&G Inc.	USA	*571*	75-29555
Explorocean Technology Ltd	UK	*502*	80-09870
Haux-Life-Support GmbH	Germany	*169*	75-11970
ISVR Consultancy Services	UK	*505*	80-13520
Rolls-Royce MatEval	UK	*480*	75-50260
Strachan & Henshaw Ltd	UK	*512*	80-47400
Vipac Engineers & Scientists Ltd	Australia	*26*	80-22580

617 50 Inspection equipment

Aero Systems Engineering Inc.	USA	*546*	75-01116
Airmatic Engineering (UK) Ltd	UK	*429*	75-01850
Baugh & Weedon Ltd	UK	*433*	75-06010
Calspan SRL Technology	USA	*540*	60-03000
Central Engineering Company	USA	*562*	75-17290
Elf Atochem	Australia	*16*	75-05630
Everest Imaging Ltd	USA	*574*	75-32657
Hocking NDT Ltd	UK	*458*	75-29580
Japan Aircraft Manufacturing Co. Ltd	Japan	*252*	75-03200
Karlheinz Hinze Optoengineering GmbH	Germany	*169*	75-12800
Meggitt Avionics (UK)	UK	*467*	75-38340
Merex Inc.	USA	*595*	75-60250
Olympus Industrial	UK	*472*	75-43020
P W Allen & Co Ltd	UK	*429*	75-02740
Quantum Inspection & Testing Ltd	Canada	*79*	80-01800
Riva Calzoni SpA	Italy	*238*	75-20500
Rolls-Royce MatEval	UK	*480*	75-50260
Santa Barbara SA	Spain	*364*	55-01000
SFIM	**France**	***144***	**75-35080**
SPM Instrument UK Ltd	UK	*486*	75-56050
Stewart R. Browne Mfg Co.Inc.	USA	*560*	75-15295
Stocker & Yale Inc.	USA	*618*	75-84250
Trimos-Sylvac Task Force	Switzerland	*387*	75-14500
Vernon Gauging Systems Ltd	UK	*491*	75-62750
Vipac Engineers & Scientists Ltd	Australia	*26*	80-22580

617 50 10 Aircraft engine health monitoring equipment

Analysis Management Systems	S. Africa	*357*	75-00770
A/S Brüel & Kjaer	Denmark	*102*	75-02790
Baganoff Engineering Inc.	USA	*557*	75-11660
Bristow Helicopters Ltd	UK	*499*	80-03950
Calspan SRL Technology	USA	*540*	60-03000
Canadian Marconi Co.	Canada	*69*	75-02600
Central Engineering Company	USA	*562*	75-17290
ELECMA	France	*136*	75-14400
Endevco Corp.	USA	*573*	75-31545
Everest Imaging Ltd	USA	*574*	75-32657
Howden Fan Company Pty Ltd	Australia	*17*	75-08740
Industrial Acoustics Company Ltd	UK	*460*	75-31300
Magnus Engineering & Maintenance Ltd	Israel	*221*	75-12750
Meggitt Avionics (UK)	UK	*467*	75-38340
Olympus Industrial	UK	*472*	75-43020
Prosig Computer Consultants Ltd	UK	*476*	75-46865
Sverdrup Corp.	USA	*643*	80-30650
Sverdrup Technology Inc.	USA	*643*	80-30680
Varian Australia Pty Ltd	Australia	*23*	75-22100

617 50 80 Inspection equipment, ultrasonic

AEA Technology plc	UK	*496*	80-00480
B&W Engineering Corp.	USA	*561*	75-16009
Baugh & Weedon Ltd	UK	*433*	75-06010
Quantum Inspection & Testing Ltd	Canada	*79*	80-01800
Rolls-Royce MatEval	UK	*480*	75-50260
SPM Instrument UK Ltd	UK	*486*	75-56050
Swiss Aircraft & Systems Co	Switzerland	*384*	55-00100

617 60 Measuring systems & equipment

AAI/ACL Technologies Inc.	USA	*545*	75-00360
Advanced Research & Applications	USA	*546*	75-00940
Advanced Technical Corp.	USA	*546*	75-01000
Aero Systems Engineering Inc.	USA	*546*	75-01116
Aeronautical & General Instr. Ltd	UK	*428*	75-01050
Airmatic Engineering (UK) Ltd	UK	*429*	75-01850
American Aerospace Controls Inc.	USA	*552*	75-05450
Angus Electronics Co.	USA	*554*	75-07630
Anritsu Wiltron	USA	*554*	75-07800
A/S Brüel & Kjaer	Denmark	*102*	75-02790
Automatic Systems Laboratories Ltd	UK	*432*	75-04470
Avery Berkel	UK	*432*	75-04560
AVO International	USA	*556*	75-11300
Bowers Group	UK	*435*	75-07770
Budenberg Gauges (Australia) Pty Ltd	Australia	*14*	75-02860
Cadex Electronics Inc.	Canada	*69*	75-01750
Calspan SRL Technology	USA	*540*	60-03000
Canadian Marconi Co.	Canada	*69*	75-02600
Compagnie Générale de Géophysique	France	*134*	75-11335
Contraves Inc	USA	*566*	75-21850
Daco Scientific Ltd	UK	*444*	75-16550
EIP Microwave Inc.	USA	*572*	75-29590
Electric Tachometer Corp.	USA	*572*	75-29680
Electro Optical Industries Inc.	USA	*573*	75-30450
ENDECO/YSI Inc.	USA	*573*	75-31540
Endevco Corp.	USA	*573*	75-31545
Faure Herman SA	France	*136*	75-17500
Fisco Products Ltd	UK	*451*	75-23670
Guildline Instruments Inc.	USA	*580*	75-40900
Hoffman Engineering Corp.	USA	*582*	75-43980
ISVR Consultancy Services	UK	*505*	80-13520
Jennings Technology Corp.	USA	*586*	75-50350
Jordan Valley Applied Radiation Ltd	Israel	*220*	75-10850
Kaman Instrumentation Corp.	USA	*588*	75-51560
Litton Industries Inc.	USA	*590*	75-55400
Meggitt Avionics (UK)	UK	*467*	75-38340
Merex Inc.	USA	*595*	75-60250
Northrop Grumman Corporation	USA	*599*	75-65450
Oerlikon-Contraves AG	Switzerland	*387*	75-09250
Philips Calibration Services	Australia	*25*	80-16490
Poli Officine Meccaniche	Italy	*238*	75-19310
Precision Filters Inc.	USA	*605*	75-71330
Quantum Inspection & Testing Ltd	Canada	*79*	80-01800
Racal Instruments Inc.	USA	*606*	75-72402
Santa Barbara SA	Spain	*364*	55-01000
Senso-Metrics Inc.	USA	*613*	75-79340
Siemens plc	UK	*483*	75-53750
Sparton of Canada Ltd	Canada	*77*	75-13100
SPM Instrument UK Ltd	UK	*486*	75-56050
Stanford Research Systems Inc.	USA	*617*	75-83750
Summit Machine Builders Corp.	USA	*618*	75-84900
Sverdrup Corp.	USA	*643*	80-30650
The Eraser Co. Inc.	USA	*574*	75-32020
THERMOCOAX	France	*147*	75-44930
Trimos-Sylvac Task Force	Switzerland	*387*	75-14500
Varian Australia Pty Ltd	Australia	*23*	75-22100
Vipac Engineers & Scientists Ltd	Australia	*26*	80-22580
Warren-Knight Instrument Co.	USA	*627*	75-95150
Wavetek Corporation	USA	*627*	75-95270
Wayne Kerr Electronics	UK	*492*	75-64860
Weinschel Corporation	USA	*627*	75-95544

617 60 20 Bathymetric survey systems

Explorocean Technology Ltd	UK	*502*	80-09870
GeoAcoustics Ltd	UK	*454*	75-25405
Litton Industries Inc.	USA	*591*	75-55400
Precision Filters Inc.	USA	*605*	75-71330

617 60 62 Oceanographic measuring instruments

Applied Microsystems Ltd	Canada	*68*	75-00720
A1Marketing	UK	*426*	75-00025
Chelsea Instruments Ltd	UK	*439*	75-12600
Compagnie Générale de Géophysique	France	*134*	75-11335
Daco Scientific Ltd	UK	*444*	75-16550
EG&G Inc.	USA	*571*	75-29555
ENDECO/YSI Inc.	USA	*573*	75-31540
Explorocean Technology Ltd	UK	*502*	80-09870
GeoAcoustics Ltd	UK	*454*	75-25405
Guildline Instruments Inc.	USA	*580*	75-40900
Hugrún h/f	Iceland	*202*	75-00500
Installocean Ltd	UK	*461*	75-31605
NEC Corporation	Japan	*253*	75-06350
Oceano Instruments (UK) Ltd	UK	*508*	80-19050
Sparton of Canada Ltd	Canada	*77*	75-13100
The Graphics Co. of America	USA	*579*	75-40360

Product–Supplier Link

617 60 62 30 Bathythermographs, expendable

Explorocean Technology Ltd	UK	502	80-09870
Sparton Electronics	USA	616	75-82900
Sparton of Canada Ltd	Canada	77	75-13100

617 60 62 45 Probes, expendable, conductivity, temperature, depth

Budenberg Gauges (Australia) Pty Ltd	Australia	14	75-02860
Calspan SRL Technology	USA	540	60-03000
EG&G Inc.	USA	571	75-29555
Hugrún h/f	Iceland	202	75-00500
Sparton Electronics	USA	616	75-82900
THERMOCOAX	France	147	75-44930

617 60 62 50 Current profilers, expendable

American Aerospace Controls Inc.	USA	552	75-05450
EG&G Inc.	USA	571	75-29555
Explorocean Technology Ltd	UK	502	80-09870

617 60 62 80 Sound velocimeters, expendable

Applied Microsystems Ltd	Canada	68	75-00720
Explorocean Technology Ltd	UK	502	80-09870
Sparton of Canada Ltd	Canada	77	75-13100

617 60 70 Topographical measuring instruments

EG&G Inc.	USA	571	75-29555
The Graphics Co. of America	USA	579	75-40360

617 61 Ballistics measurement equipment

ARSOC Sprl	Belgium	45	75-00520
Cannatech Inc.	USA	562	75-16630
Canons Delcour SA	Belgium	46	75-02200
Drello Ing P. Drewell GmbH & Co KG	Germany	167	75-06900
Hadland Photonics Ltd	UK	456	75-27200
Indústrias de Defesa	Portugal	327	55-00950
Intertechnique (Aerospace) Ltd	UK	461	75-31948
MS Instruments plc	UK	469	75-41265
Octec Ltd	UK	471	75-42950
Sabre Computers Int'l Ltd	UK	481	75-51150
Summit Machine Builders Corp.	USA	618	75-84900
WS Atkins plc.	UK	497	80-01550

617 61 30 Cameras, high speed, ballistics recording

EG&G Inc.	USA	571	75-29555
Hadland Photonics Ltd	UK	456	75-27200
Polytec-PI SA	France	141	75-30450
Visual Instrumentation Corp.	USA	626	75-94300

617 62 Muzzle velocity measuring equipment

Calspan SRL Technology	USA	540	60-03000
Cannatech Inc.	USA	562	75-16630
Canons Delcour SA	Belgium	46	75-02200
Drello Ing P. Drewell GmbH & Co KG	Germany	167	75-06900
Sabre Computers Int'l Ltd	UK	481	75-51150

617 62 60 Muzzle velocity measuring radars

Merex Inc.	USA	595	75-60250
RSL Electronics Ltd	Israel	222	75-18900

617 70 Calibration & weapon alignment equipment

Airmatic Engineering (UK) Ltd	UK	429	75-01850
Australian Defence Industries Ltd	Australia	9	55-01000
Calspan SRL Technology	USA	540	60-03000
Cannatech Inc.	USA	562	75-16630
Canons Delcour SA	Belgium	46	75-02200
Elan GmbH	Germany	168	75-07550
Hall & Watts (Defence Optics) Group	**UK**	**456**	**75-27650**
Heinsoldt Systemtechnick GmbH	Germany	171	75-17800
ISVR Consultancy Services	UK	505	80-13520
Litton Industries Inc.	USA	590	75-55400
Nanoquest Defence Products Ltd	UK	470	75-41890
Oerlikon-Contraves AG	Switzerland	387	75-09250
Pyser-SGI Ltd	UK	476	75-47300
Quantum Inspection & Testing Ltd	Canada	79	80-01800
Santa Barbara SA	Spain	364	55-01000
Siemens plc	UK	483	75-53750
SISDEF Ltda	Chile	85	80-01940
Strachan & Henshaw Ltd	UK	512	80-47400
Tamares Ltd	UK	487	75-58500
Theodor Rapp GmbH Co	Germany	173	75-24600
Tilt Measurement Ltd	UK	488	75-59930
Time Electronics Ltd	UK	488	75-59940
Vickers Shipbuilding & Eng Ltd	UK	491	75-63100
Warren-Knight Instrument Co.	USA	627	75-95150
Weinschel Corporation	USA	627	75-95544
W.F. Stanley & Co. Ltd	UK	486	75-56350

617 70 60 20 Radar alignment equipment, optical, airborne

Nanoquest Defence Products Ltd	UK	470	75-41890
Power Technology Inc.	USA	605	75-71090
Warren-Knight Instrument Co.	USA	627	75-95150

617 70 60 30 Radar alignment equipment, optical, shipborne

Calspan SRL Technology	USA	540	60-03000
Nanoquest Defence Products Ltd	UK	470	75-41890
Tilt Measurement Ltd	UK	488	75-59930
Warren-Knight Instrument Co.	USA	627	75-95150
W.F. Stanley & Co. Ltd	UK	486	75-56350

617 75 Muzzle boresight equipment

Armalon Ltd	UK	431	75-03729
Canons Delcour SA	Belgium	46	75-02200
CI Systems Ltd	Israel	218	75-03750
Graflex Inc	USA	579	75-40310
Nanoquest Defence Products Ltd	UK	470	75-41890
Pyser-SGI Ltd	UK	476	75-47300
Quantum Inspection & Testing Ltd	Canada	79	80-01800
Ring Sights Holdings Co. Ltd	UK	479	75-49700
SAGEM SA	**France**	**143**	**75-33950**
Stocker & Yale Inc.	USA	618	75-84250
Warren-Knight Instrument Co.	USA	627	75-95150
W.F. Stanley & Co. Ltd	UK	486	75-56350

617 80 Test equipment accessories

AAR Allen Group International	Netherlands	296	80-00050
Advanced Technical Corp.	USA	546	75-01000
Aero Systems Engineering Inc.	USA	546	75-01116
Air-Log Ltd	UK	429	75-01800
Airmatic Engineering (UK) Ltd	UK	429	75-01850
Angus Electronics Co.	USA	554	75-07630
Anritsu Wiltron	USA	554	75-07800
Associated Aircraft Mfg & Sales Inc.	USA	555	75-09500
California Instruments Corp.	USA	561	75-16280
Canadian Marconi Co.	Canada	69	75-02600
Cannatech Inc.	USA	562	75-16630
Canons Delcour SA	Belgium	46	75-02200
ERA Technology Ltd	UK	502	80-09600
GEC-Marconi Avionics Ltd	UK	454	75-25320
General Electrodynamics Corp.	USA	578	75-38500
Hensoldt AG	Germany	169	75-12450
Hewlett-Packard (Canada) Ltd	Canada	72	75-07520
Hewlett-Packard Company	USA	581	75-43200
Jennings Technology Corp.	USA	586	75-50350
Lloyd Instruments Ltd	UK	465	75-36150
Lucas Aftermarket Operations	UK	465	75-36655
Marconi Instruments Ltd	UK	466	75-37270
Merex Inc.	USA	595	75-60250
QES Ltd	UK	509	80-25000
Quantum Inspection & Testing Ltd	Canada	79	80-01800
Reliant Design Development Co Ltd	UK	478	75-49170

Rolls-Royce MatEval	UK	480	75-50260
SELIN Sistemi Sarl	Italy	238	75-21750
Sparton Technology Inc.	USA	616	75-82950
Techtest Ltd.	UK	484	75-54590
Tel-Instrument Electronics Corp.	USA	621	75-87860
The Graphics Co. of America	USA	579	75-40360
THERMOCOAX	France	147	75-44930
Trompeter Electronics Inc.	USA	623	75-91070
Vipac Engineers & Scientists Ltd	Australia	26	80-22580
Watkins-Johnson Co.	USA	627	75-95200

617 90 Test equipment components

Advanced Technical Corp.	USA	546	75-01000
Aero Systems Engineering Inc.	USA	546	75-01116
Air-Log Ltd	UK	429	75-01800
Airmatic Engineering (UK) Ltd	UK	429	75-01850
Anritsu Wiltron	USA	554	75-07800
Associated Aircraft Mfg & Sales Inc.	USA	555	75-09500
Bertan High Voltage	USA	558	75-13720
Cannatech Inc.	USA	562	75-16630
Canons Delcour SA	Belgium	46	75-02200
GEC-Marconi Avionics Ltd	UK	454	75-25320
General Electrodynamics Corp.	USA	578	75-38500
Hensoldt AG	Germany	169	75-12450
Hewlett-Packard (Canada) Ltd	Canada	72	75-07520
Hewlett-Packard Company	USA	581	75-43200
Hoffman Engineering Corp.	USA	582	75-43980
Interspiro Ltd	UK	461	75-31945
Litton Industries Inc.	USA	590	75-55400
Marconi Instruments Ltd	UK	466	75-37270
National Aerospace Laboratory	Netherlands	290	60-01400
Oerlikon-Contraves AG.	Switzerland	387	75-09250
Quantum Inspection & Testing Ltd	Canada	79	80-01800
Reliant Design Development Co Ltd.	UK	478	75-49170
Rolls-Royce MatEval	UK	480	75-50260
SELIN Sistemi Sarl	Italy	238	75-21750
Siemens plc	UK	483	75-53750
Sparton Technology Inc.	USA	616	75-82950
Techtest Ltd.	UK	484	75-54590
Tel-Instrument Electronics Corp.	USA	621	75-87860
The Graphics Co. of America	USA	579	75-40360
THERMOCOAX	France	147	75-44930
Turbo Lufttechnik GmbH	Germany	176	75-31900
Vipac Engineers & Scientists Ltd	Australia	26	80-22580
Watkins-Johnson Co.	USA	627	75-95200
Wavetek Corporation	USA	627	75-95270
Weinschel Corporation	USA	627	75-95544

619 Other fixed base facilities & equipment

AB Carl Munters	Sweden	377	75-08230
AEA Technology plc	UK	496	80-00480
Airmatic Engineering (UK) Ltd	UK	429	75-01850
Applied Science Laboratories	USA	554	75-08200
Automatic Systems Laboratories Ltd	UK	432	75-04470
Bauer Aerospace	USA	557	75-12250
California Instruments Corp.	USA	561	75-16280
Calspan SRL Technology	USA	540	60-03000
Columbus Jack Corp.	USA	564	75-19850
David Brown Defence Equipment Ltd	UK	437	75-10150
GEL Engineering Ltd	UK	454	75-25500
Lucas Aerospace Power Transmission	USA	592	75-56265
RFI Industries Pty Ltd	Australia	20	75-18500
Ruska Instrument Corp.	USA	611	75-77040
SPM Instrument UK Ltd.	UK	486	75-56050
Tel-Instrument Electronics Corp.	USA	621	75-87860
Teledyne Fluid Systems	USA	621	75-87570

Field Accommodation & Equipment

621 Field accommodation

Andrew Mitchell & Co. Ltd	UK	468	75-40100
Bachmann SA	France	132	75-04900
Electronic Space Systems Corp.	USA	572	75-30100
Gontrailer SL	Spain	367	75-05660
Krauss-Maffei Wehrtechnik GmbH	Germany	171	75-16345
L'hotellier Montrichard	France	138	75-24900
MFD International Ltd.	UK	506	80-16450
Morgan Marine Ltd	UK	469	75-40865
PressCut Industries Inc.	USA	605	75-71436
Rollalong	UK	479	75-50180
Sleeman Engineering	UK	484	75-54400
Sprung Instant Structures Ltd	Canada	77	75-13200
Structure-flex Ltd	UK	486	75-57525
Touringsport (Rosker Ltd)	UK	489	75-60460
Weatherhaven	Canada	77	75-15220

621 10 Buildings, lightweight demountable (see code 611.21)

Bachmann SA	France	132	75-04900
GEL Engineering Ltd	UK	454	75-25500
Gontrailer SL	Spain	367	75-05660
Metis Construction & Trade Co. Inc.	Turkey	403	80-01330
MFD International Ltd.	UK	506	80-16450
Morgan Marine Ltd	UK	469	75-40865
Rollalong	UK	479	75-50180
Rubb Buildings Ltd	UK	481	75-50750
Sleeman Engineering	UK	484	75-54400
SpanSet Inter AG	Switzerland	387	75-13550
Sprung Instant Structures Ltd	Canada	77	75-13200
Technical Products Group Inc.	USA	620	75-86764
Weatherhaven	Canada	77	75-15220

621 11 Buildings, inflatable

Andrew Mitchell & Co. Ltd	UK	468	75-40100
Bachmann SA	France	132	75-04900
Electronic Space Systems Corp.	USA	572	75-30100
MFC Survival Ltd	UK	467	75-38900

621 20 Caravans (trailer caravans)

CSC (Specialised Vehicles) Ltd	UK	444	75-16220
Genco Mobile Systems BV.	Netherlands	293	75-02340
Gontrailer SL	Spain	367	75-05660
Hunting Engineering Ltd	UK	459	75-30700
Rollalong	UK	479	75-50180

621 60 Shelters

Aeraustral Trading	Australia	11	75-00330
Bachmann SA	France	132	75-04900
Bharat Electronics Ltd	India	203	55-00200
Carrozzeria Boneschi Sarl	Italy	233	75-05000
CompAir Holman Iberica SA	Spain	366	75-01900
CSC (Specialised Vehicles) Ltd	UK	444	75-16220
Engineered Air Systems Inc.	USA	573	75-31560
Garofoli Costruzioni Meccaniche SpA	Italy	235	75-09550
GEL Engineering Ltd	UK	454	75-25500
Giat Industries	France	126	55-00100
J T Inglis & Sons Ltd	UK	461	75-31500
Krauss-Maffei Wehrtechnik GmbH	Germany	171	75-16345
L'hotellier Montrichard	France	138	75-24900
Metis Construction & Trade Co. Inc.	Turkey	403	80-01330
MFD International Ltd.	UK	506	80-16450
MSI-Defence Systems Ltd	UK	469	75-41280
Rinaldo Piaggio SpA	Italy	238	75-18790
Rubb Buildings Ltd	UK	481	75-50750
SEI Industries Ltd	Canada	76	75-12750
SITEP SpA	Italy	239	75-22595
Société des Tuyaux Bonna	France	146	75-41520
Sprung Instant Structures Ltd	Canada	77	75-13200
Technical Products Group Inc.	USA	620	75-86764
Weatherhaven	Canada	77	75-15220

621 60 30 Shelters, EMI / EMP protective

Andrew Mitchell & Co. Ltd	UK	468	75-40100
Bharat Electronics Ltd	India	203	55-00200
Carrozzeria Boneschi Sarl	Italy	233	75-05000
Daloc Sheltec AB	Sweden	376	75-04000
Giat Industries	France	126	55-00100
Hunting Engineering Ltd	UK	459	75-30700
L'hotellier Montrichard	France	138	75-24900
Metis Construction & Trade Co. Inc.	Turkey	403	80-01330
MFD International Ltd.	UK	506	80-16450
Morgan Marine Ltd	UK	469	75-40865

Presentey Engineering	Canada	*75*	75-11500
PressCut Industries Inc.	USA	*605*	75-71436
RFI Industries Pty Ltd	Australia	*20*	75-18500
Technical Products Group Inc.	USA	*620*	75-86764
Vermont Composites, Inc.	USA	*625*	75-93450

621 60 80 Shelter accessories

AGM Container Controls Inc.	USA	*547*	75-02870
Air-Log Ltd	UK	*429*	75-01800
Åkers Krutbruk Protection AB	Sweden	*375*	75-01100
Bruce Industries Inc.	USA	*561*	75-15360
Daloc Sheltec AB	Sweden	*376*	75-04000
H. Fine & Son Ltd	UK	*451*	75-23210
Hunting Engineering Ltd	UK	*459*	75-30700
L'hotellier Montrichard	France	*138*	75-24900
RFI Industries Pty Ltd	Australia	*20*	75-18500
SITEP SpA	Italy	*239*	75-22595
SpanSet Inter AG	Switzerland	*387*	75-13550
Technical Products Group Inc.	USA	*620*	75-86764
Teledyne Brown Engineering	USA	*620*	75-87508
Vermont Composites, Inc.	USA	*625*	75-93450

621 60 81 Shelter air conditioning systems

Air International Transit	Australia	*11*	75-00450
Air-Log Ltd	UK	*429*	75-01800
Aircraft Maintenance Support	UK	*428*	75-01580
A/S Dantherm	**Denmark**	***102***	**75-04100**
Daloc Sheltec AB	Sweden	*376*	75-04000
Engineered Air Systems Inc.	USA	*573*	75-31560
Giat Industries	France	*126*	55-00100
Howden Fluid Systems	USA	*582*	75-44400
Kinetics Ltd	Israel	*220*	75-11320
Technical Products Group Inc.	USA	*620*	75-86764
Teledyne Brown Engineering	USA	*620*	75-87508

621 60 83 Doors, shelter, blast-proof

Air-Log Ltd	UK	*429*	75-01800
Booth Industries Limited	UK	*435*	75-07670
Daloc Sheltec AB	Sweden	*376*	75-04000
Envirodoor Markus Ltd	UK	*449*	75-21090
Indal Technologies Inc	Canada	*72*	75-07740
L'hotellier Montrichard	France	*138*	75-24900
Metis Construction & Trade Co. Inc.	Turkey	*403*	80-01330
Overly Manufacturing Co.	USA	*602*	75-67750
Sleeman Engineering	UK	*484*	75-54400

621 61 Shelters, field, excavated

Giat Industries	France	*126*	55-00100

621 62 Shelters, modular inflatable

Bachmann SA	France	*132*	75-04900
Carrozzeria Boneschi Sarl	Italy	*233*	75-05000
Electronic Space Systems Corp.	USA	*572*	75-30100
Engineered Air Systems Inc.	USA	*573*	75-31560
Eurovinil Industries SpA	Italy	*234*	75-08400
Giat Industries	France	*126*	55-00100
Irvin Aerospace Limited	UK	*462*	75-32050
L'hotellier Montrichard	France	*138*	75-24900
MFC Survival Ltd	UK	*467*	75-38900
White Cross Rubber Products Ltd.	UK	*494*	75-65857

621 63 Shelters, portable, rigid frame

Andrew Mitchell & Co. Ltd	UK	*468*	75-40100
Bachmann SA	France	*132*	75-04900
CSC (Specialised Vehicles) Ltd	UK	*444*	75-16220
Engineered Air Systems Inc.	USA	*573*	75-31560
EPS Logistics Technology Ltd	UK	*502*	80-09550
GEL Engineering Ltd	UK	*454*	75-25500
Giat Industries	France	*126*	55-00100
Hyundai	Korea (ROK)	*265*	75-01540
L'hotellier Montrichard	France	*138*	75-24900
Presentey Engineering	Canada	*75*	75-11500
Rubb Buildings Ltd	UK	*481*	75-50750
Sprung Instant Structures Ltd	Canada	*77*	75-13200
Technical Products Group Inc.	USA	*620*	75-86764
Weatherhaven	Canada	*77*	75-15220

621 65 Shelters, airportable

Andrew Mitchell & Co. Ltd	UK	*468*	75-40100
Bharat Electronics Ltd	India	*203*	55-00200
Carrozzeria Boneschi Sarl	Italy	*233*	75-05000
CSC (Specialised Vehicles) Ltd	UK	*444*	75-16220
Engineered Air Systems Inc.	USA	*573*	75-31560
Eurovinil Industries SpA	Italy	*234*	75-08400
Giat Industries	France	*126*	55-00100
Hunting Engineering Ltd	UK	*459*	75-30700
L'hotellier Montrichard	France	*138*	75-24900
Presentey Engineering	Canada	*75*	75-11500
Sprung Instant Structures Ltd	Canada	*77*	75-13200
Technical Products Group Inc.	USA	*620*	75-86764
Weatherhaven	Canada	*77*	75-15220

621 70 Tents / marquees, canvas

Andrew Mitchell & Co. Ltd	UK	*468*	75-40100
Bachmann SA	France	*132*	75-04900
Climax Engineering Pty Ltd	Australia	*14*	75-03480
H. Fine & Son Ltd	UK	*451*	75-23210
L.N. Compton Ltd	UK	*442*	75-14600
M. Kumar Udyog	India	*205*	75-02100
Manta Industries Ltd	Canada	*73*	75-09250
Novurania SpA	Italy	*237*	75-16650
Rabintex Industries Ltd	Israel	*221*	75-17850
Research Institute for	Ukraine	*406*	60-00990
ROFi Industrier AS	Norway	*310*	75-02300
Structure-flex Ltd	UK	*486*	75-57525
Techniques Michel Brochier SA	France	*147*	75-44100
Touringsport (Rosker Ltd)	UK	*489*	75-60460
Trojan Manufacturing & Marketing Inc	Philippines	*321*	75-01350
US Army Equipment Guns and Ammo	Belgium	*47*	75-06345
Weatherhaven	Canada	*77*	75-15220

621 70 50 Tents, inflatable

Eurovinil Industries SpA	Italy	*234*	75-08400
Medesign GmbH	Germany	*172*	75-20370
MFC Survival Ltd	UK	*467*	75-38900
Novurania SpA	Italy	*237*	75-16650

622 Field service units, equipment & supplies

American Cleaning Manufacturing Co.	USA	*553*	75-05940
Andrew Mitchell & Co. Ltd	UK	*468*	75-40100
Brimotor Ltd	UK	*436*	75-08600
CMI - Ballarat	Australia	*14*	75-03600
EG&G Inc.	USA	*571*	75-29555
Exstar International Corp.	USA	*575*	75-33000
GAAM Fire & Rescue	Australia	*16*	75-07000
Gilbert Gilkes & Gordon Ltd	UK	*454*	75-25430
Haux-Life-Support GmbH	Germany	*169*	75-11970
Indústrias de Defesa	Portugal	*327*	55-00950
Karcher (UK) Ltd.	UK	*463*	75-33220
M. Paz Logistics Ltd	Israel	*224*	80-01600
Mechanical Equipment Co. Inc.	USA	*595*	75-59815
Morgan Marine Ltd	UK	*469*	75-40865
Palbam Metal Works	Israel	*221*	75-17000
Pre-Mac (Kent) Ltd	UK	*476*	75-46660
Stella-Meta	UK	*486*	75-56780
Weatherhaven	Canada	*77*	75-15220
Winco Inc.	USA	*628*	75-97050

622 05 Field kitchens

Alfred Karcher GmbH & Co	Germany	*170*	75-15650
American Cleaning Manufacturing Co.	USA	*553*	75-05940
Carlin Manufacturing Inc.	USA	*562*	75-16753
Carrozzeria Boneschi Sarl	Italy	*233*	75-05000
CSC (Specialised Vehicles) Ltd	UK	*444*	75-16220
Electrothermal Engineering Ltd	UK	*448*	75-20400
Engineered Air Systems Inc.	USA	*573*	75-31560
Equipos Móviles de Campana Arpa SA	Spain	*366*	75-02860
GEL Engineering Ltd	UK	*454*	75-25500

Indústrias de Defesa	Portugal	327	55-00950
Karcher France	France	138	75-23500
Karcher (UK) Ltd.	UK	463	75-33220
Palbam Metal Works	Israel	221	75-17000
Pland	UK	475	75-45740
Progress-Werk Oberkirch AG	Germany	173	75-24320
Stott Benham	UK	486	75-57460
Weatherhaven	Canada	77	75-15220

622 05 70 Field kitchen equipment

American Cleaning Manufacturing Co.	USA	553	75-05940
BCB International Ltd	UK	433	75-06075
Carlin Manufacturing Inc.	USA	562	75-16753
CMI - Ballarat	Australia	14	75-03600
Electrothermal Engineering Ltd	UK	448	75-20400
Equipos Móviles de Campana Arpa SA	Spain	366	75-02860
Evotec Ltd	UK	449	75-21970
Michsaf	Israel	221	75-14200
Palbam Metal Works	Israel	221	75-17000
Pland	UK	475	75-45740
Progress-Werk Oberkirch AG	Germany	173	75-24320
Stott Benham	UK	486	75-57460
Teledyne Brown Engineering	USA	620	75-87508

622 05 70 30 Field cooking ware

Evotec Ltd	UK	449	75-21970
Palbam Metal Works	Israel	221	75-17000

622 06 Mess kits

BCB International Ltd	UK	433	75-06075
Plastico Ltd	UK	475	75-45950
Weatherhaven	Canada	77	75-15220

622 08 Field rations

Atomic Energy Corp	S. Africa	357	75-01000
BCB International Ltd	UK	433	75-06075
Evotec Ltd	UK	449	75-21970
Nutrimental SA/Comexport	Brazil	57	75-14900
Plastico Ltd	UK	475	75-45950

622 08 50 Meals, ready-to-eat

BCB International Ltd	UK	433	75-06075
Evotec Ltd	UK	449	75-21970
Military Logistics MFC	USA	596	75-61550
Nutrimental SA/Comexport	Brazil	57	75-14900

622 10 Field ablution units

Carlin Manufacturing Inc.	USA	562	75-16753
CSC (Specialised Vehicles) Ltd	UK	444	75-16220
GEL Engineering Ltd	UK	454	75-25500
Karl H. Høie & Co. A/S	Norway	309	75-00970
Weatherhaven	Canada	77	75-15220

622 13 Field laundries, transportable

Alfred Karcher GmbH & Co	Germany	170	75-15650
American Cleaning Manufacturing Co.	USA	553	75-05940
Carrozzeria Boneschi Sarl	Italy	233	75-05000
Engineered Air Systems Inc.	USA	573	75-31560

622 15 Field toilets

Carlin Manufacturing Inc.	USA	562	75-16753
CSC (Specialised Vehicles) Ltd	UK	444	75-16220
Engineered Air Systems Inc.	USA	573	75-31560
Equipos Móviles de Campana Arpa SA	Spain	366	75-02860
Eurovinil Industries SpA	Italy	234	75-08400
GEL Engineering Ltd	UK	454	75-25500
Weatherhaven	Canada	77	75-15220

622 18 Water carriage packs

H. Fine & Son Ltd	UK	451	75-23210
PD Technical Mouldings plc	UK	473	75-44200
Stella-Meta	UK	486	75-56780
Weltonhurst Ltd	UK	493	75-65135

622 20 Water purification plants

Aircraft Appliances & Equipment Ltd	Canada	68	75-00430
Åkers Krutbruk Protection AB	Sweden	375	75-01100
American Cleaning Manufacturing Co.	USA	553	75-05940
Bachmann SA	France	132	75-04900
Everpure Inc.	USA	575	75-32680
Exstar International Corp.	USA	575	75-33000
GEL Engineering Ltd	UK	454	75-25500
Gilbert Gilkes & Gordon Ltd	UK	454	75-25430
IDE Technologies Ltd	Israel	219	75-09170
PD Technical Mouldings plc	UK	473	75-44200
Pre-Mac (Kent) Ltd	UK	476	75-46660
SCAM SpA	Italy	238	75-21530
SIHI Australia Pty Ltd	Australia	21	75-19505
Stella-Meta	UK	486	75-56780
Thompson Kennicott Limited	UK	513	80-55000
Weir Engineering Pty Ltd	Australia	23	75-23300

622 20 40 Desalination plants

Aircraft Appliances & Equipment Ltd	Canada	68	75-00430
Beaird Industries Inc.	USA	557	75-12540
Exstar International Corp.	USA	575	75-33000
SIHI Australia Pty Ltd	Australia	21	75-19505
Stella-Meta	UK	486	75-56780
Thompson Kennicott Limited	UK	513	80-55000
Weir Engineering Pty Ltd	Australia	23	75-23300
Weir Westgarth Ltd	UK	514	80-72250

622 20 50 Water purification plants, mobile

Åkers Krutbruk Protection AB	Sweden	375	75-01100
Engineered Air Systems Inc.	USA	573	75-31560
Exstar International Corp.	USA	575	75-33000
Gilbert Gilkes & Gordon Ltd	UK	454	75-25430
Mechanical Equipment Co. Inc.	USA	595	75-59815
PD Technical Mouldings plc	UK	473	75-44200
Pre-Mac (Kent) Ltd	UK	476	75-46660
SIHI Australia Pty Ltd	Australia	21	75-19505
Stella-Meta	UK	486	75-56780
Weir Engineering Pty Ltd	Australia	23	75-23300

622 20 80 Water purification units, portable (see code 652.90)

Åkers Krutbruk Protection AB	Sweden	375	75-01100
BCB International Ltd	UK	433	75-06075
Exstar International Corp.	USA	575	75-33000
Gilbert Gilkes & Gordon Ltd	UK	454	75-25430
PD Technical Mouldings plc	UK	473	75-44200
Pre-Mac (Kent) Ltd	UK	476	75-46660
SIHI Australia Pty Ltd	Australia	21	75-19505
Stella-Meta	UK	486	75-56780
Survival Products Inc.	USA	618	75-85200
Weir Engineering Pty Ltd	Australia	23	75-23300

622 25 Lighting systems, mobile / portable

Blakley Electrics Ltd	UK	434	75-07015
Brimotor Ltd	UK	436	75-08600
Bruce Industries Inc.	USA	561	75-15360
Carrozzeria Boneschi Sarl	Italy	233	75-05000
EG&G Inc.	USA	571	75-29555
Engineered Air Systems Inc.	USA	573	75-31560
Furneaux Riddall & Co. Ltd	UK	452	75-24890
General Electronics Communications	Greece	189	75-01300
Havis-Shields Equipment Corp.	USA	581	75-42700
Henry Cooch & Son Ltd	UK	442	75-14900
M. Paz Logistics Ltd	Israel	224	80-01600
Mitralux International AG	Switzerland	387	75-08500
Saunders-Roe Ltd	UK	482	75-51850
Smith & Prince Ltd	UK	485	75-54630

Strumech Versatower Ltd	UK	486	75-57530
Westair Ltd	UK	493	75-65410
Winco Inc.	USA	628	75-97050

622 25 60 Searchlights, mobile ground-based

BDL Systems Ltd	UK	433	75-06080
Contraves Inc	USA	566	75-21850
M. Paz Logistics Ltd	Israel	224	80-01600
OCEM SpA	Italy	237	75-16750
Saunders-Roe Ltd	UK	482	75-51850

622 30 Pumps, mobile trailer-mounted

American Cleaning Manufacturing Co.	USA	553	75-05940
Engineered Air Systems Inc.	USA	573	75-31560
GAAM Fire & Rescue	Australia	16	75-07000
Gilbert Gilkes & Gordon Ltd	UK	454	75-25430
Hamworthy Marine	UK	457	75-27850
Ingersoll-Dresser Pumps (UK) Ltd	UK	461	75-31450
PD Technical Mouldings plc	UK	473	75-44200
Plenty Ltd	UK	475	75-46050
SIHI Australia Pty Ltd	Australia	21	75-19505
Wesley Group Ltd	UK	493	75-65300

622 35 Gas generation / compression units, mobile

Heiser Josef	Austria	33	75-01230
SAS Limited	UK	482	75-51830
SNPE	France	145	75-36315
Teledyne Brown Engineering	USA	620	75-87508
Zwick Energy Research Organization	USA	629	75-98400

622 35 50 Oxygen generating systems, mobile / transportable

Litton Industries Inc.	USA	590	75-55400
Medesign GmbH	Germany	172	75-20370
SAS Limited	UK	482	75-51830
Teledyne Brown Engineering	USA	620	75-87508

622 35 55 Oxygen / nitrogen generating systems, mobile / transportable

Litton Industries Inc.	USA	590	75-55400
SAS Limited	UK	482	75-51830

622 50 Compressors, air

CompAir Reavell Ltd	UK	442	75-14530
Divex Ltd.	UK	445	75-18220
Hamworthy Marine	UK	457	75-27850
Haux-Life-Support GmbH	Germany	169	75-11970
H.C. Slingsby plc	UK	484	75-54430
HPC Engineering plc	UK	459	75-30350
Hytech BV	Netherlands	293	75-03530
SAS Limited	UK	482	75-51830
SIHI Australia Pty Ltd	Australia	21	75-19505
Weir Engineering Pty Ltd	Australia	23	75-23300

622 50 20 Charging sets, HP air

Divex Ltd.	UK	445	75-18220
Haux-Life-Support GmbH	Germany	169	75-11970
Hytech BV	Netherlands	293	75-03530
SAS Limited	UK	482	75-51830

622 50 20 25 Generating systems, pure air

Giat Industries	France	126	55-00100
Haux-Life-Support GmbH	Germany	169	75-11970
SAS Limited	UK	482	75-51830
Spar Aviation Services	Canada	76	75-13020

622 50 20 25 50 Generating systems, pure air, man-portable

Blakley Electrics Ltd	UK	434	75-07015
Giat Industries	France	126	55-00100
SAS Limited	UK	482	75-51830

622 60 Power washing equipment (see code 750)

American Cleaning Manufacturing Co.	USA	553	75-05940
H.C. Slingsby plc	UK	484	75-54430
Karcher (UK) Ltd.	UK	463	75-33220
Wesley Group Ltd	UK	493	75-65300

622 70 Field service equipment

Auto Crane Co.	USA	556	75-10360
Tecalemit (Australasia) Pty Ltd	Australia	21	75-20250

622 72 Beds, camp

Andrew Mitchell & Co. Ltd	UK	468	75-40100
BCB International Ltd	UK	433	75-06075
Progress-Werk Oberkirch AG	Germany	173	75-24320

622 90 Field service units, other types

Bombardier Inc.	Canada	69	75-01400
Systems & Electronics Inc.	USA	619	75-85650
Tecalemit (Australasia) Pty Ltd	Australia	21	75-20250
Weatherhaven	Canada	77	75-15220

623 Field medical units, equipment & supplies

Andrew Mitchell & Co. Ltd	UK	468	75-40100
Bachmann SA	France	132	75-04900
Bombardier Inc.	Canada	69	75-01400
Empl Fahrzeugwerk GesmbH	Austria	32	75-00920
Equipos Móviles de Campana Arpa SA	Spain	366	75-02860
Ferno (UK) Ltd	UK	450	75-22650
GEL Engineering Ltd	UK	454	75-25500
Mediworld Ltd.	UK	462	75-32300
Military Logistics MFC	USA	596	75-61550
Seton Healthcare Group plc	UK	483	75-53000
Weatherhaven	Canada	77	75-15220

623 10 Field medical centres / units

Carrozzeria Boneschi Sarl	Italy	233	75-05000
Equipos Móviles de Campana Arpa SA	Spain	366	75-02860
Eurovinil Industries SpA	Italy	234	75-08400
GEL Engineering Ltd	UK	454	75-25500
Giat Industries	France	126	55-00100
Lenoir Aviation AG	Switzerland	386	75-07335
Medesign GmbH	Germany	172	75-20370
Mediworld Ltd.	UK	462	75-32300
Technical Products Group Inc.	USA	620	75-86764
Weatherhaven	Canada	77	75-15220

623 10 10 Field medical centres, relocatable

Carrozzeria Boneschi Sarl	Italy	233	75-05000
CSC (Specialised Vehicles) Ltd	UK	444	75-16220
GEL Engineering Ltd	UK	454	75-25500
Giat Industries	France	126	55-00100
Lenoir Aviation AG	Switzerland	386	75-07335
Medesign GmbH	Germany	172	75-20370
Mediworld Ltd.	UK	462	75-32300
Seward Medical Limited	UK	483	75-53050
Sprung Instant Structures Ltd	Canada	77	75-13200
Technical Products Group Inc.	USA	620	75-86764
Weatherhaven	Canada	77	75-15220

623 11 Field hospitals, mobile

Andrew Mitchell & Co. Ltd	UK	468	75-40100
Bachmann SA	France	132	75-04900

Carrozzeria Boneschi Sarl	Italy	233	75-05000
CSC (Specialised Vehicles) Ltd	UK	444	75-16220
Eurovinil Industries SpA	Italy	234	75-08400
GEL Engineering Ltd	UK	454	75-25500
Giat Industries	France	126	55-00100
Glover Webb Ltd	UK	455	75-25700
Lenoir Aviation AG	Switzerland	386	75-07335
Medesign GmbH	Germany	172	75-20370
Mediworld Ltd.	UK	462	75-32300
Singapore Technologies Automotive	Singapore	348	75-04280
Sprung Instant Structures Ltd	Canada	77	75-13200
Technical Products Group Inc.	USA	620	75-86764

623 13 Medical / dental treatment units, rapid intervention

Bombardier Inc.	Canada	69	75-01400
Carrozzeria Boneschi Sarl	Italy	233	75-05000
CSC (Specialised Vehicles) Ltd	UK	444	75-16220
GEL Engineering Ltd	UK	454	75-25500
Jarvis Manufacturing Co.	UK	462	75-32330
Medesign GmbH	Germany	172	75-20370
Mediworld Ltd.	UK	462	75-32300
Technical Products Group Inc.	USA	620	75-86764

623 20 Field medical equipment

Ambu International A/S	Denmark	102	75-01170
Angénieux	France	132	75-02700
BCB International Ltd	UK	433	75-06075
Eschmann Bros. & Walsh Ltd	UK	449	75-21500
Ferno (UK) Ltd.	UK	450	75-22650
GEL Engineering Ltd	UK	454	75-25500
Jarvis Manufacturing Co.	UK	462	75-32330
Medesign GmbH	Germany	172	75-20370
Mediworld Ltd.	UK	462	75-32300
M.G. Electric (Colchester) Ltd	UK	468	75-38912
Seton Healthcare Group plc	UK	483	75-53000
STI	USA	618	75-85230
STI International Ltd	UK	486	75-57100
UO Equipment Co.	USA	625	75-92870
Urals Optical & Mechanical Plant	Russia	339	75-53000

623 20 30 First aid kits

Aimsco Inc.	USA	547	75-03050
Ambu International A/S	Denmark	102	75-01170
BCB International Ltd	UK	433	75-06075
Evotec Ltd	UK	449	75-21970
Ferno (UK) Ltd.	UK	450	75-22650
Intamar Logistics Inc.	USA	585	75-47050
Jarvis Manufacturing Co.	UK	462	75-32330
Medesign GmbH	Germany	172	75-20370
Mediworld Ltd.	UK	462	75-32300

623 70 Field medical supplies

BCB International Ltd	UK	433	75-06075
Ferno (UK) Ltd.	UK	450	75-22650
Medesign GmbH	Germany	172	75-20370
Mediworld Ltd.	UK	462	75-32300
Smith & Nephew plc	UK	484	75-54620
STI International Ltd	UK	486	75-57100

623 74 Field dressings

BCB International Ltd	UK	433	75-06075
Charcoal Cloth (International) Ltd	UK	439	75-12550
Ferno (UK) Ltd.	UK	450	75-22650
Lohmann GmbH & Co KG	Germany	171	75-18910
Medesign GmbH	Germany	172	75-20370
Mediworld Ltd.	UK	462	75-32300
Smith & Nephew plc	UK	484	75-54620

623 76 Paramedic packs

Ambu International A/S	Denmark	102	75-01170
Ferno (UK) Ltd.	UK	450	75-22650
Medesign GmbH	Germany	172	75-20370
Mediworld Ltd.	UK	462	75-32300

623 80 Field casualty transportation equipment

Bachmann SA	France	132	75-04900
Ferno (UK) Ltd.	UK	450	75-22650
Germa AB	Sweden	377	75-05700
Giat Industries	France	126	55-00100
Medesign GmbH	Germany	172	75-20370
Mediworld Ltd.	UK	462	75-32300

623 80 20 Casualty bags, field casualty transportation

BCB International Ltd	UK	433	75-06075
Ferno (UK) Ltd.	UK	450	75-22650
Germa AB	Sweden	377	75-05700
H. Fine & Son Ltd	UK	451	75-23210
Lantor (UK) Ltd	UK	464	75-34665
Manta Industries Ltd	Canada	73	75-09250
Medesign GmbH	Germany	172	75-20370
Mediworld Ltd.	UK	462	75-32300
Synrein Plastics Ltd.	UK	487	75-58325
Techniques Michel Brochier SA	France	147	75-44100

623 80 70 Stretchers, field casualty transportation

Climax Engineering Pty Ltd.	Australia	14	75-03480
Ferno (UK) Ltd.	UK	450	75-22650
Germa AB	Sweden	377	75-05700
H. Fine & Son Ltd	UK	451	75-23210
Manta Industries Ltd	Canada	73	75-09250
Medesign GmbH	Germany	172	75-20370
Mediworld Ltd.	UK	462	75-32300
ROFi Industrier AS	Norway	310	75-02300

623 90 Water systems, field hospital

Mediworld Ltd.	UK	462	75-32300

625 Field Operations Support Systems & Equipment

ADE-HML Limited	UK	427	75-00710
Advanced Technical Corp.	USA	546	75-01000
Aero Tec Laboratories Inc.	USA	546	75-01117
Aljac Engineering Ltd	UK	429	75-02650
Bombardier Inc.	Canada	69	75-01400
Brimotor Ltd	UK	436	75-08600
Clark Masts Teksam Ltd	UK	440	75-13550
Columbus Jack Corp.	USA	564	75-19850
CORIS	France	134	75-12450
CSC (Specialised Vehicles) Ltd	UK	444	75-16220
GEC-Marconi Avionics Ltd	UK	454	75-25320
GEL Engineering Ltd	UK	454	75-25500
Howden Aircontrol Ltd	UK	459	75-30240
Israel Aircraft Industries Ltd	Israel	219	75-10200
Japan Aircraft Manufacturing Co. Ltd.	Japan	252	75-03200
Leonard Farnell & Co. Ltd	UK	450	75-22400
M. Paz Logistics Ltd	Israel	224	80-01600
MAN Gutehoffnungshütte AG	Germany	172	75-19620
Marchetti Autogru SpA	Italy	236	75-14400
Metalite Aviation Lighting	UK	467	75-38750
Modra Electric Power Pty Ltd	Australia	19	75-13500
Pronal SA	France	141	75-31100
Reynolds Boughton Ltd	UK	479	75-49340
SAIT Systems SA	Belgium	47	75-05600
Schramm Inc.	USA	613	75-78460
Warren-Knight Instrument Co.	USA	627	75-95150

625 20 Field operations support equipment, air force related

Howden Aircontrol Ltd	UK	459	75-30240
M. Paz Logistics Ltd	Israel	224	80-01600
Zwick Energy Research Organization	USA	629	75-98400

Product-Supplier Link

625 20 05 Operations support systems & equipment, forward airfield, mobile / transportable

AB Carl Munters	Sweden	377	75-08230
Israel Aircraft Industries Ltd	Israel	219	75-10200
Litton FiberCom	USA	590	75-55370
Technical Products Group Inc.	USA	620	75-86764
Wesley Group Ltd	UK	493	75-65300
Zwick Energy Research Organization	USA	629	75-98400

625 20 10 Airfield runway-related systems & equipment, mobile / transportable

AAI Corp.	USA	545	75-00020
ADE-HML Limited	UK	427	75-00710
Equipos Industriales de Manutención	Spain	366	75-02850
Israel Aircraft Industries Ltd	Israel	219	75-10200
Litton FiberCom	USA	590	75-55370
M. Paz Logistics Ltd	Israel	224	80-01600
Newton Derby Ltd	UK	470	75-42520
Racal Avionics Ltd	UK	477	75-47660
Thorn DNT Airfield Lighting Pty Ltd	Australia	22	75-20540

625 20 10 05 Aircraft arresting gear, mobile tactical

Aérazur	France	131	75-00500
SpanSet Inter AG	Switzerland	387	75-13550

625 20 10 40 Helicopter pads, forward deployment

Thorn DNT Airfield Lighting Pty Ltd	Australia	22	75-20540

625 20 10 60 Airfield rapid runway repair equipment

Aveling Barford (Machines) plc	UK	432	75-04550
Glover Webb Ltd	UK	455	75-25700
Grove Europe Ltd	**UK**	**456**	**75-27030**

625 20 10 70 Runway visual range equipment, airfield

Aeronautical & General Instr. Ltd	UK	428	75-01050
M. Paz Logistics Ltd	Israel	224	80-01600
Presentey Engineering	Canada	75	75-11500
Thorn DNT Airfield Lighting Pty Ltd	Australia	22	75-20540
Vaisala Oy	Finland	119	75-02800

625 20 20 ATC systems & equipment, forward airfield, mobile / transportable

Alcatel Air Navigation Systems GmbH	Germany	164	75-00535
CORIS	France	134	75-12450
CSC (Specialised Vehicles) Ltd	UK	444	75-16220
Denro Inc.	USA	568	75-25651
Fernau Avionics Ltd	UK	450	75-22600
IDS Ingegneria dei Sistemi SpA	Italy	240	80-00920
Israel Aircraft Industries Ltd	Israel	219	75-10200
NEC Corporation	Japan	253	75-06350
Park Air Electronics Ltd	UK	472	75-43750
SAIT Systems SA	Belgium	47	75-05600
Telegenix Inc.	USA	621	75-87610

625 20 20 75 Air traffic control towers, mobile / transportable

CSC (Specialised Vehicles) Ltd	UK	444	75-16220
General Electronics Communications	Greece	189	75-01300
Glover Webb Ltd	UK	455	75-25700
Israel Aircraft Industries Ltd	Israel	219	75-10200
Park Air Electronics Ltd	UK	472	75-43750
Redifon MEL Limited	UK	478	75-48530
SAIT Systems SA	Belgium	47	75-05600

625 20 30 Meteorological systems & equipment, forward airfield, mobile / transportable

AAI Corp.	USA	545	75-00020
Adolf Thies GmbH & Co KG	Germany	175	75-30600
Advanced Technical Corp.	USA	546	75-01000
Aeronautical & General Instr. Ltd	UK	428	75-01050
Alden Electronics Inc.	USA	548	75-04350
A1Marketing	UK	426	75-00025
CORIS	France	134	75-12450
CSC (Specialised Vehicles) Ltd	UK	444	75-16220
Environmental Technologies Group Inc.	USA	574	75-31860
Glover Webb Ltd	UK	455	75-25700
Hobeco Sociedade Consignatária Ltda	Brazil	56	75-08600
Jenoptik Impulsphysik GmbH	Germany	170	75-14800
Meggitt Avionics (UK)	UK	467	75-38340
SITEP SpA.	Italy	239	75-22595
Teledyne Brown Engineering	USA	620	75-87508
Thorn DNT Airfield Lighting Pty Ltd	Australia	22	75-20540
Vaisala Artais Division	USA	625	75-93087
Vaisala Oy	Finland	119	75-02800
Warren-Knight Instrument Co.	USA	627	75-95150

625 20 30 90 Meteorological equipment, forward airfield, mobile / transportable

AAI Corp.	USA	545	75-00020
Adolf Thies GmbH & Co KG	Germany	175	75-30600
Advanced Technical Corp.	USA	546	75-01000
Alden Electronics Inc.	USA	548	75-04350
Glover Webb Ltd	UK	455	75-25700
Jenoptik Impulsphysik GmbH	Germany	170	75-14800
Meggitt Avionics (UK)	UK	467	75-38340
Thorn DNT Airfield Lighting Pty Ltd	Australia	22	75-20540
Vaisala Oy	Finland	119	75-02800
Warren-Knight Instrument Co.	USA	627	75-95150

625 20 40 Fuel storage & delivery equipment, forward airfield, mobile / transportable

ADE-HML Limited	UK	427	75-00710
Aeroquip Corp.	USA	547	75-02350
Aljac Engineering Ltd	UK	429	75-02650
Emco Wheaton Australia Pty Ltd	Australia	16	75-05760
Engineered Air Systems Inc.	USA	573	75-31560
Israel Aircraft Industries Ltd	Israel	219	75-10200
Novatraçáo Artefatos	Brazil	57	75-14780
Pronal SA	France	141	75-31100
SEI Industries Ltd	Canada	76	75-12750
SIFT SAS di Cutore & Pitotti	Italy	239	75-22150

625 20 40 60 Fuel, storage equipment, forward airfield, mobile / transportable

ADE-HML Limited	UK	427	75-00710
ContiTech Holding GmbH	Germany	166	75-05150
Engineered Air Systems Inc.	USA	573	75-31560
Israel Aircraft Industries Ltd	Israel	219	75-10200
SEI Industries Ltd	Canada	76	75-12750
The Armstrong Monitoring Corp.	Canada	68	75-00740

625 20 40 65 Fuel storage tanks, flexible

Aero Tec Laboratories Inc.	USA	546	75-01117
Continental AG Werk Uniroyal	Germany	176	75-32090
ContiTech Holding GmbH	Germany	166	75-05150
Eurovinil Industries SpA	Italy	234	75-08400
Novatraçáo Artefatos	Brazil	57	75-14780
Pronal SA	France	141	75-31100
SEI Industries Ltd	Canada	76	75-12750

625 20 45 Airfield approach & landing aids, forward airfield, mobile / transportable

Canadian Marconi Co.	Canada	69	75-02600
Fernau Avionics Ltd	UK	450	75-22600
M. Paz Logistics Ltd	Israel	224	80-01600
Metalite Aviation Lighting	UK	467	75-38750
NEC Corporation	Japan	253	75-06350
Racal Avionics Ltd	UK	477	75-47660
SEI Industries Ltd	Canada	76	75-12750
Thorn DNT Airfield Lighting Pty Ltd	Australia	22	75-20540
Wilcox Electric Inc.	USA	627	75-96730

625 20 45 45 Microwave landing systems (MLS), forward airfield, mobile / transportable

Canadian Marconi Co.	Canada	69	75-02600
GEC-Marconi Avionics Ltd	UK	454	75-25320

625 20 50 Airfield lighting equipment, forward airfield, mobile / transportable

Engineered Air Systems Inc.	USA	573	75-31560
Equipos Industriales de Manutención	Spain	366	75-02850
M. Paz Logistics Ltd	Israel	224	80-01600
Metalite Aviation Lighting	UK	467	75-38750
Thorn DNT Airfield Lighting Pty Ltd	Australia	22	75-20540

625 20 50 10 Airfield runway lighting equipment, forward airfield, mobile / transportable

M. Paz Logistics Ltd	Israel	224	80-01600
Metalite Aviation Lighting	UK	467	75-38750
Strumech Versatower Ltd	UK	486	75-57530
Thorn DNT Airfield Lighting Pty Ltd	Australia	22	75-20540

625 20 70 Maintenance facilities & equipment, aircraft, field mobile / transportable

Access International Ltd	UK	427	75-00300
ADE-HML Limited	UK	427	75-00710
Aeroquip Corp.	USA	547	75-02350
Dayton T. Brown Inc.	USA	632	80-04920
Engineered Air Systems Inc.	USA	573	75-31560
Fastening Systems International Inc.	USA	575	75-33870
GEC-Marconi Avionics Ltd	UK	454	75-25320
Lista Ltd	Switzerland	386	75-07350
M.A. Bigland (Preparations) Ltd	UK	434	75-06700
Spec Tool Co.	USA	616	75-83050
Weatherhaven	Canada	77	75-15220

625 20 75 Aircraft ground handling & loading equipment, forward airfield, mobile

Air-Log Ltd	UK	429	75-01800
Belotti SpA	Italy	232	75-02640
Equipos Industriales de Manutención	Spain	366	75-02850
Hamech Ltd	UK	456	75-27730
Indal Technologies Inc	Canada	72	75-07740
Israel Aircraft Industries Ltd	Israel	219	75-10200
Zwick Energy Research Organization	USA	629	75-98400

625 20 75 32 Cranes, mobile, off-road, aircraft salvage

Belotti SpA	Italy	232	75-02640
Grove Europe Ltd	UK	456	75-27030

625 20 75 50 Loaders, aircraft cargo, forward airfield, mobile / transportable

Equipos Industriales de Manutención	Spain	366	75-02850
Systems & Electronics Inc.	USA	619	75-85650

625 20 75 50 10 Loaders, aircraft cargo, self-propelled, forward airfield

Aveling Barford (Machines) plc	UK	432	75-04550
Equipos Industriales de Manutención	Spain	366	75-02850
Perry Engineering	Australia	20	75-16460
Systems & Electronics Inc.	USA	619	75-85650

625 20 75 70 Servicing trolleys, aircraft ground handling, forward airfield

ADE-HML Limited	UK	427	75-00710
Air-Log Ltd	UK	429	75-01800
Aircraft Maintenance Support	UK	428	75-01580
Arvico Corporation	USA	555	75-09240
Commercial Hydraulics Keelavite Ltd	UK	441	75-14410
Equipos Industriales de Manutención	Spain	366	75-02850
Howden Aircontrol Ltd	UK	459	75-30240
Israel Aircraft Industries Ltd	Israel	219	75-10200
Zwick Energy Research Organization	USA	629	75-98400

625 20 75 75 Starter trucks, aircraft engine, forward airfield, mobile / transportable

ADE-HML Limited	UK	427	75-00710
Air-Log Ltd	UK	429	75-01800
Arvico Corporation	USA	555	75-09240
Atlas Copco Airpower NV	Belgium	45	75-00900
Lear Astronics Corp.	USA	589	75-54390

625 20 75 80 Tractors, aircraft towing (see code 141.70.30.05)

Eagle Industrial Truck Mfg. Inc.	USA	570	75-28430
Schopf Maschinenbau GmbH	Germany	174	75-26600
Still GmbH	Germany	175	75-29000
Thorn DNT Airfield Lighting Pty Ltd	Australia	22	75-20540

625 40 Field operations support equipment, army

GEC-Marconi Avionics Ltd	UK	454	75-25320
Krauss-Maffei Wehrtechnik GmbH	Germany	171	75-16345
Structure-flex Ltd	UK	486	75-57525
Westford Plastics & Engineering Ltd	UK	493	75-65450

625 40 30 10 Meteorological systems & equipment, artillery, mobile / transportable

AAI Corp.	USA	545	75-00020
A1Marketing	UK	426	75-00025
Gill Instruments Ltd	UK	454	75-25450
Hobeco Sociedade Consignatária Ltda	Brazil	56	75-08600
Meggitt Avionics (UK)	UK	467	75-38340
R.M. Young Co.	USA	628	75-98020
Strumech Versatower Ltd	UK	486	75-57530
Teledyne Brown Engineering	USA	620	75-87508
Vaisala Oy	Finland	119	75-02800
Warren-Knight Instrument Co.	USA	627	75-95150

625 40 40 60 Fuel, oil & lubricants storage equipment, mobile / transportable

ADE-HML Limited	UK	427	75-00710
British Filters Ltd	UK	436	75-09030
Genco Mobile Systems BV.	Netherlands	293	75-02340
Victaulic Co. of America	USA	626	75-93840
Westford Plastics & Engineering Ltd	UK	493	75-65450

625 40 40 65 Tanks, fuel, oil & lubricants storage, flexible

Balmoral Group Ltd	UK	433	75-05420
Eurovinil Industries SpA	Italy	234	75-08400
Pronal SA	France	141	75-31100
SEI Industries Ltd	Canada	76	75-12750
Structure-flex Ltd	UK	486	75-57525

625 40 50 Lighting equipment, field mobile / transportable

Brimotor Ltd	UK	436	75-08600
Bruce Industries Inc.	USA	561	75-15360
Clark Masts Teksam Ltd	UK	440	75-13550
Engineered Air Systems Inc.	USA	573	75-31560
Equipos Industriales de Manutención	Spain	366	75-02850
Henry Cooch & Son Ltd	UK	442	75-14900
Luminescent Systems Inc.	USA	593	75-56480
M. Paz Logistics Ltd	Israel	224	80-01600
Mitralux International AG.	Switzerland	387	75-08500
Modra Electric Power Pty Ltd	Australia	19	75-13500
Pelican Products Inc.	USA	603	75-69125
Saunders-Roe Ltd.	UK	482	75-51850
Smith & Prince Ltd	UK	485	75-54630
Strumech Versatower Ltd	UK	486	75-57530

Thorn DNT Airfield Lighting Pty Ltd	Australia	22	75-20540
Westair Ltd	UK	493	75-65410

625 40 50 30 Floodlighting, emergency, mobile / transportable

Brimotor Ltd.	UK	436	75-08600
Bruce Industries Inc.	USA	561	75-15360
Clark Masts Teksam Ltd	UK	440	75-13550
Engineered Air Systems Inc.	USA	573	75-31560
Havis-Shields Equipment Corp.	USA	581	75-42700
Henry Cooch & Son Ltd	UK	442	75-14900
Modra Electric Power Pty Ltd	Australia	19	75-13500
Nitech Ltd	UK	471	75-42570
Strumech Versatower Ltd	UK	486	75-57530
The Will-Burt Co.	USA	628	75-96830
Thorn DNT Airfield Lighting Pty Ltd	Australia	22	75-20540

625 40 70 Maintenance facilities & equipment, vehicle, field mobile / transportable

ADE-HML Limited	UK	427	75-00710
Bombardier Inc.	Canada	69	75-01400
CSC (Specialised Vehicles) Ltd	UK	444	75-16220
Genco Mobile Systems BV.	Netherlands	293	75-02340
Glover Webb Ltd	UK	455	75-25700
Krauss-Maffei Wehrtechnik GmbH	Germany	171	75-16345
Sprung Instant Structures Ltd	Canada	77	75-13200
Systems & Electronics Inc.	USA	619	75-85650
Wesley Group Ltd	UK	493	75-65300
W.J. Clow (Midlands) Ltd	UK	441	75-13865

625 40 72 Maintenance workshops, electrical, mobile

GEC-Marconi Avionics Ltd	UK	454	75-25320
Genco Mobile Systems BV.	Netherlands	293	75-02340
Glover Webb Ltd	UK	455	75-25700
Systems & Electronics Inc.	USA	619	75-85650

625 40 75 32 Cranes, mobile, off-road

Auto Crane Co.	USA	556	75-10360
Belotti SpA	Italy	232	75-02640
Grove Europe Ltd	**UK**	**456**	**75-27030**
Henry Cooch & Son Ltd	UK	442	75-14900

625 40 80 Recovery equipment, vehicle-related

Empl Fahrzeugwerk GesmbH	Austria	32	75-00920
Manfred Vetter GmbH & Co	Germany	176	75-32800
Spec Tool Co.	USA	616	75-83050
Tirfor Ltd	UK	488	75-60150

625 40 80 20 Recovery cranes, vehicle-mounted

Atlas Hydraulic Loaders Ltd	UK	432	75-04300
Goussinsky B. Engineering & Mfr Ltd	Israel	219	75-07950
Grove Europe Ltd	**UK**	**456**	**75-27030**
Reynolds Boughton Ltd	UK	479	75-49340

625 40 80 30 Recovery equipment, hand-operated

Kohlbrat & Bunz GmbH	Austria	33	75-01575

625 40 80 60 Tow ropes, kinetic energy

SpanSet Inter AG	Switzerland	387	75-13550

625 40 80 90 Recovery winches, vehicle-mounted

dp Manufacturing Inc.	USA	569	75-27320
H.C. Slingsby plc	UK	484	75-54430
M.A. Bigland (Preparations) Ltd	UK	434	75-06700
Mannesmann Rexroth SA	France	139	75-25500
Ramsey Winch Co.	USA	607	75-72830

Reynolds Boughton Ltd	UK	479	75-49340
Superwinch Ltd	UK	487	75-57950

626 Construction Plant

Blakley Electrics Ltd	UK	434	75-07015
Parker Plant Limited	UK	473	75-43900

626 10 Asphalt making plant

Parker Plant Limited	UK	473	75-43900

626 20 Concrete mixing plant

EFE Endüstri ve Ticaret A.S.	Turkey	402	75-00530

626 60 Stone crushing plant

Parker Plant Limited	UK	473	75-43900

627 Combat Engineer & construction equipment (including vehicles)

AAI Corp.	USA	545	75-00020
Alvis Unipower Ltd	UK	430	75-02863
Applied Power Australia Ltd	Australia	12	75-01500
Astra	Italy	232	75-01800
Aveling Barford (Machines) plc	UK	432	75-04550
Bedek Aviation Group	Israel	218	75-02800
Case (United Kingdom) Ltd.	UK	439	75-12050
Caterpillar Inc.	USA	562	75-17100
CNIM	France	133	75-09990
Dorman Equipment Pty Ltd	Australia	15	75-04720
Israel Aircraft Industries Ltd	Israel	219	75-10200
IVECO Fiat SpA	Italy	235	75-13230
JC Bamford Excavators Ltd	UK	462	75-32297
Pearson Engineering Ltd	UK	473	75-44330
Plumett Ltd	UK	475	75-46340
Precision Machining Inc.	USA	605	75-71350
Reynolds Boughton Ltd	UK	479	75-49340
Stow Mfg Co.	USA	618	75-84280
Systems & Electronics Inc.	USA	619	75-85650
Thos. Storey (Engineers) Ltd	UK	486	75-57350
Williams Fairey Engrg	UK	494	75-66285
W.J. Clow (Midlands) Ltd	UK	441	75-13865

627 10 Bridging equipment, tactical

AAI Corp.	USA	545	75-00020
China Shipbuilding Trading Co. Ltd	China PR	88	55-00360
CNIM	France	133	75-09990
Eurobridge Mobile Brücken GmbH	Germany	168	75-08700
Indal Technologies Inc	Canada	72	75-07740
Israel Military Industries Ltd	Israel	220	75-10300
IVECO Fiat SpA	Italy	235	75-13230
Mabey & Johnson Ltd	UK	465	75-36830
MAN GHH AG	Germany	172	75-19615
MAN Gutehoffnungshütte AG	Germany	172	75-19620
Meuse & Sambre	Belgium	46	75-04971
PD Technical Mouldings plc	UK	473	75-44200
Precision Machining Inc.	USA	605	75-71350
SpanSet Inter AG	Switzerland	387	75-13550
Systems & Electronics Inc.	USA	619	75-85650
Thos. Storey (Engineers) Ltd	UK	486	75-57350
Vickers Bridging	UK	491	75-63030
Williams Fairey Engrg	UK	494	75-66285

627 10 05 Bridges, armoured vehicle launched (see code 141.10.55)

AAI Corp.	USA	545	75-00020
CNIM	France	133	75-09990
PD Technical Mouldings plc	UK	473	75-44200
Renown Engineering Ltd	UK	478	75-49227
Systems & Electronics Inc.	USA	619	75-85650
Vickers Bridging	UK	491	75-63030
ZTS TEES Martin s.p.	Slovakia	353	75-01500

627 10 10 Bridges, Bailey

Garden Reach Shipbuilders	India	*203*	55-00300
Mabey & Johnson Ltd	UK	*465*	75-36830
Thos. Storey (Engineers) Ltd	UK	*486*	75-57350

627 10 15 Bridges, tactical floating

China Shipbuilding Trading Co. Ltd	China PR	*88*	55-00360
CNIM	France	*133*	75-09990
FBM Marine Limited	UK	*450*	75-22485
MAN Gutehoffnungshütte AG	Germany	*172*	75-19620
Precision Machining Inc.	USA	*605*	75-71350
Systems & Electronics Inc.	USA	*619*	75-85650
Thos. Storey (Engineers) Ltd	UK	*486*	75-57350
Williams Fairey Engrg	UK	*494*	75-66285

627 10 20 Footbridges, tactical

Indal Technologies Inc	Canada	*72*	75-07740
SpanSet Inter AG	Switzerland	*387*	75-13550

627 10 22 Footbridges, tactical, conventional

Indal Technologies Inc	Canada	*72*	75-07740

627 10 50 Pontoons, tactical bridging

CNIM	France	*133*	75-09990
FBM Marine Limited	UK	*450*	75-22485
MAN Gutehoffnungshütte AG	Germany	*172*	75-19620
PD Technical Mouldings plc	UK	*473*	75-44200
Precision Machining Inc.	USA	*605*	75-71350
Williams Fairey Engrg	UK	*494*	75-66285

627 10 55 Pontoon pushboats, tactical bridging

FBM Marine Limited	UK	*450*	75-22485
Precision Machining Inc.	USA	*605*	75-71350
Schottel France Sarl	France	*143*	75-34450
Thos. Storey (Engineers) Ltd	UK	*486*	75-57350

627 10 80 Bridging vehicles, tactical

CNIM	France	*133*	75-09990
Eurobridge Mobile Brücken GmbH	Germany	*168*	75-08700
IVECO Fiat SpA	Italy	*235*	75-13230
Partek Cargotec Ltd	UK	*473*	75-44055
Systems & Electronics Inc.	USA	*619*	75-85650
Vickers Bridging	UK	*491*	75-63030

627 10 80 10 Bridging vehicles, tactical amphibious

Schottel France Sarl	France	*143*	75-34450

627 20 Construction tools

Atlas Copco Berema AB	Sweden	*375*	75-02500
Kango Ltd	UK	*462*	75-33210

627 20 20 Digging tools

Atlas Copco Berema AB	Sweden	*375*	75-02500
Kango Ltd	UK	*462*	75-33210

627 20 30 Drills, rock / road

Atlas Copco Berema AB	Sweden	*375*	75-02500

627 20 40 Electric power tools

Atlas Copco Berema AB	Sweden	*375*	75-02500
H.C. Slingsby plc	UK	*484*	75-54430

627 30 Construction vehicles & machines

A/S Hydrema Danmark	Denmark	*103*	75-09000
Astra	Italy	*232*	75-01800
Carleton Technologies Inc.	USA	*562*	75-16850
Case (United Kingdom) Ltd.	UK	*439*	75-12050
Caterpillar Inc.	USA	*562*	75-17100
Dorman Equipment Pty Ltd	Australia	*15*	75-04720
Israel Aircraft Industries Ltd	Israel	*219*	75-10200
JC Bamford Excavators Ltd	UK	*462*	75-32297
O&K Orenstein & Koppel AG.	Germany	*173*	75-22940
Parker Plant Limited	UK	*473*	75-43900
Pel-Job Ecomat Industrie SA	France	*136*	75-14250
Reynolds Boughton Ltd	UK	*479*	75-49340
Sandhurst Mfg Co Ltd	UK	*481*	75-51750
Terex Equipment Ltd	UK	*488*	75-59320

627 30 10 Earthmoving equipment

Aveling Barford (Machines) plc	UK	*432*	75-04550
Carleton Technologies Inc.	USA	*562*	75-16850
Caterpillar of Australia Ltd	Australia	*14*	75-03240
JC Bamford Excavators Ltd	UK	*462*	75-32297
O&K Orenstein & Koppel AG.	Germany	*173*	75-22940
Pearson Engineering Ltd	UK	*473*	75-44330
Pel-Job Ecomat Industrie SA	France	*136*	75-14250
Reynolds Boughton Ltd	UK	*479*	75-49340
Terex Equipment Ltd	UK	*488*	75-59320
Urdan Industries Ltd	Israel	*223*	75-23210
Vammas Ltd	**Finland**	***117***	**55-02220**

627 30 15 Bulldozers

Case (United Kingdom) Ltd.	UK	*439*	75-12050
Komatsu Hanomag Aktiengesellschaft	Germany	*171*	75-16185
NORINCO	China PR	*88*	55-00280
Pearson Engineering Ltd	UK	*473*	75-44330

627 30 20 Compactors, wheeled

Aveling Barford (Machines) plc	UK	*432*	75-04550
Grove Europe Ltd	**UK**	***456***	**75-27030**
JC Bamford Excavators Ltd	UK	*462*	75-32297
Komatsu Hanomag Aktiengesellschaft	Germany	*171*	75-16185

627 30 25 Compressors, air, construction type

Atlas Copco Airpower NV	Belgium	*45*	75-00900
Riva Calzoni SpA	Italy	*238*	75-20500
SAS Limited	UK	*482*	75-51830

627 30 25 50 Compressors, air, portable

Atlas Copco Airpower NV	Belgium	*45*	75-00900
CompAir Holman Iberica SA	Spain	*366*	75-01900
H.C. Slingsby plc	UK	*484*	75-54430
Riva Calzoni SpA	Italy	*238*	75-20500
SAS Limited	UK	*482*	75-51830
Zwick Energy Research Organization	USA	*629*	75-98400

627 30 25 80 Compressors, air, towed trailer

Atlas Copco Airpower NV	Belgium	*45*	75-00900
SAS Limited	UK	*482*	75-51830

627 30 30 Diggers

A.F. Trenchers Ltd	UK	*489*	75-61090
Case (United Kingdom) Ltd.	UK	*439*	75-12050
JC Bamford Excavators Ltd	UK	*462*	75-32297
O&K Orenstein & Koppel AG.	Germany	*173*	75-22940
Pearson Engineering Ltd	UK	*473*	75-44330
Pel-Job Ecomat Industrie SA	France	*136*	75-14250
Sandhurst Mfg Co Ltd	UK	*481*	75-51750

Product–Supplier Link

627 30 30 15 Diggers, backhoe

JC Bamford Excavators Ltd	UK	*462*	75-32297
O&K Orenstein & Koppel AG	Germany	*173*	75-22940

627 30 30 20 Diggers, combat

A.F. Trenchers Ltd	UK	*489*	75-61090
Case (United Kingdom) Ltd	UK	*439*	75-12050
JC Bamford Excavators Ltd	UK	*462*	75-32297
Pearson Engineering Ltd	UK	*473*	75-44330
Sandhurst Mfg Co Ltd	UK	*481*	75-51750

627 30 30 30 Digger / loaders

JC Bamford Excavators Ltd	UK	*462*	75-32297
O&K Orenstein & Koppel AG	Germany	*173*	75-22940

627 30 35 Dumpers

A/S Hydrema Danmark	Denmark	*103*	75-09000
Astra	Italy	*232*	75-01800
Aveling Barford (Machines) plc	UK	*432*	75-04550
O&K Orenstein & Koppel AG	Germany	*173*	75-22940
Thwaites Ltd	UK	*488*	75-59800

627 30 40 Excavators

A/S Hydrema Danmark	Denmark	*103*	75-09000
Atlas Hydraulic Loaders Ltd	UK	*432*	75-04300
Case (United Kingdom) Ltd	UK	*439*	75-12050
Loadlift Equipment Ltd	New Zealand	*302*	75-01250
Pel-Job Ecomat Industrie SA	France	*136*	75-14250
Sandhurst Mfg Co Ltd	UK	*481*	75-51750

627 30 40 10 Excavators, crawler type

Case (United Kingdom) Ltd	UK	*439*	75-12050
Komatsu Hanomag Aktiengesellschaft	Germany	*171*	75-16185
Pel-Job Ecomat Industrie SA	France	*136*	75-14250
Sandhurst Mfg Co Ltd	UK	*481*	75-51750

627 30 50 Loaders

Atlas Hydraulic Loaders Ltd	UK	*432*	75-04300
Howden Defense Systems Inc.	USA	*582*	75-44400
JC Bamford Excavators Ltd	UK	*462*	75-32297
Komatsu Hanomag Aktiengesellschaft	Germany	*171*	75-16185
O&K Orenstein & Koppel AG	Germany	*173*	75-22940
Terex Equipment Ltd	UK	*488*	75-59320

627 30 50 80 Loaders, shovel

JC Bamford Excavators Ltd	UK	*462*	75-32297
O&K Orenstein & Koppel AG	Germany	*173*	75-22940

627 30 60 Road rollers

Aveling Barford (Machines) plc	UK	*432*	75-04550

627 30 63 Road sweepers

Bunce (Ashbury) Ltd	UK	*438*	75-10870

627 30 65 Road surfacing machines

O&K Orenstein & Koppel AG	Germany	*173*	75-22940

627 30 70 Scrapers

Israel Aircraft Industries Ltd	Israel	*219*	75-10200
Terex Equipment Ltd	UK	*488*	75-59320

627 30 90 Welding machines, mobile

Dorman Equipment Pty Ltd	Australia	*15*	75-04720
Electrode Works Zika Ltd	Israel	*223*	75-23800
H.C. Slingsby plc	UK	*484*	75-54430
The Lincoln Electric Co.	Australia	*18*	75-12410

627 40 Combat Engineer equipment, special purpose

AAI Corp.	USA	*545*	75-00020
Applied Power Australia Ltd	Australia	*12*	75-01500
BDL Systems Ltd	UK	*433*	75-06080
Case (United Kingdom) Ltd	UK	*439*	75-12050
General Motors of Canada Ltd	Canada	*71*	75-05900
Pearson Engineering Ltd	UK	*473*	75-44330
Perry Engineering	Australia	*20*	75-16460
Urdan Industries Ltd	Israel	*223*	75-23210

627 40 10 Earth anchoring systems

Pearson Engineering Ltd	UK	*473*	75-44330
Renown Engineering Ltd	UK	*478*	75-49227

627 40 20 Cratering kits

AAI Corp.	USA	*545*	75-00020

627 40 30 Foxhole digging equipment

Case (United Kingdom) Ltd	UK	*439*	75-12050

627 40 33 Foxhole digging equipment, explosive

AAI Corp.	USA	*545*	75-00020

627 40 35 Foxhole digging machines

A.F. Trenchers Ltd	UK	*489*	75-61090
Case (United Kingdom) Ltd	UK	*439*	75-12050
Pearson Engineering Ltd	UK	*473*	75-44330
Perry Engineering	Australia	*20*	75-16460
Sandhurst Mfg Co Ltd	UK	*481*	75-51750

627 40 50 Lifting equipment

Applied Power Australia Ltd	Australia	*12*	75-01500
Grove Europe Ltd	**UK**	***456***	**75-27030**
Holmatro Inc.	USA	*636*	80-14340
Kinetics Ltd	Israel	*220*	75-11320
Logistic Projects	S. Africa	*358*	75-09800
Manfred Vetter GmbH & Co	Germany	*176*	75-32800
MFC Survival Ltd	UK	*467*	75-38900
Parsons Chain Co. Ltd	UK	*473*	75-44035
SpanSet Inter AG	Switzerland	*387*	75-13550
Strachan & Henshaw Ltd	UK	*512*	80-47400
Structure-flex Ltd	UK	*486*	75-57525

627 40 60 Prefabricated roadway surfaces

Eve Trakway Ltd	UK	*449*	75-21701

627 40 65 Portable runways, airfield

Aardvark Active Barriers Ltd.	UK	*426*	75-00032

627 60 Materials handling equipment

AAI Corp.	USA	*545*	75-00020
Aveling Barford (Machines) plc	UK	*432*	75-04550
Boss Group Limited	UK	*435*	75-07750
Carleton Technologies Inc.	USA	*562*	75-16850
Case Corporation	USA	*562*	75-16950
Caterpillar Inc.	USA	*562*	75-17100
Engineered Systems Co.	USA	*574*	75-31590
Grove Europe Ltd	**UK**	***456***	**75-27030**

Hovair Systems	UK	*459*	75-30190
JC Bamford Excavators Ltd	UK	*462*	75-32297
MIL-BASE Industries Inc.	USA	*596*	75-61530
O&K Orenstein & Koppel AG	Germany	*173*	75-22940
Parsons Chain Co. Ltd	UK	*473*	75-44035
Peco Ltd	New Zealand	*302*	75-01630
Still GmbH	Germany	*175*	75-29000
Stinar Corp.	USA	*618*	75-84220

627 60 40 Fork lift trucks (see code 613.20.50.10)

Aveling Barford (Machines) plc	UK	*432*	75-04550
Boss Group Limited	UK	*435*	75-07750
Case (United Kingdom) Ltd.	UK	*439*	75-12050
Echtman Engineering Corp. Ltd	Israel	*218*	75-05050
Hamech Ltd	UK	*456*	75-27730
JC Bamford Excavators Ltd	UK	*462*	75-32297
Polymathic Engineering Co. Ltd	UK	*475*	75-46350
Sellick Equipment Ltd.	Canada	*76*	75-12755
Still GmbH	Germany	*175*	75-29000

627 60 40 70 Fork lift trucks, rough terrain

Aveling Barford (Machines) plc	UK	*432*	75-04550
Boss Group Limited	UK	*435*	75-07750
Daewoo Heavy Industries Ltd	Korea (ROK)	*265*	75-00600
Hyundai Motor Co. Ltd	Korea (ROK)	*266*	75-01602
JC Bamford Excavators Ltd	UK	*462*	75-32297
Manitou BF	France	*139*	75-25550
O&K Orenstein & Koppel AG	Germany	*173*	75-22940
Sellick Equipment Ltd.	Canada	*76*	75-12755

627 60 50 Handlers, telescopic

Aveling Barford (Machines) plc	UK	*432*	75-04550
JC Bamford Excavators Ltd	UK	*462*	75-32297
Sellick Equipment Ltd.	Canada	*76*	75-12755
Still GmbH	Germany	*175*	75-29000
Stinar Corp.	USA	*618*	75-84220

627 60 50 70 Handlers, rough terrain

Aveling Barford (Machines) plc	UK	*432*	75-04550
Boss Group Limited	UK	*435*	75-07750
JC Bamford Excavators Ltd	UK	*462*	75-32297
Sellick Equipment Ltd.	Canada	*76*	75-12755

628 Camouflage & decoys

AAI Corp.	USA	*545*	75-00020
Barracuda Technologies AB	Sweden	*375*	75-02700
Barracuda Technologies Pty Ltd.	Australia	*12*	75-02020
Bridport Aviation Products	UK	*435*	75-08400
Bridport-Quin Pty Ltd	Australia	*13*	75-02746
Irvin Industries Inc.	USA	*586*	75-48960
J&D Wilkie Ltd	UK	*494*	75-66240
L'Garde Inc.	USA	*590*	75-55060
M. Paz Logistics Ltd	Israel	*224*	80-01600
Structure-flex Ltd	UK	*486*	75-57525
Technical Products Group Inc.	USA	*620*	75-86764
W&J Knox Ltd	UK	*463*	75-34050

628 10 Camouflage nets

Airborne Industries	UK	*428*	75-01550
Barracuda Technologies AB	Sweden	*375*	75-02700
BCB International Ltd	UK	*433*	75-06075
Bridport Aviation Products	UK	*435*	75-08400
Bridport-Quin Pty Ltd	Australia	*13*	75-02746
Climax Engineering Pty Ltd	Australia	*14*	75-03480
Deben Group Industries Ltd	UK	*444*	75-17270
Fothergill Engineered Fabrics Ltd	UK	*452*	75-24380
Irvin Aerospace Limited	UK	*462*	75-32050
Irvin Aerospace SpA	Italy	*235*	75-12200
J Blaschke Wehrtechnik GmbH	Austria	*32*	75-00560
J&D Wilkie Ltd	UK	*494*	75-66240
M. Paz Logistics Ltd	Israel	*224*	80-01600
Manta Industries Ltd	Canada	*73*	75-09250
Sala International Pty Ltd	Australia	*21*	75-18980
Technical Products Group Inc.	USA	*620*	75-86764
Valpak SA	Greece	*192*	75-03200
W&J Knox Ltd	UK	*463*	75-34050

628 20 Camouflage fabrics

Barracuda Technologies AB	Sweden	*375*	75-02700
Barracuda Technologies Pty Ltd.	Australia	*12*	75-02020
Bridport Aviation Products	UK	*435*	75-08400
Bridport-Quin Pty Ltd	Australia	*13*	75-02746
Bruck Textiles plc	Australia	*13*	75-02810
Fothergill Engineered Fabrics Ltd	UK	*452*	75-24380
H. Fine & Son Ltd	UK	*451*	75-23210
Irvin Aerospace Limited	UK	*462*	75-32050
J Blaschke Wehrtechnik GmbH	Austria	*32*	75-00560
J&D Wilkie Ltd	UK	*494*	75-66240
M. Paz Logistics Ltd	Israel	*224*	80-01600
Manta Industries Ltd	Canada	*73*	75-09250
NV Bekaert SA	Belgium	*45*	75-01300
Technical Products Group Inc.	USA	*620*	75-86764
UTEXBEL SA/NV	Belgium	*47*	75-06373

628 40 Decoys

Balmoral Group Ltd	UK	*433*	75-05420
Barracuda Technologies AB	Sweden	*375*	75-02700
Bridport Aviation Products	UK	*435*	75-08400
Chemring Countermeasures	UK	*439*	75-12800
Colebrand Ltd	UK	*441*	75-14120
Cranfield Aerospace Ltd	UK	*500*	80-07200
Denis Ferranti Meters Ltd	UK	*450*	75-22850
H. Fine & Son Ltd	UK	*451*	75-23210
Hovermarine International Ltd	UK	*459*	75-30200
Hunting Engineering Ltd	UK	*459*	75-30700
Irvin Aerospace Limited	UK	*462*	75-32050
Irvin Aerospace SpA	Italy	*235*	75-12200
Irvin Industries Inc.	USA	*586*	75-48960
J Blaschke Wehrtechnik GmbH	Austria	*32*	75-00560
L'Garde Inc.	USA	*590*	75-55060
Manta Industries Ltd	Canada	*73*	75-09250
MASS Consultants Limited	UK	*506*	80-15650
Signature Management	UK	*484*	75-53790
Technical Products Group Inc.	USA	*620*	75-86764
Teledyne Brown Engineering	USA	*620*	75-87508

628 40 30 Decoys, electronic

Flight Refuelling Limited	UK	*451*	75-23850
FR Aviation Limited	UK	*503*	80-10850
GEC-Marconi Materials Technology Ltd	UK	*419*	60-00600
Technical Products Group Inc.	USA	*620*	75-86764
Thomson Marconi Sonar Limited	UK	*426*	74-20000

628 40 40 Decoys, inflatable

AAI Corp.	USA	*545*	75-00020
Chemring Countermeasures	UK	*439*	75-12800
Crestbury Ltd	UK	*443*	75-16030
FPT Industries Ltd	UK	*452*	75-24400
Hunting Engineering Ltd	UK	*459*	75-30700
Irvin Aerospace Limited	UK	*462*	75-32050
Irvin Industries Inc.	USA	*586*	75-48960
L'Garde Inc.	USA	*590*	75-55060
ML Lifeguard Equipment Ltd	UK	*469*	75-40230
Signature Management	UK	*484*	75-53790
Technical Products Group Inc.	USA	*620*	75-86764

628 40 60 Decoy systems

AAI Corp.	USA	*545*	75-00020
Barracuda Technologies AB	Sweden	*375*	75-02700
Bridport Aviation Products	UK	*435*	75-08400
Chemring Countermeasures	UK	*439*	75-12800
Cranfield Aerospace Ltd	UK	*500*	80-07200
FPT Industries Ltd	UK	*452*	75-24400
GEC ALSTHOM Engineering Systems Ltd	UK	*453*	75-25225
Hunting Engineering Ltd	UK	*459*	75-30700
Irvin Aerospace Limited	UK	*462*	75-32050
J Blaschke Wehrtechnik GmbH	Austria	*32*	75-00560

Signature Management	UK	*484*	75-53790
Technical Products Group Inc.	USA	*620*	75-86764
Teledyne Brown Engineering	USA	*620*	75-87508

628 40 60 30 Decoy systems, naval

Barracuda Technologies AB	Sweden	*375*	75-02700
Bharat Electronics Ltd	India	*203*	55-00200
L'Garde Inc.	USA	*590*	75-55060
Technical Products Group Inc.	USA	*620*	75-86764

629 Field equipment / supplies, other types

Aero Tec Laboratories Inc.	USA	*546*	75-01117
BCB International Ltd	UK	*433*	75-06075
Carr & Day & Martin Ltd	UK	*438*	75-11750
Columbus Jack Corp.	USA	*564*	75-19850
Davies Industrial Communications Ltd	UK	*444*	75-17010
GEL Engineering Ltd	UK	*454*	75-25500
Geoquip Ltd	UK	*454*	75-25415
Giat Industries	France	*126*	55-00100
M. Paz Logistics Ltd	Israel	*224*	80-01600
MB-Microtec AG	Switzerland	*386*	75-08485
Richmond Packaging (UK) Ltd	UK	*479*	75-49630
SPL Ordnance Ltd.	Israel	*225*	80-01935
Trojan Manufacturing & Marketing Inc	Philippines	*321*	75-01350

629 30 Chemical light sources

Divex Ltd.	UK	*445*	75-18220
M. Paz Logistics Ltd	Israel	*224*	80-01600
Omniglow Corp	USA	*601*	75-66620
Saunders-Roe Ltd.	UK	*482*	75-51850

629 30 60 Route markers, luminescent

M. Paz Logistics Ltd	Israel	*224*	80-01600
Pearson Engineering Ltd	UK	*473*	75-44330
Saunders-Roe Ltd.	UK	*482*	75-51850

629 40 Flashlights, hand-held

Mag Instrument Inc.	USA	*593*	75-57020

629 60 Road barriers, flexible

Cochrane Steel Products (PTY) Ltd	S. Africa	*357*	75-03000
Elkosta VKS Dr Dörnemann GmbH & Co	Germany	*168*	75-07570
Goussinsky B. Engineering & Mfr Ltd	Israel	*219*	75-07950

629 70 Sensors, seismic, field

EG&G Inc.	USA	*571*	75-29555
Fockas Pantelis Ltd	Greece	*189*	75-01293
Geoquip Ltd	UK	*454*	75-25415

629 80 Tape / wire, barbed

Cochrane Steel Products (PTY) Ltd	S. Africa	*357*	75-03000
Eusung Industrial Co. Ltd.	Korea (ROK)	*265*	75-01200
Geoquip Ltd	UK	*454*	75-25415
SPL Ordnance Ltd.	Israel	*225*	80-01935

629 90 Water quality test sets

Katadyn Products Inc.	Switzerland	*386*	75-07000
McNab Inc.	USA	*595*	75-59720
Richmond Packaging (UK) Ltd	UK	*479*	75-49630

Power Generation

631 Power generators, fixed

A. Smith Gt Bentley Ltd	UK	*484*	75-54540
Aircraft Appliances & Equipment Ltd	Canada	*68*	75-00430
Arvico Corporation	USA	*555*	75-09240
BP Solar Limited.	UK	*435*	75-07800
Brimotor Ltd.	UK	*436*	75-08600
British International Industries Ltd.	UK	*436*	75-09150
CORIS	France	*134*	75-12450
Detroit Diesel Corp.	USA	*568*	75-25830
Elequip Projects Ltd.	UK	*448*	75-20450
Fincantieri	Italy	*234*	75-09250
Fluor Daniel Inc.	USA	*635*	80-11720
GEC ALSTHOM Paxman Diesels Ltd	UK	*453*	75-25245
Global Thermoelectric Inc.	Canada	*71*	75-06200
Hewitt Equipment Ltd	Canada	*72*	75-07500
Hitzinger Gesellschaft GmbH.	Austria	*33*	75-01280
Houchin Aerospace Ltd	UK	*459*	75-30100
Integrated Power Corp.	USA	*585*	75-47100
John Brown Engineering Ltd	UK	*437*	75-10250
Jones & Rickard Pty Ltd	Australia	*18*	75-10750
Microturbo Ltd	UK	*468*	75-39160
Microturbo SA	France	*140*	75-27000
Mirrlees Blackstone Limited	UK	*468*	75-40050
Modra Electric Power Pty Ltd	Australia	*19*	75-13500
MTU do Brasil Ltda	Brazil	*57*	75-13900
MTU Friedrichschafen GmbH	Germany	*172*	75-21802
Newton Derby Ltd	UK	*470*	75-42520
Pezetel	Poland	*325*	80-01000
Puma Manufacturing Ltd	UK	*476*	75-47100
Saft.	France	*142*	75-33920
Saft America Inc	Mexico	*282*	75-18000
Saft America Inc.	USA	*612*	75-77550
Saft Korea Co Ltd	Korea (ROK)	*266*	75-04300
Saft Nife Argentina SA	Argentina	*4*	75-09000
Saft Nife AS	Norway	*310*	75-02400
Saft Nife BV	Netherlands	*294*	75-06160
Saft Nife Corp	Canada	*75*	75-12290
Saft Nife GmbH	Austria	*34*	75-02142
Saft Nife GmbH	Germany	*174*	75-25901
Saft Nife Iberica SL	Spain	*367*	75-05100
Saft Nife M.E. Ltd	Cyprus	*96*	75-05000
Saft Nife Oy	Finland	*118*	75-01960
Saft Nife Power Systems Australia	Australia	*20*	75-18950
Saft Singapore Pte Ltd	Singapore	*348*	75-03400
SELIN Sistemi Sarl	Italy	*238*	75-21750
Solar Turbines.	USA	*616*	75-82300
Stewart & Stevenson Services Inc.	USA	*618*	75-84180
Stork-Wärtsilä Diesel BV.	Netherlands	*295*	75-07300
Technology Dynamics Inc.	USA	*620*	75-86900
Teledyne Brown Engineering	USA	*620*	75-87508
Telkoor Ltd	Israel	*223*	75-22400
THORN Microwave Devices Ltd	UK	*488*	75-59760
Van Rietschoten & Houwens	Netherlands	*296*	75-07900
VEC Srl.	Argentina	*4*	75-30000

631 10 Power generating sets

Aircraft Appliances & Equipment Ltd	Canada	*68*	75-00430
Arvico Corporation	USA	*555*	75-09240
A/S Dantherm	**Denmark**	***102***	**75-04100**
Atlas Copco Airpower NV	Belgium	*45*	75-00900
Blakley Electrics Ltd	UK	*434*	75-07015
Brimotor Ltd.	UK	*436*	75-08600
Brush Electrical Machines Ltd	UK	*437*	75-10450
C-RAN Corp.	USA	*566*	75-22400
Commercial Hydraulics Keelavite Ltd	UK	*441*	75-14410
CORIS	France	*134*	75-12450
Detroit Diesel-Allison Australia Pty	Australia	*15*	75-04450
Elequip Projects Ltd.	UK	*448*	75-20450
Fincantieri	Italy	*234*	75-09250
GEC ALSTHOM Paxman Diesels Ltd	UK	*453*	75-25245
Global Thermoelectric Inc.	Canada	*71*	75-06200
Hewitt Equipment Ltd	Canada	*72*	75-07500
Ideal Electric Co.	USA	*584*	75-45920
Integrated Power Corp.	USA	*585*	75-47100
John Brown Engineering Ltd	UK	*437*	75-10250
Lister-Petter Inc.	USA	*590*	75-55360
Lyulka-Saturn Inc	Russia	*337*	75-12000
Microturbo Ltd	UK	*468*	75-39160
Mirrlees Blackstone Limited	UK	*468*	75-40050

Modra Electric Power Pty Ltd	Australia	*19*	75-13500
MTU Friedrichschafen GmbH	Germany	*172*	75-21802
Newton Derby Ltd	UK	*470*	75-42520
NORINCO	China PR	*88*	55-00280
Normalair-Garrett Australia	Australia	*19*	75-14870
Northrop Grumman Corporation	USA	*599*	75-65450
O'Brien Energy Services Inc.	USA	*601*	75-65830
Power Spectra Inc.	USA	*605*	75-71085
SELIN Sistemi Sarl	Italy	*238*	75-21750
SEMT Pielstick	France	*144*	75-34843
Solar Turbines	USA	*616*	75-82300

631 50 Power stations, packaged

Atlas Copco Airpower NV	Belgium	*45*	75-00900
C-RAN Corp.	USA	*566*	75-22400
CORIS	France	*134*	75-12450
Detroit Diesel-Allison Australia Pty	Australia	*15*	75-04450
Fincantieri	Italy	*234*	75-09250
Hewitt Equipment Ltd	Canada	*72*	75-07500
Integrated Power Corp.	USA	*585*	75-47100
John Brown Engineering Ltd	UK	*437*	75-10250
Mirrlees Blackstone Limited	UK	*468*	75-40050
Modra Electric Power Pty Ltd	Australia	*19*	75-13500
Newton Derby Ltd	UK	*470*	75-42520
Solar Turbines	USA	*616*	75-82300
Stewart & Stevenson Services Inc.	USA	*618*	75-84180

631 60 Power generators, solar, fixed

BP Solar España SA	Spain	*365*	75-01660
BP Solar Limited	UK	*435*	75-07800
C-RAN Corp.	USA	*566*	75-22400
Integrated Power Corp.	USA	*585*	75-47100
Siemens Solar Industries	USA	*614*	75-80570
Teledyne Brown Engineering	USA	*620*	75-87508

632 Power generators, mobile

A. Smith Gt Bentley Ltd	UK	*484*	75-54540
Aermacchi SpA	Italy	*230*	75-00500
Air-Log Ltd	UK	*429*	75-01800
Aircraft Appliances & Equipment Ltd	Canada	*68*	75-00430
A/S Dantherm	**Denmark**	***102***	**75-04100**
Australian Defence Industries Ltd	Australia	*9*	55-01000
Bertel O. Steen A/S	Norway	*311*	75-02700
BP Solar Limited	UK	*435*	75-07800
Brimotor Ltd	UK	*436*	75-08600
British International Industries Ltd	UK	*436*	75-09150
Commercial Hydraulics Keelavite Ltd	UK	*441*	75-14410
Commercial Hydraulics Kontak Ltd	UK	*441*	75-14415
Daewoo Heavy Industries Ltd	Korea (ROK)	*265*	75-00600
Detroit Diesel Corp.	USA	*568*	75-25830
Elequip Projects Ltd.	UK	*448*	75-20450
Engineered Air Systems Inc.	USA	*573*	75-31560
GEL Engineering Ltd	UK	*454*	75-25500
Hewitt Equipment Ltd	Canada	*72*	75-07500
Houchin Aerospace Ltd	UK	*459*	75-30100
Hunting Engineering Ltd	UK	*459*	75-30700
Jones & Rickard Pty Ltd	Australia	*18*	75-10750
Kaman Electromagnetics Corp.	USA	*637*	80-17500
Lear Astronics Corp.	USA	*589*	75-54390
Lister-Petter Inc.	USA	*590*	75-55360
Marine Electric Systems Inc.	USA	*593*	75-58000
Microturbo Ltd	UK	*468*	75-39160
Microturbo SA	France	*140*	75-27000
Modra Electric Power Pty Ltd	Australia	*19*	75-13500
MTU Friedrichschafen GmbH	Germany	*172*	75-21802
Newage International Ltd.	UK	*470*	75-42330
Newton Derby Ltd	UK	*470*	75-42520
Pezetel	Poland	*325*	80-01000
Puma Manufacturing Ltd	UK	*476*	75-47100
Saft	France	*142*	75-33920
Saft America Inc	Mexico	*282*	75-18000
Saft America Inc.	USA	*612*	75-77550
Saft Korea Co Ltd	Korea (ROK)	*266*	75-04300
Saft Nife Argentina SA	Argentina	*4*	75-09000
Saft Nife AS	Norway	*310*	75-02400
Saft Nife BV	Netherlands	*294*	75-06160
Saft Nife Corp	Canada	*75*	75-12290
Saft Nife GmbH	Austria	*34*	75-02142
Saft Nife GmbH	Germany	*174*	75-25901
Saft Nife Iberica SL	Spain	*367*	75-05100
Saft Nife M.E. Ltd	Cyprus	*96*	75-05000
Saft Nife Oy	Finland	*118*	75-01960
Saft Nife Power Systems Australia	Australia	*20*	75-18950
Saft Singapore Pte Ltd	Singapore	*348*	75-03400
SELIN Sistemi Sarl	Italy	*238*	75-21750
Skurka Engineering Company	USA	*615*	75-81520
Solar Turbines	USA	*616*	75-82300
Stewart & Stevenson Services Inc.	USA	*618*	75-84180
Teledyne Brown Engineering	USA	*620*	75-87508
Van Rietschoten & Houwens	Netherlands	*296*	75-07900
Winco Inc.	USA	*628*	75-97050

632 10 Power generators, mobile, electric

AAI/ACL Technologies Inc.	USA	*545*	75-00360
Aermacchi SpA	Italy	*230*	75-00500
Air-Log Ltd	UK	*429*	75-01800
Arvico Corporation	USA	*555*	75-09240
Atlas Copco Airpower NV	Belgium	*45*	75-00900
Blakley Electrics Ltd	UK	*434*	75-07015
Brimotor Ltd	UK	*436*	75-08600
Commercial Hydraulics Keelavite Ltd	UK	*441*	75-14410
Commercial Hydraulics Kontak Ltd	UK	*441*	75-14415
Detroit Diesel-Allison Australia Pty	Australia	*15*	75-04450
Elequip Projects Ltd.	UK	*448*	75-20450
E.P. Barrus Ltd	UK	*433*	75-05900
EUROCONTROL SpA	Italy	*233*	75-08200
Henry Cooch & Son Ltd	UK	*442*	75-14900
Hewitt Equipment Ltd	Canada	*72*	75-07500
Hunting Engineering Ltd	UK	*459*	75-30700
Kaman Electromagnetics Corp.	USA	*637*	80-17500
Kango Ltd	UK	*462*	75-33210
Microturbo Ltd	UK	*468*	75-39160
Mirrlees Blackstone Limited	UK	*468*	75-40050
Modra Electric Power Pty Ltd	Australia	*19*	75-13500
MTU Friedrichschafen GmbH	Germany	*172*	75-21802
Newton Derby Ltd	UK	*470*	75-42520
Rantec Microwave & Elec.	USA	*607*	75-72915
SELIN Sistemi Sarl	Italy	*238*	75-21750
Skurka Engineering Company	USA	*615*	75-81520
Solar Turbines	USA	*616*	75-82300
Spar Aviation Services	Canada	*76*	75-13020
Stewart & Stevenson Services Inc.	USA	*618*	75-84180
Teledyne Brown Engineering	USA	*620*	75-87508

632 10 50 Generating sets, portable

Air-Log Ltd	UK	*429*	75-01800
Aircraft Appliances & Equipment Ltd	Canada	*68*	75-00430
Arvico Corporation	USA	*555*	75-09240
Atlas Copco Airpower NV	Belgium	*45*	75-00900
Australian Defence Industries Ltd	Australia	*9*	55-01000
Blakley Electrics Ltd	UK	*434*	75-07015
Brimotor Ltd	UK	*436*	75-08600
Commercial Hydraulics Keelavite Ltd	UK	*441*	75-14410
E.P. Barrus Ltd	UK	*433*	75-05900
G.E. Mortley, Sprague & Co. Ltd	UK	*469*	75-41000
Hewitt Equipment Ltd	Canada	*72*	75-07500
Hunting Engineering Ltd	UK	*459*	75-30700
Kango Ltd	UK	*462*	75-33210
Lister-Petter Inc.	USA	*590*	75-55360
Microturbo Ltd	UK	*468*	75-39160
Mirrlees Blackstone Limited	UK	*468*	75-40050
Modra Electric Power Pty Ltd	Australia	*19*	75-13500
Newton Derby Ltd	UK	*470*	75-42520
Normalair-Garrett Australia	Australia	*19*	75-14870
SELIN Sistemi Sarl	Italy	*238*	75-21750
Spar Aviation Services	Canada	*76*	75-13020
Stewart & Stevenson Services Inc.	USA	*618*	75-84180
Teledyne Brown Engineering	USA	*620*	75-87508
Winco Inc.	USA	*628*	75-97050

632 20 Power generators, mobile, hydraulic

AAI/ACL Technologies Inc.	USA	*545*	75-00360
Aermacchi SpA	Italy	*230*	75-00500
Aircraft Appliances & Equipment Ltd	Canada	*68*	75-00430
AV Technology LLC	USA	*556*	75-11405
Commercial Hydraulics Keelavite Ltd	UK	*441*	75-14410
Kinetics Ltd	Israel	*220*	75-11320

Lear Astronics Corp.	USA	*589*	75-54390
Mannesmann Rexroth SA	France	*139*	75-25500

632 60 Power generators, solar, mobile

AV Technology LLC	USA	*556*	75-11405
BP Solar Australia Pty Ltd	Australia	*13*	75-02630
BP Solar Limited.	UK	*435*	75-07800
Integrated Power Corp.	USA	*585*	75-47100
Siemens Solar Industries	USA	*614*	75-80570
Solarex Pty Ltd	Australia	*21*	75-19565

632 60 50 Power generators, solar, portable

BP Solar Limited.	UK	*435*	75-07800
Integrated Power Corp.	USA	*585*	75-47100
Solarex Pty Ltd	Australia	*21*	75-19565

632 90 Power generator components

Aircraft Appliances & Equipment Ltd	Canada	*68*	75-00430
Detroit Diesel-Allison Australia Pty	Australia	*15*	75-04450
Hewitt Equipment Ltd	Canada	*72*	75-07500
Integrated Power Corp.	USA	*585*	75-47100
Keltron Power Systems	USA	*588*	75-52000
Newton Derby Ltd	UK	*470*	75-42520
SELIN Sistemi Sarl	Italy	*238*	75-21750

632 93 Diesel engines, power generator

Air-Log Ltd	UK	*429*	75-01800
Aircraft Maintenance Support	UK	*428*	75-01580
Caterpillar of Australia Ltd	Australia	*14*	75-03240
Cummins Engine Co. Inc.	USA	*567*	75-22980
Detroit Diesel-Allison Australia Pty	Australia	*15*	75-04450
Deutz Corp.	USA	*569*	75-25950
Deutz Motor Industrie Motoren GmbH	Germany	*167*	75-06350
Elequip Projects Ltd.	UK	*448*	75-20450
E.P. Barrus Ltd	UK	*433*	75-05900
FP Diesel LP	USA	*576*	75-35753
GEC ALSTHOM Diesels Ltd	UK	*453*	75-25220
GEC ALSTHOM Paxman Diesels Ltd	UK	*453*	75-25245
Hercules Engines Company	USA	*581*	75-43080
Hewitt Equipment Ltd	Canada	*72*	75-07500
Lister-Petter Ltd.	UK	*465*	75-36070
Mirrlees Blackstone Limited	UK	*468*	75-40050
MTU Friedrichschafen GmbH	Germany	*172*	75-21802
Perkins Engines (Shrews.)	UK	*474*	75-44750
SEMT Pielstick	France	*144*	75-34843
Stewart & Stevenson Services Inc.	USA	*618*	75-84180
Wartsila Diesel International Sarl	France	*149*	75-47060

633 Power generators, system-integrated

Advance Hivolt	UK	*427*	75-00903
Belix Co. Ltd.	UK	*434*	75-06190
Detroit Diesel Corp.	USA	*568*	75-25830
DSB Special Batteries Limited	UK	*447*	75-18750
Eldec Corp.	USA	*572*	75-29660
Fluor Daniel Inc.	USA	*635*	80-11720
GEC ALSTHOM T&D Power Electronic	UK	*453*	75-25260
GEL Engineering Ltd	UK	*454*	75-25500
HDL Research Laboratory Inc.	USA	*581*	75-42780
Jones & Rickard Pty Ltd	Australia	*18*	75-10750
Microturbo SA	France	*140*	75-27000
Puma Manufacturing Ltd	UK	*476*	75-47100
Raytheon E-Systems	USA	*608*	75-73302
Saft.	France	*142*	75-33920
Saft America Inc	Mexico	*282*	75-18000
Saft America Inc.	USA	*612*	75-77550
Saft Korea Co Ltd	Korea (ROK)	*266*	75-04300
Saft Nife Argentina SA	Argentina	*4*	75-09000
Saft Nife AS	Norway	*310*	75-02400
Saft Nife BV	Netherlands	*294*	75-06160
Saft Nife Corp	Canada	*75*	75-12290
Saft Nife GmbH	Austria	*34*	75-02142
Saft Nife GmbH	Germany	*174*	75-25901
Saft Nife Iberica SL	Spain	*367*	75-05100
Saft Nife M.E. Ltd	Cyprus	*96*	75-05000
Saft Nife Oy	Finland	*118*	75-01960
Saft Nife Power Systems Australia	Australia	*20*	75-18950
Saft Singapore Pte Ltd	Singapore	*348*	75-03400
Stewart & Stevenson Services Inc.	USA	*618*	75-84180
Teledyne Brown Engineering	USA	*620*	75-87508

633 10 Nuclear power generators, system-integrated, spacecraft

Framatome	France	*136*	75-18405
General Atomics	USA	*578*	75-37447
Teledyne Brown Engineering	USA	*620*	75-87508

633 20 Power generators, system-integrated, aircraft

AlliedSignal Aerospace GmbH	Germany	*164*	75-00610
Belix Co. Ltd.	UK	*434*	75-06190
Century Electronics	USA	*562*	75-17385
HDL Research Laboratory Inc.	USA	*581*	75-42780
Lucas Aerospace Power Systems	USA	*592*	75-56260
Newton Derby Ltd	UK	*470*	75-42520

633 20 10 Batteries, aircraft (see code 221.40.70.30)

Aircraft Appliances & Equipment Ltd	Canada	*68*	75-00430
Chloride Industrial Batteries Ltd	UK	*457*	75-28490
Concorde Battery Corp.	USA	*566*	75-20845
Eldec Corp.	USA	*572*	75-29660
Hawker Energy Products Ltd	UK	*457*	75-28490
Hawker Eternacell Ltd	UK	*457*	75-28500
Saft.	France	*142*	75-33920
Saft America Inc	Mexico	*282*	75-18000
Saft Korea Co Ltd	Korea (ROK)	*266*	75-04300
Saft Nife AB	Sweden	*378*	75-09850
Saft Nife Argentina SA	Argentina	*4*	75-09000
Saft Nife AS	Norway	*310*	75-02400
Saft Nife BV	Netherlands	*294*	75-06160
Saft Nife Corp	Canada	*75*	75-12290
Saft Nife GmbH	Austria	*34*	75-02142
Saft Nife GmbH	Germany	*174*	75-25901
Saft Nife Iberica SL	Spain	*367*	75-05100
Saft Nife Ltd	UK	*481*	75-51470
Saft Nife M.E. Ltd	Cyprus	*96*	75-05000
Saft Nife Oy	Finland	*118*	75-01960
Saft Nife Power Systems Australia	Australia	*20*	75-18950
Saft Singapore Pte Ltd	Singapore	*348*	75-03400
Sonnenschein Lithium GmbH	Germany	*175*	75-28050
Steatite Insulations Ltd	UK	*486*	75-56450

633 20 10 50 Batteries, NiCad, aircraft

A1Marketing	UK	*426*	75-00025
Eldec Corp.	USA	*572*	75-29660
Saft Nife AB	Sweden	*378*	75-09850
Saft Nife Ltd	UK	*481*	75-51470
Saft Singapore Pte Ltd	Singapore	*348*	75-03400
VEC Srl.	Argentina	*4*	75-30000

633 30 Power generators, system-integrated, naval

Atlas Copco Airpower NV	Belgium	*45*	75-00900
Belix Co. Ltd.	UK	*434*	75-06190
Century Electronics	USA	*562*	75-17385
Detroit Diesel-Allison Australia Pty	Australia	*15*	75-04450
HDL Research Laboratory Inc.	USA	*581*	75-42780
Newton Derby Ltd	UK	*470*	75-42520
Stewart & Stevenson Services Inc.	USA	*618*	75-84180

633 30 10 Batteries, marine

Chloride Industrial Batteries Ltd	UK	*457*	75-28490
Concorde Battery Corp.	USA	*566*	75-20845
DSB Special Batteries Limited	UK	*447*	75-18750
Eldec Corp.	USA	*572*	75-29660
Hawker Energy Products Ltd	UK	*457*	75-28490
Hawker Eternacell Ltd	UK	*457*	75-28500
Saft Nife AB	Sweden	*378*	75-09850
Saft Nife Ltd	UK	*481*	75-51470
Ultralife Batteries (UK) Ltd	UK	*490*	75-61840

633 30 10 60 Batteries, submarine (see code 239.40.70.10)

Chloride Industrial Batteries Ltd	UK	*457*	75-28490
DSB Special Batteries Limited	UK	*447*	75-18750
Eldec Corp.	USA	*572*	75-29660
Hawker Energy Products Ltd	UK	*457*	75-28490
Hawker Eternacell Ltd	UK	*457*	75-28500
Saft Nife AB	Sweden	*378*	75-09850

633 30 10 70 Batteries, marine, torpedo propulsion

Chloride Industrial Batteries Ltd	UK	*457*	75-28490
Hawker Eternacell Ltd	UK	*457*	75-28500
Saft America Inc.	USA	*612*	75-77550
Saft Nife AB	Sweden	*378*	75-09850
Saft Nife Ltd	UK	*481*	75-51470
Saft Singapore Pte Ltd	Singapore	*348*	75-03400
Sonnenschein Lithium GmbH	Germany	*175*	75-28050

633 40 Power generators, system-integrated, ground-based

Century Electronics	USA	*562*	75-17385
Commercial Hydraulics Kontak Ltd	UK	*441*	75-14415
Detroit Diesel-Allison Australia Pty	Australia	*15*	75-04450
EMS-PATVAG AG	Switzerland	*386*	75-03650
Foxtronics Inc.	USA	*576*	75-35752
Global Thermoelectric Inc.	Canada	*71*	75-06200
HDL Research Laboratory Inc.	USA	*581*	75-42780
Hunting Engineering Ltd	UK	*459*	75-30700
Integrated Power Corp.	USA	*585*	75-47100
Keltron Power Systems	USA	*588*	75-52000
Newton Derby Ltd	UK	*470*	75-42520
Stewart & Stevenson Services Inc.	USA	*618*	75-84180
Teledyne Brown Engineering	USA	*620*	75-87508
The Titan Corp	USA	*622*	75-89620

633 40 10 Batteries, ground-based

Bren-Tronics Inc.	USA	*560*	75-15170
Chloride Industrial Batteries Ltd	UK	*457*	75-28490
Concorde Battery Corp.	USA	*566*	75-20845
DMS Technologies	UK	*445*	75-18270
DSB Special Batteries Limited	UK	*447*	75-18750
Exide Batteries Ltd	UK	*449*	75-22050
Fluor Daniel Inc.	USA	*635*	80-11720
Hawker Energy Products Ltd	UK	*457*	75-28490
Hawker Eternacell Ltd	UK	*457*	75-28500
Nife Brasil Sistemas Eletricos Ltda	Brazil	*57*	75-14400
Nitech Ltd	UK	*471*	75-42570
Power Conversion Inc.	USA	*604*	75-71070
Saft.	France	*142*	75-33920
Saft America Inc	Mexico	*282*	75-18000
Saft America Inc.	USA	*612*	75-77550
Saft Korea Co Ltd	Korea (ROK)	*266*	75-04300
Saft Nife AB	Sweden	*378*	75-09850
Saft Nife Argentina SA	Argentina	*4*	75-09000
Saft Nife AS	Norway	*310*	75-02400
Saft Nife BV	Netherlands	*294*	75-06160
Saft Nife Corp	Canada	*75*	75-12290
Saft Nife GmbH	Austria	*34*	75-02142
Saft Nife GmbH	Germany	*174*	75-25901
Saft Nife Iberica SL	Spain	*367*	75-05100
Saft Nife Ltd	UK	*481*	75-51470
Saft Nife M.E. Ltd	Cyprus	*96*	75-05000
Saft Nife Oy	Finland	*118*	75-01960
Saft Nife Power Systems Australia	Australia	*20*	75-18950
Saft Singapore Pte Ltd	Singapore	*348*	75-03400
Sonnenschein Lithium GmbH	Germany	*175*	75-28050
Tadiran Ltd	Israel	*222*	75-21600
Ultralife Batteries (UK) Ltd	UK	*490*	75-61840
Uniross Batteries Pty Ltd.	S. Africa	*360*	75-21500

633 40 10 30 Batteries, ground-based, communications equipment

Advanced Semiconductor Inc.	USA	*546*	75-00950
A1Marketing	UK	*426*	75-00025
Chloride Industrial Batteries Ltd	UK	*457*	75-28490
Concorde Battery Corp.	USA	*566*	75-20845
DMS Technologies	UK	*445*	75-18270
DSB Special Batteries Limited	UK	*447*	75-18750
Hagen	UK	*456*	75-27220
Hawker Energy Products Ltd	UK	*457*	75-28490
Hawker Eternacell Ltd	UK	*457*	75-28500
IBV Multipower Ltd	UK	*470*	75-41695
Saft.	France	*142*	75-33920
Saft America Inc	Mexico	*282*	75-18000
Saft America Inc.	USA	*612*	75-77550
Saft Korea Co Ltd	Korea (ROK)	*266*	75-04300
Saft Nife AB	Sweden	*378*	75-09850
Saft Nife Argentina SA	Argentina	*4*	75-09000
Saft Nife AS	Norway	*310*	75-02400
Saft Nife BV	Netherlands	*294*	75-06160
Saft Nife Corp	Canada	*75*	75-12290
Saft Nife GmbH	Austria	*34*	75-02142
Saft Nife GmbH	Germany	*174*	75-25901
Saft Nife Iberica SL	Spain	*367*	75-05100
Saft Nife Ltd	UK	*481*	75-51470
Saft Nife Oy	Finland	*118*	75-01960
Saft Nife Power Systems Australia	Australia	*20*	75-18950
Saft Singapore Pte Ltd	Singapore	*348*	75-03400
Sonnenschein Lithium GmbH	Germany	*175*	75-28050
Tadiran Ltd	Israel	*222*	75-21600
Ultralife Batteries (UK) Ltd	UK	*490*	75-61840
Uniross Batteries Pty Ltd.	S. Africa	*360*	75-21500

633 40 10 60 Batteries, ground-based, rechargeable

Advanced Semiconductor Inc.	USA	*546*	75-00950
A1Marketing	UK	*426*	75-00025
Carleton Life Support Technologies	Canada	*70*	75-02950
Concorde Battery Corp.	USA	*566*	75-20845
DMS Technologies	UK	*445*	75-18270
Hagen	UK	*456*	75-27220
IBV Multipower Ltd	UK	*470*	75-41695
Oldham France SA	France	*140*	75-28350
Power Conversion Inc.	USA	*604*	75-71070
Saft Nife Ltd	UK	*481*	75-51470
Signaal USFA	Netherlands	*295*	75-06500
Uniross Batteries Pty Ltd.	S. Africa	*360*	75-21500
VEC Srl.	Argentina	*4*	75-30000

633 40 10 80 Batteries, vehicle, ground-based (see code 241.50.40.70.30.30)

Chloride Industrial Batteries Ltd	UK	*457*	75-28490
Concorde Battery Corp.	USA	*566*	75-20845
Exide Batteries Ltd	UK	*449*	75-22050
Hawker Energy Products Ltd	UK	*457*	75-28490
Saft.	France	*142*	75-33920
Saft America Inc	Mexico	*282*	75-18000
Saft Korea Co Ltd	Korea (ROK)	*266*	75-04300
Saft Nife AB	Sweden	*378*	75-09850
Saft Nife Argentina SA	Argentina	*4*	75-09000
Saft Nife AS	Norway	*310*	75-02400
Saft Nife BV	Netherlands	*294*	75-06160
Saft Nife Corp	Canada	*75*	75-12290
Saft Nife GmbH	Austria	*34*	75-02142
Saft Nife GmbH	Germany	*174*	75-25901
Saft Nife Iberica SL	Spain	*367*	75-05100
Saft Nife Ltd	UK	*481*	75-51470
Saft Nife M.E. Ltd	Cyprus	*96*	75-05000
Saft Nife Oy	Finland	*118*	75-01960
Saft Nife Power Systems Australia	Australia	*20*	75-18950
Saft Singapore Pte Ltd	Singapore	*348*	75-03400
Vulcan Batteries Ltd	Israel	*223*	75-23290

633 50 10 Batteries, thermal, ground-based

Hawker Eternacell Ltd	UK	*457*	75-28500
Saft America Inc.	USA	*612*	75-77550
Saft Nife AB	Sweden	*378*	75-09850
Saft Nife Ltd	UK	*481*	75-51470
Saft Singapore Pte Ltd	Singapore	*348*	75-03400
Somchem	S. Africa	*359*	75-19200

633 60 Batteries, advanced technology

Chloride Industrial Batteries Ltd	UK	*457*	75-28490
Hawker Energy Products Ltd	UK	*457*	75-28490

Saft Nife Ltd	UK	*481*	75-51470
Steatite Insulations Ltd	UK	*486*	75-56450

634 Power sources, uninterruptible

Arvico Corporation	USA	*555*	75-09240
Automation NV	Belgium	*45*	75-00950
Belix Co. Ltd	UK	*434*	75-06190
CORIS	France	*134*	75-12450
Engineered Magnetics Inc.	USA	*573*	75-31568
Erskine Systems Ltd	UK	*449*	75-21470
EUROCONTROL SpA	Italy	*233*	75-08200
Fluor Daniel Inc.	USA	*635*	80-11720
GEC ALSTHOM T&D Power Electronic	UK	*453*	75-25260
Guardall Alarms New Zealand Ltd	New Zealand	*302*	75-00780
KB Electronics (1989) Ltd	Canada	*73*	75-08710
Luxtron Inc.	USA	*593*	75-56700
M Brodribb Pty Ltd	Australia	*13*	75-02760
M.J. Electronic Services	UK	*507*	80-16880
MPL Powerware Systems Ltd	UK	*469*	75-41260
Nova Electric	USA	*600*	75-65600
QES Ltd	UK	*509*	80-25000
Saft Nife Ltd	UK	*481*	75-51470
SELIN Sistemi Sarl	Italy	*238*	75-21750
Van Rietschoten & Houwens	Netherlands	*296*	75-07900
Wayne Kerr Electronics	UK	*492*	75-64860

635 Power converters

Abbott Electronics Inc.	USA	*545*	75-00150
Abbott Electronics Ltd	UK	*426*	75-00180
Advance Power	UK	*427*	75-00904
Airborne Power	USA	*548*	75-03100
Arvico Corporation	USA	*555*	75-09240
Automation NV	Belgium	*45*	75-00950
Babcock Inc.	USA	*556*	75-11600
Belix Co. Ltd	UK	*434*	75-06190
Bertan High Voltage	USA	*558*	75-13720
Canadian Marconi Co.	Canada	*69*	75-02600
Celab Ltd	UK	*439*	75-12250
Century Electronics	USA	*562*	75-17385
Circle-C Transformers	Australia	*14*	75-03450
Coutant-Lambda Ltd	UK	*443*	75-15390
Covergie SA	France	*134*	75-12400
Eldec Corp.	USA	*572*	75-29660
Elgar Corporation	USA	*573*	75-30720
Engineered Magnetics Inc.	USA	*573*	75-31568
Erskine Systems Ltd	UK	*449*	75-21470
EUROCONTROL SpA	Italy	*233*	75-08200
GEC ALSTHOM T&D Power Electronic	UK	*453*	75-25260
Grenson Electronics Ltd	UK	*456*	75-26950
ILC Data Device Corp.	USA	*584*	75-46050
Interpoint A Crane Company	USA	*585*	75-46870
Jones & Rickard Pty Ltd	Australia	*18*	75-10750
KB Electronics (1989) Ltd	Canada	*73*	75-08710
MicroSpire SA	France	*140*	75-26780
M.J. Electronic Services	UK	*507*	80-16880
Power Ten ICN	USA	*605*	75-71095
QES Ltd	UK	*509*	80-25000
Roband Electronics plc	UK	*479*	75-49880
Saft Nife Ltd	UK	*481*	75-51470
STC Keltec Corp.	USA	*588*	75-51950
Technology Dynamics Inc.	USA	*620*	75-86900
Telkoor Ltd	Israel	*223*	75-22400
The Titan Corp	USA	*622*	75-89620
Van Rietschoten & Houwens	Netherlands	*296*	75-07900
Walter Dittel GmbH	Germany	*167*	75-06510
Wayne Kerr Electronics	UK	*492*	75-64860

635 50 Power converters, nuclear thermionic

Belix Co. Ltd	UK	*434*	75-06190

Safety, Survival & Rescue

651 Safety equipment

Aircraft Materials Ltd	UK	*428*	75-01650
Altoflex	France	*132*	75-02300
Autoflug GmbH & Co	Germany	*165*	75-01400
BCB International Ltd	UK	*433*	75-06075
Beaufort Air-Sea Equipment Ltd	UK	*433*	75-06100
Bristol Uniforms Ltd	UK	*436*	75-08760
Bristow Helicopters Ltd	UK	*499*	80-03950
Capewell Aerial Delivery Systems	USA	*576*	75-36300
Checkmate Industries Ltd	Hong Kong SAR	*198*	75-00200
Checkmate Industries of EEC Ltd	Denmark	*102*	75-03300
Crestbury Ltd	UK	*443*	75-16030
Crewsaver	UK	*443*	75-16050
Dunlop Marine Safety Ltd	UK	*447*	75-19035
Elts Unlimited Inc.	USA	*573*	75-30900
Emergency Beacon Corp.	USA	*573*	75-31050
H. Koch & Sons Co.	USA	*589*	75-53030
Heckler & Koch Inc	USA	*581*	75-42860
Hutchwilco Limited	New Zealand	*302*	75-00860
Intamar Logistics Inc.	USA	*585*	75-47050
Interspiro AB	Sweden	*377*	75-06950
Interspiro Ltd	UK	*461*	75-31945
Irvin Aerospace Limited	UK	*462*	75-32050
Jetway Associates Ltd	UK	*462*	75-32520
Life Support International Inc.	USA	*590*	75-55163
Lite Flite Aps	Denmark	*103*	75-12000
Litton Industries Inc.	USA	*590*	75-55400
M. Paz Logistics Ltd	Israel	*224*	80-01600
Mullion Manufacturing Ltd	UK	*470*	75-41600
Mustang Survival (USA) Corp	USA	*597*	75-63480
National Nuclear Corporation	USA	*598*	75-63870
P.A.D. International	UK	*472*	75-43470
Pains-Wessex Ltd	UK	*472*	75-43510
Rabintex Industries Ltd	Israel	*221*	75-17850
Racal Health & Safety Group	UK	*477*	75-47675
Ruggieri Technologie	France	*142*	75-33500
Saunders-Roe Ltd	UK	*482*	75-51850
Siebe North Inc.	USA	*614*	75-80555
SNPE	France	*145*	75-36315
SpanSet Ltd	UK	*485*	75-55230
SPL Ordnance Ltd	Israel	*225*	80-01935
Strong Enterprises	USA	*618*	75-84500
Survival Products Inc.	USA	*618*	75-85200
Teufelberger GmbH	Austria	*34*	75-03150
UO Equipment Co.	USA	*625*	75-92870
Viking Life-Saving Equipment A/S	Denmark	*104*	75-22000
W.J. Clow (Midlands) Ltd	UK	*441*	75-13865
Zettler GmbH Kommunikation und	Germany	*177*	75-34930

652 Personal survival equipment

Autoflug GmbH & Co	Germany	*165*	75-01400
Avon Inflatables Ltd	UK	*432*	75-05100
BCB International Ltd	UK	*433*	75-06075
Beaufort Air-Sea Equipment Ltd	UK	*433*	75-06100
Becker Flugfunkwerk GmbH	Germany	*165*	75-02200
Capewell Aerial Delivery Systems	USA	*576*	75-36300
Crestbury Ltd	UK	*443*	75-16030
Crewsaver	UK	*443*	75-16050
Dunlop Marine Safety Ltd	UK	*447*	75-19035
EROS GIE	France	*137*	75-19500
FXC Corp.	USA	*577*	75-36700
Heckler & Koch Inc	USA	*581*	75-42860
Hutchwilco Limited	New Zealand	*302*	75-00860
Intamar Logistics Inc.	USA	*585*	75-47050
Interspiro AB	Sweden	*377*	75-06950
Irvin Aerospace Limited	UK	*462*	75-32050
Life Support International Inc.	USA	*590*	75-55163
Litton Industries Inc.	USA	*590*	75-55400
M. Paz Logistics Ltd	Israel	*224*	80-01600
Mullion Manufacturing Ltd	UK	*470*	75-41600
Mustang Survival Corp.	Canada	*74*	75-09860
Mustang Survival (USA) Corp	USA	*597*	75-63480
North American Aerodynamics Inc.	USA	*599*	75-65200
Norton Co.	USA	*600*	75-65510
Octavius Hunt Ltd	UK	*471*	75-42920
P.A.D. International	UK	*472*	75-43470
Pains-Wessex Ltd	UK	*472*	75-43510
Pointer Inc.	USA	*604*	75-70810
Pre-Mac (Kent) Ltd	UK	*476*	75-46660
Saunders-Roe Ltd	UK	*482*	75-51850
Survival Products Inc.	USA	*618*	75-85200
The North Face (Europe) Ltd	UK	*471*	75-42727
Touringsport (Rosker Ltd)	UK	*489*	75-60460
Viking Life-Saving Equipment A/S	Denmark	*104*	75-22000
Zodiac International	France	*149*	75-47310

652 05 Lifebuoys

Becker Flugfunkwerk GmbH	Germany	*165*	75-02200
Crewsaver	UK	*443*	75-16050
Eurovinil Industries SpA	Italy	*234*	75-08400
Frazier Aviation Inc.	USA	*576*	75-35970
Grabner GmbH	Austria	*33*	75-01100
Hutchwilco Limited	New Zealand	*302*	75-00860
Intamar Logistics Inc.	USA	*585*	75-47050
Life Support International Inc.	USA	*590*	75-55163
Merex Inc.	USA	*595*	75-60250
Viking Life-Saving Equipment A/S	Denmark	*104*	75-22000

652 10 Lifejackets

Air Cruisers Co	USA	*548*	75-03600
Autoflug GmbH & Co	Germany	*165*	75-01400
Avon Inflatables Ltd	UK	*432*	75-05100
Crewsaver	UK	*443*	75-16050
EROS GIE	France	*137*	75-19500
Eurovinil Industries SpA	Italy	*234*	75-08400
Frazier Aviation Inc.	USA	*576*	75-35970
Grabner GmbH	Austria	*33*	75-01100
H. Koch & Sons Co.	USA	*589*	75-53030
Hutchwilco Limited	New Zealand	*302*	75-00860
Intamar Logistics Inc.	USA	*585*	75-47050
Irvin Aerospace Canada Limited	Canada	*73*	75-08300
Irvin Aerospace Limited	UK	*462*	75-32050
Irvin Aerospace SpA	Italy	*235*	75-12200
Irvin Industries Inc.	USA	*586*	75-48960
Life Support International Inc.	USA	*590*	75-55163
Merex Inc.	USA	*595*	75-60250
MFC Survival Ltd	UK	*467*	75-38900
ML Lifeguard Equipment Ltd	UK	*469*	75-40230
Mullion Manufacturing Ltd	UK	*470*	75-41600
Mustang Survival Corp.	Canada	*74*	75-09860
Mustang Survival (USA) Corp	USA	*597*	75-63480
Musto Ltd	UK	*470*	75-41870
Pains-Wessex Ltd	UK	*472*	75-43510
Silent Partner Body Armor Inc.	USA	*615*	75-80930
Survival Products Inc.	USA	*618*	75-85200
Universal Safety Supply Pty Ltd	Australia	*23*	75-21750
Vacuum Reflex Ltd	UK	*490*	75-62600
Viking Life-Saving Equipment A/S	Denmark	*104*	75-22000
Wardle Storeys plc	UK	*492*	75-64450

652 14 Lifeboats

Balmoral Group Ltd	UK	*433*	75-05420
Cougar Limited	UK	*443*	75-15250
Emsworth Shipyard	UK	*448*	75-20750
Eurovinil Industries SpA	Italy	*234*	75-08400
Halmatic Ltd	UK	*456*	75-27700
Intamar Logistics Inc.	USA	*585*	75-47050
Life Support International Inc.	USA	*590*	75-55163
Merex Inc.	USA	*595*	75-60250
ML Lifeguard Equipment Ltd	UK	*469*	75-40230
Schat-Harding BV	Netherlands	*295*	75-06280
Sea Tiger Marine Pty Ltd	Australia	*21*	75-19350

652 15 Liferafts

Aérazur	France	*131*	75-00500
Air Cruisers Co	USA	*548*	75-03600
Autoflug GmbH & Co	Germany	*165*	75-01400
Avon Inflatables Ltd	UK	*432*	75-05100
Beaufort Air-Sea Equipment Ltd	UK	*433*	75-06100
BFGoodrich Aerospace	USA	*558*	75-13770
COES (Derby) Ltd	UK	*441*	75-14070
ContiTech Holding GmbH	Germany	*166*	75-05150
Dunlop Marine Safety Ltd	UK	*447*	75-19035
Eurovinil Industries SpA	Italy	*234*	75-08400
Frazier Aviation Inc.	USA	*576*	75-35970
H. Koch & Sons Co.	USA	*589*	75-53030
Intamar Logistics Inc.	USA	*585*	75-47050
Irvin Aerospace Limited	UK	*462*	75-32050
Irvin Aerospace SpA	Italy	*235*	75-12200
Irvin Industries Inc.	USA	*586*	75-48960
Life Support International Inc.	USA	*590*	75-55163
Merex Inc.	USA	*595*	75-60250
MFC Survival Ltd	UK	*467*	75-38900
Schat-Harding BV	Netherlands	*295*	75-06280
SMR Technologies Inc.	USA	*615*	75-82100
Survival Products Inc.	USA	*618*	75-85200
Switlik Parachute Co. Inc.	USA	*619*	75-85400
Viking Life-Saving Equipment A/S	Denmark	*104*	75-22000
Wardle Storeys plc	UK	*492*	75-64450
Zodiac International	France	*149*	75-47310

652 25 Parachutes, personnel

Aeraustral Trading	Australia	*11*	75-00330
Aérazur	France	*131*	75-00500
Frazier Aviation Inc.	USA	*576*	75-35970
FXC Corp.	USA	*577*	75-36700
Gelvenor Textiles	S. Africa	*358*	75-06100
Intamar Logistics Inc.	USA	*585*	75-47050
Irvin Aerospace Canada Limited	Canada	*73*	75-08300
Irvin Aerospace Limited	UK	*462*	75-32050
Irvin Aerospace SpA	Italy	*235*	75-12200
Irvin Industries Inc.	USA	*586*	75-48960
Life Support International Inc.	USA	*590*	75-55163
Lite Flite Aps	Denmark	*103*	75-12000
Merex Inc.	USA	*595*	75-60250
North American Aerodynamics Inc.	USA	*599*	75-65200
Nottingham International	USA	*640*	80-22000
Parachutes of Germany	Germany	*173*	75-23330
Pezetel	Poland	*325*	80-01000
Pioneer Aerospace Corp	USA	*604*	75-70200
Research Institute for	Ukraine	*406*	60-00990
Sächsische Spezialkonfektion GmbH	Germany	*174*	75-25880
Sala International Pty Ltd	Australia	*21*	75-18980
Strong Enterprises	USA	*618*	75-84500
Syndex Aerospace Inc.	USA	*619*	75-85540
Wardle Storeys plc	UK	*492*	75-64450

652 25 50 Parachutes, paratroop

Aérazur	France	*131*	75-00500
Aircraft Materials Ltd	UK	*428*	75-01650
Autoflug GmbH & Co	Germany	*165*	75-01400
Frazier Aviation Inc.	USA	*576*	75-35970
FXC Corp.	USA	*577*	75-36700
Intamar Logistics Inc.	USA	*585*	75-47050
Irvin Aerospace Canada Limited	Canada	*73*	75-08300
Irvin Aerospace Limited	UK	*462*	75-32050
Irvin Aerospace SpA	Italy	*235*	75-12200
Irvin Industries Inc.	USA	*586*	75-48960
Life Support International Inc.	USA	*590*	75-55163
Lite Flite Aps	Denmark	*103*	75-12000
Nottingham International	USA	*640*	80-22000
Pioneer Aerospace Corp	USA	*604*	75-70200
Sächsische Spezialkonfektion GmbH	Germany	*174*	75-25880
Syndex Aerospace Inc.	USA	*619*	75-85540
Wardle Storeys plc	UK	*492*	75-64450

652 25 60 Parachutes, sport

Aérazur	France	*131*	75-00500
Centro De Paracaidismo	Spain	*367*	80-00060
Gelvenor Textiles	S. Africa	*358*	75-06100
Irvin Aerospace SpA	Italy	*235*	75-12200
Irvin Industries Inc.	USA	*586*	75-48960
Life Support International Inc.	USA	*590*	75-55163
Pezetel	Poland	*325*	80-01000
Sächsische Spezialkonfektion GmbH	Germany	*174*	75-25880
Sala International Pty Ltd	Australia	*21*	75-18980
Strong Enterprises	USA	*618*	75-84500

652 25 70 Parachute systems, tactical gliding

Aérazur	France	*131*	75-00500
Irvin Aerospace Limited	UK	*462*	75-32050
Irvin Industries Inc.	USA	*586*	75-48960
Life Support International Inc.	USA	*590*	75-55163
Para-Flite Inc.	USA	*602*	75-68800
Pioneer Aerospace Corp	USA	*604*	75-70200
Research Institute for	Ukraine	*406*	60-00990
Sala International Pty Ltd	Australia	*21*	75-18980
Strong Enterprises	USA	*618*	75-84500

652 30 Survival clothing

Aérazur	France	131	75-00500
Autoflug GmbH & Co	Germany	165	75-01400
BCB International Ltd	UK	433	75-06075
Berghaus Ltd	UK	434	75-06525
Dunlop Marine Safety Ltd	UK	447	75-19035
Frazier Aviation Inc.	USA	576	75-35970
Gelvenor Textiles	S. Africa	358	75-06100
Hutchwilco Limited	New Zealand	302	75-00860
Intamar Logistics Inc.	USA	585	75-47050
Irvin Aerospace Canada Limited	Canada	73	75-08300
Life Support International Inc.	USA	590	75-55163
Merex Inc.	USA	595	75-60250
ML Lifeguard Equipment Ltd	UK	469	75-40230
Mustang Survival Corp.	Canada	74	75-09860
Mustang Survival (USA) Corp	USA	597	75-63480
Musto Ltd	UK	470	75-41870
Pains-Wessex Ltd	UK	472	75-43510
Silent Partner Body Armor Inc.	USA	615	75-80930
Survival Products Inc.	USA	618	75-85200
Trojan Manufacturing & Marketing Inc	Philippines	321	75-01350
Viking Life-Saving Equipment A/S	Denmark	104	75-22000

652 40 Survival bags

Autoflug GmbH & Co	Germany	165	75-01400
Frazier Aviation Inc.	USA	576	75-35970
H. Koch & Sons Co.	USA	589	75-53030
Intamar Logistics Inc.	USA	585	75-47050
Irvin Aerospace Canada Limited	Canada	73	75-08300
Life Support International Inc.	USA	590	75-55163
Manta Industries Ltd	Canada	73	75-09250
Merex Inc.	USA	595	75-60250
ML Lifeguard Equipment Ltd	UK	469	75-40230
PG Products Ltd.	UK	474	75-45065
Viking Life-Saving Equipment A/S	Denmark	104	75-22000

652 50 Survival packs / kits, personal

Air Cruisers Co	USA	548	75-03600
Autoflug GmbH & Co	Germany	165	75-01400
BCB International Ltd	UK	433	75-06075
Emergency Beacon Corp.	USA	573	75-31050
Evotec Ltd	UK	449	75-21970
H. Koch & Sons Co.	USA	589	75-53030
Hutchwilco Limited	New Zealand	302	75-00860
Intamar Logistics Inc.	USA	585	75-47050
Irvin Aerospace Canada Limited	Canada	73	75-08300
Life Support International Inc.	USA	590	75-55163
Manta Industries Ltd	Canada	73	75-09250
New Pac Safety AB	Sweden	378	75-08260
Ruggieri Technologie	France	142	75-33500
Saunders-Roe Ltd.	UK	482	75-51850
Survival Products Inc.	USA	618	75-85200
Trojan Manufacturing & Marketing Inc	Philippines	321	75-01350

652 50 30 First aid packs / kits, personal survival

Alfred Karcher GmbH & Co	Germany	170	75-15650
BCB International Ltd	UK	433	75-06075
Evotec Ltd	UK	449	75-21970
Ferno (UK) Ltd.	UK	450	75-22650
Giat Industries	France	126	55-00100
Hutchwilco Limited	New Zealand	302	75-00860
Life Support International Inc.	USA	590	75-55163
ML Lifeguard Equipment Ltd	UK	469	75-40230
Survival Products Inc.	USA	618	75-85200

652 50 40 Knives, survival (see code 734.45.60)

BCB International Ltd	UK	433	75-06075
Frazier Aviation Inc.	USA	576	75-35970
Life Support International Inc.	USA	590	75-55163
Nottingham International	USA	640	80-22000
Wilkinson Sword Ltd	UK	494	75-66250

652 50 60 Rations, emergency

BCB International Ltd	UK	433	75-06075
Compact A/S	Norway	309	75-00540
Evotec Ltd.	UK	449	75-21970
Hutchwilco Limited	New Zealand	302	75-00860
Life Support International Inc.	USA	590	75-55163
ML Lifeguard Equipment Ltd	UK	469	75-40230
Nottingham International	USA	640	80-22000
Nutrimental SA/Comexport	Brazil	57	75-14900

652 50 70 Shark repellant

BCB International Ltd	UK	433	75-06075
Life Support International Inc.	USA	590	75-55163
ML Lifeguard Equipment Ltd	UK	469	75-40230
Pains-Wessex Ltd	UK	472	75-43510

652 70 Rescue beacons, radio

AlliedSignal Aerospace Canada	Canada	68	75-00470
A.L.S.	UK	430	75-02770
Autoflug GmbH & Co	Germany	165	75-01400
Aviation Support Tech. Int'l Inc.	USA	556	75-11050
BCB International Ltd	UK	433	75-06075
Becker Flugfunkwerk GmbH	Germany	165	75-02200
Elts Unlimited Inc.	USA	573	75-30900
Emergency Beacon Corp.	USA	573	75-31050
Glatzer Industries Corp.	USA	579	75-39760
Hutchwilco Limited	New Zealand	302	75-00860
Intamar Logistics Inc.	USA	585	75-47050
Jotron Electronics A/S	Norway	309	75-01030
Jotron (UK) Ltd	UK	462	75-32895
Life Support International Inc.	USA	590	75-55163
Litton Industries Inc.	USA	590	75-55400
M. Paz Logistics Ltd	Israel	224	80-01600
McMurdo Ltd	UK	467	75-38170
Moonraker Australia Pty Ltd	Australia	19	75-13770
Philips Traffic & Engineering	Australia	20	75-16550
Siemens Plessey Electronic Systems	Australia	21	75-19475
Signature Industries Ltd	UK	484	75-53785
Techtest Ltd.	UK	484	75-54590
Viking Life-Saving Equipment A/S	Denmark	104	75-22000

652 90 Water desalination / purification kits, personal survival (see code 622.20.80)

Åkers Krutbruk Protection AB	Sweden	375	75-01100
BCB International Ltd	UK	433	75-06075
Bend Research	USA	540	60-02500
Life Support International Inc.	USA	590	75-55163
Pre-Mac (Kent) Ltd	UK	476	75-46660
Schering-Plough Ltd	UK	482	75-51990
Stella-Meta	UK	486	75-56780
Survival Products Inc.	USA	618	75-85200

652 90 60 Solar stills, water purification

BCB International Ltd	UK	433	75-06075
Eurovinil Industries SpA	Italy	234	75-08400
Life Support International Inc.	USA	590	75-55163

653 Diving systems & equipment

A.L.S.	UK	430	75-02770
Beaufort Air-Sea Equipment Ltd	UK	433	75-06100
Comex Pro	France	134	75-10200
Crestbury Ltd	UK	443	75-16030
EROS GIE	France	137	75-19500
Haux-Life-Support GmbH	Germany	169	75-11970
Hytech BV	Netherlands	293	75-03530
Interspiro NL.	Netherlands	293	75-03570
Nautronix Ltd	UK	470	75-41900
Perry Technologies	USA	603	75-69360
SAS Limited	UK	482	75-51830
Zodiac International	France	149	75-47310

653 30 Diving systems, shipborne

Broco Inc	USA	560	75-15260
Comex Pro	France	134	75-10200
Divex Ltd.	UK	445	75-18220

Haux-Life-Support GmbH	Germany	*169*	75-11970
Hytech BV	Netherlands	*293*	75-03530
Jotron (UK) Ltd	UK	*462*	75-32895
Merex Inc.	USA	*595*	75-60250
SAS Limited	UK	*482*	75-51830

653 30 20 Decompression chambers, shipborne

Atomic Energy Corp	S. Africa	*357*	75-01000
Comex Pro.	France	*134*	75-10200
Divex Ltd.	UK	*445*	75-18220
Fullerton, Sherwood Engineering Ltd	Canada	*71*	75-05625
Haux-Life-Support GmbH	Germany	*169*	75-11970
Hytech BV	Netherlands	*293*	75-03530
SAS Limited	UK	*482*	75-51830

653 30 55 Recompression chambers, shipborne

Comex Pro.	France	*134*	75-10200
Divex Ltd.	UK	*445*	75-18220
Fullerton, Sherwood Engineering Ltd	Canada	*71*	75-05625
Haux-Life-Support GmbH	Germany	*169*	75-11970
Hytech BV	Netherlands	*293*	75-03530
SAS Limited	UK	*482*	75-51830

653 40 Diving systems, land-based

Comex Pro.	France	*134*	75-10200
Divex Ltd.	UK	*445*	75-18220
Haux-Life-Support GmbH	Germany	*169*	75-11970
Hytech BV	Netherlands	*293*	75-03530
Interspiro Ltd	UK	*461*	75-31945
Jotron (UK) Ltd	UK	*462*	75-32895

653 40 20 Decompression chambers, land-based

Comex Pro.	France	*134*	75-10200
Divex Ltd.	UK	*445*	75-18220
Fullerton, Sherwood Engineering Ltd	Canada	*71*	75-05625
Haux-Life-Support GmbH	Germany	*169*	75-11970
Hytech BV	Netherlands	*293*	75-03530
SAS Limited	UK	*482*	75-51830

653 40 55 Recompression chambers, land-based

Atomic Energy Corp	S. Africa	*357*	75-01000
Comex Pro.	France	*134*	75-10200
Divex Ltd.	UK	*445*	75-18220
Fullerton, Sherwood Engineering Ltd	Canada	*71*	75-05625
Haux-Life-Support GmbH	Germany	*169*	75-11970
Hytech BV	Netherlands	*293*	75-03530
SAS Limited	UK	*482*	75-51830

653 40 60 Diving systems, training, land-based

Comex Pro.	France	*134*	75-10200
Fullerton, Sherwood Engineering Ltd	Canada	*71*	75-05625
Haux-Life-Support GmbH	Germany	*169*	75-11970
Hytech BV	Netherlands	*293*	75-03530
SAS Limited	UK	*482*	75-51830

653 70 Diving equipment

A.L.S.	UK	*430*	75-02770
Broco Inc	USA	*560*	75-15260
Comex Pro.	France	*134*	75-10200
Divex Ltd.	UK	*445*	75-18220
EROS GIE	France	*137*	75-19500
Haux-Life-Support GmbH	Germany	*169*	75-11970
Hytech BV	Netherlands	*293*	75-03530
Interspiro	USA	*585*	75-48680
Interspiro Ltd	UK	*461*	75-31945
Interspiro NL.	Netherlands	*293*	75-03570
JDR Cable Systems Ltd	UK	*462*	75-32480
MacArtney A/S.	Denmark	*103*	75-12800
Merex Inc.	USA	*595*	75-60250
MSI-Defence Systems Ltd	UK	*469*	75-41280
Nautronix Ltd	UK	*470*	75-41900
Pelican Products Inc.	USA	*603*	75-69125
SAS Limited	UK	*482*	75-51830

653 70 60 Scuba diving equipment, closed circuit

Comex Pro.	France	*134*	75-10200
Divex Ltd.	UK	*445*	75-18220
Fullerton, Sherwood Engineering Ltd	Canada	*71*	75-05625
Haux-Life-Support GmbH	Germany	*169*	75-11970
Hytech BV	Netherlands	*293*	75-03530
Interspiro	USA	*585*	75-48680
Merex Inc.	USA	*595*	75-60250

653 80 Ancillary equipment, diving systems

A.L.S.	UK	*430*	75-02770
Comex Pro.	France	*134*	75-10200
Divex Ltd.	UK	*445*	75-18220
EROS GIE	France	*137*	75-19500
Fullerton, Sherwood Engineering Ltd	Canada	*71*	75-05625
Haux-Life-Support GmbH	Germany	*169*	75-11970
Nautronix Ltd	UK	*470*	75-41900
SAS Limited	UK	*482*	75-51830

653 80 30 Breathing air compressors

Comex Pro.	France	*134*	75-10200
Crestbury Ltd	UK	*443*	75-16030
Divex Ltd.	UK	*445*	75-18220
EROS GIE	France	*137*	75-19500
Giat Industries	France	*126*	55-00100
Haux-Life-Support GmbH	Germany	*169*	75-11970
SAS Limited	UK	*482*	75-51830
Spar Aviation Services	Canada	*76*	75-13020

653 80 40 Helium recovery systems

Haux-Life-Support GmbH	Germany	*169*	75-11970

654 Rescue equipment

A Haberkorn GmbH & Co KG	Austria	*33*	75-01140
Aeraustral Trading	Australia	*11*	75-00330
Aircraft Materials Ltd	UK	*428*	75-01650
A.L.S.	UK	*430*	75-02770
Autoflug GmbH & Co	Germany	*165*	75-01400
Barrow Hepburn Sala Ltd.	UK	*433*	75-05830
Bristol Uniforms Ltd	UK	*436*	75-08760
Bristow Helicopters Ltd	UK	*499*	80-03950
Clark Masts Teksam Ltd	UK	*440*	75-13550
Cubic Defense Systems Inc.	USA	*567*	75-22900
Eyal Engineering & Industrial Co.	Israel	*219*	75-07200
GAAM Fire & Rescue	Australia	*16*	75-07000
Holmatro Inc.	USA	*636*	80-14340
Holmatro Industrial and Rescue	Netherlands	*293*	75-03350
Interspiro AB	Sweden	*377*	75-06950
Irvin Industries Inc.	USA	*586*	75-48960
Jason's Cradle	UK	*462*	75-32296
Life Support International Inc.	USA	*590*	75-55163
Pains-Wessex Ltd	UK	*472*	75-43510
Phillips & Smith Ltd	New Zealand	*303*	75-01650
Plumett Ltd	UK	*475*	75-46340
SPL Ordnance Ltd.	Israel	*225*	80-01935
Teufelberger GmbH.	Austria	*34*	75-03150
Wardle Storeys plc	UK	*492*	75-64450

654 05 Rescue bags

BCB International Ltd	UK	*433*	75-06075
GAAM Fire & Rescue	Australia	*16*	75-07000
Germa AB	Sweden	*377*	75-05700
Holmatro Inc.	USA	*636*	80-14340
Life Support International Inc.	USA	*590*	75-55163
SPL Ordnance Ltd.	Israel	*225*	80-01935
Woodville Environmental & Safety	UK	*494*	75-66900

654 10 Rescue floodlights

Civil Defence Supply	UK	440	75-13350
Clark Masts Teksam Ltd	UK	440	75-13550
Havis-Shields Equipment Corp.	USA	581	75-42700
Henry Cooch & Son Ltd	UK	442	75-14900
Life Support International Inc.	USA	590	75-55163
Nitech Ltd	UK	471	75-42570
Spectrolab Inc.	USA	617	75-83100
SPL Ordnance Ltd.	Israel	225	80-01935
Thorn DNT Airfield Lighting Pty Ltd	Australia	22	75-20540

654 20 Rescue harnesses, helicopter

Aircraft Materials Ltd	UK	428	75-01650
A.L.S.	UK	430	75-02770
Barrow Hepburn Sala Ltd.	UK	433	75-05830
BCB International Ltd	UK	433	75-06075
Bristow Helicopters Ltd	UK	499	80-03950
H. Fine & Son Ltd	UK	451	75-23210
Irvin Aerospace Canada Limited	Canada	73	75-08300
Life Support International Inc.	USA	590	75-55163
Sala International Pty Ltd.	Australia	21	75-18980
SpanSet Inter AG	Switzerland	387	75-13550
SPL Ordnance Ltd.	Israel	225	80-01935

654 30 Rescue lifeline, sea

Life Support International Inc.	USA	590	75-55163
Pains-Wessex Ltd	UK	472	75-43510

654 40 Rescue lifeline thrower

Life Support International Inc.	USA	590	75-55163
Pains-Wessex Ltd	UK	472	75-43510
Plumett Ltd	UK	475	75-46340

654 50 Rescue packs, air-sea

Autoflug GmbH & Co	Germany	165	75-01400
Intamar Logistics Inc.	USA	585	75-47050
Irvin Aerospace Canada Limited	Canada	73	75-08300
Irvin Aerospace Limited	UK	462	75-32050
Irvin Industries Inc.	USA	586	75-48960
Life Support International Inc.	USA	590	75-55163
Sala International Pty Ltd.	Australia	21	75-18980

654 60 Rescue stretchers

Aircraft Materials Ltd	UK	428	75-01650
Barrow Hepburn Sala Ltd.	UK	433	75-05830
BCB International Ltd	UK	433	75-06075
Ferno (UK) Ltd.	UK	450	75-22650
H. Fine & Son Ltd	UK	451	75-23210
Life Support International Inc.	USA	590	75-55163
ROFi Industrier AS	Norway	310	75-02300
Sala International Pty Ltd.	Australia	21	75-18980
SPL Ordnance Ltd.	Israel	225	80-01935
Switlik Parachute Co. Inc.	USA	619	75-85400

Salvage equipment

660 50 Lifting equipment, underwater

A1Marketing	UK	426	75-00025
Comex Pro.	France	134	75-10200
Crestbury Ltd	UK	443	75-16030
Divex Ltd.	UK	445	75-18220
MacArtney A/S	Denmark	103	75-12800
Schat-Harding BV	Netherlands	295	75-06280
SpanSet Inter AG	Switzerland	387	75-13550

General Equipment

Personal Protection

711 Overall protection (suits)

Åkers Krutbruk Protection AB	Sweden	375	75-01100
Alfred Karcher GmbH & Co	Germany	170	75-15650
Autoflug GmbH & Co	Germany	165	75-01400
Avon Technical Products	UK	432	75-05150
Bachmann SA	France	132	75-04900
Barrday Inc.	Canada	68	75-01080
Beaufort Air-Sea Equipment Ltd	UK	433	75-06100
Bondina NBC Materials	UK	435	75-07600
Calspan SRL Technology	USA	540	60-03000
Clansman Armour Systems	UK	440	75-13460
Compton Webb Ltd	UK	442	75-14550
EROS GIE	France	137	75-19500
Excalor Hitzeschutz GmbH	Germany	168	75-08750
Far West Industries Inc.	Canada	71	75-05270
Germa AB	Sweden	377	75-05700
H.R. Denne Ltd	UK	445	75-17550
ICL Technical Plastics Ltd	UK	460	75-31060
IFI Srl	Italy	235	75-11220
Meggitt Composites	UK	467	75-38345
MSA (Britain) Limited	UK	469	75-41270
Mustang Survival Corp.	Canada	74	75-09860
Mustang Survival (USA) Corp	USA	597	75-63480
NORINCO	China PR	88	55-00280
Odenwald Werke Rittersbach	Germany	173	75-22850
P.A.D. International	UK	472	75-43470
Pains-Wessex Ltd	UK	472	75-43510
Paul Boyé	France	133	75-07020
Rabintex Industries Ltd	Israel	221	75-17850
Real Santa Rita Equipamentos de	Brazil	57	75-18300
Sèkur SpA	Italy	238	75-21650
Siebe North Inc.	USA	614	75-80555
Silent Partner Body Armor Inc.	USA	615	75-80930
SPL Ordnance Ltd.	Israel	225	80-01935
Sutcliffe Speakman Carbons Ltd	UK	487	75-58000
Switlik Parachute Co. Inc.	USA	619	75-85400
Tacconi SpA	Italy	239	75-25000
Vandeputte NV	Belgium	47	75-06375
Vidaro Corp.	USA	626	75-94000
Viking Life-Saving Equipment A/S	Denmark	104	75-22000

711 10 Suits, protective, explosive ordnance disposal

Åkers Krutbruk Protection AB	Sweden	375	75-01100
Barrday Inc.	Canada	68	75-01080
BCB International Ltd	UK	433	75-06075
Beaufort Air-Sea Equipment Ltd.	UK	433	75-06100
DTC International	USA	570	75-27620
Interspiro	USA	585	75-48680
Lightweight Body Armour	UK	464	75-35650
Mechem	S. Africa	358	75-11000
Meggitt Composites	UK	467	75-38345
ML Lifeguard Equipment Ltd	UK	469	75-40230
Second Chance Body Armor Inc.	USA	613	75-79000
Silent Partner Body Armor Inc.	USA	615	75-80930
SPL Ordnance Ltd.	Israel	225	80-01935
Sutcliffe Speakman Carbons Ltd	UK	487	75-58000
Tacconi SpA	Italy	239	75-25000

711 15 Suits, protective, anti-microwave

Sutcliffe Speakman Carbons Ltd	UK	487	75-58000

711 20 Suits, protective, aircrew

Aérazur	France	131	75-00500
Autoflug GmbH & Co	Germany	165	75-01400
Beaufort Air-Sea Equipment Ltd	UK	433	75-06100
Compton Webb Ltd	UK	442	75-14550
EROS GIE	France	137	75-19500
Flight Suits Ltd	USA	575	75-34800
Germa AB	Sweden	377	75-05700
Irvin Aerospace Canada Limited	Canada	73	75-08300

Irvin Industries Inc.	USA	586	75-48960
ML Lifeguard Equipment Ltd	UK	469	75-40230
Mustang Survival Corp.	Canada	74	75-09860
Mustang Survival (USA) Corp	USA	597	75-63480
Pains-Wessex Ltd	UK	472	75-43510
Paul Boyé	France	133	75-07020
Sutcliffe Speakman Carbons Ltd	UK	487	75-58000
Viking Life-Saving Equipment A/S	Denmark	104	75-22000

711 25 Suits, protective, electro-magnetic radiation

Sutcliffe Speakman Carbons Ltd	UK	487	75-58000

711 30 Suits, protective, NBC

Aérazur	France	131	75-00500
Alfred Karcher GmbH & Co	Germany	170	75-15650
American Cleaning Manufacturing Co.	USA	553	75-05940
Autoflug GmbH & Co	Germany	165	75-01400
Avon Technical Products	UK	432	75-05150
Beaufort Air-Sea Equipment Ltd	UK	433	75-06100
Bondina NBC Materials	UK	435	75-07600
Calspan SRL Technology	USA	540	60-03000
Charcoal Cloth (International) Ltd	UK	439	75-12550
Compton Webb Ltd	UK	442	75-14550
CQC Plc	UK	443	75-15700
EROS GIE	France	137	75-19500
Excalor Hitzeschutz GmbH	Germany	168	75-08750
Germa AB	Sweden	377	75-05700
IFI Srl	Italy	235	75-11220
Interspiro	USA	585	75-48680
Interspiro Ltd	UK	461	75-31945
Intertechnique SA	France	137	75-23000
Irvin Aerospace Limited	UK	462	75-32050
Irvin Aerospace SpA	Italy	235	75-12200
J Blaschke Wehrtechnik GmbH	Austria	32	75-00560
James North & Sons Ltd	UK	471	75-42730
Kinetics Ltd	Israel	220	75-11320
Lantor (UK) Ltd	UK	464	75-34665
ML Lifeguard Equipment Ltd	UK	469	75-40230
New Pac Safety AB	Sweden	378	75-08260
Nottingham International	USA	640	80-22000
Odenwald Werke Rittersbach	Germany	173	75-22850
Paul Boyé	France	133	75-07020
Sèkur SpA	Italy	238	75-21650
Silent Partner Body Armor Inc.	USA	615	75-80930
S.M. Alexander (Plastics) Ltd	UK	429	75-02580
Sutcliffe Speakman Carbons Ltd	UK	487	75-58000
Tacconi SpA	Italy	239	75-25000
White Cross Rubber Products Ltd	UK	494	75-65857
Woodville Environmental & Safety	UK	494	75-66900

711 40 Suits, flame-resistant

Albion Clothing Ltd	New Zealand	301	75-00110
Avirex Ltd	USA	556	75-11250
BCB International Ltd	UK	433	75-06075
Beaufort Air-Sea Equipment Ltd	UK	433	75-06100
Bristol Uniforms Ltd	UK	436	75-08760
Bruck Textiles plc	Australia	13	75-02810
Excalor Hitzeschutz GmbH	Germany	168	75-08750
Flight Suits Ltd	USA	575	75-34800
GD Protective Clothing and Equipment	Ireland	214	75-00120
Germa AB	Sweden	377	75-05700
ICL Technical Plastics Ltd	UK	460	75-31060
Intamar Logistics Inc.	USA	585	75-47050
Interspiro	USA	585	75-48680
Interspiro Ltd	UK	461	75-31945
MSA (Britain) Limited	UK	469	75-41270
Mustang Survival Corp.	Canada	74	75-09860
Mustang Survival (USA) Corp	USA	597	75-63480
Pains-Wessex Ltd	UK	472	75-43510
Paul Boyé	France	133	75-07020
PG Products Ltd	UK	474	75-45065
Phillips & Smith Ltd	New Zealand	303	75-01650
Real Santa Rita Equipamentos de	Brazil	57	75-18300
Silent Partner Body Armor Inc.	USA	615	75-80930
SPL Ordnance Ltd	Israel	225	80-01935
Sutcliffe Speakman Carbons Ltd	UK	487	75-58000
Switlik Parachute Co. Inc.	USA	619	75-85400
Tacconi SpA	Italy	239	75-25000
Universal Safety Supply Pty Ltd	Australia	23	75-21750
Vacuum Reflex Ltd	UK	490	75-62600
Vandeputte NV	Belgium	47	75-06375
Vidaro Corp.	USA	626	75-94000
Viking Life-Saving Equipment A/S	Denmark	104	75-22000

711 50 Suits, protective, diving

Avon Technical Products	UK	432	75-05150
Divex Ltd.	UK	445	75-18220
Interspiro Ltd	UK	461	75-31945
Sutcliffe Speakman Carbons Ltd	UK	487	75-58000

711 51 Suits, diving, atmospheric

Plastechnol Ltd	UK	508	80-22500

711 52 Suits, diving, dry

Avon Technical Products	UK	432	75-05150
Divex Ltd.	UK	445	75-18220
Interspiro Ltd	UK	461	75-31945
White Cross Rubber Products Ltd.	UK	494	75-65857

711 53 Suits, diving, wet

Divex Ltd.	UK	445	75-18220

711 60 Suits, space

Aérazur	France	131	75-00500
David Clark Co. Inc.	USA	564	75-18850
EROS GIE	France	137	75-19500

711 70 Suits, temperature controlled

Carleton Life Support Technologies	Canada	70	75-02950
Germa AB	Sweden	377	75-05700
Interspiro Ltd	UK	461	75-31945
Survival Products Inc.	USA	618	75-85200
Vacuum Reflex Ltd	UK	490	75-62600

711 75 Suits, cooling

Carleton Life Support Technologies	Canada	70	75-02950

711 80 Suits, flotation

Beaufort Air-Sea Equipment Ltd	UK	433	75-06100
EROS GIE	France	137	75-19500
Germa AB	Sweden	377	75-05700
Irvin Aerospace Canada Limited	Canada	73	75-08300
Mullion Manufacturing Ltd	UK	470	75-41600
Mustang Survival Corp.	Canada	74	75-09860
Mustang Survival (USA) Corp	USA	597	75-63480
Silent Partner Body Armor Inc.	USA	615	75-80930
Survival Products Inc.	USA	618	75-85200
Tacconi SpA	Italy	239	75-25000
Viking Life-Saving Equipment A/S	Denmark	104	75-22000

711 90 Suits, storm-proof

Beaufort Air-Sea Equipment Ltd	UK	433	75-06100
Berghaus Ltd	UK	434	75-06525
Compton Webb Ltd	UK	442	75-14550
Mullion Manufacturing Ltd	UK	470	75-41600
Mustang Survival (USA) Corp	USA	597	75-63480
Musto Ltd	UK	470	75-41870
Pains-Wessex Ltd	UK	472	75-43510
Vacuum Reflex Ltd	UK	490	75-62600

711 95 Suits, cold weather

Aérazur	France	131	75-00500
Beaufort Air-Sea Equipment Ltd	UK	433	75-06100
Berghaus Ltd	UK	434	75-06525

Compton Webb Ltd	UK	*442*	75-14550
Germa AB	Sweden	*377*	75-05700
Irvin Aerospace Canada Limited	Canada	*73*	75-08300
Mustang Survival Corp.	Canada	*74*	75-09860
Mustang Survival (USA) Corp	USA	*597*	75-63480
Musto Ltd	UK	*470*	75-41870
Pains-Wessex Ltd	UK	*472*	75-43510
Paul Boyé	France	*133*	75-07020
Real Santa Rita Equipamentos de	Brazil	*57*	75-18300
Survival Products Inc.	USA	*618*	75-85200
Switlik Parachute Co. Inc.	USA	*619*	75-85400
Vacuum Reflex Ltd	UK	*490*	75-62600
Viking Life-Saving Equipment A/S	Denmark	*104*	75-22000

711 95 10 Suits, anti-hypothermia

Beaufort Air-Sea Equipment Ltd	UK	*433*	75-06100
Irvin Aerospace Canada Limited	Canada	*73*	75-08300
Irvin Aerospace SpA	Italy	*235*	75-12200
Mullion Manufacturing Ltd	UK	*470*	75-41600
Mustang Survival Corp.	Canada	*74*	75-09860
Mustang Survival (USA) Corp	USA	*597*	75-63480
Pains-Wessex Ltd	UK	*472*	75-43510
Survival Products Inc.	USA	*618*	75-85200
Switlik Parachute Co. Inc.	USA	*619*	75-85400
Vacuum Reflex Ltd	UK	*490*	75-62600

712 Head protection

Armourshield Limited	UK	*431*	75-03775
Autoflug GmbH & Co	Germany	*165*	75-01400
Barrday Inc.	Canada	*68*	75-01080
CGF Gallet SA	France	*133*	75-08840
Checkmate Industries of EEC Ltd	Denmark	*102*	75-03300
Colan Products	Australia	*14*	75-03650
Compton Webb Ltd	UK	*442*	75-14550
Courtaulds Aerospace	UK	*443*	75-15330
Davies Industrial Communications Ltd	UK	*444*	75-17010
DSM	Netherlands	*292*	75-01730
Excalor Hitzeschutz GmbH	Germany	*168*	75-08750
GAAM Fire & Rescue	Australia	*16*	75-07000
Helmet Integrated Systems Ltd	UK	*458*	75-29000
Hiatt & Company Ltd	UK	*458*	75-29180
ICL Technical Plastics Ltd	UK	*460*	75-31060
IFI Srl	Italy	*235*	75-11220
Indústrias de Defesa	Portugal	*327*	55-00950
M. Kumar Udyog	India	*205*	75-02100
MSA (Britain) Limited	UK	*469*	75-41270
Pacific Helmets (NZ) Ltd	New Zealand	*302*	75-01620
P.A.D. International	UK	*472*	75-43470
Pains-Wessex Ltd	UK	*472*	75-43510
Pittards plc	UK	*475*	75-45700
Protector Technologies Ltd	UK	*476*	75-46890
Rabintex Industries Ltd	Israel	*221*	75-17850
RBR Armour Ltd.	UK	*478*	75-48370
Real Santa Rita Equipamentos de	Brazil	*57*	75-18300
Schuberth Helme GmbH	Germany	*174*	75-26800
Silent Partner Body Armor Inc.	USA	*615*	75-80930
South Central Ordnance Co.	USA	*616*	75-82400
Specialty Plastic Products Inc.	USA	*616*	75-83035
SPL Ordnance Ltd.	Israel	*225*	80-01935
Universal Safety Supply Pty Ltd	Australia	*23*	75-21750
Vidaro Corp.	USA	*626*	75-94000

712 10 Helmets, protective

Åkers Krutbruk Protection AB	Sweden	*375*	75-01100
Autoflug GmbH & Co	Germany	*165*	75-01400
Barrday Inc.	Canada	*68*	75-01080
Browning SA	Belgium	*45*	75-01950
Checkmate Industries Ltd	Hong Kong SAR	*198*	75-00200
Checkmate Industries of EEC Ltd	Denmark	*102*	75-03300
Colan Products	Australia	*14*	75-03650
Courtaulds Aerospace	UK	*443*	75-15330
David Clark Co. Inc.	USA	*564*	75-18850
Davies Industrial Communications Ltd	UK	*444*	75-17010
DSM	Netherlands	*292*	75-01730
Far East Industrial Supply Co Inc.	Philippines	*320*	75-00800
Forjas Taurus SA	Brazil	*56*	75-06640
GAAM Fire & Rescue	Australia	*16*	75-07000
Helmet Integrated Systems Ltd	UK	*458*	75-29000
Hodge Clemco Ltd	UK	*458*	75-29590
H.P.P. (UK) Ltd	UK	*459*	75-30360
ICL Technical Plastics Ltd	UK	*460*	75-31060
James North & Sons Ltd	UK	*471*	75-42730
Lightweight Body Armour	UK	*464*	75-35650
M. Kumar Udyog	India	*205*	75-02100
MSA (Britain) Limited	UK	*469*	75-41270
Nottingham International	USA	*640*	80-22000
Pacific Helmets (NZ) Ltd	New Zealand	*302*	75-01620
Pains-Wessex Ltd	UK	*472*	75-43510
Real Santa Rita Equipamentos de	Brazil	*57*	75-18300
Schuberth Helme GmbH	Germany	*174*	75-26800
Silent Partner Body Armor Inc.	USA	*615*	75-80930
Specialty Plastic Products Inc.	USA	*616*	75-83035
SPL Ordnance Ltd.	Israel	*225*	80-01935
St Bernard Composites Ltd	UK	*486*	75-56222
Tacconi SpA	Italy	*239*	75-25000
Trojan Manufacturing & Marketing Inc	Philippines	*321*	75-01350
Universal Safety Supply Pty Ltd	Australia	*23*	75-21750

712 10 10 Helmets, aircrew

Autoflug GmbH & Co	Germany	*165*	75-01400
A.V. Electronics Ltd	UK	*448*	75-20200
Avirex Ltd	USA	*556*	75-11250
Barrday Inc.	Canada	*68*	75-01080
CGF Gallet SA	France	*133*	75-08840
Colan Products	Australia	*14*	75-03650
Courtaulds Aerospace	UK	*443*	75-15330
David Clark Co. Inc.	USA	*564*	75-18850
Davies Industrial Communications Ltd	UK	*444*	75-17010
EROS GIE	France	*137*	75-19500
Flight Suits Ltd	USA	*575*	75-34800
Gentex Corp.	USA	*578*	75-39300
Haynes & Cann Ltd	UK	*457*	75-28750
Helmet Integrated Systems Ltd	UK	*458*	75-29000
Intamar Logistics Inc.	USA	*585*	75-47050
Kaiser Electronics	USA	*587*	75-51350
Pacific Helmets (NZ) Ltd	New Zealand	*302*	75-01620
Scott Aviation	USA	*613*	75-78850
SPL Ordnance Ltd.	Israel	*225*	80-01935

712 10 20 Helmets, anti-riot

Barrday Inc.	Canada	*68*	75-01080
CGF Gallet SA	France	*133*	75-08840
Checkmate Industries Ltd	Hong Kong SAR	*198*	75-00200
Checkmate Industries of EEC Ltd	Denmark	*102*	75-03300
Colan Products	Australia	*14*	75-03650
Davies Industrial Communications Ltd	UK	*444*	75-17010
Defense Technology Corporation of	USA	*568*	75-24980
Forjas Taurus SA	Brazil	*56*	75-06640
Helmet Integrated Systems Ltd	UK	*458*	75-29000
Hiatt & Company Ltd	UK	*458*	75-29180
ICL Technical Plastics Ltd	UK	*460*	75-31060
Lightweight Body Armour	UK	*464*	75-35650
Nottingham International	USA	*640*	80-22000
Pacific Helmets (NZ) Ltd	New Zealand	*302*	75-01620
Pains-Wessex Ltd	UK	*472*	75-43510
PG Products Ltd.	UK	*474*	75-45065
Schuberth Helme GmbH	Germany	*174*	75-26800
Silent Partner Body Armor Inc.	USA	*615*	75-80930
South Central Ordnance Co.	USA	*616*	75-82400
SPL Ordnance Ltd.	Israel	*225*	80-01935

712 10 30 Helmets, crash tender crew

Barrday Inc.	Canada	*68*	75-01080
CGF Gallet SA	France	*133*	75-08840
Colan Products	Australia	*14*	75-03650
Davies Industrial Communications Ltd	UK	*444*	75-17010
Excalor Hitzeschutz GmbH	Germany	*168*	75-08750
GAAM Fire & Rescue	Australia	*16*	75-07000
Helmet Integrated Systems Ltd	UK	*458*	75-29000
Nottingham International	USA	*640*	80-22000
Pacific Helmets (NZ) Ltd	New Zealand	*302*	75-01620
Schuberth Helme GmbH	Germany	*174*	75-26800
SPL Ordnance Ltd.	Israel	*225*	80-01935

712 10 40 Helmets, explosive ordnance disposal

Åkers Krutbruk Protection AB	Sweden	375	75-01100
CGF Gallet SA	France	133	75-08840
Colan Products	Australia	14	75-03650
Davies Industrial Communications Ltd	UK	444	75-17010
Lightweight Body Armour	UK	464	75-35650
Pacific Helmets (NZ) Ltd	New Zealand	302	75-01620
Silent Partner Body Armor Inc.	USA	615	75-80930
South Central Ordnance Co.	USA	616	75-82400
SPL Ordnance Ltd.	Israel	225	80-01935

712 10 50 Helmets, fire protection

Bristol Uniforms Ltd	UK	436	75-08760
CGF Gallet SA	France	133	75-08840
Colan Products	Australia	14	75-03650
Davies Industrial Communications Ltd	UK	444	75-17010
Excalor Hitzeschutz GmbH	Germany	168	75-08750
GAAM Fire & Rescue	Australia	16	75-07000
GD Protective Clothing and Equipment	Ireland	214	75-00120
Helmet Integrated Systems Ltd	UK	458	75-29000
MSA (Britain) Limited	UK	469	75-41270
Pacific Helmets (NZ) Ltd	New Zealand	302	75-01620
Phillips & Smith Ltd	New Zealand	303	75-01650
Schuberth Helme GmbH	Germany	174	75-26800
SPL Ordnance Ltd.	Israel	225	80-01935

712 10 60 Helmets, military

Åkers Krutbruk Protection AB	Sweden	375	75-01100
Armourshield Limited	UK	431	75-03775
Barrday Inc.	Canada	68	75-01080
CGF Gallet SA	France	133	75-08840
Checkmate Industries Ltd	Hong Kong SAR	198	75-00200
Checkmate Industries of EEC Ltd	Denmark	102	75-03300
Colan Products	Australia	14	75-03650
Courtaulds Aerospace	UK	443	75-15330
Davies Industrial Communications Ltd	UK	444	75-17010
DSM	Netherlands	292	75-01730
Helmet Integrated Systems Ltd	UK	458	75-29000
Indústrias de Defesa	Portugal	327	55-00950
Lightweight Body Armour	UK	464	75-35650
M. Kumar Udyog	India	205	75-02100
NII Stali	Russia	338	75-14100
Nottingham International	USA	640	80-22000
Pacific Helmets (NZ) Ltd	New Zealand	302	75-01620
Schuberth Helme GmbH	Germany	174	75-26800
Silent Partner Body Armor Inc.	USA	615	75-80930
South Central Ordnance Co.	USA	616	75-82400
Specialty Plastic Products Inc.	USA	616	75-83035
SPL Ordnance Ltd.	Israel	225	80-01935

712 10 61 Helmets, military, plastic

Barrday Inc.	Canada	68	75-01080
CGF Gallet SA	France	133	75-08840
Colan Products	Australia	14	75-03650
Courtaulds Aerospace	UK	443	75-15330
Davies Industrial Communications Ltd	UK	444	75-17010
Gentex Corp.	USA	578	75-39300
M. Kumar Udyog	India	205	75-02100
NII Stali	Russia	338	75-14100
Nottingham International	USA	640	80-22000
Pacific Helmets (NZ) Ltd	New Zealand	302	75-01620
Silent Partner Body Armor Inc.	USA	615	75-80930
South Central Ordnance Co.	USA	616	75-82400
Specialty Plastic Products Inc.	USA	616	75-83035
SPL Ordnance Ltd.	Israel	225	80-01935
St Bernard Composites Ltd	UK	486	75-56222

712 10 62 Helmets, military, steel

CGF Gallet SA	France	133	75-08840
Colan Products	Australia	14	75-03650
Indústrias de Defesa	Portugal	327	55-00950
Nottingham International	USA	640	80-22000
Pacific Helmets (NZ) Ltd	New Zealand	302	75-01620
Schuberth Helme GmbH	Germany	174	75-26800
SPL Ordnance Ltd.	Israel	225	80-01935

712 10 70 Helmets, parachutist

Barrday Inc.	Canada	68	75-01080
CGF Gallet SA	France	133	75-08840
Colan Products	Australia	14	75-03650
Courtaulds Aerospace	UK	443	75-15330
Davies Industrial Communications Ltd	UK	444	75-17010
Flight Suits Ltd	USA	575	75-34800
Helmet Integrated Systems Ltd	UK	458	75-29000
Lightweight Body Armour	UK	464	75-35650
Nottingham International	USA	640	80-22000
Pacific Helmets (NZ) Ltd	New Zealand	302	75-01620
Schuberth Helme GmbH	Germany	174	75-26800
Scott Aviation	USA	613	75-78850
Silent Partner Body Armor Inc.	USA	615	75-80930
South Central Ordnance Co.	USA	616	75-82400
Specialty Plastic Products Inc.	USA	616	75-83035
SPL Ordnance Ltd.	Israel	225	80-01935

712 10 75 Helmets, police

Barrday Inc.	Canada	68	75-01080
CGF Gallet SA	France	133	75-08840
Checkmate Industries Ltd	Hong Kong SAR	198	75-00200
Checkmate Industries of EEC Ltd	Denmark	102	75-03300
Colan Products	Australia	14	75-03650
Davies Industrial Communications Ltd	UK	444	75-17010
Helmet Integrated Systems Ltd	UK	458	75-29000
ICL Technical Plastics Ltd	UK	460	75-31060
Lightweight Body Armour	UK	464	75-35650
M. Kumar Udyog	India	205	75-02100
Pacific Helmets (NZ) Ltd	New Zealand	302	75-01620
Schuberth Helme GmbH	Germany	174	75-26800
Silent Partner Body Armor Inc.	USA	615	75-80930
South Central Ordnance Co.	USA	616	75-82400
Specialty Plastic Products Inc.	USA	616	75-83035
SPL Ordnance Ltd.	Israel	225	80-01935

712 10 80 Helmets, tank crew

CGF Gallet SA	France	133	75-08840
Colan Products	Australia	14	75-03650
Courtaulds Aerospace	UK	443	75-15330
Davies Industrial Communications Ltd	UK	444	75-17010
Jarvis Manufacturing Co.	UK	462	75-32330
LARIMART SpA	Italy	236	75-13500
Nottingham International	USA	640	80-22000
Pacific Helmets (NZ) Ltd	New Zealand	302	75-01620
Specialty Plastic Products Inc.	USA	616	75-83035
SPL Ordnance Ltd.	Israel	225	80-01935
St Bernard Composites Ltd	UK	486	75-56222

713 Ear protection

A.V. Electronics Ltd	UK	448	75-20200
Bose BV	Netherlands	291	75-00790
Chapman & Smith Ltd	UK	439	75-12500
David Clark Co. Inc.	USA	564	75-18850
Davies Industrial Communications Ltd	UK	444	75-17010
Deben Group Industries Ltd	UK	444	75-17270
Ets Bollé	France	133	75-06960
IFI Srl	Italy	235	75-11220
MSA (Britain) Limited	UK	469	75-41270
P.A.D. International	UK	472	75-43470
Protector Technologies Ltd	UK	476	75-46890
Real Santa Rita Equipamentos de	Brazil	57	75-18300
Universal Safety Supply Pty Ltd	Australia	23	75-21750

713 10 Ear defenders

A.V. Electronics Ltd	UK	448	75-20200
Bose BV	Netherlands	291	75-00790
Chapman & Smith Ltd	UK	439	75-12500
David Clark Co. Inc.	USA	564	75-18850
Davies Industrial Communications Ltd	UK	444	75-17010
Deben Group Industries Ltd	UK	444	75-17270
E-A-R Specialty Composites	USA	571	75-28500
Ets Bollé	France	133	75-06960
Haynes & Cann Ltd	UK	457	75-28750
Hodge Clemco Ltd	UK	458	75-29590

James North & Sons Ltd	UK	471	75-42730
MSA (Britain) Limited	UK	469	75-41270
Real Santa Rita Equipamentos de	Brazil	57	75-18300
Universal Safety Supply Pty Ltd	Australia	23	75-21750

713 10 10 Ear defenders, gun crew

Chapman & Smith Ltd	UK	439	75-12500
David Clark Co. Inc.	USA	564	75-18850
Davies Industrial Communications Ltd	UK	444	75-17010

713 10 20 Ear defenders, aircraft ground crew

Bose BV	Netherlands	291	75-00790
Chapman & Smith Ltd	UK	439	75-12500
David Clark Co. Inc.	USA	564	75-18850
Davies Industrial Communications Ltd	UK	444	75-17010
MSA (Britain) Limited	UK	469	75-41270
Tacconi SpA	Italy	239	75-25000
Universal Safety Supply Pty Ltd	Australia	23	75-21750

713 20 Ear protection, active noise reduction

Bose BV	Netherlands	291	75-00790

714 Eye & face protection

AOtec Inc.	USA	554	75-07940
Barrday Inc.	Canada	68	75-01080
Colan Products	Australia	14	75-03650
EG&G Inc.	USA	571	75-29555
Ets Bollé	France	133	75-06960
IFI Srl	Italy	235	75-11220
Knabenhans	Switzerland	386	75-07100
Laser & Atomic Research & Devt	Croatia	94	60-00450
MSA (Britain) Limited	UK	469	75-41270
P.A.D. International	UK	472	75-43470
Pilkington Optronics Ltd	UK	426	74-10000
Protector Technologies Ltd	UK	476	75-46890
Rabintex Industries Ltd	Israel	221	75-17850
RBR Armour Ltd	UK	478	75-48370
Real Santa Rita Equipamentos de	Brazil	57	75-18300
SFIM	**France**	**144**	**75-35080**
Silent Partner Body Armor Inc.	USA	615	75-80930
South Central Ordnance Co.	USA	616	75-82400
SPL Ordnance Ltd	Israel	225	80-01935

714 10 Visors

AOtec Inc.	USA	554	75-07940
EROS GIE	France	137	75-19500
Ets Bollé	France	133	75-06960
Excalor Hitzeschutz GmbH	Germany	168	75-08750
Helmet Integrated Systems Ltd	UK	458	75-29000
Hodge Clemco Ltd	UK	458	75-29590
ICL Technical Plastics Ltd	UK	460	75-31060
James North & Sons Ltd	UK	471	75-42730
Lightweight Body Armour	UK	464	75-35650
MSA (Britain) Limited	UK	469	75-41270
PG Products Ltd	UK	474	75-45065
Racal Health & Safety Group	UK	477	75-47675
RBR Armour Ltd	UK	478	75-48370
Real Santa Rita Equipamentos de	Brazil	57	75-18300
Scott Aviation	USA	613	75-78850
Silent Partner Body Armor Inc.	USA	615	75-80930
South Central Ordnance Co.	USA	616	75-82400
SPL Ordnance Ltd	Israel	225	80-01935
Tacconi SpA	Italy	239	75-25000
Universal Safety Supply Pty Ltd	Australia	23	75-21750

714 15 Face masks, cold weather

Ets Bollé	France	133	75-06960
Ideal Knitwear Manufacturing Ltd	Canada	72	75-07720
Musto Ltd	UK	470	75-41870
Sutcliffe Speakman Carbons Ltd	UK	487	75-58000
Tacconi SpA	Italy	239	75-25000

714 20 Goggles

AOtec Inc.	USA	554	75-07940
EG&G Inc.	USA	571	75-29555
Ets Bollé	France	133	75-06960
Intertechnique SA	France	137	75-23000
James North & Sons Ltd	UK	471	75-42730
Laser & Atomic Research & Devt	Croatia	94	60-00450
MSA (Britain) Limited	UK	469	75-41270
Pilkington Optronics Ltd	UK	426	74-10000
Real Santa Rita Equipamentos de	Brazil	57	75-18300
Silent Partner Body Armor Inc.	USA	615	75-80930
SPL Ordnance Ltd	Israel	225	80-01935
Tacconi SpA	Italy	239	75-25000
Transvaro A.S.	Turkey	403	75-02080
Universal Safety Supply Pty Ltd	Australia	23	75-21750

714 30 Spectacles, safety

AOtec Inc.	USA	554	75-07940
Ets Bollé	France	133	75-06960
James North & Sons Ltd	UK	471	75-42730
Knabenhans	Switzerland	386	75-07100
Laser & Atomic Research & Devt	Croatia	94	60-00450
SPL Ordnance Ltd	Israel	225	80-01935
Tacconi SpA	Italy	239	75-25000
Universal Safety Supply Pty Ltd	Australia	23	75-21750
Vandeputte NV	Belgium	47	75-06375

715 Respiratory protection

Alfred Karcher GmbH & Co	Germany	170	75-15650
AOtec Inc.	USA	554	75-07940
Avon Technical Products	UK	432	75-05150
Chapman & Smith Ltd	UK	439	75-12500
EROS GIE	France	137	75-19500
Haux-Life-Support GmbH	Germany	169	75-11970
IFI Srl	Italy	235	75-11220
Interspiro AB	Sweden	377	75-06950
Intertechnique SA	France	137	75-23000
Israel Product Research Co. Ltd	Israel	219	75-10000
James Hardie Engineering	New Zealand	303	80-00900
Leyland & Birmingham	UK	464	75-35350
MSA (Britain) Limited	UK	469	75-41270
Ocean Technology Systems	USA	601	75-66020
Odenwald Werke Rittersbach	Germany	173	75-22850
P.A.D. International	UK	472	75-43470
Pains-Wessex Ltd	UK	472	75-43510
Pittards plc	UK	475	75-45700
Protector Technologies Ltd	UK	476	75-46890
Racal Health & Safety Inc.	USA	606	75-72400
Real Santa Rita Equipamentos de	Brazil	57	75-18300
Royal Ordnance	UK	481	75-50620
Sèkur SpA	Italy	238	75-21650
SFIM	**France**	**144**	**75-35080**
Silent Partner Body Armor Inc.	USA	615	75-80930
S.M. Alexander (Plastics) Ltd	UK	429	75-02580
Universal Safety Supply Pty Ltd	Australia	23	75-21750
Vandeputte NV	Belgium	47	75-06375

715 10 Masks, protective

Alfred Karcher GmbH & Co	Germany	170	75-15650
Avon Technical Products	UK	432	75-05150
Chapman & Smith Ltd	UK	439	75-12500
Charcoal Cloth (International) Ltd	UK	439	75-12550
EROS GIE	France	137	75-19500
Giat Industries	France	126	55-00100
IFI Srl	Italy	235	75-11220
Interspiro	USA	585	75-48680
Interspiro Ltd	UK	461	75-31945
Intertechnique SA	France	137	75-23000
James North & Sons Ltd	UK	471	75-42730
MSA (Britain) Limited	UK	469	75-41270
Nottingham International	USA	640	80-22000
Odenwald Werke Rittersbach	Germany	173	75-22850
Racal Health & Safety Group	UK	477	75-47675
Real Santa Rita Equipamentos de	Brazil	57	75-18300
Royal Ordnance	UK	481	75-50620
Scott Aviation	USA	613	75-78850
Scott Health and Safety	USA	613	75-78850
Sèkur SpA	Italy	238	75-21650

Silent Partner Body Armor Inc.	USA	615	75-80930
S.M. Alexander (Plastics) Ltd	UK	429	75-02580
Sutcliffe Speakman Carbons Ltd	UK	487	75-58000
Tacconi SpA	Italy	239	75-25000
Universal Safety Supply Pty Ltd	Australia	23	75-21750

715 10 10 Masks, gas

Avon Technical Products	UK	432	75-05150
Carleton Life Support Technologies	Canada	70	75-02950
Chapman & Smith Ltd	UK	439	75-12500
Charcoal Cloth (International) Ltd	UK	439	75-12550
Davies Industrial Communications Ltd	UK	444	75-17010
EROS GIE	France	137	75-19500
Giat Industries	France	126	55-00100
IFI Srl	Italy	235	75-11220
Interspiro Ltd	UK	461	75-31945
Intertechnique SA	France	137	75-23000
Israel Product Research Co. Ltd	Israel	219	75-10000
Leyland & Birmingham	UK	464	75-35350
MSA (Britain) Limited	UK	469	75-41270
NORINCO	China PR	88	55-00280
Nottingham International	USA	640	80-22000
Odenwald Werke Rittersbach	Germany	173	75-22850
Pains-Wessex Ltd	UK	472	75-43510
Royal Ordnance	UK	481	75-50620
Scott Aviation	USA	613	75-78850
Sèkur SpA	Italy	238	75-21650
Shalon Chemical Industries Ltd	Israel	222	75-20000
Silent Partner Body Armor Inc.	USA	615	75-80930
Sutcliffe Speakman Carbons Ltd	UK	487	75-58000
Tacconi SpA	Italy	239	75-25000

715 10 10 90 Components, gas mask

AOtec Inc.	USA	554	75-07940
Avon Technical Products	UK	432	75-05150
Chapman & Smith Ltd	UK	439	75-12500
Davies Industrial Communications Ltd	UK	444	75-17010
Dowty Seals Ltd	UK	446	75-18620
EROS GIE	France	137	75-19500
Giat Industries	France	126	55-00100
IFI Srl	Italy	235	75-11220
Interspiro	USA	585	75-48680
Intertechnique SA	France	137	75-23000
Leyland & Birmingham	UK	464	75-35350
MSA (Britain) Limited	UK	469	75-41270
Royal Ordnance	UK	481	75-50620
Tacconi SpA	Italy	239	75-25000

715 10 10 93 Filters, respirator

Avon Technical Products	UK	432	75-05150
Carleton Life Support Technologies	Canada	70	75-02950
Chapman & Smith Ltd	UK	439	75-12500
EROS GIE	France	137	75-19500
Flodins Filter AB	Sweden	377	75-05300
Giat Industries	France	126	55-00100
IFI Srl	Italy	235	75-11220
Intertechnique SA	France	137	75-23000
James North & Sons Ltd	UK	471	75-42730
Leyland & Birmingham	UK	464	75-35350
Micronel AG	Switzerland	387	75-08490
MSA (Britain) Limited	UK	469	75-41270
Odenwald Werke Rittersbach	Germany	173	75-22850
Pains-Wessex Ltd	UK	472	75-43510
Pall Land & Marine	UK	472	75-43520
Racal Filter Technologies Ltd	Canada	75	75-11600
Racal Health & Safety Inc.	USA	606	75-72400
Scott Aviation	USA	613	75-78850
Scott Health and Safety	USA	613	75-78850
Sèkur SpA	Italy	238	75-21650
Shalon Chemical Industries Ltd	Israel	222	75-20000
Sutcliffe Speakman Carbons Ltd	UK	487	75-58000
Tacconi SpA	Italy	239	75-25000

715 10 15 Masks, oxygen, aircraft

Carleton Life Support Technologies	Canada	70	75-02950
EROS GIE	France	137	75-19500
Flight Suits Ltd	USA	575	75-34800
Gentex Corp.	USA	578	75-39300
Intertechnique (Aerospace) Ltd	UK	461	75-31948
Intertechnique SA	France	137	75-23000
Meggitt Oxygen Systems	UK	467	75-38360
Puritan-Bennett Aero Systems Co.	USA	606	75-71890
Scott Aviation	USA	613	75-78850
Scott Health and Safety	USA	613	75-78850
SFIM	**France**	**144**	**75-35080**
Tokyo Aircraft Instrument Co Ltd	Japan	254	75-08600

715 10 20 Masks, diving

Divex Ltd.	UK	445	75-18220
EROS GIE	France	137	75-19500
Haux-Life-Support GmbH	Germany	169	75-11970
Interspiro	USA	585	75-48680
Interspiro Ltd	UK	461	75-31945
Interspiro NL.	Netherlands	293	75-03570
Intertechnique SA	France	137	75-23000
Ocean Technology Systems	USA	601	75-66020

715 20 Breathing apparatus

Ambu International A/S	Denmark	102	75-01170
Divex Ltd.	UK	445	75-18220
EROS GIE	France	137	75-19500
Haux-Life-Support GmbH	Germany	169	75-11970
Heiser Josef	Austria	33	75-01230
Interspiro	USA	585	75-48680
Interspiro Ltd	UK	461	75-31945
Interspiro NL.	Netherlands	293	75-03570
Jarvis Manufacturing Co.	UK	462	75-32330
MSA (Britain) Limited	UK	469	75-41270
Racal Health & Safety Group	UK	477	75-47675
Scott Aviation	USA	613	75-78850
Scott Health and Safety	USA	613	75-78850
Siebe Gorman & Co. Ltd	UK	483	75-53600
Tokyo Aircraft Instrument Co Ltd	Japan	254	75-08600

715 20 10 Breathing apparatus, fire / crash tendercrew

EROS GIE	France	137	75-19500
Interspiro	USA	585	75-48680
Interspiro Ltd	UK	461	75-31945
Jarvis Manufacturing Co.	UK	462	75-32330
MSA (Britain) Limited	UK	469	75-41270
Phillips & Smith Ltd	New Zealand	303	75-01650
Racal Health & Safety Group	UK	477	75-47675
Scott Aviation	USA	613	75-78850
Scott Health and Safety	USA	613	75-78850
Siebe Gorman & Co. Ltd	UK	483	75-53600
Universal Safety Supply Pty Ltd	Australia	23	75-21750

715 20 20 Breathing apparatus, underwater, self-contained

Divex Ltd.	UK	445	75-18220
EROS GIE	France	137	75-19500
Fullerton, Sherwood Engineering Ltd	Canada	71	75-05625
Haux-Life-Support GmbH	Germany	169	75-11970
Interspiro Ltd	UK	461	75-31945
Interspiro NL.	Netherlands	293	75-03570
Scott Aviation	USA	613	75-78850

716 Body protection

Achidatex Nazareth Elite (1977) Ltd	Israel	217	75-00200
Aeraustral Trading	Australia	11	75-00330
Airborne Industries	UK	428	75-01550
Åkers Krutbruk Protection AB	Sweden	375	75-01100
Akzo Nobel Fazer AG	Germany	164	75-00515
Ansett Australia Avionics	Australia	24	80-01010
Armour Technology International Ltd	UK	431	75-03790
Armourshield Limited	UK	431	75-03775
Avon Technical Products	UK	432	75-05150
Barrday Inc.	Canada	68	75-01080
BCB International Ltd	UK	433	75-06075
Beaufort Air-Sea Equipment Ltd	UK	433	75-06100
Bennett Safetywear Limited	UK	434	75-06200

Product-Supplier Link

Checkmate Industries Ltd	Hong Kong SAR	198	75-00200
Cochrane Steel Products (PTY) Ltd	S. Africa	357	75-03000
Colan Products	Australia	14	75-03650
Deenside Ltd	UK	445	75-17350
DSM	Netherlands	292	75-01730
DTC International	USA	570	75-27620
Excalor Hitzeschutz GmbH	Germany	168	75-08750
Hiatt & Company Ltd	UK	458	75-29180
H.P.P. (UK) Ltd	UK	459	75-30360
IFI Srl	Italy	235	75-11220
Intertechnique SA	France	137	75-23000
Kirasa Co. Ltd	Russia	337	75-11350
Meggitt Composites	UK	467	75-38345
Mehler Technische Textilien GmbH	Germany	172	75-20400
Mustang Survival (USA) Corp	USA	597	75-63480
Odenwald Werke Rittersbach	Germany	173	75-22850
P.A.D. International	UK	472	75-43470
Pains-Wessex Ltd	UK	472	75-43510
Paul Boyé	France	133	75-07020
PK Electronic International Ltd	UK	475	75-45720
Q-Systems International Inc.	USA	606	75-71960
RBR Armour Ltd	UK	478	75-48370
Real Santa Rita Equipamentos de	Brazil	57	75-18300
Second Chance Body Armor Inc.	USA	613	75-79000
Sèkur SpA	Italy	238	75-21650
Silent Partner Body Armor Inc.	USA	615	75-80930
SNPE	France	145	75-36315
South Central Ordnance Co.	USA	616	75-82400
Sparks & Ellis Uniforms	S. Africa	359	75-19390
Specialty Plastic Products Inc.	USA	616	75-83035
SPL Ordnance Ltd	Israel	225	80-01935
Tacconi SpA	Italy	239	75-25000
Vidaro Corp.	USA	626	75-94000

716 10 Vests & jackets, protective

Achidatex Nazareth Elite (1977) Ltd	Israel	217	75-00200
Aeraustral Trading	Australia	11	75-00330
Airborne Industries	UK	428	75-01550
Åkers Krutbruk Protection AB	Sweden	375	75-01100
Armour of America Inc.	USA	555	75-08850
Armour Technology International Ltd	UK	431	75-03790
Armourshield Limited	UK	431	75-03775
Avon Technical Products	UK	432	75-05150
Barrday Inc.	Canada	68	75-01080
Beaufort Air-Sea Equipment Ltd	UK	433	75-06100
Canadian Body Armour Ltd	Canada	69	75-02250
Checkmate Industries Ltd	Hong Kong SAR	198	75-00200
Checkmate Industries of EEC Ltd	Denmark	102	75-03300
Clansman Armour Systems	UK	440	75-13460
Cochrane Steel Products (PTY) Ltd	S. Africa	357	75-03000
Colan Products	Australia	14	75-03650
Deenside Ltd	UK	445	75-17350
DSM	Netherlands	292	75-01730
EROS GIE	France	137	75-19500
Excalor Hitzeschutz GmbH	Germany	168	75-08750
Flight Suits Ltd	USA	575	75-34800
H. Fine & Son Ltd	UK	451	75-23210
H.C. Slingsby plc	UK	484	75-54430
Hiatt & Company Ltd	UK	458	75-29180
Hodge Clemco Ltd	UK	458	75-29590
Intertechnique SA	France	137	75-23000
Kirasa Co. Ltd	Russia	337	75-11350
Lightweight Body Armour	UK	464	75-35650
Manta Industries Ltd	Canada	73	75-09250
Meggitt Composites	UK	467	75-38345
Mehler Technische Textilien GmbH	Germany	172	75-20400
Mustang Survival (USA) Corp	USA	597	75-63480
NII Stali	Russia	338	75-14100
Norton Co.	USA	600	75-65510
Nottingham International	USA	640	80-22000
Odenwald Werke Rittersbach	Germany	173	75-22850
Pains-Wessex Ltd	UK	472	75-43510
Paul Boyé	France	133	75-07020
PK Electronic International Ltd	UK	475	75-45720
Q-Systems International Inc.	USA	606	75-71960
RBR Armour Ltd	UK	478	75-48370
Second Chance Body Armor Inc.	USA	613	75-79000
Sèkur SpA	Italy	238	75-21650
Silent Partner Body Armor Inc.	USA	615	75-80930
SNPE	France	145	75-36315
South Central Ordnance Co.	USA	616	75-82400
Sparks & Ellis Uniforms	S. Africa	359	75-19390
Specialty Plastic Products Inc.	USA	616	75-83035
SPL Ordnance Ltd	Israel	225	80-01935
Switlik Parachute Co. Inc.	USA	619	75-85400
Tacconi SpA	Italy	239	75-25000
TBA Textiles	UK	487	75-58740
Universal Safety Supply Pty Ltd	Australia	23	75-21750

716 11 Body armour jackets, lightweight bullet-proof

Achidatex Nazareth Elite (1977) Ltd	Israel	217	75-00200
Aeraustral Trading	Australia	11	75-00330
Åkers Krutbruk Protection AB	Sweden	375	75-01100
Armour of America Inc.	USA	555	75-08850
Armour Technology International Ltd	UK	431	75-03790
Armourshield Limited	UK	431	75-03775
Barrday Inc.	Canada	68	75-01080
BCB International Ltd	UK	433	75-06075
Bennett Safetywear Limited	UK	434	75-06200
Canadian Body Armour Ltd	Canada	69	75-02250
Checkmate Industries Ltd	Hong Kong SAR	198	75-00200
Clansman Armour Systems	UK	440	75-13460
Colan Products	Australia	14	75-03650
Courtaulds Aerospace	UK	443	75-15330
DSM	Netherlands	292	75-01730
Forjas Taurus SA	Brazil	56	75-06640
Germa AB	Sweden	377	75-05700
K S I Agencies Ltd	New Zealand	303	80-01150
Kirasa Co. Ltd	Russia	337	75-11350
Lightweight Body Armour	UK	464	75-35650
Meggitt Composites	UK	467	75-38345
ML Lifeguard Equipment Ltd	UK	469	75-40230
NII Stali	Russia	338	75-14100
Norton Co.	USA	600	75-65510
Nottingham International	USA	640	80-22000
PK Electronic International Ltd	UK	475	75-45720
Plasan Sasa	Israel	221	75-17510
Q-Systems International Inc.	USA	606	75-71960
RBR Armour Ltd	UK	478	75-48370
ROFi Industrier AS	Norway	310	75-02300
Safariland Ltd Inc.	USA	612	75-77350
Second Chance Body Armor Inc.	USA	613	75-79000
Silent Partner Body Armor Inc.	USA	615	75-80930
South Central Ordnance Co.	USA	616	75-82400
Specialty Plastic Products Inc.	USA	616	75-83035
SPL Ordnance Ltd	Israel	225	80-01935
TBA Textiles	UK	487	75-58740

716 12 Vests, bullet-resistant

Achidatex Nazareth Elite (1977) Ltd	Israel	217	75-00200
Åkers Krutbruk Protection AB	Sweden	375	75-01100
Akzo Nobel Fazer AG	Germany	164	75-00515
Albion Clothing Ltd	New Zealand	301	75-00110
Armour of America Inc.	USA	555	75-08850
Armour Technology International Ltd	UK	431	75-03790
Armourshield Limited	UK	431	75-03775
Avon Technical Products	UK	432	75-05150
Barrday Inc.	Canada	68	75-01080
BCB International Ltd	UK	433	75-06075
Browning SA	Belgium	45	75-01950
Bruck Textiles plc	Australia	13	75-02810
Canadian Body Armour Ltd	Canada	69	75-02250
Checkmate Industries Ltd	Hong Kong SAR	198	75-00200
Checkmate Industries of EEC Ltd	Denmark	102	75-03300
Cochrane Steel Products (PTY) Ltd	S. Africa	357	75-03000
Colan Products	Australia	14	75-03650
DSM	Netherlands	292	75-01730
Germa AB	Sweden	377	75-05700
Guardian Technologies International	USA	580	75-40800
ITP Ltd	Israel	220	75-10800
Kirasa Co. Ltd	Russia	337	75-11350
Lightweight Body Armour	UK	464	75-35650
Meggitt Composites	UK	467	75-38345
Mehler Technische Textilien GmbH	Germany	172	75-20400
ML Lifeguard Equipment Ltd	UK	469	75-40230
NII Stali	Russia	338	75-14100
Norton Co.	USA	600	75-65510
Nottingham International	USA	640	80-22000
Paul Boyé	France	133	75-07020
PK Electronic International Ltd	UK	475	75-45720

Plasan Sasa	Israel	221	75-17510
Q-Systems International Inc.	USA	606	75-71960
Rabintex Industries Ltd	Israel	221	75-17850
RBR Armour Ltd.	UK	478	75-48370
Second Chance Body Armor Inc.	USA	613	75-79000
Silent Partner Body Armor Inc.	USA	615	75-80930
SNPE	France	145	75-36315
South Central Ordnance Co.	USA	616	75-82400
Specialty Plastic Products Inc.	USA	616	75-83035
SPL Ordnance Ltd.	Israel	225	80-01935
TBA Textiles	UK	487	75-58740

716 12 50 Combination ballistic & flotation vests

Armour of America Inc.	USA	555	75-08850
Barrday Inc.	Canada	68	75-01080
Colan Products	Australia	14	75-03650
EROS GIE	France	137	75-19500
Mustang Survival (USA) Corp	USA	597	75-63480
Q-Systems International Inc.	USA	606	75-71960
RBR Armour Ltd.	UK	478	75-48370
Silent Partner Body Armor Inc.	USA	615	75-80930
South Central Ordnance Co.	USA	616	75-82400

716 13 Vests, fragmentation

Armour of America Inc.	USA	555	75-08850
Armour Technology International Ltd.	UK	431	75-03790
Barrday Inc.	Canada	68	75-01080
Bruck Textiles plc	Australia	13	75-02810
Canadian Body Armour Ltd	Canada	69	75-02250
Colan Products	Australia	14	75-03650
CQC Plc	UK	443	75-15700
DSM	Netherlands	292	75-01730
Lightweight Body Armour	UK	464	75-35650
Meggitt Composites	UK	467	75-38345
Mehler Technische Textilien GmbH	Germany	172	75-20400
Norton Co.	USA	600	75-65510
Q-Systems International Inc.	USA	606	75-71960
RBR Armour Ltd.	UK	478	75-48370
Second Chance Body Armor Inc.	USA	613	75-79000
Silent Partner Body Armor Inc.	USA	615	75-80930
South Central Ordnance Co.	USA	616	75-82400
Specialty Plastic Products Inc.	USA	616	75-83035
SPL Ordnance Ltd.	Israel	225	80-01935
TBA Textiles	UK	487	75-58740

716 20 Gloves, protective

Airborne Industries	UK	428	75-01550
Aircraft Materials Ltd	UK	428	75-01650
Bennett Safetywear Limited	UK	434	75-06200
Burfield & Co. (Gloves) Ltd	UK	438	75-10900
Colan Products	Australia	14	75-03650
Compton Webb Ltd	UK	442	75-14550
DSM	Netherlands	292	75-01730
Excalor Hitzeschutz GmbH	Germany	168	75-08750
H.C. Slingsby plc	UK	484	75-54430
Hodge Clemco Ltd	UK	458	75-29590
Indiana Rubber	Netherlands	293	75-03540
James North & Sons Ltd	UK	471	75-42730
LRC Products Ltd	UK	465	75-36530
M. Kumar Udyog	India	205	75-02100
Mustang Survival (USA) Corp	USA	597	75-63480
New Pac Safety AB	Sweden	378	75-08260
Odenwald Werke Rittersbach	Germany	173	75-22850
Pains-Wessex Ltd	UK	472	75-43510
Real Santa Rita Equipamentos de	Brazil	57	75-18300
Southcombe Brothers Ltd	UK	485	75-55214
Tacconi SpA	Italy	239	75-25000
Universal Safety Supply Pty Ltd	Australia	23	75-21750
Vacuum Reflex Ltd	UK	490	75-62600
Vandeputte NV	Belgium	47	75-06375
Vidaro Corp.	USA	626	75-94000
Viking Life-Saving Equipment A/S	Denmark	104	75-22000

716 21 Gloves, aircrew

Beaufort Air-Sea Equipment Ltd	UK	433	75-06100
Bennett Safetywear Limited	UK	434	75-06200
Burfield & Co. (Gloves) Ltd	UK	438	75-10900
Colan Products	Australia	14	75-03650
EROS GIE	France	137	75-19500
Flight Suits Ltd	USA	575	75-34800
Southcombe Brothers Ltd	UK	485	75-55214

716 22 Gloves, combat

BCB International Ltd	UK	433	75-06075
Bennett Safetywear Limited	UK	434	75-06200
Burfield & Co. (Gloves) Ltd	UK	438	75-10900
Colan Products	Australia	14	75-03650
Southcombe Brothers Ltd	UK	485	75-55214

716 70 Trousers, protective

Airborne Industries	UK	428	75-01550
Albion Clothing Ltd	New Zealand	301	75-00110
Barrday Inc.	Canada	68	75-01080
Beaufort Air-Sea Equipment Ltd	UK	433	75-06100
Colan Products	Australia	14	75-03650
Excalor Hitzeschutz GmbH	Germany	168	75-08750
Hodge Clemco Ltd	UK	458	75-29590
Intertechnique SA.	France	137	75-23000
James North & Sons Ltd	UK	471	75-42730
Odenwald Werke Rittersbach	Germany	173	75-22850
Pains-Wessex Ltd	UK	472	75-43510
South Central Ordnance Co.	USA	616	75-82400
Tacconi SpA	Italy	239	75-25000
Universal Safety Supply Pty Ltd	Australia	23	75-21750
Vidaro Corp.	USA	626	75-94000

716 70 10 Trousers, anti-g

Beaufort Air-Sea Equipment Ltd	UK	433	75-06100
Colan Products	Australia	14	75-03650
David Clark Co. Inc.	USA	564	75-18850
EROS GIE	France	137	75-19500
Germa AB	Sweden	377	75-05700
Intertechnique SA.	France	137	75-23000
Irvin Aerospace Canada Limited	Canada	73	75-08300
Irvin Aerospace SpA	Italy	235	75-12200
ML Lifeguard Equipment Ltd	UK	469	75-40230
Mustang Survival Corp.	Canada	74	75-09860
Tacconi SpA	Italy	239	75-25000

717 Foot protection

Berghaus Ltd	UK	434	75-06525
Checkmate Industries Ltd	Hong Kong SAR	198	75-00200
DSM	Netherlands	292	75-01730
Excalor Hitzeschutz GmbH	Germany	168	75-08750
Haynes & Cann Ltd	UK	457	75-28750
IFI Srl.	Italy	235	75-11220
Pains-Wessex Ltd	UK	472	75-43510
Paul Boyé	France	133	75-07020
Pittards plc	UK	475	75-45700
Rabintex Industries Ltd	Israel	221	75-17850
Real Santa Rita Equipamentos de	Brazil	57	75-18300
SPL Ordnance Ltd.	Israel	225	80-01935
Vidaro Corp.	USA	626	75-94000

717 10 Footwear, protective

Blundstone Pty Ltd	Australia	12	75-02490
BTR Silvertown Ltd	UK	437	75-10830
Checkmate Industries Ltd	Hong Kong SAR	198	75-00200
Checkmate Industries of EEC Ltd	Denmark	102	75-03300
DSM	Netherlands	292	75-01730
Excalor Hitzeschutz GmbH	Germany	168	75-08750
G.B. Britton	UK	437	75-09900
Haynes & Cann Ltd	UK	457	75-28750
H.C. Slingsby plc	UK	484	75-54430
Leyland & Birmingham	UK	464	75-35350
Lightweight Body Armour	UK	464	75-35650
Odenwald Werke Rittersbach	Germany	173	75-22850
Pains-Wessex Ltd	UK	472	75-43510
Real Santa Rita Equipamentos de	Brazil	57	75-18300
SPL Ordnance Ltd.	Israel	225	80-01935

Product-Supplier Link

Tacconi SpA	Italy	*239*	75-25000
Tebbutt & Hall Ltd	UK	*487*	75-58800
Trojan Manufacturing & Marketing Inc	Philippines	*321*	75-01350
Universal Safety Supply Pty Ltd	Australia	*23*	75-21750
Vandeputte NV	Belgium	*47*	75-06375

719 Other personal protection equipment

Airborne Industries	UK	*428*	75-01550
Altoflex	France	*132*	75-02300
Beaufort Air-Sea Equipment Ltd	UK	*433*	75-06100
Colan Products	Australia	*14*	75-03650
CQC Plc	UK	*443*	75-15700
DSM	Netherlands	*292*	75-01730
FDB Electrical Ltd	UK	*450*	75-22490
Giat Industries	France	*126*	55-00100
IFI Srl	Italy	*235*	75-11220
Meggitt Composites	UK	*467*	75-38345
Mehler Technische Textilien GmbH	Germany	*172*	75-20400
National Nuclear Corporation	USA	*598*	75-63870
P.A.D. International	UK	*472*	75-43470
Pains-Wessex Ltd	UK	*472*	75-43510
Perimeter Products Inc.	USA	*603*	75-69290
Pittards plc	UK	*475*	75-45700
PK Electronic International Ltd	UK	*475*	75-45720
Rabintex Industries Ltd	Israel	*221*	75-17850
RBR Armour Ltd.	UK	*478*	75-48370
Second Chance Body Armor Inc.	USA	*613*	75-79000
Silent Partner Body Armor Inc.	USA	*615*	75-80930
Sleeman Engineering	UK	*484*	75-54400
South Central Ordnance Co.	USA	*616*	75-82400
SPL Ordnance Ltd.	Israel	*225*	80-01935

719 10 Armour, lightweight composite

Armour of America Inc.	USA	*555*	75-08850
Armour Technology International Ltd.	UK	*431*	75-03790
Barrday Inc.	Canada	*68*	75-01080
Colan Products	Australia	*14*	75-03650
Cyclone Aviation Products Ltd	Israel	*218*	75-04250
DSM	Netherlands	*292*	75-01730
Kirasa Co. Ltd	Russia	*337*	75-11350
Lightweight Body Armour	UK	*464*	75-35650
Meggitt Composites	UK	*467*	75-38345
Mehler Technische Textilien GmbH	Germany	*172*	75-20400
NKI	Netherlands	*294*	75-05100
Norton Co.	USA	*600*	75-65510
Nottingham International	USA	*640*	80-22000
Permali Gloucester Ltd	UK	*474*	75-44850
Plasan Sasa	Israel	*221*	75-17510
Protection Development Int'l Corp.	USA	*606*	75-71670
Q-Systems International Inc.	USA	*606*	75-71960
RBR Armour Ltd.	UK	*478*	75-48370
Silent Partner Body Armor Inc.	USA	*615*	75-80930
Simula Government Products, Inc.	USA	*615*	75-81250
South Central Ordnance Co.	USA	*616*	75-82400
SPL Ordnance Ltd.	Israel	*225*	80-01935

719 10 30 Armour, lightweight composite, helicopter crew seat

AEA Technology plc	UK	*496*	80-00480
Armour of America Inc.	USA	*555*	75-08850
Barrday Inc.	Canada	*68*	75-01080
Colan Products	Australia	*14*	75-03650
Cyclone Aviation Products Ltd	Israel	*218*	75-04250
DSM	Netherlands	*292*	75-01730
Meggitt Composites	UK	*467*	75-38345
Norton Co.	USA	*600*	75-65510
Plasan Sasa	Israel	*221*	75-17510
Protection Development Int'l Corp.	USA	*606*	75-71670
RBR Armour Ltd.	UK	*478*	75-48370
Royal Ordnance	UK	*481*	75-50620
Silent Partner Body Armor Inc.	USA	*615*	75-80930
South Central Ordnance Co.	USA	*616*	75-82400

719 60 Shields, anti-riot (see code 491.60)

Armour of America Inc.	USA	*555*	75-08850
Barrday Inc.	Canada	*68*	75-01080
Browning SA	Belgium	*45*	75-01950
Civil Defence Supply	UK	*440*	75-13350
Colan Products	Australia	*14*	75-03650
DSM	Netherlands	*292*	75-01730
Hiatt & Company Ltd	UK	*458*	75-29180
ICL Technical Plastics Ltd	UK	*460*	75-31060
Lightweight Body Armour	UK	*464*	75-35650
Pains-Wessex Ltd	UK	*472*	75-43510
PG Products Ltd.	UK	*474*	75-45065
Plasan Sasa	Israel	*221*	75-17510
Protection Development Int'l Corp.	USA	*606*	75-71670
RBR Armour Ltd.	UK	*478*	75-48370
Sas R&D Services Inc.	USA	*642*	80-26870
Silent Partner Body Armor Inc.	USA	*615*	75-80930
Sleeman Engineering	UK	*484*	75-54400
South Central Ordnance Co.	USA	*616*	75-82400
SPL Ordnance Ltd.	Israel	*225*	80-01935

719 65 Shields, nuclear

AEA Technology plc	UK	*496*	80-00480
Altoflex	France	*132*	75-02300
Colan Products	Australia	*14*	75-03650
National Nuclear Corporation	USA	*598*	75-63870
Permali Gloucester Ltd	UK	*474*	75-44850

Storage Protection, Transport Containers

721 Storage protection

AB Carl Munters	Sweden	*377*	75-08230
Achidatex Nazareth Elite (1977) Ltd	Israel	*217*	75-00200
Aerolyte Industries Pty Ltd.	Australia	*11*	75-00310
AGM Container Controls Inc.	USA	*547*	75-02870
AIM Aviation (Jecco) Ltd.	UK	*428*	75-01530
Air-Log Ltd	UK	*429*	75-01800
Andrew Mitchell & Co. Ltd	UK	*468*	75-40100
Apex Storage Systems Ltd	UK	*431*	75-03420
Avon Technical Products	UK	*432*	75-05150
Brownell Ltd.	UK	*437*	75-10385
Caledonian Mining Co. Ltd	UK	*438*	75-11250
CP Cases	UK	*443*	75-15680
EPS Logistics Technology Ltd	UK	*502*	80-09550
Esmet Inc.	USA	*574*	75-32250
GEL Engineering Ltd	UK	*454*	75-25500
J&D Wilkie Ltd	UK	*494*	75-66240
Logan Fenamec (UK) Ltd	UK	*465*	75-36270
Ludlow Corp.	USA	*592*	75-56350
Metric Systems Corp.	USA	*596*	75-60800
Mineral Derivatives Ltd	UK	*468*	75-39800
Odenwald Werke Rittersbach	Germany	*173*	75-22850
Progress-Werk Oberkirch AG	Germany	*173*	75-24320
QED Design and Manufacture Ltd	UK	*476*	75-47350
Rabintex Industries Ltd	Israel	*221*	75-17850
Sattler Texilwerke KG	Austria	*34*	75-02155
The Johnson John Company	USA	*587*	75-51000

721 10 Protective storage systems

AAR Corp	USA	*629*	80-00035
AB Carl Munters	Sweden	*377*	75-08230
Achidatex Nazareth Elite (1977) Ltd	Israel	*217*	75-00200
AGM Container Controls Inc.	USA	*547*	75-02870
Air-Log Ltd	UK	*429*	75-01800
Apex Storage Systems Ltd	UK	*431*	75-03420
Barry Controls Ltd	UK	*433*	75-05930
Brownell Ltd.	UK	*437*	75-10385
Caledonian Mining Co. Ltd	UK	*438*	75-11250
CP Cases	UK	*443*	75-15680
C.P. Mega	UK	*467*	75-38330
EPS Logistics Technology Ltd	UK	*502*	80-09550
Mineral Derivatives Ltd	UK	*468*	75-39800
Munters Cargocaire	USA	*597*	75-63445
Odenwald Werke Rittersbach	Germany	*173*	75-22850
PD Technical Mouldings plc	UK	*473*	75-44200
QED Design and Manufacture Ltd	UK	*476*	75-47350
Sattler Texilwerke KG	Austria	*34*	75-02155
3C Systems, Inc.	USA	*632*	80-05000

721 20 Protective storage equipment

AB Carl Munters	Sweden	*377*	75-08230
Achidatex Nazareth Elite (1977) Ltd	Israel	*217*	75-00200
Aerial Machine & Tool Corp.	USA	*546*	75-01105
Air-Log Ltd	UK	*429*	75-01800
Barry Controls Ltd	UK	*433*	75-05930
Chubb Safe Equipment Co.	UK	*440*	75-13205
EPS Logistics Technology Ltd	UK	*502*	80-09550
Esmet Inc.	USA	*574*	75-32250
J&D Wilkie Ltd	UK	*494*	75-66240
Munters Cargocaire	USA	*597*	75-63445
PD Technical Mouldings plc	UK	*473*	75-44200
QED Design and Manufacture Ltd	UK	*476*	75-47350
Security Engineered Machinery	USA	*613*	75-79100
Société d'Exploit Conditionnement	France	*146*	75-39455
3C Systems, Inc.	USA	*632*	80-05000

721 20 10 Bags, weapon storage

Aerial Machine & Tool Corp.	USA	*546*	75-01105
AIM Aviation (Jecco) Ltd.	UK	*428*	75-01530
Air-Log Ltd	UK	*429*	75-01800
Avon Technical Products	UK	*432*	75-05150
Barrday Inc.	Canada	*68*	75-01080
Divex Ltd.	UK	*445*	75-18220
EPS Logistics Technology Ltd	UK	*502*	80-09550
H. Fine & Son Ltd	UK	*451*	75-23210
Manta Industries Ltd	Canada	*73*	75-09250
QED Design and Manufacture Ltd	UK	*476*	75-47350
Sattler Texilwerke KG	Austria	*34*	75-02155
S.M. Alexander (Plastics) Ltd	UK	*429*	75-02580
White Cross Rubber Products Ltd.	UK	*494*	75-65857

721 80 Tarpaulins, vehicle covers

Aerial Machine & Tool Corp.	USA	*546*	75-01105
Air-Log Ltd	UK	*429*	75-01800
Andrew Mitchell & Co. Ltd.	UK	*468*	75-40100
Barrday Inc.	Canada	*68*	75-01080
H.C. Slingsby plc	UK	*484*	75-54430
J Blaschke Wehrtechnik GmbH	Austria	*32*	75-00560
J T Inglis & Sons Ltd	UK	*461*	75-31500
J&D Wilkie Ltd	UK	*494*	75-66240
L.N. Compton Ltd.	UK	*442*	75-14600
Manta Industries Ltd	Canada	*73*	75-09250
Sattler Texilwerke KG	Austria	*34*	75-02155
Stephan Wood Products Inc.	USA	*628*	75-97280
Structure-flex Ltd	UK	*486*	75-57525
The Johnson John Company	USA	*587*	75-51000

722 Transport cases & containers

AAR Corp	USA	*629*	80-00035
Air-Log Ltd	UK	*429*	75-01800
Andrew Mitchell & Co. Ltd.	UK	*468*	75-40100
Bigg & Robertson Plastics Ltd	UK	*434*	75-06680
COES (Derby) Ltd	UK	*441*	75-14070
CONCO Inc.	USA	*566*	75-20835
Container Research Corp.	USA	*566*	75-21700
CP Cases	UK	*443*	75-15680
C.P. Mega	UK	*467*	75-38330
EDAK AG	Switzerland	*385*	75-03200
EPS Logistics Technology Ltd	UK	*502*	80-09550
GEL Engineering Ltd	UK	*454*	75-25500
Genco Mobile Systems BV.	Netherlands	*293*	75-02340
Giat Industries	France	*126*	55-00100
H. Koch & Sons Co.	USA	*589*	75-53030
Hardigg Cases.	USA	*580*	75-41930
Instrument Enclosures	USA	*585*	75-46890
Irvin Industries Inc.	USA	*586*	75-48960
Knürr-Mechanik	Germany	*171*	75-16150
L'hotellier Montrichard	France	*138*	75-24900
Light Alloy Ltd.	UK	*464*	75-35560
Metric Systems Corp.	USA	*596*	75-60800
Odenwald Werke Rittersbach	Germany	*173*	75-22850
Pathfinder Fox Ltd	UK	*473*	75-44065
PD Technical Mouldings plc	UK	*473*	75-44200
Pelican Products Inc.	USA	*603*	75-69125
Polyformes Ltd	UK	*475*	75-46348
Precision Machining Inc.	USA	*605*	75-71350
Progress-Werk Oberkirch AG	Germany	*173*	75-24320
Pronal SA	France	*141*	75-31100
QED Design and Manufacture Ltd	UK	*476*	75-47350
Rowlinson Packaging Ltd	UK	*481*	75-50600
Royal Ordnance	UK	*481*	75-50620
Star Case Manufacturing Co. Inc.	USA	*617*	75-83955
Thermodyne International Ltd	USA	*622*	75-88690
Valsella Meccanotecnica SpA	Italy	*239*	75-26800
Westford Plastics & Engineering Ltd	UK	*493*	75-65450
Zero Cases (Europe)	UK	*495*	75-67700
Zero Enclosures	USA	*629*	75-98150

722 10 Protective transport cases, containers & equipment

AAR Corp	USA	*629*	80-00035
Aerial Machine & Tool Corp.	USA	*546*	75-01105
Air-Log Ltd	UK	*429*	75-01800
Balmoral Group Ltd	UK	*433*	75-05420
Barry Controls Ltd	UK	*433*	75-05930
Bigg & Robertson Plastics Ltd	UK	*434*	75-06680
Brownell Ltd.	UK	*437*	75-10385
Carrozzeria Boneschi Sarl	Italy	*233*	75-05000
C.I.A.M.-E.A. Sarl	Italy	*233*	75-05150
COES (Derby) Ltd	UK	*441*	75-14070
Container Research Corp.	USA	*566*	75-21700
CP Cases	UK	*443*	75-15680
DSM	Netherlands	*292*	75-01730
EDAK AG	Switzerland	*385*	75-03200
Engineered Air Systems Inc.	USA	*573*	75-31560
EPS Logistics Technology Ltd	UK	*502*	80-09550
FPT Inc.	USA	*576*	75-35760
H. Fine & Son Ltd	UK	*451*	75-23210
Hardigg Cases.	USA	*580*	75-41930
H.C. Slingsby plc	UK	*484*	75-54430
Instrument Enclosures	USA	*585*	75-46890
Insumat Ltd	UK	*461*	75-31660
L'hotellier Montrichard	France	*138*	75-24900
Light Alloy Ltd.	UK	*464*	75-35560
L.P. Foreman & Sons Ltd.	UK	*452*	75-24160
Metric Systems Corp.	USA	*596*	75-60800
NEFAB Sarl	France	*140*	75-27750
Odenwald Werke Rittersbach	Germany	*173*	75-22850
Pathfinder Fox Ltd	UK	*473*	75-44065
PD Technical Mouldings plc	UK	*473*	75-44200
Pelican Products Inc.	USA	*603*	75-69125
Photon Beard Ltd	UK	*474*	75-45260
Polyformes Ltd	UK	*475*	75-46348
QED Design and Manufacture Ltd	UK	*476*	75-47350
Rowlinson Packaging Ltd	UK	*481*	75-50600
Shock-Tech Inc.	USA	*614*	75-80457
Société d'Exploit Conditionnement	France	*146*	75-39455
St Bernard Composites Ltd	UK	*486*	75-56222
Star Case Manufacturing Co. Inc.	USA	*617*	75-83955
Strachan & Henshaw Ltd.	UK	*512*	80-47400
Thermodyne International Ltd	USA	*622*	75-88690
Westford Plastics & Engineering Ltd	UK	*493*	75-65450
Zero Cases (Europe)	UK	*495*	75-67700
Zero Enclosures	USA	*629*	75-98150
3C Systems, Inc.	USA	*632*	80-05000

722 10 30 Containers, transport

Air-Log Ltd	UK	*429*	75-01800
Allibert Handling	UK	*429*	75-02745
Barry Controls Ltd	UK	*433*	75-05930
Bigg & Robertson Plastics Ltd	UK	*434*	75-06680
COES (Derby) Ltd	UK	*441*	75-14070
CP Cases	UK	*443*	75-15680
EPS Logistics Technology Ltd	UK	*502*	80-09550
Hardigg Cases.	USA	*580*	75-41930
H.C. Slingsby plc	UK	*484*	75-54430
Irvin Aerospace Limited	UK	*462*	75-32050
L'hotellier Montrichard	France	*138*	75-24900
NKI	Netherlands	*294*	75-05100
PD Technical Mouldings plc	UK	*473*	75-44200
Polyformes Ltd	UK	*475*	75-46348
Precision Machining Inc.	USA	*605*	75-71350
QED Design and Manufacture Ltd	UK	*476*	75-47350
Reynolds Boughton Ltd	UK	*479*	75-49340
Thermodyne International Ltd	USA	*622*	75-88690
Valsella Meccanotecnica SpA	Italy	*239*	75-26800
Zero Cases (Europe)	UK	*495*	75-67700

Product–Supplier Link

722 10 30 10 Containers, ammunition transport

Air-Log Ltd	UK	*429*	75-01800
C.I.A.M.-E.A. Sarl	Italy	*233*	75-05150
COES (Derby) Ltd	UK	*441*	75-14070
Diehl GmbH & Co	Germany	*167*	75-06500
EPS Logistics Technology Ltd	UK	*502*	80-09550
Giat Industries	France	*126*	55-00100
H. Fine & Son Ltd	UK	*451*	75-23210
Hardigg Cases	USA	*580*	75-41930
L'hotellier Montrichard	France	*138*	75-24900
Nuova Connavi Srl	Italy	*237*	75-16700
Permali Gloucester Ltd	UK	*474*	75-44850
Polyformes Ltd	UK	*475*	75-46348
Progress-Werk Oberkirch AG	Germany	*173*	75-24320
Rowlinson Packaging Ltd	UK	*481*	75-50600
Royal Ordnance	UK	*481*	75-50620
St Bernard Composites Ltd	UK	*486*	75-56222
Star Case Manufacturing Co. Inc.	USA	*617*	75-83955
Strachan & Henshaw Ltd	UK	*512*	80-47400
Trelleborg Viking A/S	Norway	*311*	75-02950
Valsella Meccanotecnica SpA	Italy	*239*	75-26800

722 10 30 40 Jerrycans, fuel / water

BCB International Ltd	UK	*433*	75-06075
Pronal SA	France	*141*	75-31100
Weltonhurst Ltd	UK	*493*	75-65135

722 10 30 50 Missile containers, man-portable

Air-Log Ltd	UK	*429*	75-01800
COES (Derby) Ltd	UK	*441*	75-14070
CP Cases	UK	*443*	75-15680
EPS Logistics Technology Ltd	UK	*502*	80-09550
Giat Industries	France	*126*	55-00100
Hardigg Cases	USA	*580*	75-41930
L'hotellier Montrichard	France	*138*	75-24900
Nuova Connavi Srl	Italy	*237*	75-16700
QED Design and Manufacture Ltd	UK	*476*	75-47350
St Bernard Composites Ltd	UK	*486*	75-56222
Star Case Manufacturing Co. Inc.	USA	*617*	75-83955
Strachan & Henshaw Ltd	UK	*512*	80-47400
Technical Products Group Inc.	USA	*620*	75-86764

722 10 30 60 Containers, transport, refrigerated

Allibert Handling	UK	*429*	75-02745
Carrozzeria Boneschi Sarl	Italy	*233*	75-05000
VSG Verfahrenstechnik	Germany	*176*	75-33030

722 15 Protective transport covers

Aerial Machine & Tool Corp.	USA	*546*	75-01105
Air-Log Ltd	UK	*429*	75-01800
Andrew Mitchell & Co. Ltd	UK	*468*	75-40100
CP Cases	UK	*443*	75-15680
EPS Logistics Technology Ltd	UK	*502*	80-09550
H. Fine & Son Ltd	UK	*451*	75-23210
J&D Wilkie Ltd	UK	*494*	75-66240
L'hotellier Montrichard	France	*138*	75-24900
NKI	Netherlands	*294*	75-05100
PD Technical Mouldings plc	UK	*473*	75-44200
QED Design and Manufacture Ltd	UK	*476*	75-47350
Sattler Texilwerke KG	Austria	*34*	75-02155
Westford Plastics & Engineering Ltd	UK	*493*	75-65450
White Cross Rubber Products Ltd	UK	*494*	75-65857

722 60 Packing cases & containers

AAR Corp	USA	*629*	80-00035
Aerial Machine & Tool Corp.	USA	*546*	75-01105
Air-Log Ltd	UK	*429*	75-01800
Allibert Handling	UK	*429*	75-02745
Brownell Ltd	UK	*437*	75-10385
C.I.A.M.-E.A. Sarl	Italy	*233*	75-05150
CP Cases	UK	*443*	75-15680
Engineered Air Systems Inc.	USA	*573*	75-31560
EPS Logistics Technology Ltd	UK	*502*	80-09550
Instrument Enclosures	USA	*585*	75-46890
Irvin Industries Inc.	USA	*586*	75-48960
L'hotellier Montrichard	France	*138*	75-24900
NEFAB Sarl	France	*140*	75-27750
Nuova Connavi Srl	Italy	*237*	75-16700
Pathfinder Fox Ltd	UK	*473*	75-44065
PD Technical Mouldings plc	UK	*473*	75-44200
Polyformes Ltd	UK	*475*	75-46348
Polyziv	Israel	*221*	75-17580
Rowlinson Packaging Ltd	UK	*481*	75-50600
Shaw Packing Ltd	UK	*511*	80-34700
Star Case Manufacturing Co. Inc.	USA	*617*	75-83955
Thermodyne International Ltd	USA	*622*	75-88690
Zero Cases (Europe)	UK	*495*	75-67700
Zero Enclosures	USA	*629*	75-98150

722 65 Pads, energy dissipating, aerial delivery

G P Embelton & Co. Pty Ltd	Australia	*16*	75-05740
Irvin Aerospace Canada Limited	Canada	*73*	75-08300
3C Systems, Inc.	USA	*632*	80-05000

722 70 Pallets, materiel transport

AAR Corp	USA	*629*	80-00035
Air-Log Ltd	UK	*429*	75-01800
Allibert Handling	UK	*429*	75-02745
Aveling Barford (Machines) plc	UK	*432*	75-04550
EPS Logistics Technology Ltd	UK	*502*	80-09550
G P Embelton & Co. Pty Ltd	Australia	*16*	75-05740
Irvin Aerospace Canada Limited	Canada	*73*	75-08300
Irvin Aerospace Limited	UK	*462*	75-32050
Polyziv	Israel	*221*	75-17580
Shaw Packing Ltd	UK	*511*	80-34700
Technical Products Group Inc.	USA	*620*	75-86764

725 Enclosure systems & equipment

Air-Log Ltd	UK	*429*	75-01800
AMCO Engineering	USA	*552*	75-05200
A.P. Metalcraft Ltd	UK	*430*	75-03410
C & R Systems Inc.	USA	*561*	75-16016
Churchill Corp.	USA	*563*	75-18480
CP Cases	UK	*443*	75-15680
EG&G Inc.	USA	*571*	75-29555
Gichner Systems Group	USA	*579*	75-39610
Hardigg Cases	USA	*580*	75-41930
Knürr-Mechanik	Germany	*171*	75-16150
Nilsen Electric (SA) Pty Ltd.	Australia	*19*	75-14680
Optima Enclosures Ltd	UK	*472*	75-43135
ROSE Elektrotechnik GmbH & Co KG	Germany	*174*	75-25630
Semi-Alloys Co.	USA	*613*	75-79300
Shock-Tech Inc.	USA	*614*	75-80457
Zero Cases (Europe)	UK	*495*	75-67700

725 20 Consoles

Barry Controls Ltd	UK	*433*	75-05930
CP Cases	UK	*443*	75-15680
Devtek Corporation	Canada	*71*	75-04300
Knürr-Mechanik	Germany	*171*	75-16150
Lista Ltd	Switzerland	*386*	75-07350
Nilsen Electric (SA) Pty Ltd.	Australia	*19*	75-14680
Optima Enclosures Ltd	UK	*472*	75-43135
ROSE Elektrotechnik GmbH & Co KG	Germany	*174*	75-25630
Shock-Tech Inc.	USA	*614*	75-80457

725 60 Racks

Air-Log Ltd	UK	*429*	75-01800
Apex Storage Systems Ltd	UK	*431*	75-03420
Barry Controls Ltd	UK	*433*	75-05930
CP Cases	UK	*443*	75-15680
EG&G Inc.	USA	*571*	75-29555
Hardigg Cases	USA	*580*	75-41930
Knürr-Mechanik	Germany	*171*	75-16150
Lista Ltd	Switzerland	*386*	75-07350
Optima Enclosures Ltd	UK	*472*	75-43135
ROSE Elektrotechnik GmbH & Co KG	Germany	*174*	75-25630
Shock-Tech Inc.	USA	*614*	75-80457

Star Case Manufacturing Co. Inc.	USA	617	75-83955
Stop-Choc Ltd.	UK	486	75-57345

Uniforms, Clothing, Webbing & Personal Equipment

731 Uniforms, hats & caps

Albion Clothing Ltd	New Zealand	301	75-00110
Bruck Textiles plc	Australia	13	75-02810
Checkmate Industries Ltd	Hong Kong SAR	198	75-00200
Checkmate Industries of EEC Ltd	Denmark	102	75-03300
Compton Webb Ltd	UK	442	75-14550
Dynamic (India) Pvt Ltd	India	205	75-00540
Hobson & Sons (London) Ltd	UK	458	75-29550
H.R. Denne Ltd	UK	445	75-17550
Hyundai	Korea (ROK)	265	75-01540
Ideal Knitwear Manufacturing Ltd	Canada	72	75-07720
M. Kumar Udyog	India	205	75-02100
Military Logistics MFC	USA	596	75-61550
Pains-Wessex Ltd	UK	472	75-43510
Rabintex Industries Ltd	Israel	221	75-17850
Sparks & Ellis Uniforms	S. Africa	359	75-19390
SPL Ordnance Ltd	Israel	225	80-01935
Toye, Kenning & Spencer Ltd	UK	489	75-60600
Vidaro Corp.	USA	626	75-94000

732 Other clothing, workwear

Beaufort Air-Sea Equipment Ltd	UK	433	75-06100
Bruck Textiles plc	Australia	13	75-02810
CMI - Ballarat	Australia	14	75-03600
Colan Products	Australia	14	75-03650
Compton Webb Ltd	UK	442	75-14550
Dynamic (India) Pvt Ltd	India	205	75-00540
H.R. Denne Ltd	UK	445	75-17550
Hyundai	Korea (ROK)	265	75-01540
I.Spiewak & Sons Inc.	USA	617	75-83300
M. Kumar Udyog	India	205	75-02100
Mustang Survival (USA) Corp	USA	597	75-63480
P.A.D. International	UK	472	75-43470
Pains-Wessex Ltd	UK	472	75-43510
Pittards plc	UK	475	75-45700
Rabintex Industries Ltd	Israel	221	75-17850
Sächsische Spezialkonfektion GmbH	Germany	174	75-25880
Sparks & Ellis Uniforms	S. Africa	359	75-19390
SPL Ordnance Ltd	Israel	225	80-01935
Toye, Kenning & Spencer Ltd	UK	489	75-60600
T.W.Kempton (Corporate Clothing) Ltd	UK	463	75-33600
Vidaro Corp.	USA	626	75-94000

732 10 Combat clothing

Avon Technical Products	UK	432	75-05150
BCB International Ltd	UK	433	75-06075
Beaufort Air-Sea Equipment Ltd	UK	433	75-06100
Bruck Textiles plc	Australia	13	75-02810
C.I.A.M.-E.A. Sarl	Italy	233	75-05150
Colan Products	Australia	14	75-03650
Compton Webb Ltd	UK	442	75-14550
I.Spiewak & Sons Inc.	USA	617	75-83300
M. Kumar Udyog	India	205	75-02100
Nottingham International	USA	640	80-22000
Pains-Wessex Ltd	UK	472	75-43510
Radar	UK	477	75-47760
Sparks & Ellis Uniforms	S. Africa	359	75-19390
SPL Ordnance Ltd	Israel	225	80-01935
Toye, Kenning & Spencer Ltd	UK	489	75-60600

732 10 40 Flying suits (non-g)

Aeraustral Trading	Australia	11	75-00330
Avirex Ltd	USA	556	75-11250
A.W. Hainsworth & Sons Ltd	UK	456	75-27300
Beaufort Air-Sea Equipment Ltd	UK	433	75-06100
Bruck Textiles plc	Australia	13	75-02810
Colan Products	Australia	14	75-03650
EROS GIE	France	137	75-19500
Flight Suits Ltd	USA	575	75-34800
Germa AB	Sweden	377	75-05700
I.Spiewak & Sons Inc.	USA	617	75-83300
Mustang Survival Corp.	Canada	74	75-09860
Mustang Survival (USA) Corp	USA	597	75-63480
Pains-Wessex Ltd	UK	472	75-43510

732 10 50 Knitted clothing

Colan Products	Australia	14	75-03650
Ideal Knitwear Manufacturing Ltd	Canada	72	75-07720
Pains-Wessex Ltd	UK	472	75-43510
Toye, Kenning & Spencer Ltd	UK	489	75-60600
T.W.Kempton (Corporate Clothing) Ltd	UK	463	75-33600

732 10 60 Overalls, disposable

Colan Products	Australia	14	75-03650
Dynamic (India) Pvt Ltd	India	205	75-00540
H.C. Slingsby plc	UK	484	75-54430
Pains-Wessex Ltd	UK	472	75-43510
Plastico Ltd	UK	475	75-45950

732 10 70 Cold weather clothing

Albion Clothing Ltd	New Zealand	301	75-00110
BCB International Ltd	UK	433	75-06075
Berghaus Ltd	UK	434	75-06525
Bruck Textiles plc	Australia	13	75-02810
Colan Products	Australia	14	75-03650
Compton Webb Ltd	UK	442	75-14550
Dynamic (India) Pvt Ltd	India	205	75-00540
Ideal Knitwear Manufacturing Ltd	Canada	72	75-07720
I.Spiewak & Sons Inc.	USA	617	75-83300
M. Kumar Udyog	India	205	75-02100
Mountain Range	UK	469	75-41100
Mustang Survival Corp.	Canada	74	75-09860
Pains-Wessex Ltd	UK	472	75-43510
Paul Boyé	France	133	75-07020
Sparks & Ellis Uniforms	S. Africa	359	75-19390
The North Face (Europe) Ltd	UK	471	75-42727

733 Footwear, military

Blundstone Pty Ltd	Australia	12	75-02490
Checkmate Industries Ltd	Hong Kong SAR	198	75-00200
Danner Shoe Manufacturing Co.	USA	567	75-23570
G.B. Britton	UK	437	75-09900
Haynes & Cann Ltd	UK	457	75-28750
Hobson & Sons (London) Ltd	UK	458	75-29550
Hyundai	Korea (ROK)	265	75-01540
M. Kumar Udyog	India	205	75-02100
Pains-Wessex Ltd	UK	472	75-43510
Pittards plc	UK	475	75-45700
Rabintex Industries Ltd	Israel	221	75-17850
Specialty Plastic Products Inc.	USA	616	75-83035
SPL Ordnance Ltd	Israel	225	80-01935
Tebbutt & Hall Ltd	UK	487	75-58800
Toye, Kenning & Spencer Ltd	UK	489	75-60600

733 10 Boots, military

Blundstone Pty Ltd	Australia	12	75-02490
BTR Silvertown Ltd	UK	437	75-10830
Checkmate Industries of EEC Ltd	Denmark	102	75-03300
Danner Shoe Manufacturing Co.	USA	567	75-23570
M. Kumar Udyog	India	205	75-02100
Nottingham International	USA	640	80-22000
Pains-Wessex Ltd	UK	472	75-43510
Sala International Pty Ltd	Australia	21	75-18980
Specialty Plastic Products Inc.	USA	616	75-83035
SPL Ordnance Ltd	Israel	225	80-01935
Toye, Kenning & Spencer Ltd	UK	489	75-60600

733 10 10 Boots, aircrew

Haynes & Cann Ltd	UK	457	75-28750
M. Kumar Udyog	India	205	75-02100

Product-Supplier Link

Nottingham International	USA	*640*	80-22000
SPL Ordnance Ltd.	Israel	*225*	80-01935

733 10 50 Boots, parachutist

G.B. Britton	UK	*437*	75-09900
Haynes & Cann Ltd	UK	*457*	75-28750
M. Kumar Udyog	India	*205*	75-02100
Nottingham International	USA	*640*	80-22000
Sala International Pty Ltd	Australia	*21*	75-18980
SPL Ordnance Ltd.	Israel	*225*	80-01935

733 20 Shoes, military

Blundstone Pty Ltd	Australia	*12*	75-02490
G.B. Britton	UK	*437*	75-09900
M. Kumar Udyog	India	*205*	75-02100
Nottingham International	USA	*640*	80-22000
Oliver Footwear	Australia	*20*	75-15645
SPL Ordnance Ltd.	Israel	*225*	80-01935
Toye, Kenning & Spencer Ltd	UK	*489*	75-60600

733 20 50 Snow shoes

ROFi Industrier AS	Norway	*310*	75-02300

733 30 Hosiery, socks

BCB International Ltd	UK	*433*	75-06075
Dynamic (India) Pvt Ltd	India	*205*	75-00540
Pains-Wessex Ltd	UK	*472*	75-43510
Toye, Kenning & Spencer Ltd	UK	*489*	75-60600

734 Webbing & personal equipment

A Haberkorn GmbH & Co KG	Austria	*33*	75-01140
Achidatex Nazareth Elite (1977) Ltd	Israel	*217*	75-00200
Aeraustral Trading	Australia	*11*	75-00330
Airborne Industries	UK	*428*	75-01550
Aircraft Materials Ltd	UK	*428*	75-01650
Avon Technical Products	UK	*432*	75-05150
Barrday Inc.	Canada	*68*	75-01080
Barrow Hepburn Sala Ltd.	UK	*433*	75-05830
Berghaus Ltd	UK	*434*	75-06525
Checkmate Industries Ltd	Hong Kong SAR	*198*	75-00200
Checkmate Industries of EEC Ltd	Denmark	*102*	75-03300
Compton Webb Ltd	UK	*442*	75-14550
CQC Plc	UK	*443*	75-15700
C.Whitehead & Co. Ltd.	UK	*494*	75-65890
Dynamic (India) Pvt Ltd	India	*205*	75-00540
Gelvenor Textiles	S. Africa	*358*	75-06100
H. Koch & Sons Co.	USA	*589*	75-53030
Hyundai	Korea (ROK)	*265*	75-01540
Jarvis Manufacturing Co.	UK	*462*	75-32330
Kohlbrat & Bunz GmbH	Austria	*33*	75-01575
M. Kumar Udyog	India	*205*	75-02100
Military Logistics MFC	USA	*596*	75-61550
P.A.D. International	UK	*472*	75-43470
Rabintex Industries Ltd	Israel	*221*	75-17850
Shooting Systems Group Inc.	USA	*614*	75-80460
Specialty Plastic Products Inc.	USA	*616*	75-83035
Tecelagem Lady Ltda.	Brazil	*58*	75-20140
Trojan Manufacturing & Marketing Inc	Philippines	*321*	75-01350
Victrix Ltd	Canada	*77*	75-15000

734 10 Bayonets

A. Eickhorn GmbH & Co	Germany	*168*	75-07400
Giat Industries	France	*126*	55-00100

734 10 50 Bayonet scabbards

Giat Industries	France	*126*	55-00100
Manta Industries Ltd	Canada	*73*	75-09250

734 20 Canteens & mugs

Nottingham International	USA	*640*	80-22000
Trojan Manufacturing & Marketing Inc	Philippines	*321*	75-01350

734 40 Holsters, pistol

Accuracy International Ltd	UK	*427*	75-00360
Armalon Ltd	UK	*431*	75-03729
ASTRA Sport SA	Spain	*365*	75-01500
C.I.A.M.-E.A. Sarl	Italy	*233*	75-05150
H. Fine & Son Ltd	UK	*451*	75-23210
Hobson & Sons (London) Ltd	UK	*458*	75-29550
Nottingham International	USA	*640*	80-22000
Paul Evers Holsters	UK	*473*	75-44080
Price Western Leather Co. Ltd.	UK	*476*	75-46755
Safariland Ltd Inc.	USA	*612*	75-77350
Shooting Systems Group Inc.	USA	*614*	75-80460

734 43 Kit bags

Andrew Mitchell & Co. Ltd	UK	*468*	75-40100
Avon Technical Products	UK	*432*	75-05150
Barrday Inc.	Canada	*68*	75-01080
Barrow Hepburn Sala Ltd.	UK	*433*	75-05830
BCB International Ltd	UK	*433*	75-06075
CQC Plc	UK	*443*	75-15700
M. Kumar Udyog	India	*205*	75-02100
Manta Industries Ltd	Canada	*73*	75-09250
Nottingham International	USA	*640*	80-22000

734 45 Knives

A. Eickhorn GmbH & Co	Germany	*168*	75-07400
BCB International Ltd	UK	*433*	75-06075
Ets Jean Tanazacq	France	*147*	75-43670
Nottingham International	USA	*640*	80-22000
US Army Equipment Guns and Ammo	Belgium	*47*	75-06345

734 45 10 Knives, aircrew emergency

BCB International Ltd	UK	*433*	75-06075
Nottingham International	USA	*640*	80-22000

734 45 30 Knives, fighting / combat

A. Eickhorn GmbH & Co	Germany	*168*	75-07400
Ets Jean Tanazacq	France	*147*	75-43670
Nottingham International	USA	*640*	80-22000
US Army Equipment Guns and Ammo	Belgium	*47*	75-06345
Wilkinson Sword Ltd	UK	*494*	75-66250

734 45 60 Knives, survival (see code 652.50.40)

A. Eickhorn GmbH & Co	Germany	*168*	75-07400
BCB International Ltd	UK	*433*	75-06075
Ets Jean Tanazacq	France	*147*	75-43670
Nottingham International	USA	*640*	80-22000
Wilkinson Sword Ltd	UK	*494*	75-66250

734 50 Mosquito nets

Aeraustral Trading	Australia	*11*	75-00330
BCB International Ltd	UK	*433*	75-06075
C.Whitehead & Co. Ltd.	UK	*494*	75-65890
M. Kumar Udyog	India	*205*	75-02100
Manta Industries Ltd	Canada	*73*	75-09250
Nottingham International	USA	*640*	80-22000
Swiss Net UK plc	UK	*487*	75-58080

734 50 60 Ropes / cords, mountain climbing

Aircraft Materials Ltd	UK	*428*	75-01650
Roblon A/S	Denmark	*103*	75-18400

734 66 Sleeping bags

Andrew Mitchell & Co. Ltd	UK	*468*	75-40100
Avon Technical Products	UK	*432*	75-05150
BCB International Ltd	UK	*433*	75-06075
Brett Harris Ltd	UK	*435*	75-08000
Bruck Textiles plc	Australia	*13*	75-02810
CQC Plc	UK	*443*	75-15700
Gelvenor Textiles	S. Africa	*358*	75-06100
GIII Ltd	Canada	*71*	75-06050
Gold-Eck Austria	Austria	*33*	75-01060
H. Fine & Son Ltd	UK	*451*	75-23210
M. Kumar Udyog	India	*205*	75-02100
Nottingham International	USA	*640*	80-22000
Polywarm Products Ltd	UK	*475*	75-46356
Research Institute for	Ukraine	*406*	60-00990
The North Face (Europe) Ltd	UK	*471*	75-42727
US Army Equipment Guns and Ammo	Belgium	*47*	75-06345

734 90 Water bags, canvas

Aerial Machine & Tool Corp.	USA	*546*	75-01105
Andrew Mitchell & Co. Ltd	UK	*468*	75-40100
Avon Technical Products	UK	*432*	75-05150
Barrday Inc.	Canada	*68*	75-01080
H. Fine & Son Ltd	UK	*451*	75-23210
Jarvis Manufacturing Co.	UK	*462*	75-32330
Manta Industries Ltd	Canada	*73*	75-09250

734 95 Webbing, infantry

Airborne Industries	UK	*428*	75-01550
Ancra CHS BV	Netherlands	*291*	75-00510
A.W. Hainsworth & Sons Ltd	UK	*456*	75-27300
CQC Plc	UK	*443*	75-15700
C.Whitehead & Co. Ltd	UK	*494*	75-65890
Gelvenor Textiles	S. Africa	*358*	75-06100
Manta Industries Ltd	Canada	*73*	75-09250
Nottingham International	USA	*640*	80-22000
Sala International Pty Ltd	Australia	*21*	75-18980
US Army Equipment Guns and Ammo	Belgium	*47*	75-06345

734 95 10 Load carrying equipment, backpacks

Berghaus Ltd	UK	*434*	75-06525

735 Watches, chronographs

Breitling Ltd	Switzerland	*385*	75-01400
M. Paz Logistics Ltd	Israel	*224*	80-01600
Stocker & Yale Inc.	USA	*618*	75-84250

735 20 Chronographs

Avirex Ltd	USA	*556*	75-11250
Breitling Ltd	Switzerland	*385*	75-01400
Stocker & Yale Inc.	USA	*618*	75-84250

735 25 Compass watches

KVH Industries Inc.	USA	*589*	75-53670
M. Paz Logistics Ltd	Israel	*224*	80-01600

735 50 Watches, pilots

Avirex Ltd	USA	*556*	75-11250
Breitling Ltd	Switzerland	*385*	75-01400
Flight Suits Ltd	USA	*575*	75-34800
Stocker & Yale Inc.	USA	*618*	75-84250

Ceremonial equipment

741 Accoutrements, medals, insignia

A.R. Fabb Bros. Ltd.	UK	*449*	75-22160
BOMISA	Italy	*233*	75-03120
Garrard & Co Ltd	UK	*453*	75-25205
Hobson & Sons (London) Ltd	UK	*458*	75-29550
M. Kumar Udyog	India	*205*	75-02100
Rabintex Industries Ltd	Israel	*221*	75-17850
Thomas Fattorini Ltd	UK	*450*	75-22450
Toye, Kenning & Spencer Ltd	UK	*489*	75-60600

741 60 Swords & scabbards

BOMISA	Italy	*233*	75-03120
Hobson & Sons (London) Ltd	UK	*458*	75-29550
Thomas Fattorini Ltd	UK	*450*	75-22450
Toye, Kenning & Spencer Ltd	UK	*489*	75-60600
Wilkinson Sword Ltd	UK	*494*	75-66250

742 Flags, standards, pennants

A.R. Fabb Bros. Ltd.	UK	*449*	75-22160
Avirex Ltd	USA	*556*	75-11250
Bruck Textiles plc	Australia	*13*	75-02810
Hobson & Sons (London) Ltd	UK	*458*	75-29550
Hutchwilco Limited	New Zealand	*302*	75-00860
Rabintex Industries Ltd	Israel	*221*	75-17850
Toye, Kenning & Spencer Ltd	UK	*489*	75-60600
Zephyr Flags & Banners	UK	*495*	75-67680

743 Military band regalia

A.R. Fabb Bros. Ltd.	UK	*449*	75-22160
Hobson & Sons (London) Ltd	UK	*458*	75-29550
Rabintex Industries Ltd	Israel	*221*	75-17850
Toye, Kenning & Spencer Ltd	UK	*489*	75-60600

744 Musical instruments

Toye, Kenning & Spencer Ltd	UK	*489*	75-60600

Cleaning equipment, high pressure

750 20 Aircraft cleaning equipment, high pressure (see code 622.60)

Alfred Karcher GmbH & Co	Germany	*170*	75-15650
American Cleaning Manufacturing Co.	USA	*553*	75-05940
Clyde Sturtevant Ltd	UK	*441*	75-13868
IFI Srl	Italy	*235*	75-11220
Karcher France	France	*138*	75-23500
Karcher (UK) Ltd.	UK	*463*	75-33220
Vapormatt Ltd	UK	*490*	75-62630

750 80 Vehicle cleaning equipment, high pressure (see code 622.60)

Alfred Karcher GmbH & Co	Germany	*170*	75-15650
American Cleaning Manufacturing Co.	USA	*553*	75-05940
IFI Srl	Italy	*235*	75-11220
Karcher France	France	*138*	75-23500
Karcher (UK) Ltd.	UK	*463*	75-33220
Vapormatt Ltd	UK	*490*	75-62630

Railway Equipment

781 Permanent way & components

Hyundai	Korea (ROK)	*265*	75-01540
MAN Gutehoffnungshütte AG	Germany	*172*	75-19620
Permali Gloucester Ltd	UK	*474*	75-44850
Transfield Pty Ltd	Australia	*22*	75-20810

782 Railway movement control systems & equipment

Conrac Elektron GmbH	Germany	*166*	75-05100
MPE Ltd	UK	*469*	75-41210

Product-Supplier Link

SFIM	**France**	***144***	**75-35080**
Siemens Switzerland Ltd	Switzerland	*387*	75-11980

782 60 Railway signalling systems & equipment

Chrysanthacopoulos SA	Greece	*188*	75-00910
Étienne Lacroix	France	*138*	75-24050
MPE Ltd	UK	*469*	75-41210
Siemens Switzerland Ltd	Switzerland	*387*	75-11980

782 70 Railway wagon control systems & equipment

SFIM	**France**	***144***	**75-35080**

Miscellaneous Equipment & Components

791 Components

Abbott Electronics Inc.	USA	*545*	75-00150
Abbott Electronics Ltd	UK	*426*	75-00180
ABLE Corp.	USA	*545*	75-00260
Abrams Instrument Corp.	USA	*545*	75-00280
Acofab SA	France	*130*	75-00200
Advanced Semiconductor Inc.	USA	*546*	75-00950
Advanced Technical Corp.	USA	*546*	75-01000
Aeromet International plc	UK	*428*	75-01040
Aimsco Inc.	USA	*547*	75-03050
Air-Log Ltd	UK	*429*	75-01800
AIROD Sdn Bhd	Malaysia	*278*	80-00050
Alan Industries Inc.	USA	*548*	75-04205
Albacom Ltd	UK	*429*	75-02372
Amphenol Ltd	UK	*430*	75-03000
Analog Modules Inc.	USA	*553*	75-07140
A.O. Henton Engineering Co Ltd.	UK	*458*	75-29040
API-Harowe	USA	*554*	75-08015
APM Hexseal	USA	*554*	75-08030
Associated Aircraft Mfg & Sales Inc.	USA	*555*	75-09500
Australasian Steel Products Pty Ltd.	Australia	*12*	75-01720
Avica US	USA	*556*	75-11100
Aydin Corp.	USA	*556*	75-11450
BAS Components Ltd	UK	*433*	75-06000
BIW Connector Systems Inc.	USA	*559*	75-14085
Blanc Aéro Industries	France	*132*	75-06900
Bospec Neue Technologien	Germany	*166*	75-03420
Bowmar Instrument Corp.	USA	*560*	75-14860
British Sarozal Ltd.	UK	*436*	75-09570
BTR Aerospace Australia	Australia	*13*	75-02815
Busak + Shamban Inc.	USA	*561*	75-16006
Canadian Marconi Co.	Canada	*69*	75-02600
Capco Inc.	USA	*562*	75-16650
Carter Manufacturing Corp.	USA	*562*	75-16800
Cleveland Crystals Inc.	USA	*564*	75-19040
Climax Engineering Pty Ltd.	Australia	*14*	75-03480
CMI - Ballarat	Australia	*14*	75-03600
Communications Instruments Inc.	USA	*565*	75-20300
Computer Optics Inc.	USA	*565*	75-20680
Consolidated Manufacturing	Australia	*15*	75-03925
Crystallume	USA	*563*	75-18425
CTS Corporation	USA	*567*	75-22800
Curtis Instruments (UK) Ltd	UK	*444*	75-16440
Dare Electronics	USA	*567*	75-23580
Detoronics Corporation	USA	*569*	75-25850
Deutsch Relays Inc.	USA	*569*	75-25930
D.G. O'Brien UK Ltd.	UK	*471*	75-42895
Diamond H Controls Limited	UK	*445*	75-18025
Dixi-Fresard Microtechniques SA	France	*135*	75-13680
Dowty Aerospace Wolverhampton	UK	*446*	75-18589
Dunlop Aviation Division	UK	*447*	75-19000
Dynaco Corp.	USA	*570*	75-27900
Eaton Corporation	USA	*571*	75-28800
Edal Industries Inc.	USA	*571*	75-29030
EG&G Inc.	USA	*571*	75-29555
EG&G Reticon	USA	*571*	75-29562
EG&G Sealol Ltd	UK	*448*	75-19675
EG&G Sealol Pty Ltd	Australia	*15*	75-05350
Egli Fischer & Co AG	Switzerland	*386*	75-03348
Eldec Corp.	USA	*572*	75-29660
Electronic Concepts Inc.	USA	*572*	75-29850
Elettronica (UK) Ltd	UK	*448*	75-20475
Esterline Technologies	USA	*574*	75-32360
Ets Bollé	France	*133*	75-06960
Fairview Machine	USA	*575*	75-33600
Ferrite Domain Scientific Production	Russia	*333*	55-07300
FFA Flugzeugwerke Altenrhein AG	Switzerland	*386*	75-04350
Fluorocarbon Company Limited	UK	*451*	75-24020
FPI Inc.	USA	*576*	75-35755
Framatome Connectors International	France	*136*	75-18405
Gabriel Electronics Inc.	USA	*577*	75-36715
Garofoli Costruzioni Meccaniche SpA	Italy	*235*	75-09550
Gates Hydraulics Ltd	UK	*453*	75-25210
GEC-Marconi Infra-Red Limited	UK	*454*	75-25338
GEC-Marconi Materials Technology Ltd	UK	*419*	60-00600
General Pneumatics	USA	*578*	75-38950
Giat Industries	France	*126*	55-00100
Graflex Inc	USA	*579*	75-40310
G.W. Lisk Co. Inc.	USA	*590*	75-55300
Hawke Cable Glands Ltd	UK	*457*	75-28475
Herley-Vega Systems	USA	*581*	75-43090
H.I. Fraser Pty Ltd.	Australia	*16*	75-06835
Hill Industries Inc.	USA	*582*	75-43465
Holt Integrated Circuits	USA	*582*	75-44060
Honeywell Control Systems Ltd	UK	*458*	75-29730
Hughes Danbury Optical Sys.	USA	*582*	75-44857
Hy-Q International (UK) Ltd	UK	*460*	75-30925
II-VI Inc.	USA	*624*	75-91355
Industria SA	France	*137*	75-22300
ITT Cannon RF Products	UK	*462*	75-32170
Jay-El Products Inc.	USA	*586*	75-50100
Jennings Technology Corp.	USA	*586*	75-50350
JML Optical Industries Inc.	USA	*587*	75-50940
John Valves Pty Ltd	Australia	*18*	75-10400
J.T. Slocomb Company	USA	*615*	75-81700
KDI Precision Products Inc.	USA	*588*	75-51705
Kearfott Guidance & Navigation Corp.	USA	*588*	75-51760
Kern Engrng & Mfg Corp.	USA	*588*	75-52330
Kirovo Chepetsky Electro	Russia	*337*	75-11370
L&S Machine Co. Inc.	USA	*589*	75-53700
Linear Technology Corp.	USA	*590*	75-55220
Litton Industries Inc.	USA	*590*	75-55400
Lucas Aerospace FCS Paris	France	*139*	75-25390
M W Microwave Corp.	USA	*597*	75-63495
MA/COM Antenna & Cable Div.	USA	*593*	75-56953
Marconi Instruments Ltd	UK	*466*	75-37270
Matra Electronique	France	*139*	75-26005
Meller Optics Inc.	USA	*595*	75-59950
Micro Communications Inc.	USA	*596*	75-60980
Microelectronics, guidance and	Turkey	*402*	55-00050
Midcon Cables Co.	USA	*596*	75-61300
Middlesex Group Limited	UK	*468*	75-39330
MITEQ Inc.	USA	*597*	75-62000
MMR Technologies Inc.	USA	*597*	75-62200
Muller England Limited	UK	*470*	75-41550
NMB (USA) Inc	USA	*599*	75-64400
Novatronics Inc	Canada	*74*	75-10300
Nucletron Electronic Vertriebs GmbH	Germany	*179*	80-01500
Obrien D G Inc.	USA	*601*	75-65850
Omori Seikoki Co. Ltd	Japan	*254*	75-07310
Optical Cable Corp.	USA	*601*	75-66952
Oxley Developments Co. Ltd	UK	*472*	75-43400
Oxley Inc.	USA	*602*	75-67870
Parker Bertea Aerospace	USA	*602*	75-68900
Penny & Giles Blackwood Group	UK	*473*	75-44495
Penstock Inc.	USA	*603*	75-69200
Philips Components Pty Ltd	Australia	*20*	75-16490
Piezo Technology Inc.	USA	*604*	75-70000
Polytec-PI SA	France	*141*	75-30450
Power Spectra Inc.	USA	*605*	75-71085
Précitube SA	France	*141*	75-30890
Projects Unlimited Inc.	USA	*606*	75-71560
QES Ltd	UK	*509*	80-25000
Recoil Pty	Australia	*20*	75-18300
Recon/Optical Inc.	USA	*608*	75-73930
Reil Industrial Enterprises Ltd	Canada	*75*	75-12000
Richard Manufacturing Co.	USA	*609*	75-75355
Rocky Mountain Instrument Co.	USA	*611*	75-75960
Rogers Corp.	USA	*611*	75-76000
Rollet H. & Co. Ltd	UK	*479*	75-50208
Schaffner EMC Ltd	UK	*482*	75-51970
Schott Glass Technologies Inc.	USA	*613*	75-78410
Sector Microwave Industries	USA	*613*	75-79010
SFIM	**France**	***144***	**75-35080**
Sheldahl Inc.	USA	*614*	75-80150
Shipham & Co. Ltd	UK	*483*	75-53350
Siecor Corp.	USA	*614*	75-80560

Skurka Engineering Company	USA	*615*	75-81520
Société Métallurgique de Gerzat SA	France	*146*	75-40570
SP aerospace and vehicle systems bv	**Netherlands**	***295***	**75-06860**
Spaulding Composites	USA	*616*	75-82970
Spectrol Reliance Ltd	UK	*485*	75-55555
STC Keltec Corp.	USA	*588*	75-51950
Stork RMO BV	Netherlands	*295*	75-07220
Sumitomo Electric Lightwave Corp.	USA	*618*	75-84800
Taconic Plastics Ltd	USA	*619*	75-86020
Teadit Juntas SA	Brazil	*55*	75-00910
Telair International	USA	*620*	75-87400
Teltron Technologies Inc.	USA	*621*	75-87865
Tesko Engineering & Tooling	Australia	*22*	75-20400
The Barden Corporation (UK) Ltd	UK	*433*	75-05630
The Ray Engineering Company	UK	*478*	75-48350
Thomas Electronics Inc.	USA	*622*	75-88850
Timken do Brasil Com. e Ind. Ltda.	Brazil	*58*	75-20520
Torwico Electronics L.P.	USA	*623*	75-89800
Trompeter Electronics Inc.	USA	*623*	75-91070
Tungum Hydraulics Ltd	UK	*490*	75-61500
Valley Todeco	USA	*625*	75-93170
VEAM Srl	Italy	*239*	75-27100
Vectronics Microwave Corp.	USA	*625*	75-93490
Viking Ltd	UK	*491*	75-63230
Voi-Shan/Diessel GmbH	Germany	*176*	75-32870
Voltage Multipliers Inc.	USA	*626*	75-94430
Warner Electric SA	Switzerland	*388*	75-15150
Wesco Aircraft Hardware Corp.	USA	*627*	75-95700
Wessex Advanced Switching Products	UK	*493*	75-65330
Western Filament Inc.	USA	*627*	75-95930
Z. Rabiner Ltd	Israel	*221*	75-17840
Zeta	USA	*629*	75-98190

791 10 Components, electrical

AB Connectors Ltd	UK	*426*	75-00040
Abbott Electronics Inc.	USA	*545*	75-00150
Abbott Electronics Ltd	UK	*426*	75-00180
AIR-LB	France	*131*	75-01100
Air-Log Ltd	UK	*429*	75-01800
AMP of Great Britain Ltd	UK	*430*	75-03050
Amphenol Corp.	USA	*553*	75-06800
API-Harowe	USA	*554*	75-08015
AVX Ltd	UK	*433*	75-05190
BBH Windings Ltd	UK	*433*	75-06045
British Sarozal Ltd	UK	*436*	75-09570
Century Electronics	USA	*562*	75-17385
Chemring Countermeasures	UK	*439*	75-12800
Communications Instruments Inc.	USA	*565*	75-20300
Courtaulds Aerospace	UK	*443*	75-15330
Crompton Instruments Ltd	UK	*443*	75-16070
CTS Corporation	USA	*567*	75-22800
Datron/Transco Inc.	USA	*568*	75-24000
Deutsch Relays Inc.	USA	*569*	75-25930
Diamond H Controls Limited	UK	*445*	75-18025
DSM	Netherlands	*292*	75-01730
Dunlop Aviation Division	UK	*447*	75-19000
E-T-A Circuit Breakers Ltd	UK	*449*	75-21565
Eaton Corporation	USA	*571*	75-28800
EEV Inc.	USA	*571*	75-29500
EG&G Inc.	USA	*571*	75-29555
Egli Fischer & Co AG	Switzerland	*386*	75-03348
Electro Impulse Laboratory Inc.	USA	*572*	75-29700
Electro Mechanical Systems Ltd	UK	*502*	80-09380
Electro-Tec Corp.	USA	*573*	75-30700
Electronic Concepts Inc.	USA	*572*	75-29850
Electronic Specialty Corp.	USA	*572*	75-30150
Esterline Technologies	USA	*574*	75-32360
FDB Electrical Ltd	UK	*450*	75-22490
Fluorocarbon Company Limited	UK	*451*	75-24020
Framatome Connectors International	France	*136*	75-18405
GEC-Marconi Materials Technology Ltd	UK	*419*	60-00600
G.W. Lisk Co. Inc.	USA	*590*	75-55300
Holec Machines & Apparaten BV	Netherlands	*293*	75-03110
Hughes Microelectronics Europe Ltd	UK	*459*	75-30480
Hy-Q International (UK) Ltd	UK	*460*	75-30925
Industria SA	France	*137*	75-22300
ITT Cannon	USA	*586*	75-49452
Jennings Technology Corp.	USA	*586*	75-50350
Kirovo Chepetsky Electro	Russia	*337*	75-11370
Kollmorgen Artus	France	*138*	75-23820
Korry Electronics Co.	USA	*589*	75-53250
Litton Industries Inc.	USA	*590*	75-55400
Lucas Control Systems Products	UK	*465*	75-36675
Lyncole XIT Grounding	USA	*593*	75-56705
Mécanex SA	Switzerland	*386*	75-08300
MPE Ltd	UK	*469*	75-41210
Novatronics Inc	Canada	*74*	75-10300
Nucletron Electronic Vertriebs GmbH	Germany	*179*	80-01500
Omori Seikoki Co. Ltd	Japan	*254*	75-07310
Oxley Developments Co. Ltd	UK	*472*	75-43400
Park Products Ltd	UK	*473*	75-44000
Penstock Inc.	USA	*603*	75-69200
Portescap (UK) Ltd	UK	*475*	75-46420
Powerex Inc.	USA	*604*	75-71080
PowerTech Inc.	USA	*605*	75-71087
Rendar Ltd	UK	*478*	75-49226
Roband Electronics plc	UK	*479*	75-49880
Robertshaw Controls Co.	USA	*610*	75-75640
Rowe Industries Inc.	USA	*611*	75-76900
Sage Laboratories Inc.	USA	*612*	75-77610
Schaffner EMC Ltd	UK	*482*	75-51970
Sielman SA	Greece	*191*	75-02360
Skurka Engineering Company	USA	*615*	75-81520
Smith & Prince Ltd	UK	*485*	75-54630
Sparton Technology Inc.	USA	*616*	75-82950
Spectrol Reliance Ltd	UK	*485*	75-55555
TEC Electrical Components Ltd	UK	*487*	75-58840
Teledyne Electronic Technologies	USA	*620*	75-87522
Telektron Ltd	UK	*488*	75-59200
TMR-Tecnologías Metalúrgicas	Spain	*367*	75-05650
Tramec SA	France	*148*	75-45500
Trompeter Electronics Inc.	USA	*623*	75-91070
United Defense L.P.	USA	*624*	75-92050
United Moulders Ltd	UK	*490*	75-62130
VEAM Srl	Italy	*239*	75-27100
Vernitron Corp.	USA	*625*	75-93760
Viking Ltd	UK	*491*	75-63230
Western Filament Inc.	USA	*627*	75-95930
White Products BV	Netherlands	*296*	75-09100

791 10 20 Connectors, electrical

AB Connectors Ltd	UK	*426*	75-00040
AIR-LB	France	*131*	75-01100
Albacom Ltd	UK	*429*	75-02372
Amphenol Corp.	USA	*553*	75-06800
Amphenol Ltd	UK	*430*	75-03000
Australian AMP Pty Ltd	Australia	*12*	75-01780
Avica US	USA	*556*	75-11100
BIW Connector Systems Inc.	USA	*559*	75-14085
Burton Electrical Engineering	USA	*561*	75-16005
Claymount Assemblies BV	Netherlands	*292*	75-01150
Compac	USA	*565*	75-20400
Comtronic GmbH	Germany	*166*	75-05070
Connei SpA	Italy	*233*	75-05380
Detoronics Corp.	USA	*568*	75-25770
Detoronics Corporation	USA	*569*	75-25850
D.G. O'Brien UK Ltd	UK	*471*	75-42895
EG&G Inc.	USA	*571*	75-29555
Egli Fischer & Co AG	Switzerland	*386*	75-03348
Elcom Systems Inc.	USA	*572*	75-29650
Esterline Technologies	USA	*574*	75-32360
Framatome Connectors International	France	*136*	75-18405
G&H Technology Inc.	USA	*578*	75-39550
Gisma Steckverbinder GmbH	Germany	*169*	75-10875
Global Airtech International	USA	*579*	75-39769
Hawke Cable Glands Ltd	UK	*457*	75-28475
Hawke Connection Systems	UK	*457*	75-28480
HCC Industries Inc.	USA	*581*	75-42770
Hermetic Seal Corporation	USA	*581*	75-43120
Holec Machines & Apparaten BV	Netherlands	*293*	75-03110
Hughes Microelectronics Europe Ltd	UK	*459*	75-30480
Hydro-Bond Engineering Ltd	UK	*460*	75-30785
Hypertac Ltd	UK	*460*	75-30920
Icore International Ltd	UK	*460*	75-31080
ITT Cannon RF Products	UK	*462*	75-32170
JDR Cable Systems Ltd	UK	*462*	75-32480
Litton Industries Inc.	USA	*590*	75-55400
Lyncole XIT Grounding	USA	*593*	75-56705
MacArtney A/S	Denmark	*103*	75-12800
Matrix Systems Corp.	USA	*594*	75-58990
New Chapel Electronics Ltd	UK	*470*	75-42320
Obrien D G Inc.	USA	*601*	75-65850
Oceano Instruments (UK) Ltd	UK	*508*	80-19050
Oxley Developments Co. Ltd	UK	*472*	75-43400
Park Products Ltd	UK	*473*	75-44000

Penstock Inc.	USA	*603*	75-69200
Pyle-National	UK	*476*	75-47290
Radiall	France	*141*	75-31600
Radiall Nederland BV	Netherlands	*294*	75-05750
Rendar Ltd.	UK	*478*	75-49226
Reynolds Industries Ltd	UK	*479*	75-49345
Schaffner EMC Ltd	UK	*482*	75-51970
Sea Connections Systems Ltd.	UK	*482*	75-52250
Sealtron Inc.	USA	*613*	75-78917
Sielman SA	Greece	*191*	75-02360
Smith & Prince Ltd	UK	*485*	75-54630
TEC Electrical Components Ltd	UK	*487*	75-58840
Tramec SA	France	*148*	75-45500
Transradio Ltd	UK	*489*	75-60970
Trompeter Electronics Inc.	USA	*623*	75-91070
Underwater & Environmental	UK	*490*	75-61900
VEAM	USA	*625*	75-93470
VEAM Srl	Italy	*239*	75-27100
Viking Ltd.	UK	*491*	75-63230
Weald Electronics Ltd	UK	*493*	75-64880
White Products BV	Netherlands	*296*	75-09100

791 10 20 80 Connector accessories, electrical

AIR-LB	France	*131*	75-01100
Amphenol Corp.	USA	*553*	75-06800
Amphenol Ltd.	UK	*430*	75-03000
Array Connector Corp.	USA	*555*	75-09020
Claymount Assemblies BV.	Netherlands	*292*	75-01150
Dowty Seals Ltd	UK	*446*	75-18620
Egli Fischer & Co AG	Switzerland	*386*	75-03348
Framatome Connectors International	France	*136*	75-18405
Graphic Datakits International	USA	*579*	75-40355
HCC Industries Inc.	USA	*581*	75-42770
Hughes Microelectronics Europe Ltd	UK	*459*	75-30480
Icore International Ltd	UK	*460*	75-31080
ITT Cannon RF Products	UK	*462*	75-32170
KEC Limited	UK	*463*	75-33700
Litton Industries Inc.	USA	*590*	75-55400
Lyncole XIT Grounding	USA	*593*	75-56705
Oxley Developments Co. Ltd	UK	*472*	75-43400
Radiall Nederland BV	Netherlands	*294*	75-05750
Reynolds Industries Ltd	UK	*479*	75-49345
Sealtron Inc.	USA	*613*	75-78917
Sielman SA	Greece	*191*	75-02360
Smith & Prince Ltd	UK	*485*	75-54630
Transradio Ltd	UK	*489*	75-60970
Trompeter Electronics Inc.	USA	*623*	75-91070
Underwater & Environmental	UK	*490*	75-61900
VEAM Srl	Italy	*239*	75-27100
Viking Ltd.	UK	*491*	75-63230
Weald Electronics Ltd	UK	*493*	75-64880

791 10 50 Printed circuit boards, professional

Air-Log Ltd	UK	*429*	75-01800
Associated Aircraft Mfg & Sales Inc.	USA	*555*	75-09500
Associated Controls (Australia)	Australia	*12*	75-01560
Bharat Electronics Ltd	India	*203*	55-00200
Circuiti Stampati Italia	Italy	*233*	75-05183
Dynaco Corp.	USA	*570*	75-27900
EG&G Inc.	USA	*571*	75-29555
Firan Corp.	Canada	*71*	75-05400
FPE Fischer GmbH	Germany	*168*	75-09300
G.E. Mortley, Sprague & Co. Ltd	UK	*469*	75-41000
Graphico Electronics Group	Canada	*71*	75-06460
Imco Industries Ltd	Israel	*219*	75-09400
Kelan Circuits Ltd	UK	*463*	75-33400
Labtech Ltd	UK	*464*	75-34280
Litton Industries Inc.	USA	*590*	75-55400
Microelectronics, guidance and	Turkey	*402*	55-00050
Projects Unlimited Inc.	USA	*606*	75-71560
Rogers Corp.	USA	*611*	75-76000
SAGEM SA	**France**	***143***	**75-33950**
Sheldahl Inc.	USA	*614*	75-80150
Sparton Technology Inc.	USA	*616*	75-82950
Taconic Plastics Ltd	USA	*619*	75-86020
TDS Circuits Plc	UK	*487*	75-58750
Textron Marine & Land Systems	USA	*622*	75-88660
Thomson-CSF Electronics Belgium	Belgium	*47*	75-06310
TMR-Tecnologías Metalúrgicas	Spain	*367*	75-05650
VEAM Srl	Italy	*239*	75-27100
Zincolcelere Sarl	Italy	*240*	75-32000

791 10 60 Switches, electrical

Albacom Ltd	UK	*429*	75-02372
Bowmar Instrument Corp.	USA	*560*	75-14860
Cole Instrument Corp.	USA	*564*	75-19420
Communications Instruments Inc.	USA	*565*	75-20300
Communitron (Australia) Pty Ltd	Australia	*14*	75-03860
CTS Corporation	USA	*567*	75-22800
Diamond H Controls Limited	UK	*445*	75-18025
Eaton Corporation	USA	*571*	75-28800
Eaton Limited	UK	*501*	80-09050
EG&G Inc.	USA	*571*	75-29555
Egli Fischer & Co AG	Switzerland	*386*	75-03348
Esterline Technologies	USA	*574*	75-32360
Holec Machines & Apparaten BV	Netherlands	*293*	75-03110
Intertechnique (Aerospace) Ltd	UK	*461*	75-31948
Jennings Technology Corp.	USA	*586*	75-50350
Korry Electronics Co.	USA	*589*	75-53250
Litton Industries Inc.	USA	*590*	75-55400
Lucas Control Systems Products	UK	*465*	75-36675
Lyncole XIT Grounding	USA	*593*	75-56705
Meggitt Avionics (UK)	UK	*467*	75-38340
NMB (USA) Inc	USA	*599*	75-64400
Park Products Ltd	UK	*473*	75-44000
Power Spectra Inc.	USA	*605*	75-71085
Saunders-Roe Ltd.	UK	*482*	75-51850
Sector Microwave Industries	USA	*613*	75-79010
Siemens Electromechanical Components	USA	*614*	75-80565
Skurka Engineering Company	USA	*615*	75-81520
Sparton Technology Inc.	USA	*616*	75-82950
StacoSwitch	USA	*617*	75-83580
Wessex Advanced Switching Products	UK	*493*	75-65330
Whipp & Bourne	UK	*493*	75-65798

791 20 Components, electronic

AB Connectors Ltd	UK	*426*	75-00040
Abbott Electronics Inc.	USA	*545*	75-00150
Abbott Electronics Ltd	UK	*426*	75-00180
Advanced Semiconductor Inc.	USA	*546*	75-00950
Advanced Technical Corp.	USA	*546*	75-01000
AIR-LB	France	*131*	75-01100
Amphenol Corp.	USA	*553*	75-06800
Amphenol Ltd.	UK	*430*	75-03000
Amptek Inc.	USA	*553*	75-06880
Arab Organization for Ind.	Egypt	*109*	55-01700
Australian AMP Pty Ltd	Australia	*12*	75-01780
AVX Ltd	UK	*433*	75-05190
Aydin Corp.	USA	*556*	75-11450
Beck Electronics Ltd	UK	*434*	75-06125
Bharat Electronics Ltd	India	*203*	55-00200
Bowmar Instrument Corp.	USA	*560*	75-14860
British Sarozal Ltd.	UK	*436*	75-09570
Burr-Brown International Ltd.	UK	*438*	75-11032
Canadian Marconi Co.	Canada	*69*	75-02600
Capco Inc.	USA	*562*	75-16650
Carter Manufacturing Corp.	USA	*562*	75-16800
Century Electronics	USA	*562*	75-17385
Ceramic Devices Inc.	USA	*563*	75-17397
Communications & Power Industries	USA	*565*	75-20393
Compac	USA	*565*	75-20400
Crane Co.	USA	*566*	75-22452
Crompton Instruments Ltd.	UK	*443*	75-16070
Curtis Instruments (UK) Ltd	UK	*444*	75-16440
Dare Electronics	USA	*567*	75-23580
Detoronics Corporation	USA	*569*	75-25850
Dixi-Fresard Microtechniques SA	France	*135*	75-13680
DSM	Netherlands	*292*	75-01730
Eaton Corporation	USA	*571*	75-28800
Edal Industries Inc.	USA	*571*	75-29030
EEV Canada Ltd	Canada	*71*	75-04700
EEV Ltd	UK	*448*	75-19590
EG&G Inc.	USA	*571*	75-29555
EG&G Reticon	USA	*571*	75-29562
Egli Fischer & Co AG	Switzerland	*386*	75-03348
Elcom Systems Inc.	USA	*572*	75-29650
Electro Corporation	USA	*572*	75-29685
Electrodynamics Inc.	USA	*572*	75-29690
Electron Tubes Limited	UK	*448*	75-20250
Elettronica (UK) Ltd	UK	*448*	75-20475
Esterline Technologies	USA	*574*	75-32360
EuroAtlas GmbH	Germany	*168*	75-08680
Fairview Machine	USA	*575*	75-33600

Fluorocarbon Company Limited	UK	451	75-24020
Framatome Connectors International	France	136	75-18405
GEC Plessey Semiconductors	UK	454	75-25363
GEC Plessey Semiconductors	USA	578	75-37425
GEC-Marconi Infra-Red Limited	UK	454	75-25338
GEC-Marconi Materials Technology Ltd	UK	419	60-00600
Genicom Corp.	USA	578	75-39220
Giat Industries	France	126	55-00100
Hoffman Engineering Corp.	USA	582	75-43980
Honeywell Control Systems Ltd	UK	458	75-29730
Hughes Electronics Corp.	USA	583	75-44865
Hughes Microelectronics Europe Ltd	UK	459	75-30480
ILC Data Device Corp.	USA	584	75-46050
Imco Industries Ltd	Israel	219	75-09400
Industria Condensatori Applicazioni	Italy	235	75-11600
Interpoint A Crane Company	USA	585	75-46870
ITT Cannon	USA	586	75-49452
ITT Cannon RF Products	UK	462	75-32170
ITT Gallium Arsenide Tech. Center	USA	586	75-49455
Jennings Technology Corp.	USA	586	75-50350
K&L Microwave Inc.	USA	589	75-52880
Kern Engrng & Mfg Corp.	USA	588	75-52330
Kirovo Chepetsky Electro	Russia	337	75-11370
Litton Industries Inc.	USA	590	75-55400
MA/COM Antenna & Cable Div.	USA	593	75-56953
Matra Electronique	France	139	75-26005
MCL Inc.	USA	595	75-59700
Minebea Co. Ltd	Japan	253	75-05600
Monitor Products Company Inc.	USA	597	75-62450
Murata Electronics (UK) Ltd	UK	470	75-41780
Netcom Ltd	Greece	191	75-02150
Nordic VLSI A/S	Norway	310	75-01640
North Hills Signal Processing	USA	599	75-65300
Nova Electric	USA	600	75-65600
Omori Seikoki Co. Ltd	Japan	254	75-07310
Optek Technology Inc.	USA	601	75-66880
OptoElectronics Textron	USA	601	75-67100
Oxley Developments Co. Ltd	UK	472	75-43400
Oxley Inc.	USA	602	75-67870
Parker Bertea Aerospace	USA	603	75-68900
Philips Components Pty Ltd	Australia	20	75-16490
Piezo Technology Inc.	USA	604	75-70000
Powerex Inc.	USA	604	75-71080
Quartz et Silice	France	141	75-31405
Radian Technology	USA	607	75-72500
Redpoint Thermalloy Ltd	UK	478	75-48960
Richard Manufacturing Co.	USA	609	75-75355
Schaffner EMC Ltd	UK	482	75-51970
Signal Technology Corp	USA	615	75-80795
Signal Technology Corp.	USA	618	75-84240
Sorep	France	146	75-42400
Sparton Technology Inc.	USA	616	75-82950
Spectrol Reliance Ltd	UK	485	75-55555
StacoSwitch	USA	617	75-83580
Teledyne Electronic Technologies	USA	620	75-87522
Thomas Electronics Inc.	USA	622	75-88850
Tramec SA	France	148	75-45500
Transradio Ltd	UK	489	75-60970
Trompeter Electronics Inc.	USA	623	75-91070
United Moulders Ltd	UK	490	75-62130
Welwyn Components Ltd	UK	493	75-65145
Western Filament Inc.	USA	627	75-95930
Zeta	USA	629	75-98190

791 20 05 Frequency sources, very stable

EG&G Inc.	USA	571	75-29555
Elettronica (UK) Ltd	UK	448	75-20475
Frequency Electronics Inc.	USA	576	75-36190
Hewlett-Packard Company	USA	581	75-43200
Zeta	USA	629	75-98190

791 20 10 Integrated circuits

Bharat Electronics Ltd	India	203	55-00200
British Sarozal Ltd.	UK	436	75-09570
Dynaco Corp.	USA	570	75-27900
Egli Fischer & Co AG	Switzerland	386	75-03348
GEC-Marconi Materials Technology Ltd	UK	419	60-00600
Graseby Microsystems Ltd	UK	455	75-26630
Holt Integrated Circuits	USA	582	75-44060
Hughes Microelectronics Europe Ltd	UK	459	75-30480
Kirovo Chepetsky Electro	Russia	337	75-11370
Lewicki Microelectronic GmbH	Germany	171	75-18200
Linear Technology Corp.	USA	590	75-55220
Litton Industries Inc.	USA	590	75-55400
Minebea Co. Ltd	Japan	253	75-05600
Philips Components Pty Ltd	Australia	20	75-16490
Sorep	France	146	75-42400
Space Electronics Inc.	USA	642	80-29050
Teledyne Electronic Technologies	USA	620	75-87522
Vectronics Microwave Corp.	USA	625	75-93490

791 20 20 Cathode ray tubes (CRTs)

EG&G Inc.	USA	571	75-29555
Erotron	Ukraine	406	75-01500
Litton Industries Inc.	USA	590	75-55400
Teltron Technologies Inc.	USA	621	75-87865

791 20 30 Klystrons

Advanced Technical Corp.	USA	546	75-01000
Albacom Ltd	UK	429	75-02372
Bharat Electronics Ltd	India	203	55-00200
British Sarozal Ltd.	UK	436	75-09570
Communications & Power Industries	Canada	70	75-03550
Communications & Power Industries	USA	565	75-20393
EEV Canada Ltd	Canada	71	75-04700
EEV Inc.	USA	571	75-29500
EEV Ltd	UK	448	75-19590
EG&G Inc.	USA	571	75-29555
Litton Industries Inc.	USA	590	75-55400
Radio Research Instrument Co. Inc.	USA	607	75-72600
THORN Microwave Devices Ltd	UK	488	75-59760

791 20 50 Magnetrons

Advanced Technical Corp.	USA	546	75-01000
Bharat Electronics Ltd	India	203	55-00200
British Sarozal Ltd.	UK	436	75-09570
Communications & Power Industries	USA	565	75-20393
EEV Canada Ltd	Canada	71	75-04700
EEV Inc.	USA	571	75-29500
EEV Ltd	UK	448	75-19590
EG&G Inc.	USA	571	75-29555
Herley-Vega Systems	USA	581	75-43090
Litton Industries Inc.	USA	590	75-55400
Radio Research Instrument Co. Inc.	USA	607	75-72600
THORN Microwave Devices Ltd	UK	488	75-59760

791 20 53 Microwave components

Advanced Technical Corp.	USA	546	75-01000
AEA Technology plc	UK	496	80-00480
Aerospace Metal Composites Limited	UK	428	75-01180
Alan Industries Inc.	USA	548	75-04205
Albacom Ltd	UK	429	75-02372
Alcatel Bell	Belgium	44	75-00055
Amphenol Ltd	UK	430	75-03000
Andrew Canada Inc.	Canada	68	75-00600
AVX Ltd	UK	433	75-05190
Aydin Corp.	USA	556	75-11450
Beck Electronics Ltd	UK	434	75-06125
Bharat Electronics Ltd	India	203	55-00200
British Sarozal Ltd.	UK	436	75-09570
Ceramicast (P.I.C.) Ltd	UK	439	75-12410
Chelton (Electrostatics) Limited	UK	439	75-12700
Communications & Power Industries	USA	565	75-20393
Compac	USA	565	75-20400
Crystallume	USA	563	75-18425
Datron/Transco Inc.	USA	568	75-24000
EEV Canada Ltd	Canada	71	75-04700
EEV Inc.	USA	571	75-29500
EEV Ltd	UK	448	75-19590
Elettronica Aster SpA	Italy	233	75-07000
Elettronica GmbH	Germany	168	75-07565
Elettronica (UK) Ltd	UK	448	75-20475
ERA Technology Ltd	UK	502	80-09600
Evered Tube Division	UK	449	75-21800
EW Simulation Technology Ltd	UK	449	75-22150
Ferrite Domain Scientific Production	Russia	333	55-07300
FR Aviation Limited	UK	503	80-10850
Frequency Electronics Inc.	USA	576	75-36190

Gabriel Electronics Inc.	USA	*577*	75-36715
GEC-Marconi Materials Technology Ltd	UK	*419*	60-00600
GEC-Marconi Research Centre	UK	*419*	60-00700
Herley-Vega Systems	USA	*581*	75-43090
Hewlett-Packard Company	USA	*581*	75-43200
Hughes Electronics Corp.	USA	*583*	75-44865
ITT Cannon RF Products	UK	*462*	75-32170
ITT Gallium Arsenide Tech. Center	USA	*586*	75-49455
Iver Precision Engineers Ltd	UK	*462*	75-32230
James Walker & Co. Ltd	UK	*492*	75-64120
Jennings Technology Corp.	USA	*586*	75-50350
K&L Microwave Inc.	USA	*589*	75-52880
Kevlin Corporation	USA	*588*	75-52450
Labtech Ltd	UK	*464*	75-34280
Lantor (UK) Ltd	UK	*464*	75-34665
Litton Industries Inc.	USA	*590*	75-55400
M W Microwave Corp.	USA	*597*	75-63495
MA/COM Antenna & Cable Div.	USA	*593*	75-56953
Marconi Instruments Ltd	UK	*466*	75-37270
Micro Communications Inc.	USA	*596*	75-60980
Micro Metalsmiths Ltd	UK	*468*	75-39010
Microelectronics Ltd	Israel	*221*	75-14250
Microkim Ltd	Israel	*221*	75-14350
Millitech Corp.	USA	*596*	75-61665
MITEQ Inc.	USA	*597*	75-62000
MPD Technologies Inc.	USA	*597*	75-63130
Nera Telecommunications	UK	*470*	75-42318
Noise Com Inc.	USA	*599*	75-64760
Nordic VLSI A/S	Norway	*310*	75-01640
Nucletron Electronic Vertriebs GmbH	Germany	*179*	80-01500
ORITEL SA	France	*140*	75-28500
Oxley Developments Co. Ltd	UK	*472*	75-43400
Oxley Inc.	USA	*602*	75-67870
Penstock Inc.	USA	*603*	75-69200
Quasar Microwave Technology Ltd	UK	*476*	75-47450
Racal-MESL Ltd	UK	*477*	75-47682
Radiall	France	*141*	75-31600
Radiall Nederland BV	Netherlands	*294*	75-05750
Radian Technology	USA	*607*	75-72500
Radio Research Instrument Co. Inc.	USA	*607*	75-72600
Reynolds Industries Ltd	UK	*479*	75-49345
Rogers Corp.	USA	*611*	75-76000
Rollet H. & Co. Ltd	UK	*479*	75-50208
Rutherford Appleton Laboratory	UK	*420*	60-01850
Sage Laboratories Inc.	USA	*612*	75-77610
Sorep	France	*146*	75-42400
Taconic Plastics Ltd	USA	*619*	75-86020
Teledyne Electronic Technologies	USA	*620*	75-87522
THORN Microwave Devices Ltd	UK	*488*	75-59760
Transradio Ltd	UK	*489*	75-60970
Trompeter Electronics Inc.	USA	*623*	75-91070
UB Corp.	USA	*624*	75-91380
Vectronics Microwave Corp.	USA	*625*	75-93490
Weinschel Corporation	USA	*627*	75-95544
Wessex Electronics Ltd	UK	*493*	75-65333
Zeta	USA	*629*	75-98190

791 20 55 Power amplifiers

Albacom Ltd	UK	*429*	75-02372
Aydin Corp.	USA	*556*	75-11450
Bharat Electronics Ltd	India	*203*	55-00200
British Sarozal Ltd.	UK	*436*	75-09570
Century Electronics	USA	*562*	75-17385
Communications & Power Industries	Canada	*70*	75-03550
Communications & Power Industries	Switzerland	*385*	75-01900
Communications & Power Industries	USA	*565*	75-20393
EEV Ltd	UK	*448*	75-19590
EG&G Inc.	USA	*571*	75-29555
Elettronica (UK) Ltd	UK	*448*	75-20475
ERA Technology Ltd	UK	*502*	80-09600
EUROCONTROL SpA	Italy	*233*	75-08200
Hughes Electronics Corp.	USA	*583*	75-44865
Keltron Power Systems	USA	*588*	75-52000
Litton Industries Inc.	USA	*590*	75-55400
MA/COM Antenna & Cable Div.	USA	*593*	75-56953
MPD Technologies Inc.	USA	*597*	75-63130
QES Ltd	UK	*509*	80-25000
Radio Research Instrument Co. Inc.	USA	*607*	75-72600
RFI Industries Pty Ltd	Australia	*20*	75-18500
Siemens Plessey Systems	UK	*483*	75-53770
Sparton Technology Inc.	USA	*616*	75-82950
Technitrol Inc.	USA	*620*	75-86800
Teledyne Electronic Technologies	USA	*620*	75-87522
United Defense L.P.	USA	*624*	75-92050
Zeta	USA	*629*	75-98190

791 20 56 Rectifiers & Thyratrons

Automation NV	Belgium	*45*	75-00950
British Sarozal Ltd.	UK	*436*	75-09570
EEV Canada Ltd	Canada	*71*	75-04700
EEV Ltd	UK	*448*	75-19590
EG&G Inc.	USA	*571*	75-29555
Litton Industries Inc.	USA	*590*	75-55400
Polytec-PI SA	France	*141*	75-30450
Voltage Multipliers Inc.	USA	*626*	75-94430

791 20 58 Semiconductors

Advanced Semiconductor Inc.	USA	*546*	75-00950
Bharat Electronics Ltd	India	*203*	55-00200
British Sarozal Ltd.	UK	*436*	75-09570
EG&G Inc.	USA	*571*	75-29555
Egli Fischer & Co AG	Switzerland	*386*	75-03348
GEC-Marconi Materials Technology Ltd	UK	*419*	60-00600
Hughes Microelectronics Europe Ltd	UK	*459*	75-30480
International Rectifier Co.(GB) Ltd	UK	*461*	75-31860
Litton Industries Inc.	USA	*590*	75-55400
MA/COM Antenna & Cable Div.	USA	*593*	75-56953
Microsemi Corp.	USA	*596*	75-61090
Penstock Inc.	USA	*603*	75-69200
Philips Components Pty Ltd	Australia	*20*	75-16490
Rockwell Semiconductor Systems	Germany	*174*	75-25265

791 20 59 Transmitting tubes

Albacom Ltd	UK	*429*	75-02372
Bharat Electronics Ltd	India	*203*	55-00200
British Sarozal Ltd.	UK	*436*	75-09570
EEV Canada Ltd	Canada	*71*	75-04700
Litton Industries Inc.	USA	*590*	75-55400
Thomas Electronics Inc.	USA	*622*	75-88850
THORN Microwave Devices Ltd	UK	*488*	75-59760

791 20 60 Travelling-wave tubes (TWTs)

AEA Technology plc	UK	*496*	80-00480
Albacom Ltd	UK	*429*	75-02372
Bharat Electronics Ltd	India	*203*	55-00200
British Sarozal Ltd.	UK	*436*	75-09570
Communications & Power Industries	Canada	*70*	75-03550
Communications & Power Industries	USA	*565*	75-20393
EEV Canada Ltd	Canada	*71*	75-04700
EEV Inc.	USA	*571*	75-29500
EEV Ltd	UK	*448*	75-19590
EG&G Inc.	USA	*571*	75-29555
Elettronica (UK) Ltd	UK	*448*	75-20475
FR Aviation Limited	UK	*503*	80-10850
Hughes Electronics Corp.	USA	*583*	75-44865
Instrument Services Ltd	UK	*504*	80-13328
Litton Industries Inc.	USA	*590*	75-55400
Radio Research Instrument Co. Inc.	USA	*607*	75-72600
STC Keltec Corp.	USA	*588*	75-51950
Teledyne Electronic Technologies	USA	*621*	75-87534
THORN Microwave Devices Ltd	UK	*488*	75-59760

791 30 Components, hydraulic

Aeroquip Corp.	USA	*547*	75-02350
AIR-LB	France	*131*	75-01100
Associated Aircraft Mfg & Sales Inc.	USA	*555*	75-09500
Australasian Steel Products Pty Ltd.	Australia	*12*	75-01720
Blanc Aéro Industries	France	*132*	75-06900
BTR Aerospace Australia	Australia	*13*	75-02815
Busak + Shamban Inc.	USA	*561*	75-16006
Circle Seal Controls Inc.	USA	*564*	75-18680
Climax Engineering Pty Ltd.	Australia	*14*	75-03480
Crane Co.	USA	*566*	75-22452
Danfoss System Hydraulik A/S	Denmark	*102*	75-04030
Dowty Aerospace Wolverhampton	UK	*446*	75-18589
Dowty Seals Ltd	UK	*446*	75-18620
EFE Endüstri ve Ticaret A.S.	Turkey	*402*	75-00530
EG&G Inc.	USA	*571*	75-29555

Fairey Arlon Ltd	UK	*450*	75-22200
Fluorocarbon Company Limited	UK	*451*	75-24020
Forward Industries Ltd	UK	*452*	75-24200
Frederick Duffield Pty Ltd	Australia	*16*	75-06840
Gates Hydraulics Ltd	UK	*453*	75-25210
Giat Industries	France	*126*	55-00100
GKN Westland Technologies Ltd	UK	*455*	75-25690
Greene, Tweed & Co.	USA	*579*	75-40380
Greene, Tweed & Co Ltd	UK	*455*	75-26750
H.I. Fraser Pty Ltd	Australia	*16*	75-06835
Hill Industries Inc.	USA	*582*	75-43465
Howden Defense Systems Inc.	USA	*582*	75-44400
HPC Engineering plc	UK	*459*	75-30350
Hydraudyne Systems & Engineering BV	Netherlands	*293*	75-03470
Industria SA	France	*137*	75-22300
Kearfott Guidance & Navigation Corp.	USA	*588*	75-51760
Kinetics Ltd	Israel	*220*	75-11320
Lucas Aerospace Power Transmission	USA	*592*	75-56265
Lucas Aerospace Singapore Pte Ltd	Singapore	*349*	80-01200
Mannesmann Rexroth Ltd	UK	*466*	75-37100
Mannesmann Rexroth SA	France	*139*	75-25500
M.C. Aerospace Corp.	USA	*594*	75-59370
Omori Seikoki Co. Ltd	Japan	*254*	75-07310
Pall Land & Marine	UK	*472*	75-43520
Parker Bertea Aerospace	USA	*602*	75-68900
Parker Hannifin Corporation	USA	*603*	75-68910
Pezetel	Poland	*325*	80-01000
Prva Petoletka Industrie	Yugoslavia	*661*	75-00610
SP aerospace and vehicle systems bv	**Netherlands**	***295***	**75-06860**
Taylor Devices Inc.	USA	*619*	75-86420
Titeflex Europe	France	*148*	75-45300
Tungum Hydraulics Ltd	UK	*490*	75-61500
United Defense L.P.	USA	*624*	75-92050
Vickers Inc.	USA	*626*	75-93820
Vickers Systems	Germany	*176*	75-32830
Vickers Systems SA	France	*148*	75-46930
Vickers Systems Trinova SpA	Italy	*239*	75-27150

791 33 Components, electro-hydraulic

APPH (Bolton) Ltd.	UK	*431*	75-03415
Climax Engineering Pty Ltd	Australia	*14*	75-03480
Crane Co.	USA	*566*	75-22452
Dowty Aerospace Wolverhampton	UK	*446*	75-18589
EG&G Inc.	USA	*571*	75-29555
Fluorocarbon Company Limited	UK	*451*	75-24020
Forward Industries Ltd	UK	*452*	75-24200
Giat Industries	France	*126*	55-00100
GKN Westland Technologies Ltd	UK	*455*	75-25690
H.I. Fraser Pty Ltd	Australia	*16*	75-06835
Hill Industries Inc.	USA	*582*	75-43465
IFI Srl	Italy	*235*	75-11220
Kearfott Guidance & Navigation Corp.	USA	*588*	75-51760
Kinetics Ltd	Israel	*220*	75-11320
Litton Industries Inc.	USA	*590*	75-55400
Lucas Aerospace FCS Paris	France	*139*	75-25390
Lucas Aerospace Power Transmission	USA	*592*	75-56265
MacArtney A/S	Denmark	*103*	75-12800
Mannesmann Rexroth Ltd	UK	*466*	75-37100
Mannesmann Rexroth SA	France	*139*	75-25500
M.C. Aerospace Corp.	USA	*594*	75-59370
Moog GmbH	Germany	*172*	75-21650
Omori Seikoki Co. Ltd	Japan	*254*	75-07310
Parker Bertea Aerospace	USA	*602*	75-68900
Prva Petoletka Industrie	Yugoslavia	*661*	75-00610
Teledyne Electronic Technologies	USA	*620*	75-87522
United Defense L.P.	USA	*624*	75-92050
Vickers Inc.	USA	*626*	75-93820
Vickers Systems	Germany	*176*	75-32830
Vickers Systems SA	France	*148*	75-46930
Vickers Systems Trinova SpA	Italy	*239*	75-27150

791 35 Components, magnetic

Century Electronics	USA	*562*	75-17385
Electro Corporation	USA	*572*	75-29685
Litton Industries Inc.	USA	*590*	75-55400
Mecalectro SA	France	*140*	75-26040
Torwico Electronics L.P.	USA	*623*	75-89800

791 40 Components, mechanical

Air-Log Ltd	UK	*429*	75-01800
API-Harowe	USA	*554*	75-08015
Associated Aircraft Mfg & Sales Inc.	USA	*555*	75-09500
Blanc Aéro Industries	France	*132*	75-06900
Capco Inc.	USA	*562*	75-16650
Climax Engineering Pty Ltd	Australia	*14*	75-03480
Crane Co.	USA	*566*	75-22452
David Brown Gear Industries Ltd	Australia	*13*	75-02800
Deloro Stellite Ltd	UK	*445*	75-17460
Dowty Seals Ltd	UK	*446*	75-18620
Ealing Electro-Optics	UK	*447*	75-19205
Eaton Corporation	USA	*571*	75-28800
EG&G Inc.	USA	*571*	75-29555
EG&G Sealol Ltd	UK	*448*	75-19675
Eurometaal NV	Netherlands	*292*	75-02200
FFA Flugzeugwerke Altenrhein AG	Switzerland	*386*	75-04350
Fluorocarbon Company Limited	UK	*451*	75-24020
Garofoli Costruzioni Meccaniche SpA	Italy	*235*	75-09550
Giat Industries	France	*126*	55-00100
Greene, Tweed & Co Ltd	UK	*455*	75-26750
H.I. Fraser Pty Ltd	Australia	*16*	75-06835
Hill Industries Inc.	USA	*582*	75-43465
Insumat Ltd	UK	*461*	75-31660
KDI Precision Products Inc.	USA	*588*	75-51705
Kearfott Guidance & Navigation Corp.	USA	*588*	75-51760
Kinetics Ltd	Israel	*220*	75-11320
Litton Industries Inc.	USA	*590*	75-55400
Lucas Aerospace Power Transmission	USA	*592*	75-56265
McFadden Systems Inc.	USA	*594*	75-59550
Mibert Precision Ltd	UK	*468*	75-38920
Middlesex Group Limited	UK	*468*	75-39330
Nikpa Springs SA	Greece	*191*	75-02160
Normalair-Garrett Australia	Australia	*19*	75-14870
Olympic Tool Hellas SA	Greece	*191*	75-02180
Omori Seikoki Co. Ltd	Japan	*254*	75-07310
Parker Bertea Aerospace	USA	*603*	75-68900
Parker Hannifin Corporation	USA	*603*	75-68910
Portescap (UK) Ltd	UK	*475*	75-46420
Price & Orphin Limited	UK	*509*	80-24250
R&I Instrument & Gear Co. Pty Ltd	Australia	*20*	75-18650
Recoil Pty	Australia	*20*	75-18300
Reil Industrial Enterprises Ltd	Canada	*75*	75-12000
Renown Engineering Ltd	UK	*478*	75-49227
Rotek Inc.	USA	*611*	75-76750
SARMA	France	*143*	75-34330
Sielman SA	Greece	*191*	75-02360
SKF	Sweden	*378*	75-10340
Société Métallurgique de Gerzat SA	France	*146*	75-40570
SP aerospace and vehicle systems bv	**Netherlands**	***295***	**75-06860**
Stringer Welding & Machine Inc.	USA	*618*	75-84435
Teadit Juntas SA	Brazil	*55*	75-00910
The Torrington Company Ltd	UK	*489*	75-60400
Tufnol Ltd	UK	*489*	75-61410
United Defense L.P.	USA	*624*	75-92050
Vignal Artru Industries	France	*148*	75-46950
Warner & Tourco	France	*149*	75-47050
Warner Electric SA	Switzerland	*388*	75-15150
W.F. Stanley & Co. Ltd	UK	*486*	75-56350

791 40 10 Bearings

Ampep plc	UK	*430*	75-02970
Australian Timken Pty Ltd	Australia	*12*	75-01920
Barber-Colman Co.	USA	*557*	75-11960
Brassinter SA Indústria e Comércio	Brazil	*55*	75-02290
BTR Silvertown Ltd	UK	*437*	75-10830
Busak + Shamban Ltd	UK	*438*	75-11050
Deloro Stellite Ltd	UK	*445*	75-17460
Fluorocarbon Company Limited	UK	*451*	75-24020
Greene, Tweed & Co Ltd	UK	*455*	75-26750
Litton Industries Inc.	USA	*590*	75-55400
Lord SA	France	*150*	80-01350
Lucas Aerospace Power Transmission	USA	*592*	75-56265
Minebea Co. Ltd	Japan	*253*	75-05600
Railko Ltd	UK	*477*	75-47850
RBC Transport Dynamics Corp.	USA	*608*	75-73380
RHP Aerospace	UK	*479*	75-49565
Rose Bearings Ltd.	UK	*480*	75-50400
SARMA	France	*143*	75-34330
SKF	Sweden	*378*	75-10340
SKF do Brasil Ltda	Brazil	*58*	75-19680

SKF USA Inc.	USA	615	75-81500
The Barden Corporation (UK) Ltd	UK	433	75-05630
The Torrington Company Ltd	UK	489	75-60400
Timken do Brasil Com. e Ind. Ltda.	Brazil	58	75-20520
Tiodize Co. Inc.	USA	622	75-89560
Tufnol Ltd	UK	489	75-61410
United Defense L.P..	USA	624	75-92050
Valley Todeco	USA	625	75-93170

791 40 30 Fastening systems, fasteners

Abbott Interfast Corp.	USA	545	75-00180
Aimsco Inc.	USA	547	75-03050
Air-Log Ltd	UK	429	75-01800
Allfast Fastening Systems Inc.	USA	548	75-04635
APM Hexseal	USA	554	75-08030
Armstrong Fastening Systems	UK	431	75-03800
Ateliers de la Haute Garonne	France	132	75-03800
BAS Components Ltd	UK	433	75-06000
Blanc Aéro Industries	France	132	75-06900
Camloc GmbH	Germany	166	75-04580
CMI - Ballarat	Australia	14	75-03600
Consolidated Manufacturing	Australia	15	75-03925
Egli Fischer & Co AG	Switzerland	386	75-03348
Fastening Systems International Inc.	USA	575	75-33870
Hi-Shear Corp.	USA	582	75-43500
IFI Srl	Italy	235	75-11220
Lentern (Aircraft) Ltd	UK	464	75-35150
Linread Northbridge	UK	464	75-36065
Minebea Co. Ltd	Japan	253	75-05600
Permabond	UK	474	75-44800
Permali Gloucester Ltd	UK	474	75-44850
Recoil Pty	Australia	20	75-18300
Shur-Lok International SA	Belgium	47	75-05900
Tecelagem Lady Ltda	Brazil	58	75-20140
Textron Fastening Systems	USA	622	75-88656
Tiodize Co. Inc.	USA	622	75-89560
Valley Todeco	USA	625	75-93170
Victaulic Co. of America	USA	626	75-93840
Voss Industries Inc.	USA	626	75-94550
Wesco Aircraft Hardware Corp..	USA	627	75-95700

791 40 60 Seals, mechanical

Advanced Products NV	Belgium	45	75-00075
APM Hexseal	USA	554	75-08030
Bradford Cylinders Ltd	UK	435	75-07850
BTR Aerospace Australia	Australia	13	75-02815
Busak + Shamban Inc.	USA	561	75-16006
Busak + Shamban Ltd	UK	438	75-11050
Chomerics	USA	563	75-18275
ContiTech Holding GmbH	Germany	166	75-05150
Deep Sea Seals Ltd	UK	445	75-17400
Deloro Stellite Ltd	UK	445	75-17460
Dynamation Research	USA	570	75-27960
EG&G Inc.	USA	571	75-29555
EG&G Sealol Ltd	UK	448	75-19675
EG&G Sealol Pty Ltd	Australia	15	75-05350
EG&G Sealol Sarl	France	136	75-14330
Esterline Technologies	USA	574	75-32360
Firth Rixson plc	UK	451	75-23600
Flexibox Ltd	UK	451	75-23800
Fluorocarbon Company Limited	UK	451	75-24020
FPI Inc.	USA	576	75-35755
Greene, Tweed & Co.	USA	579	75-40380
Greene, Tweed & Co Ltd	UK	455	75-26750
IFI Srl	Italy	235	75-11220
James Walker & Co. Ltd	UK	492	75-64120
John Crane Inc.	USA	566	75-22500
J.T. Slocomb Company	USA	615	75-81700
Molding Solutions	USA	597	75-62400
Olympic Tool Hellas SA	Greece	191	75-02180
SpanSet Inter AG	Switzerland	387	75-13550

791 43 Components, electro-mechanical

Advanced Semiconductor Inc.	USA	546	75-00950
Aimsco Inc.	USA	547	75-03050
Associated Aircraft Mfg & Sales Inc.	USA	555	75-09500
BEI Sensors & Systems Co.	USA	557	75-12740
Bowmar Instrument Corp.	USA	560	75-14860
Chomerics	USA	563	75-18275
Circle Seal Controls Inc.	USA	564	75-18680
Comepa	France	134	75-10175
Compac	USA	565	75-20400
Control Techniques Dynamics Ltd	UK	442	75-14875
Dare Electronics	USA	567	75-23580
DRS Technologies, Inc.	USA	570	75-27400
EG&G Inc.	USA	571	75-29555
Egli Fischer & Co AG	Switzerland	386	75-03348
Electro Mechanical Systems Ltd.	UK	502	80-09380
Electrodynamics Inc.	USA	572	75-29690
Electronic Specialty Corp.	USA	572	75-30150
Fluorocarbon Company Limited	UK	451	75-24020
Garofoli Costruzioni Meccaniche SpA	Italy	235	75-09550
Hill Industries Inc.	USA	582	75-43465
Hughes Danbury Optical Sys.	USA	582	75-44857
ITT Cannon RF Products	UK	462	75-32170
Jay-El Products Inc.	USA	586	75-50100
Jennings Technology Corp.	USA	586	75-50350
KDI Precision Products Inc.	USA	588	75-51705
Kearfott Guidance & Navigation Corp.	USA	588	75-51760
Kinetics Ltd	Israel	220	75-11320
L&S Machine Co. Inc.	USA	589	75-53700
Leach International Europe SA.	France	139	75-25350
Litton Industries Inc.	USA	590	75-55400
Lucas Aerospace Power Transmission	USA	592	75-56265
Lucas Aerospace Singapore Pte Ltd	Singapore	349	80-01200
M.C. Aerospace Corp.	USA	594	75-59370
Mecalectro SA	France	140	75-26040
Mécanex SA.	Switzerland	386	75-08300
Moog GmbH.	Germany	172	75-21650
Muirhead Vactric Components Ltd	UK	470	75-41404
Murata Electronics (UK) Ltd	UK	470	75-41780
Olympic Tool Hellas SA	Greece	191	75-02180
Omori Seikoki Co. Ltd	Japan	254	75-07310
Pandect Precision Components Ltd	UK	472	75-43650
Park Products Ltd	UK	473	75-44000
Parker Bertea Aerospace	USA	602	75-68900
Penny & Giles Blackwood Group	UK	473	75-44495
Portescap (UK) Ltd	UK	475	75-46420
Radiall Nederland BV	Netherlands	294	75-05750
Sage Laboratories Inc.	USA	612	75-77610
SARMA	France	143	75-34330
Schaffner EMC Ltd	UK	482	75-51970
Sector Microwave Industries	USA	613	75-79010
Sparton Technology Inc.	USA	616	75-82950
Spectrol Reliance Ltd	UK	485	75-55555
Telair International	USA	620	75-87400
Teledyne Electronic Technologies	USA	620	75-87522
Transradio Ltd	UK	489	75-60970
United Defense L.P..	USA	624	75-92050
Warner & Tourco	France	149	75-47050
Wessex Advanced Switching Products	UK	493	75-65330

791 43 60 Slipring-brush assemblies

EUROCONTROL SpA	Italy	233	75-08200
IDM Electronics Ltd	UK	460	75-31100
Litton Industries Inc.	USA	590	75-55400
MacArtney A/S	Denmark	103	75-12800
Mécanex SA.	Switzerland	386	75-08300
Pandect Precision Components Ltd	UK	472	75-43650
Park Products Ltd	UK	473	75-44000
Rotech Components Ltd	UK	481	75-50575
United Defense L.P..	USA	624	75-92050

791 45 Components, optical

Amphenol Corp.	USA	553	75-06800
Avibras Aeroespacial SA	Brazil	55	75-00968
Avimo Electro-Optics Pte Ltd	Singapore	348	75-00800
Cleveland Crystals Inc.	USA	564	75-19040
Coherent Optics (Europe) Ltd	UK	441	75-14090
Computer Optics Inc.	USA	565	75-20680
Delft Sensor Systems	**Belgium**	**46**	**75-02500**
Delft Sensor Systems	Netherlands	292	75-01590
DRS Technologies, Inc.	USA	570	75-27400
Ealing Electro-Optics	UK	447	75-19205
EG&G Inc.	USA	571	75-29555
Egli Fischer & Co AG	Switzerland	386	75-03348
Electrodynamics Inc.	USA	572	75-29690
Elettronica (UK) Ltd	UK	448	75-20475
Eloptro	S. Africa	358	75-05680
Esterline Technologies	USA	574	75-32360

Fotona D.D.	Slovenia	*355*	75-00400
GEC-Marconi Materials Technology Ltd	UK	*419*	60-00600
Graseby Specac Ltd	UK	*455*	75-26640
Gurley Precision Instruments	USA	*580*	75-41150
Hughes Danbury Optical Sys.	USA	*582*	75-44857
Hughes Elcan Optical Technologies	Canada	*72*	75-07650
Intelect Inc.	USA	*585*	75-47160
Laser Fiber Optics Canada Ltd	Canada	*73*	75-08890
Litton Industries Inc.	USA	*590*	75-55400
Lucas Aerospace Power Transmission	USA	*592*	75-56265
MRC Precision Metal Optics	USA	*597*	75-63150
Muirhead Vactric Components Ltd	UK	*470*	75-41404
Optechs Korea Inc	Korea (ROK)	*266*	75-03750
Optical Components Inc.	USA	*601*	75-66953
Oxley Developments Co. Ltd	UK	*472*	75-43400
Oxley Inc.	USA	*602*	75-67870
Pilkington Space Technology	UK	*426*	74-10000
Power Spectra Inc.	USA	*605*	75-71085
Radiall Nederland BV	Netherlands	*294*	75-05750
Recon/Optical Inc.	USA	*608*	75-73930
Rocky Mountain Instrument Co.	USA	*611*	75-75960
Rodenstock Precision Optics Inc.	USA	*611*	75-75980
SAGEM SA	**France**	***143***	**75-33950**
Schott Glass Technologies Inc.	USA	*613*	75-78410
Urals Optical & Mechanical Plant	Russia	*339*	75-53000
Varian Australia Pty Ltd	Australia	*23*	75-22100
W.F. Stanley & Co. Ltd	UK	*486*	75-56350

791 45 20 Connectors, optical

Amphenol Corp.	USA	*553*	75-06800
Amphenol Ltd	UK	*430*	75-03000
D.G. O'Brien UK Ltd.	UK	*471*	75-42895
Egli Fischer & Co AG	Switzerland	*386*	75-03348
G&H Technology Inc.	USA	*578*	75-39550
GEC-Marconi Materials Technology Ltd	UK	*419*	60-00600
Gisma Steckverbinder GmbH	Germany	*169*	75-10875
Litton Industries Inc.	USA	*590*	75-55400
MacArtney A/S	Denmark	*103*	75-12800
Obrien D G Inc.	USA	*601*	75-65850
Oxley Developments Co. Ltd	UK	*472*	75-43400
Oxley Inc.	USA	*602*	75-67870
Radiall	France	*141*	75-31600
Reynolds Industries Ltd	UK	*479*	75-49345
Siecor Corp.	USA	*614*	75-80560
Société des Céramiques Techniques	France	*146*	75-39510
Sumitomo Electric Lightwave Corp.	USA	*618*	75-84800
Transradio Ltd	UK	*489*	75-60970
Underwater & Environmental	UK	*490*	75-61900

791 45 30 Fibre optic systems

ACAL Auriema Ltd	UK	*427*	75-00290
Alcatel Bell	Belgium	*44*	75-00055
Amphenol Ltd	UK	*430*	75-03000
Analog Modules Inc.	USA	*553*	75-07140
Andrew Ltd	UK	*430*	75-03156
Avibras Aeroespacial SA	Brazil	*55*	75-00968
British Aerospace (Systems &	UK	*436*	75-08890
CAE	UK	*438*	75-11125
Computing Devices Company Ltd	UK	*442*	75-14700
CORIS	France	*134*	75-12450
Cranfield University	UK	*419*	60-00350
Diamond Electronics Inc.	USA	*569*	75-26450
EG&G Inc.	USA	*571*	75-29555
Egli Fischer & Co AG	Switzerland	*386*	75-03348
ERA Technology Ltd	UK	*502*	80-09600
Fotec Inc.	USA	*576*	75-35680
Galileo Corp.	USA	*577*	75-36755
GEC-Marconi Materials Technology Ltd	UK	*454*	75-25340
GEC-Marconi Research Centre	UK	*419*	60-00700
General Offshore Corporation	USA	*636*	80-12550
General Offshore (UK) Ltd	UK	*503*	80-11670
Graseby Specac Ltd	UK	*455*	75-26640
ILC Technology Inc.	USA	*584*	75-46070
iMPath Networks Inc.	Canada	*77*	75-46151
Instrument Services Ltd	UK	*504*	80-13328
Laser Fiber Optics Canada Ltd	Canada	*73*	75-08890
Litton Industries Inc.	USA	*590*	75-55400
Lynwood	UK	*465*	75-36700
MacArtney A/S	Denmark	*103*	75-12800
Midcon Cables Co.	USA	*596*	75-61300
Optek Technology Inc.	USA	*601*	75-66880
Optical Cable Corp.	USA	*601*	75-66952
Oxley Developments Co. Ltd	UK	*472*	75-43400
P W Allen & Co Ltd	UK	*429*	75-02740
Pandect Precision Components Ltd	UK	*472*	75-43650
Polyot Research and Production Co.	Russia	*338*	75-14700
Pyle-National	UK	*476*	75-47290
Radiall Nederland BV	Netherlands	*294*	75-05750
Rutherford Appleton Laboratory	UK	*420*	60-01850
SAS Communications Ltd	UK	*482*	75-51800
SpecTran Corp.	USA	*616*	75-83090
Sumitomo Electric Lightwave Corp.	USA	*618*	75-84800
Teledyne Electronic Technologies	USA	*620*	75-87522
Ultrafine Technology Ltd	UK	*490*	75-61830
United Defense L.P.	USA	*624*	75-92050
Wayne Kerr Electronics	UK	*492*	75-64860

791 45 30 20 Fibre optic cable

Air Logistics Corp.	USA	*548*	75-03700
Amphenol Ltd	UK	*430*	75-03000
Computing Devices Canada	Canada	*70*	75-03600
Egli Fischer & Co AG	Switzerland	*386*	75-03348
Electric Wire & Cable Co.	Israel	*218*	75-05755
Hewlett-Packard Company	USA	*581*	75-43200
Hydrocable Systems	UK	*460*	75-30790
JDR Cable Systems BV	Netherlands	*293*	75-03600
MacArtney A/S	Denmark	*103*	75-12800
Optical Cable Corp.	USA	*601*	75-66952
Polyot Research and Production Co.	Russia	*338*	75-14700
Raydex/C.D.T. Ltd	UK	*477*	75-48340
Siecor Corp.	USA	*614*	75-80560
SpecTran Corp.	USA	*616*	75-83090
Sumitomo Electric Industries Ltd	Japan	*254*	75-07950
Sumitomo Electric Lightwave Corp.	USA	*618*	75-84800
United Defense L.P.	USA	*624*	75-92050

791 45 50 Precision optics

Bospec Neue Technologien	Germany	*166*	75-03420
Carl Zeiss Canada Ltd	Canada	*70*	75-02900
Central Design Bureau	Ukraine	*406*	75-00800
Davin Optical Ltd	UK	*444*	75-17100
Delft Sensor Systems	**Belgium**	***46***	**75-02500**
Delft Sensor Systems	Netherlands	*292*	75-01590
DRS Technologies, Inc.	USA	*570*	75-27400
Ealing Electro-Optics	UK	*447*	75-19205
Eloptro	S. Africa	*358*	75-05680
Graflex Inc	USA	*579*	75-40310
Graseby Specac Ltd	UK	*455*	75-26640
Hughes Danbury Optical Sys.	USA	*582*	75-44857
JML Optical Industries Inc.	USA	*587*	75-50940
Litton Industries Inc.	USA	*590*	75-55400
MRC Precision Metal Optics	USA	*597*	75-63150
OFC Corporation	USA	*601*	75-66220
Optical Components Inc.	USA	*601*	75-66953
Ring Sights Holdings Co. Ltd.	UK	*479*	75-49700
Rodenstock Precision Optics Inc.	USA	*611*	75-75980
SFIM	**France**	***144***	**75-35080**
Sira Ltd.	UK	*484*	75-54250
Temmek Optics Ltd.	Israel	*223*	75-22504
Urals Optical & Mechanical Plant	Russia	*339*	75-53000

791 45 55 Optical glasses

Litton Industries Inc.	USA	*590*	75-55400
Optical Components Inc.	USA	*601*	75-66953
Oxley Inc.	USA	*602*	75-67870
Quartz et Silice	France	*141*	75-31405
Schott Glass Technologies Inc.	USA	*613*	75-78410
SFIM	**France**	***144***	**75-35080**
Urals Optical & Mechanical Plant	Russia	*339*	75-53000

791 45 55 50 IR (Laser) absorbing glasses

Ets Bollé	France	*133*	75-06960
GEC-Marconi Materials Technology Ltd	UK	*419*	60-00600
Litton Industries Inc.	USA	*590*	75-55400
Oxley Inc.	USA	*602*	75-67870
SFIM	**France**	***144***	**75-35080**

791 45 55 55 IR transmitting glasses

Coherent Optics (Europe) Ltd	UK	441	75-14090
GEC-Marconi Materials Technology Ltd	UK	419	60-00600
Litton Industries Inc.	USA	590	75-55400
Oxley Inc.	USA	602	75-67870

791 45 60 Optical subassemblies

Bospec Neue Technologien	Germany	166	75-03420
Central Design Bureau	Ukraine	406	75-00800
Coherent Optics (Europe) Ltd	UK	441	75-14090
Computer Optics Inc.	USA	565	75-20680
Davin Optical Ltd	UK	444	75-17100
DRS Technologies, Inc.	USA	570	75-27400
Ealing Electro-Optics	UK	447	75-19205
Elettronica (UK) Ltd	UK	448	75-20475
Eloptro	S. Africa	358	75-05680
EUROCONTROL SpA	Italy	233	75-08200
GEC-Marconi Materials Technology Ltd	UK	419	60-00600
Graseby Specac Ltd	UK	455	75-26640
Hughes Danbury Optical Sys.	USA	582	75-44857
JML Optical Industries Inc.	USA	587	75-50940
Litton Industries Inc.	USA	590	75-55400
Meller Optics Inc.	USA	595	75-59950
MRC Precision Metal Optics	USA	597	75-63150
Optical Components Inc.	USA	601	75-66953
Ring Sights Holdings Co. Ltd.	UK	479	75-49700
Rocky Mountain Instrument Co.	USA	611	75-75960
Rodenstock Precision Optics Inc.	USA	611	75-75980
SFIM	**France**	**144**	**75-35080**
Sira Ltd.	UK	484	75-54250
Teledyne Electronic Technologies	USA	620	75-87522
Urals Optical & Mechanical Plant	Russia	339	75-53000

791 50 Components, pneumatic

Aerospace Syst. & Comp.	USA	547	75-02501
Avica US.	USA	556	75-11100
Climax Engineering Pty Ltd.	Australia	14	75-03480
Dowty Seals Ltd	UK	446	75-18620
Fluorocarbon Company Limited	UK	451	75-24020
Frederick Duffield Pty Ltd	Australia	16	75-06840
General Pneumatics.	USA	578	75-38950
GKN Westland Technologies Ltd	UK	455	75-25690
H.I. Fraser Pty Ltd.	Australia	16	75-06835
Hill Industries Inc.	USA	582	75-43465
IFI Srl	Italy	235	75-11220
Insumat Ltd	UK	461	75-31660
Interspiro Ltd	UK	461	75-31945
Lucas Aerospace Power Transmission	USA	592	75-56265
M.C. Aerospace Corp.	USA	594	75-59370
Moog Inc.	USA	597	75-62550
Parker Bertea Aerospace.	USA	602	75-68900
Parker Hannifin Corporation	USA	603	75-68910
Prva Petoletka Industrie	Yugoslavia	661	75-00610
Telektron Ltd	UK	488	75-59200
Titeflex Europe	France	148	75-45300
Tungum Hydraulics Ltd.	UK	490	75-61500
VEAM	USA	625	75-93470
Voss Industries Inc.	USA	626	75-94550

791 53 Components, electro-pneumatic

Aerospace Syst. & Comp.	USA	547	75-02501
Fluorocarbon Company Limited	UK	451	75-24020
General Pneumatics.	USA	578	75-38950
GKN Westland Technologies Ltd	UK	455	75-25690
H.I. Fraser Pty Ltd.	Australia	16	75-06835
M.C. Aerospace Corp.	USA	594	75-59370
Moog GmbH.	Germany	172	75-21650
Parker Bertea Aerospace.	USA	602	75-68900
Prva Petoletka Industrie	Yugoslavia	661	75-00610
Robertshaw Controls Co.	USA	610	75-75640
Teledyne Electronic Technologies	USA	620	75-87522
Telektron Ltd	UK	488	75-59200

791 60 Components, buoyancy

CRP Marine	UK	443	75-16140

791 90 Components, other

A.O. Henton Engineering Co Ltd.	UK	458	75-29040
Bertan High Voltage	USA	558	75-13720
Canadian Marconi Co.	Canada	69	75-02600
Climax Engineering Pty Ltd.	Australia	14	75-03480
Druck Italia Sarl	Italy	233	75-06695
EEV Ltd	UK	448	75-19590
EG&G Reticon	USA	571	75-29562
Engineered Plastics Inc.	USA	573	75-31570
Eurometaal NV	Netherlands	292	75-02200
Ferrite Domain Scientific Production	Russia	333	55-07300
Fluorocarbon Company Limited	UK	451	75-24020
Kirovo Chepetsky Electro	Russia	337	75-11370
MA/COM Antenna & Cable Div.	USA	593	75-56953
M.C. Aerospace Corp.	USA	594	75-59370
MIL-BASE Industries Inc.	USA	596	75-61530
Quasar Microwave Technology Ltd.	UK	476	75-47450
Shock-Tech Inc.	USA	614	75-80457
Société Métallurgique de Gerzat SA	France	146	75-40570
Teledyne Electronic Technologies	USA	620	75-87522
Teltron Technologies Inc.	USA	621	75-87865
Tiodize Co. Inc.	USA	622	75-89560
Transradio Ltd	UK	489	75-60970
Trompeter Electronics Inc.	USA	623	75-91070
Voss Industries Inc.	USA	626	75-94550

791 90 30 Detectors, IR

Bospec Neue Technologien	Germany	166	75-03420
Central Research Laboratories Ltd	UK	499	80-05895
Cincinnati Electronics Corp.	USA	564	75-18640
Cranfield University	UK	419	60-00350
EEV Ltd	UK	448	75-19590
EG&G Reticon	USA	571	75-29562
Electron Tubes Limited.	UK	448	75-20250
Elettronica (UK) Ltd	UK	448	75-20475
Eloptro	S. Africa	358	75-05680
Eltec Instruments Inc.	USA	573	75-30880
ERA Technology Ltd	UK	502	80-09600
GEC-Marconi Infra-Red Limited	UK	454	75-25338
GEC-Marconi Materials Technology Ltd	UK	419	60-00600
Giat Industries.	France	126	55-00100
Hewlett-Packard Company	USA	581	75-43200
Hughes Electronics Corp.	USA	583	75-44865
II-VI Inc.	USA	624	75-91355
Kidde-Graviner Ltd	UK	463	75-33880
Litton Industries Inc.	USA	590	75-55400
MMR Technologies Inc.	USA	597	75-62200
OptoElectronics Textron	USA	601	75-67100
Oxley Developments Co. Ltd	UK	472	75-43400
Polytec-PI SA	France	141	75-30450
Rutherford Appleton Laboratory	UK	420	60-01850
SAGEM SA	**France**	**143**	**75-33950**
Semi-Conductor Devices.	Israel	222	75-19500
Siemens Environmental Systems Ltd	UK	483	75-53700
Sira Ltd.	UK	484	75-54250

791 90 35 Guidance cells, silicon

EG&G Inc.	USA	571	75-29555
Electron Tubes Limited.	UK	448	75-20250
Litton Industries Inc.	USA	590	75-55400

792 Other equipment

ABLE Corp.	USA	545	75-00260
American Laser Systems Inc.	USA	553	75-06000
Amphenol Corp.	USA	553	75-06800
Avery Berkel	UK	432	75-04560
Babcock Inc.	USA	556	75-11600
Bren-Tronics Inc.	USA	560	75-15170
Busak + Shamban Ltd	UK	438	75-11050
Dowty Seals Ltd	UK	446	75-18620
EG&G Inc.	USA	571	75-29555
EG&G Reticon	USA	571	75-29562
Envirovac Inc.	USA	574	75-31920
FTL Seals Technology Limited.	UK	452	75-24870
Gates Hydraulics Ltd	UK	453	75-25210
Geier & Bluhm Inc.	USA	578	75-37427
General Electrodynamics Corp.	USA	578	75-38500
ILC Data Device Corp.	USA	584	75-46050

Infinetics Inc.	USA	*584*	75-46700
Jehier SA	France	*137*	75-23200
KEL Aerospace Pty Ltd.	Australia	*18*	75-11440
Manta Industries Ltd	Canada	*73*	75-09250
MPE Ltd	UK	*469*	75-41210
National Nuclear Corporation	USA	*598*	75-63870
NIS Ltd.	UK	*471*	75-42555
Normalair-Garrett Australia	Australia	*19*	75-14870
Pall Land & Marine	UK	*472*	75-43520
Parker Hannifin Corporation	USA	*603*	75-68910
Presentey Engineering	Canada	*75*	75-11500
Seaward International Inc.	USA	*613*	75-78940
Sheldahl Inc..	USA	*614*	75-80150
Stella-Meta	UK	*486*	75-56780
Syndex Aerospace Inc..	USA	*619*	75-85540
The Armstrong Monitoring Corp.	Canada	*68*	75-00740
The Eraser Co. Inc.	USA	*574*	75-32020
THERMOCOAX	France	*147*	75-44930
Zero Cases (Europe)	UK	*495*	75-67700

792 10 Other equipment, space-related

AEA Technology plc	UK	*496*	80-00480
AGM Container Controls Inc.	USA	*547*	75-02870
Arthur Hart Webbing	UK	*457*	75-28180
Avica Equipment	UK	*432*	75-04870
BAS Components Ltd	UK	*433*	75-06000
Bedford Trucks	UK	*434*	75-06135
Bharat Electronics Ltd	India	*203*	55-00200
Busak + Shamban Inc.	USA	*561*	75-16006
Carville Ltd.	UK	*439*	75-12000
Celab Ltd	UK	*439*	75-12250
CJB Developments Ltd.	UK	*440*	75-13430
Control Techniques Dynamics Ltd	UK	*442*	75-14875
CP Cases	UK	*443*	75-15680
Cranfield Aerospace Ltd	UK	*500*	80-07200
Crestbury Ltd	UK	*443*	75-16030
Daco Scientific Ltd	UK	*444*	75-16550
EEV Ltd	UK	*448*	75-19590
Electron Tubes Limited.	UK	*448*	75-20250
FPT Industries Ltd	UK	*452*	75-24400
Graff Electronic Machines Ltd	UK	*455*	75-26450
Howden Aircontrol Ltd.	UK	*459*	75-30240
IFI Srl.	Italy.	*235*	75-11220
ILC Technology Inc.	USA	*584*	75-46070
JDR Cable Systems Ltd	UK	*462*	75-32480
Jehier SA	France	*137*	75-23200
Labtech Ltd	UK	*464*	75-34280
Light Alloy Ltd.	UK	*464*	75-35560
Lista International Corp.	USA	*590*	75-55350
MA/COM Antenna & Cable Div.	USA	*593*	75-56953
Matra Marconi Space UK, Ltd	UK	*425*	74-06900
McMurdo Ltd	UK	*467*	75-38170
MPE Ltd	UK	*469*	75-41210
Pioneer Aerospace Corp	USA	*604*	75-70200
Reynolds Industries Ltd	UK	*479*	75-49345
Rutherford Appleton Laboratory	UK	*420*	60-01850
Safe Training Systems Ltd.	UK	*481*	75-51300
Saft Nife Ltd.	UK	*481*	75-51470
Sheldahl Inc..	USA	*614*	75-80150
Shock-Tech Inc..	USA	*614*	75-80457
SpanSet Ltd.	UK	*485*	75-55230
Stop-Choc Ltd.	UK	*486*	75-57345
Superform Aluminium	UK	*487*	75-57920
Voss Industries Inc.	USA	*626*	75-94550
Xelflex	UK	*495*	75-67400

792 20 Other equipment, air force related

AB Carl Munters	Sweden	*377*	75-08230
Aerotech International Inc	Canada	*68*	75-00300
AGM Container Controls Inc.	USA	*547*	75-02870
Avirex Ltd	USA	*556*	75-11250
Busak + Shamban Inc.	USA	*561*	75-16006
C.I.A.M.-E.A. Sarl	Italy.	*233*	75-05150
Electronic Space Systems Corp.	USA	*572*	75-30100
Ets Bollé	France	*133*	75-06960
IFI Srl.	Italy.	*235*	75-11220
Jehier SA	France	*137*	75-23200
Pioneer Aerospace Corp	USA	*604*	75-70200
Shock-Tech Inc..	USA	*614*	75-80457
Syndex Aerospace Inc..	USA	*619*	75-85540
Transradio Ltd	UK	*489*	75-60970
Victaulic Co. of America	USA	*626*	75-93840
Von Arx Ltd	Switzerland	*388*	75-14750
Voss Industries Inc.	USA	*626*	75-94550
White Products BV	Netherlands	*296*	75-09100
Zero Cases (Europe)	UK	*495*	75-67700

792 30 Other equipment, navy-related

AB Carl Munters	Sweden	*377*	75-08230
Aeroquip Corp.	USA	*547*	75-02350
AGM Container Controls Inc.	USA	*547*	75-02870
Busak + Shamban Inc.	USA	*561*	75-16006
Dynamic Controls Ltd	UK	*447*	75-19140
Fairless Engineering Supplies	UK	*450*	75-22350
Freeman Marine Equipment Inc.	USA	*576*	75-36100
Hawke Cable Glands Ltd	UK	*457*	75-28475
IFI Srl.	Italy.	*235*	75-11220
Indal Technologies Inc	Canada.	*72*	75-07740
Jehier SA	France	*137*	75-23200
New Philadelphia Fan Company	USA	*587*	75-51130
Park Products Ltd.	UK	*473*	75-44000
Pioneer Aerospace Corp	USA	*604*	75-70200
Seaward International Inc.	USA	*613*	75-78940
Shock-Tech Inc..	USA	*614*	75-80457
Telektron Ltd	UK	*488*	75-59200
Trelawny Pneumatic Tools.	UK	*489*	75-61000
Trompeter Electronics Inc..	USA	*623*	75-91070
Victaulic Co. of America	USA	*626*	75-93840
W.F. Stanley & Co. Ltd.	UK	*486*	75-56350
White Products BV	Netherlands	*296*	75-09100
Zero Cases (Europe)	UK	*495*	75-67700

792 30 50 Pressure vessels, seamless

Mannesmann Stahlflaschen GmbH	Germany	*172*	75-20006
Meggitt Oxygen Systems	UK	*467*	75-38360
Technical Products Group Inc..	USA	*620*	75-86764
W.E. Smith Engineering Pty Ltd	Australia	*21*	75-19545

792 30 90 Valves, naval

Dynamic Controls Ltd	UK	*447*	75-19140
Giat Industries.	France	*126*	55-00100
Hale Hamilton (Valves) Ltd	UK	*456*	75-27350
Haux-Life-Support GmbH	Germany	*169*	75-11970
Hopkinsons Ltd	UK	*459*	75-29800
Leafield Engineering Ltd	UK	*464*	75-34910
Orseal Ltd	UK	*472*	75-43160
Seetru Ltd	UK	*482*	75-52600
Truflo Marine Valves Ltd	UK	*489*	75-61380
Victaulic Co. of America	USA	*626*	75-93840
Worcester Controls (UK) Ltd	UK	*494*	75-67000

792 40 Other equipment, army-related

AB Carl Munters	Sweden	*377*	75-08230
Aeroquip Corp.	USA	*547*	75-02350
Aerotech International Inc	Canada	*68*	75-00300
AGM Container Controls Inc.	USA	*547*	75-02870
Avirex Ltd	USA	*556*	75-11250
Busak + Shamban Inc.	USA	*561*	75-16006
C.I.A.M.-E.A. Sarl	Italy.	*233*	75-05150
Ets Bollé	France	*133*	75-06960
Jehier SA	France	*137*	75-23200
PD Technical Mouldings plc	UK	*473*	75-44200
Saunders-Roe Ltd.	UK	*482*	75-51850
Shock-Tech Inc..	USA	*614*	75-80457
Trompeter Electronics Inc..	USA	*623*	75-91070
Victaulic Co. of America	USA	*626*	75-93840
Voss Industries Inc.	USA	*626*	75-94550
W.F. Stanley & Co. Ltd.	UK	*486*	75-56350
White Products BV	Netherlands	*296*	75-09100
Zero Cases (Europe)	UK	*495*	75-67700

792 50 Other equipment, tri-service / general

Aeroquip Corp.	USA	*547*	75-02350
Avery Berkel	UK	*432*	75-04560
Busak + Shamban Inc.	USA	*561*	75-16006
C.I.A.M.-E.A. Sarl	Italy.	*233*	75-05150

Donaldson Co. Inc.	USA	569	75-27050
Doncasters Moorside	UK	446	75-18455
Hamworthy Marine	UK	457	75-27850
Saunders-Roe Ltd.	UK	482	75-51850
Trompeter Electronics Inc.	USA	623	75-91070
Victaulic Co. of America	USA	626	75-93840
Vision Research Co.	UK	491	75-63360
Zero Cases (Europe)	UK	495	75-67700

792 50 20 Cables, defense specification

AEI Cables Limited	UK	427	75-00930
Albacom Ltd	UK	429	75-02372
Amphenol Ltd	UK	430	75-03000
Associated Aircraft Mfg & Sales Inc.	USA	555	75-09500
Axon Cable SA	France	132	75-04800
BICC Brand-Rex Limited	UK	434	75-06640
BIW Cable Systems Inc.	USA	559	75-14083
Burton Electrical Engineering.	USA	561	75-16005
Cicoil Corp.	USA	564	75-18580
Draka Kabel BV	Netherlands	292	75-01650
Dynaco Corp.	USA	570	75-27900
Dynamation Research	USA	570	75-27960
Electric Wire & Cable Co.	Israel	218	75-05755
Elfon Ltd	Greece	188	75-01150
Ez Form Cable Inc.	USA	575	75-33140
Greening Donald Co. Ltd	Canada	72	75-06500
Hydrocable Systems	UK	460	75-30790
MA/COM Antenna & Cable Div.	USA	593	75-56953
MM Cables	Australia	19	75-13495
National Wire & Cable Corp.	USA	598	75-64000
Olex Cables	Australia	19	75-15640
Optical Cable Corp.	USA	601	75-66952
Precision Tube Co. Inc.	USA	605	75-71370
Raydex/C.D.T. Ltd	UK	477	75-48340
Reynolds Industries Ltd	UK	479	75-49345
Sparton Technology Inc.	USA	616	75-82950
Teledyne Aircraft Products	USA	621	75-87580
Thermax/CDT, Inc	USA	547	75-03000
THERMOCOAX	France	147	75-44930
Thermofil-Cvalim Ltd	Israel	223	75-23000
Trompeter Electronics Inc.	USA	623	75-91070
Underwater & Environmental	UK	490	75-61900
White Products BV	Netherlands	296	75-09100
WPI-Boston Division Inc.	USA	561	75-16020

792 50 20 50 Cables, electromechanical, underwater

Associated Aircraft Mfg & Sales Inc.	USA	555	75-09500
Burton Electrical Engineering.	USA	561	75-16005
Cable and Wireless (Marine) Ltd.	UK	499	80-05280
Draka Kabel BV	Netherlands	292	75-01650
Hawke Cable Glands Ltd	UK	457	75-28475
Hawke Connection Systems.	UK	457	75-28480
Hydrocable Systems	UK	460	75-30790
JDR Cable Systems Ltd	UK	462	75-32480
NSW Technology Limited	UK	471	75-42770
Raydex/C.D.T. Ltd	UK	477	75-48340
Underwater & Environmental	UK	490	75-61900
WPI-Boston Division Inc.	USA	561	75-16020

792 50 23 Cable dispensers

Deep Sea Seals Ltd	UK	445	75-17400
Metool Products Ltd	UK	467	75-38875

792 50 35 EMP protectors, coaxial

Amphenol Ltd	UK	430	75-03000
EEV Canada Ltd	Canada	71	75-04700
EEV Ltd	UK	448	75-19590
Kopex International Ltd.	UK	463	75-34150
Presentey Engineering	Canada	75	75-11500
RFI Industries Pty Ltd	Australia	20	75-18500
Tecknit Inc.	USA	620	75-87150

792 50 40 Lightning protectors

CORIS	France	134	75-12450
EEV Canada Ltd	Canada	71	75-04700
EF International SA	Switzerland	385	75-03260
Global Atmouspherics, Inc.	USA	590	75-55175
Global Lightning Technologies.	Australia	16	75-07600
Oxley Inc.	USA	602	75-67870
Presentey Engineering	Canada	75	75-11500
RFI Industries Pty Ltd	Australia	20	75-18500
Thorn DNT Airfield Lighting Pty Ltd	Australia	22	75-20540

792 50 50 Motors, electric

Aimsco Inc.	USA	547	75-03050
Astrosyn International Technology	UK	432	75-04235
Bental Industries Ltd	Israel	218	75-02850
Combimac BV	Netherlands	292	75-01200
EG&G Inc.	USA	571	75-29555
Electro Mechanical Systems Ltd.	UK	502	80-09380
EUROCONTROL SpA	Italy	233	75-08200
G.E. Mortley, Sprague & Co. Ltd	UK	469	75-41000
GEC-Marconi Aerospace Inc.	USA	577	75-37412
Holec Machines & Apparaten BV	Netherlands	293	75-03110
Minebea Co. Ltd	Japan	253	75-05600
Park Products Ltd	UK	473	75-44000
Portescap	Switzerland	387	75-10320
Société Française de Moteurs à	France	145	75-35100
Vernitron Corp.	USA	625	75-93760

792 50 55 Storage systems & equipment (shelving & racking)

Aimsco Inc.	USA	547	75-03050
Apex Storage Systems Ltd	UK	431	75-03420
Elite Manufacturing Co Ltd	UK	448	75-20610
Lista International Corp.	USA	590	75-55350

792 50 60 Tools, maintenance

AAR Allen Group International	Netherlands	296	80-00050
Allfast Fastening Systems Inc.	USA	548	75-04635
Ametsa	Switzerland	385	75-00400
Antex (Electronics) Ltd	UK	430	75-03355
Appareillages et Matériels de	France	132	75-03050
Applied Power Australia Ltd	Australia	12	75-01500
Array Connector Corp.	USA	555	75-09020
Blanc Aéro Industries.	France	132	75-06900
Clydesdale Whitehouse Co.	UK	441	75-13890
Constructions Mécaniques des Vosges	France	134	75-12300
Desoutter UK Sales Ltd.	UK	445	75-17800
Fastening Systems International Inc.	USA	575	75-33870
Gates Hydraulics Ltd	UK	453	75-25210
Geier & Bluhm Inc.	USA	578	75-37427
Graphic Datakits International	USA	579	75-40355
Hydro-Gerätebau GmbH & Co KG	Germany	170	75-13520
IFI Srl.	Italy	235	75-11220
MHH Engineering Co. Ltd	UK	468	75-38915
Schrader do Brasil SA	Brazil	58	75-19240
Textron Marine & Land Systems	USA	622	75-88660
Tomarc Limited	UK	488	75-60363
Trompeter Electronics Inc.	USA	623	75-91070

792 50 60 20 Tools, maintenance, aircraft

Aeromax Industries Inc.	USA	547	75-01920
Appareillages et Matériels de	France	132	75-03050
Associated Aircraft Mfg & Sales Inc.	USA	555	75-09500
Desoutter UK Sales Ltd.	UK	445	75-17800
Equipment & Supply Inc.	USA	574	75-32000
Fastening Systems International Inc.	USA	575	75-33870
Graphic Datakits International	USA	579	75-40355
Hydro-Gerätebau GmbH & Co KG	Germany	170	75-13520
IFI Srl.	Italy	235	75-11220
MHH Engineering Co. Ltd	UK	468	75-38915
Normalair-Garrett Australia	Australia	19	75-14870
Société d'Études et de Fabrications	France	143	75-34810
Trompeter Electronics Inc.	USA	623	75-91070

792 90 Weighing systems & equipment

Avery Berkel	UK	432	75-04560
General Electrodynamics Corp.	USA	578	75-38500
Revere Transducers Inc.	USA	609	75-74850

Manufacturing, Semi-Finished Products, Materials

Manufacturing Facilities, Plant Equipment

811 Manufacturing facilities & plant

AAI Corp.	USA	*545*	75-00020
Ailga Rubber Works	India	*204*	75-00080
AM General Corporation	USA	*552*	75-05110
BMC Sanayi ve Ticaret AS	Turkey	*402*	75-00150
Bombardier Inc.	Canada	*69*	75-01400
BP Chemicals Ltd	UK	*435*	75-07790
Bristol Aerospace Ltd	Canada	*69*	75-01500
Canadian Marconi Co.	Canada	*69*	75-02600
Chamberlain Manufacturing Corp.	USA	*563*	75-17950
Electrothermal Engineering Ltd	UK	*448*	75-20400
Flight Suits Ltd	USA	*575*	75-34800
Fluor Daniel Inc.	USA	*635*	80-11720
FN Herstal SA	Belgium	*46*	75-03600
Haden Drysys	Australia	*17*	75-08110
Hale Group International	USA	*636*	80-13600
Hellenic Arms Industry	Greece	*187*	55-00800
Hoffman Engineering Corp.	USA	*582*	75-43980
Hunting Engineering Ltd	UK	*459*	75-30700
Hydroscience Technologies Inc.	USA	*584*	75-45550
Hyundai	Korea (ROK)	*265*	75-01540
Ika-Maschinenbau Janke & Kunkel GmbH	Germany	*170*	75-13900
Indal Technologies Inc	Canada	*72*	75-07740
Israel Military Industries Ltd	Israel	*220*	75-10300
Israel Military Industries Rocket Systems Division	**Israel**	***220***	**75-10310**
ITT Avionics	USA	*586*	75-49455
John Brown Engineering Ltd	UK	*437*	75-10250
NIS Ltd	UK	*471*	75-42555
NNC Ltd	UK	*507*	80-18500
Norton Co.	USA	*600*	75-65510
Oerlikon-Contraves AG.	Switzerland	*387*	75-09250
Perm Kirov Research Production Corp	Russia	*338*	75-15975
Perry Engineering	Australia	*20*	75-16460
Rabintex Industries Ltd	Israel	*221*	75-17850
Santa Barbara SA	Spain	*364*	55-01000
Siemens Plessey Systems	UK	*483*	75-53770
SNPE	France	*145*	75-36315
Sparton Technology Inc.	USA	*616*	75-82950
Terberg Benschop BV	Netherlands	*296*	75-07600
Tracor Inc	USA	*623*	75-90000
Tumo Ltd	Finland	*119*	75-02650
United Defense L.P.	USA	*624*	75-92050
Urdan Industries Ltd	Israel	*223*	75-23210
Vihtavuori Oy	Finland	*119*	75-02850
Wellman Furnaces Ltd	UK	*493*	75-65120
Yewlands Engineering Co. Ltd.	UK	*495*	75-67520

811 30 Explosives production plant

FN Herstal SA	Belgium	*46*	75-03600
Hale Group International	USA	*636*	80-13600
Indústrias de Defesa	Portugal	*327*	55-00950
Israel Military Industries Ltd	Israel	*220*	75-10300
Israel Military Industries Rocket Systems Division	**Israel**	***220***	**75-10310**
Perm Kirov Research Production Corp	Russia	*338*	75-15975

811 60 Explosives processing plant

CMS Inc.	USA	*564*	75-19170
Hale Group International	USA	*636*	80-13600
Israel Military Industries Ltd	Israel	*220*	75-10300
Perm Kirov Research Production Corp	Russia	*338*	75-15975

812 Production machines, special purpose

AAI Corp.	USA	*545*	75-00020
A.O. Henton Engineering Co Ltd.	UK	*458*	75-29040
Apex Broach & Machine Co.	USA	*554*	75-08000
Australian Defence Industries Ltd	Australia	*9*	55-01000
Automated Production Equipment Corp.	USA	*556*	75-10380
Benelli Armi SpA	Italy	*232*	75-02700
Chamberlain Manufacturing Corp.	USA	*563*	75-17950
Cleco Limited	UK	*440*	75-13750
Constructions Mécaniques des Vosges	France	*134*	75-12300
Fjellman Press i Mariestad AB	Sweden	*377*	75-05200
FN Herstal SA	Belgium	*46*	75-03600
GEC-Marconi Communications Ltd	UK	*454*	75-25335
Haden Drysys	Australia	*17*	75-08110
Hellenic Arms Industry	Greece	*187*	55-00800
H.I. Fraser Pty Ltd	Australia	*16*	75-06835
Hydroscience Technologies Inc.	USA	*584*	75-45550
Ika-Maschinenbau Janke & Kunkel GmbH	Germany	*170*	75-13900
Imco Industries Ltd	Israel	*219*	75-09400
Inductoheat Pty Ltd	Australia	*17*	75-09500
Israel Aircraft Industries Ltd	Israel	*219*	75-10200
ITT Avionics	USA	*586*	75-49455
L&F Industries	USA	*589*	75-53690
NIS Ltd	UK	*471*	75-42555
NNC Ltd	UK	*507*	80-18500
Oerlikon-Contraves AG.	Switzerland	*387*	75-09250
Olympic Tool Hellas SA	Greece	*191*	75-02180
Sciaky Electric Welding Machines Ltd	UK	*482*	75-52100
Swanson Erie Corp.	USA	*618*	75-85265
T S Harrison & Sons	UK	*457*	75-28150
Tracor Inc	USA	*623*	75-90000
United Defense L.P.	USA	*624*	75-92050
Urdan Industries Ltd	Israel	*223*	75-23210
Vacu-Blast Ltd	UK	*490*	75-62550
Veribest International Ltd	UK	*491*	75-62743
Yewlands Engineering Co. Ltd.	UK	*495*	75-67520

812 10 Production machines, ammunition manufacturing

Apex Broach & Machine Co.	USA	*554*	75-08000
ARSOC Sprl	Belgium	*45*	75-00520
Chamberlain Manufacturing Corp.	USA	*563*	75-17950
FN Herstal SA	Belgium	*46*	75-03600
Interstate Technical Products	UK	*461*	75-31946
NIS Ltd	UK	*471*	75-42555
Nuova Connavi Srl	Italy	*237*	75-16700
Oerlikon-Contraves AG.	Switzerland	*387*	75-09250
Swanson Erie Corp.	USA	*618*	75-85265
Vammas Ltd	**Finland**	***117***	**55-02220**

812 20 Mixing machines, explosives & propergols

Ika-Maschinenbau Janke & Kunkel GmbH	Germany	*170*	75-13900
Treu SA Máquinas e Equipamentos	Brazil	*58*	75-20780

812 50 Machine tools, workshop

AAI Corp.	USA	*545*	75-00020
AeroSpace Technologies of Australia	Australia	*11*	75-00340
A.O. Henton Engineering Co Ltd.	UK	*458*	75-29040
Chamberlain Manufacturing Corp.	USA	*563*	75-17950
Constructions Mécaniques des Vosges	France	*134*	75-12300
Cranfield Precision	UK	*443*	75-15772
FN Herstal SA	Belgium	*46*	75-03600
Frazier Aviation Inc.	USA	*576*	75-35970
Hellenic Arms Industry	Greece	*187*	55-00800
H.I. Fraser Pty Ltd	Australia	*16*	75-06835
Howa Machinery Ltd	Japan	*251*	75-02550
IFI Srl	Italy	*235*	75-11220
Imco Industries Ltd	Israel	*219*	75-09400
John Heine & Son Pty Ltd	Australia	*17*	75-08350
Radcoflex Singapore Pte Ltd	Singapore	*348*	75-03370
United Defense L.P.	USA	*624*	75-92050
Vargus Ltd	Israel	*223*	75-23238

812 70 Robotic manufacturing systems

A.O. Henton Engineering Co Ltd.	UK	*458*	75-29040
Chamberlain Manufacturing Corp.	USA	*563*	75-17950
Dassault Aviation Group	France	*135*	75-12975

Fluor Daniel Inc.	USA	*635*	80-11720
Giat Industries	France	*126*	55-00100
IFI Srl	Italy	*235*	75-11220
Israel Aircraft Industries Ltd	Israel	*219*	75-10200
Machine Technology Ltd	UK	*465*	75-36870
NIS Ltd	UK	*471*	75-42555
United Defense L.P.	USA	*624*	75-92050

812 75 30 Mixers, explosives

Treu SA Máquinas e Equipamentos	Brazil	*58*	75-20780

812 75 60 Mixers, rocket fuel

Treu SA Máquinas e Equipamentos	Brazil	*58*	75-20780

819 Manufacturing plant equipment, other

AAI Corp.	USA	*545*	75-00020
Airmatic Engineering (UK) Ltd	UK	*429*	75-01850
AMP Tools (India) PVZ Ltd.	India	*204*	75-00150
Automated Production Equipment Corp.	USA	*556*	75-10380
Belotti SpA	Italy	*232*	75-02640
Bombardier Inc.	Canada	*69*	75-01400
Chamberlain Manufacturing Corp.	USA	*563*	75-17950
Daewoo Heavy Industries Ltd	Korea (ROK)	*265*	75-00600
EL-OP Electro Optics Industries Ltd	Israel	*219*	75-06300
FN Herstal SA	Belgium	*46*	75-03600
GEC-Marconi Communications Ltd	UK	*454*	75-25335
Howa Machinery Ltd	Japan	*251*	75-02550
IMI Norgren Ltd	UK	*460*	75-31158
Inductoheat Pty Ltd	Australia	*17*	75-09500
Japan Aircraft Manufacturing Co. Ltd	Japan	*252*	75-03200
John Brown Engineering Ltd	UK	*437*	75-10250
Jordan Valley Applied Radiation Ltd	Israel	*220*	75-10850
Lista Ltd	Switzerland	*386*	75-07350
Logan Fenamec (UK) Ltd	UK	*465*	75-36270
NIS Ltd	UK	*471*	75-42555
NNC Ltd	UK	*507*	80-18500
Oerlikon-Contraves AG.	Switzerland	*387*	75-09250
Terberg Benschop BV	Netherlands	*296*	75-07600
Treu SA Máquinas e Equipamentos	Brazil	*58*	75-20780
United Defense L.P.	USA	*624*	75-92050
United Moulders Ltd	UK	*490*	75-62130
Whippendell Electrical Ltd	UK	*493*	75-65800

Semi-Finished Products, Materials

821 Semi-finished products

Agamice	France	*131*	75-00800
A.J. Worthington (Leek)	UK	*494*	75-67080
Alinabal Inc.	USA	*548*	75-04460
Allied Screw Products Inc.	USA	*549*	75-04690
AMCO Engineering	USA	*552*	75-05200
Australian Defence Industries Ltd	Australia	*9*	55-01000
Bigg & Robertson Plastics Ltd	UK	*434*	75-06680
Blanc Aéro Industries	France	*132*	75-06900
Böhler Schmiedetechnik GmbH	Austria	*32*	75-00650
Brassinter SA Indústria e Comércio	Brazil	*55*	75-02290
Brighton Sheet Metal Ltd	UK	*436*	75-08550
British Steel Engineering Steels	UK	*436*	75-09600
Cabiran (1991) Ltd	Israel	*218*	75-03400
Calif Gasket & Rubber Corp.	USA	*561*	75-16230
Chamberlain Manufacturing Corp.	USA	*563*	75-17950
Deritend Precision Castings Ltd	UK	*445*	75-17705
Doncasters plc	UK	*446*	75-18410
Doncasters Structures	UK	*446*	75-18465
Downee RMD	Australia	*15*	75-04750
Ets Robert Creuzet	France	*136*	75-16300
Hellenic Arms Industry	Greece	*187*	55-00800
Hindustan Aeronautics Ltd.	India	*204*	55-01000
Hitchiner Manufacturing Co. Inc.	USA	*582*	75-43550
Howmet Exeter Alloys	UK	*459*	75-30280
International Polymer Corp.	USA	*585*	75-48200
Israel Military Industries Ltd	Israel	*220*	75-10300
James Hardie Engineering	New Zealand	*303*	80-00900
J.S. Chinn Engineering	UK	*440*	75-13050
Magnesium Elektron	UK	*466*	75-37000
Mason & Cox Pty Ltd.	Australia	*19*	75-13165
McInnes Steel Co.	USA	*595*	75-59600
Metal Dynamics Corp.	USA	*595*	75-60400
MIFA Aluminium BV	Netherlands	*294*	75-04525
MIL-BASE Industries Inc.	USA	*596*	75-61530
National Forge Co.	USA	*598*	75-63850
Partsource Inc.	USA	*603*	75-68980
Perry Engineering	Australia	*20*	75-16460
PI Castings Ltd	UK	*474*	75-45300
Pittards plc	UK	*475*	75-45700
Precimetal Fonderie de Précision SA	Belgium	*46*	75-05500
Radcoflex Singapore Pte Ltd	Singapore	*348*	75-03370
RE Components Ltd	UK	*478*	75-48400
Rollet H. & Co. Ltd	UK	*479*	75-50208
Royal Ordnance	UK	*481*	75-50620
Santa Barbara SA	Spain	*364*	55-01000
Sheffield Forgemasters Ltd	UK	*483*	75-53200
Sleeman Engineering	UK	*484*	75-54400
Société Métallurgique de Gerzat SA	France	*146*	75-40570
Specialty Steel & Forge	USA	*616*	75-83040
Stone Foundries Ltd	UK	*486*	75-57325
Superform Aluminium	UK	*487*	75-57920
T&N Technology Ltd	UK	*513*	80-50150
Tecphy	France	*147*	75-44375
Timcast Pty Ltd	Australia	*22*	75-20680
Tiodize Co. Inc.	USA	*622*	75-89560
TMR-Tecnologías Metalúrgicas	Spain	*367*	75-05650
Tungum Hydraulics Ltd	UK	*490*	75-61500
Turbotech Precision Products Ltd	UK	*490*	75-61520
Ulbrich Stainless Steels &	USA	*624*	75-91385
UMPCO	USA	*624*	75-91505
United Defense L.P.	USA	*624*	75-92050
Urdan Industries Ltd	Israel	*223*	75-23210
Urdan Industries (USA) Inc.	USA	*625*	75-92930
Villares Metals SA	Brazil	*56*	75-05350
Wall Colomonoy Corp.	USA	*626*	75-94900
Wyman-Gordon Ltd	UK	*494*	75-67280
Yewlands Engineering Co. Ltd	UK	*495*	75-67520

821 20 Castings

AEA Technology plc	UK	*496*	80-00480
Aeromet International plc	UK	*428*	75-01040
Allied Screw Products Inc.	USA	*549*	75-04690
Aluminum Co. of America	USA	*552*	75-05050
AMCO Engineering	USA	*552*	75-05200
Arromet International plc	UK	*431*	75-04000
Bet Shemesh Engines Ltd	Israel	*218*	75-03200
British Rollmakers Corporation	UK	*436*	75-09500
Centaur Precision Limited	UK	*439*	75-12280
Cronite Precision Castings Ltd	UK	*443*	75-16136
Deloro Stellite Ltd	UK	*445*	75-17460
Doncasters plc	UK	*446*	75-18410
Finecast (Maidenhead) Ltd	UK	*451*	75-23220
Firth Rixson plc	UK	*451*	75-23600
Haley Industries Ltd.	Canada	*72*	75-06600
Hellenic Arms Industry	Greece	*187*	55-00800
Hitchiner Manufacturing Co. Inc.	USA	*582*	75-43550
Howa Machinery Ltd	Japan	*251*	75-02550
Howmet SA	France	*137*	75-21750
Hycast Metals Pty Ltd	Australia	*17*	75-08800
Indústria de Fundição Tupy Ltda.	Brazil	*56*	75-06840
Ingersoll-Dresser Pumps (UK) Ltd	UK	*461*	75-31450
John Heine & Son Pty Ltd	Australia	*17*	75-08350
John Valves Pty Ltd	Australia	*18*	75-10400
Magnesium Elektron	UK	*466*	75-37000
Mannesmann Rexroth SA	France	*139*	75-25500
Mason & Cox Pty Ltd.	Australia	*19*	75-13165
Meggitt Composites	UK	*467*	75-38345
Metal Dynamics Corp.	USA	*595*	75-60400
MIFA Aluminium BV	Netherlands	*294*	75-04525
MIL-BASE Industries Inc.	USA	*596*	75-61530
NII Stali	Russia	*338*	75-14100
Permali Gloucester Ltd	UK	*474*	75-44850
Pezetel	Poland	*325*	80-01000
PI Castings Ltd	UK	*474*	75-45300
Precimetal Fonderie de Précision SA	Belgium	*46*	75-05500
Presto Casting Co.	USA	*605*	75-71440
River Don Castings Ltd	UK	*479*	75-49830
RKB Precision Products Ltd	UK	*464*	75-36068
Robert Mitchell Inc.	Canada	*74*	75-09800
Rollet H. & Co. Ltd	UK	*479*	75-50208
Royal Ordnance	UK	*481*	75-50620
Sheffield Forgemasters Ltd	UK	*483*	75-53200

Shipham & Co. Ltd	UK	*483*	75-53350
Sleeman Engineering	UK	*484*	75-54400
Stone Foundries Ltd	UK	*486*	75-57325
Teledyne Casting Service	USA	*620*	75-87515
Timcast Pty Ltd	Australia	*22*	75-20680
Timet UK Ltd	UK	*488*	75-59950
Trucast Ltd	UK	*489*	75-61360
Turbotech Precision Products Ltd	UK	*490*	75-61520
United Defense L.P.	USA	*624*	75-92050
Urdan Industries Ltd	Israel	*223*	75-23210
Urdan Industries (USA) Inc.	USA	*625*	75-92930
Victaulic Co. of America	USA	*626*	75-93840
Westley Ltd	UK	*493*	75-65680

821 20 50 Investment castings

Aeromet International plc	UK	*428*	75-01040
AETC Ltd	UK	*428*	75-01350
Cabiran (1991) Ltd	Israel	*218*	75-03400
Ceramicast (P.I.C.) Ltd	UK	*439*	75-12410
Deloro Stellite Ltd	UK	*445*	75-17460
Deritend Precision Castings Ltd	UK	*445*	75-17705
Designed Precision Castings Inc.	Canada	*70*	75-04100
Doncasters plc	UK	*446*	75-18410
Epirus Foundries SA	Greece	*188*	75-01280
Finecast (Maidenhead) Ltd	UK	*451*	75-23220
Howmet Limited	UK	*459*	75-30280
Howmet-Cercast Group	Canada	*72*	75-07630
Hycast Metals Pty Ltd	Australia	*17*	75-08800
Israel Military Industries Ltd	Israel	*220*	75-10300
PI Castings Ltd	UK	*474*	75-45300
Rollet H. & Co. Ltd	UK	*479*	75-50208
Urdan Industries (USA) Inc.	USA	*625*	75-92930
Wyman-Gordon Co.	USA	*628*	75-97700
Wyman-Gordon Investment Castings	USA	*628*	75-97720

821 40 Forgings

Aerospace Forgings Ltd	UK	*428*	75-01160
Blanc Aéro Industries	France	*132*	75-06900
Böhler Schmiedetechnik GmbH	Austria	*32*	75-00650
Chamberlain Manufacturing Corp.	USA	*563*	75-17950
Clydesdale Engineering Ltd	UK	*441*	75-13870
Clydesdale Forge Co.	UK	*441*	75-13880
Doncasters Monk Bridge	UK	*446*	75-18450
Doncasters plc	UK	*446*	75-18410
Doncasters Structures	UK	*446*	75-18465
Ets Robert Creuzet	France	*136*	75-16300
Firth Rixson plc	UK	*451*	75-23600
Forged Rolls (UK) Ltd	UK	*452*	75-24165
Forgemasters Steel & Engineering Ltd	UK	*452*	75-24168
HDA Forgings Ltd	UK	*457*	75-28770
Indústrias de Defesa	Portugal	*327*	55-00950
Jessop Saville Ltd	UK	*462*	75-32500
King Fifth Wheel Co.	USA	*589*	75-52670
Ladish Co.Inc.	USA	*589*	75-53850
Manoir Industries	France	*139*	75-25610
McInnes Steel Co.	USA	*595*	75-59600
National Forge Co.	USA	*598*	75-63850
Otto Fuchs Metallwerke	Germany	*168*	75-09570
Overall Forge Pty Ltd	Australia	*20*	75-15890
Papadopoulos & Co Ltd	Greece	*191*	75-02195
Royal Ordnance	UK	*481*	75-50620
Saco Defense Inc.	USA	*612*	75-77340
Sheffield Forgemasters Ltd	UK	*483*	75-53200
Smith-Clayton Forge (Aerospace)	UK	*484*	75-54570
Société Métallurgique de Gerzat SA	France	*146*	75-40570
Somers Forge Ltd	UK	*485*	75-55150
Special Melted Products Ltd	UK	*485*	75-55240
Specialty Steel & Forge	USA	*616*	75-83040
Tecphy	France	*147*	75-44375
TMR-Tecnologías Metalúrgicas	Spain	*367*	75-05650
United Defense L.P.	USA	*624*	75-92050
Vammas Ltd	**Finland**	***117***	**55-02220**
Villares Metals SA	Brazil	*56*	75-05350
Wyman-Gordon Co.	USA	*628*	75-97700
Wyman-Gordon Ltd	UK	*494*	75-67280

821 40 20 Gun barrels, semi-finished

Accuracy International Ltd	UK	*427*	75-00360
Australian Defence Industries Ltd	Australia	*9*	55-01000
Böhler Schmiedetechnik GmbH	Austria	*32*	75-00650
Chamberlain Manufacturing Corp.	USA	*563*	75-17950
Indústrias de Defesa	Portugal	*327*	55-00950
National Forge Co.	USA	*598*	75-63850
Parker-Hale Ltd	UK	*473*	75-43850
Royal Ordnance	UK	*481*	75-50620
Saco Defense Inc.	USA	*612*	75-77340
Sheffield Forgemasters Ltd	UK	*483*	75-53200
Special Melted Products Ltd	UK	*485*	75-55240
Villares Metals SA	Brazil	*56*	75-05350

821 40 25 Gun breeches, semi-finished

Accuracy International Ltd	UK	*427*	75-00360
Australian Defence Industries Ltd	Australia	*9*	55-01000
Böhler Schmiedetechnik GmbH	Austria	*32*	75-00650
National Forge Co.	USA	*598*	75-63850
Royal Ordnance	UK	*481*	75-50620
Sheffield Forgemasters Ltd	UK	*483*	75-53200
Timcast Pty Ltd	Australia	*22*	75-20680
United Defense L.P.	USA	*624*	75-92050
Villares Metals SA	Brazil	*56*	75-05350

822 Metal & alloy sections, plates & extrusions, welded rings

Acieries Aubert & Duval SA	France	*132*	75-04515
Allied Screw Products Inc.	USA	*549*	75-04690
Aluminum Co. of America	USA	*552*	75-05050
AMCO Engineering	USA	*552*	75-05200
American Aluminum Co.	USA	*552*	75-05530
Böhler Schmiedetechnik GmbH	Austria	*32*	75-00650
Bosma Machine and Tool Corporation	USA	*560*	75-14750
Bremer Vulkan Marineschiffbau GmbH	Germany	*166*	75-03600
British Aluminium - Speciality and	UK	*429*	75-02455
Broken Hill Proprietary Co. Ltd	Australia	*13*	75-02770
Caledonian Mining Co. Ltd	UK	*438*	75-11250
Centaur Precision Limited	UK	*439*	75-12280
Chamberlain Manufacturing Corp.	USA	*563*	75-17950
China Steel Corporation	Taiwan	*392*	75-04700
Cold Formed Products Ltd	UK	*441*	75-14100
Delta (Manganese Bronze) Ltd	UK	*445*	75-17470
Doncasters Blaenavon	UK	*446*	75-18430
Dynamic Metal Products Co.	USA	*570*	75-28240
Fibertek	USA	*575*	75-34330
Finavicomp Oy	Finland	*118*	75-00595
Haynes International Inc.	USA	*581*	75-42740
H.C. Starck	USA	*617*	75-83957
Hellenic Arms Industry	Greece	*187*	55-00800
Hoogovens Aluminium Walzprodukte	Germany	*169*	75-13270
Howmet Exeter Alloys	UK	*459*	75-30280
Hyundai	Korea (ROK)	*265*	75-01540
Inco Alloys Ltd	UK	*460*	75-31260
J.T. Slocomb Company	USA	*615*	75-81700
L&S Machine Co. Inc.	USA	*589*	75-53700
Magnesium Elektron	UK	*466*	75-37000
McInnes Steel Co.	USA	*595*	75-59600
National Forge Co.	USA	*598*	75-63850
National Nuclear Corporation	USA	*598*	75-63870
Normalair-Garrett Australia	Australia	*19*	75-14870
Otto Fuchs Metallwerke	Germany	*168*	75-09570
Precimetal Fonderie de Précision SA	Belgium	*46*	75-05500
Radcoflex Singapore Pte Ltd	Singapore	*348*	75-03370
Rashtriya Metal Industries Ltd	India	*205*	75-01810
SAI Industries Corp.	USA	*612*	75-77700
Société Métallurgique de Gerzat SA	France	*146*	75-40570
Special Melted Products Ltd	UK	*485*	75-55240
Specialty Steel & Forge	USA	*616*	75-83040
Standard Steel	USA	*617*	75-83700
Thomas Instrument & Machine Co. Inc.	USA	*622*	75-88860
Tungum Hydraulics Ltd	UK	*490*	75-61500
Villares Metals SA	Brazil	*56*	75-05350
Wyman-Gordon Co.	USA	*628*	75-97700
Wyman-Gordon Ltd	UK	*494*	75-67280
Z. Rabiner Ltd	Israel	*221*	75-17840

822 10 Steel & alloy sections

Accles & Pollock Ltd	UK	*427*	75-00320
Acieries Aubert & Duval SA	France	*132*	75-04515
Allvac	USA	*552*	75-04800
Böhler Schmiedetechnik GmbH	Austria	*32*	75-00650

Product-Supplier Link

Bremer Vulkan Marineschiffbau GmbH	Germany	*166*	75-03600
Chamberlain Manufacturing Corp.	USA	*563*	75-17950
China Steel Corporation	Taiwan	*392*	75-04700
Designed Precision Castings Inc.	Canada	*70*	75-04100
General Motors of Canada Ltd	Canada	*71*	75-05900
Magnesium Elektron	UK	*466*	75-37000
National Forge Co.	USA	*598*	75-63850
Precimetal Fonderie de Précision SA	Belgium	*46*	75-05500
Robert Mitchell Inc.	Canada	*74*	75-09800
Special Melted Products Ltd	UK	*485*	75-55240
Specialty Steel & Forge	USA	*616*	75-83040
Spencer Clark Metal Industries plc	UK	*486*	75-55700
Tecphy	France	*147*	75-44375
Ulbrich Stainless Steels &	USA	*624*	75-91385
United Defense L.P.	USA	*624*	75-92050
Wyman-Gordon Ltd	UK	*494*	75-67280

822 11 Steel sections

Acieries Aubert & Duval SA	France	*132*	75-04515
Böhler Schmiedetechnik GmbH	Austria	*32*	75-00650
Bremer Vulkan Marineschiffbau GmbH	Germany	*166*	75-03600
British Steel Engineering Steels	UK	*436*	75-09600
Caledonian Mining Co. Ltd	UK	*438*	75-11250
Caparo Group Ltd	UK	*438*	75-11500
China Steel Corporation	Taiwan	*392*	75-04700
General Motors of Canada Ltd	Canada	*71*	75-05900
Heflin Steel Co.	USA	*581*	75-42880
National Forge Co.	USA	*598*	75-63850
Precimetal Fonderie de Précision SA	Belgium	*46*	75-05500
Robert Mitchell Inc.	Canada	*74*	75-09800
Special Melted Products Ltd	UK	*485*	75-55240
Specialty Steel & Forge	USA	*616*	75-83040
Spencer Clark Metal Industries plc	UK	*486*	75-55700
Strumech Versatower Ltd	UK	*486*	75-57530
United Defense L.P.	USA	*624*	75-92050
Wyman-Gordon Ltd	UK	*494*	75-67280

822 12 Aluminium alloy sections

Aluminum Co. of America	USA	*552*	75-05050
AMCO Engineering	USA	*552*	75-05200
British Aluminium - Speciality and	UK	*429*	75-02455
Capral Aluminium Ltd	Australia	*14*	75-03170
Chamberlain Manufacturing Corp.	USA	*563*	75-17950
China Steel Corporation	Taiwan	*392*	75-04700
Designed Precision Castings Inc.	Canada	*70*	75-04100
General Motors of Canada Ltd	Canada	*71*	75-05900
Hollow Extrusions Ltd	UK	*458*	75-29652
Hoogovens Aluminium Walzprodukte	Germany	*169*	75-13270
NII Stali	Russia	*338*	75-14100
Robert Mitchell Inc.	Canada	*74*	75-09800
Rollet H. & Co. Ltd	UK	*479*	75-50208
United Defense L.P.	USA	*624*	75-92050
Wyman-Gordon Ltd	UK	*494*	75-67280

822 12 10 Aluminium bronze wire, bar, rods

China Steel Corporation	Taiwan	*392*	75-04700
Delta (Manganese Bronze) Ltd	UK	*445*	75-17470
Primex Technologies, Inc	USA	*606*	75-71472
Rollet H. & Co. Ltd	UK	*479*	75-50208

822 20 Steel & alloy plates & sheet

Böhler Schmiedetechnik GmbH	Austria	*32*	75-00650
China Steel Corporation	Taiwan	*392*	75-04700
Dynamic Metal Products Co.	USA	*570*	75-28240
Finavicomp Oy	Finland	*118*	75-00595
General Motors of Canada Ltd	Canada	*71*	75-05900
Israel Aircraft Industries Ltd	Israel	*219*	75-10200
Sleeman Engineering	UK	*484*	75-54400
Specialty Steel & Forge	USA	*616*	75-83040
Spencer Clark Metal Industries plc	UK	*486*	75-55700
Tecphy	France	*147*	75-44375
United Defense L.P.	USA	*624*	75-92050

822 21 Steel plates

Bisalloy Steels Pty Ltd	Australia	*12*	75-02425
Böhler Schmiedetechnik GmbH	Austria	*32*	75-00650
China Steel Corporation	Taiwan	*392*	75-04700
Heflin Steel Co.	USA	*581*	75-42880
Sleeman Engineering	UK	*484*	75-54400
Spencer Clark Metal Industries plc	UK	*486*	75-55700
Strumech Versatower Ltd	UK	*486*	75-57530

822 21 10 Armour steel plates

Avesta Sheffield Ltd	UK	*432*	75-04570
Bisalloy Steels Pty Ltd	Australia	*12*	75-02425
China Steel Corporation	Taiwan	*392*	75-04700
Fibertek	USA	*575*	75-34330
General Motors of Canada Ltd	Canada	*71*	75-05900
Heflin Steel Co.	USA	*581*	75-42880
Mishra Dhatu Nigam Ltd	India	*204*	55-01300
National Forge Co.	USA	*598*	75-63850
Protection Development Int'l Corp.	USA	*606*	75-71670
Sleeman Engineering	UK	*484*	75-54400
Urdan Industries Ltd	Israel	*223*	75-23210
Urdan Industries (USA) Inc.	USA	*625*	75-92930
Villares Metals SA	Brazil	*56*	75-05350

822 22 Aluminium alloy plates

Aluminum Co. of America	USA	*552*	75-05050
AMCO Engineering	USA	*552*	75-05200
British Aluminium - Speciality and	UK	*429*	75-02455
British Aluminium Plate	UK	*436*	75-09000
Capral Aluminium Ltd	Australia	*14*	75-03170
China Steel Corporation	Taiwan	*392*	75-04700
Hoogovens Aluminium Walzprodukte	Germany	*169*	75-13270
Rollet H. & Co. Ltd	UK	*479*	75-50208

822 22 10 Aluminium armour plates

Aluminum Co. of America	USA	*552*	75-05050
British Aluminium Plate	UK	*436*	75-09000
Capral Aluminium Ltd	Australia	*14*	75-03170
China Steel Corporation	Taiwan	*392*	75-04700
General Motors of Canada Ltd	Canada	*71*	75-05900
Hoogovens Aluminium Walzprodukte	Germany	*169*	75-13270

822 30 Steel & alloy extrusions

Allvac	USA	*552*	75-04800
China Steel Corporation	Taiwan	*392*	75-04700
Crane Aluminium Extrusion	Australia	*15*	75-03954
Greening Donald Co. Ltd	Canada	*72*	75-06500
Hj. Jousi Oy	Finland	*118*	75-01350
Hollow Extrusions Ltd	UK	*458*	75-29652
Magnesium Elektron	UK	*466*	75-37000
MIFA Aluminium BV	Netherlands	*294*	75-04525
National Forge Co.	USA	*598*	75-63850
Société Métallurgique de Gerzat SA	France	*146*	75-40570
Vammas Ltd	**Finland**	***117***	**55-02220**
Villares Metals SA	Brazil	*56*	75-05350
Wyman-Gordon Ltd	UK	*494*	75-67280

822 40 Steel & alloy welded rings

China Steel Corporation	Taiwan	*392*	75-04700
Doncasters Blaenavon	UK	*446*	75-18430
Doncasters plc	UK	*446*	75-18410
Dynamic Metal Products Co.	USA	*570*	75-28240
Finavicomp Oy	Finland	*118*	75-00595
Firth Rixson plc	UK	*451*	75-23600
J.T. Slocomb Company	USA	*615*	75-81700
King Fifth Wheel Co.	USA	*589*	75-52670
McInnes Steel Co.	USA	*595*	75-59600
Standard Steel	USA	*617*	75-83700
Urdan Industries Ltd	Israel	*223*	75-23210

822 50 Honeycomb core

Chromalloy France	France	*133*	75-09650
Hexcel Composites Ltd	UK	*458*	75-29150
Israel Aircraft Industries Ltd	Israel	*219*	75-10200

822 55 Tubes, steel

Accles & Pollock Ltd	UK	*427*	75-00320
Böhler Schmiedetechnik GmbH	Austria	*32*	75-00650
Caparo Group Ltd	UK	*438*	75-11500
China Steel Corporation	Taiwan	*392*	75-04700
Fine Tubes Ltd	UK	*451*	75-23215
Précitube SA	France	*141*	75-30890
Radcoflex Singapore Pte Ltd	Singapore	*348*	75-03370
Robert Mitchell Inc.	Canada	*74*	75-09800
Rollet H. & Co. Ltd	UK	*479*	75-50208
Strumech Versatower Ltd	UK	*486*	75-57530
Tecphy	France	*147*	75-44375
Tubemakers of Australia Ltd	Australia	*22*	75-20910
Wyman-Gordon Ltd	UK	*494*	75-67280

822 60 Titanium & alloys

Allied Screw Products Inc.	USA	*549*	75-04690
Allvac	USA	*552*	75-04800
Böhler Schmiedetechnik GmbH	Austria	*32*	75-00650
Chamberlain Manufacturing Corp.	USA	*563*	75-17950
Doncasters plc	UK	*446*	75-18410
Haynes International Inc.	USA	*581*	75-42740
Haynes International Ltd	UK	*457*	75-28760
Hollow Extrusions Ltd	UK	*458*	75-29652
Israel Aircraft Industries Ltd	Israel	*219*	75-10200
King Fifth Wheel Co.	USA	*589*	75-52670
Mishra Dhatu Nigam Ltd	India	*204*	55-01300
National Nuclear Corporation	USA	*598*	75-63870
RMI Titanium Company	USA	*609*	75-75480
Rollet H. & Co. Ltd	UK	*479*	75-50208
Royal Ordnance	UK	*481*	75-50620
Smith-Clayton Forge (Aerospace)	UK	*484*	75-54570
Specialty Steel & Forge	USA	*616*	75-83040
Spencer Clark Metal Industries plc	UK	*486*	75-55700
Thomas Instrument & Machine Co. Inc.	USA	*622*	75-88860
Timet	USA	*622*	75-89400
Timet UK Ltd	UK	*488*	75-59950
Ulbrich Stainless Steels &	USA	*624*	75-91385
Urdan Industries (USA) Inc.	USA	*625*	75-92930
Wyman-Gordon Co.	USA	*628*	75-97700
Wyman-Gordon Ltd	UK	*494*	75-67280

822 65 Tantalum

H.C. Starck	USA	*617*	75-83957
National Nuclear Corporation	USA	*598*	75-63870

822 70 Zirconium

National Nuclear Corporation	USA	*598*	75-63870
Rollet H. & Co. Ltd	UK	*479*	75-50208

822 75 Niobium

H.C. Starck	USA	*617*	75-83957
King Fifth Wheel Co.	USA	*589*	75-52670
National Nuclear Corporation	USA	*598*	75-63870
Villares Metals SA	Brazil	*56*	75-05350

822 78 Magnesium alloys

Magnesium Elektron	UK	*466*	75-37000
National Nuclear Corporation	USA	*598*	75-63870
Rollet H. & Co. Ltd	UK	*479*	75-50208
Teledyne Casting Service	USA	*620*	75-87515

822 80 Superalloys

Acieries Aubert & Duval SA	France	*132*	75-04515
Allvac	USA	*552*	75-04800
Australian Hard Materials	Australia	*12*	75-01830
Böhler Schmiedetechnik GmbH	Austria	*32*	75-00650
Cannon-Muskegon Corp.	USA	*562*	75-16642
Deloro Stellite Ltd	UK	*445*	75-17460
Firth Rixson plc	UK	*451*	75-23600
Haynes International Inc.	USA	*581*	75-42740
Haynes International Ltd	UK	*457*	75-28760
Howmet Exeter Alloys	UK	*459*	75-30280
Inco Alloys Ltd	UK	*460*	75-31260
King Fifth Wheel Co.	USA	*589*	75-52670
Mishra Dhatu Nigam Ltd	India	*204*	55-01300
Rollet H. & Co. Ltd	UK	*479*	75-50208
Ross & Catherall Ltd	UK	*480*	75-50555
Sheffield Forgemasters Ltd	UK	*483*	75-53200
Smith-Clayton Forge (Aerospace)	UK	*484*	75-54570
Tecphy	France	*147*	75-44375
Ulbrich Stainless Steels &	USA	*624*	75-91385
Villares Metals SA	Brazil	*56*	75-05350
Wyman-Gordon Ltd	UK	*494*	75-67280

823 Composites

Advanced Aerodynamics	USA	*546*	75-00790
Advanced Composite Components Ltd	UK	*427*	75-00845
Advanced Composite Group Ltd	UK	*427*	75-00845
Advanced Composites Group Inc.	USA	*546*	75-00820
Advanced Composites Inc.	USA	*546*	75-00830
Aérazur	France	*131*	75-00500
Aeronca Inc.	USA	*547*	75-02200
Aerospace Metal Composites Limited	UK	*428*	75-01180
AeroSpace Technologies of Australia	Australia	*11*	75-00340
Altoflex	France	*132*	75-02300
A.T. Poeton (Gloucester Plating) Ltd	UK	*508*	80-23500
Brookhouse Paxford Ltd	UK	*437*	75-10103
Colan Products	Australia	*14*	75-03650
CTL-Aerospace Inc.	USA	*567*	75-22700
DSM	Netherlands	*292*	75-01730
Dynamic Engineering Inc.	USA	*570*	75-28100
Finavicomp Oy	Finland	*118*	75-00595
Flexibulb Plastics Inc.	Canada	*71*	75-05550
FMI Composites Ltd	UK	*451*	75-24040
Fothergill Engineered Fabrics Ltd	UK	*452*	75-24380
GEL Engineering Ltd	UK	*454*	75-25500
Herbelith NV	Belgium	*46*	75-04300
Hexcel Composites	France	*133*	75-07300
Hexcel Composites Austria	Austria	*33*	75-01240
Hexcel Composites Ltd	UK	*458*	75-29150
Hexcel Structures	USA	*582*	75-43320
House of Hardy	UK	*459*	75-30180
Intermarine SpA	Italy	*235*	75-11900
International Polymer Corp.	USA	*585*	75-48200
Israel Aircraft Industries Ltd	Israel	*219*	75-10200
Israel Military Industries Ltd	Israel	*220*	75-10300
Israel Military Industries Rocket Systems Division	**Israel**	***220***	**75-10310**
Kirasa Co. Ltd	Russia	*337*	75-11350
Knabenhans	Switzerland	*386*	75-07100
L'hotellier Montrichard	France	*138*	75-24900
Lear Astronics Corp.	USA	*589*	75-54390
M&N Aerospace	USA	*597*	75-62240
M.C. Gill Corp.	USA	*579*	75-39620
Meggitt Composites	UK	*467*	75-38345
P.A.D. International	UK	*472*	75-43470
Plastic Fabricating Co. Inc.	USA	*604*	75-70335
Royal Ordnance	UK	*481*	75-50620
Sequa Corp.	USA	*613*	75-79490
Sleeman Engineering	UK	*484*	75-54400
SNPE	France	*145*	75-36315
Société SMAC	France	*146*	75-40700
Spaulding Composites	USA	*616*	75-82970
Specialty Plastic Products Inc.	USA	*616*	75-83035
Sumitomo Electric Industries Ltd	Japan	*254*	75-07950
Superform Aluminium	UK	*487*	75-57920
Technical Products Group Inc.	USA	*620*	75-86764
Texstar Inc.	USA	*621*	75-88600
Tiodize Co. Inc.	USA	*622*	75-89560
Tufnol Ltd	UK	*489*	75-61410
United Defense L.P.	USA	*624*	75-92050
Urdan Industries (USA) Inc.	USA	*625*	75-92930
Wyman-Gordon Co.	USA	*628*	75-97700

823 10 Composite armour

AEA Technology plc	UK	*496*	80-00480
AeroSpace Technologies of Australia	Australia	*11*	75-00340
Colan Products	Australia	*14*	75-03650
Courtaulds Aerospace	UK	*443*	75-15330
CTL-Aerospace Inc.	USA	*567*	75-22700
DSM	Netherlands	*292*	75-01730
Giat Industries	France	*126*	55-00100

Israel Military Industries Ltd	Israel	220	75-10300
Kirasa Co. Ltd	Russia	337	75-11350
Lightweight Body Armour	UK	464	75-35650
Meggitt Composites	UK	467	75-38345
National Forge Co.	USA	598	75-63850
NORINCO	China PR	88	55-00280
Norton Co.	USA	600	75-65510
Permali Gloucester Ltd	UK	474	75-44850
Plastechnol Ltd	UK	508	80-22500
Protection Development Int'l Corp.	USA	606	75-71670
Roblon A/S	Denmark	103	75-18400
Royal Ordnance	UK	481	75-50620
Simula Government Products, Inc.	USA	615	75-81250
Sleeman Engineering	UK	484	75-54400
Specialty Plastic Products Inc.	USA	616	75-83035
Sumitomo Electric Industries Ltd	Japan	254	75-07950
Technical Products Group Inc.	USA	620	75-86764
Texstar Inc.	USA	621	75-88600
United Defense L.P.	USA	624	75-92050
Urdan Industries (USA) Inc.	USA	625	75-92930

824 Special materials

Agamice	France	131	75-00800
Altoflex	France	132	75-02300
Avon Technical Products	UK	432	75-05150
Bospec Neue Technologien	Germany	166	75-03420
Bullet Guard Corp.	USA	561	75-15620
Carrington	UK	438	75-11800
Cercom Inc.	USA	563	75-17400
Colan Products	Australia	14	75-03650
Crystallume	USA	563	75-18425
Diamonite Products Inc.	USA	569	75-26455
DSM	Netherlands	292	75-01730
Du Pont (UK) Limited	UK	447	75-19100
Esterline Technologies	USA	574	75-32360
Ferro-Ceramic Grinding Inc.	USA	635	80-11350
Fluorocarbon Company Limited	UK	451	75-24020
FMI Composites Ltd	UK	451	75-24040
Fothergill Engineered Fabrics Ltd	UK	452	75-24380
GEC-Marconi Materials Technology Ltd	UK	454	75-25340
Hexcel Composites	France	133	75-07300
H.R. Smith (Microwaves) Ltd	UK	484	75-54590
ICL Technical Plastics Ltd	UK	460	75-31060
Insulgard Corp.	USA	585	75-47020
J&D Wilkie Ltd	UK	494	75-66240
Jarvis Manufacturing Co.	UK	462	75-32330
Jehier SA	France	137	75-23200
Lantor (UK) Ltd	UK	464	75-34665
NanoMaterials, Inc.	USA	598	75-63590
National Forge Co.	USA	598	75-63850
Norton Co.	USA	600	75-65510
Norton Performance Plastics Corp.	USA	600	75-65520
Protechnik Laboratories	S. Africa	359	75-17000
Purification Products Ltd	UK	476	75-47120
Reade Advanced Materials	USA	608	75-73860
Richmond Technology Inc.	USA	609	75-75370
Spaulding Composites	USA	616	75-82970
Sumitomo Electric Industries Ltd	Japan	254	75-07950
Textil Tabacow SA	Brazil	58	75-20000
Thermal Structures Inc.	USA	622	75-88675
White Cross Rubber Products Ltd.	UK	494	75-65857

824 10 Materials, semiconductors

Advanced Composite Group Ltd	UK	427	75-00845
AEA Technology plc	UK	496	80-00480
Aerospace Metal Composites Limited	UK	428	75-01180
British Aluminium Plate	UK	436	75-09000
British Sarozal Ltd.	UK	436	75-09570
GEC-Marconi Materials Technology Ltd	UK	419	60-00600
Hughes Microelectronics Europe Ltd	UK	459	75-30480
RFI Shielding Ltd	UK	479	75-49500
Rockwell Semiconductor Systems	Germany	174	75-25265
Rood Technology UK Ltd	UK	510	80-29200
Ross & Catherall Ltd	UK	480	75-50555
Rutherford Appleton Laboratory	UK	420	60-01850

824 20 Ceramic powders

Capral Aluminium Ltd.	Australia	14	75-03170
NanoMaterials, Inc.	USA	598	75-63590
Norton Co.	USA	600	75-65510
Norton Performance Plastics Corp.	USA	600	75-65520
Praxair Surface Technologies Inc.	USA	640	80-24680
Reade Advanced Materials	USA	608	75-73860

824 20 50 Ceramic products

AEA Technology plc	UK	496	80-00480
Cercom Inc.	USA	563	75-17400
Channel Industries Inc.	USA	563	75-18050
Coors Ceramics Co.	USA	566	75-22010
Diamonite Products Inc.	USA	569	75-26455
EDO Corporation	USA	571	75-29250
FMI Composites Ltd	UK	451	75-24040
NanoMaterials, Inc.	USA	598	75-63590
Norton Co.	USA	600	75-65510
Norton Performance Plastics Corp.	USA	600	75-65520
Oxley Inc.	USA	602	75-67870
Reade Advanced Materials	USA	608	75-73860
Société des Céramiques Techniques	France	146	75-39510
Sumitomo Electric Industries Ltd	Japan	254	75-07950
Technical Products Group Inc.	USA	620	75-86764

824 20 50 10 Ceramic armour

AEA Technology plc	UK	496	80-00480
Cercom Inc.	USA	563	75-17400
ContiTech Holding GmbH	Germany	166	75-05150
Coors Ceramics Co.	USA	566	75-22010
Norton Co.	USA	600	75-65510
Protection Development Int'l Corp.	USA	606	75-71670
Société des Céramiques Techniques	France	146	75-39510

824 30 Special fabrics

Actil	Australia	11	75-00175
A.J. Worthington (Leek)	UK	494	75-67080
Albright & Wilson Ltd	UK	429	75-02390
A.W. Hainsworth & Sons Ltd	UK	456	75-27300
Baltex Fabrics	UK	433	75-05430
Bondina NBC Materials	UK	435	75-07600
British Vita plc.	UK	437	75-09750
Carrington	UK	438	75-11800
Colan Products	Australia	14	75-03650
DSM	Netherlands	292	75-01730
Du Pont (UK) Limited	UK	447	75-19100
Elastic Webbing (Aust) Ltd	Australia	15	75-05520
Fothergill Engineered Fabrics Ltd	UK	452	75-24380
Furon CHR	USA	576	75-36400
H. Fine & Son Ltd	UK	451	75-23210
Intermarine SpA	Italy	235	75-11900
J&D Wilkie Ltd	UK	494	75-66240
Jarvis Manufacturing Co.	UK	462	75-32330
Jehier SA	France	137	75-23200
Norwellan Ltd	Australia	19	75-14882
SpanSet Inter AG	Switzerland	387	75-13550
Swift Textile Metalizing Corp.	USA	619	75-85450
Tanner Brothers (Greenfield) Ltd	UK	487	75-58550
Textil Tabacow SA	Brazil	58	75-20000
United Electric Motor Co.	USA	624	75-92100
Universal Carbon Fibres	UK	490	75-62300
UTEXBEL SA/NV	Belgium	47	75-06373
White Cross Rubber Products Ltd.	UK	494	75-65857
Woodville Environmental & Safety	UK	494	75-66900
Woodville Polymer	UK	494	75-66900

824 30 40 Insulating materials

AIM Aviation (Jecco) Ltd.	UK	428	75-01530
Balmoral Group Ltd	UK	433	75-05420
Caledonian Mining Co. Ltd	UK	438	75-11250
Chomerics	USA	563	75-18275
Colan Products	Australia	14	75-03650
Conap Inc.	USA	565	75-20825
EDO Corporation	USA	571	75-29250
Ferro-Ceramic Grinding Inc.	USA	635	80-11350
Fluorocarbon Company Limited	UK	451	75-24020
Fothergill Engineered Fabrics Ltd	UK	452	75-24380
Furon CHR	USA	576	75-36400
Heliodinâmica SA	Brazil	56	75-08270
IMI-Tech Corp.	USA	584	75-46140

Intermarine SpA	Italy	*235*	75-11900
Jarvis Manufacturing Co.	UK	*462*	75-32330
Jehier SA	France	*137*	75-23200
Permali Gloucester Ltd	UK	*474*	75-44850
Spaulding Composites	USA	*616*	75-82970
Teledyne Aircraft Products	USA	*621*	75-87580
Thermal Structures Inc.	USA	*622*	75-88675
Tufnol Ltd	UK	*489*	75-61410
United Moulders Ltd	UK	*490*	75-62130
Vitafoam	UK	*492*	75-63390

824 30 40 10 Insulating materials, acoustic

AIM Aviation (Jecco) Ltd.	UK	*428*	75-01530
Avon Technical Products	UK	*432*	75-05150
Bestobell Aviation	UK	*434*	75-06600
Colan Products	Australia	*14*	75-03650
E-A-R Specialty Composites	USA	*571*	75-28500
EDO Corporation	USA	*571*	75-29250
Furon CHR	USA	*576*	75-36400
IMI-Tech Corp.	USA	*584*	75-46140
INC Corp. Pty Ltd	Australia	*17*	75-09250
Intermarine SpA	Italy	*235*	75-11900
James Walker & Co. Ltd	UK	*492*	75-64120
Jehier SA	France	*137*	75-23200
Spaulding Composites	USA	*616*	75-82970
Thermal Structures Inc.	USA	*622*	75-88675

824 30 40 50 Insulating materials, thermal

AIM Aviation (Jecco) Ltd.	UK	*428*	75-01530
Altoflex	France	*132*	75-02300
Balmoral Group Ltd	UK	*433*	75-05420
Bestobell Aviation	UK	*434*	75-06600
Chomerics	USA	*563*	75-18275
Colan Products	Australia	*14*	75-03650
Colebrand Ltd	UK	*441*	75-14120
EDO Corporation	USA	*571*	75-29250
FMI Composites Ltd	UK	*451*	75-24040
Fothergill Engineered Fabrics Ltd	UK	*452*	75-24380
Furon CHR	USA	*576*	75-36400
IMI-Tech Corp.	USA	*584*	75-46140
Intermarine SpA	Italy	*235*	75-11900
J&D Wilkie Ltd	UK	*494*	75-66240
Jehier SA	France	*137*	75-23200
NanoMaterials, Inc.	USA	*598*	75-63590
Parker Hannifin plc	UK	*473*	75-43880
Permali Gloucester Ltd	UK	*474*	75-44850
Reade Advanced Materials	USA	*608*	75-73860
Spaulding Composites	USA	*616*	75-82970
Thermal Structures Inc.	USA	*622*	75-88675
Tufnol Ltd	UK	*489*	75-61410
Vidco Industries Ltd	Israel	*223*	75-23240
Vitafoam	UK	*492*	75-63390

824 40 Glazing materials, bullet & attack-resistant

Bullet Guard Corp.	USA	*561*	75-15620
Flexibulb Plastics Inc.	Canada	*71*	75-05550
Insulgard Corp.	USA	*585*	75-47020
Westford Plastics & Engineering Ltd	UK	*493*	75-65450

824 42 Window materials

Alifabs Design & Construction Ltd	UK	*496*	80-01000
Avimo Ltd	UK	*432*	75-04950
Coherent Optics (Europe) Ltd	UK	*441*	75-14090
Colebrand Ltd	UK	*441*	75-14120
Cranfield University	UK	*419*	60-00350
GEC-Marconi Materials Technology Ltd	UK	*419*	60-00600
Parker Hannifin plc	UK	*473*	75-43880
Pilkington Aerospace Ltd	UK	*474*	75-45500
Pilkington Thorn Optronics	UK	*426*	74-10000
RFI Shielding Ltd	UK	*479*	75-49500
THORN Microwave Devices Ltd	UK	*488*	75-59760

824 45 Metals, powdered

Deloro Stellite Ltd.	UK	*445*	75-17460
NanoMaterials, Inc.	USA	*598*	75-63590
National Forge Co.	USA	*598*	75-63850
Reade Advanced Materials	USA	*608*	75-73860
Royal Ordnance Speciality Metals	UK	*481*	75-50630
Tecphy	France	*147*	75-44375

824 50 Radar absorbent / transparent materials

AEA Technology plc	UK	*496*	80-00480
Agamice	France	*131*	75-00800
Altoflex	France	*132*	75-02300
Avon Technical Products	UK	*432*	75-05150
Colebrand Ltd	UK	*441*	75-14120
DSM	Netherlands	*292*	75-01730
Emerson and Cuming Microwave	Belgium	*46*	75-02900
ERA Technology Ltd	UK	*502*	80-09600
Finnyards Electronics Ltd	Finland	*118*	75-00600
FMI Composites Ltd	UK	*451*	75-24040
GEC-Marconi Materials Technology Ltd	UK	*454*	75-25340
H.R. Smith (Microwaves) Ltd	UK	*484*	75-54590
Johnson Matthey Plc	UK	*462*	75-32815
Lightweight Body Armour	UK	*464*	75-35650
NanoMaterials, Inc.	USA	*598*	75-63590
Permali Gloucester Ltd	UK	*474*	75-44850
Reade Advanced Materials	USA	*608*	75-73860
RFI Industries Pty Ltd	Australia	*20*	75-18500
Signature Management	UK	*484*	75-53790
Technical Products Group Inc.	USA	*620*	75-86764
United Moulders Ltd	UK	*490*	75-62130

824 55 RF shielding materials

Altoflex	France	*132*	75-02300
Bospec Neue Technologien	Germany	*166*	75-03420
Chomerics	USA	*563*	75-18275
Doncasters plc	UK	*446*	75-18410
Dowty Seals Ltd	UK	*446*	75-18620
DSM	Netherlands	*292*	75-01730
Furon CHR	USA	*576*	75-36400
GEC-Marconi Materials Technology Ltd	UK	*454*	75-25340
Intermarine SpA	Italy	*235*	75-11900
James Walker & Co. Ltd	UK	*492*	75-64120
NanoMaterials, Inc.	USA	*598*	75-63590
Nucletron Electronic Vertriebs GmbH	Germany	*179*	80-01500
Parker Hannifin plc	UK	*473*	75-43880
Reade Advanced Materials	USA	*608*	75-73860
RFI Industries Pty Ltd	Australia	*20*	75-18500
RFI Shielding Ltd	UK	*479*	75-49500
Swift Textile Metalizing Corp.	USA	*619*	75-85450
Technical Products Group Inc.	USA	*620*	75-86764
Tecknit Inc.	USA	*620*	75-87150
Trimite Ltd.	UK	*489*	75-61150
United Defense L.P.	USA	*624*	75-92050

824 60 Coatings, metal

Acofab SA	France	*130*	75-00200
A.T. Poeton (Gloucester Plating) Ltd	UK	*508*	80-23500
Australian Defence Industries Ltd	Australia	*9*	55-01000
Bodycote Metallurgical Coatings Ltd	UK	*434*	75-07500
Bospec Neue Technologien	Germany	*166*	75-03420
Brenco Thermal Coating Pty Ltd.	Australia	*13*	75-02740
Cametoid Ltd	Canada	*78*	80-00250
CFM GmbH	Germany	*166*	75-04800
Chomerics	USA	*563*	75-18275
Chromalloy France	France	*133*	75-09650
Deloro Stellite Ltd.	UK	*445*	75-17460
Doncasters Structures	UK	*446*	75-18465
DSM	Netherlands	*292*	75-01730
Esterline Technologies	USA	*574*	75-32360
H.R. Smith (Microwaves) Ltd	UK	*484*	75-54590
Hytek Finishes Co.	USA	*584*	75-45700
Metallisation Ltd	UK	*467*	75-38760
NanoMaterials, Inc.	USA	*598*	75-63590
Praxair Surface Technologies Inc.	USA	*640*	80-24680
Reade Advanced Materials	USA	*608*	75-73860
Reime Coatection A/S	Norway	*310*	75-02250
Swift Textile Metalizing Corp.	USA	*619*	75-85450
Tiodize Co. Inc.	USA	*622*	75-89560
Transfield Corrosion Protection	Australia	*22*	75-20811
Trimite Ltd.	UK	*489*	75-61150
United Defense L.P.	USA	*624*	75-92050

Product-Supplier Link

US Paint	USA	625	75-93025
Watson & Lewis Ltd	UK	514	80-71500

824 65 Coatings, plastic, for metalwork

Balmoral Group Ltd	UK	433	75-05420
Fluorocarbon Company Limited	UK	451	75-24020
ICL Technical Plastics Ltd	UK	460	75-31060
International Polymer Corp.	USA	585	75-48200
Tiodize Co. Inc.	USA	622	75-89560
Transfield Corrosion Protection	Australia	22	75-20811

824 70 Security glazing

ICL Technical Plastics Ltd	UK	460	75-31060
Westford Plastics & Engineering Ltd	UK	493	75-65450

Chemicals, Paints, Liquids

851 Chemicals, dyes

Agro Química Maringá SA	Brazil	55	75-00650
Albright & Wilson Ltd	UK	429	75-02390
Aquarius Fabrique International Pvt	India	204	75-00160
Callington Haven Pty Ltd	Australia	14	75-03100
Carey Industries Inc.	USA	562	75-16720
DSM	Netherlands	292	75-01730
Dunlop Adhesives	UK	447	75-18990
Dynamation Research	USA	570	75-27960
Elf Atochem	Australia	16	75-05630
Holliday Dyes & Chemicals Ltd	UK	458	75-29650
Hyundai	Korea (ROK)	265	75-01540
IFI Srl	Italy	235	75-11220
Kenrich Petrochemicals Inc.	USA	588	75-52310
Merex Inc.	USA	595	75-60250
Molecular Products Ltd	UK	469	75-40540
Petrimex	Slovakia	353	55-01500
Richmond Packaging (UK) Ltd	UK	479	75-49630
Sherwin Inc.	USA	614	75-80300
Silent Partner Body Armor Inc.	USA	615	75-80930
SNPE	France	145	75-36315
Turco Italiana SpA	Italy	239	75-26600
Western Electrochemical Co.	USA	627	75-95850

851 10 Additives, ammunition

Åkers Krutbruk Protection AB	Sweden	375	75-01100
Aquarius Fabrique International Pvt	India	204	75-00160
Merex Inc.	USA	595	75-60250
Royal Ordnance	UK	481	75-50620
Uniroyal Chemical Company Inc.	USA	624	75-91800
Western Electrochemical Co.	USA	627	75-95850

851 20 CS powder

Merex Inc.	USA	595	75-60250

851 30 Decontaminating agents

Alfred Karcher GmbH & Co	Germany	170	75-15650
BCB International Ltd	UK	433	75-06075
Chemviron Carbon	Belgium	46	75-02260
Giat Industries	France	126	55-00100
IFI Srl	Italy	235	75-11220
Molecular Products Ltd	UK	469	75-40540
Richmond Packaging (UK) Ltd	UK	479	75-49630
Tiodize Co. Inc.	USA	622	75-89560
Turco-Chemie GmbH	Germany	176	75-32010

851 35 Desiccants

Crosfield Group	UK	444	75-16150
Mineral Derivatives Ltd	UK	468	75-39800
Molecular Products Ltd	UK	469	75-40540

852 Special paints & finishes

Acorn Hardas Ltd	UK	427	75-00410
AD Aerospace Finishes GmbH	Germany	164	75-00500
AD Aerospace Finishes Ltd	UK	429	75-02370
AD Aerospace Finishes SA	France	131	75-01540
AD Aerospace Finishes Srl	Italy	230	75-00020
Agamice	France	131	75-00800
Akzo Nobel Coatings SA	Spain	365	75-00440
Altoflex	France	132	75-02300
Becker Industrial Coatings	UK	434	75-06130
Bosma Machine and Tool Corporation	USA	560	75-14750
Callington Haven Pty Ltd	Australia	14	75-03100
Cametoid Ltd	Canada	78	80-00250
CFM GmbH	Germany	166	75-04800
Deft Chemical Coatings	USA	568	75-24870
DSM	Netherlands	292	75-01730
Edel Aviation	France	136	75-14280
Esterline Technologies	USA	574	75-32360
Fireseal International Ltd	UK	502	80-10450
GEC-Marconi Materials Technology Ltd	UK	454	75-25340
Grow Group Canada Ltd	Canada	72	75-06530
Hytek Finishes Co.	USA	584	75-45700
ICI Devoe Coatings	USA	584	75-45895
IFI Srl	Italy	235	75-11220
Mäder Lacke AG	Switzerland	386	75-08000
MIFA Aluminium BV	Netherlands	294	75-04525
NOF Europe NV	Belgium	46	75-05300
Perm Kirov Research Production Corp	Russia	338	75-15975
Randolph Products Co.	USA	607	75-72905
Rocol Site Safety Systems Ltd	UK	479	75-50060
Sherwin Williams do Brasil Ind.	Brazil	58	75-19540
Sterling Lacquer Manufacturing Co.	USA	617	75-84175
Tiodize Co. Inc.	USA	622	75-89560
Trimite Ltd.	UK	489	75-61150
US Paint	USA	625	75-93025

855 Liquids

Agro Química Maringá SA	Brazil	55	75-00650
Aquarius Fabrique International Pvt	India	204	75-00160
Broken Hill Proprietary Co. Ltd.	Australia	13	75-02770
Callington Haven Pty Ltd	Australia	14	75-03100
CFM GmbH	Germany	166	75-04800
Chubb Fire Products & Systems Ltd	UK	440	75-13200
Emerson and Cuming Microwave	Belgium	46	75-02900
Fireseal International Ltd	UK	502	80-10450
IFI Srl	Italy	235	75-11220
Ivar Rivenaes Ltd	UK	479	75-49820
Tiodize Co. Inc.	USA	622	75-89560

855 05 Cleaning products

Accuracy International Ltd	UK	427	75-00360
AD Aerospace Finishes Srl	Italy	230	75-00020
Alfred Karcher GmbH & Co	Germany	170	75-15650
Armalon Ltd	UK	431	75-03729
Bospec Neue Technologien	Germany	166	75-03420
Break-Free Inc.	USA	560	75-14990
Callington Haven Pty Ltd	Australia	14	75-03100
Carr & Day & Martin Ltd	UK	438	75-11750
CFM GmbH	Germany	166	75-04800
Elf Atochem	Australia	16	75-05630
MJP Corporation Ltd	Finland	119	80-01000
Parker-Hale Ltd	UK	473	75-43850
Tiodize Co. Inc.	USA	622	75-89560
Turco-Chemie GmbH	Germany	176	75-32010
Vapormatt Ltd	UK	490	75-62630

855 05 05 Cleaning fluids

AD Aerospace Finishes Srl	Italy	230	75-00020
Aquarius Fabrique International Pvt	India	204	75-00160
Aviation Fluids Service Inc.	USA	556	75-10800
Callington Haven Pty Ltd	Australia	14	75-03100
Cimcool Industrial Products BV	Netherlands	292	75-01130
Elf Atochem	Australia	16	75-05630
IFI Srl	Italy	235	75-11220
Turco-Chemie GmbH	Germany	176	75-32010

855 30 Depainting compounds, non-chemical

Callington Haven Pty Ltd	Australia	*14*	75-03100
Elf Atochem	Australia	*16*	75-05630
Maxi-Blast Inc.	USA	*594*	75-59200
Turco-Chemie GmbH	Germany	*176*	75-32010
Vacu-Blast Ltd	UK	*490*	75-62550
Vapormatt Ltd	UK	*490*	75-62630

855 40 Liquids, fire extinguishing

Chubb Australia Ltd.	Australia	*14*	75-03400
Chubb Fire Products & Systems Ltd	UK	*440*	75-13200
Heiser Josef	Austria	*33*	75-01230
Phillips & Smith Ltd	New Zealand	*303*	75-01650
Recticel NV	Belgium	*46*	75-05585
Silvani Antincendi SpA	Italy	*239*	75-22250

Fuels, oils, lubricants

860 10 Fuels

BHP Hawaii Inc.	USA	*559*	75-13774
Boulder Scientific Co.	USA	*560*	75-14780
Broken Hill Proprietary Co. Ltd.	Australia	*13*	75-02770
Nitro Nobel AB	Sweden	*378*	75-08275

860 15 Oils, lubricants, greases

American Oil & Supply Co.	USA	*553*	75-06180
Armalon Ltd	UK	*431*	75-03729
Aviation Fluids Service Inc.	USA	*556*	75-10800
Break-Free Inc.	USA	*560*	75-14990
Castrol (UK) Ltd	UK	*439*	75-12070
CFM GmbH	Germany	*166*	75-04800
Cimcool Industrial Products BV	Netherlands	*292*	75-01130
Cofran Lubrification	France	*133*	75-10100
Halocarbon Products Corp.	USA	*580*	75-41630
IFI Srl	Italy	*235*	75-11220
José Murillia Bozza SA Com. e Ind.	Brazil	*55*	75-02220
K S Paul Products Ltd	UK	*464*	75-34160
Parker-Hale Ltd	UK	*473*	75-43850
Rajkumar International	India	*205*	75-01780
Rocol Site Safety Systems Ltd	UK	*479*	75-50060
Tiodize Co. Inc.	USA	*622*	75-89560

Services

910 Advisory services

A Goninan & Co. Ltd	Australia	*16*	75-07760
Aardvark Training Schools Ltd	UK	*496*	80-00200
ACI Aviation Consultants Int'l	Australia	*23*	80-00300
Advisory, Consulting & Technical	UK	*496*	80-00450
Aerospatiale	France	*131*	75-00600
Allied Defense Industries Inc.	USA	*630*	80-02000
Aman-Management Consultants Ltd	Israel	*223*	80-00150
Analysis & Technology Inc.	USA	*631*	80-02600
Andersen Consulting	Australia	*24*	80-00910
Andrich International Ltd	UK	*497*	80-01200
Anite Systems Ltd	UK	*497*	80-01240
Armalon Ltd	UK	*431*	75-03729
Aspect Computing Pty Ltd	Australia	*24*	80-01150
Assessment Services	UK	*497*	80-01500
Atlantic Research Corp.	USA	*555*	75-10150
Babtie Group Ltd	UK	*498*	80-03050
BAeSEMA Ltd.	UK	*498*	80-03070
Beltramo & Associates	USA	*632*	80-04360
Boeing Australia Limited	Australia	*12*	75-02500
Booz, Allen & Hamilton Inc.	USA	*632*	80-04700
Bremer Vulkan Marineschiffbau GmbH	Germany	*166*	75-03600
Bristow Helicopters Ltd	UK	*499*	80-03950
Bruce D. Vallillee Electronics Ltd	Canada	*80*	80-02210
Burdeshaw Associates, Ltd	USA	*632*	80-04960
Calspan SRL Technology	USA	*540*	60-03000
Cambridge Consultants Ltd	UK	*499*	80-05300
Carmichael & Sweet Consultancy Ltd	UK	*499*	80-05500
CCS International Ltd	USA	*562*	75-17250
Cenzin Co. Ltd.	Poland	*324*	75-00200
Chamberlain Manufacturing Corp.	USA	*563*	75-17950
Chelton (Electrostatics) Limited	UK	*439*	75-12700
CIC International Ltd	USA	*563*	75-18570
CMPS&F Pty Limited	Australia	*24*	80-03600
Cogent	UK	*500*	80-06250
Cranfield Aerospace Ltd	UK	*500*	80-07200
Cranfield University	UK	*419*	60-00350
Criminal Intelligence Administration	USA	*633*	80-06730
Crompton Greaves Ltd	India	*205*	75-00280
CTA Inc.	USA	*633*	80-06810
Data Sciences Ltd.	UK	*500*	80-07740
Datel Technology Ltd.	UK	*500*	80-07750
DCS Corp.	USA	*634*	80-07400
Defence Research & Analysis	UK	*500*	80-08200
Delex Systems Inc.	USA	*634*	80-08020
Deloitte & Touche Consulting Group	UK	*501*	80-08300
DEMCO Consultants Ltd	Israel	*224*	80-00430
Design Bureau of	Russia	*336*	75-03150
Det Norske Veritas	Norway	*311*	80-01400
Divex Ltd.	UK	*445*	75-18220
DSSD plc	UK	*501*	80-08750
EASAMS Ltd	UK	*501*	80-09000
EDS Defence Ltd	UK	*501*	80-09200
EG&G Washington Analytical Svcs	USA	*634*	80-09450
Elintech GmbH	Germany	*178*	80-00550
Engineering & Professional Serv.Inc.	USA	*634*	80-10150
Fluor Daniel Inc.	USA	*635*	80-11720
Forecast International/DMS	USA	*636*	80-11760
Forges de Zeebrugge SA	Belgium	*46*	75-03750
Frazer-Nash Consultancy Limited	UK	*503*	80-10950
Garber International Associates Inc.	USA	*636*	80-12150
GEC-Marconi Avionics Ltd	UK	*454*	75-25320
George G. Sharp Inc.	USA	*642*	80-28000
Global Communications Co.	USA	*579*	75-39775
Global Defense Products Inc.	USA	*579*	75-39770
GMV SA	Spain	*367*	80-00800
Gordon Consulting Engineers Ltd	Israel	*224*	80-00750
HeavyLift Aircraft Engineering Ltd.	UK	*503*	80-12300
Horizons Technology Inc.	USA	*636*	80-14450
Horrigan Analytics	USA	*636*	80-14480
Hugo Marom Aviation Consultants Ltd	Israel	*224*	80-00880
Hyder Consulting Limited	UK	*504*	80-12900
Industrieanlagen-Betriebs GmbH	Germany	*178*	80-00945
Information Handling Services Inc.	USA	*637*	80-15530
Interactive Dev. Environments	USA	*637*	80-15950
InterCon Consultants	Canada	*78*	80-00980
International Business Consulting SA	Spain	*367*	80-00930
International Hospitals Group	UK	*505*	80-13410
Israel Aircraft Industries Ltd	Israel	*219*	75-10200
Israel Military Industries Ltd	Israel	*220*	75-10300
ISVR Consultancy Services	UK	*505*	80-13520
Jet Aviation Zürich AG	Switzerland	*388*	80-01030
John M. Davis & Associates	USA	*634*	80-07200
Kapos Associates Inc.	USA	*637*	80-17630
Keystone Helicopter Corp.	USA	*637*	80-17780
Landair International Ltd	UK	*505*	80-14200
Litton Industries Inc.	USA	*590*	75-55400
Logica SA/NV	Belgium	*48*	80-01000
M International Export	USA	*639*	80-19780
M. Paz Logistics Ltd	Israel	*224*	80-01600
M. Rosenblatt & Son Inc.	USA	*641*	80-26450
Magnesium Elektron	UK	*466*	75-37000
Management Consulting & Research	USA	*638*	80-19680
MECAR SA	Belgium	*46*	75-04930
Megkon	S. Africa	*360*	80-10000
MFD International Ltd.	UK	*506*	80-16450
MIL Systems	Canada	*79*	80-01360
Mott MacDonald	UK	*507*	80-17400
NanoMaterials, Inc.	USA	*598*	75-63590
National Systems Management Corp.	USA	*639*	80-21650
NAVFCO	France	*150*	80-01420
NEL	UK	*419*	60-01400
Nevada Automotive Test Center	USA	*540*	60-14500
Nichols Research Corp.	USA	*639*	80-21780
NIS Ltd	UK	*471*	75-42555
NNC Ltd	UK	*507*	80-18500
Noble Denton Europe Ltd	UK	*507*	80-18300
Nogay Ltd	Israel	*224*	80-01470
O.T.P. Ballistics	UK	*508*	80-19700
Pacer Systems Inc.	USA	*602*	75-68000
Pennant International Limited	UK	*473*	75-44450
Penter Associations Limited	UK	*508*	80-21000
Philips Calibration Services	Australia	*25*	80-16490

Piper Pacific International	USA	*640*	80-24300
PK Electronic International Ltd.	UK	*475*	75-45720
PKM Consultants (S) Pte Ltd	Singapore	*349*	80-01640
Planning & Research Ltd	Israel	*224*	80-01650
PRC Inc.	USA	*640*	80-24800
Programming Research Ltd	UK	*509*	80-24700
Protechnik Laboratories	S. Africa	*359*	75-17000
Public Safety Group Inc.	USA	*641*	80-24955
Reade Advanced Materials	USA	*608*	75-73860
Resource Consultants Inc.	USA	*641*	80-25920
Ricardo Hitec Ltd	UK	*509*	80-27840
Richard T. Lyons Associates Inc.	USA	*641*	80-26280
RKI/Cleanroom Services	USA	*641*	80-26200
Rollab Aktiebolag	Sweden	*380*	80-01850
Rolls-Royce (Canada) Ltd.	Canada	*75*	75-12120
Royal Ordnance	UK	*481*	75-50620
SAC Technographic	UK	*510*	80-30100
Saladin Security Ltd.	UK	*510*	80-30300
ScanAvia, Scandinavian Aviation	Sweden	*380*	80-01860
Schneider Sohn & Associates Inc.	USA	*642*	80-26890
SEAL Laboratories	USA	*642*	80-27250
Secure Consulting International Ltd	UK	*511*	80-33445
SEMCOR Inc.	USA	*642*	80-27750
Serco Aviation Services	UK	*511*	80-34150
S.G.D. Engineering Ltd	Israel	*224*	80-01930
Sidem (UK) Ltd	UK	*511*	80-35850
Siemens Plessey Systems	UK	*483*	75-53770
SIP Services Int'l	UK	*511*	80-36500
SNPE	France	*145*	75-36315
Southwest Research Institute	USA	*541*	60-19200
Space Systems Finland Ltd	Finland	*118*	75-02100
SPC International Corp.	USA	*642*	80-29200
Specialty Plastic Products Inc.	USA	*616*	75-83035
SRC (Consultants) Ltd	UK	*512*	80-46200
SRI International	USA	*643*	80-29600
Strat*X International	USA	*643*	80-30100
Strat*X Ltd	UK	*512*	80-47500
Strat*X SA	France	*151*	80-02000
Sverdrup Corp.	USA	*643*	80-30650
Sverdrup Technology Inc.	USA	*643*	80-30680
TA Group Ltd	UK	*513*	80-50110
Tarmac Professional Services	UK	*513*	80-50200
Technology Australasia Pty Ltd	Australia	*26*	80-20600
Technum Integrated Technologies Ltd	Israel	*225*	80-02025
The Titan Corp	USA	*622*	75-89620
Thomson-CSF Systems Canada Inc.	Canada	*79*	80-02020
Tracor Inc	USA	*623*	75-90000
TriLink Technologies Group Inc.	Canada	*80*	80-02100
Turbo Lufttechnik GmbH	Germany	*176*	75-31900
UBC Inc.	USA	*624*	75-91383
US Army Equipment Guns and Ammo	Belgium	*47*	75-06345
Vanguard Research Inc.	USA	*644*	80-34100
Vickers Shipbuilding & Eng Ltd	UK	*491*	75-63100
Vitro Corp.	USA	*644*	80-34400
Vitrociset SpA	Italy	*240*	80-40000
Vosper International Ltd	UK	*514*	80-67000
Wegmann & Co GmbH	Germany	*177*	75-33700
Western Tactical Systems	UK	*515*	80-73000
Wornall Electronics Inc.	USA	*645*	80-35400
Yarrow Shipbuilders Ltd	UK	*495*	75-67500
3C Systems, Inc.	USA	*632*	80-05000

911 Consultancy services

A Goninan & Co. Ltd	Australia	*16*	75-07760
A&P Consultants (Pvt) Ltd	Sri Lanka	*370*	80-00100
Aardvark Training Schools Ltd	UK	*496*	80-00200
Abacus Technology Corp.	USA	*629*	80-00050
Abbott Electronics Ltd	UK	*426*	75-00180
Absolute Security	USA	*629*	80-00080
ACI Aviation Consultants Int'l	Australia	*23*	80-00300
Ada PROS Inc.	USA	*629*	80-00150
Advanced Appl. Consultants Inc.	USA	*630*	80-00250
Advisory, Consulting & Technical	UK	*496*	80-00450
Aero International Associates	USA	*630*	80-00800
Aerospace & Defense Sales Inc	Philippines	*321*	80-00050
Aerospace Design Hugo Marom Ltd	Israel	*223*	80-00040
Air Cargo Management Group	USA	*630*	80-01450
Air-Log Ltd	UK	*429*	75-01800
Airport Engineering Hugo Marom Ltd	Israel	*223*	80-00100
AirteQ Ltd	UK	*496*	80-00830
Allied Defense Industries Inc.	USA	*630*	80-02000
Allott & Lomax	UK	*497*	80-01050
Aman-Management Consultants Ltd	Israel	*223*	80-00150
American Management Systems	USA	*630*	80-02350
AMI International	USA	*631*	80-02470
Amtac Laboratories Ltd	UK	*497*	80-01060
Analysis & Technology Australia	Australia	*24*	80-00905
Andersen Consulting	Australia	*24*	80-00910
Andrich International Ltd	UK	*497*	80-01200
Anite Systems Ltd	UK	*497*	80-01240
Arca Systems	USA	*631*	80-03100
Archer Mullins Limited	UK	*497*	80-01257
Architektur Städtebau Infrastruktur	Germany	*177*	80-00210
Armalon Ltd	UK	*431*	75-03729
Arthur D. Little Ltd	UK	*497*	80-01260
Aspect Computing Pty Ltd	Australia	*24*	80-01150
Assessment Services	UK	*497*	80-01500
Australian Technical Publications	Australia	*24*	80-01880
Aviation Support Tech. Int'l Inc.	USA	*556*	75-11050
Babtie Group Ltd	UK	*498*	80-03050
BACTEC International Ltd	**UK**	***498***	**80-03060**
BAeSEMA Ltd.	UK	*498*	80-03070
Barnes & Reinecke Inc.	USA	*631*	80-04070
Bayern-Chemie	Germany	*165*	75-02100
Beltramo & Associates	USA	*632*	80-04360
Bharat Electronics Ltd	India	*203*	55-00200
BMT Marine Information Systems Ltd	UK	*498*	80-03573
BMT Seatech Ltd	UK	*498*	80-03595
Bodycote Taussig Inc.	USA	*632*	80-04590
Boeing Australia Limited	Australia	*12*	75-02500
Booz, Allen & Hamilton Inc.	USA	*632*	80-04700
Bradley Electronics Ltd	UK	*435*	75-07860
Bristow Helicopters Ltd	UK	*499*	80-03950
British Aerospace (Consultancy	UK	*499*	80-03990
Brodoprojekt	Croatia	*95*	80-00100
Bruce D. Vallillee Electronics Ltd	Canada	*80*	80-02210
BRUHN NewTech A/S	Denmark	*104*	80-00050
Burdeshaw Associates, Ltd	USA	*632*	80-04960
Caledonian Mining Co. Ltd	UK	*438*	75-11250
Calian Technology Ltd	Canada	*69*	75-01800
Calspan SRL Technology	USA	*540*	60-03000
Cambridge Consultants Ltd	UK	*499*	80-05300
Carmichael & Sweet Consultancy Ltd	UK	*499*	80-05500
CD-ROM Systems (Europe) Ltd	UK	*499*	80-05800
Chamberlain Manufacturing Corp.	USA	*563*	75-17950
Chelton (Electrostatics) Limited	UK	*439*	75-12700
CIC International Ltd	USA	*563*	75-18570
CMPS&F Pty Limited	Australia	*24*	80-03600
Cogent	UK	*500*	80-06250
Commslogic Ltd	UK	*500*	80-06400
Computer Power Group Ltd	Australia	*24*	80-03730
COSM Ltd	Portugal	*327*	80-00400
Counterrisk Inc.	Canada	*78*	80-00385
Cranfield Aerospace Ltd	UK	*500*	80-07200
Cranfield University	UK	*419*	60-00350
Crew Services Ltd	UK	*500*	80-07370
Criminal Intelligence Administration	USA	*633*	80-06730
Crompton Greaves Ltd	India	*205*	75-00280
CSC Computer Sciences Ltd	UK	*500*	80-07500
CTA Inc.	USA	*633*	80-06810
Currie & Brown (Australia) Pty Ltd.	Australia	*24*	80-03850
Damen Shipyards	Netherlands	*292*	75-01500
Data Sciences Ltd.	UK	*500*	80-07740
Data Search Associates	USA	*633*	80-07100
Data Systems Designers	Israel	*224*	80-00400
Datel Technology Ltd.	UK	*500*	80-07750
Davis (W.R.) Engineering Ltd	Canada	*70*	75-04050
Dee Organics Laboratory	UK	*500*	80-08180
Defence Data Consulting Services	Switzerland	*388*	80-00430
Defence Research & Analysis	UK	*500*	80-08200
Deloitte & Touche Consulting Group	UK	*501*	80-08300
DEMCO Consultants Ltd	Israel	*224*	80-00430
DEMEX Consulting Engineers A/S	Denmark	*104*	80-00450
Design Bureau of	Russia	*336*	75-03150
Det Norske Veritas	Norway	*311*	80-01400
Divex Ltd.	UK	*445*	75-18220
Dopra Systems Integration Ltd	UK	*446*	75-18400
Dytecna Ltd	UK	*501*	80-08900
EASAMS Ltd	UK	*501*	80-09000
Ediconsult Internazionale Srl	Italy	*240*	80-00450
EDS Defence Ltd	UK	*501*	80-09200
Electronic Development Assoc. Inc.	USA	*634*	80-09750
Elintech GmbH	Germany	*178*	80-00550
Enator Communications AB	Sweden	*379*	80-00500
Enator Telub AB	Sweden	*379*	80-00500
Engineering & Professional Serv.Inc.	USA	*634*	80-10150
ERA Technology Ltd	UK	*502*	80-09600
ESL Services (GB) Ltd.	UK	*502*	80-09730

Evotec Ltd	UK	*449*	75-21970
Export Trading Services Sarl	France	*150*	80-00595
Eyes & Ears	Egypt	*110*	80-06000
E2S NV	Belgium	*48*	80-00600
F&F Associates Inc.	USA	*635*	80-11000
Flight Refuelling Limited	UK	*451*	75-23850
Fluid Gravity Engineering Ltd.	UK	*502*	80-10660
Forecast International/DMS	USA	*636*	80-11760
Frazer-Nash Consultancy Limited	UK	*503*	80-10950
Garber International Associates Inc.	USA	*636*	80-12150
GEC-Marconi Avionics Ltd	UK	*454*	75-25320
GEC-Marconi Communications Ltd	UK	*454*	75-25335
GED (Southampton) Ltd	UK	*503*	80-11650
Genex International Corporation	Korea (ROK)	*266*	80-00725
George G. Sharp Inc.	USA	*642*	80-28000
Global Communications Co.	USA	*579*	75-39775
Global Defense Products Inc.	USA	*579*	75-39770
GMV SA	Spain	*367*	80-00800
Goliath & Co BV	Netherlands	*297*	80-07700
Gordon Consulting Engineers Ltd	Israel	*224*	80-00750
Gutteridge Haskins & Davey Pty Ltd	Australia	*25*	80-07500
Halmatic Ltd	UK	*456*	75-27700
Hans Harboe-Hansen	Denmark	*105*	80-00800
High Integrity Systems Ltd	UK	*458*	75-29250
Hoffman Engineering Corp.	USA	*582*	75-43980
Horizons Technology Inc.	USA	*636*	80-14450
Hornmark KK	Japan	*255*	80-00180
Hugo Marom Aviation Consultants Ltd	Israel	*224*	80-00880
Hyder Consulting Limited	UK	*504*	80-12900
ICL DESC	UK	*501*	80-08400
Ilex Systems Inc.	USA	*584*	75-46080
ILS Oy	Finland	*119*	80-00900
Industrieanlagen-Betriebs GmbH	Germany	*178*	80-00945
Information Handling Services Inc.	USA	*637*	80-15530
Inovex Technology Ltd	UK	*504*	80-13310
Int'l Security & Defense Systems	Israel	*224*	80-00960
Integrated Power Corp.	USA	*585*	75-47100
Interactive Dev. Environments	USA	*637*	80-15950
InterCon Consultants	Canada	*78*	80-00980
International Business Consulting SA	Spain	*367*	80-00930
International Hospitals Group	UK	*505*	80-13410
Israel Aircraft Industries Ltd	Israel	*219*	75-10200
Israel Military Industries Ltd	Israel	*220*	75-10300
Israel Shipyards Ltd.	Israel	*220*	75-10600
ISVR Consultancy Services	UK	*505*	80-13520
ITP Ltd	Israel	*220*	75-10800
Jet Aviation Zürich AG	Switzerland	*388*	80-01030
John M. Davis & Associates	USA	*634*	80-07200
JPS Communications Inc.	USA	*587*	75-51220
Kapos Associates Inc.	USA	*637*	80-17630
Keystone Helicopter Corp.	USA	*637*	80-17780
Kinhill Engineers Pty Ltd	Australia	*25*	80-11000
La-Vey Security International Ltd	UK	*506*	80-14215
Landair International Ltd	UK	*505*	80-14200
Lansdowne Technologies Inc	Canada	*78*	80-01205
Laser-Scan	UK	*506*	80-14210
Lightweight Body Armour	UK	*464*	75-35650
Litton Industries Inc.	USA	*590*	75-55400
LMS International	Belgium	*48*	80-00900
Logica UK Limited	UK	*506*	80-14350
Logistics Pty Ltd	Australia	*25*	80-12700
M International Export	USA	*639*	80-19780
M. Paz Logistics Ltd	Israel	*224*	80-01600
M. Rosenblatt & Son Inc.	USA	*641*	80-26450
Management Consulting & Research	USA	*638*	80-19680
Management Equipment Vehicles Co	Pakistan	*315*	80-01500
Manfredi Aeronautical Industries	Panama	*317*	80-00300
Marshall of Cambridge Aerospace Ltd	UK	*466*	75-37530
Masin-Aircraft GmbH	Germany	*178*	80-01300
MASS Consultants Limited	UK	*506*	80-15650
Materialinformationszentrum Ges.	Germany	*178*	80-01310
MECAR SA	Belgium	*46*	75-04930
Metra Martech Limited	UK	*506*	80-16300
MFD International Ltd.	UK	*506*	80-16450
MIL Systems	Canada	*79*	80-01360
Milo Holdings Ltd	UK	*506*	80-16850
MJI Associates Inc.	USA	*639*	80-21320
MJP Corporation Ltd	Finland	*119*	80-01000
MMM Consultancy Group Ltd	UK	*507*	80-17000
Mott MacDonald	UK	*507*	80-17400
Mycon Marketing Services Ltd	UK	*507*	80-17900
NanoMaterials, Inc.	USA	*598*	75-63590
NAVFCO	France	*150*	80-01420
NEL	UK	*419*	60-01400
Nevada Automotive Test Center	USA	*540*	60-14500
NIS Ltd	UK	*471*	75-42555
NNC Ltd	UK	*507*	80-18500
Noble Denton Europe Ltd	UK	*507*	80-18300
Nogay Ltd	Israel	*224*	80-01470
Norsk Data Ltd	UK	*507*	80-18400
North Atlantic Consult SA	Belgium	*48*	80-01500
OMI Logistics Limited	UK	*508*	80-19450
Orlando Technology Inc.	USA	*540*	60-16000
Orman Associates Inc.	USA	*640*	80-22600
OTB	S. Africa	*357*	60-00500
O.T.P. Ballistics	UK	*508*	80-19700
Overseas Maritime Consultants EPE	Greece	*193*	80-01400
Parity Solutions Ltd	UK	*508*	80-20300
Parsons Engineering Science	USA	*640*	80-23750
Pennant International Limited	UK	*473*	75-44450
Penter Associations Limited	UK	*508*	80-21000
Philips Calibration Services	Australia	*25*	80-16490
Piper Pacific International	USA	*640*	80-24300
PK Electronic International Ltd	UK	*475*	75-45720
PKM Consultants (S) Pte Ltd	Singapore	*349*	80-01640
Planning & Research Ltd	Israel	*224*	80-01650
Praxis Critical Systems Ltd	UK	*509*	80-23930
PRC Inc.	USA	*640*	80-24800
Primary Image Ltd	UK	*476*	75-46790
Proaction	USA	*641*	80-24900
Prof Sten Luthander Ingbyra AB	Sweden	*379*	80-01280
Programming Research Ltd	UK	*509*	80-24700
Prosig Computer Consultants Ltd	UK	*476*	75-46865
Public Safety Group Inc.	USA	*641*	80-24955
Q-Systems International Inc.	USA	*606*	75-71960
Radio Holland Electronics BV	Netherlands	*297*	80-18300
Reade Advanced Materials	USA	*608*	75-73860
Redifon MEL Limited	UK	*478*	75-48530
Resource Consultants Inc.	USA	*641*	80-25920
Ricardo Aerospace Ltd	UK	*509*	80-27820
Ricardo Hitec Ltd	UK	*509*	80-27840
Richard T. Lyons Associates Inc.	USA	*641*	80-26280
Ridgeway International Ltd.	UK	*509*	80-28000
RKI/Cleanroom Services	USA	*641*	80-26200
RO Future Systems and Concepts Group	UK	*510*	80-28600
Rollab Aktiebolag	Sweden	*380*	80-01850
Rolls-Royce & Associates Ltd	UK	*510*	80-29100
Rolls-Royce (Canada) Ltd.	Canada	*75*	75-12120
Royal Ordnance	UK	*481*	75-50620
SAC Technographic	UK	*510*	80-30100
Safety Consulting Engineers Inc.	USA	*641*	80-26680
Saladin Security Ltd.	UK	*510*	80-30300
SAR-ASSIST	USA	*642*	80-26850
ScanAvia, Scandinavian Aviation	Sweden	*380*	80-01860
Schneider Sohn & Associates Inc.	USA	*642*	80-26890
Science Systems (Industrial) Ltd	UK	*510*	80-31600
S.Com Computer Systems Engineering	UK	*510*	80-31800
SEAL Laboratories	USA	*642*	80-27250
Sector Ltd	UK	*511*	80-33400
Secure Consulting International Ltd	UK	*511*	80-33445
SEMCOR Inc.	USA	*642*	80-27750
Serco Aviation Services	UK	*511*	80-34150
S.G.D. Engineering Ltd	Israel	*224*	80-01930
Shorts Support Services Division	UK	*511*	80-35600
SIP Services Int'l	UK	*511*	80-36500
Sir J.H. Biles (Naval Services) Ltd	UK	*498*	80-03350
Sira Ltd.	UK	*484*	75-54250
SISDEF Ltda	Chile	*85*	80-01940
Smith System Engineering Ltd.	UK	*512*	80-39500
Sonda Aviation Enterprises Limited	UK	*512*	80-40000
Southwest Research Institute	USA	*541*	60-19200
Space Systems Finland Ltd	Finland	*118*	75-02100
SRI Consulting	UK	*512*	80-46500
SRI International	USA	*643*	80-29600
SSPA Maritime Consulting AB	Sweden	*380*	80-01960
Strat*X International	USA	*643*	80-30100
Strat*X Ltd	UK	*512*	80-47500
Strat*X SA	France	*151*	80-02000
Studio JONA di Consulenza	Italy	*240*	80-04000
STV Group Inc.	USA	*643*	80-30400
Sverdrup Corp.	USA	*643*	80-30650
Synectics Corp.	USA	*643*	80-30750
Syscon Engineering	Israel	*225*	80-01940
Systecon AB	Sweden	*380*	80-02000
Systems Consultants Services Ltd	UK	*513*	80-49500
TA Group Ltd	UK	*513*	80-50110
Talos Engineering SA	Greece	*193*	80-02010
Tarmac Professional Services	UK	*513*	80-50200
TDSi Micro Applications Ltd	UK	*513*	80-51000

Technology Australasia Pty Ltd	Australia	26	80-20600
Teleplan A/S	Norway	311	80-30000
Thales Advanced Engineering	S. Africa	360	80-60000
The Flight Data Company Ltd	UK	502	80-10600
The Titan Corp	USA	622	75-89620
The Walters Research Company	UK	514	80-70500
Thorn DNT Airfield Lighting Pty Ltd	Australia	22	75-20540
Three Quays Marine Services Ltd	UK	514	80-55400
Top Air Concepts Sarl	France	151	80-03190
Top Line Co.	USA	644	80-33100
Toye, Kenning & Spencer Ltd	UK	489	75-60600
Tracor Inc	USA	623	75-90000
Triad Group plc	UK	514	80-58600
TriLink Technologies Group Inc.	Canada	80	80-02100
Turbo Lufttechnik GmbH	Germany	176	75-31900
UBC Inc.	USA	624	75-91383
US Army Equipment Guns and Ammo	Belgium	47	75-06345
Vanguard Research Inc.	USA	644	80-34100
Vickers Shipbuilding & Eng Ltd	UK	491	75-63100
Victoria Maritime Services	Monaco	283	80-01000
Vipac Engineers & Scientists Ltd	Australia	26	80-22580
Vitro Corp.	USA	644	80-34400
Vitrociset SpA	Italy	240	80-40000
Vosper International Ltd	UK	514	80-67000
Western Tactical Systems	UK	515	80-73000
WS Atkins plc	UK	497	80-01550
Yarrow Shipbuilders Ltd	UK	495	75-67500
3C Systems, Inc.	USA	632	80-05000

912 Evaluation services

A Goninan & Co. Ltd	Australia	16	75-07760
Advanced Appl. Consultants Inc.	USA	630	80-00250
Aero International Associates	USA	630	80-00800
American Management Systems	USA	630	80-02350
AMI International	USA	631	80-02470
Analysis & Technology Inc.	USA	631	80-02600
Andersen Consulting	Australia	24	80-00910
Australian Defence Industries Ltd	Australia	9	55-01000
BAeSEMA Ltd.	UK	498	80-03070
Bayern-Chemie	Germany	165	75-02100
Beltramo & Associates	USA	632	80-04360
Bradley Electronics Ltd	UK	435	75-07860
Bruce D. Vallillee Electronics Ltd	Canada	80	80-02210
Burdeshaw Associates, Ltd	USA	632	80-04960
Calspan SRL Technology	USA	540	60-03000
Carmichael & Sweet Consultancy Ltd	UK	499	80-05500
CIC International Ltd	USA	563	75-18570
CMPS&F Pty Limited	Australia	24	80-03600
Cranfield University	UK	419	60-00350
Data Sciences Ltd	UK	500	80-07740
DCS Corp.	USA	634	80-07400
Defence Research & Analysis	UK	500	80-08200
Delex Systems Inc.	USA	634	80-08020
Deloitte & Touche Consulting Group	UK	501	80-08300
DEMCO Consultants Ltd	Israel	224	80-00430
Design Bureau of	Russia	336	75-03150
DSD Ltd	UK	501	80-08700
EDS Defence Ltd	UK	501	80-09200
EG&G Washington Analytical Svcs	USA	634	80-09450
Elintech GmbH	Germany	178	80-00550
Engineering & Professional Serv.Inc.	USA	634	80-10150
ERA Technology Ltd	UK	502	80-09600
Frazer-Nash Consultancy Limited	UK	503	80-10950
Garber International Associates Inc.	USA	636	80-12150
George G. Sharp Inc.	USA	642	80-28000
Global Communications Co.	USA	579	75-39775
Global Defense Products Inc.	USA	579	75-39770
GMV SA	Spain	367	80-00800
Gordon Consulting Engineers Ltd	Israel	224	80-00750
Gutteridge Haskins & Davey Pty Ltd	Australia	25	80-07500
Hoffman Engineering Corp.	USA	582	75-43980
Horizons Technology Inc.	USA	636	80-14450
Hugo Marom Aviation Consultants Ltd	Israel	224	80-00880
Industrieanlagen-Betriebs GmbH	Germany	178	80-00945
Israel Aircraft Industries Ltd	Israel	219	75-10200
ISVR Consultancy Services	UK	505	80-13520
Jet Aviation Zürich AG	Switzerland	388	80-01030
John M. Davis & Associates	USA	634	80-07200
Kapos Associates Inc.	USA	637	80-17630
Keystone Helicopter Corp.	USA	637	80-17780
Landair International Ltd	UK	505	80-14200
Lightweight Body Armour	UK	464	75-35650
Litton Industries Inc.	USA	590	75-55400
Logica UK Limited	UK	506	80-14350
M. Paz Logistics Ltd	Israel	224	80-01600
M. Rosenblatt & Son Inc.	USA	641	80-26450
MECAR SA	Belgium	46	75-04930
Milo Holdings Ltd	UK	506	80-16850
National Aerospace Laboratory	Netherlands	290	60-01400
National Systems Management Corp.	USA	639	80-21650
NEL	UK	419	60-01400
NIS Ltd	UK	471	75-42555
NNC Ltd	UK	507	80-18500
Nogay Ltd	Israel	224	80-01470
OMI Logistics Limited	UK	508	80-19450
OTB	S. Africa	357	60-00500
O.T.P. Ballistics	UK	508	80-19700
Parsons Engineering Science	USA	640	80-23750
PKM Consultants (S) Pte Ltd	Singapore	349	80-01640
Programming Research Ltd	UK	509	80-24700
Protechnik Laboratories	S. Africa	359	75-17000
Public Safety Group Inc.	USA	641	80-24955
Reliant Design Development Co Ltd	UK	478	75-49170
Ricardo Hitec Ltd	UK	509	80-27840
RKI/Cleanroom Services	USA	641	80-26200
Rollab Aktiebolag	Sweden	380	80-01850
Royal Ordnance	UK	481	75-50620
SAC Technographic	UK	510	80-30100
ScanAvia, Scandinavian Aviation	Sweden	380	80-01860
SEAL Laboratories	USA	642	80-27250
Secure Consulting International Ltd	UK	511	80-33445
SEMCOR Inc.	USA	642	80-27750
Serco Aviation Services	UK	511	80-34150
Sidem (UK) Ltd	UK	511	80-35850
SIP Services Int'l	UK	511	80-36500
SISDEF Ltda	Chile	85	80-01940
Southwest Research Institute	USA	541	60-19200
Specialty Plastic Products Inc.	USA	616	75-83035
SRI International	USA	643	80-29600
SRL Operations	USA	617	75-83550
Sverdrup Corp.	USA	643	80-30650
Sverdrup Technology Inc.	USA	643	80-30680
TA Group Ltd	UK	513	80-50110
Technology Australasia Pty Ltd	Australia	26	80-20600
The Titan Corp	USA	622	75-89620
Thomson-CSF Systems Canada Inc.	Canada	79	80-02020
Top Line Co.	USA	644	80-33100
Tracor Inc	USA	623	75-90000
TriLink Technologies Group Inc.	Canada	80	80-02100
UBC Inc.	USA	624	75-91383
United Defense L.P.	USA	624	75-92050
Vickers Shipbuilding & Eng Ltd	UK	491	75-63100
Vipac Engineers & Scientists Ltd	Australia	26	80-22580
Vosper International Ltd	UK	514	80-67000
Wornall Electronics Inc.	USA	645	80-35400
Yarrow Shipbuilders Ltd	UK	495	75-67500
Zernow Technical Services Inc.	USA	645	80-36650
3C Systems, Inc.	USA	632	80-05000

912 10 Safety assessment services

Design Bureau of	Russia	336	75-03150
Frazer-Nash Consultancy Limited	UK	503	80-10950

912 40 Market research services

Aerospace & Defense Sales Inc	Philippines	321	80-00050
AirteQ Ltd	UK	496	80-00830
AMI International	USA	631	80-02470
Andersen Consulting	Australia	24	80-00910
Andrich International Ltd	UK	497	80-01200
Beltramo & Associates	USA	632	80-04360
Bruce D. Vallillee Electronics Ltd	Canada	80	80-02210
Burdeshaw Associates, Ltd	USA	632	80-04960
Cambridge Consultants Ltd	UK	499	80-05300
Carmichael & Sweet Consultancy Ltd	UK	499	80-05500
CIC International Ltd	USA	563	75-18570
CMPS&F Pty Limited	Australia	24	80-03600
Defence Data Consulting Services	Switzerland	388	80-00430
Defence Research & Analysis	UK	500	80-08200
Designers & Planners Inc.	USA	634	80-08300
Engineering & Professional Serv.Inc.	USA	634	80-10150
Forecast International/DMS	USA	636	80-11760
Frost & Sullivan Ltd	UK	503	80-10970
George G. Sharp Inc.	USA	642	80-28000
Gordon Consulting Engineers Ltd	Israel	224	80-00750

Hugo Marom Aviation Consultants Ltd	Israel	224	80-00880
John Reed (Specialist Market	UK	505	80-13900
John W. Drake	USA	634	80-08600
Landair International Ltd	UK	505	80-14200
Logica UK Limited	UK	506	80-14350
M International Export	USA	639	80-19780
M. Paz Logistics Ltd	Israel	224	80-01600
M. Rosenblatt & Son Inc.	USA	641	80-26450
Metra Martech Limited	UK	506	80-16300
Mycon Marketing Services Ltd	UK	507	80-17900
NanoMaterials, Inc.	USA	598	75-63590
Orman Associates Inc.	USA	640	80-22600
Reade Advanced Materials	USA	608	75-73860
RKI/Cleanroom Services	USA	641	80-26200
Schneider Sohn & Associates Inc.	USA	642	80-26890
Sector Ltd	UK	511	80-33400
Secure Consulting International Ltd	UK	511	80-33445
Sidem (UK) Ltd	UK	511	80-35850
SIP Services Int'l	UK	511	80-36500
Squadron International Ltd	USA	643	80-29550
Technology Australasia Pty Ltd	Australia	26	80-20600
Top Line Co.	USA	644	80-33100
TriLink Technologies Group Inc.	Canada	80	80-02100

912 50 Operations research services

American Management Systems	USA	630	80-02350
AMI International	USA	631	80-02470
Andersen Consulting	Australia	24	80-00910
Burdeshaw Associates, Ltd	USA	632	80-04960
Calspan SRL Technology	USA	540	60-03000
CIC International Ltd	USA	563	75-18570
Commslogic Ltd	UK	500	80-06400
Contahal Ltd	Israel	224	80-00300
Cranfield University	UK	419	60-00350
Delex Systems Inc.	USA	634	80-08020
Design Bureau of	Russia	336	75-03150
Dytecna Ltd	UK	501	80-08900
EASAMS Ltd	UK	501	80-09000
EDS Defence Ltd	UK	501	80-09200
Engineering & Professional Serv.Inc.	USA	634	80-10150
George G. Sharp Inc.	USA	642	80-28000
Gordon Consulting Engineers Ltd	Israel	224	80-00750
Horrigan Analytics	USA	636	80-14480
Industrieanlagen-Betriebs GmbH	Germany	178	80-00945
Israel Military Industries Ltd	Israel	220	75-10300
Kapos Associates Inc.	USA	637	80-17630
Landair International Ltd	UK	505	80-14200
Logica UK Limited	UK	506	80-14350
M. Paz Logistics Ltd	Israel	224	80-01600
Management Consulting & Research	USA	638	80-19680
MIL Systems	Canada	79	80-01360
National Aerospace Laboratory	Netherlands	290	60-01400
Operational Research Division	Canada	67	60-01500
Perceptronics Inc.	USA	603	75-69270
QUANTICS incorporated	USA	641	80-25150
Radio Holland Electronics BV	Netherlands	297	80-18300
Royal Ordnance	UK	481	75-50620
Sector Ltd	UK	511	80-33400
SEMCOR Inc.	USA	642	80-27750
SIP Services Int'l	UK	511	80-36500
Smith System Engineering Ltd.	UK	512	80-39500
Sverdrup Corp.	USA	643	80-30650
Sverdrup Technology Inc.	USA	643	80-30680
TA Group Ltd	UK	513	80-50110
The Graphics Co. of America	USA	579	75-40360
Thomson-CSF Systems Canada Inc.	Canada	79	80-02020
TriLink Technologies Group Inc.	Canada	80	80-02100
Vanguard Research Inc.	USA	644	80-34100
Wheeler Systems Inc.	USA	627	75-96400

912 60 Systems analysis services

AEA Technology plc	UK	496	80-00480
Aerospatiale	France	131	75-00600
Allott & Lomax	UK	497	80-01050
American Management Systems	USA	630	80-02350
Analysis & Technology Inc.	USA	631	80-02600
Andersen Consulting	Australia	24	80-00910
Anite Systems Ltd	UK	497	80-01240
ANSER	USA	631	80-02620
Aspect Computing Pty Ltd	Australia	24	80-01150
Avica Equipment	UK	432	75-04870
Boeing Australia Limited	Australia	12	75-02500
Burdeshaw Associates, Ltd	USA	632	80-04960
Calian Technology Ltd	Canada	69	75-01800
Calspan SRL Technology	USA	540	60-03000
Cambridge Consultants Ltd	UK	499	80-05300
Central Research Laboratories Ltd	UK	499	80-05895
Chamberlain Manufacturing Corp.	USA	563	75-17950
CIC International Ltd	USA	563	75-18570
CMPS&F Pty Limited	Australia	24	80-03600
CMS Inc.	USA	564	75-19170
Computer Power Group Ltd	Australia	24	80-03730
Computing Devices Company Ltd	UK	442	75-14700
Contahal Ltd	Israel	224	80-00300
Cranfield University	UK	419	60-00350
Crew Services Ltd	UK	500	80-07370
CSC Computer Sciences Ltd	UK	500	80-07500
DCS Corp.	USA	634	80-07400
Delex Systems Inc.	USA	634	80-08020
Deloitte & Touche Consulting Group	UK	501	80-08300
Design Bureau of	Russia	336	75-03150
DSD Ltd	UK	501	80-08700
Dytecna Ltd	UK	501	80-08900
EASAMS Ltd	UK	501	80-09000
EDS Defence Ltd	UK	501	80-09200
EG&G Washington Analytical Svcs	USA	634	80-09450
Engineering & Professional Serv.Inc.	USA	634	80-10150
ERA Technology Ltd	UK	502	80-09600
Fluid Gravity Engineering Ltd.	UK	502	80-10660
Forges de Zeebrugge SA	Belgium	46	75-03750
Frazer-Nash Consultancy Limited	UK	503	80-10950
GEC ALSTHOM Engineering Systems Ltd	UK	453	75-25225
GEC-Marconi Avionics Ltd	UK	454	75-25320
GEC-Marconi Research Centre	UK	419	60-00700
George G. Sharp Inc.	USA	642	80-28000
GMV SA	Spain	367	80-00800
Gordon Consulting Engineers Ltd	Israel	224	80-00750
Horizons Technology Inc.	USA	636	80-14450
Horrigan Analytics	USA	636	80-14480
Hunting Engineering Ltd	UK	459	75-30700
HVR Consulting Services Ltd	UK	504	80-12790
Ilex Systems Inc.	USA	584	75-46080
Industrieanlagen-Betriebs GmbH	Germany	178	80-00945
Inovex Technology Ltd	UK	504	80-13310
Israel Military Industries Ltd	Israel	220	75-10300
Kaman Sciences Corp.	USA	637	80-17600
Kapos Associates Inc.	USA	637	80-17630
Landair International Ltd	UK	505	80-14200
LMS International	Belgium	48	80-00900
Logica Pty Ltd	Australia	25	80-12500
Logica UK Limited	UK	506	80-14350
M. Paz Logistics Ltd	Israel	224	80-01600
M. Rosenblatt & Son Inc.	USA	641	80-26450
MacDonald Dettwiler and Associates	Canada	78	80-01310
Mackenzie Tribbeck Associates	UK	506	80-15150
Management Consulting & Research	USA	638	80-19680
MASS Consultants Limited	UK	506	80-15650
Matra Marconi Space UK, Ltd	UK	425	74-06900
Metra Martech Limited	UK	506	80-16300
MIL Systems	Canada	79	80-01360
Milo Holdings Ltd	UK	506	80-16850
MMM Consultancy Group Ltd	UK	507	80-17000
Mott MacDonald	UK	507	80-17400
National Aerospace Laboratory	Netherlands	290	60-01400
Octec Ltd	UK	471	75-42950
O.T.P. Ballistics	UK	508	80-19700
Parity Solutions Ltd	UK	508	80-20300
Park Air Electronics Ltd	UK	472	75-43750
Praxis Critical Systems Ltd	UK	509	80-23930
PRC Inc.	USA	640	80-24800
Primary Image Ltd	UK	476	75-46790
Programming Research Ltd	UK	509	80-24700
Public Safety Group Inc.	USA	641	80-24955
Radio Holland Electronics BV	Netherlands	297	80-18300
Resource Consultants Inc.	USA	641	80-25920
Ricardo Hitec Ltd	UK	509	80-27840
Rolls-Royce & Associates Ltd	UK	510	80-29100
Royal Ordnance	UK	481	75-50620
Rutherford Appleton Laboratory	UK	420	60-01850
SAIC Ltd	UK	510	80-30220
Science Systems (Industrial) Ltd	UK	510	80-31600
SEMCOR Inc.	USA	642	80-27750
Shorts Missile Systems Ltd	**UK**	**483**	**75-53550**
Siemens Plessey Systems	UK	483	75-53770
SIP Services Int'l	UK	511	80-36500

Smith System Engineering Ltd.	UK	*512*	80-39500
Space Systems Finland Ltd	Finland	*118*	75-02100
STASYS Ltd	UK	*512*	80-46700
Stirling Dynamics Ltd.	UK	*512*	80-47000
Sverdrup Corp.	USA	*643*	80-30650
Sverdrup Technology Inc.	USA	*643*	80-30680
Synectics Corp.	USA	*643*	80-30750
TA Group Ltd	UK	*513*	80-50110
TDSi Micro Applications Ltd.	UK	*513*	80-51000
Technology Australasia Pty Ltd	Australia	*26*	80-20600
Technology Project Services	UK	*513*	80-52800
Thomson Marconi Sonar Limited	UK	*426*	74-20000
Thomson-CSF Systems Canada Inc.	Canada	*79*	80-02020
Triad Group plc	UK	*514*	80-58600
Vanguard Research Inc.	USA	*644*	80-34100
Vickers Shipbuilding & Eng Ltd	UK	*491*	75-63100
Vitro Corp.	USA	*644*	80-34400
Wheeler Systems Inc.	USA	*627*	75-96400
WS Atkins plc.	UK	*497*	80-01550
Yarrow Shipbuilders Ltd	UK	*495*	75-67500
3C Systems, Inc.	USA	*632*	80-05000

912 70 Threat assessment services

Aerospatiale.	France	*131*	75-00600
Allott & Lomax	UK	*497*	80-01050
AMI International	USA	*631*	80-02470
Analysis & Technology Inc.	USA	*631*	80-02600
Anite Systems Ltd	UK	*497*	80-01240
BACTEC International Ltd	**UK**	***498***	**80-03060**
Burdeshaw Associates, Ltd	USA	*632*	80-04960
Calspan SRL Technology.	USA	*540*	60-03000
Carmichael & Sweet Consultancy Ltd.	UK	*499*	80-05500
Central Research Laboratories Ltd	UK	*499*	80-05895
CIC International Ltd	USA	*563*	75-18570
CMPS&F Pty Limited	Australia	*24*	80-03600
Crew Services Ltd	UK	*500*	80-07370
Criminal Intelligence Administration.	USA	*633*	80-06730
DCS Corp.	USA	*634*	80-07400
Defence Research & Analysis	UK	*500*	80-08200
Delex Systems Inc.	USA	*634*	80-08020
DEMCO Consultants Ltd	Israel	*224*	80-00430
DSD Ltd	UK	*501*	80-08700
EASAMS Ltd	UK	*501*	80-09000
EDS Defence Ltd	UK	*501*	80-09200
Engineering & Professional Serv.Inc.	USA	*634*	80-10150
Fluid Gravity Engineering Ltd.	UK	*502*	80-10660
GEC-Marconi Research Centre.	UK	*419*	60-00700
GKN Westland Systems	UK	*503*	80-11800
Gordon Consulting Engineers Ltd	Israel	*224*	80-00750
Horizons Technology Inc.	USA	*636*	80-14450
Hunting Engineering Ltd	UK	*459*	75-30700
Industrieanlagen-Betriebs GmbH	Germany	*178*	80-00945
Israel Military Industries Ltd	Israel	*220*	75-10300
Jane's Information Group Ltd	**UK**	***505***	**80-13700**
Landair International Ltd	UK	*505*	80-14200
Litton Industries Inc.	USA	*591*	75-55400
Logica UK Limited.	UK	*506*	80-14350
M. Paz Logistics Ltd	Israel	*224*	80-01600
Mackenzie Tribbeck Associates.	UK	*506*	80-15150
MASS Consultants Limited	UK	*506*	80-15650
Meggitt Composites	UK	*467*	75-38345
MFD International Ltd.	UK	*506*	80-16450
Mott MacDonald	UK	*507*	80-17400
National Aerospace Laboratory	Netherlands	*290*	60-01400
Nichols Research Corp.	USA	*639*	80-21780
O.T.P. Ballistics	UK	*508*	80-19700
Pacer Systems Inc.	USA	*602*	75-68000
Pains-Wessex Ltd	UK	*472*	75-43510
Penter Associations Limited.	UK	*508*	80-21000
Public Safety Group Inc.	USA	*641*	80-24955
Q-Systems International Inc.	USA	*606*	75-71960
Royal Ordnance	UK	*481*	75-50620
SAIC Ltd.	UK	*510*	80-30220
Sector Ltd	UK	*511*	80-33400
Security Research Ltd	UK	*482*	75-52450
SEMCOR Inc.	USA	*642*	80-27750
SIP Services Int'l	UK	*511*	80-36500
Smith System Engineering Ltd.	UK	*512*	80-39500
Sverdrup Corp.	USA	*643*	80-30650
Sverdrup Technology Inc.	USA	*643*	80-30680
Synectics Corp.	USA	*643*	80-30750
TA Group Ltd	UK	*513*	80-50110
Technology Australasia Pty Ltd	Australia	*26*	80-20600
Thomson Marconi Sonar Limited	UK	*426*	74-20000
Thomson-CSF Systems Canada Inc.	Canada	*79*	80-02020
United Defense L.P.	USA	*624*	75-92050
Vanguard Research Inc.	USA	*644*	80-34100
Vitro Corp.	USA	*644*	80-34400
Western Tactical Systems.	UK	*515*	80-73000
Wheeler Systems Inc.	USA	*627*	75-96400
Wornall Electronics Inc.	USA	*645*	80-35400

920 Management services

A Goninan & Co. Ltd	Australia	*16*	75-07760
Aerospace Project Development Group	Switzerland	*384*	60-00020
AMADEUS Inc	USA	*630*	80-02200
American Management Systems	USA	*630*	80-02350
Andersen Consulting	Australia	*24*	80-00910
Anite Systems Ltd	UK	*497*	80-01240
Arcata Associates Inc.	USA	*554*	75-08400
Aspect Computing Pty Ltd.	Australia	*24*	80-01150
Assessment Services	UK	*497*	80-01500
Atlantic Research Corp.	USA	*555*	75-10150
BMT Marine Information Systems Ltd	UK	*498*	80-03573
British Aerospace Australia Ltd	Australia	*13*	75-02750
Calspan SRL Technology.	USA	*540*	60-03000
Carmichael & Sweet Consultancy Ltd.	UK	*499*	80-05500
Cas Inc.	USA	*633*	80-05230
Chamberlain Manufacturing Corp.	USA	*563*	75-17950
CIC International Ltd	USA	*563*	75-18570
CMPS&F Pty Limited	Australia	*24*	80-03600
Cogent.	UK	*500*	80-06250
Contraves GmbH	Germany	*166*	75-05200
Cranfield University.	UK	*419*	60-00350
Crew Services Ltd	UK	*500*	80-07370
CTA Inc.	USA	*633*	80-06810
Cubic Defense Systems Inc.	USA	*567*	75-22900
Daewoo Heavy Industries Ltd	Korea (ROK)	*265*	75-00600
DCS Corp.	USA	*634*	80-07400
Delex Systems Inc.	USA	*634*	80-08020
Deloitte & Touche Consulting Group	UK	*501*	80-08300
DEMCO Consultants Ltd	Israel	*224*	80-00430
Design Bureau of	Russia	*336*	75-03150
EASAMS Ltd	UK	*501*	80-09000
EG&G Inc.	USA	*571*	75-29555
EG&G Washington Analytical Svcs	USA	*634*	80-09450
Elintech GmbH	Germany	*178*	80-00550
Engineering & Professional Serv.Inc.	USA	*634*	80-10150
Flight Refuelling Limited	UK	*451*	75-23850
Fluid Gravity Engineering Ltd.	UK	*502*	80-10660
Fluor Daniel Inc.	USA	*635*	80-11720
FR Aviation Limited.	UK	*503*	80-10850
Frazer-Nash Consultancy Limited	UK	*503*	80-10950
Garber International Associates Inc.	USA	*636*	80-12150
GEC-Marconi Avionics Ltd	UK	*454*	75-25320
George G. Sharp Inc.	USA	*642*	80-28000
Global Communications Co.	USA	*579*	75-39775
Global Defense Products Inc.	USA	*579*	75-39770
Gutteridge Haskins & Davey Pty Ltd	Australia	*25*	80-07500
Hanjin Heavy Industries Co. Ltd	Korea (ROK)	*265*	75-01505
Hunting Engineering Ltd	UK	*459*	75-30700
Indal Technologies Inc	Canada	*72*	75-07740
Industrieanlagen-Betriebs GmbH	Germany	*178*	80-00945
International Hospitals Group	UK	*505*	80-13410
Israel Aircraft Industries Ltd	Israel	*219*	75-10200
Israel Military Industries Ltd	Israel	*220*	75-10300
Keystone Helicopter Corp.	USA	*637*	80-17780
Krauss-Maffei Wehrtechnik GmbH	Germany	*171*	75-16345
Landair International Ltd	UK	*505*	80-14200
Lansdowne Technologies Inc	Canada	*78*	80-01205
Litton Industries Inc.	USA	*590*	75-55400
Lockheed Martin Aircraft and	USA	*638*	80-18530
Lockheed Martin Corp.	USA	*591*	75-55635
Logica SA/NV.	Belgium	*48*	80-01000
Logistics Services International	USA	*638*	80-18800
M. Paz Logistics Ltd	Israel	*224*	80-01600
Management Consulting & Research	USA	*638*	80-19680
Materialinformationszentrum Ges.	Germany	*178*	80-01310
MFD International Ltd.	UK	*506*	80-16450
MJI Associates Inc.	USA	*639*	80-21320
NanoMaterials, Inc.	USA	*598*	75-63590
National Systems Management Corp.	USA	*639*	80-21650
Newport News Shipbuilding	USA	*599*	75-64300
NIS Ltd	UK	*471*	75-42555
NNC Ltd	UK	*507*	80-18500
Nogay Ltd	Israel	*224*	80-01470

Northrop Grumman Corporation	USA	*599*	75-65450
OMI Logistics Limited	UK	*508*	80-19450
Pacer Systems Inc.	USA	*602*	75-68000
Parsons Power Group Inc.	USA	*640*	80-23700
Philippine Exhibition Services	Philippines	*321*	80-02000
Philips Calibration Services	Australia	*25*	80-16490
PRC Inc.	USA	*640*	80-24800
Raytheon E-Systems	USA	*608*	75-73302
Reade Advanced Materials	USA	*608*	75-73860
Resource Consultants Inc.	USA	*641*	80-25920
Ricardo Hitec Ltd	UK	*509*	80-27840
Rolls-Royce (Canada) Ltd.	Canada	*75*	75-12120
SAIT Systems SA	Belgium	*47*	75-05600
ScanAvia, Scandinavian Aviation	Sweden	*380*	80-01860
Secure Consulting International Ltd	UK	*511*	80-33445
SEMCOR Inc.	USA	*642*	80-27750
Serco Aviation Services	UK	*511*	80-34150
Shorts Support Services Division	UK	*511*	80-35600
Siemens Switzerland Ltd.	Switzerland	*387*	75-11980
SISDEF Ltda	Chile	*85*	80-01940
Spar Aerospace Ltd	Canada	*76*	75-13000
Stanley Aviation Corporation	USA	*617*	75-83900
Strachan & Henshaw Ltd	UK	*512*	80-47400
Sverdrup Corp.	USA	*643*	80-30650
Sverdrup Technology Inc.	USA	*643*	80-30680
TA Group Ltd	UK	*513*	80-50110
Taylor Woodrow Construction Ltd	UK	*513*	80-50300
Taylor Woodrow Management	UK	*513*	80-50400
The Aerospace Corp.	USA	*630*	80-01050
Thomson-CSF Systems Canada Inc.	Canada	*79*	80-02020
TriLink Technologies Group Inc.	Canada	*80*	80-02100
Vanguard Research Inc.	USA	*644*	80-34100
Vitrociset SpA	Italy	*240*	80-40000
Vosper International Ltd	UK	*514*	80-67000
Wegmann & Co GmbH	Germany	*177*	75-33700
WS Atkins plc.	UK	*497*	80-01550
Yarrow Shipbuilders Ltd	UK	*495*	75-67500

920 10 Base management services

Bruce D. Vallillee Electronics Ltd	Canada	*80*	80-02210
Calspan SRL Technology	USA	*540*	60-03000
CIC International Ltd	USA	*563*	75-18570
Design Bureau of	Russia	*336*	75-03150
DSSD plc	UK	*501*	80-08750
EG&G Inc.	USA	*571*	75-29555
Elintech GmbH	Germany	*178*	80-00550
Engineering & Professional Serv.Inc.	USA	*634*	80-10150
Flight Refuelling Limited	UK	*451*	75-23850
Fluor Daniel Inc.	USA	*635*	80-11720
FR Aviation Limited	UK	*503*	80-10850
General Motors of Canada Ltd	Canada	*71*	75-05900
George G. Sharp Inc.	USA	*642*	80-28000
Global Communications Co.	USA	*579*	75-39775
Global Defense Products Inc.	USA	*579*	75-39770
International Hospitals Group	UK	*505*	80-13410
Israel Aircraft Industries Ltd	Israel	*219*	75-10200
ITT Federal Services Corp.	USA	*586*	75-49455
Lockheed Martin Aircraft and	USA	*638*	80-18530
Logistics Services International	USA	*638*	80-18800
M. Paz Logistics Ltd	Israel	*224*	80-01600
Materialinformationszentrum Ges.	Germany	*178*	80-01310
National Aerospace Laboratory	Netherlands	*290*	60-01400
Northrop Grumman Corporation	USA	*599*	75-65450
PRC Inc.	USA	*640*	80-24800
Qualitair Engineering Services Ltd	UK	*509*	80-25200
Raytheon E-Systems	USA	*608*	75-73302
Ricardo Hitec Ltd	UK	*509*	80-27840
Rolls-Royce & Associates Ltd	UK	*510*	80-29100
ScanAvia, Scandinavian Aviation	Sweden	*380*	80-01860
Sverdrup Corp.	USA	*643*	80-30650
Sverdrup Technology Inc.	USA	*643*	80-30680
Taylor Woodrow Management	UK	*513*	80-50400
Vosper International Ltd	UK	*514*	80-67000
WS Atkins plc.	UK	*497*	80-01550

920 50 Operation services, plant / base

Calspan SRL Technology	USA	*540*	60-03000
Chamberlain Manufacturing Corp.	USA	*563*	75-17950
CIC International Ltd	USA	*563*	75-18570
Daewoo Heavy Industries Ltd	Korea (ROK)	*265*	75-00600
Design Bureau of	Russia	*336*	75-03150
EG&G Inc.	USA	*571*	75-29555
Elintech GmbH	Germany	*178*	80-00550
Engineering & Professional Serv.Inc.	USA	*634*	80-10150
Fluor Daniel Inc.	USA	*635*	80-11720
George G. Sharp Inc.	USA	*642*	80-28000
Global Communications Co.	USA	*579*	75-39775
Global Defense Products Inc.	USA	*579*	75-39770
Hanjin Heavy Industries Co. Ltd	Korea (ROK)	*265*	75-01505
International Hospitals Group	UK	*505*	80-13410
Israel Aircraft Industries Ltd	Israel	*219*	75-10200
Lockheed Martin Aircraft and	USA	*638*	80-18530
M. Paz Logistics Ltd	Israel	*224*	80-01600
Raytheon E-Systems	USA	*608*	75-73302
Santa Barbara SA	Spain	*364*	55-01000
Sonovision ITEP Technologies	France	*151*	80-01850
Sverdrup Corp.	USA	*643*	80-30650
Sverdrup Technology Inc.	USA	*643*	80-30680
Taylor Woodrow Construction Ltd	UK	*513*	80-50300
The Graphics Co. of America	USA	*579*	75-40360
United Defense L.P.	USA	*624*	75-92050

920 55 Operation services, range

AAI Corp.	USA	*545*	75-00020
Arcata Associates Inc.	USA	*554*	75-08400
Armalon Ltd	UK	*431*	75-03729
Calspan SRL Technology	USA	*540*	60-03000
CIC International Ltd	USA	*563*	75-18570
EG&G Inc.	USA	*571*	75-29555
Engineering & Professional Serv.Inc.	USA	*634*	80-10150
Fluor Daniel Inc.	USA	*635*	80-11720
GEC-Marconi Avionics Ltd	UK	*454*	75-25320
Israel Military Industries Ltd	Israel	*220*	75-10300
Lockheed Martin Aircraft and	USA	*638*	80-18530
Lockheed Martin Services Group	USA	*638*	80-18540
SEMCOR Inc.	USA	*642*	80-27750
Stanley Aviation Corporation	USA	*617*	75-83900
Sverdrup Corp.	USA	*643*	80-30650
Sverdrup Technology Inc.	USA	*643*	80-30680

920 55 10 Operation services, air-to-ground range

Arcata Associates Inc.	USA	*554*	75-08400
Calspan SRL Technology	USA	*540*	60-03000
CIC International Ltd	USA	*563*	75-18570
EG&G Inc.	USA	*571*	75-29555
Engineering & Professional Serv.Inc.	USA	*634*	80-10150
Fluor Daniel Inc.	USA	*635*	80-11720
Israel Military Industries Ltd	Israel	*220*	75-10300

920 60 Safety management services

Design Bureau of	Russia	*336*	75-03150
Frazer-Nash Consultancy Limited	UK	*503*	80-10950

922 Control, supervision services

Aerospace Design Hugo Marom Ltd	Israel	*223*	80-00040
Aviation Support Tech. Int'l Inc.	USA	*556*	75-11050
Carmichael & Sweet Consultancy Ltd	UK	*499*	80-05500
CIC International Ltd	USA	*563*	75-18570
DEMCO Consultants Ltd	Israel	*224*	80-00430
Design Bureau of	Russia	*336*	75-03150
EG&G Inc.	USA	*571*	75-29555
Elintech GmbH	Germany	*178*	80-00550
Engineering & Professional Serv.Inc.	USA	*634*	80-10150
Fluor Daniel Inc.	USA	*635*	80-11720
George G. Sharp Inc.	USA	*642*	80-28000
Global Communications Co.	USA	*579*	75-39775
Global Defense Products Inc.	USA	*579*	75-39770
Israel Military Industries Ltd	Israel	*220*	75-10300
Krauss-Maffei Wehrtechnik GmbH	Germany	*171*	75-16345
M. Paz Logistics Ltd	Israel	*224*	80-01600
Materialinformationszentrum Ges.	Germany	*178*	80-01310
Nogay Ltd	Israel	*224*	80-01470
OMI Logistics Limited	UK	*508*	80-19450
Philips Calibration Services	Australia	*25*	80-16490
Raytheon E-Systems	USA	*608*	75-73302
Ricardo Hitec Ltd	UK	*509*	80-27840
Santa Barbara SA	Spain	*364*	55-01000
SEMCOR Inc.	USA	*642*	80-27750

Product-Supplier

Serco Aviation Services	UK	511	80-34150
SISDEF Ltda	Chile	85	80-01940
Sverdrup Corp.	USA	643	80-30650
Thomson-CSF Systems Canada Inc.	Canada	79	80-02020

922 50 RAM management services

AEA Technology plc	UK	496	80-00480
American Management Systems	USA	630	80-02350
CMPS&F Pty Limited	Australia	24	80-03600
Dytecna Ltd	UK	501	80-08900
Frazer-Nash Consultancy Limited	UK	503	80-10950
Messier-Dowty Inc	Canada	74	75-09655
OMI Logistics Limited	UK	508	80-19450
SEMCOR Inc.	USA	642	80-27750
Stanley Aviation Corporation	USA	617	75-83900
Sverdrup Corp.	USA	643	80-30650
TA Group Ltd	UK	513	80-50110
Technology Australasia Pty Ltd	Australia	26	80-20600

924 Planning, scheduling services

A Goninan & Co. Ltd	Australia	16	75-07760
Aman-Management Consultants Ltd	Israel	223	80-00150
Analysis & Technology Inc.	USA	631	80-02600
Andersen Consulting	Australia	24	80-00910
Artemis Management Systems GmbH	Germany	164	75-00230
Assessment Services	UK	497	80-01500
Barton-Aschman Associates Inc.	USA	632	80-04100
Bremer Vulkan Marineschiffbau GmbH	Germany	166	75-03600
Bruce D. Vallillee Electronics Ltd	Canada	80	80-02210
Calspan SRL Technology	USA	540	60-03000
Carmichael & Sweet Consultancy Ltd	UK	499	80-05500
CIC International Ltd	USA	563	75-18570
CMPS&F Pty Limited	Australia	24	80-03600
Computer Power Group Ltd	Australia	24	80-03730
Crew Services Ltd	UK	500	80-07370
Currie & Brown (Australia) Pty Ltd.	Australia	24	80-03850
DCS Corp.	USA	634	80-07400
DEMCO Consultants Ltd	Israel	224	80-00430
Dytecna Ltd	UK	501	80-08900
EG&G Washington Analytical Svcs	USA	634	80-09450
Elintech GmbH	Germany	178	80-00550
Engineering & Professional Serv.Inc.	USA	634	80-10150
Fluor Daniel Inc.	USA	635	80-11720
Garber International Associates Inc.	USA	636	80-12150
GED (Southampton) Ltd	UK	503	80-11650
General Electronics Communications	Greece	189	75-01300
Global Defense Products Inc.	USA	579	75-39770
Horizons Technology Inc.	USA	636	80-14450
Hunting Engineering Ltd	UK	459	75-30700
Industrieanlagen-Betriebs GmbH	Germany	178	80-00945
InterCon Consultants	Canada	78	80-00980
Israel Aircraft Industries Ltd	Israel	219	75-10200
Jeppesen UK Ltd	UK	505	80-13850
Korea Express USA Inc.	USA	638	80-17900
Krauss-Maffei Wehrtechnik GmbH	Germany	171	75-16345
Lansdowne Technologies Inc	Canada	78	80-01205
Litton Industries Inc.	USA	590	75-55400
Management Consulting & Research	USA	638	80-19680
Materialinformationszentrum Ges.	Germany	178	80-01310
National Aerospace Laboratory	Netherlands	290	60-01400
National Systems Management Corp.	USA	639	80-21650
Nogay Ltd	Israel	224	80-01470
Pacer Systems Inc.	USA	602	75-68000
Planning & Research Ltd	Israel	224	80-01650
Raytheon E-Systems	USA	608	75-73302
Ricardo Hitec Ltd	UK	509	80-27840
SEMCOR Inc.	USA	642	80-27750
Serco Aviation Services	UK	511	80-34150
SISDEF Ltda	Chile	85	80-01940
Sonovision ITEP Technologies	France	151	80-01850
Stanley Aviation Corporation	USA	617	75-83900
Sverdrup Corp.	USA	643	80-30650
TA Group Ltd	UK	513	80-50110
Technology Australasia Pty Ltd	Australia	26	80-20600
The Graphics Co. of America	USA	579	75-40360
Thomson-CSF Systems Canada Inc.	Canada	79	80-02020
Tracor Inc	USA	623	75-90000
TriLink Technologies Group Inc.	Canada	80	80-02100
United Defense L.P.	USA	624	75-92050
Vitro Corp.	USA	644	80-34400

Vitrociset SpA	Italy	240	80-40000
Yarrow Shipbuilders Ltd	UK	495	75-67500

926 Logistics management services

Advanced Appl. Consultants Inc.	USA	630	80-00250
Air-Log Ltd	UK	429	75-01800
Alvis Industries Ltd	UK	430	75-02850
Amer Ind. Inc.	USA	631	80-02390
American Management Systems	USA	630	80-02350
Analysis & Technology Inc.	USA	631	80-02600
Andersen Consulting	Australia	24	80-00910
Barton-Aschman Associates Inc.	USA	632	80-04100
Bremer Vulkan Marineschiffbau GmbH	Germany	166	75-03600
Carmichael & Sweet Consultancy Ltd	UK	499	80-05500
CIC International Ltd	USA	563	75-18570
CMPS&F Pty Limited	Australia	24	80-03600
Cogent	UK	500	80-06250
Computer Power Group Ltd	Australia	24	80-03730
Contraves GmbH	Germany	166	75-05200
Cubic Defense Systems Inc.	USA	567	75-22900
Delex Systems Inc.	USA	634	80-08020
Denel Informatics	S. Africa	360	80-05000
Design Bureau of	Russia	336	75-03150
DSSD plc	UK	501	80-08750
Dytecna Ltd	UK	501	80-08900
EDS Defence Ltd	UK	501	80-09200
EG&G Inc.	USA	571	75-29555
EG&G Washington Analytical Svcs	USA	634	80-09450
Elintech GmbH	Germany	178	80-00550
Engineering & Professional Serv.Inc.	USA	634	80-10150
EPS Logistics Technology Ltd	UK	502	80-09550
Eyes & Ears	Egypt	110	80-06000
Fluor Daniel Inc.	USA	635	80-11720
Frazer-Nash Consultancy Limited	UK	503	80-10950
GEC-Marconi Avionics Ltd	UK	454	75-25320
GED (Southampton) Ltd	UK	503	80-11650
GEL Engineering Ltd	UK	454	75-25500
General Motors of Canada Ltd	Canada	71	75-05900
George G. Sharp Inc.	USA	642	80-28000
Global Communications Co.	USA	579	75-39775
Global Defense Products Inc.	USA	579	75-39770
Horizons Technology Inc.	USA	636	80-14450
Industrieanlagen-Betriebs GmbH	Germany	178	80-00945
Information Handling Services Inc.	USA	637	80-15530
InterCon Consultants	Canada	78	80-00980
Israel Aircraft Industries Ltd	Israel	219	75-10200
Israel Military Industries Ltd	Israel	220	75-10300
ITT Avionics	USA	586	75-49455
Kay & Associates Inc.	USA	637	80-17660
Keystone Helicopter Corp.	USA	637	80-17780
Kinhill Engineers Pty Ltd	Australia	25	80-11000
Krauss-Maffei Wehrtechnik GmbH	Germany	171	75-16345
Landair International Ltd	UK	505	80-14200
Lansdowne Technologies Inc	Canada	78	80-01205
LARIMART SpA	Italy	236	75-13500
Litton Industries Inc.	USA	590	75-55400
Lockheed Martin Aircraft and	USA	638	80-18530
Logistics Pty Ltd	Australia	25	80-12700
M. Paz Logistics Ltd	Israel	224	80-01600
M. Rosenblatt & Son Inc.	USA	641	80-26450
Management Consulting & Research	USA	638	80-19680
Materialinformationszentrum Ges.	Germany	178	80-01310
Messier-Dowty Inc	Canada	74	75-09655
MIL Systems	Canada	79	80-01360
National Systems Management Corp.	USA	639	80-21650
Newport News Shipbuilding	USA	599	75-64300
OMI Logistics Limited	UK	508	80-19450
Pacer Systems Inc.	USA	602	75-68000
Pilatus Aircraft Ltd	Switzerland	387	75-10100
PRC Inc.	USA	640	80-24800
Raytheon E-Systems	USA	608	75-73302
Resource Consultants Inc.	USA	641	80-25920
Ridgeway International Ltd.	UK	509	80-28000
Rolls-Royce (Canada) Ltd.	Canada	75	75-12120
SEMCOR Inc.	USA	642	80-27750
Sener Ingenieria y Sistemas SA	Spain	367	75-05200
Siemens Plessey Systems	UK	483	75-53770
Sonovision ITEP Technologies	France	151	80-01850
Stanley Aviation Corporation	USA	617	75-83900
TA Group Ltd	UK	513	80-50110
Technology Australasia Pty Ltd	Australia	26	80-20600
Textron Marine & Land Systems	USA	622	75-88660
The Graphics Co. of America	USA	579	75-40360

Thomson-CSF Systems Canada Inc.	Canada	*79*	80-02020
Tracor Inc	USA	*623*	75-90000
United Defense L.P..	USA	*624*	75-92050
Vickers Shipbuilding & Eng Ltd	UK	*491*	75-63100
Vitro Corp.	USA	*644*	80-34400
Vitrociset SpA	Italy	*240*	80-40000
Vosper International Ltd	UK	*514*	80-67000
Vosper Thornycroft	UK	*492*	75-63800
Wheeler Systems Inc.	USA	*627*	75-96400
WS Atkins plc.	UK	*497*	80-01550

927 Project management services

A Goninan & Co. Ltd	Australia	*16*	75-07760
Aeromaritime Systembau GmbH	Germany	*164*	75-00290
Aerospace Design Hugo Marom Ltd	Israel	*223*	80-00040
AeroSpace Technologies of Australia	Australia	*11*	75-00340
Air-Log Ltd	UK	*429*	75-01800
Airport Engineering Hugo Marom Ltd	Israel	*223*	80-00100
Amer Ind. Inc.	USA	*631*	80-02390
AMI International	USA	*631*	80-02470
Analysis & Technology Inc.	USA	*631*	80-02600
Andersen Consulting	Australia	*24*	80-00910
Anite Systems Ltd	UK	*497*	80-01240
Architektur Städtebau Infrastruktur	Germany	*177*	80-00210
Armalon Ltd.	UK	*431*	75-03729
Artemis Management Systems GmbH	Germany	*164*	75-00230
Artemis Management Systems Ltd	UK	*497*	80-01259
AV Seawork Limited	UK	*510*	80-33300
Aviation Support Tech. Int'l Inc..	USA	*556*	75-11050
Babtie Group Ltd	UK	*498*	80-03050
BAeSEMA Ltd.	UK	*498*	80-03070
Bradley Electronics Ltd.	UK	*435*	75-07860
British Aerospace Australia Ltd	Australia	*13*	75-02750
British Aerospace (Consultancy	UK	*499*	80-03990
Calspan SRL Technology.	USA	*540*	60-03000
Carmichael & Sweet Consultancy Ltd	UK	*499*	80-05500
CIC International Ltd	USA	*563*	75-18570
CMPS&F Pty Limited	Australia	*24*	80-03600
CMS Inc.	USA	*564*	75-19170
Cogent	UK	*500*	80-06250
Computer Power Group Ltd	Australia	*24*	80-03730
Computerised Training Systems Ltd	UK	*442*	75-14650
Contahal Ltd.	Israel	*224*	80-00300
Contraves GmbH	Germany	*166*	75-05200
Crew Services Ltd	UK	*500*	80-07370
Danzas Corporation.	USA	*633*	80-06900
Davies Industrial Communications Ltd	UK	*444*	75-17010
DCS Corp.	USA	*634*	80-07400
Delex Systems Inc.	USA	*634*	80-08020
Deloitte & Touche Consulting Group	UK	*501*	80-08300
DEMCO Consultants Ltd	Israel	*224*	80-00430
Design Bureau of	Russia	*336*	75-03150
Dytecna Ltd	UK	*501*	80-08900
EASAMS Ltd	UK	*501*	80-09000
EDS Defence Ltd	UK	*501*	80-09200
EG&G Washington Analytical Svcs	USA	*634*	80-09450
Elequip Projects Ltd.	UK	*448*	75-20450
Elintech GmbH	Germany	*178*	80-00550
Engineering & Professional Serv.Inc.	USA	*634*	80-10150
Fluor Daniel Inc.	USA	*635*	80-11720
Frazer-Nash Consultancy Limited	UK	*503*	80-10950
GEC-Marconi Avionics Ltd	UK	*454*	75-25320
GED (Southampton) Ltd	UK	*503*	80-11650
George G. Sharp Inc.	USA	*642*	80-28000
Global Communications Co.	USA	*579*	75-39775
Global Defense Products Inc.	USA	*579*	75-39770
Gutteridge Haskins & Davey Pty Ltd	Australia	*25*	80-07500
Horizons Technology Inc.	USA	*636*	80-14450
Hunting Engineering Ltd	UK	*459*	75-30700
Hyder Consulting Limited	UK	*504*	80-12900
ICL DESC	UK	*501*	80-08400
Indal Technologies Inc	Canada	*72*	75-07740
Industrieanlagen-Betriebs GmbH	Germany	*178*	80-00945
Ingenieria de Sistemas	Spain	*367*	80-00900
InterCon Consultants	Canada	*78*	80-00980
Intereurope Technology Ltd	UK	*504*	80-13340
Israel Aircraft Industries Ltd	Israel	*219*	75-10200
Israel Military Industries Ltd	Israel	*220*	75-10300
Jasmin Simtec Ltd	UK	*462*	75-32370
John Laing International Ltd	UK	*505*	80-13870
Keystone Helicopter Corp.	USA	*637*	80-17780
Kinhill Engineers Pty Ltd	Australia	*25*	80-11000
Korea Express USA Inc.	USA	*638*	80-17900
Krauss-Maffei Wehrtechnik GmbH	Germany	*171*	75-16345
Landair International Ltd	UK	*505*	80-14200
Lansdowne Technologies Inc	Canada	*78*	80-01205
Logica Pty Ltd	Australia	*25*	80-12500
Logica UK Limited.	UK	*506*	80-14350
Logicon Geodynamics Inc.	USA	*638*	80-18625
M. Paz Logistics Ltd	Israel	*224*	80-01600
M. Rosenblatt & Son Inc.	USA	*641*	80-26450
Management Consulting & Research	USA	*638*	80-19680
Marshall of Cambridge Aerospace Ltd	UK	*466*	75-37530
Martin Marietta Canada Ltd	Canada	*79*	80-01320
Messier-Dowty Inc	Canada	*74*	75-09655
MFD International Ltd.	UK	*506*	80-16450
MIL Systems	Canada	*79*	80-01360
MSI-Defence Systems Ltd	UK	*470*	75-41280
National Systems Management Corp.	USA	*639*	80-21650
NIS Ltd.	UK	*471*	75-42555
NNC Ltd	UK	*507*	80-18500
Nogay Ltd	Israel	*224*	80-01470
Northrop Grumman Corporation.	USA	*599*	75-65450
OMI Logistics Limited	UK	*508*	80-19450
Pacer Systems Inc.	USA	*602*	75-68000
Parity Solutions Ltd.	UK	*508*	80-20300
Parsons Power Group Inc.	USA	*640*	80-23700
Perry Technologies	USA	*603*	75-69360
PKM Consultants (S) Pte Ltd.	Singapore	*349*	80-01640
PRC Inc.	USA	*640*	80-24800
Prof Sten Luthander Ingbyra AB	Sweden	*379*	80-01280
Ricardo Hitec Ltd	UK	*509*	80-27840
Rolls-Royce & Associates Ltd	UK	*510*	80-29100
Rubery Owen Group Services Ltd	UK	*510*	80-29350
SAIT Systems SA	Belgium	*47*	75-05600
Science Systems (Industrial) Ltd	UK	*510*	80-31600
Scientific Management Associates	Australia	*25*	80-19100
Sector Ltd	UK	*511*	80-33400
SEMCOR Inc.	USA	*642*	80-27750
Serco Aviation Services	UK	*511*	80-34150
Siemens Plessey Systems	UK	*483*	75-53770
Siemens Switzerland Ltd.	Switzerland	*387*	75-11980
Singapore Technologies Automotive	Singapore	*348*	75-04280
SISDEF Ltda	Chile	*85*	80-01940
Skilled Engineering Ltd	Australia	*25*	80-19350
Smith System Engineering Ltd.	UK	*512*	80-39500
Sonda Aviation Enterprises Limited.	UK	*512*	80-40000
Stanley Aviation Corporation	USA	*617*	75-83900
Strachan & Henshaw Ltd.	UK	*512*	80-47400
STV Group Inc.	USA	*643*	80-30400
Supercraft Ltd.	UK	*487*	75-57910
Sverdrup Corp.	USA	*643*	80-30650
Systems Engineering & Assessment Ltd	UK	*513*	80-49600
TA Group Ltd	UK	*513*	80-50110
Talos Engineering SA	Greece	*193*	80-02010
Tarmac Professional Services	UK	*513*	80-50200
Taylor Woodrow Construction Ltd	UK	*513*	80-50300
Taylor Woodrow Management	UK	*513*	80-50400
Techmatics Inc.	USA	*644*	80-31850
Technology Australasia Pty Ltd	Australia	*26*	80-20600
Technology Project Services	UK	*513*	80-52800
The Graphics Co. of America	USA	*579*	75-40360
The Titan Corp	USA	*622*	75-89620
Thomson-CSF Systems Canada Inc.	Canada	*79*	80-02020
Thorn DNT Airfield Lighting Pty Ltd	Australia	*22*	75-20540
Three Quays Marine Services Ltd	UK	*514*	80-55400
Tracor Inc	USA	*623*	75-90000
Triad Group plc	UK	*514*	80-58600
TriLink Technologies Group Inc.	Canada	*80*	80-02100
Vanguard Research Inc.	USA	*644*	80-34100
Vitro Corp.	USA	*644*	80-34400
Vitrociset SpA	Italy	*240*	80-40000
Vosper International Ltd	UK	*514*	80-67000
Welcom Software Technology	UK	*514*	80-72500
Wheeler Systems Inc.	USA	*627*	75-96400
WS Atkins plc.	UK	*497*	80-01550

928 Procurement services

Advanced Appl. Consultants Inc.	USA	*630*	80-00250
Aero International Associates	USA	*630*	80-00800
Allied Defense Industries Inc.	USA	*630*	80-02000
Allied Ordnance of Singapore Pte Ltd.	Singapore	*347*	75-00500
Alvis Logistics Ltd	UK	*430*	75-02860
Anite Systems Ltd	UK	*497*	80-01240
Armalon Ltd.	UK	*431*	75-03729

Aspect Computing Pty Ltd	Australia	*24*	80-01150
Associated Aircraft Mfg & Sales Inc.	USA	*555*	75-09500
Atlantic Commercial (UK) Ltd	UK	*498*	80-01600
Australian Technical Publications	Australia	*24*	80-01880
BACTEC International Ltd	**UK**	***498***	**80-03060**
BAeSEMA Ltd.	UK	*498*	80-03070
Beltramo & Associates	USA	*632*	80-04360
Bremer Vulkan Marineschiffbau GmbH	Germany	*166*	75-03600
British Aerospace Australia Ltd	Australia	*13*	75-02750
CORIS	France	*134*	75-12450
Crew Services Ltd	UK	*500*	80-07370
Data Sciences Ltd.	UK	*500*	80-07740
Dytecna Ltd	UK	*501*	80-08900
EDS Defence Ltd	UK	*501*	80-09200
EG&G Washington Analytical Svcs	USA	*634*	80-09450
Elintech GmbH	Germany	*178*	80-00550
Fluor Daniel Inc.	USA	*635*	80-11720
George G. Sharp Inc.	USA	*642*	80-28000
Global Communications Co.	USA	*579*	75-39775
Global Defense Products Inc.	USA	*579*	75-39770
Hugo Marom Aviation Consultants Ltd	Israel	*224*	80-00880
IMI Trading USA Inc.	USA	*637*	80-15080
InterCon Consultants	Canada	*78*	80-00980
International Procurement Services	UK	*461*	75-31820
John M. Davis & Associates	USA	*634*	80-07200
Keystone Helicopter Corp.	USA	*637*	80-17780
Landair International Ltd	UK	*505*	80-14200
Law Enforcement International Ltd	UK	*464*	75-34870
Logistics Services International	USA	*638*	80-18800
M International Export	USA	*639*	80-19780
M. Paz Logistics Ltd	Israel	*224*	80-01600
M. Rosenblatt & Son Inc.	USA	*641*	80-26450
Merex Inc.	USA	*595*	75-60250
MIL Systems	Canada	*79*	80-01360
MTH Ltd	UK	*470*	75-41330
NanoMaterials, Inc.	USA	*598*	75-63590
Napco International Inc.	USA	*598*	75-63600
National Systems Management Corp.	USA	*639*	80-21650
NNC Ltd	UK	*507*	80-18500
Nottingham International	USA	*640*	80-22000
Orkal Industries Corp.	USA	*602*	75-67505
Parity Solutions Ltd	UK	*508*	80-20300
Parsons Power Group Inc.	USA	*640*	80-23700
PK Electronic International Ltd	UK	*475*	75-45720
Plastico Ltd	UK	*475*	75-45950
Price Western Leather Co. Ltd.	UK	*476*	75-46755
Radio Holland Electronics BV	Netherlands	*297*	80-18300
Raytheon E-Systems	USA	*608*	75-73302
Reade Advanced Materials	USA	*608*	75-73860
Saint John Shipbuilding Ltd	Canada	*75*	75-12300
SAIT Systems SA	Belgium	*47*	75-05600
Sas R&D Services Inc.	USA	*642*	80-26870
Sidem (UK) Ltd	UK	*511*	80-35850
Smith System Engineering Ltd.	UK	*512*	80-39500
Space Electronics Inc.	USA	*642*	80-29050
Stanley Aviation Corporation	USA	*617*	75-83900
Sverdrup Corp.	USA	*643*	80-30650
Sverdrup Technology Inc.	USA	*643*	80-30680
Taylor Woodrow Management	UK	*513*	80-50400
Thomson-CSF Systems Canada Inc.	Canada	*79*	80-02020
Three Quays Marine Services Ltd	UK	*514*	80-55400
United Defense L.P.	USA	*624*	75-92050
Viktor Lenac Ship Repair	Croatia	*95*	75-08000
Western Tactical Systems	UK	*515*	80-73000
Woldtare Export Limited	UK	*515*	80-76200
Yewlands Engineering Co. Ltd.	UK	*495*	75-67520

930 Data handling services

Anite Systems Ltd	UK	*497*	80-01240
Aspect Computing Pty Ltd	Australia	*24*	80-01150
ATS Aerospace Inc.	Canada	*68*	75-00752
BAeSEMA Ltd.	UK	*498*	80-03070
Calspan SRL Technology	USA	*540*	60-03000
CMPS&F Pty Limited	Australia	*24*	80-03600
Data Sciences Ltd.	UK	*500*	80-07740
EG&G Inc.	USA	*571*	75-29555
EG&G Washington Analytical Svcs	USA	*634*	80-09450
Elintech GmbH	Germany	*178*	80-00550
Engineering & Professional Serv.Inc.	USA	*634*	80-10150
Fluor Daniel Inc.	USA	*635*	80-11720
George G. Sharp Inc.	USA	*642*	80-28000
Inovex Technology Ltd	UK	*504*	80-13310
Israel Aircraft Industries Ltd	Israel	*219*	75-10200
ISVR Consultancy Services	UK	*505*	80-13520
Jeppesen UK Ltd	UK	*505*	80-13850
Litton Industries Inc.	USA	*590*	75-55400
Lockheed Martin Corp.	USA	*591*	75-55635
Logica SA/NV.	Belgium	*48*	80-01000
Materialinformationszentrum Ges.	Germany	*178*	80-01310
MTH Ltd	UK	*470*	75-41330
National Systems Management Corp.	USA	*639*	80-21650
OMI Logistics Limited	UK	*508*	80-19450
PRC Inc.	USA	*640*	80-24800
Rollab Aktiebolag	Sweden	*380*	80-01850
SEMCOR Inc.	USA	*642*	80-27750
Sonovision ITEP Technologies	France	*151*	80-01850
SOPEMEA	France	*151*	80-01997
Stanley Aviation Corporation	USA	*617*	75-83900
Storage Concepts (Europe) Ltd	UK	*486*	75-57347
Sverdrup Corp.	USA	*643*	80-30650
Sverdrup Technology Inc.	USA	*643*	80-30680
TA Group Ltd	UK	*513*	80-50110
Technical Indexes Ltd	UK	*513*	80-52400
Thomson-CSF Systems Canada Inc.	Canada	*79*	80-02020
Wheeler Systems Inc.	USA	*627*	75-96400
Yarrow Shipbuilders Ltd	UK	*495*	75-67500
3C Systems, Inc.	USA	*632*	80-05000

931 Data acquisition services

Advanced Appl. Consultants Inc.	USA	*630*	80-00250
Aspect Computing Pty Ltd.	Australia	*24*	80-01150
ATS Aerospace Inc.	Canada	*68*	75-00752
BAeSEMA Ltd.	UK	*498*	80-03070
Calspan SRL Technology	USA	*540*	60-03000
CMPS&F Pty Limited	Australia	*24*	80-03600
EG&G Inc.	USA	*571*	75-29555
EG&G Washington Analytical Svcs	USA	*634*	80-09450
George G. Sharp Inc.	USA	*642*	80-28000
GlomalMart	USA	*636*	80-13000
Israel Aircraft Industries Ltd	Israel	*219*	75-10200
ISVR Consultancy Services	UK	*505*	80-13520
Litton Industries Inc.	USA	*590*	75-55400
LMS International	Belgium	*48*	80-00900
Lockheed Martin Services Group	USA	*638*	80-18540
MTH Ltd	UK	*470*	75-41330
National Systems Management Corp.	USA	*639*	80-21650
OMI Logistics Limited	UK	*508*	80-19450
Prosig Computer Consultants Ltd	UK	*476*	75-46865
SEMCOR Inc.	USA	*642*	80-27750
Sonovision ITEP Technologies	France	*151*	80-01850
SOPEMEA	France	*151*	80-01997
Stanley Aviation Corporation	USA	*617*	75-83900
Sverdrup Corp.	USA	*643*	80-30650
Sverdrup Technology Inc.	USA	*643*	80-30680
TA Group Ltd	UK	*513*	80-50110
Technical Indexes Ltd	UK	*513*	80-52400
Tracor Inc	USA	*623*	75-90000
Vanguard Research Inc.	USA	*644*	80-34100

932 Data processing services

Analysis & Technology Inc.	USA	*631*	80-02600
Aspect Computing Pty Ltd.	Australia	*24*	80-01150
ATS Aerospace Inc.	Canada	*68*	75-00752
BAeSEMA Ltd.	UK	*498*	80-03070
Calspan SRL Technology	USA	*540*	60-03000
CMPS&F Pty Limited	Australia	*24*	80-03600
Cray Research Inc.	USA	*566*	75-22570
Dynamic Graphics Ltd	UK	*501*	80-08850
Dytecna Ltd	UK	*501*	80-08900
EASAMS Ltd	UK	*501*	80-09000
EG&G Inc.	USA	*571*	75-29555
EG&G Washington Analytical Svcs	USA	*634*	80-09450
GEC-Marconi Communications Ltd	UK	*454*	75-25335
GED (Southampton) Ltd	UK	*503*	80-11650
George G. Sharp Inc.	USA	*642*	80-28000
GlomalMart	USA	*636*	80-13000
ICL DESC	UK	*501*	80-08400
Israel Aircraft Industries Ltd	Israel	*219*	75-10200
ISVR Consultancy Services	UK	*505*	80-13520
Litton Industries Inc.	USA	*590*	75-55400
Lockheed Martin Services Group	USA	*638*	80-18540
Logica UK Limited.	UK	*506*	80-14350
Logicon Geodynamics Inc.	USA	*638*	80-18625
Materialinformationszentrum Ges.	Germany	*178*	80-01310

Matra Cap Systèmes	France	*150*	80-01405
MTH Ltd	UK	*470*	75-41330
National Systems Management Corp.	USA	*639*	80-21650
OMI Logistics Limited	UK	*508*	80-19450
PRC Inc.	USA	*640*	80-24800
Prosig Computer Consultants Ltd	UK	*476*	75-46865
SEMCOR Inc.	USA	*642*	80-27750
SISDEF Ltda	Chile	*85*	80-01940
Sonovision ITEP Technologies	France	*151*	80-01850
SOPEMEA	France	*151*	80-01997
Stanley Aviation Corporation	USA	*617*	75-83900
Sverdrup Corp.	USA	*643*	80-30650
Sverdrup Technology Inc.	USA	*643*	80-30680
Synectics Corp.	USA	*643*	80-30750
TA Group Ltd	UK	*513*	80-50110
Tracor Inc	USA	*623*	75-90000
Vanguard Research Inc.	USA	*644*	80-34100

933 Data storage services

Aspect Computing Pty Ltd.	Australia	*24*	80-01150
Calspan SRL Technology	USA	*540*	60-03000
CMPS&F Pty Limited	Australia	*24*	80-03600
EG&G Inc.	USA	*571*	75-29555
EG&G Washington Analytical Svcs	USA	*634*	80-09450
GED (Southampton) Ltd	UK	*503*	80-11650
GlomalMart	USA	*636*	80-13000
ICL DESC	UK	*501*	80-08400
Israel Aircraft Industries Ltd	Israel	*219*	75-10200
ISVR Consultancy Services	UK	*505*	80-13520
Litton Industries Inc.	USA	*590*	75-55400
Materialinformationszentrum Ges.	Germany	*178*	80-01310
MTH Ltd	UK	*470*	75-41330
National Systems Management Corp.	USA	*639*	80-21650
OMI Logistics Limited	UK	*508*	80-19450
PRC Inc.	USA	*640*	80-24800
Saint John Shipbuilding Ltd	Canada	*75*	75-12300
SEMCOR Inc.	USA	*642*	80-27750
Sonovision ITEP Technologies	France	*151*	80-01850
Stanley Aviation Corporation	USA	*617*	75-83900
Storage Concepts (Europe) Ltd	UK	*486*	75-57347
TA Group Ltd	UK	*513*	80-50110
Technical Indexes Ltd	UK	*513*	80-52400

940 Engineering services

A Goninan & Co. Ltd	Australia	*16*	75-07760
Abacus Technology Corp.	USA	*629*	80-00050
AEA Technology plc	UK	*496*	80-00480
Allied Defense Industries Inc.	USA	*630*	80-02000
Allott & Lomax	UK	*497*	80-01050
Aman-Management Consultants Ltd	Israel	*223*	80-00150
Amer Ind. Inc.	USA	*631*	80-02390
American Management Systems	USA	*630*	80-02350
Amper Programas SA	Spain	*365*	75-00480
Andersen Consulting	Australia	*24*	80-00910
Arca Systems	USA	*631*	80-03100
Armalon Ltd	UK	*431*	75-03729
Atlantic Research Corp.	USA	*555*	75-10150
ATS Aerospace Inc.	Canada	*68*	75-00752
Australian Defence Industries Ltd	Australia	*9*	55-01000
Australian Submarine Corporation	Australia	*12*	75-01910
Bachmann SA	France	*132*	75-04900
Balfour Beatty Ltd.	UK	*498*	80-03073
Barnes & Reinecke Inc.	USA	*631*	80-04070
Benthos Inc.	USA	*558*	75-13540
Birkbys Plastics Ltd.	UK	*434*	75-06720
Boeing Australia Limited	Australia	*12*	75-02500
Bofors AB	Sweden	*375*	74-00020
BP Chemicals Ltd	UK	*435*	75-07790
Bremer Vulkan Marineschiffbau GmbH	Germany	*166*	75-03600
Bristol Aerospace Ltd	Canada	*69*	75-01500
British Aerospace Systems & Services	UK	*499*	80-04000
C. Raymond Hunt Associates Inc.	USA	*637*	80-14900
C-CUBED Corp.	USA	*633*	80-05260
Calspan SRL Technology	USA	*540*	60-03000
Cambridge Consultants Ltd	UK	*499*	80-05300
Carville Ltd.	UK	*439*	75-12000
Chamberlain Manufacturing Corp.	USA	*563*	75-17950
CMPS&F Pty Limited	Australia	*24*	80-03600
CORIS	France	*134*	75-12450
Cosens Engineering Ltd	UK	*443*	75-15140
Cranfield Aerospace Ltd	UK	*500*	80-07200
Daewoo Heavy Industries Ltd	Korea (ROK)	*265*	75-00600
David Brown Defence Equipment Ltd	UK	*437*	75-10150
Davis (W.R.) Engineering Ltd	Canada	*70*	75-04050
Dayton T. Brown Inc.	USA	*632*	80-04920
DCS Corp.	USA	*634*	80-07400
De Weger Architects and Counsulting	Netherlands	*297*	80-23300
Denro Inc.	USA	*568*	75-25651
Design Bureau of	Russia	*336*	75-03150
DJH Engineering Center Inc.	USA	*634*	80-08480
Dornier GmbH	Germany	*167*	75-06700
DRS Technologies, Inc.	USA	*570*	75-27400
DUAL Incorporated	USA	*634*	80-08670
EASAMS Ltd	UK	*501*	80-09000
EG&G Inc.	USA	*571*	75-29555
EG&G Washington Analytical Svcs	USA	*634*	80-09450
Electronic Development Assoc. Inc.	USA	*634*	80-09750
Elintech GmbH	Germany	*178*	80-00550
Eurocopter Canada Ltd	Canada	*71*	75-05050
Fabbrica Italiana Apparecchiature	Italy	*234*	75-08550
FFA Flugzeugwerke Altenrhein AG	Switzerland	*386*	75-04350
Flight International Inc.	USA	*635*	80-11600
Flight Refuelling Limited	UK	*451*	75-23850
Fluid Gravity Engineering Ltd.	UK	*502*	80-10660
Fluor Daniel Inc.	USA	*635*	80-11720
Forges de Zeebrugge SA	Belgium	*46*	75-03750
Frazer-Nash Consultancy Limited	UK	*503*	80-10950
Gates Hydraulics Ltd	UK	*453*	75-25210
GEC ALSTHOM Engineering Systems Ltd	UK	*453*	75-25225
GEC-Marconi Avionics Ltd	UK	*454*	75-25320
GED (Southampton) Ltd	UK	*503*	80-11650
General Offshore Corporation	USA	*636*	80-12550
General Offshore (UK) Ltd	UK	*503*	80-11670
GENME SA Marine & Industrial	Greece	*192*	80-00700
George G. Sharp Inc.	USA	*642*	80-28000
Graseby Specac Ltd	UK	*455*	75-26640
Gutteridge Haskins & Davey Pty Ltd	Australia	*25*	80-07500
H. Koch & Sons Co.	USA	*589*	75-53030
Hellenic Arms Industry	Greece	*187*	55-00800
H.I. Fraser Pty Ltd.	Australia	*16*	75-06835
Howaldtswerke Deutsche Werft AG	Germany	*170*	75-13400
HPC Engineering plc	UK	*459*	75-30350
Indal Technologies Inc	Canada	*72*	75-07740
Innotech Aviation Ltd.	Canada	*78*	80-00960
Intereurope Technology Ltd	UK	*504*	80-13340
Israel Aircraft Industries Ltd	Israel	*219*	75-10200
Israel Shipyards Ltd.	Israel	*220*	75-10600
ISVR Consultancy Services	UK	*505*	80-13520
ITT Avionics	USA	*586*	75-49455
J & E Hofmann Engineering Pty Ltd.	Australia	*17*	75-08630
James Hardie Engineering	New Zealand	*303*	80-00900
John J. McMullen Associates Inc.	USA	*639*	80-20800
John Laing International Ltd	UK	*505*	80-13870
K S I Agencies Ltd	New Zealand	*303*	80-01150
Kaman Sciences Corp.	USA	*637*	80-17600
KJM Fabritek Inc.	USA	*589*	75-52877
KRZ Flotsky Arsenal	Bulgaria	*62*	55-02500
Kvaerner ASA	Norway	*309*	75-01090
Leo A. Daly	USA	*633*	80-06893
LIW	S. Africa	*358*	75-09500
LMS International	Belgium	*48*	80-00900
Lockheed Martin Corp.	USA	*591*	75-55635
Lockheed Martin Space Mission	USA	*638*	80-18500
Lynx Corp Pty Ltd	Australia	*25*	80-12980
M. Paz Logistics Ltd	Israel	*224*	80-01600
M. Rosenblatt & Son Inc.	USA	*641*	80-26450
Management Consulting & Research	USA	*638*	80-19680
Marshall of Cambridge Aerospace Ltd	UK	*466*	75-37530
Martin Marietta Canada Ltd	Canada	*79*	80-01320
Messier-Dowty Electronics	Canada	*74*	75-09650
Metal Improvement Company Inc.	USA	*639*	80-20950
MFD International Ltd.	UK	*506*	80-16450
Microelectronics, guidance and	Turkey	*402*	55-00050
MIL Systems	Canada	*79*	80-01360
MTH Ltd	UK	*470*	75-41330
MTL Systems Inc.	USA	*597*	75-63255
Mustang Survival (USA) Corp	USA	*597*	75-63480
NanoMaterials, Inc.	USA	*598*	75-63590
Nanoquest Defence Products Ltd	UK	*470*	75-41890
National Systems Management Corp.	USA	*639*	80-21650
Nea-Lindberg A/S	Denmark	*103*	75-14300
Newport News Shipbuilding	USA	*599*	75-64300
NNC Ltd	UK	*507*	80-18500
Nogay Ltd	Israel	*224*	80-01470
Northrop Grumman Corporation	USA	*599*	75-65450

Ogle Design Ltd	UK	508	80-19400
Omori Seikoki Co. Ltd	Japan	254	75-07310
OTO Breda	Italy	237	75-18100
Pacer Systems Inc.	USA	602	75-68000
Panhard & Levassor SA	France	140	75-28750
Parsons Engineering Science	USA	640	80-23750
Parsons Power Group Inc.	USA	640	80-23700
Penang Shipbuilding Corporation	Malaysia	277	75-00800
Per Udsen Co. Aircraft Industry A/S	Denmark	104	75-21300
Perry Technologies	USA	603	75-69360
PK Electronic International Ltd	UK	475	75-45720
Planning & Research Ltd	Israel	224	80-01650
Raytheon E-Systems	USA	608	75-73302
Reade Advanced Materials	USA	608	75-73860
Ricardo Aerospace Ltd	UK	509	80-27820
Ricardo Hitec Ltd	UK	509	80-27840
Rollab Aktiebolag	Sweden	380	80-01850
Rolls-Royce (Canada) Ltd	Canada	75	75-12120
Santa Barbara SA	Spain	364	55-01000
Sener Ingenieria y Sistemas SA	Spain	367	75-05200
Serco Aviation Services	UK	511	80-34150
S.G.D. Engineering Ltd	Israel	224	80-01930
Siecor Corp.	USA	614	75-80560
Siemens Plessey Systems	UK	483	75-53770
SIMA PERU	Peru	319	55-01950
Singapore Electronic & Engineering	Singapore	349	80-01960
Singapore Ordnance Engineering	Singapore	349	80-01970
Singapore Technologies Aerospace Ltd	Singapore	348	75-03905
SISDEF Ltda	Chile	85	80-01940
Skilled Engineering Ltd	Australia	25	80-19350
SNPE	France	145	75-36315
Southwest Research Institute	USA	541	60-19200
Spar Aerospace Ltd	Canada	76	75-13000
Spectrum Associates Inc.	USA	642	80-29350
SRL Operations	USA	617	75-83550
Stanley Aviation Corporation	USA	617	75-83900
STN ATLAS Elektronik GmbH	Germany	165	75-01150
Strachan & Henshaw Ltd	UK	512	80-47400
STV Group Inc.	USA	643	80-30400
Survival Products Inc.	USA	618	75-85200
Sverdrup Corp.	USA	643	80-30650
Sverdrup Technology Inc.	USA	643	80-30680
TA Group Ltd	UK	513	80-50110
Taylor Woodrow Management	UK	513	80-50400
Techmatics Inc.	USA	644	80-31850
Techspace Aero SA	Belgium	47	75-06200
Terberg Benschop BV	Netherlands	296	75-07600
The Aerospace Corp.	USA	630	80-01050
The Titan Corp	USA	622	75-89620
Thomson Sintra	France	148	75-45135
Thomson-CSF Systems Canada Inc.	Canada	79	80-02020
Three Quays Marine Services Ltd	UK	514	80-55400
Tracor Inc	USA	623	75-90000
Transfield Power Services	Australia	26	80-20700
TriLink Technologies Group Inc.	Canada	80	80-02100
Tufnol Ltd	UK	489	75-61410
Turbo Lufttechnik GmbH	Germany	176	75-31900
Vanguard Research Inc.	USA	644	80-34100
Varna Shipyard	Bulgaria	62	75-05000
Vazduhoplovno Remonti Zavod	Yugoslavia	661	80-02500
Verolme Elektra BV	Netherlands	296	75-08100
Vickers Shipbuilding & Eng Ltd	UK	491	75-63100
Vignal Artru Industries	France	148	75-46950
Vihtavuori Oy	Finland	119	75-02850
Vitro Corp.	USA	644	80-34400
Vitrociset SpA	Italy	240	80-40000
Vosper International Ltd	UK	514	80-67000
Wegmann & Co GmbH	Germany	177	75-33700
Yantar Shipyard	Russia	339	75-78900
Yarrow Shipbuilders Ltd	UK	495	75-67500
Yewlands Engineering Co. Ltd.	UK	495	75-67520
Zuken-Redac Ltd	UK	495	75-67800

941 Construction services

AMI International	USA	631	80-02470
ASC Shipbuilding Pty Limited	Australia	12	75-01530
Balfour Beatty Ltd.	UK	498	80-03073
Blakley Electrics Ltd	UK	434	75-07015
Bremer Vulkan Marineschiffbau GmbH	Germany	166	75-03600
Caledonian Mining Co. Ltd	UK	438	75-11250
CMPS&F Pty Limited	Australia	24	80-03600
Costain Middle East	UAE	407	75-00300
Design Bureau of	Russia	336	75-03150
Devonport Royal Dockyard Ltd	UK	501	80-08500
Flight Refuelling Limited	UK	451	75-23850
Fluor Daniel Inc.	USA	635	80-11720
George G. Sharp Inc.	USA	642	80-28000
Gutteridge Haskins & Davey Pty Ltd	Australia	25	80-07500
Hellenic Arms Industry	Greece	187	55-00800
Howaldtswerke Deutsche Werft AG	Germany	170	75-13400
Israel Aircraft Industries Ltd	Israel	219	75-10200
John Laing International Ltd	UK	505	80-13870
Kinhill Engineers Pty Ltd	Australia	25	80-11000
KRZ Flotsky Arsenal	Bulgaria	62	55-02500
Kvaerner ASA	Norway	309	75-01090
Nea-Lindberg A/S	Denmark	103	75-14300
NNC Ltd	UK	507	80-18500
Parsons Power Group Inc.	USA	640	80-23700
Planning & Research Ltd	Israel	224	80-01650
Prof Sten Luthander Ingbyra AB	Sweden	379	80-01280
Santa Barbara SA	Spain	364	55-01000
Sener Ingenieria y Sistemas SA	Spain	367	75-05200
SIMA PERU	Peru	319	55-01950
Skilled Engineering Ltd	Australia	25	80-19350
SNPE	France	145	75-36315
Spantech Pty Ltd	Australia	21	75-19573
Sverdrup Corp.	USA	643	80-30650
Sverdrup Technology Inc.	USA	643	80-30680
Taylor Woodrow Construction Ltd	UK	513	80-50300
Taylor Woodrow Management	UK	513	80-50400
Three Quays Marine Services Ltd	UK	514	80-55400
Verolme Elektra BV	Netherlands	296	75-08100
Vihtavuori Oy	Finland	119	75-02850
Vosper International Ltd	UK	514	80-67000
Yantar Shipyard	Russia	339	75-78900
Yarrow Shipbuilders Ltd	UK	495	75-67500

941 10 Airfield construction services

Aerospace Design Hugo Marom Ltd	Israel	223	80-00040
Airport Engineering Hugo Marom Ltd	Israel	223	80-00100
Architektur Städtebau Infrastruktur	Germany	177	80-00210
Blakley Electrics Ltd	UK	434	75-07015
CMPS&F Pty Limited	Australia	24	80-03600
CORIS	France	134	75-12450
Costain Civil Engineering Ltd	UK	443	75-15213
Costain Middle East	UAE	407	75-00300
Fluor Daniel Inc.	USA	635	80-11720
Israel Aircraft Industries Ltd	Israel	219	75-10200
John Laing International Ltd	UK	505	80-13870
Kinhill Engineers Pty Ltd	Australia	25	80-11000
Metis Construction & Trade Co. Inc.	Turkey	403	80-01330
Prof Sten Luthander Ingbyra AB	Sweden	379	80-01280
Sverdrup Corp.	USA	643	80-30650
Tarmac Professional Services	UK	513	80-50200
Taylor Woodrow Construction Ltd	UK	513	80-50300
Taylor Woodrow Management	UK	513	80-50400
Thorn DNT Airfield Lighting Pty Ltd	Australia	22	75-20540
Transfield Pty Ltd	Australia	22	75-20810

941 30 Harbour construction services

Blakley Electrics Ltd	UK	434	75-07015
Costain Middle East	UAE	407	75-00300
EASAMS Ltd	UK	501	80-09000
Fluor Daniel Inc.	USA	635	80-11720
John Laing International Ltd	UK	505	80-13870
Kinhill Engineers Pty Ltd	Australia	25	80-11000
Metis Construction & Trade Co. Inc.	Turkey	403	80-01330
Sverdrup Corp.	USA	643	80-30650
Tarmac Professional Services	UK	513	80-50200
Taylor Woodrow Construction Ltd	UK	513	80-50300
Taylor Woodrow Management	UK	513	80-50400
Transfield Pty Ltd	Australia	22	75-20810

941 60 Shipbuilding services

AMI International	USA	631	80-02470
ASC Shipbuilding Pty Limited	Australia	12	75-01530
Astilleros Espanoles SA	Spain	365	75-01400
Blakley Electrics Ltd	UK	434	75-07015
Bollinger Shipyard Inc.	USA	560	75-14600
Bremer Vulkan Marineschiffbau GmbH	Germany	166	75-03600
Damen Shipyards	Netherlands	292	75-01500

Elefsis Shipyards SA	Greece	188	75-01130
Frazer-Nash Consultancy Limited	UK	503	80-10950
George G. Sharp Inc.	USA	642	80-28000
Guangzhou Shipyard International Co.	China PR	89	75-00400
Hijos de J. Barreras SA	Spain	366	75-03781
Israel Shipyards Ltd.	Israel	220	75-10600
KRZ Flotsky Arsenal	Bulgaria	62	55-02500
Litton Industries Inc.	USA	590	75-55400
M. Rosenblatt & Son Inc.	USA	641	80-26450
Meuse & Sambre	Belgium	46	75-04971
MIL Systems	Canada	79	80-01360
National Steel & Shipbuilding Co.	USA	598	75-63900
Newport News Shipbuilding	USA	599	75-64300
Saint John Shipbuilding Ltd	Canada	75	75-12300
Schottel Sudamericana I.C.S.A.	Argentina	4	80-20000
Shipyard Beograd	Yugoslavia	661	75-00690
Svendborg Vaerft A/S	Denmark	104	75-19800
Three Quays Marine Services Ltd	UK	514	80-55400
Transfield Shipbuilding WA	Australia	22	75-20812
Van der Giessen-de Noord	Netherlands	296	75-07850
Varna Shipyard	Bulgaria	62	75-05000
Vickers Shipbuilding & Eng Ltd	UK	491	75-63100
Vosper International Ltd	UK	514	80-67000
Yantar Shipyard	Russia	339	75-78900

941 70 Training range construction services

Armalon Ltd	UK	431	75-03729
ATS Aerospace Inc.	Canada	68	75-00752
Blakley Electrics Ltd	UK	434	75-07015
Calspan SRL Technology	USA	540	60-03000
Costain Middle East	UAE	407	75-00300
Fluor Daniel Inc.	USA	635	80-11720
John Laing International Ltd	UK	505	80-13870
OTEC Support Services Ltd	UK	508	80-19600
Sector Ltd	UK	511	80-33400

942 Design services

ACH	France	130	75-00150
Aeromaritime Systembau GmbH	Germany	164	75-00290
Aerospace Design Hugo Marom Ltd	Israel	223	80-00040
Aerospace Project Development Group	Switzerland	384	60-00020
AeroSpace Technologies of Australia	Australia	11	75-00340
Agamice	France	131	75-00800
Aircraft Materials Ltd	UK	428	75-01650
Allied Research Corp.	USA	549	75-04670
Aman-Management Consultants Ltd	Israel	223	80-00150
Amper Programas SA	Spain	365	75-00480
Anco SA	Greece	187	75-00200
Applied Data Sciences Inc.	USA	554	75-08050
Armalon Ltd	UK	431	75-03729
Array Connector Corp.	USA	555	75-09020
Atlantic Research Corp.	USA	555	75-10150
Australian Defence Industries Ltd	Australia	9	55-01000
Babtie Group Ltd	UK	498	80-03050
BAeSEMA Ltd.	UK	498	80-03070
Becker Flugfunkwerk GmbH	Germany	165	75-02200
Birkbys Plastics Ltd	UK	434	75-06720
Bofors AB	Sweden	375	74-00020
BP Chemicals Ltd	UK	435	75-07790
Bremer Vulkan Marineschiffbau GmbH	Germany	166	75-03600
Bristol Aerospace Ltd	Canada	69	75-01500
Bristow Helicopters Ltd	UK	499	80-03950
British Aerospace Australia Ltd	Australia	13	75-02750
CAE Aviation Ltd	Canada	78	80-00245
Calspan SRL Technology	USA	540	60-03000
Cambridge Consultants Ltd	UK	499	80-05300
Capewell Aerial Delivery Systems	USA	576	75-36300
CEA Technologies Pty Ltd	Australia	24	80-03100
Commercial Hydraulics Kontak Ltd	UK	441	75-14415
Communications & Power Industries	USA	565	75-20393
Cortana Corp.	USA	540	60-03500
Costain Civil Engineering Ltd	UK	443	75-15213
Cranfield Aerospace Ltd	UK	500	80-07200
Cranfield University	UK	419	60-00350
CRL Technologies	Canada	70	75-03760
David Brown Gear Industries Ltd	Australia	13	75-02800
De Weger Architects and Counsulting	Netherlands	297	80-23300
Design Bureau of	Russia	336	75-03150
Devonport Royal Dockyard Ltd	UK	501	80-08500
Didsbury Engineering Co. Ltd	UK	445	75-18050
DJH Engineering Center Inc.	USA	634	80-08480
DRS Technologies, Inc.	USA	570	75-27400
EASAMS Ltd	UK	501	80-09000
EG&G Washington Analytical Svcs	USA	634	80-09450
Electrothermal Engineering Ltd	UK	448	75-20400
Elintech GmbH	Germany	178	80-00550
Engineering & Professional Serv.Inc.	USA	634	80-10150
ESDU International plc	UK	502	80-09700
Eurocopter Canada Ltd	Canada	71	75-05050
FFA Flugzeugwerke Altenrhein AG	Switzerland	386	75-04350
Flight Refuelling Limited	UK	451	75-23850
Fluor Daniel Inc.	USA	635	80-11720
Frazer-Nash Consultancy Limited	UK	503	80-10950
GED (Southampton) Ltd	UK	503	80-11650
General Electronics Communications	Greece	189	75-01300
George G. Sharp Inc.	USA	642	80-28000
Gordon Consulting Engineers Ltd	Israel	224	80-00750
Graseby Dynamics Ltd	UK	455	75-26590
Gutteridge Haskins & Davey Pty Ltd	Australia	25	80-07500
H. Koch & Sons Co.	USA	589	75-53030
Hale Group International	USA	636	80-13600
Helio Mirror Co. Ltd	UK	457	75-28900
Hellenic Arms Industry	Greece	187	55-00800
Hellenic Electronics Equipment	Greece	187	60-00100
Hill Industries Inc.	USA	582	75-43465
Hindustan Aeronautics Ltd.	India	204	55-01000
Hughes Elcan Optical Technologies	Canada	72	75-07650
Hughes Microelectronics Europe Ltd	UK	459	75-30480
Hyder Consulting Limited	UK	504	80-12900
ILS Oy	Finland	119	80-00900
Indal Technologies Inc	Canada	72	75-07740
Inovex Technology Ltd	UK	504	80-13310
Institute of Space Constructions	Georgia	156	60-01000
Intereurope Technology Ltd	UK	504	80-13340
Israel Aircraft Industries Ltd	Israel	219	75-10200
ISVR Consultancy Services	UK	505	80-13520
James Hardie Engineering	New Zealand	303	80-00900
John J. McMullen Associates Inc.	USA	639	80-20800
John Laing International Ltd	UK	505	80-13870
Jonathan Arthur Ciener Inc.	USA	564	75-18590
JPS Communications Inc.	USA	587	75-51220
Kaman Sciences Corp.	USA	637	80-17600
Keystone Helicopter Corp.	USA	637	80-17780
KJM Fabritek Inc.	USA	589	75-52877
Krauss-Maffei Wehrtechnik GmbH	Germany	171	75-16345
Leo A. Daly	USA	633	80-06893
Litton Industries Inc.	USA	591	75-55400
Lockheed Martin Space Mission	USA	638	80-18500
Lucas Aerospace - Australia	Australia	25	80-12850
M. Rosenblatt & Son Inc.	USA	641	80-26450
Marshall of Cambridge Aerospace Ltd	UK	466	75-37530
MECAR SA	Belgium	46	75-04930
Menasco Aerospace Ltd	Canada	74	75-09600
Messier-Dowty Electronics	Canada	74	75-09650
MFD International Ltd.	UK	506	80-16450
MIL Systems	Canada	79	80-01360
MTH Ltd	UK	470	75-41330
Mustang Survival Corp.	Canada	74	75-09860
Mustang Survival (USA) Corp	USA	597	75-63480
Nanoquest Defence Products Ltd	UK	470	75-41890
National Systems Management Corp.	USA	639	80-21650
Newport News Shipbuilding	USA	599	75-64300
Nogay Ltd	Israel	224	80-01470
NPO Molniya plc	Russia	337	75-13380
Ogle Design Ltd	UK	508	80-19400
Orlando Technology Inc.	USA	540	60-16000
Pacer Systems Inc.	USA	602	75-68000
Parsons Engineering Science	USA	640	80-23750
Parsons Power Group Inc.	USA	640	80-23700
Per Udsen Co. Aircraft Industry A/S	Denmark	104	75-21300
Pilatus Aircraft Ltd	Switzerland	387	75-10100
Pioneer Aerospace Corp	USA	604	75-70200
PK Electronic International Ltd	UK	475	75-45720
Planning & Research Ltd	Israel	224	80-01650
PRC Inc.	USA	640	80-24800
Prof Sten Luthander Ingbyra AB	Sweden	379	80-01280
QES Ltd	UK	509	80-25000
Raytheon E-Systems	USA	608	75-73302
Regent Mfg Inc.	USA	609	75-74200
Ricardo Hitec Ltd	UK	509	80-27840
Rollab Aktiebolag	Sweden	380	80-01850
Rolls-Royce (Canada) Ltd.	Canada	75	75-12120
Rotadata Ltd.	UK	481	75-50570
Rubery Owen Group Services Ltd	UK	510	80-29350
Saint John Shipbuilding Ltd	Canada	75	75-12300
Santa Barbara SA	Spain	364	55-01000

Sector Ltd	UK	511	80-33400
SEMCOR Inc.	USA	642	80-27750
Sener Ingenieria y Sistemas SA	Spain	367	75-05200
S.G.D. Engineering Ltd	Israel	224	80-01930
Shorts Support Services Division	UK	511	80-35600
Siemens Environmental Systems Ltd	UK	483	75-53700
Siemens Switzerland Ltd	Switzerland	387	75-11980
SISDEF Ltda	Chile	85	80-01940
SITEP SpA	Italy	239	75-22595
Skilled Engineering Ltd	Australia	25	80-19350
SNPE	France	145	75-36315
SONACA SA	Belgium	47	75-06100
SOPEMEA	France	151	80-01997
SP aerospace and vehicle systems bv	**Netherlands**	**295**	**75-06860**
Sparton Technology Inc.	USA	616	75-82950
Spectrum Associates Inc.	USA	642	80-29350
St Bernard Composites Ltd	UK	486	75-56222
Stanley Aviation Corporation	USA	617	75-83900
Strachan & Henshaw Ltd	UK	512	80-47400
Strategic Technologies Pte Ltd	**Singapore**	**348**	**75-04400**
STV Group Inc.	USA	643	80-30400
Supercraft Ltd	UK	487	75-57910
Sverdrup Corp.	USA	643	80-30650
Sverdrup Technology Inc.	USA	643	80-30680
T E D Engineering Australia Limited	Australia	22	75-20300
TA Group Ltd	UK	513	80-50110
Tarmac Professional Services	UK	513	80-50200
Taylor Woodrow Management	UK	513	80-50400
Teleplan A/S	Norway	311	80-30000
The Aerospace Corp.	USA	630	80-01050
Thomson-CSF Systems Canada Inc.	Canada	79	80-02020
USM, Inc.	USA	624	75-91507
Varity Perkins	UK	474	75-44750
Vickers Shipbuilding & Eng Ltd	UK	491	75-63100
Victoria Maritime Services	Monaco	283	80-01000
Vihtavuori Oy	Finland	119	75-02850
Viktor Lenac Ship Repair	Croatia	95	75-08000
Vipac Engineers & Scientists Ltd	Australia	26	80-22580
Vitro Corp.	USA	644	80-34400
Vosper International Ltd	UK	514	80-67000
Weir Westgarth Ltd	UK	514	80-72250
Wheeler Systems Inc.	USA	627	75-96400
Wornall Electronics Inc.	USA	645	80-35400
Yarrow Shipbuilders Ltd	UK	495	75-67500
Zuken-Redac Ltd	UK	495	75-67800

942 20 Design services, aircraft-related

Aerospace Project Development Group	Switzerland	384	60-00020
AeroSpace Technologies of Australia	Australia	11	75-00340
Aircraft Materials Ltd	UK	428	75-01650
Airport Engineering Hugo Marom Ltd	Israel	223	80-00100
Becker Flugfunkwerk GmbH	Germany	165	75-02200
Bristow Helicopters Ltd	UK	499	80-03950
British Aerospace Australia Ltd	Australia	13	75-02750
CAE Aviation Ltd	Canada	78	80-00245
Calspan SRL Technology	USA	540	60-03000
Cambridge Consultants Ltd	UK	499	80-05300
Capewell Aerial Delivery Systems	USA	576	75-36300
CHAM (Concentration, Heat	UK	499	80-05900
CMS Inc.	USA	564	75-19170
Cranfield Aerospace Ltd	UK	500	80-07200
Cranfield University	UK	419	60-00350
CRL Technologies	Canada	70	75-03760
Daedalus Research Inc.	USA	567	75-23360
Dassault Aviation Group	France	135	75-12975
Didsbury Engineering Co. Ltd	UK	445	75-18050
EASAMS Ltd	UK	501	80-09000
Elintech GmbH	Germany	178	80-00550
Embraer	Brazil	54	55-05000
ERA Technology Ltd	UK	502	80-09600
FLS Aerospace (Light Aircraft) Ltd	UK	451	75-23900
Frazer-Nash Consultancy Limited	UK	503	80-10950
GKN Westland Design Services	UK	503	80-11775
Hill Industries Inc.	USA	582	75-43465
Israel Aircraft Industries Ltd	Israel	219	75-10200
Keystone Helicopter Corp.	USA	637	80-17780
Litton Industries Inc.	USA	591	75-55400
LMS International	Belgium	48	80-00900
Lockridge Mapstrap Co.	USA	638	80-18612
Lucas Aerospace - Australia	Australia	25	80-12850
Marshall of Cambridge Aerospace Ltd	UK	466	75-37530
Menasco Aerospace Ltd	Canada	74	75-09600
Messier-Dowty Electronics	Canada	74	75-09650
Nanoquest Defence Products Ltd	UK	470	75-41890
National Aerospace Laboratory	Netherlands	290	60-01400
National Systems Management Corp.	USA	639	80-21650
NPO Molniya plc	Russia	337	75-13380
Nyge Aero AB	Sweden	379	80-01490
Pacer Systems Inc.	USA	602	75-68000
Pacific Aerospace Corp. Ltd	New Zealand	302	75-01600
Per Udsen Co. Aircraft Industry A/S	Denmark	104	75-21300
Pioneer Aerospace Corp	USA	604	75-70200
PRC Inc.	USA	640	80-24800
Prof Sten Luthander Ingbyra AB	Sweden	379	80-01280
Ricardo Hitec Ltd	UK	509	80-27840
Rollab Aktiebolag	Sweden	380	80-01850
Scaled Composites Inc.	USA	541	60-19160
SEMCOR Inc.	USA	642	80-27750
S.G.D. Engineering Ltd	Israel	224	80-01930
Shorts Support Services Division	UK	511	80-35600
SP aerospace and vehicle systems bv	**Netherlands**	**295**	**75-06860**
Sparton Technology Inc.	USA	616	75-82950
St Bernard Composites Ltd	UK	486	75-56222
State Research Institute	Russia	336	60-04150
STV Group Inc.	USA	643	80-30400
Supercraft Ltd	UK	487	75-57910
Sverdrup Corp.	USA	643	80-30650
Sverdrup Technology Inc.	USA	643	80-30680
The Flight Data Company Ltd	UK	502	80-10600
Thomson-CSF Systems Canada Inc.	Canada	79	80-02020
Tracor Inc	USA	623	75-90000
Transfield Pty Ltd	Australia	22	75-20810
TriLink Technologies Group Inc.	Canada	80	80-02100
Vazduhoplovno Remonti Zavod	Yugoslavia	661	80-02500
Vipac Engineers & Scientists Ltd	Australia	26	80-22580

942 30 Design services, warship-related

Analysis & Technology Inc.	USA	631	80-02600
Australian Technical Publications	Australia	24	80-01880
BAeSEMA Ltd.	UK	498	80-03070
Bremer Vulkan Marineschiffbau GmbH	Germany	166	75-03600
Brodoprojekt	Croatia	95	80-00100
Calspan SRL Technology	USA	540	60-03000
Cambridge Consultants Ltd	UK	499	80-05300
CEA Technologies Pty Ltd	Australia	24	80-03100
CHAM (Concentration, Heat	UK	499	80-05900
Crew Services Ltd	UK	500	80-07370
Daewoo Heavy Industries Ltd	Korea (ROK)	265	75-00600
Davie Industries Inc.	Canada	70	75-04040
Devonport Royal Dockyard Ltd	UK	501	80-08500
Didsbury Engineering Co. Ltd	UK	445	75-18050
Dynamic Controls Ltd	UK	447	75-19140
EASAMS Ltd	UK	501	80-09000
EG&G Washington Analytical Svcs	USA	634	80-09450
ERA Technology Ltd	UK	502	80-09600
Frazer-Nash Consultancy Limited	UK	503	80-10950
GED (Southampton) Ltd	UK	503	80-11650
George G. Sharp Inc.	USA	642	80-28000
Graseby Dynamics Ltd	UK	455	75-26590
Hans Harboe-Hansen	Denmark	105	80-00800
Indal Technologies Inc	Canada	72	75-07740
Israel Aircraft Industries Ltd	Israel	219	75-10200
Israel Shipyards Ltd	Israel	220	75-10600
John J. McMullen Associates Inc.	USA	639	80-20800
Litton Industries Inc.	USA	590	75-55400
LMS International	Belgium	48	80-00900
M. Rosenblatt & Son Inc.	USA	641	80-26450
Meggitt Avionics (UK)	UK	467	75-38340
MIL Systems	Canada	79	80-01360
Milo Holdings Ltd	UK	506	80-16850
MTG Marinetechnik GmbH	Germany	178	80-01370
Newport News Shipbuilding	USA	599	75-64300
Ricardo Hitec Ltd	UK	509	80-27840
Saint John Shipbuilding Ltd	Canada	75	75-12300
SEMCOR Inc.	USA	642	80-27750
SISDEF Ltda	Chile	85	80-01940
SITEP SpA	Italy	239	75-22595
Spectrum Associates Inc.	USA	642	80-29350
SSPA Maritime Consulting AB	Sweden	380	80-01960
Stanley Aviation Corporation	USA	617	75-83900
STN ATLAS Elektronik GmbH	Germany	165	75-01150
Strachan & Henshaw Ltd	UK	512	80-47400
STV Group Inc.	USA	643	80-30400
Supercraft Ltd	UK	487	75-57910

TA Group Ltd	UK	*513*	80-50110
Talos Engineering SA	Greece	*193*	80-02010
Thomson Sintra	France	*148*	75-45135
Thomson-CSF Systems Canada Inc.	Canada	*79*	80-02020
Three Quays Marine Services Ltd	UK	*514*	80-55400
Tracor Inc	USA	*623*	75-90000
United Defense L.P.	USA	*624*	75-92050
Van der Giessen-de Noord	Netherlands	*296*	75-07850
Vickers Shipbuilding & Eng Ltd	UK	*491*	75-63100
Vipac Engineers & Scientists Ltd	Australia	*26*	80-22580
Vitro Corp.	USA	*644*	80-34400
Vosper International Ltd	UK	*514*	80-67000

942 30 50 Naval architect services

ACH	France	*130*	75-00150
AEA Technology plc	UK	*496*	80-00480
Analysis & Technology Inc.	USA	*631*	80-02600
Bremer Vulkan Marineschiffbau GmbH	Germany	*166*	75-03600
C. Raymond Hunt Associates Inc.	USA	*637*	80-14900
CHAM (Concentration, Heat	UK	*499*	80-05900
Cortana Corp.	USA	*540*	60-03500
Damen Shipyards	Netherlands	*292*	75-01500
Designers & Planners Inc.	USA	*634*	80-08300
Devonport Royal Dockyard Ltd	UK	*501*	80-08500
Frazer-Nash Consultancy Limited	UK	*503*	80-10950
George G. Sharp Inc.	USA	*642*	80-28000
ILS Oy	Finland	*119*	80-00900
Israel Aircraft Industries Ltd	Israel	*219*	75-10200
Israel Shipyards Ltd.	Israel	*220*	75-10600
John J. McMullen Associates Inc.	USA	*639*	80-20800
Litton Industries Inc.	USA	*590*	75-55400
M. Rosenblatt & Son Inc.	USA	*641*	80-26450
Marr Technical Services Ltd	UK	*506*	80-15620
MIL Systems	Canada	*79*	80-01360
MTG Marinetechnik GmbH	Germany	*178*	80-01370
Newport News Shipbuilding	USA	*599*	75-64300
Saint John Shipbuilding Ltd	Canada	*75*	75-12300
Spectrum Associates Inc.	USA	*642*	80-29350
SSPA Maritime Consulting AB	Sweden	*380*	80-01960
Talos Engineering SA	Greece	*193*	80-02010
Three Quays Marine Services Ltd	UK	*514*	80-55400
Transfield Shipbuilding WA	Australia	*22*	75-20812
Victoria Maritime Services	Monaco	*283*	80-01000
Viktor Lenac Ship Repair	Croatia	*95*	75-08000
Vipac Engineers & Scientists Ltd	Australia	*26*	80-22580
Vosper International Ltd	UK	*514*	80-67000
Wheeler Systems Inc.	USA	*627*	75-96400

942 40 Design services, ground-based installation or equipment related

AAR Allen Group International	Netherlands	*296*	80-00050
AEA Technology plc	UK	*496*	80-00480
Aerospace Design Hugo Marom Ltd	Israel	*223*	80-00040
Airport Engineering Hugo Marom Ltd	Israel	*223*	80-00100
AlliedSignal Aerospace	USA	*551*	75-04700
Amper Programas SA	Spain	*365*	75-00480
Analysis & Technology Inc.	USA	*631*	80-02600
Applied Data Sciences Inc.	USA	*554*	75-08050
APT Radar Systems Ltd	UK	*431*	75-03560
Armalon Ltd	UK	*431*	75-03729
Array Connector Corp.	USA	*555*	75-09020
Australian Technical Publications	Australia	*24*	80-01880
Babtie Group Ltd	UK	*498*	80-03050
BAeSEMA Ltd.	UK	*498*	80-03070
Becker Flugfunkwerk GmbH	Germany	*165*	75-02200
Bremer Vulkan Marineschiffbau GmbH	Germany	*166*	75-03600
Cambridge Consultants Ltd	UK	*499*	80-05300
Continental Electronics Corp.	USA	*566*	75-21800
Costain Middle East	UAE	*407*	75-00300
De Weger Architects and Counsulting	Netherlands	*297*	80-23300
Design Bureau of	Russia	*336*	75-03150
Didsbury Engineering Co. Ltd	UK	*445*	75-18050
DRS Technologies, Inc.	USA	*570*	75-27400
Dytecna Ltd	UK	*501*	80-08900
EASAMS Ltd	UK	*501*	80-09000
Engineering & Professional Serv.Inc.	USA	*634*	80-10150
ERA Technology Ltd	UK	*502*	80-09600
Fluor Daniel Inc.	USA	*635*	80-11720
Forward Industries Ltd	UK	*452*	75-24200
Frazer-Nash Consultancy Limited	UK	*503*	80-10950
GEC-Marconi Communications Ltd	UK	*454*	75-25335
GED (Southampton) Ltd	UK	*503*	80-11650
Gordon Consulting Engineers Ltd	Israel	*224*	80-00750
Hale Group International	USA	*636*	80-13600
H.I. Fraser Pty Ltd.	Australia	*16*	75-06835
Indal Technologies Inc	Canada	*72*	75-07740
Inovex Technology Ltd	UK	*504*	80-13310
Institute of Space Constructions	Georgia	*156*	60-01000
Israel Aircraft Industries Ltd	Israel	*219*	75-10200
John Laing International Ltd	UK	*505*	80-13870
JPS Communications Inc.	USA	*587*	75-51220
Litton Industries Inc.	USA	*590*	75-55400
LMS International	Belgium	*48*	80-00900
Marine-Air Systems Ltd	New Zealand	*302*	75-01300
MFD International Ltd.	UK	*506*	80-16450
Microelectronics, guidance and	Turkey	*402*	55-00050
Milo Holdings Ltd	UK	*506*	80-16850
MTH Ltd	UK	*470*	75-41330
Mustang Survival (USA) Corp	USA	*597*	75-63480
Nanoquest Defence Products Ltd	UK	*470*	75-41890
Nogay Ltd	Israel	*224*	80-01470
Parsons Engineering Science	USA	*640*	80-23750
QES Ltd	UK	*509*	80-25000
Radio Holland Electronics BV	Netherlands	*297*	80-18300
Regent Mfg Inc.	USA	*609*	75-74200
Ricardo Hitec Ltd	UK	*509*	80-27840
Rollab Aktiebolag	Sweden	*380*	80-01850
Sector Ltd	UK	*511*	80-33400
SEMCOR Inc.	USA	*642*	80-27750
Siemens Switzerland Ltd	Switzerland	*387*	75-11980
Sira Ltd.	UK	*484*	75-54250
Spar Aviation Services	Canada	*76*	75-13020
Stanilite Electronics Pty Ltd	Australia	*21*	75-19690
Stanley Aviation Corporation	USA	*617*	75-83900
STV Group Inc.	USA	*643*	80-30400
Sverdrup Corp.	USA	*643*	80-30650
Sverdrup Technology Inc.	USA	*643*	80-30680
Syscon Engineering	Israel	*225*	80-01940
TA Group Ltd	UK	*513*	80-50110
Talos Engineering SA	Greece	*193*	80-02010
Tarmac Professional Services	UK	*513*	80-50200
Taylor Woodrow Management	UK	*513*	80-50400
Technology Australasia Pty Ltd	Australia	*26*	80-20600
Thomson-CSF Systems Canada Inc.	Canada	*79*	80-02020
Thorn DNT Airfield Lighting Pty Ltd	Australia	*22*	75-20540
Tracor Inc	USA	*623*	75-90000
United Defense L.P.	USA	*624*	75-92050
Vazduhoplovno Remonti Zavod	Yugoslavia	*661*	80-02500
Vickers Shipbuilding & Eng Ltd	UK	*491*	75-63100
Vipac Engineers & Scientists Ltd	Australia	*26*	80-22580
Vitrociset SpA	Italy	*240*	80-40000
Wornall Electronics Inc.	USA	*645*	80-35400

942 40 10 Design services, airport / airfield

Costain Civil Engineering Ltd.	UK	*443*	75-15213

942 40 70 Training range design services

Armalon Ltd	UK	*431*	75-03729
ATS Aerospace Inc.	Canada	*68*	75-00752
Bradley Electronics Ltd	UK	*435*	75-07860
Calspan SRL Technology	USA	*540*	60-03000
Costain Civil Engineering Ltd.	UK	*443*	75-15213
Costain Middle East	UAE	*407*	75-00300
DEMCO Consultants Ltd	Israel	*224*	80-00430
Engineering & Professional Serv.Inc.	USA	*634*	80-10150
Fluor Daniel Inc.	USA	*635*	80-11720
Gordon Consulting Engineers Ltd	Israel	*224*	80-00750
Lockheed Martin Services Group	USA	*638*	80-18540
OTEC Support Services Ltd	UK	*508*	80-19600
Rollab Aktiebolag	Sweden	*380*	80-01850
Sector Ltd	UK	*511*	80-33400
SEMCOR Inc.	USA	*642*	80-27750
Sverdrup Corp.	USA	*643*	80-30650
Sverdrup Technology Inc.	USA	*643*	80-30680
Thomson-CSF Systems Canada Inc.	Canada	*79*	80-02020
Tracor Inc	USA	*623*	75-90000
Vipac Engineers & Scientists Ltd	Australia	*26*	80-22580

942 40 80 Vehicle design services

Air-Log Ltd	UK	*429*	75-01800

Allied Research Corp.	USA	*549*	75-04670
Calspan SRL Technology	USA	*540*	60-03000
CHAM (Concentration, Heat	UK	*499*	80-05900
Commercial Hydraulics Kontak Ltd	UK	*441*	75-14415
Davis (W.R.) Engineering Ltd	Canada	*70*	75-04050
Dytecna Ltd	UK	*501*	80-08900
Frazer-Nash Consultancy Limited	UK	*503*	80-10950
GED (Southampton) Ltd	UK	*503*	80-11650
Grove Europe Ltd	**UK**	***456***	**75-27030**
Israel Aircraft Industries Ltd	Israel	*219*	75-10200
Krauss-Maffei Wehrtechnik GmbH	Germany	*171*	75-16345
LMS International	Belgium	*48*	80-00900
Prof Sten Luthander Ingbyra AB	Sweden	*379*	80-01280
Reumech-OMC	S. Africa	*359*	75-18265
Ricardo Hitec Ltd	UK	*509*	80-27840
Singapore Technologies Automotive	Singapore	*348*	75-04280
SP aerospace and vehicle systems bv	**Netherlands**	***295***	**75-06860**
TA Group Ltd	UK	*513*	80-50110
Terberg Benschop BV	Netherlands	*296*	75-07600
Thompson Carmichael Ltd	UK	*488*	75-59525
United Defense L.P.	USA	*624*	75-92050
Vickers Defence Systems	UK	*491*	75-63040
Vickers Shipbuilding & Eng Ltd	UK	*491*	75-63100

943 Installation services

AAI Corp.	USA	*545*	75-00020
Air Communications Pty Ltd	Australia	*23*	80-00720
AlliedSignal Aerospace	USA	*551*	75-04700
Amper Programas SA	Spain	*365*	75-00480
Analysis & Technology Inc.	USA	*631*	80-02600
APT Radar Systems Ltd	UK	*431*	75-03560
Assessment Services	UK	*497*	80-01500
Atlantic Research Corp.	USA	*555*	75-10150
Bristow Helicopters Ltd	UK	*499*	80-03950
British Aerospace Australia Ltd	Australia	*13*	75-02750
Calspan SRL Technology	USA	*540*	60-03000
Commercial Hydraulics Keelavite Ltd	UK	*441*	75-14410
Commercial Hydraulics Kontak Ltd	UK	*441*	75-14415
Computer Resources International A/S.	Denmark	*104*	80-00100
Cranfield Aerospace Ltd	UK	*500*	80-07200
Cranfield University	UK	*419*	60-00350
Design Bureau of	Russia	*336*	75-03150
Dytecna Ltd	UK	*501*	80-08900
EASAMS Ltd	UK	*501*	80-09000
EG&G Inc.	USA	*571*	75-29555
EG&G Washington Analytical Svcs	USA	*634*	80-09450
Electrothermal Engineering Ltd	UK	*448*	75-20400
Elintech GmbH	Germany	*178*	80-00550
Emprise Corp.	USA	*634*	80-09950
Engineering & Professional Serv.Inc.	USA	*634*	80-10150
Flight Refuelling Limited	UK	*451*	75-23850
FLS Aerospace (Light Aircraft) Ltd	UK	*451*	75-23900
Fluor Daniel Inc.	USA	*635*	80-11720
Forward Industries Ltd	UK	*452*	75-24200
General Electronics Communications	Greece	*189*	75-01300
Gordon Consulting Engineers Ltd	Israel	*224*	80-00750
Helio Mirror Co. Ltd.	UK	*457*	75-28900
Hellenic Arms Industry	Greece	*187*	55-00800
Hewlett-Packard (Canada) Ltd	Canada	*72*	75-07520
Hunting Engineering Ltd	UK	*459*	75-30700
Israel Aircraft Industries Ltd	Israel	*219*	75-10200
Keystone Helicopter Corp.	USA	*637*	80-17780
Lansdowne Technologies Inc	Canada	*78*	80-01205
LARIMART SpA	Italy	*236*	75-13500
Logistics Services International	USA	*638*	80-18800
M. Paz Logistics Ltd	Israel	*224*	80-01600
Marshall of Cambridge Aerospace Ltd	UK	*466*	75-37530
Microelectronics, guidance and	Turkey	*402*	55-00050
Mott MacDonald	UK	*507*	80-17400
MSL Secure Systems	UK	*507*	80-17500
National Systems Management Corp.	USA	*639*	80-21650
NEC Corporation	Japan	*253*	75-06350
Nyge Aero AB	Sweden	*379*	80-01490
Parsons Power Group Inc.	USA	*640*	80-23700
PK Electronic International Ltd	UK	*475*	75-45720
Planning & Research Ltd	Israel	*224*	80-01650
Raytheon E-Systems	USA	*608*	75-73302
SAIT Systems SA	Belgium	*47*	75-05600
Santa Barbara SA	Spain	*364*	55-01000
Schottel Sudamericana I.C.S.A.	Argentina	*4*	80-20000
SEMCOR Inc.	USA	*642*	80-27750
Sener Ingenieria y Sistemas SA	Spain	*367*	75-05200
Serco Aviation Services	UK	*511*	80-34150
S.G.D. Engineering Ltd	Israel	*224*	80-01930
Shorrock Integrated Systems	UK	*483*	75-53450
Shorts Support Services Division	UK	*511*	80-35600
Siemens Plessey Systems	UK	*483*	75-53770
Siemens Switzerland Ltd	Switzerland	*387*	75-11980
Singapore Electronic & Engineering	Singapore	*349*	80-01960
SISDEF Ltda	Chile	*85*	80-01940
SITEP SpA	Italy	*239*	75-22595
SNPE	France	*145*	75-36315
Stanley Aviation Corporation	USA	*617*	75-83900
Strategic Technologies Pte Ltd	**Singapore**	***348***	**75-04400**
Strumech Versatower Ltd	UK	*486*	75-57530
Sverdrup Corp.	USA	*643*	80-30650
Sverdrup Technology Inc.	USA	*643*	80-30680
Taylor Woodrow Management	UK	*513*	80-50400
Tracor Inc	USA	*623*	75-90000
Vickers Shipbuilding & Eng Ltd	UK	*491*	75-63100
Vihtavuori Oy	Finland	*119*	75-02850
Vitrociset SpA	Italy	*240*	80-40000
Vosper International Ltd	UK	*514*	80-67000
Wilcox Electric Inc.	USA	*627*	75-96730
Wornall Electronics Inc.	USA	*645*	80-35400
Yarrow Shipbuilders Ltd	UK	*495*	75-67500

943 60 Installation services, security equipment

AlliedSignal Aerospace	USA	*551*	75-04700
CORIS	France	*134*	75-12450
CPAD Technologies Inc.	Canada	*78*	80-00386
Dytecna Ltd	UK	*501*	80-08900
EASAMS Ltd	UK	*501*	80-09000
EG&G Inc.	USA	*571*	75-29555
Engineering & Professional Serv.Inc.	USA	*634*	80-10150
Fluor Daniel Inc.	USA	*635*	80-11720
Gordon Consulting Engineers Ltd	Israel	*224*	80-00750
Litton Industries Inc.	USA	*590*	75-55400
Marshall of Cambridge Aerospace Ltd	UK	*466*	75-37530
Metis Construction & Trade Co. Inc.	Turkey	*403*	80-01330
Microelectronics, guidance and	Turkey	*402*	55-00050
Mott MacDonald	UK	*507*	80-17400
MSL Secure Systems	UK	*507*	80-17500
NEC Corporation	Japan	*253*	75-06350
OTEC Support Services Ltd	UK	*508*	80-19600
REMOTEC Inc.	USA	*609*	75-74430
Sector Ltd	UK	*511*	80-33400
SEMCOR Inc.	USA	*642*	80-27750
S.G.D. Engineering Ltd	Israel	*224*	80-01930
Siemens Switzerland Ltd	Switzerland	*387*	75-11980
Singapore Electronic & Engineering	Singapore	*349*	80-01960
Sparton Technology Inc.	USA	*616*	75-82950
Stanley Aviation Corporation	USA	*617*	75-83900
Sverdrup Corp.	USA	*643*	80-30650
Talos Engineering SA	Greece	*193*	80-02010
Taylor Woodrow Management	UK	*513*	80-50400
Technology Australasia Pty Ltd	Australia	*26*	80-20600
Television Installation Services	UK	*488*	75-59250
Tracor Inc	USA	*623*	75-90000
Vitro Corp.	USA	*644*	80-34400
VSK Electronics	Belgium	*48*	75-08000
Wornall Electronics Inc.	USA	*645*	80-35400

943 70 Systems integration services

AlliedSignal Aerospace	USA	*551*	75-04700
Ameridata Canada Ltd	Canada	*68*	75-00480
Amper Programas SA	Spain	*365*	75-00480
Analysis & Technology Inc.	USA	*631*	80-02600
APT Radar Systems Ltd	UK	*431*	75-03560
Aspect Computing Pty Ltd	Australia	*24*	80-01150
Australian Technical Publications	Australia	*24*	80-01880
Bristow Helicopters Ltd	UK	*499*	80-03950
British Aerospace Australia Ltd	Australia	*13*	75-02750
Burdeshaw Associates, Ltd	USA	*632*	80-04960
Commercial Hydraulics Kontak Ltd	UK	*441*	75-14415
Computer Power Group Ltd	Australia	*24*	80-03730
Computer Resources International A/S.	Denmark	*104*	80-00100
Computing Devices Canada Ltd	Canada	*70*	75-03600
CORIS	France	*134*	75-12450
Cranfield Aerospace Ltd	UK	*500*	80-07200
Cranfield University	UK	*419*	60-00350
Denro Inc.	USA	*568*	75-25651
Devonport Royal Dockyard Ltd	UK	*501*	80-08500

D.E.3I Informatique	France	150	80-00400
Dytecna Ltd	UK	501	80-08900
Finavitec Oy	Finland	119	80-00580
Forward Industries Ltd	UK	452	75-24200
GEC-Marconi Communications Ltd	UK	454	75-25335
George G. Sharp Inc.	USA	642	80-28000
Information Security Systems Ltd.	UK	461	75-31430
ITT Federal Services Corp.	USA	586	75-49455
Kaman Sciences Corp.	USA	637	80-17600
Kvaerner Eureka AS	Norway	309	75-01095
Lansdowne Technologies Inc	Canada	78	80-01205
Litton Industries Inc.	USA	590	75-55400
Lockheed Martin Services Group	USA	638	80-18540
Logistics Services International	USA	638	80-18800
M. Paz Logistics Ltd	Israel	224	80-01600
MacDonald Dettwiler and Associates	Canada	78	80-01310
MIL Systems	Canada	79	80-01360
NEC Corporation	Japan	253	75-06350
Parity Solutions Ltd	UK	508	80-20300
PRC Inc.	USA	640	80-24800
Radio Holland Electronics BV	Netherlands	297	80-18300
SAIT Systems SA	Belgium	47	75-05600
SEMCOR Inc.	USA	642	80-27750
Siemens Plessey Systems	UK	483	75-53770
Siemens Switzerland Ltd	Switzerland	387	75-11980
Sierra Technologies Inc.	USA	614	75-80640
Singapore Electronic & Engineering	Singapore	349	80-01960
Singapore Engineering Software	Singapore	349	80-01965
Sonak Systems & Software SA	Greece	193	80-01970
Stanley Aviation Corporation	USA	617	75-83900
State Research Institute	Russia	336	60-04150
Steyr-Daimler-Puch Spezialfahrzeug	Austria	34	75-02810
Sverdrup Corp.	USA	643	80-30650
Sverdrup Technology Inc.	USA	643	80-30680
Systematic Software Engineering Ltd.	UK	513	80-49480
Taylor Woodrow Management	UK	513	80-50400
Technology Australasia Pty Ltd	Australia	26	80-20600
Technology Project Services	UK	513	80-52800
Tracor Inc	USA	623	75-90000
TriLink Technologies Group Inc.	Canada	80	80-02100
Van Rietschoten & Houwens	Netherlands	296	75-07900
Vanguard Research Inc.	USA	644	80-34100
Vickers Shipbuilding & Eng Ltd	UK	491	75-63100
Vosper International Ltd	UK	514	80-67000

944 Manufacturing services

A Goninan & Co. Ltd	Australia	16	75-07760
AAI Corp.	USA	545	75-00020
Aero Machining Ltd.	Canada	77	80-00020
Algram Group Ltd.	UK	429	75-02625
American Eurocopter	USA	552	75-05790
Amper Programas SA	Spain	365	75-00480
A.O. Henton Engineering Co Ltd.	UK	458	75-29040
APPH (Bolton) Ltd.	UK	431	75-03415
Armalon Ltd	UK	431	75-03729
ASCO Industries NV/SA	Belgium	45	75-00550
Associated Aircraft Mfg & Sales Inc.	USA	555	75-09500
Avionic Components Ltd.	UK	432	75-04970
Bofors AB	Sweden	375	74-00020
Bombardier Inc	Canada	69	75-01900
Brandt Fijnmechanische Industrie BV	Netherlands	292	75-00850
Bristol Aerospace Ltd	Canada	69	75-01500
British Aerospace Australia Ltd	Australia	13	75-02750
CAE Aviation Ltd	Canada	78	80-00245
Carville Ltd.	UK	439	75-12000
Chamberlain Manufacturing Corp.	USA	563	75-17950
Colledge & Morley (Gears) Ltd.	UK	441	75-14150
Communitron (Australia) Pty Ltd	Australia	14	75-03860
Constructions Mécaniques des Vosges	France	134	75-12300
Cow Industrial Polymers Ltd	UK	443	75-15650
Cranfield Aerospace Ltd	UK	500	80-07200
Cranfield University	UK	419	60-00350
Crouse Hinds Australia Pty Ltd	Australia	15	75-03963
Davis (W.R.) Engineering Ltd	Canada	70	75-04050
Devonport Royal Dockyard Ltd	UK	501	80-08500
Doncasters Structures	UK	446	75-18465
DRS Technologies, Inc.	USA	570	75-27400
DUAL Incorporated	USA	634	80-08670
Dynamit Nobel GmbH	Germany	167	75-07100
Dytecna Ltd	UK	501	80-08900
Electron Beam Processes Ltd	UK	502	80-09300
Electrothermal Engineering Ltd	UK	448	75-20400
Engineering & Professional Serv.Inc.	USA	634	80-10150
F&F Associates Inc.	USA	635	80-11000
Federal Industries Inc.	USA	575	75-34030
FFA Flugzeugwerke Altenrhein AG	Switzerland	386	75-04350
Flight Refuelling Limited	UK	451	75-23850
FLS Aerospace (Light Aircraft) Ltd	UK	451	75-23900
Fluorocarbon Company Limited	UK	451	75-24020
Forward Industries Ltd	UK	452	75-24200
Furno Co. Inc.	USA	577	75-36500
Gates Hydraulics Ltd	UK	453	75-25210
H. Fine & Son Ltd	UK	451	75-23210
Hanjin Heavy Industries Co. Ltd	Korea (ROK)	265	75-01505
Helio Mirror Co. Ltd.	UK	457	75-28900
Hellenic Aerospace Industry Ltd.	Greece	192	80-00750
Hill Industries Inc.	USA	582	75-43465
Horstman Defence Systems Ltd.	UK	459	75-29920
Howden Fan Company Pty Ltd	Australia	17	75-08740
HPC Engineering plc	UK	459	75-30350
Hughes Microelectronics Europe Ltd	UK	459	75-30480
Ingersoll-Dresser Pumps (UK) Ltd	UK	461	75-31450
Insumat Ltd	UK	461	75-31660
Israel Aircraft Industries Ltd	Israel	219	75-10200
ITT Avionics	USA	586	75-49455
J & E Hofmann Engineering Pty Ltd.	Australia	17	75-08630
Japan Aircraft Manufacturing Co. Ltd.	Japan	252	75-03200
Jaques	Australia	17	75-10200
John Heine & Son Pty Ltd	Australia	17	75-08350
Litton Industries Inc.	USA	590	75-55400
M. Paz Logistics Ltd	Israel	224	80-01600
Marshall of Cambridge Aerospace Ltd	UK	466	75-37530
MECAR SA	Belgium	46	75-04930
Menasco Aerospace Ltd	Canada	74	75-09600
Messier-Dowty Electronics	Canada	74	75-09650
Mibert Precision Ltd	UK	468	75-38920
MTH Ltd	UK	470	75-41330
NanoMaterials, Inc.	USA	598	75-63590
Northern Airborne Technology Ltd	Canada	74	75-10145
Omori Seikoki Co. Ltd	Japan	254	75-07310
Pacific Aerospace Corp. Ltd	New Zealand	302	75-01600
Per Udsen Co. Aircraft Industry A/S	Denmark	104	75-21300
Pioneer Aerospace Corp	USA	604	75-70200
PK Electronic International Ltd	UK	475	75-45720
Plastic Fabricating Co. Inc.	USA	604	75-70335
Plenty Ltd	UK	475	75-46050
Price & Orphin Limited	UK	509	80-24250
QES Ltd	UK	509	80-25000
Reade Advanced Materials	USA	608	75-73860
Renown Engineering Ltd	UK	478	75-49227
Ricardo Hitec Ltd	UK	509	80-27840
Robert Keen Ltd.	UK	505	80-14000
Robert Mitchell Inc.	Canada	74	75-09800
Rohr Inc.	USA	611	75-76150
Rolls-Royce (Canada) Ltd.	Canada	75	75-12120
SAIT Systems SA	Belgium	47	75-05600
Santa Barbara SA	Spain	364	55-01000
SEMCOR Inc.	USA	642	80-27750
Sener Ingenieria y Sistemas SA	Spain	367	75-05200
Siemens Environmental Systems Ltd	UK	483	75-53700
Siemens Switzerland Ltd	Switzerland	387	75-11980
SIMA PERU	Peru	319	55-01950
SISDEF Ltda	Chile	85	80-01940
SONACA SA	Belgium	47	75-06100
SP aerospace and vehicle systems bv	**Netherlands**	**295**	**75-06860**
St Bernard Composites Ltd	UK	486	75-56222
Strategic Technologies Pte Ltd	**Singapore**	**348**	**75-04400**
Supercraft Ltd	UK	487	75-57910
Terberg Benschop BV	Netherlands	296	75-07600
The Graphics Co. of America	USA	579	75-40360
Theodor Rapp GmbH Co	Germany	173	75-24600
Tracor Inc	USA	623	75-90000
Transfield Pty Ltd	Australia	22	75-20810
Tufnol Ltd	UK	489	75-61410
United Defense L.P.	USA	624	75-92050
United Moulders Ltd	UK	490	75-62130
Vacu-Blast Ltd	UK	490	75-62550
Vihtavuori Oy	Finland	119	75-02850
W.F. Stanley & Co. Ltd	UK	486	75-56350
Wheeler Systems Inc.	USA	627	75-96400
Wilcox Electric Inc.	USA	627	75-96730
Yarrow Shipbuilders Ltd	UK	495	75-67500
Yewlands Engineering Co. Ltd.	UK	495	75-67520

944 10 Assembly services

A Goninan & Co. Ltd	Australia	*16*	75-07760
AAI Corp.	USA	*545*	75-00020
Aero Machining Ltd.	Canada	*77*	80-00020
American Eurocopter	USA	*552*	75-05790
Amper Programas SA	Spain	*365*	75-00480
A.O. Henton Engineering Co Ltd.	UK	*458*	75-29040
Array Connector Corp.	USA	*555*	75-09020
Australian Defence Industries Ltd	Australia	*9*	55-01000
Aviation Support Tech. Int'l Inc.	USA	*556*	75-11050
Celab Ltd	UK	*439*	75-12250
Chamberlain Manufacturing Corp.	USA	*563*	75-17950
Devonport Royal Dockyard Ltd	UK	*501*	80-08500
DRS Technologies, Inc.	USA	*570*	75-27400
Dynamit Nobel GmbH	Germany	*167*	75-07100
Electron Beam Processes Ltd	UK	*502*	80-09300
Engineering & Professional Serv.Inc.	USA	*634*	80-10150
Forward Industries Ltd	UK	*452*	75-24200
GEC-Marconi Communications Ltd	UK	*454*	75-25335
GSI Corporation	USA	*580*	75-40715
Hellenic Aerospace Industry Ltd.	Greece	*192*	80-00750
Hill Industries Inc.	USA	*582*	75-43465
Horstman Defence Systems Ltd.	UK	*459*	75-29920
Howden Fan Company Pty Ltd	Australia	*17*	75-08740
Hughes Microelectronics Europe Ltd	UK	*459*	75-30480
Israel Aircraft Industries Ltd	Israel	*219*	75-10200
John Heine & Son Pty Ltd	Australia	*17*	75-08350
Lucas Aerospace - Australia	Australia	*25*	80-12850
M. Paz Logistics Ltd	Israel	*224*	80-01600
Machine Technology Ltd.	UK	*465*	75-36870
Meggitt Avionics (UK)	UK	*467*	75-38340
Menasco Aerospace Ltd	Canada	*74*	75-09600
Messier-Dowty Electronics	Canada	*74*	75-09650
Metis Construction & Trade Co. Inc.	Turkey	*403*	80-01330
Micrel Inc.	USA	*596*	75-60950
MTH Ltd	UK	*470*	75-41330
Normalair-Garrett Australia	Australia	*19*	75-14870
Northern Airborne Technology Ltd	Canada	*74*	75-10145
Omori Seikoki Co. Ltd	Japan	*254*	75-07310
Pacific Aerospace Corp. Ltd	New Zealand	*302*	75-01600
Park Products Ltd.	UK	*473*	75-44000
Per Udsen Co. Aircraft Industry A/S	Denmark	*104*	75-21300
QES Ltd	UK	*509*	80-25000
Radio Holland Electronics BV	Netherlands	*297*	80-18300
Renown Engineering Ltd	UK	*478*	75-49227
Ricardo Hitec Ltd	UK	*509*	80-27840
Robert Keen Ltd.	UK	*505*	80-14000
Robert Mitchell Inc.	Canada	*74*	75-09800
Rotadata Ltd.	UK	*481*	75-50570
SAIT Systems SA	Belgium	*47*	75-05600
Scaled Composites Inc.	USA	*541*	60-19160
SEMCOR Inc.	USA	*642*	80-27750
Siemens Environmental Systems Ltd	UK	*483*	75-53700
Siemens Switzerland Ltd.	Switzerland	*387*	75-11980
SISDEF Ltda	Chile	*85*	80-01940
SONACA SA	Belgium	*47*	75-06100
SP aerospace and vehicle systems bv	**Netherlands**	***295***	**75-06860**
Sparton Technology Inc.	USA	*616*	75-82950
Supercraft Ltd.	UK	*487*	75-57910
Terberg Benschop BV	Netherlands	*296*	75-07600
The Graphics Co. of America	USA	*579*	75-40360
Theodor Rapp GmbH Co	Germany	*173*	75-24600
Tracor Inc	USA	*623*	75-90000
Transfield Pty Ltd.	Australia	*22*	75-20810
Trompeter Electronics Inc.	USA	*623*	75-91070
United Defense L.P.	USA	*624*	75-92050
United Moulders Ltd	UK	*490*	75-62130
Vickers Defence Systems	UK	*491*	75-63040
Vickers Shipbuilding & Eng Ltd	UK	*491*	75-63100
W.F. Stanley & Co. Ltd.	UK	*486*	75-56350
WPI-Boston Division Inc.	USA	*561*	75-16020

944 15 Ammunition assembly & filling services

Allied Research Corp.	USA	*549*	75-04670
Armalon Ltd	UK	*431*	75-03729
Chamberlain Manufacturing Corp.	USA	*563*	75-17950
CMS Inc.	USA	*564*	75-19170
Giat Industries	France	*126*	55-00100
Haley & Weller Ltd	UK	*456*	75-27400
KJM Fabritek Inc.	USA	*589*	75-52877
M. Paz Logistics Ltd	Israel	*224*	80-01600
Naschem	S. Africa	*358*	75-15000
SAIT Systems SA	Belgium	*47*	75-05600

944 30 Component manufacturing services

AAI Corp.	USA	*545*	75-00020
AEA Technology plc	UK	*496*	80-00480
Algram Group Ltd.	UK	*429*	75-02625
A.O. Henton Engineering Co Ltd.	UK	*458*	75-29040
Armalon Ltd	UK	*431*	75-03729
Array Connector Corp.	USA	*555*	75-09020
Avionic Components Ltd.	UK	*432*	75-04970
Barnes Aerospace Group	USA	*557*	75-12000
Britte SA	Belgium	*45*	75-01900
CAE Aviation Ltd	Canada	*78*	80-00245
Consolidated Manufacturing	Australia	*15*	75-03925
Cranfield Aerospace Ltd	UK	*500*	80-07200
Cranfield University	UK	*419*	60-00350
Davies Industrial Communications Ltd	UK	*444*	75-17010
Diamond Electronics Inc.	USA	*569*	75-26450
Donaldson Co. Inc.	USA	*569*	75-27050
Doncasters Structures	UK	*446*	75-18465
Dynamic Controls Ltd	UK	*447*	75-19140
Electron Beam Processes Ltd	UK	*502*	80-09300
Embraer	Brazil	*54*	55-05000
Engineered Plastics Inc.	USA	*573*	75-31570
Fairview Machine	USA	*575*	75-33600
FFA Flugzeugwerke Altenrhein AG	Switzerland	*386*	75-04350
Fluorocarbon Company Limited	UK	*451*	75-24020
Gates Hydraulics Ltd	UK	*453*	75-25210
GEREP GmbH	Germany	*168*	75-10300
Grove Europe Ltd	**UK**	***456***	**75-27030**
GSI Corporation	USA	*580*	75-40715
Haley & Weller Ltd	UK	*456*	75-27400
H.I. Fraser Pty Ltd.	Australia	*16*	75-06835
Horstman Defence Systems Ltd.	UK	*459*	75-29920
HPC Engineering plc	UK	*459*	75-30350
Hughes Microelectronics Europe Ltd	UK	*459*	75-30480
John Heine & Son Pty Ltd	Australia	*17*	75-08350
Litton Industries Inc.	USA	*590*	75-55400
M. Paz Logistics Ltd	Israel	*224*	80-01600
Marshall of Cambridge Aerospace Ltd	UK	*466*	75-37530
Normalair-Garrett Australia	Australia	*19*	75-14870
Omori Seikoki Co. Ltd	Japan	*254*	75-07310
Papadopoulos & Co Ltd	Greece	*191*	75-02195
Park Products Ltd.	UK	*473*	75-44000
Parker-Hale Ltd	UK	*473*	75-43850
Permali Gloucester Ltd	UK	*474*	75-44850
Renown Engineering Ltd	UK	*478*	75-49227
Robert Keen Ltd.	UK	*505*	80-14000
SAIT Systems SA	Belgium	*47*	75-05600
Sandeep Metalcraft Pvt Ltd	India	*205*	75-01900
Scaled Composites Inc.	USA	*541*	60-19160
SONACA SA	Belgium	*47*	75-06100
SP aerospace and vehicle systems bv	**Netherlands**	***295***	**75-06860**
Sparton Technology Inc.	USA	*616*	75-82950
Supercraft Ltd.	UK	*487*	75-57910
Textron Marine & Land Systems	USA	*622*	75-88660
Theodor Rapp GmbH Co	Germany	*173*	75-24600
Tufnol Ltd	UK	*489*	75-61410
United Defense L.P.	USA	*624*	75-92050
Vickers Defence Systems	UK	*491*	75-63040
Voss Industries Inc.	USA	*626*	75-94550
Wheeler Systems Inc.	USA	*627*	75-96400
W.H.J. Fagg & Son Ltd.	UK	*450*	75-22190
Woodville Polymer	UK	*494*	75-66900
WPI-Boston Division Inc.	USA	*561*	75-16020

944 50 Micro-engineering

Aircraft Tanks Ltd	UK	*429*	75-01730
Applied Microengineering Ltd	UK	*431*	75-03530
Central Research Laboratories Ltd	UK	*499*	80-05895
ERA Technology Ltd	UK	*502*	80-09600
GEC-Marconi Research Centre	UK	*419*	60-00700
Iver Precision Engineers Ltd	UK	*462*	75-32230
Nordic VLSI A/S	Norway	*310*	75-01640

944 60 Manufacturing services, spare parts

AAI Corp.	USA	*545*	75-00020

Aero Machining Ltd.	Canada	77	80-00020
AeroSpace Technologies of Australia	Australia	11	75-00340
Amper Programas SA	Spain	365	75-00480
A.O. Henton Engineering Co Ltd.	UK	458	75-29040
APT Radar Systems Ltd	UK	431	75-03560
Armalon Ltd	UK	431	75-03729
Associated Aircraft Mfg & Sales Inc.	USA	555	75-09500
Australian Defence Industries Ltd	Australia	9	55-01000
Aviation Support Tech. Int'l Inc.	USA	556	75-11050
Avionic Components Ltd.	UK	432	75-04970
Barnes Aerospace Group	USA	557	75-12000
Bougas Panagiotis Ltd	Greece	187	75-00600
CAE Aviation Ltd	Canada	78	80-00245
Chamberlain Manufacturing Corp.	USA	563	75-17950
Chicopee Manufacturing Limited	Canada	70	75-03200
Constructions Mécaniques des Vosges	France	134	75-12300
Cranfield Aerospace Ltd	UK	500	80-07200
Cranfield University	UK	419	60-00350
Devonport Royal Dockyard Ltd	UK	501	80-08500
Donaldson Co. Inc.	USA	569	75-27050
Electron Beam Processes Ltd	UK	502	80-09300
Engineering & Professional Serv.Inc.	USA	634	80-10150
Fairview Machine	USA	575	75-33600
Federal Industries Inc.	USA	575	75-34030
FFA Flugzeugwerke Altenrhein AG	Switzerland	386	75-04350
Forward Industries Ltd	UK	452	75-24200
Frazier Aviation Inc.	USA	576	75-35970
GEREP GmbH	Germany	168	75-10300
Graseby Dynamics Ltd	UK	455	75-26590
Grove Europe Ltd	**UK**	**456**	**75-27030**
Helston Gunsmiths	UK	458	75-29010
H.I. Fraser Pty Ltd.	Australia	16	75-06835
Hill Industries Inc.	USA	582	75-43465
Horstman Defence Systems Ltd.	UK	459	75-29920
Ingersoll-Dresser Pumps (UK) Ltd	UK	461	75-31450
John Heine & Son Pty Ltd	Australia	17	75-08350
Litton Industries Inc.	USA	590	75-55400
Lucas Aerospace - Australia	Australia	25	80-12850
M. Paz Logistics Ltd	Israel	224	80-01600
Meggitt Avionics (UK)	UK	467	75-38340
Menasco Aerospace Ltd	Canada	74	75-09600
Merex Inc.	USA	595	75-60250
Normalair-Garrett Australia	Australia	19	75-14870
Omori Seikoki Co. Ltd	Japan	254	75-07310
OTEC Support Services Ltd	UK	508	80-19600
Parker-Hale Ltd	UK	473	75-43850
Regent Mfg Inc.	USA	609	75-74200
Renown Engineering Ltd	UK	478	75-49227
Robert Keen Ltd	UK	505	80-14000
SONACA SA	Belgium	47	75-06100
SP aerospace and vehicle systems bv	**Netherlands**	**295**	**75-06860**
Sparton Technology Inc.	USA	616	75-82950
Supercraft Ltd	UK	487	75-57910
Terberg Benschop BV	Netherlands	296	75-07600
The Graphics Co. of America	USA	579	75-40360
Theodor Rapp GmbH Co	Germany	173	75-24600
Thorn DNT Airfield Lighting Pty Ltd	Australia	22	75-20540
United Defense L.P.	USA	624	75-92050
Vazduhoplovno Remonti Zavod	Yugoslavia	661	80-02500
Vickers Defence Systems	UK	491	75-63040
Weir Engineering Pty Ltd	Australia	23	75-23300
Woodville Polymer	UK	494	75-66900
WPI-Boston Division Inc.	USA	561	75-16020

945 Maintenance, repair, overhaul services

AAR Corp	USA	629	80-00035
Abex GmbH Aerohydraul	Germany	164	75-00105
ACH	France	130	75-00150
Advisory, Consulting & Technical	UK	496	80-00450
Aero Corporation	USA	630	80-00650
Aero-Industrias Leaver & Cia SA	Colombia	91	80-00100
Aeromaritime Systembau GmbH	Germany	164	75-00290
Aeropartes Indústria Aeronáutica	Brazil	58	80-00100
AIM Enterprises Inc.	USA	630	80-01350
Air Communications Pty Ltd	Australia	23	80-00720
Air New Zealand Ltd	New Zealand	303	80-00100
Air Target Services Pty Ltd	Australia	24	80-00840
Aircraft Instrument & Radio Co. Inc.	USA	630	80-01600
AlliedSignal Aerospace	USA	551	75-04700
AlliedSignal Aerospace GmbH	Germany	164	75-00610
Alvis Logistics Ltd	UK	430	75-02860
Alvis Transmission Ltd.	UK	430	75-02862
Alvis Unipower Ltd	UK	430	75-02863
American Eurocopter	USA	552	75-05790
Amper Programas SA	Spain	365	75-00480
Analysis & Technology Inc.	USA	631	80-02600
Ansett Australia Avionics	Australia	24	80-01010
APT Radar Systems Ltd	UK	431	75-03560
ASCO Industries NV/SA	Belgium	45	75-00550
Associated Aircraft Mfg & Sales Inc.	USA	555	75-09500
Australian Defence Industries Ltd	Australia	9	55-01000
Aviation Spares Ltd.	UK	498	80-01850
Bachmann SA	France	132	75-04900
Balmoral Group Ltd	UK	433	75-05420
Barfield Inc.	USA	557	75-11930
Barnes Aerospace Group	USA	557	75-12000
Bavaria Keytronic Technologie GmbH	Germany	165	75-02050
Bender Shipbuilding & Repair Co.	USA	558	75-13300
BFGoodrich Aerospace	USA	558	75-13770
Billfish Inc.	USA	632	80-04510
Bofors AB	Sweden	375	74-00020
Bosman Powersource Support BV	Netherlands	292	75-00800
Bremer Vulkan Marineschiffbau GmbH	Germany	166	75-03600
Bristol Aerospace Ltd	Canada	69	75-01500
British Aerospace Australia Ltd	Australia	13	75-02750
CAE Aviation Ltd	Canada	78	80-00245
Canadian Marconi Co.	Canada	69	75-02600
Carmichael International Ltd	UK	438	75-11700
Cenzin Co. Ltd.	Poland	324	75-00200
CHC Helicopter Corporation	Canada	78	80-00280
Comaf Indústria Seronautica Ltda	Brazil	58	80-01000
Commercial Hydraulics Keelavite Ltd	UK	441	75-14410
Constructions Mécaniques des Vosges	France	134	75-12300
Corrintec Ltd	UK	442	75-15120
Cranfield Aerospace Ltd	UK	500	80-07200
Cranfield University	UK	419	60-00350
Cypher Research Laboratories	Australia	10	60-00300
Daewoo Heavy Industries Ltd	Korea (ROK)	265	75-00600
Delft Sensor Systems	**Belgium**	**46**	**75-02500**
Delft Sensor Systems	Netherlands	292	75-01590
Devonport Royal Dockyard Ltd	UK	501	80-08500
Dowty Aerospace Aviation Services	Singapore	349	80-00450
Dowty Aerospace Aviation Services	UK	501	80-08600
Dytecna Ltd	UK	501	80-08900
EG&G Washington Analytical Svcs	USA	634	80-09450
Electrothermal Engineering Ltd	UK	448	75-20400
Elintech GmbH	Germany	178	80-00550
Embraer Aviation International	France	150	80-00500
Empresa Nacional de Aeronáutica	Chile	84	55-00550
Engineering & Professional Serv.Inc.	USA	634	80-10150
Eurocopter Canada Ltd	Canada	71	75-05050
Eurodef Aerospace & Defence	Greece	189	75-01295
Fabbrica Italiana Apparecchiature	Italy	234	75-08550
FFA Flugzeugwerke Altenrhein AG	Switzerland	386	75-04350
FFV Aerotech AB	Sweden	379	80-00600
Finavitec Oy	Finland	119	80-00580
Fincantieri	Italy	234	75-09250
Flight International Inc.	USA	635	80-11600
Flight Refuelling Limited	UK	451	75-23850
Fluor Daniel Inc.	USA	635	80-11720
FN Herstal SA	Belgium	46	75-03600
FPT Industries Ltd	UK	452	75-24400
FR Aviation Limited	UK	503	80-10850
GEC-Marconi Avionics Ltd	UK	454	75-25320
GEC-Marconi Communications Ltd	UK	454	75-25335
General Motors of Canada Ltd	Canada	71	75-05900
GENME SA Marine & Industrial	Greece	192	80-00700
GEREP GmbH	Germany	168	75-10300
GKN Westland Aerospace Transmissions	UK	455	75-25630
GLS Gesellschaft für Logistischen	Germany	178	80-00750
Graseby Dynamics Ltd	UK	455	75-26590
Greenwich Air Services Inc.	USA	636	80-13170
Grove Europe Ltd	**UK**	**456**	**75-27030**
H+S Aviation Ltd	UK	503	80-12050
Halifax Shipyard Limited	Canada	78	80-00810
Hanjin Heavy Industries Co. Ltd	Korea (ROK)	265	75-01505
HeavyLift Aircraft Engineering Ltd.	UK	503	80-12300
Helio Mirror Co. Ltd	UK	457	75-28900
Hellenic Aerospace Industry Ltd.	Greece	192	80-00750
Hellenic Arms Industry	Greece	187	55-00800
Hewlett-Packard (Canada) Ltd	Canada	72	75-07520
H.I. Fraser Pty Ltd.	Australia	16	75-06835
Hindustan Aeronautics Ltd.	India	204	55-01000
Hong Leong-Lürssen Shipyard	Malaysia	277	75-00055
Horstman Defence Systems Ltd.	UK	459	75-29920
Howaldtswerke Deutsche Werft AG	Germany	170	75-13400

Howden Wade Ltd	UK	*459*	75-30260
Hunting Aviation Ltd	UK	*504*	80-12785
Hydroscience Technologies Inc.	USA	*584*	75-45550
Indivers NV/Interturbine	Netherlands	*293*	75-03550
Ingersoll-Dresser Pumps (UK) Ltd	UK	*461*	75-31450
Innotech Aviation Ltd.	Canada	*78*	80-00960
Instrument Services Ltd	UK	*504*	80-13328
Israel Aircraft Industries Ltd	Israel	*219*	75-10200
ITT Avionics	USA	*586*	75-49455
Japan Aircraft Manufacturing Co. Ltd.	Japan	*252*	75-03200
Jaques	Australia	*17*	75-10200
Jered Brown Brothers Inc.	USA	*586*	75-50380
Jet Aviation Zürich AG	Switzerland	*388*	80-01030
John Crane Inc.	USA	*566*	75-22500
John M. Piris & Sons Ltd	Gibraltar	*185*	75-01000
Jones & Rickard Pty Ltd	Australia	*18*	75-10750
Kawasaki Heavy Industries Ltd	Japan	*252*	75-04350
Keystone Helicopter Corp.	USA	*637*	80-17780
Kiev Aircraft Repair Plant	Ukraine	*406*	55-01500
Krauss-Maffei Wehrtechnik GmbH	Germany	*171*	75-16345
KUKA Wehrtechnik GmbH	Germany	*171*	75-17200
Litton Industries Inc.	USA	*590*	75-55400
Loadlift Equipment Ltd	New Zealand	*302*	75-01250
Lockheed Martin Aircraft and	USA	*638*	80-18530
Lucas Aerospace - Australia	Australia	*25*	80-12850
Lucas Aerospace Power Transmission	USA	*592*	75-56265
Lynx Corp Pty Ltd	Australia	*25*	80-12980
M International Export	USA	*639*	80-19780
M. Paz Logistics Ltd	Israel	*224*	80-01600
Maritime Contractors Inc.	USA	*639*	80-20050
Materialinformationszentrum Ges.	Germany	*178*	80-01310
Menasco Aerospace Ltd	Canada	*74*	75-09600
Merex Inc.	USA	*595*	75-60250
Messier-Dowty Limited	UK	*467*	75-38700
Microelectronics, guidance and	Turkey	*402*	55-00050
MIL Systems	Canada	*79*	80-01360
Moscow Aircraft Production	Russia	*334*	55-13050
MRF TEREM	Bulgaria	*62*	55-03900
MSI-Defence Systems Ltd	UK	*469*	75-41280
MTH Ltd	UK	*470*	75-41330
MTU München GmbH	Germany	*172*	75-21801
Newport News Shipbuilding	USA	*599*	75-64300
Nimda Co. Ltd.	Israel	*224*	80-01400
Nord-Micro Elektronik Feinmechanik	Germany	*173*	75-22700
Normalair-Garrett Australia	Australia	*19*	75-14870
Northrop Grumman Corporation	USA	*599*	75-65450
Nyge Aero AB	Sweden	*379*	80-01490
Omori Seikoki Co. Ltd	Japan	*254*	75-07310
Pacific Aerospace Corp. Ltd	New Zealand	*302*	75-01600
Pakistan Aeronautical Complex	Pakistan	*315*	55-01600
Park Products Ltd	UK	*473*	75-44000
Patria Vehicles Oy	Finland	*118*	75-01710
Pezetel	Poland	*325*	80-01000
Philips Calibration Services	Australia	*25*	80-16490
Pilatus Aircraft Ltd	Switzerland	*387*	75-10100
Pilatus Britten-Norman Ltd	UK	*474*	75-45400
Pratt & Whitney Canada Inc	Canada	*75*	75-11400
PT Industri Pesawat Terbang	Indonesia	*209*	55-00300
QES Ltd	UK	*509*	80-25000
Qualitair Engineering Services Ltd	UK	*509*	80-25200
Radio Holland Electronics BV	Netherlands	*297*	80-18300
Raytheon Aerospace	USA	*641*	80-25650
RDM Technology BV	Netherlands	*294*	75-05780
Renown Engineering Ltd	UK	*478*	75-49227
Ricardo Hitec Ltd	UK	*509*	80-27840
Robert Keen Ltd	UK	*505*	80-14000
Rohde & Schwarz France Telecom	France	*142*	75-32900
Rolls-Royce (Canada) Ltd	Canada	*75*	75-12120
SABCA	Belgium	*46*	75-05590
SAIT Systems SA	Belgium	*47*	75-05600
Santa Barbara SA	Spain	*364*	55-01000
Schottel Sudamericana I.C.S.A.	Argentina	*4*	80-20000
Sener Ingenieria y Sistemas SA	Spain	*367*	75-05200
Serck Aviation Div.	UK	*482*	75-52860
Sermatech International Inc	USA	*613*	75-79530
Sevastopol JSC	Ukraine	*407*	75-06500
Shorts Support Services Division	UK	*511*	80-35600
SIMA PERU	Peru	*319*	55-01950
Singapore Electronic & Engineering	Singapore	*349*	80-01960
Singapore Technologies Automotive	Singapore	*348*	75-04280
Société d'Applications des Machines	France	*143*	75-34300
Sonovision ITEP Technologies	France	*151*	80-01850
SP aerospace and vehicle systems bv	**Netherlands**	***295***	**75-06860**
Spar Aerospace Ltd	Canada	*76*	75-13000
ST Aerospace Engineering Pte Ltd	Singapore	*349*	80-01990
ST Aerospace Engines Pte Ltd	Singapore	*349*	80-02000
Standard Aero Ltd	Canada	*79*	80-01970
Stanley Aviation Corporation	USA	*617*	75-83900
Stocznia Szczecinska SA	Poland	*325*	75-02850
Strachan & Henshaw Ltd	UK	*512*	80-47400
Survival Products Inc.	USA	*618*	75-85200
Swiss Aircraft & Systems Co	Switzerland	*384*	55-00100
TAT Industries Ltd	Israel	*223*	75-22000
Techspace Aero SA	Belgium	*47*	75-06200
Terberg Benschop BV	Netherlands	*296*	75-07600
The NORDAM Group	USA	*599*	75-64880
Theodor Rapp GmbH Co	Germany	*173*	75-24600
Tracor Inc	USA	*623*	75-90000
Turbo Power Inc.	USA	*644*	80-33550
United Electric Motor Co.	USA	*624*	75-92100
Vacu-Blast Ltd	UK	*490*	75-62550
Vazduhoplovno Remonti Zavod	Yugoslavia	*661*	80-02500
Vestfold Bil & Karosseri A/S.	Norway	*311*	80-50000
Vickers Shipbuilding & Eng Ltd	UK	*491*	75-63100
Vinnell Corp.	USA	*644*	80-34350
Vipac Engineers & Scientists Ltd	Australia	*26*	80-22580
Vitrociset SpA	Italy	*240*	80-40000
Vosper International Ltd	UK	*514*	80-67000
Walashek Industrial & Marine Inc.	USA	*645*	80-34500
Wall Colomonoy Corp.	USA	*626*	75-94900
Wegmann & Co GmbH	Germany	*177*	75-33700
Weir Engineering Pty Ltd	Australia	*23*	75-23300
W.F. Stanley & Co. Ltd	UK	*486*	75-56350
Wood Group Aero Ltd	UK	*515*	80-76250
Yewlands Engineering Co. Ltd	UK	*495*	75-67520
ZF Luftfahrttechnik GmbH	Germany	*177*	75-36000

945 20 Maintenance, repair, overhaul services, aircraft

AAR Corp	USA	*629*	80-00035
Aero Corporation	USA	*630*	80-00650
Aeromot Group SA	Brazil	*55*	75-00630
Aeronautica Industrial SA	Spain	*365*	75-00200
Aeronavali Venezia	Italy	*240*	80-01500
Aeropartes Indústria Aeronáutica	Brazil	*58*	80-00100
AIM Enterprises Inc.	USA	*630*	80-01350
Air New Zealand Ltd	New Zealand	*303*	80-00100
Aircraft Tanks Ltd	UK	*429*	75-01730
AIROD Sdn Bhd	Malaysia	*278*	80-00050
Amper Programas SA	Spain	*365*	75-00480
Associated Aircraft Mfg & Sales Inc.	USA	*555*	75-09500
Aviation Support Tech. Int'l Inc.	USA	*556*	75-11050
CAE Aviation Ltd	Canada	*78*	80-00245
Canadian Marconi Co.	Canada	*69*	75-02600
Cenzin Co. Ltd	Poland	*324*	75-00200
Component Repair Technologies Inc.	USA	*633*	80-05896
Cranfield Aerospace Ltd	UK	*500*	80-07200
Cranfield University	UK	*419*	60-00350
Dassault Aviation Group	France	*135*	75-12975
Dowty Aerospace Aviation Services	Singapore	*349*	80-00450
Embraer	Brazil	*54*	55-05000
Embraer Aviation International	France	*150*	80-00500
Empresa Nacional de Aeronáutica	Chile	*84*	55-00550
FFA Flugzeugwerke Altenrhein AG	Switzerland	*386*	75-04350
Finavitec Oy	Finland	*119*	80-00580
Flight International Inc.	USA	*635*	80-11600
Fokker Services BV	Netherlands	*297*	80-06600
FR Aviation Limited	UK	*503*	80-10850
GEC-Marconi Avionics Ltd	UK	*454*	75-25320
Global Airtech International	USA	*579*	75-39769
Greenwich Air Services Inc.	USA	*636*	80-13170
Hellenic Aerospace Industry Ltd.	Greece	*192*	80-00750
Hindustan Aeronautics Ltd.	India	*204*	55-01000
Hong Kong Aircraft Eng. Co. Ltd	Hong Kong SAR	*198*	80-00200
Hunting Aviation Ltd	UK	*504*	80-12785
Hunting Aviation (Pty) Ltd	S. Africa	*358*	75-08000
Innotech Aviation Ltd.	Canada	*78*	80-00960
Israel Aircraft Industries Ltd	Israel	*219*	75-10200
Jet Aviation Zürich AG	Switzerland	*388*	80-01030
Kawasaki Heavy Industries Ltd	Japan	*252*	75-04350
Kay & Associates Inc.	USA	*637*	80-17660
Kiev Aircraft Repair Plant	Ukraine	*406*	55-01500
Litton Industries Inc.	USA	*590*	75-55400
Lockheed Martin Aircraft and	USA	*638*	80-18530
Lockheed Martin Aircraft Center	USA	*638*	80-18390
M International Export	USA	*639*	80-19780
Marshall of Cambridge Aerospace Ltd	UK	*466*	75-37530

Merex Inc.	USA	*595*	75-60250
Moscow Aircraft Production	Russia	*334*	55-13050
Motores Rolls-Royce Ltda	Brazil	*58*	80-20000
MRF TEREM	Bulgaria	*62*	55-03900
Nord-Micro Elektronik Feinmechanik	Germany	*173*	75-22700
Northrop Grumman Corporation	USA	*599*	75-65450
OGMA - Indústria Aeronáutica	Portugal	*327*	80-01500
Pacific Aerospace Corp. Ltd	New Zealand	*302*	75-01600
Pemco World Air Services	USA	*603*	75-69135
Pilatus Britten-Norman Ltd	UK	*474*	75-45400
Precision Standard Inc.	USA	*605*	75-71365
Qualitair Engineering Services Ltd	UK	*509*	80-25200
Raytheon Aerospace	USA	*641*	80-25650
RDM Technology BV	Netherlands	*294*	75-05780
Rohr Inc.	USA	*611*	75-76150
SABCA	Belgium	*46*	75-05590
SECA	France	*151*	80-01775
ShinMaywa Industries Ltd	Japan	*254*	75-07800
Shorts Support Services Division	UK	*511*	80-35600
Singapore Technologies Aerospace Ltd	Singapore	*348*	75-03905
Skyline Aviation BV	Netherlands	*297*	80-19000
Smiths Aerospace & Defense	USA	*615*	75-82090
Spar Aerospace Ltd	Canada	*76*	75-13000
Supercraft Ltd	UK	*487*	75-57910
Survival Products Inc.	USA	*618*	75-85200
Swiss Aircraft & Systems Co	Switzerland	*384*	55-00100
The Boeing Company	USA	*560*	75-14250
The NORDAM Group	USA	*599*	75-64880
United Beechcraft Inc.	USA	*644*	80-33600
Vazduhoplovno Remonti Zavod	Yugoslavia	*661*	80-02500

945 20 10 Maintenance, repair, overhaul services, aircraft airframe

AAR Corp	USA	*629*	80-00035
Aero Corporation	USA	*630*	80-00650
AIM Enterprises Inc.	USA	*630*	80-01350
AIROD Sdn Bhd	Malaysia	*278*	80-00050
Associated Aircraft Mfg & Sales Inc.	USA	*555*	75-09500
CAE Aviation Ltd	Canada	*78*	80-00245
Cranfield Aerospace Ltd	UK	*500*	80-07200
Cranfield University	UK	*419*	60-00350
Dowty Aerospace Aviation Services	Singapore	*349*	80-00450
Embraer Aviation International	France	*150*	80-00500
Empresa Nacional de Aeronáutica	Chile	*84*	55-00550
FFA Flugzeugwerke Altenrhein AG	Switzerland	*386*	75-04350
Finavitec Oy	Finland	*119*	80-00580
Hellenic Aerospace Industry Ltd	Greece	*192*	80-00750
Hong Kong Aircraft Eng. Co. Ltd	Hong Kong SAR	*198*	80-00200
Innotech Aviation Ltd	Canada	*78*	80-00960
Kay & Associates Inc.	USA	*637*	80-17660
Kiev Aircraft Repair Plant	Ukraine	*406*	55-01500
Lockheed Martin Aircraft and	USA	*638*	80-18530
Lockheed Martin Aircraft Center	USA	*638*	80-18390
Lucas Aerospace	Canada	*73*	75-09140
M International Export	USA	*639*	80-19780
Marshall of Cambridge Aerospace Ltd	UK	*466*	75-37530
Moscow Aircraft Production	Russia	*334*	55-13050
MTU Maintenance GmbH	Germany	*179*	80-01375
Pakistan Aeronautical Complex	Pakistan	*315*	55-01600
Pezetel	Poland	*325*	80-01000
Pilatus Britten-Norman Ltd	UK	*474*	75-45400
Qualitair Engineering Services Ltd	UK	*509*	80-25200
Shorts Support Services Division	UK	*511*	80-35600
Singapore Technologies Aerospace Ltd	Singapore	*348*	75-03905
SONACA SA	Belgium	*47*	75-06100
ST Mobile Aerospace Engineering Inc.	USA	*639*	80-21360
Supercraft Ltd	UK	*487*	75-57910
The Boeing Company	USA	*560*	75-14250
Vazduhoplovno Remonti Zavod	Yugoslavia	*661*	80-02500

945 20 11 Maintenance, repair, overhaul services, aircraft component

AAR Allen Group International	Netherlands	*296*	80-00050
AAR Corp	USA	*629*	80-00035
AAR Technical Service Center Corp	USA	*629*	80-00045
Abex GmbH Aerohydraul	Germany	*164*	75-00105
Aero Corporation	USA	*630*	80-00650
AIM Enterprises Inc.	USA	*630*	80-01350
Aircraft Equipment Overhauls & Sales	Australia	*23*	80-00650
AIROD Sdn Bhd	Malaysia	*278*	80-00050
Airscrew Howden Ltd	UK	*429*	75-02050
AlliedSignal Aerospace GmbH	Germany	*164*	75-00610
Amper Programas SA	Spain	*365*	75-00480
Associated Aircraft Mfg & Sales Inc.	USA	*555*	75-09500
Barnes Aerospace Group	USA	*557*	75-12000
Bosman Powersource Support BV	Netherlands	*292*	75-00800
Brenco Thermal Coating Pty Ltd	Australia	*13*	75-02740
CAE Aviation Ltd	Canada	*78*	80-00245
Cranfield Aerospace Ltd	UK	*500*	80-07200
Cranfield University	UK	*419*	60-00350
Dowty Aerospace Aviation Services	Singapore	*349*	80-00450
Dowty Aerospace Aviation Services	UK	*501*	80-08600
Electron Beam Processes Ltd	UK	*502*	80-09300
Embraer Aviation International	France	*150*	80-00500
Empresa Nacional de Aeronáutica	Chile	*84*	55-00550
Fabbrica Italiana Apparecchiature	Italy	*234*	75-08550
Fairey Hydraulics	UK	*450*	75-22255
FFA Flugzeugwerke Altenrhein AG	Switzerland	*386*	75-04350
FFV Aviocomp AB	Sweden	*379*	80-00620
Finavitec Oy	Finland	*119*	80-00580
FPT Industries Ltd	UK	*452*	75-24400
GEC-Marconi Avionics Ltd	UK	*454*	75-25320
Goodyear do Brasil Ltda	Brazil	*56*	75-07600
Goodyear Great Britain Ltd	UK	*455*	75-26135
Greenwich Air Services Inc.	USA	*636*	80-13170
Hellenic Aerospace Industry Ltd	Greece	*192*	80-00750
Hermetic Aircraft International Ltd	USA	*581*	75-43100
Hong Kong Aircraft Eng. Co. Ltd	Hong Kong SAR	*198*	80-00200
Howden Wade Ltd	UK	*459*	75-30260
Innotech Aviation Ltd	Canada	*78*	80-00960
Kay & Associates Inc.	USA	*637*	80-17660
Kiev Aircraft Repair Plant	Ukraine	*406*	55-01500
Litton Industries Inc.	USA	*590*	75-55400
Lockheed Martin Aircraft and	USA	*638*	80-18530
Lucas Aerospace - Australia	Australia	*25*	80-12850
Lucas Aerospace Power Transmission	USA	*592*	75-56265
M International Export	USA	*639*	80-19780
Marshall of Cambridge Aerospace Ltd	UK	*466*	75-37530
Messier-Bugatti	France	*140*	75-26400
Messier-Dowty Limited	UK	*467*	75-38700
Moscow Aircraft Production	Russia	*334*	55-13050
Omori Seikoki Co. Ltd	Japan	*254*	75-07310
Pakistan Aeronautical Complex	Pakistan	*315*	55-01600
Pilatus Britten-Norman Ltd	UK	*474*	75-45400
Qualitair Engineering Services Ltd	UK	*509*	80-25200
Repairtech International Inc.	USA	*641*	80-25880
Riva Calzoni SpA	Italy	*238*	75-20500
Singapore Technologies Aerospace Ltd	Singapore	*348*	75-03905
SONACA SA	Belgium	*47*	75-06100
SP aerospace and vehicle systems bv	**Netherlands**	***295***	**75-06860**
Spar Aviation Services	Canada	*76*	75-13020
ST Aerospace Systems Pte Ltd	Singapore	*350*	80-02005
Standard Aero Ltd	Canada	*79*	80-01970
Supercraft Ltd	UK	*487*	75-57910
The Boeing Company	USA	*560*	75-14250
Vazduhoplovno Remonti Zavod	Yugoslavia	*661*	80-02500
Vestfold Bil & Karosseri A/S	Norway	*311*	80-50000
Walter Kidde Aerospace	USA	*588*	75-52500

945 20 15 Maintenance, repair, overhaul services, aircraft instrument

AAR Allen Group International	Netherlands	*296*	80-00050
AAR Corp	USA	*629*	80-00035
AAR Technical Service Center Corp	USA	*629*	80-00045
Aero-Mach Labs Inc.	USA	*630*	80-00840
AIM Enterprises Inc.	USA	*630*	80-01350
Aircraft Instrument & Radio Co. Inc.	USA	*630*	80-01600
Amper Programas SA	Spain	*365*	75-00480
Ansett Australia Avionics	Australia	*24*	80-01010
Avionics Southwest	USA	*631*	80-03760
CAE Aviation Ltd	Canada	*78*	80-00245
Dowty Aerospace Aviation Services	Singapore	*349*	80-00450
Embraer Aviation International	France	*150*	80-00500
FFA Flugzeugwerke Altenrhein AG	Switzerland	*386*	75-04350
GEC-Marconi Avionics Ltd	UK	*454*	75-25320
Hellenic Aerospace Industry Ltd	Greece	*192*	80-00750

Hong Kong Aircraft Eng. Co. Ltd	Hong Kong SAR	198	80-00200
Innotech Aviation Ltd.	Canada	78	80-00960
Instrumentointi Oy	Finland	119	80-00950
Kay & Associates Inc.	USA	637	80-17660
Litton Industries Inc.	USA	590	75-55400
Lockheed Martin Aircraft and	USA	638	80-18530
Logistics Services International	USA	638	80-18800
M International Export	USA	639	80-19780
Marshall of Cambridge Aerospace Ltd	UK	466	75-37530
Moscow Aircraft Production	Russia	334	55-13050
Pakistan Aeronautical Complex	Pakistan	315	55-01600
Philips Calibration Services	Australia	25	80-16490
Philips Traffic & Engineering	Australia	20	75-16550
Pilatus Britten-Norman Ltd	UK	474	75-45400
SAIT Systems SA	Belgium	47	75-05600
Singapore Technologies Aerospace Ltd	Singapore	348	75-03905
ST Aerospace Systems Pte Ltd	Singapore	350	80-02005
The Boeing Company	USA	560	75-14250
Vazduhoplovno Remonti Zavod	Yugoslavia	661	80-02500

945 20 30 Maintenance, repair, overhaul services, helicopter

Aeronautica Industrial SA	Spain	365	75-00200
AIM Enterprises Inc.	USA	630	80-01350
American Eurocopter	USA	552	75-05790
Amper Programas SA	Spain	365	75-00480
Arab British Engine Co.	Egypt	109	74-00300
Aviation Support Tech. Int'l Inc.	USA	556	75-11050
CHC Helicopter Corporation	Canada	78	80-00280
Cranfield Aerospace Ltd	UK	500	80-07200
Cranfield University	UK	419	60-00350
Empresa Nacional de Aeronáutica	Chile	84	55-00550
Eurocopter Canada Ltd	Canada	71	75-05050
GEC-Marconi Avionics Ltd	UK	454	75-25320
Helitech Pty Ltd	Australia	25	80-08400
Hellenic Aerospace Industry Ltd.	Greece	192	80-00750
Hong Kong Aircraft Eng. Co. Ltd	Hong Kong SAR	198	80-00200
Innotech Aviation Ltd.	Canada	78	80-00960
Kay & Associates Inc.	USA	637	80-17660
Keystone Helicopter Corp.	USA	637	80-17780
Líder Táxi Aéreo SA	Brazil	57	75-12420
Lockheed Martin Aircraft and	USA	638	80-18530
M International Export	USA	639	80-19780
Marshall of Cambridge Aerospace Ltd	UK	466	75-37530
Merex Inc.	USA	595	75-60250
OGMA - Indústria Aeronáutica	Portugal	327	80-01500
Pezetel	Poland	325	80-01000
SABCA	Belgium	46	75-05590
ShinMaywa Industries Ltd	Japan	254	75-07800
Shorts Support Services Division	UK	511	80-35600
Standard Aero Ltd	Canada	79	80-01970
Survival Products Inc.	USA	618	75-85200
TUSAS Aerospace Industries Inc.	Turkey	403	75-02120
Vazduhoplovno Remonti Zavod	Yugoslavia	661	80-02500
ZF Luftfahrttechnik GmbH	Germany	177	75-36000

945 20 31 Maintenance, repair, overhaul services, helicopter airframe

Aeronautica Industrial SA	Spain	365	75-00200
AIM Enterprises Inc.	USA	630	80-01350
American Eurocopter	USA	552	75-05790
Associated Aircraft Mfg & Sales Inc.	USA	555	75-09500
Aviation Support Tech. Int'l Inc.	USA	556	75-11050
CHC Helicopter Corporation	Canada	78	80-00280
Cranfield Aerospace Ltd	UK	500	80-07200
Cranfield University	UK	419	60-00350
Empresa Nacional de Aeronáutica	Chile	84	55-00550
Hellenic Aerospace Industry Ltd.	Greece	192	80-00750
Innotech Aviation Ltd.	Canada	78	80-00960
Kay & Associates Inc.	USA	637	80-17660
Keystone Helicopter Corp.	USA	637	80-17780
Lockheed Martin Aircraft and	USA	638	80-18530
M International Export	USA	639	80-19780
Marshall of Cambridge Aerospace Ltd	UK	466	75-37530
Merex Inc.	USA	595	75-60250
OGMA - Indústria Aeronáutica	Portugal	327	80-01500
Shorts Support Services Division	UK	511	80-35600
SONACA SA	Belgium	47	75-06100
Vazduhoplovno Remonti Zavod	Yugoslavia	661	80-02500

945 20 39 Maintenance, repair, overhaul services, helicopter components

AAR Allen Group International	Netherlands	296	80-00050
Abex GmbH Aerohydraul	Germany	164	75-00105
Aeronautica Industrial SA	Spain	365	75-00200
AIM Enterprises Inc.	USA	630	80-01350
Airscrew Howden Ltd	UK	429	75-02050
Amper Programas SA	Spain	365	75-00480
Associated Aircraft Mfg & Sales Inc.	USA	555	75-09500
Bosman Powersource Support BV	Netherlands	292	75-00800
DRS Technologies, Inc.	USA	570	75-27400
Eurocopter Canada Ltd	Canada	71	75-05050
Fabbrica Italiana Apparecchiature	Italy	234	75-08550
Fairey Hydraulics	UK	450	75-22255
FPT Industries Ltd	UK	452	75-24400
GEC-Marconi Avionics Ltd	UK	454	75-25320
Giat Industries	France	126	55-00100
Hellenic Aerospace Industry Ltd.	Greece	192	80-00750
Innotech Aviation Ltd.	Canada	78	80-00960
Kay & Associates Inc.	USA	637	80-17660
Litton Industries Inc.	USA	590	75-55400
Lockheed Martin Aircraft and	USA	638	80-18530
Lucas Aerospace - Australia	Australia	25	80-12850
Lucas Aerospace Power Transmission	USA	592	75-56265
M International Export	USA	639	80-19780
Marshall of Cambridge Aerospace Ltd	UK	466	75-37530
Masco Service Corp.	USA	639	80-20300
Merex Inc.	USA	595	75-60250
Messier-Bugatti	France	140	75-26400
Messier-Dowty Limited	UK	467	75-38700
Omori Seikoki Co. Ltd	Japan	254	75-07310
Praxair Surface Technologies Inc.	USA	640	80-24680
Sermatech International Inc	USA	613	75-79530
SONACA SA	Belgium	47	75-06100
SP aerospace and vehicle systems bv	**Netherlands**	**295**	**75-06860**
Spar Aviation Services	Canada	76	75-13020
ST Aerospace Systems Pte Ltd	Singapore	350	80-02005
The Purdy Corp.	USA	606	75-71850
TUSAS Aerospace Industries Inc.	Turkey	403	75-02120
Vazduhoplovno Remonti Zavod	Yugoslavia	661	80-02500

945 25 Maintenance, repair, overhaul services, aircraft engine

AAR Corp	USA	629	80-00035
Aeronautica Industrial SA	Spain	365	75-00200
AIM Enterprises Inc.	USA	630	80-01350
Air New Zealand Ltd	New Zealand	303	80-00100
Aircraft Tanks Ltd	UK	429	75-01730
AIROD Sdn Bhd	Malaysia	278	80-00050
AlliedSignal Aerospace GmbH	Germany	164	75-00610
AOI Engine Factory	Egypt	109	55-01700
Bet Shemesh Engines Ltd	Israel	218	75-03200
CHC Helicopter Corporation	Canada	78	80-00280
Chromalloy Holland	Netherlands	297	80-04000
Embraer Aviation International	France	150	80-00500
Empresa Nacional de Aeronáutica	Chile	84	55-00550
Eurodef Aerospace & Defence	Greece	189	75-01295
FFA Flugzeugwerke Altenrhein AG	Switzerland	386	75-04350
Finavitec Oy	Finland	119	80-00580
GEC-Marconi Avionics Ltd	UK	454	75-25320
Greenwich Air Services Inc.	USA	636	80-13170
H+S Aviation Ltd	UK	503	80-12050
Hellenic Aerospace Industry Ltd.	Greece	192	80-00750
Hong Kong Aircraft Eng. Co. Ltd	Hong Kong SAR	198	80-00200
Interturbine Singapore Pte Ltd	Singapore	349	80-01000
Israel Aircraft Industries Ltd	Israel	219	75-10200
Jet Aviation Zürich AG	Switzerland	388	80-01030
Kay & Associates Inc.	USA	637	80-17660
Keystone Helicopter Corp.	USA	637	80-17780
Kiev Aircraft Repair Plant	Ukraine	406	55-01500
Lockheed Martin Aircraft and	USA	638	80-18530
M International Export	USA	639	80-19780
Merex Inc.	USA	595	75-60250
Northrop Grumman Corporation	USA	599	75-65450
OGMA - Indústria Aeronáutica	Portugal	327	80-01500

Pacific Aerospace Corp. Ltd	New Zealand	*302*	75-01600
Pilatus Britten-Norman Ltd	UK	*474*	75-45400
Pratt & Whitney Canada Inc	Canada	*75*	75-11400
Qualitair Engineering Services Ltd	UK	*509*	80-25200
Sabreliner Corp.	USA	*641*	80-26500
SECA	France	*151*	80-01775
Shorts Support Services Division	UK	*511*	80-35600
Singapore Technologies Aerospace Ltd	Singapore	*348*	75-03905
ST Aerospace Engines Pte Ltd	Singapore	*349*	80-02000
Standard Aero Ltd	Canada	*79*	80-01970
The Boeing Company	USA	*560*	75-14250
The NORDAM Group	USA	*599*	75-64880
Turbine Support Ltd	UK	*514*	80-58800
Vazduhoplovno Remonti Zavod	Yugoslavia	*661*	80-02500

945 25 90 Maintenance, repair, overhaul services, aircraft engine component

AAR Allen Group International	Netherlands	*296*	80-00050
AAR Corp	USA	*629*	80-00035
Aero Corporation	USA	*630*	80-00650
AIM Enterprises Inc.	USA	*630*	80-01350
Air New Zealand Ltd	New Zealand	*303*	80-00100
AIROD Sdn Bhd	Malaysia	*278*	80-00050
AlliedSignal Aerospace GmbH	Germany	*164*	75-00610
Barnes Aerospace Group	USA	*557*	75-12000
Bet Shemesh Engines Ltd	Israel	*218*	75-03200
Bosman Powersource Support BV	Netherlands	*292*	75-00800
Brenco Thermal Coating Pty Ltd	Australia	*13*	75-02740
Cadorath Aerospace Inc	Canada	*77*	80-00242
Chromalloy France	France	*133*	75-09650
CTL-Aerospace Inc.	USA	*567*	75-22700
Electron Beam Processes Ltd	UK	*502*	80-09300
Empresa Nacional de Aeronáutica	Chile	*84*	55-00550
FFA Flugzeugwerke Altenrhein AG	Switzerland	*386*	75-04350
Finavitec Oy	Finland	*119*	80-00580
GEC-Marconi Avionics Ltd	UK	*454*	75-25320
GKN Westland Industrial Products Ltd	UK	*455*	75-25675
H+S Aviation Ltd	UK	*503*	80-12050
Hellenic Aerospace Industry Ltd	Greece	*192*	80-00750
Hong Kong Aircraft Eng. Co. Ltd	Hong Kong SAR	*198*	80-00200
Interturbine Germany GmbH	Germany	*170*	75-14630
Interturbine Singapore Pte Ltd	Singapore	*349*	80-01000
Kay & Associates Inc.	USA	*637*	80-17660
Kiev Aircraft Repair Plant	Ukraine	*406*	55-01500
Lockheed Martin Aircraft and	USA	*638*	80-18530
LORI Heat Transfer Operations	USA	*592*	75-56090
Lucas Aerospace	Canada	*73*	75-09140
Lucas Aerospace - Australia	Australia	*25*	80-12850
Lucas Aerospace Power Transmission	USA	*592*	75-56265
M International Export	USA	*639*	80-19780
Merex Inc.	USA	*595*	75-60250
OGMA - Indústria Aeronáutica	Portugal	*327*	80-01500
Pilatus Britten-Norman Ltd	UK	*474*	75-45400
Praxair Surface Technologies Inc.	USA	*640*	80-24680
Riva Calzoni SpA	Italy	*238*	75-20500
Sermatech International Inc	USA	*613*	75-79530
Singapore Technologies Aerospace Ltd	Singapore	*348*	75-03905
Standard Aero Ltd	Canada	*79*	80-01970
Supercraft Ltd	UK	*487*	75-57910
The NORDAM Group	USA	*599*	75-64880
Wall Colomonoy Corp.	USA	*626*	75-94900
Walter Kidde Aerospace	USA	*588*	75-52500
Woodward Governor Company	USA	*628*	75-97290

945 30 Maintenance, repair, overhaul services, ship

Analysis & Technology Inc.	USA	*631*	80-02600
Astilleros Espanoles SA	Spain	*365*	75-01400
Australian Defence Industries Ltd	Australia	*9*	55-01000
Bender Shipbuilding & Repair Co.	USA	*558*	75-13300
Billfish Inc.	USA	*632*	80-04510
Bollinger Shipyard Inc.	USA	*560*	75-14600
Bremer Vulkan Marineschiffbau GmbH	Germany	*166*	75-03600
Cenzin Co. Ltd.	Poland	*324*	75-00200
Daewoo Heavy Industries Ltd	Korea (ROK)	*265*	75-00600
Damen Shipyards	Netherlands	*292*	75-01500
Davie Industries Inc.	Canada	*70*	75-04040
Devonport Royal Dockyard Ltd	UK	*501*	80-08500
Elefsis Shipyards SA	Greece	*188*	75-01130
Emsworth Shipyard	UK	*448*	75-20750
Fincantieri	Italy	*234*	75-09250
Garden Reach Shipbuilders	India	*203*	55-00300
GEC-Marconi Avionics Ltd	UK	*454*	75-25320
Guangzhou Shipyard International Co.	China PR	*89*	75-00400
Hellenic Shipyards Co.	Greece	*187*	55-00850
H.I. Fraser Pty Ltd	Australia	*16*	75-06835
Hijos de J. Barreras SA	Spain	*366*	75-03781
Indivers NV/Interturbine	Netherlands	*293*	75-03550
Interdyne Corp.	USA	*637*	80-16050
Israel Aircraft Industries Ltd	Israel	*219*	75-10200
Israel Shipyards Ltd.	Israel	*220*	75-10600
Jered Brown Brothers Inc.	USA	*586*	75-50380
John Crane Inc.	USA	*566*	75-22500
John M. Piris & Sons Ltd	Gibraltar	*185*	75-01000
KRZ Flotsky Arsenal	Bulgaria	*62*	55-02500
Litton Industries Inc.	USA	*590*	75-55400
Malakoff & Wm. Moore Ltd	UK	*466*	75-37065
Maritime Contractors Inc.	USA	*639*	80-20050
MIL Systems	Canada	*79*	80-01360
Morska Stocznia Remontowa SA	Poland	*324*	75-02600
MTH Ltd	UK	*470*	75-41330
National Steel & Shipbuilding Co.	USA	*598*	75-63900
Newport News Shipbuilding	USA	*599*	75-64300
Pacific Ship Repair & Fabrication	USA	*640*	80-23080
Park Products Ltd	UK	*473*	75-44000
Penang Shipbuilding Corporation	Malaysia	*277*	75-00800
Perry Engineering	Australia	*20*	75-16460
Philips Calibration Services	Australia	*25*	80-16490
Royal Schelde	Netherlands	*294*	75-06150
Saint John Shipbuilding Ltd	Canada	*75*	75-12300
Santierul Naval Constanta SA	Romania	*330*	75-00750
Schottel Sudamericana I.C.S.A.	Argentina	*4*	80-20000
Sermatech International Inc	USA	*613*	75-79530
Sevastopol JSC	Ukraine	*407*	75-06500
Shipyard Beograd	Yugoslavia	*661*	75-00690
SIRN sze	Italy	*240*	80-01950
Stanley Aviation Corporation	USA	*617*	75-83900
Stewart & Stevenson Services Inc.	USA	*618*	75-84180
Stocznia Szczecinska SA	Poland	*325*	75-02850
Strachan & Henshaw Ltd	UK	*512*	80-47400
Textron Marine & Land Systems	USA	*622*	75-88660
The Saudi Vosper Thornycroft	Saudi Arabia	*343*	80-02000
Tracor Inc	USA	*623*	75-90000
Transfield Shipbuilding WA	Australia	*22*	75-20812
Tulcea Shipyard	Romania	*331*	75-01200
UMC International plc.	UK	*514*	80-61000
United Defense L.P.	USA	*624*	75-92050
Van der Giessen-de Noord	Netherlands	*296*	75-07850
Vancouver Shipyards Co. Ltd	Canada	*77*	75-14435
Varna Shipyard	Bulgaria	*62*	75-05000
Viktor Lenac Ship Repair	Croatia	*95*	75-08000
Vipac Engineers & Scientists Ltd	Australia	*26*	80-22580
Vosper International Ltd	UK	*514*	80-67000
Walashek Industrial & Marine Inc.	USA	*645*	80-34500
Wheeler Systems Inc.	USA	*627*	75-96400
Yantar Shipyard	Russia	*339*	75-78900

945 30 39 Maintenance, repair, overhaul services, submarine

Analysis & Technology Inc.	USA	*631*	80-02600
Devonport Royal Dockyard Ltd	UK	*501*	80-08500
H.I. Fraser Pty Ltd	Australia	*16*	75-06835
MTH Ltd	UK	*470*	75-41330
Newport News Shipbuilding	USA	*599*	75-64300
Park Products Ltd	UK	*473*	75-44000
Philips Calibration Services	Australia	*25*	80-16490
Stanley Aviation Corporation	USA	*617*	75-83900
Strachan & Henshaw Ltd	UK	*512*	80-47400
Tracor Inc	USA	*623*	75-90000
UMC International plc.	UK	*514*	80-61000

945 35 Maintenance, repair, overhaul services, marine diesel engines

Detroit Diesel-Allison Australia Pty	Australia	*15*	75-04450
Devonport Royal Dockyard Ltd	UK	*501*	80-08500
Hong Leong-Lürssen Shipyard	Malaysia	*277*	75-00055
Israel Shipyards Ltd.	Israel	*220*	75-10600
John M. Piris & Sons Ltd	Gibraltar	*185*	75-01000
Merex Inc.	USA	*595*	75-60250

Mermaid Marine Engines Ltd.	UK	*467*	75-38550
MIL Systems	Canada	*79*	80-01360
Penang Shipbuilding Corporation	Malaysia	*277*	75-00800
Rolls-Royce Motor Cars Ltd	UK	*480*	75-50270
Saint John Shipbuilding Ltd	Canada	*75*	75-12300
Schottel Sudamericana I.C.S.A.	Argentina	*4*	80-20000
ShinMaywa Industries Ltd	Japan	*254*	75-07800
Stanley Aviation Corporation	USA	*617*	75-83900
Vancouver Shipyards Co. Ltd	Canada	*77*	75-14435
Viktor Lenac Ship Repair	Croatia	*95*	75-08000
Wartsila Diesel France SA	France	*149*	75-47070
Wartsila Diesel International Sarl	France	*149*	75-47060

945 40 Maintenance, repair, overhaul services, ground-based equipment

AAI Corp.	USA	*545*	75-00020
ABB Leitungsbau GmbH	Germany	*164*	75-00055
Accuracy International Ltd	UK	*427*	75-00360
Aircraft Maintenance Support	UK	*428*	75-01580
Allied Research Corp.	USA	*549*	75-04670
Aveling Barford (Machines) plc	UK	*432*	75-04550
CTL-Aerospace Inc.	USA	*567*	75-22700
Devonport Royal Dockyard Ltd	UK	*501*	80-08500
Didsbury Engineering Co. Ltd	UK	*445*	75-18050
DRS Technologies, Inc.	USA	*570*	75-27400
Dytecna Ltd	UK	*501*	80-08900
Elintech GmbH	Germany	*178*	80-00550
Fluor Daniel Inc.	USA	*635*	80-11720
FN Herstal SA	Belgium	*46*	75-03600
GEC-Marconi Communications Ltd	UK	*454*	75-25335
General Motors of Canada Ltd	Canada	*71*	75-05900
Giat Industries	France	*126*	55-00100
Grove Europe Ltd	**UK**	***456***	**75-27030**
Hellenic Aerospace Industry Ltd.	Greece	*192*	80-00750
Horstman Defence Systems Ltd.	UK	*459*	75-29920
Howard Porter Pty Ltd	Australia	*17*	75-08720
Israel Aircraft Industries Ltd	Israel	*219*	75-10200
Kay & Associates Inc.	USA	*637*	80-17660
M. Paz Logistics Ltd	Israel	*224*	80-01600
Microelectronics, guidance and	Turkey	*402*	55-00050
Nanoquest Defence Products Ltd	UK	*470*	75-41890
NEC Corporation	Japan	*253*	75-06350
Northrop Grumman Corporation.	USA	*599*	75-65450
Nottingham International	USA	*640*	80-22000
OGMA - Indústria Aeronáutica	Portugal	*327*	80-01500
Omori Seikoki Co. Ltd	Japan	*254*	75-07310
OTEC Support Services Ltd	UK	*508*	80-19600
Park Products Ltd	UK	*473*	75-44000
Philips Calibration Services	Australia	*25*	80-16490
QES Ltd	UK	*509*	80-25000
Radio Holland Electronics BV	Netherlands	*297*	80-18300
Riva Calzoni SpA	Italy	*238*	75-20500
Serco Aviation Services	UK	*511*	80-34150
Shorts Support Services Division	UK	*511*	80-35600
Singapore Electronic & Engineering	Singapore	*349*	80-01960
Spar Aviation Services	Canada	*76*	75-13020
Stanley Aviation Corporation	USA	*617*	75-83900
Stewart & Stevenson Services Inc.	USA	*618*	75-84180
Sverdrup Technology Inc.	USA	*643*	80-30680
Terberg Benschop BV	Netherlands	*296*	75-07600
The Graphics Co. of America	USA	*579*	75-40360
Theodor Rapp GmbH Co	Germany	*173*	75-24600
Tracor Inc	USA	*623*	75-90000
United Electric Motor Co.	USA	*624*	75-92100
Vazduhoplovno Remonti Zavod	Yugoslavia	*661*	80-02500
Vipac Engineers & Scientists Ltd	Australia	*26*	80-22580
Vitrociset SpA	Italy	*240*	80-40000

945 40 70 Maintenance, repair, overhaul services, armoured vehicle

A Goninan & Co. Ltd	Australia	*16*	75-07760
Allied Research Corp.	USA	*549*	75-04670
Alvis Industries Ltd	UK	*430*	75-02850
ASCO Industries NV/SA	Belgium	*45*	75-00550
Cenzin Co. Ltd.	Poland	*324*	75-00200
Diehl GmbH & Co	Germany	*167*	75-06500
General Motors of Canada Ltd	Canada	*71*	75-05900
Giat Industries	France	*126*	55-00100
Israel Aircraft Industries Ltd	Israel	*219*	75-10200
Krauss-Maffei Wehrtechnik GmbH	Germany	*171*	75-16345
KUKA Wehrtechnik GmbH	Germany	*171*	75-17200
M. Paz Logistics Ltd	Israel	*224*	80-01600
Nanoquest Defence Products Ltd	UK	*470*	75-41890
Neubrandenburger Fahrzeugwerke GmbH	Germany	*179*	80-01450
Nimda Co. Ltd.	Israel	*224*	80-01400
Norca Machinery Company	USA	*599*	75-64800
Nottingham International	USA	*640*	80-22000
Panhard & Levassor SA	France	*140*	75-28750
Patria Vehicles Oy	Finland	*118*	75-01710
RDM Technology BV	Netherlands	*294*	75-05780
Singapore Ordnance Engineering	Singapore	*349*	80-01970
Singapore Technologies Automotive	Singapore	*348*	75-04280
SP aerospace and vehicle systems bv	**Netherlands**	***295***	**75-06860**
Stanley Aviation Corporation	USA	*617*	75-83900
Vickers Defence Systems	UK	*491*	75-63040

945 40 70 10 Maintenance, repair, overhaul services, transmissions, tracked armoured vehicle

Alvis Transmission Ltd.	UK	*430*	75-02862
Aveling Barford (Machines) plc	UK	*432*	75-04550
David Brown Defence Equipment Ltd	UK	*437*	75-10150
Nottingham International	USA	*640*	80-22000
SP aerospace and vehicle systems bv	**Netherlands**	***295***	**75-06860**

945 40 70 30 Maintenance, repair, overhaul services, transmissions, wheeled armoured vehicle

Alvis Transmission Ltd.	UK	*430*	75-02862
Aveling Barford (Machines) plc	UK	*432*	75-04550
David Brown Defence Equipment Ltd	UK	*437*	75-10150
SP aerospace and vehicle systems bv	**Netherlands**	***295***	**75-06860**

945 40 70 50 Maintenance, repair, overhaul services, transmissions, non-armoured vehicle

Alvis Transmission Ltd.	UK	*430*	75-02862
Aveling Barford (Machines) plc	UK	*432*	75-04550
David Brown Defence Equipment Ltd	UK	*437*	75-10150
Thompson Carmichael Ltd	UK	*488*	75-59525

946 Reconditioning, refit, conversion services

A Goninan & Co. Ltd	Australia	*16*	75-07760
Aero Corporation	USA	*630*	80-00650
Air New Zealand Ltd	New Zealand	*303*	80-00100
Alvis Unipower Ltd	UK	*430*	75-02863
Analysis & Technology Inc.	USA	*631*	80-02600
Bender Shipbuilding & Repair Co.	USA	*558*	75-13300
C.L.S. Group	UK	*441*	75-13860
CTL-Aerospace Inc.	USA	*567*	75-22700
Dassault Aviation Group	France	*135*	75-12975
Devonport Royal Dockyard Ltd	UK	*501*	80-08500
EG&G Washington Analytical Svcs	USA	*634*	80-09450
Empresa Nacional de Aeronáutica	Chile	*84*	55-00550
Engineering & Professional Serv.Inc.	USA	*634*	80-10150
FFA Flugzeugwerke Altenrhein AG	Switzerland	*386*	75-04350
Flight Refuelling Limited	UK	*451*	75-23850
FN Herstal SA	Belgium	*46*	75-03600
FR Aviation Limited	UK	*503*	80-10850
GEC-Marconi Avionics Ltd	UK	*454*	75-25320
General Motors of Canada Ltd	Canada	*71*	75-05900
GLS Gesellschaft für Logistischen	Germany	*178*	80-00750
Hellenic Aerospace Industry Ltd.	Greece	*192*	80-00750
Israel Aircraft Industries Ltd	Israel	*219*	75-10200
IVECO Fiat SpA	Italy	*235*	75-13230
John M. Piris & Sons Ltd	Gibraltar	*185*	75-01000
Kiev Aircraft Repair Plant	Ukraine	*406*	55-01500
KUKA Wehrtechnik GmbH	Germany	*171*	75-17200
Lynx Corp Pty Ltd	Australia	*25*	80-12980
Maritime Contractors Inc.	USA	*639*	80-20050
Marshall of Cambridge Aerospace Ltd	UK	*466*	75-37530
MIL Systems	Canada	*79*	80-01360
Newport News Shipbuilding	USA	*599*	75-64300
Nimda Co. Ltd.	Israel	*224*	80-01400
Parker-Hale Ltd	UK	*473*	75-43850

Patria Vehicles Oy	Finland	*118*	75-01710
Qualitair Engineering Services Ltd	UK	*509*	80-25200
Reumech-OMC	S. Africa	*359*	75-18265
SEMCOR Inc.	USA	*642*	80-27750
Sevastopol JSC	Ukraine	*407*	75-06500
Singapore Technologies Automotive	Singapore	*348*	75-04280
SP aerospace and vehicle systems bv	**Netherlands**	***295***	**75-06860**
Stanley Aviation Corporation	USA	*617*	75-83900
Stewart & Stevenson Services Inc.	USA	*618*	75-84180
Swiss Aircraft & Systems Co	Switzerland	*384*	55-00100
Tracor Inc	USA	*623*	75-90000
Vestfold Bil & Karosseri A/S.	Norway	*311*	80-50000
Vickers Shipbuilding & Eng Ltd	UK	*491*	75-63100
Vosper International Ltd	UK	*514*	80-67000
Wegmann & Co GmbH	Germany	*177*	75-33700
Wegmann USA Inc.	USA	*627*	75-95543
Yantar Shipyard	Russia	*339*	75-78900

946 20 Reconditioning, refit, conversion services, aircraft

Aero Corporation	USA	*630*	80-00650
Aero Space Modifications	USA	*630*	80-00850
Air New Zealand Ltd	New Zealand	*303*	80-00100
Dassault Aviation Group	France	*135*	75-12975
Dowty Aerospace Aviation Services	Singapore	*349*	80-00450
Embraer Aviation International	France	*150*	80-00500
Empresa Nacional de Aeronáutica	Chile	*84*	55-00550
Engineering & Professional Serv.Inc.	USA	*634*	80-10150
European Helicopters Ltd	UK	*502*	80-09850
FFA Flugzeugwerke Altenrhein AG	Switzerland	*386*	75-04350
Flight International Inc.	USA	*635*	80-11600
FR Aviation Limited	UK	*503*	80-10850
GEC-Marconi Avionics Ltd	UK	*454*	75-25320
Hellenic Aerospace Industry Ltd	Greece	*192*	80-00750
Israel Aircraft Industries Ltd	Israel	*219*	75-10200
Kay & Associates Inc.	USA	*637*	80-17660
Keystone Helicopter Corp.	USA	*637*	80-17780
Kiev Aircraft Repair Plant	Ukraine	*406*	55-01500
Lockheed Martin Aircraft Center	USA	*638*	80-18390
Marshall of Cambridge Aerospace Ltd	UK	*466*	75-37530
Northrop Grumman Corporation	USA	*599*	75-65450
Pacific Aerospace Corp. Ltd	New Zealand	*302*	75-01600
Pemco World Air Services	USA	*603*	75-69135
Qualitair Engineering Services Ltd	UK	*509*	80-25200
SECA	France	*151*	80-01775
SEMCOR Inc.	USA	*642*	80-27750
Shorts Support Services Division	UK	*511*	80-35600
Swiss Aircraft & Systems Co	Switzerland	*384*	55-00100
TUSAS Aerospace Industries Inc.	Turkey	*403*	75-02120

946 30 Reconditioning, refit, conversion services, ship

Analysis & Technology Inc.	USA	*631*	80-02600
Astilleros Espanoles SA	Spain	*365*	75-01400
Australian Defence Industries Ltd	Australia	*9*	55-01000
Bender Shipbuilding & Repair Co.	USA	*558*	75-13300
Bollinger Shipyard Inc.	USA	*560*	75-14600
Bremer Vulkan Marineschiffbau GmbH	Germany	*166*	75-03600
Daewoo Heavy Industries Ltd	Korea (ROK)	*265*	75-00600
Damen Shipyards	Netherlands	*292*	75-01500
Devonport Royal Dockyard Ltd	UK	*501*	80-08500
EG&G Washington Analytical Svcs	USA	*634*	80-09450
Elefsis Shipyards SA	Greece	*188*	75-01130
Emsworth Shipyard	UK	*448*	75-20750
Fincantieri	Italy	*234*	75-09250
Garden Reach Shipbuilders	India	*203*	55-00300
Graseby Dynamics Ltd	UK	*455*	75-26590
Hong Leong-Lürssen Shipyard	Malaysia	*277*	75-00055
Israel Aircraft Industries Ltd	Israel	*219*	75-10200
Israel Shipyards Ltd.	Israel	*220*	75-10600
John M. Piris & Sons Ltd	Gibraltar	*185*	75-01000
KRZ Flotsky Arsenal	Bulgaria	*62*	55-02500
Litton Industries Inc.	USA	*590*	75-55400
Maritime Contractors Inc.	USA	*639*	80-20050
Marr Technical Services Ltd	UK	*506*	80-15620
MIL Systems	Canada	*79*	80-01360
Newport News Shipbuilding	USA	*599*	75-64300
Park Products Ltd	UK	*473*	75-44000
Penang Shipbuilding Corporation	Malaysia	*277*	75-00800
Saint John Shipbuilding Ltd	Canada	*75*	75-12300
Schottel Sudamericana I.C.S.A.	Argentina	*4*	80-20000
Sevastopol JSC	Ukraine	*407*	75-06500
Stanley Aviation Corporation	USA	*617*	75-83900
Strachan & Henshaw Ltd	UK	*512*	80-47400
Svendborg Vaerft A/S	Denmark	*104*	75-19800
Three Quays Marine Services Ltd	UK	*514*	80-55400
Tracor Inc	USA	*623*	75-90000
United Defense L.P.	USA	*624*	75-92050
Van der Giessen-de Noord	Netherlands	*296*	75-07850
Vancouver Shipyards Co. Ltd	Canada	*77*	75-14435
Vipac Engineers & Scientists Ltd	Australia	*26*	80-22580
Vosper International Ltd	UK	*514*	80-67000
Yantar Shipyard	Russia	*339*	75-78900

946 40 Reconditioning, refit, conversion services, ground-based equipment

Arab Organization for Ind.	Egypt	*109*	55-01700
Australian Defence Industries Ltd	Australia	*9*	55-01000
Engineering & Professional Serv.Inc.	USA	*634*	80-10150
Fabbrica Italiana Apparecchiature	Italy	*234*	75-08550
Fluor Daniel Inc.	USA	*635*	80-11720
Israel Aircraft Industries Ltd	Israel	*219*	75-10200
Kay & Associates Inc.	USA	*637*	80-17660
Marshall of Cambridge Aerospace Ltd	UK	*466*	75-37530
Park Products Ltd	UK	*473*	75-44000
Serco Aviation Services	UK	*511*	80-34150
Spar Aviation Services	Canada	*76*	75-13020
Stanley Aviation Corporation	USA	*617*	75-83900
Stewart & Stevenson Services Inc.	USA	*618*	75-84180
Theodor Rapp GmbH Co	Germany	*173*	75-24600
Tracor Inc	USA	*623*	75-90000
Vammas Ltd	**Finland**	***117***	**55-02220**
Vestfold Bil & Karosseri A/S.	Norway	*311*	80-50000
Vickers Shipbuilding & Eng Ltd	UK	*491*	75-63100
Vipac Engineers & Scientists Ltd	Australia	*26*	80-22580

946 40 70 Reconditioning, refit, conversion services, armoured vehicle

A Goninan & Co. Ltd	Australia	*16*	75-07760
Australian Defence Industries Ltd	Australia	*9*	55-01000
AV Technology LLC	USA	*556*	75-11405
Electron Beam Processes Ltd	UK	*502*	80-09300
Engineering & Professional Serv.Inc.	USA	*634*	80-10150
General Motors of Canada Ltd	Canada	*71*	75-05900
GLS Gesellschaft für Logistischen	Germany	*178*	80-00750
Israel Aircraft Industries Ltd	Israel	*219*	75-10200
Jung Jungenthal GmbH	Germany	*170*	75-15130
KUKA Wehrtechnik GmbH	Germany	*171*	75-17200
Marshall of Cambridge Aerospace Ltd	UK	*466*	75-37530
Nimda Co. Ltd.	Israel	*224*	80-01400
Panhard & Levassor SA	France	*140*	75-28750
Patria Vehicles Oy	Finland	*118*	75-01710
Singapore Technologies Automotive	Singapore	*348*	75-04280
SP aerospace and vehicle systems bv	**Netherlands**	***295***	**75-06860**
Stewart & Stevenson Services Inc.	USA	*618*	75-84180
Vammas Ltd	**Finland**	***117***	**55-02220**
Vickers Defence Systems	UK	*491*	75-63040
Wegmann USA Inc.	USA	*627*	75-95543

946 40 80 Reconditioning, refit, conversion services, weapon & weapon system

RDM Technology BV	Netherlands	*294*	75-05780

946 40 80 10 Reconditioning, refit, conversion services, weapon, ground based

RDM Technology BV	Netherlands	*294*	75-05780

946 40 90 Ordnance retrieval & recycling services

AEA Technology plc	UK	*496*	80-00480
Carolina Metals Inc.	USA	*562*	75-16760
CMS Inc.	USA	*564*	75-19170
M. Paz Logistics Ltd	Israel	*224*	80-01600
Milldale Limited	UK	*468*	75-39500
Ordnance & Tactical Systems Division	USA	*606*	75-71472

Primex Technologies, Inc	USA	*606*	75-71472
United Defense L.P..	USA	*624*	75-92050

947 Research & development, study services

Abacus Technology Corp..	USA	*629*	80-00050
Advanced Research & Applications.	USA	*546*	75-00940
Advanced Technology Ltd.	Israel	*217*	75-00400
AEA Technology plc	UK	*496*	80-00480
Aero Corporation	USA	*630*	80-00650
Aerospace Project Development Group	Switzerland	*384*	60-00020
All-Russian Institute of Aviation	Russia	*333*	55-01050
American Management Systems	USA	*630*	80-02350
AMI International	USA	*631*	80-02470
Analysis & Technology Inc.	USA	*631*	80-02600
Armalon Ltd	UK	*431*	75-03729
Atlantic Research Corp.	USA	*555*	75-10150
ATS Aerospace Inc.	Canada	*68*	75-00752
Australian Technical Publications	Australia	*24*	80-01880
bd Systems Inc..	USA	*632*	80-04140
Bend Research	USA	*540*	60-02500
BGS Systems Inc..	USA	*632*	80-04505
Bradley Electronics Ltd	UK	*435*	75-07860
CAE Aviation Ltd	Canada	*78*	80-00245
Calspan SRL Technology.	USA	*540*	60-03000
Cambridge Consultants Ltd	UK	*499*	80-05300
Cas Inc.	USA	*633*	80-05230
CEA Technologies Pty Ltd	Australia	*24*	80-03100
Central Aerohydrodynamics Institute	Russia	*335*	60-00300
Chamberlain Manufacturing Corp..	USA	*563*	75-17950
Chase Initiatives.	USA	*633*	80-05450
Chromalloy France	France	*133*	75-09650
CMPS&F Pty Limited	Australia	*24*	80-03600
Consolidated Cybernetics Controls	Canada	*78*	80-00360
Cortana Corp.	USA	*540*	60-03500
Cranfield Aerospace Ltd	UK	*500*	80-07200
CTL-Aerospace Inc.	USA	*567*	75-22700
Dassault Aviation Group	France	*135*	75-12975
Datel Technology Ltd.	UK	*500*	80-07750
David Sarnoff Research Center	USA	*540*	60-19150
Davies Industrial Communications Ltd	UK	*444*	75-17010
DCS Corp.	USA	*634*	80-07400
Defense Technology Foundation	Japan	*255*	80-00125
Delex Systems Inc.	USA	*634*	80-08020
Delfin Systems	USA	*568*	75-25390
Design Bureau of	Russia	*336*	75-03150
DRS Technologies, Inc.	USA	*570*	75-27400
Dynamit Nobel GmbH	Germany	*167*	75-07100
Dytecna Ltd	UK	*501*	80-08900
EASAMS Ltd	UK	*501*	80-09000
EDS Defence Ltd	UK	*501*	80-09200
EG&G Inc.	USA	*571*	75-29555
EG&G Washington Analytical Svcs	USA	*634*	80-09450
Eilat Consultants	Israel	*224*	80-00500
Electronic Development Assoc. Inc.	USA	*634*	80-09750
Elintech GmbH	Germany	*178*	80-00550
Emprise Corp.	USA	*634*	80-09950
Engineering & Professional Serv.Inc.	USA	*634*	80-10150
Engineering and Physical Sciences	UK	*419*	60-00400
EPS Logistics Technology Ltd	UK	*502*	80-09550
ERA Technology Ltd	UK	*502*	80-09600
Esquimalt Defence Research.	Canada	*67*	60-00500
E2S NV	Belgium	*48*	80-00600
Fincantieri	Italy	*234*	75-09250
Fluor Daniel Inc.	USA	*635*	80-11720
Frazer-Nash Consultancy Limited	UK	*503*	80-10950
GEC-Marconi Avionics Ltd	UK	*454*	75-25320
GEC-Marconi Materials Technology Ltd	UK	*419*	60-00600
General Atomics	USA	*578*	75-37447
German Aerospace Research Estab.	Germany	*161*	60-00800
Gordon Consulting Engineers Ltd	Israel	*224*	80-00750
Graseby Dynamics Ltd	UK	*455*	75-26590
Grove Europe Ltd	**UK**	***456***	**75-27030**
Halifax Shipyard Limited	Canada	*78*	80-00810
Helio Mirror Co. Ltd.	UK	*457*	75-28900
Hellenic Aerospace Industry Ltd.	Greece	*192*	80-00750
Hellenic Arms Industry	Greece	*187*	55-00800
Horizons Technology Inc.	USA	*636*	80-14450
Horrigan Analytics	USA	*636*	80-14480
Hughes Elcan Optical Technologies	Canada	*72*	75-07650
Hugo Marom Aviation Consultants Ltd	Israel	*224*	80-00880
Hunting Engineering Ltd	UK	*459*	75-30700
Imco Industries Ltd	Israel	*219*	75-09400
Industrieanlagen-Betriebs GmbH	Germany	*178*	80-00945
Inovex Technology Ltd	UK	*504*	80-13310
Institute of Space Constructions	Georgia	*156*	60-01000
Instrument Design Bureau - KBP Tula	Russia	*335*	60-02100
Interactive Dev. Environments	USA	*637*	80-15950
Interstate Technical Products	UK	*461*	75-31946
Israel Aircraft Industries Ltd	Israel	*219*	75-10200
Japan Aircraft Manufacturing Co. Ltd.	Japan	*252*	75-03200
Jered Brown Brothers Inc.	USA	*586*	75-50380
Jet Aviation Zürich AG	Switzerland	*388*	80-01030
Jones & Rickard Pty Ltd	Australia	*18*	75-10750
JPS Communications Inc.	USA	*587*	75-51220
Kaman Electromagnetics Corp.	USA	*637*	80-17500
Kaman Sciences Corp.	USA	*637*	80-17600
KUKA Wehrtechnik GmbH	Germany	*171*	75-17200
Lansdowne Technologies Inc	Canada	*78*	80-01205
Laser & Atomic Research & Devt	Croatia	*94*	60-00450
Laser-Scan	UK	*506*	80-14210
Lightweight Body Armour	UK	*464*	75-35650
LMS International	Belgium	*48*	80-00900
Logica UK Limited.	UK	*506*	80-14350
M. Paz Logistics Ltd	Israel	*224*	80-01600
M. Rosenblatt & Son Inc..	USA	*641*	80-26450
MacDonald Dettwiler and Associates.	Canada	*78*	80-01310
Management Consulting & Research	USA	*638*	80-19680
Mark Three	USA	*540*	60-13150
Maxwell Technologies Inc..	USA	*594*	75-59250
Menasco Aerospace Ltd	Canada	*74*	75-09600
MFD International Ltd.	UK	*506*	80-16450
Microelectronics, guidance and	Turkey	*402*	55-00050
Milo Holdings Ltd	UK	*506*	80-16850
MPB Technologies Inc	Canada	*74*	75-09855
MS Instruments plc	UK	*469*	75-41265
MTH Ltd	UK	*470*	75-41330
Mustang Survival Corp.	Canada	*74*	75-09860
Mustang Survival (USA) Corp	USA	*597*	75-63480
NanoMaterials, Inc.	USA	*598*	75-63590
National Aerospace Laboratory	Netherlands	*290*	60-01400
National Systems Management Corp.	USA	*639*	80-21650
NEL	UK	*419*	60-01400
Nichols Research Corp..	USA	*639*	80-21780
NNC Ltd	UK	*507*	80-18500
Office National d'Études et de	France	*127*	60-05000
Orbital Engine Company (Australia)	Australia	*25*	80-15830
Orlando Technology Inc.	USA	*540*	60-16000
Pacer Systems Inc.	USA	*602*	75-68000
Perceptronics Inc..	USA	*603*	75-69270
Pilatus Aircraft Ltd	Switzerland	*387*	75-10100
Programming Research Ltd	UK	*509*	80-24700
Prometey Central Research Institute	Russia	*335*	60-03555
PT Centronix	Indonesia	*210*	75-00200
Public Safety Group Inc.	USA	*641*	80-24955
Racal Research Ltd	UK	*509*	80-26300
Reade Advanced Materials	USA	*608*	75-73860
Research Associates of Syracuse.	USA	*641*	80-25900
Research Triangle Institute.	USA	*540*	60-18000
Ricardo Hitec Ltd	UK	*509*	80-27840
Ring Sights Ltd	UK	*509*	80-28200
RKI/Cleanroom Services	USA	*641*	80-26200
Rohde & Schwarz France Telecom	France	*142*	75-32900
Rollab Aktiebolag	Sweden	*380*	80-01850
Rolls-Royce (Canada) Ltd.	Canada	*75*	75-12120
Royal Military College of Science	UK	*420*	60-01820
Royal Ordnance	UK	*481*	75-50620
Safety Consulting Engineers Inc.	USA	*641*	80-26680
SAIT Systems SA	Belgium	*47*	75-05600
Santa Barbara SA	Spain	*364*	55-01000
Scaled Composites Inc.	USA	*541*	60-19160
SEMCOR Inc.	USA	*642*	80-27750
Sener Ingenieria y Sistemas SA	Spain	*367*	75-05200
SIMA PERU	Peru	*319*	55-01950
Sira Ltd.	UK	*484*	75-54250
SISDEF Ltda	Chile	*85*	80-01940
Smith System Engineering Ltd.	UK	*512*	80-39500
SOPEMEA	France	*151*	80-01997
Southwest Research Institute	USA	*541*	60-19200
Space Systems Finland Ltd	Finland	*118*	75-02100
Spire Corp.	USA	*643*	80-29400
SRI International	USA	*643*	80-29600
SRL Operations	USA	*617*	75-83550
ST Aerospace Engineering Pte Ltd	Singapore	*349*	80-01990
Stanley Aviation Corporation	USA	*617*	75-83900
Strachan & Henshaw Ltd	UK	*512*	80-47400
Sverdrup Corp.	USA	*643*	80-30650
Sverdrup Technology Inc.	USA	*643*	80-30680
Swiss Aircraft & Systems Co	Switzerland	*384*	55-00100

TA Group Ltd	UK	*513*	80-50110
Talos Engineering SA	Greece	*193*	80-02010
Tetra Corp.	USA	*644*	80-32700
The Titan Corp	USA	*622*	75-89620
Thomson Sintra	France	*148*	75-45135
Thomson-CSF Systems Canada Inc.	Canada	*79*	80-02020
TNO Defence Research	Netherlands	*290*	60-20100
TNO Physics & Electronics Laboratory	Netherlands	*290*	60-20100
TNO Prins Maurits Laboratory	Netherlands	*290*	60-20100
Tracor Inc	USA	*623*	75-90000
Transfield Pty Ltd	Australia	*22*	75-20810
UBC Inc.	USA	*624*	75-91383
Vanguard Research Inc.	USA	*644*	80-34100
Vickers Shipbuilding & Eng Ltd	UK	*491*	75-63100
Vipac Engineers & Scientists Ltd	Australia	*26*	80-22580
Vitro Corp.	USA	*644*	80-34400
Vitrociset SpA	Italy	*240*	80-40000
Vosper International Ltd	UK	*514*	80-67000
Wegmann & Co GmbH	Germany	*177*	75-33700
Yarrow Shipbuilders Ltd	UK	*495*	75-67500
Yewlands Engineering Co. Ltd.	UK	*495*	75-67520
3C Systems, Inc.	USA	*632*	80-05000

947 20 Software development services

Advanced System Architectures Ltd	UK	*496*	80-00420
Advanced Technology Ltd.	Israel	*217*	75-00400
AEA Technology plc	UK	*496*	80-00480
Aerobel Defence Technology Ltd	UK	*496*	80-00520
American Management Systems	USA	*630*	80-02350
Analysis & Technology Inc.	USA	*631*	80-02600
Andersen Consulting	Australia	*24*	80-00910
Anite Systems Ltd	UK	*497*	80-01240
Aonix Europe Ltd	UK	*497*	80-01251
Applied Data Sciences Inc.	USA	*554*	75-08050
Aspect Computing Pty Ltd	Australia	*24*	80-01150
ATS Aerospace Inc.	Canada	*68*	75-00752
Australian Technical Publications	Australia	*24*	80-01880
bd Systems Inc.	USA	*632*	80-04140
BGS Systems Inc.	USA	*632*	80-04505
British Aerospace Australia Ltd	Australia	*13*	75-02750
BRUHN NewTech A/S	Denmark	*104*	80-00050
CAE Invertron Ltd.	UK	*438*	75-11145
Calspan SRL Technology	USA	*540*	60-03000
Cambridge Consultants Ltd	UK	*499*	80-05300
Cas Inc.	USA	*633*	80-05230
CEA Technologies Pty Ltd	Australia	*24*	80-03100
Central Research Laboratories Ltd	UK	*499*	80-05895
Computer Resources International A/S	Denmark	*104*	80-00100
Computer Sciences Corp.	USA	*565*	75-20720
Computerised Training Systems Ltd	UK	*442*	75-14650
Computing Devices Canada Ltd	Canada	*70*	75-03600
Computing Devices Company Ltd	UK	*442*	75-14700
Contahal Ltd	Israel	*224*	80-00300
Cranfield Aerospace Ltd	UK	*500*	80-07200
Cranfield University	UK	*419*	60-00350
Cray Marine	UK	*443*	75-15915
Crew Services Ltd	UK	*500*	80-07370
DCS Corp.	USA	*634*	80-07400
Delex Systems Inc.	USA	*634*	80-08020
Delfin Systems	USA	*568*	75-25390
Deloitte & Touche Consulting Group	UK	*501*	80-08300
Designers & Planners Inc.	USA	*634*	80-08300
D.E.3I Informatique	France	*150*	80-00400
DRS Technologies, Inc.	USA	*570*	75-27400
DUAL Incorporated	USA	*634*	80-08670
EASAMS Ltd	UK	*501*	80-09000
EDS Defence Ltd	UK	*501*	80-09200
Emprise Corp.	USA	*634*	80-09950
Engineering & Professional Serv.Inc.	USA	*634*	80-10150
ERA Technology Ltd	UK	*502*	80-09600
ESDU International plc	UK	*502*	80-09700
E2S NV	Belgium	*48*	80-00600
Fluid Gravity Engineering Ltd.	UK	*502*	80-10660
Frazer-Nash Consultancy Limited	UK	*503*	80-10950
GEC-Marconi Avionics Ltd	UK	*454*	75-25320
GEC-Marconi Research Centre	UK	*419*	60-00700
GED (Southampton) Ltd	UK	*503*	80-11650
GKN Westland Systems	UK	*503*	80-11800
Horizons Technology Inc.	USA	*636*	80-14450
Horrigan Analytics	USA	*636*	80-14480
Hunterskil Howard plc	UK	*503*	80-12745
Hunting Engineering Ltd	UK	*459*	75-30700
ICL DESC	UK	*501*	80-08400
Industrieanlagen-Betriebs GmbH	Germany	*178*	80-00945
Inovex Technology Ltd	UK	*504*	80-13310
Int'l Submarine Engrg.	Canada	*73*	75-07780
Inter-National Research Institute	USA	*637*	80-16100
Interactive Dev. Environments	USA	*637*	80-15950
Intereurope Technology Ltd	UK	*504*	80-13340
Jane's Information Group Ltd	**UK**	***505***	**80-13700**
JPS Communications Inc.	USA	*587*	75-51220
Kaman Sciences Corp.	USA	*637*	80-17600
Kennedy Marketing Communications	USA	*637*	80-17680
Landair International Ltd	UK	*505*	80-14200
Lansdowne Technologies Inc	Canada	*78*	80-01205
LMS International	Belgium	*48*	80-00900
Lockheed Martin Services Group	USA	*638*	80-18540
Logica UK Limited	UK	*506*	80-14350
MacDonald Dettwiler	Malaysia	*278*	80-00270
MacDonald Dettwiler and Associates	Canada	*78*	80-01310
Mackenzie Tribbeck Associates	UK	*506*	80-15150
Management Consulting & Research	USA	*638*	80-19680
Marconi Instruments Ltd	UK	*466*	75-37270
Marine-Air Systems Ltd	New Zealand	*302*	75-01300
MASS Consultants Limited	UK	*506*	80-15650
Matra Marconi Space UK, Ltd	UK	*425*	74-06900
Medesign GmbH	Germany	*172*	75-20370
Metra Martech Limited	UK	*506*	80-16300
Microelectronics, guidance and	Turkey	*402*	55-00050
Milo Holdings Ltd	UK	*506*	80-16850
Mott MacDonald	UK	*507*	80-17400
MS Instruments plc	UK	*469*	75-41265
MTH Ltd	UK	*470*	75-41330
National Aerospace Laboratory	Netherlands	*290*	60-01400
Nichols Research Corp.	USA	*639*	80-21780
NNC Ltd	UK	*507*	80-18500
Nordic VLSI A/S	Norway	*310*	75-01640
Novamarine Instruments Pty Ltd	Australia	*19*	75-14885
Octec Ltd	UK	*471*	75-42950
OMI Logistics Limited	UK	*508*	80-19450
Optim Electronics Corp.	USA	*601*	75-66980
Oxley Developments Co. Ltd	UK	*472*	75-43400
Pacer Systems Inc.	USA	*602*	75-68000
Parity Solutions Ltd	UK	*508*	80-20300
Philips Traffic & Engineering	Australia	*20*	75-16550
Praxis Critical Systems Ltd	UK	*509*	80-23930
PRC Inc.	USA	*640*	80-24800
Primary Image Ltd	UK	*476*	75-46790
Programming Research Ltd	UK	*509*	80-24700
Prosig Computer Consultants Ltd	UK	*476*	75-46865
Public Safety Group Inc.	USA	*641*	80-24955
R-R Industrial Controls Ltd	UK	*510*	80-29340
Research Associates of Syracuse	USA	*641*	80-25900
Research Triangle Institute	USA	*540*	60-18000
Ricardo Hitec Ltd	UK	*509*	80-27840
Rollab Aktiebolag	Sweden	*380*	80-01850
Rolls-Royce & Associates Ltd	UK	*510*	80-29100
Rolls-Royce MatEval	UK	*480*	75-50260
Rood Technology UK Ltd	UK	*510*	80-29200
Rutherford Appleton Laboratory	UK	*420*	60-01850
SAGEM-B	Belgium	*47*	75-05595
SAIC Ltd	UK	*510*	80-30220
SAIT Systems SA	Belgium	*47*	75-05600
Science Systems (Industrial) Ltd	UK	*510*	80-31600
Scientific Computers Ltd	UK	*510*	80-31550
SEMCOR Inc.	USA	*642*	80-27750
Shade (Computer Services) Ltd	UK	*511*	80-34570
Shorts Missile Systems Ltd	**UK**	***483***	**75-53550**
Siemens Plessey Systems	UK	*483*	75-53770
Sigma Signals Corporation	USA	*615*	75-80790
Signal Computing Ltd	UK	*511*	80-36050
Singapore Electronic & Engineering	Singapore	*349*	80-01960
Singapore Engineering Software	Singapore	*349*	80-01965
Smith System Engineering Ltd	UK	*512*	80-39500
Sonak Systems & Software SA	Greece	*193*	80-01970
Space Systems Finland Ltd	Finland	*118*	75-02100
Spectral Intelligence Corp.	USA	*642*	80-29250
Stanley Aviation Corporation	USA	*617*	75-83900
State Research Institute	Russia	*336*	60-04150
Stirling Dynamics Ltd	UK	*512*	80-47000
Sverdrup Corp.	USA	*643*	80-30650
Sverdrup Technology Inc.	USA	*643*	80-30680
Synectics Corp.	USA	*643*	80-30750
Systematic Software Engineering Ltd.	UK	*513*	80-49480
TA Group Ltd	UK	*513*	80-50110
Tadiran Spectralink	Israel	*222*	75-21550
TDSi Micro Applications Ltd	UK	*513*	80-51000
Technology Project Services	UK	*513*	80-52800

Product-Supplier Link

Thomson Marconi Sonar Limited	UK	*426*	74-20000
Thomson Sintra	France	*148*	75-45135
Thomson-CSF Systems Canada Inc.	Canada	*79*	80-02020
Tracor Inc	USA	*623*	75-90000
Triad Group plc	UK	*514*	80-58600
Vanguard Research Inc.	USA	*644*	80-34100
Vickers Shipbuilding & Eng Ltd	UK	*491*	75-63100
Vitrociset SpA	Italy	*240*	80-40000
Vosper International Ltd	UK	*514*	80-67000
VRC Corp.	USA	*645*	80-34460
Welcom Software Technology	UK	*514*	80-72500

948 Test services

Advanced Research & Applications	USA	*546*	75-00940
AEA Technology plc	UK	*496*	80-00480
Aero Machining Ltd.	Canada	*77*	80-00020
AeroSpace Technologies of Australia	Australia	*11*	75-00340
Amtac Laboratories Ltd	UK	*497*	80-01060
Analysis & Technology Inc.	USA	*631*	80-02600
Assessment Services	UK	*497*	80-01500
ATS Aerospace Inc.	Canada	*68*	75-00752
Barnes & Reinecke Inc.	USA	*631*	80-04070
Bayern-Chemie	Germany	*165*	75-02100
Bell Technologies Inc.	USA	*631*	80-03352
Bodycote Taussig Inc.	USA	*632*	80-04590
Bofors AB	Sweden	*375*	74-00020
Calspan SRL Technology	USA	*540*	60-03000
Cas Inc.	USA	*633*	80-05230
Chamberlain Manufacturing Corp.	USA	*563*	75-17950
Dayton T. Brown Inc.	USA	*632*	80-04920
DCS Corp.	USA	*634*	80-07400
Defense Technology Foundation	Japan	*255*	80-00125
Delex Systems Inc.	USA	*634*	80-08020
Devonport Royal Dockyard Ltd	UK	*501*	80-08500
DRS Technologies, Inc.	USA	*570*	75-27400
EASAMS Ltd	UK	*501*	80-09000
EG&G Inc.	USA	*571*	75-29555
EG&G Washington Analytical Svcs	USA	*634*	80-09450
Electron Beam Processes Ltd	UK	*502*	80-09300
Emprise Corp.	USA	*634*	80-09950
Enator Driftentreprenader AB	Sweden	*379*	80-00500
Engineering & Professional Serv.Inc.	USA	*634*	80-10150
EPI Technologies Inc.	USA	*634*	80-10350
ERA Technology Ltd	UK	*502*	80-09600
Frazer-Nash Consultancy Limited	UK	*503*	80-10950
Freeman & Proctor Ltd	UK	*452*	75-24700
Gabi Shoef Ltd	Israel	*219*	75-07650
GEC-Marconi Avionics Ltd	UK	*454*	75-25320
Grove Europe Ltd	**UK**	***456***	**75-27030**
GTT Industries Inc.	USA	*636*	80-13350
Helio Mirror Co. Ltd	UK	*457*	75-28900
Hellenic Arms Industry	Greece	*187*	55-00800
Hellier Technical Training	USA	*636*	80-13930
Hewlett-Packard (Canada) Ltd	Canada	*72*	75-07520
Hi-Rel Laboratories Inc.	USA	*636*	80-13960
Hoffman Engineering Corp.	USA	*582*	75-43980
Horizons Technology Inc.	USA	*636*	80-14450
Hugo Marom Aviation Consultants Ltd	Israel	*224*	80-00880
Hunting Engineering Ltd	UK	*459*	75-30700
IDIC	Chile	*85*	60-00950
Indivers NV/Interturbine	Netherlands	*293*	75-03550
Industrieanlagen-Betriebs GmbH	Germany	*178*	80-00945
Institute of Space Constructions	Georgia	*156*	60-01000
Israel Aircraft Industries Ltd	Israel	*219*	75-10200
ITT Avionics	USA	*586*	75-49455
James Hardie Engineering	New Zealand	*303*	80-00900
Japan Aircraft Manufacturing Co. Ltd.	Japan	*252*	75-03200
Jones & Rickard Pty Ltd	Australia	*18*	75-10750
Kaman Sciences Corp.	USA	*637*	80-17600
Lightweight Body Armour	UK	*464*	75-35650
LMS International	Belgium	*48*	80-00900
Lucas Aerospace - Australia	Australia	*25*	80-12850
Lynx Corp Pty Ltd	Australia	*25*	80-12980
M. Rosenblatt & Son Inc.	USA	*641*	80-26450
Marshall of Cambridge Aerospace Ltd	UK	*466*	75-37530
Maxwell Technologies Inc.	USA	*594*	75-59250
Menasco Aerospace Ltd	Canada	*74*	75-09600
MFD International Ltd.	UK	*506*	80-16450
Micrel Inc.	USA	*596*	75-60950
MSEI Services Inc.	Canada	*79*	80-01380
MTH Ltd	UK	*470*	75-41330
NEL	UK	*419*	60-01400
Nevada Automotive Test Center	USA	*540*	60-14500
Newport News Shipbuilding	USA	*599*	75-64300
Normalair-Garrett Australia	Australia	*19*	75-14870
NPO Geofizika	Russia	*338*	75-14315
Office National d'Études et de	France	*127*	60-05000
Orlando Technology Inc.	USA	*540*	60-16000
OTB	S. Africa	*357*	60-00500
O.T.P. Ballistics	UK	*508*	80-19700
Pacer Systems Inc.	USA	*602*	75-68000
Philips Calibration Services	Australia	*25*	80-16490
Philips Traffic & Engineering	Australia	*20*	75-16550
Prototypa-ZM, s.r.o.	Czech Republic	*98*	60-01000
Public Safety Group Inc.	USA	*641*	80-24955
Q-Systems International Inc.	USA	*606*	75-71960
Quantum Inspection & Testing Ltd	Canada	*79*	80-01800
Ricardo Hitec Ltd	UK	*509*	80-27840
RKI/Cleanroom Services	USA	*641*	80-26200
Rolls-Royce (Canada) Ltd.	Canada	*75*	75-12120
Rolls-Royce MatEval	UK	*480*	75-50260
Rood Technology UK Ltd	UK	*510*	80-29200
Safety Consulting Engineers Inc.	USA	*641*	80-26680
Santa Barbara SA	Spain	*364*	55-01000
SEAL Laboratories	USA	*642*	80-27250
SEMCOR Inc.	USA	*642*	80-27750
Sener Ingenieria y Sistemas SA	Spain	*367*	75-05200
Singapore Electronic & Engineering	Singapore	*349*	80-01960
Sonovision ITEP Technologies	France	*151*	80-01850
SOPEMEA	France	*151*	80-01997
Southwest Research Institute	USA	*541*	60-19200
Space Electronics Inc.	USA	*642*	80-29050
Specialty Plastic Products Inc.	USA	*616*	75-83035
SST/Burlington Microelectronics	USA	*643*	80-29650
Standard Aero Ltd	Canada	*79*	80-01970
Stanford Technology Corp.	USA	*643*	80-29700
Stanley Aviation Corporation	USA	*617*	75-83900
Sverdrup Corp.	USA	*643*	80-30650
Sverdrup Technology Inc.	USA	*643*	80-30680
Swiss Aircraft & Systems Co	Switzerland	*384*	55-00100
Techspace Aero SA	Belgium	*47*	75-06200
The Aerospace Corp.	USA	*630*	80-01050
Thermal Processing Group Ltd	UK	*513*	80-54500
Tracor Inc	USA	*623*	75-90000
Transfield Pty Ltd	Australia	*22*	75-20810
UBC Inc.	USA	*624*	75-91383
Vipac Engineers & Scientists Ltd	Australia	*26*	80-22580
VRC Corp.	USA	*645*	80-34460
Wegmann & Co GmbH	Germany	*177*	75-33700

948 10 Ballistic test services

Accuracy International Ltd	UK	*427*	75-00360
AEA Technology plc	UK	*496*	80-00480
Aircraft Equipment International Ltd	UK	*428*	75-01560
Allott & Lomax	UK	*497*	80-01050
ATS Aerospace Inc.	Canada	*68*	75-00752
CMS Inc.	USA	*564*	75-19170
EG&G Inc.	USA	*571*	75-29555
Frazer-Nash Consultancy Limited	UK	*503*	80-10950
Hadland Photonics Ltd	UK	*456*	75-27200
Heflin Steel Co.	USA	*581*	75-42880
Helston Gunsmiths	UK	*458*	75-29010
HVR Consulting Services Ltd	UK	*504*	80-12790
Interstate Technical Products	UK	*461*	75-31946
Kaman Sciences Corp.	USA	*637*	80-17600
Lightweight Body Armour	UK	*464*	75-35650
Mark Three	USA	*540*	60-13150
MASS Consultants Limited	UK	*506*	80-15650
Meggitt Composites	UK	*467*	75-38345
Nevada Automotive Test Center	USA	*540*	60-14500
O.T.P. Ballistics	UK	*508*	80-19700
Prototypa-ZM, s.r.o.	Czech Republic	*98*	60-01000
Public Safety Group Inc.	USA	*641*	80-24955
Q-Systems International Inc.	USA	*606*	75-71960
Rollab Aktiebolag	Sweden	*380*	80-01850
Sleeman Engineering	UK	*484*	75-54400
Specialty Plastic Products Inc.	USA	*616*	75-83035
TBA Textiles	UK	*487*	75-58740
3C Systems, Inc.	USA	*632*	80-05000

949 Other engineering services

Advanced Appl. Consultants Inc.	USA	*630*	80-00250
Armalon Ltd	UK	*431*	75-03729
Atlantic Research Corp.	USA	*555*	75-10150

Australian Technical Publications	Australia	24	80-01880
Calspan SRL Technology	USA	540	60-03000
Chamberlain Manufacturing Corp.	USA	563	75-17950
DCS Corp.	USA	634	80-07400
Delex Systems Inc.	USA	634	80-08020
Design Bureau of	Russia	336	75-03150
Devonport Royal Dockyard Ltd	UK	501	80-08500
DRS Technologies, Inc.	USA	570	75-27400
Electron Beam Processes Ltd	UK	502	80-09300
Elintech GmbH	Germany	178	80-00550
Engineering & Professional Serv.Inc.	USA	634	80-10150
EPS Logistics Technology Ltd	UK	502	80-09550
FFA Flugzeugwerke Altenrhein AG	Switzerland	386	75-04350
Frazer-Nash Consultancy Limited	UK	503	80-10950
GEC-Marconi Avionics Ltd	UK	454	75-25320
GED (Southampton) Ltd	UK	503	80-11650
Indal Technologies Inc	Canada	72	75-07740
Israel Aircraft Industries Ltd	Israel	219	75-10200
Japan Aircraft Manufacturing Co. Ltd.	Japan	252	75-03200
Kaman Sciences Corp.	USA	637	80-17600
Lucas Aerospace - Australia	Australia	25	80-12850
M. Paz Logistics Ltd	Israel	224	80-01600
Medesign GmbH	Germany	172	75-20370
Menasco Aerospace Ltd	Canada	74	75-09600
MFD International Ltd.	UK	506	80-16450
Microelectronics, guidance and	Turkey	402	55-00050
MIL Systems	Canada	79	80-01360
M.J. Electronic Services	UK	507	80-16880
National Systems Management Corp.	USA	639	80-21650
NEL	UK	419	60-01400
NNC Ltd	UK	507	80-18500
Pacer Systems Inc.	USA	602	75-68000
Planning & Research Ltd	Israel	224	80-01650
PressCut Industries Inc.	USA	605	75-71436
Programming Research Ltd	UK	509	80-24700
Prototypa-ZM, s.r.o.	Czech Republic	98	60-01000
Qualitair Engineering Services Ltd	UK	509	80-25200
Radio Holland Electronics BV	Netherlands	297	80-18300
Ricardo Hitec Ltd	UK	509	80-27840
Safety Consulting Engineers Inc.	USA	641	80-26680
Santa Barbara SA	Spain	364	55-01000
Seaward International Inc.	USA	613	75-78940
SEMCOR Inc.	USA	642	80-27750
Sener Ingenieria y Sistemas SA	Spain	367	75-05200
Serco Aviation Services	UK	511	80-34150
Siemens Switzerland Ltd	Switzerland	387	75-11980
Southwest Research Institute	USA	541	60-19200
Spire Corp.	USA	643	80-29400
Standard Aero Ltd	Canada	79	80-01970
Stanley Aviation Corporation	USA	617	75-83900
Strachan & Henshaw Ltd	UK	512	80-47400
STV Group Inc.	USA	643	80-30400
Sverdrup Corp.	USA	643	80-30650
Sverdrup Technology Inc.	USA	643	80-30680
Swiss Aircraft & Systems Co	Switzerland	384	55-00100
The Titan Corp	USA	622	75-89620
Tracor Inc	USA	623	75-90000
Vickers Shipbuilding & Eng Ltd	UK	491	75-63100
Vitro Corp.	USA	644	80-34400
Wornall Electronics Inc.	USA	645	80-35400

949 20 Coating services

Acorn Hardas Ltd	UK	427	75-00410
AD Aerospace Finishes Ltd	UK	429	75-02370
AEA Technology plc	UK	496	80-00480
Armalon Ltd	UK	431	75-03729
A.T. Poeton (Gloucester Plating) Ltd	UK	508	80-23500
Avimed Ltd	UK	498	80-01900
Avimo Ltd	UK	432	75-04950
Balmoral Group Ltd	UK	433	75-05420
Barnes Aerospace Group	USA	557	75-12000
B.I. Engineering	UK	434	75-06630
Bodycote H.I.P. Ltd	UK	499	80-03650
Bodycote Metallurgical Coatings Ltd	UK	434	75-07500
Bospec Neue Technologien	Germany	166	75-03420
Brenco Thermal Coating Pty Ltd.	Australia	13	75-02740
Brighton Sheet Metal Ltd.	UK	436	75-08550
Caledonian Mining Co. Ltd	UK	438	75-11250
Calspan SRL Technology	USA	540	60-03000
Central Research Laboratories Ltd	UK	499	80-05895
Chromalloy France	France	133	75-09650
Coherent Optics (Europe) Ltd	UK	441	75-14090
Concoat Ltd	UK	442	75-14780
Cranfield University	UK	419	60-00350
Deft Chemical Coatings	USA	568	75-24870
Doncasters plc	UK	446	75-18410
Doncasters Structures	UK	446	75-18465
DRS Technologies, Inc.	USA	570	75-27400
Eloptro	S. Africa	358	75-05680
Fireseal International Ltd	UK	502	80-10450
GEC-Marconi Materials Technology Ltd	UK	419	60-00600
Hellenic Aerospace Industry Ltd.	Greece	192	80-00750
Holscot Industrial Linings Ltd	UK	458	75-29655
Hovermarine International Ltd	UK	459	75-30200
ICI Devoe Coatings	USA	584	75-45895
K S Paul Products Ltd.	UK	464	75-34160
Kaman Sciences Corp.	USA	637	80-17600
Laser & Atomic Research & Devt	Croatia	94	60-00450
Menasco Aerospace Ltd	Canada	74	75-09600
NanoMaterials, Inc.	USA	598	75-63590
Newport News Shipbuilding	USA	599	75-64300
NNC Ltd	UK	507	80-18500
Nullifire Ltd	UK	471	75-42850
OGMA - Indústria Aeronáutica	Portugal	327	80-01500
Oxley Developments Co. Ltd	UK	472	75-43400
Pilkington Thorn Optronics	UK	426	74-10000
Plasfin Ltd	UK	508	80-22350
Praxair Surface Technologies Inc.	USA	640	80-24680
Praxair Surface Technologies Ltd	UK	509	80-23925
Reade Advanced Materials	USA	608	75-73860
Renown Engineering Ltd	UK	478	75-49227
Rolls-Royce & Associates Ltd	UK	510	80-29100
Rutherford Appleton Laboratory	UK	420	60-01850
Seaward International Inc.	USA	613	75-78940
Sermatech International Inc	USA	613	75-79530
Trimite Ltd.	UK	489	75-61150
United Defense L.P.	USA	624	75-92050
W. Canning plc	UK	438	75-11480
Watson & Lewis Ltd	UK	514	80-71500
Weir Engineering Pty Ltd.	Australia	23	75-23300
Wood Group Aero Ltd	UK	515	80-76250

949 20 20 Corrosion-resistant lining services

AEA Technology plc	UK	496	80-00480
Balmoral Group Ltd	UK	433	75-05420
Brenco Thermal Coating Pty Ltd.	Australia	13	75-02740
Calspan SRL Technology	USA	540	60-03000
Hellenic Aerospace Industry Ltd.	Greece	192	80-00750
ICI Devoe Coatings	USA	584	75-45895
Kaman Sciences Corp.	USA	637	80-17600
Newport News Shipbuilding	USA	599	75-64300
OGMA - Indústria Aeronáutica	Portugal	327	80-01500
Strachan & Henshaw Ltd	UK	512	80-47400
Tecphy	France	147	75-44375
United Defense L.P.	USA	624	75-92050

949 30 Environmental testing services

AB Connectors Ltd	UK	426	75-00040
AB Electronics (Rogerstone) Ltd	UK	426	75-00080
AEA Technology plc	UK	496	80-00480
AeroSpace Technologies of Australia	Australia	11	75-00340
Airscrew Howden Ltd	UK	429	75-02050
Allott & Lomax	UK	497	80-01050
Assessment Services	UK	497	80-01500
ATS Aerospace Inc.	Canada	68	75-00752
Avica Equipment	UK	432	75-04870
Bayern-Chemie	Germany	165	75-02100
Bharat Electronics Ltd	India	203	55-00200
Chamberlain Manufacturing Corp.	USA	563	75-17950
Cincinnati Electronics Corp.	USA	564	75-18640
CMS Inc.	USA	564	75-19170
Colebrand Ltd	UK	441	75-14120
Cranfield University	UK	419	60-00350
Dayton T. Brown Inc.	USA	632	80-04920
Devonport Royal Dockyard Ltd	UK	501	80-08500
Engineering Dynamics (Southern) Ltd.	UK	449	75-20960
ERA Technology Ltd	UK	502	80-09600
Explorocean Technology Ltd	UK	502	80-09870
FHP International plc	UK	502	80-10300
Fluid Gravity Engineering Ltd.	UK	502	80-10660
Fluor Daniel Inc.	USA	635	80-11720
GEC ALSTHOM Engineering Systems Ltd	UK	453	75-25225
GEL Engineering Ltd	UK	454	75-25500

Product-Supplier Link

Haley & Weller Ltd	UK	*456*	75-27400
Hunting Engineering Ltd	UK	*459*	75-30700
HVR Consulting Services Ltd	UK	*504*	80-12790
Industrieanlagen-Betriebs GmbH	Germany	*178*	80-00945
Leafield Engineering Ltd	UK	*464*	75-34910
Litton Industries Inc.	USA	*590*	75-55400
LMS International	Belgium	*48*	80-00900
MDH Limited	UK	*467*	75-38270
Messier-Dowty Electronics	Canada	*74*	75-09650
Metra Martech Limited	UK	*506*	80-16300
MFD International Ltd.	UK	*506*	80-16450
Microelectronics, guidance and	Turkey	*402*	55-00050
Montford Instruments Ltd	UK	*469*	75-40600
Mott MacDonald	UK	*507*	80-17400
NEL.	UK	*419*	60-01400
NNC Ltd	UK	*507*	80-18500
O.T.P. Ballistics	UK	*508*	80-19700
Pains-Wessex Ltd	UK	*472*	75-43510
Parmatic Betex Limited	UK	*508*	80-20450
Parsons Engineering Science	USA	*640*	80-23750
Parsons Power Group Inc.	USA	*640*	80-23700
Philips Traffic & Engineering	Australia	*20*	75-16550
PRC Inc.	USA	*640*	80-24800
Research Triangle Institute	USA	*540*	60-18000
RKI/Cleanroom Services	USA	*641*	80-26200
Rolls-Royce & Associates Ltd	UK	*510*	80-29100
Rood Technology UK Ltd	UK	*510*	80-29200
Rutherford Appleton Laboratory	UK	*420*	60-01850
Shorts Missile Systems Ltd	**UK**	***483***	**75-53550**
Singapore Electronic & Engineering	Singapore	*349*	80-01960
Sira Ltd.	UK	*484*	75-54250
Solomat Ltd	UK	*485*	75-55000
SOPEMEA	France	*151*	80-01997
Space Electronics Inc.	USA	*642*	80-29050
Stirling Dynamics Ltd	UK	*512*	80-47000
Sverdrup Corp.	USA	*643*	80-30650
Tracor Inc	USA	*623*	75-90000

949 35 Hot isostatic pressing services

AEA Technology plc	UK	*496*	80-00480
Bodycote H.I.P. Ltd	UK	*499*	80-03650
National Forge Co.	USA	*598*	75-63850
Tecphy	France	*147*	75-44375

949 40 Optical coating services

AEA Technology plc	UK	*496*	80-00480
Bospec Neue Technologien	Germany	*166*	75-03420
Eloptro	S. Africa	*358*	75-05680
Litton Industries Inc.	USA	*590*	75-55400
Optical Components Inc.	USA	*601*	75-66953
Oxley Developments Co. Ltd	UK	*472*	75-43400
Ring Sights Holdings Co. Ltd.	UK	*479*	75-49700
Spire Corp.	USA	*643*	80-29400

949 50 RFI / EMI screening services

Bospec Neue Technologien	Germany	*166*	75-03420
Brenco Thermal Coating Pty Ltd.	Australia	*13*	75-02740
Chomerics	USA	*563*	75-18275
Cincinnati Electronics Corp.	USA	*564*	75-18640
Dayton T. Brown Inc.	USA	*632*	80-04920
Dytecna Ltd	UK	*501*	80-08900
Engineering & Professional Serv.Inc.	USA	*634*	80-10150
ERA Technology Ltd	UK	*502*	80-09600
GEC-Marconi Avionics Ltd	UK	*454*	75-25320
GEC-Marconi Materials Technology Ltd	UK	*454*	75-25340
Hellermann Electric	UK	*458*	75-28970
Hunting Engineering Ltd	UK	*459*	75-30700
Interpoint A Crane Company	USA	*585*	75-46870
Litton Industries Inc.	USA	*590*	75-55400
Oxley Developments Co. Ltd	UK	*472*	75-43400
Parker Hannifin plc	UK	*473*	75-43880
PressCut Industries Inc.	USA	*605*	75-71436
Racal-Comsec Ltd	UK	*477*	75-47665
RFI Industries Pty Ltd	Australia	*20*	75-18500
Sanders	USA	*612*	75-77850
Siemens Switzerland Ltd.	Switzerland	*387*	75-11980
Singapore Electronic & Engineering	Singapore	*349*	80-01960
SOPEMEA	France	*151*	80-01997
Technology Australasia Pty Ltd	Australia	*26*	80-20600

United Defense L.P.	USA	*624*	75-92050
Vickers Shipbuilding & Eng Ltd	UK	*491*	75-63100

949 60 Security engineering services

DEMCO Consultants Ltd	Israel	*224*	80-00430
Diamond Electronics Inc.	USA	*569*	75-26450
EG&G Inc.	USA	*571*	75-29555
Engineering & Professional Serv.Inc.	USA	*634*	80-10150
Fluor Daniel Inc.	USA	*635*	80-11720
Frazer-Nash Consultancy Limited	UK	*503*	80-10950
Industrieanlagen-Betriebs GmbH	Germany	*178*	80-00945
Mott MacDonald	UK	*507*	80-17400
Ricardo Hitec Ltd	UK	*509*	80-27840
Singapore Electronic & Engineering	Singapore	*349*	80-01960
Stanley Aviation Corporation	USA	*617*	75-83900
STV Group Inc.	USA	*643*	80-30400
Sverdrup Corp.	USA	*643*	80-30650
Talos Engineering SA	Greece	*193*	80-02010
Technology Australasia Pty Ltd	Australia	*26*	80-20600
Tracor Inc	USA	*623*	75-90000
Vision Research Co.	UK	*491*	75-63360
VSK Electronics	Belgium	*48*	75-08000
Wornall Electronics Inc.	USA	*645*	80-35400

949 70 Surface treatment services

AEA Technology plc	UK	*496*	80-00480
Armalon Ltd	UK	*431*	75-03729
Barnes Aerospace Group	USA	*557*	75-12000
Brenco Thermal Coating Pty Ltd.	Australia	*13*	75-02740
Chromalloy France	France	*133*	75-09650
Elf Atochem	Australia	*16*	75-05630
Hellenic Aerospace Industry Ltd.	Greece	*192*	80-00750
Hodge Clemco Ltd	UK	*458*	75-29590
Lucas Aerospace - Australia	Australia	*25*	80-12850
Nitram Vacuum Heat Treatment	UK	*471*	75-42580
SONACA SA	Belgium	*47*	75-06100
Thermal Processing Group Ltd	UK	*513*	80-54500

949 80 Systems validation & verification services

American Management Systems	USA	*630*	80-02350
Cas Inc.	USA	*633*	80-05230
DCS Corp.	USA	*634*	80-07400
EDS Defence Ltd	UK	*501*	80-09200
Frazer-Nash Consultancy Limited	UK	*503*	80-10950
GEC-Marconi Avionics Ltd	UK	*454*	75-25320
Mott MacDonald	UK	*507*	80-17400
MSL Secure Systems.	UK	*507*	80-17500
NNC Ltd	UK	*507*	80-18500
Pacer Systems Inc.	USA	*602*	75-68000
Programming Research Ltd	UK	*509*	80-24700
Research Triangle Institute	USA	*540*	60-18000
RKI/Cleanroom Services	USA	*641*	80-26200
SEMCOR Inc.	USA	*642*	80-27750
Stanley Aviation Corporation	USA	*617*	75-83900
State Research Institute	Russia	*336*	60-04150
Sverdrup Corp.	USA	*643*	80-30650
Sverdrup Technology Inc.	USA	*643*	80-30680
Technology Australasia Pty Ltd	Australia	*26*	80-20600
Tracor Inc	USA	*623*	75-90000
Vanguard Research Inc.	USA	*644*	80-34100

950 Logistics services

ACCO Foreign Shipping Inc.	USA	*629*	80-00100
AIM Enterprises Inc.	USA	*630*	80-01350
Air New Zealand Ltd	New Zealand	*303*	80-00100
Allied Defense Industries Inc.	USA	*630*	80-02000
Alvis Logistics Ltd	UK	*430*	75-02860
American Eurocopter	USA	*552*	75-05790
American Management Systems	USA	*630*	80-02350
Analysis & Technology Inc.	USA	*631*	80-02600
Australian Submarine Corporation	Australia	*12*	75-01910
Autoflug GmbH & Co	Germany	*165*	75-01400
AV Technology LLC	USA	*556*	75-11405
bd Systems Inc.	USA	*632*	80-04140
Bender Shipbuilding & Repair Co.	USA	*558*	75-13300
Boeing Australia Limited	Australia	*12*	75-02500
CD-ROM Systems (Europe) Ltd	UK	*499*	80-05800

CMPS&F Pty Limited	Australia	*24*	80-03600
Cogent	UK	*500*	80-06250
Contraves GmbH	Germany	*166*	75-05200
CTA Inc.	USA	*633*	80-06810
Cubic Defense Systems Inc.	USA	*567*	75-22900
Devonport Royal Dockyard Ltd	UK	*501*	80-08500
Dornier GmbH	Germany	*167*	75-06700
Dowty Aerospace Aviation Services	Singapore	*349*	80-00450
DUAL Incorporated	USA	*634*	80-08670
Dytecna Ltd	UK	*501*	80-08900
EASAMS Ltd	UK	*501*	80-09000
E.C. Transport (Wimborne) Ltd	UK	*501*	80-09100
EG&G Inc.	USA	*571*	75-29555
EG&G Washington Analytical Svcs	USA	*634*	80-09450
Embraer Aviation International	France	*150*	80-00500
Engineering & Professional Serv.Inc.	USA	*634*	80-10150
Fabbrica Italiana Apparecchiature	Italy	*234*	75-08550
FFA Flugzeugwerke Altenrhein AG	Switzerland	*386*	75-04350
FLS Aerospace (Light Aircraft) Ltd	UK	*451*	75-23900
Fluor Daniel Inc.	USA	*635*	80-11720
GEC-Marconi Avionics Ltd	UK	*454*	75-25320
GEL Engineering Ltd	UK	*454*	75-25500
General Motors of Canada Ltd	Canada	*71*	75-05900
Giat Industries	France	*126*	55-00100
Global Communications Co.	USA	*579*	75-39775
HeavyLift Cargo Airlines Ltd	UK	*503*	80-12300
Hydroscience Technologies Inc.	USA	*584*	75-45550
IAS Industrial & Airborne Systems	Germany	*178*	80-00800
Israel Aircraft Industries Ltd	Israel	*219*	75-10200
ITT Avionics	USA	*586*	75-49455
IVECO Fiat SpA	Italy	*235*	75-13230
Kay & Associates Inc.	USA	*637*	80-17660
Keystone Helicopter Corp.	USA	*637*	80-17780
L'hotellier Montrichard	France	*138*	75-24900
Leafield Logistics & Tech. Services	UK	*506*	80-14218
Litton Industries Inc.	USA	*591*	75-55400
Lockheed Martin Aircraft and	USA	*638*	80-18530
Lockheed Martin Corp.	USA	*591*	75-55635
Logtek (Logistic Technologies)	S. Africa	*360*	80-09950
M. Paz Logistics Ltd	Israel	*224*	80-01600
M. Rosenblatt & Son Inc.	USA	*641*	80-26450
Materialinformationszentrum Ges.	Germany	*178*	80-01310
Merex Inc.	USA	*595*	75-60250
Microelectronics, guidance and	Turkey	*402*	55-00050
MIL Systems	Canada	*79*	80-01360
National Systems Management Corp.	USA	*639*	80-21650
Nord-Micro Elektronik Feinmechanik	Germany	*173*	75-22700
Northrop Grumman Corporation	USA	*640*	80-21840
OMI Logistics Limited	UK	*508*	80-19450
Pilkington Optronics Ltd	UK	*426*	74-10000
R&B Aircraft Supply Inc.	USA	*641*	80-25420
Rolls-Royce (Canada) Ltd.	Canada	*75*	75-12120
RTE-Relli Engineering und Vertriebs	Germany	*179*	80-01800
Saint John Shipbuilding Ltd	Canada	*75*	75-12300
Satair A/S	Denmark	*105*	80-03500
Seamodal Transport Corp.	USA	*642*	80-27300
SEMCOR Inc.	USA	*642*	80-27750
Siemens Plessey Systems	UK	*483*	75-53770
Spar Aerospace Ltd	Canada	*76*	75-13000
SRI International	USA	*643*	80-29600
Stanley Aviation Corporation	USA	*617*	75-83900
STN ATLAS Elektronik GmbH	Germany	*165*	75-01150
STV Group Inc.	USA	*643*	80-30400
Systecon AB	Sweden	*380*	80-02000
Thomson Sintra	France	*148*	75-45135
Thomson-CSF Systems Canada Inc.	Canada	*79*	80-02020
TNT Express (UK) Ltd	UK	*514*	80-56600
Tracor Inc	USA	*623*	75-90000
United Defense L.P.	USA	*624*	75-92050
United Electric Motor Co.	USA	*624*	75-92100
Vickers Shipbuilding & Eng Ltd	UK	*491*	75-63100
Vinnell Corp.	USA	*644*	80-34350
Vitrociset SpA	Italy	*240*	80-40000
Vosper International Ltd	UK	*514*	80-67000
Wegmann & Co GmbH	Germany	*177*	75-33700
Wornall Electronics Inc.	USA	*645*	80-35400

951 Supply, provisioning, support services

AAI Corp.	USA	*545*	75-00020
Advanced Appl. Consultants Inc.	USA	*630*	80-00250
Aero Space Modifications	USA	*630*	80-00850
AIM Enterprises Inc.	USA	*630*	80-01350
Alvis Industries Ltd	UK	*430*	75-02850
Alvis Logistics Ltd	UK	*430*	75-02860
American Eurocopter	USA	*552*	75-05790
American Management Systems	USA	*630*	80-02350
AV Technology LLC	USA	*556*	75-11405
Bender Shipbuilding & Repair Co.	USA	*558*	75-13300
Cubic Defense Systems Inc.	USA	*567*	75-22900
Damen Shipyards	Netherlands	*292*	75-01500
Denro Inc.	USA	*568*	75-25651
Dowty Aerospace Aviation Services	Singapore	*349*	80-00450
Dytecna Ltd	UK	*501*	80-08900
EG&G Inc.	USA	*571*	75-29555
Embraer Aviation International	France	*150*	80-00500
Engineered Air Systems Inc.	USA	*573*	75-31560
Engineering & Professional Serv.Inc.	USA	*634*	80-10150
FLS Aerospace (Light Aircraft) Ltd	UK	*451*	75-23900
Fluor Daniel Inc.	USA	*635*	80-11720
GEC-Marconi Avionics Ltd	UK	*454*	75-25320
General Motors of Canada Ltd	Canada	*71*	75-05900
Global Airtech International	USA	*579*	75-39769
Global Communications Co.	USA	*579*	75-39775
GLS Gesellschaft für Logistischen	Germany	*178*	80-00750
Hill Industries Inc.	USA	*582*	75-43465
IAS Industrial & Airborne Systems	Germany	*178*	80-00800
Israel Aircraft Industries Ltd	Israel	*219*	75-10200
ITT Avionics	USA	*586*	75-49455
ITT Federal Services Corp.	USA	*586*	75-49455
Leafield Logistics & Tech. Services	UK	*506*	80-14218
Litton Industries Inc.	USA	*591*	75-55400
Lockheed Martin Aircraft and	USA	*638*	80-18530
M International Export	USA	*639*	80-19780
M. Paz Logistics Ltd	Israel	*224*	80-01600
M. Rosenblatt & Son Inc.	USA	*641*	80-26450
Materialinformationszentrum Ges.	Germany	*178*	80-01310
Microelectronics, guidance and	Turkey	*402*	55-00050
MIL Systems	Canada	*79*	80-01360
National Systems Management Corp.	USA	*639*	80-21650
Nord-Micro Elektronik Feinmechanik	Germany	*173*	75-22700
OMI Logistics Limited	UK	*508*	80-19450
OTO Breda	Italy	*237*	75-18100
Pilatus Aircraft Ltd	Switzerland	*387*	75-10100
Pilkington Optronics Ltd	UK	*426*	74-10000
Plastico Ltd	UK	*475*	75-45950
Ridgeway International Ltd.	UK	*509*	80-28000
Rolls-Royce (Canada) Ltd.	Canada	*75*	75-12120
Saint John Shipbuilding Ltd	Canada	*75*	75-12300
Satair A/S	Denmark	*105*	80-03500
Spar Aerospace Ltd	Canada	*76*	75-13000
SRI International	USA	*643*	80-29600
ST Aerospace Engineering Pte Ltd	Singapore	*349*	80-01990
Stanley Aviation Corporation	USA	*617*	75-83900
STV Group Inc.	USA	*643*	80-30400
Textron Marine & Land Systems	USA	*622*	75-88660
Thomson Sintra	France	*148*	75-45135
Thomson-CSF Systems Canada Inc.	Canada	*79*	80-02020
Tracor Inc	USA	*623*	75-90000
United Defense L.P.	USA	*624*	75-92050
Vosper International Ltd	UK	*514*	80-67000
Vosper Thornycroft	UK	*492*	75-63800
Wornall Electronics Inc.	USA	*645*	80-35400
Yarrow Shipbuilders Ltd	UK	*495*	75-67500

952 Storage services

AB Carl Munters	Sweden	*377*	75-08230
AIM Enterprises Inc.	USA	*630*	80-01350
Aviation Support Tech. Int'l Inc.	USA	*556*	75-11050
Bachmann SA	France	*132*	75-04900
EG&G Inc.	USA	*571*	75-29555
Fluor Daniel Inc.	USA	*635*	80-11720
GEC-Marconi Avionics Ltd	UK	*454*	75-25320
GED (Southampton) Ltd	UK	*503*	80-11650
GEL Engineering Ltd	UK	*454*	75-25500
General Motors of Canada Ltd	Canada	*71*	75-05900
Israel Aircraft Industries Ltd	Israel	*219*	75-10200
L'hotellier Montrichard	France	*138*	75-24900
Materialinformationszentrum Ges.	Germany	*178*	80-01310
National Systems Management Corp.	USA	*639*	80-21650
Ridgeway International Ltd.	UK	*509*	80-28000
Saint John Shipbuilding Ltd	Canada	*75*	75-12300
Schneider Specialized Carriers Inc.	USA	*642*	80-26900
Shaw Packing Ltd.	UK	*511*	80-34700
Textron Marine & Land Systems	USA	*622*	75-88660
United Defense L.P.	USA	*624*	75-92050

Product-Supplier Link

Vosper International Ltd	UK	514	80-67000
WTT Customs House Brokerage	USA	645	80-35300

953 Packing services

American Eurocopter	USA	552	75-05790
Aviation Support Tech. Int'l Inc.	USA	556	75-11050
Bremer Vulkan Marineschiffbau GmbH	Germany	166	75-03600
Contraves GmbH	Germany	166	75-05200
Danzas Corporation	USA	633	80-06900
Engineered Air Systems Inc.	USA	573	75-31560
GEC-Marconi Avionics Ltd	UK	454	75-25320
GED (Southampton) Ltd	UK	503	80-11650
GEL Engineering Ltd	UK	454	75-25500
General Motors of Canada Ltd	Canada	71	75-05900
Israel Aircraft Industries Ltd	Israel	219	75-10200
Korea Express USA Inc.	USA	638	80-17900
L'hotellier Montrichard	France	138	75-24900
National Systems Management Corp.	USA	639	80-21650
Ridgeway International Ltd.	UK	509	80-28000
Shaw Packing Ltd	UK	511	80-34700
Spar Aerospace Ltd	Canada	76	75-13000
United Defense L.P.	USA	624	75-92050
United Electric Motor Co.	USA	624	75-92100
Vickers Shipbuilding & Eng Ltd	UK	491	75-63100
Vosper International Ltd	UK	514	80-67000
WTT Customs House Brokerage	USA	645	80-35300

954 Export services

ACCO Foreign Shipping Inc.	USA	629	80-00100
Alvis Logistics Ltd	UK	430	75-02860
AMI International	USA	631	80-02470
Bremer Vulkan Marineschiffbau GmbH	Germany	166	75-03600
Condor Enterprises Inc.	USA	633	80-06200
CTA Inc.	USA	633	80-06810
Danzas Corporation	USA	633	80-06900
E.C. Transport (Wimborne) Ltd	UK	501	80-09100
Engineered Air Systems Inc.	USA	573	75-31560
FFA Flugzeugwerke Altenrhein AG	Switzerland	386	75-04350
GEC-Marconi Avionics Ltd	UK	454	75-25320
GEL Engineering Ltd	UK	454	75-25500
Global Airtech International	USA	579	75-39769
Kerametal Co. Ltd.	Slovakia	353	55-01000
Korea Express USA Inc.	USA	638	80-17900
LBG Agencies (1979) Ltd	Israel	224	80-00980
M. Paz Logistics Ltd	Israel	224	80-01600
Merex Inc.	USA	595	75-60250
National Systems Management Corp.	USA	639	80-21650
OTEC Support Services Ltd	UK	508	80-19600
P. Monsted Chartering A/S	Denmark	105	80-01350
Petrimex	Slovakia	353	55-01500
Ridgeway International Ltd.	UK	509	80-28000
South China Aero Technology Ltd	Hong Kong SAR	198	80-00850
United Defense L.P.	USA	624	75-92050
Vosper International Ltd	UK	514	80-67000
WTT Customs House Brokerage	USA	645	80-35300

954 10 Forwarding services

Bremer Vulkan Marineschiffbau GmbH	Germany	166	75-03600
Danzas Corporation	USA	633	80-06900
E.C. Transport (Wimborne) Ltd	UK	501	80-09100
Kay & Associates Inc.	USA	637	80-17660
Korea Express USA Inc.	USA	638	80-17900
National Systems Management Corp.	USA	639	80-21650
P. Monsted Chartering A/S	Denmark	105	80-01350
Ridgeway International Ltd.	UK	509	80-28000
Seamodal Transport Corp.	USA	642	80-27300
Transammo NV	Belgium	48	80-02000
Vinnell Corp.	USA	644	80-34350
Vosper International Ltd	UK	514	80-67000
WTT Customs House Brokerage	USA	645	80-35300

954 60 Spare parts supply services

AIM Enterprises Inc.	USA	630	80-01350
Allied Defense Industries Inc.	USA	630	80-02000
Alvis Industries Ltd	UK	430	75-02850
Alvis Logistics Ltd	UK	430	75-02860
Alvis Unipower Ltd	UK	430	75-02863
American Eurocopter	USA	552	75-05790
Armalon Ltd	UK	431	75-03729
Associated Aircraft Mfg & Sales Inc.	USA	555	75-09500
Barnes Aerospace Group	USA	557	75-12000
Bremer Vulkan Marineschiffbau GmbH	Germany	166	75-03600
CD-ROM Systems (Europe) Ltd	UK	499	80-05800
Cleveland-Guest (Engineering) Ltd	UK	440	75-13810
Continental Electronics Corp.	USA	566	75-21800
Contraves GmbH	Germany	166	75-05200
Devonport Royal Dockyard Ltd	UK	501	80-08500
Dowty Aerospace Aviation Services	Singapore	349	80-00450
DRS Technologies, Inc.	USA	570	75-27400
Dytecna Ltd	UK	501	80-08900
Econ Optics Mechanics SA	Greece	188	75-01070
Embraer Aviation International	France	150	80-00500
Engineered Air Systems Inc.	USA	573	75-31560
Engineering & Professional Serv.Inc.	USA	634	80-10150
FFA Flugzeugwerke Altenrhein AG	Switzerland	386	75-04350
FLS Aerospace (Light Aircraft) Ltd	UK	451	75-23900
GEC-Marconi Avionics Ltd	UK	454	75-25320
General Motors of Canada Ltd	Canada	71	75-05900
Holec Machines & Apparaten BV	Netherlands	293	75-03110
Howden Fan Company Pty Ltd	Australia	17	75-08740
Hydroscience Technologies Inc.	USA	584	75-45550
Indal Technologies Inc	Canada	72	75-07740
International Logistics Group	USA	585	75-48120
IVECO Fiat SpA	Italy	235	75-13230
L'hotellier Montrichard	France	138	75-24900
Leafield Logistics & Tech. Services	UK	506	80-14218
Litton Industries Inc.	USA	591	75-55400
Lucas Aerospace - Australia	Australia	25	80-12850
M International Export	USA	639	80-19780
M. Paz Logistics Ltd	Israel	224	80-01600
Merex Inc.	USA	595	75-60250
MSI-Defence Systems Ltd	UK	469	75-41280
NEC Corporation	Japan	253	75-06350
P. Monsted Chartering A/S	Denmark	105	80-01350
R&B Aircraft Supply Inc.	USA	641	80-25420
Radio Holland Electronics BV	Netherlands	297	80-18300
RTE-Relli Engineering und Vertriebs.	Germany	179	80-01800
Saint John Shipbuilding Ltd	Canada	75	75-12300
Satair A/S	Denmark	105	80-03500
Singapore Technologies Automotive	Singapore	348	75-04280
Standard Aero Ltd	Canada	79	80-01970
Stanley Aviation Corporation	USA	617	75-83900
STN ATLAS Elektronik GmbH	Germany	165	75-01150
Textron Marine & Land Systems	USA	622	75-88660
The Graphics Co. of America	USA	579	75-40360
Theodor Rapp GmbH Co	Germany	173	75-24600
Thorn DNT Airfield Lighting Pty Ltd	Australia	22	75-20540
Tracor Inc	USA	623	75-90000
United Defense L.P.	USA	624	75-92050
Vazduhoplovno Remonti Zavod	Yugoslavia	661	80-02500
Vickers Defence Systems	UK	491	75-63040
Vitrociset SpA	Italy	240	80-40000
Vosper International Ltd	UK	514	80-67000

955 Transport, delivery services

Dan-Am Shipping and Chartering Inc.	USA	633	80-06898
E.C. Transport (Wimborne) Ltd	UK	501	80-09100
EG&G Inc.	USA	571	75-29555
Engineering & Professional Serv.Inc.	USA	634	80-10150
GEL Engineering Ltd	UK	454	75-25500
Korea Express USA Inc.	USA	638	80-17900
P. Monsted Chartering A/S	Denmark	105	80-01350
Petrimex	Slovakia	353	55-01500
Ridgeway International Ltd.	UK	509	80-28000
Saint John Shipbuilding Ltd	Canada	75	75-12300
Schneider Specialized Carriers Inc.	USA	642	80-26900
Seamodal Transport Corp.	USA	642	80-27300
Skyline Aviation BV.	Netherlands	297	80-19000
Spar Aerospace Ltd	Canada	76	75-13000
Stanley Aviation Corporation	USA	617	75-83900
TNT Express (UK) Ltd	UK	514	80-56600
United Defense L.P.	USA	624	75-92050
Vinnell Corp.	USA	644	80-34350
Vosper International Ltd	UK	514	80-67000

955 10 30 Secure transportation services, dangerous / hazardous materials

Dan-Am Shipping and Chartering Inc.	USA	633	80-06898
E.C. Transport (Wimborne) Ltd	UK	501	80-09100
EG&G Inc.	USA	571	75-29555
P. Monsted Chartering A/S	Denmark	105	80-01350
Schneider Specialized Carriers Inc.	USA	642	80-26900
Transammo NV	Belgium	48	80-02000
WTT Customs House Brokerage	USA	645	80-35300

955 10 30 10 Secure transportation services, explosives

Dan-Am Shipping and Chartering Inc.	USA	633	80-06898
E.C. Transport (Wimborne) Ltd	UK	501	80-09100
EG&G Inc.	USA	571	75-29555
P. Monsted Chartering A/S	Denmark	105	80-01350
Schneider Specialized Carriers Inc.	USA	642	80-26900
Transammo NV	Belgium	48	80-02000
WTT Customs House Brokerage	USA	645	80-35300

955 20 Aircraft delivery services

Air New Zealand Ltd	New Zealand	303	80-00100
American Eurocopter	USA	552	75-05790
Aviation Support Tech. Int'l Inc.	USA	556	75-11050
FFA Flugzeugwerke Altenrhein AG	Switzerland	386	75-04350
Israel Aircraft Industries Ltd	Israel	219	75-10200
Keystone Helicopter Corp.	USA	637	80-17780
Korea Express USA Inc.	USA	638	80-17900
Marshall of Cambridge Aerospace Ltd	UK	466	75-37530
P. Monsted Chartering A/S	Denmark	105	80-01350
Skyways Inc.	USA	642	80-28250

960 Document / manual publishing & editorial services

A Goninan & Co. Ltd	Australia	16	75-07760
Advanced Appl. Consultants Inc.	USA	630	80-00250
American Eurocopter	USA	552	75-05790
Analysis & Technology Inc.	USA	631	80-02600
Armalon Ltd	UK	431	75-03729
Australian Technical Publications	Australia	24	80-01880
Bofors AB	Sweden	375	74-00020
Calspan SRL Technology	USA	540	60-03000
Chamberlain Manufacturing Corp.	USA	563	75-17950
Cincinnati Electronics Corp.	USA	564	75-18640
CMPS&F Pty Limited	Australia	24	80-03600
Contraves GmbH	Germany	166	75-05200
CTP Logistics Ltd	UK	500	80-07650
Cubic Defense Systems Inc.	USA	567	75-22900
Data Sciences Ltd.	UK	500	80-07740
Delex Systems Inc.	USA	634	80-08020
Document Engineering Co.	USA	634	80-08500
DRS Technologies, Inc.	USA	570	75-27400
Dytecna Ltd	UK	501	80-08900
EASAMS Ltd	UK	501	80-09000
EG&G Washington Analytical Svcs	USA	634	80-09450
Elintech GmbH	Germany	178	80-00550
Engineering & Professional Serv.Inc.	USA	634	80-10150
Eurocopter Canada Ltd	Canada	71	75-05050
Eurolang Services	France	150	80-00550
Forecast International/DMS	USA	636	80-11760
Gajon Institute of Technology SpA	Italy	229	60-00080
GEC-Marconi Avionics Ltd	UK	454	75-25320
GEL Engineering Ltd	UK	454	75-25500
George G. Sharp Inc.	USA	642	80-28000
IMP Aerospace Limited	Canada	72	75-07730
Intereurope Technology Ltd	UK	504	80-13340
Israel Aircraft Industries Ltd	Israel	219	75-10200
ITT Avionics	USA	586	75-49455
Jane's Information Group Ltd	**UK**	**505**	**80-13700**
Leo J. Brennan Inc.	USA	632	80-04800
Litton Industries Inc.	USA	591	75-55400
Marshall of Cambridge Aerospace Ltd	UK	466	75-37530
Materialinformationszentrum Ges.	Germany	178	80-01310
Microelectronics, guidance and	Turkey	402	55-00050
MIL Systems	Canada	79	80-01360
MSI-Defence Systems Ltd	UK	470	75-41280
National Systems Management Corp.	USA	639	80-21650
Newport News Shipbuilding	USA	599	75-64300
Nogay Ltd	Israel	224	80-01470
Nord-Micro Elektronik Feinmechanik	Germany	173	75-22700
O'Neil & Associates Inc.	USA	640	80-22372
OMI Logistics Limited	UK	508	80-19450
Pasha Publications Inc.	USA	640	80-23780
Premier Product Support Services Ltd	UK	509	80-23995
Ricardo Hitec Ltd	UK	509	80-27840
Rolls-Royce (Canada) Ltd.	Canada	75	75-12120
SEMCOR Inc.	USA	642	80-27750
Sojuzkarta - Impex	Russia	340	80-02000
Spar Aerospace Ltd	Canada	76	75-13000
Standard Aero Ltd	Canada	79	80-01970
Stanilite Electronics Pty Ltd	Australia	21	75-19690
Stanley Aviation Corporation	USA	617	75-83900
STN ATLAS Elektronik GmbH	Germany	165	75-01150
STV Group Inc.	USA	643	80-30400
Sverdrup Corp.	USA	643	80-30650
Technical Indexes Ltd	UK	513	80-52400
Thomson Sintra	France	148	75-45135
Tracor Inc	USA	623	75-90000
TriLink Technologies Group Inc.	Canada	80	80-02100
United Defense L.P.	USA	624	75-92050
Vanguard Research Inc.	USA	644	80-34100
Vanwell Publishing Ltd	Canada	80	80-02230
Vickers Shipbuilding & Eng Ltd	UK	491	75-63100
Vosper International Ltd	UK	514	80-67000
Wegmann & Co GmbH	Germany	177	75-33700

960 10 Mapping & charting services

Cartographical Services Ltd	UK	499	80-05550
Fugro-UDI Ltd	UK	452	75-24885
General Electronics Communications	Greece	189	75-01300
Instituto Geográfico Militar	Chile	85	60-00700
Jeppesen & Co GmbH	Germany	170	75-15050
Kaman Sciences Corp.	USA	637	80-17600
Laser-Scan	UK	506	80-14210
Litton Industries Inc.	USA	590	75-55400
Materialinformationszentrum Ges.	Germany	178	80-01310
SISDEF Ltda	Chile	85	80-01940
Sojuzkarta - Impex	Russia	340	80-02000
Sonar Research & Development Ltd	UK	485	75-55170
Stanley Aviation Corporation	USA	617	75-83900
Technical & Field Surveys Pty Ltd	Australia	26	80-20000

960 10 10 Airborne survey services

Calspan SRL Technology	USA	540	60-03000
Cartographical Services Ltd	UK	499	80-05550
Instituto Geográfico Militar	Chile	85	60-00700
Litton Industries Inc.	USA	590	75-55400
Technical & Field Surveys Pty Ltd	Australia	26	80-20000

960 20 Technical manuals, documentation services

A Goninan & Co. Ltd	Australia	16	75-07760
AAI Corp.	USA	545	75-00020
Advanced Appl. Consultants Inc.	USA	630	80-00250
Aeromaritime Systembau GmbH	Germany	164	75-00290
AeroSpace Technologies of Australia	Australia	11	75-00340
Allied Research Corp.	USA	549	75-04670
Alvis Industries Ltd	UK	430	75-02850
American Eurocopter	USA	552	75-05790
American Management Systems	USA	630	80-02350
Analysis & Technology Inc.	USA	631	80-02600
Armalon Ltd	UK	431	75-03729
Australian Technical Publications	Australia	24	80-01880
Barnes & Reinecke Inc.	USA	631	80-04070
Bofors AB	Sweden	375	74-00020
CAE Aviation Ltd	Canada	78	80-00245
Calspan SRL Technology	USA	540	60-03000
Chamberlain Manufacturing Corp.	USA	563	75-17950
Cincinnati Electronics Corp.	USA	564	75-18640
Contraves GmbH	Germany	166	75-05200
Cubic Defense Systems Inc.	USA	567	75-22900
Davis (W.R.) Engineering Ltd	Canada	70	75-04050
Defence Data Consulting Services	Switzerland	388	80-00430
Delex Systems Inc.	USA	634	80-08020
Design Bureau of	Russia	336	75-03150
DRS Technologies, Inc.	USA	570	75-27400
Dytecna Ltd	UK	501	80-08900
EASAMS Ltd	UK	501	80-09000
Elintech GmbH	Germany	178	80-00550
Engineered Air Systems Inc.	USA	573	75-31560

Engineering & Professional Serv.Inc.	USA	*634*	80-10150
ESDU International plc	UK	*502*	80-09700
Eurocopter Canada Ltd	Canada	*71*	75-05050
Fabbrica Italiana Apparecchiature	Italy	*234*	75-08550
Forecast International/DMS	USA	*636*	80-11760
GEC-Marconi Avionics Ltd	UK	*454*	75-25320
GED (Southampton) Ltd	UK	*503*	80-11650
George G. Sharp Inc.	USA	*642*	80-28000
GKN Westland Design Services	UK	*503*	80-11775
Global Engineering Documents	USA	*636*	80-12850
Gordon Consulting Engineers Ltd	Israel	*224*	80-00750
Hunterskil Howard plc	UK	*503*	80-12745
Information Handling Services Inc.	USA	*637*	80-15530
Israel Aircraft Industries Ltd	Israel	*219*	75-10200
Leo J. Brennan Inc.	USA	*632*	80-04800
Litton Industries Inc.	USA	*591*	75-55400
Logistics Services International	USA	*638*	80-18800
M. Rosenblatt & Son Inc.	USA	*641*	80-26450
Marshall of Cambridge Aerospace Ltd	UK	*466*	75-37530
Materialinformationszentrum Ges.	Germany	*178*	80-01310
MECAR SA	Belgium	*46*	75-04930
Medesign GmbH	Germany	*172*	75-20370
Meggitt Avionics (UK)	UK	*467*	75-38340
Microelectronics, guidance and	Turkey	*402*	55-00050
MIL Systems	Canada	*79*	80-01360
Milo Holdings Ltd	UK	*506*	80-16850
Mustang Survival Corp.	Canada	*74*	75-09860
National Systems Management Corp.	USA	*639*	80-21650
Newport Aeronautical Sales	USA	*639*	80-21750
Newport News Shipbuilding	USA	*599*	75-64300
NIS Ltd	UK	*471*	75-42555
NNC Ltd	UK	*507*	80-18500
Nord-Micro Elektronik Feinmechanik	Germany	*173*	75-22700
O'Neil & Associates Inc.	USA	*640*	80-22372
OMI Logistics Limited	UK	*508*	80-19450
Premier Product Support Services Ltd	UK	*509*	80-23995
Raytheon E-Systems	USA	*608*	75-73302
Ricardo Hitec Ltd	UK	*509*	80-27840
Rolls-Royce (Canada) Ltd.	Canada	*75*	75-12120
Saint John Shipbuilding Ltd	Canada	*75*	75-12300
Satair A/S	Denmark	*105*	80-03500
Scientific Management Associates	Australia	*25*	80-19100
SEED Technical Services Ltd	UK	*511*	80-34000
SEMCOR Inc.	USA	*642*	80-27750
SISDEF Ltda	Chile	*85*	80-01940
Sonovision ITEP Technologies	France	*151*	80-01850
Spar Aerospace Ltd	Canada	*76*	75-13000
Standard Aero Ltd	Canada	*79*	80-01970
Stanilite Electronics Pty Ltd	Australia	*21*	75-19690
Stanley Aviation Corporation	USA	*617*	75-83900
Strachan & Henshaw Ltd	UK	*512*	80-47400
STV Group Inc.	USA	*643*	80-30400
Sverdrup Corp.	USA	*643*	80-30650
TA Group Ltd	UK	*513*	80-50110
Technical Indexes Ltd	UK	*513*	80-52400
Technology Project Services	UK	*513*	80-52800
Textron Marine & Land Systems	USA	*622*	75-88660
Thomson Sintra	France	*148*	75-45135
Tracor Inc	USA	*623*	75-90000
TriLink Technologies Group Inc.	Canada	*80*	80-02100
United Defense L.P.	USA	*624*	75-92050
Vickers Shipbuilding & Eng Ltd	UK	*491*	75-63100
Vipac Engineers & Scientists Ltd	Australia	*26*	80-22580
Vitro Corp.	USA	*644*	80-34400
Vosper International Ltd	UK	*514*	80-67000
Vosper Thornycroft	UK	*492*	75-63800
Wheeler Systems Inc.	USA	*627*	75-96400

960 30 Translation services

Advanced Appl. Consultants Inc.	USA	*630*	80-00250
Elintech GmbH	Germany	*178*	80-00550
Engineering & Professional Serv.Inc.	USA	*634*	80-10150
Eurolang Services	France	*150*	80-00550
Fabbrica Italiana Apparecchiature	Italy	*234*	75-08550
General Electronics Communications	Greece	*189*	75-01300
Hunterskil Howard plc	UK	*503*	80-12745
Intereurope Technology Ltd	UK	*504*	80-13340
Israel Aircraft Industries Ltd	Israel	*219*	75-10200
Materialinformationszentrum Ges.	Germany	*178*	80-01310
OMI Logistics Limited	UK	*508*	80-19450
Premier Product Support Services Ltd	UK	*509*	80-23995
SISDEF Ltda	Chile	*85*	80-01940
Sonovision ITEP Technologies	France	*151*	80-01850
Standard Aero Ltd	Canada	*79*	80-01970

965 Marketing, advertising and public relations services

Aerospace & Defense Sales Inc	Philippines	*321*	80-00050
AirteQ Ltd	UK	*496*	80-00830
Allied Defense Industries Inc.	USA	*630*	80-02000
Alvis Logistics Ltd	UK	*430*	75-02860
American Eurocopter	USA	*552*	75-05790
AMI International	USA	*631*	80-02470
Armalon Ltd	UK	*431*	75-03729
ASTA Corp.	USA	*631*	80-03355
Bruce D. Vallillee Electronics Ltd	Canada	*80*	80-02210
Burdeshaw Associates, Ltd	USA	*632*	80-04960
Carmichael & Sweet Consultancy Ltd	UK	*499*	80-05500
CMPS&F Pty Limited	Australia	*24*	80-03600
Crew Services Ltd	UK	*500*	80-07370
CTA Inc.	USA	*633*	80-06810
Defence Research & Analysis	UK	*500*	80-08200
Engineering & Professional Serv.Inc.	USA	*634*	80-10150
Fabbrica Italiana Apparecchiature	Italy	*234*	75-08550
Forecast International/DMS	USA	*636*	80-11760
Glavcosmos	Russia	*339*	80-00500
Howden Fan Company Pty Ltd	Australia	*17*	75-08740
Hugo Marom Aviation Consultants Ltd	Israel	*224*	80-00880
InterCon Consultants	Canada	*78*	80-00980
Jotron (UK) Ltd	UK	*462*	75-32895
Kalamos International Ltd	UK	*505*	80-13950
Landair International Ltd	UK	*505*	80-14200
Lansdowne Technologies Inc	Canada	*78*	80-01205
M International Export	USA	*639*	80-19780
M. Paz Logistics Ltd	Israel	*224*	80-01600
Mycon Marketing Services Ltd	UK	*507*	80-17900
NanoMaterials, Inc.	USA	*598*	75-63590
NIS Ltd	UK	*471*	75-42555
OMI Logistics Limited	UK	*508*	80-19450
OTEC Support Services Ltd	UK	*508*	80-19600
PKM Consultants (S) Pte Ltd	Singapore	*349*	80-01640
Reade Advanced Materials	USA	*608*	75-73860
Schneider Sohn & Associates Inc.	USA	*642*	80-26890
Sector Ltd	UK	*511*	80-33400
Siemens Plessey Systems	UK	*483*	75-53770
South China Aero Technology Ltd	Hong Kong SAR	*198*	80-00850
SPC International Corp.	USA	*642*	80-29200
Spice Communications	UK	*512*	80-45800
Strat*X International	USA	*643*	80-30100
Strat*X Ltd	UK	*512*	80-47500
Strat*X SA	France	*151*	80-02000
TADS Video Production	UK	*513*	80-50100
The Graphics Co. of America	USA	*579*	75-40360
Top Line Co.	USA	*644*	80-33100
TRH Pictures	UK	*514*	80-58500

970 Medical services

International Hospitals Group	UK	*505*	80-13410
Materialinformationszentrum Ges.	Germany	*178*	80-01310
Medesign GmbH	Germany	*172*	75-20370
Mediworld Ltd.	UK	*462*	75-32300
Rolls-Royce (Canada) Ltd.	Canada	*75*	75-12120
Vessa Limited	UK	*491*	75-63010

980 Training services

Aardvark Training Schools Ltd	UK	*496*	80-00200
Aeromaritime Systembau GmbH	Germany	*164*	75-00290
Aerospace Inspection Training	UK	*496*	80-00650
AeroSpace Technologies of Australia	Australia	*11*	75-00340
Air Affairs (NZ) Limited	New Zealand	*303*	80-00050
AirteQ Ltd	UK	*496*	80-00830
Alvis Logistics Ltd	UK	*430*	75-02860
American Eurocopter	USA	*552*	75-05790
American Management Systems	USA	*630*	80-02350
AMI International	USA	*631*	80-02470
Analysis & Technology Inc.	USA	*631*	80-02600
Andrich International Ltd	UK	*497*	80-01200
Ansett Australia Avionics	Australia	*24*	80-01010
Ansul Incorporated	USA	*554*	75-07810
Arca Systems	USA	*631*	80-03100
Archer Mullins Limited	UK	*497*	80-01257

Armalon Ltd	UK	*431*	75-03729
Atlantic Research Corp.	USA	*555*	75-10150
ATS Aerospace Inc.	Canada	*68*	75-00752
Bayern-Chemie	Germany	*165*	75-02100
Becker Flugfunkwerk GmbH	Germany	*165*	75-02200
Boeing Australia Limited	Australia	*12*	75-02500
Bofors AB	Sweden	*375*	74-00020
Bristow Helicopters Ltd	UK	*499*	80-03950
British Aerospace Systems & Services	UK	*499*	80-04000
Calspan SRL Technology	USA	*540*	60-03000
Carl Cranz Gesellschaft eV	Germany	*178*	80-00350
Cenzin Co. Ltd.	Poland	*324*	75-00200
Chamberlain Manufacturing Corp.	USA	*563*	75-17950
Contraves GmbH	Germany	*166*	75-05200
Cranfield University	UK	*419*	60-00350
Crew Services Ltd	UK	*500*	80-07370
CSE Aviation Ltd	UK	*500*	80-07600
Cubic Defense Systems Inc.	USA	*567*	75-22900
Daewoo Heavy Industries Ltd	Korea (ROK)	*265*	75-00600
Datacom Inc.	USA	*567*	75-23645
Denel Informatics	S. Africa	*360*	80-05000
Designers & Planners Inc.	USA	*634*	80-08300
Dornier GmbH	Germany	*167*	75-06700
DRS Technologies, Inc.	USA	*570*	75-27400
DUAL Incorporated	USA	*634*	80-08670
Dundridge College	UK	*501*	80-08830
Dynamit Nobel GmbH	Germany	*167*	75-07100
EASAMS Ltd	UK	*501*	80-09000
EG&G Washington Analytical Svcs	USA	*634*	80-09450
Eilat Consultants	Israel	*224*	80-00500
Elettronica GmbH	Germany	*168*	75-07565
Elintech GmbH	Germany	*178*	80-00550
Embraer Aviation International	France	*150*	80-00500
Engineering & Professional Serv.Inc.	USA	*634*	80-10150
European Aviation Training Center NV	Belgium	*48*	80-00500
FFA Flugzeugwerke Altenrhein AG	Switzerland	*386*	75-04350
FFV Aerotech AB	Sweden	*379*	80-00600
Flight International Inc.	USA	*635*	80-11600
FlightSafety International Inc.	USA	*635*	80-11650
FN Herstal SA	Belgium	*46*	75-03600
FR Aviation Limited	UK	*503*	80-10850
Frost & Sullivan Ltd	UK	*503*	80-10970
Gajon Institute of Technology SpA	Italy	*229*	60-00080
GEC-Marconi Avionics Ltd	UK	*454*	75-25320
GEL Engineering Ltd	UK	*454*	75-25500
General Motors of Canada Ltd	Canada	*71*	75-05900
George G. Sharp Inc.	USA	*642*	80-28000
Global Security Complex	USA	*636*	80-12950
Hanjin Heavy Industries Co. Ltd	Korea (ROK)	*265*	75-01505
Haux-Life-Support GmbH	Germany	*169*	75-11970
HeavyLift Aircraft Engineering Ltd.	UK	*503*	80-12300
Heckler & Koch Inc	USA	*581*	75-42860
Héli Union	France	*150*	80-00820
Hellenic Aerospace Industry Ltd.	Greece	*192*	80-00750
Horizons Technology Inc.	USA	*636*	80-14450
Hydroscience Technologies Inc.	USA	*584*	75-45550
Inst. of Sound & Vibration Research	UK	*419*	60-00870
Int'l Security & Defense Systems	Israel	*224*	80-00960
Interactive Dev. Environments	USA	*637*	80-15950
Israel Aircraft Industries Ltd	Israel	*219*	75-10200
Israel Military Industries Ltd	Israel	*220*	75-10300
ITT Federal Services Corp.	USA	*586*	75-49455
IVECO Fiat SpA	Italy	*235*	75-13230
Kapos Associates Inc.	USA	*637*	80-17630
Keystone Helicopter Corp.	USA	*637*	80-17780
Krauss-Maffei Wehrtechnik GmbH	Germany	*171*	75-16345
L'hotellier Montrichard	France	*138*	75-24900
Landair International Ltd	UK	*505*	80-14200
Litton Industries Inc.	USA	*591*	75-55400
Logistics Services International	USA	*638*	80-18800
Logtek (Logistic Technologies)	S. Africa	*360*	80-09950
Marr Technical Services Ltd	UK	*506*	80-15620
Materialinformationszentrum Ges.	Germany	*178*	80-01310
Microelectronics, guidance and	Turkey	*402*	55-00050
MSI-Defence Systems Ltd	UK	*469*	75-41280
National Systems Management Corp.	USA	*639*	80-21650
NAVFCO	France	*150*	80-01420
Nogay Ltd	Israel	*224*	80-01470
Nord-Micro Elektronik Feinmechanik	Germany	*173*	75-22700
Northrop Grumman Corporation	USA	*599*	75-65450
OTEC Support Services Ltd	UK	*508*	80-19600
Pacer Systems Inc.	USA	*602*	75-68000
Pennant International Limited	UK	*473*	75-44450
PK Electronic International Ltd	UK	*475*	75-45720
Planning & Research Ltd	Israel	*224*	80-01650
Professional School of Security	Israel	*216*	60-15000
Radio Holland Electronics BV	Netherlands	*297*	80-18300
Research Associates of Syracuse	USA	*641*	80-25900
Resource Consultants Inc.	USA	*641*	80-25920
Ricardo Hitec Ltd	UK	*509*	80-27840
Royal Jordanian Air Academy	Jordan	*260*	80-29100
Santa Barbara SA	Spain	*364*	55-01000
Scientific Management Associates	Australia	*25*	80-19100
Security Industry Training	UK	*511*	80-33500
SEMCOR Inc.	USA	*642*	80-27750
Serco Aviation Services	UK	*511*	80-34150
Siecor Corp.	USA	*614*	75-80560
SISDEF Ltda	Chile	*85*	80-01940
Sonda Aviation Enterprises Limited	UK	*512*	80-40000
Sonovision ITEP Technologies	France	*151*	80-01850
Spar Aerospace Ltd	Canada	*76*	75-13000
SPL Ordnance Ltd.	Israel	*225*	80-01935
Stanley Aviation Corporation	USA	*617*	75-83900
State College Operations	USA	*617*	75-83990
STN ATLAS Elektronik GmbH	Germany	*165*	75-01150
Strat*X International	USA	*643*	80-30100
Strat*X Ltd	UK	*512*	80-47500
Strat*X SA	France	*151*	80-02000
TA Group Ltd	UK	*513*	80-50110
The Robar Companies Inc.	USA	*609*	75-75630
Thomson Sintra	France	*148*	75-45135
Thomson-CSF Systems Canada Inc.	Canada	*79*	80-02020
Tracor Inc	USA	*623*	75-90000
Transfield Pty Ltd	Australia	*22*	75-20810
TriLink Technologies Group Inc.	Canada	*80*	80-02100
UK Training Ltd	UK	*514*	80-60000
United Defense L.P.	USA	*624*	75-92050
USM, Inc.	USA	*624*	75-91507
Vinnell Corp.	USA	*644*	80-34350
Vitrociset SpA	Italy	*240*	80-40000
Wegmann & Co GmbH	Germany	*177*	75-33700
Western Tactical Systems	UK	*515*	80-73000
Wheeler Systems Inc.	USA	*627*	75-96400
Wornall Electronics Inc.	USA	*645*	80-35400

980 10 Mission & weapon training services

AAI Corp.	USA	*545*	75-00020
Analysis & Technology Inc.	USA	*631*	80-02600
Armalon Ltd	UK	*431*	75-03729
Bofors AB	Sweden	*375*	74-00020
Burdeshaw Associates, Ltd	USA	*632*	80-04960
Calspan SRL Technology	USA	*540*	60-03000
Cubic Defense Systems Inc.	USA	*567*	75-22900
Dynamit Nobel GmbH	Germany	*167*	75-07100
Elintech GmbH	Germany	*178*	80-00550
Engineering & Professional Serv.Inc.	USA	*634*	80-10150
Flight International Inc.	USA	*635*	80-11600
Gajon Institute of Technology SpA	Italy	*229*	60-00080
GEC-Marconi Avionics Ltd	UK	*454*	75-25320
General Motors of Canada Ltd	Canada	*71*	75-05900
Heckler & Koch Inc	USA	*581*	75-42860
Horizons Technology Inc.	USA	*636*	80-14450
Hydroscience Technologies Inc.	USA	*584*	75-45550
Israel Aircraft Industries Ltd	Israel	*219*	75-10200
Kapos Associates Inc.	USA	*637*	80-17630
Materialinformationszentrum Ges.	Germany	*178*	80-01310
MECAR SA	Belgium	*46*	75-04930
Northrop Grumman Corporation	USA	*599*	75-65450
Pacer Systems Inc.	USA	*602*	75-68000
Perceptronics Inc.	USA	*603*	75-69270
Saint John Shipbuilding Ltd	Canada	*75*	75-12300
Saladin Security Ltd.	UK	*510*	80-30300
SEMCOR Inc.	USA	*642*	80-27750
Smith & Wesson	USA	*615*	75-82060
SPL Ordnance Ltd.	Israel	*225*	80-01935
STN ATLAS Elektronik GmbH	Germany	*165*	75-01150
Thomson-CSF	France	*147*	75-45100
Thomson-CSF Systems Canada Inc.	Canada	*79*	80-02020
Tracor Inc	USA	*623*	75-90000
United Defense L.P.	USA	*624*	75-92050
Wheeler Systems Inc.	USA	*627*	75-96400

980 10 05 Anti-terrorist training services

Public Safety Group Inc.	USA	*641*	80-24955
Saladin Security Ltd.	UK	*510*	80-30300

SDMS Security Products Ltd.	UK	*510*	80-32400
SPL Ordnance Ltd.	Israel	*225*	80-01935

980 10 20 Bomb disposal training services

BACTEC International Ltd	**UK**	***498***	**80-03060**
Clearway Explosives Services Ltd.	UK	*500*	80-06150
CMS Inc.	USA	*564*	75-19170
Public Safety Group Inc.	USA	*641*	80-24955
SDMS Security Products Ltd.	UK	*510*	80-32400
SPL Ordnance Ltd.	Israel	*225*	80-01935

980 10 40 Law enforcement training services

DEMCO Consultants Ltd	Israel	*224*	80-00430
Global Security Complex	USA	*636*	80-12950
Night Vision Equipment Co. Inc.	USA	*599*	75-64390
SPL Ordnance Ltd.	Israel	*225*	80-01935

980 10 50 Personnel protection training services

Analysis & Technology Inc.	USA	*631*	80-02600
Calspan SRL Technology	USA	*540*	60-03000
George G. Sharp Inc.	USA	*642*	80-28000
Gordon Consulting Engineers Ltd	Israel	*224*	80-00750
Heckler & Koch Inc	USA	*581*	75-42860
Public Safety Group Inc.	USA	*641*	80-24955
Saint John Shipbuilding Ltd	Canada	*75*	75-12300
Saladin Security Ltd.	UK	*510*	80-30300
SDMS Security Products Ltd.	UK	*510*	80-32400
Sector Ltd	UK	*511*	80-33400
SPL Ordnance Ltd.	Israel	*225*	80-01935
STN ATLAS Elektronik GmbH	Germany	*165*	75-01150

980 10 70 Target presentation services

Air Affairs (NZ) Limited	New Zealand	*303*	80-00050
Air Target Services Pty Ltd	Australia	*24*	80-00840
Analysis & Technology Inc.	USA	*631*	80-02600
ATS Aerospace Inc.	Canada	*68*	75-00752
CSE Aviation Ltd	UK	*500*	80-07600
Elintech GmbH	Germany	*178*	80-00550
Flight International Inc.	USA	*635*	80-11600
Flight Operations Ltd	UK	*502*	80-10650
Flight Refuelling Limited	UK	*451*	75-23850
Hayes Targets	USA	*603*	75-69135
Lockheed Martin Services Group	USA	*638*	80-18540
Sector Ltd	UK	*511*	80-33400
Skyline Aviation BV	Netherlands	*297*	80-19000
Southwest Aerospace	USA	*616*	75-82470
STN ATLAS Elektronik GmbH	Germany	*165*	75-01150
Thomson-CSF Systems Canada Inc.	Canada	*79*	80-02020

980 10 90 Underwater training services

Haux-Life-Support GmbH	Germany	*169*	75-11970
Int'l Submarine Engrg.	Canada	*73*	75-07780
Thomson-CSF	France	*147*	75-45100

980 20 Platform operation & procedures training services

AAI Corp.	USA	*545*	75-00020
Analysis & Technology Inc.	USA	*631*	80-02600
Calspan SRL Technology	USA	*540*	60-03000
Cubic Defense Systems Inc.	USA	*567*	75-22900
Elintech GmbH	Germany	*178*	80-00550
Engineering & Professional Serv.Inc.	USA	*634*	80-10150
Gajon Institute of Technology SpA	Italy	*229*	60-00080
George G. Sharp Inc.	USA	*642*	80-28000
Israel Aircraft Industries Ltd	Israel	*219*	75-10200
Kapos Associates Inc.	USA	*637*	80-17630
Kollsman	USA	*589*	75-53150
Marr Technical Services Ltd	UK	*506*	80-15620
Materialinformationszentrum Ges.	Germany	*178*	80-01310
Pilatus Aircraft Ltd	Switzerland	*387*	75-10100
SAC Technographic	UK	*510*	80-30100
SEMCOR Inc.	USA	*642*	80-27750
The Boeing Company	USA	*559*	75-14250
Thomson-CSF	France	*147*	75-45100

Tracor Inc	USA	*623*	75-90000
Transfield Pty Ltd	Australia	*22*	75-20810
United Defense L.P.	USA	*624*	75-92050
Van Rietschoten & Houwens	Netherlands	*296*	75-07900

980 22 Aircraft flight training services

AAI Corp.	USA	*545*	75-00020
Air New Zealand Ltd	New Zealand	*303*	80-00100
American Eurocopter	USA	*552*	75-05790
Astroways/Defense Group	USA	*631*	80-03358
Bristow Helicopters Ltd	UK	*499*	80-03950
British Aerospace Flight Training	UK	*499*	80-03995
Calspan SRL Technology	USA	*540*	60-03000
Chopperline Flight Training	Australia	*24*	80-03400
CSE Aviation Ltd	UK	*500*	80-07600
EASAMS Ltd	UK	*501*	80-09000
Eilat Consultants	Israel	*224*	80-00500
Elintech GmbH	Germany	*178*	80-00550
Embraer	Brazil	*54*	55-05000
European Aviation Training Center NV	Belgium	*48*	80-00500
FFA Flugzeugwerke Altenrhein AG	Switzerland	*386*	75-04350
Flight International Inc.	USA	*635*	80-11600
Flight Operations Ltd	UK	*502*	80-10650
FlightSafety International Inc.	USA	*635*	80-11650
Frasca International Inc.	USA	*576*	75-35900
Hughes Electronics Corp.	USA	*583*	75-44865
Israel Aircraft Industries Ltd	Israel	*219*	75-10200
Keystone Helicopter Corp.	USA	*637*	80-17780
Manfredi Aeronautical Industries	Panama	*317*	80-00300
Moscow Aircraft Production	Russia	*334*	55-13050
Northrop Grumman Corporation	USA	*599*	75-65450
Oxford Air Training School	UK	*508*	80-19900
Pezetel	Poland	*325*	80-01000
Royal Jordanian Air Academy	Jordan	*260*	80-29100
SEMCOR Inc.	USA	*642*	80-27750
Shorts Support Services Division	UK	*511*	80-35600
SimuFlite Training International	USA	*642*	80-28160
Tracor Inc	USA	*623*	75-90000

980 30 Equipment operation & utilization training services

AAI Corp.	USA	*545*	75-00020
Aeroquip Corp.	USA	*547*	75-02350
Analysis & Technology Inc.	USA	*631*	80-02600
Ansul Incorporated	USA	*554*	75-07810
Bayern-Chemie	Germany	*165*	75-02100
Bofors AB	Sweden	*375*	74-00020
Calspan SRL Technology	USA	*540*	60-03000
Chamberlain Manufacturing Corp.	USA	*563*	75-17950
Continental Electronics Corp.	USA	*566*	75-21800
Contraves GmbH	Germany	*166*	75-05200
Corrintec Ltd	UK	*442*	75-15120
Cubic Defense Systems Inc.	USA	*567*	75-22900
DRS Technologies, Inc.	USA	*570*	75-27400
EASAMS Ltd	UK	*501*	80-09000
Elintech GmbH	Germany	*178*	80-00550
Engineering & Professional Serv.Inc.	USA	*634*	80-10150
FFA Flugzeugwerke Altenrhein AG	Switzerland	*386*	75-04350
Fluor Daniel Inc.	USA	*635*	80-11720
Gajon Institute of Technology SpA	Italy	*229*	60-00080
GEC-Marconi Avionics Ltd	UK	*454*	75-25320
George G. Sharp Inc.	USA	*642*	80-28000
Harris Corporation	USA	*581*	75-42050
Helio Mirror Co. Ltd	UK	*457*	75-28900
Hewlett-Packard (Canada) Ltd	Canada	*72*	75-07520
Hodge Clemco Ltd	UK	*458*	75-29590
Horizons Technology Inc.	USA	*636*	80-14450
Hydroscience Technologies Inc.	USA	*584*	75-45550
Israel Aircraft Industries Ltd	Israel	*219*	75-10200
ITT Avionics	USA	*586*	75-49455
Krauss-Maffei Wehrtechnik GmbH	Germany	*171*	75-16345
L'hotellier Montrichard	France	*138*	75-24900
Lockheed Martin Tactical Aircraft	USA	*592*	75-55650
Marr Technical Services Ltd	UK	*506*	80-15620
MECAR SA	Belgium	*46*	75-04930
Meggitt Avionics (UK)	UK	*467*	75-38340
NEC Corporation	Japan	*253*	75-06350
Newport News Shipbuilding	USA	*599*	75-64300
Night Vision Equipment Co. Inc.	USA	*599*	75-64390
Nord-Micro Elektronik Feinmechanik	Germany	*173*	75-22700
Northrop Grumman Corporation	USA	*599*	75-65450

Pennant International Limited	UK	473	75-44450
Pilatus Aircraft Ltd	Switzerland	387	75-10100
Planning & Research Ltd	Israel	224	80-01650
Raytheon E-Systems	USA	608	75-73302
Research Associates of Syracuse	USA	641	80-25900
SAC Technographic	UK	510	80-30100
Saint John Shipbuilding Ltd	Canada	75	75-12300
SEMCOR Inc.	USA	642	80-27750
Shorts Support Services Division	UK	511	80-35600
Siecor Corp.	USA	614	75-80560
Sonovision ITEP Technologies	France	151	80-01850
Spar Aerospace Ltd	Canada	76	75-13000
Stanley Aviation Corporation	USA	617	75-83900
Stork Bronswerk BV	Netherlands	295	75-07200
The Boeing Company	USA	559	75-14250
Thomson-CSF	France	147	75-45100
Tracor Inc	USA	623	75-90000
United Defense L.P.	USA	624	75-92050
Vickers Defence Systems	UK	491	75-63040
Vitro Corp.	USA	644	80-34400
Vitrociset SpA	Italy	240	80-40000
Wornall Electronics Inc.	USA	645	80-35400

980 40 Technical training services

AAI Corp.	USA	545	75-00020
Advanced Appl. Consultants Inc.	USA	630	80-00250
Air New Zealand Ltd	New Zealand	303	80-00100
AirteQ Ltd	UK	496	80-00830
Alvis Logistics Ltd	UK	430	75-02860
American Eurocopter	USA	552	75-05790
AMI International	USA	631	80-02470
Analysis & Technology Inc.	USA	631	80-02600
Anco SA	Greece	187	75-00200
Aspect Computing Pty Ltd	Australia	24	80-01150
Bayern-Chemie	Germany	165	75-02100
Becker Flugfunkwerk GmbH	Germany	165	75-02200
Bofors AB	Sweden	375	74-00020
Bristow Helicopters Ltd	UK	499	80-03950
Calspan SRL Technology	USA	540	60-03000
Commslogic Ltd	UK	500	80-06400
Computerised Training Systems Ltd	UK	442	75-14650
Continental Electronics Corp.	USA	566	75-21800
Contraves GmbH	Germany	166	75-05200
Cranfield Aerospace Ltd	UK	500	80-07200
Cranfield University	UK	419	60-00350
CSE Aviation Ltd	UK	500	80-07600
Daewoo Heavy Industries Ltd	Korea (ROK)	265	75-00600
Dundridge College	UK	501	80-08830
Dytecna Ltd	UK	501	80-08900
Elettronica GmbH	Germany	168	75-07565
Elintech GmbH	Germany	178	80-00550
Embraer	Brazil	54	55-05000
Embraer Aviation International	France	150	80-00500
Engineered Air Systems Inc.	USA	573	75-31560
Engineering & Professional Serv.Inc.	USA	634	80-10150
FFA Flugzeugwerke Altenrhein AG	Switzerland	386	75-04350
Fluor Daniel Inc.	USA	635	80-11720
FN Herstal SA	Belgium	46	75-03600
Gajon Institute of Technology SpA	Italy	229	60-00080
GEC-Marconi Avionics Ltd	UK	454	75-25320
General Motors of Canada Ltd	Canada	71	75-05900
George G. Sharp Inc.	USA	642	80-28000
Gordon Consulting Engineers Ltd	Israel	224	80-00750
Halmatic Ltd	UK	456	75-27700
Harris Corporation	USA	581	75-42050
Hellenic Aerospace Industry Ltd.	Greece	192	80-00750
Hewlett-Packard (Canada) Ltd	Canada	72	75-07520
High Integrity Systems Ltd	UK	458	75-29250
Horizons Technology Inc.	USA	636	80-14450
Hughes Electronics Corp.	USA	583	75-44865
Inst. of Sound & Vibration Research	UK	419	60-00870
Interactive Dev. Environments	USA	637	80-15950
Interspiro Ltd	UK	461	75-31945
Israel Aircraft Industries Ltd	Israel	219	75-10200
Kay & Associates Inc.	USA	637	80-17660
Kiev Aircraft Repair Plant	Ukraine	406	55-01500
Krauss-Maffei Wehrtechnik GmbH	Germany	171	75-16345
Lockheed Martin Services Group	USA	638	80-18540
Mannesmann Rexroth SA	France	139	75-25500
Marr Technical Services Ltd	UK	506	80-15620
Materialinformationszentrum Ges.	Germany	178	80-01310
Microelectronics, guidance and	Turkey	402	55-00050
National Systems Management Corp.	USA	639	80-21650
NEC Corporation	Japan	253	75-06350
Newport News Shipbuilding	USA	599	75-64300
Night Vision Equipment Co. Inc.	USA	599	75-64390
Nord-Micro Elektronik Feinmechanik	Germany	173	75-22700
Northrop Grumman Corporation	USA	599	75-65450
OMI Logistics Limited	UK	508	80-19450
OTEC Support Services Ltd	UK	508	80-19600
Oxford Air Training School	UK	508	80-19900
Pennant International Limited	UK	473	75-44450
PK Electronic International Ltd	UK	475	75-45720
Planning & Research Ltd	Israel	224	80-01650
Prosig Computer Consultants Ltd	UK	476	75-46865
Redifon MEL Limited	UK	478	75-48530
Resource Consultants Inc.	USA	641	80-25920
Ricardo Hitec Ltd	UK	509	80-27840
RKI/Cleanroom Services	USA	641	80-26200
Rollab Aktiebolag	Sweden	380	80-01850
Rolls-Royce (Canada) Ltd.	Canada	75	75-12120
Saint John Shipbuilding Ltd	Canada	75	75-12300
Santa Barbara SA	Spain	364	55-01000
Sector Ltd	UK	511	80-33400
SEMCOR Inc.	USA	642	80-27750
Serco Aviation Services	UK	511	80-34150
Shorts Support Services Division	UK	511	80-35600
Sira Ltd.	UK	484	75-54250
SISDEF Ltda	Chile	85	80-01940
Sonovision ITEP Technologies	France	151	80-01850
Spar Aerospace Ltd	Canada	76	75-13000
Standard Aero Ltd	Canada	79	80-01970
Stanley Aviation Corporation	USA	617	75-83900
State College Operations	USA	617	75-83990
Stork Bronswerk BV	Netherlands	295	75-07200
STV Group Inc.	USA	643	80-30400
TA Group Ltd	UK	513	80-50110
Technology Australasia Pty Ltd	Australia	26	80-20600
Thomson-CSF	France	147	75-45100
Thomson-CSF Systems Canada Inc.	Canada	79	80-02020
Thorn DNT Airfield Lighting Pty Ltd	Australia	22	75-20540
TriLink Technologies Group Inc.	Canada	80	80-02100
United Defense L.P.	USA	624	75-92050
Vickers Defence Systems	UK	491	75-63040
Vipac Engineers & Scientists Ltd	Australia	26	80-22580
Vitro Corp.	USA	644	80-34400
Vitrociset SpA	Italy	240	80-40000
W. Vinten Ltd	UK	491	75-63353
Yarrow Shipbuilders Ltd	UK	495	75-67500

980 50 Language training services

APT Radar Systems Ltd	UK	431	75-03560
Elintech GmbH	Germany	178	80-00550
Gajon Institute of Technology SpA	Italy	229	60-00080
Hellenic Aerospace Industry Ltd.	Greece	192	80-00750
Israel Aircraft Industries Ltd	Israel	219	75-10200
Kay & Associates Inc.	USA	637	80-17660
Materialinformationszentrum Ges.	Germany	178	80-01310
Northrop Grumman Corporation	USA	599	75-65450
OTEC Support Services Ltd	UK	508	80-19600
Oxford Air Training School	UK	508	80-19900
Sector Ltd	UK	511	80-33400
Serco Aviation Services	UK	511	80-34150

980 60 Maintenance training services

AAI Corp.	USA	545	75-00020
AB Carl Munters	Sweden	377	75-08230
Aeroquip Corp.	USA	547	75-02350
Air New Zealand Ltd	New Zealand	303	80-00100
American Eurocopter	USA	552	75-05790
Analysis & Technology Inc.	USA	631	80-02600
Ansett Australia Avionics	Australia	24	80-01010
Becker Flugfunkwerk GmbH	Germany	165	75-02200
Bofors AB	Sweden	375	74-00020
Bristow Helicopters Ltd	UK	499	80-03950
Calspan SRL Technology	USA	540	60-03000
Chamberlain Manufacturing Corp.	USA	563	75-17950
CMPS&F Pty Limited	Australia	24	80-03600
Computerised Training Systems Ltd	UK	442	75-14650
Contraves GmbH	Germany	166	75-05200
Corrintec Ltd	UK	442	75-15120
CSE Aviation Ltd	UK	500	80-07600
Cubic Defense Systems Inc.	USA	567	75-22900
EASAMS Ltd	UK	501	80-09000

Product-Supplier Link

Elettronica GmbH	Germany	*168*	75-07565
Elintech GmbH	Germany	*178*	80-00550
Embraer	Brazil	*54*	55-05000
Embraer Aviation International	France	*150*	80-00500
Engineering & Professional Serv.Inc.	USA	*634*	80-10150
FFA Flugzeugwerke Altenrhein AG	Switzerland	*386*	75-04350
Fluor Daniel Inc.	USA	*635*	80-11720
Gajon Institute of Technology SpA	Italy	*229*	60-00080
GEC-Marconi Avionics Ltd	UK	*454*	75-25320
GEL Engineering Ltd	UK	*454*	75-25500
General Motors of Canada Ltd	Canada	*71*	75-05900
Gordon Consulting Engineers Ltd	Israel	*224*	80-00750
Grove Europe Ltd	**UK**	***456***	**75-27030**
H+S Aviation Ltd	UK	*503*	80-12050
Harris Corporation	USA	*581*	75-42050
Helio Mirror Co. Ltd	UK	*457*	75-28900
Hellenic Aerospace Industry Ltd	Greece	*192*	80-00750
Hewlett-Packard (Canada) Ltd	Canada	*72*	75-07520
Hydroscience Technologies Inc.	USA	*584*	75-45550
Interspiro Ltd	UK	*461*	75-31945
Israel Aircraft Industries Ltd	Israel	*219*	75-10200
ITT Avionics	USA	*586*	75-49455
IVECO Fiat SpA	Italy	*235*	75-13230
John Crane Inc.	USA	*566*	75-22500
Kay & Associates Inc.	USA	*637*	80-17660
Keystone Helicopter Corp.	USA	*637*	80-17780
Kiev Aircraft Repair Plant	Ukraine	*406*	55-01500
Krauss-Maffei Wehrtechnik GmbH	Germany	*171*	75-16345
Marr Technical Services Ltd	UK	*506*	80-15620
Materialinformationszentrum Ges.	Germany	*178*	80-01310
Meggitt Avionics (UK)	UK	*467*	75-38340
NEC Corporation	Japan	*253*	75-06350
Night Vision Equipment Co. Inc.	USA	*599*	75-64390
Nord-Micro Elektronik Feinmechanik	Germany	*173*	75-22700
Northrop Grumman Corporation	USA	*599*	75-65450
OMI Logistics Limited	UK	*508*	80-19450
OTEC Support Services Ltd	UK	*508*	80-19600
Oxford Air Training School	UK	*508*	80-19900
Pennant International Limited	UK	*473*	75-44450
Philips Calibration Services	Australia	*25*	80-16490
Pilatus Aircraft Ltd	Switzerland	*387*	75-10100
PK Electronic International Ltd	UK	*475*	75-45720
Planning & Research Ltd	Israel	*224*	80-01650
Ricardo Hitec Ltd	UK	*509*	80-27840
RKI/Cleanroom Services	USA	*641*	80-26200
Rolls-Royce (Canada) Ltd	Canada	*75*	75-12120
Saint John Shipbuilding Ltd	Canada	*75*	75-12300
Santa Barbara SA	Spain	*364*	55-01000
Sector Ltd	UK	*511*	80-33400
SEMCOR Inc.	USA	*642*	80-27750
Serco Aviation Services	UK	*511*	80-34150
Sonovision ITEP Technologies	France	*151*	80-01850
Stanley Aviation Corporation	USA	*617*	75-83900
STN ATLAS Elektronik GmbH	Germany	*165*	75-01150
Stork Bronswerk BV	Netherlands	*295*	75-07200
STV Group Inc.	USA	*643*	80-30400
The Boeing Company	USA	*559*	75-14250
Thomson Sintra	France	*148*	75-45135
Thorn DNT Airfield Lighting Pty Ltd	Australia	*22*	75-20540
Tracor Inc	USA	*623*	75-90000
United Defense L.P.	USA	*624*	75-92050
Vickers Defence Systems	UK	*491*	75-63040
Vipac Engineers & Scientists Ltd	Australia	*26*	80-22580
Vitro Corp.	USA	*644*	80-34400
Vitrociset SpA	Italy	*240*	80-40000

980 70 Medical training services

DEMCO Consultants Ltd	Israel	*224*	80-00430
International Hospitals Group	UK	*505*	80-13410
Medesign GmbH	Germany	*172*	75-20370
Vessa Limited	UK	*491*	75-63010

980 90 Other training services

Ada PROS Inc.	USA	*629*	80-00150
Advanced Appl. Consultants Inc.	USA	*630*	80-00250
AirteQ Ltd	UK	*496*	80-00830
Analysis & Technology Inc.	USA	*631*	80-02600
Cranfield University	UK	*419*	60-00350
DEMCO Consultants Ltd	Israel	*224*	80-00430
EDS Defence Ltd	UK	*501*	80-09200
Elintech GmbH	Germany	*178*	80-00550
Engineering & Professional Serv.Inc.	USA	*634*	80-10150
Holec Machines & Apparaten BV	Netherlands	*293*	75-03110
Horizons Technology Inc.	USA	*636*	80-14450
ICL DESC	UK	*501*	80-08400
Kapos Associates Inc.	USA	*637*	80-17630
National Systems Management Corp.	USA	*639*	80-21650
Northrop Grumman Corporation	USA	*599*	75-65450
Parsons Chain Co. Ltd	UK	*473*	75-44035
PK Electronic International Ltd	UK	*475*	75-45720
Planning & Research Ltd	Israel	*224*	80-01650
Professional School of Security	Israel	*216*	60-15000
Prosig Computer Consultants Ltd	UK	*476*	75-46865
RKI/Cleanroom Services	USA	*641*	80-26200
SAC Technographic	UK	*510*	80-30100
Sector Ltd	UK	*511*	80-33400
Serco Aviation Services	UK	*511*	80-34150
Services Sound & Vision Corp.	UK	*511*	80-34200
Strat*X International	USA	*643*	80-30100
Strat*X Ltd	UK	*512*	80-47500
Strat*X SA	France	*151*	80-02000
TADS Video Production	UK	*513*	80-50100
Technology Australasia Pty Ltd	Australia	*26*	80-20600

990 Environmental control services

AB Carl Munters	Sweden	*377*	75-08230
AEA Technology plc	UK	*496*	80-00480
Carolina Metals Inc.	USA	*562*	75-16760
EG&G Washington Analytical Svcs	USA	*634*	80-09450
Engineered Air Systems Inc.	USA	*573*	75-31560
Fluor Daniel Inc.	USA	*635*	80-11720
GEL Engineering Ltd	UK	*454*	75-25500
M International Export	USA	*639*	80-19780
MIL Systems	Canada	*79*	80-01360
NIS Ltd	UK	*471*	75-42555
NNC Ltd	UK	*507*	80-18500
Ordnance & Tactical Systems Division	USA	*606*	75-71472
Parsons Engineering Science	USA	*640*	80-23750
Planning & Research Ltd	Israel	*224*	80-01650
PRC Inc.	USA	*640*	80-24800
SEMCOR Inc.	USA	*642*	80-27750
Sverdrup Corp.	USA	*643*	80-30650
Taylor Woodrow Management	UK	*513*	80-50400

991 Decontamination services

AEA Technology plc	UK	*496*	80-00480
British Nuclear Fuels plc	**UK**	***499***	**80-04175**
Chemviron Carbon	Belgium	*46*	75-02260
Engineered Air Systems Inc.	USA	*573*	75-31560
Fluor Daniel Inc.	USA	*635*	80-11720
GEL Engineering Ltd	UK	*454*	75-25500
Giat Industries	France	*126*	55-00100
NIS Ltd	UK	*471*	75-42555
NNC Ltd	UK	*507*	80-18500
Ordnance & Tactical Systems Division	USA	*606*	75-71472
Sverdrup Corp.	USA	*643*	80-30650
Vickers Shipbuilding & Eng Ltd	UK	*491*	75-63100

993 40 75 Search and Rescue (SAR) Operations / Services

Tadiran Spectralink	Israel	*222*	75-21550

995 Ground based services

Peco Ltd	New Zealand	*302*	75-01630

999 Other services

AB Carl Munters	Sweden	*377*	75-08230
Advanced Appl. Consultants Inc.	USA	*630*	80-00250
AMI International	USA	*631*	80-02470
Andersen Consulting	Australia	*24*	80-00910
Andrich International Ltd	UK	*497*	80-01200
Armalon Ltd	UK	*431*	75-03729
Artemis Management Systems Ltd	UK	*497*	80-01259
Atlantic Research Corp.	USA	*555*	75-10150
Bayern-Chemie	Germany	*165*	75-02100
Beltramo & Associates	USA	*632*	80-04360

Carleton Life Support Technologies	Canada	70	75-02950
Crompton Greaves Ltd	India	205	75-00280
Currie & Brown (Australia) Pty Ltd	Australia	24	80-03850
DCS Corp	USA	634	80-07400
Delex Systems Inc	USA	634	80-08020
Design Bureau of	Russia	336	75-03150
Dytecna Ltd	UK	501	80-08900
EDS Defence Ltd	UK	501	80-09200
EG&G Inc	USA	571	75-29555
EG&G Washington Analytical Svcs	USA	634	80-09450
Elintech GmbH	Germany	178	80-00550
Embraer Aviation International	France	150	80-00500
ENDECO/YSI Inc	USA	573	75-31540
Engineering & Professional Serv.Inc.	USA	634	80-10150
Frazer-Nash Consultancy Limited	UK	503	80-10950
Garber International Associates Inc.	USA	636	80-12150
GEL Engineering Ltd	UK	454	75-25500
GTT Industries Inc	USA	636	80-13350
Horizons Technology Inc	USA	636	80-14450
Israel Shipyards Ltd	Israel	220	75-10600
Krauss-Maffei Wehrtechnik GmbH	Germany	171	75-16345
Lansdowne Technologies Inc	Canada	78	80-01205
Lockheed Martin Corp	USA	591	75-55635
M. Paz Logistics Ltd	Israel	224	80-01600
Management Consulting & Research	USA	638	80-19680
Microelectronics, guidance and	Turkey	402	55-00050
Pioneer Aerospace Corp	USA	604	75-70200
Planning & Research Ltd	Israel	224	80-01650
Ricardo Hitec Ltd	UK	509	80-27840
SEMCOR Inc	USA	642	80-27750
Serco Aviation Services	UK	511	80-34150
Sonovision ITEP Technologies	France	151	80-01850
SOPEMEA	France	151	80-01997
SPC International Corp.	USA	642	80-29200
Stanley Aviation Corporation	USA	617	75-83900
STS Recruitment	UK	512	80-47800
Sverdrup Corp	USA	643	80-30650
Technology Australasia Pty Ltd	Australia	26	80-20600
TRH Pictures	UK	514	80-58500
TriLink Technologies Group Inc.	Canada	80	80-02100
Universal Propulsion Co. Inc.	USA	625	75-92700
Vanguard Research Inc	USA	644	80-34100
Vitro Corp	USA	644	80-34400
Vitrociset SpA	Italy	240	80-40000
Vosper International Ltd	UK	514	80-67000

999 10 Budget & marketing information services

AMI International	USA	631	80-02470
Andersen Consulting	Australia	24	80-00910
Andrich International Ltd	UK	497	80-01200
Armalon Ltd	UK	431	75-03729
Beltramo & Associates	USA	632	80-04360
Defence Research & Analysis	UK	500	80-08200
Engineering & Professional Serv.Inc.	USA	634	80-10150
Fluor Daniel Inc	USA	635	80-11720
Glavcosmos	Russia	339	80-00500
Logica UK Limited	UK	506	80-14350
M. Paz Logistics Ltd	Israel	224	80-01600
Mycon Marketing Services Ltd	UK	507	80-17900
SEMCOR Inc	USA	642	80-27750
SPC International Corp.	USA	642	80-29200
Technology Australasia Pty Ltd	Australia	26	80-20600
The Graphics Co. of America	USA	579	75-40360
Top Line Co	USA	644	80-33100

999 50 Offset fulfillment services

Australian Technical Publications	Australia	24	80-01880
Beltramo & Associates	USA	632	80-04360
CMPS&F Pty Limited	Australia	24	80-03600
Hellenic Aerospace Industry Ltd	Greece	192	80-00750
Lansdowne Technologies Inc	Canada	78	80-01205
M International Export	USA	639	80-19780
M. Paz Logistics Ltd	Israel	224	80-01600
Philips Traffic & Engineering	Australia	20	75-16550
Schneider Sohn & Associates Inc.	USA	642	80-26890
SPC International Corp.	USA	642	80-29200
Technology Australasia Pty Ltd	Australia	26	80-20600
Welt International Corp	USA	645	80-34650

999 60 Technical support services

Aero Space Modifications	USA	630	80-00850
AeroSpace Technologies of Australia	Australia	11	75-00340
Allied Research Corp	USA	549	75-04670
American Management Systems	USA	630	80-02350
AMI International	USA	631	80-02470
Analysis & Technology Inc	USA	631	80-02600
Andersen Consulting	Australia	24	80-00910
Aspect Computing Pty Ltd	Australia	24	80-01150
Atlantic Research Corp	USA	555	75-10150
Australian Technical Publications	Australia	24	80-01880
Beltramo & Associates	USA	632	80-04360
Boeing Australia Limited	Australia	12	75-02500
Bristow Helicopters Ltd	UK	499	80-03950
Burdeshaw Associates, Ltd	USA	632	80-04960
CMPS&F Pty Limited	Australia	24	80-03600
Crew Services Ltd	UK	500	80-07370
DCS Corp	USA	634	80-07400
Delex Systems Inc	USA	634	80-08020
Design Bureau of	Russia	336	75-03150
Donaldson Co. Inc	USA	569	75-27050
DRS Technologies, Inc	USA	570	75-27400
Dytecna Ltd	UK	501	80-08900
EG&G Inc	USA	571	75-29555
EG&G Washington Analytical Svcs	USA	634	80-09450
Embraer Aviation International	France	150	80-00500
Engineering & Professional Serv.Inc.	USA	634	80-10150
ERA Technology Ltd	UK	502	80-09600
FLS Aerospace (Light Aircraft) Ltd	UK	451	75-23900
Fluor Daniel Inc	USA	635	80-11720
Forges de Zeebrugge SA	Belgium	46	75-03750
Frazer-Nash Consultancy Limited	UK	503	80-10950
GEC-Marconi Avionics Ltd	UK	454	75-25320
GEC-Marconi Materials Technology Ltd	UK	454	75-25340
GED (Southampton) Ltd	UK	503	80-11650
George G. Sharp Inc	USA	642	80-28000
GTT Industries Inc	USA	636	80-13350
Harris Corporation	USA	581	75-42050
Hellenic Aerospace Industry Ltd	Greece	192	80-00750
Hewlett-Packard (Canada) Ltd	Canada	72	75-07520
Horizons Technology Inc	USA	636	80-14450
Indal Technologies Inc	Canada	72	75-07740
Infodoc Sarl	Italy	240	80-00922
ITT Federal Services Corp.	USA	586	75-49455
Kaman Sciences Corp	USA	637	80-17600
Krauss-Maffei Wehrtechnik GmbH	Germany	171	75-16345
Leafield Logistics & Tech. Services	UK	506	80-14218
Lockheed Martin Corp	USA	591	75-55635
Logica UK Limited	UK	506	80-14350
M. Paz Logistics Ltd	Israel	224	80-01600
MIL Systems	Canada	79	80-01360
Milo Holdings Ltd	UK	506	80-16850
NEC Corporation	Japan	253	75-06350
OMI Logistics Limited	UK	508	80-19450
Pioneer Aerospace Corp	USA	604	75-70200
Quantrad Sensor	USA	606	75-72250
Raytheon Aerospace	USA	641	80-25650
Ricardo Hitec Ltd	UK	509	80-27840
SAIT Systems SA	Belgium	47	75-05600
Scaled Composites Inc	USA	541	60-19160
SEMCOR Inc	USA	642	80-27750
Shorts Support Services Division	UK	511	80-35600
Sira Ltd	UK	484	75-54250
Skilled Engineering Ltd	Australia	25	80-19350
South China Aero Technology Ltd	Hong Kong SAR	198	80-00850
SPC International Corp.	USA	642	80-29200
Stanley Aviation Corporation	USA	617	75-83900
STV Group Inc	USA	643	80-30400
Sverdrup Corp	USA	643	80-30650
Sverdrup Technology Inc	USA	643	80-30680
Synectics Corp	USA	643	80-30750
Technology Australasia Pty Ltd	Australia	26	80-20600
Technology Project Services	UK	513	80-52800
Three Quays Marine Services Ltd	UK	514	80-55400
Tracor Inc	USA	623	75-90000
TriLink Technologies Group Inc.	Canada	80	80-02100
Vanguard Research Inc	USA	644	80-34100
Vitro Corp	USA	644	80-34400
Vitrociset SpA	Italy	240	80-40000
Vosper International Ltd	UK	514	80-67000
W. Vinten Ltd	UK	491	75-63353
Wheeler Systems Inc	USA	627	75-96400

999 65 Technology transfer services

Aerospace Design Hugo Marom Ltd	Israel	*223*	80-00040
AMI International	USA	*631*	80-02470
Analysis & Technology Inc.	USA	*631*	80-02600
Andersen Consulting	Australia	*24*	80-00910
Andrich International Ltd	UK	*497*	80-01200
Beltramo & Associates	USA	*632*	80-04360
Bruce D. Vallillee Electronics Ltd	Canada	*80*	80-02210
Cambridge Consultants Ltd	UK	*499*	80-05300
CIC International Ltd	USA	*563*	75-18570
Delex Systems Inc.	USA	*634*	80-08020
EDS Defence Ltd	UK	*501*	80-09200
EG&G Washington Analytical Svcs	USA	*634*	80-09450
Engineering & Professional Serv.Inc.	USA	*634*	80-10150
ERA Technology Ltd	UK	*502*	80-09600
ESDU International plc	UK	*502*	80-09700
F&F Associates Inc.	USA	*635*	80-11000
Fluor Daniel Inc.	USA	*635*	80-11720
FN Herstal SA	Belgium	*46*	75-03600
Forges de Zeebrugge SA	Belgium	*46*	75-03750
Frazer-Nash Consultancy Limited	UK	*503*	80-10950
Harris Corporation	USA	*581*	75-42050
Israel Shipyards Ltd.	Israel	*220*	75-10600
Lansdowne Technologies Inc	Canada	*78*	80-01205
Lockheed Martin Corp.	USA	*591*	75-55635
M. Paz Logistics Ltd	Israel	*224*	80-01600
Medesign GmbH	Germany	*172*	75-20370
Microelectronics, guidance and	Turkey	*402*	55-00050
OMI Logistics Limited	UK	*508*	80-19450
Philips Traffic & Engineering	Australia	*20*	75-16550
Research Triangle Institute	USA	*540*	60-18000
Ricardo Hitec Ltd	UK	*509*	80-27840
Rollab Aktiebolag	Sweden	*380*	80-01850
Schneider Sohn & Associates Inc.	USA	*642*	80-26890
SEMCOR Inc.	USA	*642*	80-27750
Sira Ltd.	UK	*484*	75-54250
SOPEMEA	France	*151*	80-01997
SPC International Corp.	USA	*642*	80-29200
Stanley Aviation Corporation	USA	*617*	75-83900
STV Group Inc.	USA	*643*	80-30400
Sverdrup Corp.	USA	*643*	80-30650
Synectics Corp.	USA	*643*	80-30750
Technology Australasia Pty Ltd	Australia	*26*	80-20600
Vosper International Ltd	UK	*514*	80-67000

Company Electronic Address

Company Electronic Address Index

Anco SA . . . Greece . . . *187* 75-00200
i4e3lis@athen.compulink.g
AN/COM Industries Inc. . . . USA . . . *553* 75-07300
ancom@ancom.com
Andrew Sensor Products . . . USA . . . *554* 75-07452
http://www.andrew.com
Andrich International Ltd . . . UK . . . *497* 80-01200
andrich@andrich.com
Angus Electronics Co. . . . USA . . . *554* 75-07630
http://www.industry.net/angus
102476.3667@compuserve.com
Anite Networks Ltd . . . Australia . . . *11* 75-01293
http://www.anite.com.au
Anite Systems Ltd . . . UK . . . *497* 80-01240
http://www.anitesystems.com
ANSER . . . USA . . . *631* 80-02620
http://www.anser.org
Aonix Europe Ltd . . . UK . . . *497* 80-01251
http://www.aonix.co.uk
AOtec Inc. . . . USA . . . *554* 75-07940
http://www.aotec.com
AP Labs . . . USA . . . *630* 80-00440
http://www.sd.aplabs.com
info@sd.aplabs.com
A.P. Publications Ltd . . . UK . . . *497* 80-01253
http://www.ap.appublications.co.uk
106142.1713@compuserve.com
Apex Broach & Machine Co. . . . USA . . . *554* 75-08000
http://www.apexbroach.com
apexl@apexbroach.com
APM Hexseal . . . USA . . . *554* 75-08030
http://www.hexseal.com
Apollo Metals (UK) Ltd . . . UK . . . *515* 85-00085
apollometals.co.uk
Applied Data Sciences Inc. . . . USA . . . *554* 75-08050
http://www.appdatsci.com
sysadm@appdatsci.com
Applied Electro Mechanics Inc. . . . USA . . . *554* 75-08130
tjenkgrp@access.digex.net
Applied Holographics plc . . . UK . . . *431* 75-03520
http://www.applied-holographics.com
100407.1630@compuserve.com
Applied Signal Technology . . . USA . . . *554* 75-08210
customer@appsig.com
Applied Technology Services . . . Israel . . . *226* 90-00135
yossi_v@netvision.net.il
APREL Laboratories . . . Canada . . . *66* 60-00050
http://www.aprel.com
info@aprel.com
Aquarius Fabrique International Pvt . . . India . . . *204* 75-00160
jopi@giaspn01.vsnl.net.in
AR Technologies Division . . . Canada . . . *77* 80-00080
http://www.orenda.com
ctrsek@orenda.com
Arabian Aviation Corp EC . . . Bahrain . . . *36* 90-00150
aviation@balelco.com.bh
Arca-Regler . . . Germany . . . *165* 75-00950
http://www.arca-valve.com
Arde Inc. . . . USA . . . *554* 75-08500
ardeinc@aol.com
Aremco Products . . . UK . . . *431* 75-03705
aremco@btinternet.com
Aries Ingenieria y Sistemas SA . . . Spain . . . *367* 80-00200
ariesing@arrakis.es
Armaz (Bagira) International . . . Israel . . . *223* 80-00180
armaz@aol.co.il
Armed Forces Communications . . . Belgium . . . *44* 65-00050
afcea@infoboard.be
Armed Forces Communications . . . USA . . . *541* 65-00130
admin@afcea.org
afcea@infoboard.be
Armed Forces University Munich . . . Germany . . . *157* 25-19100
h11cpost@rz.unibw-muenchen.de
Armour of America Inc. . . . USA . . . *555* 75-08850
http://www.armourofamerica.com
armourinc@aol.com
Array Systems Computing Inc. . . . Canada . . . *77* 80-00060
http://www.array.ca
Artemis Management Systems Ltd . . . UK . . . *497* 80-01259
http://www.artemispm.com
info@artemispm.com
Arthur Hart Webbing . . . UK . . . *457* 75-28180
http://www.ndirect.co.uk/~harts
sales.harts@ndirect.co.uk
Artisys s.r.o. . . . Czech Republic *98* 75-00080
art@traveller.cz
ASC Industries . . . USA . . . *645* 85-00180
ascind@fastlane.net
Asea Brown Boveri AB . . . Sweden . . . *374* 70-00200
http://www.abb.se
Asea Brown Boveri A/S . . . Norway . . . *308* 75-00110
http://www.abb.no
info@noabb.abb.no
Aseko Oy Electronics . . . Finland . . . *120* 90-00160
elektroniikka@aseko.fi
Asia Motors Co Ltd . . . Korea (ROK) . . *265* 75-00100
http://www.asia.co.kr
Aspen Electronics Ltd . . . UK . . . *521* 90-00155
http://www.aspelect.demon.co.uk
Assessment Services . . . UK . . . *497* 80-01500
http://www.neag.co.uk
as@as.neag.co.uk
Associated Controls (Australia) . . . Australia . . . *12* 75-01560
michael@ascom.com.au
Associated Industries . . . USA . . . *645* 85-00181
http://www.associated-ind.com
talk@associated-ind.com
Association of Consulting Engineers . . . UK . . . *420* 65-00070
http://www.acenet.co.uk
consult@acenet.co.uk
Association of the US Army . . . USA . . . *542* 65-00150
http://www.ausa.org
ausahq@aol.com
Astro-Med Inc. . . . USA . . . *555* 75-09955
http://www.astro-med.com
astro-med@astro-med.com
A.T. Poeton (Gloucester Plating) Ltd . . . UK . . . *508* 80-23500
http://www.poeton.co.uk
atp@poetong.demon.co.uk
AT-Marine Oy . . . Finland . . . *120* 90-00200
at-marine@co.inet.fi
Atlantis Aerospace Corp. . . . Canada . . . *68* 75-00750
http://www.atlantis.com
info@atlantis.com
Atlas Hydraulic Loaders Ltd . . . UK . . . *432* 75-04300
atlas@atloi.demon.co.uk
Atomic Energy Corp . . . S. Africa . . . *357* 75-01000
http://www.aec.co.za
aec.marketing@pixie.co.z
AudioPack Sound Systems . . . USA . . . *556* 75-10240
audiopack@audiopack.com
Audiotel International Ltd . . . UK . . . *432* 75-04400
sales@audiotel-international.com
Auspex Systems Limited . . . UK . . . *498* 80-01650
http:// www.auspex.com
chardy@auspex.co.uk
Austin Italia SpA . . . Italy . . . *243* 90-00185
austin@enter.it
Australian Embassy . . . Egypt . . . *108* 31-00090
austremb@sc.soficom.com.eg
Australian Embassy . . . France . . . *123* 31-00090
http://www.austgov.fr
Australian Embassy . . . Israel . . . *215* 31-00090
aussie@inter.net.il
Australian High Commission . . . Pakistan . . . *314* 31-00090
ahcisb@isb.compol.ccom
Automatic Systems Laboratories Ltd . . . UK . . . *432* 75-04470
aslco.com
Automation NV . . . Belgium . . . *45* 75-00950
automation@automation.be
Autronics Corp . . . USA . . . *556* 75-10580
http://www.autronics.com
Avalon Electronics Ltd . . . UK . . . *432* 75-04485
http://www.aval.demon.co.uk/avalon
Avesta Sheffield Ltd. . . . UK . . . *432* 75-04570
http://www.avestasheffield.com
Avex Air Training Pty Ltd . . . S. Africa . . . *360* 80-00100
avex@iafrica.com
Aviapribor Corp . . . Russia . . . *336* 75-01200
aomiea@dol.ru
Avicraft AB . . . Sweden . . . *380* 90-00180
clma@lutab.se
Avimed Ltd . . . UK . . . *498* 80-01900
ianperry@pop3.hiay.co.uk
Avimo Ltd . . . UK . . . *432* 75-04950
101317.1734@compuserve.com
Avio Consultants Pty Ltd . . . Australia . . . *28* 90-01980
avioccon@esprit.com.au

Company Electronic Address

Company Electronic Address

Company Electronic Address

DRS Laurel Technologies . . . . USA *570* 75-27380
http://www.lrltech.com
DRS Optronics, Inc. . . . . USA *572* 75-29812
http://www.drs.com
DRS Precision Echo, Inc. . . . . USA *570* 75-27360
http://www.precisionecho.com
DRS Technologies, Inc. . . . . USA *570* 75-27400
http://www.drs.com
DSO National Laboratories . . . . Singapore . . . *345* 12-10000
ltaisiew@dso.org.sg
Du Pont de Nemours & Co. . . . . USA *570* 75-27750
http://www.dupont.com
Dundridge College . . . . UK *501* 80-08830
training.dundridge@dial.pipex.com
Dutch Gas Turbine Association . . . . Netherlands . . *290* 65-00600
vgt@euronet.nl
DY-4 Systems Inc. . . . . Canada *71* 75-04650
lhall@dy4.com
Dynamic Graphics Ltd . . . . UK *501* 80-08850
dgi@dgl.co.uk
Dynamic Sales & Service . . . . India *207* 90-00450
dynaipl@gems.vsnl.net.in
Dynasound Inc. . . . . USA *570* 75-28265
http://www.dynasound.com
ltkoening@dynasound.com
Dynepco, Inc . . . . USA *570* 75-28325
http://www.dynepco.com
dynepco@nae.com
Dytec Inc. . . . . USA *652* 90-04980
dytecil@aol.com

E

E-Pool Systems . . . . Pakistan . . . *315* 85-03000
http://www.compol.com/epool
epool@isb.compol.com
E-T-A Circuit Breakers Ltd . . . . UK *449* 75-21565
http://www.etacbe.co.uk
info@etacbe.co.uk
Eaton Corporation . . . . USA *571* 75-28800
http://www.eaton.com
EBA Microwave . . . . USA *652* 90-05020
http://www.home.earthlink.net/~bdreyer/
ebell99090@aol.com
EBH Trading Corp. . . . . Canada *80* 85-00600
ebh@ican.net
ECC Simulation Ltd . . . . UK *447* 75-19280
100442.3327@compuserve.com
ED Electronics Ltd . . . . UK *522* 90-00520
electronic-disigns.com
ED.A.I. Sarl . . . . Italy *240* 80-00430
http://www.edai.it/edai
edai@edai.it
Edgcumbe Instruments Ltd . . . . UK *447* 75-19410
sales@edgcombe.co.uk
Ediconsult Internazionale Srl . . . . . . Italy *240* 80-00450
ediconsuer@iol.it
EDO Corporation . . . . USA *571* 75-29250
http://www.edocorp.com
wfrost@edocorp.com
EES Technology Ltd . . . . UK *447* 75-19500
http://www.eestech.com
sales@eestech.com
EEV France . . . . France *152* 90-00513
106173.3521@compuserve.com
EEV Inc. . . . . USA *571* 75-29500
http://www.eevinc.com
EEV Ltd . . . . UK *448* 75-19590
http://www.eev.com
info@eev.com
Efratom, Time and Frequency Products . USA *557* 75-11860
http://www.datum.com
EFW Inc. . . . . USA *571* 75-29520
efw_inc@efw.aom
EG & G Heimann Optoelectronics GmbH Germany . . . *168* 75-07255
http://www.egginc.com/heimann
EG&G Inc. . . . . USA *571* 75-29555
http://www.egginc.com
Eidsvoll Electronics AS . . . . Norway *309* 75-00850
eidel@telepost.no
Eilat Consultants . . . . Israel *224* 80-00500
zb@netvision.net.il
EIP Microwave Inc. . . . . USA *572* 75-29590
http://www.eipm.com

Eiva A/S . . . . Denmark . . . *104* 80-00500
eiva_as@pip.dknet.dk
Elbit Systems Ltd . . . . Israel *218* 75-05300
http://www.elbit.co.il
Elcom Systems Inc. . . . . USA *572* 75-29650
http://www.elcomsystems.com/home/
sales@elcomsystems.com
Electro Corporation . . . . USA *572* 75-29685
http://www.electrocorp.com
Electro Impulse Laboratory Inc. . . . . . USA *572* 75-29700
http://www.electro-impulse.com
eil@electro-impulse.com
Electro Optical Industries Inc. . . . . . . USA *573* 75-30450
eoi@sonatech.com
Electro-Tec Corp. . . . . USA *573* 75-30700
http://kaydoncorp.com
Electrode Works Zika Ltd. . . . . Israel *223* 75-23800
zika1@netvision.net.il
Electron Tubes Limited . . . . UK *448* 75-20250
http://www.electech.demon.co.uk
rmca@electron-tubes.co.uk
Electronic Concepts Inc. . . . . USA *572* 75-29850
http://www.eci-capacitors.com
sales@eci-capacitors.com
Electronic Development Assoc. Inc. . . . USA *634* 80-09750
eda@eda-design.com
Electronic Industries Association . . . . Korea (ROK) . . *264* 65-00700
eiak@soback.kornet.nm.kr
Electronic Industries Association . . . . USA *542* 65-05620
http://www.eia.org
peterm@eia.org
publicaffairs@eia.org
Electronic Space Systems Corp. . . . . . USA *572* 75-30100
essco@tiac.net
Electrophysics Corp. . . . . USA *573* 75-30500
http://www.electrophysicscorp.com
info@electrophysicscorp.com
Electrothermal Engineering Ltd . . . . . UK *448* 75-20400
http://www.electrothermal.c om
Eley Ltd . . . . UK *448* 75-20550
101622.1253@compuserve.com
Elga Ltd . . . . UK *448* 75-20600
http://www.elga.co.uk
elga@elga.demon.co.uk
Elgin Industries . . . . USA *573* 75-30800
elginind@worldnet.att.net
Eliya Travels & Tours . . . . Pakistan . . . *315* 80-01000
imranali@paknet3.ptc.pk
Eloptro . . . . S. Africa . . . *358* 75-05680
jans@eloptro.co.za
Elron Electronic Industries Ltd. . . . . . Israel *217* 70-00700
http://www.elron.com
elron@alron.net
Eltex of Sweden AB . . . . Sweden *376* 75-04820
public@eltex.se
Embassy of Austria . . . . Germany *158* 31-00100
oesterreich-botschaft-bonn@compuserve. com
Embassy of Austria . . . . Israel *215* 31-00100
avtambta@netvision.net.il
Embassy of Austria . . . . Sweden *372* 31-00100
http://www.algonet.se/~austria
austria@algonet.se
Embassy of Austria . . . . Venezuela . . . *658* 31-00100
e.austria@cantv.net
Embassy of Barbados . . . . USA *537* 31-00140
barbados@oas.org
Embassy of Belgium . . . . India *202* 31-00150
waldel@del2.vsnl.net.in
Embassy of Belgium . . . . Italy *227* 31-00150
ambelrom@pub.xplore.it
Embassy of Belgium . . . . Netherlands . . *288* 31-00150
ambel-dh@wirehub.nl
Embassy of Brazil . . . . Australia *7* 31-00230
aasbrem@interconnet.com.au
Embassy of Canada . . . . Czech Republic . *97* 31-00310
michael.jay@prgue01.x400.gc.cz
Embassy of Canada . . . . Ireland *214* 31-00310
cdnembsy@iol.ie
Embassy of Canada . . . . Venezuela . . . *658* 31-00310
crcas@crcas01.x400.gc.ca
Embassy of Chile . . . . China PR *86* 31-00350
echile@public3.bta.net.cn
Embassy of Chile . . . . Denmark . . . *100* 31-00350
chiledkáinet.uni-c.dk

Company Electronic Address

Company	Electronic address	Country	Page	Code
Embassy of Chile	echileil@inter.net.il	Israel	215	31-00350
Embassy of Chile	echilejo@go.com.jo	Jordan	259	31-00350
Embassy of Chile	echileuk@echileuk.demon.co.uk	UK	414	31-00350
Embassy of Colombia	embacolbuca@datamarkets.com	Argentina	3	31-00380
Embassy of Colombia	emcoltky@j-link.or.jp	Japan	246	31-00380
Embassy of Colombia	embcol@form-net.com	Kenya	261	31-00380
Embassy of Costa Rica	embajada.cr@supermedia.pl	Poland	322	31-00410
Embassy of Costa Rica	http:// www.embcrlon.demon.co.uk info@embcrlon.demon.co.uk	UK	414	31-00410
Embassy of Croatia	EHMADORA@WAW.PDI.NET	Poland	322	31-00425
Embassy of Cyprus	embrecyp@worldnet.fr	France	123	31-00440
Embassy of Dem.Soc.Rep. of Sri Lanka	mmenu@iway.fr	France	125	31-01690
Embassy of Dem.Soc.Rep.of Sri Lanka	lankaemba@vision.net.id	Indonesia	209	31-01690
Embassy of Dem.Soc.Rep.of Sri Lanka	ldpo3432@niftyserve.of.jp	Japan	248	31-01690
Embassy of Dem.Soc.Rep.of Sri Lanka	sloman@gto.net.om	Oman	313	31-01690
Embassy of Ecuador	mecuahon@hondutel.hn	Honduras	197	31-00500
Embassy of Ecuador	http://www.embajada-ecuador.se suecia@embajada.ecuador.se	Sweden	373	31-00500
Embassy of El Salvador	embesal@iac.co.jp	Japan	247	31-00520
Embassy of Estonia	saatkond@estemb.or.at	Austria	31	31-00535
Embassy of Estonia	saatkond@estemb.be	Belgium	41	31-00535
Embassy of Estonia	seketar@estemb.dk	Denmark	100	31-00535
Embassy of Estonia	sekretar@estemb.fi	Finland	116	31-00535
Embassy of Estonia	saadik@est-emb.fr	France	124	31-00535
Embassy of Estonia	bonn@estebonn.gihub.de	Germany	159	31-00535
Embassy of Estonia	sekretar@riia.vm.ee; sekretar@com.latnet.lv	Latvia	270	31-00535
Embassy of Estonia	sekretar@estemb.no	Norway	307	31-00535
Embassy of Estonia	mskvpost@moskva.vm.ee	Russia	332	31-00535
Embassy of Estonia	sekretar@estemb.se	Sweden	373	31-00535
Embassy of Estonia	sekretar@estonia.gov.uk	UK	415	31-00535
Embassy of Estonia	saatkond@estemb.kiev.ua	Ukraine	405	31-00535
Embassy of Estonia	73671.2064@compuserve.com loa@estemb.org	USA	538	31-00535
Embassy of Federal Rep.of Germany	http://www.docuweb.ca/Germany/ 100566.2620@compuserve.com bn555@freenet.carleton.ca	Canada	65	31-00660
Embassy of Federal Rep.of Germany	106655.2050@compuserve.com	Nigeria	305	31-00660
Embassy of Federal Rep.of Germany	106112.3427@compuserve.com	Yugoslavia	661	31-00660
Embassy of Finland	suomi@tba.com.br	Brazil	54	31-00580
Embassy of Finland	Finland@it.com.pl	Poland	322	31-00580
Embassy of Finland	finland.embassy@finland.caz	S. Africa	356	31-00580
Embassy of France	france@brnet.com.br	Brazil	54	31-00590
Embassy of France	http://www.amba-ottawa.fr presse@amba-ottawa.fr	Canada	65	31-00590
Embassy of France	ambfrajp@popmail.gol.com	Japan	247	31-00590
Embassy of France	ambafrance.presse@com.pc.net.ro	Romania	330	31-00590
Embassy of France	jeromecauchard@.diplomtic.fr	Syria	391	31-00590
Embassy of Ghana	j.ackan@flashnet.it 610270 ghanaemb i	Italy	228	31-00670
Embassy of Greece	grnewsca@globalx.net	Canada	65	31-00690
Embassy of Greece	http://www.personal.eunet.fi/pp/grembfi grembfi@grembfi.pp.fi	Finland	116	31-00690
Embassy of Greece	022883010-0001@t-online.de	Germany	159	31-00690
Embassy of Guatemala	embaguate_bonn@compuserve.com	Germany	159	31-00740
Embassy of Honduras	honduras@interlink.or.jp	Japan	247	31-00790
Embassy of Iceland	icemb.stock@utn.stjr.is	Sweden	373	31-00820
Embassy of India	indemb@batelco.com.bh	Bahrain	36	31-00830
Embassy of India	embindia@chilesat.net	Chile	84	31-00830
Embassy of India	indembprague@bohem-net.cz	Czech Republic	97	31-00830
Embassy of India	chancery@indemb.datanet.hu	Hungary	200	31-00830
Embassy of India	ambsr@indemb.nl; couns@indemb.nl	Netherlands	289	31-00830
Embassy of India	postmaster@indoperu.org.pe	Peru	319	31-00830
Embassy of Islamic Rep.of Mauritania	http://www.emb.org/mauritania werzeg@embofirm.org	USA	539	31-01170
Embassy of Israel	embassyofisrael-vienna@mailbox.telecom .at	Austria	31	31-00880
Embassy of Israel	embisrae@latino.net.co	Colombia	91	31-00880
Embassy of Israel	http://www.xs4all.nl/~ambisra ambisra@xs4all.nl	Netherlands	289	31-00880
Embassy of Israel	http://www.iconz.co.nz/~israel dimitri.geidelberg@israel. org.nz israel-ask@israel.org.nz	New Zealand	301	31-00880
Embassy of Israel	israel@singnet.com.sg	Singapore	346	31-00880
Embassy of Italy	ambcalia@cyberianet.lb	Lebanon	271	31-00890
Embassy of Italy	embitalia@tmx.com.ni	Nicaragua	304	31-00890
Embassy of Italy	ambasu@quanta.com.py	Paraguay	318	31-00890
Embassy of Italy	http://www.embital.pt/embital embital@individual.EUnet.pt	Portugal	326	31-00890
Embassy of Italy	amitalia@samara.co.zw	Zimbabwe	662	31-00890
Embassy of Jamaica	emb.jam.brussels@skynet.be	Belgium	41	31-00910
Embassy of Japan	japan-embassy@telecom.net.et	Ethiopia	114	31-00920
Embassy of Kenya	koenyla@iuol.cn.net	China PR	86	31-00960
Embassy of Latvia	latemb@public.ua.net	Ukraine	405	31-01015
Embassy of Lebanon	maalouf@iusanet.cl	Chile	84	31-01020
Embassy of Lebanon	ambaliba@japan.co.jp	Japan	247	31-01020
Embassy of Malta	maemba@bldtzheim.com	Germany	159	31-01140
Embassy of Mexico	nstn0281@fox.nstn.ca	Canada	66	31-01190
Embassy of Mexico	mexico@inf.fu-berlin.de sreale01@att.com	Germany	159	31-01190
Embassy of Mexico	embamex@twics.com	Japan	247	31-01190
Embassy of Mexico	embmexke@form-net.com	Kenya	262	31-01190
Embassy of Mexico	eumrm@k.ro	Romania	330	31-01190
Embassy of Mexico	embamex@algonet.se	Sweden	373	31-01190

Company Electronic Address

Company / Address	Country	Page	Code
Explorocean Technology Ltd	UK	502	80-09870
explorocian-technology@compuserve.com			
Expotel Hotel Reservations	UK	502	80-09890
http://www.expotel.co.uk			
lores@expotel.atfmail.com			
E2S NV	Belgium	48	80-00600
e2s@e2s.be			

F

Company / Address	Country	Page	Code
F&F Associates Inc.	USA	635	80-11000
dfowler@bmpcoe.org			
Fabrica Militar "Frey Luis Beltran"	Argentina	4	55-06160
myaflb@redigital.com.ar			
Fairchild Defense	USA	575	75-33200
http://www.orbital.com			
Fairey Microfiltrex	UK	450	75-22300
100434.2340@compuserve.com			
Fairey Microfiltrex Inc.	USA	575	75-33330
fairey.microfiltrex@infoavc.net			
Fairview Machine	USA	575	75-33600
fairview@netway.com			
Far West Industries Inc.	Canada	71	75-05270
http://www.farwestcanada.com			
farwest@junction.net			
Fastening Systems International Inc.	USA	575	75-33870
http://www.fsirivet.com			
sales@fsirivet.com			
Federal Emergency Management Agency	USA	527	20-12000
eipa@fema.gov			
Federal Institute for Materials	Germany	161	60-00600
http://www.bam.de			
Federal Ministry for Foreign Affairs	Austria	30	12-20000
http://www.gov.austria-info.at/foreign affairs			
Federal Ministry of Finance	Germany	156	12-60000
http://www.bundesfinanzministerium.de			
Federal Telecommunications Sdn Bhd.	Malaysia	278	80-00200
cstan@fedtel.po.my			
Federation of the Electronics	UK	421	65-00400
raxford@fei.org.uk			
Feedback Instruments	UK	450	75-22520
http://www.fbk.com			
feedback@fdbk.demon.co.uk			
FEI	UK	421	65-00500
http://www.fei.org.uk/eri			
feedback@fei.org.uk			
Ferro-Ceramic Grinding Inc.	USA	635	80-11350
ferroceram@aol.com			
Fesca Qualos Gears	Australia	16	75-06360
sales@fpg.com.au			
Fifth Allied Tactical Air Force	Italy	229	34-22110
http://www.facgpidiv.tin.it			
Fike Corp.	USA	575	75-34370
http://www.fike.com			
fmpsales@fike.com			
Finavitec Oy	Finland	119	80-00580
firstname.lastname@finavitec.fi			
Fine Tubes Ltd	UK	451	75-23215
http://www.finetubes.co.uk			
sales@finetubes.co.uk			
Finmeccanica	Italy	230	70-00270
news@finmeccanica.com			
Finnish Shipbuilding Surveyors Ltd	Finland	119	80-00600
birger.mickwitz@fsc.pp.fi			
Finnyards Ltd	Finland	118	75-00600
finnyards.rauma@finnyards.com			
Fireseal International Ltd	UK	502	80-10450
firesealperry@pop3.hiway.co.uk			
Fitzroy Engineering Group Limited	New Zealand	302	75-00650
greg.michetski@xttra.co.nz			
Flight International Inc.	USA	635	80-11600
http://www.fltintl.com			
flight-mktg@fltintl.com			
Flight Management International Corp	USA	635	80-11630
fmicorp@erols.com			
Flight Suits Ltd	USA	575	75-34800
http://www.flightsuits.com			
mailbox@flightsuits.com			
FN Herstal SA	Belgium	46	75-03600
info@fnherstal.com			
Foerster Instruments Inc.	USA	576	75-35330
http://www.foerstergroup.com			
foerster@usaor.net			
Fokker Elmo BV	Netherlands	292	75-02320
fo-elmo@euronet.nl			
Force Computers (UK) Ltd	UK	451	75-24155
http://www.forcecomputers.com			
Forecast International/DMS	USA	636	80-11760
http://www.forecast1.com			
fidms@ibm.net			
Foreign & Commonwealth Office	UK	409	12-20000
http://www.fco.gov.uk/			
Forrester Stephen Aviation	Australia	28	90-06830
roband@netspace.net.au			
Fotec Inc.	USA	576	75-35680
http://www.std.com/fotec			
info@fotec.com			
Fothergill Engineered Fabrics Ltd	UK	452	75-24380
106127.647@compuserve.com			
FP Diesel LP	USA	576	75-35753
marketing@fpdiesel.com			
FP Diesel Ltd	UK	516	85-00630
fpdiesel@aol.com			
FPE Fischer GmbH	Germany	168	75-09300
fpe@fischer.dbp.de			
F.R. Ordnance International Ltd	UK	452	75-24480
106427.1673@compuserve.com			
Framatome Connectors Nederland BV	Netherlands	299	90-00630
http://www.fciconnect.com			
Franz Inc.	USA	576	75-35850
http://www.franz.com			
Frasca International Inc.	USA	576	75-35900
http://wwww.frasca.com			
marketing@frasca.com			
Fraser-Volpe Corp.	USA	576	75-35950
fraservolp@aol.com			
Frazer-Nash Consultancy Limited	UK	503	80-10950
fnc@frazer.demon.co.uk			
Fred V. Fowler Co Inc.	USA	653	90-06700
http://www.fvfowler.com			
Frederick Electronics	USA	576	75-36000
http://www.sfa.com			
sales@sfa.com			
Freudenberg Technical Products LP	UK	452	75-24730
106027.1374@compuserve.com			
Froude Consine	UK	452	75-24800
http:// www.froudeconsine.co.uk/fcsales			
sales@froudeconsine.co.uk			
FT Technologies Ltd	UK	452	75-24880
http://www.fttech.co.uk			
sales@fttech.co.uk			
Fugro-UDI Ltd	UK	452	75-24885
info@fugro-udi.co.uk			
Fuji Electric Co Ltd	Japan	251	75-01050
http://www.fujieletric.co.jp			
Fuji Heavy Industries Ltd	Japan	251	75-01100
okak@sb.hq.subaru-fhi.co.jp			
Fullerton, Sherwood Engineering Ltd	Canada	71	75-05625
http://www.fseng.com			
info@fseng.com			
Futronix Inc.	USA	577	75-36650
futronix@juno.com			
FXC Corp.	USA	577	75-36700
fxc@pia.com			

G

Company / Address	Country	Page	Code
Galileo Corp.	USA	577	75-36755
http://www.galileocorp.com			
Gamatronic Electronic Industries Ltd	Israel	219	75-07680
gammaa@netvision.net.il			
GAP Instrument Corp.	USA	577	75-36900
ceo@gap.net			
Garbarino Pompe SpA	Italy	234	75-09530
boy108@acqui.alpicom.it garbarino@acqui.alpicom.it			
Garmin International	USA	577	75-36920
http://www.garmin.doc			
GEC ALSTHOM Diesels Ltd	UK	453	75-25220
http://www.gecalsthom.com			
GEC ALSTHOM Ruston Diesels Ltd	UK	453	75-25255
ruston@compuserve.com			
GEC Plessey Semiconductors	UK	454	75-25363
http://www.gpsemi.com			
GEC-Marconi Aerospace Inc.	USA	577	75-37412
76245.1237@compuserve.com			
GEC-Marconi Hazeltine Corp.	USA	577	75-37414
http://www.systems.gec.cem			

Company Electronic Address

Company	Country	Page	Code
GEC-Marconi Research Centre gmrc.enquiries@gecm.com	UK	419	60-00700
GED (Southampton) Ltd ged@hmsoton.demon.co.uk	UK	503	80-11650
Geier & Bluhm Inc. http://www.geier-bluhm.com	USA	578	75-37427
Gelvenor Textiles gelvenor@sprintlink.co.za	S. Africa	358	75-06100
GEM Elettronica Srl http://www.jth.if/gem gem@jth.it	Italy	235	75-10100
Genco Mobile Systems BV http://www.gms@gemco.nl	Netherlands	293	75-02340
General Electrodynamics Corp. http://www.gecscales.com info@gecscales.com	USA	578	75-38500
General Kinetics Engineering Corp. 70673.3204@compuserve.com	Canada	71	75-05890
General Oceanics Inc. http://www.business1.com/genocean 73410.515@compuserve.com	USA	578	75-38920
Genex International Corporation ceju@hitel.kol.co.kr	Korea (ROK)	266	80-00725
GeoAcoustics Ltd http://www.geoacoustics.co.uk sales@geoacoustics.co.uk	UK	454	75-25405
German Aerospace Research Estab. http://www.dlr.de	Germany	161	60-00800
GET Engineering Corp. http://www.getnids.com getntds@sierra.net	USA	578	75-39540
GFC Aerospaço e Equipamentos Ltda gfc@easynet.com.br	Brazil	59	90-07000
Gibbs Rifle Co. http://www.gibbsrifle.com	USA	579	75-39570
Gichner Systems Group http://www.optimaeps.com sales@optimaeps.com	USA	579	75-39610
Gidropridor gidroprb@sovam.com	Russia	333	55-07600
GIII Ltd. g3@cybersp.mb.ca	Canada	71	75-06050
Gilardoni. http://www.smartcom.gilardoni.it gx@gilardoni.inet.it	Italy	235	75-10510
Gilbert Gilkes & Gordon Ltd http://www. gilkes.u-net.com/ chris@gilkes.com	UK	454	75-25430
Gill Instruments Ltd gill@gill.co.uk	UK	454	75-25450
Global Lightning Technologies http://www.gtt.com.au info@gtt.com.au	Australia	16	75-07600
Global Security Complex global@lightspeed.net	USA	636	80-12950
Global Thermoelectric Inc. http://www.globalte.com	Canada	71	75-06200
GMC SA misios@xan.forthnet.gr	Greece	189	75-01336
GMV SA http://www.gmv.es webmaster@gmv.es	Spain	367	80-00800
GNB Technologies gnbn@xtra.co.nz	New Zealand	302	75-00750
Godiva Ltd. godivagroup@ukbusiness.com	UK	455	75-25900
Golden Company golden@qatar.net.qa	Qatar	329	85-00500
Gordon Consulting Engineers Ltd ig@netvision.net.il	Israel	224	80-00750
Gould Alloys Ltd gouldalloys@mail.att.net	UK	516	85-00700
Government Flying Service gfsbrian@hkstar.com	Hong Kong SAR	197	25-05000
Government of Luxembourg theodore.pecatore@sip .etat.lu	Luxembourg	274	11-10000
Government of Nagorno-Karabakh root@gov.nk.am	Armenia	5	15-00050
Gracemoor Consultants (UK) Ltd gracemoor@dial.pipex.com	UK	503	80-11952
Graflex Inc graflex@aol.com	USA	579	75-40310
GRC International Inc. http://www.grci.com	USA	636	80-13120
Greene Tweed & Co France SA http://www.gtweed.com	France	153	90-00775
Greene, Tweed & Co Ltd mktng@gtweed.com	UK	455	75-26750
Greenwich Air Services - Texas, LP http://www.businesswire.com/cnn/gasia. htm	USA	636	80-13170
Ground Self-Defense Force jgsdf@ppp.mindnet.or.jp	Japan	245	25-30000
Gruppo Nuova MA.NA.RO. http://www.nuovamanaro.it claudiovignoli@nuovamanro. it mailbox@nuovamanaro.it	Italy	235	75-10730
GSI Corporation 103354.3170@compuserve.com homepages/gsicable	USA	580	75-40715
GTA http://www.nsads.net/fmkb gerry@login.dkknet.dk	Denmark	105	80-00700
GTE Corp. http://www.gte.com	USA	543	70-05200
Guardian Technologies International http://www.guardiantech.com gti@interramp.com	USA	580	75-40800
GUD Manufacturing Co. Pty Ltd apaj@ozemail.com.au	Australia	17	75-07885
Gull Electronic Systems Div. http://www.parker.com markers@gull.com	USA	580	75-41130
Gurley Precision Instruments http://www.gurley.com encoders@gurley.com	USA	580	75-41150
Gutteridge Haskins & Davey Pty Ltd http://www.ghd.com.au sydneymail@ghd.com.au	Australia	25	80-07500

H

Company	Country	Page	Code
H. Fine & Son Ltd http://www.hfine.co.uk	UK	451	75-23210
H. Koch & Sons Co. koch@ape.net	USA	589	75-53030
Hadland Photonics Inc. http://www.hadlandphotonics.com	USA	653	90-08030
Hadland Photonics Ltd post@hadland.co.uk	UK	456	75-27200
Hagen arrowvale@ax.compulink.co.uk	UK	456	75-27220
Hägglunds Moelv AS mail@haggmo.no	Norway	309	75-00880
Hägglunds Vehicle AB stmp.bandvagn@haggve.se	Sweden	377	75-05800
Hägglunds Vehicle AB hagglunds.vehicle@worldnet.att.net	USA	653	90-08035
HAL Communications Corp. http://www.halcomm.com halcomm@halcomm.com	USA	580	75-41450
Hale Group International halegrp@tidalwave.net	USA	636	80-13600
Haley Industries Ltd http://www.haley.on.ca harvey.murray@haley.on.ca	Canada	72	75-06600
Halocarbon Products Corp. http://www.halocarbon.com	USA	580	75-41630
Hammonds Technical Services Inc. hammonds@worldnet.att.net	USA	580	75-41800
Hardigg Cases literature@hardigg.com	USA	580	75-41930
Hardware Specialty Co. Inc. hardspeci@juno.com	USA	647	85-00800
Harkness Hall Ltd harknesshall.com	UK	457	75-28050
Harris Brett Ltd. http://www westyorks.com/snugpak snugpak@legend.co.uk	UK	457	75-28140
Harris Corporation http://www.harris.com webmaster@harris.com	USA	581	75-42050
Hartron Corporation root@khartron.kharkov.ua	Ukraine	406	75-03000
Havis-Shields Equipment Corp. http://www.havis.com	USA	581	75-42700
Hawker Eternacell Ltd sales.eternacell@btr.inc.com	UK	457	75-28500
HDL Research Laboratory Inc. hdlpower@aol.com	USA	581	75-42780

I

Company Electronic Address Index

Ilyushin Design Bureau . . . Russia . . . 333 55-09000
ilyushin@sovam.com

ImageBase Technology Ltd . . . UK . . . 516 85-00769
imagebas@aol.com

Imgraf SA . . . Dominican R. . . . 107 90-00300
rabaez@codetel.net.do

IMI Marston Ltd . . . UK . . . 460 75-31155
aht@attmail.com

IMI Norgren Ltd . . . UK . . . 460 75-31158
http://www.norgren.com

IMI-Tech Corp. . . . USA . . . 584 75-46140
http://www.imi-tech.com

IMP Aerospace Limited . . . Canada . . . 72 75-07730
http://www.impgroup.com
impaero@impgroup.com

IMP Group Ltd . . . Canada . . . 67 70-07000
http://www.impgroup.com
impaero@impgroup.com

iMPath Networks Inc. . . . Canada . . . 77 75-46151
http://www.impathnetworks.com

Importécnica SA . . . Brazil . . . 59 90-10000
importec@wm.com.br

Impreglio SpA . . . Italy . . . 240 80-00790
impregilo@interbusiness.it

Indian Avitron . . . India . . . 206 85-01000
bmgent@giasd101.vsnl.net.in

Inductoheat Pty Ltd . . . Australia . . . 17 75-09500
http://www.ozemail.com.au/~inducto
inducto@ozemail.com.au

Industrieanlagen-Betriebs GmbH . . . Germany . . . 178 80-00945
http://www.iabg.de
iabg@iabg.de

Infodoc Sarl . . . Italy . . . 240 80-00922
http://www.infodoc.it
infodo@arcanet.it

InfoLink Technologies Inc. . . . USA . . . 585 75-46703
infolink@aol.com

Information Handling Services Inc. . . . USA . . . 637 80-15530
http://www.ihs.com
info@ihs.com

Information Security Systems Ltd . . . UK . . . 461 75-31430
infosecsys@btinternet.com

Infotel Technologies (Pte) Ltd . . . Singapore . . . 351 90-00940
general@infotel.com.sg

Inframetrics Inc. . . . USA . . . 585 75-46750
http://www.inframetrics.com

Ingenieria de Sistemas . . . Spain . . . 367 80-00900
http://www.isdefe.es
general@isdefe.es

INIMEX . . . Belgium . . . 46 75-04650
defence@inimex.be

INRI UK Ltd . . . UK . . . 504 80-13320
http://www.inri.com
sales@inri.co.uk

Inst. of Sound & Vibration Research . . . UK . . . 419 60-00870
mzs@isvr.soton.ac.uk

Institut de Sciences Criminelles . . . France . . . 127 60-04000
http://www.montesquieu.u-bordeaux.fr

Institute of Metal Science . . . Bulgaria . . . 62 60-00800
stvims@bgcict.acad.bg

Instituto Geográfico Militar . . . Chile . . . 85 60-00700
igm@reuna.al

Instituto Vasco de Criminologia . . . Spain . . . 365 60-02500
http://www.sc.ehu.es/scrwwwiv/ivac.htm l
szoivac@sz.ehu.es

Instrumentation Marketing Corp. . . . USA . . . 653 90-09030
info@photosonics.com

Insulgard Corp. . . . USA . . . 585 75-47020
http://www.insulgard.com
insulgard@aol.com

Int'l Submarine Engrg. . . . Canada . . . 73 75-07780
http://www.ise.bc.ca

Integra Consult A/S . . . Denmark . . . 105 80-00900
integra@pip.dknet.dk

Interface Flooring Systems Inc. . . . USA . . . 585 75-46847
http://www.ifsia.com

Intergraph Corp. . . . USA . . . 585 75-47400
http://www.intergraph.com/federal

Intergraph (UK) Ltd . . . UK . . . 523 90-00948
http://www.intergraph.com/uk
bfrench@ingr.com

Interhom (Pvt) Limited . . . Pakistan . . . 316 90-00950
inter@hom.1hr.erum.com.pk

Intermarine SpA . . . Italy . . . 235 75-11900
marketing@intermarine.it

Intermec UK Ltd . . . UK . . . 504 80-13345
http://www.intermec.com

InterMed Video Technologies Inc. . . . USA . . . 585 75-46855
http://www.intermedvideo.com
sales@intermedvideo.com

International Aero Engines AG . . . USA . . . 544 74-09500
http://www.v2500.com

International Atomic Energy Agency . . . Austria . . . 32 34-10000
http://www.iaca.or.at/worldatom

International Committee . . . Switzerland . . . 384 34-11000
http:www.icrc.ch

International Helicopter Parts . . . USA . . . 647 85-00950
http://www.ihpww.com
ihp@ihpww.com

International Loss Prevention . . . Canada . . . 72 75-07775
http://www.ilps.com
ilps@mindlink.bc.ca

International Maritime Organization . . . UK . . . 418 34-25000
rkohn@imo.org

International Polymer Corp. . . . USA . . . 585 75-48200
internationalpolymerhouston@postoffice .att.net

International Procurement Services . . . UK . . . 461 75-31820
100145.2747@compuserve.com

International Professional Security . . . UK . . . 421 65-00700
http://ourworld.compuserve.com/ homepages/ipsa
ipsa.secretary@dial.pipex.com

International Promotion Services Srl . . . Italy . . . 243 90-00950
md1182@mclink.it

International Tech. Applications . . . Greece . . . 193 80-00795
ita@otenet.gr

Interpoint A Crane Company . . . USA . . . 585 75-46870
http://www.interpoint.com
power@intp.com

Interpoint (UK) Ltd. . . . UK . . . 516 85-00775
http://www.interpoint.com

Interspiro AB . . . Sweden . . . 377 75-06950
info@interspiro.se

Interstate Electronics . . . USA . . . 586 75-48700
peter_rothenberg@iecorp. com

INTERTECHNIK Techn. Produktionen . . . Austria . . . 33 75-01430
info@intertechnik.at

Iscan Inc. . . . USA . . . 586 75-49100
http://www.iscaninc.com
info@iscaninc.com

Israel Aerospace and Defence . . . Israel . . . 224 80-00970
http://www.iaw.co.il
eshel@shani.net

Israel Aircraft Industries . . . Singapore . . . 350 85-01000
siairo@pacific.net.sg

Israel Product Research Co. Ltd. . . . Israel . . . 219 75-10000
ispra@netvision.net.il

ITEC Ltd . . . UK . . . 505 80-13550
itec@andrich.com

I.T.P. Ltd . . . Israel . . . 216 65-00050
ronyg@itpgroup.com

ITT Defense & Electronics . . . USA . . . 586 75-49455
http://www.ittind.com

IVEX Corp. . . . USA . . . 586 75-49950
http://www.ivex3d.com
marketing@ivex3d.com

Izhevsk Mechanical Plant . . . Russia . . . 333 55-09500
ims@shield.udmurtia.su

J

J L Weibull AB . . . Sweden . . . 379 75-12100
jlw@jlw.se

J T Inglis & Sons Ltd . . . UK . . . 461 75-31500
inglis@taynet.co.uk

J-Tec Associates Inc. . . . USA . . . 587 75-51250
http://www.j-tecassociates.com
j-tec@j-tecassociates.com

Jakab Industries Pty Ltd . . . Australia . . . 17 75-10170
jakab@mpx.com.au

Jalex BV . . . Netherlands . . . 297 80-10000
skyline.aviation@tip.nl

Jane's Information Group . . . USA . . . 637 80-17080
http://www.janes.com/janes.html
velthaus@janes.com

Jane's Information Group Ltd . . . UK . . . 505 80-13700
http://www.janes.com
info@janes.co.uk

Jaques . . . Australia . . . 17 75-10200
jaques@sprint.com.au

Company Electronic Address Index

M

M.C. Gill Corp. USA *579* 75-39620
http://www.mcgillcorp.com

McCallie Associates Inc. USA *639* 80-20500
jmaassen@mccallie.com

MCG Electronics Inc. USA *595* 75-59580
http://www.mcgsurge.com
mcginfo@specdata.com

MCL Inc. USA *595* 75-59700
http://www.mcl.com

McTurbine Inc. USA *639* 80-19780
mcturbine@intcomm.net

MDH Limited UK *467* 75-38270
101376.1130@compuserve.com

Measurement Devices Ltd UK *467* 75-38275
mdl.uk.@zetnet.co.uk

Mécanex SA Switzerland *386* 75-08300
mecanexsa@bluewin.ch

Megkon S. Africa *360* 80-10000
hennie_f@megkom.com

Mekel Engineering Inc. USA *595* 75-59850
http://www.mekel.com

Menasco Aerospace Ltd Canada *74* 75-09600
http://www.coltecind.com

Mercedes-Benz (Australia) Pty Ltd Australia *19* 75-13355
wilson.jack@mbau.mercedes-benz.com

Mercury Computer Systems Inc. USA *595* 75-60220
http://www.mc.com
info@mc.com

Metal Dynamics Corp. USA *595* 75-60400
http://www.tulsaweb.com/mdc

Meteorological Office UK *411* 20-01100
http://www.meto.gov.uk/

Metheus Corporation UK *467* 75-38800
http://www.metheus.com
europe@metheus.com

Metheus Corporation USA *595* 75-60660
info@metheus.com

Metheus Systems (S) Pte Ltd (Asia) Singapore *348* 75-02720
singapore@metheus.com

Metool Products Ltd. UK *467* 75-38875
postmaster@metool.demon.co.uk

Metra Martech Limited UK *506* 80-16300
http://www.metra-martech.com
research@metra.demon.co.uk

Metric A/S Norway *312* 90-01295
http://www.metric.no
metric@metric.no

Metric Systems Corp. USA *596* 75-60800
mscmrkt@eng-msc.com

Metrix Instrument Company UK *524* 90-01340
http://www.industry.net/metrix
bob.matrix.europe@dial.pipex.com

Metrosonics Inc USA *596* 75-60820
http://www.metrosonics.com
salesinfo@metrosonics.com

M.G. Electric (Colchester) Ltd UK *468* 75-38912
sales@mgelectric.co.uk

MHD Int'l Aviation Parts Inc. Canada *80* 85-01000
mhd_intl@sypatico.ca

Michigan Wheel Corp. USA *596* 75-60930
http://www.miwheel.com
info@miwheel.com

Micro Communications Inc. USA *596* 75-60980
dheymans@sprynet.com

Micro Data Base Systems Inc. USA *639* 80-21080
http://www.mdbs.com
info@mdbs.com

Microinfo Ltd UK *506* 80-16500
http://www.microinfo.co.uk
ops@ukminfodemon.uk

Micronic Devices India *207* 90-01340
mic.amdbt@gems.vsnl.net-in

Microsemi Corp. USA *596* 75-61090
http://www.microsemi.com
webmaster@microsemi.com

Middle East Marketing Services Ltd Jordan *260* 80-05000
mms@go.com.jo

MIL Systems Canada *79* 80-01360
http://www.milsystems.com
quality@milsystems.com

MIL-BASE Industries Inc. USA *596* 75-61530
1e0276@sshare.com

Military Logistics MFC USA *596* 75-61550
sales@millog.com

Militech International USA *647* 85-01343
oberdick@erols.com

Millitech Corp. USA *596* 75-61665
dgreenman@millitech.com

Milper Ltd Israel *226* 90-01310
ugig@milpe.co.il

Mils Elektronik GmbH & Co KG Austria *33* 75-01700
http://www.mils.com
consult@mils.com (enquiries)
support@mils.com (customers)

Mining Resource Engineering Ltd Canada *74* 75-09770
mre@limestone.kosone.com

Ministry for Foreign Affairs Iceland *201* 12-20000
postur@utn.stjr.is

Ministry of Defence Armenia *5* 12-10000
http://www.nic.nk.am/artsakh/army.html

Ministry of Defence Croatia *94* 12-10000
kopriv@zvonmir.morh.tel.hr

Ministry of Defence Estonia *112* 12-10000
info@kmin.ee
polyakoff@kmin.ee

Ministry of Defence Sweden *371* 12-10000
http://www.regeringen.se
anna.liedbergius@defence.m inistry.se
registrator@defence.ministry.se

Ministry of Defence UK *408* 12-10000
http://www.mod.uk/

Ministry of Economic Affairs Netherlands *287* 12-12000
http://www.minez.nl
voorlichting@minez.nl

Ministry of Economics Affairs Estonia *112* 12-10500
kantselei@mineco.ee

Ministry of Finance Estonia *112* 12-20550
admin@fin.ee
aili@fin.ee

Ministry of Finance Jordan *259* 12-60000
mdf@amra.nic.gov.jo

Ministry of Finance Lithuania *273* 12-61000
finmin@finmin.lt

Ministry of Foreign Affairs Estonia *112* 12-20000
vmedia@vm.ee
vminfo@vm.ee

Ministry of Foreign Affairs Finland *115* 12-20000
webmaster@formin.fi

Ministry of Foreign Affairs Latvia *270* 12-20000
http://www.mfa.gov.lv
mfa.cha@mfa.gov.lv

Ministry of Foreign Affairs Lithuania *273* 12-20000
chancery@uors.urm.it

Ministry of Foreign Affairs Morocco *284* 12-20000
http://www.minicom.gov.ma
mext@onpt.net.ma

Ministry of Foreign Affairs Seychelles *344* 12-20000
mfapesey@Seychelles

Ministry of Industry and Trade Hungary *200* 12-25000
http://www.ikm.hu

Ministry of Interior Belgium *39* 12-07000
henri.maes@cominformatics. fgov.be

Ministry of Internal Affairs Armenia *5* 12-40000
news@mia.nk.am

Ministry of Internal Affairs Estonia *112* 12-40000
piret@sisemin.gov.ee

Ministry of National Defence Belgium *39* 12-10000
sidmanager@mil.be

Ministry of National Security Bahamas *35* 12-40000
watsondpm@batelnet.bs

Ministry of Science Croatia *94* 12-80000
office@science.hr

Ministry of Science Malaysia *276* 12-60000
http://masti.gov.my

Ministry of Science and Technology Slovenia *354* 12-80000
vlado.valda@mzt.si

Ministry of Transportation and Estonia *112* 12-30500
nei@mail.kbfi.ee

Mishra Dhatu Nigam Ltd India *204* 55-01300
mdnexport@hdl.vsnl.net.in
spralloy@hdi.ysnl.net.in

MITEQ Inc. USA *597* 75-62000
http://www.miteq.com

MITM Consulting Engineers Israel *224* 80-01350
mitm_c@netvision.net.il

Mitsubishi Electric Corporation Japan *253* 75-05700
http://www.mitsubishielectric.com

Mitsubishi Precision Co. Ltd Japan *253* 75-06000
http://www.mindnet.or.jp/mpc

Company Electronic Address

N

Nichols Brothers Boat Builders Inc. USA *599* 75-64380
nichols@whidbey.com
Nichols Research Corp. USA *639* 80-21780
http://www.nichols.com
info@nichols.com
Nicotra Meccanica SpA Italy *241* 85-00800
nicotrareccanica@ga.lactica.it
Night Vision Equipment Co. Inc. USA *599* 75-64390
nvec@ptdprolog.net
Nilsen Electric (SA) Pty Ltd Australia *19* 75-14680
http://www.nilsen.com.au
nilsensa@nilsen.com.au
Noble Denton & Associates (WLL) UAE *408* 90-01200
nodent@ns2.emirates.net.ae
Noble Denton Europe Ltd UK *507* 80-18300
marketing@nodent.co.uk
ndeaberdeen@nodent.co.uk
Nobo (UK) Limited UK *471* 75-42645
info@nobo.co.uk
Nogay Ltd Israel *224* 80-01470
mogay@netvision.net.il
Noise Com Inc. USA *599* 75-64760
http://www.noisecom.com
info@noisecom.com
Noptel Oy Finland *118* 75-01650
http://www.otm.fi/noptel/
Noratech BV Netherlands *299* 90-01470
ahylkema@bart.nl
Norca Machinery Company USA *599* 75-64800
norca@idt.net
Norsam Technologies USA *639* 80-21800
http://www.norsam.com
jane@norsam.com
Norsk Data Ltd UK *507* 80-18400
http:// www.serviceteam.co.uk
kayr@serviceteam.co.uk
Norsk Jetmotor A/S Norway *310* 75-01950
office@jetmotor.no
Northern Airborne Technology Ltd Canada *74* 75-10145
johnd@natech.com
Northrop Grumman Corporation USA *599* 75-65450
http://northgrum.com
http://www.essd.com
http://www.northropgrum.com
Northrop Grumman International Inc. Belgium *50* 90-01400
t.darcy@pophost.eunet.be
Novacom Oy Finland *120* 90-01450
matti.pihlanto@novacom.inet.fi
NPO Molniya plc Russia *337* 75-13380
alex@molniya.msk.ru
NQEA Australia Pty Ltd Australia *19* 75-14890
nqea_cns@tpgi.com.au
Nuclear Associates Division USA *600* 75-65730
http://www.nucl.com
sales@nucl.com
Nuclear Metals Inc. USA *600* 75-65760
http://www.nucmet.com
sales@nucmet.com
Nuclear Research Corp. USA *600* 75-65770
http://www.nrc-ind.com
mktg@nuc.com
Numerical Algorithms Group Ltd UK *507* 80-18800
http://www.nag.co.uk/
http://www.nag.co.uk/welcome_iec.html
infodesk@nag.co.uk.
helpdesk@iec.co.uk
Nyge Aero AB Sweden *379* 80-01490
http://www.nygeaero.com
carl-gustaf.werner@nygearo.se

O

O'Gara-Hess & Eisenhardt Armoring USA *601* 75-66400
http://www.ogara.com
Ocean Data Equipment Corp. USA *601* 75-65860
mkt@seabeam.com
Ocean Technology Systems USA *601* 75-66020
ots4com@aol.com
Oceano Instruments (UK) Ltd UK *508* 80-19050
http://www.oceano.demon.co.uk
sales@oceano.demon.co.uk
Oceanographic Co Netherlands Netherlands *294* 75-05255
f.vandongen@ocn.nl
Oceanor Norway *310* 75-02125
http://www.oceanor.no
oceanor@oceanor.no
Octec Ltd UK *471* 75-42950
http://www. octec.co.uk
octec@octec.co.uk
Odetics Inc. USA *601* 75-66050
http://www.odetics.com
Odmat Rotem Israel *225* 85-01700
ad_rotem@netvision.net.il
OEA Aerospace Inc. USA *601* 75-66100
http://www.oeaa.com
OFC Corporation USA *601* 75-66220
http://www.ofccorp.com
ofccorp@ofccorp.com
OFEMA South Africa S. Africa *361* 90-14950
ofema@iafrica.com
Office of the Governor Bermuda *51* 10-10000
depgov@ibl.bm
Office of the President Estonia *112* 10-10000
http://www.ee/president
presinfo@nw.vpk.ee
sekretar@nw.vpk.ee
Office of the President Latvia *269* 10-10000
pnede@mail.parks.lv
Office of the Prime Minister Estonia *112* 11-10000
valitsus@rk.ee
Office of the Prime Minister Malaysia *276* 11-10000
http://www.mdc.com.my
Officine Panerai SpA Italy *237* 75-17500
webcom@panerai.it
O.I. Analytical USA *600* 75-65810
cms6900@aol.com
Oki Electric Industry Co. Ltd Japan *254* 75-07200
http://www.oki.co.jp
Olympus Industrial UK *472* 75-43020
106160.1063@compuserve.com
Omega DSE Ltd New Zealand *304* 90-01500
omega@sequel.co.nz
Omnipol Ltd. Czech Republic *99* 75-00750
omnipol@gls.cz
Omnisec AG Switzerland *387* 75-09330
marketing@omnisec.ch
Omnitécnica Sarl Portugal *327* 75-15000
omnitec.ca@mail.telepac.pt
Ontario Drive & Gear Ltd. Canada *74* 75-10400
http://www.argoatv.com
sales@odg.com
Ontic Engineering & Mfg Inc. USA *601* 75-66875
http://www.ontic.com
Optechs Korea Inc Korea (ROK) *266* 75-03750
http://ibk.co.kr/company/optechs.htm
Optek Technology Inc. USA *601* 75-66880
http://www.optekinc.com
optek1@optekinc.com
Optex UK *472* 75-43130
http://www.optexint.com
info@optexint.com
Optical Cable Corp. USA *601* 75-66952
http://www.occfiber.com
sadams@occfiber.com
Optical Components Inc. USA *601* 75-66953
http://www.primenet.com/~oci/
daleadv@primenet.com
Optotek Ltd Canada *74* 75-10420
http://www.optotek.com
optotek@optotek.com
Oracle Telecomputing Inc. Canada *74* 75-10430
oratel@entrenet.com
Orbital Engine Company (Australia) Australia *25* 80-15830
http://www.orbeng.com.au
oemai@orbeng.com.au
Oreet-Marketing Communications Israel *226* 90-01600
areetimic@netvision.net.il
Orion Aviation USA *602* 75-67500
whin@orionav.com
Orkal Industries Corp. USA *602* 75-67505
orkal@orkal.com
Orlando Engineering Co. Ltd Taiwan *394* 90-01480
orlando@ms7.hinet.net
Orman Associates Inc. USA *640* 80-22600
or2withdog@aol.com
Orseal Ltd UK *472* 75-43160
richard@threlfall. telme.com

Company Electronic Address

PowerTech Inc. USA 605 75-71087
http://www.power-tech.com
alexmart@ix.netcom.com
Praxis Critical Systems Ltd UK 509 80-23930
http://www.praxis-cs.co.uk
info@praxis-cs.co.uk
PRB Associates Inc. USA 640 80-24700
http://www.prb.net
Precise Flight Inc. USA 605 75-71200
http://www.preciseflight.com
preciseflight@preciseflight.com
Precision Filters Inc. USA 605 75-71330
pfinfo@pfinc.com
Precision Technologies Pte Ltd Singapore 352 90-01670
pretech@singnet.com.sg
Presentey Engineering Canada 75 75-11500
http://www.presentey.com
pepl@presentey.com
PressCut Industries Inc. USA 605 75-71436
http://www.presscut.com
info@presscut.com
Pressure Systems Inc. USA 605 75-71438
mikeh@psi-pci.com
Price & Orphin Limited UK 509 80-24250
sales@pontown.demon.co.uk
Prism Color Group USA 640 80-24875
prismcolor@earthlink.com
Proaction USA 641 80-24900
proaction@aol.com
Prof Sten Luthander Ingbyra AB Sweden 379 80-01280
info@lutab.se
Profile Security Services Ltd. UK 509 80-24500
profilesecurity@compuserve
Programming Research Ltd UK 509 80-24700
http://www.prqa.co.uk
info@prqa.co.uk
Progress-Werk Oberkirch AG Germany 173 75-24320
100517.6550@ compuserve.com
Projects Unlimited Inc. USA 606 75-71560
http://www.pui.com
conrad@pui.com
Prometey Central Research Institute Russia 335 60-03555
vvv@promety2.spb.su
Prosig Computer Consultants Ltd UK 476 75-46865
http://www.prosig.co.uk
Prosser/Enpo USA 574 75-31670
cranegov@juno.com
Protection Development Int'l Corp. USA 606 75-71670
http://www.armour-pdi.com
info@armour-pdi.com
Przemyslowy Instytut Telekomunikacji Poland 325 75-02650
http://www.pit.edu.pl
office@pit.edu.pl
PT Boma Bisma Indra Indonesia 209 55-00050
bbimark@radned.id
ptbbi@peterpetraac.id
PT PAL Indonesia (PERSERO) Indonesia 210 75-00500
paljkt@ibm.net
PT Stimec Elcom Indonesia 211 90-01970
stimec@indosat.net.id
PTI Technologies Inc. USA 606 75-71800
filters@ptitechnologies.com
Pulnix America Inc. USA 606 75-71815
http://www.pulnix.com
Pulnix Europe Ltd UK 476 75-47030
http://www.pulnix.co.uk
market@pulnix.co.uk
office@pulnix.co.uk
sales@pulnix.co.uk
PZL Mielec Poland 325 75-02670
wskpzlsa@ptcmet.mielec.pl

Q

Q-Systems International Inc. USA 606 75-71960
qsitexas@aol.com
Q-Tech Corp. USA 606 75-71970
sales @q-tech.com
QED Design and Manufacture Ltd UK 476 75-47350
gordon@tigra.demon.co.uk
Qualitair Engineering Services Ltd UK 509 80-25200
http://www.qualitair.co.uk
QUANTICS incorporated USA 641 80-25150
users@quantics.com
Quantrad Sensor USA 606 75-72250
http://www.quadrad.com
quantrad@quantrad.com

R

R&B Aircraft Supply Inc. USA 641 80-25420
rbaircraft@aol.com
R.A. Miller Industries Inc. USA 596 75-61650
http://www.rami.com
R.A. Poole and Co (Stainless) Ltd UK 517 85-01690
r.a.poole@st.st.fasteners.demon.co.uk
Racal Avionics Inc USA 655 90-17000
ravinc@clark.net
Racal Canada Inc Canada 75 75-11590
http://www.racal.ca
sales@racalcanada.com
Racal Electronics plc UK 424 70-01805
http://www.racal.com
racal_pr@compuserve.com
Racal Instruments Inc. USA 607 75-72402
http://www.racalinst.com
Racal Protection Canada Canada 81 90-01805
rpc@recorder.ca
Racal Recorders Inc. USA 655 90-18016
recorders@usa.racal.com
Racal Recorders Ltd UK 477 75-47689
recorders@racalrec.co.uk
Rada Electronic Industries Ltd. Israel 222 75-17900
http://www.rada.com
Radar/Digital Systems USA 607 75-72460
rfay@ports.com
Radiall Nederland BV Netherlands 294 75-05750
http://www.radiall.nl
info@radiall.nl
Radian Technology USA 607 75-72500
sales@radiantech.com
Radio Research Instrument Co. Inc. USA 607 75-72600
http://www.techexpo.com/www/radiores
kbps32e@prodigy.com
Radiola Corporation Ltd. New Zealand 303 85-02000
richardt@radiola.co.nz
Radstone Technology plc. UK 477 75-47810
http://www.radstone.co.uk
sales@radstone.co.uk
Radstone Technology Pty Ltd Australia 20 75-18100
unitsyd@unitronix.com.au
Rafael USA Inc. USA 655 90-18040
yudi@ix.netcom.com
RAN Maritime Headquarters Australia 7 25-41000
fpao@mhqe.mc.gov.au
RAND USA 540 60-17000
http://www.rand.org
Randolph Products Co. USA 607 75-72905
71273.3465@compuserve.com
Rapid Acquisition Targeting Systems USA 607 75-72920
http://www.ratslasers.com
Rapid Deployment Towers Inc. USA 607 75-72925
jwhii@aol.com
Raytheon Anschütz GmbH Germany 182 90-01820
http://www.raytheon.com
receptionde@raykiel.com
Raytheon E-Systems USA 608 75-73302
http://www.raytheon.com/esys
info@emed.com
Raytheon International Inc UK 524 90-01825
http://www.raytheon.com
RBR Armour Limited Hong Kong SAR 199 90-01830
100452.3251@compuserve.com
Reade Advanced Materials USA 608 75-73860
http://www.reade.com
ramsales@reade.com
Recoil Pty Australia 20 75-18300
sales@recoil.com.au
Recon/Optical Inc. USA 608 75-73930
http://www.roi.bourns.com
Recrusul SA Brazil 57 75-18340
recrusul@recrusul.com.br
Reflectone Inc USA 609 75-74150
http://www.reflectone.com
little@reflectone.com
Refractory Metals Association USA 542 65-18250
http://www.mpif.org

Company Electronic Address

Company Electronic Address Index

Senate . . . France . . . 121 15-10000
c.antonini@senat.fr
c.giulinani@senat.fr
jd.giulini@senat.fr
jm.schuh@senat.fr
mj.tulard@senat.fr
presidence@senat.fr
Sencea Ltd . . . UK . . . 482 75-52730
http://www.sencea.co.uk
sales@sencea.co.uk
Sensor Systems Inc. . . . USA . . . 613 75-79400
sensorsystems@worldnet.att.net
Senstar-Stellar Corp . . . Canada . . . 76 75-12760
http://www.senstar.ca
info@senstar.ca
SEOS Displays Ltd. . . . UK . . . 482 75-52830
info@seos.co.uk
Sermatech International Inc . . . USA . . . 613 75-79530
http://www.sermatech.com
Services Sound & Vision Corp. . . . UK . . . 511 80-34200
http://www.ssvc.com
Servogear A/S . . . Norway . . . 310 75-02435
main.office@servogear.no
Servotron SA . . . Argentina . . . 5 90-19300
75074.566@compuserve.com
servoci@microstar.com.ar
SETEC Pty Ltd . . . Australia . . . 21 75-19415
pdlloyd@compuserve.com
SEXTANT Avionique SA . . . France . . . 144 75-35000
http://www.sextant-avionique.com
Shakespeare Products Group . . . USA . . . 614 75-79800
http://www.shakespeare-ef.com
Shalon Chemical Industries Ltd . . . Israel . . . 222 75-20000
http://www.doryanet.co.il/shlon17.htm
shalon@netvision.net.il
Sheldahl Inc. . . . USA . . . 614 75-80150
http://www.sheldahl.com
Shenango Steel Buildings Inc. . . . USA . . . 614 75-80205
shenangotmc@pathway.net
Shock-Tech Inc. . . . USA . . . 614 75-80457
shocktech@worldnet.att.net
Shorrock Integrated Systems . . . UK . . . 483 75-53450
http://www/shorrock.co.uk
sis@shorrock.co.uk
Shoshin Corp . . . Japan . . . 258 90-01960
shori@shoshin.co.jp
Siemens A/S . . . Norway . . . 310 75-02450
mja@fs.siemens.no
Siemens Atea NV . . . Belgium . . . 47 75-05950
p80921@vnet.atea.be
Siemens Electromechanical Components . . . USA . . . 614 75-80565
info@sec.siemens.com
Siemens Environmental Systems Ltd . . . UK . . . 483 75-53700
radmon@poole.siemens.co.uk
Siemens Plessey Electronic Systems . . . Australia . . . 21 75-19475
spspes@siemens.spes.com.au
Sierracin/Sylmar Corporation . . . USA . . . 614 75-80650
sscmperry@aol.com
SIFT SAS di Cutore & Pitotti . . . Italy . . . 239 75-22150
sift@mbox.micanet.it
SIGARMS Inc. . . . USA . . . 614 75-80700
sai@nhultranet.com
Sigma Security Devices Ltd . . . UK . . . 483 75-53778
http:// www.nightingale.co.uk/sigma
sigma@nightingale.co.uk
Sigma Signals Corporation . . . USA . . . 615 75-80790
http://www.netcom.com/~sigmasig/info. html
sigmasignal@ix.netcom.com
Signaal USFA . . . Netherlands . . . 295 75-06500
http://www.usfa.nl
zewa@usfa.nl
Signal Processors Ltd . . . UK . . . 484 75-53780
http://www.spl.co.uk/satcomms/index. html
sales@spl.co.uk
Signal Technology Corp . . . USA . . . 615 75-80795
http://www.sigtech.com
Signal Technology Corp. . . . USA . . . 618 75-84240
http://www.sigtech.com
uffelman@sigtech.com
SIHI Australia Pty Ltd . . . Australia . . . 21 75-19505
sfsaus@ozemail.om.au
Silver Eagle Manufacturing Co. . . . USA . . . 615 75-81020
http://www.silvereaglemfg.com
SIM Security & Electronic System . . . Germany . . . 175 75-28010
sim.electronic@t-online.de
Simad SpA . . . Italy . . . 230 74-23350
http://www.amicus.it/simad
Simagine . . . Netherlands . . . 295 75-06600
simulation.systems@gw.siemens.nl
SIME s.a.s. di Paolo de Gaetano & C. . . . Italy . . . 244 90-01930
paolo@star.flashnet.it
SimuFlite Training International . . . USA . . . 642 80-28160
info@simuflite.com
Simula Government Products, Inc. . . . USA . . . 615 75-81250
isdept@simula.com
Singapore Aerospace International . . . USA . . . 656 90-19420
staerosp@aol.com
Singapore Aerospace (UK) Pte Ltd . . . UK . . . 525 90-01959
bdbc@st.com.sg
Singapore British Engineering Pte . . . Singapore . . . 350 85-04000
sbesing@mbox2.singnet.com.sg
SIP Services Int'l . . . UK . . . 511 80-36500
sipinternational@compuserve.com
Sira Ltd . . . UK . . . 484 75-54250
http://www.sira.co.uk
SIRN sze . . . Italy . . . 240 80-01950
sizn.ge@pn.itnet.it
SISDEF Ltda . . . Chile . . . 85 80-01940
sisdef@entelchile.net
SITEP SpA . . . Italy . . . 239 75-22595
sitep@mbox.vol.it
Skanacid A/S . . . Denmark . . . 105 80-03650
skanacid@post4.tele.dk
Skandinavisk Teleindustri Skanti A/S . . . Denmark . . . 106 90-19450
http://www.skanti.dk
skanti@skanti.dk
SKC Inc. . . . USA . . . 615 75-81480
http://www.skcinc.com
skcinc@skcinc.com
SKF . . . Sweden . . . 378 75-10340
http://www.skf.com
Skilled Engineering Ltd . . . Australia . . . 25 80-19350
http://www.skilled.aust.com
Skipasmidastodin hf . . . Iceland . . . 202 75-05000
skipas@sne.pa.is
Skyfox Aviation Ltd . . . Australia . . . 17 75-08300
http://www.ozemail.com.au/%7 eskyfoxpl/
skyfoxpl@ozemail.com.au
Skyline Aviation BV . . . Netherlands . . . 297 80-19000
skyline.aviation@tip.nl
Skyways Inc. . . . USA . . . 642 80-28250
intlferry@aol.com
Smith System Engineering Ltd . . . UK . . . 512 80-39500
http://www.smithsys.co.uk
Smiths Industries Aerospace . . . UK . . . 485 75-54700
http://www.smithsind.com
plc@smithsind.co.uk
Snorkel . . . USA . . . 616 75-82170
snorkel@snorkelusa.com
SNPE . . . France . . . 145 75-36315
http://www.snpe.com
Sociedade Zickermann SA . . . Portugal . . . 328 90-03850
zickermann.lis@mail.telepal.pt
Society of Flight Test Engineers . . . UK . . . 420 60-02000
http://www.hughes-ec.com/org/sfte
Sodankyla Geophysical Observatory . . . Finland . . . 117 60-03500
http://www.sgo.fi
aarne.ranta@sgo.fi
Solomat . . . USA . . . 616 75-82330
http://www.solomat.com
iaq@solomat.com
Somers Forge Ltd . . . UK . . . 485 75-55150
somers@folkes.forgings.ltd.uk
Sonar Research & Development Ltd . . . UK . . . 485 75-55170
sales@srd-ltd.demon.co.uk
Sonda Aviation Enterprises Limited . . . UK . . . 512 80-40000
office@sonda_sim.co.uk
Sonsub International Ltd . . . UK . . . 512 80-41000
http://www.sonsub.co.uk
it@sonsub.co.uk
Sony Corporation . . . Japan . . . 251 75-00975
g-huh@ccom.sony.co.jp
Sooa Corporation . . . Korea (ROK) . . . 267 90-01960
sukchul@nownuri.net
South African Embassy . . . Finland . . . 117 31-01670
saemb@icon.fi
South African Embassy . . . Japan . . . 248 31-01670
http://sunsite.sut.ac.jp/embassy/ jasanet

The Dindima Group Pty Ltd . . . . Australia 15 75-04580
dindima@ozemail.com.au
The Eraser Co. Inc. . . . . USA 574 75-32020
eraser@dreamscape.com
The Flight Data Company Ltd . . . . UK 502 80-10600
http://www.fdata.demon.co.uk
marketing@flightdata.co.uk
The Knesset . . . . Israel 215 15-10000
sifra2@netvision.net.il
The New Piper Aircraft Inc. . . . . USA 604 75-70250
http://www.newpiper.com
The Nitor Group Ltd . . . . UK 424 70-01400
nitor@compuserve.com
The OSIRIS Group . . . . Australia 25 80-16000
http://www.leadingedge.net.au
leading@magna.com.au
tscott@laurel.ocs.mq.edu. au
The Purdy Corp. . . . . USA 606 75-71850
http://www.purdytransmissions.com
sales@purdytransmissions.com
The Racal Corporation . . . . USA 544 70-18000
http://www.racal.com
The Ray Engineering Company . . . . . . UK 478 75-48350
http://www.rencol.com
sales@rencol.com
The Robar Companies Inc. . . . . USA 609 75-75630
robarguns@earthlink.com
The Seychelles High Commission . . . . UK 417 31-01620
soozyw@aol.com
The Sintef Group. . . . . Norway 308 60-01600
http://www.sintef.no/
The Titan Corp . . . . USA 622 75-89620
http://www.titan.com/
corpcomm@titan.com.
Thiokol Corpoartion . . . . USA 622 75-88740
http://www.thoikel.com
Thomas Bertie & Co. Ltda . . . . Chile 85 90-00150
tbertie@netup.cl
Thomas Electronics of Aus. Pty Ltd . . . Australia 22 75-20480
teo@thomas.com.au
Thomassen International BV . . . . Netherlands . . . 296 75-07670
thomassen@thomassen-int.nl
Thomson Marconi Sonar Limited . . . . . UK 426 74-20000
http://www.tmsluk.co.uk
Thomson Training & Simulation Ltd . . . UK 488 75-59567
http://www.ttsl.co.uk
Thomson-CSF Electronics Belgium. . . . Belgium 47 75-06310
teb@thomson-csf.fr
Thomson-CSF Inc.. . . . . USA 656 90-20300
http://www.thompson-csf.com
Thomson-CSF Systems Canada Inc . . . Canada. 80 80-02020
http://www.thomson-csf.ca
thomcsf@thomson-csf.ca
THORN Microwave Devices Ltd . . . . . . UK 488 75-59760
http://www.tmd.co.uk/
Thought Technology Ltd . . . . Canada. 77 75-13770
http://www.thoughttechnology.com
mail@thoughttechnology.com
TI Group plc. . . . . UK 424 70-02100
http://www.tigroup.com/
Time & Tune . . . . Pakistan 315 75-01980
ikram@infolink.net.pk
Tisdale Co. Inc.. . . . . USA 622 75-89580
tisdale@flex.net
TIW Systems Inc. . . . . USA 623 75-89755
info@tiw.com
Tomsed Corp. . . . . USA 623 75-89778
tomsed@bellsouth.net
Toshiba Corp . . . . Japan 250 70-05000
http://www.toshiba.co.jp
naomi.furuyu@toshiba.co.jp
Trace International Group Ltd . . . . . . USA 648 85-02350
tigl@traceww.com
TRACE Middle East Regional Liaison. . . Greece 193 85-02000
jpp@inet.gr
TRACE Moscow . . . . Russia 340 85-02000
jbarrie@bigfoot.com
TRACE Scandinavia . . . . Denmark 105 85-02000
f-vaaben@traceww.com
Tracor Aerospace Inc. . . . . USA 623 75-90010
mhk@corpl.tracor.com
Tracor Inc . . . . USA 623 75-90000
http://www.tracor.com
pr@corp1.tracor.com

Trade Mark Electronics BVBA. . . . . . . Belgium 48 85-02000
info@tme.be
Trans Security Systems & Technology . . Israel 223 75-23055
transsec@netvision.net.il
Transaero Inc. . . . . USA 656 90-20730
http://www.transaeroinc.com
Transfield Pty Ltd . . . . Australia 22 75-20810
http://www.transfield.com.au
dicksmithd@transfield.com. au
Transfield Shipbuilding WA . . . . Australia 22 75-20812
tranship@iinet.net.au
Treu SA Máquinas e Equipamentos . . . Brazil 58 75-20780
treurj01@unijet.com.br
TriLink Technologies Group Inc. . . . . Canada. 80 80-02100
mikeve@tenebris.com
Triplett Corporation . . . . USA 623 75-90970
http://www.triplett.com
Tritech International Ltd . . . . UK 489 75-61160
sales@tritech.co.uk
Triton Technology Inc. . . . . USA 623 75-91050
sales@tritontech.com
Tronic Electronic Services Ltd . . . . . . UK 489 75-61330
tronic@co.uk
TrueTime . . . . USA 624 75-91110
http://www.truetime.com
Truflo Marine Valves Ltd . . . . UK 489 75-61380
http://www.truflo.co.uk
TSM Corporation . . . . Korea (ROK) . . . 266 80-05000
tsmc@unitel.co.kr
TSS (UK) Ltd . . . . UK 489 75-61395
tssmail@tssuk.co.uk
Tubemakers of Australia Ltd. . . . . . . Australia 22 75-20910
wogig!milsonspt!bartlem@toa.corp. attmail.com
Tufnol Ltd . . . . UK 489 75-61410
http://www.tufnol.co.uk
sales@tufnol.co.uk
Turbine Support Ltd . . . . UK 514 80-58800
tur@supp.demon.co.uk
Turbomeca GmbH. . . . . Germany 183 90-02120
richard.musil@turbomeca.de
Tuthill Corporation . . . . USA 624 75-91320
cables@ix.netcom.com
Twin Disc Inc. . . . . USA 624 75-91350
http://www.twindisk.com
Twin Disc International SA . . . . Belgium 47 75-06320
http://www/twindisc.com

U

U E F Chesterfield Cylinders . . . . UK 490 75-61780
post@chestcyl.demon.co.uk
UBC Inc.. . . . . USA 624 75-91383
crane@gate.net
UBE Industries Ltd. . . . . Japan 255 75-10100
http://www.ube-ind.co.jp
UCC International Limited . . . . UK 490 75-61700
http://www.ucc.co.uk
fluidpower@ucc.co.uk
Uit (Pvt) Ltd. . . . . Pakistan 316 90-02120
uitfak@isb.comsats.net.pk
UK Hydrographic Office . . . . UK 411 20-01500
http://www.hydro.gov.uk
Ulbrich Stainless Steels & . . . . USA 624 75-91385
http://www.ulbrich.com
Ultramare Singapore Pty Ltd. . . . . . . Singapore . . . 352 90-02090
ultsin@mbox2.singnet.com.sg
Umbra Cuscinetti SpA . . . . Italy. 239 75-26630
umbraera@clilne.it
Unicon Ltd . . . . Greece 193 80-03000
unicon@acropolis.gr
Unimex Handels GmbH. . . . . Germany 183 90-02150
unimex@compuserve.com
Uniplex Ltd . . . . UK 514 80-62000
guy.norgrove@uniplex.co.uk
United Electric Motor Co. . . . . USA 624 75-92100
uemc@aol.com
United Kingdom High Commission . . . . Belize. 50 31-01890
brithicom@bti.net
United Kingdom High Commission . . . . Lesotho 271 31-01890
hcmaseru@lesoff.co.za
United Kingdom High Commission . . . . Seychelles . . . 345 31-01890
bhcsey@seychelles.net

V

W

X

Y

Z

Notes

Notes

Notes

Notes

Notes

Notes

Notes